Table of Contents

Truck and Van Sections

CHILTON'S™

TRUCK AND VAN
MANUAL 1995-99

President	Dean F. Morgantini, S.A.E.
Vice President–Finance	Barry L. Beck
Vice President–Sales	Glenn D. Potere
Executive Editor	Kevin M. G. Maher, A.S.E.
Production Manager	Ben Greisler, S.A.E.
Production Assistant	Melinda Possinger
Project Managers	George B. Heinrich III, A.S.E., S.A.E., Will Kessler, A.S.E., S.A.E., James R. Marotta, A.S.E., S.T.S., Richard Schwartz, A.S.E., Todd W. Stidham
Schematics Editor	Christopher G. Ritchie
Editors	Leonard Davis, A.S.E., S.T.S. Frank Keytanjian, A.S.E., S.A.E.

CHILTON™*Automotive Books*

PUBLISHED BY **W. G. NICHOLS, INC.**

Manufactured in USA
© 1998 W. G. Nichols
1020 Andrew Drive
West Chester, PA 19380
ISBN 0-8019-7924-2
Library of Congress Catalog Card No. 98-71357
1234567890 7654321098

Table of Contents

Truck and Van Sections

Model Index

HOW TO USE THIS MANUAL

Model Specific Sections

The model specific sections are grouped by manufacturer and arranged in alphabetical order. The text and illustrations that comprise the service procedures in each model specific section are arranged in the following order of systems and components: Engine Repair (Gasoline, then Diesel if applicable), Fuel System (Gasoline, then Diesel if applicable), Drive Train, Steering and Suspension.

All illustrations are located as close as possible to the applicable procedure. Procedures are for all models in the particular section unless specifically noted otherwise.

Unit Repair Sections

The Unit Repair Sections (URS's) are written to cover all applicable 1995-99 models for the specific URS system or component, unless specifically noted otherwise. The procedures covered in the 10 URS's are not repeated in the model specific sections; therefore, refer to the URS's for the service procedures for the applicable systems or components. Refer to the Table of Contents for URS coverage.

Locating Information

The Table of Contents, located at the front of the book, lists each Unit Repair Section (URS) and model specific section in this manual.

To find where a particular model specific section is located in the book, you need only look in the Table of Contents. Once you have found the proper section, you may wish to find where specific procedures located in that section. Turn to the Index at the front of the model specific section. At the upper left-hand side is a listing of the main topics within that section and the page number on which they may be found. Following the main topics is an alphabetical listing of all of the procedures within the section and their page numbers.

The Model Index, located just after the Table of Contents in the beginning of this manual, may also be used to locate the specific section for any vehicle model covered in this manual.

Safety Notice

Proper service and repair procedures are vital to the safe, reliable operation of all motor vehicles, as well as the personal safety of those performing the repairs. This manual outlines procedures for servicing and repairing vehicles using safe effective methods. The procedures contain many NOTES, WARNINGS and CAUTIONS which should be followed along with standard safety procedures to eliminate the possibility of personal injury or improper service which could damage the vehicle or compromise its safety.

It is important to note that repair procedures and techniques, tools and parts for servicing vehicles, as well as the skill and experience of the individual performing the work vary widely. It is not possible to anticipate all of the conceivable ways or conditions under which vehicles may be serviced, or to provide cautions as to all of the possible hazards that may result. Standard and accepted safety precautions and equipment should be used when handling toxic or flammable fluids, and safety goggles or other protection should be used during cutting, grinding, chiseling, prying, or any other process that can cause material removal or projectiles.

Some procedures require the use of tools specially designed for a specific purpose. Before substituting another tool or procedure, you must be completely satisfied that neither your personal safety, nor the performance of the vehicle will be endangered.

Although information in this manual is based on industry sources and is as complete as possible at the time of publication, the possibility exists that some vehicle manufacturers made later changes which could not be included here. Information on very late models may not be available in some circumstances. While striving for total accuracy, NP/Chilton cannot assume responsibility for any errors, changes, or omissions that may occur in the compilation of this data.

Part Numbers

Part numbers listed in this book are not recommendations by NP/Chilton for any product by brand name. They are references that can be used with interchanges manuals and aftermarket supplier catalogs to locate each brand supplier's discrete part number.

Special Tools

Special tools are recommended by the vehicle manufacturer to perform their specific job. Use has been kept to a minimum, but where absolutely necessary, they are referred to in the text by the part number of the tool manufacturer. These tools may be purchased, under the appropriate part number, from your local dealer or regional distributor, or an equivalent tool can be purchased locally from a tool supplier or parts outlet. Before substituting any tool for the one recommended, read the previous Safety Notice.

Copyright Notice

NP/Chilton would like to thank all manufacturer's involved for their generous assistance.

Get ready for ASE testing with Motor Age Self Study Guides.

Each training unit contains a complete description of the ASE Task Analysis and Test Specifications, and covers the subject areas of the corresponding ASE test question group. Also included are sample ASE test questions. In addition, each book includes a special glossary, sample questions and an expanded answer analysis to increase your knowledge of the subject.

AA Car & Light Truck
A1 Engine Repair
A2 Automatic
Transmission/Transaxle
A3 Manual Drive Train & Axles
A4 Suspension & Steering
A5 Brakes
A6 Electrical/Electronic Systems
A7 Heating & A/C
A8 Engine Performance

Parts Specialist
P1 Medium/Heavy Parts Specialist
P2 Automobile Parts Specialist

Advanced Level
L1 Advanced Engine Performance Specialist
L2 Med/Hvy Vehicle Electronic Diesel Engine Diagnosis Specialist
F1 Light Vehicle Compressed Natural Gas

ALSO AVAILABLE:
TT Medium/Heavy Truck Service: T1 Gasoline Engines, T2 Diesel Engines, T3 Drive Train, T4 Brakes, T5 Suspension & Steering, T6 Electrical/Electronic Systems, T7 Heating, Ventilation & A/C, T8
Preventive Maintenance Inspection (PMI), & MM Engine Machinist (M1, M2, M3),
BB Collision Repair/Paint & Refinish: B2 Paint & Refinishing, B3 Non-Structural Analysis & Damage Repair, B4 Structural Analysis & Damage Repair, B5 Mechanical & Electrical Components, B6 Damage Analysis & Estimating

Name: _____
first   middle   last

Company _____

Address: _____ Apt. # _____

City: _____ State: _____ Zip: _____

Phone (DAYTIME): _____ Fax _____

**For pricing and shipping information,
fax this form to Trudy Kolb, 610-964-4251**

SPECIFICATIONS

1

ACURA
SLX

VEHICLE IDENTIFICATION CHART

Engine						Model Year	
Code	Liters	Cu. In. (cc)	Cyl.	Fuel Sys.	Eng. Mfg.	Code	Year
6VD1	3.2	193 (3165)	6	MFI	Isuzu	T	1996
6VE1	3.5	213 (3494)	6	MFI	Isuzu	V	1997
						W	1998
						X	1999

MFI - Multi-port Fuel Injection

79241C01

ENGINE IDENTIFICATION

Year	Model	Engine Displacement Liters (cc)	Engine Series (ID/VIN)	Fuel System	No. of Cylinders	Engine Type
1996	SLX	3.2 (3165)	6VD1	MFI	6	SOHC
1997	SLX	3.2 (3165)	6VD1	MFI	6	SOHC
1998-99	SLX	3.5 (3494)	6VE1	MFI	6	DOHC

MFI - Multi-port Fuel Injection
SOHC - Single Overhead Camshaft
DOHC - Double Overhead Camshaft

79241C02

GENERAL ENGINE SPECIFICATIONS

Year	Engine ID/VIN	Engine Displacement Liters (cc)	Fuel System Type	Net Horsepower @ rpm	Net Torque @ rpm (ft. lbs.)	Bore x Stroke (in.)	Compression Ratio	Oil Pressure @ rpm
1996	6VD1	3.2 (3165)	MFI	190@5600	188@4000	3.68x3.03	9.1:1	57-80@3000
1997	6VD1	3.2 (3165)	MFI	190@5600	188@4000	3.68x3.03	9.3:1	57-80@3000
1998-99	6VE1	3.5 (3494)	MFI	215@5400	230@3000	3.68x3.35	9.1:1	57-80@3000

MFI - Multi-port Fuel Injection

79241C03

GASOLINE ENGINE TUNE-UP SPECIFICATIONS

Year	Engine ID/VIN	Engine Displacement Liters (cc)	Spark Plug Gap (in.)	Ignition Timing (deg.) MT	Ignition Timing (deg.) AT	Fuel Pump (psi)	Idle Speed (rpm) MT	Idle Speed (rpm) AT	Valve Clearance In.	Valve Clearance Ex.
1996	6VD1	3.2 (3165)	0.040	①	①	41-46	①	①	HYD	HYD
1997	6VD1	3.2 (3165)	0.040	①	①	48-55	①	①	HYD	HYD
1998-99	6VE1	3.5 (3494)	0.040-0.043	①	①	48-55	①	①	HYD	HYD

NOTE: The Vehicle Emission Control Information (VECI) label often reflects specification changes made during production. The label figures must be used if they differ from those in this chart.

HYD - Hydraulic (No periodic valve lash adjustments are necessary.)

① Controlled by the ECM and is not adjustable

79241C04

CAPACITIES

Year	Model	Engine ID/VIN	Engine Displacement Liters (cc)	Engine Oil with Filter (qts.)	Transmission (pts.) 4-Spd	Transmission (pts.) 5-Spd	Transmission (pts.) Auto.	Transfer Case (pts.)	Drive Axle Front (pts.)	Drive Axle Rear (pts.)	Fuel Tank (gal.)	Cooling System (qts.)
1996	SLX	6VD1	3.2 (3165)	5.7	—	6.2	18.2	3.0	3.2	3.8	22.5	①
1997	SLX	6VD1	3.2 (3165)	5.7	—	6.2	18.2	3.0	3.0	3.8	22.5	①
1998-99	SLX	6VE1	3.5 (3494)	5.0	—	5.8	18.2	②	3.0	6.4	22.5	③

NOTE: All capacities are approximate. Add fluid gradually and check to be sure a proper fluid level is obtained.

① Manual transmission: 9.3   ② Without TOD: 3.0   ③ Manual transmission: 7.0
   Automatic transmission: 9.0   With TOD: 4.0   Automatic transmission: 7.4

79241C05

VALVE SPECIFICATIONS

Year	Engine ID/VIN	Engine Displacement Liters (cc)	Seat Angle (deg.)	Face Angle (deg.)	Spring Test Pressure (lbs. @ in.)	Spring Installed Height (in.)	Stem-to-Guide Clearance (in.) Intake	Stem-to-Guide Clearance (in.) Exhaust	Stem Diameter (in.) Intake	Stem Diameter (in.) Exhaust
1996	6VD1	3.2 (3165)	45	45	45-55@1.54	1.54	0.0009-0.0079	0.0012-0.0079	0.2323-0.2353	0.2323-0.2350
1997	6VD1	3.2 (3165)	45	45	45-55@1.54	1.54	0.0009-0.0079	0.0012-0.0079	0.2323-0.2353	0.2323-0.2350
1998-99	6VE1	3.5 (3494)	45	45	41-44@1.38	1.38	0.0009-0.0079	0.0012-0.0079	0.2323-0.2353	0.2323-0.2350

79241C06

TORQUE SPECIFICATIONS
All readings in ft. lbs.

Year	Engine ID/VIN	Engine Displacement Liters (cc)	Cylinder Head Bolts	Main Bearing Bolts	Rod Bearing Bolts	Crankshaft Damper Bolts	Flywheel Bolts	Manifold Intake	Manifold Exhaust	Spark Plugs	Lug Nut
1996	6VD1	3.2 (3165)	①	②	40	123	40	17	42	13	87
1997	6VD1	3.2 (3165)	①	②	40	123	40	17	42	13	87
1998-99	6VE1	3.5 (3494)	47	③	40	123	40	18	38	13	87

① 8x1.25 bolts: 15 ft. lbs.
  11x1.5 bolts: 47 ft. lbs.

② Main bearing cap bolts: 29 ft. lbs.
  Oil gallery bolts: 29 ft. lbs. plus 55-65 degrees
  Buttress bolts: 29 ft. lbs.

③ 22 ft. lbs. plus an additional 60 degrees (1/6 turn).

79241C07

1995-96 SCHEDULED MAINTENANCE INTERVALS
(ACURA SLX)

TO BE SERVICED	TYPE OF SERVICE	VEHICLE MILEAGE INTERVAL (x1000)												
		7.5	15	22.5	30	37.5	45	52.5	60	67.5	75	82.5	90	97.5
Engine oil & filter	R	✓	✓	✓	✓	✓	✓	✓	✓	✓	✓	✓	✓	✓
Battery fluid level	S/I	✓	✓	✓	✓	✓	✓	✓	✓	✓	✓	✓	✓	✓
Body & chassis lubrication	S/I	✓	✓	✓	✓	✓	✓	✓	✓	✓	✓	✓	✓	✓
Brake & clutch fluid level	S/I	✓	✓	✓	✓	✓	✓	✓	✓	✓	✓	✓	✓	✓
Brake lines & hoses	S/I	✓	✓	✓	✓	✓	✓	✓	✓	✓	✓	✓	✓	✓
Check & rotate tires	S/I	✓	✓	✓	✓	✓	✓	✓	✓	✓	✓	✓	✓	✓
Engine coolant	S/I	✓	✓	✓	✓	✓	✓	✓	✓	✓	✓	✓	✓	✓
Exhaust system	S/I	✓	✓	✓	✓	✓	✓	✓	✓	✓	✓	✓	✓	✓
Lubricate accelerator linkage	S/I	✓	✓	✓	✓	✓	✓	✓	✓	✓	✓	✓	✓	✓
Starter safety switch	S/I	✓	✓	✓	✓	✓	✓	✓	✓	✓	✓	✓	✓	✓
Steering operation	S/I	✓	✓	✓	✓	✓	✓	✓	✓	✓	✓	✓	✓	✓
Lubricate front & rear propeller shaft	S/I	✓		✓		✓		✓		✓		✓		✓
Propeller shaft flange torque 46 lb. ft. (63 Nm)	S/I	✓		✓		✓		✓		✓		✓		✓
Accelerator linkage	S/I		✓		✓		✓		✓		✓		✓	
Auto cruise control linkage & hose	S/I		✓		✓		✓		✓		✓		✓	
Brake pedal play	S/I		✓		✓		✓		✓		✓		✓	
Clutch pedal play	S/I		✓		✓		✓		✓		✓		✓	
Cooling & heater hoses	S/I		✓		✓		✓		✓		✓		✓	
Disc brakes	S/I		✓		✓		✓		✓		✓		✓	
Front & rear axle oil	R		✓		✓				✓				✓	
Lubricate clutch pedal spring, bushing & clevis pin	S/I		✓		✓		✓		✓		✓		✓	
Lubricate key lock cylinder	S/I		✓		✓		✓		✓		✓		✓	
Parking brake	S/I		✓		✓		✓		✓		✓		✓	
Suspension & steering	S/I		✓		✓		✓		✓		✓		✓	
Manual transmission & transfer case oil	R		✓		✓				✓				✓	
Air cleaner filter	R				✓				✓				✓	

79241C08

1995-96 SCHEDULED MAINTENANCE INTERVALS
(ACURA SLX) (Cont.)

TO BE SERVICED	TYPE OF SERVICE	VEHICLE MILEAGE INTERVAL (x1000)												
		7.5	15	22.5	30	37.5	45	52.5	60	67.5	75	82.5	90	97.5
Engine coolant	R				✓				✓				✓	
Power steering fluid	R				✓				✓				✓	
Clutch lines & hose	S/I				✓				✓				✓	
Engine drive belt	S/I				✓				✓				✓	
Front wheel bearings	S/I				✓				✓				✓	
Clean radiator core & A/C condenser	S/I								✓				✓	
Spark plugs	R								✓					
Timing belt	R								✓					
Fuel tank, cap & lines	S/I								✓					

R – Replace   S/I – Service or Inspect

FREQUENT OPERATION MAINTENANCE (SEVERE SERVICE)
If a vehicle is operated under any of the following conditions it is considered severe service:
- Extremely dusty areas.
- 50% or more of the vehicle operation is in 32°C (90°F) or higher temperatures, or constant operation in temperatures below 0°C (32°F).
- Prolonged idling (vehicle operation in stop and go traffic).
- Frequent short running periods (engine does not warm to normal operating temperatures).
- Police, taxi, delivery usage or trailer towing usage.

Oil & oil filter change – change every 3000 miles.
Front & rear axle oil – change every 15,000 miles.
Automatic transmission fluid & filter – change every 20,000 miles.

79241C09

1997-99 SCHEDULED MAINTENANCE INTERVALS
(ACURA SLX)

TO BE SERVICED	TYPE OF SERVICE	VEHICLE MILEAGE INTERVAL (x1000)															
		7.5	15	22.5	30	37.5	45	52.5	60	67.5	75	82.5	90	97.5	105	112.5	120
Accelerator linkage ①	L	✓	✓	✓	✓	✓	✓	✓	✓	✓	✓	✓	✓	✓	✓	✓	✓
Accessory drive belts ②	S/I				✓				✓				✓				✓
Air cleaner filter	R				✓				✓				✓				✓
Auto cruise control linkage & hose ③	S/I		✓		✓		✓		✓		✓		✓		✓		✓
Automatic transmission fluid level ③	S/I	✓		✓		✓		✓		✓		✓		✓		✓	
Battery fluid level ③	S/I	✓	✓	✓	✓	✓	✓	✓	✓	✓	✓	✓	✓	✓	✓	✓	✓
Body and chassis ①	L	✓	✓	✓	✓	✓	✓	✓	✓	✓	✓	✓	✓	✓	✓	✓	✓
Brake fluid level ③	S/I	✓	✓	✓	✓	✓	✓	✓	✓	✓	✓	✓	✓	✓	✓	✓	✓
Brake lines & hoses ③	S/I	✓	✓	✓	✓	✓	✓	✓	✓	✓	✓	✓	✓	✓	✓	✓	✓
Brake pedal play ③	S/I		✓		✓		✓		✓		✓		✓		✓		✓
Clutch fluid level ③	S/I	✓	✓	✓	✓	✓	✓	✓	✓	✓	✓	✓	✓	✓	✓	✓	✓
Clutch lines & hose ③	S/I				✓				✓				✓				✓
Clutch pedal free-play ③	S/I		✓		✓		✓		✓		✓		✓		✓		✓
Clutch pedal spring, bushing and clevis pin ①	S/I		✓		✓		✓		✓		✓		✓		✓		✓
Cooling and heating system hoses ③	S/I		✓		✓		✓		✓		✓		✓		✓		✓
Driveshaft flange torque ③	S/I	✓		✓		✓		✓		✓		✓		✓		✓	
Drum and disc brakes ③	S/I		✓		✓		✓		✓		✓		✓		✓		✓
Engine coolant	R				✓				✓				✓				✓
Engine coolant level ③	S/I	✓	✓	✓	✓	✓	✓	✓	✓	✓	✓	✓	✓	✓	✓	✓	✓
Engine oil & filter ③	R	✓	✓	✓	✓	✓	✓	✓	✓	✓	✓	✓	✓	✓	✓	✓	✓
Exhaust system ③	S/I	✓	✓	✓	✓	✓	✓	✓	✓	✓	✓	✓	✓	✓	✓	✓	✓
Front and rear axle lubricant	R		✓		✓				✓				✓				✓
Front and rear driveshafts ①	S/I	✓	✓	✓	✓	✓	✓	✓	✓	✓	✓	✓	✓	✓	✓	✓	✓
Front wheel bearings	S/I & L				✓				✓				✓				✓
Fuel lines & tank cap ③	S/I								✓								✓
Inspect for fluid leaks ③	S/I	✓	✓	✓	✓	✓	✓	✓	✓	✓	✓	✓	✓	✓	✓	✓	✓
Key lock cylinder ③	L		✓		✓		✓		✓		✓		✓		✓		✓
Manual transmission and transfer case fluid	R		✓		✓				✓				✓				✓
Parking brake system ③	S/I		✓		✓		✓		✓		✓		✓		✓		✓

79241C10

1997-99 SCHEDULED MAINTENANCE INTERVALS
(ACURA SLX)

TO BE SERVICED	TYPE OF SERVICE	VEHICLE MILEAGE INTERVAL (x1000)															
		7.5	15	22.5	30	37.5	45	52.5	60	67.5	75	82.5	90	97.5	105	112.5	120
Power steering fluid	R				✓				✓				✓				✓
Radiator core and A/C condenser	S/I & C								✓								✓
Rotate tires	S/I	✓	✓	✓	✓	✓	✓	✓	✓	✓	✓	✓	✓	✓	✓	✓	✓
Shift-on-the-fly system gear fluid ②	S/I		✓		✓		✓		✓		✓		✓		✓		✓
Spark plugs	R	Every 100,000 miles															
Starter safety switch ③	S/I	✓	✓	✓	✓	✓	✓	✓	✓	✓	✓	✓	✓	✓	✓	✓	✓
Steering operation ③	S/I	✓	✓	✓	✓	✓	✓	✓	✓	✓	✓	✓	✓	✓	✓	✓	✓
Suspension & steering ③	S/I	✓	✓	✓	✓	✓	✓	✓	✓	✓	✓	✓	✓	✓	✓	✓	✓
Throttle linkage ③	S/I		✓		✓		✓		✓		✓		✓		✓		✓
Timing belt	R									✓							
Tires and wheels ③	S/I	✓	✓	✓	✓	✓	✓	✓	✓	✓	✓	✓	✓	✓	✓	✓	✓
Valve clearance	A								✓								✓

① Perform this at the mileage indicated or every 6 months, whichever occurs first.
② Perform this at the mileage indicated or every 24 months, whichever occurs first.
③ Perform this at the mileage indicated or every 12 months, whichever occurs first.

R - Replace   S/I - Service or Inspect   L - Lubricate   A - Adjust   C - Clean

FREQUENT OPERATION MAINTENANCE (SEVERE SERVICE)

If a vehicle is operated under any of the following conditions it is considered severe service:

- Towing a trailer or using a camper or car-top carrier.
- Repeated short trips of less than 5 miles in temperatures below freezing, or trips of less than 10 miles in any temperature.
- Extensive idling or low-speed driving for long distances as in heavy commercial use, such as delivery, taxi or police cars.
- Operating on rough, muddy or salt-covered roads.
- Operating on unpaved or dusty roads.
- Frequent operation in temperatures above 90°F.

Air cleaner element - replace every 15,000 miles.

Engine oil and filter - replace every 3,000 miles or 3 months, whichever occurs first.

Automatic transmission fluid - replace every 20,000 miles.

Rear axle lubricant - replace every 15,000 miles.

79241C11

SCHEDULED MAINTENANCE
ACURA
SLX

The following should be used as a guide when determining the amount of work required for a particular service if taken to a repair shop.
In estimating how long a particular Scheduled Maintenance Service should take, please observe the following:

- **Chilton Time** is time based on field research and data supplied by the vehicle manufacturer.
- Labor time operations are given in hours and tenths of an hour.
- All labor operations, are to be used as a guide.

Mechanic Skill Level Codes:
(G) GENERAL: Normally skilled with certification.
(M) MAINTENANCE: Semi-skilled working on certication.
(P) PRECISION: Really skilled with multiple certification.

	Chilton Time		Chilton Time		Chilton Time
M) 7500 Mile Service		**(G) 45000 Mile Service**		**(G) 90000 Mile Service**	
1996-97	2.0	1996-97	3.2	1996-97	7.8
1998-99	2.8	1998-99	3.9	1998-99	8.5
G) 15000 Mile Service		**(M) 52500 Mile Service**		**(M) 97500 Mile Service**	
1996-97	4.0	1996-97	2.0	1996-97	2.0
1998-99	4.5	1998-99	2.8	1998-99	2.8
M) 22500 Mile Service		**(G) 60000 Mile Service**		**(G) 100000 Mile Service**	
1996-97	2.0	1996-97	12.6	1998-99	
1998-99	2.8	1998-99	11.1	Replace spark plugs	.8
G) 30000 Mile Service		**(M) 67500 Mile Service**		**(G) 105000 Mile Service**	
1996-97	7.8	1996-97	2.0	1998-99	3.9
1998-99	8.5	1998-99	2.8		
M) 37500 Mile Service		**(G) 75000 Mile Service**		**(M) 112500 Mile Service**	
1996-97	2.0	1996-97	3.1	1998-99	2.8
1998-99	2.8	1998-99	7.3		
		(M) 82500 Mile Service		**(G) 120000 Mile Service**	
		1996-97	2.0	1998-99	11.1
		1998-99	2.8		

79241C12

CHRYSLER CORP.
Chrysler Town & Country • Dodge Caravan • Plymouth Voyager

VEHICLE IDENTIFICATION CHART

Engine Code						Model Year	
Code	Liters	Cu. In. (cc)	Cyl.	Fuel Sys.	Eng. Mfg.	Code	Year
B	2.4	148 (2429)	I4	SMFI	Chrysler	S	1995
K	2.5	153 (2507)	I4	TFI	Chrysler	T	1996
3	3.0	181 (2972)	V6	SMFI	Mitsubishi	V	1997
R	3.3	201 (3300)	V6	SMFI	Chrysler	W	1998
L	3.8	231 (3785)	V6	SMFI	Chrysler	X	1999

SMFI - Sequential Multiport Fuel Injection
TFI - Throttle body Fuel Injection

79241C57

ENGINE IDENTIFICATION

Year	Model	Engine Displacement Liters (cc)	Engine Series (ID/VIN)	Fuel System	No. of Cylinders	Engine Type
1995	Caravan	3.0 (2972)	3	SMFI	6	SOHC
	Caravan	3.3 (3300)	R	SMFI	6	OHV
	Caravan	3.8 (3785)	L	SMFI	6	OHV
	Town & Country	3.8 (3785)	L	SMFI	6	OHV
	Voyager	2.5 (2507)	K	TFI	4	SOHC
	Voyager	3.0 (2972)	3	SMFI	6	SOHC
	Voyager	3.3 (3300)	R	SMFI	6	OHV
	Voyager	3.8 (3785)	L	SMFI	6	OHV
1996	Caravan	2.4 (2429)	B	SMFI	4	DOHC
	Caravan	3.0 (2972)	3	SMFI	6	SOHC
	Caravan	3.3 (3300)	R	SMFI	6	OHV
	Caravan	3.8 (3785)	L	SMFI	6	OHV
	Town & Country	3.3 (3301)	R	SMFI	6	OHV
	Town & Country	3.8 (3778)	L	SMFI	6	OHV
	Voyager	2.4 (2429)	B	SMFI	4	DOHC
	Voyager	3.0 (2972)	3	SMFI	6	SOHC
	Voyager	3.3 (3301)	R	SMFI	6	OHV
	Voyager	3.8 (3785)	L	SMFI	6	OHV
1997	Caravan	2.4 (2429)	B	SMFI	4	DOHC
	Caravan	3.0 (2972)	3	SMFI	6	SOHC
	Caravan	3.3 (3300)	R	SMFI	6	OHV
	Caravan	3.8 (3785)	L	SMFI	6	OHV
	Town & Country	3.3 (3301)	R	SMFI	6	OHV
	Town & Country	3.8 (3778)	L	SMFI	6	OHV
	Voyager	2.4 (2429)	B	SMFI	4	DOHC
	Voyager	3.0 (2972)	3	SMFI	6	SOHC
	Voyager	3.3 (3301)	R	SMFI	6	OHV
	Voyager	3.8 (3785)	L	SMFI	6	OHV
1998-99	Caravan	2.4 (2429)	B	SMFI	4	DOHC
	Caravan	3.0 (2972)	3	SMFI	6	SOHC
	Caravan	3.3 (3300)	R	SMFI	6	OHV
	Caravan	3.8 (3785)	L	SMFI	6	OHV
	Town & Country	3.3 (3301)	R	SMFI	6	OHV
	Town & Country	3.8 (3778)	L	SMFI	6	OHV
	Voyager	2.4 (2429)	B	SMFI	4	DOHC
	Voyager	3.0 (2972)	3	SMFI	6	SOHC
	Voyager	3.3 (3301)	R	SMFI	6	OHV
	Voyager	3.8 (3785)	L	SMFI	6	OHV

DSL - Diesel
SMFI - Sequential Multiport Fuel Injection
TFI - Throttle body Fuel Injection
OHV - Overhead Valve
SOHC - Single Overhead Camshaft
DOHC - Double Overhead Camshaft

79241C58

GENERAL ENGINE SPECIFICATIONS

Year	Engine ID/VIN	Engine Displacement Liters (cc)	Fuel System Type	Net Horsepower @ rpm	Net Torque @ rpm (ft. lbs.)	Bore x Stroke (in.)	Compression Ratio	Oil Pressure @ rpm
1995	K	2.5 (2507)	TFI	100@4800	135@2800	3.44x4.09	8.9:1	25-80@3000
	3	3.0 (2972)	SMFI	143@5000	170@2800	3.59x2.99	8.9:1	30-80@3000
	R	3.3 (3300)	SMFI	162@4800	194@3600	3.66x3.19	8.9:1	30-80@3000
	L	3.8 (3785)	SMFI	162@4400	213@3300	3.78x3.43	9.0:1	30-80@3000
1996	B	2.4 (2429)	SMFI	150@5200	167@4000	3.44x3.98	9.4:1	25-80@3000
	3	3.0 (2972)	SMFI	143@5000	170@2800	3.59x2.99	8.9:1	30-80@3000
	R	3.3 (3300)	SMFI	162@4800	194@3600	3.66x3.19	8.9:1	30-80@3000
	L	3.8 (3785)	SMFI	162@4400	213@3300	3.78x3.43	9.0:1	30-80@3000
1997	B	2.4 (2429)	SMFI	150@5200	167@4000	3.44x3.98	9.4:1	25-80@3000
	3	3.0 (2972)	SMFI	150@5200	176@4000	3.59x2.99	8.9:1	35-75@3000
	R	3.3 (3300)	SMFI	158@4850	203@3250	3.66x3.19	8.9:1	30-80@3000
	L	3.8 (3785)	SMFI	166@4300	227@3100	3.78x3.43	8.9:1	30-80@3000
1998-99	B	2.4 (2429)	SMFI	150@5200	167@4000	3.44x3.98	9.4:1	25-80@3000
	3	3.0 (2972)	SMFI	150@5200	176@4000	3.59x2.99	8.9:1	45-75@3000
	R	3.3 (3300)	SMFI	158@4850	203@3250	3.66x3.19	8.9:1	30-80@3000
	L	3.8 (3785)	SMFI	180@4400	240@3200	3.78x3.43	9.6:1	30-80@3000

SMFI - Sequential Multiport Fuel Injection

TFI - Throttle body Fuel Injection

79241C59

ENGINE TUNE-UP SPECIFICATIONS

Year	Engine ID/VIN	Engine Displacement Liters (cc)	Spark Plugs Gap (in.)	Ignition Timing (deg.) MT	Ignition Timing (deg.) AT	Fuel Pump (psi)	Idle Speed (rpm) MT	Idle Speed (rpm) AT	Valve Clearance In.	Valve Clearance Ex.
1995	K	2.5 (2507)	0.035	—	①	39	—	②	HYD	HYD
	3	3.0 (2972)	0.035	—	①	48	—	②	HYD	HYD
	R	3.3 (3300)	0.050	—	③	48	—	②	HYD	HYD
	L	3.8 (3785)	0.050	—	③	48	—	②	HYD	HYD
1996	B	2.4 (2429)	0.050	③	③	49	②	②	HYD	HYD
	3	3.0 (2972)	0.035	—	③	48	—	②	HYD	HYD
	R	3.3 (3300)	0.050	—	③	49	—	②	HYD	HYD
	L	3.8 (3785)	0.050	—	③	49	—	②	HYD	HYD
1997	B	2.4 (2429)	0.048-0.053	—	③	49	—	②	HYD	HYD
	3	3.0 (2972)	0.039-0.044	—	③	48	—	②	HYD	HYD
	R	3.3 (3300)	0.048-0.053	—	③	49	—	②	HYD	HYD
	L	3.8 (3785)	0.048-0.053	—	③	49	—	②	HYD	HYD
1998-99	B	2.4 (2429)	0.048-0.053	—	③	49	—	②	HYD	HYD
	3	3.0 (2972)	0.039-0.044	—	③	48	—	②	HYD	HYD
	R	3.3 (3300)	0.048-0.053	—	③	55	—	②	HYD	HYD
	L	3.8 (3785)	0.048-0.053	—	③	49	—	②	HYD	HYD

NOTE: The Vehicle Emission Control Information label often reflects specification changes made during production. The label figures must be used if they differ from those in this chart.

HYD - Hydraulic

① Refer to the Vehicle Emission Control Information label for correct timing specification with a range of +/- 2.

② Refer to the Vehicle Emission Control Information (VECI) label for correct specification.

③ Ignition timing is regulated by the Powertrain Control Module (PCM), and cannot be adjusted.

79241C60

CAPACITIES

Year	Model	Engine ID/VIN	Engine Displacement Liters (cc)	Engine Oil with Filter (qts.)	Transmission (pts.)			Transfer Case (pts.)	Drive Axle		Fuel Tank (gal.)	Cooling System (qts.)
					4-Spd	5-Spd	Auto.		Front (pts.)	Rear (pts.)		
1995	Caravan	R	3.3 (3300)	4.5	—	—	18.0 ⑥	2.4	—	4.0 ⑤	③	10.5
	Caravan	L	3.8 (3785)	4.5	—	—	18.0 ⑥	2.4	—	4.0 ⑤	③	10.5
	Town & Country	L	3.8 (3785)	4.5	—	—	③	2.4	4.0	4.0 ④	②	10.5
	Voyager	K	2.5 (2507)	4.5	—	—	①	—	—	-	20.0	9.5
	Voyager	3	3.0 (2972)	4.5	—	—	①	—	—	-	20.0	10.5
	Voyager	R	3.3 (3300)	4.5	—	—	①	2.4	—	4.0 ②	③	10.5
	Voyager	L	3.8 (3785)	4.5	—	—	①	2.4	—	4.0 ②	③	10.5
1996	Caravan	B	2.4 (2429)	4.5	—	—	18.0 ⑥	—	—	—	20.0	9.5
	Caravan	3	3.0 (2972)	4.5	—	—	18.0 ⑥	—	—	—	20.0	10.5
	Caravan	R	3.3 (3300)	4.5	—	—	18.0 ⑥	2.4	—	4.0 ⑤	③	10.5
	Caravan	L	3.8 (3785)	4.5	—	—	18.0 ⑥	2.4	—	4.0 ⑤	③	10.5
	Town & Country	R	3.3 (3301)	4.5	—	—	18.2 ⑥	NA	NA	NA	20.0	10.5
	Town & Country	L	3.8 (3778)	4.5	—	—	18.2 ⑥	NA	NA	NA	20.0	10.5
	Voyager	B	2.4 (2429)	4.5	—	—	8.0 ⑦	—	—	—	20.0	9.5
	Voyager	3	3.0 (2972)	4.5	—	—	8.0 ⑦	—	—	—	20.0	10.5
	Voyager	R	3.3 (3301)	4.5	—	—	8.0 ⑦	—	—	—	20.0	10.5
	Voyager	L	3.8 (3785)	4.5	—	—	8.0 ⑦	—	—	4.0 ⑤	20.0	10.5
1997	Caravan	B	2.4 (2429)	4.5	—	—	8.0 ⑦	—	—	4.0 ⑤	20.0	9.5
	Caravan	3	3.0 (2972)	4.5	—	—	8.0 ⑦	—	—	4.0 ⑤	20.0	10.5
	Caravan	R	3.3 (3300)	4.5	—	—	8.0 ⑦	—	—	4.0 ⑤	20.0	10.5
	Caravan	L	3.8 (3785)	4.5	—	—	8.0 ⑦	—	—	4.0 ⑤	20.0	10.5
	Town & Country	R	3.3 (3301)	4.5	—	—	8.0 ⑦	—	—	4.0 ⑤	20.0	10.5
	Town & Country	L	3.8 (3778)	4.5	—	—	8.0 ⑦	—	—	4.0 ⑤	20.0	10.5
	Voyager	B	2.4 (2429)	4.5	—	—	8.0 ⑦	—	—	4.0 ⑤	20.0	9.5
	Voyager	3	3.0 (2972)	4.5	—	—	8.0 ⑦	—	—	4.0 ⑤	20.0	10.5
	Voyager	R	3.3 (3301)	4.5	—	—	8.0 ⑦	—	—	4.0 ⑤	20.0	10.5
	Voyager	L	3.8 (3785)	4.5	—	—	8.0 ⑦	—	—	4.0 ⑤	20.0	10.5
1998-99	Caravan	B	2.4 (2429)	4.5	—	—	8.0 ⑦	—	—	4.0 ⑤	20.0	9.5
	Caravan	3	3.0 (2972)	4.5	—	—	8.0 ⑦	—	—	4.0 ⑤	20.0	10.5
	Caravan	R	3.3 (3300)	4.5	—	—	8.0 ⑦	—	—	4.0 ⑤	20.0	10.5
	Caravan	L	3.8 (3785)	4.5	—	—	8.0 ⑦	—	—	4.0 ⑤	20.0	10.5
	Town & Country	R	3.3 (3301)	4.5	—	—	8.0 ⑦	—	—	4.0 ⑤	20.0	10.5
	Town & Country	L	3.8 (3778)	4.5	—	—	8.0 ⑦	—	—	4.0 ⑤	20.0	10.5
	Voyager	B	2.4 (2429)	4.5	—	—	8.0 ⑦	—	—	4.0 ⑤	20.0	9.5
	Voyager	3	3.0 (2972)	4.5	—	—	8.0 ⑦	—	—	4.0 ⑤	20.0	10.5
	Voyager	R	3.3 (3301)	4.5	—	—	8.0 ⑦	—	—	4.0 ⑤	20.0	10.5
	Voyager	L	3.8 (3785)	4.5	—	—	8.0 ⑦	—	—	4.0 ⑤	20.0	10.5

NOTE: All capacities are approximate. Add fluid gradually and check to be sure a proper fluid level is obtained.

① Fleet vehicles: 19.0 pts.
  Non-fleet vehicles: 18.0 pts.
② Overrunning clutch: 0.78 pts.

③ FWD: 20.0 gals.
  AWD: 18.0 gals.
④ With 7.25 in. rear: 2.5 pts.
  With 8.25 in. rear: 4.4 pts.

⑤ Overrunning clutch: 0.75 pts.
⑥ Overhaul fill capacity with torque converter empty
⑦ 31TH overhaul fill capacity with torque converter empty: 17.0
  41TE overhaul fill capacity with torque converter empty: 18.2

79241C61

VALVE SPECIFICATIONS

Year	Engine ID/VIN	Engine Displacement Liters (cc)	Seat Angle (deg.)	Face Angle (deg.)	Spring Test Pressure (lbs. @ in.)	Spring Installed Height (in.)	Stem-to-Guide Clearance (in.)		Stem Diameter (in.)	
							Intake	Exhaust	Intake	Exhaust
1995	K	2.5 (2507)	45	45	115@1.65	1.65	0.001-0.003	0.003-0.005	0.3124	0.3103
	3	3.0 (2972)	44.5	45.5	73@1.59	1.59	0.001-0.002	0.002-0.003	0.3130-0.3140	0.3120-0.3130
	R	3.3 (3300)	45	44.5	95@1.57	1.62-1.68	0.001-0.003	0.002-0.006	0.3120-0.3130	0.3110-0.3120
	L	3.8 (3785)	45	44.5	95@1.57	1.62-1.68	0.001-0.003	0.002-0.006	0.3120-0.3130	0.3110-0.3120
1996	B	2.4 (2429)	45	44.5-45.0	129-143@1.17	1.50	0.0018-0.0025	0.0029-0.0037	0.234	0.233
	3	3.0 (2972)	44.5	45.5	73@1.59	1.59	0.001-0.002	0.002-0.003	0.313-0.314	0.312-0.313
	R	3.3 (3300)	45	44.5	207-229@1.17	1.62-1.68	0.001-0.003	0.002-0.006	0.312-0.313	0.311-0.312
	L	3.8 (3785)	45	44.5	207-229@1.17	1.62-1.68	0.001-0.003	0.002-0.006	0.312-0.313	0.311-0.312
1997	B	2.4 (2429)	45	44.5-45.0	129-143@1.17	1.50	0.0018-0.0025	0.0029-0.0037	0.234	0.233
	3	3.0 (2972)	44.0-44.3	45.0-45.3	73@1.59	1.59	0.001-0.004	0.002-0.006	0.313-0.314	0.3120-0.3125
	R	3.3 (3300)	45.0-45.5	②	207-229@1.169	1.622-1.681	0.001-0.003	0.002-0.006	0.312-0.313	0.3112-0.3119
	L	3.8 (3785)	45.0-45.5	②	207-229@1.169	1.622-1.681	0.001-0.003	0.002-0.006	0.312-0.313	0.3112-0.3119
1998-99	B	2.4 (2429)	45	44.5-45.0	129-143@1.17	1.50	0.0018-0.0025	0.0029-0.0037	0.234	0.233
	3	3.0 (2972)	44.0-44.3	45.0-45.3	73@1.59	1.59	0.001-0.004	0.002-0.006	0.313-0.314	0.3120-0.3125
	R	3.3 (3300)	45.0-45.5	②	207-229@1.169	1.622-1.681	0.001-0.003	0.002-0.006	0.312-0.313	0.3112-0.3119
	L	3.8 (3785)	45.0-45.5	②	207-229@1.169	1.622-1.681	0.001-0.003	0.002-0.006	0.312-0.313	0.3112-0.3119

① Town & Country
② Intake valve: 44.5 degrees
  Exhaust valve: 45 degrees

79241C62

TORQUE SPECIFICATIONS
All readings in ft. lbs.

Year	Engine ID/VIN	Engine Displacement Liters (cc)	Cylinder Head Bolts	Main Bearing Bolts	Rod Bearing Bolts	Crankshaft Damper Bolts	Flywheel Bolts	Manifold Intake ①	Exhaust	Spark Plugs	Lug Nut
1995	K	2.5 (2507)	②	③	④	85	70	17	17	20	95
	3	3.0 (2972)	80	60	38	112	70	15	16	20	95
	R	3.3 (3300)	②	③	④	40	70	17	17	20	95
	L	3.8 (3785)	②	③	④	40	70	17	17	20	95
1996	B	2.4 (2429)	⑤	⑥	⑦	100	70	20	17	20	95
	3	3.0 (2972)	80	60	38	112	70	15	16	20	95
	R	3.3 (3300)	②	③	④	40	70	17	17	20	95
	L	3.8 (3785)	②	③	④	40	70	17	17	20	95
1997	B	2.4 (2429)	⑤	⑥	⑦	100	70	20	17	20	85-115
	3	3.0 (2972)	80	⑧	⑨	100	70	20	17	20	85-115
	R	3.3 (3300)	②	③	④	40	70	17	17	20	85-115
	L	3.8 (3785)	②	③	④	40	70	17	17	20	85-115
1998-99	B	2.4 (2429)	⑤	⑥	⑦	100	70	20	17	20	85-115
	3	3.0 (2972)	80	⑧	⑨	100	70	20	17	20	85-115
	R	3.3 (3300)	②	③	④	40	70	17	17	20	85-115
	L	3.8 (3785)	②	③	④	40	70	17	17	20	85-115

① This applies to the lower intake manifold only.
② Step 1: 45 ft. lbs. / Step 2: 65 ft. lbs. / Step 3: 65 ft. lbs. / Step 4: Plus 1/4 turn
③ Step 1: 30 ft. lbs. / Step 2: Plus 1/4 turn
④ Step 1: 40 ft. lbs. / Step 2: Plus 1/4 turn
⑤ Step 1: 25 ft. lbs. / Step 2: 50 ft. lbs. / Step 3: 50 ft. lbs. / Step 4: Plus 1/4 turn
⑥ M8 bolts: 250 in. lbs. / M11 bolts: 30 ft. lbs. plus 1/4 turn
⑦ Step 1: 20 ft. lbs. / Step 2: Plus 1/4 turn

7924AC63

SCHEDULED MAINTENANCE INTERVALS
(CHRYSLER CARAVAN, TOWN & COUNTRY, VOYAGER)

TO BE SERVICED	TYPE OF SERVICE	VEHICLE MILEAGE INTERVAL (x1000)												
		7.5	15	22.5	30	37.5	45	52.5	60	67.5	75	82.5	90	97.5
Engine oil & filter	R	✓	✓	✓	✓	✓	✓	✓	✓	✓	✓	✓	✓	✓
Driveshaft boots	S/I	✓	✓	✓	✓	✓	✓	✓	✓	✓	✓	✓	✓	✓
Exhaust system	S/I	✓	✓	✓	✓	✓	✓	✓	✓	✓	✓	✓	✓	✓
Engine coolant level, hoses & clamps	S/I	✓	✓	✓	✓	✓	✓	✓	✓	✓	✓	✓	✓	✓
Rotate tires	S/I	✓	✓	✓	✓	✓	✓	✓	✓	✓	✓	✓	✓	✓
Drive belts	S/I		✓		✓		✓		✓		✓		✓	
Brake hoses & linings	S/I			✓			✓			✓			✓	
Automatic transaxle fluid & filter	R				✓				✓				✓	
Air filter	R				✓				✓				✓	
Spark plugs① (2.5L & 3.0L)	R				✓				✓				✓	
Serpentine belts (3.0L & 3.3L)	S/I								✓		✓		✓	
Lubricate tie rod ends	S/I				✓				✓				✓	
PCV valve	S/I				✓				✓				✓	
Engine Coolant	R							✓				✓		
Timing belt (2.5L)	R												✓	
Timing belt (3.0L)	R								✓					
Distributor cap & rotor	R							✓						
Ignition cables (2.5L & 3.0L)	R								✓					
Ignition timing	S/I								✓					

① Platinum tip spark plugs & ignition cables (3.3L & 3.8L) - replace every 100,000 miles.

R – Replace   S/I – Service or Inspect

FREQUENT OPERATION MAINTENANCE (SEVERE SERVICE)

If a vehicle is operated under any of the following conditions it is considered severe service:
- Extremely dusty areas.
- 50% or more of the vehicle operation is in 32°C (90°F) or higher temperatures, or constant operation in temperatures below 0°C (32°F).
- Prolonged idling (vehicle operation in stop and go traffic).
- Frequent short running periods (engine does not warm to normal operating temperatures).
- Police, taxi, delivery usage or trailer towing usage.

Oil & oil filter change – change every 3000 miles.
Automatic transaxle fluid & filter - change every 15,000 miles.
Brake hoses & linings - check every 9000 miles.
CV joints & front suspension ball joints - check every 3000 miles.
Tie rod ends & steering linkage - check every 15,000 miles.
Air filter - change every 15,000 miles.

79231C64

SCHEDULED MAINTENANCE
CHRYSLER CORPORATION
CHRYSLER TOWN & COUNTRY, DODGE CARAVAN, PLYMOUTH VOYAGER

**The following should be used as a guide when determining the amount of work required for a particular service if taken to a repair shop.
In estimating how long a particular Scheduled Maintenance Service should take, please observe the following:**

- **Chilton Time** is time based on field research and data supplied by the vehicle manufacturer.
- Labor time operations are given in hours and tenths of an hour.
- All labor operations, are to be used as a guide.

Mechanic Skill Level Codes:

        (G) GENERAL: Normally skilled with certification.
        (M) MAINTENANCE: Semi-skilled working on certicication.
        (P) PRECISION: Really skilled with multiple certification.

	Chilton Time
(M) 7500 Mile Service	
1995-99	1.1
(M) 15000 Mile Service	
1995-99	1.2
(M) 22500 Mile Service	
1995-99	1.3
(G) 30000 Mile Service	
1995-99	2.3
Renew spark plugs (2.4L, 3.0L) add	1.0
(M) 37500 Mile Service	
1995-99	1.1

	Chilton Time
(G) 45000 Mile Service	
1995-99	1.4
(G) 52500 Mile Service	
1995-99	1.6
Renew dist. cap & rotor add	.3
(G) 60000 Mile Service	
1995-99	
2.5L	2.6
3.0L	8.4
3.3L, 3.8L	2.7
(M) 67500 Mile Service	
1995-99	1.3

	Chilton Time
(M) 75000 Mile Service	
1995-99	
2.5L	1.2
3.0L, 3.3L, 3.8L	1.3
(M) 82500 Mile Service	
1995-99	1.6
(G) 90000 Mile Service	
1995-99	
2.5L	5.9
3.0L	7.4
3.3L, 3.8L	2.7
(M) 97500 Mile Service	
1995-99	1.1

7924AC65

CHRYSLER CORP.
Dakota • Durango • RAM Trucks • RAM Vans

VEHICLE IDENTIFICATION

Engine Code						Model Year	
Code	Liters	Cu. In. (cc)	Cyl.	Fuel Sys.	Eng. Mfg.	Code	Year
5	5.9	360 (5899)	V8	SMFI	Chrysler	S	1995
6	5.9	359 (5882)	I6	DSL-24V	Cummins	T	1996
C	5.9	359 (5882)	I6	DSL-12V	Cummins	V	1997
D	5.9	359 (5882)	I6	DSL-12V	Cummins	W	1998
G	2.5	153 (2507)	I4	TFI	Chrysler	X	1999
P	2.5	153 (2507)	I4	SMFI	Chrysler		
W	8.0	488 (7994)	V10	SMFI	Chrysler		
X	3.9	238 (3916)	V6	SMFI	Chrysler		
Y	5.2	318 (5208)	V8	SMFI	Chrysler		
Z	5.9	360 (5899)	V8	SMFI	Chrysler		

DSL-12V - Diesel with 12 valve cylinder head
DSL-24V - Diesel with 24 valve cylinder head
SMFI - Sequential Multi-port Fuel Injection
TFI - Throttle body Fuel Injection

79241CA0

ENGINE IDENTIFICATION

Year	Model	Engine Displacement Liters (cc)	Engine Series (ID/VIN)	Fuel System	No. of Cylinders	Engine Type
1995	B150 Van	3.9 (3916)	X	SMFI	6	OHV
		5.2 (5211)	Y	SMFI	8	OHV
	B250 Van	3.9 (3916)	X	SMFI	6	OHV
		5.2 (5211)	Y	SMFI	8	OHV
		5.9 (5899)	Z	SMFI	8	OHV
	B350 Van	5.2 (5211)	Y	SMFI	8	OHV
		5.9 (5899)	Z	SMFI	8	OHV
	C 1500 Pick-up	3.9 (3916)	X	SMFI	6	OHV
	C/F 1500 Pick-up	5.2 (5211)	Y	SMFI	8	OHV
		5.9 (5899)	Z	SMFI	8	OHV
	C/F 2500 Pick-up	5.2 (5211)	Y	SMFI	8	OHV
		5.9 (5882)	C	DSL	6	OHV
		5.9 (5899)	Z	SMFI	8	OHV
		8.0 (7997)	W	SMFI	10	OHV
	C/F 3500 Pick-up	5.9 (5882)	C	DSL	6	OHV
		5.9 (5899)	5	SMFI	8	OHV
		8.0 (7997)	W	SMFI	10	OHV
	Dakota	2.5 (2507)	G	TFI	4	SOHC
		3.9 (3916)	X	SMFI	6	OHV
		5.2 (5211)	Y	SMFI	8	OHV
1996	B1500 Van	3.9 (3916)	X	SMFI	6	OHV
		5.2 (5211)	Y	SMFI	8	OHV
	B2500 Van	3.9 (3916)	X	SMFI	6	OHV
		5.2 (5211)	Y	SMFI	8	OHV
		5.9 (5899)	Z	SMFI	8	OHV
	B3500 Van	5.2 (5211)	Y	SMFI	8	OHV
		5.9 (5899)	Z	SMFI	8	OHV
	C/F 1500 Pick-up	3.9 (3916)	X	SMFI	6	OHV
		5.2 (5211)	Y	SMFI	8	OHV
		5.9 (5899)	Z	SMFI	8	OHV
	C/F 2500 Pick-up	5.2 (5211)	Y	SMFI	8	OHV
		5.9 (5882)	C	DSL	6	OHV
		5.9 (5899)	Z	SMFI	8	OHV
		8.0 (7997)	W	SMFI	10	OHV
	C/F 3500 Pick-up	5.9 (5882)	C	DSL	6	OHV
		5.9 (5899)	5	SMFI	8	OHV
		8.0 (7997)	W	SMFI	10	OHV
	Dakota	2.5 (2458)	P	SMFI	4	OHV
		3.9 (3916)	X	SMFI	6	OHV
		5.2 (5211)	Y	SMFI	8	OHV
1997	Dakota	2.5 (2458)	P	SMFI	4	OHV
		3.9 (3916)	X	SMFI	6	OHV
		5.2 (5211)	Y	SMFI	8	OHV
	Ram Truck 1500	3.9 (3916)	X	SMFI	6	OHV
		5.2 (5211)	Y	SMFI	8	OHV
		5.9 (5899)	Z	SMFI	8	OHV
	Ram Truck 2500	5.2 (5211)	Y	SMFI	8	OHV
		5.9 (5882)	C	DSL	6	OHV
		5.9 (5899)	Z	SMFI	8	OHV
		8.0 (7997)	W	SMFI	10	OHV

79241CB1

ENGINE IDENTIFICATION

Year	Model	Engine Displacement Liters (cc)	Engine Series (ID/VIN)	Fuel System	No. of Cylinders	Engine Type
1997 cont.	Ram Truck 3500	5.9 (5882)	C	DSL	6	OHV
		5.9 (5899)	5	SMFI	8	OHV
		8.0 (7997)	W	SMFI	10	OHV
	Ram Van 1500	3.9 (3916)	X	SMFI	6	OHV
		5.2 (5211)	Y	SMFI	8	OHV
	Ram Van 2500	3.9 (3916)	X	SMFI	6	OHV
		5.2 (5211)	Y	SMFI	8	OHV
		5.9 (5899)	Z	SMFI	8	OHV
	Ram Van 3500	5.2 (5211)	Y	SMFI	8	OHV
		5.9 (5899)	Z	SMFI	8	OHV
1998-99	Dakota	2.5 (2464)	P	SMFI	4	OHV
		3.9 (3906)	X	SMFI	6	OHV
		5.2 (5208)	Y	SMFI	8	OHV
		5.9 (5825)	Z	SMFI	8	OHV
	Durango	3.9 (3906)	X	SMFI	6	OHV
		5.2 (5208)	Y	SMFI	8	OHV
		5.9 (5825)	Z	SMFI	8	OHV
	Ram Truck 1500	3.9 (3906)	X	SMFI	6	OHV
		5.2 (5208)	Y	SMFI	8	OHV
		5.9 (5825)	Z	SMFI	8	OHV
	Ram Truck 2500	5.9 (5825)	D	DSL-12V	6	OHV
		5.9 (5825)	6	DSL-24V	6	OHV
		5.9 (5825)	Z	SMFI	8	OHV
		8.0 (7994)	W	SMFI	10	OHV
	Ram Truck 3500	5.9 (5825)	D	DSL-12V	6	OHV
		5.9 (5825)	6	DSL-24V	6	OHV
		5.9 (5825)	5	SMFI	8	OHV
		8.0 (7994)	W	SMFI	10	OHV
	Ram Van 1500	3.9 (3916)	X	SMFI	6	OHV
		5.2 (5208)	Y	SMFI	8	OHV
	Ram Van 2500	3.9 (3916)	X	SMFI	6	OHV
		5.2 (5208)	Y	SMFI	8	OHV
		5.9 (5825)	Z	SMFI	8	OHV
	Ram Van 3500	5.2 (5208)	Y	SMFI	8	OHV
		5.9 (5825)	Z	SMFI	8	OHV

DOHC - Double Overhead Camshaft
DSL-12V - Diesel engine with 12 valve cylinder head
DSL-24V - Diesel engine with 24 valve cylinder head
OHV - Overhead Valve
SMFI - Sequential Multi-port Fuel Injection
SOHC - Single Overhead Camshaft
TFI - Throttle body Fuel Injection

79241CB2

GENERAL ENGINE SPECIFICATIONS

Year	Engine ID/VIN	Engine Displacement Liters (cc)	Fuel System Type	Net Horsepower @ rpm	Net Torque @ rpm (ft. lbs.)	Bore x Stroke (in.)	Compression Ratio	Oil Pressure @ rpm
1995	5	5.9 (5899)	SMFI	230@4000	330@2800	4.00x3.58	8.9:1	30-80@3000
	C	5.9 (5882)	DSL	160@2500 ①	400@1600 ②	4.02x4.72	17.5:1	30@2500
	G	2.5 (2507)	TFI	100@4800	135@2800	3.45x4.09	8.9:1	25-80@3000
	W	8.0 (7997)	SMFI	300@4000	450@2400	4.00x3.88	8.6:1	50-60@3000
	X	3.9 (3916)	SMFI	175@4800	220@3200	3.91x3.31	9.1:1	30-80@3000
	Y	5.2 (5211)	SMFI	220@4400	300@3200	3.91x3.31	9.1:1	30-80@3000
	Z	5.9 (5899)	SMFI	230@4000	330@3200	4.00x3.58	8.9:1	30-80@3000
1996	5	5.9 (5899)	SMFI	230@4000	330@2800	4.00x3.58	8.9:1	30-80@3000
	C	5.9 (5882)	DSL	180@2500 ⑤	420@1600 ⑥	4.02x4.72	17.5:1	30@2500
	P	2.5 (2458)	SMFI	120@5200	145@3400	3.88x3.19	9.2:1	25-80@3000
	W	8.0 (7997)	SMFI	300@4000	450@2400	4.00x3.58	8.4:1	50-60@3000
	X	3.9 (3916)	SMFI	175@4800	220@3200	3.91x3.31	9.1:1	30-80@3000
	Y	5.2 (5211)	SMFI	220@4400 ③	300@3200 ④	3.91x3.31	9.1:1	30-80@3000
	Z	5.9 (5899)	SMFI	230@4000	330@3250	4.00x3.58	9.1:1	30-80@3000
1997	5	5.9 (5899)	SMFI	230@4000	330@2800	4.00x3.58	8.9:1	30-80@3000
	C	5.9 (5882)	DSL	180@2500 ⑤	420@1600 ⑥	4.02x4.72	17.5:1	30@2500
	P	2.5 (2458)	SMFI	120@5200	145@3400	3.88x3.19	9.2:1	25-80@3000
	W	8.0 (7997)	SMFI	300@4000	450@2400	4.00x3.58	8.4:1	50-60@3000
	X	3.9 (3916)	SMFI	175@4800	220@3200	3.91x3.31	9.1:1	30-80@3000
	Y	5.2 (5211)	SMFI	220@4400 ③	300@3200 ④	3.91x3.31	9.1:1	30-80@3000
	Z	5.9 (5899)	SMFI	230@4000	330@3250	4.00x3.58	9.1:1	30-80@3000
1998-99	5	5.9 (5825)	SMFI	250@4400	345@3200	4.00x3.58	9.1:1	30-80@3000
	6	5.9 (5825)	DSL-24V	NA	NA	4.02x4.72	NA	30@2500
	D	5.9 (5825)	DSL-12V	⑦	⑧	4.02x4.72	17.5:1	30-80@3000
	P	2.5 (2464)	SMFI	120@5200	145@3250	3.88x3.19	9.2:1	37-75 above 1600
	W	8.0 (7994)	SMFI	300@4000	⑨	4.00x3.88	8.6:1	30-80@3000
	X	3.9 (3906)	SMFI	175@4800	225@3200	3.91x3.31	9.1:1	30-80@3000
	Y	5.2 (5208)	SMFI	230@4400	300@3200	3.91x3.31	9.1:1	30-80@3000
	Z	5.9 (5825)	SMFI	250@4400	345@3200	4.00x3.58	9.1:1	30-80@3000

MFI - Multi-port Fuel Injection
TFI - Throttle body fuel injection
DSL-12V - Diesel with 12 valve cylinder head
DSL-24V - Diesel with 24 valve cylinder head
① Manual transmission: 430@1600
② Manual transmission: 175@2500
③ Compressed natural gas engine: 200@4400
④ Compressed natural gas engine: 250@3600

⑤ Manual transmission: 200@2500
⑥ Manual transmission: 440@1600
⑦ Federal models w/manual transmission: 215@2600
Except Federal models w/manual transmissions: 180@2500
⑧ Federal models w/manual transmission: 440@1600
Except Federal models w/manual transmissions: 420@1500
⑨ Federal models: 450@2800
California models: 440@2800

79241CB3

GASOLINE ENGINE TUNE-UP SPECIFICATIONS

Year	Engine ID/VIN	Engine Displacement Liters (cc)	Spark Plugs Gap (in.)	Ignition Timing (deg.)		Fuel Pump (psi)	Idle Speed (rpm)		Valve Clearance	
				MT	AT		MT	AT	In.	Ex.
1995	G	2.5 (2507)	0.035	①	①	14.5	②	②	HYD	HYD
	X	3.9 (3916)	0.030	①	①	35-45	②	②	HYD	HYD
	Y	5.2 (5211)	0.030	①	①	35-45	②	②	HYD	HYD
	5	5.9 (5899)	0.030	①	①	35-45	②	②	HYD	HYD
	Z	5.9 (5899)	0.030	①	①	35-45	②	②	HYD	HYD
	W	8.0 (7997)	0.030	①	①	35-45	②	②	HYD	HYD
1996	P	2.5 (2458)	0.035	—	②	49.2	②	②	HYD	HYD
	X	3.9 (3916)	0.035	①	①	49.2	②	②	HYD	HYD
	Y	5.2 (5211)	0.035	①	①	49.2	②	②	HYD	HYD
	5	5.9 (5899)	0.035	①	①	49.2	②	②	HYD	HYD
	Z	5.9 (5899)	0.035	①	①	49.2	②	②	HYD	HYD
	W	8.0 (7997)	0.045	①	①	49.2	②	②	HYD	HYD
1997	P	2.5 (2458)	0.035	—	②	49.2	②	②	HYD	HYD
	X	3.9 (3916)	0.035	①	①	49.2	②	②	HYD	HYD
	Y	5.2 (5211)	0.035	①	①	49.2	②	②	HYD	HYD
	5	5.9 (5899)	0.035	①	①	49.2	②	②	HYD	HYD
	Z	5.9 (5899)	0.035	①	①	49.2	②	②	HYD	HYD
	W	8.0 (7997)	0.045	①	①	49.2	②	②	HYD	HYD
1998-99	5	5.9 (5825)	0.040	①	①	44.2-54.2	②	②	HYD	HYD
	P	2.5 (2464)	0.035	①	①	44.2-54.2	②	②	HYD	HYD
	W	8.0 (7994)	0.045	①	①	44.2-54.2	②	②	HYD	HYD
	X	3.9 (3916)	0.040	①	①	44.2-54.2	②	②	HYD	HYD
	Y	5.2 (5208)	0.040	①	①	44.2-54.2	②	②	HYD	HYD
	Z	5.9 (5899)	0.040	①	①	44.2-54.2	②	②	HYD	HYD

NOTE: The Vehicle Emission Control Information (VECI) label often reflects specification changes made during production. The label figures must be used if they differ from those in this chart.

HYD - Hydraulic

① Ignition timing cannot be adjusted. Base engine timing is set during assembly, and controlled by the PCM.

② Refer to the VECI label for correct specification.

79241CB4

DIESEL ENGINE TUNE-UP SPECIFICATIONS

Year	Engine ID/VIN	Engine Displacement cu. in. (cc)	Valve Clearance		Intake Valve Opens (deg.)	Injection Pump Setting (deg.)	Injection Nozzle Pressure (psi)		Idle Speed (rpm)	Cranking Compression Pressure (psi)
			Intake (in.)	Exhaust (in.)			New	Used		
1995	C	5.9 (5882)	0.010	0.020	NA	①	3822	NA	②	NA
1996	C	5.9 (5882)	0.010	0.020	NA	①	3822	NA	②	NA
1997	C	5.9 (5882)	0.010	0.020	NA	①	3822	NA	②	NA
1998-99	6	5.9 (5825)	0.006-0.015	0.0015-0.0300	NA	③	4250-4750	NA	④	NA
	D	5.9 (5825)	0.010	0.020	NA	⑤	3394-3887	NA	⑥	NA

NOTE: The Vehicle Emission Control Information label often reflects specification changes made during production. The label figures must be used if they differ from those in this chart

NA - Not Available

① Align marks on pump flange and gear housing

② Automatic transmission with A/C: 750-800 rpm
   Manual transmission with A/C: 780 rpm

③ Align the marks on the crankshaft, camshaft and pump sprockets.

④ The idle speed is computer-controlled and cannot be adjusted.

⑤ Federal models with manual transmissions: 13.5 degrees BTDC
   Except Federal models with manual transmissions: 14.0 degrees BTDC

⑥ Automatic transmission: 750-800 rpm with trans. in drive and A/C on.
   Manual transmission: 780 rpm with trans. in Neutral and A/C on.

79241CB5

CAPACITIES

Year	Model	Engine ID/VIN	Engine Displacement Liters (cc)	Engine Oil with Filter (qts.)	Transmission (pts.)			Transfer Case (pts.)	Drive Axle		Fuel Tank (gal.)	Cooling System (qts.)
					4-Spd	5-Spd	Auto.		Front (pts.)	Rear (pts.)		
1995	B150 Van	X	3.9 (3916)	4.0	—	—	①	—	—	②	22.0 ③	14.6
		Y	5.2 (5211)	5.0	—	—	①	—	—	②	22.0 ③	16.5
	B250 Van	X	3.9 (3916)	4.0	—	—	①	—	—	②	22.0 ③	14.6
		Y	5.2 (5211)	5.0	—	—	①	—	—	②	22.0 ③	16.5
		Z	5.9 (5899)	5.0	—	—	①	—	—	②	35.0	15.0 ④
	B350 Van	Y	5.2 (5211)	5.0	—	—	①	—	—	②	22.0 ③	16.5
		Z	5.9 (5899)	5.0	—	—	①	—	—	②	35.0	15.0 ④
	C 1500 Pick-up	X	3.9 (3916)	4.0	—	⑤	①	—	—	②	26.0 ③	20.0
	C/F 1500 Pick-up	Y	5.2 (5211)	5.0	—	⑤	①	⑥	⑦	②	26.0 ③	20.0
		Z	5.9 (5899)	5.0	—	⑤	①	⑥	⑦	②	26.0 ③	20.0
	C/F 2500 Pick-up	Y	5.2 (5211)	5.0	—	⑤	①	⑥	⑦	②	26.0 ③	20.0
		C	5.9 (5882)	11.0	—	⑤	①	⑥	⑦	②	26.0 ③	26.0
		Z	5.9 (5899)	5.0	—	⑤	①	⑥	⑦	②	26.0 ③	20.0
		W	8.0 (7997)	7.0	—	⑤	①	⑥	⑦	②	26.0 ③	24.0
	C/F 3500 Pick-up	C	5.9 (5882)	11.0	—	⑤	①	⑥	⑦	②	26.0 ③	26.0
		5	5.9 (5899)	5.0	—	⑤	①	⑥	⑦	②	26.0 ③	20.0
		W	8.0 (7997)	7.0	—	⑤	①	⑥	⑦	②	26.0 ③	24.0
	Dakota	G	2.5 (2507)	4.5	—	⑤	①	2.5	3.0	⑧	15.0 ⑨	9.8
		X	3.9 (3916)	4.5	—	⑤	①	2.5	3.0	⑧	15.0 ⑨	14.0
		Y	5.2 (5211)	5.0	—	⑤	①	2.5	3.0	⑧	15.0 ⑨	14.3
1996	B1500 Van	X	3.9 (3916)	4.0	—	—	①	—	—	②	22.0 ③	14.6
		Y	5.2 (5211)	5.0	—	—	①	—	—	②	22.0 ③	16.5
	B2500 Van	X	3.9 (3916)	4.0	—	—	①	—	—	②	22.0 ③	14.6
		Y	5.2 (5211)	5.0	—	—	①	—	—	②	22.0 ③	16.5
		Z	5.9 (5899)	5.0	—	—	①	—	—	②	35.0	15.0 ④
	B3500 Van	Y	5.2 (5211)	5.0	—	—	①	—	—	②	22.0 ③	16.5
		Z	5.9 (5899)	5.0	—	—	①	—	—	②	35.0	15.0 ④
	C/F 1500 Pick-up	X	3.9 (3916)	4.0	—	⑤	①	—	—	⑩	26.0 ③	20.0
		Y	5.2 (5211)	5.0	—	⑤	①	⑪	⑦	⑩	26.0 ③	20.0
		Z	5.9 (5899)	5.0	—	⑤	①	⑪	⑦	⑩	26.0 ③	20.0
	C/F 2500 Pick-up	Y	5.2 (5211)	5.0	—	⑤	①	⑪	⑦	⑩	26.0 ③	20.0
		C	5.9 (5882)	11.0	—	⑤	①	⑪	⑦	⑩	26.0 ③	26.0
		Z	5.9 (5899)	5.0	—	⑤	①	⑪	⑦	⑩	26.0 ③	20.0
		W	8.0 (7997)	7.0	—	⑤	①	⑪	⑦	⑩	26.0 ③	24.0
	C/F 3500 Pick-up	C	5.9 (5882)	11.0	—	⑤	①	⑪	⑦	⑩	26.0 ③	26.0
		5	5.9 (5899)	5.0	—	⑤	①	⑪	⑦	⑩	26.0 ③	20.0
		W	8.0 (7997)	7.0	—	⑤	①	⑪	⑦	⑩	26.0 ③	24.0
	Dakota	P	2.5 (2458)	4.5	—	⑤	①	2.5	3.0	⑧	⑫	9.8
		X	3.9 (3916)	4.5	—	⑤	①	2.5	3.0	⑧	⑫	14.0
		Y	5.2 (5211)	5.0	—	⑤	①	2.5	3.0	⑧	⑫	14.3
1997	Dakota	P	2.5 (2458)	4.5	—	⑤	①	2.5	3.0	⑧	⑫	9.8
		X	3.9 (3916)	4.5	—	⑤	①	2.5	3.0	⑧	⑫	14.0
		Y	5.2 (5211)	5.0	—	⑤	①	2.5	3.0	⑧	⑫	14.3
	Ram Truck 1500	X	3.9 (3916)	4.0	—	⑤	①	—	—	⑩	26.0 ③	20.0
		Y	5.2 (5211)	5.0	—	⑤	①	⑪	⑦	⑩	26.0 ③	20.0
		Z	5.9 (5899)	5.0	—	⑤	①	⑪	⑦	⑩	26.0 ③	20.0
	Ram Truck 2500	Y	5.2 (5211)	5.0	—	⑤	①	⑪	⑦	⑩	26.0 ③	20.0
		C	5.9 (5882)	11.0	—	⑤	①	⑪	⑦	⑩	26.0 ③	26.0
		Z	5.9 (5899)	5.0	—	⑤	①	⑪	⑦	⑩	26.0 ③	20.0
		W	8.0 (7997)	7.0	—	⑤	①	⑪	⑦	⑩	26.0 ③	24.0
	Ram Truck 3500	C	5.9 (5882)	11.0	—	⑤	①	⑪	⑦	⑩	26.0 ③	26.0
		5	5.9 (5899)	5.0	—	⑤	①	⑪	⑦	⑩	26.0 ③	20.0

79241CB6

CAPACITIES

Year	Model	Engine ID/VIN	Engine Displacement Liters (cc)	Engine Oil with Filter (qts.)	Transmission (pts.) 4-Spd	5-Spd	Auto.	Transfer Case (pts.)	Drive Axle Front (pts.)	Rear (pts.)	Fuel Tank (gal.)	Cooling System (qts.)
1997 (cont.)	Ram Truck 3500	W	8.0 (7997)	7.0	—	⑤	①	⑪	⑦	⑩	26.0 ③	24.0
	Ram Van 1500	X	3.9 (3916)	4.0	—	—	①	—	—	②	22.0 ③	14.6
		Y	5.2 (5211)	5.0	—	—	①	—	—	②	22.0 ③	16.5
	Ram Van 2500	X	3.9 (3916)	4.0	—	—	①	—	—	②	22.0 ③	14.6
		Y	5.2 (5211)	5.0	—	—	①	—	—	②	22.0 ③	16.5
		Z	5.9 (5899)	5.0	—	—	①	—	—	②	35.0	15.0 ④
	Ram Van 3500	Y	5.2 (5211)	5.0	—	—	①	—	—	②	22.0	16.5
		Z	5.9 (5899)	5.0	—	—	①	—	—	②	35.0	15.0 ④
1998-99	Dakota	P	2.5 (2458)	4.5	—	⑬	—	—	—	⑭	⑫	9.8
		X	3.9 (3916)	4.0	—	⑬	⑮	2.5	3.0	⑭	⑫	14.0
		Y	5.2 (5211)	5.0	—	⑬	⑮	2.5	3.0	⑭	⑫	14.3
		Z	5.9 (5899)	5.0	—	—	⑮	2.5	3.0	⑭	⑫	14.3
	Durango	X	3.9 (3916)	4.0	—	—	⑮	⑯	3.0	⑭	25.0	14.0
		Y	5.2 (5211)	5.0	—	—	⑮	⑯	3.0	⑭	25.0	14.3
		Z	5.9 (5899)	5.0	—	—	⑮	⑯	3.0	⑭	25.0	14.3
	Ram Truck 1500	X	3.9 (3916)	4.0	—	4.2 ⑰	3.0 ⑱	2.5	⑲	⑳	21	20.0
		Y	5.2 (5211)	5.0	—	4.2 ⑰	3.0 ⑱	2.5	⑲	⑳	21	20.0
		Z	5.9 (5899)	5.0	—	—	3.0 ⑱	2.5	⑲	⑳	21	20.0
	Ram Truck 2500	D	5.9 (5882)	11.0	—	8.0 22	3.0 ⑱	23	⑲	⑳	21	26.0
		6	5.9 (5882)	11.0	—	8.0 22	3.0 ⑱	23	⑲	⑳	21	26.0
		Z	5.9 (5899)	5.0	—	8.0 24	3.0 ⑱	23	⑲	⑳	21	20.0
		W	8.0 (7997)	7.0	—	8.0 22	3.0 ⑱	23	⑲	⑳	21	24.0
	Ram Truck 3500	D	5.9 (5882)	11.0	—	8.0 22	3.0 ⑱	23	⑲	⑳	21	26.0
		6	5.9 (5882)	11.0	—	8.0 22	3.0 ⑱	23	⑲	⑳	21	26.0
		5	5.9 (5899)	5.0	—	8.0 24	3.0 ⑱	23	⑲	⑳	21	20.0
		W	8.0 (7997)	7.0	—	8.0 22	3.0 ⑱	23	⑲	⑳	21	24.0
	Ram Van	X	3.9 (3916)	4.0	—	4.2 ⑰	3.0 ⑱	2.5	⑲	⑳	21	20.0
		Y	5.2 (5211)	5.0	—	4.2 ⑰	3.0 ⑱	2.5	⑲	⑳	21	20.0
		Z	5.9 (5899)	5.0	—	—	3.0 ⑱	2.5	⑲	⑳	21	20.0

NOTE: All capacities are approximate. Add fluid gradually and check to be sure a proper fluid level is obtained.

① 32RH: 17.0 pts.
  36RH: 16.6 pts.
  42RH: 20.2 pts.
  32RH: 17.0 pts.
  36RH: 16.6 pts.
  42RH: 20.2 pts.
② Chrysler 8.25 in.: 4.4 pts.
  Chrysler 9.25 in.: 4.8 pts.
  Dana 60: 6.3 pts.
③ Optional fuel tank: 35 gals.
④ With rear heater: 16.0 qts.
⑤ NV3500: 4.2 pts.
  NV4500: 8.0 pts.
  AX15: 6.6 pts.
  Getrag: 7.0 pts.
⑥ NP231HD: 2.5 pts.
  NP241: 4.7 pts.
  NP241HD: 13 pts.
⑦ 7.25 in.: 3 pts.
  Dana 44: 5.6 pts.
  Dana 60: 6.5 pts.
⑧ 7.25 in.: 2.9 pts.
  8.25 in.: 4.4 pts.

⑨ Optional fuel tank: 22 gals.
⑩ Chrysler 7.25 in.: 3 pts.
  Chrysler 8.25 in. and 9.25 in.: 4.8 pts.
  Spicer and Dana 60: 6.0 pts.
  Dana 70 and 80: 7.0 pts.
⑪ NP231HD: 2.5 pts.
  NP241: 4.7 pts.
  NP241HD: 6.5 pts.
⑫ Standard fuel tank: 15 gal.
  Optional fuel tank: 22 gal.
⑬ AX15 Transmission: 6.6 pts.
  NV3500 Transmission: 4.2 pts.
⑭ The following values include 0.25 pt. of friction modifier for LSD axles.
  8.25 axle: 4.4 pts.
  9.25 axle: 4.9 pts.
⑮ Fluid drain/filter service: 8.0 pts.
  Overhaul dry fill: 20-23 pts.
⑯ NV231: 2.5 pts.
  NV231-HD: 2.5 pts.
  NV242: 3.0 pts.
⑰ NV 3500 transmission

⑱ For fluid drain and filter replacement only.
  For complete overhaul, or dry fill-
  42 RE: 17-22 pts.
  46 RE: 19-23 pts.
  47 RE: 29-33 pts.
⑲ 216-FBI front axle: 4.8
  248-FBI front axle: 7.6
⑳ - 9.25 in. axle: 4.9; includes 0.25 pts. of friction modifier for LSD axles.
  - 248-RBI axle: 6.3; includes 0.25 pts. of friction modifier for LSD axles.
  - 267-RBI axle: 7.0; includes 0.25 pts. of friction modifier for LSD axles.
  - 286-RBI (2WD): 6.8; includes 0.25 pts. of friction modifier for LSD axles.
  - 286-RBI (4WD): 10.1; includes 0.4 pts. of friction modifier for LSD axles.
21 119 in. wheel base models: 26 gal.
  135 in. wheel base models: 26 gal.
  All other models: 35 gal.
22 NV 4500 HD transmission
23 NV241: 5.0
  NV241 HD: 6.5
  NV241 HD w/PTO: 9.0
24 NV 4500 transmission

79241CB7

VALVE SPECIFICATIONS

Year	Engine ID/VIN	Engine Displacement Liters (cc)	Seat Angle (deg.)	Face Angle (deg.)	Spring Test Pressure (lbs. @ in.)	Spring Installed Height (in.)	Stem-to-Guide Clearance (in.)		Stem Diameter (in.)	
							Intake	Exhaust	Intake	Exhaust
1995	G	2.5 (2507)	45.0	45.0	115@1.65	1.65	0.001-0.003	0.003-0.005	0.3124	0.3103
	X	3.9 (3916)	44.25-44.75	43.25-43.75	85@1.64	1.64	0.001-0.003	0.001-0.003	0.3110-0.3120	0.3110-0.3120
	Y	5.2 (5211)	44.25-44.75	43.25-43.75	85@1.64	1.64	0.001-0.003	0.001-0.003	0.3110-0.3120	0.3110-0.3120
	C	5.9 (5882)	①	①	81@1.94	2.36	0.002-0.006	0.002-0.006	0.3130-0.3140	0.3130-0.3140
	5	5.9 (5899)	44.25-44.75	43.25-43.75	85@1.64	1.64	0.001-0.003	0.002-0.004	0.3720-0.3730	0.3710-0.3720
	Z	5.9 (5899)	44.25-44.75	43.25-43.75	85@1.64	1.64	0.001-0.003	0.002-0.004	0.3720-0.3730	0.3710-0.3720
	W	8.0 (7997)	44.5	45.0	81-89@1.64	1.64	0.001-0.003	0.001-0.003	0.3110-0.3120	0.3110-0.3120
1996	P	2.5 (2507)	45.0	45.0	184-196@1.22	1.65	0.001-0.003	0.001-0.003	0.311-0.312	0.311-0.312
	X	3.9 (3916)	44.25-44.75	43.25-43.75	200@1.21	1.64	0.001-0.003	0.001-0.003	0.311-0.312	0.311-0.312
	Y	5.2 (5211)	44.25-44.75	43.25-43.75	200@1.21	1.64	0.001-0.003	0.001-0.003	0.311-0.312	0.311-0.312
	C	5.9 (5882)	①	①	81@1.94	2.36	0.002-0.006	0.002-0.006	0.313-0.314	0.313-0.314
	5	5.9 (5899)	44.25-44.75	43.25-43.75	200@1.21	1.64	0.001-0.003	0.002-0.004	0.372-0.373	0.371-0.372
	Z	5.9 (5899)	44.25-44.75	43.25-43.75	200@1.21	1.64	0.001-0.003	0.002-0.004	0.372-0.373	0.371-0.372
	W	8.0 (7997)	44.5	45.0	190-210@1.22	1.64	0.001-0.003	0.001-0.003	0.311-0.312	0.311-0.312
1997	P	2.5 (2507)	45.0	45.0	184-196@1.22	1.65	0.001-0.003	0.001-0.003	0.311-0.312	0.311-0.312
	X	3.9 (3916)	44.25-44.75	43.25-43.75	200@1.21	1.64	0.001-0.003	0.001-0.003	0.311-0.312	0.311-0.312
	Y	5.2 (5211)	44.25-44.75	43.25-43.75	200@1.21	1.64	0.001-0.003	0.001-0.003	0.311-0.312	0.311-0.312
	C	5.9 (5882)	①	①	81@1.94	2.36	0.002-0.006	0.002-0.006	0.313-0.314	0.313-0.314
	5	5.9 (5899)	44.25-44.75	43.25-43.75	200@1.21	1.64	0.001-0.003	0.002-0.004	0.372-0.373	0.371-0.372
	Z	5.9 (5899)	44.25-44.75	43.25-43.75	200@1.21	1.64	0.001-0.003	0.002-0.004	0.372-0.373	0.371-0.372
	W	8.0 (7997)	44.5	45.0	190-210@1.22	1.64	0.001-0.003	0.001-0.003	0.311-0.312	0.311-0.312
1998-99	P	2.5 (2458)	44.5	45.0	184-196@1.216	1.64	0.001-0.003	0.001-0.003	0.311-0.312	0.311-0.312
	X	3.9 (3916)	44.25-44.75	43.25-43.75	200@1.21	1.64	0.001-0.003	0.001-0.003	0.311-0.312	0.311-0.312
	Y	5.2 (5208)	44.25-44.75	43.25-43.75	200@1.21	1.64	0.001-0.003	0.001-0.003	0.311-0.312	0.311-0.312

79241CB8

VALVE SPECIFICATIONS

Year	Engine ID/VIN	Engine Displacement Liters (cc)	Seat Angle (deg.)	Face Angle (deg.)	Spring Test Pressure (lbs. @ in.)	Spring Installed Height (in.)	Stem-to-Guide Clearance (in.)		Stem Diameter (in.)	
							Intake	Exhaust	Intake	Exhaust
1998-99 cont.	Z	5.9 (5899)	44.25-44.75	43.25-43.75	200@1.21	1.64	0.001-0.003	0.002-0.004	0.372-0.373	0.371-0.372
	D	5.9 (5825)	①	①	81@1.94	1.94	0.0031-0.0051	0.0031-0.0051	0.3126-0.3134	0.3126-0.3134
	6	5.9 (5825)	①	①	76.4@1.39	1.39	0.002	0.002	0.2752-0.2760	0.2752-0.2760
	W	8.0 (7994)	44.5	45.0	200@1.212	1.64	0.001-0.003	0.001-0.003	0.311-0.312	0.311-0.312
	5	5.9 (5899)	44.25-44.75	43.25-43.75	200@1.21	1.64	0.001-0.003	0.002-0.004	0.372-0.373	0.371-0.372

① Intake: 30 degrees
   Exhaust: 45 degrees

79241CB9

TORQUE SPECIFICATIONS
All readings in ft. lbs.

Year	Engine ID/VIN	Engine Displacement Liters (cc)	Cylinder Head Bolts	Main Bearing Bolts	Rod Bearing Bolts	Crankshaft Damper Bolts	Flywheel Bolts	Manifold Intake ①	Exhaust	Spark Plugs	Lug Nut
1995	5	5.9 (5899)	②	85	45	135	55	③	25	30	④
	C	5.9 (5882)	⑤	⑥	⑦	92	101	⑧	32	—	④
	G	2.5 (2507)	⑨	⑩	⑪	85	70	17	17	20	95
	W	8.0 (7997)	②	85	45	135	55	⑫	16	30	④
	X	3.9 (3916)	②	85	45	135	55	③	25	30	④
	Y	5.2 (5211)	②	85	45	135	55	③	25	30	④
	Z	5.9 (5899)	②	85	45	135	55	③	25	30	④
1996	5	5.9 (5899)	②	85	45	135	55	③	25	30	④
	C	5.9 (5882)	⑤	⑥	⑦	92	101	⑧	32	—	④
	P	2.5 (2507)	⑬	80	33	80	105	17	17	27	95
	W	8.0 (7997)	②	85	45	135	55	⑫	16	30	④
	X	3.9 (3916)	②	85	45	135	55	③	25	30	④
	Y	5.2 (5211)	②	85	45	135	55	③	25	30	④
	Z	5.9 (5899)	②	85	45	135	55	③	25	30	④
1997	5	5.9 (5899)	②	85	45	135	55	③	25	30	④
	C	5.9 (5882)	⑤	⑥	⑦	92	101	⑧	32	—	④
	P	2.5 (2507)	⑬	80	33	80	105	17	17	27	95
	W	8.0 (7997)	②	85	45	135	55	⑫	16	30	④
	X	3.9 (3916)	②	85	45	135	55	③	25	30	④
	Y	5.2 (5211)	②	85	45	135	55	③	25	30	④
	Z	5.9 (5899)	②	85	45	135	55	③	25	30	④
1998-99	5	5.9 (5899)	②	85	45	18	㉔	③	25	30	④
	6	5.9 (5825)	⑭	⑮	⑯	92	⑰	18	32	—	④
	D	5.9 (5825)	⑱	⑲	⑯	92	⑰	18	32	—	④
	P	2.5 (2458)	⑳	80	33	80	105	㉑	㉑	27	④
	W	8.0 (7994)	㉒	㉓	45	135	55	40	16	30	④
	X	3.9 (3916)	②	85	45	18	23	⑥	25	30	④
	Y	5.2 (5211)	②	85	45	18	㉔	③	25	30	④
	Z	5.9 (5899)	②	85	45	18	㉔	③	25	30	④

① Applies to the lower intake manifold only.
② Step 1: 50 ft. lbs.
Step 2: 105 ft. lbs.
③ For V6 engines:
Step 1: Bolts 1-2: 72 inch lbs. in sequence and in 12 inch lbs. steps.
Step 2: Bolts 3-12: 72 inch lbs.
Step 3: Check that all bolts are tightened to 72 inch lbs.
Step 4: Tighten all bolts in sequence to 12 ft. lbs.
Step 5: Check that all bolts are tightened to 12 ft. lbs.
For V8 Engines:
Step 1: Bolts 1-4: 72 inch lbs. in sequence and in 12 inch lbs. steps.
Step 2: Bolts 5-12: 72 inch lbs.
Step 3: Check that all bolts are tightened to 72 inch lbs.
Step 4: Tighten all bolts in sequence to 12 ft. lbs.
Step 5: Check that all bolts are tightened to 12 ft. lbs.
④ 5 stud wheel: 95 ft. lbs.
8 stud wheel: 135 ft. lbs.
8 stud dual wheel: 145 ft. lbs.
⑤ All bolts: 66 ft. lbs.
Long bolts: 89 ft. lbs.
All bolts and additional 1/4 turn

⑥ Step 1: 45 ft. lbs.
Step 2: 88 ft. lbs.
Step 3: 129 ft. lbs.
⑦ Step 1: 26 ft. lbs.
Step 2: 51 ft. lbs.
⑧ Intake manifold cover bolts: 18 ft. lbs.
⑨ Step 1: 45 ft. lbs.
Step 2: 65 ft. lbs.
Step 3: 65 ft. lbs.
Step 4: Plus 1/4 turn
⑩ Step 1: 30 ft. lbs.
Step 2: Plus 1/4 turn
⑪ Step 1: 40 ft. lbs.
Step 2: Plus 1/4 turn
⑫ Lower intake manifold: 40 ft. lbs.
Upper intake manifold: 16 ft. lbs.
⑬ Step 1: 25 ft. lbs.
Step 2: 50 ft. lbs.
Step 3: 50 ft. lbs.
Step 4: Plus 1/4 turn
⑭ Step 1: 66 ft. lbs.
Step 2: recheck at 66 ft. lbs.
Step 3: additional 1/4 turn (90 degrees)
⑮ Step 1: 44 ft. lbs.
Step 2: 88 ft. lbs.
Step 3: 129 ft. lbs.

⑯ Step 1: 26 ft. lbs.
Step 2: 51 ft. lbs.
Step 3: 73 ft. lbs.
⑰ Manual transmission: 101 ft. lbs.
Automatci transmission: 32 ft. lbs.
⑱ Step 1: 66 ft. lbs.
Step 2: recheck at 66 ft. lbs.
Step 3: 90 ft. lbs. (long bolts only)
Step 4: recheck at 90 ft. lbs. (long bolts only)
Step 5: all bolts an additional 1/4 turn (90 degrees)
⑲ Step 1: 45 ft. lbs.
Step 2: 60 ft. lbs.
Step 3: additional 1/4 turn (90 degrees)
⑳ Bolts 1-10 and 12-14: 110 ft. lbs.
Bolt 11: 100 ft. lbs.
㉑ Exhaust manifold bolt 1: 30 ft. lbs.
Intake/exhaust manifold bolts 2-5: 23 ft. lbs.
Exhaust manifold nuts 6 & 7: 23 ft. lbs.
㉒ Step 1: 43 ft. lbs.
Step 2: 105 ft. lbs.
㉓ Step 1: 20 ft. lbs.
Step 2: 85 ft. lbs.
㉔ Flywheel bolts: 55 ft. lbs.
Torque converter driveplate: 23 ft. lbs

79241CB0

SCHEDULED MAINTENANCE INTERVALS
(DODGE RAM VAN, DAKOTA, DURANGO & RAM TRUCK—LIGHT DUTY)

TO BE SERVICED	TYPE OF SERVICE	VEHICLE MILEAGE INTERVAL (x1000)												
		7.5	15	22.5	30	37.5	45	52.5	60	67.5	75	82.5	90	97.5
Engine oil & filter	R	✓	✓	✓	✓	✓	✓	✓	✓	✓	✓	✓	✓	✓
Exhaust system	S/I	✓	✓	✓	✓	✓	✓	✓	✓	✓	✓	✓	✓	✓
Engine coolant level, hoses & clamps	S/I	✓	✓	✓	✓	✓	✓	✓	✓	✓	✓	✓	✓	✓
Rotate tires	S/I	✓		✓		✓		✓		✓		✓		✓
Drive belts	S/I		✓		✓		✓		✓		✓		✓	
Brake booster bellcrank pivot	S/I		✓		✓		✓		✓		✓		✓	
Steering linkage	S/I		✓		✓		✓		✓		✓		✓	
Brake hoses & linings	S/I			✓			✓			✓			✓	
Front suspension ball joints	S/I			✓			✓			✓			✓	
Front wheel bearings	S/I			✓			✓			✓			✓	
Steering linkage	S/I			✓			✓			✓			✓	
Spark plugs	R				✓				✓				✓	
Engine air cleaner element	R				✓				✓				✓	
Automatic transmission fluid, filter & adjust bands	S/I					✓					✓			
Manual transmission fluid	R					✓					✓			
Transfer case fluid	R					✓					✓			
Engine Coolant	R						✓				✓			
PCV valve	R								✓				✓	
Battery①	R								✓					
Fuel filter	R								✓					
Ignition cables, distributor cap & rotor	R								✓					
Timing belt (2.5L)	R								✓					

① Replace at 60,000 miles, if not replaced previously.
R – Replace   S/I – Service or Inspect

79241CC3

SCHEDULED MAINTENANCE INTERVALS
(DODGE RAM VAN, DAKOTA, DURANGO & RAM TRUCK—LIGHT DUTY Cont.)

FREQUENT OPERATION MAINTENANCE (SEVERE SERVICE)

If a vehicle is operated under any of the following conditions it is considered severe service:
- **Extremely dusty areas.**
- 50% or more of the vehicle operation is in 32°C (90°F) or higher temperatures, or constant operation in temperatures below 0°C (32°F).
- **Prolonged idling (vehicle operation in stop and go traffic).**
- **Frequent short running periods (engine does not warm to normal operating temperatures).**
- **Police, taxi, delivery usage or trailer towing usage.**

Oil & oil filter change – change every 3000 miles.
Air filter/air pump air filter – change every 24,000 miles.
Engine coolant level, hoses & clamps - check every 6000 miles.
Exhaust system - check every 6000 miles.
Drive belts - check every 18,000 miles; replace every 24,000 miles.
Crankcase inlet air filter (6 & 8 cyl.) - clean every 24,000 miles.
Ignition timing (1995 2.5L) - adjust every 60,000 miles.
Oxygen sensor - replace every 82,500 miles.
Automatic transmission fluid, filter & bands - change & adjust every 12,000 miles.
Steering linkage - lubricate every 6000 miles.
Rear axle fluid - change every 12,000 miles.

SCHEDULED MAINTENANCE INTERVALS
(DODGE RAM TRUCK—GASOLINE MEDIUM & HEAVY DUTY)

TO BE SERVICED	TYPE OF SERVICE	VEHICLE MILEAGE INTERVAL (x1000)												
		6	12	18	24	30	36	42	48	54	60	66	72	78
Engine oil & filter	R	✓	✓	✓	✓	✓	✓	✓	✓	✓	✓	✓	✓	✓
Exhaust system	S/I	✓	✓	✓	✓	✓	✓	✓	✓	✓	✓	✓	✓	✓
Engine coolant level, hoses & clamps	S/I	✓	✓	✓	✓	✓	✓	✓	✓	✓	✓	✓	✓	✓
Rotate tires	S/I	✓		✓		✓		✓		✓		✓		✓
Drive belts	S/I		✓		✓		✓		✓		✓		✓	
Brake hoses & linings	S/I			✓			✓			✓			✓	
Engine air cleaner element & air pump filter (heavy duty)	R				✓				✓				✓	
Automatic transmission fluid, filter & adjust bands	S/I				✓				✓				✓	
Crankcase inlet air filter (5.9L) (heavy duty)	S/I				✓				✓				✓	
Front wheel bearings (4x2)	S/I				✓				✓				✓	

79241CC1

SCHEDULED MAINTENANCE INTERVALS
(DODGE RAM TRUCK—GASOLINE MEDIUM & HEAVY DUTY Cont.)

TO BE SERVICED	TYPE OF SERVICE	VEHICLE MILEAGE INTERVAL (x1000)												
		6	12	18	24	30	36	42	48	54	60	66	72	78
Engine air cleaner element (medium duty)	R					✓					✓			
Engine Coolant	R						✓					✓		
Spark plugs	R					✓					✓			
Transfer case fluid	R						✓						✓	
Battery①	R										✓			
Distributor cap & rotor (5.9L) (heavy duty)	R										✓			
EGR valve (5.9L) (heavy duty)	R										✓			
Ignition cables	R										✓			
Oxygen sensor (5.9L) (heavy duty)	R										✓			
PCV valve (5.9L) (heavy duty)	R										✓			
EGR passages (5.9L) (heavy duty)	S/I										✓			

① Replace at 60,000 miles, if not replaced previously.
R – Replace   S/I – Service or Inspect

FREQUENT OPERATION MAINTENANCE (SEVERE SERVICE)
If a vehicle is operated under any of the following conditions it is considered severe service:
- Extremely dusty areas.
- 50% or more of the vehicle operation is in 32°C (90°F) or higher temperatures, or constant operation in temperatures below 0°C (32°F).
- Prolonged idling (vehicle operation in stop and go traffic).
- Frequent short running periods (engine does not warm to normal operating temperatures).
- Police, taxi, delivery usage or trailer towing usage.

Oil & oil filter change – change every 3000 miles.
Automatic transmission fluid, filter & bands - change & adjust every 12,000 miles.
Rear axle fluid - change every 12,000 miles.
Brake hoses & linings - check every 12,000 miles.
Front axle fluid (4x4) - change every 24,000 miles.
Engine air cleaner element & air pump filter - replace every 12,000 miles.
Crankcase inlet air filter (5.9L) (heavy duty) - clean & relubricate every 12,000 miles.
PCV valve (5.9L) (heavy duty) - check every 30,000 miles.

79241CC2

SCHEDULED MAINTENANCE INTERVALS
(DODGE RAM TRUCK—DIESEL)

TO BE SERVICED	TYPE OF SERVICE	VEHICLE MILEAGE INTERVAL (x1000)												
		6	12	18	24	30	36	42	48	54	60	66	72	78
Engine oil & filter	R	✓	✓	✓	✓	✓	✓	✓	✓	✓	✓	✓	✓	✓
Brake hoses	S/I	✓	✓	✓	✓	✓	✓	✓	✓	✓	✓	✓	✓	✓
Exhaust system	S/I	✓	✓	✓	✓	✓	✓	✓	✓	✓	✓	✓	✓	✓
Engine coolant level, hoses & clamps	S/I	✓	✓	✓	✓	✓	✓	✓	✓	✓	✓	✓	✓	✓
Steering linkage	S/I	✓	✓	✓	✓	✓	✓	✓	✓	✓	✓	✓	✓	✓
Rotate tires	S/I	✓		✓		✓		✓		✓		✓		✓
Fuel filter	R		✓		✓		✓		✓		✓		✓	
Water pump weep hole	S/I		✓		✓		✓		✓		✓		✓	
Drive belts	S/I			✓			✓			✓			✓	
Brake linings	S/I			✓			✓			✓			✓	
Automatic transmission fluid, filter & adjust bands	S/I				✓				✓				✓	
Damper	S/I				✓				✓				✓	
Fan hub	S/I				✓				✓				✓	
Front wheel bearings	S/I				✓				✓				✓	
Valve lash clearance	S/I				✓				✓				✓	
Air filter	R					✓					✓			
Engine Coolant	R						✓					✓		
Transfer case fluid	R						✓						✓	

R – Replace   S/I – Service or Inspect

FREQUENT OPERATION MAINTENANCE (SEVERE SERVICE)

If a vehicle is operated under any of the following conditions it is considered severe service:
- **Extremely dusty areas.**
- 50% or more of the vehicle operation is in 32°C (90°F) or higher temperatures, or constant operation in temperatures below 0°C (32°F).
- Prolonged idling (vehicle operation in stop and go traffic).
- Frequent short running periods (engine does not warm to normal operating temperatures).
- Police, taxi, delivery usage or trailer towing usage.

Oil & oil filter change – change every 3000 miles.
Automatic transmission fluid, filter & adjust bands - change & adjust every 12,000 miles.
Rear axle fluid - change every 12,000 miles.
Brake linings - check every 12,000 miles.
Front axle fluid (4x4) - change every 24,000 miles.

79241CC4

SCHEDULED MAINTENANCE
CHRYSLER CORPORATION
DODGE RAM VAN, RAM TRUCK

The following should be used as a guide when determining the amount of work required for a particular service if taken to a repair shop. In estimating how long a particular Scheduled Maintenance Service should take, please observe the following:

- **Chilton Time** is time based on field research and data supplied by the vehicle manufacturer.
- Labor time operations are given in hours and tenths of an hour.
- All labor operations, are to be used as a guide.

Mechanic Skill Level Codes:
(G) GENERAL: Normally skilled with certification.
(M) MAINTENANCE: Semi-skilled working on certicication.
(P) PRECISION: Really skilled with multiple certification.

	Chilton Time
(M) 6000 Mile Service	
1995-99	
Gas	1.0
Diesel	1.2
(M) 7500 Mile Service	
1995-99	1.0
(M) 12000 Mile Service	
1995-99	
Gas	.6
Diesel	1.1
(M) 15000 Mile Service	
1995-99	.8
(M) 18000 Mile Service	
1995-99	
Gas	1.2
Diesel	1.5
(M) 22500 Mile Service	
1995-99	1.5
(G) 24000 Mile Service	
1995-99	
Gas	
medium duty	3.0
heavy duty	3.1
Diesel	3.4
Inspect wheel bearings 2WD	
add	.1
(G) 30000 Mile Service	
1995-99	
Gas	
light duty	2.1
medium/heavy duty	2.3
Diesel	1.5

	Chilton Time
(M) 36000 Mile Service	
1995-99	
Gas	1.5
Diesel	2.2
(G) 37500 Mile Service	
1995-99	3.6
(M) 42000 Mile Service	
1995-99	
Gas	1.0
Diesel	1.2
(M) 45000 Mile Service	
1995-99	1.6
(G) 48000 Mile Service	
1995-99	
Gas	
medium duty	3.0
heavy duty	3.1
Diesel	3.4
Inspect wheel bearings 2WD	
add	.1
(M) 52500 Mile Service	
1995-99	1.0
(M) 54000 Mile Service	
1995-99	
Gas	1.2
Diesel	1.5

	Chilton Time
(G) 60000 Mile Service	
1995-99	
Gas	
light duty	4.9
medium duty	3.1
heavy duty	5.0
Diesel	1.4
(M) 66000 Mile Service	
1995-99	
Gas	1.0
Diesel	1.7
(M) 67500 Mile Service	
1995-99	1.5
(G) 72000 Mile Service	
1995-99	
Gas	
medium duty	3.6
heavy duty	3.7
Diesel	4.0
Inspect wheel bearings 2WD	
add	.1
(G) 75000 Mile Service	
1995-99	3.9
(M) 78000 Mile Service	
1995-99	
Gas	1.0
Diesel	1.2
(M) 82500 Mile Service	
1995-99	1.0
(G) 90000 Mile Service	
1995-99	2.8
(M) 97500 Mile Service	
1995-99	1.0

79241CC5

CHRYSLER CORP.
Jeep Cherokee • Grand Cherokee • Wrangler

VEHICLE IDENTIFICATION CHART

Engine Code						Model Year	
Code	Liters	Cu. In. (cc)	Cyl.	Fuel Sys.	Eng. Mfg.	Code	Year
N	4.7	287 (4701)	8	MFI	Chrysler	S	1995
P	2.5	150 (2458)	4	MFI	Chrysler	T	1996
S	4.0	242 (3966)	6	MFI	Chrysler	V	1997
Y	5.2	318 (5211)	8	MFI	Chrysler	W	1998
Z	5.9	360 (5899)	8	MFI	Chrysler	X	1999

MFI - Multi-port Fuel Injection

79241CF3

ENGINE IDENTIFICATION

Year	Model	Engine Displacement Liters (cc)	Engine Series (ID/VIN)	Fuel System	No. of Cylinders	Engine Type
1995	Cherokee	2.5 (2458)	P	MFI	4	OHV
	Cherokee	4.0 (3966)	S	MFI	6	OHV
	Grand Cherokee	4.0 (3966)	S	MFI	6	OHV
	Grand Cherokee	5.2 (5211)	Y	MFI	8	OHV
	Wrangler	2.5 (2458)	P	MFI	4	OHV
	Wrangler	4.0 (3966)	S	MFI	6	OHV
1996	Cherokee	2.5 (2458)	P	MFI	4	OHV
	Cherokee	4.0 (3966)	S	MFI	6	OHV
	Grand Cherokee	4.0 (3966)	S	MFI	6	OHV
	Grand Cherokee	5.2 (5211)	Y	MFI	8	OHV
	Wrangler	2.5 (2464)	P	MFI	4	OHV
	Wrangler	4.0 (3958)	S	MFI	6	OHV
1997	Cherokee	2.5 (2458)	P	MFI	4	OHV
	Cherokee	4.0 (3966)	S	MFI	6	OHV
	Grand Cherokee	4.0 (3966)	S	MFI	6	OHV
	Grand Cherokee	5.2 (5211)	Y	MFI	8	OHV
	Wrangler	2.5 (2464)	P	MFI	4	OHV
	Wrangler	4.0 (3958)	S	MFI	6	OHV
1998	Cherokee	2.5 (2458)	P	MFI	4	OHV
	Cherokee	4.0 (3966)	S	MFI	6	OHV
	Grand Cherokee	4.0 (3966)	S	MFI	6	OHV
	Grand Cherokee	5.2 (5211)	Y	MFI	8	OHV
	Grand Cherokee	5.9 (5899)	Z	MFI	8	OHV
	Wrangler	2.5 (2464)	P	MFI	4	OHV
	Wrangler	4.0 (3958)	S	MFI	6	OHV
1999	Cherokee	2.5 (2458)	P	MFI	4	OHV
	Cherokee	4.0 (3966)	S	MFI	6	OHV
	Grand Cherokee	4.0 (3966)	S	MFI	6	OHV
	Grand Cherokee	4.7 (4701)	N	MFI	8	SOHC
	Wrangler	2.5 (2464)	P	MFI	4	OHV
	Wrangler	4.0 (3958)	S	MFI	6	OHV

MFI - Multi-port Fuel Injection
OHV - Overhead Valve
SOHC - Single Overhead Camshaft

79241CF4

GENERAL ENGINE SPECIFICATIONS

Year	Engine ID/VIN	Engine Displacement Liters (cc)	Fuel System Type	Net Horsepower @ rpm	Net Torque @ rpm (ft. lbs.)	Bore x Stroke (in.)	Compression Ratio	Oil Pressure @ rpm
1995	P	2.5 (2458)	MFI	130@5250	139@3250 ①	3.88x3.19	9.1:1	37@1600 ②
	S	4.0 (3966)	MFI	180@4750 ③	220@4000 ④	3.88x3.44	8.7:1	37@1600 ②
	Y	5.2 (5211)	MFI	220@4400	285@3600	3.91x3.31	9.1:1	30@3000 ②
1996	P	2.5 (2458)	MFI	130@5250	149@3250	3.88x3.19	9.1:1	37@1600 ②
	S	4.0 (3966)	MFI	190@4750	225@4000	3.88x3.44	8.7:1	37@1600 ②
	Y	5.2 (5211)	MFI	220@4400	285@3600	3.91x3.31	9.1:1	30@3000 ②
1997	P	2.5 (2458)	MFI	130@5250	149@3250	3.88x3.19	9.1:1	37@1600 ②
	S	4.0 (3966)	MFI	190@4750	225@4000	3.88x3.44	8.7:1	37@1600 ②
	Y	5.2 (5211)	MFI	220@4400	285@3600	3.91x3.31	9.1:1	30@3000 ②
1998-99	P	2.5 (2458)	MFI	125@5400	140@3500 ①	3.88x3.19	9.2:1	37@1600 ②
	S	4.0 (3966)	MFI	181@4600 ⑤	222@2800 ⑥	3.88x3.41	8.8:1	37@1600 ②
	Y	5.2 (5211)	MFI	220@4400	300@2800	3.91x3.31	9.1:1	30@3000 ②
	N	4.7 (4701)	MFI	230@4600	300@3200	3.66x3.40	9.0:1	25@3000 ⑦
	Z	5.9 (5899)	MFI	245@4000	345@3200	4.00x3.58	9.1:1	30@3000 ⑦

MFI - Multi-port Fuel Injection
① Cherokee: 149@3250
② Above 3000 rpm, pressure can vary to a maximum of 75 psi, except on 5.2L.
   On 5.2L, pressure can vary to 80 psi maximum
③ Cherokee and Grand Cherokee: 190@4750
④ Cherokee and Grand Wagoneer: 225@4000
⑤ Cherokee: 190@4600
   Grand Cherokee: 185@4600
⑥ Cherokee: 225@3000
   Grand Cherokee: 220@2400

79241CF5

GASOLINE ENGINE TUNE-UP SPECIFICATIONS

Year	Engine ID/VIN	Engine Displacement Liters (cc)	Spark Plugs Gap (in.)	Ignition Timing (deg.) MT	AT	Fuel Pump (psi)	Idle Speed (rpm) MT	AT	Valve Clearance In.	Ex.
1995	P	2.5 (2458)	0.035	①	①	39-41 ②	①	①	HYD	HYD
	S	4.0 (3966)	0.035	①	①	39-41 ②	①	①	HYD	HYD
	Y	5.2 (5211)	0.035	①	①	39-41 ②	①	①	HYD	HYD
1996	P	2.5 (2458)	0.035	①	①	47-51 ②	①	①	HYD	HYD
	S	4.0 (3966)	0.035	①	①	47-51 ②	①	①	HYD	HYD
	Y	5.2 (5211)	0.035	①	①	47-51 ②	①	①	HYD	HYD
1997	P	2.5 (2458)	0.035	①	①	47-51 ②	①	①	HYD	HYD
	S	4.0 (3966)	0.035	①	①	47-51 ②	①	①	HYD	HYD
	Y	5.2 (5211)	0.035	①	①	47-51 ②	①	①	HYD	HYD
1998-99	N ③	4.7 (4701)	0.040	—	①	47-51 ②	—	①	HYD	HYD
	P	2.5 (2458)	0.035	①	①	47-51 ②	①	①	HYD	HYD
	S	4.0 (3966)	0.035	①	①	47-51 ②	①	①	HYD	HYD
	Y ④	5.2 (5211)	0.040	—	①	47-51 ②	—	①	HYD	HYD
	Z ④	5.9 (5899)	0.040	—	①	47-51 ②	—	①	HYD	HYD

NOTE: The Vehicle Emission Control Information label often reflects specification changes made during production. The label figures must be used if they differ from those in this chart.
HYD - Hydraulic
① Not adjustable
② With the vacuum line disconnected from the fuel pressure regulator (if equipped).
   Fuel pressure is measured at the test port pressure fitting on fuel rail.
③ New for 1999.
④ 1998 only.

79241CF6

CAPACITIES

Year	Model	Engine ID/VIN	Engine Displacement Liters (cc)	Engine Oil with Filter	Transmission (pts.) 4-Spd	5-Spd	Auto.	Transfer Case (pts.)	Drive Axle Front (pts.)	Rear (pts.)	Fuel Tank (gal.)	Cooling System (qts.)
1995	Cherokee	P	2.5 (2468)	4.0	—	6.6 [1]	17.0	3.0 [2]	3.1	3.5 [3]	20.2	10.0
		S	4.0 (3966)	6.0	—	6.6	17.0	3.0 [2]	3.1	3.5 [3]	20.2	12.0
	Grand Cherokee	S	4.0 (3966)	6.0	—	6.5	17.0	3.2 [4]	3.1	3.4	23.0	12.0
		Y	5.2 (5211)	5.0	—	6.5	19.5	3.2 [4]	3.1	3.4	23.0	14.9
	Wrangler	P	2.5 (2468)	4.0	—	6.6	17.5	[5]	3.7	3.5 [3]	15.0 [6]	9.0
		S	4.0 (3966)	6.0	—	6.6	17.5	[5]	3.7	3.5 [3]	15.0 [6]	10.5
1996	Cherokee	P	2.5 (2468)	4.0	—	6.6 [7]	17.0	3.0 [2]	3.1	3.5 [3]	20.2	10.0
		S	4.0 (3966)	6.0	—	6.6	17.0	3.0 [2]	3.1	3.5 [3]	20.2	12.0
	Grand Cherokee	S	4.0 (3966)	6.0	—	6.5	17.0	3.2 [8]	3.1	3.4	23.0	12.0
		Y	5.2 (5211)	5.0	—	6.5	19.5	3.2 [8]	3.1	3.4	23.0	14.9
	Wrangler	P	2.5 (2464)	4.0	—	6.6	17.5	[9]	3.7	3.5	[10]	9.0
		S	4.0 (3958)	6.0	—	6.6	17.5	[9]	3.7	3.5	[10]	10.5
1997	Cherokee	P	2.5 (2468)	4.0	—	6.6 [7]	17.0	3.0 [2]	3.1	3.5 [3]	20.2	10.0
		S	4.0 (3966)	6.0	—	6.6	17.0	3.0 [2]	3.1	3.5 [3]	20.2	12.0
	Grand Cherokee	S	4.0 (3966)	6.0	—	6.5	17.0	3.2 [8]	3.1	3.4	23.0	12.0
		Y	5.2 (5211)	5.0	—	6.5	19.5	3.2 [8]	3.1	3.4	23.0	14.9
	Wrangler	P	2.5 (2464)	4.0	—	6.6	17.5	[9]	3.7	3.5	[10]	9.0
		S	4.0 (3958)	6.0	—	6.6	17.5	[9]	3.7	3.5	[10]	10.5
1998-99	Cherokee	P	2.5 (2468)	4.0	—	6.6 [7]	17.0	3.0 [2]	3.1	3.5 [3]	20.2	10.0
		S	4.0 (3966)	6.0	—	6.6	17.0	3.0 [2]	3.1	3.5 [3]	20.2	12.0
	Grand Cherokee	N	4.7 (4701)	6.0	—	—	[11]	[12]	2.5	[13]	20.5	13.0
		S	4.0 (3966)	6.0	—	—	[11]	[12]	3.1	[13]	[14]	12.0
		Y	5.2 (5211)	5.0	—	—	[11]	[12]	3.1	[13]	23.0	14.9
		Z	5.9 (5899)	5.0	—	—	[11]	[12]	2.5	[13]	23.0	14.9
	Wrangler	P	2.5 (2464)	4.0	—	6.6	17.5	[9]	3.7	3.5	[10]	9.0
		S	4.0 (3958)	6.0	—	6.6	17.5	[9]	3.7	3.5	[10]	10.5

[1] 2WD: 7.0 pts.
[2] Command-Trac - 2.2 pts.
[3] 8 1/4 axle: 4.4 pts.
  When equipped with TRAC-LOK,
  include 2 oz. of friction modifier additive
[4] NP242: 2.9
  NP249: 2.5
[5] NP242: 2.9
  NP249: 3.0
[6] Optional - 20 gal.
[7] 2WD: 7.0 pts.
[8] NP242: 2.9 pts.
  NP249: 2.5 pts.
[9] Command-Trac:
  Automatic: 2.2 pts.
  Manual: 3.3 pts.
[10] Standard: 15.0 gals.
  Optional: 19.6 gals.

[11] 42RE: 19-22 pts.
  44RE: 19-22 pts.
  45RFE: 28.0 pts.
  46RE: 19.5-28.0 pts.
  The amounts vary depending on type and size
  of internal cooler, length and inside diameter of cooler
  lines, or auxiliary cooler.
[12] 242 NVG: 3.0 pts.
  247 NVG: 2.5 pts.
  249 NVG: 2.5 pts.
[13] 194 RBI: 3.5 pts.
  198 RBI: 3.75 pts.
  216 RBA: 4.75 pts.
  226 RBA: 4.75 pts.
  If equipped with VARI-LOCK or TRAC-LOCK,
  include approx. 0.25 pt. of Friction Modifier.
[14] 1998: 23 gallons
  1999: 20.5 gallons

79241CF7

VALVE SPECIFICATIONS

Year	Engine ID/VIN	Engine Displacement Liters (cc)	Seat Angle (deg.)	Face Angle (deg.)	Spring Test Pressure (lbs. @ in.)	Spring Installed Height (in.)	Stem-to-Guide Clearance (in.)		Stem Diameter (in.)	
							Intake	Exhaust	Intake	Exhaust
1995	P	2.5 (2458)	44.5	45	200@1.216	1.640	0.0010-0.0030	0.0010-0.0030	0.3110-0.3120	0.3110-0.3120
	S	4.0 (3966)	44.5	45	200@1.216	1.640	0.0010-0.0030	0.0010-0.0030	0.3110-0.3120	0.3110-0.3120
	Y	5.2 (5211)	44.25-44.75	43.25-43.75	200@1.212	1.640	0.0010-0.0030	0.0010-0.0030	0.3110-0.3120	0.3110-0.3120
1996	P	2.5 (2458)	44.5	45	184-196@1.216	1.640	0.0010-0.0030	0.0010-0.0030	0.3110-0.3120	0.3110-0.3120
	S	4.0 (3966)	44.5	45	184-196@1.216	1.640	0.0010-0.0030	0.0010-0.0030	0.3110-0.3120	0.3110-0.3120
	Y	5.2 (5211)	44.25-44.75	43.25-43.75	200@1.212	1.640	0.0010-0.0030	0.0010-0.0030	0.3110-0.3120	0.3110-0.3120
1997	P	2.5 (2458)	44.5	45	184-196@1.216	1.640	0.0010-0.0030	0.0010-0.0030	0.3110-0.3120	0.3110-0.3120
	S	4.0 (3966)	44.5	45	184-196@1.216	1.640	0.0010-0.0030	0.0010-0.0030	0.3110-0.3120	0.3110-0.3120
	Y	5.2 (5211)	44.25-44.75	43.25-43.75	200@1.212	1.640	0.0010-0.0030	0.0010-0.0030	0.3110-0.3120	0.3110-0.3120
1998-99	N	4.7 (4701)	44.50-45	45.0-45.5	176.2-192.4@1.1532	1.602	0.0011-0.0017	0.0029	0.2728-0.2739	0.2717-0.2728
	P	2.5 (2458)	44.5	45	184-196@1.216	1.640	0.0010-0.0030	0.0010-0.0030	0.3110-0.3120	0.3110-0.3120
	S	4.0 (3966)	44.5	45	184-196@1.216	1.640	0.0010-0.0030	0.0010-0.0030	0.3110-0.3120	0.3110-0.3120
	Y	5.2 (5211)	44.25-44.75	43.25-43.75	200@1.212	1.640	0.0010-0.0030	0.0010-0.0030	0.3110-0.3120	0.3110-0.3120
	Z	5.9 (5899)	44.25-44.75	43.25-43.75	200@1.212	1.640	0.0010-0.0030	0.0020-0.0040	0.3720-0.3730	0.3710-0.3720

79241CF8

TORQUE SPECIFICATIONS
All readings in ft. lbs.

Year	Engine ID/VIN		Engine Displacement Liters (cc)	Cylinder Head Bolts	Main Bearing Bolts	Rod Bearing Bolts	Crankshaft Damper Bolts	Flywheel Bolts	Manifold Intake *	Manifold Exhaust	Spark Plugs	Lug Nut
1995	P		2.5 (2458)	①	80	33	80	105	②	②	27	80-110
	S		4.0 (3966)	③	80	33	80	105	④	8	27	80-110
	Y		5.2 (5211)	⑤	85	45	135	105	⑥	20	30	80-110
1996	P	⑦	2.5 (2458)	①	80	33	80	105	⑧	⑧	27	80-110
	P	⑨	2.5 (2464)	⑩	80	33	80	105	⑧	20	27	95
	S	⑦	4.0 (3966)	③	80	33	80	105	④	④	27	80-110
	S	⑨	4.0 (3958)	⑩	80	33	80	105	④	20	27	95
	Y		5.2 (5211)	⑤	85	45	135	105	⑥	20	30	80-110
1997	P	⑦	2.5 (2458)	①	80	33	80	105	⑧	⑧	27	80-110
	P	⑨	2.5 (2464)	⑩	80	33	80	105	⑧	20	27	95
	S	⑦	4.0 (3966)	③	80	33	80	105	④	④	27	80-110
	S	⑨	4.0 (3958)	⑩	80	33	80	105	④	20	27	95
	Y		5.2 (5211)	⑤	85	45	135	105	⑥	20	30	80-110
1998-99	N		4.7 (4701)	⑪	⑫	15 ⑬	130	45	9	18	27	85-115
	P	⑦	2.5 (2458)	①	80	33	80	105	②	②	27	85-115
	P	⑨	2.5 (2464)	⑩	80	33	80	105	②	②	27	85-115
	S	⑦	4.0 (3966)	③	80	33	80	105	④	④	27	85-115
	S	⑨	4.0 (3958)	③	80	33	80	105	④	④	27	85-115
	Y		5.2 (5211)	⑤	85	45	135	105	⑥	20	30	85-115
	Z		5.9 (5899)	⑤	85	45	135	55	⑥	20	30	85-115

* NOTE: Applies to Lower Manifold only.

① Step 1: 22 ft. lbs.
Step 2: 45 ft. lbs.
Step 3: Bolts 1-6: 110 ft. lbs.
Step 4: Bolt 7: 100 ft. lbs.
Step 5: Bolts 8-10: 110 ft. lbs.

② Bolt 1: 30 ft. lbs.
Bolts 2-7: 23 ft. lbs.

③ Step 1: 22 ft. lbs.
Step 2: 45 ft. lbs.
Step 3: Bolts 1-10 and 12-14: 110 ft. lbs.

④ Bolts 1-5 and 8-11: 24 ft. lbs.
Bolts 6-7: 23 ft. lbs.

⑤ Step 1: 50 ft. lbs.
Step 2: 105 ft. lbs.

⑥ Step 1: Torque bolts 1-4 to 72 in. lbs.
Step 2: Torque bolts 5-12 to 72 in. lbs.
Step 3: Torque all bolts to 12 ft. lbs.

⑦ Cherokee and Grand Cherokee

⑧ Bolts 1, 6 & 7: 30 ft. lbs.
Bolts 2-5: 23 ft. lbs.

⑨ Wrangler

⑩ Bolts 1-10 and 12-14: 110 ft. lbs.
Bolt 11: 100 ft. lbs.

⑪ M11 bolts: 60 ft. lbs.
M8 bolts: 250 inch lbs.

⑫ Step 1: Bolts 1-10 - 25 inch lbs.
Step 2: Bolts 1-10 - additional 90°
Step 3: Bolts A-K - 40 ft. lbs.
Step 4: Bolts A1-A5 - 20 ft. lbs.

⑬ Plus an additional 110°

79241CF9

SCHEDULED MAINTENANCE INTERVALS
(JEEP CHEROKEE, GRAND CHEROKEE & WRANGLER)

TO BE SERVICED	TYPE OF SERVICE	VEHICLE MILEAGE INTERVAL (x1000)												
		7.5	15	22.5	30	37.5	45	52.5	60	67.5	75	82.5	90	97.5
Engine oil & filter	R	✓	✓	✓	✓	✓	✓	✓	✓	✓	✓	✓	✓	✓
Brake hoses & linings	S/I	✓	✓	✓	✓	✓	✓	✓	✓	✓	✓	✓	✓	✓
Engine coolant level, hoses & clamps	S/I	✓	✓	✓	✓	✓	✓	✓	✓	✓	✓	✓	✓	✓
Exhaust system	S/I	✓	✓	✓	✓	✓	✓	✓	✓	✓	✓	✓	✓	✓
Lubricate steering linkage (4x2)	S/I	✓	✓	✓	✓	✓	✓	✓	✓	✓	✓	✓	✓	✓
Lubricate steering linkage (4x4)	S/I	✓		✓		✓		✓		✓		✓		✓
Air filter	R				✓				✓				✓	
Automatic transmission fluid & filter	R				✓				✓				✓	
Spark plugs	R				✓				✓				✓	
Transfer case oil	R				✓				✓				✓	
Drive belts	S/I				✓				✓				✓	
Front & rear axle oil	R				✓				✓				✓	
Prop shaft universal joints	S/I				✓				✓				✓	
Rotate tires	S/I				✓				✓				✓	
Engine Coolant	R						✓				✓			
Manual transmission fluid	R					✓					✓			
Distributor cap & rotor	R								✓					
Fuel filter	R								✓					
Ignition cables	R								✓					

R – Replace   S/I – Service or Inspect

FREQUENT OPERATION MAINTENANCE (SEVERE SERVICE)

If a vehicle is operated under any of the following conditions it is considered severe service:
- **Extremely dusty areas.**
- **50% or more of the vehicle operation is in 32°C (90°F) or higher temperatures, or constant operation in temperatures below 0°C (32°F).**
- **Prolonged idling (vehicle operation in stop and go traffic).**
- **Frequent short running periods (engine does not warm to normal operating temperatures).**
- **Police, taxi, delivery usage or trailer towing usage.**

Oil & oil filter change – change every 3000 miles.
Automatic transmission fluid & filter - change every 12,000 miles.
Brake hoses & linings - check every 12,000 miles.
Lubricate steering linkage - check every 3000 miles.
Manual transmission fluid - change every 18,000 miles.
Prop shaft universal joints - lubricate every 3000 miles.
Front & rear axle oil - change every 12,000 miles.

79241CF0

SCHEDULED MAINTENANCE
CHRYSLER CORPORATION
JEEP CHEROKEE, GRAND CHEROKEE, WRANGLER

The following should be used as a guide when determining the amount of work required for a particular service if taken to a repair shop.
In estimating how long a particular Scheduled Maintenance Service should take, please observe the following:

- **Chilton Time** is time based on field research and data supplied by the vehicle manufacturer.

- Labor time operations are given in hours and tenths of an hour.

- All labor operations, are to be used as a guide.

Mechanic Skill Level Codes:
- **(G) GENERAL:** Normally skilled with certification.
- **(M) MAINTENANCE:** Semi-skilled working on certicication.
- **(P) PRECISION:** Really skilled with multiple certification.

	Chilton Time
(M) 7500 Mile Service	
1995-99	.8
w/4WD add	.1
(M) 15000 Mile Service	
1995-99	.8
(M) 22500 Mile Service	
1995-99	.8
w/4WD add	.1
(G) 30000 Mile Service	
1995-99	2.6
w/AT add	.5
w/4WD add	.4

	Chilton Time
(G) 37500 Mile Service	
1995-99	1.1
w/4WD add	.1
(G) 45000 Mile Service	
1995-99	1.3
(M) 52500 Mile Service	
1995-99	.8
w/4WD add	.1
(G) 60000 Mile Service	
1995-99	3.8
w/AT add	.5
w/4WD add	.4
(M) 67500 Mile Service	
1995-99	.8
w/4WD add	.1

	Chilton Time
(G) 75000 Mile Service	
1995-99	1.6
(M) 82500 Mile Service	
1995-99	.8
w/4WD add	.1
(G) 90000 Mile Service	
1995-99	2.6
w/AT add	.5
w/4WD add	.4
(M) 97500 Mile Service	
1995-99	.8
w/4WD add	.1

79241CG1

FORD MOTOR CO.
Ford F-Series • E-Series • Aerostar • Bronco • Club Wagon
Expedition • Explorer • Ranger • Super-Duty • Lincoln Navigator

VEHICLE IDENTIFICATION CHART

Engine Code						Model Year	
Code	Liters	Cu. In. (cc)	Cyl.	Fuel Sys.	Eng. Mfg.	Code	Year
2	4.2	256 (4195)	6	MFI	Ford	S	1995
5	6.8	415 (6802)	10	MFI	Ford	T	1996
6	4.6	280 (4588)	8	MFI	Ford	V	1997
A	2.3	140 (2294)	4	MFI	Ford	W	1998
C	2.5	152 (2500)	4	MFI	Ford	X	1999
E	4.0	244 (4000)	6	MFI	Ford		
F	7.3	445 (7292)	8	DI	Navistar		
G	7.5	460 (7538)	8	MFI	Ford		
H	5.8	351 (5752)	8	MFI	Ford		
L	5.4	330 (5409)	8	MFI	Ford		
N	5.0	302 (4949)	8	MFI	Ford		
P	5.0	302 (4949)	8	MFI	Ford		
R	5.8	351 (5752)	8	MFI	Ford		
U	3.0	183 (2999)	6	MFI	Ford		
W	4.6	280 (4588)	8	MFI	Ford		
X	4.0	244 (3998)	6	MFI	Ford		
Y	4.9	300 (4916)	6	MFI	Ford		

MFI - Multi-port Fuel Injection
DSL - Diesel
DI - Direct Injection Turbo

79241C66

ENGINE IDENTIFICATION

Year	Model	Engine Displacement Liters (cc)	Engine Series (ID/VIN)	Fuel System	No. of Cylinders	Engine Type
1995	Aerostar	3.0 (2999)	U	MFI	6	OHV
	Aerostar	4.0 (3998)	X	MFI	6	OHV
	Bronco	5.0 (4949)	N	MFI	8	OHV
	Bronco	5.8 (5752)	H	MFI	8	OHV
	E-150	4.9 (4916)	Y	MFI	6	OHV
	E-150	5.0 (4949)	N	MFI	8	OHV
	E-150	5.8 (5752)	H	MFI	8	OHV
	E-250	4.9 (4916)	Y	MFI	6	OHV
	E-250	5.0 (4949)	N	MFI	8	OHV
	E-250	5.8 (5752)	H	MFI	8	OHV
	E-350	4.9 (4916)	Y	MFI	6	OHV
	E-350	5.8 (5752)	H	MFI	8	OHV
	E-350	7.3 (7292)	F	DI	8	OHV
	E-350	7.3 (7292)	K	IDI	8	OHV
	E-350	7.3 (7292)	M	IDI	8	OHV
	E-350	7.5 (7538)	G	MFI	8	OHV
	Explorer	4.0 (3998)	X	MFI	6	OHV
	F-150	4.9 (4916)	Y	MFI	6	OHV
	F-150	5.0 (4949)	N	MFI	8	OHV
	F-150	5.8 (5752)	H	MFI	8	OHV
	F-250	4.9 (4916)	Y	MFI	6	OHV
	F-250	5.0 (4949)	N	MFI	8	OHV
	F-250	5.8 (5752)	H	MFI	8	OHV
	F-250	7.3 (7292)	M	IDI	8	OHV
	F-250	7.5 (7538)	G	MFI	8	OHV
	F-350	4.9 (4916)	Y	MFI	6	OHV
	F-350	5.8 (5752)	H	MFI	8	OHV
	F-350	7.3 (7292)	K	IDI	8	OHV
	F-350	7.3 (7292)	F	DI	8	OHV
	F-350	7.3 (7292)	M	IDI	8	OHV
	F-350	7.5 (7538)	G	MFI	8	OHV
	F-Super Duty	7.3 (7292)	F	DI	8	OHV
	F-Super Duty	7.3 (7292)	M	DDI	8	OHV
	F-Super Duty	7.3 (7292)	K	IDI	8	OHV
	F-Super Duty	7.5 (7538)	G	EFI	8	OHV
	Lightning Pick-up	5.8 (5752)	R	EFI	8	OHV
	Ranger	2.3 (2294)	A	MFI	4	SOHC
	Ranger	3.0 (2999)	U	MFI	6	OHV
	Ranger	4.0 (3998)	X	MFI	6	OHV
1996	Aerostar	3.0 (2982)	U	MFI	6	OHV
	Aerostar	4.0 (3950)	X	MFI	6	OHV
	Bronco	5.0 (4949)	N	MFI	8	OHV
	Bronco	5.8 (5752)	H	MFI	8	OHV
	E-150	4.9 (4916)	Y	MFI	6	OHV
	E-150	5.0 (4949)	N	MFI	8	OHV
	E-150	5.8 (5752)	H	MFI	8	OHV
	E-250	4.9 (4916)	Y	MFI	6	OHV
	E-250	5.0 (4949)	N	MFI	8	OHV
	E-250	5.8 (5752)	H	MFI	8	OHV
	E-350	4.9 (4916)	Y	MFI	6	OHV
	E-350	5.8 (5752)	H	MFI	8	OHV
	E-350	7.3 (7292)	F	DI	8	OHV
	E-350	7.5 (7538)	G	MFI	8	OHV

79241C67

ENGINE IDENTIFICATION

Year	Model	Engine Displacement Liters (cc)	Engine Series (ID/VIN)	Fuel System	No. of Cylinders	Engine Type
1996 (cont.)	Explorer	4.0 (3950)	X	MFI	6	OHV
	Explorer	5.0 (4949)	P	MFI	8	OHV
	F-150	4.2 (4195)	2	MFI	6	OHV
	F-150	4.6 (4588)	W	MFI	8	SOHC
	F-150	4.6 (4588)	6	MFI	8	SOHC
	F-150	4.6 (4588)	9	NA	8	SOHC
	F-150	4.9 (4916)	Y	MFI	6	OHV
	F-150	5.0 (4949)	N	MFI	8	OHV
	F-150	5.8 (5752)	H	MFI	8	OHV
	F-250	4.9 (4916)	Y	MFI	6	OHV
	F-250	5.0 (4949)	N	MFI	8	OHV
	F-250	5.8 (5752)	H	MFI	8	OHV
	F-250	7.3 (7292)	F	DI	8	OHV
	F-250	7.5 (7538)	G	MFI	8	OHV
	F-350	4.9 (4916)	Y	MFI	6	OHV
	F-350	5.8 (5752)	H	MFI	8	OHV
	F-350	7.3 (7292)	F	DI	8	OHV
	F-350	7.5 (7538)	G	MFI	8	OHV
	F-Super Duty	7.3 (7292)	F	DI	8	OHV
	F-Super Duty	7.5 (7538)	G	MFI	8	OHV
	Mountaineer	5.0 (4949)	P	MFI	8	OHV
	Ranger	2.3 (2300)	A	MFI	4	SOHC
	Ranger	3.0 (2982)	U	MFI	6	OHV
	Ranger	4.0 (3950)	X	MFI	6	OHV
1997	Aerostar	3.0 (2982)	U	MFI	6	OHV
	Aerostar	4.0 (3950)	X	MFI	6	OHV
	E-150	4.2 (4195)	2	MFI	6	OHV
	E-150	4.6 (4588)	6/W	MFI	8	OHV
	E-150	5.4 (5409)	L	MFI	8	OHV
	E-250	4.2 (4195)	2	MFI	6	OHV
	E-250	5.4 (5409)	L	MFI	8	OHV
	E-350	5.4 (5409)	L	MFI	8	OHV
	E-350	6.8 (6802)	5	MFI	10	OHV
	E-350	7.3 (7292)	F	DI	8	OHV
	Expedition	4.6 (4588)	6	MFI	8	SOHC
	Expedition	5.4 (5409)	L	MFI	8	SOHC
	Explorer	4.0 (4000)	X	FUEL	6	SOHC
	Explorer	5.0 (4949)	P	MFI	8	OHV
	F-150	4.2 (4195)	2	MFI	6	OHV
	F-150	4.6 (4588)	W	MFI	8	SOHC
	F-150	4.6 (4588)	6	MFI	8	SOHC
	F-150	5.4 (5409)	L	MFI	8	SOHC
	F-250	5.8 (5752)	H	MFI	8	OHV
	F-250	7.3 (7292)	F	DI	8	OHV
	F-250	7.5 (7538)	G	MFI	8	OHV
	F-350	5.8 (5752)	H	MFI	8	OHV
	F-350	7.3 (7292)	F	DI	8	OHV
	F-350	7.5 (7538)	G	MFI	8	OHV
	F-Super Duty	5.8 (5752)	H	MFI	8	OHV
	F-Super Duty	7.3 (7292)	F	DI	8	OHV
	F-Super Duty	7.5 (7538)	G	MFI	8	OHV
	Mountaineer	5.0 (4949)	P	MFI	8	OHV
	Ranger	2.3 (2300)	A	MFI	4	SOHC

79241C68

ENGINE IDENTIFICATION

Year	Model	Engine Displacement Liters (cc)	Engine Series (ID/VIN)	Fuel System	No. of Cylinders	Engine Type
1997 (cont.)	Ranger	3.0 (2982)	U	MFI	6	OHV
	Ranger	4.0 (3950)	X	MFI	6	OHV
1998-99	E-150	4.2 (4195)	2	MFI	6	OHV
	E-150	4.6 (4588)	6/W	MFI	8	OHV
	E-150	5.4 (5409)	L	MFI	8	OHV
	E-250	4.2 (4195)	2	MFI	6	OHV
	E-250	5.4 (5409)	L	MFI	8	OHV
	E-350	5.4 (5409)	L	MFI	8	OHV
	E-350	6.8 (6802)	5	MFI	10	OHV
	E-350	7.3 (7292)	F	DI	8	OHV
	Expedition	4.6 (4588)	6/W	MFI	8	SOHC
	Expedition	5.4 (5409)	L	MFI	8	SOHC
	Explorer	4.0 (4000)	X	MFI	6	SOHC
	Explorer	5.0 (4949)	P	MFI	8	OHV
	F-150	4.2 (4195)	2	MFI	6	OHV
	F-150	4.6 (4588)	6/W	MFI	8	SOHC
	F-150	5.4 (5409)	L	MFI	8	SOHC
	F-250	4.6 (4588)	6/W	MFI	8	SOHC
	F-250	5.4 (5409)	L	MFI	8	SOHC
	F-350	5.4 (5409)	L	MFI	8	SOHC
	F-350	5.8 (5752)	H	MFI	8	OHV
	F-350	6.8 (6802)	5	MFI	10	OHV
	F-350	7.3 (7292)	F	DI	8	OHV
	F-350	7.5 (7538)	G	MFI	8	OHV
	F-Super Duty	5.4 (5409)	L	MFI	8	SOHC
	F-Super Duty	5.8 (5752)	H	MFI	8	OHV
	F-Super Duty	6.8 (6802)	5	MFI	10	OHV
	F-Super Duty	7.3 (7292)	F	DI	8	OHV
	F-Super Duty	7.5 (7538)	G	MFI	8	OHV
	Mountaineer	5.0 (4949)	P	MFI	8	OHV
	Navigator	5.4 (5409)	L	MFI	8	SOHC
	Ranger	2.5 (2500)	C	MFI	4	OHV
	Ranger	3.0 (2982)	U	MFI	6	OHV
	Ranger	4.0 (3950)	X	MFI	6	OHV

MFI - Multi-port Fuel Injection
DSL - Diesel
DI - Direct Injection Turbo
OHV- Overhead Valve
SOHC- Single Overhead Camshaft

79241C69

GENERAL ENGINE SPECIFICATIONS

Year	Engine ID/VIN	Engine Displacement Liters (cc)	Fuel System Type	Net Horsepower @ rpm	Net Torque @ rpm (ft. lbs.)	Bore x Stroke (In.)	Compression Ratio	Oil Pressure @ rpm
1995	A	2.3 (2294)	EFI	100@4600	133@2600	3.78x3.13	9.2:1	40-60@2000
	F	7.3 (7292)	DI	210@3000	425@2000	4.11x4.18	17.5:1	40-70@3000
	G	7.5 (7538)	EFI	245@4000	400@2200	4.36x3.85	8.5:1	40-88@2000
	H	5.8 (5752)	EFI	210@3600	325@2800	4.00x3.50	8.8:1	40-65@2000
	M	7.3 (7292)	IDI	185@3000 ①	360@1400 ②	4.11x4.18	21.5:1	40-70@2000
	N	5.0 (4949)	EFI	205@4000	275@3000	4.00x3.00	9.0:1	40-60@2000
	R	5.8 (5752)	EFI	240@4200	340@3200	4.00x3.50	8.8:1	40-65@2000
	U	3.0 (2999)	MFI	135@4600	160@2800	3.50x3.14	9.3:1	40-60@2500
	X	4.0 (3998)	EFI	160@4000	225@2500	3.81x3.39	9.0:1	40-60@2000
	Y	4.9 (4916)	EFI	145@3400 ③	265@2000 ③	4.00x3.98	8.8:1	40-60@2000
1996	2	4.2 (4195)	MFI	205@4400	255@3000	3.81x3.74	9.3:1	50@2000
	6	4.6 (4588)	MFI	210@4400	290@3250	3.55x3.54	9.0:1	20-45@1500
	9	4.6 (4588)	NA	NA	NA	3.55x3.54	9.0:1	20-45@1500
	A	2.3 (2300)	MFI	112@4800	135@2400	3.78x3.13	9.4:1	40-60@2000
	F	7.3 (7292)	DI	210@3000	425@2000	4.11x4.18	17.5:1	40-70@3000
	G	7.5 (7538)	MFI	245@4000	400@2200	4.36x3.85	8.5:1	40-88@2000
	H	5.8 (5752)	MFI	210@3600	325@2800	4.00x3.50	8.8:1	40-65@2000
	N	5.0 (4949)	MFI	199@4200	270@2400	4.00x3.00	9.0:1	40-60@2000
	P	5.0 (4949)	MFI	210@4500	280@3500	4.00x3.00	9.0:1	40-60@2500
	U	3.0 (2982)	MFI	147@5000	162@3250	3.50x3.14	9.3:1	40-60@2500
	W	4.6 (4588)	MFI	210@4400	290@3250	3.55x3.54	9.0:1	20-45@1500
	X	4.0 (3950)	MFI	160@4000	225@2500	3.81x3.39	9.0:1	40-60@2000
	Y	4.9 (4916)	MFI	145@3400 ③	265@2000 ③	4.00x3.98	8.8:1	40-60@2000
1997	2	4.2 (4195)	MFI	205@4400	255@3000	3.81x3.74	9.3:1	50@2000
	5	6.8 (6802)	MFI	265@4250	410@2750	4.09X4.17	9.0:1	40-70@1500
	6	4.6 (4588)	MFI	210@4400	290@3250	3.55x3.54	9.0:1	20-45@1500
	A	2.3 (2300)	MFI	112@4800	135@2400	3.78x3.13	9.4:1	40-60@2000
	E	4.0 (3950)	MFI	205@5000	250@3000	3.95X3.31	9.7:1	40-60@2000
	F	7.3 (7292)	DI	210@3000	425@2000	4.11x4.18	17.5:1	40-70@3000
	G	7.5 (7538)	MFI	245@4000	400@2200	4.36x3.85	8.5:1	40-88@2000
	H	5.8 (5752)	MFI	210@3600	325@2800	4.00x3.50	8.8:1	40-65@2000
	L	5.4 (5409)	MFI	235@4250	330@3000	3.55X4.17	9.0:1	40-70@1500
	N	5.0 (4949)	MFI	199@4200	270@2400	4.00x3.00	9.0:1	40-60@2000
	P	5.0 (4949)	MFI	210@4500	280@3500	4.00x3.00	9.0:1	40-60@2500
	U	3.0 (2982)	MFI	147@5000	162@3250	3.50x3.14	9.3:1	40-60@2500
	W	4.6 (4588)	MFI	210@4400	290@3250	3.55x3.54	9.0:1	20-45@1500
	X	4.0 (3950)	MFI	160@4000	225@2500	3.81x3.39	9.0:1	40-60@2000
1998-99	5	6.8 (6802)	MFI	265@4250	410@2750	4.09X4.17	9.0:1	40-70@1500
	6/W	4.6 (4588)	MFI	210@4400	290@3250	3.55x3.54	9.0:1	20-45@1500
	C	2.5 (2500)	MFI	119@5000	146@3000	3.78X3.90	9.1:1	40-60@2000
	E	4.0 (3950)	MFI	205@5000	250@3000	3.95X3.31	9.7:1	40-60@2000
	G	7.5 (7538)	MFI	245@4000	400@2200	4.36x3.85	8.5:1	40-88@2000
	L	5.4 (5409)	MFI	235@4250	330@3000	3.55X4.17	9.0:1	40-70@1500
	P	5.0 (4949)	MFI	210@4500	280@3500	4.00x3.00	9.0:1	40-60@2500
	P	5.0 (4949)	MFI	210@4500	280@3500	4.00x3.00	9.0:1	40-60@2500
	R	5.8 (5752)	MFI	210@3600	325@2800	4.00x3.50	8.8:1	40-65@2000
	U	3.0 (2982)	MFI	147@5000	162@3250	3.50x3.14	9.3:1	40-60@2500
	X	4.0 (3950)	MFI	160@4000	225@2500	3.81x3.39	9.0:1	40-60@2000

NA - Not Available
MFI - Multiport fuel injection
EFI - Electronic fuel injection
IDI - Indirect diesel injection
DI - Direct injection Turbo

① High altitude: 165@3000
② High altitude: 325@1600
③ Ratings are for E150-250 Van and regular Wagon with 4 spd automatic OD (E40D). Use 150hp@3400 rpm and 260 ft. lbs. @2000 rpm for all other applications

79241C

GASOLINE ENGINE TUNE-UP SPECIFICATIONS

Year	Engine ID/VIN	Engine Displacement Liters (cc)	Spark Plugs Gap (in.)	Ignition Timing (deg.) MT	AT	Fuel Pump (psi)	Idle Speed (rpm) MT	AT	Valve Clearance In.	Ex.
1995	A	2.3 (2294)	0.044	10B	10B	35–45	725	675	HYD	HYD
	G	7.5 (7538)	0.044	10B	10B	35–45	775	675	HYD	HYD
	H	5.8 (5752)	0.044	10B	10B	35–45	775	675	HYD	HYD
	N	5.0 (4949)	0.044	10B	10B	35–45	775	675	HYD	HYD
	R	5.8 (5752)	0.044	10B	10B	35–45	775	675	HYD	HYD
	U	3.0 (2999)	0.044	10B	10B	35–45	①	①	HYD	HYD
	X	4.0 (3998)	0.054	10B	10B	35–45	①	①	HYD	HYD
	Y	4.9 (4916)	0.044	10B	10B	50–60	700	575	HYD	HYD
1996	2	4.2 (4195)	0.052-0.056	10B ②	10B ②	30–45 ③	NA	NA	HYD	HYD
	6	4.6 (4588)	0.052-0.056	8-12B ②	8-12B ②	30–45 ③	NA	NA	HYD	HYD
	9	4.6 (4588)	NA	NA	NA	30–45 ③	NA	NA	HYD	HYD
	A	2.3 (2300)	0.044	10B	10B	35–45	725	675	HYD	HYD
	G	7.5 (7538)	0.044	10B	10B	35–45	775	675	HYD	HYD
	H	5.8 (5752)	0.044	10B	10B	35–45	775	675	HYD	HYD
	N	5.0 (4949)	0.044	10B	10B	35–45	775	675	HYD	HYD
	P	5.0 (4949)	0.044	—	10B	35–45	—	①	HYD	HYD
	U	3.0 (2982)	0.044	10B	10B	35–45	①	①	HYD	HYD
	W	4.6 (4588)	0.052-0.056	8-12B ②	8-12B ②	30–45 ③	NA	NA	HYD	HYD
	X	4.0 (3950)	0.054	10B	10B	35–45	①	①	HYD	HYD
	Y	4.9 (4916)	0.044	10B	10B	50–60	700	575	HYD	HYD
1997	2	4.2 (4195)	0.052-0.056	10B ②	10B ②	30–45 ③	NA	NA	HYD	HYD
	6	4.6 (4588)	0.052-0.056	8-12B ②	8-12B ②	30–45 ③	NA	NA	HYD	HYD
	A	2.3 (2300)	0.044	10B	10B	35–45	725	675	HYD	HYD
	G	7.5 (7538)	0.044	10B	10B	35–45	775	675	HYD	HYD
	H	5.8 (5752)	0.044	10B	10B	35–45	775	675	HYD	HYD
	N	5.0 (4949)	0.044	10B	10B	35–45	775	675	HYD	HYD
	P	5.0 (4949)	0.044	—	10B	35–45	—	①	HYD	HYD
	U	3.0 (2982)	0.044	10B	10B	35–45	①	①	HYD	HYD
	W	4.6 (4588)	0.052-0.056	8-12B ②	8-12B ②	30–45 ③	NA	NA	HYD	HYD
	X	4.0 (3950)	0.054	10B	10B	35–45	①	①	HYD	HYD
1998-99	2	4.2 (4195)	0.052-0.056	10B ②	10B ②	30–45 ③	NA	NA	HYD	HYD
	5	6.8 (6802)	0.052-0.055	10B ②	10B ②	28–45	①	①	HYD	HYD
	6/W	4.6 (4588)	0.052-0.056	8-12B ②	8-12B ②	30–45 ③	NA	NA	HYD	HYD
	C	2.5 (2500)	0.044	10B ②	10B ②	56–72	①	①	HYD	HYD
	E	4.0 (3950)	0.052-0.056	10B ②	10B ②	35–45	①	①	HYD	HYD
	G	7.5 (7538)	0.044	10B	10B	35–45	775	675	HYD	HYD
	H	5.8 (5752)	0.044	10B	10B	35–45	775	675	HYD	HYD
	L	5.4 (5409)	0.052-0.056	10B ②	10B ②	28–45	①	①	HYD	HYD
	P	5.0 (4949)	0.044	—	10B	35–45	—	①	HYD	HYD
	U	3.0 (2982)	0.044	10B	10B	35–45	①	①	HYD	HYD
	X	4.0 (3950)	0.054	10B	10B	35–45	①	①	HYD	HYD

NOTE: The Vehicle Emission Control Information label often reflects specification changes changes made during production. The label figures must be used if they differ from those in this chart.

B - Before top dead center

HYD - Hydraulic

NA - Not Available

① Idle speed is electronically controlled and cannot be adjusted

② Ignition timing is preset and cannot be adjusted

③ With engine running

79241C71

DIESEL ENGINE TUNE-UP SPECIFICATIONS

Year	Engine ID/VIN	Engine Displacement cu. in. (cc)	Valve Clearance		Intake Valve Opens (deg.)	Injection Pump Setting (deg.)		Injection Nozzle Pressure (psi)		Idle Speed (rpm)	Cranking Compression Pressure (psi)
			Intake (in.)	Exhaust (in.)				New	Used		
1995	F	7.3 (7292)	HYD	HYD	—		①	1875	1425	②	③
	M	7.3 (7292)	HYD	HYD	—	8.5B	④	NA	NA	②	③
1996	F	7.3 (7292)	HYD	HYD	—		①	1875	1425	②	③
1997	F	7.3 (7292)	HYD	HYD	—		①	1875	1425	②	③
1998-99	F	7.3 (7292)	HYD	HYD	—		①	1875	1425	②	③

NOTE: The Vehicle Emission Control Information label often reflects specification changes made during production. The label figures must be used if they differ from those in this chart

HYD - Hydraulic

B - Before top dead center

NA - Not Available

① PCM controlled

② See underhood emission label

③ Compression pressure in the lowest cylinder must be at least 75%
   of the highest cylinder
   Minimum pressure: 195 psi
   Maximum pressure: 440 psi

④ At 2000 rpm

79241C72

CAPACITIES

Year	Model	Engine ID/VIN	Engine Displacement Liters (cc)	Engine Oil with Filter (qts.)	Transmission (pts.) 4-Spd	5-Spd	Auto.	Transfer Case (pts.)	Drive Axle Front (pts.)	Rear (pts.)	Fuel Tank (gal.)	Cooling System (qts.)
1995	Aerostar	U	3.0 (2999)	4.5	—	5.6	19.0	2.5	3.0	[11]	21.0	11.8
	Aerostar	X	4.0 (3998)	5.0	—	5.6	19.0	2.5	3.0	[11]	21.0	12.6
	Bronco	Y	4.9 (4916)	6.0	7.0	7.0	24.0	[1]	5.5	5.5	32.0	14.0
	Bronco	N	5.0 (4949)	6.0	7.0 [2]	7.0	24.0	[1]	5.5	5.5	32.0	14.0
	Bronco	H	5.8 (5752)	6.0	7.0 [2]	7.0	24.0	[1]	5.5	5.5	32.0	15.0
	E-150	Y	4.9 (4916)	6.0	7.0 [2]	7.0	24.0	—	—	6.0 [3]	[4]	14.0
	E-150	N	5.0 (4949)	6.0	7.0 [2]	7.0	24.0	—	—	6.0 [3]	6.0 [4]	15.0
	E-150	H	5.8 (5752)	6.0	7.0 [2]	7.0	24.0	—	—	6.0 [3]	[4]	14.0
	E-250	Y	4.9 (4916)	6.0	7.0 [2]	7.0	24.0	—	—	6.0 [3]	[4]	14.0
	E-250	N	5.0 (4949)	6.0	7.0 [2]	7.0	24.0	—	—	6.0 [3]	[4]	15.0
	E-250	H	5.8 (5752)	6.0	7.0 [2]	7.0	24.0	—	—	6.0 [3]	[4]	15.0
	E-250	M	7.3 (7292)	10.0	7.0 [2]	7.0	24.0	—	—	6.0 [3]	[4]	20.0
	E-250	L	7.5 (7538)	6.0	7.0 [2]	7.0	24.0	—	—	6.0 [3]	[4]	10.9
	E-350	Y	4.9 (4916)	6.0	7.0 [2]	7.0	24.0	—	—	6.0 [3]	[4]	17.5
	E-350	H	5.8 (5752)	6.0	7.0 [2]	7.0	24.0	—	—	6.0 [3]	[4]	15.0
	E-350	M	7.3 (7292)	10.0	7.0 [2]	7.0	24.0	—	—	6.0 [3]	[4]	20.0
	E-350	L	7.5 (7538)	6.0	7.0 [2]	7.0	24.0	—	—	6.0 [3]	[4]	19.8
	Explorer	X	4.0 (3998)	5.0	—	5.6	[14]	3.0	[11]	[11]	19.3	[18]
	F-150	Y	4.9 (4916)	6.0	7.0 [2]	7.0	24.0 [10]	[1]	6.0	6.0 [3]	[4]	[5]
	F-150	N	5.0 (4949)	6.0	7.0 [2]	7.0	24.0 [10]	[1]	6.0	6.0 [3]	[4]	[6]
	F-150	H	5.8 (5752)	6.0	7.0 [2]	7.0	24.0 [10]	[1]	6.0	6.0 [3]	[4]	[7]
	F-250	Y	4.9 (4916)	6.0	7.0 [2]	7.0	24.0 [10]	[1]	6.0	6.0 [3]	[4]	[5]
	F-250	N	5.0 (4949)	6.0	7.0 [2]	7.0	24.0 [10]	[1]	6.0	6.0 [3]	[4]	[5]
	F-250	H	5.8 (5752)	6.0	7.0 [2]	7.0	24.0 [10]	[1]	6.0	6.0 [3]	[4]	[7]
	F-250	L	7.5 (7538)	6.0	7.0 [2]	7.0	24.0 [10]	[1]	6.0	6.0 [3]	19.0	19.8
	F-350	Y	4.9 (4916)	6.0	7.0 [2]	7.0	24.0 [10]	[1]	6.0	6.0 [3]	19.0	[5]
	F-350	H	5.8 (5752)	6.0	7.0 [2]	7.0	24.0 [10]	[1]	6.0	6.0 [3]	19.0	[7]
	F-350	M	7.3 (7292)	10.0	7.0 [2]	7.0	24.0 [10]	[1]	6.0	6.0 [3]	19.0	20.0
	F-350	L	7.5 (7538)	6.0	7.0 [2]	7.0	24.0 [10]	[1]	6.0	6.0 [3]	19.0	19.8
	F-Super Duty	C	7.3 (7292)	10.0	7.0 [2]	7.0	24.0 [10]	[1]	6.0	6.0 [3]	19.0	20.0
	Ranger	A	2.3 (2294)	5.0	—	[12]	[14]	3.0	[11]	5.5	[17]	[19]
	Ranger	U	3.0 (2999)	4.5	—	3.0	[14]	[15]	[16]	[16]	[17]	[20]
	Ranger	X	4.0 (3998)	5.0	—	3.0	[14]	[15]	[16]	[16]	[17]	[18]
1996	Aerostar	U	3.0 (2982)	5.0	—	5.6	19.0	2.5	3.0	[11]	21.0	11.8
	Aerostar	X	4.0 (3950)	5.0	—	5.6	19.0	2.5	3.0	[11]	21.0	12.6
	Bronco	Y	4.9 (4916)	6.0	[9]	7.6	24.0 [10]	[8]	5.5	5.5	32.0	14.0
	Bronco	N	5.0 (4949)	6.0	[9]	7.6	24.0	[8]	5.5	5.5	32.0	14.0
	Bronco	H	5.8 (5752)	6.0	[9]	7.6	24.0 [9]	[8]	5.5	5.5	32.0	15.0
	E-150	Y	4.9 (4916)	6.0	[9]	7.6	24.0 [9]	—	—	6.0 [3]	[4]	14.0
	E-150	N	5.0 (4949)	6.0	[9]	7.6	24.0 [9]	—	—	6.0 [3]	[4]	15.0
	E-150	H	5.8 (5752)	6.0	[9]	7.6	24.0 [9]	—	—	6.0 [3]	[4]	14.0
	E-250	Y	4.9 (4916)	6.0	[9]	7.6	24.0 [9]	—	—	6.0 [3]	[4]	14.0
	E-250	N	5.0 (4949)	6.0	[9]	7.6	24.0 [9]	—	—	6.0 [3]	[4]	15.0
	E-250	H	5.8 (5752)	6.0	[9]	7.6	24.0 [9]	—	—	6.0 [3]	[4]	15.0
	E-250	L	7.5 (7538)	6.0	[9]	7.6	24.0 [9]	—	—	6.0 [3]	[4]	10.9
	E-350	Y	4.9 (4916)	6.0	[9]	7.6	24.0 [9]	—	—	6.0 [3]	[4]	17.5
	E-350	H	5.8 (5752)	6.0	[9]	7.6	24.0 [9]	—	—	6.0 [3]	[4]	15.0
	E-350	F	7.3 (7292)	14.0	[9]	7.6	24.0 [9]	—	—	6.0 [3]	[4]	23.0
	E-350	L	7.5 (7538)	6.0	[9]	7.6	24.0 [9]	—	—	6.0 [3]	[4]	19.8
	Explorer	X	4.0 (3950)	5.0	—	5.6	[14]	3.0	[11]	[11]	19.3	[18]
	Explorer	P	5.0 (4949)	5.0	—	—	13.9	—	—	5.5	19.0	12.8
	F-150	2	4.2 (4195)	6.0	—	7.6	26.0	4.0	3.7	5.5	24.5 [21]	15.7 [22]

79241C73

CAPACITIES

Year	Model	Engine ID/VIN	Engine Displacement Liters (cc)	Engine Oil with Filter (qts.)	Transmission (pts.) 4-Spd	5-Spd	Auto.	Transfer Case (pts.)	Drive Axle Front (pts.)	Rear (pts.)	Fuel Tank (gal.)	Cooling System (qts.)
1996 (cont.)	F-150	W	4.6 (4588)	6.0	—	7.6	26.0	4.0	3.7	5.5	24.5 [21]	17.9
	F-150	6	4.6 (4588)	6.0	—	7.6	26.0	4.0	3.7	5.5	24.5 [21]	17.9
	F-150	9	4.6 (4588)	6.0	—	7.6	26.0	4.0	3.7	5.5	24.5 [21]	17.9
	F-150	Y	4.9 (4916)	6.0	[9]	7.6	24.0 [9]	[8]	6.0	6.0 [3]	[10]	[5]
	F-150	N	5.0 (4949)	6.0	[9]	7.6	24.0 [9]	[8]	6.0	6.0 [3]	[10]	[6]
	F-150	H	5.8 (5752)	6.0	[9]	7.6	24.0 [9]	[8]	6.0	6.0 [3]	[10]	[7]
	F-250	Y	4.9 (4916)	6.0	[9]	7.6	24.0 [9]	[8]	6.0	6.0 [3]	[10]	[5]
	F-250	N	5.0 (4949)	6.0	[9]	7.6	24.0 [9]	[8]	6.0	6.0 [3]	[10]	[6]
	F-250	H	5.8 (5752)	6.0	[9]	7.6	24.0 [9]	[8]	6.0	6.0 [3]	[10]	[7]
	F-250	F	7.3 (7292)	14.0	[9]	7.6	24.0 [9]	[8]	6.0	6.0 [3]	[10]	23.0
	F-250	L	7.5 (7538)	6.0	[9]	7.6	24.0 [9]	[8]	6.0	6.0 [3]	[10]	19.8
	F-350	Y	4.9 (4916)	6.0	[9]	7.6	24.0 [9]	[8]	6.0	6.0 [3]	[10]	[5]
	F-350	H	5.8 (5752)	6.0	[9]	7.6	24.0 [9]	[8]	6.0	6.0 [3]	[10]	[7]
	F-350	F	7.3 (7292)	14.0	—	7.6	24.0 [9]	[8]	6.0	6.0 [3]	[10]	23.0
	F-350	L	7.5 (7538)	6.0	[9]	7.6	24.0 [9]	[8]	6.0	6.0 [3]	[10]	19.8
	F-Super Duty	C	7.3 (7292)	10.0	7.0 [2]	7.0	24.0 [10]	[1]	6.0	6.0 [3]	19.0	20.0
	Mountaineer	P	5.0 (4949)	5.0	—	—	13.9	—	—	5.5		12.8
	Ranger	A	2.3 (2300)	5.0	—	[12]	[14]	3.0	[11]	5.5	[17]	[19]
	Ranger	U	3.0 (2982)	4.5	—	3.0	[14]	[15]	[16]	[16]	[17]	[20]
	Ranger	X	4.0 (3950)	5.0	—	3.0	[14]	[15]	[16]	[16]	[17]	[18]
1997	Aerostar	U	3.0 (2982)	5.0	—	5.6	19.0	2.5	3.0	[11]	21.0	11.8
	Aerostar	X	4.0 (3950)	5.0	—	5.6	19.0	2.5	3.0	[11]	21.0	12.6
	E-150	2	4.2 (4195)	6.0	—	—	[9]	—	—	6.0	35.0	15.7
	E-150	6/W	4.6 (4588)	6.0	—	—	[9]	—	—	6.0	35.0	17.9
	E-150	L	5.4 (5409)	7.0	—	—	[9]	—	—	6.0	35.0	19.8
	E-250	2	4.2 (4195)	6.0	—	—	[9]	—	—	6.0 [3]	35.0	15.7
	E-250	L	5.4 (5409)	7.0	—	—	[9]	—	—	6.0 [3]	35.0	19.8
	E-350	L	5.4 (5409)	7.0	—	—	[9]	—	—	6.0 [3]	35.0	19.8
	E-350	5	6.8 (6802)	7.0	—	—	[9]	—	—	6.0 [3]	35.0	23.0
	E-350	F	7.3 (7292)	7.0	—	—	[9]	—	—	6.0 [3]	35.0	23.0
	Explorer	X	4.0 (3950)	5.0	—	5.6	[14]	3.0	[11]	[11]	19.3	[18]
	Explorer	P	5.0 (4949)	5.0	—	—	13.9	—	—	5.5	19.0	12.8
	Explorer	E	4.0 (4000)	7.0	—	—	[9]	—	—	6.0 [3]	[23]	23.0
	F-150	2	4.2 (4195)	6.0	—	7.6	26.0	4.0	3.7	5.5	24.5 [21]	15.7 [22]
	F-150	W	4.6 (4588)	6.0	—	7.6	26.0	4.0	3.7	5.5	24.5 [21]	17.9
	F-150	6	4.6 (4588)	6.0	—	7.6	26.0	4.0	3.7	5.5	24.5 [21]	17.9
	F-150	L	5.4 (5409)	7.0	—	7.6	[9]	4.2	6.0	6.0	30.0	19.8
	F-250	H	5.8 (5752)	6.0	[9]	7.6	24.0 [9]	[8]	6.0	6.0 [3]	[10]	[7]
	F-250	F	7.3 (7292)	14.0	[9]	7.6	24.0 [9]	[8]	6.0	6.0 [3]	[10]	23.0
	F-250	L	7.5 (7538)	6.0	[9]	7.6	24.0 [9]	[8]	6.0	6.0 [3]	[10]	19.8
	F-350	H	5.8 (5752)	6.0	[9]	7.6	24.0 [9]	[8]	6.0	6.0 [3]	[10]	[7]
	F-350	F	7.3 (7292)	14.0	—	7.6	24.0 [9]	[8]	6.0	6.0 [3]	[10]	23.0
	F-350	L	7.5 (7538)	6.0	[9]	7.6	24.0 [9]	[8]	6.0	6.0 [3]	[10]	19.8
	F-Super Duty	C	7.3 (7292)	10.0	7.0 [2]	7.0	24.0 [10]	[1]	6.0	6.0 [3]	19.0	20.0
	F-Super Duty	R	5.8 (5758)	7.0	7.0 [2]	7.0	24.0	[1]	6.0 [3]	6.0 [3]	35.0	23.0
	Mountaineer	P	5.0 (4949)	5.0	—	—	13.9	—	—	5.5		12.8
	Ranger	A	2.3 (2300)	5.0	—	[12]	[14]	3.0	[11]	5.5	[17]	[19]
	Ranger	U	3.0 (2982)	4.5	—	3.0	[14]	[15]	[16]	[16]	[17]	[20]
	Ranger	X	4.0 (3950)	5.0	—	3.0	[14]	[15]	[16]	[16]	[17]	[18]
	Expedition	6/W	4.6 (4588)	6.0	—	—	[9]	4.0	3.7	5.5	24.5 [21]	17.9
	Expedition	L	5.4 (5409)	7.0	—	—	[9]	4.0	3.7	5.5	24.5 [21]	20.8

79241C74

CAPACITIES

Year	Model	Engine ID/VIN	Engine Displacement Liters (cc)	Engine Oil with Filter (qts.)	Transmission (pts.) 4-Spd	5-Spd	Auto.	Transfer Case (pts.)	Drive Axle Front (pts.)	Rear (pts.)	Fuel Tank (gal.)	Cooling System (qts.)
1998-99	E-150	2	4.2 (4195)	6.0	—	—	⑨	—	—	6.0	35.0	15.7
	E-150	6/W	4.6 (4588)	6.0	—	—	⑨	—	—	6.0	35.0	17.9
	E-150	L	5.4 (5409)	7.0	—	—	⑨	—	—	6.0	35.0	19.8
	E-250	2	4.2 (4195)	6.0	—	—	⑨	—	—	6.0 ③	35.0	15.7
	E-250	L	5.4 (5409)	7.0	—	—	⑨	—	—	6.0 ③	35.0	19.8
	E-350	L	5.4 (5409)	7.0	—	—	⑨	—	—	6.0 ③	35.0	19.8
	E-350	5	6.8 (6802)	7.0	—	—	⑨	—	—	6.0 ③	35.0	23.0
	E-350	F	7.3 (7292)	7.0	—	—	⑨	—	—	6.0 ③	35.0	23.0
	Explorer	X	4.0 (3950)	5.0	—	5.6	⑭	3.0	⑪	⑪	㉓	⑱
	Explorer	P	5.0 (4949)	5.0	—	—	13.9	—	—	5.5	㉓	12.8
	Explorer	E	4.0 (4000)	7.0	—	—	⑨	—	—	6.0 ③	㉓	23.0
	F-150	2	4.2 (4195)	6.0	—	7.6	26.0	4.0	3.7	5.5	24.5 ㉑	15.7 ㉒
	F-150	6/W	4.6 (4588)	6.0	—	7.6	26.0	4.0	3.7	5.5	24.5 ㉑	17.9
	F-150	L	5.4 (5409)	7.0	—	7.6	⑨	4.2	6.0	6.0	30.0	19.8
	F-250	6/W	4.6 (4588)	6.0	—	7.6	26.0	4.0	3.7	5.5	24.5 ㉑	17.9
	F-250	L	5.4 (5409)	7.0	—	7.6	⑨	4.2	6.0	6.0	30.0	19.8
	F-350	L	5.4 (5409)	7.0	—	7.6	⑨	4.2	6.0	6.0	30.0	19.8
	F-350	R	5.8 (5758)	7.0	7.0 ②	7.0	24.0	①	6.0 ③	6.0 ③	35.0	23.0
	F-350	5	6.8 (6802)	7.0	—	7.6	⑨	4.2	6.0 ③	6.0 ③	35.0	23.0
	F-350	F	7.3 (7292)	7.0	—	7.6	⑨	4.2	6.0 ③	6.0 ③	35.0	23.0
	F-350	G	7.5 (7538)	7.0	—	7.6	⑨	4.2	6.0 ③	6.0 ③	35.0	23.0
	F-Super Duty	L	5.4 (5409)	7.0	—	7.6	⑨	4.2	6.0	6.0	30.0	19.8
	F-Super Duty	5	6.8 (6802)	7.0	—	7.6	⑨	4.2	6.0 ③	6.0 ③	35.0	23.0
	F-Super Duty	F	7.3 (7292)	7.0	—	7.6	⑨	4.2	6.0 ③	6.0 ③	35.0	23.0
	F-Super Duty	G	7.5 (7538)	7.0	—	7.6	⑨	4.2	6.0 ③	6.0 ③	35.0	23.0
	F-Super Duty	R	5.8 (5758)	7.0	7.0 ②	7.0	24.0	①	6.0 ③	6.0 ③	35.0	23.0
	Expedition	6/W	4.6 (4588)	6.0	—	—	⑨	4.0	3.7	5.5	24.5 ㉔	17.9
	Expedition	L	5.4 (5409)	7.0	—	—	⑨	4.0	3.7	5.5	24.5 ㉔	20.8
	Navigator	L	5.4 (5409)	7.0	—	—	⑨	4.0	3.7	5.5	24.5 ㉔	20.8
	Mountaineer	P	5.0 (4949)	5.0	—	—	13.9	—	—	5.5	㉓	12.8
	Ranger	C	2.5 (2500)	5.0	—	⑫	⑭	3.0	⑪	5.5	⑰	⑲
	Ranger	U	3.0 (2982)	4.5	—	3.0	⑭	⑮	⑯	⑯	⑰	⑳
	Ranger	X	4.0 (3950)	5.0	—	3.0	⑭	⑮	⑮	⑯	⑰	⑱

NOTE: All capacities are approximate. Add fluid gradually and check to be sure a proper fluid level is obtained.

① New Process: 9 pts. Dexron II
BW 1345: 6.5 pts. Dexron II
BW 1356: 4 pts. Mercon

② With OD: 4.5 pts.

③ Heavy duty: 7.5 pts.

④ 124" Wheelbase: 18 gals.
138", 158" and 176" Wheelbase and front-mounted tank: 22 gals.
138", 158" and 176" Wheelbase and rear-mounted tank: 16 gals.

⑤ 4.9L without AC: 13.0 qts.
4.9L with AC or supercooling: 14.0 qts.
4.9L with AC and supercooling: 15.6 qts.

⑥ 5.0L with manual trans. and standard cooling system: 15.7 qts.
5.0L with automatic trans. and standard cooling: 16.5 qts.
5.0L with manual/automatic trans. and AC: 16.4 qts.
5.0L with manual/automatic trans. with supercooling/AC: 18.3 qts.

⑦ Manual trans. with standard cooling: 15.7 qts.
Automatic trans. with standard cooling: 16.4 qts.
Manual/automatic trans. with AC: 16.4 qts.
Manual/automatic trans. with supercooling and AC: 18.0 qts.

⑧ Without PTO: 4.2 pts.
With PTO: 12.0 pts.

⑨ With 4R70W: 28 pts.
With E40D: 32.0 pts.

⑩ 4WD: 27 pts.

⑪ Front axle Dana 28: 1.1 pts.
Front axle Dana 35: 3.5 pts.
Rear axle: 5.5 pts.

⑫ Mazda trans.: 3 pts.
Mitsubishi trans.: 4.8 pts.

⑬ Short wheelbase: 17 gals.
Long wheelbase: 17 or 21 gals.
Ranger Supercab: 17 or 21 gals.

⑭ 2WD: 19.4 pts.
4WD: 20.0 pts.

⑮ BW 13-50 manual shift: 3.0 pts.
BW 13-50 electric shift: 6.5 pts.
BW 13-54 mechanical shift: 3.0 pts.
BW 13-50 contains no lubricant
and none should be added

⑯ 6.75" ring gear: 3 pts.
7.50" ring gear: 5 pts.

⑰ Short wheelbase: 16.3
Long wheelbase: 19.6
Ranger Supercab: 19.6

⑱ 4.0L without AC: 7.8 qts.
4.0L with AC: 8.6 qts.

⑲ 2.3L without AC: 6.5 qts.
2.3L with AC: 7.2 qts.

⑳ 3.0L without AC: 9.5 qts.
3.0L with AC: 10.2 qts.

㉑ Also available with a 30 gallon
tank with 8 ft. box

㉒ Includes recovery reservoir

㉓ 2-door: 17.5
4-door: 21.0

㉔ 4.6L: 26.0
5.4L: 30.0

79241C75

VALVE SPECIFICATIONS

Year	Engine ID/VIN	Engine Displacement Liters (cc)	Seat Angle (deg.)	Face Angle (deg.)	Spring Test Pressure (lbs. @ in.)	Spring Installed Height (in.)	Stem-to-Guide Clearance (in.)		Stem Diameter (in.)	
							Intake	Exhaust	Intake	Exhaust
1995	A	2.3 (2300)	45	44	126-142@1.12	1.53-1.59	0.0010-0.0027	0.0015-0.0032	0.3416-0.3423	0.3411-0.3418
	U	3.0 (2999)	45	44	185@1.16	1.58-1.61	0.0010-0.0027	0.0015-0.0032	0.3126-0.3134	0.3121-0.3129
	X	4.0 (3998)	45	44	138@1.22	1.58-1.61	0.0008-0.0025	0.0018-0.0035	0.3159-0.3167	0.3149-0.3156
	Y	4.9 (4916)	45	44	⑧	①	0.0010-0.0027	0.0010-0.0027	0.3415-0.3423	0.3415-0.3423
	N	5.0 (4949)	45	44	200@1.20	⑥	0.0010-0.0027	0.0015-0.0032	0.3415-0.3423	0.3415-0.3423
	H	5.8 (5752)	45	44	200@1.20	③	0.0010-0.0027	0.0010-0.0027	0.3415-0.3420	0.3415-0.3420
	R	5.8 (5752)	45	44	200@1.20	③	0.0010-0.0027	0.0010-0.0027	0.3415-0.3420	0.3415-0.3420
	F	7.3 (7292)	④	④	200@1.38	⑤	0.0055	0.0055	0.3119-0.3126	0.3119-0.3126
	M	7.3 (7292)	④	④	200@1.40	⑤	0.0055	0.0055	0.3716-0.3723	0.3716-0.3723
	G	7.5 (7538)	45	44	220@1.33	1.83	0.0010-0.0027	0.0010-0.0027	0.3415-0.3423	0.3415-0.3423
1996	A	2.3 (2300)	45	44	126-142@1.12	1.53-1.59	0.0010-0.0027	0.0015-0.0032	0.3416-0.3423	0.3411-0.3418
	U	3.0 (2982)	45	44	185@1.16	1.58-1.61	0.0010-0.0027	0.0015-0.0032	0.3126-0.3134	0.3121-0.3129
	X	4.0 (3950)	45	44	138@1.22	1.58-1.61	0.0008-0.0025	0.0018-0.0035	0.3159-0.3167	0.3149-0.3156
	2	4.2 (4195)	44.75	NA	NA	1.566-1.637	0.0008-0.0027	0.0018-0.0037	0.3423-0.3415	0.3418-0.3410
	W	4.6 (4588)	45	45.5	132@1.100	1.570	0.0008-0.0027	0.0018-0.0037	0.2750-0.2746	0.2740-0.2736
	6	4.6 (4588)	45	45.5	132@1.100	1.570	0.0008-0.0027	0.0018-0.0037	0.2750-0.2746	0.2740-0.2736
	9	4.6 (4588)	45	45.5	132@1.100	1.570	0.0008-0.0027	0.0018-0.0037	0.2750-0.2746	0.2740-0.2736
	Y	4.9 (4916)	45	44	⑧	①	0.0010-0.0027	0.0010-0.0027	0.3415-0.3423	0.3415-0.3423
	N	5.0 (4949)	45	44	200@1.20	⑥	0.0010-0.0027	0.0015-0.0032	0.3415-0.3423	0.3415-0.3423
	P	5.0 (4949)	45	44	200@1.20	⑥	0.0010-0.0027	0.0015-0.0032	0.3415-0.3423	0.3415-0.3410
	H	5.8 (5752)	45	44	200@1.20	③	0.0010-0.0027	0.0010-0.0027	0.3415-0.3420	0.3415-0.3420
	F	7.3 (7292)	④	④	200@1.38	⑤	0.0055	0.0055	0.3119-0.3126	0.3119-0.3126
	G	7.5 (7538)	45	44	220@1.33	1.83	0.0010-0.0027	0.0010-0.0027	0.3415-0.3423	0.3415-0.3423
1997	A	2.3 (2300)	45	44	126-142@1.12	1.53-1.59	0.0010-0.0027	0.0015-0.0032	0.3416-0.3423	0.3411-0.3418
	U	3.0 (2982)	45	44	185@1.16	1.58-1.61	0.0010-0.0027	0.0015-0.0032	0.3126-0.3134	0.3121-0.3129
	X	4.0 (3950)	45	44	138@1.22	1.58-1.61	0.0008-0.0025	0.0018-0.0035	0.3159-0.3167	0.3149-0.3156

VALVE SPECIFICATIONS

Year	Engine ID/VIN	Engine Displacement Liters (cc)	Seat Angle (deg.)	Face Angle (deg.)	Spring Test Pressure (lbs. @ in.)	Spring Installed Height (in.)	Stem-to-Guide Clearance (in.)		Stem Diameter (in.)	
							Intake	Exhaust	Intake	Exhaust
1997 (cont.)	2	4.2 (4195)	44.75	NA	NA	1.566-1.637	0.0008-0.0027	0.0018-0.0037	0.3423-0.3415	0.3418-0.3410
	W	4.6 (4588)	45	45.5	132@1.100	1.570	0.0008-0.0027	0.0018-0.0037	0.2750-0.2746	0.2740-0.2736
	6	4.6 (4588)	45	45.5	132@1.100	1.570	0.0008-0.0027	0.0018-0.0037	0.2750-0.2746	0.2740-0.2736
	N	5.0 (4949)	45	44	200@1.20	⑥	0.0010-0.0027	0.0015-0.0032	0.3415-0.3423	0.3415-0.3423
	P	5.0 (4949)	45	44	200@1.20	⑥	0.0010-0.0027	0.0015-0.0032	0.3415-0.3423	0.3410-0.3418
	H	5.8 (5752)	45	44	200@1.20	③	0.0010-0.0027	0.0010-0.0027	0.3415-0.3420	0.3415-0.3420
	F	7.3 (7292)	④	④	200@1.38	⑤	0.0055	0.0055	0.3119-0.3126	0.3119-0.3126
	G	7.5 (7538)	45	44	220@1.33	1.83	0.0010-0.0027	0.0010-0.0027	0.3415-0.3423	0.3415-0.3423
	L	5.4 (5409)	45	45.5	150@1.10	1.570	0.0008-0.0027	0.0018-0.0037	0.275-0.2746	0.274-0.2736
	5	6.8 (6802)	44.50-45.25	45.25-45.75	150@1.10	1.570	0.0008-0.0027	0.0018-0.0037	0.275-0.2746	0.274-0.2735
	E	4.0 (4000)	45	45	202-225@1.413-1.445	1.569-1.601	0.0010-0.0020	0.0010-0.0020	0.274-0.2748	0.273-0.2740
1998-99	E	4.0 (4000)	45	45	202-225@1.413-1.445	1.569-1.601	0.0010-0.0020	0.0010-0.0020	0.274-0.2748	0.273-0.2740
	X	4.0 (3950)	45	44	138@1.22	1.58-1.61	0.0008-0.0025	0.0018-0.0035	0.3159-0.3167	0.3149-0.3156
	P	5.0 (4949)	45	44	200@1.20	⑥	0.0010-0.0027	0.0015-0.0032	0.3415-0.3423	0.3410-0.3418
	6/W	4.6 (4588)	45	45.5	132@1.100	1.570	0.0008-0.0027	0.0018-0.0037	0.2750-0.2746	0.2740-0.2736
	2	4.2 (4195)	44.75	NA	NA	1.566-1.637	0.0008-0.0027	0.0018-0.0037	0.3423-0.3415	0.3418-0.3410
	L	5.4 (5409)	45	45.5	150@1.10	1.570	0.0008-0.0027	0.0018-0.0037	0.275-0.2746	0.274-0.2736
	R	5.8 (5752)	45	44	200@1.20	③	0.0010-0.0027	0.0010-0.0027	0.3415-0.3420	0.3415-0.3420
	5	6.8 (6802)	44.50-45.25	45.25-45.75	150@1.10	1.570	0.0008-0.0027	0.0018-0.0037	0.275-0.2746	0.274-0.2735
	G	7.5 (7538)	45	44	220@1.33	1.83	0.0010-0.0027	0.0010-0.0027	0.3415-0.3423	0.3415-0.3423
	F	7.3 (7292)	④	④	200@1.38	⑤	0.0055	0.0055	0.3119-0.3126	0.3119-0.3126
	U	3.0 (2982)	45	44	185@1.16	1.58-1.61	0.0010-0.0027	0.0015-0.0032	0.3126-0.3134	0.3121-0.3129
	C	2.5 (2500)	45	44	57-63@1.56	1.540-1.580	0.0008-0.0025	0.0018-0.0037	0.2746-0.2754	0.2736-0.2744

① Intake: 1.64 in.
Exhaust: 1.47 in.
② Intake: 1.68 in.
Exhaust: 1.59 in.
③ Intake: 1.78 in.
Exhaust: 1.59 in.
④ Intake: 30
Exhaust: 37.5
⑤ Intake: 1.767 in.
Exhaust: 1.833 in.
⑥ Intake: 1.75-1.81 in.
Exhaust: 1.59 in.
⑦ Intake: 166@1.240
Exhaust: 166@1.070
⑧ Intake: 166-184@1.240
Exhaust: 1.66-184@1.070

79241C77

TORQUE SPECIFICATIONS
All readings in ft. lbs.

	Engine ID/VIN	Engine Displacement Liters (cc)	Cylinder Head Bolts	Main Bearing Bolts	Rod Bearing Bolts	Crankshaft Damper Bolts	Flywheel Bolts	Manifold Intake *	Manifold Exhaust	Spark Plugs	Lug Nut
1995	A	2.3 (2294)	51	75-85	30-36	103-133	54-64	19-28	14-21	5-10	⑦
	U	3.0 (2999)	⑫	60	26	107	54-64	24	25	8-10	⑦
	X	4.0 (3998)	⑨	66-77	19-24	⑩	59	⑪	19	10-15	⑦
	Y	4.9 (4916)	①	60-70	40-45	130-150	75-85	22-32	22-32	10-15	②
	N	5.0 (4949)	③	60-70	19-24	70-90	75-85	23-25	18-24	10-15	②
	H	5.8 (5752)	④	95-105	40-45	70-90	75-90	23-25	20-24	10-15	②
	R	5.8 (5752)	④	95-105	40-45	70-90	75-90	23-25	20-24	10-15	②
	F	7.3 (7292)	⑤	95	70	90	89	18	45	—	—
	M	7.3 (7292)	⑤	13	14	90	47	23-25	20-24	—	②
	G	7.5 (7538)	⑥	95-105	45-40	70-90	75-85	22-32	28-33	5-10	②
1996	A	2.3 (2300)	51	75-85	30-36	103-133	54-64	19-28	14-21	5-10	100
	U	3.0 (2982)	⑫	60	26	107	54-64	24	25	8-10	100
	X	4.0 (3950)	⑬	66-77	19-24	⑩	59	⑪	19	10-15	100
	2	4.2 (4195)	⑭	81-88	⑮	103-117	54-63	⑯	15-22	8-14	83-113
	W	4.6 (4588)	⑰	⑱	29-33	⑲	54-64	⑳	15	7-14	83-113
	6	4.6 (4588)	⑱	㉑	29-33	⑲	54-64	⑳	15	7-14	83-113
	9	4.6 (4588)	⑰	NA	29-33	⑲	54-64	⑳	15	7-14	83-113
	P	5.0 (4949)	③	60-70	19-24	110-130	75-85	12-18	26-32	7-15	100
	Y	4.9 (4916)	①	60-70	40-45	130-150	75-85	22-32	22-32	10-15	②
	N	5.0 (4949)	③	60-70	19-24	70-90	75-85	23-25	18-24	10-15	②
	H	5.8 (5752)	④	95-105	40-45	70-90	75-90	23-25	20-24	10-15	②
	F	7.3 (7292)	⑤	95	70	90	89	18	45	—	—
	G	7.5 (7538)	⑥	95-105	40-45	70-90	75-85	22-32	28-33	5-10	—
1997	A	2.3 (2300)	51	75-85	30-36	103-133	54-64	19-28	14-21	5-10	100
	U	3.0 (2982)	⑫	60	26	107	54-64	24	25	8-10	100
	X	4.0 (3950)	⑬	66-77	19-24	⑩	59	⑪	19	10-15	100
	2	4.2 (4195)	⑭	81-88	⑮	103-117	54-63	⑯	15-22	8-14	83-113
	W	4.6 (4588)	⑰	⑱	29-33	⑲	54-64	⑳	15	7-14	83-113
	6	4.6 (4588)	⑰	㉑	29-33	⑲	54-64	⑳	15	7-14	83-113
	P	5.0 (4949)	③	60-70	19-24	110-130	75-85	12-18	26-32	7-15	100
	N	5.0 (4949)	③	60-70	19-24	70-90	75-85	23-25	18-24	10-15	②
	H	5.8 (5752)	④	95-105	40-45	70-90	75-90	23-25	20-24	10-15	②
	F	7.3 (7292)	⑤	95	70	90	89	18	45	—	—
	G	7.5 (7538)	⑥	95-105	40-45	70-90	75-85	22-32	28-33	5-10	—
	L	5.4 (5409)	㉔	㉕	㉖	⑲	54-64	㉓	17-19	9-20	100
	5	6.8 (6802)	㉔	㉕	㉖	⑲	54-64	㉓	17-20	7-14	140
	E	4.0 (4000)	㉒	67-74	19-24	⑧	54-64	9-10	15-18	7-14	100
1998-99	U	3.0 (2982)	⑫	60	26	107	54-64	24	25	8-10	100
	X	4.0 (3950)	⑬	66-77	19-24	⑩	59	⑪	19	10-15	100
	2	4.2 (4195)	⑭	81-88	⑮	103-117	54-63	⑯	15-22	8-14	83-113
	6/W	4.6 (4588)	⑰	㉑	29-33	⑲	54-64	⑳	15	7-14	83-113
	P	5.0 (4949)	③	60-70	19-24	110-130	75-85	12-18	26-32	7-15	100
	H	5.8 (5752)	④	95-105	40-45	70-90	75-90	23-25	20-24	10-15	②
	F	7.3 (7292)	⑤	95	70	90	89	18	45	—	—
	G	7.5 (7538)	⑥	95-105	40-45	70-90	75-85	22-32	28-33	5-10	—
	L	5.4 (5409)	㉔	㉕	㉖	⑲	54-64	㉓	17-19	9-20	100

79241C78

TORQUE SPECIFICATIONS
All readings in ft. lbs.

	Engine ID/VIN	Engine Displacement Liters (cc)	Cylinder Head Bolts	Main Bearing Bolts	Rod Bearing Bolts	Crankshaft Damper Bolts	Flywheel Bolts	Manifold		Spark Plugs	Lug Nut
								Intake *	Exhaust		
1998-99	C	2.5 (2500)	51	75-85	30-36	103-133	54-64	19-28	14-21	5-10	100
	5	6.8 (6802)	㉔	㉕	㉖		⑲	54-64	㉓	17-20	7-14	140
	E	4.0 (4000)	㉒	67-74	19-24		⑧	54-64	9-10	15-18	7-14	100

* NOTE: Applies to Lower Manifold only.

① Step 1: 55 ft. lbs.
   Step 2: 65 ft. lbs.
   Step 3: 85 ft. lbs.
② E-F100, E-F150, E-F250: 90 ft. lbs.
   E-F350 with single rear wheels: 135 ft. lbs.
   F350 with dual rear wheels: 210 ft. lbs.
③ With flanged head bolts:
   Step 1: 25-35 ft. lbs.
   Step 2: 40-55 ft. lbs.
   Step 3: Turn an additional 1/4 turn
   With hex head bolts:
   Step 1: 55-65 ft. lbs.
   Step 2: 65-72 ft. lbs.
④ Step 1: 95-105 ft. lbs.
   Step 2: 105-112 ft. lbs.
⑤ Step 1: 65 ft. lbs.
   Step 2: 85 ft. lbs.
   Step 3: 105 ft. lbs.
⑥ Step 1: 70-80 ft. lbs.
   Step 2: 100-110 ft. lbs.
   Step 3: 130-140 ft. lbs.
⑦ Aerostar, Explorer and Ranger: 100 ft. lbs.
⑧ Step 1: 20-28 ft. lbs.
   Step 2: Back off a minimum of two turns
   Step 3: 20-25 ft. lbs.
⑨ Tighten cylinder head bolts to 44 ft. lbs.
   Tighten intake manifold bolts to 3-6 ft. lbs.
   Tighten cylinder head bolts to 59 ft. lbs.
   Tighten intake manifold bolts to 6-11 ft. lbs.
   Tighten cylinder head bolts 85 degrees
   Tighten intake manifold bolts to 11-15 ft. lbs.
   Tighten intake manifold bolts to 15-18 ft. lbs.
⑩ Step 1: 30-37 ft. lbs.
   Step 2: Turn 90 degrees
⑪ Step 1: 3-6 ft. lbs.
   Step 2: 6-11 ft. lbs.
   Step 3: 11-15 ft. lbs.
   Step 4: 15-18 ft. lbs.
⑫ Step 1: 37 ft. lbs.
   Step 2: 68 ft. lbs.
⑬ Cylinder head bolts:
   Step 1: 44 ft. lbs.
   Step 2: 59 ft. lbs.
   Step 3: Plus 85 degrees
   Intake manifold bolts:
   Step 1: 3-6 ft. lbs.
   Step 2: 6-11 ft. lbs.
   Step 3: 11-15 ft. lbs.
   Step 4: 15-18 ft. lbs.

⑭ Step 1: 29 ft. lbs.
   Step 2: 36 ft. lbs.
   Step 3: Loosen each bolt and tighten one at a time
   Short bolts to 32 ft. lbs.
   Long bolts to 36 ft. lbs.
   Step 4: Turn each bolt an additional 135 degrees
⑮ Step 1: 29 ft. lbs.
   Step 2: Plus 90 degrees
⑯ Tighten bolts in sequence to 71-101 in. lbs.
⑰ Step 1: 31 ft. lbs.
   Step 2: Plus 90 degrees
   Step 3: Plus 90 degrees
⑱ Tighten main bearing jack screws in sequence as follows:
   Step 1: 45 in. lbs.
   Step 2: 98 in. lbs.
   Tighten cross-mounted cap bolts in sequence as follows:
   Step 1: 89 in. lbs.
   Step 2: 17 ft. lbs.
⑲ Step 1: 88 ft. lbs.
   Step 2: Loosen bolt
   Step 3: 39 ft. lbs.
   Step 4: Plus 90 degrees
⑳ Step 1: 18 in. lbs.
   Step 2: 8 ft. lbs.
㉑ Tighten jack screws in sequence as follows:
   Step 1: 45 in. lbs.
   Step 2: 98 in. lbs.
   Tighten cross-mounted cap bolts in sequence as follows:
   Step 1: 24 ft. lbs.
   Step 2: Plus 90 degrees
㉒ Step 1: 28 ft. lbs.
   Step 2: Plus 90 degrees
   Step 3: Plus another 90 degrees
㉓ Step 1: 18 inch lbs.
   Step 2: 71-106 inch lbs.
㉔ Step 1: 27-32 inch lbs.
   Step 2: Plus 90 degrees
   Step 3: Plus another 90 degrees
㉕ Step 1: 27-32 ft. lbs.
   Step 2: Plus 90 degrees
㉖ Step 1: 30-33 ft. lbs.
   Step 2: 90-120 degrees

79241C79

SCHEDULED MAINTENANCE INTERVALS
(FORD AEROSTAR, RANGER, EXPLORER, & MOUNTAINEER)

TO BE SERVICED	TYPE OF SERVICE	VEHICLE MILEAGE INTERVAL (x1000)												
		5	10	15	20	25	30	35	40	45	50	55	60	65
Engine oil & filter	R	✓	✓	✓	✓	✓	✓	✓	✓	✓	✓	✓	✓	✓
Automatic transmission shift linkage (Bell crank system)	S/I	✓	✓	✓	✓	✓	✓	✓	✓	✓	✓	✓	✓	✓
Exhaust system & heat shields	S/I	✓		✓		✓		✓		✓		✓		✓
Rotate tires	S/I	✓		✓		✓		✓		✓		✓		✓
Steering linkage & driveshaft U-joint (if equipped with fitting)	S/I	✓		✓		✓		✓		✓		✓		✓
Clutch reservoir fluid level (Ranger, Explorer & Mountaineer)	S/I	✓		✓		✓		✓		✓		✓		✓
Rear driveshaft double cardan joint centering ball (Ranger, Explorer & Mountaineer SWB 4x4)	S/I	✓		✓		✓		✓		✓		✓		✓
Disc brake system & caliper slide rails	S/I			✓			✓			✓			✓	
Drum brake systems, hoses & lines	S/I			✓			✓			✓			✓	
Engine coolant strength, hoses & clamps	S/I			✓			✓			✓			✓	
Transfer case shift lever pivot bolt & control rod connecting pins (Ranger, Explorer & Mountaineer 4x4)	S/I			✓			✓			✓			✓	
Air cleaner filter	R						✓						✓	
Automatic transmission fluid & filter	R						✓						✓	
Engine Coolant①	R						✓						✓	
Fuel filter	R						✓						✓	
Front axle RH axle shaft slip yoke (Ranger, Explorer & Mountaineer 4x4)	S/I						✓						✓	

79241C80

SCHEDULED MAINTENANCE INTERVALS
(FORD AEROSTAR, RANGER, EXPLORER & MOUNTAINER Cont.)

TO BE SERVICED	TYPE OF SERVICE	VEHICLE MILEAGE INTERVAL (x1000)												
		5	10	15	20	25	30	35	40	45	50	55	60	65
Front suspension ball joints, bushings, arms, springs & rear jounce bumpers (Aerostar)	S/I						✓						✓	
Front wheel bearings (4x2)	S/I						✓						✓	
Hub lock (Ranger, Explorer & Mountaineer 4x4)	S/I						✓						✓	
Parking brake system	S/I						✓						✓	
Spindle needle bearing (Ranger, Explorer & Mountaineer 4x4)	S/I						✓						✓	
Throttle or TV lever ball studs	S/I						✓						✓	
Front axle & transfer case oil (E-4WD)	R												✓	
Manual transmission	R												✓	
PCV valve	R												✓	
Spark plugs (2.3L)②	R												✓	
Transfer case oil (Ranger, Explorer & Mountaineer 4x4)	R												✓	
Drive belts	S/I						✓						✓	

① Engine coolant - change initially at 50,000 miles and every 30,000 miles thereafter.
② Spark plugs (3.0L & 4.0L) - replace every 100,000 miles.
R – Replace   S/I – Service or Inspect

FREQUENT OPERATION MAINTENANCE (SEVERE SERVICE)
If a vehicle is operated under any of the following conditions it is considered severe service:
- **Extremely dusty areas.**
- **50% or more of the vehicle operation is in 32°C (90°F) or higher temperatures, or constant operation in temperatures below 0°C (32°F).**
- **Prolonged idling (vehicle operation in stop and go traffic).**
- **Frequent short running periods (engine does not warm to normal operating temperatures).**
- **Police, taxi, delivery usage or trailer towing usage.**
Oil & oil filter change – change every 3000 miles.
Air cleaner filter - service or inspect every 6000 miles.
Exhaust system - check every 6000 miles.
Rotate tires every 9000 miles. (City delivery vehicles & other unique applications that require constant turning may need frequent tire rotation.)
Automatic transmission fluid & filter - change every 21,000 miles.

79241C81

1995–96 SCHEDULED MAINTENANCE INTERVALS
(FORD F-150/250/250HD/350 & SUPER DUTY/E-150/250/350 & CLUB WAGON)

TO BE SERVICED	TYPE OF SERVICE	VEHICLE MILEAGE INTERVAL (x1000)												
		5	10	15	20	25	30	35	40	45	50	55	60	65
Engine oil & filter	R	✓	✓	✓	✓	✓	✓	✓	✓	✓	✓	✓	✓	✓
Automatic transmission shift linkage	S/I	✓		✓		✓		✓		✓		✓		✓
Clutch reservoir fluid level	S/I	✓		✓		✓		✓		✓		✓		✓
Exhaust system & heat shields	S/I	✓		✓		✓		✓		✓		✓		✓
Rotate tires⑥	S/I	✓		✓		✓		✓		✓		✓		✓
Steering linkage suspension, driveshaft U-joint, & slip-yoke (if equipped)	S/I	✓		✓		✓		✓		✓		✓		✓
Clutch release lever (7.3L diesel & 7.5L)	S/I	✓			✓			✓			✓			✓
Fuel filter③	R			✓			✓			✓			✓	
Disc brake system & caliper slide rails	S/I			✓			✓			✓			✓	
Drum brake systems, hoses & lines	S/I			✓			✓			✓			✓	
Parking brake fluid level (F-Super Duty)	S/I			✓			✓			✓			✓	
Spring U bolts (F-Super Duty)	S/I			✓			✓			✓			✓	
Transfer case shift lever pivot bolt & control rod connecting pins (4x4)	S/I			✓			✓			✓			✓	
Engine coolant strength, hoses & clamps	S/I			✓			✓			✓			✓	
Air cleaner filter④	R						✓						✓	
Automatic transmission fluid & filter⑤	R						✓						✓	
Front axle RH axle slip yoke (4x4)	S/I						✓						✓	
Front & rear driveshaft slip-yoke	S/I						✓						✓	
Front wheel bearings	S/I						✓						✓	
Hub lock (4x4)	S/I						✓						✓	

79241C82

1995-96 SCHEDULED MAINTENANCE INTERVALS
(FORD F-150/250/250HD/350 & SUPER DUTY/E-150/250/350 & CLUB WAGON Cont.)

TO BE SERVICED	TYPE OF SERVICE	VEHICLE MILEAGE INTERVAL (x1000)												
		5	10	15	20	25	30	35	40	45	50	55	60	65
Parking brake system	S/I						✓						✓	
Spindle needle bearing (4x4)	S/I						✓						✓	
Throttle & TV lever ball studs	S/I						✓						✓	
Crankcase emission air filter	R						✓						✓	
Engine coolant②	R										✓			
Manual transmission & rear axle oil①	R												✓	
PCV valve	R												✓	
Spark plugs	R												✓	
Transfer case oil	R												✓	
Drive belts	S/I												✓	
Thermactor hoses & clamps	S/I												✓	

① Rear axle oil - change every 100,000 miles.
② Engine coolant - change at 50,000 miles:
      Gasoline - change every 30,000 miles thereafter.
      Diesel - add 8-10 oz. FW-15 every 15,000 miles & 4 pints FW-15 every time coolant is changed.
③ Fuel filter (7.3L diesel) - change filter at 15,000 miles & when ever fuel restriction lamp is illuminated.
④ Air cleaner filter (7.3L diesel) - change filter when restriction gauge is in red zone.
⑤ Automatic transmission fluid & filter - C6 & E40D transmissions do not require regular fluid changes under normal operating conditions.
⑥ Rotate front tires for dual rear wheel vehicles from side to side only.
R – Replace   S/I – Service or Inspect

FREQUENT OPERATION MAINTENANCE (SEVERE SERVICE)
   If a vehicle is operated under any of the following conditions it is considered severe service:
- Extremely dusty areas.
- 50% or more of the vehicle operation is in 32°C (90°F) or higher temperatures, or constant operation in temperatures below 0°C (32°F).
- Prolonged idling (vehicle operation in stop and go traffic).
- Frequent short running periods (engine does not warm to normal operating temperatures).
- Police, taxi, delivery usage or trailer towing usage.
Air cleaner filter - check every 3000 miles.
Oil & oil filter change – change every 3000 miles.
Rear axle oil (F-Super Duty) - change every 3000 miles.
Automatic transmission shift linkage - lubricate every 6000 miles.
Clutch reservoir fluid level - check every 6000 miles.
Exhaust system - check every 6000 miles.
Steering linkage suspension, driveshaft U-joint & slip-yoke (if equipped) - lubricate every 6000 miles.
Rotate tires every 9000 miles. (City delivery vehicles & other unique applications that require constant turning, may need more frequent tire rotation.)
Clutch release lever (7.3L diesel & 7.5L) - lubricate every 15,000 miles.
Automatic transmission fluid & filter - change every 21,000 miles.

79241C83

1997-99 SCHEDULED MAINTENANCE INTERVALS
(FORD F-150, LIGHT-DUTY F-250, EXPEDITION & NAVIGATOR)

TO BE SERVICED	TYPE OF SERVICE	VEHICLE MILEAGE INTERVAL (x1000)																							
		5	10	15	20	25	30	35	40	45	50	55	60	65	70	75	80	85	90	95	100	105	110	115	120
Accessory drive belt	S/I												✓												✓
Air cleaner filter ①	R						✓						✓						✓			✓			✓
Automatic transmission fluid	R						✓						✓						✓			✓			✓
Automatic transmission shift linkage	S/I & L	✓	✓	✓	✓	✓	✓	✓	✓	✓	✓	✓	✓	✓	✓	✓	✓	✓	✓	✓	✓	✓	✓	✓	✓
Brake caliper slide rails	L			✓			✓			✓			✓			✓			✓			✓			✓
Brake system, hoses & lines	S/I			✓			✓			✓			✓			✓			✓			✓			✓
Clutch reservoir fluid level	S/I	✓	✓	✓	✓	✓	✓	✓	✓	✓	✓	✓	✓	✓	✓	✓	✓	✓	✓	✓	✓	✓	✓	✓	✓
Engine coolant ②	R										✓						✓						✓		
Engine cooling system hoses, clamps & coolant	S/I			✓			✓			✓			✓			✓			✓			✓			✓
Engine oil & filter	R	✓	✓	✓	✓	✓	✓	✓	✓	✓	✓	✓	✓	✓	✓	✓	✓	✓	✓	✓	✓	✓	✓	✓	✓
Exhaust system	S/I	✓	✓	✓	✓	✓	✓	✓	✓	✓	✓	✓	✓	✓	✓	✓	✓	✓	✓	✓	✓	✓	✓	✓	✓
Front wheel bearings	S/I & L						✓						✓						✓						✓
Front/rear axle driveshaft slip yoke	L						✓						✓						✓						✓
Front/rear axle fluid ③	R																				✓				
Fuel filter	R			✓			✓			✓			✓			✓			✓			✓			✓
Manual transmission fluid	R												✓												✓
Parking brake system	S/I						✓						✓						✓						✓
PCV valve	R												✓												✓
Rotate tires	S/I	✓	✓	✓	✓	✓	✓	✓	✓	✓	✓	✓	✓	✓	✓	✓	✓	✓	✓	✓	✓	✓	✓	✓	✓
Spark plugs	R																				✓				
Steering linkage, suspension, driveshaft U joints	S/I & L	✓	✓	✓	✓	✓	✓	✓	✓	✓	✓	✓	✓	✓	✓	✓	✓	✓	✓	✓	✓	✓	✓	✓	✓

① Perform this at the mileage shown or every 30 months, whichever occurs first.

② Drain, flush and refill the cooling system initially at 50,000 miles or 48 months, whichever occurs first, then every 30,000 miles or 30 months thereafter.

③ The axle lubricant must be replaced every 100,000 miles or if the axle has been submerged under water. Otherwise the lube should not be checked or changed unless a repair is required.

R - Replace   S/I - Inspect and service, if needed   L - Lubricate   A - Adjust   C - Clean

FREQUENT OPERATION MAINTENANCE (SEVERE SERVICE)

If a vehicle is operated under any of the following conditions it is considered severe service:

- Towing a trailer or using a camper or car-top carrier.

- Repeated short trips of less than 5 miles in temperatures below freezing, or trips of less than 10 miles in any temperature.

79241C84

1997-99 SCHEDULED MAINTENANCE INTERVALS
(FORD F-150, LIGHT-DUTY F-250, EXPEDITION & NAVIGATOR)

FREQUENT OPERATION MAINTENANCE (SEVERE SERVICE)—continued

- Extensive idling or low-speed driving for long distances as in heavy commercial use, such as delivery, taxi or police cars.
- Operating on rough, muddy or salt-covered roads.
- Operating on unpaved or dusty roads.
- Driving in extremely hot (over 90°) conditions.

Engine oil & filter - replace every 3,000 miles.

Tires - rotate and inspect every 6,000 miles.

Clutch reservoir fluid level - inspect every 6,000 miles.

Automatic transmission shift linkage - lubricate every 6,000 miles.

Steering linkage, suspension, U-joints - lubricate every 6,000 miles.

Exhaust system - inspect for leaks or damage every 6,000 miles.

Fuel filter - replace every 15,000 miles.

Automatic transmission fluid - change every 21,000 miles.

Crankcase emission air filter - replace every 60,000 miles.

PCV valve - replace every 60,000 miles.

Accessory drive belt - inspect every 60,000 miles.

Spark plugs - replace every 99,000 miles.

79241C85

1997-99 SCHEDULED MAINTENANCE INTERVALS
(FORD ECONOLINE, CLUB WAGON, HEAVY-DUTY F-250 & F-350)

TO BE SERVICED	TYPE OF SERVICE	VEHICLE MILEAGE INTERVAL (x1000)																			
		5	10	15	20	25	30	35	40	45	50	55	60	65	70	75	80	85	90	95	100
Accessory drive belt	S/I																				✓
Air cleaner filter ①②	R						✓						✓						✓		
Automatic transmission fluid ③	R						✓						✓						✓		
Engine coolant ④⑤	R										✓					✓					
Engine cooling system hoses, clamps & coolant ⑥	S/I			✓			✓			✓			✓			✓			✓		
Engine oil & filter	R	✓	✓	✓	✓	✓	✓	✓	✓	✓	✓	✓	✓	✓	✓	✓	✓	✓	✓	✓	✓
Exhaust system	S/I			✓			✓			✓			✓			✓			✓		
Front wheel bearings	S/I & L																		✓		
Front/rear axle lubricant ⑦	R																				✓
Fuel filter ⑧	R						✓						✓						✓		
PCV valve	R	Every 120,000 miles																			
Rotate tires	S/I	✓	✓	✓	✓	✓	✓	✓	✓	✓	✓	✓	✓	✓	✓	✓	✓	✓	✓	✓	✓
Spark plugs	R																				✓
Steering linkage, suspension, driveshaft U joints	S/I & L	✓	✓	✓	✓	✓	✓	✓	✓	✓	✓	✓	✓	✓	✓	✓	✓	✓	✓	✓	✓

① Perform this at the mileage shown or every 30 months, whichever occurs first.

② 7.3L DIT Diesel engine: the air filter should be replaced when the restriction gauge is in the red zone.

③ Except the E40D transmission.

④ Drain, flush and refill the cooling system initially at 50,000 miles or 48 months, whichever occurs first, then every 30,000 miles or 30 months thereafter.

⑤ 7.3L DIT Diesel engine: add 4 pints of FW-15 each time the coolant is replaced.

⑥ 7.3L DIT Diesel engine: add 8-10 oz. of FW-15 to the engine coolant every 15,000 miles.

⑦ The axle lubricant must be replaced every 100,000 miles or if the axle has been submerged under water. Otherwise the lube should not be checked or changed unless a repair is required.

⑧ 7.3L DIT Diesel engine: the fuel filter should be replaced when the restriction lamp is illuminated.

R - Replace   S/I - Inspect and service, if needed   L - Lubricate   A - Adjust   C - Clean

FREQUENT OPERATION MAINTENANCE (SEVERE SERVICE)

If a vehicle is operated under any of the following conditions it is considered severe service:

- Towing a trailer or using a camper or car-top carrier.
- Repeated short trips of less than 5 miles in temperatures below freezing, or trips of less than 10 miles in any temperature.
- Extensive idling or low-speed driving for long distances as in heavy commercial use, such as delivery, taxi or police cars.
- Operating on rough, muddy or salt-covered roads.
- Operating on unpaved or dusty roads.
- Driving in extremely hot (over 90°) conditions.

Engine oil & filter - replace every 3,000 miles.

Tires - rotate and inspect every 6,000 miles.

Steering linkage, suspension, U-joints - lubricate every 6,000 miles.

Exhaust system - inspect for leaks or damage every 12,000 miles.

79241C86

1997-99 SCHEDULED MAINTENANCE INTERVALS
(FORD ECONOLINE, CLUB WAGON, HEAVY-DUTY F-250 & F-350)

FREQUENT OPERATION MAINTENANCE (SEVERE SERVICE)—continued

Fuel filter - replace every 15,000 miles.

Automatic transmission fluid - change every 21,000 miles.

Front wheel bearings (2WD) - inspect and repack every 30,000 miles.

Rear axle lubricant (E-Super Duty only) - replace every 30,000 miles.

Spark plugs (except 4.2L engine) - replace every 60,000 miles.

PCV valve - replace every 60,000 miles.

Accessory drive belt - inspect every 60,000 miles.

Spark plugs (4.2L engine only) - replace every 100,000 miles.

79241C87

SCHEDULED MAINTENANCE
FORD MOTOR COMPANY
FORD F-SERIES, E-SERIES, AEROSTAR, BRONCO, CLUB WAGON, EXPEDITION RANGER, SUPER-DUTY, LINCOLN NAVIGATOR

The following should be used as a guide when determining the amount of work required for a particular service if taken to a repair shop. In estimating how long a particular Scheduled Maintenance Service should take, please observe the following:

● **Chilton Time** is time based on field research and data supplied by the vehicle manufacturer.
● Labor time operations are given in hours and tenths of an hour.
● All labor operations, are to be used as a guide.

Mechanic Skill Level Codes:
(G) GENERAL: Normally skilled with certification.
(M) MAINTENANCE: Semi-skilled working on certication.
(P) PRECISION: Really skilled with multiple certification.

	Chilton Time
(M) 5000 Mile Service	
1995-97 E-Series, F-Series (thru 96) Bronco, Club Wagon	1.1
w/AT add	.1
w/7.5L add	.1
w/7.3L Diesel add	.1
1997-99 E-Series, F-Series, Club Wagon, Expedition, Navigator	7
Rotate tires add	.6
1995-99 Aerostar	1.1
1995-99 Ranger	1.1
w/AT add	.1
w/4WD add	.1
(M) 10000 Mile Service	
1995-97 E-Series, F-Series (thru 96) Bronco, Club Wagon	.3
1997-99 E-Series, F-Series, Club Wagon, Expedition, Navigator	.7
Rotate tires add	.6
1995-99 Aerostar	.4
1995-99 Ranger	.3
w/AT add	.1
(G) 15000 Mile Service	
1995-97 E-Series, F-Series (thru 96) Bronco, Club Wagon	1.9
w/AT add	.1
w/4WD add	.1
w/Super Duty add	.2
1997-99 E-Series, F-Series, Club Wagon, Expedition, Navigator	1.4
Rotate tires add	.6
1995-99 Aerostar	1.5
1995-99 Ranger	1.5
w/AT add	.1
w/4WD add	.2

	Chilton Time
(M) 20000 Mile Service	
1995-97 E-Series, F-Series (thru 96) Bronco, Club Wagon	.3
w/7.5L add	.1
w/7.3L Diesel add	.1
1997-99 E-Series, F-Series, Club Wagon, Expedition, Navigator	.7
Rotate tires add	.6
1995-99 Aerostar	.4
1995-99 Ranger	.3
w/AT add	.1
(M) 25000 Mile Service	
1995-97 E-Series, F-Series (thru 96) Bronco, Club Wagon	1.1
w/AT add	.1
1997-99 E-Series, F-Series, Club Wagon, Expedition, Navigator	.7
Rotate tires add	.6
1995-99 Aerostar	1.1
1995-99 Ranger	1.1
w/AT add	.1
w/4WD add	.1
(G) 30000 Mile Service	
1995-97 E-Series, F-Series (thru 96) Bronco, Club Wagon	2.5
w/AT add	.6
w/4WD add	.5
w/Super Duty add	.2
1997-99 E-Series, F-Series, Club Wagon, Expedition, Navigator	2.0
w/AT add	.9
Rotate tires add	.6
1995-99 Aerostar	2.8
1995-99 Ranger	2.8
w/AT add	.6
w/4WD add	.4

	Chilton Time
(M) 35000 Mile Service	
1996-97 E-Series, F-Series (thru 96) Bronco, Club Wagon	1.1
w/AT add	.1
w/7.5L add	.1
w/7.3L Diesel add	.1
1997-99 E-Series, F-Series, Club Wagon, Expedition, Navigator	.7
Rotate tires add	.6
1995-99 Aerostar	1.1
1995-99 Ranger	1.1
w/AT add	.1
w/4WD add	.1
(M) 40000 Mile Service	
1995-97 E-Series, F-Series (thru 96) Bronco, Club Wagon	.3
1997-99 E-Series, F-Series, Club Wagon, Expedition, Navigator	.7
Rotate tires add	.6
1995-99 Aerostar	.4
1995-99 Ranger	.3
w/AT add	.1
(G) 45000 Mile Service	
1995-97 E-Series, F-Series (thru 96) Bronco, Club Wagon	2.0
w/AT add	.1
w/4WD add	.1
w/Super Duty add	.2
1997-99 E-Series, F-Series, Club Wagon, Expedition, Navigator	1.4
Rotate tires add	.6
1995-99 Aerostar	1.5
1995-99 Ranger	1.5
w/AT add	.1
w/4WD add	.2

79241C88

FORD MOTOR CO.
Ford Windstar • Mercury Villager

VEHICLE IDENTIFICATION CHART

		Engine Code						Model Year
Code	Liters	Cu. In. (cc)	Cyl.	Fuel Sys.	Eng. Mfg.		Code	Year
U	3.0	183 (2999)	6	SEFI	Ford		S	1995
W	3.0	183 (2999)	6	MFI	Nissan		T	1996
V	3.0	182 (2986)	6	SEFI	Ford		V	1997
1	3.0	182 (2960)	6	SEFI	Nissan		W	1998
4	3.8	231 (3800)	6	SEFI	Ford		Y	1999

MFI - Multi-port Fuel Injection

SEFI - Sequential Multi-port Fuel Injection

79241C89

ENGINE IDENTIFICATION

Year	Model	Engine Displacement Liters (cc)	Engine Series (ID/VIN)	Fuel System	No. of Cylinders	Engine Type
1995	Windstar	3.8 (3802)	4	MFI	6	OHV
1996	Windstar	3.0 (2982)	U	MFI	6	OHV
	Windstar	3.8 (3802)	4	SPI	6	OHV
	Villager	3.0 (2966)	W	MFI	6	SOHC
1997	Windstar	3.0 (2982)	U	SEFI	6	OHV
	Windstar	3.8 (3802)	4	SEFI	6	OHV
	Villager	3.0 (2966)	W	MFI	6	SOHC
1998-99	Windstar	3.0 (2986)	V	SEFI	6	OHV
	Windstar	3.8 (3800)	4	SEFI	6	OHV
	Villager	3.0 (2960)	1	SEFI	6	SOHC

MFI - Multiport fuel injection

SPI - Split port injection

OHV - Overhead valve

SOHC - Single overhead camshaft

79241C90

GENERAL ENGINE SPECIFICATIONS

Year	Engine ID/VIN	Engine Displacement Liters (cc)	Fuel System Type	Net Horsepower @ rpm	Net Torque @ rpm (ft. lbs.)	Bore x Stroke (in.)	Compression Ratio	Oil Pressure @ rpm
1995	4	3.8 (3802)	MFI	140@3800	215@2400	3.81x3.39	9.0:1	40-60@2500
1996	U	3.0 (2982)	MFI	147@5000	162@3250	3.50x3.14	9.3:1	40-60@2500
	W	3.0 (2966)	MFI	151@4800	174@4400	3.43x3.27	9.0:1	40-60@2500
	4	3.8 (3802)	MFI	200@5000	230@3000	3.81x3.39	9.3:1	40-60@2500
1997	U	3.0 (2982)	SEFI	147@5000	162@3250	3.50x3.14	9.3:1	40-60@2500
	W	3.0 (2966)	MFI	151@4800	174@4400	3.43x3.27	9.0:1	40-60@2500
	4	3.8 (3802)	SEFI	200@5000	230@3000	3.81x3.39	9.3:1	40-60@2500
1998-99	V	3.0 (2986)	SEFI	150@5000	172@3300	3.50x3.14	9.3:1	40-60@2500
	1	3.0 (2960)	SEFI	151@4800	174@4400	3.43x3.27	9.0:1	40-60@2500
	4	3.8 (3800)	SEFI	200@5000	225@3000	3.81x3.39	9.3:1	40-60@2500

MFI - Multiport fuel injection

SEFI - Sequential Multi-port Fuel Injection

79241C91

GASOLINE ENGINE TUNE-UP SPECIFICATIONS

Year	Engine ID/VIN	Engine Displacement Liters (cc)	Spark Plugs Gap (in.)	Ignition Timing (deg.) MT	AT	Fuel Pump (psi)	Idle Speed (rpm) MT	AT	Valve Clearance In.	Ex.
1995	4	3.8 (3802)	0.054	—	10B	30-45	①	①	HYD	HYD
1996	U	3.0 (2982)	0.044	—	10B	35-45	①	①	HYD	HYD
	W	3.0 (2966)	0.033	—	15B	36-38 ②	—	①	HYD	HYD
	4	3.8 (3802)	0.054	—	10B	30-45	①	①	HYD	HYD
1997	U	3.0 (2982)	0.044	—	10B	35-45	①	①	HYD	HYD
	W	3.0 (2966)	0.033	—	15B	36-38 ②	—	①	HYD	HYD
	4	3.8 (3802)	0.054	—	10B	30-45	①	①	HYD	HYD
1998-99	V	3.0 (2986)	0.042-0.046	—	①	③	—	①	HYD	HYD
	1	3.0 (2960)	0.031-0.035	—	13-17B	④	—	①	HYD	HYD
	4	3.8 (3800)	0.052-0.056	—	①	③	—	①	HYD	HYD

NOTE: The Vehicle Emission Control Information label often reflects specification changes changes made during production. The label figures must be used if they differ from those in this chart.

B - Before top dead center

HYD - Hydraulic

① Controlled by the Powertrain Control Module (PCM) and cannot be manually adjusted.

② Fuel pressure is with engine running, pressure regulator vacuum hose disconnected.

③ Engine running: 28-45 psi
Key On, Engine Off (KOEO): 35-45 psi

④ Engine running: 30-38 psi
Key On, Engine Off (KOEO): 37-43 psi

79241C92

CAPACITIES

Year	Model	Engine ID/VIN	Engine Displacement Liters (cc)	Engine Oil with Filter (qts.)	Transmission (pts.) 4-Spd	5-Spd	Auto.	Transfer Case (pts.)	Drive Axle Front (pts.)	Rear (pts.)	Fuel Tank (gal.)	Cooling System (qts.)
1995	Windstar	4	3.8 (3802)	4.5	—	—	24.5	—	①	—	20.0	12.1
1996	Windstar	U	3.0 (2966)	4.2	—	—	16.5	—	①	—	20.0	②
	Windstar	4	3.8 (3802)	4.5	—	—	24.5	—	①	—	20.0 ③	12.1
	Villager	W	3.0 (2966)	4.2	—	—	16.5	—	①	—	20.0	②
1997	Windstar	U	3.0 (2966)	4.2	—	—	16.5	—	①	—	20.0	②
	Windstar	4	3.8 (3802)	4.5	—	—	24.5	—	①	—	20.0 ③	12.1
	Villager	W	3.0 (2966)	4.2	—	—	16.5	—	①	—	20.0	②
1998-99	Windstar	V	3.0 (2986)	4.5	—	—	24.5	—	①	—	20.0 ③	12.1
	Windstar	4	3.8 (3800)	4.5	—	—	24.5	—	①	—	20.0 ③	12.1
	Villager	1	3.0 (2960)	4.2	—	—	17.4	—	①	—	20.0	④

NOTE: All capacities are approximate. Add fluid gradually and check to be sure a proper fluid level is obtained.

① Included in transaxle capacity

② With coolant recovery reservoir: 7.9
Without coolant recovery reservoir: 5.8

③ Optional: 26 gals.

④ With rear heater: 12.7
Without rear heater: 11.4

79241C93

VALVE SPECIFICATIONS

Year	Engine ID/VIN	Engine Displacement Liters (cc)	Seat Angle (deg.)	Face Angle (deg.)	Spring Test Pressure (lbs. @ in.)	Spring Installed Height (in.)	Stem-to-Guide Clearance (in.)		Stem Diameter (in.)	
							Intake	Exhaust	Intake	Exhaust
1995	4	3.8 (3802)	44.5	45.8	220@1.18	1.97	0.0010-0.0028	0.0015-0.0033	0.3423-0.3415	0.3418-0.3410
1996	U	3.0 (2982)	45	44	185@1.16	1.58-1.61	0.0010-0.0027	0.0015-0.0032	0.3126-0.3134	0.3121-0.3129
	W	3.0 (2966)	45	45	①	②	0.0008-0.0021	0.0016-0.0029	0.2742-0.2748	0.3136-0.3138
	4	3.8 (3802)	44.75	45.8	220@1.18	1.97	0.0010-0.0028	0.0015-0.0033	0.3423-0.3415	0.3410-0.3418
1997	U	3.0 (2982)	45	44	185@1.16	1.58-1.61	0.0010-0.0027	0.0015-0.0032	0.3126-0.3134	0.3121-0.3129
	W	3.0 (2966)	45	45	①	②	0.0008-0.0021	0.0016-0.0029	0.2742-0.2748	0.3136-0.3138
	4	3.8 (3802)	44.75	45.8	220@1.18	1.97	0.0010-0.0028	0.0015-0.0033	0.3423-0.3415	0.3410-0.3418
1998-99	V	3.0 (2986)	45	44	180@1.16	1.650-1.736	0.0010-0.0027	0.0015-0.0032	0.3126-0.3134	0.3121-0.3129
	1	3.0 (2960)	45	45	③	②	④	⑤	0.2742-0.2748	0.3136-0.3138
	4	3.8 (3800)	44.75	45.8	198-220@1.18	1.970	0.0010-0.0028	0.0015-0.0033	0.3415-0.3423	0.3410-0.3418

① Outer spring: 118@1.81
   Inner spring: 57.3@0.984
② Spring height measured unloaded
   Minimum length, outer spring: 2.016
   Minimum length, inner spring: 1.736
③ Outer spring: 118@1.18
   Inner spring: 57.3@0.984
④ Nominal: 0.0008-0.0021
   Maximum: 0.0039
⑤ Nominal: 0.0016-0.0029
   Maximum: 0.0039

79241C94

TORQUE SPECIFICATIONS
All readings in ft. lbs.

	Engine ID/VIN	Engine Displacement Liters (cc)	Cylinder Head Bolts	Main Bearing Bolts	Rod Bearing Bolts	Crankshaft Damper Bolts	Flywheel Bolts	Manifold		Spark Plugs	Lug Nut
								Intake	Exhaust		
1995	4	3.8 (3802)	①	65-81	31-36	103-132	54-64	②	19	8-10	85-105
1996	U	3.0 (2982)	③	60	26	107	54-64	24	25	8-10	100
	W	3.0 (2966)	④	67-74	⑤	141-156	61-69	⑥	13-16	14-22	80
	4	3.8 (3802)	①	65-81	31-36	103-132	54-64	②	19	8-10	85-105
1997	U	3.0 (2982)	③	60	26	107	54-64	24	25	8-10	100
	W	3.0 (2966)	④	67-74	⑤	141-156	61-69	⑥	13-16	14-22	80
	4	3.8 (3802)	①	65-81	31-36	103-132	54-64	②	19	8-10	85-105
1998-99	V	3.0 (2982)	⑦	56-62	23-28	93-121	54-64	⑧	15-22	7-14	85-104
	1	3.0 (2960)	⑨	⑩	⑤	141-156	61-69	⑪	13-16	14-22	72-87
	4	3.8 (3802)	⑫	82-88	⑬	104-132	54-64	71-106	15-22	7-15	85-104

① Step 1: 15 ft. lbs.
Step 2: 29 ft. lbs.
Step 3: 37 ft. lbs.
Step 4: Loosen each bolt one at a time
Step 5: Long bolts to 11-19 ft. lbs. plus 1/4 turn
Step 6: Short bolts to 7-15 ft. lbs. plus 1/4 turn

② Lower intake manifold:
Step 1: 13 ft. lbs.
Step 2: 16 ft. lbs.
Upper intake manifold:
Step 1: 8 ft. lbs.
Step 2: 15 ft. lbs.
Step 3: 24 ft. lbs.

③ Step 1: 37 ft. lbs.
Step 2: 68 ft. lbs.

④ Step 1: 22 ft. lbs.
Step 2: 43 ft. lbs.
Step 3: Loosen bolts one turn
Step 4: 22 ft. lbs.
Step 5: Rotate 60-65 degrees or 40-47 ft. lbs.
Step 6: Small cylinder head bolt located outside of valve cover: 6 ft. lbs.

⑤ Step 1: 10-12 ft. lbs.
Step 2: 28-33 ft. lbs.

⑥ Step 1: Nuts and bolts: 3 ft. lbs.
Step 2: Nuts: 17-20 ft. lbs., Bolts: 12-14 ft. lbs.
Step 3: Repeat Step 2

⑦ Step 1: 59 ft. lbs.
Step 2: Loosen all bolts 360 degrees
Step 3: 36 ft. lbs.
Step 4: 67 ft. lbs.

⑧ Step 1: 15-22 ft. lbs.
Step 2: 20-23 ft. lbs.

⑨ Step 1: 22 ft. lbs.
Step 2: 43 ft. lbs.
Step 3: Loosen all bolts completely
Step 4: 22 ft. lbs.
Step 5: Tighten an additional 60-65 degrees or to 40-47 ft. lbs.
Step 6: Head bolt A to 80-104 inch lbs.

⑩ Step 1: 34-37 ft. lbs.
Step 2: 67-74 ft. lbs.

⑪ Step 1: 26-44 inch lbs.
Step 2: bolts to 9-12 ft. lbs.; nuts to 17-20 ft. lbs.
Step 3: Repeat Step 2

⑫ Step 1: 15 ft. lbs.
Step 2: 29 ft. lbs.
Step 3: 37 ft. lbs.
Step 4: Loosen all bolts in sequence
Step 5: Long bolts to 30-36 ft. lbs. plus 1/2 turn
Short bolts to 7-15 ft. lbs. plus 1/4 turn
Move on to next bolt in sequence.

⑬ Step 1: 30-34 ft. lbs.
Step 2: Tighten an additional 90-120 degrees

79241C95

SCHEDULED MAINTENANCE INTERVALS
(FORD WINDSTAR, MERCURY VILLAGER)

TO BE SERVICED	TYPE OF SERVICE	VEHICLE MILEAGE INTERVAL (x1000)												
		5	10	15	20	25	30	35	40	45	50	55	60	65
Engine oil & filter	R	✓	✓	✓	✓	✓	✓	✓	✓	✓	✓	✓	✓	✓
Rotate tires	S/I	✓		✓		✓				✓		✓		✓
Engine coolant strength, hoses & clamps	S/I			✓			✓			✓			✓	
Air cleaner filter	R						✓						✓	
Automatic transmission fluid & filter	R						✓						✓	
Engine Coolant①	R						✓						✓	
PCV valve	R												✓	
Spark plugs②	R						✓						✓	
Drive belts	S/I						✓						✓	
Exhaust system & heat shields	S/I						✓						✓	
Front & rear brakes	S/I						✓						✓	

① Engine coolant - change initially at 50,000 miles and every 30,000 miles thereafter.
② Spark plugs (Windstar) - replace every 100,000 miles.
R – Replace   S/I – Service or Inspect

FREQUENT OPERATION MAINTENANCE (SEVERE SERVICE)

If a vehicle is operated under any of the following conditions it is considered severe service:
- **Extremely dusty areas.**
- **50% or more of the vehicle operation is in 32°C (90°F) or higher temperatures, or constant operation in temperatures below 0°C (32°F).**
- **Prolonged idling (vehicle operation in stop and go traffic).**
- **Frequent short running periods (engine does not warm to normal operating temperatures).**
- **Police, taxi, delivery usage or trailer towing usage.**
Oil & oil filter change – change every 3000 miles.
Rotate tires initially at 6000 miles, and every 9000 miles thereafter.
Air cleaner filter - change every 15,000 miles.
Engine coolant strength, hoses & clamps - check every 15,000 miles.
Exhaust system - check every 15,000 miles.
Automatic transmission fluid & filter - change every 21,000 miles.

79241C96

SCHEDULED MAINTENANCE
FORD MOTOR COMPANY
FORD WINDSTAR

The following should be used as a guide when determining the amount of work required for a particular service if taken to a repair shop. In estimating how long a particular Scheduled Maintenance Service should take, please observe the following:

- **Chilton Time** is time based on field research and data supplied by the vehicle manufacturer.
- Labor time operations are given in hours and tenths of an hour.
- All labor operations, are to be used as a guide.

Mechanic Skill Level Codes:
- **(G) GENERAL:** Normally skilled with certification.
- **(M) MAINTENANCE:** Semi-skilled working on certicication.
- **(P) PRECISION:** Really skilled with multiple certification.

	Chilton Time		Chilton Time		Chilton Time
(G) 5000 Mile Service		**(G) 25000 Mile Service**		**(M) 50000 Mile Service**	
1995-99	.8	1995-99	.8	1995-99	.4
(M) 10000 Mile Service		**(G) 30000 Mile Service**		**(G) 55000 Mile Service**	
1995-99	.4	1995-99	3.4	1995-99	.8
(G) 15000 Mile Service		**(G) 35000 Mile Service**		**(G) 60000 Mile Service**	
1995-99	.9	1995-99	.8	1995-99	3.5
(M) 20000 Mile Service		**(M) 40000 Mile Service**		**(G) 65000 Mile Service**	
1995-99	.4	1995-99	.4	1995-99	.8
		(G) 45000 Mile Service			
		1995-99	.9		

79241C97

SCHEDULED MAINTENANCE
FORD MOTOR COMPANY
MERCURY VILLAGER

The following should be used as a guide when determining the amount of work required for a particular service if taken to a repair shop. In estimating how long a particular Scheduled Maintenance Service should take, please observe the following:

- **Chilton Time** is time based on field research and data supplied by the vehicle manufacturer.
- Labor time operations are given in hours and tenths of an hour.
- All labor operations, are to be used as a guide.

Mechanic Skill Level Codes:
- **(G) GENERAL:** Normally skilled with certification.
- **(M) MAINTENANCE:** Semi-skilled working on certicication.
- **(P) PRECISION:** Really skilled with multiple certification.

	Chilton Time		Chilton Time		Chilton Time
(G) 5000 Mile Service		**(G) 25000 Mile Service**		**(M) 50000 Mile Service**	
1995-99	.8	1995-99	.8	1995-99	.4
(M) 10000 Mile Service		**(G) 30000 Mile Service**		**(G) 55000 Mile Service**	
1995-99	.4	1995-99	3.2	1995-99	.8
(G) 15000 Mile Service		**(G) 35000 Mile Service**		**(G) 60000 Mile Service**	
1995-99	.9	1995-99	.8	1995-99	3.4
(M) 20000 Mile Service		**(M) 40000 Mile Service**		**(G) 65000 Mile Service**	
1995-99	.4	1995-99	.4	1995-99	.8
		(G) 45000 Mile Service			
		1995-99	.9		

79241C98

GENERAL MOTORS
Cadillac Escalade • Chevrolet Astro • Blazer • C/K Pick-Ups Express • G/P Vans • S10 • Suburban • Tahoe • Venture GMC C/K Pick-Ups • Denali • Envoy • G/P Vans • Jimmy • S15 Safari • Savana • Sonoma • Yukon • Oldsmobile Bravada

VEHICLE IDENTIFICATION CHART

Engine Code							Model Year	
Code	Liters	Cu. In. (cc)	Cyl.	Fuel Sys.	Eng. Mfg.		Code	Year
4	2.2	134 (2189)	4	MFI	CPC		S	1995
A	2.5	151 (2474)	4	TFI	CPC		T	1996
C	6.2	379 (6210)	8	DSL	CPC		V	1997
F	6.5	395 (6473)	8	DSL	CPC		W	1998
H	5.0	305 (4999)	8	TFI	CPC		X	1999
J	6.2	379 (6210)	8	DSL	CPC			
J	7.4	454 (7440)	8	MFI	CPC			
K	5.7	350 (5735)	8	TFI	CPC			
M	5.0	305 (4999)	8	MFI	CPC			
N	7.4	454 (7440)	8	TFI	CPC			
P	6.5	395 (6473)	8	DSL	CPC			
R	2.8	173 (2835)	6	TFI	CPC			
R	5.7	350 (5735)	8	MFI	CPC			
S	6.5	395 (6473)	8	DSL	CPC			
W	4.3	263 (4293)	6	TFI	CPC			
W	4.3	263 (4293)	6	MFI	CPC			
X	4.3	263 (4293)	6	MFI	CPC			
Y	6.5	395 (6473)	8	DSL	CPC			
Z	4.3	263 (4293)	6	TFI	CPC			
Z	4.3	263 (4293)	6	MFI	CPC			

CPC - Chevrolet/Pontiac/Canada

DSL - Diesel

MFI - Multi-port Fuel Injection

TFI - Throttle body Fuel Injection

79241C13

ENGINE IDENTIFICATION

All measurements are given in inches.

Year	Model	Engine Displacement Liters (cc)	Engine Series (ID/VIN)	Fuel System	No. of Cylinders	Engine Type
1995	Astro	4.3 (4293)	W	TFI	6	OHV
	Bravada	4.3 (4293)	W	MFI	6	OHV
	C1500	4.3 (4293)	Z	TFI	6	OHV
	C1500	5.0 (4999)	H	TFI	8	OHV
	C1500	5.7 (5735)	K	TFI	8	OHV
	C1500	6.5 (6505)	P	DSL	8	OHV
	C1500	6.5 (6505)	S	DSL	8	OHV
	C2500	4.3 (4293)	Z	TFI	6	OHV
	C2500	5.0 (4999)	H	TFI	8	OHV
	C2500	5.7 (5735)	K	TFI	8	OHV
	C2500	6.5 (6505)	F	DSL	8	OHV
	C2500	6.5 (6505)	P	DSL	8	OHV
	C2500	6.5 (6505)	S	DSL	8	OHV
	C2500	7.4 (7440)	N	TFI	8	OHV
	C3500	5.7 (5735)	K	TFI	8	OHV
	C3500	6.5 (6505)	F	DSL	8	OHV
	C3500	6.5 (6505)	S	DSL	8	OHV
	G/P20	4.3 (4293)	Z	TFI	6	OHV
	G/P20	5.0 (4999)	H	TFI	8	OHV
	G/P20	5.7 (5735)	K	TFI	8	OHV
	G/P20	6.5 (6505)	P	DSL	8	OHV
	G/P30	4.3 (4293)	Z	TFI	6	OHV
	G/P30	5.7 (5735)	K	TFI	8	OHV
	G/P30	6.5 (6505)	Y	DSL	8	OHV
	G/P30	7.4 (7440)	N	TFI	8	OHV
	Jimmy	4.3 (4293)	W	MFI	6	OHV
	K1500	4.3 (4293)	Z	TFI	6	OHV
	K1500	5.0 (4999)	H	TFI	8	OHV
	K1500	5.7 (5735)	K	TFI	8	OHV
	K1500	6.5 (6505)	P	DSL	8	OHV
	K1500	6.5 (6505)	S	DSL	8	OHV
	K1500	7.4 (7440)	N	TFI	8	OHV
	K2500	4.3 (4293)	Z	TFI	6	OHV
	K2500	5.0 (4999)	H	TFI	8	OHV
	K2500	5.7 (5735)	K	TFI	8	OHV
	K2500	6.5 (6505)	F	DSL	8	OHV
	K2500	6.5 (6505)	P	DSL	8	OHV
	K2500	6.5 (6505)	S	DSL	8	OHV
	K2500	7.4 (7440)	N	TFI	8	OHV
	K3500	5.7 (5735)	K	TFI	8	OHV
	K3500	6.5 (6505)	F	DSL	8	OHV
	K3500	7.4 (7440)	N	TFI	8	OHV
	S10 Blazer	4.3 (4293)	W	MFI	6	OHV
	S10 Pick-up	2.2 (2189)	4	MFI	4	OHV
	S10 Pick-up	4.3 (4293)	W	MFI	6	OHV
	S10 Pick-up	4.3 (4293)	Z	MFI	6	OHV

79241C14

ENGINE IDENTIFICATION

All measurements are given in inches.

Year	Model	Engine Displacement Liters (cc)	Engine Series (ID/VIN)	Fuel System	No. of Cylinders	Engine Type
1995 (cont.)	S15 Pick-up	2.2 (2189)	4	MFI	4	OHV
	S15 Pick-up	4.3 (4293)	W	MFI	6	OHV
	S15 Pick-up	4.3 (4293)	Z	MFI	6	OHV
	Safari	4.3 (4293)	W	TFI	6	OHV
	Sonoma	2.2 (2189)	4	MFI	4	OHV
	Sonoma	4.3 (4293)	W	MFI	6	OHV
	Sonoma	4.3 (4293)	Z	MFI	6	OHV
	Suburban	5.7 (5735)	K	TFI	8	OHV
	Suburban	6.5 (6505)	F	DSL	8	OHV
	Suburban	7.4 (7440)	N	TFI	8	OHV
	Tahoe	5.7 (5735)	K	MFI	8	OHV
	Tahoe	6.5 (6505)	S	DSL	8	OHV
	Yukon	5.7 (5735)	K	MFI	8	OHV
	Yukon	6.5 (6505)	S	DSL	8	OHV
1996	Astro	4.3 (4293)	W	MFI	6	OHV
	Bravada	4.3 (4293)	W	MFI	6	OHV
	Bravada	4.3 (4293)	X	MFI	6	OHV
	C1500	4.3 (4293)	W	MFI	6	OHV
	C1500	5.0 (4999)	M	MFI	8	OHV
	C1500	5.7 (5735)	R	MFI	8	OHV
	C2500	4.3 (4293)	W	MFI	6	OHV
	C2500	5.0 (4999)	M	MFI	8	OHV
	C2500	5.7 (5735)	R	MFI	8	OHV
	C2500	6.5 (6374)	F	DSL	8	OHV
	C3500	6.5 (6374)	F	DSL	8	OHV
	C3500	7.4 (7440)	J	MFI	8	OHV
	G/P1500	4.3 (4293)	W	MFI	6	OHV
	G/P1500	5.0 (4999)	M	MFI	8	OHV
	G/P1500	5.7 (5735)	R	MFI	8	OHV
	G/P2500	5.0 (4999)	M	MFI	8	OHV
	G/P2500	5.7 (5735)	R	MFI	8	OHV
	G/P3500	6.5 (6374)	F	DSL	8	OHV
	G/P3500	7.4 (7440)	J	MFI	8	OHV
	Jimmy	4.3 (4293)	W	MFI	6	OHV
	Jimmy	4.3 (4293)	X	MFI	6	OHV
	K1500	4.3 (4293)	W	MFI	6	OHV
	K1500	5.0 (4999)	M	MFI	8	OHV
	K1500	5.7 (5735)	R	MFI	8	OHV
	K1500	6.5 (6374)	F	DSL	8	OHV
	K2500	5.0 (4999)	M	MFI	8	OHV
	K2500	5.7 (5735)	R	MFI	8	OHV
	K2500	6.5 (6374)	F	DSL	8	OHV
	K3500	5.7 (5735)	R	MFI	8	OHV
	K3500	6.5 (6374)	F	DSL	8	OHV
	K3500	7.4 (7440)	J	MFI	8	OHV

79241C15

ENGINE IDENTIFICATION

All measurements are given in inches.

Year	Model	Engine Displacement Liters (cc)	Engine Series (ID/VIN)	Fuel System	No. of Cylinders	Engine Type
1996 (cont.)	S10 Blazer	4.3 (4293)	W	MFI	6	OHV
	S10 Blazer	4.3 (4293)	X	MFI	6	OHV
	S10 Pick-up	2.2 (2189)	4	MFI	4	OHV
	S10 Pick-up	4.3 (4293)	W	MFI	6	OHV
	S10 Pick-up	4.3 (4293)	X	MFI	6	OHV
	S15 Pick-up	2.2 (2189)	4	MFI	4	OHV
	S15 Pick-up	4.3 (4293)	W	MFI	6	OHV
	S15 Pick-up	4.3 (4293)	X	MFI	6	OHV
	Safari	4.3 (4293)	W	MFI	6	OHV
	Sonoma	2.2 (2189)	4	MFI	4	OHV
	Sonoma	4.3 (4293)	W	MFI	6	OHV
	Sonoma	4.3 (4293)	X	MFI	6	OHV
	Suburban	5.7 (5735)	R	MFI	8	OHV
	Suburban	7.4 (7440)	J	MFI	8	OHV
	Tahoe	5.7 (5735)	R	MFI	8	OHV
	Tahoe	6.5 (6374)	S	DSL	8	OHV
	Yukon	5.7 (5735)	R	MFI	8	OHV
	Yukon	6.5 (6374)	S	DSL	8	OHV
1997	Astro	4.3 (4293)	W	MFI	6	OHV
	Bravada	4.3 (4293)	W	MFI	6	OHV
	Bravada	4.3 (4293)	X	MFI	6	OHV
	C1500	4.3 (4293)	W	MFI	6	OHV
	C1500	5.0 (4999)	M	MFI	8	OHV
	C1500	5.7 (5735)	R	MFI	8	OHV
	C2500	4.3 (4293)	W	MFI	6	OHV
	C2500	5.0 (4999)	M	MFI	8	OHV
	C2500	5.7 (5735)	R	MFI	8	OHV
	C2500	6.5 (6374)	F	DSL	8	OHV
	C3500	6.5 (6374)	F	DSL	8	OHV
	C3500	7.4 (7440)	J	MFI	8	OHV
	G/P1500	4.3 (4293)	W	MFI	6	OHV
	G/P1500	5.0 (4999)	M	MFI	8	OHV
	G/P1500	5.7 (5735)	R	MFI	8	OHV
	G/P2500	5.0 (4999)	M	MFI	8	OHV
	G/P2500	5.7 (5735)	R	MFI	8	OHV
	G/P3500	6.5 (6374)	F	DSL	8	OHV
	G/P3500	7.4 (7440)	J	MFI	8	OHV
	Jimmy	4.3 (4293)	W	MFI	6	OHV
	Jimmy	4.3 (4293)	X	MFI	6	OHV
	K1500	4.3 (4293)	W	MFI	6	OHV
	K1500	5.0 (4999)	M	MFI	8	OHV
	K1500	5.7 (5735)	R	MFI	8	OHV
	K1500	6.5 (6374)	F	DSL	8	OHV
	K2500	5.0 (4999)	M	MFI	8	OHV
	K2500	5.7 (5735)	R	MFI	8	OHV
	K2500	6.5 (6374)	F	DSL	8	OHV

79241C16

ENGINE IDENTIFICATION

All measurements are given in inches.

Year	Model	Engine Displacement Liters (cc)	Engine Series (ID/VIN)	Fuel System	No. of Cylinders	Engine Type
1997 (cont.)	K3500	5.7 (5735)	R	MFI	8	OHV
	K3500	6.5 (6374)	F	DSL	8	OHV
	K3500	7.4 (7440)	J	MFI	8	OHV
	S10 Blazer	4.3 (4293)	W	MFI	6	OHV
	S10 Blazer	4.3 (4293)	X	MFI	6	OHV
	S10 Pick-up	2.2 (2189)	4	MFI	4	OHV
	S10 Pick-up	4.3 (4293)	W	MFI	6	OHV
	S10 Pick-up	4.3 (4293)	X	MFI	6	OHV
	S15 Pick-up	2.2 (2189)	4	MFI	4	OHV
	S15 Pick-up	4.3 (4293)	W	MFI	6	OHV
	S15 Pick-up	4.3 (4293)	X	MFI	6	OHV
	Safari	4.3 (4293)	W	MFI	6	OHV
	Sonoma	2.2 (2189)	4	MFI	4	OHV
	Sonoma	4.3 (4293)	W	MFI	6	OHV
	Sonoma	4.3 (4293)	X	MFI	6	OHV
	Suburban	5.7 (5735)	R	MFI	8	OHV
	Suburban	7.4 (7440)	J	MFI	8	OHV
	Tahoe	5.7 (5735)	R	MFI	8	OHV
	Tahoe	6.5 (6374)	S	DSL	8	OHV
	Yukon	5.7 (5735)	R	MFI	8	OHV
	Yukon	6.5 (6374)	S	DSL	8	OHV
1998-99	Astro	4.3 (4293)	W	MFI	6	OHV
	Bravada	4.3 (4293)	W	MFI	6	OHV
	Bravada	4.3 (4293)	X	MFI	6	OHV
	C1500	4.3 (4293)	W	MFI	6	OHV
	C1500	5.0 (4999)	M	MFI	8	OHV
	C1500	5.7 (5735)	R	MFI	8	OHV
	C2500	4.3 (4293)	W	MFI	6	OHV
	C2500	5.0 (4999)	M	MFI	8	OHV
	C2500	5.7 (5735)	R	MFI	8	OHV
	C2500	6.5 (6374)	F	DSL	8	OHV
	C3500	6.5 (6374)	F	DSL	8	OHV
	C3500	7.4 (7440)	J	MFI	8	OHV
	Denali	5.7 (5735)	R	MFI	8	OHV
	Denali	6.5 (6374)	S	DSL	8	OHV
	Envoy	4.3 (4293)	W	MFI	6	OHV
	Envoy	4.3 (4293)	X	MFI	6	OHV
	Escalade	5.7 (5735)	R	MFI	8	OHV
	Escalade	6.5 (6374)	S	DSL	8	OHV
	G/P1500	4.3 (4293)	W	MFI	6	OHV
	G/P1500	5.0 (4999)	M	MFI	8	OHV
	G/P1500	5.7 (5735)	R	MFI	8	OHV
	G/P2500	5.0 (4999)	M	MFI	8	OHV
	G/P2500	5.7 (5735)	R	MFI	8	OHV
	G/P3500	6.5 (6374)	F	DSL	8	OHV
	G/P3500	7.4 (7440)	J	MFI	8	OHV

79241C17

ENGINE IDENTIFICATION

All measurements are given in inches.

Year	Model	Engine Displacement Liters (cc)	Engine Series (ID/VIN)	Fuel System	No. of Cylinders	Engine Type
1998-99 (cont.)	Jimmy	4.3 (4293)	W	MFI	6	OHV
	Jimmy	4.3 (4293)	X	MFI	6	OHV
	K1500	4.3 (4293)	W	MFI	6	OHV
	K1500	5.0 (4999)	M	MFI	8	OHV
	K1500	5.7 (5735)	R	MFI	8	OHV
	K1500	6.5 (6374)	F	DSL	8	OHV
	K2500	5.0 (4999)	M	MFI	8	OHV
	K2500	5.7 (5735)	R	MFI	8	OHV
	K2500	6.5 (6374)	F	DSL	8	OHV
	K3500	5.7 (5735)	R	MFI	8	OHV
	K3500	6.5 (6374)	F	DSL	8	OHV
	K3500	7.4 (7440)	J	MFI	8	OHV
	S10 Blazer	4.3 (4293)	W	MFI	6	OHV
	S10 Blazer	4.3 (4293)	X	MFI	6	OHV
	S10 Pick-up	2.2 (2189)	4	MFI	4	OHV
	S10 Pick-up	4.3 (4293)	W	MFI	6	OHV
	S10 Pick-up	4.3 (4293)	X	MFI	6	OHV
	S15 Pick-up	2.2 (2189)	4	MFI	4	OHV
	S15 Pick-up	4.3 (4293)	W	MFI	6	OHV
	S15 Pick-up	4.3 (4293)	X	MFI	6	OHV
	Safari	4.3 (4293)	W	MFI	6	OHV
	Sonoma	2.2 (2189)	4	MFI	4	OHV
	Sonoma	4.3 (4293)	W	MFI	6	OHV
	Sonoma	4.3 (4293)	X	MFI	6	OHV
	Suburban	5.7 (5735)	R	MFI	8	OHV
	Suburban	6.5 (6374)	S	DSL	8	OHV
	Suburban	7.4 (7440)	J	MFI	8	OHV
	Tahoe	5.7 (5735)	R	MFI	8	OHV
	Yukon	5.7 (5735)	R	MFI	8	OHV

DSL - Diesel

MFI - Multi-port Fuel Injection

OHV - Overhead Valve

TFI - Throttle body Fuel Injection

79241C18

GENERAL ENGINE SPECIFICATIONS

Year	Engine ID/VIN	Engine Displacement Liters (cc)	Fuel System Type	Net Horsepower @ rpm	Net Torque @ rpm (ft. lbs.)	Bore x Stroke (in.)	Compression Ratio	Oil Pressure @ rpm
1995	4	2.2 (2189)	MFI	118@5200	130@2800	3.50x3.46	9.0:1	56@3000
	F	6.5 (6473)	DSL	190@3400	⑤	4.06x3.82	21.5:1	40-45@2000
	H	5.0 (4999)	TFI	175@4200	265@2800	3.74x3.48	9.1:1	18@2000
	K	5.7 (5735)	MFI	③	④	4.00x3.48	9.1:1	18@2000
	K	5.7 (5735)	TFI	③	④	4.00x3.48	9.1:1	18@2000
	N	7.4 (7440)	TFI	230@3600	385@1600	4.25x4.00	7.9:1	25@2000
	P	6.5 (6473)	DSL	155@3600	275@1700	4.06x3.82	21.5:1	40-45@2000
	S	6.5 (6473)	DSL	155@3600	275@1700	4.06x3.82	21.5:1	40-45@2000
	W	4.3 (4293)	TFI	191@4500	260@3600	4.00x3.48	9.1:1	18@2000
	Y	6.5 (6473)	DSL	⑥	⑦	4.06x3.82	21.5:1	40-45@2000
	Z	4.3 (4293)	MFI	①	②	4.00x3.48	9.1:1	18@2000
	Z	4.3 (4293)	TFI	①	②	4.00x3.48	9.1:1	18@2000
1996	4	2.2 (2189)	MFI	118@5200	130@2800	3.50x3.46	9.0:1	56@3000
	F	6.5 (6473)	DSL	⑫	⑤	4.05x3.80	21.5:1	40-45@2000
	J	7.4 (7440)	MFI	290@4200	410@3200	4.25x4.00	9.0:1	40@2000
	M	5.0 (4999)	MFI	220@4600	285@2800	3.74x3.48	9.4:1	18@2000
	R	5.7 (5735)	MFI	250@4600	335@2800	4.00x3.48	9.4:1	18@2000
	S	6.5 (6473)	DSL	180@3400	360@1700	4.06x3.82	21.5:1	40-45@2000
	W	4.3 (4293)	MFI	⑧	⑨	4.00x3.48	9.2:1	18@2000
	X	4.3 (4293)	MFI	⑩	⑪	4.00x3.48	9.2:1	18@2000
1997	4	2.2 (2189)	MFI	118@5200	130@2800	3.50x3.46	9.0:1	56@3000
	F	6.5 (6473)	DSL	⑫	⑤	4.05x3.80	21.5:1	40-45@2000
	J	7.4 (7440)	MFI	290@4200	410@3200	4.25x4.00	9.0:1	40@2000
	M	5.0 (4999)	MFI	220@4600	285@2800	3.74x3.48	9.4:1	18@2000
	R	5.7 (5735)	MFI	250@4600	335@2800	4.00x3.48	9.4:1	18@2000
	S	6.5 (6473)	DSL	180@3400	360@1700	4.06x3.82	21.5:1	40-45@2000
	W	4.3 (4293)	MFI	⑧	⑨	4.00x3.48	9.2:1	18@2000
	X	4.3 (4293)	MFI	⑩	⑪	4.00x3.48	9.2:1	18@2000
1998-99	4	2.2 (2189)	MFI	120@5000	140@3600	3.50x3.46	9.0:1	56@3000
	F	6.5 (6473)	DSL	195@3400	430@1800	4.05x3.80	21.5:1	30-43@2000
	J	7.4 (7440)	MFI	290@4200	410@3200	4.25x4.00	9.0:1	40@2000
	M	5.0 (4999)	MFI	230@4600	285@2800	3.74x3.48	9.4:1	18@2000
	R	5.7 (5735)	MFI	255@4600	335@2800	4.00x3.48	9.4:1	18@2000
	S	6.5 (6473)	DSL	180@3400	360@1700	4.06x3.82	21.5:1	30-43@2000
	W	4.3 (4293)	MFI	⑬	240@2800	4.00x3.48	9.2:1	18@2000
	X	4.3 (4293)	MFI	⑭	⑮	4.00x3.48	9.2:1	18@2000

TFI - Throttle body fuel injection
MFI - Multiport fuel injection
DSL - Diesel

① Below 15,000 GVWR: 385@1700
　Above 15,000 GVWR: 380@1700
② Below 8500 GVWR: 210@4000
　Above 8500 GVWR: 190@4000
③ Below 8500 GVWR: 385@1700
　Above 8500 GVWR: 300@2400
④ Below 8500 GVWR: 155@3600
　Above 8500 GVWR: 160@3600
⑤ Below 8500 GVWR: 275@1700
　Above 8500 GVWR: 290@1700

⑥ S10: 155@4000
　C/K: 160@4000
　C/K HD: 155@4000
　G Van: 165@4000
⑦ S10: 230@2800
　C/K Pick-up: 235@2400
　C/K HD Pick-up and
　G-Van: 230@2400
⑧ Below 15,000 GVWR: 180@3400
　Above 15,000 GVWR: 190@3400
⑨ 2WD: 180@4400
　4WD: 190@4400

⑩ 2WD: 245@2800
　4WD: 250@2800
⑪ 2WD: 170@4400
　4WD: 180@4400
⑫ 2WD: 235@2800
　4WD: 240@2800
⑬ 2WD: 175@4400
　4WD: 180@4400
⑭ 2WD: 180@4400
　4WD: 190@4400
⑮ 2WD: 245@2800
　4WD: 250@2800

79241C19

GASOLINE ENGINE TUNE-UP SPECIFICATIONS

Year	Engine ID/VIN	Engine Displacement Liters (cc)	Spark Plugs Gap (in.)	Ignition Timing (deg.) MT	AT	Fuel Pump (psi)	Idle Speed (rpm) MT	AT	Valve Clearance In.	Ex.
1995	4	2.2 (2189)	0.060	①	①	41-47 ②	950	890	HYD	HYD
	H	5.0 (4999)	0.035	③	③	9-13	③	③	HYD	HYD
	K	5.7 (5735)	0.035	③	③	9-13	③	③	HYD	HYD
	N	7.4 (7440)	0.035	③	③	26-32	③	③	HYD	HYD
	W	4.3 (4293)	0.035	④	④	55-61 ②		625	HYD	HYD
	Z	4.3 (4293)	0.035	⑤	⑤	9-13	650	725	HYD	HYD
1996	4	2.2 (2189)	0.060	①	①	41-47	⑥	⑥	HYD	HYD
	J	7.4 (7440)	0.060	④	④	60-66 ②	750 ⑦	675 ⑦	HYD	HYD
	M	5.0 (4999)	0.060	④	④	60-66 ②	650 ⑧	550	HYD	HYD
	R	5.7 (5735)	0.060	④	④	60-66 ②	660 ⑨	525	HYD	HYD
	W	4.3 (4293)	0.060	④	④	58-64 ②	600	625	HYD	HYD
	X	4.3 (4293)	0.045	①	①	41-47	⑥	⑥	HYD	HYD
1997	4	2.2 (2189)	0.060	①	①	41-47	⑥	⑥	HYD	HYD
	J	7.4 (7440)	0.060	④	④	60-66 ②	750 ⑦	675 ⑦	HYD	HYD
	M	5.0 (4999)	0.060	④	④	60-66 ②	650 ⑧	550	HYD	HYD
	R	5.7 (5735)	0.060	④	④	60-66 ②	660 ⑨	525	HYD	HYD
	W	4.3 (4293)	0.060	④	④	58-64 ②	600	625	HYD	HYD
	X	4.3 (4293)	0.045	①	①	41-47	⑥	⑥	HYD	HYD
1998-99	4	2.2 (2189)	0.060	①	①	41-47	⑥	⑥	HYD	HYD
	J	7.4 (7440)	0.060	④	④	60-66 ②	750 ⑦	675 ⑦	HYD	HYD
	M	5.0 (4999)	0.060	④	④	60-66 ②	650 ⑧	550	HYD	HYD
	R	5.7 (5735)	0.060	④	④	60-66 ②	660 ⑨	525	HYD	HYD
	W	4.3 (4293)	0.060	④	④	58-64 ②	600	625	HYD	HYD
	X	4.3 (4293)	0.045	①	①	41-47	⑥	⑥	HYD	HYD

NOTE: The Vehicle Emission Control Information label often reflects specification changes made during production. The label figures must be used if they differ from those in this chart.

HYD - Hydraulic

① Distributorless ignition
② With key ON and engine OFF.
③ Refer to underhood label for exact setting
④ Ignition timing is preset and cannot be adjusted
⑤ 0°, disconnect set timing connector (tan with black striped wire taped to engine harness near distributor)
⑥ Idle speed is maintained by the PCM
⑦ Over 8500 GVW
⑧ Under 8500 GVW
⑨ Over 8500:
   Manual: 565-615
   Automatic: 525-575

79241C20

DIESEL ENGINE TUNE-UP SPECIFICATIONS

Year	Engine ID/VIN	Engine Displacement cu. in. (cc)	Valve Clearance Intake (in.)	Exhaust (in.)	Intake Valve Opens (deg.)	Injection Pump Setting (deg.)	Injection Nozzle Pressure (psi) New	Used	Idle Speed (rpm)	Cranking Compression Pressure (psi)
1995	F	6.5 (6473)	HYD	HYD	①	①	1600	1500	①	NA
	P	6.5 (6473)	HYD	HYD	①	①	1800	1700	①	NA
	S	6.5 (6473)	HYD	HYD	①	①	1800	1700	①	NA
	Y	6.5 (6473)	HYD	HYD	①	①	1600	1500	①	NA
1996	F	6.5 (6473)	HYD	HYD	①	①	1800	1700	①	NA
	S	6.5 (6473)	HYD	HYD	①	①	1800	1700	①	NA
1997	F	6.5 (6473)	HYD	HYD	①	①	1800	1700	①	NA
	S	6.5 (6473)	HYD	HYD	①	①	1800	1700	①	NA
1998-99	F	6.5 (6473)	HYD	HYD	①	①	1800	1700	①	380-400
	S	6.5 (6473)	HYD	HYD	①	①	1800	1700	①	380-400

NOTE: The Vehicle Emission Control Information label often reflects specification changes made during production. The label figures must be used if they differ from those in this chart

HYD - Hydraulic

NA - Not Available

① Refer to Vehicle Emission Control Information label

79241C21

CAPACITIES

Year	Model	Engine ID/VIN	Engine Displacement Liters (cc)	Engine Oil with Filter (qts.)	Transmission (pts.) 5-Spd	Transmission (pts.) Auto.	Transfer Case (pts.)	Drive Axle Front (pts.)	Drive Axle Rear (pts.)	Fuel Tank (gal.)	Cooling System (qts.)
1995	Astro	W	4.3 (4293)	5.0	4.4	10.0	3.0	2.6	3.8	27.0	12.8 [1]
	Astro	Z	4.3 (4293)	5.0	4.4	10.0	—	—	3.8	27.0	12.8 [1]
	Safari	W	4.3 (4293)	5.0	4.4	10.0	3.0	2.6	3.8	27.0	12.8 [1]
	Safari	Z	4.3 (4293)	5.0	4.4	10.0	—	—	3.8	27.0	12.8 [1]
	C1500	Z	4.3 (4293)	5.0	[2]	[3]	—	—	[4]	[5]	11.0
	C1500	H	5.0 (4999)	5.0	[2]	[3]	—	—	[4]	[5]	18.0
	C1500	K	5.7 (5735)	5.0	[2]	[3]	—	—	[4]	[5]	18.0
	C1500	F	6.5 (6473)	7.0	[2]	[3]	—	—	[4]	[5]	26.5
	C1500	N	7.4 (7440)	7.0	[2]	[3]	—	—	[4]	[5]	25.0
	C2500	Z	4.3 (4293)	5.0	[2]	[3]	—	—	[4]	[5]	11.0
	C2500	H	5.0 (4999)	5.0	[2]	[3]	—	—	[4]	[5]	18.0
	C2500	K	5.7 (5735)	5.0	[2]	[3]	—	—	[4]	[5]	18.0
	C2500	P	6.5 (6473)	7.0	[2]	[3]	—	—	[4]	[5]	26.5
	C2500	S	6.5 (6473)	7.0	[2]	[3]	—	—	[4]	[5]	26.5
	C3500	H	5.0 (4999)	5.0	[2]	[3]	—	—	[4]	[5]	18.0
	C3500	K	5.7 (5735)	5.0	[2]	[3]	—	—	[4]	[5]	18.0 [6]
	C3500	F	6.5 (6473)	7.0	[2]	[3]	—	—	[4]	[5]	26.5
	C3500	P	6.5 (6473)	7.0	[2]	[3]	—	—	[4]	[5]	26.5
	C3500	S	6.5 (6473)	7.0	[2]	[3]	—	—	[4]	[5]	26.5
	C3500	N	7.4 (7440)	6.0	[2]	[3]	—	—	[4]	[5]	25.0 [7]
	G/P10	Z	4.3 (4293)	5.0	[2]	[3]	—	—	[4]	22.0 [8]	11.0 [9]
	G/P10	H	5.0 (4999)	5.0	[2]	[3]	—	—	[4]	22.0 [8]	17.0 [9]
	G/P10	K	5.7 (5735)	5.0	[2]	[3]	—	—	[4]	[10]	18.0 [9]
	G/P10	P	6.5 (6505)	7.0	[2]	[3]	—	—	[4]	22.0 [8]	24.0 [9]
	G/P20	Z	4.3 (4293)	5.0	[2]	[3]	—	—	[4]	22.0 [8]	11.0 [9]
	G/P20	H	5.0 (4999)	5.0	[2]	[3]	—	—	[4]	22.0 [8]	17.0 [9]
	G/P20	K	5.7 (5735)	5.0	[2]	[3]	—	—	[4]	[10]	18.0 [9]
	G/P20	P	6.5 (6505)	7.0	[2]	[3]	—	—	[4]	22.0 [8]	24.0 [9]
	G/P20	Y	6.5 (6505)	7.0	[2]	[3]	—	—	[4]	22.0 [8]	24.0 [9]
	G/P30	Z	4.3 (4293)	5.0	[2]	[3]	—	—	[4]	22.0 [8]	11.0 [9]
	G/P30	K	5.7 (5735)	5.0	[2]	[3]	—	—	[4]	[10]	18.0 [9]
	G/P30	P	6.5 (6505)	7.0	[2]	[3]	—	—	[4]	22.0 [8]	24.0 [9]
	G/P30	Y	6.5 (6505)	7.0	[2]	[3]	—	—	[4]	22.0 [8]	24.0 [9]
	G/P30	N	7.4 (7440)	6.0	[2]	[3]	—	—	[4]	[11]	24.5 [9]
	K1500	Z	4.3 (4293)	5.0	[2]	[3]	[12]	[13]	[4]	[5]	11.0
	K1500	H	5.0 (4999)	5.0	[2]	[3]	[12]	[13]	[4]	[5]	18.0
	K1500	K	5.7 (5735)	5.0	[2]	[3]	[12]	[13]	[4]	[5]	18.0
	K1500	F	6.5 (6505)	7.0	[2]	[3]	[12]	[13]	[4]	[5]	25.0
	K1500	N	7.4 (7440)	7.0	[2]	[3]	[12]	[13]	[4]	[5]	25.0 [7]
	K2500	Z	4.3 (4293)	5.0	[2]	[3]	[12]	[13]	[4]	[5]	11.0
	K2500	H	5.0 (4999)	5.0	[2]	[3]	[12]	[13]	[4]	[5]	18.0
	K2500	K	5.7 (5735)	5.0	[2]	[3]	[12]	[13]	[4]	[5]	18.0
	K2500	P	6.5 (6505)	7.0	[2]	[3]	[12]	[13]	[4]	[5]	25.0
	K2500	S	6.5 (6505)	7.0	[2]	[3]	[12]	[13]	[4]	[5]	25.0
	K2500	F	6.5 (6473)	7.0	[2]	[3]	[12]	[13]	[4]	[5]	26.5
	K2500	N	7.4 (7440)	7.0	[2]	[3]	[12]	[13]	[4]	[5]	25.0 [7]
	K3500	H	5.0 (4999)	5.0	[2]	[3]	[12]	[13]	[4]	[5]	18.0
	K3500	K	5.7 (5735)	5.0	[2]	[3]	[12]	[13]	[4]	[5]	18.0

79241C22

CAPACITIES

Year	Model	Engine ID/VIN	Engine Displacement Liters (cc)	Engine Oil with Filter (qts.)	Transmission (pts.) 5-Spd	Transmission (pts.) Auto.	Transfer Case (pts.)	Drive Axle Front (pts.)	Drive Axle Rear (pts.)	Fuel Tank (gal.)	Cooling System (qts.)
1995 (cont.)	K3500	P	6.5 (6505)	7.0	②	③	⑫	⑬	④	⑤	25.0
	K3500	S	6.5 (6505)	7.0	②	③	⑫	⑬	④	⑤	25.0
	K3500	F	6.5 (6473)	7.0	②	③	⑫	⑬	④	⑤	26.5
	K3500	N	7.4 (7440)	7.0	②	③	⑫	⑬	④	⑤	25.0
	Bravada	W	4.3 (4293)	4.5	4.4	10.0	—	3.5	3.5	20.0	12.1
	Jimmy	W	4.3 (4293)	4.5	4.4	10.0	—	3.5	3.5	20.0	12.1
	S10 Blazer	W	4.3 (4293)	4.5	4.4	10.0	—	3.5	3.5	20.0	12.1
	S10 Pick-up	4	2.2 (2189)	4.0	4.4	10.0	4.6	2.6	4.0	13.0 ⑭	11.5
	S10 Pick-up	W	4.3 (4293)	4.5	4.4	10.0	4.6	2.6	4.0	20.0	12.0
	S10 Pick-up	Z	4.3 (4293)	5.0	4.4	10.0	4.6	2.6	4.0	20.0	12.0
	S15 Pick-up	4	2.2 (2189)	4.0	4.4	10.0	4.6	2.6	4.0	13.0 ⑭	11.5
	S15 Pick-up	W	4.3 (4293)	4.5	4.4	10.0	4.6	2.6	4.0	20.0	12.0
	S15 Pick-up	Z	4.3 (4293)	5.0	4.4	10.0	4.6	2.6	4.0	20.0	12.0
	Sonoma	4	2.2 (2189)	4.0	4.4	10.0	4.6	2.6	4.0	13.0 ⑭	11.5
	Sonoma	W	4.3 (4293)	4.5	4.4	10.0	4.6	2.6	4.0	20.0	12.0
	Sonoma	Z	4.3 (4293)	5.0	4.4	10.0	4.6	2.6	4.0	20.0	12.0
	Suburban	K	5.7 (5735)	5.0	—	③	—	—	④	⑤	18.0
	Suburban	P	6.5 (6505)	7.0	②	③	—	—	④	⑤	25.0
	Suburban	S	6.5 (6505)	7.0	②	③	—	—	④	⑤	25.0
	Suburban	N	7.4 (7440)	6.0	—	③	10.0	4.0	④	25.0 ⑤	24.5
	Tahoe	K	5.7 (5735)	5.0	—	③	10.0	4.0	④	25.0 ⑯	18.0
	Tahoe	F	6.5 (6473)	7.0	②	③	—	—	④	⑤	26.5
	Yukon	K	5.7 (5735)	5.0	—	③	10.0	4.0	④	25.0 ⑯	18.0
	Yukon	F	6.5 (6473)	7.0	②	③	—	—	④	⑤	26.5
1996	Astro	W	4.3 (4293)	5.0	—	10.0	3.0	2.6	3.8	27.0	⑰
	Safari	W	4.3 (4293)	5.0	—	10.0	3.0	2.6	3.8	27.0	⑰
	C1500	W	4.3 (4293)	5.0	②	⑱	—	—	④	⑤	13.0
	C1500	M	5.0 (4999)	5.0	②	⑱	—	—	④	⑤	18.0
	C1500	R	5.7 (5735)	5.0	②	⑱	—	—	④	⑤	18.0
	C2500	W	4.3 (4293)	5.0	②	⑱	—	—	④	⑤	13.0
	C2500	M	5.0 (4999)	5.0	②	⑱	—	—	④	⑤	18.0
	C2500	R	5.7 (5735)	5.0	②	⑱	—	—	④	⑤	18.0
	C2500	F	6.5 (6473)	7.0	②	⑱	—	—	④	⑤	27.5
	C2500	S	6.5 (6473)	7.0	②	⑱	—	—	④	⑤	27.5
	C3500	F	6.5 (6473)	7.0	②	⑱	—	—	④	⑤	27.5
	C3500	J	7.4 (7440)	6.0	②	⑱	—	—	④	⑤	25.0 ⑦
	G/P1500	W	4.3 (4293)	5.0	②	⑱	—	—	④	22.0 ⑤	13.0 ⑨
	G/P1500	M	5.0 (4999)	5.0	②	⑱	—	—	④	22.0 ⑤	17.0 ⑨
	G/P1500	R	5.7 (5735)	5.0	②	⑱	—	—	④	⑤	18.0 ⑨
	G/P2500	M	5.0 (4999)	5.0	②	⑱	—	—	④	22.0 ⑤	17.0 ⑨
	G/P2500	R	5.7 (5735)	5.0	②	⑱	—	—	④	⑤	18.0 ⑨
	G/P3500	F	6.5 (6473)	7.0	②	⑱	—	—	④	22.0 ⑤	27.5 ⑨
	G/P3500	J	7.4 (7440)	6.0	②	⑱	—	—	④	⑪	24.5 ⑨
	K1500	W	4.3 (4293)	5.0	②	⑱	⑲	⑬	④	⑤	11.0
	K1500	M	5.0 (4999)	5.0	②	⑱	⑲	⑬	④	⑤	18.0
	K1500	R	5.7 (5735)	5.0	②	⑱	⑲	⑬	④	⑤	18.0
	K1500	F	6.5 (6473)	7.0	②	⑱	⑲	⑬	④	⑤	27.5
	K2500	M	5.0 (4999)	5.0	②	⑱	⑲	⑬	④	⑤	18.0

79241C23

CAPACITIES

Year	Model	Engine ID/VIN	Engine Displacement Liters (cc)	Engine Oil with Filter (qts.)	Transmission (pts.) 5-Spd	Transmission (pts.) Auto.	Transfer Case (pts.)	Drive Axle Front (pts.)	Drive Axle Rear (pts.)	Fuel Tank (gal.)	Cooling System (qts.)
1996 (cont.)	K2500	R	5.7 (5735)	5.0	[2]	[18]	[19]	[13]	[4]	[5]	18.0
	K2500	F	6.5 (6473)	7.0	[2]	[18]	[19]	[13]	[4]	[5]	27.5
	K3500	F	6.5 (6473)	7.0	[2]	[18]	[19]	[13]	[4]	[5]	27.5
	K3500	J	7.4 (7440)	6.0	[2]	[18]	[19]	[13]	[4]	[5]	25.0
	S10 Blazer	W	4.3 (4293)	5.0	4.4	10.0	2.6	2.6	3.9	20.0	11.9
	S10 Blazer	X	4.3 (4293)	5.0	4.4	10.0	2.6	2.6	3.9	20.0	11.9
	Jimmy	W	4.3 (4293)	5.0	4.4	10.0	2.6	2.6	3.9	20.0	11.9
	Jimmy	X	4.3 (4293)	5.0	4.4	10.0	2.6	2.6	3.9	20.0	11.9
	Bravada	W	4.3 (4293)	5.0	4.4	10.0	2.6	2.6	3.9	20.0	11.9
	Bravada	X	4.3 (4293)	5.0	4.4	10.0	2.6	2.6	3.9	20.0	11.9
	Sonoma	4	2.2 (2189)	4.0	4.4	10.0	2.6	2.6	3.9	13.0 [14]	11.5
	Sonoma	W	4.3 (4293)	5.0	4.4	10.0	2.6	2.6	3.9	20.0	11.9
	Sonoma	X	4.3 (4293)	5.0	4.4	10.0	2.6	2.6	3.9	20.0	11.9
	S15 Pick-up	4	2.2 (2189)	4.0	4.4	10.0	2.6	2.6	3.9	13.0 [14]	11.5
	S15 Pick-up	W	4.3 (4293)	5.0	4.4	10.0	2.6	2.6	3.9	20.0	11.9
	S15 Pick-up	X	4.3 (4293)	5.0	4.4	10.0	2.6	2.6	3.9	20.0	11.9
	S10 Pick-up	4	2.2 (2189)	4.0	4.4	10.0	2.6	2.6	3.9	13.0 [14]	11.5
	S10 Pick-up	W	4.3 (4293)	5.0	4.4	10.0	2.6	2.6	3.9	20.0	11.9
	S10 Pick-up	X	4.3 (4293)	5.0	4.4	10.0	2.6	2.6	3.9	20.0	11.9
	Suburban	R	5.7 (5735)	5.0	—	[18]	—	—	[4]	[5]	18.0
	Suburban	F	7.4 (7440)	6.0	—	[18]	[19]	[13]	[4]	25.0 [5]	24.5
	Tahoe	R	5.7 (5735)	5.0	—	[18]	[19]	[13]	[4]	[10]	18.0
	Tahoe	S	6.5 (6473)	7.0	—	[18]	—	—	[4]	[10]	23.8
	Yukon	R	5.7 (5735)	5.0	—	[18]	[19]	[13]	[4]	[10]	18.0
	Yukon	S	6.5 (6473)	7.0	—	[18]	—	—	[4]	[10]	23.8
1997	Astro	W	4.3 (4293)	5.0	—	10.0	3.0	2.6	3.8	27.0	[17]
	Safari	W	4.3 (4293)	5.0	—	10.0	3.0	2.6	3.8	27.0	[17]
	C1500	W	4.3 (4293)	5.0	[2]	[18]	—	—	[4]	[5]	13.0
	C1500	M	5.0 (4999)	5.0	[2]	[18]	—	—	[4]	[5]	18.0
	C1500	R	5.7 (5735)	5.0	[2]	[18]	—	—	[4]	[5]	18.0
	C2500	W	4.3 (4293)	5.0	[2]	[18]	—	—	[4]	[5]	13.0
	C2500	M	5.0 (4999)	5.0	[2]	[18]	—	—	[4]	[5]	18.0
	C2500	R	5.7 (5735)	5.0	[2]	[18]	—	—	[4]	[5]	18.0
	C2500	F	6.5 (6473)	7.0	[2]	[18]	—	—	[4]	[5]	27.5
	C2500	S	6.5 (6473)	7.0	[2]	[18]	—	—	[4]	[5]	27.5
	C3500	F	6.5 (6473)	7.0	[2]	[18]	—	—	[4]	[5]	27.5
	C3500	J	7.4 (7440)	6.0	[2]	[18]	—	—	[4]	[5]	25.0 [7]
	G/P1500	W	4.3 (4293)	5.0	[2]	[18]	—	—	[4]	22.0 [5]	13.0 [9]
	G/P1500	M	5.0 (4999)	5.0	[2]	[18]	—	—	[4]	22.0 [5]	17.0 [9]
	G/P1500	R	5.7 (5735)	5.0	[2]	[18]	—	—	[4]	[5]	18.0 [9]
	G/P2500	M	5.0 (4999)	5.0	[2]	[18]	—	—	[4]	22.0 [5]	17.0 [9]
	G/P2500	R	5.7 (5735)	5.0	[2]	[18]	—	—	[4]	[5]	18.0 [9]
	G/P3500	F	6.5 (6473)	7.0	[2]	[18]	—	—	[4]	22.0 [5]	27.5 [9]
	G/P3500	J	7.4 (7440)	6.0	[2]	[18]	—	—	[4]	[11]	24.5 [9]
	K1500	W	4.3 (4293)	5.0	[2]	[18]	[19]	[13]	[4]	[5]	11.0
	K1500	M	5.0 (4999)	5.0	[2]	[18]	[19]	[13]	[4]	[5]	18.0
	K1500	R	5.7 (5735)	5.0	[2]	[18]	[19]	[13]	[4]	[5]	18.0
	K1500	F	6.5 (6473)	7.0	[2]	[18]	[19]	[13]	[4]	[5]	27.5

79241C24

CAPACITIES

Year	Model	Engine ID/VIN	Engine Displacement Liters (cc)	Engine Oil with Filter (qts.)	Transmission (pts.) 5-Spd	Transmission (pts.) Auto.	Transfer Case (pts.)	Drive Axle Front (pts.)	Drive Axle Rear (pts.)	Fuel Tank (gal.)	Cooling System (qts.)
1997 (cont.)	K2500	M	5.0 (4999)	5.0	[2]	[18]	[19]	[13]	[4]	[5]	18.0
	K2500	R	5.7 (5735)	5.0	[2]	[18]	[19]	[13]	[4]	[5]	18.0
	K2500	F	6.5 (6473)	7.0	[2]	[18]	[19]	[13]	[4]	[5]	27.5
	K3500	F	6.5 (6473)	7.0	[2]	[18]	[19]	[13]	[4]	[5]	27.5
	K3500	J	7.4 (7440)	6.0	[2]	[18]	[19]	[13]	[4]	[5]	25.0
	S10 Blazer	W	4.3 (4293)	5.0	4.4	10.0	2.6	2.6	3.9	20.0	11.9
	S10 Blazer	X	4.3 (4293)	5.0	4.4	10.0	2.6	2.6	3.9	20.0	11.9
	Jimmy	W	4.3 (4293)	5.0	4.4	10.0	2.6	2.6	3.9	20.0	11.9
	Jimmy	X	4.3 (4293)	5.0	4.4	10.0	2.6	2.6	3.9	20.0	11.9
	Bravada	W	4.3 (4293)	5.0	4.4	10.0	2.6	2.6	3.9	20.0	11.9
	Bravada	X	4.3 (4293)	5.0	4.4	10.0	2.6	2.6	3.9	20.0	11.9
	Sonoma	4	2.2 (2189)	4.0	4.4	10.0	2.6	2.6	3.9	13.0 [14]	11.5
	Sonoma	W	4.3 (4293)	5.0	4.4	10.0	2.6	2.6	3.9	20.0	11.9
	Sonoma	X	4.3 (4293)	5.0	4.4	10.0	2.6	2.6	3.9	20.0	11.9
	S15 Pick-up	4	2.2 (2189)	4.0	4.4	10.0	2.6	2.6	3.9	13.0 [14]	11.5
	S15 Pick-up	W	4.3 (4293)	5.0	4.4	10.0	2.6	2.6	3.9	20.0	11.9
	S15 Pick-up	X	4.3 (4293)	5.0	4.4	10.0	2.6	2.6	3.9	20.0	11.9
	S10 Pick-up	4	2.2 (2189)	4.0	4.4	10.0	2.6	2.6	3.9	13.0 [14]	11.5
	S10 Pick-up	W	4.3 (4293)	5.0	4.4	10.0	2.6	2.6	3.9	20.0	11.9
	S10 Pick-up	X	4.3 (4293)	5.0	4.4	10.0	2.6	2.6	3.9	20.0	11.9
	Suburban	R	5.7 (5735)	5.0	—	[18]	—	—	[4]	[5]	18.0
	Suburban	F	7.4 (7440)	6.0	—	[18]	[19]	[13]	[4]	25.0 [5]	24.5
	Tahoe	R	5.7 (5735)	5.0	—	[18]	[19]	[13]	[4]	[10]	18.0
	Tahoe	S	6.5 (6473)	7.0	—	[18]	—	—	[4]	[10]	23.8
	Yukon	R	5.7 (5735)	5.0	—	[18]	[19]	[13]	[4]	[10]	18.0
	Yukon	S	6.5 (6473)	7.0	—	[18]	—	—	[4]	[10]	23.8
1998-99	Astro	W	4.3 (4293)	5.0	—	10.0	3.0	2.6	3.8	27.0	[17]
	Bravada	W	4.3 (4293)	5.0	[20]	11.0	2.6	2.6	3.9	20.0	11.9
	Bravada	X	4.3 (4293)	5.0	[20]	11.0	2.6	2.6	3.9	20.0	11.9
	C1500	M	5.0 (4999)	5.0	[2]	[18]	—	—	[4]	[5]	18.0
	C1500	R	5.7 (5735)	5.0	[2]	[18]	—	—	[4]	[5]	18.0
	C1500	W	4.3 (4293)	5.0	[2]	[18]	—	—	[4]	[5]	13.0
	C2500	F	6.5 (6473)	7.0	[2]	[18]	—	—	[4]	[5]	27.5
	C2500	M	5.0 (4999)	5.0	[2]	[18]	—	—	[4]	[5]	18.0
	C2500	R	5.7 (5735)	5.0	[2]	[18]	—	—	[4]	[5]	18.0
	C2500	S	6.5 (6473)	7.0	[2]	[18]	—	—	[4]	[5]	27.5
	C2500	W	4.3 (4293)	5.0	[2]	[18]	—	—	[4]	[5]	13.0
	C3500	F	6.5 (6473)	7.0	[2]	[18]	—	—	[4]	[5]	27.5
	C3500	J	7.4 (7440)	6.0	[2]	[18]	—	—	[4]	[5]	25.0 [7]
	Denali	R	5.7 (5735)	5.0	—	[18]	[19]	[13]	[4]	[10]	18.0
	Denali	S	6.5 (6473)	7.0	—	[18]	-	—	[4]	[10]	23.8
	Envoy	W	4.3 (4293)	5.0	[20]	11.0	2.6	2.6	3.9	20.0	11.9
	Envoy	X	4.3 (4293)	5.0	[20]	11.0	2.6	2.6	3.9	20.0	11.9
	Escalade	R	5.7 (5735)	5.0	—	[18]	[19]	[13]	[4]	[10]	18.0
	Escalade	S	6.5 (6473)	7.0	—	[18]	-	—	[4]	[10]	23.8
	G/P1500	M	5.0 (4999)	5.0	[2]	[18]	—	—	[4]	22.0 [5]	17.0 [9]
	G/P1500	R	5.7 (5735)	5.0	[2]	[18]	—	—	[4]	[5]	18.0 [9]
	G/P1500	W	4.3 (4293)	5.0	[2]	[18]	—	—	[4]	22.0 [5]	13.0 [9]

79241C25

CAPACITIES

Year	Model	Engine ID/VIN	Engine Displacement Liters (cc)	Engine Oil with Filter (qts.)	Transmission (pts.) 5-Spd	Auto.	Transfer Case (pts.)	Drive Axle Front (pts.)	Rear (pts.)	Fuel Tank (gal.)	Cooling System (qts.)
1998-99 (cont.)	G/P2500	M	5.0 (4999)	5.0	②	⑱	—	—	④	22.0 ⑮	17.0 ⑨
	G/P2500	R	5.7 (5735)	5.0	②	⑱	—	—	④	⑮	18.0 ⑨
	G/P3500	F	6.5 (6473)	7.0	②	⑱	—	—	④	22.0 ⑮	27.5 ⑨
	G/P3500	J	7.4 (7440)	6.0	②	⑱	—	—	④	⑪	24.5 ⑨
	Jimmy	W	4.3 (4293)	5.0	⑳	11.0	2.6	2.6	3.9	20.0	11.9
	Jimmy	X	4.3 (4293)	5.0	⑳	11.0	2.6	2.6	3.9	20.0	11.9
	K1500	F	6.5 (6473)	7.0	②	⑱	⑲	⑬	④	⑤	27.5
	K1500	M	5.0 (4999)	5.0	②	⑱	⑲	⑬	④	⑤	18.0
	K1500	R	5.7 (5735)	5.0	②	⑱	⑲	⑬	④	⑤	18.0
	K1500	W	4.3 (4293)	5.0	②	⑱	⑲	⑬	④	⑤	11.0
	K2500	F	6.5 (6473)	7.0	②	⑱	⑲	⑬	④	⑤	27.5
	K2500	M	5.0 (4999)	5.0	②	⑱	⑲	⑬	④	⑤	18.0
	K2500	R	5.7 (5735)	5.0	②	⑱	⑲	⑬	④	⑤	18.0
	K3500	F	6.5 (6473)	7.0	②	⑱	⑲	⑬	④	⑤	27.5
	K3500	J	7.4 (7440)	6.0	②	⑱	⑲	⑬	④	⑤	25.0
	S10 Blazer	W	4.3 (4293)	5.0	⑳	11.0	2.6	2.6	3.9	20.0	11.9
	S10 Blazer	X	4.3 (4293)	5.0	⑳	11.0	2.6	2.6	3.9	20.0	11.9
	S10 Pick-up	4	2.2 (2189)	4.0	⑳	11.0	2.6	2.6	3.9	13.0 ⑭	11.5
	S10 Pick-up	W	4.3 (4293)	5.0	⑳	11.0	2.6	2.6	3.9	20.0	11.9
	S10 Pick-up	X	4.3 (4293)	5.0	⑳	11.0	2.6	2.6	3.9	20.0	11.9
	S15 Pick-up	4	2.2 (2189)	4.0	⑳	11.0	2.6	2.6	3.9	13.0 ⑭	11.5
	S15 Pick-up	W	4.3 (4293)	5.0	⑳	11.0	2.6	2.6	3.9	20.0	11.9
	S15 Pick-up	X	4.3 (4293)	5.0	⑳	11.0	2.6	2.6	3.9	20.0	11.9
	Safari	W	4.3 (4293)	5.0	—	10.0	3.0	2.6	3.8	27.0	⑰
	Sonoma	4	2.2 (2189)	4.0	⑳	11.0	2.6	2.6	3.9	13.0 ⑭	11.5
	Sonoma	W	4.3 (4293)	5.0	⑳	11.0	2.6	2.6	3.9	20.0	11.9
	Sonoma	X	4.3 (4293)	5.0	⑳	11.0	2.6	2.6	3.9	20.0	11.9
	Suburban	F	7.4 (7440)	6.0	—	⑱	⑲	⑬	④	25.0 ⑮	24.5
	Suburban	R	5.7 (5735)	5.0	—	⑱	—	—	④	⑤	18.0
	Tahoe	R	5.7 (5735)	5.0	—	⑱	⑲	⑬	④	⑩	18.0
	Tahoe	S	6.5 (6473)	7.0	—	⑱	-	—	④	⑩	23.8
	Yukon	R	5.7 (5735)	5.0	—	⑱	⑲	⑬	④	⑩	18.0
	Yukon	S	6.5 (6473)	7.0	—	⑱	-	—	④	⑩	23.8

NOTE: All capacities are approximate. Add fluid gradually and check to be sure a proper fluid level is obtained.

① 16.5 qts. with rear heater
② New Venture gear 4500: 8.0 pts.
New Venture gear 5LM60: 4.4 pts.
③ 350C trans.: 6.3 pts.
THM400 and 4L80 trans.: 9.0 pts.
THM700 R4 and 4L60 trans.: 10.0 pts.
THM700 R4 and 4L80-E trans.: 14.3 pts.
④ 8.5" ring gear: 4.2 pts.
9.5" ring gear: 6.5 pts.
9.75" ring gear: 6.0 pts.
10.5" ring gear: 6.5 pts.

⑤ Std. available with 25 and 34 gallon tanks
Chassis cab available with 22, 30 and 34 gallon tanks
⑥ HD: 27.0 qts.
⑦ 3500HD: 28.5 qts. capacity
⑧ Available 32 and 41 gallon tanks
⑨ Add three qts. with rear heater
⑩ Short bed: 26 gals.
Long bed: 34 gals.
⑪ Available with a variety of fuel tanks
⑫ Fill to bottom of filler plug hole
⑬ K2 models: 1.75 qts.
K3 models: 2.25 qts.

⑭ Available with 20 gallon tank
⑮ Available 31 and 40 gallon tanks
⑯ Available with optional 31 gallon tank
⑰ With rear heater: 16.5 qts.
Without rear heater: 14.3 qts.
⑱ 4L60E trans.: 10.0 pts.
4L80E trans.: 14.5 pts.
⑲ NV241 and NV243: 4.5 pts.
4401 and 4470: 6.6 pts.
⑳ NV1500: 6 pts.
NV3500: 4.5 pts.

79241C26

VALVE SPECIFICATIONS

Year	Engine ID/VIN	Engine Displacement Liters (cc)	Seat Angle (deg.)	Face Angle (deg.)	Spring Test Pressure (lbs. @ in.)	Spring Installed Height (in.)	Stem-to-Guide Clearance (in.)		Stem Diameter (in.)	
							Intake	Exhaust	Intake	Exhaust
1995	4	2.2 (2189)	46	45	228@1.27	1.71	0.0010-0.0020	0.0010-0.0030	NA	NA
	W	4.3 (4293)	46	45	194-206@1.25	1.69-1.71	0.0011-0.0027	0.0011-0.0027	NA	NA
	Z	4.3 (4293)	46	45	194-206@1.25	1.69-1.71	0.0010-0.0027	0.0010-0.0027	NA	NA
	H	5.0 (4999)	46	45	76-84@1.70	1.72	0.0010-0.0027	0.0010-0.0027	NA	NA
	K	5.7 (5735)	46	45	76-84@1.70	1.72	0.0010-0.0027	0.0010-0.0027	NA	NA
	F	6.5 (6473)	46	45	230@1.39	1.81	0.0010-0.0027	0.0010-0.0027	NA	NA
	P	6.5 (6473)	46	45	230@1.39	1.81	0.0010-0.0027	0.0010-0.0027	NA	NA
	S	6.5 (6473)	46	45	230@1.39	1.81	0.0010-0.0027	0.0010-0.0027	NA	NA
	Y	6.5 (6473)	46	45	230@1.39	1.81	0.0010-0.0027	0.0010-0.0027	NA	NA
	N	7.4 (7440)	46	45	205-225@1.40	1.80	0.0010-0.0027	0.0012-0.0029	NA	NA
1996	4	2.2 (2189)	46	45	228@1.28	1.71	0.0010-0.0020	0.0010-0.0030	NA	NA
	W	4.3 (4293)	46	45	187-203@1.27	1.69-1.71	0.0010	0.0020	NA	NA
	X	4.3 (4293)	46	45	187-203@1.27	1.69-1.71	0.0010	0.0020	NA	NA
	M	5.0 (4999)	46	45	187-203@1.27	1.69-1.71	0.0010-0.0027	0.0010-0.0027	NA	NA
	R	5.7 (5735)	46	45	187-203@1.27	1.69-1.71	0.0010-0.0027	0.0010-0.0027	NA	NA
	F	6.5 (6473)	46	45	230@1.40	1.80	0.0010-0.0027	0.0010-0.0027	NA	NA
	S	6.5 (6473)	46	45	230@1.40	1.80	0.0010-0.0027	0.0010-0.0027	NA	NA
	J	7.4 (7440)	46	45	238-262@1.34	1.83	0.0010-0.0029 ①	0.0012-0.0031 ①	NA	NA
1997	4	2.2 (2189)	46	45	228@1.28	1.71	0.0010-0.0020	0.0010-0.0030	NA	NA
	W	4.3 (4293)	46	45	187-203@1.27	1.69-1.71	0.0010	0.0020	NA	NA
	X	4.3 (4293)	46	45	187-203@1.27	1.69-1.71	0.0010	0.0020	NA	NA
	M	5.0 (4999)	46	45	187-203@1.27	1.69-1.71	0.0010-0.0027	0.0010-0.0027	NA	NA
	R	5.7 (5735)	46	45	187-203@1.27	1.69-1.71	0.0010-0.0027	0.0010-0.0027	NA	NA
	F	6.5 (6473)	46	45	230@1.40	1.80	0.0010-0.0027	0.0010-0.0027	NA	NA

79241C27

VALVE SPECIFICATIONS

Year	Engine ID/VIN	Engine Displacement Liters (cc)	Seat Angle (deg.)	Face Angle (deg.)	Spring Test Pressure (lbs. @ in.)	Spring Installed Height (in.)	Stem-to-Guide Clearance (in.)		Stem Diameter (in.)	
							Intake	Exhaust	Intake	Exhaust
1997 (cont.)	S	6.5 (6473)	46	45	230@1.40	1.80	0.0010-0.0027	0.0010-0.0027	NA	NA
	J	7.4 (7440)	46	45	238-262@1.34	1.83	0.0010-0.0029 ①	0.0012-0.0031 ①	NA	NA
1998-99	4	2.2 (2189)	46	45	228@1.28	1.71	0.0010-0.0020	0.0010-0.0030	NA	NA
	W	4.3 (4293)	46	45	187-203@1.27	1.69-1.71	0.0010	0.0020	NA	NA
	X	4.3 (4293)	46	45	187-203@1.27	1.69-1.71	0.0010	0.0020	NA	NA
	M	5.0 (4999)	46	45	187-203@1.27	1.69-1.71	0.0010-0.0027	0.0010-0.0027	NA	NA
	R	5.7 (5735)	46	45	187-203@1.27	1.69-1.71	0.0010-0.0027	0.0010-0.0027	NA	NA
	F	6.5 (6473)	46	45	230@1.40	1.80	0.0010-0.0027	0.0010-0.0027	NA	NA
	S	6.5 (6473)	46	45	230@1.40	1.80	0.0010-0.0027	0.0010-0.0027	NA	NA
	J	7.4 (7440)	46	45	238-262@1.34	1.83	0.0010-0.0029 ①	0.0012-0.0031 ①	NA	NA

NA - Not Available

① Service limit:
  Intake: 0.0037 MAX
  Exhaust: 0.0049 MAX

79241C2

TORQUE SPECIFICATIONS
All readings in ft. lbs.

Year	Engine ID/VIN	Engine Displacement Liters (cc)	Cylinder Head Bolts	Main Bearing Bolts	Rod Bearing Bolts	Crankshaft Damper Bolts	Flywheel Bolts	Manifold Intake *	Exhaust	Spark Plugs	Lug Nut
1995	4	2.2 (2189)	(1)	70	38	77	55	(2)	10	(3)	100
	W	4.3 (4293)	65	81	(4)	70	74	35	(5)	11	90
	Z	4.3 (4293)	65	81	(4)	70	74	35	(5)	11	90
	H	5.0 (4999)	65	(6)	45	70	75	35	(5)	15	(7)
	K	5.7 (5735)	65	(6)	45	70	75	35	(5)	15	(7)
	F	6.5 (6473)	(8)	(9)	48	200	66	31	26	—	(7)
	P	6.5 (6473)	(8)	(9)	48	200	66	31	26	—	(7)
	S	6.5 (6473)	(8)	(9)	48	200	66	31	26	—	(7)
	Y	6.5 (6473)	(8)	(9)	48	200	66	31	26	—	(7)
	N	7.4 (7440)	80	100	48	85	65	35	40	22	(7)
1996	4	2.2 (2189)	(1)	70	38	77	55	(2)	10	11	100
	W	4.3 (4293)	(10)	77	(4)	74	74	(11)	(12)	11	90
	X	4.3 (4293)	(10)	77	(4)	74	74	(11)	(12)	11	90
	M	5.0 (4999)	(13)	(14)	(15)	74	74	(11)	(12)	15	(7)
	R	5.7 (5735)	(13)	(14)	(15)	74	74	(11)	(12)	15	(7)
	F	6.5 (6473)	(8)	(9)	48	200	65	31	26	—	(7)
	S	6.5 (6473)	(8)	(9)	48	200	65	31	26	—	(7)
	J	7.4 (7440)	85	100	45	110	67	30	22	15	(7)
1997	4	2.2 (2189)	(1)	70	38	77	55	(2)	10	11	100
	W	4.3 (4293)	(10)	77	(4)	74	74	(11)	(12)	11	90
	X	4.3 (4293)	(10)	77	(4)	74	74	(11)	(12)	11	90
	M	5.0 (4999)	(13)	(14)	(15)	74	74	(11)	(12)	15	(7)
	R	5.7 (5735)	(13)	(14)	(15)	74	74	(11)	(12)	15	(7)
	F	6.5 (6473)	(8)	(9)	48	200	65	31	26	—	(7)
	S	6.5 (6473)	(8)	(9)	48	200	65	31	26	—	(7)
	J	7.4 (7440)	85	100	45	110	67	30	22	15	(7)
1998-99	4	2.2 (2189)	(1)	70	38	77	55	(2)	10	11	100
	W	4.3 (4293)	(10)	77	(4)	74	74	(11)	(12)	11	90
	X	4.3 (4293)	(10)	77	(4)	74	74	(11)	(12)	11	90
	M	5.0 (4999)	(8)	(14)	(15)	74	74	(11)	(12)	15	(7)
	R	5.7 (5735)	(8)	(14)	(15)	74	74	(11)	(12)	15	(7)
	F	6.5 (6473)	(8)	(9)	48	200	65	31	26	—	(7)
	S	6.5 (6473)	(8)	(9)	48	200	65	31	26	—	(7)
	J	7.4 (7440)	85	100	45	110	67	30	22	15	(7)

* NOTE: Applies to Lower Manifold only.

(1) Short bolts: 43 ft. lbs. plus 90 degrees
Long bolts: 46 ft. lbs. plus 90 degrees

(2) Lower intake manifold nuts: 24 ft. lbs.
Lower intake manifold studs: 22 ft. lbs.
Upper intake manifold bolts: 22 ft. lbs.

(3) 1st-time installation (new head): 22 ft. lbs.
All other installations: 12 ft. lbs.

(4) 20 ft. lbs. plus 70 degrees

(5) Two center bolts: 26 ft. lbs.
All others: 20 ft. lbs.

(6) Outer bolts on caps 2-4: 70 ft. lbs
All others: 80 ft. lbs.

(7) All 5 & 6 stud single rear wheels: 110 ft. lbs.
All 8 stud single rear wheels: 120 ft. lbs.
All 8 stud dual rear wheels: 140 ft. lbs.
All 10 stud dual wheels: 175 ft. lbs.

(8) Apply sealer
Step 1: 20 ft. lbs.
Step 2: 50 ft. lbs.
Step 3: 50 ft. lbs.
Step 4: Plus 90-100 degrees

(9) Outer bolts: 100 ft. lbs.
Inner bolts: 111 ft. lbs.

(10) 1st pass: 22 ft. lbs.
2nd pass:
Short bolt: Plus 55 degrees
Medium bolt: Plus 65 degrees
Long bolt: Plus 75 degrees

(11) Lower intake manifold:
1st pass: 27 in. lbs.
2nd pass: 106 in. lbs.
Final pass: 11 ft. lbs.
Upper manifold bolts:
1st pass: 44 in. lbs.
2nd pass: 88 in. lbs.

(12) Tighten bolts to 12 ft. lbs.
Retorque to 22 ft. lbs.

(13) Step 1: 22 ft. lbs.
Step 2:
Short bolt: Plus 55 degrees
Medium bolt: Plus 65 degrees
Long bolt: Plus 75 degrees

(14) Outer bolts on caps 2-4: 67 ft. lbs.
All others: 74 ft. lbs.

(15) Coat threads with sealant
Tighten all bolts to 20 ft. lbs.
Retorque to 50 ft. lbs.

79241C29

1995–96 SCHEDULED MAINTENANCE INTERVALS
(GENERAL MOTORS G SERIES VANS, C/K SERIES PICK UP, TAHOE, YUKON & SUBURBAN—DIESEL

TO BE SERVICED	TYPE OF SERVICE	VEHICLE MILEAGE INTERVAL (x1000)												
		5	10	15	20	25	30	35	40	45	50	55	60	65
Engine oil & filter	R	✓	✓	✓	✓	✓	✓	✓	✓	✓	✓	✓	✓	✓
Brake pedal springs	S/I	✓	✓	✓	✓	✓	✓	✓	✓	✓	✓	✓	✓	✓
Chassis lubrication	S/I	✓	✓	✓	✓	✓	✓	✓	✓	✓	✓	✓	✓	✓
CV joints & axle seals	S/I	✓	✓	✓	✓	✓	✓	✓	✓	✓	✓	✓	✓	✓
Front axle propshaft splines	S/I	✓	✓	✓	✓	✓	✓	✓	✓	✓	✓	✓	✓	✓
Front suspension	S/I	✓	✓	✓	✓	✓	✓	✓	✓	✓	✓	✓	✓	✓
Kingpin bushings	S/I	✓	✓	✓	✓	✓	✓	✓	✓	✓	✓	✓	✓	✓
Parking brake cable guides	S/I	✓	✓	✓	✓	✓	✓	✓	✓	✓	✓	✓	✓	✓
Rear driveline center splines	S/I	✓	✓	✓	✓	✓	✓	✓	✓	✓	✓	✓	✓	✓
Rear axle fluid level	S/I	✓	✓	✓	✓	✓	✓	✓	✓	✓	✓	✓	✓	✓
Steering linkage	S/I	✓	✓	✓	✓	✓	✓	✓	✓	✓	✓	✓	✓	✓
Transfer case shift linkage (4WD)	S/I	✓	✓	✓	✓	✓	✓	✓	✓	✓	✓	✓	✓	✓
Transmission shift linkage	S/I	✓	✓	✓	✓	✓	✓	✓	✓	✓	✓	✓	✓	✓
Rotate tires	S/I	✓		✓		✓		✓		✓		✓		✓
Air intake system	S/I		✓		✓		✓		✓		✓		✓	
Engine coolant hoses, ducts & valves①	S/I		✓		✓		✓		✓		✓		✓	
Exhaust system & shields	S/I		✓		✓		✓		✓		✓		✓	
Clutch fork ball stud	S/I				✓				✓				✓	
Front wheel bearings (2 wheel drive)	S/I				✓				✓				✓	
Air cleaner filter	R						✓						✓	
Engine idle speed adjustment	S/I						✓						✓	
Fuel filter	R						✓						✓	
Automatic transmission fluid & filter②	S/I						✓						✓	
Crankcase depression regulator valve	S/I												✓	
EGR system	S/I												✓	

79241C30

1995-96 SCHEDULED MAINTENANCE INTERVALS
(GENERAL MOTORS G SERIES VANS, C/K SERIES PICK UP, TAHOE, YUKON & SUBURBAN—DIESEL Cont.)

TO BE SERVICED	TYPE OF SERVICE	VEHICLE MILEAGE INTERVAL (x1000)												
		5	10	15	20	25	30	35	40	45	50	55	60	65
Fuel tank, cap & lines	S/I												✓	
Serpentine belt	S/I												✓	
Engine Coolant③	R													

① Only for thermostatically controlled cooling fan.
② Under 8600 GVWR shown; over 8600 GVWR change every 24,000 miles.
③ Engine coolant (1996) - replace every 100,000 miles. Use O.E. specified (DEX-COOL™) coolant only. If any silicate coolant is used, the service interval is every 30,000 miles.

R – Replace   S/I – Service or Inspect

FREQUENT OPERATION MAINTENANCE (SEVERE SERVICE)
   If a vehicle is operated under any of the following conditions it is considered severe service:
- Extremely dusty areas.
- 50% or more of the vehicle operation is in 32°C (90°F) or higher temperatures, or constant operation in temperatures below 0°C (32°F).
- Prolonged idling (vehicle operation in stop and go traffic).
- Frequent short running periods (engine does not warm to normal operating temperatures).
- Police, taxi, delivery usage or trailer towing usage.
Oil & oil filter change – change every 2500 miles.
Lubricate chassis every 2500 miles.
Drive axle - check every 2500 miles.
Rotate tires every 7500 miles.
Exhaust system & shields - check every 10,000 miles.
Automatic transmission fluid & filter (1995 over 8600 GVWR) - change every 12,000 miles.
Automatic transmission fluid & filter (1995 under 8600 GVWR) - change every 15,000 miles.
Air cleaner filter - change every 15,000 miles.
Front wheel bearings - repack every 15,000 miles.

79241C31

1995–96 SCHEDULED MAINTENANCE INTERVALS
(GENERAL MOTORS G SERIES VANS, C/K SERIES PICK UP, TAHOE, YUKON & SUBURBAN—GASOLINE 1995 LIGHT DUTY EMISSIONS & 1996)

TO BE SERVICED	TYPE OF SERVICE	VEHICLE MILEAGE INTERVAL (x1000)												
		7.5	15	22.5	30	37.5	45	52.5	60	67.5	75	82.5	90	97.5
Engine oil & filter	R	✓	✓	✓	✓	✓	✓	✓	✓	✓	✓	✓	✓	✓
Chassis lubrication	S/I	✓	✓	✓	✓	✓	✓	✓	✓	✓	✓	✓	✓	✓
CV joints & axle seals	S/I	✓	✓	✓	✓	✓	✓	✓	✓	✓	✓	✓	✓	✓
Front axle propshaft splines	S/I	✓	✓	✓	✓	✓	✓	✓	✓	✓	✓	✓	✓	✓
Front suspension	S/I	✓	✓	✓	✓	✓	✓	✓	✓	✓	✓	✓	✓	✓
Kingpin bushings	S/I	✓	✓	✓	✓	✓	✓	✓	✓	✓	✓	✓	✓	✓
Parking brake cable guides	S/I	✓	✓	✓	✓	✓	✓	✓	✓	✓	✓	✓	✓	✓
Rear driveline center splines	S/I	✓	✓	✓	✓	✓	✓	✓	✓	✓	✓	✓	✓	✓
Rear axle fluid level	S/I	✓	✓	✓	✓	✓	✓	✓	✓	✓	✓	✓	✓	✓
Steering linkage	S/I	✓	✓	✓	✓	✓	✓	✓	✓	✓	✓	✓	✓	✓
Transfer case shift linkage (4WD)	S/I	✓	✓	✓	✓	✓	✓	✓	✓	✓	✓	✓	✓	✓
Transmission shift linkage	S/I	✓	✓	✓	✓	✓	✓	✓	✓	✓	✓	✓	✓	✓
Rotate tires	S/I	✓			✓		✓		✓		✓		✓		✓
Engine coolant hoses, ducts & valves①	S/I		✓		✓		✓		✓		✓		✓		
Air cleaner filter	R				✓				✓				✓		
Automatic transmission fluid & filter	R②				✓				✓				✓		
Engine Coolant③	R				✓				✓				✓		
Fuel filter	R				✓				✓				✓		
Spark plugs④	R				✓				✓				✓		
Clutch fork ball stud	S/I				✓				✓				✓		
Exhaust system & shields	S/I				✓				✓				✓		
Front wheel bearings (2 wheel drive)	S/I				✓				✓				✓		
Serpentine belt	S/I				✓				✓						
EGR system	S/I								✓						
Engine timing check	S/I								✓						
EVAP system	S/I								✓						

79241C34

1995–96 SCHEDULED MAINTENANCE INTERVALS
(GENERAL MOTORS G SERIES VANS, C/K SERIES PICK UP, TAHOE, YUKON & SUBURBAN—GASOLINE 1995 LIGHT DUTY EMISSIONS & 1996 Cont.)

TO BE SERVICED	TYPE OF SERVICE	VEHICLE MILEAGE INTERVAL (x1000)												
		7.5	15	22.5	30	37.5	45	52.5	60	67.5	75	82.5	90	97.5
Fuel tank, cap & lines	S/I								✓					

① Only for thermostatically controlled cooling fan.
② Under 8600 GVWR shown; over 8600 GVWR change every 24,000 miles.
③ Engine coolant (1996-97) - replace every 100,000 miles. Use O.E. specified (DEX-COOL™) coolant only. If any silicate coolant is used, the service interval is every 30,000 miles.
④ Spark plugs (1996-97) - replace every 100,000 miles.
R – Replace   S/I – Service or Inspect

FREQUENT OPERATION MAINTENANCE (SEVERE SERVICE)
   If a vehicle is operated under any of the following conditions it is considered severe service:
- **Extremely dusty areas.**
- **50% or more of the vehicle operation is in 32°C (90°F) or higher temperatures, or constant operation in temperatures below 0°C (32°F).**
- **Prolonged idling (vehicle operation in stop and go traffic).**
- **Frequent short running periods (engine does not warm to normal operating temperatures).**
- **Police, taxi, delivery usage or trailer towing usage.**
Oil & oil filter change – change every 3000 miles.
Lubricate chassis every 3000 miles.
Drive axle - check every 3000 miles.
Rotate tires every 6000 miles.
Automatic transmission fluid & filter (1993-95 over 8600 GVWR) - change every 12,000 miles.
Automatic transmission fluid & filter (1993-95 under 8600 GVWR) - change every 15,000 miles.
Exhaust system & shields - check every 15,000 miles.
Front wheel bearings - repack every 15,000 miles.
Air cleaner filter - change every 24,000 miles.

1995–96 SCHEDULED MAINTENANCE INTERVALS
(GENERAL MOTORS G SERIES VANS, C/K SERIES PICK UP, TAHOE, YUKON & SUBURBAN—GASOLINE 1995 HEAVY DUTY EMISSIONS)

TO BE SERVICED	TYPE OF SERVICE	VEHICLE MILEAGE INTERVAL (x1000)												
		6	12	18	24	30	36	42	48	54	60	66	72	78
Engine oil & filter	R	✓	✓	✓	✓	✓	✓	✓	✓	✓	✓	✓	✓	✓
Chassis lubrication	S/I	✓	✓	✓	✓	✓	✓	✓	✓	✓	✓	✓	✓	✓
Drive axle	S/I	✓	✓	✓	✓	✓	✓	✓	✓	✓	✓	✓	✓	✓
Rotate tires	S/I	✓	✓	✓	✓	✓	✓	✓	✓	✓	✓	✓	✓	✓
Engine coolant hoses, ducts & valves	S/I		✓		✓		✓		✓		✓		✓	
Exhaust system & shields	S/I		✓		✓		✓		✓		✓		✓	
Serpentine belt	S/I		✓		✓		✓		✓		✓		✓	

79241C33

1995–96 SCHEDULED MAINTENANCE INTERVALS
(GENERAL MOTORS G SERIES VANS, C/K SERIES PICK UP, TAHOE, YUKON & SUBURBAN—GASOLINE 1995 HEAVY DUTY EMISSIONS Cont.)

TO BE SERVICED	TYPE OF SERVICE	VEHICLE MILEAGE INTERVAL (x1000)												
		6	12	18	24	30	36	42	48	54	60	66	72	78
Air cleaner filter	R				✓				✓				✓	
Engine Coolant	R				✓				✓				✓	
Fuel filter	R				✓				✓				✓	
Spark plugs	R				✓				✓				✓	
Front wheel bearings (2 wheel drive)	S/I				✓				✓				✓	
Front wheel bearings (2 wheel drive)	S/I				✓				✓				✓	
Thermostatically controlled air cleaner	S/I				✓				✓				✓	
Automatic transmission fluid & filter	R②					✓					✓			
Clutch fork ball stud	S/I					✓					✓			
EGR system	S/I										✓			
Engine timing check	S/I										✓			
EVAP system	S/I										✓			
EVRV system	S/I										✓			
Fuel tank, cap & lines	S/I										✓			
Ignition wires	S/I										✓			

① Only for thermostatically controlled cooling fan.
② Under 8600 GVWR shown; over 8600 GVWR change every 24,000 miles.
R – Replace   S/I – Service or Inspect

FREQUENT OPERATION MAINTENANCE (SEVERE SERVICE)
If a vehicle is operated under any of the following conditions it is considered severe service:
- Extremely dusty areas.
- 50% or more of the vehicle operation is in 32°C (90°F) or higher temperatures, or constant operation in temperatures below 0°C (32°F).
- Prolonged idling (vehicle operation in stop and go traffic).
- Frequent short running periods (engine does not warm to normal operating temperatures).
- Police, taxi, delivery usage or trailer towing usage.
Oil & oil filter change – change every 3000 miles.
Lubricate chassis every 3000 miles.
Drive axle - check every 3000 miles.
Automatic transmission fluid & filter (over 8600 GVWR) - change every 12,000 miles.
Automatic transmission fluid & filter (under 8600 GVWR) - change every 15,000 miles.

79241C32

1995–96 SCHEDULED MAINTENANCE INTERVALS
(GENERAL MOTORS ASTRO, SAFARI, BLAZER, JIMMY, BRAVADA, SONOMA & S SERIES)

TO BE SERVICED	TYPE OF SERVICE	VEHICLE MILEAGE INTERVAL (x1000)												
		7.5	15	22.5	30	37.5	45	52.5	60	67.5	75	82.5	90	97.5
Engine oil & filter	R	✓	✓	✓	✓	✓	✓	✓	✓	✓	✓	✓	✓	✓
Brake/clutch pedal springs & parking brake cable guides	S/I	✓	✓	✓	✓	✓	✓	✓	✓	✓	✓	✓	✓	✓
Chassis lubrication & CV joints, ball joints & axle seals	S/I	✓	✓	✓	✓	✓	✓	✓	✓	✓	✓	✓	✓	✓
Drive axle, steering linkage & front suspension	S/I	✓	✓	✓	✓	✓	✓	✓	✓	✓	✓	✓	✓	✓
Front/rear axle fluid & transfer case shift linkage (4WD)	S/I	✓	✓	✓	✓	✓	✓	✓	✓	✓	✓	✓	✓	✓
Rotate tires	S/I	✓		✓		✓		✓		✓		✓		✓
Engine coolant strength, hoses & clamps	S/I		✓		✓		✓		✓		✓		✓	
Air cleaner filter & fuel filter	R				✓				✓				✓	
Automatic transmission fluid & filter (Astro & Safari)	R				✓				✓				✓	
Automatic transmission fluid & filter (1995 Blazer, Jimmy, Bravada, Sonoma, S Series (Pick-up) & Typhoon②	R				✓				✓				✓	
Engine coolant③	R				✓				✓				✓	
Spark plugs①	R				✓				✓				✓	
Accessory drive belt	S/I				✓				✓				✓	
Exhaust system	S/I				✓				✓				✓	
Front wheel bearings	S/I				✓				✓				✓	
Fuel tank, cap & lines	S/I								✓				✓	
Engine timing check	S/I								✓					

① Check ignition wires. Spark plugs (1996) - replace every 100,000 miles.
② 1995 Blazer, Jimmy, Bravada, Sonoma, S Series (Pick-up) & Typhoon - change every 30,000 miles. 1996 Blazer, Jimmy, Bravada, Sonoma, S Series (Pick-up) & Typhoon - change only when necessary.
③ Engine coolant (1996) - replace every 100,000 miles. Use O.E. specified (DEX-COOL™) coolant only. If any silicate coolant is used, the service interval is every 30,000 miles.

R – Replace   S/I – Service or Inspect

79241C35

1995–96 SCHEDULED MAINTENANCE INTERVALS
(GENERAL MOTORS ASTRO, SAFARI, BLAZER, JIMMY, BRAVADA, SONOMA & S SERIES Cont.)

FREQUENT OPERATION MAINTENANCE (SEVERE SERVICE)

If a vehicle is operated under any of the following conditions it is considered severe service:

- Extremely dusty areas.
- 50% or more of the vehicle operation is in 32°C (90°F) or higher temperatures, or constant operation in temperatures below 0°C (32°F).
- Prolonged idling (vehicle operation in stop and go traffic).
- Frequent short running periods (engine does not warm to normal operating temperatures).
- Police, taxi, delivery usage or trailer towing usage.

Oil & oil filter change – change every 3000 miles.
Lubricate chassis every 3000 miles.
Rotate tires at 6000 miles, then every 15,000 miles.
Drive axle - check every 15,000 miles.
Automatic transmission fluid & filter (Astro & Safari) - change every 15,000 miles.
Automatic transmission fluid & filter (1995 Blazer, Jimmy, Bravada, Sonoma, S Series (Pick-up) & Typhoon - change every 15,000 miles.
Automatic transmission fluid & filter (1996 Blazer, Jimmy, Bravada, Sonoma, S Series (Pick-up) & Typhoon) - change every 50,000 miles.
Air cleaner filter - change every 15,000 miles.

79241C36

1997-99 SCHEDULED MAINTENANCE INTERVALS
(GENERAL MOTORS S/T SERIES ASTRO, BLAZER, BRAVADA, ENVOY, PICK-UP, JIMMY, SAFARI, & SONOMA)

TO BE SERVICED	TYPE OF SERVICE	VEHICLE MILEAGE INTERVAL (x1000)															
		7.5	15	22.5	30	37.5	45	52.5	60	67.5	75	82.5	90	97.5	105	112.5	120
Accessory drive belt	S/I								✓								✓
Air cleaner filter	R				✓				✓				✓				✓
Automatic transmission fluid	R	Every 50,000 miles															
Brake system ①	S/I	✓	✓	✓	✓	✓	✓	✓	✓	✓	✓	✓	✓	✓	✓	✓	✓
Chassis & suspension grease points	L	✓	✓	✓	✓	✓	✓	✓	✓	✓	✓	✓	✓	✓	✓	✓	✓
CV-joint boots & axle seals	S/I	✓	✓	✓	✓	✓	✓	✓	✓	✓	✓	✓	✓	✓	✓	✓	✓
Engine coolant system ②	S/I	Every 150,000 miles															
Engine oil & filter	R	✓	✓	✓	✓	✓	✓	✓	✓	✓	✓	✓	✓	✓	✓	✓	✓
Front wheel bearings ③	S/I & L				✓				✓				✓				✓
Fuel filter	R				✓				✓				✓				✓
Fuel tank, cap & lines	S/I								✓								✓
PCV valve	S/I	Every 100,000 miles															
Rear/front axle fluid level	S/I	✓	✓	✓	✓	✓	✓	✓	✓	✓	✓	✓	✓	✓	✓	✓	✓
Rotate tires	S/I	✓	✓	✓	✓	✓	✓	✓	✓	✓	✓	✓	✓	✓	✓	✓	✓
Spark plug wires	S/I	Every 100,000 miles															
Spark plugs	R	Every 100,000 miles															

① This should be performed when the tires are removed for rotation.

② Drain, flush and refill the cooling system, inspect the system hoses, and clean the radiator and condenser.

③ 2-wheel drive models only.

R - Replace   S/I - Inspect, and service if necessary   L - Lubricate   A - Adjust   C - Clean

FREQUENT OPERATION MAINTENANCE (SEVERE SERVICE)

 If a vehicle is operated under any of the following conditions it is considered severe service:

- Towing a trailer or using a camper or car-top carrier.

- Repeated short trips of less than 5 miles in temperatures below freezing, or trips of less than 10 miles in any temperature.

- Extensive idling or low-speed driving for long distances as in heavy commercial use, such as delivery, taxi or police cars.

- Operating on rough, muddy or salt-covered roads.

- Operating on unpaved or dusty roads.

- Driving in extremely hot (over 90°) conditions.

Engine oil & filter - replace every 3,000 miles or 3 months, whichever occurs first.

Chassis and suspension grease points - lubricate every 3,000 miles.

Rear/front axle fluid level - inspect every 3,000 miles.

Rotate the tires every 6,000 miles.

Brake system components - inspect every 6,000 miles.

Front wheel bearings (2-wheel drive only) - clean, inspect and repack every 15,000 miles.

Air cleaner filter - inspect every 15,000 miles.

Automatic transmission fluid & filter - replace every 15,000 miles.

79241C37

1997-99 SCHEDULED MAINTENANCE INTERVALS
(GENERAL MOTORS C/K SERIES PICK-UP, DENALI, ESCALADE, SIERRA, SUBURBAN, TAHOE AND YUKON—GASOLINE)

TO BE SERVICED	TYPE OF SERVICE	VEHICLE MILEAGE INTERVAL (x1000)															
		7.5	15	22.5	30	37.5	45	52.5	60	67.5	75	82.5	90	97.5	105	112.5	120
Accessory drive belt	S/I								✓								✓
Automatic transmission fluid ①	R	Every 50,000 miles															
Brake system	S/I	✓	✓	✓	✓	✓	✓	✓	✓	✓	✓	✓	✓	✓	✓	✓	✓
Chassis & suspension grease points	L	✓	✓	✓	✓	✓	✓	✓	✓	✓	✓	✓	✓	✓	✓	✓	✓
Cooling fan operation	S/I		✓		✓		✓				✓			✓			✓
CV-joint boots & axle seals	S/I	✓			✓		✓		✓	✓	✓	✓	✓	✓	✓	✓	✓
EGR system	S/I								✓								✓
Engine coolant	R	Every 150,000 miles															
Engine oil & filter	R	✓	✓	✓	✓	✓	✓	✓	✓	✓	✓	✓	✓	✓	✓	✓	✓
EVAP system	S/I								✓								✓
Front wheel bearings ②	S/I & L				✓				✓				✓				✓
Fuel filter	R								✓								✓
Fuel system	S/I								✓								✓
Rear/Front axle fluid level	S/I	✓	✓	✓	✓	✓	✓	✓	✓	✓	✓	✓	✓	✓	✓	✓	✓
Rotate tires	S/I	✓	✓	✓	✓	✓	✓	✓	✓	✓	✓	✓	✓	✓	✓	✓	✓
Shields & underhood insulation ①	S/I		✓		✓		✓		✓		✓		✓		✓		✓
Spark plugs	R	Every 100,000 miles															
Spark plug wires	S/I	Every 100,000 miles															

① Vehicles with a GVWR of 8,500 lbs. or more only.

② 2-wheel drive models only.

R - Replace   S/I - Inspect, and service if necessary   L - Lubricate   A - Adjust   C - Clean

FREQUENT OPERATION MAINTENANCE (SEVERE SERVICE)

If a vehicle is operated under any of the following conditions it is considered severe service:

- Towing a trailer or using a camper or car-top carrier.
- Repeated short trips of less than 5 miles in temperatures below freezing, or trips of less than 10 miles in any temperature.
- Extensive idling or low-speed driving for long distances as in heavy commercial use, such as delivery, taxi or police cars.
- Operating on rough, muddy or salt-covered roads.
- Operating on unpaved or dusty roads.
- Driving in extremely hot (over 90°) conditions.

Engine oil & filter - replace every 3,000 miles or 3 months, whichever occurs first.

Chassis and suspension grease points - lubricate every 3,000 miles.

Rear/front axle fluid level - inspect every 3,000 miles.

Rotate the tires every 6,000 miles.

Brake system components - inspect every 6,000 miles.

Front wheel bearings (2-wheel drive only) - clean, inspect and repack every 15,000 miles.

79241C3

1997-99 SCHEDULED MAINTENANCE INTERVALS
(GENERAL MOTORS C/K SERIES PICK-UP, DENALI, ESCALADE, SIERRA, SUBURBAN, TAHOE AND YUKON—GASOLINE)

FREQUENT OPERATION MAINTENANCE (SEVERE SERVICE)

Shields & underhood insulation (vehicles w/GVWR over 8,500 lbs. only) - inspect every 15,000 miles.

Cooling fan system hoses & connections - inspect every 15,000 miles.

Fuel filter - replace every 30,000 miles.

Air cleaner filter - inspect every 45,000 miles.

Automatic transmission fluid & filter - replace every 50,000 miles.

Accessory drive belt - inspect every 60,000 miles.

Fuel system tank, cap and lines - inspect every 60,000 miles.

EVAP system - inspect every 60,000 miles.

EGR system - inspect every 60,000 miles.

PCV system - inspect every 100,000 miles.

Engine cooling system components - inspect and clean every 150,000 miles.

79241C39

1997-99 SCHEDULED MAINTENANCE INTERVALS
(GENERAL MOTORS C/K SERIES PICK-UP, DENALI, ESCALADE, SIERRA, SUBURBAN, TAHOE AND YUKON—DIESEL)

TO BE SERVICED	TYPE OF SERVICE	VEHICLE MILEAGE INTERVAL (x1000)																									
		5	8	10	15	20	23	25	30	35	38	40	45	50	53	55	60	65	68	70	75	80	83	85	90	95	98
Air intake system	S/I			✓		✓			✓			✓		✓			✓			✓	✓				✓		
Automatic transmission fluid ①	R	Every 50,000 miles																									
Brake system	S/I	✓	✓		✓			✓		✓		✓		✓		✓	✓		✓		✓		✓		✓		✓
Chassis & suspension grease points	L	✓		✓	✓	✓			✓	✓	✓		✓	✓	✓		✓	✓	✓		✓	✓	✓			✓	✓	✓
Cooling fan operation	S/I			✓		✓			✓			✓		✓			✓			✓	✓				✓		
Crankcase depression regulator valve system hoses	S/I																✓										
CV-joint boots & axle seals	S/I	✓			✓	✓	✓		✓	✓	✓		✓	✓	✓		✓	✓	✓		✓	✓	✓			✓	✓	✓
EGR system ②	S/I																✓										
Engine coolant	R	Every 150,000 miles																									
Engine cooling system hoses & radiator	S/I & C	Initially at 100,000 miles, then every 50,000 miles																									
Engine oil & filter	R	✓			✓	✓	✓		✓	✓	✓		✓	✓	✓		✓	✓	✓		✓	✓	✓			✓	✓	✓
Front wheel bearings ③	S/I & L								✓								✓								✓		
Fuel filter	R								✓								✓								✓		
Rear/Front axle fluid level	S/I	✓			✓	✓	✓		✓	✓	✓		✓	✓	✓		✓	✓	✓		✓	✓	✓			✓	✓	✓
Rotate tires	S/I	✓	✓		✓		✓		✓		✓		✓		✓		✓		✓		✓		✓		✓		✓
Shields & underhood insulation ①	S/I			✓		✓			✓			✓		✓			✓			✓	✓				✓		

① Vehicles with a GVWR of 8,500 lbs. or more only.

② If equipped.

③ 2-wheel drive models only.

R - Replace   S/I - Inspect, and service if necessary   L - Lubricate   A - Adjust   C - Clean

FREQUENT OPERATION MAINTENANCE (SEVERE SERVICE)

If a vehicle is operated under any of the following conditions it is considered severe service:

- Towing a trailer or using a camper or car-top carrier.
- Repeated short trips of less than 5 miles in temperatures below freezing, or trips of less than 10 miles in any temperature.
- Extensive idling or low-speed driving for long distances as in heavy commercial use, such as delivery, taxi or police cars.
- Operating on rough, muddy or salt-covered roads.
- Operating on unpaved or dusty roads.
- Driving in extremely hot (over 90°) conditions.

Engine oil & filter - replace every 2,500 miles.

Chassis and suspension grease points - lubricate every 2,500 miles.

Rear/front axle fluid level - inspect initially at 5,000 miles, then every 2,500 miles thereafter.

Rotate tires - every 7,500 miles.

Brake system - inspect every 7,500 miles.

Air cleaner filter - inspect every 15,000 miles.

Front wheel bearings (2-wheel drive only) - clean, inspect and repack every 15,000 miles.

79241C

1997-99 SCHEDULED MAINTENANCE INTERVALS
(GENERAL MOTORS G/P SERIES VAN, EXPRESS AND SAVANA—GASOLINE)

TO BE SERVICED	TYPE OF SERVICE	VEHICLE MILEAGE INTERVAL (x1000)															
		7.5	15	22.5	30	37.5	45	52.5	60	67.5	75	82.5	90	97.5	105	112.5	120
Accessory drive belt	S/I								✓								✓
Air cleaner filter	R				✓				✓				✓				✓
Automatic transmission fluid ①	R	Every 50,000 miles															
Chassis & suspension grease points	L	✓	✓	✓	✓	✓	✓	✓	✓	✓	✓	✓	✓	✓	✓	✓	✓
CV-joint boots & axle seals	S/I	✓	✓	✓	✓	✓	✓	✓	✓	✓	✓	✓	✓	✓	✓	✓	✓
EGR system	S/I								✓								✓
Engine coolant	R	Every 150,000 miles															
Engine oil & filter	R	✓	✓	✓	✓	✓	✓	✓	✓	✓	✓	✓	✓	✓	✓	✓	✓
EVAP system	S/I								✓								✓
Front wheel bearings	S/I & L				✓				✓				✓				✓
Fuel filter	R				✓				✓				✓				✓
Fuel system	S/I								✓								✓
PCV system	S/I	Every 100,000 miles															
Rear axle fluid level	S/I	✓	✓	✓	✓	✓	✓	✓	✓	✓	✓	✓	✓	✓	✓	✓	✓
Rotate tires	S/I	✓	✓	✓	✓	✓	✓	✓	✓	✓	✓	✓	✓	✓	✓	✓	✓
Shields & underhood insulation ①	S/I		✓		✓		✓		✓		✓		✓		✓		✓
Spark plugs	R	Every 100,000 miles															
Spark plug wires	S/I	Every 100,000 miles															

① Vehicles with a GVWR of 8,500 lbs. or more only.

R - Replace   S/I - Inspect, and service if necessary   L - Lubricate   A - Adjust   C - Clean

FREQUENT OPERATION MAINTENANCE (SEVERE SERVICE)

If a vehicle is operated under any of the following conditions it is considered severe service:
- Towing a trailer or using a camper or car-top carrier.
- Repeated short trips of less than 5 miles in temperatures below freezing, or trips of less than 10 miles in any temperature.
- Extensive idling or low-speed driving for long distances as in heavy commercial use, such as delivery, taxi or police cars.
- Operating on rough, muddy or salt-covered roads.
- Operating on unpaved or dusty roads.
- Driving in extremely hot (over 90°) conditions.

Engine oil & filter - replace every 3,000 miles or 3 months, whichever occurs first.

Chassis and suspension grease points - lubricate every 3,000 miles.

Rear/front axle fluid level - inspect every 3,000 miles.

Rotate the tires every 6,000 miles.

Brake system components - inspect every 6,000 miles.

Front wheel bearings (2-wheel drive only) - clean, inspect and repack every 15,000 miles.

Shields & underhood insulation (vehicles w/GVWR over 8,500 lbs. only) - inspect every 15,000 miles.

Cooling fan system hoses & connections - inspect every 15,000 miles.

79241C41

1997-99 SCHEDULED MAINTENANCE INTERVALS
(GENERAL MOTORS G/P SERIES VAN, EXPRESS AND SAVANA—GASOLINE)

FREQUENT OPERATION MAINTENANCE (SEVERE SERVICE)—continued

Fuel filter - replace every 30,000 miles.

Air cleaner filter - inspect every 45,000 miles.

Automatic transmission fluid & filter - replace every 50,000 miles.

Accessory drive belt - inspect every 60,000 miles.

Fuel system tank, cap and lines - inspect every 60,000 miles.

EVAP system - inspect every 60,000 miles.

EGR system - inspect every 60,000 miles.

PCV system - inspect every 100,000 miles.

Engine cooling system components - inspect and clean every 150,000 miles.

79241C

1997-99 SCHEDULED MAINTENANCE INTERVALS
(GENERAL MOTORS G/P SERIES VAN, EXPRESS AND SAVANA—DIESEL)

TO BE SERVICED	TYPE OF SERVICE	5	10	15	20	25	30	35	40	45	50	55	60	65	70	75	80	85	90	95	100	105	110	115	120
													VEHICLE MILEAGE INTERVAL (x1000)												
Air cleaner filter	R						✓						✓						✓						✓
Air intake system	S/I		✓		✓		✓		✓		✓		✓		✓		✓		✓		✓		✓		✓
Automatic transmission fluid ①	R										✓										✓				
Chassis & suspension grease points	L	✓	✓	✓	✓	✓	✓	✓	✓	✓	✓	✓	✓	✓	✓	✓	✓	✓	✓	✓	✓	✓	✓	✓	✓
Cooling fan ducts & hoses	S/I												✓												✓
Crankcase depression regulator valve system hoses	S/I												✓												✓
CV-joint boots & axle seals	S/I	✓	✓	✓	✓	✓	✓	✓	✓	✓	✓	✓	✓	✓	✓	✓	✓	✓	✓	✓	✓	✓	✓	✓	✓
EGR system ②	S/I												✓												
Engine coolant	R	Every 100,000 miles																							
Engine cooling system hoses & radiator	S/I & C	Initially at 100,000 miles, then every 50,000 miles																							
Engine oil & filter ③	R	✓	✓	✓	✓	✓	✓	✓	✓	✓	✓	✓	✓	✓	✓	✓	✓	✓	✓	✓	✓	✓	✓	✓	✓
Front wheel bearings	S/I & L						✓						✓						✓						✓
Fuel filter	R												✓												✓
Rear axle fluid level	S/I	✓	✓	✓	✓	✓	✓	✓	✓	✓	✓	✓	✓	✓	✓	✓	✓	✓	✓	✓	✓	✓	✓	✓	✓
Rotate tires	S/I	✓	✓	✓	✓	✓	✓	✓	✓	✓	✓	✓	✓	✓	✓	✓	✓	✓	✓	✓	✓	✓	✓	✓	✓
Shields & underhood insulation	S/I						✓						✓						✓						✓

① For vehicles with a GVWR of 8,500 lbs. or more.

② If equipped.

③ Perform at the mileage specified or every 3 months, whichever occurs first.

R - Replace   S/I - Inspect, and service if necessary   L - Lubricate   A - Adjust   C - Clean

FREQUENT OPERATION MAINTENANCE (SEVERE SERVICE)

If a vehicle is operated under any of the following conditions it is considered severe service:

- **Towing a trailer or using a camper or car-top carrier.**
- **Repeated short trips of less than 5 miles in temperatures below freezing, or trips of less than 10 miles in any temperature.**
- **Extensive idling or low-speed driving for long distances as in heavy commercial use, such as delivery, taxi or police cars.**
- **Operating on rough, muddy or salt-covered roads.**
- **Operating on unpaved or dusty roads.**
- **Driving in extremely hot (over 90°) conditions.**

Engine oil & filter - replace every 2,500 miles.

Chassis and suspension grease points - lubricate every 2,500 miles.

Rear axle fluid level - inspect every 2,500 miles.

CV-joint boots and axle seals - inspect for leakage every 2,500 miles.

Rotate tires - every 7,500 miles.

Air cleaner filter - inspect every 15,000 miles.

Front wheel bearings (2-wheel drive only) - clean, inspect and repack every 15,000 miles.

Automatic transmission fluid & filter - replace every 50,000 miles.

79241C43

SCHEDULED MAINTENANCE
GENERAL MOTORS
C/K SERIES PICK-UP, DENALI, ESCALADE, EXPRESS, G SERIES VANS SAVANA, SIERRA, SUBURBAN, TAHOE, YUKON

The following should be used as a guide when determining the amount of work required for a particular service if taken to a repair shop. In estimating how long a particular Scheduled Maintenance Service should take, please observe the following:

- **Chilton Time** is time based on field research and data supplied by the vehicle manufacturer.
- Labor time operations are given in hours and tenths of an hour.
- All labor operations, are to be used as a guide.

Mechanic Skill Level Codes:
- **(G)** GENERAL: Normally skilled with certification.
- **(M)** MAINTENANCE: Semi-skilled working on certication.
- **(P)** PRECISION: Really skilled with multiple certification.

	Chilton Time
(M) 5000 Mile Service	
1995-99 Diesel	2.0
w/4WD add	.1
(M) 7500 Mile Service	
1995-96 G Vans, C/K Pick-up, Tahoe, Yukon, Suburban	1.8
w/4WD add	.1
1997-99 G Vans, Express, Savava, C/K Pick-up, Denali, Escalade, Sierra, Suburban, Tahoe, Yukon	1.2
Rotate Tires add	.6
(M) 8000 Mile Service	
1997-99 Diesel	.8
(M) 10000 Mile Service	
1995-99 Diesel	1.8
w/4WD add	.1
(M) 12000 Mile Service	
1995 Diesel	1.3
(M) 15000 Mile Service	
1995-96 G Vans, C/K Pick-up, Tahoe, Yukon, Suburban	
Gas	1.4
Diesel	2.0
w/4WD add	.1
1997-99 G Vans, Express, Savava, C/K Pick-up, Denali, Escalade, Sierra, Suburban, Tahoe, Yukon	1.3
Rotate Tires add	.6
1997-99 Diesel	2.0
(M) 18000 Mile Service	
1995 Diesel	1.0
(M) 20000 Mile Service	
1995-99 Diesel	1.8
w/4WD add	.1

	Chilton Time
(M) 22500 Mile Service	
1995-96 G Vans, C/K Pick-up, Tahoe, Yukon, Suburban	1.8
w/4WD add	.1
1997-99 G Vans, Express, Savava, C/K Pick-up, Denali, Escalade, Sierra, Suburban, Tahoe, Yukon	1.2
Rotate Tires add	.6
(M) 23000 Mile Service	
1997-99 Diesel	.8
(M) 25000 Mile Service	
1995-99 Diesel	1.0
w/4WD add	.1
(G) 30000 Mile Service	
1995 G Vans, C/K Pick-up, Tahoe, Yukon, Suburban	
Gas	
light duty	4.0
w/AT add	.5
w/4WD add	.1
heavy duty	.6
w/AT add	.5
Diesel	2.6
w/AT add	.5
w/4WD add	.1
1996 G Vans, C/K Pick-up, Tahoe, Yukon, Suburban	
Gas	4.0
Diesel	2.6
w/AT add	.5
w/4WD add	.1
1997-99 G Vans, Express, Savava, C/K Pick-up, Denali, Escalade, Sierra, Suburban, Tahoe, Yukon	1.3
Rotate Tires add	.6
Pack Ft. Brgs 2WD add	1.3
1997-99 Diesel	2.3
Pack Ft Brgs add	1.3
(M) 35000 Mile Service	
1995-99 Diesel	1.0
w/4WD add	.1

	Chilton Time
(M) 37500 Mile Service	
1995-96 G Vans, C/K Pick-up, Tahoe, Yukon, Suburban	1.8
w/4WD add	.1
1997-99 G Vans, Express, Savava, C/K Pick-up, Denali, Escalade, Sierra, Suburban, Tahoe, Yukon	1.2
Rotate Tires add	.6
(M) 38000 Mile Service	
1997-99 Diesel	.8
(M) 45000 Mile Service	
1995-96 G Vans, C/K Pick-up, Tahoe, Yukon, Suburban	
Gas	1.4
Diesel	1.8
w/4WD add	.1
1997-99 G Vans, Express, Savava, C/K Pick-up, Denali, Escalade, Sierra, Suburban, Tahoe, Yukon	1.3
Rotate Tires add	.6
1997-99 Diesel	1.8
(M) 50000 Mile Service	
1995-96 Diesel	2.1
w/4WD add	.1
1997-99 Diesel	1.2
(M) 52500 Mile Service	
1995-96 G Vans, C/K Pick-up, Tahoe, Yukon, Suburban	1.8
w/4WD add	.1
1997-99 G Vans, Express, Savava, C/K Pick-up, Denali, Escalade, Sierra, Suburban, Tahoe, Yukon	1.2
Rotate Tires add	.6
w/4WD add	.2
(M) 53000 Mile Service	
1997-99 Diesel	.8

79241C44

SCHEDULED MAINTENANCE
GENERAL MOTORS
C/K SERIES PICK-UP, DENALI, ESCALADE, EXPRESS, G SERIES VANS
SAVANA, SIERRA, SUBURBAN, TAHOE, YUKON

The following should be used as a guide when determining the amount of work required for a particular service if taken to a repair shop. In estimating how long a particular Scheduled Maintenance Service should take, please observe the following:

- **Chilton Time** is time based on field research and data supplied by the vehicle manufacturer.
- Labor time operations are given in hours and tenths of an hour.
- All labor operations, are to be used as a guide.

Mechanic Skill Level Codes:
        (G) GENERAL: Normally skilled with certification.
        (M) MAINTENANCE: Semi-skilled working on certification.
        (P) PRECISION: Really skilled with multiple certification.

	Chilton Time
(M) 55000 Mile Service	
1995-96 Diesel	1.8
1997-99 Diesel	1.0
(G) 60000 Mile Service	
1995 G Vans, C/K Pick-up, Tahoe, Yukon, Suburban	
Gas	
light duty	4.4
w/AT add	.5
w/4WD add	.1
heavy duty	2.4
w/AT add	.5
Diesel	3.4
w/AT add	.5
w/4WD add	.1
1996 G Vans, C/K Pick-up, Tahoe, Yukon, Suburban	
Gas	4.4
Diesel	3.4
w/AT add	.5
w/4WD add	.1
1997-99 G Vans, Express, Savava, C/K Pick-up, Denali, Escalade, Sierra, Suburban, Tahoe, Yukon	2.1
Rotate Tires add	.6
Pack Ft Brgs 2WD add	1.3
1997-98 Diesel	2.7
Pack Ft Brgs add	1.3
(M) 65000 Mile Service	
1995-96 Diesel	1.8
w/4WD add	.1
1997-99 Diesel	1.0
(M) 67500 Mile Service	
1995-96 G Vans, C/K Pick-up, Tahoe, Yukon, Suburban	1.8
w/4WD add	.1
1997-99 G Vans, Express, Savava, C/K Pick-up, Denali, Escalade, Sierra, Suburban, Tahoe, Yukon	1.2
Rotate Tires add	.6
(M) 68000 Mile Service	
1997-99 Diesel	.8

	Chilton Time
(M) 75000 Mile Service	
1995-96 Vans, C/K Pick-up, Tahoe, Yukon, Suburban	1.4
w/4WD add	.1
1997-99 G Vans, Express, Savava, C/K Pick-up, Denali, Escalade, Sierra, Suburban, Tahoe, Yukon	1.3
Rotate Tires add	.6
1997-99 Diesel	1.8
(M) 82500 Mile Service	
1995-96 G Vans, C/K Pick-up, Tahoe, Yukon, Suburban	1.8
w/4WD add	.1
1997-99 G Vans, Express, Savava, C/K Pick-up, Denali, Escalade, Sierra, Suburban, Tahoe, Yukon	1.2
Rotate Tires add	.6
(M) 83000 Mile Service	
1997-99 Diesel	.8
(M) 85000 Mile Service	
1997-99 Diesel	1.0
(G) 90000 Mile Service	
1995-96 G Vans, C/K Pick-up, Tahoe, Yukon, Suburban	3.9
w/AT add	.5
w/4WD add	.1
1997-99 G Vans, Express, Savava, C/K Pick-up, Denali, Escalade, Sierra, Suburban, Tahoe, Yukon	1.3
Rotate Tires add	.6
Pack Ft Brgs 2WD add	1.3
1997-99 Diesel	2.7
Pack Ft Brgs add	1.3
(M) 90000 Mile Service	
1997-99 Diesel	1.0
(M) 97500 Mile Service	
1995-96 G Vans, C/K Pick-up, Tahoe, Yukon, Suburban	1.8
w/4WD add	.1

	Chilton Time
1997-99 G Vans, Express, Savava, C/K Pick-up, Denali, Escalade, Sierra, Suburban, Tahoe, Yukon	1.2
Rotate Tires add	.6
(M) 98000 Mile Service	
1997-99 Diesel	.8
(M) 100000 Mile Service	
1997-99 G Vans, Express, Savava, C/K Pick-up, Denali, Escalade, Sierra, Suburban, Tahoe, Yukon	
Replace spark plugs	
4.3L	.8
5.0L, 5.7L	.8
7.4L	1.0
Replace spark plug wires	
4.3L	1.2
5.0L, 5.7L	1.2
7.4L	1.0
(M) 105000 Mile Service	
1995-96 G Vans, C/K Pick-up, Tahoe, Yukon, Suburban	1.8
w/4WD add	.1
1997-99 G Vans, Express, Savava, C/K Pick-up, Denali, Escalade, Sierra, Suburban, Tahoe, Yukon	1.3
Rotate Tires add	.6
(M) 112500 Mile Service	
1997-99 G Vans, Express, Savava, C/K Pick-up, Denali, Escalade, Sierra, Suburban, Tahoe, Yukon	1.2
Rotate Tires add	.6
(M) 120000 Mile Service	
1997-99 G Vans, Express, Savava, C/K Pick-up, Denali, Escalade, Sierra, Suburban, Tahoe, Yukon	2.1
Rotate Tires add	.6
Pack Ft Brgs 2WD add	1.3

79241C45

SCHEDULED MAINTENANCE
GENERAL MOTORS
ASTRO, S/T SERIES, BLAZER, BRAVADA, ENVOY, PICK-UP
JIMMY, SAFARI, SONOMA

The following should be used as a guide when determining the amount of work required for a particular service if taken to a repair shop. In estimating how long a particular Scheduled Maintenance Service should take, please observe the following:

- **Chilton Time** is time based on field research and data supplied by the vehicle manufacturer.
- Labor time operations are given in hours and tenths of an hour.
- All labor operations, are to be used as a guide.

Mechanic Skill Level Codes:
  (G) GENERAL: Normally skilled with certification.
  (M) MAINTENANCE: Semi-skilled working on certicication.
  (P) PRECISION: Really skilled with multiple certification.

	Chilton Time
(M) 7500 Mile Service	
1995-96	1.3
w/4WD add	.2
1997-99	1.1
Rotate tires add	.4
(M) 15000 Mile Service	
1995-96	.9
w/4WD add	.2
1997-99	1.1
Rotate tires add	.4
(M) 22500 Mile Service	
1995-96	1.3
w/4WD add	.2
1997-99	1.1
Rotate tires add	.4
(G) 30000 Mile Service	
1995-96	3.5
w/AT add	.5
w/4WD add	.2
1997-99	3.0
Rotate tires add	.4
Pack Ft Brgs 2WD add	1.1
(M) 37500 Mile Service	
1995-96	1.3
w/4WD add	.2
1997-99	1.1
Rotate tires add	.4
(M) 45000 Mile Service	
1995-96	.9
w/4WD add	.2
1997-99	1.1
Rotate tires add	.4

	Chilton Time
(G) 50000 Mile Service	
1997-99	
Change AT fluid & filter	.9
(M) 52500 Mile Service	
1995-96	1.3
w/4WD add	.2
1997-99	1.1
Rotate tires add	.4
(G) 60000 Mile Service	
1995-96	3.8
w/AT add	.5
w/4WD add	.2
1997-99	3.2
Rotate tires add	.4
Pack Ft Brgs 2WD add	1.1
(M) 67500 Mile Service	
1995-96	1.3
w/4WD add	.2
1997-99	1.1
Rotate tires add	.4
(M) 75000 Mile Service	
1995-96	.9
w/4WD add	.2
1997-99	1.1
Rotate tires add	.4
(M) 82500 Mile Service	
1995-96	1.3
w/4WD add	.2
1997-99	1.1
Rotate tires add	.4
(G) 90000 Mile Service	
1995-96	3.6

	Chilton Time
w/AT add	.5
w/4WD add	.2
1997-99	1.8
Rotate tires add	.4
Pack Ft Brgs 2WD add	1.1
(M) 97500 Mile Service	
1995-96	1.3
w/4WD add	.2
1997-99	1.1
Rotate tires add	.4
(G) 100000 Mile Service	
1997-99	
Replace PCV Valve	.5
Replace Spark Plugs	
2.2L	.8
4.3L	1.5
Replace Spark Plug Wires	
2.2L	.5
4.3L	.9
(G) 120000 Mile Service	
1997-99	3.2
(M) 105500 Mile Service	
1997-99	1.1
Rotate tires add.	.4
(M) 112500 Mile Service	
1997-99	1.1
Rotate tires add.	.4
(G) 120000 Mile Service	
1997-99	3.2
Rotate tires add.	.4
Pack Ft Brgs 2WD add	1.1

79241C46

GENERAL MOTORS
Chevrolet Lumina APV • Venture • Oldsmobile Silhouette
Pontiac Trans Sport

VEHICLE IDENTIFICATION CHART

Engine Code						Model Year	
Code	Liters	Cu. In. (cc)	Cyl.	Fuel Sys.	Eng. Mfg.	Code	Year
D	3.1	191 (3130)	6	TFI	CPC	S	1995
E	3.4	207 (3350)	6	MFI/SFI	CPC	T	1996
L	3.8	231 (3785)	6	MFI	CPC	V	1997
						W	1998
						X	1999

TFI - Throttle body Fuel Injection

MFI - Multi-port Fuel Injection

SFI - Sequential Fuel Injection

CPC - Chevrolet/Pontiac/Canada

79241C47

ENGINE IDENTIFICATION
All measurements are given in inches.

Year	Model	Engine Displacement Liters (cc)	Engine Series (ID/VIN)	Fuel System	No. of Cylinders	Engine Type
1995	Lumina APV	3.1 (3097)	D	TFI	6	OHV
	Lumina APV	3.8 (3785)	L	MFI	6	OHV
	Silhouette	3.1 (3097)	D	TFI	6	OHV
	Silhouette	3.8 (3785)	L	MFI	6	OHV
	Trans Sport	3.1 (3097)	D	TFI	6	OHV
	Trans Sport	3.8 (3785)	L	MFI	6	OHV
1996	Lumina APV	3.4 (3350)	E	MFI	6	OHV
	Silhouette	3.4 (3350)	E	MFI	6	OHV
	Trans Sport	3.4 (3350)	E	MFI	6	OHV
1997	Silhouette	3.4 (3350)	E	MFI	6	OHV
	Trans Sport	3.4 (3350)	E	MFI	6	OHV
	Venture	3.4 (3350)	E	MFI	6	OHV
1998-99	Silhouette	3.4 (3350)	E	SFI	6	OHV
	Trans Sport	3.4 (3350)	E	SFI	6	OHV
	Venture	3.4 (3350)	E	SFI	6	OHV

TFI - Throttle body Fuel Injection

MFI - Multi-port Fuel Injection

SFI - Sequential Fuel Injection

79241C48

GENERAL ENGINE SPECIFICATIONS

Year	Engine ID/VIN	Engine Displacement Liters (cc)	Fuel System Type	Net Horsepower @ rpm	Net Torque @ rpm (ft. lbs.)	Bore x Stroke (in.)	Compression Ratio	Oil Pressure @ rpm
1995	D	3.1 (3130)	TFI	120@4400	175@2200	3.50x3.31	8.5:1	15@1100
	L	3.8 (3785)	MFI	170@4300	225@3200	3.80x3.40	9.0:1	60@1850
1996	E	3.4 (3350)	MFI	180@5200	205@4000	3.62x3.31	9.5:1	15@1100
1997	E	3.4 (3350)	MFI	180@5200	205@4000	3.62x3.31	9.5:1	15@1100
1998-99	E	3.4 (3350)	SFI	180@5200	205@4000	3.62x3.31	9.5:1	15@1100

TFI - Throttle body Fuel Injection

MFI - Multi-port Fuel Injection

SFI - Sequential Fuel Injection

79241C49

GASOLINE ENGINE TUNE-UP SPECIFICATIONS

Year	Engine ID/VIN	Engine Displacement Liters (cc)	Spark Plugs Gap (in.)	Ignition Timing (deg.) MT	Ignition Timing (deg.) AT	Fuel Pump (psi)	Idle Speed (rpm) MT	Idle Speed (rpm) AT	Valve Clearance In.	Valve Clearance Ex.
1995	D	3.1 (3130)	0.045	①	①	9-13	—	725	HYD	HYD
	L	3.8 (3785)	0.060	①	①	41-47 ②	—	725	HYD	HYD
1996	E	3.4 (3350)	0.060	①	①	41-47	—	③	HYD	HYD
1997	E	3.4 (3350)	0.060	①	①	41-47	—	③	HYD	HYD
1998-99	E	3.4 (3350)	0.060	①	①	41-47	—	③	HYD	HYD

NOTE: The Vehicle Emission Control Information label often reflects specification changes made during production. The label figures must be used if they differ from those in this chart.

HYD - Hydraulic

① Refer to underhood label for exact setting.

② With key ON and engine OFF.

③ Idle speed is maintained by the PCM.

79241C50

CAPACITIES

Year	Model	Engine ID/VIN	Engine Displacement Liters (cc)	Engine Oil with Filter (qts.)	Transmission (pts.) 4-Spd	Transmission (pts.) 5-Spd	Transmission (pts.) Auto.	Fuel Tank (gal.)	Cooling System (qts.)
1995	Lumina APV	D	3.1 (3130)	4.5	—	—	8.0	20.0	13.4
	Lumina APV	L	3.8 (3785)	4.5	—	—	12.0	20.0	11.4
	Silhouette	D	3.1 (3130)	4.5	—	—	8.0	20.0	13.4
	Silhouette	L	3.8 (3785)	4.5	—	—	12.0	20.0	11.4
	Trans Sport	D	3.1 (3130)	4.5	—	—	8.0	20.0	13.4
	Trans Sport	L	3.8 (3785)	4.5	—	—	12.0	20.0	11.4
1996	Lumina APV	E	3.4 (3350)	4.5	—	—	12.0	20.0	①
	Silhouette	E	3.4 (3350)	4.5	—	—	12.0	20.0	①
	Trans Sport	E	3.4 (3350)	4.5	—	—	12.0	20.0	①
1997	Silhouette	E	3.4 (3350)	4.5	—	—	12.0	20.0	①
	Trans Sport	E	3.4 (3350)	4.5	—	—	12.0	20.0	①
	Venture	E	3.4 (3350)	4.5	—	—	12.0	20.0	①
1998-99	Silhouette	E	3.4 (3350)	4.5	—	—	12.0	20.0	①
	Trans Sport	E	3.4 (3350)	4.5	—	—	12.0	20.0	①
	Venture	E	3.4 (3350)	4.5	—	—	12.0	20.0	①

NOTE: All capacities are approximate. Add fluid gradually and check to be sure a proper fluid level is obtained.

① With rear heater: 13.5 qts.

Without rear heater: 11.8 qts.

79241C51

VALVE SPECIFICATIONS

Year	Engine ID/VIN	Engine Displacement Liters (cc)	Seat Angle (deg.)	Face Angle (deg.)	Spring Test Pressure (lbs. @ in.)	Spring Installed Height (in.)	Stem-to-Guide Clearance (in.)		Stem Diameter (in.)	
							Intake	Exhaust	Intake	Exhaust
1995	D	3.1 (3130)	46	45	191@1.18	1.58	0.0010-0.0027	0.0010-0.0027	NA	NA
	L	3.8 (3785)	45	45	210@1.32	1.69-1.72	0.0015-0.0035	0.0015-0.0032	NA	NA
1996	E	3.4 (3350)	46	45	230@1.26	1.70	0.0010-0.0027	0.0010-0.0027	NA	NA
1997	E	3.4 (3350)	46	45	230@1.26	1.70	0.0010-0.0027	0.0010-0.0027	NA	NA
1998-99	E	3.4 (3350)	46	45	230@1.26	1.70	0.0010-0.0027	0.0010-0.0027	NA	NA

NA - Not Avaliable

79241C52

TORQUE SPECIFICATIONS
All readings in ft. lbs.

Year	Engine ID/VIN	Engine Displacement Liters (cc)	Cylinder Head Bolts	Main Bearing Bolts	Rod Bearing Bolts	Crankshaft Damper Bolts	Flywheel Bolts	Manifold		Spark Plugs	Lug Nut
								Intake	Exhaust		
1995	D	3.1 (3130)	①	78	37	76	57	②	25	③	100
	L	3.8 (3785)	④	⑤	⑥	⑦	⑧	7	38	11	100
1996	E	3.4 (3350)	①	⑨	⑩	76	61	⑪	12	11	100
1997	E	3.4 (3350)	①	⑨	⑩	76	61	⑪	12	11	100
1998-99	E	3.4 (3350)	①	⑨	⑩	76	61	⑪	12	11	100

① Coat threads with sealer
   Tighten all bolts to 33 ft. lbs.
   Tighten all an additional 90 degrees (1/4 turn)
② Tighten bolts to 12 ft. lbs.
   Retorque to 22 ft. lbs.
③ 1st-time installation (new head): 22 ft. lbs.
   All other installations: 12 ft. lbs.
④ Tighten all bolts to 35 ft. lbs. then rotate 130 degrees
   Tighten four center bolts an additional 30 degrees
⑤ 26 ft. lbs. plus 50 degrees
⑥ 20 ft. lbs. plus 50 degrees
⑦ 110 ft. lbs. plus 76 degrees
⑧ 11 ft. lbs. plus 50 degrees
⑨ 37 ft. lbs. plus 77 degrees
⑩ Step 1: 15 ft. lbs.
   Step 2: Plus 75 degrees
⑪ Lower manifold: 10 ft. lbs.
   Upper manifold: 18 ft. lbs.

79241C53

1995–96 SCHEDULED MAINTENANCE INTERVALS
(GENERAL MOTORS CHEVROLET LUMINA APV, OLDSMOBILE SILHOUETTE & PONTIAC TRANS SPORT)

TO BE SERVICED	TYPE OF SERVICE	VEHICLE MILEAGE INTERVAL (x1000)												
		7.5	15	22.5	30	37.5	45	52.5	60	67.5	75	82.5	90	97.5
Engine oil & filter	R	✓	✓	✓	✓	✓	✓	✓	✓	✓	✓	✓	✓	✓
Automatic transaxle fluid & filter③	S/I	✓	✓	✓	✓	✓	✓	✓	✓	✓	✓	✓	✓	✓
Chassis lubrication	S/I	✓	✓	✓	✓	✓	✓	✓	✓	✓	✓	✓	✓	✓
Throttle body mount bolt torque	S/I	✓												
Disc brake pads, rear drum brake linings, wheel cylinders & parking brake	S/I	✓		✓		✓		✓		✓		✓		✓
Rotate tires	S/I	✓		✓		✓		✓		✓		✓		✓
Air cleaner filter & PCV valve & filter	R			✓					✓				✓	
Engine coolant②	R				✓				✓				✓	
Spark plugs①	R				✓				✓				✓	
Accessory drive belt	S/I				✓				✓				✓	
Engine coolant strength, hoses & clamps	S/I				✓				✓				✓	
EGR system	S/I				✓				✓				✓	
Exhaust system & shields	S/I				✓				✓				✓	
Ignition wires	S/I				✓				✓				✓	
Fuel tank, cap & lines	S/I								✓				✓	

① Replace platinum tip spark plugs every 100,000 miles.
② Engine coolant (1996) - replace every 100,000 miles. Use O.E. specified (DEX-COOL™) coolant only. If any silicate coolant is used, the service interval is every 30,000 miles.
③ (1995) - change every 100,000 miles.
R – Replace   S/I – Service or Inspect

FREQUENT OPERATION MAINTENANCE (SEVERE SERVICE)
If a vehicle is operated under any of the following conditions it is considered severe service:
- Extremely dusty areas.
- 50% or more of the vehicle operation is in 32°C (90°F) or higher temperatures, or constant operation in temperatures below 0°C (32°F).
- Prolonged idling (vehicle operation in stop and go traffic).
- Frequent short running periods (engine does not warm to normal operating temperatures).
- Police, taxi, delivery usage or trailer towing usage.
Oil & oil filter change – change every 3000 miles.
Lubricate parking brake cable guides, underbody contact points & linkage every 6000 miles.
Rotate tires at 6000 miles, then every 15,000 miles (1995) or every 12,000 miles (1996)
Throttle body mounting torque – check at 6000 miles.
Air cleaner filter - service or inspect every 15,000 miles.
Automatic transaxle - change fluid & filter every 15,000 miles (1995) or every 50,000 miles (1996)

1997-99 SCHEDULED MAINTENANCE INTERVALS
(GENERAL MOTORS CHEVROLET VENTURE, OLDSMOBILE SILHOUETTE & PONTIAC TRANS SPORT)

TO BE SERVICED	TYPE OF SERVICE	VEHICLE MILEAGE INTERVAL (x1000)															
		7.5	15	22.5	30	37.5	45	52.5	60	67.5	75	82.5	90	97.5	105	112.5	120
Accessory drive belt	I								✓								✓
Air cleaner filter	R								✓								✓
Air distributor air filter	R		✓		✓		✓		✓		✓		✓		✓		✓
Brake system	I	✓	✓	✓	✓	✓	✓	✓	✓	✓	✓	✓	✓	✓	✓	✓	✓
Engine coolant	R	Every 150,000 miles															
Engine oil & filter ①	R	✓	✓	✓	✓	✓	✓	✓	✓	✓	✓	✓	✓	✓	✓	✓	✓
Fuel system tank, cap & lines	I								✓								✓
Rotate Tires	S/I	✓	✓	✓	✓	✓	✓	✓	✓	✓	✓	✓	✓	✓	✓	✓	✓
Spark plug wires	S/I	Every 100,000 miles															
Spark plugs	R	Every 100,000 miles															

① Perform this at the mileage indicated or every 12 months, whichever occurs first.

R - Replace   I - Inspect   L - Lubricate   A - Adjust   C - Clean

FREQUENT OPERATION MAINTENANCE (SEVERE SERVICE)

If a vehicle is operated under any of the following conditions it is considered severe service:
- Towing a trailer or using a camper or car-top carrier.
- Repeated short trips of less than 5 miles in temperatures below freezing, or trips of less than 10 miles in any temperature.
- Extensive idling or low-speed driving for long distances as in heavy commercial use, such as delivery, taxi or police cars.
- Operating on rough, muddy or salt-covered roads.
- Operating on unpaved or dusty roads.
- Driving in extremely hot (over 90°) conditions.

Automatic transaxle fluid and filter - replace every 50,000 miles.

Engine oil and filter - replace every 3,000 miles or 3 months, whichever occurs first.

Tires - rotate every 6,000 miles.

Brake system - inspect every 6,000 miles.

Air distributor air filter - replace every 12,000 miles.

Air cleaner filter - inspect and replace (if necessary) every 15,000 miles.

79241C55

SCHEDULED MAINTENANCE
GENERAL MOTORS
CHEVROLET LUMINA APV, VENTURE, OLDSMOBILE SILHOUETTE
PONTIAC TRANS SPORT

The following should be used as a guide when determining the amount of work required for a particular service if taken to a repair shop. In estimating how long a particular Scheduled Maintenance Service should take, please observe the following:

- **Chilton Time** is time based on field research and data supplied by the vehicle manufacturer.
- Labor time operations are given in hours and tenths of an hour.
- All labor operations, are to be used as a guide.

Mechanic Skill Level Codes:
- **(G) GENERAL:** Normally skilled with certification.
- **(M) MAINTENANCE:** Semi-skilled working on certicication.
- **(P) PRECISION:** Really skilled with multiple certification.

	Chilton Time
(M) 7500 Mile Service	
1995-99	1.5
w/AT add	.1
(M) 15000 Mile Service	
1995-99	.4
w/AT add (1995)	.4
w/AT add (1996-99)	.1
(M) 22500 Mile Service	
1995-99	1.3
w/AT add	.1
(G) 30000 Mile Service	
1995-99	3.8
w/AT add (1995)	.4
w/AT add (1996-99)	.1
(M) 37500 Mile Service	
1995-99	1.3
w/AT add	.1
(M) 45000 Mile Service	
1995-99	.4
w/AT add (1995)	.4
w/AT add (1996-99)	.1

	Chilton Time
(M) 52500 Mile Service	
1995-99	1.3
w/AT add	.1
(G) 60000 Mile Service	
1995-99	3.8
w/AT add (1995)	.4
w/AT add (1996-99)	.1
(M) 67500 Mile Service	
1995-99	1.3
w/AT add	.1
(M) 75000 Mile Service	
1995-99	.4
w/AT add (1995)	.4
w/AT add (1996-99)	.1
(M) 82500 Mile Service	
1995-99	1.3
w/AT add	.1
(G) 90000 Mile Service	
1995-99	3.8
w/AT add (1995)	.4
w/AT add (1996-99)	.1

	Chilton Time
(M) 97500 Mile Service	
1995-99	1.3
w/AT add	.1
(G) 100000 Mile Service	
1997-99	
Replace Spark Plugs	1.2
Replace Spark Plug Wires	1.9
(G) 105500 Mile Service	
1997-99	1.3
w/AT add	.1
(M) 112500 Mile Service	
1997-99	1.3
w/AT add	.1
(M) 120000 Mile Service	
1997-99	3.8
w/AT add	.1
(G) 150000 Mile Service	
1997-99	
ReplaceEngine Coolant	1.5

79241C56

GEO CHEVROLET
Tracker

VEHICLE IDENTIFICATION CHART

		Engine Code					Model Year	
Code	Liters	Cu. In. (cc)	Cyl.	Fuel Sys.	Eng. Mfg.		Code	Year
U	1.6	98 (1590)	4	TFI	Suzuki		S	1995
6	1.6	98 (1590)	4	MFI	Suzuki		T	1996
							V	1997
MFI - Multi-port Fuel Injection							W	1998
TFI - Throttle Body Fuel Injection							X	1999

79241C99

ENGINE IDENTIFICATION

Year	Model		Engine Displacement Liters (cc)	Engine Series (ID/VIN)	Fuel System	No. of Cylinders	Engine Type
1995	Tracker		1.6 (1590)	U	TFI	4	SOHC
	Tracker	①	1.6 (1590)	6	MFI	4	SOHC
1996	Tracker		1.6 (1590)	6	MFI	4	SOHC
1997	Tracker		1.6 (1590)	6	MFI	4	SOHC
1998	Tracker		1.6 (1590)	6	MFI	4	SOHC

TFI - Throttle body fuel injection
MFI - Multiport fuel injection
SOHC - Single overhead camshaft
① California and New York models

79241CA1

GENERAL ENGINE SPECIFICATIONS

Year	Engine ID/VIN		Engine Displacement Liters (cc)	Fuel System Type	Net Horsepower @ rpm	Net Torque @ rpm (ft. lbs.)	Bore x Stroke (in.)	Com-pression Ratio	Oil Pressure @ rpm
1995	U		1.6 (1590)	TFI	80@5400	94@3000	2.95x3.54	8.9:1	47-61@4000
	6	①	1.6 (1590)	MFI	95@5600	98@4000	2.95x3.54	9.5:1	47-61@4000
1996	6		1.6 (1590)	MFI	95@5600	98@4000	2.95x3.54	9.5:1	47-61@4000
1997	6		1.6 (1590)	MFI	95@5600	98@4000	2.95x3.54	9.5:1	47-61@4000
1998	6		1.6 (1590)	MFI	95@5600	98@4000	2.95x3.54	9.5:1	47-61@4000

TFI - Throttle body fuel injection
MFI - Multiport fuel injection
① California and New York models

79241CA2

GASOLINE ENGINE TUNE-UP SPECIFICATIONS

Year	Engine ID/VIN	Engine Displacement Liters (cc)	Spark Plugs Gap (in.)	Ignition Timing (deg.) MT	AT	Fuel Pump (psi)	Idle Speed (rpm) MT	AT	Valve Clearance In.	Ex.
1995	U	1.6 (1590)	0.030	8B ⑤	8B ⑤	34-41	800-850	800-850	0.0090-0.0110 ①	0.0102-0.0114 ①
	6 ②	1.6 (1590)	0.030	8B ⑤	8B ⑤	30-37	800-850	800-850	0.0050-0.0070	0.0050-0.0070
1996	6	1.6 (1590)	0.030	5B ③	5B ③	30-37	800-850	800-850	0.0050-0.0070 ④	0.0050-0.0070 ④
1997	6	1.6 (1590)	0.030	5B ③	5B ③	30-37	800-850	800-850	0.0050-0.0070 ④	0.0050-0.0070
1998	6	1.6 (1590)	0.030	5B ③	5B ③	30-37	800-850	800-850	0.0050-0.0070 ④	0.0050-0.0070 ④

NOTE: The Vehicle Emission Control Information label often reflects specification changes made during production. The label figures must be used if they differ from those in this chart.

B - Before top dead center

HYD - Hydraulic

① Specifications for hot engine. Cold adjustment set valve lash:
Intake: 0.0051-0.0067
Exhaust: 0.0063-0.0073

② California and New York models

③ Connect fused jumper from Duty Check Cavity 4 to cavity 5 for fixed timing (DLC connector located at left strut tower)

④ Cold settings

⑤ Connect jumper wire from Duty Check DLC cavity 3 to 4 (DLC connector located next to battery)

79241CA3

CAPACITIES

Year	Model	Engine ID/VIN	Engine Displacement Liters (cc)	Engine Oil with Filter (qts.)	Transmission (pts.) 4-Spd	5-Spd	Auto.	Transfer Case (pts.)	Drive Axle Front (pts.)	Rear (pts.)	Fuel Tank (gal.)	Cooling System (qts.)
1995	Tracker	U	1.6 (1590)	4.5	—	3.2	9.2 ①	1.8	2.4	2.2	11.1	②
	Tracker ③	6	1.6 (1590)	4.5	—	3.2	9.2 ①	3.6	2.4	2.2	11.1	②
1996	Tracker	6	1.6 (1590)	4.5	—	3.2	10.6	3.6	2.4	4.6	11.0	②
1997	Tracker	6	1.6 (1590)	4.5	—	3.2	10.6	3.6	2.4	4.6	11.0	②
1998	Tracker	6	1.6 (1590)	4.5	—	3.2	10.6	3.6	2.4	4.6	11.0	②

NOTE: All capacities are approximate. Add fluid gradually and check to be sure a proper fluid level is obtained.

① Automatic transmission - Specification is after complete overhaul. Drain and fill will be less

② Manual transmission: 5.6 qts.
Automatic transmission: 5.5 qts.

③ California and New York

79241CA4

VALVE SPECIFICATIONS

Year	Engine ID/VIN	Engine Displacement Liters (cc)	Seat Angle (deg.)	Face Angle (deg.)	Spring Test Pressure (lbs. @ in.)	Spring Installed Height (in.)	Stem-to-Guide Clearance (in.)		Stem Diameter (in.)	
							Intake	Exhaust	Intake	Exhaust
1995	U	1.6 (1590)	45	45	50.2-64.3@ 1.63	1.63	0.0008- 0.0019	0.0014- 0.0025	0.2742- 0.2748	0.2737- 0.2742
	6 ①	1.6 (1590)	45	45	23.6-27.5@ 1.24	1.24	0.0008- 0.0018	0.0018- 0.0028	0.2152- 0.2157	0.2142- 0.2148
1996	6	1.6 (1590)	45	45	23.6-27.5@ 1.24	1.24	0.0008- 0.0018	0.0018- 0.0028	0.2152- 0.2157	0.2142- 0.2148
1997	6	1.6 (1590)	45	45	23.6-27.5@ 1.24	1.24	0.0008- 0.0018	0.0018- 0.0028	0.2152- 0.2157	0.2142- 0.2148
1998	6	1.6 (1590)	45	45	23.6-27.5@ 1.24	1.24	0.0008- 0.0018	0.0018- 0.0028	0.2152- 0.2157	0.2142- 0.2148

① California and New York models

79241CA5

TORQUE SPECIFICATIONS
All readings in ft. lbs.

Year	Engine ID/VIN	Engine Displacement Liters (cc)	Cylinder Head Bolts	Main Bearing Bolts	Rod Bearing Bolts	Crankshaft Damper Bolts	Flywheel Bolts	Manifold		Spark Plugs	Lug Nut
								Intake	Exhaust		
1995	6 ③	1.8 (1803)	④	44	⑥	87 ①	⑤	14	25	21	76
	U	1.6 (1590)	②	40	40	81 ①	58	17	17	21	70
1996	6	1.6 (1590)	②	40	26	44	58	17	17	21	70
1997	6	1.6 (1590)	②	40	26	44	58	17	17	21	70
1998	6	1.6 (1590)	②	40	26	44	58	17	17	21	70

① Crankshaft timing belt sprocket
② Step 1: 26 ft. lbs.
   Step 2: 41 ft. lbs.
   Step 3: 52 ft. lbs.

③ California and New York
④ Step 1: 22 ft. lbs.
   Step 2: Plus two additional steps of 90 degrees

⑤ Manual transaxle: 58 ft. lbs.
   Automatic transaxle: 47 ft. lbs.
⑥ Step 1: 18 ft. lbs.
   Step 2: Plus 90 degrees

79241CA6

SCHEDULED MAINTENANCE INTERVALS
(GEO TRACKER)

TO BE SERVICED	TYPE OF SERVICE	VEHICLE MILEAGE INTERVAL (x1000)												
		7.5	15	22.5	30	37.5	45	52.5	60	67.5	75	82.5	90	97.5
Engine oil & filter	R	✓	✓	✓	✓	✓	✓	✓	✓	✓	✓	✓	✓	✓
Automatic transmission fluid⑤	S/I	✓	✓	✓	✓	✓	✓	✓	✓	✓	✓	✓	✓	✓
Disc brake pads, rotors, drum brake linings, drums, wheel cylinders & parking brake	S/I	✓	✓	✓	✓	✓	✓	✓	✓	✓	✓	✓	✓	✓
Exhaust system	S/I	✓	✓	✓	✓	✓	✓	✓	✓	✓	✓	✓	✓	✓
Free-wheeling hubs	S/I	✓	✓	✓	✓	✓	✓	✓	✓	✓	✓	✓	✓	✓
Locking front hubs	S/I	✓	✓	✓	✓	✓	✓	✓	✓	✓	✓	✓	✓	✓
Manual transmission/ transfer case fluids①	S/I	✓	✓	✓	✓	✓	✓	✓	✓	✓	✓	✓	✓	✓
Rotate tires	S/I	✓	✓	✓	✓	✓	✓	✓	✓	✓	✓	✓	✓	✓
Steering & suspension	S/I	✓	✓	✓	✓	✓	✓	✓	✓	✓	✓	✓	✓	✓
Throttle linkage	S/I	✓	✓	✓	✓	✓	✓	✓	✓	✓	✓	✓	✓	✓
Adjust valve lash	S/I		✓		✓		✓		✓		✓		✓		
Engine idle speed	S/I		✓		✓		✓		✓		✓		✓		
Propeller shafts & U-joints	S/I		✓		✓		✓		✓		✓		✓		
Air cleaner filter	R				✓				✓					✓	
Camshaft timing belt	R								✓						
Fuel filter	R				✓				✓					✓	
Spark plugs	R				✓				✓					✓	
Engine coolant	R				✓				✓					✓	
Engine accessory drive belt③	S/I				✓				✓					✓	
Front wheel bearings	S/I				✓				✓					✓	
Fuel tank, cap & lines	S/I				✓				✓					✓	
Brake fluid	R								✓						
Fuel tank cap gasket	R								✓						
Ignition wires	R								✓						
Emission system hoses	S/I								✓						
Engine timing	S/I								✓						

79241CA7

SCHEDULED MAINTENANCE INTERVALS
(GEO TRACKER) (Cont.)

TO BE SERVICED	TYPE OF SERVICE	VEHICLE MILEAGE INTERVAL (x1000)												
		7.5	15	22.5	30	37.5	45	52.5	60	67.5	75	82.5	90	97.5
EVAP canister④	R													
PCV valve②	R													
Fuel injectors⑥	S/I													

① Replace every 30,000 miles.
② Replace every 50,000 miles.
③ Replace every 60,000 miles.
③ Replace every 100,000 miles.
⑤ Replace every 100,000 miles (1993-94)
⑥ Service or inspect every 100,000 miles.
R – Replace   S/I – Service or Inspect

FREQUENT OPERATION MAINTENANCE (SEVERE SERVICE)
If a vehicle is operated under any of the following conditions it is considered severe service:
- Extremely dusty areas.
- 50% or more of the vehicle operation is in 32°C (90°F) or higher temperatures, or constant operation in temperatures below 0°C (32°F).
- Prolonged idling (vehicle operation in stop and go traffic).
- Frequent short running periods (engine does not warm to normal operating temperatures).
- Police, taxi, delivery usage or trailer towing usage.
Oil & oil filter change – change every 3000 miles.
Free-wheeling hubs - service or inspect every 3000 miles.
Rotate tires every 6000 miles.
Air cleaner filter - service or inspect every 15,000 miles.
Repack front wheel bearings every 15,000 miles.
Manual transmission fluid - change every 15,000 miles.
Engine idle speed - check every 15,000 miles.
Propeller shafts & U-joints - service or inspect every 15,000 miles.
Automatic transmission fluid & filter - change every 15,000 miles (1993-94) or every 50,000 miles (1995-97).

79241CA8

SCHEDULED MAINTENANCE
GENERAL MOTORS CORPORATION
GEO/CHEVROLET
TRACKER

The following should be used as a guide when determining the amount of work required for a particular service if taken to a repair shop. In estimating how long a particular Scheduled Maintenance Service should take, please observe the following:

- **Chilton Time** is time based on field research and data supplied by the vehicle manufacturer.
- Labor time operations are given in hours and tenths of an hour.
- All labor operations, are to be used as a guide.

Mechanic Skill Level Codes:
(G) GENERAL: Normally skilled with certification.
(M) MAINTENANCE: Semi-skilled working on certication.
(P) PRECISION: Really skilled with multiple certification.

	Chilton Time
(M) 7500 Mile Service	
1995-99	1.1
(G) 15000 Mile Service	
1995-99	1.3
Adjust valves add	1.1
(M) 22500 Mile Service	
1995-99	1.1
(G) 30000 Mile Service	
1995-99	3.9
Adjust valves add	1.1
w/AT, add	1.0
(M) 37500 Mile Service	
1995-99	1.1

	Chilton Time
(G) 45000 Mile Service	
1995-99	1.3
Adjust valves add	1.1
(M) 52500 Mile Service	
1995-99	1.1
(G) 60000 Mile Service	
1995-99	8.6
w/AT, add	1.0

	Chilton Time
(M) 67500 Mile Service	
1995-99	1.1
(G) 75000 Mile Service	
1995-99	1.3
Adjust valves add	1.1
(M) 82500 Mile Service	
1995-99	1.1
(G) 90000 Mile Service	
1995-99	8.6
w/AT, add	1.0
(M) 97500 Mile Service	
1995-99	1.1

79241CA9

HONDA
CR-V • Odyssey • Passport

VEHICLE IDENTIFICATION CHART

		Engine Code					Model Year	
Code	Liters	Cu. In. (cc)	Cyl.	Fuel Sys.	Eng. Mfg.		Code	Year
B20B4	2.0	120 (1973)	4	SFI	Honda		S	1995
X22SE/D	2.2	132 (2198)	4	SFI	Isuzu		T	1996
F22B6	2.2	132 (2156)	4	MFI	Honda		V	1997
F23A7	2.3	137 (2254)	4	SFI	Honda		W	1998
4ZE1/E	2.6	156 (2559)	6	MFI	Isuzu		X	1999
6VD1/V	3.2	193 (3165)	6	MFI	Isuzu			
6VD1/W	3.2	193 (3165)	6	SFI	Isuzu			

SFI - Sequential Fuel Injection

MFI - Multi-port Fuel Injection

79241CC6

ENGINE IDENTIFICATION
All measurements are given in inches.

Year	Model	Engine Displacement Liters (cc)	Engine Series (ID/VIN)	Fuel System	No. of Cylinders	Engine Type
1995	Odyssey	2.2 (2156)	F22B6	MFI	4	SOHC 16V
	Passport	2.6 (2559)	4ZE1/E	MFI	4	SOHC 8V
		3.2 (3165)	6VD1/V	MFI	6	SOHC 24V
1996	Odyssey	2.2 (2156)	F22B6	MFI	4	SOHC 16V
	Passport	2.6 (2559)	4ZE1/E	MFI	4	SOHC 8V
		3.2 (3165)	6VD1/V	MFI	6	SOHC 24V
1997	CR-V	2.0 (1973)	B20B4	SFI	4	DOHC 16V
	Odyssey	2.2 (2156)	F22B6	MFI	4	SOHC 16V
	Passport	2.6 (2559)	4ZE1/E	MFI	4	SOHC 8V
		3.2 (3165)	6VD1/V	MFI	6	SOHC 24V
1998-99	CR-V	2.0 (1973)	B20B4	SFI	4	DOHC 16V
	Odyssey	2.3 (2254)	F23A7	SFI	4	SOHC 16V
	Passport	2.2 (2198)	X22SE/D	SFI	4	DOHC 16V
		3.2 (3165)	6VD1/W	SFI	6	DOHC 24V

MFI - Multipoint fuel injection

SFI - Sequential Fuel Injection

SOHC - Single overhead camshaft

DOHC - Double overhead camshaft

8V - 8 Valve cylinder heads

16V - 16 Valve cylinder heads

79241CC7

GENERAL ENGINE SPECIFICATIONS

Year	Engine ID/VIN	Engine Displacement Liters (cc)	Fuel System Type	Net Horsepower @ rpm	Net Torque @ rpm (ft. lbs.)	Bore x Stroke (in.)	Compression Ratio	Oil Pressure @ rpm
1995	F22B6	2.2 (2156)	MFI	140@5600	145@4600	3.35x3.74	8.8:1	50@3000
	4ZE1	2.6 (2559)	MFI	120@4600	150@2600	3.65x3.74	8.6:1	57-71@4000
	6VD1	3.2 (3165)	MFI	175@5200	188@4000	3.68x3.03	9.3:1	57-80@3000
1996	F22B6	2.2 (2156)	MFI	140@5600	145@4600	3.35x3.74	8.8:1	50@3000
	4ZE1	2.6 (2559)	MFI	120@4600	150@2600	3.65x3.74	8.6:1	57-80@3000
	6VD1	3.2 (3165)	MFI	190@5600	188@4000	3.68x3.03	9.0:1	57-80@3000
1997	B20B4	2.0 (1973)	SFI	126@5400	133@4300	3.31X3.50	9.2:1	50@3000
	F22B6	2.2 (2156)	MFI	140@5600	145@4600	3.35x3.74	8.8:1	50@3000
	4ZE1	2.6 (2559)	MFI	120@4600	150@2600	3.65x3.74	8.6:1	57-80@3000
	6VD1	3.2 (3165)	MFI	190@5600	188@4000	3.68x3.03	9.0:1	57-80@3000
1998-99	B20B4	2.0 (1973)	SFI	126@5400	133@4300	3.31X3.50	9.2:1	50@3000
	F23A7	2.3 (2254)	SFI	150@5600	152@4700	3.39X3.82	9.3:1	50@3000
	X22SE	2.2 (2198)	SFI	130@5200	144@4000	NA	9.6:1	22@800
	6VD1	3.2 (3165)	SFI	205@5400	214@3000	3.68X3.03	9.1:1	57-80@3000

MFI - Multi-port Fuel Injection
SFI - Sequential Fuel Injection

79241CC8

GASOLINE ENGINE TUNE-UP SPECIFICATIONS

Year	Engine ID/VIN	Engine Displacement Liters (cc)	Spark Plugs Gap (in.)	Ignition Timing (deg.) MT	Ignition Timing (deg.) AT	Fuel Pump (psi)	Idle Speed (rpm) MT	Idle Speed (rpm) AT	Valve Clearance (in.) In.	Valve Clearance (in.) Ex.
1995	F22B6	2.2 (2156)	0.039-0.043	—	15B	36-43	—	650-750	0.009-0.010	0.011-0.013
	4ZE1	2.6 (2559)	0.039-0.043	12B ①	—	35	850-950	—	0.006	0.001
	6VD1	3.2 (3156)	0.040	NA	NA	25-30	750	750	NA	NA
1996	F22B6	2.2 (2156)	0.039-0.043	—	15B	30-37	—	650-750	0.009-0.011	0.011-0.013
	4ZE1	2.6 (2559)	0.040	12B ①	—	35	900 ②	—	0.006	0.010
	6VD1	3.2 (3156)	0.040	5B ②	5B ②	41-46	750 ②	750 ②	HLA	HLA
1997	B20B4	2.0 (1973)	0.043	—	14-18B	38-46	—	700-800	③	③
	F22B6	2.2 (2156)	0.039-0.043	—	15B	30-37	—	650-750	0.009-0.011	0.011-0.013
	4ZE1	2.6 (2559)	0.040	12B ①	—	35	900 ②	—	0.006	0.010
	6VD1	3.2 (3156)	0.040	5B ②	5B ②	41-46	750 ②	750 ②	HLA	HLA
1998-99	B20B4	2.0 (1973)	0.043	—	14-18B	38-46	—	700-800	③	③
	F23A7	2.3 (2254)	0.039-0.043	—	10-14B	47-54	—	650-750	0.009-0.011	0.011-0.013
	X22SE	2.2 (2198)	0.027-0.031	②	②	41-55	B	B	HLA	HLA
	6VD1	3.2 (3165)	0.040-0.043	16B ②	16B ②	48-55	750 ②	750 ②	0.009-0.013	0.010-0.014

NOTE: The Vehicle Emission Control Information label often reflects specification changes made during production. The label figures must be used if they differ from those in this chart.
NOTE: The fuel pressure readings are given with the vacuum hose connected to the regulator and the engine running
B - Before top dead center
HLA - Hydraulic Lash Adjuster
① Measured at 900 rpm
② Controlled by ECM and is not adjustable
③ Measured between the rocker arm and valve:
  - Intake: 0.13-0.17 in.
  - Exhaust: 0.26-0.30 in.
  Measured between the rocker arm and the camshaft:
  - Intake: 0.003-0.005 in.
  - Exhaust: 0.006-0.008 in.

79241CC9

CAPACITIES

Year	Model	Engine ID/VIN	Engine Displacement Liters (cc)	Engine Oil with Filter (qts.)	Transmission (pts.)			Transfer Case (pts.)	Drive Axle		Fuel Tank (gal.)	Cooling System (qts.)
					4-Spd	5-Spd	Auto.		Front (pts.)	Rear (pts.)		
1995	Odyssey	F22B6	2.2 (2156)	4.0	—	—	5.0	—	—	—	17.2	6.7
	Passport	4ZE1	2.6 (2559)	4.4	—	4.5	—	3.0	3.2	①	21.9	9.5
		6VD1	3.2 (3165)	5.7	—	5.9	9.1	3.0	3.2	①	21.9	②
1996-97	Odyssey	F22B6	2.2 (2156)	4.0	—	—	5.0	—	—	—	17.2	6.7
	Passport	4ZE1	2.6 (2559)	4.4	—	4.5	—	3.0	3.2	①	21.9	9.5
		6VD1	3.2 (3165)	5.7	—	5.9	18.2	3.0	3.2	①	21.9	②
1996-97	Odyssey	F22B6	2.2 (2156)	4.0	—	—	5.0	—	—	—	17.2	6.7
	Passport	4ZE1	2.6 (2559)	4.4	—	4.5	—	3.0	3.2	①	21.9	9.5
		6VD1	3.2 (3165)	5.7	—	5.9	18.2	3.0	3.2	①	21.9	②
	CR-V	B20B4	2.0 (1973)	4.0	—	—	6.2	③	—	2.2	15.3	4.1
1996-97	Odyssey	F23A7	2.3 (2254)	4.5	—	—	5.8	—	—	—	17.2	6.7
	Passport	X22SE	2.2 (2198)	4.8	—	4.5	18.2	3.0	3.0	3.74	21.1	④
		6VD1	3.2 (3165)	5.0	—	6.2	18.2	3.0	3.0	3.74	21.1	⑤
	CR-V	B20B4	2.0 (1973)	4.0	—	—	6.2	③	—	2.2	15.3	4.1

NOTE: All capacities are approximate. Add fluid gradually and check to be sure a proper fluid level is obtained.

NOTE: Capacities given are service, not overhaul capacities

① Saginaw: 4.0
  Dana: 3.8
② Automatic transmission: 9.3
  Manual transmission: 9.7
③ The transfer case fluid is shared with the transaxle assembly.
④ M/T: 14.6 pts.
  A/T: 22.2 pts.
⑤ M/T: 22.4 pts.
  A/T: 22.2 pts.

79241CC0

VALVE SPECIFICATIONS

Year	Engine ID/VIN	Engine Displacement Liters (cc)	Seat Angle (deg.)	Face Angle (deg.)	Spring Test Pressure (lbs. @ in.)	Spring Installed Height (in.)	Stem-to-Guide Clearance (in.)		Stem Diameter (in.)	
							Intake	Exhaust	Intake	Exhaust
1995	F22B6	2.2 (2156)	45	45	NA	NA	0.0008-0.0030	0.0022-0.0050	0.2148-0.2163	0.2134-0.2150
	4ZE1	2.6 (2559)	45	45	45-55@1.61	1.610	0.0009-0.0079	0.0015-0.0098	0.3120-0.3134	0.3091-0.3124
	6VD1	3.2 (3165)	45	45	45-55@1.54	1.540	0.0009-0.0079	0.0012-0.0079	0.2323-0.2353	0.2323-0.2350
1996	F22B6	2.2 (2156)	45	45	NA	NA	0.0008-0.0018	0.0022-0.0050	0.2159-0.2163	0.2146-0.2150
	4ZE1	2.6 (2559)	45	45	45-55@1.61	1.610	0.0009-0.0079	0.0015-0.0098	0.3120-0.3134	0.3091-0.3124
	6VD1	3.2 (3165)	45	45	45-55@1.54	1.540	0.0009-0.0079	0.0012-0.0079	0.2323-0.2353	0.2323-0.2350
1997	F22B6	2.2 (2156)	45	45	NA	NA	0.0008-0.0018	0.0022-0.0050	0.2159-0.2163	0.2146-0.2150
	4ZE1	2.6 (2559)	45	45	45-55@1.61	1.610	0.0009-0.0079	0.0015-0.0098	0.3120-0.3134	0.3091-0.3124
	B20B4	2.0 (1973)	45	45	NA	①	0.001-0.002	0.002-0.0030	0.2591-0.2594	0.2579-0.2583
	6VD1	3.2 (3165)	45	45	45-55@1.54	1.540	0.0009-0.0079	0.0012-0.0079	0.2323-0.2353	0.2323-0.2350
1998-99	B20B4	2.0 (1973)	45	45	NA	①	0.001-0.002	0.002-0.0030	0.2591-0.2594	0.2579-0.2583
	F23A7	2.3 (2254)	45	45	NA	②	0.008-0.0018	0.0022-0.0031	0.2159-0.2163	0.2146-0.2150
	X22SE	2.2 (2198)	NA	NA	NA	NA	0.0012-0.0022	0.0016-0.0026	NA	NA
	6VD1	3.2 (3165)	45	45	44@1.38	1.380	0.0009-0.0079	0.0012-0.0079	0.2323-0.2353	0.2323-0.2350

NA - Not Available

① Valve spring free length:
  - Intake: 1.668 in.
  - Exhaust: 1.745 in.

② Valve spring free length:
  - Intake: 2.011 in.
  - Exhaust: 2.188 in.

79241CD1

TORQUE SPECIFICATIONS
All readings in ft. lbs.

Year	Engine ID/VIN	Engine Displacement Liters (cc)	Cylinder Head Bolts	Main Bearing Bolts	Rod Bearing Bolts	Crankshaft Damper Bolts	Flywheel Bolts	Manifold Intake *	Manifold Exhaust	Spark Plugs	Lug Nut
1995	F22B6	2.2 (2156)	①	②	34	181	54	16	23	13	80
	4ZE1/E	2.6 (2559)	③	72	43	87	40	16	33	14	④
	6VD1/V	3.2 (3165)	⑤	29	49	123	40	17	42	13	④
1996	F22B6	2.2 (2156)	①	②	34	181	54	16	23	13	80
	4ZE1/E	2.6 (2559)	③	72	43	87	40	16	33	14	④
	6VD1/V	3.2 (3165)	⑤	⑥	40	123	40	17	42	13	87
1997	B20B4	2.0 (1973)	⑦	⑧	23	130	54	17	23	13	80
	F22B6	2.2 (2156)	①	②	34	181	54	16	23	13	80
	4ZE1/E	2.6 (2559)	③	72	43	87	40	16	33	14	④
	6VD1/V	3.2 (3165)	⑤	⑥	40	123	40	17	42	13	87
1998-99	B20B4	2.0 (1973)	⑦	⑧	23	130	54	17	23	13	80
	X22SE	2.2 (2198)	⑩	⑪	⑪	14	⑫	16	⑬	18	87
	F23A7	2.3 (2254)	⑨	51	14	181	54	16	23	13	80
	6VD1/V	3.2 (3165)	⑭	29	40	123	40	17	42	13	87

NOTE: Dip main bearing bolts and crankshaft damper bolt in clean engine oil prior to tightening.

* NOTE: Applies to Lower Manifold only.

① Step 1: 29 ft. lbs.
  Step 2: 51 ft. lbs.
  Step 3: 72 ft. lbs.
② Step 1: 22 ft. lbs.
  Step 2: 54 ft. lbs.
③ Step 1: 58 ft. lbs.
  Step 2: 72 ft. lbs.
④ Steel wheels: 66 ft. lbs.
  Aluminum wheels: 87 ft. lbs.
⑤ M8x1.25 bolts: 15 ft. lbs.
  M11x1.50 bolts: 47 ft. lbs.
⑥ Main bearing cap bolts: 29 ft. lbs.
  Oil gallery bolts: 29 ft. lbs. plus 55-65 degrees
  Buttress bolts: 29 ft. lbs.

⑦ Step 1: 22 ft. lbs.
  Step 2: 63 ft. lbs.
⑧ Step 1: 18 ft. lbs.
  Step 2: 56 ft. lbs.
⑨ Step 1: 22 ft. lbs.
  Step 2: 90°
  Step 3: New bolts - 90°
⑩ Step 1: 18 ft. lbs.
  Step 2: 90°
  Step 3: 90°
  Step 4: 90°
⑪ Step 1: 37 ft. lbs.
  Step 2: 45°
  Step 3: 15°

⑫ Step 1: 48 ft. lbs.
  Step 2: 30°
  Step 3: 15°
⑬ Step 1: 10 ft. lbs.
  Step 2: 14 ft. lbs.
  Step 3: 14 ft. lbs.
⑭ Step 1: 21 ft. lbs.
  Step 2: 47 ft. lbs.

79241CD2

1995–96 SCHEDULED MAINTENANCE INTERVALS
(HONDA PASSPORT & ODYSSEY)

TO BE SERVICED	TYPE OF SERVICE	VEHICLE MILEAGE INTERVAL (x1000)												
		7.5	15	22.5	30	37.5	45	52.5	60	67.5	75	82.5	90	97.5
Engine oil & filter	R	✓	✓	✓	✓	✓	✓	✓	✓	✓	✓	✓	✓	✓
Suspension & steering	S/I	✓	✓	✓	✓	✓	✓	✓	✓	✓	✓	✓	✓	✓
Automatic transmission fluid	S/I	✓	✓	✓	✓	✓	✓	✓	✓	✓	✓	✓	✓	✓
Body & chassis lubrication	S/I	✓	✓	✓	✓	✓	✓	✓	✓	✓	✓	✓	✓	✓
Brake & clutch fluid level	S/I	✓	✓	✓	✓	✓	✓	✓	✓	✓	✓	✓	✓	✓
Brake lines & hoses	S/I	✓	✓	✓	✓	✓	✓	✓	✓	✓	✓	✓	✓	✓
Check & rotate tires	S/I	✓	✓	✓	✓	✓	✓	✓	✓	✓	✓	✓	✓	✓
Driveshaft boots	S/I	✓	✓	✓	✓	✓	✓	✓	✓	✓	✓	✓	✓	✓
Engine coolant	S/I	✓	✓	✓	✓	✓	✓	✓	✓	✓	✓	✓	✓	✓
Exhaust system	S/I	✓	✓	✓	✓	✓	✓	✓	✓	✓	✓	✓	✓	✓
Lubricate accelerator linkage	S/I	✓	✓	✓	✓	✓	✓	✓	✓	✓	✓	✓	✓	✓
Starter safety switch	S/I	✓	✓	✓	✓	✓	✓	✓	✓	✓	✓	✓	✓	✓
Parking brake	S/I	✓	✓	✓	✓	✓	✓	✓	✓	✓	✓	✓	✓	✓
Lubricate front & rear propeller shaft	S/I	✓		✓		✓				✓		✓		
Propeller shaft flange torque 46 lb. ft. (63 Nm)	S/I	✓		✓		✓		✓		✓		✓		✓
Auto cruise control linkage & hose	S/I		✓		✓		✓		✓		✓		✓	
Brake & clutch pedal play	S/I		✓		✓		✓		✓		✓		✓	
Cooling & heater hoses	S/I		✓		✓		✓		✓		✓		✓	
Disc & drum brakes	S/I		✓		✓		✓		✓		✓		✓	
Lubricate clutch pedal spring, bushing & clevis pin	S/I		✓		✓		✓		✓		✓		✓	
Shift on-the-fly system gear fluid	S/I		✓		✓		✓		✓		✓		✓	
Throttle linkage	S/I		✓		✓		✓		✓		✓		✓	
Valve clearance (2.2L & 2.6L)	S/I		✓		✓		✓		✓		✓		✓	
Automatic transmission fluid & filter (Odyssey)	R				✓				✓				✓	

79241CD3

1995–96 SCHEDULED MAINTENANCE INTERVALS
(HONDA PASSPORT & ODYSSEY Cont.)

TO BE SERVICED	TYPE OF SERVICE	VEHICLE MILEAGE INTERVAL (x1000)												
		7.5	15	22.5	30	37.5	45	52.5	60	67.5	75	82.5	90	97.5
Manual transmission & transfer case oil	R		✓		✓				✓					
Air cleaner filter	R				✓				✓				✓	
Power steering fluid	R				✓				✓				✓	
Clutch lines & hose	S/I				✓				✓				✓	
Engine drive belts	S/I				✓				✓				✓	
Engine idle speed (2.6L)	S/I	✓			✓				✓				✓	
Front wheel bearings & free wheeling hubs	S/I				✓				✓				✓	
Front & rear axle oil	R		✓		✓				✓					
Steering gear play	S/I				✓				✓				✓	
Brake fluid (include. ABS) (Odyssey)	R				✓				✓				✓	
Spark plugs (2.2L & 2.6L)	R				✓				✓				✓	
Spark plugs (3.2L)	R								✓					
Engine coolant	R				✓				✓				✓	
Distributor cap, rotor & ignition wires (2.2L & 2.6L)	S/I								✓				✓	
Oxygen sensor (1993-95)	R												✓	
Timing belt①	R													
PCV valve (Odyssey)	S/I								✓					
Clean radiator core & A/C condenser	S/I								✓					
Fuel tank, cap & lines	S/I								✓					

① Replace timing belt for Odyssey at 90,000 miles or for Passport at 60,000 miles.
R – Replace   S/I – Service or Inspect

FREQUENT OPERATION MAINTENANCE (SEVERE SERVICE)
If a vehicle is operated under any of the following conditions it is considered severe service:
- Extremely dusty areas.
- 50% or more of the vehicle operation is in 32°C (90°F) or higher temperatures, or constant operation in temperatures below 0°C (32°F).
- Prolonged idling (vehicle operation in stop and go traffic).
- Frequent short running periods (engine does not warm to normal operating temperatures).
- Police, taxi, delivery usage or trailer towing usage.

Oil & oil filter change – change every 3000 miles.
Front & rear axle oil - change every 15,000 miles.
Automatic transmission fluid & filter - change every 20,000 miles.

79241CD4

1997-99 SCHEDULED MAINTENANCE INTERVALS
(HONDA PASSPORT)

TO BE SERVICED	TYPE OF SERVICE	VEHICLE MILEAGE INTERVAL (x1000)															
		7.5	15	22.5	30	37.5	45	52.5	60	67.5	75	82.5	90	97.5	105	112.5	120
Accelerator linkage ①	L	✓	✓	✓	✓	✓	✓	✓	✓	✓	✓	✓	✓	✓	✓	✓	✓
Accessory drive belts ②	S/I				✓				✓				✓				✓
Air cleaner filter	R				✓				✓				✓				✓
Auto cruise control linkage & hose ③	S/I		✓		✓		✓		✓		✓		✓		✓		✓
Automatic transmission fluid level ③	S/I	✓		✓		✓		✓		✓		✓		✓		✓	
Battery fluid level ③	S/I	✓	✓	✓	✓	✓	✓	✓	✓	✓	✓	✓	✓	✓	✓	✓	✓
Body and chassis ①	L	✓	✓	✓	✓	✓	✓	✓	✓	✓	✓	✓	✓	✓	✓	✓	✓
Brake fluid level ③	S/I	✓	✓	✓	✓	✓	✓	✓	✓	✓	✓	✓	✓	✓	✓	✓	✓
Brake lines & hoses ③	S/I	✓	✓	✓	✓	✓	✓	✓	✓	✓	✓	✓	✓	✓	✓	✓	✓
Brake pedal play ③	S/I		✓		✓		✓		✓		✓		✓		✓		✓
Clutch fluid level ③	S/I	✓	✓	✓	✓	✓	✓	✓	✓	✓	✓	✓	✓	✓	✓	✓	✓
Clutch lines & hose ③	S/I				✓				✓				✓				✓
Clutch pedal free-play ③	S/I		✓		✓		✓		✓		✓		✓		✓		✓
Clutch pedal spring, bushing and clevis pin ①	S/I		✓		✓		✓		✓		✓		✓		✓		✓
Cooling and heating system hoses ③	S/I		✓		✓		✓		✓		✓		✓		✓		✓
Driveshaft flange torque ③	S/I	✓		✓		✓		✓		✓		✓		✓		✓	
Drum and disc brakes ③	S/I		✓		✓		✓		✓		✓		✓		✓		✓
Engine coolant	R				✓				✓				✓				✓
Engine coolant level ③	S/I	✓	✓	✓	✓	✓	✓	✓	✓	✓	✓	✓	✓	✓	✓	✓	✓
Engine oil & filter ③	R	✓	✓	✓	✓	✓	✓	✓	✓	✓	✓	✓	✓	✓	✓	✓	✓
Exhaust system ③	S/I	✓	✓	✓	✓	✓	✓	✓	✓	✓	✓	✓	✓	✓	✓	✓	✓
Front and rear axle lubricant	R		✓		✓				✓				✓				✓
Front and rear driveshafts ①	S/I	✓	✓	✓	✓	✓	✓	✓	✓	✓	✓	✓	✓	✓	✓	✓	✓
Front wheel bearings	S/I & L				✓				✓				✓				✓
Fuel lines & tank cap ③	S/I								✓								✓
Inspect for fluid leaks ③	S/I	✓	✓	✓	✓	✓	✓	✓	✓	✓	✓	✓	✓	✓	✓	✓	✓
Key lock cylinder ③	L		✓		✓		✓		✓		✓		✓		✓		✓
Manual transmission and transfer case fluid ④	R		✓		✓				✓				✓				✓
Parking brake system ③	S/I		✓		✓		✓		✓		✓		✓		✓		✓

79241CD5

1997-99 SCHEDULED MAINTENANCE INTERVALS
(HONDA PASSPORT)

TO BE SERVICED	TYPE OF SERVICE	VEHICLE MILEAGE INTERVAL (x1000)															
		7.5	15	22.5	30	37.5	45	52.5	60	67.5	75	82.5	90	97.5	105	112.5	120
Power steering fluid	R				✓				✓				✓				✓
Radiator core and A/C condenser	S/I & C								✓								✓
Rotate tires	S/I	✓	✓	✓	✓	✓	✓	✓	✓	✓	✓	✓	✓	✓	✓	✓	✓
Shift-on-the-fly system gear fluid ③	S/I		✓		✓			✓	✓		✓		✓		✓		✓
Spark plugs ⑤	R				✓				✓				✓				✓
Spark plugs ④	R								✓								✓
Starter safety switch ③	S/I	✓	✓	✓	✓	✓	✓	✓	✓	✓	✓	✓	✓	✓	✓	✓	✓
Steering operation ③	S/I	✓	✓	✓	✓	✓	✓	✓	✓	✓	✓	✓	✓	✓	✓	✓	✓
Suspension & steering ③	S/I	✓	✓	✓	✓	✓	✓	✓	✓	✓	✓	✓	✓	✓	✓	✓	✓
Throttle linkage ③	S/I		✓		✓			✓	✓		✓		✓		✓		✓
Timing belt	R								✓								✓
Tires and wheels ③	S/I	✓	✓	✓	✓	✓	✓	✓	✓	✓	✓	✓	✓	✓	✓	✓	✓
Valve clearance ⑤	A		✓		✓			✓	✓		✓		✓		✓		✓

① Perform this at the mileage indicated or every 6 months, whichever occurs first.
② Perform this at the mileage indicated or every 24 months, whichever occurs first.
③ Perform this at the mileage indicated or every 12 months, whichever occurs first.
④ 3.2L V6 engine.
⑤ 2.6L I4 engine.

R - Replace   S/I - Service or Inspect   L - Lubricate   A - Adjust   C - Clean

FREQUENT OPERATION MAINTENANCE (SEVERE SERVICE)

If a vehicle is operated under any of the following conditions it is considered severe service:
- Towing a trailer or using a camper or car-top carrier.
- Repeated short trips of less than 5 miles in temperatures below freezing.
- Extensive idling or low-speed driving for long distances as in heavy commercial use, such as delivery, taxi or police cars.
- Operating on rough, muddy or salt-covered roads.
- Operating on unpaved or dusty roads.

Air cleaner element - replace every 15,000 miles.
Engine oil and filter - replace every 3,000 miles or 3 months, whichever occurs first.
Automatic transmission fluid - replace every 20,000 miles.
Rear axle lubricant - replace every 15,000 miles.

79241CD6

1997-99 SCHEDULED MAINTENANCE INTERVALS
(HONDA CR-V & ODYSSEY)

TO BE SERVICED	TYPE OF SERVICE	VEHICLE MILEAGE INTERVAL (x1000)															
		7.5	15	22.5	30	37.5	45	52.5	60	67.5	75	82.5	90	97.5	105	112.5	120
Accessory drive belts	I & A				✓				✓				✓				✓
Air cleaner element	R				✓				✓				✓				✓
Air conditioning filter	R				✓				✓				✓				✓
Brake fluid	R						✓						✓				
Brake hoses & lines (including ABS)	I		✓		✓		✓		✓		✓		✓		✓		✓
Cooling system hoses & connections	I		✓		✓		✓		✓		✓		✓		✓		✓
Engine coolant	R						✓						✓				
Engine oil	R	✓	✓	✓	✓	✓	✓	✓	✓	✓	✓	✓	✓	✓	✓	✓	✓
Engine oil and coolant levels	I	Inspect at each fuel stop															
Engine oil filter	R		✓		✓		✓		✓		✓		✓		✓		✓
Exhaust system	I		✓		✓		✓		✓		✓		✓		✓		✓
Fluid levels and condition	I		✓		✓		✓		✓		✓		✓		✓		✓
Front and rear brakes	I		✓		✓		✓		✓		✓		✓		✓		✓
Fuel lines & connections	I		✓		✓		✓		✓		✓		✓		✓		✓
Halfshaft boots	I		✓		✓		✓		✓		✓		✓		✓		✓
Idle speed	I & A														✓		
Parking brake system	I & A		✓		✓		✓		✓		✓		✓		✓		✓
Rear differential fluid	R												✓				
Rotate and inspect tires	I	✓	✓	✓	✓	✓	✓	✓	✓	✓	✓	✓	✓	✓	✓	✓	✓
Spark plugs	R				✓				✓				✓				✓
Supplemental Restrain System (SRS) ①	I	Inspect the SRS 10 years after production															
Suspension components	I		✓		✓		✓		✓		✓		✓		✓		✓
Tie rod ends, steering gear box & boots	I		✓		✓		✓		✓		✓		✓		✓		✓
Timing balancer belt ②	R														✓		
Timing belt	R														✓		
Transmission fluid ③	R												✓				
Transmission fluid ②	R						✓				✓				✓		
Valve clearance ③	I	Adjust if valves are noisy															
Valve clearance ②	I				✓				✓				✓				✓

79241CD7

1997-99 SCHEDULED MAINTENANCE INTERVALS
(HONDA CR-V & ODYSSEY)

TO BE SERVICED	TYPE OF SERVICE	VEHICLE MILEAGE INTERVAL (x1000)															
		7.5	15	22.5	30	37.5	45	52.5	60	67.5	75	82.5	90	97.5	105	112.5	120
Water pump	S/I														✓		

① The SRS should only be inspected or serviced by a qualified automotive technician—this system can cause severe personal injury.

② Odyssey

③ CR-V

R - Replace   I - Inspect   L - Lubricate   A - Adjust   C - Clean

FREQUENT OPERATION MAINTENANCE (SEVERE SERVICE)

If a vehicle is operated under any of the following conditions it is considered severe service:

- Towing a trailer or using a camper or car-top carrier.
- Repeated short trips of less than 5 miles in temperatures below freezing, or trips of less than 10 miles in any temperature.
- Extensive idling or low-speed driving for long distances as in heavy commercial use, such as delivery, taxi or police cars.
- Operating on rough, muddy or salt-covered roads.
- Operating on unpaved or dusty roads.
- Driving in extremely hot (over 90°) conditions.

Air cleaner element - replace every 15,000 miles.

Engine oil and filter - replace every 3,750 miles or 6 months, whichever occurs first.

Timing belt - replace every 60,000 miles if the vehicle is regularly driven in temperatures above 110°F or below -20°F.

Transmission fluid - replace every 30,000 miles.

Rear differential fluid - replace every 60,000 miles.

Front and rear brakes - inspect every 7,500 miles or 6 months, whichever occurs first.

Locks and hinges - lubricate every 15,000 miles.

Tie rods, steering gear box, and boots - inspect every 7,500 miles or 6 months, whichever occurs first.

Suspension components - inspect every 7,500 miles or 6 months, whichever occurs first.

Halfshaft boots - inspect every 7,500 miles or 6 months, whichever occurs first.

79241CD8

SCHEDULED MAINTENANCE
HONDA
CR-V, ODYSSEY, PASSPORT

The following should be used as a guide when determining the amount of work required for a particular service if taken to a repair shop. In estimating how long a particular Scheduled Maintenance Service should take, please observe the following:

- **Chilton Time** is time based on field research and data supplied by the vehicle manufacturer.
- Labor time operations are given in hours and tenths of an hour.
- All labor operations, are to be used as a guide.

Mechanic Skill Level Codes:
- **(G) GENERAL:** Normally skilled with certification.
- **(M) MAINTENANCE:** Semi-skilled working on certicication.
- **(P) PRECISION:** Really skilled with multiple certification.

	Chilton Time
(M) 7500 Mile Service	
1995-99 Passport	1.9
1995-99 Odyssey, CR-V	1.4
(G) 15000 Mile Service	
1995-99 Passport	
2.6L	4.8
3.2L	3.7
1995-99 Odyssey, CR-V	1.5
(M) 22500 Mile Service	
1995-99 Passport	1.9
1995-99 Odyssey, CR-V	1.4
(G) 30000 Mile Service	
1995-99 Passport	
2.6L	7.4
3.2L	3.7
1995-99 Odyssey, CR-V	4.1
(M) 37500 Mile Service	
1995-99 Passport	1.9
1995-99 Odyssey, CR-V	1.4

	Chilton Time
(G) 45000 Mile Service	
1995-99 Passport	4.2
2.6L	2.9
3.2L	2.0
1995-99 Odyssey, CR-V	1.9
(M) 52500 Mile Service	
1995-99 Passport	1.9
1995-99 Odyssey, CR-V	1.4
(G) 60000 Mile Service	
1995-99 Passport	
2.6L	8.9
3.2L	8.2
1995-99 Odyssey, CR-V	4.0
(M) 67500 Mile Service	
1995-99 Passport	1.9
1995-99 Odyssey, CR-V	1.4
(G) 75000 Mile Service	
1995-99 Passport	
2.6L	2.9
3.2L	2.0
1995-99 Odyssey, CR-V	1.9
(M) 82500 Mile Service	
1995-99 Passport	1.9
1995-99 Odyssey, CR-V	1.4

	Chilton Time
(G) 90000 Mile Service	
1995-99 Passport	
2.6L	7.4
3.2L	6.7
1995-99 Odyssey, CR-V	7.0
(M) 97500 Mile Service	
1995-99 Passport	1.9
1995-99 Odyssey, CR-V	1.4
(G) 105000 Mile Service	
1995-99 Passport	4.2
1995-99 Odyssey, CR-V	1.9
Replace timing belt add	2.5
(M) 112500 Mile Service	
1995-99 Passport	1.9
1995-99 Odyssey, CR-V	1.4
(G) 120000 Mile Service	
1995-99 Passport	
2.6L	7.4
3.2L	6.7
1995-99 Odyssey, CR-V	4.0

79241CD9

ISUZU
Amigo • Hombre • Pick-Up • Rodeo • Trooper

VEHICLE IDENTIFICATION CHART

Engine Code						Model Year	
Code	Liters	Cu. In. (cc)	Cyl.	Fuel Sys.	Eng. Mfg.	Code	Year
4	2.2	134 (2189)	4	MFI	GM	S	1995
D	2.2	134 (2198)	4	MFI	Isuzu	T	1996
E	2.6	156 (2559)	4	MFI	Isuzu	V	1997
F22B6	2.2	134 (2156)	4	MFI	Honda	W	1998
F23A7	2.3	138 (2253)	4	MFI	Honda	X	1999
L	2.3	137 (2254)	4	MFI	Isuzu		
V	3.2	193 (3165)	6	MFI	Isuzu		
W	2.2	134 (2156)	4	MFI	GM		
W	3.2	193 (3165)	6	MFI	Isuzu		
X	3.5	213 (3494)	6	MFI	Isuzu		
X	4.3	262 (4300)	6	CPI	GM		

MFI - Multi-port Fuel Injection
CPI - Central Port Injection

79241CD0

ENGINE IDENTIFICATION

Year	Model	Engine Displacement Liters (cc)	Engine Series (ID/VIN)	Fuel System	No. of Cylinders	Engine Type
1995	Pick-up	2.3 (2254)	L	MFI	4	SOHC
		2.6 (2559)	E	MFI	4	SOHC
	Rodeo	2.6 (2559)	E	MFI	4	SOHC
		3.2 (3165)	V	MFI	6	SOHC
	Trooper	3.2 (3165)	V	MFI	6	SOHC
		3.2 (3165)	V	MFI	6	DOHC
1996	Hombre	2.2 (2189)	4	MFI	4	OHV
	Oasis	2.2 (2156)	F22B6	MFI	4	SOHC
	Rodeo	2.6 (2559)	E	MFI	4	SOHC
		3.2 (3165)	V	MFI	6	SOHC
	Trooper	3.2 (3165)	V	MFI	6	SOHC
1997	Hombre	2.2 (2189)	4	MFI	4	OHV
		4.3 (4300)	X	CPI	6	OHV
	Oasis	2.2 (2156)	F22B6	MFI	4	SOHC
	Rodeo	2.6 (2559)	E	MFI	4	SOHC
		3.2 (3165)	V	MFI	6	SOHC
	Trooper	3.2 (3165)	V	MFI	6	SOHC
1998-99	Amigo	2.2 (2198)	D	MFI	4	DOHC
		3.2 (3165)	W	MFI	6	DOHC
	Hombre	2.2 (2189)	4	MFI	4	OHV
		4.3 (4300)	X	CPI	6	OHV
	Oasis	2.3 (2253)	F23A7	MFI	4	SOHC
	Rodeo	2.2 (2198)	D	MFI	4	DOHC
		3.2 (3165)	W	MFI	6	DOHC
	Trooper	3.5 (3494)	X	MFI	6	DOHC

CPI - Central Port Injection
DOHC - Double overhead camshaft
MFI - Multiport fuel injection
OHV - Overhead valve
SOHC - Single overhead camshaft

79241CE1

GENERAL ENGINE SPECIFICATIONS

Year	Engine ID/VIN		Engine Displacement Liters (cc)	Fuel System Type	Net Horsepower @ rpm	Net Torque @ rpm (ft. lbs.)	Bore x Stroke (in.)	Compression Ratio	Oil Pressure @ rpm	
1995	E		2.6 (2559)	MFI	120@4600	150@2600	3.65x3.74	8.6:1	57-71@4000	
	L		2.3 (2254)	MFI	100@4600	125@2600	3.52x3.54	8.3:1	57@3000	
	V	①	3.2 (3165)	MFI	175@5200	188@4000	3.67x3.03	9.3:1	57-80@3000	
	V	②	3.2 (3165)	MFI	190@5600	195@3800	3.67x3.03	9.8:1	57-80@3000	
1996	4		2.2 (2189)	MFI	118@5200	130@2800	3.50x3.46	9.0:1	56@3000	
	E		2.6 (2559)	MFI	120@4600	150@2600	3.65x3.74	8.6:1	57-80@3000	
	F22B6		2.2 (2156)	MFI	140@5600	145@4500	3.35x3.74	8.8:1	50@3000	
	V		3.2 (3165)	MFI	190@5600	188@4000	3.68x3.03	9.1:1	57-80@3000	
1997	4		2.2 (2189)	MFI	118@5200	130@2800	3.50x3.46	9.0:1	56@3000	
	E		2.6 (2559)	MFI	120@4600	150@2600	3.65x3.74	8.6:1	57-80@3000	
	F22B6		2.2 (2156)	MFI	140@5600	145@4500	3.35x3.74	8.8:1	50@3000	
	V		3.2 (3165)	MFI	190@5600	188@4000	3.68x3.03	9.1:1	57-80@3000	
	X		4.3 (4300)	CPI	180@4400	240@2800	4.00x3.48	9.2:1	24@4000	③
1998-99	4		2.2 (2189)	MFI	120@5000	140@3600	3.50x3.46	8.9:1	56@3000	
	D		2.2 (2198)	MFI	130@5200	144@4000	NA	9.6:1	22@800	
	F23A7		2.3 (2253)	MFI	150@5600	152@4700	3.39x3.82	9.3:1	50@3000	③
	W		3.2 (3165)	MFI	205@5400	214@3000	3.68x3.03	9.1:1	60-80@3000	
	X		3.5 (3494)	MFI	215@5400	230@3000	3.68x3.35	9.1:1	60-80@3000	
	X		4.3 (4300)	CPI	180@4400	240@2800	4.00x3.48	9.2:1	24@4000	③

CPI - Central Port Injection
MFI - Multi-port Fuel Injection
① Single Overhead Camshaft
② Double Overhead Camshaft
③ Minimum value

79241CE2

GASOLINE ENGINE TUNE-UP SPECIFICATIONS

Year	Engine ID/VIN	Engine Displacement Liters (cc)	Spark Plugs Gap (in.)	Ignition Timing (deg.) MT	AT	Fuel Pump (psi)	Idle Speed (rpm) MT	AT	Valve Clearance In.	Ex.
1995	E	2.6 (2559)	0.040	12B	12B	35	850	950	0.008	0.008
	L	2.3 (2254)	0.040	12B	—	35	850	950	0.008	0.008
	V	3.2 (3165)	0.040-0.043	5B	5B	41-46	750	750	HYD	HYD
1996	4	2.2 (2189)	0.060	①	—	41-47	①	—	HYD	HYD
	E	2.6 (2559)	0.040	12B ⑤	12B ⑤	35	900 ⑤	900 ⑤	0.006	0.010
	F22B6	2.2 (2156)	0.039-0.043	—	15B	38-46	—	650-750	0.009-0.011	0.011-0.013
	V	3.2 (3165)	0.040	5B ⑤	5B ⑤'	41-46	750 ⑤	750 ⑤	HYD	HYD
1997	4	2.2 (2189)	0.060	①	①	41-47	①	①	HYD	HYD
	E	2.6 (2559)	0.040	12B ⑤	12B ⑤	35	900 ⑤	900 ⑤	0.006	0.010
	F22B6	2.2 (2156)	0.039-0.043	—	15B	38-46	—	650-750	0.009-0.011	0.011-0.013
	V	3.2 (3165)	0.040	5B ⑤	5B ⑤	41-46	750 ⑤	750 ⑤	HYD	HYD
	X	4.3 (4300)	0.060	①	①	60-66	①	①	HYD	HYD
1998-99	4	2.2 (2189)	④	①	①	41-47	①	①	HYD	HYD
	D	2.2 (2198)	0.040	⑤	⑤	41-55	800 ⑤	800 ⑤	HYD	HYD
	F23A7	2.3 (2253)	0.039-0.043	—	10-14 B	47-54 ②	—	650-750 ⑥	0.009-0.011 ③	0.011-0.013 ③
	W	3.2 (3165)	0.040	⑤	⑤	48-55	750 ⑤	750 ⑤	0.009-0.013 ③	0.010-0.014 ③
	X	3.5 (3494)	0.040	⑤	⑤	48-55	750 ⑤	750 ⑤	0.009-0.013 ③	0.010-0.014 ③
	X	4.3 (4300)	0.060	①	①	60-66	①	①	HYD	HYD

NOTE: The Vehicle Emission Control Information label often reflects specification changes made during production. The label figures must be used if they differ from those in this chart.

B - Before top dead center

HYD - Hydraulic

NA - Non-adjustable

① PCM controlled, varies with calibration.

② With pressure regulator hose disconnected.

③ Cold

④ See underhood sticker

⑤ Controlled by the ECM and is not adjustable.

⑥ Slow idle with the headlights and cooling fan OFF.

79241C

CAPACITIES

Year	Model	Engine ID/VIN	Engine Displacement Liters (cc)	Engine Oil with Filter (qts.)	Transmission (pts.)			Transfer Case (pts.)	Drive Axle		Fuel Tank (gal.)	Cooling System (qts.)
					4-Spd	5-Spd	Auto.		Front (pts.)	Rear (pts.)		
1995	Pick-up	L	2.3 (2254)	3.7	—	3.2	—	—	—	3.2	①	9.5
		E	2.6 (2559)	4.4	—	6.2	—	3.0	3.2	3.8	①	9.5
	Rodeo	E	2.6 (2559)	4.4	—	②	—	—	—	⑤	21.9	9.5
		V	3.2 (3165)	6.2	—	②	18.2	3.0	3.2	⑤	21.9	③
	Trooper	V	3.2 (3165)	5.7	—	6.2	18.2	3.0	3.2	3.8	22.5	④
		V	3.2 (3165)	5.7	—	6.2	18.2	3.0	3.2	3.8	22.6	④
1996	Hombre	4	2.2 (2189)	4.5	—	4.4	—	—	—	3.9	19	11.5
	Oasis	F22B6	2.2 (2156)	3.3	—	—	10.6	—	—	—	14.3	6.9
	Rodeo	E	2.6 (2559)	4.4	—	②	—	—	—	⑤	21.9	9.5
		V	3.2 (3165)	5.7	—	②	18.2	3.0	3.2	⑤	21.9	③
	Trooper	V	3.2 (3165)	5.7	—	6.2	18.2	3.0	3.2	3.8	22.5	④
1997	Hombre	4	2.2 (2189)	4.5	—	4.4	—	—	—	3.9	19	11.5
		X	4.3 (4300)	4.5	—	5.0 ⑪	10 ⑨	4.4	3.0	4.0	19.0 ⑩	12.1
	Oasis	F22B6	2.2 (2156)	3.3	—	—	10.6	—	—	—	14.3	6.9
	Rodeo	E	2.6 (2559)	4.4	—	②	—	—	—	⑤	21.9	9.5
		V	3.2 (3165)	5.7	—	②	18.2	3.0	3.2	⑤	21.9	③
	Trooper	V	3.2 (3165)	5.7	—	6.2	18.2	3.0	3.2	3.8	22.5	④
1998-99	Amigo	D	2.2 (2198)	4.8	—	4.5	—	3.0	3.0	3.7	17.7	7.3
		W	3.2 (3165)	5.0	—	6.2	—	3.0	3.0	3.7	17.7	11.2
	Hombre	4	2.2 (2189)	4.5	—	5.8	10 ⑨	—	—	4.0	19.0 ⑩	11.5
		X	4.3 (4300)	4.5	—	5.0 ⑪	10 ⑨	4.4	3.0	4.0	19.0 ⑩	12.1
	Oasis	F23A7	2.3 (2253)	4.8 ⑥	—	—	5.8 ⑧	—	—	—	17.2	6.7 ⑦
	Rodeo	D	2.2 (2198)	4.8	—	4.5	18.2	3.0	3.0	3.7	21.1	7.3
		W	3.2 (3165)	5.0	—	6.2	18.2	3.0	3.0	3.7	21.1	11.2
	Trooper	X	3.5 (3494)	5.0	—	5.8	18.2	3.0 ⑫	3.0	6.4	22.5	7.4 ⑬

NOTE: All capacities are approximate. Add fluid gradually and check to ensure a proper level has been reached.
① Standard bed: 14.0
  Spacecab and long bed: 19.8
② MUA transmission: 6.2
  Borg-Warner transmission: 4.8
③ Manual transmission: 9.7
  Automatic transmission: 9.3
④ Manual transmission: 9.3
  Automatic transmission: 9.0
⑤ Saginaw: 4.0
  Dana: 3.8
⑥ 5.9 after overhaul
⑦ 8.2 after overhaul
⑧ 12.6 after overhaul
⑨ 22 after overhaul
⑩ Specification is for steel tank,
  Plastic tank: 18.0
⑪ Specification is for RWD,
  4WD: 4.4
⑫ 4.0 if equipped with Torque On Demand (TOD)
⑬ Specification is for A/T,
  M/T: 7.0

79241CE4

VALVE SPECIFICATIONS

Year	Engine ID/VIN	Engine Displacement Liters (cc)	Seat Angle (deg.)	Face Angle (deg.)	Spring Test Pressure (lbs. @ in.)	Spring Installed Height (in.)	Stem-to-Guide Clearance (in.)		Stem Diameter (in.)	
							Intake	Exhaust	Intake	Exhaust
1995	E	2.6 (2559)	45	45	45-55@1.61	1.61	0.0009-0.0080	0.0015-0.0098	0.3102-0.3134	0.3091-0.3124
	L	2.3 (2254)	45	45	49-56@1.61	1.61	0.0009-0.0080	0.0015-0.0098	0.3102-0.3134	0.3091-0.3124
	V	3.2 (3165)	45	45	45-55@1.54	1.54	0.0009-0.0078	0.0012-0.0078	0.2323-0.2353	0.2323-0.2350
1996	4	2.2 (2189)	45	45	75-81@1.71	1.71	0.0010-0.0027	0.0014-0.0031	NA	NA
	E	2.6 (2559)	45	45	45-55@1.61	1.61	0.0009-0.0079	0.0015-0.0098	0.3102-0.3134	0.3091-0.3124
	F22B6	2.2 (2156)	45	45	NA	NA	0.0008-0.0018	0.0022-0.0031	0.2159-0.2163	0.2146-0.2150
	V	3.2 (3165)	45	45	45-55@1.54	1.54	0.0009-0.0079	0.0012-0.0079	0.2323-0.2353	0.2323-0.2350
1997	4	2.2 (2189)	45	45	75-81@1.71	1.71	0.0010-0.0027	0.0014-0.0031	NA	NA
	E	2.6 (2559)	45	45	45-55@1.61	1.61	0.0009-0.0079	0.0015-0.0098	0.3102-0.3134	0.3091-0.3124
	F22B6	2.2 (2156)	45	45	NA	NA	0.0008-0.0018	0.0022-0.0031	0.2159-0.2163	0.2146-0.2150
	V	3.2 (3165)	45	45	45-55@1.54	1.54	0.0009-0.0079	0.0012-0.0079	0.2323-0.2353	0.2323-0.2350
1998-99	4	2.2 (2189)	45	45	201-215@1.175	①	0.0007-0.0020	0.0014-0.0029	NA	NA
	D	2.2 (2198)	NA	NA	NA	NA	0.0012-0.0022	0.0016-0.0026	NA	NA
	F23A7	2.3 (2253)	45	45	NA	②	0.0008-0.0018	0.0022-0.0031	0.2159-0.2163	0.2146-0.2150
	W	3.2 (3165)	45	45	41-44@1.38	1.380	0.0002-0.0009	0.0012-0.0025	0.2346-0.2353	0.2343-0.2350
	X	4.3 (4300)	46	45	187-203@1.27	1.78 ③	0.0011-0.0027	0.0011-0.0027	NA	NA
	X	3.5 (3494)	45	45	41-44@1.38	1.380	0.0002-0.0009	0.0012-0.0025	0.2346-0.2353	0.2343-0.2350

NA - Not Available
① Free length: 1.91
② Free length
   Intake: 2.011
   Exhaust: 2.188
③ Specification is for intake,
   Exhaust: 1.69-1.71

79241CE5

TORQUE SPECIFICATIONS
All readings in ft. lbs.

Year	Engine ID/VIN	Engine Displacement Liters (cc)	Cylinder Head Bolts	Main Bearing Bolts	Rod Bearing Bolts	Crankshaft Damper Bolts	Flywheel Bolts	Manifold Intake	Manifold Exhaust	Spark Plugs	Lug Nut
1995	E	2.6 (2559)	①	72	43	87	40	16	16	14	②
	L	2.3 (2254)	①	72	43	87	40	16	16	14	②
	V	3.2 (3165)	④	29	40	123	40	17	42	13	87
1996	4	2.2 (2189)	⑥	70	38	77	55	22	10	13	95
	E	2.6 (2559)	①	72	43	87	40	16	33	14	②
	F22B6	2.2 (2156)	⑤	54	34	181	54	16	23	13	80
	V	3.2 (3165)	④	⑦	40	123	40	17	42	13	87
1997	4	2.2 (2189)	⑥	70	38	77	55	22	10	13	95
	E	2.6 (2559)	①	72	43	87	40	16	33	14	②
	F22B6	2.2 (2156)	⑤	54	34	181	54	16	23	13	80
	V	3.2 (3165)	④	⑦	40	123	40	17	42	13	87
	X	4.3 (4300)	⑫	⑬	⑭	74	74	⑮	⑯	⑰	95
1998-99	4	2.2 (2189)	⑩	70	38	77	55	K	10	13	95
	D	2.2 (2198)	⑱	⑱	⑱	⑱	⑱	16	R	18	87
	F23A7	2.3 (2253)	③	⑧	⑨	181	54	16	23	13	80
	W	3.2 (3165)	⑲	29	40	123	⑲	18	42*	13	87
	X	3.5 (3494)	⑳	⑳	40	123	40	18	38*	13	87
	X	4.3 (4300)	⑫	⑬	⑭	74	74	⑮	⑯	⑰	95

* Use new bolts during assembly.

① Step 1: 58 ft. lbs.
  Step 2: 72 ft. lbs.
② Steel wheels: 58-72 ft. lbs.
  Aluminum wheels: 80-94 ft. lbs.
③ Step 1: 22 ft. lbs.
  Step 2: 180°
  Step 3: For new bolts, tighten an additional 90°
④ 8x1.25 bolts: 15 ft. lbs.
  11x1.5 bolts: 47 ft. lbs.
⑤ Step 1: 29 ft. lbs.
  Step 2: 51 ft. lbs.
  Step 3: 72 ft. lbs.
⑥ Long bolts: 46 ft. lbs.
  Short bolts: 43 ft. lbs.
⑦ Main bearing cap bolts: 29 ft. lbs.
  Oil gallery bolts: 29 ft. lbs. plus 55-65 degrees
  Buttress bolts: 29 ft. lbs.
⑧ 11 mm bolts:
  Step 1: 22 ft. lbs.
  Step 2: 58 ft. lbs.
  6 mm bolts: 8.7 ft. lbs.
  NOTE: If the bearings were replaced, the engine should be run to normal operating temperature at idle and be continued to run for 15 minutes.
⑨ Step 1: 14 ft. lbs.
  Step 2: an additional 90°
  NOTE: If the bearings were replaced, the engine should be run to normal operating temperature at idle and be continued to run for 15 minutes.
⑩ Step 1:
  Long bolts: 46 ft. lbs.
  Short bolts: 43 ft. lbs.
  Step 2: All bolts an additional 90°
⑪ Nuts: 17 ft. lbs.
  Studs: 105 inch lbs.

⑫ Step 1: 22 ft. lbs.
  Step 2:
  Long bolts: an additional 75°
  Medium bolts: an additional 65°
  Short bolts: an additional 55°
  Or, if an angle gauge is not avaliable:
  Step 1: 24 ft. lbs.
  Step 2: 45 ft. lbs.
  Step 3: 65 ft. lbs.
⑬ Step 1: 15 ft. lbs.
  Step 2: an additional 73° or 77 ft. lbs.
⑭ Step 1: 20 ft. lbs.
  Step 2: an additional 70°
⑮ Lower intake manifold-
  Step 1: 27 inch lbs.
  Step 2: 106 inch lbs.
  Step 3: 11 ft. lbs.
  Upper intake manifold-
  Step 1: 44 inch lbs.
  Step 2: 88 inch lbs.
⑯ Step 1: 11 ft. lbs.
  Step 2: 22 ft. lbs.
⑰ New cylinder head only: 22 ft. lbs.
  All subsequent plug installations: 15 ft. lbs.
⑱ Cylinder head-
  Step 1: 18 ft. lbs.
  Step 2: an additional 90°
  Step 3: an additional 90°
  Step 4: an additional 90°
  Main bearings-
  Step 1: 37 ft. lbs.
  Step 2: an additional 45°
  Step 3: an additional 15°
  Connecting rod bearing bolts-
  Step 1: 25 ft. lbs.
  Step 2: an additional 45°
  Step 3: an additional 15°

⑱ Continued-
  Crankshaft sprocket-
  Step 1: 94 ft. lbs.
  Step 2: an additional 45°
  Crankshaft balancer-
  Step 1: 14 ft. lbs.
  Step 2: an additional 45°
  Flywheel bolts-
  Step 1: 48 ft. lbs.
  Step 2: an additional 30°
  Step 3: an additional 15°
  Exhaust manifold-
  Step 1: 112 inch lbs.
  Step 2: 14 ft. lbs.
  Step 3: 14 ft. lbs. again
⑲ Cylinder head bolts-
  Step 1: 21 ft. lbs.
  Step 2: 47 ft. lbs.
  Flywheel bolts: 40 ft. lbs.
  NOTE: Always use new flywheel bolts during assembly, and do not lubricate the bolt threads.
⑳ Cylinder head bolts-
  Step 1: 21 ft. lbs.
  Step 47 ft. lbs.
  NOTE: Always use new cylinder head bolts during assembly, and do not lubricate the bolt threads.
  Main bearing/oil galley bolts-
  Step 1: 22 ft. lbs.
  Step 2: an additional 55-65°
  Crankcase side bolts: 29 ft. lbs.
  Flywheel bolts: 40 ft. lbs.
  NOTE: Always use new flywheel bolts during assembly, and do not use threadlocker on the bolts.

79241CE6

1995-97 SCHEDULED MAINTENANCE INTERVALS
(ISUZU HOMBRE, OASIS, RODEO & TROOPER)

TO BE SERVICED	TYPE OF SERVICE	VEHICLE MILEAGE INTERVAL (x1000)												
		7.5	15	22.5	30	37.5	45	52.5	60	67.5	75	82.5	90	97.5
Engine oil & filter	R	✓	✓	✓	✓	✓	✓	✓	✓	✓	✓	✓	✓	✓
Automatic transmission fluid	S/I	✓	✓	✓	✓	✓	✓	✓	✓	✓	✓	✓	✓	✓
Battery fluid level	S/I	✓	✓	✓	✓	✓	✓	✓	✓	✓	✓	✓	✓	✓
Body & chassis lubrication	S/I	✓	✓	✓	✓	✓	✓	✓	✓	✓	✓	✓	✓	✓
Brake & clutch fluid level	S/I	✓	✓	✓	✓	✓	✓	✓	✓	✓	✓	✓	✓	✓
Brake lines & hoses	S/I	✓	✓	✓	✓	✓	✓	✓	✓	✓	✓	✓	✓	✓
Check & rotate tires	S/I	✓	✓	✓	✓	✓	✓	✓	✓	✓	✓	✓	✓	✓
Engine coolant strength, hoses & clamps	S/I	✓	✓	✓	✓	✓	✓	✓	✓	✓	✓	✓	✓	✓
Exhaust system	S/I	✓	✓	✓	✓	✓	✓	✓	✓	✓	✓	✓	✓	✓
Front suspension, ball joints, steering linkage, parking brake cable guides, propeller shaft splines, universal joints, brake & clutch pedal springs (Hombre)	S/I	✓	✓	✓	✓	✓	✓	✓	✓	✓	✓	✓	✓	✓
Lubricate accelerator linkage (except Rodeo)	S/I	✓	✓	✓	✓	✓	✓	✓	✓	✓	✓	✓	✓	✓
Rear axle seals (Hombre)	S/I	✓	✓	✓	✓	✓	✓	✓	✓	✓	✓	✓	✓	✓
Starter safety switch	S/I	✓	✓	✓	✓	✓	✓	✓	✓	✓	✓	✓	✓	✓
Suspension & steering (Rodeo)	S/I	✓	✓	✓	✓	✓	✓	✓	✓	✓	✓	✓	✓	✓
Suspension & steering (Trooper)	S/I		✓		✓		✓		✓		✓		✓	
Lubricate front & rear propeller shaft	S/I	✓		✓		✓		✓		✓		✓		✓
Propeller shaft flange torque 46 lb. ft. (63 Nm)	S/I	✓		✓		✓		✓		✓		✓		✓
Auto cruise control linkage & hose	S/I		✓		✓		✓		✓		✓		✓	
Brake & clutch pedal play	S/I		✓		✓		✓		✓		✓		✓	
Cooling & heater hoses	S/I		✓		✓		✓		✓		✓		✓	

79241CE7

1995-97 SCHEDULED MAINTENANCE INTERVALS
(ISUZU HOMBRE, OASIS, RODEO & TROOPER Cont.)

TO BE SERVICED	TYPE OF SERVICE	VEHICLE MILEAGE INTERVAL (x1000)												
		7.5	15	22.5	30	37.5	45	52.5	60	67.5	75	82.5	90	97.5
Disc brakes (Trooper)	S/I		✓		✓		✓		✓		✓		✓	
Disc & drum brakes (Amigo, Pick Up & Rodeo)	S/I		✓		✓		✓		✓		✓		✓	
Lubricate clutch pedal spring, bushing & clevis pin	S/I		✓		✓		✓		✓		✓		✓	
Lubricate key lock cylinder	S/I		✓		✓		✓		✓		✓		✓	
Parking brake	S/I		✓		✓		✓		✓		✓		✓	
Shift on-the-fly system gear fluid (Rodeo)	S/I		✓		✓		✓		✓		✓		✓	
Clutch control cable (2.3L)	S/I		✓		✓		✓		✓		✓		✓	
Throttle linkage (Rodeo)	S/I		✓		✓		✓		✓		✓		✓	
Valve clearance (Rodeo 2.6L)	S/I		✓		✓		✓		✓		✓		✓	
Manual transmission & transfer case oil	R		✓		✓				✓				✓	
Front & rear axle oil (Trooper)	R		✓		✓				✓				✓	
Engine idle speed (Amigo & Pick Up 2.3L & 2.6L)	S/I	✓							✓		✓		✓	
Air cleaner filter	R				✓				✓				✓	
Automatic transmission fluid & filter (Oasis)	R				✓				✓				✓	
Brake fluid (include ABS) (Oasis)	R				✓				✓				✓	
Engine coolant (Trooper)	R				✓				✓				✓	
Engine coolant (Rodeo)	R				✓								✓	
Front & rear axle oil	R		✓		✓				✓				✓	
Fuel filter (Hombre)	R				✓				✓				✓	
Power steering fluid	R				✓				✓				✓	
Spark plugs (Rodeo 2.6L, Amigo & Pick Up)①	R				✓				✓				✓	

79241CE8

1995-97 SCHEDULED MAINTENANCE INTERVALS
(ISUZU HOMBRE, OASIS, RODEO & TROOPER Cont.)

TO BE SERVICED	TYPE OF SERVICE	VEHICLE MILEAGE INTERVAL (x1000)												
		7.5	15	22.5	30	37.5	45	52.5	60	67.5	75	82.5	90	97.5
Spark plugs (Rodeo 3.2L & Trooper)①	R								✓					
Carburetor choke (2.3L)	S/I				✓				✓				✓	
Clutch lines & hose	S/I				✓				✓				✓	
Engine drive belts	S/I				✓				✓				✓	
Engine idle speed (Rodeo 2.6L)	S/I	✓							✓				✓	
Front wheel bearings & free wheeling hubs	S/I				✓				✓				✓	
Steering gear play (Amigo & Pick Up)	S/I				✓				✓				✓	
Thermostatically controlled air cleaner (3.1L)	S/I				✓				✓				✓	
Clean radiator core & A/C condenser	S/I				✓				✓				✓	
Ignition wires (Amigo, Hombre, Pick Up & Rodeo 2.6L)	S/I								✓				✓	
Oxygen sensor (Amigo, Pick Up & 1993-95 Rodeo)	R												✓	
Timing belt (Oasis)	R												✓	
Timing belt (Amigo, Rodeo, Pick Up 2.3L & 2.6L)	R								✓					
Engine timing (Amigo, Hombre & Pick Up)	S/I								✓					
Fuel tank, cap & lines	S/I								✓					
PCV valve (Oasis)	S/I								✓					

① Spark plugs (Hombre) - replace every 100,000 miles.
R – Replace   S/I – Service or Inspect

FREQUENT OPERATION MAINTENANCE (SEVERE SERVICE)
If a vehicle is operated under any of the following conditions it is considered severe service:
- Extremely dusty areas.
- 50% or more of the vehicle operation is in 32°C (90°F) or higher temperatures, or constant operation in temperatures below 0°C (32°F).
- Prolonged idling (vehicle operation in stop and go traffic).
- Frequent short running periods (engine does not warm to normal operating temperatures).
- Police, taxi, delivery usage or trailer towing usage.
Oil & oil filter change, body & chassis lubrication – every 3000 miles.
Rotate tires every 6000 miles.
Front & rear axle oil - change every 15,000 miles. Automatic transmission fluid & filter - change every 20,000 miles.

79241CE9

1998-99 SCHEDULED MAINTENANCE INTERVALS
(ISUZU AMIGO, HOMBRE, OASIS, RODEO & TROOPER)

TO BE SERVICED	TYPE OF SERVICE	VEHICLE MILEAGE INTERVAL (x1000)															
		7.5	15	22.5	30	37.5	45	52.5	60	67.5	75	82.5	90	97.5	105	112.5	120
Accelerator linkage ①	L	✓	✓	✓	✓	✓	✓	✓	✓	✓	✓	✓	✓	✓	✓	✓	✓
Accessory drive belts ②	S/I				✓				✓				✓				✓
Air cleaner filter	R				✓				✓				✓				✓
Auto cruise control linkage & hose ③	S/I		✓		✓		✓		✓		✓		✓		✓		✓
Automatic transmission fluid level ③	S/I	✓		✓		✓		✓		✓		✓		✓		✓	
Battery fluid level ③	S/I	✓	✓	✓	✓	✓	✓	✓	✓	✓	✓	✓	✓	✓	✓	✓	✓
Body and chassis ①	L	✓	✓	✓	✓	✓	✓	✓	✓	✓	✓	✓	✓	✓	✓	✓	✓
Brake fluid level ③	S/I	✓	✓	✓	✓	✓	✓	✓	✓	✓	✓	✓	✓	✓	✓	✓	✓
Brake lines & hoses ③	S/I	✓	✓	✓	✓	✓	✓	✓	✓	✓	✓	✓	✓	✓	✓	✓	✓
Brake pedal play ③	S/I		✓		✓		✓		✓		✓		✓		✓		✓
Clutch fluid level ③	S/I	✓	✓	✓	✓	✓	✓	✓	✓	✓	✓	✓	✓	✓	✓	✓	✓
Clutch lines & hose ③	S/I				✓				✓				✓				✓
Clutch pedal free-play ③	S/I		✓		✓		✓		✓		✓		✓		✓		✓
Clutch pedal spring, bushing and clevis pin ①	S/I		✓		✓		✓		✓		✓		✓		✓		✓
Cooling and heating system hoses ③	S/I		✓		✓		✓		✓		✓		✓		✓		✓
Driveshaft flange torque ③	S/I	✓		✓		✓		✓		✓		✓		✓		✓	
Drum and disc brakes ③	S/I		✓		✓		✓		✓		✓		✓		✓		✓
Engine coolant	R				✓				✓				✓				✓
Engine coolant level ③	S/I	✓	✓	✓	✓	✓	✓	✓	✓	✓	✓	✓	✓	✓	✓	✓	✓
Engine oil & filter ③	R	✓	✓	✓	✓	✓	✓	✓	✓	✓	✓	✓	✓	✓	✓	✓	✓
Exhaust system ③	S/I	✓	✓	✓	✓	✓	✓	✓	✓	✓	✓	✓	✓	✓	✓	✓	✓
Front and rear axle lubricant	R		✓		✓				✓				✓				✓
Front and rear driveshafts ①	S/I	✓	✓	✓	✓	✓	✓	✓	✓	✓	✓	✓	✓	✓	✓	✓	✓
Front wheel bearings	S/I & L				✓				✓				✓				✓
Fuel lines & tank cap ③	S/I								✓								✓
Inspect for fluid leaks ③	S/I	✓	✓	✓	✓	✓	✓	✓	✓	✓	✓	✓	✓	✓	✓	✓	✓
Key lock cylinder ③	L		✓		✓		✓		✓		✓		✓		✓		✓
Manual transmission and transfer case fluid ④	R		✓		✓				✓				✓				✓
Parking brake system ③	S/I		✓		✓		✓		✓		✓		✓		✓		✓

79241CE0

1998-99 SCHEDULED MAINTENANCE INTERVALS
(ISUZU AMIGO, HOMBRE, OASIS, RODEO & TROOPER)

TO BE SERVICED	TYPE OF SERVICE	VEHICLE MILEAGE INTERVAL (x1000)															
		7.5	15	22.5	30	37.5	45	52.5	60	67.5	75	82.5	90	97.5	105	112.5	120
Power steering fluid	R				✓				✓				✓				✓
Radiator core and A/C condenser	S/I & C								✓								✓
Rotate tires	S/I	✓	✓	✓	✓	✓	✓	✓	✓	✓	✓	✓	✓	✓	✓	✓	✓
Shift-on-the-fly system gear fluid ③	S/I		✓		✓		✓		✓		✓		✓		✓		✓
Spark plug wires ⑤	S/I								✓								✓
Spark plugs	R	Every 100,000 miles															
Starter safety switch ③	S/I	✓	✓	✓	✓	✓	✓	✓	✓	✓	✓	✓	✓	✓	✓	✓	✓
Steering operation ③	S/I	✓	✓	✓	✓	✓	✓	✓	✓	✓	✓	✓	✓	✓	✓	✓	✓
Suspension & steering ③	S/I	✓	✓	✓	✓	✓	✓	✓	✓	✓	✓	✓	✓	✓	✓	✓	✓
Throttle linkage ③	S/I		✓		✓		✓		✓		✓		✓		✓		✓
Timing belt	R										✓						
Tires and wheels ③	S/I	✓	✓	✓	✓	✓	✓	✓	✓	✓	✓	✓	✓	✓	✓	✓	✓
Valve clearance ④	A								✓								✓

① Perform this at the mileage indicated or every 6 months, whichever occurs first.
② Perform this at the mileage indicated or every 24 months, whichever occurs first.
③ Perform this at the mileage indicated or every 12 months, whichever occurs first.
④ 3.2L V6 engine.
⑤ 2.2L I4 engine.

R - Replace   S/I - Service or Inspect   L - Lubricate   A - Adjust   C - Clean

FREQUENT OPERATION MAINTENANCE (SEVERE SERVICE)

If a vehicle is operated under any of the following conditions it is considered severe service:

- Towing a trailer or using a camper or car-top carrier.
- Repeated short trips of less than 5 miles in temperatures below freezing.
- Extensive idling or low-speed driving for long distances as in heavy commercial use, such as delivery, taxi or police cars.
- Operating on rough, muddy or salt-covered roads.
- Operating on unpaved or dusty roads.

Air cleaner element - replace every 15,000 miles.
Engine oil and filter - replace every 3,000 miles or 3 months, whichever occurs first.
Automatic transmission fluid - replace every 20,000 miles.
Rear axle lubricant - replace every 15,000 miles.

79241CF1

SCHEDULED MAINTENANCE
ISUZU
AMIGO, HOMBRE, OASIS, RODEO, TROOPER

The following should be used as a guide when determining the amount of work required for a particular service if taken to a repair shop. In estimating how long a particular Scheduled Maintenance Service should take, please observe the following:

- **Chilton Time** is time based on field research and data supplied by the vehicle manufacturer.
- Labor time operations are given in hours and tenths of an hour.
- All labor operations, are to be used as a guide.

Mechanic Skill Level Codes:
  (G) GENERAL: Normally skilled with certification.
  (M) MAINTENANCE: Semi-skilled working on certicication.
  (P) PRECISION: Really skilled with multiple certification.

	Chilton Time
(M) 7500 Mile Service	
1995-99 Trooper	2.9
1995-99 Rodeo	1.9
w/AT add	.3
1996-99 Hombre	2.0
1996-99 Oasis	2.1
1998-99 Amigo	2.9
(G) 15000 Mile Service	
1995-99 Trooper	5.0
1995-97 Rodeo	5.5
w/AT add	.3
1996-99 Hombre	4.9
1996-99 Oasis	5.1
1998-99 Amigo	5.0
(M) 22500 Mile Service	
1995-99 Trooper	2.9
1995-99 Rodeo	2.8
w/AT add	.3
1996-99 Hombre	3.0
1996-99 Oasis	3.1
1998-99 Amigo	2.9
(G) 30000 Mile Service	
1995-99 Trooper	9.8
1995-99 Rodeo	7.2
w/AT add	.3
1996-99 Hombre	7.1
1996-99 Oasis	7.3
1998-99 Amigo	9.8
(M) 37500 Mile Service	
1995-99 Trooper	2.9
1995-99 Rodeo	2.0
w/AT add	.3
1996-99 Hombre	3.0
1996-99 Oasis	3.1
1998-99 Amigo	2.9

	Chilton Time
(G) 45000 Mile Service	
1995-99 Trooper	5.5
1995-99 Rodeo	4.5
w/AT add	.3
1996-99 Hombre	5.0
1996-99 Oasis	5.1
1998-99 Amigo	5.5
(M) 52500 Mile Service	
1995-99 Trooper	3.0
1995-99 Rodeo	3.0
w/AT add	.3
1996-99 Hombre	3.1
1996-99 Oasis	3.3
1998-99 Amigo	3.0
(G) 60000 Mile Service	
1995-99 Trooper	11.0
1995-99 Rodeo	10.8
w/AT add	.3
1996-99 Hombre	10.7
1996-99 Oasis	10.9
1998-99 Amigo	11.0
(M) 67500 Mile Service	
1995-99 Trooper	2.9
1995-99 Rodeo	2.0
w/AT add	.3
1996-99 Hombre	1.9
1996-99 Oasis	2.1
1998-99 Amigo	2.0
(G) 75000 Mile Service	
1995-99 Trooper	5.0
1995-99 Rodeo	4.5
w/AT add	.3
Replace timing belt add	2.7
1996-99 Hombre	4.9
1996-99 Oasis	5.1
1998-99 Amigo	5.0
(M) 82500 Mile Service	
1995-99 Trooper	2.9
1995-99 Rodeo	2.0
w/AT add	.3
1996-99 Hombre	1.9
1996-99 Oasis	2.1
1998-99 Amigo	2.9

	Chilton Time
(G) 90000 Mile Service	
1995-99 Trooper	11.0
1995-99 Rodeo	7.2
w/AT add	.3
1996-99 Hombre	10.7
1996-99 Oasis	10.9
Replace timing belt add	2.5
1998-99 Amigo	11.0
(M) 97500 Mile Service	
1995-99 Trooper	2.9
1995-99 Rodeo	2.0
w/AT add	.3
1996-99 Hombre	1.9
1996-99 Oasis	2.1
1998-99 Amigo	2.9
(G) 105000 Mile Service	
1995-99 Trooper	5.5
1995-99 Rodeo	4.5
w/AT add	.3
1996-99 Hombre	4.9
1996-99 Oasis	5.1
1998-99 Amigo	5.5
(G) 112500 Mile Service	
1995-99 Trooper	3.0
1995-99 Rodeo	3.0
w/AT add	.3
1996-99 Hombre	3.1
1996-99 Oasis	3.3
1998-99 Amigo	3.0
(G) 120000 Mile Service	
1995-99 Trooper	11.0
1995-99 Rodeo	10.8
w/AT add	.3
1996-99 Hombre	10.7
1996-99 Oasis	10.9
1998-99 Amigo	11.0

79241CF2

KIA
Sportage

VEHICLE IDENTIFICATION CHART

		Engine Code					Model Year	
Code	Liters	Cu. In. (cc)	Cyl.	Fuel Sys.	Eng. Mfg.		Code	Year
1	2.0	122 (1998)	4	MFI	KIA		S	1995
3	2.0	122 (1998)	4	MFI	KIA		T	1996
							V	1997
							W	1998
							X	1999

MFI - Multi-port Fuel Injection

79241CG2

ENGINE IDENTIFICATION

Year	Model	Engine Displacement Liters (cc)	Engine Series (ID/VIN)	Fuel System	No. of Cylinders	Engine Type
1995	Sportage	2.0 (1998)	1	MFI	4	SOHC
		2.0 (1998)	3	MFI	4	DOHC
1996	Sportage	2.0 (1998)	3	MFI	4	DOHC
1997	Sportage	2.0 (1998)	3	MFI	4	DOHC
1998-99	Sportage	2.0 (1998)	3	MFI	4	DOHC

MFI - Multi-port Fuel Injection
SOHC - Single Overhead Camshaft
DOHC - Double Overhead Camshafts

79241CG3

GENERAL ENGINE SPECIFICATIONS

Year	Engine ID/VIN	Engine Displacement Liters (cc)	Fuel System Type	Net Horsepower @ rpm	Net Torque @ rpm (ft. lbs.)	Bore x Stroke (in.)	Com- pression Ratio	Oil Pressure @ rpm
1995	1	2.0 (1998)	MFI	94@5000	114@2500	3.39x3.39	9.2:1	43-57 ①
	3	2.0 (1998)	MFI	139@6000	134@4000	3.39x3.39	9.2:1	43-57 ①
1996	3	2.0 (1998)	MFI	130@5500	127@4000	3.39x3.39	9.2:1	43-57 ①
1997	3	2.0 (1998)	MFI	130@5500	127@4000	3.39x3.39	9.2:1	43-57 ①
1998-99	3	2.0 (1998)	MFI	130@5500	127@4000	3.39x3.39	9.2:1	43-57 ①

MFI - Multi-port Fuel Injection
① The manufacturer does not provide an engine speed specification for oil pump pressure.

79241C

GASOLINE ENGINE TUNE-UP SPECIFICATIONS

Year	Engine ID/VIN	Engine Displacement Liters (cc)	Spark Plugs Gap (in.)	Ignition Timing (deg.) MT	AT	Fuel Pump (psi)	Idle Speed (rpm) MT	AT	Valve Clearance In.	Ex.
1995	1	2.0 (1998)	0.028-0.032	4° BTDC	—	50	750-850	—	0.012 in.	0.012 in.
	3	2.0 (1998)	0.039-0.043	4° BTDC	—	50	750-850	—	HYD.	HYD.
1996	3	2.0 (1998)	0.039-0.043	4° BTDC	—	50	750-850	—	HYD.	HYD.
1997	3	2.0 (1998)	0.039-0.043	4° BTDC	—	50	750-850	—	HYD.	HYD.
1998-99	3	2.0 (1998)	0.039-0.043	4° BTDC	—	50	750-850	—	HYD.	HYD.

BTDC - Before Top Dead Center
HYD. - Hydraulic lash adjusters

79241CG5

CAPACITIES

Year	Model	Engine ID/VIN	Engine Displacement Liters (cc)	Engine Oil with Filter (qts.)	Transmission (pts.) 4-Spd	5-Spd	Auto.	Transfer Case (pts.)	Drive Axle Front (pts.)	Rear (pts.)	Fuel Tank (gal.)	Cooling System (qts.)
1995	Sportage	1	2.0 (1998)	4.4	—	2.6	5.4	2.8	2.6	2.6	15.8	7.9
		3	2.0 (1998)	4.4	—	2.6	5.4	2.8	2.6	2.6	15.8	7.9
1996	Sportage	3	2.0 (1998)	4.4	—	2.6	5.4	2.8	2.6	2.6	15.8	7.9
1997	Sportage	3	2.0 (1998)	4.4	—	2.6	5.4	2.8	2.6	2.6	15.8	7.9
1998-99	Sportage	3	2.0 (1998)	4.4	—	2.6	5.4	2.8	2.6	2.6	15.8	7.9

NOTE: All capacities are approximate. Add fluid gradually and check to be sure a proper fluid level is obtained.

79241CG6

VALVE SPECIFICATIONS

Year	Engine ID/VIN	Engine Displacement Liters (cc)	Seat Angle (deg.)	Face Angle (deg.)	Spring Test Pressure (lbs. @ in.)	Spring Installed Height (in.)	Stem-to-Guide Clearance (in.)		Stem Diameter (in.)	
							Intake	Exhaust	Intake	Exhaust
1995	1	2.0 (1998)	45	45	①	①	0.0079	0.0079	0.3142	0.3139
	3	2.0 (1998)	45	45	②	②	0.0010-0.0024	0.0012-0.0026	0.2350-0.2356	0.2348-0.2354
1996	3	2.0 (1998)	45	45	②	②	0.0010-0.0024	0.0012-0.0026	0.2350-0.2356	0.2348-0.2354
1997	3	2.0 (1998)	45	45	②	②	0.0010-0.0024	0.0012-0.0026	0.2350-0.2356	0.2348-0.2354
1998-99	3	2.0 (1998)	45	45	②	②	0.0010-0.0024	0.0012-0.0026	0.2350-0.2356	0.2348-0.2354

① Spring test pressure or installed height not provided by the manufacturer.
   Valve Spring Free Length:
   -Outer spring: 2.047 in.
   -Inner spring: 1.732 in.
② Spring test pressure or installed height not provided by the manufacturer.
   Valve Spring Free Length:
   -Outer spring: 1.524-1.539 in.
   -Inner spring: 1.484-1.496 in.

79241CG7

TORQUE SPECIFICATIONS
All readings in ft. lbs.

Year	Engine ID/VIN	Engine Displacement Liters (cc)	Cylinder Head Bolts	Main Bearing Bolts	Rod Bearing Bolts	Crankshaft Damper Bolts	Flywheel Bolts	Manifold		Spark Plugs	Lug Nuts
								Intake	Exhaust		
1995	1	2.0 (1998)	62	63	50	11	73	16	31	11-17	73
	3	2.0 (1998)	62	63	50	11	73	16	31	11-17	73
1996	3	2.0 (1998)	62	63	50	11	73	16	31	11-17	73
1997	3	2.0 (1998)	62	63	50	11	73	16	31	11-17	73
1998-99	3	2.0 (1998)	62	63	50	11	73	16	31	11-17	73

79241CG8

SCHEDULED MAINTENANCE INTERVALS
(KIA SPORTAGE)

TO BE SERVICED	TYPE OF SERVICE	VEHICLE MILEAGE INTERVAL (x1000)															
		7.5	15	22.5	30	37.5	45	52.5	60	67.5	75	82.5	90	97.5	105	112.5	120
Accessory drive belt	S/I				✓				✓				✓				✓
Air cleaner filter	R				✓				✓				✓				✓
Automatic transmission fluid	R				✓		✓		✓		✓		✓		✓		✓
Ball joints	S/I				✓				✓				✓				✓
Brake lines & connections	S/I				✓				✓				✓				✓
Chassis/body fasteners	S/I				✓				✓				✓				✓
Cooling system	S/I				✓				✓				✓				✓
CV-joint boots	S/I		✓		✓		✓		✓		✓		✓		✓		✓
Disc brakes	S/I		✓		✓		✓		✓		✓		✓		✓		✓
Driveshaft U-joints	L		✓		✓		✓		✓		✓		✓		✓		✓
Drum brakes	S/I				✓								✓				✓
Emission hoses & tubes	S/I								✓								✓
Emission hoses & tubes (California)	R														✓		
Engine coolant	R				✓				✓				✓				✓
Engine oil & filter	R	✓	✓	✓	✓	✓	✓	✓	✓	✓	✓	✓	✓	✓	✓	✓	✓
Exhaust system heat shields	S/I				✓				✓				✓				✓
Front differential fluid ①	R				✓				✓				✓				✓
	S/I	✓	✓	✓	✓	✓	✓	✓	✓	✓	✓	✓	✓	✓	✓	✓	✓
Fuel filter	R				✓				✓				✓				✓
Fuel lines & hoses	S/I				✓				✓				✓				✓
Idle speed	S/I				✓				✓				✓				✓
Locks & hinges	L	✓	✓	✓	✓	✓	✓	✓	✓	✓	✓	✓	✓	✓	✓	✓	✓
Manual transmission fluid	R				✓				✓				✓				✓
PCV valve	S/I								✓								✓
Rear differential fluid	R				✓				✓				✓				✓
	S/I	✓	✓	✓	✓	✓	✓	✓	✓	✓	✓	✓	✓	✓	✓	✓	✓
Spark plug wires	S/I								✓								✓
Spark plugs	R				✓				✓				✓				✓
Steering operation & linkage	S/I				✓				✓				✓				✓

79241CG9

SCHEDULED MAINTENANCE INTERVALS
(KIA SPORTAGE)

Timing belt (California)	R														✓		
	S/I						✓				✓						
Timing belt (non-California)	R						✓										✓
Transfer case fluid ①	R			✓			✓				✓						✓
Transfer case fluid ①	S/I		✓		✓		✓		✓		✓		✓		✓		
Transmission fluid	S/I	✓	✓	✓	✓	✓	✓	✓	✓	✓	✓	✓	✓	✓	✓	✓	✓

① If equipped.

R - Replace   S/I - Inspect and service, if needed   L - Lubricate   A - Adjust   C - Clean

FREQUENT OPERATION MAINTENANCE (SEVERE SERVICE)

If a vehicle is operated under any of the following conditions it is considered severe service:

- Towing a trailer or using a camper or car-top carrier.
- Repeated short trips of less than 5 miles in temperatures below freezing, or trips of less than 10 miles in any temperature.
- Extensive idling or low-speed driving for long distances as in heavy commercial use, such as delivery, taxi or police cars.
- Operating on rough, muddy or salt-covered roads.
- Operating on unpaved or dusty roads.
- Driving in extremely hot (over 90°) conditions.

Engine oil & filter - replace every 5,000 miles or 5 months, whichever occurs first.

Air cleaner filter - inspect, and replace if necessary, every 15,000 miles or 15 months, whichever occurs first.

Transfer case fluid - inspect the level every 5,000 miles or 5 months, and replace every 15,000 miles or 15 months, whichever occurs first.

Transmission fluid - inspect the level every 5,000 miles or 5 months, and replace every 15,000 miles or 15 months, whichever occurs first.

Front differential fluid - inspect the level every 5,000 miles or 5 months, and replace every 15,000 miles or 15 months, whichever occurs first.

Rear differential fluid - inspect the level every 5,000 miles or 5 months, and replace every 15,000 miles or 15 months, whichever occurs first.

79241CG0

SCHEDULED MAINTENANCE
KIA
SPORTAGE

The following should be used as a guide when determining the amount of work required for a particular service if taken to a repair shop.
In estimating how long a particular Scheduled Maintenance Service should take, please observe the following:

- **Chilton Time** is time based on field research and data supplied by the vehicle manufacturer.
- Labor time operations are given in hours and tenths of an hour.
- All labor operations, are to be used as a guide.

Mechanic Skill Level Codes:
    (G) GENERAL: Normally skilled with certification.
    (M) MAINTENANCE: Semi-skilled working on certicication.
    (P) PRECISION: Really skilled with multiple certification.

	Chilton Time			Chilton Time			Chilton Time
(M) 7500 Mile Service 1995-99	.5		**(M) 45000 Mile Service** 1995-99	1.0		**(M) 75000 Mile Service** 1995-99	1.0
(M) 15000 Mile Service 1995-99	1.0		**(M) 52500 Mile Service** 1995-99	.5		**(M) 82500 Mile Service** 1995-99	.5
(M) 22500 Mile Service 1995-99	.5		**(G) 60000 Mile Service** 1995-99	5.2		**(G) 90000 Mile Service** 1995-99	5.2
(G) 30000 Mile Service 1995-99	3.0		**(M) 67500 Mile Service** 1995-99	.5		**(M) 97500 Mile Service** 1995-99	.5
(M) 37500 Mile Service 1995-99	.5					**(M) 105000 Mile Service** 1995-99	1.0

79241CH1

LAND ROVER
Defender 90 • Discovery • Range Rover

VEHICLE IDENTIFICATION CHART

Engine Code						Model Year	
Code	Liters	Cu. In. (cc)	Cyl.	Fuel Sys.	Eng. Mfg.	Code	Year
2	4.0	241 (3950)	8	MFI	Land Rover	S	1995
J	4.6	278 (4554)	8	MFI	Land Rover	T	1996
V	3.9	241 (3949)	8	MFI	Land Rover	V	1997
V	4.2	261 (4278)	8	MFI	Land Rover	W	1998
						X	1999

MFI - Multi-port Fuel Injection

79241CH2

ENGINE IDENTIFICATION

Year	Model	Engine Displacement Liters (cc)	Engine Series (ID/VIN)	Fuel System	No. of Cylinders	Engine Type
1995	Defender 90	3.9 (3949)	V	MFI	8	OHV
	Discovery	3.9 (3949)	V	MFI	8	OHV
	Range Rover	4.2 (4278)	V	MFI	8	OHV
1996	Discovery	4.0 (3950)	V	MFI	8	OHV
	Range Rover	4.0 (3950)	2	MFI	8	OHV
		4.6 (4554)	J	MFI	8	OHV
1997	Defender 90	4.0 (3950)	V	MFI	8	OHV
	Discovery	4.0 (3950)	V	MFI	8	OHV
	Range Rover	4.0 (3950)	2	MFI	8	OHV
		4.6 (4554)	J	MFI	8	OHV
1998-99	Defender 90	4.0 (3950)	V	MFI	8	OHV
	Discovery	4.0 (3950)	V	MFI	8	OHV
	Range Rover	4.0 (3950)	2	MFI	8	OHV
		4.6 (4554)	J	MFI	8	OHV

MFI - Multi-port Fuel Injection
OHV - Overhead Valve

79241CH3

GENERAL ENGINE SPECIFICATIONS

Year	Engine ID/VIN	Engine Displacement Liters (cc)	Fuel System Type	Net Horsepower @ rpm	Net Torque @ rpm (ft. lbs.)	Bore x Stroke (in.)	Compression Ratio	Oil Pressure @ rpm
1995	V	3.9 (3949)	MFI	182@4750	232@3100	3.7x2.8	9.35:1	40@2500
	V	4.2 (4278)	MFI	200@4750	251@3250	3.7x3.3	8.9:1	40@2400
1996	2	4.0 (3950)	MFI	190@4750	236@3000	3.7x2.8	9.34:1	35@2400
	J	4.6 (4554)	MFI	225@4750	280@3000	3.7x3.2	9.34:1	35@2400
	V	4.0 (3950)	MFI	182@4750	233@3000	3.7x2.8	9.34:1	35@2400
1997	2	4.0 (3950)	MFI	190@4750	236@3000	3.7x2.8	9.34:1	35@2400
	J	4.6 (4554)	MFI	225@4750	280@3000	3.7x3.2	9.34:1	35@2400
	V	4.0 (3950)	MFI	182@4750	233@3000	3.7x2.8	9.34:1	35@2400
1998-99	2	4.0 (3950)	MFI	190@4750	236@3000	3.7x2.8	9.34:1	35@2400
	J	4.6 (4554)	MFI	225@4750	280@3000	3.7x3.2	9.34:1	35@2400
	V	4.0 (3950)	MFI	182@4750	233@3000	3.7x2.8	9.34:1	35@2400

79241CH4

GASOLINE ENGINE TUNE-UP SPECIFICATIONS

Year	Engine ID/VIN	Engine Displacement Liters (cc)	Spark Plugs Gap (in.)	Ignition Timing (deg.) MT	Ignition Timing (deg.) AT	Fuel Pump (psi)	Idle Speed (rpm) MT	Idle Speed (rpm) AT	Valve Clearance In.	Valve Clearance Ex.
1995	V	3.9 (3949)	0.032-0.039	3-5BTDC	3-5BTDC	34-37	675-725	575-625 ①	HYD	HYD
	V	4.2 (4278)	0.032-0.039	3-5BTDC	3-5BTDC	34-37	675-725	575-625	HYD	HYD
1996	2	4.0 (3950)	0.032-0.039	②	②	34-37	675-725	675-725	HYD	HYD
	J	4.6 (4554)	0.035-0.040	②	②	34-37	680-720	680-720	HYD	HYD
	V	4.0 (3950)	0.035-0.040	②	②	34-37	675-725	675-725	HYD	HYD
1997	2	4.0 (3950)	0.032-0.039	②	②	34-37	675-725	675-725	HYD	HYD
	J	4.6 (4554)	0.035-0.040	②	②	34-37	680-720	680-720	HYD	HYD
	V	4.0 (3950)	0.035-0.040	②	②	34-37	675-725	675-725	HYD	HYD
1998-99	2	4.0 (3950)	0.032-0.039	②	②	34-37	675-725	675-725	HYD	HYD
	J	4.6 (4554)	0.035-0.040	②	②	34-37	680-720	680-720	HYD	HYD
	V	4.0 (3950)	0.035-0.040	②	②	34-37	675-725	675-725	HYD	HYD

① With the A/C turned ON.
② Automatically controlled by the Powertrain Control Module (PCM).

79241CH5

CAPACITIES

Year	Model	Engine ID/VIN	Engine Displacement Liters (cc)	Engine Oil with Filter (qts.)	Transmission (pts.) 4-Spd	Transmission (pts.) 5-Spd	Transmission (pts.) Auto.	Transfer Case (pts.)	Drive Axle Front (pts.)	Drive Axle Rear (pts.)	Fuel Tank (gal.)	Cooling System (qts.)
1995	Defender 90	V	3.9 (3949)	7	—	—	20	4.8	3.6	4.8	15.6	13.5
	Discovery	V	3.9 (3949)	7	—	5.7	19.2	4.9	3.6	3.6	23.0	23.9
	Range Rover	V	4.2 (4278)	7	—	NA	NA	NA	NA	NA	NA	NA
1996	Discovery	V	4.0 (3950)	5.6	—	5.7	19.2	4.9	3.6	3.6	23	23.9
	Range Rover	2	4.0 (3950)	7	—	5.7	20.5	5.0	3.6	3.6	26.4	24
		J	4.6 (4554)	7	—	5.7	23.2	5.0	3.6	3.6	26.4	24
1997	Defender 90	V	4.0 (3950)	5.6	—	—	19.2	4.8	3.6	4.8	15.6	13.5
	Discovery	V	4.0 (3950)	5.6	—	5.7	19.2	4.9	3.6	3.6	23	23.9
	Range Rover	2	4.0 (3950)	7	—	5.7	20.5	5.0	3.6	3.6	26.4	24
		J	4.6 (4554)	7	—	5.7	23.2	5.0	3.6	3.6	26.4	24
1998-99	Defender 90	V	4.0 (3950)	5.6	—	—	19.2	4.8	3.6	4.8	15.6	13.5
	Discovery	V	4.0 (3950)	5.6	—	5.7	19.2	4.9	3.6	3.6	23	23.9
	Range Rover	2	4.0 (3950)	7	—	5.7	20.5	5.0	3.6	3.6	26.4	24
		J	4.6 (4554)	7	—	5.7	23.2	5.0	3.6	3.6	26.4	24

NOTE: All capacities are approximate. Add fluid gradually and check to be sure a proper fluid level is obtained.
NA - Not Available

79241CH

VALVE SPECIFICATIONS

Year	Engine ID/VIN	Engine Displacement Liters (cc)	Seat Angle (deg.)	Face Angle (deg.)	Spring Test Pressure (lbs. @ in.)	Spring Installed Height (in.)	Stem-to-Guide Clearance (in.)		Stem Diameter (in.)	
							Intake	Exhaust	Intake	Exhaust
1995	V	3.9 (3949)	46°-46°25'	46°	65@1.59	1.59	0.0010-0.0026	0.0015-0.0031	0.3411-0.3417	0.3406-0.3412
	V	4.2 (4278)	46°-46°25'	46°	65@1.59	1.59	0.0010-0.0026	0.0015-0.0031	0.3411-0.3417	0.3406-0.3412
1996	2	4.0 (3950)	46°-46°25'	46°	65@1.59	1.59	0.0010-0.0026	0.0015-0.0031	0.3411-0.3417	0.3406-0.3412
	J	4.6 (4554)	46°-46°25'	46°	65@1.59	1.59	0.0010-0.0026	0.0015-0.0031	0.3411-0.3417	0.3406-0.3412
	V	4.0 (3950)	46°-46°25'	46°	65@1.59	1.59	0.0010-0.0026	0.0015-0.0031	0.3411-0.3417	0.3406-0.3412
1997	2	4.0 (3950)	46°-46°25'	46°	65@1.59	1.59	0.0010-0.0026	0.0015-0.0031	0.3411-0.3417	0.3406-0.3412
	J	4.6 (4554)	46°-46°25'	46°	65@1.59	1.59	0.0010-0.0026	0.0015-0.0031	0.3411-0.3417	0.3406-0.3412
	V	4.0 (3950)	46°-46°25'	46°	65@1.59	1.59	0.0010-0.0026	0.0015-0.0031	0.3411-0.3417	0.3406-0.3412
1998-99	2	4.0 (3950)	46°-46°25'	46°	65@1.59	1.59	0.0010-0.0026	0.0015-0.0031	0.3411-0.3417	0.3406-0.3412
	J	4.6 (4554)	46°-46°25'	46°	65@1.59	1.59	0.0010-0.0026	0.0015-0.0031	0.3411-0.3417	0.3406-0.3412
	V	4.0 (3950)	46°-46°25'	46°	65@1.59	1.59	0.0010-0.0026	0.0015-0.0031	0.3411-0.3417	0.3406-0.3412

79241CH7

TORQUE SPECIFICATIONS
All readings in ft. lbs.

Year	Engine ID/VIN	Engine Displacement Liters (cc)	Cylinder Head Bolts	Main Bearing Bolts	Rod Bearing Bolts	Crankshaft Damper Bolts	Flywheel Bolts	Manifold		Spark Plugs	Lug Nut
								Intake	Exhaust		
1995	V	3.9 (3949)	①	52	37	200	58	②	40	12	80
	V	4.2 (4278)	①	52	37	200	58	②	40	12	80
1996	2	4.0 (3950)	③	52	37	200	58	②	40	12	80
	J	4.6 (4554)	①	52	37	200	58	②	40	14	80
	V	4.0 (3950)	③	52	37	200	58	②	40	12	80
1997	2	4.0 (3950)	③	52	37	200	58	②	40	12	80
	J	4.6 (4554)	①	52	37	200	58	②	40	14	80
	V	4.0 (3950)	③	52	37	200	58	②	40	12	80
1998-99	2	4.0 (3950)	③	52	37	200	58	②	40	12	80
	J	4.6 (4554)	①	52	37	200	58	②	40	14	80
	V	4.0 (3950)	③	52	37	200	58	②	40	12	80

① Step 1: 15 ft. lbs.
  Step 2: additional 90°
  Step 3: additional 90°
② Step 1: 84 inch lbs.
  Step 2: 38 ft. lbs.
③ Step 1: Outer row - 44 ft. lbs.
  Step 2: Center row - 67 ft. lbs.
  Step 3: Inner row - 67 ft. lbs.

79241CH8

SCHEDULED MAINTENANCE INTERVALS
(LAND ROVER DISCOVERY, DEFENDER 90 & RANGE ROVER)

TO BE SERVICED	TYPE OF SERVICE	VEHICLE MILEAGE INTERVAL (x1000)															
		7.5	15	22.5	30	37.5	45	52.5	60	67.5	75	82.5	90	97.5	105	112.5	120
Air cleaner filter	R				✔				✔				✔				✔
Brake fluid	R				✔				✔				✔				✔
Brake fluid level	S/I		✔				✔				✔				✔		
Brake lines	S/I	✔	✔	✔	✔	✔	✔	✔	✔	✔	✔	✔	✔	✔	✔	✔	✔
Brake pads, calipers & rotors	S/I	✔	✔	✔	✔	✔	✔	✔	✔	✔	✔	✔	✔	✔	✔	✔	✔
Coolant hoses	S/I	✔	✔	✔	✔	✔	✔	✔	✔	✔	✔	✔	✔	✔	✔	✔	✔
Door locks & hinges	L		✔		✔		✔		✔		✔		✔		✔		✔
Driveshafts & U-joints	L		✔		✔		✔		✔		✔		✔		✔		✔
Engine & transmission mounts	S/I						✔						✔				
Engine coolant	R				✔				✔				✔				✔
Engine oil & filter	R	✔	✔	✔	✔	✔	✔	✔	✔	✔	✔	✔	✔	✔	✔	✔	✔
Exhaust system & heat shields	S/I	✔	✔	✔	✔	✔	✔	✔	✔	✔	✔	✔	✔	✔	✔	✔	✔
Front & rear axle oil	R				✔				✔				✔				✔
Fuel filter	R								✔								✔
Fuel lines	S/I		✔		✔		✔		✔		✔		✔		✔		✔
Hood latch, safety catch & fuel door hinges	L		✔		✔		✔		✔		✔		✔		✔		✔
Oxygen sensors	R											✔					
Parking brake	S/I		✔		✔		✔		✔		✔		✔		✔		✔
Power steering fluid level	S/I		✔		✔		✔		✔		✔		✔		✔		✔
Radiator & A/C condenser	S/I		✔		✔		✔		✔		✔		✔		✔		✔
Seat belts	S/I		✔		✔		✔		✔		✔		✔		✔		✔
Serpentine drive belt	R										✔						
Serpentine drive belt	S/I				✔				✔				✔				✔
Shock absorbers	S/I		✔		✔		✔		✔		✔		✔		✔		✔
Spark plugs	R				✔				✔				✔				✔
Steering box	S/I & A						✔						✔				
Steering rods, joints & dust covers	S/I		✔		✔		✔		✔		✔		✔		✔		✔
Supplemental Restraint System (SRS) ①	S/I	Every 10 years															

79241CH9

SCHEDULED MAINTENANCE INTERVALS
(LAND ROVER DISCOVERY, DEFENDER 90 & RANGE ROVER)

TO BE SERVICED	TYPE OF SERVICE	VEHICLE MILEAGE INTERVAL (x1000)															
		7.5	15	22.5	30	37.5	45	52.5	60	67.5	75	82.5	90	97.5	105	112.5	120
Suspension links & mountings	S/I		✓		✓		✓		✓		✓		✓		✓		✓
Tires	S/I	✓	✓	✓	✓	✓	✓	✓	✓	✓	✓	✓	✓	✓	✓	✓	✓
Transfer gearbox oil	R				✓				✓				✓				✓
Transmission fluid	R				✓				✓				✓				✓
Transmission fluid filter	R				✓								✓				
Wheel speed sensor wiring	S/I		✓		✓		✓		✓		✓		✓		✓		✓
Wiper blades	S/I	✓	✓	✓	✓	✓	✓	✓	✓	✓	✓	✓	✓	✓	✓	✓	✓

① The Supplemental Restraint System (SRS) must only be serviced by a qualified automotive service technician familiar with the system, otherwise severe personal injury may be the result.

R - Replace   S/I - Inspect and service, if needed   L - Lubricate   A - Adjust   C - Clean

FREQUENT OPERATION MAINTENANCE (SEVERE SERVICE)

If a vehicle is operated under any of the following conditions it is considered severe service:

- **Towing a trailer or using a camper or car-top carrier.**
- **Repeated short trips of less than 5 miles in temperatures below freezing, or trips of less than 10 miles in any temperature.**
- **Extensive idling or low-speed driving for long distances as in heavy commercial use, such as delivery, taxi or police cars.**
- **Operating on rough, muddy or salt-covered roads, or used frequently off-road.**
- **Operating on unpaved or dusty roads.**
- **Driving in extremely hot (over 90°) conditions.**

Air cleaner filter - replace every 15,000 miles.

Brake fluid - replace every 15,000 miles.

Brake fluid level - inspect initially at 7,500 miles, then every 15,000 miles.

Brake pads, calipers & rotors - inspect every 3,750 miles.

Driveshafts & U-joints - lubricate every 7,500 miles.

Engine & transmission mounts - inspect every 22,500 miles.

Engine coolant - replace every 15,000 miles.

Engine oil & filter - replace every 3,750 miles.

Front & rear axle oil - replace every 15,000 miles.

Fuel filter - replace every 30,000 miles.

Power steering fluid level - inspect every 7,500 miles.

Serpentine drive belt - inspect every 15,000 miles, and replace every 30,000 miles.

Shock absorbers - inspect every 7,500 miles.

Spark plugs - replace every 15,000 miles.

Steering rods, joints & dust covers - inspect every 7,500 miles.

Suspension links & mountings - inspect every 7,500 miles.

Tires - inspect every 3,750 miles.

Transfer gearbox oil - replace every 15,000 miles.

Transmission fluid & filter - replace every 15,000 miles.

79241CH0

SCHEDULED MAINTENANCE
LAND ROVER
DISCOVERY, DEFENDER 90, RANGE ROVER

The following should be used as a guide when determining the amount of work required for a particular service if taken to a repair shop. In estimating how long a particular Scheduled Maintenance Service should take, please observe the following:

- **Chilton Time** is time based on field research and data supplied by the vehicle manufacturer.
- Labor time operations are given in hours and tenths of an hour.
- All labor operations, are to be used as a guide.

Mechanic Skill Level Codes:
- **(G)** GENERAL: Normally skilled with certification.
- **(M)** MAINTENANCE: Semi-skilled working on certication.
- **(P)** PRECISION: Really skilled with multiple certification.

	Chilton Time		Chilton Time		Chilton Time
(M) 7500 Mile Service		**(M) 52500 Mile Service**		**(M) 82500 Mile Service**	
1995-99	2.3	1995-99	2.6	1995-99	2.3
(G) 15000 Mile Service		**(G) 60000 Mile Service**		**(G) 90000 Mile Service**	
1995-99	4.9	1995-99	5.8	1995-99	5.9
(M) 22500 Mile Service		**(M) 67500 Mile Service**		**(M) 97500 Mile Service**	
1995-99	2.9	1995-99	2.5	1995-99	2.3
(G) 30000 Mile Service		**(G) 75000 Mile Service**		**(G) 105000 Mile Service**	
1995-99	5.8	1995-99	5.1	1995-99	5.1
(M) 37500 Mile Service				**(M) 112500 Mile Service**	
1995-99	2.3			1995-99	2.5
(M) 45000 Mile Service				**(G) 120000 Mile Service**	
1995-99	1.0			1995-99	5.8

79241CI1

LEXUS
LX450 • RX300

VEHICLE IDENTIFICATION CHART

		Engine Code					Model Year	
Code	Liters	Cu. In. (cc)	Cyl.	Fuel Sys.	Eng. Mfg.		Code	Year
1FZ-FE	4.5	273 (4477)	6	MFI/SFI	Toyota		T	1996
1MZ-FE	3.0	183 (2995)	6	SFI	Toyota		V	1997
							W	1998
							X	1999

MFI - Multi-port Fuel Injection

SFI - Sequential Fuel Injection

79241CI2

ENGINE IDENTIFICATION

Year	Model	Engine Displacement Liters (cc)	Engine Series (ID/VIN)	Fuel System	No. of Cylinders	Engine Type
1996	LX450	4.5 (4477)	1FZ-FE	MFI	6	DOHC
1997	LX450	4.5 (4477)	1FZ-FE	MFI	6	DOHC
1998-99	LX450	4.5 (4477)	1FZ-FE	SFI	6	DOHC
	RX300	3.0 (2995)	1MZ-FE	SFI	6	DOHC

MFI - Multi-port Fuel Injection

SFI - Sequential Fuel Injection

DOHC - Double Overhead Camshaft

79241CI3

GENERAL ENGINE SPECIFICATIONS

Year	Engine ID/VIN	Engine Displacement Liters (cc)	Fuel System Type	Net Horsepower @ rpm	Net Torque @ rpm (ft. lbs.)	Bore x Stroke (in.)	Com-pression Ratio	Oil Pressure @ rpm
1996	1FZ-FE	4.5 (4477)	MFI	212@4600	275@3200	3.94x3.74	9.0:1	36-71@3000
1997	1FZ-FE	4.5 (4477)	MFI	212@4600	275@3200	3.94x3.74	9.0:1	36-71@3000
1998-99	1FZ-FE	4.5 (4477)	SFI	212@4600	275@3200	3.94x3.74	9.0:1	36-71@3000
	1MZ-FE	3.0 (2995)	SFI	194@5200	209@4400	3.44x3.27	10.5:1	43-78@3000

MFI - Multi-port Fuel Injection

SFI - Sequential Fuel Injection

79241CI4

GASOLINE ENGINE TUNE-UP SPECIFICATIONS

Year	Engine ID/VIN	Engine Displacement Liters (cc)	Spark Plugs Gap (in.)	Ignition Timing (deg.) MT	Ignition Timing (deg.) AT	Fuel Pump (psi)	Idle Speed (rpm) MT	Idle Speed (rpm) AT	Valve Clearance In.	Valve Clearance Ex.
1996	1FZ-FE	4.5 (4477)	0.031	—	3B ①	38-44	—	600-700	0.006-0.010	0.010-0.014
1997	1FZ-FE	4.5 (4477)	0.031	—	3B ①	38-44	—	600-700	0.006-0.010	0.010-0.014
1998-99	1FZ-FE	4.5 (4477)	0.031	—	3B ①	38-44	—	600-700	0.006-0.010	0.010-0.014
	1MZ-FE	3.0 (2995)	0.043	—	8-12B ①	44-50	—	650-750	0.006-0.010	0.010-0.014

NOTE: The Vehicle Emission Control Information label often reflects specification changes made during production. The label figures must be used if they differ from those in this chart.

B - Before top dead center

① Terminals TE1 and E1 check connector must be connected.

79241CI5

CAPACITIES

Year	Model	Engine ID/VIN	Engine Displacement Liters (cc)	Engine Oil with Filter (qts.)	Transmission (pts.) 4-Spd	Transmission (pts.) 5-Spd	Transmission (pts.) Auto.	Transfer Case (pts.)	Drive Axle Front (pts.)	Drive Axle Rear (pts.)	Fuel Tank (gal.)	Cooling System (qts.)
1996	LX450	1FZ-FE	4.5 (4477)	7.8	—	—	4.0	3.6	①	6.8	25.1	②
1997	LX450	1FZ-FE	4.5 (4477)	7.8	—	—	4.0	3.6	①	6.8	25.1	②
1998-99	LX450	1FZ-FE	4.5 (4477)	7.8	—	—	4.0	3.6	①	6.8	25.1	②
	RX300	1MZ-FE	3.0 (2995)	5	—	—	③	—	1.9	1.9	NA	8.5

NOTE: All capacities are approximate. Add fluid gradually and check to be sure a proper fluid level is obtained.

① With differential lock: 5.6
Without differential lock: 5.8

② With rear heater: 14.2
Without rear heater: 13.2

③ U140E Transaxle
Dry Fill: 17.44 pts.
Drain and Refill: 7.4 pts.
U140F Transaxle
Dry Fill: 19.34 pts.
Drain and Refill: 8.6 pts.

79241CI6

VALVE SPECIFICATIONS

Year	Engine ID/VIN	Engine Displacement Liters (cc)	Seat Angle (deg.)	Face Angle (deg.)	Spring Test Pressure (lbs. @ in.)	Spring Installed Height (in.)	Stem-to-Guide Clearance (in.) Intake	Stem-to-Guide Clearance (in.) Exhaust	Stem Diameter (in.) Intake	Stem Diameter (in.) Exhaust
1996	1FZ-FE	4.5 (4477)	45	44.5	48.1-53.4@ 1.437	1.437	0.0010-0.0024	0.0012-0.0026	0.2744-0.2750	0.2742-0.2748
1997	1FZ-FE	4.5 (4477)	45	44.5	48.1-53.4@ 1.437	1.437	0.0010-0.0024	0.0012-0.0026	0.2744-0.2750	0.2742-0.2748
1998-99	1FZ-FE	4.5 (4477)	45	44.5	48.1-53.4@ 1.437	1.437	0.0010-0.0024	0.0012-0.0026	0.2744-0.2750	0.2742-0.2748
	1MZ-FE	3.0 (2995)	45	40.5	41.9-46.3@ 1.437	1.331	0.0010-0.0024	0.0012-0.0026	0.2154-0.2159	0.2152-0.2156

79241C

TORQUE SPECIFICATIONS
All readings in ft. lbs.

Year	Engine ID/VIN	Engine Displacement Liters (cc)	Cylinder Head Bolts	Main Bearing Bolts	Rod Bearing Bolts	Crankshaft Damper Bolts	Flywheel Bolts	Manifold Intake	Manifold Exhaust	Spark Plugs	Lug Nut
1996	1FZ-FE	4.5 (4477)	①	②	③	304	74	15	29	14	④
1997	1FZ-FE	4.5 (4477)	①	②	③	304	74	15	29	14	④
1998-99	1FZ-FE	4.5 (4477)	①	②	③	304	74	15	29	14	④
	1MZ-FE	3.0 (2995)	⑤	⑥	⑦	159	61	32	36	13	70

① Step 1: 29 ft. lbs.
   Step 2: Plus 90 degrees
   Step 3: Plus 90 degrees
② Step 1: 54 ft. lbs.
   Step 2: Plus 90 degrees
③ Step 1: 35 ft. lbs.
   Step 2: Plus 90 degrees
④ Steel wheels: 109 ft. lbs.
   Aluminum wheels: 76 ft. lbs.
⑤ Step 1: 40 ft. lbs.
   Step 2: Plus 90 degrees
   Step 3: 13 ft. lbs.
⑥ Step 1: 16 ft. lbs.
   Step 2: Plus 90 degrees
   Step 3: 20 ft. lbs.
⑦ Step 1: 18 ft. lbs.
   Step 2: Plus 90 degrees

79241CI8

SCHEDULED MAINTENANCE INTERVALS
(LEXUS LX450, LX470 & RX300)

TO BE SERVICED	TYPE OF SERVICE	VEHICLE MILEAGE INTERVAL (x1000) 7.5	15	22.5	30	37.5	45	52.5	60	67.5	75	82.5	90	97.5
Engine oil & filter	R	✓	✓	✓	✓	✓	✓	✓	✓	✓	✓	✓	✓	✓
Automatic transmission fluid & filter	S/I		✓		✓		✓		✓		✓		✓	
Ball joints & dust covers	S/I		✓		✓		✓		✓		✓		✓	
Bolts & nuts on chassis & body	S/I		✓		✓		✓		✓		✓		✓	

79241CI9

SCHEDULED MAINTENANCE INTERVALS
(LEXUS LX450, LX470 & RX300)

TO BE SERVICED	TYPE OF SERVICE	VEHICLE MILEAGE INTERVAL (x1000)												
		7.5	15	22.5	30	37.5	45	52.5	60	67.5	75	82.5	90	97.5
Brake linings & drums	S/I		✓		✓		✓		✓		✓		✓	
Brake line pipes & hoses	S/I		✓		✓		✓		✓		✓		✓	
Brake pads discs (front & rear)	S/I		✓		✓		✓		✓		✓		✓	
Propeller shaft grease	S/I		✓		✓		✓		✓		✓		✓	
Steering knuckle & chassis grease	S/I		✓		✓		✓		✓		✓		✓	
Steering linkage	S/I		✓		✓		✓		✓		✓		✓	
Transfer, differential & steering gear box oil	S/I		✓		✓		✓		✓		✓		✓	
Air cleaner filter	R				✓				✓				✓	
Front wheel bearing & thrust bush grease	R				✓				✓				✓	
Spark plugs	R				✓				✓				✓	
Drive belts	S/I				✓				✓				✓	
Exhaust pipes & mountings	S/I				✓				✓				✓	
Fuel lines & connections	S/I				✓				✓				✓	
Engine Coolant	R						✓				✓			
Charcoal canister (Calif.)	R								✓					
Fuel tank cap gasket	R								✓					
Heated oxygen sensors (except Calif.)①	R													

① Heated oxygen sensors (except Calif.) - replace every 80,000 miles.

R – Replace   S/I – Service or Inspect

FREQUENT OPERATION MAINTENANCE (SEVERE SERVICE)

If a vehicle is operated under any of the following conditions it is considered severe service:
- Extremely dusty areas.
- 50% or more of the vehicle operation is in 32°C (90°F) or higher temperatures, or constant operation in temperatures below 0°C (32°F).
- Prolonged idling (vehicle operation in stop and go traffic).
- Frequent short running periods (engine does not warm to normal operating temperatures).
- Police, taxi, delivery usage or trailer towing usage.

Air cleaner filter - service or inspect every 3750 miles.
Oil & oil filter change – change every 3750 miles.
Ball joints & dust covers - service or inspect every 7500 miles.
Bolts & nuts on body & chassis - service or inspect every 7500 miles.
Brake linings & drums - service or inspect every 7500 miles.
Brake pads & discs (front & rear) - service or inspect every 7500 miles.
Steering knuckle & chassis grease - service or inspect every 7500 miles.
Steering linkage - service or inspect every 7500 miles.
Propeller shaft grease - service or inspect every 7500 miles.
Exhaust pipes & mountings - service or inspect every 15,000 miles.

SCHEDULED MAINTENANCE
LEXUS
LX450, LX470, RX300

The following should be used as a guide when determining the amount of work required for a particular service if taken to a repair shop.
In estimating how long a particular Scheduled Maintenance Service should take, please observe the following:

- **Chilton Time** is time based on field research and data supplied by the vehicle manufacturer.
- Labor time operations are given in hours and tenths of an hour.
- All labor operations, are to be used as a guide.

Mechanic Skill Level Codes:
    (G) GENERAL: Normally skilled with certification.
    (M) MAINTENANCE: Semi-skilled working on certication.
    (P) PRECISION: Really skilled with multiple certification.

	Chilton Time		Chilton Time		Chilton Time
(M) 7500 Mile Service		**(M) 37500 Mile Service**		**(G) 75000 Mile Service**	
1995-99	.7	1995-99	.7	1995-99	5.1
(M) 15000 Mile Service		**(M) 45000 Mile Service**		**(M) 82500 Mile Service**	
1995-99	2.1	1995-99	2.1	1995-99	.7
(M) 22500 Mile Service		**(M) 52500 Mile Service**		**(G) 90000 Mile Service**	
1995-99	.7	1995-99	.7	1995-99	6.9
(G) 30000 Mile Service		**(G) 60000 Mile Service**		**(M) 97500 Mile Service**	
1995-99	6.9	1995-99	6.9	1995-99	.7
		(M) 67500 Mile Service			
		1995-99	.7		

79241CJ1

MAZDA
B2300 • B2500 • B3000 • B4000 • MPV

VEHICLE IDENTIFICATION CHART

	Engine Code					Model Year	
Code	Liters	Cu. In. (cc)	Cyl.	Fuel Sys.	Eng. Mfg.	Code	Year
A	2.3	140 (2298)	4	MFI	Ford	S	1995
C	2.5	152 (2500)	4	MFI	Ford	T	1996
JE	3.0	180 (2954)	6	MFI	Mazda	V	1997
U	3.0	182 (2968)	6	MFI	Ford	W	1998
X	4.0	245 (4016)	6	MFI	Ford	X	1999

MFI - Multi-port Fuel Injection

79241CJ2

ENGINE IDENTIFICATION

Year	Model	Engine Displacement Liters (cc)	Engine Series (ID/VIN)	Fuel System	No. of Cylinders	Engine Type
1995	B2300	2.3 (2298	A	EFI	4	SOHC
	B3000	3.0 (2968)	U	EFI	6	OHV
	B4000	4.0 (4016)	X	EFI	6	OHV
	MPV	3.0 (2954)	JE	EFI	6	SOHC
1996	B2300	2.3 (2298)	A	EFI	4	SOHC
	B3000	3.0 (2968)	U	EFI	6	OHV
	B4000	4.0 (4016)	X	EFI	6	OHV
	MPV	3.0 (2954)	JE	EFI	6	SOHC
1997	B2300	2.3 (2298)	A	EFI	4	SOHC
	B3000	3.0 (2968)	U	EFI	6	OHV
	B4000	4.0 (4016)	X	EFI	6	OHV
	MPV	3.0 (2954)	JE	EFI	6	SOHC
1998-99	B2500	2.5 (2500)	C	MFI	4	SOHC
	B3000	3.0 (2968)	U	MFI	6	OHV
	B4000	4.0 (4016)	X	MFI	6	OHV
	MPV	3.0 (2954)	JE	MFI	6	SOHC

EFI - Electronic Fuel Injection
MFI - Multi-port Fuel Injection
SOHC - Single Overhead Camshaft
OHV - Overhead Valve

79241CJ3

GENERAL ENGINE SPECIFICATIONS

Year	Engine ID/VIN	Engine Displacement Liters (cc)	Fuel System Type	Net Horsepower @ rpm	Net Torque @ rpm (ft. lbs.)	Bore x Stroke (in.)	Compression Ratio	Oil Pressure @ rpm
1995	A	2.3 (2298)	EFI	112@4800	135@2400	3.78x3.13	9.2:1	36-71@3000
	JE	3.0 (2954)	EFI	155@5000	169@4000	3.50x3.00	8.5:1	53-75@3000
	U	3.0 (2968)	EFI	145@4800	165@3000	3.50x3.14	9.3:1	36-71@3000
	X	4.0 (4016)	EFI	160@4200	220@3000	3.95x3.32	9.0:1	36-71@3000
1996	A	2.3 (2298)	EFI	112@4800	135@2400	3.78x3.13	9.2:1	40-60@2000
	JE	3.0 (2954)	EFI	155@5000	169@4000	3.54x3.05	8.5:1	53-75@3000
	U	3.0 (2968)	EFI	145@4800	165@3000	3.50x3.14	9.3:1	40-60@2500
	X	4.0 (4016)	EFI	160@4200	220@3000	3.94x3.31	9.0:1	40-60@2000
1997	A	2.3 (2298)	EFI	112@4800	135@2400	3.78x3.13	9.2:1	40-60@2000
	JE	3.0 (2954)	EFI	155@5000	169@4000	3.54x3.05	8.5:1	53-75@3000
	U	3.0 (2968)	EFI	145@4800	165@3000	3.50x3.14	9.3:1	40-60@2500
	X	4.0 (4016)	EFI	160@4200	220@3000	3.94x3.31	9.0:1	40-60@2000
1998-99	C	2.5 (2500)	MFI	119@5000	146@3000	3.78x3.40	9.4:1	40-60@2000
	JE	3.0 (2954)	MFI	155@5000	169@4000	3.54x3.05	8.5:1	53–75@3000
	U	3.0 (2968)	MFI	150@5000	185@3750	3.95x3.32	9.1:1	40–60@2500
	X	4.0 (4016)	MFI	160@4200	225@3000	3.95x3.32	9.0:1	40–60@2000

EFI - Electronic fuel injection
MFI - Multi-port Fuel Injection

79241CJ4

GASOLINE ENGINE TUNE-UP SPECIFICATIONS

Year	Engine ID/VIN	Engine Displacement Liters (cc)	Spark Plugs Gap (in.)	Ignition Timing (deg.) MT	Ignition Timing (deg.) AT	Fuel Pump (psi)	Idle Speed (rpm) MT	Idle Speed (rpm) AT	Valve Clearance In.	Valve Clearance Ex.
1995	A	2.3 (2298)	0.042-0.046	8-12B ①	8-12B ①	35-45 ②	475-575	475-575	HYD	HYD
	JE	3.0 (2954)	0.039-0.043	—	10-12B	30-37 ②	—	780-820 ③	HYD	HYD
	U	3.0 (2968)	0.042-0.046	8-12B ④	8-12B ④	35-45 ②	⑤	⑤	HYD	HYD
	X	4.0 (4016)	0.052-0.056	8-12B ①	8-12B ①	35-45 ②	⑤	⑤	HYD	HYD
1996	A	2.3 (2298)	⑥	8-12B ⑦	8-12B ⑦	35-45 ②	475-575	475-575	HYD	HYD
	JE	3.0 (2954)	0.040-0.043	—	10-12B	30-37 ②	—	780-820	HYD	HYD
	U	3.0 (2968)	⑥	8-12B ⑦	8-12B ⑦	35-45 ②	⑤	⑤	HYD	HYD
	X	4.0 (4016)	⑥	8-12B ⑦	8-12B ⑦	35-45 ②	⑤	⑤	HYD	HYD
1997	A	2.3 (2298)	⑥	8-12B ⑦	8-12B ⑦	35-45 ②	475-575	475-575	HYD	HYD
	JE	3.0 (2954)	0.040-0.043	—	10-12B	30-37 ②	—	780-820	HYD	HYD
	U	3.0 (2968)	⑥	8-12B ⑦	8-12B ⑦	35-45 ②	⑤	⑤	HYD	HYD
	X	4.0 (4016)	⑥	8-12B ⑦	8-12B ⑦	35-45 ②	⑤	⑤	HYD	HYD
1998-99	C	2.5 (2500)	0.044	10B	10B	56-72	⑧	⑧	HYD	HYD
	JE	3.0 (2954)	0.041	—	10-12B	31-38	—	⑧	HYD	HYD
	U	3.0 (2968)	0.044	10B	10B	56-72	⑧	⑧	HYD	HYD
	X	4.0 (4016)	0.054	NA	NA	56-72	⑧	⑧	HYD	HYD

NOTE: The Vehicle Emission Control Information label often reflects specification changes made during production. The label figures must be used if they differ from those in this chart.

B - Before top dead center
HYD - Hydraulic
① With "SPOUT" shorting bar disconnected
② Pressure indicated is with gauge in-line, regulator vacuum hose connected and engine idling
③ Data link connector terminal 10 grounded and transmission in park
④ With "SPOUT" shorting bar disconnected
⑤ Not adjustable
⑥ Refer to Vehicle's Emission Control Information label
⑦ Base timing, not adjustable
⑧ Automatically controlled by the Powertrain Control Module (PCM).

79241CJ5

CAPACITIES

Year	Model	Engine ID/VIN	Engine Displacement Liters (cc)	Engine Oil with Filter (qts.)	Transmission (pts.)			Transfer Case (pts.)	Drive Axle		Fuel Tank (gal.)	Cooling System (qts.)
					4-Spd	5-Spd	Auto.		Front (pts.)	Rear (pts.)		
1995	B2300	A	2.3 (2298)	5.0	—	5.6	19.4	2.5	3.5	5.0	16.3 ①	②
	B3000	U	3.0 (2968)	5.0	—	5.6	③	2.5	3.5	5.0	16.3 ①	④
	B4000	X	4.0 (4016)	5.0	—	5.6	③	2.5	3.5	5.0	16.3 ①	⑤
	MPV	JE	3.0 (2954)	5.0	—	—	18.2	3.2	3.6	3.2	⑥	10.3
1996	B2300	A	2.3 (2298)	5.0	—	5.6	19.0	2.5	⑦	5.0	16.3 ①	⑧
	B3000	U	3.0 (2968)	4.5	—	5.6	③	2.5	⑦	5.0	16.3 ①	⑨
	B4000	X	4.0 (4016)	5.0	—	5.6	③	2.5	⑦	5.0	16.3 ①	⑤
	MPV	JE	3.0 (2954)	5.0	—	—	18.2	3.2	3.6	3.2	⑥	10.3
1997	B2300	A	2.3 (2298)	5.0	—	5.6	19.0	2.5	⑦	5.0	16.3 ①	⑧
	B3000	U	3.0 (2968)	4.5	—	5.6	③	2.5	⑦	5.0	16.3 ①	⑨
	B4000	X	4.0 (4016)	5.0	—	5.6	③	2.5	⑦	5.0	16.3 ①	⑤
	MPV	JE	3.0 (2954)	5.0	—	—	18.2	3.2	3.6	3.2	⑥	10.3
1998-99	B2500	C	2.5 (2500)	4.5	5.6	9.5	—	—	5.0	⑩	⑪	⑪
	B3000	U	3.0 (2968)	4.5	5.6	9.5 ⑫	2.5	3.25	5.0	⑩	⑪	⑪
	B4000	X	4.0 (4016)	5.0	5.6	9.5 ⑫	2.5	3.25	5.0	⑩	⑪	⑪
	MPV	JE	3.0 (2954)	5.0	—	9.1	3.2	3.6	3.2	⑬	10.3	10.3

NOTE: All capacities are approximate. Add fluid gradually and ensure a proper fluid level is obtained.

① Long bed and Supercab: 19.6 gals.

② Without A/C: 6.5 qts.
With A/C: 7.2 qts.

③ 2WD: 19.4 pts.
4WD: 20.0 pts.

④ Without A/C: 9.5 qts.
With A/C: 10.2 qts.

⑤ Without A/C: 7.8 qts.
With A/C: 8.6 qts.

⑥ 2WD: 19.6 gals.
4WD: 19.8 gals.

⑦ Dana 28: 3.0 pts.
Dana 35: 3.5 pts.

⑧ Without A/C: 6.5 qts.
With A/C:

⑨ Without A/C: 9.5 qts.
With A/C: 10.2 qts.

⑩ Regular cab/short bed: 16.3
Regular cab/long bed: 19.6
Cab Plus: 20.0

⑪ 2.5L engine with manual transmission: 10.5
2.5L engine with automatic transmission: 10.2
3.0L engine with manual transmission: 15.2
3.0L engine with automatic transmission: 14.8
4.0L engine with manual transmission: 13.5
4.0L engine with automatic transmission: 13.2

⑫ Measurement given is for 4x2 vehicles
for 4x4 vehicles, add 0.3 pts.

⑬ 19.6 gallon tank on 2wd models
19.8 gallon tank on 4wd models

79241CJ6

VALVE SPECIFICATIONS

Year	Engine ID/VIN	Engine Displacement Liters (cc)	Seat Angle (deg.)	Face Angle (deg.)	Spring Test Pressure (lbs. @ in.)	Spring Installed Height (in.)	Stem-to-Guide Clearance (in.)		Stem Diameter (in.)	
							Intake	Exhaust	Intake	Exhaust
1995	A	2.3 (2299)	45	44	57-63@1.56	1.540-1.580	0.0010-0.0027	0.0015-0.0032	0.2746-0.2754	0.2736-0.2744
	JE	3.0 (2954)	45	45	①	②	0.0010-0.0023	0.0012-0.0025	0.2745-0.2750	0.3160-0.3165
	U	3.0 (2968)	45	44	65@1.58	1.736-1.650	0.0010-0.0027	0.0015-0.0032	0.3134-0.3126	0.3129-0.3121
	X	4.0 (4016)	45	44	60-68@1.59	1.641-1.729	0.0008-0.0025	0.0018-0.0035	0.3159-0.3167	0.3149-0.3156
1996	A	2.3 (2299)	45	44	57-63@1.56	1.540-1.580	0.0010-0.0027	0.0015-0.0032	0.2746-0.2754	0.2736-0.2744
	JE	3.0 (2954)	45	45	③	④	0.0010-0.0023	0.0012-0.0025	0.2745-0.2750	0.3160-0.3165
	U	3.0 (2968)	45	44	⑤	1.736-1.650	0.0010-0.0017	0.0015-0.0032	0.3134-0.3126	0.3129-0.3121
	X	4.0 (4016)	45	44	60-68@1.59	1.910 ⑥	0.0008-0.0025	0.0018-0.0035	0.3159-0.3167	0.3149-0.3156
1997	A	2.3 (2299)	45	44	57-63@1.56	1.540-1.580	0.0010-0.0027	0.0015-0.0032	0.2746-0.2754	0.2736-0.2744
	JE	3.0 (2954)	45	45	③	④	0.0010-0.0023	0.0012-0.0025	0.2745-0.2750	0.3160-0.3165
	U	3.0 (2968)	45	44	⑤	1.736-1.650	0.0010-0.0017	0.0015-0.0032	0.3134-0.3126	0.3129-0.3121
	X	4.0 (4016)	45	44	60-68@1.59	1.910 ⑥	0.0008-0.0025	0.0018-0.0035	0.3159-0.3167	0.3149-0.3156
1998-99	C	2.5 (2500)	45	44	57-63@1.56	1.540-1.580	0.0008-0.0027	0.0018-0.0037	0.2746-0.2754	0.2736-0.2744
	JE	3.0 (2954)	45	45	③	④	0.0010-0.0023	0.0012-0.0025	0.2745-0.2750	0.3160-0.3165
	U	3.0 (2968)	45	44	⑤	1.736-1.650	0.0010-0.0017	0.0015-0.0032	0.3134-0.3126	0.3129-0.3121
	X	4.0 (4016)	45	44	60-68@1.59	1.910 ⑥	0.0008-0.0025	0.0018-0.0035	0.3159-0.3167	0.3149-0.3156

NA - Not Available

① Intake:
　Inner: 21-22@1.56
　Outer: 31-33@1.73
　Exhaust:
　Inner: 33-37@1.59
　Outer: 52-58@1.77

② Intake:
　Inner: 1.555
　Outer: 1.732
　Exhaust:
　Inner: 1.594
　Outer: 1.772

③ Intake:
　Inner: 21-22@1.77
　Outer: 31-33@1.73
　Exhaust:
　Inner: 33-37@1.59
　Outer: 21-22@1.56

④ Intake:
　Inner: 1.840
　Outer: 2.004
　Exhaust:
　Inner: 2.092
　Outer: 2.296

⑤ Loaded: 180@1.16
　Unloaded: 65@1.58

⑥ Free length only

79241CJ7

TORQUE SPECIFICATIONS
All readings in ft. lbs.

Year	Engine ID/VIN	Engine Displacement Liters (cc)	Cylinder Head Bolts	Main Bearing Bolts	Rod Bearing Bolts	Crankshaft Damper Bolts	Flywheel Bolts	Manifold Intake	Manifold Exhaust	Spark Plugs	Lug Nut
1995	A	2.3 (2298)	①	②	30-36	103-133	56-64	19-28	③	7-15	100
	JE	3.0 (2954)	④	⑤	⑥	116-122	76-81	14-19	16-21	10-13	65-87
	U	3.0 (2968)	⑦	55-62	23-28	92-122	54-64	⑧	15-22	7-14	100
	X	4.0 (4016)	⑨	66-77	18-24	30-37	⑩	⑪	18	7-15	100
1996	A	2.3 (2298)	⑫	②	⑬	92-121	56-64	19-28	③	7-15	100
	JE	3.0 (2954)	④	⑤	⑥	116-122	76-81	14-19	16-21	11-16	65-87
	U	3.0 (2968)	⑭	55-62	23-28	92-122	54-64	H	15-22	7-14	100
	X	4.0 (4016)	⑨	66-77	18-24	30-37	60	⑮	18	7-15	100
1997	A	2.3 (2298)	⑫	②	⑬	92-121	56-64	19-28	③	7-15	100
	JE	3.0 (2954)	④	⑤	⑥	116-122	76-81	14-19	16-21	11-16	65-87
	U	3.0 (2968)	⑭	55-62	23-28	92-122	54-64	H	15-22	7-14	100
	X	4.0 (4016)	⑨	66-77	18-24	30-37	60	⑮	18	7-15	100
1998-99	C	2.5 (2500)	⑫	②	⑬	92-121	56-64	19-28	③	7-15	100
	JE	3.0 (2954)	④	⑤	⑥	116-122	76-81	14-19	16-21	11-16	65-87
	U	3.0 (2968)	⑭	55-62	23-28	92-122	54-64	H	15-22	7-14	100
	X	4.0 (4016)	⑨	66-77	18-24	30-37	60	⑮	18	7-15	100

① Step 1: 51 ft. lbs.
 Step 2: Plus 90-100 degrees
② Step 1: Tighten by hand until seated
 Step 2: 50-60 ft. lbs.
 Step 3: 75-85 ft. lbs.
③ Step 1: 15-22 ft. lbs.
 Step 2: 45-59 ft. lbs.
④ Step 1: 12.7-16.2 ft. lbs.
 Step 2: Turn each bolt, in order, 90 degrees
 Step 3: Repeat Step 2
⑤ Step 1: 12.7-16.2 ft. lbs.
 Step 2: Turn each bolt, in order, 90 degrees
 Step 3: Turn each bolt, in order, 45 degrees
⑥ Step 1: 20-23.5 ft. lbs.
 Step 2: Plus 90 degrees
⑦ Step 1: 33-41 ft. lbs.
 Step 2: 63-73 ft. lbs.
⑧ Step 1: 11 ft. lbs.
 Step 2: 19-24 ft. lbs.

⑨ Step 1: 22-26 ft. lbs.
 Step 2: 52-56 ft. lbs.
 Step 3: Plus 90 degrees
⑩ Step 1: 9-11 ft. lbs.
 Step 2: 50-55 ft. lbs.
⑪ Step 1: 6 ft. lbs.
 Step 2: 11 ft. lbs.
 Step 3: 16 ft. lbs.
⑬ Step 1: 25-30 ft. lbs.
 Step 2: 30-36 ft. lbs.
⑫ Step 1: 52 ft. lbs.
 Step 2: 52 ft. lbs.
 Step 3: Plus 90-100 degrees
⑭ Step 1: 59 ft. lbs.
 Step 2: Loosen all bolts 1 turn
 Step 3: 33-41 ft. lbs.
 Step 4: 63-73 ft. lbs.
⑮ Step 1: 6 ft. lbs.
 Step 2: 11 ft. lbs.

79241CJ8

SCHEDULED MAINTENANCE INTERVALS
(MAZDA MPV)

TO BE SERVICED	TYPE OF SERVICE	VEHICLE MILEAGE INTERVAL (x1000)												
		7.5	15	22.5	30	37.5	45	52.5	60	67.5	75	82.5	90	97.5
Engine oil & filter	R	✓	✓	✓	✓	✓	✓	✓	✓	✓	✓	✓	✓	✓
Air cleaner filter	R				✓				✓				✓	
Brake fluid	R				✓				✓				✓	
Spark plugs	R				✓				✓				✓	
Bolts & nuts on chassis & body	S/I				✓				✓				✓	
Cooling system	S/I				✓				✓				✓	
Disc brakes, brake lines, hoses & connections	S/I				✓				✓				✓	
Drive belt(s)	S/I				✓				✓				✓	
Drive shaft dust boots (4WD)	S/I				✓				✓				✓	
Exhaust system heat shields	S/I				✓				✓				✓	
Front suspension ball joints	S/I				✓				✓				✓	
Fuel lines & hoses	S/I				✓				✓				✓	
Idle speed	S/I		✓				✓				✓			
Steering operation & linkages	S/I				✓				✓				✓	
Engine coolant	R						✓				✓			
Timing belt (except Calif.)	R								✓					
Timing belt (Calif.)①	S/I								✓				✓	
Automatic transmission fluid & filter	R								✓					
Front & rear axle oil	R								✓					
Fuel filter & PCV valve	R								✓					
Transfer case oil (4WD)	R								✓					
Emission hoses & tubes②	S/I								✓					
Ignition timing	S/I								✓					

① Timing belt (Calif.) - replace at 105,000 miles, unless previously replaced.
② Emission hoses & tubes - replace at 80,000 miles.
R – Replace   S/I – Service or Inspect

79241CJ9

SCHEDULED MAINTENANCE INTERVALS
(MAZDA MPV) (Cont.)

FREQUENT OPERATION MAINTENANCE (SEVERE SERVICE)
   If a vehicle is operated under any of the following conditions it is considered severe service:
- Extremely dusty areas.
- 50% or more of the vehicle operation is in 32°C (90°F) or higher temperatures, or constant operation in temperatures below 0°C (32°F).
- Prolonged idling (vehicle operation in stop and go traffic).
- Frequent short running periods (engine does not warm to normal operating temperatures).
- Police, taxi, delivery usage or trailer towing usage.
Oil & oil filter - change every 5000 miles.
Air cleaner filter - service or inspect every 15,000 miles.
Bolts & nuts on chassis & body - tighten every 15,000 miles.
Spark plugs - replace every 15,000 miles.
Automatic transmission fluid & filter - replace every 30,000 miles.
Front & rear axle oil - replace every 30,000 miles.
Transfer case oil (4WD) - replace every 30,000 miles.

79241CK1

1995-97 SCHEDULED MAINTENANCE INTERVALS
(MAZDA B SERIES)

TO BE SERVICED	TYPE OF SERVICE	VEHICLE MILEAGE INTERVAL (x1000)												
		5	10	15	20	25	30	35	40	45	50	55	60	65
Engine oil & filter	R	✓	✓	✓	✓	✓	✓	✓	✓	✓	✓	✓	✓	✓
Automatic transmission shift linkage	S/I	✓		✓		✓		✓		✓		✓		✓
Clutch reservoir fluid	S/I	✓		✓		✓		✓		✓		✓		✓
Exhaust system	S/I	✓		✓		✓		✓		✓		✓		✓
Propeller shaft slip yoke (B Series)	S/I	✓		✓		✓		✓		✓		✓		✓
Propeller shaft U-joints	S/I	✓		✓		✓		✓		✓		✓		✓
Rear propeller shaft double cardan joint centering ball (B Series short bed 4x4)	S/I	✓		✓		✓		✓		✓		✓		✓
Rotate tires	S/I	✓		✓		✓		✓		✓		✓		✓
Slip yoke (Navajo) (if equipped)	S/I	✓		✓		✓		✓		✓		✓		✓
Steering linkage suspension	S/I	✓		✓		✓		✓		✓		✓		✓
Disc brake system & caliper slide rails	S/I			✓			✓			✓	✓			
Drum brake linings, lines & hoses	S/I			✓			✓			✓		✓		
Engine cooling hoses, clamps & coolant condition	S/I			✓			✓			✓			✓	
Transfer case shift lever pivot bolt & control rod connecting pins (4x4)	S/I			✓			✓			✓			✓	
Air cleaner filter	R						✓						✓	
Automatic transmission fluid & filter (B Series)	R						✓						✓	
Engine coolant (Navajo)	R						✓						✓	
Engine coolant (B Series)③	R										✓			
Fuel filter	R						✓						✓	
Accessory drive belts	S/I						✓						✓	
Front axle R.H. axle - shaft slip yoke (4x4)	S/I						✓						✓	

79241CJ0

1995-97 SCHEDULED MAINTENANCE INTERVALS
(MAZDA B SERIES) (Cont.)

TO BE SERVICED	TYPE OF SERVICE	VEHICLE MILEAGE INTERVAL (x1000)												
		5	10	15	20	25	30	35	40	45	50	55	60	65
Front wheel bearings	S/I						✓						✓	
Hub lock (4x4)	S/I						✓						✓	
Parking brake system	S/I						✓						✓	
Spindle needle bearing spindle thrust bearing (4x4)	S/I						✓						✓	
Throttle linkage & kick down cable ball studs (Navajo)	S/I						✓						✓	
Manual transmission oil	R												✓	
PCV valve	R												✓	
Spark plugs (1993-95)	R						✓						✓	
Spark plugs (1996-97 Calif.)	R												✓	
Spark plugs (1996-97) (exc. Calif.)①	R													
Timing belt (B Series 2.2L) ②	R												✓	
Timing belt (B Series 2.3L) ②	S/I													
Transfer case oil (4x4)	R												✓	

① Replace every 100,000 miles.
② Timing belt (B Series 2.3L) - service or inspect at 120,000 miles.
③ Engine coolant (B Series) - replace initially at 50,000 miles, and every 30,000 miles thereafter.
R – Replace   S/I – Service or Inspect

FREQUENT OPERATION MAINTENANCE (SEVERE SERVICE)

If a vehicle is operated under any of the following conditions it is considered severe service:
- Extremely dusty areas.
- 50% or more of the vehicle operation is in 32°C (90°F) or higher temperatures, or constant operation in temperatures below 0°C (32°F).
- Prolonged idling (vehicle operation in stop and go traffic).
- Frequent short running periods (engine does not warm to normal operating temperatures).
- Police, taxi, delivery usage or trailer towing usage.

Oil & oil filter - replace every 3000 miles.
Automatic transmission shift linkage - lubricate every 6000 miles.
Exhaust system - service or inspect every 6000 miles.
Clutch reservoir fluid - service or inspect every 6000 miles.
Propeller shaft slip yoke (B Series) - lubricate every 6000 miles.
Propeller shaft U-joints - lubricate every 6000 miles.
Rear propeller shaft double cardan joint centering ball (B Series short bed 4x4) - lubricate every 6000 miles.
Rotate tires (B Series) - rotate every 6000 miles.
Slip yoke (Navajo if equipped) - lubricate every 6000 miles.
Steering linkage suspension - lubricate every 6000 miles.
Automatic transmission fluid & filter (B Series) - replace every 21,000 miles.
Automatic transmission fluid & filter (Navajo) - replace every 24,000 miles.
Throttle linkage & kick down cable ball (Navajo) - lubricate every 24,000 miles.
Manual transmission oil - replace every 30,000 miles.
Rear axle oil - replace at 99,000 miles.

1998-99 SCHEDULED MAINTENANCE INTERVALS
(MAZDA B SERIES)

TO BE SERVICED	TYPE OF SERVICE	VEHICLE MILEAGE INTERVAL (x1000)																		
		5	10	15	20	25	30	35	40	45	50	55	60	65	70	75	80	85	90	95
Accessory drive belts	S/I												✔						✔	
Air cleaner element	R						✔						✔						✔	
Caliper slide rails	L			✔			✔			✔			✔			✔			✔	
Clutch reservoir fluid level	S/I	✔		✔		✔		✔		✔		✔		✔		✔		✔		✔
Cooling system (hoses, clamps, coolant)	S/I			✔			✔			✔			✔						✔	
Disc brake system	S/I			✔			✔			✔			✔			✔			✔	
Driveshaft slip joint (if equipped)	L	✔		✔		✔		✔		✔		✔		✔		✔		✔		✔
Driveshaft U-joints (if equipped w/grease fittings)	L	✔		✔		✔		✔		✔		✔		✔		✔		✔		✔
Drum brake system, lines & hoses	S/I			✔			✔			✔			✔			✔			✔	
Engine coolant	R										✔				✔					
Engine oil & filter	R	✔	✔	✔	✔	✔	✔	✔	✔	✔	✔	✔	✔	✔	✔	✔	✔	✔	✔	✔
Exhaust system	S/I	✔		✔		✔		✔		✔		✔		✔		✔		✔		✔
Exhaust system shielding	S/I	✔		✔		✔		✔		✔		✔		✔		✔		✔		✔
Front axle RH axle shaft slip yoke (4x4)	L						✔						✔						✔	
Front wheel bearings	S/I & L						✔						✔						✔	
Fuel filter	R						✔						✔						✔	
Manual transmission fluid	R												✔							
Parking brake system	S/I						✔						✔						✔	
PCV valve	R												✔							
Rear axle lubricant ③	R																			
Rear driveshaft double cardan joint centering ball (short bed 4x4)	L	✔		✔		✔		✔		✔		✔		✔		✔		✔		✔
Rotate tires & check air pressure	S/I	✔		✔		✔		✔		✔		✔		✔		✔		✔		✔
Spark plugs (2.5L)	R												✔							
Spark plugs (3.0L/4.0L) ②	R																			
Timing belt ①	S/I																			
Transfer case fluid (4x4)	R												✔							

79241CK3

1998-99 SCHEDULED MAINTENANCE INTERVALS
(MAZDA B SERIES)

TO BE SERVICED	TYPE OF SERVICE	VEHICLE MILEAGE INTERVAL (x1000)																		
		5	10	15	20	25	30	35	40	45	50	55	60	65	70	75	80	85	90	95
Wheel lug nut torque	S/I	✓	✓	✓	✓	✓	✓	✓	✓	✓	✓	✓	✓	✓	✓	✓	✓	✓	✓	✓

① On the 2.5L engine, the timing belt tension and condition should be inspected at the 120,000 mile mark.

② The spark plugs should be replaced every 100,000 miles.

③ The rear axle lubricant should be replaced every 100,000 miles or whenever the axle housing is submerged beneath water. Otherwise, the lube should not be checked or changed unless there is a leak or service is required.

R - Replace   S/I - Service or Inspect   L - Lubricate

FREQUENT OPERATION MAINTENANCE (SEVERE SERVICE)

If a vehicle is operated under any of the following conditions it is considered severe service:

- Towing a trailer or using a camper or car-top carrier.
- Repeated short trips of less than 5 miles in temperatures below freezing.
- Extensive idling or low-speed driving for long distances as in heavy commercial use, such as delivery, taxi or police cars.
- Operating on rough, muddy or salt-covered roads.
- Operating on unpaved or dusty roads.

Engine oil and filter - replace every 3,000 miles or 6 months, whichever occurs first.

Spark plugs - replace every 60,000 miles.

Engine coolant - replace initially at the 48,000 mile mark then every 30,000 miles, or at 48 months then every 30 months.

Engine cooling system - inspect and service every 15,000 miles or every 12 months, whichever occurs first.

Air cleaner filter - replace every 30,000 miles.

PCV valve - replace every 60,000 miles.

Camshaft drive belt tension (2.5L) - inspect and service every 120,000 miles.

Accessory drive belt (2.5L/4.0L) - inspect and service initially at the 60,000 mile mark, then every 30,000 miles.

Accessory drive belt (3.0L) - inspect and service initially at the 90,000 mile mark, then every 30,000 miles.

Wheel lug nut torque - inspect initially at the 500 mile mark, then every 6,000 miles.

Rotate tires and adjust tire pressure - perform initially at the 6,000 mile mark, then every 9,000 miles.

Clutch reservoir level - inspect every 6,000 miles.

Front wheel bearings (4x2 only) - inspect and repack every 30,000 miles.

Disc brake system - inspect every 15,000 miles.

Caliper slide rails - lubricate every 15,000 miles.

Brake lines, hoses and linings - inspect every 15,000 miles.

Exhaust system - inspect for leaks, damage or looseness every 6,000 miles.

Driveshaft U-joints (if equipped with grease fittings) - lubricate every 6,000 miles.

Parking brake system - inspect every 30,000 miles.

Rear driveshaft double cardan joint centering ball (short bed 4x4) - lubricate every 6,000 miles.

Transfer case fluid - replace every 30,000 miles.

Automatic transmission fluid - replace every 51,000 miles.

Fuel filter - replace every 51,000 miles.

Rear axle lubricant - replace every 102,000 miles.

79241CK4

SCHEDULED MAINTENANCE
MAZDA
B SERIES, MPV

The following should be used as a guide when determining the amount of work required for a particular service if taken to a repair shop.
In estimating how long a particular Scheduled Maintenance Service should take, please observe the following:

- **Chilton Time** is time based on field research and data supplied by the vehicle manufacturer.
- Labor time operations are given in hours and tenths of an hour.
- All labor operations, are to be used as a guide.

Mechanic Skill Level Codes:
- **(G)** GENERAL: Normally skilled with certification.
- **(M)** MAINTENANCE: Semi-skilled working on certicication.
- **(P)** PRECISION: Really skilled with multiple certification.

	Chilton Time
(M) 5000 Mile Service	
1995-97 B Series	1.2
w/AT add	.1
w/4WD short bed add	.1
1998-99 B Series	1.0
(M) 7500 Mile Service	
1995-99 MPV	.3
(M) 10000 Mile Service	
1995-97 B Series	.3
1998-99 B Series	.7
(M) 15000 Mile Service	
1995-97 B Series	1.8
w/AT add	.1
w/4WD add	.1
1998-99 B Series	1.3
1995-99 MPV	.5
(M) 20000 Mile Service	
1995-97 B Series	.3
1998-99 B Series	.7
(M) 22500 Mile Service	
1995-99 MPV	.3
(M) 25000 Mile Service	
1995-97 B Series	1.2
w/AT add	.1
w/4WD add	.1
1998-99 B Series	1.0
(G) 30000 Mile Service	
1995 B Series	2.3
1996-97 B Series	1.7
w/AT add	.5
w/4WD add	.6
1998-99 B Series	
MT 2WD	2.3
4WD	3.1
AT 2WD	2.9
4WD	3.8
1995-99 MPV	2.3

	Chilton Time
(M) 35000 Mile Service	
1995-97 B Series	1.2
w/AT add	.1
w/4WD add	.1
1998-99 B Series	1.0
(M) 37500 Mile Service	
1995-99 MPV	.3
(M) 40000 Mile Service	
1995-97 B Series	.3
1998-99 B Series	.7
(M) 45000 Mile Service	
1995-99 B Series	1.8
w/AT add	.1
w/4WD add	.1
1998-99 B Series	1.3
1995-99 MPV	1.0
(M) 50000 Mile Service	
1995-97 B Series	.8
1998-99 B Series	1.0
(M) 52500 Mile Service	
1995-99 MPV	.3
(M) 55000 Mile Service	
1995-97 B Series	1.2
w/AT add	.1
w/4WD add	.1
1998-99 B Series	1.0
(G) 60000 Mile Service	
1995 B Series	2.8
1996-97 B Series	2.2
w/AT add	.6
w/4WD add	.6
Renew spark plugs Calif. engs. add	.8
Renew timing belt, 2.3L add	2.3

	Chilton Time
1998-99 B Series	
MT 2WD	2.7
4WD	4.0
AT 2WD	3.0
4WD	4.1
1995-99 MPV	6.3
w/AT add	.5
w/4WD add	.6
(M) 65000 Mile Service	
1995-97 B Series	1.2
w/AT add	.1
w/4WD add	.1
1998-99 B Series	1.0
(M) 67500 Mile Service	
1995-99 MPV	.3
(M) 75000 Mile Service	
1995-99 MPV	1.0
(M) 80000 Mile Service	
1995-99 B Series	1.0
(M) 82500 Mile Service	
1995-99 MPV	.3
(M) 85000 Mile Service	
1995-99 B Series	1.0
(G) 90000 Mile Service	
1995-99 MPV	2.3
Renew timing belt (Calif.) add	3.0
w/4WD add	.1
1998-99 B Series	
MT 2WD	2.3
4WD	2.5
AT 2WD	3.0
4WD	3.7
(M) 95000 Mile Service	
1998-99 B Series	1.0
(M) 97500 Mile Service	
1995-99 MPV	.3

79241CK5

MERCEDES-BENZ
ML320

VEHICLE IDENTIFICATION CHART

		Engine Code					Model Year	
Code	Liters	Cu. In. (cc)	Cyl.	Fuel Sys.	Eng. Mfg.		Code	Year
M112	3.2	195	6	SFI	MB		W	1998
							X	1999

SFI - Sequential Fuel Injection

MB - Mercedes-Benz

79241CK6

ENGINE IDENTIFICATION

Year	Model	Engine Displacement Liters (cc)	Engine Series (ID/VIN)	Fuel System	No. of Cylinders	Engine Type
1998-99	ML320	3.2 (3199)	M112	SFI	6	SOHC

SFI - Sequential Fuel Injection

SOHC - Single Overhead Camshaft

79241CK7

GENERAL ENGINE SPECIFICATIONS

Year	Engine ID/VIN	Engine Displacement Liters (cc)	Fuel System Type	Net Horsepower @ rpm	Net Torque @ rpm (ft. lbs.)	Bore x Stroke (in.)	Compression Ratio	Oil Pressure @ rpm
1998-99	M112	3.2 (3199)	SFI	215@5600	233@3000	3.54x3.31	10.0:1	①

SFI - Sequential Fuel Injection

① 43.5@3000

　10@700

79241CK8

GASOLINE ENGINE TUNE-UP SPECIFICATIONS

Year	Engine ID/VIN	Engine Displacement Liters (cc)	Spark Plugs Gap (in.)	Ignition Timing (deg.) MT	Ignition Timing (deg.) AT	Fuel Pump (psi)	Idle Speed (rpm) MT	Idle Speed (rpm) AT	Valve Clearance In.	Valve Clearance Ex.
1998-99	M112	3.2 (3199)	0.032	—	①	55	—	700	HYD	HYD

HYD - Hydraulic
① ECM controlled

79241CK9

CAPACITIES

Year	Model	Engine ID/VIN	Engine Displacement Liters (cc)	Engine Oil with Filter	Transmission (pts.) 4-Spd	Transmission (pts.) 5-Spd	Transmission (pts.) Auto.	Transfer Case (pts.)	Drive Axle Front (pts.)	Drive Axle Rear (pts.)	Fuel Tank (gal.)	Cooling System (qts.)
1998-99	ML320	M112	3.2 (3199)	8.5	—	—	20	3.0	2.4	3.2	19.0	9

NOTE: All capacities are approximate. Add fluid gradually and check to be sure a proper fluid level is obtained.

79241CK0

TORQUE SPECIFICATIONS
All readings in ft. lbs.

Year	Engine ID/VIN	Engine Displacement Liters (cc)	Cylinder Head Bolts	Main Bearing Bolts	Rod Bearing Bolts	Crankshaft Damper Bolts	Flywheel Bolts	Manifold Intake	Manifold Exhaust	Spark Plugs	Lug Nut
1998-99	M112	3.2 (3199)	①	②	③	④	⑤	15	12	21	111

① Step 1: 15 ft. lbs.
Step 2: 37 ft. lbs.
Step 3: 65 degrees
Step 4: 65 degrees
② M8x40: 18 ft. lbs.
M8x75
Step 1: 10 ft. lbs.
Step 2: 90-100 degrees
M10x90
Step 1: 15 ft. lbs.
Step 2: 90-100 degrees
③ Step 1: 44 inch lbs.
Step 2: 18 ft. lbs
Step 3: 90 degrees
④ Step 1: 148 ft. lbs.
Step 2: 95 degrees
⑤ Step 1: 33 ft. lbs.
Step 2: 90 degrees

79241CL1

SCHEDULED MAINTENANCE INTERVALS
(MERCEDES BENZ ML320)

TO BE SERVICED	TYPE OF SERVICE	VEHICLE MILEAGE INTERVAL (x1000)															
		7.5	15	22.5	30	37.5	45	52.5	60	67.5	75	82.5	90	97.5	105	112.5	120
Accessory drive belt ①	S/I		✓		✓		✓		✓		✓		✓		✓		✓
Air filter element ②	R								✓								
Body for paint damage ③	S/I				✓				✓				✓				✓
Brake fluid ③	R				✓				✓				✓				✓
Engine coolant ④	R						✓						✓				
Engine oil & filter ①	R		✓		✓		✓		✓		✓		✓		✓		✓
Fuel filter ②	R								✓								
Spark plugs ②	R	Every 100,000 miles															
Suspension & body structure ③	S/I				✓				✓				✓				✓
Underside of vehicle ②	S/I								✓								

① Perform this at the mileage shown or once a year, whichever occurs first.

② Perform this at the mileage shown, or every 4 years, whichever occurs first.

③ Perform this at the mileage shown or every 2 years, whichever occurs first.

④ Perform this at the mileage shown or every 3 years, whichever occurs first.

R - Replace   S/I - Inspect and service, if needed   L - Lubricate   A - Adjust   C - Clean

79241CL2

SCHEDULED MAINTENANCE
MERCEDES BENZ
ML320

The following should be used as a guide when determining the amount of work required for a particular service if taken to a repair shop. In estimating how long a particular Scheduled Maintenance Service should take, please observe the following:

- **Chilton Time** is time based on field research and data supplied by the vehicle manufacturer.
- Labor time operations are given in hours and tenths of an hour.
- All labor operations, are to be used as a guide.

    Mechanic Skill Level Codes:
        (G) GENERAL: Normally skilled with certification.
        (M) MAINTENANCE: Semi-skilled working on certicication.
        (P) PRECISION: Really skilled with multiple certification.

	Chilton Time			Chilton Time			Chilton Time
(G) 15000 Mile Service 1998-99	1.2		**(G) 60000 Mile Service** 1998-99	5.8		**(G) 90000 Mile Service** 1998-99	4.1
(G) 30000 Mile Service 1998-99	2.5		**(G) 75000 Mile Service** 1998-99	1.2		**(G) 105000 Mile Service** 1998-99	1.2
(G) 45000 Mile Service 1998-99	2.7					**(G) 120000 Mile Service** 1998-99	2.6

79241CL3

MITSUBISHI
Mighty Max • Montero • Montero Sport

VEHICLE IDENTIFICATION CHART

	Engine Code						Model Year	
Code	Liters	Cu. In. (cc)	Cyl.	Fuel Sys.	Eng. Mfg.		Code	Year
G	2.4	143.4(2351)	4	MFI	Mitsubishi		S	1995
M	3.5	213.4(3479)	6	MFI	Mitsubishi		T	1996
P	3.0	181.4(2972)	6	MFI	Mitsubishi		V	1997
							W	1998
							X	1999

MFI - Multi-port Fuel Injection

79241CL4

ENGINE IDENTIFICATION

Year	Model	Engine Displacement Liters (cc)	Engine Series (ID/VIN)	Fuel System	No. of Cylinders	Engine Type
1995	Mighty Max	2.4 (2350)	G	MFI	4	SOHC
	Montero	3.0 (2972)	H	MFI	6	SOHC
	Montero	3.5 (3497)	M	MFI	6	DOHC
1996	Mighty Max	2.4 (2350)	G	MFI	4	SOHC
	Montero	3.0 (2972)	H	MFI	6	SOHC
	Montero	3.5 (3497)	M	MFI	6	DOHC
1997	Montero	3.0 (2972)	H	MFI	6	SOHC
	Montero	3.5 (3497)	M	MFI	6	DOHC
	Montero Sport	2.4 (2351)	G	MFI	4	SOHC
1998-99	Montero	3.5 (3497)	M	MFI	6	SOHC
	Montero Sport	2.4 (2351)	G	MFI	4	SOHC
	Montero Sport	3.0 (2972)	P	MFI	6	SOHC

MFI - Multi-port Fuel Injection
SOHC - Single Overhead Camshaft
DOHC - Double Overhead Camshaft

79241CL5

GENERAL ENGINE SPECIFICATIONS

Year	Engine ID/VIN	Engine Displacement Liters (cc)	Fuel System Type	Net Horsepower @ rpm	Net Torque @ rpm (ft. lbs.)	Bore x Stroke (in.)	Compression Ratio	Oil Pressure @ rpm
1995	G	2.4 (2350)	MFI	①	148@3000	3.41x3.94	9.5:1	41@2000
	H	3.0 (2972)	MFI	②	③	3.59x2.99	9.0:1	30-80@2000
	M	3.5 (3497)	MFI	214@5000	228@3000	3.66x3.38	9.5:1	30-80@2000
1996	G	2.4 (2350)	MFI	①	148@3000	3.41x3.94	9.5:1	41@2000
	H	3.0 (2972)	MFI	②	③	3.59x2.99	9.0:1	30-80@2000
	M	3.5 (3497)	MFI	214@5000	228@3000	3.66x3.38	9.5:1	30-80@2000
1997	G	2.4(2351)	MFI	134@5500	148@3000	3.41x3.94	9.5:1	41@2000
	H	3.0 (2972)	MFI	②	③	3.59x2.99	9.0:1	30-80@2000
	M	3.5 (3497)	MFI	214@5000	228@3000	3.66x3.38	9.5:1	30-80@2000
1998-99	G	2.4(2351)	MFI	134@5500	148@3000	3.41x3.94	9.5:1	41@2000
	M	3.5 (3497)	MFI	④	⑤	3.66x3.38	9.0:1	30-80@2000
	P	3.0 (2972)	MFI	173@5500	188@4500	3.59x2.99	9.0:1	30-80@2000

MFI - Multi-port Fuel Injection
① California: 138@5500
    Except California: 141@5500
② California: 168@5500
    Except California: 177@5500
③ California: 183@4500
    Except California: 188@4500
④ California: 197@5000
    Except California: 200@5000
⑤ California: 221@3500
    Except California: 225@3500

79241CL6

GASOLINE ENGINE TUNE-UP SPECIFICATIONS

Year	Engine ID/VIN	Engine Displacement Liters (cc)	Spark Plugs Gap (in.)	Ignition Timing (deg.) MT	Ignition Timing (deg.) AT	Fuel Pump (psi)	Idle Speed (rpm) MT	Idle Speed (rpm) AT	Valve Clearance In.	Valve Clearance Ex.
1995	G	2.4 (2350)	0.039-0.043	5B	5B	38	800	800	HYD	HYD
	H	3.0 (2972)	0.039-0.043	5B	5B	38	700	700	HYD	HYD
	M	3.5 (3497)	0.039-0.043	5B	5B	38	700	700	HYD	HYD
1996	G	2.4 (2350)	0.039-0.043	5B	5B	38 ①	650-850	650-850	HYD	HYD
	H	3.0 (2972)	0.039-0.043	5B	5B	38 ①	600-800	600-800	HYD	HYD
	M	3.5 (3497)	0.039-0.043	5B	5B	47-53 ②	600-800	600-800	HYD	HYD
1997	H	3.0 (2972)	0.039-0.043	5B	5B	38 ①	600-800	600-800	HYD	HYD
	M	3.5 (3497)	0.039-0.043	5B	5B	47-53 ②	600-800	600-800	HYD	HYD
	P	3.0 (2972)	0.039-0.043	5B	5B	38 ①	750	750	HYD	HYD
1998-99	G	2.4 (2351)	0.039-0.043	5B	5B	38 ①	750	750	HYD	HYD
	M	3.5 (3479)	0.039-0.043	5B	5B	38 ①	700	700	HYD	HYD
	P	3.0 (2972)	0.039-0.043	5B	5B	38 ①	750	750	HYD	HYD

NOTE: The Vehicle Emission Control Information label often reflects specification changes made during production. The label figures must be used if they differ from those in this chart.
B - Before top dead center
HYD - Hydraulic
① With vacuum hose connected
② With vacuum hose disconnected

79241CL7

CAPACITIES

Year	Model	Engine ID/VIN	Engine Displacement Liters (cc)	Engine Oil with Filter (qts.)	Transmission (pts.) 4-Spd	Transmission (pts.) 5-Spd	Transmission (pts.) Auto.	Transfer Case (pts.)	Drive Axle Front (pts.)	Drive Axle Rear (pts.)	Fuel Tank (gal.)	Cooling System (qts.)
1995	Mighty Max	G	2.4 (2350)	4.2	—	4.9	14.8	NA	NA	3.2	13.7	6.3
	Montero	H	3.0 (2972)	5.1	—	5.3	15.2	4.8	2.6	5.5	24.3	10.0
	Montero	M	3.5 (3497)	5.1	—	5.3	15.2	5.2	2.6	5.5	24.3	10.0
1996	Mighty Max	G	2.4 (2350)	4.2	—	4.9	14.8	NA	NA	3.2	①	6.3
	Montero	H	3.0 (2972)	5.0	—	5.3	15.2	4.8	2.6	5.5	24.3	10.0
	Montero	M	3.5 (3497)	5.0	—	5.3	15.2	5.2	2.5	5.5	24.3	10.0
1997	Montero	H	3.0 (2972)	5.0	—	5.3	15.2	4.8	2.6	5.5	24.3	10.0
	Montero	M	3.5 (3497)	5.0	—	5.3	15.2	5.2	2.5	5.5	24.3	10.0
	Montero Sport	G	2.4 (2351)	4.5	—	4.8	20.8	4.8	2.4	3.2	19.5	8.5
1998-99	Montero	M	3.5 (3497)	5.5	—	6.6	17.8	5.2	2.4	6.6	24.3	10.0
	Montero Sport	G	2.4 (2351)	4.5	—	4.8	20.8	4.8	2.4	3.2	19.5	8.5
	Montero Sport	P	3.0 (2972)	5.2	—	4.8	20.8	4.8	2.4	5.5	19.5	9.5

NOTE: All capacities are approximate. Add fluid gradually and ensure a proper fluid level is obtained.

NA - Not Available

① Std. body: 13.7 gals.
   Long body: 18.2 gals.

79241CL8

VALVE SPECIFICATIONS

Year	Engine ID/VIN	Engine Displacement Liters (cc)	Seat Angle (deg.)	Face Angle (deg.)	Spring Test Pressure (lbs. @ in.)	Spring Installed Height (in.)	Stem-to-Guide Clearance (in.) Intake	Stem-to-Guide Clearance (in.) Exhaust	Stem Diameter (in.) Intake	Stem Diameter (in.) Exhaust
1995	G	2.4 (2350)	44-44.5	45-45.5	73@1.59	1.59	0.0010-0.0020	0.0020-0.0035	0.315	0.311
	H	3.0 (2972)	44-44.5	45-45.5	72.5@1.59	1.59	0.0010-0.0020	0.0020-0.0035	0.315	0.311
	M	3.5 (3497)	44-44.5	45-45.5	52.9@1.49	1.49	0.0010-0.0020	0.0020-0.0035	0.260	0.256
1996	G	2.4 (2350)	44-44.5	45-45.5	73@1.59	1.59	0.0008-0.0020	0.0020-0.0035	0.315	0.311
	H	3.0 (2972)	44-44.5	45-45.5	72.5@1.59	1.59	0.0012-0.0024	0.0020-0.0035	0.315	0.311
	M	3.5 (3497)	44-44.5	45-45.5	52.9@1.49	1.49	0.0008-0.0020	0.0020-0.0035	0.260	0.256
1997	G	2.4 (2351)	44-44.5	45-45.5	60@1.74	1.740	0.0008-0.0020	0.0012-0.0027	0.240	0.230
	H	3.0 (2972)	44-44.5	45-45.5	72.5@1.59	1.59	0.0012-0.0024	0.0020-0.0035	0.315	0.311
	M	3.5 (3497)	44-44.5	45-45.5	52.9@1.49	1.49	0.0008-0.0020	0.0020-0.0035	0.260	0.256
1998-99	G	2.4 (2351)	44-44.5	45-45.5	60@1.74	1.740	0.0008-0.0020	0.0012-0.0027	0.240	0.230
	M	3.5 (3497)	44-44.5	45-45.5	60@1.74	1.740	0.0008-0.0020	0.0016-0.0028	0.236	0.236
	P	3.0 (2972)	44-44.5	45-45.5	60@1.74	1.740	0.0008-0.0020	0.0016-0.0028	0.240	0.240

79241CL9

TORQUE SPECIFICATIONS
All readings in ft. lbs.

Year	Engine ID/VIN	Engine Displacement Liters (cc)	Cylinder Head Bolts	Main Bearing Bolts	Rod Bearing Bolts	Crankshaft Damper Bolts	Flywheel Bolts	Manifold Intake *	Manifold Exhaust	Spark Plugs	Lug Nut
1995	G	2.4 (2350)	①	②	③	87	98	13	13	18	87-101
	H	3.0 (2972)	80	47	38	136	54	10	14	18	72-87
	M	3.5 (3497)	80	54	NA	134	54	10	33	18	72-87
1996	G	2.4 (2350)	①	③	③	87	98	13	④	18	87-101
	H	3.0 (2972)	80	57	38	136	54	10	14	18	72-87
	M	3.5 (3497)	80	54	25 ⑤	134	54	10	33	18	72-87
1997	G	2.4 (2351)	⑥	②	⑦	87	98	14	⑧	18	⑨
	H	3.0 (2972)	80	57	38	136	54	10	14	18	72-87
	M	3.5 (3497)	80	54	25 ⑤	134	54	10	33	18	72-87
1998-99	G	2.4 (2351)	⑥	②	⑦	87	98	14	⑧	18	⑨
	M	3.5 (3497)	80	54	38	134	54	16	22	18	72-87
	P	3.0 (2972)	80	69	37	134	55	16	33	18	⑨

* NOTE: Applies to Lower Manifold only.

① Step 1: 54 ft. lbs.
   Step 2: 14.5 ft. lbs. plus 1/4 turn
   Step 3: Plus an additional 1/4 turn

② Step 1: 18 ft. lbs.
   Step 2: Plus 1/4 turn

③ Step 1: 14.5 ft. lbs.
   Step 2: Plus 1/4 turn

④ Mighty Max: 87-101 ft. lbs.
   Montero: 72-87 ft. lbs.

⑤ Torque to value plus an additional 1/4 turn

⑥ Step 1: 58 ft. lbs.
   Step 2: 14 ft. lbs. plus 1/4 turn
   Step 3: Plus an additional 1/4 turn

⑦ Step 1: 14 ft. lbs.
   Step 2: Plus 1/4 turn

⑧ M8 fasteners: 22 ft. lbs.
   M10 fasteners: 36 ft. lbs.

⑨ Size 15 x 6 JJ: 87-101 ft. lbs.
   Size 15 x 7 JJ: 73-86 ft. lbs.

79241CL0

SCHEDULED MAINTENANCE INTERVALS
(MITSUBISHI MONTERO, MONTERO SPORT, & PICK UP)

TO BE SERVICED	TYPE OF SERVICE	VEHICLE MILEAGE INTERVAL (x1000)												
		7.5	15	22.5	30	37.5	45	52.5	60	67.5	75	82.5	90	97.5
Engine oil & filter	R	✓	✓	✓	✓	✓	✓	✓	✓	✓	✓	✓	✓	✓
Automatic transmission & transfer oil	S/I		✓		✓		✓		✓		✓		✓	
Brake hoses	S/I		✓		✓		✓		✓		✓		✓	
Disc brake pads & rotors	S/I		✓		✓		✓		✓		✓		✓	
Drive shaft boots	S/I		✓		✓		✓		✓		✓		✓	
Air cleaner filter	R				✓				✓				✓	
Automatic transmission & transfer oil (4WD)	R				✓				✓				✓	
Engine coolant	R				✓				✓				✓	
Ball joints & steering linkage seals	S/I				✓				✓				✓	
Drive belt(s)	S/I				✓				✓				✓	
Drum brake linings & wheel cylinders	S/I				✓				✓				✓	
Exhaust system	S/I				✓				✓				✓	
Front & rear axle	S/I				✓				✓				✓	
Fuel hoses	S/I				✓				✓				✓	
Manual transmission & transfer oil (4WD)	S/I				✓				✓				✓	
Propeller shaft joint	S/I				✓				✓				✓	
Spark plugs (Montero)	R				✓				✓				✓	
Spark plugs (Pick Up or Montero Sport w/platinum tip)	R								✓					
Ignition cables	R								✓					
Timing belt	R								✓					
Distributor cap & rotor	S/I								✓					
EVAP system (except EVAP canister)	S/I								✓					
Fuel system (tank, pipe line connection & fuel tank filler tube cap)	S/I								✓					

79241CM0

SCHEDULED MAINTENANCE INTERVALS
(MITSUBISHI MONTERO, MONTERO SPORT, & PICK UP) (Cont.)

TO BE SERVICED	TYPE OF SERVICE	VEHICLE MILEAGE INTERVAL (x1000)												
		7.5	15	22.5	30	37.5	45	52.5	60	67.5	75	82.5	90	97.5
EGR valve②	S/I													
EVAP canister②	S/I													
PCV system①	S/I													

① PCV system (except EVAP canister) - service or inspect at 100,000 miles.
② Replace at 100,000 miles.
R – Replace   S/I – Service or Inspect

FREQUENT OPERATION MAINTENANCE (SEVERE SERVICE)
If a vehicle is operated under any of the following conditions it is considered severe service:
- Extremely dusty areas.
- 50% or more of the vehicle operation is in 32°C (90°F) or higher temperatures, or constant operation in temperatures below 0°C (32°F).
- Prolonged idling (vehicle operation in stop and go traffic).
- Frequent short running periods (engine does not warm to normal operating temperatures).
- Police, taxi, delivery usage or trailer towing usage.
Oil & oil filter - change every 3000 miles.
Front disc brake pads (dusty or salty conditions) - service or inspect every 6000 miles.
Front disc brake pads - service or inspect every 7500 miles.
Air cleaner filter - service or inspect every 15,000 miles.
Rear drum brake linings & rear wheel cylinders - service or inspect every 15,000 miles.
Spark plugs (except platinum tip) - replace every 15,000 miles.
PCV system - service or inspect every 60,000 miles.

79241CN1

SCHEDULED MAINTENANCE
MITSUBISHI
MONTERO, MONTERO SPORT, PICK-UP

The following should be used as a guide when determining the amount of work required for a particular service if taken to a repair shop.
In estimating how long a particular Scheduled Maintenance Service should take, please observe the following:

- **Chilton Time** is time based on field research and data supplied by the vehicle manufacturer.
- Labor time operations are given in hours and tenths of an hour.
- All labor operations, are to be used as a guide.

Mechanic Skill Level Codes:
- **(G)** GENERAL: Normally skilled with certification.
- **(M)** MAINTENANCE: Semi-skilled working on certicication.
- **(P)** PRECISION: Really skilled with multiple certification.

	Chilton Time		Chilton Time		Chilton Time
(M) 7500 Mile Service		**(M) 45000 Mile Service**		**(M) 67500 Mile Service**	
1995-99	.3	1995-99	.6	1995-99	.3
(M) 15000 Mile Service		w/AT add	.1	**(M) 75000 Mile Service**	
1995-99	.6	w/4WD add	.1	1995-99	.6
w/AT add	.1	**(M) 52500 Mile Service**		w/AT add	.1
w/4WD add	.1	1995-99	.3	w/4WD add	.1
(M) 22500 Mile Service		**(G) 60000 Mile Service**		**(M) 82500 Mile Service**	
1995-99	.3	1995-99	6.1	1995-99	.3
(G) 30000 Mile Service		Renew spark plugs (Montero)		**(G) 90000 Mile Service**	
1995-99	2.1	add	.6	1995-99	2.2
Renew spark plugs (Montero)		w/AT add	.6	Renew spark plugs (Montero)	
add	.6	w/4WD add	.5	add	.6
w/AT add	.2			w/AT add	.6
w/4WD add	1.0			w/4WD add	.4
(M) 37500 Mile Service				**(M) 97500 Mile Service**	
1995-99	.3			1995-99	.3

79241CN2

NISSAN/INFINITI
Nissan Frontier • Pathfinder • Pick-Up • Quest • Infiniti QX4

VEHICLE IDENTIFICATION CHART

Engine Code							Model Year	
Code	Liters	Cu. In. (cc)	Cyl.	Fuel Sys.	Eng. Mfg.		Code	Year
KA24E	2.4	146 (2389)	4	MFI	Nissan		S	1995
KA24DE	2.4	146 (2389)	4	MFI	Nissan		T	1996
VG30E	3.0	181 (2960)	6	MFI	Nissan		V	1997
VG33E	3.3	199 (3277)	6	MFI	Nissan		W	1998
							X	1999

MFI - Multi-port Fuel Injection

79241CM1

ENGINE IDENTIFICATION

Year	Model	Engine Displacement Liters (cc)	Engine Series (ID/VIN)	Fuel System	No. of Cylinders	Engine Type
1995	Pick-up	2.4 (2389)	KA24E (S)	MFI	4	SOHC
	Pick-up	3.0 (2960)	VG30E (H)	MFI	6	SOHC
	Quest	3.0 (2960)	VG30E (W)	MFI	6	SOHC
	Pathfinder	3.0 (2960)	VG30E (H)	MFI	6	SOHC
1996	Pick-up	2.4 (2389)	KA24E (S)	MFI	4	SOHC
	Pick-up	3.0 (2960)	VG30E (H)	MFI	6	SOHC
	Quest	3.0 (2960)	VG30E (W)	MFI	6	SOHC
	Pathfinder	3.3 (3277)	VG33E	MFI	6	SOHC
1997	Pathfinder	3.3 (3277)	VG33E	MFI	6	SOHC
	Pick-up	2.4 (2389)	KA24E (S)	MFI	4	SOHC
	Pick-up	3.0 (2960)	VG30E (H)	MFI	6	SOHC
	Quest	3.0 (2960)	VG30E (W)	MFI	6	SOHC
	QX4	3.3 (3277)	VG33E	MFI	6	SOHC
1998-99	Frontier	2.4 (2389)	KA24DE	MFI	4	DOHC
	Pathfinder	3.3 (3277)	VG33E	MFI	6	SOHC
	Quest	3.0 (2960)	VG30E	MFI	6	SOHC
	QX4	3.3 (3277)	VG33E	MFI	6	SOHC

MFI - Multi-port Fuel Injection
SOHC - Single Overhead Camshaft
DOHC - Double Overhead Camshafts

79241CM2

GENERAL ENGINE SPECIFICATIONS

Year	Engine ID/VIN	Engine Displacement Liters (cc)	Fuel System Type	Net Horsepower @ rpm	Net Torque @ rpm (ft. lbs.)	Bore x Stroke (in.)	Com-pression Ratio	Oil Pressure @ rpm
1995	KA24E	2.4 (2389)	MFI	134@5200	154@3600	3.50X3.78	8.6:1	60@3000
	VG30E	3.0 (2960)	MFI	153@4800	180@4000	3.43X3.27	9.0:1	53@3200
1996	KA24E	2.4 (2389)	MFI	134@5200	154@3600	3.50x3.78	8.6:1	60@3000
	VG30E	3.0 (2960)	MFI	①	②	3.43x3.27	9.0:1	53@3200
	VG33E	3.3 (3277)	MFI	168@4800	196@2800	3.60X3.27	8.9:1	53@3200
1997	KA24E	2.4 (2389)	MFI	134@5200	154@3600	3.50x3.78	8.6:1	60@3000
	VG30E	3.0 (2960)	MFI	①	②	3.43x3.27	9.0:1	53@3200
	VG33E	3.3 (3277)	MFI	168@4800	196@2800	3.60X3.27	8.9:1	53@3200
1998-99	KA24DE	2.4 (2389)	MFI	143@5200	154@4000	3.50x3.78	9.2:1	60-70@3000
	VG30E	3.0 (2960)	MFI	151@4800	174@4400	3.43x3.27	9.0:1	57-70@3200
	VG33E	3.3 (3277)	MFI	168@4800	196@2800	3.60X3.27	8.9:1	60-65@2000

MFI - Multi-port Fuel Injection
NA - Not Available
① Quest: 151@4800
   Pick-up and Pathfinder: 153@4800
② Quest: 174@4400
   Pick-up and Pathfinder: 180@4000

79241CM3

GASOLINE ENGINE TUNE-UP SPECIFICATIONS

Year	Engine ID/VIN	Engine Displacement Liters (cc)	Spark Plugs Gap (in.)	Ignition Timing (deg.) MT	Ignition Timing (deg.) AT	Fuel Pump (psi)		Idle Speed (rpm) MT	Idle Speed (rpm) AT		Valve Clearance In.	Valve Clearance Ex.
1995	KA24E	2.4 (2389)	0.033	10B	10B	36	①	800	800	②	HYD	HYD
	VG30E	3.0 (2960)	③	15B	15B	34	①	750	750	②	HYD	HYD
1996	KA24E	2.4 (2389)	0.033	10B	10B	36	①	800	800	②	HYD	HYD
	VG30E	3.0 (2960)	③	15B	15B	34	①	700	700	②	HYD	HYD
	VG33E	3.3 (3277)	0.041	15B	15B	34	①	750	750	②	HYD	HYD
1997	KA24E	2.4 (2389)	0.033	10B	10B	36	①	800	800	②	HYD	HYD
	VG30E	3.0 (2960)	③	15B	15B	34	①	700	700	②	HYD	HYD
	VG33E	3.3 (3277)	0.041	15B	15B	34	①	750	750	②	HYD	HYD
1998-99	KA24DE	2.4 (2389)	0.039-0.043	18-22B	18-22B	34	①	750-850	750-850	②	HYD	HYD
	VG30E	3.0 (2960)	0.031-0.035	—	13-17B	34	①	—	700-800	②	HYD	HYD
	VG33E	3.3 (3277)	0.039-0.043	13-17B	13-17B	34	①	700-800	700-800	②	HYD	HYD

NOTE: The Vehicle Emission Control Information label often reflects specification changes made during production. The label figures must be used if they differ from those in this chart.
B - Before top dead center
HYD - Hydraulic
① System pressure at idle with vacuum hose connected
   Should increase to 43 psi when disconnected
② Automatic transmission in Neutral
③ Quest: 0.033
   Pick-up and Pathfinder: 0.041

79241CM4

VALVE SPECIFICATIONS

Year	Engine ID/VIN	Engine Displacement Liters (cc)	Seat Angle (deg.)	Face Angle (deg.)	Spring Test Pressure (lbs. @ in.)	Spring Installed Height (in.)	Stem-to-Guide Clearance (in.)		Stem Diameter (in.)	
							Intake	Exhaust	Intake	Exhaust
1995	KA24E	2.4 (2389)	45	45.5	①	NA	0.0008-0.0021	0.0016-0.0028	0.2742-0.2748	0.3129-0.3134
	VG30E	3.0 (2960)	45	45.25-45.75	②	NA	0.0008-0.0021	0.0016-0.0029	0.2742-0.2748	0.3136-0.3138
1996	KA24E	2.4 (2389)	45	45.5	①	NA	0.0008-0.0021	0.0016-0.0029	0.2742-0.2748	0.3129-0.3134
	VG30E	3.0 (2960)	45	45.25-45.75	②	NA	0.0008-0.0021	0.0016-0.0029	0.2742-0.2748	0.3136-0.3138
	VG33E	3.3 (3277)	45	45.25-46.75	②	NA	0.0008-0.0021	0.0016-0.0029	0.2742-0.2748	0.3136-0.3138
1997	KA24E	2.4 (2389)	45	45.5	①	NA	0.0008-0.0021	0.0016-0.0029	0.2742-0.2748	0.3129-0.3134
	VG30E	3.0 (2960)	45	45.25-45.75	②	NA	0.0008-0.0021	0.0016-0.0029	0.2742-0.2748	0.3136-0.3138
	VG33E	3.3 (3277)	45	45.25-46.75	②	NA	0.0008-0.0021	0.0016-0.0029	0.2742-0.2748	0.3136-0.3138
1998-99	KA24DE	2.4 (2389)	45	45.5	93.9@1.15	NA	0.0008-0.0021	0.0016-0.0029	0.2742-0.2748	0.2734-0.2740
	VG30E	3.0 (2960)	45	45.25-45.75	②	NA	0.0008-0.0021	0.0016-0.0029	0.2742-0.2748	0.3136-0.3138
	VG33E	3.3 (3277)	45	45.25-46.75	②	NA	0.0008-0.0021	0.0016-0.0029	0.2742-0.2748	0.3135-0.3138

NA - Not Available
① Intake:
  Inner: 63.9 @ 1.28
  Outer: 135.2 @ 1.48
  Exhaust:
  Inner: 74 @ 1.15
  Outer: 144 @ 1.34
② Inner: 57.3 @ 0.984
  Outer: 117.7 @ 1.181

79241CM

TORQUE SPECIFICATIONS
All readings in ft. lbs.

Year	Engine ID/VIN	Engine Displacement Liters (cc)	Cylinder Head Bolts	Main Bearing Bolts	Rod Bearing Bolts	Crankshaft Damper Bolts	Flywheel Bolts	Manifold		Spark Plugs	Lug Nut
								Intake	Exhaust		
1995	KA24E	2.4 (2389)	①	34-38	②	87-116	③	14	14	18	④
	VG30E	3.0 (2960)	⑤	67-74	②	90-98	72-80	⑥	15	18	④
1996	KA24E	2.4 (2389)	①	34-38	②	87-116	③	14	14	18	④
	VG30E	3.0 (2960)	⑤	67-74	②	90-98	72-80	⑥	15	18	④
	VG33E	3.3 (3277)	⑤	67-74	②	141-156	61-69	⑥	21-25	18	④
1997	KA24E	2.4 (2389)	①	34-38	②	87-116	③	14	14	18	④
	VG30E	3.0 (2960)	⑤	67-74	②	90-98	72-80	⑥	15	18	④
	VG33E	3.3 (3277)	⑤	67-74	②	141-156	61-69	⑥	21-25	18	④
1998-99	KA24DE	2.4 (2389)	①	34-41	②	105-112	105-112	⑥	27-35	14-22	87-108
	VG30E	3.0 (2960)	⑤	67-74	②	141-156	61-69	⑥	13-16	14-22	72-87
	VG33E	3.3 (3277)	⑤	67-74	②	141-156	61-69	⑥	21-25	14-22	87-108

① Step 1: 22 ft. lbs.
　Step 2: 58 ft. lbs.
　Step 3: Loosen completely then retorque to 22 ft. lbs.
　Step 4: 58 ft. lbs. or an additional 80-85 degrees
② 10-12 ft. lbs. plus 60-65 degrees or 28-33 ft. lbs.
③ Manual transmission: 105-112 ft. lbs.
　Automatic transmission: 69-76 ft. lbs.
④ Quest: 80 ft. lbs.
　Pick-up with single wheels: 87-108 ft. lbs.
　Pick-up with dual wheels: 166-203 ft. lbs.
⑤ Step 1: 22 ft. lbs.
　Step 2: 43 ft. lbs.
　Step 3: Loosen completely then retorque to 22 ft. lbs.
　Step 4: 40-47 ft. lbs. or an additional 60-65 degrees
⑥ Step 1: Tighten nuts and bolts to 3 ft. lbs.
　Step 2: Tighten bolts to 12-14 ft. lbs.; nuts to 17-20 ft. lbs.
　Step 3: Repeat Step 2

79241CM6

SCHEDULED MAINTENANCE INTERVALS
(NISSAN FRONTIER, PATHFINDER, PICK UP & QUEST)

TO BE SERVICED	TYPE OF SERVICE	VEHICLE MILEAGE INTERVAL (x1000)												
		7.5	15	22.5	30	37.5	45	52.5	60	67.5	75	82.5	90	97.5
Engine oil & filter	R	✓	✓	✓	✓	✓	✓	✓	✓	✓	✓	✓	✓	✓
Brake lines & cables	S/I		✓		✓		✓		✓		✓		✓	
Brake pads, discs, drums & linings	S/I		✓		✓		✓		✓		✓		✓	
Drive shaft boots (Quest & 1993 Pathfinder & Pick Up)	S/I		✓		✓		✓		✓		✓		✓	
Drive shaft boots & propeller shaft (1994-97 4x4 Pathfinder & Pick Up)	S/I		✓		✓		✓		✓		✓		✓	
Front wheel bearings (4x2 Pathfinder & Pick Up)	S/I				✓				✓				✓	
Automatic transaxle oil (Quest)	S/I		✓		✓		✓		✓		✓		✓	
Automatic & manual transmission, transfer & differential gear oil (Pathfinder & Pick Up)②	S/I		✓		✓		✓		✓		✓		✓	
Front wheel bearings (1996-97 4x4 Pathfinder)	S/I		✓		✓		✓		✓		✓		✓	
Front wheel bearings & free running hubs (4x4 Pick Up & 1993-95 4x4 Pathfinder)	S/I		✓		✓		✓		✓		✓		✓	
Propeller shaft (1996-97 Pathfinder)	S/I		✓		✓		✓		✓		✓		✓	
Air cleaner filter	R				✓				✓				✓	
Engine coolant (Quest)	R				✓				✓				✓	
Engine coolant (Pathfinder & Pick Up)	R								✓					
PCV filter (Pathfinder & Pick Up KA24E)	R				✓				✓				✓	
Spark plugs	R				✓				✓				✓	
Drive belt(s) (Pathfinder & Pick Up)	S/I				✓				✓				✓	
Exhaust system	S/I				✓				✓				✓	

79241CM7

SCHEDULED MAINTENANCE INTERVALS
(NISSAN FRONTIER, PATHFINDER, PICK UP & QUEST) (Cont.)

TO BE SERVICED	TYPE OF SERVICE	VEHICLE MILEAGE INTERVAL (x1000)												
		7.5	15	22.5	30	37.5	45	52.5	60	67.5	75	82.5	90	97.5
Drive belt(s) (Quest)	S/I								✓		✓		✓	
Fuel lines	S/I				✓				✓				✓	
Steering gear (box) & linkage, (steering damper-4x4), axle & suspension parts (Pathfinder & Pick Up)	S/I				✓				✓				✓	
Steering gear linkage, axle & suspension parts (Quest)	S/I				✓				✓				✓	
Vapor lines	S/I				✓				✓				✓	
Steering linkage ball joints & front suspension ball joints (Pathfinder & Pick Up)	S/I								✓					
Timing belt (1993 VG30E)	R								✓					
Timing belt (1994-97)①	R													

① Timing belt - replace at 105,000 miles.
② Differential (w/limited-slip differential) oil - replace oil every 30,000 miles.
R – Replace   S/I – Service or Inspect

FREQUENT OPERATION MAINTENANCE (SEVERE SERVICE)

If a vehicle is operated under any of the following conditions it is considered severe service:
- Extremely dusty areas.
- 50% or more of the vehicle operation is in 32°C (90°F) or higher temperatures, or constant operation in temperatures below 0°C (32°F).
- Prolonged idling (vehicle operation in stop and go traffic).
- Frequent short running periods (engine does not warm to normal operating temperatures).
- Police, taxi, delivery usage or trailer towing usage.
Oil & oil filter change – change every 3750 miles.
Brake pads, discs, drums & linings - service or inspect every 7500 miles.
Drive shaft boots (Quest) - service or inspect every 7500 miles.
Drive shaft boots & propeller shaft (Pathfinder & Pick Up) - service or inspect every 7500 miles.
Exhaust system - service or inspect every 7500 miles.
Propeller shaft (1996-97 Pathfinder) - service or inspect every 7500 miles. (if immersed in water, grease daily.)
Steering gear (box) & linkage, (steering damper-4x4), axle & suspension parts (Pathfinder & Pick Up) - service or inspect every 7500 miles.
Steering gear linkage, axle & suspension parts (Quest) - service or inspect every 7500 miles.
Steering linkage ball joints & front suspension ball joints - service or inspect every 7500 miles.

79241CM8

SCHEDULED MAINTENANCE
NISSAN
PATHFINDER, FRONTIER, PICK-UP, QUEST

The following should be used as a guide when determining the amount of work required for a particular service if taken to a repair shop. In estimating how long a particular Scheduled Maintenance Service should take, please observe the following:

- **Chilton Time** is time based on field research and data supplied by the vehicle manufacturer.
- Labor time operations are given in hours and tenths of an hour.
- All labor operations, are to be used as a guide.

Mechanic Skill Level Codes:
- **(G) GENERAL:** Normally skilled with certification.
- **(M) MAINTENANCE:** Semi-skilled working on certication.
- **(P) PRECISION:** Really skilled with multiple certification.

	Chilton Time
(M) 7500 Mile Service	
1995-99	.5
(M) 15000 Mile Service	
1995-99 Pick-up, Pathfinder Frontier	.9
w/4WD add	.3
1995-99 Quest	.9
(M) 22500 Mile Service	
1995-99	.5
(G) 30000 Mile Service	
1995-99 Pick-up, Pathfinder Frontier	2.9
w/4WD add	.5
1995-99 Quest	3.1
w/KA24E engine add	.2
(M) 37500 Mile Service	
1994-99	.5

	Chilton Time
(M) 45000 Mile Service	
1995-99 Pick-up, Pathfinder Frontier	.9
w/4WD add	.3
1995-99 Quest	.9
(M) 52500 Mile Service	
1995-99	.5
(G) 60000 Mile Service	
1995-99 Pick-up, Pathfinder Frontier	3.7
w/4WD add	.5
1995-99 Quest	3.2
w/KA24E engine add	.2

	Chilton Time
(M) 67500 Mile Service	
1995-99	
(M) 75000 Mile Service	
1995-99 Pick-up, Pathfinder Frontier	
w/4WD add	
1995-99 Quest	1.
(M) 82500 Mile Service	
1995-99	.
(G) 90000 Mile Service	
1995-99 Pick-up, Pathfinder Frontier	3.
w/4WD add	
1995-99 Quest	3.
(M) 97500 Mile Service	
1995-99	

792410

SUBARU
Forester

VEHICLE IDENTIFICATION CHART

| | | Engine Code | | | | Model Year | |
Code	Liters	Cu. In. (cc)	Cyl.	Fuel Sys.	Eng. Mfg.	Code	Year
6	2.5	150 (2457)	4	MFI	Subaru	W	1998
						X	1999

MFI - Multiport Fuel Injection

79241CN3

ENGINE IDENTIFICATION

Year	Model	Engine Displacement Liters (cc)	Engine Series (ID/VIN)	Fuel System	No. of Cylinders	Engine Type
1998-99	Forester	2.5 (2457)	6	MFI	4	DOHC 16V

MFI - Multiport Fuel Injection
DOHC - Double Overhead Camshafts
16V - 16 Valve cylinder heads

79241CN4

GENERAL ENGINE SPECIFICATIONS

Year	Engine ID/VIN	Engine Displacement Liters (cc)	Fuel System Type	Net Horsepower @ rpm	Net Torque @ rpm (ft. lbs.)	Bore x Stroke (in.)	Compression Ratio	Oil Pressure @ rpm
1998-99	6	2.5 (2457)	MFI	165@5600	162@4000	3.92x3.11	9.7:1	14 psi @ 800

MFI - Multi-port Fuel Injection

79241CN5

GASOLINE ENGINE TUNE-UP SPECIFICATIONS

Year	Engine ID/VIN	Engine Displacement Liters (cc)	Spark Plugs Gap (in.)	Ignition Timing (deg.) ① MT	Ignition Timing (deg.) ① AT	Fuel Pump (psi)	Idle Speed (rpm) ② MT	Idle Speed (rpm) ② AT	Valve Clearance ③ In.	Valve Clearance ③ Ex.
1998-99	6	2.5 (2457)	0.039-0.043	7-23° BTDC	7-23° BTDC	34-38	600-800	600-800	0.0071-0.0087	0.0090-0.0106

BTDC - Before Top Dead Center.
  ① At idle speed.
  ② With engine under no load.
  ③ With engine cold.

79241CN6

CAPACITIES

Year	Model	Engine ID/VIN	Engine Displacement Liters (cc)	Engine Oil with Filter (qts.)	Transmission (pts.) 4-Spd	Transmission (pts.) 5-Spd	Transmission (pts.) Auto.	Transfer Case (pts.)	Drive Axle Front (pts.)	Drive Axle Rear (pts.)	Fuel Tank (gal.)	Cooling System (qts.)
1998-99	Forester	6	2.5 (2457)	4.7	—	7.4	20	—	2.6 ①	1.6	15.9	6.3

  ① A/T differential only.

79241CN7

VALVE SPECIFICATIONS

Year	Engine ID/VIN	Engine Displacement Liters (cc)	Seat Angle (deg.)	Face Angle (deg.)	Spring Test Pressure (lbs. @ in.)	Spring Installed Height (in.)	Stem-to-Guide Clearance (in.) Intake	Stem-to-Guide Clearance (in.) Exhaust	Stem Diameter (in.) Intake	Stem Diameter (in.) Exhaust
1998-99	6	2.5 (2457)	①	①	33-38@ 1.654 ②	③	0.0014-0.0024 ④	0.0016-0.0026 ④	0.2343-0.2348	0.2343-0.2348

  ① Refacing angle: 90°
  ② Wear limit: 0.0059 in.
  ③ 102-118 lbs. @ 1.315 in.
  ④ Free length: 1.8913 in.

79241CN8

TORQUE SPECIFICATIONS
All readings in ft. lbs.

Year	Engine ID/VIN	Engine Displacement Liters (cc)	Cylinder Head Bolts	Main Bearing Bolts	Rod Bearing Bolts	Crankshaft Damper Bolts	Flywheel Bolts	Manifold		Spark Plugs	Lug Nut
								Intake	Exhaust		
1998-99	6	2.5 (2457)	①	②	31-34	123-137	51-55	14-17	19-26 ③	13-17	58-72

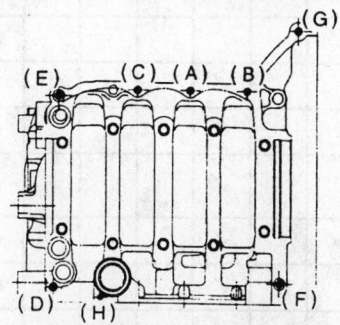

8 mm case bolt torque sequence

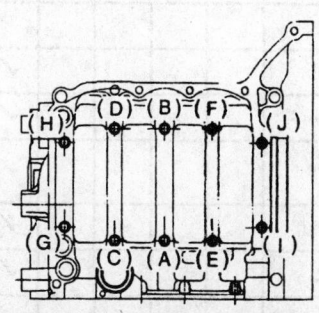

10 mm case bolt torque sequence

① Step 1: Oil all bolt threads.
Step 2: Tighten all bolts, in sequence, to 22 ft. lbs.
Step 3: Tighten all bolts, in sequence, to 51 ft. lbs.
Step 4: Loosen all bolts 180° (one-half turn).
Step 5: Loosen all bolts another 180° (one-half turn).
Step 6: Tighetn bolts A and B, in sequence, to 25 ft. lbs.
Step 7: Tighten bolts C, D, E and F, in sequence, to 11 ft. lbs.
Step 8: Tighten all bolts, in sequence, 80-90°.
Step 9: Tighten all bolts, in sequence, another 80-90°.
(Refer to the cylinder head procedure for sequences and bolt locations.)

② Split Engine Case Bolts:
10mm bolts: 33-37 ft. lbs.
8mm bolts: A thru G, 17-20 ft. lbs.; H, 5 ft. lbs.
③ No separate exhaust manifold is used, the front pipe bolts directly to the cylinder heads.

79241CN9

SCHEDULED MAINTENANCE INTERVALS
(SUBARU FORESTER)

TO BE SERVICED	TYPE OF SERVICE	VEHICLE MILEAGE INTERVAL (x1000) OR MONTHS																
		3	7.5	15	22.5	30	37.5	45	52.5	60	67.5	75	82.5	90	97.5	105	112.5	120
Accessory drive belts	R									✓								✓
Accessory drive belts	S/I					✓								✓				
Air cleaner filter	R					✓				✓				✓				✓
Automatic transmission fluid	S/I					✓				✓				✓				✓
Axle shaft joints	S/I			✓		✓		✓		✓		✓		✓		✓		✓
Brake fluid	R					✓				✓				✓				✓
Brake system lines	S/I			✓		✓		✓		✓		✓		✓		✓		✓
Clutch operation	S/I			✓		✓		✓		✓		✓		✓		✓		✓
Disc brake pads & rotors	S/I			✓		✓		✓		✓		✓		✓		✓		✓
Drums brake linings & drums	S/I					✓				✓				✓				✓
Engine coolant	R					✓				✓				✓				✓
Engine cooling system, hoses & connections	S/I					✓				✓				✓				✓
Engine oil & filter	R	✓	✓	✓	✓	✓	✓	✓	✓	✓	✓	✓	✓	✓	✓	✓	✓	✓
Front & rear axle boots	S/I			✓		✓		✓		✓		✓		✓		✓		✓
Front & rear wheel bearings	S/I & L									✓								✓
Fuel filter	R					✓				✓				✓				✓
Fuel system, hoses & connections	S/I					✓				✓				✓				✓
Parking & service brake systems' operation	S/I			✓		✓		✓		✓		✓		✓		✓		✓
Spark plugs	R									✓								✓
Steering & suspension	S/I			✓		✓		✓		✓		✓		✓		✓		✓
Supplemental Restraint System (SRS) ①	S/I	Every 10 years																
Timing belt	R															✓		
Timing belt	S/I					✓				✓				✓				
Transmission & differential fluid levels	S/I					✓				✓				✓				✓
Valve clearance	S/I															✓		

① The Supplemental Restraint System (SRS) must only be serviced by a qualified automotive technician familiar with the system, otherwise severe personal injury may be the result.

R - Replace   S/I - Inspect and service, if needed   L - Lubricate   A - Adjust   C - Clean

79241C

SCHEDULED MAINTENANCE INTERVALS
(SUBARU FORESTER)

FREQUENT OPERATION MAINTENANCE (SEVERE SERVICE)

If a vehicle is operated under any of the following conditions it is considered severe service:

Towing a trailer or using a camper or car-top carrier.

Repeated short trips of less than 5 miles in temperatures below freezing, or trips of less than 10 miles in any temperature.

Extensive idling or low-speed driving for long distances as in heavy commercial use, such as delivery, taxi or police cars.

Operating on rough, muddy or salt-covered roads, or extensive mountain driving.

Operating on unpaved or dusty roads.

Driving in extremely hot (over 90°) conditions.

Engine oil & filter - replace every 3,000 miles or 3 months, whichever occurs first.

Fuel filter - replace every 7,500 miles or 7.5 months, whichever occurs first.

Fuel system, hoses & connections - inspect every 7,500 miles or 7.5 months, whichever occurs first.

Transmission & differential fluid - replace every 15,000 miles.

Automatic transmission fluid - replace every 15,000 miles.

Brake fluid - replace every 15,000 miles.

Disc brake pads & rotors - inspect every 7,500 miles or 7.5 months, whichever occurs first.

Front & rear axle boots - inspect every 7,500 miles or 7.5 months, whichever occurs first.

Axle shaft boots - inspect every 7,500 miles or 7.5 months, whichever occurs first.

Drum brake linings & drums - inspect every 7,500 miles or 7.5 months, whichever occurs first.

Brake lines - inspect every 7,500 miles or 7.5 months, whichever occurs first.

Parking & service brake system operation - inspect every 7,500 miles or 7.5 months, whichever occurs first.

Clutch operation - inspect every 7,500 miles or 7.5 months, whichever occurs first.

Steering.& suspension - inspect every 7,500 miles or 7.5 months, whichever occurs first.

Valve clearance - inspect every 7,500 miles or 7.5 months, whichever occurs first.

79241C01

SCHEDULED MAINTENANCE
SUBARU
FORESTER

The following should be used as a guide when determining the amount of work required for a particular service if taken to a repair shop. In estimating how long a particular Scheduled Maintenance Service should take, please observe the following:

- **Chilton Time** is time based on field research and data supplied by the vehicle manufacturer.
- Labor time operations are given in hours and tenths of an hour.
- All labor operations, are to be used as a guide.

Mechanic Skill Level Codes:
        (G) GENERAL: Normally skilled with certification.
        (M) MAINTENANCE: Semi-skilled working on certicication.
        (P) PRECISION: Really skilled with multiple certification.

	Chilton Time			Chilton Time			Chil Tim
(M) 3000 Mile Service		**(M) 45000 Mile Service**		**(G) 90000 Mile Service**			
1997-99	.4	1997-99	.7	1997-99		2	
(M) 7500 Mile Service		**(M) 52500 Mile Service**		Replace timing belt add		3	
1997-99	.4	1997-99	.4	**(M) 97500 Mile Service**			
(M) 15000 Mile Service		**(G) 60000 Mile Service**		1997-99			
1997-99	.7	1997-99	8.0	**(M) 10500 Mile Service**			
(M) 22500 Mile Service		**(M) 67500 Mile Service**		1997-99			
1997-99	.4	1997-99	.4	**(M) 112500 Mile Service**			
(G) 30000 Mile Service		**(M) 75000 Mile Service**		1997-99			
1997-99	2.6	1997-99	.7	**(G) 120000 Mile Service**			
(M) 37500 Mile Service		**(M) 82500 Mile Service**		1997-99		8	
1997-99	.4	1997-99	.4				

79241

SUZUKI
Samurai • Sidekick • Sidekick Sport • X90

VEHICLE IDENTIFICATION CHART

Engine Code						Model Year	
Code	Liters	Cu. In. (cc)	Cyl.	Fuel Sys.	Eng. Mfg.	Code	Year
0	1.6 (1590)	97	4	TFI	Suzuki	S	1995
0	1.6 (1590)	97	4	MFI	Suzuki	T	1996
2	1.8 (1843)	112.5	4	MFI	Suzuki	V	1997
3	1.3 (1298)	79.2	4	TFI	Suzuki	W	1998

79241C03

ENGINE IDENTIFICATION

Year	Model	Engine Displacement Liters (cc)	Engine Series (ID/VIN)	Fuel System	No. of Cylinders	Engine Type
1995	Samurai	1.3 (1298)	3	TFI	4	8V-SOHC
	Sidekick	1.6 (1590)	0	TFI	4	8V-SOHC
		1.6 (1590)	0	MFI	4	16V-SOHC
1996	X90	1.6 (1590)	0	MFI	4	16V-SOHC
	Sidekick	1.6 (1590)	0	MFI	4	16V-SOHC
	Sport	1.8 (1843)	2	MFI	4	16V-DOHC
1997	X90	1.6 (1590)	0	MFI	4	16V-SOHC
	Sidekick	1.6 (1590)	0	MFI	4	16V-SOHC
	Sport	1.8 (1843)	2	MFI	4	16V-DOHC
1998	X90	1.6 (1590)	0	MFI	4	16V-SOHC
	Sidekick	1.6 (1590)	0	MFI	4	16V-SOHC
	Sport	1.8 (1843)	2	MFI	4	16V-DOHC

TFI – Throttle-body Fuel Injection
MFI – Multi-port Fuel Injection
8V-SOHC – 8 Valve Single Overhead Camshaft
16V-SOHC – 16 Valve Single Overhead Camshaft
16V-DOHC – 16 Valve Double Overhead Camshafts

79241C04

GENERAL ENGINE SPECIFICATIONS

Year	Engine ID/VIN	Engine Displacement Liters (cc)	Fuel System Type	Net Horsepower @ rpm	Net Torque @ rpm (ft. lbs.)	Bore x Stroke (in.)	Compression Ratio	Oil Pressure @ rpm
1995	3 ①	1.3 (1298)	TFI	66@6000	76@3500	2.91x2.97	9.5:1	43-60@3000
	0	1.6 (1590)	TFI	80@5400	94@3000	2.95x3.54	8.9:1	51-63@3000
	0 ②	1.6 (1590)	MFI	95@5600	98@4000	2.95x3.54	9.5:1	47-61@3000
1996	0 ③	1.6 (1590)	MFI	95@5600	98@4000	2.95x3.54	9.5:1	47-61@3000
	2	1.8 (1843)	MFI	120@6500	114@3500	3.31x3.27	9.8:1	55-67@4000
1997	0 ③	1.6 (1590)	MFI	95@5600	98@4000	2.95x3.54	9.5:1	47-61@3000
	2	1.8 (1843)	MFI	120@6500	114@3500	3.31x3.27	9.8:1	55-67@4000
1998	0 ③	1.6 (1590)	MFI	95@5600	98@4000	2.95x3.54	9.5:1	47-61@3000
	2	1.8 (1843)	MFI	120@6500	114@3500	3.31x3.27	9.8:1	55-67@4000

MFI - Multi-port Fuel Injection
TFI - Throttle body Fuel Injection
① Samurai
② Sidekick 16 valve engine
③ X90 and Sidekick

79241C05

GASOLINE ENGINE TUNE-UP SPECIFICATIONS

Year	Model	Engine ID/VIN	Engine Displacement Liters (cc)	Spark Plugs Gap (in.)	Ignition Timing (deg.) MT	Ignition Timing (deg.) AT	Fuel Pump (psi)	Idle Speed (rpm) MT	Idle Speed (rpm) AT	Valve Clearance In.		Valve Clearance Ex.	
1995	Samurai	3	1.3 (1298)	0.029	8B	—	34-40	800	—	0.0051-0.0067	①	0.0063-0.0079	①
	Sidekick ②	0	1.6 (1590)	0.029	8B	8B	34-40	800	800	0.005-0.007	③	0.0050-0.0070	③
	④	0	1.6 (1590)	0.029	5B	5B	34-40	800	800	0.005-0.007	①	0.006-0.008	⑤
1996	X90	0	1.6 (1590)	0.029	5B	5B	28-37	800	800	0.0050-0.0070	⑥	0.0050-0.0070	⑥
	Sidekick	0	1.6 (1590)	0.029	5B	5B	28-37	800	800	0.0050-0.0070	⑥	0.0050-0.0070	⑥
	Sidekick Sport	2	1.8 (1843)	0.029	5B	5B	31-37	750-800	750-800	HYD.		HYD.	
1997	X90	0	1.6 (1590)	0.029	5B	5B	28-37	800	800	0.0050-0.0070	⑥	0.0050-0.0070	⑥
	Sidekick	0	1.6 (1590)	0.029	5B	5B	28-37	800	800	0.0050-0.0070	⑥	0.0050-0.0070	⑥
	Sidekick Sport	2	1.8 (1843)	0.029	5B	5B	31-37	750-800	750-800	HYD.		HYD.	
1998	X90	0	1.6 (1590)	0.029	5B	5B	28-37	800	800	0.0050-0.0070	⑥	0.0050-0.0070	⑥
	Sidekick	0	1.6 (1590)	0.029	5B	5B	28-37	800	800	0.0050-0.0070	⑥	0.0050-0.0070	⑥
	Sidekick Sport	2	1.8 (1843)	0.029	5B	5B	31-37	750-800	750-800	HYD.		HYD.	

① 2-door model / 8 valve engine
② 4-door model / 16 valve engine
③ Cold engine specifications.
④ California and New York models.
⑤ Cold engine specifications. Hot engine specifications are:
  Intake: 0.0090-0.0110 in.
  Exhaust: 0.0102-0.0118 in.
⑥ Cold engine specifications. Hot engine specifications are 0.010-0.012 in.

79241CO

CAPACITIES

Year	Model	Engine ID/VIN	Engine Displacement Liters (cc)	Engine Oil with Filter (qts.)	Transmission (pts.) 5-Spd	Transmission (pts.) Auto.	Transfer Case (pts.)	Drive Axle Front (pts.)	Drive Axle Rear (pts.)	Fuel Tank (gal.)	Cooling System (qts.)
1995	Samurai	3	1.3 (1298)	3.9	2.75	—	1.7	4.2	3.2	10.6	5.3
	Sidekick	0	1.6 (1590)	4.75	①	②	3.6	2.1	4.6	11	5.5
1996	X90	0	1.6 (1590)	4.75	①	②	3.6	2.1	4.6	11	5.5
	Sidekick	0	1.6 (1590)	4.75	①	②	3.6	2.1	4.6	③	5.5
	Sidekick Sport	2	1.8 (1843)	5.5	①	②	3.6	2.1	4.6	18.5	5.5
1997	X90	0	1.6 (1590)	4.75	①	②	3.6	2.1	4.6	11	5.5
	Sidekick	0	1.6 (1590)	4.75	①	②	3.6	2.1	4.6	③	5.5
	Sidekick Sport	2	1.8 (1843)	5.5	①	②	3.6	2.1	4.6	18.5	5.5
1998	X90	0	1.6 (1590)	4.75	①	②	3.6	2.1	4.6	11	5.5
	Sidekick	0	1.6 (1590)	4.75	①	②	3.6	2.1	4.6	③	5.5
	Sidekick Sport	2	1.8 (1843)	5.5	①	②	3.6	2.1	4.6	18.5	5.5

Note: All capacities are approximate. Add fluid gradually and check to be sure a proper fluid level is obtained.

① 2-wheel drive model: 4.0 pts.
   4-wheel drive model: 3.2 pts.
② 3-speed transmission –
   – Fluid drain, and filter and pan removal only: 5.9 pts.
   – After complete transmission overhaul: 10.8 pts.
   4-speed overdrive transmission –
   – Fluid drain, and filter and pan removal only: 5.3 pts.
   – After complete transmission overhaul: 14.6 pts.
③ 2-door model: 11 gals.
   4-door model: 14.5 gals.

79241C08

VALVE SPECIFICATIONS

Year	Engine ID/VIN	Engine Displacement Liters (cc)	Seat Angle (deg.)	Face Angle (deg.)	Spring Test Pressure (lbs. @ in.)	Spring Installed Height (in.)	Stem-to-Guide Clearance (in.) Intake	Stem-to-Guide Clearance (in.) Exhaust	Stem Diameter (in.) Intake	Stem Diameter (in.) Exhaust
1995	3 ①	1.3 (1298)	45	45	61-74@1.67	2.008	0.0008-0.0018	0.0014-0.0024	0.2152-0.2157	0.2146-0.2151
	0 ②	1.6 (1590)	45	45	55-64@1.63	1.986	0.0008-0.0019	0.0014-0.0025	0.2742-0.2748	0.2737-0.2742
	0 ③	1.6 (1590)	45	45	50-57@1.28	1.450	0.0008-0.0018	0.0018-0.0028	0.2152-0.2157	0.2142-0.2148
1996	0 ④	1.6 (1590)	45	45	24-28@1.24	1.245	0.0008-0.0018	0.0018-0.0028	0.2152-0.2157	0.2142-0.2148
	2	1.8 (1843)	45	45	50-57@1.28	1.280	0.0008-0.0018	0.0018-0.0028	0.2348-0.2354	0.2339-0.2344
1997	0 ④	1.6 (1590)	45	45	24-28@1.24	1.245	0.0008-0.0018	0.0018-0.0028	0.2152-0.2157	0.2142-0.2148
	2	1.8 (1843)	45	45	50-57@1.28	1.280	0.0008-0.0018	0.0018-0.0028	0.2348-0.2354	0.2339-0.2344
1998	0 ④	1.6 (1590)	45	45	24-28@1.24	1.245	0.0008-0.0018	0.0018-0.0028	0.2152-0.2157	0.2142-0.2148
	2	1.8 (1843)	45	45	50-57@1.28	1.280	0.0008-0.0018	0.0018-0.0028	0.2348-0.2354	0.2339-0.2344

① Samurai
② Sidekick 8 valve engine
③ Sidekick 16 valve engine
④ X90 and Sidekick

79241C09

TORQUE SPECIFICATIONS
All readings in ft. lbs.

Year	Engine ID/VIN	Engine Displacement Liters (cc)	Cylinder Head Bolts	Main Bearing Bolts	Rod Bearing Bolts	Crankshaft Damper Bolts	Flywheel Bolts	Manifold Intake	Manifold Exhaust	Spark Plugs	Lug Nut
1995	3 ①	1.3 (1298)	51-54	36-41	24-26	76-83 ②	50-52	13-20	13-20	15-22	36-50
	0 ③	1.6 (1590)	48-51	36-41	24-26	76-83 ②	55-58	13-20	13-20	14-21	58-80
	0 ④	1.6 (1590)	51-54	36-41	24-26	76-83 ②	55-58	13-20	13-20	14-21	58-80
1996	0 ⑤	1.6 (1590)	48-51	36-41	24-26	76-83 ②	55-58	13-20	13-20	14-21	58-80
	2	1.8 (1843)	⑥	⑦	33	108.5	51	13-20	13-20	14-21	58-80
1997	0 ⑤	1.6 (1590)	48-51	36-41	24-26	76-83 ②	55-58	13-20	13-20	14-21	58-80
	2	1.8 (1843)	⑥	⑦	33	108.5	51	13-20	13-20	14-21	58-80
1998	0 ⑤	1.6 (1590)	48-51	36-41	24-26	76-83 ②	55-58	13-20	13-20	14-21	58-80
	2	1.8 (1843)	⑥	⑦	33	108.5	51	13-20	13-20	14-21	58-80

① Samurai
② Specification shown is for crankshaft timing sprocket bolt
③ Sidekick 8 valve engine
④ Sidekick 16 valve engine
⑤ X90 and Sidekick
⑥ M10: 76 ft. lbs.
   M6: 8 ft. lbs.
⑦ 10mm threaded diameter: 42 ft. lbs.
   8mm threaded diameter: 19.5 ft. lbs.

79241C0

SCHEDULED MAINTENANCE INTERVALS
(SUZUKI SAMURAI, SIDEKICK & X90)

TO BE SERVICED	TYPE OF SERVICE	VEHICLE MILEAGE INTERVAL (x1000)												
		7.5	15	22.5	30	37.5	45	52.5	60	67.5	75	82.5	90	97.5
Engine oil & filter	R	✓	✓	✓	✓	✓	✓	✓	✓	✓	✓	✓	✓	✓
Automatic transmission fluid (Sidekick) ④	S/I	✓	✓	✓	✓	✓	✓	✓	✓	✓	✓	✓	✓	✓
Manual transmission oil (Sidekick) ⑧	S/I	✓	✓	✓	✓	✓	✓	✓	✓	✓	✓	✓	✓	✓
Power steering system (Sidekick)	S/I	✓	✓	✓	✓	✓	✓	✓	✓	✓	✓	✓	✓	✓
Steering system	S/I	✓	✓	✓	✓	✓	✓	✓	✓	✓	✓	✓	✓	✓
Transfer & differential oil (Sidekick) ⑧	S/I	✓	✓	✓	✓	✓	✓	✓	✓	✓	✓	✓	✓	✓
Transmission, transfer & differential oil (Samurai) ⑦	S/I	✓	✓	✓	✓	✓	✓	✓	✓	✓	✓	✓	✓	✓
Wheel discs & free wheeling hubs	S/I	✓	✓	✓	✓	✓	✓	✓	✓	✓	✓	✓	✓	✓
Shock absorbers (Samurai)	S/I	✓	✓		✓		✓		✓		✓		✓	
Suspension bolts & nuts (Samurai)	S/I	✓	✓		✓		✓		✓		✓		✓	
Suspension system (Sidekick)	S/I	✓	✓		✓		✓		✓		✓		✓	
Brake discs & pads (front)	S/I		✓		✓		✓		✓		✓		✓	
Brake drums & shoes (rear)	S/I		✓		✓		✓		✓		✓		✓	
Brake fluid ⑥	S/I		✓		✓		✓		✓		✓		✓	
Brake hoses & pipes	S/I		✓		✓		✓		✓		✓		✓	
Brake pedal	S/I		✓		✓		✓		✓		✓		✓	
Brake lever & cable	S/I		✓		✓		✓		✓		✓		✓	
Clutch	S/I		✓		✓		✓		✓		✓		✓	
Idle speed	S/I		✓		✓		✓		✓		✓		✓	
Propeller shafts	S/I		✓		✓		✓		✓		✓		✓	
Valve lash (clearance)	S/I		✓		✓		✓		✓		✓		✓	
Wheel bearings	S/I		✓		✓		✓		✓		✓		✓	
Steering knuckle oil seals (Samurai)	R			✓			✓			✓			✓	

79241CP1

SCHEDULED MAINTENANCE INTERVALS
(SUZUKI SAMURAI, SIDEKICK & X90) (Cont.)

TO BE SERVICED	TYPE OF SERVICE	VEHICLE MILEAGE INTERVAL (x1000)												
		7.5	15	22.5	30	37.5	45	52.5	60	67.5	75	82.5	90	97.5
Air cleaner filter element	R				✓				✓				✓	
Engine Coolant	R				✓				✓				✓	
Fuel filter	R				✓				✓				✓	
Spark plugs	R				✓				✓				✓	
Cooling system hoses & connections	S/I				✓				✓				✓	
Drive belt(s)	S/I				✓				✓				✓	
Exhaust pipes & mountings	S/I				✓				✓				✓	
Fuel lines & connections	S/I				✓				✓				✓	
Fuel tank cap gasket	S/I				✓				✓				✓	
Leaf springs (Samurai)	S/I				✓				✓				✓	
Camshaft timing belt	R								✓				✓	
Distributor cap & rotor	S/I								✓					
Emission-related hoses & tubes	S/I								✓					
Oxygen sensor or heated oxygen sensor②	R													
EVAP canister④	R													
PCV valve①	R													
EGR system⑤	S/I													
Fuel Injectors③	S/I													
TWC converter③	S/I													

① PCV valve - replace every 50,000 miles.
② Oxygen sensor or heated oxygen sensor - service or inspect at 80,000 miles.
③ Service or inspect at 100,000 miles.
④ Replace at 100,000 miles.
⑤ EGR system - service or inspect every 50,000 miles.
⑥ Replace every 60,000 miles.
⑦ Replace oil at 7500 miles and every 30,000 miles thereafter.
⑧ Replace oil every 30,000 miles.
R – Replace   S/I – Service or Inspect

79241CP2

SCHEDULED MAINTENANCE INTERVALS
(SUZUKI SAMURAI, SIDEKICK & X90) (Cont.)

FREQUENT OPERATION MAINTENANCE (SEVERE SERVICE)

If a vehicle is operated under any of the following conditions it is considered severe service:

- Extremely dusty areas.
- 50% or more of the vehicle operation is in 32°C (90°F) or higher temperatures, or constant operation in temperatures below 0°C (32°F).
- Prolonged idling (vehicle operation in stop and go traffic).
- Frequent short running periods (engine does not warm to normal operating temperatures).
- Police, taxi, delivery usage or trailer towing usage.

Oil & oil filter change – change every 3000 miles.
Air cleaner filter element - service or inspect every 3000 miles & replace every 15,000 miles.
Steering wheel free play, gear box oil & linkage - service or inspect every 3000 miles.
Bolts & nuts on chassis - tighten every 6000 miles.
Brake discs & pads (front) - service or inspect every 6000 miles.
Brake drums & shoes (rear) - service or inspect every 6000 miles.
Exhaust pipes & mountings - tighten every 6000 miles.
Propeller shafts - service or inspect every 6000 miles.
Automatic transmission fluid & filter - replace every 15,000 miles.
Distributor cap & ignition wires - service or inspect every 15,000 miles.
Drive belt(s) - service or inspect every 15,000 miles.
Manual transmission oil - replace every 15,000 miles.
Transfer & differential oil - replace every 15,000 miles.

79241CP3

SCHEDULED MAINTENANCE
SUZUKI
SAMURAI, SIDEKICK, SIDEKICK SPORT, X90

The following should be used as a guide when determining the amount of work required for a particular service if taken to a repair shop.
In estimating how long a particular Scheduled Maintenance Service should take, please observe the following:

- **Chilton Time** is time based on field research and data supplied by the vehicle manufacturer.
- Labor time operations are given in hours and tenths of an hour.
- All labor operations, are to be used as a guide.

Mechanic Skill Level Codes:
(G) GENERAL: Normally skilled with certification.
(M) MAINTENANCE: Semi-skilled working on certication.
(P) PRECISION: Really skilled with multiple certification.

	Chilton Time		Chilton Time		Chilton Time
(M) 7500 Mile Service		**(M) 37500 Mile Service**		**(G) 75000 Mile Service**	
1995 Samurai	.8	1995 Samurai	.8	1995 Samurai	2.3
1995-98 Sidekick, Sidekick Sport, X-90	.7	1995-98 Sidekick, Sidekick Sport, X-90	.7	1995-98 Sidekick, Sidekick Sport, X-90	1.9
w/AT add	.1	w/AT add	.1	w/AT add	.1
w/4WD add	.3	w/4WD add	.3	w/4WD add	.3
(G) 15000 Mile Service		**(G) 45000 Mile Service**		**(M) 82500 Mile Service**	
1995 Samurai	2.3	1995 Samurai	4.4	1995 Samurai	.8
1995-98 Sidekick, Sidekick Sport, X-90	2.0	1995-98 Sidekick, Sidekick Sport, X-90	2.0	1995-98 Sidekick, Sidekick Sport, X-90	.6
w/AT add	.1	w/AT add	.1	w/AT add	.1
w/4WD add	.3	w/4WD add	.3	w/4WD add	.3
(G) 22500 Mile Service		**(M) 52500 Mile Service**		**(G) 90000 Mile Service**	
1995 Samurai	2.9	1995 Samurai	.8	1995 Samurai	9.8
1995-98 Sidekick, Sidekick Sport, X-90	.6	1995-98 Sidekick, Sidekick Sport, X-90	.6	1995-98 Sidekick, Sidekick Sport, X-90	7.3
w/AT add	.1	w/AT add	.1	w/AT add	.1
w/4WD add	.3	w/4WD add	.3	w/4WD add	.3
(G) 30000 Mile Service		**(G) 60000 Mile Service**		**(M) 97500 Mile Service**	
1995 Samurai	4.8	1995 Samurai	8.1	1995 Samurai	.8
1995-98 Sidekick, Sidekick Sport, X-90	4.4	1995-98 Sidekick, Sidekick Sport, X-90	7.6	1995-98 Sidekick, Sidekick Sport, X-90	.6
w/AT add	.1	w/AT add	.1	w/AT add	.1
w/4WD add	.3	w/4WD add	.3	w/4WD add	.3
		(M) 67500 Mile Service			
		1995 Samurai	.8		
		1995-98 Sidekick, Sidekick Sport, X-90	.6		
		w/AT add	.1		
		w/4WD add	.3		

79241CP4

TOYOTA
4Runner • Land Cruiser • Pick-Up • Previa • RAV4 • Sienna • T100 • Tacoma

VEHICLE IDENTIFICATION CHART

| Engine Code | | | | | | Model Year | |
Code	Liters	Cu. In. (cc)	Cyl.	Fuel Sys.	Eng. Mfg.	Code	Year
1FZ-FE	4.5	273 (4477)	8	MFI	Toyota	S	1995
1MZ-FE	3.0	183 (2995)	6	MFI	Toyota	T	1996
2RZ-FE	2.4	149 (2438)	4	MFI	Toyota	V	1997
2UZ-FE	4.7	285 (4664)	8	MFI	Toyota	W	1998
3RZ-FE	2.7	164 (2693)	4	MFI	Toyota	X	1999
3S-FE	2.0	122 (1998)	4	MFI	Toyota		
5VZ-FE	3.4	206 (3378)	6	MFI	Toyota		

MFI - Multi-port Fuel Injection

79241CP5

ENGINE IDENTIFICATION

Year	Model	Engine Displacement Liters (cc)	Engine Series (ID/VIN)	Fuel System	No. of Cylinders	Engine Type
1995	Previa ①	2.4 (2438)	2TZ-FE	MFI	4	DOHC
	4Runner	2.4 (2366)	22R-E	MFI	4	SOHC
	4Runner	3.0 (2959)	3VZ-E	MFI	6	SOHC
	Land Cruiser	4.5 (4477)	1FZ-FE	MFI	6	DOHC
	Pick-up	2.4 (2366)	22R-E	MFI	4	SOHC
	Pick-up	3.0 (2959)	3VZ-E	MFI	6	SOHC
	Previa	2.4 (2438)	2TZ-FE	MFI	4	DOHC
	T100	2.7 (2693)	3RZ-FE	MFI	4	DOHC
	T100	3.4 (3378)	5VZ-FE	MFI	6	DOHC
	Tacoma	2.4 (2438)	2RZ-FE	MFI	4	DOHC
	Tacoma	2.7 (2693)	3RZ-FE	MFI	4	DOHC
	Tacoma	3.4 (3378)	5VZ-FE	MFI	6	DOHC
1996	Previa ①	2.4 (2438)	2TZ-FZE	MFI	4	DOHC
	4Runner	2.7 (2693)	3RZ-FE	MFI	4	DOHC
	4Runner	3.4 (3378)	5VZ-FE	MFI	6	DOHC
	Land Cruiser	4.5 (4477)	1FZ-FE	MFI	6	DOHC
	RAV4	2.0 (1998)	3S-FE	MFI	4	DOHC
	T100	2.7 (2693)	3RZ-FE	MFI	4	DOHC
	T100	3.4 (3378)	5VZ-FE	MFI	6	DOHC
	Tacoma	2.4 (2438)	2RZ-FE	MFI	4	DOHC
	Tacoma	2.7 (2693)	3RZ-FE	MFI	4	DOHC
	Tacoma	3.4 (3378)	5VZ-FE	MFI	4	DOHC
1997	Previa ①	2.4 (2438)	2TZ-FZE	MFI	4	DOHC
	4Runner	2.7 (2693)	3RZ-FE	MFI	4	DOHC
	4Runner	3.4 (3378)	5VZ-FE	MFI	6	DOHC
	Land Cruiser	4.5 (4477)	1FZ-FE	MFI	6	DOHC
	RAV4	2.0 (1998)	3S-FE	MFI	4	DOHC
	T100	2.7 (2693)	3RZ-FE	MFI	4	DOHC
	T100	3.4 (3378)	5VZ-FE	MFI	6	DOHC
	Tacoma	2.4 (2438)	2RZ-FE	MFI	4	DOHC
	Tacoma	2.7 (2693)	3RZ-FE	MFI	4	DOHC
	Tacoma	3.4 (3378)	5VZ-FE	MFI	4	DOHC
1998-99	4Runner	2.7 (2693)	3RZ-FE	MFI	4	DOHC
	4Runner	3.4 (3378)	5VZ-FE	MFI	6	DOHC
	Land Cruiser	4.7 (4664)	2UZ-FE	MFI	8	DOHC
	RAV4	2.0 (1998)	3S-FE	MFI	4	DOHC
	Sienna	3.0 (2995)	1MZ-FE	MFI	6	DOHC
	T100	2.7 (2693)	3RZ-FE	MFI	4	DOHC
	T100	3.4 (3378)	5VZ-FE	MFI	6	DOHC
	Tacoma	2.4 (2438)	2RZ-FE	MFI	4	DOHC
	Tacoma	2.7 (2693)	3RZ-FE	MFI	4	DOHC
	Tacoma	3.4 (3378)	5VZ-FE	MFI	4	DOHC

MFI - Multi-port Fuel Injection
SOHC - Single Overhead Camshaft
DOHC - Double Overhead Camshaft
OHV - Overhead Valve
① Supercharged

79241CP6

GENERAL ENGINE SPECIFICATIONS

Year	Engine ID/VIN	Engine Displacement Liters (cc)	Fuel System Type	Net Horsepower @ rpm	Net Torque @ rpm (ft. lbs.)	Bore x Stroke (in.)	Com-pression Ratio	Oil Pressure @ rpm
1995	1FZ-FE	4.5 (4477)	MFI	212@4600	275@3200	3.94x3.64	9.0:1	36-71@3000
	22R-E	2.4 (2366)	MFI	116@4800	140@2800	3.62x3.50	9.3:1	36-71@3000
	2RZ-FE	2.4 (2438)	MFI	142@5000	160@4000	3.74x3.38	9.5:1	36-71@3000
	2TZ-FE	2.4 (2438)	MFI	161@5000	201@3000	3.74x3.39	8.9:1	36-71@3000
	2TZ-FE	2.4 (2438)	MFI	138@5000	154@4000	3.74x3.39	9.1:1	36-71@3000
	3RZ-FE	2.7 (2693)	MFI	150@4800	177@4000	3.74x3.74	9.5:1	36-71@3000
	3VZ-E	3.0 (2959)	MFI	150@4800	180@3400	3.44x3.23	9.0:1	36-71@3000
	5VZ-FE	3.4 (3378)	MFI	190@4800	220@3400	3.68x3.23	9.6:1	NA
1996	1FZ-FE	4.5 (4477)	MFI	212@4600	275@3200	3.94x3.74	9.0:1	36-71@3000
	2RZ-FE	2.4 (2438)	MFI	142@5000	160@4000	3.74x3.38	9.5:1	36-71@3000
	2TZ-FZE	2.4 (2438)	MFI	161@5000	201@3600	3.74x3.39	8.9:1	36@3000
	3RZ-FE	2.7 (2693)	MFI	150@4800	177@4000	3.74x3.74	9.5:1	36-71@3000
	3S-FE	2.0 (1998)	MFI	120@5400	125@4600	3.40x3.40	9.5:1	NA
	5VZ-FE	3.4 (3378)	MFI	190@4800	220@3600	3.68x3.23	9.6:1	NA
1997	1FZ-FE	4.5 (4477)	MFI	212@4600	275@3200	3.94x3.74	9.0:1	36-71@3000
	2RZ-FE	2.4 (2438)	MFI	142@5000	160@4000	3.74x3.38	9.5:1	36-71@3000
	2TZ-FZE	2.4 (2438)	MFI	161@5000	201@3600	3.74x3.39	8.9:1	36@3000
	3RZ-FE	2.7 (2693)	MFI	150@4800	177@4000	3.74x3.74	9.5:1	36-71@3000
	3S-FE	2.0 (1998)	MFI	120@5400	125@4600	3.40x3.40	9.5:1	NA
	5VZ-FE	3.4 (3378)	MFI	190@4800	220@3600	3.68x3.23	9.6:1	NA
1998-99	1FZ-FE	4.5 (4477)	MFI	212@4600	275@3200	3.94x3.74	9.0:1	36-71@3000
	1MZ-FE	3.0 (2995)	MFI	188@5200	203@4400	3.44x3.27	10.5:1	43-78@3000
	2RZ-FE	2.4 (2438)	MFI	142@5000	160@4000	3.74x3.38	9.5:1	36-71@3000
	2UZ-FE	4.7 (4664)	MFI	230@4800	320@3400	3.70x3.30	9.6:1	45-65@3000
	3RZ-FE	2.7 (2693)	MFI	150@4800	177@4000	3.74x3.74	9.5:1	36-71@3000
	3S-FE	2.0 (1998)	MFI	120@5400	125@4600	3.40x3.40	9.5:1	NA
	5VZ-FE	3.4 (3378)	MFI	190@4800	220@3600	3.68x3.23	9.6:1	NA

MFI - Multi-port Fuel Injection

NA - Not Available

79241CP7

GASOLINE ENGINE TUNE-UP SPECIFICATIONS

Year	Engine ID/VIN	Engine Displacement Liters (cc)	Spark Plugs Gap (in.)	Ignition Timing (deg.) MT	AT	Fuel Pump (psi)	Idle Speed (rpm) MT	AT	Valve Clearance In.	Ex.
1995	1FZ-FE	4.5 (4477)	0.031	—	3B	38-44	—	600-700	0.006-	0.010-
	22R-E	2.4 (2366)	0.031	5B	5B	38-44	750	850	0.008	0.012
	2RZ-FE	2.4 (2438)	0.031	①	①	38-44	650-750	650-750	0.006-0.010	0.010-0.014
	2TZ-FE	2.4 (2438)	0.043	5B	5B	38-44	—	750	②	③
	3RZ-FE	2.7 (2693)	0.031	①	①	38-44	750	—	0.008	0.012
	3VZ-E	3.0 (2959)	0.031	10B	10B	38-44	800	800	④	⑤
	5VZ-FE	3.4 (3378)	0.043	⑥	⑥	38-44	650-750	650-750	0.006-0.009	0.011-0.014
1996	1FZ-FE	4.5 (4477)	0.031	—	3B	38-44	—	600-700	0.006-0.010	0.010-0.014
	2RZ-FE	2.4 (2438)	0.031	5B ⑥	5B ⑥	38-44	650-750	—	0.006-0.010	0.010-0.014
	2TZ-FZE	2.4 (2438)	0.043	—	5B ⑥	38-44	700-800	700-800	0.006-0.010	0.010-0.014
	3RZ-FE	2.7 (2693)	0.031	5B ⑥	5B ⑥	38-44	650-750	650-750	0.006-0.010	0.010-0.014
	3S-FE	2.0 (1998)	0.043	10B ①	5B ①	44-50	700-800	700-800	0.007-0.011	0.011-0.015
	5VZ-FE	3.4 (3378)	0.043	10B ①	10B ①	38-44	650-750	650-750	0.006-0.009	0.011-0.014
1997	1FZ-FE	4.5 (4477)	0.031	—	3B	38-44	—	600-700	0.006-0.010	0.010-0.014
	2RZ-FE	2.4 (2438)	0.031	5B ⑥	5B ⑥	38-44	650-750	—	0.006-0.010	0.010-0.014
	2TZ-FZE	2.4 (2438)	0.043	—	5B ⑥	38-44	700-800	700-800	0.006-0.010	0.010-0.014
	3RZ-FE	2.7 (2693)	0.031	5B ⑥	5B ⑥	38-44	650-750	650-750	0.006-0.010	0.010-0.014
	3S-FE	2.0 (1998)	0.043	10B ①	5B ①	44-50	700-800	700-800	0.007-0.011	0.011-0.015
	5VZ-FE	3.4 (3378)	0.043	10B ①	10B ①	38-44	650-750	650-750	0.006-0.009	0.011-0.014
1998-99	1FZ-FE	4.5 (4477)	0.031	—	3B	38-44	—	600-700	0.006-0.010	0.010-0.014
	1MZ-FE	3.0 (2995)	0.043	⑥	⑥	38-44	—	650-750	0.006-0.010	0.010-0.014
	2RZ-FE	2.4 (2438)	0.031	5B ⑥	5B ⑥	38-44	650-750	—	0.006-0.010	0.010-0.014
	2UZ-FE	4.7 (4664)	0.043	—	—	38-44	—	650-750	0.006-0.010	0.010-0.014
	3RZ-FE	2.7 (2693)	0.031	5B ⑥	5B ⑥	38-44	650-750	650-750	0.006-0.010	0.010-0.014

79241CP8

GASOLINE ENGINE TUNE-UP SPECIFICATIONS

Year	Engine ID/VIN	Engine Displacement Liters (cc)	Spark Plugs Gap (in.)	Ignition Timing (deg.)		Fuel Pump (psi)	Idle Speed (rpm)		Valve Clearance	
				MT	AT		MT	AT	In.	Ex.
1998-99 (cont.)	3S-FE	2.0 (1998)	0.043	10B ①	5B ①	44-50	700-800	700-800	0.007-0.011	0.011-0.015
	5VZ-FE	3.4 (3378)	0.043	10B ①	10B ①	38-44	650-750	650-750	0.006-0.009	0.011-0.014

NOTE: The Vehicle Emission Control Information label often reflects specification changes made during production. The label figures must be used if they differ from those in this chart.

B - Before top dead center

① 5B at idle, with terminal TE1 and E1 connected of DLC1
② Intake: 0.006-0.010 (cold)
③ Exhaust: 0.010-0.014 (cold)
④ Intake: 0.007-0.011 (cold)
⑤ Exhaust: 0.009-0.013 (cold
⑥ 10B at idle, with terminal TE1 and E1 connected of DLC1

79241CP9

CAPACITIES

Year	Model	Engine ID/VIN	Engine Displacement Liters (cc)	Engine Oil with Filter (qts.)	Transmission (pts.)			Transfer Case (pts.)	Drive Axle		Fuel Tank (gal.)	Cooling System (qts.)
					4-Spd	5-Spd	Auto.		Front (pts.)	Rear (pts.)		
1995	4Runner	22R-E	2.4 (2366)	4.5	—	①	②	③	④	⑤	⑥	⑦
	4Runner	3VZ-E	3.0 (2959)	4.8	—	①	②	③	④	⑤	⑥	⑦
	Land Cruiser	1FZ-FE	4.5 (4477)	7.8	—	—	12.6	⑧	⑨	6.8	25.1	14.8
	Pick-up	22R-E	2.4 (2366)	4.5	—	①	②	③	④	⑤	⑥	⑩
	Pick-up	3VZ-E	3.0 (2959)	4.8	—	①	②	③	④	⑤	⑥	⑦
	Previa	2TZ-FE	2.4 (2438)	6.1	—	—	5.0	3.0	2.2	3.2	19.8	13.0
	Previa	2TZ-FZE	2.4 (2438)	6.1	—	—	5.0	3.0	2.2	3.2	19.8	13.0
	T100	3RZ-FE	2.7 (2693)	5.6	—	2.7	—	—	—	3.8	19.8	9.2
	T100	5VZ-FE	3.4 (3378)	⑪	⑫	②	2.4	3.9	⑤		24.0	⑦
	Tacoma	2RZ-FE	2.4 (2438)	5.8	—	⑬	⑭	2.2	⑮	2.9	15.1	⑯
	Tacoma	3RZ-FE	2.7 (2693)	5.8	—	⑬	⑭	2.2	⑮	⑰	18.0	⑯
	Tacoma	5VZ-FE	3.4 (3378)	R	—	⑬	⑭	2.2	⑮	⑰	18.0	⑲
1996	4Runner	3RZ-FE	2.7 (2693)	5.8	—	⑬	⑭	2.4	⑮	⑳	18.0	㉑
	4Runner	5VZ-FE	3.4 (3378)	5.5	—	⑬	⑭	2.4	⑮	⑳	18.0	㉒
	Land Cruiser	1FZ-FE	4.5 (4477)	7.8	—	—	4.0	3.6	⑨	6.8	25.1	38.0
	Previa	2TZ-FZE	2.4 (2438)	6.1	—	—	3.4	2.8	㉓	3.2	19.8	13.0
	RAV4	3S-FE	2.0 (1998)	4.1	—	㉔	7.0	—	—	—	15.3	㉕
	T100	3RZ-FE	2.7 (2693)	5.8	—	2.7	—	—	—	3.8	24.0	9.2
	T100	5VZ-FE	3.4 (3378)	⑪	⑪	②	2.4	3.9	⑤	—	24.0	⑦
	Tacoma	2RZ-FE	2.4 (2438)	5.8	—	⑬	⑭	2.4	⑮	2.9	15.1	⑯
	Tacoma	3RZ-FE	2.7 (2693)	5.8	—	⑬	⑭	2.4	⑮	⑰	18.0	⑯
	Tacoma	5VZ-FE	3.4 (3378)	⑱	—	⑬	⑭	2.4	⑮	⑰	18.0	⑲
1997	4Runner	3RZ-FE	2.7 (2693)	5.8	—	⑬	⑭	2.4	⑮	⑳	18.0	㉑
	4Runner	5VZ-FE	3.4 (3378)	5.5	—	⑬	⑭	2.4	⑮	⑳	18.0	㉒
	Land Cruiser	1FZ-FE	4.5 (4477)	7.8	—	—	4.0	3.6	⑨	6.8	25.1	38.0
	Previa	2TZ-FZE	2.4 (2438)	6.1	—	—	3.4	2.8	㉓	3.2	19.8	13.0
	RAV4	3S-FE	2.0 (1998)	4.1	—	㉔	7.0	—	—	—	15.3	㉕
	T100	3RZ-FE	2.7 (2693)	5.8	—	2.7	—	—	—	3.8	24.0	9.2
	T100	5VZ-FE	3.4 (3378)	⑪	⑪	②	2.4	3.9	⑤	—	24.0	⑦
	Tacoma	2RZ-FE	2.4 (2438)	5.8	—	⑬	⑭	2.4	⑮	2.9	15.1	⑯
	Tacoma	3RZ-FE	2.7 (2693)	5.8	—	⑬	⑭	2.4	⑮	⑰	18.0	⑯
	Tacoma	5VZ-FE	3.4 (3378)	R	—	⑬	⑭	2.4	⑮	⑰	18.0	⑲
1998-99	4Runner	3RZ-FE	2.7 (2693)	5.8	—	⑬	⑭	2.4	⑮	⑳	18.0	㉑
	4Runner	5VZ-FE	3.4 (3378)	5.5	—	⑬	⑭	2.4	⑮	⑳	18.0	㉒
	Land Cruiser	1FZ-FE	4.5 (4477)	7.8	—	—	4.0	3.6	⑨	6.8	25.1	38
	Land Cruiser	2UZ-FE	4.7 (4664)	5.5	—	—	4	3.6	⑨	6.8	25.1	38
	RAV4	3S-FE	2.0 (1998)	4.1	—	㉔	7.0	—	—	—	15.3	㉕
	Sienna	1MZ-FE	3.0 (2995)	5	—	—	3.7	—	—	—	13.0	㉖
	T100	3RZ-FE	2.7 (2693)	5.8	—	2.7	—	—	—	3.8	24.0	9.2
	T100	5VZ-FE	3.4 (3378)	⑪	⑪	B	2.4	3.9	⑤	—	24.0	⑦

79241CP(

CAPACITIES

Year	Model	Engine ID/VIN	Engine Displacement Liters (cc)	Engine Oil with Filter (qts.)	Transmission (pts.)			Transfer Case (pts.)	Drive Axle		Fuel Tank (gal.)	Cooling System (qts.)
					4-Spd	5-Spd	Auto.		Front (pts.)	Rear (pts.)		
1998-99 (cont.)	Tacoma	2RZ-FE	2.4 (2438)	5.8	—	⑬	⑭	2.4	⑮	2.9	15.1	⑯
	Tacoma	3RZ-FE	2.7 (2693)	5.8	—	⑬	⑭	2.4	⑮	⑰	18.0	⑯
	Tacoma	5VZ-FE	3.4 (3378)	R	—	⑬	⑭	2.4	⑮	⑰	18.0	⑲

NOTE: All capacities are approximate. Add fluid gradually and check to be sure a proper fluid level is obtained.

① G58: 8.2
R150F: 6.4
② Drain and refill:
A340E: 3.4
A340F: 4.2
③ Except 3VZ-E AT (VF1A type): 2.4
3VZ-E AT (A340H): 1.6
④ Standard: 3.4; ADD: 4.0
⑤ 2WD: 4.4
4WD: 4.3
⑥ With standard tires: 17.2
With optional 31x10.5 tires: 18.8
⑦ 2WD M/T: 10.6
2WD A/T: 10.5
4WD M/T: 10.6
4WD A/T: 10.8
⑧ With ABS: 2.8
Without ABS: 3.6
⑨ With differential lock: 5.6
Without differential lock: 5.8
⑩ With rear heater: 9.2
Without rear heater: 8.9

⑪ 2WD: 5.5
4WD: 5.0
⑫ 2WD: 5.4
4WD: 4.6
All others: 5.4
⑬ W59:
2WD: 5.4
4WD: 5.2
R150, R150F:
2WD: 5.4
4WD: 4.6
⑭ A43D: 5.0
A340E: 3.4
A340F: 4.2
⑮ Without ADD: 2.32
With ADD: 2.44
⑯ 2WD M/T: 8.5
2WD A/T: 8.2
4WD M/T: 8.8
4WD A/T: 8.7
⑰ Extra long: 4.4
All others: 5.4

⑱ 2WD: 5.7
4WD: 5.5
⑲ With rear heater: 11.0
M/T: 10.7
A/T: 10.5
⑳ 2WD: 5.8
4WD with differential locks: 5.8
4WD without differential locks: 5.2
㉑ With rear heater: 11.6
Without rear heater: 10.6
㉒ With rear heater: 9.5
Without rear heater: 8.5
㉓ 2WD: 3.2
4WD: 2.2
㉔ 2WD: 8.2
4WD: 10.6
㉕ M/T: 8.5
A/T: 8.1
㉖ Without rear heater: 10.0

79241CQ1

VALVE SPECIFICATIONS

Year	Engine ID/VIN	Engine Displacement Liters (cc)	Seat Angle (deg.)		Face Angle (deg.)	Spring Test Pressure (lbs. @ in.)	Spring Installed Height (in.)	Stem-to-Guide Clearance (in.)		Stem Diameter (in.)	
								Intake	Exhaust	Intake	Exhaust
1995	1FZ-FE	4.5 (4477)	45		44.5	53.4@1.437	1.437	0.0010-0.0024	0.0012-0.0026	0.2744-0.2750	0.2742-0.2748
	22R-E	2.4 (2366)	45	①	44.5	66.1@1.594	1.594	0.0010-0.0024	0.0012-0.0026	0.3138-0.3144	0.3136-0.3142
	2RZ-FE	2.4 (2438)	45	①	44.5	40-46@1.406	1.406	0.0010-0.0024	0.0012-0.0026	0.2350-0.2356	0.2348-0.2354
	2TZ-FE	2.4 (2438)	45	①	44.5	57-63@1.406	1.406	0.0010-0.0024	0.0012-0.0026	0.2350-0.2356	0.2348-0.2354
	3RZ-FE	2.7 (2693)	45	①	44.5	57-63@1.406	1.406	0.0010-0.0024	0.0012-0.0026	0.2350-0.2356	0.2348-0.2354
	3VZ-E	3.0 (2959)	45	①	44.5	54-57@1.575	1.575	0.0010-0.0024	0.0012-0.0026	0.3138-0.3144	0.3136-0.3142
	5VZ-FE	3.4 (3378)	45		44.5	41.9-46.3@1.311	1.311	0.0010-0.0024	0.0012-0.0026	0.2350-0.2356	0.2348-0.2354
1996	1FZ-FE	4.5 (4477)	45		44.5	48.1-53.4@1.437	1.437	0.0010-0.0024	0.0012-0.0026	0.2744-0.2750	0.2742-0.2748
	2RZ-FE	2.4 (2438)	45		44.5	40-46@1.406	1.406	0.0010-0.0024	0.0012-0.0026	0.2350-0.2356	0.2348-0.2354
	2TZ-FZE	2.4 (2438)	45		44.5	38.7-42.8@1.594	1.406	0.0010-0.0024	0.0012-0.0026	0.2350-0.2356	0.2348-0.2354
	3RZ-FE	2.7 (2693)	45		44.5	40-46@1.406	1.406	0.0010-0.0024	0.0012-0.0026	0.2350-0.2356	0.2348-0.2354
	3S-FE	2.0 (1998)	45		44.5	36.8-42.5@1.366	1.366	0.0010-0.0024	0.0012-0.0026	0.2350-0.2356	0.2348-0.2354
	5VZ-FE	3.4 (3378)	45		44.5	41.9-46.3@1.311	1.311	0.0010-0.0024	0.0012-0.0026	0.2350-0.2356	0.2348-0.2354
1997	1FZ-FE	4.5 (4477)	45		44.5	48.1-53.4@1.437	1.437	0.0010-0.0024	0.0012-0.0026	0.2744-0.2750	0.2742-0.2748
	2RZ-FE	2.4 (2438)	45		44.5	40-46@1.406	1.406	0.0010-0.0024	0.0012-0.0026	0.2350-0.2356	0.2348-0.2354
	2TZ-FZE	2.4 (2438)	45		44.5	38.7-42.8@1.594	1.406	0.0010-0.0024	0.0012-0.0026	0.2350-0.2356	0.2348-0.2354
	3RZ-FE	2.7 (2693)	45		44.5	40-46@1.406	1.406	0.0010-0.0024	0.0012-0.0026	0.2350-0.2356	0.2348-0.2354
	3S-FE	2.0 (1998)	45		44.5	36.8-42.5@1.366	1.366	0.0010-0.0024	0.0012-0.0026	0.2350-0.2356	0.2348-0.2354
	5VZ-FE	3.4 (3378)	45		44.5	41.9-46.3@1.311	1.311	0.0010-0.0024	0.0012-0.0026	0.2350-0.2356	0.2348-0.2354
1998-99	1FZ-FE	4.5 (4477)	45		44.5	48.1-53.4@1.437	1.437	0.0010-0.0024	0.0012-0.0026	0.2744-0.2750	0.2742-0.2748
	1MZ-FE	3.0 (2995)	45		44.5	41.9-46.3@1.33	1.791	0.0010-0.0024	0.0012-0.0026	0.2154-0.2159	0.2152-0.2157
	2RZ-FE	2.4 (2438)	45		44.5	40-46@1.406	1.406	0.0010-0.0024	0.0012-0.0026	0.2350-0.2356	0.2348-0.2354
	2UZ-FE	4.7 (4664)	30, 45, 60		44.5	45.9-50.7@1.378	1.380	0.0010-0.0024	0.0012-0.0026	0.2154-0.2159	0.2152-0.2157
	3RZ-FE	2.7 (2693)	45		44.5	40-46@1.406	1.406	0.0010-0.0024	0.0012-0.0026	0.2350-0.2356	0.2348-0.2354
	3S-FE	2.0 (1998)	45		44.5	36.8-42.5@1.366	1.366	0.0010-0.0024	0.0012-0.0026	0.2350-0.2356	0.2348-0.2354
	5VZ-FE	3.4 (3378)	45		44.5	41.9-46.3@1.311	1.311	0.0010-0.0024	0.0012-0.0026	0.2350-0.2356	0.2348-0.2354

① Blend seat with 30 and 60 degree cutters to center the 45 degree portion on valve face

79241CQ2

TORQUE SPECIFICATIONS
All readings in ft. lbs.

Year	Engine ID/VIN	Engine Displacement Liters (cc)	Cylinder Head Bolts	Main Bearing Bolts	Rod Bearing Bolts	Crankshaft Damper Bolts	Flywheel Bolts	Manifold Intake	Manifold Exhaust	Spark Plugs	Lug Nut
1995	2TZ-FE	2.4 (2438)	①	②	③	192	④	15	36	11-15	—
	22R-E	2.4 (2366)	53-63	69-83	40-47	120-130	73-86	13-19	26-36	11-15	—
	2RZ-FE	2.4 (2438)	①	⑬	⑭	193	65	22	36	14	83
	3RZ-FE	2.7 (2693)	①	⑬	⑭	⑮	⑮	22	36	14	—
	3VZ-E	3.0 (2959)	⑤	⑥	⑦	176-186	63-67	11-15	25-33	11-15	—
	5VZ-FE	3.4 (3378)	⑤	⑦	⑮	176-186	63-67	11-15	25-33	11-15	76
	1FZ-FE	4.5 (4477)	⑩	⑪	⑫	304	74	15	29	15	—
1996	3S-FE	2.0 (1998)	⑱	43	⑰	80	⑲	14	36	13	76
	2RZ-FE	2.4 (2438)	⑧	⑨	⑯	193	⑳	22	36	14	83
	2TZ-FZE	2.4 (2438)	⑧	⑨	⑯	192	54	15	36	14	76
	3RZ-FE	2.7 (2693)	⑧	⑨	⑯	⑰	㉑	22	36	14	83
	5VZ-FE	3.4 (3378)	㉒	㉓	⑰	184	63-67	13	30	13	76
	1FZ-FE	4.5 (4477)	⑧	㉔	㉕	304	74	15	29	14	㉖
1997	3S-FE	2.0 (1998)	⑱	43	⑰	80	⑲	14	36	13	76
	2RZ-FE	2.4 (2438)	⑧	⑨	⑯	193	⑳	22	36	14	83
	3RZ-FE	2.7 (2693)	⑧	⑨	⑯	⑰	㉑	22	36	14	83
	5VZ-FE	3.4 (3378)	㉒	㉓	⑰	184	63-67	13	30	13	76
	1FZ-FE	4.5 (4477)	⑧	24	㉕	304	74	15	29	14	㉖
1998-99	1MZ-FE	3.0 (2995)	㉗	㉘	㉙	159	61	11	36	13	76
	3S-FE	2.0 (1998)	⑱	43	⑰	80	⑲	14	36	13	76
	2RZ-FE	2.4 (2438)	⑧	⑨	⑯	193	⑳	22	36	14	83
	3RZ-FE	2.7 (2693)	⑧	⑨	⑯	⑰	㉑	22	36	14	83
	5VZ-FE	3.4 (3378)	㉒	㉓	⑰	184	63-67	13	30	13	76
	1FZ-FE	4.5 (4477)	⑧	24	㉕	304	74	15	29	14	㉖
	2UZ-FE	4.7 (4664)	㉚	㉛	⑰	181	⑱	13	33	13	㉖

① Step 1: 29 ft. lbs.
Step 2: 90 degree turn
Step 3: 90 degree turn
② Step 1: 20 ft. lbs.
Step 2: 35 ft. lbs.
Step 3: 58 ft. lbs.
③ Step 1: 22 ft. lbs.
Step 2: 90 degree turn
④ Manual transmission: 65 ft. lbs.
Automatic transmission: 54 ft. lbs.
⑤ Step 1: 27 ft. lbs.
Step 2: 33 ft. lbs.
Step 3: 90 degree turn
Step 4: 90 degree turn
⑥ Step 1: 18 ft. lbs.
Step 2: 90 degree turn
⑦ Step 1: 45 ft. lbs.
Step 2: 90 degree turn

⑧ Step 1: 29 ft. lbs.
Step 2: 90 degree turn
Step 3: 90 degree turn
⑨ Step 1: 29 ft. lbs.
Step 2: 90 degree turn
⑩ Step 1: 27 ft. lbs.
Step 2: 90 degree turn
Step 3: 90 degree turn
⑪ Step 1: 54 ft. lbs.
Step 2: 90 degree turn
⑫ Step 1: 35 ft. lbs.
Step 2: 90 degree turn
⑬ Step 1: 29 ft. lbs.
Step 2: 90 degree turn
⑭ Step 1: 33 ft. lbs.
Step 2: 90 degree turn
⑮ Step 1: 19 ft. lbs.
Step 2: 90 degree turn

⑯ Step 1: 33 ft. lbs.
Step 2: 90 degree turn
⑰ Step 1: 18 ft. lbs.
Step 2: 90 degree turn
⑱ Step 1: 36 ft. lbs.
Step 2: 90 degree turn
⑲ Manual transmission: 65 ft. lbs.
Automatic transmission: 61 ft. lbs.
⑳ Manual transmission: 65 ft. lbs.
Automatic transmission: 54 ft. lbs.
㉑ Manual transmission: 19 ft. lbs. + 90°
Automatic transmission: 54 ft. lbs.
㉒ Step 1: 25 ft. lbs.
Step 2: 90 degree turn
Recessed head: 13 ft. lbs.
㉓ Step 1: 45 ft. lbs.
Step 2: 90 degree turn
㉔ Step 1: 54 ft. lbs.
Step 2: 90 degree turn

㉕ Step 1: 35 ft. lbs.
Step 2: 90 degree turn
㉖ Steel wheel: 109 ft. lbs.
Aluminum wheel: 76 ft. lbs.
㉗ Step 1: 40 ft. lbs.
Step 2: an additional 90°
Recessed bolt: 13 ft. lbs.
㉘ 6-point bolts: 20 ft. lbs.
12-point bolts:
Step 1: 16 ft. lbs.
Step 2: an additional 90°
㉙ Step 1: 29 ft. lbs.
Step 2: an additional 90°
㉚ Step 1: 24 ft. lbs.
Step 2: an additional 90°
Step 3: another 90°
㉛ Step 1: 20 ft. lbs.
Step 2: an additional 90°

79241CQ3

1995 SCHEDULED MAINTENANCE INTERVALS
(TOYOTA LAND CRUISER, PICK UP, PREVIA, RAV4, T100, TACOMA & 4RUNNER)

TO BE SERVICED	TYPE OF SERVICE	VEHICLE MILEAGE INTERVAL (x1000)												
		7.5	15	22.5	30	37.5	45	52.5	60	67.5	75	82.5	90	97.5
Engine oil & filter	R	✓	✓	✓	✓	✓	✓	✓	✓	✓	✓	✓	✓	✓
Rotate tires②	S/I	✓	✓	✓	✓	✓	✓	✓	✓	✓	✓	✓	✓	✓
Driveshaft boots (Pick Up 4WD)	S/I	✓	✓	✓	✓	✓	✓	✓	✓	✓	✓	✓	✓	✓
Driveshaft boots (T100 & Tacoma 4WD)	S/I		✓			✓		✓	✓		✓		✓	
Idle speed (T100, 4Runner & 1995-97 Pick Up)	S/I	✓		✓		✓		✓		✓		✓		✓
Drive belts (T100, 4Runner & 1995-97 Pick Up)	S/I				✓				✓	✓	✓	✓	✓	✓
Drive belts (Previa & Tacoma)	S/I								✓	✓	✓	✓	✓	✓
Drive belts (Land Cruiser & 1993-94 Pick Up)	S/I				✓				✓				✓	
Automatic transmission fluid & filter	S/I		✓		✓		✓		✓		✓		✓	
Ball joints & dust covers	S/I		✓		✓		✓		✓		✓		✓	
Bolts & nuts on chassis & body	S/I		✓		✓		✓		✓		✓		✓	
Brake linings & drums	S/I		✓		✓		✓		✓		✓		✓	
Brake line pipes & hoses	S/I		✓		✓		✓		✓		✓		✓	
Brake pads discs (front & rear)	S/I		✓		✓		✓		✓		✓		✓	
Manual transmission fluid (Tacoma, 4Runner & Pick Up)	S/I		✓		✓		✓		✓		✓		✓	
Propeller shaft grease	S/I		✓		✓		✓		✓		✓		✓	
Steering knuckle & chassis grease	S/I		✓		✓		✓		✓		✓		✓	
Steering linkage	S/I		✓		✓		✓		✓		✓		✓	
Transfer, differential & steering gear box oil	S/I		✓		✓		✓		✓		✓		✓	
Air cleaner filter	R				✓				✓				✓	

79241CQ4

1995 SCHEDULED MAINTENANCE INTERVALS
(TOYOTA LAND CRUISER, PICK UP, PREVIA, RAV4, T100, TACOMA & 4RUNNER) (Cont.)

TO BE SERVICED	TYPE OF SERVICE	VEHICLE MILEAGE INTERVAL (x1000)												
		7.5	15	22.5	30	37.5	45	52.5	60	67.5	75	82.5	90	97.5
Front wheel bearing & thrust bushing grease	R				✓				✓				✓	
Spark plugs (except Previa)	R				✓				✓				✓	
Spark plugs (Previa)	R								✓					
Exhaust pipes & mountings	S/I				✓				✓				✓	
Fuel lines & connections	S/I				✓				✓				✓	
Valve clearance (4Runner & Pick Up 22R-E)	S/I				✓				✓				✓	
Valve clearance (Previa, Tacoma, T100, 4Runner & Pick Up 3VZ-E)	S/I								✓					
Engine Coolant	R						✓				✓			
Charcoal canister (Calif.)	R								✓					
Fuel tank cap gasket	R								✓					
Oxygen sensor or heated oxygen sensors①	R													

① Oxygen sensor or heated oxygen sensors - replace every 80,000 miles.
② 4WD vehicles - rotate tires every 5000 miles.
R – Replace   S/I – Service or Inspect

FREQUENT OPERATION MAINTENANCE (SEVERE SERVICE)

If a vehicle is operated under any of the following conditions it is considered severe service:
- Extremely dusty areas.
- 50% or more of the vehicle operation is in 32°C (90°F) or higher temperatures, or constant operation in temperatures below 0°C (32°F).
- Prolonged idling (vehicle operation in stop and go traffic).
- Frequent short running periods (engine does not warm to normal operating temperatures).
- Police, taxi, delivery usage or trailer towing usage.
Air cleaner filter - service or inspect every 3750 miles.
Oil & oil filter change – change every 3750 miles.
Ball joints & dust covers - service or inspect every 7500 miles.
Bolts & nuts on body & chassis - service or inspect every 7500 miles.
Brake linings & drums - service or inspect every 7500 miles.
Brake pads & discs (front & rear) - service or inspect every 7500 miles.
Drivebelts (T100 & Tacoma) - service or inspect initially at 45,000 miles and every 7500 miles thereafter.
Steering knuckle & chassis grease - service or inspect every 7500 miles.
Steering linkage - service or inspect every 7500 miles.
Propeller shaft grease - service or inspect every 7500 miles.
Driveshaft boots (Previa, T100 & Tacoma 4WD) - service or inspect every 7500 miles.
Driveshaft boots (4Runner 4WD) - service or inspect every 15,000 miles.
Exhaust pipes & mountings - service or inspect every 15,000 miles.
Timing belt (T100, Tacoma 5VZ-FE & Pick Up 3VZ-E) - replace every 60,000 miles.

79241CQ5

1996-99 SCHEDULED MAINTENANCE INTERVALS
(TOYOTA LAND CRUISER, PREVIA, RAV4, SIENNA, T100, TACOMA & 4RUNNER)

TO BE SERVICED	TYPE OF SERVICE	VEHICLE MILEAGE INTERVAL (x1000)																		
		5	10	15	20	25	30	35	40	45	50	55	60	65	70	75	80	85	90	95
Automatic transmission and differential fluid	S/I			✓			✓			✓			✓			✓			✓	
Ball joints and boots	S/I			✓			✓			✓			✓			✓			✓	
Brake linings, discs/drums, lines & hoses	S/I			✓			✓			✓			✓			✓			✓	
Charcoal canister	S/I												✓							
Drive belts	S/I						✓						✓						✓	
Driveshaft bushing (4WD except Previa)	L						✓						✓						✓	
Engine coolant	R						✓						✓						✓	
Engine oil & filter	R	✓	✓	✓	✓	✓	✓	✓	✓	✓	✓	✓	✓	✓	✓	✓	✓	✓	✓	✓
Exhaust pipes & mounts	S/I			✓			✓			✓			✓			✓			✓	
Fuel lines & connections, fuel tank vapor vent system hoses, fuel tank band	S/I						✓						✓						✓	
Fuel tank cap gasket	S/I						✓						✓						✓	
Halfshaft boots & flange bolts	S/I			✓			✓			✓			✓			✓			✓	
Limited slip differential fluid	R						✓						✓						✓	
Manual transmission and differential fluid	S/I						✓						✓						✓	
Non-platinum spark plugs	R						✓						✓						✓	
Platinum spark plugs	R												✓							
Propeller shaft (4WD models except RAV4)	L			✓			✓			✓			✓			✓			✓	
Propeller shaft bolts	S/I			✓			✓			✓			✓			✓			✓	
Rack and pinion assembly	S/I			✓			✓			✓			✓			✓			✓	
Rear wheel bearing (T100, Tacoma, 4Runner and Land Cruiser)	L						✓						✓						✓	
Steering Knuckle (Land Cruiser)	L			✓			✓			✓			✓			✓			✓	
Steering linkage	S/I			✓			✓			✓			✓			✓			✓	
Supercharger gear oil (Previa)	S/I						✓						✓						✓	

79241CQ6

1996-99 SCHEDULED MAINTENANCE INTERVALS
(TOYOTA LAND CRUISER, PREVIA, RAV4, SIENNA, T100, TACOMA & 4RUNNER)

TO BE SERVICED	TYPE OF SERVICE	VEHICLE MILEAGE INTERVAL (x1000)																		
		5	10	15	20	25	30	35	40	45	50	55	60	65	70	75	80	85	90	95
Transfer case and differential fluid (RAV4)	S/I			✓			✓			✓			✓			✓			✓	
Valves	S/I												✓							

R - Replace   S/I - Service or Inspect   L - Lubricate

FREQUENT OPERATION MAINTENANCE (SEVERE SERVICE)

If a vehicle is operated under any of the following conditions it is considered severe service:
- Towing a trailer or using a camper or car-top carrier.
- Repeated short trips of less than 5 miles in temperatures below freezing.
- Extensive idling or low-speed driving for long distances as in heavy commercial use, such as delivery, taxi or police cars.
- Operating on rough, muddy or salt-covered roads.
- Operating on unpaved or dusty roads.

Air filter - service or inspect every 5,000 miles or 4 months, whichever occurs first.

Brake linings and discs or drums - service or inspect every 5,000 miles or 4 months, whichever occurs first.

Steering linkage - service or inspect every 5,000 miles or 4 months, whichever occurs first.

Ball joints and boots - service or inspect every 5,000 miles or 4 months, whichever occurs first.

Halfshaft boots - service or inspect every 5,000 miles or 4 months. Retighten the flange bolts, whichever occurs first.

Body/chassis bolts and nuts - service or inspect every 5,000 miles or 4 months, whichever occurs first.

Transmission and differential fluid - replace every 15,000 miles or 12 months, whichever occurs first.

Transfer case and differential fluid (Tacoma, T100, 4Runner and Land Cruiser) - replace every 15,000 miles or 12 months, whichever occurs first.

Timing belt - replace every 60,000 miles or 48 months, whichever occurs first.

79241CQ7

SCHEDULED MAINTENANCE
TOYOTA
LAND CRUISER, PICK-UP, PREVIA, RAV4
SIENNA, T100, TACOMA, 4RUNNER

**The following should be used as a guide when determining the amount of work required for a particular service if taken to a repair shop.
In estimating how long a particular Scheduled Maintenance Service should take, please observe the following:**

- **Chilton Time** is time based on field research and data supplied by the vehicle manufacturer.
- Labor time operations are given in hours and tenths of an hour.
- All labor operations, are to be used as a guide.

Mechanic Skill Level Codes:
(G) GENERAL: Normally skilled with certification.
(M) MAINTENANCE: Semi-skilled working on certication.
(P) PRECISION: Really skilled with multiple certification.

	Chilton Time
(G) 60000 Mile Service	
1995-99 Land Cruiser	3.7
1995 Pick-up	3.8
w/AT add	.1
w/4WD add	.1
w/22RE engine add	.2
1995-97 Previa	2.8
1998-99 Sienna	2.6
1996-99 RAV4	3.6
1995-99 T100	3.6
1995-99 Tacoma	3.7
1995-99 4Runner	4.0
(M) 65000 Mile Service	
1996-99	.3
(M) 67500 Mile Service	
1995 Land Cruiser	.8
1995 Pick-up	1.0
w/4WD add	.2
1995 Previa	.8
1995 T100	1.0
w/4WD add	.1
1995 Tacoma	.8
w/4WD add	.1
(M) 70000 Mile Service	
1996-99	.3
(M) 75000 Mile Service	
1995 Land Cruiser	1.8
1996-99 Land Cruiser	2.0

	Chilton Time
1995 Pick-up	1.9
w/AT add	.1
w/4WD add	.1
1995 Previa	1.8
1996-97 Previa	
2WD	1.5
4WD	1.8
1998-99 Sienna	1.6
1996-99 RAV4	1.8
1995 T100	1.8
1996-99 T100	
4 cyl	1.7
6 cyl	
2WD	1.2
4WD	1.7
1995 Tacoma	1.9
1996-99 Tacoma	
2WD	1.7
4WD	1.8
1995 4Runner	1.9
1996-99 4Runner	
2WD	1.2
4WD	1.7
(M) 80000 Mile Service	
1996-99	.3
(M) 82500 Mile Service	
1995 Land Cruiser	.8
1995 Pick-up	1.1
w/4WD add	.1
1995 Previa	.9

	Chilton Time
1996-97 RAV4	.8
1995 T100	.9
1995 Tacoma	.8
1995 4Runner	1.1
(M) 85000 Mile Service	
1996-99	.3
(G) 90000 Mile Service	
1995-99 Land Cruiser	3.7
1995 Pick-up	3.8
w/AT add	.1
w/4WD add	.1
w/22RE engine add	.2
1995-97 Previa	2.8
1998-99 Sienna	2.6
1996-99 RAV4	3.6
1995-99 T100	3.6
1995-99 Tacoma	3.7
1995-99 4Runner	4.0
(G) 95000 Mile Service	
1996-99	.3
(M) 97500 Mile Service	
1995 Land Cruiser	.8
1995 Pick-up	1.0
w/4WD add	.2
1995 Previa	.8
1995 T100	1.0
1995 Tacoma	.8
1995 4Runner	1.0

79241CQ8

MAINTENANCE LIGHT RESETTING AND DTC RETRIEVAL

2

MAINTENANCE LIGHT RESETTING

This section describes reset procedures for maintenance lights. Maintenance lights are used to indicate to the operator of the vehicle that some type of routine maintenance should be performed. Unlike a Check Engine light that will be displayed when there is a fault with the engine management system, the maintenance light will be displayed when an engine or transmission oil change is recommended according to driving conditions. Also, the light will be displayed to indicate when the emission control system needs to be serviced.

Chrysler Corp.

RESETTING

The Emission Maintenance Reminder (EMR) light is now referred to as a Service Reminder Indicator (SRI) lamp. It is located in the dash and is labeled MAINT REQD. It is used on 5.9L V8 HDC engine and 8.0L V10 gas powered engine vehicles only. The SRI lamp will illuminate at the 60,000 and 82,000 mileage (96,000 and 131,000 km) marks and will remain ON until it is reset. Perform the required maintenance before resetting the lamp. Failure to adhere to part replacement or service required may be a violation of federal law. Resetting the SRI lamp requires the use of the second generation Digital Readout Box (DRB-II) scan tool or equivalent. Consult the scan tool's instruction guide for this procedure.

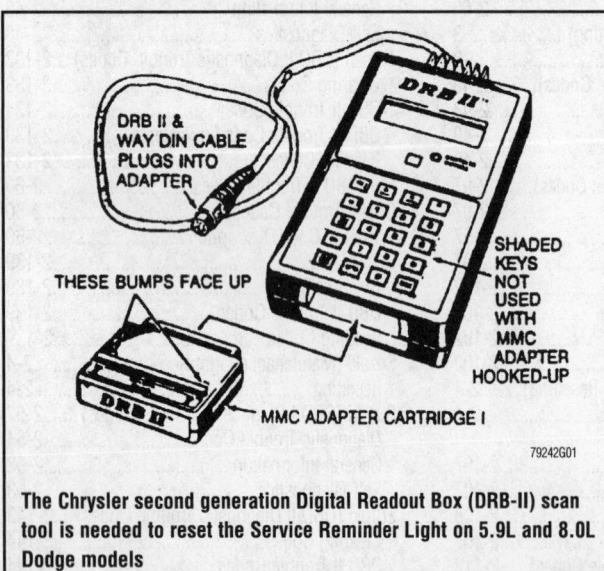

The Chrysler second generation Digital Readout Box (DRB-II) scan tool is needed to reset the Service Reminder Light on 5.9L and 8.0L Dodge models

Ford Motor Co.

RESETTING

1996–97 Explorer

The 1996–97 Explorer is equipped with the CHANGE OIL SOON or OIL CHANGE REQUIRED light. When oil life left is between five percent and zero percent, CHANGE OIL SOON will be displayed on the message center. When oil life reaches zero percent, the OIL CHANGE REQUIRED message will be displayed. The message center indicator will indicate the percent of oil life left during the System Check. This percentage is based on the driver's driving history and the time since the last oil change. In order to ensure oil life left indications, the driver should only perform the OIL CHANGE RESET procedure after every oil change. Reset the system by pressing the OIL CHANGE RESET switch and holding it for 5 seconds. After a successful reset the Message Center will display oil life indicators. The CHANGE OIL SOON or OIL CHANGE REQUIRED message will disappear after the 5 second interval.

Geo/Chevrolet

RESETTING

1995 Tracker

The CHECK ENGINE light is utilized to indicate periodic service intervals or to indicate there is a problem in the engine management system. On federal Tracker vehicles, the CHECK ENGINE light will come ON while the engine is running at the 50,000 mile (80,000 km), 80,000 mile (128,000 km), and 100,000 mile (160,000 km) marks. This alerts the driver that it is time for a scheduled service. When servicing is completed the light must be reset by turning the cancel switch, located behind lower trim panel near the steering column, off.

On 1995 California Tracker vehicles, the CHECK ENGINE light will come ON only when a fault code is sensed in the engine management system. When the problem has been corrected, clear the codes and the light should go out. A cancel switch is not used on these vehicles.

On all vehicles equipped with a CHECK ENGINE light, the light will illuminate when a fault is sensed in the engine management system. The fault will be memorized by the ECU for recall later. Once the fault has been recalled and the cause of the fault repaired, the memory should be erased. Disconnect the negative battery cable for 60 seconds or more. Reconnect the cable and start the engine. Allow the engine to reach operating temperature and ground the diagnostic switch. Check for a Code 12 and disconnect the diagnostic switch. The codes will be reset at this point.

Honda

RESETTING

1995 Passport

The 1995 Passport DX model is equipped with an Oxygen Sensor Life Indicator light. It is no longer used in 1996–99 models. The oxygen sensor must be replaced after 90,000 miles (144,000 km) of vehicle operation. When the odometer reading reaches 90,000 miles (144,000 km), the oxygen sensor life indicator light (02) will illuminate to remind the driver to change the

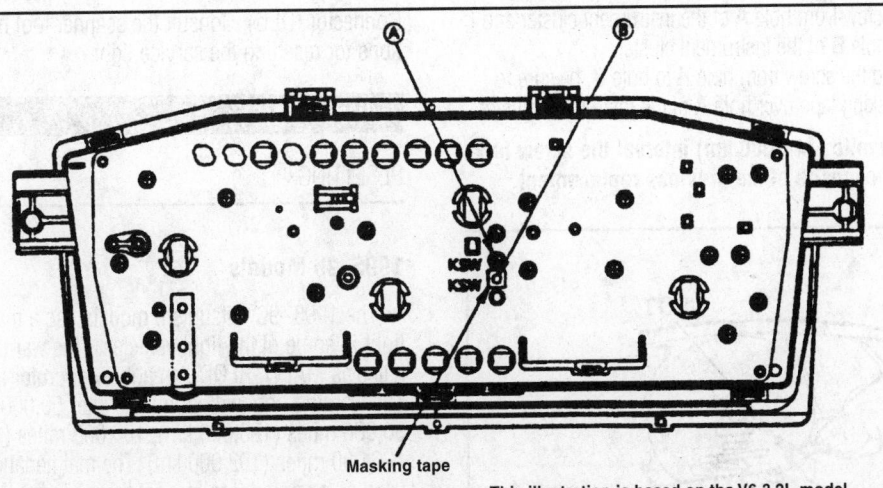

Masking tape

This illustration is based on the V6-3.2L model.

79242G02

Oxygen Sensor Life Indicator reset screw and hole locations—1995 Honda Passport DX

oxygen sensor. After replacing the oxygen sensor, the oxygen sensor life indicator light must be reset to remind the driver to replace the oxygen sensor after the next 90,000 miles (144,000 km).

➡**The reset screw is located in the back of the instrument cluster.**

1. Perform the reset procedure as follows:
   a. Remove the instrument panel cluster assembly.
   b. Remove the masking tape from hole **B**.
   c. Remove the screw from hole **A** and install it in hole **B**.
   d. Apply new masking tape over hole **A**.

➡**The above procedure assumes that the oxygen sensor is being replaced for the first time (after 90,000 miles/144,000 km). For subsequent reset procedures (at the next 90,000 miles/144,000 km), hole positions will alternate accordingly from the procedure presented here.**

Isuzu

RESETTING

1995 Amigo, Pick-Up, Trooper and Rodeo

The 1995 Amigo, Pick-Up, Trooper and Rodeo are equipped with an oxygen sensor maintenance light, located in the dash panel, that will illuminate every 90,000 miles (144,000 km), indicating that the oxygen sensor must be replaced. This light was used up through 1995, but was dropped for 1996 models. After the sensor has been replaced and the emission system checked, reset the maintenance light as follows:

1. To reset the reminder light, proceed as follows:
   a. Remove the instrument cluster assembly.
   b. Working on the backside of the instrument cluster, remove the masking tape over hole **B** in the instrument cluster.

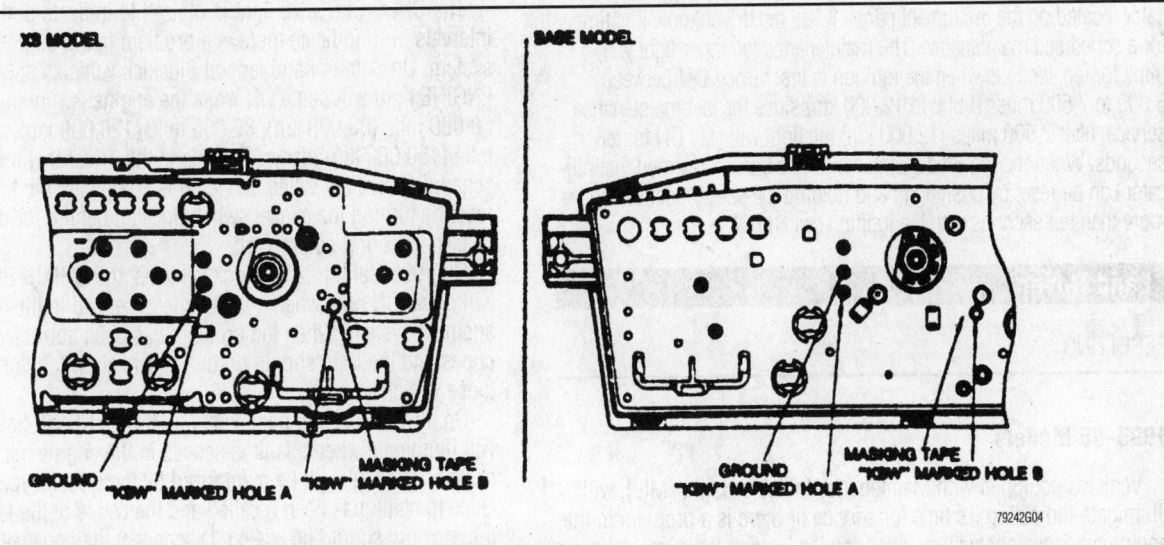

XS MODEL

BASE MODEL

GROUND "K3W" MARKED HOLE A MASKING TAPE "K3W" MARKED HOLE B

GROUND "K3W" MARKED HOLE A MASKING TAPE "K3W" MARKED HOLE B

79242G04

Oxygen sensor reminder light reset screw and hole locations—1995 Isuzu models

c. Remove the screw from hole **A** of the instrument cluster and place that screw in hole **B** of the instrument cluster.

d. After switching the screw from hole **A** to hole **B**, be sure to place a piece of masking tape over hole **A** of the instrument cluster.

➡**At the next 90,000 mile (144,000 km) interval the screw hole positions will be the opposite of the previous replacement.**

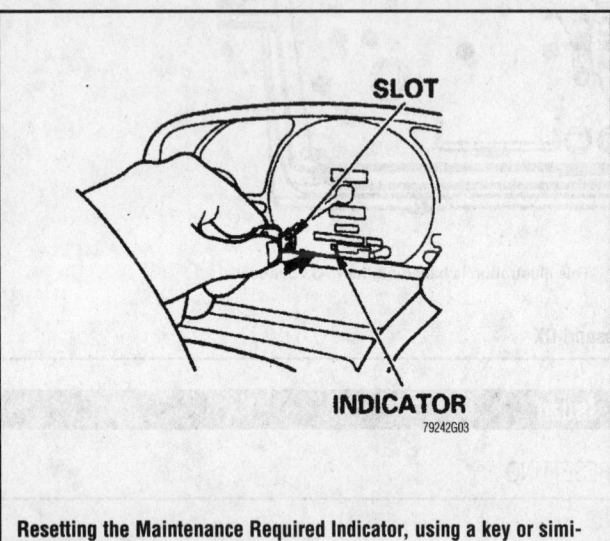

Resetting the Maintenance Required Indicator, using a key or similar object—1996 Isuzu Oasis

Oasis

The 1996–97 Oasis is equipped with a maintenance reminder indicator located on the instrument panel. It lets the driver know it is time for a scheduled maintenance. When it is near 7,500 miles (12,000 km) since the last maintenance, the indicator will turn yellow. If you exceed 7,500 miles (12,000 km), the indicator will turn red. When the required maintenance has been performed the indicator can be reset by inserting the ignition key or other similar object into the slot below the indicator. This will extinguish the indicator for the next 7500 miles (12,000 km).

The 1998–99 Oasis is equipped with a maintenance reminder indicator located on the instrument panel. It lets the driver know it is time for a scheduled maintenance. The maintenance reminder light will blink for ten seconds when the ignition is first turned **ON** between 6,000 to 7,500 miles (9500 to 12,000 km) since the last maintenance service, after 7,500 miles (12,000 km) the light will stay ON for ten seconds. When the required maintenance has been performed the indicator can be reset by pushing in and holding the select/reset switch for more than ten seconds with the ignition switch **ON**.

Land Rover

RESETTING

1995–96 Models

Vehicles equipped with Maintenance Indicator Light (MIL), will illuminate indicating it's time for service or there is a problem in the engine management system. After service is completed, reset the maintenance light.

The service reminder light can only be only reset using a special scanner tool. The scanner tool is connected to the Diagnostic Link Connector (DLC). Consult the scanner tool manufacturer's instructions for resetting the service light.

Mitsubishi

RESETTING

1995–96 Models

The 1995–96 Mitsubishi models use a maintenance warning light in some of the light vehicles. This warning light now illuminates as MAINT. REQD., an abbreviation for maintenance required. The warning light will illuminate after 50,000 miles (80,000 km), 80,000 miles (128,000 km), 100,000 miles (160,000 km) and 120,000 miles (192,000 km). The maintenance required light is used as a reminder to inspect the emission control system and to perform the required emission service related to the mileage interval. After completing the necessary emission service, the maintenance light can be reset as follows:

The reset procedure and reset switch locations have remained the same as previous vehicles. On some models, the reset switch is located on the back of the instrument cluster near the speedometer cable junction. On other models, the reset switch is located on the lower right-hand corner of the instrument cluster, behind the face panel. After the switch is located, slide the switch knob to the other side to reset the maintenance warning light.

➡**After 120,000 miles (192,000 km), the warning light bulb should be removed to prevent it from continuing to illuminate.**

Suzuki

RESETTING

1995 Sidekick and Samurai

The CHECK ENGINE light is utilized to indicate periodic service intervals or to indicate there is a problem in the engine management system. On Samurai and federal Sidekick vehicles only, the CHECK ENGINE light will come ON while the engine is running at the 50,000 mile (80,000 km), 80,000 mile (128,000 km), and 100,000 mile (160,000 km) marks. This alerts the driver that it is time for a scheduled service. When servicing is completed the light must be reset by turning the cancel switch, located behind lower trim panel near the steering column, off.

On 1995 California Sidekick vehicles, the CHECK ENGINE light will come ON only when a fault code is sensed in the engine management system. When the problem has been corrected, clear the codes and the light should go out. A cancel switch is not used on these vehicles.

On all vehicles equipped with a CHECK ENGINE light, the light will illuminate when a fault is sensed in the engine management system. The fault will be memorized by the ECU for recall later. Once the fault has been recalled and the cause of the fault repaired, the memory should be erased. Disconnect the negative battery cable for 60 seconds or more. Reconnect the cable and start the engine. Allow the engine to reach operating temperature and ground the diagnostic switch. Check for a Code 12 and disconnect the diagnostic switch. The codes will be reset at this point.

OBD I DIAGNOSTIC TROUBLE CODES

General Information

For years, vehicles have been capable of storing diagnostic trouble codes. Codes prior to the 1996 OBD II legislation have been proprietary to the vehicle manufacturer. In some cases, the codes are specific to the individual make and model.

Furthermore, some manufacturers have developed specialized devices to read their codes. This complicates code reading and clearing.

Chrysler Corp.

SELF-DIAGNOSTICS

The Chrysler fuel injection systems combine electronic spark advance and fuel control. At the center of these systems is a digital, pre-programmed computer, known as an Powertrain Control Module (PCM). The PCM can also be referred to as the Single Module Engine Controller (SMEC) or as the Single Board Engine Controller (SBEC). The PCM regulates ignition timing, air-fuel ratio, emission control devices, cooling fan, charging system idle speed and speed control. It has the ability to update and revise its commands to meet changing operating conditions.

Various sensors provide the input necessary for PCM to correctly regulate fuel flow at the injectors. These include the Manifold Absolute Pressure (MAP), Throttle Position Sensor (TPS), oxygen sensor, coolant temperature sensor, charge temperature sensor, and vehicle speed sensors.

In addition to the sensors, various switches are used to provide important information to the PCM. These include the neutral safety switch, air conditioning clutch switch, brake switch and speed control switch. These signals cause the PCM to change either the fuel flow at the injectors or the ignition timing or both.

The PCM is designed to test it's own input and output circuits, If a fault is found in a major system, this information is stored in the PCM for eventual display to the technician. Information on this fault can be displayed to the technician by means of the instrument panel CHECK ENGINE light or by connecting a diagnostic read-out tester and reading a numbered display code, which directly relates to a general fault. Some inputs and outputs are checked continuously and others are checked under certain conditions. If the problem is repaired or no longer exists, the PCM cancels the fault code after approximately 50 key **ON/OFF** cycles.

When a fault code is detected, it appears as either a flash of the CHECK ENGINE light on the instrument panel or by watching the Diagnostic Readout Box version II (ORB-II). This indicates that an abnormal signal in the system has been recognized by the PCM. Fault codes do indicate the presence of a failure but they don't identify the failed component directly.

Fault Codes

Fault codes are 2 digit numbers that tell the technician which circuit is bad. Fault codes do indicate the presence of a failure but they don't identify the failed component directly. Therefore a fault code a result and not always the reason for the problem.

Indicator Codes

Indicator codes are 2 digit numbers that tell the technician if particular sequences or conditions have occurred. Such a condition where the indicator code will be displayed is at the beginning or the end of a diagnostic test. Indicator codes will not generate a CHECK ENGINE light or engine running test code.

Actuator Test Mode (ATM) Codes

Starting in 1985, ATM test codes are 2 digit numbers that identify the various circuits used by the technician during the diagnosis procedure. In 1989 the PCM and test equipment changed design. The actuator test functions where expanded, but access to these functions may have changed, dependent on vehicle or test equipment being used.

Engine Running Test Codes

Engine running test codes where introduced on fuel injected vehicles. These are 2 digit numbers. The codes are used to access sensor readouts while the engine is running and place the engine in particular operating conditions for diagnosis. Feedback carburetor system does not offer engine running sensor test mode.

Check Engine Light

This is possibly the most critical step of diagnosis. A detailed examination of connectors, wiring and vacuum hoses can Offen lead to a repair without further diagnosis. A careful inspector will check the undersides of hoses as well as the integrity of hard-to-reach hoses blocked by the air cleaner or other component. Wiring should be checked carefully for any sign of strain, burning, crimping, or terminals pulled-out from a connector. Checking connectors at components or in harnesses is required; usually, pushing them together will reveal a loose fit.

The CHECK ENGINE or Maintenance Indicator Lamp (MIL) light has 2 modes of operation: diagnostic mode and switch test mode.

If a ORB-II diagnostic tester is not available, the PCM can show the technician fault codes by flashing the CHECK ENGINE light on the instrument panel in the diagnostic mode. In the switch test mode, after all codes are displayed, switch function can be confirmed. The light will turn on and off when a switch is turned ON and OFF.

Even though the light can be used as a diagnostic tool, it cannot do the following:

Once the light starts to display fault codes, it cannot be stopped. If the technician loses count, he must start the test procedure again. The light cannot display all of the codes or any blank displays.

The light cannot tell the technician if the oxygen feed-back system is lean or rich and if the idle motor and detonation systems are operational. The light cannot perform the actuation test mode, sensor test mode or engine running test mode.

➡ Be advised that the CHECK ENGINE light can only perform a limited amount of functions and is not to be used as a substitute for a diagnostic tester. All diagnostic procedure described herein are intended for use with a Diagnostic Readout Box II (DRB-II) or equivalent tool.

Limp-In Mode

The limp-in mode is the attempt by the PCM to compensate for the failure of certain components by substituting information from other sources. If the PCM senses incorrect data or no data at all from the MAP sensor, throttle position sensor or coolant temperature sensor, the system is placed into limp-in mode and the CHECK ENGINE light on the instrument panel is activated. This mode will keep the vehicle drive able until the customer can get it to a service facility.

Test Modes

There are 5 modes of testing required for the proper diagnosis of the system. They are as follows:

Diagnostic Test Mode This mode is used to access the fault codes from the PCM's memory.

Circuit Actuation Test Mode (ATM Test) This mode is used to turn a certain circuit on and off in order to test it. ATM test codes are used in this mode.

Switch Test Mode This mode is used to determine if specific switch inputs are being received by the PCM.

Sensor Test Mode This mode looks at the output signals of certain sensors as they are received by the PCM when the engine is not running. Sensor access codes are read in this mode. Also this mode is used to clear the PCM memory of stored codes.

Engine Running Test Mode This mode looks at sensor output signals as seen by the PCM when the engine is running. Also this mode is used to determine some specific running conditions necessary for diagnosis.

READING CODES

Obtaining Trouble Codes

Entering the Jeep or Eagle self-diagnostic system requires the use of a special adapter that connects with the Diagnostic Readout Box II (DRB-II). These systems require the adapter because all of the system diagnosis is done Off-Board instead of On-Board like most vehicles. The adapter, which is a computer module itself, measures signals at the diagnostic connector and converts the signals into a form which the DRB can use to perform tests. On vehicles other than Jeep and Eagle the following procedures will obtain stored Diagnostic Trouble Codes (DTC).

USING THE CHECK ENGINE LAMP

Codes display on vehicles built before 1989 are displayed in numerical order, after 1989 codes are displayed in order of occurrence.

1. Connect the readout box to the diagnostic connector located in the engine compartment near PCM.

2. Start the engine, if possible, cycle the transmission selector and the A/C switch if applicable. Shut off the engine.

3. Turn the ignition switch ON—OFF, ON—OFF, ON—OFF, ON within 5 seconds.

4. Observe the CHECK ENGINE light on the instrument panel.

5. Just after the last ON cycle, the dash warning (MIL) lamp will begin flashing the stored codes.

6. The codes are transmitted as two digit flashes.

7. Example would be Code 21 will be displayed as a FLASH FLASH pause FLASH.

8. Be ready to write down the codes as they appear; the only way to repeat the codes is to start over at the beginning.

SCAN TOOL

The scan tool is the preferred choice for fault recovery and system diagnosis. Some hints on using the ORB-II include:

• To use the HELP screen, press and hold F3 at any time.

• To restart the ORB-II at any time, hold the MODE button and press ATM at the same time.

• Pressing the up or down arrows will move forward or backward one item within a menu.

• To select an item, either press the number of the item or move the cursor arrow to the selection, then press ENTER.

• To return to the previous display (screen), press ATM.

• Some test screens display multiple items. To view only one, move the cursor arrow to the desired item, then press ENTER.

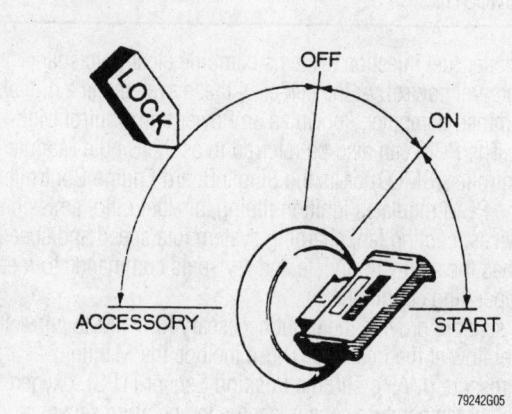

Cycle the ignition switch ON-OFF three times to enter the diagnostic mode

To read stored faults with the ORB-II:

1. With the ignition switch OFF, connect the tool to the diagnostic connector near the engine controller under the hood. On some 1988 and earlier models, cycling the ignition key ON-OFF three times may be necessary to enter the diagnostics. On 1989 and newer models, simply turn the ignition switch ON to access the read fault code data.

2. Start the engine if possible. Cycle the transmission from Park to a forward gear, then back to Park. Cycle the air conditioning ON and OFF. Turn the ignition switch OFF.

3. Turn the ignition switch ON but do not start the engine. The ORB-II will begin its power-up sequence; do not touch any keys on the scan tool during this sequence.

4. Reading faults must be selected from the FUEL/IGN MENU. To reach this menu on the ORB-II:

a. When the initial menu is displayed after the power-up sequence, use the down arrow to display choice 4) SELECT SYSTEM and select this choice.

b. Once on the—SELECT SYSTEM—screen, choose 1) ENGINE. This will enter the engine diagnostics section of the program.

c. The screen will momentarily display the engine family and SBEC identification numbers. After a few seconds the screen displays the choices 1) With A/C and 2) Without A/C. Select and enter the correct choice for the vehicle.

d. When the—ENGINE SYSTEM—screen appears, select 1) FUEL/IGNITION from the menu.

e. On the next screen, select 2) READ FAULTS.

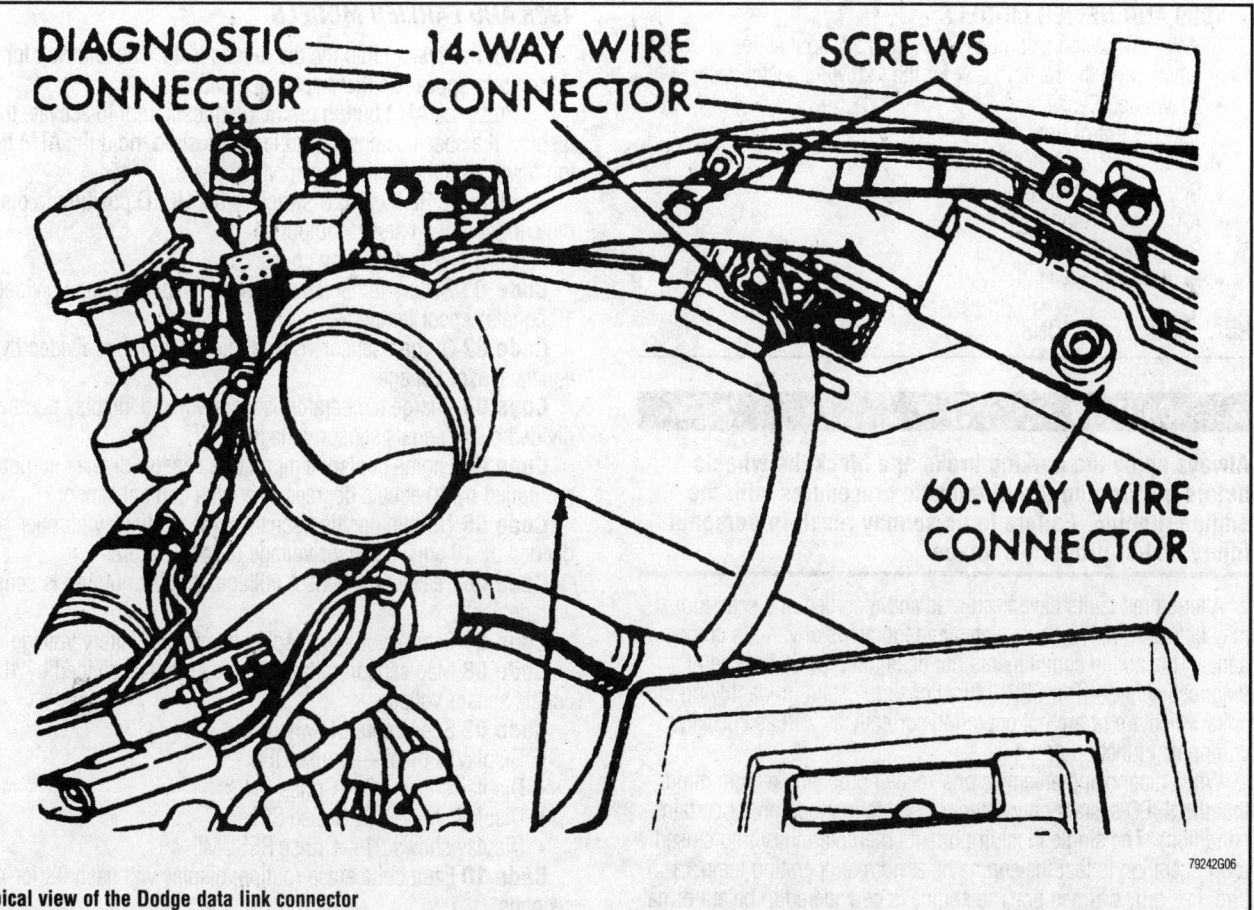

Typical view of the Dodge data link connector

5. If any faults are stored, the display will show how many are stored (1 of 4 faults, etc.) and issue a text description of the problem, such as COOLANT SENSOR VOLTAGE TOO LOW. The last line of the display shows the number of engine starts since the code was set. If the number displayed is 0 starts, this indicates a hard or current fault. Faults are displayed in reverse order of occurrence; the first fault shown is the most current and the last fault shown is the oldest.

6. Press the down arrow to read each fault after the first. Record the screen data carefully for easy reference.

7. If no faults are stored in the controller, the display will state NO FAULTS DETECTED and show the number of starts since the system memory was last erased.

8. After all faults have been read and recorded, press ATM.

9. Refer to the appropriate diagnostic chart for a diagnostic path. Remember that the fault message identifies a circuit problem, not a component. Use of the charts is required to sequentially test a circuit and identify the fault.

SWITCH TEST

The PCM only recognizes 2 switch input states—HI and LOW. For this reason the PCM cannot tell the difference between a selected switch position and an open circuit, short circuit or an open switch. However, if one of the switches is toggled, the controller does have the ability to respond to the change of state in the switch. If the change is displayed, it can be assumed that entire switch circuit to the PCM is operational.

1988 AND EARLIER MODELS:

After all codes have been shown and has indicated Code 55 end of message, actuate the following component switches. The digital display must change its numbers between 00 and 88 and the CHECK ENGINE light will blink when the following switches are activated and released:

- Brake pedal
- Gear shift selector
- A/C switch
- Electric defogger switch (1984)

```
----- FUEL/IGN FAULTS -----
   NO FAULTS DETECTED

   X STARTS SINCE ERS
```

```
   1 OF X FAULTS
      [message
    appears here]
   X STARTS SINCE SET
```

79242G07

Example of the DRB-II display screen while reading the trouble codes

1989 AND NEWER MODELS:

To enter the switch test mode, activate read input states or equivalent function on the readout box for the following switch tests:

- Z1 Voltage Sense
- Speed Control Set
- Speed Control ON/OFF
- Speed Control Resume
- A/C Switch Sense
- Brake Switch
- Park/neutral Switch

SCAN TOOL FUNCTIONS

✳✳ CAUTION

Always apply the parking brake and block the wheels before performing any diagnostic procedures with the engine running. Failure to do so may result in personal injury and/or property damage.

After stored faults have been read and recorded, the scan tool may be used to investigate states and functions of various components. This ability compliments but does not replace the use of diagnostic charts. The ORB-II functions are useful in identifying circuits which are or are not operating correctly as well as checking component function or signal.

When diagnosing an emissions-related problem, keep in mind that the SBEC system only enters closed loop mode under certain conditions. The single most important criteria for entry into closed loop operation is that the engine be at normal operating temperature; i.e., fully warmed up. The engine is considered to be at normal operating temperature if any of the following are true: the electric cooling fan cycles on at least once or the upper radiator hose is hot to the touch or the heater is able to deliver hot air.

In open loop operation, the signal from the oxygen sensor is ignored by the engine controller and the fuel injection is controlled by pre-programmed values within the computer. Once closed loop operation is begun, the signal from the oxygen sensor is used by the engine controller to constantly adjust the fuel injection to maintain the proper air/fuel ratio. The system will switch in and out of closed loop operation depending on sensor signals and driver input. In most cases, the system will be in closed loop operation during normal driving, acceleration or deceleration and idle. Wide open throttle will cause the system to momentarily switch to open loop operation. Additionally, some engine control systems will momentarily switch to open loop under hard acceleration or deceleration until the MAP sensor signal stabilizes.

The ORB-II may be operated in the following diagnostic modes from the FUEL/IGN MENU screen.

Sensors

This function displays current data being transmitted from the fuel and ignition sensors to the engine controller. Examples of sensor data available include MAP voltage, throttle position sensor voltage and percentage, RPM, coolant temperature, voltage sensor, total spark advance, and vehicle speed. Many other sensors may be monitored depending on engine/transmission combinations.

Data for each sensor is displayed in the appropriate units, such as volts, mph, in. Hg, degrees F, etc.

1988 AND EARLIER MODELS

1. Put the system into the diagnostic test mode and wait for Code 55 to appear on the display screen.

2. Press the ATM button on the diagnostic tool to activate the display. If a specific sensor read test is desired, hold the ATM button down until the desired test code appears.

3. Slide the READ/HOLD switch to the HOLD position to display the corresponding sensor output.

Sensor Read Test Display codes:

Code 01 Battery temperature sensor; display voltage divided by 10 equals sensor temperature

Code 02 Oxygen sensor voltage; display number divided by 10 equals sensor voltage

Code 03 Charge temperature sensor voltage; display number divided by 10 equals sensor voltage

Code 04 Engine coolant temperature sensor; display number multiplied by 10 equals degrees of engine coolant sensor

Code 05 Throttle position sensor voltage; display number divided by 10 equals sensor voltage or temperature

Code 06 Peak knock sensor voltage; display number is sensor voltage

Code 07 Battery voltage; display number is battery voltage

Code 08 Map sensor voltage; display number divided by 10 equals sensor voltage

Code 09 Speed control switches:
- Display is blank—Cruise OFF
- Display shows 00—Cruise ON
- Display shows 10- Cruise SET
- Display shows 01—Cruise RESUME

Code 10 Fault code erase routine; display will flash 0's for 4 seconds

The State Display programs allow the operator to view the present conditions in the SBEC system. These choices are displayed on the FUEL IGN STATE screen and offer the choices of MODULE INFO, SENSORS, INPUTS/OUTPUTS or MONITORS. Viewing system data through these windows can be helpful in observing the effects of repairs or to compare the problem vehicle to a known-good vehicle.

1989 AND NEWER MODELS

To enter the sensor test mode, activate read sensor voltage or read sensor values or equivalent functions on the readout box for the following sensor displays:

Read Sensor Voltage
- Battery temperature sensor
- Oxygen sensor input
- Throttle body temperature sensor
- Coolant temperature sensor
- Throttle position
- Minimum throttle
- Battery voltage
- MAP sensor voltage

Read Sensor Values
- Throttle body temperature
- Coolant temperature
- MAP gauge reading
- AIS motor position
- Added adaptive fuel
- Adaptive fuel factor

- Barometric pressure
- Engine speed
- Module spark advance
- Vehicle speed
- Oxygen sensor state

Engine Running Test Mode

1988 AND EARLIER MODELS

The Engine Running Test Mode monitors the sensors on the vehicle which check operating conditions while the engine is running. The engine running test mode can be performed with the engine idling in NEUTRAL and with parking brake set or under actual driving conditions. With the diagnostic readout box READ/HOLD switch in the READ position, the engine running test mode is initiated after the engine is started.

Select a test code by switching the READ/HOLD switch to the READ position and pressing the actuator button until the desired code appears. Release actuator button and switch the READ/HOLD switch to the HOLD position. The logic module will monitor that system test and results will be displayed.

Only fuel injected engines offer this function. The Feedback carburetor system does not offer engine running sensor test mode.

ENGINE RUNNING TEST DISPLAY CODES:

Code 61 Battery temperature sensor; display number divided by 10 equals voltage

Code 62 Oxygen sensor; display number divided by 10 equals voltage

Code 63 Fuel injector temperature sensor; display number divided by 10 equals voltage

Code 64 Engine coolant temperature sensor; display number multiplied by 10 equals degrees F

Code 65 Throttle position sensor; display number divided by 10 equals voltage

Code 67 Battery voltage sensor; display is voltage

Code 68 Manifold vacuum sensor; display is in. Hg code 69—Minimum throttle position sensor; display number divided by 10 equals voltage

Code 70 Minimum airflow idle speed sensor; display number multiplied by 10 equals rpm (see minimum air flow check procedure)

Code 71 Vehicle speed sensor; display is mph Code 72—Engine speed sensor; display number multiplied by 10 equals rpm

FUEL/IGNITION INPUT/OUTPUT:

The engine controller recognizes only two states of electrical signals, voltage high or low. In some cases this corresponds to a switch or circuit being on or off; in other circuits a voltage signal may change from low voltage to higher voltage as a sensor opens. The controller cannot recognize the difference between a selected switch position and an open or shorted circuit.

In this test mode, the change in the circuit may be viewed as the switch is operated. For example, if the BRAKE SWITCH state is selected, the display should change from Low to High as the brake pedal is pressed. If a change in a circuit is displayed as the switch is used, it may be reasonably assumed that the entire switch circuit into the engine controller is operating correctly.

Depending on the engine/transmission in the vehicle, some of the switch states which may be checked include the air conditioning switch, brake switch, park/neutral switch, fuel flow signal, air conditioning clutch relay, radiator fan relay, CHECK ENGINE lamp, overdrive solenoid(s), lock-up solenoid and the speed control vent or vacuum solenoids. The scan tool will recognize the correct choices for each vehicle and only offer the appropriate systems on the screen.

MONITORS

On vehicles built before 1991, this display is called ENGINE PARAMETERS. 1991 and newer vehicles name the screen MONITORS. This display allows close observation of groups of related signals. For example, if RPM is chosen, the screen will display data for many of the factors affecting the rpm such as throttle position sensor, advance, air conditioning status, park/neutral status, AIS status and coolant temperature.

One of the screens within this test is NO START. When this display is selected, the screen shows the initial data sent to the engine controller during cranking. Using this screen to identify missing or unusual signals can shorten diagnostic time.

Actuator Tests

The purpose of the circuit actuation mode test is to check for proper operation of the output circuits that the PCM cannot internally recognize. The PCM can attempt to activate these outputs and allow the technician to affirm proper operation. Most of the tests performed in this mode issue an audible click or visual indication of component operation (click of relay contacts, injector spray, etc.). Except for intermittent conditions, if a component functions properly when it is tested, it can be assumed that the component, attendant wiring and driving circuit are functioning properly.

1988 AND EARLIER MODELS

The Actuator Test Mode 10 Code number was introduced in 1985. In 1983–84 ATM function only provided 3 ignition sparks, 2 AIS motor cycles and 1 injector pulse.

1. Put the system into the diagnostic test mode and wait for Code 55 to appear on the display screen.

2. Press ATM button on the tool to activate the display. If a specific ATM test is desired, hold the ATM button down until the desired test code appears.

3. The computer will continue to turn the selected circuit on and off for as long as 5 minutes or until the ATM button is pressed again or the ignition switch is turned to the OFF position.

4. If the ATM button is not pressed again, the computer will continue to cycle the selected circuit for 5 minutes and then shut the system off. Turning the ignition to the OFF position will also turn the test mode off.

ACTUATOR TEST DISPLAY CODES:

Code 01 Spark activation—once every 2 seconds

Code 02 Injector activation—once every 2 seconds

Code 03 AIS activation—one step open, one step closed every 4 seconds

Code 04 Radiator fan relay—once every 2 seconds code 05—A/C WOT cutout relay—once every 2 seconds

Code 06 ASD relay activation—once every 2 seconds code 07—Purge solenoid activation—one toggle every 2 seconds (The A/C fan will run continuously and the A/C switch must be in the ON position to allow for actuation)

Code 08 Speed control activation—speed control vent and vacuum every 2 seconds (Speed control switch must be in the ON position to allow for activation)

Code 09 Alternator control field activation—one toggle every 2 seconds

Code 10 Shift indicator activation—one toggle every 2 seconds

Code 11 EGR diagnosis solenoid activation—one toggle every 2 seconds

1989 AND NEWER MODELS

This family of tests is chosen from the FUEL/IGN MENU screen. The actuator tests allow the operation of the output circuits not recognized by the engine controller to be checked by energizing them on command. Testing in this fashion is necessary because the controller does not recognize the function of all the external components. If an output to a relay is triggered, and the relay is heard to click, it may be reasonably assumed that both the output circuit and the relay are operating properly. In this mode, most of the tests cause a response that may be seen or heard, although close attention may be necessary to notice the change.

Once selected, the ACTUATOR TEST screen offers a choice of items to be activated. Depending on engine and fuel system, some of the choices include:

- Stop all tests
- Engine rpm
- Ignition coil
- Fuel injector
- Fuel system
- Solenoid/relay
- AIS motor

The engine speed may be set to a desired level through the ENGINE RPM screen. Once a system is chosen, related screens will appear allowing detailed selection of which relay, injector or component is to be operated.

Exiting Diagnostic Test

By turning the ignition switch to the OFF position, the test mode system is exited. With a Diagnostic Readout Box attached to the system and the ATM control button not pressed, the computer will continue to cycle the selected circuits for 5 minutes and then automatically shut the system down.

Clearing Codes

Stored faults should only be cleared by use of the ORB-II or similar scan tool. Disconnecting the battery will clear codes but is not recommended as doing so will also clear all other memories on the vehicle and may affect drive ability. Disconnecting the PCM connector will also clear codes, but on newer models it may store a power loss code and will affect driveability until the vehicle is driven and the PCM can relearn it's drive ability memory.

The—ERASE—screen will appear when ATM is pressed at the end of the stored faults. Select the desired action from ERASE or DON'T ERASE. If ERASE is chosen, the display asks ARE YOU SURE? Pressing ENTER erases stored faults and displays the message FAULTS ERASED. After the faults are erased, press ATM to return the FUEL/IGN MENU.

VEHICLE SELF-DIAGNOSTICS

Dodge Built Fuel Injection System

Code 88 Display used for start of test

Code 11 Camshaft signal or Ignition signal—no reference signal detected during engine cranking

Code 12 Memory to controller has been cleared within 50-100 engine starts

Code 13 MAP sensor pneumatic signal—no variation in MAP sensor signal is detected or no difference is recognized between the engine MAP reading and the stored barometric pressure reading

Code 14 MAP voltage too high or too low

Code 15 Vehicle speed sensor signal—no distance sensor signal detected during road load conditions

Code 16 Knock sensor circuit—Open or short has been detected in the knock sensor circuit

Code 16 Battery input sensor—battery voltage sensor input below 4 volts with engine running

Code 17 Low engine temperature—engine coolant temperature remains below normal operating temperature during vehicle travel; possible thermostat problem

Code 21 Oxygen sensor signal—neither rich or lean condition is detected from the oxygen sensor input

Code 22 Coolant voltage low—coolant temperature sensor input below the minimum acceptable voltage/Coolant voltage high—coolant temperature sensor input above the maximum acceptable voltage

Code 23 Air Charge or Throttle Body temperature voltage HIGH/LOW—charge air temperature sensor input is above or below the acceptable voltage limits

Code 24 Throttle Position sensor voltage high or low.

Code 25 Automatic Idle Speed (AIS) motor driver circuit—short or open detected in 1 or more of the AIS control circuits

Code 26 Injectors No. 1, 2, or 3 peak current not reached, high resistance in circuit

Code 27 Injector control circuit—bank output driver stage does not respond properly to the control signal

Code 27 Injectors No. 1, 2, or 3 control circuit and peak current not reached

Code 31 Purge solenoid circuit—open or short detected in the purge solenoid circuit

Code 32 Exhaust Gas Recirculation (EGR) solenoid circuit—open or short detected in the EGR solenoid circuit EGR system failure—required change in fuel/air ratio not detected during diagnostic test

Code 32 Surge valve solenoid—open or short in turbocharger surge valve circuit—some 1993 vehicles

Code 33 Air conditioner clutch relay circuit—open or short detected in the air conditioner clutch relay circuit. If vehicle doesn't have air conditioning ignore this code

Code 34 Speed control servo solenoids or MUX speed control circuit HIGH/LOW—open or short detected in the vacuum or vent solenoid circuits or speed control switch input above or below allowable voltage

Code 35 Radiator fan control relay circuit—open or short
etected in the radiator fan relay circuit
Code 35 Idle switch shorted—switch input shorted to
ound—some 1993 vehicles
Code 36 Wastegate solenoid—open or short detected in the
rbocharger wastegate control solenoid circuit
Code 37 Part Throttle Unlock (PTU) circuit for torque converter
utch—open or short detected in the torque converter part throttle
nlock solenoid circuit
Code 37 Baro Reed Solenoid—solenoid does not turn off when
should
Code 37 Shift indicator circuit (manual transaxle)
Code 37 Transaxle temperature out of range—some 1993 models
Code 41 Charging system circuit—output driver stage for gen-
ator field does not respond properly to the voltage regulator con-
ol signal
Code 42 Fuel pump or no Auto shut-down (ASD) relay voltage
nse at controller
Code 43 Ignition control circuit—peak primary circuit current
ot respond properly with maximum dwell time
Code 43 Ignition coil #1, 2, or 3 primary circuits—peak pri-
ary was not achieved within the maximum allowable dwell time
Code 44 Battery temperature voltage—problem exists in the
CM battery temperature circuit or there is an open or short in the
ngine coolant temperature circuit
Code 44 Fused J2 circuit in not present in the logic board; used
n the single engine module controller system
Code 45 Turbo boost limit exceeded—MAP sensor detects
verboost
Code 44 Overdrive solenoid circuit—open or short in overdrive
olenoid circuit
Code 46 Battery voltage too high—battery voltage sense input
bove target charging voltage during engine operation
Code 47 Battery voltage too low—battery voltage sense input
elow target charging voltage
Code 51 Air/fuel at limit—oxygen sensor signal input indicates
EAN air/fuel ratio condition during engine operation
Code 52 Air/fuel at limit—oxygen sensor signal input indicates
CH air/fuel ratio condition during engine operation
Code 52 Logic module fault—1984 vehicles. Code 53—Inter-
al controller failure—internal engine controller fault condition
etected during self test
Code 54 Camshaft or (distributor sync.) reference circuit—No
amshaft position sensor signal detected during engine rotation
Code 55 End of message
Code 61 Baro read solenoid—open or short detected in the
aro read solenoid circuit
Code 62 EMR mileage not stored—unsuccessful attempt to
pdate EMR mileage in the controller EEPROM
Code 63 EEPROM write denied—unsuccessful attempt to write
 an EEPROM location by the controller
Code 64 Flex fuel sensor—Flex fuel sensor signal out of
nge—(new in 1993)- CNG Temperature voltage out of range—CN
as pressure out of range
Code 65 Manifold tuning valve—an open or short has been
etected in the manifold tuning valve solenoid circuit (3.3L and
SL LH-Platform)

Code 66 No CCO messages or no BODY CCD messages or no
EATX CCD messages—messages from the CCD bus or the BODY
CCD or the EATX CCD were not received by the PCM
Code 76 Ballast bypass relay—open or short in fuel pump relay
circuit
Code 77 Speed control relay—an open or short has been
detected in the speed control relay
Code 88 Display used for start of test
Code Error Fault code error—Unrecognized fault 10 received
by DRB

➡️**This list is for reference and does not mean that a component is defective. The code identifies the circuit and component that require further testing.**

Dodge Feedback Carburetor System

Code 88 Display used for start of test—must appear or other
codes aren't valid
Code 11 Carburetor oxygen solenoid
Code 12 Transmission unlock relay—3.7L and 5.2L
Code 13 Air switching solenoid—3.7L and 5.2L—or Vacuum
operated secondary control solenoid—2.2L
Code 14 Battery feed to computer disconnected with 20–40
engine starts
Code 16 Ignore
Code 17 Electronic throttle control solenoid
Code 18 EGR or Purge control solenoid
Code 21 Distributor pick-up signal
Code 22 Oxygen feedback stays rich or lean too long—3.7L
and 5.2L—or Oxygen feedback is LEAN too long—2.2L
Code 23 Oxygen feedback is RICH too long—2.2L
Code 24 Vacuum transducer signal problem
Code 25 Charge temperature switch signal—3.7L and
5.2L engine—or Radiator fan temperature switch signal—2.2L
engine
Code 26 Charge temperature sensor signal—3.7L and 5.2L
engine—or Engine temperature sensor signal—2.2L
Code 28 Speed sensor circuit (if equipped)
Code 31 Battery feed to computer
Code 32 Computer can't enter diagnostics
Code 33 Computer can't enter diagnostics
Code 55 End of message
Code 88 Display used for start of test
Code 00 Diagnostic readout box is powered up and waiting for
codes

➡️**This list is for reference and does not mean that a component is defective. The code identifies the circuit and component that require further testing.**

Jeep and Eagle Built Fuel Systems

1988–90 2.5L, 3.0L AND 4.0L ENGINE

Code 1000 Ignition line low
Code 1001 Ignition line high
Code 1002 Oxygen heater line
Code 1004 Battery voltage low
Code 1005 Sensor ground line out of limits

Code 1010 Diagnostic enable line low
Code 1011 Diagnostic enable line high
Code 1012 MAP line low
Code 1013 MAP line high
Code 1014 Fuel pump line low
Code 1015 Fuel pump line high
Code 1016 Charge air temperature sensor low
Code 1017 Charge air temperature sensor high
Code 1018 No serial data from the ECU
Code 1021 Engine failed to start due to mechanical, fuel, or ignition problem
Code 1022 Start line low
Code 1024 ECU does not see start signal
Code 1025 Wide open throttle circuit low
Code 1027 ECU sees wide open throttle
Code 1028 ECU does not see wide open throttle
Code 1031 ECU sees closed throttle
Code 1032 ECU does not see closed throttle
Code 1033 Idle speed increase line low
Code 1034 Idle speed increase line high
Code 1035 Idle speed decrease line low
Code 1036 Idle speed decrease line high
Code 1037 Throttle position sensor reads low
Code 1038 Park/Neutral line high
Code 1040 Latched B+ line low
Code 1041 Latched B+ line high
Code 1042 No Latched B+ 1/2 volt drop
Code 1047 Wrong ECU
Code 1048 Manual vehicle equipped with automatic ECU
Code 1949 Automatic vehicle equipped with manual ECU
Code 1050 Idle RPM less than 500
Code 1051 Idle RPM greater than 2000
Code 1052 MAP sensor out of limits
Code 1053 Change in MAP reading out of limits
Code 1054 Coolant temperature sensor line low
Code 1055 Coolant temperature sensor line high
Code 1056 Inactive coolant temperature sensor
Code 1057 Knock circuit shorted
Code 1058 Knock value out of limits
Code 1059 A/C request line low
Code 1060 A/C request line high
Code 1061 A/C select line low
Code 1062 A/C select line high
Code 1063 A/C clutch line low
Code 1064 A/C clutch line high
Code 1065 Oxygen reads rich
Code 1066 Oxygen reads lean
Code 1067 Latch relay line low
Code 1068 Latch relay line high
Code 1070 A/C cutout line low
Code 1071 A/C cutout line high
Code 1073 ECU does not see speed sensor signal
Code 1200 ECU defective
Code 1202 Injector shorted to ground
Code 1209 Injector open
Code 1218 No voltage at ECU from power latch relay
Code 1220 No voltage at ECU from EGR solenoid
Code 1221 No injector voltage

Code 1222 MAP not grounded
Code 1223 No ECU tests run

➡**Prior to 1988 vehicles used an Off-Board Diagnostic system which required special diagnostic equipment to read codes. After 1991 Jeep and Eagle vehicles used the Chrysler Domestic Built Engine Control system. The code list for Chrysler Built Domestic Fuel injection System also covers 1991 and newer Jeep and Eagle vehicles.**

Ford Motor Co.

INTRODUCTION TO FORD SELF-DIAGNOSTICS

The engine control systems are used in conjunction with either throttle body (CFI) injection or multi-point (EFI and SEFI) injection fuel delivery system or feedback carburetor systems depending on the year, model and powertrain. Although the individual system components vary slightly, the electronic control system operation i basically the same. The major difference is the number and type of output devices being controlled by the ECA.

Automotive manufacturers have developed on-board computers to control engines, transmissions and many other components. These on-board computers with dozens of sensors and actuators have become almost impossible to test without the help of electronic test equipment.

One of these electronic test devices has become the on-board computer itself. The Powertrain Control Modules (PCM), sometime called the Electronic Control Assembly (ECA), used on toadies veh cles has a built in self testing system. This self test ability is called self-diagnosis. The self-diagnosis system will test many or all of th sensors and controlled devices for proper function. When a malfunction is detected this system will store a fault code in memory that's related to that specific circuit. You can access the computer t obtain fault codes recorded in memory by using an analog voltmet or special diagnostic scan tool. This will help narrow down what area to begin testing.

Fault code meanings can vary from year to year even on the same model. It is extremely important after retrieving a fault code t verify its meaning with a proper manual. Servicing a fault code incorrectly will not only lead to the wrong conclusion but could als cause damage if tested or serviced incorrectly. There is a list of ge eral code descriptions provide later in this manual.

What System Is On My Vehicle?

There are 3 electronic fuel control systems used by Ford Motor Company. These systems all operate using similar components an on-board computers. Self-Diagnostic on these systems will vary, but, the basic fuel control operation is the same. Ford uses the following systems:

• **EEC-IV and EEC-V** engine control system: used on most domestic built Ford vehicles since 1984.
• **Non-NAAO EEC** engine control system: used on import buil Ford vehicles, referred to as Non-NAAO cars.
• **MCU** feedback carburetor system: used on most Ford vehicle before 1984 and some later model vehicles equipped with a V8 engine and feedback carburetor.

Most Ford vehicles made after 1983 use the 4th generation electronic Engine Control system, commonly designated EEC-IV.

If you own a vehicle with a 2.0L, 2.2L, or 2.5L engine, then the fuel control system is referred to as NON-NAAO (Not North American Automotive Operations produced vehicles) system. The fuel system used on these vehicles is called Electronic Engine Control (EEC). This Non-NAAO EEC system components and operation are basically the same as the EEC-IV system. The self-diagnostic function on the EEC system differs from the EEC-VI system and is covered under NON-NAAO vehicle.

Most 1984–94 Ford domestic built vehicles employ the 4th generation Electronic Engine Control system, commonly called EEC-IV, to manage fuel, ignition and emissions on vehicle engines. In 1994 the EEC-V system was introduced on some models. The diagnostic system on EEC-V provides 3 digit codes in place of 2 digit codes, and it is capable of monitoring more inputs and outputs.

If your vehicle was made before 1984, or has a feedback carburetor equipped V8 engine, then it probably uses the Microprocessor Control Unit (MCU). The MCU system was used on most 1981-83 carburetor equipped vehicles, and 1984 and newer V8 engines with feedback carburetors. The MCU system uses a large six sided connector, identical to the one used with EEC-IV systems. The MCU system does NOT use the small single wire connector, like the EEC-IV system. The MCU system is covered in greater detail later in this manual.

EEC-IV & EEC-V DIAGNOSTIC SYSTEMS

Most 1984–94 Ford domestic built vehicles employ the 4th generation Electronic Engine Control system, commonly designated EEC-IV, to manage fuel, ignition and emissions on vehicle engines. In 1994 the EEC-V system was introduced on some models. The diagnostic system on EEC-V provides 3 digit codes in place of 2 digit codes and monitors more components.

Engine Control System

The Powertrain Control Modules (PCM), usually referred to as the Electronic Control Assembly (ECA) by Ford, is given responsibility for the operation of the emission control devices, cooling fans, ignition and advance and in some cases, automatic transmission functions. Because the EEC-IV oversees both the ignition timing and the fuel injector operation, a precise air/fuel ratio will be maintained under all operating conditions. The ECA is a microprocessor or small computer which receives electrical inputs from several sensors, switches and relays on and around the engine.

Based on combinations of these inputs, the ECA controls outputs to various devices concerned with engine operation and emissions. The engine control assembly relies on the signals to form a correct picture of current vehicle operation. If any of the input signals is incorrect, the ECA reacts to whatever picture is painted for it. For example, if the coolant temperature sensor is inaccurate and reads too low, the ECA may see a picture of the engine never warming up. Consequently, the engine settings will be maintained as if the engine were cold. Because so many

inputs can affect one output, correct diagnostic procedures are essential on these systems.

One part of the ECA is devoted to monitoring both input and output functions within the system. This ability forms the core of the self-diagnostic system. If a problem is detected within a circuit, the controller will recognize the fault, assign it an identification code, and store the code in a memory section. Depending on the year and model, the fault code(s) may be represented by two or three digit numbers. The stored code(s) may be retrieved during diagnosis.

When the term Powertrain Control Module (PCM) is used in this manual it will refer to the engine control computer regardless that it may also be called an Electronic Control Assembly (ECA).

While the EEC-IV system is capable of recognizing many internal faults, certain faults will not be recognized. Because the computer system sees only electrical signals, it cannot sense or react to mechanical or vacuum faults affecting engine operation. Some of these faults may affect another component which will set a code. For example, the ECA monitors the output signal to the fuel injectors, but cannot detect a partially clogged injector. As long as the output driver responds correctly, the computer will read the system as functioning correctly. However, the improper flow of fuel may result in a lean mixture. This would, in turn, be detected by the oxygen sensor and noticed as a constantly lean signal by the ECA. Once the signal falls outside the pre-programmed limits, the engine control assembly would notice the fault and set an identification code.

Additionally, the EEC-IV system employs adaptive fuel logic. This process is used to compensate for normal wear and variability within the fuel system. Once the engine enters steady-state operation, the engine control assembly watches the oxygen sensor signal for a bias or tendency to run slightly rich or lean. If such a bias is detected, the adaptive logic corrects the fuel delivery to bring the air/fuel mixture towards a centered or 14.7:1 ratio. This compensating shift is stored in a non-volatile memory which is retained by battery power even with the ignition switched off. The correction factor is then available the next time the vehicle is operated.

➡ **If the battery is disconnected for longer than 5 minutes, the adaptive fuel factor will be lost. After repair it will be necessary to drive the car at least 10 miles to allow the processor to relearn the correct factors. The driving period should include steady-throttle open road driving if possible. During the drive, the vehicle may exhibit driveability symptoms not noticed before. These symptoms should clear as the ECA computes the correction factor. The ECA will also store Code 19 indicating loss of power to the controller.**

FAILURE MODE EFFECTS MANAGEMENT (FMEM)

The engine controller assembly contains back-up programs which allow the engine to operate if a sensor signal is lost. If a sensor input is seen to be out of range—either high or low—the FMEM program is used. The processor substitutes a fixed value for the missing sensor signal. The engine will continue to operate, although performance and driveability may be noticeably reduced. This function of the controller is sometimes referred to as the limp-in or fail-safe mode. If the missing sensor signal is restored, the FMEM system immediately returns the system to normal operation. The dashboard warning lamp will be lit when FMEM is in effect.

HARDWARE LIMITED OPERATION STRATEGY (HLOS)

This mode is only used if the fault is too extreme for the FMEM circuit to handle. In this mode, the processor has ceased all computation and control; the entire system is run on fixed values. The vehicle may be operated but performance and driveability will be greatly reduced. The fixed or default settings provide minimal calibration, allowing the vehicle to be carefully driven in for service. The dashboard warning lamp will be lit when HLOS is engaged. Codes cannot be read while the system is operating in this mode.

Dashboard Warning Lamp

The CHECK ENGINE or SERVICE ENGINE SOON dashboard warning lamp is referred to as the Malfunction Indicator Lamp (MIL). The lamp is connected to the engine control assembly and will alert the driver to certain malfunctions within the EEC-IV system. When the lamp is lit, the ECA has detected a fault and stored an identity code in memory. The engine control system will usually enter either FMEM or HLOS mode and driveability will be impaired.

The light will stay on as long as the fault causing it is present. Should the fault self-correct, the MIL will extinguish but the stored code will remain in memory.

Under normal operating conditions, the MIL should light briefly when the ignition key is turned ON. As soon as the ECA receives a signal that the engine is cranking, the lamp will be extinguished. The dash warning lamp should remain out during the entire operating cycle.

➡**On Continental, the CHECK ENGINE message is displayed on the message center. When a fault is detected, the message is accompanied by a 1 second tone every 5 seconds. The tone stops after 1 minute. When the Continental system enters HLOS, the additional message CHECK DCL is displayed. DCL refers to the Data Communications Link running between the engine controller and the message center.**

EEC-IV & EEC-V SCAN TOOL FUNCTIONS

Although stored codes may be read by using a analog voltmeter, the use of hand-held scan tools such as Ford's Self-Test Automatic Readout (STAR) tester or the second generation SUPER STAR II tester or their equivalent is recommended. There are many manufacturers of these tools; the purchaser must be certain that the tool is proper for the intended use.

Both the STAR and SUPER STAR testers are designed to communicate directly with the EEC-IV system and interpret the electrical signals. The SUPER STAR tester may be used to read either 2 or 3 digit codes; the original STAR tester will not read the 3 digit codes used on many 1990 and newer vehicles.

The scan tool allows any stored faults to be read from the engine controller memory. Use of the scan tool provides additional data during troubleshooting but does not eliminate the use of the charts. The scan tool makes collecting information easier; the data must be correctly interpreted by an operator familiar with the system.

Electrical Tools

The most commonly required electrical diagnostic tool is the Digital Multimeter, allowing voltage, resistance and amperage to be read by one instrument. Many of the diagnostic charts require the use of a volt or ohmmeter during diagnosis.

The multimeter must be a high impedance unit, with 10 megohms of impedance in the voltmeter. This type of meter will not place an additional load on the circuit it is testing; this is extremely important in low voltage circuits. The multimeter must be of high quality in all respects. It should be handled carefully and protected from impact or damage. Replace the batteries frequently in the unit.

Additionally, an analog (needle type) voltmeter may be used to read stored fault codes if the STAR tester is not available. The codes are transmitted as visible needle sweeps on the face of the instrument.

Almost all diagnostic procedures will require the use of the Breakout Box, a device which connects into the EEC-IV harness and provides testing ports for the 60 wires in the harness. Direct testing of the harness connectors at the terminals or by back-probing is not recommended; damage to the wiring and terminal is almost certain to occur.

Other necessary tools include a quality tachometer with inductive (clip-on) pickup, a fuel pressure gauge with system adapters and a vacuum gauge with an auxiliary source of vacuum.

EEC-IV & EEC-V SELF-DIAGNOSTICS

Diagnosis of a driveability problem requires attention to detail and following the diagnostic procedures in the correct order. Resist the temptation to begin extensive testing before completing the preliminary diagnostic steps. The preliminary or visual inspection must be completed in detail before diagnosis begins. In many cases this will shorten diagnostic time and Offen cure the problem without electronic testing.

Visual Inspection

This is possibly the most critical step of diagnosis. A detailed examination of all connectors, wiring and vacuum hoses can Offen lead to a repair without further diagnosis. Performance of this step relies on the skill of the technician performing it; a careful inspector will check the undersides of hoses as well as the integrity of hard-to-reach hoses blocked by the air cleaner or other components. Wiring should be checked carefully for any sign of strain , burning, crimping or terminal pull-out from a connector.

Checking connectors at components or in harnesses is required usually, pushing them together will reveal a loose fit. Pay particular attention to ground circuits, making sure they are not loose or corroded. Remember to inspect connectors and hose fittings at components not mounted on the engine, such as the evaporative canister or relays mounted on the fender aprons. Any component or wiring in the vicinity of a fluid leak or spillage should be given extra attention during inspection.

Additionally, inspect maintenance items such as belt condition and tension, battery charge and condition and the radiator cap carefully. Any of these very simple items may affect the system enough to set a fault.

Diagnostic Connector Location

The Diagnostic Link Connectors (DLC) are located a 6 basic locations:

- Near the bulkhead (right or left side of vehicle)
- Near the wheel well (right or left side of vehicle)
- Near the front corner of the engine compartment (right or left side of vehicle)

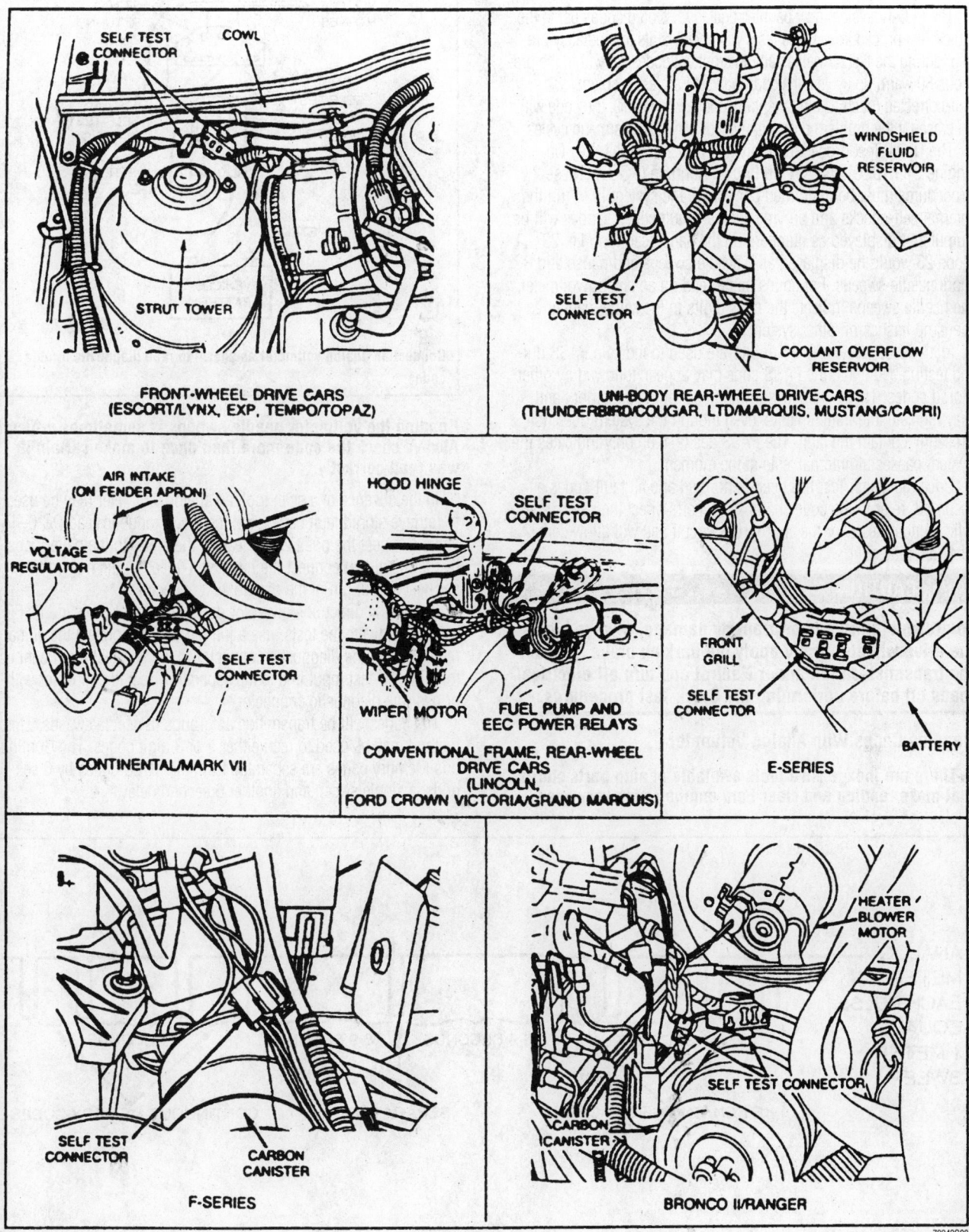

FRONT-WHEEL DRIVE CARS
(ESCORT/LYNX, EXP, TEMPO/TOPAZ)

UNI-BODY REAR-WHEEL DRIVE-CARS
(THUNDERBIRD/COUGAR, LTD/MARQUIS, MUSTANG/CAPRI)

CONTINENTAL/MARK VII

CONVENTIONAL FRAME, REAR-WHEEL DRIVE CARS
(LINCOLN, FORD CROWN VICTORIA/GRAND MARQUIS)

E-SERIES

F-SERIES

BRONCO II/RANGER

79242G08

Typical diagnostic link connector locations. Locations will vary depending on the year or model

EEC-IV & EEC-V READING CODES

The EEC-IV system may be interrogated for stored codes using the Quick Test procedures. These tests will reveal faults immediately present during the test as well as any intermittent codes set within the previous 80 warm up cycles. If a code was set before a problem self-corrected (such as a momentarily loose connector), the code will be erased if the problem does not reoccur within 80 warm-up cycles.

The Quick Test procedure is divided into 2 sections, Key On Engine Off (KOEO) and Key On Engine Running (KOER). These 2 procedures must be performed correctly if the system is to run the internal self-checks and provide accurate fault codes. Codes will be output and displayed as numbers on the hand scan tool, i.e. 23. Code 23 would be displayed as 2 needle sweeps and pause and 3 more needle sweeps. For codes being read on an analog voltmeter, the needle sweeps indicate the code digits in the same manner as the lamp flashes on other systems.

In all cases, the codes 11 or 111 are used to indicate PASS during testing. Note that the PASS code may appear, followed by other stored codes. These are codes from the continuous memory and may indicate intermittent faults, even though the system does not presently contain the fault. The PASS designation only indicates the system passes all internal tests at the moment.

Once the Quick Test has been performed and all fault codes recorded, refer to the code charts. The charts direct the use of specific pinpoint tests for the appropriate circuit and will allow complete circuit testing.

✳✳ CAUTION

To prevent injury and/or property damage, always block the drive wheels, firmly apply the parking brake, place the transmission in Park or Neutral and turn all electrical loads off before performing the Quick Test procedures.

Reading Codes With Analog Voltmeter

➥**There are inexpensive tools available at auto parts stores that make reading and clear Ford engine codes very easy.**

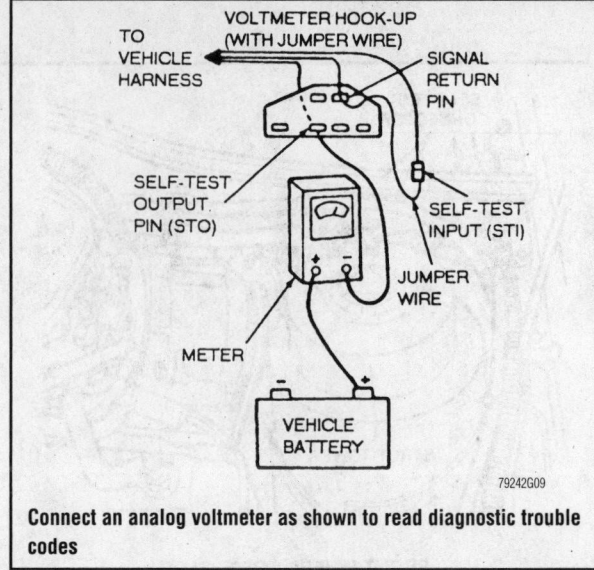

Connect an analog voltmeter as shown to read diagnostic trouble codes

Reading the voltmeter needle sweeps is sometimes difficult. Always check the code more than once to make certain it was read correctly.

In the absence of a scan tool, an analog voltmeter may be used to retrieve stored fault codes. Set the meter range to read DC 0–15 volts. Connect the positive (+) lead of the meter to the battery positive terminal and connect the negative (-) lead of the meter to the self-test output pin of the diagnostic connector.

Follow the directions given for performing the KOEO and KOER tests. To activate the tests, use a jumper wire to connect the signal return pin on the diagnostic connector to the self-test input connector. The self-test input line is the separate wire and connector with or near the diagnostic connector.

The codes will be transmitted as groups of needle sweeps. This method may be used to read either 2 or 3 digit codes. The Continuous Memory codes are separated from the KOEO codes by 6 seconds, a single sweep and another 6 second delay.

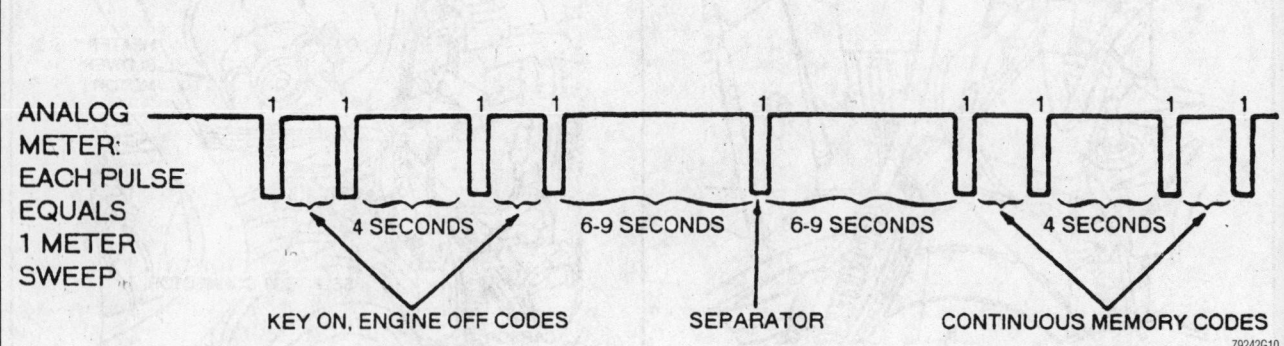

Code transmission during KOEO test. Note the continuous memory codes are transmitted after a pause and a separator pulse

KEY ON ENGINE OFF (KOEO) TEST

1. Connect the scan tool to the self-test connectors. Make certain the test button is unlatched or up.

2. Start the engine and run it until normal operating temperature is reached.

3. Turn the engine OFF for 10 seconds.

4. Activate the test button on the STAR tester.

5. Turn the ignition switch ON but do not start the engine. For vehicles with 4.9L engines, depress the clutch during the entire test. For vehicles with the 7.3L diesel engine, hold the accelerator to the floor during the test.

6. The KOEO codes will be transmitted. Six to nine seconds after the last KOEO code, a single separator pulse will be transmitted. Six to nine seconds after this pulse, the codes from the Continuous Memory will be transmitted.

7. Record all service codes displayed. Do not depress the throttle on gasoline engines during the test.

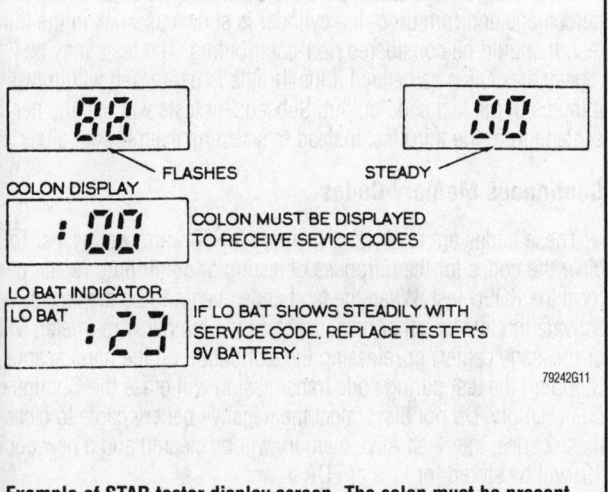

Example of STAR tester display screen. The colon must be present before the codes can be retrieved

KEY ON ENGINE RUNNING (KOER) TEST

1. Make certain the self-test button is released or de-activated on the STAR tester.

2. Start the engine and run it at 2000 rpm for two minutes. This action warms up the oxygen sensor.

3. Turn the ignition switch OFF for 10 seconds.

4. Activate or latch the self-test button on the scan tool.

5. Start the engine. The engine identification code will be transmitted. This is a single digit number representing ½ the number of cylinders in a gasoline engine. On the STAR tester, this number may appear with a zero, i.e., 20 = 2. For 7.3L diesel engines, the 10 code is 5. The code is used to confirm that the correct processor is installed and that the self-test has begun.

6. If the vehicle is equipped with a Brake On/Off (BOO) switch, the brake pedal must be depressed and released after the 10 code is transmitted.

7. If the vehicle is equipped with a Power Steering Pressure Switch (PSPS), the steering wheel must be turned at least `/2 turn and released within 2 seconds after the engine 10 code is transmitted.

8. If the vehicle is equipped with the E400 transmission, the Overdrive Cancel Switch (OCS) must be cycled after the engine 10 code is transmitted.

9. Certain Ford vehicles will display a Dynamic Response code 6–20 seconds after the engine 10 code. This will appear as one pulse on a meter or as a 10 on the STAR tester. When this code appears, briefly take the engine to wide open throttle. This allows the system to test the throttle position, MAF and MAP sensors.

10. All relevant codes will be displayed and should be recorded. Remember that the codes refer only to faults present during this test cycle. Codes stored in Continuous Memory are not displayed in this test mode.

11. Do not depress the throttle during testing unless a dynamic response code is displayed.

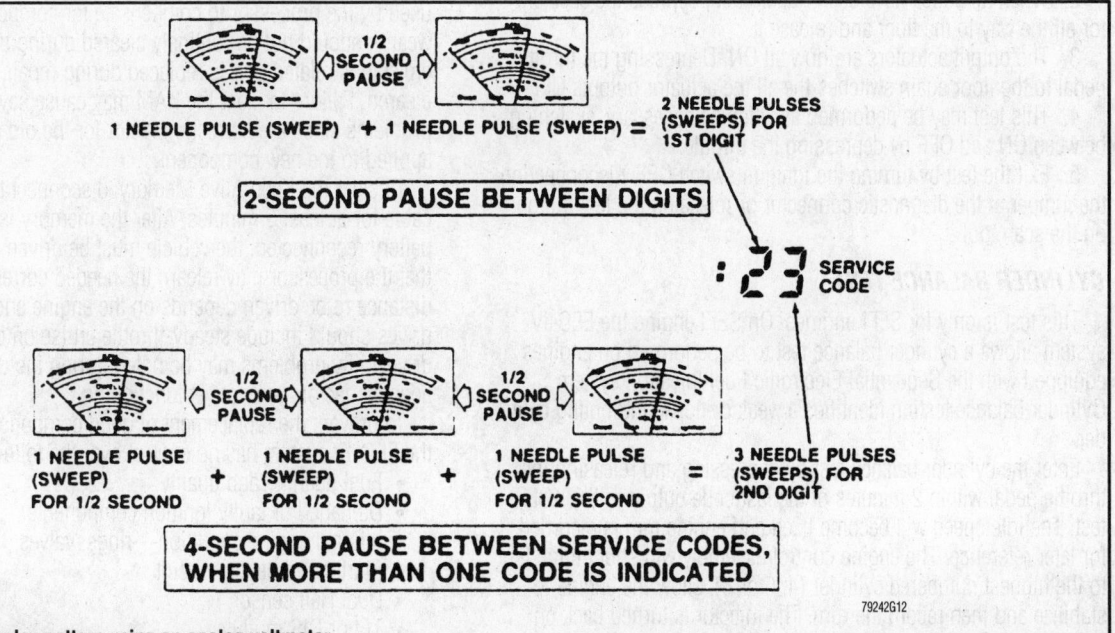

Example of code display pattern using an analog voltmeter

Advanced Test Modes

CONTINUOUS MONITOR OR WIGGLE TEST MODE

Once entered, this mode allows the technician to attempt to recreate intermittent faults by wiggling or tapping components, wiring or connectors. The test may be performed during either KOEO or KOER procedures. The test requires the use of either an analog voltmeter or a hand scan tool.

To enter the continuous monitor mode during KOEO testing, turn the ignition switch ON. Activate the test, wait 10 seconds, then deactivate and reactivate the test; the system will enter the continuous monitor mode. Tap, move or wiggle the harness, component or connector suspected of causing the problem; if a fault is detected, the code will store in the memory. When the fault occurs, the dash warning lamp will illuminate, the STAR tester will light a red indicator (and possibly beep) and the analog meter needle will sweep once.

To enter this mode in the KOER test:

1. Start the engine and run it at 2000 rpm for two minutes. This action warms up the oxygen sensor.
2. Turn the ignition switch OFF for 10 seconds.
3. Start the engine.
4. Activate the test, wait 10 seconds, then deactivate and reactivate the test; the system will enter the continuous monitor mode.
5. Tap, move or wiggle the harness, component or connector suspected of causing the problem; if a fault is detected, the code will store in the memory.
6. When the fault occurs, the dash warning lamp will illuminate, the STAR tester will light a red indicator (and possibly beep) and the analog meter needle will sweep once.

OUTPUT STATE CHECK

This testing mode allows the operator to energize and de-energize most of the outputs controlled by the EEC-IV system. Many of the outputs may be checked at the component by listening for a click or feeling the item move or engage by a hand placed on the case. To enter this check:

1. Enter the KOEO test mode.
2. When all codes have been transmitted, depress the accelerator all the way to the floor and release it.
3. The output actuators are now all ON. Depressing the throttle pedal to the floor again switches the all the actuator outputs OFF.
4. This test may be performed as Offen as necessary, switching between ON and OFF by depressing the throttle.
5. Exit the test by turning the ignition switch OFF, disconnecting the jumper at the diagnostic connector or releasing the test button on the scan tool.

CYLINDER BALANCE TEST

This test is only for SEFI engines. On SEFI engine the EEC-IV system allows a cylinder balance test to be performed on engines equipped with the Sequential Electronic Fuel Injection system. Cylinder balance testing identifies a weak or non-contributing cylinder.

Enter the cylinder balance test by depressing and releasing the throttle pedal within 2 minutes of the last code output in the KOER test. The idle speed will become fixed and engine mm is recorded for later reference. The engine control assembly will shut off the fuel to the highest numbered cylinder (4, 6 or 8), allow the engine to stabilize and then record the rpm. The injector is turned back on and the next one shut off and the process continues through cylinder No. 1.

The controller selects the highest rpm drop from all the cylinders tested, multiplies it by a percentage and arrives at an rpm drop value for all cylinders. For example, if the greatest drop for any cylinder was 150 rpm, the processor applies a multiple of 65% and arrives at 98 mm. The processor then checks the recorded rpm drops, checking that each was at least 98 rpm. If all cylinders meet the criteria, the test is complete and the ECA outputs Code 90 indicating PASS.

If one cylinder did not drop at least this amount, then the cylinder number is output instead of the 90 code. The cylinder number will be followed by a zero, so 30 indicates cylinder No. 3 did not meet the minimum rpm drop.

The test may be repeated a second time by depressing and releasing the throttle pedal within 2 minutes of the last code output. For the second test, the controller uses a lower percentage (and thus a lower rpm) to determine the minimum acceptable rpm drop. Again, either Code 90 or the number of the weak cylinder will be output.

Performing a third test causes the ECA to select an even lower percentage and rpm drop. If a cylinder is shown as weak in the third test, it should be considered non-contributing. The tests may be repeated as Offen as needed if the throttle is depressed within two minutes of the last code output. Subsequent tests will use the percentage from the third test instead of selecting even lower values.

Continuous Memory Codes

These codes are retained in memory for 80 warm-up cycles. To clear the codes for the purposes of testing or confirming repair, perform the KOEO test. When the fault codes begin to be displayed, deactivate the test by either disconnecting the jumper wire (meter, MIL or message center) or releasing the test button on the hand scanner. Stopping the test during code transmission will erase the Continuous Memory. Do not disconnect the negative battery cable to clear these codes; the Keep Alive memory will be cleared and a new code 19, will be stored for loss of ECA power.

KEEP ALIVE MEMORY

The Keep Alive Memory (KAM) contains the adaptive factors used by the processor to compensate for component tolerances and wear. It should not be routinely cleared during diagnosis. If an emissions related part is replaced during repair, the KAM must be cleared. Failure to clear the KAM may cause severe driveability problems since the correction factor for the old component will be applied to the new component.

To clear the Keep Alive Memory, disconnect the negative battery cable for at least 5 minutes. After the memory is cleared and the battery reconnected, the vehicle must be driven at least 10 miles so that the processor may relearn the needed correction factors. The distance to be driven depends on the engine and vehicle, but all drives should include steady-throttle cruise on open roads. Certain driveability problems may be noted during the drive because the adaptive factors are not yet functioning.

To prevent the replacement of good components, remember that the EEC-IV system has no control over the following items:

- Fuel quantity and quality
- Damaged or faulty ignition components
- Internal engine condition—rings, valves, timing belt, etc.
- Starter and battery circuit
- Dual Hall sensor
- TFI or DIS module
- Distributor condition or function

- Camshaft sensor
- Crankshaft sensor
- Ignition or DIS coil
- Engine governor module

Any of these systems can cause erratic engine behavior easily mistaken for an EEC-IV problem.

NON-NAAO DIAGNOSTIC SYSTEM

The 2.0L, 2.2L and 2.5L engines are referred to by Ford Motor Company as NON-NAAO, indicating the vehicles and/or their engines originate outside North American Automotive Operations.

Although these vehicles share many similarities in their engine control systems, differences must also be considered. While the fault codes are almost standardized (i.e., Code 14 indicates the barometric pressure sensor), not all engines use the same components so a code may be unique to a particular engine or family. These procedures encompass both turbocharged .and non-turbocharged engines.

Beside the engine diagnostic function, these procedures will also display codes related to the 4-speed Electronically-controlled Automatic Transaxle (4EAT) used in these vehicles. Note that the 4EAT codes are displayed by these procedures even though retrieving the engine fault codes may require the North American procedures described at the beginning of this section.

Engine Control System

These vehicles employ the Electronic Engine Control system, commonly designated EEC, to manage fuel, ignition and emissions on vehicle engines. This system is not EEC-IV, but does share some similarities.

The engine control assembly (ECA) is given responsibility for the operation of the emission control devices, cooling fans, ignition and advance and in some cases, automatic transmission functions. Because the EEC oversees both the ignition timing and the fuel injector operation, a precise air/fuel ratio will be maintained under all operating conditions. The ECA is a microprocessor or small computer which receives electrical in-puts from several sensors, switches and relays on and around the engine.

Based on combinations of these inputs, the ECA controls outputs to various devices concerned with engine operation and emissions. The engine control assembly relies on the signals to form a correct picture of current vehicle operation. If any of the input signals is incorrect, the ECA reacts to what ever picture is painted for it. For example, if the coolant temperature sensor is inaccurate and reads too low, the ECA may see a picture of the engine never warming up. Consequently, the engine settings will be maintained as if the engine were cold. Because so many inputs can affect one output, correct diagnostic procedures are essential on these systems.

One part of the ECA is devoted to monitoring both input and output functions within the system. This ability forms the core of the self-diagnostic system. If a problem is detected within a circuit, the controller will recognize the fault, assign it an identification code, and store the code in a memory section. Most NON-NAAO vehicles use two-digit codes for both engine and 4EAT transaxle faults. The stored code(s) may be retrieved during diagnosis.

➡When the term **Powertrain Control Module (PCM)** is used in this manual it will refer to the engine control computer regardless that it may also be called an **Electronic Control Assembly (ECA).**

While the EEC system is capable of recognizing many internal faults, certain faults will not be recognized. Because the computer system sees only electrical signals, it cannot sense or react to mechanical or vacuum faults affecting engine operation. Some of these faults may affect another component which will set a code. For example, the ECA monitors the output signal to the fuel injectors, but cannot detect a partially clogged injector. As long as the output driver responds correctly, the computer will read the system as functioning correctly. However, the improper flow of fuel may result in a lean mixture. This would, in turn, be detected by the oxygen sensor and noticed as a constantly lean signal by the ECA. Once the signal falls outside the pre-programmed limits, the engine control assembly would notice the fault and set an identification code.

Dashboard Warning Lamp

The CHECK ENGINE dashboard warning lamp is referred to as the Malfunction Indicator Lamp (MIL). The lamp is connected to the engine control assembly and will alert the driver to certain malfunctions within the EEC system. When the lamp is lit, the ECA has detected a fault and stored an identity code in memory.

The light will stay on as long as the fault causing it is present. Should the fault self-correct, the MIL will extinguish but the stored code will remain in memory.

Under normal operating conditions, the MIL should light briefly when the ignition key is turned ON. As soon as the ECA receives a signal that the engine is running, the lamp will be extinguished. The dash warning lamp should remain out during the entire operating cycle.

Vehicles with a 4EAT transaxle also provide a manual shift light, indicating when the transmission is in manual shift mode.

NON-NAAO SCAN TOOL FUNCTIONS

Although stored codes may be read by using an analog voltmeter by counting the needle sweeps, the use of hand-held scan tools such as Ford's second generation SUPER STAR II tester or equivalent is recommended. There are many manufacturers of these tools; the purchaser must be certain that the tool is proper for the intended use.

➡**The engine and 4EAT fault codes on NON-NAAO vehicles may only be read with the SUPER STAR II or its equivalent. The regular STAR tester or voltmeter may be capable not retrieve the stored codes.**

The SUPER STAR II tester is designed to communicate directly with the EEC system and interpret the electrical signals. The scan tool allows any stored faults to be read from the engine controller memory. Use of the scan tool provides additional data during troubleshooting but does not eliminate the use of the charts. The scan tool makes collecting information easier; the data must be correctly interpreted by an operator familiar with the system.

An adapter cable will be required to connect the scan tool to the vehicle; the adapter(s) may differ depending on the vehicle being tested.

Electrical Tools

The most commonly required electrical diagnostic tool is the Digital Multimeter, allowing voltage, resistance and amperage to be read by one instrument. Many of the diagnostic charts require the use of a voltmeter or ohmmeter during diagnosis.

The multimeter must be a high impedance unit, with 10 megohms of impedance in the voltmeter. This type of meter will not place an additional load on the circuit it is testing; this is extremely important in low voltage circuits. The multimeter must be of high quality in all respects. It should be handled carefully and protected from impact or damage. Replace the batteries frequently in the unit.

Additionally, an analog (needle type) voltmeter may be used to read stored fault codes if the SUPER STAR II tester is not available. The codes are transmitted as visible needle sweeps on the face of the instrument.

Almost all diagnostic procedures will require the use of the Breakout Box, a device which connects into the EEC harness and provides testing ports for the 60 wires in the harness. Direct testing of the harness connectors at the terminals or by backprobing is not recommended; damage to the wiring and terminals is almost certain to occur.

Other necessary tools include a quality tachometer with inductive (clip-on) pickup, a fuel pressure gauge with system adapters and a vacuum gauge with an auxiliary source of vacuum.

NON-NAAO SELF-DIAGNOSTICS

Diagnosis of a driveability problem requires attention to detail and following the diagnostic procedures in the correct order. Resist the temptation to begin extensive testing before completing the preliminary diagnostic steps. The preliminary or visual inspection must be completed in detail before diagnosis begins. In many cases this will shorten diagnostic time and Offen cure the problem without electronic testing.

Keep in mind that all the things that previously went wrong with vehicles, before the age of electronics, can still go wrong and are still the cause of the majority of the driveability problems. The best diagnosis starts with a list of symptoms and possible causes, followed by careful checking of those causes in the most likely order. Eliminate all the possible mechanical causes before considering electrical faults.

Visual Inspection

This is possibly the most critical step of diagnosis. A detailed examination of all connectors, wiring and vacuum hoses can Offen lead to a repair without further diagnosis. Performance of this step relies on the skill of the technician performing it; a careful inspector will check the undersides of hoses as well as the integrity of hard-to-reach hoses blocked by the air cleaner or other components. Wiring should be checked carefully for any sign of strain , burning, crimping or terminal pull-out from a connector.

Checking connectors at components or in harnesses is required; usually, pushing them together will reveal a loose fit. Pay particular attention to ground circuits, making sure they are not loose or corroded. Remember to inspect connectors and hose fittings at components not mounted on the engine, such as the evaporative canister or relays mounted on the fender aprons. Any component or wiring in the vicinity of a fluid leak or spillage should be given extra attention during inspection.

Additionally, inspect maintenance items such as belt condition and tension, battery charge and condition and the radiator cap carefully. Any of these very simple items may affect the system enough to set a fault.

NON-NAAO READING CODES

The EEC system may be interrogated for stored codes using the Quick Test procedures. If a code was set before a problem self-corrected (such as a momentarily loose connector), the code will remain in memory until cleared.

The Quick Test procedure is divided into 3 sections, Key On Engine Off (KOEO), Key On Engine Running (KOER) and the Switch Monitor test. These 3 procedures must be performed correctly if the system is to run the internal self-checks and provide accurate fault codes. Codes will be output and displayed as numbers on the hand scan tool, i.e. 23. If the codes are being read by an analog voltmeter, the codes will be displayed as groups of needle sweeps separated by pauses.

Code 23 would be shown as two sweeps, a pause and three more sweeps. A longer pause will occur between codes. Unlike the EEC-IV system, the EEC system does not broadcast a PASS designator or code. If no fault codes are stored, the display screen of the hand scanner will remain blank. Additionally, the EEC system does not operate switches or sensors during KOEO or KOER testing.

Once the Quick Test has been performed and all fault codes recorded, refer to the service code charts. The charts direct the use of specific pinpoint tests for the appropriate circuit and will allow complete circuit testing.

The EEC diagnostic connector is located at the left rear corner of the engine compartment on most vehicles. When connecting the test equipment and adapters, note that the Self-Test Input (STI) connector is separate from the main diagnostic connector on all NON-NAAO engines except for the 1.8L engine. The Self-Test Output (STO) connector is contained within the main diagnostic connector.

✳✳ CAUTION

To prevent injury and/or property damage, always block the drive wheels, firmly apply the parking brake, place the transmission in Park or Neutral and turn all electrical loads off before performing the Quick Test procedures.

Reading Codes With Analog Voltmeter

In the absence of a scan tool, an analog voltmeter may be used to retrieve stored fault codes. Set the meter range to read DC 0–20 volts. Connect the + lead of the meter to the STO pin in the diagnostic connector and connect the—lead of the meter to the negative battery terminal or a good engine ground.

Follow the directions given for performing the KOEO and KOER tests. To activate the tests, use a jumper wire to connect the STI connector to ground. The codes will be transmitted as groups of needle sweeps.

1 NEEDLE PULSE (SWEEP) + 1 NEEDLE PULSE (SWEEP) = 2 NEEDLE PULSES (SWEEPS) FOR 1ST DIGIT

1 6-SECOND PULSE BETWEEN DIGITS

:*23* SERVICE CODE

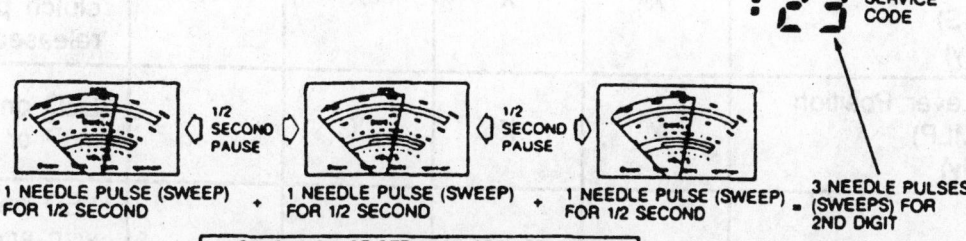

1 NEEDLE PULSE (SWEEP) FOR 1/2 SECOND + 1 NEEDLE PULSE (SWEEP) FOR 1/2 SECOND + 1 NEEDLE PULSE (SWEEP) FOR 1/2 SECOND = 3 NEEDLE PULSES (SWEEPS) FOR 2ND DIGIT

1/2 SECOND PAUSE   1/2 SECOND PAUSE

4-SECOND PAUSE BETWEEN SERVICE CODES, WHEN MORE THAN ONE CODE IS INDICATED

79242G13

Code display patterns on an analog voltmeter

KEY ON ENGINE OFF (KOEO) TEST

1. Make certain the scan tool is OFF; connect it to the self-test connectors. Switch the scan tool to the MECS position. Except on 1 8L engines, make certain the adapter ground cable is connected to the negative battery terminal. On the 1.8L engine, make certain the switch on the adapter is set to EEC or ECA if engine codes are to be retrieved. The other switch position will retrieve codes from the 4EAT.

2. Make certain the scan tool test button is ON or latched down.

3. For all engine or 4EAT codes except 1.8L and 1.9L engines, turn the ignition switch ON but do not start the engine, then turn the scan tool ON. On 1.8L and 1.9L engines, turn the scan tool ON first, then turn the ignition switch ON.

4. Once energized, the tester should display 888 and beep for 2 seconds. Release the test button; 00 should appear, signifying the tool is ready to read codes.

5. Re-engage the test button.

6. The KOEO codes will be transmitted.

7. Record all service codes displayed.

8. After all codes are received, release the test button to review all the codes retained in tester memory.

9. Make sure all codes displayed are recorded. Clear the ECA memory and perform the KOEO test again. This will isolate hard faults from intermittent ones. Any hard faults will cause the code(s) to be repeated in the 2nd test. An intermittent which is not now present will not set a new code.

10. Record all codes from the 2nd test. After repairs are made on hard fault items, the intermittent ones must be recreated by tapping suspect sensors, wiggling wires or connectors or reproducing circumstances on a test drive.

➤For both KOEO and KOER tests, the message STO LO always displayed on the screen indicates that the system cannot initiate the Self-Test. The message STI LO displayed with an otherwise blank screen indicates Pass or No Codes Stored.

KEY ON ENGINE RUNNING (KOER) TEST

1. Make certain the self-test button is released or de-activated on the SUPER STAR II tester and that the tester is properly connected.

2. Start the engine and run it at 2000 rpm for 2 minutes. This action warms up the oxygen sensor.

3. Turn the ignition switch OFF.

4. Turn the ignition switch ON for 10 seconds but do not start the engine.

5. Start the engine and run it at idle.

6. Activate or latch the self-test button on the scan tool.

7. All relevant codes will be displayed and should be recorded.

SWITCH MONITOR TESTS

This test mode allows the operator to check the input signal from individual switches to the ECA. All switches to be tested must be OFF at the time the test begins; if one switch is on, it will affect the testing of another. The test must begin with the engine cool. The tests may be performed with either the SUPER STAR II tester or an analog voltmeter. When using the scan tool, the small LED on the adapter cable will light to show that the ECA has received the switch signal. If the voltmeter is used, the voltage will change when the switch is engaged or disengaged.

1. The engine must be off and cooled. Place the transmission in Park or Neutral.

2. Turn all accessories OFF.

3. If using the SUPER STAR II, connect it properly. If using an analog voltmeter, use a jumper to ground the STI terminal. Connect the positive (+) voltmeter lead to the SML terminal of the diagnostic connector and connect the negative (-) lead to a good engine ground.

4. Turn the ignition switch **ON**. Engage the center button on the SUPER STAR II. Most switches can be exercised without starting the engine.

5. Operate each switch according to the test chart and note the response either on the LED or the volt scale. Remember that an improper response means the ECA did not see the switch operation; check circuitry and connectors before assuming the switch is faulty.

6. Turn the ignition switch **OFF** when testing is complete.

Switch	1.3L	1.8L	2.2L	2.2L Turbo	SUPER STAR II Tester LED or Analog VOM Indications
Clutch engage Switch/ Neutral Gear Switch (CES/NGS) (MTX only)	X	X	X	X	LED on or 12V in gear and clutch pedal released
Manual Lever Position Switch (MLP) (ATX Only)	X	X	X		LED on or 12V in P or N
Idle Switch (IDL)	X	X	X	X	LED on or 12V with accelerator pedal depressed
Brake On-Off Switch (BOO)	X	X MTX	X	X	LED on or 12V with brake pedal depressed
Headlamps Switch (HLDT)	X	X	X	X	LED on or 12V with headlamp switch on
Blower Motor Switch (BLMT)	X	X	X	X	LED on or 12V with blower switch at 2nd or above position
A/C Switch (ACS)	X	X			LED on or 12V with A/C switch on and blower on
Defrost Switch (DEF)	X	X	X	X	LED on or 12V with defrost switch on
Coolant Temperature Switch (CTS)	X	X		X	LED on or 12V with cooling fan on
Wide Open Throttle Switch (WOT)	X	X MTX			LED off or 0V with accelerator pedal fully depressed

Switch tests for 1990 and early Ford Non-NAAO vehicles

79242G14

Switch	1.3L	1.6L	1.8L	2.2L	2.2L Turbo	SUPER STAR II Tester LED or Analog VOM Indications
Clutch engage Switch/ Neutral Gear Switch (CES/NGS) (MTX only)	X	X	X	X	X	LED on or less than 1.5V in gear and clutch pedal released
Manual Lever Position Switch (MLP) (ATX Only)	X	X	X	X	X	LED on or less than 1.5V in P or N
Idle Switch (IDL)	X	X	X	X	X	LED on or less than 1.5V with accelerator pedal depressed
Brake On-Off Switch (BOO)	X	X	X MTX	X	X	LED on or less than 1.5V with brake pedal depressed (not fully)
Headlamps Switch (HLDT)	X	X	X	X	X	LED on or less than 1.5V with headlamp switch on
Blower Motor Switch (BLMT)	X	X	X	X	X	LED on or less than 1.5V with blower switch at 2nd or above position
A/C Switch (ACS)	X	X	X	X	X	LED on or less than 1.5V with A/C switch on and blower on
Defrost Switch (DEF)	X	X	X	X	X	LED on or less than 1.5V with defrost switch on
Coolant Temperature Switch (CTS)	X	X	X	X	X	LED on or less than 1.5V with cooling fan on
Wide Open Throttle Switch (WOT)	X		X			LED off or 0V with accelerator pedal fully depressed
Knock Control (KC)					X	LED on or less than 1.5V while tapping on engine

79242G15

Switch tests for 1991 and newer Ford Non-NAAO vehicles

Clearing Codes

Codes stored within the memory must be erased when repairs are completed. Additionally, erasing codes during diagnosis can separate hard faults from intermittent ones.

To erase stored codes, disconnect the negative battery cable, then depress the brake pedal for at least 10 seconds. Reconnect the battery cable and recheck the system for any remaining or newly-set codes.

EEC-IV System

The code definitions listed general 2-digit codes for Ford Vehicles using the Ford EEC-IV engine control system. In 1991 Ford started introducing vehicles that use 3-digit codes. The code definitions for both the 2 and 3-digit codes are found in this section. For a specific code definition or component test procedure consult your 'Chilton Total Car Care' manual for your vehicle. A diagnostic code does not mean the component is defective. For example a Code 29 is a vehicle speed sensor code. This does not mean the sensor is defective, but to check the sensor and related components. A defective speedometer cable or transmission problem will also set this code.

➡**When the term Powertrain Control Module (PCM) is used in this manual it will refer to the engine control computer regardless that it may be a PCM or Electronic Control Module (ECM) or Electronic Control Assembly (ECA).**

2-DIGIT DTC'S

1984–94 LIGHT TRUCKS:

Code 11 System Pass

Code 12 (R) Idle control fault—RPM Unable To Reach Upper Limit Self-Test

Code 13 (C) DC Motor Did Follow Dashpot

Code 13 (O) DC Motor Did Not Move

Code 13 (R) Idle control fault—Cannot control RPM during Self-Test low RPM check

Code 14(C)—Engine RPM signal fault—Profile Ignition Pickup (PIP) circuit failure or RPM sensor.

Code 15 (C) EEC Processor, power to Keep Alive Memory (KAM) interrupted or test failed

Code 15 (O) Power Interrupted To Processor or EEC Processor ROM Test failure

Code 16 (O,R) RPM too low to perform Exhaust Gas Oxygen (EGO) sensor test or fuel control error.

Code 1 (O)7 CFI Fuel Control System fault—Rich/Lean condition indicated; 3.8L V-6/5.0LV-8 (1984).

Code 17 (R) RPM Below Self-Test Limit, Set Too Low. Code 18 (C)—Ignition diagnostic monitor (IDM) circuit failure, loss of RPM signal or SPOUT circuit grounded

Code 18 (O)—Ignition Diagnostic Monitor (IDM) circuit

Code 18 (R)—SPOUT or SAW circuit open

Code 19 (C) Cylinder Identification (CID) Sensor Input failure

Code 19 (O) Failure in EEC Processor internal voltage. Code 19 (R)—Erratic RPM During EGR Test or RPM Too Low During ISC Off Test

Code 21 Engine Coolant Temperature (ECT) out of Self-Test range

Code 22 (O, R) Manifold Absolute Pressure (MAP)/Barometric Pressure (BP/BARO) Sensor circuit out of Self-Test range

Code 23 Throttle Position (TP) Sensor out of Self-Test range

Code 24 (O, R) Air Charge (ACT) or Intake Air (IAT) Temperature out of Self-Test range

Code 25 (R) Knock not sensed during dynamic response test

Code 26 (O, R) Transmission Fluid Temp (TFT) out of Self-Test range

Code 26 (O, R) Vane Air (VAF) or Mass Air (MAF) sensor out of self-test range

Code 28 (C) Loss Of Primary Tach, Right Side.

Code 29 (C) Insufficient input from Vehicle Speed Sensor (VSS) or Programmable Speedometer/Odometer Module (PSOM)

Code 31 EGR valve position sensor circuit below minimum voltage

Code 32 EGR Valve Position (EVP) sensor circuit voltage below closed limit

Code 33 (C) Throttle Position (TP) sensor noisy/harsh on line

Code 33 (R, C) EGR valve position sensor circuit, EGR valve opening not detected

Code 34 EGR valve circuit out of self-test range or valve not closing

Code 35 EGR valve circuit above maximum voltage -except 2.3L HSC with Feedback Carburetor System—or—Throttle Kicker on 2.3L HSC with Feedback Carburetor System.

Code 38 (C) Idle Track Switch Circuit Open.

Code 39 (C) AXOD Torque Converter or Bypass Clutch Not Applying Properly

Code 41 (R,C) Oxygen Sensor circuit indicates system always lean

Code 42 (R,C) Oxygen Sensor circuit indicates system always rich, right side if 2 sensors used

Code 43 (C) Oxygen Sensor Out Of Test Range—on 1992 and earlier vehicles—or—Throttle Position Sensor failure—on 1993 and newer vehicles

Code 43 (R) Exhaust Gas Oxygen (EGO) sensor cool down has occurred during testing—2.3L HSC and 2.8L FBC truck

Code 44 (R) Air injection control system failure (right side cylinders, if a split system)

Code 45 (C) Coil 1 primary circuit failure

Code 45 (R) Air injection control system air flow misdirected

Code 46 (C) Coil Primary Circuit failure

Code 46 (R) Thermactor air not bypassed during Self-Test

Code 47 (C) 4x4 switch is closed—on Truck.

Code 47 (R) Airflow low at idle—on fuel injected engines—or—4 x 4 switch is closed—on Truck—or—Fuel control system/Exhaust Gas Oxygen (EGO) Sensor fault—on 2.3L HSC and 2.8L FBC truck

Code 48 (C) Coil Primary Circuit failure; Except 2.3L Truck—or—Loss Of Secondary Tach, Left Side—with 2.3L Truck engine

Code 48 Airflow high at base idle

Code 49 (C) El electronic Transmission Shift Error—on Truck and 1992 and later cars—or—SPOUT Signal Defaulted To 10 Degrees BTDC or SPOUT Open—Up to 1991 passenger cars

Code 51 (O, C) Engine Coolant Temperature (ECT) circuit open or out of range during self-test

Code 52 (O) Power Steering Pressure Switch (PSPS) circuit open

Code 52 (R) Power Steering Pressure Switch (PSPS) circuit did not change states

Code 53 (O, C) Throttle Position (TP) circuit above maximum voltage

Code 54 (O, C) Air Charge (ACT) or Intake Air (IAT) Temperature circuit open

Code 55 (R) Key Power Input To Processor—open circuit

Code 56 (O, C) Mass Air (MAF) or Vane Air (VAF) Flow circuit above maximum voltage—Port fuel injected engines—or—Transmission oil temperature (TOT) circuit open—on vehicles with automatic transaxle

Code 57 (C) AXOD Circuit failure—on vehicles with automatic overdrive transaxle—or—Octane Adjust Circuit failure—on some 1992 and newer cars

Code 58 (R) Idle Tracking Switch circuit fault.

Code 59 (C) Automatic Transmission Shift Error—on 1991 and newer—or—AXOD 4/3 or Neutral Pressure Switch Failed Open—on 3.0L EFI and 3.8L AXOD—vehicles with automatic overdrive transaxle

Code 59 (O) AXOD 4/3 Pressure Switch Failed Closed—on 3.8L engine AXOD—vehicles with automatic transaxle—or—Idle Adjust Service Pin In Use—on 2.9L EFI engine—or—Low Speed Fuel Pump Circuit failure—on 3.0L SHO engine

Code 61 (O, C) Engine Coolant Temperature (ECT) circuit grounded

Code 62 (C) Converter clutch error

Code 62 (O) Electronic Transmission Shift Error.

Code 63 (O, C) Throttle Position (TP) circuit below minimum voltage

Code 64 (O, C) Air Charge (ACT) or Intake Air (IAT) Temperature circuit grounded

Code 65 (C) Fuel System Failed To Enter Closed Loop Mode or key power

Code 65 (O) Key Power Check—Possible Charging System overvoltage condition

Code 65 (R) Overdrive Cancel Switch (OCS) circuit did not switch

Code 66 (C) Mass Air (MAF) or Vane Air (VAF) Flow circuit below minimum voltage—engine with Port fuel injection—or—Transmission Oil Temperature (TOT) circuit grounded—vehicles with automatic transaxle

Code 67 (O, C) Manual Lever Position (MLP) sensor out of range and A/C ON

Code 67 (O, C) Neutral/Drive Switch (NDS) circuit open/A/C on during Self-Test

Code 67 (O, R) Neutral Drive Circuit Failed or A/C Input High or—Clutch Switch Circuit failed—on vehicles with manual transaxle -or—Manual Lever Position Sensor out of range—on vehicles with automatic transaxle

Code 68 (C) Transmission Fluid Temp (TFT) transmission over temp (over heated)

Code 68 (O) Idle Tracking Switch circuit—on 2.8L FBC truck only—or—Air temperature sensor—except FBC truck.

Code 68 (R, C) Air Temperature Sensor Circuit failure—on 2.9L EFI engine—or—Idle Tracking Switch Circuit failure—on CFI engine—or—Transmission Temperature Circuit

Code 69 (O, C) Transmission Shift Error

Code 70 (C) Data Communications Link Circuit failure

Code 71(C) SOffware Re-Initialization Detected—on 1 .9L EFI and 2.3L Turbo—or—Idle Tracking Switch failure—on CFI engine—or -Message Center Control Circuit failure—on vehicles with Message Center Control Center—or—Power Interrupt Detected—except vehicles with 3.8L AXOD (automatic overdrive transaxle)

Code 72 (R) Insufficient Manifold Absolute Pressure (MAP) change during Dynamic Response Test

Code 73 (R) Insufficient Throttle Position (TP) change during Dynamic Response Test

Code 74 (R, C) Brake On/Off (BOO) circuit open/not actuated during Self-Test

Code 75 (R) Brake On/Off (BOO) circuit closed/EEC processor input open

Code 76 (R) Insufficient Airflow Output Change During Test

Code 77 (R) Brief Wide Open Throttle (WOT) not sensed during Self-Test/operator error (Dynamic Response/Cylinder Balance Tests)

Code 78 (C) Power Interrupt Detected

Code 79 (O) A/C on/Defrost on during Self-Test

Code 81(C) MAP Sensor Has Not Changing Normally

Code 81(O) Air Management Circuit failure

Code 82 (O) Supercharger Bypass Circuit failure, 3.8L SC engine—or—Air Management Circuit failure, Except 3.8L SC engine—or -EGR Solenoid Circuit failure, 2.3L OHC engine

Code 83 OIC—Low speed fuel pump relay circuit failure

Code 83 (O) High Speed Electro Drive Fan Circuit failure, Except 2.3L OHC and 3.0L SHO engine—or—Low Speed Fuel Pump Relay Circuit failure, 3.0L SHO engine

Code 84 (O) EGR Vacuum Regulator (EVR) circuit failure

Code 84 (R) EGR Solenoid Circuit failure

Code 85 (C) Adaptive Lean Limit Reached

Code 85 (O) Canister Purge (CANP) circuit failure

Code 86 (C) Adaptive Rich Limit Reached

Code 86 (O) Shift Solenoid (SS) circuit failure—or—Wide Open Throttle (WOT) A/C Cutoff Solenoid circuit—on Carbureted engine

Code 87 Fuel Pump circuit fault

Code 88 (C) Loss Of Dual Plug Input control

Code 88 (O) Electro Drive Fan Circuit failure—fuel injected engine—or—Throttle Kicker, feedback carburetor system

Code 89 (O) Transmission solenoid circuit failure.

Code 89 (O) Clutch Converter Override (CCO) circuit failure—or—Exhaust Heat Control (EHC) Solenoid circuit—3.8L CFI engine

Code 91(C) No Heated Exhaust Gas Oxygen (HEGO) sensor switching detected—left HEGO

Code 91(O) Shift Solenoid 1 (SS1) circuit failure.

Code 91(R) Heated Exhaust Gas Oxygen (H EGO) sensor circuit indicates system lean—left HEGO

Code 92 (O) Shift Solenoid Circuit failure

Code 92 (R) Oxygen Sensor Circuit failure

Code 93 (O) Throttle Position Sensor (TPS) input low at maximum DC motor extension—OR—Shift solenoid circuit failure

Code 93 (O) Coast Clutch Solenoid (CCS) circuit failure

Code 94 (O) Torque Converter Clutch (TCC) solenoid circuit failure

Code 94 (O) Converter Clutch Control (CCC) Solenoid circuit failure

Code 94 (R) Thermactor Air System inoperative, left side

Code 95 (O, C) Fuel Pump secondary circuit failure/Fuel Pump circuit open—EEC processor to motor ground

Code 96 (O, C) Fuel Pump secondary circuit failure/Fuel Pump circuit open—battery to EEC processor

Code 97 (O) Overdrive Cancel Indicator Light (OCIL) circuit failure

Code 98 (R) Electronic control assembly failure

Code 98 (O) Electronic Pressure Control (EPC) Driver open in EEC processor

Code 98 (R) Hard fault is present—FMEM mode

Code 99 (O,C) Electronic Pressure Control (EPC) circuit failure

Code 92 (O) Shift Solenoid 2 (SS2) circuit failure

Code 92 (R) Heated Exhaust Gas Oxygen (HEGO) sensor circuit indicates system rich—left HEGO

Code 93 (O) Throttle Position Sensor Input Low At Max DC Motor Extension, CFI engine—or—Shift Solenoid Circuit failure—Except CFI engine

Code 94 (O) Converter Clutch Solenoid Circuit failure

Code 94 (R) Thermactor Air System Inoperative

Code 95 (O, C) Fuel Pump Circuit failure, ECA To ground

Code 96 (O, C) Fuel Pump Circuit failure

Code 97 (O) Transmission Indicator Circuit failure

Code 98 (O) Electronic Pressure Control Circuit failure

Code 98 (R) Electronic Control Assembly failure

Code 99 (O, C) Electronic Pressure Control Circuit or Transmission Shift failure

Code 99 (R) EEC System Has Not Learned To Control Idle: Ignore Codes 12 & 13

No Code–Unable to Run Self Test or Output Codes, or list does not apply to vehicle tested, refer to service manual.

➡**This list is to be used as a reference for testing and does not mean a specific component Is defective.**

(O)—Key On, Engine Off
(R)—Engine running
(C)—Continuous Memory

3-DIGIT DTC'S

1991–95 VEHICLES:

Code 111 System pass

Code 112 Intake Air Temperature (IAT) Sensor circuit below minimum voltage

Code 113 Intake Air Temperature (IAT) Sensor circuit above maximum voltage

Code 114 Intake Air Temperature (IAT) higher or lower than expected

Code 116 Engine Coolant Temperature (ECT) higher or lower than expected

Code 117 Engine Coolant Temperature (ECT) Sensor circuit below minimum voltage

Code 118 Engine Coolant Temperature (ECT) Sensor circuit above maximum voltage

Code 121 Closed throttle voltage higher or lower than expected

Code 121 Indicates Throttle Position voltage inconsistent with Mass Air Flow (MAF) Sensor

Code 122 Throttle Position (TP) Sensor circuit below minimum voltage

Code 123 Throttle Position (TP) Sensor circuit above maximum voltage

Code 124 Throttle Position (TP) Sensor circuit voltage higher than expected

Code 125 Throttle Position (TP) Sensor circuit voltage lower than expected

Code 126 Manifold Absolute Pressure/Barometric Pressure (MAP/BARO) Sensor higher or lower than expected

Code 128 Manifold Absolute Pressure (MAP) Sensor vacuum hose damaged/disconnected

Code 129 Insufficient Manifold Absolute Pressure (MAP)/Mass Air Flow (MAF) change during Dynamic Response Test-KOER

Code 136 Lack of Heated Oxygen Sensor (HO2S-2) switches during KOER, indicates lean—Bank # 2

Code 137 Lack of Heated Oxygen Sensor (HO2S-2) switches during KOER, indicates rich—Bank # 2

Code 138 Cold Start Injector (CSI) flow insufficient—KOER

Code 139 No Heated Oxygen Sensor (HO2S-2) switches detected—Bank # 2

Code 141 Fuel system indicates lean

Code 144 No Heated Oxygen Sensor (HO2S-1) switches detected—Bank # 1

Code 157 Mass Air Flow (MAF) Sensor circuit below minimum voltage

Code 158 Mass Air Flow (MAF) Sensor circuit above maximum voltage

Code 159 Mass Air Flow (MAF) higher or lower than expected

Code 167 Insufficient Throttle Position (TP) change during Dynamic Response Test—KOER

Code 171 Fuel system at adaptive limits, Heated Oxygen Sensor (HO2S-I) unable to switch—Bank # 1

Code 172 Lack of Heated Oxygen Sensor (HO2S-1) switches, indicates lean—Bank # 1

Code 173 Lack of Heated Oxygen Sensor (HO2S-1) switches, indicates rich—Bank # 1

Code 174 Heated Oxygen Sensor (HO2S) switching time is slow—Right side—1992 vehicles only

Code 175 Fuel system at adaptive limits, Heated Oxygen Sensor (HO2S-2) unable to switch—Bank # 2

Code 176 Lack of Heated Oxygen Sensor (HO2S-2) switches, indicates lean—Bank # 2

Code 177 Lack of Heated Oxygen Sensor (HO2S-2) switches, indicates rich—Bank # 2

Code 178 Heated Oxygen Sensor (HO2S) switching time is slow—Left side—1992 vehicles only

Code 179 Fuel system at lean adaptive limit at part throttle, system rich—Bank # 1

Code 181 Fuel system at rich adaptive limit at part throttle, system lean—Bank # 1

Code 182 Fuel system at lean adaptive limit at idle, system rich—Right side—1992 vehicles only

Code 183 Fuel system at rich adaptive limit at idle, system lean—Right side—1992 vehicles only

Code 184 Mass Air Flow (MAF) higher than expected

Code 185 Mass Air Flow (MAF) lower than expected

Code 186 Injector pulse width higher or Mass Air Flow (MAF) lower than expected (without BARO Sensor)

Code 187 Injector pulse width lower than expected (with BARO Sensor)

Code 187 Injector pulse width lower or Mass Air Flow (MAF) higher than expected (without BARO Sensor)

Code 188 Fuel system at lean adaptive limit at part throttle, system rich—Bank # 2

Code 189 Fuel system at rich adaptive limit at part throttle, system lean—Bank # 2

Code 191 Adaptive fuel lean limit is reached at idle -Left side-1992 vehicles only

Code 192 Adaptive fuel rich limit is reached at idle—Left side—1992 vehicles only

Code 193 Flexible Fuel (FF) Sensor circuit failure

Code 211 Profile Ignition Pickup (PIP) circuit failure

Code 212 Loss of Ignition Diagnostic monitor (1DM) input to Powertrain Control Module (PCM)/SPOUT circuit grounded

Code 213 SPOUT circuit open

Code 214 Cylinder Identification (CID) circuit failure

Code 215 Powertrain Control Module (PCM) detected Coil 1 Primary circuit failure (EI)

Code 216 Powertrain Control Module (PCM) detected Coil 2 Primary circuit failure (EI)

Code 217 Powertrain Control Module (PCM) detected Coil 3 Primary circuit failure (EI)

Code 218 Loss of Ignition Diagnostic Monitor (1DM) signal left side (dual plug EI)

Code 219 Spark Timing defaulted to 10 degrees -SPOUT circuit open (EI)

Code 221 Spark Timing error (EI)

Code 222 Loss of Ignition Diagnostic Monitor (1DM) signal - right side (dual plug EI)

Code 223 Loss of Dual Plug Inhibit (DPI) control (Dual Plug EI)

Code 224 Powertrain Control Module (PCM) detected Coil 1, 2, or 4 Primary circuit failure (Dual Plug EI)

Code 225 Knock not sensed during Dynamic Response Test—KOER

Code 226 Ignition Diagnostic Monitor (1DM) signal not received (EI)

Code 232 Powertrain Control Module (PCM) detected Coil 1, 2, or 4 Primary circuit failure (EI)

Code 238 Powertrain Control Module (PCM) detected Coil 4 Primary circuit failure (EI)

Code 241 Ignition Control Module (1CM) to Powertrain Control Module (PCM) Ignition Diagnostic Monitor (1DM) Pulse Width Transmission error (EI)

Code 244 Cylinder Identification (CID) circuit fault present when Cylinder Balance Test requested

Code 311 Secondary Air Injection (AIR) system inoperative during KOER Bank # 1 with dual HO2S

Code 312 Secondary Air Injection (AIR) misdirected during KOER

Code 313 Secondary Air Injection (AIR) not bypassed during KOER

Code 314 Secondary Air Injection (AIR) system inoperative during KOER—Bank # 2 with dual HO2S

Code 326 EGR (PFE/DPFE) circuit voltage lower than expected

Code 327 EGR (EVP/PFE/DPFE) circuit below minimum voltage

Code 328 EGR (EVP) closed valve voltage lower than expected

Code 332 Insufficient EGR flow detected/EGR Valve opening not detected (EVP/PFE/DPFE)

Code 334 EGR (EVP) closed valve voltage higher than expected

Code 335 EGR (PFE/DPFE) Sensor voltage higher or lower than expected during KOEO

Code 336 Exhaust pressure high/EGR (PFE/DPFE) circuit voltage higher than expected

Code 337 EGR (EVP/PFE/DPFE) circuit above maximum voltage

Code 338 Engine Coolant Temperature (ECT) lower than expected (thermostat test)

Code 339 Engine Coolant Temperature (ECT) higher than expected (thermostat test)

Code 341 Octane Adjust service pin open

Code 411 Cannot control RPM during KOER low rpm check

Code 412 Cannot control RPM during KOER high rpm check

Code 415 Idle Air Control (IAC) system at maximum adaptive lower limit

Code 416 Idle Air Control (IAC) system at upper adaptive learning limit

Code 452 Insufficient input from Vehicle Speed Sensor (VSS) to PCM

Code 453 Servo leaking down (KOER IVSC test)

Code 454 Servo leaking up (KOER IVSC test)

Code 455 Insufficient RPM increase (KOER IVSC test)

Code 456 Insufficient RPM decrease (KOER IVSC test)

Code 457 Speed Control Command Switch(s) circuit not functioning (KOEO IVSC test)

Code 458 Speed Control Command Switch(s) stuck/circuit grounded (KOEO IVSC test)

Code 459 Speed Control ground circuit open (KOEO IVSC test)

Code 511 Powertrain Control Module (PCM) Read Only Memory (ROM) test failure (KOEO)

Code 512 Powertrain Control Module (PCM) Keep Alive Memory (KAM) test failure

Code 513 Powertrain Control Module (PCM) internal voltage failure (KOEO)

Code 519 Power Steering Pressure (PSP) Switch circuit open—KOEO

Code 521 Power Steering Pressure (PSP) Switch circuit did not change states—KOER

Code 522 Vehicle not in park or neutral during KOEO/Park/Neutral Position (PNP) Switch circuit open

Code 524 Low speed Fuel Pump circuit open—battery to PCM

Code 525 Indicates vehicle in gear/A/C on

Code 526 Neutral Pressure Switch (NPS) circuit closed; A/C on -1992 vehicles only

Code 527 Park/Neutral Position (PNP) Switch open—A/C on, KOEO

Code 528 Clutch Pedal Position (CPP) switch circuit failure

Code 529 Data Communications Link (DCL) or PCM circuit failure

Code 532 Cluster Control Assembly (CCA) circuit failure

Code 533 Data Communications Link (DCL) or Electronic Instrument Cluster (EIC) circuit failure

Code 536 Brake On/Off (BOO) circuit failure/not actuated during KOER

Code 538 Insufficient RPM change during KOER Dynamic Response Test

Code 538 Invalid Cylinder Balance Test due to throttle movement during test—SFI only

Code 538 Invalid Cylinder Balance test due to Cylinder Identification (CID) circuit failure

Code 539 A/C on/Defrost on during Self-Test

Code 542 Fuel Pump secondary circuit failure

Code 543 Fuel Pump secondary circuit failure

Code 551 Idle Air Control (IAC) circuit failure—KOEO

Code 552 Secondary Air Injection Bypass (AIRB) circuit failure -KOEO

Code 553 Secondary Air Injection Diverter (AIRD) circuit failure—KOEO

Code 554 Fuel Pressure Regulator Control (FPRC) circuit failure

Code 556 Fuel Pump Relay primary circuit failure

Code 557 Low speed Fuel Pump primary circuit failure

Code 558 EGR Vacuum Regulator (EVR) circuit failure -KOEO

Code 559 Air Conditioning On (ACON) Relay circuit failure-KOEO

Code 563 High Fan Control (HFC) circuit failure—KOEO

Code 564 Fan Control (FC) circuit failure—KOEO

Code 565 Canister Purge (CANP) circuit failure—KOEO

Code 566 3–4 Shift Solenoid circuit failure, A4LD transmission—KOEO

Code 567 Speed Control Vent (SCVNT) circuit failure -KOEO IVSC test

Code 568 Speed Control Vacuum (SCVAC) circuit failure—KOEO IVSC test

Code 569 Auxiliary Canister Purge (CANP2) circuit failure-KOEO

Code 571 EGRA solenoid circuit failure KOEO

Code 572 EGRV solenoid circuit failure KOEO

Code 578 A/C Pressure Sensor circuit shorted (VCRM) mode

Code 579 Insufficient A/C pressure change (VCRM) mode
Code 581 Power to fan circuit over current (VCRM) mode
Code 582 Fan circuit open (VCRM) mode
Code 583 Power to Fuel Pump over current (VCRM) mode
Code 584 Power ground circuit open (Pin 1) (VCRM) mode
Code 585 Power to A/C Clutch over current (VCRM) mode
Code 586 A/C Clutch circuit open (VCRM) mode
Code 587 Variable Control Relay Module (VCRM) communication failure
Code 593 Heated Oxygen Sensor Heater (HO2S HTR)
Code 617 1–2 Shift error
Code 618 2–3 Shift error
Code 619 3–4 Shift error
Code 621 Shift Solenoid 1 (SS1) circuit failure—KOEO
Code 622 Shift Solenoid 2 (SS2) circuit failure—KOEO
Code 623 Transmission Control Indicator Lamp (TCIL) circuit failure
Code 624 Electronic Pressure Control (EPC) circuit failure
Code 625 Electronic Pressure Control (EPC) driver open in PCM
Code 626 Coast Clutch Solenoid (CCS) circuit failure—KOEO
Code 627 Torque Converter Clutch (TCC) solenoid circuit failure
Code 628 Excessive Converter Clutch slippage
Code 629 Torque Converter Clutch (TCC) solenoid circuit failure
Code 631 Transmission Control Indicator Lamp (TCIL) circuit failure—KOEO
Code 632 Transmission Control Switch (TCS) circuit did not change states during KOER
Code 633 4 x 4L Switch closed during KOEO
Code 634 Manual Lever Position (MLP) voltage higher or lower than expected/ error in Transmission Select Switch (TSS) circuit(s)
Code 636 Transmission Oil Temperature (TOT) higher or lower than expected
Code 637 Transmission Oil Temperature (TOT) Sensor circuit above maximum voltage/circuit open
Code 638 Transmission Oil Temperature (TOT) Sensor circuit below minimum voltage/circuit shorted
Code 639 Insufficient input from Transmission Speed Sensor (TSS)
Code 641 Shift Solenoid 3 (SS3) circuit failure
Code 643 Torque Converter Clutch (TCC) circuit failure
Code 645 Incorrect gear ratio obtained for first gear
Code 646 Incorrect gear ratio obtained for second gear
Code 647 Incorrect gear ratio obtained for third gear
Code 648 Incorrect gear ratio obtained for fourth gear
Code 649 Electronic Pressure Control (EPC) higher or lower than expected
Code 651 Electronic Pressure Control (EPC) circuit failure
Code 652 Torque Converter Clutch (TCC) Solenoid circuit failure
Code 654 Manual Lever Position (MLP) Sensor not indicating park during KOEO
Code 655 Manual Lever Position (MLP) Sensor indicating not in neutral during Self-Test
Code 656 Torque Converter Clutch (TCC) continuous slip error
Code 657 Transmission Over Temperature condition occurred
Code 659 High vehicle speed in park indicated
Code 667 Transmission Range sensor circuit voltage below minimum voltage
Code 668 Transmission Range sensor circuit voltage above maximum voltage

Code 675 Transmission Range sensor circuit voltage out of range
Code 691 4x4 Low switch open or short circuit
Code 692 Transmission state does not match calculated ratio
Code 998 Hard fault present—FMEM Mode

➡**If specific cylinder banks or sides are referred to in any of the above codes, but the vehicle code is being obtained from has a 4 cylinder engine, or only one Oxygen Sensor, disregard the bank side reference, but the code definition and components it pertains to is always the same.**

EEC-V System

1994 LIGHT TRUCKS

DTC P0102 Mass Air Flow (MAF) Sensor circuit low input
DTC P0103 Mass Air Flow (MAF) Sensor circuit high input
DTC P0112 Intake Air Temperature (IAT) Sensor circuit low input
DTC P0113 Intake Air Temperature (IAT) Sensor high input
DTC P0117 Engine Coolant Temperature (ECT) low input
DTC P0118 Engine Coolant Temperature (ECT) Sensor circuit high input
DTC P0122 Throttle Position (TP) Sensor circuit low input
DTC P0123 Throttle Position (TP) Sensor high input
DTC P0125 Insufficient coolant temperature to enter closed loop fuel control
DTC P0132 Upstream Heated Oxygen Sensor (HO2S 11) circuit high voltage (Bank #1)
DTC P0135 Heated Oxygen Sensor Heater (HTR 11) circuit malfunction
DTC P0138 Downstream Heated Oxygen Sensor (HO2S 12) circuit high voltage (Bank #1)
DTC P0140 Heated Oxygen Sensor (HO2S 12) circuit no activity detected (Bank #1)
DTC P0141 Heated Oxygen Sensor Heater (HTR 12) circuit malfunction
DTC P0152 Upstream Heated Oxygen Sensor (HO2S 21) circuit high voltage (Bank #2)
DTC P0155 Heated Oxygen Sensor Heater (HTR 21) circuit malfunction
DTC P0158 Downstream Heated Oxygen Sensor (HO2S 22) circuit high voltage (Bank #2)
DTC P0160 Heated Oxygen Sensor (HO2S 12) circuit no activity detected (Bank #2)
DTC P0161 Heated Oxygen Sensor Heater (HTR 22) circuit malfunction
DTC P0171 System (adaptive fuel) too lean (Bank #1)
DTC P0172 System (adaptive fuel) too lean (Bank #1)
DTC P0174 System (adaptive fuel) too lean (Bank #1)
DTC P0175 System (adaptive fuel) too lean (Bank #1)
DTC P0300 Random misfire detected
DTC P0301 Cylinder #1 misfire detected
DTC P0302 Cylinder #2 misfire detected
DTC P0303 Cylinder #3 misfire detected
DTC P0304 Cylinder #4 misfire detected
DTC P0305 Cylinder #5 misfire detected
DTC P0306 Cylinder #6 misfire detected
DTC P0307 Cylinder #7 misfire detected
DTC P0308 Cylinder #8 misfire detected
DTC P0320 Ignition engine speed (Profile Ignition Pickup) input circuit malfunction

DTC P0340 Camshaft Position (CMP) sensor circuit malfunction (CID)

DTC P0402 Exhaust Gas Recirculation (EGR) excess flow detected (valve open at idle)

DTC P0420 Catalyst system efficiency below threshold (Bank #1)

DTC P0430 Catalyst system efficiency below threshold (Bank #2)

DTC P0443 Evaporative emission control system Canister Purge (CANP) Control Valve circuit malfunction

DTC P0500 Vehicle Speed Sensor (VSS) malfunction

DTC P0505 Idle Air Control (IAC) system malfunction

DTC P0605 Powertrain Control Module (PCM)—Read Only Memory (ROM) test error

DTC P0703 Brake On/Off (BOO) switch input malfunction

DTC P0707 Manual Lever Position (MLP) sensor circuit low input

DTC P0708 Manual Lever Position (MLP) sensor circuit high input

DTC P0720 Output Shaft Speed (OSS) sensor circuit malfunction

DTC P0741 Torque Converter Clutch (TCC) system incorrect mechanical performance

DTC P0743 Torque Converter Clutch (TCC) system electrical failure

DTC P0750 Shift Solenoid #1(SS1) circuit malfunction

DTC P0751 Shift Solenoid #1(SS1) performance

DTC P0755 Shift Solenoid #2 (SS2) circuit malfunction

DTC P0756 Shift Solenoid #2 (SS2) performance

DTC P1000 OBD II Monitor Testing not complete

DTC P1100 Mass Air Flow (MAF) sensor intermittent

DTC P1101 Mass Air Flow (MAF) sensor out of Self-Test range

DTC P1112 Intake Air Temperature (IAT) sensor intermittent

DTC P1116 Engine Coolant Temperature (ECT) sensor out of Self-Test range

DTC P1117 Engine Coolant Temperature (ECT) sensor intermittent

DTC P1120 Throttle Position (TP) sensor out of range low

DTC P1121 Throttle Position (TP) sensor inconsistent with MAF sensor

DTC P1124 Throttle Position (TP) sensor out of Self-Test range

DTC P1125 Throttle Position (TP) sensor circuit intermittent

DTC P1130 Lack of HO2S 11 switch, adaptive fuel at limit

DTC P1131 Lack of HO2S 11 switch, sensor indicates lean Bank #1)

DTC P1132 Lack of HO2S 11 switch, sensor indicates rich Bank #1)

DTC P1137 Lack of HO2S 12 switch, sensor indicates lean Bank #1)

DTC P1138 Lack of HO2S 12 switch, sensor indicates rich Bank #1)

DTC P1150 Lack of HO2S 21 switch, adaptive fuel at limit

DTC P1151 Lack of HO2S 21 switch, sensor indicates lean Bank #2)

DTC P1152 Lack of HO2S 21 switch, sensor indicates rich Bank #2)

DTC P1157 Lack of HO2S 22 switch, sensor indicates lean Bank #2)

DTC P1158 Lack of HO2S 22 switch, sensor indicates rich Bank #2)

DTC P1351 Ignition Diagnostic Monitor (IDM) circuit input malfunction

DTC P1352 Ignition coil A primary circuit malfunction

DTC P1353 Ignition coil B primary circuit malfunction

DTC P1354 Ignition coil C primary circuit malfunction

DTC P1355 Ignition coil D primary circuit malfunction

DTC P1364 Ignition coil primary circuit malfunction

DTC P1390 Octane Adjust (OCT ADJ) out of Self-Test range

DTC P1400 Differential Pressure Feedback Electronic (DPFE) sensor circuit low voltage detected

DTC P1401 Differential Pressure Feedback Electronic (DPFE) sensor circuit high voltage detected

DTC P1403 Differential Pressure Feedback Electronic (DPFE) sensor hoses reversed

DTC P1405 Differential Pressure Feedback Electronic (DPFE) sensor upstream hose off or plugged

DTC P1406 Differential Pressure Feedback Electronic (DPFE) sensor downstream hose off or plugged

DTC P1407 Exhaust Gas Recirculation (EGR) no flow detected (valve stuck closed or inoperative)

DTC P1408 Exhaust Gas Recirculation (EGR) flow out of Self-Test range

DTC P1473 Fan Secondary High with fan(s) off

DTC P1474 Low Fan Control primary circuit malfunction

DTC P1479 High Fan Control primary circuit malfunction

DTC P1480 Fan Secondary low with low fan on

DTC P1481 Fan Secondary low with high fan on

DTC P1500 Vehicle Speed Sensor (VSS) circuit intermittent

DTC P1505 Idle Air Control (IAC) system at adaptive clip

DTC P1605 Powertrain Control Module (PCM)—Keep Alive Memory (KAM) test error

DTC P1703 Brake On/Off (BOO) switch out of Self-Test range

DTC P1705 Manual Lever Position (MLP) sensor out of Self-Test range

DTC P1711 Transmission Fluid Temperature (TFT) sensor out of Self-Test range

DTC P1742 Torque Converter Clutch (TCC) solenoid mechanically failed (turns MIL on)

DTC P1743 Torque Converter Clutch (TCC) solenoid mechanically failed (turns TCIL on)

DTC P1744 Torque Converter Clutch (TCC) system mechanically stuck in off position

DTC P1746 Electronic Pressure Control (EPC) solenoid circuit low input (open circuit)

DTC P1747 Electronic Pressure Control (EPC) solenoid circuit high input (short circuit)

DTC P1751 Shift Solenoid #1(SS1) performance

DTC P1756 Shift Solenoid #2 (SS2) performance

DTC P1780 Transmission Control Switch (TCS) circuit out of Self-Test range

1995 LIGHT TRUCKS

DTC P0102 Mass Air Flow (MAF) Sensor circuit low input

DTC P0103 Mass Air Flow (MAF) Sensor circuit high input

DTC P0112 Intake Air Temperature (IAT) Sensor circuit low input

DTC P0113 Intake Air Temperature (IAT) Sensor high input

DTC P0117 Engine Coolant Temperature (ECT) low input

DTC P0118 Engine Coolant Temperature (ECT) Sensor circuit high input

DTC P0121 In range operating Throttle Position (TP) sensor circuit failure

DTC P0122 Throttle Position (TP) Sensor circuit low input

DTC P0123 Throttle Position (TP) Sensor high input

DTC P0125 Insufficient coolant temperature to enter closed loop fuel control

DTC P0126 Insufficient coolant temperature for stable operation

DTC P0131 Upstream Heated Oxygen Sensor (HO₂S 11) circuit out of range low voltage (Bank #1)

DTC P0132 Upstream Heated Oxygen Sensor (HO₂S 11) circuit high voltage (Bank #1)

DTC P0133 Upstream Heated Oxygen Sensor (HO₂S 11) circuit slow response (Bank #1)

DTC P0135 Heated Oxygen Sensor Heater (HTR 11) circuit malfunction

DTC P0136 Downstream Heated Oxygen Sensor (HO₂S 12) circuit malfunction (Bank #1

DTC P0138 Downstream Heated Oxygen Sensor (HO₂S 12) circuit high voltage (Bank #1)

DTC P0140 Heated Oxygen Sensor (HO₂S 12) circuit no activity detected (Bank #1)

DTC P0141 Heated Oxygen Sensor Heater (HTR 12) circuit malfunction

DTC P0151 Upstream Heated Oxygen Sensor (HO₂S 21) circuit out of range low voltage (Bank #2)

DTC P0152 Upstream Heated Oxygen Sensor (HO₂S 21) circuit high voltage (Bank #2)

DTC P0153 Upstream Heated Oxygen Sensor (HO₂S 21) circuit slow response (Bank #2)

DTC P0155 Heated Oxygen Sensor Heater (HTR 21) circuit malfunction

DTC P0156 Downstream Heated Oxygen Sensor (HO₂S 22) circuit malfunction (Bank #2)

DTC P0158 Downstream Heated Oxygen Sensor (HO₂S 22) circuit high voltage (Bank #2)

DTC P0160 Heated Oxygen Sensor (HO₂S 12) circuit no activity detected (Bank #2)

DTC P0161 Heated Oxygen Sensor Heater (HTR 22) circuit malfunction

DTC P0171 System (adaptive fuel) too lean (Bank #1)

DTC P0172 System (adaptive fuel) too rich (Bank #1)

DTC P0174 System (adaptive fuel) too lean (Bank #2)

DTC P0175 System (adaptive fuel) too rich (Bank #2)

DTC P0222 Throttle Position Sensor B (TP-B) circuit low input

DTC P0223 Throttle Position Sensor B (TP-B) circuit high input

DTC P0230 Fuel pump primary circuit malfunction

DTC P0231 Fuel pump secondary circuit low

DTC P0232 Fuel pump secondary circuit high

DTC P0300 Random misfire detected

DTC P0301 Cylinder #1 misfire detected

DTC P0302 Cylinder #2 misfire detected

DTC P0303 Cylinder #3 misfire detected

DTC P0304 Cylinder #4 misfire detected

DTC P0305 Cylinder #5 misfire detected

DTC P0306 Cylinder #6 misfire detected

DTC P0307 Cylinder #7 misfire detected

DTC P0308 Cylinder #8 misfire detected

DTC P0320 Ignition engine speed (Profile Ignition Pickup) input circuit malfunction

DTC P0340 Camshaft Position (CMP) sensor circuit malfunction (CID)

DTC P0350 Ignition Coil primary circuit malfunction

DTC P0351 Ignition Coil A primary circuit malfunction

DTC P0352 Ignition Coil B primary circuit malfunction

DTC P0353 Ignition Coil C primary circuit malfunction

DTC P0354 Ignition Coil D primary circuit malfunction

DTC P0400 Exhaust Gas Recirculation (EGR) flow malfunction

DTC P0401 Exhaust Gas Recirculation (EGR) flow insufficient detected

DTC P0402 Exhaust Gas Recirculation (EGR) excess flow detected (valve open at idle)

DTC P0411 Secondary Air Injection system incorrect flow detected

DTC P0412 Secondary Air Injection system control valve malfunction

DTC P0420 Catalyst system efficiency below threshold (Bank #1)

DTC P0430 Catalyst system efficiency below threshold (Bank #2)

DTC P0443 Evaporative emission control system Canister Purge (CANP) Control Valve circuit malfunction

DTC P0500 Vehicle Speed Sensor (VSS) malfunction

DTC P0505 Idle Air Control (IAC) system malfunction

DTC P0603 Powertrain Control Module (PCM)—Keep Alive Memory (KAM) test error

DTC P0605 Powertrain Control Module (PCM)—Read Only Memory (ROM) test error

DTC P0704 Clutch Pedal Position (CPP) switch input circuit malfunction

DTC P0703 Brake On/Off (BOO) switch input malfunction

DTC P0707 Manual Lever Position (MLP) sensor circuit low input

DTC P0708 Manual Lever Position (MLP) sensor circuit high input

DTC P0712 Transmission Fluid Temperature (TFT) sensor circuit low input

DTC P0713 Transmission Fluid Temperature (TFT) sensor circuit high input

DTC P0715 Turbine Shaft Speed (TSS) sensor circuit malfunction

DTC P0720 Output Shaft Speed (OSS) sensor circuit malfunction

DTC P0731 Incorrect ratio for first gear

DTC P0732 Incorrect ratio for second gear

DTC P0733 Incorrect ratio for third gear

DTC P0734 Incorrect ratio for fourth gear

DTC P0736 Reverse incorrect gear

DTC P0741 Torque Converter Clutch (TCC) system incorrect mechanical performance

DTC P0746 Electronic Pressure Control (EPC) solenoid performance

DTC P0743 Torque Converter Clutch (TCC) system electrical failure

DTC P0750 Shift Solenoid #1(SS1) circuit malfunction

DTC P0751 Shift Solenoid #1(SS1) performance

DTC P0755 Shift Solenoid #2 (SS2) circuit malfunction

DTC P0756 Shift Solenoid #2 (SS2) performance

DTC P0760 Shift Solenoid #3 (SS3) circuit malfunction

DTC P0761 Shift Solenoid #3 (SS3) performance

DTC P0781 1 to 2 shift error

DTC P0782 2 to 3 shift error

DTC P0783 3 to 4 shift error

DTC P0784 4 to 5 shift error

DTC P1000 OBD II Monitor Testing not complete

DTC U1039 OBD II Monitor not complete

DTC UIOS1 Brake switch signal missing or incorrect

DTC P1100 Mass Air Flow (MAF) sensor intermittent

DTC P1101 Mass Air Flow (MAF) sensor out of Self-Test range

DTC P1112 Intake Air Temperature (IAT) sensor intermittent

DTC P1116 Engine Coolant Temperature (ECT) sensor out of Self-Test range

DTC P1117 Engine Coolant Temperature (ECT) sensor intermittent

DTC P1120 Throttle Position (TP) sensor out of range low

DTC P1121 Throttle Position (TP) sensor inconsistent with MAF sensor

DTC P1124 Throttle Position (TP) sensor out of Self-Test range

DTC P1125 Throttle Position (TP) sensor circuit intermittent

DTC P1130 Lack of HO2S 11 switch, adaptive fuel at limit

DTC P1131 Lack of HO2S 11 switch, sensor indicates lean (Bank #1)

DTC P1132 Lack of HO2S 11 switch, sensor indicates rich (Bank #1)

DTC U1135 Ignition switch signal missing or incorrect

DTC P1137 Lack of HO2S 12 switch, sensor indicates lean (Bank #1)

DTC P1138 Lack of HO2S 12 switch, sensor indicates rich (Bank #1)

DTC P1150 Lack of HO2S 21 switch, adaptive fuel at limit

DTC P1151 Lack of HO2S 21 switch, sensor indicates lean (Bank #2)

DTC P1152 Lack of HO2S 21 switch, sensor indicates rich (Bank #2)

DTC P1157 Lack of HO2S 22 switch, sensor indicates lean (Bank #2)

DTC P1158 Lack of HO2S 22 switch, sensor indicates rich (Bank #2)

DTC P1220 Series Throttle Control malfunction

DTC P1224 Throttle Position Sensor (TP-B) out of Self-test range

DTC P1233 Fuel Pump driver Module off-line

DTC P1234 Fuel Pump driver Module off-line

DTC P1235 Fuel Pump control out of range

DTC P1236 Fuel Pump control out of range

DTC P1237 Fuel Pump secondary circuit malfunction

DTC P1238 Fuel Pump secondary circuit malfunction

DTC P1260 THEFT detected—engine disabled

DTC P1270 Engine RPM or vehicle speed limiter reached

DTC P1351 Ignition Diagnostic Monitor (1DM) circuit input malfunction

DTC P1352 Ignition coil A primary circuit malfunction

DTC P1353 Ignition coil B primary circuit malfunction

DTC P1354 Ignition coil C primary circuit malfunction

DTC P1355 Ignition coil D primary circuit malfunction

DTC P1358 Ignition Diagnostic Monitor (1DM) signal out of Self-Test range

DTC P1359 Spark output circuit malfunction

DTC P1364 Ignition coil primary circuit malfunction

DTC P1390 Octane Adjust (OCT ADJ) out of Self-Test range

DTC P1400 Differential Pressure Feedback Electronic (DPFE) sensor circuit low voltage detected

DTC P1401 Differential Pressure Feedback Electronic (DPFE) sensor circuit high voltage detected

DTC P1403 Differential Pressure Feedback Electronic (DPFE) sensor hoses reversed

DTC P1405 Differential Pressure Feedback Electronic (DPFE) sensor upstream hose off or plugged

DTC P1406 Differential Pressure Feedback Electronic (DPFE) sensor downstream hose off or plugged

DTC P1407 Exhaust Gas Recirculation (EGR) no flow detected (valve stuck closed or inoperative)

DTC P1408 Exhaust Gas Recirculation (EGR) flow out of Self-Test range

DTC P1409 Electronic Vacuum Regulator (EVR) control circuit malfunction

DTC P1414 Secondary Air Injection system monitor circuit high voltage

DTC P1443 Evaporative emission control system—vacuum system purge control solenoid or purge control valve malfunction

DTC P1444 4- Purge Flow Sensor (PFS) circuit low input

DTC P1445 Purge Flow Sensor (PFS) circuit high input

DTC U1451 Lack of response from Passive Anti-Theft system (PATS) module—engine disabled

DTC P1460 Wide Open Throttle Air Conditioning Cut-off (WAC) circuit malfunction

DTC P1461 Air Conditioning Pressure (ACP) sensor circuit low input

DTC P1462 Air Conditioning Pressure (ACP) sensor circuit high input

DTC P1463 Air Conditioning Pressure (ACP) sensor insufficient pressure change

DTC P1469 Low air conditioning cycling period

DTC P1473 Fan Secondary High with fan(s) off

DTC P1474 Low Fan Control primary circuit malfunction

DTC P1479 High Fan Control primary circuit malfunction

DTC P1480 Fan Secondary low with low fan on

DTC P1481 Fan Secondary low with high fan on

DTC P1500 Vehicle Speed Sensor (VSS) circuit intermittent

DTC P1505 Idle Air Control (IAC) system at adaptive clip

DTC P1506 Idle Air control (IAC) over speed error

DTC P1518 Intake Manifold Runner Control (IMRC) malfunction (stuck open)

DTC P1519 Intake Manifold Runner Control (IMRC) malfunction (stuck closed)

DTC P1520 Intake Manifold Runner Control (IMRC) circuit malfunction

DTC P1507 Idle Air control (IAC) under speed error

DTC P1605 Powertrain Control Module (PCM)—Keep Alive Memory (KAM) test error

DTC P1650 Power steering Pressure (PSP) switch out of Self-Test range

DTC P1651 Power steering Pressure (PSP) switch input malfunction

DTC P1701 Reverse engagement error

DTC P1703 Brake On/Off (BOO) switch out of Self-Test range

DTC P1705 Manual Lever Position (MLP) sensor out of Self-Test range

DTC P1709 Park or Neutral Position (PNP) switch out of Self-test range

DTC P1729 4X4 Low switch error

DTC P1711 Transmission Fluid Temperature (TFT) sensor out of Self-Test range

DTC P1741 Torque Converter Clutch (TOC) control error

DTC P1742 Torque Converter Clutch (TCC) solenoid mechanically failed (turns MIL on)

DTC P1743 Torque Converter Clutch (TCC) solenoid mechanically failed (turns TOIL on)

DTC P1744 Torque Converter Clutch (TCC) system mechanically stuck in off position

DTC P1748 Electronic Pressure Control (EPC) solenoid circuit low input (open circuit)

DTC P1747 Electronic Pressure Control (EPC) solenoid circuit high input (short circuit)

DTC P1749 Electric Pressure Control (EPC) solenoid failed low

DTC P1751 Shift Solenoid #1(SS1) performance

DTC P1756 Shift Solenoid #2 (SS2) performance

DTC P1780 Transmission Control Switch (TCS) circuit out of Self-Test range

General Motors

SELF-DIAGNOSTICS

Automotive manufacturers have developed on-board computers to control engines, transmissions and many other components. These on-board computers with dozens of sensors and actuators have become almost impossible to test without the help of electronic test equipment.

One of these electronic test devices has become the on-board computer itself. The Powertrain Control Modules (PCM), sometimes called the Electronic Control Module (ECM), used on toadies vehicles has a built in self testing system. This self test ability is called self-diagnosis. The self-diagnosis system will test many or all of the sensors and controlled devices for proper function. When a malfunction is detected this system will store a code in memory that's related to that specific circuit. The computer can later be accessed to obtain fault codes recorded in memory using the procedures for Reading Codes. This helps narrow down what area to begin testing.

Fault code meanings can vary from year to year even on the same model. It is extremely important after retrieving a fault code to verify its meaning with a proper manual. Servicing a code incorrectly will not only lead to the wrong conclusion but could also cause damage if tested or serviced incorrectly.

Since the control module is programmed to recognize the presence and value of electrical inputs, it will also note the lack of a signal or a radical change in values. It will, for example, react to the loss of signal from the vehicle speed sensor or note that engine coolant temperature has risen beyond acceptable (programmed) limits. Once a fault is recognized, a numeric code is assigned and held in memory. The dashboard warning lamp—CHECK ENGINE or SERVICE ENGINE SOON—will illuminate to advise the operator that the system has detected a fault.

More than one code may be stored. Although not every engine uses every code and the same code may carry different meanings relative to each engine or engine family. For example, on the 3.3L (VIN N), Code 46 indicates a fault found in the power steering pressure switch circuit. The same code on the 5.7L (VIN F) engine indicates a fault in the VATS anti-theft system. The list of codes and descriptions can be found in the 'Code Descriptions' section of the manual.

In the event of an PCM failure, the system will default to a pre-programmed set of values. These are compromise values which allow the engine to operate, although possibly at reduced efficiency. This is also known as the default, limp-in or back-up mode. Driveability is almost always affected when the PCM enters this mode.

Service Precautions

• Protect the on-board solid-state components from rough handling or extremes of temperature.

• Always turn the ignition OFF when connecting or disconnecting battery cables, jumper cables, or a battery charger. Failure to do this can result in PCM or other electronic component damage.

• Remove the PCM before any arc welding is performed to the vehicle

• Electronic components are very susceptible to damage caused by electrostatic discharge (static electricity). To prevent electronic component damage, do not touch the control module connector pins or soldered components on the control module circuit board.

Visual Inspection

This is possibly the most critical step of diagnosis. A detailed examination of all connectors, wiring and vacuum hoses can Offen lead to a repair without further diagnosis. Also, take into consideration if the vehicle has been serviced recently? Sometimes things get reconnected in the wrong place, or not at all. A careful inspector will check the undersides of hoses as well as the integrity of hard-to-reach hoses blocked by the air cleaner or other components. Correct routing for vacuum hoses can be obtained from your specific vehicle service manual or Vehicle Emission Control Information (VECI) label in the engine compartment of the vehicle. Wiring should be checked carefully for any sign of strain, burning, crimping or terminals pulled-out from a connector.

Checking connectors at components or in harnesses is required; usually, pushing them together will reveal a loose fit. Also, check electrical connectors for corroded, bent, damaged, improperly seated pins, and bad wire crimps to terminals. Pay particular attention to ground circuits, making sure they are not loose or corroded. Remember to inspect connectors and hose fittings at components not mounted on the engine, such as the evaporative canister or relays mounted on the fender aprons. Any component or wiring in the vicinity of a fluid leak or spillage should be given extra attention during inspection.

➡**There are many problems with connectors on electronic engine control systems. Due to the low voltage signals that these systems use any dirt, corrosion or damage will affect their operation. Note that some connectors use a special grease on the contacts to prevent corrosion. Do not wipe this grease off, it is a special type for this purpose. You can obtain this grease from your vehicle dealer.**

Additionally, inspect maintenance items such as belt condition and tension, battery charge and condition and the radiator cap carefully. Any of these very simple items may affect the system enough to set a fault.

Dashboard Warning Lamp

The primary function of the dash warning lamp is to advise the operator that a fault has been detected, and, in most cases, a code stored. Under normal conditions, the dash warning lamp will illuminate when the ignition is turned ON. Once the engine is started and running, the PCM will perform a system check and extinguish the warning lamp if no fault is found.

"SERVICE ENGINE SOON"

79242G16

The Check Engine or Service Engine Soon (MIL) lamp is used for reading trouble codes

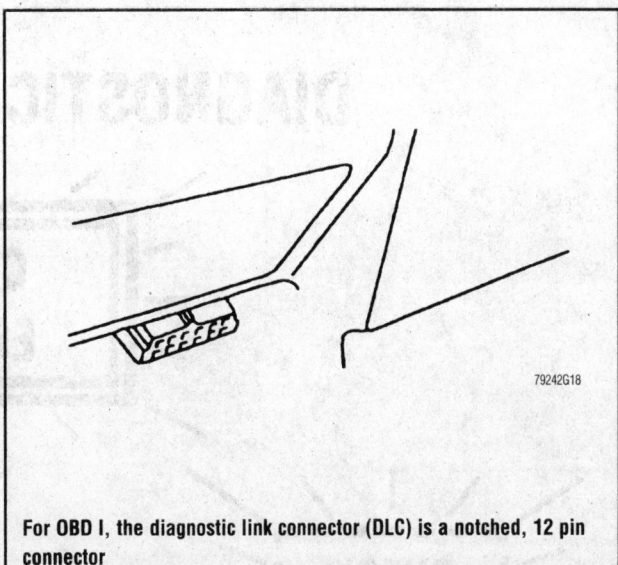

79242G18

For OBD I, the diagnostic link connector (DLC) is a notched, 12 pin connector

Additionally, the dash warning lamp can be used to retrieve stored codes after the system is placed in the Diagnostic Mode. Codes are transmitted as a series of flashes with short or long pauses. When the system is placed in the Field Service Mode (available on fuel injected model), the dash lamp will indicate open loop or closed loop function.

Intermittent Problems

If a fault occurs intermittently, such as a loose connector pin breaking contact as the vehicle hits a bump, the PCM will note the fault as it occurs and energize the dash warning lamp. If the problem self-corrects, as with the terminal pin again making contact, the dash lamp will extinguish after 10 seconds but a code will remain stored in the PCM memory. When an unexpected code appears during an intermittent failure that self-corrected; the codes are still useful in diagnosis and should not be discounted.

Diagnostic Connector Location

The Assembly Line Communication Link (ALCL) or Assembly Line Diagnostic Link (ALDL) is a Diagnostic Link Connector (DLC)

79242G17

On most vehicles, the diagnostic link connector is located under the instrument panel

located in the passenger compartment. It has terminals which are used in the assembly plant to check that the engine is operating properly before it leaves the plant.

This DLC is where you connect you jumper the terminals to place the engine control computer into self-diagnostic mode. The standard term DLC is sometimes referred to as the ALCL or the ALDL in different manuals. Either way it is referred to, they all still perform the same function.

READING & CLEARING CODES

Reading Codes

Since the inception of electronic engine management systems on General Motors vehicles, there has been a variety of connectors provided to the technician for retrieving Diagnostic Trouble Codes (DTC)s. Additionally, there have been a number of different names given to these connectors over the years; Assembly Line Communication Link (ALCL), Assembly Line Diagnostic Link (ALDL), Data Link Connector (DLC). Actually when the system was initially introduced to the 49 states in 1979, early 1980, there was no connector used at all. On these early vehicles there was a green spade terminal taped to the ECM harness and connected to the diagnostic enable line at the computer. When this terminal was grounded with the key ON, the system would flash any stored diagnostic trouble codes. The introduction of the ALDL was found to be a much more convenient way of retrieving fault codes. This connector was located underneath the instrument panel on most GM vehicles, however on some models it will not be found there. On early Corvettes the ALDL is located underneath the ashtray, it can be found in the glove compartment of some early FWD Oldsmobiles, and between the seats in the Pontiac Fiero. The connector was first introduced as a square connector with four terminals, then progressed to a flat five terminal connector, and finally to what is still used in 1993, a 12 terminal double row connector. To access stored Diagnostic Trouble Codes (DTC) from the square connector, turn the ignition ON and identify the diagnostic enable terminal (usually a white wire with a black tracer) and ground it. The flat five terminal connector is identified from left to right as A, B, C, D, and E. There is a space between terminal D and E which permits a spade to be inserted for the purposes of diagnostics when the ignition key is ON. On this connector

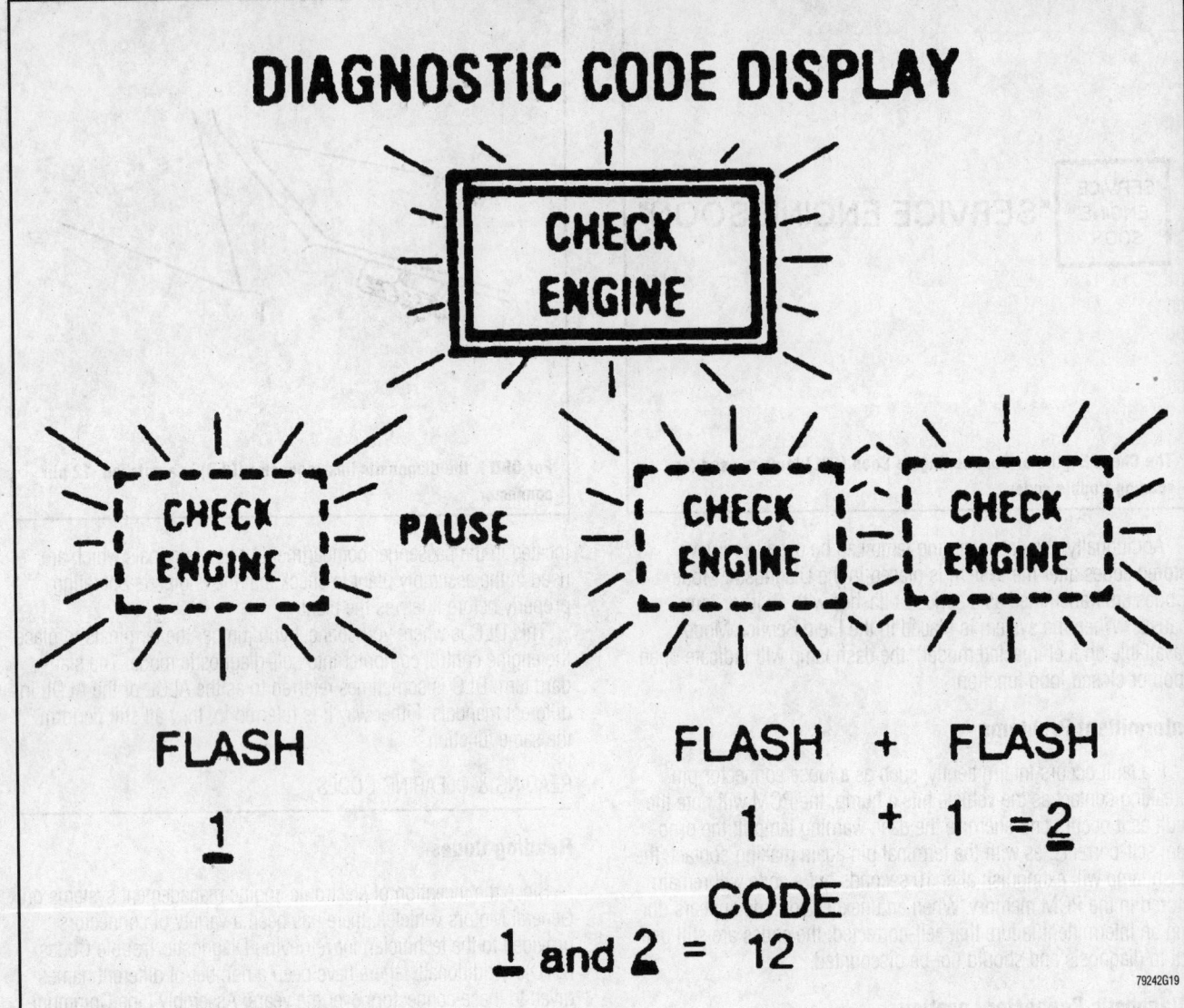

DIAGNOSTIC CODE DISPLAY

Example of a code 12 displayed on the check engine lamp

terminal D is the diagnostic enable line, and E is a ground. The 12 terminal double row connector has been continually expanded through the years as vehicles acquired more on-board electronic systems such as Anti-lock Brakes. Despite this the terminals used for engine code retrieval have remained the same. The 12 terminal connector is identified from right-to-left on the top row A-F, and on the bottom row from left-to-right, G-L. To access engine codes turn the ignition ON and insert a jumper between terminals A and B. Terminal A is a ground, and terminal B is the diagnostic request line. Stored trouble codes can be read through the flashing of the Check Engine Light or on later vehicles the Service Engine Soon lamp. Trouble codes are identified by the timed flash of the indicator light. When diagnostics are first entered the light will flash once, pause; then two quick flashes.

This reads as DTC 12 which indicates that the diagnostic system is working. This code will flash indefinitely if there are no stored trouble codes. If codes are stored in memory, Code 12 will flash three times before the next code appears. Codes are displayed in the next highest numerical sequence. For example, Code 13 would be displayed next if it was stored in memory and would read as follow:

flash, pause, flash, flash, flash, long pause, repeat twice. This sequence will continue until all codes have been displayed, and then start all over again with Code 12.

Clearing Codes

To clear any Diagnostic Trouble Codes (DTC's) from the PCM memory, either to determine if the malfunction will occur again or because repair has been completed, power feed must be disconnected for at least 30 seconds. Depending on how the vehicle is equipped, the system power feed can be disconnected at the positive battery terminal pigtail, the inline fuseholder that originates at the positive connection at the battery, or the ECM/PCM fuse in the fuse block. The negative battery terminal may be disconnected but other on-board memory data such as preset radio tuning will also be lost. To prevent system damage, the ignition switch must be in the OFF position when disconnecting or reconnecting power.

When using a Diagnostic Computer such as Tech 1, or equivalent scan tool to read the diagnostic trouble codes, clearing the codes is done in the same manner. On some systems, DTC's may be cleared through the Tech 1, or equivalent scan tool.

DIAGNOSTIC TROUBLE CODES

Code 12 No engine RPM reference pulses—System Normal

Code 13 Oxygen Sensor (02S) circuit open—left side on 2 sensor system

Code 14 Engine Coolant Temperature (ECT) sensor—possible circuit high or shorted sensor

Code 15 Engine Coolant Temperature (ECT) sensor—circuit low or open circuit

Code 16 Direct ignition system (DIS), fault line circuit or Distributor ignition system (low resolution pulse) or Missing 2x reference circuit or OPTI-Spark ignition timing system (low resolution pulse) or System voltage out of range

Code 17 Camshaft Position Sensor (CPS) or spark reference circuit error

Code 18 Crank/Cam error

Code 19 Crankshaft Position Sensor (CPS) circuit

Code 21 Throttle Position (TP) sensor circuit—signal voltage out of range, probably high

Code 22 Throttle Position (TP) sensor circuit—signal voltage low

Code 23 Intake Air Temperature (IAT or MAT) sensor circuit temperature out of range, low or Open or grounded M/C solenoid Feedback Carburetor system

Code 24 Vehicle Speed Sensor (VSS) circuit

Code 25 Intake Air Temperature (IAT or MAT) sensor circuit temperature out of range, high

Code 26 Quad-Driver Module #1 circuit or Transaxle gear switch circuit

Code 27 Quad-Driver Module circuit or Transaxle gear switch, probably 2nd gear switch circuit

Code 28 Quad-Driver Module (QDM) #2 circuit or Transaxle gear switch, probably 3rd gear switch circuit

Code 29 Transaxle gear switch, probably 4th gear switch circuit

Code 31 Camshaft sensor circuit fault or Park/Neutral Position (PNP) switch circuit or Wastegate circuit signal

Code 32 Exhaust Gas Recirculation (EGR) circuit fault or Barometric Pressure Sensor circuit low Feedback Carburetor system

Code 33 Manifold Absolute Pressure (MAP) sensor—signal voltage out of range, high or Mass Air Flow (MAF) sensor—signal voltage out of range, probably high

Code 34—Manifold Absolute Pressure (MAP) sensor—circuit out of range voltage, low or Mass Air Flow (MAF) sensor circuit (gm/sec low)

Code 35—Idle Air Control (IAC) or idle speed error or Idle Speed Control (ISO) circuit throttle switch shorted Feedback Carburetor system

Code 36 Ignition system circuit error or Transaxle shift problem—4T60E Transaxle

Code 38 Brake input circuit fault—Torque converter clutch signal

Code 39 Clutch input circuit fault—Torque converter clutch signal

Code 41 Cam sensor or cylinder select circuit fault ignition control (IC) reference pulse system fault or Electronic Spark Timing (EST) circuit open or shorted

Code 42 Electronic Spark Timing (EST) circuit grounded or Ignition Control (IC) circuit grounded or faulty bypass line

Code 43 Knock Sensor (KS) or Electronic Spark Control (ESC) circuit fault

Code 44 Oxygen Sensor (02S), left side on 2 sensor system lean exhaust indicated

Code 45 Oxygen Sensor (02S), left side on 2 sensor system rich exhaust indicated

Code 46 Personal Automotive Security System (PASSKey II) circuit or Power Steering Pressure Switch (PSPS) circuit

Code 47 PCM-BCM data circuit

Code 48 Misfire diagnosis

Code 51 Calibration error, faulty MEM-CAL, ECM or EEPROM failure

Code 52 Engine oil temperature sensor circuit, low temperature indicated or Fuel Calpac missing or Over voltage condition or EGR Circuit fault

Code 53 Battery voltage error or EGR problem or Personal Automotive Security System (PASS-Key) circuit

Code 54 EGR #2 problem or Fuel pump circuit (low voltage) or Shorted mixture control solenoid circuit Feedback Carburetor system

Code 55 A/D Converter error, PCM error or not grounded, EGR #3 problem, Fuel lean monitor, Grounded voltage reference, faulty oxygen sensor or fuel lean Feedback Carburetor system

Code 56 Quad-Driver Module (QDM) #2 circuit or Secondary air inlet valve actuator vacuum sensor circuit signal high 5.7L (VIN J)

Code 57 Boost control problem

Code 58 Vehicle Anti-theft System fuel enable circuit

Code 61 A/C system performance or Cruise vent solenoid circuit fault or Oxygen Sensor (02S) degraded signal or Secondary port throttle valve system fault 5.7L (VIN J) or Transaxle gear switch signal

Code 62 Cruise vacuum solenoid circuit fault or Engine oil temperature sensor, high temperature indicated or Transaxle gear switch signal circuit fault

Code 63 Oxygen Sensor (02S), right side circuit open or Cruise system problem (speed error) or Manifold Absolute Pressure (MAP) sensor circuit out of range

Code 64 Oxygen Sensor (02S), right side—lean exhaust indicated

Code 65 Oxygen Sensor (02S), right side—rich exhaust indicated or Cruise servo position circuit or Fuel injector circuit low current

Code 66 A/C pressure sensor circuit fault, probably low pressure or Engine power switch, voltage high or low or PCM fault 5.7L (VIN J)

Code 67 A/C pressure sensor circuit, sensor or A/C clutch circuit failure or Cruise switch circuit fault

Code 68 A/C compressor relay (shorted circuit) or Cruise system fault

Code 69 A/C clutch circuit or head pressure high

Code 70 A/C refrigerant pressure sensor circuit (high pressure)

Code 71 A/C evaporator temperature sensor circuit (low temperature)

Code 72 Gear selector switch circuit

Code 73 A/C evaporator temperature sensor circuit (high temperature)

Code 75 Digital EGR #1 solenoid error

Code 76 Digital EGR #2 solenoid error

Code 77 Digital EGR #3 solenoid error

Code 79 Vehicle Speed Sensor (VSS) circuit signal high

Code 80 Vehicle Speed Sensor (VSS) circuit signal low

Code 81 Brake input circuit fault—Torque converter clutch signal

Code 82 Ignition Control (IC) 3X signal error

Code 85 PROM error

Code 86 Analog/Digital ECM error

Code 87 EEPROM error

Code 99 Power management

➡This list is for reference and does not mean a specific component is defective.

Honda

GENERAL INFORMATION

Honda utilizes 2 types of fuel systems. The first is the feedback carburetor system of which there are 2 types; a 2 barrel down draft-fixed venturi type, and 2 side draft carburetors variable venturi type. The feedback carburetor was in use up to 1991 in Honda vehicles.

The second type fuel system is Programmed Fuel Injection (PGM-FI) system. As of 1992 all Hondas are fuel injected.

SELF-DIAGNOSTICS

Service Precautions

- Make sure all ECM harness connectors are fastened securely. A poor connection can cause an extremely high voltage surge and result in damage to integrated circuits.
- Keep all ECM parts and harnesses dry during service. Protect the ECM and all solid-state components from rough handling or extremes of temperature.
- Use extreme care when working around the ECM or other components. The airbag or SRS wiring may be in the vicinity. On these vehicles, the SRS wiring and connectors are yellow. Do not cut or test these circuits.
- Before attempting to remove any parts, turn the ignition switch OFF and disconnect the battery ground cable.
- Always use a 12 volt battery as a power source for the engine, never a booster or high-voltage charging unit.
- Do not disconnect the battery cables with the engine running.
- Do not disconnect any wiring connector with the engine running or the ignition ON unless specifically instructed.
- Do not apply battery power directly to injectors.
- Whenever possible, use a flashlight instead of a drop light.
- Relieve fuel system pressure before servicing any fuel system component.
- Always use eye or full-face protection when working around fuel lines, fittings or components.

Reading Codes

1985–89 VEHICLES

When a fault is noted, the ECU stores an identifying code and illuminates the CHECK ENGINE light. The code will remain in memory until cleared; the dashboard warning lamp may not illuminate during the next ignition cycle if the fault is no longer present. Not all faults noted by the ECU will trigger the dashboard warning lamp although the fault code will be set in memory. For this reason, troubleshooting should be based on the presence of stored codes, not the illumination of the warning lamp while the car is operating.

Stored codes are displayed by either a single flashing LED (Light Emitting Diode) light, or an illuminated light pattern of 4 LED lights on the ECU. When the CHECK ENGINE warning lamp has been on or reported on, check the ECU LED for presence of codes.

The location of the malfunction is determined by observing the LED display. Earlier Hondas used 2 types of LED displays: a single LED and a 4 LED display. After 1987 all models use the single LED display.

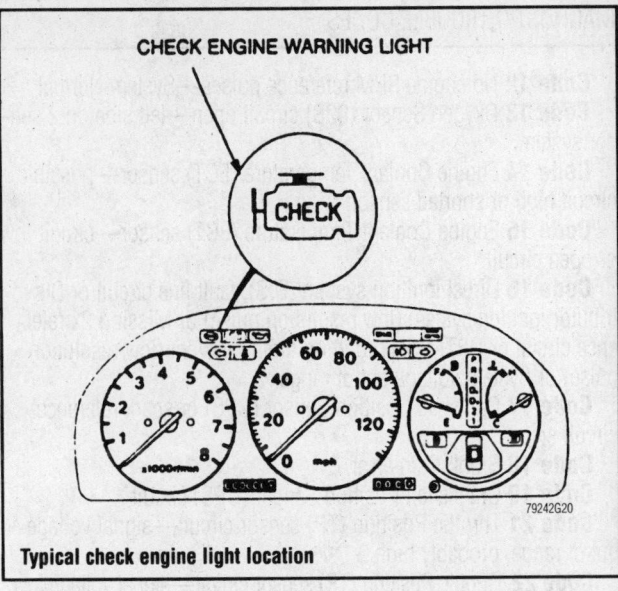

CHECK ENGINE WARNING LIGHT

Typical check engine light location

Systems with a single LED indicate the malfunction with a series of flashes. The number of flashes indicates a code which identifies the location of the component or system malfunction. The code will flash, followed by a 2 second pause, repeat, followed by another 2 second pause, then move to the next code.

On systems with 4 LED's a display pattern identifies the malfunction. The LED's are numbered 1, 2, 4 and 8 as counted from right-to-left. The code is determined by observing which LED's are lit on the display. Each code is displayed once, followed by a 2 second pause, then the next code is displayed.

The LED's are part of the Electronic Control Module (ECM).

Turn the ignition switch ON; the LED will display any stored codes.

When counting flashes to determine codes, a code not valid for the vehicle may be found. In this case, first recount the flashes to confirm an accurate count. If necessary, turn the ignition switch OFF, then recycle the system and begin the count again. If the Code is not valid for the vehicle, the ECU must be replaced.

➡On vehicles with electronically controlled automatic transaxles, the 5, D or D4 lamp may flash with the CHECK ENGINE lamp if certain codes are stored. If this does occur, proceed with the diagnosis based on the engine code shown. After repairs, recheck the lamp. If the additional warning lamp is still lit, proceed with diagnosis for that system.

1990–95 VEHICLES

When a fault is noted, the ECM stores an identifying code and illuminates the CHECK ENGINE light. The code will remain in memory until cleared. The dashboard warning lamp may not illuminate during the next ignition cycle if the fault is no longer present. Not all faults noted by the ECM will trigger the dashboard warning lamp although the fault code will be set in memory. For this reason, troubleshooting should be based on the presence of stored codes, not the illumination of the warning lamp while the car is operating.

In 1990, the Accord and Prelude were equipped with a 2-pin service connector in addition to the LED. if the service connector is umpered, with the ignition key in the ON position, the CHECK ENGINE lamp will display the stored codes in a series of flashes. The 2-pin service connector is located under the passenger side of dash on the Accord and behind the center console on the Prelude.

As of 1992, the LED on the ECU was eliminated and all vehicles obtain codes by jumping the 2-pin connector when the ignition switch is ON. The CHECK ENGINE light will then flash codes present in the ECU memory.

Diagnostic Codes 1-9 are indicated by a series of short flashes; two-digit codes use a number of long flashes for the first digit followed by the appropriate number of short flashes. For example, Code 43 would be indicated by 4 long flashes followed by 3 short flashes. Codes are separated by a longer pause between transmissions. The position of the codes during output can be helpful in diagnostic work. Multiple codes transmitted in isolated order indicate unique occurrences; a display of showing 1-1-1 pause 9-9-9 indicates two problems or problems occurring at different times. An alternating display, such as 1-9-1-9-1, indicates simultaneous occurrences of the faults.

When counting flashes to determine codes, a code not valid for the vehicle may be found. In this case, first recount the flashes to confirm an accurate count. If necessary, turn the ignition switch OFF, then recycle the system and begin the count again. If the code is not valid for the vehicle, the ECM must be replaced.

➡**On vehicles with electronically controlled automatic transaxles, the D4 lamp may flash with the CHECK ENGINE lamp if certain codes are stored. If this does occur, proceed with the diagnosis based on the engine code shown. After repairs, recheck the lamp. If the additional warning lamp is still lit, proceed with diagnosis for that system.**

Clearing Codes

1985–87 VEHICLES

The memory for the PGM-FI CHECK ENGINE lamp on the dashboard will be erased when the ignition switch is turned OFF; however, the memory for the LED display will not be canceled. Thus, the CHECK ENGINE lamp will not come on when the ignition switch is again turned ON unless the trouble is once more detected. Troubleshooting should be done according to the LED display even if the CHECK ENGINE lamp is off.

After making repairs, disconnect the battery negative cable from the battery negative terminal for at least 10 seconds and reset the ECU memory. After reconnecting the cable, check that the LED display is turned off.

Turn the ignition switch ON. The PGM-FI CHECK ENGINE lamp should come on for about 2 seconds. If the CHECK ENGINE lamp won't come on, check for:—Blown CHECK ENGINE lamp bulb—Blown fuse (causing faulty back up light, seat belt alarm, clock, memory function of the car radio) -Open circuit in Yellow wire—Open circuit in wiring and control unit.

After the PGM-FI CHECK ENGINE lamp and self-diagnosis indicators have been turned on, turn the ignition switch OFF. If the LED display fails to come on when the ignition switch is turned ON again, check for:—Blown fuses, especially No. 10 fuse—Open circuit in wire between ECU fuse.

Replace the ECU only after making sure that all couplers and connectors are connected securely.

1988–90 VEHICLES

The memory for the PGM-CARB and PGM-FI CHECK ENGINE lamp on the dashboard will be erased when the ignition switch is turned OFF; however, the memory for the LED display will not be canceled. Thus, the CHECK ENGINE lamp will not come on when the ignition switch is again turned ON unless the trouble is once more detected. Troubleshooting should be done according to the LED display even if the CHECK ENGINE lamp is off.

To clear the ECU trouble code memory, remove the ECU memory power fuse for at least 10 seconds.

➡**Removing this fuse will also erase the clock, radio station presets, and the radio anti-theft codes. Make sure you have the anti-theft code and station presets before removing fuse so they may be reset when repairs are complete.**

1990–95 VEHICLES

Stored codes are removed from memory by removing power to the ECU. Disconnecting the power may also clear the memories used for other solid-state equipment such as the clock and radio. For this reason, always make note of the radio presets before clearing the system. Additionally, some radios contain anti-theft programming; obtain the owner's code number before clearing the codes.

While disconnecting the battery will clear the memory, this is not the recommended procedure. The memory should be cleared after the ignition is switched OFF by removing the appropriate fuse for at least 10 seconds.

➡**Removing this fuse will also erase the clock, radio station presets, and the radio anti-theft codes. Make sure you have the anti-theft code and station presets before removing fuse so they may be reset when repairs are complete.**

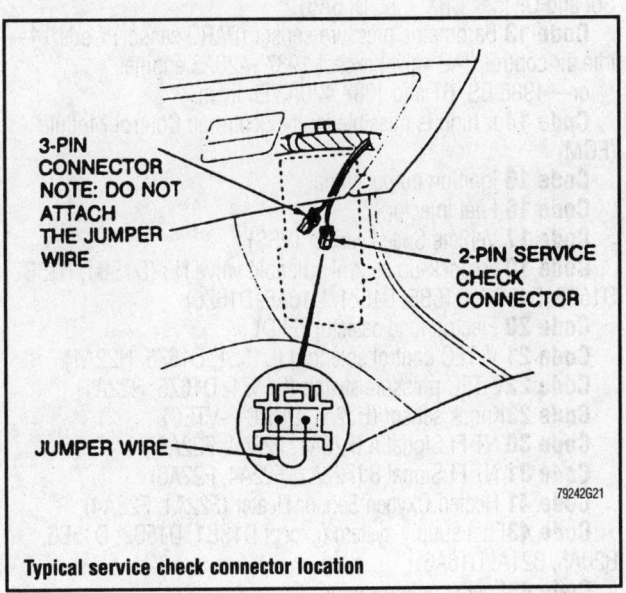

Typical service check connector location

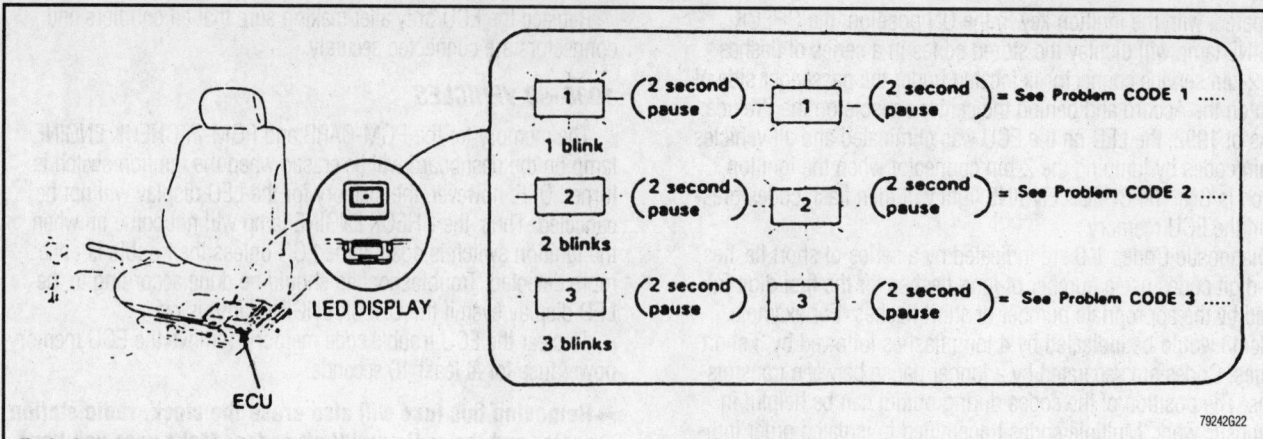

Typical electronic control module location

DIAGNOSTIC TROUBLE CODES

Code 0 Electronic Control Module (ECM)

Code 1 Heated oxygen sensor (or Oxygen content) or—Oxygen content A (A20A3, B20A5)

Code 2 Oxygen content B (A20A3, B20A5) or—Electronic Control Module (ECM) (BS, BT—1986 only) and (A20A3—1987 only)

Code 3 Manifold Absolute Pressure (MAP)

Code 4 Crankshaft position sensor or—Faulty ECU (BS, BT—1986 only) and (A20A3 -1987 only, B20A5, B21A1)

Code 5 Manifold Absolute Pressure (MAP)

Code 6 Engine coolant temperature (ECT)

Code 7 Throttle position sensor (TP sensor)

Code 8 Top dead center sensor (TDC sensor)

Code 9 No. 1 cylinder position sensor

Code 10 - Intake air temperature sensor (IAT sensor)

Code 11 - Electronic Control Module (ECM) (BS, BT -1986 only) and (A20A3—1987 only)

Code 12 - Exhaust Gas Recirculation (EGR) System (except Del Sol and Civic & CRX 1 .6L DI 6A6)

Code 13 Barometric pressure sensor (BARO sensor) Code 14— Idle air control (IAC valve) except 1987 -A20A3 engine.

or—1986 BS, BT and 1987 A20A3 Engines,

Code 14 or high is possible faulty Electronic Control Module (ECM)

Code 15 Ignition output signal

Code 16 Fuel Injector

Code 17 Vehicle Speed sensor (VSS)

Code 19 NT lock-up control solenoid valve NB (D1SB1, D15B2, D15B6, D15B7, D15B8, D15Z1, D16A6, D16Z6)

Code 20 Electric load detector (ELD)

Code 21 V-TEC control solenoid (D15Z1, D16Z6, H22A1)

Code 22 V-TEC pressure switch (D15Z1, D16Z6, H22A1)

Code 23 Knock sensor (H22A1-DOHC—VTEC)

Code 30 NT FI Signal A (F22A1, F22A4, F22A6)

Code 31 NT FI Signal B (F22A1, F22A4, F22A6)

Code 41 Heated Oxygen Sensor Heater (F22A1, F22A4)

Code 43 Fuel supply system (except D1SB1, D15B2, D15B6, B20A5, B21A, D16A6)

Code 45 Fuel supply metering

Code 48 Heated oxygen sensor (D15Z1 engine only, except Calif. emission)

Code 61 Front Heated Oxygen Sensor

Code 63 Rear Heated Oxygen Sensor

Code 65 Rear Heated Oxygen Sensor Heater

Code 67 Catalytic Converter System

Code 70 Automatic Transaxle or A/F FI Data line

Code 71 Misfire detected; cylinder No. 1 or random misfire

Code 72 Misfire detected; cylinder No. 2 or random misfire

Code 73 Misfire detected; cylinder No. 3 or random misfire

Code 74 Misfire detected; cylinder No. 4 or random misfire

Code 75 Misfire detected; cylinder No. 5 or random misfire

Code 76 Misfire detected; cylinder No. 6 or random misfire

Code 80 Exhaust Gas Recirculation (EGR) system

Code 86 Coolant temperature

Code 92 Evaporative Emission Control System

Infiniti

GENERAL INFORMATION

The Infiniti Electronic Concentrated Control System (ECCS) is an air flow controlled, sequential port fuel injection and engine control system. It is used on all models equipped with 2.0L, 3.0L and 4.5L engines. The ECCS electronic control unit consists of a microcomputer, an inspection lamp, a diagnostic mode selector and connectors for signal input and output, powers and grounds.

The safety relay prevents electrical damage to the electronic control unit, or ECU, and the injectors in case the battery terminals are accidentally connected in reverse. The safety relay is built into the fuel pump control circuit.

Ignition timing is controlled in response to engine operating conditions. The optimum ignition timing in each driving condition is pre-programmed in the computer. The signal from the control unit is transmitted to the power transistor and this signal controls when the transistor turns the ignition coil primary circuit on and off (hence, the ignition timing). The idle speed is also controlled according to engine operating conditions, temperature and gear position. On manual transmission models, if battery voltage is less than 12 volts for a few seconds, a higher idle speed will be maintained by the control unit to improve charging function.

There is a fail-safe system built into the ECCS control unit. This system makes engine starting possible if a portion of the ECU's central processing unit circuit fails. Also, if a major component such as the crank angle sensor or the air flow meter were to malfunction, the ECU substitutes or borrows data to compensate for the fault. For example, if the output voltage of the air flow meter is extremely low, the ECU will substitute a pre-programmed value for the air flow meter signal and allows the vehicle to be driven as long as the engine speed is kept below 2000 rpm. Or, if the cylinder head temperature sensor circuit is open, the control unit clamps the warm-up enrichment at a certain amount. This amount is almost the same as that when the cylinder head temperature is between 68–176°F (20–80°C).

If the fuel pump circuit malfunctions, the fuel pump relay comes on until the engine stops. This allows the fuel pump to receive power from the relay.

The electronic control unit controls the following functions:

- Injector pulse width
- Ignition timing
- Intake valve timing control (045)
- Air regulator control (G20)
- Exhaust gas recirculation (EGR) solenoid valve operation
- Exhaust gas sensor heater operation
- Idle speed
- FICD solenoid valve operation (G20 and M30)
- Fuel pump relay operation
- Fuel pump voltage (M30 and 045)
- Fuel pressure regulator control (M30)
- AIV control (G20)
- Carbon canister control solenoid valve operation
- Air conditioner relay operation (During early wide-open throttle)
- Radiator fan operation (G20)
- Traction control system (TCS) operation (045, if equipped)
- Self-diagnosis
- Fail-safe mode operation

SELF-DIAGNOSTICS

Service Precautions

- Do not disconnect the injector harness connectors with the engine running.
- Do not apply battery power directly to the injectors.
- Do not disconnect the ECU harness connectors before the battery ground cable has been disconnected.
- Make sure all ECU connectors are fastened securely. A poor connection can cause an extremely high surge voltage in the coil and condenser and result in damage to integrated circuits.
- When testing the ECU with a DVOM make sure that the probes of the tester never touch each other as this will result in damage to a transistor in the ECU.
- Keep the ECCS harness at least 4 in. away from adjacent harnesses to prevent an ECCS system malfunction due to external electronic noise.
- Keep all parts and harnesses dry during service.
- Before attempting to remove any parts, turn OFF the ignition switch and disconnect the battery ground cable.
- Always use a 12 volt battery as a power source.
- Do not attempt to disconnect the battery cables with the engine running or the ignition key ON.
- Do not clean the air flow meter with any type of detergent.
- Do not attempt to disassemble the ECCS control unit under any circumstances.
- Avoid static electricity build-up by properly grounding yourself prior to handling any ECU or related parts.

Reading Codes

➡Diagnostic codes may be retrieved by observing code flashes through the LED lights located on the Electronic Control Module (ECM). A special Nissan Consult monitor tool can be used, but is not required- When using special diagnostic equipment, always observe the tool manufacturer's instructions.

1990–95 VEHICLES

2-Mode Diagnostic System

Infiniti vehicles use a 2-mode diagnostic system incorporated in the ECU which uses inputs from various sensors to determine the correct air/fuel ratio. If any of the sensors malfunction the ECU will store the code in memory.

An Infiniti/Nissan Consult monitor may be used to retrieve these codes by simply connecting the monitor to the diagnosis connector located on the driver's side near the hood release.

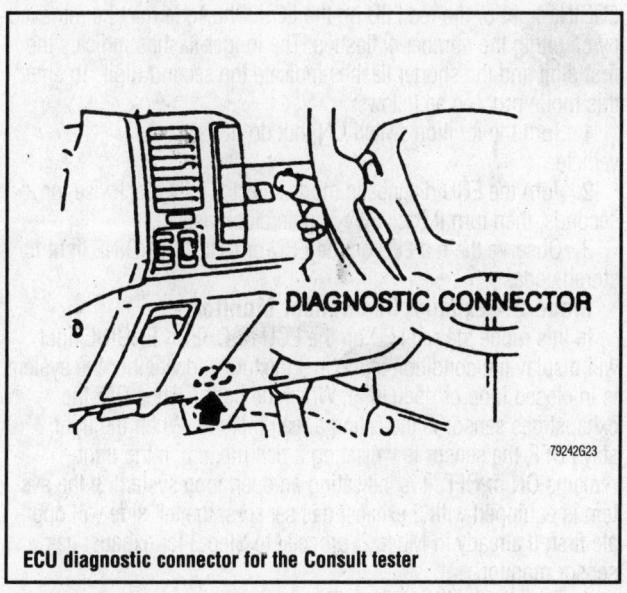

ECU diagnostic connector for the Consult tester

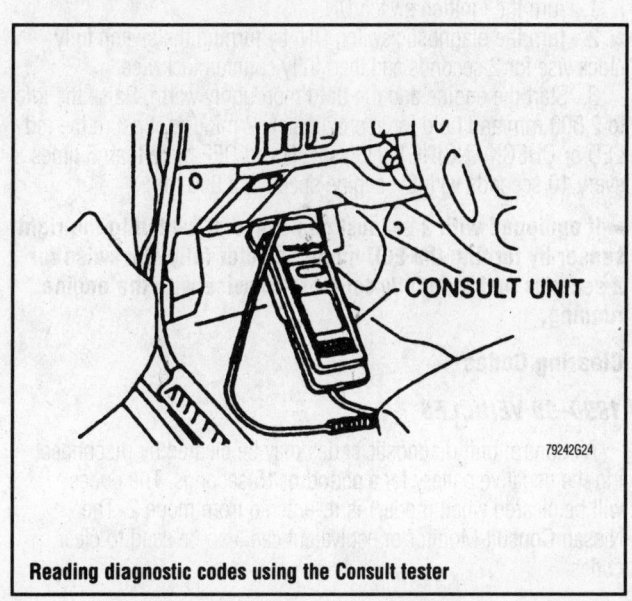

Reading diagnostic codes using the Consult tester

Turn the ignition switch ON and press START, ENGINE and then SELF-DIAG RESULTS, the results will then be output to the monitor.

The conventional CHECK ENGINE or red-LED ECU light may be used for self-diagnostics. The conventional 2-mode diagnostic system is broken into 2 separate modes each capable of 2 tests, an ignition switch ON or engine running test as outlined below:

Mode I—Bulb Check

In this mode the RED indicator light on the ECU and the CHECK ENGINE light should be ON. To enter this mode simply turn the ignition switch ON and observe the light.

Mode 1—Malfunction Warning

In this mode the ECU is acknowledging if there is a malfunction by illuminating the RED indicator light on the ECU and the CHECK ENGINE light. If the light turns OFF, the system is normal. To enter this mode, simply start the engine and observe the light.

Mode 2—Self-Diagnostic Codes

In this mode the ECU will output all malfunctions via the CHECK ENGINE light or the red LED on the ECU. The code may be retrieved by counting the number of flashes. The longer flashes indicate the first digit and the shorter flashes indicate the second digit. To enter this mode proceed as follows:

1. Turn the ignition switch ON, but do not start the vehicle.
2. Turn the ECU diagnostic mode selector fully clockwise for 2 seconds, then turn it back fully counterclockwise.
3. Observe the red LED on the ECU or CHECK ENGINE light for stored codes.

Mode 2—Exhaust Gas Sensor Monitor

In this mode the red LED on the ECU or CHECK ENGINE light will display the condition of the fuel mixture and whether the system is in closed loop or open loop. When the light flashes ON, the exhaust gas sensor is indicating a lean mixture. When the light stays OFF, the sensor is indicating a rich mixture. If the light remains ON or OFF, it is indicating an open loop system. If the system is equipped with 2 exhaust gas sensors, the left side will operate first. If already in Mode 2, proceed to Step 3 for exhaust gas sensor monitor.

1. Turn the ignition switch ON.
2. Turn the diagnostic switch ON, by turning the switch fully clockwise for 2 seconds and then fully counterclockwise.
3. Start the engine and run until thoroughly warm. Raise the idle to 2,000 rpm and hold for approximately 2 minutes. Ensure the red LED or CHECK ENGINE light flash ON and OFF more than 5 times every 10 seconds with the engine speed at 2,000 rpm.

➡If equipped with 2 exhaust gas sensors, switch to the right sensor by turning the ECU mode selector fully clockwise for 2 seconds and then fully counterclockwise with the engine running.

Clearing Codes

1990–95 VEHICLES

All control unit diagnostic codes may be cleared by disconnecting the negative battery for a period of 15 seconds. The codes will be cleared when mode 1 is re-entered from mode 2. The Nissan Consult Monitor or equivalent can also be used to clear codes.

DIAGNOSTIC TROUBLE CODES

1990–95 Vehicles

Code 16 TCS Signal
Code 21 Ignition signal missing in primary coil
Code 31 ECM (engine ECCS control unit)
Code 32 EGR circuit
Code 33 Heated oxygen sensor circuit
Code 34 Knock Sensor (KS) circuit
Code 35 EGR temperature sensor circuit
Code 42 Fuel temperature sensor circuit
Code 43 Throttle sensor circuit
Code 45 Injector leak
Code 46 Secondary throttle sensor circuit
Code 51 Injector circuit
Code 53 Heated oxygen sensor circuit (right bank)
Code 54 NT controller circuit
Code 55 No malfunctioning in the above circuit
Code 11 Crankshaft position sensor
Code 12 Mass Air flow sensor
Code 13 Engine coolant temperature sensor circuit
Code 14 Vehicle speed sensor

Isuzu

GENERAL INFORMATION

Isuzu vehicles may be fitted with either a Feedback Carburetor (FBC), a Throttle Body Fuel Injection (TBI) system or a Multi-port Fuel Injection System (MFI).

The Feedback Carburetor System (FBC) is primarily used on 1985 and earlier normally aspirated engines, although some engines may use this system up to 1993. The system Electronic Control Module (ECM) constantly monitors and controls engine operation by reading data from various sensors and outputting signals to the carburetor. This helps lower emissions while maintaining the fuel economy, driveability and performance of the vehicle.

The Throttle Body (TBI) fuel injection system was put into production in 1989. It is used on 2.8L and 3.1L engines. The system functions much the same as a multi-port fuel injection system but with one exception—fuel is injected into the intake manifold rather than into each individual cylinder. This system may also control the ignition system.

The 1-TEC Multi-port Fuel Injection (MFI) system was first used in 1985 and continues to be used today. The system constantly monitors and controls engine operation through the use of data sensors, an Electronic Control Module (ECM) and other components. Individual fuel injectors are mounted at each cylinder and provide a metered amount of fuel as required by current operating conditions. This system may also control the ignition system and, as equipped, the turbocharger system.

SELF-DIAGNOSTICS

All vehicles covered in this section have self-diagnostic capabilities. The ECM diagnostics are in the form of trouble codes stored in the system's memory. When a trouble code is detected by the control module, it will turn the malfunction indicator lamp ON until the code is cleared. An intermittent problem will set a code. The lamp will turn OFF if the problem goes away, but the trouble code will stay in memory until ECM power is interrupted.

Service Precautions

• Keep all ECM parts and harnesses dry during service. Protect the ECM and all solid-state components from rough handling or extremes of temperature.

• Use extreme care when working around the ECM or other solid-state components. Do not allow any open circuit to short or ground in the ECM circuit. Voltage spikes may cause damage to solid-state components.

• Before attempting to remove any parts, turn the ignition switch OFF or disconnect the negative battery cable.

• Remove the ECM before any arc welding is performed to the vehicle.

• Electronic components are very susceptible to damage caused by electrostatic discharge (static electricity). To prevent electronic component damage, do not touch the control module connector pins or soldered components on the control module circuit board.

Reading Codes

➡**Diagnostic codes may be retrieve through the use of the CHECK ENGINE light or Malfunction Indicator Lamp (MIL). A special Scan tool can be used, but is not required. When using special diagnostic equipment, always observe the tool manufacturer's instructions.**

1982–86 VEHICLES

The trouble code system is actuated by connecting a diagnostic lead to ground. The location of the diagnostic lead differs from model-to-model and, in some cases from year-to-year within the same model.

Trooper II for 1985; connect the diagnostic lead terminals together (1 male and 1 female). The terminals are located under dash, on the passenger's side, behind the radio. The terminal leads for 1986 models are located near the ALDL connector, under dash, on the driver's side behind the cigarette lighter.

Pick-up truck for 1982; connect the diagnostic lead terminals together (1 male and 1 female). The terminals are branched from the harness near the ECM, under the dash on the driver's side behind the hood release.

The trouble code is determined by counting the flashes of the Check Engine lamp. Trouble Code 12 will flash first, indicating that the self-diagnostic system is working. Code 12 consists of 1 flash, short pause, then 2 flashes. There will be a longer pause and Code 12 will repeat 2 more times. Each code flashes 3 times. The cycle will then repeat itself until the engine is started or the ignition switch is turned OFF. In most cases, the codes will be checked with the engine running since no codes other than 12 or 51 will be present on initial key ON.

1987–94 VEHICLES

With Scan Tool

1. Turn the ignition key to the OFF position.
2. Connect the scan tool to the Assembly Line Diagnostic Link (ALDL).
3. Turn ON the ignition for scan tool to access engine computer.

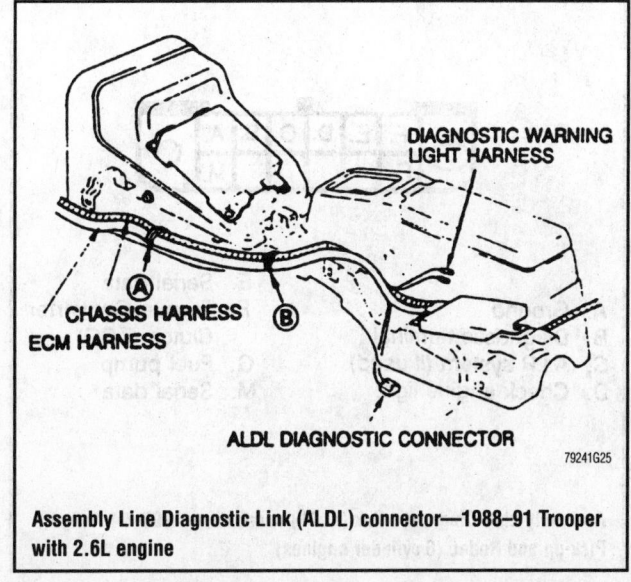

Assembly Line Diagnostic Link (ALDL) connector—1988–91 Trooper with 2.6L engine

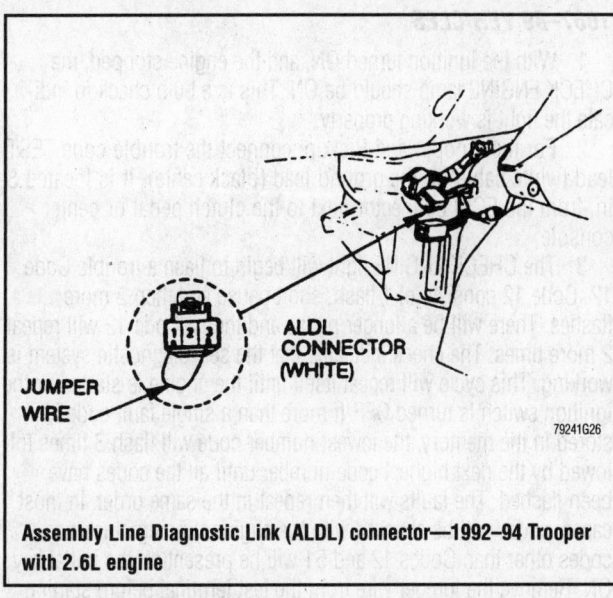

Assembly Line Diagnostic Link (ALDL) connector—1992–94 Trooper with 2.6L engine

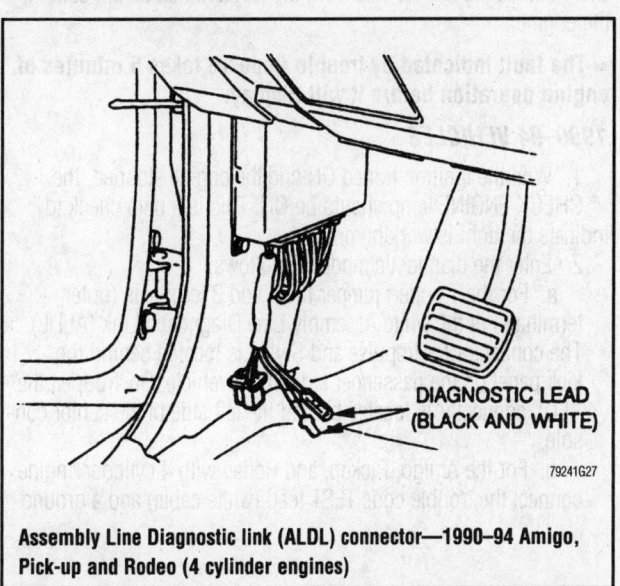

Assembly Line Diagnostic link (ALDL) connector—1990–94 Amigo, Pick-up and Rodeo (4 cylinder engines)

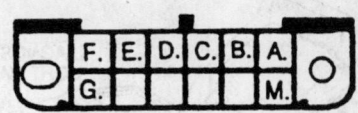

A. Ground
B. Diagnostic terminal
C. A.I.R system (if used)
D. Check engine light
E. Serial data
F. Torque Converter Clutch (TCC)
G. Fuel pump
M. Serial data

79241G28

Assembly Line Diagnostic link (ALDL) connector—1990–94 Amigo, Pick-up and Rodeo (6 cylinder engines)

1987–89 VEHICLES

1. With the ignition turned ON, and the engine stopped, the CHECK ENGINE lamp should be ON. This is a bulb check to indicate the light is working properly.

2. For the Trooper and Pickup; connect the trouble code TEST lead (white cable) to the ground lead (black cable). It is located 8 in. from the ECM connector next to the clutch pedal or center console.

3. The CHECK ENGINE light will begin to flash a trouble Code 12. Code 12 consists of 1 flash, short pause and then 2 more flashes. There will be a longer pause and then a Code 12 will repeat 2 more times. The check indicates that the self-diagnostic system is working. This cycle will repeat itself until the engine is started or the ignition switch is turned OFF. If more than a single fault code is stored in the memory, the lowest number code will flash 3 times followed by the next highest code number until all the codes have been flashed. The faults will then repeat in the same order. In most cases, codes will be checked with the engine running since no codes other than Codes 12 and 51 will be present on the initial key ON. Remove the jumper wire from the test terminal before starting the engine.

➡ The fault indicated by trouble Code 15 takes 5 minutes of engine operation before it will display.

1990–94 VEHICLES

1. With the ignition turned ON and the engine stopped, the CHECK ENGINE lamp should be ON. This is a bulb check to indicate the light is working properly.

2. Enter the diagnostic modes as follows:
   a. For the Trooper; jumper the 1 and 3 terminals (outer terminals) of the white Assembly Line Diagnostic Link (ALDL). The connector for Impulse and Stylus is located behind the kick panel on the passenger side of the vehicle. On Trooper, the ALDL connector is located behind the left side of the center console.
   b. For the Amigo, Pickup, and Rodeo with 4 cylinder engine; connect the trouble code TEST lead (white cable) and a ground

lead (black cable) together. It is located 8 in. from the ECM connector (next to the clutch pedal or brake pedal).
   c. For the Amigo, Pickup, and Rodeo with 6 cylinder engine; jumper wire the A and B terminals together of the Assembly Line Diagnostic Link (ALDL). The ALDL is located in the center console and is sometimes covered by a plastic cover labeled DIAGNOSTIC CONNECTOR. Read the trouble codes with the ignition switch ON and the engine OFF.

3. The CHECK ENGINE light will begin to flash a trouble Code 12. Code 12 consists of 1 flash, a short pause and then 2 more flashes. There will be a longer pause and a Code 12 will repeat 2 more times. Code 12 indicates that the self-diagnostic system is working. If any other faults are present, the faults will be displayed 3 times each in the same fashion. Fault codes are flashed from lowest to highest after the Code 12. Remember to remove the jumper wire from the ALDL connector before starting the engine. After all codes have been displayed, the cycle will repeat itself until the engine is started or the ignition switch is turned OFF.

➡ The fault indicated by trouble Code 15 takes 5 minutes of engine operation before it will display (4 cylinder engine only).

Clearing Codes

1982–86 VEHICLES

The trouble code memory is fed a continuous 12 volts even with the ignition switch in the OFF position. After a fault has been corrected, it will be necessary to remove the voltage for 10 seconds to clear any stored codes. Voltage can be removed by disconnecting the 14 pin ECM connector or by removing the fuse marked 'ECM' or fuse No. 4 on some models. Since all memory will be lost when removing the fuse, it will be necessary to reset the clock and other electrical equipment.

1987–94 VEHICLES

The trouble code memory is fed a continuous 12 volts even with the ignition switch in the OFF position. After a fault has been corrected, it will be necessary to remove the voltage for 30 seconds to clear any stored codes. The quickest way to remove the voltage is to remove the ECM fuse from the fuse block or the MAIN 60A fuse for 10 seconds. The voltage can also be removed by disconnecting the negative battery cable. This will mean electronic instrumentation, such as a clock and radio, would have to be reset.

1987–89 Pickup and Trooper; turn the ignition switch OFF and disconnect the ECM 13-pin connector or remove the No. 4 fuse from the fuse block for 30 seconds.

The 60 amp slow blow fuse may be removed from the fuse block in the engine compartment. However, the electronic functions with memory have to be reset after removing the No. 4 fuse for 30 seconds.

1992–94 Amigo, Pickup, Rodeo and Trooper; To clear the trouble codes; turn the ignition switch OFF and remove the ECM fuse from the under-dash fuse block for 30 seconds. Removing the number 3 fuse from the under dash fuse panel will result in having to reset all the electronic functions with memory in the vehicle. This applies to trucks with 4-cylinder engines.

Removing the 60 amp slow blow fuse from the fuse block in the engine compartment will also erase codes.

DIAGNOSTIC TROUBLE CODES

1982–94 Vehicles

1982 1.8L FBC TRUCK ENGINE

Code 12 Idle switch is not turned ON
Code 13 Idle switch is not turned OFF
Code 14 Wide Open Throttle (WOT) switch is not turned ON
Code 15 Wide Open Throttle (WOT) switch is not turned ON
Code 21 Output transistor is not turned ON
Code 22 Output transistor is not turned OFF
Code 23 Abnormal oxygen sensor
Code 24 Abnormal Water Temperature Sensor (WTS) switch
Code 25 Abnormal Random Access Memory (RAM)
Code 12, 13, 14 and 15 Check Engine lamp not ON
Code 21, 22, 23, 24 and 25 Check Engine lamp ON

1985–89 1.SL FBC (VIN 7) ENGINE

1983 1.8L FBC ENGINE

1983–86 2.0L FBC (VIN A) ENGINE

1986–94 2.3L FBC (VIN L) ENGINE

Code 12 Normal
Code 13 Oxygen sensor circuit
Code 14 Coolant Temperature Sensor (CTS)—circuit shorted
Code 15 Coolant Temperature Sensor (CTS)—circuit open
Code 21 Idle switch—circuit open or Wide Open Throttle (WOT) switch—circuit shorted
Code 22 Fuel Cut Solenoid (FCS)—circuit open or grounded
Code 23 Mixture Control (M/C) solenoid—circuit open or grounded, or Vacuum Control Solenoid (VCS)—circuit open or grounded (1983 1.8L Truck, 1983–86 2.0L Truck, 1986–88 2.3L Truck)
Code 24 Vehicle Speed Sensor (VSS) circuit
Code 25 Air Switching Solenoid (ASS)—circuit open or grounded
Code 26 Vacuum Switching Valve (VSV) system for canister purge -circuit open or grounded
Code 27 Vacuum Switching Valve (VSV)-constant high voltage to ECM
Code 31 No ignition reference pulses to ECM
Code 32 EGR temperature sensor—system malfunction
Code 34 EGR temperature sensor—circuit failure electronic idle control
Code 42 Fuel Cut Relay and/or circuit shorted
Code 44 Oxygen Sensor circuit—lean indication
Code 45 Oxygen Sensor circuit—rich indication
Code 52 Faulty Electronic Control Module (ECM)—Random Access Memory (RAM) problem in ECM
Code 53 Shorted Air Switching Solenoid (ASS) or Air Injection System and/or faulty Electronic Control Module (ECM)
Code 54 Shorted Vacuum Control Solenoid (VCS) and/or faulty Electronic Control Module (ECM)
Code 55 Faulty Electronic Control Module (ECM)

1985–87 2.0L TURBO EFI (VIN F) ENGINE

1983–89 2.0L EFI (VIN A) ENGINE

1988–89 2.3L EFI (VIN L) ENGINE

1988–94 2.6L EFI (VIN E) ENGINE

Code 12 Normal
Code 13 Oxygen sensor circuit

Code 14 Engine Coolant Temperature (ECT) sensor -grounded
Code 15 Engine Coolant Temperature (ECT) sensor—incorrect signal (open circuit on 1988-94 2.6L)
Code 16 Engine Coolant Temperature (ECT) sensor -open circuit
Code 21 Throttle Valve Switch (TVS) system—idle contact and full contact made simultaneously
Code 22 Starter—no signal input
Code 23 Ignition power transistor—output terminal grounded
Code 25 Vacuum Switching Valve (VSV)—output terminal grounded or open
Code 26 Canister purge Vacuum Switching Valve (VSV)—open or grounded
Code 27 Canister purge Vacuum Switching Valve (VSV)—faulty transistor or bad ground circuit
Code 32 EGR temperature sensor—faulty sensor or harness
Code 33 Fuel injector system—output terminal grounded or open
Code 34 EGR Nacuum switching valve—output terminal grounded or open
Code 35 Ignition power transistor—open circuit
Code 41 Crank Angle sensor (CAS)—no signal or faulty signal
Code 43 Throttle Valve Switch—idle contact closed continuously
Code 44 Fuel metering system—lean signal (Oxygen sensor-low voltage)
Code 45 Fuel metering system—rich signal (Oxygen sensor-high voltage)
Code 51 Faulty ECM
Code 52 Faulty ECM
Code 53 Vacuum Switching Valve (VSV)—grounded or faulty power transistor
Code 54 Ignition power transistor—grounded or faulty power transistor
Code 55 Faulty ECM
Code 61 Air Flow Sensor (AFS)—grounded, shorted, open or broken HOT wire
Code 62 Air Flow Sensor (AFS)—broken COLD wire
Code 63 Vehicle Speed Sensor (VSS)—no signal input
Code 64 Fuel injector system—grounded or faulty transistor
Code 65 Throttle Valve Switch (TVS)—full contact closed continuously
Code 66 Knock sensor—grounded or open circuit
Code 71 Throttle Position Sensor (TPS)—turbo control system—abnormal signal
Code 72 EGR Vacuum switching valve—output terminal grounded or open
Code 73 EGR Vacuum switching valve—faulty transistor or grounded system

1989–91 2.8L TBI (VIN R) ENGINE

1991–94 3.1L TBI (VIN Z) ENGINE

Code 12 Normal
Code 13 Oxygen sensor circuit
Code 14 Engine Coolant Temperature (ECT) sensor—high temperature indicated
Code 15 Engine Coolant Temperature (ECT) sensor—low temperature indicated
Code 21 Throttle Position Sensor (TPS)—voltage high
Code 22 Throttle Position Sensor (TPS)—voltage low
Code 23 Intake Air Temperature (IAT)—low temperature indicated

Code 24 Vehicle Speed Sensor (VSS)—no input signal Code 25—Intake Air Temperature (IAT)—high temperature indicated

Code 31 Turbocharger wastegate control

Code 32 EGR system fault

Code 33 Manifold Absolute Pressure (MAP) sensor—voltage high

Code 34 Manifold Absolute Pressure (MAP) sensor—voltage low

Code 42 Electronic Spark Timing (EST) circuit fault

Code 43 Electronic Spark Control (ESC)—knock failure circuit

Code 44 Oxygen sensor circuit—lean exhaust

Code 45 Oxygen sensor circuit—rich exhaust

Code 51 PROM error—faulty or incorrect PROM

Code 52 CALPAK error—faulty or incorrect CALPAK

Code 54 Fuel Pump Circuit—low voltage

Code 55 ECM error

1991–94 2.3L EFI (VIN 5/6) ENGINE

1992–94 3.2L EFI (VIN V/W) ENGINE

Code 13 Oxygen sensor circuit

Code 14 Engine Coolant Temperature (ECT) sensor—out of range

Code 21 Throttle Position Sensor (TPS)—out of range

Code 23 Intake Air Temperature (IAT)—out of range

Code 24 Vehicle Speed Sensor (VSS)—no input signal

Code 32 EGR system fault

Code 33 Manifold Absolute Pressure (MAP) sensor—out of range

Code 44 Oxygen sensor circuit—lean exhaust

Code 45 Oxygen sensor circuit—rich exhaust

Code 51 ECM failure

Mazda

GENERAL INFORMATION

Mazda utilizes 2 types of fuel systems between the years 1984–94. The Feedback Carburetor (FBC) system and Electronic Gas Injection (EGI) system, or fuel injection system. The feedback carburetor system was used in the B2000, B2200 and B2600 pickup trucks between years 1984–92.

Electronic Gas Injection (EGI) was first available in the 1984 RX-7. 626 picked it up in 1986, 323 in 1987. In 1988 all models except B2200 and B2600 pickup trucks came equipped with fuel injection. Mazda uses various variations of EGI. Navajo uses the Ford EEC-IV system. However, the EEC-IV system will not be covered in this section.

SELF-DIAGNOSTICS

Service Precautions

• Before connecting or disconnecting the ECU harness connectors, make sure the ignition switch is OFF and the negative battery cable is disconnected to avoid the possibility of damage to the control unit.

• When performing ECU input/output signal diagnosis, remove the pin terminal retainer from the connectors to make it easier to insert tester probes into the connector.

• When connecting or disconnecting pin connectors from the ECU, take care not to bend or break any pin terminals. Check that there are no bends or breaks on ECU pin terminals before attempting any connections.

• Before replacing any ECU, perform the ECU input/output signal diagnosis to make sure the ECU is functioning properly or not.

• After checking through EGI troubleshooting, perform the EFI self-diagnosis and driving test.

• When measuring supply voltage of ECU controlled components with a circuit tester, separate 1 tester probe from another. If the 2 tester probes accidentally make contact with each other during measurement, a short circuit will result and may damage the ECU.

Reading Codes

➡**Diagnostic codes may be retrieved through the use of the CHECK ENGINE light or Malfunction Indicator Lamp (MIL). Special System Checker No. 83, Digital Code Checker and a Self-diagnosis Checker are all special diagnostic equipment used to retrieve codes, however these tools are not required. When using special diagnostic equipment, always observe the tool manufacturer's instructions.**

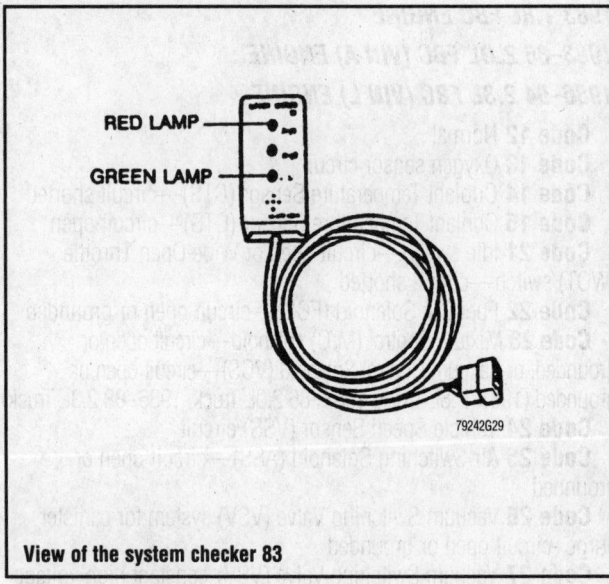

RED LAMP

GREEN LAMP

79242G29

View of the system checker 83

1984–86 VEHICLES WITH SYSTEM CHECKER TOOL 83

On 1986 B2000 Pick-up, the System Checker No. 83 (tool No. 49-G030-920), is used to detect and indicate any problems of each sensor, damaged wiring, poor contact or a short circuit between each of the sensor control units. Trouble is indicated by a red lamp and a buzzer. If there are more than 2 problems at a time, the indicator lamp turns ON in the numerical order of the code number. Even if the problem is corrected during indication, 1 cycle will be indicated, If after a malfunction has occurred and the ignition key is switched OFF, the malfunction indicator for the feedback system will not be displayed on the checker.

Read engine trouble codes using the following procedures:

1984–85 B2200

1. Operate the engine until normal operating temperatures are reached. Allow the engine to run at idle.

2. Connect System Checker tool No. 83 (49-G030-920) to the check connector, located near the ECU.

3. Check whether the trouble indication light turns ON.

If there is more than 2 problems at the same time, the indicator lamp lights on in the numerical order of the code number. Even if the problem is corrected during indication, 1 cycle will be Indicated. If after a malfunction has occurred the Ignition key is switched off, the malfunction indicator for the feedback system will not be displayed on the checker. The control unit has a built In fall-safe mechanism. If a malfunction occurs during driving, the control unit will on its own initiative, send out a command and driving performance will be affected. The commands are as follows:

    a. Water Thermo-Sensor—the control unit outputs a constant 176°F (80°C) command.

    b. Feed-Back Sensor—the control unit holds air/fuel solenoid to dwell meter reading 27° (duty 30%) for B2200.

    c. Vacuum Sensor—the control unit prevents operation of the EGR valve, and holds the air/fuel solenoid to a duty of 0%.

    d. EGR Position Sensor—the control unit prevents operation of the EGR valve.

➡️If the trouble code is code number 3 (feedback system), proceed as follows:

4. Start the engine, letting it run until it reaches normal operating temperature. Connect a tachometer to the engine.

5. Connect a dwell meter (90 degrees, 4 cylinder) to the yellow wire in the service (check) connector of the air/fuel solenoid valve.

6. Run the engine at idle and note the reading on the dwell meter.

7. If the dwell meter reading is 0° degrees, the probable causes are as follows:

    a. The wiring harness from the IG to the check connector BrY terminal is open.

    b. The wiring harness from the check connector Y terminal to the control unit (F) terminal is grounded.

    c. The transistor in the control unit for the air/fuel solenoid is open.

8. If the dwell meter reading is 90°, the probable causes are as follows:

    a. The wiring harness from the IG to the check connector BrY terminal is open.

    b. The wiring harness from the check connector BrY terminal to the control unit (F) terminal is grounded.

    c. The transistor in the control unit for the air/fuel solenoid is short circuited.

9. If the dwell meter reading is 18°, check whether the green lamp (feedback signal) illuminates or does not illuminate.

10. If the oxygen sensor signal lamp does not illuminate, proceed as follows:

    a. If the green lamp does not illuminate, the air is sucked from the intake system or the air is sucked from the exhaust manifold.

    b. Carburetor jets are clogged.

    c. The valve of the air/fuel solenoid is stuck to the lower position, giving a lean air/fuel mixture condition.

11. If the oxygen sensor signal lamp illuminates, proceed as follows:

    a. If green lamp turns ON, the mixture is richer than stoichiometric air/fuel ratio.

    b. If the green lamp turns ON and OFF, the O2 sensor signal is fed to the control unit.

    c. If the green lamp turns OFF, the mixture is leaner than stoichiometric.

WITH SELF-DIAGNOSIS CHECKER

The self-diagnosis checker (49-HOl 8-9A1) and System Selector (49-BOl 9-9A0), are used to retrieve code numbers of malfunctions which have happened and were memorized or are continuing. The malfunction is indicated by a code number.

If there is more than 1 malfunction, the code numbers will display on the self-diagnosis checker in numerical order. The ECU has a built in fail-safe mechanism for the main input sensors. If a malfunction occurs, the emission control unit will substitute values. This will affect driving performance, but the vehicle may still be driven.

The ECU continuously checks for malfunctions of the input devices. But the ECU checks for malfunctions of the output devices within 3 seconds after the green (1 pin) test connector or TEN terminal of the diagnosis connector is grounded and the ignition switch is turned to the ON position.

Read engine trouble codes using the following procedures:

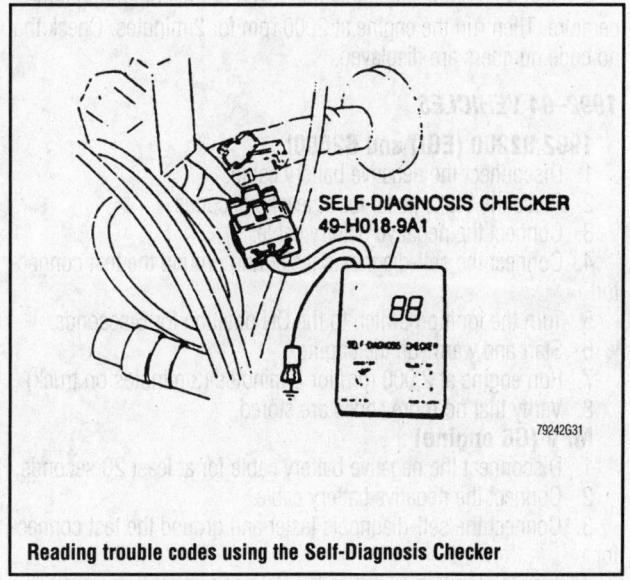

Reading trouble codes using the Self-Diagnosis Checker

1992 MPV, B2200 and B2600i

The check connector is located at the rear of the left side wheel house on 626/MX-6, front of the left side wheel house on MPV, above the right side wheel house on B2200 and near the fuel filter on B2600i.

1. Connect the tester to the check connector and to ground.

2. Set the tester select switch to the A setting.

3. With a jumper wire, ground the 1 pin test connector.

4. Turn the ignition switch ON.

5. Make sure that 88 flashes on the monitor and that the audible buzzer sounds for 3 seconds after turning the ignition switch ON.

6. If 88 does not flash, check the main relay, power supply circuit and the check connector wiring.

7. If 88 flashes and the buzzer sounds for more than 20 seconds, replace the engine control unit and perform steps number 1 through 6 again.

➡️Before replacing the ECU on the MPV or B2600i, check for a short circuit between ECU terminal IB for JE engine and 1F for G6 engine and the 6 pin check connector.

8. Note and record any other code numbers that are present.

Clearing Codes

1984–86 VEHICLES

1. Turn the ignition switch OFF.
2. Disconnect the negative battery cable.
3. Depress the brake pedal for at least 5 seconds.
4. Reconnect the negative battery cable.

1987–91 VEHICLES

1. Cancel the memory of the malfunction by disconnecting the negative battery cable and depressing the brake pedal for at least 20 seconds, then reconnect the negative battery cable.
2. Connect the Self-Diagnosis Checker 49-H018-9A1 to the check connector. Ground the test connector (green: 1 pin) using a jumper wire.
3. Turn the ignition switch ON, but do not start the engine for approximately 6 seconds.
4. Start the engine and allow it to reach normal operating temperature. Then run the engine at 2000 rpm for 2 minutes. Check that no code numbers are displayed.

1992–94 VEHICLES

1992 B2200 (EGI) and B2600i

1. Disconnect the negative battery cable.
2. Press the brake pedal for at least 5 seconds.
3. Connect the negative battery cable.
4. Connect the self-diagnosis tester and ground the test connector.
5. Turn the ignition switch to the ON position for 6 seconds.
6. Start and warm-up the engine.
7. Run engine at 2,000 rpm for 2 minutes (3 minutes on truck).
8. Verify that no more codes are stored.

MPV (G6 engine)

1. Disconnect the negative battery cable for at least 20 seconds.
2. Connect the negative battery cable.
3. Connect the self-diagnosis tester and ground the test connector.
4. Turn the ignition switch to the ON position for 6 seconds.
5. Start and warm-up the engine.
6. Run engine at 2,000 rpm for 3 minutes.
7. Verify that no more codes are stored.

B2200 (FOC engine)

For the pickup with the feedback carburetor fuel system disconnect the negative battery cable for at least 5 seconds.

DIAGNOSTIC TROUBLE CODES

1984–94

1984–85 VEHICLES

2.0L ENGINE (CODE FE)

Code 01 Engine speed
Code 02 Water thermosensor
Code 03 Oxygen sensor
Code 04 Vacuum sensor
Code 05 EGR position sensor

1986–87 VEHICLES

1.6L, 2.0L AND Z2L ENGINES

Code 01 Ignition pulse
Code 02 Air flow meter

Code 03 Water thermosensor
Code 04 Intake air thermo or Temperature sensor
Code 05 Feedback system
Code 06 Atmospheric pressure sensor (1986 1.6L)
Code 08 EGR position sensor
Code 09 Atmospheric pressure sensor
Code 22 No. 1 Cylinder sensor (2.2L turbocharged)

1988–94 VEHICLES

2.0L, 2.2L, 2.5L, 2.6L AND 3.0L ENGINES

Code 01 Ignition pulse
Code 02 Ne signal—distributor
Code 02 NE 2 signal—crankshaft (1994 2.0L, 1992–94 3.0L)
Code 03 Gi signal—distributor (1988–91 3.0L)
Code 03 G signal—distributor
Code 04 G2 signal—distributor (1988–91 3.0L);NE 1 signal—distributor (1994 2.0L, 1992–93 3.0L)
Code 05 Knock sensor and control unit (Left side on 1992-94 3.0L)
Code 06 Speed signal
Code 07 Knock sensor; right side (1992-94 3.0L)
Code 08 Air flow meter
Code 09 Engine coolant temperature sensor (C IS)
Code 10 Intake air temperature sensor
Code 11 Intake air thermosensor—dynamic chamber (3.0L, 2.6L)
Code 13 Intake manifold pressure sensor (1.3L)
Code 14 Atmospheric pressure sensor (in ECU on 2.6L and 1994 2.5L)
Code 15 Oxygen sensor
Code 15 Oxygen sensor; left side on 1994 2.5L, 1990-94 3.0L
Code 16 EGR position sensor
Code 17 Closed loop system
Code 17 Closed loop system; left side on 1993-94 2.5L 1990-94 3.0L
Code 23 Heated oxygen sensor; right side on 1992-94 1.8L V6, 1994 2.5L 1990-91 3.0L
Code 24 Closed loop system; right side on 1992-94 1.8L V6, 1993 2.5L 1990-91 3.0L
Code 25 Solenoid valve—pressure regulator
Code 26 Solenoid valve—purge control
Code 26 Solenoid valve—purge control No. 2 (1988-89 3.0L)
Code 27 Solenoid valve—purge control No. 1 (1988-89 3.0L)
Code 27 Solenoid valve—No. 2 purge control (1989 1.6L)
Code 28 Solenoid valve—EGR vacuum
Code 29 Solenoid valve—EGR vent
Code 30 Relay (cold start injector 3.0L)
Code 34 ISC valve
Code 34 Idle air control valve (1993-94 2.0L and 2.5L, 1.6L, 1.8L, 2.6L, 3.1L)
Code 36 Oxygen sensor heater relay (1990 3.0L)
Code 36 Right side oxygen sensor heater (1992-94 3.0L)
Code 37 Left side oxygen sensor heater (1992-94 3.0L)
Code 37 Coolant fan relay
Code 40 Oxygen sensor heater relay (1991 3.0L)
Code 40 Solenoid (triple induction system) and oxygen sensor relay (1988-89 3.0L)
Code 41 Solenoid valve—VRIS (1989-94 MPV 3.0L)
Code 41—Solenoid valve—VRIS 1 (1992-94 1.8L V6, 1993 2.5L)
Code 41 Solenoid valve—VICS (3.0L)

Code 42 Solenoid valve—Waste gate (turbocharged)
Code 46 Solenoid valve—VRIS 2 (1992–94 1.8L V6, 1993
.5L)
Code 65 C signal—PCMT (1 992–94 3.0L)
Code 67 Coolant fan relay No. 1 (1993 2.5L)
Code 67 Coolant fan relay No. 2 (1992–94 1.8L V6)
Code 68 Coolant fan relay No. 2, No.3 with ATX (1993 2.5L)
Code 69 Engine coolant temperature sensor—fan (1992–94
.8L V6, 1993 2.0L and 2.5L)

Mitsubishi

GENERAL INFORMATION

Mitsubishi uses 2 types of fuel systems. Feedback carburetor
system and fuel injection. The type of fuel injection system is known
is Electronic Controlled Injection (ECI).

Mitsubishi uses a conventional downdraft two-barrel compound
ype carburetor which incorporates an automatic choke, accelerator
ump, and enrichment system. In addition, a deceleration device is
provided.

The Electronic Fuel Injection (EFI) system, used on Mitsubishi
vehicles, is classified as a Multi-Point Injection (MPI) system.
The MPI system controls the fuel flow, idle speed, and ignition
timing. The basic function of the MPI system is to control the
air/fuel ratio in accordance with all engine operating conditions.
An Electronic Control Unit (ECU) is the heart of the MPI system.
Based on data from various sensors, the ECU computes the
desired air/fuel ratio.

SELF-DIAGNOSTICS

Service Precautions

• Before connecting or disconnecting the ECU harness connectors, make sure the ignition switch is OFF and the negative battery
cable is disconnected to avoid the possibility of damage to the control unit.

• When performing ECU input/output signal diagnosis, remove
the pin terminal retainer from the connectors to make it easier to
insert tester probes into the connector.

• When connecting or disconnecting pin connectors from the
ECU, take care not to bend or break any pin terminals. Check that
there are no bends or breaks on ECU pin terminals before attempting any connections.

• Before replacing any ECU, perform the ECU input/output signal diagnosis to make sure the ECU is functioning properly.

• When measuring supply voltage of ECU-controlled components with a circuit tester, separate 1 tester probe from another. If
the 2 tester probes accidentally make contact with each other during
measurement, a short circuit will result and damage the ECU.

Reading Codes

➡**All though the CHECK ENGINE light or Malfunction Indicator Lamp (MIL) will illuminate when there is trouble
detected, diagnostic codes can only be retrieved with the
use of either a analog voltmeter or a Multi-use Tester. When
using diagnostic equipment, always observe the tool manufacturer's instructions.**

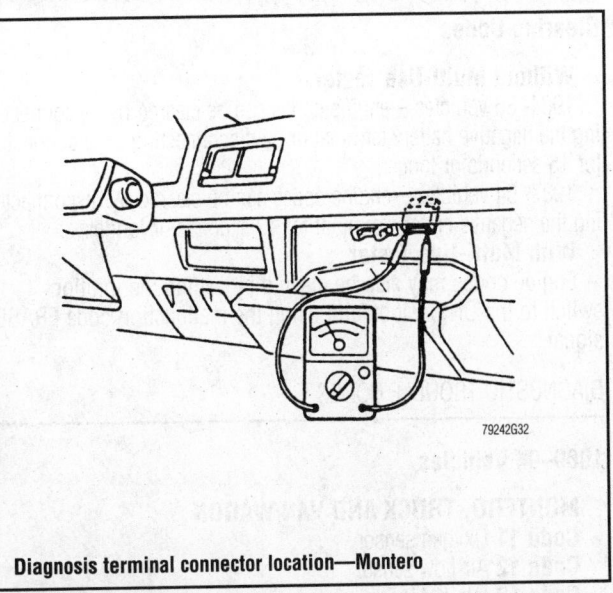

Diagnosis terminal connector location—Montero

1984–86 VEHICLES

With ECI/MPI Tester

Refer to manufacturer's tester manual regarding diagnosis with
this tester.

1985–94 VEHICLES

With Analog Voltmeter

The voltmeter can be used to retrieve code numbers of malfunctions which have happened and were memorized or are continuing to happen. On the voltmeter, the malfunction is indicated
by a sweep of the needle. The voltmeter should be connected to
the data link connector located under the driver side dashboard.
Connect the voltmeter between the Multi-Point Injection (MPI)
terminal and the ground terminal. Turn the ignition switch ON if
the normal condition exists, the voltmeter pointer will indicate a
normal pattern. A normal pattern is indicated by constant needle
sweeps. If a problem exists in the system the voltmeter pointer
will indicate it in a series of pointer sweeps. For example, a
Code 3 would be 3 consecutive short sweeps of the voltmeter
needle.

If there is more than 1 malfunction, the low code numbers will
first be indicated and after a 2 second pause (no code indication)
the higher code will be indicated.

With Multi-Use Tester

To read the trouble codes using the Multi-Use Tester (MB991341
or equivalent) follow the steps below:

1. Turn the ignition switch OFF.
2. Insert the power supply terminal to the cigarette lighter
socket.
3. Connect the tester connector to the diagnosis connector in
the glove compartment, under the `hood or under the driver side
dashboard.
4. Turn the ignition switch ON and push the DIAG key.
5. Observe the trouble code and make the necessary repairs.

On most models the CHECK ENGINE malfunction indicator light
will light up and remain illuminated to indicate that there is a problem in the system. After this light has been reported to be ON, the
system should be checked for malfunction codes.

Clearing Codes

Without Multi-Use Tester

1984–86 vehicles—engine codes can be cleared by disconnecting the negative battery terminal or by disconnecting ECU connector for 15 seconds or longer.

1987–94 vehicles—engine codes can be cleared by disconnecting the negative battery terminal for 10 seconds or longer.

With Multi-Use Tester

Engine codes may also be cleared by setting the ignition switch to the ON position and using the malfunction code ERASE signal.

DIAGNOSTIC TROUBLE CODES

1989–94 Vehicles

MONTERO, TRUCK AND VAN/WAGON

Code 11 Oxygen sensor

Code 12 Air flow sensor

Code 13 Intake Air Temperature Sensor

Code 14 Throttle Position Sensor (TPS)

Code 15 SC Motor Position Sensor (MPS)

Code 21 Engine Coolant Temperature Sensor

Code 22 Crank angle sensor

Code 23 No. 1 cylinder TDC (Camshaft position) Sensor

Code 24 Vehicle speed sensor

Code 25 Barometric pressure sensor

Code 31 Knock (KS) sensor

Code 32 Manifold pressure sensor

Code 36 Ignition timing adjustment signal

Code 39 Oxygen sensor (rear—turbocharged)

Code 41 Injector

Code 42 Fuel pump

Code 43 EGR-California

Code 44 Ignition Coil; power transistor unit (No. 1 and No. 4 cylinders) on 3.0L

Code 52 Ignition Coil; power transistor unit (No. 2 and No. 5 cylinders) on 3.0L

Code 53 Ignition Coil; power transistor unit (No. 3 and No. 6 cylinders)

Code 55 AC valve position sensor

Code 59 Heated oxygen sensor

Code 61 Transaxle control unit cable (automatic transmission)

Code 62 Warm-up control valve position sensor (non-turbo)

Nissan

GENERAL INFORMATION

Nissan uses 2 types of fuel systems. Electronic Control Carburetor (ECC) system and Electronic Concentrated Control System (ECCS). The ECC system is a Feedback carburetor system. The ECCS is a fuel injected system which may be either throttle body injection or Multi-port injection. Both ECC and ECCS systems were available as of 1984.

SELF-DIAGNOSTICS

Service Precautions

• Do not disconnect the injector harness connectors with the engine running.

• Do not apply battery power directly to the injectors.

• Do not disconnect the ECU harness connectors before the battery ground cable has been disconnected.

• Make sure all ECU connectors are fastened securely. A poor connection can cause an extremely high surge voltage in the coil and condenser and result in damage to integrated circuits.

• When testing the ECU with a DVOM make sure that the probes of the tester never touch each other as this will result in damage to a transistor in the ECU.

• Keep the ECCS harness at least 4 in. away from adjacent harnesses to prevent an ECCS system malfunction due to external electronic noise.

• Keep all parts and harnesses dry during service.

• Before attempting to remove any parts, turn OFF the ignition switch and disconnect the battery ground cable.

• Always use a 12 volt battery as a power source.

• Do not attempt to disconnect the battery cables with the engine running or the ignition key ON.

• Do not clean the air flow meter with any type of detergent.

• Do not attempt to disassemble the ECCS control unit under any circumstances.

• Avoid static electricity build-up by properly grounding yourself prior to handling any ECU or related parts.

Reading Codes

➡**Diagnostic codes may be retrieved by observing the code flashes through the LED lights located on the Electronic Control Module (ECM). A special Nissan Consult monitor tool can be used, but is not required. When using special diagnostic equipment, always observe the tool manufacturer's instructions.**

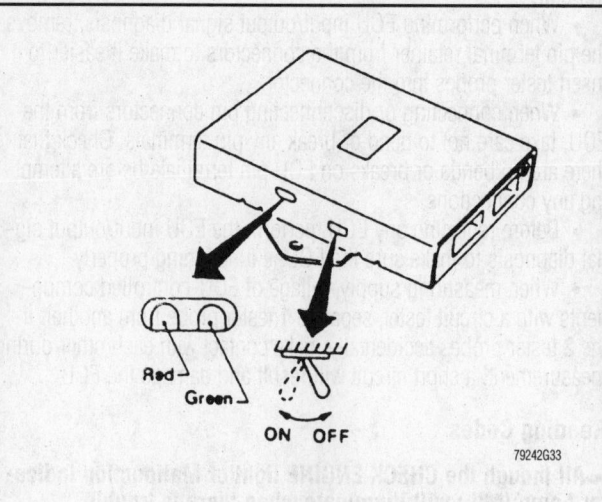

Entering the self-diagnostics using the ON/OFF mode switch

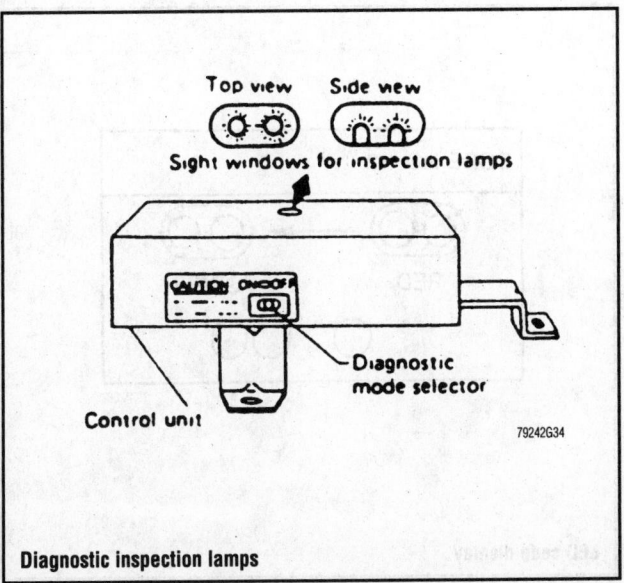

Diagnostic inspection lamps

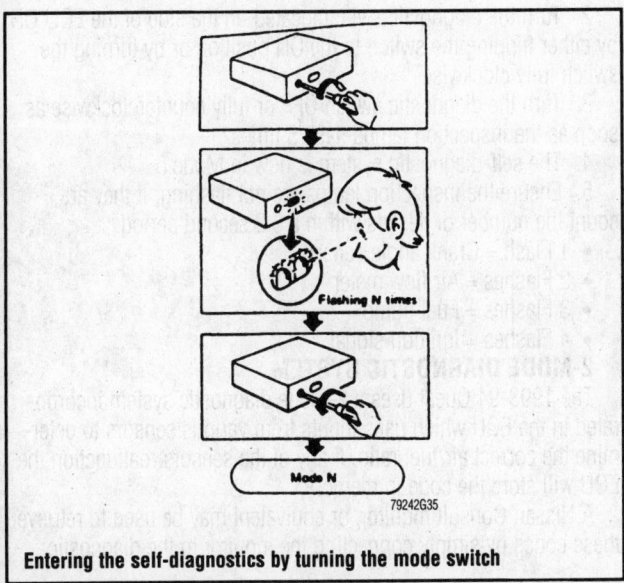

Entering the self-diagnostics by turning the mode switch

ELECTRONIC CONTROLLED CARBURETOR

1987–88 VEHICLES

The 1 .6L (EI 6S) carbureted engine utilizes a duty-controlled solenoid valve for fuel enrichment and an Idle Speed Control (ISC) actuator for basic controls instead of the conventional choke valve plate and fast idle cam. There are several other inputs which further affect the air/fuel ratio. The system is controlled in 2 ways: open or closed loop. To inspect the system for malfunctions, proceed as follows:

1. Position the ECU so the red and green LED's are visible.
2. Run the engine until it is at normal operating temperature.
3. Verify the diagnosis switch on the ECU is OFF.
4. Run the engine 2000 rpm for 5 minutes. After 5 minutes, observe the green LED light while maintaining 2000 rpm. The light should be blinking ON and OFF at least 5 times in 10 seconds. If not as specified, inspect the exhaust gas sensor.
5. Turn the engine OFF and turn the ECU diagnosis switch ON.

6. Turn the ignition switch ON. The green LED on the ECU should stay ON and the red LED will either flash for a short period indicating a malfunctioning input sensor or for a longer time indicating a malfunctioning output sensor.

ELECTRONIC FUEL INJECTION

1984–94 VEHICLES

Two types of diagnostic systems are used in Nissan vehicles: the 2-mode diagnostic system and the 5-mode diagnostic system. The 2 mode system is used in some vehicles starting in 1990, ultimately, all vehicles used the 2-mode system after 1991 with the exception of 1991-94 Maxima (VG3OE engine), Pathfinder and Truck. These vehicles continued to use the 5-mode system. The 5-mode system began in 1984.

The 5-mode diagnostic system is incorporated in the ECU which uses inputs from various sensors to determine the correct air/fuel ratio. If any of the sensors malfunction the ECU will store the code in memory. The 5-mode diagnostic system is capable of various tests as outlined below. When using these modes, the ECM may have to be removed from its mounting bracket to better access the mode selector switch.

➡ Vehicles are equipped with a CHECK ENGINE light on the instrument panel. If any systems are malfunctioning, the light will illuminate the same time as the red lamp while the engine is running and the system is in Mode 1.

Mode 1—Heated Oxygen Sensor

During closed loop operation the green lamp turns ON when a lean condition is detected and turns OFF under a rich condition. During open loop the green lamp remains ON or OFF. This mode is used to check Heated Oxygen sensor functions for correct operation. To enter Mode 1, proceed as follows:

1. Turn the ignition switch ON.
2. Turn the diagnostic switch located on the side of the ECU ON by either flipping the switch to the ON position or turning the screw switch fully clockwise.
3. Turn the diagnostic switch OFF or fully counterclockwise as soon as the inspection lamps flash 1 time.
4. The self-diagnostic system is now in Mode 1.

Mode 2—Mixture Ratio Feedback Control Monitor

The green inspection lamp is operating in the same manner as in Mode 1. During closed loop operation the red inspection lamp turns ON and OFF simultaneously with the green lamp when the mixture ratio is controlled within the specified value. During open loop the red lamp remains ON or OFF. Mode 2 is used for checking that optimum control of the fuel mixture is obtained. To enter Mode 2, proceed as follows:

1. Turn the ignition switch ON.
2. Turn the diagnostic switch ON, by either flipping the switch to the ON position or use a screwdriver and turn the switch fully clockwise.
3. Turn the diagnostic switch OFF or fully counterclockwise as soon as the inspection lamps flash 2 times.
4. The self-diagnostic system is now in Mode 2.

Mode 3—Self-Diagnosis System

This mode of the self-diagnostics is for stored code retrieval. To enter Mode 3, proceed as follows:

1. Thoroughly warm the engine before proceeding. With the engine OFF, turn the ignition switch ON. 2. Turn the diagnostic switch located on the side of the ECU ON by either flipping the switch to the ON position or using a screwdriver, turn the switch fully clockwise.

2. Turn the diagnostic switch OFF or fully counterclockwise as soon as the inspection lamps flash 3 times.

3. The self-diagnostic system is now in Mode 3.

➡When the battery is disconnected or self-diagnostic Mode 4 is selected after using Mode 3, all stored codes will be cleared. However if the ignition key is turned OFF and then the procedure is followed to enter Mode 4 directly, the stored codes will not be cleared.

4. The codes will now be displayed by the red and green inspection lamps flashing. The red lamp will flash first and the green lamp will follow. The red lamp is the tens and the green lamp is the units, that is, the red lamp flashes 1 time and the green lamp flashes 2 times, this would indicate a Code 12.

Mode 4—On/Off Switches

This mode checks the operation of the Vehicle Speed Sensor (VSS), Closed Throttle Position (CTP) and starter switches. Entering this mode will also clear all stored codes in the ECU. To enter Mode 4, proceed as follows:

1. Turn the ignition switch ON.

2. Turn the diagnostic switch located on the side of the ECU ON by either flipping the switch to the ON position or turning the mode switch fully clockwise.

3. Turn the diagnostic switch OFF or fully counterclockwise as soon as the inspection lamps flash 4 times.

4. The self-diagnostic system is now in Mode 4.

5. Turn the ignition switch to the START position and verify the red inspection lamp illuminates. This verifies that the starter switch is working.

6. Depress the accelerator and verify the red inspection lamp goes OFF. This verifies that the CTP switch is working.

7. Raise and properly support the vehicle and verify the lamp goes ON when the vehicle speed is above 12 mph (20 km/h). This verifies that the VSS is working

8. Turn the ignition switch OFF.

Mode 5—Real Time Diagnostics

In this mode the ECU is capable of detecting and alerting the technician the instant a malfunction in the crank angle sensor, air flow meter, ignition signal or the fuel pump occurs while operating/driving the vehicle. Items which are noted to be malfunctioning are not stored in the ECU's memory. To enter Mode 5, proceed as follows:

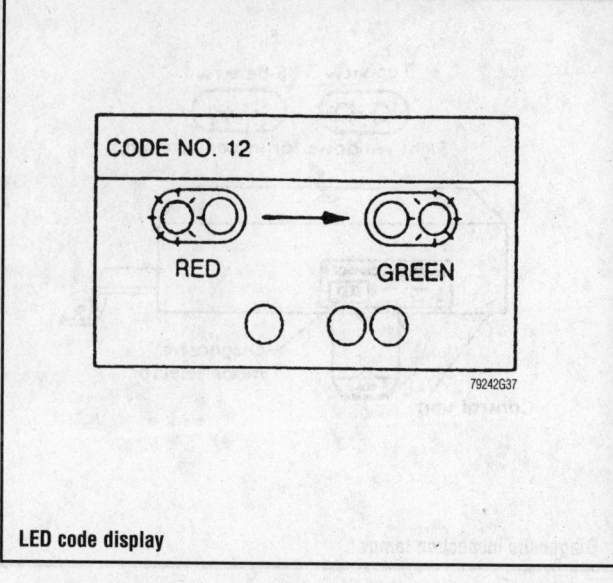

LED code display

1. Turn the ignition switch ON.

2. Turn the diagnostic switch located on the side of the ECU ON by either flipping the switch to the ON position or by turning the switch fully clockwise.

3. Turn the diagnostic switch OFF or fully counterclockwise as soon as the inspection lamps flash 5 times.

4. The self-diagnostic system is now in Mode 5.

5. Ensure the inspection lamps are not flashing. If they are, count the number of flashes within a 3.2 second period:

• 1 Flash = Crank angle sensor
• 2 Flashes = Air flow meter
• 3 Flashes = Fuel pump
• 4 Flashes = Ignition signal

2-MODE DIAGNOSTIC SYSTEM

The 1993-94 Quest uses a 2-mode diagnostic system incorporated in the ECU which uses inputs from various sensors to determine the correct air/fuel ratio. If any of the sensors malfunction the ECU will store the code in memory.

A Nissan Consult monitor, or equivalent may be used to retrieve these codes by simply connecting the monitor to the diagnostic

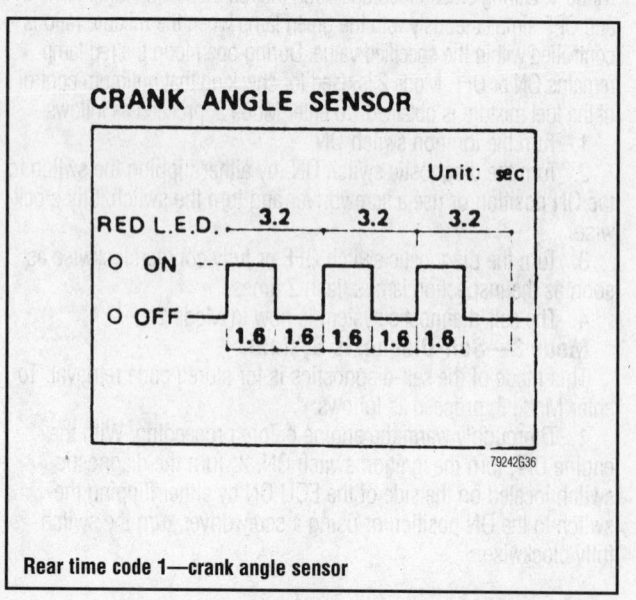

Rear time code 1—crank angle sensor

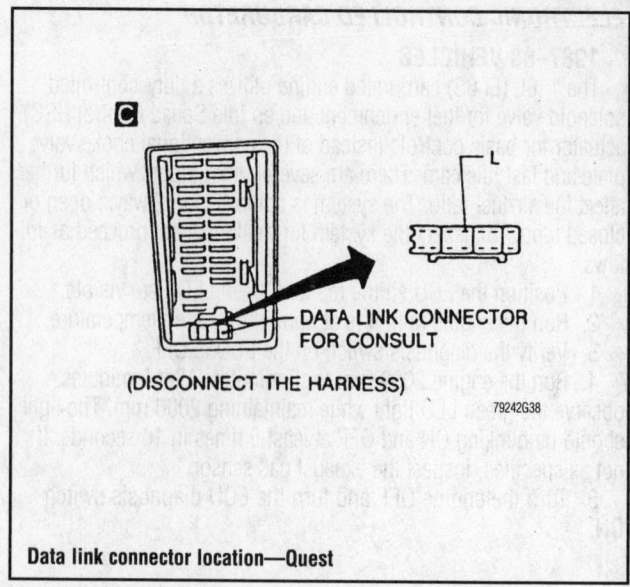

Data link connector location—Quest

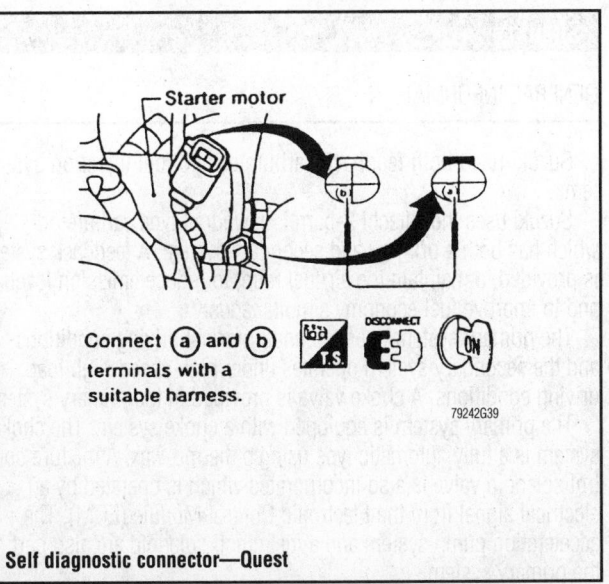

Connect (a) and (b) terminals with a suitable harness.

79242G39

Self diagnostic connector—Quest

connector located on the driver's side near the hood release. Turn the ignition switch to ON and press START, ENGINE and then SELF-DIAG RESULTS, the results will then be output to the monitor.

The conventional CHECK ENGINE or red LED ECU light may also be used for self-diagnostics. The conventional 2-Mode diagnostic system is broken into 2 separate modes each capable of 2 tests, an ignition switch ON or engine running test as outlined below:

Mode 1—Bulb Check
In this mode the RED indicator light on the ECU and the CHECK ENGINE light should be ON. To enter this mode simply turn the ignition switch ON and observe the light.

Mode I—Malfunction Warning
In this mode the ECU is acknowledging if there is a malfunction by illuminating the RED indicator light on the ECU and the CHECK ENGINE light. If the light turns OFF, the system is normal. To enter this mode, simply start the engine and observe the light.

Mode 2—Self-Diagnostic Codes—Quest
In this mode the ECU will output all malfunctions via the CHECK ENGINE light or the red LED on the ECU. The code may be retrieved by counting the number of flashes. The longer flashes indicate the first digit and the shorter flashes indicate the second digit. To enter this mode proceed as follows:
1. Turn the ignition switch ON, but do not start the vehicle.
2. Disconnect harness connectors and connect terminals A and B with a jumper wire.
3. Wait 2 seconds, remove the jumper wire and reconnect the harness connector.
4. Observe the CHECK ENGINE light for stored codes.

Mode 2—Exhaust Gas Sensor Monitor
In this mode the red LED on the ECU or CHECK ENGINE light will display the condition of the fuel mixture and whether the system is in closed loop or open loop. When the light flashes ON, the exhaust gas sensor is indicating a lean mixture. When the light stays OFF, the sensor is indicating a rich mixture. If the light remains ON or OFF, it is indicating an open loop system. If the system is equipped with 2 exhaust gas sensors, the left side will operate first. If already in Mode 2, proceed to Step C to enter the exhaust gas sensor monitor.

1. On all models except Quest, perform the following steps:
   a. Turn the ignition switch ON.
   b. Turn the diagnostic switch ON, by turning the switch fully clockwise for 2 seconds and then fully counterclockwise.
   c. Start the engine and run until thoroughly warm. Raise the idle to 2,000 rpm and hold for approximately 2 minutes. Ensure the red LED or CHECK ENGINE light flashes ON and OFF more than 5 times every 10 seconds with the engine speed at 2,000 rpm.

➡**If equipped with 2 exhaust gas sensors, switch to the right sensor by turning the ECU mode selector fully clockwise for 2 seconds and then fully counterclockwise with the engine running.**

2. On Quest models, perform the following steps:
   a. Turn the ignition switch ON.
   b. Disconnect harness connectors and connect terminals A and B with a jumper wire.
   c. Wait 2 seconds, remove the jumper wire and reconnect the harness connectors.
   d. Start the engine and run until thoroughly warm. Raise the idle to 2,000 rpm and hold for approximately 2 minutes. Ensure the red LED or CHECK ENGINE light flashes ON and OFF more than 5 times every 10 seconds with the engine speed at 2,000 rpm.

Clearing Codes

ENGINE CODES

Except Mode 5, 3 and 2 Systems
All control unit diagnostic codes may be cleared by disconnecting the negative battery cable for a period of 15 seconds. Entering Mode 4 of the Electronic Fuel Injection system diagnostics will also clear stored ECM engine codes.

Mode 5, 3 and 2 Systems
On 5-mode systems, enter mode 4 immediately after using mode 3 and the codes will be cleared. On 2-mode systems, the codes will be cleared when mode 1 is re-entered from mode 2. The Nissan Consult Monitor or equivalent can also be used to clear codes on 2-mode systems.

DIAGNOSTIC TROUBLE CODES

1984–87 Vehicles

Code 11 Crankshaft position sensor circuit
Code 12 Mass Air flow sensor circuit
Code 13 Engine coolant temperature sensor circuit
Code 21 Ignition signal circuit
Code 22 Fuel pump circuit
Code 23 Idle switch circuit
Code 24 Transmission switch
Code 31 AC switch, fast idle control of load signal
Code 32 Starter signal
Code 33 EGR gas sensor
Code 34 Detonation (Knock) sensor
Code 41 Air or Fuel temperature sensor
Code 42 Throttle sensor (or BP sensor in Canada)
Code 43 Mixture feedback control slips out (or low battery in Canada)
Code 44 No Malfunctioning circuits

1988–94 Vehicles

Code 11 Crankshaft position sensor circuit
Code 12 Mass Air flow sensor circuit
Code 13 Engine coolant temperature sensor circuit
Code 14 Vehicle speed sensor circuit
Code 15 Mixture ratio feedback control slips out (1988)
Code 21 Ignition signal circuit
Code 22 Fuel pump circuit (to 1991)
Code 23 Idle switch circuit (to 1991)
Code 24 Fuel Switch circuit or OD. switch circuit (to 1990)
Code 25 AAC valve circuit (to 1991)
Code 31 Electronic Control Module (ECM) or A/C circuit
Code 32 Exhaust Gas Recirculation (EGR) function
Code 33 Oxygen sensor circuit (left side, if two)
Code 34 Knock sensor circuit
Code 35 Exhaust gas temperature sensor circuit
Code 41 Air temperature sensor circuit
Code 42 Fuel temperature sensor circuit
Code 43 Throttle position sensor circuit
Code 44 No malfunctioning circuits
Code 45 Injector leak
Code 51 Injector circuit
Code 53 Heated oxygen sensor circuit (right side)
Code 54 Signal circuit from NT control unit to ECM
Code 55 No malfunctioning in the above circuits

Suzuki

GENERAL INFORMATION

Suzuki used both feedback carburetor and fuel injection systems.

Suzuki uses the Hitachi 2-barrel, downdraft type carburetor, which has both a primary and secondary system. A feedback system is provided to maintain the air/fuel ratio, to reduce emission levels and to improve fuel economy simultaneously.

The primary system operates under normal driving conditions and the secondary system operates under high speed-high load driving conditions. A choke valve is provided in the primary system.

The primary system is equipped with a choke system. The choke system is a fully automatic type using a thermo-wax. A mixture control solenoid valve is also incorporated which is operated by an electrical signal from the Electronic Control Module (ECM). The acceleration pump system and a fuel cutoff solenoid are also part of the primary system.

The secondary system is equipped with a secondary diaphragm through which vacuum is supplied from the primary side, via a Vacuum Switching Valve (VSV) and a Vacuum Transmitting Valve (VTV), to operate the secondary throttle vavle. The VSV and VTV are used only on the Suzuki Samurai and have been eliminated on the Suzuki Sidekick 1300cc.

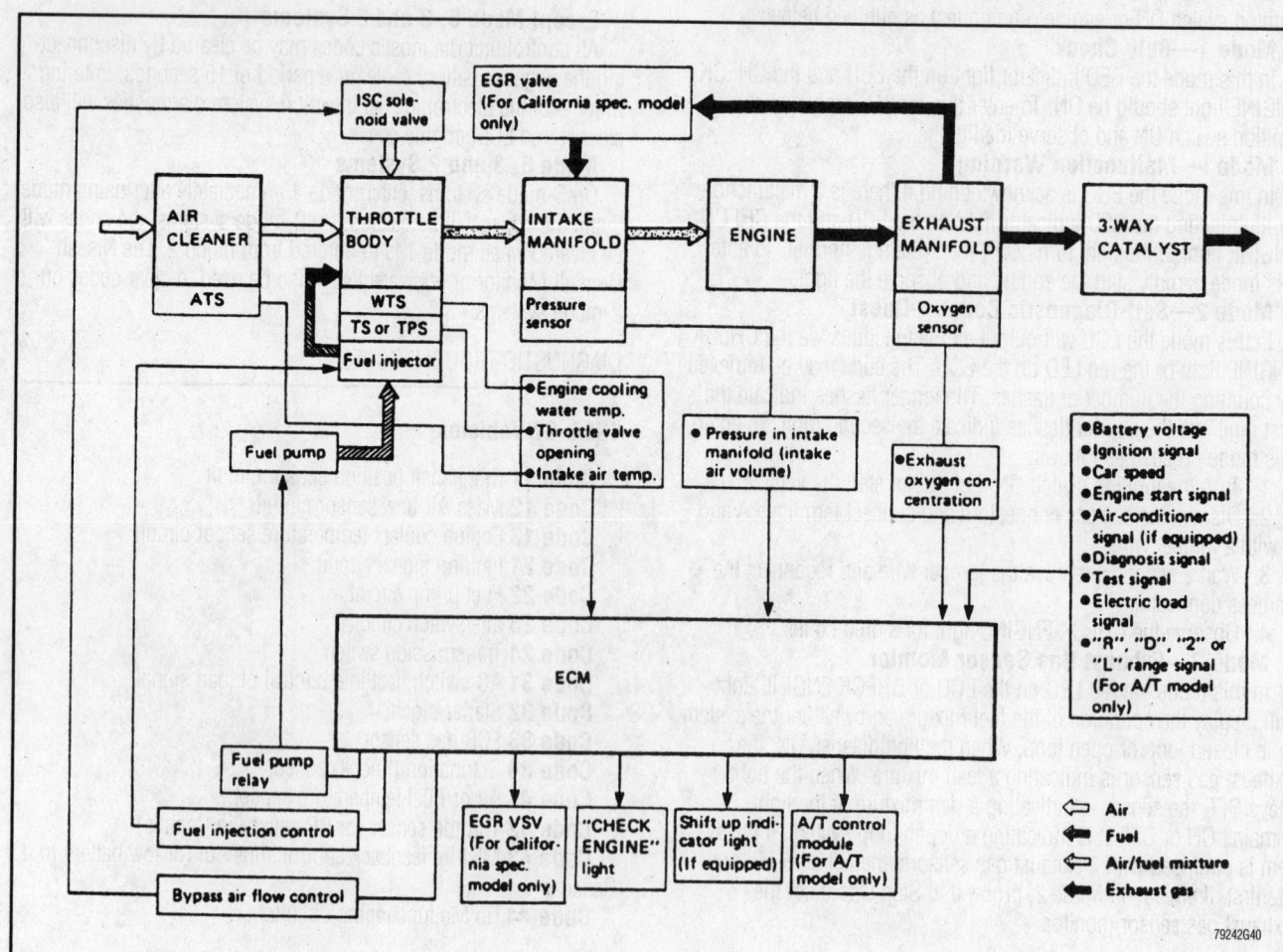

Example of input and output strategy used by Suziki

79242G40

A Switch vent solenoid valve is provided on the top of the float chamber. Its purpose is to reduce the evaporative emissions. The 2-barrel, downdraft type carburetor is also equipped with a idle-up system.

This system operates at idle and compensates the idle speed when any of the following conditions exist:

- When any electrical load (lights, rear defogger, heater fan, etc.) is operating.
- When the vehicle is at a high altitude.
- When the engine temperature is below 44°F (7°C).
- When the engine speed is lower than 1500 rpm after the engine is started.

The Electronic Fuel Injection (EFI) system supplies the vehicle's combustion chambers with air/fuel mixture of optimized ratio under varying driving conditions. Fuel delivery through the injector is controlled electrically by the Electronic Control Module (ECM).

SELF-DIAGNOSTICS

Service Precautions

- Keep the ECM parts and harnesses dry during service. Protect the ECM and all solid-state components from rough handling or temperature extremes.
- Use extreme care when working around the ECM or other components.
- Disconnect the negative battery cable before attempting to disconnect or remove any parts.
- Disconnect the negative battery cable and ECM connector before performing arc welding on the vehicle.
- Disconnect and remove the ECM from the vehicle before subjecting the vehicle to the temperatures experienced in a heated paint booth.

Reading Codes

1989–95 VEHICLES

On Swift and Samurai, the ECM memory is activated by connecting the spare fuse to the diagnosis switch terminal and turning the ignition switch ON. The fuse panel is located under the instrument panel, near the driver's side kick panel. On Sidekick models the

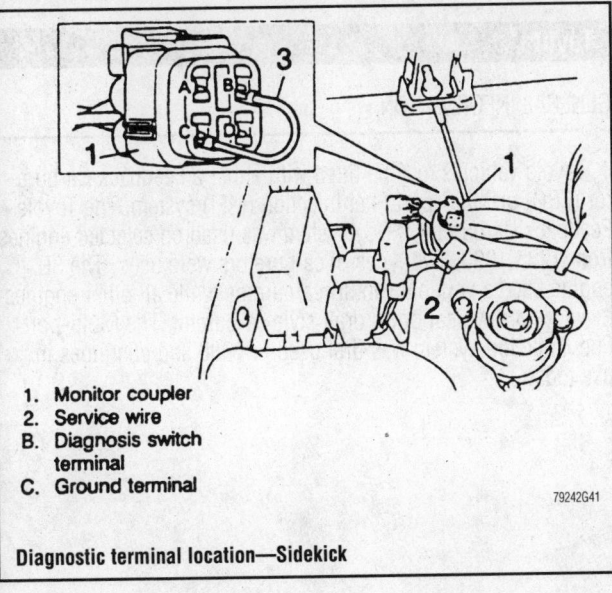

1. Monitor coupler
2. Service wire
B. Diagnosis switch terminal
C. Ground terminal

79242G41

Diagnostic terminal location—Sidekick

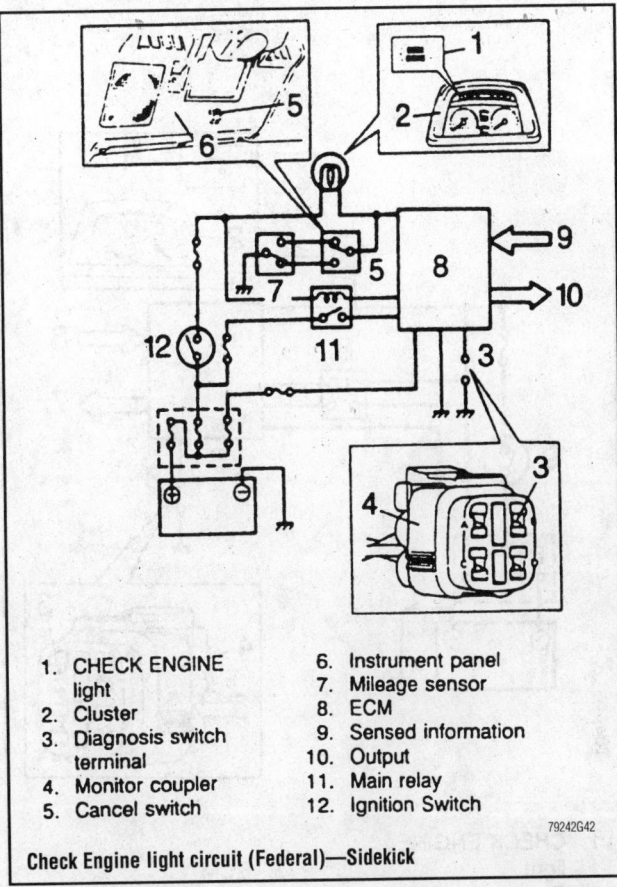

1. CHECK ENGINE light	6. Instrument panel
2. Cluster	7. Mileage sensor
3. Diagnosis switch terminal	8. ECM
4. Monitor coupler	9. Sensed information
5. Cancel switch	10. Output
	11. Main relay
	12. Ignition Switch

79242G42

Check Engine light circuit (Federal)—Sidekick

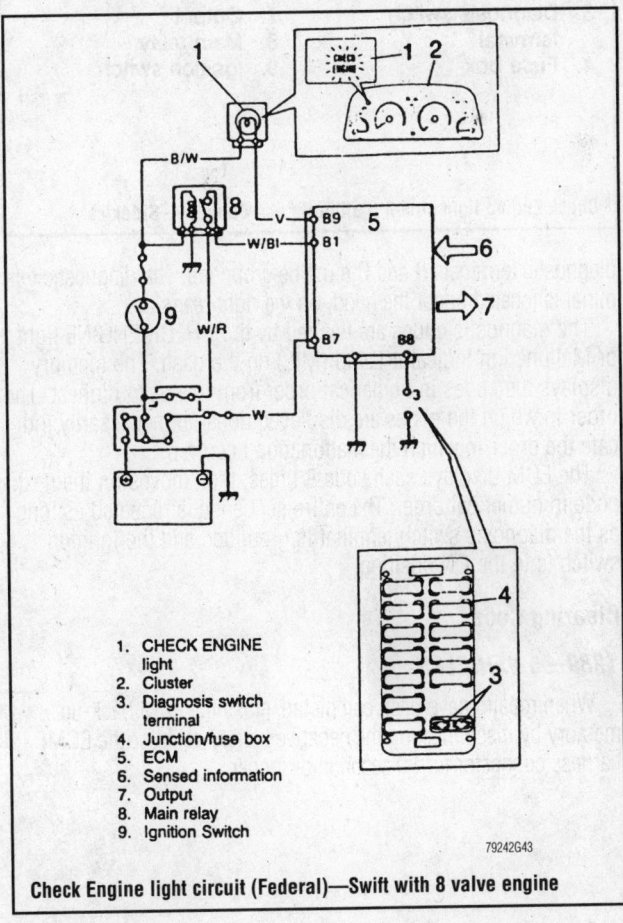

1. CHECK ENGINE light
2. Cluster
3. Diagnosis switch terminal
4. Junction/fuse box
5. ECM
6. Sensed information
7. Output
8. Main relay
9. Ignition Switch

79242G43

Check Engine light circuit (Federal)—Swift with 8 valve engine

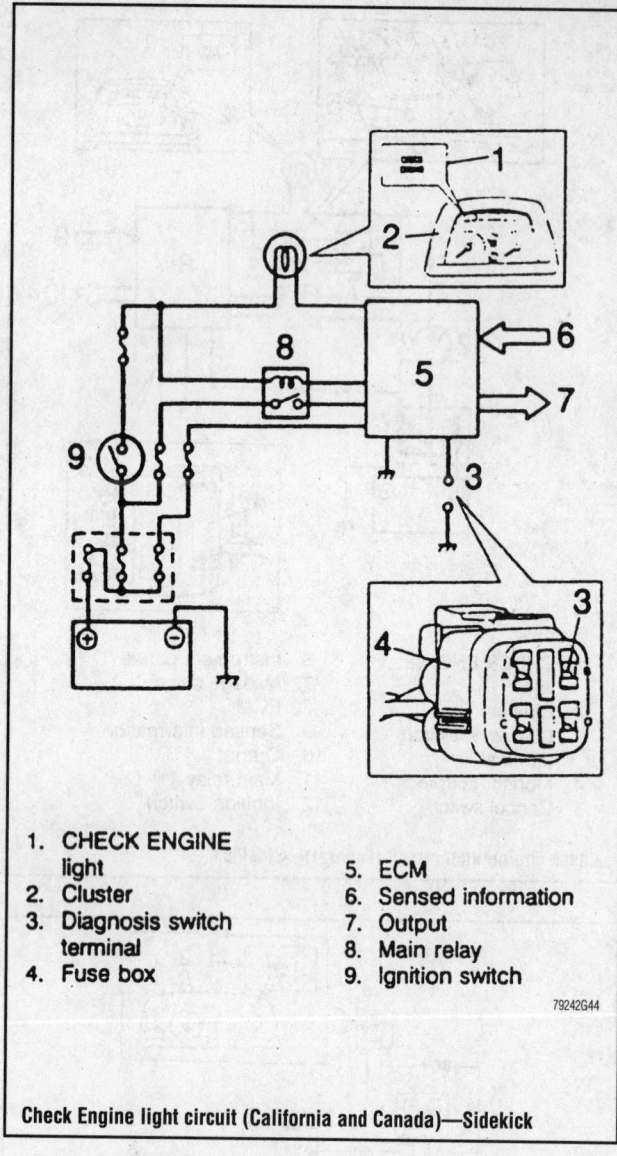

1. CHECK ENGINE light
2. Cluster
3. Diagnosis switch terminal
4. Fuse box
5. ECM
6. Sensed information
7. Output
8. Main relay
9. Ignition switch

79242G44

Check Engine light circuit (California and Canada)—Sidekick

diagnostic terminals B and C must be grounded. The diagnostic terminal is located under the hood, on the right rear side.

The diagnostic codes are flashed by the CHECK ENGINE light or Malfunction Indicator Lamp (MIL) on the dash. The memory displays the codes in numerical order from lowest to highest. The order in which the codes are displayed does not necessarily indicate the order in which the malfunction occurred.

The ECM displays each code 3 times, then moves on the next code in numerical order. The entire sequence is repeated as long as the diagnosis switch terminal is grounded and the ignition switch is in the ON position.

Clearing Codes

1989–95 VEHICLES

When repairs have been completed, erase the ECM back-up memory by disconnecting the negative battery cable or the ECM harness connector for 30 seconds or longer.

DIAGNOSTIC TROUBLE CODES

1986–95 Vehicles

Code 12 Normal
Code 13 Oxygen sensor circuit
Code 14 Engine Coolant Temperature (ECT) Sensor circuit—low temperature indicated, signal voltage high
Code 15 Engine Coolant Temperature (ECT) Sensor circuit—high temperature indicated, signal voltage low
Code 21 Throttle Position Sensor (TPS) circuit—signal voltage high
Code 22 Throttle Position Sensor (TPS) circuit—signal voltage low
Code 23 Air Temperature Sensor (ATS) circuit—low temperature indicated, signal voltage high
Code 24 Vehicle Speed Sensor (VSS) circuit
Code 25 Air Temperature Sensor (ATS) circuit—high temperature indicated, signal voltage low
Code 31 Pressure Sensor (PS) circuit—high pressure indicated, signal voltage high
Code 32 Pressure Sensor (PS) circuit—low pressure indicated, signal voltage low
Code 33 Mass Air Flow Sensor (MAS) circuit—signal voltage high
Code 34 Mass Air Flow Sensor (MAS) circuit—signal voltage low
Code 41 Ignition signal
Code 42 Crank Angle Sensor (CAS) circuit (except 1989–90 Sidekick) or Fifth switch circuit, Lock-up signal circuit (1989–90 Sidekick)
Code 44 Idle switch of Throttle Position Sensor (TPS) —open circuit
Code 45 Idle switch of Throttle Position Sensor (TPS) — shorted circuit
Code 51 Exhaust Gas Recirculation (EGR) system and/or Recirculated Exhaust Gas Temperature Sensor (REGTS) system—California vehicle
Code 52 Fuel Injector—California vehicle
Code 53 Ground circuit—California vehicle
Code 54 Fifth gear switch circuit
Code 71 Test switch circuit

Toyota

GENERAL INFORMATION

Toyota vehicles may be fitted with either a Feedback Carburetor (FBC) or Multi-port Fuel Injection (MEI) system. The Toyota Feedback Carburetor (FBC) system was used on selected engines from 1983 1990. Two types of carburetors were used. The 3E engine used a variable venturi carburetor, while all other engines used a more typical down draft style carburetor. The Multi-port Fuel Injection system was first used in 1980 and continues in use today.

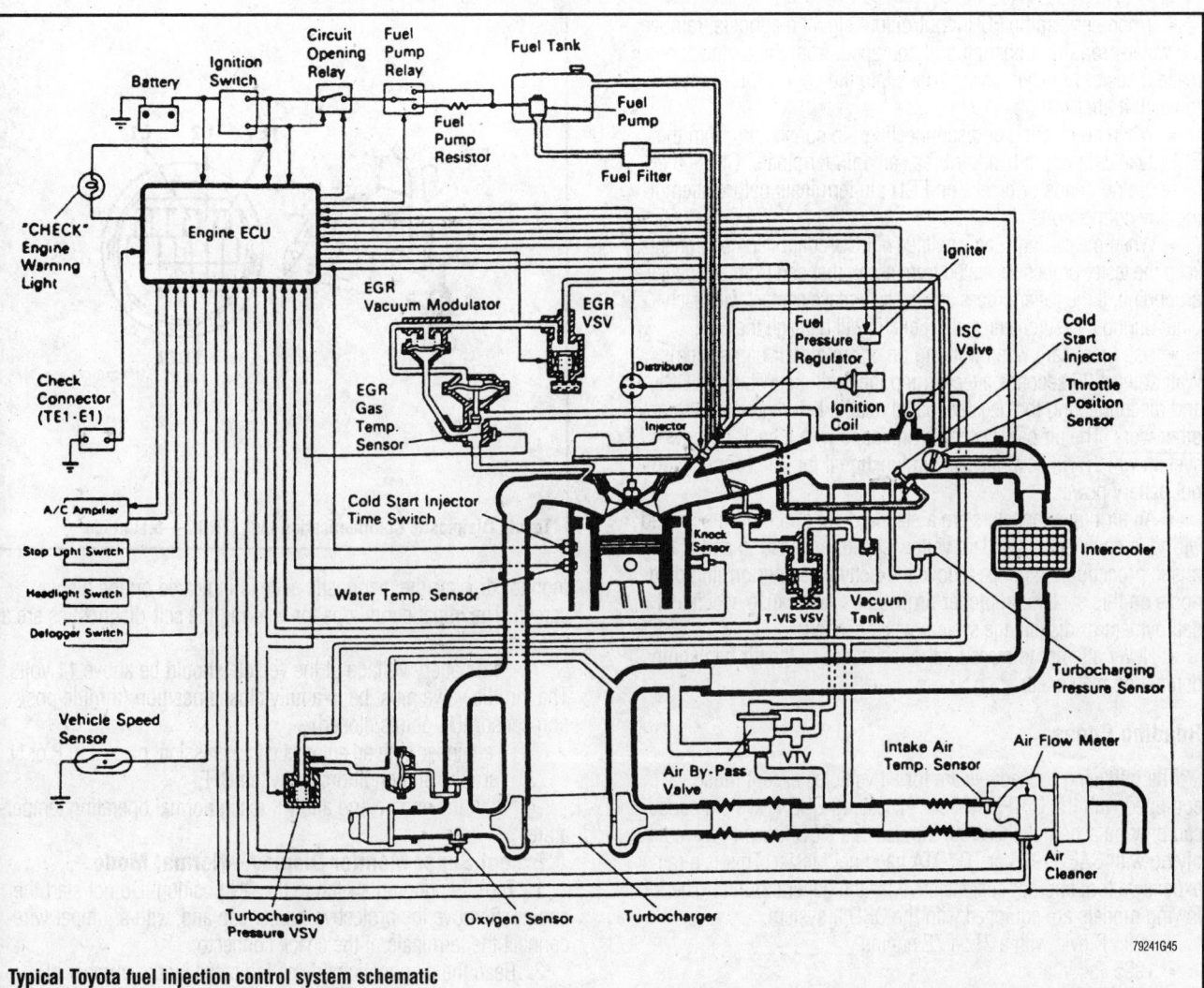

Typical Toyota fuel injection control system schematic

SELF-DIAGNOSTICS

As the engine control computers became capable of more functions, self-diagnostic and memory circuits were added.

These systems allow the ECU to note a fault, assign an identity code and store the code in memory for later retrieval.

All fuel injected control engine units possess the ability to provide fault codes during diagnosis. The number, type and meaning of engine codes vary by year and model.

While most fault codes are held in an electronic memory and are retained even after the ignition is switched OFF, certain codes are only held or displayed as long as the ignition is ON. If the fault is present at the next restart, the code will reset.

When a controller or ECU notes a fault, the dash warning lamp for the appropriate system will be lit to advise the operator. If the dash lamp is normally lit during system operation, as in the case of cruise control, the lamp will flash when a fault is found. The illumination or flashing of the dash lamp indicates that the controller has detected a fault and placed itself into the back-up or default mode.

Beginning in 1995 some models were equipped with an on board diagnostic system known as OBD II. To diagnose this system an OBD II scan tool, complying with SAE J1978 or TOYOTA hand held tester is necessary to access codes and read data output from the ECM. This is a rather expensive tool and not cost effective for the general public. The following model and engine applications are equipped with the OBD II system:

- 1995 Previa with a 2TZ-FZE engine
- 1995 Tacoma
- 1995 T100

Service Precautions

- Keep all ECU parts and harnesses dry during service. Protect the ECU and all solid-state components from rough handling or extremes of temperature.
- Before attempting to remove any parts, turn the ignition switch OFF and disconnect the battery ground cable.
- Make sure all harness connectors are fastened securely. A poor connection can cause an extremely high voltage surge, resulting in damage to integrated circuits.
- Always use a 12 volt battery as a power source.
- Do not attempt to disconnect the battery cables with the engine running.
- Do not attempt to disassemble the ECU unit under any circumstances.
- If installing a 2-way or CB radio, mobile phone or other radio equipment, keep the antenna as far as possible away from the electronic control unit. Keep the antenna feeder line at least 8 in. away from the EEI harness and do not run the lines parallel for a long distance. Be sure to ground the radio to the vehicle body.

• When performing ECU input/output signal diagnosis, remove the water-proofing rubber plug, if equipped, from the connectors to make it easier to insert tester probes into the connector. Always reinstall it after testing.

• When connecting or disconnecting pin connectors from the ECU, take care not to bend or break any pin terminals. Check that there are no bends or breaks on ECU pin terminals before attempting any connections.

• When measuring supply voltage of ECU-controlled components, keep the tester probes separated from each other and from accidental grounding. If the tester probes accidentally make contact with each other during measurement, a short circuit will damage the ECU.

• Use great care when working on or around air bag systems. Wait at least 20 seconds after turning the ignition switch to LOCK and disconnecting the negative battery cable before performing any other work. The air bag system is equipped with a back-up power system which will keep the system functional for 20 seconds without battery power.

• All air bag connectors are a standard yellow color; the related wiring is encased in standard yellow sheathing. Testing and diagnostic procedures must be followed exactly when performing diagnosis on this system. Improper procedures may cause accidental deployment or disable the system when needed.

• Never attempt to measure the resistance of the air bag squib; detonation may occur.

Reading Codes

The following procedures are for all vehicles except those equipped with the OBD II system. Accessing OBD II system codes can only be accomplished with the use of a OBD II scan tool, complying with SAE J1978 or TOYOTA hand held tester. This is a rather expensive tool and not cost effective for the general public. The following models are equipped with the OBD II system:

• 1995 Previa with a 2TZ-FZE engine
• 1995 Tacoma
• 1995 T100

1983–86 VEHICLES

The diagnostic codes can be read by the number of blinks of the 'Check Engine' warning light when the proper terminals of the check connector are short-circuited. If the vehicle is equipped with a super

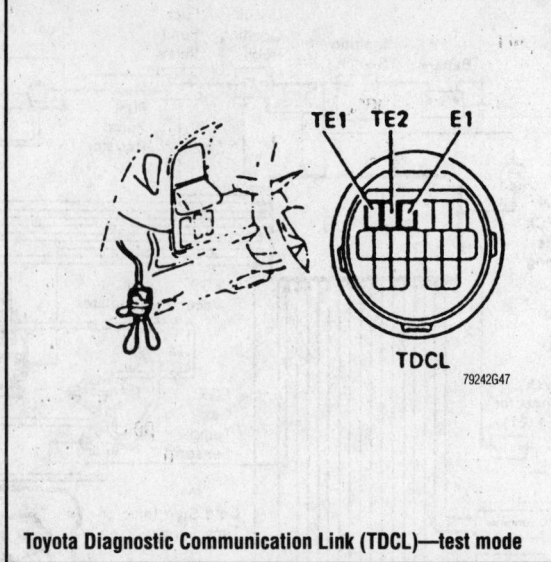

Toyota Diagnostic Communication Link (TDCL)—test mode

monitor display, the diagnostic code is indicated on the display screen. The initial conditions for entering the self-diagnostics are as follows:

1. The battery voltage of the vehicle should be above 11 volts. The throttle valve must be in a fully closed position (throttle position sensor IDL points closed).
2. If equipped with an automatic transmission, place it in P or N.
3. Turn the air conditioning switch OFF.
4. Start the engine and allow it reach normal operating temperature.

Except Super Monitor Display—Normal Mode

1. Turn the ignition switch to the ON position. Do not start the engine. Remove the protective rubber cap and, with a jumper wire connect the terminals of the check connector.
2. Read the diagnostic code as indicated by the number of flashes of the 'Check Engine' warning light.

➡**On some early models, install an analog voltmeter to the EFI service connector. Read diagnostic codes by voltmeter needle deflection between OV-2.5V-5V. The voltmeter needle will fluctuate between 5V and 2.5V every 0.6 seconds.**

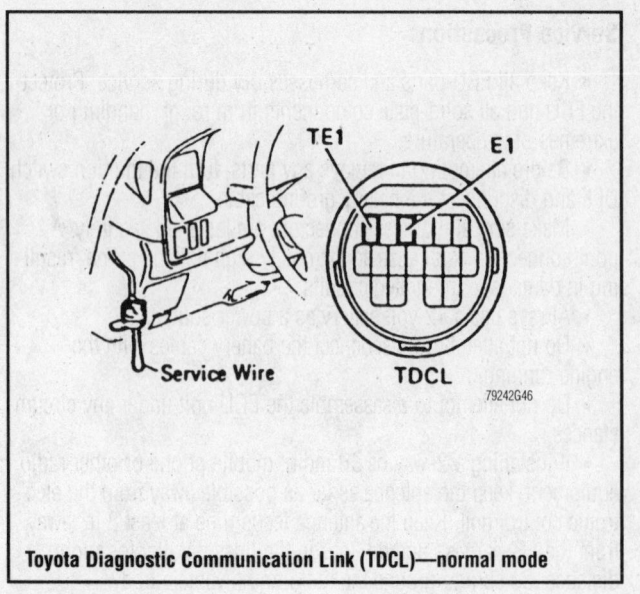

Toyota Diagnostic Communication Link (TDCL)—normal mode

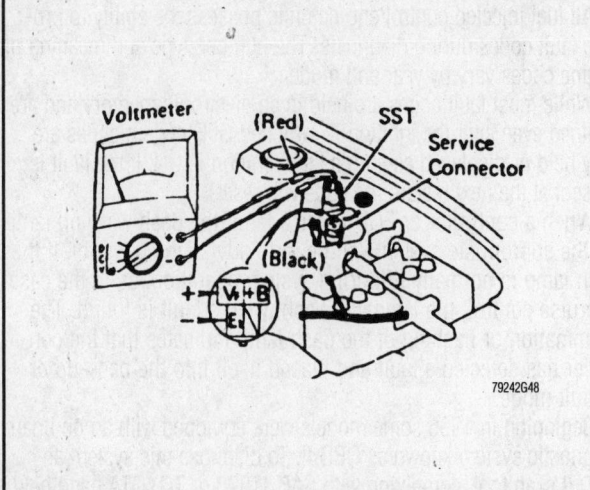

Installing an analog voltmeter to the EFI service connector—early models

3. If the system is operating normally (no malfunction), the light ll blink once every ¼ second. On single digit code number sys- ms, the light will blink once every 3 or 4.5 seconds.

4. In the event of a malfunction, the light will blink once every ½ cond (on some models it may be 1, 2 or 3 seconds). The 1st mber of blinks will equal the 1st digit of a 2-digit diagnostic de. After a 1.5 second pause, the 2nd number of blinks will equal e 2nd number of a 2-digit diagnostic code. If there are 2 or more des, there will be a 2.5 second pause between each. On single git code number systems the light will blink a number of times qual to the malfunction code indication every 2 or 4.5 seconds.

5. After all the codes have been output, there will be a 4.5 sec- nd pause and they will be repeated as long as the terminals of the neck connector are shorted.

▶In event of multiple trouble codes, indication will begin om the smaller value and continue to the larger in order.

6. After the diagnosis check, remove the jumper wire from the neck connector and install the protective rubber cap.

Test Mode

1. Using a jumper wire, connect the TE2 and E1 terminals of the oyota Diagnostic Communication Link (TDCL), then turn the igni- on switch ON to begin the diagnostic test mode.

2. Start the engine and drive the vehicle at a speed of 10 mph or ore. Simulate the conditions where the malfunction has been ported to happen.

3. Using a jumper wire, connect the TE2 and E1 terminals of the DCL connector.

4. Read the diagnosis code as indicated by the number of Check Engine' light flashes.

5. After diagnosis check remove the jumper wires.

Super Monitor Display

The super monitor display system was offered as an option on ome late model Toyota vehicles.

1. Turn the ignition switch ON but do not start the engine.

2. Simultaneously push and hold in the SELECT and INPUT M eys for at least 3 seconds. The letters DIAG will appear on the creen.

3. After a short pause, hold the SET key in for at least 3 sec- nds. If the system is normal (no malfunctions), ENG-OK will ppear on the screen.

4. If there is a malfunction, the code number for it will appear on the screen. In event of 2 or more numbers, there will be a 3 second pause between each (example:EN-42).

1987–89 VEHICLES

Stored fault codes are transmitted through the blinking of the CHECK engine warning lamp. This occurs only when the system is placed into the diagnostic mode; it does not occur while the vehicle is being driven.

To read the fault codes:
1. The following initial conditions must be met:
   a. Battery voltage at or above 11 volts.
   b. Throttle fully closed.
   c. Transmission in N or P.
   d. All electrical systems and accessories OFF.
2. Turn the ignition ON but do not start the engine.
3. Use a jumper wire to connect terminals T and E1 at the diag- nostic connector. On all 1989 vehicles except Corolla, MR2 and Ter- cel, connect terminals TE1 and Et. For 1988–89 Vans, jumper the 2 pins of the service connector. On 1989 Corolla, MR2 and Tercel, connect terminals T and E1.

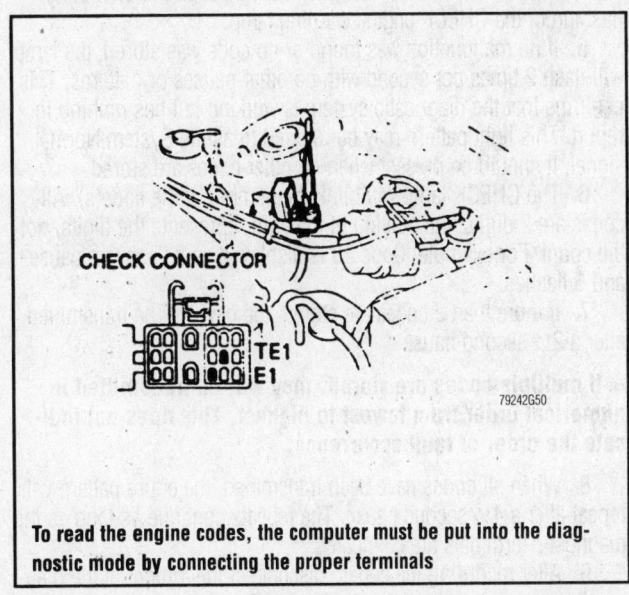

To read the engine codes, the computer must be put into the diag- nostic mode by connecting the proper terminals

Example of the Super Monitor display. When correctly engaged, the system will provide system identifiers such as ENG, ABS or ECT.

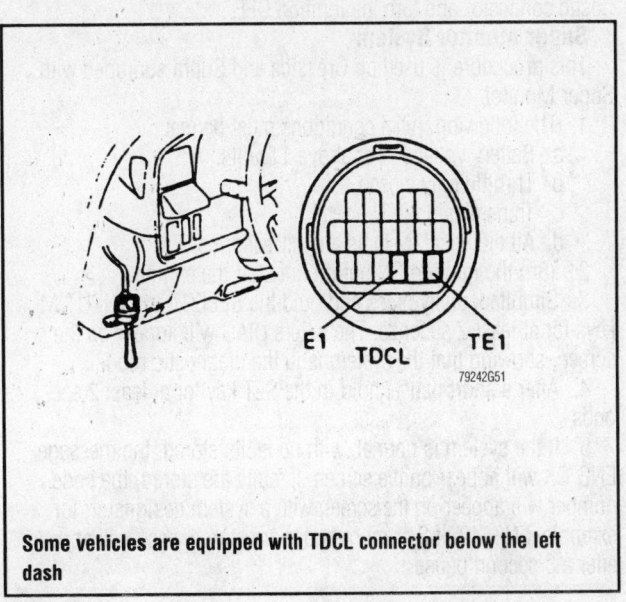

Some vehicles are equipped with TDCL connector below the left dash

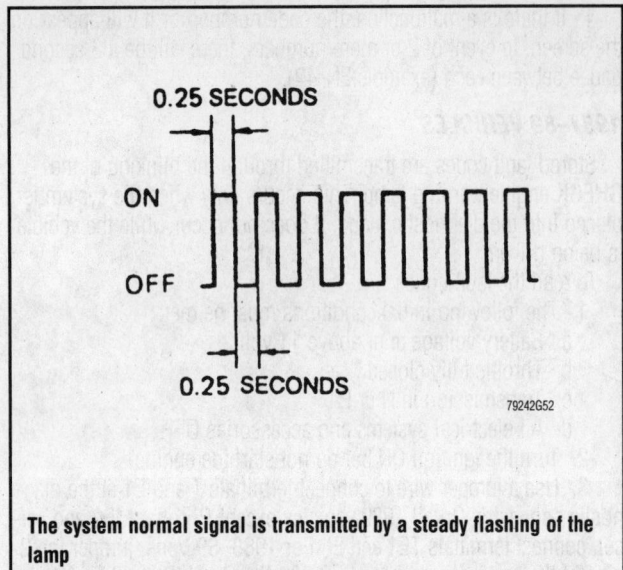

0.25 SECONDS

ON

OFF

0.25 SECONDS

79242G52

The system normal signal is transmitted by a steady flashing of the lamp

4. The fault codes will be transmitted through the controlled flashing of the CHECK engine warning lamp.

5. If no malfunction was found or no code was stored, the lamp will flash 2 times per second with no other pauses or patterns. This confirms that the diagnostic system is working but has nothing to report. This light pattern may be referred to as the System Normal signal; it should be present when no other codes are stored.

6. The CHECK lamp will blink the number of the code(s). All codes are 2 digits; the pulsing of the light represents the digits, not the count. For example, Code 25 is displayed as 2 flashes a pause and 5 flashes.

7. If more than 2 codes are stored, the next will be transmitted after a 2½ second pause.

➡**If multiple codes are stored, they will be transmitted in numerical order from lowest to highest. This does not indicate the order of fault occurrence.**

8. When all codes have been transmitted, the entire pattern will repeat after a 4½ second pause. The repeats continue as long as the diagnostic terminals are connected.

9. After recording the codes, disconnect the jumper at the diagnostic connector and turn the ignition OFF.

Super Monitor System

This procedure is used on Cressida and Supra equipped with Super Monitor.

1. The following initial conditions must be met:
   a. Battery voltage at or above 11 volts.
   b. Throttle fully closed.
   c. Transmission in N or P.
   d. All electrical systems and accessories OFF.
2. Turn the ignition ON but do not start the engine.
3. Simultaneously press and hold the SELECT and INPUT M keys for at least 2 seconds. The letters DIAG will appear on the screen, showing that the system is in the diagnostic mode.
4. After a short pause, hold in the SET key for at least 2 seconds.
5. If the system is normal, with no faults stored, the message ENG OK will appear on the screen. If faults are stored, the code number will appear on the screen with a system designator; for example, ENG-42. If 2 or more codes are stored, each will appear after a 3 second pause.

1990–95 VEHICLES

Stored fault codes are transmitted through the blinking of the CHECK engine warning lamp. This occurs only when the system is placed into the diagnostic mode; it does not occur while the vehicle is being driven.

To read the fault codes:

1. The following initial conditions must be met:
   a. Battery voltage at or above 11 volts.
   b. Throttle fully closed.
   c. Transmission in N or P.
   d. All electrical systems and accessories OFF.
2. Turn the ignition ON but do not start the engine.
3. Use a jumper wire to connect terminals TE1 and E1 at the diagnostic connector in the engine compartment or at the TDCL connector below the left dashboard if so equipped.
4. The fault codes will be transmitted through the controlled flashing of the CHECK ENGINE warning lamp.
5. If no malfunction was found or no code was stored, the lamp will flash 2 times per second with no other pauses or patterns. This confirms that the diagnostic system is working but has nothing to report. This light pattern may be referred to as the System Normal signal; it should be present when no other codes are stored.
6. The CHECK lamp will blink the number of the code(s). All codes are 2-digit; the pulsing of the light represents the digits, not the count. For example, Code 25 is displayed as 2 flashes, a pause and 5 flashes.
7. If more than 1 code is stored, the next will be transmitted after a 2½ second pause.

➡**If multiple codes are stored, they will be transmitted in numerical order from lowest to highest. This does not indicate the order of fault occurrence.**

8. When all codes have been transmitted, the entire pattern will repeat after a 4½ second pause. The repeats continue as long as the diagnostic terminals are connected.
9. After recording the codes, disconnect the jumper at the diagnostic connector and turn the ignition OFF.

Clearing Codes

1986–95

Stored codes will remain in memory until cleared. The correct method of clearing codes is to turn the ignition switch OFF, then remove the proper fuse. On all vehicles except as noted below, remove the EEI fuse. Each fuse must be removed for at least 10 seconds. The time required may be longer in cold weather.

Disconnecting the negative battery cable will also clear the memory but is not recommended due to other on-board memories being cleared as well. Once the system power is restored, re-check for stored codes. Only the System Normal indication should be present. If any other code is stored, the clearing procedure must be repeated or additional repairs performed; the old code will remain stored along with any new ones.

After repairs, it is recommended to clear the memory before test driving the vehicle. Upon returning from the drive, interrogate the memory; if the original code is again present, the repair was unsuccessful.

DIAGNOSTIC TROUBLE CODES

1983–1984 Engine

Code 1 Normal operation
Code 2 Open or shorted air flow meter circuit—defective air flow meter or Electronic Control Unit (ECU)

Code 3 Open or shorted air flow meter circuit—defective air flow meter or Electronic Control Unit (ECU)

Code 4 Open Water Thermosnsor (THW) circuit—defective Water Thermosensor (THW) or Electronic Control Unit (ECU)

Code 5 Open or shorted oxygen sensor circuit—lean or rich indication—defective oxygen sensor or Electronic Control Unit (ECU)

Code 6 No ignition signal—defective ignition system circuit, Integrated Ignition Assembly (IIA) or Electronic Control Unit (ECU)

Code 7 Defective Throttle Position Sensor (TPS) circuit, Throttle Position Sensor (TPS) or Electronic Control Unit (ECU)

1985–1987 Engines

➡ The 1985 2.0L (25-E and 3Y-EC) engines use 1984 Codes

Code 1 Normal operation

Code 2 Open or shorted air flow meter circuit—defective air flow meter or Electronic Control Unit (ECU)

Code 3 No signal from igniter 4 times in succession -defective igniter or main relay circuit, igniter or Electronic Control Unit (ECU)

Code 4 Open Water Thermosensor (THW) circuit—defective Water Thermosensor (THW) or Electronic Control Unit (ECU)

Code 5 Open or shorted oxygen sensor circuit—lean or rich indication—defective oxygen sensor or Electronic Control Unit (ECU)

Code 6 No engine revolution sensor (Ne) signal to Electronic Control Unit (ECU) or Ne value being over 1000 rpm in spite of no Ne signal to ECU—defective igniter circuit, igniter, distributor or Electronic Cont rol Unit (ECU)

Code 7 Open or shorted Throttle Position Sensor (TPS) circuit, Throttle Position Sensor (TPS) or Electronic Control Unit (ECU)

Code 8 Open or shorted intake air thermosensor circuit -defective intake air thermosensor circuit or Electronic Control Unit (ECU)

Code 10 No starter switch signal to Electronic Control Unit (ECU) with vehicle speed at 0 and engine speed over 800 rpm—defective speed sensor circuit, main relay circuit, igniter switch to starter circuit, igniter switch or Electronic Control Unit (ECU)

Code 11 Short circuit in check connector terminal T with the air conditioning switch ON or throttle switch (IDL) contact point OFF—defective air conditioner switch, Throttle Position Sensor (TPS) circuit, Throttle Position Sensor (TPS) or Electronic Control Unit (ECU)

Code 12 Knock control sensor signal has not reached judgement level in succession—defective knock control sensor circuit, knock control sensor or Electronic Control Unit (ECU)

Code 13 Knock CPU faulty

1988–95 Engines except 1995 OBD II

Constant blinking of indicator light: No faults detected

Code 11 Momentary interruption in power supply to ECU; up to 1991

Code 12 Engine revolution (NE or G) signal to ECU; missing within several seconds after engine is cranked

Code 13 Rpm NE signal to ECU; missing when engine speed is above 1000 rpm

Code 14 Igniter (IGE) signal to ECU; missing 4–11 times in succession

Code 16 ECT control signal—normal signal missing from ECT CPU (1990–94)

Code 16 A/T control system—normal signal missing from between the engine CPU and A/T CPU in the ECM (1995)

Code 21 Main oxygen sensor signal; voltage output does not exceed a set value on the lean and rich sides continuously for a certain period of time or open/short sensor heater circuit

Code 22 Water temperature sensor circuit (THW); open/short for 500 msec. or more

Code 23 Intake air temperature signal (THA)

Code 24 Intake air temperature sensor circuit (THA); open/short for 500 msec. or more

Code 25 Air/fuel ratio LEAN malfunction; Oxygen sensor output is less than 0.45 V for at least 90 seconds when oxygen sensor is warmed up (engine racing at 2000 rpm). California only: air/fuel ratio feedback compensat ion/adaptive control:

feedback value continues at upper (LEAN) limit, or is not renewed, for a certain period of time.

Code 26 Air/fuel ratio RICH malfunction; California only:

Air/fuel ratio feedback compensation/adaptive control: feedback value continues at lower (RICH) limit, or is not renewed, for a certain period of time.

Code 27 Sub-oxygen sensor signal; detection of sensor/signal deterioration or open/short sensor heater circuit (California only)

Code 28 No. 2 oxygen sensor signal/heater signal

Code 31 Air flow meter circuit; open or shorted when idle contacts are closed

Code 31 Vacuum (Manifold absolute pressure) sensor signal; open/short circuit

Code 32 Air flow meter circuit; circuit open or shorted when idling

Code 34 Turbocharging pressure signal; excessive pressure

Code 35 Altitude compensation (HAC) sensor signal; open/short

Code 35 Turbocharging pressure sensor signal; open/short

Code 36 Turbocharging pressure sensor signal; open or short detected for 0.5 sec or more in the turbocharging pressure sensor signal circuit; 1992–94

Code 41 Throttle position sensor circuit (VTA); open/short

Code 42 Vehicle speed sensor circuit

Code 43 No starter switch (STA) signal to ECU until engine speed reaches 800 rpm when cranking

Code 51 NC signal ON, DL contact OFF, or shift position in R, D, 2 or 1 range; with check terminals T and El connected

Code 52 Knock sensor signal (KNK); open/short

Code 53 Knock control signal in ECU; ECU knock control faulty

Code 55 Knock sensor (rear side) signal in ECU; ECU knock control faulty

Code 71 EGR system malfunction; EGR gas temperature signal (THG) is below water temperature sensor signal or below intake air temperature sensor signal plus 86°F (30°C), after driving for 240 seconds in EGR operation range (California only)

Code 72 Fuel cut solenoid signal circuit (FCS) open; up to 1991

Code 78 Fuel pump control signal input circuit to pump (FPC) open

Code 81 TCM communication; open detected in ECT1 circuit for 2 or more seconds

Code 83 TCM communication; open detected in ESA1 circuit 0.5 sec after idle

Code 84 TCM communication; open in ESA2 circuit for 0.5 seconds after idle

Code 85 TCM communication; open in ESA3 circuit for more than 0.5 seconds after idle

OBD II DIAGNOSTIC TROUBLE CODES

Introduction

To comply with OBD II Regulations, the Control Module is equipped with sOffware designed to allow it to monitor vehicle emission control systems and components. Once the ignition is turned on or the engine is started, and certain test conditions are met, the PCM runs a series of monitors to test the emission control systems and components. Test conditions include different inputs such as time since startup, run-time, engine speed and temperature, transaxle gear position, and the engine open or closed loop status. Once the monitor is started, the control module attempts to run it to completion. If a particular monitor fails a test, a code is set and operating conditions at that time are recorded in memory. If the same component or system fails twice in succession, the Malfunction Indicator Lamp (MIL) is activated.

Monitors are divided into two types: Main Monitors and the Comprehensive Component Monitors.

- Catalyst Monitor
- EGR Monitor
- EVAP Monitor
- Fuel System Monitor
- Misfire Monitor
- Oxygen Sensor Monitor
- Oxygen Sensor Heater Monitor

Certain monitors, in particular the fuel system and misfire monitors, have limitations that are different from any of the other monitors. The first time either of these monitors fail, the MIL is activated, and engine conditions at the time of the fault are recorded. In order for the control module to turn off an MIL related to these two monitors, it must determine that no faults are present with engine operating conditions similar to when it detected the fault. To qualify, the engine must be operated within a specified speed range, engine load range and temperature range.

A warm-up cycle is considered to be vehicle operation after the engine has been turned off for a period of time, with the ECT input rising a specified amount and reaching normal operating temperature. When the MIL is turned off because a fault is no longer present, most OBD II codes will be erased after a minimum of 40 warm-up cycles. Misfire and fuel system codes require a minimum of 80 warm-up cycles before they clear.

OBD II Systems use a standardized test connector, called the Data Link Connector (DLC). It is located beneath the left side of the instrument panel. The DLC is located out of the line of sight of vehicle passengers, but is easily viewable from a kneeling position outside the vehicle. The connector is rectangular in design and contains up to I6 terminals. It has keying features to allow for easy connection. Both the DLC and Scan Tool connectors have latching features that ensure the scan tool will remain properly connected.

Some common uses of the Scan Tool are to identify and clear Diagnostic Trouble Codes (DTCs) and to read control module freeze frame.

The Malfunction Indicator Lamp (MIL) looks similar to the "Check Engine" lamp. However, on OBD II Systems, it is controlled under a strict set of guidelines that dictate when the MIL is illuminated. If any of the control module monitors detects a fault that could impact vehicle emissions, a fault code is set. A One-Trip Monitor requires that a test fail once, a Two-Trip Monitor requires a test fail twice in succession, and a Three-Trip Monitor requires that a test fail three times in succession to activate the MIL.

The MIL is mounted in the instrument panel and has two functions: To act as a bulb check at key On and to inform the driver tha an emissions fault has occurred.

Once the engine is started, if no faults are detected, the control module should extinguish the MIL after a few seconds. If the MIL remains On or flashes with the engine running a driveability symptom is present.

Non-OBD II Trouble Codes

For years, vehicles have been capable of storing diagnostic trou ble codes. Codes prior to the 1996 OBD II legislation have been proprietary to the vehicle manufacturer. In some cases, the codes are specific to the individual make and model.

Furthermore, some manufacturers have developed specialized devices to read their codes. This complicates code reading and clearing.

For further information on Non-OBD II trouble codes, please ref to the "Chilton Total vehicle Care" manual for your particular make and model of vehicle.

OBD II Trouble Codes

Federal law required all vehicle manufacturers to meet On Board Diagnostics, Second Generation or OBD II standards by 1996. In order to meet this standard, the automobile's on-board computer must monitor and perform diagnostic tests on vehicle emissions to ensure that the vehicle is operating at an acceptable (legal) emission level. The maximum allowable emission level is set by the Federal Test Procedure (FTP).

Some 1995 and all 1996–99 vehicles are OBD II compliant. All OBD II vehicles have the same 16 pin diagnostic connector or DLC This eliminates the need to have a manufacturer specific connector to plug a scan tool into your vehicle.

➡**Many 1995 vehicles have a 16 pin OBD II connector, however, this does not mean that the vehicle is OBD II compliant.**

TROUBLE CODE DESCRIPTION

In the past, trouble code numbers varied between manufacturers years, makes and models. OBD II requires that all vehicle manufacturers use a common Diagnostic Trouble Code (DTC) numbering system. Since the generic listing was not specific enough, most manufacturers came up with their own DTC listings which are calle manufacturer specific codes. Both generic and manufacturer specifi codes are 5 digits. The numbers can be decoded as follows:

The first digit is a letter which identifies the function of the devic or circuit which has the fault. This digit can be either:

- P—Powertrain
- B—Body
- C—Chassis
- U—Network or data link code

The second digit is either a 0 or 1 and indicates whether the cod is generic or manufacturer specific.

- 0—Generic
- 1—Manufacturer Specific

The third digit represents the specific vehicle circuit or system that has the fault. Listed below are the number identifiers for the powertrain system.

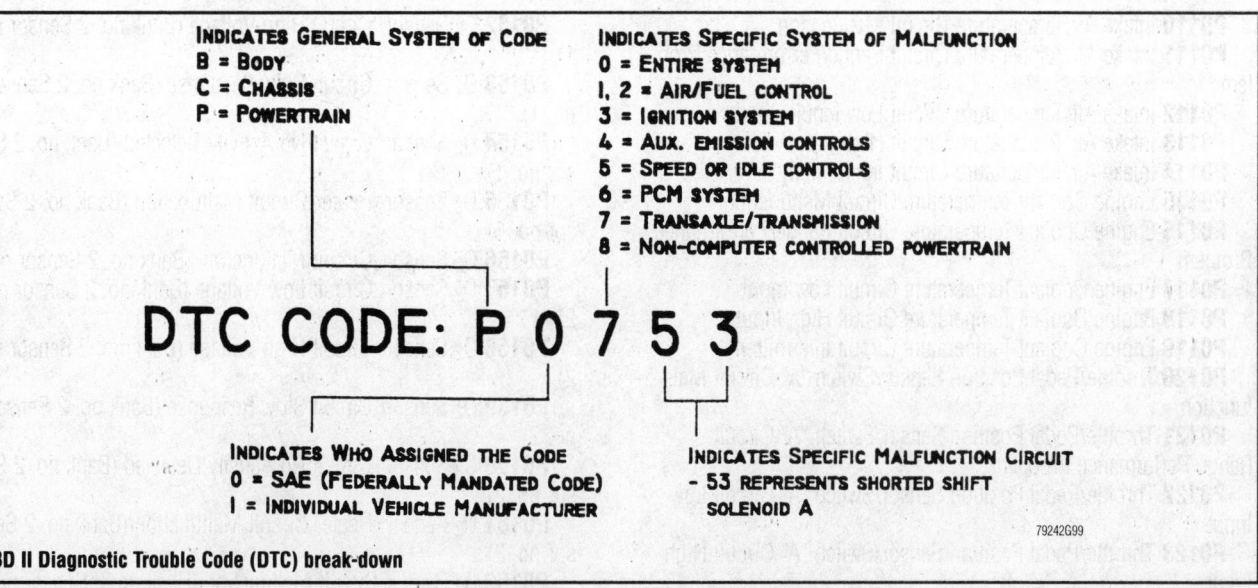

OBD II Diagnostic Trouble Code (DTC) break-down

- 1—Fuel and Air Metering
- 2—Fuel and Air Metering (Injector Circuit Malfunctions Only)
- 3—Ignition System or Misfire
- 4—Auxiliary Emission Control
- 5—Vehicle Speed Control and Idle Control System
- 6—Computer and Auxiliary Outputs
- 7—Transmission
- 8—Transmission

The last two digits indicate the specific trouble code.

On OBD II vehicles there are two different types of DTCs: Stored and Pending. For a DTC to become Stored, certain malfunction conditions must occur. The condition(s) required to Store codes are different for every DTC and vary by vehicle manufacturer.

In order for some DTCs to become Stored, a malfunction condition has to happen more than once. If the malfunction conditions are required to occur more than once, the potential malfunction is called a Pending DTC. The DTC remains pending until the malfunction condition occurs the required number of times to make the code stored. If the malfunction condition does not occur again after a set time the pending DTC will be cleared.

Acura

READING CODES

Reading the control module memory is one of the first steps in OBD II system diagnostics. This step should be initially performed to determine the general nature of the fault. Subsequent readings will determine if the fault has been cleared.

Reading codes can be performed by any of the methods below:
- Read the control module memory with the Generic Scan Tool (GST)
- Read the control module memory with the vehicle manufacturer's specific tester

To read the fault codes, connect the scan tool or tester according to the manufacturer's instructions. Follow the manufacturer's specified procedure for reading the codes.

CLEARING CODES

Control module reset procedures are a very important part of OBD II system diagnostics. This step should be done at the end of any fault code repair and at the end of any driveability repair.

Clearing codes can be performed by any of the methods below:
- Clear the control module memory with the Generic Scan Tool (GST)
- Clear the control module memory with the vehicle manufacturer's specific tester
- Turn the ignition off and remove the negative battery cable for at least 1 minute.

Removing the negative battery cable may cause other systems in the vehicle to loose their memory. Prior to removing the cable, ensure you have the proper reset codes for radios and alarms.

➡The MIL will may also be de-activated for some codes if the vehicle completes three consecutive trips without a fault detected with vehicle conditions similar to those present during the fault.

OBD II TROUBLE CODES

P0100 Mass or Volume Air Flow Circuit Malfunction

P0101 Mass or Volume Air Flow Circuit Range/Performance Problem

P0102 Mass or Volume Air Flow Circuit Low Input

P0103 Mass or Volume Air Flow Circuit High Input

P0104 Mass or Volume Air Flow Circuit Intermittent

P0105 Manifold Absolute Pressure/Barometric Pressure Circuit Malfunction

P0106 Manifold Absolute Pressure/Barometric Pressure Circuit Range/Performance Problem

P0107 Manifold Absolute Pressure/Barometric Pressure Circuit Low Input

P0108 Manifold Absolute Pressure/Barometric Pressure Circuit High Input

P0109 Manifold Absolute Pressure/Barometric Pressure Circuit Intermittent

P0110 Intake Air Temperature Circuit Malfunction

P0111 Intake Air Temperature Circuit Range/Performance Problem

P0112 Intake Air Temperature Circuit Low Input

P0113 Intake Air Temperature Circuit High Input

P0114 Intake Air Temperature Circuit Intermittent

P0115 Engine Coolant Temperature Circuit Malfunction

P0116 Engine Coolant Temperature Circuit Range/Performance Problem

P0117 Engine Coolant Temperature Circuit Low Input

P0118 Engine Coolant Temperature Circuit High Input

P0119 Engine Coolant Temperature Circuit Intermittent

P0120 Throttle/Pedal Position Sensor/Switch "A" Circuit Malfunction

P0121 Throttle/Pedal Position Sensor/Switch "A" Circuit Range/Performance Problem

P0122 Throttle/Pedal Position Sensor/Switch "A" Circuit Low Input

P0123 Throttle/Pedal Position Sensor/Switch "A" Circuit High Input

P0124 Throttle/Pedal Position Sensor/Switch "A" Circuit Intermittent

P0125 Insufficient Coolant Temperature For Closed Loop Fuel Control

P0126 Insufficient Coolant Temperature For Stable Operation

P0130 O_2 Circuit Malfunction (Bank no. 1 Sensor no. 1)

P0131 O_2 Sensor Circuit Low Voltage (Bank no. 1 Sensor no. 1)

P0132 O_2 Sensor Circuit High Voltage (Bank no. 1 Sensor no. 1)

P0133 O_2 Sensor Circuit Slow Response (Bank no. 1 Sensor no. 1)

P0134 O_2 Sensor Circuit No Activity Detected (Bank no. 1 Sensor no. 1)

P0135 O_2 Sensor Heater Circuit Malfunction (Bank no. 1 Sensor no. 1)

P0136 O_2 Sensor Circuit Malfunction (Bank no. 1 Sensor no. 2)

P0137 O_2 Sensor Circuit Low Voltage (Bank no. 1 Sensor no. 2)

P0138 O_2 Sensor Circuit High Voltage (Bank no. 1 Sensor no. 2)

P0139 O_2 Sensor Circuit Slow Response (Bank no. 1 Sensor no. 2)

P0140 O_2 Sensor Circuit No Activity Detected (Bank no. 1 Sensor no. 2)

P0141 O_2 Sensor Heater Circuit Malfunction (Bank no. 1 Sensor no. 2)

P0142 O_2 Sensor Circuit Malfunction (Bank no. 1 Sensor no. 3)

P0143 O_2 Sensor Circuit Low Voltage (Bank no. 1 Sensor no. 3)

P0144 O_2 Sensor Circuit High Voltage (Bank no. 1 Sensor no. 3)

P0145 O_2 Sensor Circuit Slow Response (Bank no. 1 Sensor no. 3)

P0146 O_2 Sensor Circuit No Activity Detected (Bank no. 1 Sensor no. 3)

P0147 O_2 Sensor Heater Circuit Malfunction (Bank no. 1 Sensor no. 3)

P0150 O_2 Sensor Circuit Malfunction (Bank no. 2 Sensor no. 1)

P0151 O_2 Sensor Circuit Low Voltage (Bank no. 2 Sensor no. 1)

P0152 O_2 Sensor Circuit High Voltage (Bank no. 2 Sensor no. 1)

P0153 O_2 Sensor Circuit Slow Response (Bank no. 2 Sensor no. 1)

P0154 O_2 Sensor Circuit No Activity Detected (Bank no. 2 Sensor no. 1)

P0155 O_2 Sensor Heater Circuit Malfunction (Bank no. 2 Sensor no. 1)

P0156 O_2 Sensor Circuit Malfunction (Bank no. 2 Sensor no. 2)

P0157 O_2 Sensor Circuit Low Voltage (Bank no. 2 Sensor no. 2)

P0158 O_2 Sensor Circuit High Voltage (Bank no. 2 Sensor no. 2)

P0159 O_2 Sensor Circuit Slow Response (Bank no. 2 Sensor no. 2)

P0160 O_2 Sensor Circuit No Activity Detected (Bank no. 2 Sensor no. 2)

P0161 O_2 Sensor Heater Circuit Malfunction (Bank no. 2 Sensor no. 2)

P0162 O_2 Sensor Circuit Malfunction (Bank no. 2 Sensor no. 3)

P0163 O_2 Sensor Circuit Low Voltage (Bank no. 2 Sensor no. 3)

P0164 O_2 Sensor Circuit High Voltage (Bank no. 2 Sensor no. 3)

P0165 O_2 Sensor Circuit Slow Response (Bank no. 2 Sensor no. 3)

P0166 O_2 Sensor Circuit No Activity Detected (Bank no. 2 Sensor no. 3)

P0167 O_2 Sensor Heater Circuit Malfunction (Bank no. 2 Sensor no. 3)

P0170 Fuel Trim Malfunction (Bank no. 1)

P0171 System Too Lean (Bank no. 1)

P0172 System Too Rich (Bank no. 1)

P0173 Fuel Trim Malfunction (Bank no. 2)

P0174 System Too Lean (Bank no. 2)

P0175 System Too Rich (Bank no. 2)

P0176 Fuel Composition Sensor Circuit Malfunction

P0177 Fuel Composition Sensor Circuit Range/Performance

P0178 Fuel Composition Sensor Circuit Low Input

P0179 Fuel Composition Sensor Circuit High Input

P0180 Fuel Temperature Sensor "A" Circuit Malfunction

P0181 Fuel Temperature Sensor "A" Circuit Range/Performance

P0182 Fuel Temperature Sensor "A" Circuit Low Input

P0183 Fuel Temperature Sensor "A" Circuit High Input

P0184 Fuel Temperature Sensor "A" Circuit Intermittent

P0185 Fuel Temperature Sensor "B" Circuit Malfunction

P0186 Fuel Temperature Sensor "B" Circuit Range/Performance

P0187 Fuel Temperature Sensor "B" Circuit Low Input

P0188 Fuel Temperature Sensor "B" Circuit High Input

P0189 Fuel Temperature Sensor "B" Circuit Intermittent

P0190 Fuel Rail Pressure Sensor Circuit Malfunction

P0191 Fuel Rail Pressure Sensor Circuit Range/Performance

P0192 Fuel Rail Pressure Sensor Circuit Low Input

P0193 Fuel Rail Pressure Sensor Circuit High Input

P0194 Fuel Rail Pressure Sensor Circuit Intermittent

P0195 Engine Oil Temperature Sensor Malfunction

P0196 Engine Oil Temperature Sensor Range/Performance

P0197 Engine Oil Temperature Sensor Low

P0198 Engine Oil Temperature Sensor High

P0199 Engine Oil Temperature Sensor Intermittent

P0200 Injector Circuit Malfunction

P0201 Injector Circuit Malfunction—Cylinder no. 1

P0202 Injector Circuit Malfunction—Cylinder no. 2
P0203 Injector Circuit Malfunction—Cylinder no. 3
P0204 Injector Circuit Malfunction—Cylinder no. 4
P0205 Injector Circuit Malfunction—Cylinder no. 5
P0206 Injector Circuit Malfunction—Cylinder no. 6
P0207 Injector Circuit Malfunction—Cylinder no. 7
P0208 Injector Circuit Malfunction—Cylinder no. 8
P0209 Injector Circuit Malfunction—Cylinder no. 9
P0210 Injector Circuit Malfunction—Cylinder no. 10
P0211 Injector Circuit Malfunction—Cylinder no. 11
P0212 Injector Circuit Malfunction—Cylinder no. 12
P0213 Cold Start Injector no. 1 Malfunction
P0214 Cold Start Injector no. 2 Malfunction
P0215 Engine Shutoff Solenoid Malfunction
P0216 Injection Timing Control Circuit Malfunction
P0217 Engine Over Temperature Condition
P0218 Transmission Over Temperature Condition
P0219 Engine Over Speed Condition
P0220 Throttle/Pedal Position Sensor/Switch "B" Circuit Malfunction
P0221 Throttle/Pedal Position Sensor/Switch "B" Circuit Range/Performance Problem
P0222 Throttle/Pedal Position Sensor/Switch "B" Circuit Low Input
P0223 Throttle/Pedal Position Sensor/Switch "B" Circuit High Input
P0224 Throttle/Pedal Position Sensor/Switch "B" Circuit Intermittent
P0225 Throttle/Pedal Position Sensor/Switch "C" Circuit Malfunction
P0226 Throttle/Pedal Position Sensor/Switch "C" Circuit Range/Performance Problem
P0227 Throttle/Pedal Position Sensor/Switch "C" Circuit Low Input
P0228 Throttle/Pedal Position Sensor/Switch "C" Circuit High Input
P0229 Throttle/Pedal Position Sensor/Switch "C" Circuit Intermittent
P0230 Fuel Pump Primary Circuit Malfunction
P0231 Fuel Pump Secondary Circuit Low
P0232 Fuel Pump Secondary Circuit High
P0233 Fuel Pump Secondary Circuit Intermittent
P0234 Engine Over Boost Condition
P0261 Cylinder no. 1 Injector Circuit Low
P0262 Cylinder no. 1 Injector Circuit High
P0263 Cylinder no. 1 Contribution/Balance Fault
P0264 Cylinder no. 2 Injector Circuit Low
P0265 Cylinder no. 2 Injector Circuit High
P0266 Cylinder no. 2 Contribution/Balance Fault
P0267 Cylinder no. 3 Injector Circuit Low
P0268 Cylinder no. 3 Injector Circuit High
P0269 Cylinder no. 3 Contribution/Balance Fault
P0270 Cylinder no. 4 Injector Circuit Low
P0271 Cylinder no. 4 Injector Circuit High
P0272 Cylinder no. 4 Contribution/Balance Fault
P0273 Cylinder no. 5 Injector Circuit Low
P0274 Cylinder no. 5 Injector Circuit High
P0275 Cylinder no. 5 Contribution/Balance Fault
P0276 Cylinder no. 6 Injector Circuit Low
P0277 Cylinder no. 6 Injector Circuit High
P0278 Cylinder no. 6 Contribution/Balance Fault

P0279 Cylinder no. 7 Injector Circuit Low
P0280 Cylinder no. 7 Injector Circuit High
P0281 Cylinder no. 7 Contribution/Balance Fault
P0282 Cylinder no. 8 Injector Circuit Low
P0283 Cylinder no. 8 Injector Circuit High
P0284 Cylinder no. 8 Contribution/Balance Fault
P0285 Cylinder no. 9 Injector Circuit Low
P0286 Cylinder no. 9 Injector Circuit High
P0287 Cylinder no. 9 Contribution/Balance Fault
P0288 Cylinder no. 10 Injector Circuit Low
P0289 Cylinder no. 10 Injector Circuit High
P0290 Cylinder no. 10 Contribution/Balance Fault
P0291 Cylinder no. 11 Injector Circuit Low
P0292 Cylinder no. 11 Injector Circuit High
P0293 Cylinder no. 11 Contribution/Balance Fault
P0294 Cylinder no. 12 Injector Circuit Low
P0295 Cylinder no. 12 Injector Circuit High
P0296 Cylinder no. 12 Contribution/Balance Fault
P0300 Random/Multiple Cylinder Misfire Detected
P0301 Cylinder no. 1—Misfire Detected
P0302 Cylinder no. 2—Misfire Detected
P0303 Cylinder no. 3—Misfire Detected
P0304 Cylinder no. 4—Misfire Detected
P0305 Cylinder no. 5—Misfire Detected
P0306 Cylinder no. 6—Misfire Detected
P0307 Cylinder no. 7—Misfire Detected
P0308 Cylinder no. 8—Misfire Detected
P0309 Cylinder no. 9—Misfire Detected
P0310 Cylinder no. 10—Misfire Detected
P0311 Cylinder no. 11—Misfire Detected
P0312 Cylinder no. 12—Misfire Detected
P0320 Ignition/Distributor Engine Speed Input Circuit Malfunction
P0321 Ignition/Distributor Engine Speed Input Circuit Range/Performance
P0322 Ignition/Distributor Engine Speed Input Circuit No Signal
P0323 Ignition/Distributor Engine Speed Input Circuit Intermittent
P0325 Knock Sensor no. 1—Circuit Malfunction (Bank no. 1 or Single Sensor)
P0326 Knock Sensor no. 1—Circuit Range/Performance (Bank no. 1 or Single Sensor)
P0327 Knock Sensor no. 1—Circuit Low Input (Bank no. 1 or Single Sensor)
P0328 Knock Sensor no. 1—Circuit High Input (Bank no. 1 or Single Sensor)
P0329 Knock Sensor no. 1—Circuit Input Intermittent (Bank no. 1 or Single Sensor)
P0330 Knock Sensor no. 2—Circuit Malfunction (Bank no. 2)
P0331 Knock Sensor no. 2—Circuit Range/Performance (Bank no. 2)
P0332 Knock Sensor no. 2—Circuit Low Input (Bank no. 2)
P0333 Knock Sensor no. 2—Circuit High Input (Bank no. 2)
P0334 Knock Sensor no. 2—Circuit Input Intermittent (Bank no. 2)
P0335 Crankshaft Position Sensor "A" Circuit Malfunction
P0336 Crankshaft Position Sensor "A" Circuit Range/Performance
P0337 Crankshaft Position Sensor "A" Circuit Low Input
P0338 Crankshaft Position Sensor "A" Circuit High Input
P0339 Crankshaft Position Sensor "A" Circuit Intermittent

P0340 Camshaft Position Sensor Circuit Malfunction
P0341 Camshaft Position Sensor Circuit Range/Performance
P0342 Camshaft Position Sensor Circuit Low Input
P0343 Camshaft Position Sensor Circuit High Input
P0344 Camshaft Position Sensor Circuit Intermittent
P0350 Ignition Coil Primary/Secondary Circuit Malfunction
P0351 Ignition Coil "A" Primary/Secondary Circuit Malfunction
P0352 Ignition Coil "B" Primary/Secondary Circuit Malfunction
P0353 Ignition Coil "C" Primary/Secondary Circuit Malfunction
P0354 Ignition Coil "D" Primary/Secondary Circuit Malfunction
P0355 Ignition Coil "E" Primary/Secondary Circuit Malfunction
P0356 Ignition Coil "F" Primary/Secondary Circuit Malfunction
P0357 Ignition Coil "G" Primary/Secondary Circuit Malfunction
P0358 Ignition Coil "H" Primary/Secondary Circuit Malfunction
P0359 Ignition Coil "I" Primary/Secondary Circuit Malfunction
P0360 Ignition Coil "J" Primary/Secondary Circuit Malfunction
P0361 Ignition Coil "K" Primary/Secondary Circuit Malfunction
P0362 Ignition Coil "L" Primary/Secondary Circuit Malfunction
P0370 Timing Reference High Resolution Signal "A" Malfunction
P0371 Timing Reference High Resolution Signal "A" Too Many Pulses
P0372 Timing Reference High Resolution Signal "A" Too Few Pulses
P0373 Timing Reference High Resolution Signal "A" Intermittent/Erratic Pulses
P0374 Timing Reference High Resolution Signal "A" No Pulses
P0375 Timing Reference High Resolution Signal "B" Malfunction
P0376 Timing Reference High Resolution Signal "B" Too Many Pulses
P0377 Timing Reference High Resolution Signal "B" Too Few Pulses
P0378 Timing Reference High Resolution Signal "B" Intermittent/Erratic Pulses
P0379 Timing Reference High Resolution Signal "B" No Pulses
P0380 Glow Plug/Heater Circuit "A" Malfunction
P0381 Glow Plug/Heater Indicator Circuit Malfunction
P0382 Glow Plug/Heater Circuit "B" Malfunction
P0385 Crankshaft Position Sensor "B" Circuit Malfunction
P0386 Crankshaft Position Sensor "B" Circuit Range/Performance
P0387 Crankshaft Position Sensor "B" Circuit Low Input
P0388 Crankshaft Position Sensor "B" Circuit High Input
P0389 Crankshaft Position Sensor "B" Circuit Intermittent
P0400 Exhaust Gas Recirculation Flow Malfunction
P0401 Exhaust Gas Recirculation Flow Insufficient Detected
P0402 Exhaust Gas Recirculation Flow Excessive Detected
P0403 Exhaust Gas Recirculation Circuit Malfunction
P0404 Exhaust Gas Recirculation Circuit Range/Performance
P0405 Exhaust Gas Recirculation Sensor "A" Circuit Low
P0406 Exhaust Gas Recirculation Sensor "A" Circuit High
P0407 Exhaust Gas Recirculation Sensor "B" Circuit Low
P0408 Exhaust Gas Recirculation Sensor "B" Circuit High
P0410 Secondary Air Injection System Malfunction
P0411 Secondary Air Injection System Incorrect Flow Detected
P0412 Secondary Air Injection System Switching Valve "A" Circuit Malfunction
P0413 Secondary Air Injection System Switching Valve "A" Circuit Open
P0414 Secondary Air Injection System Switching Valve "A" Circuit Shorted

P0415 Secondary Air Injection System Switching Valve "B" Circuit Malfunction
P0416 Secondary Air Injection System Switching Valve "B" Circuit Open
P0417 Secondary Air Injection System Switching Valve "B" Circuit Shorted
P0418 Secondary Air Injection System Relay "A" Circuit Malfunction
P0419 Secondary Air Injection System Relay "B" Circuit Malfunction
P0420 Catalyst System Efficiency Below Threshold (Bank no. 1
P0421 Warm Up Catalyst Efficiency Below Threshold (Bank no. 1)
P0422 Main Catalyst Efficiency Below Threshold (Bank no. 1)
P0423 Heated Catalyst Efficiency Below Threshold (Bank no. 1)
P0424 Heated Catalyst Temperature Below Threshold (Bank no. 1)
P0430 Catalyst System Efficiency Below Threshold (Bank no. 2
P0431 Warm Up Catalyst Efficiency Below Threshold (Bank no. 2)
P0432 Main Catalyst Efficiency Below Threshold (Bank no. 2)
P0433 Heated Catalyst Efficiency Below Threshold (Bank no. 2)
P0434 Heated Catalyst Temperature Below Threshold (Bank no. 2)
P0440 Evaporative Emission Control System Malfunction
P0441 Evaporative Emission Control System Incorrect Purge Flow
P0442 Evaporative Emission Control System Leak Detected (Small Leak)
P0443 Evaporative Emission Control System Purge Control Valve Circuit Malfunction
P0444 Evaporative Emission Control System Purge Control Valve Circuit Open
P0445 Evaporative Emission Control System Purge Control Valve Circuit Shorted
P0446 Evaporative Emission Control System Vent Control Circuit Malfunction
P0447 Evaporative Emission Control System Vent Control Circuit Open
P0448 Evaporative Emission Control System Vent Control Circuit Shorted
P0449 Evaporative Emission Control System Vent Valve/Solenoid Circuit Malfunction
P0450 Evaporative Emission Control System Pressure Sensor Malfunction
P0451 Evaporative Emission Control System Pressure Sensor Range/Performance
P0452 Evaporative Emission Control System Pressure Sensor Low Input
P0453 Evaporative Emission Control System Pressure Sensor High Input
P0454 Evaporative Emission Control System Pressure Sensor Intermittent
P0455 Evaporative Emission Control System Leak Detected (Gross Leak)
P0460 Fuel Level Sensor Circuit Malfunction
P0461 Fuel Level Sensor Circuit Range/Performance
P0462 Fuel Level Sensor Circuit Low Input
P0463 Fuel Level Sensor Circuit High Input
P0464 Fuel Level Sensor Circuit Intermittent
P0465 Purge Flow Sensor Circuit Malfunction
P0466 Purge Flow Sensor Circuit Range/Performance

P0467 Purge Flow Sensor Circuit Low Input
P0468 Purge Flow Sensor Circuit High Input
P0469 Purge Flow Sensor Circuit Intermittent
P0470 Exhaust Pressure Sensor Malfunction
P0471 Exhaust Pressure Sensor Range/Performance
P0472 Exhaust Pressure Sensor Low
P0473 Exhaust Pressure Sensor High
P0474 Exhaust Pressure Sensor Intermittent
P0475 Exhaust Pressure Control Valve Malfunction
P0476 Exhaust Pressure Control Valve Range/Performance
P0477 Exhaust Pressure Control Valve Low
P0478 Exhaust Pressure Control Valve High
P0479 Exhaust Pressure Control Valve Intermittent
P0480 Cooling Fan no. 1 Control Circuit Malfunction
P0481 Cooling Fan no. 2 Control Circuit Malfunction
P0482 Cooling Fan no. 3 Control Circuit Malfunction
P0483 Cooling Fan Rationality Check Malfunction
P0484 Cooling Fan Circuit Over Current
P0485 Cooling Fan Power/Ground Circuit Malfunction
P0500 Vehicle Speed Sensor Malfunction
P0501 Vehicle Speed Sensor Range/Performance
P0502 Vehicle Speed Sensor Circuit Low Input
P0503 Vehicle Speed Sensor Intermittent/Erratic/High
P0505 Idle Control System Malfunction
P0506 Idle Control System RPM Lower Than Expected
P0507 Idle Control System RPM Higher Than Expected
P0510 Closed Throttle Position Switch Malfunction
P0520 Engine Oil Pressure Sensor/Switch Circuit Malfunction
P0521 Engine Oil Pressure Sensor/Switch Range/Performance
P0522 Engine Oil Pressure Sensor/Switch Low Voltage
P0523 Engine Oil Pressure Sensor/Switch High Voltage
P0530 A/C Refrigerant Pressure Sensor Circuit Malfunction
P0531 A/C Refrigerant Pressure Sensor Circuit Range/Performance
P0532 A/C Refrigerant Pressure Sensor Circuit Low Input
P0533 A/C Refrigerant Pressure Sensor Circuit High Input
P0534 A/C Refrigerant Charge Loss
P0550 Power Steering Pressure Sensor Circuit Malfunction
P0551 Power Steering Pressure Sensor Circuit Range/Performance
P0552 Power Steering Pressure Sensor Circuit Low Input
P0553 Power Steering Pressure Sensor Circuit High Input
P0554 Power Steering Pressure Sensor Circuit Intermittent
P0560 System Voltage Malfunction
P0561 System Voltage Unstable
P0562 System Voltage Low
P0563 System Voltage High
P0565 Cruise Control On Signal Malfunction
P0566 Cruise Control Off Signal Malfunction
P0567 Cruise Control Resume Signal Malfunction
P0568 Cruise Control Set Signal Malfunction
P0569 Cruise Control Coast Signal Malfunction
P0570 Cruise Control Accel Signal Malfunction
P0571 Cruise Control/Brake Switch "A" Circuit Malfunction
P0572 Cruise Control/Brake Switch "A" Circuit Low
P0573 Cruise Control/Brake Switch "A" Circuit High
P0574 **Through P0580** Reserved for Cruise Codes
P0600 Serial Communication Link Malfunction
P0601 Internal Control Module Memory Check Sum Error
P0602 Control Module Programming Error
P0603 Internal Control Module Keep Alive Memory (KAM) Error

P0604 Internal Control Module Random Access Memory (RAM) Error
P0605 Internal Control Module Read Only Memory (ROM) Error
P0606 PCM Processor Fault
P0608 Control Module VSS Output "A" Malfunction
P0609 Control Module VSS Output "B" Malfunction
P0620 Generator Control Circuit Malfunction
P0621 Generator Lamp "L" Control Circuit Malfunction
P0622 Generator Field "F" Control Circuit Malfunction
P0650 Malfunction Indicator Lamp (MIL) Control Circuit Malfunction
P0654 Engine RPM Output Circuit Malfunction
P0655 Engine Hot Lamp Output Control Circuit Malfunction
P0656 Fuel Level Output Circuit Malfunction
P0700 Transmission Control System Malfunction
P0701 Transmission Control System Range/Performance
P0702 Transmission Control System Electrical
P0703 Torque Converter/Brake Switch "B" Circuit Malfunction
P0704 Clutch Switch Input Circuit Malfunction
P0705 Transmission Range Sensor Circuit Malfunction (PRNDL Input)
P0706 Transmission Range Sensor Circuit Range/Performance
P0707 Transmission Range Sensor Circuit Low Input
P0708 Transmission Range Sensor Circuit High Input
P0709 Transmission Range Sensor Circuit Intermittent
P0710 Transmission Fluid Temperature Sensor Circuit Malfunction
P0711 Transmission Fluid Temperature Sensor Circuit Range/Performance
P0712 Transmission Fluid Temperature Sensor Circuit Low Input
P0713 Transmission Fluid Temperature Sensor Circuit High Input
P0714 Transmission Fluid Temperature Sensor Circuit Intermittent
P0715 Input/Turbine Speed Sensor Circuit Malfunction
P0716 Input/Turbine Speed Sensor Circuit Range/Performance
P0717 Input/Turbine Speed Sensor Circuit No Signal
P0718 Input/Turbine Speed Sensor Circuit Intermittent
P0719 Torque Converter/Brake Switch "B" Circuit Low
P0720 Output Speed Sensor Circuit Malfunction
P0721 Output Speed Sensor Circuit Range/Performance
P0722 Output Speed Sensor Circuit No Signal
P0723 Output Speed Sensor Circuit Intermittent
P0724 Torque Converter/Brake Switch "B" Circuit High
P0725 Engine Speed Input Circuit Malfunction
P0726 Engine Speed Input Circuit Range/Performance
P0727 Engine Speed Input Circuit No Signal
P0728 Engine Speed Input Circuit Intermittent
P0730 Incorrect Gear Ratio
P0731 Gear no. 1 Incorrect Ratio
P0732 Gear no. 2 Incorrect Ratio
P0733 Gear no. 3 Incorrect Ratio
P0734 Gear no. 4 Incorrect Ratio
P0735 Gear no. 5 Incorrect Ratio
P0736 Reverse Incorrect Ratio
P0740 Torque Converter Clutch Circuit Malfunction
P0741 Torque Converter Clutch Circuit Performance or Stuck Off
P0742 Torque Converter Clutch Circuit Stuck On
P0743 Torque Converter Clutch Circuit Electrical

P0744 Torque Converter Clutch Circuit Intermittent
P0745 Pressure Control Solenoid Malfunction
P0746 Pressure Control Solenoid Performance or Stuck Off
P0747 Pressure Control Solenoid Stuck On
P0748 Pressure Control Solenoid Electrical
P0749 Pressure Control Solenoid Intermittent
P0750 Shift Solenoid "A" Malfunction
P0751 Shift Solenoid "A" Performance or Stuck Off
P0752 Shift Solenoid "A" Stuck On
P0753 Shift Solenoid "A" Electrical
P0754 Shift Solenoid "A" Intermittent
P0755 Shift Solenoid "B" Malfunction
P0756 Shift Solenoid "B" Performance or Stuck Off
P0757 Shift Solenoid "B" Stuck On
P0758 Shift Solenoid "B" Electrical
P0759 Shift Solenoid "B" Intermittent
P0760 Shift Solenoid "C" Malfunction
P0761 Shift Solenoid "C" Performance Or Stuck Off
P0762 Shift Solenoid "C" Stuck On
P0763 Shift Solenoid "C" Electrical
P0764 Shift Solenoid "C" Intermittent
P0765 Shift Solenoid "D" Malfunction
P0766 Shift Solenoid "D" Performance Or Stuck Off
P0767 Shift Solenoid "D" Stuck On
P0768 Shift Solenoid "D" Electrical
P0769 Shift Solenoid "D" Intermittent
P0770 Shift Solenoid "E" Malfunction
P0771 Shift Solenoid "E" Performance Or Stuck Off
P0772 Shift Solenoid "E" Stuck On
P0773 Shift Solenoid "E" Electrical
P0774 Shift Solenoid "E" Intermittent
P0780 Shift Malfunction
P0781 1–2 Shift Malfunction
P0782 2–3 Shift Malfunction
P0783 3–4 Shift Malfunction
P0784 4–5 Shift Malfunction
P0785 Shift/Timing Solenoid Malfunction
P0786 Shift/Timing Solenoid Range/Performance
P0787 Shift/Timing Solenoid Low
P0788 Shift/Timing Solenoid High
P0789 Shift/Timing Solenoid Intermittent
P0790 Normal/Performance Switch Circuit Malfunction
P0801 Reverse Inhibit Control Circuit Malfunction
P0803 1–4 Upshift (Skip Shift) Solenoid Control Circuit Malfunction
P0804 1–4 Upshift (Skip Shift) Lamp Control Circuit Malfunction
P1106 Map Sensor Circuit Intermittent High Voltage
P1107 MAP Sensor Circuit Intermittent Low Voltage
P1111 IAT Sensor Circuit Intermittent High Voltage
P1112 IAT Sensor Circuit Intermittent Low Voltage
P1114 ECT Sensor Circuit Intermittent Low Voltage
P1115 ECT Sensor Circuit Intermittent High Voltage
P1121 TP Sensor Circuit Intermittent High Voltage
P1122 TP Sensor Circuit Intermittent Low Voltage
P1133 HO2S-11 Insufficient Switching (Bank 1 Sensor 1)
P1134 HO2S-11 Transition Time Ratio (Bank 1 Sensor 1)
P1153 HO2S-21 Insufficient Switching (Bank 2 Sensor I)
P1154 HO2S-21 Transition Time Ratio (Bank 2 Sensor 1)
P1171 Fuel System Lean During Acceleration
P1391 G-Acceleration Sensor Intermittent Low Voltage
P1390 G-Acceleration (Low G) Sensor Performance

P1392 Rough Road G-Sensor Circuit Low Voltage
P1393 Rough Road G-Sensor Circuit High Voltage
P1394 G-Acceleration Sensor Intermittent High Voltage
P1406 EGR Valve Pintle Position Sensor Circuit Fault
P1441 EVAP System Flow During Non-Purge
P1442 EVAP System Flow During Non-Purge
P1508 Idle Speed Control System-Low
P1509 Idle Speed Control System-High
P1618 Serial Peripheral Interface Communication Error
P1640 Output Driver Module `A' Fault
P1790 PCM ROM (Transmission Side) Check Sum Error
P1792 PCM EEPROM (Transmission Side) Check Sum Error
P1835 Kick Down Switch Always On
P1850 Brake Band Apply Solenoid Electrical Fault
P1860 TCC PWM Solenoid Electrical Fault
P1870 Transmission Component Slipping

Chrysler Corp.

READING CODES

Reading the control module memory is one of the first steps in OBD II system diagnostics. This step should be initially performed to determine the general nature of the fault. Subsequent readings will determine if the fault has been cleared.

Reading codes can be performed by any of the methods below:
• Read the control module memory with the Generic Scan Tool (GST)
• Read the control module memory with the vehicle manufacturer's specific tester

To read the fault codes, connect the scan tool or tester according to the manufacturer's instructions. Follow the manufacturer's specified procedure for reading the codes.

CLEARING CODES

Control module reset procedures are a very important part of OBD II System diagnostics. This step should be done at the end of any fault code repair and at the end of any driveability repair.

Clearing codes can be performed by any of the methods below:
• Clear the control module memory with the Generic Scan Tool (GST)
• Clear the control module memory with the vehicle manufacturer's specific tester
• Turn the ignition off and remove the negative battery cable for at least 1 minute.

Removing the negative battery cable may cause other systems in the vehicle to loose their memory. Prior to removing the cable, ensure you have the proper reset codes for radios and alarms.

➡**The MIL will may also be de-activated for some codes if the vehicle completes three consecutive trips without a fault detected with vehicle conditions similar to those present during the fault.**

OBD II TROUBLE CODES

P0100 Mass or Volume Air Flow Circuit Malfunction
P0101 Mass or Volume Air Flow Circuit Range/Performance Problem
P0102 Mass or Volume Air Flow Circuit Low Input

P0103 Mass or Volume Air Flow Circuit High Input

P0104 Mass or Volume Air Flow Circuit Intermittent

P0105 Manifold Absolute Pressure/Barometric Pressure Circuit Malfunction

P0106 Manifold Absolute Pressure/Barometric Pressure Circuit Range/Performance Problem

P0107 Manifold Absolute Pressure/Barometric Pressure Circuit Low Input

P0108 Manifold Absolute Pressure/Barometric Pressure Circuit High Input

P0109 Manifold Absolute Pressure/Barometric Pressure Circuit Intermittent

P0110 Intake Air Temperature Circuit Malfunction

P0111 Intake Air Temperature Circuit Range/Performance Problem

P0112 Intake Air Temperature Circuit Low Input

P0113 Intake Air Temperature Circuit High Input

P0114 Intake Air Temperature Circuit Intermittent

P0115 Engine Coolant Temperature Circuit Malfunction

P0116 Engine Coolant Temperature Circuit Range/Performance Problem

P0117 Engine Coolant Temperature Circuit Low Input

P0118 Engine Coolant Temperature Circuit High Input

P0119 Engine Coolant Temperature Circuit Intermittent

P0120 Throttle/Pedal Position Sensor/Switch "A" Circuit Malfunction

P0121 Throttle/Pedal Position Sensor/Switch "A" Circuit Range/Performance Problem

P0122 Throttle/Pedal Position Sensor/Switch "A" Circuit Low Input

P0123 Throttle/Pedal Position Sensor/Switch "A" Circuit High Input

P0124 Throttle/Pedal Position Sensor/Switch "A" Circuit Intermittent

P0125 Insufficient Coolant Temperature For Closed Loop Fuel Control

P0126 Insufficient Coolant Temperature For Stable Operation

P0130 02 Circuit Malfunction (Bank no. 1 Sensor no. 1)

P0131 02 Sensor Circuit Low Voltage (Bank no. 1 Sensor no. 1)

P0132 02 Sensor Circuit High Voltage (Bank no. 1 Sensor no. 1)

P0133 02 Sensor Circuit Slow Response (Bank no. 1 Sensor no. 1)

P0134 02 Sensor Circuit No Activity Detected (Bank no. 1 Sensor no. 1)

P0135 02 Sensor Heater Circuit Malfunction (Bank no. 1 Sensor no. 1)

P0136 02 Sensor Circuit Malfunction (Bank no. 1 Sensor no. 2)

P0137 02 Sensor Circuit Low Voltage (Bank no. 1 Sensor no. 2)

P0138 02 Sensor Circuit High Voltage (Bank no. 1 Sensor no. 2)

P0139 02 Sensor Circuit Slow Response (Bank no. 1 Sensor no. 2)

P0140 02 Sensor Circuit No Activity Detected (Bank no. 1 Sensor no. 2)

P0141 02 Sensor Heater Circuit Malfunction (Bank no. 1 Sensor no. 2)

P0142 02 Sensor Circuit Malfunction (Bank no. 1 Sensor no. 3)

P0143 02 Sensor Circuit Low Voltage (Bank no. 1 Sensor no. 3)

P0144 02 Sensor Circuit High Voltage (Bank no. 1 Sensor no. 3)

P0145 02 Sensor Circuit Slow Response (Bank no. 1 Sensor no. 3)

P0146 02 Sensor Circuit No Activity Detected (Bank no. 1 Sensor no. 3)

P0147 02 Sensor Heater Circuit Malfunction (Bank no. 1 Sensor no. 3)

P0150 02 Sensor Circuit Malfunction (Bank no. 2 Sensor no. 1)

P0151 02 Sensor Circuit Low Voltage (Bank no. 2 Sensor no. 1)

P0152 02 Sensor Circuit High Voltage (Bank no. 2 Sensor no. 1)

P0153 02 Sensor Circuit Slow Response (Bank no. 2 Sensor no. 1)

P0154 02 Sensor Circuit No Activity Detected (Bank no. 2 Sensor no. 1)

P0155 02 Sensor Heater Circuit Malfunction (Bank no. 2 Sensor no. 1)

P0156 02 Sensor Circuit Malfunction (Bank no. 2 Sensor no. 2)

P0157 02 Sensor Circuit Low Voltage (Bank no. 2 Sensor no. 2)

P0158 02 Sensor Circuit High Voltage (Bank no. 2 Sensor no. 2)

P0159 02 Sensor Circuit Slow Response (Bank no. 2 Sensor no. 2)

P0160 02 Sensor Circuit No Activity Detected (Bank no. 2 Sensor no. 2)

P0161 02 Sensor Heater Circuit Malfunction (Bank no. 2 Sensor no. 2)

P0162 02 Sensor Circuit Malfunction (Bank no. 2 Sensor no. 3)

P0163 02 Sensor Circuit Low Voltage (Bank no. 2 Sensor no. 3)

P0164 02 Sensor Circuit High Voltage (Bank no. 2 Sensor no. 3)

P0165 02 Sensor Circuit Slow Response (Bank no. 2 Sensor no. 3)

P0166 02 Sensor Circuit No Activity Detected (Bank no. 2 Sensor no. 3)

P0167 02 Sensor Heater Circuit Malfunction (Bank no. 2 Sensor no. 3)

P0170 Fuel Trim Malfunction (Bank no. 1)

P0171 System Too Lean (Bank no. 1)

P0172 System Too Rich (Bank no. 1)

P0173 Fuel Trim Malfunction (Bank no. 2)

P0174 System Too Lean (Bank no. 2)

P0175 System Too Rich (Bank no. 2)

P0176 Fuel Composition Sensor Circuit Malfunction

P0177 Fuel Composition Sensor Circuit Range/Performance

P0178 Fuel Composition Sensor Circuit Low Input

P0179 Fuel Composition Sensor Circuit High Input

P0180 Fuel Temperature Sensor "A" Circuit Malfunction

P0181 Fuel Temperature Sensor "A" Circuit Range/Performance

P0182 Fuel Temperature Sensor "A" Circuit Low Input

P0183 Fuel Temperature Sensor "A" Circuit High Input

P0184 Fuel Temperature Sensor "A" Circuit Intermittent

P0185 Fuel Temperature Sensor "B" Circuit Malfunction

P0186 Fuel Temperature Sensor "B" Circuit Range/Performance

P0187 Fuel Temperature Sensor "B" Circuit Low Input

P0188 Fuel Temperature Sensor "B" Circuit High Input

P0189 Fuel Temperature Sensor "B" Circuit Intermittent

P0190 Fuel Rail Pressure Sensor Circuit Malfunction

P0191 Fuel Rail Pressure Sensor Circuit Range/Performance

P0192 Fuel Rail Pressure Sensor Circuit Low Input

P0193 Fuel Rail Pressure Sensor Circuit High Input

P0194 Fuel Rail Pressure Sensor Circuit Intermittent

P0195 Engine Oil Temperature Sensor Malfunction

P0196 Engine Oil Temperature Sensor Range/Performance

P0197 Engine Oil Temperature Sensor Low

P0198 Engine Oil Temperature Sensor High
P0199 Engine Oil Temperature Sensor Intermittent
P0200 Injector Circuit Malfunction
P0201 Injector Circuit Malfunction—Cylinder no. 1
P0202 Injector Circuit Malfunction—Cylinder no. 2
P0203 Injector Circuit Malfunction—Cylinder no. 3
P0204 Injector Circuit Malfunction—Cylinder no. 4
P0205 Injector Circuit Malfunction—Cylinder no. 5
P0206 Injector Circuit Malfunction—Cylinder no. 6
P0207 Injector Circuit Malfunction—Cylinder no. 7
P0208 Injector Circuit Malfunction—Cylinder no. 8
P0209 Injector Circuit Malfunction—Cylinder no. 9
P0210 Injector Circuit Malfunction—Cylinder no. 10
P0211 Injector Circuit Malfunction—Cylinder no. 11
P0212 Injector Circuit Malfunction—Cylinder no. 12
P0213 Cold Start Injector no. 1 Malfunction
P0214 Cold Start Injector no. 2 Malfunction
P0215 Engine Shutoff Solenoid Malfunction
P0216 Injection Timing Control Circuit Malfunction
P0217 Engine Over Temperature Condition
P0218 Transmission Over Temperature Condition
P0219 Engine Over Speed Condition
P0220 Throttle/Pedal Position Sensor/Switch "B" Circuit Malfunction
P0221 Throttle/Pedal Position Sensor/Switch "B" Circuit Range/Performance Problem
P0222 Throttle/Pedal Position Sensor/Switch "B" Circuit Low Input
P0223 Throttle/Pedal Position Sensor/Switch "B" Circuit High Input
P0224 Throttle/Pedal Position Sensor/Switch "B" Circuit Intermittent
P0225 Throttle/Pedal Position Sensor/Switch "C" Circuit Malfunction
P0226 Throttle/Pedal Position Sensor/Switch "C" Circuit Range/Performance Problem
P0227 Throttle/Pedal Position Sensor/Switch "C" Circuit Low Input
P0228 Throttle/Pedal Position Sensor/Switch "C" Circuit High Input
P0229 Throttle/Pedal Position Sensor/Switch "C" Circuit Intermittent
P0230 Fuel Pump Primary Circuit Malfunction
P0231 Fuel Pump Secondary Circuit Low
P0232 Fuel Pump Secondary Circuit High
P0233 Fuel Pump Secondary Circuit Intermittent
P0234 Engine Over Boost Condition
P0261 Cylinder no. 1 Injector Circuit Low
P0262 Cylinder no. 1 Injector Circuit High
P0263 Cylinder no. 1 Contribution/Balance Fault
P0264 Cylinder no. 2 Injector Circuit Low
P0265 Cylinder no. 2 Injector Circuit High
P0266 Cylinder no. 2 Contribution/Balance Fault
P0267 Cylinder no. 3 Injector Circuit Low
P0268 Cylinder no. 3 Injector Circuit High
P0269 Cylinder no. 3 Contribution/Balance Fault
P0270 Cylinder no. 4 Injector Circuit Low
P0271 Cylinder no. 4 Injector Circuit High
P0272 Cylinder no. 4 Contribution/Balance Fault
P0273 Cylinder no. 5 Injector Circuit Low
P0274 Cylinder no. 5 Injector Circuit High
P0275 Cylinder no. 5 Contribution/Balance Fault

P0276 Cylinder no. 6 Injector Circuit Low
P0277 Cylinder no. 6 Injector Circuit High
P0278 Cylinder no. 6 Contribution/Balance Fault
P0279 Cylinder no. 7 Injector Circuit Low
P0280 Cylinder no. 7 Injector Circuit High
P0281 Cylinder no. 7 Contribution/Balance Fault
P0282 Cylinder no. 8 Injector Circuit Low
P0283 Cylinder no. 8 Injector Circuit High
P0284 Cylinder no. 8 Contribution/Balance Fault
P0285 Cylinder no. 9 Injector Circuit Low
P0286 Cylinder no. 9 Injector Circuit High
P0287 Cylinder no. 9 Contribution/Balance Fault
P0288 Cylinder no. 10 Injector Circuit Low
P0289 Cylinder no. 10 Injector Circuit High
P0290 Cylinder no. 10 Contribution/Balance Fault
P0291 Cylinder no. 11 Injector Circuit Low
P0292 Cylinder no. 11 Injector Circuit High
P0293 Cylinder no. 11 Contribution/Balance Fault
P0294 Cylinder no. 12 Injector Circuit Low
P0295 Cylinder no. 12 Injector Circuit High
P0296 Cylinder no. 12 Contribution/Balance Fault
P0300 Random/Multiple Cylinder Misfire Detected
P0301 Cylinder no. 1—Misfire Detected
P0302 Cylinder no. 2—Misfire Detected
P0303 Cylinder no. 3—Misfire Detected
P0304 Cylinder no. 4—Misfire Detected
P0305 Cylinder no. 5—Misfire Detected
P0306 Cylinder no. 6—Misfire Detected
P0307 Cylinder no. 7—Misfire Detected
P0308 Cylinder no. 8—Misfire Detected
P0309 Cylinder no. 9—Misfire Detected
P0310 Cylinder no. 10—Misfire Detected
P0311 Cylinder no. 11—Misfire Detected
P0312 Cylinder no. 12—Misfire Detected
P0320 Ignition/Distributor Engine Speed Input Circuit Malfunction
P0321 Ignition/Distributor Engine Speed Input Circuit Range/Performance
P0322 Ignition/Distributor Engine Speed Input Circuit No Signal
P0323 Ignition/Distributor Engine Speed Input Circuit Intermittent
P0325 Knock Sensor no. 1—Circuit Malfunction (Bank no. 1 or Single Sensor)
P0326 Knock Sensor no. 1—Circuit Range/Performance (Bank no. 1 or Single Sensor)
P0327 Knock Sensor no. 1—Circuit Low Input (Bank no. 1 or Single Sensor)
P0328 Knock Sensor no. 1—Circuit High Input (Bank no. 1 or Single Sensor)
P0329 Knock Sensor no. 1—Circuit Input Intermittent (Bank no. 1 or Single Sensor)
P0330 Knock Sensor no. 2—Circuit Malfunction (Bank no. 2)
P0331 Knock Sensor no. 2—Circuit Range/Performance (Bank no. 2)
P0332 Knock Sensor no. 2—Circuit Low Input (Bank no. 2)
P0333 Knock Sensor no. 2—Circuit High Input (Bank no. 2)
P0334 Knock Sensor no. 2—Circuit Input Intermittent (Bank no. 2)
P0335 Crankshaft Position Sensor "A" Circuit Malfunction
P0336 Crankshaft Position Sensor "A" Circuit Range/Performance
P0337 Crankshaft Position Sensor "A" Circuit Low Input

P0338 Crankshaft Position Sensor "A" Circuit High Input
P0339 Crankshaft Position Sensor "A" Circuit Intermittent
P0340 Camshaft Position Sensor Circuit Malfunction
P0341 Camshaft Position Sensor Circuit Range/Performance
P0342 Camshaft Position Sensor Circuit Low Input
P0343 Camshaft Position Sensor Circuit High Input
P0344 Camshaft Position Sensor Circuit Intermittent
P0350 Ignition Coil Primary/Secondary Circuit Malfunction
P0351 Ignition Coil "A" Primary/Secondary Circuit Malfunction
P0352 Ignition Coil "B" Primary/Secondary Circuit Malfunction
P0353 Ignition Coil "C" Primary/Secondary Circuit Malfunction
P0354 Ignition Coil "D" Primary/Secondary Circuit Malfunction
P0355 Ignition Coil "E" Primary/Secondary Circuit Malfunction
P0356 Ignition Coil "F" Primary/Secondary Circuit Malfunction
P0357 Ignition Coil "G" Primary/Secondary Circuit Malfunction
P0358 Ignition Coil "H" Primary/Secondary Circuit Malfunction
P0359 Ignition Coil "I" Primary/Secondary Circuit Malfunction
P0360 Ignition Coil "J" Primary/Secondary Circuit Malfunction
P0361 Ignition Coil "K" Primary/Secondary Circuit Malfunction
P0362 Ignition Coil "L" Primary/Secondary Circuit Malfunction
P0370 Timing Reference High Resolution Signal "A" Malfunction
P0371 Timing Reference High Resolution Signal "A" Too Many Pulses
P0372 Timing Reference High Resolution Signal "A" Too Few Pulses
P0373 Timing Reference High Resolution Signal "A" Intermittent/Erratic Pulses
P0374 Timing Reference High Resolution Signal "A" No Pulses
P0375 Timing Reference High Resolution Signal "B" Malfunction
P0376 Timing Reference High Resolution Signal "B" Too Many Pulses
P0377 Timing Reference High Resolution Signal "B" Too Few Pulses
P0378 Timing Reference High Resolution Signal "B" Intermittent/Erratic Pulses
P0379 Timing Reference High Resolution Signal "B" No Pulses
P0380 Glow Plug/Heater Circuit "A" Malfunction
P0381 Glow Plug/Heater Indicator Circuit Malfunction
P0382 Glow Plug/Heater Circuit "B" Malfunction
P0385 Crankshaft Position Sensor "B" Circuit Malfunction
P0386 Crankshaft Position Sensor "B" Circuit Range/Performance
P0387 Crankshaft Position Sensor "B" Circuit Low Input
P0388 Crankshaft Position Sensor "B" Circuit High Input
P0389 Crankshaft Position Sensor "B" Circuit Intermittent
P0400 Exhaust Gas Recirculation Flow Malfunction
P0401 Exhaust Gas Recirculation Flow Insufficient Detected
P0402 Exhaust Gas Recirculation Flow Excessive Detected
P0403 Exhaust Gas Recirculation Circuit Malfunction
P0404 Exhaust Gas Recirculation Circuit Range/Performance
P0405 Exhaust Gas Recirculation Sensor "A" Circuit Low
P0406 Exhaust Gas Recirculation Sensor "A" Circuit High
P0407 Exhaust Gas Recirculation Sensor "B" Circuit Low
P0408 Exhaust Gas Recirculation Sensor "B" Circuit High
P0410 Secondary Air Injection System Malfunction
P0411 Secondary Air Injection System Incorrect Flow Detected
P0412 Secondary Air Injection System Switching Valve "A" Circuit Malfunction
P0413 Secondary Air Injection System Switching Valve "A" Circuit Open

P0414 Secondary Air Injection System Switching Valve "A" Circuit Shorted
P0415 Secondary Air Injection System Switching Valve "B" Circuit Malfunction
P0416 Secondary Air Injection System Switching Valve "B" Circuit Open
P0417 Secondary Air Injection System Switching Valve "B" Circuit Shorted
P0418 Secondary Air Injection System Relay "A" Circuit Malfunction
P0419 Secondary Air Injection System Relay "B" Circuit Malfunction
P0420 Catalyst System Efficiency Below Threshold (Bank no. 1)
P0421 Warm Up Catalyst Efficiency Below Threshold (Bank no. 1)
P0422 Main Catalyst Efficiency Below Threshold (Bank no. 1)
P0423 Heated Catalyst Efficiency Below Threshold (Bank no. 1)
P0424 Heated Catalyst Temperature Below Threshold (Bank no. 1)
P0430 Catalyst System Efficiency Below Threshold (Bank no. 2)
P0431 Warm Up Catalyst Efficiency Below Threshold (Bank no. 2)
P0432 Main Catalyst Efficiency Below Threshold (Bank no. 2)
P0433 Heated Catalyst Efficiency Below Threshold (Bank no. 2)
P0434 Heated Catalyst Temperature Below Threshold (Bank no. 2)
P0440 Evaporative Emission Control System Malfunction
P0441 Evaporative Emission Control System Incorrect Purge Flow
P0442 Evaporative Emission Control System Leak Detected (Small Leak)
P0443 Evaporative Emission Control System Purge Control Valve Circuit Malfunction
P0444 Evaporative Emission Control System Purge Control Valve Circuit Open
P0445 Evaporative Emission Control System Purge Control Valve Circuit Shorted
P0446 Evaporative Emission Control System Vent Control Circuit Malfunction
P0447 Evaporative Emission Control System Vent Control Circuit Open
P0448 Evaporative Emission Control System Vent Control Circuit Shorted
P0449 Evaporative Emission Control System Vent Valve/Solenoid Circuit Malfunction
P0450 Evaporative Emission Control System Pressure Sensor Malfunction
P0451 Evaporative Emission Control System Pressure Sensor Range/Performance
P0452 Evaporative Emission Control System Pressure Sensor Low Input
P0453 Evaporative Emission Control System Pressure Sensor High Input
P0454 Evaporative Emission Control System Pressure Sensor Intermittent
P0455 Evaporative Emission Control System Leak Detected (Gross Leak)
P0460 Fuel Level Sensor Circuit Malfunction
P0461 Fuel Level Sensor Circuit Range/Performance
P0462 Fuel Level Sensor Circuit Low Input
P0463 Fuel Level Sensor Circuit High Input
P0464 Fuel Level Sensor Circuit Intermittent

P0465 Purge Flow Sensor Circuit Malfunction
P0466 Purge Flow Sensor Circuit Range/Performance
P0467 Purge Flow Sensor Circuit Low Input
P0468 Purge Flow Sensor Circuit High Input
P0469 Purge Flow Sensor Circuit Intermittent
P0470 Exhaust Pressure Sensor Malfunction
P0471 Exhaust Pressure Sensor Range/Performance
P0472 Exhaust Pressure Sensor Low
P0473 Exhaust Pressure Sensor High
P0474 Exhaust Pressure Sensor Intermittent
P0475 Exhaust Pressure Control Valve Malfunction
P0476 Exhaust Pressure Control Valve Range/Performance
P0477 Exhaust Pressure Control Valve Low
P0478 Exhaust Pressure Control Valve High
P0479 Exhaust Pressure Control Valve Intermittent
P0480 Cooling Fan no. 1 Control Circuit Malfunction
P0481 Cooling Fan no. 2 Control Circuit Malfunction
P0482 Cooling Fan no. 3 Control Circuit Malfunction
P0483 Cooling Fan Rationality Check Malfunction
P0484 Cooling Fan Circuit Over Current
P0485 Cooling Fan Power/Ground Circuit Malfunction
P0500 Vehicle Speed Sensor Malfunction
P0501 Vehicle Speed Sensor Range/Performance
P0502 Vehicle Speed Sensor Circuit Low Input
P0503 Vehicle Speed Sensor Intermittent/Erratic/High
P0505 Idle Control System Malfunction
P0506 Idle Control System RPM Lower Than Expected
P0507 Idle Control System RPM Higher Than Expected
P0510 Closed Throttle Position Switch Malfunction
P0520 Engine Oil Pressure Sensor/Switch Circuit Malfunction
P0521 Engine Oil Pressure Sensor/Switch Range/Performance
P0522 Engine Oil Pressure Sensor/Switch Low Voltage
P0523 Engine Oil Pressure Sensor/Switch High Voltage
P0530 A/C Refrigerant Pressure Sensor Circuit Malfunction
P0531 A/C Refrigerant Pressure Sensor Circuit Range/Performance
P0532 A/C Refrigerant Pressure Sensor Circuit Low Input
P0533 A/C Refrigerant Pressure Sensor Circuit High Input
P0534 A/C Refrigerant Charge Loss
P0550 Power Steering Pressure Sensor Circuit Malfunction
P0551 Power Steering Pressure Sensor Circuit Range/Performance
P0552 Power Steering Pressure Sensor Circuit Low Input
P0553 Power Steering Pressure Sensor Circuit High Input
P0554 Power Steering Pressure Sensor Circuit Intermittent
P0560 System Voltage Malfunction
P0561 System Voltage Unstable
P0562 System Voltage Low
P0563 System Voltage High
P0565 Cruise Control On Signal Malfunction
P0566 Cruise Control Off Signal Malfunction
P0567 Cruise Control Resume Signal Malfunction
P0568 Cruise Control Set Signal Malfunction
P0569 Cruise Control Coast Signal Malfunction
P0570 Cruise Control Accel Signal Malfunction
P0571 Cruise Control/Brake Switch "A" Circuit Malfunction
P0572 Cruise Control/Brake Switch "A" Circuit Low
P0573 Cruise Control/Brake Switch "A" Circuit High
P0574 Through P0580 Reserved for Cruise Codes
P0600 Serial Communication Link Malfunction
P0601 Internal Control Module Memory Check Sum Error
P0602 Control Module Programming Error

P0603 Internal Control Module Keep Alive Memory (KAM) Error
P0604 Internal Control Module Random Access Memory (RAM) Error
P0605 Internal Control Module Read Only Memory (ROM) Error
P0606 PCM Processor Fault
P0608 Control Module VSS Output "A" Malfunction
P0609 Control Module VSS Output "B" Malfunction
P0620 Generator Control Circuit Malfunction
P0621 Generator Lamp "L" Control Circuit Malfunction
P0622 Generator Field "F" Control Circuit Malfunction
P0650 Malfunction Indicator Lamp (MIL) Control Circuit Malfunction
P0654 Engine RPM Output Circuit Malfunction
P0655 Engine Hot Lamp Output Control Circuit Malfunction
P0656 Fuel Level Output Circuit Malfunction
P0700 Transmission Control System Malfunction
P0701 Transmission Control System Range/Performance
P0702 Transmission Control System Electrical
P0703 Torque Converter/Brake Switch "B" Circuit Malfunction
P0704 Clutch Switch Input Circuit Malfunction
P0705 Transmission Range Sensor Circuit Malfunction (PRNDL Input)
P0706 Transmission Range Sensor Circuit Range/Performance
P0707 Transmission Range Sensor Circuit Low Input
P0708 Transmission Range Sensor Circuit High Input
P0709 Transmission Range Sensor Circuit Intermittent
P0710 Transmission Fluid Temperature Sensor Circuit Malfunction
P0711 Transmission Fluid Temperature Sensor Circuit Range/Performance
P0712 Transmission Fluid Temperature Sensor Circuit Low Input
P0713 Transmission Fluid Temperature Sensor Circuit High Input
P0714 Transmission Fluid Temperature Sensor Circuit Intermittent
P0715 Input/Turbine Speed Sensor Circuit Malfunction
P0716 Input/Turbine Speed Sensor Circuit Range/Performance
P0717 Input/Turbine Speed Sensor Circuit No Signal
P0718 Input/Turbine Speed Sensor Circuit Intermittent
P0719 Torque Converter/Brake Switch "B" Circuit Low
P0720 Output Speed Sensor Circuit Malfunction
P0721 Output Speed Sensor Circuit Range/Performance
P0722 Output Speed Sensor Circuit No Signal
P0723 Output Speed Sensor Circuit Intermittent
P0724 Torque Converter/Brake Switch "B" Circuit High
P0725 Engine Speed Input Circuit Malfunction
P0726 Engine Speed Input Circuit Range/Performance
P0727 Engine Speed Input Circuit No Signal
P0728 Engine Speed Input Circuit Intermittent
P0730 Incorrect Gear Ratio
P0731 Gear no. 1 Incorrect Ratio
P0732 Gear no. 2 Incorrect Ratio
P0733 Gear no. 3 Incorrect Ratio
P0734 Gear no. 4 Incorrect Ratio
P0735 Gear no. 5 Incorrect Ratio
P0736 Reverse Incorrect Ratio
P0740 Torque Converter Clutch Circuit Malfunction
P0741 Torque Converter Clutch Circuit Performance or Stuck Off
P0742 Torque Converter Clutch Circuit Stuck On
P0743 Torque Converter Clutch Circuit Electrical

P0744 Torque Converter Clutch Circuit Intermittent
P0745 Pressure Control Solenoid Malfunction
P0746 Pressure Control Solenoid Performance or Stuck Off
P0747 Pressure Control Solenoid Stuck On
P0748 Pressure Control Solenoid Electrical
P0749 Pressure Control Solenoid Intermittent
P0750 Shift Solenoid "A" Malfunction
P0751 Shift Solenoid "A" Performance or Stuck Off
P0752 Shift Solenoid "A" Stuck On
P0753 Shift Solenoid "A" Electrical
P0754 Shift Solenoid "A" Intermittent
P0755 Shift Solenoid "B" Malfunction
P0756 Shift Solenoid "B" Performance or Stuck Off
P0757 Shift Solenoid "B" Stuck On
P0758 Shift Solenoid "B" Electrical
P0759 Shift Solenoid "B" Intermittent
P0760 Shift Solenoid "C" Malfunction
P0761 Shift Solenoid "C" Performance Or Stuck Off
P0762 Shift Solenoid "C" Stuck On
P0763 Shift Solenoid "C" Electrical
P0764 Shift Solenoid "C" Intermittent
P0765 Shift Solenoid "D" Malfunction
P0766 Shift Solenoid "D" Performance Or Stuck Off
P0767 Shift Solenoid "D" Stuck On
P0768 Shift Solenoid "D" Electrical
P0769 Shift Solenoid "D" Intermittent
P0770 Shift Solenoid "E" Malfunction
P0771 Shift Solenoid "E" Performance Or Stuck Off
P0772 Shift Solenoid "E" Stuck On
P0773 Shift Solenoid "E" Electrical
P0774 Shift Solenoid "E" Intermittent
P0780 Shift Malfunction
P0781 1–2 Shift Malfunction
P0782 2–3 Shift Malfunction
P0783 3–4 Shift Malfunction
P0784 4–5 Shift Malfunction
P0785 Shift/Timing Solenoid Malfunction
P0786 Shift/Timing Solenoid Range/Performance
P0787 Shift/Timing Solenoid Low
P0788 Shift/Timing Solenoid High
P0789 Shift/Timing Solenoid Intermittent
P0790 Normal/Performance Switch Circuit Malfunction
P0801 Reverse Inhibit Control Circuit Malfunction
P0803 1–4 Upshift (Skip Shift) Solenoid Control Circuit Malfunction
P0804 1–4 Upshift (Skip Shift) Lamp Control Circuit Malfunction
P1290 CNG Fuel System Pressure Too High (3.3L CNG vehicles only)
P1291 No Temp Rise Seen From Intake Air Heaters
P1292 CNG Pressure Sensor Voltage Too High (3.3L CNG vehicles only)
P1293 CNG Pressure Sensor Voltage Too Low (3.3L CNG vehicles only)
P1294 Target Idle Not Reached
P1296 No 5-Volts To MAP Sensor
P1297 No Change In MAP From Start To Run
P1391 Intermittent Loss Of CMP Or CKP

P1398 Misfire Adaptive Numerator At Limit
P1486 EVAP Leak Monitor Pinched Hose Or Obstruction Found
P1491 Radiator Fan Control Relay Circuit
P1492 Battery Temp Sensor Voltage Too High
P1493 Battery Temp Sensor Voltage Too Low
P1494 Leak Detection Pump Pressure Switch Or Mechanical Fault
P1495 Leak Detection Pump Solenoid Circuit
P1498 Auxiliary 5-Volt Supply Output Too Low
P1697 PCM Failure SRI Mile Not Stored
P1698 PCM Failure EEPROM Write Denied
P1756 Governor Pressure Not Equal To Target @ 15–20 PSI
P1757 Governor Pressure Above 3 PSI In Gear With 0 MPH
P1762 Governor Pressure Sensor Offset Volts Too Low Or High
P1763 Governor Pressure Sensor Volts Too High
P1764 Governor Pressure Sensor Volts Too Low
P1765 Trans 12-Volt Supply Relay Control Circuit
P1899 P/N Switch Stuck In Park Or In Gear

Ford Motor Co.

➡**The Mercury Villager is covered under the Nissan section since it shares a platform with the Nissan Quest.**

READING CODES

Reading the control module memory is one of the first steps in OBD II system diagnostics. This step should be initially performed to determine the general nature of the fault. Subsequent readings will determine if the fault has been cleared.

Reading codes can be performed by any of the methods below:
- Read the control module memory with the Generic Scan Tool (GST)
- Read the control module memory with the vehicle manufacturer's specific tester

To read the fault codes, connect the scan tool or tester according to the manufacturer's instructions. Follow the manufacturer's specified procedure for reading the codes.

CLEARING CODES

Control module reset procedures are a very important part of OBD II System diagnostics. This step should be done at the end of any fault code repair and at the end of any driveability repair.

Clearing codes can be performed by any of the methods below:
- Clear the control module memory with the Generic Scan Tool (GST)
- Clear the control module memory with the vehicle manufacturer's specific tester
- Turn the ignition off and remove the negative battery cable for at least 1 minute.

Removing the negative battery cable may cause other systems in the vehicle to loose their memory. Prior to removing the cable, ensure you have the proper reset codes for radios and alarms.

➡**The MIL will may also be de-activated for some codes if the vehicle completes three consecutive trips without a fault detected with vehicle conditions similar to those present during the fault.**

OBD II TROUBLE CODES

1995 Models

P0102 Mass Air Flow (MAF) Sensor circuit low input
P0103 Mass Air Flow (MAF) Sensor circuit high input
P0112 Intake Air Temperature (IAT) Sensor circuit low input
P0113 Intake Air Temperature (IAT) Sensor high input
P0117 Engine Coolant Temperature (ECT) low input
P0118 Engine Coolant Temperature (ECT) Sensor circuit high input
P0121 In range operating Throttle Position (TP) sensor circuit failure
P0122 Throttle Position (TP) Sensor circuit low input
P0123 Throttle Position (TP) Sensor high input
P0125 Insufficient coolant temperature to enter closed loop fuel control
P0126 Insufficient coolant temperature for stable operation
P0131 Upstream Heated Oxygen Sensor (HO2S 11) circuit out of range low voltage (bank #1)
P0132 Upstream Heated Oxygen Sensor (HO2S 11) circuit high voltage (Bank #1)
P0133 Upstream Heated Oxygen Sensor (HO2S 11) circuit slow response (Bank #1)
P0135 Heated Oxygen Sensor Heater (HTR 11) circuit malfunction
P0136 Downstream Heated Oxygen Sensor (HO2S 12) circuit malfunction (Bank #1
P0138 Downstream Heated Oxygen Sensor (HO2S 12) circuit high voltage (Bank #1)
P0140 Heated Oxygen Sensor (HO2S 12) circuit no activity detected (Bank #1)
P0141 Heated Oxygen Sensor Heater (HTR 12) circuit malfunction
P0151 Upstream Heated Oxygen Sensor (HO2S 21) circuit out of range low voltage (Bank #2)
P0152 Upstream Heated Oxygen Sensor (HO2S 21) circuit high voltage (Bank #2)
P0153 Upstream Heated Oxygen Sensor (HO2S 21) circuit slow response (Bank #2)
P0155 Heated Oxygen Sensor Heater (HTR 21) circuit malfunction
P0156 Downstream Heated Oxygen Sensor (HO2S 22) circuit malfunction (Bank #2)
P0158 Downstream Heated Oxygen Sensor (HO2S 22) circuit high voltage (Bank #2)
P0160 Heated Oxygen Sensor (HO2S 12) circuit no activity detected (Bank #2)
P0161 Heated Oxygen Sensor Heater (HTR 22) circuit malfunction
P0171 System (adaptive fuel) too lean (Bank #1)
P0172 System (adaptive fuel) too rich (Bank #1)
P0174 System (adaptive fuel) too lean (Bank #2)
P0175 System (adaptive fuel) too rich (Bank #2)
P0222 Throttle Position Sensor B (TP-B) circuit low input
P0223 Throttle Position Sensor B (TP-B) circuit high input
P0230 Fuel pump primary circuit malfunction
P0231 Fuel pump secondary circuit low
P0232 Fuel pump secondary circuit high
P0300 Random misfire detected
P0301 Cylinder #1 misfire detected
P0302 Cylinder #2 misfire detected

P0303 Cylinder #3 misfire detected
P0304 Cylinder #4 misfire detected
P0305 Cylinder #5 misfire detected
P0306 Cylinder #6 misfire detected
P0307 Cylinder #7 misfire detected
P0308 Cylinder #8 misfire detected
P0320 Ignition engine speed (Profile Ignition Pick-up) input circuit malfunction
P0340 Camshaft Position (CMP) sensor circuit malfunction (CID)
P0350 Ignition Coil primary circuit malfunction
P0351 Ignition Coil A primary circuit malfunction
P0352 Ignition Coil B primary circuit malfunction
P0353 Ignition Coil C primary circuit malfunction
P0354 Ignition Coil D primary circuit malfunction
P0400 Exhaust Gas Recirculation (EGR) flow malfunction
P0401 Exhaust Gas Recirculation (EGR) flow insufficient detected
P0402 Exhaust Gas Recirculation (EGR) excess flow detected (valve open at idle)
P0411 Secondary Air Injection system incorrect flow detected
P0412 Secondary Air Injection system control valve malfunction
P0420 Catalyst system efficiency below threshold (Bank #1)
P0430 Catalyst system efficiency below threshold (Bank #2)
P0443 Evaporative emission control system Canister Purge (CANP) Control Valve circuit malfunction
P0500 Vehicle Speed Sensor (VSS) malfunction
P0505 Idle Air Control (IAC) system malfunction
P0603 Powertrain Control Module (PCM)—Keep Alive Memory (KAM) test error
P0605 Powertrain Control Module (PCM)—Read Only Memory (ROM) test error
P0704 Clutch Pedal Position (CPP) switch input circuit malfunction
P0703 Brake On/Off (BOO) switch input malfunction
P0707 Manual Lever Position (MLP) sensor circuit low input
P0708 Manual Lever Position (MLP) sensor circuit high input
P0712 Transmission Fluid Temperature (TFT) sensor circuit low input
P0713 Transmission Fluid Temperature (TFT) sensor circuit high input
P0715 Turbine Shaft Speed (TSS) sensor circuit malfunction
P0720 Output Shaft Speed (OSS) sensor circuit malfunction
P0731 Incorrect ratio for first gear
P0732 Incorrect ratio for second gear
P0733 Incorrect ratio for third gear
P0734 Incorrect ratio for fourth gear
P0736 Reverse incorrect gear
P0741 Torque Converter Clutch (TCC) system incorrect mechanical performance
P0746 Electronic Pressure Control (EPC) solenoid performance
P0743 Torque Converter Clutch (TCC) system electrical failure
P0750 Shift Solenoid #1 (SS1) circuit malfunction
P0751 Shift Solenoid #1 (SS1) performance
P0755 Shift Solenoid #2 (SS2) circuit malfunction
P0756 Shift Solenoid #2 (SS2) performance
P0760 Shift Solenoid #3 (SS3) circuit malfunction
P0761 Shift Solenoid #3 (SS3) performance
P0781 1 to 2 shift error
P0782 2 to 3 shift error
P0783 3 to 4 shift error
P0784 4 to 5 shift error

P1000 OBD II Monitor Testing not complete
U1039 OBD II Monitor not complete
UI051 Brake switch signal missing or incorrect
P1100 Mass Air Flow (MAF) sensor intermittent
P1101 Mass Air Flow (MAF) sensor out of Self-Test range
P1112 Intake Air Temperature (IAT) sensor intermittent
P1116 Engine Coolant Temperature (ECT) sensor out of Self-Test range
P1117 Engine Coolant Temperature (ECT) sensor intermittent
P1120 Throttle Position (TP) sensor out of range low
P1121 Throttle Position (TP) sensor inconsistent with MAF sensor
P1124 Throttle Position (TP) sensor out of Self-Test range
P1125 Throttle Position (TP) sensor circuit intermittent
P1130 Lack of HO2S 11 switch, adaptive fuel at limit
P1131 Lack of HO2S 11 switch, sensor indicates lean (Bank #1)
P1132 Lack of HO2S 11 switch, sensor indicates rich (Bank #1)
U1135 Ignition switch signal missing or incorrect
P1137 Lack of HO2S 12 switch, sensor indicates lean (Bank #1)
P1138 Lack of HO2S 12 switch, sensor indicates rich (Bank #1)
P1150 Lack of HO2S 21 switch, adaptive fuel at limit
P1151 Lack of HO2S 21 switch, sensor indicates lean (Bank #2)
P1152 Lack of HO2S 21 switch, sensor indicates rich (Bank #2)
P1157 Lack of HO2S 22 switch, sensor indicates lean (Bank #2)
P1158 Lack of HO2S 22 switch, sensor indicates rich (Bank #2)
P1220 Series Throttle Control malfunction
P1224 Throttle Position Sensor (TP-B) out of Self-test range
P1233 Fuel Pump driver Module off-line
P1234 Fuel Pump driver Module off-line
P1235 Fuel Pump control out of range
P1236 Fuel Pump control out of range
P1237 Fuel Pump secondary circuit malfunction
P1238 Fuel Pump secondary circuit malfunction
P1260 THEFT detected—engine disabled
P1270 Engine RPM or vehicle speed limiter reached
P1351 Ignition Diagnostic Monitor (IDM) circuit input malfunction
P1352 Ignition coil A primary circuit malfunction
P1353 Ignition coil B primary circuit malfunction
P1354 Ignition coil C primary circuit malfunction
P1355 Ignition coil D primary circuit malfunction
P1358 Ignition Diagnostic Monitor (IDM) signal out of Self-Test range
P1359 Spark output circuit malfunction
P1364 Ignition coil primary circuit malfunction
P1390 Octane Adjust (OCT ADJ) out of Self-Test range
P1400 Differential Pressure Feedback Electronic (DPFE) sensor circuit low voltage detected
P1401 Differential Pressure Feedback Electronic (DPFE) sensor circuit high voltage detected
P1403 Differential Pressure Feedback Electronic (DPFE) sensor hoses reversed
P1405 Differential Pressure Feedback Electronic (DPFE) sensor upstream hose off or plugged
P1406 Differential Pressure Feedback Electronic (DPFE) sensor downstream hose off or plugged
P1407 Exhaust Gas Recirculation (EGR) no flow detected (valve stuck closed or inoperative)
P1408 Exhaust Gas Recirculation (EGR) flow out of Self-Test range
P1409 Electronic Vacuum Regulator (EVR) control circuit malfunction

P1414 Secondary Air Injection system monitor circuit high voltage
P1443 Evaporative emission control system—vacuum system purge control solenoid or purge control valve malfunction
P1444 Purge Flow Sensor (PFS) circuit low input
P1445 Purge Flow Sensor (PFS) circuit high input
U1451 Lack of response from Passive Anti-Theft system (PATS) module—engine disabled
P1460 Wide Open Throttle Air Conditioning Cut-off (WAC) circuit malfunction
P1461 Air Conditioning Pressure (ACP) sensor circuit low input
P1462 Air Conditioning Pressure (ACP) sensor circuit high input
P1463 Air Conditioning Pressure (ACP) sensor insufficient pressure change
P1469 Low air conditioning cycling period
P1473 Fan Secondary High with fan(s) off
P1474 Low Fan Control primary circuit malfunction
P1479 High Fan Control primary circuit malfunction
P1480 Fan Secondary low with low fan on
P1481 Fan Secondary low with high fan on
P1500 Vehicle Speed Sensor (VSS) circuit intermittent
P1505 Idle Air Control (IAC) system at adaptive clip
P1506 Idle Air control (IAC) overspeed error
P1518 Intake Manifold Runner Control (IMRC) malfunction (stuck open)
P1519 Intake Manifold Runner Control (IMRC) malfunction (stuck closed)
P1520 Intake Manifold Runner Control (IMRC) circuit malfunction
P1507 Idle Air control (IAC) under speed error
P1605 Powertrain Control Module (POM)—Keep Alive Memory (KAM) test error
P1650 Power steering Pressure (PSP) switch out of Self-Test range
P1651 Power steering Pressure (PSP) switch input malfunction
P1701 Reverse engagement error
P1703 Brake On/Off (BOO) switch out of Self-Test range
P1705 Manual Lever Position (MLP) sensor out of Self-Test range
P1709 Park or Neutral Position (PNP) switch out of Self-test range
P1729 4X4 Low switch error
P1711 Transmission Fluid Temperature (TFT) sensor out of Self-Test range
P1741 Torque Converter Clutch (TCC) control error
P1742 Torque Converter Clutch (TCC) solenoid mechanically failed (turns MIL on)
P1743 Torque Converter Clutch (TCC) solenoid mechanically failed (turns TOIL on)
P1744 Torque Converter Clutch (TCC) system mechanically stuck in off position
P1748 Electronic Pressure Control (EPC) solenoid circuit low input (open circuit)
P1747 Electronic Pressure Control (EPC) solenoid circuit high input (short circuit)
P1749 Electric Pressure Control (EPC) solenoid failed low
P1751 Shift Solenoid #1(SS1) performance
P1756 Shift Solenoid #2 (SS2) performance
P1780 Transmission Control Switch (TCS) circuit out of Self-Test range

1996–99 Models

P0000 No Failures
P0100 Mass or Volume Air Flow Circuit Malfunction
P0101 Mass or Volume Air Flow Circuit Range/Performance Problem
P0102 Mass or Volume Air Flow Circuit Low Input
P0103 Mass or Volume Air Flow Circuit High Input
P0104 Mass or Volume Air Flow Circuit Intermittent
P0105 Manifold Absolute Pressure/Barometric Pressure Circuit Malfunction
P0106 Manifold Absolute Pressure/Barometric Pressure Circuit Range/Performance Problem
P0107 Manifold Absolute Pressure/Barometric Pressure Circuit Low Input
P0108 Manifold Absolute Pressure/Barometric Pressure Circuit High Input
P0109 Manifold Absolute Pressure/Barometric Pressure Circuit Intermittent
P0110 Intake Air Temperature Circuit Malfunction
P0111 Intake Air Temperature Circuit Range/Performance Problem
P0112 Intake Air Temperature Circuit Low Input
P0113 Intake Air Temperature Circuit High Input
P0114 Intake Air Temperature Circuit Intermittent
P0115 Engine Coolant Temperature Circuit Malfunction
P0116 Engine Coolant Temperature Circuit Range/Performance Problem
P0117 Engine Coolant Temperature Circuit Low Input
P0118 Engine Coolant Temperature Circuit High Input
P0119 Engine Coolant Temperature Circuit Intermittent
P0120 Throttle/Pedal Position Sensor/Switch "A" Circuit Malfunction
P0121 Throttle/Pedal Position Sensor/Switch "A" Circuit Range/Performance Problem
P0122 Throttle/Pedal Position Sensor/Switch "A" Circuit Low Input
P0123 Throttle/Pedal Position Sensor/Switch "A" Circuit High Input
P0124 Throttle/Pedal Position Sensor/Switch "A" Circuit Intermittent
P0125 Insufficient Coolant Temperature For Closed Loop Fuel Control
P0126 Insufficient Coolant Temperature For Stable Operation
P0130 O2 Circuit Malfunction (Bank no. 1 Sensor no. 1)
P0131 O2 Sensor Circuit Low Voltage (Bank no. 1 Sensor no. 1)
P0132 O2 Sensor Circuit High Voltage (Bank no. 1 Sensor no. 1)
P0133 O2 Sensor Circuit Slow Response (Bank no. 1 Sensor no. 1)
P0134 O2 Sensor Circuit No Activity Detected (Bank no. 1 Sensor no. 1)
P0135 O2 Sensor Heater Circuit Malfunction (Bank no. 1 Sensor no. 1)
P0136 O2 Sensor Circuit Malfunction (Bank no. 1 Sensor no. 2)
P0137 O2 Sensor Circuit Low Voltage (Bank no. 1 Sensor no. 2)
P0138 O2 Sensor Circuit High Voltage (Bank no. 1 Sensor no. 2)
P0139 O2 Sensor Circuit Slow Response (Bank no. 1 Sensor no. 2)
P0140 O2 Sensor Circuit No Activity Detected (Bank no. 1 Sensor no. 2)

P0141 O2 Sensor Heater Circuit Malfunction (Bank no. 1 Sensor no. 2)
P0142 O2 Sensor Circuit Malfunction (Bank no. 1 Sensor no. 3)
P0143 O2 Sensor Circuit Low Voltage (Bank no. 1 Sensor no. 3)
P0144 O2 Sensor Circuit High Voltage (Bank no. 1 Sensor no. 3)
P0145 O2 Sensor Circuit Slow Response (Bank no. 1 Sensor no. 3)
P0146 O2 Sensor Circuit No Activity Detected (Bank no. 1 Sensor no. 3)
P0147 O2 Sensor Heater Circuit Malfunction (Bank no. 1 Sensor no. 3)
P0150 O2 Sensor Circuit Malfunction (Bank no. 2 Sensor no. 1)
P0151 O2 Sensor Circuit Low Voltage (Bank no. 2 Sensor no. 1)
P0152 O2 Sensor Circuit High Voltage (Bank no. 2 Sensor no. 1)
P0153 O2 Sensor Circuit Slow Response (Bank no. 2 Sensor no. 1)
P0154 O2 Sensor Circuit No Activity Detected (Bank no. 2 Sensor no. 1)
P0155 O2 Sensor Heater Circuit Malfunction (Bank no. 2 Sensor no. 1)
P0156 O2 Sensor Circuit Malfunction (Bank no. 2 Sensor no. 2)
P0157 O2 Sensor Circuit Low Voltage (Bank no. 2 Sensor no. 2)
P0158 O2 Sensor Circuit High Voltage (Bank no. 2 Sensor no. 2)
P0159 O2 Sensor Circuit Slow Response (Bank no. 2 Sensor no. 2)
P0160 O2 Sensor Circuit No Activity Detected (Bank no. 2 Sensor no. 2)
P0161 O2 Sensor Heater Circuit Malfunction (Bank no. 2 Sensor no. 2)
P0162 O2 Sensor Circuit Malfunction (Bank no. 2 Sensor no. 3)
P0163 O2 Sensor Circuit Low Voltage (Bank no. 2 Sensor no. 3)
P0164 O2 Sensor Circuit High Voltage (Bank no. 2 Sensor no. 3)
P0165 O2 Sensor Circuit Slow Response (Bank no. 2 Sensor no. 3)
P0166 O2 Sensor Circuit No Activity Detected (Bank no. 2 Sensor no. 3)
P0167 O2 Sensor Heater Circuit Malfunction (Bank no. 2 Sensor no. 3)
P0170 Fuel Trim Malfunction (Bank no. 1)
P0171 System Too Lean (Bank no. 1)
P0172 System Too Rich (Bank no. 1)
P0173 Fuel Trim Malfunction (Bank no. 2)
P0174 System Too Lean (Bank no. 2)
P0175 System Too Rich (Bank no. 2)
P0176 Fuel Composition Sensor Circuit Malfunction
P0177 Fuel Composition Sensor Circuit Range/Performance
P0178 Fuel Composition Sensor Circuit Low Input
P0179 Fuel Composition Sensor Circuit High Input
P0180 Fuel Temperature Sensor "A" Circuit Malfunction
P0181 Fuel Temperature Sensor "A" Circuit Range/Performance
P0182 Fuel Temperature Sensor "A" Circuit Low Input
P0183 Fuel Temperature Sensor "A" Circuit High Input
P0184 Fuel Temperature Sensor "A" Circuit Intermittent
P0185 Fuel Temperature Sensor "B" Circuit Malfunction
P0186 Fuel Temperature Sensor "B" Circuit Range/Performance
P0187 Fuel Temperature Sensor "B" Circuit Low Input
P0188 Fuel Temperature Sensor "B" Circuit High Input
P0189 Fuel Temperature Sensor "B" Circuit Intermittent

P0190 Fuel Rail Pressure Sensor Circuit Malfunction
P0191 Fuel Rail Pressure Sensor Circuit Range/Performance
P0192 Fuel Rail Pressure Sensor Circuit Low Input
P0193 Fuel Rail Pressure Sensor Circuit High Input
P0194 Fuel Rail Pressure Sensor Circuit Intermittent
P0195 Engine Oil Temperature Sensor Malfunction
P0196 Engine Oil Temperature Sensor Range/Performance
P0197 Engine Oil Temperature Sensor Low
P0198 Engine Oil Temperature Sensor High
P0199 Engine Oil Temperature Sensor Intermittent
P0200 Injector Circuit Malfunction
P0201 Injector Circuit Malfunction—Cylinder no. 1
P0202 Injector Circuit Malfunction—Cylinder no. 2
P0203 Injector Circuit Malfunction—Cylinder no. 3
P0204 Injector Circuit Malfunction—Cylinder no. 4
P0205 Injector Circuit Malfunction—Cylinder no. 5
P0206 Injector Circuit Malfunction—Cylinder no. 6
P0207 Injector Circuit Malfunction—Cylinder no. 7
P0208 Injector Circuit Malfunction—Cylinder no. 8
P0209 Injector Circuit Malfunction—Cylinder no. 9
P0210 Injector Circuit Malfunction—Cylinder no. 10
P0211 Injector Circuit Malfunction—Cylinder no. 11
P0212 Injector Circuit Malfunction—Cylinder no. 12
P0213 Cold Start Injector no. 1 Malfunction
P0214 Cold Start Injector no. 2 Malfunction
P0215 Engine Shutoff Solenoid Malfunction
P0216 Injection Timing Control Circuit Malfunction
P0217 Engine Over Temperature Condition
P0218 Transmission Over Temperature Condition
P0219 Engine Over Speed Condition
P0220 Throttle/Pedal Position Sensor/Switch "B" Circuit Malfunction
P0221 Throttle/Pedal Position Sensor/Switch "B" Circuit Range/Performance Problem
P0222 Throttle/Pedal Position Sensor/Switch "B" Circuit Low Input
P0223 Throttle/Pedal Position Sensor/Switch "B" Circuit High Input
P0224 Throttle/Pedal Position Sensor/Switch "B" Circuit Intermittent
P0225 Throttle/Pedal Position Sensor/Switch "C" Circuit Malfunction
P0226 Throttle/Pedal Position Sensor/Switch "C" Circuit Range/Performance Problem
P0227 Throttle/Pedal Position Sensor/Switch "C" Circuit Low Input
P0228 Throttle/Pedal Position Sensor/Switch "C" Circuit High Input
P0229 Throttle/Pedal Position Sensor/Switch "C" Circuit Intermittent
P0230 Fuel Pump Primary Circuit Malfunction
P0231 Fuel Pump Secondary Circuit Low
P0232 Fuel Pump Secondary Circuit High
P0233 Fuel Pump Secondary Circuit Intermittent
P0234 Engine Over Boost Condition
P0261 Cylinder no. 1 Injector Circuit Low
P0262 Cylinder no. 1 Injector Circuit High
P0263 Cylinder no. 1 Contribution/Balance Fault
P0264 Cylinder no. 2 Injector Circuit Low
P0265 Cylinder no. 2 Injector Circuit High
P0266 Cylinder no. 2 Contribution/Balance Fault
P0267 Cylinder no. 3 Injector Circuit Low

P0268 Cylinder no. 3 Injector Circuit High
P0269 Cylinder no. 3 Contribution/Balance Fault
P0270 Cylinder no. 4 Injector Circuit Low
P0271 Cylinder no. 4 Injector Circuit High
P0272 Cylinder no. 4 Contribution/Balance Fault
P0273 Cylinder no. 5 Injector Circuit Low
P0274 Cylinder no. 5 Injector Circuit High
P0275 Cylinder no. 5 Contribution/Balance Fault
P0276 Cylinder no. 6 Injector Circuit Low
P0277 Cylinder no. 6 Injector Circuit High
P0278 Cylinder no. 6 Contribution/Balance Fault
P0279 Cylinder no. 7 Injector Circuit Low
P0280 Cylinder no. 7 Injector Circuit High
P0281 Cylinder no. 7 Contribution/Balance Fault
P0282 Cylinder no. 8 Injector Circuit Low
P0283 Cylinder no. 8 Injector Circuit High
P0284 Cylinder no. 8 Contribution/Balance Fault
P0285 Cylinder no. 9 Injector Circuit Low
P0286 Cylinder no. 9 Injector Circuit High
P0287 Cylinder no. 9 Contribution/Balance Fault
P0288 Cylinder no. 10 Injector Circuit Low
P0289 Cylinder no. 10 Injector Circuit High
P0290 Cylinder no. 10 Contribution/Balance Fault
P0291 Cylinder no. 11 Injector Circuit Low
P0292 Cylinder no. 11 Injector Circuit High
P0293 Cylinder no. 11 Contribution/Balance Fault
P0294 Cylinder no. 12 Injector Circuit Low
P0295 Cylinder no. 12 Injector Circuit High
P0296 Cylinder no. 12 Contribution/Balance Fault
P0300 Random/Multiple Cylinder Misfire Detected
P0301 Cylinder no. 1—Misfire Detected
P0302 Cylinder no. 2—Misfire Detected
P0303 Cylinder no. 3—Misfire Detected
P0304 Cylinder no. 4—Misfire Detected
P0305 Cylinder no. 5—Misfire Detected
P0306 Cylinder no. 6—Misfire Detected
P0307 Cylinder no. 7—Misfire Detected
P0308 Cylinder no. 8—Misfire Detected
P0309 Cylinder no. 9—Misfire Detected
P0310 Cylinder no. 10—Misfire Detected
P0311 Cylinder no. 11—Misfire Detected
P0312 Cylinder no. 12—Misfire Detected
P0320 Ignition/Distributor Engine Speed Input Circuit Malfunction
P0321 Ignition/Distributor Engine Speed Input Circuit Range/Performance
P0322 Ignition/Distributor Engine Speed Input Circuit No Signal
P0323 Ignition/Distributor Engine Speed Input Circuit Intermittent
P0325 Knock Sensor no. 1—Circuit Malfunction (Bank no. 1 or Single Sensor)
P0326 Knock Sensor no. 1—Circuit Range/Performance (Bank no. 1 or Single Sensor)
P0327 Knock Sensor no. 1—Circuit Low Input (Bank no. 1 or Single Sensor)
P0328 Knock Sensor no. 1—Circuit High Input (Bank no. 1 or Single Sensor)
P0329 Knock Sensor no. 1—Circuit Input Intermittent (Bank no. 1 or Single Sensor)
P0330 Knock Sensor no. 2—Circuit Malfunction (Bank no. 2)
P0331 Knock Sensor no. 2—Circuit Range/Performance (Bank no. 2)

P0332 Knock Sensor no. 2—Circuit Low Input (Bank no. 2)

P0333 Knock Sensor no. 2—Circuit High Input (Bank no. 2)

P0334 Knock Sensor no. 2—Circuit Input Intermittent (Bank no. 2)

P0335 Crankshaft Position Sensor "A" Circuit Malfunction

P0336 Crankshaft Position Sensor "A" Circuit Range/Performance

P0337 Crankshaft Position Sensor "A" Circuit Low Input

P0338 Crankshaft Position Sensor "A" Circuit High Input

P0339 Crankshaft Position Sensor "A" Circuit Intermittent

P0340 Camshaft Position Sensor Circuit Malfunction

P0341 Camshaft Position Sensor Circuit Range/Performance

P0342 Camshaft Position Sensor Circuit Low Input

P0343 Camshaft Position Sensor Circuit High Input

P0344 Camshaft Position Sensor Circuit Intermittent

P0350 Ignition Coil Primary/Secondary Circuit Malfunction

P0351 Ignition Coil "A" Primary/Secondary Circuit Malfunction

P0352 Ignition Coil "B" Primary/Secondary Circuit Malfunction

P0353 Ignition Coil "C" Primary/Secondary Circuit Malfunction

P0354 Ignition Coil "D" Primary/Secondary Circuit Malfunction

P0355 Ignition Coil "E" Primary/Secondary Circuit Malfunction

P0356 Ignition Coil "F" Primary/Secondary Circuit Malfunction

P0357 Ignition Coil "G" Primary/Secondary Circuit Malfunction

P0358 Ignition Coil "H" Primary/Secondary Circuit Malfunction

P0359 Ignition Coil "I" Primary/Secondary Circuit Malfunction

P0360 Ignition Coil "J" Primary/Secondary Circuit Malfunction

P0361 Ignition Coil "K" Primary/Secondary Circuit Malfunction

P0362 Ignition Coil "L" Primary/Secondary Circuit Malfunction

P0370 Timing Reference High Resolution Signal "A" Malfunction

P0371 Timing Reference High Resolution Signal "A" Too Many Pulses

P0372 Timing Reference High Resolution Signal "A" Too Few Pulses

P0373 Timing Reference High Resolution Signal "A" Intermittent/Erratic Pulses

P0374 Timing Reference High Resolution Signal "A" No Pulses

P0375 Timing Reference High Resolution Signal "B" Malfunction

P0376 Timing Reference High Resolution Signal "B" Too Many Pulses

P0377 Timing Reference High Resolution Signal "B" Too Few Pulses

P0378 Timing Reference High Resolution Signal "B" Intermittent/Erratic Pulses

P0379 Timing Reference High Resolution Signal "B" No Pulses

P0380 Glow Plug/Heater Circuit "A" Malfunction

P0381 Glow Plug/Heater Indicator Circuit Malfunction

P0382 Glow Plug/Heater Circuit "B" Malfunction

P0385 Crankshaft Position Sensor "B" Circuit Malfunction

P0386 Crankshaft Position Sensor "B" Circuit Range/Performance

P0387 Crankshaft Position Sensor "B" Circuit Low Input

P0388 Crankshaft Position Sensor "B" Circuit High Input

P0389 Crankshaft Position Sensor "B" Circuit Intermittent

P0400 Exhaust Gas Recirculation Flow Malfunction

P0401 Exhaust Gas Recirculation Flow Insufficient Detected

P0402 Exhaust Gas Recirculation Flow Excessive Detected

P0403 Exhaust Gas Recirculation Circuit Malfunction

P0404 Exhaust Gas Recirculation Circuit Range/Performance

P0405 Exhaust Gas Recirculation Sensor "A" Circuit Low

P0406 Exhaust Gas Recirculation Sensor "A" Circuit High

P0407 Exhaust Gas Recirculation Sensor "B" Circuit Low

P0408 Exhaust Gas Recirculation Sensor "B" Circuit High

P0410 Secondary Air Injection System Malfunction

P0411 Secondary Air Injection System Incorrect Flow Detected

P0412 Secondary Air Injection System Switching Valve "A" Circuit Malfunction

P0413 Secondary Air Injection System Switching Valve "A" Circuit Open

P0414 Secondary Air Injection System Switching Valve "A" Circuit Shorted

P0415 Secondary Air Injection System Switching Valve "B" Circuit Malfunction

P0416 Secondary Air Injection System Switching Valve "B" Circuit Open

P0417 Secondary Air Injection System Switching Valve "B" Circuit Shorted

P0418 Secondary Air Injection System Relay "A" Circuit Malfunction

P0419 Secondary Air Injection System Relay "B" Circuit Malfunction

P0420 Catalyst System Efficiency Below Threshold (Bank no. 1)

P0421 Warm Up Catalyst Efficiency Below Threshold (Bank no. 1)

P0422 Main Catalyst Efficiency Below Threshold (Bank no. 1)

P0423 Heated Catalyst Efficiency Below Threshold (Bank no. 1)

P0424 Heated Catalyst Temperature Below Threshold (Bank no. 1)

P0430 Catalyst System Efficiency Below Threshold (Bank no. 2)

P0431 Warm Up Catalyst Efficiency Below Threshold (Bank no. 2)

P0432 Main Catalyst Efficiency Below Threshold (Bank no. 2)

P0433 Heated Catalyst Efficiency Below Threshold (Bank no. 2)

P0434 Heated Catalyst Temperature Below Threshold (Bank no. 2)

P0440 Evaporative Emission Control System Malfunction

P0441 Evaporative Emission Control System Incorrect Purge Flow

P0442 Evaporative Emission Control System Leak Detected (Small Leak)

P0443 Evaporative Emission Control System Purge Control Valve Circuit Malfunction

P0444 Evaporative Emission Control System Purge Control Valve Circuit Open

P0445 Evaporative Emission Control System Purge Control Valve Circuit Shorted

P0446 Evaporative Emission Control System Vent Control Circuit Malfunction

P0447 Evaporative Emission Control System Vent Control Circuit Open

P0448 Evaporative Emission Control System Vent Control Circuit Shorted

P0449 Evaporative Emission Control System Vent Valve/Solenoid Circuit Malfunction

P0450 Evaporative Emission Control System Pressure Sensor Malfunction

P0451 Evaporative Emission Control System Pressure Sensor Range/Performance

P0452 Evaporative Emission Control System Pressure Sensor Low Input

P0453 Evaporative Emission Control System Pressure Sensor High Input

P0454 Evaporative Emission Control System Pressure Sensor Intermittent

P0455 Evaporative Emission Control System Leak Detected (Gross Leak)
P0460 Fuel Level Sensor Circuit Malfunction
P0461 Fuel Level Sensor Circuit Range/Performance
P0462 Fuel Level Sensor Circuit Low Input
P0463 Fuel Level Sensor Circuit High Input
P0464 Fuel Level Sensor Circuit Intermittent
P0465 Purge Flow Sensor Circuit Malfunction
P0466 Purge Flow Sensor Circuit Range/Performance
P0467 Purge Flow Sensor Circuit Low Input
P0468 Purge Flow Sensor Circuit High Input
P0469 Purge Flow Sensor Circuit Intermittent
P0470 Exhaust Pressure Sensor Malfunction
P0471 Exhaust Pressure Sensor Range/Performance
P0472 Exhaust Pressure Sensor Low
P0473 Exhaust Pressure Sensor High
P0474 Exhaust Pressure Sensor Intermittent
P0475 Exhaust Pressure Control Valve Malfunction
P0476 Exhaust Pressure Control Valve Range/Performance
P0477 Exhaust Pressure Control Valve Low
P0478 Exhaust Pressure Control Valve High
P0479 Exhaust Pressure Control Valve Intermittent
P0480 Cooling Fan no. 1 Control Circuit Malfunction
P0481 Cooling Fan no. 2 Control Circuit Malfunction
P0482 Cooling Fan no. 3 Control Circuit Malfunction
P0483 Cooling Fan Rationality Check Malfunction
P0484 Cooling Fan Circuit Over Current
P0485 Cooling Fan Power/Ground Circuit Malfunction
P0500 Vehicle Speed Sensor Malfunction
P0501 Vehicle Speed Sensor Range/Performance
P0502 Vehicle Speed Sensor Circuit Low Input
P0503 Vehicle Speed Sensor Intermittent/Erratic/High
P0505 Idle Control System Malfunction
P0506 Idle Control System RPM Lower Than Expected
P0507 Idle Control System RPM Higher Than Expected
P0510 Closed Throttle Position Switch Malfunction
P0520 Engine Oil Pressure Sensor/Switch Circuit Malfunction
P0521 Engine Oil Pressure Sensor/Switch Range/Performance
P0522 Engine Oil Pressure Sensor/Switch Low Voltage
P0523 Engine Oil Pressure Sensor/Switch High Voltage
P0530 A/C Refrigerant Pressure Sensor Circuit Malfunction
P0531 A/C Refrigerant Pressure Sensor Circuit Range/Performance
P0532 A/C Refrigerant Pressure Sensor Circuit Low Input
P0533 A/C Refrigerant Pressure Sensor Circuit High Input
P0534 A/C Refrigerant Charge Loss
P0550 Power Steering Pressure Sensor Circuit Malfunction
P0551 Power Steering Pressure Sensor Circuit Range/Performance
P0552 Power Steering Pressure Sensor Circuit Low Input
P0553 Power Steering Pressure Sensor Circuit High Input
P0554 Power Steering Pressure Sensor Circuit Intermittent
P0560 System Voltage Malfunction
P0561 System Voltage Unstable
P0562 System Voltage Low
P0563 System Voltage High
P0565 Cruise Control On Signal Malfunction
P0566 Cruise Control Off Signal Malfunction
P0567 Cruise Control Resume Signal Malfunction
P0568 Cruise Control Set Signal Malfunction
P0569 Cruise Control Coast Signal Malfunction
P0570 Cruise Control Accel Signal Malfunction

P0571 Cruise Control/Brake Switch "A" Circuit Malfunction
P0572 Cruise Control/Brake Switch "A" Circuit Low
P0573 Cruise Control/Brake Switch "A" Circuit High
P0574 **Through P0580** Reserved for Cruise Codes
P0600 Serial Communication Link Malfunction
P0601 Internal Control Module Memory Check Sum Error
P0602 Control Module Programming Error
P0603 Internal Control Module Keep Alive Memory (KAM) Error
P0604 Internal Control Module Random Access Memory (RAM) Error
P0605 Internal Control Module Read Only Memory (ROM) Error
P0606 PCM Processor Fault
P0608 Control Module VSS Output "A" Malfunction
P0609 Control Module VSS Output "B" Malfunction
P0620 Generator Control Circuit Malfunction
P0621 Generator Lamp "L" Control Circuit Malfunction
P0622 Generator Field "F" Control Circuit Malfunction
P0650 Malfunction Indicator Lamp (MIL) Control Circuit Malfunction
P0654 Engine RPM Output Circuit Malfunction
P0655 Engine Hot Lamp Output Control Circuit Malfunction
P0656 Fuel Level Output Circuit Malfunction
P0700 Transmission Control System Malfunction
P0701 Transmission Control System Range/Performance
P0702 Transmission Control System Electrical
P0703 Torque Converter/Brake Switch "B" Circuit Malfunction
P0704 Clutch Switch Input Circuit Malfunction
P0705 Transmission Range Sensor Circuit Malfunction (PRNDL Input)
P0706 Transmission Range Sensor Circuit Range/Performance
P0707 Transmission Range Sensor Circuit Low Input
P0708 Transmission Range Sensor Circuit High Input
P0709 Transmission Range Sensor Circuit Intermittent
P0710 Transmission Fluid Temperature Sensor Circuit Malfunction
P0711 Transmission Fluid Temperature Sensor Circuit Range/Performance
P0712 Transmission Fluid Temperature Sensor Circuit Low Input
P0713 Transmission Fluid Temperature Sensor Circuit High Input
P0714 Transmission Fluid Temperature Sensor Circuit Intermittent
P0715 Input/Turbine Speed Sensor Circuit Malfunction
P0716 Input/Turbine Speed Sensor Circuit Range/Performance
P0717 Input/Turbine Speed Sensor Circuit No Signal
P0718 Input/Turbine Speed Sensor Circuit Intermittent
P0719 Torque Converter/Brake Switch "B" Circuit Low
P0720 Output Speed Sensor Circuit Malfunction
P0721 Output Speed Sensor Circuit Range/Performance
P0722 Output Speed Sensor Circuit No Signal
P0723 Output Speed Sensor Circuit Intermittent
P0724 Torque Converter/Brake Switch "B" Circuit High
P0725 Engine Speed Input Circuit Malfunction
P0726 Engine Speed Input Circuit Range/Performance
P0727 Engine Speed Input Circuit No Signal
P0728 Engine Speed Input Circuit Intermittent
P0730 Incorrect Gear Ratio
P0731 Gear no. 1 Incorrect Ratio
P0732 Gear no. 2 Incorrect Ratio
P0733 Gear no. 3 Incorrect Ratio
P0734 Gear no. 4 Incorrect Ratio

P0735 Gear no. 5 Incorrect Ratio
P0736 Reverse Incorrect Ratio
P0740 Torque Converter Clutch Circuit Malfunction
P0741 Torque Converter Clutch Circuit Performance or Stuck Off
P0742 Torque Converter Clutch Circuit Stuck On
P0743 Torque Converter Clutch Circuit Electrical
P0744 Torque Converter Clutch Circuit Intermittent
P0745 Pressure Control Solenoid Malfunction
P0746 Pressure Control Solenoid Performance or Stuck Off
P0747 Pressure Control Solenoid Stuck On
P0748 Pressure Control Solenoid Electrical
P0749 Pressure Control Solenoid Intermittent
P0750 Shift Solenoid "A" Malfunction
P0751 Shift Solenoid "A" Performance or Stuck Off
P0752 Shift Solenoid "A" Stuck On
P0753 Shift Solenoid "A" Electrical
P0754 Shift Solenoid "A" Intermittent
P0755 Shift Solenoid "B" Malfunction
P0756 Shift Solenoid "B" Performance or Stuck Off
P0757 Shift Solenoid "B" Stuck On
P0758 Shift Solenoid "B" Electrical
P0759 Shift Solenoid "B" Intermittent
P0760 Shift Solenoid "C" Malfunction
P0761 Shift Solenoid "C" Performance Or Stuck Off
P0762 Shift Solenoid "C" Stuck On
P0763 Shift Solenoid "C" Electrical
P0764 Shift Solenoid "C" Intermittent
P0765 Shift Solenoid "D" Malfunction
P0766 Shift Solenoid "D" Performance Or Stuck Off
P0767 Shift Solenoid "D" Stuck On
P0768 Shift Solenoid "D" Electrical
P0769 Shift Solenoid "D" Intermittent
P0770 Shift Solenoid "E" Malfunction
P0771 Shift Solenoid "E" Performance Or Stuck Off
P0772 Shift Solenoid "E" Stuck On
P0773 Shift Solenoid "E" Electrical
P0774 Shift Solenoid "E" Intermittent
P0780 Shift Malfunction
P0781 1–2 Shift Malfunction
P0782 2–3 Shift Malfunction
P0783 3–4 Shift Malfunction
P0784 4–5 Shift Malfunction
P0785 Shift/Timing Solenoid Malfunction
P0786 Shift/Timing Solenoid Range/Performance
P0787 Shift/Timing Solenoid Low
P0788 Shift/Timing Solenoid High
P0789 Shift/Timing Solenoid Intermittent
P0790 Normal/Performance Switch Circuit Malfunction
P0801 Reverse Inhibit Control Circuit Malfunction
P0803 1–4 Upshift (Skip Shift) Solenoid Control Circuit Malfunction
P0804 1–4 Upshift (Skip Shift) Lamp Control Circuit Malfunction
P1000 OBD II Monitor Testing Not Complete More Driving Required
P1001 Key On Engine Running (KOER) Self-Test Not Able To Complete, KOER Aborted
P1100 Mass Air Flow (MAF) Sensor Intermittent
P1101 Mass Air Flow (MAF) Sensor Out Of Self-Test Range
P1111 System Pass 49 State Except Econoline
P1112 Intake Air Temperature (IAT) Sensor Intermittent

P1116 Engine Coolant Temperature (ECT) Sensor Out Of Self-Test Range
P1117 Engine Coolant Temperature (ECT) Sensor Intermittent
P1120 Throttle Position (TP) Sensor Out Of Range (Low)
P1121 Throttle Position (TP) Sensor Inconsistent With MAF Sensor
P1124 Throttle Position (TP) Sensor Out Of Self-Test Range
P1125 Throttle Position (TP) Sensor Circuit Intermittent
P1127 Exhaust Not Warm Enough, Downstream Heated Oxygen Sensors (HO2S) Not Tested
P1128 Upstream Heated Oxygen Sensors (HO2S) Swapped From Bank To Bank
P1129 Downstream Heated Oxygen Sensors (HO2S) Swapped From Bank To Bank
P1130 Lack Of Upstream Heated Oxygen Sensor (HO2S 11) Switch, Adaptive Fuel At Limit (Bank #1)
P1131 Lack Of Upstream Heated Oxygen Sensor (HO2S 11) Switch, Sensor Indicates Lean (Bank #1)
P1132 Lack Of Upstream Heated Oxygen Sensor (HO2S 11) Switch, Sensor Indicates Rich (Bank#1)
P1137 Lack Of Downstream Heated Oxygen Sensor (HO2S 12) Switch, Sensor Indicates Lean (Bank#1)
P1138 Lack Of Downstream Heated Oxygen Sensor (HO2S 12) Switch, Sensor Indicates Rich (Bank#1)
P1150 Lack Of Upstream Heated Oxygen Sensor (HO2S 21) Switch, Adaptive Fuel At Limit (Bank #2)
P1151 Lack Of Upstream Heated Oxygen Sensor (HO2S 21) Switch, Sensor Indicates Lean (Bank#2)
P1152 Lack Of Upstream Heated Oxygen Sensor (HO2S 21) Switch, Sensor Indicates Rich (Bank #2)
P1157 Lack Of Downstream Heated Oxygen Sensor (HO2S 22) Switch, Sensor Indicates Lean (Bank #2)
P1158 Lack Of Downstream Heated Oxygen Sensor (HO2S 22) Switch, Sensor Indicates Rich (Bank#2)
P1169 (HO2S 12) Signal Remained Unchanged For More Than 20 Seconds After Closed Loop
P1170 (HO2S 11) Signal Remained Unchanged For More Than 20 Seconds After Closed Loop
P1173 Feedback A/F Mixture Control (HO2S 21) Signal Remained Unchanged For More Than 20 Seconds After Closed Loop
P1184 Engine Oil Temp Sensor Circuit Performance
P1195 Barometric (BARO) Pressure Sensor Circuit Malfunction (Signal Is From EGR Boost Sensor)
P1196 Starter Switch Circuit Malfunction
P1209 Injection Control Pressure (ICP) Peak Fault
P1210 Injection Control Pressure (ICP) Above Expected Level
P1211 Injection Control Pressure (ICP) Not Controllable— Pressure Above/Below Desired
P1212 Injection Control Pressure (ICP) Voltage Not At Expected Level
P1218 Cylinder Identification (CID) Stuck High
P1219 Cylinder Identification (CID) Stuck Low
P1220 Series Throttle Control Malfunction (Traction Control System)
P1224 Throttle Position Sensor "B" (TP-B) Out Of Self-Test Range (Traction Control System)
P1230 Fuel Pump Low Speed Malfunction
P1231 Fuel Pump Secondary Circuit Low With High Speed Pump On
P1232 Low Speed Fuel Pump Primary Circuit Malfunction
P1233 Fuel Pump Driver Module Off-line (MIL DTC)

P1234 Fuel Pump Driver Module Disabled Or Off-line (No MIL)
P1235 Fuel Pump Control Out Of Range (MIL DTC)
P1236 Fuel Pump Control Out Of Range (No MIL)
P1237 Fuel Pump Secondary Circuit Malfunction (MIL DTC)
P1238 Fuel Pump Secondary Circuit Malfunction (No DMIL)
P1250 Fuel Pressure Regulator Control (FPRC) Solenoid Malfunction
P1260 THEFT Detected—Engine Disabled
P1261 High To Low Side Short—Cylinder #1 (Indicates Low Side Circuit Is Shorted To B+ Or To The High Side Between The IDM And The Injector)
P1262 High To Low Side Short—Cylinder #2 (Indicates Low Side Circuit Is Shorted To B+ Or To The High Side Between The IDM And The Injector)
P1263 High To Low Side Short—Cylinder #3 (Indicates Low Side Circuit Is Shorted To B+ Or To The High Side Between The IDM And The Injector)
P1264 High To Low Side Short—Cylinder #4 (Indicates Low Side Circuit Is Shorted To B+ Or To The High Side Between The IDM And The Injector)
P1265 High To Low Side Short—Cylinder #5 (Indicates Low Side Circuit Is Shorted To B+ Or To The High Side Between The IDM And The Injector)
P1266 High To Low Side Short—Cylinder #6 (Indicates Low Side Circuit Is Shorted To B+ Or To The High Side Between The IDM And The Injector)
P1267 High To Low Side Short—Cylinder #7 (Indicates Low Side Circuit Is Shorted To B+ Or To The High Side Between The IDM And The Injector)
P1268 High To Low Side Short—Cylinder #8 (Indicates Low Side Circuit Is Shorted To B+ Or To The High Side Between The IDM And The Injector)
P1270 Engine RPM Or Vehicle Speed Limiter Reached
P1271 High To Low Side Open—Cylinder #1 (Indicates A High To Low Side Open Between The Injector And The IDM)
P1272 High To Low Side Open—Cylinder #2 (Indicates A High To Low Side Open Between The Injector And The IDM)
P1273 High To Low Side Open—Cylinder #3 (Indicates A High To Low Side Open Between The Injector And The IDM)
P1274 High To Low Side Open—Cylinder #4 (Indicates A High To Low Side Open Between The Injector And The IDM)
P1275 High To Low Side Open—Cylinder #5 (Indicates A High To Low Side Open Between The Injector And The IDM)
P1276 High To Low Side Open—Cylinder #6 (Indicates A High To Low Side Open Between The Injector And The IDM)
P1277 High To Low Side Open—Cylinder #7 (Indicates A High To Low Side Open Between The Injector And The IDM)
P1278 High To Low Side Open—Cylinder #8 (Indicates A High To Low Side Open Between The Injector And The IDM)
P1280 Injection Control Pressure (ICP) Circuit Out Of Range Low
P1281 Injection Control Pressure (ICP) Circuit Out Of Range High
P1282 Injection Control Pressure (ICP) Excessive
P1283 Injection Pressure Regulator (IPR) Circuit Failure
P1284 Injection Control Pressure (ICP) Failure—Aborts KOER Or CCT Test
P1285 Cylinder Head Temperature (CHT) Over Temperature Sensed
P1288 Cylinder Head Temperature (CHT) Sensor Out Of Self-Test Range
P1289 Cylinder Head Temperature (CHT) Sensor Circuit Low Input

P1290 Cylinder Head Temperature (CHT) Sensor Circuit High Input
P1291 IDM To Injector High Side Circuit #1 (Right Bank) Short To GND Or B+
P1292 IDM To Injector High Side Circuit #2 (Right Bank) Short To GND Or B+
P1293 IDM To Injector High Side Circuit Open Bank #1 (Right Bank)
P1294 IDM To Injector High Side Circuit Open Bank #2 (Left Bank)
P1295 Multiple IDM/Injector Circuit Faults On Bank #1 (Right)
P1296 Multiple IDM/Injector Circuit Faults On Bank#2 (Left)
P1297 High Sides Shorted Together
P1298 IDM Failure
P1299 Engine Over Temperature Condition
P1309 Misfire Detection Monitor Is Not Enabled
P1316 Injector Circuit/IDM Codes Detected
P1320 Distributor Signal Interrupt
P1336 Crankshaft Position Sensor (Gear)
P1345 No Camshaft Position Sensor Signal
P1351 Ignition Diagnostic Monitor (IDM) Circuit Input Malfunction
P1351 Indicates Ignition System Malfunction
P1352 Indicates Ignition System Malfunction
P1353 Indicates Ignition System Malfunction
P1354 Indicates Ignition System Malfunction
P1355 Indicates Ignition System Malfunction
P1356 PIPs Occurred While IDM Pulse width Indicates Engine Not Turning
P1357 Ignition Diagnostic Monitor (IDM) Pulse width Not Defined
P1358 Ignition Diagnostic Monitor (IDM) Signal Out Of Self-Test Range
P1359 Spark Output Circuit Malfunction
P1364 Spark Output Circuit Malfunction
P1390 Octane Adjust (OCT ADJ) Out Of Self-Test Range
P1391 Glow Plug Circuit Low Input Bank #1 (Right)
P1392 Glow Plug Circuit High Input Bank #1 (Right)
P1393 Glow Plug Circuit Low Input Bank #2 (Left)
P1394 Glow Plug Circuit High Input Bank #2 (Left)
P1395 Glow Plug Monitor Fault Bank #1
P1396 Glow Plug Monitor Fault Bank #2
P1397 System Voltage Out Of Self Test Range
P1400 Differential Pressure Feedback EGR (DPFE) Sensor Circuit Low Voltage Detected
P1401 Differential Pressure Feedback EGR (DPFE) Sensor Circuit High Voltage Detected/EGR Temperature Sensor
P1402 EGR Valve Position Sensor Open Or Short
P1403 Differential Pressure Feedback EGR (DPFE) Sensor Hoses Reversed
P1405 Differential Pressure Feedback EGR (DPFE) Sensor Upstream Hose Off Or Plugged
P1406 Differential Pressure Feedback EGR (DPFE) Sensor Downstream Hose Off Or Plugged
P1407 Exhaust Gas Recirculation (EGR) No Flow Detected (Valve Stuck Closed Or Inoperative)
P1408 Exhaust Gas Recirculation (EGR) Flow Out Of Self-Test Range
P1409 Electronic Vacuum Regulator (EVR) Control Circuit Malfunction
P1410 Check That Fuel Pressure Regulator Control Solenoid And The EGR Check Solenoid Connectors Are Not Swapped

P1411 Secondary Air Injection System Incorrect Downstream Flow Detected

P1413 Secondary Air Injection System Monitor Circuit Low Voltage

P1414 Secondary Air Injection System Monitor Circuit High Voltage

P1442 Evaporative Emission Control System Small Leak Detected

P1443 Evaporative Emission Control System—Vacuum System, Purge Control Solenoid Or Purge Control Valve Malfunction

P1444 Purge Flow Sensor (PFS) Circuit Low Input

P1445 Purge Flow Sensor (PFS) Circuit High Input

P1449 Evaporative Emission Control System Unable To Hold Vacuum

P1450 Unable To Bleed Up Fuel Tank Vacuum

P1455 Evaporative Emission Control System Control Leak Detected (Gross Leak)

P1460 Wide Open Throttle Air Conditioning Cut-Off Circuit Malfunction

P1461 Air Conditioning Pressure (ACP) Sensor Circuit Low Input

P1462 Air Conditioning Pressure (ACP) Sensor Circuit High Input

P1463 Air Conditioning Pressure (ACP) Sensor Insufficient Pressure Change

P1464 Air Conditioning (A/C) Demand Out Of Self-Test Range/A/C On During KOER Or CCT Test

P1469 Low Air Conditioning Cycling Period

P1473 Fan Secondary High, With Fan(s) Off

P1474 Low Fan Control Primary Circuit Malfunction

P1479 High Fan Control Primary Circuit Malfunction

P1480 Fan Secondary Low, With Low Fan On

P1481 Fan Secondary Low, With High Fan On

P1483 Power To Fan Circuit Over current

P1484 Open Power/Ground To Variable Load Control Module (VLCM)

P1485 EGR Control Solenoid Open Or Short

P1486 EGR Vent Solenoid Open Or Short

P1487 EGR Boost Check Solenoid Open Or Short

P1500 Vehicle Speed Sensor (VSS) Circuit Intermittent

P1501 Vehicle Speed Sensor (VSS) Out Of Self-Test Range/Vehicle Moved During Test

P1502 Invalid Self Test—Auxiliary Powertrain Control Module (APCM) Functioning

P1504 Idle Air Control (IAC) Circuit Malfunction

P1505 Idle Air Control (IAC) System At Adaptive Clip

P1506 Idle Air Control (IAC) Overspeed Error

P1507 Idle Air Control (IAC) Underspeed Error

P1512 Intake Manifold Runner Control (IMRC) Malfunction (Bank#1 Stuck Closed)

P1513 Intake Manifold Runner Control (IMRC) Malfunction (Bank#2 Stuck Closed)

P1516 Intake Manifold Runner Control (IMRC) Input Error (Bank #1)

P1517 Intake Manifold Runner Control (IMRC) Input Error (Bank #2)

P1518 Intake Manifold Runner Control (IMRC) Malfunction (Stuck Open)

P1519 Intake Manifold Runner Control (IMRC) Malfunction (Stuck Closed)

P1520 Intake Manifold Runner Control (IMRC) Circuit Malfunction

P1521 Variable Resonance Induction System (VRIS) Solenoid #1 Open Or Short

P1522 Variable Resonance Induction System (VRIS) Solenoid #2 Open Or Short

P1523 High Speed Inlet Air (HSIA) Solenoid Open Or Short

P1530 Air Condition (A/C) Clutch Circuit Malfunction

P1531 Invalid Test—Accelerator Pedal Movement

P1536 Parking Brake Applied Failure

P1537 Intake Manifold Runner Control (IMRC) Malfunction (Bank#1 Stuck Open)

P1538 Intake Manifold Runner Control (IMRC) Malfunction (Bank#2 Stuck Open)

P1539 Power To Air Condition (A/C) Clutch Circuit Overcurrent

P1549 Problem In Intake Manifold Tuning (IMT) Valve System

P1550 Power Steering Pressure (PSP) Sensor Out Of Self-Test Range

P1601 Serial Communication Error

P1605 Powertrain Control Module (PCM)—Keep Alive Memory (KAM) Test Error

P1608 PCM Internal Circuit Malfunction

P1609 PCM Internal Circuit Malfunction (2.5L Only)

P1625 B+ Supply To Variable Load Control Module (VLCM) Fan Circuit Malfunction

P1626 B+ Supply To Variable Load Control Module (VLCM) Air Conditioning (A/C) Circuit

P1650 Power Steering Pressure (PSP) Switch Out Of Self-Test Range

P1651 Power Steering Pressure (PSP) Switch Input Malfunction

P1660 Output Circuit Check Signal High

P1661 Output Circuit Check Signal Low

P1662 Injection Driver Module Enable (IDM EN) Circuit Failure

P1663 Fuel Delivery Command Signal (FDCS) Circuit Failure

P1667 Cylinder Identification (CID) Circuit Failure

P1668 PCM—IDM Diagnostic Communication Error

P1670 EF Feedback Signal Not Detected

P1701 Reverse Engagement Error

P1701 Fuel Trim Malfunction (Villager)

P1703 Brake On/Off (BOO) Switch Out Of Self-Test Range

P1704 Digital Transmission Range (TR) Sensor Failed To Transition State

P1705 Transmission Range (TR) Sensor Out Of Self-Test Range

P1705 TP Sensor (AT) Villager

P1705 Clutch Pedal Position (CPP) Or Park Neutral Position (PNP) Problem

P1706 High Vehicle Speed In Park

P1709 Park Or Neutral Position (PNP) Or Clutch Pedal Position (CPP) Switch Out Of Self-Test Range

P1709 Throttle Position (TP) Sensor Malfunction (Aspire 1.3L, Escort/ Tracer 1.8L, Probe 2.5L)

P1711 Transmission Fluid Temperature (TFT) Sensor Out Of Self-Test Range

P1714 Shift Solenoid "A" Inductive Signature Malfunction

P1715 Shift Solenoid "B" Inductive Signature Malfunction

P1716 Transmission Malfunction

P1717 Transmission Malfunction

P1719 Transmission Malfunction

P1720 Vehicle Speed Sensor (VSS) Circuit Malfunction

P1727 Coast Clutch Solenoid Inductive Signature Malfunction

P1728 Transmission Slip Error—Converter Clutch Failed

P1729 4x4 Low Switch Error

P1731 Improper 1–2 Shift

P1732 Improper 2–3 Shift

P1733 Improper 3–4 Shift

P1734 Improper 4–5 Shift

P1740 Torque Converter Clutch (TCC) Inductive Signature Malfunction

P1741 Torque Converter Clutch (TCC) Control Error

P1742 Torque Converter Clutch (TCC) Solenoid Failed On (Turns On MIL)

P1743 Torque Converter Clutch (TCC) Solenoid Failed On (Turns On TCIL)

P1744 Torque Converter Clutch (TCC) System Mechanically Stuck In Off Position

P1744 Torque Converter Clutch (TCC) Solenoid Malfunction (2.5L Only)

P1746 Electronic Pressure Control (EPC) Solenoid Open Circuit (Low Input)

P1747 Electronic Pressure Control (EPC) Solenoid Short Circuit (High Input)

P1748 Electronic Pressure Control (EPC) Malfunction

P1749 Electronic Pressure Control (EPC) Solenoid Failed Low

P1751 Shift Solenoid#1 (SS1) Performance

P1754 Coast Clutch Solenoid (CCS) Circuit Malfunction

P1756 Shift Solenoid#2 (SS2) Performance

P1760 Overrun Clutch SN

P1761 Shift Solenoid #(SS2) Performance

P1762 Transmission Malfunction

P1765 3–2 Timing Solenoid Malfunction (2.5L Only)

P1779 TCIL Circuit Malfunction

P1780 Transmission Control Switch (TCS) Circuit Out Of Self-Test Range

P1781 4x4 Low Switch, Out Of Self-Test Range

P1783 Transmission Over Temperature Condition

P1784 Transmission Malfunction

P1785 Transmission Malfunction

P1786 Transmission Malfunction

P1787 Transmission Malfunction

P1788 3–2 Timing/Coast Clutch Solenoid (3–2/CCS) Circuit Open

P1789 3–2 Timing/Coast Clutch Solenoid (3–2/CCS) Circuit Shorted

P1792 Idle (IDL) Switch (Closed Throttle Position Switch) Malfunction

P1794 Loss Of Battery Voltage Input

P1795 EGR Boost Sensor Malfunction

P1797 Clutch Pedal Position (CPP) Switch Or Neutral Switch Circuit Malfunction

P1900 Cooling Fan

U1021 SCP Indicating The Lack Of Air Conditioning (A/C) Clutch Status Response

U1039 Vehicle Speed Signal (VSS) Missing Or Incorrect

U1051 Brake Switch Signal Missing Or Incorrect

U1073 SCP Indicating The Lack Of Engine Coolant Fan Status Response

U1131 SCP Indicating The Lack Of Fuel Pump Status Response

U1135 SCP Indicating The Ignition Switch Signal Missing Or Incorrect

U1256 SCP Indicating A Communications Error

U1451 Lack Of Response From Passive Anti-Theft System (PATS) Module—Engine Disabled

General Motors

READING CODES

Reading the control module memory is one of the first steps in OBD II system diagnostics. This step should be initially performed to determine the general nature of the fault. Subsequent readings will determine if the fault has been cleared.

Reading codes can be performed by any of the methods below:
- Read the control module memory with the Generic Scan Tool (GST)
- Read the control module memory with the vehicle manufacturer's specific tester

To read the fault codes, connect the scan tool or tester according to the manufacturer's instructions. Follow the manufacturer's specified procedure for reading the codes.

CLEARING CODES

Control module reset procedures are a very important part of OBD II System diagnostics. This step should be done at the end of any fault code repair and at the end of any driveability repair.

Clearing codes can be performed by any of the methods below:
- Clear the control module memory with the Generic Scan Tool (GST)
- Clear the control module memory with the vehicle manufacturer's specific tester
- Turn the ignition off and remove the negative battery cable for at least 1 minute.

Removing the negative battery cable may cause other systems in the vehicle to loose their memory. Prior to removing the cable, ensure you have the proper reset codes for radios and alarms.

➡**The MIL will may also be de-activated for some codes if the vehicle completes three consecutive trips without a fault detected with vehicle conditions similar to those present during the fault.**

OBD II TROUBLE CODES

P0100 Mass or Volume Air Flow Circuit Malfunction

P0101 Mass or Volume Air Flow Circuit Range/Performance Problem

P0102 Mass or Volume Air Flow Circuit Low Input

P0103 Mass or Volume Air Flow Circuit High Input

P0104 Mass or Volume Air Flow Circuit Intermittent

P0105 Manifold Absolute Pressure/Barometric Pressure Circuit Malfunction

P0106 Manifold Absolute Pressure/Barometric Pressure Circuit Range/Performance Problem

P0107 Manifold Absolute Pressure/Barometric Pressure Circuit Low Input

P0108 Manifold Absolute Pressure/Barometric Pressure Circuit High Input

P0109 Manifold Absolute Pressure/Barometric Pressure Circuit Intermittent

P0110 Intake Air Temperature Circuit Malfunction

P0111 Intake Air Temperature Circuit Range/Performance Problem

P0112 Intake Air Temperature Circuit Low Input

P0113 Intake Air Temperature Circuit High Input

P0114 Intake Air Temperature Circuit Intermittent

P0115 Engine Coolant Temperature Circuit Malfunction

P0116 Engine Coolant Temperature Circuit Range/Performance Problem

P0117 Engine Coolant Temperature Circuit Low Input

P0118 Engine Coolant Temperature Circuit High Input

P0119 Engine Coolant Temperature Circuit Intermittent

P0120 Throttle/Pedal Position Sensor/Switch "A" Circuit Malfunction

P0121 Throttle/Pedal Position Sensor/Switch "A" Circuit Range/Performance Problem

P0122 Throttle/Pedal Position Sensor/Switch "A" Circuit Low Input

P0123 Throttle/Pedal Position Sensor/Switch "A" Circuit High Input

P0124 Throttle/Pedal Position Sensor/Switch "A" Circuit Intermittent

P0125 Insufficient Coolant Temperature For Closed Loop Fuel Control

P0126 Insufficient Coolant Temperature For Stable Operation

P0130 O2 Circuit Malfunction (Bank no. 1 Sensor no. 1)

P0131 O2 Sensor Circuit Low Voltage (Bank no. 1 Sensor no. 1)

P0132 O2 Sensor Circuit High Voltage (Bank no. 1 Sensor no. 1)

P0133 O2 Sensor Circuit Slow Response (Bank no. 1 Sensor no. 1)

P0134 O2 Sensor Circuit No Activity Detected (Bank no. 1 Sensor no. 1)

P0135 O2 Sensor Heater Circuit Malfunction (Bank no. 1 Sensor no. 1)

P0136 O2 Sensor Circuit Malfunction (Bank no. 1 Sensor no. 2)

P0137 O2 Sensor Circuit Low Voltage (Bank no. 1 Sensor no. 2)

P0138 O2 Sensor Circuit High Voltage (Bank no. 1 Sensor no. 2)

P0139 O2 Sensor Circuit Slow Response (Bank no. 1 Sensor no. 2)

P0140 O2 Sensor Circuit No Activity Detected (Bank no. 1 Sensor no. 2)

P0141 O2 Sensor Heater Circuit Malfunction (Bank no. 1 Sensor no. 2)

P0142 O2 Sensor Circuit Malfunction (Bank no. 1 Sensor no. 3)

P0143 O2 Sensor Circuit Low Voltage (Bank no. 1 Sensor no. 3)

P0144 O2 Sensor Circuit High Voltage (Bank no. 1 Sensor no. 3)

P0145 O2 Sensor Circuit Slow Response (Bank no. 1 Sensor no. 3)

P0146 O2 Sensor Circuit No Activity Detected (Bank no. 1 Sensor no. 3)

P0147 O2 Sensor Heater Circuit Malfunction (Bank no. 1 Sensor no. 3)

P0150 O2 Sensor Circuit Malfunction (Bank no. 2 Sensor no. 1)

P0151 O2 Sensor Circuit Low Voltage (Bank no. 2 Sensor no. 1)

P0152 O2 Sensor Circuit High Voltage (Bank no. 2 Sensor no. 1)

P0153 O2 Sensor Circuit Slow Response (Bank no. 2 Sensor no. 1)

P0154 O2 Sensor Circuit No Activity Detected (Bank no. 2 Sensor no. 1)

P0155 O2 Sensor Heater Circuit Malfunction (Bank no. 2 Sensor no. 1)

P0156 O2 Sensor Circuit Malfunction (Bank no. 2 Sensor no. 2)

P0157 O2 Sensor Circuit Low Voltage (Bank no. 2 Sensor no. 2)

P0158 O2 Sensor Circuit High Voltage (Bank no. 2 Sensor no. 2)

P0159 O2 Sensor Circuit Slow Response (Bank no. 2 Sensor no. 2)

P0160 O2 Sensor Circuit No Activity Detected (Bank no. 2 Sensor no. 2)

P0161 O2 Sensor Heater Circuit Malfunction (Bank no. 2 Sensor no. 2)

P0162 O2 Sensor Circuit Malfunction (Bank no. 2 Sensor no. 3)

P0163 O2 Sensor Circuit Low Voltage (Bank no. 2 Sensor no. 3)

P0164 O2 Sensor Circuit High Voltage (Bank no. 2 Sensor no. 3)

P0165 O2 Sensor Circuit Slow Response (Bank no. 2 Sensor no. 3)

P0166 O2 Sensor Circuit No Activity Detected (Bank no. 2 Sensor no. 3)

P0167 O2 Sensor Heater Circuit Malfunction (Bank no. 2 Sensor no. 3)

P0170 Fuel Trim Malfunction (Bank no. 1)

P0171 System Too Lean (Bank no. 1)

P0172 System Too Rich (Bank no. 1)

P0173 Fuel Trim Malfunction (Bank no. 2)

P0174 System Too Lean (Bank no. 2)

P0175 System Too Rich (Bank no. 2)

P0176 Fuel Composition Sensor Circuit Malfunction

P0177 Fuel Composition Sensor Circuit Range/Performance

P0178 Fuel Composition Sensor Circuit Low Input

P0179 Fuel Composition Sensor Circuit High Input

P0180 Fuel Temperature Sensor "A" Circuit Malfunction

P0181 Fuel Temperature Sensor "A" Circuit Range/Performance

P0182 Fuel Temperature Sensor "A" Circuit Low Input

P0183 Fuel Temperature Sensor "A" Circuit High Input

P0184 Fuel Temperature Sensor "A" Circuit Intermittent

P0185 Fuel Temperature Sensor "B" Circuit Malfunction

P0186 Fuel Temperature Sensor "B" Circuit Range/Performance

P0187 Fuel Temperature Sensor "B" Circuit Low Input

P0188 Fuel Temperature Sensor "B" Circuit High Input

P0189 Fuel Temperature Sensor "B" Circuit Intermittent

P0190 Fuel Rail Pressure Sensor Circuit Malfunction

P0191 Fuel Rail Pressure Sensor Circuit Range/Performance

P0192 Fuel Rail Pressure Sensor Circuit Low Input

P0193 Fuel Rail Pressure Sensor Circuit High Input

P0194 Fuel Rail Pressure Sensor Circuit Intermittent

P0195 Engine Oil Temperature Sensor Malfunction

P0196 Engine Oil Temperature Sensor Range/Performance

P0197 Engine Oil Temperature Sensor Low

P0198 Engine Oil Temperature Sensor High

P0199 Engine Oil Temperature Sensor Intermittent

P0200 Injector Circuit Malfunction

P0201 Injector Circuit Malfunction—Cylinder no. 1

P0202 Injector Circuit Malfunction—Cylinder no. 2

P0203 Injector Circuit Malfunction—Cylinder no. 3

P0204 Injector Circuit Malfunction—Cylinder no. 4

P0205 Injector Circuit Malfunction—Cylinder no. 5

P0206 Injector Circuit Malfunction—Cylinder no. 6

P0207 Injector Circuit Malfunction—Cylinder no. 7

P0208 Injector Circuit Malfunction—Cylinder no. 8

P0209 Injector Circuit Malfunction—Cylinder no. 9

P0210 Injector Circuit Malfunction—Cylinder no. 10
P0211 Injector Circuit Malfunction—Cylinder no. 11
P0212 Injector Circuit Malfunction—Cylinder no. 12
P0213 Cold Start Injector no. 1 Malfunction
P0214 Cold Start Injector no. 2 Malfunction
P0215 Engine Shutoff Solenoid Malfunction
P0216 Injection Timing Control Circuit Malfunction
P0217 Engine Over Temperature Condition
P0218 Transmission Over Temperature Condition
P0219 Engine Over Speed Condition
P0220 Throttle/Pedal Position Sensor/Switch "B" Circuit Malfunction
P0221 Throttle/Pedal Position Sensor/Switch "B" Circuit Range/Performance Problem
P0222 Throttle/Pedal Position Sensor/Switch "B" Circuit Low Input
P0223 Throttle/Pedal Position Sensor/Switch "B" Circuit High Input
P0224 Throttle/Pedal Position Sensor/Switch "B" Circuit Intermittent
P0225 Throttle/Pedal Position Sensor/Switch "C" Circuit Malfunction
P0226 Throttle/Pedal Position Sensor/Switch "C" Circuit Range/Performance Problem
P0227 Throttle/Pedal Position Sensor/Switch "C" Circuit Low Input
P0228 Throttle/Pedal Position Sensor/Switch "C" Circuit High Input
P0229 Throttle/Pedal Position Sensor/Switch "C" Circuit Intermittent
P0230 Fuel Pump Primary Circuit Malfunction
P0231 Fuel Pump Secondary Circuit Low
P0232 Fuel Pump Secondary Circuit High
P0233 Fuel Pump Secondary Circuit Intermittent
P0234 Engine Over Boost Condition
P0261 Cylinder no. 1 Injector Circuit Low
P0262 Cylinder no. 1 Injector Circuit High
P0263 Cylinder no. 1 Contribution/Balance Fault
P0264 Cylinder no. 2 Injector Circuit Low
P0265 Cylinder no. 2 Injector Circuit High
P0266 Cylinder no. 2 Contribution/Balance Fault
P0267 Cylinder no. 3 Injector Circuit Low
P0268 Cylinder no. 3 Injector Circuit High
P0269 Cylinder no. 3 Contribution/Balance Fault
P0270 Cylinder no. 4 Injector Circuit Low
P0271 Cylinder no. 4 Injector Circuit High
P0272 Cylinder no. 4 Contribution/Balance Fault
P0273 Cylinder no. 5 Injector Circuit Low
P0274 Cylinder no. 5 Injector Circuit High
P0275 Cylinder no. 5 Contribution/Balance Fault
P0276 Cylinder no. 6 Injector Circuit Low
P0277 Cylinder no. 6 Injector Circuit High
P0278 Cylinder no. 6 Contribution/Balance Fault
P0279 Cylinder no. 7 Injector Circuit Low
P0280 Cylinder no. 7 Injector Circuit High
P0281 Cylinder no. 7 Contribution/Balance Fault
P0282 Cylinder no. 8 Injector Circuit Low
P0283 Cylinder no. 8 Injector Circuit High
P0284 Cylinder no. 8 Contribution/Balance Fault
P0285 Cylinder no. 9 Injector Circuit Low
P0286 Cylinder no. 9 Injector Circuit High
P0287 Cylinder no. 9 Contribution/Balance Fault

P0288 Cylinder no. 10 Injector Circuit Low
P0289 Cylinder no. 10 Injector Circuit High
P0290 Cylinder no. 10 Contribution/Balance Fault
P0291 Cylinder no. 11 Injector Circuit Low
P0292 Cylinder no. 11 Injector Circuit High
P0293 Cylinder no. 11 Contribution/Balance Fault
P0294 Cylinder no. 12 Injector Circuit Low
P0295 Cylinder no. 12 Injector Circuit High
P0296 Cylinder no. 12 Contribution/Balance Fault
P0300 Random/Multiple Cylinder Misfire Detected
P0301 Cylinder no. 1—Misfire Detected
P0302 Cylinder no. 2—Misfire Detected
P0303 Cylinder no. 3—Misfire Detected
P0304 Cylinder no. 4—Misfire Detected
P0305 Cylinder no. 5—Misfire Detected
P0306 Cylinder no. 6—Misfire Detected
P0307 Cylinder no. 7—Misfire Detected
P0308 Cylinder no. 8—Misfire Detected
P0309 Cylinder no. 9—Misfire Detected
P0310 Cylinder no. 10—Misfire Detected
P0311 Cylinder no. 11—Misfire Detected
P0312 Cylinder no. 12—Misfire Detected
P0320 Ignition/Distributor Engine Speed Input Circuit Malfunction
P0321 Ignition/Distributor Engine Speed Input Circuit Range/Performance
P0322 Ignition/Distributor Engine Speed Input Circuit No Signal
P0323 Ignition/Distributor Engine Speed Input Circuit Intermittent
P0325 Knock Sensor no. 1—Circuit Malfunction (Bank no. 1 or Single Sensor)
P0326 Knock Sensor no. 1—Circuit Range/Performance (Bank no. 1 or Single Sensor)
P0327 Knock Sensor no. 1—Circuit Low Input (Bank no. 1 or Single Sensor)
P0328 Knock Sensor no. 1—Circuit High Input (Bank no. 1 or Single Sensor)
P0329 Knock Sensor no. 1—Circuit Input Intermittent (Bank no. 1 or Single Sensor)
P0330 Knock Sensor no. 2—Circuit Malfunction (Bank no. 2)
P0331 Knock Sensor no. 2—Circuit Range/Performance (Bank no. 2)
P0332 Knock Sensor no. 2—Circuit Low Input (Bank no. 2)
P0333 Knock Sensor no. 2—Circuit High Input (Bank no. 2)
P0334 Knock Sensor no. 2—Circuit Input Intermittent (Bank no. 2)
P0335 Crankshaft Position Sensor "A" Circuit Malfunction
P0336 Crankshaft Position Sensor "A" Circuit Range/Performance
P0337 Crankshaft Position Sensor "A" Circuit Low Input
P0338 Crankshaft Position Sensor "A" Circuit High Input
P0339 Crankshaft Position Sensor "A" Circuit Intermittent
P0340 Camshaft Position Sensor Circuit Malfunction
P0341 Camshaft Position Sensor Circuit Range/Performance
P0342 Camshaft Position Sensor Circuit Low Input
P0343 Camshaft Position Sensor Circuit High Input
P0344 Camshaft Position Sensor Circuit Intermittent
P0350 Ignition Coil Primary/Secondary Circuit Malfunction
P0351 Ignition Coil "A" Primary/Secondary Circuit Malfunction
P0352 Ignition Coil "B" Primary/Secondary Circuit Malfunction
P0353 Ignition Coil "C" Primary/Secondary Circuit Malfunction
P0354 Ignition Coil "D" Primary/Secondary Circuit Malfunction

P0355 Ignition Coil "E" Primary/Secondary Circuit Malfunction
P0356 Ignition Coil "F" Primary/Secondary Circuit Malfunction
P0357 Ignition Coil "G" Primary/Secondary Circuit Malfunction
P0358 Ignition Coil "H" Primary/Secondary Circuit Malfunction
P0359 Ignition Coil "I" Primary/Secondary Circuit Malfunction
P0360 Ignition Coil "J" Primary/Secondary Circuit Malfunction
P0361 Ignition Coil "K" Primary/Secondary Circuit Malfunction
P0362 Ignition Coil "L" Primary/Secondary Circuit Malfunction
P0370 Timing Reference High Resolution Signal "A" Malfunction
P0371 Timing Reference High Resolution Signal "A" Too Many Pulses
P0372 Timing Reference High Resolution Signal "A" Too Few Pulses
P0373 Timing Reference High Resolution Signal "A" Intermittent/Erratic Pulses
P0374 Timing Reference High Resolution Signal "A" No Pulses
P0375 Timing Reference High Resolution Signal "B" Malfunction
P0376 Timing Reference High Resolution Signal "B" Too Many Pulses
P0377 Timing Reference High Resolution Signal "B" Too Few Pulses
P0378 Timing Reference High Resolution Signal "B" Intermittent/Erratic Pulses
P0379 Timing Reference High Resolution Signal "B" No Pulses
P0380 Glow Plug/Heater Circuit "A" Malfunction
P0381 Glow Plug/Heater Indicator Circuit Malfunction
P0382 Glow Plug/Heater Circuit "B" Malfunction
P0385 Crankshaft Position Sensor "B" Circuit Malfunction
P0386 Crankshaft Position Sensor "B" Circuit Range/Performance
P0387 Crankshaft Position Sensor "B" Circuit Low Input
P0388 Crankshaft Position Sensor "B" Circuit High Input
P0389 Crankshaft Position Sensor "B" Circuit Intermittent
P0400 Exhaust Gas Recirculation Flow Malfunction
P0401 Exhaust Gas Recirculation Flow Insufficient Detected
P0402 Exhaust Gas Recirculation Flow Excessive Detected
P0403 Exhaust Gas Recirculation Circuit Malfunction
P0404 Exhaust Gas Recirculation Circuit Range/Performance
P0405 Exhaust Gas Recirculation Sensor "A" Circuit Low
P0406 Exhaust Gas Recirculation Sensor "A" Circuit High
P0407 Exhaust Gas Recirculation Sensor "B" Circuit Low
P0408 Exhaust Gas Recirculation Sensor "B" Circuit High
P0410 Secondary Air Injection System Malfunction
P0411 Secondary Air Injection System Incorrect Flow Detected
P0412 Secondary Air Injection System Switching Valve "A" Circuit Malfunction
P0413 Secondary Air Injection System Switching Valve "A" Circuit Open
P0414 Secondary Air Injection System Switching Valve "A" Circuit Shorted
P0415 Secondary Air Injection System Switching Valve "B" Circuit Malfunction
P0416 Secondary Air Injection System Switching Valve "B" Circuit Open
P0417 Secondary Air Injection System Switching Valve "B" Circuit Shorted
P0418 Secondary Air Injection System Relay "A" Circuit Malfunction
P0419 Secondary Air Injection System Relay "B" Circuit Malfunction

P0420 Catalyst System Efficiency Below Threshold (Bank no. 1
P0421 Warm Up Catalyst Efficiency Below Threshold (Bank no. 1)
P0422 Main Catalyst Efficiency Below Threshold (Bank no. 1)
P0423 Heated Catalyst Efficiency Below Threshold (Bank no. 1)
P0424 Heated Catalyst Temperature Below Threshold (Bank no. 1)
P0430 Catalyst System Efficiency Below Threshold (Bank no. 2
P0431 Warm Up Catalyst Efficiency Below Threshold (Bank no. 2)
P0432 Main Catalyst Efficiency Below Threshold (Bank no. 2)
P0433 Heated Catalyst Efficiency Below Threshold (Bank no. 2)
P0434 Heated Catalyst Temperature Below Threshold (Bank no. 2)
P0440 Evaporative Emission Control System Malfunction
P0441 Evaporative Emission Control System Incorrect Purge Flow
P0442 Evaporative Emission Control System Leak Detected (Small Leak)
P0443 Evaporative Emission Control System Purge Control Valve Circuit Malfunction
P0444 Evaporative Emission Control System Purge Control Valve Circuit Open
P0445 Evaporative Emission Control System Purge Control Valve Circuit Shorted
P0446 Evaporative Emission Control System Vent Control Circuit Malfunction
P0447 Evaporative Emission Control System Vent Control Circuit Open
P0448 Evaporative Emission Control System Vent Control Circuit Shorted
P0449 Evaporative Emission Control System Vent Valve/Solenoid Circuit Malfunction
P0450 Evaporative Emission Control System Pressure Sensor Malfunction
P0451 Evaporative Emission Control System Pressure Sensor Range/Performance
P0452 Evaporative Emission Control System Pressure Sensor Low Input
P0453 Evaporative Emission Control System Pressure Sensor High Input
P0454 Evaporative Emission Control System Pressure Sensor Intermittent
P0455 Evaporative Emission Control System Leak Detected (Gross Leak)
P0460 Fuel Level Sensor Circuit Malfunction
P0461 Fuel Level Sensor Circuit Range/Performance
P0462 Fuel Level Sensor Circuit Low Input
P0463 Fuel Level Sensor Circuit High Input
P0464 Fuel Level Sensor Circuit Intermittent
P0465 Purge Flow Sensor Circuit Malfunction
P0466 Purge Flow Sensor Circuit Range/Performance
P0467 Purge Flow Sensor Circuit Low Input
P0468 Purge Flow Sensor Circuit High Input
P0469 Purge Flow Sensor Circuit Intermittent
P0470 Exhaust Pressure Sensor Malfunction
P0471 Exhaust Pressure Sensor Range/Performance
P0472 Exhaust Pressure Sensor Low
P0473 Exhaust Pressure Sensor High
P0474 Exhaust Pressure Sensor Intermittent
P0475 Exhaust Pressure Control Valve Malfunction
P0476 Exhaust Pressure Control Valve Range/Performance

P0477 Exhaust Pressure Control Valve Low
P0478 Exhaust Pressure Control Valve High
P0479 Exhaust Pressure Control Valve Intermittent
P0480 Cooling Fan no. 1 Control Circuit Malfunction
P0481 Cooling Fan no. 2 Control Circuit Malfunction
P0482 Cooling Fan no. 3 Control Circuit Malfunction
P0483 Cooling Fan Rationality Check Malfunction
P0484 Cooling Fan Circuit Over Current
P0485 Cooling Fan Power/Ground Circuit Malfunction
P0500 Vehicle Speed Sensor Malfunction
P0501 Vehicle Speed Sensor Range/Performance
P0502 Vehicle Speed Sensor Circuit Low Input
P0503 Vehicle Speed Sensor Intermittent/Erratic/High
P0505 Idle Control System Malfunction
P0506 Idle Control System RPM Lower Than Expected
P0507 Idle Control System RPM Higher Than Expected
P0510 Closed Throttle Position Switch Malfunction
P0520 Engine Oil Pressure Sensor/Switch Circuit Malfunction
P0521 Engine Oil Pressure Sensor/Switch Range/Performance
P0522 Engine Oil Pressure Sensor/Switch Low Voltage
P0523 Engine Oil Pressure Sensor/Switch High Voltage
P0530 A/C Refrigerant Pressure Sensor Circuit Malfunction
P0531 A/C Refrigerant Pressure Sensor Circuit Range/Performance
P0532 A/C Refrigerant Pressure Sensor Circuit Low Input
P0533 A/C Refrigerant Pressure Sensor Circuit High Input
P0534 A/C Refrigerant Charge Loss
P0550 Power Steering Pressure Sensor Circuit Malfunction
P0551 Power Steering Pressure Sensor Circuit Range/Performance
P0552 Power Steering Pressure Sensor Circuit Low Input
P0553 Power Steering Pressure Sensor Circuit High Input
P0554 Power Steering Pressure Sensor Circuit Intermittent
P0560 System Voltage Malfunction
P0561 System Voltage Unstable
P0562 System Voltage Low
P0563 System Voltage High
P0565 Cruise Control On Signal Malfunction
P0566 Cruise Control Off Signal Malfunction
P0567 Cruise Control Resume Signal Malfunction
P0568 Cruise Control Set Signal Malfunction
P0569 Cruise Control Coast Signal Malfunction
P0570 Cruise Control Accel Signal Malfunction
P0571 Cruise Control/Brake Switch "A" Circuit Malfunction
P0572 Cruise Control/Brake Switch "A" Circuit Low
P0573 Cruise Control/Brake Switch "A" Circuit High
P0574 **Through P0580** Reserved for Cruise Codes
P0600 Serial Communication Link Malfunction
P0601 Internal Control Module Memory Check Sum Error
P0602 Control Module Programming Error
P0603 Internal Control Module Keep Alive Memory (KAM) Error
P0604 Internal Control Module Random Access Memory (RAM) Error
P0605 Internal Control Module Read Only Memory (ROM) Error
P0606 PCM Processor Fault
P0608 Control Module VSS Output "A" Malfunction
P0609 Control Module VSS Output "B" Malfunction
P0620 Generator Control Circuit Malfunction
P0621 Generator Lamp "L" Control Circuit Malfunction
P0622 Generator Field "F" Control Circuit Malfunction
P0650 Malfunction Indicator Lamp (MIL) Control Circuit Malfunction

P0654 Engine RPM Output Circuit Malfunction
P0655 Engine Hot Lamp Output Control Circuit Malfunction
P0656 Fuel Level Output Circuit Malfunction
P0700 Transmission Control System Malfunction
P0701 Transmission Control System Range/Performance
P0702 Transmission Control System Electrical
P0703 Torque Converter/Brake Switch "B" Circuit Malfunction
P0704 Clutch Switch Input Circuit Malfunction
P0705 Transmission Range Sensor Circuit Malfunction (PRNDL Input)
P0706 Transmission Range Sensor Circuit Range/Performance
P0707 Transmission Range Sensor Circuit Low Input
P0708 Transmission Range Sensor Circuit High Input
P0709 Transmission Range Sensor Circuit Intermittent
P0710 Transmission Fluid Temperature Sensor Circuit Malfunction
P0711 Transmission Fluid Temperature Sensor Circuit Range/Performance
P0712 Transmission Fluid Temperature Sensor Circuit Low Input
P0713 Transmission Fluid Temperature Sensor Circuit High Input
P0714 Transmission Fluid Temperature Sensor Circuit Intermittent
P0715 Input/Turbine Speed Sensor Circuit Malfunction
P0716 Input/Turbine Speed Sensor Circuit Range/Performance
P0717 Input/Turbine Speed Sensor Circuit No Signal
P0718 Input/Turbine Speed Sensor Circuit Intermittent
P0719 Torque Converter/Brake Switch "B" Circuit Low
P0720 Output Speed Sensor Circuit Malfunction
P0721 Output Speed Sensor Circuit Range/Performance
P0722 Output Speed Sensor Circuit No Signal
P0723 Output Speed Sensor Circuit Intermittent
P0724 Torque Converter/Brake Switch "B" Circuit High
P0725 Engine Speed Input Circuit Malfunction
P0726 Engine Speed Input Circuit Range/Performance
P0727 Engine Speed Input Circuit No Signal
P0728 Engine Speed Input Circuit Intermittent
P0730 Incorrect Gear Ratio
P0731 Gear no. 1 Incorrect Ratio
P0732 Gear no. 2 Incorrect Ratio
P0733 Gear no. 3 Incorrect Ratio
P0734 Gear no. 4 Incorrect Ratio
P0735 Gear no. 5 Incorrect Ratio
P0736 Reverse Incorrect Ratio
P0740 Torque Converter Clutch Circuit Malfunction
P0741 Torque Converter Clutch Circuit Performance or Stuck Off
P0742 Torque Converter Clutch Circuit Stuck On
P0743 Torque Converter Clutch Circuit Electrical
P0744 Torque Converter Clutch Circuit Intermittent
P0745 Pressure Control Solenoid Malfunction
P0746 Pressure Control Solenoid Performance or Stuck Off
P0747 Pressure Control Solenoid Stuck On
P0748 Pressure Control Solenoid Electrical
P0749 Pressure Control Solenoid Intermittent
P0750 Shift Solenoid "A" Malfunction
P0751 Shift Solenoid "A" Performance or Stuck Off
P0752 Shift Solenoid "A" Stuck On
P0753 Shift Solenoid "A" Electrical
P0754 Shift Solenoid "A" Intermittent
P0755 Shift Solenoid "B" Malfunction

P0756 Shift Solenoid "B" Performance or Stuck Off

P0757 Shift Solenoid "B" Stuck On

P0758 Shift Solenoid "B" Electrical

P0759 Shift Solenoid "B" Intermittent

P0760 Shift Solenoid "C" Malfunction

P0761 Shift Solenoid "C" Performance Or Stuck Off

P0762 Shift Solenoid "C" Stuck On

P0763 Shift Solenoid "C" Electrical

P0764 Shift Solenoid "C" Intermittent

P0765 Shift Solenoid "D" Malfunction

P0766 Shift Solenoid "D" Performance Or Stuck Off

P0767 Shift Solenoid "D" Stuck On

P0768 Shift Solenoid "D" Electrical

P0769 Shift Solenoid "D" Intermittent

P0770 Shift Solenoid "E" Malfunction

P0771 Shift Solenoid "E" Performance Or Stuck Off

P0772 Shift Solenoid "E" Stuck On

P0773 Shift Solenoid "E" Electrical

P0774 Shift Solenoid "E" Intermittent

P0780 Shift Malfunction

P0781 1–2 Shift Malfunction

P0782 2–3 Shift Malfunction

P0783 3–4 Shift Malfunction

P0784 4–5 Shift Malfunction

P0785 Shift/Timing Solenoid Malfunction

P0786 Shift/Timing Solenoid Range/Performance

P0787 Shift/Timing Solenoid Low

P0788 Shift/Timing Solenoid High

P0789 Shift/Timing Solenoid Intermittent

P0790 Normal/Performance Switch Circuit Malfunction

P0801 Reverse Inhibit Control Circuit Malfunction

P0803 1–4 Upshift (Skip Shift) Solenoid Control Circuit Malfunction

P0804 1–4 Upshift (Skip Shift) Lamp Control Circuit Malfunction

P1106 MAP Sensor Voltage Intermittently High (Except 2.2L)

P1107 MAP Sensor Voltage Intermittently Low (Except 2.2L)

P1111 IAT Sensor Circuit Intermittent High Voltage (Except 2.2L)

P1112 IAT Sensor Circuit Intermittent Low Voltage (Except 2.2L)

P1114 ECT Sensor Circuit Intermittent Low Voltage (Except 2.2L)

P1115 ECT Sensor Circuit Intermittent High Voltage (Except 2.2L)

P1121 TP Sensor Voltage Intermittently High (Except 2.2L)

P1122 TP Sensor Voltage Intermittently Low (Except 2.2L)

P1133 HO2S Insufficient Switching Sensor (3.4L)

P1133 HO2S Insufficient Switching Bank #1, Sensor #1 (Except 3.4L & 4.3L)

P1134 HO2S #1 Transition Time Ratio (3.4L)

P1134 HO2S Transition Time Ratio Bank #1, Sensor #1 (4.3L, 5.0L, 5.7L & 7.4L)

P1153 HO2S Insufficient Switching Sensor Bank #2, Sensor #1 (4.3L, 5.0L, 5.7L & 7.4L)

P1154 HO2S Transition Time Ratio Bank #2, Sensor #1 (4.3L, 5.0L, 5.7L & 7.4L)

P1345 Crankshaft/Camshaft (CKP/CMP) Correlation (4.3L, 5.0L, 5.7L & 7.4L)

P1350 Ignition Control (IC) Circuit Malfunction (3.4L)

P1351 Ignition Control (IC) Circuit High Voltage (4.3L, 5.0L, 5.7L & 7.4L)

P1361 Ignition Control (IC) Circuit Not Toggling (3.4L)

P1361 Ignition Control (IC) Circuit Low Voltage (4.3L, 5.0L, 5.7L & 7.4L)

P1380 Electronic Brake Control Module (EBCM) DTC Detected Rough Road Data Unusable

P1381 Misfire Detected, No EBCM/PCM/VCM Serial Data (Except "P" Series)

P1406 EGR Pintle Position Circuit Fault (Except "P" Series)

P1415 AIR System Bank #1 (Except "P" Series)

P1416 AIR System Bank #2 (Except "P" Series)

P1441 EVAP Control System Flow During Non-Purge

P1442 EVAP Vacuum Switch Circuit (3.4L)

P1508 IAC System Low RPM (4.3L, 5.0L, 5.7L & 7.4L)

P1509 IAC System High RPM (4.3L, 5.7L & 7.4L)

P1520 PNP Circuit (2.2L)

P1635 5-Volt Reference "A" Circuit (3.4L)

P1639 5-Volt Reference "B" Circuit (3.4L)

P1641 MIL Control Circuit (3.4L)

P1651 Fan #1 Relay Control Circuit (3.4L)

P1652 Fan #2 Relay Control Circuit (3.4L)

P1654 A/C Relay Control (3.4L)

P1655 EVAP Purge Solenoid Control Circuit (3.4L)

P1672 Low Engine Oil Level Light Control Circuit (3.4L)

Geo/Chevrolet

READING CODES

Reading the control module memory is one of the first steps in OBD II system diagnostics. This step should be initially performed to determine the general nature of the fault. Subsequent readings will determine if the fault has been cleared.

Reading codes can be performed by any of the methods below:
- Read the control module memory with the Generic Scan Tool (GST)
- Read the control module memory with the vehicle manufacturer's specific tester

To read the fault codes, connect the scan tool or tester according to the manufacturer's instructions. Follow the manufacturer's specified procedure for reading the codes.

CLEARING CODES

Control module reset procedures are a very important part of OBD II System diagnostics. This step should be done at the end of any fault code repair and at the end of any driveability repair.

Clearing codes can be performed by any of the methods below:
- Clear the control module memory with the Generic Scan Tool (GST)
- Clear the control module memory with the vehicle manufacturer's specific tester
- Turn the ignition off and remove the negative battery cable for at least 1 minute.

Removing the negative battery cable may cause other systems in the vehicle to loose their memory. Prior to removing the cable, ensure you have the proper reset codes for radios and alarms.

➡**The MIL will may also be de-activated for some codes if the vehicle completes three consecutive trips without a fault detected with vehicle conditions similar to those present during the fault.**

OBD II TROUBLE CODES

P0100 Mass or Volume Air Flow Circuit Malfunction

P0101 Mass or Volume Air Flow Circuit Range/Performance Problem

P0102 Mass or Volume Air Flow Circuit Low Input

P0103 Mass or Volume Air Flow Circuit High Input

P0104 Mass or Volume Air Flow Circuit Intermittent

P0105 Manifold Absolute Pressure/Barometric Pressure Circuit Malfunction

P0106 Manifold Absolute Pressure/Barometric Pressure Circuit Range/Performance Problem

P0107 Manifold Absolute Pressure/Barometric Pressure Circuit Low Input

P0108 Manifold Absolute Pressure/Barometric Pressure Circuit High Input

P0109 Manifold Absolute Pressure/Barometric Pressure Circuit Intermittent

P0110 Intake Air Temperature Circuit Malfunction

P0111 Intake Air Temperature Circuit Range/Performance Problem

P0112 Intake Air Temperature Circuit Low Input

P0113 Intake Air Temperature Circuit High Input

P0114 Intake Air Temperature Circuit Intermittent

P0115 Engine Coolant Temperature Circuit Malfunction

P0116 Engine Coolant Temperature Circuit Range/Performance Problem

P0117 Engine Coolant Temperature Circuit Low Input

P0118 Engine Coolant Temperature Circuit High Input

P0119 Engine Coolant Temperature Circuit Intermittent

P0120 Throttle/Pedal Position Sensor/Switch "A" Circuit Malfunction

P0121 Throttle/Pedal Position Sensor/Switch "A" Circuit Range/Performance Problem

P0122 Throttle/Pedal Position Sensor/Switch "A" Circuit Low Input

P0123 Throttle/Pedal Position Sensor/Switch "A" Circuit High Input

P0124 Throttle/Pedal Position Sensor/Switch "A" Circuit Intermittent

P0125 Insufficient Coolant Temperature For Closed Loop Fuel Control

P0126 Insufficient Coolant Temperature For Stable Operation

P0130 O2 Circuit Malfunction (Bank no. 1 Sensor no. 1)

P0131 O2 Sensor Circuit Low Voltage (Bank no. 1 Sensor no. 1)

P0132 O2 Sensor Circuit High Voltage (Bank no. 1 Sensor no. 1)

P0133 O2 Sensor Circuit Slow Response (Bank no. 1 Sensor no. 1)

P0134 O2 Sensor Circuit No Activity Detected (Bank no. 1 Sensor no. 1)

P0135 O2 Sensor Heater Circuit Malfunction (Bank no. 1 Sensor no. 1)

P0136 O2 Sensor Circuit Malfunction (Bank no. 1 Sensor no. 2)

P0137 O2 Sensor Circuit Low Voltage (Bank no. 1 Sensor no. 2)

P0138 O2 Sensor Circuit High Voltage (Bank no. 1 Sensor no. 2)

P0139 O2 Sensor Circuit Slow Response (Bank no. 1 Sensor no. 2)

P0140 O2 Sensor Circuit No Activity Detected (Bank no. 1 Sensor no. 2)

P0141 O2 Sensor Heater Circuit Malfunction (Bank no. 1 Sensor no. 2)

P0142 O2 Sensor Circuit Malfunction (Bank no. 1 Sensor no. 3)

P0143 O2 Sensor Circuit Low Voltage (Bank no. 1 Sensor no. 3)

P0144 O2 Sensor Circuit High Voltage (Bank no. 1 Sensor no. 3)

P0145 O2 Sensor Circuit Slow Response (Bank no. 1 Sensor no. 3)

P0146 O2 Sensor Circuit No Activity Detected (Bank no. 1 Sensor no. 3)

P0147 O2 Sensor Heater Circuit Malfunction (Bank no. 1 Sensor no. 3)

P0150 O2 Sensor Circuit Malfunction (Bank no. 2 Sensor no. 1)

P0151 O2 Sensor Circuit Low Voltage (Bank no. 2 Sensor no. 1)

P0152 O2 Sensor Circuit High Voltage (Bank no. 2 Sensor no. 1)

P0153 O2 Sensor Circuit Slow Response (Bank no. 2 Sensor no. 1)

P0154 O2 Sensor Circuit No Activity Detected (Bank no. 2 Sensor no. 1)

P0155 O2 Sensor Heater Circuit Malfunction (Bank no. 2 Sensor no. 1)

P0156 O2 Sensor Circuit Malfunction (Bank no. 2 Sensor no. 2)

P0157 O2 Sensor Circuit Low Voltage (Bank no. 2 Sensor no. 2)

P0158 O2 Sensor Circuit High Voltage (Bank no. 2 Sensor no. 2)

P0159 O2 Sensor Circuit Slow Response (Bank no. 2 Sensor no. 2)

P0160 O2 Sensor Circuit No Activity Detected (Bank no. 2 Sensor no. 2)

P0161 O2 Sensor Heater Circuit Malfunction (Bank no. 2 Sensor no. 2)

P0162 O2 Sensor Circuit Malfunction (Bank no. 2 Sensor no. 3)

P0163 O2 Sensor Circuit Low Voltage (Bank no. 2 Sensor no. 3)

P0164 O2 Sensor Circuit High Voltage (Bank no. 2 Sensor no. 3)

P0165 O2 Sensor Circuit Slow Response (Bank no. 2 Sensor no. 3)

P0166 O2 Sensor Circuit No Activity Detected (Bank no. 2 Sensor no. 3)

P0167 O2 Sensor Heater Circuit Malfunction (Bank no. 2 Sensor no. 3)

P0170 Fuel Trim Malfunction (Bank no. 1)

P0171 System Too Lean (Bank no. 1)

P0172 System Too Rich (Bank no. 1)

P0173 Fuel Trim Malfunction (Bank no. 2)

P0174 System Too Lean (Bank no. 2)

P0175 System Too Rich (Bank no. 2)

P0176 Fuel Composition Sensor Circuit Malfunction

P0177 Fuel Composition Sensor Circuit Range/Performance

P0178 Fuel Composition Sensor Circuit Low Input

P0179 Fuel Composition Sensor Circuit High Input

P0180 Fuel Temperature Sensor "A" Circuit Malfunction

P0181 Fuel Temperature Sensor "A" Circuit Range/Performance

P0182 Fuel Temperature Sensor "A" Circuit Low Input

P0183 Fuel Temperature Sensor "A" Circuit High Input

P0184 Fuel Temperature Sensor "A" Circuit Intermittent

P0185 Fuel Temperature Sensor "B" Circuit Malfunction

P0186 Fuel Temperature Sensor "B" Circuit Range/Performance

P0187 Fuel Temperature Sensor "B" Circuit Low Input

P0188 Fuel Temperature Sensor "B" Circuit High Input

P0189 Fuel Temperature Sensor "B" Circuit Intermittent
P0190 Fuel Rail Pressure Sensor Circuit Malfunction
P0191 Fuel Rail Pressure Sensor Circuit Range/Performance
P0192 Fuel Rail Pressure Sensor Circuit Low Input
P0193 Fuel Rail Pressure Sensor Circuit High Input
P0194 Fuel Rail Pressure Sensor Circuit Intermittent
P0195 Engine Oil Temperature Sensor Malfunction
P0196 Engine Oil Temperature Sensor Range/Performance
P0197 Engine Oil Temperature Sensor Low
P0198 Engine Oil Temperature Sensor High
P0199 Engine Oil Temperature Sensor Intermittent
P0200 Injector Circuit Malfunction
P0201 Injector Circuit Malfunction—Cylinder no. 1
P0202 Injector Circuit Malfunction—Cylinder no. 2
P0203 Injector Circuit Malfunction—Cylinder no. 3
P0204 Injector Circuit Malfunction—Cylinder no. 4
P0205 Injector Circuit Malfunction—Cylinder no. 5
P0206 Injector Circuit Malfunction—Cylinder no. 6
P0207 Injector Circuit Malfunction—Cylinder no. 7
P0208 Injector Circuit Malfunction—Cylinder no. 8
P0209 Injector Circuit Malfunction—Cylinder no. 9
P0210 Injector Circuit Malfunction—Cylinder no. 10
P0211 Injector Circuit Malfunction—Cylinder no. 11
P0212 Injector Circuit Malfunction—Cylinder no. 12
P0213 Cold Start Injector no. 1 Malfunction
P0214 Cold Start Injector no. 2 Malfunction
P0215 Engine Shutoff Solenoid Malfunction
P0216 Injection Timing Control Circuit Malfunction
P0217 Engine Over Temperature Condition
P0218 Transmission Over Temperature Condition
P0219 Engine Over Speed Condition
P0220 Throttle/Pedal Position Sensor/Switch "B" Circuit Malfunction
P0221 Throttle/Pedal Position Sensor/Switch "B" Circuit Range/Performance Problem
P0222 Throttle/Pedal Position Sensor/Switch "B" Circuit Low Input
P0223 Throttle/Pedal Position Sensor/Switch "B" Circuit High Input
P0224 Throttle/Pedal Position Sensor/Switch "B" Circuit Intermittent
P0225 Throttle/Pedal Position Sensor/Switch "C" Circuit Malfunction
P0226 Throttle/Pedal Position Sensor/Switch "C" Circuit Range/Performance Problem
P0227 Throttle/Pedal Position Sensor/Switch "C" Circuit Low Input
P0228 Throttle/Pedal Position Sensor/Switch "C" Circuit High Input
P0229 Throttle/Pedal Position Sensor/Switch "C" Circuit Intermittent
P0230 Fuel Pump Primary Circuit Malfunction
P0231 Fuel Pump Secondary Circuit Low
P0232 Fuel Pump Secondary Circuit High
P0233 Fuel Pump Secondary Circuit Intermittent
P0234 Engine Over Boost Condition
P0261 Cylinder no. 1 Injector Circuit Low
P0262 Cylinder no. 1 Injector Circuit High
P0263 Cylinder no. 1 Contribution/Balance Fault
P0264 Cylinder no. 2 Injector Circuit Low
P0265 Cylinder no. 2 Injector Circuit High
P0266 Cylinder no. 2 Contribution/Balance Fault

P0267 Cylinder no. 3 Injector Circuit Low
P0268 Cylinder no. 3 Injector Circuit High
P0269 Cylinder no. 3 Contribution/Balance Fault
P0270 Cylinder no. 4 Injector Circuit Low
P0271 Cylinder no. 4 Injector Circuit High
P0272 Cylinder no. 4 Contribution/Balance Fault
P0273 Cylinder no. 5 Injector Circuit Low
P0274 Cylinder no. 5 Injector Circuit High
P0275 Cylinder no. 5 Contribution/Balance Fault
P0276 Cylinder no. 6 Injector Circuit Low
P0277 Cylinder no. 6 Injector Circuit High
P0278 Cylinder no. 6 Contribution/Balance Fault
P0279 Cylinder no. 7 Injector Circuit Low
P0280 Cylinder no. 7 Injector Circuit High
P0281 Cylinder no. 7 Contribution/Balance Fault
P0282 Cylinder no. 8 Injector Circuit Low
P0283 Cylinder no. 8 Injector Circuit High
P0284 Cylinder no. 8 Contribution/Balance Fault
P0285 Cylinder no. 9 Injector Circuit Low
P0286 Cylinder no. 9 Injector Circuit High
P0287 Cylinder no. 9 Contribution/Balance Fault
P0288 Cylinder no. 10 Injector Circuit Low
P0289 Cylinder no. 10 Injector Circuit High
P0290 Cylinder no. 10 Contribution/Balance Fault
P0291 Cylinder no. 11 Injector Circuit Low
P0292 Cylinder no. 11 Injector Circuit High
P0293 Cylinder no. 11 Contribution/Balance Fault
P0294 Cylinder no. 12 Injector Circuit Low
P0295 Cylinder no. 12 Injector Circuit High
P0296 Cylinder no. 12 Contribution/Balance Fault
P0300 Random/Multiple Cylinder Misfire Detected
P0301 Cylinder no. 1—Misfire Detected
P0302 Cylinder no. 2—Misfire Detected
P0303 Cylinder no. 3—Misfire Detected
P0304 Cylinder no. 4—Misfire Detected
P0305 Cylinder no. 5—Misfire Detected
P0306 Cylinder no. 6—Misfire Detected
P0307 Cylinder no. 7—Misfire Detected
P0308 Cylinder no. 8—Misfire Detected
P0309 Cylinder no. 9—Misfire Detected
P0310 Cylinder no. 10—Misfire Detected
P0311 Cylinder no. 11—Misfire Detected
P0312 Cylinder no. 12—Misfire Detected
P0320 Ignition/Distributor Engine Speed Input Circuit Malfunction
P0321 Ignition/Distributor Engine Speed Input Circuit Range/Performance
P0322 Ignition/Distributor Engine Speed Input Circuit No Signal
P0323 Ignition/Distributor Engine Speed Input Circuit Intermittent
P0325 Knock Sensor no. 1—Circuit Malfunction (Bank no. 1 or Single Sensor)
P0326 Knock Sensor no. 1—Circuit Range/Performance (Bank no. 1 or Single Sensor)
P0327 Knock Sensor no. 1—Circuit Low Input (Bank no. 1 or Single Sensor)
P0328 Knock Sensor no. 1—Circuit High Input (Bank no. 1 or Single Sensor)
P0329 Knock Sensor no. 1—Circuit Input Intermittent (Bank no. 1 or Single Sensor)
P0330 Knock Sensor no. 2—Circuit Malfunction (Bank no. 2)
P0331 Knock Sensor no. 2—Circuit Range/Performance (Bank no. 2)

P0332 Knock Sensor no. 2—Circuit Low Input (Bank no. 2)

P0333 Knock Sensor no. 2—Circuit High Input (Bank no. 2)

P0334 Knock Sensor no. 2—Circuit Input Intermittent (Bank no.)

P0335 Crankshaft Position Sensor "A" Circuit Malfunction

P0336 Crankshaft Position Sensor "A" Circuit Range/Performance

P0337 Crankshaft Position Sensor "A" Circuit Low Input

P0338 Crankshaft Position Sensor "A" Circuit High Input

P0339 Crankshaft Position Sensor "A" Circuit Intermittent

P0340 Camshaft Position Sensor Circuit Malfunction

P0341 Camshaft Position Sensor Circuit Range/Performance

P0342 Camshaft Position Sensor Circuit Low Input

P0343 Camshaft Position Sensor Circuit High Input

P0344 Camshaft Position Sensor Circuit Intermittent

P0350 Ignition Coil Primary/Secondary Circuit Malfunction

P0351 Ignition Coil "A" Primary/Secondary Circuit Malfunction

P0352 Ignition Coil "B" Primary/Secondary Circuit Malfunction

P0353 Ignition Coil "C" Primary/Secondary Circuit Malfunction

P0354 Ignition Coil "D" Primary/Secondary Circuit Malfunction

P0355 Ignition Coil "E" Primary/Secondary Circuit Malfunction

P0356 Ignition Coil "F" Primary/Secondary Circuit Malfunction

P0357 Ignition Coil "G" Primary/Secondary Circuit Malfunction

P0358 Ignition Coil "H" Primary/Secondary Circuit Malfunction

P0359 Ignition Coil "I" Primary/Secondary Circuit Malfunction

P0360 Ignition Coil "J" Primary/Secondary Circuit Malfunction

P0361 Ignition Coil "K" Primary/Secondary Circuit Malfunction

P0362 Ignition Coil "L" Primary/Secondary Circuit Malfunction

P0370 Timing Reference High Resolution Signal "A" Malfunction

P0371 Timing Reference High Resolution Signal "A" Too Many Pulses

P0372 Timing Reference High Resolution Signal "A" Too Few Pulses

P0373 Timing Reference High Resolution Signal "A" Intermittent/Erratic Pulses

P0374 Timing Reference High Resolution Signal "A" No Pulses

P0375 Timing Reference High Resolution Signal "B" Malfunction

P0376 Timing Reference High Resolution Signal "B" Too Many Pulses

P0377 Timing Reference High Resolution Signal "B" Too Few Pulses

P0378 Timing Reference High Resolution Signal "B" Intermittent/Erratic Pulses

P0379 Timing Reference High Resolution Signal "B" No Pulses

P0380 Glow Plug/Heater Circuit "A" Malfunction

P0381 Glow Plug/Heater Indicator Circuit Malfunction

P0382 Glow Plug/Heater Circuit "B" Malfunction

P0385 Crankshaft Position Sensor "B" Circuit Malfunction

P0386 Crankshaft Position Sensor "B" Circuit Range/Performance

P0387 Crankshaft Position Sensor "B" Circuit Low Input

P0388 Crankshaft Position Sensor "B" Circuit High Input

P0389 Crankshaft Position Sensor "B" Circuit Intermittent

P0400 Exhaust Gas Recirculation Flow Malfunction

P0401 Exhaust Gas Recirculation Flow Insufficient Detected

P0402 Exhaust Gas Recirculation Flow Excessive Detected

P0403 Exhaust Gas Recirculation Circuit Malfunction

P0404 Exhaust Gas Recirculation Circuit Range/Performance

P0405 Exhaust Gas Recirculation Sensor "A" Circuit Low

P0406 Exhaust Gas Recirculation Sensor "A" Circuit High

P0407 Exhaust Gas Recirculation Sensor "B" Circuit Low

P0408 Exhaust Gas Recirculation Sensor "B" Circuit High

P0410 Secondary Air Injection System Malfunction

P0411 Secondary Air Injection System Incorrect Flow Detected

P0412 Secondary Air Injection System Switching Valve "A" Circuit Malfunction

P0413 Secondary Air Injection System Switching Valve "A" Circuit Open

P0414 Secondary Air Injection System Switching Valve "A" Circuit Shorted

P0415 Secondary Air Injection System Switching Valve "B" Circuit Malfunction

P0416 Secondary Air Injection System Switching Valve "B" Circuit Open

P0417 Secondary Air Injection System Switching Valve "B" Circuit Shorted

P0418 Secondary Air Injection System Relay "A" Circuit Malfunction

P0419 Secondary Air Injection System Relay "B" Circuit Malfunction

P0420 Catalyst System Efficiency Below Threshold (Bank no. 1)

P0421 Warm Up Catalyst Efficiency Below Threshold (Bank no. 1)

P0422 Main Catalyst Efficiency Below Threshold (Bank no. 1)

P0423 Heated Catalyst Efficiency Below Threshold (Bank no. 1)

P0424 Heated Catalyst Temperature Below Threshold (Bank no. 1)

P0430 Catalyst System Efficiency Below Threshold (Bank no. 2)

P0431 Warm Up Catalyst Efficiency Below Threshold (Bank no. 2)

P0432 Main Catalyst Efficiency Below Threshold (Bank no. 2)

P0433 Heated Catalyst Efficiency Below Threshold (Bank no. 2)

P0434 Heated Catalyst Temperature Below Threshold (Bank no. 2)

P0440 Evaporative Emission Control System Malfunction

P0441 Evaporative Emission Control System Incorrect Purge Flow

P0442 Evaporative Emission Control System Leak Detected (Small Leak)

P0443 Evaporative Emission Control System Purge Control Valve Circuit Malfunction

P0444 Evaporative Emission Control System Purge Control Valve Circuit Open

P0445 Evaporative Emission Control System Purge Control Valve Circuit Shorted

P0446 Evaporative Emission Control System Vent Control Circuit Malfunction

P0447 Evaporative Emission Control System Vent Control Circuit Open

P0448 Evaporative Emission Control System Vent Control Circuit Shorted

P0449 Evaporative Emission Control System Vent Valve/Solenoid Circuit Malfunction

P0450 Evaporative Emission Control System Pressure Sensor Malfunction

P0451 Evaporative Emission Control System Pressure Sensor Range/Performance

P0452 Evaporative Emission Control System Pressure Sensor Low Input

P0453 Evaporative Emission Control System Pressure Sensor High Input

P0454 Evaporative Emission Control System Pressure Sensor Intermittent

P0455 Evaporative Emission Control System Leak Detected (Gross Leak)
P0460 Fuel Level Sensor Circuit Malfunction
P0461 Fuel Level Sensor Circuit Range/Performance
P0462 Fuel Level Sensor Circuit Low Input
P0463 Fuel Level Sensor Circuit High Input
P0464 Fuel Level Sensor Circuit Intermittent
P0465 Purge Flow Sensor Circuit Malfunction
P0466 Purge Flow Sensor Circuit Range/Performance
P0467 Purge Flow Sensor Circuit Low Input
P0468 Purge Flow Sensor Circuit High Input
P0469 Purge Flow Sensor Circuit Intermittent
P0470 Exhaust Pressure Sensor Malfunction
P0471 Exhaust Pressure Sensor Range/Performance
P0472 Exhaust Pressure Sensor Low
P0473 Exhaust Pressure Sensor High
P0474 Exhaust Pressure Sensor Intermittent
P0475 Exhaust Pressure Control Valve Malfunction
P0476 Exhaust Pressure Control Valve Range/Performance
P0477 Exhaust Pressure Control Valve Low
P0478 Exhaust Pressure Control Valve High
P0479 Exhaust Pressure Control Valve Intermittent
P0480 Cooling Fan no. 1 Control Circuit Malfunction
P0481 Cooling Fan no. 2 Control Circuit Malfunction
P0482 Cooling Fan no. 3 Control Circuit Malfunction
P0483 Cooling Fan Rationality Check Malfunction
P0484 Cooling Fan Circuit Over Current
P0485 Cooling Fan Power/Ground Circuit Malfunction
P0500 Vehicle Speed Sensor Malfunction
P0501 Vehicle Speed Sensor Range/Performance
P0502 Vehicle Speed Sensor Circuit Low Input
P0503 Vehicle Speed Sensor Intermittent/Erratic/High
P0505 Idle Control System Malfunction
P0506 Idle Control System RPM Lower Than Expected
P0507 Idle Control System RPM Higher Than Expected
P0510 Closed Throttle Position Switch Malfunction
P0520 Engine Oil Pressure Sensor/Switch Circuit Malfunction
P0521 Engine Oil Pressure Sensor/Switch Range/Performance
P0522 Engine Oil Pressure Sensor/Switch Low Voltage
P0523 Engine Oil Pressure Sensor/Switch High Voltage
P0530 A/C Refrigerant Pressure Sensor Circuit Malfunction
P0531 A/C Refrigerant Pressure Sensor Circuit Range/Performance
P0532 A/C Refrigerant Pressure Sensor Circuit Low Input
P0533 A/C Refrigerant Pressure Sensor Circuit High Input
P0534 A/C Refrigerant Charge Loss
P0550 Power Steering Pressure Sensor Circuit Malfunction
P0551 Power Steering Pressure Sensor Circuit Range/Performance
P0552 Power Steering Pressure Sensor Circuit Low Input
P0553 Power Steering Pressure Sensor Circuit High Input
P0554 Power Steering Pressure Sensor Circuit Intermittent
P0560 System Voltage Malfunction
P0561 System Voltage Unstable
P0562 System Voltage Low
P0563 System Voltage High
P0565 Cruise Control On Signal Malfunction
P0566 Cruise Control Off Signal Malfunction
P0567 Cruise Control Resume Signal Malfunction
P0568 Cruise Control Set Signal Malfunction
P0569 Cruise Control Coast Signal Malfunction
P0570 Cruise Control Accel Signal Malfunction

P0571 Cruise Control/Brake Switch "A" Circuit Malfunction
P0572 Cruise Control/Brake Switch "A" Circuit Low
P0573 Cruise Control/Brake Switch "A" Circuit High
P0574 **Through P0580** Reserved for Cruise Codes
P0600 Serial Communication Link Malfunction
P0601 Internal Control Module Memory Check Sum Error
P0602 Control Module Programming Error
P0603 Internal Control Module Keep Alive Memory (KAM) Error
P0604 Internal Control Module Random Access Memory (RAM) Error
P0605 Internal Control Module Read Only Memory (ROM) Error
P0606 PCM Processor Fault
P0608 Control Module VSS Output "A" Malfunction
P0609 Control Module VSS Output "B" Malfunction
P0620 Generator Control Circuit Malfunction
P0621 Generator Lamp "L" Control Circuit Malfunction
P0622 Generator Field "F" Control Circuit Malfunction
P0650 Malfunction Indicator Lamp (MIL) Control Circuit Malfunction
P0654 Engine RPM Output Circuit Malfunction
P0655 Engine Hot Lamp Output Control Circuit Malfunction
P0656 Fuel Level Output Circuit Malfunction
P0700 Transmission Control System Malfunction
P0701 Transmission Control System Range/Performance
P0702 Transmission Control System Electrical
P0703 Torque Converter/Brake Switch "B" Circuit Malfunction
P0704 Clutch Switch Input Circuit Malfunction
P0705 Transmission Range Sensor Circuit Malfunction (PRNDL Input)
P0706 Transmission Range Sensor Circuit Range/Performance
P0707 Transmission Range Sensor Circuit Low Input
P0708 Transmission Range Sensor Circuit High Input
P0709 Transmission Range Sensor Circuit Intermittent
P0710 Transmission Fluid Temperature Sensor Circuit Malfunction
P0711 Transmission Fluid Temperature Sensor Circuit Range/Performance
P0712 Transmission Fluid Temperature Sensor Circuit Low Input
P0713 Transmission Fluid Temperature Sensor Circuit High Input
P0714 Transmission Fluid Temperature Sensor Circuit Intermittent
P0715 Input/Turbine Speed Sensor Circuit Malfunction
P0716 Input/Turbine Speed Sensor Circuit Range/Performance
P0717 Input/Turbine Speed Sensor Circuit No Signal
P0718 Input/Turbine Speed Sensor Circuit Intermittent
P0719 Torque Converter/Brake Switch "B" Circuit Low
P0720 Output Speed Sensor Circuit Malfunction
P0721 Output Speed Sensor Circuit Range/Performance
P0722 Output Speed Sensor Circuit No Signal
P0723 Output Speed Sensor Circuit Intermittent
P0724 Torque Converter/Brake Switch "B" Circuit High
P0725 Engine Speed Input Circuit Malfunction
P0726 Engine Speed Input Circuit Range/Performance
P0727 Engine Speed Input Circuit No Signal
P0728 Engine Speed Input Circuit Intermittent
P0730 Incorrect Gear Ratio
P0731 Gear no. 1 Incorrect Ratio
P0732 Gear no. 2 Incorrect Ratio
P0733 Gear no. 3 Incorrect Ratio
P0734 Gear no. 4 Incorrect Ratio

P0735 Gear no. 5 Incorrect Ratio
P0736 Reverse Incorrect Ratio
P0740 Torque Converter Clutch Circuit Malfunction
P0741 Torque Converter Clutch Circuit Performance or Stuck Off
P0742 Torque Converter Clutch Circuit Stuck On
P0743 Torque Converter Clutch Circuit Electrical
P0744 Torque Converter Clutch Circuit Intermittent
P0745 Pressure Control Solenoid Malfunction
P0746 Pressure Control Solenoid Performance or Stuck Off
P0747 Pressure Control Solenoid Stuck On
P0748 Pressure Control Solenoid Electrical
P0749 Pressure Control Solenoid Intermittent
P0750 Shift Solenoid "A" Malfunction
P0751 Shift Solenoid "A" Performance or Stuck Off
P0752 Shift Solenoid "A" Stuck On
P0753 Shift Solenoid "A" Electrical
P0754 Shift Solenoid "A" Intermittent
P0755 Shift Solenoid "B" Malfunction
P0756 Shift Solenoid "B" Performance or Stuck Off
P0757 Shift Solenoid "B" Stuck On
P0758 Shift Solenoid "B" Electrical
P0759 Shift Solenoid "B" Intermittent
P0760 Shift Solenoid "C" Malfunction
P0761 Shift Solenoid "C" Performance Or Stuck Off
P0762 Shift Solenoid "C" Stuck On
P0763 Shift Solenoid "C" Electrical
P0764 Shift Solenoid "C" Intermittent
P0765 Shift Solenoid "D" Malfunction
P0766 Shift Solenoid "D" Performance Or Stuck Off
P0767 Shift Solenoid "D" Stuck On
P0768 Shift Solenoid "D" Electrical
P0769 Shift Solenoid "D" Intermittent
P0770 Shift Solenoid "E" Malfunction
P0771 Shift Solenoid "E" Performance Or Stuck Off
P0772 Shift Solenoid "E" Stuck On
P0773 Shift Solenoid "E" Electrical
P0774 Shift Solenoid "E" Intermittent
P0780 Shift Malfunction
P0781 1–2 Shift Malfunction
P0782 2–3 Shift Malfunction
P0783 3–4 Shift Malfunction
P0784 4–5 Shift Malfunction
P0785 Shift/Timing Solenoid Malfunction
P0786 Shift/Timing Solenoid Range/Performance
P0787 Shift/Timing Solenoid Low
P0788 Shift/Timing Solenoid High
P0789 Shift/Timing Solenoid Intermittent
P0790 Normal/Performance Switch Circuit Malfunction
P0801 Reverse Inhibit Control Circuit Malfunction
P0803 1–4 Upshift (Skip Shift) Solenoid Control Circuit Malfunction
P0804 1–4 Upshift (Skip Shift) Lamp Control Circuit Malfunction
P1450 Barometric Pressure Sensor Circuit Fault**P1451** Barometric Pressure Sensor Performance
P1460 Cooling Fan Control System Fault
P1500 Starter Signal Circuit Fault
P1510 Back-up Power Supply Fault
P1530 Ignition Timing Adjustment Switch Circuit
P1600 PCM Battery Circuit Fault

Honda

➡**The Honda Passport is covered in the Isuzu section since it shares a platform with the Isuzu Rodeo.**

READING CODES

With Scan Tool

Reading the control module memory is one of the first steps in OBD II system diagnostics. This step should be initially performed to determine the general nature of the fault. Subsequent readings will determine if the fault has been cleared.

Reading codes can be performed by any of the methods below:
• Read the control module memory with the Generic Scan Tool (GST)
• Read the control module memory with the vehicle manufacturer's specific tester

To read the fault codes, connect the scan tool or tester according to the manufacturer's instructions. Follow the manufacturer's specified procedure for reading the codes.

Without Scan Tool

Honda also provides a way of reading OBD II trouble code equivalents using a service connector and viewing the MIL. This method is similar to the flash codes from non-OBD II vehicles.

To read codes, plug the service connector into the service check connector and turn the ignition on. The MIL will flash any stored trouble codes.

CLEARING CODES

Control module reset procedures are a very important part of OBD II System diagnostics. This step should be done at the end of any fault code repair and at the end of any driveability repair.

Clearing codes can be performed by any of the methods below:
• Clear the control module memory with the Generic Scan Tool (GST)
• Clear the control module memory with the vehicle manufacturer's specific tester
• Turn the ignition off and remove the negative battery cable for at least 1 minute.

Removing the negative battery cable may cause other systems in the vehicle to loose their memory. Prior to removing the cable, ensure you have the proper reset codes for radios and alarms.

➡**The MIL will may also be de-activated for some codes if the vehicle completes three consecutive trips without a fault detected with vehicle conditions similar to those present during the fault.**

OBD II TROUBLE CODES

P0100 Mass or Volume Air Flow Circuit Malfunction
P0101 Mass or Volume Air Flow Circuit Range/Performance Problem
P0102 Mass or Volume Air Flow Circuit Low Input
P0103 Mass or Volume Air Flow Circuit High Input
P0104 Mass or Volume Air Flow Circuit Intermittent

P0105 Manifold Absolute Pressure/Barometric Pressure Circuit Malfunction

P0106 Manifold Absolute Pressure/Barometric Pressure Circuit Range/Performance Problem

P0107 Manifold Absolute Pressure/Barometric Pressure Circuit Low Input

P0108 Manifold Absolute Pressure/Barometric Pressure Circuit High Input

P0109 Manifold Absolute Pressure/Barometric Pressure Circuit Intermittent

P0110 Intake Air Temperature Circuit Malfunction

P0111 Intake Air Temperature Circuit Range/Performance Problem

P0112 Intake Air Temperature Circuit Low Input

P0113 Intake Air Temperature Circuit High Input

P0114 Intake Air Temperature Circuit Intermittent

P0115 Engine Coolant Temperature Circuit Malfunction

P0116 Engine Coolant Temperature Circuit Range/Performance Problem

P0117 Engine Coolant Temperature Circuit Low Input

P0118 Engine Coolant Temperature Circuit High Input

P0119 Engine Coolant Temperature Circuit Intermittent

P0120 Throttle/Pedal Position Sensor/Switch "A" Circuit Malfunction

P0121 Throttle/Pedal Position Sensor/Switch "A" Circuit Range/Performance Problem

P0122 Throttle/Pedal Position Sensor/Switch "A" Circuit Low Input

P0123 Throttle/Pedal Position Sensor/Switch "A" Circuit High Input

P0124 Throttle/Pedal Position Sensor/Switch "A" Circuit Intermittent

P0125 Insufficient Coolant Temperature For Closed Loop Fuel Control

P0126 Insufficient Coolant Temperature For Stable Operation

P0130 O2 Circuit Malfunction (Bank no. 1 Sensor no. 1)

P0131 O2 Sensor Circuit Low Voltage (Bank no. 1 Sensor no. 1)

P0132 O2 Sensor Circuit High Voltage (Bank no. 1 Sensor no. 1)

P0133 O2 Sensor Circuit Slow Response (Bank no. 1 Sensor no. 1)

P0134 O2 Sensor Circuit No Activity Detected (Bank no. 1 Sensor no. 1)

P0135 O2 Sensor Heater Circuit Malfunction (Bank no. 1 Sensor no. 1)

P0136 O2 Sensor Circuit Malfunction (Bank no. 1 Sensor no. 2)

P0137 O2 Sensor Circuit Low Voltage (Bank no. 1 Sensor no. 2)

P0138 O2 Sensor Circuit High Voltage (Bank no. 1 Sensor no. 2)

P0139 O2 Sensor Circuit Slow Response (Bank no. 1 Sensor no. 2)

P0140 O2 Sensor Circuit No Activity Detected (Bank no. 1 Sensor no. 2)

P0141 O2 Sensor Heater Circuit Malfunction (Bank no. 1 Sensor no. 2)

P0142 O2 Sensor Circuit Malfunction (Bank no. 1 Sensor no. 3)

P0143 O2 Sensor Circuit Low Voltage (Bank no. 1 Sensor no. 3)

P0144 O2 Sensor Circuit High Voltage (Bank no. 1 Sensor no. 3)

P0145 O2 Sensor Circuit Slow Response (Bank no. 1 Sensor no. 3)

P0146 O2 Sensor Circuit No Activity Detected (Bank no. 1 Sensor no. 3)

P0147 O2 Sensor Heater Circuit Malfunction (Bank no. 1 Sensor no. 3)

P0150 O2 Sensor Circuit Malfunction (Bank no. 2 Sensor no. 1)

P0151 O2 Sensor Circuit Low Voltage (Bank no. 2 Sensor no. 1)

P0152 O2 Sensor Circuit High Voltage (Bank no. 2 Sensor no. 1)

P0153 O2 Sensor Circuit Slow Response (Bank no. 2 Sensor no. 1)

P0154 O2 Sensor Circuit No Activity Detected (Bank no. 2 Sensor no. 1)

P0155 O2 Sensor Heater Circuit Malfunction (Bank no. 2 Sensor no. 1)

P0156 O2 Sensor Circuit Malfunction (Bank no. 2 Sensor no. 2)

P0157 O2 Sensor Circuit Low Voltage (Bank no. 2 Sensor no. 2)

P0158 O2 Sensor Circuit High Voltage (Bank no. 2 Sensor no. 2)

P0159 O2 Sensor Circuit Slow Response (Bank no. 2 Sensor no. 2)

P0160 O2 Sensor Circuit No Activity Detected (Bank no. 2 Sensor no. 2)

P0161 O2 Sensor Heater Circuit Malfunction (Bank no. 2 Sensor no. 2)

P0162 O2 Sensor Circuit Malfunction (Bank no. 2 Sensor no. 3)

P0163 O2 Sensor Circuit Low Voltage (Bank no. 2 Sensor no. 3)

P0164 O2 Sensor Circuit High Voltage (Bank no. 2 Sensor no. 3)

P0165 O2 Sensor Circuit Slow Response (Bank no. 2 Sensor no. 3)

P0166 O2 Sensor Circuit No Activity Detected (Bank no. 2 Sensor no. 3)

P0167 O2 Sensor Heater Circuit Malfunction (Bank no. 2 Sensor no. 3)

P0170 Fuel Trim Malfunction (Bank no. 1)

P0171 System Too Lean (Bank no. 1)

P0172 System Too Rich (Bank no. 1)

P0173 Fuel Trim Malfunction (Bank no. 2)

P0174 System Too Lean (Bank no. 2)

P0175 System Too Rich (Bank no. 2)

P0176 Fuel Composition Sensor Circuit Malfunction

P0177 Fuel Composition Sensor Circuit Range/Performance

P0178 Fuel Composition Sensor Circuit Low Input

P0179 Fuel Composition Sensor Circuit High Input

P0180 Fuel Temperature Sensor "A" Circuit Malfunction

P0181 Fuel Temperature Sensor "A" Circuit Range/Performance

P0182 Fuel Temperature Sensor "A" Circuit Low Input

P0183 Fuel Temperature Sensor "A" Circuit High Input

P0184 Fuel Temperature Sensor "A" Circuit Intermittent

P0185 Fuel Temperature Sensor "B" Circuit Malfunction

P0186 Fuel Temperature Sensor "B" Circuit Range/Performance

P0187 Fuel Temperature Sensor "B" Circuit Low Input

P0188 Fuel Temperature Sensor "B" Circuit High Input

P0189 Fuel Temperature Sensor "B" Circuit Intermittent

P0190 Fuel Rail Pressure Sensor Circuit Malfunction

P0191 Fuel Rail Pressure Sensor Circuit Range/Performance

P0192 Fuel Rail Pressure Sensor Circuit Low Input

P0193 Fuel Rail Pressure Sensor Circuit High Input

P0194 Fuel Rail Pressure Sensor Circuit Intermittent

P0195 Engine Oil Temperature Sensor Malfunction

P0196 Engine Oil Temperature Sensor Range/Performance

P0197 Engine Oil Temperature Sensor Low

P0198 Engine Oil Temperature Sensor High

P0199 Engine Oil Temperature Sensor Intermittent

P0200 Injector Circuit Malfunction
P0201 Injector Circuit Malfunction—Cylinder no. 1
P0202 Injector Circuit Malfunction—Cylinder no. 2
P0203 Injector Circuit Malfunction—Cylinder no. 3
P0204 Injector Circuit Malfunction—Cylinder no. 4
P0205 Injector Circuit Malfunction—Cylinder no. 5
P0206 Injector Circuit Malfunction—Cylinder no. 6
P0207 Injector Circuit Malfunction—Cylinder no. 7
P0208 Injector Circuit Malfunction—Cylinder no. 8
P0209 Injector Circuit Malfunction—Cylinder no. 9
P0210 Injector Circuit Malfunction—Cylinder no. 10
P0211 Injector Circuit Malfunction—Cylinder no. 11
P0212 Injector Circuit Malfunction—Cylinder no. 12
P0213 Cold Start Injector no. 1 Malfunction
P0214 Cold Start Injector no. 2 Malfunction
P0215 Engine Shutoff Solenoid Malfunction
P0216 Injection Timing Control Circuit Malfunction
P0217 Engine Over Temperature Condition
P0218 Transmission Over Temperature Condition
P0219 Engine Over Speed Condition
P0220 Throttle/Pedal Position Sensor/Switch "B" Circuit Malfunction
P0221 Throttle/Pedal Position Sensor/Switch "B" Circuit Range/Performance Problem
P0222 Throttle/Pedal Position Sensor/Switch "B" Circuit Low Input
P0223 Throttle/Pedal Position Sensor/Switch "B" Circuit High Input
P0224 Throttle/Pedal Position Sensor/Switch "B" Circuit Intermittent
P0225 Throttle/Pedal Position Sensor/Switch "C" Circuit Malfunction
P0226 Throttle/Pedal Position Sensor/Switch "C" Circuit Range/Performance Problem
P0227 Throttle/Pedal Position Sensor/Switch "C" Circuit Low Input
P0228 Throttle/Pedal Position Sensor/Switch "C" Circuit High Input
P0229 Throttle/Pedal Position Sensor/Switch "C" Circuit Intermittent
P0230 Fuel Pump Primary Circuit Malfunction
P0231 Fuel Pump Secondary Circuit Low
P0232 Fuel Pump Secondary Circuit High
P0233 Fuel Pump Secondary Circuit Intermittent
P0234 Engine Over Boost Condition
P0261 Cylinder no. 1 Injector Circuit Low
P0262 Cylinder no. 1 Injector Circuit High
P0263 Cylinder no. 1 Contribution/Balance Fault
P0264 Cylinder no. 2 Injector Circuit Low
P0265 Cylinder no. 2 Injector Circuit High
P0266 Cylinder no. 2 Contribution/Balance Fault
P0267 Cylinder no. 3 Injector Circuit Low
P0268 Cylinder no. 3 Injector Circuit High
P0269 Cylinder no. 3 Contribution/Balance Fault
P0270 Cylinder no. 4 Injector Circuit Low
P0271 Cylinder no. 4 Injector Circuit High
P0272 Cylinder no. 4 Contribution/Balance Fault
P0273 Cylinder no. 5 Injector Circuit Low
P0274 Cylinder no. 5 Injector Circuit High
P0275 Cylinder no. 5 Contribution/Balance Fault
P0276 Cylinder no. 6 Injector Circuit Low
P0277 Cylinder no. 6 Injector Circuit High

P0278 Cylinder no. 6 Contribution/Balance Fault
P0279 Cylinder no. 7 Injector Circuit Low
P0280 Cylinder no. 7 Injector Circuit High
P0281 Cylinder no. 7 Contribution/Balance Fault
P0282 Cylinder no. 8 Injector Circuit Low
P0283 Cylinder no. 8 Injector Circuit High
P0284 Cylinder no. 8 Contribution/Balance Fault
P0285 Cylinder no. 9 Injector Circuit Low
P0286 Cylinder no. 9 Injector Circuit High
P0287 Cylinder no. 9 Contribution/Balance Fault
P0288 Cylinder no. 10 Injector Circuit Low
P0289 Cylinder no. 10 Injector Circuit High
P0290 Cylinder no. 10 Contribution/Balance Fault
P0291 Cylinder no. 11 Injector Circuit Low
P0292 Cylinder no. 11 Injector Circuit High
P0293 Cylinder no. 11 Contribution/Balance Fault
P0294 Cylinder no. 12 Injector Circuit Low
P0295 Cylinder no. 12 Injector Circuit High
P0296 Cylinder no. 12 Contribution/Balance Fault
P0300 Random/Multiple Cylinder Misfire Detected
P0301 Cylinder no. 1—Misfire Detected
P0302 Cylinder no. 2—Misfire Detected
P0303 Cylinder no. 3—Misfire Detected
P0304 Cylinder no. 4—Misfire Detected
P0305 Cylinder no. 5—Misfire Detected
P0306 Cylinder no. 6—Misfire Detected
P0307 Cylinder no. 7—Misfire Detected
P0308 Cylinder no. 8—Misfire Detected
P0309 Cylinder no. 9—Misfire Detected
P0310 Cylinder no. 10—Misfire Detected
P0311 Cylinder no. 11—Misfire Detected
P0312 Cylinder no. 12—Misfire Detected
P0320 Ignition/Distributor Engine Speed Input Circuit Malfunction
P0321 Ignition/Distributor Engine Speed Input Circuit Range/Performance
P0322 Ignition/Distributor Engine Speed Input Circuit No Signal
P0323 Ignition/Distributor Engine Speed Input Circuit Intermittent
P0325 Knock Sensor no. 1—Circuit Malfunction (Bank no. 1 or Single Sensor)
P0326 Knock Sensor no. 1—Circuit Range/Performance (Bank no. 1 or Single Sensor)
P0327 Knock Sensor no. 1—Circuit Low Input (Bank no. 1 or Single Sensor)
P0328 Knock Sensor no. 1—Circuit High Input (Bank no. 1 or Single Sensor)
P0329 Knock Sensor no. 1—Circuit Input Intermittent (Bank no. 1 or Single Sensor)
P0330 Knock Sensor no. 2—Circuit Malfunction (Bank no. 2)
P0331 Knock Sensor no. 2—Circuit Range/Performance (Bank no. 2)
P0332 Knock Sensor no. 2—Circuit Low Input (Bank no. 2)
P0333 Knock Sensor no. 2—Circuit High Input (Bank no. 2)
P0334 Knock Sensor no. 2—Circuit Input Intermittent (Bank no. 2)
P0335 Crankshaft Position Sensor "A" Circuit Malfunction
P0336 Crankshaft Position Sensor "A" Circuit Range/Performance
P0337 Crankshaft Position Sensor "A" Circuit Low Input
P0338 Crankshaft Position Sensor "A" Circuit High Input
P0339 Crankshaft Position Sensor "A" Circuit Intermittent

P0340 Camshaft Position Sensor Circuit Malfunction
P0341 Camshaft Position Sensor Circuit Range/Performance
P0342 Camshaft Position Sensor Circuit Low Input
P0343 Camshaft Position Sensor Circuit High Input
P0344 Camshaft Position Sensor Circuit Intermittent
P0350 Ignition Coil Primary/Secondary Circuit Malfunction
P0351 Ignition Coil "A" Primary/Secondary Circuit Malfunction
P0352 Ignition Coil "B" Primary/Secondary Circuit Malfunction
P0353 Ignition Coil "C" Primary/Secondary Circuit Malfunction
P0354 Ignition Coil "D" Primary/Secondary Circuit Malfunction
P0355 Ignition Coil "E" Primary/Secondary Circuit Malfunction
P0356 Ignition Coil "F" Primary/Secondary Circuit Malfunction
P0357 Ignition Coil "G" Primary/Secondary Circuit Malfunction
P0358 Ignition Coil "H" Primary/Secondary Circuit Malfunction
P0359 Ignition Coil "I" Primary/Secondary Circuit Malfunction
P0360 Ignition Coil "J" Primary/Secondary Circuit Malfunction
P0361 Ignition Coil "K" Primary/Secondary Circuit Malfunction
P0362 Ignition Coil "L" Primary/Secondary Circuit Malfunction
P0370 Timing Reference High Resolution Signal "A" Malfunction
P0371 Timing Reference High Resolution Signal "A" Too Many Pulses
P0372 Timing Reference High Resolution Signal "A" Too Few Pulses
P0373 Timing Reference High Resolution Signal "A" Intermittent/Erratic Pulses
P0374 Timing Reference High Resolution Signal "A" No Pulses
P0375 Timing Reference High Resolution Signal "B" Malfunction
P0376 Timing Reference High Resolution Signal "B" Too Many Pulses
P0377 Timing Reference High Resolution Signal "B" Too Few Pulses
P0378 Timing Reference High Resolution Signal "B" Intermittent/Erratic Pulses
P0379 Timing Reference High Resolution Signal "B" No Pulses
P0380 Glow Plug/Heater Circuit "A" Malfunction
P0381 Glow Plug/Heater Indicator Circuit Malfunction
P0382 Glow Plug/Heater Circuit "B" Malfunction
P0385 Crankshaft Position Sensor "B" Circuit Malfunction
P0386 Crankshaft Position Sensor "B" Circuit Range/Performance
P0387 Crankshaft Position Sensor "B" Circuit Low Input
P0388 Crankshaft Position Sensor "B" Circuit High Input
P0389 Crankshaft Position Sensor "B" Circuit Intermittent
P0400 Exhaust Gas Recirculation Flow Malfunction
P0401 Exhaust Gas Recirculation Flow Insufficient Detected
P0402 Exhaust Gas Recirculation Flow Excessive Detected
P0403 Exhaust Gas Recirculation Circuit Malfunction
P0404 Exhaust Gas Recirculation Circuit Range/Performance
P0405 Exhaust Gas Recirculation Sensor "A" Circuit Low
P0406 Exhaust Gas Recirculation Sensor "A" Circuit High
P0407 Exhaust Gas Recirculation Sensor "B" Circuit Low
P0408 Exhaust Gas Recirculation Sensor "B" Circuit High
P0410 Secondary Air Injection System Malfunction
P0411 Secondary Air Injection System Incorrect Flow Detected
P0412 Secondary Air Injection System Switching Valve "A" Circuit Malfunction
P0413 Secondary Air Injection System Switching Valve "A" Circuit Open
P0414 Secondary Air Injection System Switching Valve "A" Circuit Shorted

P0415 Secondary Air Injection System Switching Valve "B" Circuit Malfunction
P0416 Secondary Air Injection System Switching Valve "B" Circuit Open
P0417 Secondary Air Injection System Switching Valve "B" Circuit Shorted
P0418 Secondary Air Injection System Relay "A" Circuit Malfunction
P0419 Secondary Air Injection System Relay "B" Circuit Malfunction
P0420 Catalyst System Efficiency Below Threshold (Bank no. 1)
P0421 Warm Up Catalyst Efficiency Below Threshold (Bank no. 1)
P0422 Main Catalyst Efficiency Below Threshold (Bank no. 1)
P0423 Heated Catalyst Efficiency Below Threshold (Bank no. 1)
P0424 Heated Catalyst Temperature Below Threshold (Bank no. 1)
P0430 Catalyst System Efficiency Below Threshold (Bank no. 2)
P0431 Warm Up Catalyst Efficiency Below Threshold (Bank no. 2)
P0432 Main Catalyst Efficiency Below Threshold (Bank no. 2)
P0433 Heated Catalyst Efficiency Below Threshold (Bank no. 2)
P0434 Heated Catalyst Temperature Below Threshold (Bank no. 2)
P0440 Evaporative Emission Control System Malfunction
P0441 Evaporative Emission Control System Incorrect Purge Flow
P0442 Evaporative Emission Control System Leak Detected (Small Leak)
P0443 Evaporative Emission Control System Purge Control Valve Circuit Malfunction
P0444 Evaporative Emission Control System Purge Control Valve Circuit Open
P0445 Evaporative Emission Control System Purge Control Valve Circuit Shorted
P0446 Evaporative Emission Control System Vent Control Circuit Malfunction
P0447 Evaporative Emission Control System Vent Control Circuit Open
P0448 Evaporative Emission Control System Vent Control Circuit Shorted
P0449 Evaporative Emission Control System Vent Valve/Solenoid Circuit Malfunction
P0450 Evaporative Emission Control System Pressure Sensor Malfunction
P0451 Evaporative Emission Control System Pressure Sensor Range/Performance
P0452 Evaporative Emission Control System Pressure Sensor Low Input
P0453 Evaporative Emission Control System Pressure Sensor High Input
P0454 Evaporative Emission Control System Pressure Sensor Intermittent
P0455 Evaporative Emission Control System Leak Detected (Gross Leak)
P0460 Fuel Level Sensor Circuit Malfunction
P0461 Fuel Level Sensor Circuit Range/Performance
P0462 Fuel Level Sensor Circuit Low Input
P0463 Fuel Level Sensor Circuit High Input
P0464 Fuel Level Sensor Circuit Intermittent
P0465 Purge Flow Sensor Circuit Malfunction
P0466 Purge Flow Sensor Circuit Range/Performance

P0467 Purge Flow Sensor Circuit Low Input
P0468 Purge Flow Sensor Circuit High Input
P0469 Purge Flow Sensor Circuit Intermittent
P0470 Exhaust Pressure Sensor Malfunction
P0471 Exhaust Pressure Sensor Range/Performance
P0472 Exhaust Pressure Sensor Low
P0473 Exhaust Pressure Sensor High
P0474 Exhaust Pressure Sensor Intermittent
P0475 Exhaust Pressure Control Valve Malfunction
P0476 Exhaust Pressure Control Valve Range/Performance
P0477 Exhaust Pressure Control Valve Low
P0478 Exhaust Pressure Control Valve High
P0479 Exhaust Pressure Control Valve Intermittent
P0480 Cooling Fan no. 1 Control Circuit Malfunction
P0481 Cooling Fan no. 2 Control Circuit Malfunction
P0482 Cooling Fan no. 3 Control Circuit Malfunction
P0483 Cooling Fan Rationality Check Malfunction
P0484 Cooling Fan Circuit Over Current
P0485 Cooling Fan Power/Ground Circuit Malfunction
P0500 Vehicle Speed Sensor Malfunction
P0501 Vehicle Speed Sensor Range/Performance
P0502 Vehicle Speed Sensor Circuit Low Input
P0503 Vehicle Speed Sensor Intermittent/Erratic/High
P0505 Idle Control System Malfunction
P0506 Idle Control System RPM Lower Than Expected
P0507 Idle Control System RPM Higher Than Expected
P0510 Closed Throttle Position Switch Malfunction
P0520 Engine Oil Pressure Sensor/Switch Circuit Malfunction
P0521 Engine Oil Pressure Sensor/Switch Range/Performance
P0522 Engine Oil Pressure Sensor/Switch Low Voltage
P0523 Engine Oil Pressure Sensor/Switch High Voltage
P0530 A/C Refrigerant Pressure Sensor Circuit Malfunction
P0531 A/C Refrigerant Pressure Sensor Circuit Range/Performance
P0532 A/C Refrigerant Pressure Sensor Circuit Low Input
P0533 A/C Refrigerant Pressure Sensor Circuit High Input
P0534 A/C Refrigerant Charge Loss
P0550 Power Steering Pressure Sensor Circuit Malfunction
P0551 Power Steering Pressure Sensor Circuit Range/Performance
P0552 Power Steering Pressure Sensor Circuit Low Input
P0553 Power Steering Pressure Sensor Circuit High Input
P0554 Power Steering Pressure Sensor Circuit Intermittent
P0560 System Voltage Malfunction
P0561 System Voltage Unstable
P0562 System Voltage Low
P0563 System Voltage High
P0565 Cruise Control On Signal Malfunction
P0566 Cruise Control Off Signal Malfunction
P0567 Cruise Control Resume Signal Malfunction
P0568 Cruise Control Set Signal Malfunction
P0569 Cruise Control Coast Signal Malfunction
P0570 Cruise Control Accel Signal Malfunction
P0571 Cruise Control/Brake Switch "A" Circuit Malfunction
P0572 Cruise Control/Brake Switch "A" Circuit Low
P0573 Cruise Control/Brake Switch "A" Circuit High
P0574 Through P0580 Reserved for Cruise Codes
P0600 Serial Communication Link Malfunction
P0601 Internal Control Module Memory Check Sum Error
P0602 Control Module Programming Error
P0603 Internal Control Module Keep Alive Memory (KAM) Error

P0604 Internal Control Module Random Access Memory (RAM) Error
P0605 Internal Control Module Read Only Memory (ROM) Error
P0606 PCM Processor Fault
P0608 Control Module VSS Output "A" Malfunction
P0609 Control Module VSS Output "B" Malfunction
P0620 Generator Control Circuit Malfunction
P0621 Generator Lamp "L" Control Circuit Malfunction
P0622 Generator Field "F" Control Circuit Malfunction
P0650 Malfunction Indicator Lamp (MIL) Control Circuit Malfunction
P0654 Engine RPM Output Circuit Malfunction
P0655 Engine Hot Lamp Output Control Circuit Malfunction
P0656 Fuel Level Output Circuit Malfunction
P0700 Transmission Control System Malfunction
P0701 Transmission Control System Range/Performance
P0702 Transmission Control System Electrical
P0703 Torque Converter/Brake Switch "B" Circuit Malfunction
P0704 Clutch Switch Input Circuit Malfunction
P0705 Transmission Range Sensor Circuit Malfunction (PRNDL Input)
P0706 Transmission Range Sensor Circuit Range/Performance
P0707 Transmission Range Sensor Circuit Low Input
P0708 Transmission Range Sensor Circuit High Input
P0709 Transmission Range Sensor Circuit Intermittent
P0710 Transmission Fluid Temperature Sensor Circuit Malfunction
P0711 Transmission Fluid Temperature Sensor Circuit Range/Performance
P0712 Transmission Fluid Temperature Sensor Circuit Low Input
P0713 Transmission Fluid Temperature Sensor Circuit High Input
P0714 Transmission Fluid Temperature Sensor Circuit Intermittent
P0715 Input/Turbine Speed Sensor Circuit Malfunction
P0716 Input/Turbine Speed Sensor Circuit Range/Performance
P0717 Input/Turbine Speed Sensor Circuit No Signal
P0718 Input/Turbine Speed Sensor Circuit Intermittent
P0719 Torque Converter/Brake Switch "B" Circuit Low
P0720 Output Speed Sensor Circuit Malfunction
P0721 Output Speed Sensor Circuit Range/Performance
P0722 Output Speed Sensor Circuit No Signal
P0723 Output Speed Sensor Circuit Intermittent
P0724 Torque Converter/Brake Switch "B" Circuit High
P0725 Engine Speed Input Circuit Malfunction
P0726 Engine Speed Input Circuit Range/Performance
P0727 Engine Speed Input Circuit No Signal
P0728 Engine Speed Input Circuit Intermittent
P0730 Incorrect Gear Ratio
P0731 Gear no. 1 Incorrect Ratio
P0732 Gear no. 2 Incorrect Ratio
P0733 Gear no. 3 Incorrect Ratio
P0734 Gear no. 4 Incorrect Ratio
P0735 Gear no. 5 Incorrect Ratio
P0736 Reverse Incorrect Ratio
P0740 Torque Converter Clutch Circuit Malfunction
P0741 Torque Converter Clutch Circuit Performance or Stuck Off
P0742 Torque Converter Clutch Circuit Stuck On
P0743 Torque Converter Clutch Circuit Electrical

P0744 Torque Converter Clutch Circuit Intermittent
P0745 Pressure Control Solenoid Malfunction
P0746 Pressure Control Solenoid Performance or Stuck Off
P0747 Pressure Control Solenoid Stuck On
P0748 Pressure Control Solenoid Electrical
P0749 Pressure Control Solenoid Intermittent
P0750 Shift Solenoid "A" Malfunction
P0751 Shift Solenoid "A" Performance or Stuck Off
P0752 Shift Solenoid "A" Stuck On
P0753 Shift Solenoid "A" Electrical
P0754 Shift Solenoid "A" Intermittent
P0755 Shift Solenoid "B" Malfunction
P0756 Shift Solenoid "B" Performance or Stuck Off
P0757 Shift Solenoid "B" Stuck On
P0758 Shift Solenoid "B" Electrical
P0759 Shift Solenoid "B" Intermittent
P0760 Shift Solenoid "C" Malfunction
P0761 Shift Solenoid "C" Performance Or Stuck Off
P0762 Shift Solenoid "C" Stuck On
P0763 Shift Solenoid "C" Electrical
P0764 Shift Solenoid "C" Intermittent
P0765 Shift Solenoid "D" Malfunction
P0766 Shift Solenoid "D" Performance Or Stuck Off
P0767 Shift Solenoid "D" Stuck On
P0768 Shift Solenoid "D" Electrical
P0769 Shift Solenoid "D" Intermittent
P0770 Shift Solenoid "E" Malfunction
P0771 Shift Solenoid "E" Performance Or Stuck Off
P0772 Shift Solenoid "E" Stuck On
P0773 Shift Solenoid "E" Electrical
P0774 Shift Solenoid "E" Intermittent
P0780 Shift Malfunction
P0781 1–2 Shift Malfunction
P0782 2–3 Shift Malfunction
P0783 3–4 Shift Malfunction
P0784 4–5 Shift Malfunction
P0785 Shift/Timing Solenoid Malfunction
P0786 Shift/Timing Solenoid Range/Performance
P0787 Shift/Timing Solenoid Low
P0788 Shift/Timing Solenoid High
P0789 Shift/Timing Solenoid Intermittent
P0790 Normal/Performance Switch Circuit Malfunction
P0801 Reverse Inhibit Control Circuit Malfunction
P0803 1–4 Upshift (Skip Shift) Solenoid Control Circuit Malfunction
P0804 1–4 Upshift (Skip Shift) Lamp Control Circuit Malfunction
P0505 Idle Control System Malfunction
P0700 Automatic Transaxle
P0715 Automatic Transaxle
P0720 Automatic Transaxle
P0730 Automatic Transaxle
P0740 Automatic Transaxle
P0753 Automatic Transaxle
P0758 Automatic Transaxle
P1106 Barometric Pressure Circuit Range/Performance Problem
P1107 Barometric Pressure Circuit Low Input
P1108 Barometric Pressure Circuit High Input
P1121 Throttle Position Lower Than Expected
P1122 Throttle Position Higher Than Expected
P1128 Manifold Absolute Pressure Lower Than Expected
P1129 Manifold Absolute Pressure Higher Than Expected

P1259 VTEC System Malfunction
P1297 Electrical Load Detector Circuit Low Input
P1298 Electrical Load Detector Circuit High Input
P1297 Electrical Load Detector Circuit Low Input
P1298 Electrical Load Detector Circuit High Input
P1336 Crankshaft Speed Fluctuation Sensor Intermittent Interruption
P1337 Crankshaft Speed Fluctuation Sensor No Signal
P1359 Crankshaft Position Top Dead Center Sensor/Cylinder Position Connector Disconnection
P1361 Top Dead Center Sensor Intermittent Interruption
P1362 Top Dead Center Sensor No Signal
P1381 Cylinder Position Sensor Intermittent Interruption
P1382 Cylinder Position Sensor No Signal
P1456 Evaporative Emission Control System Leak Detected (Fuel Tank System)
P1457 Evaporative Emission Control System Leak Detected (EVAP Control Canister Leak)
P1491 EGR Valve Lift Insufficient Detected
P1498 EGR Valve Lift Sensor High Voltage
P1519 Idle Air Control Valve Circuit Failure
P1508 Idle Air Control Valve Circuit Failure
P1607 Powertrain Control Module Internal Circuit Failure A
P1705 Automatic Transaxle
P1706 Automatic Transaxle
P1753 Automatic Transaxle
P1768 Automatic Transaxle
P1790 Automatic Transaxle
P1791 Automatic Transaxle

OBD II TROUBLE CODE EQUIVALENTS

If a scan tool is not available for code retrieval, the following codes may be retrieved without one.

1 O2 Sensor Circuit High Voltage (Bank no. 1 Sensor no. 1)
1 O2 Sensor Circuit Low Voltage (Bank no. 1 Sensor no. 1)
3 Manifold Absolute Pressure/Barometric Pressure Circuit Low Input
3 Manifold Absolute Pressure/Barometric Pressure Circuit High Input
4 Crankshaft Position Sensor "A" Circuit Malfunction
4 Crankshaft Position Sensor "A" Circuit Range/Performance
5 Manifold Absolute Pressure Higher Than Expected
5 Manifold Absolute Pressure Lower Than Expected
6 Engine Coolant Temperature Circuit High Input
6 Engine Coolant Temperature Circuit Low Input
7 Throttle Position Higher Than Expected
7 Throttle Position Lower Than Expected
7 Throttle/Pedal Position Sensor/Switch "A" Circuit High Input
7 Throttle/Pedal Position Sensor/Switch "A" Circuit Low Input
8 Crankshaft Position Top Dead Center Sensor/Cylinder Position Connector Disconnection
8 Top Dead Center Sensor Intermittent Interruption
8 Top Dead Center Sensor No Signal
9 Cylinder Position Sensor Intermittent Interruption
9 Cylinder Position Sensor No Signal
10 Intake Air Temperature Circuit High Input
10 Intake Air Temperature Circuit Low Input
12 EGR Valve Lift Insufficient Detected
12 EGR Valve Lift Sensor High Voltage
13 Barometric Pressure Circuit High Input
13 Barometric Pressure Circuit Low Input

13 Barometric Pressure Circuit Range/Performance Problem
14 Idle Air Control Valve Circuit Failure
14 Idle Air Control Valve Circuit Failure
14 Idle Control System Malfunction
20 Electrical Load Detector Circuit High Input
20 Electrical Load Detector Circuit High Input
20 Electrical Load Detector Circuit Low Input
20 Electrical Load Detector Circuit Low Input
22 VTEC System Malfunction
23 Knock Sensor no. 1—Circuit Malfunction (Bank no. 1 or Single Sensor) **80** Exhaust Gas Recirculation Flow Insufficient Detected
41 O2 Sensor Heater Circuit Malfunction (Bank no. 1 Sensor no. 1)
45 System Too Lean (Bank no. 1)
45 System Too Rich (Bank no. 1)
54 Crankshaft Speed Fluctuation Sensor Intermittent Interruption
54 Crankshaft Speed Fluctuation Sensor No Signal
61 O2 Sensor Circuit Slow Response (Bank no. 1 Sensor no. 1)
63 O2 Sensor Circuit High Voltage (Bank no. 1 Sensor no. 2)
63 O2 Sensor Circuit Low Voltage (Bank no. 1 Sensor no. 2)
63 O2 Sensor Circuit Slow Response (Bank no. 1 Sensor no. 2)
65 O2 Sensor Heater Circuit Malfunction (Bank no. 1 Sensor no. 2)
67 Catalyst System Efficiency Below Threshold (Bank no. 1)
70 Automatic Transaxle
70 Transmission Control System Malfunction
70 Input/Turbine Speed Sensor Circuit Malfunction
70 Output Speed Sensor Circuit Malfunction
70 Incorrect Gear Ratio
70 Torque Converter Clutch Circuit Malfunction
70 Shift Solenoid "A" Electrical
70 Shift Solenoid "B" Electrical
71 Cylinder no. 1—Misfire Detected
72 Cylinder no. 2—Misfire Detected
73 Cylinder no. 3—Misfire Detected
74 Cylinder no. 4—Misfire Detected
86 Engine Coolant Temperature Circuit Range/Performance Problem
90 Evaporative Emission Control System Leak Detected (EVAP Control Canister Leak)
90 Evaporative Emission Control System Leak Detected (Fuel Tank System)
91 Evaporative Emission Control System Pressure Sensor Low Input
91 Evaporative Emission Control System Pressure Sensor High Input

Isuzu

➡**This section also provides coverage for the Honda Passport since it shares a platform with the Isuzu Rodeo.**

READING CODES

Reading the control module memory is one of the first steps in OBD II system diagnostics. This step should be initially performed to determine the general nature of the fault. Subsequent readings will determine if the fault has been cleared.

Reading codes can be performed by any of the methods below:
• Read the control module memory with the Generic Scan Tool (GST)
• Read the control module memory with the vehicle manufacturer's specific tester

To read the fault codes, connect the scan tool or tester according to the manufacturer's instructions. Follow the manufacturer's specified procedure for reading the codes.

CLEARING CODES

Control module reset procedures are a very important part of OBD II System diagnostics. This step should be done at the end of any fault code repair and at the end of any driveability repair.

Clearing codes can be performed by any of the methods below:
• Clear the control module memory with the Generic Scan Tool (GST)
• Clear the control module memory with the vehicle manufacturer's specific tester
• Turn the ignition off and remove the negative battery cable for at least 1 minute.

Removing the negative battery cable may cause other systems in the vehicle to loose their memory. Prior to removing the cable, ensure you have the proper reset codes for radios and alarms.

➡**The MIL will may also be de-activated for some codes if the vehicle completes three consecutive trips without a fault detected with vehicle conditions similar to those present during the fault.**

OBD II TROUBLE CODES

P0100 Mass or Volume Air Flow Circuit Malfunction
P0101 Mass or Volume Air Flow Circuit Range/Performance Problem
P0102 Mass or Volume Air Flow Circuit Low Input
P0103 Mass or Volume Air Flow Circuit High Input
P0104 Mass or Volume Air Flow Circuit Intermittent
P0105 Manifold Absolute Pressure/Barometric Pressure Circuit Malfunction
P0106 Manifold Absolute Pressure/Barometric Pressure Circuit Range/Performance Problem
P0107 Manifold Absolute Pressure/Barometric Pressure Circuit Low Input
P0108 Manifold Absolute Pressure/Barometric Pressure Circuit High Input
P0109 Manifold Absolute Pressure/Barometric Pressure Circuit Intermittent
P0110 Intake Air Temperature Circuit Malfunction
P0111 Intake Air Temperature Circuit Range/Performance Problem
P0112 Intake Air Temperature Circuit Low Input
P0113 Intake Air Temperature Circuit High Input
P0114 Intake Air Temperature Circuit Intermittent
P0115 Engine Coolant Temperature Circuit Malfunction
P0116 Engine Coolant Temperature Circuit Range/Performance Problem
P0117 Engine Coolant Temperature Circuit Low Input
P0118 Engine Coolant Temperature Circuit High Input
P0119 Engine Coolant Temperature Circuit Intermittent
P0120 Throttle/Pedal Position Sensor/Switch "A" Circuit Malfunction
P0121 Throttle/Pedal Position Sensor/Switch "A" Circuit Range/Performance Problem

P0122 Throttle/Pedal Position Sensor/Switch "A" Circuit Low Input

P0123 Throttle/Pedal Position Sensor/Switch "A" Circuit High Input

P0124 Throttle/Pedal Position Sensor/Switch "A" Circuit Intermittent

P0125 Insufficient Coolant Temperature For Closed Loop Fuel Control

P0126 Insufficient Coolant Temperature For Stable Operation

P0130 O2 Circuit Malfunction (Bank no. 1 Sensor no. 1)

P0131 O2 Sensor Circuit Low Voltage (Bank no. 1 Sensor no. 1)

P0132 O2 Sensor Circuit High Voltage (Bank no. 1 Sensor no. 1)

P0133 O2 Sensor Circuit Slow Response (Bank no. 1 Sensor no. 1)

P0134 O2 Sensor Circuit No Activity Detected (Bank no. 1 Sensor no. 1)

P0135 O2 Sensor Heater Circuit Malfunction (Bank no. 1 Sensor no. 1)

P0136 O2 Sensor Circuit Malfunction (Bank no. 1 Sensor no. 2)

P0137 O2 Sensor Circuit Low Voltage (Bank no. 1 Sensor no. 2)

P0138 O2 Sensor Circuit High Voltage (Bank no. 1 Sensor no. 2)

P0139 O2 Sensor Circuit Slow Response (Bank no. 1 Sensor no. 2)

P0140 O2 Sensor Circuit No Activity Detected (Bank no. 1 Sensor no. 2)

P0141 O2 Sensor Heater Circuit Malfunction (Bank no. 1 Sensor no. 2)

P0142 O2 Sensor Circuit Malfunction (Bank no. 1 Sensor no, 3)

P0143 O2 Sensor Circuit Low Voltage (Bank no. 1 Sensor no. 3)

P0144 O2 Sensor Circuit High Voltage (Bank no. 1 Sensor no. 3)

P0145 O2 Sensor Circuit Slow Response (Bank no. 1 Sensor no. 3)

P0146 O2 Sensor Circuit No Activity Detected (Bank no. 1 Sensor no. 3)

P0147 O2 Sensor Heater Circuit Malfunction (Bank no. 1 Sensor no. 3)

P0150 O2 Sensor Circuit Malfunction (Bank no. 2 Sensor no. 1)

P0151 O2 Sensor Circuit Low Voltage (Bank no. 2 Sensor no. 1)

P0152 O2 Sensor Circuit High Voltage (Bank no. 2 Sensor no. 1)

P0153 O2 Sensor Circuit Slow Response (Bank no. 2 Sensor no. 1)

P0154 O2 Sensor Circuit No Activity Detected (Bank no. 2 Sensor no. 1)

P0155 O2 Sensor Heater Circuit Malfunction (Bank no. 2 Sensor no. 1)

P0156 O2 Sensor Circuit Malfunction (Bank no. 2 Sensor no. 2)

P0157 O2 Sensor Circuit Low Voltage (Bank no. 2 Sensor no. 2)

P0158 O2 Sensor Circuit High Voltage (Bank no. 2 Sensor no. 2)

P0159 O2 Sensor Circuit Slow Response (Bank no. 2 Sensor no. 2)

P0160 O2 Sensor Circuit No Activity Detected (Bank no. 2 Sensor no. 2)

P0161 O2 Sensor Heater Circuit Malfunction (Bank no. 2 Sensor no. 2)

P0162 O2 Sensor Circuit Malfunction (Bank no. 2 Sensor no. 3)

P0163 O2 Sensor Circuit Low Voltage (Bank no. 2 Sensor no. 3)

P0164 O2 Sensor Circuit High Voltage (Bank no. 2 Sensor no. 3)

P0165 O2 Sensor Circuit Slow Response (Bank no. 2 Sensor no. 3)

P0166 O2 Sensor Circuit No Activity Detected (Bank no. 2 Sensor no. 3)

P0167 O2 Sensor Heater Circuit Malfunction (Bank no. 2 Sensor no. 3)

P0170 Fuel Trim Malfunction (Bank no. 1)

P0171 System Too Lean (Bank no. 1)

P0172 System Too Rich (Bank no. 1)

P0173 Fuel Trim Malfunction (Bank no. 2)

P0174 System Too Lean (Bank no. 2)

P0175 System Too Rich (Bank no. 2)

P0176 Fuel Composition Sensor Circuit Malfunction

P0177 Fuel Composition Sensor Circuit Range/Performance

P0178 Fuel Composition Sensor Circuit Low Input

P0179 Fuel Composition Sensor Circuit High Input

P0180 Fuel Temperature Sensor "A" Circuit Malfunction

P0181 Fuel Temperature Sensor "A" Circuit Range/Performance

P0182 Fuel Temperature Sensor "A" Circuit Low Input

P0183 Fuel Temperature Sensor "A" Circuit High Input

P0184 Fuel Temperature Sensor "A" Circuit Intermittent

P0185 Fuel Temperature Sensor "B" Circuit Malfunction

P0186 Fuel Temperature Sensor "B" Circuit Range/Performance

P0187 Fuel Temperature Sensor "B" Circuit Low Input

P0188 Fuel Temperature Sensor "B" Circuit High Input

P0189 Fuel Temperature Sensor "B" Circuit Intermittent

P0190 Fuel Rail Pressure Sensor Circuit Malfunction

P0191 Fuel Rail Pressure Sensor Circuit Range/Performance

P0192 Fuel Rail Pressure Sensor Circuit Low Input

P0193 Fuel Rail Pressure Sensor Circuit High Input

P0194 Fuel Rail Pressure Sensor Circuit Intermittent

P0195 Engine Oil Temperature Sensor Malfunction

P0196 Engine Oil Temperature Sensor Range/Performance

P0197 Engine Oil Temperature Sensor Low

P0198 Engine Oil Temperature Sensor High

P0199 Engine Oil Temperature Sensor Intermittent

P0200 Injector Circuit Malfunction

P0201 Injector Circuit Malfunction—Cylinder no. 1

P0202 Injector Circuit Malfunction—Cylinder no. 2

P0203 Injector Circuit Malfunction—Cylinder no. 3

P0204 Injector Circuit Malfunction—Cylinder no. 4

P0205 Injector Circuit Malfunction—Cylinder no. 5

P0206 Injector Circuit Malfunction—Cylinder no. 6

P0207 Injector Circuit Malfunction—Cylinder no. 7

P0208 Injector Circuit Malfunction—Cylinder no. 8

P0209 Injector Circuit Malfunction—Cylinder no. 9

P0210 Injector Circuit Malfunction—Cylinder no. 10

P0211 Injector Circuit Malfunction—Cylinder no. 11

P0212 Injector Circuit Malfunction—Cylinder no. 12

P0213 Cold Start Injector no. 1 Malfunction

P0214 Cold Start Injector no. 2 Malfunction

P0215 Engine Shutoff Solenoid Malfunction

P0216 Injection Timing Control Circuit Malfunction

P0217 Engine Over Temperature Condition

P0218 Transmission Over Temperature Condition

P0219 Engine Over Speed Condition

P0220 Throttle/Pedal Position Sensor/Switch "B" Circuit Malfunction

P0221 Throttle/Pedal Position Sensor/Switch "B" Circuit Range/Performance Problem

P0222 Throttle/Pedal Position Sensor/Switch "B" Circuit Low Input

P0223 Throttle/Pedal Position Sensor/Switch "B" Circuit High Input

P0224 Throttle/Pedal Position Sensor/Switch "B" Circuit Intermittent

P0225 Throttle/Pedal Position Sensor/Switch "C" Circuit Malfunction

P0226 Throttle/Pedal Position Sensor/Switch "C" Circuit Range/Performance Problem

P0227 Throttle/Pedal Position Sensor/Switch "C" Circuit Low Input

P0228 Throttle/Pedal Position Sensor/Switch "C" Circuit High Input

P0229 Throttle/Pedal Position Sensor/Switch "C" Circuit Intermittent

P0230 Fuel Pump Primary Circuit Malfunction
P0231 Fuel Pump Secondary Circuit Low
P0232 Fuel Pump Secondary Circuit High
P0233 Fuel Pump Secondary Circuit Intermittent
P0234 Engine Over Boost Condition /Injector)
P0261 Cylinder no. 1 Injector Circuit Low
P0262 Cylinder no. 1 Injector Circuit High
P0263 Cylinder no. 1 Contribution/Balance Fault
P0264 Cylinder no. 2 Injector Circuit Low
P0265 Cylinder no. 2 Injector Circuit High
P0266 Cylinder no. 2 Contribution/Balance Fault
P0267 Cylinder no. 3 Injector Circuit Low
P0268 Cylinder no. 3 Injector Circuit High
P0269 Cylinder no. 3 Contribution/Balance Fault
P0270 Cylinder no. 4 Injector Circuit Low
P0271 Cylinder no. 4 Injector Circuit High
P0272 Cylinder no. 4 Contribution/Balance Fault
P0273 Cylinder no. 5 Injector Circuit Low
P0274 Cylinder no. 5 Injector Circuit High
P0275 Cylinder no. 5 Contribution/Balance Fault
P0276 Cylinder no. 6 Injector Circuit Low
P0277 Cylinder no. 6 Injector Circuit High
P0278 Cylinder no. 6 Contribution/Balance Fault
P0279 Cylinder no. 7 Injector Circuit Low
P0280 Cylinder no. 7 Injector Circuit High
P0281 Cylinder no. 7 Contribution/Balance Fault
P0282 Cylinder no. 8 Injector Circuit Low
P0283 Cylinder no. 8 Injector Circuit High
P0284 Cylinder no. 8 Contribution/Balance Fault
P0285 Cylinder no. 9 Injector Circuit Low
P0286 Cylinder no. 9 Injector Circuit High
P0287 Cylinder no. 9 Contribution/Balance Fault
P0288 Cylinder no. 10 Injector Circuit Low
P0289 Cylinder no. 10 Injector Circuit High
P0290 Cylinder no. 10 Contribution/Balance Fault
P0291 Cylinder no. 11 Injector Circuit Low
P0292 Cylinder no. 11 Injector Circuit High
P0293 Cylinder no. 11 Contribution/Balance Fault
P0294 Cylinder no. 12 Injector Circuit Low
P0295 Cylinder no. 12 Injector Circuit High
P0296 Cylinder no. 12 Contribution/Balance Fault
P0300 Random/Multiple Cylinder Misfire Detected
P0301 Cylinder no. 1—Misfire Detected
P0302 Cylinder no. 2—Misfire Detected
P0303 Cylinder no. 3—Misfire Detected
P0304 Cylinder no. 4—Misfire Detected
P0305 Cylinder no. 5—Misfire Detected
P0306 Cylinder no. 6—Misfire Detected

P0307 Cylinder no. 7—Misfire Detected
P0308 Cylinder no. 8—Misfire Detected
P0309 Cylinder no. 9—Misfire Detected
P0310 Cylinder no. 10—Misfire Detected
P0311 Cylinder no. 11—Misfire Detected
P0312 Cylinder no. 12—Misfire Detected
P0320 Ignition/Distributor Engine Speed Input Circuit Malfunction

P0321 Ignition/Distributor Engine Speed Input Circuit Range/Performance

P0322 Ignition/Distributor Engine Speed Input Circuit No Signal
P0323 Ignition/Distributor Engine Speed Input Circuit Intermittent

P0325 Knock Sensor no. 1—Circuit Malfunction (Bank no. 1 or Single Sensor)

P0326 Knock Sensor no. 1—Circuit Range/Performance (Bank no. 1 or Single Sensor)

P0327 Knock Sensor no. 1—Circuit Low Input (Bank no. 1 or Single Sensor)

P0328 Knock Sensor no. 1—Circuit High Input (Bank no. 1 or Single Sensor)

P0329 Knock Sensor no. 1—Circuit Input Intermittent (Bank no. 1 or Single Sensor)

P0330 Knock Sensor no. 2—Circuit Malfunction (Bank no. 2)
P0331 Knock Sensor no. 2—Circuit Range/Performance (Bank no. 2)

P0332 Knock Sensor no. 2—Circuit Low Input (Bank no. 2)
P0333 Knock Sensor no. 2—Circuit High Input (Bank no. 2)
P0334 Knock Sensor no. 2—Circuit Input Intermittent (Bank no. 2)

P0335 Crankshaft Position Sensor "A" Circuit Malfunction
P0336 Crankshaft Position Sensor "A" Circuit Range/Performance

P0337 Crankshaft Position Sensor "A" Circuit Low Input
P0338 Crankshaft Position Sensor "A" Circuit High Input
P0339 Crankshaft Position Sensor "A" Circuit Intermittent
P0340 Camshaft Position Sensor Circuit Malfunction
P0341 Camshaft Position Sensor Circuit Range/Performance
P0342 Camshaft Position Sensor Circuit Low Input
P0343 Camshaft Position Sensor Circuit High Input
P0344 Camshaft Position Sensor Circuit Intermittent
P0350 Ignition Coil Primary/Secondary Circuit Malfunction
P0351 Ignition Coil "A" Primary/Secondary Circuit Malfunction
P0352 Ignition Coil "B" Primary/Secondary Circuit Malfunction
P0353 Ignition Coil "C" Primary/Secondary Circuit Malfunction
P0354 Ignition Coil "D" Primary/Secondary Circuit Malfunction
P0355 Ignition Coil "E" Primary/Secondary Circuit Malfunction
P0356 Ignition Coil "F" Primary/Secondary Circuit Malfunction
P0357 Ignition Coil "G" Primary/Secondary Circuit Malfunction
P0358 Ignition Coil "H" Primary/Secondary Circuit Malfunction
P0359 Ignition Coil "I" Primary/Secondary Circuit Malfunction
P0360 Ignition Coil "J" Primary/Secondary Circuit Malfunction
P0361 Ignition Coil "K" Primary/Secondary Circuit Malfunction
P0362 Ignition Coil "L" Primary/Secondary Circuit Malfunction
P0370 Timing Reference High Resolution Signal "A" Malfunction

P0371 Timing Reference High Resolution Signal "A" Too Many Pulses

P0372 Timing Reference High Resolution Signal "A" Too Few Pulses

P0373 Timing Reference High Resolution Signal "A" Intermittent/Erratic Pulses

P0374 Timing Reference High Resolution Signal "A" No Pulses

P0375 Timing Reference High Resolution Signal "B" Malfunction

P0376 Timing Reference High Resolution Signal "B" Too Many Pulses

P0377 Timing Reference High Resolution Signal "B" Too Few Pulses

P0378 Timing Reference High Resolution Signal "B" Intermittent/Erratic Pulses

P0379 Timing Reference High Resolution Signal "B" No Pulses

P0380 Glow Plug/Heater Circuit "A" Malfunction

P0381 Glow Plug/Heater Indicator Circuit Malfunction

P0382 Glow Plug/Heater Circuit "B" Malfunction

P0385 Crankshaft Position Sensor "B" Circuit Malfunction

P0386 Crankshaft Position Sensor "B" Circuit Range/Performance

P0387 Crankshaft Position Sensor "B" Circuit Low Input

P0388 Crankshaft Position Sensor "B" Circuit High Input

P0389 Crankshaft Position Sensor "B" Circuit Intermittent

P0400 Exhaust Gas Recirculation Flow Malfunction

P0401 Exhaust Gas Recirculation Flow Insufficient Detected

P0402 Exhaust Gas Recirculation Flow Excessive Detected

P0403 Exhaust Gas Recirculation Circuit Malfunction

P0404 Exhaust Gas Recirculation Circuit Range/Performance

P0405 Exhaust Gas Recirculation Sensor "A" Circuit Low

P0406 Exhaust Gas Recirculation Sensor "A" Circuit High

P0407 Exhaust Gas Recirculation Sensor "B" Circuit Low

P0408 Exhaust Gas Recirculation Sensor "B" Circuit High

P0410 Secondary Air Injection System Malfunction

P0411 Secondary Air Injection System Incorrect Flow Detected

P0412 Secondary Air Injection System Switching Valve "A" Circuit Malfunction

P0413 Secondary Air Injection System Switching Valve "A" Circuit Open

P0414 Secondary Air Injection System Switching Valve "A" Circuit Shorted

P0415 Secondary Air Injection System Switching Valve "B" Circuit Malfunction

P0416 Secondary Air Injection System Switching Valve "B" Circuit Open

P0417 Secondary Air Injection System Switching Valve "B" Circuit Shorted

P0418 Secondary Air Injection System Relay "A" Circuit Malfunction

P0419 Secondary Air Injection System Relay "B" Circuit Malfunction

P0420 Catalyst System Efficiency Below Threshold (Bank no. 1)

P0421 Warm Up Catalyst Efficiency Below Threshold (Bank no. 1)

P0422 Main Catalyst Efficiency Below Threshold (Bank no. 1)

P0423 Heated Catalyst Efficiency Below Threshold (Bank no. 1)

P0424 Heated Catalyst Temperature Below Threshold (Bank no. 1)

P0430 Catalyst System Efficiency Below Threshold (Bank no. 2)

P0431 Warm Up Catalyst Efficiency Below Threshold (Bank no. 2)

P0432 Main Catalyst Efficiency Below Threshold (Bank no. 2)

P0433 Heated Catalyst Efficiency Below Threshold (Bank no. 2)

P0434 Heated Catalyst Temperature Below Threshold (Bank no. 2)

P0440 Evaporative Emission Control System Malfunction

P0441 Evaporative Emission Control System Incorrect Purge Flow

P0442 Evaporative Emission Control System Leak Detected (Small Leak)

P0443 Evaporative Emission Control System Purge Control Valve Circuit Malfunction

P0444 Evaporative Emission Control System Purge Control Valve Circuit Open

P0445 Evaporative Emission Control System Purge Control Valve Circuit Shorted

P0446 Evaporative Emission Control System Vent Control Circuit Malfunction

P0447 Evaporative Emission Control System Vent Control Circuit Open

P0448 Evaporative Emission Control System Vent Control Circuit Shorted

P0449 Evaporative Emission Control System Vent Valve/Solenoid Circuit Malfunction

P0450 Evaporative Emission Control System Pressure Sensor Malfunction

P0451 Evaporative Emission Control System Pressure Sensor Range/Performance

P0452 Evaporative Emission Control System Pressure Sensor Low Input

P0453 Evaporative Emission Control System Pressure Sensor High Input

P0454 Evaporative Emission Control System Pressure Sensor Intermittent

P0455 Evaporative Emission Control System Leak Detected (Gross Leak)

P0460 Fuel Level Sensor Circuit Malfunction

P0461 Fuel Level Sensor Circuit Range/Performance

P0462 Fuel Level Sensor Circuit Low Input

P0463 Fuel Level Sensor Circuit High Input

P0464 Fuel Level Sensor Circuit Intermittent

P0465 Purge Flow Sensor Circuit Malfunction

P0466 Purge Flow Sensor Circuit Range/Performance

P0467 Purge Flow Sensor Circuit Low Input

P0468 Purge Flow Sensor Circuit High Input

P0469 Purge Flow Sensor Circuit Intermittent

P0470 Exhaust Pressure Sensor Malfunction

P0471 Exhaust Pressure Sensor Range/Performance

P0472 Exhaust Pressure Sensor Low

P0473 Exhaust Pressure Sensor High

P0474 Exhaust Pressure Sensor Intermittent

P0475 Exhaust Pressure Control Valve Malfunction

P0476 Exhaust Pressure Control Valve Range/Performance

P0477 Exhaust Pressure Control Valve Low

P0478 Exhaust Pressure Control Valve High

P0479 Exhaust Pressure Control Valve Intermittent

P0480 Cooling Fan no. 1 Control Circuit Malfunction

P0481 Cooling Fan no. 2 Control Circuit Malfunction

P0482 Cooling Fan no. 3 Control Circuit Malfunction

P0483 Cooling Fan Rationality Check Malfunction

P0484 Cooling Fan Circuit Over Current

P0485 Cooling Fan Power/Ground Circuit Malfunction

P0500 Vehicle Speed Sensor Malfunction

P0501 Vehicle Speed Sensor Range/Performance

P0502 Vehicle Speed Sensor Circuit Low Input

P0503 Vehicle Speed Sensor Intermittent/Erratic/High

P0505 Idle Control System Malfunction

P0506 Idle Control System RPM Lower Than Expected
P0507 Idle Control System RPM Higher Than Expected
P0510 Closed Throttle Position Switch Malfunction
P0520 Engine Oil Pressure Sensor/Switch Circuit Malfunction
P0521 Engine Oil Pressure Sensor/Switch Range/Performance
P0522 Engine Oil Pressure Sensor/Switch Low Voltage
P0523 Engine Oil Pressure Sensor/Switch High Voltage
P0530 A/C Refrigerant Pressure Sensor Circuit Malfunction
P0531 A/C Refrigerant Pressure Sensor Circuit Range/Performance
P0532 A/C Refrigerant Pressure Sensor Circuit Low Input
P0533 A/C Refrigerant Pressure Sensor Circuit High Input
P0534 A/C Refrigerant Charge Loss
P0550 Power Steering Pressure Sensor Circuit Malfunction
P0551 Power Steering Pressure Sensor Circuit Range/Performance
P0552 Power Steering Pressure Sensor Circuit Low Input
P0553 Power Steering Pressure Sensor Circuit High Input
P0554 Power Steering Pressure Sensor Circuit Intermittent
P0560 System Voltage Malfunction
P0561 System Voltage Unstable
P0562 System Voltage Low
P0563 System Voltage High
P0565 Cruise Control On Signal Malfunction
P0566 Cruise Control Off Signal Malfunction
P0567 Cruise Control Resume Signal Malfunction
P0568 Cruise Control Set Signal Malfunction
P0569 Cruise Control Coast Signal Malfunction
P0570 Cruise Control Accel Signal Malfunction
P0571 Cruise Control/Brake Switch "A" Circuit Malfunction
P0572 Cruise Control/Brake Switch "A" Circuit Low
P0573 Cruise Control/Brake Switch "A" Circuit High
P0574 Through P0580 Reserved for Cruise Codes
P0600 Serial Communication Link Malfunction
P0601 Internal Control Module Memory Check Sum Error
P0602 Control Module Programming Error
P0603 Internal Control Module Keep Alive Memory (KAM) Error
P0604 Internal Control Module Random Access Memory (RAM) Error
P0605 Internal Control Module Read Only Memory (ROM) Error
P0606 PCM Processor Fault
P0608 Control Module VSS Output "A" Malfunction
P0609 Control Module VSS Output "B" Malfunction
P0620 Generator Control Circuit Malfunction
P0621 Generator Lamp "L" Control Circuit Malfunction
P0622 Generator Field "F" Control Circuit Malfunction
P0650 Malfunction Indicator Lamp (MIL) Control Circuit Malfunction
P0654 Engine RPM Output Circuit Malfunction
P0655 Engine Hot Lamp Output Control Circuit Malfunction
P0656 Fuel Level Output Circuit Malfunction
P0700 Transmission Control System Malfunction
P0701 Transmission Control System Range/Performance
P0702 Transmission Control System Electrical
P0703 Torque Converter/Brake Switch "B" Circuit Malfunction
P0704 Clutch Switch Input Circuit Malfunction
P0705 Transmission Range Sensor Circuit Malfunction (PRNDL Input)
P0706 Transmission Range Sensor Circuit Range/Performance
P0707 Transmission Range Sensor Circuit Low Input
P0708 Transmission Range Sensor Circuit High Input
P0709 Transmission Range Sensor Circuit Intermittent

P0710 Transmission Fluid Temperature Sensor Circuit Malfunction
P0711 Transmission Fluid Temperature Sensor Circuit Range/Performance
P0712 Transmission Fluid Temperature Sensor Circuit Low Input
P0713 Transmission Fluid Temperature Sensor Circuit High Input
P0714 Transmission Fluid Temperature Sensor Circuit Intermittent
P0715 Input/Turbine Speed Sensor Circuit Malfunction
P0716 Input/Turbine Speed Sensor Circuit Range/Performance
P0717 Input/Turbine Speed Sensor Circuit No Signal
P0718 Input/Turbine Speed Sensor Circuit Intermittent
P0719 Torque Converter/Brake Switch "B" Circuit Low
P0720 Output Speed Sensor Circuit Malfunction
P0721 Output Speed Sensor Circuit Range/Performance
P0722 Output Speed Sensor Circuit No Signal
P0723 Output Speed Sensor Circuit Intermittent
P0724 Torque Converter/Brake Switch "B" Circuit High
P0725 Engine Speed Input Circuit Malfunction
P0726 Engine Speed Input Circuit Range/Performance
P0727 Engine Speed Input Circuit No Signal
P0728 Engine Speed Input Circuit Intermittent
P0730 Incorrect Gear Ratio
P0731 Gear no. 1 Incorrect Ratio
P0732 Gear no. 2 Incorrect Ratio
P0733 Gear no. 3 Incorrect Ratio
P0734 Gear no. 4 Incorrect Ratio
P0735 Gear no. 5 Incorrect Ratio
P0736 Reverse Incorrect Ratio
P0740 Torque Converter Clutch Circuit Malfunction
P0741 Torque Converter Clutch Circuit Performance or Stuck Off
P0742 Torque Converter Clutch Circuit Stuck On
P0743 Torque Converter Clutch Circuit Electrical
P0744 Torque Converter Clutch Circuit Intermittent
P0745 Pressure Control Solenoid Malfunction
P0746 Pressure Control Solenoid Performance or Stuck Off
P0747 Pressure Control Solenoid Stuck On
P0748 Pressure Control Solenoid Electrical
P0749 Pressure Control Solenoid Intermittent
P0750 Shift Solenoid "A" Malfunction
P0751 Shift Solenoid "A" Performance or Stuck Off
P0752 Shift Solenoid "A" Stuck On
P0753 Shift Solenoid "A" Electrical
P0754 Shift Solenoid "A" Intermittent
P0755 Shift Solenoid "B" Malfunction
P0756 Shift Solenoid "B" Performance or Stuck Off
P0757 Shift Solenoid "B" Stuck On
P0758 Shift Solenoid "B" Electrical
P0759 Shift Solenoid "B" Intermittent
P0760 Shift Solenoid "C" Malfunction
P0761 Shift Solenoid "C" Performance Or Stuck Off
P0762 Shift Solenoid "C" Stuck On
P0763 Shift Solenoid "C" Electrical
P0764 Shift Solenoid "C" Intermittent
P0765 Shift Solenoid "D" Malfunction
P0766 Shift Solenoid "D" Performance Or Stuck Off
P0767 Shift Solenoid "D" Stuck On
P0768 Shift Solenoid "D" Electrical
P0769 Shift Solenoid "D" Intermittent

P0770 Shift Solenoid "E" Malfunction
P0771 Shift Solenoid "E" Performance Or Stuck Off
P0772 Shift Solenoid "E" Stuck On
P0773 Shift Solenoid "E" Electrical
P0774 Shift Solenoid "E" Intermittent
P0780 Shift Malfunction
P0781 1–2 Shift Malfunction
P0782 2–3 Shift Malfunction
P0783 3–4 Shift Malfunction
P0784 4–5 Shift Malfunction
P0785 Shift/Timing Solenoid Malfunction
P0786 Shift/Timing Solenoid Range/Performance
P0787 Shift/Timing Solenoid Low
P0788 Shift/Timing Solenoid High
P0789 Shift/Timing Solenoid Intermittent
P0790 Normal/Performance Switch Circuit Malfunction
P0801 Reverse Inhibit Control Circuit Malfunction
P0803 1–4 Upshift (Skip Shift) Solenoid Control Circuit Malfunction
P0804 1–4 Upshift (Skip Shift) Lamp Control Circuit Malfunction
P1106 Map Sensor Circuit Intermittent High Voltage
P1107 MAP Sensor Circuit Intermittent Low Voltage
P1111 IAT Sensor Circuit Intermittent High Voltage
P1112 IAT Sensor Circuit Intermittent Low Voltage
P1114 ECT Sensor Circuit Intermittent Low Voltage
P1115 ECT Sensor Circuit Intermittent High Voltage
P1121 TP Sensor Circuit Intermittent High Voltage
P1122 TP Sensor Circuit Intermittent Low Voltage
P1133 HO2S-11 Insufficient Switching (Bank 1 Sensor 1)
P1134 HO2S-11 Transition Time Ratio (Bank 1 Sensor 1)
P1153 HO2S-21 Insufficient Switching (Bank 2 Sensor I)
P1154 HO2S-21 Transition Time Ratio (Bank 2 Sensor 1)
P1171 Fuel System Lean During Acceleration
P1391 G-Acceleration Sensor Intermittent Low Voltage
P1390 G-Acceleration (Low G) Sensor Performance
P1392 Rough Road G-Sensor Circuit Low Voltage
P1393 Rough Road G-Sensor Circuit High Voltage
P1394 G-Acceleration Sensor Intermittent High Voltage
P1406 EGR Valve Pintle Position Sensor Circuit Fault
P1441 EVAP System Flow During Non-Purge
P1442 EVAP System Flow During Non-Purge
P1508 Idle Speed Control System-Low
P1509 Idle Speed Control System-High
P1618 Serial Peripheral Interface Communication Error
P1640 Output Driver Module `A' Fault
P1790 PCM ROM (Transmission Side) Check Sum Error
P1792 PCM EEPROM (Transmission Side) Check Sum Error
P1835 Kick Down Switch Always On
P1850 Brake Band Apply Solenoid Electrical Fault
P1860 TCC PWM Solenoid Electrical Fault
P1870 Transmission Component Slipping

KIA

READING CODES

Reading the control module memory is one of the first steps in OBD II system diagnostics. This step should be initially performed to determine the general nature of the fault. Subsequent readings will determine if the fault has been cleared.

Reading codes can be performed by any of the methods below:
- Read the control module memory with the Generic Scan Tool (GST)
- Read the control module memory with the vehicle manufacturer's specific tester

To read the fault codes, connect the scan tool or tester according to the manufacturer's instructions. Follow the manufacturer's specified procedure for reading the codes.

CLEARING CODES

Control module reset procedures are a very important part of OBD II System diagnostics. This step should be done at the end of any fault code repair and at the end of any driveability repair.

Clearing codes can be performed by any of the methods below:
- Clear the control module memory with the Generic Scan Tool (GST)
- Clear the control module memory with the vehicle manufacturer's specific tester
- Turn the ignition off and remove the negative battery cable for at least 1 minute.

Removing the negative battery cable may cause other systems in the vehicle to loose their memory. Prior to removing the cable, ensure you have the proper reset codes for radios and alarms.

➡**The MIL will may also be de-activated for some codes if the vehicle completes three consecutive trips without a fault detected with vehicle conditions similar to those present during the fault.**

OBD II TROUBLE CODES

P0100 Mass or Volume Air Flow Circuit Malfunction
P0101 Mass or Volume Air Flow Circuit Range/Performance Problem
P0102 Mass or Volume Air Flow Circuit Low Input
P0103 Mass or Volume Air Flow Circuit High Input
P0104 Mass or Volume Air Flow Circuit Intermittent
P0105 Manifold Absolute Pressure/Barometric Pressure Circuit Malfunction
P0106 Manifold Absolute Pressure/Barometric Pressure Circuit Range/Performance Problem
P0107 Manifold Absolute Pressure/Barometric Pressure Circuit Low Input
P0108 Manifold Absolute Pressure/Barometric Pressure Circuit High Input
P0109 Manifold Absolute Pressure/Barometric Pressure Circuit Intermittent
P0110 Intake Air Temperature Circuit Malfunction
P0111 Intake Air Temperature Circuit Range/Performance Problem
P0112 Intake Air Temperature Circuit Low Input
P0113 Intake Air Temperature Circuit High Input
P0114 Intake Air Temperature Circuit Intermittent
P0115 Engine Coolant Temperature Circuit Malfunction
P0116 Engine Coolant Temperature Circuit Range/Performance Problem
P0117 Engine Coolant Temperature Circuit Low Input
P0118 Engine Coolant Temperature Circuit High Input
P0119 Engine Coolant Temperature Circuit Intermittent
P0120 Throttle/Pedal Position Sensor/Switch "A" Circuit Malfunction

P0121 Throttle/Pedal Position Sensor/Switch "A" Circuit Range/Performance Problem

P0122 Throttle/Pedal Position Sensor/Switch "A" Circuit Low Input

P0123 Throttle/Pedal Position Sensor/Switch "A" Circuit High Input

P0124 Throttle/Pedal Position Sensor/Switch "A" Circuit Intermittent

P0125 Insufficient Coolant Temperature For Closed Loop Fuel Control

P0126 Insufficient Coolant Temperature For Stable Operation

P0130 O2 Circuit Malfunction (Bank no. 1 Sensor no. 1)

P0131 O2 Sensor Circuit Low Voltage (Bank no. 1 Sensor no. 1)

P0132 O2 Sensor Circuit High Voltage (Bank no. 1 Sensor no. 1)

P0133 O2 Sensor Circuit Slow Response (Bank no. 1 Sensor no. 1)

P0134 O2 Sensor Circuit No Activity Detected (Bank no. 1 Sensor no. 1)

P0135 O2 Sensor Heater Circuit Malfunction (Bank no. 1 Sensor no. 1)

P0136 O2 Sensor Circuit Malfunction (Bank no. 1 Sensor no. 2)

P0137 O2 Sensor Circuit Low Voltage (Bank no. 1 Sensor no. 2)

P0138 O2 Sensor Circuit High Voltage (Bank no. 1 Sensor no. 2)

P0139 O2 Sensor Circuit Slow Response (Bank no. 1 Sensor no. 2)

P0140 O2 Sensor Circuit No Activity Detected (Bank no. 1 Sensor no. 2)

P0141 O2 Sensor Heater Circuit Malfunction (Bank no. 1 Sensor no. 2)

P0142 O2 Sensor Circuit Malfunction (Bank no. 1 Sensor no. 3)

P0143 O2 Sensor Circuit Low Voltage (Bank no. 1 Sensor no. 3)

P0144 O2 Sensor Circuit High Voltage (Bank no. 1 Sensor no. 3)

P0145 O2 Sensor Circuit Slow Response (Bank no. 1 Sensor no. 3)

P0146 O2 Sensor Circuit No Activity Detected (Bank no. 1 Sensor no. 3)

P0147 O2 Sensor Heater Circuit Malfunction (Bank no. 1 Sensor no. 3)

P0150 O2 Sensor Circuit Malfunction (Bank no. 2 Sensor no. 1)

P0151 O2 Sensor Circuit Low Voltage (Bank no. 2 Sensor no. 1)

P0152 O2 Sensor Circuit High Voltage (Bank no. 2 Sensor no. 1)

P0153 O2 Sensor Circuit Slow Response (Bank no. 2 Sensor no. 1)

P0154 O2 Sensor Circuit No Activity Detected (Bank no. 2 Sensor no. 1)

P0155 O2 Sensor Heater Circuit Malfunction (Bank no. 2 Sensor no. 1)

P0156 O2 Sensor Circuit Malfunction (Bank no. 2 Sensor no. 2)

P0157 O2 Sensor Circuit Low Voltage (Bank no. 2 Sensor no. 2)

P0158 O2 Sensor Circuit High Voltage (Bank no. 2 Sensor no. 2)

P0159 O2 Sensor Circuit Slow Response (Bank no. 2 Sensor no. 2)

P0160 O2 Sensor Circuit No Activity Detected (Bank no. 2 Sensor no. 2)

P0161 O2 Sensor Heater Circuit Malfunction (Bank no. 2 Sensor no. 2)

P0162 O2 Sensor Circuit Malfunction (Bank no. 2 Sensor no. 3)

P0163 O2 Sensor Circuit Low Voltage (Bank no. 2 Sensor no. 3)

P0164 O2 Sensor Circuit High Voltage (Bank no. 2 Sensor no. 3)

P0165 O2 Sensor Circuit Slow Response (Bank no. 2 Sensor no. 3)

P0166 O2 Sensor Circuit No Activity Detected (Bank no. 2 Sensor no. 3)

P0167 O2 Sensor Heater Circuit Malfunction (Bank no. 2 Sensor no. 3)

P0170 Fuel Trim Malfunction (Bank no. 1)

P0171 System Too Lean (Bank no. 1)

P0172 System Too Rich (Bank no. 1)

P0173 Fuel Trim Malfunction (Bank no. 2)

P0174 System Too Lean (Bank no. 2)

P0175 System Too Rich (Bank no. 2)

P0176 Fuel Composition Sensor Circuit Malfunction

P0177 Fuel Composition Sensor Circuit Range/Performance

P0178 Fuel Composition Sensor Circuit Low Input

P0179 Fuel Composition Sensor Circuit High Input

P0180 Fuel Temperature Sensor "A" Circuit Malfunction

P0181 Fuel Temperature Sensor "A" Circuit Range/Performance

P0182 Fuel Temperature Sensor "A" Circuit Low Input

P0183 Fuel Temperature Sensor "A" Circuit High Input

P0184 Fuel Temperature Sensor "A" Circuit Intermittent

P0185 Fuel Temperature Sensor "B" Circuit Malfunction

P0186 Fuel Temperature Sensor "B" Circuit Range/Performance

P0187 Fuel Temperature Sensor "B" Circuit Low Input

P0188 Fuel Temperature Sensor "B" Circuit High Input

P0189 Fuel Temperature Sensor "B" Circuit Intermittent

P0190 Fuel Rail Pressure Sensor Circuit Malfunction

P0191 Fuel Rail Pressure Sensor Circuit Range/Performance

P0192 Fuel Rail Pressure Sensor Circuit Low Input

P0193 Fuel Rail Pressure Sensor Circuit High Input

P0194 Fuel Rail Pressure Sensor Circuit Intermittent

P0195 Engine Oil Temperature Sensor Malfunction

P0196 Engine Oil Temperature Sensor Range/Performance

P0197 Engine Oil Temperature Sensor Low

P0198 Engine Oil Temperature Sensor High

P0199 Engine Oil Temperature Sensor Intermittent

P0200 Injector Circuit Malfunction

P0201 Injector Circuit Malfunction—Cylinder no. 1

P0202 Injector Circuit Malfunction—Cylinder no. 2

P0203 Injector Circuit Malfunction—Cylinder no. 3

P0204 Injector Circuit Malfunction—Cylinder no. 4

P0205 Injector Circuit Malfunction—Cylinder no. 5

P0206 Injector Circuit Malfunction—Cylinder no. 6

P0207 Injector Circuit Malfunction—Cylinder no. 7

P0208 Injector Circuit Malfunction—Cylinder no. 8

P0209 Injector Circuit Malfunction—Cylinder no. 9

P0210 Injector Circuit Malfunction—Cylinder no. 10

P0211 Injector Circuit Malfunction—Cylinder no. 11

P0212 Injector Circuit Malfunction—Cylinder no. 12

P0213 Cold Start Injector no. 1 Malfunction

P0214 Cold Start Injector no. 2 Malfunction

P0215 Engine Shutoff Solenoid Malfunction

P0216 Injection Timing Control Circuit Malfunction

P0217 Engine Over Temperature Condition

P0218 Transmission Over Temperature Condition

P0219 Engine Over Speed Condition

P0220 Throttle/Pedal Position Sensor/Switch "B" Circuit Malfunction

P0221 Throttle/Pedal Position Sensor/Switch "B" Circuit Range/Performance Problem

P0222 Throttle/Pedal Position Sensor/Switch "B" Circuit Low Input

P0223 Throttle/Pedal Position Sensor/Switch "B" Circuit High Input

P0224 Throttle/Pedal Position Sensor/Switch "B" Circuit Intermittent

P0225 Throttle/Pedal Position Sensor/Switch "C" Circuit Malfunction

P0226 Throttle/Pedal Position Sensor/Switch "C" Circuit Range/Performance Problem

P0227 Throttle/Pedal Position Sensor/Switch "C" Circuit Low Input

P0228 Throttle/Pedal Position Sensor/Switch "C" Circuit High Input

P0229 Throttle/Pedal Position Sensor/Switch "C" Circuit Intermittent

P0230 Fuel Pump Primary Circuit Malfunction

P0231 Fuel Pump Secondary Circuit Low

P0232 Fuel Pump Secondary Circuit High

P0233 Fuel Pump Secondary Circuit Intermittent

P0234 Engine Over Boost Condition

P0261 Cylinder no. 1 Injector Circuit Low

P0262 Cylinder no. 1 Injector Circuit High

P0263 Cylinder no. 1 Contribution/Balance Fault

P0264 Cylinder no. 2 Injector Circuit Low

P0265 Cylinder no. 2 Injector Circuit High

P0266 Cylinder no. 2 Contribution/Balance Fault

P0267 Cylinder no. 3 Injector Circuit Low

P0268 Cylinder no. 3 Injector Circuit High

P0269 Cylinder no. 3 Contribution/Balance Fault

P0270 Cylinder no. 4 Injector Circuit Low

P0271 Cylinder no. 4 Injector Circuit High

P0272 Cylinder no. 4 Contribution/Balance Fault

P0273 Cylinder no. 5 Injector Circuit Low

P0274 Cylinder no. 5 Injector Circuit High

P0275 Cylinder no. 5 Contribution/Balance Fault

P0276 Cylinder no. 6 Injector Circuit Low

P0277 Cylinder no. 6 Injector Circuit High

P0278 Cylinder no. 6 Contribution/Balance Fault

P0279 Cylinder no. 7 Injector Circuit Low

P0280 Cylinder no. 7 Injector Circuit High

P0281 Cylinder no. 7 Contribution/Balance Fault

P0282 Cylinder no. 8 Injector Circuit Low

P0283 Cylinder no. 8 Injector Circuit High

P0284 Cylinder no. 8 Contribution/Balance Fault

P0285 Cylinder no. 9 Injector Circuit Low

P0286 Cylinder no. 9 Injector Circuit High

P0287 Cylinder no. 9 Contribution/Balance Fault

P0288 Cylinder no. 10 Injector Circuit Low

P0289 Cylinder no. 10 Injector Circuit High

P0290 Cylinder no. 10 Contribution/Balance Fault

P0291 Cylinder no. 11 Injector Circuit Low

P0292 Cylinder no. 11 Injector Circuit High

P0293 Cylinder no. 11 Contribution/Balance Fault

P0294 Cylinder no. 12 Injector Circuit Low

P0295 Cylinder no. 12 Injector Circuit High

P0296 Cylinder no. 12 Contribution/Balance Fault

P0300 Random/Multiple Cylinder Misfire Detected

P0301 Cylinder no. 1—Misfire Detected

P0302 Cylinder no. 2—Misfire Detected

P0303 Cylinder no. 3—Misfire Detected

P0304 Cylinder no. 4—Misfire Detected

P0305 Cylinder no. 5—Misfire Detected

P0306 Cylinder no. 6—Misfire Detected

P0307 Cylinder no. 7—Misfire Detected

P0308 Cylinder no. 8—Misfire Detected

P0309 Cylinder no. 9—Misfire Detected

P0310 Cylinder no. 10—Misfire Detected

P0311 Cylinder no. 11—Misfire Detected

P0312 Cylinder no. 12—Misfire Detected

P0320 Ignition/Distributor Engine Speed Input Circuit Malfunction

P0321 Ignition/Distributor Engine Speed Input Circuit Range/Performance

P0322 Ignition/Distributor Engine Speed Input Circuit No Signal

P0323 Ignition/Distributor Engine Speed Input Circuit Intermittent

P0325 Knock Sensor no. 1—Circuit Malfunction (Bank no. 1 or Single Sensor)

P0326 Knock Sensor no. 1—Circuit Range/Performance (Bank no. 1 or Single Sensor)

P0327 Knock Sensor no. 1—Circuit Low Input (Bank no. 1 or Single Sensor)

P0328 Knock Sensor no. 1—Circuit High Input (Bank no. 1 or Single Sensor)

P0329 Knock Sensor no. 1—Circuit Input Intermittent (Bank no. 1 or Single Sensor)

P0330 Knock Sensor no. 2—Circuit Malfunction (Bank no. 2)

P0331 Knock Sensor no. 2—Circuit Range/Performance (Bank no. 2)

P0332 Knock Sensor no. 2—Circuit Low Input (Bank no. 2)

P0333 Knock Sensor no. 2—Circuit High Input (Bank no. 2)

P0334 Knock Sensor no. 2—Circuit Input Intermittent (Bank no. 2)

P0335 Crankshaft Position Sensor "A" Circuit Malfunction

P0336 Crankshaft Position Sensor "A" Circuit Range/Performance

P0337 Crankshaft Position Sensor "A" Circuit Low Input

P0338 Crankshaft Position Sensor "A" Circuit High Input

P0339 Crankshaft Position Sensor "A" Circuit Intermittent

P0340 Camshaft Position Sensor Circuit Malfunction

P0341 Camshaft Position Sensor Circuit Range/Performance

P0342 Camshaft Position Sensor Circuit Low Input

P0343 Camshaft Position Sensor Circuit High Input

P0344 Camshaft Position Sensor Circuit Intermittent

P0350 Ignition Coil Primary/Secondary Circuit Malfunction

P0351 Ignition Coil "A" Primary/Secondary Circuit Malfunction

P0352 Ignition Coil "B" Primary/Secondary Circuit Malfunction

P0353 Ignition Coil "C" Primary/Secondary Circuit Malfunction

P0354 Ignition Coil "D" Primary/Secondary Circuit Malfunction

P0355 Ignition Coil "E" Primary/Secondary Circuit Malfunction

P0356 Ignition Coil "F" Primary/Secondary Circuit Malfunction

P0357 Ignition Coil "G" Primary/Secondary Circuit Malfunction

P0358 Ignition Coil "H" Primary/Secondary Circuit Malfunction

P0359 Ignition Coil "I" Primary/Secondary Circuit Malfunction

P0360 Ignition Coil "J" Primary/Secondary Circuit Malfunction

P0361 Ignition Coil "K" Primary/Secondary Circuit Malfunction

P0362 Ignition Coil "L" Primary/Secondary Circuit Malfunction

P0370 Timing Reference High Resolution Signal "A" Malfunction

P0371 Timing Reference High Resolution Signal "A" Too Many Pulses

P0372 Timing Reference High Resolution Signal "A" Too Few Pulses

P0373 Timing Reference High Resolution Signal "A" Intermittent/Erratic Pulses

P0374 Timing Reference High Resolution Signal "A" No Pulses

P0375 Timing Reference High Resolution Signal "B" Malfunction

P0376 Timing Reference High Resolution Signal "B" Too Many Pulses

P0377 Timing Reference High Resolution Signal "B" Too Few Pulses

P0378 Timing Reference High Resolution Signal "B" Intermittent/Erratic Pulses

P0379 Timing Reference High Resolution Signal "B" No Pulses

P0380 Glow Plug/Heater Circuit "A" Malfunction

P0381 Glow Plug/Heater Indicator Circuit Malfunction

P0382 Glow Plug/Heater Circuit "B" Malfunction

P0385 Crankshaft Position Sensor "B" Circuit Malfunction

P0386 Crankshaft Position Sensor "B" Circuit Range/Performance

P0387 Crankshaft Position Sensor "B" Circuit Low Input

P0388 Crankshaft Position Sensor "B" Circuit High Input

P0389 Crankshaft Position Sensor "B" Circuit Intermittent

P0400 Exhaust Gas Recirculation Flow Malfunction

P0401 Exhaust Gas Recirculation Flow Insufficient Detected

P0402 Exhaust Gas Recirculation Flow Excessive Detected

P0403 Exhaust Gas Recirculation Circuit Malfunction

P0404 Exhaust Gas Recirculation Circuit Range/Performance

P0405 Exhaust Gas Recirculation Sensor "A" Circuit Low

P0406 Exhaust Gas Recirculation Sensor "A" Circuit High

P0407 Exhaust Gas Recirculation Sensor "B" Circuit Low

P0408 Exhaust Gas Recirculation Sensor "B" Circuit High

P0410 Secondary Air Injection System Malfunction

P0411 Secondary Air Injection System Incorrect Flow Detected

P0412 Secondary Air Injection System Switching Valve "A" Circuit Malfunction

P0413 Secondary Air Injection System Switching Valve "A" Circuit Open

P0414 Secondary Air Injection System Switching Valve "A" Circuit Shorted

P0415 Secondary Air Injection System Switching Valve "B" Circuit Malfunction

P0416 Secondary Air Injection System Switching Valve "B" Circuit Open

P0417 Secondary Air Injection System Switching Valve "B" Circuit Shorted

P0418 Secondary Air Injection System Relay "A" Circuit Malfunction

P0419 Secondary Air Injection System Relay "B" Circuit Malfunction

P0420 Catalyst System Efficiency Below Threshold (Bank no. 1)

P0421 Warm Up Catalyst Efficiency Below Threshold (Bank no. 1)

P0422 Main Catalyst Efficiency Below Threshold (Bank no. 1)

P0423 Heated Catalyst Efficiency Below Threshold (Bank no. 1)

P0424 Heated Catalyst Temperature Below Threshold (Bank no. 1)

P0430 Catalyst System Efficiency Below Threshold (Bank no. 2)

P0431 Warm Up Catalyst Efficiency Below Threshold (Bank no. 2)

P0432 Main Catalyst Efficiency Below Threshold (Bank no. 2)

P0433 Heated Catalyst Efficiency Below Threshold (Bank no. 2)

P0434 Heated Catalyst Temperature Below Threshold (Bank no. 2)

P0440 Evaporative Emission Control System Malfunction

P0441 Evaporative Emission Control System Incorrect Purge Flow

P0442 Evaporative Emission Control System Leak Detected (Small Leak)

P0443 Evaporative Emission Control System Purge Control Valve Circuit Malfunction

P0444 Evaporative Emission Control System Purge Control Valve Circuit Open

P0445 Evaporative Emission Control System Purge Control Valve Circuit Shorted

P0446 Evaporative Emission Control System Vent Control Circuit Malfunction

P0447 Evaporative Emission Control System Vent Control Circuit Open

P0448 Evaporative Emission Control System Vent Control Circuit Shorted

P0449 Evaporative Emission Control System Vent Valve/Solenoid Circuit Malfunction

P0450 Evaporative Emission Control System Pressure Sensor Malfunction

P0451 Evaporative Emission Control System Pressure Sensor Range/Performance

P0452 Evaporative Emission Control System Pressure Sensor Low Input

P0453 Evaporative Emission Control System Pressure Sensor High Input

P0454 Evaporative Emission Control System Pressure Sensor Intermittent

P0455 Evaporative Emission Control System Leak Detected (Gross Leak)

P0460 Fuel Level Sensor Circuit Malfunction

P0461 Fuel Level Sensor Circuit Range/Performance

P0462 Fuel Level Sensor Circuit Low Input

P0463 Fuel Level Sensor Circuit High Input

P0464 Fuel Level Sensor Circuit Intermittent

P0465 Purge Flow Sensor Circuit Malfunction

P0466 Purge Flow Sensor Circuit Range/Performance

P0467 Purge Flow Sensor Circuit Low Input

P0468 Purge Flow Sensor Circuit High Input

P0469 Purge Flow Sensor Circuit Intermittent

P0470 Exhaust Pressure Sensor Malfunction

P0471 Exhaust Pressure Sensor Range/Performance

P0472 Exhaust Pressure Sensor Low

P0473 Exhaust Pressure Sensor High

P0474 Exhaust Pressure Sensor Intermittent

P0475 Exhaust Pressure Control Valve Malfunction

P0476 Exhaust Pressure Control Valve Range/Performance

P0477 Exhaust Pressure Control Valve Low

P0478 Exhaust Pressure Control Valve High

P0479 Exhaust Pressure Control Valve Intermittent

P0480 Cooling Fan no. 1 Control Circuit Malfunction

P0481 Cooling Fan no. 2 Control Circuit Malfunction

P0482 Cooling Fan no. 3 Control Circuit Malfunction

P0483 Cooling Fan Rationality Check Malfunction

P0484 Cooling Fan Circuit Over Current

P0485 Cooling Fan Power/Ground Circuit Malfunction

P0500 Vehicle Speed Sensor Malfunction

P0501 Vehicle Speed Sensor Range/Performance

P0502 Vehicle Speed Sensor Circuit Low Input

P0503 Vehicle Speed Sensor Intermittent/Erratic/High

P0505 Idle Control System Malfunction

P0506 Idle Control System RPM Lower Than Expected
P0507 Idle Control System RPM Higher Than Expected
P0510 Closed Throttle Position Switch Malfunction
P0520 Engine Oil Pressure Sensor/Switch Circuit Malfunction
P0521 Engine Oil Pressure Sensor/Switch Range/Performance
P0522 Engine Oil Pressure Sensor/Switch Low Voltage
P0523 Engine Oil Pressure Sensor/Switch High Voltage
P0530 A/C Refrigerant Pressure Sensor Circuit Malfunction
P0531 A/C Refrigerant Pressure Sensor Circuit Range/Performance
P0532 A/C Refrigerant Pressure Sensor Circuit Low Input
P0533 A/C Refrigerant Pressure Sensor Circuit High Input
P0534 A/C Refrigerant Charge Loss
P0550 Power Steering Pressure Sensor Circuit Malfunction
P0551 Power Steering Pressure Sensor Circuit Range/Performance
P0552 Power Steering Pressure Sensor Circuit Low Input
P0553 Power Steering Pressure Sensor Circuit High Input
P0554 Power Steering Pressure Sensor Circuit Intermittent
P0560 System Voltage Malfunction
P0561 System Voltage Unstable
P0562 System Voltage Low
P0563 System Voltage High
P0565 Cruise Control On Signal Malfunction
P0566 Cruise Control Off Signal Malfunction
P0567 Cruise Control Resume Signal Malfunction
P0568 Cruise Control Set Signal Malfunction
P0569 Cruise Control Coast Signal Malfunction
P0570 Cruise Control Accel Signal Malfunction
P0571 Cruise Control/Brake Switch "A" Circuit Malfunction
P0572 Cruise Control/Brake Switch "A" Circuit Low
P0573 Cruise Control/Brake Switch "A" Circuit High
P0574 Through P0580 Reserved for Cruise Codes
P0600 Serial Communication Link Malfunction
P0601 Internal Control Module Memory Check Sum Error
P0602 Control Module Programming Error
P0603 Internal Control Module Keep Alive Memory (KAM) Error
P0604 Internal Control Module Random Access Memory (RAM) Error
P0605 Internal Control Module Read Only Memory (ROM) Error
P0606 PCM Processor Fault
P0608 Control Module VSS Output "A" Malfunction
P0609 Control Module VSS Output "B" Malfunction
P0620 Generator Control Circuit Malfunction
P0621 Generator Lamp "L" Control Circuit Malfunction
P0622 Generator Field "F" Control Circuit Malfunction
P0650 Malfunction Indicator Lamp (MIL) Control Circuit Malfunction
P0654 Engine RPM Output Circuit Malfunction
P0655 Engine Hot Lamp Output Control Circuit Malfunction
P0656 Fuel Level Output Circuit Malfunction
P0700 Transmission Control System Malfunction
P0701 Transmission Control System Range/Performance
P0702 Transmission Control System Electrical
P0703 Torque Converter/Brake Switch "B" Circuit Malfunction
P0704 Clutch Switch Input Circuit Malfunction
P0705 Transmission Range Sensor Circuit Malfunction (PRNDL Input)
P0706 Transmission Range Sensor Circuit Range/Performance
P0707 Transmission Range Sensor Circuit Low Input
P0708 Transmission Range Sensor Circuit High Input
P0709 Transmission Range Sensor Circuit Intermittent

P0710 Transmission Fluid Temperature Sensor Circuit Malfunction
P0711 Transmission Fluid Temperature Sensor Circuit Range/Performance
P0712 Transmission Fluid Temperature Sensor Circuit Low Input
P0713 Transmission Fluid Temperature Sensor Circuit High Input
P0714 Transmission Fluid Temperature Sensor Circuit Intermittent
P0715 Input/Turbine Speed Sensor Circuit Malfunction
P0716 Input/Turbine Speed Sensor Circuit Range/Performance
P0717 Input/Turbine Speed Sensor Circuit No Signal
P0718 Input/Turbine Speed Sensor Circuit Intermittent
P0719 Torque Converter/Brake Switch "B" Circuit Low
P0720 Output Speed Sensor Circuit Malfunction
P0721 Output Speed Sensor Circuit Range/Performance
P0722 Output Speed Sensor Circuit No Signal
P0723 Output Speed Sensor Circuit Intermittent
P0724 Torque Converter/Brake Switch "B" Circuit High
P0725 Engine Speed Input Circuit Malfunction
P0726 Engine Speed Input Circuit Range/Performance
P0727 Engine Speed Input Circuit No Signal
P0728 Engine Speed Input Circuit Intermittent
P0730 Incorrect Gear Ratio
P0731 Gear no. 1 Incorrect Ratio
P0732 Gear no. 2 Incorrect Ratio
P0733 Gear no. 3 Incorrect Ratio
P0734 Gear no. 4 Incorrect Ratio
P0735 Gear no. 5 Incorrect Ratio
P0736 Reverse Incorrect Ratio
P0740 Torque Converter Clutch Circuit Malfunction
P0741 Torque Converter Clutch Circuit Performance or Stuck Off
P0742 Torque Converter Clutch Circuit Stuck On
P0743 Torque Converter Clutch Circuit Electrical
P0744 Torque Converter Clutch Circuit Intermittent
P0745 Pressure Control Solenoid Malfunction
P0746 Pressure Control Solenoid Performance or Stuck Off
P0747 Pressure Control Solenoid Stuck On
P0748 Pressure Control Solenoid Electrical
P0749 Pressure Control Solenoid Intermittent
P0750 Shift Solenoid "A" Malfunction
P0751 Shift Solenoid "A" Performance or Stuck Off
P0752 Shift Solenoid "A" Stuck On
P0753 Shift Solenoid "A" Electrical
P0754 Shift Solenoid "A" Intermittent
P0755 Shift Solenoid "B" Malfunction
P0756 Shift Solenoid "B" Performance or Stuck Off
P0757 Shift Solenoid "B" Stuck On
P0758 Shift Solenoid "B" Electrical
P0759 Shift Solenoid "B" Intermittent
P0760 Shift Solenoid "C" Malfunction
P0761 Shift Solenoid "C" Performance Or Stuck Off
P0762 Shift Solenoid "C" Stuck On
P0763 Shift Solenoid "C" Electrical
P0764 Shift Solenoid "C" Intermittent
P0765 Shift Solenoid "D" Malfunction
P0766 Shift Solenoid "D" Performance Or Stuck Off
P0767 Shift Solenoid "D" Stuck On
P0768 Shift Solenoid "D" Electrical
P0769 Shift Solenoid "D" Intermittent

P0770 Shift Solenoid "E" Malfunction
P0771 Shift Solenoid "E" Performance Or Stuck Off
P0772 Shift Solenoid "E" Stuck On
P0773 Shift Solenoid "E" Electrical
P0774 Shift Solenoid "E" Intermittent
P0780 Shift Malfunction
P0781 1–2 Shift Malfunction
P0782 2–3 Shift Malfunction
P0783 3–4 Shift Malfunction
P0784 4–5 Shift Malfunction
P0785 Shift/Timing Solenoid Malfunction
P0786 Shift/Timing Solenoid Range/Performance
P0787 Shift/Timing Solenoid Low
P0788 Shift/Timing Solenoid High
P0789 Shift/Timing Solenoid Intermittent
P0790 Normal/Performance Switch Circuit Malfunction
P0801 Reverse Inhibit Control Circuit Malfunction
P0803 1–4 Upshift (Skip Shift) Solenoid Control Circuit Malfunction
P0804 1–4 Upshift (Skip Shift) Lamp Control Circuit Malfunction
P0740 Torque Converter Clutch System Fault
P0750 TCM Shift Solenoid `A' Electrical Fault
P0755 TCM Shift Solenoid `B' Electrical Fault
P0760 TCM Shift Solenoid `C' Electrical Fault
P1102 H02S-11 Heater Circuit High Voltage
P1105 H02S-12 Heater Circuit High Voltage
P1115 H02S-11 Heater Circuit Low Voltage
P1117 H02S-12 Heater Circuit Low Voltage
P1123 Long Term Fuel Trim Adaptive Air System Low
P1124 Long Term Fuel Trim Adaptive Air System High
P1127 Long Term Fuel Trim Multiplicative Air System Low
P1128 Long Term Fuel Trim Multiplicative Air System High
P1140 Load Calculation Cross Check
P1170 H02S-11 Circuit Voltage Stuck At Mid-Range
P1195 EGR Boost Or Pressure Sensor Circuit Fault
P1196 Ignition Switch Start Circuit Fault
P1213 Fuel Injector 1, 2, 3 Or 4 Circuit High Voltage
P1214 Fuel Injector 1, 2, 3 Or 4 Circuit High Voltage
P1215 Fuel Injector 1, 2, 3 Or 4 Circuit High Voltage
P1216 Fuel Injector 1, 2, 3 Or 4 Circuit High Voltage
P1225 Fuel Injector 1, 2, 5 Or 4 Circuit Low Voltage
P1226 Fuel Injector 1, 2, 5 Or 4 Circuit Low Voltage
P1227 Fuel Injector 1, 2, 5 Or 4 Circuit Low Voltage
P1228 Fuel Injector 1, 2, 5 Or 4 Circuit Low Voltage
P1250 Pressure Regulator Control Solenoid Circuit Fault
P1345 No SGC (CMP) Signal To PCM
P1386 Knock Sensor Control Zero Test
P1401 EGR Control Solenoid Circuit Signal Low
P1402 EGR Control Solenoid Circuit Signal High
P1402 EGR Valve Position Sensor Circuit Fault
P1410 EVAP Purge Control Solenoid Circuit High Voltage
P1412 EGR Differential Pressure Sensor Signal Low
P1413 EGR Differential Pressure Sensor Signal High
P1425 EVAP Purge Control Solenoid Circuit Low Voltage
P1449 Canister Drain Cut Valve Solenoid Circuit Fault
P1455 Fuel Tank Sending Unit Circuit Fault
P1458 Air Conditioning Compressor Clutch Signal Fault
P1485 EGR Vent Control Solenoid Circuit Fault
P1486 EGR Vacuum Control Solenoid Circuit Fault

P1487 EGR Boost Sensor Solenoid Circuit Fault
P1510 Idle Air Control Valve Closing Coil High Voltage
P1513 Idle Air Control Valve Closing Coil Low Voltage
P1515 A/T To M/T Codification
P1523 VICS Solenoid Valve Circuit Fault
P1552 Idle Air Control Valve Opening Coil Low Voltage
P1553 Idle Air Control Valve Opening Coil High Voltage
P1606 Chassis Accelerator Sensor Signal Circuit Fault
P1608 PCM Internal Fault
P1611 MIL Request Circuit Low Voltage
P1614 MIL Request Circuit High Voltage
P1616 Chassis Accelerator Sensor Signal Low Voltage
P1617 Chassis Accelerator Sensor Signal High Voltage
P1624 TCM to PCM MIL Request Circuit Fault
P1655 Unused Power Stage "B"
P1660 Unused Power Stage `A'
P1660 Unused Power Stage `B'
P1665 Power Stage Group `A'
P1743 Torque Converter Clutch Solenoid Circuit Fault
P1794 Battery Or Circuit Fault
P1797 Clutch Pedal Switch (MT) Or PIN Switch Circuit Fault

Land Rover

READING CODES

Reading the control module memory is one of the first steps in OBD II system diagnostics. This step should be initially performed to determine the general nature of the fault. Subsequent readings will determine if the fault has been cleared.

Reading codes can be performed by any of the methods below:
• Read the control module memory with the Generic Scan Tool (GST)
• Read the control module memory with the vehicle manufacturer's specific tester

To read the fault codes, connect the scan tool or tester according to the manufacturer's instructions. Follow the manufacturer's specified procedure for reading the codes.

CLEARING CODES

Control module reset procedures are a very important part of OBD II System diagnostics. This step should be done at the end of any fault code repair and at the end of any driveability repair.

Clearing codes can be performed by any of the methods below:
• Clear the control module memory with the Generic Scan Tool (GST)
• Clear the control module memory with the vehicle manufacturer's specific tester
• Turn the ignition off and remove the negative battery cable for at least 1 minute.

Removing the negative battery cable may cause other systems in the vehicle to loose their memory. Prior to removing the cable, ensure you have the proper reset codes for radios and alarms.

➡**The MIL will may also be de-activated for some codes if the vehicle completes three consecutive trips without a fault detected with vehicle conditions similar to those present during the fault.**

OBD II TROUBLE CODES

P0100 Mass or Volume Air Flow Circuit Malfunction
P0101 Mass or Volume Air Flow Circuit Range/Performance Problem
P0102 Mass or Volume Air Flow Circuit Low Input
P0103 Mass or Volume Air Flow Circuit High Input
P0104 Mass or Volume Air Flow Circuit Intermittent
P0105 Manifold Absolute Pressure/Barometric Pressure Circuit Malfunction
P0106 Manifold Absolute Pressure/Barometric Pressure Circuit Range/Performance Problem
P0107 Manifold Absolute Pressure/Barometric Pressure Circuit Low Input
P0108 Manifold Absolute Pressure/Barometric Pressure Circuit High Input
P0109 Manifold Absolute Pressure/Barometric Pressure Circuit Intermittent
P0110 Intake Air Temperature Circuit Malfunction
P0111 Intake Air Temperature Circuit Range/Performance Problem
P0112 Intake Air Temperature Circuit Low Input
P0113 Intake Air Temperature Circuit High Input
P0114 Intake Air Temperature Circuit Intermittent
P0115 Engine Coolant Temperature Circuit Malfunction
P0116 Engine Coolant Temperature Circuit Range/Performance Problem
P0117 Engine Coolant Temperature Circuit Low Input
P0118 Engine Coolant Temperature Circuit High Input
P0119 Engine Coolant Temperature Circuit Intermittent
P0120 Throttle/Pedal Position Sensor/Switch "A" Circuit Malfunction
P0121 Throttle/Pedal Position Sensor/Switch "A" Circuit Range/Performance Problem
P0122 Throttle/Pedal Position Sensor/Switch "A" Circuit Low Input
P0123 Throttle/Pedal Position Sensor/Switch "A" Circuit High Input
P0124 Throttle/Pedal Position Sensor/Switch "A" Circuit Intermittent
P0125 Insufficient Coolant Temperature For Closed Loop Fuel Control
P0126 Insufficient Coolant Temperature For Stable Operation
P0130 O2 Circuit Malfunction (Bank no. 1 Sensor no. 1)
P0131 O2 Sensor Circuit Low Voltage (Bank no. 1 Sensor no. 1)
P0132 O2 Sensor Circuit High Voltage (Bank no. 1 Sensor no. 1)
P0133 O2 Sensor Circuit Slow Response (Bank no. 1 Sensor no. 1)
P0134 O2 Sensor Circuit No Activity Detected (Bank no. 1 Sensor no. 1)
P0135 O2 Sensor Heater Circuit Malfunction (Bank no. 1 Sensor no. 1)
P0136 O2 Sensor Circuit Malfunction (Bank no. 1 Sensor no. 2)
P0137 O2 Sensor Circuit Low Voltage (Bank no. 1 Sensor no. 2)
P0138 O2 Sensor Circuit High Voltage (Bank no. 1 Sensor no. 2)
P0139 O2 Sensor Circuit Slow Response (Bank no. 1 Sensor no. 2)
P0140 O2 Sensor Circuit No Activity Detected (Bank no. 1 Sensor no. 2)
P0141 O2 Sensor Heater Circuit Malfunction (Bank no. 1 Sensor no. 2)
P0142 O2 Sensor Circuit Malfunction (Bank no. 1 Sensor no. 3)
P0143 O2 Sensor Circuit Low Voltage (Bank no. 1 Sensor no. 3)
P0144 O2 Sensor Circuit High Voltage (Bank no. 1 Sensor no. 3)
P0145 O2 Sensor Circuit Slow Response (Bank no. 1 Sensor no. 3)
P0146 O2 Sensor Circuit No Activity Detected (Bank no. 1 Sensor no. 3)
P0147 O2 Sensor Heater Circuit Malfunction (Bank no. 1 Sensor no. 3)
P0150 O2 Sensor Circuit Malfunction (Bank no. 2 Sensor no. 1)
P0151 O2 Sensor Circuit Low Voltage (Bank no. 2 Sensor no. 1)
P0152 O2 Sensor Circuit High Voltage (Bank no. 2 Sensor no. 1)
P0153 O2 Sensor Circuit Slow Response (Bank no. 2 Sensor no. 1)
P0154 O2 Sensor Circuit No Activity Detected (Bank no. 2 Sensor no. 1)
P0155 O2 Sensor Heater Circuit Malfunction (Bank no. 2 Sensor no. 1)
P0156 O2 Sensor Circuit Malfunction (Bank no. 2 Sensor no. 2)
P0157 O2 Sensor Circuit Low Voltage (Bank no. 2 Sensor no. 2)
P0158 O2 Sensor Circuit High Voltage (Bank no. 2 Sensor no. 2)
P0159 O2 Sensor Circuit Slow Response (Bank no. 2 Sensor no. 2)
P0160 O2 Sensor Circuit No Activity Detected (Bank no. 2 Sensor no. 2)
P0161 O2 Sensor Heater Circuit Malfunction (Bank no. 2 Sensor no. 2)
P0162 O2 Sensor Circuit Malfunction (Bank no. 2 Sensor no. 3)
P0163 O2 Sensor Circuit Low Voltage (Bank no. 2 Sensor no. 3)
P0164 O2 Sensor Circuit High Voltage (Bank no. 2 Sensor no. 3)
P0165 O2 Sensor Circuit Slow Response (Bank no. 2 Sensor no. 3)
P0166 O2 Sensor Circuit No Activity Detected (Bank no. 2 Sensor no. 3)
P0167 O2 Sensor Heater Circuit Malfunction (Bank no. 2 Sensor no. 3)
P0170 Fuel Trim Malfunction (Bank no. 1)
P0171 System Too Lean (Bank no. 1)
P0172 System Too Rich (Bank no. 1)
P0173 Fuel Trim Malfunction (Bank no. 2)
P0174 System Too Lean (Bank no. 2)
P0175 System Too Rich (Bank no. 2)
P0176 Fuel Composition Sensor Circuit Malfunction
P0177 Fuel Composition Sensor Circuit Range/Performance
P0178 Fuel Composition Sensor Circuit Low Input
P0179 Fuel Composition Sensor Circuit High Input
P0180 Fuel Temperature Sensor "A" Circuit Malfunction
P0181 Fuel Temperature Sensor "A" Circuit Range/Performance
P0182 Fuel Temperature Sensor "A" Circuit Low Input
P0183 Fuel Temperature Sensor "A" Circuit High Input
P0184 Fuel Temperature Sensor "A" Circuit Intermittent
P0185 Fuel Temperature Sensor "B" Circuit Malfunction
P0186 Fuel Temperature Sensor "B" Circuit Range/Performance
P0187 Fuel Temperature Sensor "B" Circuit Low Input
P0188 Fuel Temperature Sensor "B" Circuit High Input

P0189 Fuel Temperature Sensor "B" Circuit Intermittent
P0190 Fuel Rail Pressure Sensor Circuit Malfunction
P0191 Fuel Rail Pressure Sensor Circuit Range/Performance
P0192 Fuel Rail Pressure Sensor Circuit Low Input
P0193 Fuel Rail Pressure Sensor Circuit High Input
P0194 Fuel Rail Pressure Sensor Circuit Intermittent
P0195 Engine Oil Temperature Sensor Malfunction
P0196 Engine Oil Temperature Sensor Range/Performance
P0197 Engine Oil Temperature Sensor Low
P0198 Engine Oil Temperature Sensor High
P0199 Engine Oil Temperature Sensor Intermittent
P0200 Injector Circuit Malfunction
P0201 Injector Circuit Malfunction—Cylinder no. 1
P0202 Injector Circuit Malfunction—Cylinder no. 2
P0203 Injector Circuit Malfunction—Cylinder no. 3
P0204 Injector Circuit Malfunction—Cylinder no. 4
P0205 Injector Circuit Malfunction—Cylinder no. 5
P0206 Injector Circuit Malfunction—Cylinder no. 6
P0207 Injector Circuit Malfunction—Cylinder no. 7
P0208 Injector Circuit Malfunction—Cylinder no. 8
P0209 Injector Circuit Malfunction—Cylinder no. 9
P0210 Injector Circuit Malfunction—Cylinder no. 10
P0211 Injector Circuit Malfunction—Cylinder no. 11
P0212 Injector Circuit Malfunction—Cylinder no. 12
P0213 Cold Start Injector no. 1 Malfunction
P0214 Cold Start Injector no. 2 Malfunction
P0215 Engine Shutoff Solenoid Malfunction
P0216 Injection Timing Control Circuit Malfunction
P0217 Engine Over Temperature Condition
P0218 Transmission Over Temperature Condition
P0219 Engine Over Speed Condition
P0220 Throttle/Pedal Position Sensor/Switch "B" Circuit Malfunction
P0221 Throttle/Pedal Position Sensor/Switch "B" Circuit Range/Performance Problem
P0222 Throttle/Pedal Position Sensor/Switch "B" Circuit Low Input
P0223 Throttle/Pedal Position Sensor/Switch "B" Circuit High Input
P0224 Throttle/Pedal Position Sensor/Switch "B" Circuit Intermittent
P0225 Throttle/Pedal Position Sensor/Switch "C" Circuit Malfunction
P0226 Throttle/Pedal Position Sensor/Switch "C" Circuit Range/Performance Problem
P0227 Throttle/Pedal Position Sensor/Switch "C" Circuit Low Input
P0228 Throttle/Pedal Position Sensor/Switch "C" Circuit High Input
P0229 Throttle/Pedal Position Sensor/Switch "C" Circuit Intermittent
P0230 Fuel Pump Primary Circuit Malfunction
P0231 Fuel Pump Secondary Circuit Low
P0232 Fuel Pump Secondary Circuit High
P0233 Fuel Pump Secondary Circuit Intermittent
P0234 Engine Over Boost Condition
P0261 Cylinder no. 1 Injector Circuit Low
P0262 Cylinder no. 1 Injector Circuit High
P0263 Cylinder no. 1 Contribution/Balance Fault
P0264 Cylinder no. 2 Injector Circuit Low
P0265 Cylinder no. 2 Injector Circuit High
P0266 Cylinder no. 2 Contribution/Balance Fault

P0267 Cylinder no. 3 Injector Circuit Low
P0268 Cylinder no. 3 Injector Circuit High
P0269 Cylinder no. 3 Contribution/Balance Fault
P0270 Cylinder no. 4 Injector Circuit Low
P0271 Cylinder no. 4 Injector Circuit High
P0272 Cylinder no. 4 Contribution/Balance Fault
P0273 Cylinder no. 5 Injector Circuit Low
P0274 Cylinder no. 5 Injector Circuit High
P0275 Cylinder no. 5 Contribution/Balance Fault
P0276 Cylinder no. 6 Injector Circuit Low
P0277 Cylinder no. 6 Injector Circuit High
P0278 Cylinder no. 6 Contribution/Balance Fault
P0279 Cylinder no. 7 Injector Circuit Low
P0280 Cylinder no. 7 Injector Circuit High
P0281 Cylinder no. 7 Contribution/Balance Fault
P0282 Cylinder no. 8 Injector Circuit Low
P0283 Cylinder no. 8 Injector Circuit High
P0284 Cylinder no. 8 Contribution/Balance Fault
P0285 Cylinder no. 9 Injector Circuit Low
P0286 Cylinder no. 9 Injector Circuit High
P0287 Cylinder no. 9 Contribution/Balance Fault
P0288 Cylinder no. 10 Injector Circuit Low
P0289 Cylinder no. 10 Injector Circuit High
P0290 Cylinder no. 10 Contribution/Balance Fault
P0291 Cylinder no. 11 Injector Circuit Low
P0292 Cylinder no. 11 Injector Circuit High
P0293 Cylinder no. 11 Contribution/Balance Fault
P0294 Cylinder no. 12 Injector Circuit Low
P0295 Cylinder no. 12 Injector Circuit High
P0296 Cylinder no. 12 Contribution/Balance Fault
P0300 Random/Multiple Cylinder Misfire Detected
P0301 Cylinder no. 1—Misfire Detected
P0302 Cylinder no. 2—Misfire Detected
P0303 Cylinder no. 3—Misfire Detected
P0304 Cylinder no. 4—Misfire Detected
P0305 Cylinder no. 5—Misfire Detected
P0306 Cylinder no. 6—Misfire Detected
P0307 Cylinder no. 7—Misfire Detected
P0308 Cylinder no. 8—Misfire Detected
P0309 Cylinder no. 9—Misfire Detected
P0310 Cylinder no. 10—Misfire Detected
P0311 Cylinder no. 11—Misfire Detected
P0312 Cylinder no. 12—Misfire Detected
P0320 Ignition/Distributor Engine Speed Input Circuit Malfunction
P0321 Ignition/Distributor Engine Speed Input Circuit Range/Performance
P0322 Ignition/Distributor Engine Speed Input Circuit No Signal
P0323 Ignition/Distributor Engine Speed Input Circuit Intermittent
P0325 Knock Sensor no. 1—Circuit Malfunction (Bank no. 1 or Single Sensor)
P0326 Knock Sensor no. 1—Circuit Range/Performance (Bank no. 1 or Single Sensor)
P0327 Knock Sensor no. 1—Circuit Low Input (Bank no. 1 or Single Sensor)
P0328 Knock Sensor no. 1—Circuit High Input (Bank no. 1 or Single Sensor)
P0329 Knock Sensor no. 1—Circuit Input Intermittent (Bank no. 1 or Single Sensor)
P0330 Knock Sensor no. 2—Circuit Malfunction (Bank no. 2)
P0331 Knock Sensor no. 2—Circuit Range/Performance (Bank no. 2)

P0332 Knock Sensor no. 2—Circuit Low Input (Bank no. 2)
P0333 Knock Sensor no. 2—Circuit High Input (Bank no. 2)
P0334 Knock Sensor no. 2—Circuit Input Intermittent (Bank no. 2)
P0335 Crankshaft Position Sensor "A" Circuit Malfunction
P0336 Crankshaft Position Sensor "A" Circuit Range/Performance
P0337 Crankshaft Position Sensor "A" Circuit Low Input
P0338 Crankshaft Position Sensor "A" Circuit High Input
P0339 Crankshaft Position Sensor "A" Circuit Intermittent
P0340 Camshaft Position Sensor Circuit Malfunction
P0341 Camshaft Position Sensor Circuit Range/Performance
P0342 Camshaft Position Sensor Circuit Low Input
P0343 Camshaft Position Sensor Circuit High Input
P0344 Camshaft Position Sensor Circuit Intermittent
P0350 Ignition Coil Primary/Secondary Circuit Malfunction
P0351 Ignition Coil "A" Primary/Secondary Circuit Malfunction
P0352 Ignition Coil "B" Primary/Secondary Circuit Malfunction
P0353 Ignition Coil "C" Primary/Secondary Circuit Malfunction
P0354 Ignition Coil "D" Primary/Secondary Circuit Malfunction
P0355 Ignition Coil "E" Primary/Secondary Circuit Malfunction
P0356 Ignition Coil "F" Primary/Secondary Circuit Malfunction
P0357 Ignition Coil "G" Primary/Secondary Circuit Malfunction
P0358 Ignition Coil "H" Primary/Secondary Circuit Malfunction
P0359 Ignition Coil "I" Primary/Secondary Circuit Malfunction
P0360 Ignition Coil "J" Primary/Secondary Circuit Malfunction
P0361 Ignition Coil "K" Primary/Secondary Circuit Malfunction
P0362 Ignition Coil "L" Primary/Secondary Circuit Malfunction
P0370 Timing Reference High Resolution Signal "A" Malfunction
P0371 Timing Reference High Resolution Signal "A" Too Many Pulses
P0372 Timing Reference High Resolution Signal "A" Too Few Pulses
P0373 Timing Reference High Resolution Signal "A" Intermittent/Erratic Pulses
P0374 Timing Reference High Resolution Signal "A" No Pulses
P0375 Timing Reference High Resolution Signal "B" Malfunction
P0376 Timing Reference High Resolution Signal "B" Too Many Pulses
P0377 Timing Reference High Resolution Signal "B" Too Few Pulses
P0378 Timing Reference High Resolution Signal "B" Intermittent/Erratic Pulses
P0379 Timing Reference High Resolution Signal "B" No Pulses
P0380 Glow Plug/Heater Circuit "A" Malfunction
P0381 Glow Plug/Heater Indicator Circuit Malfunction
P0382 Glow Plug/Heater Circuit "B" Malfunction
P0385 Crankshaft Position Sensor "B" Circuit Malfunction
P0386 Crankshaft Position Sensor "B" Circuit Range/Performance
P0387 Crankshaft Position Sensor "B" Circuit Low Input
P0388 Crankshaft Position Sensor "B" Circuit High Input
P0389 Crankshaft Position Sensor "B" Circuit Intermittent
P0400 Exhaust Gas Recirculation Flow Malfunction
P0401 Exhaust Gas Recirculation Flow Insufficient Detected
P0402 Exhaust Gas Recirculation Flow Excessive Detected
P0403 Exhaust Gas Recirculation Circuit Malfunction
P0404 Exhaust Gas Recirculation Circuit Range/Performance
P0405 Exhaust Gas Recirculation Sensor "A" Circuit Low
P0406 Exhaust Gas Recirculation Sensor "A" Circuit High

P0407 Exhaust Gas Recirculation Sensor "B" Circuit Low
P0408 Exhaust Gas Recirculation Sensor "B" Circuit High
P0410 Secondary Air Injection System Malfunction
P0411 Secondary Air Injection System Incorrect Flow Detected
P0412 Secondary Air Injection System Switching Valve "A" Circuit Malfunction
P0413 Secondary Air Injection System Switching Valve "A" Circuit Open
P0414 Secondary Air Injection System Switching Valve "A" Circuit Shorted
P0415 Secondary Air Injection System Switching Valve "B" Circuit Malfunction
P0416 Secondary Air Injection System Switching Valve "B" Circuit Open
P0417 Secondary Air Injection System Switching Valve "B" Circuit Shorted
P0418 Secondary Air Injection System Relay "A" Circuit Malfunction
P0419 Secondary Air Injection System Relay "B" Circuit Malfunction
P0420 Catalyst System Efficiency Below Threshold (Bank no. 1)
P0421 Warm Up Catalyst Efficiency Below Threshold (Bank no. 1)
P0422 Main Catalyst Efficiency Below Threshold (Bank no. 1)
P0423 Heated Catalyst Efficiency Below Threshold (Bank no. 1)
P0424 Heated Catalyst Temperature Below Threshold (Bank no. 1)
P0430 Catalyst System Efficiency Below Threshold (Bank no. 2)
P0431 Warm Up Catalyst Efficiency Below Threshold (Bank no. 2)
P0432 Main Catalyst Efficiency Below Threshold (Bank no. 2)
P0433 Heated Catalyst Efficiency Below Threshold (Bank no. 2)
P0434 Heated Catalyst Temperature Below Threshold (Bank no. 2)
P0440 Evaporative Emission Control System Malfunction
P0441 Evaporative Emission Control System Incorrect Purge Flow
P0442 Evaporative Emission Control System Leak Detected (Small Leak)
P0443 Evaporative Emission Control System Purge Control Valve Circuit Malfunction
P0444 Evaporative Emission Control System Purge Control Valve Circuit Open
P0445 Evaporative Emission Control System Purge Control Valve Circuit Shorted
P0446 Evaporative Emission Control System Vent Control Circuit Malfunction
P0447 Evaporative Emission Control System Vent Control Circuit Open
P0448 Evaporative Emission Control System Vent Control Circuit Shorted
P0449 Evaporative Emission Control System Vent Valve/Solenoid Circuit Malfunction
P0450 Evaporative Emission Control System Pressure Sensor Malfunction
P0451 Evaporative Emission Control System Pressure Sensor Range/Performance
P0452 Evaporative Emission Control System Pressure Sensor Low Input
P0453 Evaporative Emission Control System Pressure Sensor High Input
P0454 Evaporative Emission Control System Pressure Sensor Intermittent

P0455 Evaporative Emission Control System Leak Detected (Gross Leak)
P0460 Fuel Level Sensor Circuit Malfunction
P0461 Fuel Level Sensor Circuit Range/Performance
P0462 Fuel Level Sensor Circuit Low Input
P0463 Fuel Level Sensor Circuit High Input
P0464 Fuel Level Sensor Circuit Intermittent
P0465 Purge Flow Sensor Circuit Malfunction
P0466 Purge Flow Sensor Circuit Range/Performance
P0467 Purge Flow Sensor Circuit Low Input
P0468 Purge Flow Sensor Circuit High Input
P0469 Purge Flow Sensor Circuit Intermittent
P0470 Exhaust Pressure Sensor Malfunction
P0471 Exhaust Pressure Sensor Range/Performance
P0472 Exhaust Pressure Sensor Low
P0473 Exhaust Pressure Sensor High
P0474 Exhaust Pressure Sensor Intermittent
P0475 Exhaust Pressure Control Valve Malfunction
P0476 Exhaust Pressure Control Valve Range/Performance
P0477 Exhaust Pressure Control Valve Low
P0478 Exhaust Pressure Control Valve High
P0479 Exhaust Pressure Control Valve Intermittent
P0480 Cooling Fan no. 1 Control Circuit Malfunction
P0481 Cooling Fan no. 2 Control Circuit Malfunction
P0482 Cooling Fan no. 3 Control Circuit Malfunction
P0483 Cooling Fan Rationality Check Malfunction
P0484 Cooling Fan Circuit Over Current
P0485 Cooling Fan Power/Ground Circuit Malfunction
P0500 Vehicle Speed Sensor Malfunction
P0501 Vehicle Speed Sensor Range/Performance
P0502 Vehicle Speed Sensor Circuit Low Input
P0503 Vehicle Speed Sensor Intermittent/Erratic/High
P0505 Idle Control System Malfunction
P0506 Idle Control System RPM Lower Than Expected
P0507 Idle Control System RPM Higher Than Expected
P0510 Closed Throttle Position Switch Malfunction
P0520 Engine Oil Pressure Sensor/Switch Circuit Malfunction
P0521 Engine Oil Pressure Sensor/Switch Range/Performance
P0522 Engine Oil Pressure Sensor/Switch Low Voltage
P0523 Engine Oil Pressure Sensor/Switch High Voltage
P0530 A/C Refrigerant Pressure Sensor Circuit Malfunction
P0531 A/C Refrigerant Pressure Sensor Circuit Range/Performance
P0532 A/C Refrigerant Pressure Sensor Circuit Low Input
P0533 A/C Refrigerant Pressure Sensor Circuit High Input
P0534 A/C Refrigerant Charge Loss
P0550 Power Steering Pressure Sensor Circuit Malfunction
P0551 Power Steering Pressure Sensor Circuit Range/Performance
P0552 Power Steering Pressure Sensor Circuit Low Input
P0553 Power Steering Pressure Sensor Circuit High Input
P0554 Power Steering Pressure Sensor Circuit Intermittent
P0560 System Voltage Malfunction
P0561 System Voltage Unstable
P0562 System Voltage Low
P0563 System Voltage High
P0565 Cruise Control On Signal Malfunction
P0566 Cruise Control Off Signal Malfunction
P0567 Cruise Control Resume Signal Malfunction
P0568 Cruise Control Set Signal Malfunction
P0569 Cruise Control Coast Signal Malfunction
P0570 Cruise Control Accel Signal Malfunction

P0571 Cruise Control/Brake Switch "A" Circuit Malfunction
P0572 Cruise Control/Brake Switch "A" Circuit Low
P0573 Cruise Control/Brake Switch "A" Circuit High
P0574 **Through P0580** Reserved for Cruise Codes
P0600 Serial Communication Link Malfunction
P0601 Internal Control Module Memory Check Sum Error
P0602 Control Module Programming Error
P0603 Internal Control Module Keep Alive Memory (KAM) Error
P0604 Internal Control Module Random Access Memory (RAM) Error
P0605 Internal Control Module Read Only Memory (ROM) Error
P0606 PCM Processor Fault
P0608 Control Module VSS Output "A" Malfunction
P0609 Control Module VSS Output "B" Malfunction
P0620 Generator Control Circuit Malfunction
P0621 Generator Lamp "L" Control Circuit Malfunction
P0622 Generator Field "F" Control Circuit Malfunction
P0650 Malfunction Indicator Lamp (MIL) Control Circuit Malfunction
P0654 Engine RPM Output Circuit Malfunction
P0655 Engine Hot Lamp Output Control Circuit Malfunction
P0656 Fuel Level Output Circuit Malfunction
P0700 Transmission Control System Malfunction
P0701 Transmission Control System Range/Performance
P0702 Transmission Control System Electrical
P0703 Torque Converter/Brake Switch "B" Circuit Malfunction
P0704 Clutch Switch Input Circuit Malfunction
P0705 Transmission Range Sensor Circuit Malfunction (PRNDL Input)
P0706 Transmission Range Sensor Circuit Range/Performance
P0707 Transmission Range Sensor Circuit Low Input
P0708 Transmission Range Sensor Circuit High Input
P0709 Transmission Range Sensor Circuit Intermittent
P0710 Transmission Fluid Temperature Sensor Circuit Malfunction
P0711 Transmission Fluid Temperature Sensor Circuit Range/Performance
P0712 Transmission Fluid Temperature Sensor Circuit Low Input
P0713 Transmission Fluid Temperature Sensor Circuit High Input
P0714 Transmission Fluid Temperature Sensor Circuit Intermittent
P0715 Input/Turbine Speed Sensor Circuit Malfunction
P0716 Input/Turbine Speed Sensor Circuit Range/Performance
P0717 Input/Turbine Speed Sensor Circuit No Signal
P0718 Input/Turbine Speed Sensor Circuit Intermittent
P0719 Torque Converter/Brake Switch "B" Circuit Low
P0720 Output Speed Sensor Circuit Malfunction
P0721 Output Speed Sensor Circuit Range/Performance
P0722 Output Speed Sensor Circuit No Signal
P0723 Output Speed Sensor Circuit Intermittent
P0724 Torque Converter/Brake Switch "B" Circuit High
P0725 Engine Speed Input Circuit Malfunction
P0726 Engine Speed Input Circuit Range/Performance
P0727 Engine Speed Input Circuit No Signal
P0728 Engine Speed Input Circuit Intermittent
P0730 Incorrect Gear Ratio
P0731 Gear no. 1 Incorrect Ratio
P0732 Gear no. 2 Incorrect Ratio
P0733 Gear no. 3 Incorrect Ratio
P0734 Gear no. 4 Incorrect Ratio

P0735 Gear no. 5 Incorrect Ratio
P0736 Reverse Incorrect Ratio
P0740 Torque Converter Clutch Circuit Malfunction
P0741 Torque Converter Clutch Circuit Performance or Stuck Off
P0742 Torque Converter Clutch Circuit Stuck On
P0743 Torque Converter Clutch Circuit Electrical
P0744 Torque Converter Clutch Circuit Intermittent
P0745 Pressure Control Solenoid Malfunction
P0746 Pressure Control Solenoid Performance or Stuck Off
P0747 Pressure Control Solenoid Stuck On
P0748 Pressure Control Solenoid Electrical
P0749 Pressure Control Solenoid Intermittent
P0750 Shift Solenoid "A" Malfunction
P0751 Shift Solenoid "A" Performance or Stuck Off
P0752 Shift Solenoid "A" Stuck On
P0753 Shift Solenoid "A" Electrical
P0754 Shift Solenoid "A" Intermittent
P0755 Shift Solenoid "B" Malfunction
P0756 Shift Solenoid "B" Performance or Stuck Off
P0757 Shift Solenoid "B" Stuck On
P0758 Shift Solenoid "B" Electrical
P0759 Shift Solenoid "B" Intermittent
P0760 Shift Solenoid "C" Malfunction
P0761 Shift Solenoid "C" Performance Or Stuck Off
P0762 Shift Solenoid "C" Stuck On
P0763 Shift Solenoid "C" Electrical
P0764 Shift Solenoid "C" Intermittent
P0765 Shift Solenoid "D" Malfunction
P0766 Shift Solenoid "D" Performance Or Stuck Off
P0767 Shift Solenoid "D" Stuck On
P0768 Shift Solenoid "D" Electrical
P0769 Shift Solenoid "D" Intermittent
P0770 Shift Solenoid "E" Malfunction
P0771 Shift Solenoid "E" Performance Or Stuck Off
P0772 Shift Solenoid "E" Stuck On
P0773 Shift Solenoid "E" Electrical
P0774 Shift Solenoid "E" Intermittent
P0780 Shift Malfunction
P0781 1–2 Shift Malfunction
P0782 2–3 Shift Malfunction
P0783 3–4 Shift Malfunction
P0784 4–5 Shift Malfunction
P0785 Shift/Timing Solenoid Malfunction
P0786 Shift/Timing Solenoid Range/Performance
P0787 Shift/Timing Solenoid Low
P0788 Shift/Timing Solenoid High
P0789 Shift/Timing Solenoid Intermittent
P0790 Normal/Performance Switch Circuit Malfunction
P0801 Reverse Inhibit Control Circuit Malfunction
P0803 1–4 Upshift (Skip Shift) Solenoid Control Circuit Malfunction
P0804 1–4 Upshift (Skip Shift) Lamp Control Circuit Malfunction

Lexus

READING CODES

Reading the control module memory is one of the first steps in OBD II system diagnostics.

This step should be initially performed to determine the general nature of the fault. Subsequent readings will determine if the fault has been cleared.

Reading codes can be performed by any of the methods below:

• Read the control module memory with the Generic Scan Tool (GST)

• Read the control module memory with the vehicle manufacturer's specific tester

To read the fault codes, connect the scan tool or tester according to the manufacturer's instructions. Follow the manufacturer's specified procedure for reading the codes.

CLEARING CODES

Control module reset procedures are a very important part of OBD II System diagnostics. This step should be done at the end of any fault code repair and at the end of any driveability repair.

Clearing codes can be performed by any of the methods below:

• Clear the control module memory with the Generic Scan Tool (GST)

• Clear the control module memory with the vehicle manufacturer's specific tester

• Turn the ignition off and remove the negative battery cable for at least 1 minute.

Removing the negative battery cable may cause other systems in the vehicle to loose their memory. Prior to removing the cable, ensure you have the proper reset codes for radios and alarms.

➡**The MIL will may also be de-activated for some codes if the vehicle completes three consecutive trips without a fault detected with vehicle conditions similar to those present during the fault.**

OBD II TROUBLE CODES

P0100 Mass or Volume Air Flow Circuit Malfunction
P0101 Mass or Volume Air Flow Circuit Range/Performance Problem
P0102 Mass or Volume Air Flow Circuit Low Input
P0103 Mass or Volume Air Flow Circuit High Input
P0104 Mass or Volume Air Flow Circuit Intermittent
P0105 Manifold Absolute Pressure/Barometric Pressure Circuit Malfunction
P0106 Manifold Absolute Pressure/Barometric Pressure Circuit Range/Performance Problem
P0107 Manifold Absolute Pressure/Barometric Pressure Circuit Low Input
P0108 Manifold Absolute Pressure/Barometric Pressure Circuit High Input
P0109 Manifold Absolute Pressure/Barometric Pressure Circuit Intermittent
P0110 Intake Air Temperature Circuit Malfunction
P0111 Intake Air Temperature Circuit Range/Performance Problem
P0112 Intake Air Temperature Circuit Low Input
P0113 Intake Air Temperature Circuit High Input
P0114 Intake Air Temperature Circuit Intermittent
P0115 Engine Coolant Temperature Circuit Malfunction
P0116 Engine Coolant Temperature Circuit Range/Performance Problem
P0117 Engine Coolant Temperature Circuit Low Input

P0118 Engine Coolant Temperature Circuit High Input
P0119 Engine Coolant Temperature Circuit Intermittent
P0120 Throttle/Pedal Position Sensor/Switch "A" Circuit Malfunction
P0121 Throttle/Pedal Position Sensor/Switch "A" Circuit Range/Performance Problem
P0122 Throttle/Pedal Position Sensor/Switch "A" Circuit Low Input
P0123 Throttle/Pedal Position Sensor/Switch "A" Circuit High Input
P0124 Throttle/Pedal Position Sensor/Switch "A" Circuit Intermittent
P0125 Insufficient Coolant Temperature For Closed Loop Fuel Control
P0126 Insufficient Coolant Temperature For Stable Operation
P0130 O2 Circuit Malfunction (Bank no. 1 Sensor no. 1)
P0131 O2 Sensor Circuit Low Voltage (Bank no. 1 Sensor no. 1)
P0132 O2 Sensor Circuit High Voltage (Bank no. 1 Sensor no. 1)
P0133 O2 Sensor Circuit Slow Response (Bank no. 1 Sensor no. 1)
P0134 O2 Sensor Circuit No Activity Detected (Bank no. 1 Sensor no. 1)
P0135 O2 Sensor Heater Circuit Malfunction (Bank no. 1 Sensor no. 1)
P0136 O2 Sensor Circuit Malfunction (Bank no. 1 Sensor no. 2)
P0137 O2 Sensor Circuit Low Voltage (Bank no. 1 Sensor no. 2)
P0138 O2 Sensor Circuit High Voltage (Bank no. 1 Sensor no. 2)
P0139 O2 Sensor Circuit Slow Response (Bank no. 1 Sensor no. 2)
P0140 O2 Sensor Circuit No Activity Detected (Bank no. 1 Sensor no. 2)
P0141 O2 Sensor Heater Circuit Malfunction (Bank no. 1 Sensor no. 2)
P0142 O2 Sensor Circuit Malfunction (Bank no. 1 Sensor no. 3)
P0143 O2 Sensor Circuit Low Voltage (Bank no. 1 Sensor no. 3)
P0144 O2 Sensor Circuit High Voltage (Bank no. 1 Sensor no. 3)
P0145 O2 Sensor Circuit Slow Response (Bank no. 1 Sensor no. 3)
P0146 O2 Sensor Circuit No Activity Detected (Bank no. 1 Sensor no. 3)
P0147 O2 Sensor Heater Circuit Malfunction (Bank no. 1 Sensor no. 3)
P0150 O2 Sensor Circuit Malfunction (Bank no. 2 Sensor no. 1)
P0151 O2 Sensor Circuit Low Voltage (Bank no. 2 Sensor no. 1)
P0152 O2 Sensor Circuit High Voltage (Bank no. 2 Sensor no. 1)
P0153 O2 Sensor Circuit Slow Response (Bank no. 2 Sensor no. 1)
P0154 O2 Sensor Circuit No Activity Detected (Bank no. 2 Sensor no. 1)
P0155 O2 Sensor Heater Circuit Malfunction (Bank no. 2 Sensor no. 1)
P0156 O2 Sensor Circuit Malfunction (Bank no. 2 Sensor no. 2)
P0157 O2 Sensor Circuit Low Voltage (Bank no. 2 Sensor no. 2)
P0158 O2 Sensor Circuit High Voltage (Bank no. 2 Sensor no. 2)
P0159 O2 Sensor Circuit Slow Response (Bank no. 2 Sensor no. 2)

P0160 O2 Sensor Circuit No Activity Detected (Bank no. 2 Sensor no. 2)
P0161 O2 Sensor Heater Circuit Malfunction (Bank no. 2 Sensor no. 2)
P0162 O2 Sensor Circuit Malfunction (Bank no. 2 Sensor no. 3)
P0163 O2 Sensor Circuit Low Voltage (Bank no. 2 Sensor no. 3)
P0164 O2 Sensor Circuit High Voltage (Bank no. 2 Sensor no. 3)
P0165 O2 Sensor Circuit Slow Response (Bank no. 2 Sensor no. 3)
P0166 O2 Sensor Circuit No Activity Detected (Bank no. 2 Sensor no. 3)
P0167 O2 Sensor Heater Circuit Malfunction (Bank no. 2 Sensor no. 3)
P0170 Fuel Trim Malfunction (Bank no. 1)
P0171 System Too Lean (Bank no. 1)
P0172 System Too Rich (Bank no. 1)
P0173 Fuel Trim Malfunction (Bank no. 2)
P0174 System Too Lean (Bank no. 2)
P0175 System Too Rich (Bank no. 2)
P0176 Fuel Composition Sensor Circuit Malfunction
P0177 Fuel Composition Sensor Circuit Range/Performance
P0178 Fuel Composition Sensor Circuit Low Input
P0179 Fuel Composition Sensor Circuit High Input
P0180 Fuel Temperature Sensor "A" Circuit Malfunction
P0181 Fuel Temperature Sensor "A" Circuit Range/Performance
P0182 Fuel Temperature Sensor "A" Circuit Low Input
P0183 Fuel Temperature Sensor "A" Circuit High Input
P0184 Fuel Temperature Sensor "A" Circuit Intermittent
P0185 Fuel Temperature Sensor "B" Circuit Malfunction
P0186 Fuel Temperature Sensor "B" Circuit Range/Performance
P0187 Fuel Temperature Sensor "B" Circuit Low Input
P0188 Fuel Temperature Sensor "B" Circuit High Input
P0189 Fuel Temperature Sensor "B" Circuit Intermittent
P0190 Fuel Rail Pressure Sensor Circuit Malfunction
P0191 Fuel Rail Pressure Sensor Circuit Range/Performance
P0192 Fuel Rail Pressure Sensor Circuit Low Input
P0193 Fuel Rail Pressure Sensor Circuit High Input
P0194 Fuel Rail Pressure Sensor Circuit Intermittent
P0195 Engine Oil Temperature Sensor Malfunction
P0196 Engine Oil Temperature Sensor Range/Performance
P0197 Engine Oil Temperature Sensor Low
P0198 Engine Oil Temperature Sensor High
P0199 Engine Oil Temperature Sensor Intermittent
P0200 Injector Circuit Malfunction
P0201 Injector Circuit Malfunction—Cylinder no. 1
P0202 Injector Circuit Malfunction—Cylinder no. 2
P0203 Injector Circuit Malfunction—Cylinder no. 3
P0204 Injector Circuit Malfunction—Cylinder no. 4
P0205 Injector Circuit Malfunction—Cylinder no. 5
P0206 Injector Circuit Malfunction—Cylinder no. 6
P0207 Injector Circuit Malfunction—Cylinder no. 7
P0208 Injector Circuit Malfunction—Cylinder no. 8
P0209 Injector Circuit Malfunction—Cylinder no. 9
P0210 Injector Circuit Malfunction—Cylinder no. 10
P0211 Injector Circuit Malfunction—Cylinder no. 11
P0212 Injector Circuit Malfunction—Cylinder no. 12
P0213 Cold Start Injector no. 1 Malfunction
P0214 Cold Start Injector no. 2 Malfunction
P0215 Engine Shutoff Solenoid Malfunction
P0216 Injection Timing Control Circuit Malfunction

P0217 Engine Over Temperature Condition
P0218 Transmission Over Temperature Condition
P0219 Engine Over Speed Condition
P0220 Throttle/Pedal Position Sensor/Switch "B" Circuit Malfunction
P0221 Throttle/Pedal Position Sensor/Switch "B" Circuit Range/Performance Problem
P0222 Throttle/Pedal Position Sensor/Switch "B" Circuit Low Input
P0223 Throttle/Pedal Position Sensor/Switch "B" Circuit High Input
P0224 Throttle/Pedal Position Sensor/Switch "B" Circuit Intermittent
P0225 Throttle/Pedal Position Sensor/Switch "C" Circuit Malfunction
P0226 Throttle/Pedal Position Sensor/Switch "C" Circuit Range/Performance Problem
P0227 Throttle/Pedal Position Sensor/Switch "C" Circuit Low Input
P0228 Throttle/Pedal Position Sensor/Switch "C" Circuit High Input
P0229 Throttle/Pedal Position Sensor/Switch "C" Circuit Intermittent
P0230 Fuel Pump Primary Circuit Malfunction
P0231 Fuel Pump Secondary Circuit Low
P0232 Fuel Pump Secondary Circuit High
P0233 Fuel Pump Secondary Circuit Intermittent
P0234 Engine Over Boost Condition
P0261 Cylinder no. 1 Injector Circuit Low
P0262 Cylinder no. 1 Injector Circuit High
P0263 Cylinder no. 1 Contribution/Balance Fault
P0264 Cylinder no. 2 Injector Circuit Low
P0265 Cylinder no. 2 Injector Circuit High
P0266 Cylinder no. 2 Contribution/Balance Fault
P0267 Cylinder no. 3 Injector Circuit Low
P0268 Cylinder no. 3 Injector Circuit High
P0269 Cylinder no. 3 Contribution/Balance Fault
P0270 Cylinder no. 4 Injector Circuit Low
P0271 Cylinder no. 4 Injector Circuit High
P0272 Cylinder no. 4 Contribution/Balance Fault
P0273 Cylinder no. 5 Injector Circuit Low
P0274 Cylinder no. 5 Injector Circuit High
P0275 Cylinder no. 5 Contribution/Balance Fault
P0276 Cylinder no. 6 Injector Circuit Low
P0277 Cylinder no. 6 Injector Circuit High
P0278 Cylinder no. 6 Contribution/Balance Fault
P0279 Cylinder no. 7 Injector Circuit Low
P0280 Cylinder no. 7 Injector Circuit High
P0281 Cylinder no. 7 Contribution/Balance Fault
P0282 Cylinder no. 8 Injector Circuit Low
P0283 Cylinder no. 8 Injector Circuit High
P0284 Cylinder no. 8 Contribution/Balance Fault
P0285 Cylinder no. 9 Injector Circuit Low
P0286 Cylinder no. 9 Injector Circuit High
P0287 Cylinder no. 9 Contribution/Balance Fault
P0288 Cylinder no. 10 Injector Circuit Low
P0289 Cylinder no. 10 Injector Circuit High
P0290 Cylinder no. 10 Contribution/Balance Fault
P0291 Cylinder no. 11 Injector Circuit Low
P0292 Cylinder no. 11 Injector Circuit High
P0293 Cylinder no. 11 Contribution/Balance Fault
P0294 Cylinder no. 12 Injector Circuit Low

P0295 Cylinder no. 12 Injector Circuit High
P0296 Cylinder no. 12 Contribution/Balance Fault
P0300 Random/Multiple Cylinder Misfire Detected
P0301 Cylinder no. 1—Misfire Detected
P0302 Cylinder no. 2—Misfire Detected
P0303 Cylinder no. 3—Misfire Detected
P0304 Cylinder no. 4—Misfire Detected
P0305 Cylinder no. 5—Misfire Detected
P0306 Cylinder no. 6—Misfire Detected
P0307 Cylinder no. 7—Misfire Detected
P0308 Cylinder no. 8—Misfire Detected
P0320 Ignition/Distributor Engine Speed Input Circuit Malfunction
P0321 Ignition/Distributor Engine Speed Input Circuit Range/Performance
P0322 Ignition/Distributor Engine Speed Input Circuit No Signal
P0323 Ignition/Distributor Engine Speed Input Circuit Intermittent
P0325 Knock Sensor no. 1—Circuit Malfunction (Bank no. 1 or Single Sensor)
P0326 Knock Sensor no. 1—Circuit Range/Performance (Bank no. 1 or Single Sensor)
P0327 Knock Sensor no. 1—Circuit Low Input (Bank no. 1 or Single Sensor)
P0328 Knock Sensor no. 1—Circuit High Input (Bank no. 1 or Single Sensor)
P0329 Knock Sensor no. 1—Circuit Input Intermittent (Bank no. 1 or Single Sensor)
P0330 Knock Sensor no. 2—Circuit Malfunction (Bank no. 2)
P0331 Knock Sensor no. 2—Circuit Range/Performance (Bank no. 2)
P0332 Knock Sensor no. 2—Circuit Low Input (Bank no. 2)
P0333 Knock Sensor no. 2—Circuit High Input (Bank no. 2)
P0334 Knock Sensor no. 2—Circuit Input Intermittent (Bank no. 2)
P0335 Crankshaft Position Sensor "A" Circuit Malfunction
P0336 Crankshaft Position Sensor "A" Circuit Range/Performance
P0337 Crankshaft Position Sensor "A" Circuit Low Input
P0338 Crankshaft Position Sensor "A" Circuit High Input
P0339 Crankshaft Position Sensor "A" Circuit Intermittent
P0340 Camshaft Position Sensor Circuit Malfunction
P0341 Camshaft Position Sensor Circuit Range/Performance
P0342 Camshaft Position Sensor Circuit Low Input
P0343 Camshaft Position Sensor Circuit High Input
P0344 Camshaft Position Sensor Circuit Intermittent
P0350 Ignition Coil Primary/Secondary Circuit Malfunction
P0351 Ignition Coil "A" Primary/Secondary Circuit Malfunction
P0352 Ignition Coil "B" Primary/Secondary Circuit Malfunction
P0353 Ignition Coil "C" Primary/Secondary Circuit Malfunction
P0354 Ignition Coil "D" Primary/Secondary Circuit Malfunction
P0355 Ignition Coil "E" Primary/Secondary Circuit Malfunction
P0356 Ignition Coil "F" Primary/Secondary Circuit Malfunction
P0357 Ignition Coil "G" Primary/Secondary Circuit Malfunction
P0358 Ignition Coil "H" Primary/Secondary Circuit Malfunction
P0359 Ignition Coil "I" Primary/Secondary Circuit Malfunction
P0360 Ignition Coil "J" Primary/Secondary Circuit Malfunction
P0361 Ignition Coil "K" Primary/Secondary Circuit Malfunction
P0362 Ignition Coil "L" Primary/Secondary Circuit Malfunction
P0370 Timing Reference High Resolution Signal "A" Malfunction
P0371 Timing Reference High Resolution Signal "A" Too Many Pulses

P0372 Timing Reference High Resolution Signal "A" Too Few Pulses

P0373 Timing Reference High Resolution Signal "A" Intermittent/Erratic Pulses

P0374 Timing Reference High Resolution Signal "A" No Pulses

P0375 Timing Reference High Resolution Signal "B" Malfunction

P0376 Timing Reference High Resolution Signal "B" Too Many Pulses

P0377 Timing Reference High Resolution Signal "B" Too Few Pulses

P0378 Timing Reference High Resolution Signal "B" Intermittent/Erratic Pulses

P0379 Timing Reference High Resolution Signal "B" No Pulses

P0380 Glow Plug/Heater Circuit "A" Malfunction

P0381 Glow Plug/Heater Indicator Circuit Malfunction

P0382 Glow Plug/Heater Circuit "B" Malfunction

P0385 Crankshaft Position Sensor "B" Circuit Malfunction

P0386 Crankshaft Position Sensor "B" Circuit Range/Performance

P0387 Crankshaft Position Sensor "B" Circuit Low Input

P0388 Crankshaft Position Sensor "B" Circuit High Input

P0389 Crankshaft Position Sensor "B" Circuit Intermittent

P0400 Exhaust Gas Recirculation Flow Malfunction

P0401 Exhaust Gas Recirculation Flow Insufficient Detected

P0402 Exhaust Gas Recirculation Flow Excessive Detected

P0403 Exhaust Gas Recirculation Circuit Malfunction

P0404 Exhaust Gas Recirculation Circuit Range/Performance

P0405 Exhaust Gas Recirculation Sensor "A" Circuit Low

P0406 Exhaust Gas Recirculation Sensor "A" Circuit High

P0407 Exhaust Gas Recirculation Sensor "B" Circuit Low

P0408 Exhaust Gas Recirculation Sensor "B" Circuit High

P0410 Secondary Air Injection System Malfunction

P0411 Secondary Air Injection System Incorrect Flow Detected

P0412 Secondary Air Injection System Switching Valve "A" Circuit Malfunction

P0413 Secondary Air Injection System Switching Valve "A" Circuit Open

P0414 Secondary Air Injection System Switching Valve "A" Circuit Shorted

P0415 Secondary Air Injection System Switching Valve "B" Circuit Malfunction

P0416 Secondary Air Injection System Switching Valve "B" Circuit Open

P0417 Secondary Air Injection System Switching Valve "B" Circuit Shorted

P0418 Secondary Air Injection System Relay "A" Circuit Malfunction

P0419 Secondary Air Injection System Relay "B" Circuit Malfunction

P0420 Catalyst System Efficiency Below Threshold (Bank no. 1)

P0421 Warm Up Catalyst Efficiency Below Threshold (Bank no. 1)

P0422 Main Catalyst Efficiency Below Threshold (Bank no. 1)

P0423 Heated Catalyst Efficiency Below Threshold (Bank no. 1)

P0424 Heated Catalyst Temperature Below Threshold (Bank no. 1)

P0430 Catalyst System Efficiency Below Threshold (Bank no. 2)

P0431 Warm Up Catalyst Efficiency Below Threshold (Bank no. 2)

P0432 Main Catalyst Efficiency Below Threshold (Bank no. 2)

P0433 Heated Catalyst Efficiency Below Threshold (Bank no. 2)

P0434 Heated Catalyst Temperature Below Threshold (Bank no. 2)

P0440 Evaporative Emission Control System Malfunction

P0441 Evaporative Emission Control System Incorrect Purge Flow

P0442 Evaporative Emission Control System Leak Detected (Small Leak)

P0443 Evaporative Emission Control System Purge Control Valve Circuit Malfunction

P0444 Evaporative Emission Control System Purge Control Valve Circuit Open

P0445 Evaporative Emission Control System Purge Control Valve Circuit Shorted

P0446 Evaporative Emission Control System Vent Control Circuit Malfunction

P0447 Evaporative Emission Control System Vent Control Circuit Open

P0448 Evaporative Emission Control System Vent Control Circuit Shorted

P0449 Evaporative Emission Control System Vent Valve/Solenoid Circuit Malfunction

P0450 Evaporative Emission Control System Pressure Sensor Malfunction

P0451 Evaporative Emission Control System Pressure Sensor Range/Performance

P0452 Evaporative Emission Control System Pressure Sensor Low Input

P0453 Evaporative Emission Control System Pressure Sensor High Input

P0454 Evaporative Emission Control System Pressure Sensor Intermittent

P0455 Evaporative Emission Control System Leak Detected (Gross Leak)

P0460 Fuel Level Sensor Circuit Malfunction

P0461 Fuel Level Sensor Circuit Range/Performance

P0462 Fuel Level Sensor Circuit Low Input

P0463 Fuel Level Sensor Circuit High Input

P0464 Fuel Level Sensor Circuit Intermittent

P0465 Purge Flow Sensor Circuit Malfunction

P0466 Purge Flow Sensor Circuit Range/Performance

P0467 Purge Flow Sensor Circuit Low Input

P0468 Purge Flow Sensor Circuit High Input

P0469 Purge Flow Sensor Circuit Intermittent

P0470 Exhaust Pressure Sensor Malfunction

P0471 Exhaust Pressure Sensor Range/Performance

P0472 Exhaust Pressure Sensor Low

P0473 Exhaust Pressure Sensor High

P0474 Exhaust Pressure Sensor Intermittent

P0475 Exhaust Pressure Control Valve Malfunction

P0476 Exhaust Pressure Control Valve Range/Performance

P0477 Exhaust Pressure Control Valve Low

P0478 Exhaust Pressure Control Valve High

P0479 Exhaust Pressure Control Valve Intermittent

P0480 Cooling Fan no. 1 Control Circuit Malfunction

P0481 Cooling Fan no. 2 Control Circuit Malfunction

P0482 Cooling Fan no. 3 Control Circuit Malfunction

P0483 Cooling Fan Rationality Check Malfunction

P0484 Cooling Fan Circuit Over Current

P0485 Cooling Fan Power/Ground Circuit Malfunction

P0500 Vehicle Speed Sensor Malfunction

P0501 Vehicle Speed Sensor Range/Performance

P0502 Vehicle Speed Sensor Circuit Low Input

P0503 Vehicle Speed Sensor Intermittent/Erratic/High
P0505 Idle Control System Malfunction
P0506 Idle Control System RPM Lower Than Expected
P0507 Idle Control System RPM Higher Than Expected
P0510 Closed Throttle Position Switch Malfunction
P0520 Engine Oil Pressure Sensor/Switch Circuit Malfunction
P0521 Engine Oil Pressure Sensor/Switch Range/Performance
P0522 Engine Oil Pressure Sensor/Switch Low Voltage
P0523 Engine Oil Pressure Sensor/Switch High Voltage
P0530 A/C Refrigerant Pressure Sensor Circuit Malfunction
P0531 A/C Refrigerant Pressure Sensor Circuit Range/Performance
P0532 A/C Refrigerant Pressure Sensor Circuit Low Input
P0533 A/C Refrigerant Pressure Sensor Circuit High Input
P0534 A/C Refrigerant Charge Loss
P0550 Power Steering Pressure Sensor Circuit Malfunction
P0551 Power Steering Pressure Sensor Circuit Range/Performance
P0552 Power Steering Pressure Sensor Circuit Low Input
P0553 Power Steering Pressure Sensor Circuit High Input
P0554 Power Steering Pressure Sensor Circuit Intermittent
P0560 System Voltage Malfunction
P0561 System Voltage Unstable
P0562 System Voltage Low
P0563 System Voltage High
P0565 Cruise Control On Signal Malfunction
P0566 Cruise Control Off Signal Malfunction
P0567 Cruise Control Resume Signal Malfunction
P0568 Cruise Control Set Signal Malfunction
P0569 Cruise Control Coast Signal Malfunction
P0570 Cruise Control Accel Signal Malfunction
P0571 Cruise Control/Brake Switch "A" Circuit Malfunction
P0572 Cruise Control/Brake Switch "A" Circuit Low
P0573 Cruise Control/Brake Switch "A" Circuit High
P0574 Through P0580 Reserved for Cruise Codes
P0600 Serial Communication Link Malfunction
P0601 Internal Control Module Memory Check Sum Error
P0602 Control Module Programming Error
P0603 Internal Control Module Keep Alive Memory (KAM) Error
P0604 Internal Control Module Random Access Memory (RAM) Error
P0605 Internal Control Module Read Only Memory (ROM) Error
P0606 PCM Processor Fault
P0608 Control Module VSS Output "A" Malfunction
P0609 Control Module VSS Output "B" Malfunction
P0620 Generator Control Circuit Malfunction
P0621 Generator Lamp "L" Control Circuit Malfunction
P0622 Generator Field "F" Control Circuit Malfunction
P0650 Malfunction Indicator Lamp (MIL) Control Circuit Malfunction
P0654 Engine RPM Output Circuit Malfunction
P0655 Engine Hot Lamp Output Control Circuit Malfunction
P0656 Fuel Level Output Circuit Malfunction
P0700 Transmission Control System Malfunction
P0701 Transmission Control System Range/Performance
P0702 Transmission Control System Electrical
P0703 Torque Converter/Brake Switch "B" Circuit Malfunction
P0704 Clutch Switch Input Circuit Malfunction
P0705 Transmission Range Sensor Circuit Malfunction (PRNDL Input)
P0706 Transmission Range Sensor Circuit Range/Performance
P0707 Transmission Range Sensor Circuit Low Input

P0708 Transmission Range Sensor Circuit High Input
P0709 Transmission Range Sensor Circuit Intermittent
P0710 Transmission Fluid Temperature Sensor Circuit Malfunction
P0711 Transmission Fluid Temperature Sensor Circuit Range/Performance
P0712 Transmission Fluid Temperature Sensor Circuit Low Input
P0713 Transmission Fluid Temperature Sensor Circuit High Input
P0714 Transmission Fluid Temperature Sensor Circuit Intermittent
P0715 Input/Turbine Speed Sensor Circuit Malfunction
P0716 Input/Turbine Speed Sensor Circuit Range/Performance
P0717 Input/Turbine Speed Sensor Circuit No Signal
P0718 Input/Turbine Speed Sensor Circuit Intermittent
P0719 Torque Converter/Brake Switch "B" Circuit Low
P0720 Output Speed Sensor Circuit Malfunction
P0721 Output Speed Sensor Circuit Range/Performance
P0722 Output Speed Sensor Circuit No Signal
P0723 Output Speed Sensor Circuit Intermittent
P0724 Torque Converter/Brake Switch "B" Circuit High
P0725 Engine Speed Input Circuit Malfunction
P0726 Engine Speed Input Circuit Range/Performance
P0727 Engine Speed Input Circuit No Signal
P0728 Engine Speed Input Circuit Intermittent
P0730 Incorrect Gear Ratio
P0731 Gear no. 1 Incorrect Ratio
P0732 Gear no. 2 Incorrect Ratio
P0733 Gear no. 3 Incorrect Ratio
P0734 Gear no. 4 Incorrect Ratio
P0735 Gear no. 5 Incorrect Ratio
P0736 Reverse Incorrect Ratio
P0740 Torque Converter Clutch Circuit Malfunction
P0741 Torque Converter Clutch Circuit Performance or Stuck Off
P0742 Torque Converter Clutch Circuit Stuck On
P0743 Torque Converter Clutch Circuit Electrical
P0744 Torque Converter Clutch Circuit Intermittent
P0745 Pressure Control Solenoid Malfunction
P0746 Pressure Control Solenoid Performance or Stuck Off
P0747 Pressure Control Solenoid Stuck On
P0748 Pressure Control Solenoid Electrical
P0749 Pressure Control Solenoid Intermittent
P0750 Shift Solenoid "A" Malfunction
P0751 Shift Solenoid "A" Performance or Stuck Off
P0752 Shift Solenoid "A" Stuck On
P0753 Shift Solenoid "A" Electrical
P0754 Shift Solenoid "A" Intermittent
P0755 Shift Solenoid "B" Malfunction
P0756 Shift Solenoid "B" Performance or Stuck Off
P0757 Shift Solenoid "B" Stuck On
P0758 Shift Solenoid "B" Electrical
P0759 Shift Solenoid "B" Intermittent
P0760 Shift Solenoid "C" Malfunction
P0761 Shift Solenoid "C" Performance Or Stuck Off
P0762 Shift Solenoid "C" Stuck On
P0763 Shift Solenoid "C" Electrical
P0764 Shift Solenoid "C" Intermittent
P0765 Shift Solenoid "D" Malfunction
P0766 Shift Solenoid "D" Performance Or Stuck Off
P0767 Shift Solenoid "D" Stuck On

P0768 Shift Solenoid "D" Electrical
P0769 Shift Solenoid "D" Intermittent
P0770 Shift Solenoid "E" Malfunction
P0771 Shift Solenoid "E" Performance Or Stuck Off
P0772 Shift Solenoid "E" Stuck On
P0773 Shift Solenoid "E" Electrical
P0774 Shift Solenoid "E" Intermittent
P0780 Shift Malfunction
P0781 1–2 Shift Malfunction
P0782 2–3 Shift Malfunction
P0783 3–4 Shift Malfunction
P0784 4–5 Shift Malfunction
P0785 Shift/Timing Solenoid Malfunction
P0786 Shift/Timing Solenoid Range/Performance
P0787 Shift/Timing Solenoid Low
P0788 Shift/Timing Solenoid High
P0789 Shift/Timing Solenoid Intermittent
P0790 Normal/Performance Switch Circuit Malfunction
P0801 Reverse Inhibit Control Circuit Malfunction
P0803 1–4 Upshift (Skip Shift) Solenoid Control Circuit Malfunction
P0804 1–4 Upshift (Skip Shift) Lamp Control Circuit Malfunction
P1100 Barometric Pressure Sensor Circuit Fault
P1200 Fuel Pump Relay Circuit Fault
P1300 Igniter Circuit Fault (Bank 1)
P1305 Igniter Circuit Fault (Bank 2)
P1335 Crankshaft Position Sensor Circuit Fault
P1400 Sub-Throttle Position Sensor Circuit Fault
P1401 Sub-Throttle Position Sensor Performance
P1500 Starter Signal Circuit Fault
P1510 Air Volume Too Low With Supercharger On
P1600 PCM Battery Back-up Circuit Fault
P1605 Knock Control CPU Fault
P1700 Vehicle Speed Sensor Circuit Fault
P1705 Direct Clutch Speed Sensor Circuit Fault
P1765 Linear Shift Solenoid Circuit Fault
P1780 Park Neutral Position Switch Fault

Mazda

READING CODES

Reading the control module memory is one of the first steps in OBD II system diagnostics. This step should be initially performed to determine the general nature of the fault. Subsequent readings will determine if the fault has been cleared.

Reading codes can be performed by any of the methods below:
• Read the control module memory with the Generic Scan Tool (GST)
• Read the control module memory with the vehicle manufacturer's specific tester

To read the fault codes, connect the scan tool or tester according to the manufacturer's instructions. Follow the manufacturer's specified procedure for reading the codes.

CLEARING CODES

Control module reset procedures are a very important part of OBD II System diagnostics. This step should be done at the end of any fault code repair and at the end of any driveability repair.

Clearing codes can be performed by any of the methods below:
• Clear the control module memory with the Generic Scan Tool (GST)
• Clear the control module memory with the vehicle manufacturer's specific tester
• Turn the ignition off and remove the negative battery cable for at least 1 minute.

Removing the negative battery cable may cause other systems in the vehicle to loose their memory. Prior to removing the cable, ensure you have the proper reset codes for radios and alarms.

➡The MIL will may also be de-activated for some codes if the vehicle completes three consecutive trips without a fault detected with vehicle conditions similar to those present during the fault.

OBD II TROUBLE CODES

P0100 Mass or Volume Air Flow Circuit Malfunction
P0101 Mass or Volume Air Flow Circuit Range/Performance Problem
P0102 Mass or Volume Air Flow Circuit Low Input
P0103 Mass or Volume Air Flow Circuit High Input
P0104 Mass or Volume Air Flow Circuit Intermittent
P0105 Manifold Absolute Pressure/Barometric Pressure Circuit Malfunction
P0106 Manifold Absolute Pressure/Barometric Pressure Circuit Range/Performance Problem
P0107 Manifold Absolute Pressure/Barometric Pressure Circuit Low Input
P0108 Manifold Absolute Pressure/Barometric Pressure Circuit High Input
P0109 Manifold Absolute Pressure/Barometric Pressure Circuit Intermittent
P0110 Intake Air Temperature Circuit Malfunction
P0111 Intake Air Temperature Circuit Range/Performance Problem
P0112 Intake Air Temperature Circuit Low Input
P0113 Intake Air Temperature Circuit High Input
P0114 Intake Air Temperature Circuit Intermittent
P0115 Engine Coolant Temperature Circuit Malfunction
P0116 Engine Coolant Temperature Circuit Range/Performance Problem
P0117 Engine Coolant Temperature Circuit Low Input
P0118 Engine Coolant Temperature Circuit High Input
P0119 Engine Coolant Temperature Circuit Intermittent
P0120 Throttle/Pedal Position Sensor/Switch "A" Circuit Malfunction
P0121 Throttle/Pedal Position Sensor/Switch "A" Circuit Range/Performance Problem
P0122 Throttle/Pedal Position Sensor/Switch "A" Circuit Low Input
P0123 Throttle/Pedal Position Sensor/Switch "A" Circuit High Input
P0124 Throttle/Pedal Position Sensor/Switch "A" Circuit Intermittent
P0125 Insufficient Coolant Temperature For Closed Loop Fuel Control
P0126 Insufficient Coolant Temperature For Stable Operation
P0130 O2 Circuit Malfunction (Bank no. 1 Sensor no. 1)
P0131 O2 Sensor Circuit Low Voltage (Bank no. 1 Sensor no. 1)
P0132 O2 Sensor Circuit High Voltage (Bank no. 1 Sensor no. 1)

P0133 O2 Sensor Circuit Slow Response (Bank no. 1 Sensor no. 1)

P0134 O2 Sensor Circuit No Activity Detected (Bank no. 1 Sensor no. 1)

P0135 O2 Sensor Heater Circuit Malfunction (Bank no. 1 Sensor no. 1)

P0136 O2 Sensor Circuit Malfunction (Bank no. 1 Sensor no. 2)

P0137 O2 Sensor Circuit Low Voltage (Bank no. 1 Sensor no. 2)

P0138 O2 Sensor Circuit High Voltage (Bank no. 1 Sensor no. 2)

P0139 O2 Sensor Circuit Slow Response (Bank no. 1 Sensor no. 2)

P0140 O2 Sensor Circuit No Activity Detected (Bank no. 1 Sensor no. 2)

P0141 O2 Sensor Heater Circuit Malfunction (Bank no. 1 Sensor no. 2)

P0142 O2 Sensor Circuit Malfunction (Bank no. 1 Sensor no. 3)

P0143 O2 Sensor Circuit Low Voltage (Bank no. 1 Sensor no. 3)

P0144 O2 Sensor Circuit High Voltage (Bank no. 1 Sensor no. 3)

P0145 O2 Sensor Circuit Slow Response (Bank no. 1 Sensor no. 3)

P0146 O2 Sensor Circuit No Activity Detected (Bank no. 1 Sensor no. 3)

P0147 O2 Sensor Heater Circuit Malfunction (Bank no. 1 Sensor no. 3)

P0150 O2 Sensor Circuit Malfunction (Bank no. 2 Sensor no. 1)

P0151 O2 Sensor Circuit Low Voltage (Bank no. 2 Sensor no. 1)

P0152 O2 Sensor Circuit High Voltage (Bank no. 2 Sensor no. 1)

P0153 O2 Sensor Circuit Slow Response (Bank no. 2 Sensor no. 1)

P0154 O2 Sensor Circuit No Activity Detected (Bank no. 2 Sensor no. 1)

P0155 O2 Sensor Heater Circuit Malfunction (Bank no. 2 Sensor no. 1)

P0156 O2 Sensor Circuit Malfunction (Bank no. 2 Sensor no. 2)

P0157 O2 Sensor Circuit Low Voltage (Bank no. 2 Sensor no. 2)

P0158 O2 Sensor Circuit High Voltage (Bank no. 2 Sensor no. 2)

P0159 O2 Sensor Circuit Slow Response (Bank no. 2 Sensor no. 2)

P0160 O2 Sensor Circuit No Activity Detected (Bank no. 2 Sensor no. 2)

P0161 O2 Sensor Heater Circuit Malfunction (Bank no. 2 Sensor no. 2)

P0162 O2 Sensor Circuit Malfunction (Bank no. 2 Sensor no. 3)

P0163 O2 Sensor Circuit Low Voltage (Bank no. 2 Sensor no. 3)

P0164 O2 Sensor Circuit High Voltage (Bank no. 2 Sensor no. 3)

P0165 O2 Sensor Circuit Slow Response (Bank no. 2 Sensor no. 3)

P0166 O2 Sensor Circuit No Activity Detected (Bank no. 2 Sensor no. 3)

P0167 O2 Sensor Heater Circuit Malfunction (Bank no. 2 Sensor no. 3)

P0170 Fuel Trim Malfunction (Bank no. 1)

P0171 System Too Lean (Bank no. 1)

P0172 System Too Rich (Bank no. 1)

P0173 Fuel Trim Malfunction (Bank no. 2)

P0174 System Too Lean (Bank no. 2)

P0175 System Too Rich (Bank no. 2)

P0176 Fuel Composition Sensor Circuit Malfunction

P0177 Fuel Composition Sensor Circuit Range/Performance

P0178 Fuel Composition Sensor Circuit Low Input

P0179 Fuel Composition Sensor Circuit High Input

P0180 Fuel Temperature Sensor "A" Circuit Malfunction

P0181 Fuel Temperature Sensor "A" Circuit Range/Performance

P0182 Fuel Temperature Sensor "A" Circuit Low Input

P0183 Fuel Temperature Sensor "A" Circuit High Input

P0184 Fuel Temperature Sensor "A" Circuit Intermittent

P0185 Fuel Temperature Sensor "B" Circuit Malfunction

P0186 Fuel Temperature Sensor "B" Circuit Range/Performance

P0187 Fuel Temperature Sensor "B" Circuit Low Input

P0188 Fuel Temperature Sensor "B" Circuit High Input

P0189 Fuel Temperature Sensor "B" Circuit Intermittent

P0190 Fuel Rail Pressure Sensor Circuit Malfunction

P0191 Fuel Rail Pressure Sensor Circuit Range/Performance

P0192 Fuel Rail Pressure Sensor Circuit Low Input

P0193 Fuel Rail Pressure Sensor Circuit High Input

P0194 Fuel Rail Pressure Sensor Circuit Intermittent

P0195 Engine Oil Temperature Sensor Malfunction

P0196 Engine Oil Temperature Sensor Range/Performance

P0197 Engine Oil Temperature Sensor Low

P0198 Engine Oil Temperature Sensor High

P0199 Engine Oil Temperature Sensor Intermittent

P0200 Injector Circuit Malfunction

P0201 Injector Circuit Malfunction—Cylinder no. 1

P0202 Injector Circuit Malfunction—Cylinder no. 2

P0203 Injector Circuit Malfunction—Cylinder no. 3

P0204 Injector Circuit Malfunction—Cylinder no. 4

P0205 Injector Circuit Malfunction—Cylinder no. 5

P0206 Injector Circuit Malfunction—Cylinder no. 6

P0207 Injector Circuit Malfunction—Cylinder no. 7

P0208 Injector Circuit Malfunction—Cylinder no. 8

P0209 Injector Circuit Malfunction—Cylinder no. 9

P0210 Injector Circuit Malfunction—Cylinder no. 10

P0211 Injector Circuit Malfunction—Cylinder no. 11

P0212 Injector Circuit Malfunction—Cylinder no. 12

P0213 Cold Start Injector no. 1 Malfunction

P0214 Cold Start Injector no. 2 Malfunction

P0215 Engine Shutoff Solenoid Malfunction

P0216 Injection Timing Control Circuit Malfunction

P0217 Engine Over Temperature Condition

P0218 Transmission Over Temperature Condition

P0219 Engine Over Speed Condition

P0220 Throttle/Pedal Position Sensor/Switch "B" Circuit Malfunction

P0221 Throttle/Pedal Position Sensor/Switch "B" Circuit Range/Performance Problem

P0222 Throttle/Pedal Position Sensor/Switch "B" Circuit Low Input

P0223 Throttle/Pedal Position Sensor/Switch "B" Circuit High Input

P0224 Throttle/Pedal Position Sensor/Switch "B" Circuit Intermittent

P0225 Throttle/Pedal Position Sensor/Switch "C" Circuit Malfunction

P0226 Throttle/Pedal Position Sensor/Switch "C" Circuit Range/Performance Problem

P0227 Throttle/Pedal Position Sensor/Switch "C" Circuit Low Input

P0228 Throttle/Pedal Position Sensor/Switch "C" Circuit High Input

P0229 Throttle/Pedal Position Sensor/Switch "C" Circuit Intermittent

P0230 Fuel Pump Primary Circuit Malfunction

P0231 Fuel Pump Secondary Circuit Low

P0232 Fuel Pump Secondary Circuit High

P0233 Fuel Pump Secondary Circuit Intermittent

P0234 Engine Over Boost Condition

P0261 Cylinder no. 1 Injector Circuit Low

P0262 Cylinder no. 1 Injector Circuit High

P0263 Cylinder no. 1 Contribution/Balance Fault

P0264 Cylinder no. 2 Injector Circuit Low

P0265 Cylinder no. 2 Injector Circuit High

P0266 Cylinder no. 2 Contribution/Balance Fault

P0267 Cylinder no. 3 Injector Circuit Low

P0268 Cylinder no. 3 Injector Circuit High

P0269 Cylinder no. 3 Contribution/Balance Fault

P0270 Cylinder no. 4 Injector Circuit Low

P0271 Cylinder no. 4 Injector Circuit High

P0272 Cylinder no. 4 Contribution/Balance Fault

P0273 Cylinder no. 5 Injector Circuit Low

P0274 Cylinder no. 5 Injector Circuit High

P0275 Cylinder no. 5 Contribution/Balance Fault

P0276 Cylinder no. 6 Injector Circuit Low

P0277 Cylinder no. 6 Injector Circuit High

P0278 Cylinder no. 6 Contribution/Balance Fault

P0279 Cylinder no. 7 Injector Circuit Low

P0280 Cylinder no. 7 Injector Circuit High

P0281 Cylinder no. 7 Contribution/Balance Fault

P0282 Cylinder no. 8 Injector Circuit Low

P0283 Cylinder no. 8 Injector Circuit High

P0284 Cylinder no. 8 Contribution/Balance Fault

P0285 Cylinder no. 9 Injector Circuit Low

P0286 Cylinder no. 9 Injector Circuit High

P0287 Cylinder no. 9 Contribution/Balance Fault

P0288 Cylinder no. 10 Injector Circuit Low

P0289 Cylinder no. 10 Injector Circuit High

P0290 Cylinder no. 10 Contribution/Balance Fault

P0291 Cylinder no. 11 Injector Circuit Low

P0292 Cylinder no. 11 Injector Circuit High

P0293 Cylinder no. 11 Contribution/Balance Fault

P0294 Cylinder no. 12 Injector Circuit Low

P0295 Cylinder no. 12 Injector Circuit High

P0296 Cylinder no. 12 Contribution/Balance Fault

P0300 Random/Multiple Cylinder Misfire Detected

P0301 Cylinder no. 1—Misfire Detected

P0302 Cylinder no. 2—Misfire Detected

P0303 Cylinder no. 3—Misfire Detected

P0304 Cylinder no. 4—Misfire Detected

P0305 Cylinder no. 5—Misfire Detected

P0306 Cylinder no. 6—Misfire Detected

P0307 Cylinder no. 7—Misfire Detected

P0308 Cylinder no. 8—Misfire Detected

P0309 Cylinder no. 9—Misfire Detected

P0310 Cylinder no. 10—Misfire Detected

P0311 Cylinder no. 11—Misfire Detected

P0312 Cylinder no. 12—Misfire Detected

P0320 Ignition/Distributor Engine Speed Input Circuit Malfunction

P0321 Ignition/Distributor Engine Speed Input Circuit Range/Performance

P0322 Ignition/Distributor Engine Speed Input Circuit No Signal

P0323 Ignition/Distributor Engine Speed Input Circuit Intermittent

P0325 Knock Sensor no. 1—Circuit Malfunction (Bank no. 1 or Single Sensor)

P0326 Knock Sensor no. 1—Circuit Range/Performance (Bank no. 1 or Single Sensor)

P0327 Knock Sensor no. 1—Circuit Low Input (Bank no. 1 or Single Sensor)

P0328 Knock Sensor no. 1—Circuit High Input (Bank no. 1 or Single Sensor)

P0329 Knock Sensor no. 1—Circuit Input Intermittent (Bank no. 1 or Single Sensor)

P0330 Knock Sensor no. 2—Circuit Malfunction (Bank no. 2)

P0331 Knock Sensor no. 2—Circuit Range/Performance (Bank no. 2)

P0332 Knock Sensor no. 2—Circuit Low Input (Bank no. 2)

P0333 Knock Sensor no. 2—Circuit High Input (Bank no. 2)

P0334 Knock Sensor no. 2—Circuit Input Intermittent (Bank no. 2)

P0335 Crankshaft Position Sensor "A" Circuit Malfunction

P0336 Crankshaft Position Sensor "A" Circuit Range/Performance

P0337 Crankshaft Position Sensor "A" Circuit Low Input

P0338 Crankshaft Position Sensor "A" Circuit High Input

P0339 Crankshaft Position Sensor "A" Circuit Intermittent

P0340 Camshaft Position Sensor Circuit Malfunction

P0341 Camshaft Position Sensor Circuit Range/Performance

P0342 Camshaft Position Sensor Circuit Low Input

P0343 Camshaft Position Sensor Circuit High Input

P0344 Camshaft Position Sensor Circuit Intermittent

P0350 Ignition Coil Primary/Secondary Circuit Malfunction

P0351 Ignition Coil "A" Primary/Secondary Circuit Malfunction

P0352 Ignition Coil "B" Primary/Secondary Circuit Malfunction

P0353 Ignition Coil "C" Primary/Secondary Circuit Malfunction

P0354 Ignition Coil "D" Primary/Secondary Circuit Malfunction

P0355 Ignition Coil "E" Primary/Secondary Circuit Malfunction

P0356 Ignition Coil "F" Primary/Secondary Circuit Malfunction

P0357 Ignition Coil "G" Primary/Secondary Circuit Malfunction

P0358 Ignition Coil "H" Primary/Secondary Circuit Malfunction

P0359 Ignition Coil "I" Primary/Secondary Circuit Malfunction

P0360 Ignition Coil "J" Primary/Secondary Circuit Malfunction

P0361 Ignition Coil "K" Primary/Secondary Circuit Malfunction

P0362 Ignition Coil "L" Primary/Secondary Circuit Malfunction

P0370 Timing Reference High Resolution Signal "A" Malfunction

P0371 Timing Reference High Resolution Signal "A" Too Many Pulses

P0372 Timing Reference High Resolution Signal "A" Too Few Pulses

P0373 Timing Reference High Resolution Signal "A" Intermittent/Erratic Pulses

P0374 Timing Reference High Resolution Signal "A" No Pulses

P0375 Timing Reference High Resolution Signal "B" Malfunction

P0376 Timing Reference High Resolution Signal "B" Too Many Pulses

P0377 Timing Reference High Resolution Signal "B" Too Few Pulses

P0378 Timing Reference High Resolution Signal "B" Intermittent/Erratic Pulses

P0379 Timing Reference High Resolution Signal "B" No Pulses

P0380 Glow Plug/Heater Circuit "A" Malfunction
P0381 Glow Plug/Heater Indicator Circuit Malfunction
P0382 Glow Plug/Heater Circuit "B" Malfunction
P0385 Crankshaft Position Sensor "B" Circuit Malfunction
P0386 Crankshaft Position Sensor "B" Circuit Range/Performance
P0387 Crankshaft Position Sensor "B" Circuit Low Input
P0388 Crankshaft Position Sensor "B" Circuit High Input
P0389 Crankshaft Position Sensor "B" Circuit Intermittent
P0400 Exhaust Gas Recirculation Flow Malfunction
P0401 Exhaust Gas Recirculation Flow Insufficient Detected
P0402 Exhaust Gas Recirculation Flow Excessive Detected
P0403 Exhaust Gas Recirculation Circuit Malfunction
P0404 Exhaust Gas Recirculation Circuit Range/Performance
P0405 Exhaust Gas Recirculation Sensor "A" Circuit Low
P0406 Exhaust Gas Recirculation Sensor "A" Circuit High
P0407 Exhaust Gas Recirculation Sensor "B" Circuit Low
P0408 Exhaust Gas Recirculation Sensor "B" Circuit High
P0410 Secondary Air Injection System Malfunction
P0411 Secondary Air Injection System Incorrect Flow Detected
P0412 Secondary Air Injection System Switching Valve "A" Circuit Malfunction
P0413 Secondary Air Injection System Switching Valve "A" Circuit Open
P0414 Secondary Air Injection System Switching Valve "A" Circuit Shorted
P0415 Secondary Air Injection System Switching Valve "B" Circuit Malfunction
P0416 Secondary Air Injection System Switching Valve "B" Circuit Open
P0417 Secondary Air Injection System Switching Valve "B" Circuit Shorted
P0418 Secondary Air Injection System Relay "A" Circuit Malfunction
P0419 Secondary Air Injection System Relay "B" Circuit Malfunction
P0420 Catalyst System Efficiency Below Threshold (Bank no. 1)
P0421 Warm Up Catalyst Efficiency Below Threshold (Bank no. 1)
P0422 Main Catalyst Efficiency Below Threshold (Bank no. 1)
P0423 Heated Catalyst Efficiency Below Threshold (Bank no. 1)
P0424 Heated Catalyst Temperature Below Threshold (Bank no. 1)
P0430 Catalyst System Efficiency Below Threshold (Bank no. 2)
P0431 Warm Up Catalyst Efficiency Below Threshold (Bank no. 2)
P0432 Main Catalyst Efficiency Below Threshold (Bank no. 2)
P0433 Heated Catalyst Efficiency Below Threshold (Bank no. 2)
P0434 Heated Catalyst Temperature Below Threshold (Bank no. 2)
P0440 Evaporative Emission Control System Malfunction
P0441 Evaporative Emission Control System Incorrect Purge Flow
P0442 Evaporative Emission Control System Leak Detected (Small Leak)
P0443 Evaporative Emission Control System Purge Control Valve Circuit Malfunction
P0444 Evaporative Emission Control System Purge Control Valve Circuit Open
P0445 Evaporative Emission Control System Purge Control Valve Circuit Shorted
P0446 Evaporative Emission Control System Vent Control Circuit Malfunction

P0447 Evaporative Emission Control System Vent Control Circuit Open
P0448 Evaporative Emission Control System Vent Control Circuit Shorted
P0449 Evaporative Emission Control System Vent Valve/Solenoid Circuit Malfunction
P0450 Evaporative Emission Control System Pressure Sensor Malfunction
P0451 Evaporative Emission Control System Pressure Sensor Range/Performance
P0452 Evaporative Emission Control System Pressure Sensor Low Input
P0453 Evaporative Emission Control System Pressure Sensor High Input
P0454 Evaporative Emission Control System Pressure Sensor Intermittent
P0455 Evaporative Emission Control System Leak Detected (Gross Leak)
P0460 Fuel Level Sensor Circuit Malfunction
P0461 Fuel Level Sensor Circuit Range/Performance
P0462 Fuel Level Sensor Circuit Low Input
P0463 Fuel Level Sensor Circuit High Input
P0464 Fuel Level Sensor Circuit Intermittent
P0465 Purge Flow Sensor Circuit Malfunction
P0466 Purge Flow Sensor Circuit Range/Performance
P0467 Purge Flow Sensor Circuit Low Input
P0468 Purge Flow Sensor Circuit High Input
P0469 Purge Flow Sensor Circuit Intermittent
P0470 Exhaust Pressure Sensor Malfunction
P0471 Exhaust Pressure Sensor Range/Performance
P0472 Exhaust Pressure Sensor Low
P0473 Exhaust Pressure Sensor High
P0474 Exhaust Pressure Sensor Intermittent
P0475 Exhaust Pressure Control Valve Malfunction
P0476 Exhaust Pressure Control Valve Range/Performance
P0477 Exhaust Pressure Control Valve Low
P0478 Exhaust Pressure Control Valve High
P0479 Exhaust Pressure Control Valve Intermittent
P0480 Cooling Fan no. 1 Control Circuit Malfunction
P0481 Cooling Fan no. 2 Control Circuit Malfunction
P0482 Cooling Fan no. 3 Control Circuit Malfunction
P0483 Cooling Fan Rationality Check Malfunction
P0484 Cooling Fan Circuit Over Current
P0485 Cooling Fan Power/Ground Circuit Malfunction
P0500 Vehicle Speed Sensor Malfunction
P0501 Vehicle Speed Sensor Range/Performance
P0502 Vehicle Speed Sensor Circuit Low Input
P0503 Vehicle Speed Sensor Intermittent/Erratic/High
P0505 Idle Control System Malfunction
P0506 Idle Control System RPM Lower Than Expected
P0507 Idle Control System RPM Higher Than Expected
P0510 Closed Throttle Position Switch Malfunction
P0520 Engine Oil Pressure Sensor/Switch Circuit Malfunction
P0521 Engine Oil Pressure Sensor/Switch Range/Performance
P0522 Engine Oil Pressure Sensor/Switch Low Voltage
P0523 Engine Oil Pressure Sensor/Switch High Voltage
P0530 A/C Refrigerant Pressure Sensor Circuit Malfunction
P0531 A/C Refrigerant Pressure Sensor Circuit Range/Performance
P0532 A/C Refrigerant Pressure Sensor Circuit Low Input
P0533 A/C Refrigerant Pressure Sensor Circuit High Input
P0534 A/C Refrigerant Charge Loss

P0550 Power Steering Pressure Sensor Circuit Malfunction
P0551 Power Steering Pressure Sensor Circuit Range/Performance
P0552 Power Steering Pressure Sensor Circuit Low Input
P0553 Power Steering Pressure Sensor Circuit High Input
P0554 Power Steering Pressure Sensor Circuit Intermittent
P0560 System Voltage Malfunction
P0561 System Voltage Unstable
P0562 System Voltage Low
P0563 System Voltage High
P0565 Cruise Control On Signal Malfunction
P0566 Cruise Control Off Signal Malfunction
P0567 Cruise Control Resume Signal Malfunction
P0568 Cruise Control Set Signal Malfunction
P0569 Cruise Control Coast Signal Malfunction
P0570 Cruise Control Accel Signal Malfunction
P0571 Cruise Control/Brake Switch "A" Circuit Malfunction
P0572 Cruise Control/Brake Switch "A" Circuit Low
P0573 Cruise Control/Brake Switch "A" Circuit High
P0574 **Through P0580** Reserved for Cruise Codes
P0600 Serial Communication Link Malfunction
P0601 Internal Control Module Memory Check Sum Error
P0602 Control Module Programming Error
P0603 Internal Control Module Keep Alive Memory (KAM) Error
P0604 Internal Control Module Random Access Memory (RAM) Error
P0605 Internal Control Module Read Only Memory (ROM) Error
P0606 PCM Processor Fault
P0608 Control Module VSS Output "A" Malfunction
P0609 Control Module VSS Output "B" Malfunction
P0620 Generator Control Circuit Malfunction
P0621 Generator Lamp "L" Control Circuit Malfunction
P0622 Generator Field "F" Control Circuit Malfunction
P0650 Malfunction Indicator Lamp (MIL) Control Circuit Malfunction
P0654 Engine RPM Output Circuit Malfunction
P0655 Engine Hot Lamp Output Control Circuit Malfunction
P0656 Fuel Level Output Circuit Malfunction
P0700 Transmission Control System Malfunction
P0701 Transmission Control System Range/Performance
P0702 Transmission Control System Electrical
P0703 Torque Converter/Brake Switch "B" Circuit Malfunction
P0704 Clutch Switch Input Circuit Malfunction
P0705 Transmission Range Sensor Circuit Malfunction (PRNDL Input)
P0706 Transmission Range Sensor Circuit Range/Performance
P0707 Transmission Range Sensor Circuit Low Input
P0708 Transmission Range Sensor Circuit High Input
P0709 Transmission Range Sensor Circuit Intermittent
P0710 Transmission Fluid Temperature Sensor Circuit Malfunction
P0711 Transmission Fluid Temperature Sensor Circuit Range/Performance
P0712 Transmission Fluid Temperature Sensor Circuit Low Input
P0713 Transmission Fluid Temperature Sensor Circuit High Input
P0714 Transmission Fluid Temperature Sensor Circuit Intermittent
P0715 Input/Turbine Speed Sensor Circuit Malfunction
P0716 Input/Turbine Speed Sensor Circuit Range/Performance
P0717 Input/Turbine Speed Sensor Circuit No Signal

P0718 Input/Turbine Speed Sensor Circuit Intermittent
P0719 Torque Converter/Brake Switch "B" Circuit Low
P0720 Output Speed Sensor Circuit Malfunction
P0721 Output Speed Sensor Circuit Range/Performance
P0722 Output Speed Sensor Circuit No Signal
P0723 Output Speed Sensor Circuit Intermittent
P0724 Torque Converter/Brake Switch "B" Circuit High
P0725 Engine Speed Input Circuit Malfunction
P0726 Engine Speed Input Circuit Range/Performance
P0727 Engine Speed Input Circuit No Signal
P0728 Engine Speed Input Circuit Intermittent
P0730 Incorrect Gear Ratio
P0731 Gear no. 1 Incorrect Ratio
P0732 Gear no. 2 Incorrect Ratio
P0733 Gear no. 3 Incorrect Ratio
P0734 Gear no. 4 Incorrect Ratio
P0735 Gear no. 5 Incorrect Ratio
P0736 Reverse Incorrect Ratio
P0740 Torque Converter Clutch Circuit Malfunction
P0741 Torque Converter Clutch Circuit Performance or Stuck Off
P0742 Torque Converter Clutch Circuit Stuck On
P0743 Torque Converter Clutch Circuit Electrical
P0744 Torque Converter Clutch Circuit Intermittent
P0745 Pressure Control Solenoid Malfunction
P0746 Pressure Control Solenoid Performance or Stuck Off
P0747 Pressure Control Solenoid Stuck On
P0748 Pressure Control Solenoid Electrical
P0749 Pressure Control Solenoid Intermittent
P0750 Shift Solenoid "A" Malfunction
P0751 Shift Solenoid "A" Performance or Stuck Off
P0752 Shift Solenoid "A" Stuck On
P0753 Shift Solenoid "A" Electrical
P0754 Shift Solenoid "A" Intermittent
P0755 Shift Solenoid "B" Malfunction
P0756 Shift Solenoid "B" Performance or Stuck Off
P0757 Shift Solenoid "B" Stuck On
P0758 Shift Solenoid "B" Electrical
P0759 Shift Solenoid "B" Intermittent
P0760 Shift Solenoid "C" Malfunction
P0761 Shift Solenoid "C" Performance Or Stuck Off
P0762 Shift Solenoid "C" Stuck On
P0763 Shift Solenoid "C" Electrical
P0764 Shift Solenoid "C" Intermittent
P0765 Shift Solenoid "D" Malfunction
P0766 Shift Solenoid "D" Performance Or Stuck Off
P0767 Shift Solenoid "D" Stuck On
P0768 Shift Solenoid "D" Electrical
P0769 Shift Solenoid "D" Intermittent
P0770 Shift Solenoid "E" Malfunction
P0771 Shift Solenoid "E" Performance Or Stuck Off
P0772 Shift Solenoid "E" Stuck On
P0773 Shift Solenoid "E" Electrical
P0774 Shift Solenoid "E" Intermittent
P0780 Shift Malfunction
P0781 1–2 Shift Malfunction
P0782 2–3 Shift Malfunction
P0783 3–4 Shift Malfunction
P0784 4–5 Shift Malfunction
P0785 Shift/Timing Solenoid Malfunction
P0786 Shift/Timing Solenoid Range/Performance
P0787 Shift/Timing Solenoid Low

P0788 Shift/Timing Solenoid High

P0789 Shift/Timing Solenoid Intermittent

P0790 Normal/Performance Switch Circuit Malfunction

P0801 Reverse Inhibit Control Circuit Malfunction

P0803 1–4 Upshift (Skip Shift) Solenoid Control Circuit Malfunction

P0804 1–4 Upshift (Skip Shift) Lamp Control Circuit Malfunction

P1000 OBD II Monitor Testing Not Complete More Driving Required

P1001 Key On Engine Running (KOER) Self-Test Not Able To Complete, KOER Aborted

P1100 Mass Air Flow (MAF) Sensor Intermittent

P1101 Mass Air Flow (MAF) Sensor Out Of Self-Test Range

P1110 Intake Air Temperature (IAT) Sensor Signal Circuit Fault

P1112 Intake Air Temperature (IAT) Sensor Intermittent

P1113 Intake Air Temperature (IAT) Sensor Intermittent

P1116 Engine Coolant Temperature (ECT) Sensor Out Of Self-Test Range

P1117 Engine Coolant Temperature (ECT) Sensor Intermittent

P1120 Throttle Position (TP) Sensor Out Of Range (Low)

P1121 Throttle Position (TP) Sensor Inconsistent With MAF Sensor

P1124 Throttle Position (TP) Sensor Out Of Self-Test Range

P1125 Throttle Position (TP) Sensor Circuit Intermittent

P1127 Exhaust Not Warm Enough, Downstream Heated Oxygen Sensors (HO2S) Not Tested

P1128 Upstream Heated Oxygen Sensors (HO2S) Swapped From Bank To Bank

P1129 Downstream Heated Oxygen Sensors (HO2S) Swapped From Bank To Bank

P1130 Lack Of Upstream Heated Oxygen Sensor (HO2S 11) Switch, Adaptive Fuel At Limit (Bank #1)

P1131 Lack Of Upstream Heated Oxygen Sensor (HO2S 11) Switch, Sensor Indicates Lean (Bank #1)

P1132 Lack Of Upstream Heated Oxygen Sensor (HO2S 11) Switch, Sensor Indicates Rich (Bank#1)

P1137 Lack Of Downstream Heated Oxygen Sensor (HO2S 12) Switch, Sensor Indicates Lean (Bank#1)

P1138 Lack Of Downstream Heated Oxygen Sensor (HO2S 12) Switch, Sensor Indicates Rich (Bank#1)

P1150 Lack Of Upstream Heated Oxygen Sensor (HO2S 21) Switch, Adaptive Fuel At Limit (Bank #2)

P1151 Lack Of Upstream Heated Oxygen Sensor (HO2S 21) Switch, Sensor Indicates Lean (Bank#2)

P1152 Lack Of Upstream Heated Oxygen Sensor (HO2S 21) Switch, Sensor Indicates Rich (Bank #2)

P1170 (HO2S 11) Signal Remained Unchanged For More Than 20 Seconds After Closed Loop

P1173 Feedback A/F Mixture Control (HO2S 21) Signal Remained Unchanged For More Than 20 Seconds After Closed Loop

P1195 Barometric (BARO) Pressure Sensor Circuit Malfunction (Signal Is From EGR Boost Sensor)

P1196 Starter Switch Circuit Malfunction

P1235 Fuel Pump Control Out Of Range (MIL DTC)

P1236 Fuel Pump Control Out Of Range (No MIL)

P1250 Fuel Pressure Regulator Control (FPRC) Solenoid Malfunction

P1252 Fuel Pressure Regulator Control (FPRC) Solenoid Malfunction

P1260 THEFT Detected—Engine Disabled

P1270 Engine RPM Or Vehicle Speed Limiter Reached

P1345 No Camshaft Position Sensor Signal

P1351 Ignition Diagnostic Monitor (IDM) Circuit Input Malfunction

P1351 Indicates Ignition System Malfunction

P1352 Indicates Ignition System Malfunction

P1353 Indicates Ignition System Malfunction

P1354 Indicates Ignition System Malfunction

P1358 Ignition Diagnostic Monitor (IDM) Signal Out Of Self-Test Range

P1359 Spark Output Circuit Malfunction

P1360 Ignition Coil "A" Secondary Circuit Fault

P1361 Ignition Coil "A" Secondary Circuit Fault

P1362 Ignition Coil "A" Secondary Circuit Fault

P1364 Spark Output Circuit Malfunction

P1365 Ignition Coil Secondary Circuit Fault

P1390 Octane Adjust (OCT ADJ) Out Of Self-Test Range

P1400 Differential Pressure Feedback EGR (DPFE) Sensor Circuit Low Voltage Detected

P1401 Differential Pressure Feedback EGR (DPFE) Sensor Circuit High Voltage Detected/EGR Temperature Sensor

P1402 EGR Valve Position Sensor Open Or Short

P1405 Differential Pressure Feedback EGR (DPFE) Sensor Upstream Hose Off Or Plugged

P1406 Differential Pressure Feedback EGR (DPFE) Sensor Downstream Hose Off Or Plugged

P1407 Exhaust Gas Recirculation (EGR) No Flow Detected (Valve Stuck Closed Or Inoperative)

P1408 Exhaust Gas Recirculation (EGR) Flow Out Of Self-Test Range

P1409 Electronic Vacuum Regulator (EVR) Control Circuit Malfunction

P1443 Evaporative Emission Control System—Vacuum System Purge Control Solenoid Or Purge Control Valve Malfunction

P1444 Purge Flow Sensor (PFS) Circuit Low Input

P1445 Purge Flow Sensor (PFS) Circuit High Input

P1449 Evaporative Emission Control System Unable To Hold Vacuum

P1455 Evaporative Emission Control System Control Leak Detected (Gross Leak)

P1460 Wide Open Throttle Air Conditioning Cut-Off Circuit Malfunction

P1464 Air Conditioning (A/C) Demand Out Of Self-Test Range/A/C On During KOER Or CCT Test

P1474 Low Fan Control Primary Circuit Malfunction

P1485 EGR Control Solenoid Open Or Short

P1486 EGR Vent Solenoid Open Or Short

P1487 EGR Boost Check Solenoid Open Or Short

P1500 Vehicle Speed Sensor (VSS) Circuit Intermittent

P1501 Vehicle Speed Sensor (VSS) Out Of Self-Test Range/Vehicle Moved During Test

P1502 Invalid Self Test—Auxiliary Powertrain Control Module (APCM) Functioning

P1504 Idle Air Control (IAC) Circuit Malfunction

P1505 Idle Air Control (IAC) System At Adaptive Clip

P1506 Idle Air Control (IAC) Overspeed Error

P1507 Idle Air Control (IAC) Underspeed Error

P1508 Bypass Air Solenoid "1" Circuit Fault

P1509 Bypass Air Solenoid "2" Circuit Fault

P1521 Variable Resonance Induction System (VRIS) Solenoid #1 Open Or Short

P1522 Variable Resonance Induction System (VRIS) Solenoid #2 Open Or Short

P1523 High Speed Inlet Air (HSIA) Solenoid Open Or Short

P1524 Charge Air Cooler Bypass Solenoid Circuit Fault

P1525 ABV Vacuum Solenoid Circuit Fault

P1526 ABV Vent Solenoid Circuit Fault

P1529 Atmospheric balance Air Control Valve Fault

P1540 ABV System Fault

P1601 Serial Communication Error

P1602 Serial Communication Error

P1605 Powertrain Control Module (PCM)—Keep Alive Memory (KAM) Test Error

P1608 PCM Internal Circuit Malfunction

P1609 PCM Internal Circuit Malfunction

P1627 Serial Communication Error

P1628 Serial Communication Error

P1650 Power Steering Pressure (PSP) Switch Out Of Self-Test Range

P1651 Power Steering Pressure (PSP) Switch Input Malfunction

P1701 Reverse Engagement Error

P1703 Brake On/Off (BOO) Switch Out Of Self-Test Range

P1705 Transmission Range (TR) Sensor Out Of Self-Test Range

P1706 High Vehicle Speed In Park

P1709 Park Or Neutral Position (PNP) Or Clutch Pedal Position (CPP) Switch Out Of Self-Test Range

P1711 Transmission Fluid Temperature (TFT) Sensor Out Of Self-Test Range

P1720 Vehicle Speed Sensor (VSS) Circuit Malfunction

P1729 4x4 Low Switch Error

P1741 Torque Converter Clutch (TCC) Control Error

P1742 Torque Converter Clutch (TCC) Solenoid Failed On (Turns On MIL)

P1743 Torque Converter Clutch (TCC) Solenoid Failed On (Turns On TCIL)

P1746 Electronic Pressure Control (EPC) Solenoid Open Circuit (Low Input)

P1747 Electronic Pressure Control (EPC) Solenoid Short Circuit (High Input)

P1749 Electronic Pressure Control (EPC) Solenoid Failed Low

P1751 Shift Solenoid#1 (SS1) Performance

P1754 Coast Clutch Solenoid (CCS) Circuit Malfunction

P1756 Shift Solenoid#2 (SS2) Performance

P1761 Shift Solenoid #(SS2) Performance

P1780 Transmission Control Switch (TCS) Circuit Out Of Self-Test Range

P1781 4x4 Low Switch, Out Of Self-Test Range

P1783 Transmission Over Temperature Condition

P1794 PCM Battery Direct Power Circuit Fault

P1797 P/N Switch Open or Short Circuit Fault

Mercedes-Benz

READING CODES

Reading the control module memory is one of the first steps in OBD II system diagnostics. This step should be initially performed to determine the general nature of the fault. Subsequent readings will determine if the fault has been cleared.

Reading codes can be performed by any of the methods below:

• Read the control module memory with the Generic Scan Tool (GST)

• Read the control module memory with the vehicle manufacturer's specific tester

To read the fault codes, connect the scan tool or tester according to the manufacturer's instructions. Follow the manufacturer's specified procedure for reading the codes.

CLEARING CODES

Control module reset procedures are a very important part of OBD II System diagnostics. This step should be done at the end of any fault code repair and at the end of any driveability repair.

Clearing codes can be performed by any of the methods below:

• Clear the control module memory with the Generic Scan Tool (GST)

• Clear the control module memory with the vehicle manufacturer's specific tester

• Turn the ignition off and remove the negative battery cable for at least 1 minute.

Removing the negative battery cable may cause other systems in the vehicle to loose their memory. Prior to removing the cable, ensure you have the proper reset codes for radios and alarms.

➡**The MIL will may also be de-activated for some codes if the vehicle completes three consecutive trips without a fault detected with vehicle conditions similar to those present during the fault.**

OBD II TROUBLE CODES

P0100 Mass or Volume Air Flow Circuit Malfunction

P0101 Mass or Volume Air Flow Circuit Range/Performance Problem

P0102 Mass or Volume Air Flow Circuit Low Input

P0103 Mass or Volume Air Flow Circuit High Input

P0104 Mass or Volume Air Flow Circuit Intermittent

P0105 Manifold Absolute Pressure/Barometric Pressure Circuit Malfunction

P0106 Manifold Absolute Pressure/Barometric Pressure Circuit Range/Performance Problem

P0107 Manifold Absolute Pressure/Barometric Pressure Circuit Low Input

P0108 Manifold Absolute Pressure/Barometric Pressure Circuit High Input

P0109 Manifold Absolute Pressure/Barometric Pressure Circuit Intermittent

P0110 Intake Air Temperature Circuit Malfunction

P0111 Intake Air Temperature Circuit Range/Performance Problem

P0112 Intake Air Temperature Circuit Low Input

P0113 Intake Air Temperature Circuit High Input

P0114 Intake Air Temperature Circuit Intermittent

P0115 Engine Coolant Temperature Circuit Malfunction

P0116 Engine Coolant Temperature Circuit Range/Performance Problem

P0117 Engine Coolant Temperature Circuit Low Input

P0118 Engine Coolant Temperature Circuit High Input

P0119 Engine Coolant Temperature Circuit Intermittent

P0120 Throttle/Pedal Position Sensor/Switch "A" Circuit Malfunction

P0121 Throttle/Pedal Position Sensor/Switch "A" Circuit Range/Performance Problem

P0122 Throttle/Pedal Position Sensor/Switch "A" Circuit Low Input

P0123 Throttle/Pedal Position Sensor/Switch "A" Circuit High Input

P0124 Throttle/Pedal Position Sensor/Switch "A" Circuit Intermittent

P0125 Insufficient Coolant Temperature For Closed Loop Fuel Control

P0126 Insufficient Coolant Temperature For Stable Operation

P0130 O2 Circuit Malfunction (Bank no. 1 Sensor no. 1)

P0131 O2 Sensor Circuit Low Voltage (Bank no. 1 Sensor no. 1)

P0132 O2 Sensor Circuit High Voltage (Bank no. 1 Sensor no. 1)

P0133 O2 Sensor Circuit Slow Response (Bank no. 1 Sensor no. 1)

P0134 O2 Sensor Circuit No Activity Detected (Bank no. 1 Sensor no. 1)

P0135 O2 Sensor Heater Circuit Malfunction (Bank no. 1 Sensor no. 1)

P0136 O2 Sensor Circuit Malfunction (Bank no. 1 Sensor no. 2)

P0137 O2 Sensor Circuit Low Voltage (Bank no. 1 Sensor no. 2)

P0138 O2 Sensor Circuit High Voltage (Bank no. 1 Sensor no. 2)

P0139 O2 Sensor Circuit Slow Response (Bank no. 1 Sensor no. 2)

P0140 O2 Sensor Circuit No Activity Detected (Bank no. 1 Sensor no. 2)

P0141 O2 Sensor Heater Circuit Malfunction (Bank no. 1 Sensor no. 2)

P0142 O2 Sensor Circuit Malfunction (Bank no. 1 Sensor no. 3)

P0143 O2 Sensor Circuit Low Voltage (Bank no. 1 Sensor no. 3)

P0144 O2 Sensor Circuit High Voltage (Bank no. 1 Sensor no. 3)

P0145 O2 Sensor Circuit Slow Response (Bank no. 1 Sensor no. 3)

P0146 O2 Sensor Circuit No Activity Detected (Bank no. 1 Sensor no. 3)

P0147 O2 Sensor Heater Circuit Malfunction (Bank no. 1 Sensor no. 3)

P0150 O2 Sensor Circuit Malfunction (Bank no. 2 Sensor no. 1)

P0151 O2 Sensor Circuit Low Voltage (Bank no. 2 Sensor no. 1)

P0152 O2 Sensor Circuit High Voltage (Bank no. 2 Sensor no. 1)

P0153 O2 Sensor Circuit Slow Response (Bank no. 2 Sensor no. 1)

P0154 O2 Sensor Circuit No Activity Detected (Bank no. 2 Sensor no. 1)

P0155 O2 Sensor Heater Circuit Malfunction (Bank no. 2 Sensor no. 1)

P0156 O2 Sensor Circuit Malfunction (Bank no. 2 Sensor no. 2)

P0157 O2 Sensor Circuit Low Voltage (Bank no. 2 Sensor no. 2)

P0158 O2 Sensor Circuit High Voltage (Bank no. 2 Sensor no. 2)

P0159 O2 Sensor Circuit Slow Response (Bank no. 2 Sensor no. 2)

P0160 O2 Sensor Circuit No Activity Detected (Bank no. 2 Sensor no. 2)

P0161 O2 Sensor Heater Circuit Malfunction (Bank no. 2 Sensor no. 2)

P0162 O2 Sensor Circuit Malfunction (Bank no. 2 Sensor no. 3)

P0163 O2 Sensor Circuit Low Voltage (Bank no. 2 Sensor no. 3)

P0164 O2 Sensor Circuit High Voltage (Bank no. 2 Sensor no. 3)

P0165 O2 Sensor Circuit Slow Response (Bank no. 2 Sensor no. 3)

P0166 O2 Sensor Circuit No Activity Detected (Bank no. 2 Sensor no. 3)

P0167 O2 Sensor Heater Circuit Malfunction (Bank no. 2 Sensor no. 3)

P0170 Fuel Trim Malfunction (Bank no. 1)

P0171 System Too Lean (Bank no. 1)

P0172 System Too Rich (Bank no. 1)

P0173 Fuel Trim Malfunction (Bank no. 2)

P0174 System Too Lean (Bank no. 2)

P0175 System Too Rich (Bank no. 2)

P0176 Fuel Composition Sensor Circuit Malfunction

P0177 Fuel Composition Sensor Circuit Range/Performance

P0178 Fuel Composition Sensor Circuit Low Input

P0179 Fuel Composition Sensor Circuit High Input

P0180 Fuel Temperature Sensor "A" Circuit Malfunction

P0181 Fuel Temperature Sensor "A" Circuit Range/Performance

P0182 Fuel Temperature Sensor "A" Circuit Low Input

P0183 Fuel Temperature Sensor "A" Circuit High Input

P0184 Fuel Temperature Sensor "A" Circuit Intermittent

P0185 Fuel Temperature Sensor "B" Circuit Malfunction

P0186 Fuel Temperature Sensor "B" Circuit Range/Performance

P0187 Fuel Temperature Sensor "B" Circuit Low Input

P0188 Fuel Temperature Sensor "B" Circuit High Input

P0189 Fuel Temperature Sensor "B" Circuit Intermittent

P0190 Fuel Rail Pressure Sensor Circuit Malfunction

P0191 Fuel Rail Pressure Sensor Circuit Range/Performance

P0192 Fuel Rail Pressure Sensor Circuit Low Input

P0193 Fuel Rail Pressure Sensor Circuit High Input

P0194 Fuel Rail Pressure Sensor Circuit Intermittent

P0195 Engine Oil Temperature Sensor Malfunction

P0196 Engine Oil Temperature Sensor Range/Performance

P0197 Engine Oil Temperature Sensor Low

P0198 Engine Oil Temperature Sensor High

P0199 Engine Oil Temperature Sensor Intermittent

P0200 Injector Circuit Malfunction

P0201 Injector Circuit Malfunction—Cylinder no. 1

P0202 Injector Circuit Malfunction—Cylinder no. 2

P0203 Injector Circuit Malfunction—Cylinder no. 3

P0204 Injector Circuit Malfunction—Cylinder no. 4

P0205 Injector Circuit Malfunction—Cylinder no. 5

P0206 Injector Circuit Malfunction—Cylinder no. 6

P0207 Injector Circuit Malfunction—Cylinder no. 7

P0208 Injector Circuit Malfunction—Cylinder no. 8

P0209 Injector Circuit Malfunction—Cylinder no. 9

P0210 Injector Circuit Malfunction—Cylinder no. 10

P0211 Injector Circuit Malfunction—Cylinder no. 11

P0212 Injector Circuit Malfunction—Cylinder no. 12

P0213 Cold Start Injector no. 1 Malfunction

P0214 Cold Start Injector no. 2 Malfunction

P0215 Engine Shutoff Solenoid Malfunction

P0216 Injection Timing Control Circuit Malfunction

P0217 Engine Over Temperature Condition

P0218 Transmission Over Temperature Condition

P0219 Engine Over Speed Condition

P0220 Throttle/Pedal Position Sensor/Switch "B" Circuit Malfunction

P0221 Throttle/Pedal Position Sensor/Switch "B" Circuit Range/Performance Problem

P0222 Throttle/Pedal Position Sensor/Switch "B" Circuit Low Input

P0223 Throttle/Pedal Position Sensor/Switch "B" Circuit High Input

P0224 Throttle/Pedal Position Sensor/Switch "B" Circuit Intermittent

P0225 Throttle/Pedal Position Sensor/Switch "C" Circuit Malfunction

P0226 Throttle/Pedal Position Sensor/Switch "C" Circuit Range/Performance Problem

P0227 Throttle/Pedal Position Sensor/Switch "C" Circuit Low Input

P0228 Throttle/Pedal Position Sensor/Switch "C" Circuit High Input

P0229 Throttle/Pedal Position Sensor/Switch "C" Circuit Intermittent

P0230 Fuel Pump Primary Circuit Malfunction
P0231 Fuel Pump Secondary Circuit Low
P0232 Fuel Pump Secondary Circuit High
P0233 Fuel Pump Secondary Circuit Intermittent
P0234 Engine Over Boost Condition
P0261 Cylinder no. 1 Injector Circuit Low
P0262 Cylinder no. 1 Injector Circuit High
P0263 Cylinder no. 1 Contribution/Balance Fault
P0264 Cylinder no. 2 Injector Circuit Low
P0265 Cylinder no. 2 Injector Circuit High
P0266 Cylinder no. 2 Contribution/Balance Fault
P0267 Cylinder no. 3 Injector Circuit Low
P0268 Cylinder no. 3 Injector Circuit High
P0269 Cylinder no. 3 Contribution/Balance Fault
P0270 Cylinder no. 4 Injector Circuit Low
P0271 Cylinder no. 4 Injector Circuit High
P0272 Cylinder no. 4 Contribution/Balance Fault
P0273 Cylinder no. 5 Injector Circuit Low
P0274 Cylinder no. 5 Injector Circuit High
P0275 Cylinder no. 5 Contribution/Balance Fault
P0276 Cylinder no. 6 Injector Circuit Low
P0277 Cylinder no. 6 Injector Circuit High
P0278 Cylinder no. 6 Contribution/Balance Fault
P0279 Cylinder no. 7 Injector Circuit Low
P0280 Cylinder no. 7 Injector Circuit High
P0281 Cylinder no. 7 Contribution/Balance Fault
P0282 Cylinder no. 8 Injector Circuit Low
P0283 Cylinder no. 8 Injector Circuit High
P0284 Cylinder no. 8 Contribution/Balance Fault
P0285 Cylinder no. 9 Injector Circuit Low
P0286 Cylinder no. 9 Injector Circuit High
P0287 Cylinder no. 9 Contribution/Balance Fault
P0288 Cylinder no. 10 Injector Circuit Low
P0289 Cylinder no. 10 Injector Circuit High
P0290 Cylinder no. 10 Contribution/Balance Fault
P0291 Cylinder no. 11 Injector Circuit Low
P0292 Cylinder no. 11 Injector Circuit High
P0293 Cylinder no. 11 Contribution/Balance Fault
P0294 Cylinder no. 12 Injector Circuit Low
P0295 Cylinder no. 12 Injector Circuit High
P0296 Cylinder no. 12 Contribution/Balance Fault
P0300 Random/Multiple Cylinder Misfire Detected
P0301 Cylinder no. 1—Misfire Detected
P0302 Cylinder no. 2—Misfire Detected
P0303 Cylinder no. 3—Misfire Detected
P0304 Cylinder no. 4—Misfire Detected
P0305 Cylinder no. 5—Misfire Detected
P0306 Cylinder no. 6—Misfire Detected
P0307 Cylinder no. 7—Misfire Detected

P0308 Cylinder no. 8—Misfire Detected
P0309 Cylinder no. 9—Misfire Detected
P0310 Cylinder no. 10—Misfire Detected
P0311 Cylinder no. 11—Misfire Detected
P0312 Cylinder no. 12—Misfire Detected
P0320 Ignition/Distributor Engine Speed Input Circuit Malfunction

P0321 Ignition/Distributor Engine Speed Input Circuit Range/Performance

P0322 Ignition/Distributor Engine Speed Input Circuit No Signal

P0323 Ignition/Distributor Engine Speed Input Circuit Intermittent

P0325 Knock Sensor no. 1—Circuit Malfunction (Bank no. 1 or Single Sensor)

P0326 Knock Sensor no. 1—Circuit Range/Performance (Bank no. 1 or Single Sensor)

P0327 Knock Sensor no. 1—Circuit Low Input (Bank no. 1 or Single Sensor)

P0328 Knock Sensor no. 1—Circuit High Input (Bank no. 1 or Single Sensor)

P0329 Knock Sensor no. 1—Circuit Input Intermittent (Bank no. 1 or Single Sensor)

P0330 Knock Sensor no. 2—Circuit Malfunction (Bank no. 2)

P0331 Knock Sensor no. 2—Circuit Range/Performance (Bank no. 2)

P0332 Knock Sensor no. 2—Circuit Low Input (Bank no. 2)
P0333 Knock Sensor no. 2—Circuit High Input (Bank no. 2)
P0334 Knock Sensor no. 2—Circuit Input Intermittent (Bank no. 2)

P0335 Crankshaft Position Sensor "A" Circuit Malfunction

P0336 Crankshaft Position Sensor "A" Circuit Range/Performance

P0337 Crankshaft Position Sensor "A" Circuit Low Input
P0338 Crankshaft Position Sensor "A" Circuit High Input
P0339 Crankshaft Position Sensor "A" Circuit Intermittent
P0340 Camshaft Position Sensor Circuit Malfunction
P0341 Camshaft Position Sensor Circuit Range/Performance
P0342 Camshaft Position Sensor Circuit Low Input
P0343 Camshaft Position Sensor Circuit High Input
P0344 Camshaft Position Sensor Circuit Intermittent
P0350 Ignition Coil Primary/Secondary Circuit Malfunction
P0351 Ignition Coil "A" Primary/Secondary Circuit Malfunction
P0352 Ignition Coil "B" Primary/Secondary Circuit Malfunction
P0353 Ignition Coil "C" Primary/Secondary Circuit Malfunction
P0354 Ignition Coil "D" Primary/Secondary Circuit Malfunction
P0355 Ignition Coil "E" Primary/Secondary Circuit Malfunction
P0356 Ignition Coil "F" Primary/Secondary Circuit Malfunction
P0357 Ignition Coil "G" Primary/Secondary Circuit Malfunction
P0358 Ignition Coil "H" Primary/Secondary Circuit Malfunction
P0359 Ignition Coil "I" Primary/Secondary Circuit Malfunction
P0360 Ignition Coil "J" Primary/Secondary Circuit Malfunction
P0361 Ignition Coil "K" Primary/Secondary Circuit Malfunction
P0362 Ignition Coil "L" Primary/Secondary Circuit Malfunction
P0370 Timing Reference High Resolution Signal "A" Malfunction

P0371 Timing Reference High Resolution Signal "A" Too Many Pulses

P0372 Timing Reference High Resolution Signal "A" Too Few Pulses

P0373 Timing Reference High Resolution Signal "A" Intermittent/Erratic Pulses

P0374 Timing Reference High Resolution Signal "A" No Pulses

P0375 Timing Reference High Resolution Signal "B" Malfunction

P0376 Timing Reference High Resolution Signal "B" Too Many Pulses

P0377 Timing Reference High Resolution Signal "B" Too Few Pulses

P0378 Timing Reference High Resolution Signal "B" Intermittent/Erratic Pulses

P0379 Timing Reference High Resolution Signal "B" No Pulses

P0380 Glow Plug/Heater Circuit "A" Malfunction

P0381 Glow Plug/Heater Indicator Circuit Malfunction

P0382 Glow Plug/Heater Circuit "B" Malfunction

P0385 Crankshaft Position Sensor "B" Circuit Malfunction

P0386 Crankshaft Position Sensor "B" Circuit Range/Performance

P0387 Crankshaft Position Sensor "B" Circuit Low Input

P0388 Crankshaft Position Sensor "B" Circuit High Input

P0389 Crankshaft Position Sensor "B" Circuit Intermittent

P0400 Exhaust Gas Recirculation Flow Malfunction

P0401 Exhaust Gas Recirculation Flow Insufficient Detected

P0402 Exhaust Gas Recirculation Flow Excessive Detected

P0403 Exhaust Gas Recirculation Circuit Malfunction

P0404 Exhaust Gas Recirculation Circuit Range/Performance

P0405 Exhaust Gas Recirculation Sensor "A" Circuit Low

P0406 Exhaust Gas Recirculation Sensor "A" Circuit High

P0407 Exhaust Gas Recirculation Sensor "B" Circuit Low

P0408 Exhaust Gas Recirculation Sensor "B" Circuit High

P0410 Secondary Air Injection System Malfunction

P0411 Secondary Air Injection System Incorrect Flow Detected

P0412 Secondary Air Injection System Switching Valve "A" Circuit Malfunction

P0413 Secondary Air Injection System Switching Valve "A" Circuit Open

P0414 Secondary Air Injection System Switching Valve "A" Circuit Shorted

P0415 Secondary Air Injection System Switching Valve "B" Circuit Malfunction

P0416 Secondary Air Injection System Switching Valve "B" Circuit Open

P0417 Secondary Air Injection System Switching Valve "B" Circuit Shorted

P0418 Secondary Air Injection System Relay "A" Circuit Malfunction

P0419 Secondary Air Injection System Relay "B" Circuit Malfunction

P0420 Catalyst System Efficiency Below Threshold (Bank no. 1)

P0421 Warm Up Catalyst Efficiency Below Threshold (Bank no. 1)

P0422 Main Catalyst Efficiency Below Threshold (Bank no. 1)

P0423 Heated Catalyst Efficiency Below Threshold (Bank no. 1)

P0424 Heated Catalyst Temperature Below Threshold (Bank no. 1)

P0430 Catalyst System Efficiency Below Threshold (Bank no. 2)

P0431 Warm Up Catalyst Efficiency Below Threshold (Bank no. 2)

P0432 Main Catalyst Efficiency Below Threshold (Bank no. 2)

P0433 Heated Catalyst Efficiency Below Threshold (Bank no. 2)

P0434 Heated Catalyst Temperature Below Threshold (Bank no. 2)

P0440 Evaporative Emission Control System Malfunction

P0441 Evaporative Emission Control System Incorrect Purge Flow

P0442 Evaporative Emission Control System Leak Detected (Small Leak)

P0443 Evaporative Emission Control System Purge Control Valve Circuit Malfunction

P0444 Evaporative Emission Control System Purge Control Valve Circuit Open

P0445 Evaporative Emission Control System Purge Control Valve Circuit Shorted

P0446 Evaporative Emission Control System Vent Control Circuit Malfunction

P0447 Evaporative Emission Control System Vent Control Circuit Open

P0448 Evaporative Emission Control System Vent Control Circuit Shorted

P0449 Evaporative Emission Control System Vent Valve/Solenoid Circuit Malfunction

P0450 Evaporative Emission Control System Pressure Sensor Malfunction

P0451 Evaporative Emission Control System Pressure Sensor Range/Performance

P0452 Evaporative Emission Control System Pressure Sensor Low Input

P0453 Evaporative Emission Control System Pressure Sensor High Input

P0454 Evaporative Emission Control System Pressure Sensor Intermittent

P0455 Evaporative Emission Control System Leak Detected (Gross Leak)

P0460 Fuel Level Sensor Circuit Malfunction

P0461 Fuel Level Sensor Circuit Range/Performance

P0462 Fuel Level Sensor Circuit Low Input

P0463 Fuel Level Sensor Circuit High Input

P0464 Fuel Level Sensor Circuit Intermittent

P0465 Purge Flow Sensor Circuit Malfunction

P0466 Purge Flow Sensor Circuit Range/Performance

P0467 Purge Flow Sensor Circuit Low Input

P0468 Purge Flow Sensor Circuit High Input

P0469 Purge Flow Sensor Circuit Intermittent

P0470 Exhaust Pressure Sensor Malfunction

P0471 Exhaust Pressure Sensor Range/Performance

P0472 Exhaust Pressure Sensor Low

P0473 Exhaust Pressure Sensor High

P0474 Exhaust Pressure Sensor Intermittent

P0475 Exhaust Pressure Control Valve Malfunction

P0476 Exhaust Pressure Control Valve Range/Performance

P0477 Exhaust Pressure Control Valve Low

P0478 Exhaust Pressure Control Valve High

P0479 Exhaust Pressure Control Valve Intermittent

P0480 Cooling Fan no. 1 Control Circuit Malfunction

P0481 Cooling Fan no. 2 Control Circuit Malfunction

P0482 Cooling Fan no. 3 Control Circuit Malfunction

P0483 Cooling Fan Rationality Check Malfunction

P0484 Cooling Fan Circuit Over Current

P0485 Cooling Fan Power/Ground Circuit Malfunction

P0500 Vehicle Speed Sensor Malfunction

P0501 Vehicle Speed Sensor Range/Performance

P0502 Vehicle Speed Sensor Circuit Low Input

P0503 Vehicle Speed Sensor Intermittent/Erratic/High

P0505 Idle Control System Malfunction

P0506 Idle Control System RPM Lower Than Expected
P0507 Idle Control System RPM Higher Than Expected
P0510 Closed Throttle Position Switch Malfunction
P0520 Engine Oil Pressure Sensor/Switch Circuit Malfunction
P0521 Engine Oil Pressure Sensor/Switch Range/Performance
P0522 Engine Oil Pressure Sensor/Switch Low Voltage
P0523 Engine Oil Pressure Sensor/Switch High Voltage
P0530 A/C Refrigerant Pressure Sensor Circuit Malfunction
P0531 A/C Refrigerant Pressure Sensor Circuit Range/Performance
P0532 A/C Refrigerant Pressure Sensor Circuit Low Input
P0533 A/C Refrigerant Pressure Sensor Circuit High Input
P0534 A/C Refrigerant Charge Loss
P0550 Power Steering Pressure Sensor Circuit Malfunction
P0551 Power Steering Pressure Sensor Circuit Range/Performance
P0552 Power Steering Pressure Sensor Circuit Low Input
P0553 Power Steering Pressure Sensor Circuit High Input
P0554 Power Steering Pressure Sensor Circuit Intermittent
P0560 System Voltage Malfunction
P0561 System Voltage Unstable
P0562 System Voltage Low
P0563 System Voltage High
P0565 Cruise Control On Signal Malfunction
P0566 Cruise Control Off Signal Malfunction
P0567 Cruise Control Resume Signal Malfunction
P0568 Cruise Control Set Signal Malfunction
P0569 Cruise Control Coast Signal Malfunction
P0570 Cruise Control Accel Signal Malfunction
P0571 Cruise Control/Brake Switch "A" Circuit Malfunction
P0572 Cruise Control/Brake Switch "A" Circuit Low
P0573 Cruise Control/Brake Switch "A" Circuit High
P0574 **Through P0580** Reserved for Cruise Codes
P0600 Serial Communication Link Malfunction
P0601 Internal Control Module Memory Check Sum Error
P0602 Control Module Programming Error
P0603 Internal Control Module Keep Alive Memory (KAM) Error
P0604 Internal Control Module Random Access Memory (RAM) Error
P0605 Internal Control Module Read Only Memory (ROM) Error
P0606 PCM Processor Fault
P0608 Control Module VSS Output "A" Malfunction
P0609 Control Module VSS Output "B" Malfunction
P0620 Generator Control Circuit Malfunction
P0621 Generator Lamp "L" Control Circuit Malfunction
P0622 Generator Field "F" Control Circuit Malfunction
P0650 Malfunction Indicator Lamp (MIL) Control Circuit Malfunction
P0654 Engine RPM Output Circuit Malfunction
P0655 Engine Hot Lamp Output Control Circuit Malfunction
P0656 Fuel Level Output Circuit Malfunction
P0700 Transmission Control System Malfunction
P0701 Transmission Control System Range/Performance
P0702 Transmission Control System Electrical
P0703 Torque Converter/Brake Switch "B" Circuit Malfunction
P0704 Clutch Switch Input Circuit Malfunction
P0705 Transmission Range Sensor Circuit Malfunction (PRNDL Input)
P0706 Transmission Range Sensor Circuit Range/Performance
P0707 Transmission Range Sensor Circuit Low Input
P0708 Transmission Range Sensor Circuit High Input

P0709 Transmission Range Sensor Circuit Intermittent
P0710 Transmission Fluid Temperature Sensor Circuit Malfunction
P0711 Transmission Fluid Temperature Sensor Circuit Range/Performance
P0712 Transmission Fluid Temperature Sensor Circuit Low Input
P0713 Transmission Fluid Temperature Sensor Circuit High Input
P0714 Transmission Fluid Temperature Sensor Circuit Intermittent
P0715 Input/Turbine Speed Sensor Circuit Malfunction
P0716 Input/Turbine Speed Sensor Circuit Range/Performance
P0717 Input/Turbine Speed Sensor Circuit No Signal
P0718 Input/Turbine Speed Sensor Circuit Intermittent
P0719 Torque Converter/Brake Switch "B" Circuit Low
P0720 Output Speed Sensor Circuit Malfunction
P0721 Output Speed Sensor Circuit Range/Performance
P0722 Output Speed Sensor Circuit No Signal
P0723 Output Speed Sensor Circuit Intermittent
P0724 Torque Converter/Brake Switch "B" Circuit High
P0725 Engine Speed Input Circuit Malfunction
P0726 Engine Speed Input Circuit Range/Performance
P0727 Engine Speed Input Circuit No Signal
P0728 Engine Speed Input Circuit Intermittent
P0730 Incorrect Gear Ratio
P0731 Gear no. 1 Incorrect Ratio
P0732 Gear no. 2 Incorrect Ratio
P0733 Gear no. 3 Incorrect Ratio
P0734 Gear no. 4 Incorrect Ratio
P0735 Gear no. 5 Incorrect Ratio
P0736 Reverse Incorrect Ratio
P0740 Torque Converter Clutch Circuit Malfunction
P0741 Torque Converter Clutch Circuit Performance or Stuck Off
P0742 Torque Converter Clutch Circuit Stuck On
P0743 Torque Converter Clutch Circuit Electrical
P0744 Torque Converter Clutch Circuit Intermittent
P0745 Pressure Control Solenoid Malfunction
P0746 Pressure Control Solenoid Performance or Stuck Off
P0747 Pressure Control Solenoid Stuck On
P0748 Pressure Control Solenoid Electrical
P0749 Pressure Control Solenoid Intermittent
P0750 Shift Solenoid "A" Malfunction
P0751 Shift Solenoid "A" Performance or Stuck Off
P0752 Shift Solenoid "A" Stuck On
P0753 Shift Solenoid "A" Electrical
P0754 Shift Solenoid "A" Intermittent
P0755 Shift Solenoid "B" Malfunction
P0756 Shift Solenoid "B" Performance or Stuck Off
P0757 Shift Solenoid "B" Stuck On
P0758 Shift Solenoid "B" Electrical
P0759 Shift Solenoid "B" Intermittent
P0760 Shift Solenoid "C" Malfunction
P0761 Shift Solenoid "C" Performance Or Stuck Off
P0762 Shift Solenoid "C" Stuck On
P0763 Shift Solenoid "C" Electrical
P0764 Shift Solenoid "C" Intermittent
P0765 Shift Solenoid "D" Malfunction
P0766 Shift Solenoid "D" Performance Or Stuck Off
P0767 Shift Solenoid "D" Stuck On

P0768 Shift Solenoid "D" Electrical
P0769 Shift Solenoid "D" Intermittent
P0770 Shift Solenoid "E" Malfunction
P0771 Shift Solenoid "E" Performance Or Stuck Off
P0772 Shift Solenoid "E" Stuck On
P0773 Shift Solenoid "E" Electrical
P0774 Shift Solenoid "E" Intermittent
P0780 Shift Malfunction
P0781 1–2 Shift Malfunction
P0782 2–3 Shift Malfunction
P0783 3–4 Shift Malfunction
P0784 4–5 Shift Malfunction
P0785 Shift/Timing Solenoid Malfunction
P0786 Shift/Timing Solenoid Range/Performance
P0787 Shift/Timing Solenoid Low
P0788 Shift/Timing Solenoid High
P0789 Shift/Timing Solenoid Intermittent
P0790 Normal/Performance Switch Circuit Malfunction
P0801 Reverse Inhibit Control Circuit Malfunction
P0803 1–4 Upshift (Skip Shift) Solenoid Control Circuit Malfunction
P0804 1–4 Upshift (Skip Shift) Lamp Control Circuit Malfunction

Mitsubishi

READING CODES

Reading the control module memory is one of the first steps in OBD II system diagnostics. This step should be initially performed to determine the general nature of the fault. Subsequent readings will determine if the fault has been cleared.

Reading codes can be performed by any of the methods below:
• Read the control module memory with the Generic Scan Tool (GST)
• Read the control module memory with the vehicle manufacturer's specific tester

To read the fault codes, connect the scan tool or tester according to the manufacturer's instructions. Follow the manufacturer's specified procedure for reading the codes.

CLEARING CODES

Control module reset procedures are a very important part of OBD II System diagnostics. This step should be done at the end of any fault code repair and at the end of any driveability repair.

Clearing codes can be performed by any of the methods below:
• Clear the control module memory with the Generic Scan Tool (GST)
• Clear the control module memory with the vehicle manufacturer's specific tester
• Turn the ignition off and remove the negative battery cable for at least 1 minute.

Removing the negative battery cable may cause other systems in the vehicle to loose their memory. Prior to removing the cable, ensure you have the proper reset codes for radios and alarms.

➡The MIL will may also be de-activated for some codes if the vehicle completes three consecutive trips without a fault detected with vehicle conditions similar to those present during the fault.

OBD II TROUBLE CODES

P0100 Mass or Volume Air Flow Circuit Malfunction
P0101 Mass or Volume Air Flow Circuit Range/Performance Problem
P0102 Mass or Volume Air Flow Circuit Low Input
P0103 Mass or Volume Air Flow Circuit High Input
P0104 Mass or Volume Air Flow Circuit Intermittent
P0105 Manifold Absolute Pressure/Barometric Pressure Circuit Malfunction
P0106 Manifold Absolute Pressure/Barometric Pressure Circuit Range/Performance Problem
P0107 Manifold Absolute Pressure/Barometric Pressure Circuit Low Input
P0108 Manifold Absolute Pressure/Barometric Pressure Circuit High Input
P0109 Manifold Absolute Pressure/Barometric Pressure Circuit Intermittent
P0110 Intake Air Temperature Circuit Malfunction
P0111 Intake Air Temperature Circuit Range/Performance Problem
P0112 Intake Air Temperature Circuit Low Input
P0113 Intake Air Temperature Circuit High Input
P0114 Intake Air Temperature Circuit Intermittent
P0115 Engine Coolant Temperature Circuit Malfunction
P0116 Engine Coolant Temperature Circuit Range/Performance Problem
P0117 Engine Coolant Temperature Circuit Low Input
P0118 Engine Coolant Temperature Circuit High Input
P0119 Engine Coolant Temperature Circuit Intermittent
P0120 Throttle/Pedal Position Sensor/Switch "A" Circuit Malfunction
P0121 Throttle/Pedal Position Sensor/Switch "A" Circuit Range/Performance Problem
P0122 Throttle/Pedal Position Sensor/Switch "A" Circuit Low Input
P0123 Throttle/Pedal Position Sensor/Switch "A" Circuit High Input
P0124 Throttle/Pedal Position Sensor/Switch "A" Circuit Intermittent
P0125 Insufficient Coolant Temperature For Closed Loop Fuel Control
P0126 Insufficient Coolant Temperature For Stable Operation
P0130 O2 Circuit Malfunction (Bank no. 1 Sensor no. 1)
P0131 O2 Sensor Circuit Low Voltage (Bank no. 1 Sensor no. 1)
P0132 O2 Sensor Circuit High Voltage (Bank no. 1 Sensor no. 1)
P0133 O2 Sensor Circuit Slow Response (Bank no. 1 Sensor no. 1)
P0134 O2 Sensor Circuit No Activity Detected (Bank no. 1 Sensor no. 1)
P0135 O2 Sensor Heater Circuit Malfunction (Bank no. 1 Sensor no. 1)
P0136 O2 Sensor Circuit Malfunction (Bank no. 1 Sensor no. 2)
P0137 O2 Sensor Circuit Low Voltage (Bank no. 1 Sensor no. 2)
P0138 O2 Sensor Circuit High Voltage (Bank no. 1 Sensor no. 2)
P0139 O2 Sensor Circuit Slow Response (Bank no. 1 Sensor no. 2)
P0140 O2 Sensor Circuit No Activity Detected (Bank no. 1 Sensor no. 2)

P0141 O2 Sensor Heater Circuit Malfunction (Bank no. 1 Sensor no. 2)

P0142 O2 Sensor Circuit Malfunction (Bank no. 1 Sensor no. 3)

P0143 O2 Sensor Circuit Low Voltage (Bank no. 1 Sensor no. 3)

P0144 O2 Sensor Circuit High Voltage (Bank no. 1 Sensor no. 3)

P0145 O2 Sensor Circuit Slow Response (Bank no. 1 Sensor no. 3)

P0146 O2 Sensor Circuit No Activity Detected (Bank no. 1 Sensor no. 3)

P0147 O2 Sensor Heater Circuit Malfunction (Bank no. 1 Sensor no. 3)

P0150 O2 Sensor Circuit Malfunction (Bank no. 2 Sensor no. 1)

P0151 O2 Sensor Circuit Low Voltage (Bank no. 2 Sensor no. 1)

P0152 O2 Sensor Circuit High Voltage (Bank no. 2 Sensor no. 1)

P0153 O2 Sensor Circuit Slow Response (Bank no. 2 Sensor no. 1)

P0154 O2 Sensor Circuit No Activity Detected (Bank no. 2 Sensor no. 1)

P0155 O2 Sensor Heater Circuit Malfunction (Bank no. 2 Sensor no. 1)

P0156 O2 Sensor Circuit Malfunction (Bank no. 2 Sensor no. 2)

P0157 O2 Sensor Circuit Low Voltage (Bank no. 2 Sensor no. 2)

P0158 O2 Sensor Circuit High Voltage (Bank no. 2 Sensor no. 2)

P0159 O2 Sensor Circuit Slow Response (Bank no. 2 Sensor no. 2)

P0160 O2 Sensor Circuit No Activity Detected (Bank no. 2 Sensor no. 2)

P0161 O2 Sensor Heater Circuit Malfunction (Bank no. 2 Sensor no. 2)

P0162 O2 Sensor Circuit Malfunction (Bank no. 2 Sensor no. 3)

P0163 O2 Sensor Circuit Low Voltage (Bank no. 2 Sensor no. 3)

P0164 O2 Sensor Circuit High Voltage (Bank no. 2 Sensor no. 3)

P0165 O2 Sensor Circuit Slow Response (Bank no. 2 Sensor no. 3)

P0166 O2 Sensor Circuit No Activity Detected (Bank no. 2 Sensor no. 3)

P0167 O2 Sensor Heater Circuit Malfunction (Bank no. 2 Sensor no. 3)

P0170 Fuel Trim Malfunction (Bank no. 1)

P0171 System Too Lean (Bank no. 1)

P0172 System Too Rich (Bank no. 1)

P0173 Fuel Trim Malfunction (Bank no. 2)

P0174 System Too Lean (Bank no. 2)

P0175 System Too Rich (Bank no. 2)

P0176 Fuel Composition Sensor Circuit Malfunction

P0177 Fuel Composition Sensor Circuit Range/Performance

P0178 Fuel Composition Sensor Circuit Low Input

P0179 Fuel Composition Sensor Circuit High Input

P0180 Fuel Temperature Sensor "A" Circuit Malfunction

P0181 Fuel Temperature Sensor "A" Circuit Range/Performance

P0182 Fuel Temperature Sensor "A" Circuit Low Input

P0183 Fuel Temperature Sensor "A" Circuit High Input

P0184 Fuel Temperature Sensor "A" Circuit Intermittent

P0185 Fuel Temperature Sensor "B" Circuit Malfunction

P0186 Fuel Temperature Sensor "B" Circuit Range/Performance

P0187 Fuel Temperature Sensor "B" Circuit Low Input

P0188 Fuel Temperature Sensor "B" Circuit High Input

P0189 Fuel Temperature Sensor "B" Circuit Intermittent

P0190 Fuel Rail Pressure Sensor Circuit Malfunction

P0191 Fuel Rail Pressure Sensor Circuit Range/Performance

P0192 Fuel Rail Pressure Sensor Circuit Low Input

P0193 Fuel Rail Pressure Sensor Circuit High Input

P0194 Fuel Rail Pressure Sensor Circuit Intermittent

P0195 Engine Oil Temperature Sensor Malfunction

P0196 Engine Oil Temperature Sensor Range/Performance

P0197 Engine Oil Temperature Sensor Low

P0198 Engine Oil Temperature Sensor High

P0199 Engine Oil Temperature Sensor Intermittent

P0200 Injector Circuit Malfunction

P0201 Injector Circuit Malfunction—Cylinder no. 1

P0202 Injector Circuit Malfunction—Cylinder no. 2

P0203 Injector Circuit Malfunction—Cylinder no. 3

P0204 Injector Circuit Malfunction—Cylinder no. 4

P0205 Injector Circuit Malfunction—Cylinder no. 5

P0206 Injector Circuit Malfunction—Cylinder no. 6

P0207 Injector Circuit Malfunction—Cylinder no. 7

P0208 Injector Circuit Malfunction—Cylinder no. 8

P0209 Injector Circuit Malfunction—Cylinder no. 9

P0210 Injector Circuit Malfunction—Cylinder no. 10

P0211 Injector Circuit Malfunction—Cylinder no. 11

P0212 Injector Circuit Malfunction—Cylinder no. 12

P0213 Cold Start Injector no. 1 Malfunction

P0214 Cold Start Injector no. 2 Malfunction

P0215 Engine Shutoff Solenoid Malfunction

P0216 Injection Timing Control Circuit Malfunction

P0217 Engine Over Temperature Condition

P0218 Transmission Over Temperature Condition

P0219 Engine Over Speed Condition

P0220 Throttle/Pedal Position Sensor/Switch "B" Circuit Malfunction

P0221 Throttle/Pedal Position Sensor/Switch "B" Circuit Range/Performance Problem

P0222 Throttle/Pedal Position Sensor/Switch "B" Circuit Low Input

P0223 Throttle/Pedal Position Sensor/Switch "B" Circuit High Input

P0224 Throttle/Pedal Position Sensor/Switch "B" Circuit Intermittent

P0225 Throttle/Pedal Position Sensor/Switch "C" Circuit Malfunction

P0226 Throttle/Pedal Position Sensor/Switch "C" Circuit Range/Performance Problem

P0227 Throttle/Pedal Position Sensor/Switch "C" Circuit Low Input

P0228 Throttle/Pedal Position Sensor/Switch "C" Circuit High Input

P0229 Throttle/Pedal Position Sensor/Switch "C" Circuit Intermittent

P0230 Fuel Pump Primary Circuit Malfunction

P0231 Fuel Pump Secondary Circuit Low

P0232 Fuel Pump Secondary Circuit High

P0233 Fuel Pump Secondary Circuit Intermittent

P0234 Engine Over Boost Condition

P0261 Cylinder no. 1 Injector Circuit Low

P0262 Cylinder no. 1 Injector Circuit High

P0263 Cylinder no. 1 Contribution/Balance Fault

P0264 Cylinder no. 2 Injector Circuit Low

P0265 Cylinder no. 2 Injector Circuit High

P0266 Cylinder no. 2 Contribution/Balance Fault
P0267 Cylinder no. 3 Injector Circuit Low
P0268 Cylinder no. 3 Injector Circuit High
P0269 Cylinder no. 3 Contribution/Balance Fault
P0270 Cylinder no. 4 Injector Circuit Low
P0271 Cylinder no. 4 Injector Circuit High
P0272 Cylinder no. 4 Contribution/Balance Fault
P0273 Cylinder no. 5 Injector Circuit Low
P0274 Cylinder no. 5 Injector Circuit High
P0275 Cylinder no. 5 Contribution/Balance Fault
P0276 Cylinder no. 6 Injector Circuit Low
P0277 Cylinder no. 6 Injector Circuit High
P0278 Cylinder no. 6 Contribution/Balance Fault
P0279 Cylinder no. 7 Injector Circuit Low
P0280 Cylinder no. 7 Injector Circuit High
P0281 Cylinder no. 7 Contribution/Balance Fault
P0282 Cylinder no. 8 Injector Circuit Low
P0283 Cylinder no. 8 Injector Circuit High
P0284 Cylinder no. 8 Contribution/Balance Fault
P0285 Cylinder no. 9 Injector Circuit Low
P0286 Cylinder no. 9 Injector Circuit High
P0287 Cylinder no. 9 Contribution/Balance Fault
P0288 Cylinder no. 10 Injector Circuit Low
P0289 Cylinder no. 10 Injector Circuit High
P0290 Cylinder no. 10 Contribution/Balance Fault
P0291 Cylinder no. 11 Injector Circuit Low
P0292 Cylinder no. 11 Injector Circuit High
P0293 Cylinder no. 11 Contribution/Balance Fault
P0294 Cylinder no. 12 Injector Circuit Low
P0295 Cylinder no. 12 Injector Circuit High
P0296 Cylinder no. 12 Contribution/Balance Fault
P0300 Random/Multiple Cylinder Misfire Detected
P0301 Cylinder no. 1—Misfire Detected
P0302 Cylinder no. 2—Misfire Detected
P0303 Cylinder no. 3—Misfire Detected
P0304 Cylinder no. 4—Misfire Detected
P0305 Cylinder no. 5—Misfire Detected
P0306 Cylinder no. 6—Misfire Detected
P0307 Cylinder no. 7—Misfire Detected
P0308 Cylinder no. 8—Misfire Detected
P0309 Cylinder no. 9—Misfire Detected
P0310 Cylinder no. 10—Misfire Detected
P0311 Cylinder no. 11—Misfire Detected
P0312 Cylinder no. 12—Misfire Detected
P0320 Ignition/Distributor Engine Speed Input Circuit Malfunction
P0321 Ignition/Distributor Engine Speed Input Circuit Range/Performance
P0322 Ignition/Distributor Engine Speed Input Circuit No Signal
P0323 Ignition/Distributor Engine Speed Input Circuit Intermittent
P0325 Knock Sensor no. 1—Circuit Malfunction (Bank no. 1 or Single Sensor)
P0326 Knock Sensor no. 1—Circuit Range/Performance (Bank no. 1 or Single Sensor)
P0327 Knock Sensor no. 1—Circuit Low Input (Bank no. 1 or Single Sensor)
P0328 Knock Sensor no. 1—Circuit High Input (Bank no. 1 or Single Sensor)
P0329 Knock Sensor no. 1—Circuit Input Intermittent (Bank no. 1 or Single Sensor)

P0330 Knock Sensor no. 2—Circuit Malfunction (Bank no. 2)
P0331 Knock Sensor no. 2—Circuit Range/Performance (Bank no. 2)
P0332 Knock Sensor no. 2—Circuit Low Input (Bank no. 2)
P0333 Knock Sensor no. 2—Circuit High Input (Bank no. 2)
P0334 Knock Sensor no. 2—Circuit Input Intermittent (Bank no. 2)
P0335 Crankshaft Position Sensor "A" Circuit Malfunction
P0336 Crankshaft Position Sensor "A" Circuit Range/Performance
P0337 Crankshaft Position Sensor "A" Circuit Low Input
P0338 Crankshaft Position Sensor "A" Circuit High Input
P0339 Crankshaft Position Sensor "A" Circuit Intermittent
P0340 Camshaft Position Sensor Circuit Malfunction
P0341 Camshaft Position Sensor Circuit Range/Performance
P0342 Camshaft Position Sensor Circuit Low Input
P0343 Camshaft Position Sensor Circuit High Input
P0344 Camshaft Position Sensor Circuit Intermittent
P0350 Ignition Coil Primary/Secondary Circuit Malfunction
P0351 Ignition Coil "A" Primary/Secondary Circuit Malfunction
P0352 Ignition Coil "B" Primary/Secondary Circuit Malfunction
P0353 Ignition Coil "C" Primary/Secondary Circuit Malfunction
P0354 Ignition Coil "D" Primary/Secondary Circuit Malfunction
P0355 Ignition Coil "E" Primary/Secondary Circuit Malfunction
P0356 Ignition Coil "F" Primary/Secondary Circuit Malfunction
P0357 Ignition Coil "G" Primary/Secondary Circuit Malfunction
P0358 Ignition Coil "H" Primary/Secondary Circuit Malfunction
P0359 Ignition Coil "I" Primary/Secondary Circuit Malfunction
P0360 Ignition Coil "J" Primary/Secondary Circuit Malfunction
P0361 Ignition Coil "K" Primary/Secondary Circuit Malfunction
P0362 Ignition Coil "L" Primary/Secondary Circuit Malfunction
P0370 Timing Reference High Resolution Signal "A" Malfunction
P0371 Timing Reference High Resolution Signal "A" Too Many Pulses
P0372 Timing Reference High Resolution Signal "A" Too Few Pulses
P0373 Timing Reference High Resolution Signal "A" Intermittent/Erratic Pulses
P0374 Timing Reference High Resolution Signal "A" No Pulses
P0375 Timing Reference High Resolution Signal "B" Malfunction
P0376 Timing Reference High Resolution Signal "B" Too Many Pulses
P0377 Timing Reference High Resolution Signal "B" Too Few Pulses
P0378 Timing Reference High Resolution Signal "B" Intermittent/Erratic Pulses
P0379 Timing Reference High Resolution Signal "B" No Pulses
P0380 Glow Plug/Heater Circuit "A" Malfunction
P0381 Glow Plug/Heater Indicator Circuit Malfunction
P0382 Glow Plug/Heater Circuit "B" Malfunction
P0385 Crankshaft Position Sensor "B" Circuit Malfunction
P0386 Crankshaft Position Sensor "B" Circuit Range/Performance
P0387 Crankshaft Position Sensor "B" Circuit Low Input
P0388 Crankshaft Position Sensor "B" Circuit High Input
P0389 Crankshaft Position Sensor "B" Circuit Intermittent
P0400 Exhaust Gas Recirculation Flow Malfunction
P0401 Exhaust Gas Recirculation Flow Insufficient Detected
P0402 Exhaust Gas Recirculation Flow Excessive Detected

P0403 Exhaust Gas Recirculation Circuit Malfunction
P0404 Exhaust Gas Recirculation Circuit Range/Performance
P0405 Exhaust Gas Recirculation Sensor "A" Circuit Low
P0406 Exhaust Gas Recirculation Sensor "A" Circuit High
P0407 Exhaust Gas Recirculation Sensor "B" Circuit Low
P0408 Exhaust Gas Recirculation Sensor "B" Circuit High
P0410 Secondary Air Injection System Malfunction
P0411 Secondary Air Injection System Incorrect Flow Detected
P0412 Secondary Air Injection System Switching Valve "A" Circuit Malfunction
P0413 Secondary Air Injection System Switching Valve "A" Circuit Open
P0414 Secondary Air Injection System Switching Valve "A" Circuit Shorted
P0415 Secondary Air Injection System Switching Valve "B" Circuit Malfunction
P0416 Secondary Air Injection System Switching Valve "B" Circuit Open
P0417 Secondary Air Injection System Switching Valve "B" Circuit Shorted
P0418 Secondary Air Injection System Relay "A" Circuit Malfunction
P0419 Secondary Air Injection System Relay "B" Circuit Malfunction
P0420 Catalyst System Efficiency Below Threshold (Bank no. 1)
P0421 Warm Up Catalyst Efficiency Below Threshold (Bank no. 1)
P0422 Main Catalyst Efficiency Below Threshold (Bank no. 1)
P0423 Heated Catalyst Efficiency Below Threshold (Bank no. 1)
P0424 Heated Catalyst Temperature Below Threshold (Bank no. 1)
P0430 Catalyst System Efficiency Below Threshold (Bank no. 2)
P0431 Warm Up Catalyst Efficiency Below Threshold (Bank no. 2)
P0432 Main Catalyst Efficiency Below Threshold (Bank no. 2)
P0433 Heated Catalyst Efficiency Below Threshold (Bank no. 2)
P0434 Heated Catalyst Temperature Below Threshold (Bank no. 2)
P0440 Evaporative Emission Control System Malfunction
P0441 Evaporative Emission Control System Incorrect Purge Flow
P0442 Evaporative Emission Control System Leak Detected (Small Leak)
P0443 Evaporative Emission Control System Purge Control Valve Circuit Malfunction
P0444 Evaporative Emission Control System Purge Control Valve Circuit Open
P0445 Evaporative Emission Control System Purge Control Valve Circuit Shorted
P0446 Evaporative Emission Control System Vent Control Circuit Malfunction
P0447 Evaporative Emission Control System Vent Control Circuit Open
P0448 Evaporative Emission Control System Vent Control Circuit Shorted
P0449 Evaporative Emission Control System Vent Valve/Solenoid Circuit Malfunction
P0450 Evaporative Emission Control System Pressure Sensor Malfunction
P0451 Evaporative Emission Control System Pressure Sensor Range/Performance

P0452 Evaporative Emission Control System Pressure Sensor Low Input
P0453 Evaporative Emission Control System Pressure Sensor High Input
P0454 Evaporative Emission Control System Pressure Sensor Intermittent
P0455 Evaporative Emission Control System Leak Detected (Gross Leak)
P0460 Fuel Level Sensor Circuit Malfunction
P0461 Fuel Level Sensor Circuit Range/Performance
P0462 Fuel Level Sensor Circuit Low Input
P0463 Fuel Level Sensor Circuit High Input
P0464 Fuel Level Sensor Circuit Intermittent
P0465 Purge Flow Sensor Circuit Malfunction
P0466 Purge Flow Sensor Circuit Range/Performance
P0467 Purge Flow Sensor Circuit Low Input
P0468 Purge Flow Sensor Circuit High Input
P0469 Purge Flow Sensor Circuit Intermittent
P0470 Exhaust Pressure Sensor Malfunction
P0471 Exhaust Pressure Sensor Range/Performance
P0472 Exhaust Pressure Sensor Low
P0473 Exhaust Pressure Sensor High
P0474 Exhaust Pressure Sensor Intermittent
P0475 Exhaust Pressure Control Valve Malfunction
P0476 Exhaust Pressure Control Valve Range/Performance
P0477 Exhaust Pressure Control Valve Low
P0478 Exhaust Pressure Control Valve High
P0479 Exhaust Pressure Control Valve Intermittent
P0480 Cooling Fan no. 1 Control Circuit Malfunction
P0481 Cooling Fan no. 2 Control Circuit Malfunction
P0482 Cooling Fan no. 3 Control Circuit Malfunction
P0483 Cooling Fan Rationality Check Malfunction
P0484 Cooling Fan Circuit Over Current
P0485 Cooling Fan Power/Ground Circuit Malfunction
P0500 Vehicle Speed Sensor Malfunction
P0501 Vehicle Speed Sensor Range/Performance
P0502 Vehicle Speed Sensor Circuit Low Input
P0503 Vehicle Speed Sensor Intermittent/Erratic/High
P0505 Idle Control System Malfunction
P0506 Idle Control System RPM Lower Than Expected
P0507 Idle Control System RPM Higher Than Expected
P0510 Closed Throttle Position Switch Malfunction
P0520 Engine Oil Pressure Sensor/Switch Circuit Malfunction
P0521 Engine Oil Pressure Sensor/Switch Range/Performance
P0522 Engine Oil Pressure Sensor/Switch Low Voltage
P0523 Engine Oil Pressure Sensor/Switch High Voltage
P0530 A/C Refrigerant Pressure Sensor Circuit Malfunction
P0531 A/C Refrigerant Pressure Sensor Circuit Range/Performance
P0532 A/C Refrigerant Pressure Sensor Circuit Low Input
P0533 A/C Refrigerant Pressure Sensor Circuit High Input
P0534 A/C Refrigerant Charge Loss
P0550 Power Steering Pressure Sensor Circuit Malfunction
P0551 Power Steering Pressure Sensor Circuit Range/Performance
P0552 Power Steering Pressure Sensor Circuit Low Input
P0553 Power Steering Pressure Sensor Circuit High Input
P0554 Power Steering Pressure Sensor Circuit Intermittent
P0560 System Voltage Malfunction
P0561 System Voltage Unstable
P0562 System Voltage Low

P0563 System Voltage High
P0565 Cruise Control On Signal Malfunction
P0566 Cruise Control Off Signal Malfunction
P0567 Cruise Control Resume Signal Malfunction
P0568 Cruise Control Set Signal Malfunction
P0569 Cruise Control Coast Signal Malfunction
P0570 Cruise Control Accel Signal Malfunction
P0571 Cruise Control/Brake Switch "A" Circuit Malfunction
P0572 Cruise Control/Brake Switch "A" Circuit Low
P0573 Cruise Control/Brake Switch "A" Circuit High
P0574 Through P0580 Reserved for Cruise Codes
P0600 Serial Communication Link Malfunction
P0601 Internal Control Module Memory Check Sum Error
P0602 Control Module Programming Error
P0603 Internal Control Module Keep Alive Memory (KAM) Error
P0604 Internal Control Module Random Access Memory (RAM) Error
P0605 Internal Control Module Read Only Memory (ROM) Error
P0606 PCM Processor Fault
P0608 Control Module VSS Output "A" Malfunction
P0609 Control Module VSS Output "B" Malfunction
P0620 Generator Control Circuit Malfunction
P0621 Generator Lamp "L" Control Circuit Malfunction
P0622 Generator Field "F" Control Circuit Malfunction
P0650 Malfunction Indicator Lamp (MIL) Control Circuit Malfunction
P0654 Engine RPM Output Circuit Malfunction
P0655 Engine Hot Lamp Output Control Circuit Malfunction
P0656 Fuel Level Output Circuit Malfunction
P0700 Transmission Control System Malfunction
P0701 Transmission Control System Range/Performance
P0702 Transmission Control System Electrical
P0703 Torque Converter/Brake Switch "B" Circuit Malfunction
P0704 Clutch Switch Input Circuit Malfunction
P0705 Transmission Range Sensor Circuit Malfunction (PRNDL Input)
P0706 Transmission Range Sensor Circuit Range/Performance
P0707 Transmission Range Sensor Circuit Low Input
P0708 Transmission Range Sensor Circuit High Input
P0709 Transmission Range Sensor Circuit Intermittent
P0710 Transmission Fluid Temperature Sensor Circuit Malfunction
P0711 Transmission Fluid Temperature Sensor Circuit Range/Performance
P0712 Transmission Fluid Temperature Sensor Circuit Low Input
P0713 Transmission Fluid Temperature Sensor Circuit High Input
P0714 Transmission Fluid Temperature Sensor Circuit Intermittent
P0715 Input/Turbine Speed Sensor Circuit Malfunction
P0716 Input/Turbine Speed Sensor Circuit Range/Performance
P0717 Input/Turbine Speed Sensor Circuit No Signal
P0718 Input/Turbine Speed Sensor Circuit Intermittent
P0719 Torque Converter/Brake Switch "B" Circuit Low
P0720 Output Speed Sensor Circuit Malfunction
P0721 Output Speed Sensor Circuit Range/Performance
P0722 Output Speed Sensor Circuit No Signal
P0723 Output Speed Sensor Circuit Intermittent
P0724 Torque Converter/Brake Switch "B" Circuit High
P0725 Engine Speed Input Circuit Malfunction
P0726 Engine Speed Input Circuit Range/Performance

P0727 Engine Speed Input Circuit No Signal
P0728 Engine Speed Input Circuit Intermittent
P0730 Incorrect Gear Ratio
P0731 Gear no. 1 Incorrect Ratio
P0732 Gear no. 2 Incorrect Ratio
P0733 Gear no. 3 Incorrect Ratio
P0734 Gear no. 4 Incorrect Ratio
P0735 Gear no. 5 Incorrect Ratio
P0736 Reverse Incorrect Ratio
P0740 Torque Converter Clutch Circuit Malfunction
P0741 Torque Converter Clutch Circuit Performance or Stuck Off
P0742 Torque Converter Clutch Circuit Stuck On
P0743 Torque Converter Clutch Circuit Electrical
P0744 Torque Converter Clutch Circuit Intermittent
P0745 Pressure Control Solenoid Malfunction
P0746 Pressure Control Solenoid Performance or Stuck Off
P0747 Pressure Control Solenoid Stuck On
P0748 Pressure Control Solenoid Electrical
P0749 Pressure Control Solenoid Intermittent
P0750 Shift Solenoid "A" Malfunction
P0751 Shift Solenoid "A" Performance or Stuck Off
P0752 Shift Solenoid "A" Stuck On
P0753 Shift Solenoid "A" Electrical
P0754 Shift Solenoid "A" Intermittent
P0755 Shift Solenoid "B" Malfunction
P0756 Shift Solenoid "B" Performance or Stuck Off
P0757 Shift Solenoid "B" Stuck On
P0758 Shift Solenoid "B" Electrical
P0759 Shift Solenoid "B" Intermittent
P0760 Shift Solenoid "C" Malfunction
P0761 Shift Solenoid "C" Performance Or Stuck Off
P0762 Shift Solenoid "C" Stuck On
P0763 Shift Solenoid "C" Electrical
P0764 Shift Solenoid "C" Intermittent
P0765 Shift Solenoid "D" Malfunction
P0766 Shift Solenoid "D" Performance Or Stuck Off
P0767 Shift Solenoid "D" Stuck On
P0768 Shift Solenoid "D" Electrical
P0769 Shift Solenoid "D" Intermittent
P0770 Shift Solenoid "E" Malfunction
P0771 Shift Solenoid "E" Performance Or Stuck Off
P0772 Shift Solenoid "E" Stuck On
P0773 Shift Solenoid "E" Electrical
P0774 Shift Solenoid "E" Intermittent
P0780 Shift Malfunction
P0781 1–2 Shift Malfunction
P0782 2–3 Shift Malfunction
P0783 3–4 Shift Malfunction
P0784 4–5 Shift Malfunction
P0785 Shift/Timing Solenoid Malfunction
P0786 Shift/Timing Solenoid Range/Performance
P0787 Shift/Timing Solenoid Low
P0788 Shift/Timing Solenoid High
P0789 Shift/Timing Solenoid Intermittent
P0790 Normal/Performance Switch Circuit Malfunction
P0801 Reverse Inhibit Control Circuit Malfunction
P0803 1–4 Upshift (Skip Shift) Solenoid Control Circuit Malfunction
P0804 1–4 Upshift (Skip Shift) Lamp Control Circuit Malfunction
P1100 Induction Control Motor Position Sensor Fault

P1101 Traction Control Vacuum Solenoid Circuit Fault
P1102 Traction Control Ventilation Solenoid Circuit Fault
P1103 Turbocharger Waste Gate Actuator Circuit Fault
P1104 Turbocharger Waste Gate Solenoid Circuit Fault
P1105 Fuel Pressure Solenoid Circuit Fault
P1294 Target Idle Speed Not Reached
P1295 No 5-Volt Supply To TP Sensor
P1296 No 5-Volt Supply To MAP Sensor
P1297 No Change In MAP From Start To Run
P1300 Ignition Timing Adjustment Circuit
P1390 Timing Belt Skipped One Tooth Or More
P1391 Intermittent Loss Of CMP Or CKP Sensor Signals
P1400 Manifold Differential Pressure Sensor Fault
P1443 EVAP Purge Control Solenoid "2" Circuit Fault
P1486 EVAP Leak Monitor Pinched Hose Detected
P1487 High Speed Radiator Fan Control Relay Circuit Fault
P1989 High Speed Condenser Fan Control Relay Fault
P1490 Low Speed Fan Control Relay Fault
P1492 Battery Temperature Sensor High Voltage
P1494 EVAP Ventilation Switch Or Mechanical Fault
P1495 EVAP Ventilation Solenoid Circuit Fault
P1496 5-Volt Supply Output Too Low
P1500 Generator FR Terminal Circuit Fault
P1600 PCM-TCM Serial Communication Link Circuit Fault
P1696 PCM Failure- EEPROM Write Denied
P1715 No CCD Messages From TCM
P1750 TCM Pulse Generator Circuit Fault
P1791 Pressure Control, Shift Control, TCC Solenoid Fault
P1899 PCM ECT Level Signal to TCM Circuit Fault

Nissan

➡ This section also provides coverage for the Mercury Villager since it shares a platform with the Nissan Quest

READING CODES

With Scan Tool

Reading the control module memory is one of the first steps in OBD II system diagnostics. This step should be initially performed to determine the general nature of the fault. Subsequent readings will determine if the fault has been cleared.

Reading codes can be performed by any of the methods below:
• Read the control module memory with the Generic Scan Tool (GST)
• Read the control module memory with the vehicle manufacturer's specific tester

To read the fault codes, connect the scan tool or tester according to the manufacturer's instructions. Follow the manufacturer's specified procedure for reading the codes.

Without Scan Tool

The ECM is capable of outputting data in four different modes, depending on the position of the mode switch and the ignition key. Modes are switched by turning the mode screw on the side of the ECM, near the red LED. Additional modes are accessed by turning the ignition key on or off.

The ECM is located forward of the center console, behind an access panel on the Altima, and in the passenger's side kick panel on the 240SX.

With the ECM set in Mode 1 and the ignition in the **ON** position, a malfunction indicator lamp bulb check may be performed. When the engine is started, the ECM will illuminate the indicator lamps as a warning of a fault in the system.

Mode 2 is set by turning the mode selector screw fully clockwise, waiting 2 seconds, then turning the screw fully counterclockwise. With the ignition in the **ON** position, self-diagnostic results will be output as a series of lamp flashes. When the engine is started, the oxygen sensor monitor function is enabled and the red LED on the ECM is used to determine proper oxygen sensor function.

1. Remove the access cover and locate the mode adjusting screw and LED on the ECM.

2. Turn the ignition switch **ON**, but do not start the engine. Both the LED and the malfunction indicator lamp on the instrument panel should be illuminated. This is a bulb check.

3. Start the engine.

➡ Switching modes is not possible while the engine is running.

4. If the LED or malfunction indicator lamp illuminates, there is a fault in the system.

5. Turn the mode selector screw fully clockwise. Wait 2 seconds, then turn the screw fully counterclockwise.

6. The diagnostic trouble codes will now be read from the ECM memory. They will appear as flashes of the malfunction indicator lamp, or the ECM's LED.

7. After all codes have been read, turn the mode selector screw fully clockwise to erase the codes.

➡ Turn the mode adjusting screw to the fully counterclockwise position whenever the vehicle is in use.

8. Turn the ignition **OFF**.

➡ When the ignition switch is turned OFF during diagnosis, power to the ECM will drop after approximately 5 seconds. The diagnosis will automatically return to Mode 1 at this time.

CLEARING CODES

With Scan Tool

Control module reset procedures are a very important part of OBD II System diagnostics. This step should be done at the end of any fault code repair and at the end of any driveability repair.

Clearing codes can be performed by any of the methods below:
• Clear the control module memory with the Generic Scan Tool (GST)
• Clear the control module memory with the vehicle manufacturer's specific tester

➡ The MIL will may also be de-activated for some codes if the vehicle completes three consecutive trips without a fault detected with vehicle conditions similar to those present during the fault.

Without Scan Tool

The easiest way to clear trouble codes without a scan tool is to turn the mode selector screw fully clockwise after all codes have been read.

➡ Turn the mode adjusting screw to the fully counterclockwise position whenever the vehicle is in use.

Codes may also be erased by turning the ignition off and remove the negative battery cable for at least 1 minute. However, removing the negative battery cable may cause other systems in the vehicle to loose their memory. Prior to removing the cable, ensure you have the proper reset codes for radios and alarms.

OBD II TROUBLE CODES

P0000 No Self Diagnostic Failure Indicated
P0100 Mass or Volume Air Flow Circuit Malfunction
P0101 Mass or Volume Air Flow Circuit Range/Performance Problem
P0102 Mass or Volume Air Flow Circuit Low Input
P0103 Mass or Volume Air Flow Circuit High Input
P0104 Mass or Volume Air Flow Circuit Intermittent
P0105 Manifold Absolute Pressure/Barometric Pressure Circuit Malfunction
P0106 Manifold Absolute Pressure/Barometric Pressure Circuit Range/Performance Problem
P0107 Manifold Absolute Pressure/Barometric Pressure Circuit Low Input
P0108 Manifold Absolute Pressure/Barometric Pressure Circuit High Input
P0109 Manifold Absolute Pressure/Barometric Pressure Circuit Intermittent
P0110 Intake Air Temperature Circuit Malfunction
P0111 Intake Air Temperature Circuit Range/Performance Problem
P0112 Intake Air Temperature Circuit Low Input
P0113 Intake Air Temperature Circuit High Input
P0114 Intake Air Temperature Circuit Intermittent
P0115 Engine Coolant Temperature Circuit Malfunction
P0116 Engine Coolant Temperature Circuit Range/Performance Problem
P0117 Engine Coolant Temperature Circuit Low Input
P0118 Engine Coolant Temperature Circuit High Input
P0119 Engine Coolant Temperature Circuit Intermittent
P0120 Throttle/Pedal Position Sensor/Switch "A" Circuit Malfunction
P0121 Throttle/Pedal Position Sensor/Switch "A" Circuit Range/Performance Problem
P0122 Throttle/Pedal Position Sensor/Switch "A" Circuit Low Input
P0123 Throttle/Pedal Position Sensor/Switch "A" Circuit High Input
P0124 Throttle/Pedal Position Sensor/Switch "A" Circuit Intermittent
P0125 Insufficient Coolant Temperature For Closed Loop Fuel Control
P0126 Insufficient Coolant Temperature For Stable Operation
P0130 O2 Circuit Malfunction (Bank no. 1 Sensor no. 1)
P0131 O2 Sensor Circuit Low Voltage (Bank no. 1 Sensor no. 1)
P0132 O2 Sensor Circuit High Voltage (Bank no. 1 Sensor no. 1)
P0133 O2 Sensor Circuit Slow Response (Bank no. 1 Sensor no. 1)
P0134 O2 Sensor Circuit No Activity Detected (Bank no. 1 Sensor no. 1)
P0135 O2 Sensor Heater Circuit Malfunction (Bank no. 1 Sensor no. 1)
P0136 O2 Sensor Circuit Malfunction (Bank no. 1 Sensor no. 2)
P0137 O2 Sensor Circuit Low Voltage (Bank no. 1 Sensor no. 2)

P0138 O2 Sensor Circuit High Voltage (Bank no. 1 Sensor no. 2)
P0139 O2 Sensor Circuit Slow Response (Bank no. 1 Sensor no. 2)
P0140 O2 Sensor Circuit No Activity Detected (Bank no. 1 Sensor no. 2)
P0141 O2 Sensor Heater Circuit Malfunction (Bank no. 1 Sensor no. 2)
P0142 O2 Sensor Circuit Malfunction (Bank no. 1 Sensor no. 3)
P0143 O2 Sensor Circuit Low Voltage (Bank no. 1 Sensor no. 3)
P0144 O2 Sensor Circuit High Voltage (Bank no. 1 Sensor no. 3)
P0145 O2 Sensor Circuit Slow Response (Bank no. 1 Sensor no. 3)
P0146 O2 Sensor Circuit No Activity Detected (Bank no. 1 Sensor no. 3)
P0147 O2 Sensor Heater Circuit Malfunction (Bank no. 1 Sensor no. 3)
P0150 O2 Sensor Circuit Malfunction (Bank no. 2 Sensor no. 1)
P0151 O2 Sensor Circuit Low Voltage (Bank no. 2 Sensor no. 1)
P0152 O2 Sensor Circuit High Voltage (Bank no. 2 Sensor no. 1)
P0153 O2 Sensor Circuit Slow Response (Bank no. 2 Sensor no. 1)
P0154 O2 Sensor Circuit No Activity Detected (Bank no. 2 Sensor no. 1)
P0155 O2 Sensor Heater Circuit Malfunction (Bank no. 2 Sensor no. 1)
P0156 O2 Sensor Circuit Malfunction (Bank no. 2 Sensor no. 2)
P0157 O2 Sensor Circuit Low Voltage (Bank no. 2 Sensor no. 2)
P0158 O2 Sensor Circuit High Voltage (Bank no. 2 Sensor no. 2)
P0159 O2 Sensor Circuit Slow Response (Bank no. 2 Sensor no. 2)
P0160 O2 Sensor Circuit No Activity Detected (Bank no. 2 Sensor no. 2)
P0161 O2 Sensor Heater Circuit Malfunction (Bank no. 2 Sensor no. 2)
P0162 O2 Sensor Circuit Malfunction (Bank no. 2 Sensor no. 3)
P0163 O2 Sensor Circuit Low Voltage (Bank no. 2 Sensor no. 3)
P0164 O2 Sensor Circuit High Voltage (Bank no. 2 Sensor no. 3)
P0165 O2 Sensor Circuit Slow Response (Bank no. 2 Sensor no. 3)
P0166 O2 Sensor Circuit No Activity Detected (Bank no. 2 Sensor no. 3)
P0167 O2 Sensor Heater Circuit Malfunction (Bank no. 2 Sensor no. 3)
P0170 Fuel Trim Malfunction (Bank no. 1)
P0171 System Too Lean (Bank no. 1)
P0172 System Too Rich (Bank no. 1)
P0173 Fuel Trim Malfunction (Bank no. 2)
P0174 System Too Lean (Bank no. 2)
P0175 System Too Rich (Bank no. 2)
P0176 Fuel Composition Sensor Circuit Malfunction
P0177 Fuel Composition Sensor Circuit Range/Performance
P0178 Fuel Composition Sensor Circuit Low Input
P0179 Fuel Composition Sensor Circuit High Input
P0180 Fuel Temperature Sensor "A" Circuit Malfunction
P0181 Fuel Temperature Sensor "A" Circuit Range/Performance
P0182 Fuel Temperature Sensor "A" Circuit Low Input
P0183 Fuel Temperature Sensor "A" Circuit High Input

P0184 Fuel Temperature Sensor "A" Circuit Intermittent
P0185 Fuel Temperature Sensor "B" Circuit Malfunction
P0186 Fuel Temperature Sensor "B" Circuit Range/Performance
P0187 Fuel Temperature Sensor "B" Circuit Low Input
P0188 Fuel Temperature Sensor "B" Circuit High Input
P0189 Fuel Temperature Sensor "B" Circuit Intermittent
P0190 Fuel Rail Pressure Sensor Circuit Malfunction
P0191 Fuel Rail Pressure Sensor Circuit Range/Performance
P0192 Fuel Rail Pressure Sensor Circuit Low Input
P0193 Fuel Rail Pressure Sensor Circuit High Input
P0194 Fuel Rail Pressure Sensor Circuit Intermittent
P0195 Engine Oil Temperature Sensor Malfunction
P0196 Engine Oil Temperature Sensor Range/Performance
P0197 Engine Oil Temperature Sensor Low
P0198 Engine Oil Temperature Sensor High
P0199 Engine Oil Temperature Sensor Intermittent
P0200 Injector Circuit Malfunction
P0201 Injector Circuit Malfunction—Cylinder no. 1
P0202 Injector Circuit Malfunction—Cylinder no. 2
P0203 Injector Circuit Malfunction—Cylinder no. 3
P0204 Injector Circuit Malfunction—Cylinder no. 4
P0205 Injector Circuit Malfunction—Cylinder no. 5
P0206 Injector Circuit Malfunction—Cylinder no. 6
P0207 Injector Circuit Malfunction—Cylinder no. 7
P0208 Injector Circuit Malfunction—Cylinder no. 8
P0209 Injector Circuit Malfunction—Cylinder no. 9
P0210 Injector Circuit Malfunction—Cylinder no. 10
P0211 Injector Circuit Malfunction—Cylinder no. 11
P0212 Injector Circuit Malfunction—Cylinder no. 12
P0213 Cold Start Injector no. 1 Malfunction
P0214 Cold Start Injector no. 2 Malfunction
P0215 Engine Shutoff Solenoid Malfunction
P0216 Injection Timing Control Circuit Malfunction
P0217 Engine Over Temperature Condition
P0218 Transmission Over Temperature Condition
P0219 Engine Over Speed Condition
P0220 Throttle/Pedal Position Sensor/Switch "B" Circuit Malfunction
P0221 Throttle/Pedal Position Sensor/Switch "B" Circuit Range/Performance Problem
P0222 Throttle/Pedal Position Sensor/Switch "B" Circuit Low Input
P0223 Throttle/Pedal Position Sensor/Switch "B" Circuit High Input
P0224 Throttle/Pedal Position Sensor/Switch "B" Circuit Intermittent
P0225 Throttle/Pedal Position Sensor/Switch "C" Circuit Malfunction
P0226 Throttle/Pedal Position Sensor/Switch "C" Circuit Range/Performance Problem
P0227 Throttle/Pedal Position Sensor/Switch "C" Circuit Low Input
P0228 Throttle/Pedal Position Sensor/Switch "C" Circuit High Input
P0229 Throttle/Pedal Position Sensor/Switch "C" Circuit Intermittent
P0230 Fuel Pump Primary Circuit Malfunction
P0231 Fuel Pump Secondary Circuit Low
P0232 Fuel Pump Secondary Circuit High
P0233 Fuel Pump Secondary Circuit Intermittent
P0234 Engine Over Boost Condition
P0261 Cylinder no. 1 Injector Circuit Low

P0262 Cylinder no. 1 Injector Circuit High
P0263 Cylinder no. 1 Contribution/Balance Fault
P0264 Cylinder no. 2 Injector Circuit Low
P0265 Cylinder no. 2 Injector Circuit High
P0266 Cylinder no. 2 Contribution/Balance Fault
P0267 Cylinder no. 3 Injector Circuit Low
P0268 Cylinder no. 3 Injector Circuit High
P0269 Cylinder no. 3 Contribution/Balance Fault
P0270 Cylinder no. 4 Injector Circuit Low
P0271 Cylinder no. 4 Injector Circuit High
P0272 Cylinder no. 4 Contribution/Balance Fault
P0273 Cylinder no. 5 Injector Circuit Low
P0274 Cylinder no. 5 Injector Circuit High
P0275 Cylinder no. 5 Contribution/Balance Fault
P0276 Cylinder no. 6 Injector Circuit Low
P0277 Cylinder no. 6 Injector Circuit High
P0278 Cylinder no. 6 Contribution/Balance Fault
P0279 Cylinder no. 7 Injector Circuit Low
P0280 Cylinder no. 7 Injector Circuit High
P0281 Cylinder no. 7 Contribution/Balance Fault
P0282 Cylinder no. 8 Injector Circuit Low
P0283 Cylinder no. 8 Injector Circuit High
P0284 Cylinder no. 8 Contribution/Balance Fault
P0285 Cylinder no. 9 Injector Circuit Low
P0286 Cylinder no. 9 Injector Circuit High
P0287 Cylinder no. 9 Contribution/Balance Fault
P0288 Cylinder no. 10 Injector Circuit Low
P0289 Cylinder no. 10 Injector Circuit High
P0290 Cylinder no. 10 Contribution/Balance Fault
P0291 Cylinder no. 11 Injector Circuit Low
P0292 Cylinder no. 11 Injector Circuit High
P0293 Cylinder no. 11 Contribution/Balance Fault
P0294 Cylinder no. 12 Injector Circuit Low
P0295 Cylinder no. 12 Injector Circuit High
P0296 Cylinder no. 12 Contribution/Balance Fault
P0300 Random/Multiple Cylinder Misfire Detected
P0301 Cylinder no. 1—Misfire Detected
P0302 Cylinder no. 2—Misfire Detected
P0303 Cylinder no. 3—Misfire Detected
P0304 Cylinder no. 4—Misfire Detected
P0305 Cylinder no. 5—Misfire Detected
P0306 Cylinder no. 6—Misfire Detected
P0307 Cylinder no. 7—Misfire Detected
P0308 Cylinder no. 8—Misfire Detected
P0309 Cylinder no. 9—Misfire Detected
P0310 Cylinder no. 10—Misfire Detected
P0311 Cylinder no. 11—Misfire Detected
P0312 Cylinder no. 12—Misfire Detected
P0320 Ignition/Distributor Engine Speed Input Circuit Malfunction
P0321 Ignition/Distributor Engine Speed Input Circuit Range/Performance
P0322 Ignition/Distributor Engine Speed Input Circuit No Signal
P0323 Ignition/Distributor Engine Speed Input Circuit Intermittent
P0325 Knock Sensor no. 1—Circuit Malfunction (Bank no. 1 or Single Sensor)
P0326 Knock Sensor no. 1—Circuit Range/Performance (Bank no. 1 or Single Sensor)
P0327 Knock Sensor no. 1—Circuit Low Input (Bank no. 1 or Single Sensor)
P0328 Knock Sensor no. 1—Circuit High Input (Bank no. 1 or Single Sensor)

P0329 Knock Sensor no. 1—Circuit Input Intermittent (Bank no. 1 or Single Sensor)

P0330 Knock Sensor no. 2—Circuit Malfunction (Bank no. 2)

P0331 Knock Sensor no. 2—Circuit Range/Performance (Bank no. 2)

P0332 Knock Sensor no. 2—Circuit Low Input (Bank no. 2)

P0333 Knock Sensor no. 2—Circuit High Input (Bank no. 2)

P0334 Knock Sensor no. 2—Circuit Input Intermittent (Bank no. 2)

P0335 Crankshaft Position Sensor "A" Circuit Malfunction

P0336 Crankshaft Position Sensor "A" Circuit Range/Performance

P0337 Crankshaft Position Sensor "A" Circuit Low Input

P0338 Crankshaft Position Sensor "A" Circuit High Input

P0339 Crankshaft Position Sensor "A" Circuit Intermittent

P0340 Camshaft Position Sensor Circuit Malfunction

P0341 Camshaft Position Sensor Circuit Range/Performance

P0342 Camshaft Position Sensor Circuit Low Input

P0343 Camshaft Position Sensor Circuit High Input

P0344 Camshaft Position Sensor Circuit Intermittent

P0350 Ignition Coil Primary/Secondary Circuit Malfunction

P0351 Ignition Coil "A" Primary/Secondary Circuit Malfunction

P0352 Ignition Coil "B" Primary/Secondary Circuit Malfunction

P0353 Ignition Coil "C" Primary/Secondary Circuit Malfunction

P0354 Ignition Coil "D" Primary/Secondary Circuit Malfunction

P0355 Ignition Coil "E" Primary/Secondary Circuit Malfunction

P0356 Ignition Coil "F" Primary/Secondary Circuit Malfunction

P0357 Ignition Coil "G" Primary/Secondary Circuit Malfunction

P0358 Ignition Coil "H" Primary/Secondary Circuit Malfunction

P0359 Ignition Coil "I" Primary/Secondary Circuit Malfunction

P0360 Ignition Coil "J" Primary/Secondary Circuit Malfunction

P0361 Ignition Coil "K" Primary/Secondary Circuit Malfunction

P0362 Ignition Coil "L" Primary/Secondary Circuit Malfunction

P0370 Timing Reference High Resolution Signal "A" Malfunction

P0371 Timing Reference High Resolution Signal "A" Too Many Pulses

P0372 Timing Reference High Resolution Signal "A" Too Few Pulses

P0373 Timing Reference High Resolution Signal "A" Intermittent/Erratic Pulses

P0374 Timing Reference High Resolution Signal "A" No Pulses

P0375 Timing Reference High Resolution Signal "B" Malfunction

P0376 Timing Reference High Resolution Signal "B" Too Many Pulses

P0377 Timing Reference High Resolution Signal "B" Too Few Pulses

P0378 Timing Reference High Resolution Signal "B" Intermittent/Erratic Pulses

P0379 Timing Reference High Resolution Signal "B" No Pulses

P0380 Glow Plug/Heater Circuit "A" Malfunction

P0381 Glow Plug/Heater Indicator Circuit Malfunction

P0382 Glow Plug/Heater Circuit "B" Malfunction

P0385 Crankshaft Position Sensor "B" Circuit Malfunction

P0386 Crankshaft Position Sensor "B" Circuit Range/Performance

P0387 Crankshaft Position Sensor "B" Circuit Low Input

P0388 Crankshaft Position Sensor "B" Circuit High Input

P0389 Crankshaft Position Sensor "B" Circuit Intermittent

P0400 Exhaust Gas Recirculation Flow Malfunction

P0401 Exhaust Gas Recirculation Flow Insufficient Detected

P0402 Exhaust Gas Recirculation Flow Excessive Detected

P0403 Exhaust Gas Recirculation Circuit Malfunction

P0404 Exhaust Gas Recirculation Circuit Range/Performance

P0405 Exhaust Gas Recirculation Sensor "A" Circuit Low

P0406 Exhaust Gas Recirculation Sensor "A" Circuit High

P0407 Exhaust Gas Recirculation Sensor "B" Circuit Low

P0408 Exhaust Gas Recirculation Sensor "B" Circuit High

P0410 Secondary Air Injection System Malfunction

P0411 Secondary Air Injection System Incorrect Flow Detected

P0412 Secondary Air Injection System Switching Valve "A" Circuit Malfunction

P0413 Secondary Air Injection System Switching Valve "A" Circuit Open

P0414 Secondary Air Injection System Switching Valve "A" Circuit Shorted

P0415 Secondary Air Injection System Switching Valve "B" Circuit Malfunction

P0416 Secondary Air Injection System Switching Valve "B" Circuit Open

P0417 Secondary Air Injection System Switching Valve "B" Circuit Shorted

P0418 Secondary Air Injection System Relay "A" Circuit Malfunction

P0419 Secondary Air Injection System Relay "B" Circuit Malfunction

P0420 Catalyst System Efficiency Below Threshold (Bank no. 1

P0421 Warm Up Catalyst Efficiency Below Threshold (Bank no. 1)

P0422 Main Catalyst Efficiency Below Threshold (Bank no. 1)

P0423 Heated Catalyst Efficiency Below Threshold (Bank no. 1

P0424 Heated Catalyst Temperature Below Threshold (Bank no. 1)

P0430 Catalyst System Efficiency Below Threshold (Bank no. 2

P0431 Warm Up Catalyst Efficiency Below Threshold (Bank no. 2)

P0432 Main Catalyst Efficiency Below Threshold (Bank no. 2)

P0433 Heated Catalyst Efficiency Below Threshold (Bank no. 2)

P0434 Heated Catalyst Temperature Below Threshold (Bank no. 2)

P0440 Evaporative Emission Control System Malfunction

P0441 Evaporative Emission Control System Incorrect Purge Flow

P0442 Evaporative Emission Control System Leak Detected (Small Leak)

P0443 Evaporative Emission Control System Purge Control Valve Circuit Malfunction

P0444 Evaporative Emission Control System Purge Control Valve Circuit Open

P0445 Evaporative Emission Control System Purge Control Valve Circuit Shorted

P0446 Evaporative Emission Control System Vent Control Circuit Malfunction

P0447 Evaporative Emission Control System Vent Control Circuit Open

P0448 Evaporative Emission Control System Vent Control Circuit Shorted

P0449 Evaporative Emission Control System Vent Valve/Solenoid Circuit Malfunction

P0450 Evaporative Emission Control System Pressure Sensor Malfunction

P0451 Evaporative Emission Control System Pressure Sensor Range/Performance

P0452 Evaporative Emission Control System Pressure Sensor Low Input

P0453 Evaporative Emission Control System Pressure Sensor High Input

P0454 Evaporative Emission Control System Pressure Sensor Intermittent

P0455 Evaporative Emission Control System Leak Detected (Gross Leak)

P0460 Fuel Level Sensor Circuit Malfunction

P0461 Fuel Level Sensor Circuit Range/Performance

P0462 Fuel Level Sensor Circuit Low Input

P0463 Fuel Level Sensor Circuit High Input

P0464 Fuel Level Sensor Circuit Intermittent

P0465 Purge Flow Sensor Circuit Malfunction

P0466 Purge Flow Sensor Circuit Range/Performance

P0467 Purge Flow Sensor Circuit Low Input

P0468 Purge Flow Sensor Circuit High Input

P0469 Purge Flow Sensor Circuit Intermittent

P0470 Exhaust Pressure Sensor Malfunction

P0471 Exhaust Pressure Sensor Range/Performance

P0472 Exhaust Pressure Sensor Low

P0473 Exhaust Pressure Sensor High

P0474 Exhaust Pressure Sensor Intermittent

P0475 Exhaust Pressure Control Valve Malfunction

P0476 Exhaust Pressure Control Valve Range/Performance

P0477 Exhaust Pressure Control Valve Low

P0478 Exhaust Pressure Control Valve High

P0479 Exhaust Pressure Control Valve Intermittent

P0480 Cooling Fan no. 1 Control Circuit Malfunction

P0481 Cooling Fan no. 2 Control Circuit Malfunction

P0482 Cooling Fan no. 3 Control Circuit Malfunction

P0483 Cooling Fan Rationality Check Malfunction

P0484 Cooling Fan Circuit Over Current

P0485 Cooling Fan Power/Ground Circuit Malfunction

P0500 Vehicle Speed Sensor Malfunction

P0501 Vehicle Speed Sensor Range/Performance

P0502 Vehicle Speed Sensor Circuit Low Input

P0503 Vehicle Speed Sensor Intermittent/Erratic/High

P0505 Idle Control System Malfunction

P0506 Idle Control System RPM Lower Than Expected

P0507 Idle Control System RPM Higher Than Expected

P0510 Closed Throttle Position Switch Malfunction

P0520 Engine Oil Pressure Sensor/Switch Circuit Malfunction

P0521 Engine Oil Pressure Sensor/Switch Range/Performance

P0522 Engine Oil Pressure Sensor/Switch Low Voltage

P0523 Engine Oil Pressure Sensor/Switch High Voltage

P0530 A/C Refrigerant Pressure Sensor Circuit Malfunction

P0531 A/C Refrigerant Pressure Sensor Circuit Range/Performance

P0532 A/C Refrigerant Pressure Sensor Circuit Low Input

P0533 A/C Refrigerant Pressure Sensor Circuit High Input

P0534 A/C Refrigerant Charge Loss

P0550 Power Steering Pressure Sensor Circuit Malfunction

P0551 Power Steering Pressure Sensor Circuit Range/Performance

P0552 Power Steering Pressure Sensor Circuit Low Input

P0553 Power Steering Pressure Sensor Circuit High Input

P0554 Power Steering Pressure Sensor Circuit Intermittent

P0560 System Voltage Malfunction

P0561 System Voltage Unstable

P0562 System Voltage Low

P0563 System Voltage High

P0565 Cruise Control On Signal Malfunction

P0566 Cruise Control Off Signal Malfunction

P0567 Cruise Control Resume Signal Malfunction

P0568 Cruise Control Set Signal Malfunction

P0569 Cruise Control Coast Signal Malfunction

P0570 Cruise Control Accel Signal Malfunction

P0571 Cruise Control/Brake Switch "A" Circuit Malfunction

P0572 Cruise Control/Brake Switch "A" Circuit Low

P0573 Cruise Control/Brake Switch "A" Circuit High

P0574 **Through P0580** Reserved for Cruise Codes

P0600 Serial Communication Link Malfunction

P0601 Internal Control Module Memory Check Sum Error

P0602 Control Module Programming Error

P0603 Internal Control Module Keep Alive Memory (KAM) Error

P0604 Internal Control Module Random Access Memory (RAM) Error

P0605 Internal Control Module Read Only Memory (ROM) Error

P0606 PCM Processor Fault

P0608 Control Module VSS Output "A" Malfunction

P0609 Control Module VSS Output "B" Malfunction

P0620 Generator Control Circuit Malfunction

P0621 Generator Lamp "L" Control Circuit Malfunction

P0622 Generator Field "F" Control Circuit Malfunction

P0650 Malfunction Indicator Lamp (MIL) Control Circuit Malfunction

P0654 Engine RPM Output Circuit Malfunction

P0655 Engine Hot Lamp Output Control Circuit Malfunction

P0656 Fuel Level Output Circuit Malfunction

P0700 Transmission Control System Malfunction

P0701 Transmission Control System Range/Performance

P0702 Transmission Control System Electrical

P0703 Torque Converter/Brake Switch "B" Circuit Malfunction

P0704 Clutch Switch Input Circuit Malfunction

P0705 Transmission Range Sensor Circuit Malfunction (PRNDL Input)

P0706 Transmission Range Sensor Circuit Range/Performance

P0707 Transmission Range Sensor Circuit Low Input

P0708 Transmission Range Sensor Circuit High Input

P0709 Transmission Range Sensor Circuit Intermittent

P0710 Transmission Fluid Temperature Sensor Circuit Malfunction

P0711 Transmission Fluid Temperature Sensor Circuit Range/Performance

P0712 Transmission Fluid Temperature Sensor Circuit Low Input

P0713 Transmission Fluid Temperature Sensor Circuit High Input

P0714 Transmission Fluid Temperature Sensor Circuit Intermittent

P0715 Input/Turbine Speed Sensor Circuit Malfunction

P0716 Input/Turbine Speed Sensor Circuit Range/Performance

P0717 Input/Turbine Speed Sensor Circuit No Signal

P0718 Input/Turbine Speed Sensor Circuit Intermittent

P0719 Torque Converter/Brake Switch "B" Circuit Low

P0720 Output Speed Sensor Circuit Malfunction

P0721 Output Speed Sensor Circuit Range/Performance

P0722 Output Speed Sensor Circuit No Signal

P0723 Output Speed Sensor Circuit Intermittent

P0724 Torque Converter/Brake Switch "B" Circuit High

P0725 Engine Speed Input Circuit Malfunction

P0726 Engine Speed Input Circuit Range/Performance

P0727 Engine Speed Input Circuit No Signal

P0728 Engine Speed Input Circuit Intermittent
P0730 Incorrect Gear Ratio
P0731 Gear no. 1 Incorrect Ratio
P0732 Gear no. 2 Incorrect Ratio
P0733 Gear no. 3 Incorrect Ratio
P0734 Gear no. 4 Incorrect Ratio
P0735 Gear no. 5 Incorrect Ratio
P0736 Reverse Incorrect Ratio
P0740 Torque Converter Clutch Circuit Malfunction
P0741 Torque Converter Clutch Circuit Performance or Stuck Off
P0742 Torque Converter Clutch Circuit Stuck On
P0743 Torque Converter Clutch Circuit Electrical
P0744 Torque Converter Clutch Circuit Intermittent
P0745 Pressure Control Solenoid Malfunction
P0746 Pressure Control Solenoid Performance or Stuck Off
P0747 Pressure Control Solenoid Stuck On
P0748 Pressure Control Solenoid Electrical
P0749 Pressure Control Solenoid Intermittent
P0750 Shift Solenoid "A" Malfunction
P0751 Shift Solenoid "A" Performance or Stuck Off
P0752 Shift Solenoid "A" Stuck On
P0753 Shift Solenoid "A" Electrical
P0754 Shift Solenoid "A" Intermittent
P0755 Shift Solenoid "B" Malfunction
P0756 Shift Solenoid "B" Performance or Stuck Off
P0757 Shift Solenoid "B" Stuck On
P0758 Shift Solenoid "B" Electrical
P0759 Shift Solenoid "B" Intermittent
P0760 Shift Solenoid "C" Malfunction
P0761 Shift Solenoid "C" Performance Or Stuck Off
P0762 Shift Solenoid "C" Stuck On
P0763 Shift Solenoid "C" Electrical
P0764 Shift Solenoid "C" Intermittent
P0765 Shift Solenoid "D" Malfunction
P0766 Shift Solenoid "D" Performance Or Stuck Off
P0767 Shift Solenoid "D" Stuck On
P0768 Shift Solenoid "D" Electrical
P0769 Shift Solenoid "D" Intermittent
P0770 Shift Solenoid "E" Malfunction
P0771 Shift Solenoid "E" Performance Or Stuck Off
P0772 Shift Solenoid "E" Stuck On
P0773 Shift Solenoid "E" Electrical
P0774 Shift Solenoid "E" Intermittent
P0780 Shift Malfunction
P0781 1–2 Shift Malfunction
P0782 2–3 Shift Malfunction
P0783 3–4 Shift Malfunction
P0784 4–5 Shift Malfunction
P0785 Shift/Timing Solenoid Malfunction
P0786 Shift/Timing Solenoid Range/Performance
P0787 Shift/Timing Solenoid Low
P0788 Shift/Timing Solenoid High
P0789 Shift/Timing Solenoid Intermittent
P0790 Normal/Performance Switch Circuit Malfunction
P0801 Reverse Inhibit Control Circuit Malfunction
P0803 1–4 Upshift (Skip Shift) Solenoid Control Circuit Malfunction
P0804 1–4 Upshift (Skip Shift) Lamp Control Circuit Malfunction
P1120 Secondary Throttle Position Sensor Circuit Fault
P1125 Tandem Throttle Position Sensor Circuit Fault

P1210 Traction Control System Signal Fault
P1220 Fuel Pump Control Module Fault
P1320 Ignition Control Signal Fault
P1336 Crankshaft Position Sensor Circuit Fault
P1400 EGR/EVAP Control Solenoid Circuit Fault
P1401 EGR Temperature Sensor Circuit Fault
P1443 EVAP Canister Control Vacuum Switch Circuit Fault
P1445 EVAP Purge Volume Control Valve Circuit Fault
P1605 TCM A~T Diagnosis Communication Line Fault
P1705 Throttle Position Sensor (Switch) Circuit Fault
P1760 Overrun Clutch Solenoid Valve Circuit Fault
P1900 Cooling Fan Control Circuit Fault

OBD II TROUBLE CODE EQUIVALENTS

If a scan tool is not available for code retrieval, the following codes may be retrieved without one.

0505 No Self Diagnostic Failure Indicated
0102 Mass or Volume Air Flow Circuit Malfunction
0401 Intake Air Temperature Circuit Malfunction
0103 Engine Coolant Temperature Circuit Malfunction
0403 Throttle/Pedal Position Sensor/Switch "A" Circuit Malfunction
0908 Insufficient Coolant Temperature For Closed Loop Fuel Control
0303 O2 Circuit Malfunction
0307 Closed Loop Control
0901 O2 Sensor Heater Circuit Malfunction (Bank no. 1 Sensor no. 1)
0707 O2 Sensor Circuit Malfunction (Bank no. 1 Sensor no. 2)
0902 O2 Sensor Heater Circuit Malfunction (Bank no. 1 Sensor no. 2)
0115 System Too Lean (Bank no. 1)
0114 System Too Rich (Bank no. 1)
0701 Random/Multiple Cylinder Misfire Detected
0608 Cylinder no. 1—Misfire Detected
0607 Cylinder no. 2—Misfire Detected
0606 Cylinder no. 3—Misfire Detected
0605 Cylinder no. 4—Misfire Detected
0304 Knock Sensor no. 1—Circuit Malfunction (Bank no. 1 or Single Sensor)
0802 Crankshaft Position Sensor "A" Circuit Malfunction
0101 Camshaft Position Sensor Circuit Malfunction
0302 Exhaust Gas Recirculation Flow Malfunction
0306 Exhaust Gas Recirculation Flow Excessive Detected
0702 Catalyst System Efficiency Below Threshold (Bank no. 1)
0104 Vehicle Speed Sensor Malfunction
0205 Idle Control System Malfunction
0301 Internal Control Module Read Only Memory (ROM) Error
1003 Transmission Range Sensor Circuit Malfunction (PRNDL Input)
1101 Inhibitor Switch Circuit
1208 Transmission Fluid Temperature Sensor Circuit Malfunction
1102 Output Speed Sensor Circuit Malfunction
1207 Engine Speed Input Circuit Malfunction
1103 Gear no. 1 Incorrect Ratio
1104 Gear no. 2 Incorrect Ratio
1105 Gear no. 3 Incorrect Ratio
1106 Gear no. 4 Incorrect Ratio
1204 Torque Converter Clutch Circuit Malfunction
1205 Pressure Control Solenoid Malfunction

1108 Shift Solenoid "A" Malfunction
1201 Shift Solenoid "B" Malfunction
0201 Ignition Control Signal Fault
0905 Crankshaft Position Sensor Circuit Fault
1005 EGR/EVAP Control Solenoid Circuit Fault
0305 EGR Temperature Sensor Circuit Fault
0804 TCM A~T Diagnosis Communication Line Fault
1206 Throttle Position Sensor (Switch) Circuit Fault
1203 Overrun Clutch Solenoid Valve Circuit Fault
1308 Cooling Fan Control Circuit Fault

Subaru

READING CODES

Reading the control module memory is one of the first steps in OBD II system diagnostics. This step should be initially performed to determine the general nature of the fault. Subsequent readings will determine if the fault has been cleared.

Reading codes can be performed by any of the methods below:
• Read the control module memory with the Generic Scan Tool (GST)
• Read the control module memory with the vehicle manufacturer's specific tester

To read the fault codes, connect the scan tool or tester according to the manufacturer's instructions. Follow the manufacturer's specified procedure for reading the codes.

CLEARING CODES

Control module reset procedures are a very important part of OBD II System diagnostics. This step should be done at the end of any fault code repair and at the end of any driveability repair.

Clearing codes can be performed by any of the methods below:
• Clear the control module memory with the Generic Scan Tool (GST)
• Clear the control module memory with the vehicle manufacturer's specific tester
• Turn the ignition off and remove the negative battery cable for at least 1 minute.

Removing the negative battery cable may cause other systems in the vehicle to loose their memory. Prior to removing the cable, ensure you have the proper reset codes for radios and alarms.

➡**The MIL will may also be de-activated for some codes if the vehicle completes three consecutive trips without a fault detected with vehicle conditions similar to those present during the fault.**

OBD II TROUBLE CODES

P0100 Mass or Volume Air Flow Circuit Malfunction
P0101 Mass or Volume Air Flow Circuit Range/Performance Problem
P0102 Mass or Volume Air Flow Circuit Low Input
P0103 Mass or Volume Air Flow Circuit High Input
P0104 Mass or Volume Air Flow Circuit Intermittent
P0105 Manifold Absolute Pressure/Barometric Pressure Circuit Malfunction

P0106 Manifold Absolute Pressure/Barometric Pressure Circuit Range/Performance Problem
P0107 Manifold Absolute Pressure/Barometric Pressure Circuit Low Input
P0108 Manifold Absolute Pressure/Barometric Pressure Circuit High Input
P0109 Manifold Absolute Pressure/Barometric Pressure Circuit Intermittent
P0110 Intake Air Temperature Circuit Malfunction
P0111 Intake Air Temperature Circuit Range/Performance Problem
P0112 Intake Air Temperature Circuit Low Input
P0113 Intake Air Temperature Circuit High Input
P0114 Intake Air Temperature Circuit Intermittent
P0115 Engine Coolant Temperature Circuit Malfunction
P0116 Engine Coolant Temperature Circuit Range/Performance Problem
P0117 Engine Coolant Temperature Circuit Low Input
P0118 Engine Coolant Temperature Circuit High Input
P0119 Engine Coolant Temperature Circuit Intermittent
P0120 Throttle/Pedal Position Sensor/Switch "A" Circuit Malfunction
P0121 Throttle/Pedal Position Sensor/Switch "A" Circuit Range/Performance Problem
P0122 Throttle/Pedal Position Sensor/Switch "A" Circuit Low Input
P0123 Throttle/Pedal Position Sensor/Switch "A" Circuit High Input
P0124 Throttle/Pedal Position Sensor/Switch "A" Circuit Intermittent
P0125 Insufficient Coolant Temperature For Closed Loop Fuel Control
P0126 Insufficient Coolant Temperature For Stable Operation
P0130 O2 Circuit Malfunction (Bank no. 1 Sensor no. 1)
P0131 O2 Sensor Circuit Low Voltage (Bank no. 1 Sensor no. 1)
P0132 O2 Sensor Circuit High Voltage (Bank no. 1 Sensor no. 1)
P0133 O2 Sensor Circuit Slow Response (Bank no. 1 Sensor no. 1)
P0134 O2 Sensor Circuit No Activity Detected (Bank no. 1 Sensor no. 1)
P0135 O2 Sensor Heater Circuit Malfunction (Bank no. 1 Sensor no. 1)
P0136 O2 Sensor Circuit Malfunction (Bank no. 1 Sensor no. 2)
P0137 O2 Sensor Circuit Low Voltage (Bank no. 1 Sensor no. 2)
P0138 O2 Sensor Circuit High Voltage (Bank no. 1 Sensor no. 2)
P0139 O2 Sensor Circuit Slow Response (Bank no. 1 Sensor no. 2)
P0140 O2 Sensor Circuit No Activity Detected (Bank no. 1 Sensor no. 2)
P0141 O2 Sensor Heater Circuit Malfunction (Bank no. 1 Sensor no. 2)
P0142 O2 Sensor Circuit Malfunction (Bank no. 1 Sensor no. 3)
P0143 O2 Sensor Circuit Low Voltage (Bank no. 1 Sensor no. 3)
P0144 O2 Sensor Circuit High Voltage (Bank no. 1 Sensor no. 3)
P0145 O2 Sensor Circuit Slow Response (Bank no. 1 Sensor no. 3)
P0146 O2 Sensor Circuit No Activity Detected (Bank no. 1 Sensor no. 3)

P0147 O2 Sensor Heater Circuit Malfunction (Bank no. 1 Sensor no. 3)
P0150 O2 Sensor Circuit Malfunction (Bank no. 2 Sensor no. 1)
P0151 O2 Sensor Circuit Low Voltage (Bank no. 2 Sensor no. 1)
P0152 O2 Sensor Circuit High Voltage (Bank no. 2 Sensor no. 1)
P0153 O2 Sensor Circuit Slow Response (Bank no. 2 Sensor no. 1)
P0154 O2 Sensor Circuit No Activity Detected (Bank no. 2 Sensor no. 1)
P0155 O2 Sensor Heater Circuit Malfunction (Bank no. 2 Sensor no. 1)
P0156 O2 Sensor Circuit Malfunction (Bank no. 2 Sensor no. 2)
P0157 O2 Sensor Circuit Low Voltage (Bank no. 2 Sensor no. 2)
P0158 O2 Sensor Circuit High Voltage (Bank no. 2 Sensor no. 2)
P0159 O2 Sensor Circuit Slow Response (Bank no. 2 Sensor no. 2)
P0160 O2 Sensor Circuit No Activity Detected (Bank no. 2 Sensor no. 2)
P0161 O2 Sensor Heater Circuit Malfunction (Bank no. 2 Sensor no. 2)
P0162 O2 Sensor Circuit Malfunction (Bank no. 2 Sensor no. 3)
P0163 O2 Sensor Circuit Low Voltage (Bank no. 2 Sensor no. 3)
P0164 O2 Sensor Circuit High Voltage (Bank no. 2 Sensor no. 3)
P0165 O2 Sensor Circuit Slow Response (Bank no. 2 Sensor no. 3)
P0166 O2 Sensor Circuit No Activity Detected (Bank no. 2 Sensor no. 3)
P0167 O2 Sensor Heater Circuit Malfunction (Bank no. 2 Sensor no. 3)
P0170 Fuel Trim Malfunction (Bank no. 1)
P0171 System Too Lean (Bank no. 1)
P0172 System Too Rich (Bank no. 1)
P0173 Fuel Trim Malfunction (Bank no. 2)
P0174 System Too Lean (Bank no. 2)
P0175 System Too Rich (Bank no. 2)
P0176 Fuel Composition Sensor Circuit Malfunction
P0177 Fuel Composition Sensor Circuit Range/Performance
P0178 Fuel Composition Sensor Circuit Low Input
P0179 Fuel Composition Sensor Circuit High Input
P0180 Fuel Temperature Sensor "A" Circuit Malfunction
P0181 Fuel Temperature Sensor "A" Circuit Range/Performance
P0182 Fuel Temperature Sensor "A" Circuit Low Input
P0183 Fuel Temperature Sensor "A" Circuit High Input
P0184 Fuel Temperature Sensor "A" Circuit Intermittent
P0185 Fuel Temperature Sensor "B" Circuit Malfunction
P0186 Fuel Temperature Sensor "B" Circuit Range/Performance
P0187 Fuel Temperature Sensor "B" Circuit Low Input
P0188 Fuel Temperature Sensor "B" Circuit High Input
P0189 Fuel Temperature Sensor "B" Circuit Intermittent
P0190 Fuel Rail Pressure Sensor Circuit Malfunction
P0191 Fuel Rail Pressure Sensor Circuit Range/Performance
P0192 Fuel Rail Pressure Sensor Circuit Low Input
P0193 Fuel Rail Pressure Sensor Circuit High Input
P0194 Fuel Rail Pressure Sensor Circuit Intermittent
P0195 Engine Oil Temperature Sensor Malfunction
P0196 Engine Oil Temperature Sensor Range/Performance
P0197 Engine Oil Temperature Sensor Low
P0198 Engine Oil Temperature Sensor High
P0199 Engine Oil Temperature Sensor Intermittent

P0200 Injector Circuit Malfunction
P0201 Injector Circuit Malfunction—Cylinder no. 1
P0202 Injector Circuit Malfunction—Cylinder no. 2
P0203 Injector Circuit Malfunction—Cylinder no. 3
P0204 Injector Circuit Malfunction—Cylinder no. 4
P0205 Injector Circuit Malfunction—Cylinder no. 5
P0206 Injector Circuit Malfunction—Cylinder no. 6
P0207 Injector Circuit Malfunction—Cylinder no. 7
P0208 Injector Circuit Malfunction—Cylinder no. 8
P0209 Injector Circuit Malfunction—Cylinder no. 9
P0210 Injector Circuit Malfunction—Cylinder no. 10
P0211 Injector Circuit Malfunction—Cylinder no. 11
P0212 Injector Circuit Malfunction—Cylinder no. 12
P0213 Cold Start Injector no. 1 Malfunction
P0214 Cold Start Injector no. 2 Malfunction
P0215 Engine Shutoff Solenoid Malfunction
P0216 Injection Timing Control Circuit Malfunction
P0217 Engine Over Temperature Condition
P0218 Transmission Over Temperature Condition
P0219 Engine Over Speed Condition
P0220 Throttle/Pedal Position Sensor/Switch "B" Circuit Malfunction
P0221 Throttle/Pedal Position Sensor/Switch "B" Circuit Range/Performance Problem
P0222 Throttle/Pedal Position Sensor/Switch "B" Circuit Low Input
P0223 Throttle/Pedal Position Sensor/Switch "B" Circuit High Input
P0224 Throttle/Pedal Position Sensor/Switch "B" Circuit Intermittent
P0225 Throttle/Pedal Position Sensor/Switch "C" Circuit Malfunction
P0226 Throttle/Pedal Position Sensor/Switch "C" Circuit Range/Performance Problem
P0227 Throttle/Pedal Position Sensor/Switch "C" Circuit Low Input
P0228 Throttle/Pedal Position Sensor/Switch "C" Circuit High Input
P0229 Throttle/Pedal Position Sensor/Switch "C" Circuit Intermittent
P0230 Fuel Pump Primary Circuit Malfunction
P0231 Fuel Pump Secondary Circuit Low
P0232 Fuel Pump Secondary Circuit High
P0233 Fuel Pump Secondary Circuit Intermittent
P0261 Cylinder no. 1 Injector Circuit Low
P0262 Cylinder no. 1 Injector Circuit High
P0263 Cylinder no. 1 Contribution/Balance Fault
P0264 Cylinder no. 2 Injector Circuit Low
P0265 Cylinder no. 2 Injector Circuit High
P0266 Cylinder no. 2 Contribution/Balance Fault
P0267 Cylinder no. 3 Injector Circuit Low
P0268 Cylinder no. 3 Injector Circuit High
P0269 Cylinder no. 3 Contribution/Balance Fault
P0270 Cylinder no. 4 Injector Circuit Low
P0271 Cylinder no. 4 Injector Circuit High
P0272 Cylinder no. 4 Contribution/Balance Fault
P0273 Cylinder no. 5 Injector Circuit Low
P0274 Cylinder no. 5 Injector Circuit High
P0275 Cylinder no. 5 Contribution/Balance Fault
P0276 Cylinder no. 6 Injector Circuit Low
P0277 Cylinder no. 6 Injector Circuit High
P0278 Cylinder no. 6 Contribution/Balance Fault

P0279 Cylinder no. 7 Injector Circuit Low
P0280 Cylinder no. 7 Injector Circuit High
P0281 Cylinder no. 7 Contribution/Balance Fault
P0282 Cylinder no. 8 Injector Circuit Low
P0283 Cylinder no. 8 Injector Circuit High
P0284 Cylinder no. 8 Contribution/Balance Fault
P0285 Cylinder no. 9 Injector Circuit Low
P0286 Cylinder no. 9 Injector Circuit High
P0287 Cylinder no. 9 Contribution/Balance Fault
P0288 Cylinder no. 10 Injector Circuit Low
P0289 Cylinder no. 10 Injector Circuit High
P0290 Cylinder no. 10 Contribution/Balance Fault
P0291 Cylinder no. 11 Injector Circuit Low
P0292 Cylinder no. 11 Injector Circuit High
P0293 Cylinder no. 11 Contribution/Balance Fault
P0294 Cylinder no. 12 Injector Circuit Low
P0295 Cylinder no. 12 Injector Circuit High
P0296 Cylinder no. 12 Contribution/Balance Fault
P0300 Random/Multiple Cylinder Misfire Detected
P0301 Cylinder no. 1—Misfire Detected
P0302 Cylinder no. 2—Misfire Detected
P0303 Cylinder no. 3—Misfire Detected
P0304 Cylinder no. 4—Misfire Detected
P0305 Cylinder no. 5—Misfire Detected
P0306 Cylinder no. 6—Misfire Detected
P0307 Cylinder no. 7—Misfire Detected
P0308 Cylinder no. 8—Misfire Detected
P0309 Cylinder no. 9—Misfire Detected
P0310 Cylinder no. 10—Misfire Detected
P0311 Cylinder no. 11—Misfire Detected
P0312 Cylinder no. 12—Misfire Detected
P0320 Ignition/Distributor Engine Speed Input Circuit Malfunction
P0321 Ignition/Distributor Engine Speed Input Circuit Range/Performance
P0322 Ignition/Distributor Engine Speed Input Circuit No Signal
P0323 Ignition/Distributor Engine Speed Input Circuit Intermittent
P0325 Knock Sensor no. 1—Circuit Malfunction (Bank no. 1 or Single Sensor)
P0326 Knock Sensor no. 1—Circuit Range/Performance (Bank no. 1 or Single Sensor)
P0327 Knock Sensor no. 1—Circuit Low Input (Bank no. 1 or Single Sensor)
P0328 Knock Sensor no. 1—Circuit High Input (Bank no. 1 or Single Sensor)
P0329 Knock Sensor no. 1—Circuit Input Intermittent (Bank no. 1 or Single Sensor)
P0330 Knock Sensor no. 2—Circuit Malfunction (Bank no. 2)
P0331 Knock Sensor no. 2—Circuit Range/Performance (Bank no. 2)
P0332 Knock Sensor no. 2—Circuit Low Input (Bank no. 2)
P0333 Knock Sensor no. 2—Circuit High Input (Bank no. 2)
P0334 Knock Sensor no. 2—Circuit Input Intermittent (Bank no. 2)
P0335 Crankshaft Position Sensor "A" Circuit Malfunction
P0336 Crankshaft Position Sensor "A" Circuit Range/Performance
P0337 Crankshaft Position Sensor "A" Circuit Low Input
P0338 Crankshaft Position Sensor "A" Circuit High Input
P0339 Crankshaft Position Sensor "A" Circuit Intermittent
P0340 Camshaft Position Sensor Circuit Malfunction

P0341 Camshaft Position Sensor Circuit Range/Performance
P0342 Camshaft Position Sensor Circuit Low Input
P0343 Camshaft Position Sensor Circuit High Input
P0344 Camshaft Position Sensor Circuit Intermittent
P0350 Ignition Coil Primary/Secondary Circuit Malfunction
P0351 Ignition Coil "A" Primary/Secondary Circuit Malfunction
P0352 Ignition Coil "B" Primary/Secondary Circuit Malfunction
P0353 Ignition Coil "C" Primary/Secondary Circuit Malfunction
P0354 Ignition Coil "D" Primary/Secondary Circuit Malfunction
P0355 Ignition Coil "E" Primary/Secondary Circuit Malfunction
P0356 Ignition Coil "F" Primary/Secondary Circuit Malfunction
P0357 Ignition Coil "G" Primary/Secondary Circuit Malfunction
P0358 Ignition Coil "H" Primary/Secondary Circuit Malfunction
P0359 Ignition Coil "I" Primary/Secondary Circuit Malfunction
P0360 Ignition Coil "J" Primary/Secondary Circuit Malfunction
P0361 Ignition Coil "K" Primary/Secondary Circuit Malfunction
P0362 Ignition Coil "L" Primary/Secondary Circuit Malfunction
P0370 Timing Reference High Resolution Signal "A" Malfunction
P0371 Timing Reference High Resolution Signal "A" Too Many Pulses
P0372 Timing Reference High Resolution Signal "A" Too Few Pulses
P0373 Timing Reference High Resolution Signal "A" Intermittent/Erratic Pulses
P0374 Timing Reference High Resolution Signal "A" No Pulses
P0375 Timing Reference High Resolution Signal "B" Malfunction
P0376 Timing Reference High Resolution Signal "B" Too Many Pulses
P0377 Timing Reference High Resolution Signal "B" Too Few Pulses
P0378 Timing Reference High Resolution Signal "B" Intermittent/Erratic Pulses
P0379 Timing Reference High Resolution Signal "B" No Pulses
P0380 Glow Plug/Heater Circuit "A" Malfunction
P0381 Glow Plug/Heater Indicator Circuit Malfunction
P0382 Glow Plug/Heater Circuit "B" Malfunction
P0385 Crankshaft Position Sensor "B" Circuit Malfunction
P0386 Crankshaft Position Sensor "B" Circuit Range/Performance
P0387 Crankshaft Position Sensor "B" Circuit Low Input
P0388 Crankshaft Position Sensor "B" Circuit High Input
P0389 Crankshaft Position Sensor "B" Circuit Intermittent
P0400 Exhaust Gas Recirculation Flow Malfunction
P0401 Exhaust Gas Recirculation Flow Insufficient Detected
P0402 Exhaust Gas Recirculation Flow Excessive Detected
P0403 Exhaust Gas Recirculation Circuit Malfunction
P0404 Exhaust Gas Recirculation Circuit Range/Performance
P0405 Exhaust Gas Recirculation Sensor "A" Circuit Low
P0406 Exhaust Gas Recirculation Sensor "A" Circuit High
P0407 Exhaust Gas Recirculation Sensor "B" Circuit Low
P0408 Exhaust Gas Recirculation Sensor "B" Circuit High
P0410 Secondary Air Injection System Malfunction
P0411 Secondary Air Injection System Incorrect Flow Detected
P0412 Secondary Air Injection System Switching Valve "A" Circuit Malfunction
P0413 Secondary Air Injection System Switching Valve "A" Circuit Open
P0414 Secondary Air Injection System Switching Valve "A" Circuit Shorted
P0415 Secondary Air Injection System Switching Valve "B" Circuit Malfunction

P0416 Secondary Air Injection System Switching Valve "B" Circuit Open

P0417 Secondary Air Injection System Switching Valve "B" Circuit Shorted

P0418 Secondary Air Injection System Relay "A" Circuit Malfunction

P0419 Secondary Air Injection System Relay "B" Circuit Malfunction

P0420 Catalyst System Efficiency Below Threshold (Bank no. 1)

P0421 Warm Up Catalyst Efficiency Below Threshold (Bank no. 1)

P0422 Main Catalyst Efficiency Below Threshold (Bank no. 1)

P0423 Heated Catalyst Efficiency Below Threshold (Bank no. 1)

P0424 Heated Catalyst Temperature Below Threshold (Bank no. 1)

P0430 Catalyst System Efficiency Below Threshold (Bank no. 2)

P0431 Warm Up Catalyst Efficiency Below Threshold (Bank no. 2)

P0432 Main Catalyst Efficiency Below Threshold (Bank no. 2)

P0433 Heated Catalyst Efficiency Below Threshold (Bank no. 2)

P0434 Heated Catalyst Temperature Below Threshold (Bank no. 2)

P0440 Evaporative Emission Control System Malfunction

P0441 Evaporative Emission Control System Incorrect Purge Flow

P0442 Evaporative Emission Control System Leak Detected (Small Leak)

P0443 Evaporative Emission Control System Purge Control Valve Circuit Malfunction

P0444 Evaporative Emission Control System Purge Control Valve Circuit Open

P0445 Evaporative Emission Control System Purge Control Valve Circuit Shorted

P0446 Evaporative Emission Control System Vent Control Circuit Malfunction

P0447 Evaporative Emission Control System Vent Control Circuit Open

P0448 Evaporative Emission Control System Vent Control Circuit Shorted

P0449 Evaporative Emission Control System Vent Valve/Solenoid Circuit Malfunction

P0450 Evaporative Emission Control System Pressure Sensor Malfunction

P0451 Evaporative Emission Control System Pressure Sensor Range/Performance

P0452 Evaporative Emission Control System Pressure Sensor Low Input

P0453 Evaporative Emission Control System Pressure Sensor High Input

P0454 Evaporative Emission Control System Pressure Sensor Intermittent

P0455 Evaporative Emission Control System Leak Detected (Gross Leak)

P0460 Fuel Level Sensor Circuit Malfunction

P0461 Fuel Level Sensor Circuit Range/Performance

P0462 Fuel Level Sensor Circuit Low Input

P0463 Fuel Level Sensor Circuit High Input

P0464 Fuel Level Sensor Circuit Intermittent

P0465 Purge Flow Sensor Circuit Malfunction

P0466 Purge Flow Sensor Circuit Range/Performance

P0467 Purge Flow Sensor Circuit Low Input

P0468 Purge Flow Sensor Circuit High Input

P0469 Purge Flow Sensor Circuit Intermittent

P0470 Exhaust Pressure Sensor Malfunction

P0471 Exhaust Pressure Sensor Range/Performance

P0472 Exhaust Pressure Sensor Low

P0473 Exhaust Pressure Sensor High

P0474 Exhaust Pressure Sensor Intermittent

P0475 Exhaust Pressure Control Valve Malfunction

P0476 Exhaust Pressure Control Valve Range/Performance

P0477 Exhaust Pressure Control Valve Low

P0478 Exhaust Pressure Control Valve High

P0479 Exhaust Pressure Control Valve Intermittent

P0480 Cooling Fan no. 1 Control Circuit Malfunction

P0481 Cooling Fan no. 2 Control Circuit Malfunction

P0482 Cooling Fan no. 3 Control Circuit Malfunction

P0483 Cooling Fan Rationality Check Malfunction

P0484 Cooling Fan Circuit Over Current

P0485 Cooling Fan Power/Ground Circuit Malfunction

P0500 Vehicle Speed Sensor Malfunction

P0501 Vehicle Speed Sensor Range/Performance

P0502 Vehicle Speed Sensor Circuit Low Input

P0503 Vehicle Speed Sensor Intermittent/Erratic/High

P0505 Idle Control System Malfunction

P0506 Idle Control System RPM Lower Than Expected

P0507 Idle Control System RPM Higher Than Expected

P0510 Closed Throttle Position Switch Malfunction

P0520 Engine Oil Pressure Sensor/Switch Circuit Malfunction

P0521 Engine Oil Pressure Sensor/Switch Range/Performance

P0522 Engine Oil Pressure Sensor/Switch Low Voltage

P0523 Engine Oil Pressure Sensor/Switch High Voltage

P0530 A/C Refrigerant Pressure Sensor Circuit Malfunction

P0531 A/C Refrigerant Pressure Sensor Circuit Range/Performance

P0532 A/C Refrigerant Pressure Sensor Circuit Low Input

P0533 A/C Refrigerant Pressure Sensor Circuit High Input

P0534 A/C Refrigerant Charge Loss

P0550 Power Steering Pressure Sensor Circuit Malfunction

P0551 Power Steering Pressure Sensor Circuit Range/Performance

P0552 Power Steering Pressure Sensor Circuit Low Input

P0553 Power Steering Pressure Sensor Circuit High Input

P0554 Power Steering Pressure Sensor Circuit Intermittent

P0560 System Voltage Malfunction

P0561 System Voltage Unstable

P0562 System Voltage Low

P0563 System Voltage High

P0565 Cruise Control On Signal Malfunction

P0566 Cruise Control Off Signal Malfunction

P0567 Cruise Control Resume Signal Malfunction

P0568 Cruise Control Set Signal Malfunction

P0569 Cruise Control Coast Signal Malfunction

P0570 Cruise Control Accel Signal Malfunction

P0571 Cruise Control/Brake Switch "A" Circuit Malfunction

P0572 Cruise Control/Brake Switch "A" Circuit Low

P0573 Cruise Control/Brake Switch "A" Circuit High

P0574 Through P0580 Reserved for Cruise Codes

P0600 Serial Communication Link Malfunction

P0601 Internal Control Module Memory Check Sum Error

P0602 Control Module Programming Error

P0603 Internal Control Module Keep Alive Memory (KAM) Error

P0604 Internal Control Module Random Access Memory (RAM) Error

P0605 Internal Control Module Read Only Memory (ROM) Error

P0606 PCM Processor Fault

P0608 Control Module VSS Output "A" Malfunction
P0609 Control Module VSS Output "B" Malfunction
P0620 Generator Control Circuit Malfunction
P0621 Generator Lamp "L" Control Circuit Malfunction
P0622 Generator Field "F" Control Circuit Malfunction
P0650 Malfunction Indicator Lamp (MIL) Control Circuit Malfunction
P0654 Engine RPM Output Circuit Malfunction
P0655 Engine Hot Lamp Output Control Circuit Malfunction
P0656 Fuel Level Output Circuit Malfunction
P0700 Transmission Control System Malfunction
P0701 Transmission Control System Range/Performance
P0702 Transmission Control System Electrical
P0703 Torque Converter/Brake Switch "B" Circuit Malfunction
P0704 Clutch Switch Input Circuit Malfunction
P0705 Transmission Range Sensor Circuit Malfunction (PRNDL Input)
P0706 Transmission Range Sensor Circuit Range/Performance
P0707 Transmission Range Sensor Circuit Low Input
P0708 Transmission Range Sensor Circuit High Input
P0709 Transmission Range Sensor Circuit Intermittent
P0710 Transmission Fluid Temperature Sensor Circuit Malfunction
P0711 Transmission Fluid Temperature Sensor Circuit Range/Performance
P0712 Transmission Fluid Temperature Sensor Circuit Low Input
P0713 Transmission Fluid Temperature Sensor Circuit High Input
P0714 Transmission Fluid Temperature Sensor Circuit Intermittent ·
P0715 Input/Turbine Speed Sensor Circuit Malfunction
P0716 Input/Turbine Speed Sensor Circuit Range/Performance
P0717 Input/Turbine Speed Sensor Circuit No Signal
P0718 Input/Turbine Speed Sensor Circuit Intermittent
P0719 Torque Converter/Brake Switch "B" Circuit Low
P0720 Output Speed Sensor Circuit Malfunction
P0721 Output Speed Sensor Circuit Range/Performance
P0722 Output Speed Sensor Circuit No Signal
P0723 Output Speed Sensor Circuit Intermittent
P0724 Torque Converter/Brake Switch "B" Circuit High
P0725 Engine Speed Input Circuit Malfunction
P0726 Engine Speed Input Circuit Range/Performance
P0727 Engine Speed Input Circuit No Signal
P0728 Engine Speed Input Circuit Intermittent
P0730 Incorrect Gear Ratio
P0731 Gear no. 1 Incorrect Ratio
P0732 Gear no. 2 Incorrect Ratio
P0733 Gear no. 3 Incorrect Ratio
P0734 Gear no. 4 Incorrect Ratio
P0735 Gear no. 5 Incorrect Ratio
P0736 Reverse Incorrect Ratio
P0740 Torque Converter Clutch Circuit Malfunction
P0741 Torque Converter Clutch Circuit Performance or Stuck Off
P0742 Torque Converter Clutch Circuit Stuck On
P0743 Torque Converter Clutch Circuit Electrical
P0744 Torque Converter Clutch Circuit Intermittent
P0745 Pressure Control Solenoid Malfunction
P0746 Pressure Control Solenoid Performance or Stuck Off
P0747 Pressure Control Solenoid Stuck On
P0748 Pressure Control Solenoid Electrical

P0749 Pressure Control Solenoid Intermittent
P0750 Shift Solenoid "A" Malfunction
P0751 Shift Solenoid "A" Performance or Stuck Off
P0752 Shift Solenoid "A" Stuck On
P0753 Shift Solenoid "A" Electrical
P0754 Shift Solenoid "A" Intermittent
P0755 Shift Solenoid "B" Malfunction
P0756 Shift Solenoid "B" Performance or Stuck Off
P0757 Shift Solenoid "B" Stuck On
P0758 Shift Solenoid "B" Electrical
P0759 Shift Solenoid "B" Intermittent
P0760 Shift Solenoid "C" Malfunction
P0761 Shift Solenoid "C" Performance Or Stuck Off
P0762 Shift Solenoid "C" Stuck On
P0763 Shift Solenoid "C" Electrical
P0764 Shift Solenoid "C" Intermittent
P0765 Shift Solenoid "D" Malfunction
P0766 Shift Solenoid "D" Performance Or Stuck Off
P0767 Shift Solenoid "D" Stuck On
P0768 Shift Solenoid "D" Electrical
P0769 Shift Solenoid "D" Intermittent
P0770 Shift Solenoid "E" Malfunction
P0771 Shift Solenoid "E" Performance Or Stuck Off
P0772 Shift Solenoid "E" Stuck On
P0773 Shift Solenoid "E" Electrical
P0774 Shift Solenoid "E" Intermittent
P0780 Shift Malfunction
P0781 1–2 Shift Malfunction
P0782 2–3 Shift Malfunction
P0783 3–4 Shift Malfunction
P0784 4–5 Shift Malfunction
P0785 Shift/Timing Solenoid Malfunction
P0786 Shift/Timing Solenoid Range/Performance
P0787 Shift/Timing Solenoid Low
P0788 Shift/Timing Solenoid High
P0789 Shift/Timing Solenoid Intermittent
P0790 Normal/Performance Switch Circuit Malfunction
P0801 Reverse Inhibit Control Circuit Malfunction
P0803 1–4 Upshift (Skip Shift) Solenoid Control Circuit Malfunction
P0804 1–4 Upshift (Skip Shift) Lamp Control Circuit Malfunction
P1100 Starter Switch Circuit Fault
P1101 Neutral Position Switch Circuit Fault (MT)

Suzuki

READING CODES

Reading the control module memory is one of the first steps in OBD II system diagnostics. This step should be initially performed to determine the general nature of the fault. Subsequent readings will determine if the fault has been cleared.

Reading codes can be performed by any of the methods below:

• Read the control module memory with the Generic Scan Tool (GST)

• Read the control module memory with the vehicle manufacturer's specific tester

To read the fault codes, connect the scan tool or tester according to the manufacturer's instructions. Follow the manufacturer's specified procedure for reading the codes.

CLEARING CODES

Control module reset procedures are a very important part of OBD II System diagnostics. This step should be done at the end of any fault code repair and at the end of any driveability repair.

Clearing codes can be performed by any of the methods below:

• Clear the control module memory with the Generic Scan Tool (GST)

• Clear the control module memory with the vehicle manufacturer's specific tester

• Turn the ignition off and remove the negative battery cable for at least 1 minute.

Removing the negative battery cable may cause other systems in the vehicle to loose their memory. Prior to removing the cable, ensure you have the proper reset codes for radios and alarms.

➡**The MIL will may also be de-activated for some codes if the vehicle completes three consecutive trips without a fault detected with vehicle conditions similar to those present during the fault.**

OBD II TROUBLE CODES

P0100 Mass or Volume Air Flow Circuit Malfunction
P0101 Mass or Volume Air Flow Circuit Range/Performance Problem
P0102 Mass or Volume Air Flow Circuit Low Input
P0103 Mass or Volume Air Flow Circuit High Input
P0104 Mass or Volume Air Flow Circuit Intermittent
P0105 Manifold Absolute Pressure/Barometric Pressure Circuit Malfunction
P0106 Manifold Absolute Pressure/Barometric Pressure Circuit Range/Performance Problem
P0107 Manifold Absolute Pressure/Barometric Pressure Circuit Low Input
P0108 Manifold Absolute Pressure/Barometric Pressure Circuit High Input
P0109 Manifold Absolute Pressure/Barometric Pressure Circuit Intermittent
P0110 Intake Air Temperature Circuit Malfunction
P0111 Intake Air Temperature Circuit Range/Performance Problem
P0112 Intake Air Temperature Circuit Low Input
P0113 Intake Air Temperature Circuit High Input
P0114 Intake Air Temperature Circuit Intermittent
P0115 Engine Coolant Temperature Circuit Malfunction
P0116 Engine Coolant Temperature Circuit Range/Performance Problem
P0117 Engine Coolant Temperature Circuit Low Input
P0118 Engine Coolant Temperature Circuit High Input
P0119 Engine Coolant Temperature Circuit Intermittent
P0120 Throttle/Pedal Position Sensor/Switch "A" Circuit Malfunction
P0121 Throttle/Pedal Position Sensor/Switch "A" Circuit Range/Performance Problem
P0122 Throttle/Pedal Position Sensor/Switch "A" Circuit Low Input
P0123 Throttle/Pedal Position Sensor/Switch "A" Circuit High Input
P0124 Throttle/Pedal Position Sensor/Switch "A" Circuit Intermittent

P0125 Insufficient Coolant Temperature For Closed Loop Fuel Control
P0126 Insufficient Coolant Temperature For Stable Operation
P0130 O2 Circuit Malfunction (Bank no. 1 Sensor no. 1)
P0131 O2 Sensor Circuit Low Voltage (Bank no. 1 Sensor no. 1)
P0132 O2 Sensor Circuit High Voltage (Bank no. 1 Sensor no. 1)
P0133 O2 Sensor Circuit Slow Response (Bank no. 1 Sensor no. 1)
P0134 O2 Sensor Circuit No Activity Detected (Bank no. 1 Sensor no. 1)
P0135 O2 Sensor Heater Circuit Malfunction (Bank no. 1 Sensor no. 1)
P0136 O2 Sensor Circuit Malfunction (Bank no. 1 Sensor no. 2)
P0137 O2 Sensor Circuit Low Voltage (Bank no. 1 Sensor no. 2)
P0138 O2 Sensor Circuit High Voltage (Bank no. 1 Sensor no. 2)
P0139 O2 Sensor Circuit Slow Response (Bank no. 1 Sensor no. 2)
P0140 O2 Sensor Circuit No Activity Detected (Bank no. 1 Sensor no. 2)
P0141 O2 Sensor Heater Circuit Malfunction (Bank no. 1 Sensor no. 2)
P0142 O2 Sensor Circuit Malfunction (Bank no. 1 Sensor no. 3)
P0143 O2 Sensor Circuit Low Voltage (Bank no. 1 Sensor no. 3)
P0144 O2 Sensor Circuit High Voltage (Bank no. 1 Sensor no. 3)
P0145 O2 Sensor Circuit Slow Response (Bank no. 1 Sensor no. 3)
P0146 O2 Sensor Circuit No Activity Detected (Bank no. 1 Sensor no. 3)
P0147 O2 Sensor Heater Circuit Malfunction (Bank no. 1 Sensor no. 3)
P0150 O2 Sensor Circuit Malfunction (Bank no. 2 Sensor no. 1)
P0151 O2 Sensor Circuit Low Voltage (Bank no. 2 Sensor no. 1)
P0152 O2 Sensor Circuit High Voltage (Bank no. 2 Sensor no. 1)
P0153 O2 Sensor Circuit Slow Response (Bank no. 2 Sensor no. 1)
P0154 O2 Sensor Circuit No Activity Detected (Bank no. 2 Sensor no. 1)
P0155 O2 Sensor Heater Circuit Malfunction (Bank no. 2 Sensor no. 1)
P0156 O2 Sensor Circuit Malfunction (Bank no. 2 Sensor no. 2)
P0157 O2 Sensor Circuit Low Voltage (Bank no. 2 Sensor no. 2)
P0158 O2 Sensor Circuit High Voltage (Bank no. 2 Sensor no. 2)
P0159 O2 Sensor Circuit Slow Response (Bank no. 2 Sensor no. 2)
P0160 O2 Sensor Circuit No Activity Detected (Bank no. 2 Sensor no. 2)
P0161 O2 Sensor Heater Circuit Malfunction (Bank no. 2 Sensor no. 2)
P0162 O2 Sensor Circuit Malfunction (Bank no. 2 Sensor no. 3)
P0163 O2 Sensor Circuit Low Voltage (Bank no. 2 Sensor no. 3)

P0164 O2 Sensor Circuit High Voltage (Bank no. 2 Sensor no. 3)

P0165 O2 Sensor Circuit Slow Response (Bank no. 2 Sensor no. 3)

P0166 O2 Sensor Circuit No Activity Detected (Bank no. 2 Sensor no. 3)

P0167 O2 Sensor Heater Circuit Malfunction (Bank no. 2 Sensor no. 3)

P0170 Fuel Trim Malfunction (Bank no. 1)

P0171 System Too Lean (Bank no. 1)

P0172 System Too Rich (Bank no. 1)

P0173 Fuel Trim Malfunction (Bank no. 2)

P0174 System Too Lean (Bank no. 2)

P0175 System Too Rich (Bank no. 2)

P0176 Fuel Composition Sensor Circuit Malfunction

P0177 Fuel Composition Sensor Circuit Range/Performance

P0178 Fuel Composition Sensor Circuit Low Input

P0179 Fuel Composition Sensor Circuit High Input

P0180 Fuel Temperature Sensor "A" Circuit Malfunction

P0181 Fuel Temperature Sensor "A" Circuit Range/Performance

P0182 Fuel Temperature Sensor "A" Circuit Low Input

P0183 Fuel Temperature Sensor "A" Circuit High Input

P0184 Fuel Temperature Sensor "A" Circuit Intermittent

P0185 Fuel Temperature Sensor "B" Circuit Malfunction

P0186 Fuel Temperature Sensor "B" Circuit Range/Performance

P0187 Fuel Temperature Sensor "B" Circuit Low Input

P0188 Fuel Temperature Sensor "B" Circuit High Input

P0189 Fuel Temperature Sensor "B" Circuit Intermittent

P0190 Fuel Rail Pressure Sensor Circuit Malfunction

P0191 Fuel Rail Pressure Sensor Circuit Range/Performance

P0192 Fuel Rail Pressure Sensor Circuit Low Input

P0193 Fuel Rail Pressure Sensor Circuit High Input

P0194 Fuel Rail Pressure Sensor Circuit Intermittent

P0195 Engine Oil Temperature Sensor Malfunction

P0196 Engine Oil Temperature Sensor Range/Performance

P0197 Engine Oil Temperature Sensor Low

P0198 Engine Oil Temperature Sensor High

P0199 Engine Oil Temperature Sensor Intermittent

P0200 Injector Circuit Malfunction

P0201 Injector Circuit Malfunction—Cylinder no. 1

P0202 Injector Circuit Malfunction—Cylinder no. 2

P0203 Injector Circuit Malfunction—Cylinder no. 3

P0204 Injector Circuit Malfunction—Cylinder no. 4

P0205 Injector Circuit Malfunction—Cylinder no. 5

P0206 Injector Circuit Malfunction—Cylinder no. 6

P0207 Injector Circuit Malfunction—Cylinder no. 7

P0208 Injector Circuit Malfunction—Cylinder no. 8

P0209 Injector Circuit Malfunction—Cylinder no. 9

P0210 Injector Circuit Malfunction—Cylinder no. 10

P0211 Injector Circuit Malfunction—Cylinder no. 11

P0212 Injector Circuit Malfunction—Cylinder no. 12

P0213 Cold Start Injector no. 1 Malfunction

P0214 Cold Start Injector no. 2 Malfunction

P0215 Engine Shutoff Solenoid Malfunction

P0216 Injection Timing Control Circuit Malfunction

P0217 Engine Over Temperature Condition

P0218 Transmission Over Temperature Condition

P0219 Engine Over Speed Condition

P0220 Throttle/Pedal Position Sensor/Switch "B" Circuit Malfunction

P0221 Throttle/Pedal Position Sensor/Switch "B" Circuit

Range/Performance Problem

P0222 Throttle/Pedal Position Sensor/Switch "B" Circuit Low Input

P0223 Throttle/Pedal Position Sensor/Switch "B" Circuit High Input

P0224 Throttle/Pedal Position Sensor/Switch "B" Circuit Intermittent

P0225 Throttle/Pedal Position Sensor/Switch "C" Circuit Malfunction

P0226 Throttle/Pedal Position Sensor/Switch "C" Circuit Range/Performance Problem

P0227 Throttle/Pedal Position Sensor/Switch "C" Circuit Low Input

P0228 Throttle/Pedal Position Sensor/Switch "C" Circuit High Input

P0229 Throttle/Pedal Position Sensor/Switch "C" Circuit Intermittent

P0230 Fuel Pump Primary Circuit Malfunction

P0231 Fuel Pump Secondary Circuit Low

P0232 Fuel Pump Secondary Circuit High

P0233 Fuel Pump Secondary Circuit Intermittent

P0234 Engine Over Boost Condition

P0261 Cylinder no. 1 Injector Circuit Low

P0262 Cylinder no. 1 Injector Circuit High

P0263 Cylinder no. 1 Contribution/Balance Fault

P0264 Cylinder no. 2 Injector Circuit Low

P0265 Cylinder no. 2 Injector Circuit High

P0266 Cylinder no. 2 Contribution/Balance Fault

P0267 Cylinder no. 3 Injector Circuit Low

P0268 Cylinder no. 3 Injector Circuit High

P0269 Cylinder no. 3 Contribution/Balance Fault

P0270 Cylinder no. 4 Injector Circuit Low

P0271 Cylinder no. 4 Injector Circuit High

P0272 Cylinder no. 4 Contribution/Balance Fault

P0273 Cylinder no. 5 Injector Circuit Low

P0274 Cylinder no. 5 Injector Circuit High

P0275 Cylinder no. 5 Contribution/Balance Fault

P0276 Cylinder no. 6 Injector Circuit Low

P0277 Cylinder no. 6 Injector Circuit High

P0278 Cylinder no. 6 Contribution/Balance Fault

P0279 Cylinder no. 7 Injector Circuit Low

P0280 Cylinder no. 7 Injector Circuit High

P0281 Cylinder no. 7 Contribution/Balance Fault

P0282 Cylinder no. 8 Injector Circuit Low

P0283 Cylinder no. 8 Injector Circuit High

P0284 Cylinder no. 8 Contribution/Balance Fault

P0285 Cylinder no. 9 Injector Circuit Low

P0286 Cylinder no. 9 Injector Circuit High

P0287 Cylinder no. 9 Contribution/Balance Fault

P0288 Cylinder no. 10 Injector Circuit Low

P0289 Cylinder no. 10 Injector Circuit High

P0290 Cylinder no. 10 Contribution/Balance Fault

P0291 Cylinder no. 11 Injector Circuit Low

P0292 Cylinder no. 11 Injector Circuit High

P0293 Cylinder no. 11 Contribution/Balance Fault

P0294 Cylinder no. 12 Injector Circuit Low

P0295 Cylinder no. 12 Injector Circuit High

P0296 Cylinder no. 12 Contribution/Balance Fault

P0300 Random/Multiple Cylinder Misfire Detected

P0301 Cylinder no. 1—Misfire Detected

P0302 Cylinder no. 2—Misfire Detected

P0303 Cylinder no. 3—Misfire Detected
P0304 Cylinder no. 4—Misfire Detected
P0305 Cylinder no. 5—Misfire Detected
P0306 Cylinder no. 6—Misfire Detected
P0307 Cylinder no. 7—Misfire Detected
P0308 Cylinder no. 8—Misfire Detected
P0309 Cylinder no. 9—Misfire Detected
P0310 Cylinder no. 10—Misfire Detected
P0311 Cylinder no. 11—Misfire Detected
P0312 Cylinder no. 12—Misfire Detected
P0320 Ignition/Distributor Engine Speed Input Circuit Malfunction
P0321 Ignition/Distributor Engine Speed Input Circuit Range/Performance
P0322 Ignition/Distributor Engine Speed Input Circuit No Signal
P0323 Ignition/Distributor Engine Speed Input Circuit Intermittent
P0325 Knock Sensor no. 1—Circuit Malfunction (Bank no. 1 or Single Sensor)
P0326 Knock Sensor no. 1—Circuit Range/Performance (Bank no. 1 or Single Sensor)
P0327 Knock Sensor no. 1—Circuit Low Input (Bank no. 1 or Single Sensor)
P0328 Knock Sensor no. 1—Circuit High Input (Bank no. 1 or Single Sensor)
P0329 Knock Sensor no. 1—Circuit Input Intermittent (Bank no. 1 or Single Sensor)
P0330 Knock Sensor no. 2—Circuit Malfunction (Bank no. 2)
P0331 Knock Sensor no. 2—Circuit Range/Performance (Bank no. 2)
P0332 Knock Sensor no. 2—Circuit Low Input (Bank no. 2)
P0333 Knock Sensor no. 2—Circuit High Input (Bank no. 2)
P0334 Knock Sensor no. 2—Circuit Input Intermittent (Bank no. 2)
P0335 Crankshaft Position Sensor "A" Circuit Malfunction
P0336 Crankshaft Position Sensor "A" Circuit Range/Performance
P0337 Crankshaft Position Sensor "A" Circuit Low Input
P0338 Crankshaft Position Sensor "A" Circuit High Input
P0339 Crankshaft Position Sensor "A" Circuit Intermittent
P0340 Camshaft Position Sensor Circuit Malfunction
P0341 Camshaft Position Sensor Circuit Range/Performance
P0342 Camshaft Position Sensor Circuit Low Input
P0343 Camshaft Position Sensor Circuit High Input
P0344 Camshaft Position Sensor Circuit Intermittent
P0350 Ignition Coil Primary/Secondary Circuit Malfunction
P0351 Ignition Coil "A" Primary/Secondary Circuit Malfunction
P0352 Ignition Coil "B" Primary/Secondary Circuit Malfunction
P0353 Ignition Coil "C" Primary/Secondary Circuit Malfunction
P0354 Ignition Coil "D" Primary/Secondary Circuit Malfunction
P0355 Ignition Coil "E" Primary/Secondary Circuit Malfunction
P0356 Ignition Coil "F" Primary/Secondary Circuit Malfunction
P0357 Ignition Coil "G" Primary/Secondary Circuit Malfunction
P0358 Ignition Coil "H" Primary/Secondary Circuit Malfunction
P0359 Ignition Coil "I" Primary/Secondary Circuit Malfunction
P0360 Ignition Coil "J" Primary/Secondary Circuit Malfunction
P0361 Ignition Coil "K" Primary/Secondary Circuit Malfunction
P0362 Ignition Coil "L" Primary/Secondary Circuit Malfunction
P0370 Timing Reference High Resolution Signal "A" Malfunction
P0371 Timing Reference High Resolution Signal "A" Too Many Pulses

P0372 Timing Reference High Resolution Signal "A" Too Few Pulses
P0373 Timing Reference High Resolution Signal "A" Intermittent/Erratic Pulses
P0374 Timing Reference High Resolution Signal "A" No Pulses
P0375 Timing Reference High Resolution Signal "B" Malfunction
P0376 Timing Reference High Resolution Signal "B" Too Many Pulses
P0377 Timing Reference High Resolution Signal "B" Too Few Pulses
P0378 Timing Reference High Resolution Signal "B" Intermittent/Erratic Pulses
P0379 Timing Reference High Resolution Signal "B" No Pulses
P0380 Glow Plug/Heater Circuit "A" Malfunction
P0381 Glow Plug/Heater Indicator Circuit Malfunction
P0382 Glow Plug/Heater Circuit "B" Malfunction
P0385 Crankshaft Position Sensor "B" Circuit Malfunction
P0386 Crankshaft Position Sensor "B" Circuit Range/Performance
P0387 Crankshaft Position Sensor "B" Circuit Low Input
P0388 Crankshaft Position Sensor "B" Circuit High Input
P0389 Crankshaft Position Sensor "B" Circuit Intermittent
P0400 Exhaust Gas Recirculation Flow Malfunction
P0401 Exhaust Gas Recirculation Flow Insufficient Detected
P0402 Exhaust Gas Recirculation Flow Excessive Detected
P0403 Exhaust Gas Recirculation Circuit Malfunction
P0404 Exhaust Gas Recirculation Circuit Range/Performance
P0405 Exhaust Gas Recirculation Sensor "A" Circuit Low
P0406 Exhaust Gas Recirculation Sensor "A" Circuit High
P0407 Exhaust Gas Recirculation Sensor "B" Circuit Low
P0408 Exhaust Gas Recirculation Sensor "B" Circuit High
P0410 Secondary Air Injection System Malfunction
P0411 Secondary Air Injection System Incorrect Flow Detected
P0412 Secondary Air Injection System Switching Valve "A" Circuit Malfunction
P0413 Secondary Air Injection System Switching Valve "A" Circuit Open
P0414 Secondary Air Injection System Switching Valve "A" Circuit Shorted
P0415 Secondary Air Injection System Switching Valve "B" Circuit Malfunction
P0416 Secondary Air Injection System Switching Valve "B" Circuit Open
P0417 Secondary Air Injection System Switching Valve "B" Circuit Shorted
P0418 Secondary Air Injection System Relay "A" Circuit Malfunction
P0419 Secondary Air Injection System Relay "B" Circuit Malfunction
P0420 Catalyst System Efficiency Below Threshold (Bank no. 1)
P0421 Warm Up Catalyst Efficiency Below Threshold (Bank no. 1)
P0422 Main Catalyst Efficiency Below Threshold (Bank no. 1)
P0423 Heated Catalyst Efficiency Below Threshold (Bank no. 1)
P0424 Heated Catalyst Temperature Below Threshold (Bank no. 1)
P0430 Catalyst System Efficiency Below Threshold (Bank no. 2)
P0431 Warm Up Catalyst Efficiency Below Threshold (Bank no. 2)
P0432 Main Catalyst Efficiency Below Threshold (Bank no. 2)

P0433 Heated Catalyst Efficiency Below Threshold (Bank no. 2)

P0434 Heated Catalyst Temperature Below Threshold (Bank no. 2)

P0440 Evaporative Emission Control System Malfunction

P0441 Evaporative Emission Control System Incorrect Purge Flow

P0442 Evaporative Emission Control System Leak Detected (Small Leak)

P0443 Evaporative Emission Control System Purge Control Valve Circuit Malfunction

P0444 Evaporative Emission Control System Purge Control Valve Circuit Open

P0445 Evaporative Emission Control System Purge Control Valve Circuit Shorted

P0446 Evaporative Emission Control System Vent Control Circuit Malfunction

P0447 Evaporative Emission Control System Vent Control Circuit Open

P0448 Evaporative Emission Control System Vent Control Circuit Shorted

P0449 Evaporative Emission Control System Vent Valve/Solenoid Circuit Malfunction

P0450 Evaporative Emission Control System Pressure Sensor Malfunction

P0451 Evaporative Emission Control System Pressure Sensor Range/Performance

P0452 Evaporative Emission Control System Pressure Sensor Low Input

P0453 Evaporative Emission Control System Pressure Sensor High Input

P0454 Evaporative Emission Control System Pressure Sensor Intermittent

P0455 Evaporative Emission Control System Leak Detected (Gross Leak)

P0460 Fuel Level Sensor Circuit Malfunction

P0461 Fuel Level Sensor Circuit Range/Performance

P0462 Fuel Level Sensor Circuit Low Input

P0463 Fuel Level Sensor Circuit High Input

P0464 Fuel Level Sensor Circuit Intermittent

P0465 Purge Flow Sensor Circuit Malfunction

P0466 Purge Flow Sensor Circuit Range/Performance

P0467 Purge Flow Sensor Circuit Low Input

P0468 Purge Flow Sensor Circuit High Input

P0469 Purge Flow Sensor Circuit Intermittent

P0470 Exhaust Pressure Sensor Malfunction

P0471 Exhaust Pressure Sensor Range/Performance

P0472 Exhaust Pressure Sensor Low

P0473 Exhaust Pressure Sensor High

P0474 Exhaust Pressure Sensor Intermittent

P0475 Exhaust Pressure Control Valve Malfunction

P0476 Exhaust Pressure Control Valve Range/Performance

P0477 Exhaust Pressure Control Valve Low

P0478 Exhaust Pressure Control Valve High

P0479 Exhaust Pressure Control Valve Intermittent

P0480 Cooling Fan no. 1 Control Circuit Malfunction

P0481 Cooling Fan no. 2 Control Circuit Malfunction

P0482 Cooling Fan no. 3 Control Circuit Malfunction

P0483 Cooling Fan Rationality Check Malfunction

P0484 Cooling Fan Circuit Over Current

P0485 Cooling Fan Power/Ground Circuit Malfunction

P0500 Vehicle Speed Sensor Malfunction

P0501 Vehicle Speed Sensor Range/Performance

P0502 Vehicle Speed Sensor Circuit Low Input

P0503 Vehicle Speed Sensor Intermittent/Erratic/High

P0505 Idle Control System Malfunction

P0506 Idle Control System RPM Lower Than Expected

P0507 Idle Control System RPM Higher Than Expected

P0510 Closed Throttle Position Switch Malfunction

P0520 Engine Oil Pressure Sensor/Switch Circuit Malfunction

P0521 Engine Oil Pressure Sensor/Switch Range/Performance

P0522 Engine Oil Pressure Sensor/Switch Low Voltage

P0523 Engine Oil Pressure Sensor/Switch High Voltage

P0530 A/C Refrigerant Pressure Sensor Circuit Malfunction

P0531 A/C Refrigerant Pressure Sensor Circuit Range/Performance

P0532 A/C Refrigerant Pressure Sensor Circuit Low Input

P0533 A/C Refrigerant Pressure Sensor Circuit High Input

P0534 A/C Refrigerant Charge Loss

P0550 Power Steering Pressure Sensor Circuit Malfunction

P0551 Power Steering Pressure Sensor Circuit Range/Performance

P0552 Power Steering Pressure Sensor Circuit Low Input

P0553 Power Steering Pressure Sensor Circuit High Input

P0554 Power Steering Pressure Sensor Circuit Intermittent

P0560 System Voltage Malfunction

P0561 System Voltage Unstable

P0562 System Voltage Low

P0563 System Voltage High

P0565 Cruise Control On Signal Malfunction

P0566 Cruise Control Off Signal Malfunction

P0567 Cruise Control Resume Signal Malfunction

P0568 Cruise Control Set Signal Malfunction

P0569 Cruise Control Coast Signal Malfunction

P0570 Cruise Control Accel Signal Malfunction

P0571 Cruise Control/Brake Switch "A" Circuit Malfunction

P0572 Cruise Control/Brake Switch "A" Circuit Low

P0573 Cruise Control/Brake Switch "A" Circuit High

P0574 Through P0580 Reserved for Cruise Codes

P0600 Serial Communication Link Malfunction

P0601 Internal Control Module Memory Check Sum Error

P0602 Control Module Programming Error

P0603 Internal Control Module Keep Alive Memory (KAM) Error

P0604 Internal Control Module Random Access Memory (RAM) Error

P0605 Internal Control Module Read Only Memory (ROM) Error

P0606 PCM Processor Fault

P0608 Control Module VSS Output "A" Malfunction

P0609 Control Module VSS Output "B" Malfunction

P0620 Generator Control Circuit Malfunction

P0621 Generator Lamp "L" Control Circuit Malfunction

P0622 Generator Field "F" Control Circuit Malfunction

P0650 Malfunction Indicator Lamp (MIL) Control Circuit Malfunction

P0654 Engine RPM Output Circuit Malfunction

P0655 Engine Hot Lamp Output Control Circuit Malfunction

P0656 Fuel Level Output Circuit Malfunction

P0700 Transmission Control System Malfunction

P0701 Transmission Control System Range/Performance

P0702 Transmission Control System Electrical

P0703 Torque Converter/Brake Switch "B" Circuit Malfunction

P0704 Clutch Switch Input Circuit Malfunction

P0705 Transmission Range Sensor Circuit Malfunction (PRNDL Input)

P0706 Transmission Range Sensor Circuit Range/Performance

P0707 Transmission Range Sensor Circuit Low Input
P0708 Transmission Range Sensor Circuit High Input
P0709 Transmission Range Sensor Circuit Intermittent
P0710 Transmission Fluid Temperature Sensor Circuit Malfunction
P0711 Transmission Fluid Temperature Sensor Circuit Range/Performance
P0712 Transmission Fluid Temperature Sensor Circuit Low Input
P0713 Transmission Fluid Temperature Sensor Circuit High Input
P0714 Transmission Fluid Temperature Sensor Circuit Intermittent
P0715 Input/Turbine Speed Sensor Circuit Malfunction
P0716 Input/Turbine Speed Sensor Circuit Range/Performance
P0717 Input/Turbine Speed Sensor Circuit No Signal
P0718 Input/Turbine Speed Sensor Circuit Intermittent
P0719 Torque Converter/Brake Switch "B" Circuit Low
P0720 Output Speed Sensor Circuit Malfunction
P0721 Output Speed Sensor Circuit Range/Performance
P0722 Output Speed Sensor Circuit No Signal
P0723 Output Speed Sensor Circuit Intermittent
P0724 Torque Converter/Brake Switch "B" Circuit High
P0725 Engine Speed Input Circuit Malfunction
P0726 Engine Speed Input Circuit Range/Performance
P0727 Engine Speed Input Circuit No Signal
P0728 Engine Speed Input Circuit Intermittent
P0730 Incorrect Gear Ratio
P0731 Gear no. 1 Incorrect Ratio
P0732 Gear no. 2 Incorrect Ratio
P0733 Gear no. 3 Incorrect Ratio
P0734 Gear no. 4 Incorrect Ratio
P0735 Gear no. 5 Incorrect Ratio
P0736 Reverse Incorrect Ratio
P0740 Torque Converter Clutch Circuit Malfunction
P0741 Torque Converter Clutch Circuit Performance or Stuck Off
P0742 Torque Converter Clutch Circuit Stuck On
P0743 Torque Converter Clutch Circuit Electrical
P0744 Torque Converter Clutch Circuit Intermittent
P0745 Pressure Control Solenoid Malfunction
P0746 Pressure Control Solenoid Performance or Stuck Off
P0747 Pressure Control Solenoid Stuck On
P0748 Pressure Control Solenoid Electrical
P0749 Pressure Control Solenoid Intermittent
P0750 Shift Solenoid "A" Malfunction
P0751 Shift Solenoid "A" Performance or Stuck Off
P0752 Shift Solenoid "A" Stuck On
P0753 Shift Solenoid "A" Electrical
P0754 Shift Solenoid "A" Intermittent

P0755 Shift Solenoid "B" Malfunction
P0756 Shift Solenoid "B" Performance or Stuck Off
P0757 Shift Solenoid "B" Stuck On
P0758 Shift Solenoid "B" Electrical
P0759 Shift Solenoid "B" Intermittent
P0760 Shift Solenoid "C" Malfunction
P0761 Shift Solenoid "C" Performance Or Stuck Off
P0762 Shift Solenoid "C" Stuck On
P0763 Shift Solenoid "C" Electrical
P0764 Shift Solenoid "C" Intermittent
P0765 Shift Solenoid "D" Malfunction
P0766 Shift Solenoid "D" Performance Or Stuck Off
P0767 Shift Solenoid "D" Stuck On
P0768 Shift Solenoid "D" Electrical
P0769 Shift Solenoid "D" Intermittent
P0770 Shift Solenoid "E" Malfunction
P0771 Shift Solenoid "E" Performance Or Stuck Off
P0772 Shift Solenoid "E" Stuck On
P0773 Shift Solenoid "E" Electrical
P0774 Shift Solenoid "E" Intermittent
P0780 Shift Malfunction
P0781 1–2 Shift Malfunction
P0782 2–3 Shift Malfunction
P0783 3–4 Shift Malfunction
P0784 4–5 Shift Malfunction
P0785 Shift/Timing Solenoid Malfunction
P0786 Shift/Timing Solenoid Range/Performance
P0787 Shift/Timing Solenoid Low
P0788 Shift/Timing Solenoid High
P0789 Shift/Timing Solenoid Intermittent
P0790 Normal/Performance Switch Circuit Malfunction
P0801 Reverse Inhibit Control Circuit Malfunction
P0803 1–4 Upshift (Skip Shift) Solenoid Control Circuit Malfunction
P0804 1–4 Upshift (Skip Shift) Lamp Control Circuit Malfunction
P1250 EFI Heater Circuit Fault
P1408 Manifold Differential Pressure Sensor Circuit Fault
P1410 Fuel Tank Pressure Control Solenoid Circuit Fault
P1450 Barometric Pressure Sensor Circuit Fault
P1451 Barometric Pressure Sensor Performance
P1460 Cooling Fan Control System Fault
P1500 Starter Signal Circuit Fault
P1510 Back-up Power Supply Fault
P1530 Ignition Timing Adjustment Switch Circuit
P1600 PCM Battery Circuit Fault
P1700 TCM Throttle Position Sensor Circuit Fault
P1705 TCM ECT Circuit Fault
P1715 PNP Switch Circuit Fault
P1717 AT Drive Range Signal Circuit Fault

FIRING ORDERS

3

FIRING ORDERS

On every vehicle manufactured between 1995 and 1999, there are essentially only two basic methods for distributing the ignition system spark to the spark plugs: distributor system and Distributorless Ignition System (DIS). The distributor system uses a rotating rotor within a distributor cap to dispense the system's spark to the applicable spark plug. DIS systems use one of three general set-ups for spark distribution: remote coil pack(s), waste spark system, and direct ignition system (also often referred to as DIS). All DIS systems are controlled by the engine control computer, which computes the proper ignition timing based upon incoming reference signals from engine sensors.

The remote coil pack set-up uses one or more coil packs connected to the spark plugs via plug wires. The waste spark system is actually a sub-type of the remote coil pack system. The only difference being that two spark plugs are fired simultaneously because they share one coil. Many waste spark systems are designed as a hybrid of a direct ignition system and a remote coil pack system, because a coil is mounted directly on top of one spark plug and attached to another spark plug via a plug wire. Direct ignition does away with the spark plug wires completely and uses a single coil pack mounted directly on top of the spark plug for each cylinder.

Firing orders are most important for vehicles equipped with distributor ignition systems because the distributor can be rotated (which can lead to confusion as to which plug tower is what). If the distributor is rotated and the spark plug wires are installed on the original cap towers, the ignition timing will be adversely affected. DIS systems are not adjustable in the same manner as distributor systems. Therefore, if you connect the wires (when applicable) to the proper coil pack towers, the ignition timing will always be correct. Thus, if your vehicle is equipped with a DIS system and the firing order illustration does not contain a specific firing order, simply attach the wires to the proper coil pack towers and the ignition system will function properly.

➡**The coil packs used on DIS systems are often labeled with the cylinder number of their corresponding spark plugs.**

If your vehicle is equipped with a distributor which is not keyed for installation with only one orientation, it could have been removed previously and rewired. The resultant wiring would hold the correct firing order, but could change the relative placement of the plug towers in relation to the engine. For this reason it is imperative that you label all wires before disconnecting any of them. Also, before removal, compare the current wiring with the accompanying illustrations. If the current wiring does not match, make notes in your book to reflect how your engine is wired.

➡**To avoid confusion, remove and tag the spark plug wires one at a time, for replacement.**

FIRING ORDER INDEX

MANUFACTURER ENGINE	FIGURE
Acura	
3.2L (VIN V and W) Engines	1
3.5L Engines	2
Chrysler	
2.4L Engines	3
2.5L Engines	4
3.0L Engines	5
3.3L and 3.8L Engines	6
Dodge	
2.5L (VIN G) Engines	7
2.5L (VIN P) Engines	8
3.9L (VIN X) Engines	9
5.2L and 5.9L Gasoline Engines	10
8.0L Engines	11
Ford	
2.3L Engines	22
2.5L Engines	23
3.0L (except Aerostar & Windstar) Engines	
1995-97 Models	25
1998-99 Models	26
3.0L (VIN U) Windstar Engines	12
1995-97 Models	12
1998-99 Models	13
3.0L Aerostar Engines	24
3.8L (VIN 4) Engines	14
4.0L (except Aerostar) Engines	
1995-97 Models	25
1998-99 Models	26
4.2L (VIN 2) Engines	15
4.6L (VIN W, 6, and 9) Engines	16
4.9L (VIN Y) Engines	17
5.0L (VIN N) and 7.5L (VIN G) Engines	18
5.0L (VIN P) Engines	27
5.4L Engines	19
5.8L (VIN H and R) Engines	20
6.8L Engines	21
General Motors	
2.2L (VIN W and 4) Engines	28
3.1L Engines	29
3.4L Engines	30
3.8L Engines	31
4.3L (VIN W, X and Z) Engines	
1995-97 Models	32
1998-99 Models	33

79243C01

FIRING ORDER INDEX

79243C02

FIRING ORDER INDEX

MANUFACTURER ENGINE	FIGURE
Land Rover	
3.9L Engines	52
4.0L and 4.6L Engines	53
Lexus	
4.5L (1FZ-FE) Engines	54
Mazda	
2.3L Engines	22
2.5L Engines	23
3.0L Aerostar Engines	24
3.0L and 4.0L B-Series Truck Engines	
1995-97 Models	25
1998-99 Models	26
3.0L MPV Engines	
1995-97 Models	55
1998-99 Models	56
5.0L Engines	27
Mercedes-Benz	
3.2L Engines	57
Mitsubishi	
2.4L Engines	58
3.0L Engines without DIS	59
3.0L Engines with DIS	60
3.5L Engines	61
Nissan	
2.4L Engines	62
3.0L (VIN W) Engines	63
3.0L (VG30E) and 3.3L (VG33E) Engines	
1995-97 Models	64
1998-99 Models	65
Mercury	
3.0L (VIN 1) Engines	66
Subaru	
2.5L Engines	67
Suzuki	
1.6L Engines	
1995-97 Models	36
1998-99 Models	37
1.8L Engines	38
Toyota	
2.0L (3SFE) Engines without DIS	68
2.0L (3SFE) Engines with DIS	69
2.4L (22R-E and 2TZ-FE) Engines	70
2.4L (2RZ-FE) and 2.7L (3RZ-FE) Engines without DIS	71
2.4L (2RZ-FE) and 2.7L (3RZ-FE) Engines with DIS	72
3.4L (5VZ-FE) Engines	73
4.5L (1FZ-FE) Engines	54

79243C03

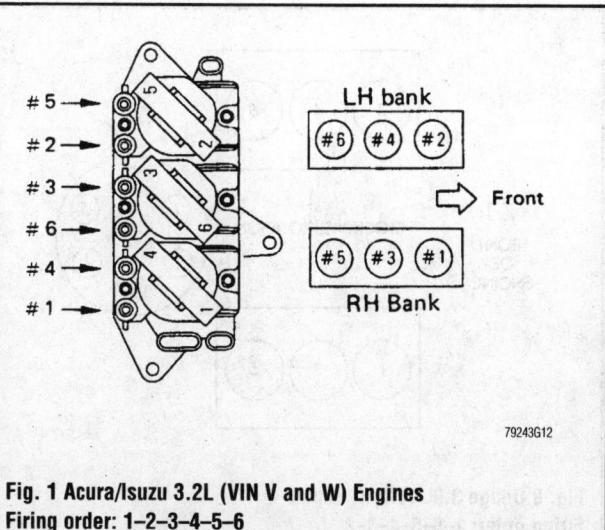

Fig. 1 Acura/Isuzu 3.2L (VIN V and W) Engines
Firing order: 1–2–3–4–5–6
Distributorless ignition system

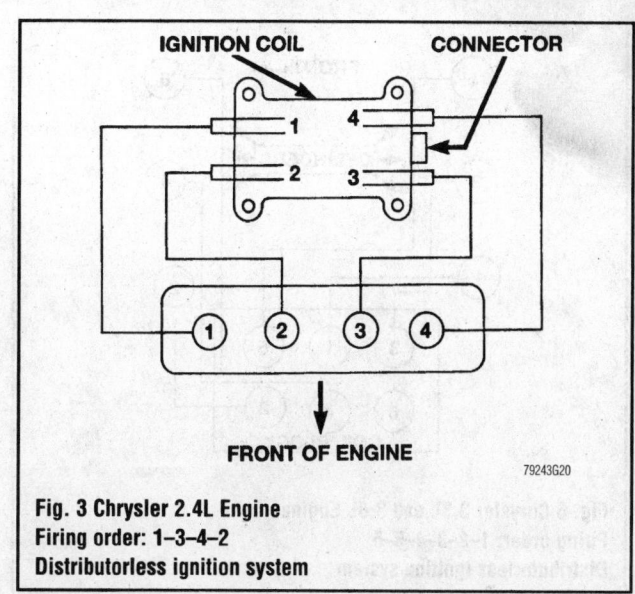

Fig. 3 Chrysler 2.4L Engine
Firing order: 1–3–4–2
Distributorless ignition system

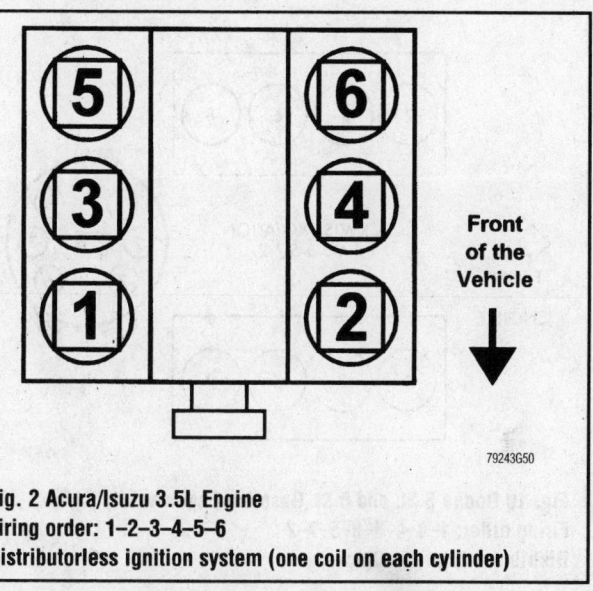

Fig. 2 Acura/Isuzu 3.5L Engine
Firing order: 1–2–3–4–5–6
Distributorless ignition system (one coil on each cylinder)

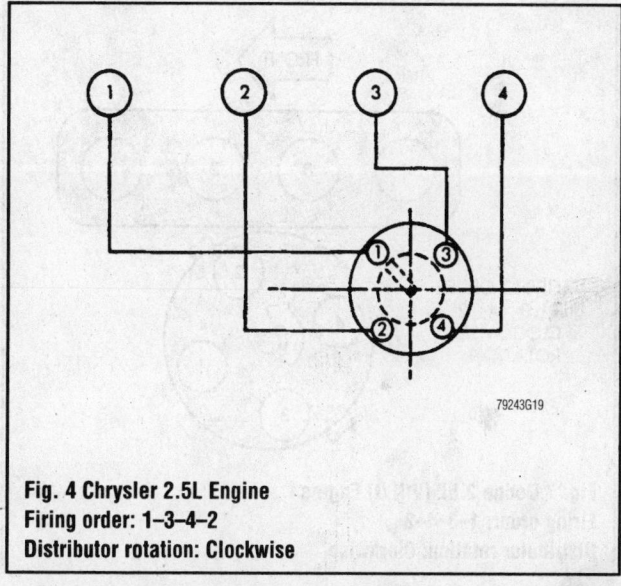

Fig. 4 Chrysler 2.5L Engine
Firing order: 1–3–4–2
Distributor rotation: Clockwise

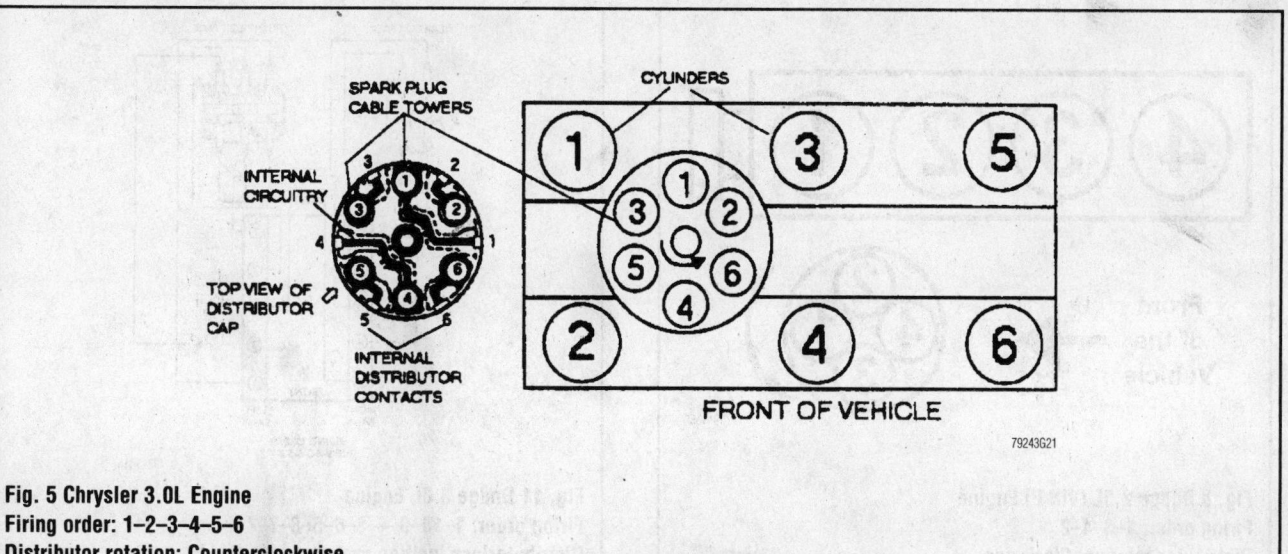

Fig. 5 Chrysler 3.0L Engine
Firing order: 1–2–3–4–5–6
Distributor rotation: Counterclockwise

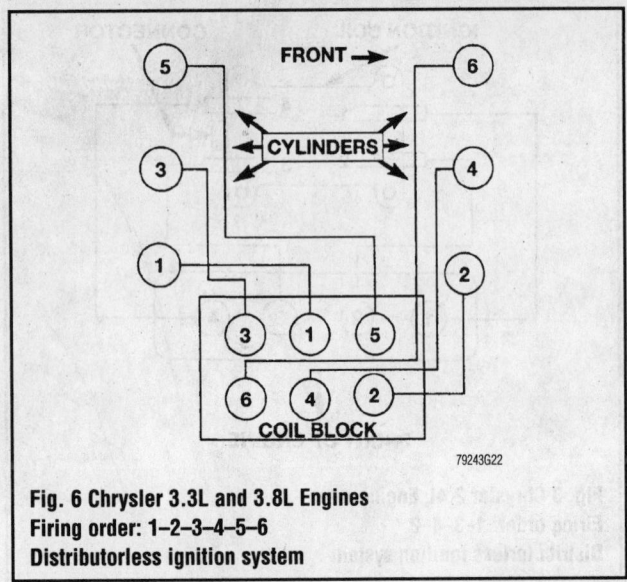

Fig. 6 Chrysler 3.3L and 3.8L Engines
Firing order: 1–2–3–4–5–6
Distributorless ignition system

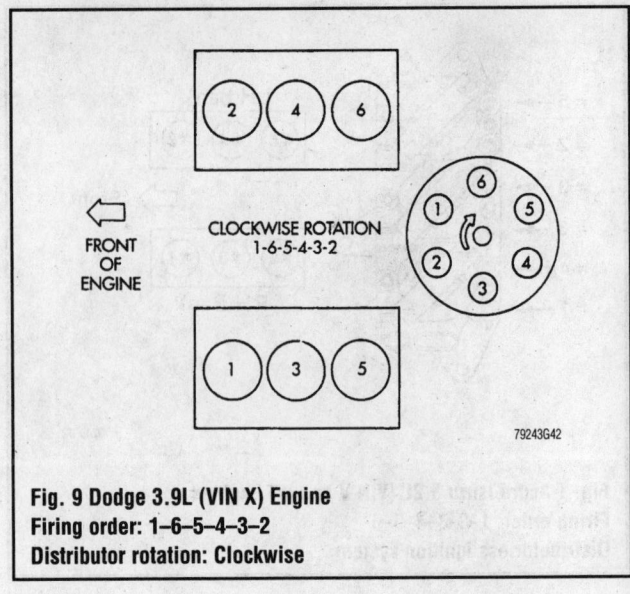

Fig. 9 Dodge 3.9L (VIN X) Engine
Firing order: 1–6–5–4–3–2
Distributor rotation: Clockwise

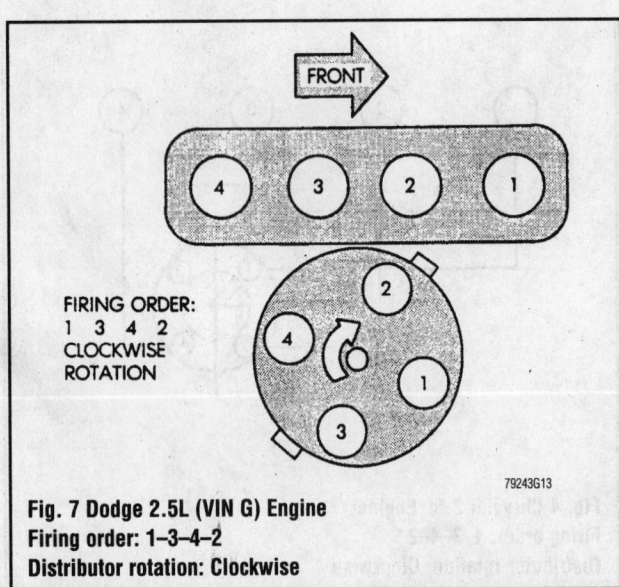

Fig. 7 Dodge 2.5L (VIN G) Engine
Firing order: 1–3–4–2
Distributor rotation: Clockwise

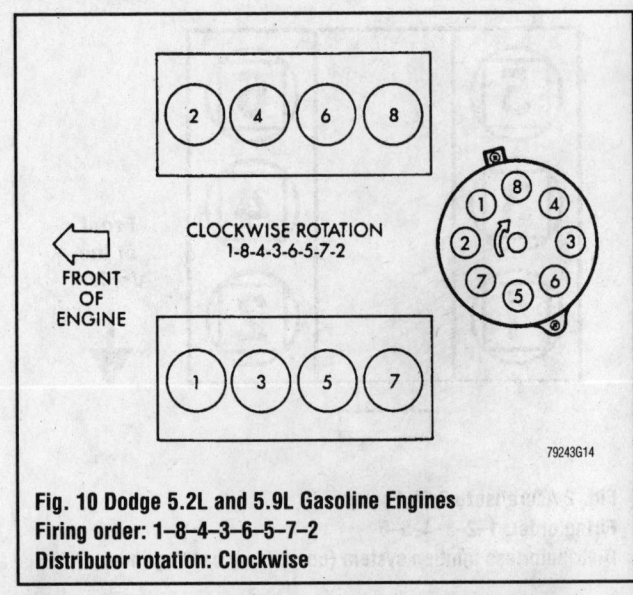

Fig. 10 Dodge 5.2L and 5.9L Gasoline Engines
Firing order: 1–8–4–3–6–5–7–2
Distributor rotation: Clockwise

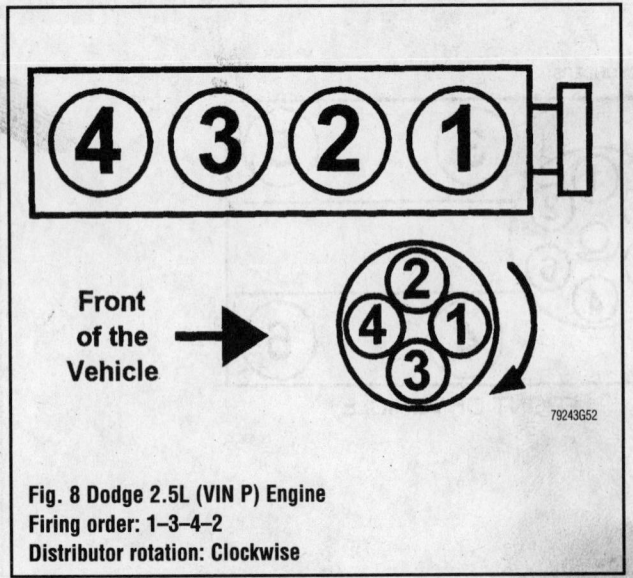

Fig. 8 Dodge 2.5L (VIN P) Engine
Firing order: 1–3–4–2
Distributor rotation: Clockwise

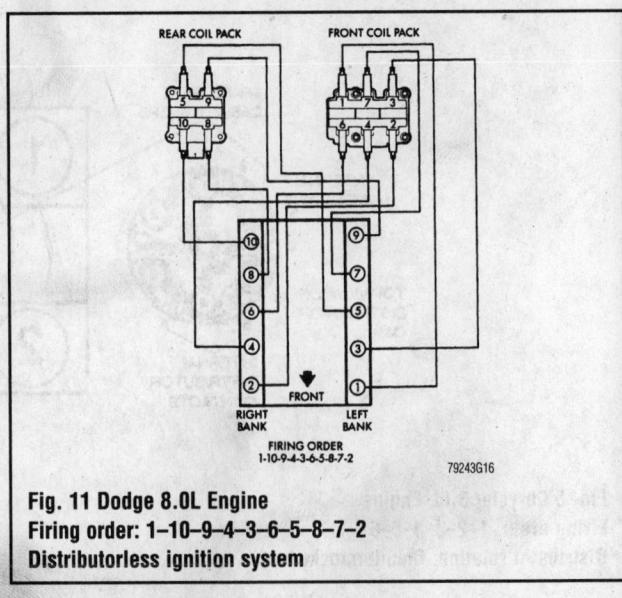

Fig. 11 Dodge 8.0L Engine
Firing order: 1–10–9–4–3–6–5–8–7–2
Distributorless ignition system

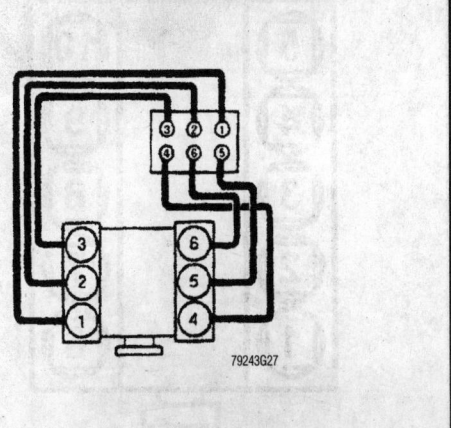

Fig. 12 Ford 1995–97 3.0L (VIN U) Engine
Firing order: 1–4–2–5–3–6
Distributorless ignition system

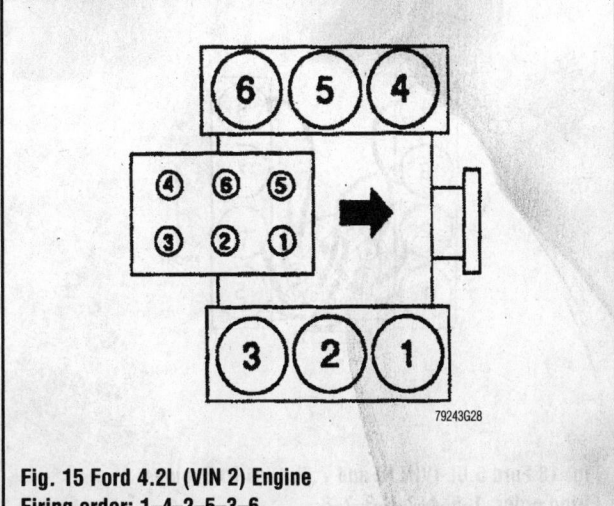

Fig. 15 Ford 4.2L (VIN 2) Engine
Firing order: 1–4–2–5–3–6
Distributorless ignition system

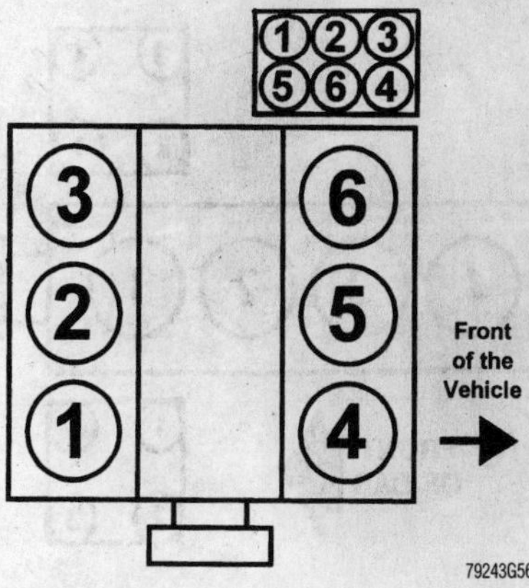

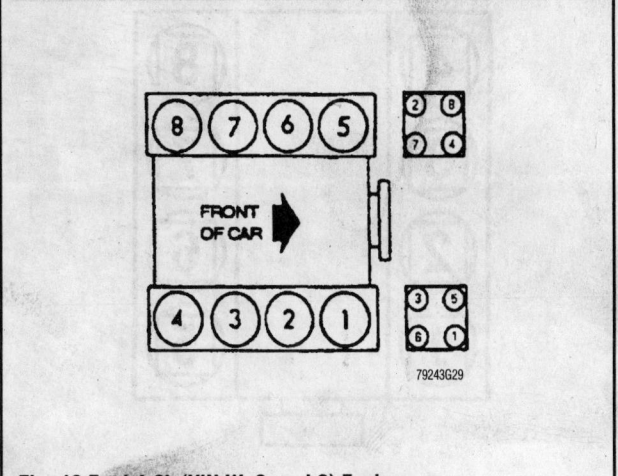

Fig. 13 Ford 1998–99 3.0L (VIN U) Engine
Firing order: 1–4–2–5–3–6
Distributorless ignition system

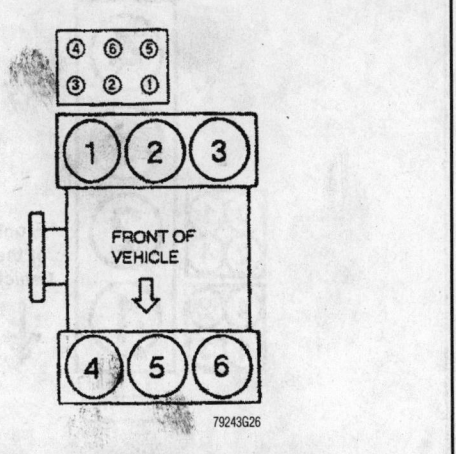

Fig. 16 Ford 4.6L (VIN W, 6, and 9) Engines
Firing order: 1–3–7–2–6–5–4–8
Distributorless ignition system

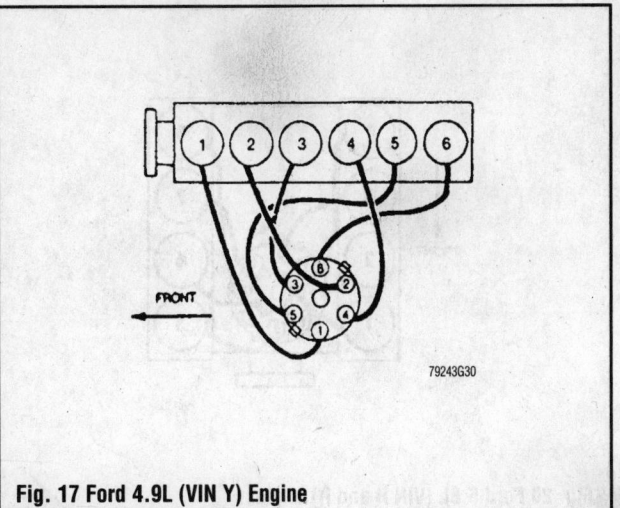

Fig. 14 Ford 3.8L (VIN 4) Engine
Firing order: 1–4–2–5–3–6
Distributorless ignition system

Fig. 17 Ford 4.9L (VIN Y) Engine
Firing order: 1–5–3–6–2–4
Distributor rotation: Clockwise

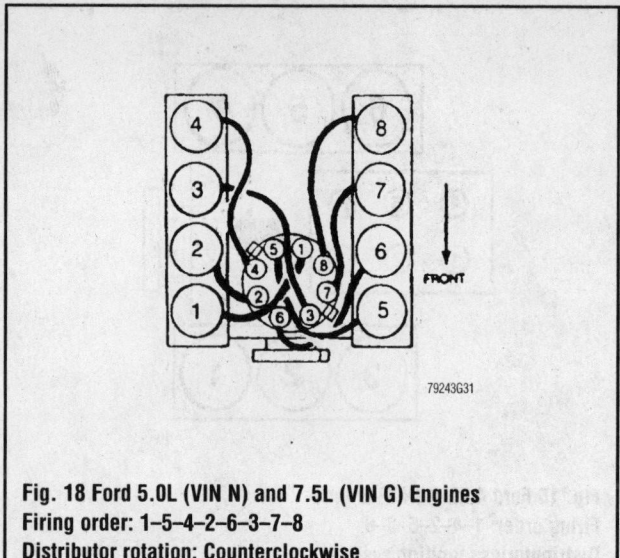

Fig. 18 Ford 5.0L (VIN N) and 7.5L (VIN G) Engines
Firing order: 1–5–4–2–6–3–7–8
Distributor rotation: Counterclockwise

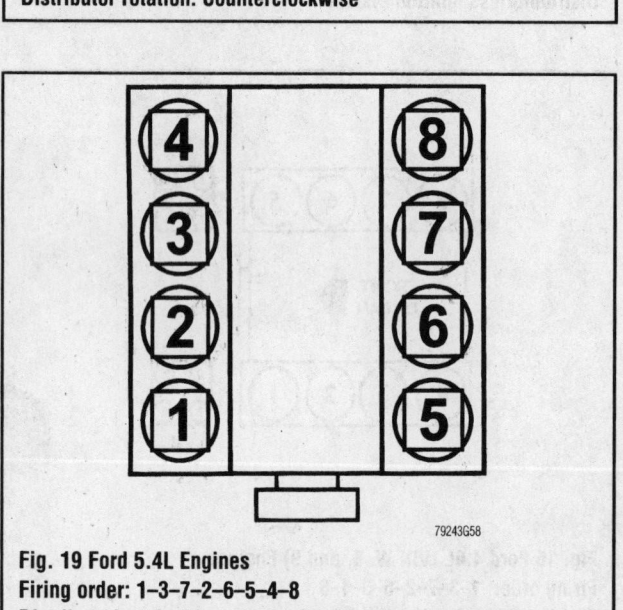

Fig. 19 Ford 5.4L Engines
Firing order: 1–3–7–2–6–5–4–8
Distributorless ignition system (one coil on each cylinder)

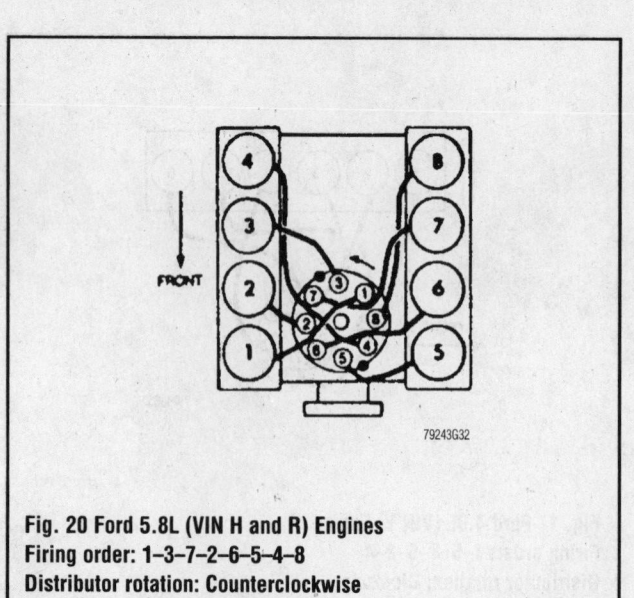

Fig. 20 Ford 5.8L (VIN H and R) Engines
Firing order: 1–3–7–2–6–5–4–8
Distributor rotation: Counterclockwise

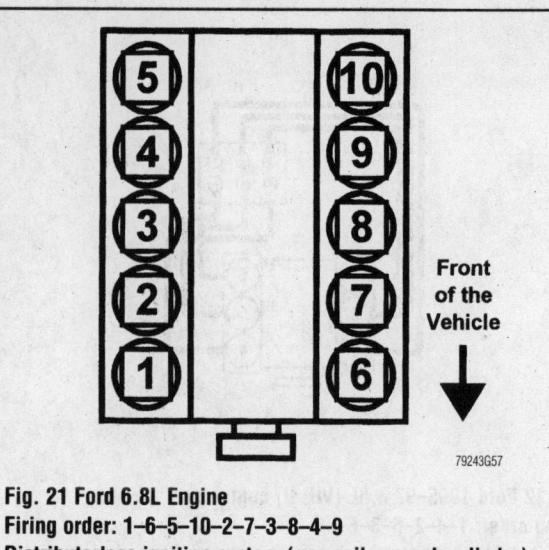

Fig. 21 Ford 6.8L Engine
Firing order: 1–6–5–10–2–7–3–8–4–9
Distributorless ignition system (one coil on each cylinder)

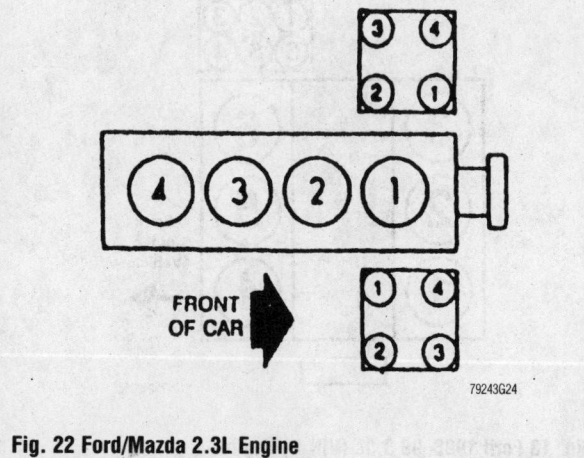
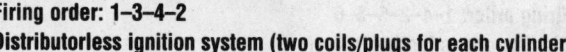

Fig. 22 Ford/Mazda 2.3L Engine
Firing order: 1–3–4–2
Distributorless ignition system (two coils/plugs for each cylinder)

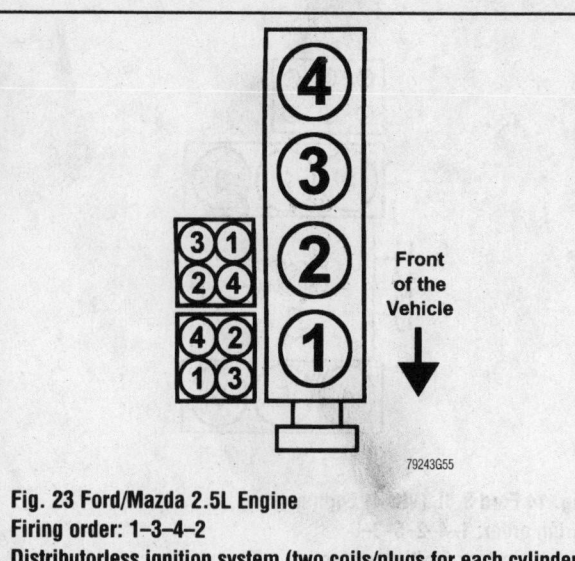

Fig. 23 Ford/Mazda 2.5L Engine
Firing order: 1–3–4–2
Distributorless ignition system (two coils/plugs for each cylinder)

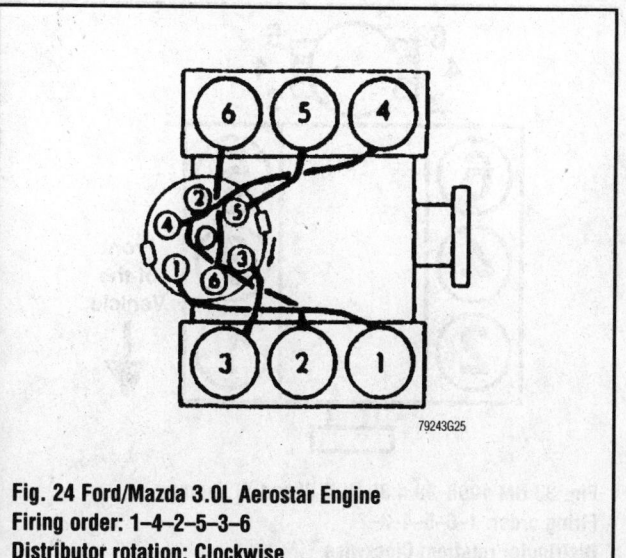

Fig. 24 Ford/Mazda 3.0L Aerostar Engine
Firing order: 1–4–2–5–3–6
Distributor rotation: Clockwise

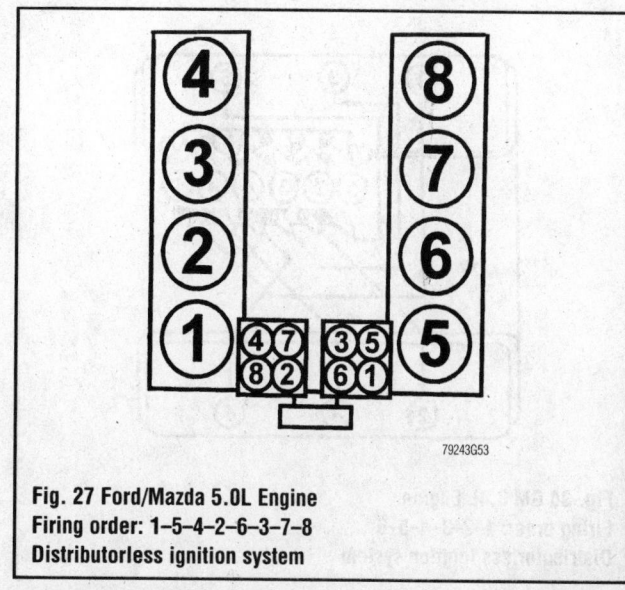

Fig. 27 Ford/Mazda 5.0L Engine
Firing order: 1–5–4–2–6–3–7–8
Distributorless ignition system

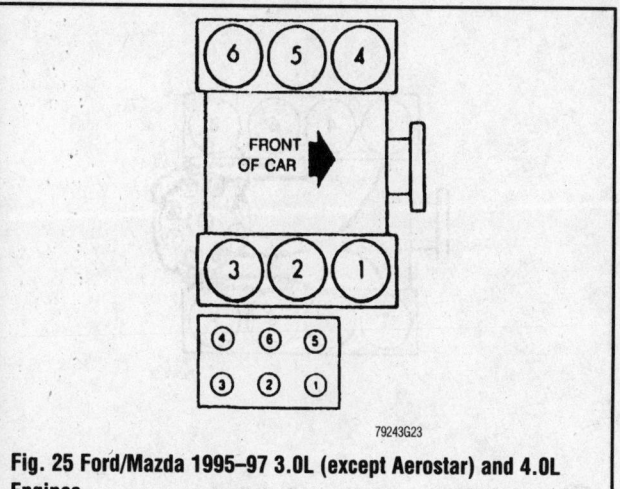

Fig. 25 Ford/Mazda 1995–97 3.0L (except Aerostar) and 4.0L Engines
Firing order: 1–4–2–5–3–6
Distributorless ignition system

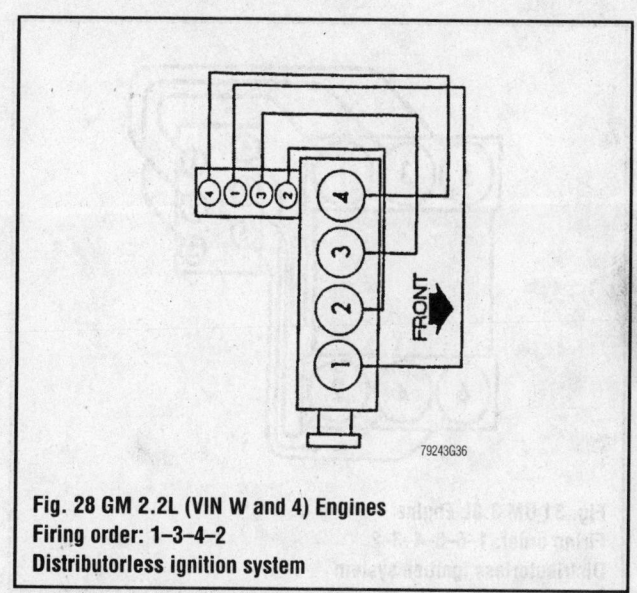

Fig. 28 GM 2.2L (VIN W and 4) Engines
Firing order: 1–3–4–2
Distributorless ignition system

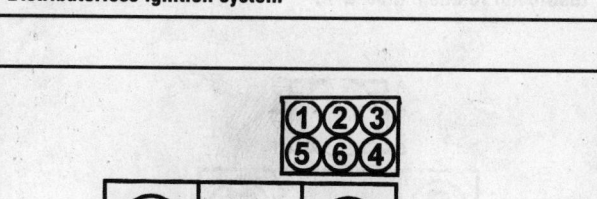

Fig. 26 Ford/Mazda 1998–99 3.0L and 4.0L Engines
Firing order: 1–4–2–5–3–6
Distributorless ignition system

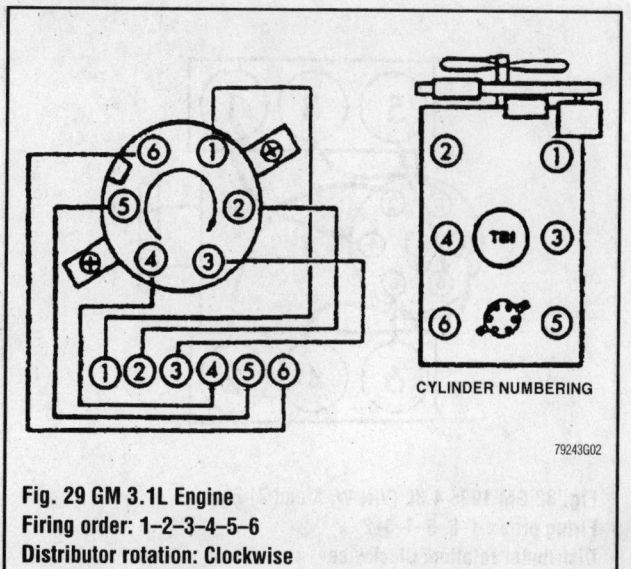

Fig. 29 GM 3.1L Engine
Firing order: 1–2–3–4–5–6
Distributor rotation: Clockwise

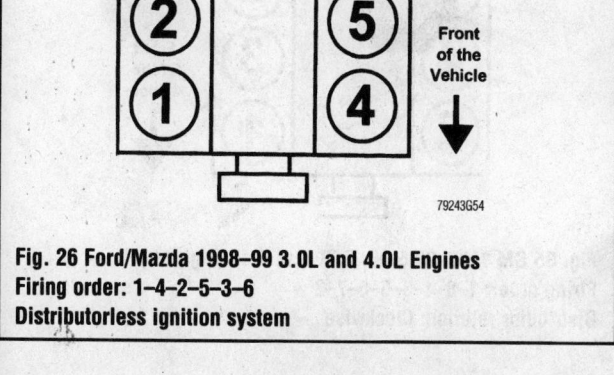

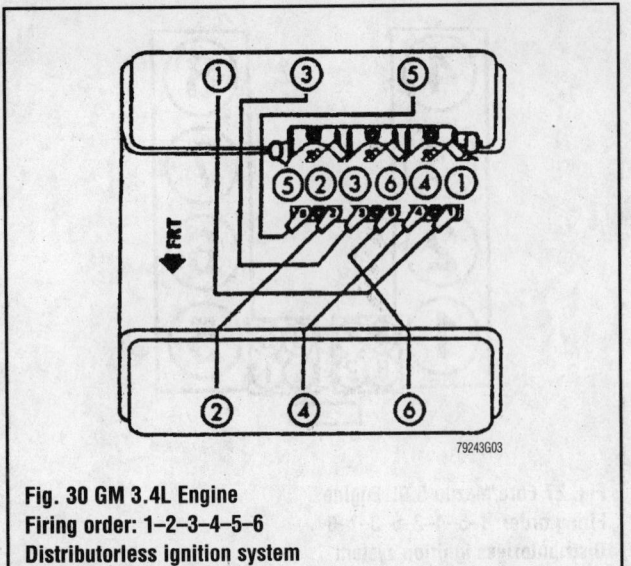

Fig. 30 GM 3.4L Engine
Firing order: 1–2–3–4–5–6
Distributorless ignition system

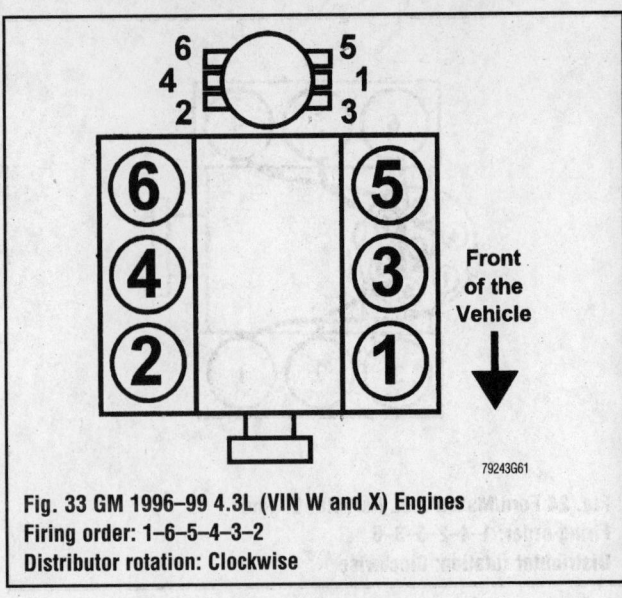

Fig. 33 GM 1996–99 4.3L (VIN W and X) Engines
Firing order: 1–6–5–4–3–2
Distributor rotation: Clockwise

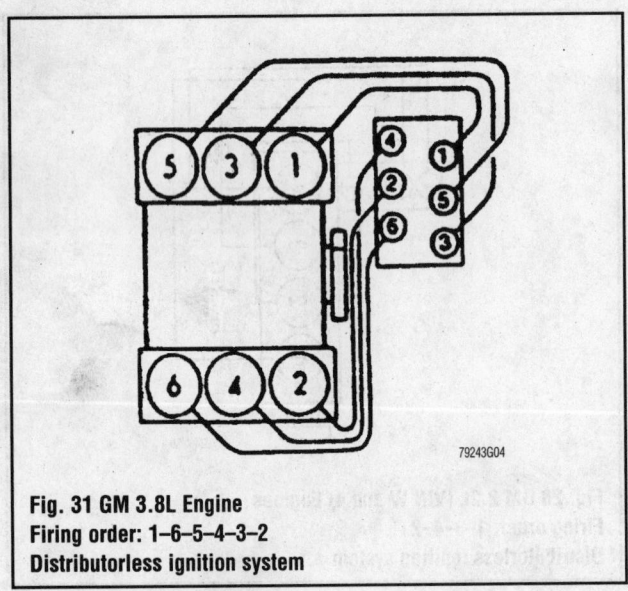

Fig. 31 GM 3.8L Engine
Firing order: 1–6–5–4–3–2
Distributorless ignition system

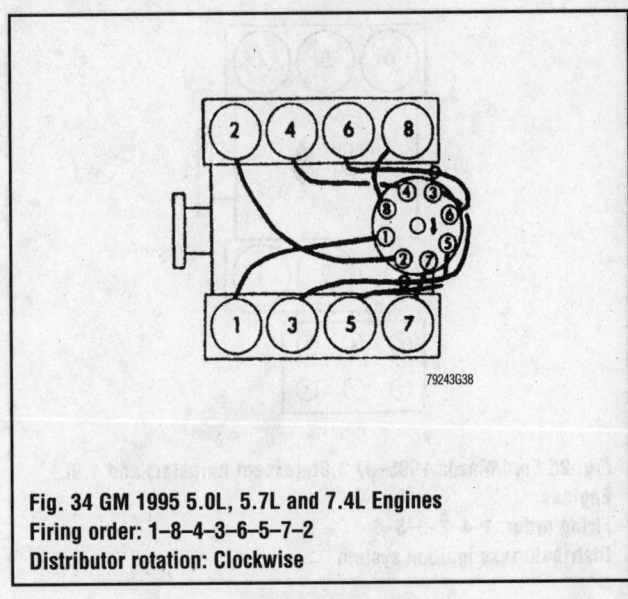

Fig. 34 GM 1995 5.0L, 5.7L and 7.4L Engines
Firing order: 1–8–4–3–6–5–7–2
Distributor rotation: Clockwise

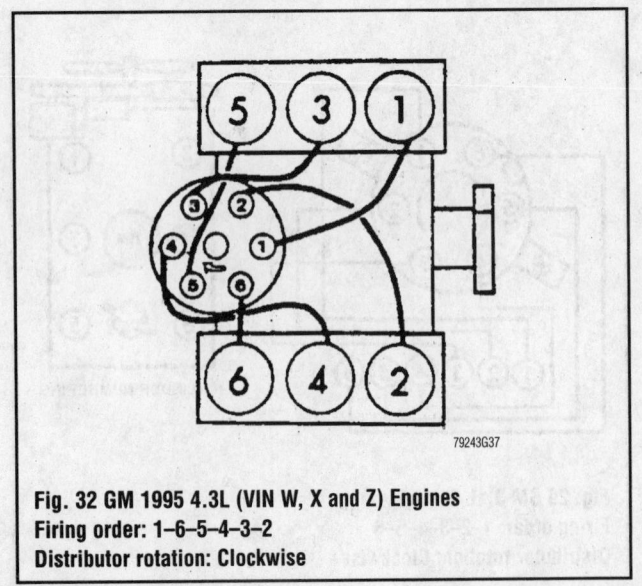

Fig. 32 GM 1995 4.3L (VIN W, X and Z) Engines
Firing order: 1–6–5–4–3–2
Distributor rotation: Clockwise

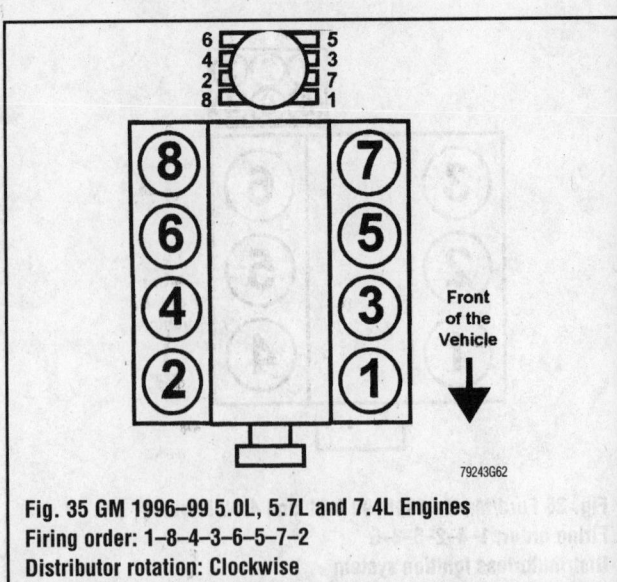

Fig. 35 GM 1996–99 5.0L, 5.7L and 7.4L Engines
Firing order: 1–8–4–3–6–5–7–2
Distributor rotation: Clockwise

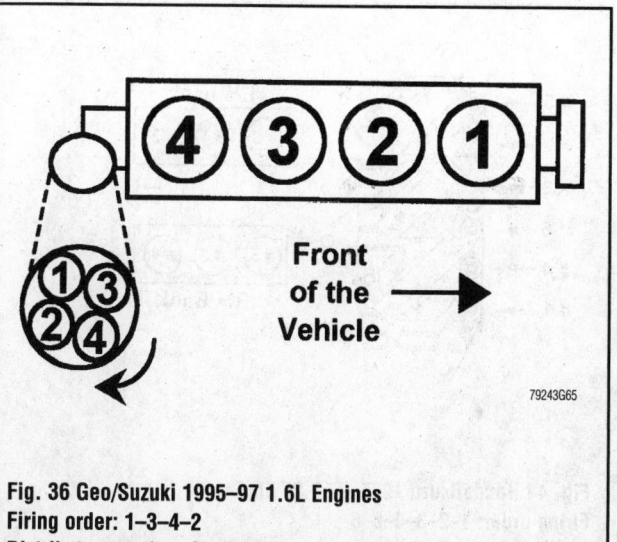

Fig. 36 Geo/Suzuki 1995–97 1.6L Engines
Firing order: 1–3–4–2
Distributor rotation: Clockwise

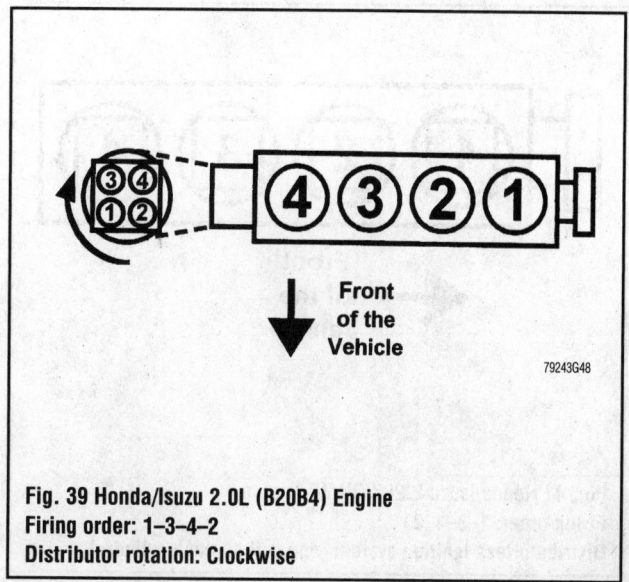

Fig. 39 Honda/Isuzu 2.0L (B20B4) Engine
Firing order: 1–3–4–2
Distributor rotation: Clockwise

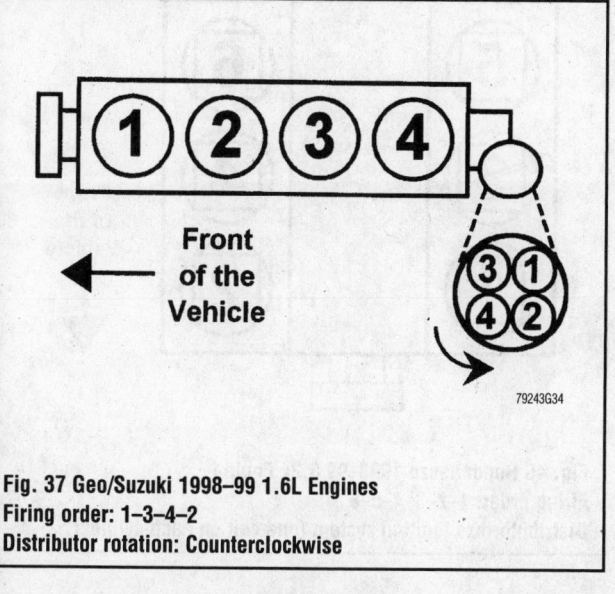

Fig. 37 Geo/Suzuki 1998–99 1.6L Engines
Firing order: 1–3–4–2
Distributor rotation: Counterclockwise

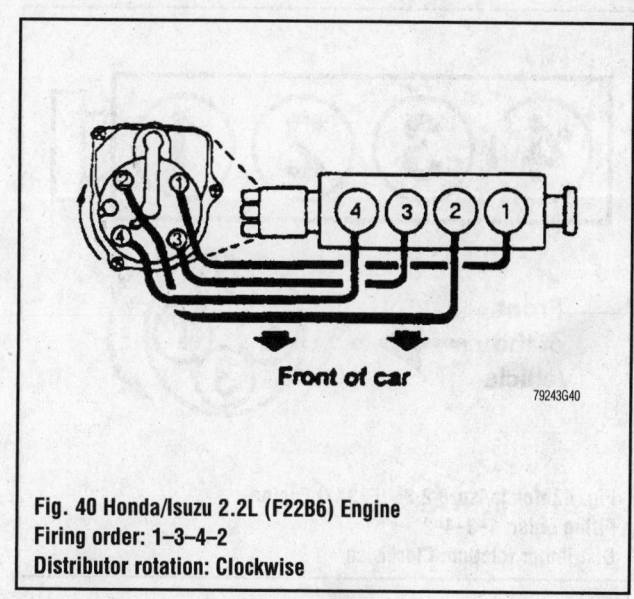

Fig. 40 Honda/Isuzu 2.2L (F22B6) Engine
Firing order: 1–3–4–2
Distributor rotation: Clockwise

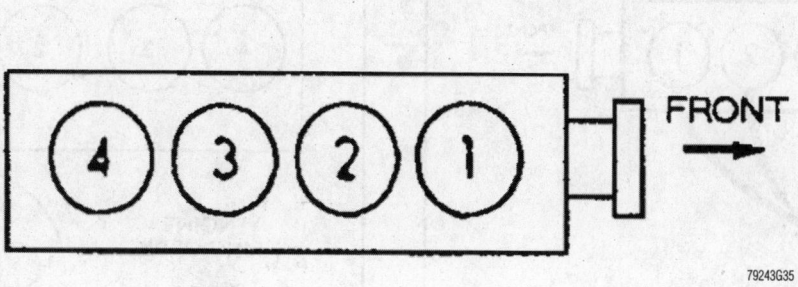

Fig. 38 Geo/Suzuki 1.8L Engine
Firing order: 1–3–4–2
Distributorless ignition system

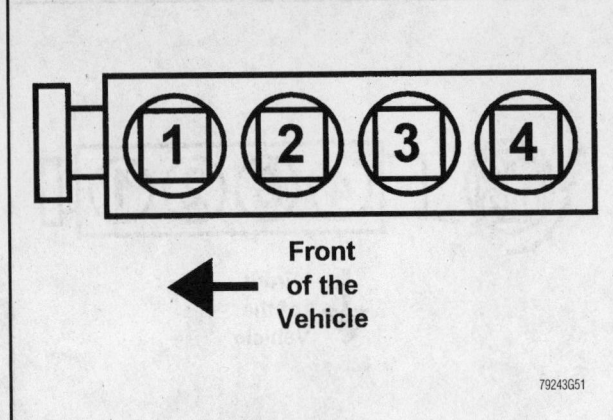

Fig. 41 Honda/Isuzu 2.2L (VIN W) Engine
Firing order: 1–3–4–2
Distributorless ignition system (one coil on each cylinder)

79243G51

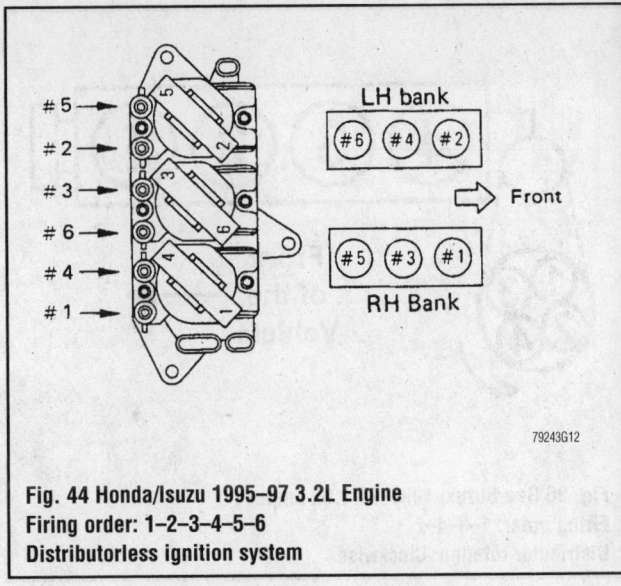

Fig. 44 Honda/Isuzu 1995–97 3.2L Engine
Firing order: 1–2–3–4–5–6
Distributorless ignition system

79243G12

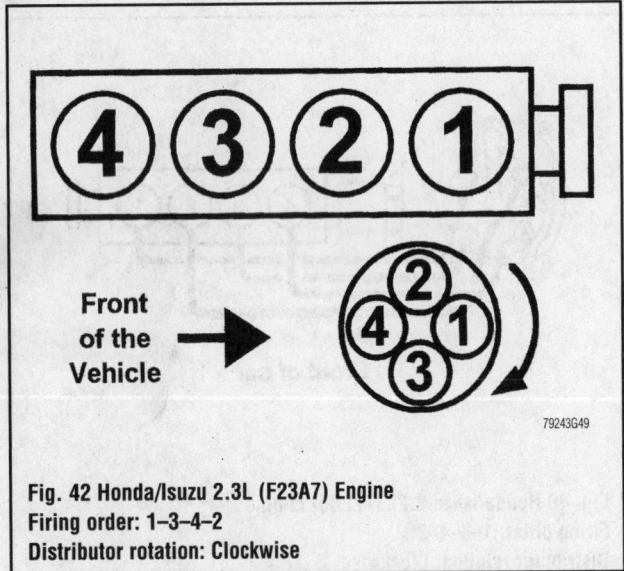

Fig. 42 Honda/Isuzu 2.3L (F23A7) Engine
Firing order: 1–3–4–2
Distributor rotation: Clockwise

79243G49

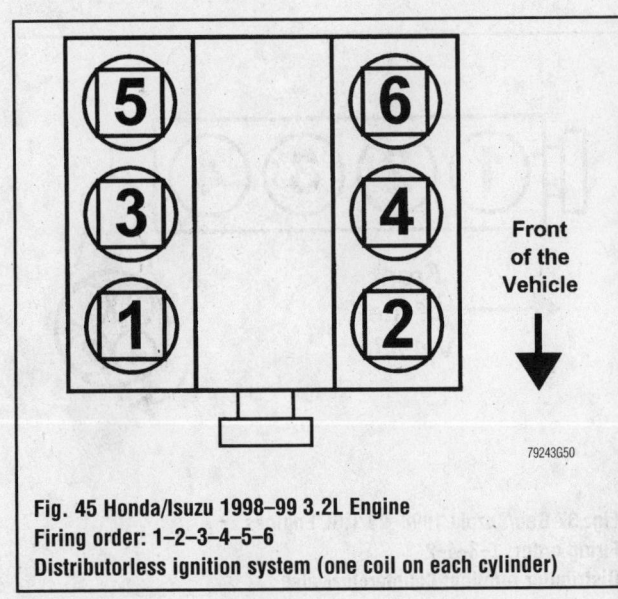

Fig. 45 Honda/Isuzu 1998–99 3.2L Engine
Firing order: 1–2–3–4–5–6
Distributorless ignition system (one coil on each cylinder)

79243G50

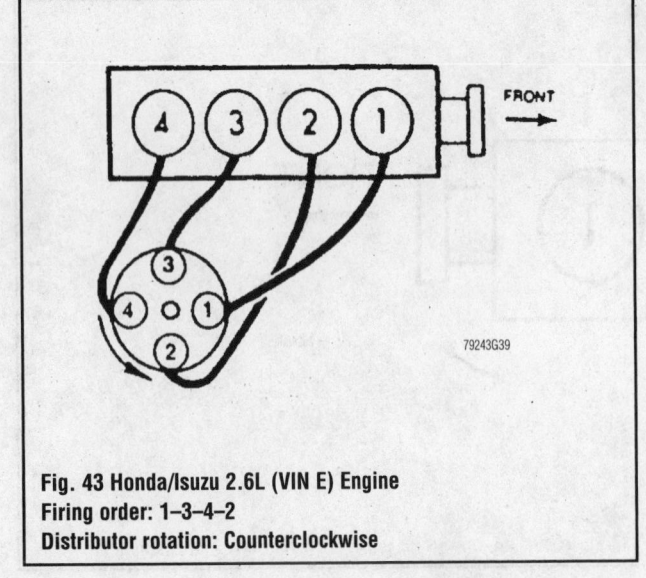

Fig. 43 Honda/Isuzu 2.6L (VIN E) Engine
Firing order: 1–3–4–2
Distributor rotation: Counterclockwise

79243G39

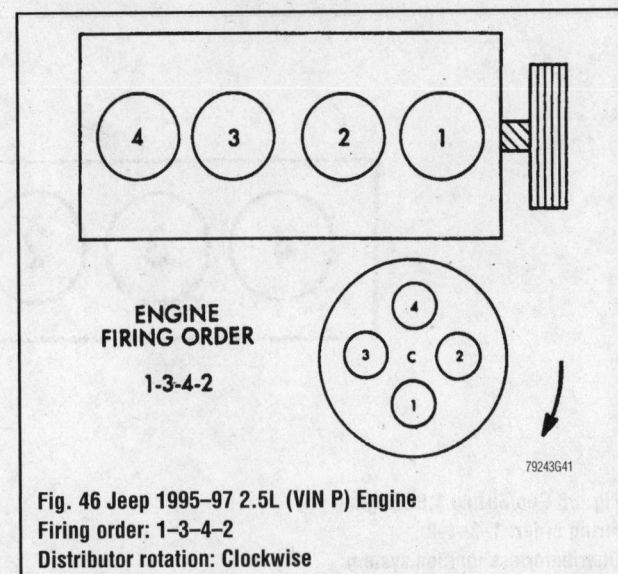

Fig. 46 Jeep 1995–97 2.5L (VIN P) Engine
Firing order: 1–3–4–2
Distributor rotation: Clockwise

79243G41

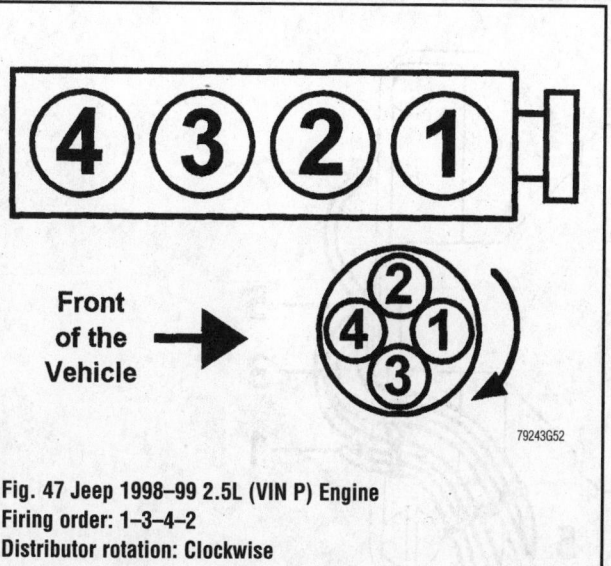

Fig. 47 Jeep 1998–99 2.5L (VIN P) Engine
Firing order: 1–3–4–2
Distributor rotation: Clockwise

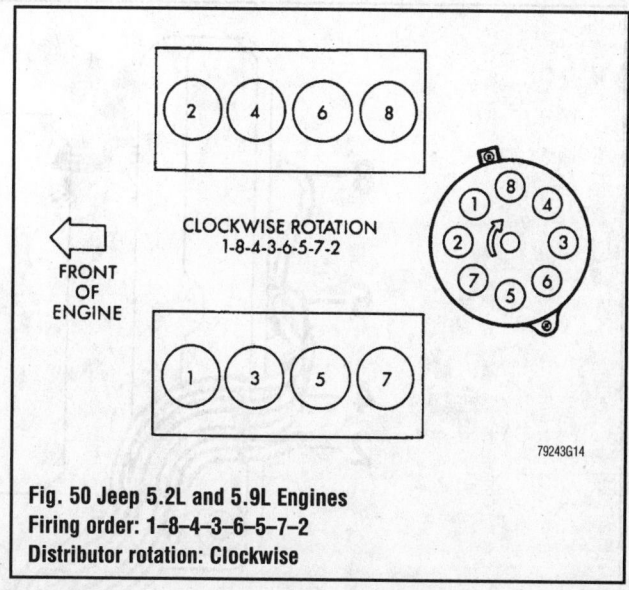

Fig. 50 Jeep 5.2L and 5.9L Engines
Firing order: 1–8–4–3–6–5–7–2
Distributor rotation: Clockwise

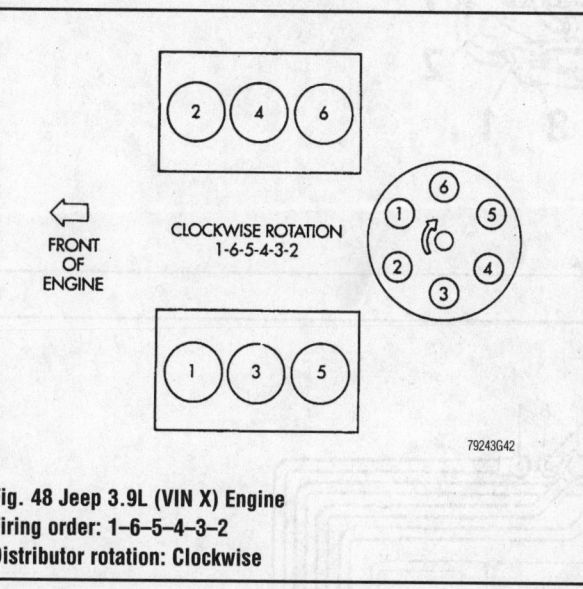

Fig. 48 Jeep 3.9L (VIN X) Engine
Firing order: 1–6–5–4–3–2
Distributor rotation: Clockwise

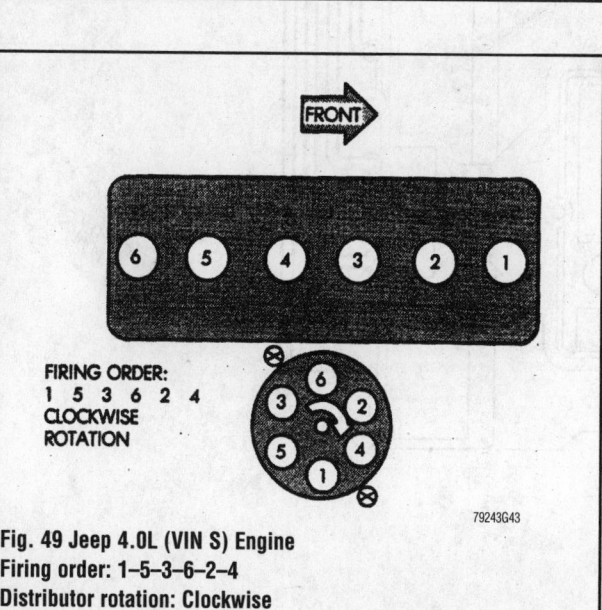

Fig. 49 Jeep 4.0L (VIN S) Engine
Firing order: 1–5–3–6–2–4
Distributor rotation: Clockwise

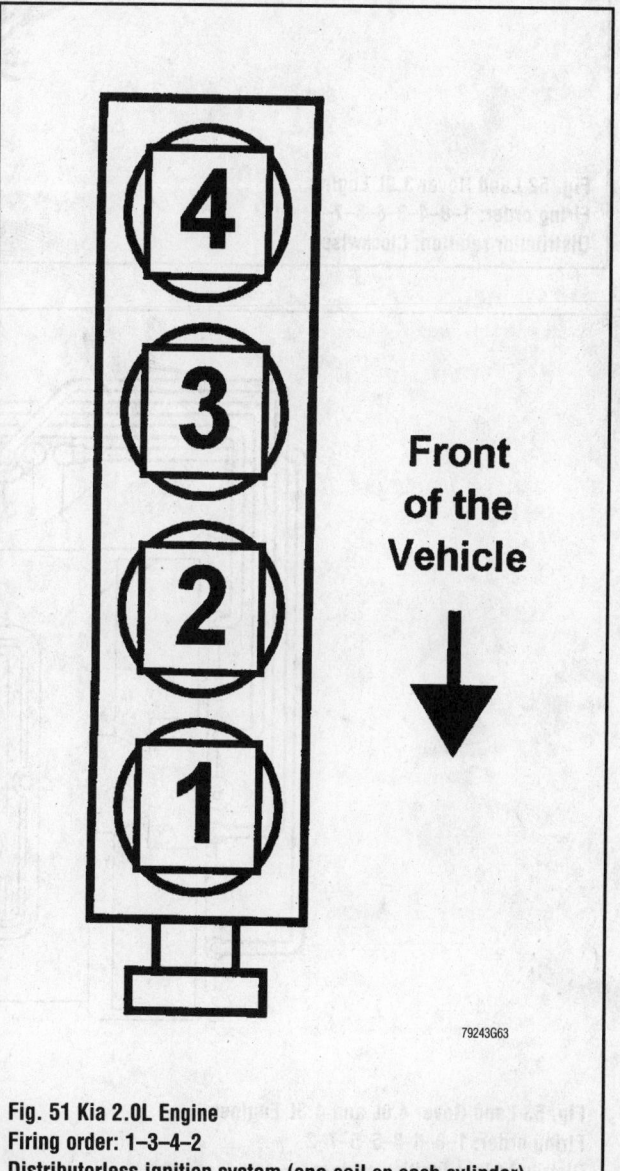

Fig. 51 Kia 2.0L Engine
Firing order: 1–3–4–2
Distributorless ignition system (one coil on each cylinder)

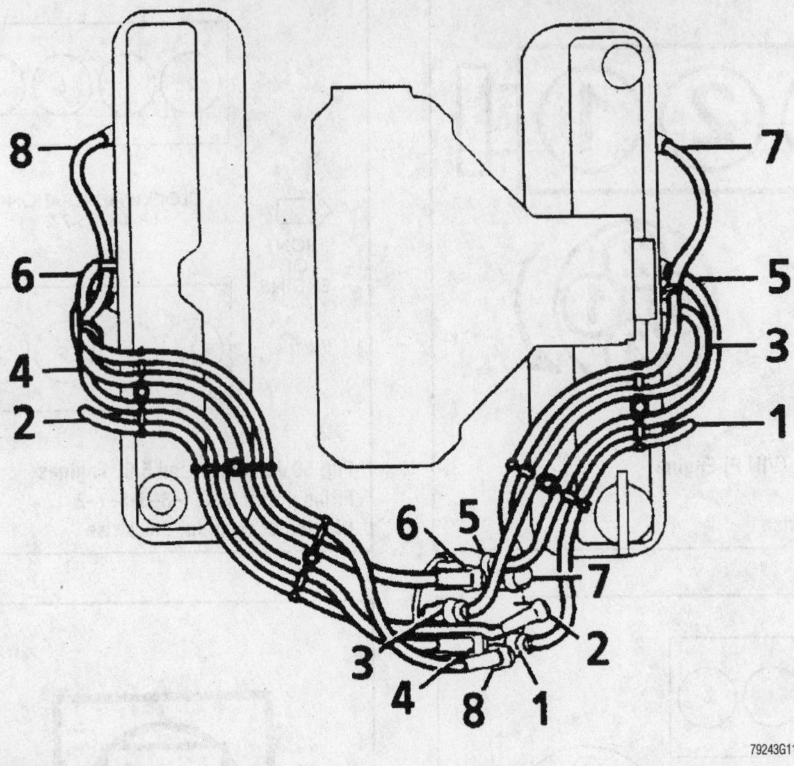

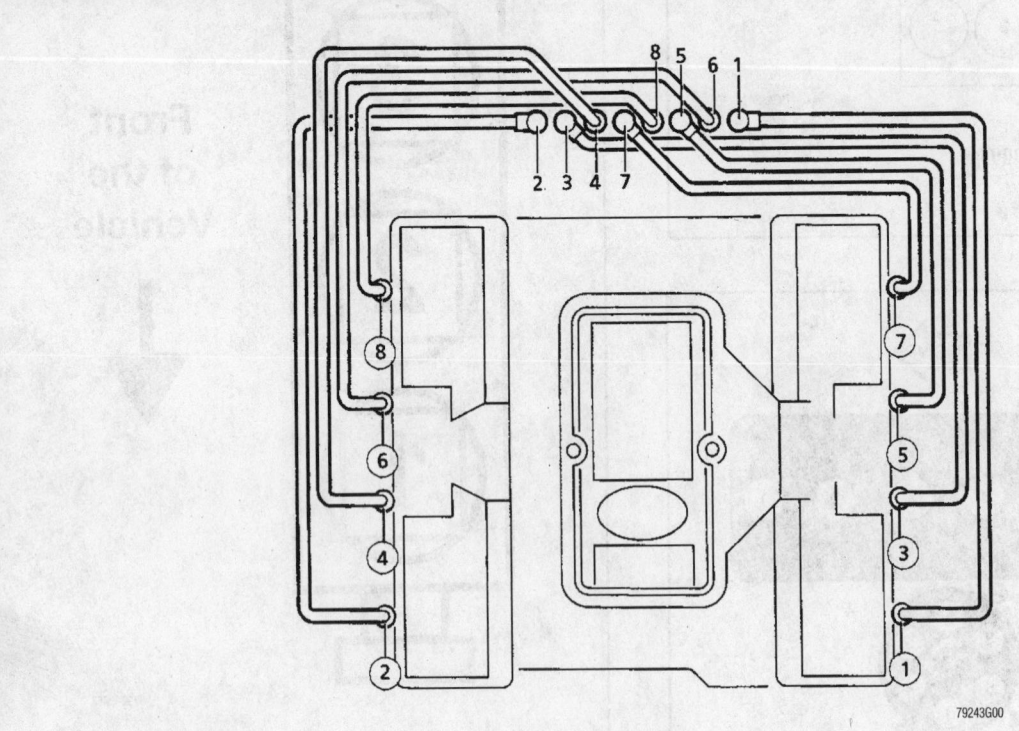

Fig. 52 Land Rover 3.9L Engine
Firing order: 1–8–4–3–6–5–7–2
Distributor rotation: Clockwise

79243G11

Fig. 53 Land Rover 4.0L and 4.6L Engines
Firing order: 1–8–4–3–6–5–7–2
Distributorless ignition system

79243G00

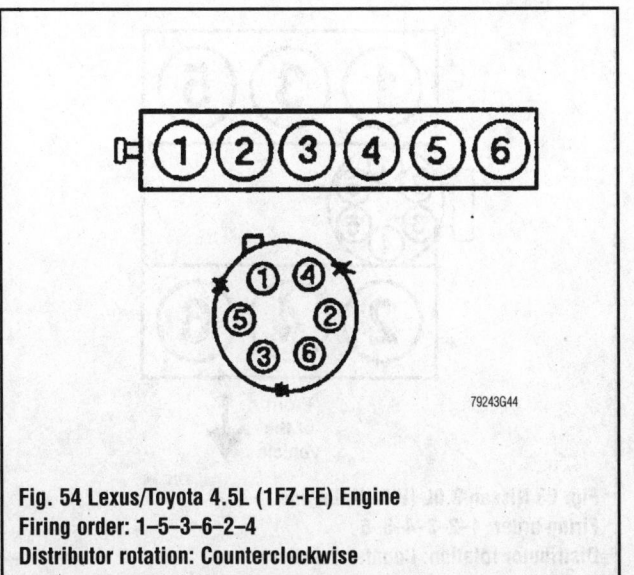

Fig. 54 Lexus/Toyota 4.5L (1FZ-FE) Engine
Firing order: 1–5–3–6–2–4
Distributor rotation: Counterclockwise

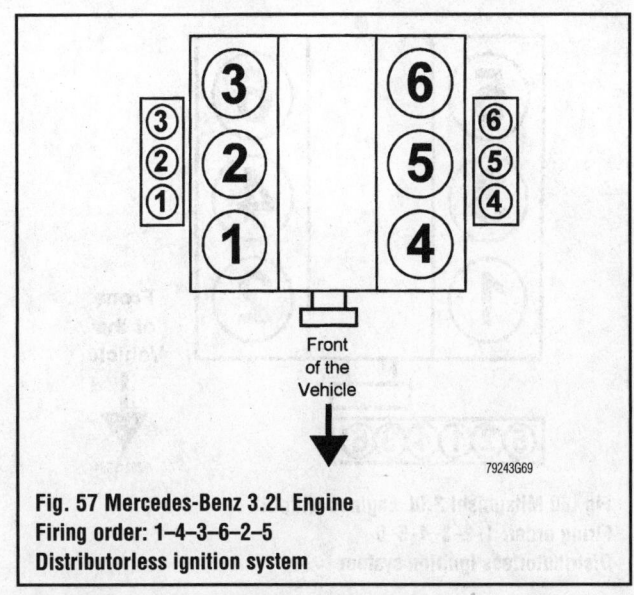

Fig. 57 Mercedes-Benz 3.2L Engine
Firing order: 1–4–3–6–2–5
Distributorless ignition system

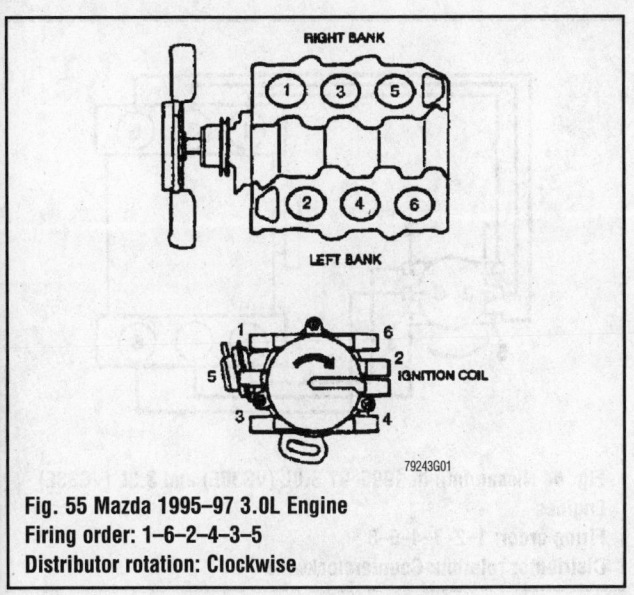

Fig. 55 Mazda 1995–97 3.0L Engine
Firing order: 1–6–2–4–3–5
Distributor rotation: Clockwise

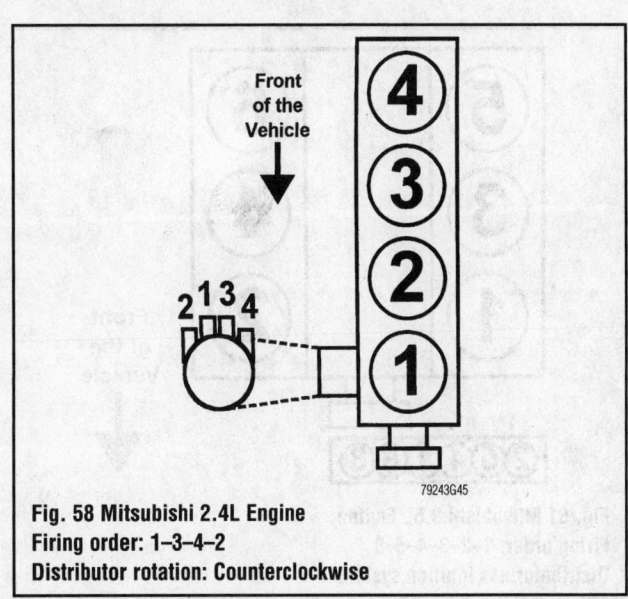

Fig. 58 Mitsubishi 2.4L Engine
Firing order: 1–3–4–2
Distributor rotation: Counterclockwise

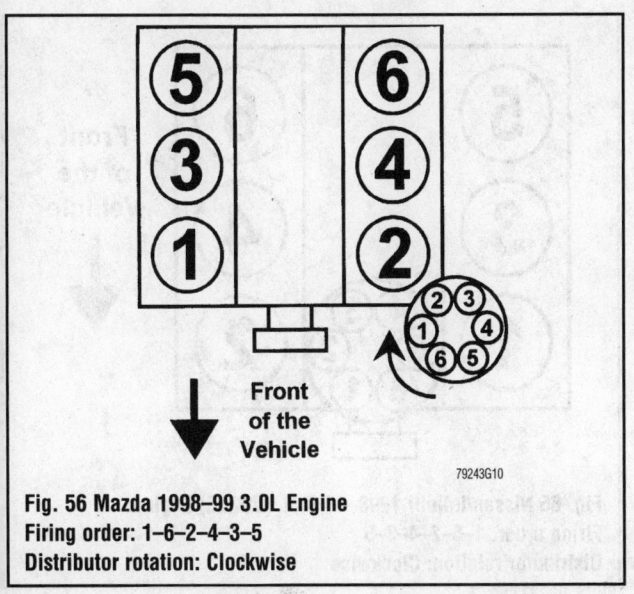

Fig. 56 Mazda 1998–99 3.0L Engine
Firing order: 1–6–2–4–3–5
Distributor rotation: Clockwise

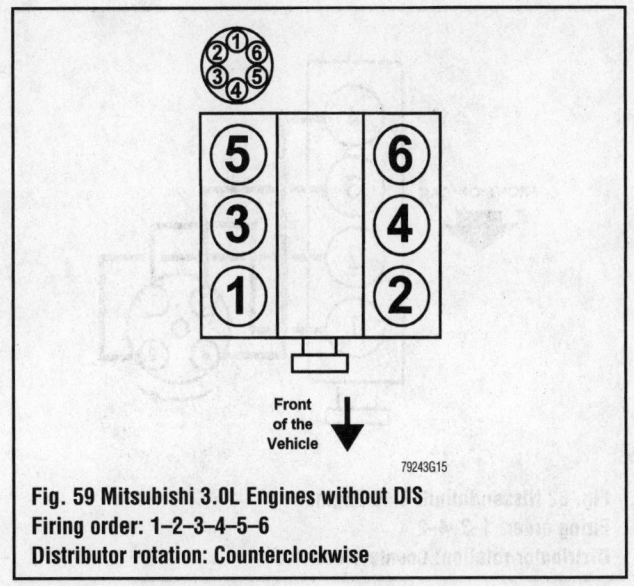

Fig. 59 Mitsubishi 3.0L Engines without DIS
Firing order: 1–2–3–4–5–6
Distributor rotation: Counterclockwise

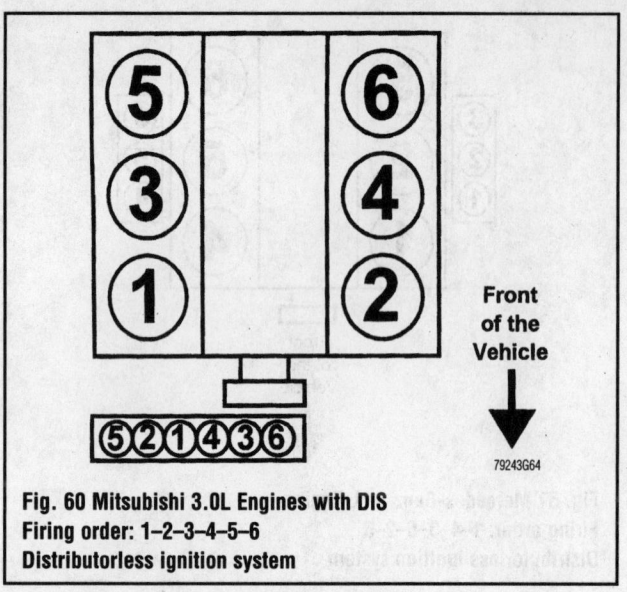

Fig. 60 Mitsubishi 3.0L Engines with DIS
Firing order: 1–2–3–4–5–6
Distributorless ignition system

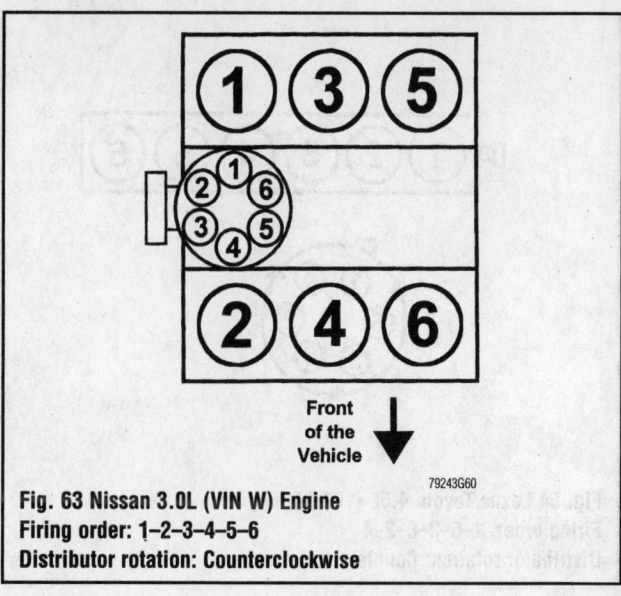

Fig. 63 Nissan 3.0L (VIN W) Engine
Firing order: 1–2–3–4–5–6
Distributor rotation: Counterclockwise

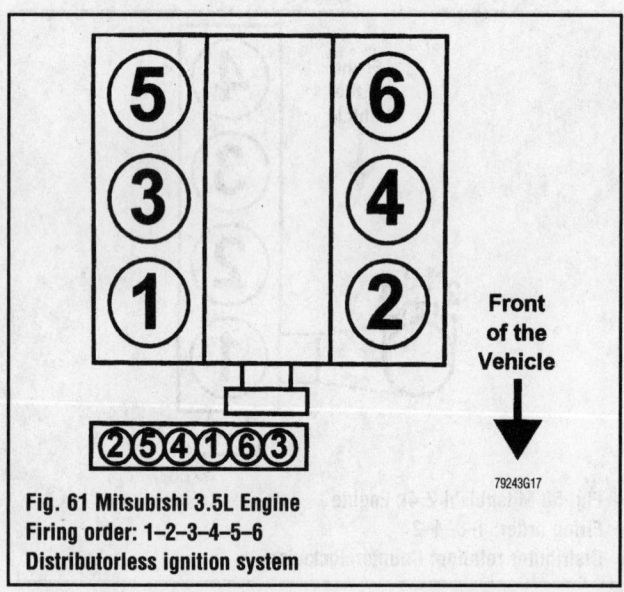

Fig. 61 Mitsubishi 3.5L Engine
Firing order: 1–2–3–4–5–6
Distributorless ignition system

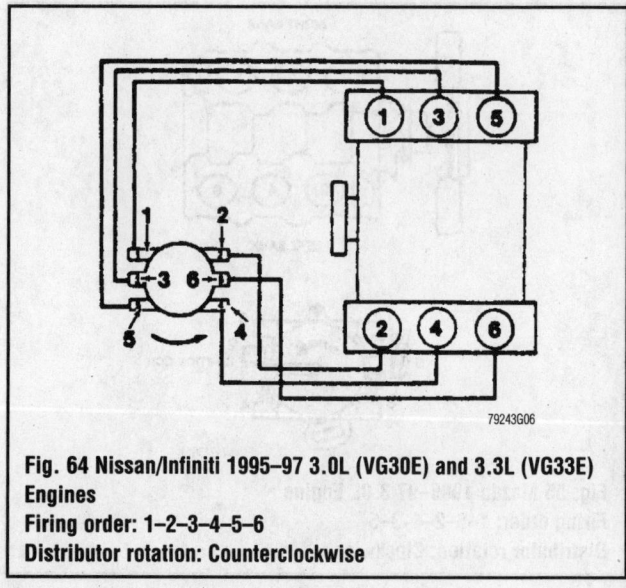

Fig. 64 Nissan/Infiniti 1995–97 3.0L (VG30E) and 3.3L (VG33E)
Engines
Firing order: 1–2–3–4–5–6
Distributor rotation: Counterclockwise

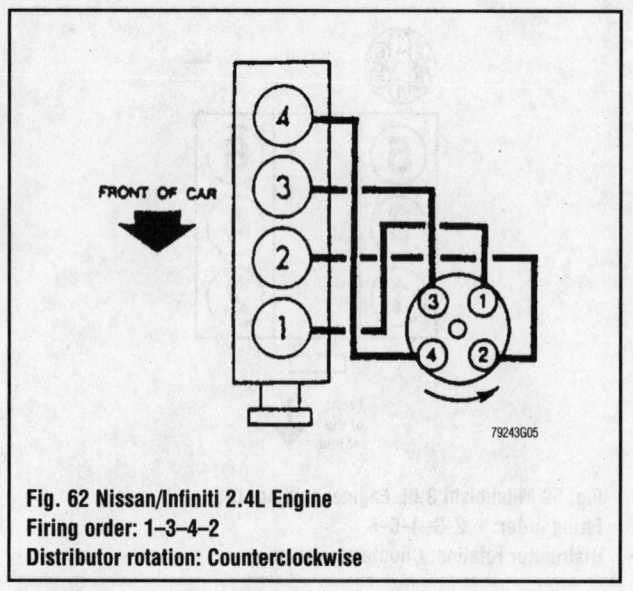

Fig. 62 Nissan/Infiniti 2.4L Engine
Firing order: 1–3–4–2
Distributor rotation: Counterclockwise

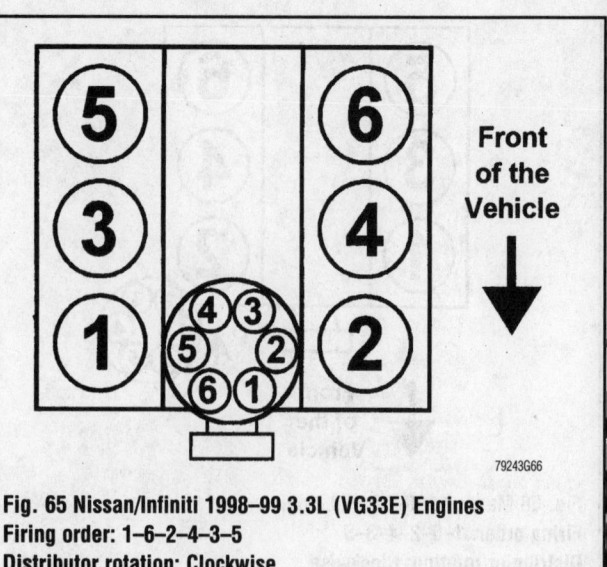

Fig. 65 Nissan/Infiniti 1998–99 3.3L (VG33E) Engines
Firing order: 1–6–2–4–3–5
Distributor rotation: Clockwise

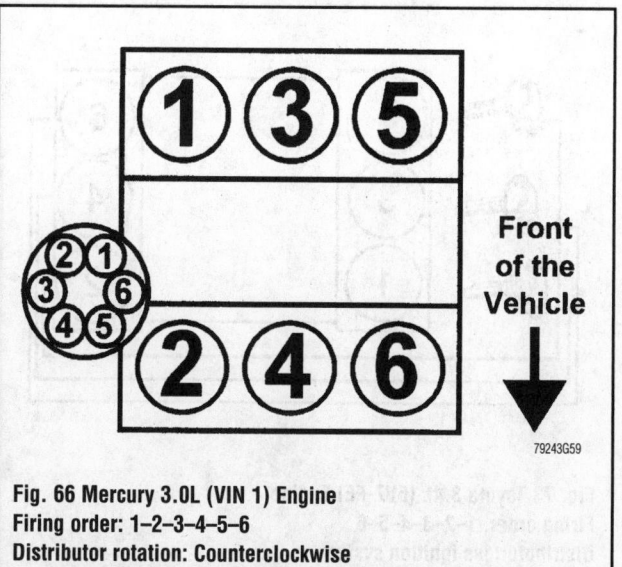

Fig. 66 Mercury 3.0L (VIN 1) Engine
Firing order: 1–2–3–4–5–6
Distributor rotation: Counterclockwise

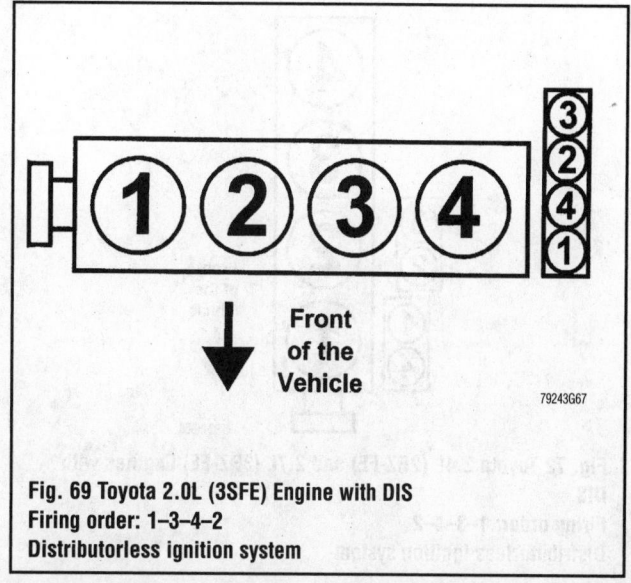

Fig. 69 Toyota 2.0L (3SFE) Engine with DIS
Firing order: 1–3–4–2
Distributorless ignition system

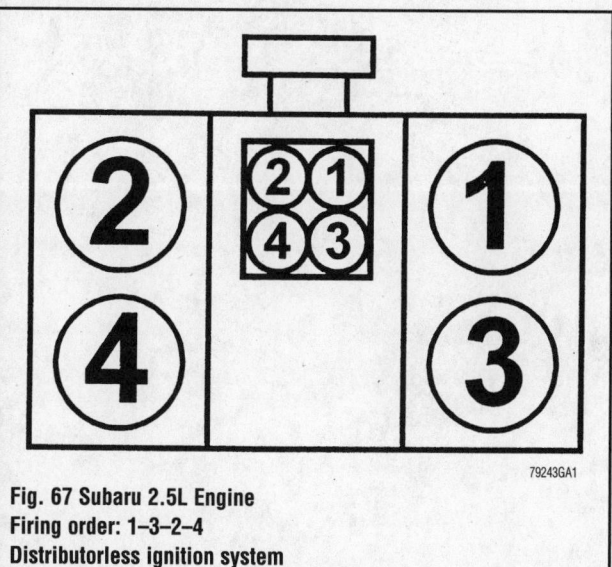

Fig. 67 Subaru 2.5L Engine
Firing order: 1–3–2–4
Distributorless ignition system

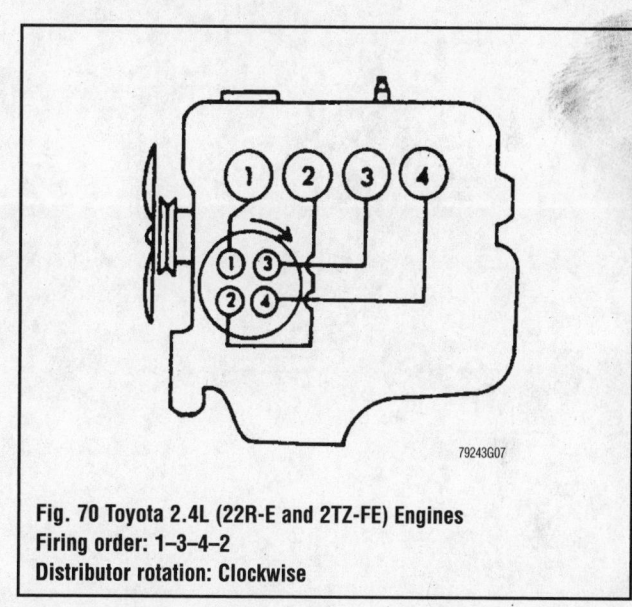

Fig. 70 Toyota 2.4L (22R-E and 2TZ-FE) Engines
Firing order: 1–3–4–2
Distributor rotation: Clockwise

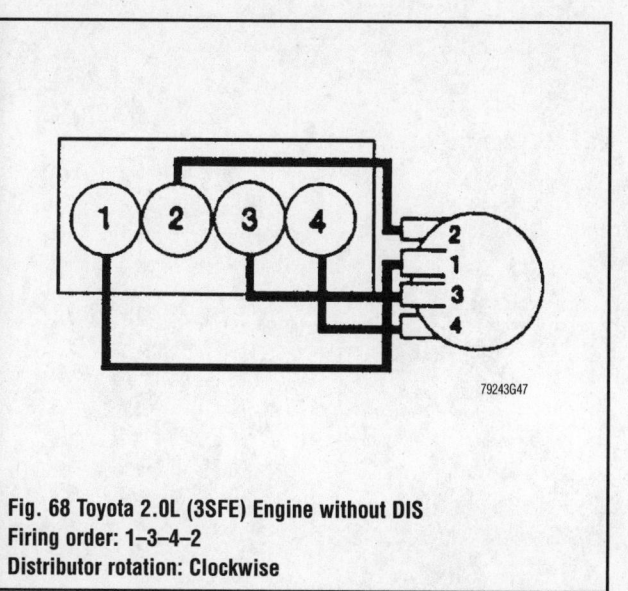

Fig. 68 Toyota 2.0L (3SFE) Engine without DIS
Firing order: 1–3–4–2
Distributor rotation: Clockwise

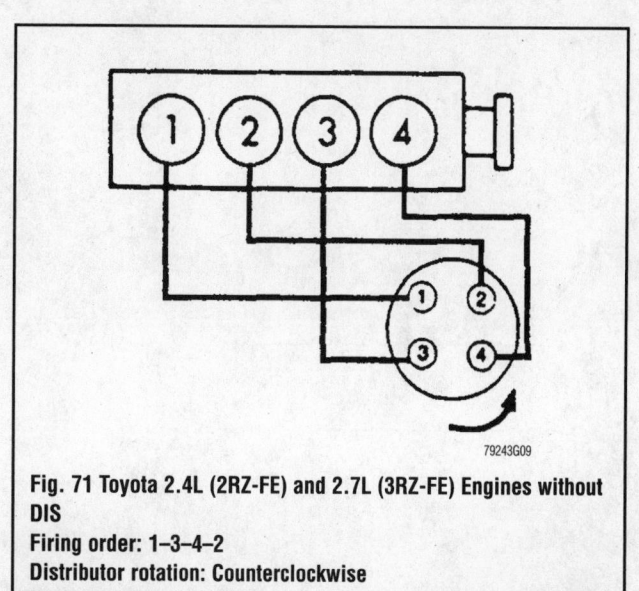

Fig. 71 Toyota 2.4L (2RZ-FE) and 2.7L (3RZ-FE) Engines without DIS
Firing order: 1–3–4–2
Distributor rotation: Counterclockwise

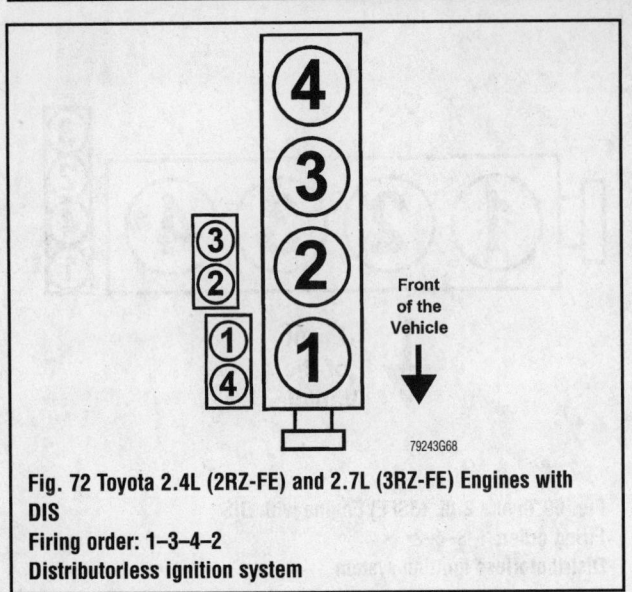

Fig. 72 Toyota 2.4L (2RZ-FE) and 2.7L (3RZ-FE) Engines with DIS
Firing order: 1–3–4–2
Distributorless ignition system

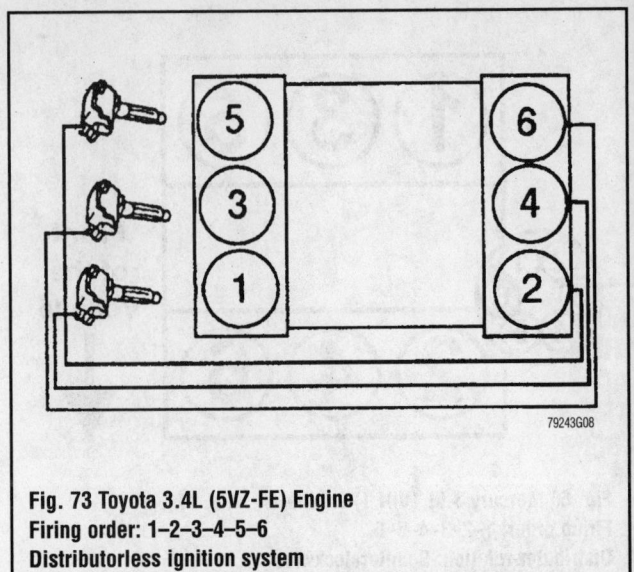

Fig. 73 Toyota 3.4L (5VZ-FE) Engine
Firing order: 1–2–3–4–5–6
Distributorless ignition system

ACCESSORY DRIVE BELTS 4

ACCESSORY DRIVE BELTS

Accessory drive belts are usually divided into two basic types: V-belts (conventional, cogged, and flat multi-ribbed) and serpentine (multi-ribbed) belts. The flat multi-ribbed V-belt actually resembles a serpentine belt, however, unlike a serpentine belt, only the inner surface of the belt makes contact with the components' pulleys. (Rarely, the back of multi-ribbed V-belts may ride against an idler or tensioner pulley, however.) V-belts ride in pulleys with V-shaped groove(s) to rotate various accessories, such as the power steering pump, air conditioner compressor, alternator/generator, water pump, and air pump. Only the inside of a V-belt is used, unlike a serpentine belt which utilizes both sides. V-belts typically operate one or two accessories per belt, whereas a single serpentine belt can drive all of the accessories. V-belts and a few serpentine belts require periodic adjustment because the belts are under tension and stretch over time. Most serpentine belts utilize an automatic belt tensioner that constantly provides the proper tension to the belt.

V-Belts

INSPECTION

Although different maintenance intervals are given by each manufacturer, it is a good rule of thumb to inspect the drive belts every 15,000 miles (24,000 km) or 12 months (whichever occurs first). Determine the belt tension at a point half-way between the pulleys by pressing on the belt with moderate thumb pressure. The belt should deflect about ¼–½ in. (6–13mm) at this point. Note that "deflection" is not play, but the ability of the belt, under actual tension, to stretch slightly and give.

Inspect the belts for the following signs of damage or wear: glazing, cracking, fraying, crumbling or missing chunks. A glazed belt will be perfectly smooth from slippage, while a good belt will have a slight texture of fabric visible. Cracks will usually start at the inner edge of the belt and run outward. A belt that is fraying will have the fabric backing de-laminating its self from the belt. A belt that is crumbling or missing chunks will have voids in the cross-section of the belt, some times the section missing chunks will be in the pulley groove and not easily seen. All worn or damaged drive belts

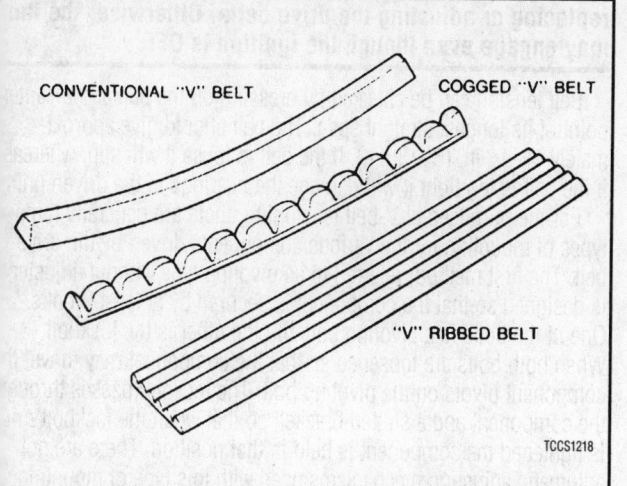

Typical accessory drive belts found on vehicles today

CONVENTIONAL "V" BELT   COGGED "V" BELT

"V" RIBBED BELT

TCCS1218

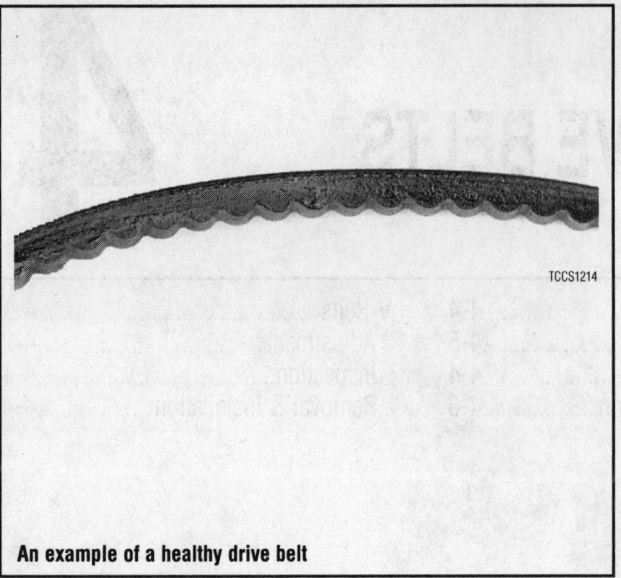

TCCS1214

An example of a healthy drive belt

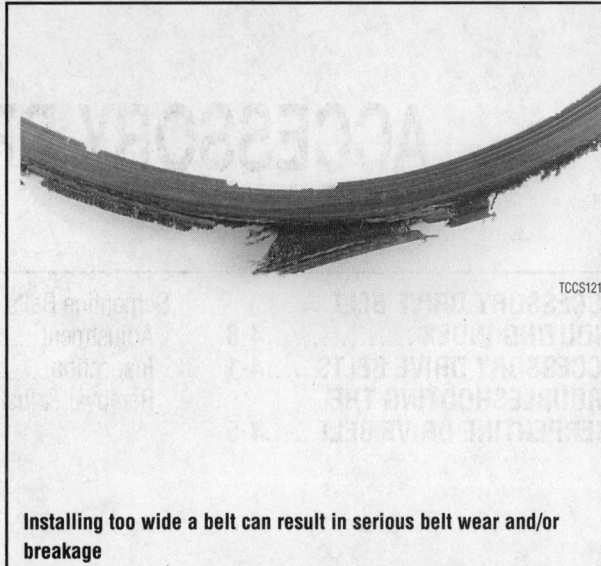

TCCS1217

Installing too wide a belt can result in serious belt wear and/or breakage

TCCS1215

Deep cracks in this belt will cause flex, building up heat that will eventually lead to belt failure

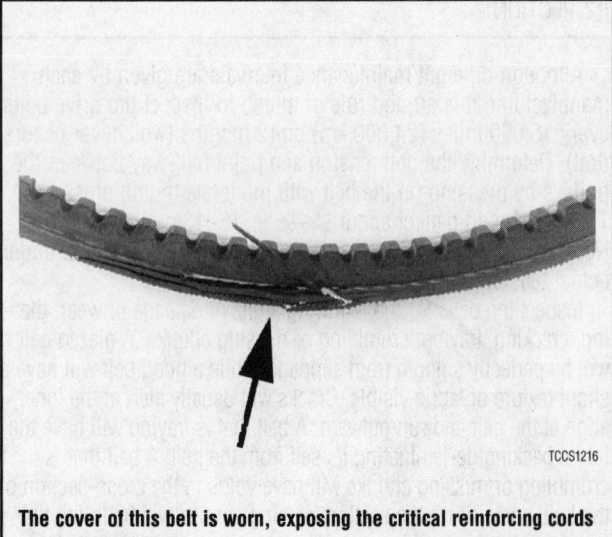

TCCS1216

The cover of this belt is worn, exposing the critical reinforcing cords to excessive wear

should be replaced immediately. It is best to replace all drive belts at one time, as a preventive maintenance measure.

Although it is generally easier on the component to have the belt too loose than too tight, a very loose belt may place a high impact load on a bearing due to the whipping or snapping action of the belt. A belt that is slightly loose may slip, especially when component loads are high. This slippage may be hard to identify. For example, the generator belt may run okay during the day, and then slip at night when headlights are turned on. Slipping belts wear quickly not only due to the direct effect of slippage but also because of the heat the slippage generates. Extreme slippage may even cause a belt to burn. A very smooth, glazed appearance on the belt's sides, as opposed to the obvious pattern of a fabric cover, indicates that the belt has been slipping.

ADJUSTMENT

> ※※ **CAUTION**
>
> **On vehicles with an electric cooling fan, disable the power to the fan by disengaging the fan motor wiring connector or removing the negative battery cable before replacing or adjusting the drive belts. Otherwise, the fan may engage even though the ignition is OFF.**

Belt tension can be checked by pressing on the belt at the center point of its longest straight span. The belt should give approximately ¼–½ in. (6–13mm). If the belt is loose it will slip, whereas if it is too tight it will damage the bearings in the driven unit.

For the purposes of V-belt tensioning, there are generally three types of mounting for the various components driven by the drive belt. The first method, referred to as pivoting type without adjuster, is designed so that the component is secured by at least 2 bolts. One of the bolts is a pivoting bolt and the other is the lockbolt. When both bolts are loosened so that the component may move, the component pivots on the pivoting bolt. The lockbolt passes through the component and a slotted bracket, so that when the lockbolt's nut is tightened the component is held in that position. There are not automatic adjusting mechanisms used with this type of mounting.

The second method of component mounting, referred to as pivoting type with adjuster, is almost identical except for the addition of

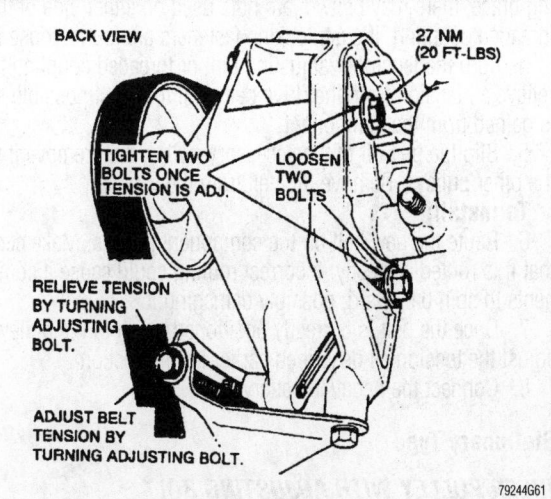

A typical pivoting accessory with an adjusting bolt

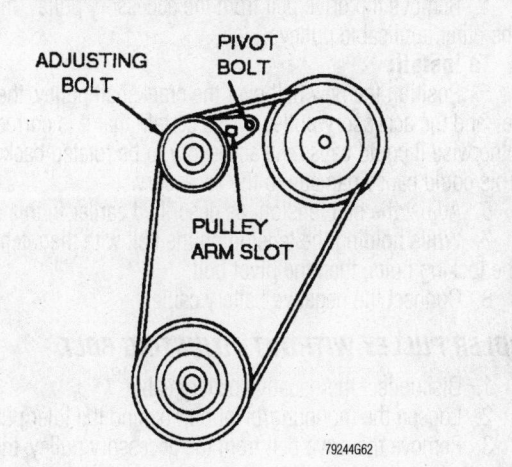

An accessory that is fixed will have an adjustable pulley—notice the square slot to aid the adjustment

an adjuster of some sort. Usually the adjuster is composed of a bracket attached to the component and a threaded adjusting bolt. After loosening the pivoting and lockbolts, the adjusting bolt can be tightened or loosened to increase or decrease the drive belt's tension. With this type of mounting, you do not have to hold the component in a tensioned position and tighten the pivoting and lockbolts; the adjusting bolt does the job for you.

Some versions of this method of mounting use an adjuster which is built into one of the components mounting braces. The brace attaches the component to the engine and incorporates a threaded adjuster in its mid-span, so that when the threaded adjuster is turned the brace shortens or lengthens. This in turn increases or decreases the amount of tension on the component.

The third type of mounting, referred to as stationary type, is designed so that the component is mounted on its brackets. There are no pivoting or lockbolts, and the component is not designed to be moved. Rather, this type of mounting uses an extra tensioner idler pulley assembly. The drive belt is tensioned by adjusting the position of the idler pulley, usually accomplished by turning the adjuster bolt on the idler mechanism.

Pivoting Type

WITHOUT ADJUSTER

1. Disconnect the negative battery cable.
2. Loosen the component's lockbolt and pivoting bolt only enough for the component to move.
3. Using a strong wooden, plastic or metal prytool, move the component either closer to, or farther away from, the engine to provide the correct tension on the belt.

✳✳ WARNING

If using a metal prytool, always wrap the end with a rag or towel to prevent accidentally damaging the component from undue stress.

4. Once the proper amount of tension is applied to the drive belt, hold the prytool with one hand while tightening the lockbolt securely with the other hand.
5. Release the pressure from the prytool and tighten the pivoting bolt securely.
6. Double check the drive belt's tension, in case the component moved slightly while tightening the bolts.
7. Connect the negative battery cable.

WITH ADJUSTER

This type of drive belt is tensioned by a tensioner, which makes precise tension adjustment easy.

1. Disconnect the negative battery cable.
2. Loosen the component's pivot and lockbolts.
3. Inspect the tensioner assembly on the component; the tensioner adjusting bolt may use a locknut or screw to prevent it from loosening over time. On the type of adjuster with a threaded mounting brace, there may be two jam nuts used on either side of the threaded coupling. If such locking fasteners are found, loosen them.
4. Turn the tensioner adjusting bolt or threaded coupling to increase or decrease the amount of tension on the drive belt, as necessary.
5. When the belt tension is correct, tighten the lockbolt and the pivot bolt.
6. If equipped, tighten the tension adjusting bolt locknut or screw to prevent the adjuster from slowly loosening over time. If equipped, tighten the two jam nuts.
7. Connect the negative battery cable.

Stationary Type

IDLER PULLEY WITH ADJUSTING BOLT

1. Loosen the idler bracket pivot bolt and locking bolts.
2. Adjust the belt tension by inserting the proper size ratchet in the square slot of the idler bracket and rotating the bracket until tension is applied.
3. While holding the tension on the belt with the ratchet, tighten the locking bolts, then the pivot bolt.

IDLER PULLEY WITHOUT ADJUSTING BOLT

1. Loosen the mounting/pivot bolt behind the idler pulley.
2. Swivel the idler pulley with a pair of pliers or a wrench on the bearing mounting until the proper tension is achieved.
3. While holding the idler pulley, at the proper tension, tighten the mounting/pivot bolt.

REMOVAL & INSTALLATION

If a belt must be replaced, the driven unit or idler pulley must be loosened and moved to its extreme loosest position, generally by moving it toward the center of the motor. After removing the old belt, check the pulleys for dirt or built-up material which could affect belt contact. Carefully install the new belt, remembering that it is new and unused; it may appear to be just a little too small to fit over the pulley flanges. Fit the belt over the largest pulley (usually the crankshaft pulley at the bottom center of the motor) first, then work on the smaller one(s). Gentle pressure in the direction of rotation is helpful. Some belts run around a third, or idler pulley, which acts as an additional pivot in the belt's path. It may be possible to loosen the idler pulley as well as the main component, making your job much easier. Depending on which belt(s) you are changing, it may be necessary to loosen or remove other interfering belts to get at the one(s) you want.

When buying replacement belts, remember that the fit is critical according to the length of the belt ("diameter"), the width of the belt, the depth of the belt and the angle or profile of the V shape or the ribs. The belt shape should match the shape of the pulley exactly; belts that are not an exact match can cause noise, slippage and premature failure.

After the new belt is installed, draw tension on it by moving the driven unit or idler pulley away from the motor and tighten its mounting bolts. This is sometimes a three or four-handed job; you may find an assistant helpful. Make sure that all the bolts you loosened get retightened and that any other loosened belts also have the correct tension. A new belt can be expected to stretch a bit after installation so be prepared to readjust your new belt, if needed, within the first two hundred miles of use.

Pivoting Type

> ⁕⁕ **CAUTION**
>
> **On vehicles with an electric cooling fan, disable the power to the fan by disengaging the fan motor wiring connector or removing the negative battery cable before replacing or adjusting the drive belts. Otherwise, the fan may engage even though the ignition is OFF.**

WITHOUT ADJUSTER

1. Disconnect the negative battery cable.
2. Loosen the accessory's slotted adjusting bracket bolt. If the hinge bolt is excessively tight, it too will have to be loosened.
3. Push the component toward the engine to provide enough slack in the belt so that it will slide over one of the accessory drive pulleys. Remove the drive belt from the accessory drive pulleys and from the vehicle.

To install:

4. Position the new drive belt over the component pulleys. Make sure that it is routed correctly.
5. Adjust the tension of the belt, as described earlier in this section.
6. Connect the negative battery cable.

WITH ADJUSTER

1. Disconnect the negative battery cable.
2. Loosen the component's pivot and lockbolts.
3. Inspect the tensioner assembly on the component; the tensioner adjusting bolt may use a locknut or screw to prevent it from loosening over time. On the type of adjuster with a threaded mount-

ing brace, there may be two jam nuts used on either side of the threaded coupling. If such locking fasteners are found, loosen them.

4. Turn the tensioner adjusting bolt or threaded coupling to relieve all tension from the drive belt until the most possible slack is gained from the component.
5. Slip the belt off of the accessory pulley, then remove it from the other pulleys. Remove the belt from the vehicle.

To install:

6. Route the new belt on the component pulleys. Make certain that it is routed correctly; incorrect routing could cause a components to spin backward, possibly damaging it.
7. Once the belt is correctly positioned on all of the pulleys, adjust the tension as described earlier in this section.
8. Connect the negative battery cable.

Stationary Type

IDLER PULLEY WITH ADJUSTING BOLT

1. Disconnect the negative battery cable.
2. Loosen the idler bracket pivot bolt and locking bolts.
3. Move the idler pulley until the most amount of slack is gained.
4. Remove the drive belt from the accessory pulley, then from the other applicable pulleys.

To install:

5. Position the new belt over the crankshaft pulley, the idler pulley and the accessory pulley. Make certain that it is correctly routed otherwise it could cause the accessory to be rotated backwards. This could cause damage to the accessory.
6. Adjust the belt tension, as described earlier in this section.
7. While holding the tension on the belt with the ratchet, tighten the locking bolts, then the pivot bolt.
8. Connect the negative battery cable.

IDLER PULLEY WITHOUT ADJUSTING BOLT

1. Disconnect the negative battery cable.
2. Loosen the mounting/pivot bolt behind the idler pulley.
3. Remove the drive belt from the accessory pulley, then from the other applicable pulleys.

To install:

4. Position the new belt over the crankshaft pulley, the idler pulley and the accessory pulley. Make certain that it is correctly routed otherwise it could cause the accessory to be rotated backwards. This could cause damage to the accessory.
5. Swivel the idler pulley with a pair of pliers or a wrench on the bearing mounting until the proper tension is achieved.
6. While holding the idler pulley, at the proper tension, tighten the mounting/pivot bolt.
7. Connect the negative battery cable.

Serpentine Belts

INSPECTION

Although many manufacturers recommend that the drive belt(s) be inspected every 30,000 miles (48,000 km) or more, it is really a good idea to check them at least once a year, or at every major fluid change. Whichever interval you choose, the belts should be checked for wear or damage. Obviously, a damaged drive belt can cause problems should give way while the vehicle is in operation. But, improper length belts (too short or long), as well as excessively worn belts, can also cause problems. Loose accessory drive belts can lead to poor engine cooling and diminished output from the alternator, air conditioning compressor

Troubleshooting the Serpentine Drive Belt

Problem	Cause	Solution
Tension sheeting fabric failure (woven fabric on outside circumference of belt has cracked or separated from body of belt)	• Grooved or backside idler pulley diameters are less than minimum recommended • Tension sheeting contacting (rubbing) stationary object • Excessive heat causing woven fabric to age • Tension sheeting splice has fractured	• Replace pulley(s) not conforming to specification • Correct rubbing condition • Replace belt • Replace belt
Noise (objectional squeal, squeak, or rumble is heard or felt while drive belt is in operation)	• Belt slippage • Bearing noise • Belt misalignment • Belt-to-pulley mismatch • Driven component inducing vibration • System resonant frequency inducing vibration	• Adjust belt • Locate and repair • Align belt/pulley(s) • Install correct belt • Locate defective driven component and repair • Vary belt tension within specifications. Replace belt.
Rib chunking (one or more ribs has separated from belt body)	• Foreign objects imbedded in pulley grooves • Installation damage • Drive loads in excess of design specifications • Insufficient internal belt adhesion	• Remove foreign objects from pulley grooves • Replace belt • Adjust belt tension • Replace belt
Rib or belt wear (belt ribs contact bottom of pulley grooves)	• Pulley(s) misaligned • Mismatch of belt and pulley groove widths • Abrasive environment • Rusted pulley(s) • Sharp or jagged pulley groove tips • Rubber deteriorated	• Align pulley(s) • Replace belt • Replace belt • Clean rust from pulley(s) • Replace pulley • Replace belt
Longitudinal belt cracking (cracks between two ribs)	• Belt has mistracked from pulley groove • Pulley groove tip has worn away rubber-to-tensile member	• Replace belt • Replace belt
Belt slips	• Belt slipping because of insufficient tension • Belt or pulley subjected to substance (belt dressing, oil, ethylene glycol) that has reduced friction • Driven component bearing failure • Belt glazed and hardened from heat and excessive slippage	• Adjust tension • Replace belt and clean pulleys • Replace faulty component bearing • Replace belt
"Groove jumping" (belt does not maintain correct position on pulley, or turns over and/or runs off pulleys)	• Insufficient belt tension • Pulley(s) not within design tolerance • Foreign object(s) in grooves	• Adjust belt tension • Replace pulley(s) • Remove foreign objects from grooves

TCCS3C09

Troubleshooting the Serpentine Drive Belt

Problem	Cause	Solution
"Groove jumping" (belt does not maintain correct position on pulley, or turns over and/or runs off pulleys)	• Excessive belt speed • Pulley misalignment • Belt-to-pulley profile mismatched • Belt cordline is distorted	• Avoid excessive engine acceleration • Align pulley(s) • Install correct belt • Replace belt
Belt broken (Note: identify and correct problem before replacement belt is installed)	• Excessive tension • Tensile members damaged during belt installation • Belt turnover • Severe pulley misalignment • Bracket, pulley, or bearing failure	• Replace belt and adjust tension to specification • Replace belt • Replace belt • Align pulley(s) • Replace defective component and belt
Cord edge failure (tensile member exposed at edges of belt or separated from belt body)	• Excessive tension • Drive pulley misalignment • Belt contacting stationary object • Pulley irregularities • Improper pulley construction • Insufficient adhesion between tensile member and rubber matrix	• Adjust belt tension • Align pulley • Correct as necessary • Replace pulley • Replace pulley • Replace belt and adjust tension to specifications
Sporadic rib cracking (multiple cracks in belt ribs at random intervals)	• Ribbed pulley(s) diameter less than minimum specification • Backside bend flat pulley(s) diameter less than minimum • Excessive heat condition causing rubber to harden • Excessive belt thickness • Belt overcured • Excessive tension	• Replace pulley(s) • Replace pulley(s) • Correct heat condition as necessary • Replace belt • Replace belt • Adjust belt tension

TCCS3C10

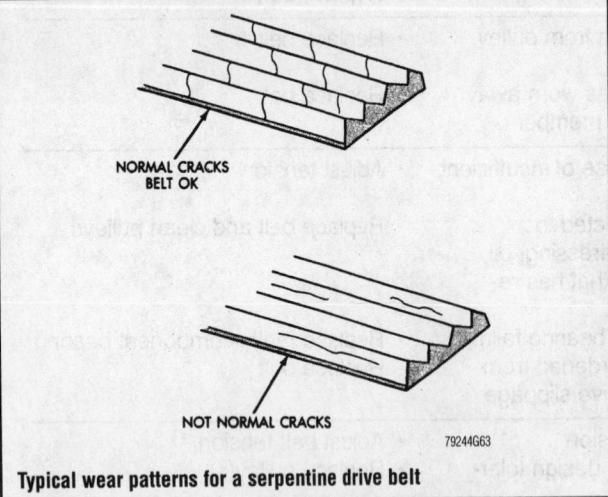

NORMAL CRACKS
BELT OK

NOT NORMAL CRACKS

79244G63

Typical wear patterns for a serpentine drive belt

or power steering pump. A belt that is too tight places a severe strain on the driven unit and can wear out bearings quickly.

Serpentine drive belts should be inspected for rib chunking (pieces of the ribs breaking off), severe glazing, frayed cords or other visible damage. Any belt which is missing sections of 2 or more adjacent ribs which are ½ in. (13mm) or longer must be replaced. You might want to note that serpentine belts do tend to form small cracks across the backing. If the only wear you find is in the form of one or more cracks are across the backing and NOT parallel to the ribs, the belt is still good and does not need to be replaced.

ADJUSTMENT

Periodic drive belt tensioning is not necessary, because an automatic spring-loaded tensioner is used with these belts to maintain proper adjustment at all times. The tensioner is also useful as a wear indicator. When the belt is properly installed, the arrow on the tensioner housing must point within the acceptable range lines on the tensioner's face. If the arrow falls outside the range, either an improper belt has been installed or the belt is worn beyond its useful life span. In either case, a new belt must be installed immediately to assure proper engine operation and to prevent possible accessory damage.

REMOVAL & INSTALLATION

Because serpentine belts use a spring loaded tensioner for adjustment, belt replacement tends to be somewhat easier than it used to be on engines where accessories were pivoted and bolted in place for tension adjustment. Basically, all belt replacement involves is to pivot the tensioner to loosen the belt, then slide the belt off of the pulleys. The two most important points are to pay CLOSE attention to the proper belt routing (since serpentine belts tend to be "snaked" all different ways through the pulleys) and to make sure the V-ribs are properly seated in all the pulleys.

Although belt routing diagrams have been included in this section, the first places you should check for proper belt routing are the labels in your engine compartment. These should include a belt routing diagram which may reflect changes made during a production run.

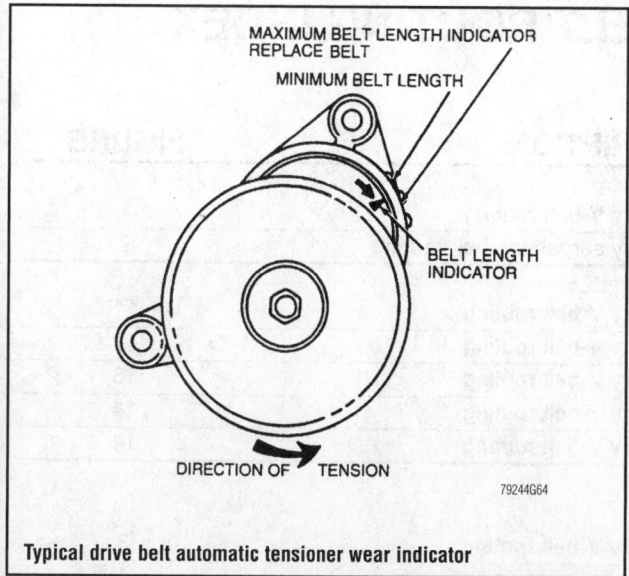

Typical drive belt automatic tensioner wear indicator

Often the underhood label will display the serpentine drive belt routing

1. Disconnect the negative battery cable for safety. This will help assure that no one mistakenly cranks the engine over with your hands between the pulleys, and that the cooling fan cannot activate while servicing the belt(s).

➡Take a good look at the installed belt and make a note of the routing. Before removing the belt, make sure the routing matches that of the belt routing label or one of the diagrams in this book. If for some reason a diagram does not match (you may not have the original engine or it may have been modified), carefully note the changes on a piece of paper.

2. For tensioners equipped with a ½ in. (13mm) square hole, insert the drive end of a large breaker bar into the hole. Use the breaker bar to pivot the tensioner away from the drive belt. For tensioners not equipped with this hole, use the proper-sized socket and breaker bar (or a large handled wrench) on the tensioner idler pulley center bolt to pivot the tensioner away from the belt. This will loosen the belt sufficiently that it can be pulled off of one or more of the pulleys. It is usually easiest to carefully pull the belt out from underneath the tensioner pulley itself.

3. Once the belt is off one of the pulleys, gently pivot the tensioner back into position. DO NOT allow the tensioner to snap back, as this could damage the tensioner's internal parts.

4. Now finish removing the belt from the other pulleys and remove it from the engine.

To install:

5. While referring to the proper routing diagram (which you identified earlier), begin to route the belt over the pulleys, leaving whichever pulley you first released it from for last.

6. Once the belt is mostly in place, carefully pivot the tensioner and position the belt over the final pulley. As you begin to allow the tensioner back into contact with the belt, run your hand around the pulleys and make sure the belt is properly seated in the ribs. If not, release the tension and seat the belt.

7. Once the belt is installed, take another look at all the pulleys to double check your installation.

8. Connect the negative battery cable, then start and run the engine to check belt operation.

9. Once the engine has reached normal operating temperature, turn the ignition OFF and check that the belt tensioner arrow is within the proper adjustment range.

Relieve the belt tension by pivoting the automatic tensioner away from the belt, then remove the belt

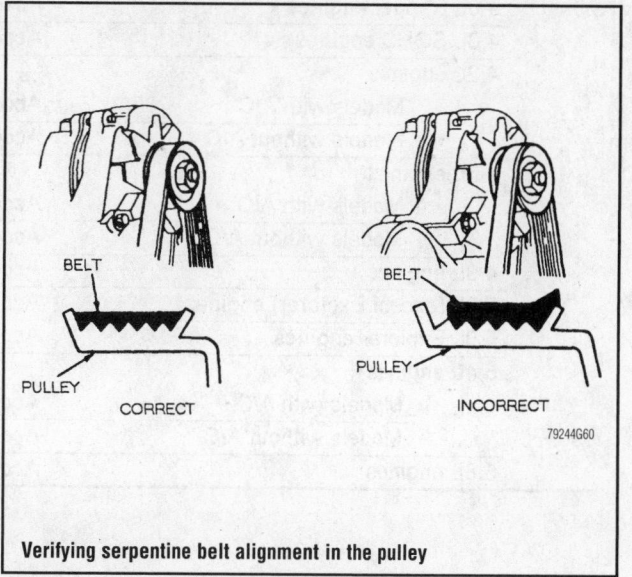

Verifying serpentine belt alignment in the pulley

ACCESSORY DRIVE BELT ROUTING INDEX

MANUFACTURER

ENGINES	DESCRIPTION	FIGURE
Acura		
3.2L SLX 3.2L engines	Accessory V-belt routing	1
3.5L SLX engines	Accessory serpentine belt routing	2
Chrysler		
2.4L engines	Accessory V-belt routing	11
2.5L engines	Accessory V-belt routing	12
3.0L engines	Accessory V-belt routing	13
3.3L engines	Accessory V-belt routing	14
3.8L engines	Accessory V-belt routing	14
Dodge		
2.5L engines		
1995 Models	Accessory V-belt routing	3
1996-99 Models with A/C	Accessory serpentine belt routing	4
1996-99 Models without A/C	Accessory serpentine belt routing	5
3.9L, 5.2L and 5.9L LDC gasoline engines	Accessory serpentine belt routing	6
5.9L Diesel engines		
Models with A/C	Accessory serpentine belt routing	9
Models without A/C	Accessory serpentine belt routing	10
5.9L HDC gasoline engines		
Models with A/C	Accessory serpentine belt routing	7
Models without A/C	Accessory serpentine belt routing	8
8.0L engines with A/C		
Models with A/C	Accessory serpentine belt routing	7
Models without A/C	Accessory serpentine belt routing	8
Ford		
2.3L Ranger engines	Accessory serpentine belt routing	15
3.0L Aerostar engines		
Models with A/C	Accessory serpentine belt routing	17
Models without A/C	Accessory serpentine belt routing	18
3.0L Ranger engines	Accessory serpentine belt routing	16
3.0L Windstar engines	Accessory serpentine belt routing	31
3.8L Windstar engines	Accessory serpentine belt routing	32
4.0L Ranger engines	Accessory serpentine belt routing	19
4.0L SOHC engines	Accessory serpentine belt routing	29
4.2L engines		
Models with A/C	Accessory serpentine belt routing	27
Models without A/C	Accessory serpentine belt routing	28
4.6L engines		
Models with A/C	Accessory serpentine belt routing	25
Models without A/C	Accessory serpentine belt routing	26
4.9L engines	Accessory serpentine belt routing	20
5.0L (except Explorer) engines	Accessory serpentine belt routing	21
5.0L Explorer engines	Accessory serpentine belt routing	30
5.4L engines		
Models with A/C	Accessory serpentine belt routing	25
Models without A/C	Accessory serpentine belt routing	26
5.8L engines	Accessory serpentine belt routing	21

79244C01

ACCESSORY DRIVE BELT ROUTING INDEX

MANUFACTURER

ENGINES	DESCRIPTION	FIGURE
Ford (cont.)		
6.8L engines		
Models with A/C	Accessory serpentine belt routing	25
Models without A/C	Accessory serpentine belt routing	26
7.3L turbo Diesel engines	Accessory serpentine belt routing	23
7.5L (except Motorhome) engines	Accessory serpentine belt routing	22
7.5L F-Super Duty Motorhome engines	Accessory serpentine belt routing	24
General Motors		
2.2L engines with A/C	Accessory serpentine belt routing	35
2.2L engines without A/C	Accessory serpentine belt routing	34
3.1L and 3.4L engines	Accessory serpentine belt routing	47
3.8L engines	Accessory serpentine belt routing	46
4.3L (VIN W and X) engines		
1995 Models	Accessory serpentine belt routing	36
1996-99 Models	Accessory serpentine belt routing	37
4.3L (except Full-size) engines		
1996-99 Models with A/C	Accessory serpentine belt routing	43
1996-99 Models without A/C	Accessory serpentine belt routing	42
4.3L Full-size Truck engines		
1995 Models with A.I.R.	Accessory serpentine belt routing	39
1995 Models without A.I.R.	Accessory serpentine belt routing	38
5.0L engines		
1995 Models with A.I.R.	Accessory serpentine belt routing	39
1996-99 Models with A/C	Accessory serpentine belt routing	43
1996-99 Models without A/C	Accessory serpentine belt routing	42
1995 Models without A.I.R.	Accessory serpentine belt routing	38
5.7L engines		
1995 Models with A.I.R.	Accessory serpentine belt routing	39
1996-99 Models with A/C	Accessory serpentine belt routing	43
1995 Models without A.I.R.	Accessory serpentine belt routing	38
1996-99 Models without A/C	Accessory serpentine belt routing	42
7.4L engines		
1995 Models	Accessory serpentine belt routing	41
1996-99 Models	Accessory serpentine belt routing	44
Diesel engines		
1995 Models	Accessory serpentine belt routing	40
1996-99 Models	Accessory serpentine belt routing	45
Honda		
2.0L, 2.2L, 2.3L, and 2.6L engines	Accessory V-belt routing	48
3.2L engines		
1995-97 Models	Accessory V-belt routing	49
1998-99 Models	Accessory serpentine belt routing	50
Isuzu		
2.2L (except Hombre) and 2.6L engines	Accessory V-belt routing	48
2.2L Hombre engines		
Models with A/C	Accessory serpentine belt routing	35
Models without A/C	Accessory serpentine belt routing	34

79244C02

ACCESSORY DRIVE BELT ROUTING INDEX

MANUFACTURER

ENGINES	DESCRIPTION	FIGURE
Isuzu (cont.)		
3.2L Trooper engines	Accessory V-belt routing	1
3.5L Trooper engines	Accessory serpentine belt routing	2
3.2L Amigo and Rodeo engines		
1995-97 Models	Accessory V-belt routing	49
1998-99 Models	Accessory serpentine belt routing	50
Jeep		
2.5L Cherokee engines		
1995 Models with A/C	Accessory serpentine belt routing	55
1995 Models without A/C	Accessory serpentine belt routing	56
1996-99 Models with A/C	Accessory serpentine belt routing	66
1996-99 Models without A/C	Accessory serpentine belt routing	67
2.5L Wrangler engines		
1997-99 Models with A/C	Accessory serpentine belt routing	61
1997-99 Models without A/C	Accessory serpentine belt routing	60
4.0L Cherokee (right-hand drive) engines		
1996-99 Models with A/C	Accessory serpentine belt routing	62
1996-99 Models without A/C	Accessory serpentine belt routing	63
1995 Models	Accessory serpentine belt routing	59
4.0L Cherokee engines		
1995 Models with A/C	Accessory serpentine belt routing	58
1995 Models without A/C	Accessory serpentine belt routing	57
1996-99 Models with A/C	Accessory serpentine belt routing	65
1996-99 Models without A/C	Accessory serpentine belt routing	64
4.0L Grand Cherokee engines		
1995 Models	Accessory serpentine belt routing	51
1996-99 Models	Accessory serpentine belt routing	53
4.0L Wrangler engines		
1995 Models with A/C	Accessory serpentine belt routing	55
1997-99 Models with A/C	Accessory serpentine belt routing	61
1995 Models without A/C	Accessory serpentine belt routing	56
1997-99 Models without A/C	Accessory serpentine belt routing	60
5.2L Grand Cherokee engines		
1995 Models	Accessory serpentine belt routing	52
1996-99 Models	Accessory serpentine belt routing	54
	Accessory serpentine belt routing	54
Kia		
1.6L engines	Accessory V-belt routing	68
Land Rover		
3.9L engines		
Models with A/C	Accessory serpentine belt routing	69
Models without A/C	Accessory serpentine belt routing	70
4.0L engines		
Models with A/C	Accessory serpentine belt routing	69
Models without A/C	Accessory serpentine belt routing	70
4.6L engines		
Models with A/C	Accessory serpentine belt routing	69
Models without A/C	Accessory serpentine belt routing	70

ACCESSORY DRIVE BELT ROUTING INDEX

MANUFACTURER

ENGINES	DESCRIPTION	FIGURE
Lexus		
4.5L engines	Accessory V-belt routing	87
Mazda		
2.3L B-series truck engines	Accessory serpentine belt routing	15
3.0L B-series truck engines	Accessory serpentine belt routing	16
3.0L MPV engines	Accessory V-belt routing (alternator)	71
3.0L MPV engines	Accessory V-belt routing (power steering)	72
3.0L MPV engines	Accessory V-belt routing (A/C compressor)	73
4.0L B-series truck engines	Accessory serpentine belt routing	19
Mercedes-Benz		
3.2L engines	Accessory serpentine belt routing	74
Mercury		
3.0L Villager engines	Accessory V-belt routing	33
5.0L Mountaineer engines	Accessory serpentine belt routing	30
Mitsubishi		
2.4L engines	Accessory serpentine belt routing	75
3.0L engines	Accessory serpentine belt routing	76
3.5L engines	Accessory serpentine belt routing	77
Nissan		
2.4L (KA24E) engines	Accessory V-belt routing	78
3.0L (VG30E) engines	Accessory V-belt routing	79
3.3L (VG33E) engines	Accessory V-belt routing	80
3.0L Quest engines	Accessory V-belt routing	33
Subaru		
2.5L engines	Accessory V-belt routing	81
Toyota		
2.0L (3SFE) engines	Accessory V-belt routing	82
2.4L (2RZFE) engines	Accessory V-belt routing	83
2.4L (2TZFZE) engines	Accessory V-belt routing	84
2.7L engines	Accessory V-belt routing	85
3.4L engines	Accessory V-belt routing	86
4.5L engines	Accessory V-belt routing	87

79244C04

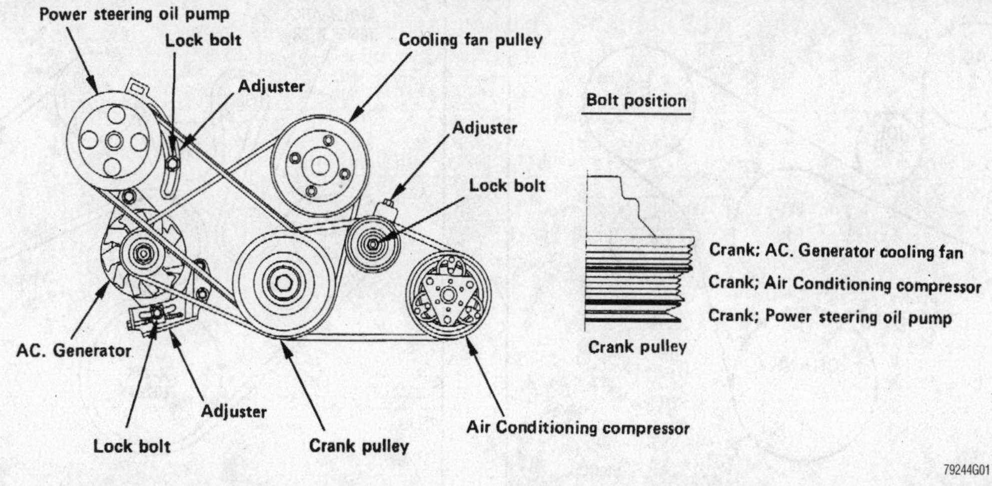

Fig. 1 Accessory V-belt routing—Acura/Isuzu 3.2L engines

79244G01

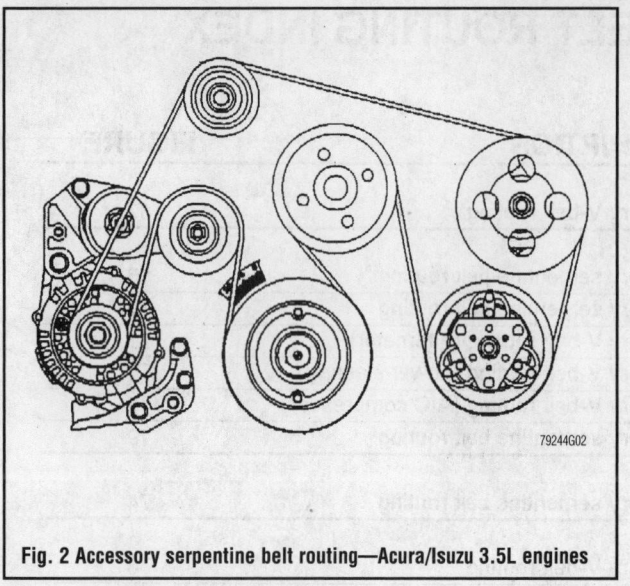

Fig. 2 Accessory serpentine belt routing—Acura/Isuzu 3.5L engines

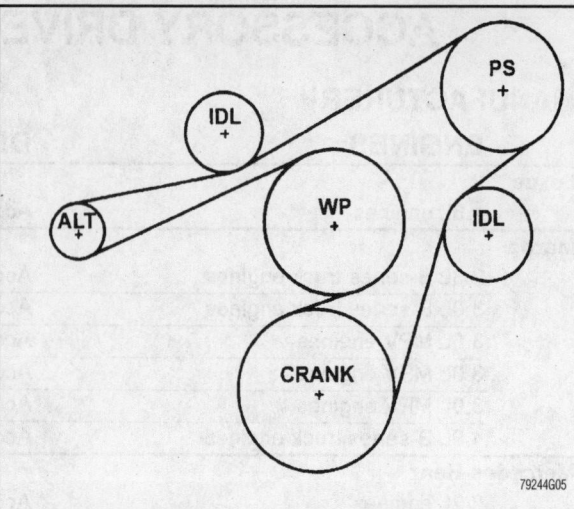

Fig. 5 Accessory serpentine belt routing—Dodge 1996–99 2.5L engines without A/C

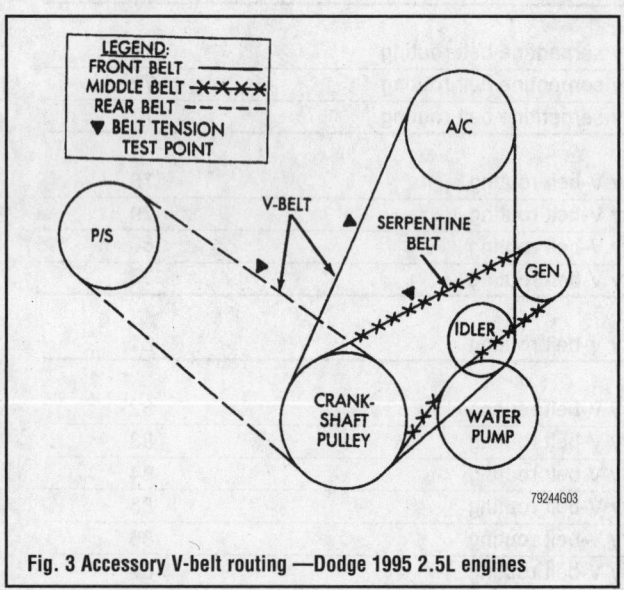

Fig. 3 Accessory V-belt routing —Dodge 1995 2.5L engines

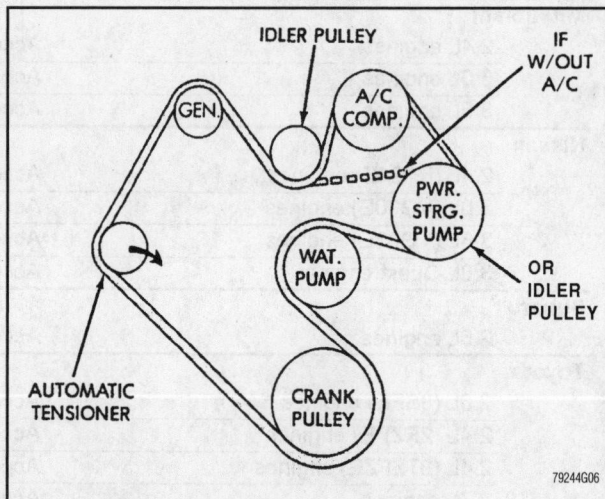

Fig. 6 Accessory serpentine belt routing—Dodge 3.9L, 5.2L and 5.9L LDC gasoline engines

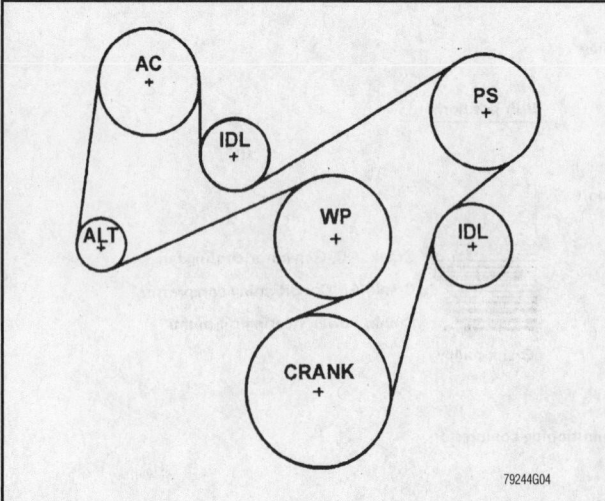

Fig. 4 Accessory serpentine belt routing—Dodge 1996–99 2.5L engines with A/C

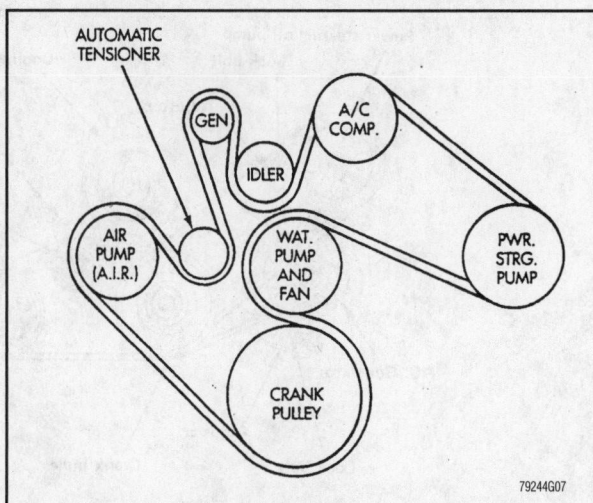

Fig. 7 Accessory serpentine belt routing—Dodge 5.9L HDC and 8.0L gasoline engines with A/C

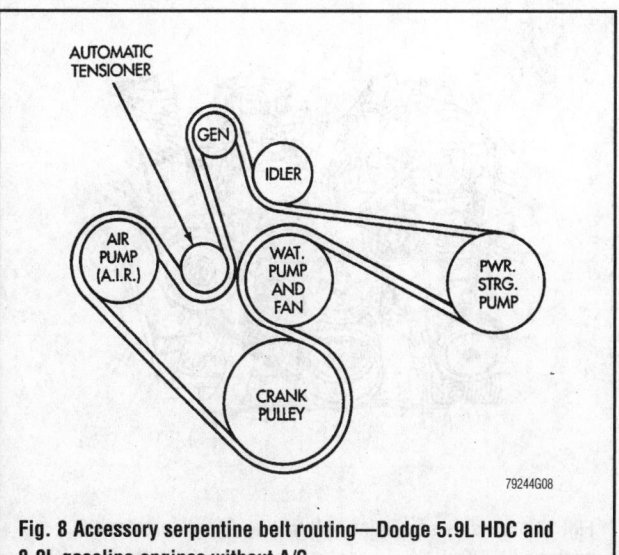

Fig. 8 Accessory serpentine belt routing—Dodge 5.9L HDC and 8.0L gasoline engines without A/C

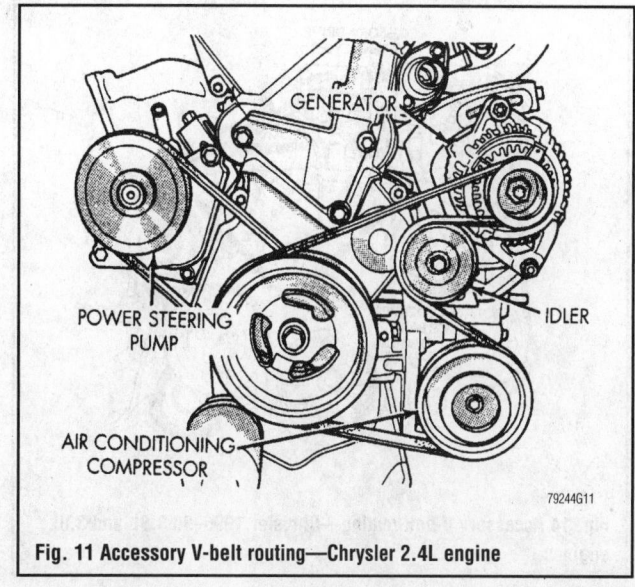

Fig. 11 Accessory V-belt routing—Chrysler 2.4L engine

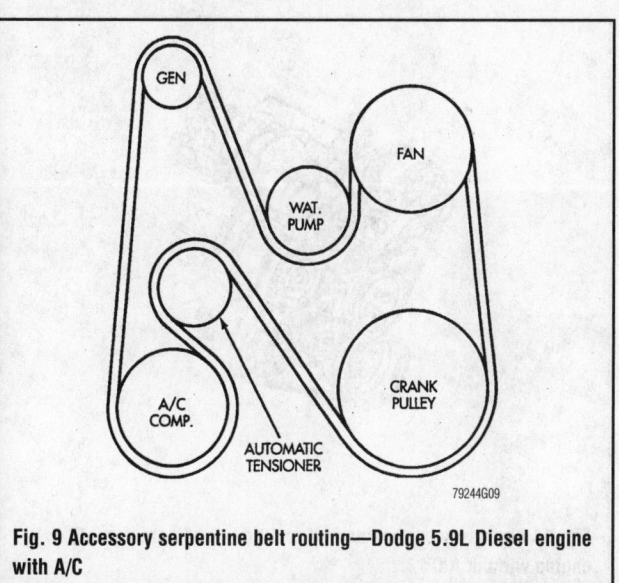

Fig. 9 Accessory serpentine belt routing—Dodge 5.9L Diesel engine with A/C

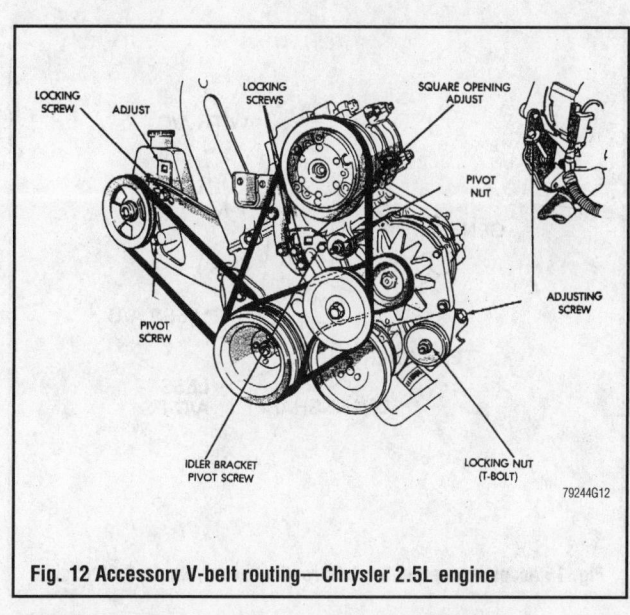

Fig. 12 Accessory V-belt routing—Chrysler 2.5L engine

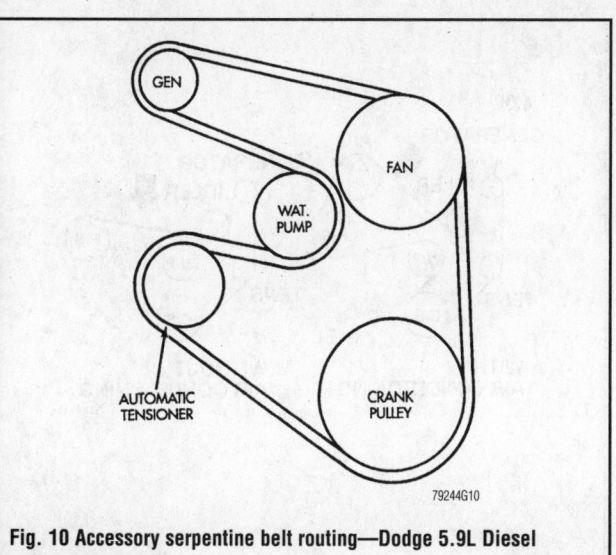

Fig. 10 Accessory serpentine belt routing—Dodge 5.9L Diesel engine without A/C

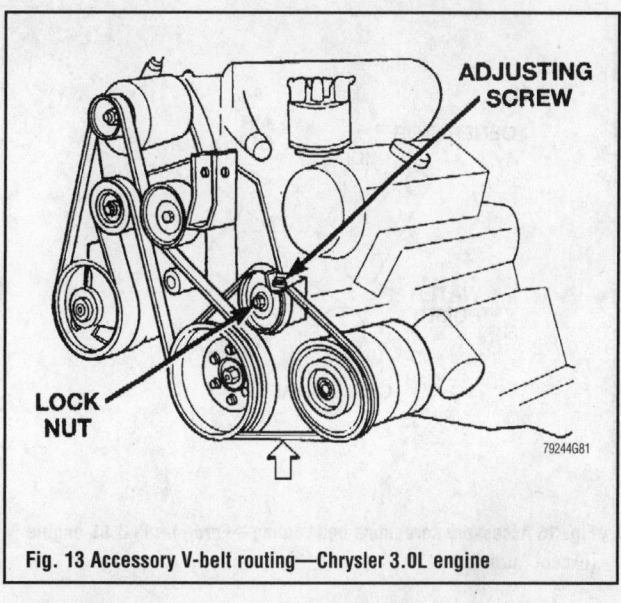

Fig. 13 Accessory V-belt routing—Chrysler 3.0L engine

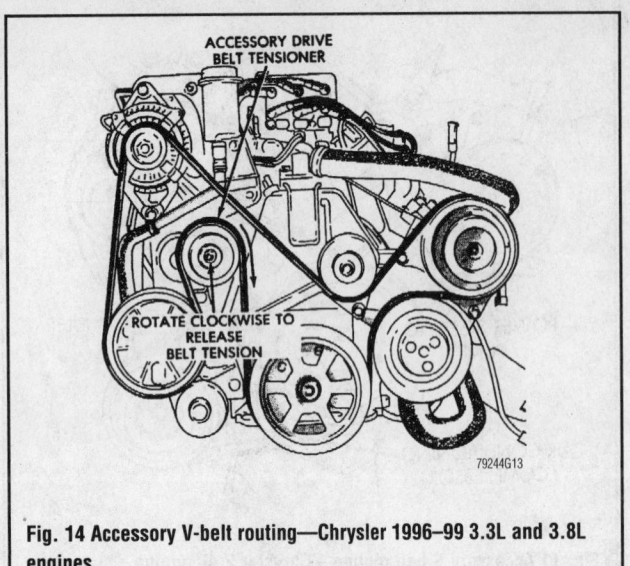

Fig. 14 Accessory V-belt routing—Chrysler 1996–99 3.3L and 3.8L engines

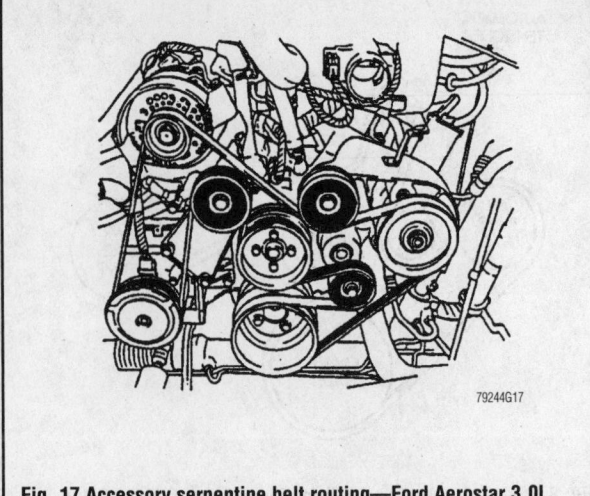

Fig. 17 Accessory serpentine belt routing—Ford Aerostar 3.0L engine with A/C

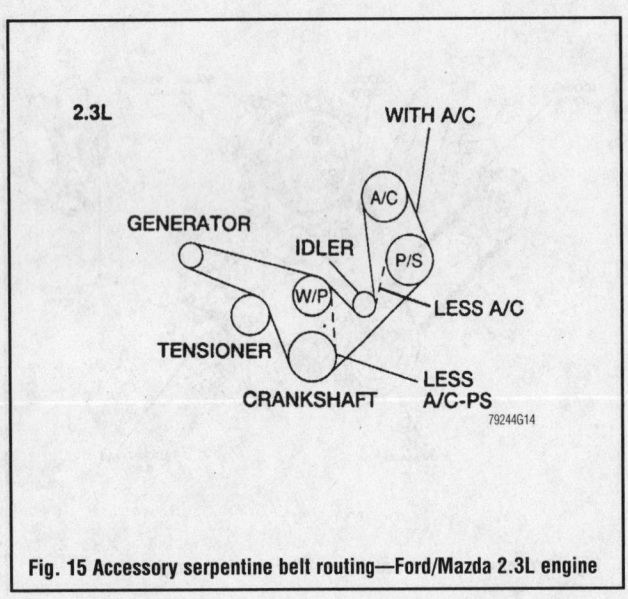

Fig. 15 Accessory serpentine belt routing—Ford/Mazda 2.3L engine

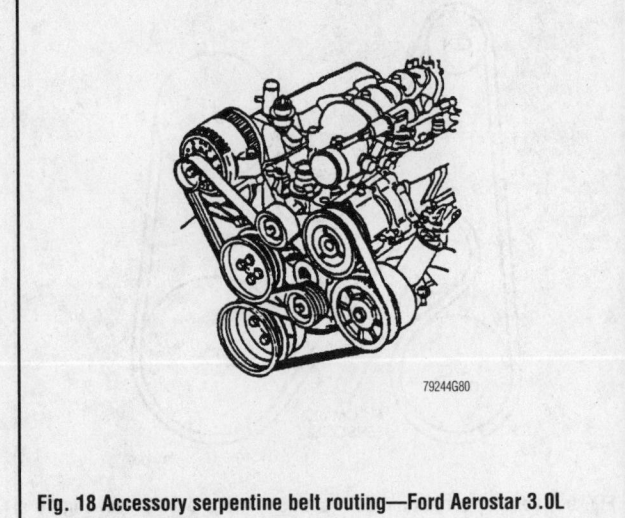

Fig. 18 Accessory serpentine belt routing—Ford Aerostar 3.0L engine without A/C

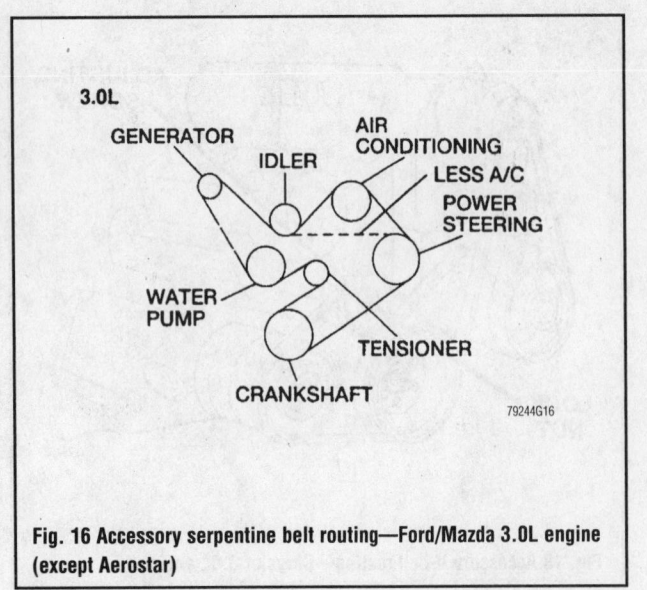

Fig. 16 Accessory serpentine belt routing—Ford/Mazda 3.0L engine (except Aerostar)

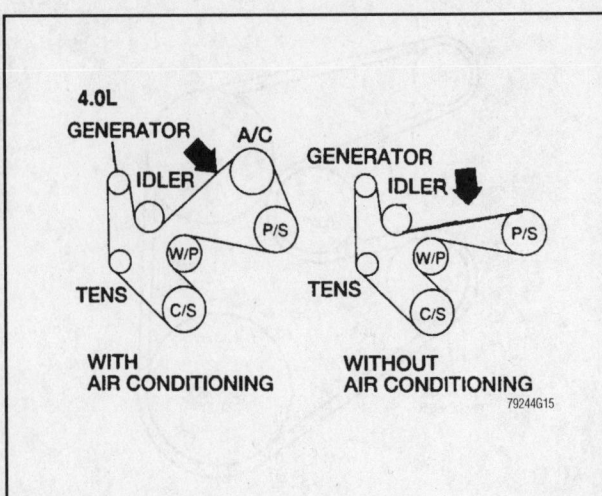

Fig. 19 Accessory serpentine belt routing—Ford/Mazda 4.0L engine

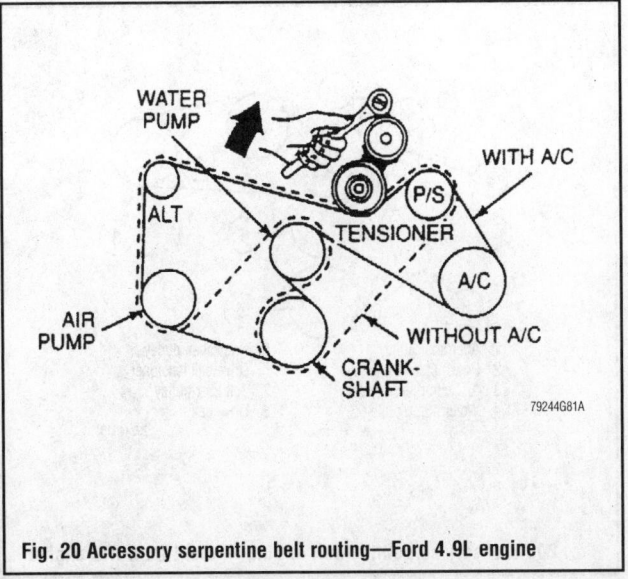

Fig. 20 Accessory serpentine belt routing—Ford 4.9L engine

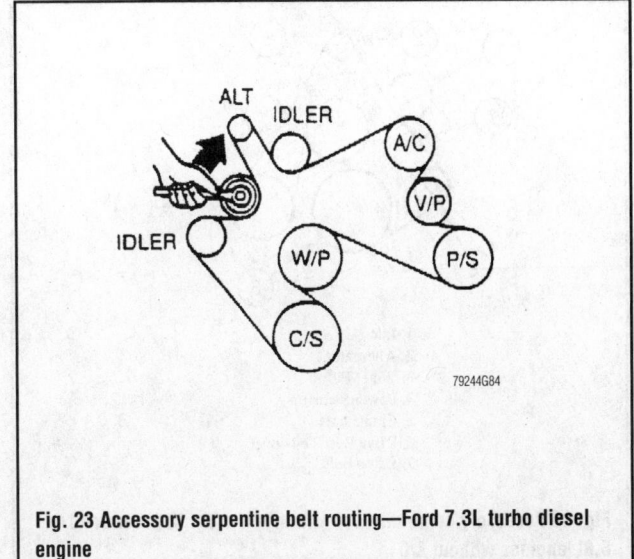

Fig. 23 Accessory serpentine belt routing—Ford 7.3L turbo diesel engine

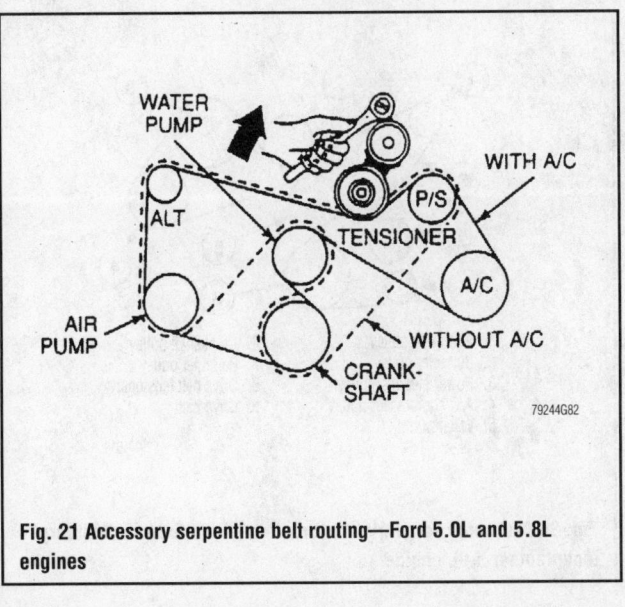

Fig. 21 Accessory serpentine belt routing—Ford 5.0L and 5.8L engines

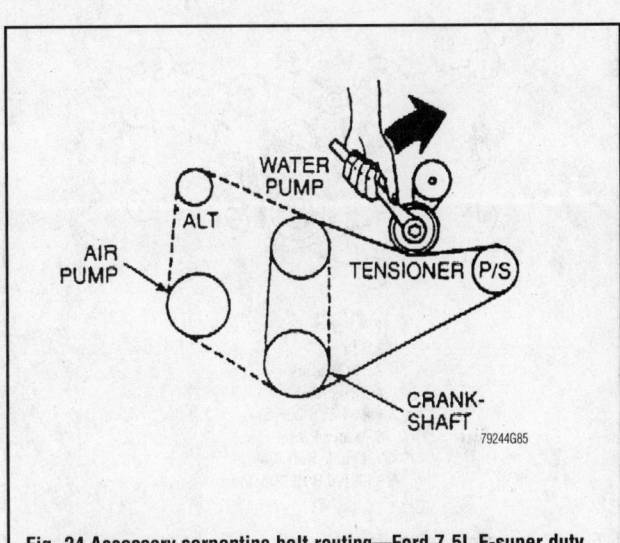

Fig. 24 Accessory serpentine belt routing—Ford 7.5L F-super duty motorhome engine

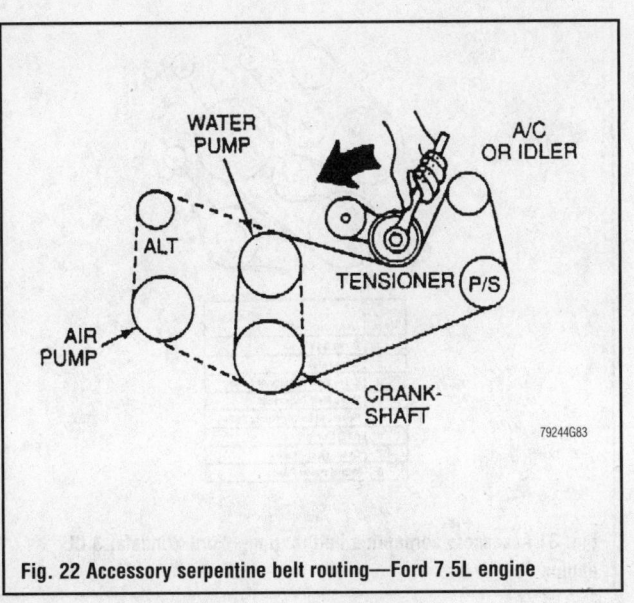

Fig. 22 Accessory serpentine belt routing—Ford 7.5L engine

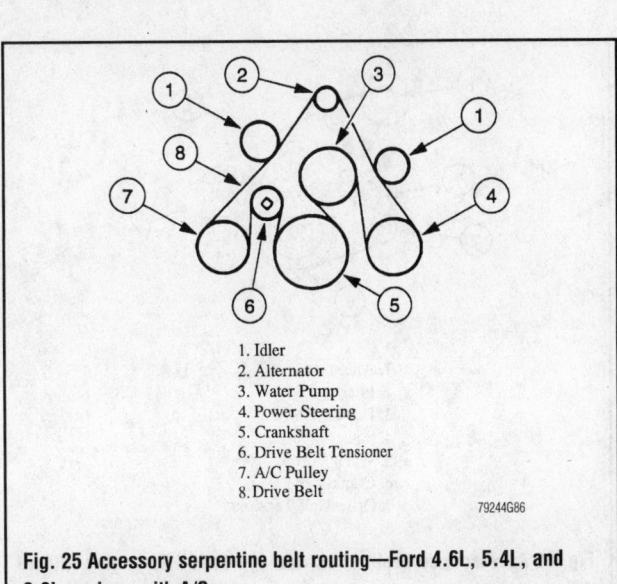

1. Idler
2. Alternator
3. Water Pump
4. Power Steering
5. Crankshaft
6. Drive Belt Tensioner
7. A/C Pulley
8. Drive Belt

Fig. 25 Accessory serpentine belt routing—Ford 4.6L, 5.4L, and 6.8L engines with A/C

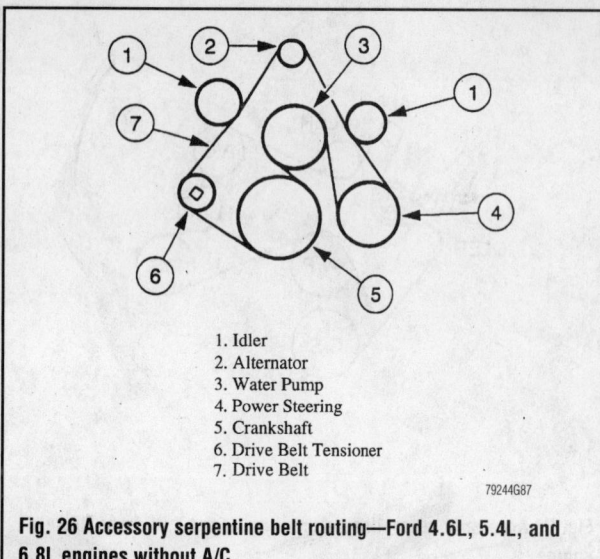

1. Idler
2. Alternator
3. Water Pump
4. Power Steering
5. Crankshaft
6. Drive Belt Tensioner
7. Drive Belt

79244G87

Fig. 26 Accessory serpentine belt routing—Ford 4.6L, 5.4L, and 6.8L engines without A/C

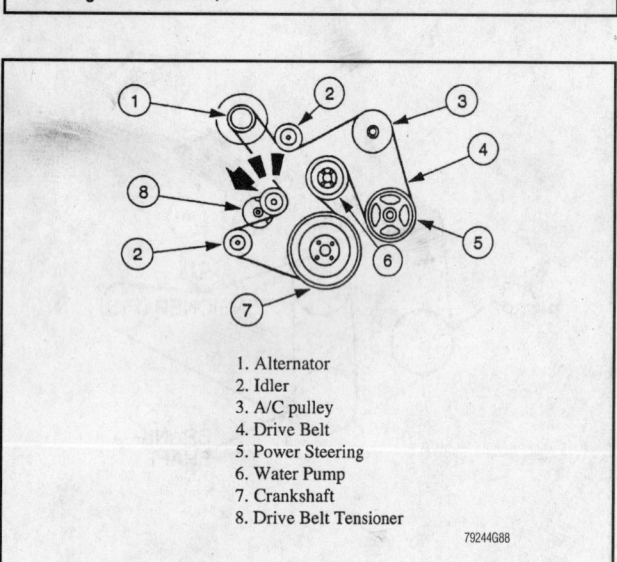

1. Alternator
2. Idler
3. A/C pulley
4. Drive Belt
5. Power Steering
6. Water Pump
7. Crankshaft
8. Drive Belt Tensioner

79244G88

Fig. 27 Accessory serpentine belt routing—Ford 4.2L engine with A/C

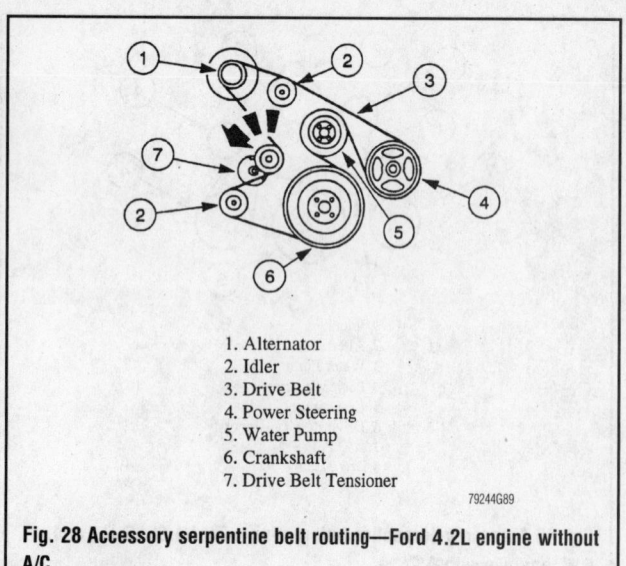

1. Alternator
2. Idler
3. Drive Belt
4. Power Steering
5. Water Pump
6. Crankshaft
7. Drive Belt Tensioner

79244G89

Fig. 28 Accessory serpentine belt routing—Ford 4.2L engine without A/C

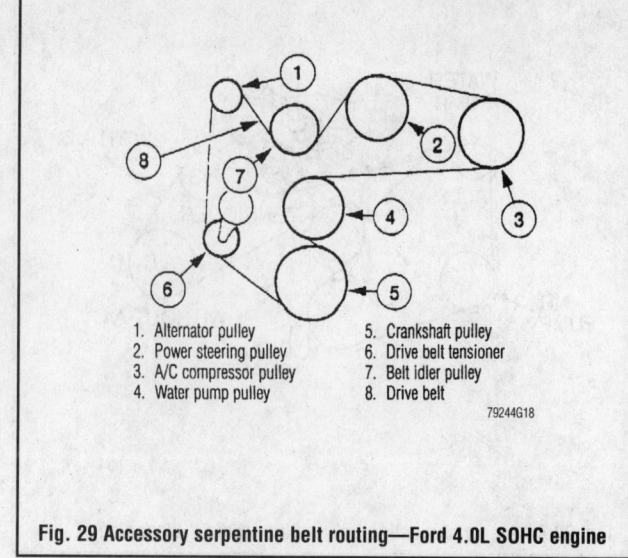

1. Alternator pulley
2. Power steering pulley
3. A/C compressor pulley
4. Water pump pulley
5. Crankshaft pulley
6. Drive belt tensioner
7. Belt idler pulley
8. Drive belt

79244G18

Fig. 29 Accessory serpentine belt routing—Ford 4.0L SOHC engine

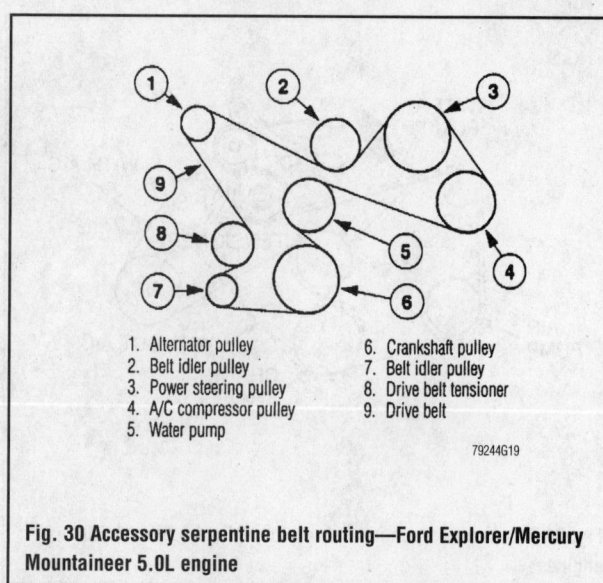

1. Alternator pulley
2. Belt idler pulley
3. Power steering pulley
4. A/C compressor pulley
5. Water pump
6. Crankshaft pulley
7. Belt idler pulley
8. Drive belt tensioner
9. Drive belt

79244G19

Fig. 30 Accessory serpentine belt routing—Ford Explorer/Mercury Mountaineer 5.0L engine

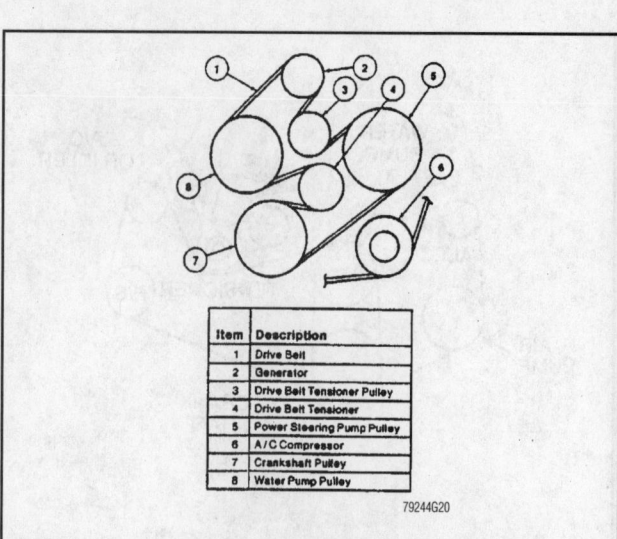

Item	Description
1	Drive Belt
2	Generator
3	Drive Belt Tensioner Pulley
4	Drive Belt Tensioner
5	Power Steering Pump Pulley
6	A/C Compressor
7	Crankshaft Pulley
8	Water Pump Pulley

79244G20

Fig. 31 Accessory serpentine belt routing—Ford Windstar 3.0L engine

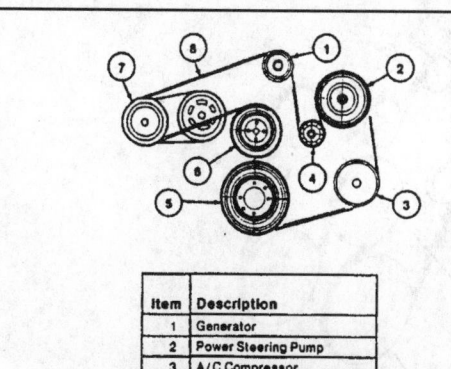

Item	Description
1	Generator
2	Power Steering Pump
3	A/C Compressor
4	Drive Belt Tensioner Pulley
5	Crankshaft Pulley
6	Water Pump Pulley
7	Drive Belt Tensioner
8	Drive Belt

79244G21

Fig. 32 Accessory serpentine belt routing—Ford Windstar 3.8L engine

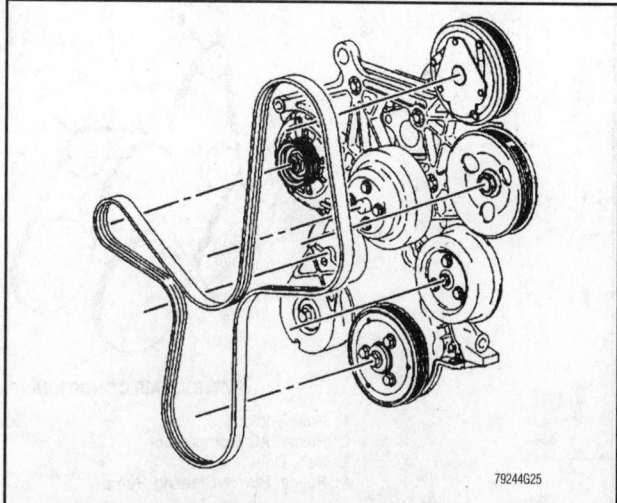

79244G25

Fig. 35 Accessory serpentine belt routing—GM/Isuzu 2.2L engine with A/C

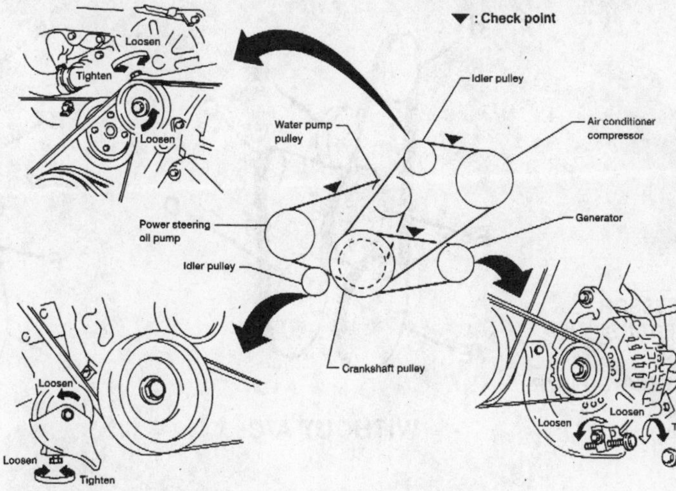

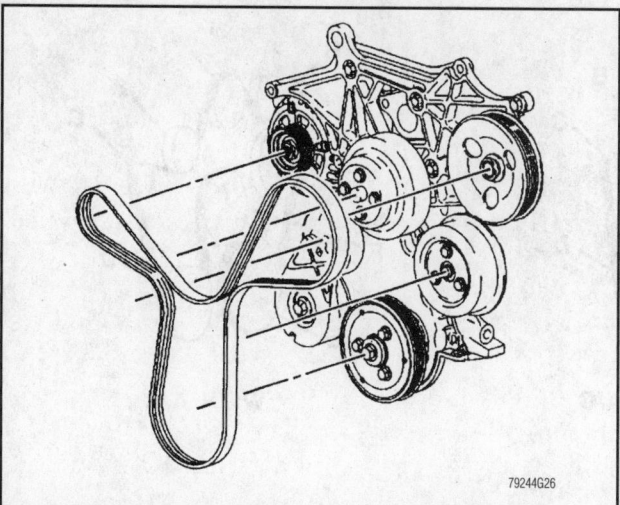

79244G22

Fig. 33 Accessory V-belt routing—Nissan Quest/Mercury Villager 3.0L engine

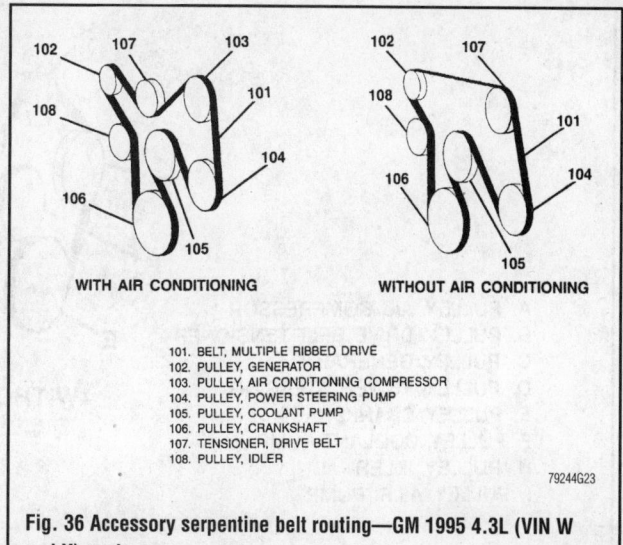

101. BELT, MULTIPLE RIBBED DRIVE
102. PULLEY, GENERATOR
103. PULLEY, AIR CONDITIONING COMPRESSOR
104. PULLEY, POWER STEERING PUMP
105. PULLEY, COOLANT PUMP
106. PULLEY, CRANKSHAFT
107. TENSIONER, DRIVE BELT
108. PULLEY, IDLER

79244G23

Fig. 34 Accessory serpentine belt routing—GM/Isuzu 2.2L engine without A/C

Fig. 36 Accessory serpentine belt routing—GM 1995 4.3L (VIN W and X) engines

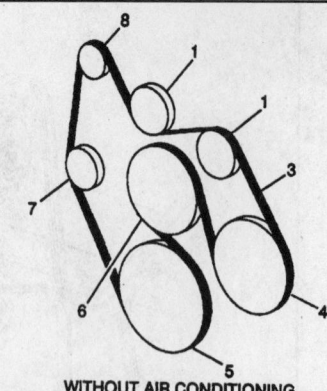
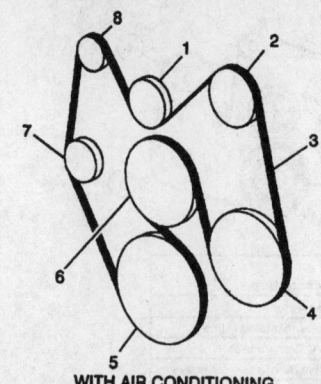

WITHOUT AIR CONDITIONING

WITH AIR CONDITIONING

1. Pulley, Idler
2. Pulley, AC Compressor
3. Belt, Drive
4. Pulley. Power Steering Pump

5. Pulley, Crankshaft
6. Pulley, Water Pump
7. Pulley, Drive Belt Tensioner
8. Pulley, Generator

79244G24

Fig. 37 Accessory serpentine belt routing—GM 1996–99 4.3L (VIN W and X) engines

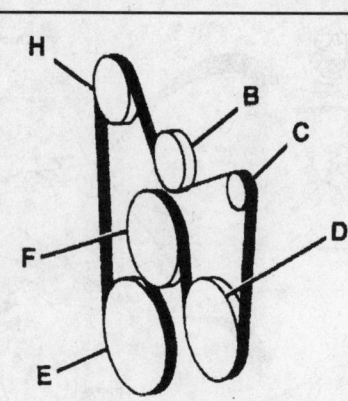
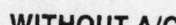

A. PULLEY, A/C COMPRESSOR
B. PULLEY, DRIVE BELT TENSIONER
C. PULLEY, GENERATOR
D. PULLEY, POWER STEERING PUMP
E. PULLEY, CRANKSHAFT
F. PULLEY, COOLANT PUMP
H. PULLEY, IDLER

WITHOUT A/C

WITH A/C

79244G27

Fig. 38 Accessory serpentine belt routing—GM 1995 Full-size Truck 4.3L, 5.0L, and 5.7L engines without A.I.R.

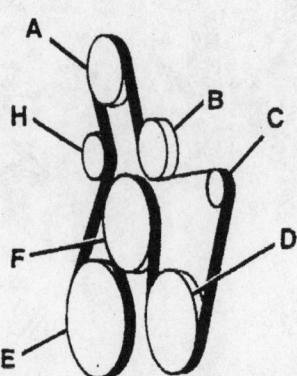

A. PULLEY, A/C COMPRESSOR
B. PULLEY, DRIVE BELT TENSIONER
C. PULLEY, GENERATOR
D. PULLEY, POWER STEERING
E. PULLEY, CRANKSHAFT
F. PULLEY, COOLANT PUMP
H. PULLEY, IDLER
I. PULLEY, A.I.R. PUMP

WITHOUT A/C

WITH A/C

79244G28

Fig. 39 Accessory serpentine belt routing—GM 1995 Full-size Truck 4.3L and 5.7L engines with A.I.R.

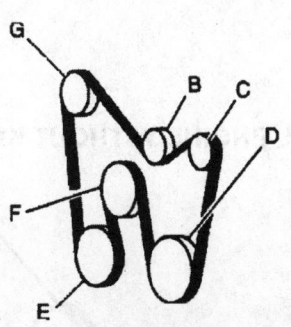
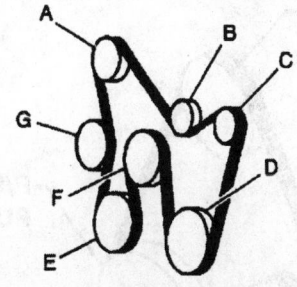

WITHOUT A/C
A. PULLEY, A/C COMPRESSOR
B. PULLEY, DRIVE BELT
   TENSIONER
C. PULLEY, GENERATOR

WITH A/C
D. PULLEY, POWER STEERING PUMP
E. PULLEY, CRANKSHAFT
F. PULLEY, COOLANT PUMP
G. PULLEY, VACUUM PUMP

79244G30

Fig. 40 Accessory serpentine belt routing—GM 1995 Diesel engine

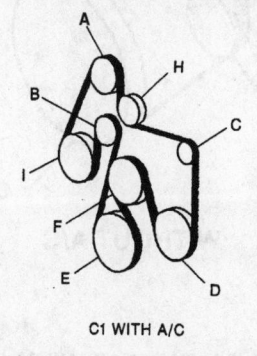
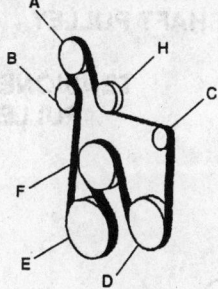
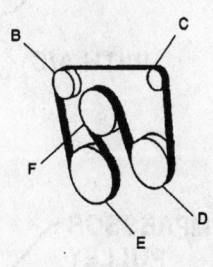

C1 WITH A/C

CK 2,3 WITH A/C

CK 2,3 WITHOUT A/C

A. PULLEY, A/C COMPRESSOR
B. PULLEY, DRIVE BELT TENSIONER
C. PULLEY, GENERATOR
D. PULLEY, POWER STEERING PUMP

E. PULLEY, CRANKSHAFT
F. PULLEY, COOLANT PUMP
H. PULLEY, IDLER
I. PULLEY, A.I.R. PUMP

79244G29

Fig. 41 Accessory serpentine belt routing—GM 1995 7.4L engine

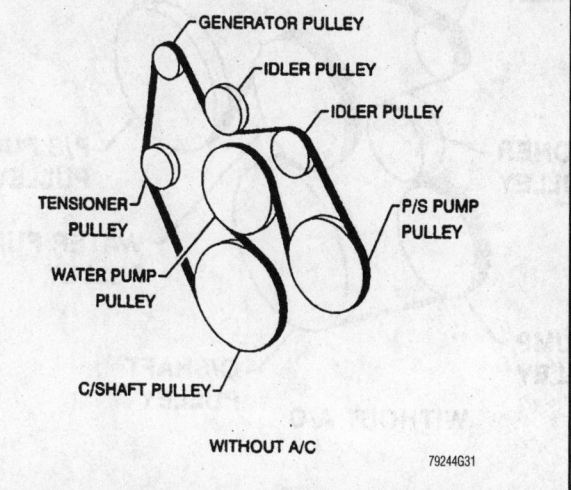

WITHOUT A/C

79244G31

Fig. 42 Accessory serpentine belt routing—GM 1996–99 4.3L, 5.0L and 5.7L engines without A/C

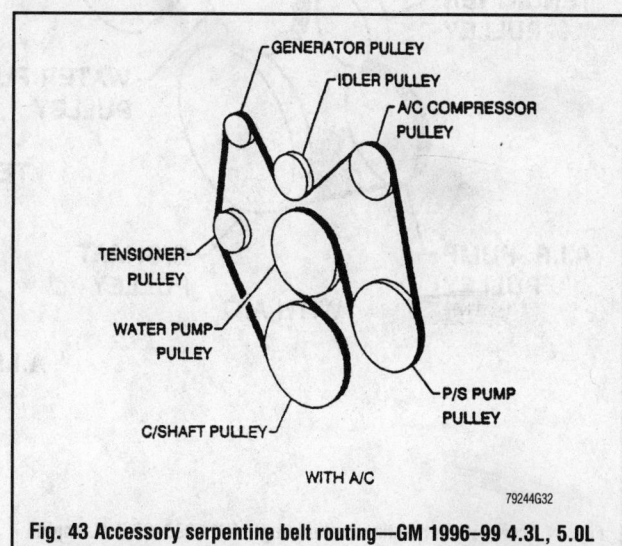

WITH A/C

79244G32

Fig. 43 Accessory serpentine belt routing—GM 1996–99 4.3L, 5.0L and 5.7L engines with A/C

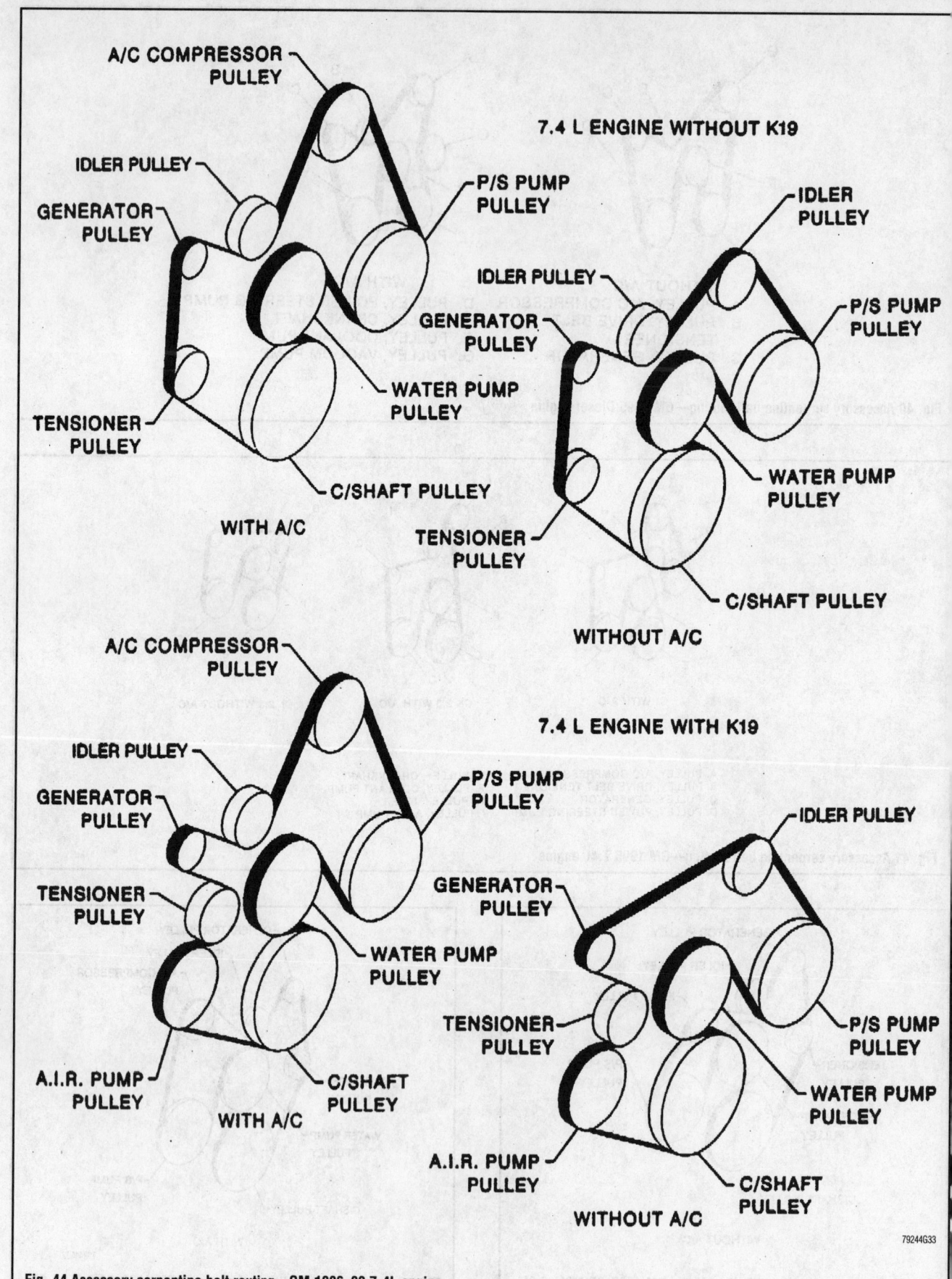

A/C COMPRESSOR PULLEY

IDLER PULLEY

GENERATOR PULLEY

P/S PUMP PULLEY

WATER PUMP PULLEY

TENSIONER PULLEY

C/SHAFT PULLEY

WITH A/C

7.4 L ENGINE WITHOUT K19

IDLER PULLEY

IDLER PULLEY

GENERATOR PULLEY

P/S PUMP PULLEY

WATER PUMP PULLEY

TENSIONER PULLEY

C/SHAFT PULLEY

WITHOUT A/C

A/C COMPRESSOR PULLEY

IDLER PULLEY

GENERATOR PULLEY

P/S PUMP PULLEY

TENSIONER PULLEY

WATER PUMP PULLEY

A.I.R. PUMP PULLEY

C/SHAFT PULLEY

WITH A/C

7.4 L ENGINE WITH K19

IDLER PULLEY

GENERATOR PULLEY

P/S PUMP PULLEY

TENSIONER PULLEY

WATER PUMP PULLEY

A.I.R. PUMP PULLEY

C/SHAFT PULLEY

WITHOUT A/C

79244G33

Fig. 44 Accessory serpentine belt routing—GM 1996–99 7.4L engine

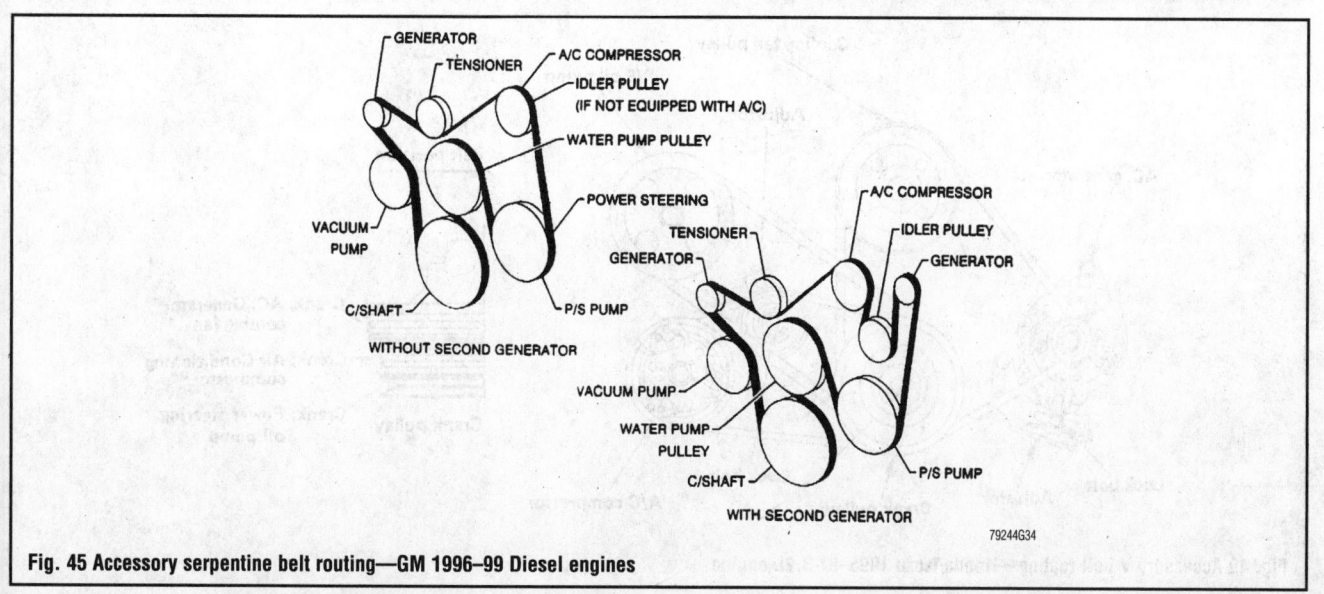

Fig. 45 Accessory serpentine belt routing—GM 1996–99 Diesel engines

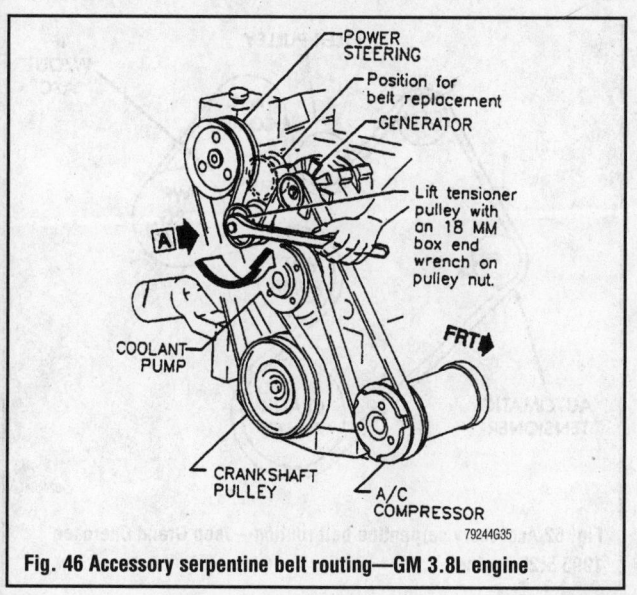

Fig. 46 Accessory serpentine belt routing—GM 3.8L engine

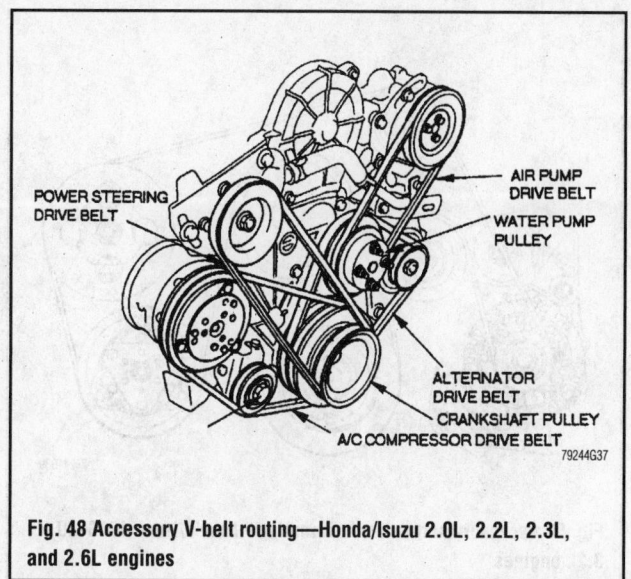

Fig. 48 Accessory V-belt routing—Honda/Isuzu 2.0L, 2.2L, 2.3L, and 2.6L engines

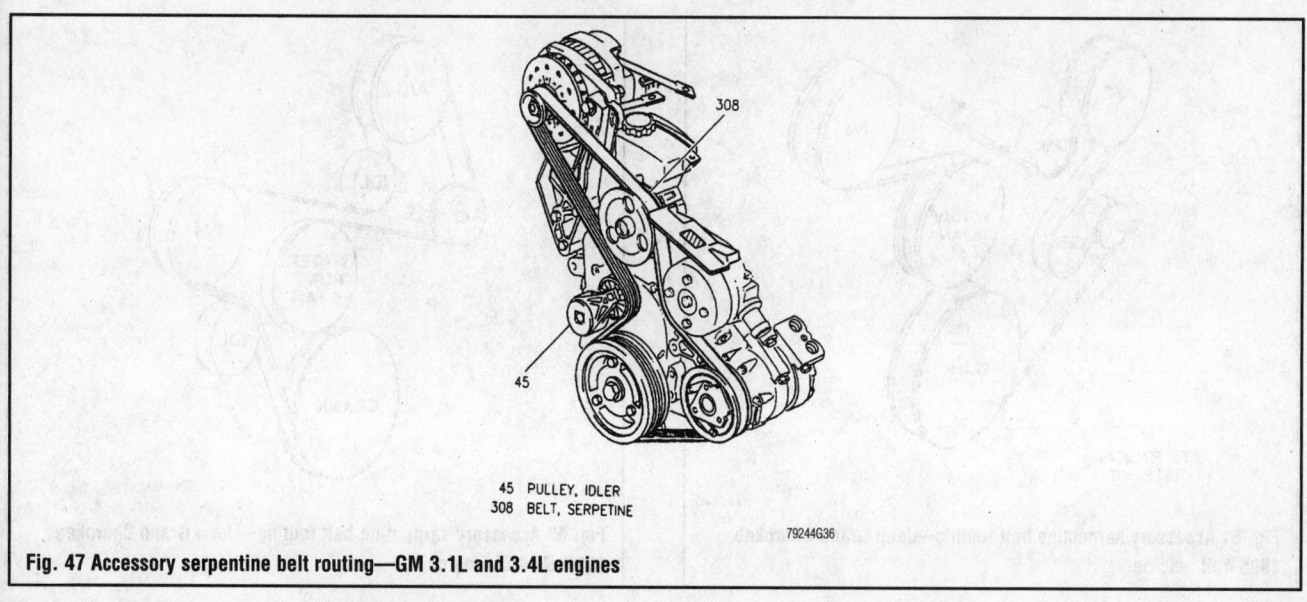

45 PULLEY, IDLER
308 BELT, SERPETINE

Fig. 47 Accessory serpentine belt routing—GM 3.1L and 3.4L engines

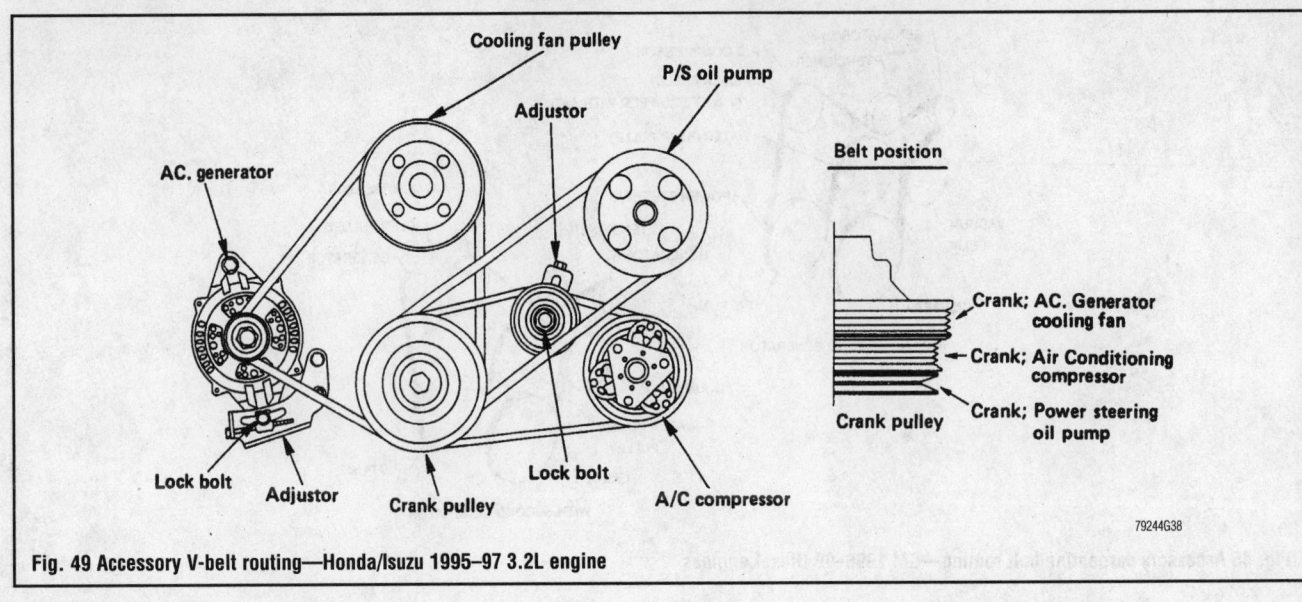

Fig. 49 Accessory V-belt routing—Honda/Isuzu 1995–97 3.2L engine

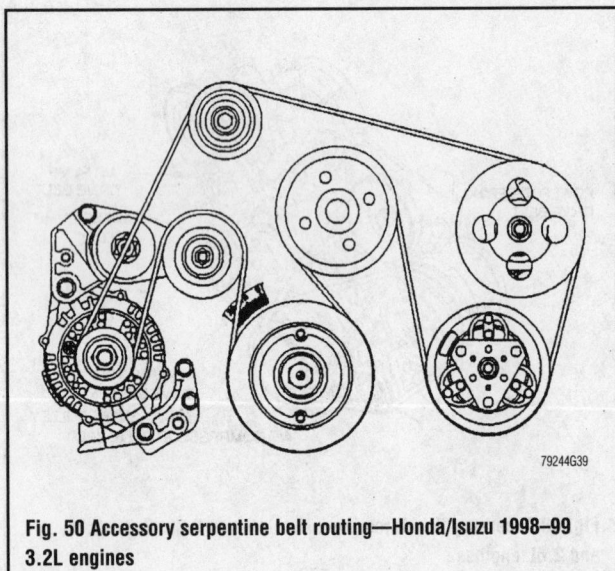

Fig. 50 Accessory serpentine belt routing—Honda/Isuzu 1998–99 3.2L engines

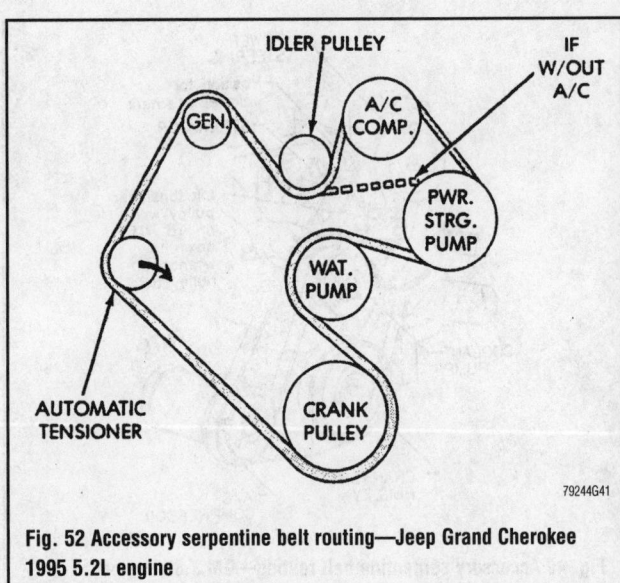

Fig. 52 Accessory serpentine belt routing—Jeep Grand Cherokee 1995 5.2L engine

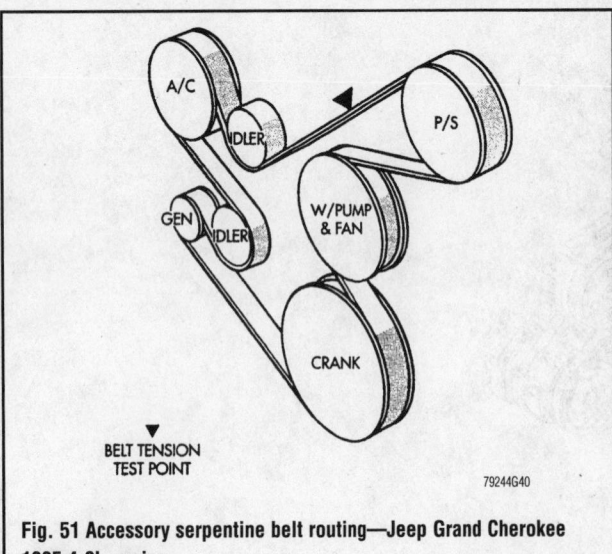

Fig. 51 Accessory serpentine belt routing—Jeep Grand Cherokee 1995 4.0L engine

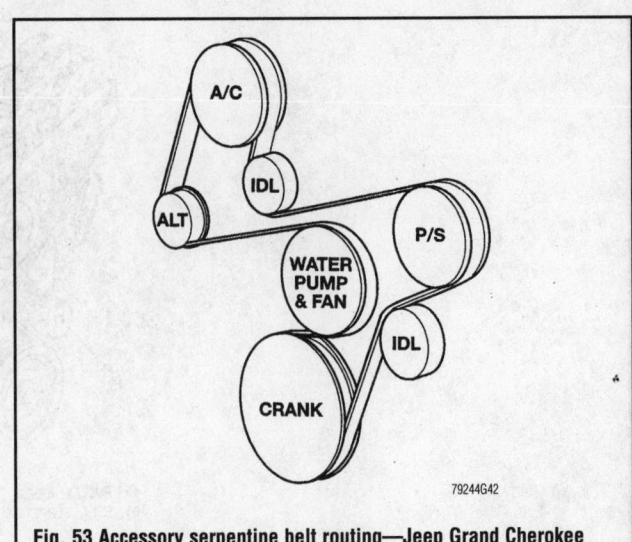

Fig. 53 Accessory serpentine belt routing—Jeep Grand Cherokee 1996–99 4.0L engine

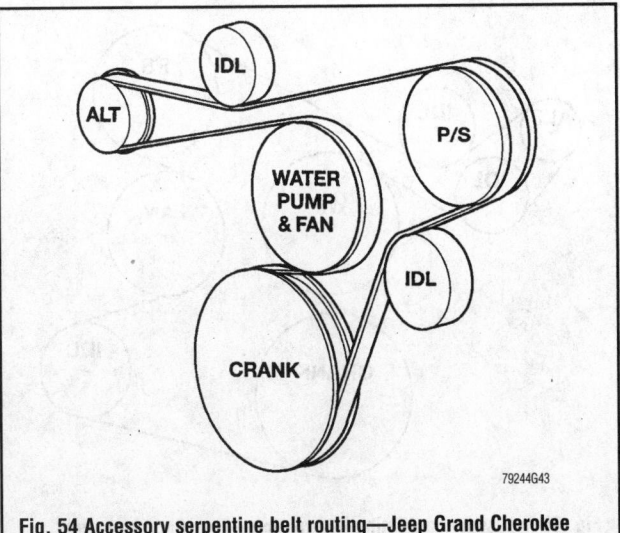

Fig. 54 Accessory serpentine belt routing—Jeep Grand Cherokee 1996–99 5.2L and 5.9L engines

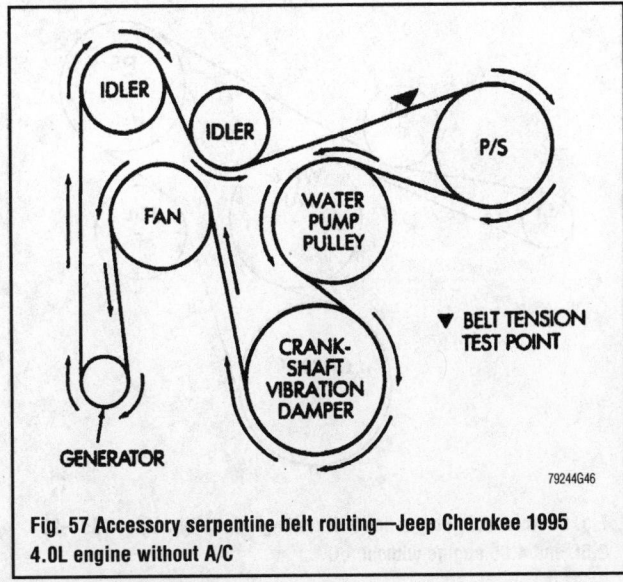

Fig. 57 Accessory serpentine belt routing—Jeep Cherokee 1995 4.0L engine without A/C

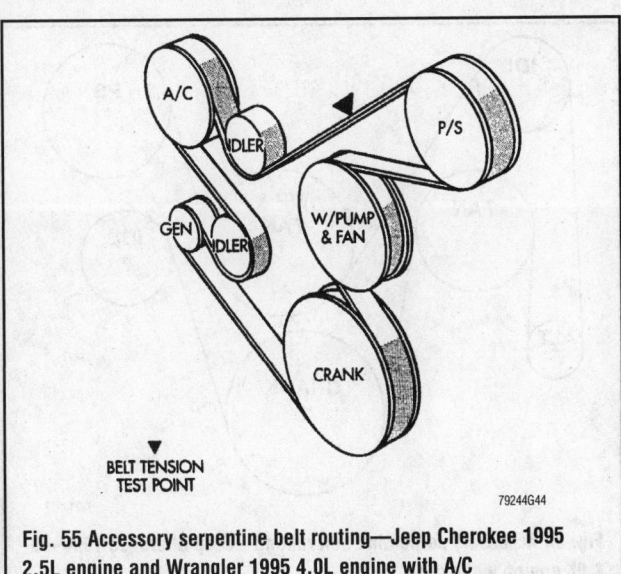

Fig. 55 Accessory serpentine belt routing—Jeep Cherokee 1995 2.5L engine and Wrangler 1995 4.0L engine with A/C

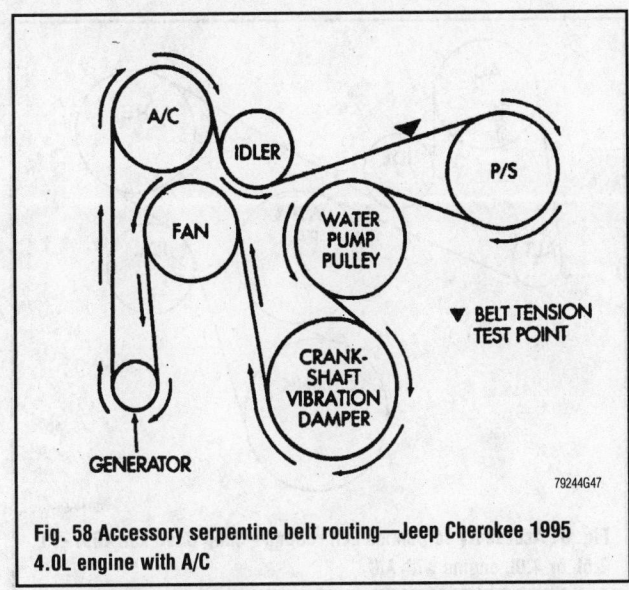

Fig. 58 Accessory serpentine belt routing—Jeep Cherokee 1995 4.0L engine with A/C

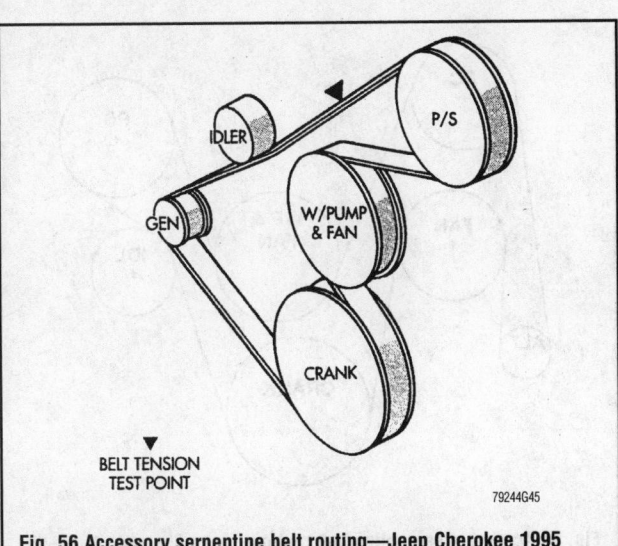

Fig. 56 Accessory serpentine belt routing—Jeep Cherokee 1995 2.5L engine and Wrangler 1995 4.0L engine without A/C

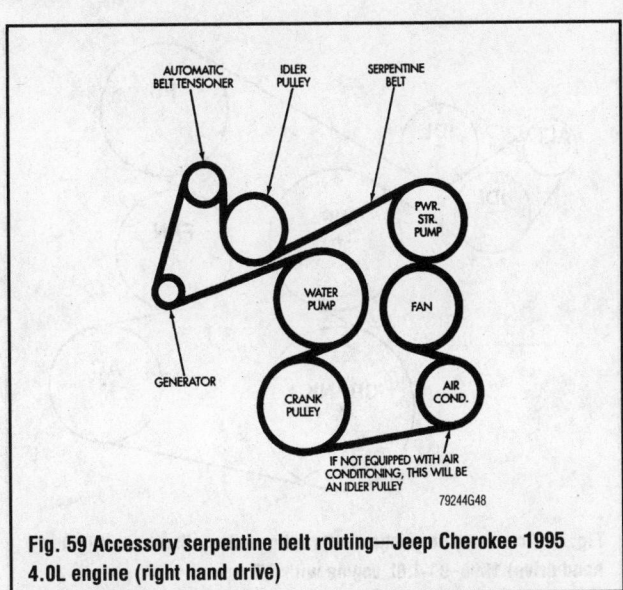

Fig. 59 Accessory serpentine belt routing—Jeep Cherokee 1995 4.0L engine (right hand drive)

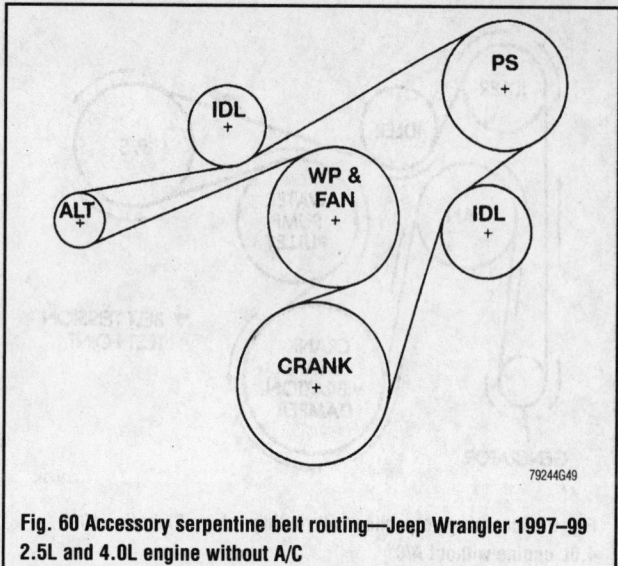

Fig. 60 Accessory serpentine belt routing—Jeep Wrangler 1997–99 2.5L and 4.0L engine without A/C

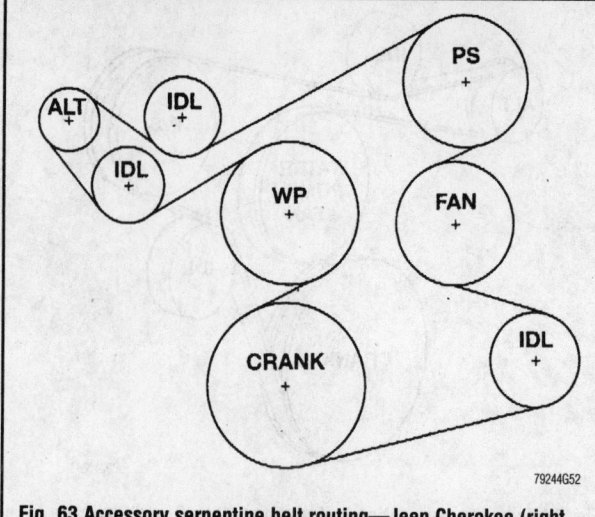

Fig. 63 Accessory serpentine belt routing—Jeep Cherokee (right hand drive) 1996–99 4.0L engine without A/C

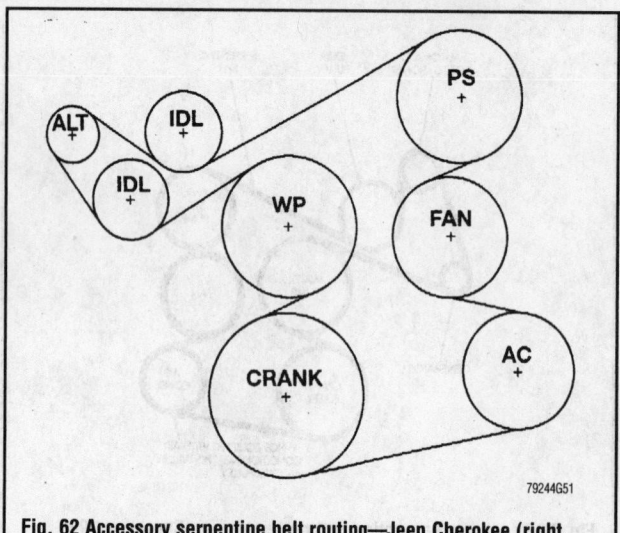

Fig. 61 Accessory serpentine belt routing—Jeep Cherokee 1997–99 2.5L or 4.0L engine with A/C

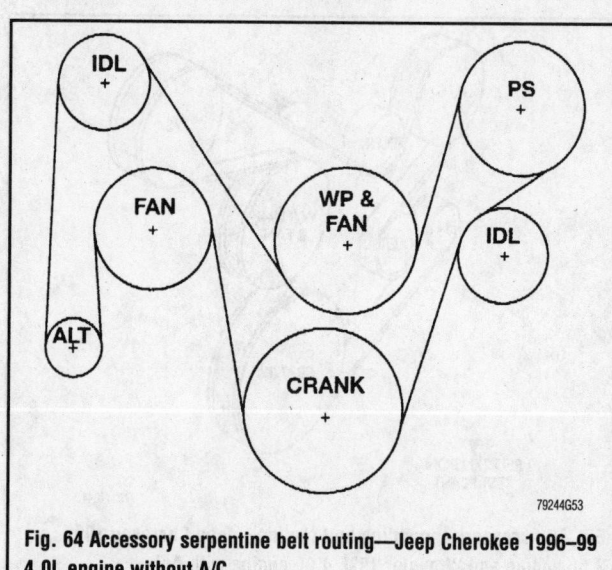

Fig. 64 Accessory serpentine belt routing—Jeep Cherokee 1996–99 4.0L engine without A/C

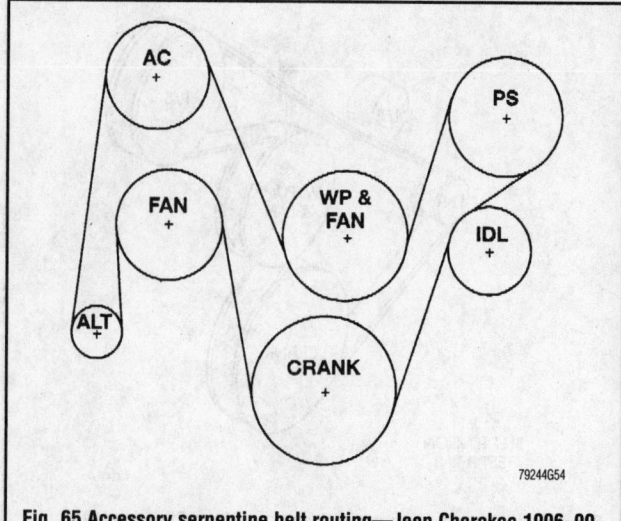

Fig. 62 Accessory serpentine belt routing—Jeep Cherokee (right hand drive) 1996–99 4.0L engine with A/C

Fig. 65 Accessory serpentine belt routing—Jeep Cherokee 1996–99 4.0L engine with A/C

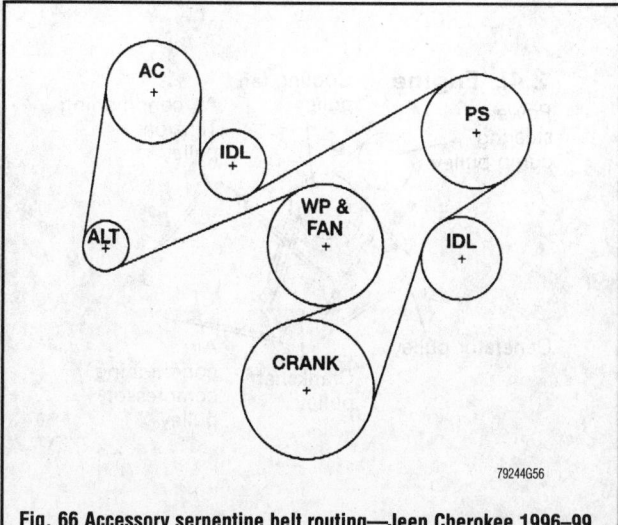

Fig. 66 Accessory serpentine belt routing—Jeep Cherokee 1996–99 2.5L engine with A/C

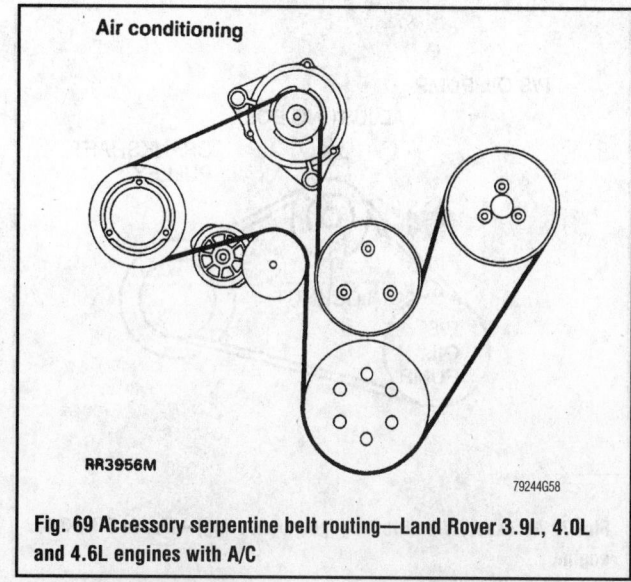

Fig. 69 Accessory serpentine belt routing—Land Rover 3.9L, 4.0L and 4.6L engines with A/C

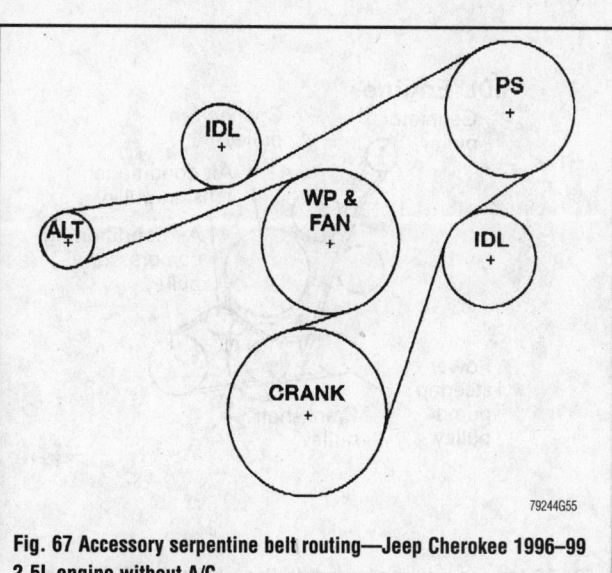

Fig. 67 Accessory serpentine belt routing—Jeep Cherokee 1996–99 2.5L engine without A/C

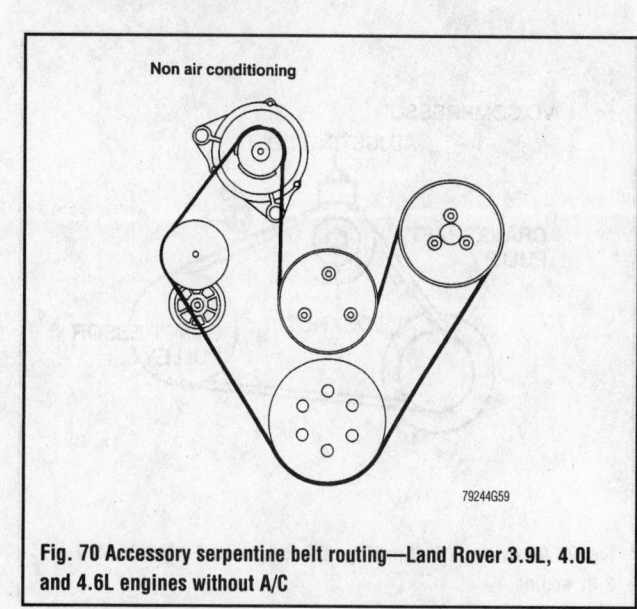

Fig. 70 Accessory serpentine belt routing—Land Rover 3.9L, 4.0L and 4.6L engines without A/C

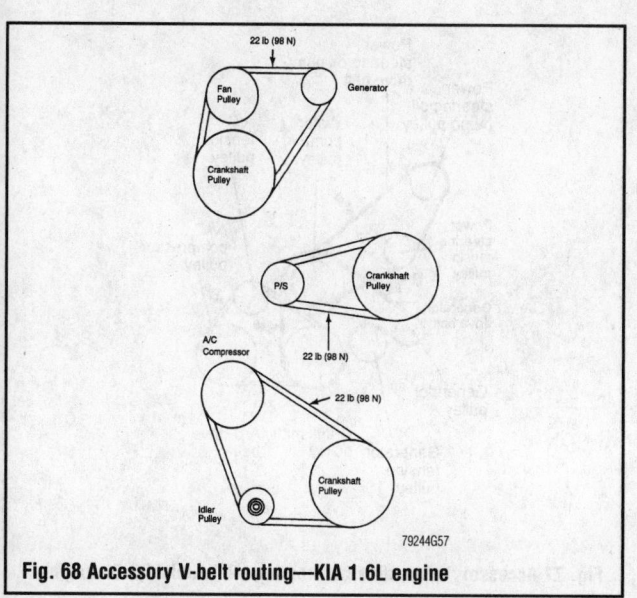

Fig. 68 Accessory V-belt routing—KIA 1.6L engine

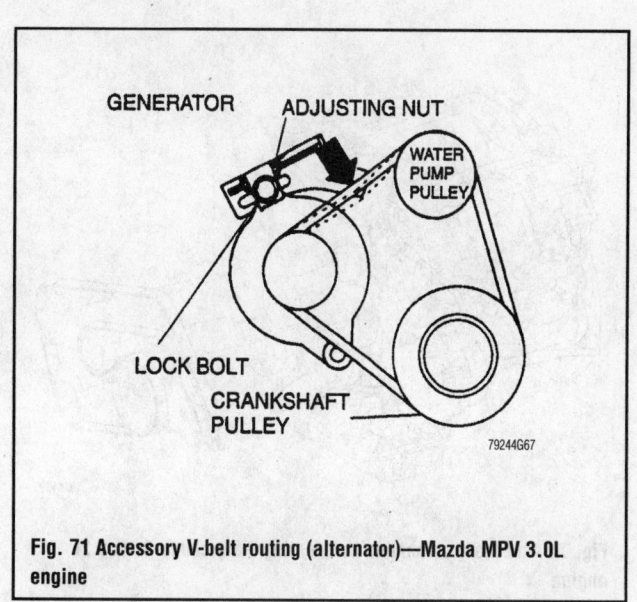

Fig. 71 Accessory V-belt routing (alternator)—Mazda MPV 3.0L engine

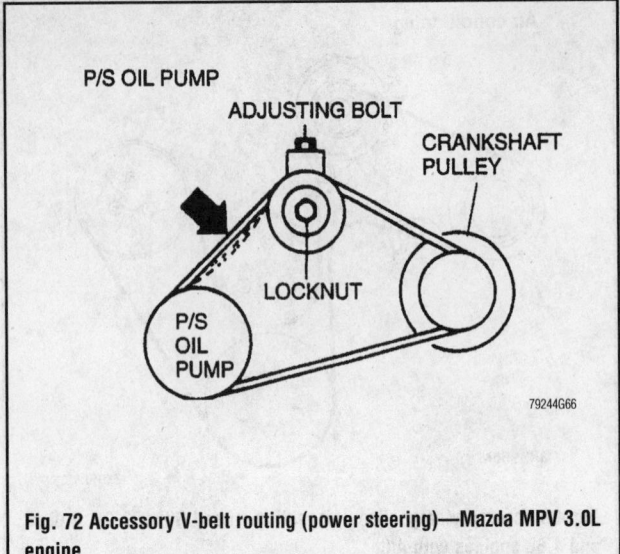

Fig. 72 Accessory V-belt routing (power steering)—Mazda MPV 3.0L engine

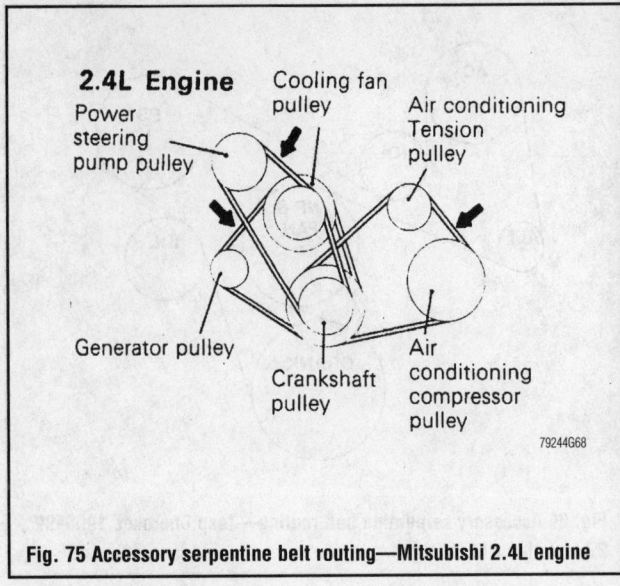

Fig. 75 Accessory serpentine belt routing—Mitsubishi 2.4L engine

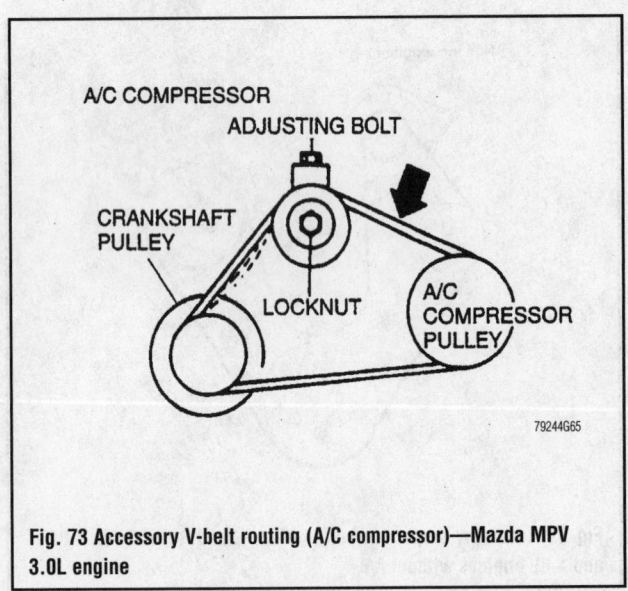

Fig. 73 Accessory V-belt routing (A/C compressor)—Mazda MPV 3.0L engine

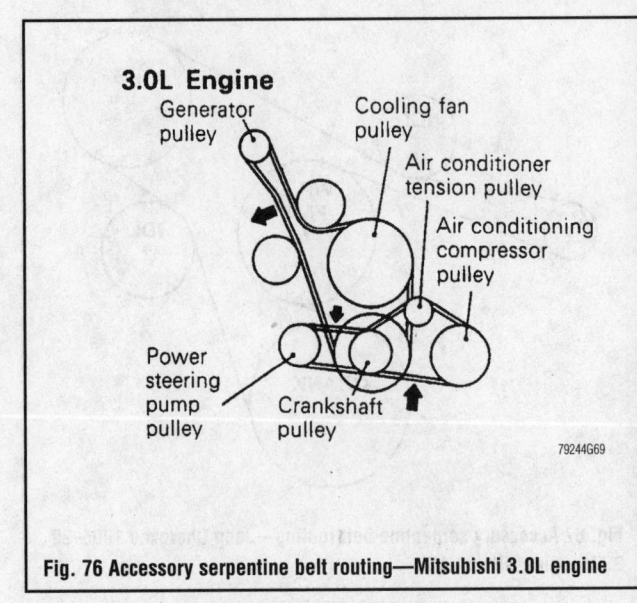

Fig. 76 Accessory serpentine belt routing—Mitsubishi 3.0L engine

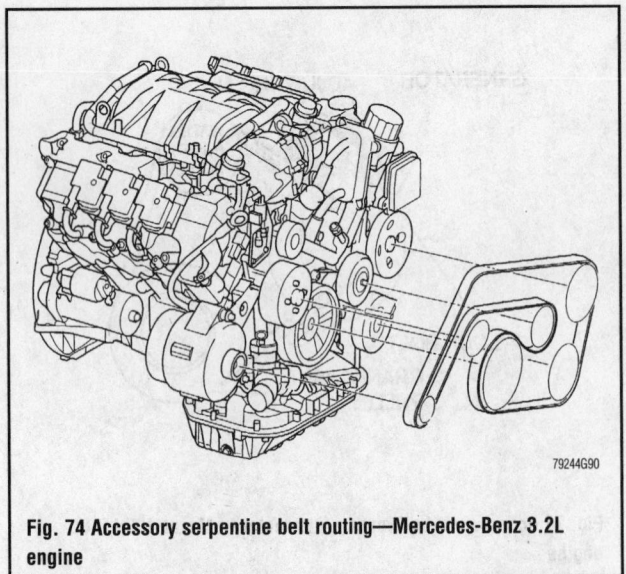

Fig. 74 Accessory serpentine belt routing—Mercedes-Benz 3.2L engine

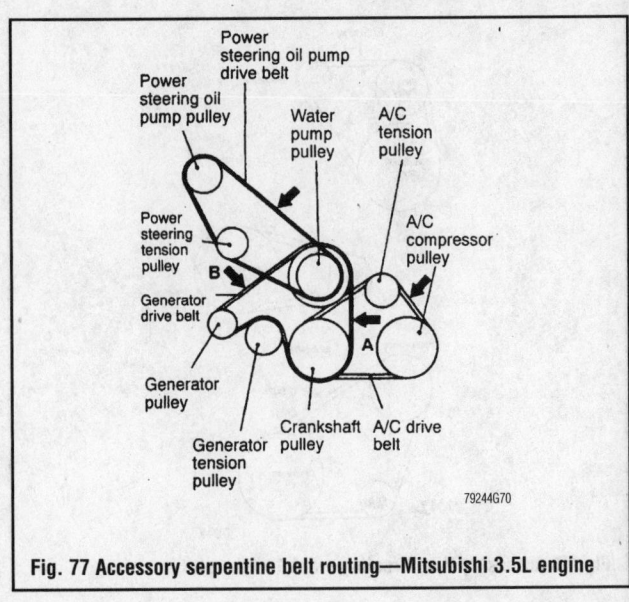

Fig. 77 Accessory serpentine belt routing—Mitsubishi 3.5L engine

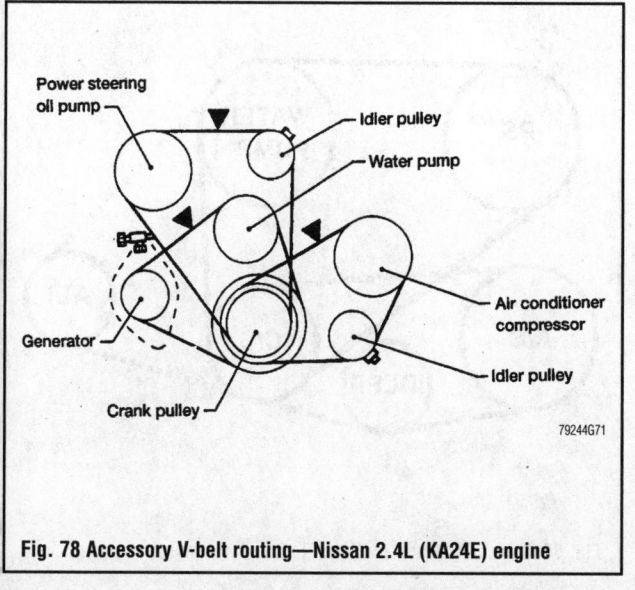

Fig. 78 Accessory V-belt routing—Nissan 2.4L (KA24E) engine

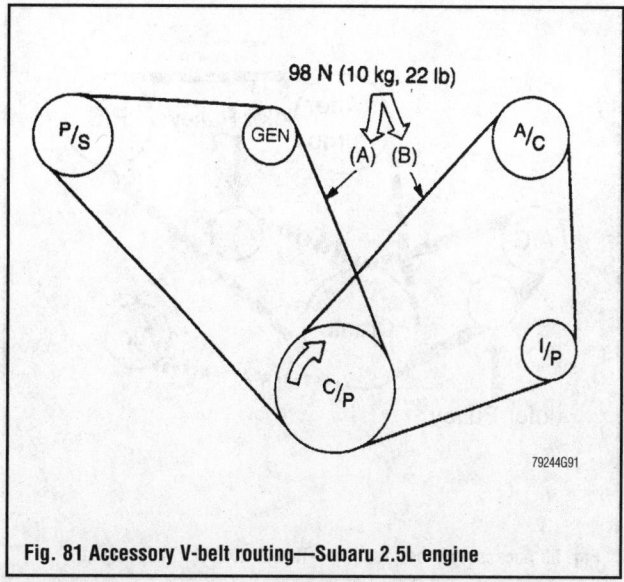

Fig. 81 Accessory V-belt routing—Subaru 2.5L engine

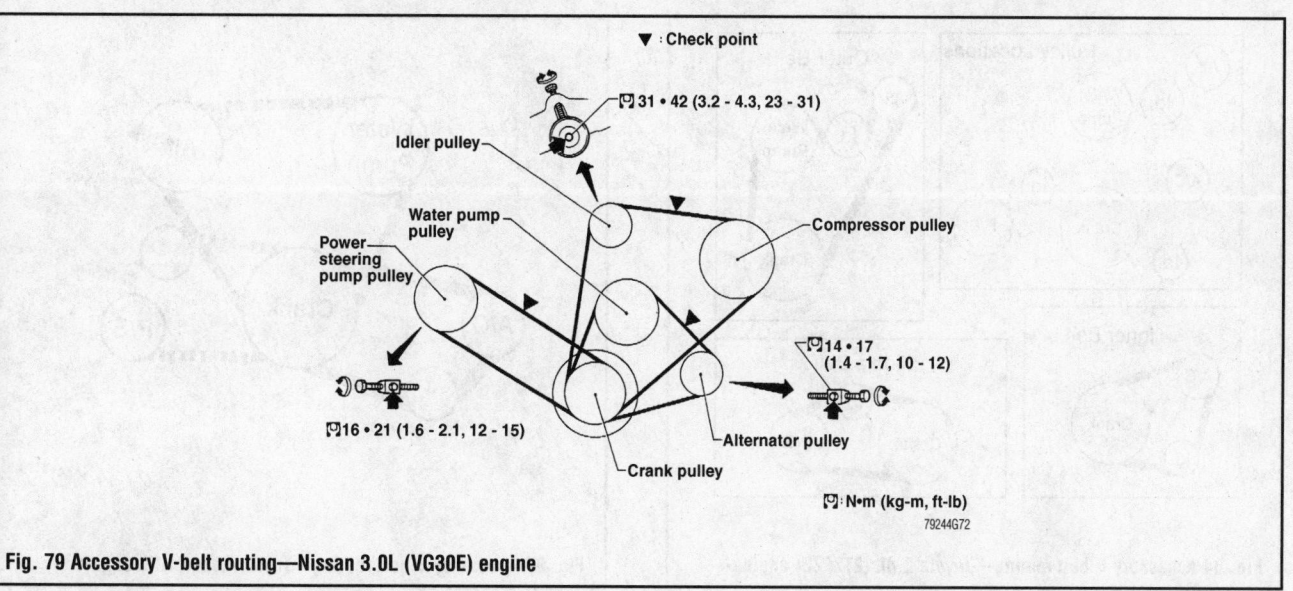

Fig. 79 Accessory V-belt routing—Nissan 3.0L (VG30E) engine

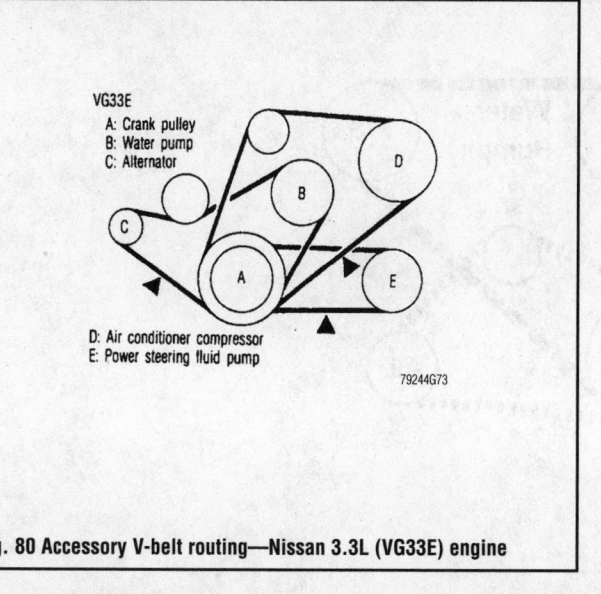

Fig. 80 Accessory V-belt routing—Nissan 3.3L (VG33E) engine

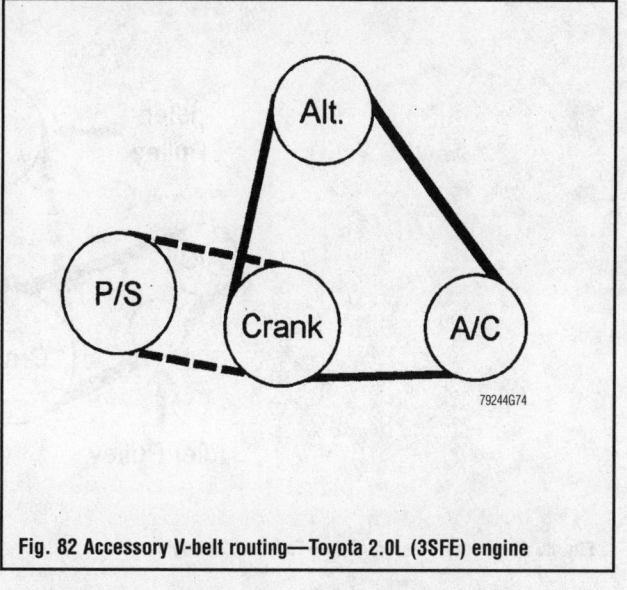

Fig. 82 Accessory V-belt routing—Toyota 2.0L (3SFE) engine

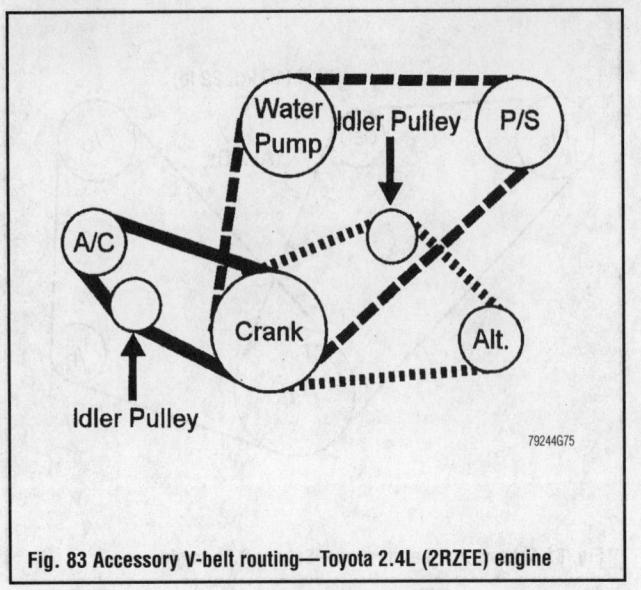

Fig. 83 Accessory V-belt routing—Toyota 2.4L (2RZFE) engine

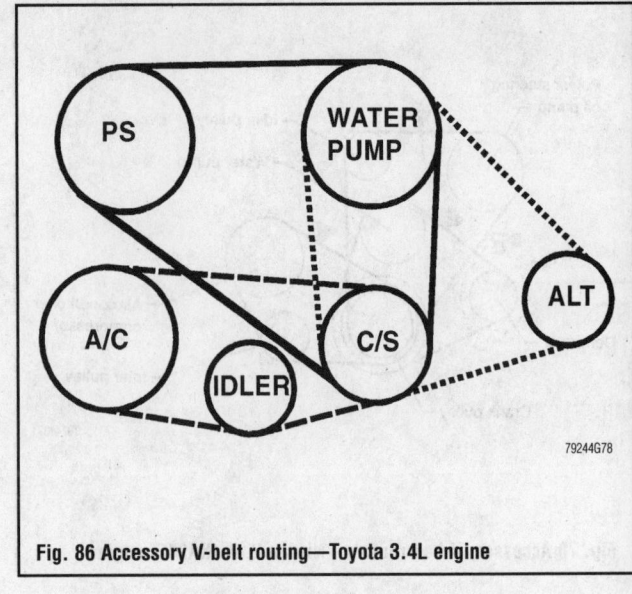

Fig. 86 Accessory V-belt routing—Toyota 3.4L engine

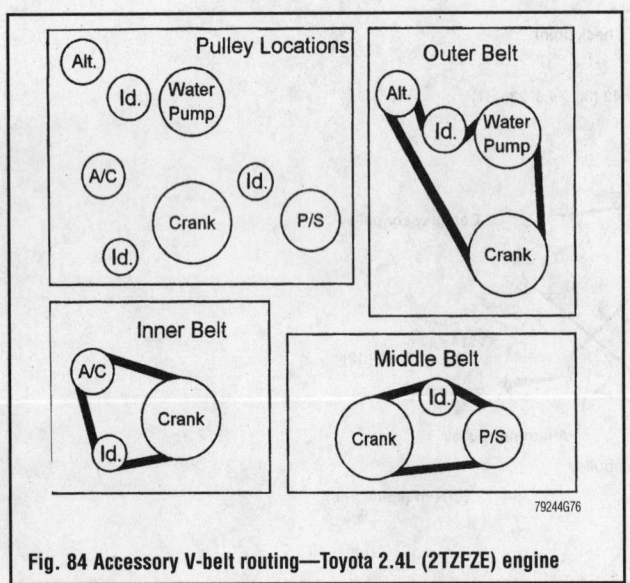

Fig. 84 Accessory V-belt routing—Toyota 2.4L (2TZFZE) engine

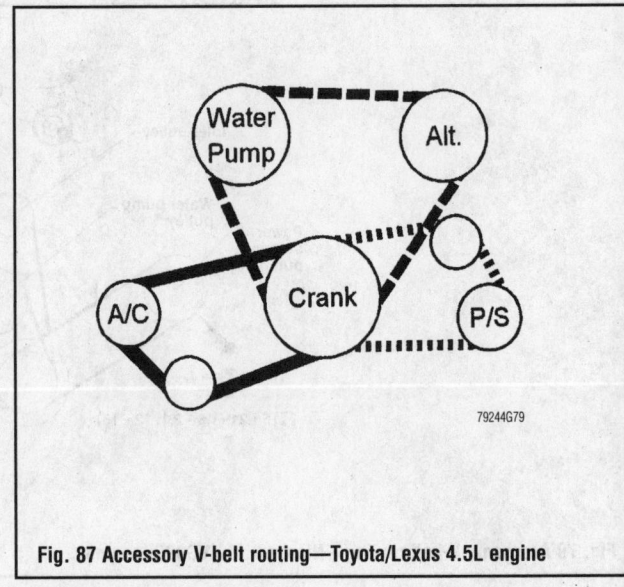

Fig. 87 Accessory V-belt routing—Toyota/Lexus 4.5L engine

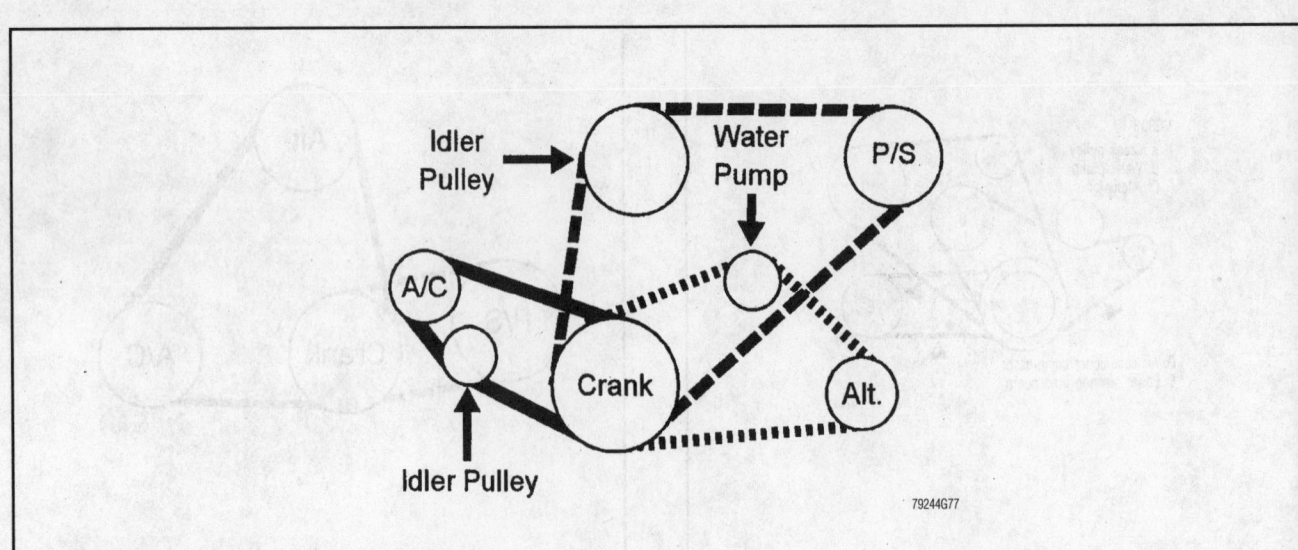

Fig. 85 Accessory V-belt routing—Toyota 2.7L engine

TIMING BELTS

5

GENERAL INFORMATION

Timing belts are typically only used on overhead camshaft engines. Timing belts are used to synchronize the crankshaft with the camshaft, similar to a timing chain on a overhead valve (pushrod) engine. Unlike a timing belt, a timing chain will normally last the life of the engine without needing service or replacement. Timing belts use raised teeth to mesh with sprockets to operate the valve train of an overhead camshaft engine.

Whenever a vehicle with an unknown service history comes into your repair facility or is recently purchased, here are some points that should be asked to help prevent costly engine d amage:

• Does the owner know if, or when the belt was replaced?
• If the vehicle purchased is used, or the condition and mileage of the last timing belt replacement are unknown, it is recommended to inspect, replace, or at least inform the owner that the vehicle is equipped with a timing belt.
• Note the mileage of the vehicle. The average replacement interval for a timing belt is approximately 60,000 miles (96,000 km).

Interference Engines

Engines, chain- or belt-driven, can be classified as either free-running or interference, depending on what would happen if the piston-to-valve timing is disrupted. A free-running engine is designed with enough clearance between the pistons and valves to allow the crankshaft to rotate (pistons still moving) while the camshaft stays in one position (several valves fully open). If this condition occurs normally, no internal engine damage will result. In an interference engine, there is not enough clearance between the pistons and valves to allow the crankshaft to turn without the camshaft being in time.

An interference engine can suffer extensive internal damage if a timing belt fails. The piston design does not allow clearance for the valve to be fully open and the piston to be at the top of its stroke. If the belt fails, the piston will collide with the valve and will bend or break the valve, damage the piston, and/or bend a connecting rod. When this type of failure occurs, the engine will need to be replaced or disassembled for further internal inspection; either choice costing many times that of replacing the timing belt.

TIMING BELT SERVICE

Inspection

→**For manufacturer's recommended service interval, refer to the maintenance interval chart located in this manual.**

The average replacement interval for a timing belt is approximately 60,000 miles (96,000km). If, however, the timing belt is inspected earlier or more frequently than suggested, and shows signs of wear or defects, the belt should be replaced at that time.

✳✳ WARNING

Never allow antifreeze, oil or solvents to come into with a timing belt. If this occurs immediately wash the solution from the timing belt. Also, never excessive bend or twist the timing belt; this can damage the belt so that its lifetime is severely shortened.

Inspect both sides of the timing belt. Replace the belt with a new one if any of the following conditions exist:

- Hardening of the rubber—back side is glossy without resilience and leaves no indentation when pressed with a fingernail
- Cracks on the rubber backing
- Cracks or peeling of the canvas backing
- Cracks on rib root
- Cracks on belt sides
- Missing teeth or chunks of teeth
- Abnormal wear of belt sides—the sides are normal if they are sharp, as if cut by a knife.

If none of these conditions exist, the belt does not need replacement unless it is at the recommended interval. The belt MUST be replaced at the recommended interval.

✳✳ WARNING

On interference engines, it is very important to replace the timing belt at the recommended intervals, otherwise expensive engine damage will likely result if the belt fails.

Never bend or twist a timing belt excessively, and do not allow solvents, antifreeze, gasoline, acid or oil to come into contact with the belt

TCCS1242

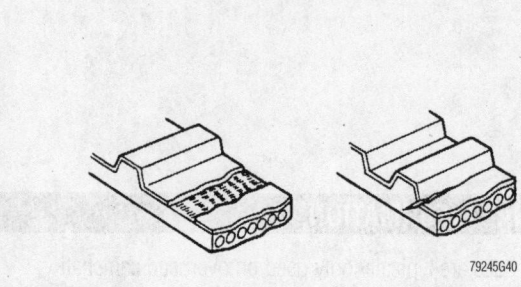

79245G40

Inspect the timing belt for damage, such as a broken or missing tooth, which may be due to a damaged pulley

TCCS1301

Clean the timing belt before inspection so that imperfections or defects are easier to recognize

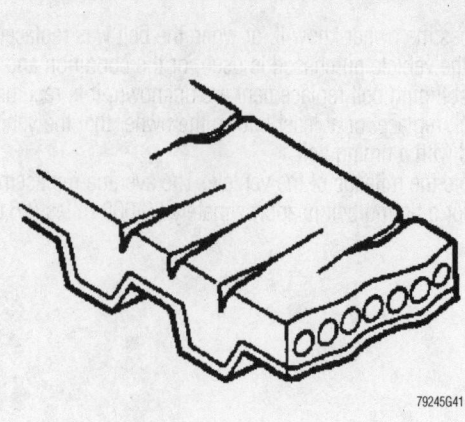

79245G41

Back surface worn or cracked from a possible overheated engine or interference with the belt cover

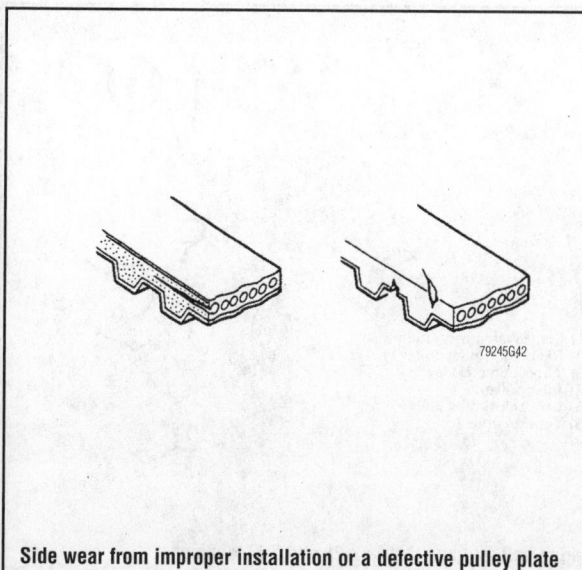

Side wear from improper installation or a defective pulley plate

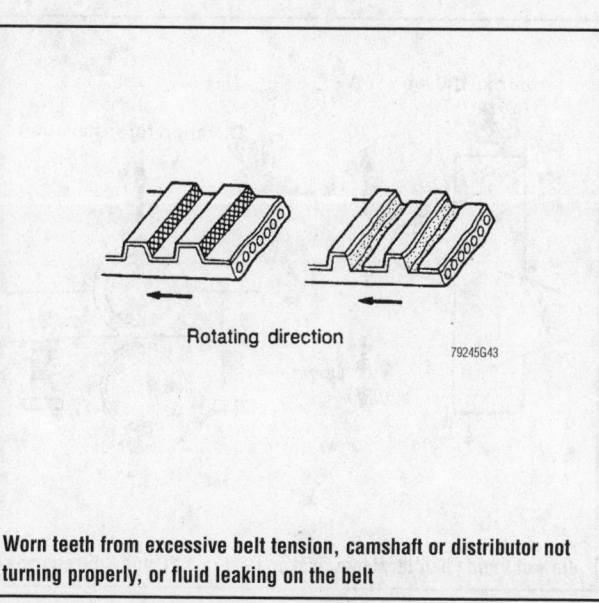

Rotating direction

Worn teeth from excessive belt tension, camshaft or distributor not turning properly, or fluid leaking on the belt

Removal & Installation

ACURA

SLX

3.2L AND 3.5L ENGINES

1. Disconnect the negative battery cable.
2. Drain the engine coolant into a sealable container.
3. Remove the air cleaner assembly and intake air duct.
4. Disconnect the upper radiator hose from the coolant inlet.
5. Remove the upper fan shroud from the radiator.
6. Remove the four nuts retaining the cooling fan assembly. Remove the cooling fan from the fan pulley.
7. Loosen and remove the drive belts.
8. Remove the upper timing belt covers.
9. Remove the fan pulley assembly.

10. Rotate the crankshaft to align the camshaft timing marks with the pointer dots on the back covers. Verify that the pointer on the crankshaft aligns with the mark on the lower timing cover.

➡ **When the timing marks are aligned on 1995 vehicles, no pistons will be at TDC/compression. When the timing marks are aligned on 1995½–99 vehicles, the No. 2 piston is at TDC/compression.**

❋❋ WARNING

Align the camshaft and crankshaft sprockets with their alignment marks before removing the timing belt. Failure to align the belt and sprocket marks may result in valve damage.

11. Use tool No. J-8614–01, or a suitable pulley holding tool to remove the crankshaft pulley center bolt. Remove the crankshaft pulley.

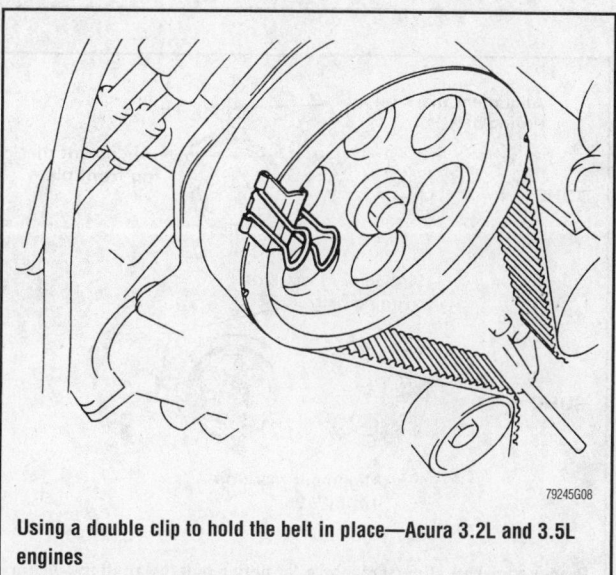

Using a double clip to hold the belt in place—Acura 3.2L and 3.5L engines

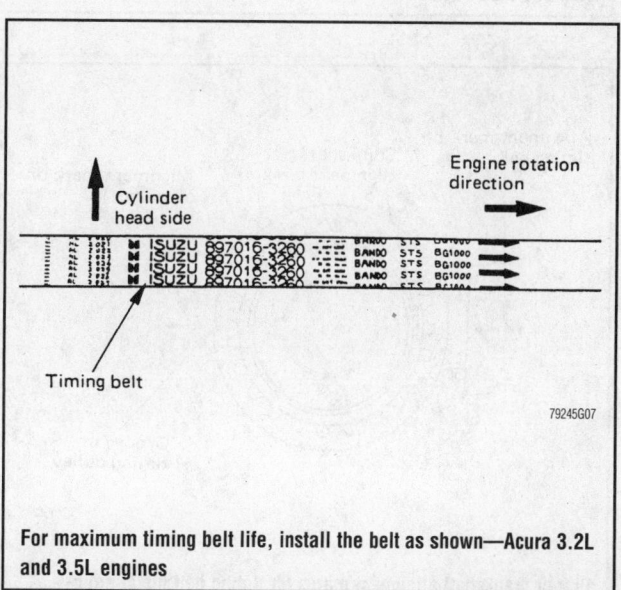

For maximum timing belt life, install the belt as shown—Acura 3.2L and 3.5L engines

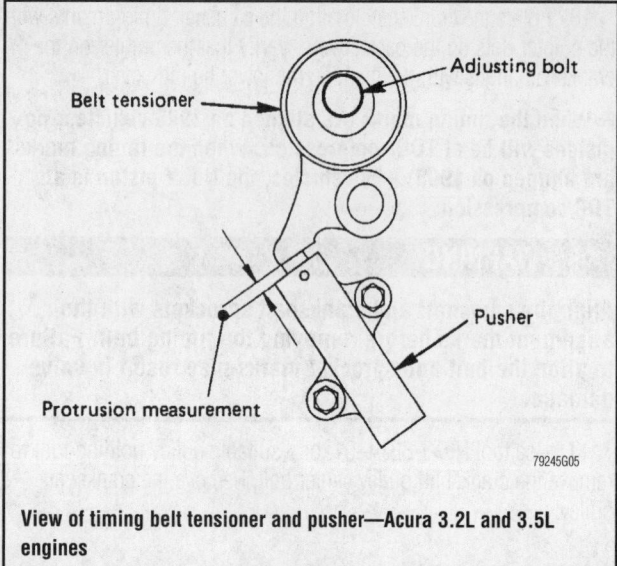

View of timing belt tensioner and pusher—Acura 3.2L and 3.5L engines

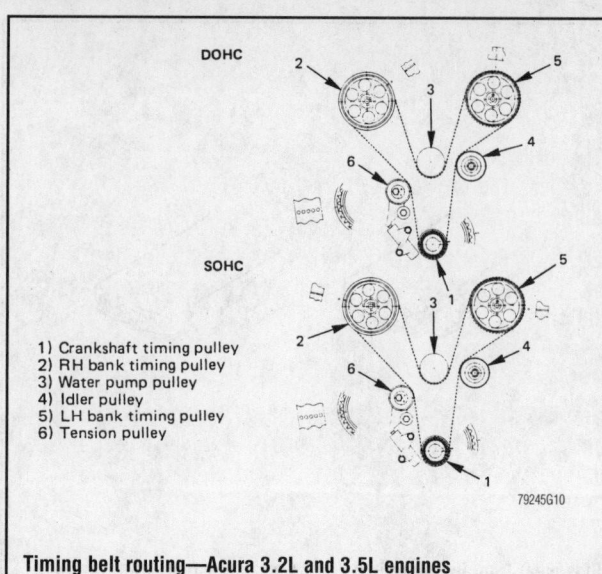

1) Crankshaft timing pulley
2) RH bank timing pulley
3) Water pump pulley
4) Idler pulley
5) LH bank timing pulley
6) Tension pulley

Timing belt routing—Acura 3.2L and 3.5L engines

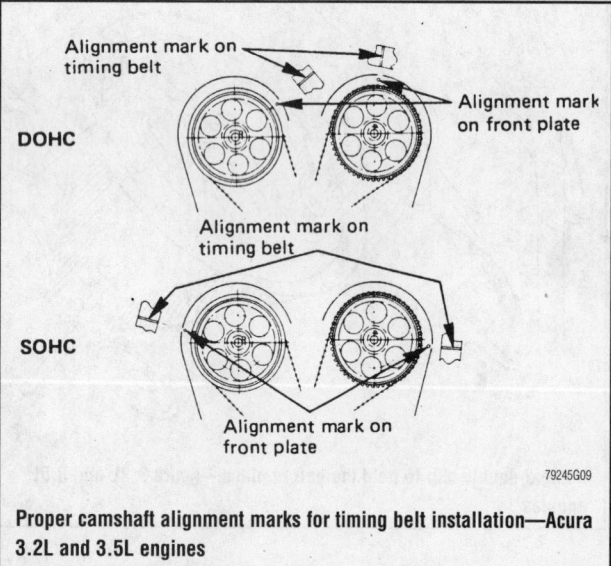

Proper camshaft alignment marks for timing belt installation—Acura 3.2L and 3.5L engines

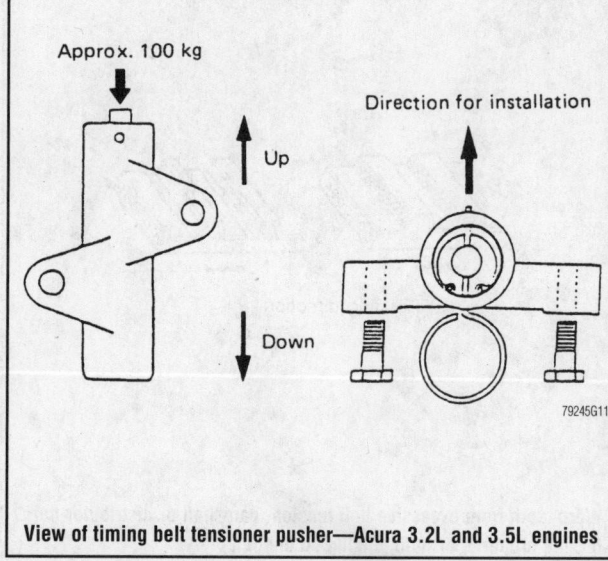

View of timing belt tensioner pusher—Acura 3.2L and 3.5L engines

12. If present, disconnect the two oil cooler hose bracket bolts on the timing cover. Move the oil cooler hoses and bracket off of the lower timing cover.

13. Remove the lower timing belt cover.

14. Remove the pusher assembly (tensioner) from below the belt tensioner pulley. The pusher rod must always face upward to prevent oil leakage. Depress the pusher rod, and insert a wire pin into the hole to keep the pusher rod retracted.

15. Remove the timing belt.

16. Inspect the water pump and replace it if there is any doubt about its condition.

17. Repair any oil or coolant leaks before installing a new timing belt. If the timing belt has been contaminated with oil or coolant, or is damaged, it must be replaced.

To install:

18. Verify that the sprocket timing marks are still aligned and that the groove and the keyway on the crankshaft timing sprocket align with the mark on the oil pump. The white pointers on the camshaft timing sprockets should align with the dots on the front plate.

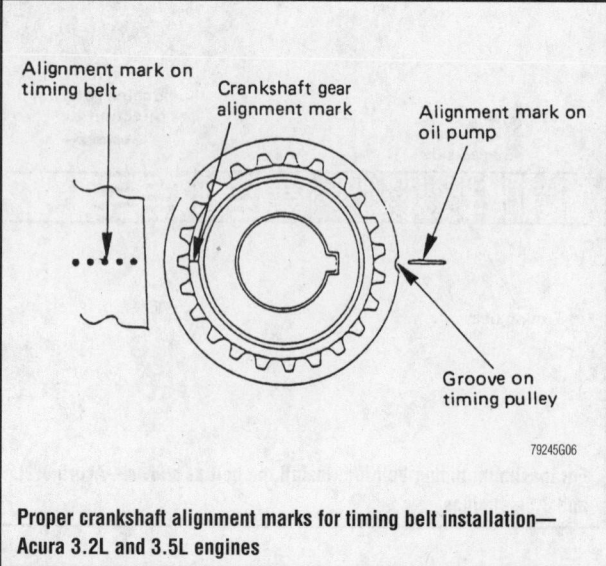

Proper crankshaft alignment marks for timing belt installation—Acura 3.2L and 3.5L engines

19. Install the timing belt. Use clips to secure the belt onto each sprocket until the installation is complete. Align the dotted marks on the timing belt with the timing mark opposite the groove on the crankshaft sprocket.

➡The arrows on the timing belt must follow the belt's direction of rotation. The manufacturer's trademark on the belt's spine should be readable left-to-right when the belt is installed.

20. Align the white line on the timing belt with the alignment mark on the right bank camshaft timing pulley. Secure the belt with a clip.

❋❋ WARNING

If any binding is felt when adjusting the timing belt tension by turning the crankshaft, STOP turning the engine, because the pistons may be hitting the valves.

21. Rotate the crankshaft counterclockwise to remove the slack between the crankshaft sprocket and the right camshaft timing belt sprocket.
22. Install the belt around the water pump pulley.
23. Install the belt on the idler pulley.
24. Align the white alignment mark on the timing belt with the alignment mark on the left bank camshaft timing belt sprocket.
25. Install the crankshaft pulley and tighten the center bolt by hand. Rotate the crankshaft pulley clockwise to give slack between the crankshaft timing belt pulley and the right bank camshaft timing belt pulley.
26. Insert a 1.4mm piece of wire through the hole in the pusher to hold the rod in. Install the pusher assembly while pushing the tension pulley toward the belt.
27. Pull the pin out from the pusher to release the rod.
28. Remove the clamps from the sprockets. Rotate the crankshaft pulley clockwise two turns. Measure the rod protrusion to ensure it is between 0.16–0.24 in. (4–6mm).
29. If the tensioner pulley bracket pivot bolt was removed, tighten it to 31 ft. lbs. (42 Nm).
30. Tighten the pusher bolts to 14 ft. lbs. (19 Nm).
31. Remove the crankshaft pulley. Install the lower and upper timing belt covers and tighten their bolts to 12 ft. lbs. (17 Nm).
32. Fit the oil cooler hose onto the timing cover and tighten its mounting bracket bolts to 16 ft. lbs. (22 Nm).
33. Install the crankshaft pulley and tighten the pulley bolt to 123 ft. lbs. (167 Nm).
34. Install fan pulley assembly and tighten the bolts to 16 ft. lbs. (22 Nm).
35. Install and adjust the accessory drive belts.
36. Install the cooling fan assembly and tighten the bolts to 6 ft. lbs. (8 Nm).
37. Install the upper fan shroud.
38. Install the air cleaner assembly and intake air duct.
39. Connect the upper radiator hose to the coolant inlet.
40. Refill and bleed the cooling system.
41. Connect the negative battery cable.

CHRYSLER

Minivans

This procedure applies only to the Dodge Caravan and Plymouth Voyager. The Chrysler Town & Country is only available with a 3.3L or 3.8L engine, both of which are not designed with timing belts.

2.4L (VIN B) ENGINE

➡You may need DRB scan tool to perform the crankshaft and camshaft relearn alignment procedure.

1. Disconnect the negative battery cable remote connection, located on the left strut tower.
2. Remove the right inner splash shield.
3. Remove the accessory drive belts.
4. Remove the crankshaft damper.
5. Remove the right engine mount.
6. Place a floor jack under the engine to support it while the engine mount is removed.
7. Remove the engine mount bracket.
8. Remove the timing belt cover.

➡This is an interference engine. Do not rotate the crankshaft or the camshafts after the timing belt has been removed. Damage to the valve components may occur. Before removing the timing belt, always align the timing marks.

9. Align the timing marks of the timing belt sprockets to the timing marks on the rear timing belt cover and oil pump cover. Loosen the timing belt tensioner bolts.
10. Remove the timing belt and the tensioner.
11. If necessary, remove the camshaft timing belt sprockets.
12. If necessary, remove the crankshaft timing belt sprocket using removal tool No. 6793 or equivalent.
13. Place the tensioner into a soft-jawed vise to compress the tensioner.
14. After compressing the tensioner, insert a pin (5/64 in. Allen wrench will also work) into the plunger side hole to retain the plunger until installation.

To install:
15. If necessary, using tool No. 6792 or equivalent, to install the crankshaft timing belt sprocket onto the crankshaft.
16. If necessary, install the camshaft sprockets onto the camshafts. Install and tighten the camshaft sprocket bolts to 75 ft. lbs. (101 Nm).
17. Set the crankshaft sprocket to Top Dead Center (TDC) by aligning the notch on the sprocket with the arrow on the oil pump housing.

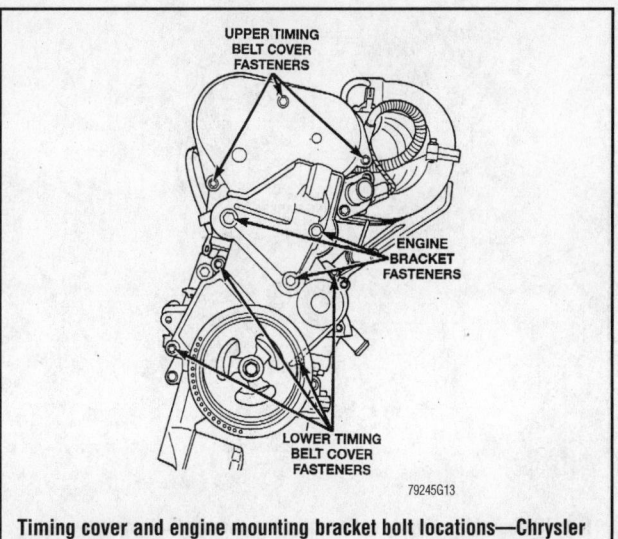

79245G13

Timing cover and engine mounting bracket bolt locations—Chrysler 2.4L (VIN B) engine

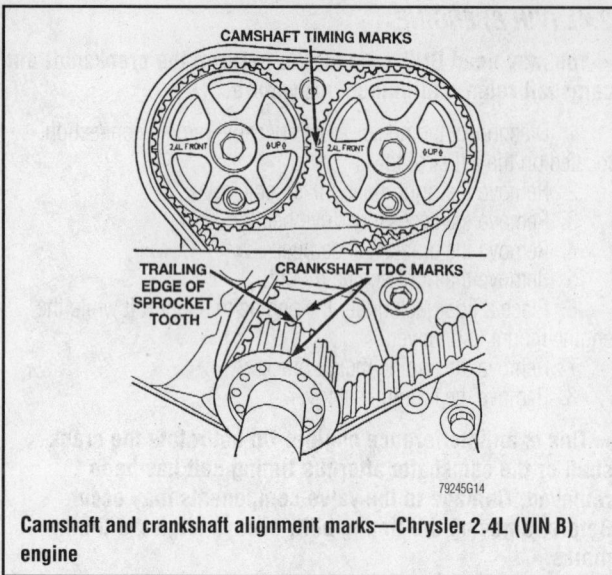

Camshaft and crankshaft alignment marks—Chrysler 2.4L (VIN B) engine

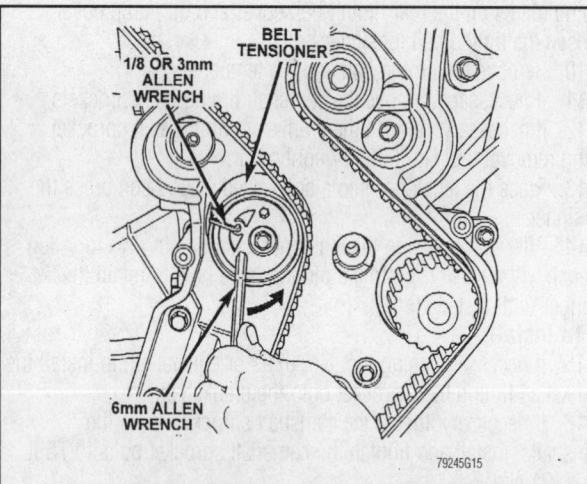

To lock the timing belt tensioner, be sure to fully insert the smaller Allen wrench into the tensioner as shown—Chrysler 2.4L (VIN B) engine

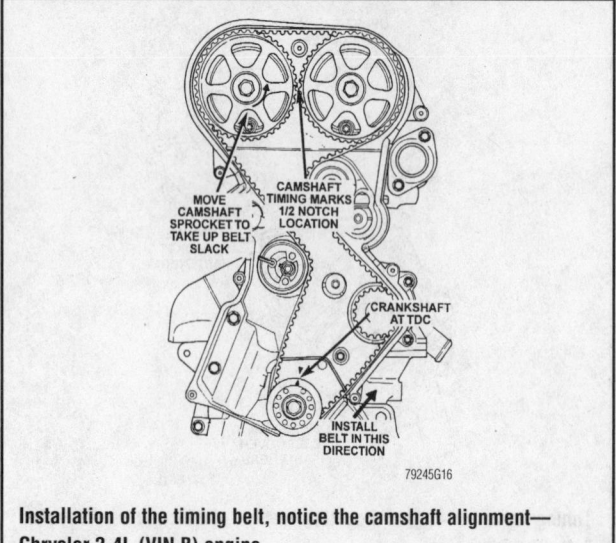

Installation of the timing belt, notice the camshaft alignment—Chrysler 2.4L (VIN B) engine

18. Set the camshafts to align the timing marks on the sprockets.

19. Move the crankshaft to ½ notch before TDC.

20. Install the timing belt starting at the crankshaft, then around the water pump sprocket, idler pulley, camshaft sprockets and around the tensioner pulley.

21. Move the crankshaft sprocket to TDC to take up the belt slack.

22. Install the tensioner on the engine block but do not tighten.

23. Using a torque wrench on the tensioner pulley, apply 250 inch lbs.(28 Nm) of torque to the tensioner pulley.

24. With torque being applied to the tensioner pulley, move the tensioner up against the tensioner pulley bracket and tighten the fasteners to 23 ft. lbs.(31 Nm).

25. Remove the tensioner plunger pin, the tension is correct when the plunger pin can be removed and reinserted easily.

✳✳ WARNING

If any binding is felt when adjusting the timing belt tension by turning the crankshaft, STOP turning the engine, because the pistons may be hitting the valves.

26. Rotate the crankshaft two revolutions and recheck the timing marks. Wait several minutes and then recheck that the plunger pin can easily be removed and installed.

27. Install the front timing belt cover.

28. Install the engine mount bracket.

29. Install the right engine mount.

30. Remove the floor jack from under the vehicle.

31. Install the crankshaft damper and tighten it to 105 ft. lbs. (142 Nm).

32. Install the accessory drive belts and adjust to the proper tension.

33. Install the right inner splash shield.

34. Reconnect the negative battery cable.

35. Perform the crankshaft and camshaft relearn alignment procedure using the DRB scan tool or equivalent.

2.5L (VIN K) ENGINE

Working on any engine (especially overhead camshaft engines) requires much care be given to valve timing. It is good practice to set the engine up to TDC No. 1 cylinder firing position. Verify that all timing marks on the crankshaft and camshaft sprockets are properly aligned before removing the timing belt. This serves as a point of reference for all work that follows. Valve timing is most important and engine damage will result if the work is incorrect.

➡This is an interference engine. Do not rotate the crankshaft or the camshafts after the timing belt has been removed. Damage to the valve components may occur. Before removing the timing belt, always align the timing marks.

1. Disconnect the negative battery cable.

2. Remove the accessory drive belts.

3. Remove the right engine mount yoke screw.

4. Remove the air conditioning compressor and set it aside. Remove the solid mount compressor bracket mounting bolts.

5. Turn the solid mount bracket away from the engine and slide it on the No. 2 stud until it is free. The front bolt and spacer will be removed with the bracket.

6. Remove the alternator and the drive belt idler.

7. Raise and safely support the vehicle. Remove the right inner fender splash shield.

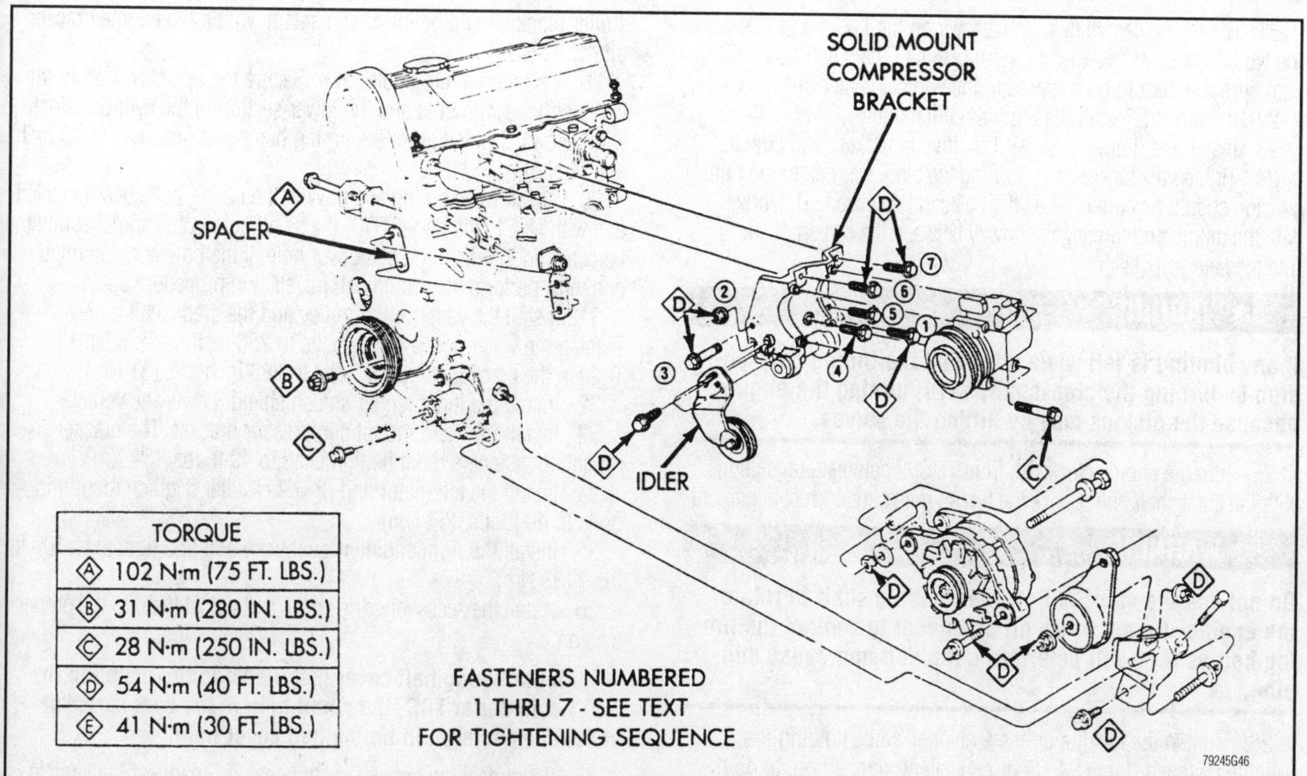

SOLID MOUNT
COMPRESSOR
BRACKET

SPACER

Ⓐ

Ⓑ

Ⓒ

Ⓓ

Ⓓ ②

③

Ⓓ

IDLER

⑦

⑥

⑤ ①

④

Ⓓ

Ⓒ

Ⓓ Ⓓ

Ⓓ

Ⓓ

Ⓓ

TORQUE	
Ⓐ	102 N·m (75 FT. LBS.)
Ⓑ	31 N·m (280 IN. LBS.)
Ⓒ	28 N·m (250 IN. LBS.)
Ⓓ	54 N·m (40 FT. LBS.)
Ⓔ	41 N·m (30 FT. LBS.)

FASTENERS NUMBERED
1 THRU 7 - SEE TEXT
FOR TIGHTENING SEQUENCE

79245G46

Exploded view of the solid mount compressor bracket—Chrysler 2.5L (VIN K) engine

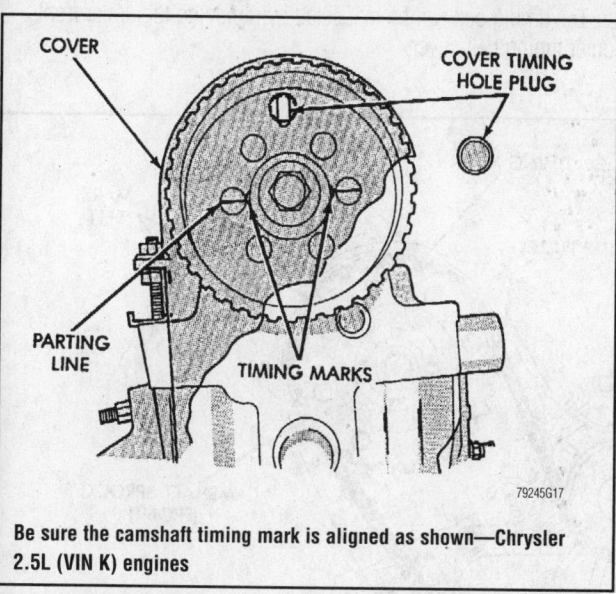

COVER

COVER TIMING
HOLE PLUG

PARTING
LINE

TIMING MARKS

79245G17

Be sure the camshaft timing mark is aligned as shown—Chrysler 2.5L (VIN K) engines

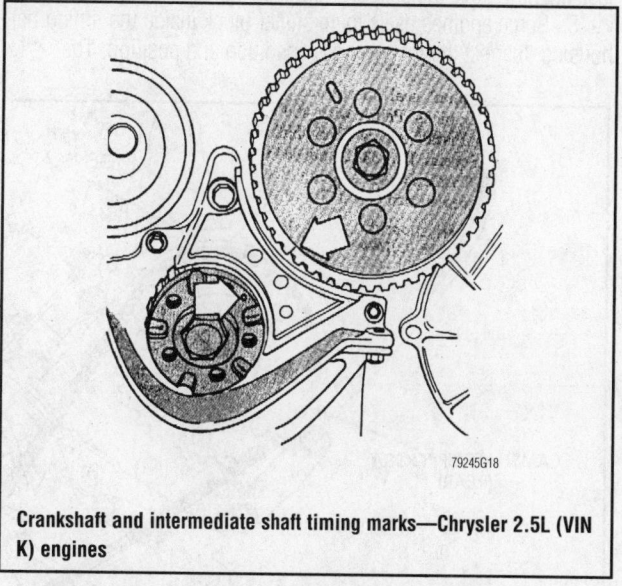

79245G18

Crankshaft and intermediate shaft timing marks—Chrysler 2.5L (VIN K) engines

8. Loosen and remove the three water pump pulley mounting bolts and remove the pulley.

9. Remove the four crankshaft pulley retaining bolts and the crankshaft pulley.

10. Remove the nuts at the upper portion of the timing cover and the bolts from the lower portion, then remove both halves of the cover.

11. Remove the timing belt covers.

12. Position a jack under the engine.

13. Separate the right engine mount and raise the engine slightly.

14. Loosen the timing belt tensioner bolt, rotate the hex nut, and remove timing belt.

15. Remove the timing belt tensioner, if necessary.

16. Remove the crankshaft sprocket with a suitable puller tool and a bolt approximately 6 in.(15 cm) long.

17. Remove the camshaft sprocket and intermediate shaft sprocket, if necessary.

To install:

18. Clean all parts well. A small amount of white paint on the sprocket timing marks may make alignment easier.

19. Install the crankshaft sprocket. Tighten the crankshaft sprocket bolt to 85 ft. lbs. (115 Nm).

20. If necessary, turn the crankshaft and intermediate shaft until markings on both sprockets are aligned.

21. Rotate the camshaft so the arrows on the hub are in line with the No. 1 camshaft cap-to-cylinder head line. The small hole in the cam sprocket should be centered in the vertical center line.

22. If removed, install the timing belt tensioner.

23. Install the timing belt over the drive sprockets and adjust.

24. Tighten the tensioner by turning the tensioner hex to the right. Tension should be correct when the belt can be twisted 90 degrees with the thumb and forefinger, midway between the camshaft and intermediate sprocket.

�֎ WARNING

If any binding is felt when adjusting the timing belt tension by turning the crankshaft, STOP turning the engine, because the pistons may be hitting the valves.

25. Turn the engine clockwise from TDC, 2 complete revolutions with the crankshaft bolt. Check the timing marks for correct alignment.

✖ WARNING

Do not use the camshaft or intermediate shaft to rotate the engine. Do not allow oil or solvent to contact the timing belt as they will deteriorate the belt and cause slipping.

26. Tighten the locknut on the tensioner, while holding the weighted wrench (tool C-4503 or equivalent) in position, to 45 ft. lbs. (61 Nm).

27. Lower the engine onto the right engine mount and install the fasteners. Remove the support from the engine.

28. Some engines use a foam stuffer block inside the timing belt housing. Inspect the foam block's condition and position. The stuffer block should be intact and secure within the engine bracket tunnel.

29. Install the timing belt cover. Secure the upper section to the cylinder head with nuts and the lower section to the cylinder block with screws. Tighten all of the timing belt cover fasteners to 40 inch lbs. (4 Nm).

30. Check valve timing again. With the timing belt cover installed, and with No. 1 cylinder at TDC, the small hole in the sprocket must be centered in the timing belt cover hole. If the hole is not aligned correctly, perform the timing belt installation procedure again.

31. Install the water pump pulley and the crankshaft pulley. Tighten the water pump pulley bolts to 250 inch lbs. (28 Nm). Tighten the crankshaft pulley bolts to 280 inch lbs. (31 Nm).

32. Install the inner fender splash shield. Lower the vehicle.

33. Install the solid mount compressor bracket. The bracket mounting fasteners must be tightened to 40 ft. lbs. (54 Nm).

34. Install the alternator and drive belt idler. Tighten mounting bolts to 40 ft. lbs. (54 Nm).

35. Install the right engine mount yoke bolt and tighten to 100 ft. lbs. (133 Nm).

36. Install the accessory drive belts and adjust them to the proper tension.

➡ **With the timing belt cover installed and the piston in the No. 1 cylinder at TDC, the small hole in the cam sprocket should be centered in timing belt cover hole.**

37. Reconnect the negative battery cable. Road test the vehicle.

3.0L (VIN 3) ENGINE

The timing belt can be inspected by removing the upper front outer timing belt cover.

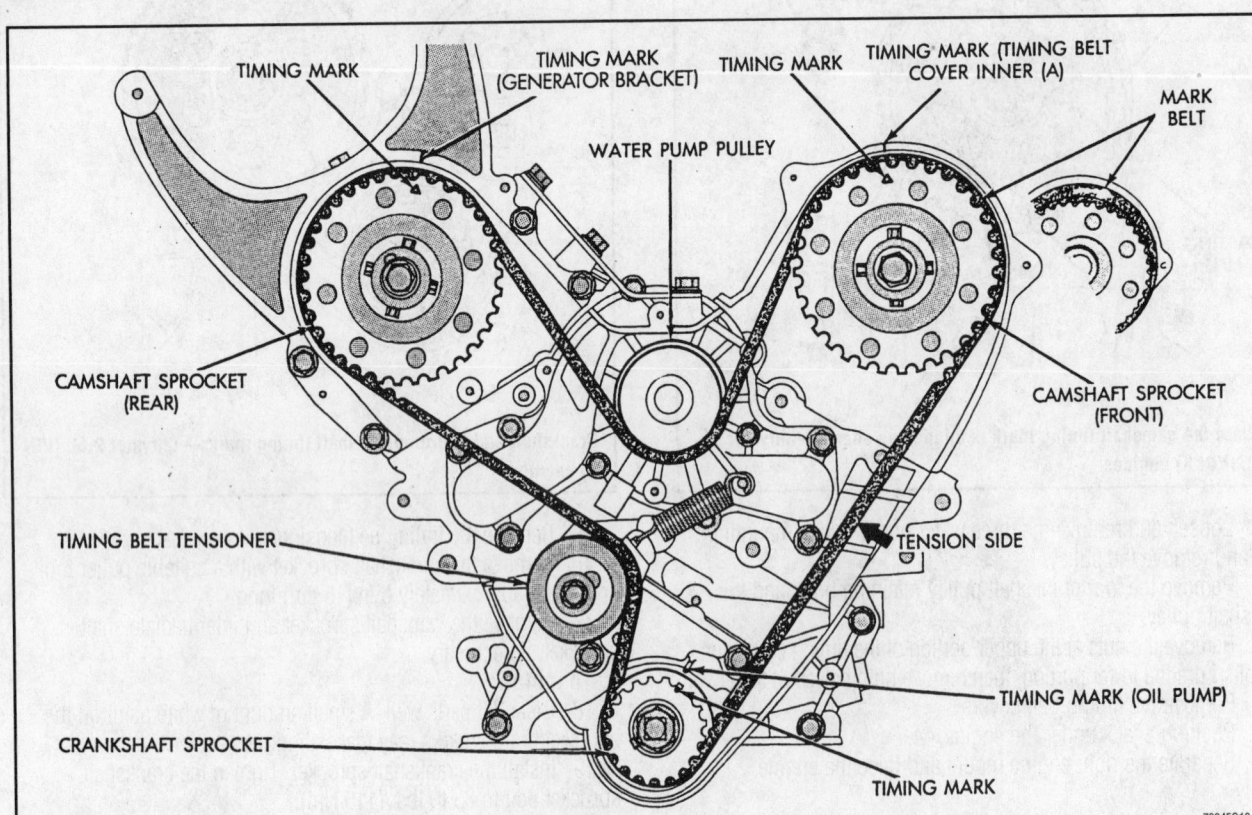

Timing belt sprocket timing marks for proper timing belt installation—Chrysler 3.0L (VIN 3) engine

79245G19

Working on any engine (especially overhead camshaft engines) requires much care be given to valve timing. It is good practice to set the engine up at TDC No. 1 cylinder firing position before beginning work. Verify that all timing marks on the crankshaft and camshaft sprockets are properly aligned before removing the timing belt and starting camshaft service. This serves as a point of reference for all work that follows. Valve timing is very important and engine damage will result if the work is incorrect.

1. Disconnect the negative battery cable.
2. Remove the accessory drive belts. Remove the engine mount insulator from the engine support bracket.
3. Remove the engine support bracket. Remove the crankshaft pulleys and torsional damper. Remove the timing belt covers.
4. Rotate the crankshaft until the sprocket timing marks are aligned. The crankshaft sprocket timing mark should align with the oil pump timing mark. The rear camshaft sprocket timing mark should align with the generator bracket timing mark and the front camshaft sprocket timing mark should align with the inner timing belt cover timing mark.
5. If the belt is to be reused, mark the direction of rotation on the belt for installation reference.
6. Loosen the timing belt tensioner bolt and remove the timing belt.
7. If necessary, remove the timing belt tensioner.
8. Remove the crankshaft sprocket flange shield and crankshaft sprocket.
9. Hold the camshaft sprocket using spanner tool MB990775 or equivalent, and remove the camshaft sprocket bolt and washer. Remove the camshaft sprocket.

To install:
10. Install the camshaft sprocket on the camshaft with the retaining bolt and washer. Hold the camshaft sprocket using spanner tool MB990775 or equivalent, and tighten the bolt to 70 ft. lbs. (95 Nm).
11. Install the crankshaft sprocket.
12. If removed, install the timing belt tensioner and tensioner spring. Hook the spring upper end to the water pump pin and the lower end to the tensioner bracket with the hook out.
13. Turn the timing belt tensioner counterclockwise full travel in the adjustment slot and tighten the bolt to temporarily hold it in this position.
14. Rotate the crankshaft sprocket until its timing mark is aligned with the oil pump timing mark.
15. Rotate the rear camshaft sprocket until its timing mark is aligned with the timing mark on the generator bracket.
16. Rotate the front (radiator side) camshaft sprocket until its mark is aligned with the timing mark on the inner timing belt cover.
17. Install the timing belt on the crankshaft sprocket while keeping the belt tight on the tension side.

➡**If the original belt is being reused, be sure to install it in the same rotational direction.**

18. Position the timing belt over the front camshaft sprocket (radiator side). Next, position the belt under the water pump pulley, then over the rear camshaft sprocket and finally over the tensioner.

✳✳ WARNING

f any binding is felt when adjusting the timing belt tension by turning the crankshaft, STOP turning the engine, because the pistons may be hitting the valves.

19. Apply rotating force in the opposite direction to the front camshaft sprocket (radiator side) to create tension on the timing belt tension side. Check that all timing marks are aligned.
20. Install the crankshaft sprocket flange.
21. Loosen the tensioner bolt and allow the tensioner spring to tension the belt.
22. Rotate the crankshaft two full turns in a clockwise direction. Turn the crankshaft smoothly and in a clockwise direction only.
23. Again line up the timing marks. If all marks are aligned, tighten the tensioner bolt to 250 inch lbs. (28 Nm). Otherwise repeat the installation procedure.
24. Install the timing belt covers. Install the engine support bracket. Tighten the support bracket mounting bolts to 35 ft. lbs. (47 Nm).
25. Install the engine mount insulator, torsional damper and crankshaft pulleys. Tighten the crankshaft pulley bolt to 112 ft. lbs. (151 Nm).
26. Install the accessory drive belts and adjust them to the proper tension.
27. Reconnect the negative battery cable.
28. Run the engine and check for proper operation. Road test the vehicle.

DODGE

Dakota

2.5L (VIN G) ENGINE

1. Position the engine so that the No. 1 piston is at TDC.
2. Disconnect the negative battery cable.
3. Remove the timing belt covers.
4. Remove the timing belt tensioner and allow the belt to hang free.
5. Remove the air conditioning compressor belt idler pulley, if equipped, and remove the mounting stud. Unbolt the compressor/alternator bracket and position it to the side.
6. Remove the timing belt from the vehicle.

To install:
7. Turn the crankshaft sprocket and intermediate shaft sprocket until the marks are aligned. Use a straightedge from bolt-to-bolt to confirm alignment.
8. Turn the camshaft until the small hole in the sprocket is at the top and rows on the hub are aligned with the camshaft cap-to-cylinder head mounting lines. Use a mirror to see the alignment so it is viewed straight on and not at an angle from above. Install the belt, but let it hang free at this point.
9. Install the air conditioning compressor/alternator bracket, idler pulley and motor mount. Raise the vehicle and support safely. Have the tensioner within an arm's reach because the timing belt will have to be held in position with one hand.
10. To properly install the timing belt, reach up and engage it with the camshaft sprocket. Turn the intermediate shaft counterclockwise slightly, then engage the belt with the intermediate shaft sprocket. Hold the belt against the intermediate shaft sprocket and turn the sprocket clockwise to take up all tension; if the timing marks are out of alignment, repeat the installation until alignment is correct.
11. Using a 13mm wrench, turn the crankshaft sprocket counterclockwise slightly and wrap the belt around it. Turn the sprocket clockwise so there is no slack in the belt between the sprockets, if the timing marks are out of alignment, repeat the installation until alignment is correct.

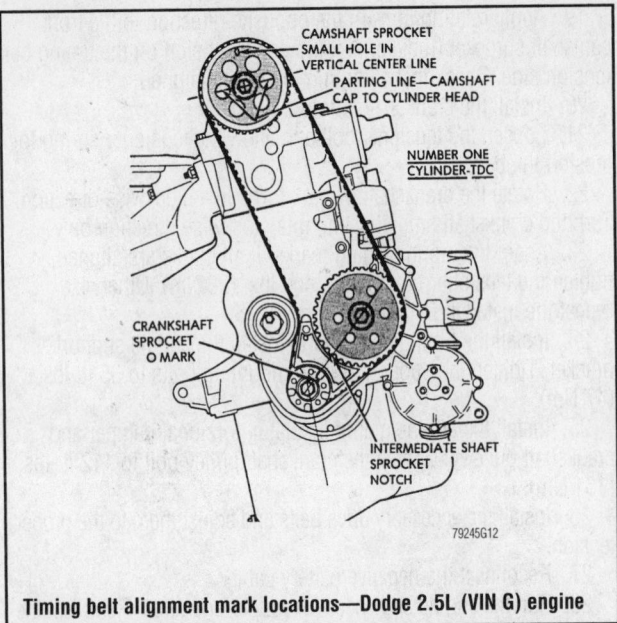

Timing belt alignment mark locations—Dodge 2.5L (VIN G) engine

✱✱ WARNING

If any binding is felt when adjusting the timing belt tension by turning the crankshaft, STOP turning the engine, because the pistons may be hitting the valves.

➡ **If the timing marks are aligned, but slack exists in the belt between either the camshaft and intermediate shaft sprockets or the intermediate and crankshaft sprockets, the timing will be incorrect when the belt is tensioned. All slack must be between the crankshaft and camshaft sprockets only.**

12. Install the tensioner and install the mounting bolt loosely. Place special tensioning tool C-4703, or equivalent, on the hex of the tensioner so the weight is approximately at the 9 o'clock position (parallel to the ground, hanging to the left) plus or minus 15°.

13. Hold the tool in position and tighten the bolt to 45 ft. lbs. (61 Nm). Do not pull the tool past the 9 o'clock position; this will make the belt too tight and will cause it to howl or possibly break during engine use.

14. Lower the vehicle and recheck the camshaft sprocket positioning. If it is correct, install the timing belt covers and all related parts.

15. Connect the negative battery cable and road test the vehicle.

FORD

Ranger

2.3L (VIN A) AND 2.5L (VIN C) ENGINES

1. Rotate the engine so that No. 1 cylinder is at TDC on the compression stroke. Check that the timing marks are aligned on the camshaft and crankshaft pulleys. An access plug is provided in the cam belt cover so that the camshaft timing can be checked without removal of the cover or any other parts. Set the crankshaft to TDC by aligning the timing mark on the crank pulley with the TDC mark on the belt cover. Look through the access hole in the belt cover to be sure that the timing mark on the cam drive sprocket is lined up with the pointer on the inner belt cover.

➡ **Always turn the engine in the normal direction of rotation. Backward rotation may cause the timing belt to jump time, due to the arrangement of the belt tensioner.**

2. Drain cooling system. Remove the upper radiator hose as necessary. Remove the fan blade and water pump pulley bolts.

✱✱ CAUTION

When draining the coolant, keep in mind that cats and dogs are attracted by ethylene glycol antifreeze, and are quite likely to drink any that is left in an uncovered container or in puddles on the ground. This will prove fatal in sufficient quantity. Always drain the coolant into a sealable container. Coolant should be reused unless it is contaminated or several years old.

3. Loosen the alternator retaining bolts and remove the drive belt from the pulleys. Remove the water pump pulley.

4. Remove the power steering pump and set it aside.

5. Remove the four timing belt outer cover retaining bolts and remove the cover. Remove the crankshaft pulley and belt guide.

6. Loosen the belt tensioner pulley assembly, then position a camshaft belt adjuster tool (T74P-6254-A or equivalent) on the tension spring rollpin and retract the belt tensioner away from the timing belt. Tighten the adjustment bolt to lock the tensioner in the retracted posi-tion.

7. If the belt is to be reused, mark the direction of rotation on the belt for installation reference.

8. Remove the timing belt.

To install:

9. Install the new belt over the crankshaft sprocket and then counterclockwise over the auxiliary and camshaft sprockets, making sure the lugs on the belt properly engage the sprocket teeth on the pulleys. Be careful not to rotate the pulleys when installing the belt.

10. Release the timing belt tensioner pulley, allowing the tensioner to take up the belt slack. If the spring does not have enough tension to move the roller against the belt (belt hangs loose), it might be necessary to manually push the roller against the belt and tighten the bolt.

➡ **The spring cannot be used to set belt tension; a wrench must be used on the tensioner assembly.**

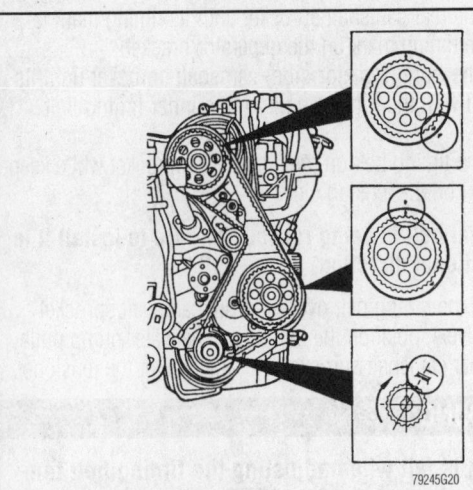

Camshaft, auxiliary shaft and crankshaft timing belt sprocket alignment mark locations—Ford 2.3L and 2.5L engines

⁕⁕ WARNING

If any binding is felt when adjusting the timing belt tension by turning the crankshaft, STOP turning the engine, because the pistons may be hitting the valves.

11. Rotate the crankshaft two complete turns by hand (in the normal direction of rotation) to remove slack from the belt, then tighten the tensioner adjustment to 26–33 ft. lbs. (35–45 Nm) and pivot bolts to 30–40 ft. lbs. (40–55 Nm). Be sure the belt is seated properly on the pulleys and that the timing marks are still in alignment when No. 1 cylinder is again at TDC/compression.

12. Install the crankshaft pulley and belt guide.

13. Install the timing belt cover.

14. Install the water pump pulley and fan blades. Install the upper radiator hose if necessary. Refill the cooling system.

15. Install the accessory drive belts.

16. Start the engine and check the ignition timing. Adjust the timing, if necessary.

GEO/CHEVROLET

Tracker

➡ **During these procedures, identify all components removed from the engine so that they may be reinstalled in their original positions. If discarding the old components so that new components can be installed, identifying the old items is not necessary.**

1.6L 8-VALVE ENGINE

➡ **Do not rotate the crankshaft counterclockwise or attempt to rotate the crankshaft by turning the camshaft sprocket.**

1. Remove the timing belt cover.

2. Remove rocker arm cover.

3. If the timing belt is not already marked with a directional arrow, use white paint, a grease pencil or correction fluid to do so.

4. Disconnect one end of the tensioner spring. Loosen the timing belt tensioner bolt and stud, then, using your finger, press the tensioner plate up and remove the timing belt from the crankshaft and camshaft sprockets.

5. Remove the timing belt tensioner, tensioner plate and spring from the engine.

6. Insert a metal rod through the hole in the camshaft to lock the camshaft from rotating. Loosen the camshaft sprocket retaining bolt, then pull the camshaft sprocket off of the end of the camshaft.

7. Remove the crankshaft timing belt sprocket by loosening the center bolt, while preventing the crankshaft from rotating. To hold the crankshaft from turning, you can use Suzuki Tool 09927–56010 (or equivalent), or a large prybar inserted in the transmission housing slot and the flywheel teeth. Pull the sprocket off of the end of the crankshaft. Be sure to retain the crankshaft sprocket key and belt guide for assem-bly.

8. If necessary, remove the timing belt inside cover from the cylinder head.

To install:

9. If necessary, install the timing belt inside cover.

10. Slide the timing belt guide on the crankshaft so that the concave side faces the oil pump, then install the sprocket key in the groove in the crankshaft.

11. Slide the pulley onto the crankshaft, and install the center retaining bolt. Tighten the center bolt to 58–65 ft. lbs. (80–90 Nm).

To hold the crankshaft from turning, you can use Suzuki Tool 09927–56010 (or equivalent), or a large prybar inserted in the transmission housing slot and the flywheel teeth.

12. Install the timing belt camshaft sprocket, ensuring that the slot in the sprocket engages the camshaft (pulley) pin; this ensures that the sprocket is properly positioned on the end of the camshaft. Secure the camshaft with the metal rod used during removal, then tighten the sprocket bolt to 41–46 ft. lbs. (56–64 Nm).

13. Assemble the timing belt tensioner plate and the tensioner, making sure that the lug of the tensioner plate engages the tensioner.

14. Install the timing belt tensioner, tensioner plate and spring on the engine. Tighten the mounting bolt and stud only finger-tight at this time. Ensure that when the tensioner is moved in a counterclockwise direction, the tensioner moves in the same direction. If the tensioner does not move, remove it and the tensioner plate to reassemble them properly.

15. Loosen all rocker arm valve lash locknuts and adjusting screws. This will permit movement of the camshaft without any rocker arm associated drag, which is essential for proper timing belt tensioning. If the camshaft does not rotate freely (free of rocker arm drag), the belt will not be properly tensioned.

⁕⁕ WARNING

If any binding is felt when adjusting the timing belt tension by turning the crankshaft, STOP turning the engine, because the pistons may be hitting the valves.

16. Rotate the camshaft sprocket clockwise until the timing mark on the sprocket and the V mark on the timing belt inside cover are aligned.

17. Using a 17mm wrench, or socket and breaker bar, on the crankshaft sprocket center bolt, turn the crankshaft clockwise until the punch mark on the sprocket is aligned with the arrow mark on the oil pump.

18. With the camshaft and crankshaft marks properly aligned, push the tensioner up with your finger and install the timing belt on the two sprockets, ensuring that the drive side of the belt is free of all slack. Release your finger from the tensioner. Be sure to install the timing belt so that the directional arrow is pointing in the appropriate direction.

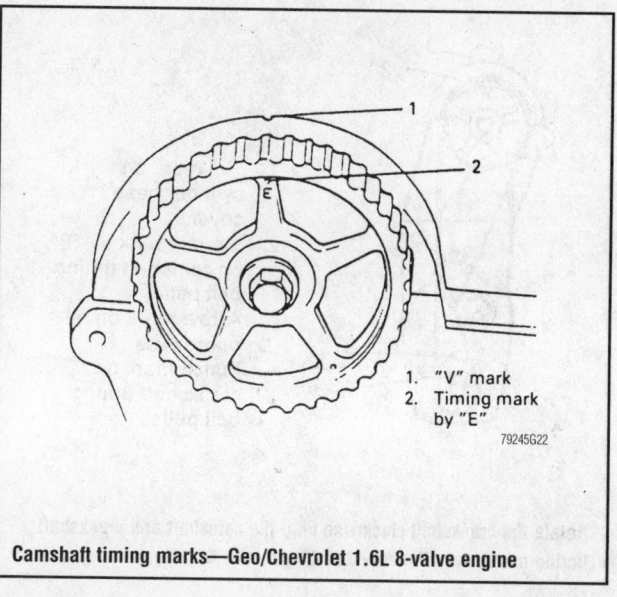

1. "V" mark
2. Timing mark by "E"

79245G22

Camshaft timing marks—Geo/Chevrolet 1.6L 8-valve engine

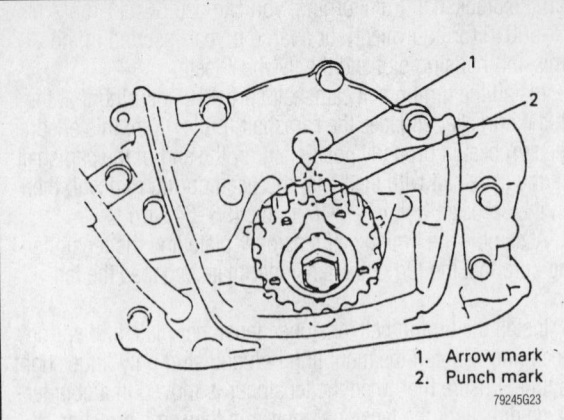

1. Arrow mark
2. Punch mark

79245G23

Align the punch mark with the arrow for proper timing belt installation—Geo/Chevrolet 1.6L 8-valve engine

➡**In this position, the No. 4 cylinder is at Top Dead Center (TDC) on the compression stroke.**

19. Rotate the crankshaft clockwise two full revolutions, then tighten the tensioner stud to 80–106 inch lbs. (9–12 Nm). Then, tighten the tensioner bolt to 18–21 ft. lbs. (24–30 Nm).

20. Ensure that all four timing marks are still aligned as before; if they are not, remove the timing belt, and install and tension it again.

21. Install the timing belt cover and all related components.

1.6L 16-VALVE ENGINE

The 1.6L 16-valve engine is known as an interference motor, because it is fabricated with such close tolerances between the pistons and valves that, if the timing belt is incorrectly positioned, jumps teeth on one of the sprockets or breaks, the valve and pistons will come into contact. This can cause severe internal engine damage

➡**Do not rotate the crankshaft counterclockwise or attempt to rotate the crankshaft by turning the camshaft sprocket.**

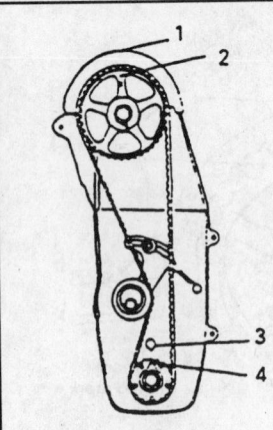

1. "V" mark on cylinder head cover
2. Timing mark by "E" on camshaft timing belt pulley
3. Arrow mark on oil pump case
4. Punch mark on crankshaft timing belt pulley

79245G47

Rotate the crankshaft clockwise until the camshaft and crankshaft timing marks are aligned—Geo/Chevrolet 1.6L 16-valve engine

1. Remove the timing belt cover.

2. If the timing belt is not already marked with a directional arrow, use white paint, a grease pencil or correction fluid to do so.

3. Rotate the crankshaft clockwise until the timing mark on the camshaft sprocket and the V mark on the timing belt inside cover are aligned, and the punch mark on the crankshaft sprocket is aligned with the mark on the engine.

❊❊ WARNING

Do not rotate the crankshaft or camshaft once the timing belt is removed, because the valves and pistons can come into contact, which may cause internal engine damage.

4. Disconnect one end of the tensioner spring. Loosen the timing belt tensioner bolt and stud, then, using your finger, press the tensioner plate up and remove the timing belt from the crankshaft and camshaft sprockets.

5. Remove the timing belt tensioner, tensioner plate and spring from the engine.

6. Install Suzuki Tool 09917–68220, or equivalent, onto the camshaft sprocket to hold the camshaft from rotating. Loosen the camshaft sprocket retaining bolt, then pull the camshaft sprocket off of the end of the camshaft.

7. Remove the crankshaft timing belt sprocket by loosening the center bolt, while preventing the crankshaft from rotating. To hold the crankshaft from turning, you can use Suzuki Tool 09927–56010 (or equivalent), or a large prybar inserted in the transmission housing slot and the flywheel teeth. Pull the sprocket off of the end of the crankshaft. Be sure to retain the crankshaft sprocket key and belt guide for assembly.

8. If necessary, remove the timing belt inside cover from the cylinder head.

To install:

9. If necessary, install the timing belt inside cover.

10. Slide the timing belt guide on the crankshaft so that the concave side faces the oil pump, then install the sprocket key in the groove in the crankshaft.

11. Slide the pulley onto the crankshaft, and install the center retaining bolt. Tighten the center bolt to 80 ft. lbs. (110 Nm). To hold the crankshaft from turning, you can use Suzuki Tool 09927–56010 (or equivalent), or a large prybar inserted in the transmission housing slot and the flywheel teeth.

12. Install the timing belt camshaft sprocket, ensuring that the slot in the sprocket engages the camshaft (pulley) pin; this ensures that the sprocket is properly positioned on the end of the camshaft. Secure the camshaft with the holding tool used during removal, then tighten the sprocket bolt to 44 ft. lbs. (60 Nm).

13. Assemble the timing belt tensioner plate and the tensioner, making sure that the lug of the tensioner plate engages the tensioner.

❊❊ WARNING

If any binding is felt when adjusting the timing belt tension by turning the crankshaft, STOP turning the engine, because the pistons may be hitting the valves.

14. Install the timing belt tensioner, tensioner plate and spring on the engine. Tighten the mounting bolt and stud only finger-tight at this time. Ensure that when the tensioner is moved in a counterclockwise direction, the tensioner moves in the same direction. If

the tensioner does not move, remove it and the tensioner plate to reassemble them properly.

15. Loosen all rocker arm valve lash locknuts and adjusting screws. This will permit movement of the camshaft without any rocker arm associated drag, which is essential for proper timing belt tensioning. If the camshaft does not rotate freely (free of rocker arm drag), the belt will not be properly tensioned.

16. Rotate the camshaft sprocket clockwise until the timing mark on the sprocket and the V mark on the timing belt inside cover are aligned.

17. Using a wrench, or socket and breaker bar, on the crankshaft sprocket center bolt, turn the crankshaft clockwise until the punch mark on the sprocket is aligned with the arrow mark on the oil pump.

18. With the camshaft and crankshaft marks properly aligned, push the tensioner up with your finger and install the timing belt on the two sprockets, ensuring that the drive side of the belt is free of all slack. Release your finger from the tensioner. Be sure to install the timing belt so that the directional arrow is pointing in the appropriate direction.

➡ **In this position, the No. 4 cylinder is at Top Dead Center (TDC) on the compression stroke.**

19. Rotate the crankshaft clockwise two full revolutions, then tighten the tensioner stud to 97 inch lbs. (11 Nm). Then, tighten the tensioner bolt to 18 ft. lbs. (24 Nm).

20. Ensure that all four timing marks are still aligned as before; if they are not, remove the timing belt, and install and tension it again.

21. Install the timing belt cover and all related components.

HONDA

CR-V and Odyssey

2.0L (B20B4) ENGINE

1. Disconnect the negative battery cable.
2. Position crankshaft so that No. 1 piston is at TDC.
3. Remove the splash guard.
4. Remove the accessory drive belts.
5. If equipped, remove the cruise control actuator.
6. Place a piece of wood between the oil pan and the jack, support the engine with a jack.
7. Remove upper engine bracket.
8. Remove the valve cover.
9. Remove the timing belt covers.
10. Loosen the adjusting bolt 180°. Release the tension from the belt by pushing on the tensioner, then retighting the adjusting bolt.
11. Remove the belt.
To install:
12. Be sure the timing marks are properly aligned.
13. Install the timing belt on the pulleys following this sequence:
  a. Crankshaft pulley.
  b. Adjusting pulley.
  c. Water pump pulley.
  d. Exhaust camshaft pulley.
  e. Intake camshaft pulley.
14. Loosen and retighten the adjusting bolt to allow tension to be applied to the belt.
15. Install the lower and middle timing covers .
16. Install the crankshaft pulley and tighten the bolt to 130 ft. lbs.(177 Nm).

✷✷ WARNING

If any binding is felt when adjusting the timing belt tension by turning the crankshaft, STOP turning the engine, because the pistons may be hitting the valves.

17. Rotate the crankshaft about five or six times counterclockwise to seat the timing belt.
18. Position the No. 1 piston to TDC.
19. Loosen the adjusting bolt ½ turn.
20. Rotate the crankshaft counterclockwise three teeth on the camshaft pulley.
21. Tighten the adjusting bolt to 40 ft. lbs.(54Nm).
22. Retighten the crankshaft pulley bolt to 130 ft. lbs.(177 Nm).
23. Install valve cover .
24. Install the engine mounting bracket, then remove the jack.
25. If removed, install the cruise control actuator.
26. Install the accessory drive belts.
27. Install the splash guard.

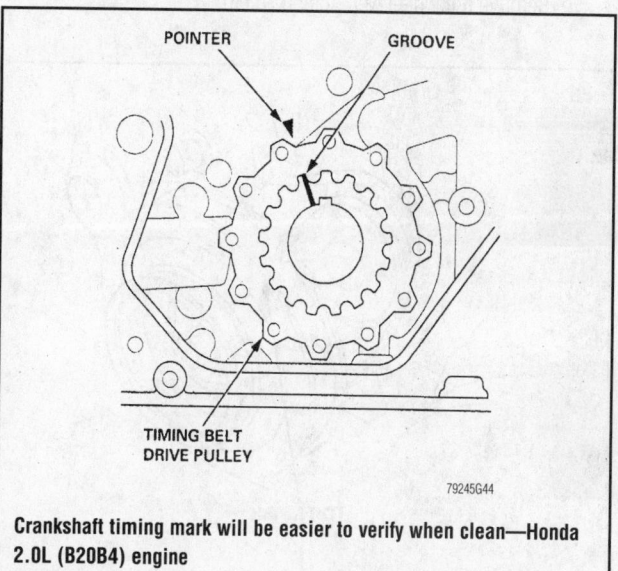

Crankshaft timing mark will be easier to verify when clean—Honda 2.0L (B20B4) engine

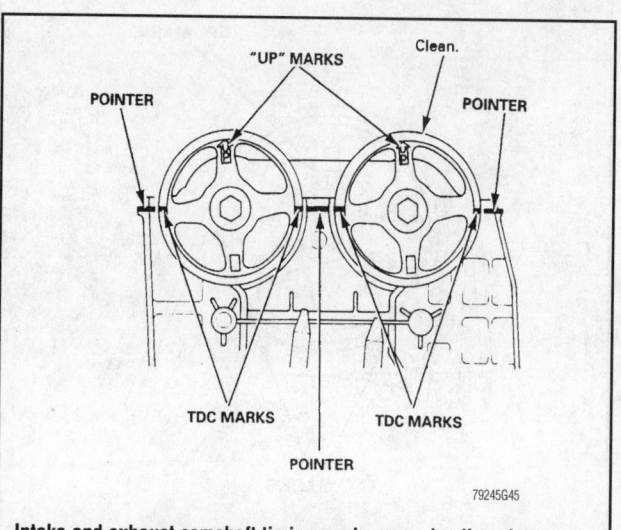

Intake and exhaust camshaft timing marks properly aligned at TDC—Honda 2.0L (B20B4) engine

28. Connect the negative battery cable.
29. Check the engine operation and road test.

2.2L (F22B6) AND 2.3L (F23A7) ENGINES

➡**The radio may contain a coded theft protection circuit. Always make note of your code number before disconnecting the battery.**

1. Disconnect the negative and positive battery cables.
2. Remove the valve cover.
3. Remove the upper timing belt cover.
4. Turn the engine to align the timing marks and set cylinder No.1 to TDC for the compression stroke. The white mark on the crankshaft pulley should align with the pointer on the timing belt cover. The words **UP** embossed on the camshaft pulley should be aligned in the upward position. The marks on the edge of the pulley should be aligned with the cylinder head or the back cover upper edge. Once in this position, the engine must NOT be turned or disturbed.
5. Remove the splash shield from below the engine.
6. Remove the wheel well splash shield.

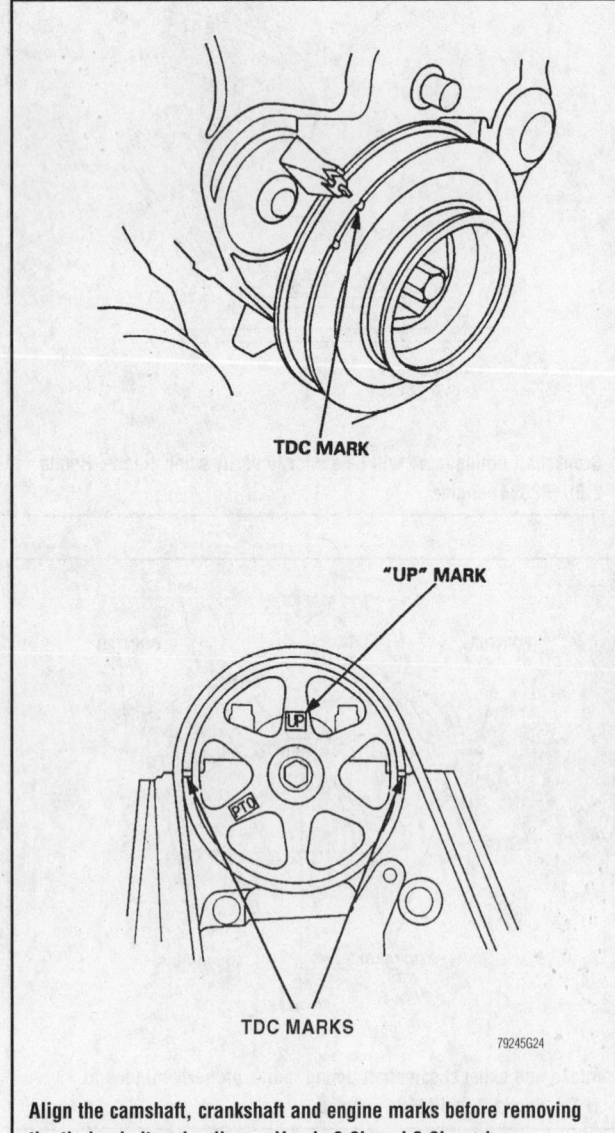

TDC MARK

"UP" MARK

PTO

TDC MARKS

79245G24

Align the camshaft, crankshaft and engine marks before removing the timing belt and pulleys—Honda 2.2L and 2.3L engines

7. Loosen and remove the power steering pump belt. Remove the power steering pump.
8. Loosen the adjusting and mounting bolts for the alternator and remove the drive belt.
9. Support the engine with a floor jack cushioned with a piece of wood.
10. Remove the through-bolt for the side engine mount and remove the mount.
11. Remove the crankshaft pulley bolt and remove the crankshaft pulley. Use a crank pulley holder (part No. 07MAB-PY3010A) and holder handle (part No. 07JAB-001020A),or there equivalents, to hold the crankshaft pulley in place while removing the bolt.
12. Remove the lower timing belt cover.
13. Remove the balancer shaft belt and its drive pulley.
14. Insert a suitable tool into the maintenance hole in the front balancer shaft. Unbolt and remove the balancer driven pulley.

➡**For servicing the balance shafts, front refers to the side of the engine facing the radiator. Rear refers to the side of the engine facing the firewall.**

15. Remove the timing belt.
16. If equipped with a TDC sensor assembly at the crankshaft sprocket, unbolt the assembly and move it to the side before removing the sprocket.
17. Remove the key and the spacers to remove the crankshaft timing sprocket.
18. Unbolt and remove the camshaft timing sprocket.

To install:

19. Install the camshaft timing sprocket so that the **UP** mark is up and the TDC marks are parallel to the cylinder head gasket surface. Install the key and tighten the bolt to 27 ft. lbs. (37 Nm).
20. Install the crankshaft timing sprocket so that the TDC mark aligns with the pointer on the oil pump. Install the spacers with the concave surfaces facing in. Install the key. Install the TDC sensor assembly back into position before installing the timing belt.
21. Install and tension the timing belt.
22. Rotate the crankshaft counterclockwise five to six turns to be sure the belt is properly seated.
23. Set the No. 1 piston at TDC for its compression stroke.

✳✳ WARNING

If any binding is felt when adjusting the timing belt tension by turning the crankshaft, STOP turning the engine, because the pistons may be hitting the valves.

24. Rotate the crankshaft counterclockwise so that the camshaft pulley moves only three teeth beyond its TDC mark.
25. Tighten the tensioner adjusting nut to 33 ft. lbs. (45 Nm).
26. Tighten the crankshaft pulley bolt to 181 ft. lbs. (245 Nm).
27. Install the balancer shaft belt drive pulley.
28. Align the groove on the pulley edge to the pointer on the balancer gear case.
29. Check the alignment of the pointer on the balancer pulley to the pointer on the oil pump.
30. Install and tension the balancer shaft belt.
31. Be sure the timing belts have been tensioned correctly and that all TDC and alignment marks are in their proper positions.
32. Install the lower timing cover and the crankshaft pulley. Apply engine oil to the pulley bolt threads and washer surface. Install the pulley bolt and tighten it to 181 ft. lbs. (245 Nm).

33. Install the upper timing cover and the valve cover. Be sure the seals are properly seated.

34. Install the side engine mount. Tighten the through-bolt to 47 ft. lbs. (64 Nm). Tighten the mount nut and bolt to 40 ft. lbs. (55 Nm) each.

35. Remove the floor jack.

36. Install and tension the alternator belt.

37. Install the power steering pump and tension its belt.

38. Install the splash shields.

39. Reconnect the positive and negative battery cables. Enter the radio security code.

40. Check engine operation.

Passport

2.6L ENGINE

1. Disconnect the negative battery cable.

2. Loosen and remove the engine accessory drive belts.

3. Remove the cooling fan assembly and the water pump pulley.

4. Drain the fluid from the power steering reservoir.

5. Unbolt and remove the power steering pump. Unbolt the hydraulic line brackets from the upper timing cover and move the pump out of the work area without disconnecting the hydraulic lines.

6. Disconnect and remove the starter motor if a flywheel holder (part No. J–38674 or equivalent) is to be used.

7. Remove the upper timing belt cover.

8. Rotate the crankshaft to set the engine at TDC/compression for the No. 1 cylinder. The arrow mark on the camshaft sprocket will be aligned with the mark on the rear timing cover.

9. Remove the crankshaft pulley.

10. Remove the lower timing belt cover.

11. Verify that the engine is set at TDC/compression for the No. 1 cylinder. The notch on the crankshaft sprocket will be aligned with the pointer on the oil seal retainer.

12. Release and remove the tensioner spring to release the timing belt's tension.

13. Remove the timing belt.

14. Unbolt the tensioner pulley bracket from the engine's front cover.

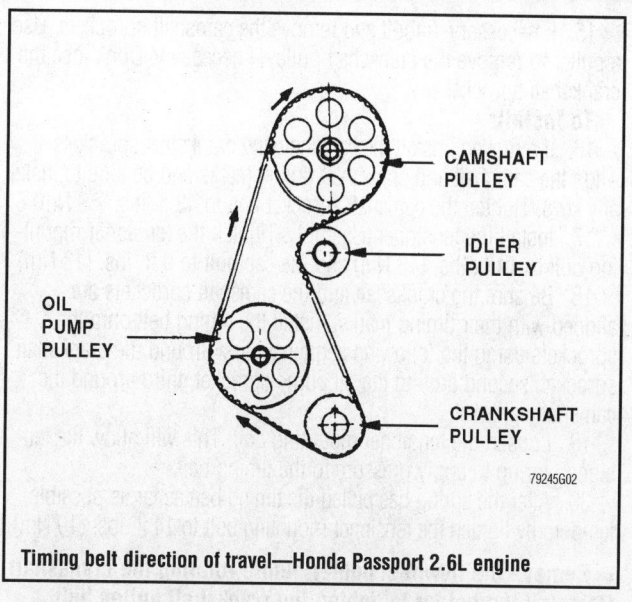

Timing belt direction of travel—Honda Passport 2.6L engine

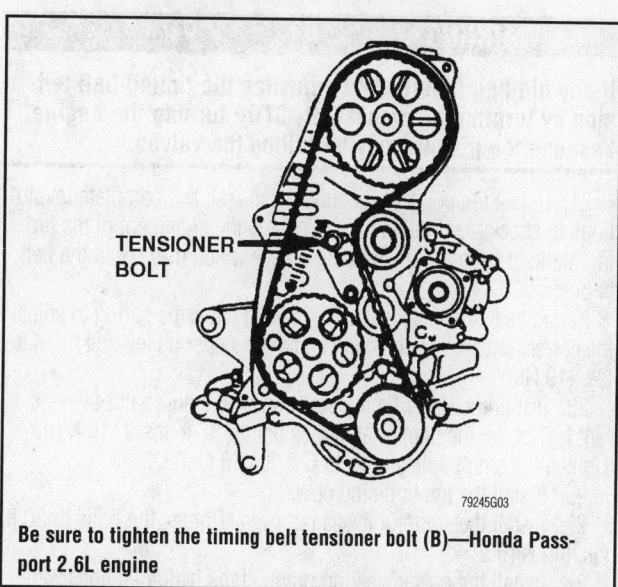

Be sure to tighten the timing belt tensioner bolt (B)—Honda Passport 2.6L engine

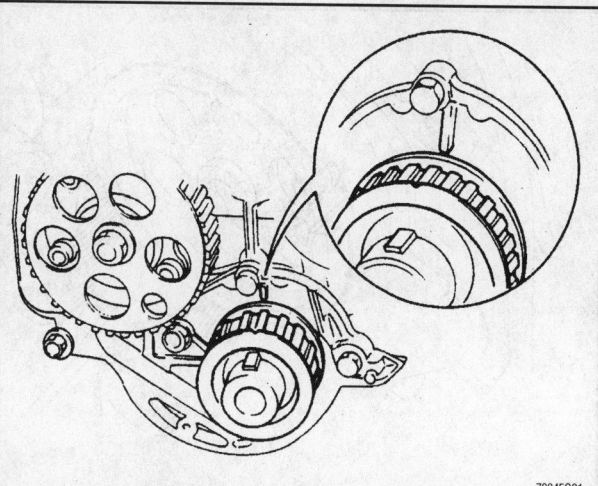

Align the crankshaft pulley timing mark the with oil retainer setting mark—Honda Passport 2.6L engine

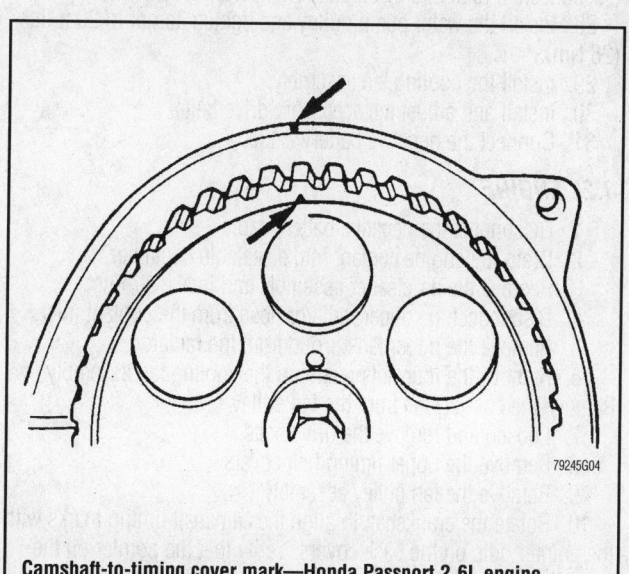

Camshaft-to-timing cover mark—Honda Passport 2.6L engine

15. If necessary, unbolt and remove the camshaft sprockets. Use a puller to remove the crankshaft pulley if necessary. Don't lose the crankshaft sprocket key.

To install:

16. If removed, install the camshaft and crankshaft sprockets. Align the camshaft and crankshaft timing marks and be sure to install any keys. Tighten the camshaft sprocket bolt to 43 ft. lbs. (59 Nm).

17. Install the tensioner assembly. Tighten the tensioner mounting bolt to 14 ft. lbs. (19 Nm) and the cap bolt to 9 ft. lbs. (13 Nm).

18. Be sure the crankshaft and the camshaft sprockets are aligned with their timing marks. Install the timing belt onto the sprockets using the following sequence: first around the crankshaft sprocket; second around the oil pump sprocket; third around the camshaft sprocket.

19. Loosen the tensioner mounting bolt. This will allow the tensioner spring to apply pressure to the timing belt.

20. After the spring has pulled the timing belt as far as possible, temporarily tighten the tensioner mounting bolt to 14 ft. lbs. (19 Nm).

➥**Remove the flywheel holder before rotating the crankshaft. Reinstall the holder to tighten the crankshaft pulley bolt.**

✳✳ WARNING

If any binding is felt when adjusting the timing belt tension by turning the crankshaft, STOP turning the engine, because the pistons may be hitting the valves.

21. Rotate the crankshaft counterclockwise two complete revolutions to check the rotation of the belt and the alignment of the timing marks. Listen for any rubbing noises which may mean the belt is binding.

22. Loosen the tensioner pulley bolt to allow the spring to adjust the correct tension. Then, retighten the tensioner pulley bolt to 14 ft. lbs. (19 Nm).

23. Install the lower timing cover and the crankshaft pulley.

24. Tighten the crankshaft pulley bolt to 87 ft. lbs. (118 Nm). Tighten the small pulley bolts to 6 ft. lbs. (8 Nm).

25. Install the upper timing cover.

26. Install the starter if it was removed. Tighten the bolts to 30 ft. lbs. (40 Nm).

27. Install the power steering pump. If the hydraulic lines were disconnected, refill and bleed the power steering system.

28. Install the water pump pulley and tighten its nut to 20 ft. lbs. (26 Nm).

29. Install the cooling fan assembly.

30. Install and adjust the accessory drive belts.

31. Connect the negative battery cable.

3.2L ENGINE

1. Disconnect the negative battery cable.
2. Drain the engine coolant into a sealable container.
3. Remove the air cleaner assembly and intake air duct.
4. Disconnect the upper radiator hose from the coolant inlet.
5. Remove the upper fan shroud from the radiator.
6. Remove the four nuts retaining the cooling fan assembly. Remove the cooling fan from the fan pulley.
7. Loosen and remove the drive belts.
8. Remove the upper timing belt covers.
9. Remove the fan pulley assembly.
10. Rotate the crankshaft to align the camshaft timing marks with the pointer dots on the back covers. Verify that the pointer on the crankshaft aligns with the mark on the lower timing cover.

➥**When the timing marks are aligned on 1995 vehicles, no pistons will be at TDC/compression. When the timing marks are aligned on 1995½–99 vehicles, the No. 2 piston is at TDC/compression.**

✳✳ WARNING

Align the camshaft and crankshaft sprockets with their alignment marks before removing the timing belt. Failure to align the belt and sprocket marks may result in valve damage.

11. Use tool No. J-8614–01, or a suitable pulley holding tool to remove the crankshaft pulley center bolt. Remove the crankshaft pulley.

12. If present, disconnect the two oil cooler hose bracket bolts on the timing cover. Move the oil cooler hoses and bracket off of the lower timing cover.

13. Remove the lower timing belt cover.

14. Remove the pusher assembly (tensioner) from below the belt tensioner pulley. The pusher rod must always face upward to prevent oil leakage. Depress the pusher rod, and insert a wire pin into the hole to keep the pusher rod retracted.

15. Remove the timing belt.

16. Inspect the water pump and replace it if there is any doubt about its condition.

17. Repair any oil or coolant leaks before installing a new timing belt. If the timing belt has been contaminated with oil or coolant, or is damaged, it must be replaced.

To install:

18. Verify that the sprocket timing marks are still aligned and that the groove and the keyway on the crankshaft timing sprocket align with the mark on the oil pump. The white pointers on the camshaft timing sprockets should align with the dots on the front plate.

19. Install the timing belt. Use clips to secure the belt onto each sprocket until the installation is complete. Align the dotted marks on the timing belt with the timing mark opposite the groove on the crankshaft sprocket.

➥**The arrows on the timing belt must follow the belt's direction of rotation. The manufacturer's trademark on the belt's spine should be readable left-to-right when the belt is installed.**

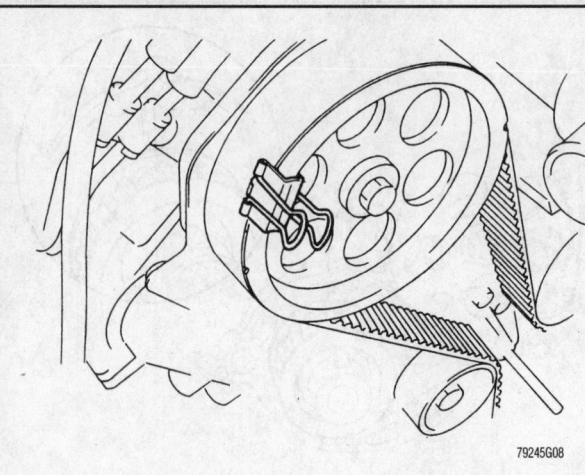

79245G08

Using a double clip to hold the belt in place—Honda Passport 3.2L engine

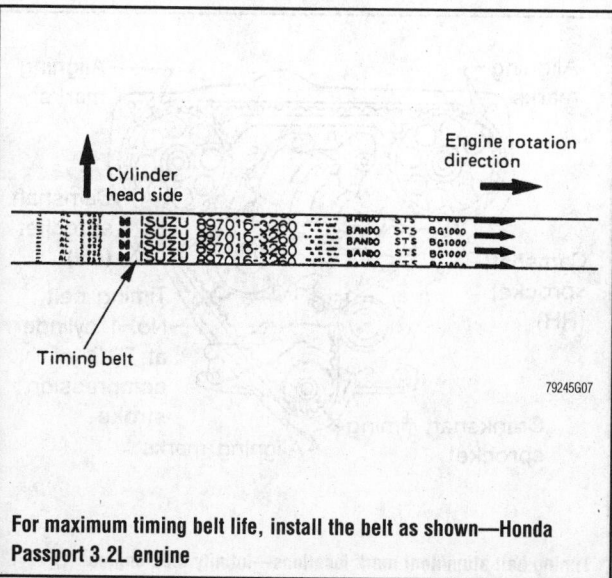

For maximum timing belt life, install the belt as shown—Honda Passport 3.2L engine

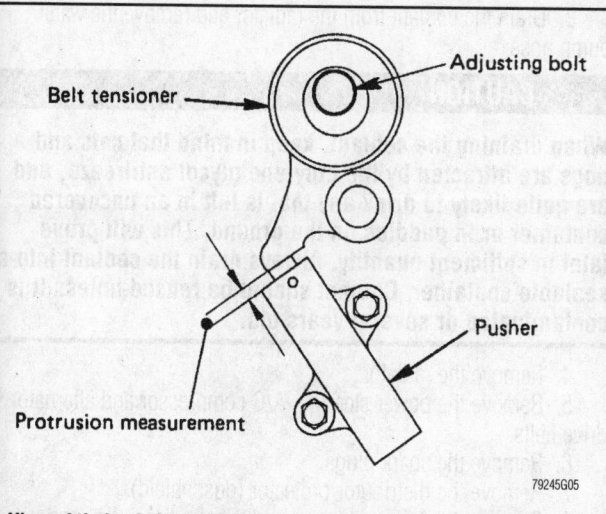

View of timing belt tensioner and pusher—Honda Passport 3.2L engine

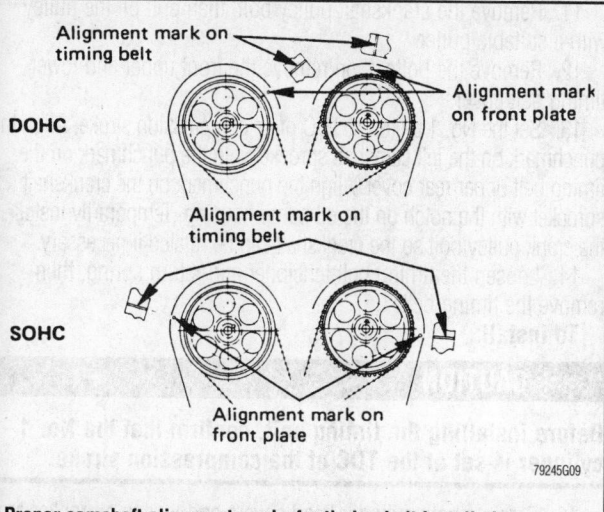

Proper camshaft alignment marks for timing belt installation—Honda Passport 3.2L engine

20. Align the white line on the timing belt with the alignment mark on the right bank camshaft timing pulley. Secure the belt with a clip.

❊❊ WARNING

If any binding is felt when adjusting the timing belt tension by turning the crankshaft, STOP turning the engine, because the pistons may be hitting the valves.

21. Rotate the crankshaft counterclockwise to remove the slack between the crankshaft sprocket and the right camshaft timing belt sprocket.

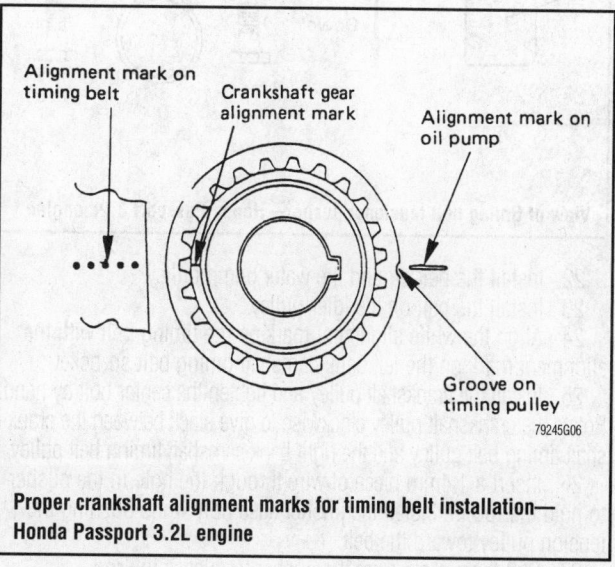

Proper crankshaft alignment marks for timing belt installation—Honda Passport 3.2L engine

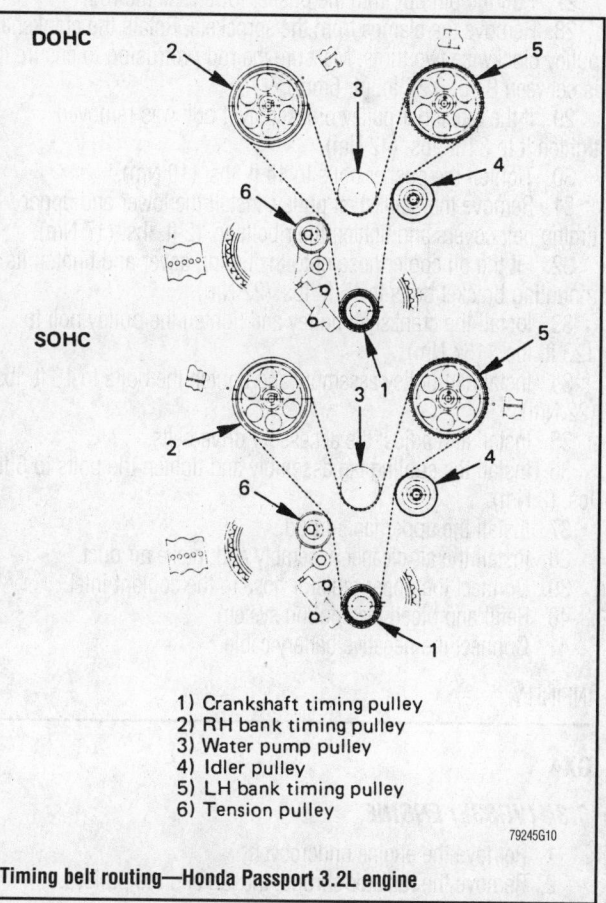

1) Crankshaft timing pulley
2) RH bank timing pulley
3) Water pump pulley
4) Idler pulley
5) LH bank timing pulley
6) Tension pulley

Timing belt routing—Honda Passport 3.2L engine

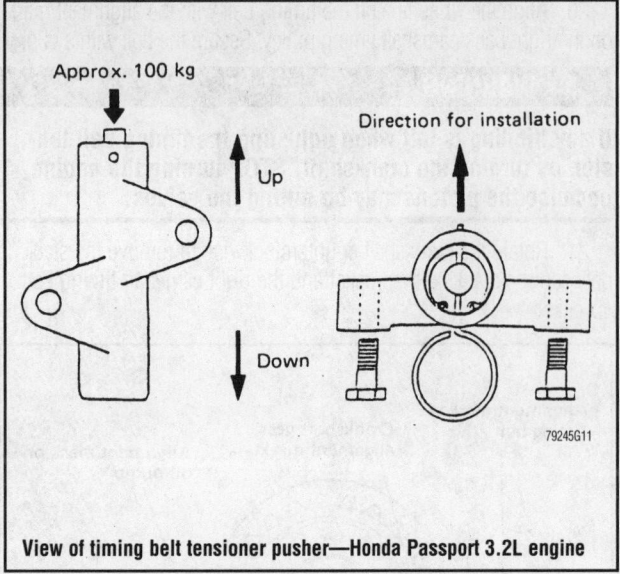

View of timing belt tensioner pusher—Honda Passport 3.2L engine

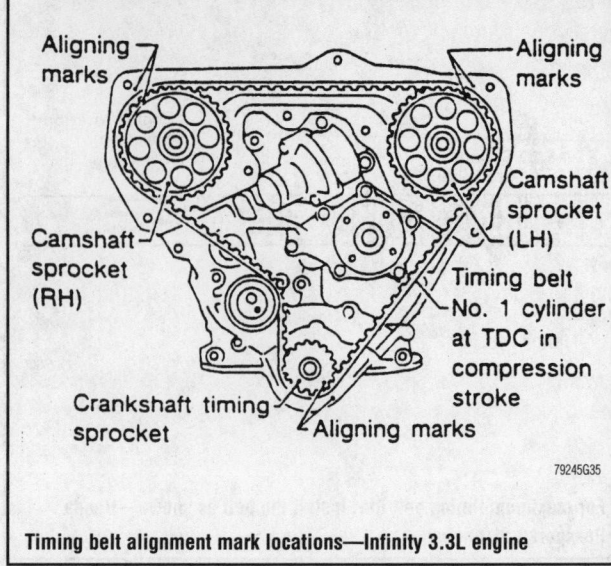

Timing belt alignment mark locations—Infinity 3.3L engine

22. Install the belt around the water pump pulley.

23. Install the belt on the idler pulley.

24. Align the white alignment mark on the timing belt with the alignment mark on the left bank camshaft timing belt sprocket.

25. Install the crankshaft pulley and tighten the center bolt by hand. Rotate the crankshaft pulley clockwise to give slack between the crankshaft timing belt pulley and the right bank camshaft timing belt pulley.

26. Insert a 1.4mm piece of wire through the hole in the pusher to hold the rod in. Install the pusher assembly while pushing the tension pulley toward the belt.

27. Pull the pin out from the pusher to release the rod.

28. Remove the clamps from the sprockets. Rotate the crankshaft pulley clockwise two turns. Measure the rod protrusion to ensure it is between 0.16–0.24 in. (4–6mm).

29. If the tensioner pulley bracket pivot bolt was removed, tighten it to 31 ft. lbs. (42 Nm).

30. Tighten the pusher bolts to 14 ft. lbs. (19 Nm).

31. Remove the crankshaft pulley. Install the lower and upper timing belt covers and tighten their bolts to 12 ft. lbs. (17 Nm).

32. Fit the oil cooler hose onto the timing cover and tighten its mounting bracket bolts to 16 ft. lbs. (22 Nm).

33. Install the crankshaft pulley and tighten the pulley bolt to 123 ft. lbs. (167 Nm).

34. Install fan pulley assembly and tighten the bolts to 16 ft. lbs. (22 Nm).

35. Install and adjust the accessory drive belts.

36. Install the cooling fan assembly and tighten the bolts to 6 ft. lbs. (8 Nm).

37. Install the upper fan shroud.

38. Install the air cleaner assembly and intake air duct.

39. Connect the upper radiator hose to the coolant inlet.

40. Refill and bleed the cooling system.

41. Connect the negative battery cable.

INFINITY

QX4

3.3L (VG33E) ENGINE

1. Remove the engine undercover.

2. Remove the radiator shroud, the fan and the pulleys.

3. Drain the coolant from the radiator and remove the water pump hose.

✳✳ CAUTION

When draining the coolant, keep in mind that cats and dogs are attracted by the ethylene glycol antifreeze, and are quite likely to drink any that is left in an uncovered container or in puddles on the ground. This will prove fatal in sufficient quantity. Always drain the coolant into a sealable container. Coolant should be reused unless it is contaminated or several years old.

4. Remove the radiator.

5. Remove the power steering, A/C compressor and alternator drive belts.

6. Remove the spark plugs.

7. Remove the distributor protector (dust shield).

8. Remove the A/C compressor drive belt idler pulley and bracket.

9. Remove the fresh air intake tube at the cylinder head cover.

10. Disconnect the radiator hose at the thermostat housing.

11. Remove the crankshaft pulley bolt, then pull off the pulley with a suitable puller.

12. Remove the bolts, then remove the front upper and lower timing belt covers.

13. Set the No. 1 piston at TDC of its compression stroke. Align the punchmark on the left camshaft sprocket with the punchmark on the timing belt upper rear cover. Align the punchmark on the crankshaft sprocket with the notch on the oil pump housing. Temporarily install the crank pulley bolt so the crankshaft can be rotated if necessary.

14. Loosen the timing belt tensioner and return spring, then remove the timing belt.

To install:

✳✳ CAUTION

Before installing the timing belt, confirm that the No. 1 cylinder is set at the TDC of the compression stroke.

15. Remove both cylinder head covers and loosen all rocker arm shaft retaining bolts.

➡The rocker arm shaft bolts MUST be loosened so that the correct belt tension can be obtained.

16. Install the tensioner and the return spring. Using a hexagon wrench, turn the tensioner clockwise and temporarily tighten the locknut.

17. Be sure that the timing belt is clean and free from oil or water.

18. When installing the timing belt align the white lines on the belt with the punchmarks on the camshaft and crankshaft sprockets. Have the arrow on the timing belt pointing toward the front belt covers.

➡A good way (although rather tedious!) to check for proper timing belt installation is to count the number of belt teeth between the timing marks. There are 133 teeth on the belt; there should be 40 teeth between the timing marks on the left and right side camshaft sprockets, and 43 teeth between the timing marks on the left side camshaft sprocket and the crankshaft sprocket.

19. While keeping the tensioner steady, loosen the locknut with a hex wrench.

20. Turn the tensioner approximately 70–80 degrees clockwise with the wrench, then tighten the locknut.

☀ WARNING

If any binding is felt when adjusting the timing belt tension by turning the crankshaft, STOP turning the engine, because the pistons may be hitting the valves.

21. Turn the crankshaft in a clockwise direction several times, then **slowly** set the No. 1 piston to TDC of the compression stroke.

22. Apply 22 lbs. of pressure (push it in!) to the center span of the timing belt between the right side camshaft sprocket and the tensioner pulley, then loosen the tensioner locknut.

23. Using a 0.0138 in. (0.35mm) thick feeler gauge (the actual width of the blade **must** be ½ in. or 13mm!), turn the crankshaft clockwise (**slowly!**). The timing belt should move approximately 2½ teeth. Tighten the tensioner locknut, turn the crankshaft slightly and remove the feeler gauge.

24. Slowly rotate the crankshaft clockwise several more times, then set the No. 1 piston to TDC of the compression stroke.

25. Position the two timing covers on the block, then tighten the mounting bolts to 24 ft. lbs. (35 Nm).

26. Press the crankshaft pulley onto the shaft, then tighten the bolt to 90–98 ft. lbs. (123–132 Nm).

27. Connect the radiator hose to the thermostat housing.

28. Reconnect the fresh air intake tube at the cylinder head cover.

29. Install the A/C compressor drive belt idler pulley and bracket.

30. Install the distributor protector (dust shield).

31. Install the spark plugs.

32. Install the power steering, A/C compressor and alternator drive belts.

33. Install the radiator.

34. Reconnect the water pump hose and fill the engine with coolant. Install the fan shroud and pulleys.

35. Install the engine undercover.

36. Start the engine and check for any leaks.

ISUZU

Amigo, Rodeo, and Trooper

2.6L ENGINE

1. Disconnect the negative battery cable.
2. Loosen and remove the engine accessory drive belts.

3. Remove the cooling fan assembly and the water pump pulley.

4. Drain the fluid from the power steering reservoir.

5. Unbolt and remove the power steering pump. Unbolt the hydraulic line brackets from the upper timing cover and move the pump out of the work area without disconnecting the hydraulic lines.

6. Disconnect and remove the starter motor if a flywheel holder (part No. J-38674 or equivalent) is to be used.

7. Remove the upper timing belt cover.

8. Rotate the crankshaft to set the engine at TDC/compression for the No. 1 cylinder. The arrow mark on the camshaft sprocket will be aligned with the mark on the rear timing cover.

9. Remove the crankshaft pulley.

10. Remove the lower timing belt cover.

11. Verify that the engine is set at TDC/compression for the No. 1 cylinder. The notch on the crankshaft sprocket will be aligned with the pointer on the oil seal retainer.

12. Release and remove the tensioner spring to release the timing belt's tension.

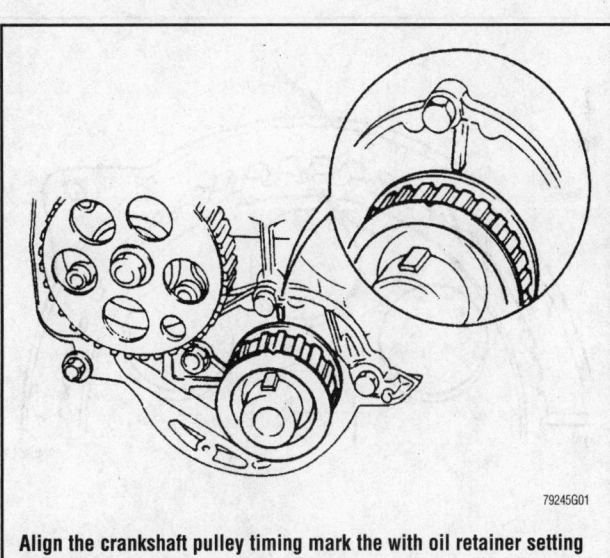

Align the crankshaft pulley timing mark the with oil retainer setting mark—Isuzu 2.6L engine

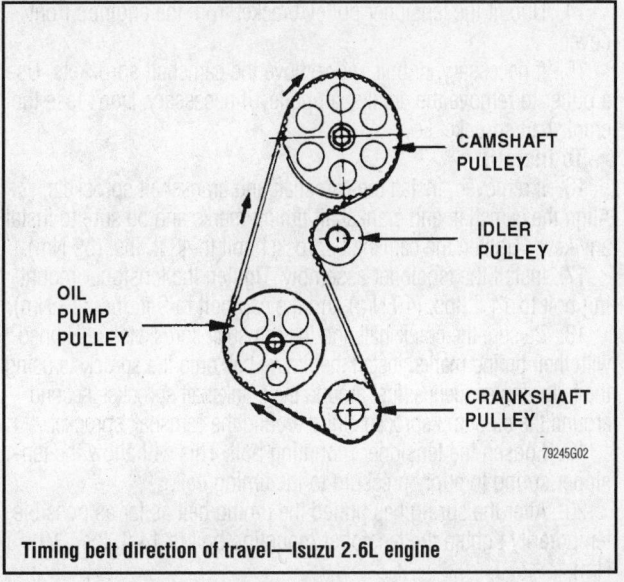

Timing belt direction of travel—Isuzu 2.6L engine

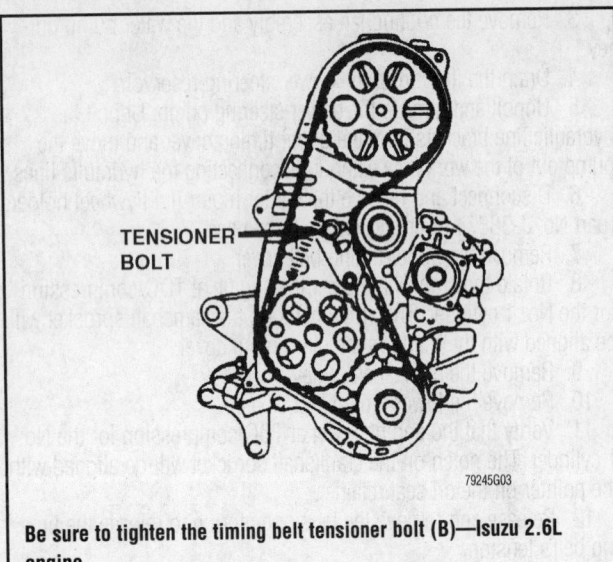

79245G03

Be sure to tighten the timing belt tensioner bolt (B)—Isuzu 2.6L engine

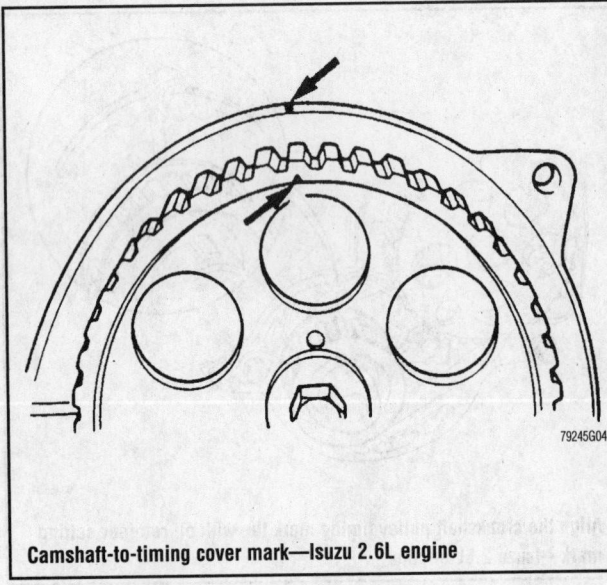

79245G04

Camshaft-to-timing cover mark—Isuzu 2.6L engine

13. Remove the timing belt.

14. Unbolt the tensioner pulley bracket from the engine's front cover.

15. If necessary, unbolt and remove the camshaft sprockets. Use a puller to remove the crankshaft pulley if necessary. Don't lose the crankshaft sprocket key.

To install:

16. If removed, install the camshaft and crankshaft sprockets. Align the camshaft and crankshaft timing marks and be sure to install any keys. Tighten the camshaft sprocket bolt to 43 ft. lbs. (59 Nm).

17. Install the tensioner assembly. Tighten the tensioner mounting bolt to 14 ft. lbs. (19 Nm) and the cap bolt to 9 ft. lbs. (13 Nm).

18. Be sure the crankshaft and the camshaft sprockets are aligned with their timing marks. Install the timing belt onto the sprockets using the following sequence: first around the crankshaft sprocket; second around the oil pump sprocket; third around the camshaft sprocket.

19. Loosen the tensioner mounting bolt. This will allow the tensioner spring to apply pressure to the timing belt.

20. After the spring has pulled the timing belt as far as possible, temporarily tighten the tensioner mounting bolt to 14 ft. lbs. (19 Nm).

➡**Remove the flywheel holder before rotating the crankshaft. Reinstall the holder to tighten the crankshaft pulley bolt.**

✳✳ **WARNING**

If any binding is felt when adjusting the timing belt tension by turning the crankshaft, STOP turning the engine, because the pistons may be hitting the valves.

21. Rotate the crankshaft counterclockwise two complete revolutions to check the rotation of the belt and the alignment of the timing marks. Listen for any rubbing noises which may mean the belt is binding.

22. Loosen the tensioner pulley bolt to allow the spring to adjust the correct tension. Then, retighten the tensioner pulley bolt to 14 ft. lbs. (19 Nm).

23. Install the lower timing cover and the crankshaft pulley.

24. Tighten the crankshaft pulley bolt to 87 ft. lbs. (118 Nm). Tighten the small pulley bolts to 6 ft. lbs. (8 Nm).

25. Install the upper timing cover.

26. Install the starter if it was removed. Tighten the bolts to 30 ft. lbs. (40 Nm).

27. Install the power steering pump. If the hydraulic lines were disconnected, refill and bleed the power steering system.

28. Install the water pump pulley and tighten its nut to 20 ft. lbs. (26 Nm).

29. Install the cooling fan assembly.

30. Install and adjust the accessory drive belts.

31. Connect the negative battery cable.

3.2L AND 3.5L ENGINES

1. Disconnect the negative battery cable.
2. Drain the engine coolant into a sealable container.
3. Remove the air cleaner assembly and intake air duct.
4. Disconnect the upper radiator hose from the coolant inlet.
5. Remove the upper fan shroud from the radiator.
6. Remove the four nuts retaining the cooling fan assembly. Remove the cooling fan from the fan pulley.
7. Loosen and remove the drive belts.
8. Remove the upper timing belt covers.
9. Remove the fan pulley assembly.
10. Rotate the crankshaft to align the camshaft timing marks with the pointer dots on the back covers. Verify that the pointer on the crankshaft aligns with the mark on the lower timing cover.

➡**When the timing marks are aligned on 1995 vehicles, no pistons will be at TDC/compression. When the timing marks are aligned on 1995½–99 vehicles, the No. 2 piston is at TDC/compression.**

✳✳ **WARNING**

Align the camshaft and crankshaft sprockets with their alignment marks before removing the timing belt. Failure to align the belt and sprocket marks may result in valve damage.

11. Use tool No. J-8614–01, or a suitable pulley holding tool to remove the crankshaft pulley center bolt. Remove the crankshaft pulley.

12. If present, disconnect the two oil cooler hose bracket bolts on the timing cover. Move the oil cooler hoses and bracket off of the lower timing cover.

13. Remove the lower timing belt cover.

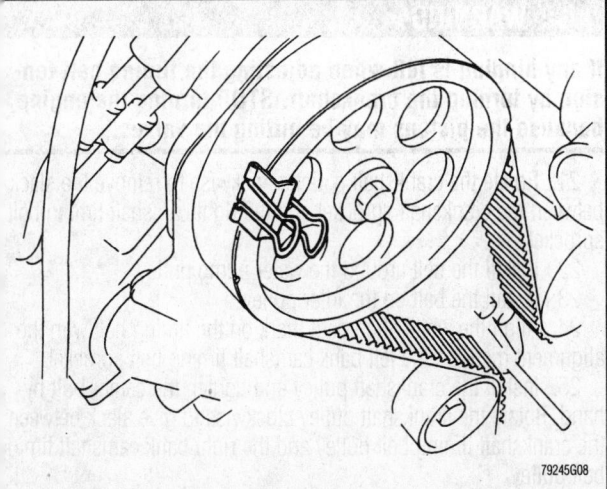

Using a double clip to hold the belt in place—Isuzu 3.2L and 3.5L engines

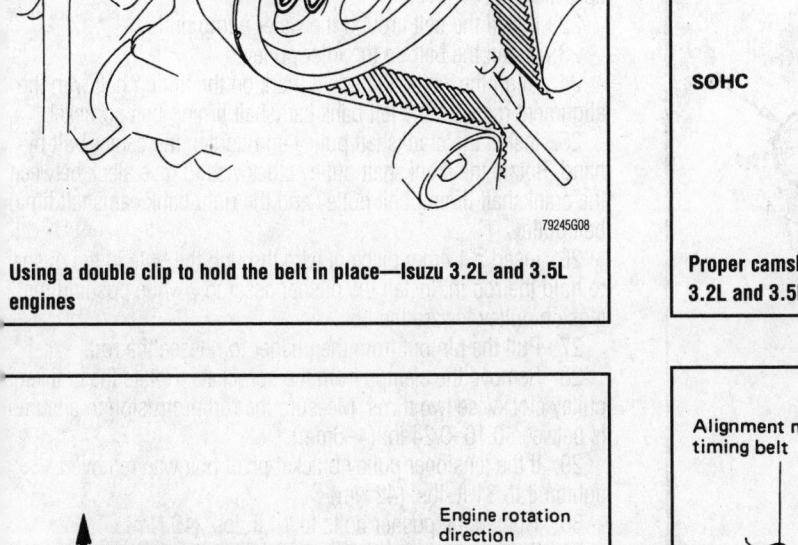

For maximum timing belt life, install the belt as shown—Isuzu 3.2L and 3.5L engines

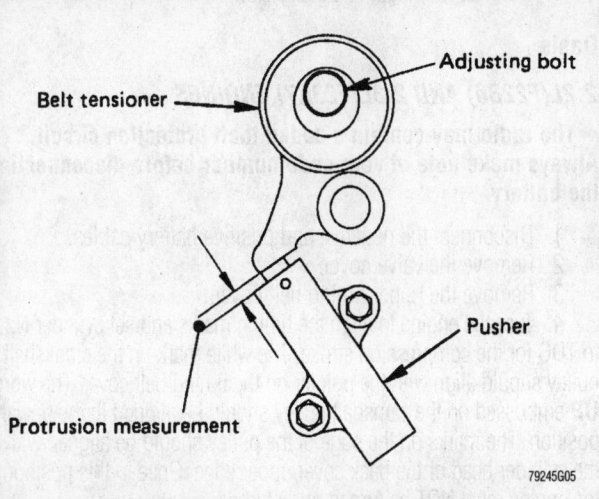

View of timing belt tensioner and pusher—Isuzu 3.2L and 3.5L engines

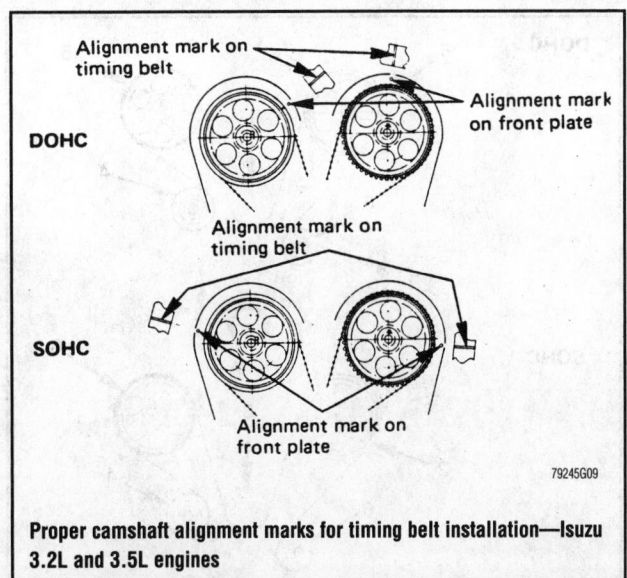

Proper camshaft alignment marks for timing belt installation—Isuzu 3.2L and 3.5L engines

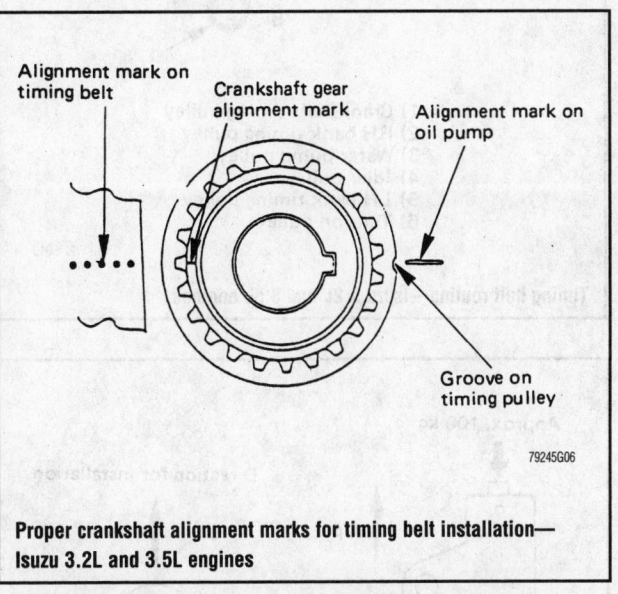

Proper crankshaft alignment marks for timing belt installation—Isuzu 3.2L and 3.5L engines

14. Remove the pusher assembly (tensioner) from below the belt tensioner pulley. The pusher rod must always face upward to prevent oil leakage. Depress the pusher rod, and insert a wire pin into the hole to keep the pusher rod retracted.

15. Remove the timing belt.

16. Inspect the water pump and replace it if there is any doubt about its condition.

17. Repair any oil or coolant leaks before installing a new timing belt. If the timing belt has been contaminated with oil or coolant, or is damaged, it must be replaced.

To install:

18. Verify that the sprocket timing marks are still aligned and that the groove and the keyway on the crankshaft timing sprocket align with the mark on the oil pump. The white pointers on the camshaft timing sprockets should align with the dots on the front plate.

19. Install the timing belt. Use clips to secure the belt onto each sprocket until the installation is complete. Align the dotted marks on the timing belt with the timing mark opposite the groove on the crankshaft sprocket.

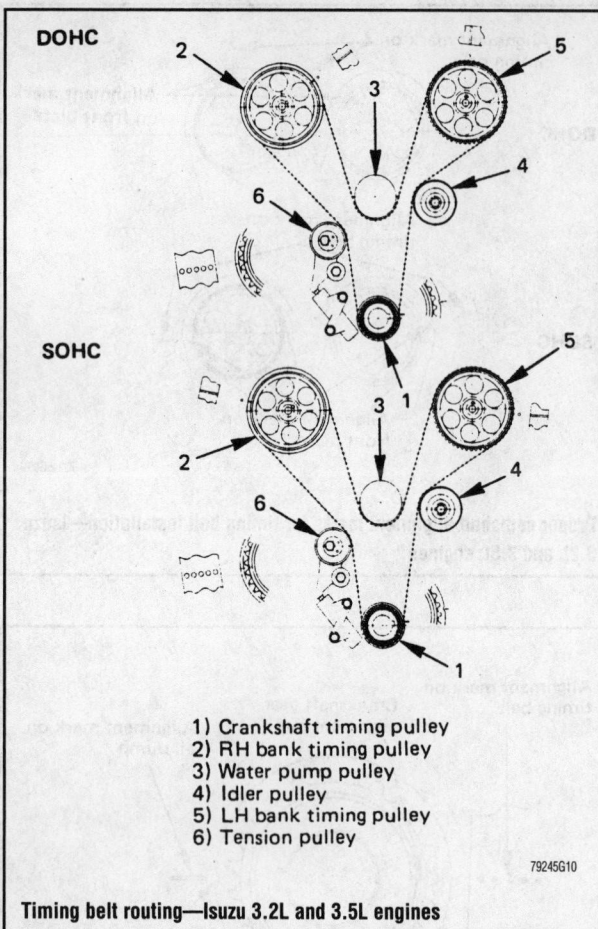

1) Crankshaft timing pulley
2) RH bank timing pulley
3) Water pump pulley
4) Idler pulley
5) LH bank timing pulley
6) Tension pulley

79245G10

Timing belt routing—Isuzu 3.2L and 3.5L engines

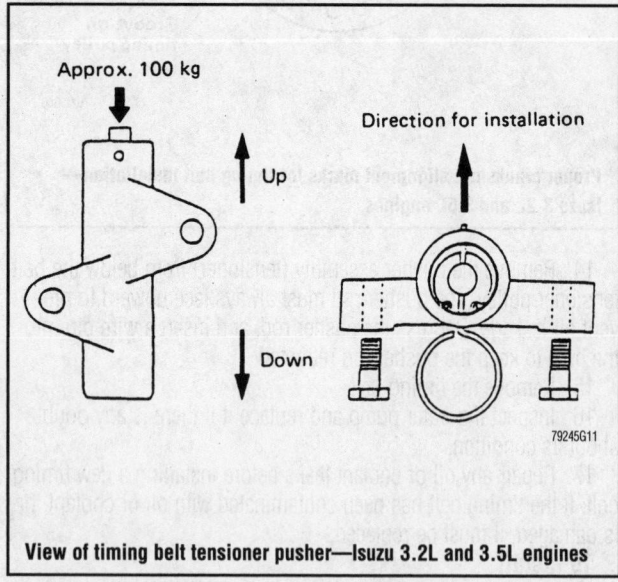

Approx. 100 kg

Up

Direction for installation

Down

79245G11

View of timing belt tensioner pusher—Isuzu 3.2L and 3.5L engines

➡️The arrows on the timing belt must follow the belt's direction of rotation. The manufacturer's trademark on the belt's spine should be readable left-to-right when the belt is installed.

20. Align the white line on the timing belt with the alignment mark on the right bank camshaft timing pulley. Secure the belt with a clip.

If any binding is felt when adjusting the timing belt tension by turning the crankshaft, STOP turning the engine, because the pistons may be hitting the valves.

21. Rotate the crankshaft counterclockwise to remove the slack between the crankshaft sprocket and the right camshaft timing belt sprocket.
22. Install the belt around the water pump pulley.
23. Install the belt on the idler pulley.
24. Align the white alignment mark on the timing belt with the alignment mark on the left bank camshaft timing belt sprocket.
25. Install the crankshaft pulley and tighten the center bolt by hand. Rotate the crankshaft pulley clockwise to give slack between the crankshaft timing belt pulley and the right bank camshaft timing belt pulley.
26. Insert a 1.4mm piece of wire through the hole in the pusher to hold the rod in. Install the pusher assembly while pushing the tension pulley toward the belt.
27. Pull the pin out from the pusher to release the rod.
28. Remove the clamps from the sprockets. Rotate the crankshaft pulley clockwise two turns. Measure the rod protrusion to ensure it is between 0.16–0.24 in. (4–6mm).
29. If the tensioner pulley bracket pivot bolt was removed, tighten it to 31 ft. lbs. (42 Nm).
30. Tighten the pusher bolts to 14 ft. lbs. (19 Nm).
31. Remove the crankshaft pulley. Install the lower and upper timing belt covers and tighten their bolts to 12 ft. lbs. (17 Nm).
32. Fit the oil cooler hose onto the timing cover and tighten its mounting bracket bolts to 16 ft. lbs. (22 Nm).
33. Install the crankshaft pulley and tighten the pulley bolt to 123 ft. lbs. (167 Nm).
34. Install fan pulley assembly and tighten the bolts to 16 ft. lbs. (22 Nm).
35. Install and adjust the accessory drive belts.
36. Install the cooling fan assembly and tighten the bolts to 6 ft. lbs. (8 Nm).
37. Install the upper fan shroud.
38. Install the air cleaner assembly and intake air duct.
39. Connect the upper radiator hose to the coolant inlet.
40. Refill and bleed the cooling system.
41. Connect the negative battery cable.

Oasis

2.2L(F22B6) AND 2.3L(F23A7) ENGINES

➡️The radio may contain a coded theft protection circuit. Always make note of your code number before disconnecting the battery.

1. Disconnect the negative and positive battery cables.
2. Remove the valve cover.
3. Remove the upper timing belt cover.
4. Turn the engine to align the timing marks and set cylinder No. 1 to TDC for the compression stroke. The white mark on the crankshaft pulley should align with the pointer on the timing belt cover. The word **UP** embossed on the camshaft pulley should be aligned in the upward position. The marks on the edge of the pulley should be aligned with the cylinder head or the back cover upper edge. Once in this position, the engine must NOT be turned or disturbed.

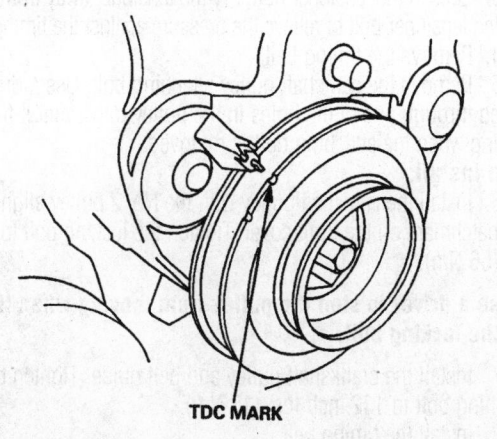

TDC MARK

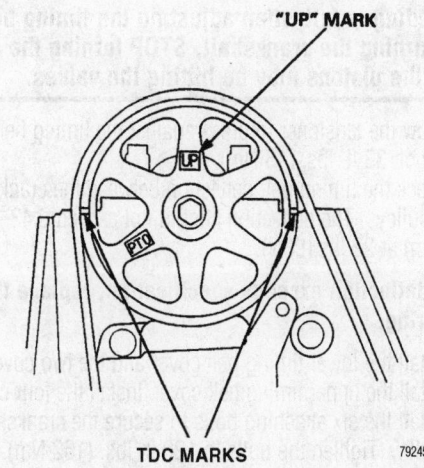

"UP" MARK

TDC MARKS

79245G24

Align the camshaft, crankshaft and engine marks before removing the timing belt and pulleys—Isuzu 2.2L and 2.3L engines

5. Remove the splash shield from below the engine.
6. Remove the wheel well splash shield.
7. Loosen and remove the power steering pump belt. Remove the power steering pump.
8. Loosen the adjusting and mounting bolts for the alternator and remove the drive belt.
9. Support the engine with a floor jack cushioned with a piece of wood.
10. Remove the through-bolt for the side engine mount and remove the mount.
11. Remove the crankshaft pulley bolt and remove the crankshaft pulley. Use a crank pulley holder (part No. 07MAB-PY3010A) and holder handle (part No. 07JAB-001020A),or there equivalents, to hold the crankshaft pulley in place while removing the bolt.
12. Remove the lower timing belt cover.
13. Remove the balancer shaft belt and its drive pulley.
14. Insert a suitable tool into the maintenance hole in the front balancer shaft. Unbolt and remove the balancer driven pulley.

➡**For servicing the balance shafts, front refers to the side of the engine facing the radiator. Rear refers to the side of the engine facing the firewall.**

15. Remove the timing belt.
16. If equipped with a TDC sensor assembly at the crankshaft sprocket, unbolt the assembly and move it to the side before removing the sprocket.
17. Remove the key and the spacers to remove the crankshaft timing sprocket.
18. Unbolt and remove the camshaft timing sprocket.
   To install:
19. Install the camshaft timing sprocket so that the **UP** mark is up and the TDC marks are parallel to the cylinder head gasket surface. Install the key and tighten the bolt to 27 ft. lbs. (37 Nm).
20. Install the crankshaft timing sprocket so that the TDC mark aligns with the pointer on the oil pump. Install the spacers with their concave surfaces facing in. Install the key. Install the TDC sensor assembly back into position before installing the timing belt.
21. Install and tension the timing belt.
22. Rotate the crankshaft counterclockwise five to six turns to be sure the belt is properly seated.
23. Set the No. 1 piston at TDC for its compression stroke.

✳✳ WARNING

If any binding is felt when adjusting the timing belt tension by turning the crankshaft, STOP turning the engine, because the pistons may be hitting the valves.

24. Rotate the crankshaft counterclockwise so that the camshaft pulley moves only three teeth beyond its TDC mark.
25. Tighten the tensioner adjusting nut to 33 ft. lbs. (45 Nm).
26. Tighten the crankshaft pulley bolt to 181 ft. lbs. (245 Nm).
27. Install the balancer shaft belt drive pulley.
28. Align the groove on the pulley edge to the pointer on the balancer gear case.
29. Check the alignment of the pointer on the balancer pulley to the pointer on the oil pump.
30. Install and tension the balancer shaft belt.
31. Be sure the timing belts have been tensioned correctly and that all TDC and alignment marks are in their proper positions.
32. Install the lower timing cover and the crankshaft pulley. Apply engine oil to the pulley bolt threads and washer surface. Install the pulley bolt and tighten it to 181 ft. lbs. (245 Nm).
33. Install the upper timing cover and the valve cover. Be sure the seals are properly seated.
34. Install the side engine mount. Tighten the through-bolt to 47 ft. lbs. (64 Nm). Tighten the mount nut and bolt to 40 ft. lbs. (55 Nm) each.
35. Remove the floor jack.
36. Install and tension the alternator belt.
37. Install the power steering pump and tension its belt.
38. Install the splash shields.
39. Reconnect the positive and negative battery cables. Enter the radio security code.
40. Check engine operation.

KIA

Sportage

SOHC ENGINE

1. Disconnect the negative battery cable.
2. Properly relieve the fuel system pressure.
3. Remove the alternator, power steering and A/C drive belts.

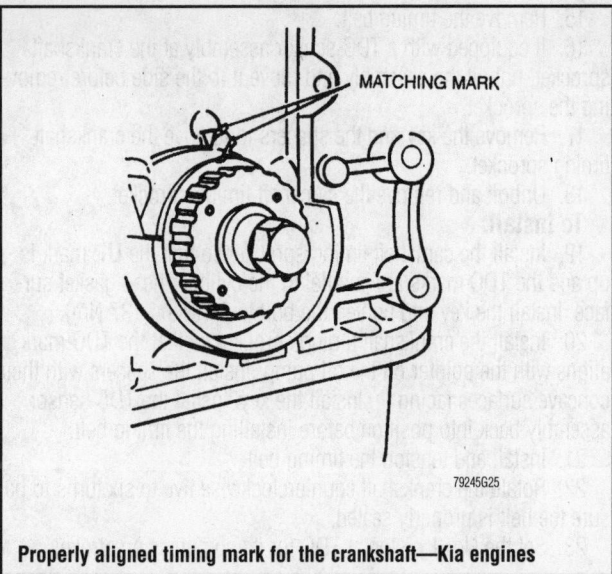

79245G25

Properly aligned timing mark for the crankshaft—Kia engines

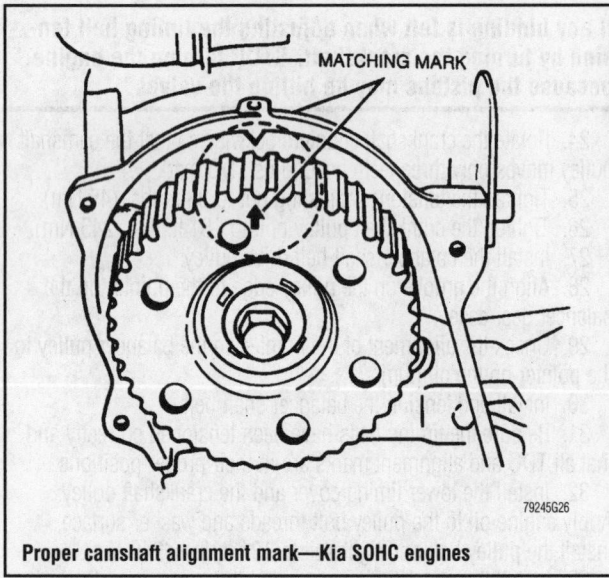

79245G26

Proper camshaft alignment mark—Kia SOHC engines

4. Remove the fresh air duct from the top of the radiator.

5. Remove the upper radiator hose.

6. Remove the four attaching nuts to the clutch fan.

7. Remove the five fan shroud bolts. Remove the fan and shroud as an assembly.

8. Remove the six attaching bolts to the crankshaft pulley.

9. Remove the damper pulley.

10. Remove the four upper timing belt cover bolts and remove the cover.

11. Remove the two lower timing belt cover bolts and remove the cover.

12. Align the timing marks. Align the camshaft pulley No. 2 with the matchmark on the front cover and the crankshaft pulley with the matchmark on the oil pump housing.

✳✳ CAUTION

When aligning the timing marks, do not turn the timing gear counterclockwise. Damage to the engine will occur.

13. Loosen the tensioner bolt. Pry the tensioner away from the belt. Tighten tensioner bolt to relieve the pressure against the timing belt.

14. Remove the timing belt.

15. Remove the camshaft pulley attaching bolt. Use a driver placed through one of the holes in the pulley to prevent it from moving when the attaching bolt is removed.

To install:

16. Install the camshaft pulley with the No. 2 pulley aligned with the matchmark on the front cover. Tighten the locking bolt to 42 ft. lbs. (56 Nm).

➡**Use a driver to stop the pulley from moving when tightening the locking bolt.**

17. Install the crankshaft pulley and belt guide. Tighten the attaching bolt to 132 inch lbs. (15 Nm).

18. Install the timing belt.

✳✳ WARNING

If any binding is felt when adjusting the timing belt tension by turning the crankshaft, STOP turning the engine, because the pistons may be hitting the valves.

19. Allow the tensioner to press against the timing belt. Tighten tensioner bolt 33 ft. lbs. (45 Nm).

20. Check the timing belt deflection between the crankshaft and camshaft pulleys. The deflection should not exceed 0.43–0.51 in. (11–13 mm) at 22 lbs.(98N).

➡ **If the deflection exceeds specification, replace the tensioner spring.**

21. Install the lower timing belt cover and the two cover bolts.

22. Install the upper timing belt cover. Install the four cover bolts.

23. Install the six attaching bolts to secure the crankshaft damper pulley. Tighten the bolts to 120 ft. lbs. (162 Nm).

24. Install the fan and shroud as an assembly.

25. Install the four attaching nuts to the clutch fan.

26. Install the five fan shroud bolts.

27. Install the upper radiator hose.

28. Install the fresh air duct to the top of the radiator.

29. Install the alternator, power steering and A/C drive belts.

30. Connect the negative battery cable.

31. Start the engine and check for leaks.

32. Road test the vehicle.

DOHC ENGINE

1. Disconnect the negative battery cable.

2. Properly relieve the fuel system pressure.

3. Remove the alternator drive belt.

4. Remove the fresh air duct from the top of the radiator.

5. Remove the upper radiator hose.

6. Remove the four attaching nuts to the clutch fan.

7. Remove the five fan shroud bolts. Remove the fan and shroud as an assembly.

8. Remove the four splash guard mounting bolts and the splash guard.

9. Loosen the lockbolts and loosen the A/C drive belt.

10. Loosen the power steering lock and mounting bolt. Remove the power steering belt.

11. Remove the five upper timing belt cover bolts and remove the cover.

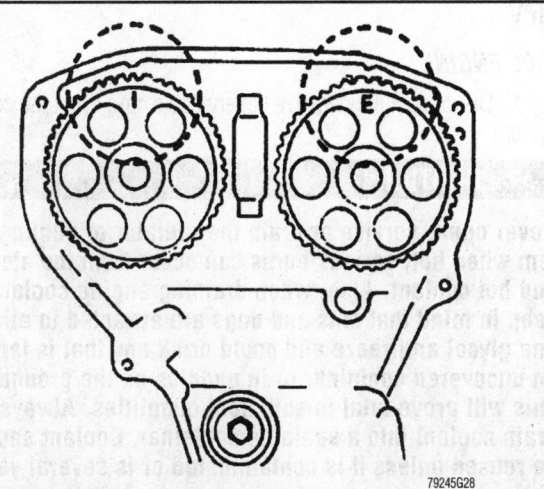

Proper alignment of the intake and exhaust camshaft pulley timing marks—Kia DOHC engine

12. Remove the two lower timing belt cover bolts and remove the cover.

13. Align the timing marks.

➡When aligning the cam pulleys with the seal plate marks, align the left cam pulley "I" mark and the right cam pulley on the "E" mark.

✳✳ **WARNING**

When aligning the timing marks, do not turn the timing gear counterclockwise. Damage to the engine will occur.

14. Loosen the tensioner bolt. Pry the tensioner away from the belt. Tighten the tensioner bolt to relieve the pressure against the timing belt.

15. Remove the timing belt.

16. Remove the camshaft pulley attaching bolts. Use a driver placed through one of the holes in the pulley to prevent it from moving when the attaching bolt is removed. Remove and mark the pulleys.

17. Remove the lower timing belt pulley and locking bolt.

To install:

18. Install the camshaft pulleys. Tighten the bolts to 35–48 ft. lbs. (47–65 Nm).

19. Install the lower timing belt pulley and locking bolt. Tighten the bolt to 120 ft. lbs. (162 Nm).

20. If necessary, align the timing marks.

➡When aligning the cam pulleys with the seal plate marks, align the left cam pulley "I" mark and the right cam pulley on the "E" mark.

✳✳ **WARNING**

When aligning the timing marks, do not turn the timing gear counterclockwise. Damage to the engine will occur.

21. Loosen the tensioner bolt. Pry the tensioner away from the belt. Tighten tensioner bolt to relieve the pressure against the timing belt.

22. Install the timing belt.

✳✳ **WARNING**

If any binding is felt when adjusting the timing belt tension by turning the crankshaft, STOP turning the engine, because the pistons may be hitting the valves.

23. Loosen the tensioner bolt and allow the tensioner to tighten the timing belt. Tighten the tensioner bolt 27–38 ft. lbs. (37–52 Nm).

24. Check the timing belt deflection. If there is more than 0.30–0.33 in. (7.5–8.5 mm) replace the tensioner spring.

25. Install the two lower timing belt cover bolts to the cover.

26. Install the five upper timing belt cover bolts to the cover.

27. Install and adjust the A/C and power steering drive belts.

28. Install the splash guard.

29. Install and tighten the alternator belt.

30. Install the upper radiator hose.

31. Install the fan and shroud as an assembly.

32. Install the four attaching nuts to the clutch fan.

33. Install the five fan shroud bolts.

34. Install the fresh air duct to the top of the radiator.

35. Properly fill the cooling system.

36. Connect the negative battery cable.

37. Start the engine and check for leaks.

38. Road test the vehicle.

MAZDA

B-series Trucks

2.3L (VIN A) AND 2.5L (VIN C) ENGINES

1. Rotate the engine so that No. 1 cylinder is at TDC on the compression stroke. Check that the timing marks are aligned on the camshaft and crankshaft pulleys. An access plug is provided in the cam belt cover so that the camshaft timing can be checked without removal of the cover or any other parts. Set the crankshaft to TDC by aligning the timing mark on the crank pulley with the TDC mark on the belt cover. Look through the access hole in the belt cover to be sure that the timing mark on the cam drive sprocket is lined up with the pointer on the inner belt cover.

➡Always turn the engine in the normal direction of rotation. Backward rotation may cause the timing belt to jump time, due to the arrangement of the belt tensioner.

2. Drain cooling system. Remove the upper radiator hose as necessary. Remove the fan blade and water pump pulley bolts.

✳✳ **CAUTION**

When draining the coolant, keep in mind that cats and dogs are attracted by ethylene glycol antifreeze, and are quite likely to drink any that is left in an uncovered container or in puddles on the ground. This will prove fatal in sufficient quantity. Always drain the coolant into a sealable container. Coolant should be reused unless it is contaminated or several years old.

3. Loosen the alternator retaining bolts and remove the drive belt from the pulleys. Remove the water pump pulley.

4. Remove the power steering pump and set it aside.

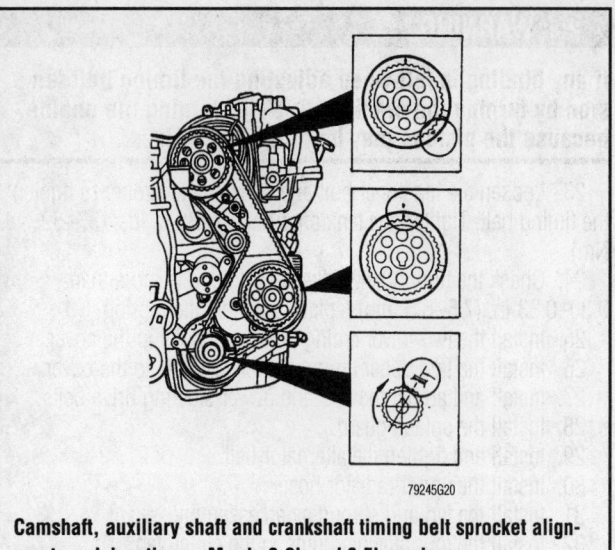

79245G20

Camshaft, auxiliary shaft and crankshaft timing belt sprocket alignment mark locations—Mazda 2.3L and 2.5L engines

5. Remove the four timing belt outer cover retaining bolts and remove the cover. Remove the crankshaft pulley and belt guide.

6. Loosen the belt tensioner pulley assembly, then position a camshaft belt adjuster tool (T74P-6254-A or equivalent) on the tension spring rollpin and retract the belt tensioner away from the timing belt. Tighten the adjustment bolt to lock the tensioner in the retracted position.

7. If the belt is to be reused, mark the direction of rotation on the belt for installation reference.

8. Remove the timing belt.

To install:

9. Install the new belt over the crankshaft sprocket and then counterclockwise over the auxiliary and camshaft sprockets, making sure the lugs on the belt properly engage the sprocket teeth on the pulleys. Be careful not to rotate the pulleys when installing the belt.

10. Release the timing belt tensioner pulley, allowing the tensioner to take up the belt slack. If the spring does not have enough tension to move the roller against the belt (belt hangs loose), it might be necessary to manually push the roller against the belt and tighten the bolt.

➡The spring cannot be used to set belt tension; a wrench must be used on the tensioner assembly.

✳ WARNING

If any binding is felt when adjusting the timing belt tension by turning the crankshaft, STOP turning the engine, because the pistons may be hitting the valves.

11. Rotate the crankshaft two complete turns by hand (in the normal direction of rotation) to remove slack from the belt, then tighten the tensioner adjustment to 26–33 ft. lbs. (35–45 Nm) and pivot bolts to 30–40 ft. lbs. (40–55 Nm). Be sure the belt is seated properly on the pulleys and that the timing marks are still in alignment when No. 1 cylinder is again at TDC/compression.

12. Install the crankshaft pulley and belt guide.

13. Install the timing belt cover.

14. Install the water pump pulley and fan blades. Install the upper radiator hose if necessary. Refill the cooling system.

15. Install the accessory drive belts.

16. Start the engine and check the ignition timing. Adjust the timing, if necessary.

MPV

3.0L ENGINE

1. Disconnect the negative battery cable, and drain the cooling system.

✳ CAUTION

Never open, service or drain the radiator or cooling system when hot; serious burns can occur from the steam and hot coolant. Also, when draining engine coolant, keep in mind that cats and dogs are attracted to ethylene glycol antifreeze and could drink any that is left in an uncovered container or in puddles on the ground. This will prove fatal in sufficient quantities. Always drain coolant into a sealable container. Coolant should be reused unless it is contaminated or is several years old.

2. Remove the timing belt covers.

3. Remove the upper idler pulley.

4. Turn the crankshaft to align the matching marks on the sprockets. If the timing belt is to be reused, draw an arrow on the belt to indicate rotation direction.

5. Remove the timing belt and automatic tensioner.

6. Using SST 49 H012 010 tool or its equivalent, loosen and remove the camshaft sprocket lockbolts. Remove the camshaft sprockets.

7. Using a suitable puller, remove the crankshaft sprocket.

To install:

8. Install the crankshaft sprocket on the crankshaft.

9. Install the camshaft sprockets with the lockbolts and tighten the bolts to 52–59 ft. lbs. (71–80 Nm).

10. Set a plain washer at the bottom of the tensioner body to prevent damage to the body plug. Press in the tensioner rod slowly, using a press or a vise.

➡**Do not press the tensioner rod with more than 2200 lbs. (9800N).**

11. Insert a pin to hold the tensioner rod in the body. Install the automatic tensioner and tighten the mounting bolts to 14–19 ft. lbs. (19–25 Nm).

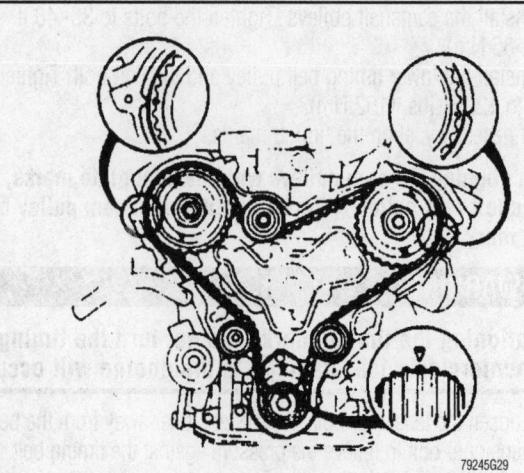

79245G29

Timing belt routing and timing mark locations—Mazda MPV 3.0L engine

12. Install the crankshaft pulley lockbolt and loosely tighten. Check the alignment of the matching marks on the sprockets.

13. With the upper idler pulley removed, install the timing belt, making sure there is no slack between the crankshaft and camshaft sprockets. If the timing belt is being reused, it must be installed in the same direction of rotation.

14. Install the upper idler pulley and tighten the attaching bolt to 27–38 ft. lbs. (37–52 Nm).

15. Turn the crankshaft twice in the direction of rotation (clockwise) and align the matching marks. If the marks do not align, repeat the previous three steps.

✳✳ WARNING

If any binding is felt when adjusting the timing belt tension by turning the crankshaft, STOP turning the engine, because the pistons may be hitting the valves.

16. Remove the pin from the automatic tensioner. Turn the crankshaft twice and align the matching marks. Be sure the marks are aligned.

17. Check the timing belt deflection. The deflection should be 0.20–0.28 in. (5–7mm). Do not apply tension other than that of the automatic tensioner.

18. If the deflection is not correct, repeat the previous three steps.

19. Remove the crankshaft pulley lockbolt.

20. Install the timing belt cover, along with the related components.

21. Reinstall the crankshaft pulley and tighten the lockbolt to 116–123 ft. lbs. (37–52 Nm).

22. Fill and bleed the cooling system.

23. Run the engine and check for leaks and proper operation. Check the idle speed and the ignition timing.

MERCURY

Villager

3.0L (VIN W) ENGINE

On this vehicle, right side refers to the "rear" components (near the firewall) and left side refers to the "front" components (near the radiator).

1. If the timing belt is to be removed, it is good practice to turn the crankshaft until the engine is at Top Dead Center (TDC) of the number one cylinder, compression stroke (firing position), before beginning work. This should align all timing marks and serve as a reference for all work that follows. After verifying that the engine is at TDC for the number one cylinder, do not crank the engine or allow the crankshaft or camshaft sprockets to be turned otherwise engine timing will be lost.

2. Drain the cooling system.

3. Disconnect the negative battery cable.

4. Remove the alternator drive belt, water pump and power steering pump belt and the A/C compressor belt (if equipped), using the recommended drive belt removal procedure.

5. If equipped with A/C, remove the three A/C compressor drive belt idler pulley bolts and remove the idler pulley.

6. Remove the upper radiator hose bracket bolt. Remove the upper hose with the bracket from the vehicle.

7. Remove the water bypass hose from between the thermostat housing and the lower water hose connection.

8. Remove the main wiring harness from the upper engine front cover.

9. Remove the eight upper engine front cover bolts and remove the upper cover.

10. Raise and safely support the vehicle.

11. Remove the right side front wheel and tire assembly.

12. Remove the four right side engine and transmission splash shield bolts and two screws, and remove the right side outer engine and transaxle splash shield.

13. Use a strap wrench to hold the water pump pulley. Remove the four pulley bolts, and the water pump pulley.

14. Use a strap wrench to hold the crankshaft pulley. Remove the center pulley bolt, and the crankshaft pulley using a harmonic balancer (damper) puller to draw the pulley from the front of the crankshaft.

15. Remove the five lower engine front cover bolts, then remove the lower engine front cover.

16. Be sure that the timing marks between the crankshaft sprocket and the oil pump housing line up.

17. If the timing belt is to be reused, mark an arrow on the belt indicating the direction of rotation. The directional arrow is necessary to ensure that the timing belt, if it to be reused, can reinstalled in the same direction.

18. Loosen the timing belt tensioner nut and slip the timing belt off of the sprockets.

19. If necessary, the camshaft sprockets can be removed. A special spanner tool is designed to hold the sprocket to keep it from turning while the center bolt is being loosened. Use care if using substitutes.

➡**The sprockets are not interchangeable.**

20. If necessary, the crankshaft sprocket can be removed. The outer timing belt guide (looks like a large washer) and the crankshaft sprocket simply pull off the front of the crankshaft.

➡**Be careful, there are two crankshaft keys. Use care not to loose them.**

To install:

21. Clean all parts well. If removed, inspect the crankshaft sprocket for warping or abnormal wear. Check the sprocket teeth for

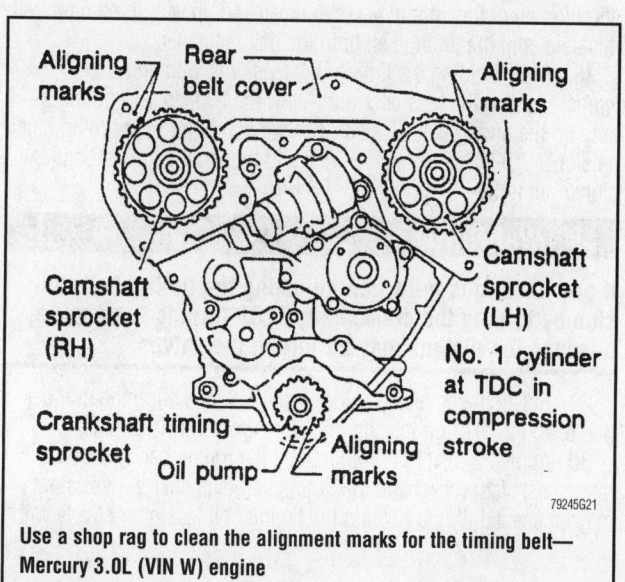

Use a shop rag to clean the alignment marks for the timing belt—Mercury 3.0L (VIN W) engine

wear, deformation, chipping or other damage. Replace as necessary. Clean the sprocket mounting surface to ease installation. Install the key. Slip the sprocket onto the crankshaft. Tap it in place with a suitably-sized socket.

22. If removed, inspect the camshaft sprockets for damage and wear. Replace as required. The sprockets should be marked **L3** to designate the front, or left side camshaft and **R3** to designate the rear, or right side camshaft. Use care to install the sprockets properly. A special spanner tool is designed to hold the sprocket to keep it from turning while the center bolt is being tightened. Use care if using a substitute. Tighten the camshaft sprocket center bolts to 58–65 ft. lbs. (78–88 Nm). Verify that the timing marks on the camshaft sprockets and the timing marks on the rear cover (called the seal plate) are aligned.

23. Use an Allen wrench to turn the timing belt tensioner clockwise until the belt tensioner spring is fully extended. Temporarily tighten the tensioner nut to 32–43 ft. lbs. (43–58 Nm).

24. If a new timing belt is to be installed, look for a printed arrow on the belt. Be sure the arrow is pointing away from the engine. If the original timing belt is to be reused, be sure that the directional arrow that was marked at disassembly is facing the correct direction.

25. A new Original Equipment Manufacture (OEM) timing belt should have three white timing marks on it that indicate the correct timing positions of the camshafts and the crankshaft. These marks are to help ensure that the engine is properly timed. When the engine is properly timed, each white timing mark on the timing belt will be aligned with the corresponding camshaft and crankshaft timing mark on the sprocket. Because the white timing marks are not evenly spaced, the technician needs to use care in installing the belt. There should be 40 timing belt teeth between the timing marks on the front and rear camshaft sprockets and 43 teeth between the timing mark on the front camshaft sprocket and the timing mark on the crankshaft sprocket.

26. Verify that the camshaft timing marks are aligned with the timing marks on the rear cover (seal plate) and that the crankshaft sprocket timing mark is aligned with the timing mark on the oil pump housing.

27. Install the timing belt starting at the crankshaft sprocket and moving around the camshaft sprockets following a counterclockwise path. Do not allow any slack in the timing belt between the sprockets. After all of the timing marks are matched up with the timing belt installed, slip the timing belt onto the belt tensioner.

28. While holding the timing belt tensioner with an Allen wrench, loosen the tensioner nut. Allow the tensioner to put pressure on the timing belt. Use an Allen wrench to turn the timing belt tensioner 70–80 degrees clockwise and tighten the timing belt tensioner nut to 32–43 ft. lbs. (43–58 Nm).

⁂ WARNING

If any binding is felt when adjusting the timing belt tension by turning the crankshaft, STOP turning the engine, because the pistons may be hitting the valves.

29. Rotate the crankshaft clockwise twice and align the number one piston to TDC on the compression stroke (firing position).

30. Apply 22 lbs. (10kg) of force on the timing belt between the rear camshaft sprocket and the timing belt tensioner. An assistant may be needed. While holding the timing belt tensioner steady with

an Allen wrench, loosen the timing belt tensioner nut. Remove the Allen wrench and adjust the timing belt tensioner using the following procedure:

   a. Install a 0.0138 in. (0.35mm) thick and 0.500 in. (12.7mm) wide feeler gauge where the timing belt just starts to go around the tensioner (approximately the four o'clock position, looking at the tensioner).

   b. Turn the crankshaft sprocket clockwise, which should force the feeler gauge between the timing belt and the tensioner, up to a position on the tensioner of about 1 o'clock.

   c. Tighten the timing belt tensioner nut to 32–43 ft. lbs. (43–58 Nm).

   d. Turn the crankshaft clockwise to rotate the feeler gauge out from between the timing belt tensioner and the timing belt.

31. Rotate the crankshaft clockwise twice, and once again align the number one piston to TDC on the compression stroke (firing position).

32. Apply 22 lbs. (10kg) of force on the timing belt between the front and rear camshaft sprockets. Measure the amount of belt deflection. Belt deflection should be between 0.51–0.59 in. (13–15mm). If belt deflection is out of specification, repeat Steps 29 through 33. If the timing belt deflection cannot be adjusted into specification, the timing belt will have to be replaced.

33. Position the lower engine front cover and install the five lower cover bolts. Do not overtighten. Tighten to 27–44 inch lbs. (3–5 Nm).

34. Install the outer timing belt guide next to the crankshaft sprocket with the dished side facing away from the cylinder block. Install the crankshaft pulley. Use a strap wrench to keep the crankshaft pulley from turning and tighten the center bolt to 90–98 ft. lbs. (123–132 Nm).

35. Position the water pump pulley on the pump. Install the four bolts. Use a strap wrench to keep the water pump pulley from turning and tighten the four water pump pulley bolts to 12–15 ft. lbs. (16–21 Nm).

36. Position the right side outer engine and transaxle splash shield, and secure with the four bolts and two screws.

37. Install the right side front wheel and tire assembly. Tighten the lug nuts to 72–87 ft. lbs. (98–118 Nm).

38. Lower the vehicle.

39. Position the upper engine timing belt front cover, and tighten the eight bolts to 27–44 inch lbs. (3–5 Nm).

40. Install the main wiring harness on the upper engine front cover.

41. Position the water bypass hose between the thermostat housing and water connection. Install the upper radiator hose between the radiator and the water hose connection. Secure the hoses with clamps. Install the upper radiator hose bracket. Tighten the bracket bolt to 34–58 ft. lbs. (46–65 Nm).

42. If equipped, position the A/C compressor drive belt idler pulley and install the three bolts. Tighten to 15 ft. lbs. (21 Nm).

43. Install and adjust the alternator drive belt, the water pump and power steering pump drive belt and the A/C compressor drive belt (of equipped).

44. Connect the battery cable.

45. Fill the cooling system.

46. Start the engine and allow it to warm to operating temperature. Check and adjust the ignition timing. Road test to verify correct engine operation.

MITSUBISHI

Mighty Max, Montero, and Montero Sport

2.4L (VIN G) ENGINE

1. Be sure that the engine's No. 1 piston is at TDC in the compression stroke.

✳✳ CAUTION

Wait at least 90 seconds after the negative battery cable is disconnected to prevent possible deployment of the air bag.

2. Disconnect the negative battery cable.
3. Remove the spark plug wires from the tree on the upper cover.
4. Drain the cooling system.
5. Remove the shroud, fan and accessory drive belts.
6. Remove the radiator as required.
7. Remove the power steering pump, alternator, air conditioning compressor, tension pulley and accompanying brackets, as required.
8. Remove the upper front timing belt cover.
9. Remove the water pump pulley and the crankshaft pulley(s).
10. Remove the lower timing belt cover mounting screws and remove the cover.
11. If the belt(s) are to be reused, mark the direction of rotation on the belt.
12. Remove the timing (outer) belt tensioner and remove the belt. Unbolt the tensioner from the block and remove.
13. Remove the outer crankshaft sprocket and flange.
14. Remove the silent shaft (inner) belt tensioner and remove the inner belt. Unbolt the tensioner from the block and remove it.
15. To remove the camshaft sprockets, use SST MB990767–01 and MIT308239, or their equivalents.

To install:

16. Install the camshaft sprockets and tighten the center bolt to 65 ft. lbs. (90 Nm).
17. Align the timing mark of the silent shaft belt sprockets on the crankshaft and silent shaft with the marks on the front case. Wrap the silent shaft belt around the sprockets so there is no slack in the upper span of the belt and the timing marks are still in line.

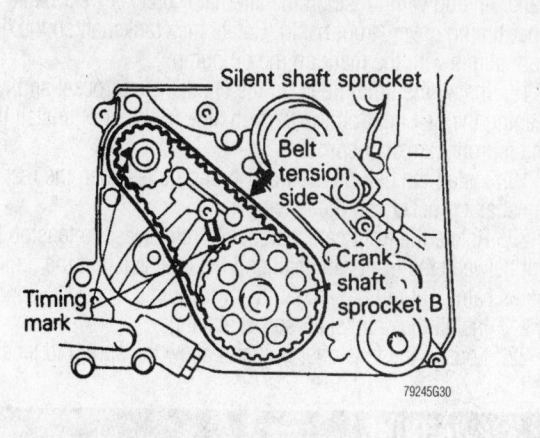

Silent shaft alignment marks. Notice the tension side of the inner (silent shaft) belt—Mitsubishi 2.4L engine

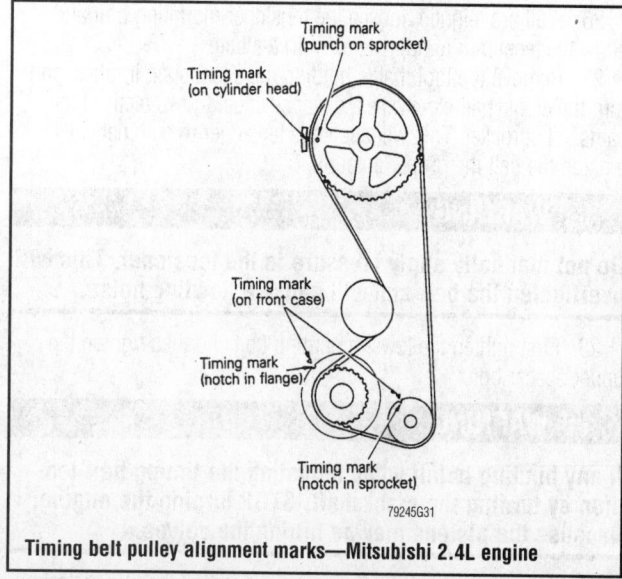

Timing belt pulley alignment marks—Mitsubishi 2.4L engine

18. Install the tensioner initially so the actual center of the pulley is above and to the left of the installation bolt.
19. Move the pulley up by hand so the center span of the long side of the belt deflects about ¼ in. (6mm).
20. Hold the pulley tightly so it does not rotate when the bolt is tightened. Tighten the bolt to 15 ft. lbs. (20 Nm). If the pulley has moved, the belt will be too tight.
21. Install the timing belt tensioner fully toward the water pump and temporarily tighten the bolts. Place the upper end of the spring against the water pump body. Align the timing marks of the cam, crankshaft and oil pump sprockets with the corresponding marks on the front case or head.

➡**If the following steps are not followed exactly, there is a chance that the silent shaft alignment will be 180 degrees off. This will cause a noticeable vibration in the engine and the entire procedure will have to be repeated.**

22. Before installing the timing belt, ensure that the left side silent shaft is in the correct position.

➡**It is possible to align the timing marks on the camshaft sprocket, crankshaft sprocket and the oil pump sprocket with the left balance shaft out of alignment.**

23. With the timing mark on the oil pump pulley aligned with the mark on the front case, check the alignment of the left balance shaft to assure correct shaft timing.
   a. Remove the plug located on the left side of the block in the area of the starter.
   b. Insert a tool having a shaft diameter of 0.3 in. (8mm) into the hole.
   c. With the timing marks still aligned, the tool must be able to go in at least 2⅓ in. If it can only go in about 1 in. turn the oil pump sprocket one complete revolution.
   d. Recheck the position of the balance shaft with the timing marks realigned. Leave the tool in place to hold the silent shaft while continuing.
24. Install the belt to the crankshaft sprocket, oil pump sprocket and the camshaft sprocket, in that order. While doing so, be sure there is no slack between the sprockets except where the tensioner will take it up when released.
25. Recheck the timing marks' alignment.

26. If all are aligned, loosen the tensioner mounting bolt and allow the tensioner to apply tension to the belt.

27. Remove the tool that is holding the silent shaft in place and turn the crankshaft clockwise a distance equal to two teeth of the camshaft sprocket. This will allow the tensioner to automatically tension the belt the proper amount.

✷✷ WARNING

Do not manually apply pressure to the tensioner. This will overtighten the belt and will cause a howling noise.

28. First tighten the lower mounting bolt and then tighten the upper spacer bolt.

✷✷ WARNING

If any binding is felt when adjusting the timing belt tension by turning the crankshaft, STOP turning the engine, because the pistons may be hitting the valves.

29. To verify that belt tension is correct, check that the deflection of the longest span (between the camshaft and oil pump sprockets) is ½ in. (13mm).

30. Install the lower timing belt cover. Be sure the packing is properly positioned in the inner grooves of the covers when installing.

31. Install the water pump pulley and the crankshaft pulley(s).

32. Install the upper front timing belt cover.

33. Install the power steering pump, alternator, air conditioning compressor, tension pulley and accompanying brackets, as required.

34. Install the radiator, shroud, fan and accessory drive belts.

35. Install the spark plug wires to the tree on the upper cover.

36. Refill the cooling system.

37. Connect the negative battery cable. Start the engine and check for leaks.

1995–96 3.0L (VIN H) ENGINE

1. Position engine with No. 1 cylinder at TDC.
2. Disconnect the negative battery cable.
3. Drain the cooling system. Remove the drive belts.

✷✷ CAUTION

Never open, service or drain the radiator or cooling system when hot; serious burns can occur from the steam and hot coolant. Also, when draining engine coolant, keep in mind that cats and dogs are attracted to ethylene glycol antifreeze and could drink any that is left in an uncovered container or in puddles on the ground. This will prove fatal in sufficient quantities. Always drain coolant into a sealable container. Coolant should be reused unless it is contaminated or is several years old.

4. Remove the upper radiator shroud.
5. Remove the fan and fan pulley.
6. Without disconnecting the lines, remove the power steering pump from its bracket and position it to the side. Remove the pump brackets.
7. Remove the belt tensioner pulley bracket.
8. Without releasing the refrigerant, remove the air conditioning compressor from its bracket and position it to the side. Remove the bracket.

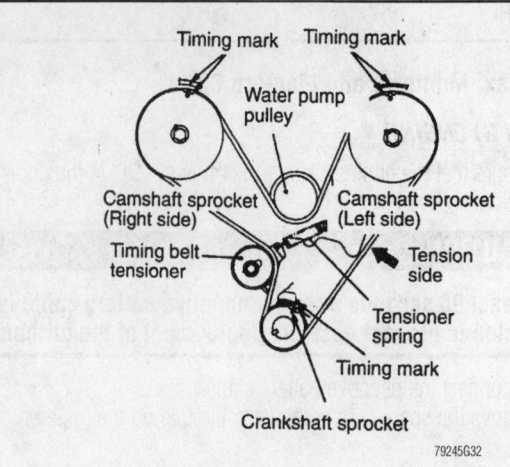

Timing belt routing and alignment mark locations for 1995–96 Mitsubishi 3.0L 12-valve engine—note the tension side of the belt

9. Remove the cooling fan bracket.

10. On some vehicles it may be necessary to remove the pulley from the crankshaft to access the lower cover bolts.

11. Remove the timing belt cover bolts and the upper and lower covers from the engine.

12. If the same timing belt will be reused, mark the direction of timing belt's rotation, for installation in the same direction. Be sure the engine is positioned so that the No. 1 cylinder is at TDC and the sprockets timing marks are aligned with the engine's timing mark indicators.

13. Loosen the timing belt tensioner bolt and remove the belt. If not removing the tensioner, position it as far away from the center of the engine as possible and tighten the bolt.

14. If the tensioner is being removed, mark the outside of the spring to ensure that it is not installed backwards. Unbolt the tensioner and remove it along with the spring.

15. Slide the timing belt off of the sprockets.

To install:

16. Install the tensioner, if removed, and hook the upper end of the spring to the water pump pin. Install the lower end of the spring to the tensioner in exactly the same position as originally installed.

17. If not already done, position both camshafts so the timing marks line up with those on the alternator bracket (rear bank) and inner timing cover (front bank). Rotate the crankshaft so the timing mark aligns with the mark on the oil pump.

18. Install the timing belt on the crankshaft sprocket and, while keeping the belt tight on the tension side (right side), install the belt on the front camshaft sprocket.

19. Install the belt on the water pump pulley, then the rear camshaft sprocket and the tensioner.

20. Rotate the front camshaft counterclockwise to tension the belt between the front camshaft and the crankshaft. If the timing marks came out of line, repeat the procedure.

21. Install the crankshaft sprocket flange.

22. Loosen the tensioner bolt and allow the spring to tension the belt.

✷✷ WARNING

If any binding is felt when adjusting the timing belt tension by turning the crankshaft, STOP turning the engine, because the pistons may be hitting the valves.

23. Slowly turn the crankshaft two full turns in the clockwise direction until the timing marks align. Now that the belt is properly tensioned, tighten the tensioner lockbolt to 35 ft. lbs. (48 Nm).

24. Install the upper and lower covers to the engine and secure with the retaining screws. Be sure the packing is positioned in the inner grooves of the covers properly when installing.

25. Install the crankshaft pulley if it was removed. Tighten the bolt to 110 ft. lbs. (150 Nm).

26. Install the air conditioning bracket and compressor to the engine. Install the belt tensioner.

27. Install the power steering pump in position. Install the fan pulley and fan.

28. Install the fan shroud to the radiator.

29. Refill the cooling system.

30. Connect the negative battery cable. Start the engine and check for fluid leaks.

1997–99 3.0L (VIN H) ENGINE

1. Position the engine with No. 1 cylinder at TDC.
2. Disconnect the negative battery cable.
3. Drain the cooling system. Remove the drive belts.

✷✷ CAUTION

Never open, service or drain the radiator or cooling system when hot; serious burns can occur from the steam and hot coolant. Also, when draining engine coolant, keep in mind that cats and dogs are attracted to ethylene glycol antifreeze and could drink any that is left in an uncovered container or in puddles on the ground. This will prove fatal in sufficient quantities. Always drain coolant into a sealable container. Coolant should be reused unless it is contaminated or is several years old.

4. Remove the upper radiator shroud.
5. Remove the fan and fan pulley.
6. Without disconnecting the lines, remove the power steering pump from its bracket and position it to the side. Remove the pump brackets.
7. Remove the belt tensioner pulley bracket.

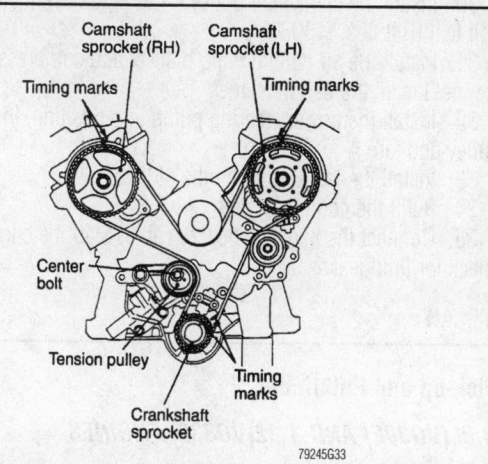

Be sure to line up the timing marks before removing or installing the timing belt—1997–99 Mitsubishi 3.0L 12-valve and all 3.0L 24-valve engines

79245G33

8. Without releasing the refrigerant, remove the air conditioning compressor from its bracket and position it to the side. Remove the bracket.

9. Remove the cooling fan bracket.

10. On some vehicles it may be necessary to remove the pulley from the crankshaft to access the lower cover bolts.

11. Remove the timing belt cover bolts, and the upper and lower covers from the engine.

12. If the same timing belt will be reused, mark the direction of the timing belt's rotation, for installation in the same direction. Be sure engine is positioned so that the No. 1 cylinder is at TDC of its compression stroke and the sprockets' timing marks are aligned with the engine's timing mark indicators.

13. Loosen the timing belt tensioner bolt and remove the belt. If not removing the tensioner, position it as far away from the center of the engine as possible and tighten the bolt.

14. If tensioner is being removed, mark outside of the spring to ensure that it is not installed backwards. Unbolt the tensioner and remove it along with the spring.

15. Using SST MB990767–01 and MIT308239, or their equivalents, remove the camshaft sprockets.

To install:

16. Hold the hexagonal portion of the camshaft with a wrench when tightening the camshaft sprocket bolt and tighten to:
- 3.0L 12-valve engines—66 ft. lbs. (90 Nm)
- 3.0L 24-valve engines—64 ft. lbs. (88 Nm)

17. If removed, install the tensioner and hook the upper end of the spring to the water pump pin. Install the lower end of the spring to the tensioner in exactly the same position as originally installed.

18. Position both camshafts so the timing marks line up with those on the alternator bracket (rear bank) and inner timing cover (front bank). Rotate the crankshaft so the timing mark aligns with the mark on the oil pump.

19. Install the timing belt on the crankshaft sprocket, and while keeping the belt tight on the tension side (right side), install the belt on the front camshaft sprocket.

20. Install the belt on the water pump pulley, then the rear camshaft sprocket and the tensioner.

21. Rotate the front camshaft counterclockwise to tension the belt between the front camshaft and the crankshaft. If the timing marks came out of line, repeat the procedure.

22. Install the crankshaft sprocket flange.

23. Loosen the tensioner bolt and allow the spring to tension the belt.

✷✷ WARNING

If any binding is felt when adjusting the timing belt tension by turning the crankshaft, STOP turning the engine, because the pistons may be hitting the valves.

24. Slowly turn the crankshaft two full turns in the clockwise direction until the timing marks align. Now that the belt is properly tensioner, tighten the tensioner lockbolt to 35 ft. lbs. (48 Nm).

25. Install the upper and lower covers to the engine and secure with the retaining screws. Be sure the packing is positioned in the inner grooves of the covers properly when installing.

26. Install the crankshaft pulley if it was removed. Tighten the bolt to 110 ft. lbs. (150 Nm).

27. Install the air conditioning bracket and compressor on the engine. Install the belt tensioner.

28. Install the power steering pump into position. Install the fan pulley and fan.

29. Install the fan shroud to the radiator.
30. Refill the cooling system.
31. Connect the negative battery cable. Start the engine and check for fluid leaks.

3.5L (VIN M) ENGINE

1. Disconnect the negative battery cable.
2. Drain the cooling system. Remove the drive belts.

⁂ CAUTION

Never open, service or drain the radiator or cooling system when hot; serious burns can occur from the steam and hot coolant. Also, when draining engine coolant, keep in mind that cats and dogs are attracted to ethylene glycol antifreeze and could drink any that is left in an uncovered container or in puddles on the ground. This will prove fatal in sufficient quantities. Always drain coolant into a sealable container. Coolant should be reused unless it is contaminated or is several years old.

3. Remove the upper radiator shroud.
4. Remove the fan and fan pulley.
5. Without disconnecting the lines, remove the power steering pump from its bracket and position it to the side. Remove the pump brackets.
6. Remove the belt tensioner pulley bracket.
7. Without releasing the refrigerant, remove the air conditioning compressor from its bracket and position it to the side. Remove the bracket.
8. Remove the cooling fan bracket.
9. On some vehicles it may be necessary to remove the pulley from the crankshaft to access the lower cover bolts.
10. Remove the timing belt cover bolts and the upper and lower covers from the engine.
11. Remove the crankshaft position sensor connector.
12. Using SST MB990767–01 and MD998754, or their equivalents, remove the crankshaft pulley from the crankshaft.
13. Use a shop rag to clean the timing marks to assist in properly aligning the timing marks.
14. Loosen the center bolt on the tension pulley and remove the timing belt.

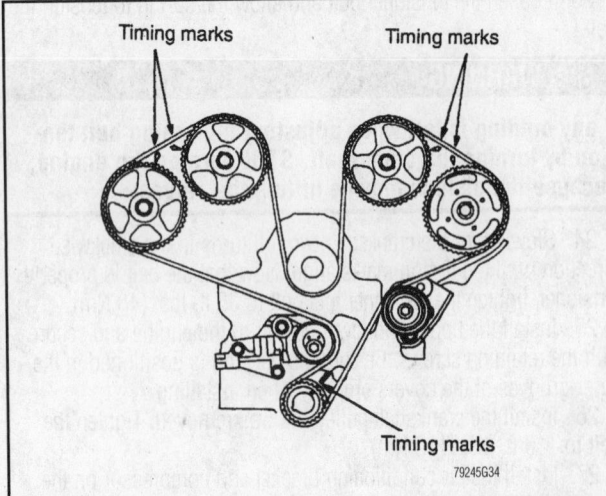

Clean the timing marks to verify their position when aligning—Mitsubishi 3.5L engine

➡ If the same timing belt will be reused, mark the direction of timing belt's rotation, for installation in the same direction. Be sure engine is positioned so No. 1 cylinder is at the TDC of it's compression stroke and the sprockets timing marks are aligned with the engine's timing mark indicators.

15. Remove the auto-tensioner, the tension pulley and the tension arm assembly.
16. Remove the sprockets by holding the hexagonal portion of the camshaft with a wrench while removing the sprocket bolt.

To install:

17. Install the crankshaft pulley and turn the crankshaft sprocket timing mark forward (clockwise) three teeth to move the piston slightly past No. 1 cylinder top dead center.
18. If removed, install the camshaft sprockets and tighten the bolts to 64 ft. lbs. (88 Nm).
19. Align the timing mark of the left bank side camshaft sprocket.
20. Align the timing mark of the right bank side camshaft sprocket, and hold the sprocket with a wrench so that it doesn't turn.
21. Set the timing belt onto the water pump pulley.
22. Check that the camshaft sprocket timing mark of the left bank side is aligned and clamp the timing belt with double clips.
23. Set the timing belt onto the idler pulley.

⁂ WARNING

If any binding is felt when adjusting the timing belt tension by turning the crankshaft, STOP turning the engine, because the pistons may be hitting the valves.

24. Turn the crankshaft one turn counterclockwise and set the timing belt onto the crankshaft sprocket.
25. Set the timing belt on the tension pulley.
26. Place the tension pulley pin hole so that it is towards the top. Press the tension pulley onto the timing belt, and then provisionally tighten the fixing bolt. Tighten the bolt to 35 ft. lbs. (48 Nm).
27. Slowly turn the crankshaft two full turns in the clockwise direction until the timing marks align. Remove the four double clips.
28. Install the crankshaft position sensor connector.
29. Install the upper and lower covers on the engine and secure them with the retaining screws. Be sure the packing is properly positioned in the inner grooves of the covers when installing.
30. Install the crankshaft pulley if it was removed. Tighten the bolt to 110 ft. lbs. (150 Nm).
31. Install the air conditioning bracket and compressor on the engine. Install the belt tensioner.
32. Install the power steering pump into position. Install the fan pulley and fan.
33. Install the fan shroud on the radiator.
34. Refill the cooling system.
35. Connect the negative battery cable. Start the engine and check for fluid leaks.

NISSAN

Pick-up and Pathfinder

3.0L (VG30E) AND 3.3L (VG33E) ENGINES

1. Remove the engine undercover.
2. Remove the radiator shroud, the fan and the pulleys.
3. Drain the coolant from the radiator and remove the water pump hose.

When draining the coolant, keep in mind that cats and dogs are attracted by the ethylene glycol antifreeze, and are quite likely to drink any that is left in an uncovered container or in puddles on the ground. This will prove fatal in sufficient quantity. Always drain the coolant into a sealable container. Coolant should be reused unless it is contaminated or several years old.

4. Remove the radiator.
5. Remove the power steering, A/C compressor and alternator drive belts.
6. Remove the spark plugs.
7. Remove the distributor protector (dust shield).
8. Remove the A/C compressor drive belt idler pulley and bracket.
9. Remove the fresh air intake tube at the cylinder head cover.
10. Disconnect the radiator hose at the thermostat housing.
11. Remove the crankshaft pulley bolt, then pull off the pulley with a suitable puller.
12. Remove the bolts, then remove the front upper and lower timing belt covers.
13. Set the No. 1 piston at TDC of its compression stroke. Align the punchmark on the left camshaft sprocket with the punchmark on the timing belt upper rear cover. Align the punchmark on the crankshaft sprocket with the notch on the oil pump housing. Temporarily install the crank pulley bolt so the crankshaft can be rotated if necessary.
14. Loosen the timing belt tensioner and return spring, then remove the timing belt.

To install:

❊❊ **CAUTION**

Before installing the timing belt, confirm that the No. 1 cylinder is set at the TDC of the compression stroke.

15. Remove both cylinder head covers and loosen all rocker arm shaft retaining bolts.

➡The rocker arm shaft bolts MUST be loosened so that the correct belt tension can be obtained.

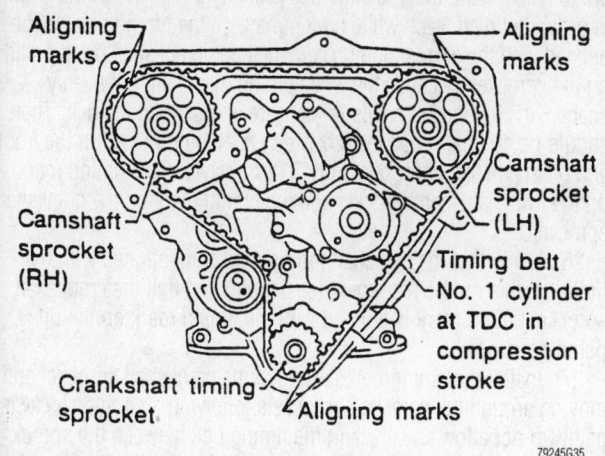

Timing belt alignment mark locations—Nissan 3.0L and 3.3L engines

16. Install the tensioner and the return spring. Using a hexagon wrench, turn the tensioner clockwise and temporarily tighten the locknut.
17. Be sure that the timing belt is clean and free from oil or water.
18. When installing the timing belt align the white lines on the belt with the punchmarks on the camshaft and crankshaft sprockets. Have the arrow on the timing belt pointing toward the front belt covers.

➡A good way (although rather tedious!) to check for proper timing belt installation is to count the number of belt teeth between the timing marks. There are 133 teeth on the belt; there should be 40 teeth between the timing marks on the left and right side camshaft sprockets, and 43 teeth between the timing marks on the left side camshaft sprocket and the crankshaft sprocket.

19. While keeping the tensioner steady, loosen the locknut with a hex wrench.
20. Turn the tensioner approximately 70–80 degrees clockwise with the wrench, then tighten the locknut.

❊❊ **WARNING**

If any binding is felt when adjusting the timing belt tension by turning the crankshaft, STOP turning the engine, because the pistons may be hitting the valves.

21. Turn the crankshaft in a clockwise direction several times, then **slowly** set the No. 1 piston to TDC of the compression stroke.
22. Apply 22 lbs. of pressure (push it in!) to the center span of the timing belt between the right side camshaft sprocket and the tensioner pulley, then loosen the tensioner locknut.
23. Using a 0.0138 in. (0.35mm) thick feeler gauge (the actual width of the blade **must** be ½ in. or 13mm!), turn the crankshaft clockwise (**slowly!**). The timing belt should move approximately 2 ½ teeth. Tighten the tensioner locknut, turn the crankshaft slightly and remove the feeler gauge.
24. Slowly rotate the crankshaft clockwise several more times, then set the No. 1 piston to TDC of the compression stroke.
25. Position the two timing covers on the block, then tighten the mounting bolts to 24 ft. lbs. (35 Nm).
26. Press the crankshaft pulley onto the shaft, then tighten the bolt to 90–98 ft. lbs. (123–132 Nm).
27. Connect the radiator hose to the thermostat housing.
28. Reconnect the fresh air intake tube at the cylinder head cover.
29. Install the A/C compressor drive belt idler pulley and bracket.
30. Install the distributor protector (dust shield).
31. Install the spark plugs.
32. Install the power steering, A/C compressor and alternator drive belts.
33. Install the radiator.
34. Reconnect the water pump hose and fill the engine with coolant. Install the fan shroud and pulleys.
35. Install the engine undercover.
36. Start the engine and check for any leaks.

Quest

3.0L (VIN W) ENGINE

On this vehicle, right side refers to the "rear" components (near the firewall) and left side refers to the "front" components (near the radiator).

1. If the timing belt is to be removed, it is good practice to turn the crankshaft until the engine is at Top Dead Center (TDC) of the number one cylinder, compression stroke (firing position), before beginning work. This should align all timing marks and serve as a reference for all work that follows. After verifying that the engine is at TDC for the number one cylinder, do not crank the engine or allow the crankshaft or camshaft sprockets to be turned otherwise engine timing will be lost.

2. Drain the cooling system.

3. Disconnect the negative battery cable.

4. Remove the alternator drive belt, water pump and power steering pump belt and the A/C compressor belt (if equipped), using the recommended drive belt removal procedure.

5. If equipped with A/C, remove the three A/C compressor drive belt idler pulley bolts and remove the idler pulley.

6. Remove the upper radiator hose bracket bolt. Remove the upper hose with the bracket from the vehicle.

7. Remove the water bypass hose from between the thermostat housing and the lower water hose connection.

8. Remove the main wiring harness from the upper engine front cover.

9. Remove the eight upper engine front cover bolts and remove the upper cover.

10. Raise and safely support the vehicle.

11. Remove the right side front wheel and tire assembly.

12. Remove the four right side engine and transmission splash shield bolts and two screws, and remove the right side outer engine and transaxle splash shield.

13. Use a strap wrench to hold the water pump pulley. Remove the four pulley bolts, and the water pump pulley.

14. Use a strap wrench to hold the crankshaft pulley. Remove the center pulley bolt, and the crankshaft pulley using a harmonic balancer (damper) puller to draw the pulley from the front of the crankshaft.

15. Remove the five lower engine front cover bolts, then remove the lower engine front cover.

16. Be sure that the timing marks between the crankshaft sprocket and the oil pump housing line up.

17. If the timing belt is to be reused, mark an arrow on the belt indicating the direction of rotation. The directional arrow is necessary to ensure that the timing belt, if it to be reused, can reinstalled in the same direction.

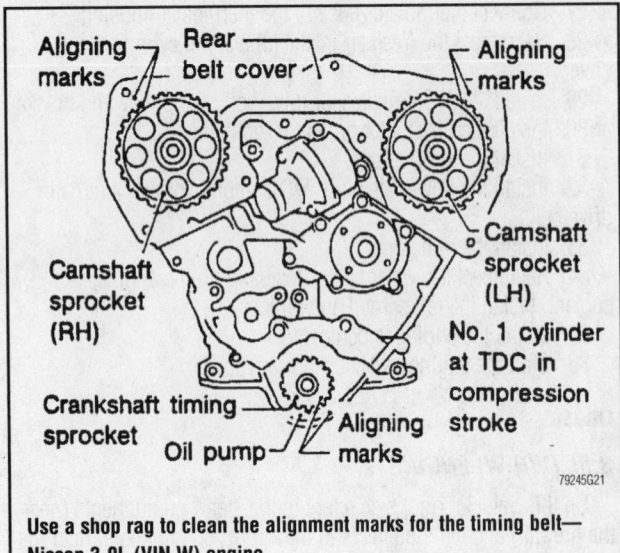

Use a shop rag to clean the alignment marks for the timing belt—Nissan 3.0L (VIN W) engine

18. Loosen the timing belt tensioner nut and slip the timing belt off of the sprockets.

19. If necessary, the camshaft sprockets can be removed. A special spanner tool is designed to hold the sprocket to keep it from turning while the center bolt is being loosened. Use care if using substitutes.

➡**The sprockets are not interchangeable.**

20. If necessary, the crankshaft sprocket can be removed. The outer timing belt guide (looks like a large washer) and the crankshaft sprocket simply pull off the front of the crankshaft.

➡**Be careful, there are two crankshaft keys. Use care not to loose them.**

To install:

21. Clean all parts well. If removed, inspect the crankshaft sprocket for warping or abnormal wear. Check the sprocket teeth for wear, deformation, chipping or other damage. Replace as necessary. Clean the sprocket mounting surface to ease installation. Install the key. Slip the sprocket onto the crankshaft. Tap it in place with a suitably-sized socket.

22. If removed, inspect the camshaft sprockets for damage and wear. Replace as required. The sprockets should be marked **L3** to designate the front, or left side camshaft and **R3** to designate the rear, or right side camshaft. Use care to install the sprockets properly. A special spanner tool is designed to hold the sprocket to keep it from turning while the center bolt is being tightened. Use care if using a substitute. Tighten the camshaft sprocket center bolts to 58–65 ft. lbs. (78–88 Nm). Verify that the timing marks on the camshaft sprockets and the timing marks on the rear cover (called the seal plate) are aligned.

23. Use an Allen wrench to turn the timing belt tensioner clockwise until the belt tensioner spring is fully extended. Temporarily tighten the tensioner nut to 32–43 ft. lbs. (43–58 Nm).

24. If a new timing belt is to be installed, look for a printed arrow on the belt. Be sure the arrow is pointing away from the engine. If the original timing belt is to be reused, be sure that the directional arrow that was marked at disassembly is facing the correct direction.

25. A new Original Equipment Manufacture (OEM) timing belt should have three white timing marks on it that indicate the correct timing positions of the camshafts and the crankshaft. These marks are to help ensure that the engine is properly timed. When the engine is properly timed, each white timing mark on the timing belt will be aligned with the corresponding camshaft and crankshaft timing mark on the sprocket. Because the white timing marks are not evenly spaced, the technician needs to use care in installing the belt. There should be 40 timing belt teeth between the timing marks on the front and rear camshaft sprockets and 43 teeth between the timing mark on the front camshaft sprocket and the timing mark on the crankshaft sprocket.

26. Verify that the camshaft timing marks are aligned with the timing marks on the rear cover (seal plate) and that the crankshaft sprocket timing mark is aligned with the timing mark on the oil pump housing.

27. Install the timing belt starting at the crankshaft sprocket and moving around the camshaft sprockets following a counterclockwise path. Do not allow any slack in the timing belt between the sprockets. After all of the timing marks are matched up with the timing belt installed, slip the timing belt onto the belt tensioner.

28. While holding the timing belt tensioner with an Allen wrench, loosen the tensioner nut. Allow the tensioner to put pres-

sure on the timing belt. Use an Allen wrench to turn the timing belt tensioner 70–80 degrees clockwise and tighten the timing belt tensioner nut to 32–43 ft. lbs. (43–58 Nm).

✳✳ WARNING

If any binding is felt when adjusting the timing belt tension by turning the crankshaft, STOP turning the engine, because the pistons may be hitting the valves.

29. Rotate the crankshaft clockwise twice and align the number one piston to TDC on the compression stroke (firing position).

30. Apply 22 lbs.(10kg) of force on the timing belt between the rear camshaft sprocket and the timing belt tensioner. An assistant may be needed. While holding the timing belt tensioner steady with an Allen wrench, loosen the timing belt tensioner nut. Remove the Allen wrench and adjust the timing belt tensioner using the following procedure:

    a. Install a 0.0138 in. (0.35mm) thick and 0.500 in. (12.7mm) wide feeler gauge where the timing belt just starts to go around the tensioner (approximately the four o'clock position, looking at the tensioner).

    b. Turn the crankshaft sprocket clockwise, which should force the feeler gauge between the timing belt and the tensioner, up to a position on the tensioner of about 1 o'clock.

    c. Tighten the timing belt tensioner nut to 32–43 ft. lbs. (43–58 Nm).

    d. Turn the crankshaft clockwise to rotate the feeler gauge out from between the timing belt tensioner and the timing belt.

31. Rotate the crankshaft clockwise twice, and once again align the number one piston to TDC on the compression stroke (firing position).

32. Apply 22lbs.(10kg) of force on the timing belt between the front and rear camshaft sprockets. Measure the amount of belt deflection. Belt deflection should be between 0.51–0.59 in. (13–15mm). If belt deflection is out of specification, repeat Steps 29 through 33. If the timing belt deflection cannot be adjusted into specification, the timing belt will have to be replaced.

33. Position the lower engine front cover and install the five lower cover bolts. Do not overtighten. Tighten to 27–44 inch lbs. (3–5 Nm).

34. Install the outer timing belt guide next to the crankshaft sprocket with the dished side facing away from the cylinder block. Install the crankshaft pulley. Use a strap wrench to keep the crankshaft pulley from turning and tighten the center bolt to 90–98 ft. lbs. (123–132 Nm).

35. Position the water pump pulley on the pump. Install the four bolts. Use a strap wrench to keep the water pump pulley from turning and tighten the four water pump pulley bolts to 12–15 ft. lbs. (16–21 Nm).

36. Position the right side outer engine and transaxle splash shield, and secure with the four bolts and two screws.

37. Install the right side front wheel and tire assembly. Tighten the lug nuts to 72–87 ft. lbs. (98–118 Nm).

38. Lower the vehicle.

39. Position the upper engine timing belt front cover, and tighten the eight bolts to 27–44 inch lbs. (3–5 Nm).

40. Install the main wiring harness on the upper engine front cover.

41. Position the water bypass hose between the thermostat housing and water connection. Install the upper radiator hose between the radiator and the water hose connection. Secure the hoses with clamps. Install the upper radiator hose bracket. Tighten the bracket bolt to 34–58 ft. lbs. (46–65 Nm).

42. If equipped, position the A/C compressor drive belt idler pulley and install the three bolts. Tighten to 15 ft. lbs. (21 Nm).

43. Install and adjust the alternator drive belt, the water pump and power steering pump drive belt and the A/C compressor drive belt (of equipped).

44. Connect the battery cable.

45. Fill the cooling system.

46. Start the engine and allow it to warm to operating temperature. Check and adjust the ignition timing. Road test to verify correct engine operation.

SUBARU

Forester

When servicing the timing belt heed the following:

• The intake and exhaust camshafts for the 2.5L DOHC engine can be rotated independently when the timing belt is removed. If the intake and exhaust valves are lifted off of their seats simultaneously, their heads will contact each other, possibly causing damage.

• When the timing belt is removed, the camshafts are positioned so that none of the valves are lifted off of their seats, resulting in a "zero-lift" position.

• The left-hand cylinder head camshafts must be rotated from the "zero-lift" position as little as possible when orienting it for timing belt installation, otherwise possible valve head interference may occur.

• Never allow the camshafts to rotate in the direction shown in the accompanying illustration, which would cause both the intake and exhaust valves to lift simultaneously, causing interference.

1. Remove all necessary components to gain access to the timing belt.

2. On models equipped with manual transmissions, loosen the two timing belt guide mounting bolts, then separate the guide from the engine block.

3. If the directional arrow and alignment marks on the timing belt are faded, and the belt is to be reused, remark the belt with white paint or a grease pencil as follows:

    a. Using a specific socket (Subaru part No. ST 499987500 Crankshaft Socket) installed on the crankshaft sprocket, rotate the crank-shaft until the crankshaft sprocket, left-hand exhaust camshaft sprocket, left-hand intake camshaft sprocket, right-hand

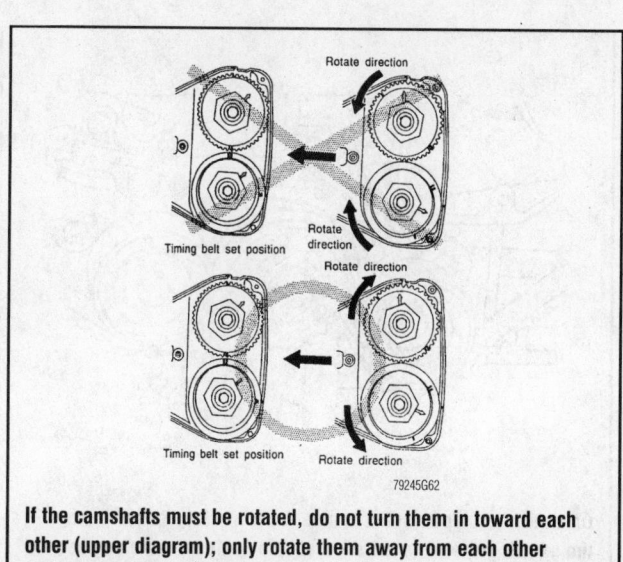

If the camshafts must be rotated, do not turn them in toward each other (upper diagram); only rotate them away from each other (lower diagram)—Subaru Forester

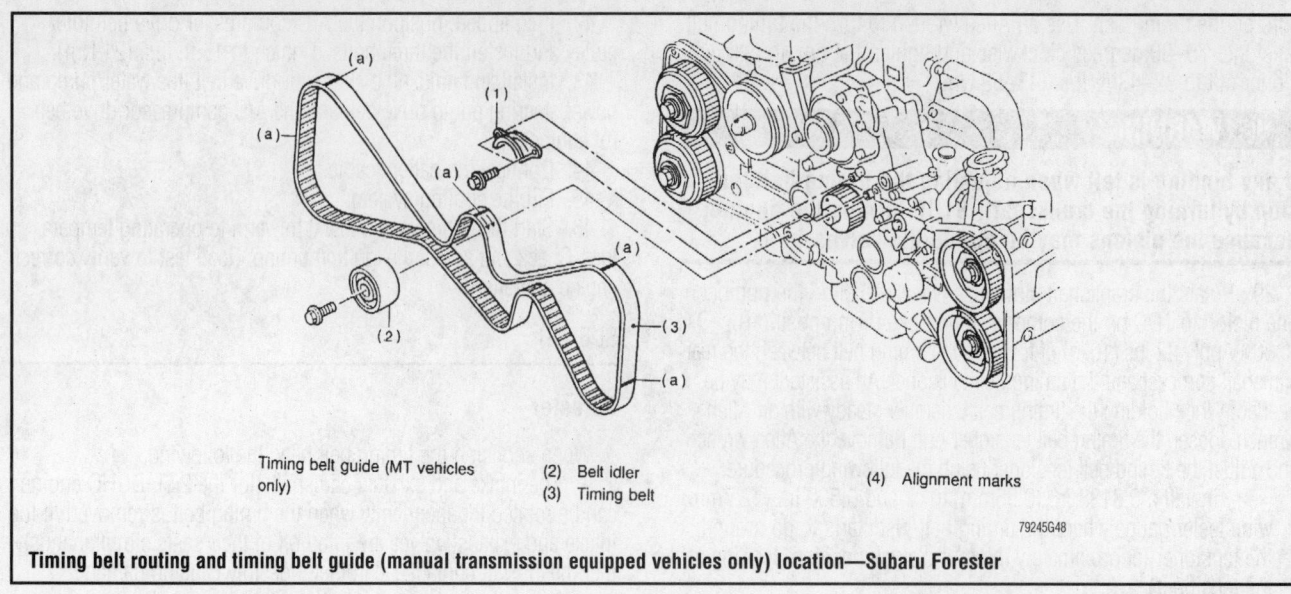

Timing belt guide (MT vehicles only)
(2) Belt idler
(3) Timing belt
(4) Alignment marks

79245G48

Timing belt routing and timing belt guide (manual transmission equipped vehicles only) location—Subaru Forester

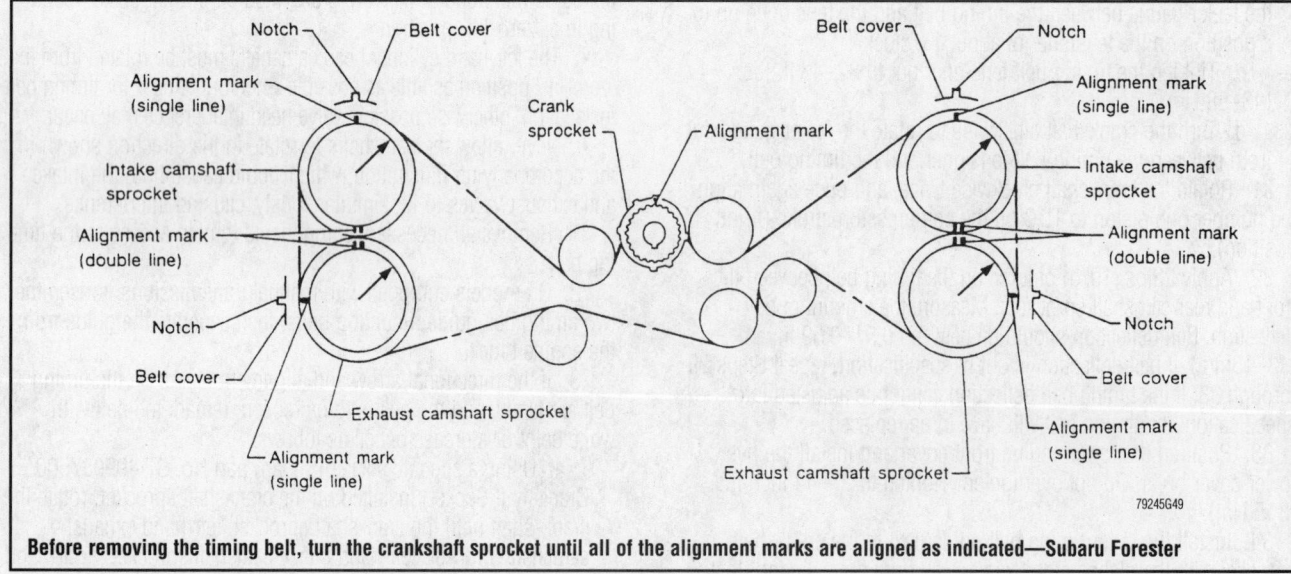

79245G49

Before removing the timing belt, turn the crankshaft sprocket until all of the alignment marks are aligned as indicated—Subaru Forester

79245G50

On models equipped with a manual transmission, loosen the two timing belt guide bolts and separate the guide from the engine block—Subaru Forester

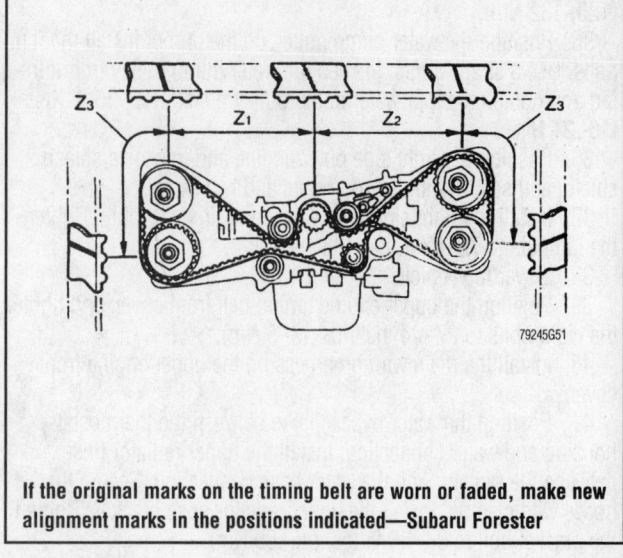

79245G51

If the original marks on the timing belt are worn or faded, make new alignment marks in the positions indicated—Subaru Forester

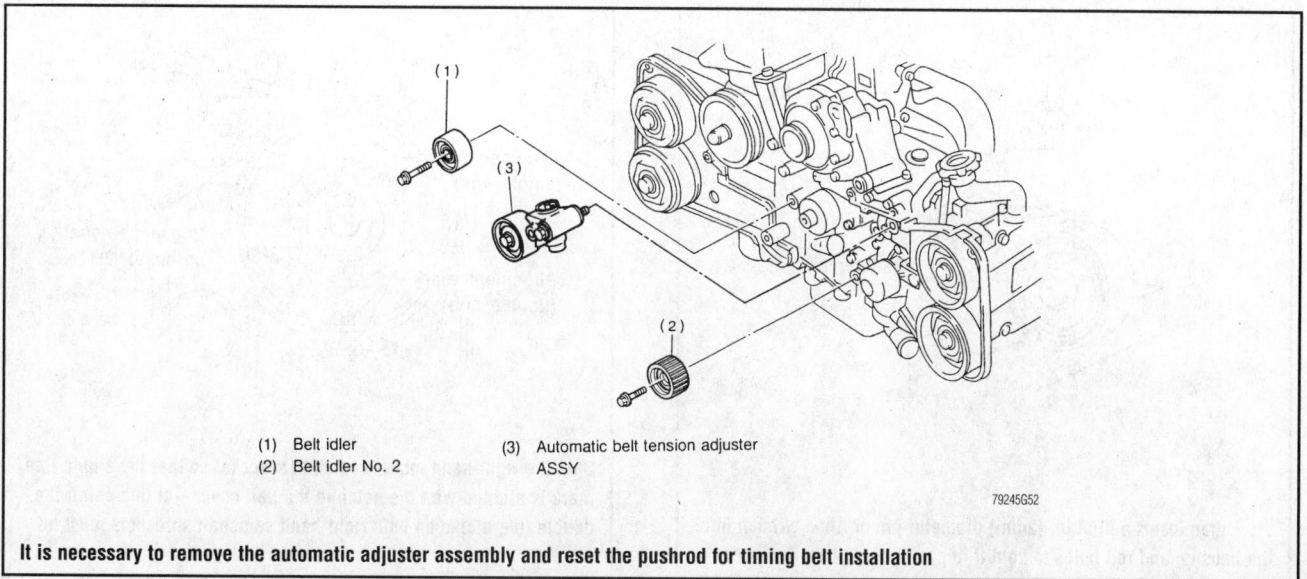

(1) Belt idler
(2) Belt idler No. 2
(3) Automatic belt tension adjuster
ASSY

79245G52

It is necessary to remove the automatic adjuster assembly and reset the pushrod for timing belt installation

intake camshaft sprocket and right-hand exhaust camshaft sprocket timing mark notches are aligned with the respective marks on the belt cover and engine block.

b. Make alignment and/or arrow marks on the timing belt in relation to the sprockets as indicated in the accompanying illustration.

- Z1—54.5 tooth length
- Z2—51 tooth length
- Z3—28 tooth length

4. Loosen the center bolt from the timing belt idler pulley, then remove the idler pulley from the engine block.

❊❊ WARNING

After removing the timing belt, DO NOT rotate the camshafts. Damage to the valves may occur.

5. Carefully remove the timing belt from all of the sprockets.
6. Remove the automatic belt tension adjuster assembly as follows:

a. Remove the two timing belt idler pulleys, as indicated in the accompanying illustration.

b. Loosen the automatic tension adjuster assembly mounting bolts, then separate the adjuster assembly from the engine block.

To install:

❊❊ WARNING

Do not allow oil, grease, or coolant to come in contact with the timing belt. If this occurs, quickly and thoroughly remove all traces of the compound. Also, never bend the timing belt sharply; the minimum bending radius is 2.36 in. (60mm).

7. Inspect the camshaft and crankshaft sprocket teeth for abnormal or excessive wear or scratches. Ensure there is no free-play between the sprocket and the key. Inspect the crankshaft sprocket sensor notch for damage or contamination with debris or dirt.

➡ **When preparing the automatic tension adjuster assembly for installation, adhere to the following points:**

- Always use a vertical press, rather than a horizontal press or vise, to depress the adjuster assembly rod

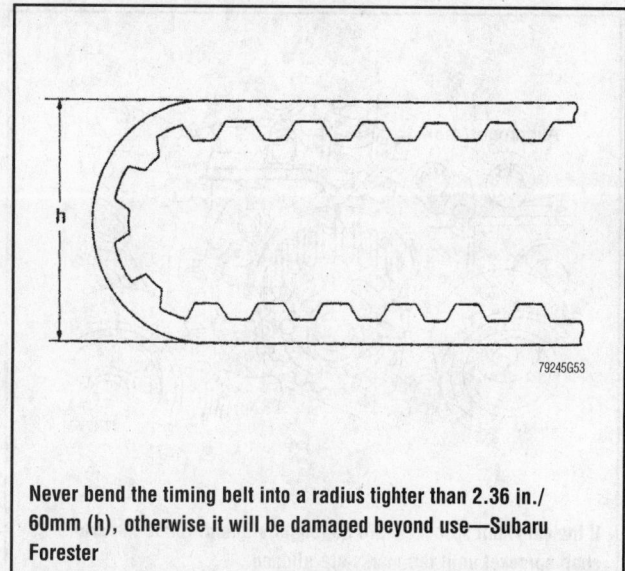

79245G53

Never bend the timing belt into a radius tighter than 2.36 in./ 60mm (h), otherwise it will be damaged beyond use—Subaru Forester

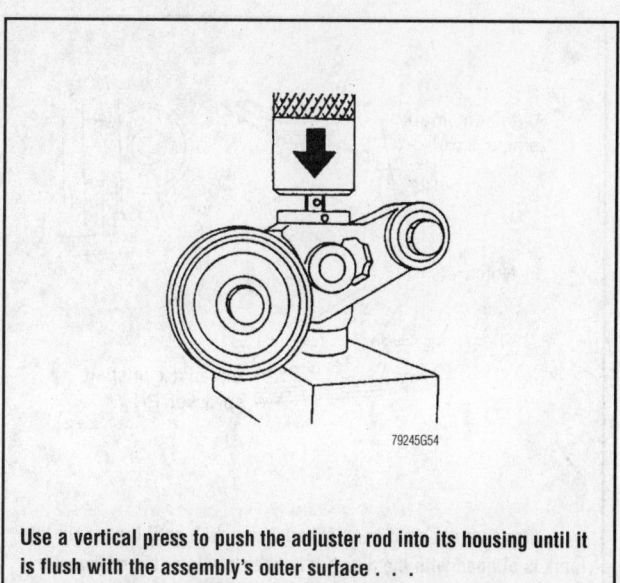

79245G54

Use a vertical press to push the adjuster rod into its housing until it is flush with the assembly's outer surface . . .

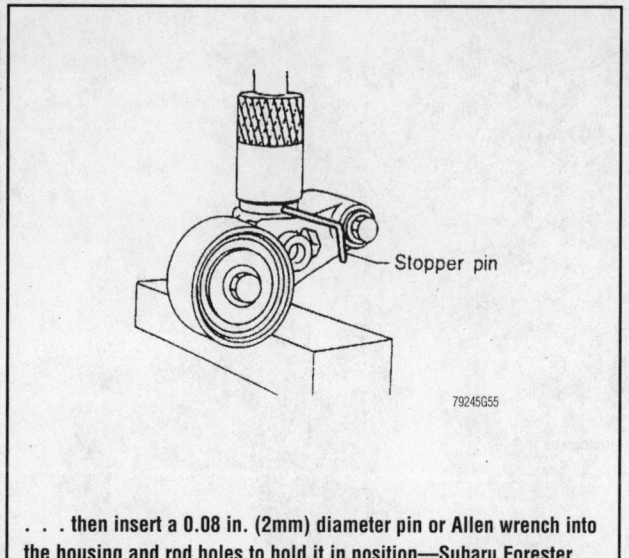

79245G55

. . . then insert a 0.08 in. (2mm) diameter pin or Allen wrench into the housing and rod holes to hold it in position—Subaru Forester

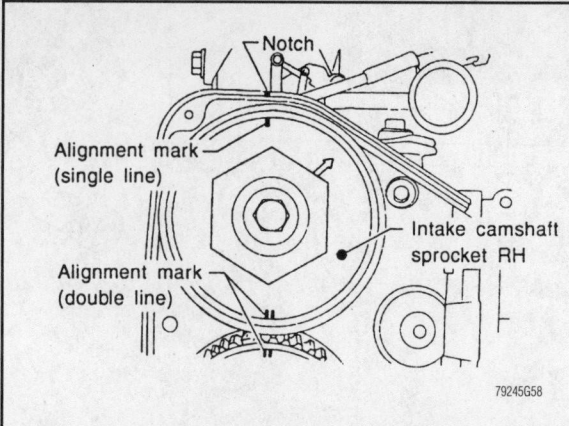

Notch

Alignment mark (single line)

Intake camshaft sprocket RH

Alignment mark (double line)

79245G58

Spin the right-hand intake camshaft sprocket so that the single line mark is aligned with the notch in the belt cover—at this point, the double line marks on both right-hand camshaft sprockets must be aligned

Alignment mark

79245G56

If the camshaft sprockets are no longer aligned, rotate the crankshaft sprocket until the marks are aligned . . .

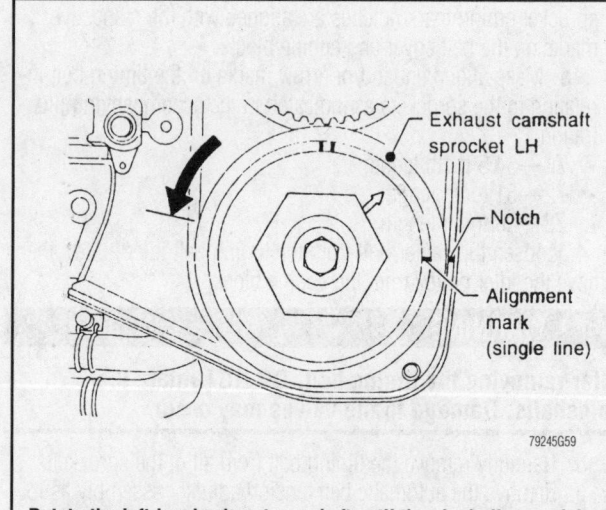

Exhaust camshaft sprocket LH

Notch

Alignment mark (single line)

79245G59

Rotate the left-hand exhaust camshaft until the single line mark is aligned with the notch in the belt cover . . .

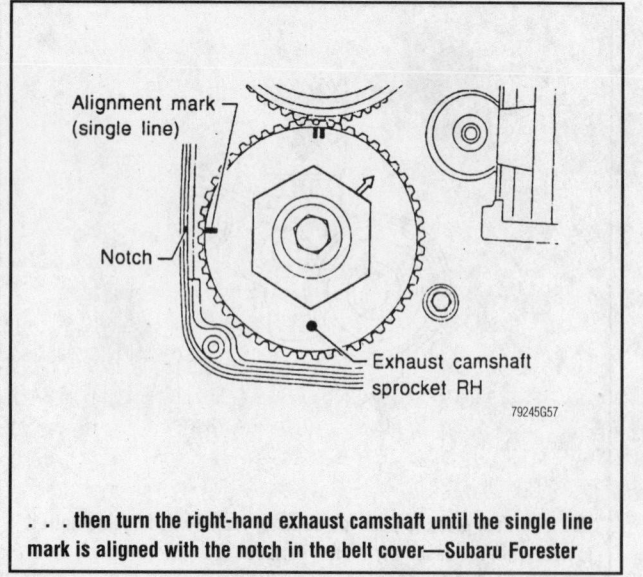

Alignment mark (single line)

Notch

Exhaust camshaft sprocket RH

79245G57

. . . then turn the right-hand exhaust camshaft until the single line mark is aligned with the notch in the belt cover—Subaru Forester

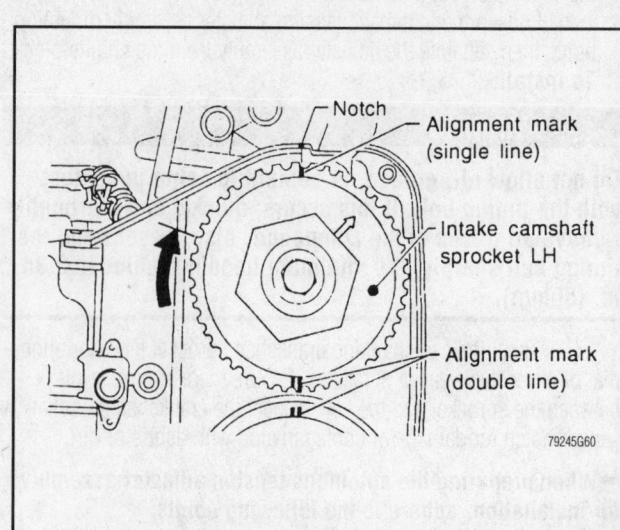

Notch

Alignment mark (single line)

Intake camshaft sprocket LH

Alignment mark (double line)

79245G60

. . . then align the single line mark on the left-hand intake sprocket with the belt cover notch—Subaru Forester

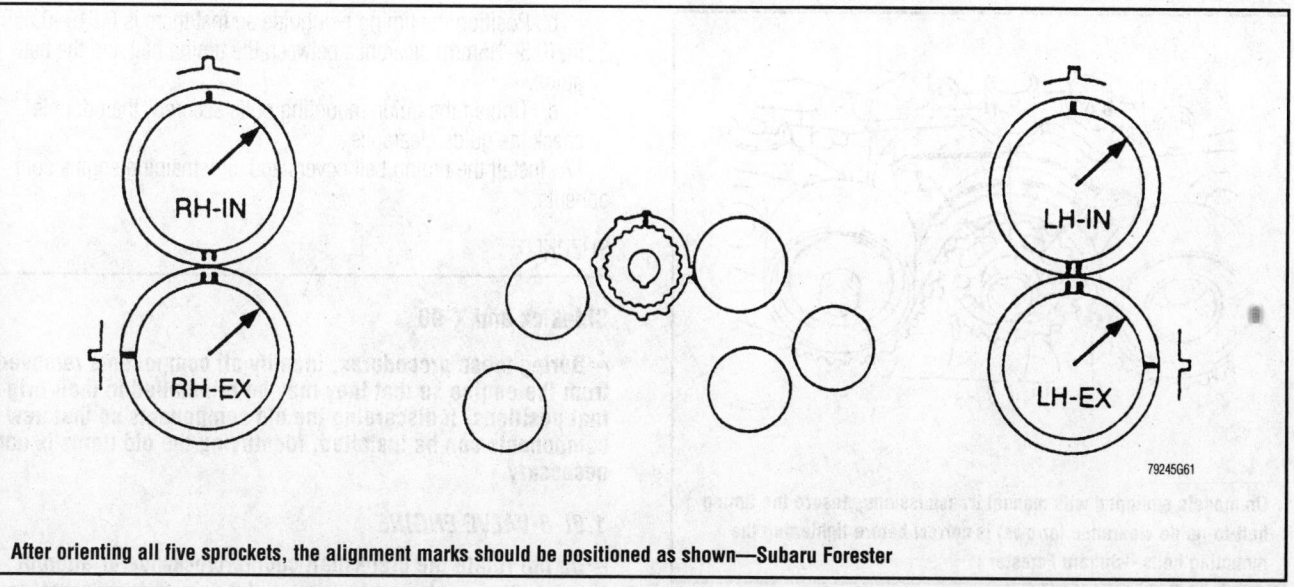

After orienting all five sprockets, the alignment marks should be positioned as shown—Subaru Forester

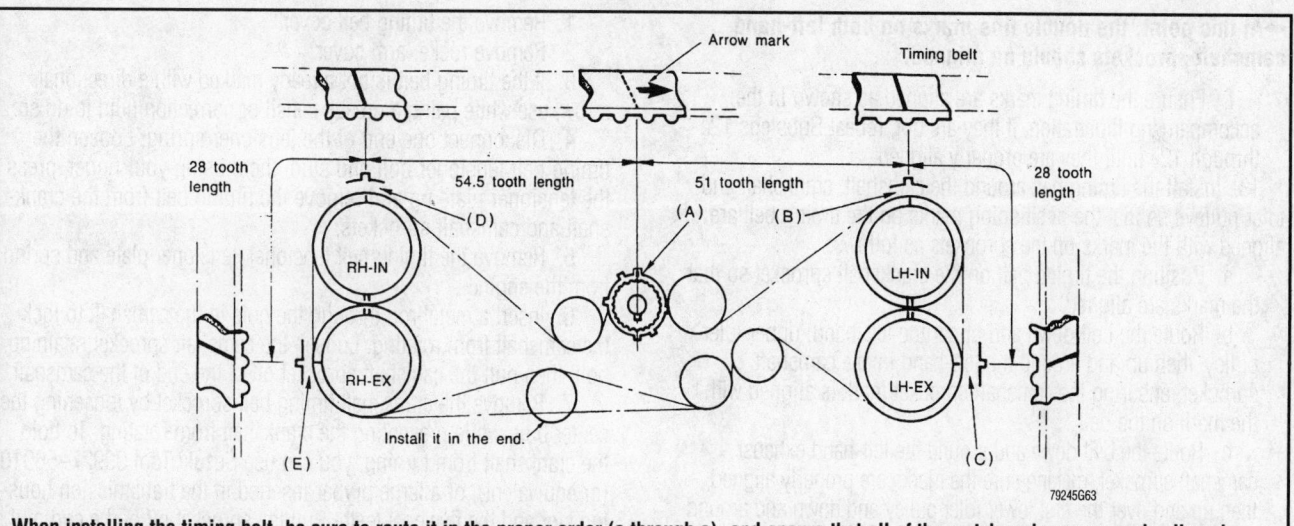

When installing the timing belt, be sure to route it in the proper order (a through e), and ensure that all of the matchmarks are properly aligned—Subaru Forester

- Depress the adjuster rod in a vertical position ONLY
- Depress the adjuster rod slowly (taking more than 3 minutes) with a force of 66 lbs. (30 kg)
- Do not allow the press force to exceed 2205 lbs. (1000 kg)
- Press the adjuster rod in as far as the end surface of the cylinder—do not press the rod into the cylinder, which may cause damage to the assembly.
- Do not release the press force from the rod until the stopper pin is completely inserted in the cylinder

8. Prepare the automatic timing belt tension adjuster assembly for installation as follows:

a. Position the adjuster assembly in a vertical press.

b. Slowly depress the adjuster rod with a force of 66 lbs. (30 kg) until the hole in the rod is aligned with the hole in the adjuster cylinder housing.

c. Insert a 0.08 in. (2mm) diameter stopper pin or Allen wrench through the hole in the cylinder housing and rod, then slowly release the press force from the adjuster rod.

9. Install the adjuster assembly onto the engine block.

10. Install timing belt idler pulley No. 2 on the engine block.

11. Install the timing belt idler pulley No. 1 on the engine block.

12. If the camshaft and crankshaft timing marks are no longer aligned, perform the following:

a. Position the crankshaft sprocket so that its mark is aligned with the mark on the oil pump cover on the engine block.

b. Align the single line mark on the right-hand exhaust camshaft sprocket with the notch on the belt cover.

c. Rotate the right-hand intake camshaft so that the single line mark is aligned with the notch on the belt cover.

➡ **At this point, the double line marks on both right-hand camshaft sprockets should be aligned.**

d. Turn the left-hand exhaust (lower) camshaft counterclockwise (as viewed from the front of the engine) until the single line mark is aligned with the notch on the belt cover.

e. Position the single line mark on the left-hand intake camshaft sprocket so that it is aligned with the notch on the belt cover. When rotating the camshaft, do so only in a clockwise direction (as viewed from the front of the engine).

On models equipped with manual transmissions, ensure the timing belt-to-guide clearance (arrows) is correct before tightening the mounting bolts—Subaru Forester

➡At this point, the double line marks on both left-hand camshaft sprockets should be aligned.

f. Ensure the timing marks are aligned as shown in the accompanying illustration. If they are not, repeat Substeps 12a through 12e until they are properly aligned.

13. Install the timing belt around the camshaft, crankshaft and idler pulleys so that the positioning marks on the timing belt are aligned with the marks on the sprockets as follows:

a. Position the timing belt on the crankshaft sprocket so that the marks are aligned.

b. Route the belt down and under the left-hand, upper idler pulley, then up and around the left-hand intake camshaft sprocket, ensuring the camshaft sprocket mark is aligned with the mark on the belt.

c. Route the belt down and around the left-hand exhaust camshaft sprocket, making sure the marks are properly aligned, then up and over the first lower idler pulley and down and around the second lower idler pulley.

d. While holding the timing belt on the inner, left-hand, lower idler pulley, route the other side of the timing belt (from the crankshaft sprocket) down and under the right-hand upper idler pulley.

e. Route the timing belt up and around the right-hand intake camshaft sprocket so that the belt and sprocket marks are aligned.

f. Position the belt down and around the right-hand exhaust camshaft sprocket, ensuring the positioning marks are aligned.

14. Install the right-hand lower idler pulley so that the timing belt is routed over the top side of it.

➡Once the belt is completely installed on all of the pulleys and sprockets, ensure that the positioning marks are still all aligned.

15. After ensuring all of the marks are still aligned, use a pair of pliers to withdraw the stopper pin or Allen wrench from the adjuster assembly housing.

16. On models with manual transmissions, perform the following:

a. Install the timing belt guide by temporarily tightening the mounting bolts.

b. Position the timing belt guide so that there is 0.019–0.059 in. (0.5–1.5mm) clearance between the timing belt and the belt guide.

c. Tighten the guide mounting bolts securely, then double check the guide clearance.

17. Install the timing belt covers and all remaining engine components.

SUZUKI

Sidekick and X-90

➡During these procedures, identify all components removed from the engine so that they may be reinstalled in their original positions. If discarding the old components so that new components can be installed, identifying the old items is not necessary.

1.6L 8-VALVE ENGINE

➡Do not rotate the crankshaft counterclockwise or attempt to rotate the crankshaft by turning the camshaft sprocket.

1. Remove the timing belt cover.
2. Remove rocker arm cover.
3. If the timing belt is not already marked with a directional arrow, use white paint, a grease pencil or correction fluid to do so.
4. Disconnect one end of the tensioner spring. Loosen the timing belt tensioner bolt and stud, then, using your finger, press the tensioner plate up and remove the timing belt from the crankshaft and camshaft sprockets.
5. Remove the timing belt tensioner, tensioner plate and spring from the engine.
6. Insert a metal rod through the hole in the camshaft to lock the camshaft from rotating. Loosen the camshaft sprocket retaining bolt, then pull the camshaft sprocket off of the end of the camshaft.
7. Remove the crankshaft timing belt sprocket by loosening the center bolt, while preventing the crankshaft from rotating. To hold the crankshaft from turning, you can use Suzuki Tool 09927–56010 (or equivalent), or a large prybar inserted in the transmission housing slot and the flywheel teeth. Pull the sprocket off of the end of the crankshaft. Be sure to retain the crankshaft sprocket key and belt guide for assembly.
8. If necessary, remove the timing belt inside cover from the cylinder head.

To install:

9. If necessary, install the timing belt inside cover.
10. Slide the timing belt guide on the crankshaft so that the concave side faces the oil pump, then install the sprocket key in the groove in the crankshaft.
11. Slide the pulley onto the crankshaft, and install the center retaining bolt. Tighten the center bolt to 58–65 ft. lbs. (80–90 Nm). To hold the crankshaft from turning, you can use Suzuki Tool 09927–56010 (or equivalent), or a large prybar inserted in the transmission housing slot and the flywheel teeth.
12. Install the timing belt camshaft sprocket, ensuring that the slot in the sprocket engages the camshaft (pulley) pin; this ensures that the sprocket is properly positioned on the end of the camshaft. Secure the camshaft with the metal rod used during removal, then tighten the sprocket bolt to 41–46 ft. lbs. (56–64 Nm).
13. Assemble the timing belt tensioner plate and the tensioner, making sure that the lug of the tensioner plate engages the tensioner.
14. Install the timing belt tensioner, tensioner plate and spring on the engine. Tighten the mounting bolt and stud only finger-

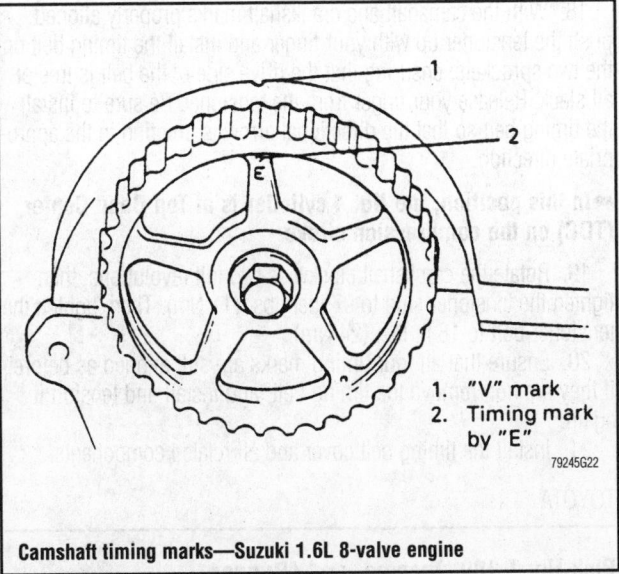

1. "V" mark
2. Timing mark by "E"

79245G22

Camshaft timing marks—Suzuki 1.6L 8-valve engine

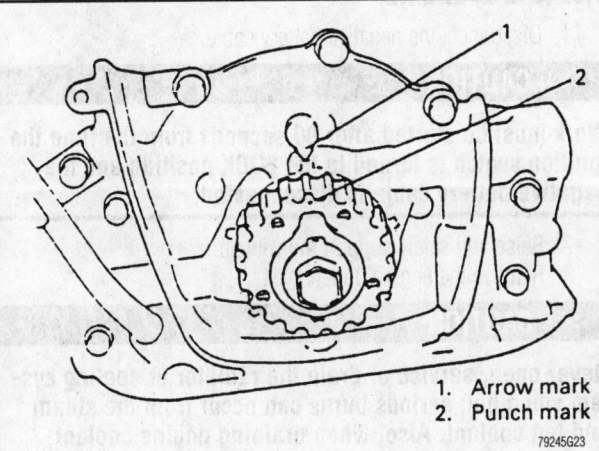

1. Arrow mark
2. Punch mark

79245G23

Align the punch mark with the arrow for proper timing belt installation—Suzuki 1.6L 8-valve engine

tight at this time. Ensure that when the tensioner is moved in a counterclockwise direction, the tensioner moves in the same direction. If the tensioner does not move, remove it and the tensioner plate to reassemble them properly.

15. Loosen all rocker arm valve lash locknuts and adjusting screws. This will permit movement of the camshaft without any rocker arm associated drag, which is essential for proper timing belt tensioning. If the camshaft does not rotate freely (free of rocker arm drag), the belt will not be properly tensioned.

✳✳ WARNING

If any binding is felt when adjusting the timing belt tension by turning the crankshaft, STOP turning the engine, because the pistons may be hitting the valves.

16. Rotate the camshaft sprocket clockwise until the timing mark on the sprocket and the V mark on the timing belt inside cover are aligned.

17. Using a 17mm wrench, or socket and breaker bar, on the crankshaft sprocket center bolt, turn the crankshaft clockwise until

the punch mark on the sprocket is aligned with the arrow mark on the oil pump.

18. With the camshaft and crankshaft marks properly aligned, push the tensioner up with your finger and install the timing belt on the two sprockets, ensuring that the drive side of the belt is free of all slack. Release your finger from the tensioner. Be sure to install the timing belt so that the directional arrow is pointing in the appropriate direction.

➡ **In this position, the No. 4 cylinder is at Top Dead Center (TDC) on the compression stroke.**

19. Rotate the crankshaft clockwise two full revolutions, then tighten the tensioner stud to 80–106 inch lbs. (9–12 Nm). Then, tighten the tensioner bolt to 18–21 ft. lbs. (24–30 Nm).

20. Ensure that all four timing marks are still aligned as before; if they are not, remove the timing belt, and install and tension it again.

21. Install the timing belt cover and all related components.

1.6L 16-VALVE ENGINE

The 1.6L 16-valve engine is known as an interference motor, because it is fabricated with such close tolerances between the pistons and valves that, if the timing belt is incorrectly positioned, jumps teeth on one of the sprockets or breaks, the valve and pistons will come into contact. This can cause severe internal engine damage

➡ **Do not rotate the crankshaft counterclockwise or attempt to rotate the crankshaft by turning the camshaft sprocket.**

1. Remove the timing belt cover.
2. If the timing belt is not already marked with a directional arrow, use white paint, a grease pencil or correction fluid to do so.
3. Rotate the crankshaft clockwise until the timing mark on the camshaft sprocket and the V mark on the timing belt inside cover are aligned, and the punch mark on the crankshaft sprocket is aligned with the mark on the engine.

✳✳ WARNING

Do not rotate the crankshaft or camshaft once the timing belt is removed, because the valves and pistons can come into contact, which may cause internal engine damage.

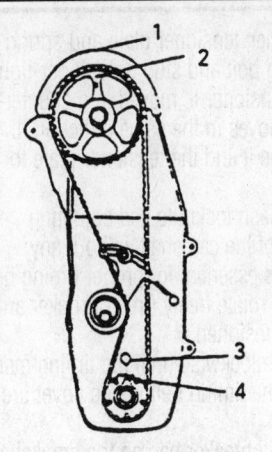

1. "V" mark on cylinder head cover
2. Timing mark by "E" on camshaft timing belt pulley
3. Arrow mark on oil pump case
4. Punch mark on crankshaft timing belt pulley

79245G47

Rotate the crankshaft clockwise until the camshaft and crankshaft timing marks are aligned—Suzuki 1.6L 16-valve engine

4. Disconnect one end of the tensioner spring. Loosen the timing belt tensioner bolt and stud, then, using your finger, press the tensioner plate up and remove the timing belt from the crankshaft and camshaft sprockets.

5. Remove the timing belt tensioner, tensioner plate and spring from the engine.

6. Install Suzuki Tool 09917–68220, or equivalent, onto the camshaft sprocket to hold the camshaft from rotating. Loosen the camshaft sprocket retaining bolt, then pull the camshaft sprocket off of the end of the camshaft.

7. Remove the crankshaft timing belt sprocket by loosening the center bolt, while preventing the crankshaft from rotating. To hold the crankshaft from turning, you can use Suzuki Tool 09927–56010 (or equivalent), or a large prybar inserted in the transmission housing slot and the flywheel teeth. Pull the sprocket off of the end of the crankshaft. Be sure to retain the crankshaft sprocket key and belt guide for assembly.

8. If necessary, remove the timing belt inside cover from the cylinder head.

To install:

9. If necessary, install the timing belt inside cover.

10. Slide the timing belt guide on the crankshaft so that the concave side faces the oil pump, then install the sprocket key in the groove in the crankshaft.

11. Slide the pulley onto the crankshaft, and install the center retaining bolt. Tighten the center bolt to 80 ft. lbs. (110 Nm). To hold the crankshaft from turning, you can use Suzuki Tool 09927–56010 (or equivalent), or a large prybar inserted in the transmission housing slot and the flywheel teeth.

12. Install the timing belt camshaft sprocket, ensuring that the slot in the sprocket engages the camshaft (pulley) pin; this ensures that the sprocket is properly positioned on the end of the camshaft. Secure the camshaft with the holding tool used during removal, then tighten the sprocket bolt to 44 ft. lbs. (60 Nm).

13. Assemble the timing belt tensioner plate and the tensioner, making sure that the lug of the tensioner plate engages the tensioner.

✳✳ WARNING

If any binding is felt when adjusting the timing belt tension by turning the crankshaft, STOP turning the engine, because the pistons may be hitting the valves.

14. Install the timing belt tensioner, tensioner plate and spring on the engine. Tighten the mounting bolt and stud only finger-tight at this time. Ensure that when the tensioner is moved in a counterclockwise direction, the tensioner moves in the same direction. If the tensioner does not move, remove it and the tensioner plate to reassemble them properly.

15. Loosen all rocker arm valve lash locknuts and adjusting screws. This will permit movement of the camshaft without any rocker arm associated drag, which is essential for proper timing belt tensioning. If the camshaft does not rotate freely (free of rocker arm drag), the belt will not be properly tensioned.

16. Rotate the camshaft sprocket clockwise until the timing mark on the sprocket and the V mark on the timing belt inside cover are aligned.

17. Using a wrench, or socket and breaker bar, on the crankshaft sprocket center bolt, turn the crankshaft clockwise until the punch mark on the sprocket is aligned with the arrow mark on the oil pump.

18. With the camshaft and crankshaft marks properly aligned, push the tensioner up with your finger and install the timing belt on the two sprockets, ensuring that the drive side of the belt is free of all slack. Release your finger from the tensioner. Be sure to install the timing belt so that the directional arrow is pointing in the appropriate direction.

➡ **In this position, the No. 4 cylinder is at Top Dead Center (TDC) on the compression stroke.**

19. Rotate the crankshaft clockwise two full revolutions, then tighten the tensioner stud to 97 inch lbs. (11 Nm). Then, tighten the tensioner bolt to 18 ft. lbs. (24 Nm).

20. Ensure that all four timing marks are still aligned as before; if they are not, remove the timing belt, and install and tension it again.

21. Install the timing belt cover and all related components.

TOYOTA

Pick-Up, T-100, Tacoma, and 4Runner

3.0L (3VZ-E) ENGINE

1. Disconnect the negative battery cable.

✳✳ CAUTION

Work must be started after 90 seconds from the time the ignition switch is turned to the LOCK position and the negative battery cable is disconnected.

2. Raise and safely support the vehicle.
3. Remove the engine undercover.

✳✳ CAUTION

Never open, service or drain the radiator or cooling system when hot; serious burns can occur from the steam and hot coolant. Also, when draining engine coolant, keep in mind that cats and dogs are attracted to ethylene glycol antifreeze and could drink any that is left in an uncovered container or in puddles on the ground. This will prove fatal in sufficient quantities. Always drain

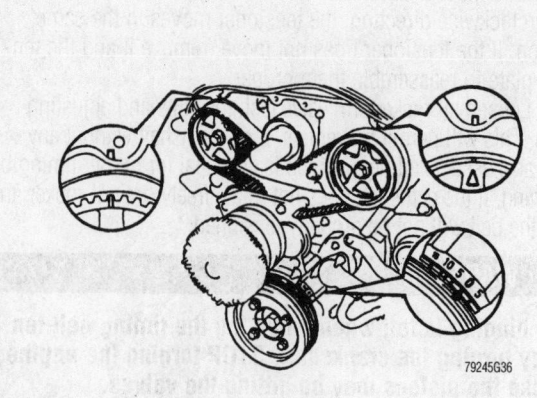

79245G36

Be sure to properly align the crankshaft timing mark before removing the belt—Toyota 3.0L (3VZE) engine

coolant into a sealable container. Coolant should be reused unless it is contaminated or is several years old.

4. Drain the engine coolant.
5. Disconnect the hoses and remove the radiator.
6. Disconnect the No. 2 and No. 3 air hoses from the air pipe.
7. Disconnect the wires from the spark plugs.
8. Remove the spark plugs.
9. Remove the accessory drive belts.
10. Disconnect the power steering pump from the engine.
11. Remove the cooling fan.
12. Remove the water outlet.
13. Disconnect the spark plug wire clamps from the mounting bolts and remove the No. 2 timing belt cover.
14. If reusing the timing belt, check that there are installation marks on the timing belt.
15. Set the No. 1 cylinder at TDC of the compression stroke, as follows:

a. Turn the crankshaft pulley and align its groove with the timing mark **0** of the No. 1 timing belt cover.

b. Check that the timing marks of the camshaft timing pulleys and the No. 3 timing belt cover are aligned. If not, turn the crankshaft pulley one revolution (360°).

16. Remove the timing belt tensioner.
17. Remove the fan bracket.
18. Disconnect the timing belt from the camshaft timing pulleys.

➡If reusing the timing belt, be sure that you can still read the installation marks. If not, place new installation marks on the timing belt to match the timing marks of the camshaft timing pulleys.

19. Remove the camshaft timing pulleys, as follows:

a. Using SST 09960–10010, or equivalent, remove the pulley bolt, the timing pulley, and the knock pin. Remove the two timing pulleys.

20. Remove the crankshaft pulley, as follows:

a. Using SST 09213–58012 and 09330–00021, or their equivalents, loosen the pulley bolt.

b. Remove the SST and the pulley bolt. Remove the crankshaft pulley.

21. Remove the No. 1 timing belt cover.
22. Remove the timing belt guide and remove the timing belt.
23. Remove the pivot bolt, the No. 1 idler pulley, and the plate washer.
24. Remove the crankshaft timing pulley.

To install:
25. Install the crankshaft timing pulley, as follows:

a. Align the timing pulley set key with the key groove of the pulley.

b. Using SST 09214–60010, or equivalent, and a hammer, tap in the timing pulley with the flange side facing inward.

26. Install the plate washer and the No. 1 idler pulley with the pivot bolt and tighten to 25 ft. lbs. (34 Nm). Check that the pulley bracket moves smoothly.

27. Temporarily install the timing belt, as follows:

a. Using the crankshaft pulley bolt, turn the crankshaft and align the timing marks of the crankshaft timing pulley and the oil pump body.

b. Align the installation mark on the timing belt with the dot mark on the crankshaft timing pulley.

c. Install the timing belt on the crankshaft timing pulley and the water pump pulley.

28. Install the timing belt guide with the cup side facing outward.
29. Install the No. 1 timing belt cover.
30. Install the crankshaft pulley, as follows:

a. Align the pulley set key with the groove of the crankshaft pulley.

b. Install the pulley bolt and tighten it to 181 ft. lbs. (245 Nm).

31. Install the left camshaft timing pulley.

a. Install the knock pin to the camshaft.

b. Align the knock pin hole of the camshaft with the knock pin groove of the timing pulley.

c. Slide the timing pulley on the camshaft with the flange side facing outward. Tighten the pulley bolt to 80 ft. lbs. (108 Nm).

32. Set the No. 1 cylinder to TDC of the compression stroke, as follows:

a. Turn the crankshaft pulley, and align its groove with the timing mark **0** of the No. 1 timing belt cover.

b. Turn the camshaft to align the knock pin hole of the camshaft with the timing mark of the No. 3 timing belt cover.

c. Turn the camshaft timing pulley and align the timing marks of the camshaft timing pulley and the No. 3 timing belt cover.

33. Connect the timing belt to the left camshaft timing pulley. Check that the installation mark on the timing belt is aligned with the end of the No. 1 timing belt cover.

a. Using SST 09960–01000, or equivalent, slightly turn the left camshaft timing pulley clockwise. Align the installation mark on the timing belt with the timing mark of the camshaft timing pulley, and hang the timing belt on the left camshaft timing pulley.

b. Align the timing marks of the left camshaft pulley and the No. 3 timing belt cover.

c. Check that the timing belt has tension between the crankshaft timing pulley and the left camshaft timing pulley.

34. Install the right camshaft timing pulley and the timing belt, as follows:

a. Align the installation mark on the timing belt with the timing mark of the right camshaft timing pulley, and hang the timing belt on the right camshaft timing pulley with the flange side facing inward.

b. Slide the right camshaft timing pulley on the camshaft. Align the timing marks on the right camshaft timing pulley and the No. 3 timing belt cover.

c. Align the knock pin hole of the camshaft with the knock pin groove of the pulley and install the knock pin. Install the bolt and tighten it to 80 ft. lbs. (108 Nm).

35. Install the fan bracket and tighten the bolts to 30 ft. lbs. (41 Nm).

❄❄ WARNING

If any binding is felt when adjusting the timing belt tension by turning the crankshaft, STOP turning the engine, because the pistons may be hitting the valves.

36. Set the timing belt tensioner, as follows:

a. Using a press, slowly depress in the pushrod using 200–2,205 lbs. (981–9,807 N) of force.

b. Align the holes of the pushrod and housing, pass a 1.5 mm hexagon wrench through the holes to keep the setting position of the pushrod.

c. Release the press and install the dust boot on the tensioner.

37. Install the timing belt tensioner and alternately tighten the bolts to 20 ft. lbs. (28 Nm). Using pliers, remove the 1.5 mm hexagon wrench from the belt tensioner.

38. Check the valve timing, as follows:

a. Slowly turn the crankshaft pulley two revolutions from TDC to TDC. Always turn the crankshaft pulley clockwise.

b. Check that each pulley aligns with the timing marks. If the timing marks do not align, remove the timing belt and reinstall it.

39. Install the No. 2 timing belt cover. Connect the four clamps on the spark plug wires to the mounting bolts of the No. 2 timing belt cover.

40. Remove the old packing material from the water outlet and apply new seal packing before installation. Tighten the nuts to 11 ft. lbs. (15 Nm).

41. Install the alternator drive belt.

42. Install the cooling fan and tighten the nuts to 48 inch lbs. (5.4 Nm).

43. Install the air conditioning drive belt.

44. Install the power steering pump, pump pulley, and the drive belt.

45. Install the spark plugs and tighten them to 13 ft. lbs. (18 Nm).

46. Connect the spark plug wires to the spark plugs.

47. Connect the No. 2 and the No. 3 air hoses to the air pipe.

48. Install the radiator and connect the hoses.

49. Fill the engine with coolant.

50. Connect the negative battery cable.

51. Start the engine and check for leaks.

52. Check the ignition timing and install the engine undercover.

3.4L (5VZ-FE) ENGINE—T-100 AND TACOMA

1. Disconnect the negative battery cable.

✳✳ CAUTION

Work must be started after 90 seconds from the time the ignition switch is turned to the LOCK position and the negative battery cable is disconnected.

2. Raise and safely support the vehicle.
3. Remove the engine undercover.
4. Drain the engine coolant.

✳✳ CAUTION

Never open, service or drain the radiator or cooling system when hot; serious burns can occur from the steam and hot coolant. Also, when draining engine coolant, keep in mind that cats and dogs are attracted to ethylene glycol antifreeze and could drink any that is left in an uncovered container or in puddles on the ground. This will prove fatal in sufficient quantities. Always drain coolant into a sealable container. Coolant should be reused unless it is contaminated or is several years old.

5. Disconnect the upper radiator hose from the engine.
6. Remove the power steering drive belt.
7. Remove the air conditioning drive belt by loosening the idler pulley nut and the adjusting bolt.
8. Loosen the lockbolt, pivot bolt, and the adjusting bolt and the alternator drive belt.
9. Remove the No. 2 fan shroud by removing the two clips.

Turn the crankwise clockwise to line up the timing marks before removing the timing belt—Toyota T-100 and Tacoma 3.4L (5VZ-FE) engine

10. Remove the fan with the fluid coupling and fan pulleys.

11. Disconnect the power steering pump from the engine and set aside. Do not disconnect the lines from the pump.

12. If equipped with air conditioning, disconnect the compressor from the engine and set aside. Do not disconnect the lines from the compressor.

13. If equipped with air conditioning, disconnect the air conditioning bracket.

14. Remove the No. 2 timing belt cover, as follows:

a. Detach the camshaft position sensor connector from the No. 2 timing belt cover.

b. Disconnect the three spark plug wire clamps from the No. 2 timing belt cover.

c. Remove the six bolts and remove the timing belt cover.

15. Remove the fan bracket, as follows:

a. Remove the power steering adjusting strut by removing the nut.

b. Remove the fan bracket by removing the bolt and nut.

16. Set the No. 1 cylinder at TDC of the compression stroke, as follows:

a. Turn the crankshaft pulley and align its groove with the timing mark **0** of the No. 1 timing belt cover.

b. Check that the timing marks of the camshaft timing pulleys and the No. 3 timing belt cover are aligned. If not, turn the crankshaft pulley one revolution (360°).

➡️**If reusing the timing belt, be sure that you can still read the installation marks. If not, place new installation marks on the timing belt to match the timing marks of the camshaft timing pulleys.**

17. Remove the timing belt tensioner by alternately loosening the two bolts.

18. Remove the camshaft timing pulleys, as follows:

a. Using SST 09960–10010 or equivalent, remove the pulley bolt, the timing pulley and the knock pin. Remove the two timing pulleys with the timing belt.

19. Remove the crankshaft pulley, as follows:

a. Using SST 09213–54015 and 09330–00021, or their equivalents, loosen the pulley bolt.

b. Remove the SST tool, the pulley bolt, and the pulley.

20. Remove the starter wire bracket and the No. 1 timing belt cover.

21. Remove the timing belt guide and remove the timing belt.

22. Remove the bolt and the No. 2 idler pulley.

23. Remove the pivot bolt, the No. 1 idler pulley, and the plate washer.

24. Remove the crankshaft gear.

To install:

25. Install the crankshaft timing gear.

   a. Align the timing pulley set key with the key groove of the gear.

   b. Using SST 09214–60010, or equivalent, and a hammer, tap in the timing gear with the flange side facing inward.

26. Install the plate washer and the No. 1 idler pulley with the pivot bolt and tighten it to 26 ft. lbs. (35 Nm). Check that the pulley bracket moves smoothly.

27. Install the No. 2 timing belt idler with the bolt. Tighten the bolt to 30 ft. lbs. (40 Nm). Check that the pulley bracket moves smoothly.

28. Temporarily install the timing belt, as follows:

   a. Using the crankshaft pulley bolt, turn the crankshaft and align the timing marks of the crankshaft timing pulley and the oil pump body.

   b. Align the installation mark on the timing belt with the dot mark of the crankshaft timing pulley.

   c. Install the timing belt on the crankshaft timing pulley, No. 1 idler pulley, and the water pump pulleys.

29. Install the timing belt guide with the cup side facing outward.

30. Install the No. 1 timing belt cover and starter wire bracket. Tighten the timing belt cover bolts to 80 inch lbs. (9 Nm).

✸✸ WARNING

If any binding is felt when adjusting the timing belt tension by turning the crankshaft, STOP turning the engine, because the pistons may be hitting the valves.

31. Install the crankshaft pulley, as follows:

   a. Align the pulley set key with the key groove of the crankshaft pulley.

   b. Install the pulley bolt and tighten it to 184 ft. lbs. (250 Nm).

32. Install the left camshaft timing pulley.

   a. Install the knock pin to the camshaft.

   b. Align the knock pin hose of the camshaft with the knock pin groove of the timing pulley.

   c. Slide the timing belt pulley on the camshaft with the flange side facing outward. Tighten the pulley bolt to 81 ft. lbs. (110 Nm).

33. Set the No. 1 cylinder to TDC of the compression stroke, as follows:

   a. Turn the crankshaft pulley, and align its groove with the timing mark **0** of the No. 1 timing belt cover.

   b. Turn the camshaft to align the knock pin hole of the camshaft with the timing mark of the No. 3 timing belt cover.

   c. Turn the camshaft timing pulley, and align the timing marks of the camshaft timing pulley and the No. 3 timing belt cover.

34. Connect the timing belt to the left camshaft timing pulley, as follows:

➡ **Check that the installation mark on the timing belt is aligned with the end of the No. 1 timing belt cover.**

   a. Using SST 09960–01000 or equivalent, slightly turn the left camshaft timing pulley clockwise. Align the installation mark on the timing belt with the timing mark of the camshaft timing pulley, and hang the timing belt on the left camshaft timing pulley.

   b. Align the timing marks of the left camshaft pulley and the No. 3 timing belt cover.

   c. Check that the timing belt has tension between the crankshaft timing pulley and the left camshaft timing pulley.

35. Install the right camshaft timing pulley and the timing belt, as follows:

   a. Align the installation mark on the timing belt with the timing mark of the right camshaft timing pulley, and hang the timing belt on the right camshaft timing pulley with the flange side facing inward.

   b. Slide the right camshaft timing pulley on the camshaft. Align the timing marks on the right camshaft timing pulley and the No. 3 timing belt cover.

   c. Align the knock pin hole of the camshaft with the knock pin groove of the pulley and install the knock pin. Install the bolt and tighten it to 81 ft. lbs. (110 Nm).

36. Set the timing belt tensioner, as follows:

   a. Using a press, slowly press in the pushrod using 220–2,205 lbs. (981–9,807 N) of force.

   b. align the holes of the pushrod and housing, pass a 1.5 mm hex wrench through the holes to keep the setting position of the pushrod.

   c. Release the press and install the dust boot on the tensioner.

37. Install the timing belt tensioner and alternately tighten the bolts to 20 ft. lbs. (28 Nm). Using pliers, remove the 1.5 mm hex wrench from the belt tensioner.

38. Check the valve timing, as follows:

   a. Slowly turn the crankshaft pulley two revolutions from TDC to TDC. Always turn the crankshaft pulley clockwise.

   b. Check that each pulley aligns with the timing marks. If the timing marks do not align, remove the timing belt and reinstall it.

39. Install the fan bracket with the bolt and nut.

40. Install the power steering adjusting strut with the nut.

41. Install the No. 2 timing belt cover. Tighten the bolts to 80 inch lbs. (9 Nm). Install the remaining components.

42. Fill the cooling system with coolant.

43. Connect the negative battery cable.

44. Start the engine and check for leaks.

3.4L (5VZFE) ENGINE—4RUNNER

1. Disconnect the negative battery cable.

✸✸ CAUTION

Wait 90 seconds from the time the key is turned to LOCK and the negative battery cable is disconnected to begin work. This allows the SRS capacitor to discharge and prevent deployment of the air bag(s).

2. Raise and safely support the vehicle.

3. Remove the engine undercover.

4. Drain the engine coolant.

✳✳ CAUTION

Never open, service or drain the radiator or cooling system when hot; serious burns can occur from the steam and hot coolant. Also, when draining engine coolant, keep in mind that cats and dogs are attracted to ethylene glycol antifreeze and could drink any that is left in an uncovered container or in puddles on the ground. This will prove fatal in sufficient quantities. Always drain coolant into a sealable container. Coolant should be reused unless it is contaminated or is several years old.

5. Disconnect the upper radiator hose from the engine.

6. Remove the power steering drive belt.

7. Remove the air conditioning drive belt by loosening the idler pulley nut and the adjusting bolt.

8. If equipped with air conditioning, disconnect the compressor from the engine and set aside. Do not disconnect the lines from the compressor.

9. If equipped with air conditioning, disconnect the air conditioning bracket.

10. Remove the fan with the fluid coupling and fan pulleys.

11. Loosen the lockbolt, pivot bolt, and the adjusting bolt and the alternator drive belt.

12. Remove the No. 2 fan shroud by removing the two clips.

13. Disconnect the power steering pump from the engine and set aside. Do not disconnect the lines from the pump.

14. Remove the oil dipstick and the guide.

15. Remove the No. 2 timing belt cover as follows:

a. Detach the camshaft position sensor connector from the No. 2 timing belt cover.

b. Disconnect the four spark plug wire clamps from the No. 2 timing belt cover.

c. Remove the six bolts and remove the timing belt cover.

16. Remove the fan bracket as follows:

a. Remove the power steering adjusting strut by removing the nut.

b. Remove the fan bracket by removing the bolt and nut.

17. Using SST 09213–54015 or equivalent, remove the crankshaft pulley.

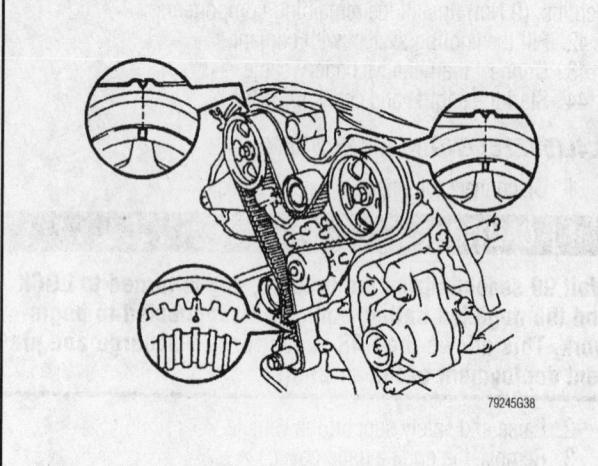

Crankshaft and camshaft timing mark locations—Toyota 4Runner 3.4L (5VZFE) engine

18. Remove the starter wire bracket and the No. 1 timing belt cover.

19. Remove the timing belt guide.

20. Set the No. 1 cylinder at TDC of the compression stroke, as follows:

a. Temporarily install the crankshaft pulley bolt to the crankshaft.

b. Turn the crankshaft and align the timing marks of the crankshaft timing pulley and the oil pump body.

c. Check that the timing marks of the camshaft timing pulleys and the No. 3 timing belt cover are aligned. If not, turn the crankshaft pulley one revolution (360°).

➡**If reusing the timing belt, be sure that you can still read the installation marks. If not, place new installation marks on the timing belt to match the timing marks of the camshaft timing pulleys.**

21. Remove the timing belt tensioner by alternately loosening the two bolts.

22. Remove the right and left camshaft pulleys.

23. Remove the No. 2 idler pulley.

24. Using a 10mm hex wrench, remove the pivot bolt, No.1 idler pulley and the plate washer.

25. Remove the timing belt guide and remove the timing belt.

26. Remove the crankshaft timing pulley.

To install:

27. Install the crankshaft timing belt pulley, as follows:

a. Align the timing belt pulley set key with the key groove of the timing pulley and slide on the timing pulley.

b. Slide on the timing belt pulley with the flange side facing inward.

28. Install the plate washer and the No. 1 idler pulley with the pivot bolt and tighten it to 26 ft. lbs. (35 Nm). Check that the pulley bracket moves smoothly.

29. Install the No. 2 timing belt idler with the bolt. Tighten the bolt to 30 ft. lbs. (40 Nm). Check that the pulley bracket moves smoothly.

30. Install the left and right camshaft timing pulleys.

31. Set the No. 1 cylinder to TDC of the compression stroke, as follows:

a. Using the crankshaft pulley bolt, turn the crankshaft and align the timing marks of the crankshaft timing pulley and the oil pump body.

b. Using SST 09960–10010 or equivalent to turn the camshaft pulley to align the marks of the camshaft timing belt pulley and the No. 3 timing belt cover.

32. Install the timing belt, as follows:

➡**The engine should be cold.**

a. Face the front mark on the timing belt forward.

b. Align the installation mark on the timing belt with the timing mark of the crankshaft timing pulley.

c. Align the installation marks on the timing belt with the timing marks of the camshaft pulleys.

33. Install the timing belt in the following order:

- Left camshaft pulley
- No. 2 idler pulley
- Right camshaft pulley
- Water pump pulley
- Crankshaft pulley
- No. 1 idler pulley

If any binding is felt when adjusting the timing belt tension by turning the crankshaft, STOP turning the engine, because the pistons may be hitting the valves.

34. Set the timing belt tensioner as follows:

a. Using a press, slowly press in the pushrod using 220–2,205 lbs. (981–9,807 N) of force.

b. Align the holes of the pushrod and housing, pass a 1.27mm wrench through the holes to keep the setting position of the pushrod.

c. Release the press and install the dust boot to the tensioner.

35. Install the timing belt tensioner and alternately tighten the bolts to 20 ft. lbs. (27 Nm). Using pliers, remove the 1.27mm wrench from the belt tensioner.

36. Check the valve timing, as follows:

a. Slowly turn the crankshaft and align the timing marks of the crankshaft timing pulley and the oil pump body. Always turn the crankshaft pulley clockwise.

b. Check that the timing marks of the right and left timing pulleys align with the timing marks of the No. 3 timing belt cover. If the marks do not align, remove the timing belt and reinstall it.

37. Install the timing belt guide with the cup side facing outward.

38. Install the No. 1 timing belt cover and starter wire bracket. Tighten the timing belt cover fasteners to 80 inch lbs. (9 Nm).

39. Install the crankshaft pulley, as follows:

a. Align the pulley set key with the key groove of the pulley and slide the pulley.

b. Using SST 09213–54014 or equivalent, tighten the bolt to 184 ft. lbs. (250 Nm).

40. Install the fan bracket with the bolt and nut.

41. Install the No. 2 timing belt cover, and tighten the bolts to 80 inch lbs. (9 Nm). Install the remaining components.

42. Fill the cooling system with coolant.

43. Connect the negative battery cable.

44. Start the engine and check for leaks.

45. Check the ignition timing.

RAV4

2.0L (VIN P) ENGINE

The timing belt is not adjustable.

1. Disconnect the negative battery cable.

To avoid air bag deployment, if equipped, work must be started after approximately 90 seconds or longer from the time the ignition switch is turned to the LOCK position and the negative battery cable is disconnected from the battery.

2. Disconnect the power steering reservoir tank and remove the reservoir bracket.

3. Detach the wiring harness bracket for the Data Link Connector 1(DLC1).

4. Remove the alternator and alternator bracket.

5. If equipped with ABS brakes, remove the ABS actuator.

6. Remove the right front wheel and the fender apron seal.

7. Remove the power steering drive belt.

8. Slightly raise the engine using a block of wood and floor jack under the oil pan to prevent damage.

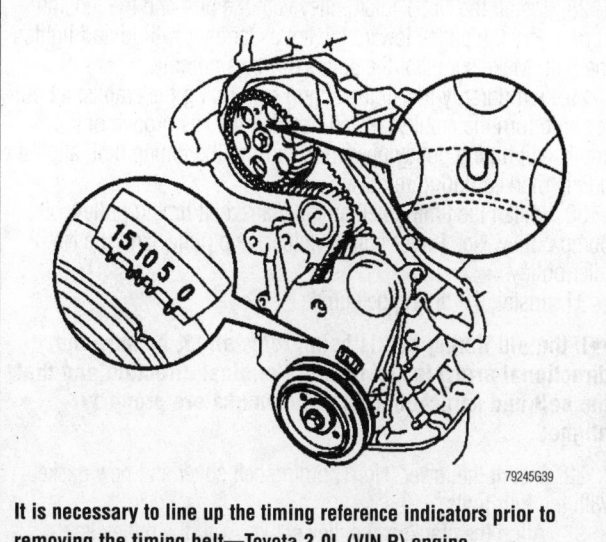

It is necessary to line up the timing reference indicators prior to removing the timing belt—Toyota 2.0L (VIN P) engine

9. Remove the four bolts, two nuts, and right-hand mounting bracket.

10. Remove the spark plugs.

11. Using SST 09213–54015 or equivalent, loosen the crankshaft pulley bolt and remove it by pulling it straight off the crankshaft.

12. Using SST 09249–63010 or equivalent, loosen the retaining bolts and remove the right engine mounting bracket.

13. Remove the upper (No. 2) timing belt cover.

14. Install the crankshaft pulley to the crankshaft and temporarily install the retaining bolt.

15. Turn the crankshaft pulley and align its groove with the timing mark **0** of the No. 1 timing belt cover. Check that the hole of the camshaft timing pulley is aligned with the timing mark of the bearing cap. If not, turn the crankshaft 360° and align the marks.

➡**If the timing belt is to be reused, matchmark the timing belt to the timing pulleys and timing belt covers so the belt can be reinstalled in its original position. Also, be sure to mark an arrow on the belt to indicate which direction it was turning.**

16. Remove the timing belt from the camshaft timing pulley.

17. Hold the camshaft sprocket with a spanner wrench and remove the mounting bolt. Remove the camshaft pulley.

18. Remove the crankshaft pulley bolt and remove the crankshaft pulley.

19. Remove the No. 1 timing belt cover.

20. Remove the timing belt guide and the timing belt.

21. Remove the No. 1 idler pulley and tension spring.

22. Remove the No. 2 idler pulley.

23. Remove the crankshaft timing pulley.

24. Support the oil pump sprocket with a spanner wrench, then remove the mounting bolt and remove the sprocket.

To install:

25. Install the oil pump pulley. Tighten the nut to 18 ft. lbs. (24 Nm).

26. Install the crankshaft timing pulley. Align the pulley set key with the key groove of the pulley. Slide on the pulley facing the flange side inward.

27. Install the No. 2 idler pulley and tighten the mounting bolt to 31 ft. lbs. (42 Nm). Be sure that the pulley moves smoothly.

28. Install the No. 1 idler pulley with the bolt and the tension spring. Pry the pulley toward the left as far as it will go and tighten the bolt. Make sure that the pulley moves smoothly.

29. Temporarily install the timing belt. Using the crankshaft pulley bolt, turn the crankshaft and position the key groove of the crankshaft timing pulley upward. If reusing the timing belt, align the points marked during removal.

30. Install the timing belt on the crankshaft timing pulley, oil pump pulley, No. 1 idler pulley, water pump pulley and the No. 2 idler pulley.

31. Install the timing belt guide.

➡ **If the old timing belt is being reinstalled, be sure the directional arrow is facing in the original direction and that the belt and sprocket/cover matchmarks are properly aligned.**

32. Install the lower (No. 1) timing belt cover and new gasket with the four bolts.

33. Align the crankshaft pulley set key with the pulley key groove. Temporarily install the crankshaft pulley and bolt.

34. Align the camshaft knock pin with the groove of the pulley, and slide the timing pulley onto the camshaft with the plate washer and set bolt.

35. Tighten the pulley set bolt to 40 ft. lbs. (54 Nm).

✲✲ WARNING

If any binding is felt when adjusting the timing belt tension by turning the crankshaft, STOP turning the engine, because the pistons may be hitting the valves.

36. Turn the crankshaft pulley and align the **0** mark on the lower (No. 1) timing belt cover.

37. Finish installing the timing belt and check the valve timing, as follows:

    a. If reusing the old timing belt, align the matchmarks that you made previously, and install the timing belt onto the camshaft pulley.

    b. Align the marks on the timing belt with the marks on the camshaft pulley.

    c. Loosen the No. 1 idler pulley set bolt ½ turn.

    d. Turn the crankshaft pulley two complete revolutions TDC to TDC. ALWAYS turn the crankshaft CLOCKWISE. Check that the pulleys are still in alignment with the timing marks.

    e. If the No. 1 idler pulley uses a green tension spring, slowly turn the crankshaft pulley 1⅞ revolutions, and align its groove with the mark at 45° BTDC (for the No. 1 cylinder) of the No. 1 timing belt cover.

    f. Tighten the No. 1 idler pulley set bolt to 31 ft. lbs. (42 Nm).

    g. Be sure there is belt tension between the crankshaft and camshaft timing pulleys.

38. Place the right-hand engine mounting bracket in position but do not install the bolts.

39. Install the upper (No. 2) timing cover with a new gasket(s).

40. Remove the engine crankshaft pulley bolt and pulley.

41. Using SST 09249–63010 or equivalent, install the mounting bolts for the right-hand mounting bracket. Tighten the mounting bolts to 38 ft. lbs. (52 Nm).

42. Align the crankshaft pulley set key with the pulley key groove. Install the pulley. Tighten the pulley bolt to 80 ft. lbs. (108 Nm).

43. Install the spark plugs.

44. Install the right-hand mounting insulator, as follows:

    a. Attach the mounting insulator to the body and mounting bracket with the four bolts and two nuts.

    b. Tighten the three bolts to hold the mounting insulator to the body. Tighten the bolts to 47 ft. lbs. (64 Nm).

    c. Tighten the two nuts and bolt to hold the mounting insulator to the mounting bracket. Tighten the bolt to 27 ft. lbs. (37 Nm) and the nut to 38 ft. lbs. (52 Nm).

45. Install and adjust the power steering pump drive belt.

46. Install the right-hand engine undercover.

47. Install the right front wheel.

48. Lower the engine.

49. If equipped, install the ABS actuator.

50. Install the alternator and alternator bracket.

51. Install the wiring harness bracket for the DLC1.

52. Install the power steering reservoir bracket and reservoir.

53. Connect the negative battery cable.

54. Start the engine and check the timing.

BRAKES

6

BRAKE OPERATING SYSTEM

Basic Operating Principles

Hydraulic systems are used to actuate the brakes of all modern automobiles. The system transports the power required to force the frictional surfaces of the braking system together from the pedal to the individual brake units at each wheel. A hydraulic system is used for two reasons.

First, fluid under pressure can be carried to all parts of an automobile by small pipes and flexible hoses without taking up a significant amount of room or posing routing problems.

Second, a great mechanical advantage can be given to the brake pedal end of the system, and the foot pressure required to actuate the brakes can be reduced by making the surface area of the master cylinder pistons smaller than that of any of the pistons in the wheel cylinders or calipers.

The master cylinder consists of a fluid reservoir along with a double cylinder and piston assembly. Double type master cylinders are designed to separate the front and rear braking systems hydraulically in case of a leak. The master cylinder coverts mechanical motion from the pedal into hydraulic pressure within the lines. This pressure is translated back into mechanical motion at the wheels by either the wheel cylinder (drum brakes) or the caliper (disc brakes).

Steel lines carry the brake fluid to a point on the vehicle's frame near each of the vehicle's wheels. The fluid is, then carried to the calipers and wheel cylinders by flexible tubes in order to allow for suspension and steering movements.

In drum brake systems, each wheel cylinder contains two pistons, one at either end, which push outward in opposite directions and force the brake shoe into contact with the drum.

In disc brake systems, the cylinders are part of the calipers. At least one cylinder in each caliper is used to force the brake pads against the disc.

All pistons employ some type of seal, usually made of rubber, to minimize fluid leakage. A rubber dust boot seals the outer end of the cylinder against dust and dirt. The boot fits around the outer end of the piston on disc brake calipers, and around the brake actuating rod on wheel cylinders.

The hydraulic system operates as follows: When at rest, the entire system, from the piston(s) in the master cylinder to those in the wheel cylinders or calipers, is full of brake fluid. Upon application of the brake pedal, fluid trapped in front of the master cylinder piston(s) is forced through the lines to the wheel cylinders. Here, it forces the pistons outward, in the case of drum brakes, and inward toward the disc, in the case of disc brakes. The motion of the pistons is opposed by return springs mounted outside the cylinders in drum brakes, and by spring seals, in disc brakes.

Upon release of the brake pedal, a spring located inside the master cylinder immediately returns the master cylinder pistons to the normal position. The pistons contain check valves and the master cylinder has compensating ports drilled in it. These are uncovered as the pistons reach their normal position. The piston check valves allow fluid to flow toward the wheel cylinders or calipers as the pistons withdraw. Then, as the return springs force the brake pads or shoes into the released position, the excess fluid reservoir through the compensating ports. It is during the time the pedal is in the released position that any fluid that has leaked out of the system will be replaced through the compensating ports.

Dual circuit master cylinders employ two pistons, located one behind the other, in the same cylinder. The primary piston is actuated directly by mechanical linkage from the brake pedal through the power booster. The secondary piston is actuated by fluid trapped between the two pistons. If a leak develops in front of the secondary piston, it moves forward until it bottoms against the front of the master cylinder, and the fluid trapped between the pistons will operate the rear brakes. If the rear brakes develop a leak, the primary piston will move forward until direct contact with the secondary piston takes place, and it will force the secondary piston to actuate the front brakes. In either case, the brake pedal moves farther when the brakes are applied, and less braking power is available.

All dual circuit systems use a switch to warn the driver when only half of the brake system is operational. This switch is usually located in a valve body which is mounted on the firewall or the frame below the master cylinder. A hydraulic piston receives pressure from both circuits, each circuit's pressure being applied to one end of the piston. When the pressures are in balance, the piston remains stationary. When one circuit has a leak, however, the greater pressure in that circuit during application of the brakes will push the piston to one side, closing the switch and activating the brake warning light.

In disc brake systems, this valve body also contains a metering valve, in some cases, a proportioning valve. The metering valve keeps pressure from traveling to the disc brakes on the front wheels until the brake shoes on the rear wheels have contacted the drums, ensuring that the front brakes will never be used alone. The proportioning valve controls the pressure to the rear brakes to lessen the chance of rear wheel lock-up during very hard braking.

Warning lights may be tested by depressing the brake pedal and holding it while opening one of the wheel cylinder bleeder screws. If this does not cause the light to go on, substitute a new lamp, make continuity checks, finally, replace the switch as necessary.

The hydraulic system may be checked for leaks by applying pressure to the pedal gradually and steadily. If the pedal sinks very slowly to the floor, the system has a leak. This is not to be confused with a springy or spongy feel due to the compression of air within the lines. If the system leaks, there will be a gradual change in the position of the pedal with a constant pressure.

Check for leaks along all lines and at wheel cylinders. If no external leaks are apparent, the problem is inside the master cylinder.

DISC BRAKES

Instead of the traditional expanding brakes that press outward against a circular drum, disc brake systems utilize a disc (rotor) with brake pads positioned on either side of it. An easily-seen analogy is the hand brake arrangement on a bicycle. The pads squeeze onto the rim of the bike wheel, slowing its motion. Automobile disc brakes use the identical principle but apply the braking effort to a separate disc instead of the wheel.

The disc (rotor) is a casting, usually equipped with cooling fins between the two braking surfaces. This enables air to circulate between the braking surfaces making them less sensitive to heat buildup and more resistant to fade. Dirt and water do not drastically affect braking action since contaminants are thrown off by the centrifugal action of the rotor or scraped off the by the pads. Also, the equal clamping action of the two brake pads tends to ensure uniform, straight line stops. Disc brakes are inherently self-adjusting. There are three general types of disc brake:

1. Fixed calipers.
2. Floating calipers.
3. Sliding calipers.

The fixed caliper design uses one or two pistons mounted on each side of the rotor (in each side of the caliper). The caliper is mounted rigidly and does not move.

The sliding and floating designs are quite similar. In fact, these two types are often lumped together. In both designs, the pad on the inside of the rotor is moved into contact with the rotor by hydraulic force. The caliper, which is not held in a fixed position, moves slightly, bringing the outside pad into contact with the rotor.

Floating calipers use threaded guide pins and bushings, or sleeves to allow the caliper to slide and apply the brake pads.

There are typically three methods of securing a sliding caliper to its mounting bracket: with a retaining pin, with a key and bolt, or with a wedge and pin. On calipers which use the retaining pin method, you will find pins driven into the slot between the caliper and the caliper mount. On calipers which use the bolt and key method, a key is used between the caliper and the mounting bracket to allow the caliper to slide. The key is held in position by a lockbolt. On calipers which use the pin and wedge method, a wedge, retained by a pin, is used between the caliper and the mounting bracket.

For pad removal purposes, fixed calipers are usually not removed, floating calipers are either removed or flipped (hinged up or down on one pin), and sliding calipers are removed.

DRUM BRAKES

Drum brakes employ two brake shoes mounted on a stationary backing plate. These shoes are positioned inside a circular drum which rotates with the wheel assembly. The shoes are held in place by springs. This allows them to slide toward the drums (when they are applied) while keeping the linings and drums in alignment. The shoes are actuated by a wheel cylinder which is mounted at the top of the backing plate. When the brakes are applied, hydraulic pressure forces the wheel cylinder's actuating links outward. Since these links bear directly against the top of the brake shoes, the tops of the shoes are, then forced against the inner side of the drum. This action forces the bottoms of the two shoes to contact the brake drum by rotating the entire assembly slightly (known as servo action). When pressure within the wheel cylinder is relaxed, return springs pull the shoes back away from the drum.

Most modern drum brakes are designed to self-adjust themselves during application when the vehicle is moving in reverse. This motion causes both shoes to rotate very slightly with the drum, rocking an adjusting lever, thereby causing rotation of the adjusting screw. Some drum brake systems are designed to self-adjust during application whenever the brakes are applied. This on-board adjustment system reduces the need for maintenance adjustments and keeps both the brake function and pedal feel satisfactory.

POWER BOOSTERS

Virtually all modern vehicles use a power assisted brake system to multiply the braking force and reduce pedal effort. There are two types of power assist used. The most widely used, by far, is the vacuum assist booster. The other is the hydraulically assisted booster.

Vacuum-Assisted Boosters

Most modern vehicles use a vacuum assisted power brake. This system was likely developed, since on all internal combustion engines, except diesels, vacuum is always available when the engine is operating, making the system is simple and efficient.

With diesel engines, vacuum is created and stored by way of a belt-driven vacuum pump and reservoir. In either case, the operation of the vacuum assist is the same.

A vacuum diaphragm is located on the front of the master cylinder and assists the driver in applying the brakes, reducing both the effort and travel one must put into moving the brake pedal. The vacuum diaphragm housing is normally connected to the intake manifold by a vacuum hose. A check valve is placed at the point where the hose enters the diaphragm housing, so that during periods of low manifold vacuum brake assist will not be lost.

Depressing the brake pedal closes off the vacuum source and allows atmospheric pressure to enter on one side of the diaphragm. This causes the master cylinder pistons to move and apply the brakes. When the brake pedal is released, vacuum is applied to both sides of the diaphragm and springs return the diaphragm and master cylinder pistons to the released position.

If the vacuum supply fails, the brake pedal rod will contact the end of the master cylinder actuator rod and the system will apply the brakes without any power assistance. The driver will notice that much higher pedal effort is needed to stop the vehicle and that the pedal feels harder than usual.

If you think this is the case you can check it as follows:

VACUUM LEAK TEST

1. Operate the engine at idle without touching the brake pedal for at least one minute.
2. Turn off the engine and wait one minute.
3. Test for the presence of assist vacuum by depressing the brake pedal and releasing it several times. If vacuum is present in the system, light application will produce less and less pedal travel. If there is no vacuum, air is leaking into the system.

SYSTEM OPERATION TEST

1. With the engine **OFF**, pump the brake pedal until the supply vacuum is entirely gone.
2. Apply light, steady pressure to the brake pedal.
3. Start the engine and let it idle. If the system is operating correctly, the brake pedal should fall toward the floor if constant pressure is maintained.

Power brake systems may be tested for hydraulic leaks just as ordinary systems are tested.

❄ WARNING

Clean, high quality brake fluid is essential to the safe and proper operation of the brake system. You should always buy the highest quality brake fluid that is available. If the brake fluid becomes contaminated, drain and flush the system, then refill the master cylinder with new fluid. Never reuse any brake fluid. Any brake fluid that is removed from the system should be discarded.

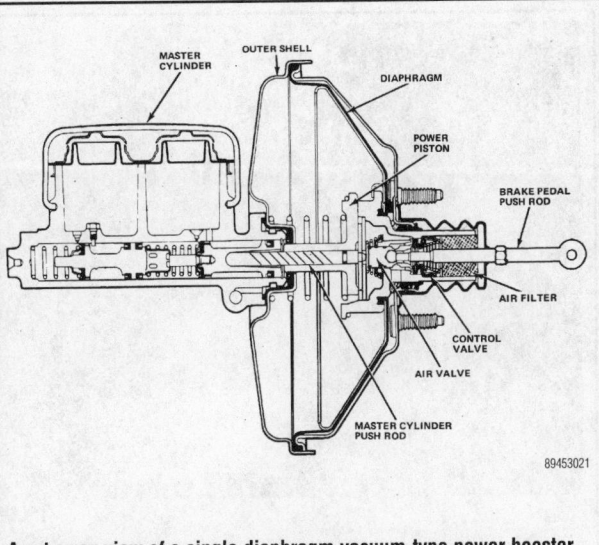

A cut away view of a single diaphragm vacuum-type power booster

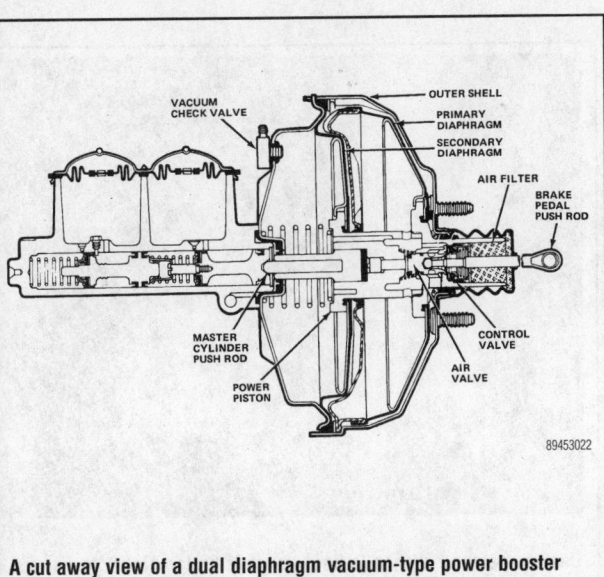

A cut away view of a dual diaphragm vacuum-type power booster

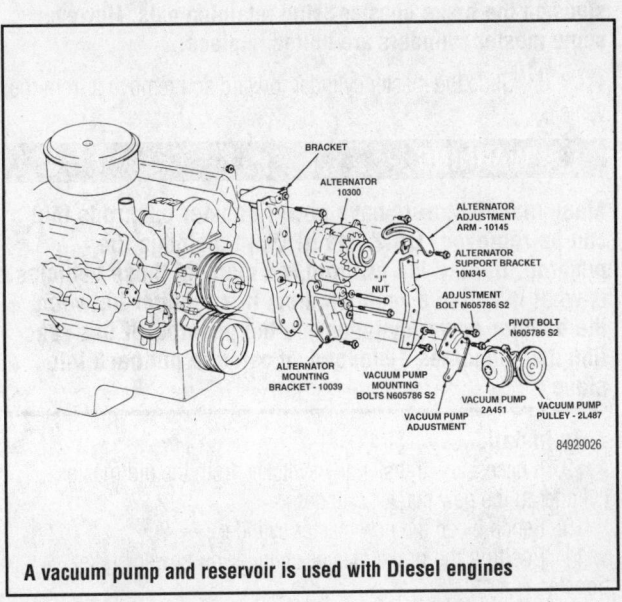

A vacuum pump and reservoir is used with Diesel engines

Hydraulically-Assisted Boosters

Used on some light vehicles, the unit is fed hydraulic fluid through the power steering system. The booster assembly, sometimes known generically by the brand name Hydro-Boost, contains a valve which controls pump pressure while braking, a lever to control the position of the valve and a boost piston to provide the force to operate the master cylinder attached to the front of the booster. The unit has a reserve system designed to store pressurized fluid to provide at least 2 brake applications in the event of hydraulic supply system failure, such as a broken power steering belt. The brakes can also be applied unassisted in the event of system depletion.

Master Cylinder

➡The following procedures apply to non-ABS systems and ABS system master cylinders that are separate from other ABS system components. ABS systems with integral master cylinder components often require special tools and model-specific procedures.

REMOVAL & INSTALLATION

With Power-Assisted Brakes

1. Disconnect the negative battery cable.
2. If applicable, apply the brake pedal several times to exhaust all vacuum from the power boost system.
3. Remove any components in the engine compartment which may interfere with master cylinder removal.
4. Disengage any electrical connectors from any switches mounted in the master cylinder.
5. Place absorbent rags under the points at which the brake pipes connect to the master cylinder.
6. Remove the brake lines from the primary and secondary outlet ports of the master cylinder. Cap or plug the lines to prevent fluid loss and contamination.
7. Remove the fasteners retaining the master cylinder to the power brake booster.

➡Most master cylinder assemblies are secured to mounting studs on the brake booster using retaining nuts. However, some master cylinders are bolted in place.

8. Slide the master cylinder forward and remove it from the vehicle.

❊❊ WARNING

Many manufacturers have power booster pushrods that can be removed. DON'T do it! Don't dislodge the pushrod. Behind the pushrod, on many of these vehicles, is what is called a reaction disc. It is a buffer between the booster power cylinder and the pushrod. If this reaction disc becomes dislodged, it can't be put back into place.

To install:
9. If necessary, transfer any switches from the old master cylinder to the new master cylinder.
10. Bench bleed the new master cylinder.
11. Position the brake master cylinder on power brake booster.

Disconnect any electrical connectors at . . .

. . . or near the master cylinder

Use an open-end wrench (a line wrench is preferable) to loosen the brake pipe fittings . . .

. . . the disconnect the pipes from the master cylinder assembly

Most master cylinders are secured to the brake booster using 2 retaining nuts

Loosen the master cylinder retainers . . .

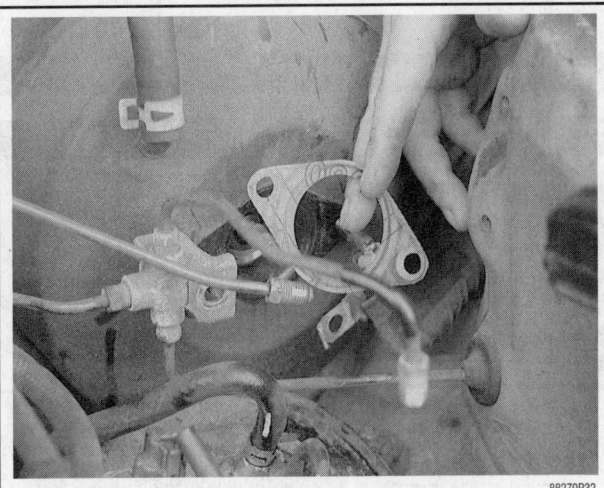

Some vehicles have gaskets between the master cylinder and booster

12. Install the retaining nuts or bolts and tighten them securely.

13. Install both the primary and secondary brake lines at the master cylinder.

14. When both brake lines are installed, tighten them securely.

15. Reattach any electrical connectors.

16. Fill the master cylinder with the proper brake fluid.

17. Bleed the brake system. Top off the master cylinder when complete.

18. Connect the negative battery cable.

19. Road test the vehicle and check for proper brake system operation.

Without Power-Assisted Brakes

1. Disconnect the master cylinder pushrod from the brake pedal linkage. This connection can be either inside the passenger compartment or inside the engine compartment. The connection is usually by way of a rod fitting over a stud on the pedal arm, retained by washers and a cotter pin or clip.

![photo caption] . . . then slide the master cylinder assembly from the mount

2. Place absorbent rags under the points at which the brake pipes connect to the master cylinder.

3. Remove the brake lines from the primary and secondary outlet ports of the master cylinder. Cap or plug the lines to prevent fluid loss and contamination.

4. Remove the fasteners retaining the master cylinder to the firewall.

5. Slide the master cylinder forward and remove it from the vehicle.

To install:

6. Bench bleed the new master cylinder.

7. Place the brake master cylinder onto the firewall.

8. Install the fasteners and tighten them securely.

9. Install both the primary and secondary brake lines at the master cylinder.

10. When both brake lines are installed, tighten them securely.

11. Fill the master cylinder with the proper brake fluid.

12. Bleed the brake system. Top off the master cylinder when complete.

13. Road test the vehicle and check for proper brake system operation.

Brake System Bleeding

✳✳ CAUTION

Brake fluid contains polyglycol ethers and polyglycols. Avoid contact with the eyes and wash your hands thoroughly after handling brake fluid. If you do get brake fluid in your eyes, flush your eyes with clean, running water for 15 minutes. If eye irritation persists, or if you have taken brake fluid internally, IMMEDIATELY seek medical assistance.

The hydraulic brake system must be bled any time any of the lines is disconnected or any time air enters the system. If a point in the system, such as a wheel cylinder or caliper brake line is the only point which was opened, the bleeder screws down stream in the hydraulic system are the only ones which must be bled. If however, the master cylinder fittings are opened, or if the reservoir level drops sufficiently that air is drawn into the system, air must be bled from the entire hydraulic system. If the brake pedal feels spongy upon application and travels almost to the floor but regains height when pumped, air has entered the system. It must be bled out. If no fittings were recently opened for service, check for leaks that would have allowed the entry of air and repair them before attempting to bleed the system.

As a general rule, once the master cylinder (and the brake pressure modulator valve or combination valve on ABS systems) is bled, the remainder of the hydraulic system should be bled in the proper sequence.

The hydraulic system can be bled in one of two ways: manual bleeding and bleeding using a pressure bleeder.

MODELS WITHOUT ABS

Manual Bleeding

MASTER CYLINDER

If the unit is removed from the vehicle, there are 2 ways to "bench-bleed" a master cylinder.

One method is with a large, clear plastic syringe made for the purpose. They are usually available at auto parts stores. In this procedure, the master cylinder is clamped in a soft-jawed vise and filled with fluid. The outlet ports are capped or plugged. Then, uncap each port, place the syringe securely in the outlet port and draw fluid into the syringe until no air is left in the master cylinder, capping the ports when done.

The other is with 2 lengths of hose or pipe (to use as bleeder tubes). Plastic hoses, made for the purpose, are available at most auto parts stores. These hoses have threaded ends for attachment to the outlet ports. Otherwise, you'll have to make your own bleeder pipes from 2 lengths of brake pipe equipped with threaded ends. Try to get the plastic ones. In this procedure, clamp the master cylinder in a soft-jawed vise. Connect the pieces of brake pipe or the plastic hoses to the outlet fittings, bend them until the free end is the master cylinder reservoir. Fill the reservoir with fresh DOT 3, or equivalent, brake fluid from a closed container, completely covering the tube ends. Pump the piston slowly until no more air bubbles appear in the reservoir. Remove the tubes, refill the brake master cylinder and securely install the caps or plugs in the ports.

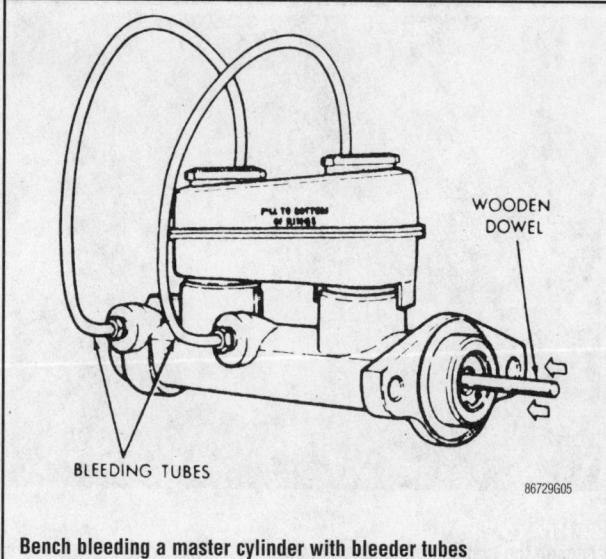

Bench bleeding a master cylinder with bleeder tubes

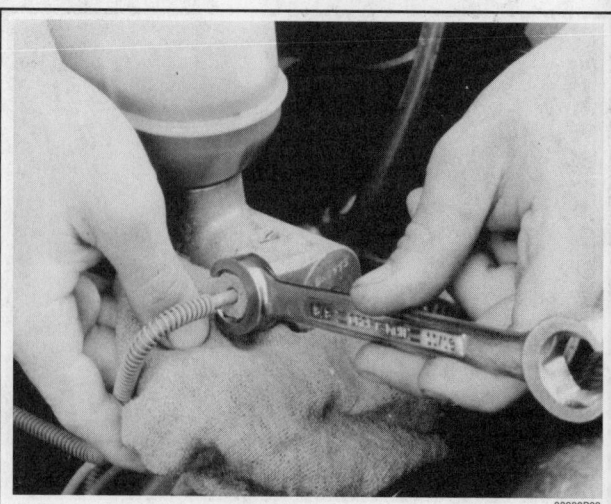

Bleeding the master cylinder by cracking open the fittings

If the brake master cylinder is on the vehicle, place a large, absorbent rag under the fittings. Open the brake lines slightly with the flare nut wrench while pressure is applied to the brake pedal by a helper inside the vehicle. Be sure to tighten the line before the brake pedal is released. Repeat the process with both lines until no air bubbles come out.

In both cases, the rest of the brake system must be bled to assure that all trapped has been removed and that the system will operate properly.

CALIPERS AND WHEEL CYLINDERS

We recommend that the brake system be bled using the jar and tube method. We know some people just let the fluid spray all over the place from the nipple. This is not only unprofessional, but it's messy and potentially dangerous. Brake fluid damages paint, concrete, your clothes, your skin, most importantly, your eyes.

➡ **Hydraulic brake systems must be totally flushed if the fluid becomes contaminated with water, dirt or other corrosive chemicals. Also, many manufacturers recommend that the system be flushed routinely, every 2 years or so. To flush, bleed the entire system until all fluid has been replaced and the new brake fluid runs clear.**

The hydraulic system on vehicles with a split system—a 2-chambered master cylinder—can be split either into front/rear or diagonally. In the diagonally split system there is one front and one rear component in each circuit. If you are in doubt as to the design of your vehicle's system, you can check the brake lines. Follow them to each wheel and see which are paired.

➡ **If, during the bleeding procedure, you can't get a good flow of fluid from the front brakes, the problem is with the metering part of the combination valve. Check the valve and you'll see a small stem sticking out of one end. You'll have to fabricate a little clip to hold the stem out as far as it will go. This will allow a full flow to the front brakes. Also, when using this clip on vehicles with power brakes, try bleeding with the engine running. The greater pressure allowed by the power booster will aid in purging the system.**

1. Fill the brake master cylinder with the fluid recommended for your vehicle. Check the level often during the procedure. Never let the master cylinder go dry or the procedure will have to be performed again.

2. Raise and support the vehicle safely.

3. If necessary for better access, remove the wheels.

4. On vehicles with a single chamber system or dual chambered systems split front/rear, you can bleed the system in the following order:
- Right rear
- Left rear
- Right front
- Left front

5. On vehicles with a dual chambered system split diagonally, the usual bleeding order is:
- Right rear
- Left front
- Left rear
- Right front

6. Find a wrench, a box wrench if possible, of the right size for the bleeder screw and place it on the nipple of the first cylinder to be bled.

7. Connect a clear, vinyl tube to the bleeder nipple. Place the other end of the tube in a clear glass jar of at least 8 oz. (237mL) capacity. The jar should be about ½ full of clean brake fluid. Submerge the end of the tube in the brake fluid.

8. Have an assistant pump the brake pedal, then hold it down. Slowly open the bleeder screw. When the brake pedal reaches the floor, close the bleeder and have the helper slowly release the pedal. Wait 15 seconds, then repeat the procedure until no more air comes out of the bleeder.

9. Repeat the procedure on the remaining calipers or wheel cylinders in the appropriate order.

10. If the brake pedal has a spongy feel, the brake system must be bled again to remove air still trapped in the system.

11. Install the bleeder caps to keep dirt out.

12. If removed for access, install the wheels.

13. Lower the vehicle.

14. Road test the vehicle and check for proper brake system operation.

Bleeding the calipers

Bleeding the wheel cylinders

Pressure Bleeding

A pressure bleeder is a device that uses compressed air and a series of adapters to forcibly expel air from the hydraulic system. When using a pressure bleeder, always follow the manufacturer's instructions. What we've given you here are general instructions.

When using pressure bleeding equipment, it's best to use a bladder-type bleeder tank. In this type of bleeder, the brake fluid is separated from the air by a rubber diaphragm. The bleeder tank must contain enough brake fluid to complete the bleeding operation and should be charged with only 10–30 psi (69–207 kPa). Never exceed 50 psi (345 kPa).

1. Clean all dirt from the master cylinder fluid reservoir filler cap.

➡**The reservoir must be at least ¾ -full during the bleeding procedure. Fill the reservoir as necessary. Use only clean, fresh brake fluid from a sealed container. Fill to the MAX level line on the reservoir.**

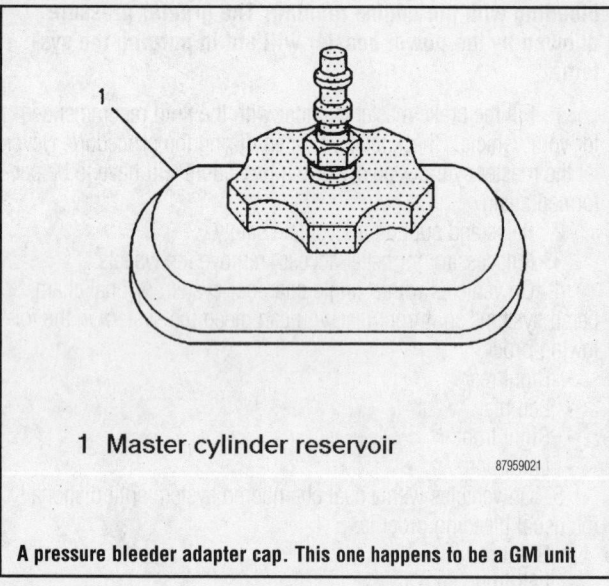

1 Master cylinder reservoir

87959021

A pressure bleeder adapter cap. This one happens to be a GM unit

2. Install the bleeder adapter tool on the master cylinder and attach the hose from the bleeder tank to the fitting on the adapter. Follow the manufacturer's instructions when installing and connecting the master cylinder adapter.

3. Open the valve on the bleeder tank.

MASTER CYLINDER

1. If the master cylinder is known or suspected to contain air, it must be bled before the wheel cylinders or calipers. Place a large, absorbent rag under the pipe fittings. Beginning at the front of the master cylinder, alternately loosen and tighten the brake line flare nuts. Allow the fluid to flow for several seconds before tightening the flare nut. Repeat this operation several times to be sure all air has been removed from the master cylinder.

CALIPERS AND WHEEL CYLINDERS

Pressure system bleeding must be performed in the correct order. Refer to the manual bleeding procedure for proper bleeding sequences.

1. Raise and safely support the vehicle.

2. Remove the protective bleeder screw cap from the caliper or wheel cylinder and clean the nipple.

3. Place a wrench, preferably a box end wrench, on the bleeder screw.

4. Attach a length of clear vinyl hose onto the bleeder nipple. The hose must fit tightly around the bleeder screw.

5. Submerge the free end of the hose in a large (approximately 16 oz./475mL) clean glass jar about half filled with clean brake fluid.

6. Loosen the bleeder screw approximately ¾ of a turn. When the fluid entering the jar is completely free of bubbles, tighten the bleeder screw.

7. Remove the bleeder hose and attach the protective screw cap.

8. Repeat the bleeding procedure at each brake.

9. Close the valve at the bleeder tank, disconnect the hose from the master cylinder adapter and remove the master cylinder adapter.

10. Check the fluid level in the remote reservoir, refilling with clean, fresh brake fluid, as necessary.

11. Check the brake pedal feel. If spongy, repeat the bleeding process and/or look for defective system components.

MODELS WITH ABS

There are 2 potential problems with attempting to bleed an ABS system. The first is that many use control valves and pressure modulators which might trap air if they are not opened and closed during the procedure using a scan tool. The second potential problem is that some ABS systems operate under extremely high pressure (making bleeding dangerous at worst or messy at best).

With this said, there are still many systems which can be bled with common tools. Many of the control valves have pressure relief knobs at one end of the valve which can be held open using a small tool (or pair of locking pliers)., just about all systems can be bled at the wheels provided that the openings are capped immediately during service. The caps keep enough fluid in the lines to prevent air from working its way back to the control or modulator valves.

Before starting, remember that many manufacturers require the use of special scan tools to bleed any part of the system other than the caliper or wheel cylinders. Some manufacturers recommend the scan tool be used when bleeding any part of the system on some of their models. All manufacturers recommend the use of pressure bleeding equipment for ABS systems, especially when bleeding the rear brakes even though manual bleeding can be done successfully in most cases.

If you decide to attempt bleeding the calipers or wheel cylinders, and you are sure that any residual high pressure is depleted, use the same procedure for bleeding as described for non-ABS systems. During the bleeding procedure, wait 10–15 seconds after closing the bleeder screw before reopening it each time. This is recommended by most manufacturers due to the number of valved components in the system.

Once the procedure is complete, start the engine and allow it to run for 15–30 seconds. Depress the brake pedal. The ABS light should not be **ON**. If the light is **ON**, there is a system problem, probably air still trapped somewhere. At this point, you can try the bleeding procedure again or have the vehicle towed to a dealer or repair shop for system bleeding.

As in all bleeding procedures, DO NOT attempt to move the vehicle unless a firm brake pedal feel has been obtained.

DISC BRAKES

Brake Pads

INSPECTION

To inspect the brake pads, remove the wheel. It is usually possible to view the pad thickness through a large hole in the caliper, or by looking at the side of the pad. However, on a few models, it may be necessary to remove the pads for inspection.

As a rule of thumb, the brake pad lining material should be worn no more than ⅛ in. (3mm). On brake pads glued to the backing material, the pad material can be measured from the edge of the backing material. However, on pads which are riveted to the backing material, the lining should be measured from the rivet heads (in the holes in the lining material)

The brake lining material should not exhibit any dampness, crumbling or cracking. If any such damage is evident the pads must be replaced. If the pads showed evidence of dampness, locate the source of the fluid leak and repair it before installing the new pads. If the brake pads exhibit uneven wear, (such as, one pair of pads is worn more on one side of the vehicle than the other pair of pads on the other side; the inner pad is worn more than the outer pad, or vice versa, on one wheel; the pad lining material is worn more on the front edge of a pad, or more on the rear edge of a pad) the disc brake caliper is either defective or mounted improperly.

※ WARNING

Never polish the pad lining with sandpaper, because hard particles from the sandpaper will become imbedded in the lining, which will damage the brake rotor. If the pad lining is damaged or worn excessively or unevenly, replace the pads with new ones.

REMOVAL & INSTALLATION

※ CAUTION

Brake dust may contain asbestos! Asbestos is harmful to your health. Never use compressed air to clean any brake component. A filtering mask should be worn during any brake repair.

Brake pad replacement should always be performed on both front or rear wheels at the same time. Never replace pads on only one wheel. When servicing any brakes use only OEM or better quality pads and parts. When the caliper is removed some brake pads stay with the caliper, others remain on the caliper mounting bracket. Use new pad mounting hardware (springs, anti-rattle clips, or shims) whenever possible to ensure a better repair.

Sliding and Floating Calipers

➡**On certain floating calipers it may be possible to remove one of the guide pins and pivot the caliper up or down to gain access to the brake pads. If you decide to do this, be sure that pivoting the caliper will not damage the flexible brake hose.**

1. Open the hood and locate the master brake cylinder fluid reservoir. Clean the area surrounding the reservoir cap, then remove the cap. Remove some of the brake fluid from the reservoir.

2. Loosen the lug nuts on the applicable wheels.
3. Raise and safely support the vehicle.
4. Remove the wheels.
5. Disconnect any electrical brake pad wear sensors.

➡**It is not necessary, and actually discouraged, to detach the brake hose from the caliper during this procedure. If you decide to detach the hose, it will be necessary for you to bleed your brake system.**

6. Remove and suspend the caliper with a piece of wire, cord or strong string. Be sure that it is not placing any stress on the brake hose.
7. For caliper bracket-mounted pads, perform the following:
   a. If present, remove any anti-squeal shims, noting their positions.
   b. Also, remove any anti-rattle springs that may be present. If these springs don't provide good tension, then replace them.

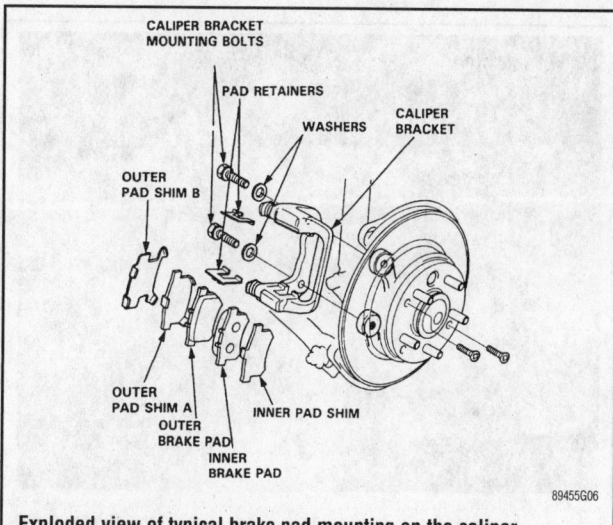

Exploded view of typical brake pad mounting on the caliper bracket—sliding and floating calipers

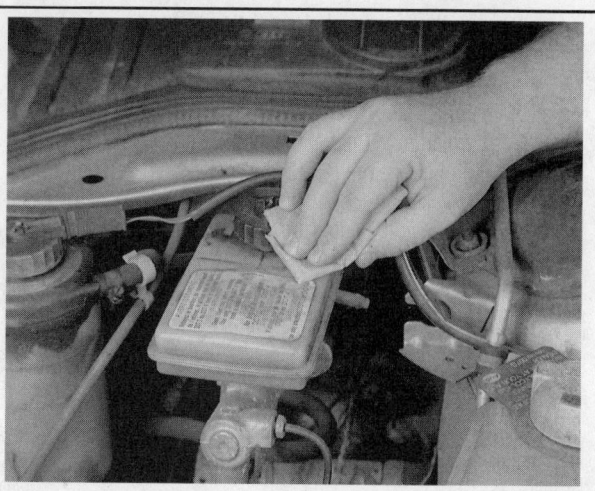

To remove the brake pads, first clean the brake master cylinder reservoir cap . . .

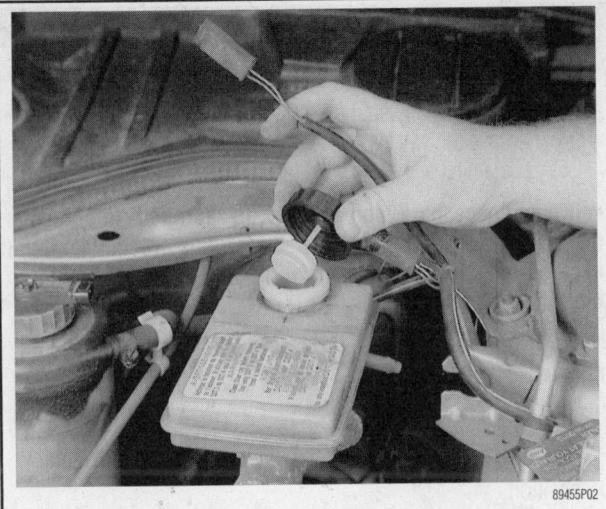

. . . then remove it

Remove the disc brake caliper from the rotor—sliding and floating calipers

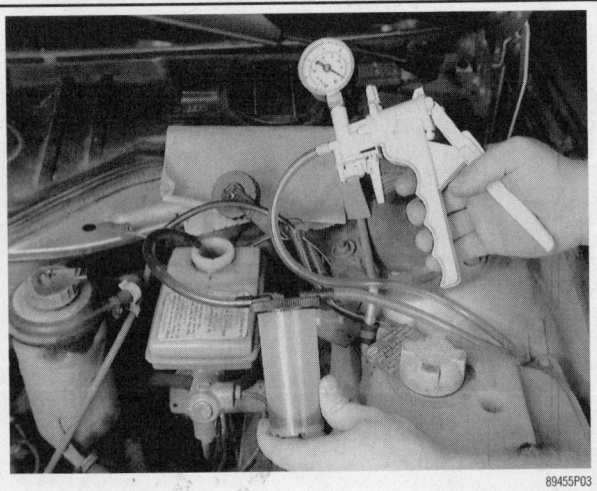

Using a vacuum pump, or some other method, remove some of the brake fluid from the reservoir

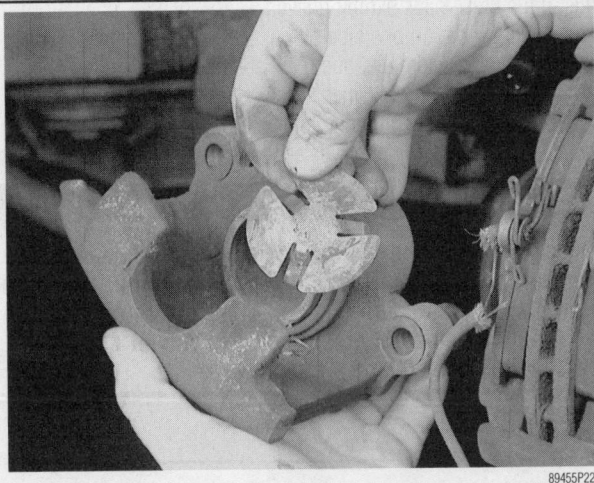

Be sure to note the positions of any clips or springs on the caliper—sliding and floating calipers

   c. Remove the brake pads from the caliper bracket by lifting the pad out by hand or with a slight tap of a hammer to help.

   8. For caliper mounted pads, perform the following:

   a. Some outer pads have tabs that are bent over the edge of the caliper, which hold the pads tight in the caliper. Straighten the tabs with pliers before trying to remove the brake pad from the caliper.

   b. Then, remove the outer brake pad with a slight tap to the back of the pad with a hammer.

   c. Other outer pads use a spring-clip to mount to the caliper. To remove this type of pad, press the pad towards the center of the caliper and slide it off. It may be helpful to use a small pry-bar.

   d. Remove the inner pad by pulling it out of the piston.

To install:

   9. Clean the caliper sliding area using a wire brush and spray brake cleaner.

   10. Lubricate the sliding area of the caliper and the pins with high temperature brake grease.

Remove the outboard pad from the mounting bracket . . .

... then remove the inboard pad—sliding and floating calipers

Apply a thin coat of high-temperature brake grease to the sliding surfaces of the bracket and caliper

11. Apply anti-squeal compound to the back side of both brake pads. Allow the compound to set-up according to the instructions on the package.

12. Install one of the old brake pads against the caliper piston, then use a large C-clamp to press the piston back into its bore.

13. Install any new hardware provided with the new pads.

14. For bracket-mounted pads, perform the following steps:

   a. Install the pads onto the caliper bracket. Some pads are marked for position.

   b. Be sure that the notches or ears of the brake pads are properly engaged on the bracket.

   c. Place the caliper over the pads and onto the caliper mounting bracket.

   d. Install the caliper mounting hardware and anti-rattle clips. Tighten the guide pins or lockbolt to the proper specification.

➡**It is a good idea to use some thread-locking compound (removable type) to the threaded fasteners of the caliper.**

15. For caliper mounted pads, perform the following:

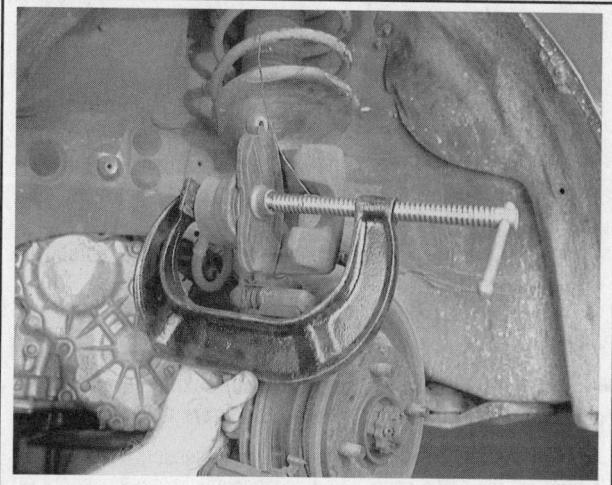

On calipers without integral parking brake mechanisms, a C-clamp can seat the piston in the caliper bore

Clean the caliper and mounting bracket with spray brake solvent and a wire brush

Install all of the springs and clips in their original positions—sliding and floating calipers

When installing the caliper and pads, make sure not to pinch the sensor wire (if equipped)—sliding and floating calipers

a. Install the inner pad by pushing the retaining fingers of the pad into the piston of the caliper.

b. If the outer pad has a spring-clip, slide the pad over the edge of the caliper into the caliper frame.

c. If you have the bent-tab style outer brake pad, then test fit the pad; it should fit tight. If the tabs do not secure the pad snugly in the caliper, place the pad on a piece of wood and tap the tab with a hammer to adjust it. It may take a few tries to get it right.

d. Place the caliper with the pads onto the rotor, if equipped, caliper bracket.

e. Install the caliper mounting hardware and anti-rattle clips. Tighten the guide pins or lockbolt(s) securely.

➡**It is a good idea to use some thread-locking compound (removable type) on the threaded fasteners of the caliper.**

16. Connect any electrical brake pad wear sensors.

17. Seat the brake pads, otherwise the vehicle may coast out of the work area and into traffic before the brakes become effective. It will take several pumps of the brake pedal to seat the pads against the rotor.

18. If a firm pedal is not achieved, it may be necessary to bleed the brakes.

19. Check the brake fluid level in the reservoir and top off as needed.

20. Install the wheels and tighten the lug nuts.

21. Road test the vehicle.

Fixed Calipers

➡**It is usually not necessary to remove the caliper to replace the brake pads on a fixed caliper.**

1. Loosen the lug nuts on the applicable wheels.

2. Raise and safely support the vehicle.

3. Remove the wheels.

4. Disconnect any electrical brake pad ware sensors.

5. Remove the pad retaining pins by pulling out the spring-clip or cotter pin, then use a punch and hammer to drive the pin out. Pins without a spring-clip or cotter pin, may be equipped with a spring steel collar on the head of the pin. To remove this style pin, just drive the pin out with a punch and hammer.

6. On calipers with hold-down clips, remove the bolt that holds the clip down.

7. Remove the pads from the caliper with a pair of pliers.

8. To seat the pistons of a fixed caliper, use a piece of wood or a prybar with a rag wrapped around the end, then wedge it between the rotor and the piston and slide the piston into its seat.

➡**It is helpful to replace one pad at a time, to reduce the risk of a piston coming out of its bore, which would lead to the caliper needing to be rebuilt.**

9. Lubricate the sliding area of the caliper and the brake pads with high temperature brake grease.

10. Apply anti-squeal compound to the back side of both brake pads. Allow the compound to set-up according to the instructions on the product.

11. Insert the new pads into the caliper.

12. If equipped, install the anti-rattle clip or retaining pin spring-clip or cotter pin. On pins with a spring steel collar, you must knock them in until seated against the shoulder in the caliper.

➡**It is a good idea to use some thread-locking compound (removable type) to the threaded fasteners of the caliper.**

13. Connect any electrical brake pad wear sensors.

14. Seat the brake pads, otherwise the vehicle may coast out of the work area and into traffic before the brakes become effective. It will take several pumps of the brake pedal to seat the pads against the rotor.

➡**If a firm pedal is not achieved, it may be necessary to bleed the brakes.**

15. Check the brake fluid level in the reservoir and top off as needed.

16. Install the wheels and tighten the lug nuts.

17. Road test the vehicle.

Brake Calipers

REMOVAL & INSTALLATION

Calipers without Integral Parking Brake Mechanisms

SLIDING CALIPERS

⁂ **CAUTION**

Brake dust may contain asbestos! Asbestos is harmful to your health. Never use compressed air to clean any brake component. A filtering mask should be worn during any brake repair.

There are typically three methods of securing a sliding caliper to its mounting bracket: with a retaining pin, with a key and bolt, or with a wedge and pin. On calipers which use the retaining pin method, you will find pins driven into the slot between the caliper and the caliper mount. On calipers which use the bolt and key method, a key (small piece of metal) is used between the caliper and the mounting bracket to allow the caliper to slide. The key is held in position by a lockbolt. On calipers which use the pin and wedge method, a wedge, retained by a pin, is used between the caliper and the mounting bracket in much the same manner as with the key and bolt method.

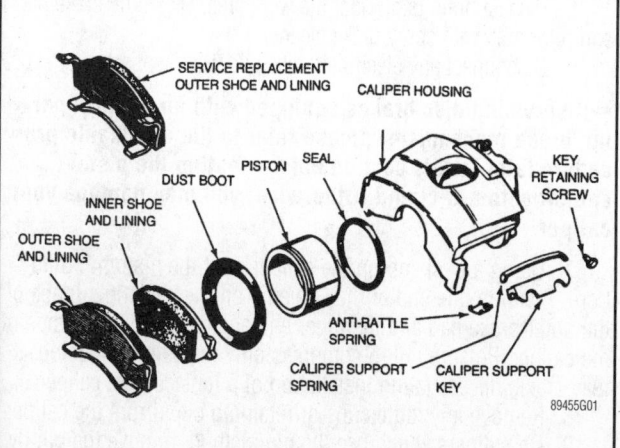

Exploded view of a typical sliding caliper, showing the key and bolt (retaining screw)

1. Loosen the lug nuts on the applicable wheels.
2. Raise and safely support the vehicle.
3. Remove the wheels.

※※ CAUTION

Any brake fluid that is removed from the system should be discarded. Also, do not allow any brake fluid to come in contact with a painted surface; it will damage the paint. Also, brake fluid contains polyglycol ethers and polyglycols. Avoid contact with the eyes and wash your hands thoroughly after handling brake fluid. If you do get brake fluid in your eyes, flush your eyes with clean, running water for 15 minutes. If eye irritation persists, or if you have taken brake fluid internally, IMMEDIATELY seek medical assistance.

4. Remove some brake fluid from the brake fluid reservoir. Use a clean suction pump, a turkey baster, or an absorbent pad to do so. Never reuse any brake fluid.

To remove a typical sliding caliper, remove the anti-rattle clips (if equipped)

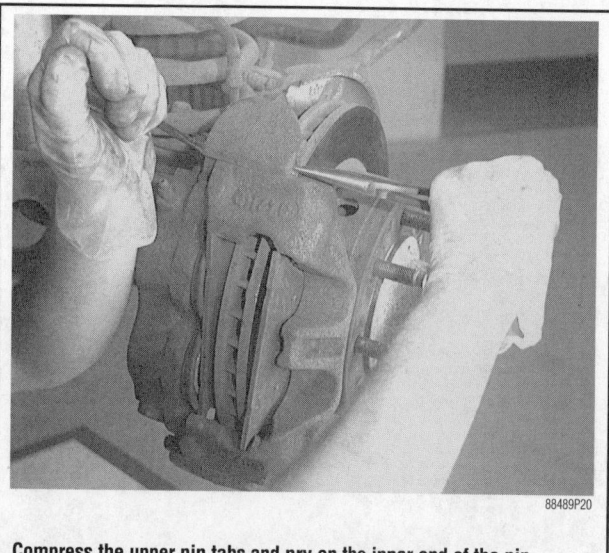

Compress the upper pin tabs and pry on the inner end of the pin . . .

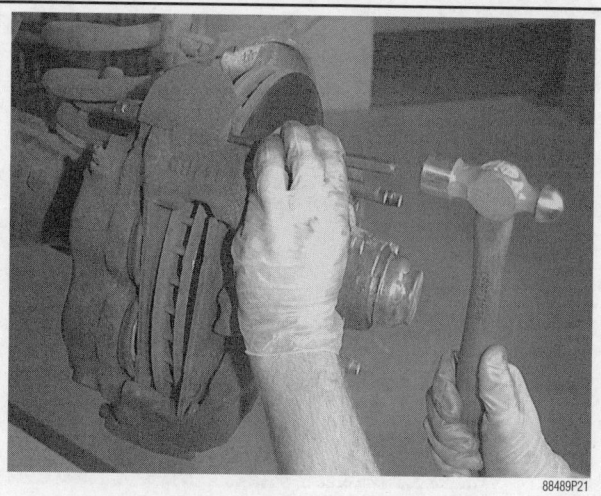

. . . then use a hammer and punch to drive the pin out of the groove . . .

. . . until it can be removed by hand

Perform the same for the lower pin as well . . .

. . . then pull the caliper off of the rotor and bracket

Once the caliper is removed, the brake pads can be removed

5. Place a drain pan under the work area. Clean the brake pad and rotor area with spray brake cleaner.

6. Disconnect any electrical brake pad wear sensor.

→**If servicing disc brakes equipped with an integral parking brake mechanism, please refer to the applicable procedure later in this section before seating the piston caliper with a C-clamp. Otherwise, you may damage your caliper.**

7. Using a C-clamp on the caliper, seat the piston into its bore. Position one end of the C-clamp on the backing surface of the outer brake pad and the other end against the inboard side of the caliper. Be sure not to compress only the caliper housing; it may crack, necessitating installation of a replacement caliper.

8. Remove any rattle clips or retaining clips from the caliper.

9. On calipers which use the pin method, remove the pin by squeezing the outboard end of the lower pin with a pair of pliers while prying out on the inboard end with a prybar. Once the pin retaining tabs are positioned in the caliper/bracket groove, use a punch and hammer to knock the lower pin the rest of the way

A vacuum pump setup can be used to draw brake fluid from the reservoir

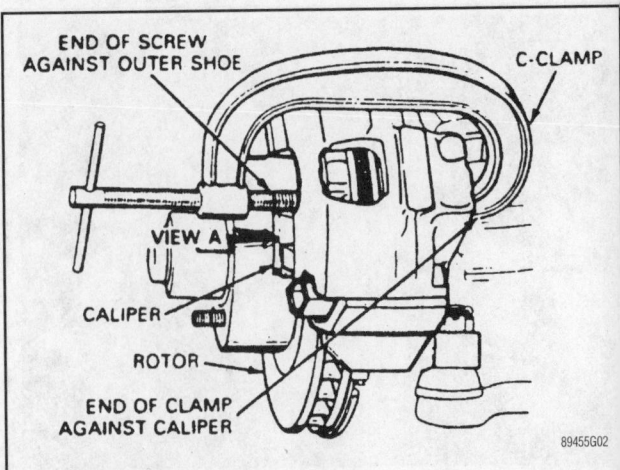

Use a large C-clamp to seat the caliper piston, make sure that one end of the clamp is positioned against the outer shoe—calipers without integral parking brake mechanisms

out of the groove. Repeat this step for the upper pin. Inspect the pins for damage, wear, and rust. Replace as needed in pairs.

10. On calipers which use the bolt and key method, remove the retaining bolt, then use a hammer and punch to drive the key out. (Be careful not to lose the caliper support spring, if equipped.) Check parts for wear and replace as necessary.

11. On calipers which use the wedge and pin method, remove the retaining pin from the guide plate, then use a punch and hammer to tap out the guide plate. Inspect parts for wear and replace as necessary.

12. If the caliper is going to be removed for overhaul or replacement, loosen the brake hose, lift off the caliper and remove the brake hose completely. Immediately plug the open end of the rubber brake hose to prevent contamination of the brake fluid. If the brake hose was attached to the caliper with a banjo connection, be sure to remove and discard the two copper washers.

13. If the caliper does not require overhaul or replacement, prepare a length of wire (a coat hanger works well), cord, or a length of strong string to support the caliper. DO NOT let the caliper hang from the brake hose; it may be damaged.

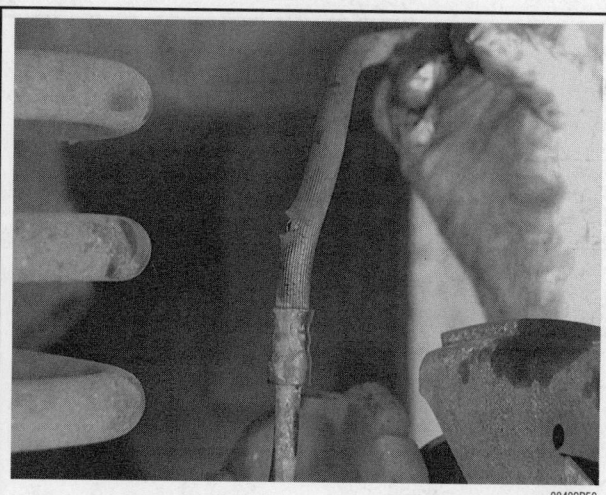

When inspecting the flexible brake hoses, check for rips (as shown), tears and cracks

14. Remove the caliper and suspend it from the wire.

15. If the brake pads came off the rotor with the caliper, remove them by prying the pads out of the caliper piston.

16. Inspect the caliper for fluid leakage, torn dust boots, or missing parts. Rebuild or replace the caliper if a problem is found.

17. Inspect the rubber brake hose for cracks or signs of rubbing against the body or steering components. Also, it is a good idea to replace them if they are over 10 years old to maintain proper brake operation.

18. Inspect metal lines for corrosion and kinks from road debris kicked up under the vehicle. If a problem is found, replace the line.

19. Inspect the rotor for non-machine grooves, heat stress cracks, glazing, minimum wear thickness, and disk run-out. Replace the rotor or have it machined to repair the damage.

20. Inspect the brake pads for minimum thickness, loose rivets, or glazing. Install new brake pads if any such problems exist.

To install:

21. Clean the sliding surfaces of the caliper and mounting bracket with spray brake cleaner and a small wire brush, then lubricate them with high temperature brake grease.

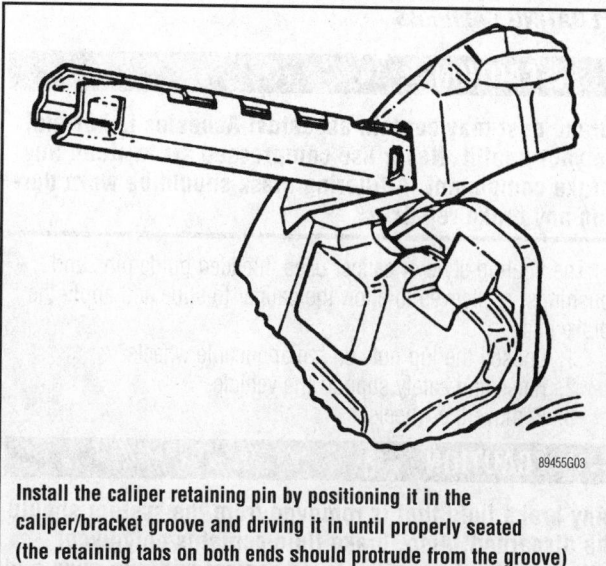

Install the caliper retaining pin by positioning it in the caliper/bracket groove and driving it in until properly seated (the retaining tabs on both ends should protrude from the groove)

22. If necessary, place the pad(s) back onto the caliper or mounting bracket.

23. If the brake hose was removed, reattach it to the caliper. If so equipped, use two new copper washers for the banjo fitting.

24. Install the caliper onto its mounting bracket.

25. For calipers which use the pin retaining method, use a hammer to tap the pins back into position, then install any anti-rattle clips.

26. For calipers which use the bolt and key method, use a pry-bar to lift the caliper up to create a gap into which the key and spring can slide. Tap the key and spring into position, then install the locking bolt and any anti-rattle clips. Tighten the locking bolt securely.

27. For calipers which use the wedge and pin method, slide the guide plates (wedge) between the gaps of the caliper and mounting bracket, then install the retaining pin. Tighten the retaining pin securely.

28. Reattach any electrical brake pad sensors.

✳✳ WARNING

Clean, high quality brake fluid is essential to the safe and proper operation of the brake system. You should always buy the highest quality brake fluid that is available. If the brake fluid becomes contaminated, drain and flush the system, then refill the master cylinder with new fluid. Never reuse any brake fluid. Any brake fluid that is removed from the system should be discarded. Also, do not allow any brake fluid to come in contact with a painted surface; it will damage the paint.

29. Bleed the brakes if a brake line was replaced, or the caliper was detached from a brake line.

30. Seat the brake pads, otherwise the vehicle may coast out of the work area and into traffic before the brakes become effective. It will take several pumps of the brake pedal to seat the pads against the rotor.

31. Check the brake fluid level in the reservoir and top off as needed.

32. Install the wheels and tighten the lug nuts.

33. Road test the vehicle.

FLOATING CALIPERS

> ✳✳ **CAUTION**
>
> Brake dust may contain asbestos! Asbestos is harmful to your health. Never use compressed air to clean any brake component. A filtering mask should be worn during any brake repair.

The floating style of caliper uses threaded guide pins and bushings, or sleeves to allow the caliper to slide and apply the brake pads.

1. Loosen the lug nuts on the applicable wheels.
2. Raise and safely support the vehicle.
3. Remove the wheels.

> ✳✳ **CAUTION**
>
> Any brake fluid that is removed from the system should be discarded. Also, brake fluid contains polyglycol ethers and polyglycols. Avoid contact with the eyes and wash your hands thoroughly after handling brake fluid. If you do get brake fluid in your eyes, flush your eyes with clean, running water for 15 minutes. If eye irritation persists, or if you have taken brake fluid internally, IMMEDIATELY seek medical assistance.

4. Remove some brake fluid from the brake fluid reservoir. Use a clean suction pump, a turkey baster (not to be returned to the kitchen), or an absorbent pad to do so. Never reuse any brake fluid.
5. Place a drain pan under the work area. Clean the brake pad and rotor area with spray brake cleaner.
6. Disconnect any electrical brake pad wear sensor.
7. If an anti-rattle spring is used and is not part of the brake pad, it can usually be pried off or pulled out.

➡ If servicing disc brakes equipped with an integral parking brake mechanism, please refer to the applicable procedure later in this section before seating the piston caliper with a C-clamp. Otherwise, you may damage the caliper.

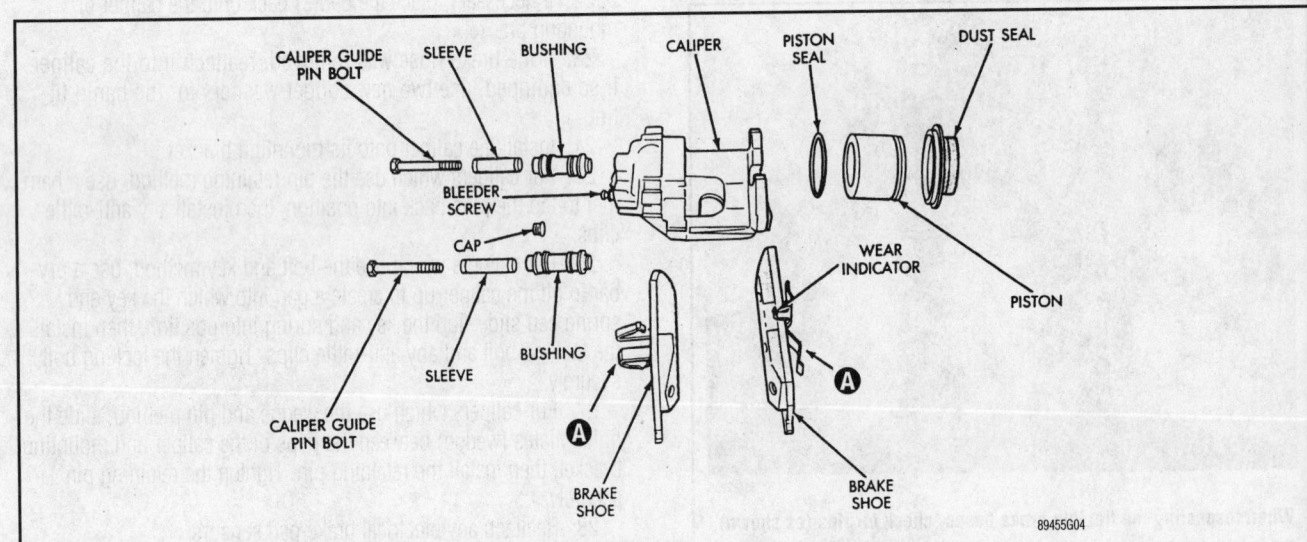

Exploded view of a typical floating caliper—when installing the brake pads, ensure that the retaining clips (A) are properly engaged in the caliper

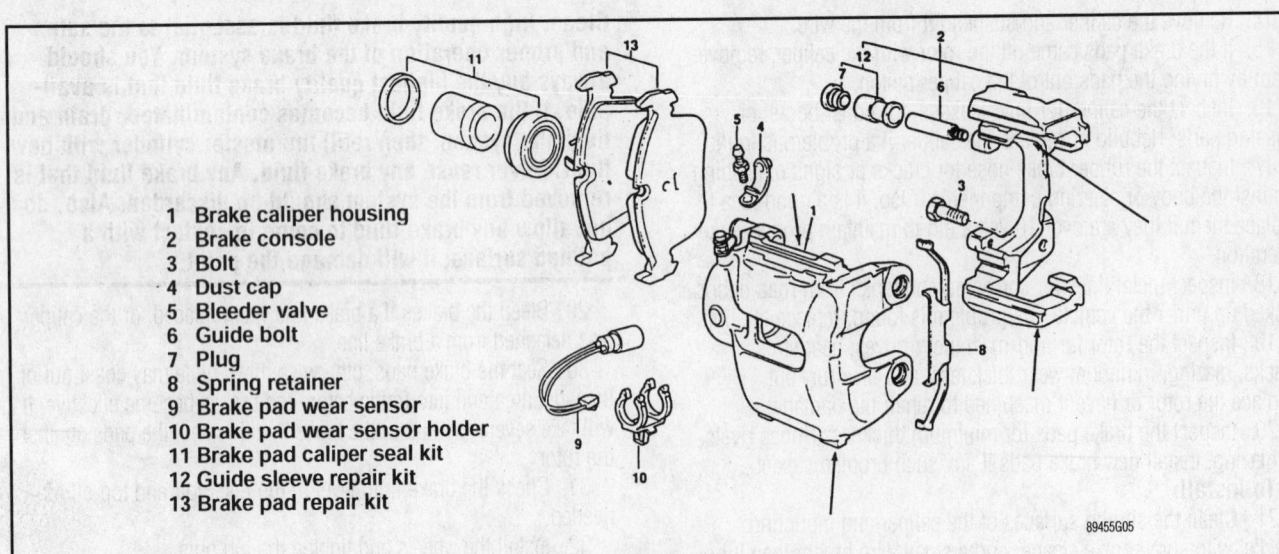

1. Brake caliper housing
2. Brake console
3. Bolt
4. Dust cap
5. Bleeder valve
6. Guide bolt
7. Plug
8. Spring retainer
9. Brake pad wear sensor
10. Brake pad wear sensor holder
11. Brake pad caliper seal kit
12. Guide sleeve repair kit
13. Brake pad repair kit

Exploded view of another floating caliper—note that this vehicle is equipped with a pad wear sensor

Some vehicles are equipped with brake pad wear sensors, indicated by the wire leading to the pad

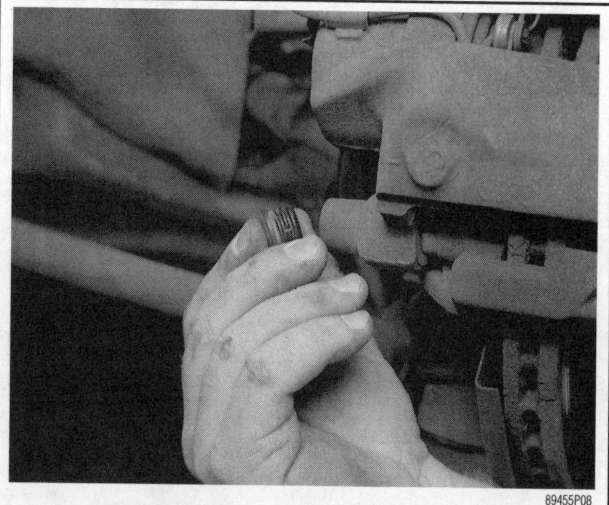

If equipped, remove the sliding pin covers . . .

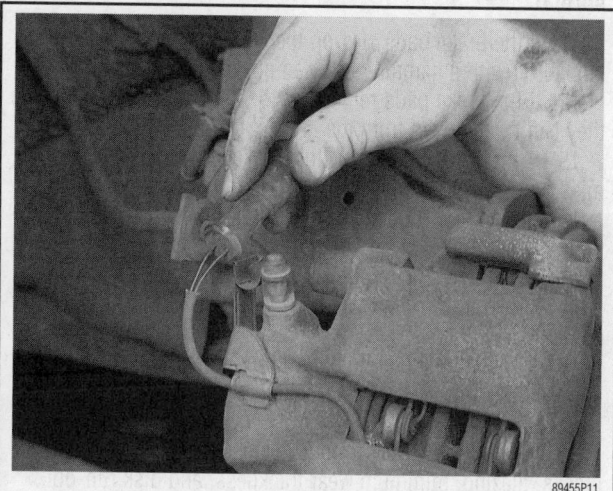

To remove a typical floating caliper, disengage the sensor wire connector from its mounting clip (if equipped) . . .

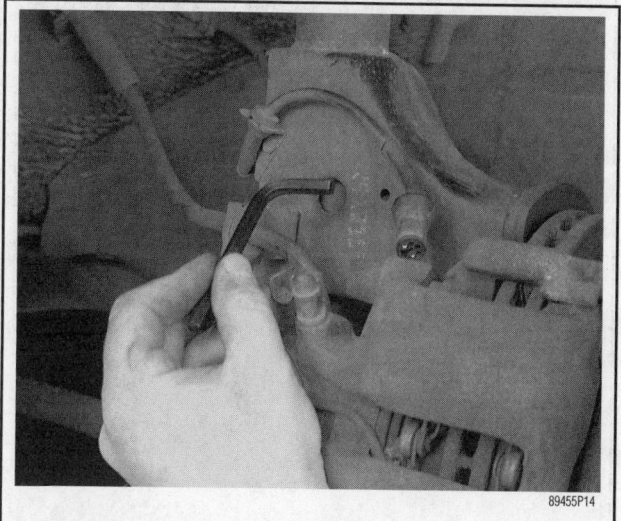

. . . then loosen . . .

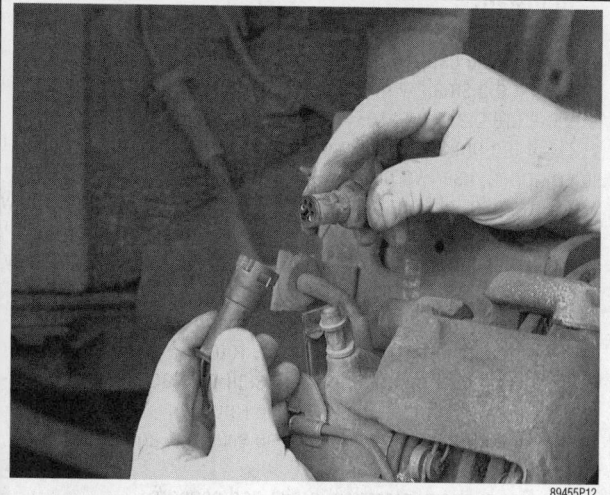

. . . then separate the two connector halves

. . . and remove the sliding pins from the floating caliper

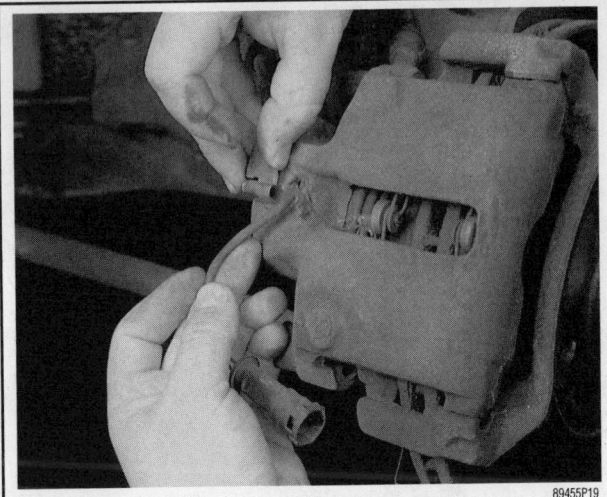

If equipped, unfasten the sensor wire from its retaining clip . . .

. . . then lift the caliper up and off of the mounting bracket

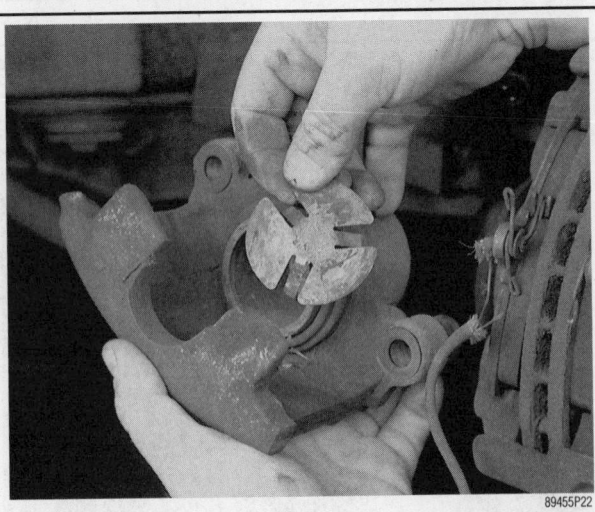

Be sure to note the positions of any clips or springs on the caliper

8. Using a C-clamp on the caliper, seat the piston into its bore. Position one end of the C-clamp on the backing surface of the outer brake pad and the other end against the inboard side of the caliper. Be sure not to compress only the caliper; it may crack, necessitating installation of a replacement caliper.

9. Loosen and remove the guide pins from the caliper.

10. If the caliper is going to be removed for overhaul or replacement, loosen the brake hose, lift off the caliper and remove the brake hose completely. Immediately plug the open end of the rubber brake hose to prevent contamination of the brake fluid. If the brake hose was attached to the caliper via a banjo connection, be sure to remove and discard the two copper washers.

11. If the caliper does not require overhaul or replacement, prepare a length of wire (a coat hanger works well), cord, or a length of strong string to support the caliper. DO NOT let the caliper hang from the brake hose; it may be damaged.

12. Remove the caliper from the rotor, if equipped, the mounting bracket.

➡The pads may or may not come off with the caliper; this is normal.

13. If the brake pads stay on the caliper, they can usually be tapped off with a hammer, or pried out by hand or with prytool.

14. If the brake pads remain on the bracket, when applicable, they can be removed from the bracket by hand.

15. Inspect the caliper for fluid leakage, torn dust boot, or missing parts. Rebuild or replace if a problem is found.

16. Inspect the rubber brake hose for cracks or signs of rubbing against the body or steering components. Install a new rubber hose if any such conditions exist. Also, it is a good idea to replace them if over 10 years old to maintain proper brake operation.

17. Inspect the metal lines for corrosion and kinks from road debris kicked up under the vehicle. If a problem is found, replace the line.

18. Inspect the rotor for non-machine grooves, heat stress cracks, glazing, minimum wear thickness, and disk run-out. Replace the rotor or have it machined to repair the damage.

19. Inspect the brake pads for minimum thickness, loose rivets, or glazing. If such a problem is found, new pads must be installed.

To install:

20. If equipped with a mounting bracket, clean the sliding surfaces of the caliper and mounting bracket with spray brake cleaner and a small wire brush, then lubricate them with high temperature brake grease.

21. If the brake hose was removed, reattach it to the caliper. so equipped, use two new copper washers for the banjo fitting.

22. Transfer old pad hardware to the new pads, or install new hardware.

23. Clean and inspect the caliper guide pins, if they are okay, then lubricate them with high temperature brake grease.

24. On caliper mounted pads, position the pads on the caliper, then install the caliper on the rotor.

25. On bracket mounted pads, install the pads on the mounting bracket. Install the caliper on the rotor.

26. Tighten the caliper guide pins securely, and replace any anti rattle clips.

27. Connect any electrical brake pad sensors.

✳✳ WARNING

Clean, high quality brake fluid is essential to the safe and proper operation of the brake system. You should always buy the highest quality brake fluid that is available. If the brake fluid becomes contaminated, drain and flush the system, then refill the master cylinder with new fluid. Never reuse any brake fluid. Any brake fluid that is removed from the system should be discarded. Also, do not allow any brake fluid to come in contact with a painted surface; it will damage the paint.

28. Bleed the brakes, if a brake line was replaced or the caliper was detached from a brake line.

29. Seat the brake pads, otherwise the vehicle may coast out of the work area and into traffic before the brakes become effective. It will take several pumps of the brake pedal to seat the pads against the rotor.

30. Check the brake fluid level in the reservoir and top off as needed.

31. Install the wheels and tighten the lug nuts.

32. Road test the vehicle.

FIXED CALIPERS

✳✳ CAUTION

Brake dust may contain asbestos! Asbestos is harmful to your health. Never use compressed air to clean any brake component. A filtering mask should be worn during any brake repair.

The fixed type caliper is bolted to the steering knuckle. The brake pads on this style of caliper are typically held in place by one or two retaining pins. Some other pads use hold down clips. It may not be necessary to remove the brake pads in order to remove the caliper.

1. Loosen the lug nuts on the applicable wheels.
2. Raise and safely support the vehicle.

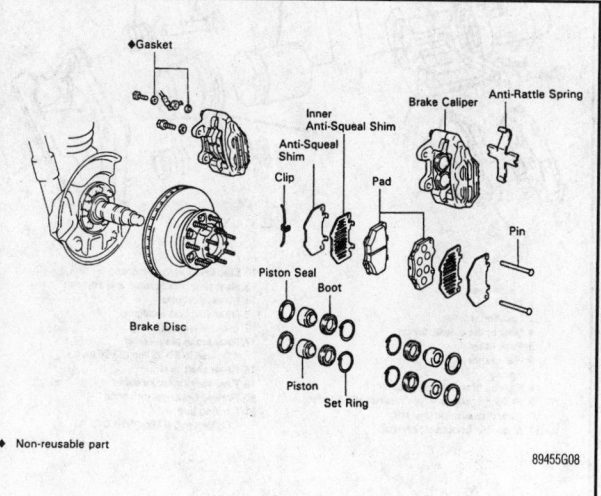

◆ Non-reusable part

89455G08

Exploded view of a common four piston fixed caliper

✳✳ WARNING

Any brake fluid that is removed from the system should be discarded. Also, do not allow any brake fluid to come in contact with a painted surface; it will damage the paint.

3. Remove the wheels.

✳✳ CAUTION

Brake fluid contains polyglycol ethers and polyglycols. Avoid contact with the eyes and wash your hands thoroughly after handling brake fluid. If you do get brake fluid in your eyes, flush your eyes with clean, running water for 15 minutes. If eye irritation persists, or if you have taken brake fluid internally, IMMEDIATELY seek medical assistance.

4. Remove some brake fluid from the brake fluid reservoir. Use a clean suction pump, a turkey baster, or an absorbent pad to do so. Never reuse any brake fluid.

5. Place a drain pan under the work area. Clean the brake pad and rotor area with spray brake cleaner.

6. If equipped, disconnect any electrical brake pad wear sensor.

7. Although not necessary for caliper removal, the brake pads can now be removed from the caliper.

8. Loosen the caliper mounting bolts.

9. If the caliper is going to be removed for overhaul or replacement purposes, loosen the brake hose, remove the caliper bolts, and disconnect the brake line.

10. If the caliper does not require overhaul or replacement (in other words, you only need to remove it for access to some other component), prepare a length of wire (coat hanger), cord, or a length of strong string from which the caliper can be hung. DO NOT let the caliper hang from the brake hose; it may be damaged and need to be replaced. Remove the caliper and hang it from the wire.

11. Inspect the caliper for fluid leakage, torn dust boot, or missing parts. Rebuild or replace if a problem is found.

12. Inspect the rubber brake hose for cracks or signs of rubbing against the body or steering components. Install a new brake hose if any such damage is evident. Also, it is a good idea to replace them if over 10 years old to maintain proper brake operation.

13. Inspect the metal brake lines for corrosion and kinks from road debris kicked up under the vehicle. If a problem is found replace the line.

14. Inspect the rotor for non-machine grooves, heat stress cracks, glazing, minimum wear thickness, and disk run-out. Replace the rotor or have it machined to repair the damage.

15. Inspect the brake pads for minimum thickness, loose rivets, or glazing. If any such problem is found, new pads must be installed.

To install:

16. Install the caliper and tighten the mounting bolts securely.

17. If the brake hose was removed, reattach it to the caliper. If so equipped, use two new copper washers for the banjo fitting.

18. If removed, install the brake pads.

19. Reconnect any electrical brake pad sensors.

20. Bleed the brakes, if a brake line was replaced or the caliper was detached from a brake line.

21. Seat the brake pads, otherwise the vehicle may coast out of the work area and into traffic before the brakes become effective. It will take several pumps of the brake pedal to seat the pads against the rotor.

22. Check the brake fluid level in the reservoir and top off as needed.

23. Install the wheels and tighten the lug nuts.

24. Road test the vehicle.

Calipers with Integral Parking Brake Mechanisms

The procedure to remove or replace the caliper and/or pads on vehicles equipped with disc brakes designed with integral parking brake mechanisms is essentially the same as disc brake calipers without integral parking brakes. There are usually two major differences between these two disc brake caliper designs.

➡ **For the actual caliper removal and installation process, refer to the applicable procedure earlier in this section. Read the following two procedures, and perform them in conjunction with the caliper procedures.**

REMOVING THE PARKING BRAKE CABLE

The first, and most obvious, difference is that, in one fashion or another, the parking brake cable is attached to the caliper. Before removing the caliper from the rotor, you must first disengage the parking brake cable from the caliper. To detach the parking brake cable from the caliper, perform the following:

➡ **This is a general procedure and may need slight alteration to apply fully to your specific vehicle. The most important thing to remember is to carefully inspect your caliper to identify the applicable parking brake cable components before disconnecting anything.**

1. Loosen the lug nuts on the applicable wheels.
2. Raise and safely support the vehicle.

➡ **Some vehicles, in fact, may be designed with front parking brake assemblies.**

3. Remove the wheels for easier access to the brake assembly.

4. Relieve the parking brake cable tension.

5. Carefully inspect the parking brake cable mounting and attaching (to the caliper) points. Most parking brake cable conduits are retained to a mounting bracket either by a jam nut and locknut setup, or by a retaining clip. Either remove the jam and locknuts, or pull the retaining clip off of the bracket, then disengage the cable conduit from the mounting bracket. If your vehicle utilizes jam and locknuts to secure the conduit onto the bracket, matchmark the

nuts' locations on the cable conduit threads for reinstallation; if marking the threads is not possible, measure (and note the measurements) from the end of the cable conduit to the jam nut and to the locknut.

➡ **With the conduit detached from its mounting bracket, there should be enough slack to disengage the parking brake cable end from the caliper lever, or similar linkage. On some models, there may be a cable end retaining fastener (clip, bolt, etc.), which must be removed before the cable can be detached from the caliper.**

6. Detach the parking brake cable end from the caliper lever, or linkage. Often, the cable end must be twisted up and around (or some similar manipulation) to disengage it from the caliper lever.

7. Remove the caliper, as described earlier in this section. Be sure to read the following procedure on seating the caliper piston before commencing with the caliper removal procedure.

To install:

8. After installing the brake caliper, as described earlier in this section, reattach the parking brake cable end to the caliper lever. If equipped, install the cable end securing fastener.

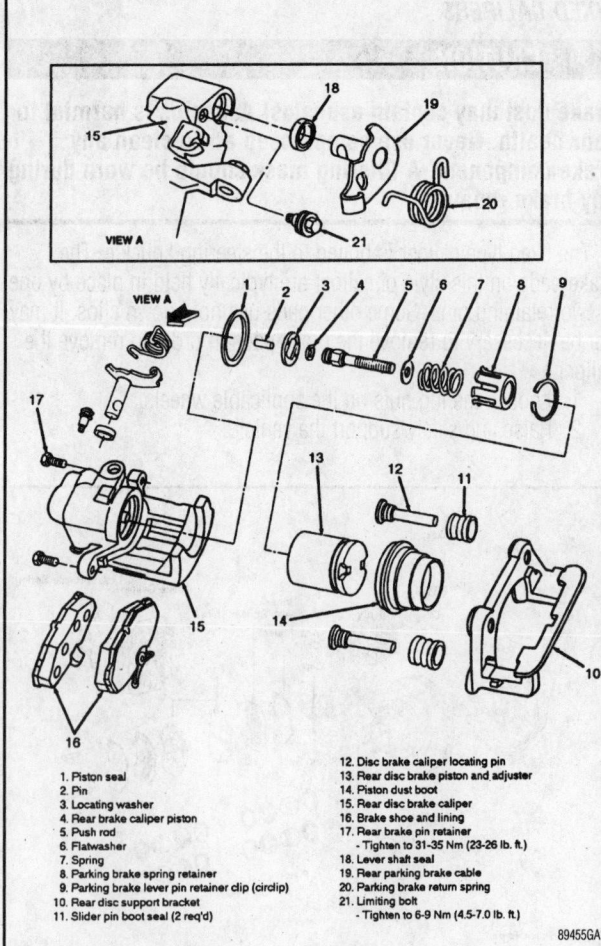

1. Piston seal
2. Pin
3. Locating washer
4. Rear brake caliper piston
5. Push rod
6. Flatwasher
7. Spring
8. Parking brake spring retainer
9. Parking brake lever pin retainer clip (circlip)
10. Rear disc support bracket
11. Slider pin boot seal (2 req'd)
12. Disc brake caliper locating pin
13. Rear disc brake piston and adjuster
14. Piston dust boot
15. Rear disc brake caliper
16. Brake shoe and lining
17. Rear brake pin retainer
   - Tighten to 31-35 Nm (23-26 lb. ft.)
18. Lever shaft seal
19. Rear parking brake cable
20. Parking brake return spring
21. Limiting bolt
   - Tighten to 6-9 Nm (4.5-7.0 lb. ft.)

89455GA1

Exploded view of a typical rear brake caliper with integral parking brake—note the wedge-shaped notches on the face of the piston (13)

9. Position the cable conduit in the mounting bracket, then either install the retaining clip, or the jam and locknuts. If equipped with jam and locknuts, position the nuts on the cable conduit so that the nuts are positioned as before (using the marks on the threads or a ruler).

10. Adjust the parking brake cable tension, as described in Section 3.

11. Install the wheels and snug the lug nuts.

12. Lower the vehicle.

13. Tighten the lug nuts fully.

14. Depress the brake pedal a few times to ensure that the brake pads are fully seated.

✳✳ CAUTION

If you do not seat the pads before driving the vehicle, the first few times you apply the brake pedal the vehicle may not stop as anticipated; this could lead to an accident with a telephone pole or one of your neighbors' cars.

SEATING THE CALIPER PISTON

Be sure to read this entirely before commencing with caliper service.

The second difference between brake calipers with and without integral parking brake mechanisms is in how the caliper pistons should be seated into their bores.

Whereas most pistons on calipers which are not equipped with integral parking brake mechanisms can be seated by using a large C-clamp, this is USUALLY not the case with calipers designed with integral parking brake mechanisms. Most integral parking brake calipers apply parking brake pressure to the rotor as follows: when the parking brake is applied, the cable pulls on the caliper lever. The lever, in turn, applies a rotational (spinning) movement to the caliper piston. The piston is designed much like an ordinary screw, so that when a rotational movement is applied to the piston, it slowly presses in against the rotor. To prevent having to constantly adjust the parking brake cable tension as the brake pads slowly wear down, the internal parking brake mechanism is designed with a ratcheting apparatus, which automatically readjusts the parking brake tension.

Since the caliper is designed to protrude from its bore when turned, usually, it cannot be seated in its bore in a conventional manner (with a large C-clamp).

✳✳ WARNING

On most of these calipers, if you use a C-clamp, or similar method, to seat the piston in its bore, you will damage the caliper beyond use. A new caliper will have to be purchased.

To seat the piston in the caliper, a spanner wrench or other model-specific tool must be used to turn the piston back into its bore. However (and to complicate things), a few of the integral parking brake calipers utilize an internal cam and/or lever type device that applies parking brake pressure to the rotor by pushing the caliper piston outward rather than turning it. On these uncommon type of calipers, you use a C-clamp to seat the piston into the caliper bore, just like the non-integral parking brake calipers. Unfortunately, the only way to tell which style of caliper you have is to remove it and inspect it.

✳✳ WARNING

When removing a caliper equipped with an integral parking brake mechanism, DO NOT seat the pads with a C-clamp.

Once the caliper and pads are removed, examine the caliper piston to determine how the piston is to be seated back into the caliper bore. All pistons which are rotated into the caliper will have some type of notch, slot or hexagonal depression or protrusion on its face, to which a tool can be attached and rotational force applied. To determine in which direction the piston must be rotated, SLOWLY turn the piston in one direction, and watch the piston's movement. Ensure that the piston moves inward in the bore. If the piston moves outward, reverse the direction of rotation and fully seat the piston. If the piston does not seem to be moving in or out, apply slight inward pressure by hand and continue turning the piston. Some models may have an adjuster or lockbolt on the back of the caliper which must be loosened or removed in order for the piston to rotate in.

➡**On some vehicles, namely some GM models, the pistons in the calipers on both sides of the vehicle must be turned in opposite directions. That means that, if the right-hand caliper piston must be turned clockwise, then the left-hand caliper piston must be turned counterclockwise (this is ONLY an example).**

If the piston does not seem to move in or out while rotating, if it moves in while rotating it in BOTH directions, or if there is no visible depressions or protrusions to which a tool could be attached, you may have a press-in style of caliper. Place an old brake pad against the piston face and install a C-clamp on the caliper. Slowly, and gently, press the piston into the caliper. If the piston does not move inward, DO NOT force it! Damage to the caliper can occur.

Once the caliper piston is fully seated in its bore, install the caliper (depending on its type: sliding, floating or fixed) as described earlier in this section.

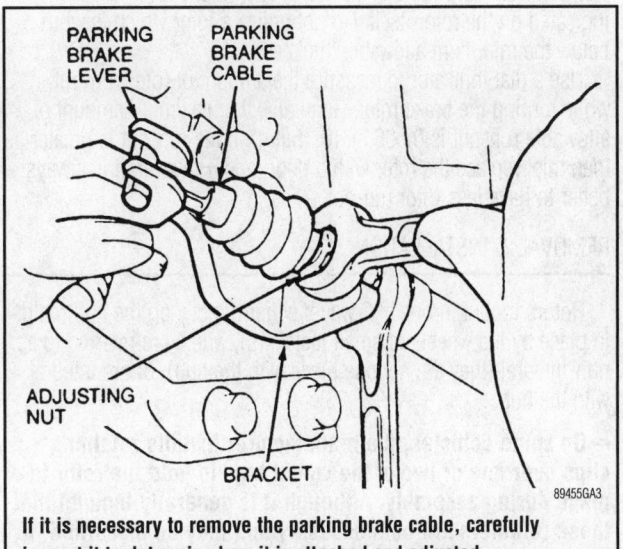

If it is necessary to remove the parking brake cable, carefully inspect it to determine how it is attached and adjusted

Brake Rotors

INSPECTION

To inspect the brake rotor, remove the caliper (without disconnecting the flexible brake hose) and the pads. The rotor should be machined or replaced with a new one, if it exhibits any of the following conditions:

- Bluing or excessive discoloration due to heat
- Cracks, or missing chunks
- Excessive scoring (run your fingernail over the rotor—if it snags any of the scores, it should be machined)
- Excessive run-out

Glaze on the rotor can be removed by hand-sanding it with medium grit garnet paper or aluminum oxide sandpaper.

Use a micrometer to measure the rotor thickness, and replace it if it is below specifications

Use a micrometer to measure the thickness of the brake rotor. The minimum allowable thickness of each brake rotor is usually indicated on the rotor itself. Do not utilize a rotor which is worn below the minimum allowable thickness

Use a dial indicator to measure the amount of rotor run-out, while turning the brake rotor. Generally, the maximum amount of allowable run-out is 0.006 in. (0.15mm); if the run-out is greater than this, replace the rotor with a good one. However, it is always better to have less rotor run-out.

REMOVAL & INSTALLATION

Rotors mount in one of 2 ways: either directly on the hub (held in place by the wheels or small fasteners), which are referred to as non-integral (they are not one piece with the hub), or are integral with the hub.

➡**On some vehicles, the manufacturer installs retaining clips over one or two of the wheel lugs to hold the rotor in place during assembly. Although it is generally thought that these retainers are not necessary and may be discarded, it** is a good idea to reinstall them anyway (better safe than sorry). Other manufacturers use one or two small machine screws to hold the rotor in place on the hub; these screws MUST be reinstalled.

Non-Integral Rotors

1. Loosen the lug nuts on the applicable wheels.
2. Raise and safely support the vehicle.
3. Remove the wheels.
4. Clean the brake assembly thoroughly with spray brake cleaner.
5. Remove the caliper.
6. If any rotor retainers are present, remove them. The push-nut type of retainer is usually damaged during removal; discard the old ones and purchase new ones.

To remove the rotor, remove the disc brake caliper . . .

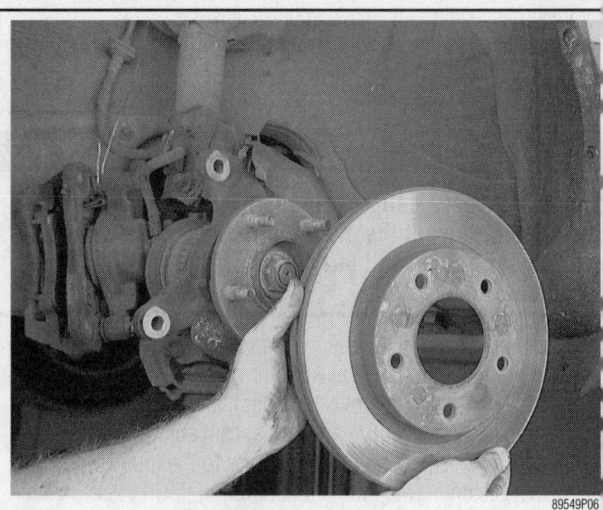

. . . then pull the rotor off by hand—non-integral rotors

If the rotor is equipped with holes and is difficult to remove, it can be loosened using two small bolts . . .

. . . then removed by hand—non-integral rotors

7. Remove the rotor. On some vehicles, the rotor simply slides off the wheel studs. However, some rotors are pressed into place and must be removed by screwing bolts in the threaded holes provided, thereby forcing the rotor off the hub. Other rotors, not equipped with the threaded holes for press-off bolts, may require the use of a puller to dislodge them from the hub.

➡The rotor may be rusted in place. Spray the area liberally with WD-40®, Liquid Wrench® or equivalent and tap the rotor loose.

To install:

➡New rotors come with an oily, rust-preventive coating on the braking surface. This coating can be removed with brake parts cleaner or most cleaners which are good for oil removal. Be sure that all traces of the coating are removed. Allow the rotor to dry before installation.

8. Position the rotor on the hub and install any retainers.
9. Install the caliper.
10. Install the wheels.
11. Lower the vehicle.

12. Seat the brake pads, otherwise the vehicle may coast out of the work area and into traffic before the brakes become effective. It will take several pumps of the brake pedal to seat the pads against the rotor.
13. Check the brake system for proper operation.

Integral Rotor/Hub Assemblies

EXCEPT 4WD VEHICLE FRONT ROTORS—NON-SEALED HUB/BEARING ASSEMBLIES

1. Loosen the lug nuts on the applicable wheels.
2. Raise and safely support the vehicle.
3. Remove the wheels.
4. Clean the brake assembly thoroughly with spray brake cleaner.
5. Remove the caliper and suspend it out of the way with wire.
6. Remove the hub grease cap.
7. Remove the cotter pin and wheel bearing nut locking cap. Discard the cotter pin.
8. Remove the wheel bearing nut.

Pry loose the grease cap . . .

. . . then remove it from the rotor hub

Remove the cotter pin . . .

. . . then pull the outer bearing out of the hub

. . . then pull the locking cap off of the end of the axle shaft

Remove the hub/rotor assembly—integral non-sealed hub/bearing (except 4WD vehicle front rotors)

Remove the adjusting nut . . .

➥On some vehicles, a left-hand threaded nut is used on the right wheel spindle. Turn this locknut clockwise to loosen.

9. Remove the brake rotor/hub, washer and bearings as an assembly. Be careful not to let the outer wheel bearing fall out of the hub during removal.

10. If the brake rotor is to be machined or replaced, remove the wheel bearings and grease seal.

To install:

➥New rotors come with an oily, rust-preventive coating on the braking surface. This coating can be removed with brake parts cleaner or most cleaners which are good for oil removal. Be sure that all traces of the coating are removed. Allow the rotor to dry before installation.

11. If removed, install the inner wheel bearing and a new grease seal.

12. Be sure the bearings and hub contain an adequate amount of clean wheel bearing grease.

13. Position the rotor/hub assembly on the spindle. Keep the hub centered on the spindle to prevent damage to the grease seal and spindle threads.

14. Install the outer wheel bearing, washer and wheel bearing nut.

15. Properly adjust the wheel bearing. On most vehicles (those with tapered roller bearing) this is done by tightening the adjusting nut until drag is felt on the bearing while rotating the rotor, then, back off the nut about ¼ turn (90°). The rotor/hub should turn freely with no end-play. If you are in any doubt about the proper adjustment procedure, refer to the appropriate model-specific section in this manual.

➡**On some vehicles (those with ball bearings), the nut is not so much an adjuster as a locknut. Tighten this nut to the manufacturer's specifications.**

16. Install the wheel bearing nut cover and a new cotter pin.
17. Install the caliper.
18. Install the wheels and snug the lug nuts.
19. Lower the vehicle.
20. Tighten the lug nuts fully.
21. Seat the brake pads, otherwise the vehicle may coast out of the work area and into traffic before the brakes become effective. It will take several pumps of the brake pedal to seat the pads against the rotor.
22. Check the brake system for proper operation.

EXCEPT 4WD VEHICLE FRONT ROTORS—SEALED HUB/BEARING ASSEMBLIES

These are unitized hubs that contain the bearing assembly. The hub/bearing unit is replaced as an assembly.

1. Loosen the lug nuts on the applicable wheels.
2. Raise and safely support the vehicle.
3. Remove the wheels.
4. Clean the brake assembly thoroughly with spray brake cleaner.
5. Remove the caliper and suspend it out of the way with wire.
6. On models so equipped, disconnect the ABS sensor wire.
7. Working through the hole provided in the rotor, or working from behind the rotor, remove the hub retaining bolts or nuts.
8. Remove the hub assembly.

To install:

➡**New rotors come with an oily, rust-preventive coating on the braking surface. This coating can be removed with brake parts cleaner or most cleaners which are good for oil removal. Be sure that all traces of the coating are removed. Allow the rotor to dry before installation.**

9. Clean the mounting surfaces of the hub and spindle.
10. Install the hub assembly and tighten the bolts/nuts securely.
11. Connect the ABS wire on models so equipped.
12. Install the caliper.
13. Install the wheels and snug the lug nuts.
14. Lower the vehicle.
15. Tighten the lug nuts fully.
16. Seat the brake pads, otherwise the vehicle may coast out of the work area and into traffic before the brakes become effective. It will take several pumps of the brake pedal to seat the pads against the rotor.
17. Check the brake system for proper operation.

4WD VEHICLE FRONT ROTORS

The following procedure applies to the front rotors on most 4WD vehicles.

1. Loosen the lug nuts on the front wheels.
2. Apply the parking brake, block the rear wheels, then raise and safely support the front of the vehicle securely.
3. Remove the wheels.
4. Remove the locking hub assemblies.
5. On most vehicles, remove the outer locknut. This requires a socket made expressly for that purpose, available at many auto parts stores. Then, remove the lockring from the bearing adjusting nut. This can be done with your finger tips or a prytool. Use the locknut socket to remove the bearing adjusting nut.

On some vehicles, a self-locking adjusting nut is used. Remove the self-locking nut with a hub nut wrench, applying inward pressure to disengage the adjusting nut locking splines, while turning it counterclockwise to remove it.

6. Remove the caliper and suspend it out of the way.
7. Slide the hub assembly off of the knuckle. The outer wheel bearing will slide out as the hub is removed, so be prepared to catch it.

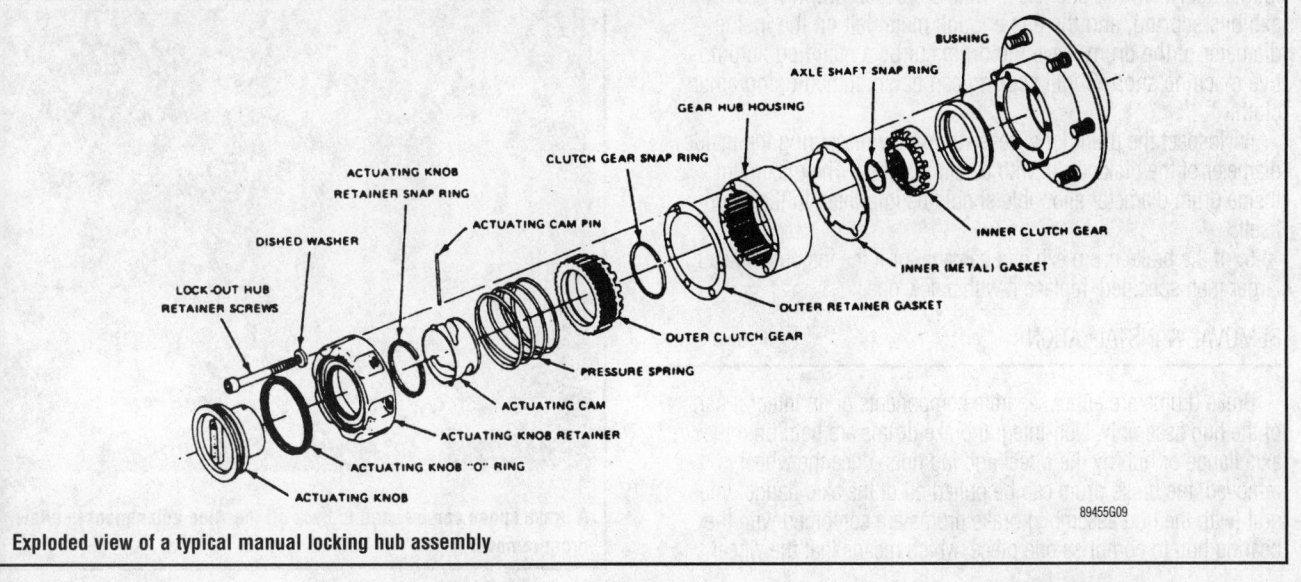

Exploded view of a typical manual locking hub assembly

89455G09

To install:

⁂ WARNING

Install the hub on the knuckle. To prevent damage to the seal and spindle threads, keep the hub centered on the spindle.

8. Carefully position the hub assembly on the spindle.
9. Install the outer bearing cone and roller, and the adjusting nut. Adjust the wheel bearings as described in the model-specific sections of this manual.

DRUM BRAKES

Brake Drums

➡Most vehicles have rubber plugs in the backing plates that are removed to access the brake adjusters. However, some vehicles are built with what are called knock-out plugs. These are areas in the backing plate that are made to be knocked out with a hammer and punch. Once the drum is off, the knock-out plug is removed and a rubber plug used in its place.

INSPECTION

➡While the brake drum is removed from the vehicle, inspect the wheel cylinder for damage and leakage.

1. Remove the brake drum from the vehicle.

⁂ CAUTION

Older brake pads or shoes may contain asbestos, which has been determined to be a cancer causing agent. Never clean the brake surfaces with compressed air! Avoid inhaling any dust from any brake surface! When cleaning brake surfaces, use a commercially available brake cleaning fluid.

2. Thoroughly clean the brake drum.
3. Inspect the brake drum for cracks, scores deep grooves, etc. A damaged drum is unsafe for use, and should be replaced immediately. Do not attempt to weld a cracked drum. If the drum exhibits scoring, and there is enough metal left on the inside diameter of the drum, have the drum cut by a qualified automotive machine shop. Slight scoring can be smoothed using emery cloth.
4. Inspect the drum for excessive wear by measuring the inside diameter of the brake drum with a caliper gauge. The maximum inside drum diameter allowable should be imprinted in the drum itself.
5. If the brake drum exhibits damage, or if the inside diameter is larger than specified, replace it with a new one.

REMOVAL & INSTALLATION

Brake drums are either separate components or an integral part of the hub assembly. Non-integral brake drums are held onto the axle flange or hub by the wheel and lug nuts; once the wheel is removed, the brake drum can be pulled off of the axle flange. Integral (with the hub assembly) brake drums are combined with the bearing hub to comprise one piece, which means that the wheel

10. Reassemble the hub parts.
11. Install the caliper(s).
12. Install the wheels and snug the lug nuts.
13. Lower the vehicle.
14. Tighten the lug nuts fully.
15. Seat the brake pads, otherwise the vehicle may coast out of the work area and into traffic before the brakes become effective. It will take several pumps of the brake pedal to seat the pads against the rotor.
16. Check the brake system for proper operation.

bearings must be disturbed (loosened or removed) in one way or another to remove the drum/hub assembly.

⁂ WARNING

If the drum is excessively difficult to remove, loosen the brake pads by adjusting their position with a brake spoon. Access for adjusting the brake pads is often gained through a small hole in the backing plate. If a brake drum is forced off of an axle flange without loosening the brake pads, damage can occur to the brake or axle components.

Non-integral drums (those that are not part of the hub) are usually fairly easy to remove. There are always exceptions to the rule, however. There are drums that are retained to the hub with one or two small bolts. Some drums can be drawn off of the hub by installing two small bolts into threaded holes in the drum; as these bolts are tightened, they slowly press the drum off of the hub. Occasionally a drum is difficult to remove because it binds on the hub flange; these must be worked off by prying gently between the drum and backing plate while applying penetrating oil to the drum/flange contact point. Some older vehicles have a drum assembly that fits over splines on the end of the axle shaft. Others just rust in place. If this occurs, just spray the area around each lug stud and the hub flange with a penetrant such as WD-40®, Liquid Wrench® or equivalent. Let the stuff work for a while, then try pulling or prying the drum off.

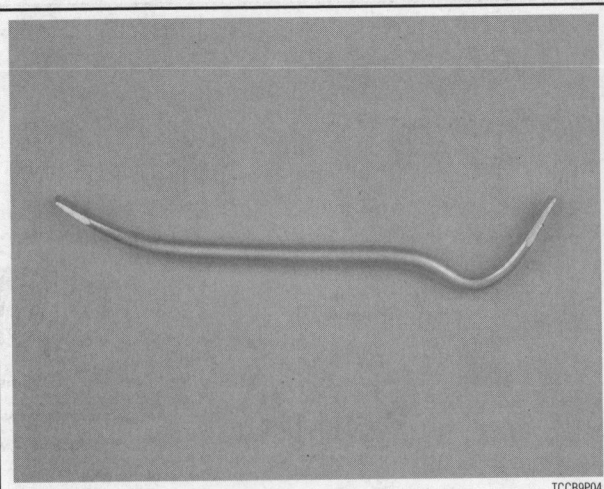

TCCB9P04

A brake spoon can be used to back off the shoe adjustment to allow drum removal

Non-Integral Drums

FREE-MOUNTED TYPE

➡Some vehicles are built with retainers threaded over 2 or more lug studs to hold the drum in place during assembly. Although these retainers may not be necessary (according to the manufacturer), it may be a good idea to reinstall new retainers anyhow.

1. Loosen the lug nuts on the applicable wheels.
2. Raise and safely support the vehicle.
3. Remove the wheels.
4. If necessary, remove and discard the retainers holding the drum to the hub.
5. If applicable, back off the parking brake adjustment.
6. Back off the brake adjustment until the wheels rotate freely, as follows:

   a. On vehicles with a starwheel-type adjuster: Remove the plug on the backing plate, then insert a thin prytool and a brake spoon into the slot. Hold the adjuster lever away from the adjuster wheel with the thin prytool and back of the adjuster wheel with the brake spoon.

   b. On vehicles with an expanding-type adjuster, remove the plug and rotate the adjuster screw (usually in an upward motion).

   c. On vehicles with ratcheting-type adjusters, remove the plug and insert a thin punch in the hole until it contacts the

To remove a free-mounted brake drum, first safely raise the vehicle and remove the wheel . . .

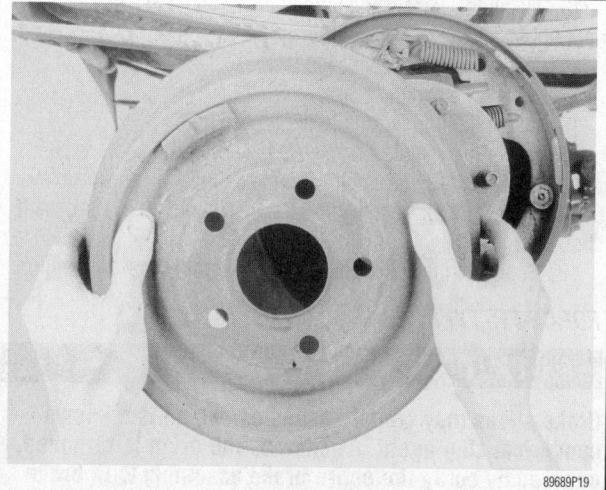

. . . then grasp hold of the drum and pull it from the axle flange and brake shoes

adjuster assembly pivot. Apply side pressure on this pivot point to allow the adjuster quadrant to ratchet and release the brake adjustment.

   d. Some vehicles, notably with manual adjusters, use adjusting cams. On these vehicles, the cam can be turned back from behind the backing plate.

7. Grasp the drum and pull it off the hub.

➡On some vehicles, the drum won't come off even with the shoes completely backed off. This is due to the drum binding on the hub boss. The safest way to remove the drum when this happens, is to spray the binding point with lubricant and to carefully pry between the hub and backing plate. Use a small prybar and pry at various points while rotating the drum. It helps to occasionally tap the hub with a deadblow, or brass mallet.

8. Spray the brake shoe assembly thoroughly with brake parts cleaner and let it dry. Similarly, spray the inside of the drum.

9. Inspect the drum for wear and/or damage, such as deep grooves, excessive thinness, cracks, etc. Machine or replace the drum as necessary. When machining, observe the maximum diameter specification. The maximum machining diameter is stamped into the drum. If the drum braking surface shows signs of blue discoloration, overheating is indicated. If the bluing is extensive the drum must be replaced. Extensive bluing indicates a weakening of the metal.

To install:

➡New brake drums come with an oily, rust-preventive coating on the braking surface. This coating can be removed with brake parts cleaner or most cleaners which are good for oil removal. Be sure that all traces of the coating are removed. Allow the drum to dry before installation.

10. If a new brake drum is being installed, remove the protective coating from the inner braking surface.
11. Adjust the brake shoes to just smaller than the inside diameter of the brake drum.
12. Slide the brake drum onto the hub. Be sure that the brake shoes are not dragging on the brake drum. Install new brake drum retainers.

13. Install the wheels and tighten the lug nuts in a star pattern until tight.

14. Adjust the brakes shoes.

15. Adjust the parking brake.

16. Install the rubber plug in the access hole.

17. Lower the vehicle. To activate the adjusters, some vehicles require you to make several quick pulls on the parking brake lever. On most, however, several short back-ups, about 10 ft. (3m) each, should do it.

18. Road test the vehicle and check for proper brake operation.

FORCE-FIT TYPE

✻✻ CAUTION

Brake shoes may contain asbestos, which is a known cancer-causing agent. As soon as the drum is removed, generously spray the entire brake assembly with brake parts cleaner. Let it dry before proceeding. It's a good idea to wear a filter mask when doing brake work.

1. Loosen the lug nuts on the applicable wheels.

2. Raise and safely support the vehicle.

3. Remove the wheels.

4. If necessary, remove and discard the retainers holding the drum to the hub.

5. If applicable, back off the parking brake adjustment.

6. Back off the brake adjustment until the wheels rotate freely, as follows:

    a. On vehicles with a starwheel-type adjuster: Remove the plug on the backing plate, then insert a thin prytool and a brake spoon into the slot. Hold the adjuster lever away from the adjuster wheel with the thin prytool and back of the adjuster wheel with the brake spoon.

    b. On vehicles with an expanding-type adjuster, remove the plug and rotate the adjuster screw (usually in an upward motion).

    c. On vehicles with ratcheting-type adjusters, remove the plug and insert a thin punch in the hole until it contacts the adjuster assembly pivot. Apply side pressure on this pivot point to allow the adjuster quadrant to ratchet and release the brake adjustment.

88279P13

If equipped with threaded holes, it is possible to press a force-fit type drum off of the hub using bolts, as shown

    d. Some vehicles, notably with manual adjusters, use adjusting cams. On these vehicles, the cam can be turned back from behind the backing plate.

7. Thread the proper size bolts into the holes provided in the drum until each contacts the hub. Turn the bolts evenly, a little at a time, until the drum slides free.

8. Grasp the drum and remove it from the axle flange or hub assembly. Remove the forcing bolts.

9. Spray the brake assembly thoroughly with brake parts cleaner and let it dry. Similarly, spray the inside of the drum.

10. Inspect the drum for wear and/or damage, such as deep grooves, excessive thinness, cracks, etc. Machine or replace the drum as necessary. When machining, observe the maximum diameter specification. The maximum machining diameter is stamped into the drum. If the drum braking surface shows signs of blue discoloration, overheating is indicated. If the bluing is extensive the drum must be replaced. Extensive bluing indicates a weakening of the metal.

To install:

➡**New brake drums come with an oily, rust-preventive coating on the braking surface. This coating can be removed with brake parts cleaner or most cleaners which are good for oil removal. Be sure that all traces of the coating are removed. Allow the drum to dry before installation.**

11. If a new brake drum is being installed, remove the protective coating from the inner braking surface.

12. Adjust the brake shoes to match the inside diameter of the brake drum.

13. Slide the brake drum onto the hub. Install 2 wheel lug nuts and tighten them, forcing the drum into place on the hub. Remove the lug nuts, then, if equipped, install new drum retainers.

14. Install the wheels.

15. Adjust the brake shoes.

16. Adjust the parking brake.

17. Install the rubber plug in the access hole.

18. Lower the vehicle. To activate the adjusters, some vehicles require you to make several quick pulls on the parking brake lever. On most, however, several short back-ups, about 10 ft. (3m) each, should do it.

19. Road test the vehicle and check for proper brake operation.

BOLTED-IN-PLACE TYPE

✻✻ CAUTION

Brake shoes may contain asbestos, which is a known cancer-causing agent. As soon as the drum is removed, generously spray the entire brake assembly with brake parts cleaner. Let it dry before proceeding. It's a good idea to wear a filter mask when doing brake work.

1. Loosen the lug nuts on the applicable wheels.

2. Raise and safely support the vehicle.

3. Remove the wheels.

4. If applicable, back off the parking brake adjustment.

5. Back off the brake adjustment until the wheels rotate freely, as follows:

    a. On vehicles with a starwheel-type adjuster: Remove the plug on the backing plate, then insert a thin prytool and a brake spoon into the slot. Hold the adjuster lever away from the adjuster wheel with the thin prytool and back of the adjuster wheel with the brake spoon.

b. On vehicles with an expanding-type adjuster, remove the plug and rotate the adjuster screw (usually in an upward motion).

c. On vehicles with ratcheting-type adjusters, remove the plug and insert a thin punch in the hole until it contacts the adjuster assembly pivot. Apply side pressure on this pivot point to allow the adjuster quadrant to ratchet and release the brake adjustment.

d. Some vehicles, notably with manual adjusters, use adjusting cams. On these vehicles, the cam can be turned back from behind the backing plate.

6. Remove the drum-to-hub attaching bolts.

7. Grasp the drum and remove it from the axle flange or hub assembly.

8. Spray the brake assembly thoroughly with brake parts cleaner and let it dry. Similarly, spray the inside of the drum.

➡**On some vehicles, the drum won't come off even with the shoes completely backed off. This is due to the drum binding on the hub boss. The safest way to remove the drum when this happens, is to spray the binding point with lubricant and pry, carefully between the hub and backing plate. Use a small prybar and pry at various points while rotating the drum. It helps to occasionally rap the hub with a dead-blow, or brass mallet.**

9. Inspect the drum for wear and/or damage, such as deep grooves, excessive thinness, cracks, etc. Machine or replace the drum as necessary. When machining, observe the maximum diameter specification. The maximum machining diameter is stamped into the drum. If the drum braking surface shows signs of blue discoloration, overheating is indicated. If the bluing is extensive the drum must be replaced. Extensive bluing indicates a weakening of the metal.

To install:

➡**New brake drums come with an oily, rust-preventive coating on the braking surface. This coating can be removed with brake parts cleaner or most cleaners which are good for oil removal. Be sure that all traces of the coating are removed. Allow the drum to dry before installation.**

10. If a new brake drum is being installed, remove the protective coating from the inner braking surface.

11. Adjust the brake shoes to match the inside diameter of the brake drum.

12. Slide the brake drum onto the hub. Be sure that the brake shoes are not dragging on the brake drum.

13. Install the drum-to-hub attaching bolts and tighten them securely.

14. Install the wheels.

15. Adjust the brakes as follows:

a. Adjust the brake shoes so that you can feel a slight drag on, or hear a scraping noise coming from the wheel when you spin it.

b. Back the shoes off just until the drag is no longer felt, or the rasping noise is no longer heard.

16. Adjust the parking brake.

17. Install the rubber plug in the access hole.

18. Lower the vehicle. To activate the adjusters, some vehicles require you to make several quick pulls on the parking brake lever. On most, however, several short back-ups, about 10 ft. (3m) each, should do it.

19. Road test the vehicle and check for proper brake operation.

Integral Drum/Hub Assemblies

> ✳✳ **CAUTION**
>
> **It is always a good idea to wear eye protection when working on brake components, especially drum brakes. Drum brakes often use powerful springs which could cause severe eye injury if they accidentally break.**

EXCEPT 4WD VEHICLE FRONT DRUM/HUB ASSEMBLIES

> ✳✳ **CAUTION**
>
> **Brake shoes may contain asbestos, which is a known cancer-causing agent. As soon as the drum is removed, generously spray the entire brake assembly with brake parts cleaner. Let it dry before proceeding. It's a good idea to wear a filter mask when doing brake work.**

Some Rear Wheel Drive (RWD) front drums and some Front Wheel Drive (FWD) rear drums are designed with the bearing hub as an integral assembly with the drum.

1. Raise and safely support the vehicle.

2. Remove the wheels.

3. If applicable, back off the parking brake adjustment.

4. Back off the brake adjustment until the wheels rotate freely.

a. On vehicles with a starwheel-type adjuster: Remove the plug on the backing plate, then insert a thin prytool and a brake spoon into the slot. Hold the adjuster lever away from the adjuster wheel with the thin prytool and back of the adjuster wheel with the brake spoon.

b. On vehicles with an expanding-type adjuster, remove the plug and rotate the adjuster screw (usually in an upward motion).

c. On vehicles with ratcheting-type adjusters, remove the plug and insert a thin punch in the hole until it contacts the adjuster assembly pivot. Apply side pressure on this pivot point to allow the adjuster quadrant to ratchet and release the brake adjustment.

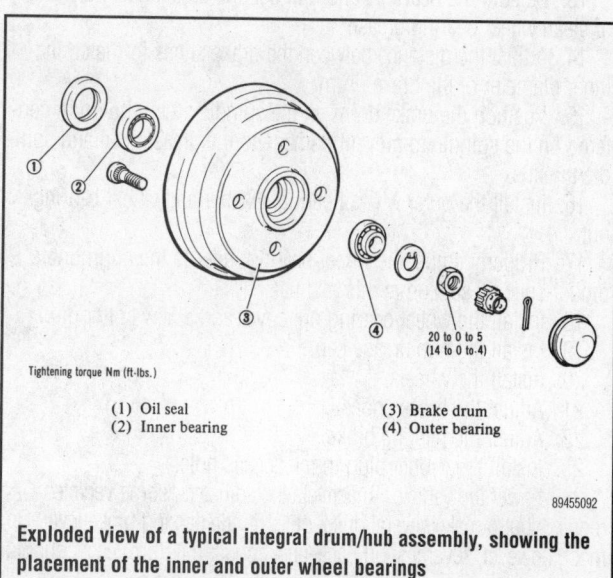

Tightening torque Nm (ft-lbs.)

20 to 0 to 5
(14 to 0 to.4)

(1) Oil seal
(2) Inner bearing
(3) Brake drum
(4) Outer bearing

89455092

Exploded view of a typical integral drum/hub assembly, showing the placement of the inner and outer wheel bearings

d. Some vehicles, notably with manual adjusters, use adjusting cams. On these vehicles, the cam can be turned back from behind the backing plate.

5. Remove the hub grease cap.

6. Remove the cotter pin and wheel bearing adjusting nut cover. Discard the cotter pin.

7. Remove the wheel bearing nut.

✳✳ WARNING

On some vehicles, a left-hand threaded nut is used on the right wheel spindle. Turn this locknut clockwise to loosen, otherwise damage to the spindle threads will occur.

8. Remove the brake drum, washer and bearings as an assembly. Be careful not to let the outer wheel bearing fall out of the hub during removal.

9. Spray the brake assembly thoroughly with brake parts cleaner and let it dry. Similarly, spray the inside of the drum.

10. Remove the brake drum/hub assembly. Inspect the drum for wear and/or damage. Machine or replace as necessary. When machining, observe the maximum diameter specification. The maximum machining diameter is stamped into the drum. If the drum braking surface shows signs of blue discoloration, overheating is indicated. If the bluing is extensive the drum/hub assembly must be replaced. Extensive bluing indicates a weakening of the metal.

➡ **If the brake drum is to be machined or replaced, remove the inner wheel bearing and grease seal.**

To install:

➡ **New brake drums come with an oily, rust-preventive coating on the braking surface. This coating can be removed with brake parts cleaner or most cleaners which are good for oil removal. Be sure that all traces of the coating are removed. Allow the drum to dry before installation.**

11. If a new brake drum is being installed, remove the protective coating from the inner braking surface.

12. If removed, install the inner wheel bearing and a new grease seal.

13. Be sure the bearings and hub contain an adequate amount of clean wheel bearing grease.

14. Adjust the distance between the brake shoes to match the inner diameter of the brake drum.

15. Position the brake drum on the spindle. Keep the drum centered on the spindle to prevent damage to the grease seal and spindle threads.

16. Install the outer wheel bearing, washer and wheel bearing nut.

17. Properly adjust the wheel bearing; refer to the appropriate model-specific section in this manual.

18. Install the wheel bearing nut cover and a new cotter pin.

19. Install the hub grease cap.

20. Install the wheels.

21. Adjust the brake shoes.

22. Adjust the parking brake.

23. Install the rubber plug in the access hole.

24. Lower the vehicle. To activate the adjusters, some vehicles require you to make several quick pulls on the parking brake lever. On most, however, several short back-ups, about 10 ft. (3m) each, should do it.

25. Road test the vehicle and check for proper brake operation.

4WD VEHICLE FRONT DRUM/HUB ASSEMBLIES

✳✳ CAUTION

Brake shoes may contain asbestos, which is a known cancer-causing agent. As soon as the drum is removed, generously spray the entire brake assembly with brake parts cleaner. Let it dry before proceeding. It's a good idea to wear a filter mask when doing brake work.

1. These vehicles may be equipped with locking hubs. If so, the hub mechanism will have to be disassembled to gain access to the bearing assembly. Once access is gained, the locknut, locking ring and adjusting nut are removed and the hub/drum assembly can be slid off the axle shaft.

2. Spray the brake assembly thoroughly with brake parts cleaner and let it dry. Similarly, spray the inside of the drum.

3. Inspect the drum for wear and/or damage. Machine or replace as necessary. When machining, observe the maximum diameter specification. The maximum machining diameter is stamped into the drum. If the drum braking surface shows signs of blue discoloration, overheating is indicated. If the bluing is extensive the drum/hub assembly must be replaced. Extensive bluing indicates a weakening of the metal.

➡ **New brake drums come with an oily, rust-preventive coating on the braking surface. This coating can be removed with brake parts cleaner or most cleaners which are good for oil removal. Be sure that all traces of the coating are removed. Allow the drum to dry before installation.**

4. If no locking hub is present, simply remove the grease cap and locknut and washer assembly, then slide the hub/drum assembly from the axle shaft.

Full Floating Axle Drums

✳✳ CAUTION

It is always a good idea to wear eye protection when working on brake components, especially drum brakes. Drum brakes often use powerful springs which could cause severe eye injury if they accidentally break.

NON-SPLINED TYPE

✳✳ CAUTION

Brake shoes may contain asbestos, which is a known cancer-causing agent. As soon as the drum is removed, generously spray the entire brake assembly with brake parts cleaner. Let it dry before proceeding. It's a good idea to wear a filter mask when doing brake work.

To remove the drums from full floating rear axles, the axle shaft will have to be removed. The bearing housing protruding through the center of the wheel can readily identify full floating rear axles.

1. Loosen the lug nuts on the applicable wheels.

2. Raise and safely support the vehicle.

3. Remove the wheels.

4. If applicable, back off the parking brake adjustment.

5. Back off the brake adjustment until the wheels rotate freely.

a. On vehicles with a starwheel-type adjuster, remove the plug on the backing plate, then insert a thin prytool and a brake spoon into the slot. Hold the adjuster lever away from the adjuster wheel with the thin prytool and back of the adjuster wheel with the brake spoon.

b. On vehicles with an expanding-type adjuster, remove the plug and rotate the adjuster screw (usually in an upward motion).

c. On vehicles with ratcheting-type adjusters, remove the plug and insert a thin punch in the hole until it contacts the adjuster assembly pivot. Apply side pressure on this pivot point to allow the adjuster quadrant to ratchet and release the brake adjustment.

d. Some vehicles, notably with manual adjusters, use adjusting cams. On these vehicles, the cam can be turned back from behind the backing plate.

6. Remove the axle shaft.

7. Remove the retaining ring, key and axle shaft nut.

8. Remove the hub and drum.

9. Spray the brake assembly thoroughly with brake parts cleaner and let it dry. Similarly, spray the inside of the drum.

10. Inspect the drum for wear and/or damage. Machine or replace as necessary. When machining, observe the maximum diameter specification. The maximum machining diameter is stamped into the drum. If the drum braking surface shows signs of blue discoloration, overheating is indicated. If the bluing is extensive the drum assembly must be replaced. Extensive bluing indicates a weakening of the metal.

To install:

➡ **New brake drums come with an oily, rust-preventive coating on the braking surface. This coating can be removed with brake parts cleaner or most cleaners which are good for oil removal. Be sure that all traces of the coating are removed. Allow the drum to dry before installation.**

11. If a new brake drum is being installed, remove the protective coating from the inner braking surface.

12. Install the hub and drum to the tube.

13. Install the axle shaft nut and tighten it to the manufacturer's specification. Refer to the appropriate model-specific section in this manual for hub removal and installation.

14. Install the key and retaining ring.

15. Install the axle shaft and wheel.

16. Adjust the brake shoes, as described in Section 3 of this manual.

17. Adjust the parking brake.

18. Install the rubber plug in the access hole.

19. Lower the vehicle. To activate the adjusters, some vehicles require you to make several quick pulls on the parking brake lever. On most, however, several short back-ups, about 10 ft. (3m) each, should do it.

20. Road test the vehicle and check for proper brake operation.

SPLINED TYPE

✴✴ CAUTION

Brake shoes may contain asbestos, which is a known cancer-causing agent. As soon as the drum is removed, generously spray the entire brake assembly with brake parts cleaner. Let it dry before proceeding. It's a good idea to wear a filter mask when doing brake work.

Splines are raised ridges on a shaft or in a bore. The axle shaft on these vehicles has splines cast into the end which mesh with corresponding splines cast into the center hole of the drum. A large, 3-jawed puller is necessary for this job.

1. Loosen the lug nuts on the applicable wheels.

2. Raise and safely support the vehicle.

3. Remove the wheels.

4. If applicable, back off the parking brake adjustment.

5. Back off the brake adjustment until the wheels rotate freely.

a. On vehicles with a starwheel-type adjuster, remove the plug on the backing plate, then insert a thin prytool and a brake spoon into the slot. Hold the adjuster lever away from the adjuster wheel with the thin prytool and back of the adjuster wheel with the brake spoon.

b. On vehicles with an expanding-type adjuster, remove the plug and rotate the adjuster screw (usually in an upward motion).

c. On vehicles with ratcheting-type adjusters, remove the plug and insert a thin punch in the hole until it contacts the adjuster assembly pivot. Apply side pressure on this pivot point to allow the adjuster quadrant to ratchet and release the brake adjustment.

d. Some vehicles, notably with manual adjusters, use adjusting cams. On these vehicles, the cam can be turned back from behind the backing plate.

6. Remove the axle shaft nut.

➡ **Some of these nuts may be left-hand threads. That is, you turn them right to loosen.**

7. Grasp the drum and try pulling it off the shaft. If it comes off, great! If not, place the puller screw on the end of the shaft, with the jaws evenly spaced around the drum rim. Slowly tighten the puller screw until the drum slide free. It may be helpful to spray the splines with a penetrant such as WD-40®, Liquid Wrench® or equivalent.

8. Remove the brake drum.

9. Spray the brake assembly thoroughly with brake parts cleaner and let it dry. Similarly, spray the inside of the drum.

10. Inspect the drum for wear and/or damage. Machine or replace as necessary. When machining, observe the maximum diameter specification. The maximum machining diameter is stamped into the drum. If the drum braking surface shows signs of blue discoloration, overheating is indicated. If the bluing is extensive the drum must be replaced. Extensive bluing indicates a weakening of the metal.

To install:

➡ **New brake drums come with an oily, rust-preventive coating on the braking surface. This coating can be removed with brake parts cleaner or most cleaners which are good for oil removal. Be sure that all traces of the coating are removed. Allow the drum to dry before installation.**

11. If a new brake drum is being installed, remove the protective coating from the inner braking surface.

12. Adjust the brake shoes to match the inside diameter of the brake drum.

13. Slide the brake drum onto the shaft splines as far as you can by hand.

14. Install the axle shaft nut and tighten it until the drum seats completely. Tighten the nut to specification. Check the appropriate model-specific section in this manual for hub removal and installation.

15. Install the wheels.

16. Adjust the brakes as follows:

a. Adjust the brake shoes so that you can feel a slight drag on, or hear a scraping noise coming from the wheel when you spin it.

b. Back the shoes off just until the drag is no longer felt, or the rasping noise is no longer heard.

17. Adjust the parking brake.
18. Install the rubber plug in the access hole.
19. Lower the vehicle. To activate the adjusters, some vehicles require you to make several quick pulls on the parking brake lever. On most, however, several short back-ups, about 10 ft. (3m) each, should do it.
20. Road test the vehicle and check for proper brake operation.

Brake Shoes

GENERAL INFORMATION

Most vehicles use a 2-shoe leading/trailing, internal expanding type of drum brake with automatic self-adjuster mechanisms. The automatic self-adjuster mechanisms can take several forms, but the overwhelming majority utilize the starwheel-type, located between the bottom ends of the two shoes, or the ratcheting type, located directly below the wheel cylinder. When the ratcheting type of adjuster is used, the lower ends of the brake shoes usually rest on an anchor plate.

➡On some vehicles, notably those with unitized rear hubs, and some vehicles with full-floating axles, not only does the brake drum have to be removed, but the hub assembly must be removed as well.

✱✱ CAUTION

Brake shoes must always be replaced as an axle set. That is, do not just replace the shoes on one side of the vehicle. Replace them on both sides. Replacing shoes on only one side will result in poor braking performance. Besides, if the shoes wore out on one side faster than the other side, there is a malfunction in the brake system. Inspect the brake system, if necessary, repair the problem before proceeding.

➡It is not a good idea to disassemble the brakes on both sides at the same time. There are a lot of parts involved which must be replaced in a certain way. Work on one side at a time, only. If you become confused as to the particular position of the various brake parts during the brake shoe replacement, refer to the other side. Remember, however, the other side is a mirror image (everything is reversed).

INSPECTION

1. Remove the brake drum.
2. Inspect the brake shoe lining material for cracks, crumbling or evidence of wetness. Replace the shoes with new ones if any such damage is found. If evidence of wetness is evident, repair the leaking component prior to installing the new shoes.
3. Measure the thickness of the brake shoe lining (not including the shoe backing). Generally, the minimum allowable lining thickness is either 1/16 in. (1.6mm) above the head of the rivet (for rivet mounted linings), or 3/32 in. (2.4mm) from the shoe backing (for glued linings).
4. If one of the brake linings is worn to or beyond the allowable limit, all four of the rear brake shoes must be replaced.

✱✱ WARNING

Never polish the shoe lining with sandpaper, because hard particles from the sandpaper will become imbedded in the lining, which will damage the brake drum. If the shoe lining is damaged or worn excessively or unevenly, replace the shoes with new ones.

5. Install the brake drum.

REMOVAL & INSTALLATION

✱✱ CAUTION

Brake shoes may contain asbestos, which is a known cancer-causing agent. As soon as the drum is removed, generously spray the entire brake assembly with brake parts cleaner. Let it dry before proceeding. It's a good idea to wear a filter mask when doing brake work.

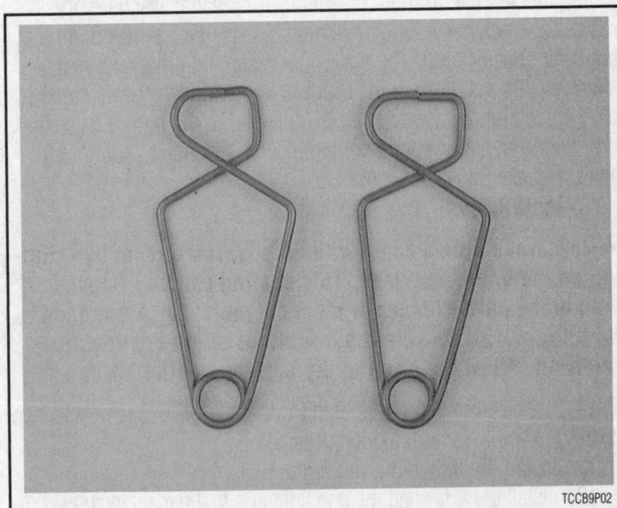

TCCB9P02

Spring clamp tools, such as those shown, can hold the wheel cylinder pistons in while servicing the shoes

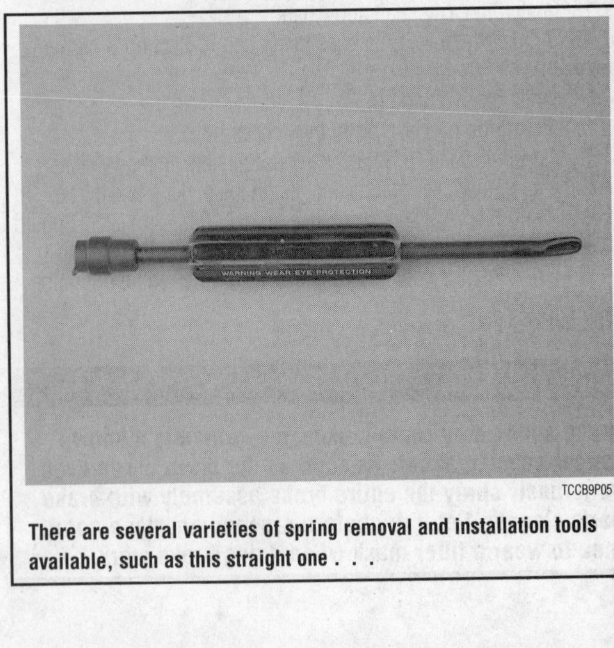

TCCB9P05

There are several varieties of spring removal and installation tools available, such as this straight one . . .

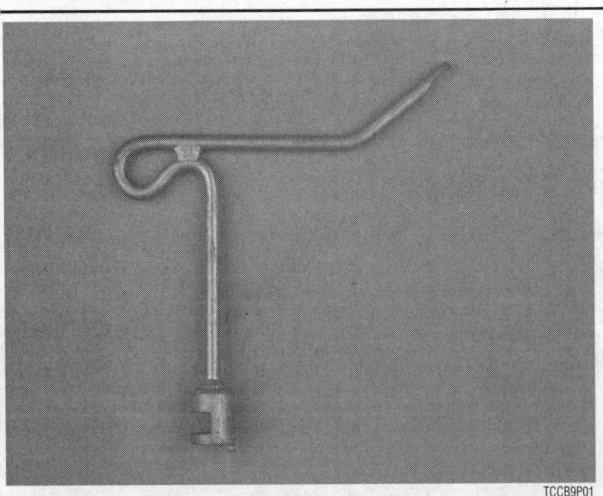

TCCB9P01

. . . and this curved one—The shape of this tool is designed to provide more leverage during use

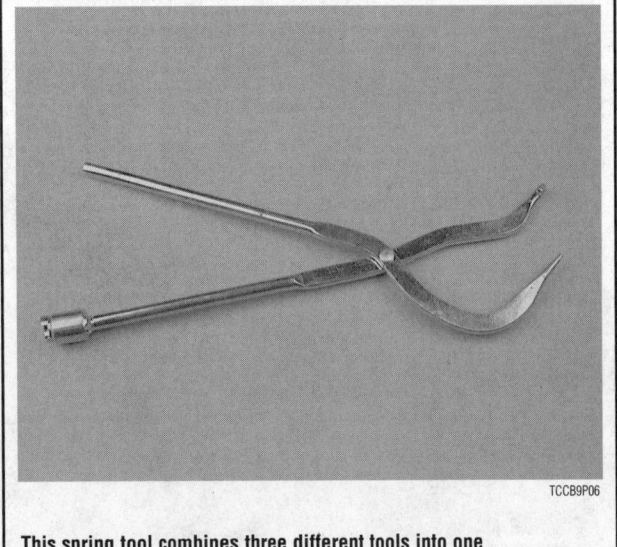

TCCB9P06

This spring tool combines three different tools into one

Models with Dual Return Springs and a Starwheel-type Adjuster

✳✳ CAUTION

It is always a good idea to wear eye protection when working on brake components, especially drum brakes. Drum brakes often use powerful springs which could cause severe eye injury if they accidentally break.

1. Remove the brake drum.
2. Spray the brake assembly thoroughly with brake parts cleaner and let it dry. Similarly, spray the inside of the drum.

3. Inspect the drum for wear and/or damage. Machine or replace as necessary. When machining, observe the maximum diameter specification. The maximum machining diameter is stamped into the drum. If the drum braking surface shows signs of blue discoloration, overheating is indicated. If the bluing is extensive the drum/hub assembly must be replaced. Extensive bluing indicates a weakening of the metal.

➡Note the location of all springs and clips for proper assembly. If an instant camera is handy, it may be a good idea to take a picture of the brake assembly with the brake drum removed. This will make reassembly much easier.

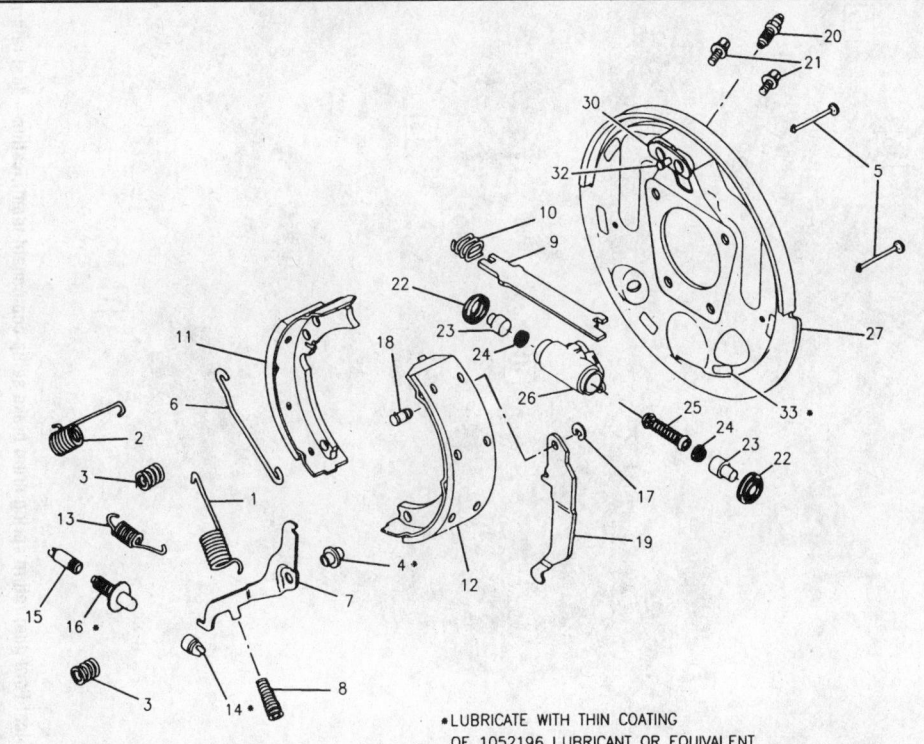

1	RETURN SPRING
2	RETURN SPRING
3	HOLD DOWN SPRING
4	BEARING SLEEVE
5	HOLD-DOWN PIN
6	ACTUATOR LINK
7	ACTUATOR LEVER
8	LEVER RETURN SPRING
9	PARKING BRAKE STRUT
10	STRUT SPRING
11	PRIMARY SHOE AND LINING
12	SECONDARY SHOE AND LINING
13	ADJUSTING SCREW SPRING
14	SOCKET
15	PIVOT NUT
16	ADJUSTING SCREW
17	RETAINING RING
18	PIN
19	PARKING BRAKE LEVER
20	BLEEDER VALVE
21	BOLT
22	BOOT
23	PISTON
24	SEAL
25	SPRING ASSEMBLY
26	WHEEL CYLINDER
27	BACKING PLATE
30	SHOE RETAINER
32	ANCHOR PIN
33	SHOE PADS (6 PLACES)

•LUBRICATE WITH THIN COATING OF 1052196 LUBRICANT OR EQUIVALENT

87959042

Exploded view of the most common GM rear drum brake setup—dual return spring and a starwheel-type adjuster type

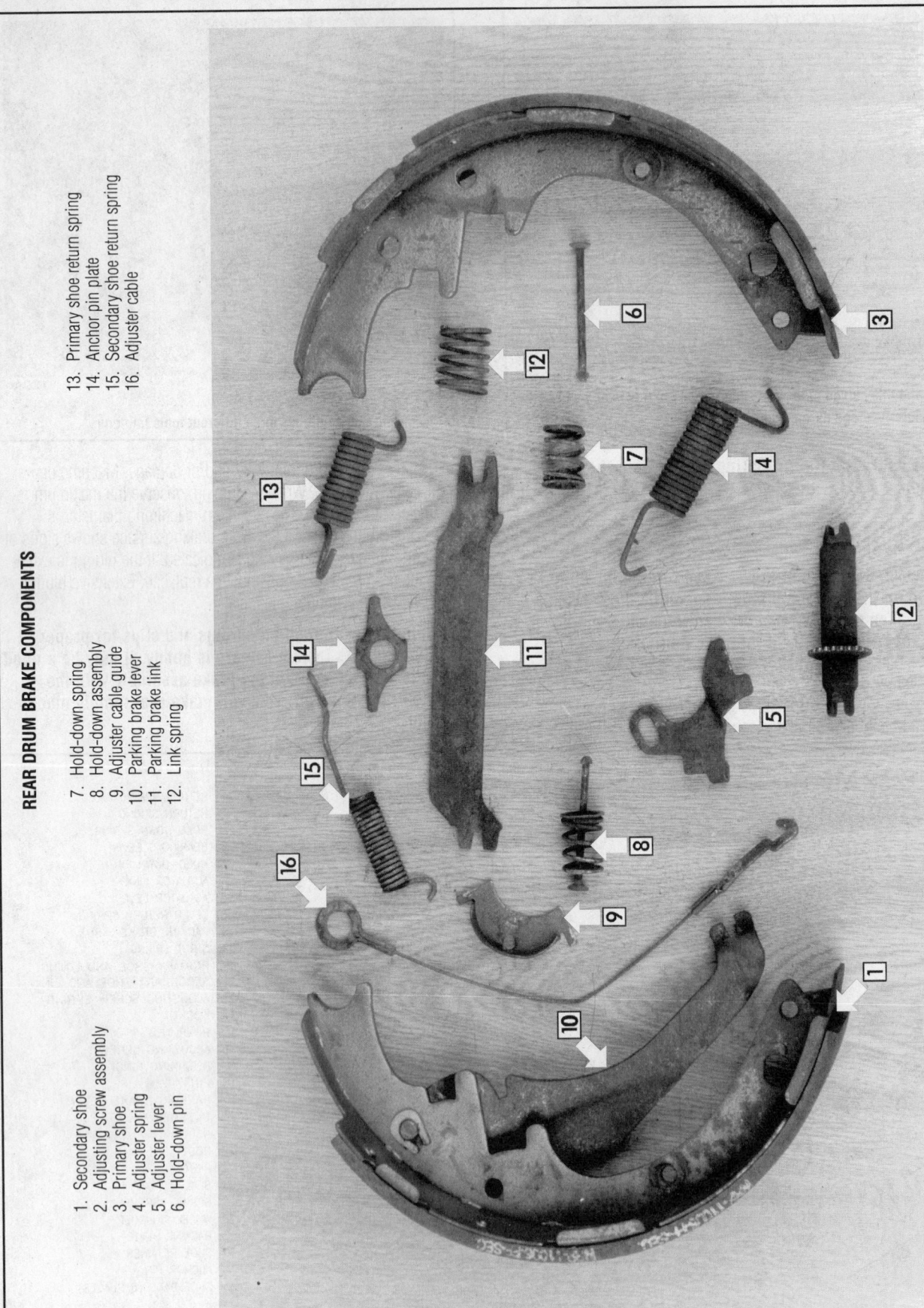

REAR DRUM BRAKE COMPONENTS

1. Secondary shoe
2. Adjusting screw assembly
3. Primary shoe
4. Adjuster spring
5. Adjuster lever
6. Hold-down pin
7. Hold-down spring
8. Hold-down assembly
9. Adjuster cable guide
10. Parking brake lever
11. Parking brake link
12. Link spring
13. Primary shoe return spring
14. Anchor pin plate
15. Secondary shoe return spring
16. Adjuster cable

Typical Ford dual return spring drum brake setup component identification—dual return spring and a starwheel-type adjuster type

4. Completely retract the adjuster by rotating the starwheel to relieve tension on the lower spring.

5. Remove the starwheel assembly and adjuster lever from between the two brake shoes.

6. Using a brake spring tool, remove the 2 upper return springs.

7. Remove the adjuster cable and cable guide.

8. Remove the anchor block plate.

9. Using a hold-down spring tool or pliers, while holding the back of the spring mounting pin with one hand, press inward on the hold-down spring plate, turn it slightly to align the notches and pin ears, then remove the hold-down spring assembly with your other hand. Remove the other hold-down spring in the same manner.

10. Lift the shoes off the pins and remove the pins from the backing plate.

11. Remove the parking brake link.

12. Pull back on the parking brake cable spring and twist the cable out of the parking brake lever.

13. The parking brake lever is held onto the rear shoe with a horseshoe clip. Spread the clip and remove the lever and washer.

A specially-designed brake tool can make disconnecting the upper return springs much easier—dual return spring and a starwheel-type adjuster type

Clean the brake shoe assemblies with a liquid cleaning solution, NEVER with compressed air

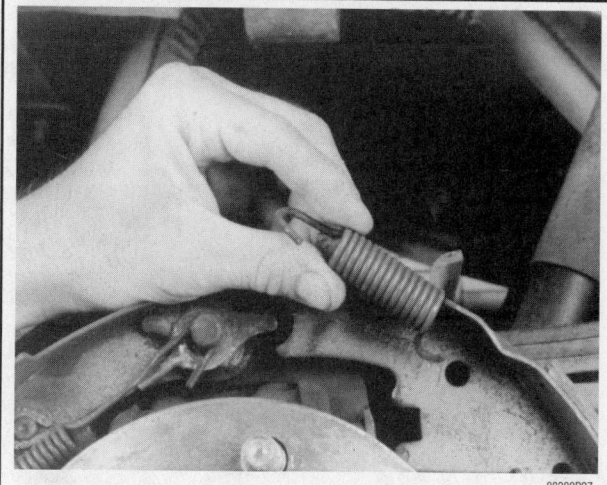

Detach the upper return springs first from the anchor bolt, then from the brake shoes . . .

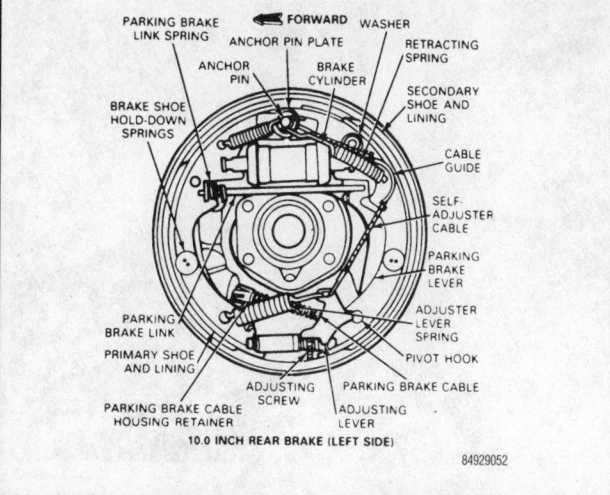

PARKING BRAKE
LINK SPRING
ANCHOR PIN PLATE
FORWARD
WASHER
RETRACTING
SPRING
ANCHOR
PIN
BRAKE
CYLINDER
SECONDARY
SHOE AND
LINING
BRAKE SHOE
HOLD-DOWN
SPRINGS
CABLE
GUIDE
SELF-
ADJUSTER
CABLE
PARKING
BRAKE
LEVER
PARKING
BRAKE LINK
ADJUSTER
LEVER
SPRING
PRIMARY SHOE
AND LINING
PIVOT HOOK
ADJUSTING
SCREW
PARKING BRAKE CABLE
PARKING BRAKE CABLE
HOUSING RETAINER
ADJUSTING
LEVER
10.0 INCH REAR BRAKE (LEFT SIDE)

Identify the brake components and note their locations prior to disassembling the brake assembly

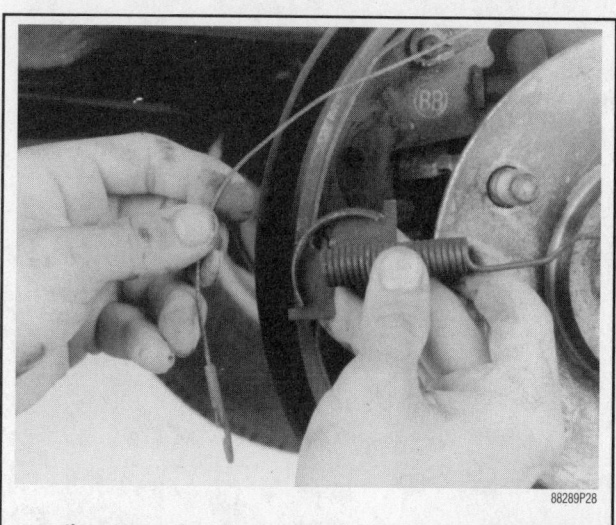

. . . then remove the adjusting cable from the guide, and the guide from the brake shoe

Remove the anchor block plate . . .

. . . then detach the parking brake cable from the lever—dual return spring and a starwheel-type adjuster type

. . . then remove the hold-down springs, retainers and pins from both shoes—dual return spring and a starwheel-type adjuster type

Another way to remove the shoes for a dual spring setup is to pull the adjuster cable toward the shoe . . .

Lift the brake shoes off of the backing plate . . .

. . . and disconnect the pivot hook from the adjusting lever. Wind the starwheel all the way in

Disconnect the adjuster lever return spring from the lever . . .

. . . disconnect the primary brake shoe return spring from the anchor pin

. . . and remove the spring and the lever—dual return spring and a starwheel-type adjuster type

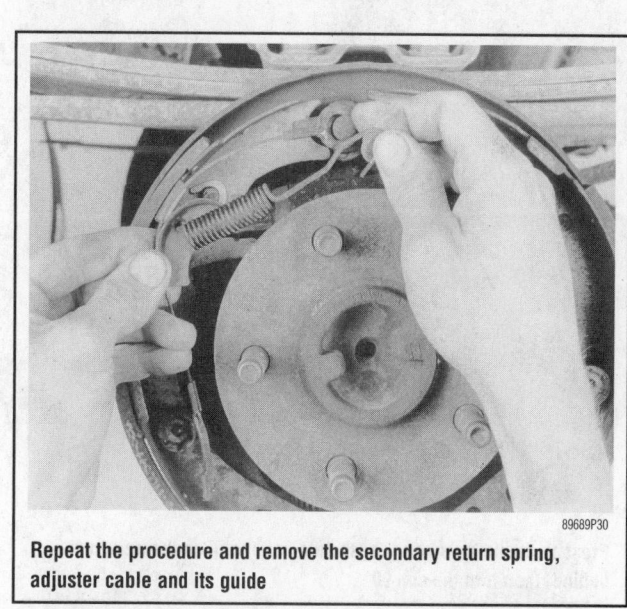

Repeat the procedure and remove the secondary return spring, adjuster cable and its guide

Next, using a brake spring removal tool . . .

Also remove the anchor pin plate—dual return spring and a starwheel-type adjuster type

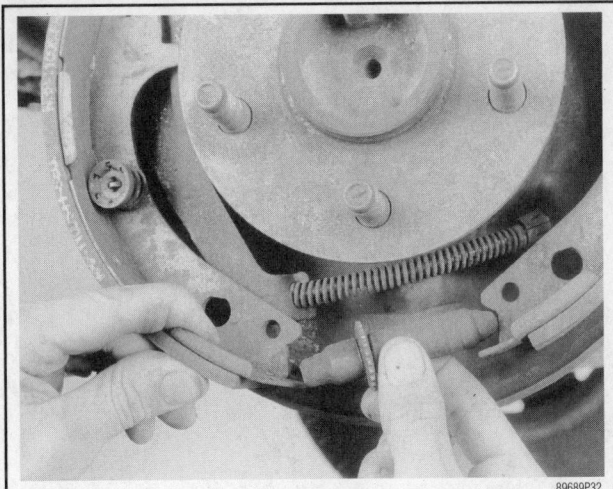

Pull the bottoms of the shoes apart and remove the adjuster screw assembly

Remove the primary (front) brake shoe from the backing plate . . .

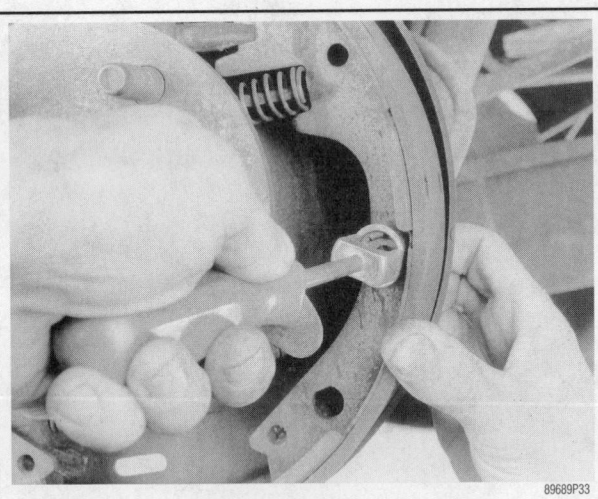

Press in the hold-down springs while holding in on the nail from behind, then turn the cup 90° . . .

. . . and the parking brake strut as well—dual return spring and a starwheel-type adjuster type

. . . and release to remove the hold-down spring. Pull the nail out from the backing plate

Remove the secondary shoe hold-down, pull the shoe out then press up on the cable spring . . .

. . . and disconnect the parking brake cable from its lever by pulling it from the slot

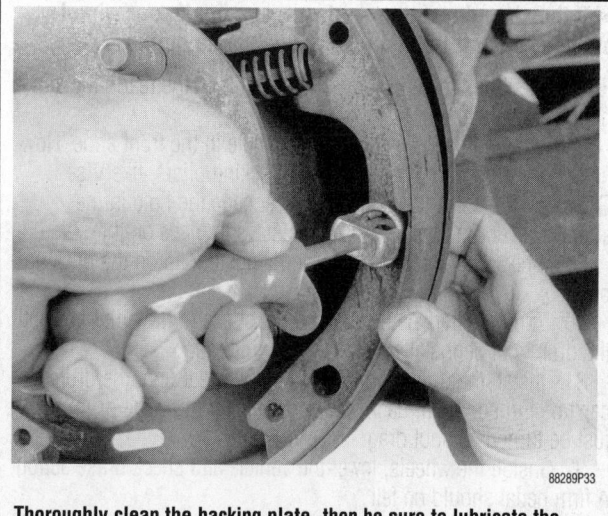

Thoroughly clean the backing plate, then be sure to lubricate the brake shoe bosses on the backing plate

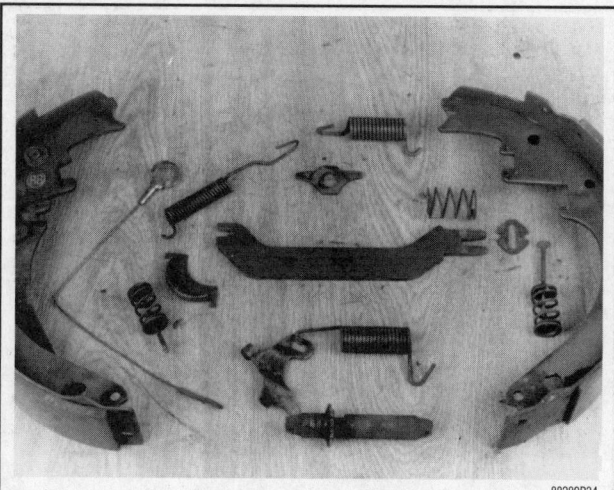

It's a good idea to arrange all the parts in their approximate installed positions on a clean work surface

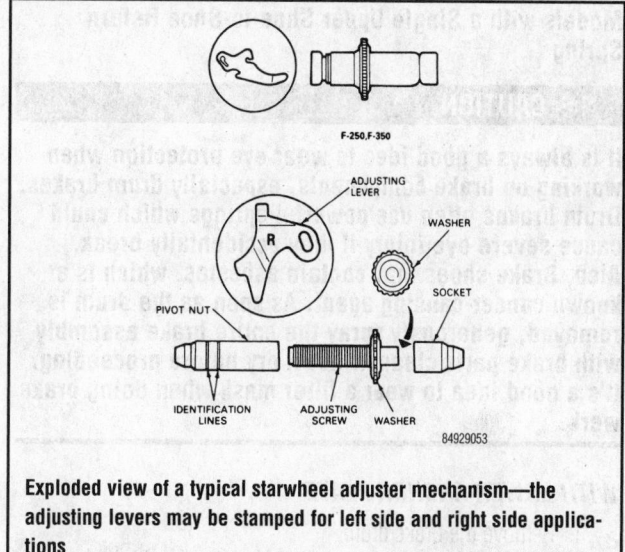

Exploded view of a typical starwheel adjuster mechanism—the adjusting levers may be stamped for left side and right side applications

To install:

14. Thoroughly clean and dry the backing plate and starwheel assembly.

15. Lubricate the backing plate bosses, anchor plate surfaces, and starwheel threads and contact points with silicone grease. High-temperature wheel bearing grease or synthetic brake grease also work well for this application.

✳✳ CAUTION

When applying lubricant to the backing plate and other components, do not use so much grease that it may get spread onto the new brake shoes' friction material; this can adversely affect the performance of the new brake shoes, therefore, increase vehicle stopping distance.

16. Insert the parking brake lever pivot stud through the applicable hole in the rear shoe, then install a new wave washer and horseshoe clip. Squeeze the clip ends until the clip cannot be pulled from the lever pivot stud.

17. Connect the parking brake cable to the lever.

18. Position the rear shoe assembly on the backing plate and install the hold-down pin and spring assembly.

19. Install the front shoe and secure it with the hold-down spring assembly.

20. Position the parking brake link and spring between the front shoe and parking brake lever.

21. Position the adjuster cable on the anchor plate pin, install the cable guide and lay the cable across the guide.

22. Be sure that the notch in the upper end of the shoe is engaging the wheel cylinder piston or piston pin.

23. Position the rear shoe return spring into the guide and shoe hole, using a brake spring tool, stretch the spring onto the anchor plate pin. Be sure that the cable guide remained in place.

24. Position the front shoe return spring in its hole in the shoe.

25. Be sure that the parking brake link is properly positioned and that the upper end of the shoe will enter the wheel cylinder or engage the wheel cylinder piston.

26. Using the spring tool, stretch the spring into position on the anchor plate pin.

➡If the shoe doesn't properly engage the link or wheel cylinder piston, try again by removing the spring.

27. Position the adjuster lever in its hole in the rear shoe and hook the cable to it.

28. Position the lower spring in its hole in the front shoe. Now comes the hard part. Clamp a pair of locking pliers, like Vise Grips® on the spring and stretch it to engage the hole in the adjuster lever. Be sure that the cable stays in place on the guide.

29. Check that the shoes are evenly positioned on the backing plate.

30. Turn the starwheel to spread the shoes to the point at which the drum can be installed with very slight drag.

31. Install the drum and adjust the starwheel until the drum can't be turned. Then, back off the adjustment until the drum can just be turned without drag.

32. Install the wheels, lower the vehicle and check brake action. A firm pedal should be felt.

33. To activate the adjusters, some vehicles require you to make several quick pulls on the parking brake lever. On most, however, several short back-ups, about 10 ft. (3m) each, should do it.

Models with a Single Upper Shoe-to-Shoe Return Spring

※※ CAUTION

It is always a good idea to wear eye protection when working on brake components, especially drum brakes. Drum brakes often use powerful springs which could cause severe eye injury if they accidentally break. Also, Brake shoes may contain asbestos, which is a known cancer-causing agent. As soon as the drum is removed, generously spray the entire brake assembly with brake parts cleaner. Let it dry before proceeding. It's a good idea to wear a filter mask when doing brake work.

WITH LOWER ANCHOR PLATE

1. Remove the brake drum.

Clean the brake assembly and drum thoroughly with brake parts cleaner and let it dry.

Inspect the drum for wear and/or damage. Machine or replace as necessary. When machining, observe the maximum diameter specification. The maximum machining diameter is stamped into the drum. If the drum braking surface shows signs of blue discoloration, overheating is indicated. If the bluing is extensive the drum/hub assembly must be replaced. Extensive bluing indicates a weakening of the metal.

➡Note the location of all springs and clips for proper assembly. If you own an instant camera, to make installation easier it may be a good idea to take a picture of your brake assembly with the brake drum removed.

2. Remove the shoe-to-lever spring and remove the adjuster lever.
3. Remove the auto-adjuster assembly.
4. Remove the retainer spring.
5. Using a hold-down spring tool or pliers, while holding the back of the spring mounting pin with one hand, press inward on the hold-down spring plate, turn it slightly to align the notches and pin ears, then remove the hold-down spring assemblies with your other hand.

Pliers can be used to disengage the hold-down spring retainer by rotating it until aligned with the pin tabs . . .

. . . then remove the retainer, spring and pin from the shoe and backing plate—models with a single upper shoe-to-shoe return spring and lower anchor plate

Use a pair of needlenose pliers, or similar tool, to detach the upper return spring from both shoes . . .

. . . then remove the brake shoes from the backing plate . . .

. . . and detach the parking brake cable from the applicable brake shoe—models with a single upper shoe-to-shoe return spring and lower anchor plate

6. Remove the shoe-to-shoe spring.
7. Remove the brake shoes from the backing plate.
8. Using a flat-tipped tool, pry open the parking brake lever retaining clip. Remove the clip and washer from the pin on the shoe assembly and remove the shoe from the lever assembly.

➡On some vehicles, the parking brake actuating lever is permanently attached to the trailing brake shoe assembly. Do not attempt to remove it from the original brake shoe assembly or reuse the original actuating lever on a replacement brake shoe assembly. All replacement brake shoe assemblies for these vehicles must come with the actuating lever as part of the trailing brake shoe assembly.

To install:
9. Thoroughly clean all parts.
10. On vehicles with the ratcheting upper mounted adjuster, clean and inspect the brake support plate and the automatic

adjuster mechanism. Be sure the quadrant (toothed part) of the adjuster is free to rotate throughout its entire tooth contact range and is free to slide the full length of its mounting slot. Check the knurled pin. It should be securely attached to the adjuster mechanism and its teeth should be in good condition. If the adjuster is worn or damaged, replace it. If the adjuster is serviceable, lubricate lightly with high-temperature grease between the strut and the quadrant.

✳✳ CAUTION

The trailing brake shoe assemblies used on the rear brakes of these vehicles are different for the left and right side of the vehicle. Care must be taken to ensure the brake shoes are properly installed in their correct side of the vehicle. Otherwise the brakes will probably malfunction, thereby creating a very dangerous condition. When the trailing shoes are properly installed on their correct side of the vehicle, the park brake actuating lever will be positioned under the brake shoe web.

11. Thoroughly clean and dry the backing plate. Lubricate the backing plate at the brake shoe contact points. Also, lubricate backing plate bosses, anchor pin, and parking brake actuating mechanism with silicone grease. High-temperature wheel bearing grease or synthetic brake grease also work well for this application.
12. Install the parking brake lever assembly on the lever pin. Install the wave washer and a new retaining clip. Use pliers, or the like, to install the retainer on the pin. If removed, connect the parking brake lever to the parking brake cable and verify that the cable is properly routed.
13. Clean and lubricate the adjuster assembly. Be sure the nut-adjuster is drawn all the way to the stop, but the nut must NOT lock firmly at the end of the assembly.
14. Install the brake shoes on the backing plate with the hold-down springs, washers and pins.
15. Install the shoe-to-shoe spring.
16. Install the retainer spring.
17. Install the auto-adjuster assembly and install the adjuster lever and the shoe-to-lever spring.
18. Pre-adjust the shoes so the drum slides on with a light drag and install the brake drum.
19. Adjust the brake shoes.
20. Install the rear wheels.
21. To activate the adjusters, some vehicles require you to make several quick pulls on the parking brake lever. On most, however, several short back-ups, about 10 ft. (3m) each, should do it.
22. Adjust the parking brake cable.
23. Lower the vehicle and check for proper brake operation.

WITH LOWER STARWHEEL-TYPE ADJUSTER

1. Loosen the lug nuts on the applicable wheels.
2. If servicing the front brakes, apply the parking brake, block the rear wheels, then raise and safely support the front of the vehicle securely.
3. If servicing the rear brakes, block the front wheels, then raise and safely support the rear of the vehicle securely.
4. Remove the wheels.
5. Remove the drums.

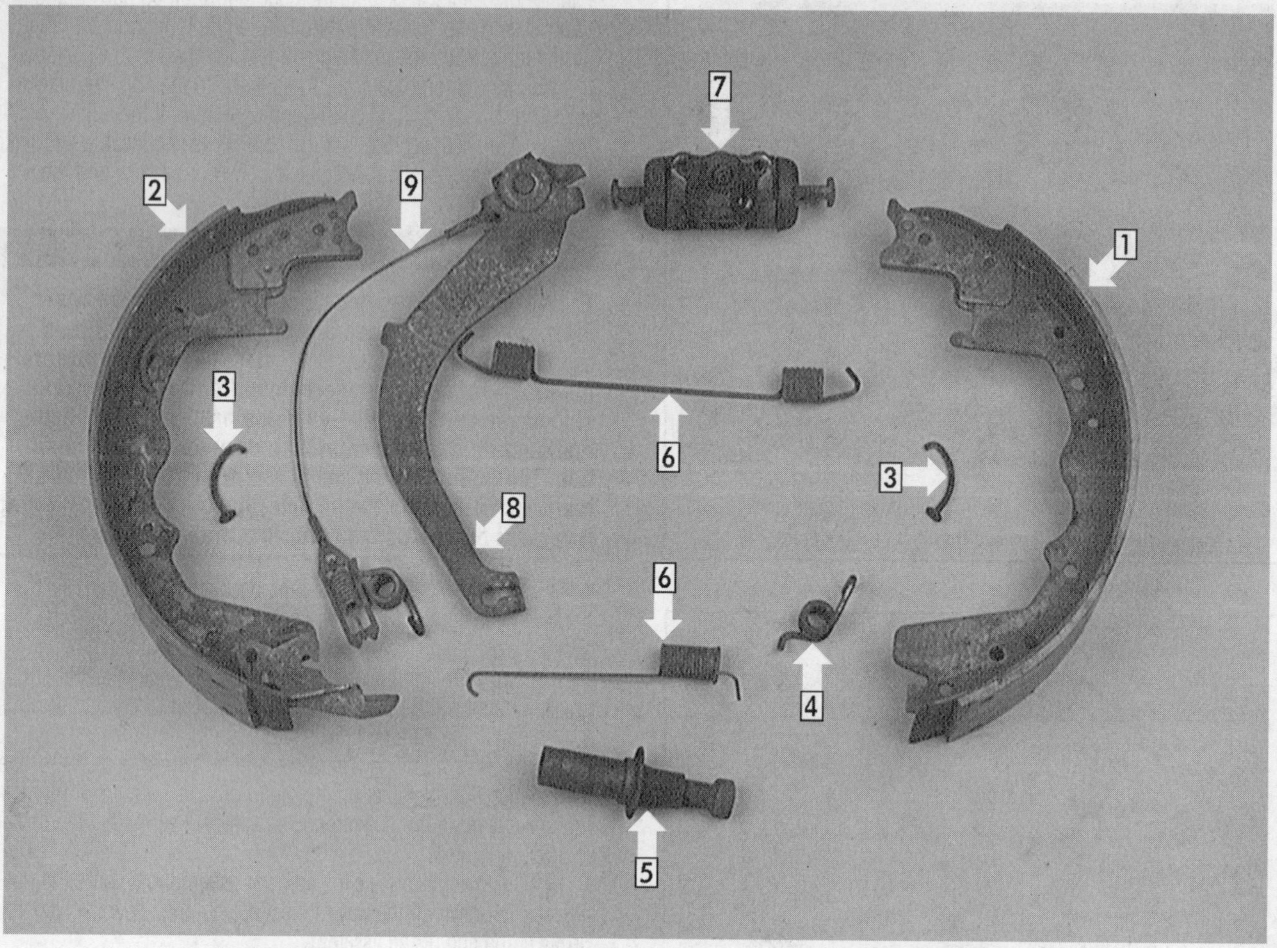

1. Front brake shoe
2. Rear brake shoe
3. Hold-down pin
4. Shoe hold-down spring
5. Adjuster
6. Return spring
7. Wheel cylinder
8. Parking brake lever
9. Parking brake adjuster cable

88489P44

It is a good idea to lay the brake parts out in their positions on a clean work surface as they are removed

88489P31

Remove the brake drum from the rear axle

88489P32

Remove the parking brake lever retaining nut which is located behind the backing plate

Disconnect the adjusting cable from the anchor pin, guide and lever—models with a single upper shoe-to-shoe return spring and starwheel adjuster

Disconnect the parking brake cable from the lever

6. Spray the brake assembly thoroughly with brake parts cleaner and let it dry. Similarly, spray the inside of the drum.

7. Inspect the drum for wear and/or damage. Machine or replace as necessary. When machining, observe the maximum diameter specification. The maximum machining diameter is stamped into the drum. If the drum braking surface shows signs of blue discoloration, overheating is indicated. If the bluing is extensive the drum/hub assembly must be replaced. Extensive bluing indicates a weakening of the metal.

8. Remove the parking brake lever assembly from the backing plate.

9. Remove the adjusting cable assembly from the anchor pin, cable guide and adjusting lever.

10. Remove the brake shoe retracting springs.

11. Remove the brake shoe hold-down spring from each shoe.

12. Remove the brake shoes and adjusting screw assembly.

13. Disassemble the adjusting screw assembly.

➡It's a good idea to arrange all the parts in the approximate installed positions as a guide for reassembly.

Use an appropriate tool to disconnect the return springs from their retaining holes

Slide the parking brake lever out from its mounting—models with a single upper shoe-to-shoe return spring and starwheel adjuster

Disengage the hold-down springs from the retaining clips on the backing plate—models with a single upper shoe-to-shoe return spring and starwheel adjuster

Back off the adjusting screw and remove it from the brake assembly

Connecting the lower retracting spring can often be difficult—be careful and have patience

Spread the shoes apart and remove them from the backing plate

This is how everything should look after assembly—models with a single upper shoe-to-shoe return spring and starwheel adjuster

To install:

14. Clean the ledge pads on the backing plate. Apply a light coat of silicone grease to the ledge pads (where the brake shoes rub the backing plate). High-temperature wheel bearing grease or synthetic brake grease (designed specifically for this) also work well. Also, apply grease to the adjusting screw assembly and the hold-down and retracting spring contacts on the brake shoes.

15. Install the upper retracting spring on the primary and secondary shoes, then position the shoe assembly on the backing plate with the wheel cylinder pistons engaged with the shoes.

16. Install the brake shoe hold-down springs.

17. Install the brake shoe adjustment screw assembly so that the slot in the head of the adjusting screw is toward the primary (leading) shoe, along with the lower retracting spring, adjusting lever spring, adjusting lever assembly and connect the adjusting cable to the adjusting lever. Position the cable in the cable guide and install the cable anchor fitting on the anchor pin.

18. Install the adjusting screw assemblies in the same locations from which they were removed.

✱✱ CAUTION

Interchanging the brake shoe adjusting screws from one side of the vehicle to the other will cause the brake shoes to retract rather than expand each time the automatic adjusting mechanism is operated; this will create an extremely dangerous condition when driving the vehicle. To prevent incorrect installation, the socket end of each adjusting screw is usually stamped with an R or an L to indicate their installation on the right or left side of the vehicle. In some cases, the adjusting pivot nuts can be distinguished by the number of lines machined around the body of the nut. Two lines indicate a nut which should be installed on the right side of the vehicle; one line indicates a nut that must be installed on the left side of the vehicle.

19. Install the parking brake assembly in the anchor pin and secure with the retaining nut behind the backing plate.

20. Adjust the brakes before installing the brake drums and wheels. Install the brake drums and wheels.

21. To activate the adjusters, some vehicles require you to make several quick pulls on the parking brake lever. On most, however, several short back-ups, about 10 ft. (3m) each, should do it.

22. Lower the vehicle and road test the brakes. New brakes may pull to one side or the other before they are seated. Continued pulling or erratic braking should not occur.

Models with a Single U-Shaped Return Spring

✷✷ CAUTION

It is always a good idea to wear eye protection when working on brake components, especially drum brakes. Drum brakes often use powerful springs which could cause severe eye injury if they accidentally break. Also, brake shoes may contain asbestos, which is a known cancer-causing agent. As soon as the drum is removed, generously spray the entire brake assembly with brake parts cleaner. Let it dry before proceeding. It's a good idea to wear a filter mask when doing brake work.

1. Loosen the lug nuts on the applicable wheels.
2. If servicing the front brakes, apply the parking brake, block the rear wheels, then raise and safely support the front of the vehicle securely.
3. If servicing the rear brakes, block the front wheels, then raise and safely support the rear of the vehicle securely.
4. Remove the wheels.
5. Remove the brake drum.
6. Spray the brake assembly thoroughly with brake parts cleaner and let it dry. Similarly, spray the inside of the drum.
7. Inspect the drum for wear and/or damage. Machine or replace as necessary. When machining, observe the maximum diameter specification. The maximum machining diameter is stamped into the drum. If the drum braking surface shows signs of blue discoloration, overheating is indicated. If the bluing is extensive the drum/hub assembly must be replaced. Extensive bluing indicates a weakening of the metal.
8. Remove the return spring clip from the lower anchor block.
9. Squeeze the upper ends of the return spring slightly and remove it from the shoes.

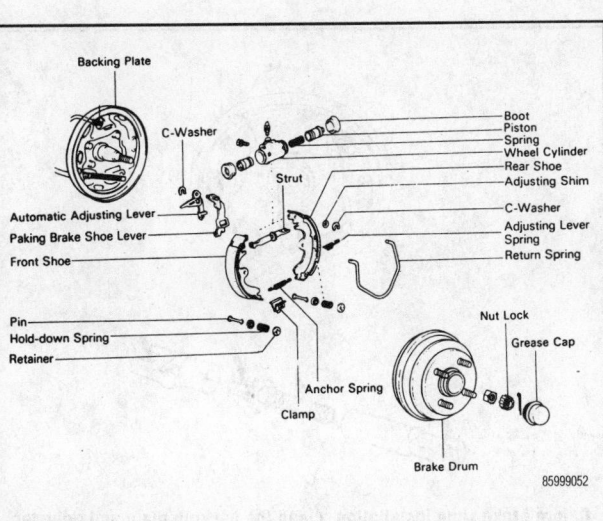

Exploded view of a typical single U-shaped return spring drum brake setup

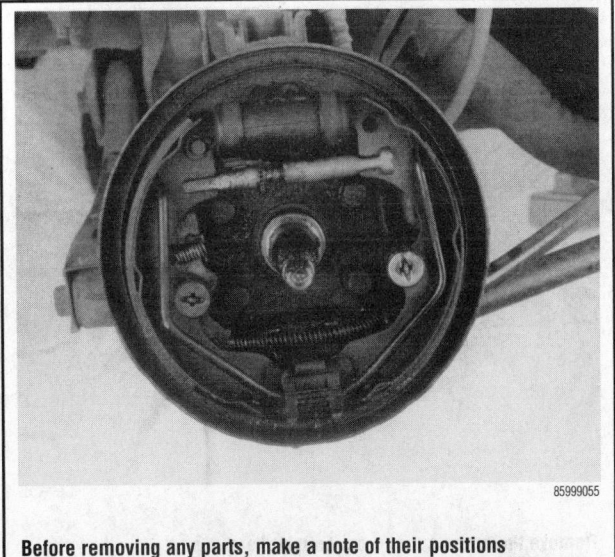

Before removing any parts, make a note of their positions

For models with a single U-shaped return spring, depress and rotate the hold-down spring retainer . . .

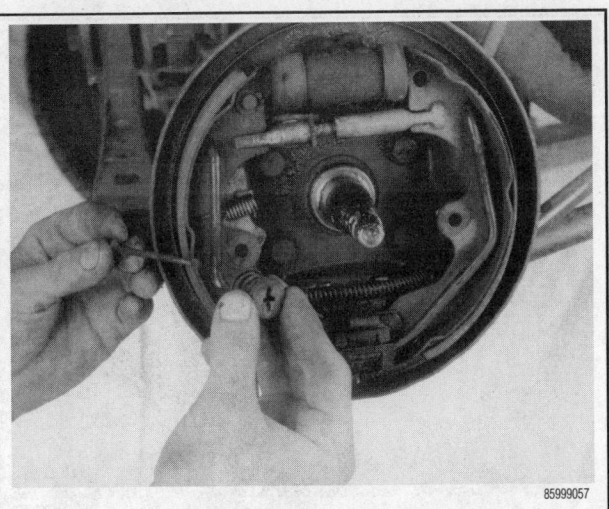

. . . then remove the spring, retainer and pin from the backing plate and shoes

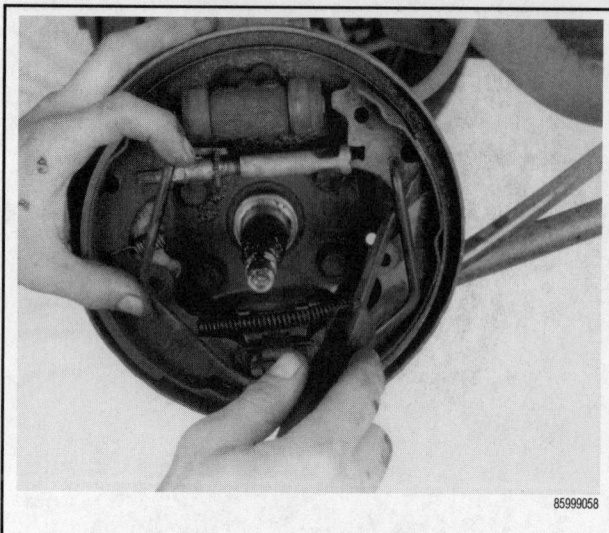

Remove the return spring from both brake shoes . . .

. . . then separate the shoes from the backing plate

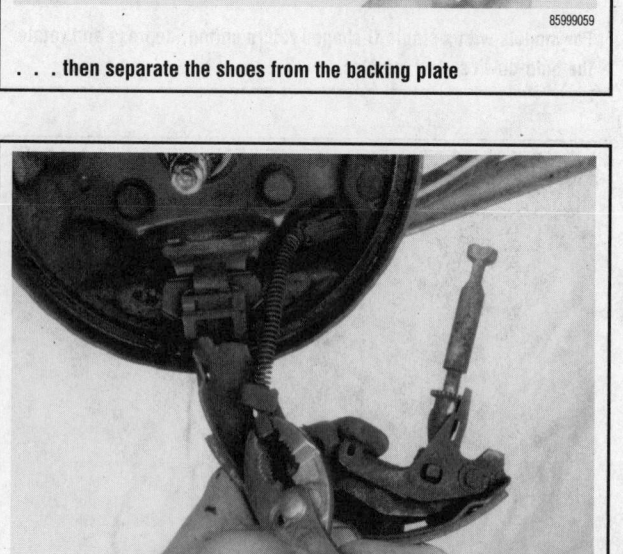

A large pair of pliers can be used to disconnect the parking brake cable from the lever

10. Using a hold-down spring tool or pliers, remove the hold-down springs. While holding the back of the spring mounting pin with one hand, press inward on the hold-down spring plate, turn it slightly to align the notches and pin ears, then remove the hold-down spring assemblies with your other hand.

11. Lift the shoes off of the pins, then remove the pins from the backing plate.

12. Remove the shoes and adjuster as an assembly.

13. Pull back on the parking brake cable spring and twist the cable out of the parking brake lever.

14. The parking brake lever is held onto the rear shoe with a horseshoe clip. Spread the clip and detach the lever and washer from the shoe.

To install:

15. Thoroughly clean and dry the backing plate assembly.

16. Lubricate the backing plate bosses, anchor plate surfaces, and all contact points with silicone grease. High-temperature wheel bearing grease or synthetic brake grease (designed specifically for this) also work well.

17. Lubricate the parking brake lever pivot stud, then insert the pivot stud through the applicable hole in the rear shoe, then install a new wave washer and horseshoe clip. Squeeze the clip ends until the clip cannot be pulled from the lever pivot stud.

18. Connect the parking brake cable to the lever.

19. Position the front and rear shoe assemblies and adjuster on the backing plate, then install the hold-down pin and spring assemblies.

20. Position the return spring in the shoes, rotate it down into position on the anchor block, and install the retaining clip.

21. Turn the strut adjusting screw to spread the shoes to the point at which the drum can just be installed without drag.

22. Install the drum.

23. Adjust the brake shoes.

24. Install the wheels, lower the vehicle and check brake action. A firm pedal should be felt.

25. To activate the adjusters, some vehicles require you to make several quick pulls on the parking brake lever. On most, however, several short back-ups, about 10 ft. (3m) each, should do it.

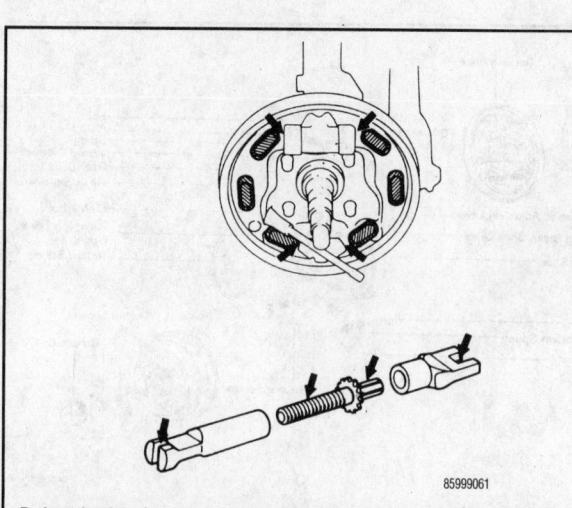

Before brake shoe installation, clean the backing plate and adjuster mechanism, then apply high temperature grease at all shoe-to-backing plate points (arrows)

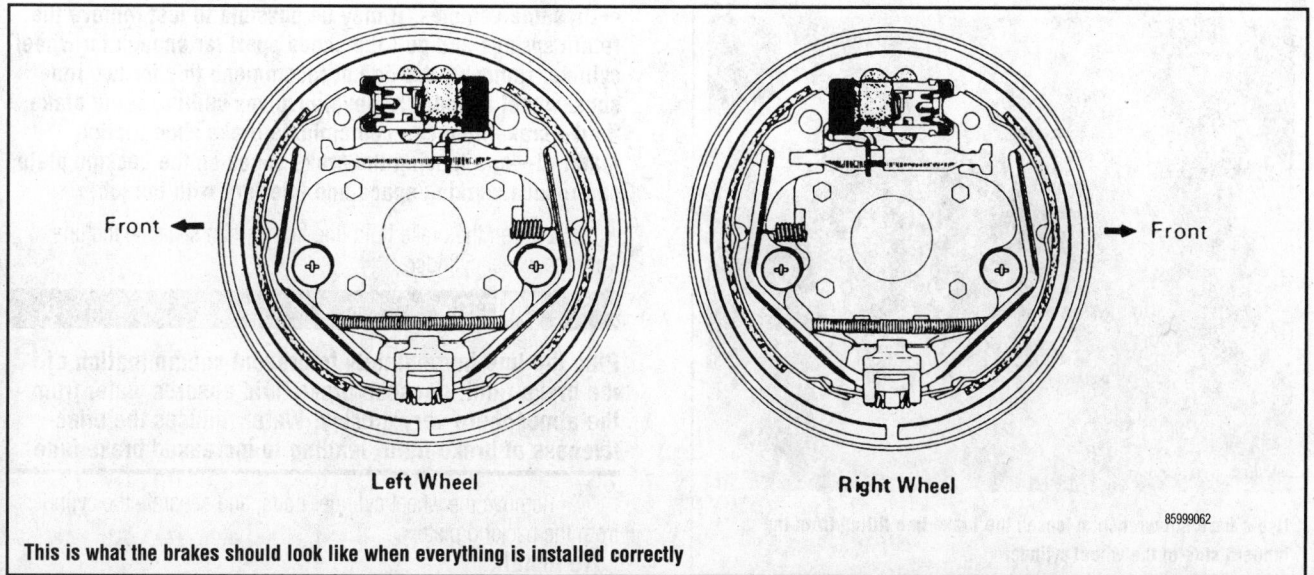

Left Wheel   Right Wheel

85999062

This is what the brakes should look like when everything is installed correctly

ADJUSTMENT

Drum brakes on all modern vehicles are self-adjusting, however, when the shoes are replaced, a preliminary adjustment makes the job easier.

On most vehicles, the adjustment is made with an expanding adjuster that is a threaded sleeve/stud assembly. Turning the knurled nut or starwheel expands or contracts the spring-loaded brake shoes. On most vehicles, this adjuster can be accessed without removing the drum, or, for that matter, the wheel.

Raise the vehicle and support it safely. Release the parking brake. Put the transmission in neutral. All this allows the wheels to turn freely. Remove the rubber plug in the brake backing plate and insert a brake adjusting tool. If you're applying brake pressure, that is, expanding the brakes, just turn the starwheel or knurled adjuster until the brake shoes lock the drum; meaning you can't turn it. Then, back off the adjustment until the drum can JUST turn freely without any drag. Some manufacturers even say it's okay to have a SLIGHT amount of drag. If the vehicle at hand is equipped with self-adjusters, you'll find that the adjuster can't be backed off. That's because the adjusting lever is holding it in place. You'll have to insert a thin punch or similar device in the hole with the brake adjusting tool. Just push slightly on the adjusting lever. That'll free the adjuster.

There are a few vehicle models that use cam-type adjusters. With these, a hex or square headed stud protrudes through the backing plate. Turning this stud rotates an eccentric cam that contacts the brake shoe. Turning it one way pushes the shoe outward; turning it the other way rotates the cam away from the shoe allowing the springs to pull the shoe away from the drum.

Wheel Cylinders

REMOVAL & INSTALLATION

Wheel cylinders are held in place on the backing plate with either bolts or spring clips. A first glance, this looks like a fairly easy job, and it can be. However, a lot can go wrong. If the wheel cylinder has been there a long time, the bolts or clips can be rusted in place. Worse, the brake line flare nut may be rusted in place. The flats on the nut are easily rounded off. Also, the flare nut can be rusted to the line, meaning the line will twist when the nut

is turned. So, before starting, it's best to thoroughly soak the area with penetrating oil where the brake line threads into the wheel cylinder. Also, apply penetrating oil to the mounting bolts or clips.

If you run into problems, here are some general tips:
- Use a flare nut wrench on the flare nuts. Sounds logical, doesn't it? Flare nut wrenches are designed to reduce the possibility of rounding-off.
- Use a box end wrench, or, if room permits, a socket on the bolts. The better grip of a box end wrench or socket will help prevent rounding off the bolt head(s).
- If you round off a bolt head, you'll have to try using Vise-Grips® (or equivalent), one of those wrenches designed for rounded-off bolts (space permitting), a nut splitter (again, space permitting), or grind off the bolt head.
- If the brake line won't budge, you fear kinking or twisting the line, or you rounded off the flare nut, try this: remove the wheel cylinder bolts or clips and pull the wheel cylinder, line attached, away from the backing plate. Usually, there is enough play in the brake line. Hold the flare nut with Vise-Grips® or equivalent, and try turning the wheel cylinder. The wheel cylinder gives you greater mechanical advantage than the flare nut. If nothing works, disconnect the line at the junction box. You'll have to install a new line.

Bolt-on Type

❊❊ CAUTION

It is always a good idea to wear eye protection when working on brake components, especially drum brakes. Drum brakes often use powerful springs which could cause severe eye injury if they accidentally break. Also, brake shoes may contain asbestos, which is a known cancer-causing agent. As soon as the drum is removed, generously spray the entire brake assembly with brake parts cleaner. Let it dry before proceeding. It's a good idea to wear a filter mask when doing brake work.

1. Loosen the lug nuts on the applicable wheels.
2. Raise and safely support the vehicle.
3. Remove the wheels.
4. Remove the drum.
5. Remove the brake shoes.

Use a flare nut wrench to loosen the brake line fitting from the inboard side of the wheel cylinder

When the brake line is disconnected there will be some fluid leakage—plug the line to avoid contamination

Remove the wheel cylinder retaining bolts, then separate the cylinder from the backing plate—bolt-on type

➡On some vehicles, it may be possible to just remove the return springs and pull the shoes apart far enough for wheel cylinder removal. We do not recommend this for two reasons: wheel cylinder removal involves spilling some brake fluid—brake fluid can contaminate brake shoe friction material—and leaving the brake shoes on the backing plate can reduce working space and interfere with the job.

6. Loosen the brake fluid line fitting, then separate the line from the wheel cylinder.

✳✳ CAUTION

Plug the line immediately to prevent contamination of the brake fluid, because brake fluid absorbs water from the atmosphere very quickly. Water reduces the effectiveness of brake fluid, leading to increased brake fade.

7. Remove the wheel cylinder bolts, and separate the cylinder from the backing plate.

To install:

8. Clean the backing plate thoroughly.

9. Apply a very thin coating of RTV silicone sealer to the cylinder mounting surface. This will aid in keeping moisture and dirt out of the brakes.

10. Position the cylinder on the backing plate, then install the retaining bolts.

11. Reattach the brake line to the wheel cylinder.

12. Install the brake shoes.

13. Install the drum.

14. Bleed the brake system.

15. Adjust the brake shoes.

16. Install the wheels and tighten the lug nuts.

Spring Clip Type

✳✳ CAUTION

It is always a good idea to wear eye protection when working on brake components, especially drum brakes. Drum brakes often use powerful springs which could cause severe eye injury if they accidentally break. Also, brake shoes may contain asbestos, which is a known cancer-causing agent. As soon as the drum is removed, generously spray the entire brake assembly with brake parts cleaner. Let it dry before proceeding. It's a good idea to wear a filter mask when doing brake work.

1. Loosen the lug nuts on the applicable wheels.

2. Raise and safely support the vehicle.

3. Remove the wheels.

4. Remove the brake drum.

5. Remove the brake shoes.

➡On some vehicles, it may be possible to just remove the return springs and pull the shoes apart far enough for wheel cylinder removal. We do not recommend this for two reasons: wheel cylinder removal involves spilling some brake fluid—brake fluid can contaminate brake shoe friction material—and leaving the brake shoes on the backing plate can reduce working space and interfere with the job.

6. Disconnect and cap the brake line at the wheel cylinder.

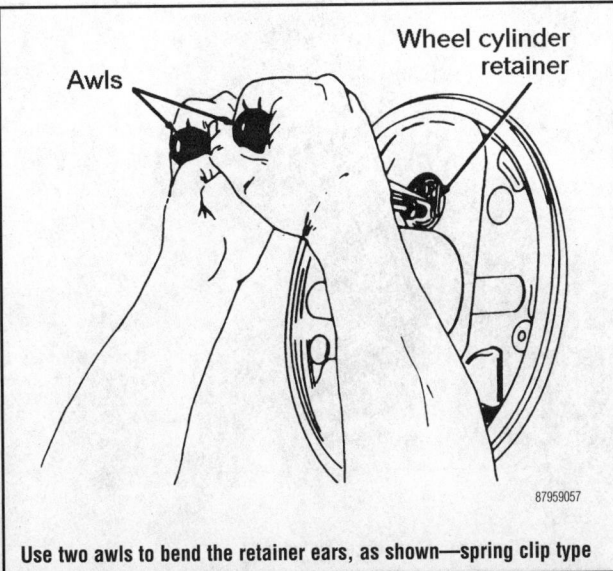

Use two awls to bend the retainer ears, as shown—spring clip type

---

☀☀ CAUTION

Plug the line immediately to prevent contamination of the brake fluid, because brake fluid absorbs water from the atmosphere very quickly. Water reduces the effectiveness of brake fluid, leading to increased brake fade.

---

7. Using two awls, release the spring clip securing the wheel cylinder to the backing plate.

8. Remove the wheel cylinder from the vehicle.

➡**On some GM vehicles it may be necessary to remove the bleeder screw from the wheel cylinder to remove it from the backing plate.**

To install:

9. If you are installing a new wheel cylinder, remove the bleeder screw from the wheel cylinder, then position the cylinder in

the backing plate. Removing the bleeder screw will keep it out of harm's way when installing the retaining clip.

10. Hold the wheel cylinder in place with a small prybar, using a socket (usually 1⅛ in./28.5mm on domestic vehicles) on the end of an extension, push the spring clip into place. Be sure both spring clip ears are seated correctly.

11. Connect the brake line to the wheel cylinder.

12. Install the bleeder screw and temporarily tighten it.

13. Install the brake shoes.

14. Install the brake drum.

15. Bleed the brake system.

16. Adjust the brake shoes.

17. Install the wheels and tighten the lug nuts.

OVERHAUL

Wheel cylinders can be overhauled, although most people do not bother. Replacing the wheel cylinder is much easier and requires no special tools or experience. If the cost difference between a rebuilding kit and new cylinder is not great, it's much safer to install the new cylinder.

If you decide to overhaul your wheel cylinder(s), you will need a wheel cylinder hone and a rebuild parts kit.

➡**It is possible to rebuild the wheel cylinder while still in place on the backing plate. There is no good reason to do so other than that, for some reason, you can't remove the cylinder. If you choose to do this, it is of the UTMOST importance that all material be flushed out of the bore before installing new parts. We DO NOT recommend rebuilding a wheel cylinder while it is installed on the backing plate.**

1. Remove the old wheel cylinder.

2. Thoroughly clean the outside of the unit with brake parts cleaner.

3. Place the cylinder on a clean work surface.

4. Remove the boots, then use a finger to push the pistons, cups and spring out of the bore.

---

1 Socket extension
2 1-1/8 in., 12 pt. socket

If the wheel cylinder uses a round type retainer, a socket and extension can be used to seat the retainer—spring clip type

To disassemble the wheel cylinder, first remove the outer boots . . .

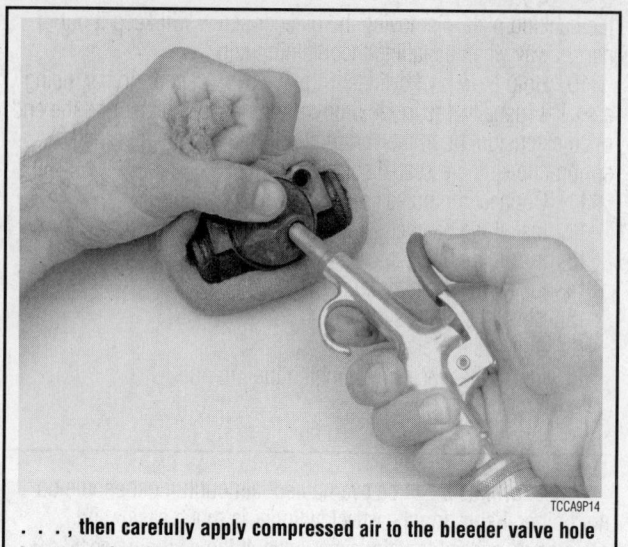

. . . , then carefully apply compressed air to the bleeder valve hole to extract the pistons and seals

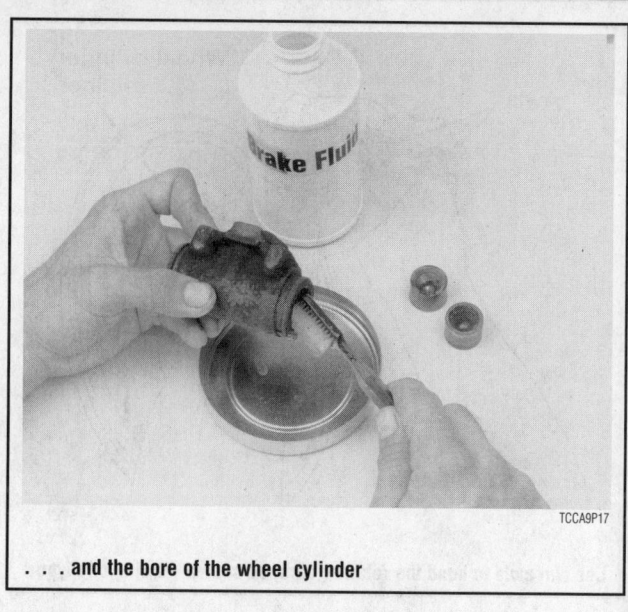

. . . and the bore of the wheel cylinder

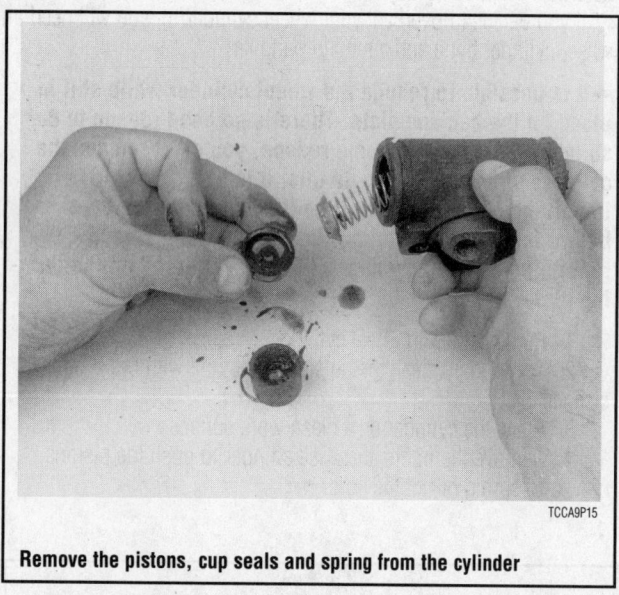

Remove the pistons, cup seals and spring from the cylinder

Once cleaned and inspected, the wheel cylinder is ready for assembly

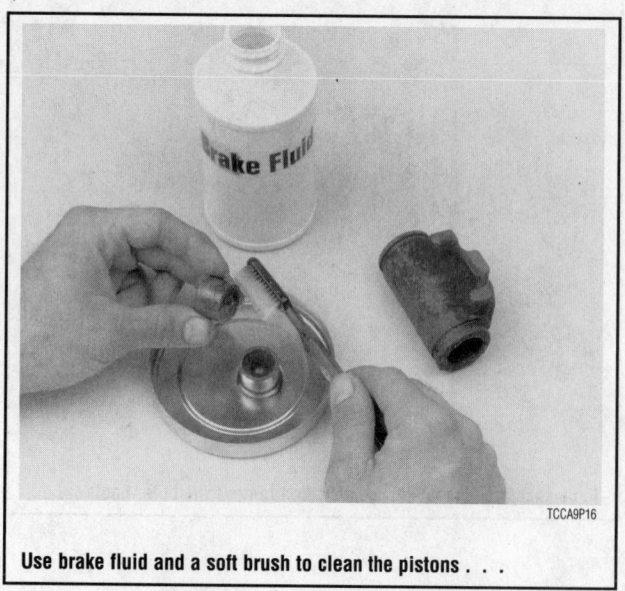

Use brake fluid and a soft brush to clean the pistons . . .

Lubricate the cup seals with brake fluid . . .

. . . , then install the spring and cup seals in the bore

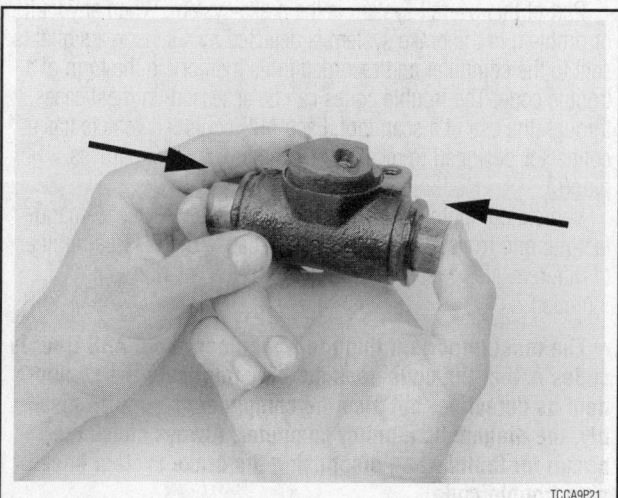

Lightly lubricate the pistons, then insert them into the wheel cylinder bore

5. Inspect the inner bore surface. If it is not badly pitted, rusted or scored, it can be rebuilt.

6. Remove the bleeder screw.

ANTI-LOCK BRAKE SYSTEMS

General Information

The purpose of the Anti-lock Brake System (ABS) is to prevent wheel lock-up under hard braking conditions. This is especially critical on wet or slippery surfaces. ABS is desirable because a vehicle that is stopped without locking one or more wheels, can stop with more control and in a shorter distance than a vehicle with locked wheels.

Under normal braking conditions, the ABS system operates just like a standard system. When one or more wheels shows a tendency to lock during braking, the ABS computer detects this and puts the system into the anti-lock mode. In this mode, hydraulic pressure is modulated to each wheel, preventing any one wheel from locking. The system can hold or reduce pressure at each

Finally, install the boots over the wheel cylinder piston ends

7. Install a wheel cylinder hone into a low-speed drill, and coat the inside of the cylinder with clean brake fluid.

8. Make several passes through the cylinder bore with the hone, never stopping in one place or passing completely through the bore.

9. Remove just enough material to establish a clean, cross-hatched inner surface.

10. Thoroughly clean the wheel cylinder bore with alcohol and let it dry. Blow out all passages with compressed air, including the bleeder screw area.

11. Coat the bore with clean brake fluid.

✳✳ WARNING

Be sure to use all of the replacement parts which come with the rebuild kit you purchased, otherwise the rebuilt wheel cylinder may not function properly.

12. Coat all replacement parts with clean brake fluid.

13. Install a cup and piston in one side, place the spring into the other side, followed by the other cup and piston. Push the pistons in until both are within the bore.

14. Install the end caps.

15. Loosely install the bleeder screw.

16. Install the rebuilt wheel cylinder.

wheel as necessary, depending on the signal received by the computer.

The effect is sort of like pumping your brakes, although it's done hundreds of time faster. In fact, when driving an ABS vehicle on ice or snow, a driver must overcome the urge to pump the brake during a stop. Let the ABS system work. Pumping the pedal on an ABS equipped vehicle will defeat the system.

PRECAUTIONS

• Do not use rubber hoses or other parts not specifically designed for the ABS system used by your vehicle. When using repair kits, replace all parts included in the kit. Partial or incorrect repair may lead to functional problems and require the replacement of components. NEVER fabricate your own replacement parts!

• Lubricate rubber parts with clean, fresh brake fluid to ease assembly. Do not use lubricated shop air to clean parts; damage to rubber components may result.

• Use only specified brake fluid from an unopened container.

• If any hydraulic component or line is removed or replaced, it may be necessary to bleed the entire system. This is always true when any upper end component (master cylinder, accumulator, control unit, etc.) is opened. It is also true when any lower end component (caliper or wheel cylinder) is opened and too much brake fluid has been lost; this does not happen often. If simply servicing a brake caliper, wheel cylinder, etc. and the line was adequately plugged after it was disconnected, the entire system will not need bleeding; only the component which was serviced. However, when in doubt, play it safe and bleed the entire system.

• A clean repair area is essential. Always clean the reservoir and cap thoroughly before removing the cap. The slightest amount of dirt in the fluid may plug an orifice and impair system function. Perform repairs after components have been thoroughly cleaned; use only denatured alcohol to clean components. Do not allow ABS components to come into contact with any substance containing mineral oil; this includes used shop rags.

• The anti-lock control unit is a microprocessor similar to other computer units in the vehicle. Ensure that the ignition switch is **OFF** before removing or installing controller wiring harnesses. Avoid static electricity discharge at or near the controller.

• If any arc welding is to be done on the vehicle, the control unit should be unplugged before welding operations begin.

ABS DEPRESSURIZING

Some ABS systems store the brake fluid at high pressures, which must be released before any service is attempted.

On these systems, the hydraulic accumulator contains brake fluid and nitrogen gas at extremely high pressures. Certain other system components may also contain brake fluid at high pressure. It is mandatory that the system pressure is relieved before disconnecting any hoses, lines or fittings, otherwise personal injury may result.

✳✳ CAUTION

On ABS systems designed to store brake fluid at high pressures, it is necessary to depressurize the system before disconnecting any hoses, lines or fittings. Otherwise, personal injury may result.

On most vehicles, ABS pressure can be depleted simply by pumping the brake pedal 20–30 times with the ignition switch **OFF**.

On some systems, particularly some GM systems, pressure should be bled using a specific, expensive scan tool. For this reason, we recommend that when in doubt, all ABS system service be referred to a professional, qualified technician.

DIAGNOSTIC TROUBLE CODES

The on-board computer system receives input from sensors all over the vehicle. The sensors signal the operating condition of every controlled component from the engine on down to the wheels.

Part of this overall system is the brake system. When any fault or problem in the brake system is detected by a sensor, a signal is sent to the computer and recorded in its memory in the form of a trouble code. The trouble codes can be accessed, in most cases, through the use of a scan tool. Each ABS equipped vehicle has a connector designed to receive the scan tools wiring harness plug(s).

Vehicle computer systems vary from manufacturer-to-manufacturer and from model-to-model. Because of the large number of different ABS systems, code retrieval information is not included.

➥**The most important thing to remember about ABS trouble codes is that the code does not only implicate the component as defective, but also the component's circuit, possibly, the diagnostic monitor computer. Always check the circuit for faults when diagnosing the brake system based on a trouble code.**

BRAKE SPECIFICATIONS
ACURA SLX
All measurements in inches unless noted

Year	Model		Master Cylinder Bore	Brake Disc Original Thickness	Brake Disc Minimum Thickness	Maximum Run-out	Brake Drum Diameter Original Inside Diameter	Brake Drum Diameter Max. Wear Limit	Brake Drum Diameter Maximum Machine Diameter	Minimum Lining Thickness Front	Minimum Lining Thickness Rear
1996	SLX	F	1.000	1.024	0.969 ①	0.005	—	—	—	0.039	—
		R	—	0.710	0.654 ②	0.005	8.27 ③	8.32 ③	8.32 ③	—	0.039 ④
1997	SLX	F	1.000	1.024	0.969 ①	0.005	—	—	—	0.039	—
		R	—	0.710	0.654 ②	0.005	8.27 ③	8.32 ③	8.32 ③	—	0.039 ④
1998-99	SLX	F	1.000	1.020	0.969 ①	0.005	—	—	—	0.039	—
		R	—	0.710	0.654 ②	0.005	8.27 ③	8.32 ③	8.32 ③	—	0.039 ④

① Minimum machine diameter: 0.983
② Minimum machine diameter: 0.668
③ Emergency brake drum surface
④ Specification includes disc pads and parking brake shoes

79246C01

BRAKE SPECIFICATIONS
CADILLAC ESCALADE, CHEVROLET ASTRO, BLAZER, C/K PICK-UPS, EXPRESS, G/P VANS, S10, SUBURBAN, TAHOE, VENTURE, GMC C/K PICK-UPS, DENALI, ENVOY, G/P VANS, JIMMY, S15, SAFARI, SAVANA, SONOMA, YUKON, OLDSMOBILE BRAVADA
All measurements in inches unless noted

Year	Model	Master Cylinder Bore	Brake Disc			Brake Drum Diameter			Minimum Lining Thickness		Brake Caliper	
			Original Thickness	Minimum Thickness	Maximum Runout	Original Inside Diameter	Max. Wear Limit	Maximum Machine Diameter	Front	Rear	Bracket Bolts (ft. lbs.)	Mounting Bolts (ft. lbs.)
1995	Astro	NA	①	②	0.004	9.50	9.59	9.56	0.030	0.030	—	38
	Bravada	NA	1.030	0.980	0.002	9.50	9.59	9.56	0.030	0.030	—	38
	C1500	NA	1.250	1.230	0.004	③	④	⑤	0.030	0.030	—	38
	C2500	NA	1.500	1.480	0.004	③	④	⑤	0.030	0.030	—	38
	C3500	NA	1.500	1.480	0.004	③	④	⑤	0.030	0.030	—	—
	G/P20	NA	⑥	⑦	0.004	③	④	⑤	0.030	0.030	—	38
	G/P30	NA	⑥	⑦	0.004	③	④	⑤	0.030	0.030	—	38
	Jimmy	NA	1.030	0.980	0.002	9.50	9.59	9.56	0.030	0.030	—	38
	K1500	NA	1.500	1.480	0.004	③	④	⑤	0.030	0.030	—	38
	K2500	NA	1.500	1.480	0.004	③	④	⑤	0.030	0.030	—	38
	K3500	NA	1.500	1.480	0.004	③	④	⑤	0.030	0.030	—	38
	S10 Blazer	NA	1.030	0.980	0.002	9.50	9.59	9.56	0.030	0.030	—	38
	S10 Pick-up	NA	1.030	0.980	0.002	9.50	9.59	9.56	0.030	0.030	—	38
	S15 Pick-up	NA	1.030	0.980	0.002	9.50	9.59	9.56	0.030	0.030	—	38
	Safari	NA	①	②	0.004	9.50	9.59	9.56	0.030	0.030	—	38
	Sonoma	NA	1.030	0.980	0.002	9.50	9.59	9.56	0.030	0.030	—	38
	Suburban	NA	1.500	1.480	0.004	③	④	⑤	0.030	0.030	—	38
	Tahoe	NA	1.500	1.480	0.004	③	④	⑤	0.030	0.030	—	38
	Yukon	NA	1.500	1.480	0.004	③	④	⑤	0.030	0.030	—	38
1996	Astro	NA	①	②	0.004	9.50	9.59	9.56	0.030	0.030	—	38
	Bravada	NA	1.030	0.965	0.003	9.50	9.59	9.56	0.030	0.030	—	38
	C1500	NA	1.250	1.230	0.004	③	④	⑤	0.030	0.030	—	38
	C2500	NA	1.500	1.480	0.004	③	④	⑤	0.030	0.030	—	38
	C3500	NA	1.500	1.480	0.004	③	④	⑤	0.030	0.030	—	—
	G/P1500	NA	⑥	⑦	0.004	③	④	⑤	0.030	0.030	—	38
	G/P2500	NA	⑥	⑦	0.004	③	④	⑤	0.030	0.030	—	38
	G/P3500	NA	⑥	⑦	0.004	③	④	⑤	0.030	0.030	—	38
	Jimmy	NA	1.030	0.965	0.003	9.50	9.59	9.56	0.030	0.030	—	38
	K1500	NA	1.500	1.480	0.004	③	④	⑤	0.030	0.030	—	38
	K2500	NA	1.500	1.480	0.004	③	④	⑤	0.030	0.030	—	38
	K3500	NA	1.500	1.480	0.004	③	④	⑤	0.030	0.030	—	38
	S10 Blazer	NA	1.030	0.965	0.003	9.50	9.59	9.56	0.030	0.030	—	38
	S10 Pick-up	NA	1.030	0.965	0.003	9.50	9.59	9.56	0.030	0.030	—	38
	S15 Pick-up	NA	1.030	0.965	0.003	9.50	9.59	9.56	0.030	0.030	—	38
	Safari	NA	①	②	0.004	9.50	9.59	9.56	0.030	0.030	—	38
	Sonoma	NA	1.030	0.965	0.003	9.50	9.59	9.56	0.030	0.030	—	38
	Suburban	NA	1.500	1.480	0.004	③	④	⑤	0.030	0.030	—	38
	Tahoe	NA	1.500	1.480	0.004	③	④	⑤	0.030	0.030	—	38
	Yukon	NA	1.500	1.480	0.004	③	④	⑤	0.030	0.030	—	38
1997	Astro	NA	①	②	0.004	9.50	9.59	9.56	0.030	0.030	—	38
	Bravada	NA	1.030	0.965	0.003	9.50	9.59	9.56	0.030	0.030	52	⑧
	C1500	NA	1.250	1.230	0.004	③	④	⑤	0.030	0.030	—	38
	C2500	NA	1.500	1.480	0.004	③	④	⑤	0.030	0.030	—	38
	C3500	NA	1.500	1.480	0.004	③	④	⑤	0.030	0.030	—	—

79246C02

BRAKE SPECIFICATIONS
CADILLAC ESCALADE, CHEVROLET ASTRO, BLAZER, C/K PICK-UPS, EXPRESS, G/P VANS, S10, SUBURBAN, TAHOE, VENTURE, GMC C/K PICK-UPS, DENALI, ENVOY, G/P VANS, JIMMY, S15, SAFARI, SAVANA, SONOMA, YUKON, OLDSMOBILE BRAVADA
All measurements in inches unless noted

Year	Model	Master Cylinder Bore	Brake Disc		Maximum Runout	Brake Drum Diameter			Minimum Lining Thickness		Brake Caliper	
			Original Thickness	Minimum Thickness		Original Inside Diameter	Max. Wear Limit	Maximum Machine Diameter	Front	Rear	Bracket Bolts (ft. lbs.)	Mounting Bolts (ft. lbs.)
1997 (cont.)	G/P1500	NA	⑥	⑦	0.004	③	④	⑤	0.030	0.030	—	38
	G/P2500	NA	⑥	⑦	0.004	③	④	⑤	0.030	0.030	—	38
	G/P3500	NA	⑥	⑦	0.004	③	④	⑤	0.030	0.030	—	38
	Jimmy	NA	1.030	0.965	0.003	9.50	9.59	9.56	0.030	0.030	52	⑧
	K1500	NA	1.500	1.480	0.004	③	④	⑤	0.030	0.030	—	38
	K2500	NA	1.500	1.480	0.004	③	④	⑤	0.030	0.030	—	38
	K3500	NA	1.500	1.480	0.004	③	④	⑤	0.030	0.030	—	38
	S10 Blazer	NA	1.030	0.965	0.003	9.50	9.59	9.56	0.030	0.030	52	⑧
	S10 Pick-up	NA	1.030	0.965	0.003	9.50	9.59	9.56	0.030	0.030	—	38
	S15 Pick-up	NA	1.030	0.965	0.003	9.50	9.59	9.56	0.030	0.030	—	38
	Safari	NA	①	②	0.004	9.50	9.59	9.56	0.030	0.030	—	38
	Sonoma	NA	1.030	0.965	0.003	9.50	9.59	9.56	0.030	0.030	—	38
	Suburban	NA	1.500	1.480	0.004	③	④	⑤	0.030	0.030	—	38
	Tahoe	NA	1.500	1.480	0.004	③	④	⑤	0.030	0.030	—	38
	Yukon	NA	1.500	1.480	0.004	③	④	⑤	0.030	0.030	—	38
1998-99	Astro	NA	①	②	0.004	9.50	9.59	9.56	0.030	0.030	—	38
	Bravada	NA	1.030	0.965	0.003	9.50	9.59	9.56	0.030	0.030	52	⑧
	C1500	NA	1.250	1.230	0.004	③	④	⑤	0.030	0.030	—	38
	C2500	NA	1.500	1.480	0.004	③	④	⑤	0.030	0.030	—	38
	C3500	NA	1.500	1.480	0.004	③	④	⑤	0.030	0.030	—	—
	Denali	NA	1.500	1.480	0.004	③	④	⑤	0.030	0.030	—	38
	Envoy	NA	1.030	0.965	0.003	9.50	9.59	9.56	0.030	0.030	52	⑧
	Escalade	NA	1.500	1.480	0.004	③	④	⑤	0.030	0.030	—	38
	G/P1500	NA	⑥	⑦	0.004	③	④	⑤	0.030	0.030	—	38
	G/P2500	NA	⑥	⑦	0.004	③	④	⑤	0.030	0.030	—	38
	G/P3500	NA	⑥	⑦	0.004	③	④	⑤	0.030	0.030	—	38
	Jimmy	NA	1.030	0.965	0.003	9.50	9.59	9.56	0.030	0.030	52	⑧
	K1500	NA	1.500	1.480	0.004	③	④	⑤	0.030	0.030	—	38
	K2500	NA	1.500	1.480	0.004	③	④	⑤	0.030	0.030	—	38
	K3500	NA	1.500	1.480	0.004	③	④	⑤	0.030	0.030	—	38
	S10 Blazer	NA	1.030	0.965	0.003	9.50	9.59	9.56	0.030	0.030	52	⑧
	S10 Pick-up	NA	1.030	0.965	0.003	9.50	9.59	9.56	0.030	0.030	—	38
	S15 Pick-up	NA	1.030	0.965	0.003	9.50	9.59	9.56	0.030	0.030	—	38
	Safari	NA	①	②	0.004	9.50	9.59	9.56	0.030	0.030	—	38
	Sonoma	NA	1.030	0.965	0.003	9.50	9.59	9.56	0.030	0.030	—	38
	Suburban	NA	1.500	1.480	0.004	③	④	⑤	0.030	0.030	—	38
	Tahoe	NA	1.500	1.480	0.004	③	④	⑤	0.030	0.030	—	38
	Yukon	NA	1.500	1.480	0.004	③	④	⑤	0.030	0.030	—	38

NA - Not Available
① Available with 1.040" and 1.250"rotors
② 1.040" rotors: 0.980
   1.250" rotors" 1.230
③ Available with 10", 11.15" and 13" drums

④ 10" drum: 10.05
   11.15" drum: 11.24
   13" drum: 13.09
⑤ 10" drum: 10.09
   11.15" drum: 11.21
   13" drum: 13.06

⑥ Available with 1.280 and 1.540 discs
⑦ 1.28" disc: 1.230
   1.54" disc: 1.480
⑧ 2WD: 38 ft. lbs.
   4WD: 77 ft. lbs.

BRAKE SPECIFICATIONS
CHEVROLET LUMINA APV, OLDSMOBILE SILHOUETTE, PONTIAC TRANS SPORT
All measurements in inches unless noted

Year	Model	Master Cylinder Bore	Brake Disc			Brake Drum Diameter			Minimum Lining Thickness		Brake Caliper	
			Original Thickness	Minimum Thickness	Maximum Runout	Original Inside Diameter	Max. Wear Limit	Maximum Machine Diameter	Front	Rear	Bracket Bolts (ft. lbs.)	Mounting Bolts (ft. lbs.)
1995	Lumina APV	0.944	1.260	1.209	0.002	8.86	8.92	8.90	0.030	0.030	—	38
	Silhouette	0.944	1.260	1.209	0.002	8.86	8.92	8.91	0.030	0.030	—	38
	Trans Sport	0.944	1.260	1.209	0.002	8.86	8.92	8.92	0.030	0.030	—	38
1996	Lumina APV	0.944	1.260	1.209	0.002	8.86	8.92	8.90	0.030	0.030	—	38
	Silhouette	0.944	1.260	1.209	0.002	8.86	8.92	8.91	0.030	0.030	—	38
	Trans Sport	0.944	1.260	1.209	0.002	8.86	8.92	8.92	0.030	0.030	—	38
1997	Silhouette	0.944	1.260	1.209	0.002	8.86	8.92	8.91	0.030	0.030	137	63
	Trans Sport	0.944	1.260	1.209	0.002	8.86	8.92	8.92	0.030	0.030	137	63
	Venture	0.944	1.260	1.209	0.002	8.86	8.92	8.90	0.030	0.030	137	63
1998-99	Silhouette	0.944	1.260	1.209	0.002	8.86	8.92	8.91	0.030	0.030	NA	NA
	Trans Sport	0.944	1.260	1.209	0.002	8.86	8.92	8.92	0.030	0.030	NA	NA
	Venture	0.944	1.260	1.209	0.002	8.86	8.92	8.90	0.030	0.030	NA	NA

NA - Not Available

79246C04

BRAKE SPECIFICATIONS
CHRYSLER TOWN & COUNTRY, DODGE CARAVAN, PLYMOUTH VOYAGER
All measurements in inches unless noted

Year	Model	Master Cylinder Bore	Brake Disc			Brake Drum Diameter			Minimum Lining Thickness	
			Original Thickness	Minimum Thickness	Maximum Run-out	Original Inside Diameter	Max. Wear Limit	Maximum Machine Diameter	Front	Rear
1995	Caravan	0.940	0.940	0.880	0.005	9.00	9.09	9.06	0.060	0.060 ①
	Town & Country	0.940	0.940	0.880	0.005	9.00	9.09	9.06	0.062	0.062 ①
	Voyager	0.940	0.940	0.880	0.005	9.00	9.09	9.06	0.060	0.060 ①
1996	Caravan	0.940	0.940	0.880	0.005	9.00	9.09	9.06	0.060	0.060 ①
	Town & Country	0.937	0.940	0.881	0.005	9.84	9.93	9.90	0.062	0.062 ①
	Voyager	0.937	0.939	0.881	0.005	9.84	9.93	9.90	0.060	0.060 ①
1997	Caravan	0.937	②	③	0.005	9.84	9.93	9.90	0.313	④
	Town & Country	0.937	②	③	0.005	9.84	9.93	9.90	0.313	④
	Voyager	0.937	②	③	0.005	9.84	9.93	9.90	0.313	④
1998-99	Caravan	⑤	⑥	⑦	0.005	9.84	9.93	9.90	0.313	④
	Town & Country	⑤	⑥	⑦	0.005	9.84	9.93	9.90	0.313	④
	Voyager	⑤	⑥	⑦	0.005	9.84	9.93	9.90	0.313	④

① For riveted brake shoes: 0.031
② Front rotor: 0.939-0.949
   Rear rotor: 0.458-0.478
③ Front rotor: 0.881
   Rear rotor: 0.409
④ Rear drum brake shoes: 0.031
   Rear disc brake pads: 0.281
⑤ FWD models: 0.937
   AWD models: 1.000
⑥ Front rotor: 0.939-0.949
   Rear rotor: 0.482-0.502
⑦ Front rotor: 0.881
   Rear rotor: 0.443

79246C05

BRAKE SPECIFICATIONS
FORD AEROSTAR, BRONCO, E-SERIES, EXPEDITION, EXPLORER, F-SERIES, RANGER
LINCOLN NAVIGATOR, MERCURY MOUNTAINEER
All measurements in inches unless noted

Year	Model		Master Cyl. Bore	Brake Disc Original Thickness	Brake Disc Minimum Thickness	Brake Disc Maximum Runout	Brake Drum Diameter Original Inside Diameter	Max. Wear Limit	Maximum Machine Diameter	Minimum Lining Thickness Front	Rear	Brake Caliper Bracket Bolts (ft. lbs.)	Mounting Bolts (ft. lbs.)
1995	Aerostar	④	0.938	0.850	0.810	0.003	9.00	9.09	9.06	0.030	0.030	72-97	21-26
		⑤	0.938	0.850	0.810	0.003	10.00	10.09	10.06	0.030	0.030	72-97	21-26
	Bronco		1.000	1.160	0.960	0.003	11.03	11.09	11.06	0.030	0.030	125-169	21-26
	E-150		1.000	1.160	1.120	0.003	11.03	11.09	11.06	0.030	0.030	141-191	22-36
	E-250		①	1.220	1.180	0.003	12.00	12.09	12.06	0.030	0.030	141-191	22-36
	E-350		NA	1.220	1.180	0.003	12.00	12.09	12.06	0.030	0.030	141-191	22-36
	Explorer	④	0.938	0.850	0.810	0.003	9.00	9.09	9.06	0.030	0.030	72-97	21-26
		⑥	0.938	0.850	0.810	0.003	10.00	10.09	10.06	0.030	0.030	72-97	21-26
		⑦	0.938	0.850	0.810	0.003	10.00	10.09	10.06	0.030	0.030	72-97	21-26
	F-150		1.000	1.160	0.960	0.003	11.03	11.09	11.06	0.030	0.030	125-169	21-26
	F-250		①	1.220	②	0.003	12.00	12.09	12.06	0.030	0.030	125-169	21-26
	F-350		1.125	1.220	②	③	12.00	12.09	12.06	0.030	0.030	125-169	21-26
	F-Super Duty	F	NA	1.220	1.180	0.008	—	—	—	0.030	—	166	42
		R	—	NA	1.430	0.008	—	—	—	—	0.030	—	—
	Ranger	④	0.938	0.850	0.810	0.003	9.00	9.09	9.06	0.030	0.030	72-97	21-26
		⑥	0.938	0.850	0.810	0.003	10.00	10.09	10.06	0.030	0.030	72-97	21-26
		⑦	0.938	0.850	0.810	0.003	10.00	10.09	10.06	0.030	0.030	72-97	21-26
1996	Aerostar	④	0.938	0.850	0.810	0.003	9.00	9.09	9.06	0.030	0.030	72-97	21-26
		⑤	0.938	0.850	0.810	0.003	10.00	10.09	10.06	0.030	0.030	72-97	21-26
	Bronco		1.000	1.160	0.960	0.003	11.03	11.09	11.06	0.030	0.030	125-169	21-26
	E-150		1.000	1.160	1.120	0.003	11.03	11.09	11.06	0.030	0.030	141-191	22-36
	E-250		①	1.220	1.180	0.003	12.00	12.09	12.06	0.030	0.030	141-191	22-36
	E-350		NA	1.220	1.180	0.003	12.00	12.09	12.06	0.030	0.030	141-191	22-36
	Explorer	④	0.938	0.850	0.810	0.003	9.00	9.09	9.06	0.030	0.030	72-97	21-26
		⑥	0.938	0.850	0.810	0.003	10.00	10.09	10.06	0.030	0.030	72-97	21-26
		⑦	0.938	0.850	0.810	0.003	10.00	10.09	10.06	0.030	0.030	72-97	21-26
	F-150	⑧	1.062	NA	0.972	NA	11.03	11.12	NA	0.156	0.030	125-169	21-26
	F-150		1.000	1.160	0.960	0.003	11.03	11.09	11.06	0.030	0.030	125-169	21-26
	F-250		①	1.220	②	0.003	12.00	12.09	12.06	0.030	0.030	125-169	21-26
	F-350		1.125	1.220	②	③	12.00	12.09	12.06	0.030	0.030	125-169	21-26
	Mountaineer	④	0.938	0.850	0.810	0.003	9.00	9.09	9.06	0.030	0.030	72-97	21-26
		⑥	0.938	0.850	0.810	0.003	10.00	10.09	10.06	0.030	0.030	72-97	21-26
		⑦	0.938	0.850	0.810	0.003	10.00	10.09	10.06	0.030	0.030	72-97	21-26
	Ranger	④	0.938	0.850	0.810	0.003	9.00	9.09	9.06	0.030	0.030	72-97	21-26
		⑥	0.938	0.850	0.810	0.003	10.00	10.09	10.06	0.030	0.030	72-97	21-26
		⑦	0.938	0.850	0.810	0.003	10.00	10.09	10.06	0.030	0.030	72-97	21-26
	F-Super Duty	F	NA	1.220	1.180	0.008	—	—	—	0.030	—	166	42
		R	—	NA	1.430	0.008	—	—	—	—	0.030	—	—
1997	Aerostar	④	0.938	0.850	0.810	0.003	9.00	9.09	9.06	0.030	0.030	72-97	21-26
		⑤	0.938	0.850	0.810	0.003	10.00	10.09	10.06	0.030	0.030	72-97	21-26
	E-150		1.000	1.160	1.120	0.003	11.03	11.09	11.06	0.030	0.030	141-191	22-36
	E-250		①	1.220	1.180	0.003	12.00	12.09	12.06	0.030	0.030	141-191	22-36
	E-350		NA	1.220	1.180	0.003	12.00	12.09	12.06	0.030	0.030	141-191	22-36

79246C06

BRAKE SPECIFICATIONS
FORD AEROSTAR, BRONCO, E-SERIES, EXPEDITION, EXPLORER, F-SERIES, RANGER
LINCOLN NAVIGATOR, MERCURY MOUNTAINEER
All measurements in inches unless noted

Year	Model			Master Cyl. Bore	Brake Disc			Brake Drum Diameter			Minimum Lining Thickness		Brake Caliper	
					Original Thickness	Minimum Thickness	Maximum Runout	Original Inside Diameter	Max. Wear Limit	Maximum Machine Diameter	Front	Rear	Bracket Bolts (ft. lbs.)	Mounting Bolts (ft. lbs.)
1997 (cont.)	Explorer	④		0.938	0.850	0.810	0.003	9.00	9.09	9.06	0.030	0.030	72-97	21-26
		⑥		0.938	0.850	0.810	0.003	10.00	10.09	10.06	0.030	0.030	72-97	21-26
		⑦		0.938	0.850	0.810	0.003	10.00	10.09	10.06	0.030	0.030	72-97	21-26
	F-150	⑧		1.062	NA	0.972	NA	11.03	11.12	NA	0.156	0.030	125-169	21-26
	F-150			1.000	1.160	0.960	0.003	11.03	11.09	11.06	0.030	0.030	125-169	21-26
	F-250			①	1.220	②	0.003	12.00	12.09	12.06	0.030	0.030	125-169	21-26
	F-350			1.125	1.220	②	③	12.00	12.09	12.06	0.030	0.030	125-169	21-26
	Mountaineer	④		0.938	0.850	0.810	0.003	9.00	9.09	9.06	0.030	0.030	72-97	21-26
		⑥		0.938	0.850	0.810	0.003	10.00	10.09	10.06	0.030	0.030	72-97	21-26
		⑦		0.938	0.850	0.810	0.003	10.00	10.09	10.06	0.030	0.030	72-97	21-26
	Ranger	④		0.938	0.850	0.810	0.003	9.00	9.09	9.06	0.030	0.030	72-97	21-26
		⑥		0.938	0.850	0.810	0.003	10.00	10.09	10.06	0.030	0.030	72-97	21-26
		⑦		0.938	0.850	0.810	0.003	10.00	10.09	10.06	0.030	0.030	72-97	21-26
	F-Super Duty		F	NA	1.220	1.180	0.008	—	—	—	0.030	—	166	42
			R	—	NA	1.430	0.008	—	—	—	—	0.030	—	—
	Expedition		F	1.000	1.023	0.964	0.0025	—	—	—	0.030	0.030	125-168	21-26
			R	—	0.700	0.657	0.025	—	—	—	0.030	0.030	120	20
1998-99	E-150			0.938	1.160	0.960	0.0025	11.03	11.09	11.06	0.030	0.030	141-191	22-26
	E-250			1.000	1.30	1.100	0.0003	12.00	12.09	12.06	0.030	0.030	141-191	22-26
	E-350			1.125	1.30	1.100	0.0003	12.00	12.09	12.06	0.030	0.030	141-191	22-26
	F-150			1.000	⑩	⑨	0.0025	11.03	11.09	11.06	0.030	0.030	125-169	21-26
	F-250			1.062	⑩	⑨	0.0025	12.00	12.09	12.06	0.030	0.030	125-169	21-26
	F-350			1.125	⑩	⑨	0.0025	12.00	12.09	12.06	0.030	0.030	125-169	21-26
	F-Super Duty			1.125	1.220	1.180	0.0025	12.00	12.09	12.06	0.030	0.030	166	42
	Expedition		F	1.000	1.023	0.964	0.0025	—	—	—	0.030	0.030	125-168	21-26
			R	—	0.700	0.657	0.025	—	—	—	0.030	0.030	120	20
	Navigator		F	1.000	1.023	0.964	0.0025	—	—	—	0.030	0.030	125-168	21-26
			R	—	0.700	0.657	0.025	—	—	—	0.030	0.030	120	20
	Explorer	④		0.938	0.850	0.810	0.003	9.00	9.09	9.06	0.030	0.030	72-97	21-26
		⑥		0.938	0.850	0.810	0.003	10.00	10.09	10.06	0.030	0.030	72-97	21-26
		⑦		0.938	0.850	0.810	0.003	10.00	10.09	10.06	0.030	0.030	72-97	21-26
	Mountaineer	④		0.938	0.850	0.810	0.003	9.00	9.09	9.06	0.030	0.030	72-97	21-26
		⑥		0.938	0.850	0.810	0.003	10.00	10.09	10.06	0.030	0.030	72-97	21-26
		⑦		0.938	0.850	0.810	0.003	10.00	10.09	10.06	0.030	0.030	72-97	21-26
	Ranger	④		0.938	0.850	0.810	0.003	9.00	9.09	9.06	0.030	0.030	72-97	21-26
		⑥		0.938	0.850	0.810	0.003	10.00	10.09	10.06	0.030	0.030	72-97	21-26
		⑦		0.938	0.850	0.810	0.003	10.00	10.09	10.06	0.030	0.030	72-97	21-26

NOTE: Due to changes made during production, refer to manufacturer's specifications if they differ from those in this chart

NA - Not Available

F - Front

R - Rear

① Under 6900 GVW: 1.062
   Over 6900 GVW: 1.125
② 4x2: 1.100
   4x4: 1.120
③ Except F-350 4x2 with dual rear wheel and 2-piece rotor/hub:0.003 in.
   F-350 4x2 with dual rear wheel and 2-piece rotor/hub: 0.010 in.
④ With 9 inch brakes
⑤ With 10 inch brakes

⑥ 4x2 with 10 inch brakes
⑦ 4x4 with 10 inch brakes
⑧ 1997 only
⑨ 0.972 for 4x2
   1.09 for 4x4
⑩ 1.020 for 4x2
   1.220 for 4x4

79246C07

BRAKE SPECIFICATIONS
FORD WINDSTAR, MERCURY VILLAGER
All measurements in inches unless noted

Year	Model		Master Cyl. Bore	Brake Disc Original Thickness	Brake Disc Minimum Thickness	Brake Disc Maximum Runout	Brake Drum Diameter Original Inside Diameter	Brake Drum Diameter Max. Wear Limit	Brake Drum Diameter Maximum Machine Diameter	Minimum Lining Thickness Front	Minimum Lining Thickness Rear	Brake Caliper Bracket-to-Hub Bolt (ft. lbs.)	Brake Caliper Mounting Pin or Bolt (ft. lbs.)
1995	Windstar		NA	1.020	0.097	0.003	9.84	9.90	9.90	0.040	0.590	84	25
1996	Windstar	F	NA	1.020	0.907	0.003	—	—	—	0.125	—	84	25
		R	NA	0.472	0.409	0.003	9.84	9.90	9.90	—	①	—	11-14
	Villager	F	1.000	1.005	0.945	0.003	—	—	—	0.079	—	—	18-25
		R	—	—	—	—	9.84	9.90	9.86	—	0.059	—	—
1997	Windstar	F	NA	1.020	0.907	0.003	—	—	—	0.125	—	84	25
		R	NA	0.472	0.409	0.003	9.84	9.90	9.90	—	①	—	11-14
	Villager	F	1.000	1.005	0.945	0.003	—	—	—	0.079	—	—	18-25
		R	—	—	—	—	9.84	9.90	9.86	—	0.059	—	—
1998-99	Windstar	F	NA	1.020	0.974	0.003 ②	—	—	—	0.125	—	73-97	25
		R	—	0.472	0.409	0.003 ②	9.84	9.90	9.90	—	③	—	11-14
	Villager	F	1.000	1.005	④	0.0028	—	—	—	0.080	—	—	18-25
		R	—	—	—	—	9.84	9.90	9.86	—	0.059	—	—

NOTE: Due to changes made during production, refer to the manufacturer's specifications if they differ from those in this chart

NA - Not Available

F - Front

R - Rear

① With drum brakes: 0.030
 With disc brakes: 0.125

② On vehicle.

③ With drum brakes: 0.059
 With disc brakes: 0.125

④ Discard thickness: 0.945
 Machining limit: 0.974

79246C08

BRAKE SPECIFICATIONS
GEO/CHEVROLET TRACKER
All measurements in inches unless noted

Year	Model	Master Cylinder Bore	Brake Disc Original Thickness	Brake Disc Minimum Thickness	Brake Disc Maximum Runout	Brake Drum Diameter Original Inside Diameter	Brake Drum Diameter Max. Wear Limit	Brake Drum Diameter Maximum Machine Diameter	Minimum Lining Thickness Front	Minimum Lining Thickness Rear	Brake Caliper Bracket Bolts (ft. lbs.)	Brake Caliper Mounting Bolts (ft. lbs.)
1995	Tracker	NA	0.394	0.315	0.006	8.66	8.74	8.74	0.236 ①	0.210 ①	—	20
1996	Tracker	NA	②	0.315	0.006	③	8.74	8.74	0.236 ①	0.210 ①	—	20
1997	Tracker	NA	②	0.315	0.006	③	8.74	8.74	0.236 ①	0.210 ①	—	20
1998	Tracker	NA	②	0.315	0.006	③	8.74	8.74	0.236 ①	0.210 ①	—	20

NA - Not Available

① Minimum lining thickness includes pad/shoe backing

② 2 door: 0.394
 4 door: 0.670 (service limit 0.590)

③ 2 door: 8.66 (service limit 0.874)
 4 door: 10.00 (service limit 10.07)

79246C09

BRAKE SPECIFICATIONS
DODGE DAKOTA, DURANGO, RAM TRUCKS, RAM VANS
All measurements in inches unless noted

Year	Model	Master Cylinder Bore	Brake Disc Original Thickness	Brake Disc Minimum Thickness	Maximum Run-out	Brake Drum Diameter Original Inside Diameter	Brake Drum Diameter Max. Wear Limit	Brake Drum Diameter Maximum Machine Diameter	Minimum Lining Thickness Front	Minimum Lining Thickness Rear
1995	B150 Van	1.125	—	①	0.004	11.00	11.09	11.06	0.125	0.062 ②
	B250 Van	1.125	—	①	0.004	11.00	11.09	11.06	0.125	0.062 ②
	B350 Van	1.125	—	①	0.004	12.00	12.09	12.06	0.125	0.062 ②
	Dakota	NA	0.861	0.810	0.004	9.00	9.09	9.06	0.060	0.060 ②
	Dakota	NA	0.861	0.810	0.004	10.00	10.09	10.06	0.060	0.060 ②
	D1500 Pick-up	1.125	1.260	①	0.004	11.00	11.09	11.06	0.062	0.062 ②
	D2500 Pick-up	1.250	1.500	①	0.005	13.00	13.09	13.06	0.062	0.062 ②
	D3500 Pick-up	1.250	1.500	①	0.005	13.00	13.09	13.06	0.062	0.062 ②
	W1500 Pick-up	1.125	1.260	①	0.004	11.00	11.09	11.06	0.062	0.062 ②
	W2500 Pick-up	1.250	1.500	①	0.005	13.00	13.09	13.06	0.062	0.062 ②
	W3500 Pick-up	1.250	1.500	①	0.005	13.00	13.09	13.06	0.062	0.062 ②
1996	Ram 1500 Pick-up	1.125	1.260	①	0.004	11.00	11.09	11.06	0.062	0.062 ②
	Ram 2500 Pick-up	1.250	1.500	①	0.005	13.00	13.09	13.06	0.062	0.062 ②
	Ram 3500 Pick-up	1.250	1.500	①	0.005	13.00	13.09	13.06	0.062	0.062 ②
	B1500 Van	1.125	—	①	0.004	11.00	11.09	11.06	0.125	0.062 ②
	B2500 Van	1.125	—	①	0.004	11.00	11.09	11.06	0.125	0.062 ②
	B3500 Van	1.125	—	①	0.004	12.00	12.09	12.06	0.125	0.062 ②
	Dakota ③	NA	0.861	0.810	0.004	9.00	9.09	9.06	0.060	0.060 ②
	Dakota ④	NA	0.861	0.810	0.004	10.00	10.09	10.06	0.060	0.060 ②
1997	Ram 1500 Pick-up	1.125	1.260	①	0.004	11.00	11.09	11.06	0.062	0.062 ②
	Ram 2500 Pick-up	1.250	1.500	①	0.005	13.00	13.09	13.06	0.062	0.062 ②
	Ram 3500 Pick-up	1.250	1.500	①	0.005	13.00	13.09	13.06	0.062	0.062 ②
	B1500 Van	1.125	—	①	0.004	11.00	11.09	11.06	0.125	0.062 ②
	B2500 Van	1.125	—	①	0.004	11.00	11.09	11.06	0.125	0.062 ②
	B3500 Van	1.125	—	①	0.004	12.00	12.09	12.06	0.125	0.062 ②
	Dakota ③	NA	0.861	0.810	0.004	9.00	9.09	9.06	0.060	0.060 ②
	Dakota ④	NA	0.861	0.810	0.004	10.00	10.09	10.06	0.060	0.060 ②
1998-99	Ram 1500 Pick-up	1.25	⑤	⑥	0.005	2.0	2.060	2.060	⑦	⑧
	Ram 2500 Pick-up	1.25	⑨	⑩	0.005	2.5	2.560	2.560	⑦	⑧
	Ram 3500 Pick-up	1.25	⑪	⑫	0.005	3.5	3.560	3.560	⑦	⑧
	B1500 Van	NA	1.26	1.181	0.004	11.03	⑬	⑬	⑦	⑧
	B2500 Van	NA	1.26	1.181	0.004	11.03	⑬	⑬	⑦	⑧
	B3500 Van	NA	1.26	1.181	0.004	12.125	⑬	⑬	⑦	⑧
	Dakota ③	NA	0.944	0.890	0.004	9.00	⑬	⑬	⑦	⑧
	Dakota ④	NA	.0944	0.890	0.004	10.00	⑬	⑬	⑦	⑧
	Durango	1.06	0.900	0.890	0.004	11.00	⑬	⑬	⑦	⑧

① Minimum thickness indicated on rotor hub
② For riveted brake shoes: 0.031
③ With 9 inch rear brakes
④ With 10 inch rear brakes
⑤ 2WD: 1.26 in.
   4WD: 1.5 in.
⑥ 2WD: 1.215 in.
   4WD: 1.269 in.
⑦ Riveted brake pads: 0.0625 in.
   Bonded brake pads: 0.1875 in.

⑧ Riveted brake shoes: 0.031 in.
   Bonded brake shoes: 0.0625 in.
⑨ 2WD: 1.5 in.
   4WD LD: 1.5 in.
   4WD HD: 1.75 in.
⑩ 2WD: 1.269 in.
   4WD LD: 1.269 in.
   4WD HD: 1.521

⑪ 2WD: 1.75 in.
   4WD: 1.75 in.
⑫ 2WD: 1.518 in.
   4WD: 1.521 in.
⑬ Maximum allowable drum diameter, either from wear or machining, is stamped on the drum.

79246C10

BRAKE SPECIFICATIONS
HONDA CR-V, ODYSSEY, PASSPORT
All measurements in inches unless noted

Year	Model		Master Cylinder Bore	Brake Disc Original Thickness	Brake Disc Minimum Thickness	Maximum Runout	Brake Drum Diameter Original Inside Diameter	Brake Drum Diameter Max. Wear Limit	Brake Drum Diameter Maximum Machine Diameter	Minimum Lining Thickness Front	Minimum Lining Thickness Rear	Brake Caliper Bracket Bolts (ft. lbs.)	Brake Caliper Mounting Bolts (ft. lbs.)
1995	Odyssey	F	NA	0.930	0.830	0.004	—	—	—	0.060	—	—	36
		R	—	0.350	0.300	0.004	6.69	6.73	6.73	—	0.040 ①	28	17
	Passport	F	1.000	②	③	0.005	—	—	—	0.039	—	115	54
		R	—	0.710	0.654	0.005	10.00 ④	10.06 ④	10.06	—	0.039 ⑤	76	32
1996	Odyssey	F	NA	0.937	0.830	0.004	—	—	—	0.060	—	—	36
		R	—	0.358	0.300	0.004	6.69	6.73	6.73	—	0.040	28	17
	Passport	F	1.000	1.020	0.983	0.005	—	—	—	0.039	—	115	54
		R	—	0.710	0.654	0.005	10.00 ④	10.06 ④	10.06	—	0.039 ⑤	76	32
1997	Odyssey	F	NA	0.937	0.830	0.004	—	—	—	0.060	—	—	36
		R	—	0.358	0.300	0.004	6.69	6.73	6.73	—	0.040	28	17
	CR-V	F	NA	0.910	0.830	0.004	—	—	—	0.060		83	36
		R	—	—	—	—	8.66	8.70	NA	—	0.080	—	—
	Passport	F	1.000	1.020	0.983	0.005	—	—	—	0.039	—	115	54
		R	—	0.710	0.654	0.005	10.00 ④	10.06 ④	10.06	·	0.039 ⑤	76	32
1998-99	Odyssey	F	NA	0.933	0.830	0.004	—	—	—	0.060	—	—	36
		R	—	0.354	0.300	0.004	—	—	—	—	0.060	28	17
	CR-V	F	NA	0.910	0.830	0.004	—	—	—	0.060	—	83	36
		R	—	—	—	—	8.66	8.70	NA	—	0.080	—	—
	Passport	F	1.000	1.020	0.983	0.005	—	—	—	0.039	—	115	54
		R	—	0.710	0.654	0.005	10.00 ④	10.06 ④	10.06	—	0.039 ⑤	76	32

NA - Not Available

F - Front

R - Rear

① Rear disc brakes: 0.060

② 3.2L engine: 1.020
   2.6L engine: 0.866

③ 3.2L engine: 0.983
   2.6L engine: 0.811

④ Parking brake drum:
   Original inside diameter: 8.27
   Maximum wear limit: 8.32

⑤ Specifications include disc pads
   and parking brake shoes

79246C11

BRAKE SPECIFICATIONS
INFINITI QX4, NISSAN FRONTIER, PATHFINDER, PICK-UP, QUEST
All measurements in inches unless noted

Year	Model	Master Cylinder Bore	Brake Disc			Brake Drum Diameter			Minimum Lining Thickness		Brake Caliper	
			Original Thickness	Minimum Thickness	Maximum Runout	Original Inside Diameter	Max. Wear Limit	Maximum Machine Diameter	Front	Rear	Bracket Bolts (ft. lbs.)	Mounting Bolts (ft. lbs.)
1995	Pathfinder	1.000	①	②	0.003	③	NA	④	0.079	0.059	53-72	24-31
	Pick-up	1.000	⑤	⑥	0.003	⑥	NA	⑦	0.079	0.059	53-72	16-23
	Quest	1.000	1.020	0.945	0.003	9.84	NA	9.90	0.079	0.079	—	12-14
1996	Pathfinder	1.000	①	②	0.003	③	NA	④	0.079	0.059	53-72	24-31
	Pick-up	1.000	⑤	⑥	0.003	⑥	NA	⑦	0.079	0.059	53-72	16-23
	Quest	1.000	1.020	0.945	0.003	9.84	NA	9.90	0.079	0.079	—	12-14
1997	Pathfinder	1.000	①	②	0.003	③	NA	④	0.079	0.059	53-72	24-31
	Pick-up	1.000	⑤	⑥	0.003	⑥	NA	⑦	0.079	0.059	53-72	16-23
	Quest	1.000	1.020	0.945	0.003	9.84	NA	9.90	0.079	0.079	—	12-14
	QX4	1.000	1.100	1.024	0.004	11.61	NA	11.67	0.079	0.059	53-72	24-31
1998-99	Frontier	1.000	⑤	0.945	0.003	⑧	NA	⑦	0.079	0.059	53-72	24-31
	Pathfinder	1.000	1.100	1.024	0.004	11.61	NA	11.67	0.079	0.059	53-72	16-23
	Quest	1.000	1.020	0.945	0.003	9.84	NA	9.90	0.079	0.079	—	12-14
	QX4	1.000	1.100	1.024	0.004	11.61	NA	11.67	0.079	0.059	53-72	24-31

NA - Not Available

① Front: 1.020
   Rear: 0.710
② Front: 0.945
   Rear: 0.630
③ Rear drum: 10.24
   Parking drum: 7.48
④ Rear drum brake: 10.30
   Rear disc parking brake drum: 7.52
⑤ 2WD KA24E: 0.870
   2WD VG30E: 1.020
   4WD: 1.020
⑥ 2WD KA24E: 0.787
   2WD/4WD VG30E: 0.945
⑦ 2WD: 10.30
   4WD: 11.67
   Parking brake drum: 7.52
⑧ 2WD: 0.787
   4WD: 0.945

79246C19

BRAKE SPECIFICATIONS
ISUZU HOMBRE, PICK-UP, RODEO, TROOPER
All measurements in inches unless noted

Year	Model		Master Cylinder Bore	Brake Disc Original Thickness	Brake Disc Minimum Thickness	Brake Disc Maximum Runout	Brake Drum Diameter Original Inside Diameter	Brake Drum Diameter Max. Wear Limit	Brake Drum Diameter Maximum Machine Diameter	Minimum Lining Thickness Front	Minimum Lining Thickness Rear	Brake Caliper Bracket Bolts (ft. lbs.)	Brake Caliper Mounting Bolts (ft. lbs.)
1995	Pick-up	F	0.938	0.886	0.811	0.0050	—	—	—	0.039	—	103-126	24
		R	—	0.472	0.417	0.0051	10.01	10.06	10.06	—	①	69-84	12-17
	Rodeo ②		1.000	0.866	NA	0.005	10.00	10.06	10.06	0.039	0.039	115	54
	Rodeo ③	F	1.000	1.026	0.969	0.005	—	—	—	0.039	—	115	54
		R	—	0.709	0.654	0.005	8.27 ④	8.32 ④	8.32 ④	—	0.039 ⑤	76	32
	Trooper		1.000	⑥	⑦	0.005	8.27 ④	8.32 ④	8.32 ④	0.039	0.039 ⑤	115	54
1996	Hombre		NA	1.040	0.980	0.004	9.50	9.59	9.56	0.030	0.030	—	38
	Oasis	F	NA	0.930	0.830	0.004	—	—	—	0.060	—	—	36
		R	—	0.350	0.300	0.004	6.69 ④	6.73 ④	6.73 ④	—	0.040	—	17
	Rodeo ②	F	1.000	1.020	0.983	0.005	—	—	—	0.039	—	115	54
		R	—	—	—	0.005	10.00	10.06	10.06	—	0.039	76	32
	Rodeo ③	F	1.000	1.024	0.969 ⑧	0.005	—	—	—	0.039	—	115	54
		R	—	0.710	0.654 ⑨	0.005	8.27 ④	8.32 ④	8.32 ④	—	0.039 ⑤	76	32
	Trooper	F	1.000	1.024	0.969 ⑧	0.005	—	—	—	0.039	—	115	54
		R	—	0.710	0.654 ⑨	0.005	8.27 ④	8.32 ④	8.32 ④	—	0.039 ⑤	76	32
1997	Hombre	F	NA	1.027 ⑩	0.965 ⑪	0.004	—	—	—	0.030	—	—	38
		R	—	0.787	0.728 ⑫	0.004	9.50	9.59	9.56	—	0.030	52	23 ⑬
	Oasis	F	NA	0.930	0.830	0.004	—	—	—	0.060	—	—	36
		R	—	0.350	0.300	0.004	6.69 ④	6.73 ④	6.73 ④	—	0.040	—	17
	Rodeo ②	F	1.000	1.020	0.983	0.005	—	—	—	0.039	—	115	54
		R	—	—	—	0.005	10.00	10.06	10.06	—	0.039	76	32
	Rodeo ③	F	1.000	1.024	0.969 ⑧	0.005	—	—	—	0.039	—	115	54
		R	—	0.710	0.654 ⑨	0.005	8.27 ④	8.32 ④	8.32 ④	—	0.039 ⑤	76	32
	Trooper	F	1.000	1.024	0.969 ⑧	0.005	—	—	—	0.039	—	115	54
		R	—	0.710	0.654 ⑨	0.005	8.27 ④	8.32 ④	8.32 ④	—	0.039 ⑤	76	32
1998-99	Amigo	F	1.000	1.020	0.969 ⑭	0.005	—	—	—	0.039	—	115	54
		R	—	0.710	0.654 ⑮	0.005	11.60 ⑯	11.67 ⑯	NA ⑯	—	0.039 ⑤	76	32
	Hombre	F	NA	1.027 ⑩	0.965 ⑪	0.004	—	—	—	0.030	—	—	38
		R	—	0.787	0.728 ⑫	0.004	9.50	9.59	9.56	—	0.030	52	23 ⑬
	Oasis	F	⑰	0.929-0.937	0.830	0.004	—	—	—	0.060	—	—	36
		R	—	0.350-0.358	0.300	0.004	6.69 ④	6.73 ④	NA ④	—	0.060 ⑱	—	17
	Rodeo	F	1.000	1.020	0.969 ⑭	0.005	—	—	—	0.039	—	115	54
		R	—	0.710	0.654 ⑮	0.005	11.60 ⑯	11.67 ⑯	NA ⑯	—	0.039 ⑤	76	32
	Trooper	F	1.000	1.020	0.969 ⑭	0.005	—	—	—	0.039	—	115	54
		R	—	0.710	0.654 ⑮	0.005	8.27 ④	8.32 ④	NA ④	—	0.039	76	32

NA - Not Available
① Disc: 0.040
   Drum: 0.039
② 2.6L engine
③ 3.2L engine
④ Emergency brake drum surface
⑤ Specification includes disc pads or shoes
⑥ Front: 1.026
   Rear: 0.709
⑦ Front: 0.969 (Minimum machine diameter: 0.983)
   Rear: 0.654 (Minimum machine diameter: 0.668)

⑧ Minimum machine diameter: 0.983
⑨ Minimum machine diameter: 0.668
⑩ Heavy duty models: 1.140
⑪ Specification is discard for light duty models.
   Discard for heavy duty models: 1.080
   Refinish for light duty models: 0.980
   Refinish for heavy duty models: 1.130
⑫ Specification is for discard.
   Refinish: 0.735
⑬ Use new bolts during assembly.

⑭ Specification is for discard.
   Refinish: 0.9.83
⑮ Specification is for discard.
   Refinish: 0.668
⑯ Parking brake drum original diameter: 8.27
   Maximum limit is 8.32
⑰ Piston-to-bore clearance: 0.00-0.02 in.
⑱ Parking brake lining: 0.040

79246C

BRAKE SPECIFICATIONS
JEEP CHEROKEE, GRAND CHEROKEE, WRANGLER
All measurements in inches unless noted

Year	Model	Master Cylinder Bore	Brake Disc Original Thickness	Brake Disc Minimum Thickness	Brake Disc Maximum Runout	Brake Drum Diameter Original Inside Diameter	Brake Drum Diameter Max. Wear Limit	Brake Drum Diameter Maximum Machine Diameter	Minimum Lining Thickness Front	Minimum Lining Thickness Rear	Brake Caliper Bracket Bolts (ft. lbs.)	Brake Caliper Mounting Bolts (ft. lbs.)
1995	Cherokee	NA	0.94	0.89	0.005	9	①	9.06	0.030	0.030	—	11
	Grand Cherokee	0.99	0.94 ②	0.89 ③	0.005	—	—	—	0.030	0.030	—	7-15
	Wrangler	NA	0.94	0.89	0.005	9	①	9.06	0.030	0.030	—	11
1996	Cherokee	NA	0.94	0.89	0.005	9	①	9.06	0.030	0.030	—	11
	Grand Cherokee	0.99	0.94 ②	0.89 ④	0.005	—	—	—	0.030	0.030	—	7-15
	Wrangler	NA	0.94	0.89	0.005	9	①	9.06	0.030	0.030	—	11
1997	Cherokee	NA	0.94	0.89	0.005	9	①	9.06	0.030	0.030	—	11
	Grand Cherokee	0.99	0.94 ②	0.89 ④	0.005	—	—	—	0.030	0.030	—	7-15
	Wrangler	NA	0.94	0.89	0.005	9	①	9.06	0.030	0.030	—	11
1998	Cherokee	NA	0.94	0.89	0.005	9	①	9.06	0.030	⑤	—	11
	Grand Cherokee	0.99	0.94 ②	0.89 ④	0.005	—	—	—	0.030	0.030	—	7-15
	Wrangler	NA	0.94	0.89	0.005	9	①	9.06	0.030	⑤	—	11
1999	Cherokee	NA	0.94	0.89	0.005	⑥	①	9.06 ⑦	0.030	⑤	—	11
	Grand Cherokee	NA	1.01 ⑧	0.96 ⑨	0.003	—	—	—	0.030	0.030	—	⑩
	Wrangler	NA	0.94	0.89	0.005	9	①	9.06	0.030	⑤	—	11

NA - Not Available

① Maximum diameter is listed on outside of drum
② Rear rotor original thickness: 0.440 in.
③ Rear rotors have minimum allowable thickness listed 0.370 in.
④ Rear rotors have minimum allowable thickness listed: 0.370 in.
⑤ Riveted brake shoes: 0.030 in.
   Bonded brake shoes: 0.060 in.
⑥ Standard: 9.00 in.
   Optional 10.00 in.
⑦ Maximum machine diameter is 10.06 if equipped with 10.00 in. drums.
⑧ Rear rotor original thickness: 0.3850 in.
⑨ Rear rotor minimum allowable thickness: 0.335 in.
⑩ Slide pin bolts: 21-30 ft. lbs.
   Anchor bolts: 66-85 ft. lbs.

79246C13

BRAKE SPECIFICATIONS
KIA SPORTAGE
All measurements in inches unless noted

Year	Model	Master Cylinder Bore	Brake Disc Original Thickness	Brake Disc Minimum Thickness	Brake Disc Maximum Runout	Brake Drum Diameter Original Inside Diameter	Brake Drum Diameter Max. Wear Limit	Brake Drum Diameter Maximum Machine Diameter	Minimum Lining Thickness Front	Minimum Lining Thickness Rear	Brake Caliper Bracket Bolts (ft. lbs.)	Brake Caliper Mounting Bolts (ft. lbs.)
1995	Sportage	NA	0.940	0.880	0.004	NA	9.89	NA	0.080	0.060	—	72
1996	Sportage	NA	0.940	0.880	0.004	NA	9.89	NA	0.080	0.060	—	72
1997	Sportage	NA	0.940	0.880	0.004	NA	9.89	NA	0.080	0.060	—	72
1998-99	Sportage	NA	0.940	0.880	0.004	NA	9.89	NA	0.080	0.060	—	72

NA - Not Available

79246C14

BRAKE SPECIFICATIONS
LAND ROVER DEFENDER 90, DISCOVERY, RANGE ROVER
All measurements in inches unless noted

Year	Model	Front Brake Disc			Rear Brake Disc			Minimum Lining Thickness		Brake Caliper ①	
		Original Thickness	Minimum Thickness	Maximum Runout	Original Thickness	Minimum Thickness	Maximum Runout	Front	Rear	Bracket Bolts (ft. lbs.)	Mounting Bolts (ft. lbs.)
1995	Defender 90	0.945	0.080	0.006	0.490	0.030	0.006	0.120	0.120	—	60
	Discovery	1.000	0.870	0.006	0.500	0.460	0.006	0.080	0.080	74	19
	Range Rover	1.000	0.870	0.006	0.500	0.460	0.006	0.080	0.080	74	19
1996	Defender 90	0.945	0.080	0.006	0.490	0.030	0.006	0.120	0.120	—	60
	Discovery	1.000	0.870	0.006	0.500	0.460	0.006	0.080	0.080	74	19
	Range Rover	1.000	0.870	0.006	0.500	0.460	0.006	0.080	0.080	74	19
1997	Defender 90	0.945	0.080	0.006	0.490	0.030	0.006	0.120	0.120	—	60
	Discovery	1.000	0.870	0.006	0.500	0.460	0.006	0.080	0.080	74	19
	Range Rover	1.000	0.870	0.006	0.500	0.460	0.006	0.080	0.080	74	19
1998-99	Defender 90	0.945	0.080	0.006	0.490	0.030	0.006	0.120	0.120	—	60
	Discovery	1.000	0.870	0.006	0.500	0.460	0.006	0.080	0.080	74	19
	Range Rover	1.000	0.870	0.006	0.500	0.460	0.006	0.080	0.080	74	19

① Both front and rear calipers.

79246C1

BRAKE SPECIFICATIONS
LEXUS LX450, RX300
All measurements in inches unless noted

Year	Model		Master Cylinder Bore	Brake Disc			Brake Drum Diameter			Minimum Lining Thickness		Brake Caliper	
				Original Thickness	Minimum Thickness	Maximum Runout	Original Inside Diameter	Max. Wear Limit	Maximum Machine Diameter	Front	Rear	Bracket Bolts (ft. lbs.)	Mounting Bolts (ft. lbs.)
1996	LX450	F	NA	1.260	1.181	0.0059	—	—	—	0.039	—	—	90
		R	NA	0.709	0.630	0.0059	—	—	—	—	0.039	76	65
1997	LX450	F	NA	1.260	1.181	0.0059	—	—	—	0.039	—	—	90
		R	NA	0.709	0.630	0.0059	—	—	—	—	0.039	76	65
1998-99	LX450	F	NA	1.260	1.181	0.0059	—	—	—	0.039	—	—	90
		R	NA	0.709	0.630	0.0059	—	—	—	—	0.039	76	65
	RX300	F	NA	1.02	1.024	0.0020	—	—	—	0.039	—	79	25
		R	NA	0.394	0.354	0.0059	—	—	—	—	0.039	34	14

NA - Not Available
F - Front
R - Rear

79246C1

BRAKE SPECIFICATIONS
MAZDA B2300, B2500, B3000, B4000, MPV
All measurements in inches unless noted

Year	Model		Master Cylinder Bore	Brake Disc Original Thickness	Brake Disc Minimum Thickness	Brake Disc Maximum Runout	Brake Drum Diameter Original Inside Diameter	Brake Drum Diameter Max. Wear Limit	Brake Drum Diameter Maximum Machine Diameter	Minimum Lining Thickness Front	Minimum Lining Thickness Rear	Brake Caliper Bracket Bolts (ft. lbs.)	Brake Caliper Mounting Bolts (ft. lbs.)
1995	B2300		NA	NA	0.810	0.003	NA	①	0.003	0.012	0.003	72-97	21-26
	B3000		NA	NA	0.810	0.003	NA	①	0.003	0.012	0.003	72-97	21-26
	B4000		NA	NA	0.810	0.003	NA	①	0.003	0.012	0.003	72-97	21-26
	MPV		0.940	②	③	0.004	NA	NA	NA	0.080	0.080	66-79	62-68
1996	B2300		NA	NA	0.810	0.003	NA	①	0.003	0.012	0.003	72-97	21-26
	B3000		NA	NA	0.810	0.003	NA	①	0.003	0.012	0.003	72-97	21-26
	B4000		NA	NA	0.810	0.003	NA	①	0.003	0.012	0.003	72-97	21-26
	MPV	F	0.940	1.100	1.020	0.004	NA	NA	NA	0.080	NA	66-79	62-68
		R	NA	0.710	0.630	0.004	NA	NA	NA	NA	0.040	66-79	62-68
1997	B2300		NA	NA	0.810	0.003	NA	①	0.003	0.012	0.003	72-97	21-26
	B3000		NA	NA	0.810	0.003	NA	①	0.003	0.012	0.003	72-97	21-26
	B4000		NA	NA	0.810	0.003	NA	①	0.003	0.012	0.003	72-97	21-26
	MPV	F	0.940	1.100	1.020	0.004	NA	NA	NA	0.080	NA	66-79	62-68
		R	NA	0.710	0.630	0.004	NA	NA	NA	NA	0.040	66-79	62-68
1998-99	B2500		NA	NA	0.810	0.003	NA	①	0.003	0.012	0.003	72-97	21-26
	B3000		NA	NA	0.810	0.003	NA	①	0.003	0.012	0.003	72-97	21-26
	B4000		NA	NA	0.810	0.003	NA	①	0.003	0.012	0.003	72-97	21-26
	MPV	F	0.940	1.100	1.020	0.004	NA	NA	NA	0.080	NA	66-79	62-68
		R	NA	0.710	0.630	0.004	NA	NA	NA	NA	0.040	66-79	62-68

NA - Not Available
① Refer to the maximum diameter stamped on drum
② Front 4x2: 1.180
Front 4x4: 1.100
Rear: 0.710
③ Front 4x2: 1.100
Front 4x4: 1.020
Rear: 0.630

79246C17

BRAKE SPECIFICATIONS
MERCEDES-BENZ ML320
All measurements in inches unless noted

Year	Model	Master Cylinder Bore	Front Brake Disc Original Thickness	Front Brake Disc Minimum Thickness	Front Brake Disc Maximum Runout	Rear Brake Disc Original Thickness	Rear Brake Disc Minimum Thickness	Rear Brake Disc Maximum Runout	Minimum Lining Thickness Front	Minimum Lining Thickness Rear	Brake Caliper Bracket Bolts (ft. lbs.)	Brake Caliper Mounting Bolts (ft. lbs.)
1998-99	ML320	NA	1.02	0.91	NA	0.59	0.49	NA	①	②	26	18

NA - Not Available
① New: Inner pad w/backing plate .65 in.
Wear limit: .47 in
New: Outter pad w/backing plate .61 in.
Wear limit: .43 in.
② New: .61 in.
Wear limit: .43 in.

79246C18

BRAKE SPECIFICATIONS
MITSUBISHI MIGHTY MAX, MONTERO, MONTERO SPORT
All measurements in inches unless noted

Year	Model		Master Cylinder Bore	Brake Disc			Brake Drum Diameter			Minimum Lining Thickness		Brake Caliper	
				Original Thickness	Minimum Thickness	Maximum Runout	Original Inside Diameter	Max. Wear Limit	Maximum Machine Diameter	Front	Rear	Bracket Bolts (ft. lbs.)	Mounting Bolts (ft. lbs.)
1995	Mighty Max		0.938	0.866	0.803	0.006	10.00	10.08	—	0.079	0.039	58-72	29-36
	Montero		0.938	①	②	0.003	—	7.80	—	0.079	0.079 ③	58-72	29-36
1996	Mighty Max		0.938	0.866	0.803	0.004	10.00	—	10.08	0.079	0.039	58-72	29-36
	Montero	F	0.938	④	⑤	0.004	—	—	—	0.079	—	65	54
		R	—	0.710	0.646	0.003	—	—	—	—	0.040	65	32
1997	Montero	F	0.938	④	⑤	0.004	—	—	—	0.079	—	65	54
		R	-	0.710	0.646	0.003	—	—	—	—	0.040	65	32
	Montero Sport	F	0.938	0.940	0.880	0.002	—	—	—	—	0.040	65	55
		R	—	0.700	0.650	0.003	10.63	10.71	10.71	—	0.080	94	32
1998-99	Montero	F	0.938	1.060	1.000	0.002	—	—	—	0.079	—	65	54
		R	—	0.710	0.646	0.003	—	—	—	—	0.040	65	32
	Montero Sport	F	0.938	0.940	0.880	0.002	—	—	—	0.080	0.040	65	55
		R	—	0.700	0.650	0.003	10.63	10.71	10.71	—	0.080	94	32

① Front: 0.940
  Rear: 0.710
② Front: 0.880
  Rear: 0.330
③ Drum shoe: 0.040
④ 3.0L engine: 0.940
  3.5L engine: 1.060
⑤ 3.0L engine: 0.880
  3.5L engine: 1.000

79246C20

BRAKE SPECIFICATIONS
SUBARU FORESTER
All measurements in inches unless noted

Year	Model		Master Cylinder Bore	Brake Disc			Brake Drum Diameter			Minimum Lining Thickness		Brake Caliper	
				Original Thickness	Minimum Thickness	Maximum Runout	Original Inside Diameter	Max. Wear Limit	Maximum Machine Diameter	Front	Rear	Bracket Bolts (ft. lbs.)	Mounting Bolts (ft. lbs.)
1998-99	Forester	F	1.0625	0.940	0.870	0.003	—	—	—	0.059	—	51-65	25-31
		R	—	0.390	0.340	0.004	9.00 ①	9.079 ②	NA	—	0.059	—	25-31

① Parking brake drum on vehicles with rear disc brakes: 6.69 in.
② Parking brake drum on vehicles with rear disc brakes: 6.73 in.

79246C21

BRAKE SPECIFICATIONS
SUZIKI SAMURAI, SIDEKICK, SIDEKICK SPORT, X-90
All measurements in inches unless noted

Year	Model	Master Cylinder Bore	Brake Disc			Brake Drum Diameter			Minimum Lining Thickness		Brake Caliper	
			Original Thickness	Minimum Thickness	Maximum Runout	Original Inside Diameter	Max. Wear Limit	Maximum Machine Diameter	Front	Rear	Bracket Bolts (ft. lbs.)	Mounting Bolts (ft. lbs.)
1995	Samurai	NA		0.334	0.006	8.66	8.74	8.74	0.236	0.120	—	19-21
	Sidekick ①	NA	0.669	0.591	0.006	10.00	10.07	10.07	0.315	0.120	51-72	19-21
	Sidekick ②	NA	0.669	0.591	0.006	10.00	10.07	10.07	0.315	0.120	51-72	19-21
1996	X-90	NA	0.394	0.315	0.006	8.66	8.74	8.74	0.240	0.120	51-72	19-21
	Sidekick ①	NA	0.394	0.315	0.006	8.66	8.74	8.74	0.240	0.120	51-72	19-21
	Sidekick ②	NA	0.670	0.590	0.006	10.00	10.07	10.07	0.295	0.120	51-72	19-21
	Sidekick Sport	NA	0.866	0.787	0.006	10.00	10.07	10.07	0.275	0.120	51-72	③
1997	X-90	NA	0.394	0.315	0.006	8.66	8.74	8.74	0.240	0.120	51-72	19-21
	Sidekick ①	NA	0.394	0.315	0.006	8.66	8.74	8.74	0.240	0.120	51-72	19-21
	Sidekick ②	NA	0.670	0.590	0.006	10.00	10.07	10.07	0.295	0.120	51-72	19-21
	Sidekick Sport	NA	0.866	0.787	0.006	10.00	10.07	10.07	0.275	0.120	51-72	③
1998	X-90	NA	0.394	0.315	0.006	8.66	8.74	8.74	0.240	0.120	51-72	19-21
	Sidekick ①	NA	0.394	0.315	0.006	8.66	8.74	8.74	0.240	0.120	51-72	19-21
	Sidekick ②	NA	0.670	0.590	0.006	10.00	10.07	10.07	0.295	0.120	51-72	19-21
	Sidekick Sport	NA	0.866	0.787	0.006	10.00	10.07	10.07	0.275	0.120	51-72	③

① 2-door models only
② 4-door models only
③ Bottom bolt: 37 ft. lbs.
　Top bolt: 42 ft. lbs.

79246C22

BRAKE SPECIFICATIONS
TOYOTA 4RUNNER, LAND CRUISER, PICK-UP, PREVIA, RAV4, SIENNA, T100, TACOMA
All measurements in inches unless noted

Year	Model		Master Cylinder Bore	Brake Disc Original Thickness	Brake Disc Minimum Thickness	Brake Disc Maximum Runout	Brake Drum Diameter Original Inside Diameter	Max. Wear Limit	Maximum Machine Diameter	Minimum Lining Thickness Front	Minimum Lining Thickness Rear	Brake Caliper Bracket Bolts (ft. lbs.)	Brake Caliper Mounting Bolts (ft. lbs.)
1995	4Runner		NA	0.984	0.906	0.0035	11.61	—	11.69	0.059	0.039	—	90
	Land Cruiser		NA	1.260 ①	1.181 ②	0.0059	11.61	—	11.69	0.059	③	—	90
	Pick-up		NA	④	①	0.0035	⑤	—	⑥	⑦	0.039	65	80
	Previa		NA	⑧	⑨	0.0028	10.00	—	10.08	0.039	0.039	—	65
	Tacoma		NA	0.866	0.787	0.0028	⑤	—	⑥	0.039	0.039	65	90
1996	4Runner		NA	0.866	0.787	0.0028	11.61	—	11.69	0.039	0.039	—	90
	Land Cruiser	F	NA	1.260	1.181	0.0059	—	—	—	0.039	—	—	90
		R	NA	0.709	0.630	—	—	—	—	—	③	65	76
	Previa	F	NA	⑩	⑪	0.0028	—	—	—	0.039	—	—	65
		R	NA	0.709	0.630	0.0039	10.00	—	10.18	—	0.039	—	—
	RAV4		NA	0.709	0.630	0.0020	9.00	—	9.08	0.039	0.039	—	79
	T100 2WD		NA	0.984	0.906	⑫	11.61	—	11.69	0.039	0.039	29	80
	T100 4WD		NA	0.984	0.984	0.0028	11.61	—	11.69	0.039	0.039	90	80
	Tacoma 2WD		NA	0.866	0.787	0.0028	10.00	—	10.08	0.039	—	65	90
	Tacoma 4WD		NA	0.866	0.787	0.0028	11.61	—	11.69	—	0.039	—	90
1997	4Runner		NA	0.866	0.787	0.0028	11.61	—	11.69	0.039	0.039	—	90
	Land Cruiser	F	NA	1.260	1.181	0.0059	—	—	—	0.039	—	—	90
		R	NA	0.709	0.630	—	—	—	—	—	③	65	76
	Previa	F	NA	⑩	⑪	0.0028	—	—	—	0.039	—	—	65
		R	NA	0.709	0.630	0.0039	10.00	—	10.18	—	0.039	—	—
	RAV4		NA	0.709	0.630	0.0020	9.00	—	9.08	0.039	0.039	—	79
	T100 2WD		NA	0.984	0.906	⑫	11.61	—	11.69	0.039	0.039	29	80
	T100 4WD		NA	0.984	0.984	0.0028	11.61	—	11.69	0.039	0.039	90	80
	Tacoma 2WD		NA	0.866	0.787	0.0028	10.00	—	10.08	0.039	—	65	90
	Tacoma 4WD		NA	0.866	0.787	0.0028	11.61	—	11.69	—	0.039	—	90
1998-99	4Runner		NA	0.866	0.787	0.0028	11.61	—	11.69	0.039	0.039	—	90
	Land Cruiser	F	NA	1.260	1.181	0.0059	—	—	—	0.039	—	—	90
		R	NA	0.709	0.630	—	—	—	—	—	③	65	76
	RAV4		NA	0.709	0.630	0.0020	9.00	—	9.08	0.039	0.039	—	79
	Sienna	F	NA	1.102	1.024	0.002	—	—	—	0.039	—	—	79
		R	NA	—	—	—	9.84	—	9.921	—	0.039	—	—
	T100 2WD		NA	0.984	0.906	⑫	11.61	—	11.69	0.039	0.039	29	80
	T100 4WD		NA	0.984	0.984	0.0028	11.61	—	11.69	0.039	0.039	90	80
	Tacoma 2WD		NA	0.984	0.906	0.0028	10.00	—	10.08	0.039	—	65	90
	Tacoma 4WD		NA	0.866	0.787	0.0028	11.61	—	11.69	—	0.039	—	90

NA - Not Available
① Rear disc: 0.630
② Rear disc: 0.709
③ Brake shoe lining: 0.059
  Disc pad lining: 0.039
④ 2WD: 0.866; 4WD: 0.787

⑤ 2WD: 10.00; 4WD: 11.61
⑥ 2WD: 10.08; 4WD: 11.69
⑦ 2WD: 0.059; 4WD: 0.039
⑧ Front with rear drum brake: 0.984
  Front with rear disc brake: 0.866
  Rear disc: 0.709

⑨ Front with rear drum brake: 0.906
  Front with rear disc brake: 0.787
  Rear disc: 0.669
⑩ With rear drum brake: 0.985
  With rear disc brake: 0.866

⑪ With rear drum brake: 0.906
  With rear disc brake: 0.787
⑫ 1 ton: 0.0035
  1/2 ton: 0.0028

79246C

DRIVESHAFTS, U-JOINTS AND CV-JOINT BOOTS

7

DRIVESHAFTS, U-JOINTS AND CV-JOINT BOOTS

Driveshafts

➡**The term driveshaft does not refer to halfshaft (often termed driveshaft by various manufacturers), which are used on front wheel drive vehicles.**

GENERAL INFORMATION

The driveshaft is a long steel tube used to transmit power from the transmission to the rear differential, and on 4-Wheel Drive (4WD) vehicles, from the transfer case to the front differential. Located at either end of the driveshaft is a universal joint (U-joint), which is designed to transmit torsional power at many different angles (within designed limits) in order to match the motion of the rear axle. A slip joint is attached to the U-joint closest to the transmission or transfer case. The shaft is designed with yokes at each end that are inline with each other in order to produce the smoothest possible running shaft.

Since the vehicles can be obtained in either 2WD or 4WD, and in various combinations (two and four door models), various types of driveshafts may be employed. Some models will be equipped with a one-piece rear driveshaft, while others have a two-piece rear driveshaft which uses a center support bearing. Most 4WD vehicles are equipped with a front shaft which is a two-piece telescopic type with internal splines. This joint allows the length of the shaft to change as the vehicle is driven over bumps or the load in the vehicle changes.

Because some vehicles covered by this manual utilize 2-piece shafts and splined yokes, it may be possible to reinstall the shaft incorrectly or "out of phase" which would cause vibration. Many of these vehicles utilize a keyed slip yoke to prevent this, but DO NOT risk improper installation. ALWAYS matchmark the shaft ends to the yokes before removal.

At the front of the rear one or two-piece driveshafts, the U-joint connects the driveshaft to a slip-jointed yoke. This yoke is internally splined and allows the driveshaft to move in and out on the transmission splines (one-piece) or the shaft splines (two-piece). The rear of the one or two-piece driveshaft is attached to the differential, on some vehicles, the U-joint may be clamped or bolted to the yoke

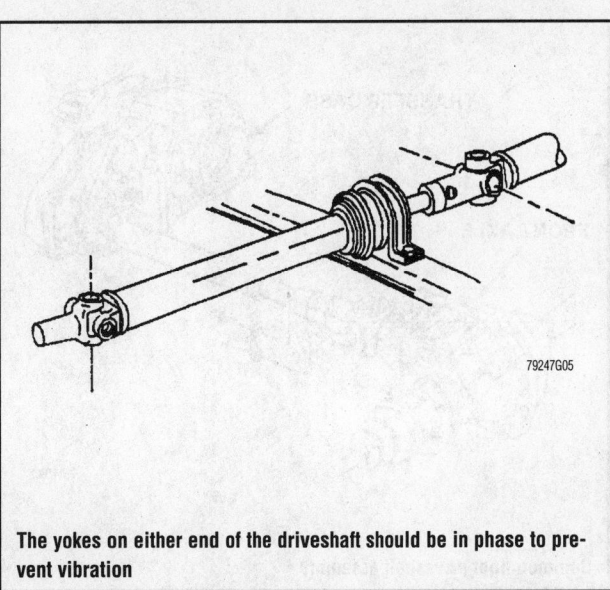

79247G05

The yokes on either end of the driveshaft should be in phase to prevent vibration

Make alignment marks on the U-joint and shaft before disassembly to prevent possible vibration when assembled

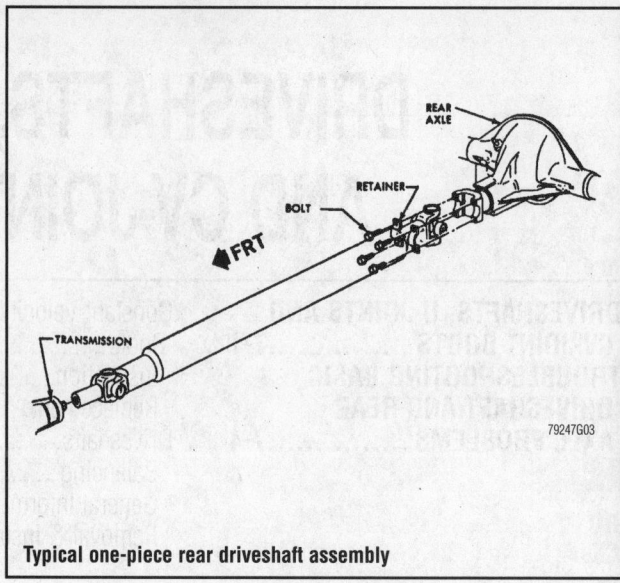

Typical one-piece rear driveshaft assembly

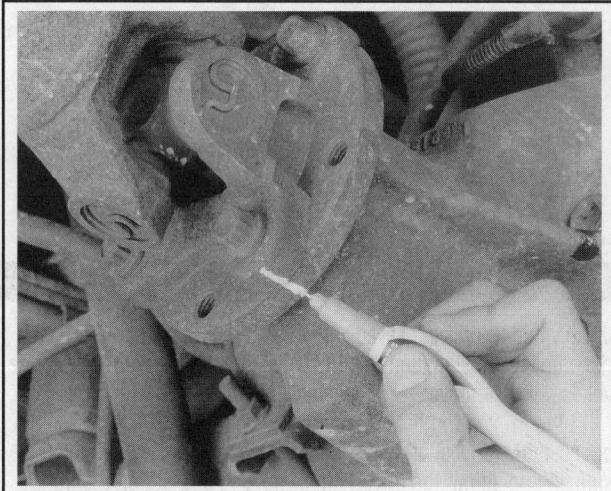

On this type of driveshaft, make alignment marks across the two flanges so it can be installed in the same position

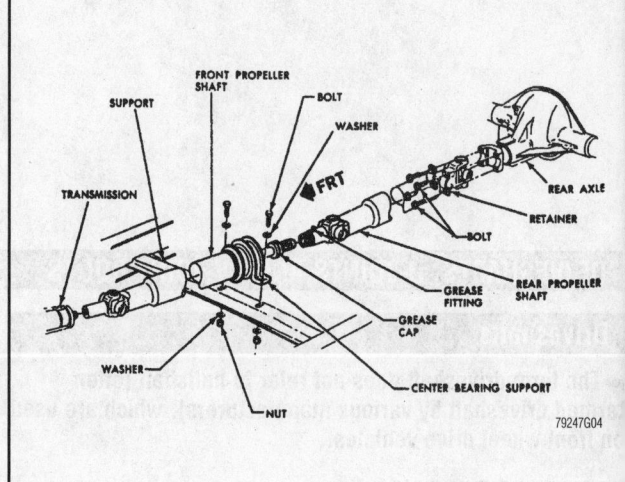

Typical two-piece rear driveshaft assembly with center support bearing

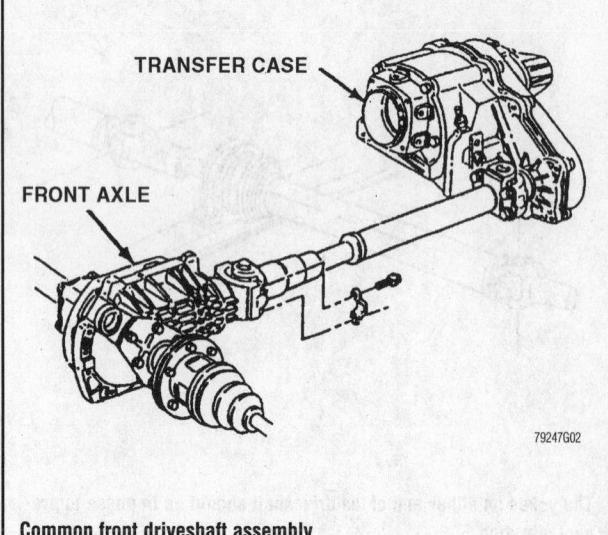

Common front driveshaft assembly

on the pinion shaft, on other vehicles the driveshaft terminates with a flange that gets bolted to the pinion flange on the differential.

REMOVAL & INSTALLATION

Rear Driveshaft

1. Mark the relationship of the driveshaft-to-pinion flange or yoke and disconnect the rear universal joint from the differential by removing the bolts. If the bearing cups are loose, wrap tape around the universal joint to prevent the cups from falling off.

2. If equipped with a one-piece driveshaft, perform the following steps:

   a. Slide the rear driveshaft forward to disengage it from the rear axle flange or yoke.

   b. Move the driveshaft rearward to disengage it from the transmission slip-joint, passing it under the axle housing.

3. If equipped with a two-piece driveshaft, perform the following steps:

a. Slide the driveshaft forward to disengage it from the rear axle flange or yoke.

b. Slide the driveshaft rearward to disengage it from the slip-joint of the front driveshaft, passing it under the axle housing.

c. Remove the center bearing-to-support nuts and bolts.

d. Slide the front driveshaft rearward to disengage it from the transfer case slip-joint.

➡**DO NOT allow the driveshaft to hang by the U-joint or bend to extreme angles, as damage to the U-joint may occur. Support the driveshaft with wire as needed during removal.**

To install:

4. Inspect the slip-joint for damage, burrs or wear, for these can damage the transmission seal. Apply engine oil to all splined driveshaft joints.

❊❊ WARNING

DO NOT use a hammer to force the driveshaft into place. Check for burrs on the transmission output shaft spline, twisted slip yoke splines or possibly the wrong U-joint.

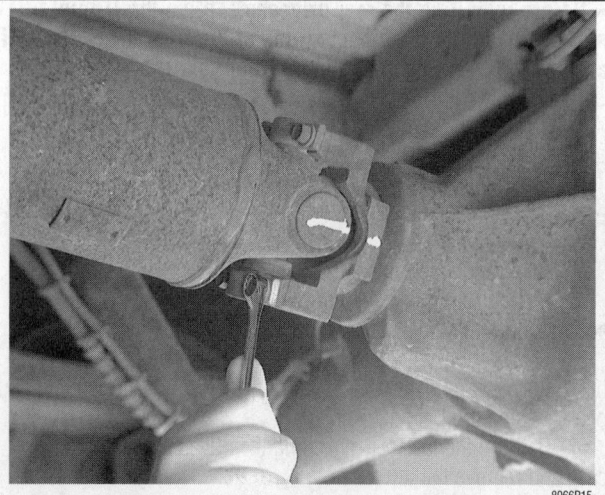

The U-joint on this driveshaft is attached to the pinion yoke by two small brackets and four bolts

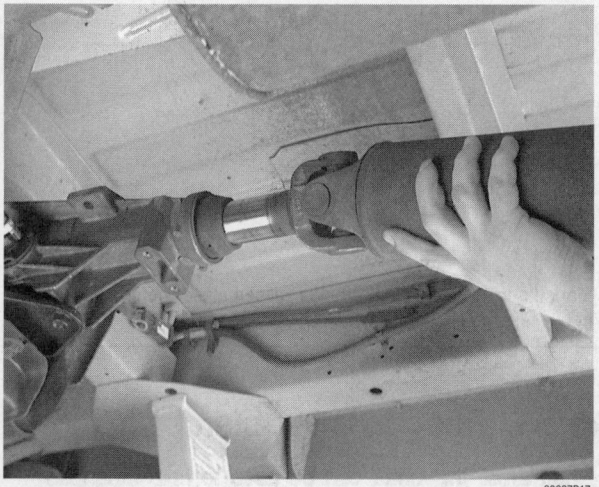

This type of driveshaft is removed from the transmission by simply sliding it out

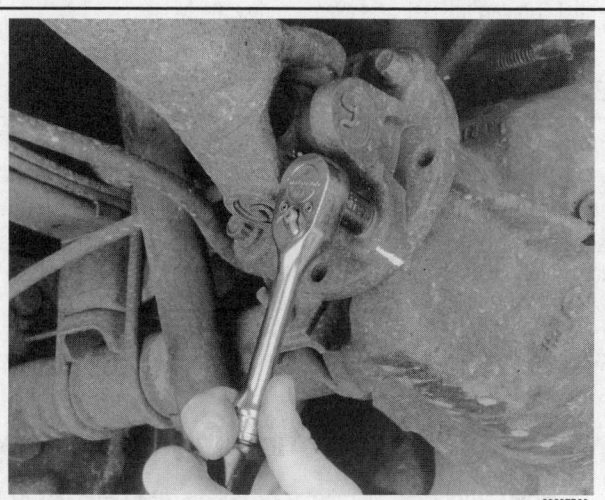

This type of driveshaft is attached to the pinion flange with four bolts and nuts

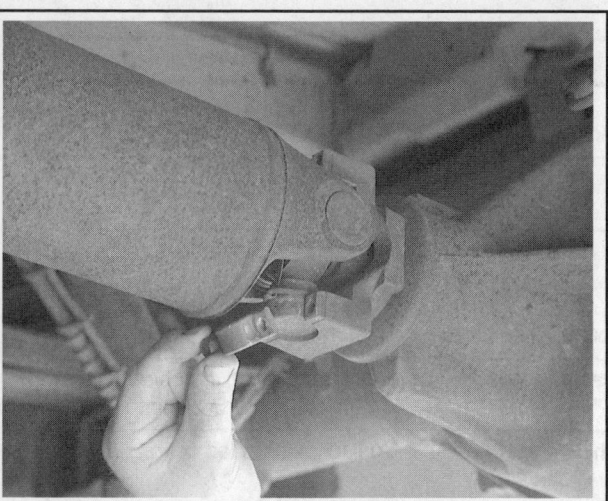

The driveshaft can be detached from the pinion shaft yoke after removing the four bolts and two brackets

Make sure the splines agree in number and fit. To prevent trunnion seal damage, **DO NOT** place any tool between the yoke and splines.

5. If installing a one-piece driveshaft, perform the following steps:

a. Attach the driveshaft to the transmission.

b. Align the rear universal joint-to-rear axle pinion flange, making sure the bearings are properly seated in the pinion flange yoke.

c. Install the rear driveshaft-to-pinion flange. Tighten the fasteners securely.

6. If installing a two-piece driveshaft, perform the following steps:

a. Install the front driveshaft to the transmission and bolt the center bearing-to-support. Tighten the nuts and bolts securely.

➡**The front driveshaft yoke must be bottomed out in the transmission (fully forward) before being installed to the support.**

b. Rotate the shaft so that the front U-joint trunnion is in the correct position.

➡**Before installing the rear driveshaft, align the U-joint trunnions (a "key" in the output spline of the front driveshaft will align with a missing spline in the rear yoke).**

   c. Attach the rear U-joint to the axle flange. Tighten the retainers securely.

7. Road test the vehicle.

Front Driveshaft

1. Raise and safely support the vehicle.
2. Place matchmarks on the front and rear flanges (if so equipped) so the driveshaft can be installed in its original position.
3. Remove the fasteners attaching the driveshaft to the differential. Suspend the driveshaft with wire to prevent it from hanging and overstressing the U-joint.
4. Remove the fasteners attaching the driveshaft to the transfer case and remove the driveshaft.

To install:

5. Align the matchmarks and install the driveshaft to the transfer case. Tighten the fasteners evenly and securely.
6. Align the matchmarks and connect the driveshaft to the differential. Tighten the fasteners evenly and securely.
7. Lower the vehicle to the floor.

BALANCING

The following procedure is used to help eliminate minor driveshaft vibration of an otherwise good driveshaft.

Before attempting this, carefully examine the driveshaft for damage such as dents and deformations. Driveshafts are subjected to large amounts of twisting force which can literally twist the driveshaft. Also check for missing weights that may have been knocked off of the shaft. If the driveshaft is deformed, replace it. If any weights appear to be missing, take the driveshaft to a machine shop that is equipped to balance the shaft and have it repaired. Driveshafts typically turn at speeds 2½ to 4 or more times faster than the rear axle; don't use a damaged driveshaft.

This type of balancing is performed by installing one or two hose clamps near the end of the driveshaft closest to the drive axle. The trial and error method is used to determine the best position of the clamp(s).

➡**Removing and turning the driveshaft 180° relative to the yoke may reduce some vibration. This should be done prior to the hose clamp method.**

1. Mark the rear of the driveshaft in four equal sections. Number the marks 1 through 4.

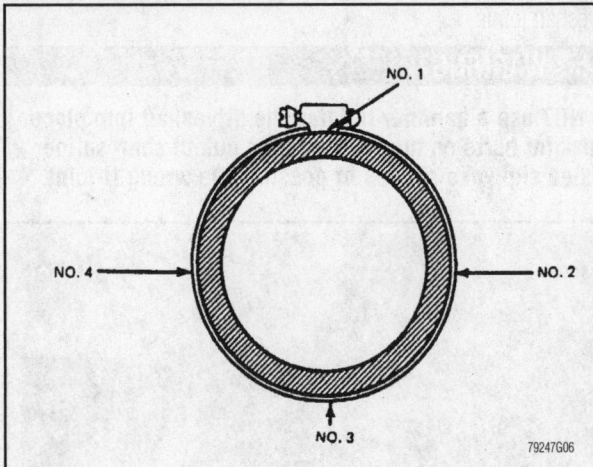

Mark the driveshaft in four equally spaced places before starting the balancing procedure for hose clamp positioning

Troubleshooting Basic Driveshaft and Rear Axle Problems

When abnormal vibrations or noises are detected in the driveshaft area, this chart can be used to help diagnose possible causes. Remember that other components such as wheels, tires, rear axle and suspension can also produce similar conditions.

BASIC DRIVESHAFT PROBLEMS

Problem	Cause	Solution
Shudder as car accelerates from stop or low speed	• Loose U-joint • Defective center bearing	• Replace U-joint • Replace center bearing
Loud clunk in driveshaft when shifting gears	• Worn U-joints	• Replace U-joints
Roughness or vibration at any speed	• Out-of-balance, bent or dented driveshaft • Worn U-joints • U-joint clamp bolts loose	• Balance or replace driveshaft • Replace U-joints • Tighten U-joint clamp bolts
Squeaking noise at low speeds	• Lack of U-joint lubrication	• Lubricate U-joint; if problem persists, replace U-joint
Knock or clicking noise	• U-joint or driveshaft hitting frame tunnel • Worn CV joint	• Correct overloaded condition • Replace CV joint

79247C01

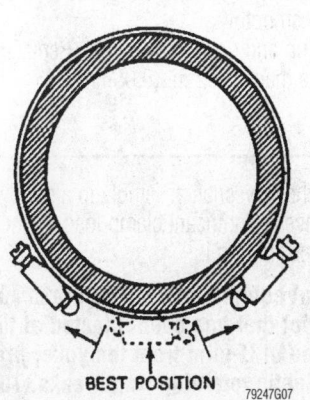

BEST POSITION

Move the hose clamp heads an equal distance from the best position a little at a time until the vibration is reduced to an acceptable level

2. Install a hose clamp with the screw portion of the clamp on the No. 1 mark.

3. Test drive the vehicle to see if the vibration condition has improved.

4. Recheck the vibration with the clamp positioned at the remaining three positions. If the vibration is equally reduced at, for example, position number 2 and position number 3, then position the screw portion of the clamp halfway between the marks.

5. Test drive the vehicle. If the vibration is still apparent, install another clamp in the same position as the first.

6. Test drive the vehicle. If the vibration is the same, move both clamps an equal distance from the point determined to be the best position. At first, position the clamps approximately ½ in. (12mm) apart.

7. Continue the process until the vibration is reduced to an acceptable level.

➡**If the vibration cannot be reduced to an acceptable level, take the driveshaft to a qualified machine shop for balancing.**

Universal Joints (U-Joints)

GENERAL INFORMATION

The universal joint (U-Joint) is used to provide a strong and flexible connection between the driveshaft and axle assembly. A flexible joint is necessary because of the constant movement of the axle assembly relative to the body of the vehicle. A U-Joint consists of the spider (trunnion), needle (roller) bearings, bearing cups, seals and snaprings. In most cases, U-Joints will last the life of the vehicle. The life of the U-Joint may decrease significantly if the operating angle has been changed or exceeded. This occurs when the vehicle ride height is changed. Vehicles that have been lifted will benefit by using a Double Cardon type joint. The Double Cardon type joint has a greater operating angle than the single U-joint.

When two components are connected by a conventional U-joint, the bend that is formed is called the operating angle. The larger the angle, the larger the amount of angular acceleration and deceleration of the joint. In other words, when the driveshaft is turning at a

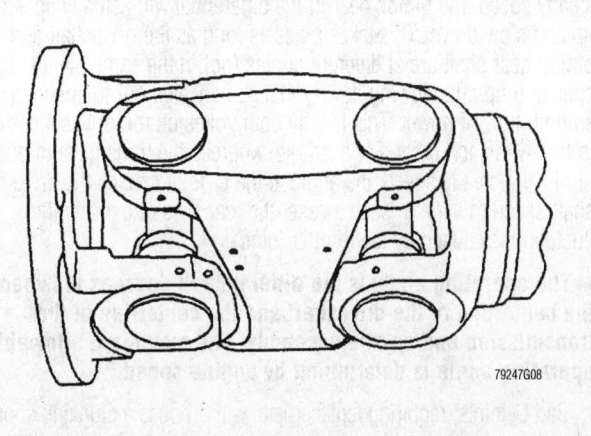

A Double Cardon universal joint has a greater operating angle than a single joint. This joint has been punch marked before disassembly so the components can be reassembled in their original positions

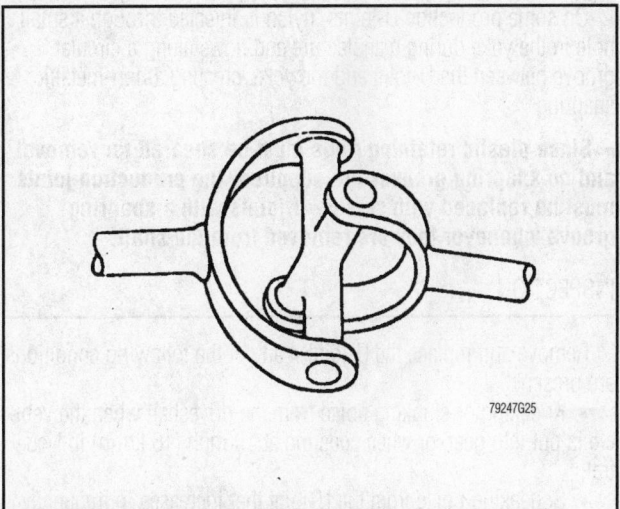

This simplified version of a universal joint shows how the angles can change while still transmitting power

PROPELLER SHAFT R.P.M.	MAX. NORMAL OPERATING ANGLES
5000	3°
4500	3°
4000	4°
3500	5°
3000	5°
2500	7°
2000	8°
1500	11°

Maximum normal operating angle between the driveshaft and transmission and/or axle assembly

steady speed, the pinion gear in the differential will actually speed up and slow down. This takes place as long as the driveshaft and pinion gear shaft are at different angles (not in the same plane). The speeding up and slowing down must be canceled out to ensure a smooth flow of power. This is why both yokes on the driveshaft are in line with each other. For example, whereas the transmission output is at a steady speed, the angle at the U-joint causes the driveshaft speed to vary. In such a case, the rear U-joint cancels the fluctuations caused by the front U-joint.

➡**The operating angle is the difference in degrees between the centerline of the driveshaft and the centerline of the transmission and/or axle assembly. The maximum allowable operating angle is determined by engine speed.**

Bad U-joints, requiring replacement, will produce a clunking sound when the vehicle is put into gear and when the transmission shifts from gear-to-gear. This is due to worn needle bearings or scored trunnion ends. Most U-joints are permanently lubricated at the factory and require no periodic maintenance. Those that do have grease fittings should be lubricated at every oil change. Clean the fitting with a shop rag before pumping grease to avoid forcing dirt into the joint.

On some production U-joints, nylon is injected through a small hole in the yoke during manufacture and flows along a circular groove between the U-joint and the yoke, creating a non-metallic snapring.

➡**Since plastic retaining rings must be sheared for removal and no snapring grooves are supplied, the production joints must be replaced with service U-joints with a snapring groove whenever they are removed from the shaft.**

INSPECTION

Remove and replace the U-joint if any of the following conditions are present:
- Knocking or clunking noise from the driveshaft when the vehicle is put into gear, or when coasting at 10 mph (16 km/h) in Neutral.
- Squeaking noise from the U-joint that increases in frequency as the speed of the vehicle increases.

- Roughness in the U-joint bearing when felt by hand. The U-joint should turn smoothly.
- Axial play (up and down movement). Replace the U-joint if the axial play is more than 0.002 in. (0.05mm).

OVERHAUL

1. Position the driveshaft assembly in a sturdy soft-jawed vise, BUT DO NOT place a significant clamp load on the shaft or you will risk deforming and ruining it.

➡**Some original equipment U-joints are secured in the yoke by nylon (plastic) that has been injected at the factory. To remove this type of U-joint from the yoke, press the bearing cup until the plastic retaining ring breaks. The replacement U-joint will have a snapring groove like a conventional joint.**

2. If applicable, remove the snaprings which retain the bearings in the yoke.

SNAP RING

79247G09

Use a pair of snapring pliers, or similar tool, to remove the outer snapring which retains the bearing in the yoke

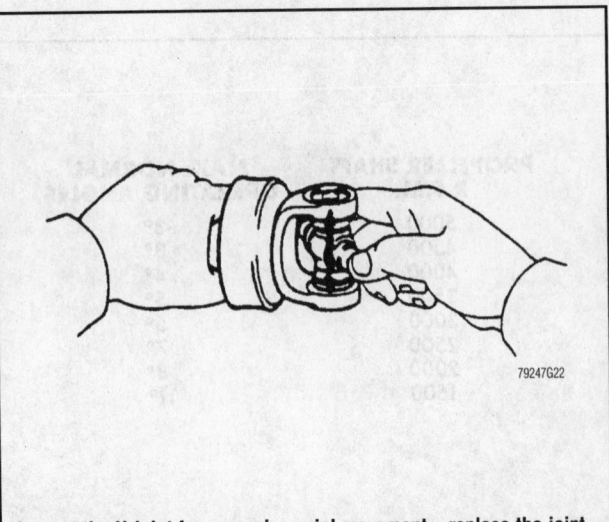

79247G22

Inspect the U-joint for excessive axial movement—replace the joint if the play is more than 0.002 in. (0.05mm).

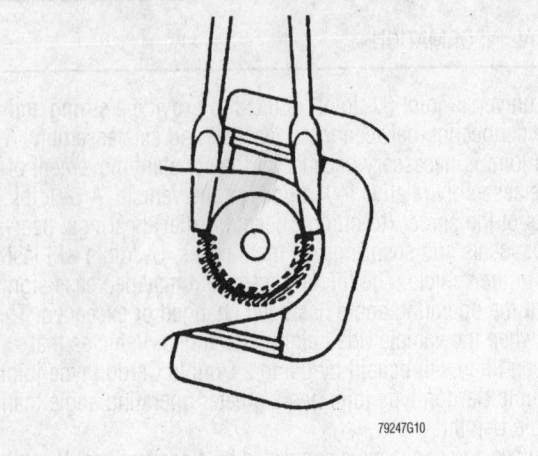

79247G10

Using two thin prytools is one method for removing the inner snaprings

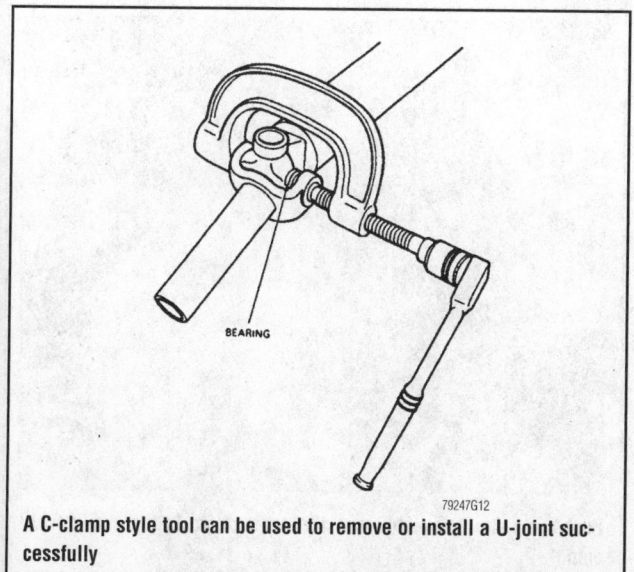

A C-clamp style tool can be used to remove or install a U-joint successfully

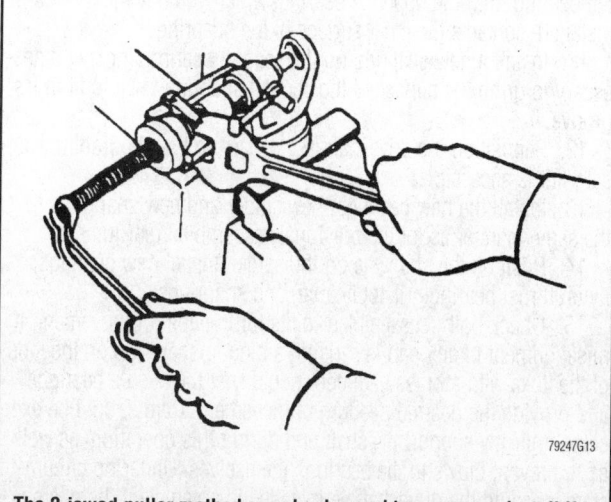

The 2-jawed puller method can also be used to remove or install U-joints

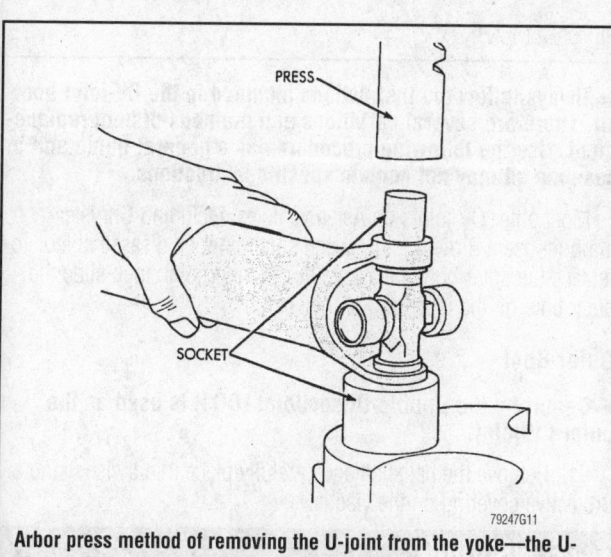

Arbor press method of removing the U-joint from the yoke—the U-joint can also be installed in a similar fashion

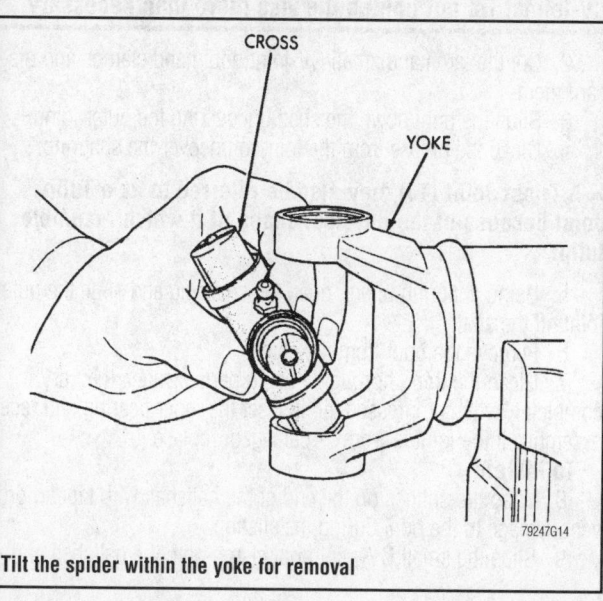

Tilt the spider within the yoke for removal

➡A U-joint removal and installation tool (which looks like a large C-clamp) is available to significantly ease the task, but it is very possible to replace the U-joints using an arbor press or a large vise and a variety of sockets.

3. Using a large C-clamp, vise or an arbor press, along with a socket smaller than the bearing cap (on one side) and a socket larger than the bearing cap (on the other side), drive one of the bearings in toward the center of the universal joint, which will force the opposite bearing out.

➡The smaller socket is used as a driver here, as it can pass through the opening of the U-joint or slip yoke flange. The larger socket is used to support the other side of the flange so that the bearing cap has room to exit the flange (into the socket).

4. As each bearing is forced far enough out of the universal joint to be accessible, grip it with a pair of pliers and pull it from the driveshaft yoke. Drive the spider in the opposite direction in order to make the opposite bearing accessible and pull it free with a pair of pliers. Use this procedure to remove all the bearings from both universal joints.

5. After removing the bearings, lift the spider from the yoke.

6. Thoroughly clean all dirt and foreign matter from the yokes on both ends of the driveshaft.

To assemble:

✳✳ WARNING

When installing new bearings in the yokes, it is advisable to use an arbor press or the special C-clamp tool. If this tool is not available, the bearings should be pressed into position with extreme care, as a heavy jolt on the needle bearings can easily damage or misalign them. This will greatly shorten their life and hamper their efficiency.

7. Start a new bearing into the yoke at the rear of the driveshaft.

8. Position a new spider in the rear yoke and press the new bearing ¼ in. (6mm) below the outer surface of the yoke.

9. With the bearing in position, install a new snapring.

10. Start a new bearing into the opposite side of the yoke. Press

the bearing until the opposite bearing, which you have just installed, contacts the inner surface of the snapring.

11. Install a new snapring on the second bearing. It may be necessary to grind the surface of the second snapring for it to fit in it's groove.

12. Reposition the driveshaft in the vise, so that the front universal joint is accessible.

13. Install the new bearings, new spider and new snaprings in the same manner as for the previously assembled rear joint.

14. Position the slip yoke on the spider. Install new bearings, nylon thrust bearings (if applicable) and snaprings.

15. Check both reassembled joints for freedom of movement, If misalignment of any part is causing a bind, a sharp rap on the side of the yoke with a brass hammer should seat the needle bearings and provide the desired freedom of movement. Care should be exercised to firmly support the shaft end during this operation, as well as to prevent blows to the bearings themselves. Under no circumstance should the driveshaft be installed in a vehicle if there is any binding in the universal joints.

16. Grease the U-joint fittings, if equipped.

Constant Velocity Joint (CV-Joint) Boots

INSPECTION

Whenever undercarriage work is performed, such as brakes, exhaust or suspension work, the Constant Velocity (CV) joint boots should be inspected for breaks and tears. The first sign of boot damage will be dark spots (grease) on the inside of the tire and wheel. If boot damage is caught early enough, the joint can be saved by cleaning, regreasing and replacing the boot. If the boot is left unrepaired, damage to the bearing will occur and replacement of the CV-joint is required. In most cases, it may be more economical to replace the entire halfshaft with a remanufactured one.

➡**Check with your parts supplier for price and availability to determine weather you should replace the entire halfshaft or separate components.**

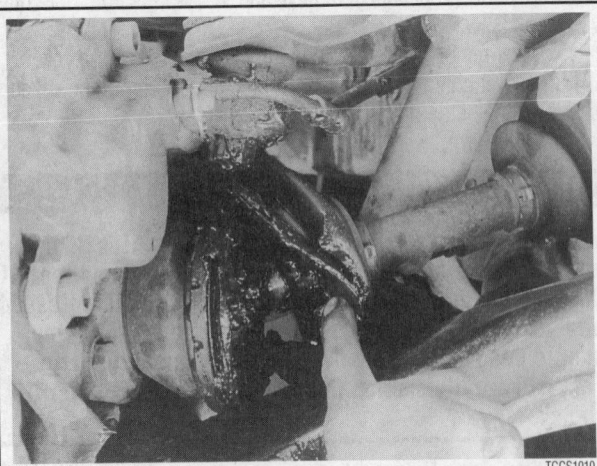

Remove, clean and inspect this CV joint—it may be possible to save the joint by replacing the boot as long as the joint is not beyond repair

Push apart the bellows to inspect for tears or cracks that may developing

REPLACEMENT

➡**Always follow the instructions included in the CV-joint boot kit. There are several variations and methods of boot replacement. Use the following procedures as a general guide and in case the kit may not contain specific instructions.**

Most outer CV joints on Asian vehicles, including Chrysler imports, use a Birfield joint, which should not be disassembled. To replace the outer boot, disassemble the inner joint, then slide the outer boot off the inner end of the shaft.

Outer Boot

➡**Generally the Double Offset Joint (DOJ) is used as the outer CV-joint.**

1. Remove the halfshaft and carefully place it in a vise using a protective covering on the vise jaws.

✳✳ WARNING

Some halfshafts may utilize hollow shafts between the CV-joints. Do not tighten the vise more than necessary.

2. Cut the large and small CV-joint boot band clamps and discard them.

3. Slide the boot down the shaft uncovering the outer joint.

4. Clean the grease from the joint to uncover the snapring.

➡**A Tripot Joint (TJ) may also be referred to as a Tulip Joint because of the physical shape of it which resembles a tulip.**

5. Using snapring pliers, open the snapring and slide the outer joint off the shaft.

6. Remove the boot from the shaft.

7. Clean the joint thoroughly using parts cleaner, then dry it completely with compressed air. Inspect the inner bearing and race assembly. If the joint is worn or damaged, replace it.

To install:

8. Wrap the splines on the end of the halfshaft with tape to prevent damage to the boot during installation.

9. Slide the small CV-joint boot clamp onto the halfshaft and

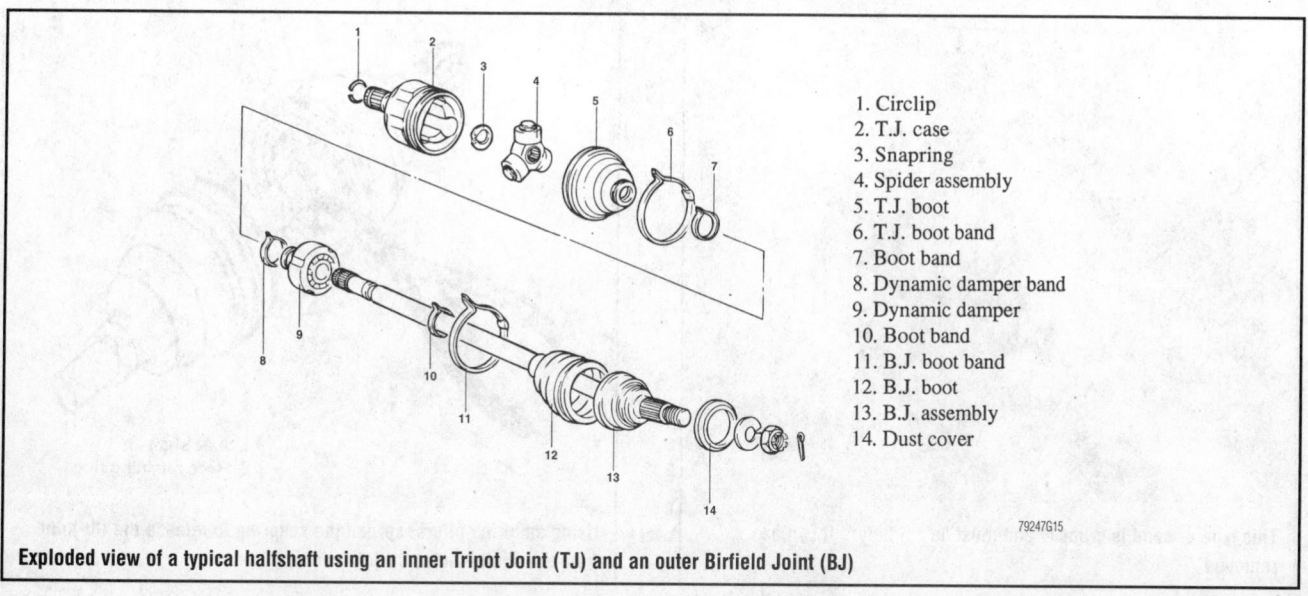

1. Circlip
2. T.J. case
3. Snapring
4. Spider assembly
5. T.J. boot
6. T.J. boot band
7. Boot band
8. Dynamic damper band
9. Dynamic damper
10. Boot band
11. B.J. boot band
12. B.J. boot
13. B.J. assembly
14. Dust cover

79247G15

Exploded view of a typical halfshaft using an inner Tripot Joint (TJ) and an outer Birfield Joint (BJ)

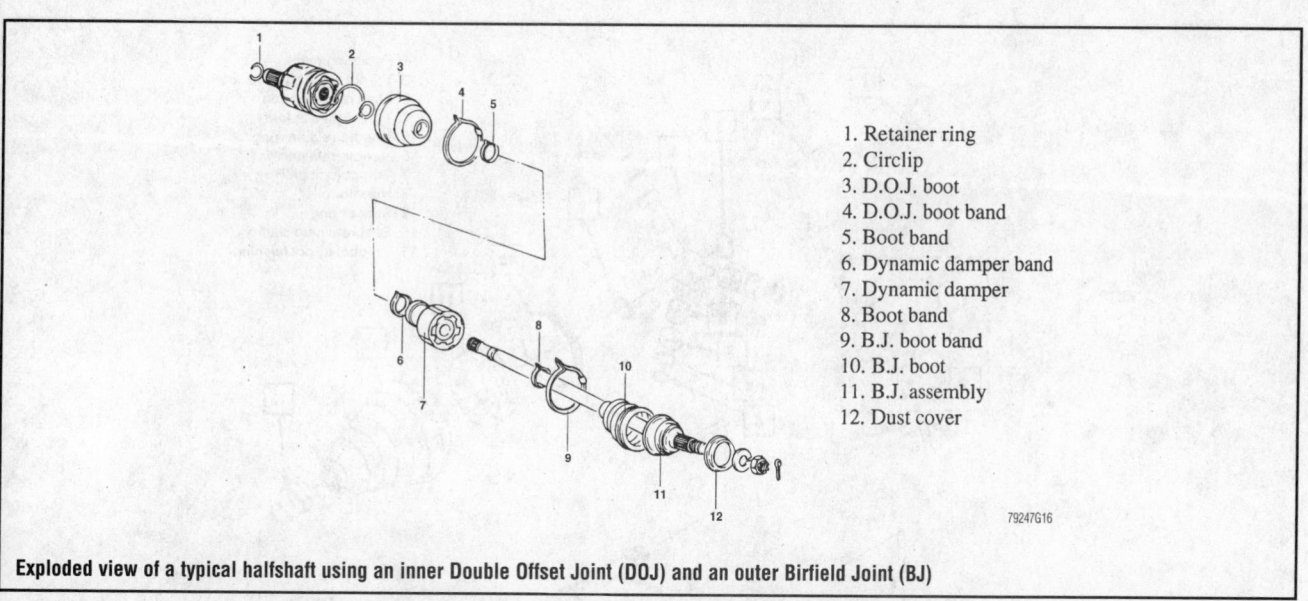

1. Retainer ring
2. Circlip
3. D.O.J. boot
4. D.O.J. boot band
5. Boot band
6. Dynamic damper band
7. Dynamic damper
8. Boot band
9. B.J. boot band
10. B.J. boot
11. B.J. assembly
12. Dust cover

79247G16

Exploded view of a typical halfshaft using an inner Double Offset Joint (DOJ) and an outer Birfield Joint (BJ)

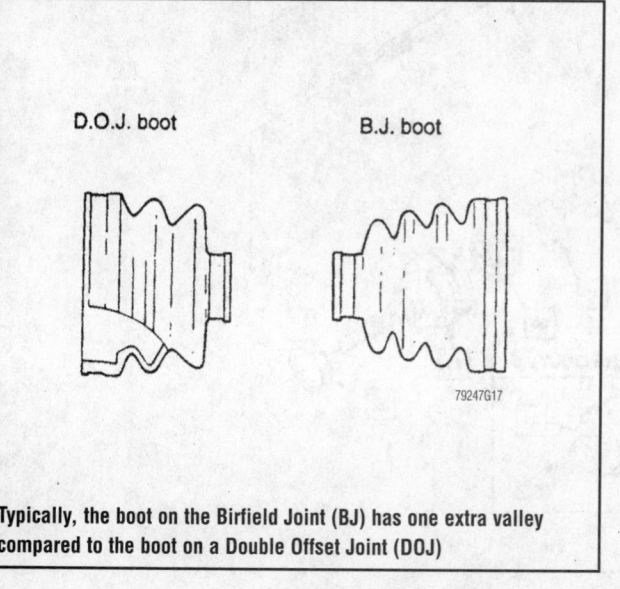

D.O.J. boot   B.J. boot

79247G17

Typically, the boot on the Birfield Joint (BJ) has one extra valley compared to the boot on a Double Offset Joint (DOJ)

TCCS7031

Pry under the hook to remove this type of band from the CV-joint

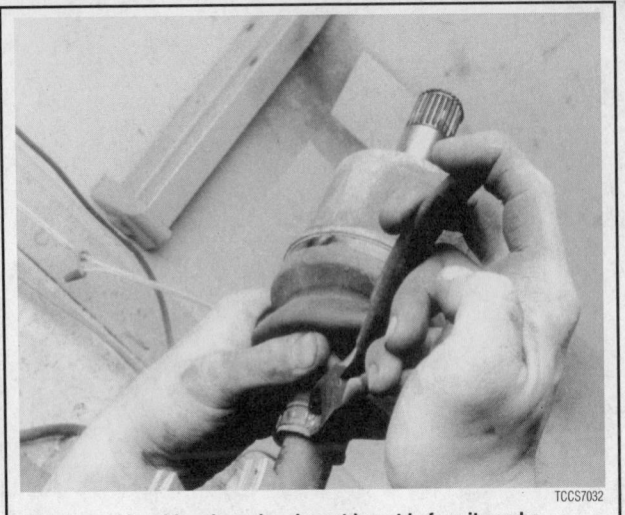

This type of band is crimped and must be cut before it can be removed

TCCS7032

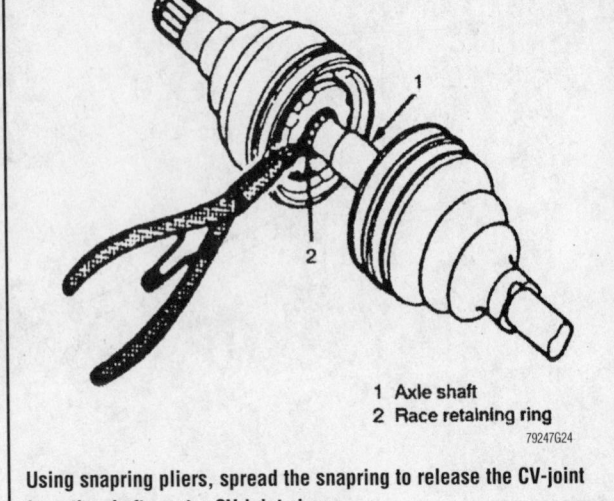

1 Axle shaft
2 Race retaining ring

79247G24

Using snapring pliers, spread the snapring to release the CV-joint from the shaft—outer CV-joint shown

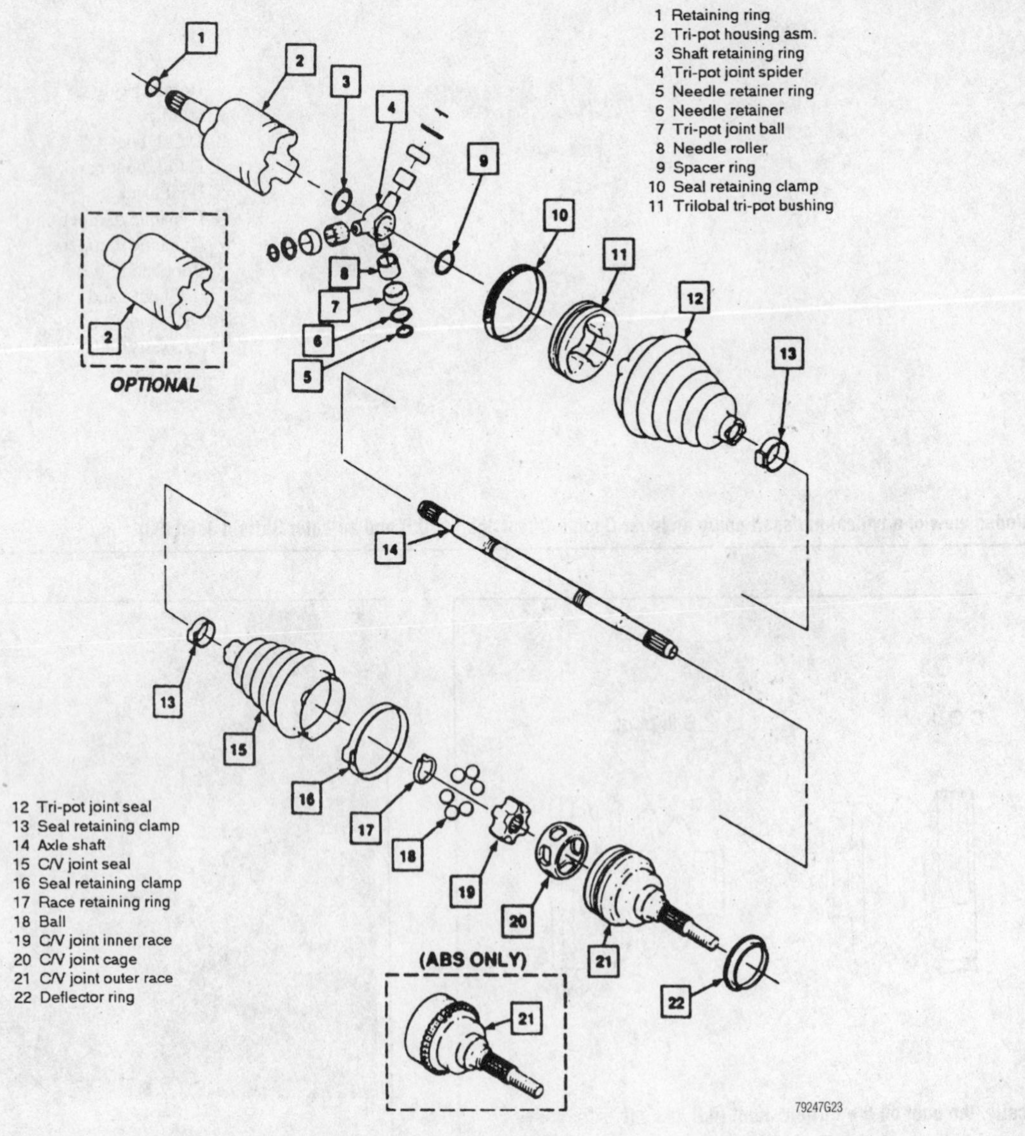

1 Retaining ring
2 Tri-pot housing asm.
3 Shaft retaining ring
4 Tri-pot joint spider
5 Needle retainer ring
6 Needle retainer
7 Tri-pot joint ball
8 Needle roller
9 Spacer ring
10 Seal retaining clamp
11 Trilobal tri-pot bushing

OPTIONAL

12 Tri-pot joint seal
13 Seal retaining clamp
14 Axle shaft
15 C/V joint seal
16 Seal retaining clamp
17 Race retaining ring
18 Ball
19 C/V joint inner race
20 C/V joint cage
21 C/V joint outer race
22 Deflector ring

(ABS ONLY)

79247G23

Exploded view of a halfshaft with TJ inner and DOJ outer joints

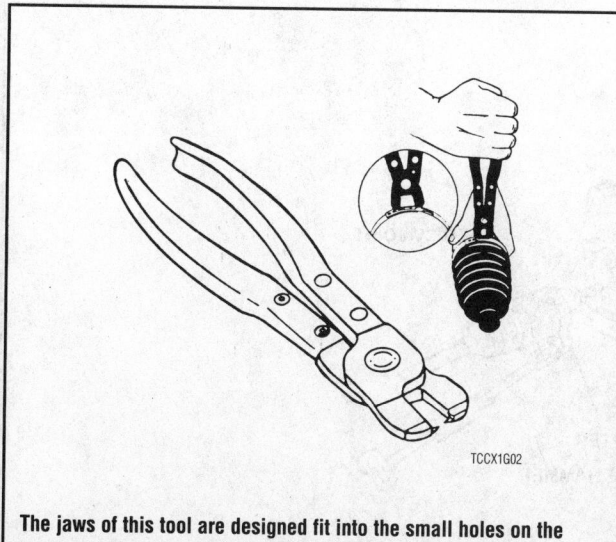

The jaws of this tool are designed fit into the small holes on the band and allow it to be tightened

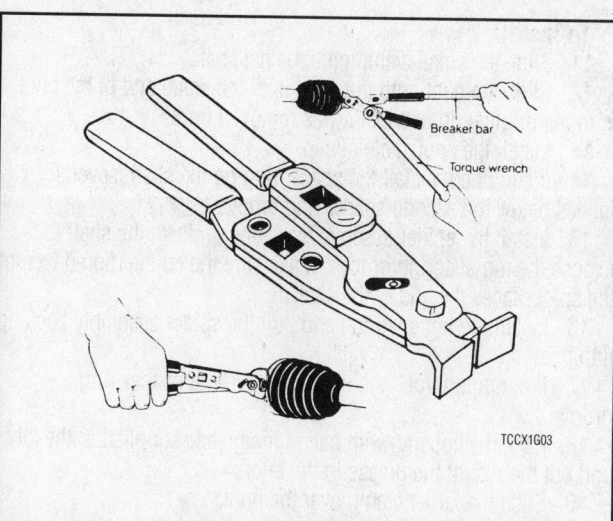

This tool allows a torque wrench to be used when the manufacturer specifies that a certain pressure is required to crimp the band

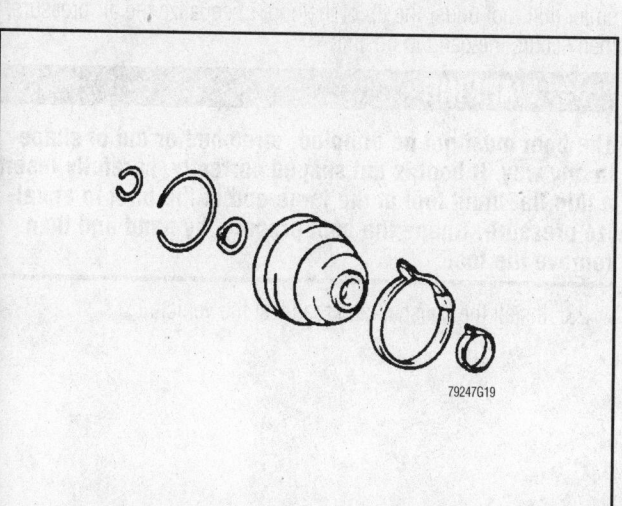

The typical boot replacement kit contains a new boot, two clamps and special grease—new circlips may also be included is some kits

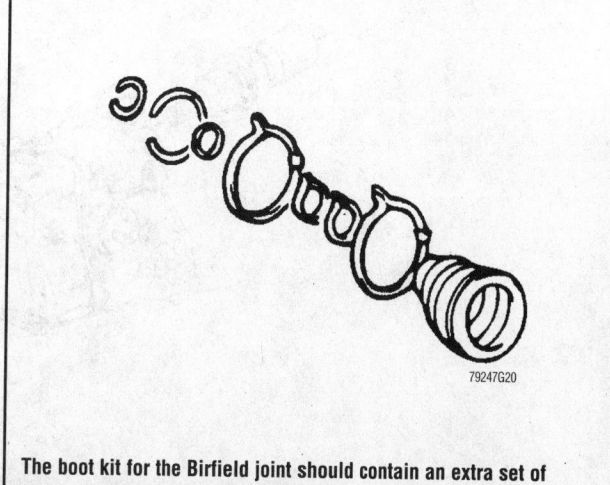

The boot kit for the Birfield joint should contain an extra set of bands, because the inner joint must be removed in order to install a new boot on the outer (Birfield) joint

push the boot down several inches past the seal mounting area. Remove the tape from the halfshaft splines.

10. Check the snapring in the outer joint for damage or excessive wear and replace as necessary. Pack the joint with half of the grease supplied in the boot kit and install it on the shaft.

11. Insert the shaft into the joint until the splines engage. With a brass drift, lightly tap the joint down until the snapring engages.

12. Pack the remaining grease from the kit into the boot, then pull the large side of the boot over the CV-joint. Seat the small end of the boot on the seal mounting area.

➡ **Some CV-joint boot bands require the use of special pliers that are designed to grip the band and allow it to be tightened.**

13. Slide the small clamp into position and secure it.

14. Install the large clamp in the proper position. Slide a small, dull tool under the lip of the boot to equalize the air pressure, then secure the band.

❋❋ WARNING

Incorrect CV-joint boot installation may lead to early failure of the boot. The boot must not be dimpled, stretched or out of shape in any way when installed. If the boot is not shaped correctly, carefully insert a thin, blunt tool under the large end of the boot to equalize air pressure. Shape the boot properly by hand, then remove the tool.

15. Install the halfshaft. Road test the vehicle to check for abnormal noise or vibration.

Inner Boot

1. Remove the halfshaft and carefully place it in a vise using a protective covering on the vise jaws.

2. Cut the large and small boot clamps and discard.

❋❋ WARNING

Do not cut through the boot and damage the sealing surface of the CV-joint housing.

3. Pull the boot down the shaft to expose the joint.

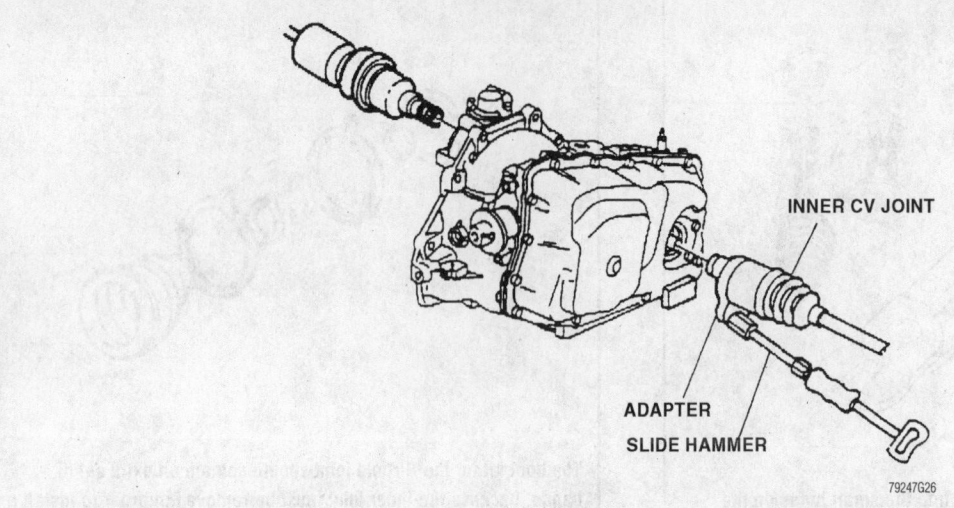

A special adapter for a slide hammer is available to assist in removing the halfshaft from the transaxle

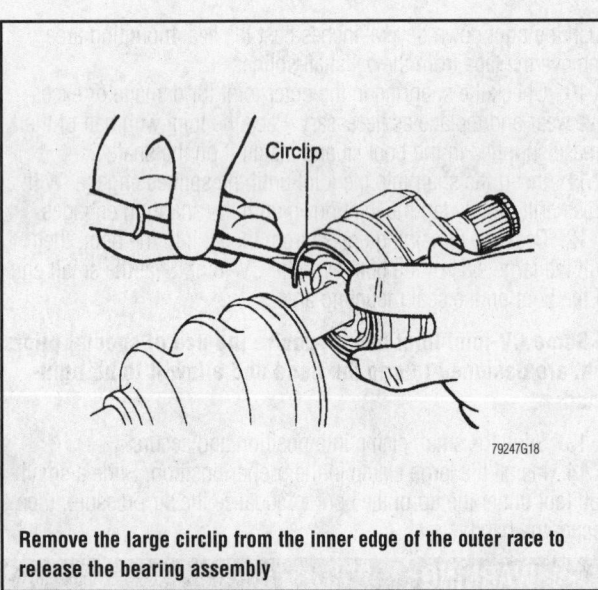

Remove the large circlip from the inner edge of the outer race to release the bearing assembly

4. If equipped, remove the large circlip from the inner edge of the outer bearing race.

5. Matchmark the bearing and outer case so they can be installed in their original positions.

6. Remove the housing from the spider and axle. Clean and dry all components thoroughly. Replace any parts that show signs of wear.

7. Push the spider assembly down the shaft to uncover the snapring on the end of the shaft. Remove the snapring and slide the spider assembly off the end of the shaft.

8. If equipped, remove the spacer ring from the shaft.

9. Remove the remaining circlip and slide the boot off the shaft.

10. Clean and dry all components thoroughly. Replace any parts that show signs of wear.

To install:

11. Slide the small clamp onto the halfshaft.

12. Slide the boot onto the shaft until the small end of the boot is in the original groove that it was removed from.

13. Install the applicable circlip.

14. If equipped, install the spacer ring on the shaft, several inches below the second spacer ring groove.

15. Install the spider assembly far enough down the shaft to expose the top snapring groove. Make sure the counterbored face of the spider faces the end of the shaft.

16. Install the top snapring and pull the spider assembly back up into position.

17. If equipped, lock the spacer ring in the spacer ring groove.

18. Pack the housing with half of the grease supplied in the kit and put the rest of the grease in the boot.

19. Slide the larger clamp over the boot.

20. Push the housing over the spider assembly.

21. If equipped, install the large circlip in the outer race.

22. Slide the larger diameter of the boot into position. Slide a small dull tool under the lip of the boot to equalize the air pressure, then secure the band in position.

✳✳ WARNING

The boot must not be dimpled, stretched or out of shape in any way. If boot is not shaped correctly, carefully insert a thin flat blunt tool at the large end of the boot to equalize pressure. Shape the boot properly by hand and then remove the tool.

23. Install the halfshaft and road test the vehicle.

OXYGEN (O₂) SENSORS

8

OXYGEN (O₂) SENSORS

General Information

An Oxygen (O₂) sensor is an input device used by the engine control computer to monitor the amount of oxygen in the exhaust gas stream. This information is used by the computer, along with other inputs, to fine-tune the air/fuel mixture so that the engine can run with the greatest efficiency in all conditions. The O₂ sensor sends this information to the computer in the form of a 100–900 millivolt (mV) reference signal, which is actually created by the O₂ sensor itself through chemical interactions between the sensor tip material (zirconium dioxide in almost all cases) and the oxygen levels in the exhaust gas stream and ambient atmosphere gas. At operating temperatures, approximately 1100°F (600°C), the element becomes a semiconductor. Essentially, through the differing levels of oxygen in the exhaust gas stream and in the surrounding atmosphere, the sensor creates a voltage signal which is directly and consistently related to the concentration of oxygen in the exhaust stream. Typically, a higher than normal amount of oxygen

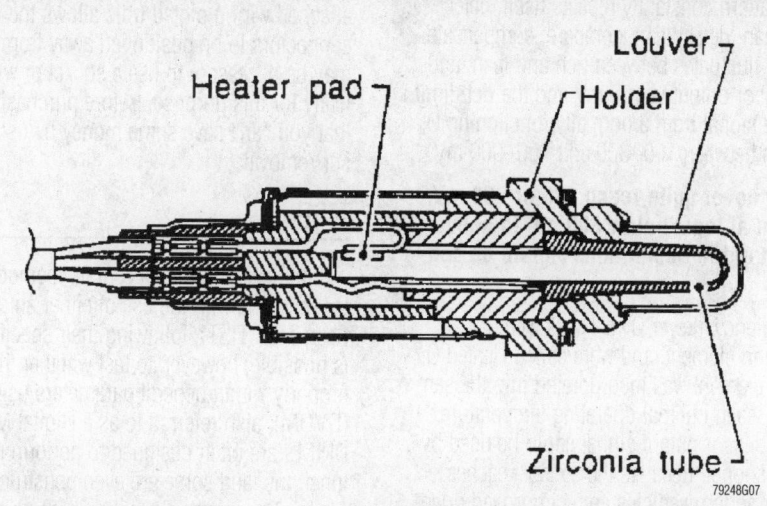

A cut away view of a heated oxygen sensor

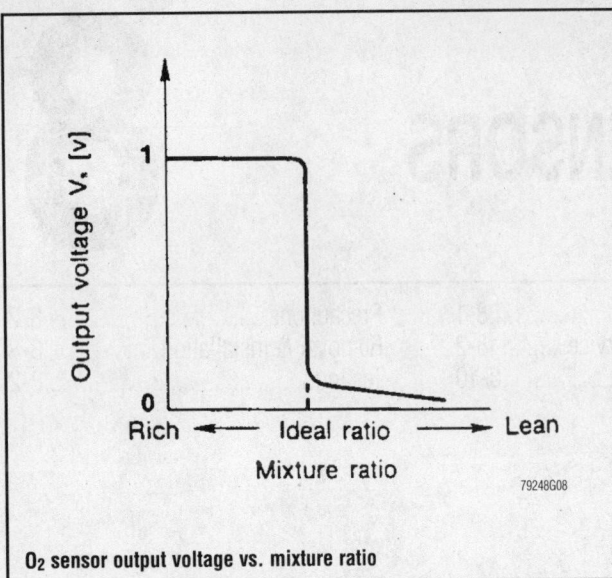

O₂ sensor output voltage vs. mixture ratio

in the exhaust stream indicates that not all of the available oxygen was used in the combustion process, because there was not enough fuel (lean condition) present. Inversely, a lower than normal concentration of oxygen in the exhaust stream indicates that a large amount was used in the combustion process, because a larger than necessary amount of fuel was present (rich condition). Thus, the engine control computer can correct the amount of fuel introduced into the combustion chambers.

Since the control computer uses the O₂ sensor output voltage as an indication of the oxygen concentration, and the oxygen concentration directly affects O₂ sensor output, the signal voltage from the sensor to the computer fluctuates constantly. This fluctuation is caused by the nature of the interaction between the computer and the O₂ sensor, which follows a general pattern: detect, compare, compensate, detect, compare, compensate, etc. This means that when the computer detects a lean signal from the O₂ sensor, it compares the reading with known parameters stored within its memory. It calculates that there is too much oxygen present in the exhaust gases, so it compensates by adding more fuel to the air/fuel mixture. This, in turn, causes the O₂ sensor to send a rich signal to the computer, which then compares this new signal, and adjusts the air/fuel mixture again. This pattern constantly repeats itself: detect rich, compare, compensate lean, detect lean, compare, compensate rich, etc. Since the O₂ sensor fluctuates between rich and lean, and because the lean limit for sensor output is 100 mV and the rich limit is 900 mV, the proper voltage signal from a normally functioning O₂ sensor consistently fluctuates between 100–300 and 700–900 mV.

➡**The sensor voltage may never quite reach 100 or 900 mV, but it should fluctuate from at least below 300 mV to above 700 mV, and the mid-point of the fluctuations should be centered around 500 mV.**

To improve O₂ sensor efficiency, newer O₂ sensors were designed with a built-in heating element, and were called Heated O₂ (HO₂) sensors. This heating element was incorporated into the sensor so that the sensor would reach optimal operating temperature quicker, meaning that the O₂ sensor output signal could be used by the engine control computer sooner. Because the sensor reaches optimal temperature quicker, modern vehicles enjoy improved driveability and fuel economy even before the engine reaches normal operating temperature.

Although a few manufacturers changed earlier, in 1995 all vehicles were required to implement a new set of engine control parameters, referred to as On-Board Diagnostics second generation (OBD-II). This updated system (based on the former OBD-I), called for additional O₂ sensors to be used after the catalytic converter, so that catalytic converter efficiency could be measured by the vehicle's engine control computer. The O₂ sensors mounted in the exhaust system after the catalytic converters are not used to affect air/fuel mixture; they are used solely to monitor catalytic converter efficiency.

O₂ (Oxygen) Sensor Service

PRECAUTIONS

When testing or servicing an O₂ sensor you will need to start and warm the engine to operating temperature in order to either perform the necessary testing procedures or to easily remove the sensor from its fitting. This will create a situation in which you will be working around a **HOT** exhaust system. The following is a list of precautions to consider during this service:

• Do not pierce any wires when testing an O₂ sensor, as this can lead to wiring harness damage. Backprobe the connector, when necessary.

• While testing the sensor, be sure to keep out of the way of moving engine components, such as the cooling fan. Refrain from wearing loose clothing which may become tangled in moving engine components.

• Safety glasses must be worn at all times when working on, or near, the exhaust system. Older exhaust systems may be covered with loose rust particles which can shower you when disturbed. These particles are more than a nuisance and can injure your eye.

• Be cautious when working on and around the hot exhaust system. Painful burns will result if skin is exposed to the exhaust system pipes or manifolds.

• The O₂ sensor may be difficult to remove when the engine temperature is below 120°F (48°C). Excessive force may damage the threads in the exhaust manifold or pipe, therefore always start the engine and allow it to reach normal operating temperature prior to removal.

• Since O₂ sensors are usually designed with a permanently-attached wiring pigtail (this allows the wiring harness and sensor connectors to be positioned away from the hot exhaust system), it may be necessary to use a socket or wrench that is designed specifically for this purpose. Before purchasing such a socket, be sure that you can't save some money by using a box end wrench for sensor removal.

TESTING

The best, and most accurate method to test the operation of an O₂ sensor is with the use of either an oscilloscope or a Diagnostic Scan Tool (DST), following their specific instructions for testing. It is possible, however, to test whether the O₂ sensor is functioning properly within general parameters using a Digital Volt-Ohmmeter (DVOM), also referred to as a Digital Multi-Meter (DMM). Newer DMM's are often designed to perform many advanced diagnostic functions, and some are even constructed to be used as an oscilloscope. Two in-vehicle testing procedures, and one bench test procedure, will be provided for the common zirconium dioxide oxygen sensor. The first in-vehicle test makes use of a standard DVOM with a 10 megohm impedance, whereas the second in-vehicle test presented necessitates the usage of an advanced DMM with MIN/MAX/Average functions. Both of these in-vehicle test procedures are likely to set Diagnostic Trouble Codes (DTC's) in the engine control computer. Therefore, after testing, be sure to clear all DTC's before retesting the sensor, if necessary.

These are some of the common DTC's which may be set during testing:

- Open in the O$_2$ sensor circuit
- Constant low voltage in the O$_2$ sensor circuit
- Constant high voltage in the O$_2$ sensor circuit
- Other fuel system problems could set a O$_2$ sensor code

→**Because an improperly functioning fuel delivery and/or control system can adversely affect the O$_2$ sensor voltage output signal, testing only the O$_2$ sensor is an inaccurate method for diagnosing an engine driveability problem.**

If after testing the sensor, the sensor is thought to be defective because of high or low readings, be sure to check that the fuel delivery and engine management system is working properly before condemning the O$_2$ sensor. Otherwise, the new O$_2$ sensor may continue to register the same high or low readings.

Often, by testing the O$_2$ sensor, another problem in the engine control management system can be diagnosed. If the sensor appears to be defective while installed in the vehicle, perform the bench test. If the sensor functions properly during the bench test, chances are that there may be a larger problem in the vehicle's fuel delivery and/or control system.

Many things can cause an O$_2$ sensor to fail, including old age, antifreeze contamination, physical damage, prolonged exposure to overly-rich exhaust gases, and exposure to silicone sealant fumes. Be sure to remedy any such condition prior to installing a new sensor, otherwise the new sensor may be damaged as well.

→**Perform a visual inspection of the sensor. Black sooty deposits may indicate a rich air/fuel mixture, brown deposits may indicate an oil consumption problem, and white gritty deposits may indicate an internal coolant leak. All of these conditions can destroy a new sensor if not corrected before installation.**

O$_2$ Sensor Terminal Identification

The easiest method for determining sensor terminal identification is to use a wiring diagram for the vehicle and engine in question. However, if a wiring diagram is not available there is a method for determining terminal identification. Throughout the testing procedures, the following terms will be used for clarity:

- Vehicle harness connector—this refers to the connector on the wires which are attached to the vehicle; NOT the connector at the end of the sensor pigtail.
- Sensor pigtail connector—this refers to the connector attached to the sensor itself.
- O$_2$ circuit—this refers to the circuit in a Heated O$_2$ (HO$_2$) sensor which corresponds to the oxygen-sensing function of the sensor; NOT the heating element circuit.
- Heating circuit—this refers to the circuit in a HO$_2$ sensor which is designed to warm the HO$_2$ sensor quickly to improve driveability.
- Sensor Output (SOUT) terminal—this is the terminal which corresponds to the O$_2$ circuit output. This is the terminal which will

register the millivolt signals created by the sensor based upon the amount of oxygen in the exhaust gas stream.

- Sensor Ground (SGND) terminal—when a sensor is so equipped, this refers to the O$_2$ circuit ground terminal. Many O$_2$ sensors are not equipped with a ground wire, rather they utilize the exhaust system for the ground circuit.
- Heating Power (HPWR) terminal—this terminal corresponds to the circuit which provides the O$_2$ sensor heating circuit with power when the ignition key is turned to the **ON** or **RUN** positions.
- Heating Ground (HGND) terminal—this is the terminal connected to the heating circuit ground wire.

ONE WIRE SENSOR

One wire sensors are by far the easiest to determine sensor terminal identification, but this is self-evident. On one wire O$_2$ sensors, the single wire terminal is the SOUT and the exhaust system is used to provide the sensor ground pathway. Proceed to the test procedures.

TWO WIRE SENSOR

On two wire sensors, one of the connector terminals is the SOUT and the other is the SGND. To determine which one is which, perform the following:

1. Locate the O$_2$ sensor and its pigtail connector. It may be necessary to raise and safely support the vehicle to gain access to the connector.
2. Start the engine and allow it to warm up to normal operating temperature, then turn the engine **OFF**.
3. Using a DVOM set to read 100–900 mV (millivolts) DC, backprobe the positive DVOM lead to one of the unidentified terminals and attach the negative lead to a good engine ground.

✳✳ CAUTION

While the engine is running, keep clear of all moving and hot components. Do not wear loose clothing. Otherwise severe personal injury or death may occur.

4. Have an assistant restart the engine and allow it to idle.
5. Check the DVOM for voltage.
6. If no voltage is evident, check your DVOM leads to ensure that they are properly connected to the terminal and engine ground. If still no voltage is evident at the first terminal, move the positive meter lead to backprobe the second terminal.
7. If voltage is now present, the positive meter lead is attached to the SOUT terminal. The remaining terminal is the SGND terminal. If still no voltage is evident, either the O$_2$ sensor is defective or the meter leads are not making adequate contact with the engine ground and terminal contacts; clean the contacts and retest. If still no voltage is evident, the sensor is defective.
8. Have your assistant turn the engine **OFF**.
9. Label the sensor pigtail SOUT and SGND terminals.
10. Proceed to the test procedures.

THREE WIRE SENSOR

→**Three wire sensors are HO$_2$ sensors.**

On three wire sensors, one of the connector terminals is the SOUT, one of the terminals is the HPWR and the other is the HGND. The SGND is achieved through the exhaust system, as with the one wire O$_2$ sensor. To identify the three terminals, perform the following:

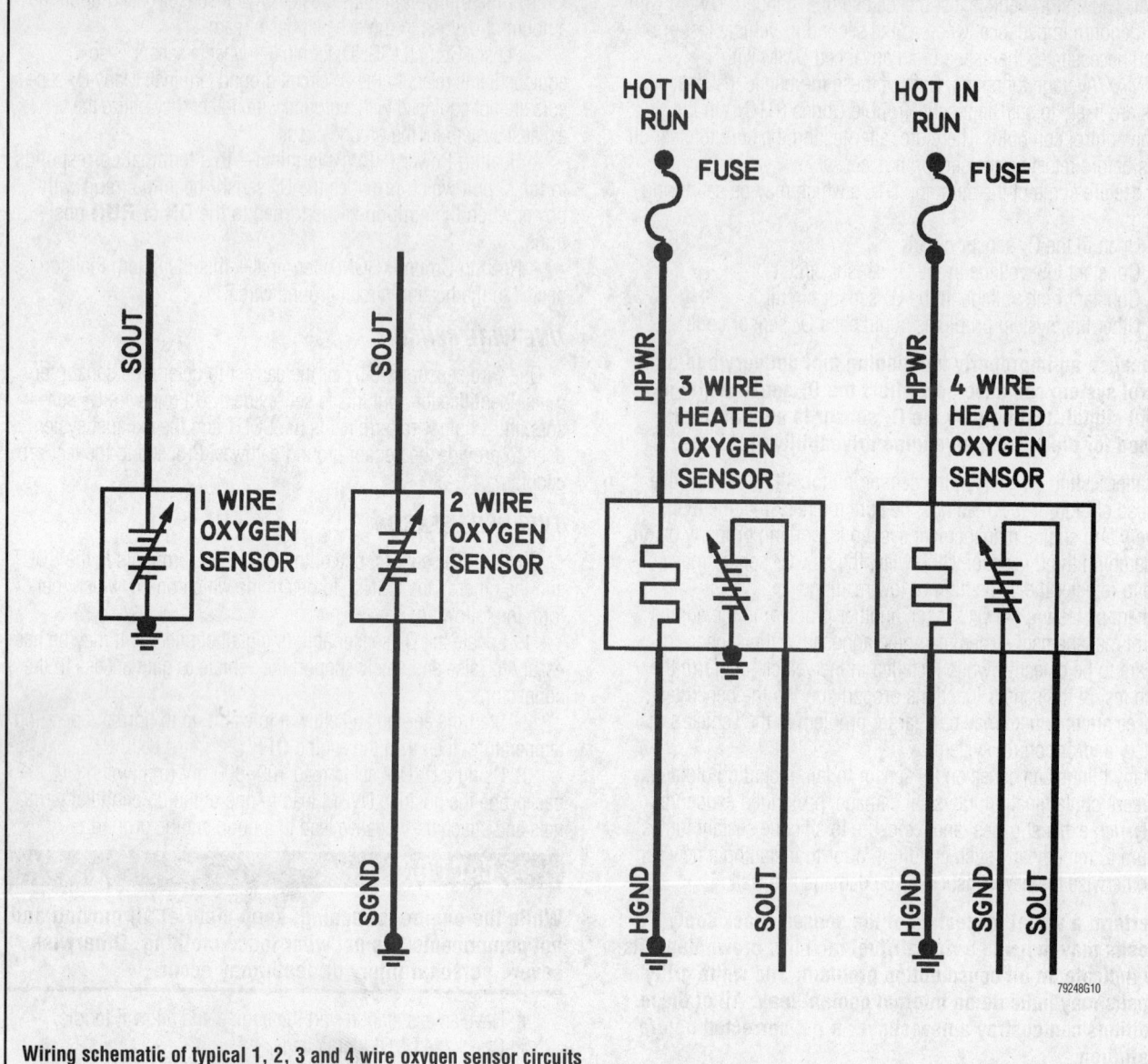

Wiring schematic of typical 1, 2, 3 and 4 wire oxygen sensor circuits

1. Locate the O₂ sensor and its pigtail connector. It may be necessary to raise and safely support the vehicle to gain access to the connector.

2. Disengage the sensor pigtail connector from the vehicle harness connector.

3. Using a DVOM set to read 12 volts, attach the DVOM ground lead to a good engine ground.

4. Have an assistant turn the ignition switch **ON** without actually starting the engine.

5. Probe all three terminals in the vehicle harness connector. One of the terminals should exhibit 12 volts of power with the ignition key **ON**; this is the HPWR terminal.

    a. If the HPWR terminal was identified, note which of the sensor harness connector terminals is the HPWR, then match the vehicle harness connector to the sensor pigtail connector. Label the corresponding sensor pigtail connector terminal with HPWR.

    b. If none of the terminals showed 12 volts of power, locate and test the heater relay or fuse. Then, perform Steps 3–6 again.

6. Start the engine and allow it to warm up to normal operating temperature, then turn the engine **OFF**.

7. Have your assistant turn the ignition **OFF**.

8. Using the DVOM set to measure resistance (ohms), attach one of the leads to the HPWR terminal of the sensor pigtail connector. Use the other lead to probe the two remaining terminals of the sensor pigtail connector, one at a time. The DVOM should show continuity with only one of the remaining unidentified terminals; this is the HGND terminal. The remaining terminal is the SOUT.

    a. If continuity was found with only one of the two unidentified terminals, label the HGND and SOUT terminals on the sensor pigtail connector.

    b. If no continuity was evident, or if continuity was evident from both unidentified terminals, the O₂ sensor is defective.

9. All three wire terminals should now be labeled on the sensor pigtail connector. Proceed with the test procedures.

FOUR WIRE SENSOR

→**Four wire sensors are HO$_2$ sensors.**

On four wire sensors, one of the connector terminals is the SOUT, one of the terminals is the SGND, one of the terminals is the HPWR and the other is the HGND. To identify the four terminals, perform the following:

1. Locate the O$_2$ sensor and its pigtail connector. It may be necess-ary to raise and safely support the vehicle to gain access to the connector.

2. Disengage the sensor pigtail connector from the vehicle harness connector.

3. Using a DVOM set to read 12 volts, attach the DVOM ground lead to a good engine ground.

4. Have an assistant turn the ignition switch **ON** without actually starting the engine.

5. Probe all four terminals in the vehicle harness connector. One of the terminals should exhibit 12 volts of power with the ignition key **ON**; this is the HPWR terminal.

   a. If the HPWR terminal was identified, note which of the sensor harness connector terminals is the HPWR, then match the vehicle harness connector to the sensor pigtail connector. Label the corresponding sensor pigtail connector terminal with HPWR.

   b. If none of the terminals showed 12 volts of power, locate and test the heater relay or fuse. Then, perform Steps 2–6 again.

6. Have your assistant turn the ignition **OFF**.

7. Using the DVOM set to measure resistance (ohms), attach one of the leads to the HPWR terminal of the sensor pigtail connector. Use the other lead to probe the three remaining terminals of the sensor pigtail connector, one at a time. The DVOM should show continuity with only one of the remaining unidentified terminals; this is the HGND terminal.

   a. If continuity was found with only one of the two unidentified terminals, label the HGND terminal on the sensor pigtail connector.

   b. If no continuity was evident, or if continuity was evident from all unidentified terminals, the O$_2$ sensor is defective.

   c. If continuity was found at two of the other terminals, the sensor is probably defective. However, the sensor may not necessarily be defective, because it may have been designed with the two ground wires joined inside the sensor in case one of the ground wires is damaged; the other circuit could still function properly. Though, this is highly unlikely. A wiring diagram is necessary in this particular case to know whether the sensor was so designed.

8. Reattach the sensor pigtail connector to the vehicle harness connector.

9. Start the engine and allow it to warm up to normal operating temperature, then turn the engine **OFF**.

10. Using a DVOM set to read 100–900 mV (millivolts) DC, backprobe the negative DVOM lead to one of the unidentified terminals and the positive lead to the other unidentified terminal.

✳✳ CAUTION

While the engine is running, keep clear of all moving and hot components. Do not wear loose clothing. Otherwise severe personal injury or death may occur.

11. Have an assistant restart the engine and allow it to idle.

12. Check the DVOM for voltage.

   a. If no voltage is evident, check your DVOM leads to ensure that they are properly connected to the terminals. If still no voltage is evident at either of the terminals, either the terminals were accidentally marked incorrectly or the sensor is defective.

   b. If voltage is present, but the polarity is reversed (the DVOM will show a negative voltage amount), turn the engine **OFF** and swap the two DVOM leads on the terminals. Start the engine and ensure that the voltage now shows the proper polarity.

   c. If voltage is evident and is the proper polarity, the positive DVOM lead is attached to the SOUT and the negative lead to the SGND terminals.

13. Have your assistant turn the engine **OFF**.

14. Label the sensor pigtail SOUT and SGND terminals.

In-Vehicle Tests

✳✳ WARNING

Never apply voltage to the O$_2$ circuit of the sensor, otherwise it may be damaged. Also, never connect an ohmmeter (or a DVOM set on the ohm function) to both of the O$_2$ circuit terminals (SOUT and SGND) of the sensor pigtail connector; it may damage the sensor.

Test 1 makes use of a standard DVOM with a 10 megohm impedance, whereas Test 2 necessitates the usage of an advanced Digital Multi-Meter (DMM) with MIN/MAX/Average functions or a sliding bar graph function. Both of these in-vehicle test procedures are likely to set Diagnostic Trouble Codes (DTC's) in the engine control computer. Therefore, after testing, be sure to clear all DTC's before retesting the sensor, if necessary. The third in-vehicle test is designed for the use of a scan tool or oscilloscope. The fourth test (Heating Circuit Test) is designed to check the function of the heating circuit in a HO$_2$ sensor.

→**If the O$_2$ sensor being tested is designed to use the exhaust system for the SGND, excessive corrosion between the exhaust and the O$_2$ sensor may affect sensor functioning.**

The in-vehicle tests may be performed for O$_2$ sensors located in the exhaust system after the catalytic converter. However, the O$_2$ sensors located behind the catalytic converter will not fluctuate like

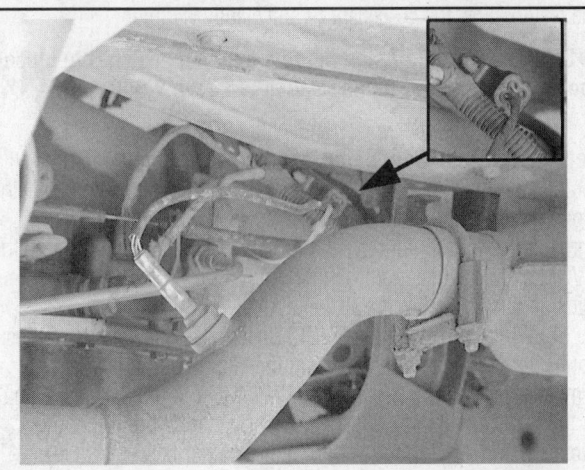

89664P30

To test the O$_2$ sensor, locate it and its connector (inset), which should be positioned away from the exhaust system to prevent heat damage

the sensors mounted before the converter, because the converter, when functioning properly, emits a steady amount of oxygen. If the O₂ sensor mounted after the catalytic converter exhibits a fluctuating signal (like other O₂ sensors), the catalytic converter is most likely defective.

TEST 1—DIGITAL VOLT-OHMMETER

This test will not only verify proper sensor functioning, but is also designed to ensure the engine control computer and associated wiring is functioning properly as well.

1. Start the engine and allow it to warm up to normal operating temperature.

➡**If you are using the opening of the thermostat to gauge normal operating temperature, be forewarned: a defective thermostat can open too early and prevent the engine from reaching normal operating temperature. This can cause a slightly rich condition in the exhaust, which can throw the O₂ sensor readings off slightly.**

2. Turn the ignition switch **OFF**, then locate the O₂ sensor pigtail connector.

3. Perform a visual inspection of the connector to ensure it is properly engaged and all terminals are straight, tight and free from corrosion or damage.

4. Disengage the sensor pigtail connector from the vehicle harness connector.

5. On sensors equipped with a SGND terminal (sensors which do not use the exhaust system for the sensor ground pathway), connect a jumper wire to the SGND terminal and to a good, clean engine ground (preferably the negative terminal of the battery).

6. Using a DVOM set to read DC voltage, attach the positive lead to the SOUT terminal of the sensor pigtail connector, and the DVOM negative lead to a good engine ground.

✷✷ CAUTION

While the engine is running, keep clear of all moving and hot components. Do not wear loose clothing. Otherwise severe personal injury or death may occur.

7. Have an assistant start the engine and hold it at approximately 2,000 rpm. Wait at least 1 minute before commencing with the test to allow the O₂ sensor to sufficiently warm up.

➡**Some carbureted Asian models may not switch into closed loop operation until engine speed is above 2,500 rpm.**

8. Using a jumper wire, connect the SOUT terminal of the **vehicle harness connector** to a good engine ground. This will fool the engine control computer into thinking it is receiving a lean signal from the O₂ sensor, and, therefore, the computer will enrichen the air/fuel ratio. With the SOUT terminal so grounded, the DVOM should register at least 800 mV, as the control computer adds additional fuel to the air/fuel ratio.

9. While observing the DVOM, disconnect the vehicle harness connector SOUT jumper wire from the engine ground. Use the jumper wire to apply slightly less than 1 volt to the SOUT terminal of the vehicle harness connector. One method to do this is by grasping and squeezing the end of the jumper between your forefinger and thumb of one hand while touching the positive terminal of the battery post with your other hand. This allows your body to act as a resistor for the battery positive voltage, and fools the engine control computer into thinking it is receiving a rich signal.

Or, use a mostly-drained AA battery by connecting the positive terminal of the AA battery to the jumper wire and the negative terminal of the battery to a good engine ground. (Another jumper wire may be necessary to do this.) The computer should lean the air/fuel mixture out. This lean mixture should register as 150 mV or less on the DVOM.

10. If the DVOM did not register millivoltages as indicated, the problem may be either the sensor, the engine control computer or the associated wiring. Perform the following to determine which is the defective component:

    a. Remove the vehicle harness connector SOUT jumper wire.

    b. While observing the DVOM, artificially enrich the air/fuel charge using propane. The DVOM reading should register higher than normal millivoltages. (Normal voltage for an ideal air/fuel mixture is approximately 450–550 mV DC). Then, lean the air/fuel intake charger by either disconnecting one of the fuel injector wiring harness connectors (to prevent the injector from delivering fuel) or by detaching one or two vacuum lines (to add additional non-metered air into the engine). The DVOM should now register lower than normal millivoltages. If the DVOM functioned as indicated, the problem lies elsewhere in the fuel delivery and control system. If the DVOM readings were still unresponsive, the O₂ sensor is defective; replace the sensor and retest.

➡**Poor wire connections and/or ground circuits may shift a normal O₂ sensor's millivoltage readings up into the rich range or down into the lean range. It is a good idea to check the wire condition and continuity before replacing a component which will not fix the problem. A voltage drop test between the sensor case and ground which reveals 14–16 mV, or more, indicates a probable bad ground.**

11. Turn the engine **OFF**, remove the DVOM and all associated jumper wires. Reattach the vehicle harness connector to the sensor pigtail connector. If applicable, reattach the fuel injector wiring connector and/or the vacuum line(s).

12. Clear any DTC's present in the engine control computer memory, as necessary.

TEST 2—DIGITAL MULTI-METER

This test method is a more straight forward O₂ sensor test, and does not test the engine control computer's response to the O₂ sensor signal. The use of a DMM with the MIN/MAX/Average function or sliding bar graph/wave function is necessary for this test. Don't forget that the O₂ sensor mounted after the catalytic converter (if equipped) will not fluctuate like the other O₂ sensor(s) will.

1. Start the engine and allow it to warm up to normal operating temperature.

➡**If you are using the opening of the thermostat to gauge normal operating temperature, be forewarned: a defective thermostat can open too early and prevent the engine from reaching normal operating temperature. This can cause a slightly rich condition in the exhaust, which can throw the O₂ sensor readings off slightly.**

2. Turn the ignition switch **OFF**, then locate the O₂ sensor pigtail connector.

3. Perform a visual inspection of the connector to ensure it is properly engaged and all terminals are straight, tight and free from corrosion or damage.

4. Backprobe the O₂ sensor connector terminals. Attach the DMM positive test lead to the SOUT terminal of the sensor pigtail

connector and the negative lead to either the SGND terminal of the sensor pigtail connector (if equipped—refer to the terminal identification procedures earlier in this section for clarification) or to a good, clean engine ground.

5. Activate the MIN/MAX/Average or sliding bar graph/wave function on the DMM.

✳✳ CAUTION

While the engine is running, keep clear of all moving and hot components. Do not wear loose clothing. Otherwise severe personal injury or death may occur.

6. Have an assistant start the engine and wait a few minutes before commencing with the test to allow the O₂ sensor to sufficiently warm up.

7. Read the minimum, maximum and average readings exhibited by the O₂ sensor, or observe the bar graph/wave form. The average reading for a properly functioning O₂ sensor is be approximately 450–550 mV DC. The minimum and maximum readings should vary more than 300–600 mV. A typical O₂ sensor can fluctuate from as low as 100 mV to as high as 900 mV; if the sensor range of fluctuation is not large enough, the sensor is defective. Also, if the fluctuation range is biased up or down in the scale. For example, if the fluctuation range is 400 mV to 900 mV the sensor is defective, because the readings are pushed up into the rich range (as long as the fuel delivery system is functioning properly). The same goes for a fluctuation range pushed down into the lean range. The mid-point of the fluctuation range should be around 400–500 mV. Finally, if the O₂ sensor voltage fluctuates too slowly (usually the voltage wave should oscillate past the mid-way point of 500 mV

several times per second) the sensor is defective. (Technician's refer to this state as "lazy.")

➡**Poor wire connections and/or ground circuits may shift a normal O₂ sensor's millivoltage readings up into the rich range or down into the lean range. It is a good idea to check the wire condition and continuity before replacing a component which will not fix the problem. A voltage drop test between the sensor case and ground which reveals 14–16 mV, or more, indicates a probable bad ground.**

8. Using the propane method, enrichen the air/fuel mixture and observe the DMM readings. The average O₂ sensor output signal voltage should rise into the rich range.

9. Lean the air/fuel mixture by either disconnecting a fuel injector wiring harness connector or by disconnecting a vacuum line. The O₂ sensor average output signal voltage should drop into the lean range.

10. If the O₂ sensor did not react as indicated, the sensor is defective and should be replaced.

11. Turn the engine **OFF**, remove the DMM and all associated jumper wires. Reattach the vehicle harness connector to the sensor pigtail connector. If applicable, reattach the fuel injector wiring connector and/or the vacuum line(s).

12. Clear any DTC's present in the engine control computer memory, as necessary.

TEST 3—OSCILLOSCOPE

This test is designed for the use of an oscilloscope to test the functioning of an O₂ sensor.

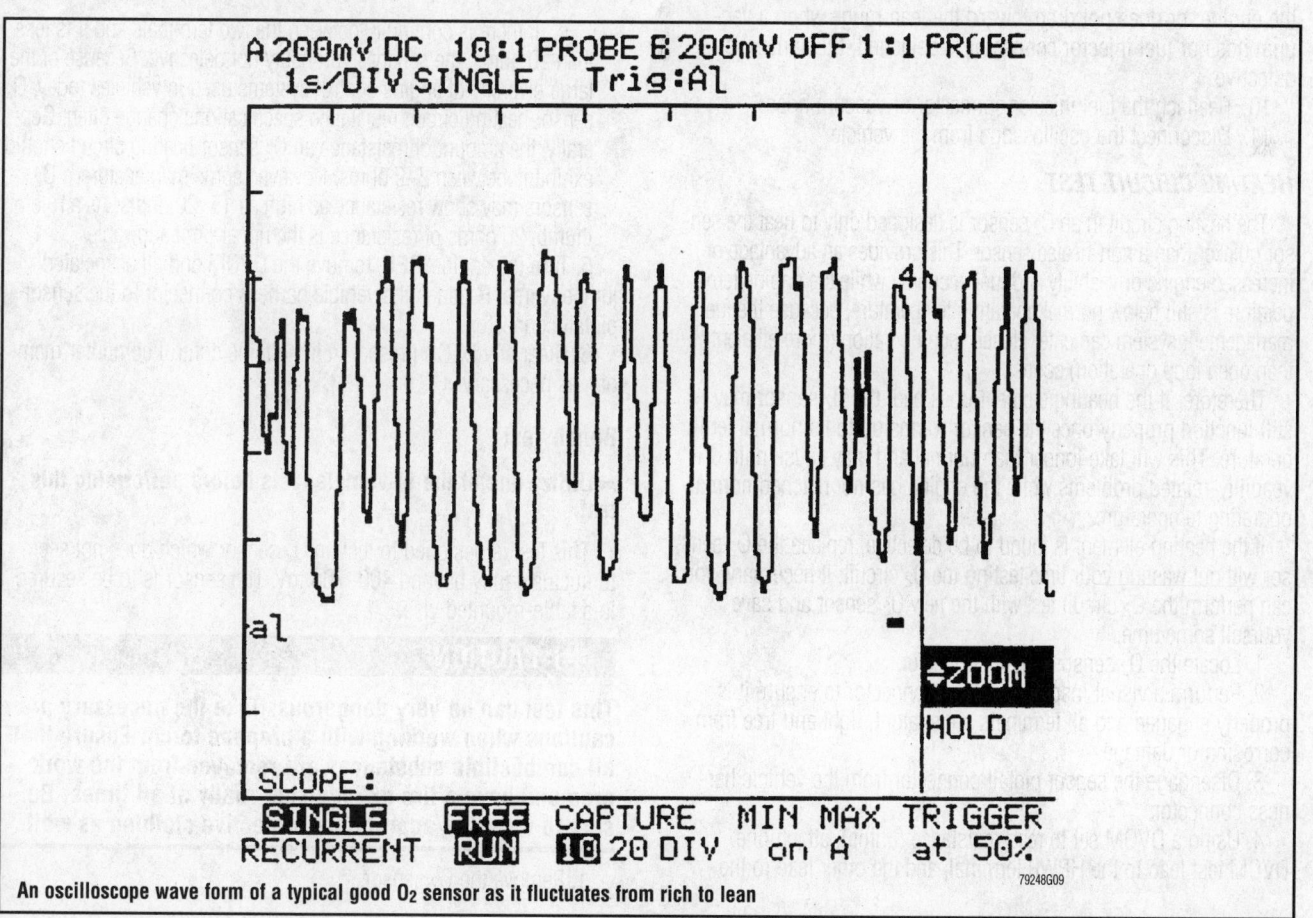

A 200mV DC 10:1 PROBE B 200mV OFF 1:1 PROBE
1s/DIV SINGLE   Trig:A1

al

⇕ZOOM
HOLD

SCOPE:
SINGLE   FREE   CAPTURE   MIN MAX TRIGGER
RECURRENT RUN 10 20 DIV on A at 50%

An oscilloscope wave form of a typical good O₂ sensor as it fluctuates from rich to lean

79248G09

➡This test is only applicable for O₂ sensors mounted in the exhaust system before the catalytic converter.

1. Start the engine and allow it to reach normal operating temperature.

2. Turn the engine **OFF**, and locate the O₂ sensor connector. Backprobe the scope lead to the O₂ sensor connector SOUT terminal. Refer to the manufacturer's instructions for more information on attaching the scope to the vehicle.

3. Turn the scope ON.

4. Set the oscilloscope amplitude to 200 mV per division, and the time to 1 second per division. Use the 1:1 setting of the probe, and be sure to connect the scope's ground lead to a good, clean engine ground. Set the signal function to automatic or internal triggering.

5. Start the engine and run it at 2,000 rpm.

6. The oscilloscope should display a wave form, representative of the O₂ sensor switching between lean (100–300 mV) and rich (700–900 mV). The sensor should switch between rich and lean, or lean and rich (crossing the mid-point of 500 mV) several times per second. Also, the range of each wave should reach at least above 700 mV and below 300 mV. However, an occasional low peak is acceptable.

7. Force the air/fuel mixture rich by introducing propane into the engine, then observe the oscilloscope readings. The fluctuating range of the O₂ sensor should climb into the rich range.

8. Lean the air/fuel mixture out by either detaching a vacuum line or by disengaging one of the fuel injector's wiring connectors. Watch the scope readings; the O₂ sensor wave form should drop toward the lean range.

9. If the O₂ sensor's wave form does not fluctuate adequately, is not centered around 500 mV during normal engine operation, does not climb toward the rich range when propane is added to the engine, or does not drop toward the lean range when a vacuum hose or fuel injector connector is detached, the sensor is defective.

10. Reattach the fuel injector connector or vacuum hose.

11. Disconnect the oscilloscope from the vehicle.

HEATING CIRCUIT TEST

The heating circuit in an O₂ sensor is designed only to heat the sensor quicker than a non-heated sensor. This provides an advantage of increased engine driveability and fuel economy while the engine temperature is still below normal operating temperature, because the fuel management system can enter closed loop operation (more efficient than open loop operation) sooner.

Therefore, if the heating element goes bad, the O₂ sensor may still function properly once the sensor warms up to its normal temperature. This will take longer than normal and may cause mild driveability-related problems while the engine has not reached normal operating temperature.

If the heating element is found to be defective, replace the O₂ sensor without wasting your time testing the O₂ circuit; if necessary, you can perform the O₂ circuit test with the new O₂ sensor and save yourself some time.

1. Locate the O₂ sensor pigtail connector.

2. Perform a visual inspection of the connector to ensure it is properly engaged and all terminals are straight, tight and free from corrosion or damage.

3. Disengage the sensor pigtail connector from the vehicle harness connector.

4. Using a DVOM set to read resistance (ohms), attach one DVOM test lead to the HPWR terminal, and the other lead to the

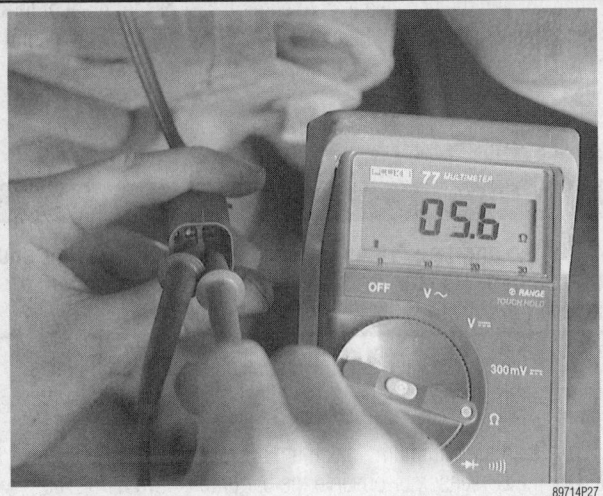

The heating circuit of the O₂ sensor can be tested with a DMM set to measure resistance

HGND terminal, of the sensor pigtail connector, then observe the resistance readings.

a. If there is no continuity between the HPWR and HGND terminals, the sensor is defective. Replace it with a new one and retest.

b. If there is continuity between the two terminals, but the resistance is greater than approximately 20 ohms, the sensor is defective. Replace it with a new one and retest.

➡For the following step, the HO₂ sensor should be approximately 75°F (23°C) for the proper resistance values.

c. If there is continuity between the two terminals and it is less than 20 ohms, the sensor is probably not defective. Because of the large diversity of engine control systems used in vehicles today, O₂ sensor heating circuit resistance specifications change often. Generally, the amount of resistance an O₂ sensor heating circuit should exhibit is between 2–9 ohms. However, some manufacturer's O₂ sensors may show resistance as high as 15–20 ohms. As a rule of thumb, 20 ohms of resistance is the upper limit allowable.

5. Turn the engine **OFF**, remove the DVOM and all associated jumper wires. Reattach the vehicle harness connector to the sensor pigtail connector.

6. Clear any DTC's present in the engine control computer memory, as necessary.

Bench Test

➡Utilize one of the in-vehicle tests before performing this test.

This test is designed to test an O₂ sensor which does not seem to fluctuate fully beyond 400–700 mV. The sensor is to be secured in a table-mounted vise.

✳✳ CAUTION

This test can be very dangerous. Take the necessary precautions when working with a propane torch. Ensure that all combustible substances are removed from the work area and have a fire extinguisher ready at all times. Be sure to wear the appropriate protective clothing as well.

1. Remove the O₂ sensor.

➡Perform a visual inspection of the sensor. Black sooty deposits may indicate a rich air/fuel mixture, brown deposits may indicate an oil consumption problem, and white gritty deposits may indicate an internal coolant leak. All of these conditions can destroy a new sensor if not corrected before installation.

2. Position the sensor in a vise so that the vise holds the sensor by the hex portion of its case.

3. Attach one lead of a DVOM set to read DC millivoltages to the sensor case and the other lead to the SOUT terminal of the sensor pigtail connector.

4. Carefully use a propane torch to heat the tip (and ONLY the tip) of the sensor. Once the sensor reaches close to normal operating temperature range, alternately heat the sensor up and allow it to cool down; the sensor output voltage signal should change with the temperature change.

➡This may also clean a sensor covered with a heavy coat of carbon.

5. If the sensor voltage does not change with the fluctuation in temperature, replace the sensor with a new one. Install the new sensor and perform one of the in-vehicle tests to rule out additional fuel management system faults.

REMOVAL & INSTALLATION

1. Start the engine and allow it to reach normal operating temperature, then turn the ignition switch **OFF**.

2. Disconnect the negative battery cable.

3. Open the hood and locate the O₂ sensor connector. It may be necessary to raise and safely support the vehicle for access to the sensor and its connector.

➡On a few models, it may be necessary to remove the passenger seat and lift the carpeting in order to access the connector for a downstream O₂ sensor.

4. Disengage the O₂ sensor pigtail connector from the vehicle harness connector.

➡There are generally two methods used to mount an O₂ sensor in the exhaust system: either the O₂ sensor is threaded directly into the exhaust component (screw-in type), or the O₂ sensor is retained by a flange and two nuts or bolts (flange type).

✳✳ WARNING

To prevent damaging a screw-in type O₂ sensor, if excessive force is needed to remove the sensor lubricate it with penetrating oil prior to removal. Also, be sure to protect the tip of the sensor; O₂ sensor tips are very sensitive and may be easily damaged if allowed to strike or come in contact with other objects.

5. Remove the sensor, as follows:

• For screw-in type sensors—Since O₂ sensors are usually designed with a permanently-attached wiring pigtail (this allows the wiring harness and sensor connectors to be positioned away from the hot exhaust system), it may be necessary to use a socket or wrench that is designed specifically for this purpose. Before purchasing such a socket, be sure that you can't save some money by using a box end wrench for sensor removal.

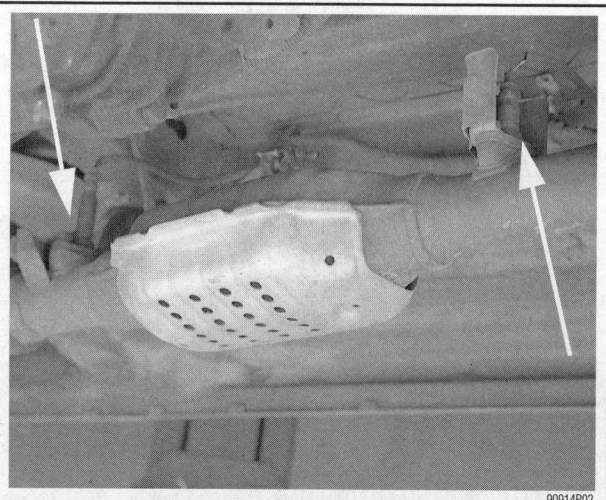

Since sensor locations vary between vehicles, the first step in removal is to locate the O₂ sensors (arrows) . . .

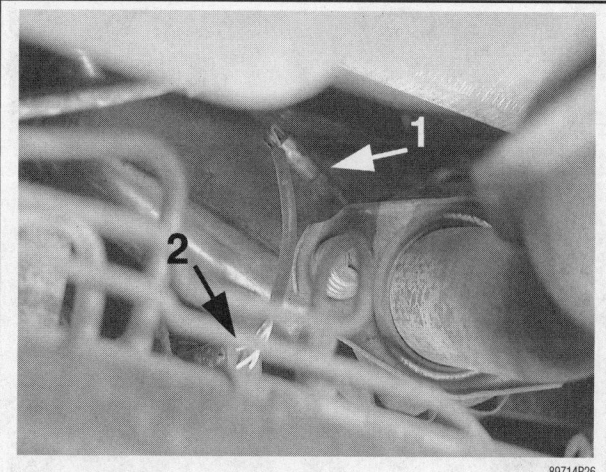

. . . and the sensor connector (2), which is usually near the O₂ sensor (1), but removed enough from the heat of the exhaust system

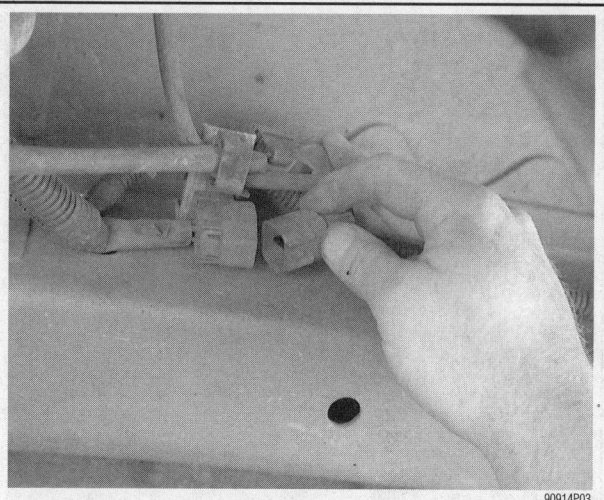

Disengage the sensor pigtail connector half from the vehicle harness connector half

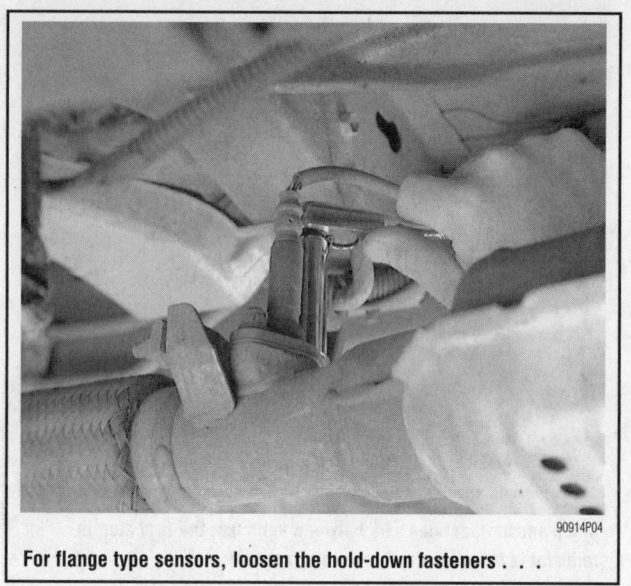

For flange type sensors, loosen the hold-down fasteners . . .

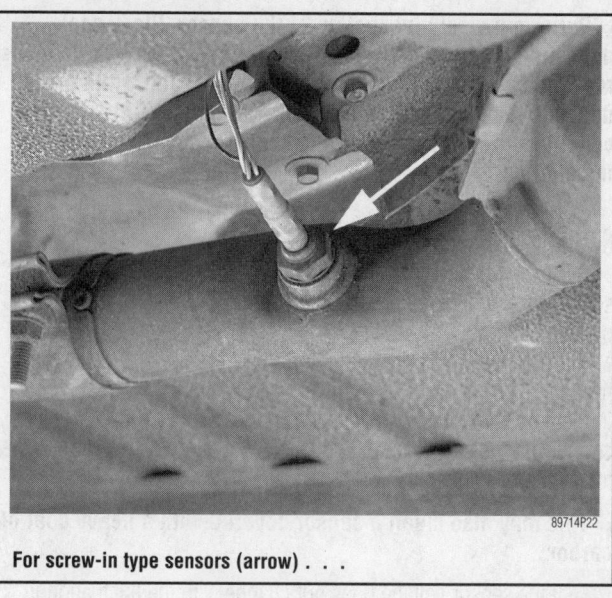

For screw-in type sensors (arrow) . . .

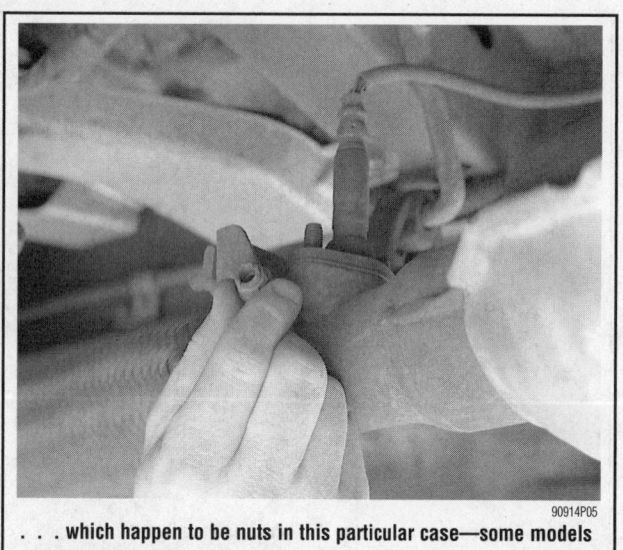

. . . which happen to be nuts in this particular case—some models may use bolts rather than nuts

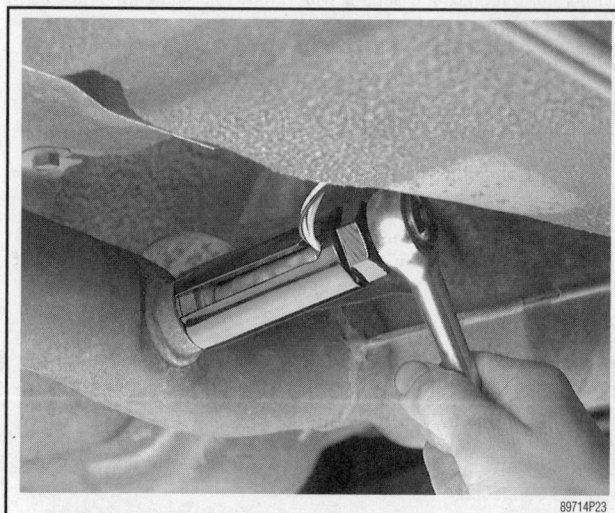

. . . either use a box end wrench to loosen the sensor or a socket designed expressly for this purpose . . .

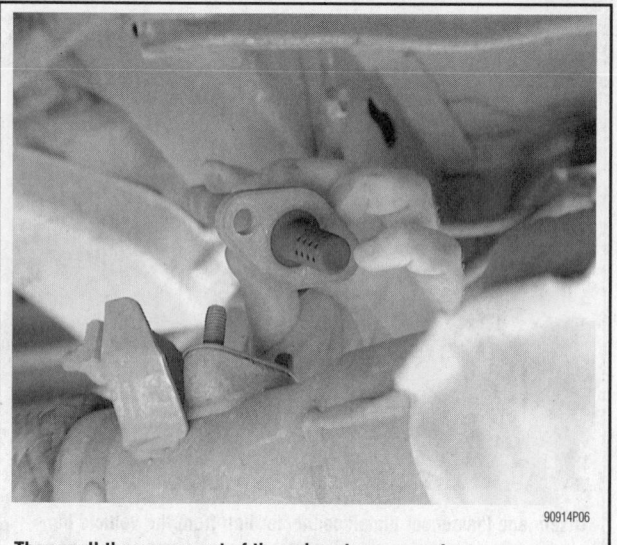

Then, pull the sensor out of the exhaust component

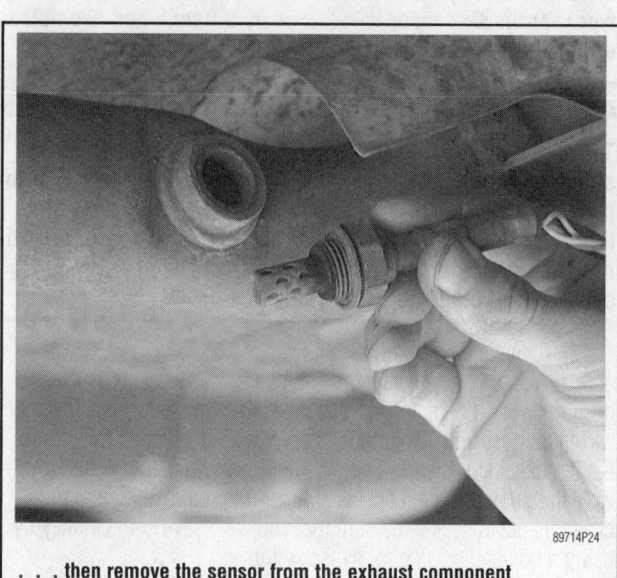

. . . then remove the sensor from the exhaust component

- For flange type sensors—Loosen the hold-down nuts or bolts and pull the sensor out of the exhaust component. Be sure to remove and discard the old sensor gasket, if equipped. You will need a new gasket for installation.

6. Perform a visual inspection of the sensor. Black sooty deposits may indicate a rich air/fuel mixture, brown deposits may indicate an oil consumption problem, and white gritty deposits may indicate an internal coolant leak. All of these conditions can destroy a new sensor if not corrected before installation.

To install:

7. Install the sensor, as follows:

➡A special anti-seize compound is used on most screw-in type O₂ sensor threads, and is designed to ease O₂ sensor removal. New sensors usually have the compound already applied to the threads. However, if installing the old O₂ sensor or the new sensor did not come with compound, apply a thin coating of electrically-conductive anti-seize compound to the sensor threads.

❊❊ WARNING

Be sure to prevent any of the anti-seize compound from coming in contact with the O₂ sensor tip. Also, take precautions to protect the sensor tip from physical damage during installation.

- For screw-in type sensors—Install the sensor in the mounting boss, then tighten it securely.
- For flange type sensors—Position a new sensor gasket on the exhaust component and insert the sensor. Tighten the hold-down fasteners securely and evenly.

8. Reattach the sensor pigtail connector to the vehicle harness connector.

9. Lower the vehicle.

10. Connect the negative battery cable.

11. Start the engine and ensure no Diagnostic Trouble Codes (DTC's) are set.

LOCATIONS

Generally, there are only five different locations in the exhaust system where O₂ sensors are positioned. The five locations have been given numbers and will be used in the accompanying charts to identify the positions of O₂ sensors in most 1995–99 vehicles.

Due to mid-year production changes or factory inconsistencies, all models may not be covered. If a vehicle you are servicing is not covered in the charts, inspect the exhaust system (while cold!) in the five general locations to find the applicable O₂ sensors.

➡On models equipped with dual exhaust systems, there may be up to 4 or 5 O₂ sensors in the exhaust system. Be sure to locate all of them before commencing with any testing or service.

The five locations are as follows:

- Location No.1—exhaust manifold or down pipe.
- Location No.2—both exhaust manifolds or down pipes of a V-type engine.
- Location No.3—exhaust collector.
- Location No.4—outlet of the catalytic converter.
- Location No.5—both the inlet and outlet of catalytic converter. This location is used to monitor the efficiency of the catalytic converter.

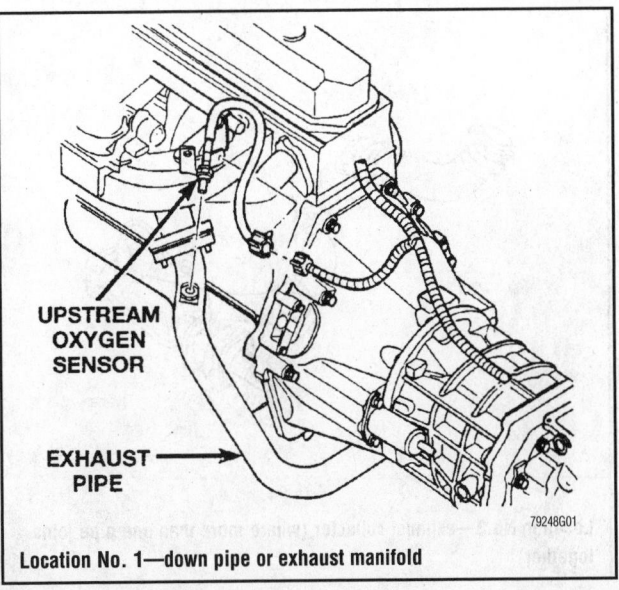

Location No. 1—down pipe or exhaust manifold

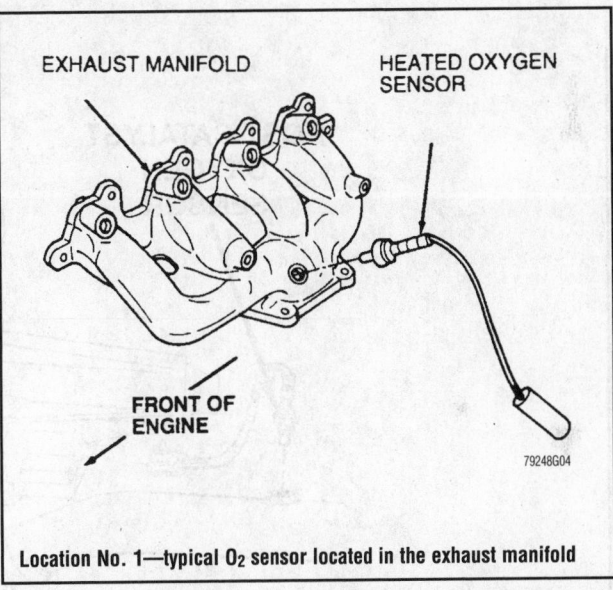

Location No. 1—typical O₂ sensor located in the exhaust manifold

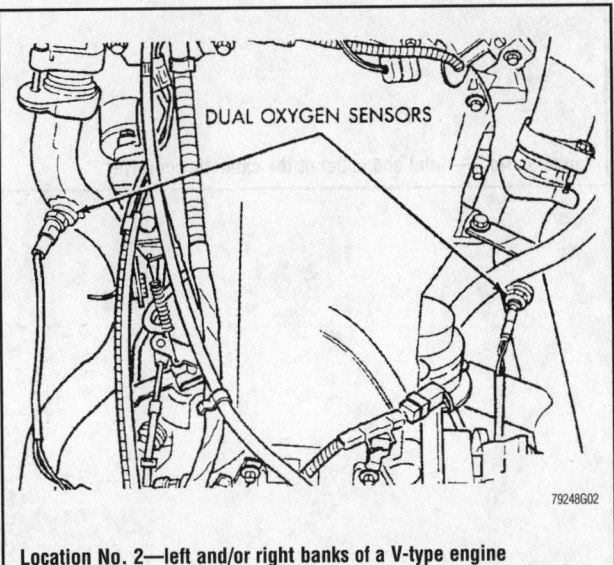

Location No. 2—left and/or right banks of a V-type engine

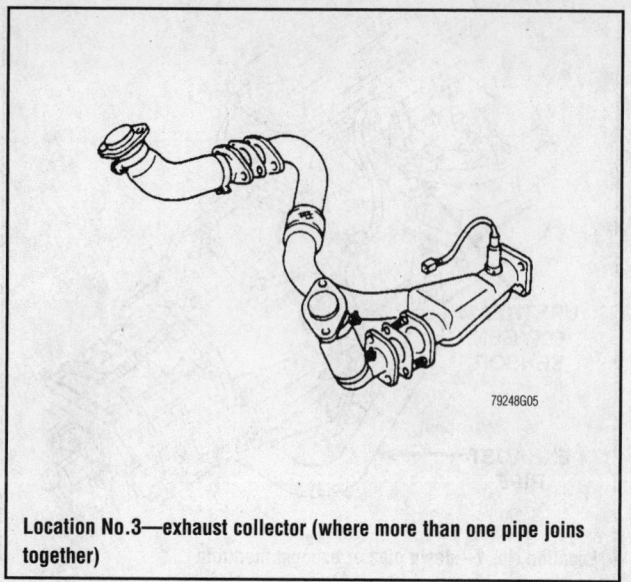

Location No.3—exhaust collector (where more than one pipe joins together)

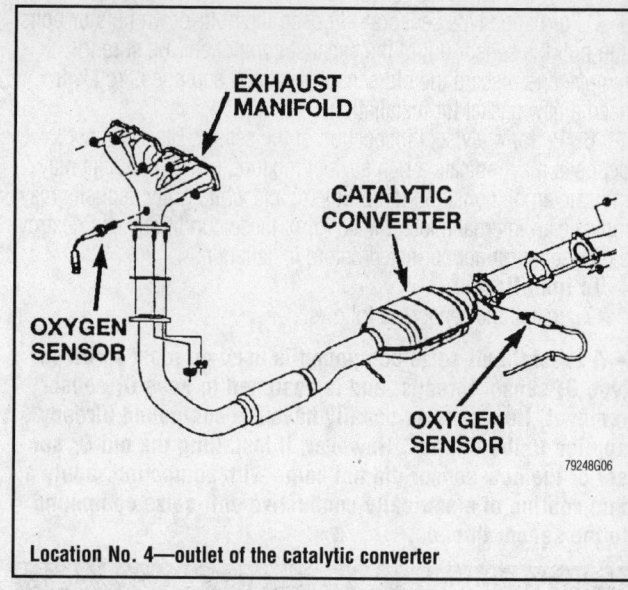

Location No. 4—outlet of the catalytic converter

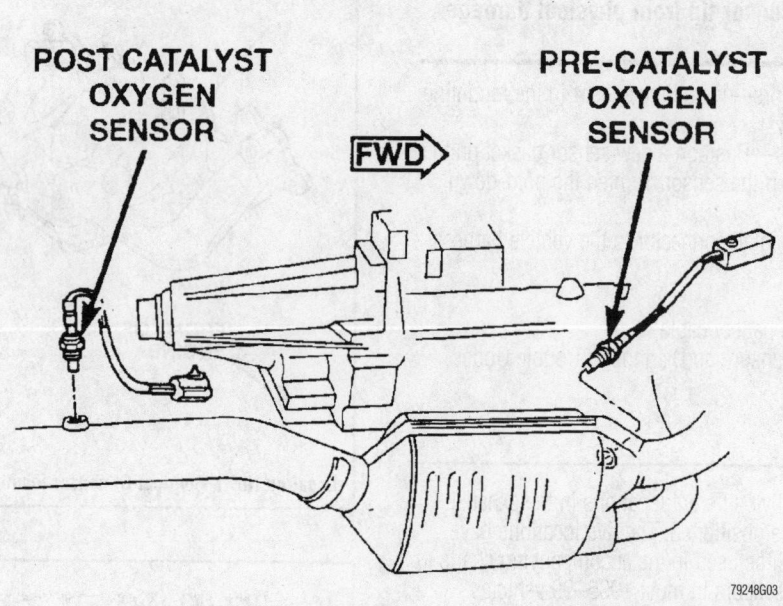

Location No. 5—inlet and outlet of the catalytic converter

OXYGEN SENSOR LOCATIONS

Manufacturer Year	Engines	No. of Sensors	Location
Acura/Isuzu Sport Utility Vehicles			
1995	3.2L	1	3
1996-99	3.2L/3.5L	4	2, 5
Dodge Trucks and Vans			
1995	all	1	1
1996-99	2.5L	2	1, 4
	3.9L	2	1, 4
	5.2L	2	1, 4
	5.9L	2	1, 4
	5.9L HDC	4	2, 5
	8.0L HDC	4	2, 5
Chrysler Minivans			
1995	all	1	1
1996-99	all	2	1, 4
Ford/Mazda Small Trucks and Vans			
1995	all	1	1
1996-99	2.3L	2	1, 4
	2.5L	2	1, 4
	3.0L	3	2, 4
	4.0L (E)	4	2, 4
	4.0L (X)	4	2, 4
	5.0L	4	2, 4
Ford Minivans			
1995-99	3.0L	4	2, 4
	3.8L	4	2, 4
Ford Full-size Trucks and Vans			
1995-96	4.9L	1	1
	5.0L	3	2, 4
	5.8L	3	2, 4
1997-99	4.2L	2	3, 4
	4.6L	4	2, 4
	5.4L	4	2, 4
	6.8L	4	2, 4
Mercury/Nissan Sport Utility Vehicles			
1995	3.0L	1	1
1996-99	3.0L	2	1, 4
GEO/Suzuki Sport Utility Vehicles			
1995	1.3L	1	1
1996-99	1.6L	2	1, 4
	1.8L	2	1, 4
GM/Isuzu Small Trucks and Vans			
1995	2.2L	2	1, 4
	4.3L	4	2, 5
1996-99	2.2L	2	1, 4
	4.3L	4	2, 5

Manufacturer Year	Engines	No. of Sensors	Location
GM Full-size Trucks and Vans			
1995-99	4.3L	4	2, 5
	5.0L	4	2, 5
	5.7L	4	2, 5
	7.4L	4	2, 5
GM Minivans			
1995	3.1L	1	1
	3.8L	1	1
1996-99	3.4L	2	1, 4
Honda/Isuzu Trucks and Sport Utility Vehicles			
1995	2.2L	2	3, 4
	2.3L	1	1
	2.6L	1	1
	3.2L	1	3
1996-99	2.0L	2	3, 4
	2.2L	2	3, 4
	2.6L	2	4, 3
	3.2L	4	2, 5
Jeep Sport Utility Vehicles			
1995	2.5L	1	1
	4.0L	1	1
	5.2L	1	1
1996-99	2.5L	2	1, 4
	4.0L	2	1, 4
	5.2L	3	1, 5
	5.9L	3	1, 5
Kia Sport Utility Vehicles			
1995	2.0L	1	1
1996-99	2.0L	2	1, 4
Land Rover Sport Utility Vehicles			
1995	3.9L	2	2
	4.0L	2	2
1996-99	4.0L	4	2, 4
	4.6L	4	2, 4
Lexus/Toyota Sport Utility Vehicles			
1995-99	4.5L	2	1,4
Mazda Minivans			
1995-99	3.0L	2	1, 4
Mercedes-Benz Sport Utility Vehicles			
1997-99	3.2L	4	5
Mitsubishi Trucks and Sport Utility Vehicles			
1995-99	2.4L	2	1, 4
	3.0L	3	2, 4
	3.5L	3	2, 4

OXYGEN SENSOR LOCATIONS

Manufacturer Year	Engines	No. of Sensors	Location	Manufacturer Year	Engines	No. of Sensors	Location
Nissan/Infinity Trucks and Sport Utility Vehicles				**Toyota Trucks and Sport Utility Vehicles (cont.)**			
1995-99	2.4L	2	1, 4	1995	3.0L	2	3, 4
	3.0L	4	2, 4		3.4L	2	3, 4
	3.3L	4	2, 4	1996-99	2.0L	2	1, 4
	4.3L	4	2, 4		2.4L	2	3, 4
Subaru Sport Utility Vehicles					2.7L	2	3, 4
1997-99	2.5L	3	3, 4		3.0L	2	3, 4
Toyota Trucks and Sport Utility Vehicles					3.4L	2	3, 4
1995	2.4L	①	②		4.7L	2	3, 4
	2.7L	2	3, 4				

① Federal models use one O₂ sensor, whereas California models use two O₂ sensors.

② Federal models: location No. 1.
California models: location Nos. 1 and 4.

79248C02

ELECTRIC COOLING FANS

9

ELECTRIC COOLING FANS

General Information

A basic vehicle cooling system consists of a radiator, water pump, thermostat, electric or engine-driven cooling fan, and hoses. Electric cooling fans are common on today's vehicles due to engine compartment space limitations or engine layout. Electric cooling fans operate in either a pusher or a puller capacity. A pusher type fan is typically mounted on the front of the radiator assembly and forces air through the radiator, whereas a puller type fan is mounted on the engine side of the radiator and draws air through the grill and radiator assembly. Vehicles that utilize a transversely-mounted engine will always be equipped with at least one electric cooling fan (most having two), because none of the engine pulleys are in-line with the radiator air-flow.

There are generally two types of electric cooling fans: primary cooling fans and secondary cooling fans. Primary cooling fans are typically of the puller style. Vehicles that do not incorporate an engine-driven mechanical cooling fan will utilize a primary cooling fan. The secondary cooling fan, also known as a A/C condenser fan or auxiliary cooling fan by certain manufacturers, could be of either a pusher or a puller style. Vehicles equipped with A/C will either utilize the radiator cooling fan or a separate fan as the A/C condenser cooling fan (which performs the same function as an auxiliary cooling fan on vehicles with a primary mechanical fan). The engine control computer that receives inputs from various sensors in the engine compartment commonly controls electric cooling fans. The engine control computer receives inputs from the engine coolant temperature sensors and A/C system pressure switches, then actuates the necessary cooling fan relays to engage the applicable cooling fan for the condition. On models equipped with only one electric primary cooling fan, the fan can operate at two speeds: low speed and high speed. The low speed condition is enabled when the engine begins to heat up or when the A/C is engaged. As the engine demands more cooling, the cooling fan will be stepped-up to high speed.

Electric Cooling Fan Service

Due to the wide variety of vehicle manufacturers and suppliers of electric cooling fans it is almost impossible to cover every specific combination of cooling fan and model. The following procedures will cover the most common types of mountings and troubleshooting techniques.

REMOVAL & INSTALLATION

Puller Type

➡**It may be simpler to remove the cooling fan(s) with the radiator as an assembly.**

1. Disconnect the negative battery cable.
2. Inspect the cooling fan and take note of any wires, hoses or A/C lines which may hamper fan removal. Also at this time, decide whether it is necessary to remove the fan along with the radiator or not.
3. Position aside all wires, hoses and A/C lines for fan removal. It may not always be possible to create enough clearance for fan removal by simply moving these obstructions aside; often they must be disconnected. If any cooling system lines must be disconnected, drain and recycle the engine coolant. If any of the A/C lines must be disconnected, the A/C system will need to be discharged and evacuated by a MVAC-trained technician using an approved recovery machine.

4. Disengage the cooling fan wiring harness connector.

5. If the fan can be removed without the radiator, perform the following:

    a. Loosen the mounting fasteners. Usually there are two nuts or bolts along the top edge of the cooling fan shroud and either two retaining clips or bolts along the bottom edge.

    b. Carefully lift the fan up and out of the engine compartment, making sure that no wires or hoses get hung up on it.

6. If it is necessary to remove the radiator for fan removal, perform the following:

    a. Disconnect all cooling system hoses from it after draining the cooling system.

    b. Locate all of the radiator mounting fasteners (usually two or more nuts or bolts along the top and, possibly two along the bottom).

➡**Quite a few radiators are secured along the bottom by two posts which fit into rubber grommets. The rubber grommets help isolate the radiator from harsh vibrations in the frame. If no nuts or bolts can be located along the bottom of the radiator, chances are that the radiator is secured with the posts and grommets.**

    c. Lift the radiator and cooling fan up and out of the engine compartment together.

    d. Separate the cooling fan from the radiator by removing the attaching fasteners.

To install:

7. If applicable, install the cooling fan on the radiator.

8. Install the cooling fan and shroud assembly (also the radiator if necessary). Tighten the fan shroud mounting bolts.

9. Reattach all wires, hoses and A/C lines as applicable. If the A/C lines were detached, the system must be evacuated and recharged by a MVAC-trained technician.

10. If drained, refill and bleed the cooling system.

11. Reattach the cooling fan electrical harness connector.

12. Connect the negative battery cable.

13. Start the engine and check for leaks.

14. Verify the operation of the cooling fan(s).

To remove a typical puller type cooling fan, first detach any braces (1), wires (2) or other obstructions . . .

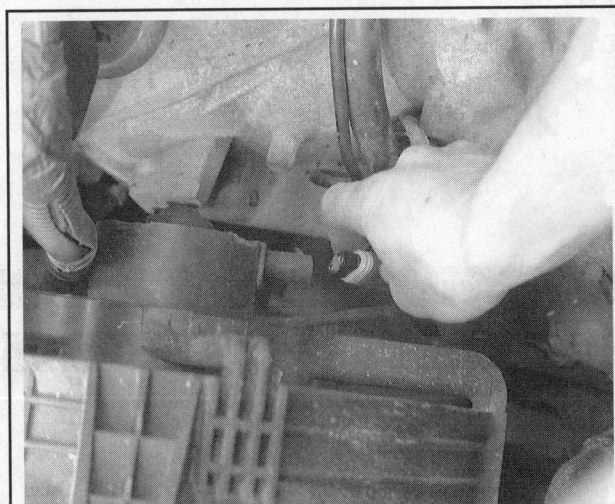

Disengage the fan wiring connector(s) . . .

. . . including cooling system hoses, to allow fan removal

. . . and loosen all fan mounting fasteners

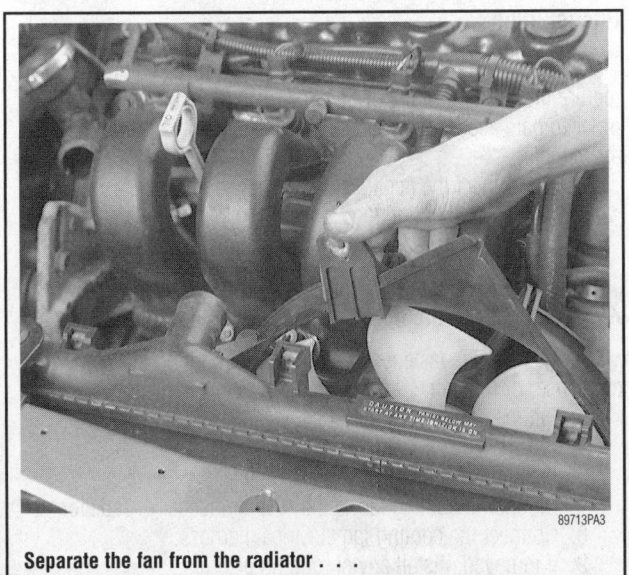

Separate the fan from the radiator . . .

. . . then lift the fan up and out of the engine compartment

This fan mounts to the fan shroud, then the shroud mounts to the radiator—molded clips in the radiator hold the bottom in place and screws at the top

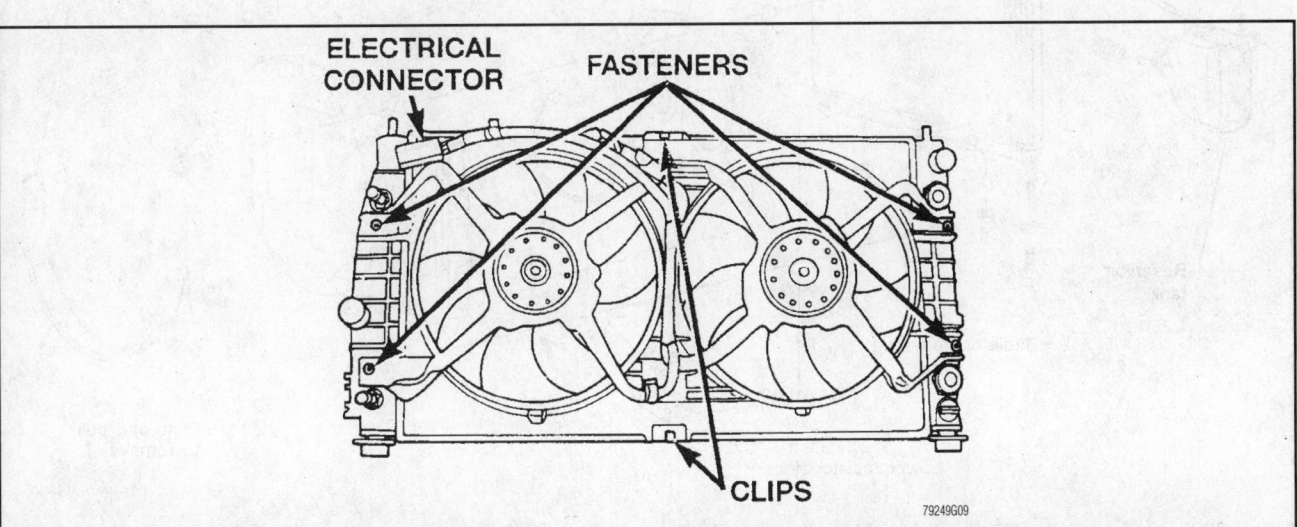

ELECTRICAL CONNECTOR   **FASTENERS**

CLIPS

Typical mounting of a puller type cooling fan assembly utilizing retaining clips and screws—note that this particular model uses a dual puller fan setup

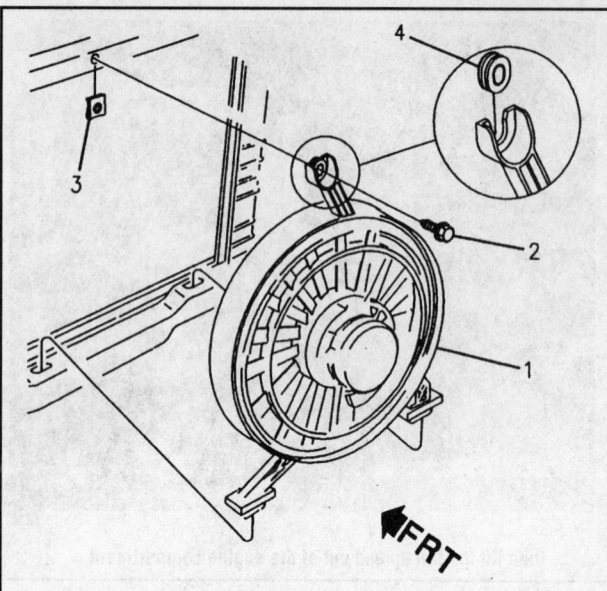

1 FAN ASSEMBLY
2 BOLT
3 CLIP
4 INSULATOR, ENGINE COOLING FAN

79249G10

Notice the slots in the bottom of the radiator, in which the fan housing posts rest—common mounting of a puller type cooling fan.

Pusher Type

Vehicles that utilize the pusher type of electric cooling fan, may require the removal of the grilles and/or upper radiator shroud in order to gain access the fasteners that mount the fan assembly in the vehicle.

1. Disconnect the negative battery cable.
2. Access the cooling fan.
3. Label and disconnect the cooling fan electrical harness.

➡️**It may be necessary to loosen the mounting bolts for the A/C condenser to the body**

4. Remove the fasteners that mount the cooling fan to the A/C condenser or radiator.
5. Lift the cooling fan out of the vehicle.

To install:

6. Insert the cooling fan into the vehicle.
7. Mount the cooling fan to the A/C condenser or radiator
8. Connect the cooling fan electrical harness.
9. If removed, install any shrouding or grills.
10. Connect the negative battery cable.

TROUBLESHOOTING

When diagnosing an inoperative cooling fan it may be necessary to use a diagnostic scan tool to monitor engine coolant temperature and the engine control computer.

1. Perform a visual inspection of the cooling fan. If the fan does not turn with ease, the fan motor is seized and needs to be replaced.

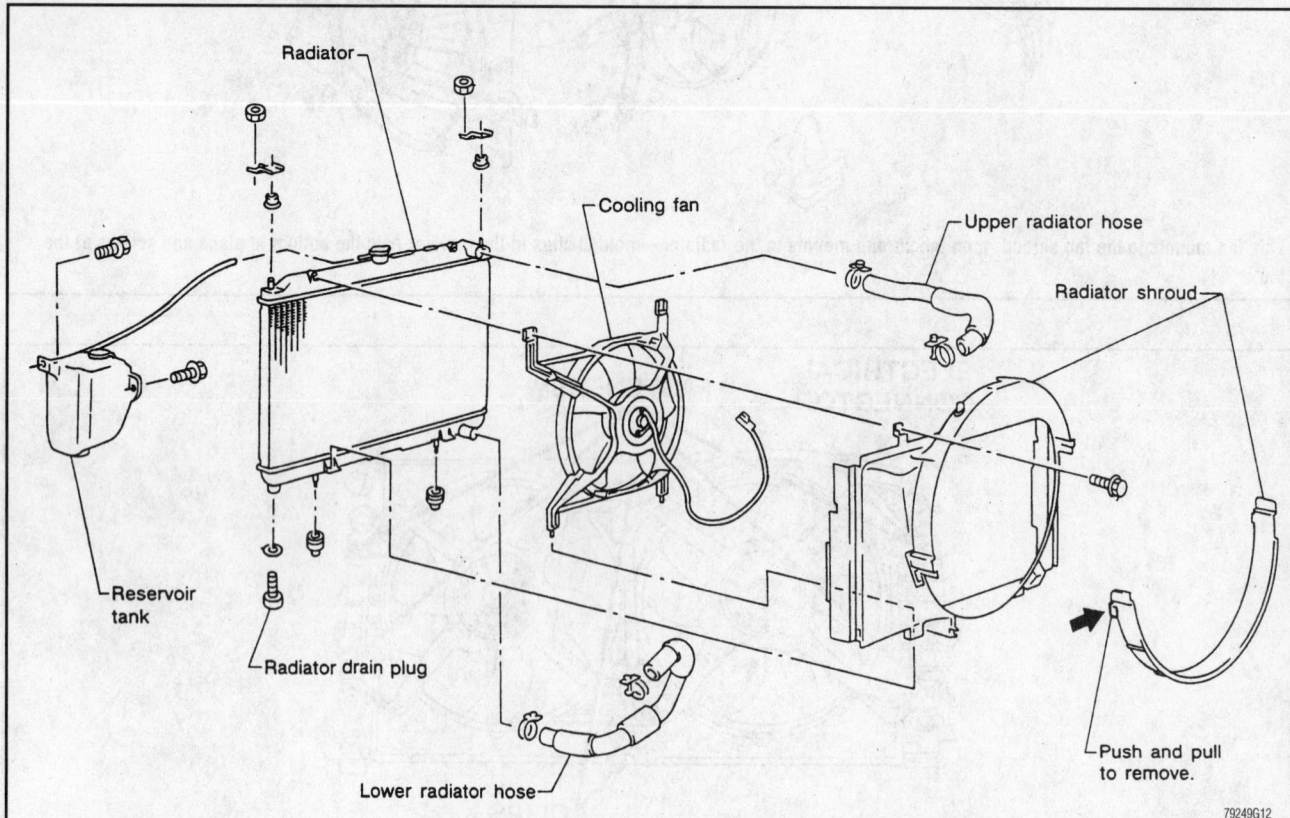

79249G12

Typically the cooling fan is rubber mounted to isolate vibration and noise—usually the rubber grommets are located at the mount, verify their position before installation

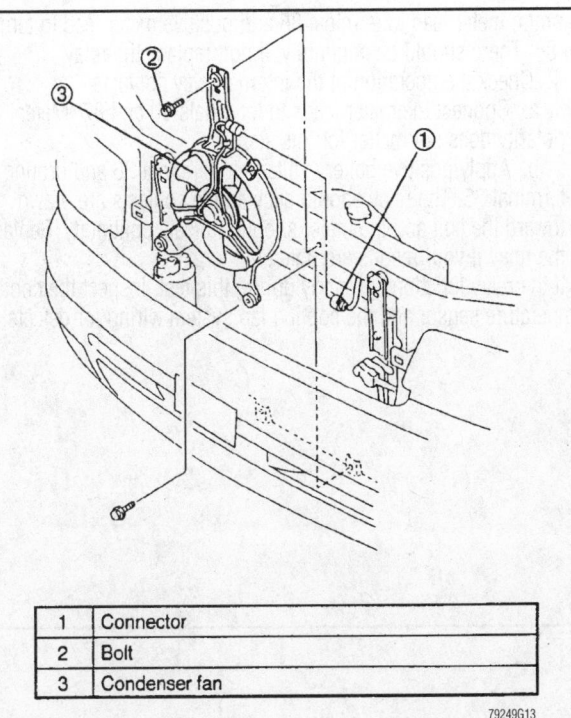

1	Connector
2	Bolt
3	Condenser fan

79249G13

After removal of the grill assembly, the pusher type of cooling fan can be removed

2. Check all the fuses and fusible links related to the cooling fan circuit.

3. Check the integrity of the electrical connections related to the cooling fan circuit.

4. Check the cooling fan motor.

5. Check the relays associated with the cooling fan circuit.

6. Using a scan tool, determine if the engine control computer is calling for the fan to activate.

Cooling Fan Motor

1. Disconnect the negative battery cable.
2. Disengage the cooling fan motor connector.
3. Identify and label the ground and the power terminals of the cooling fan connector using the wiring diagrams provided.
4. Using jumper leads with a fuse in series, apply battery voltage to the appropriate terminals of the cooling fan.
5. The cooling fan should operate. If not, replace the cooling fan.

If the cooling fan functions properly during this test, proceed to the cooling fan relay test.

Cooling Fan Relay

1. Turn the ignition **OFF**.
2. Remove the relay.
3. Locate the two terminals on the relay, which are connected to the coil windings. Check the relay coil for continuity. Connect the

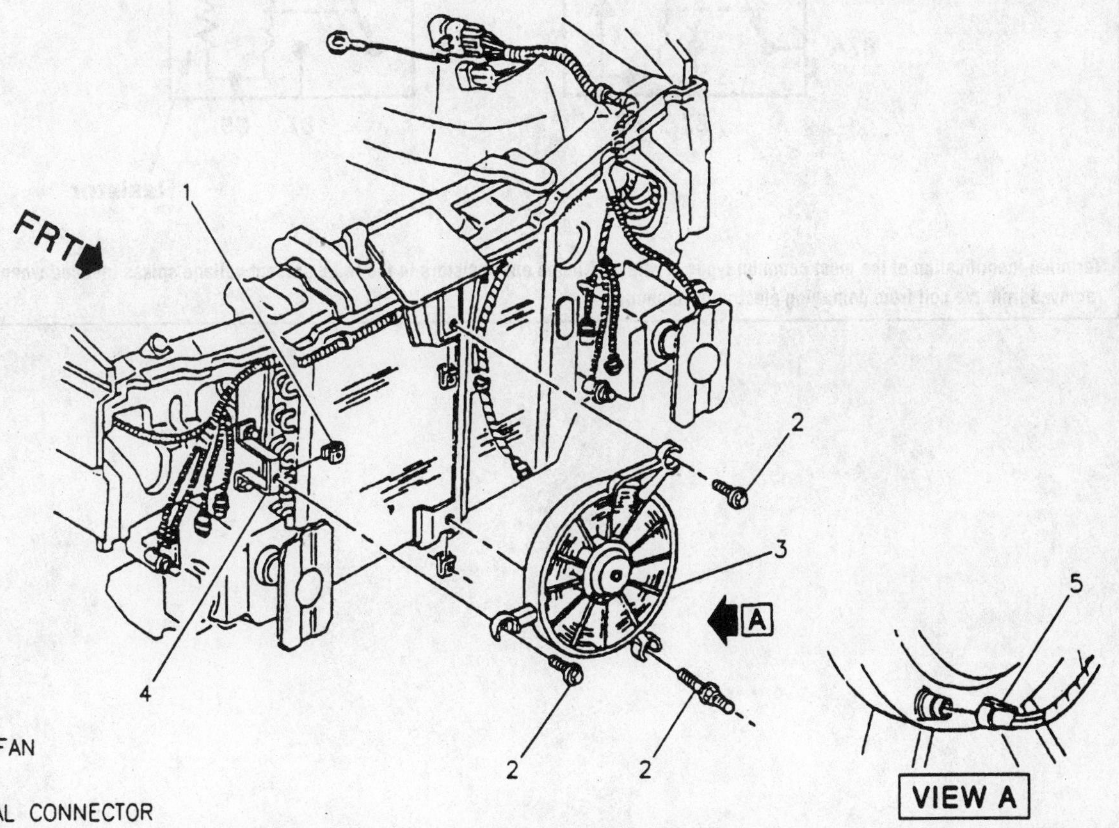

FRT

1 J–CLIP
2 BOLT
3 COOLING FAN
4 BRACKET
5 ELECTRICAL CONNECTOR

A

VIEW A

79249G14

It may be necessary to remove the grill assembly to access the A/C condenser cooling fan—pusher type

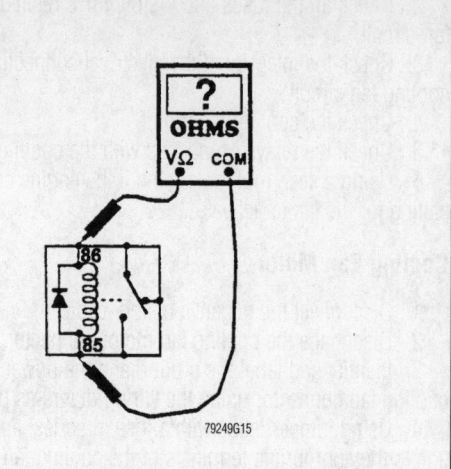

Use an ohmmeter to check for circuit continuity of the coil in the relay

common meter lead to terminal 85 and positive meter lead to terminal 86. There should be continuity. If not, replace the relay.

4. Check the operation of the internal relay contacts.

a. Connect the meter leads to terminals 30 and 87. Meter polarity does not matter for this step.

b. Apply positive battery voltage to terminal 86 and ground to terminal 85. The relay should click as the contacts are drawn toward the coil and the meter should indicate continuity. Replace the relay if your results are different.

If the relay functions properly during this test, inspect the coolant temperature sensor and the cooling fan system wiring for defects.

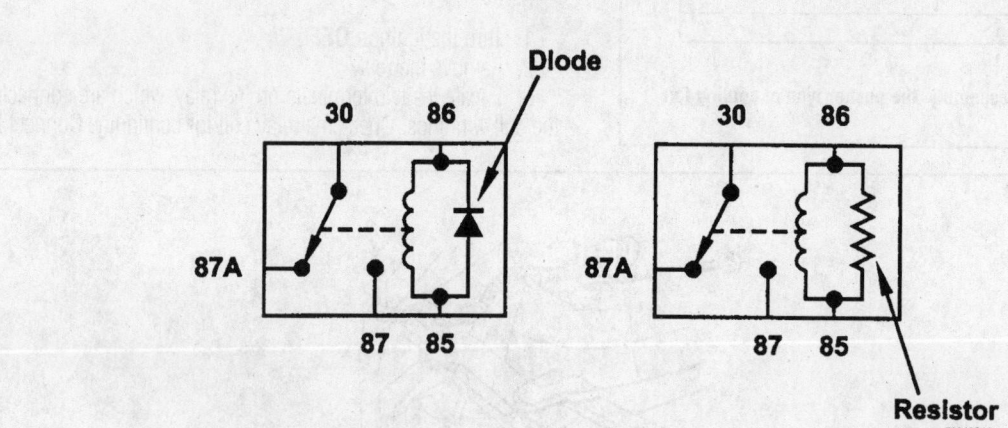

Terminal identification of the most common types of relays. Diodes and resistors in the relay prevent voltage spikes induced when the current is removed from the coil from damaging electronic components

COOLING FAN DIAGRAM INDEX

MANUFACTURER MODEL AND ENGINE	DIAGRAM
Chrysler	
Town & Country/Caravan/Voyager 2.4L/2.5L/3.0L/3.3L/3.8L	
1995 Models	1
1996-99 Models	2
Dodge	
Dakota 2.5L	
1995-96 Models	3
1997-99 Models	4
Ford	
Windstar 3.0L/3.8L	
1995-96 Models	5
1997-99 Models	6
General Motors	
Lumina APV/Silhouette/Trans Sport 3.1L	7
Lumina APV/Silhouette/Trans Sport/Venture 3.4L	
1996 Models	9
1997-99 Models	10
Lumina APV/Silhouette/Trans Sport 3.8L	8
Geo/Chevrolet	
Tracker 1.6L	11
Honda	
CR-V 2.0L	15
Oasis 2.2L	17
Odyssey 2.0L/2.2L	16
Kia	
Sportage 2.0L	18
Land Rover	
Discovery 3.9L	19
Range Rover 4.0L	20
Mazda	
MPV 3.0L	
1995 Models	21
1996-99 Models	22
Mercury	
Villager 3.0L	23
Mitsubishi	
Montero 3.0L/3.5L	24
Montero Sport 2.4L/3.0L	25
Nissan	
Quest 3.0L	
1995 Models	26
1996-97 Models	27
1998-99 Models	28
Suzuki	
Sidekick 1.6L	
1995-96 Models	12
1997-98 Models	13
X90 1.6L, Sport 1.8L	14
Toyota	
RAV4 2.0L	29

79249C01

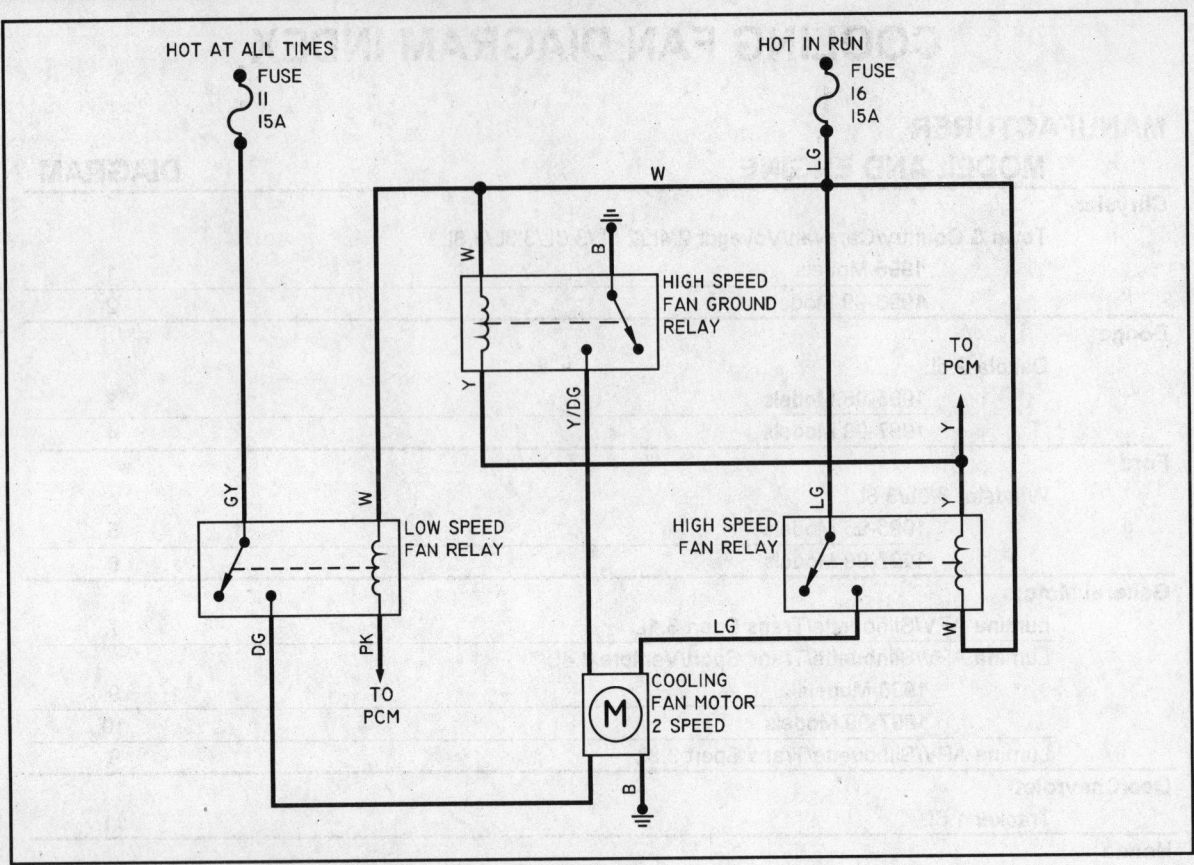

DIA. 1- 1995 Chrysler Town & Country/Caravan/Voyager 2.5L/3.0L/3.3L/3.8L

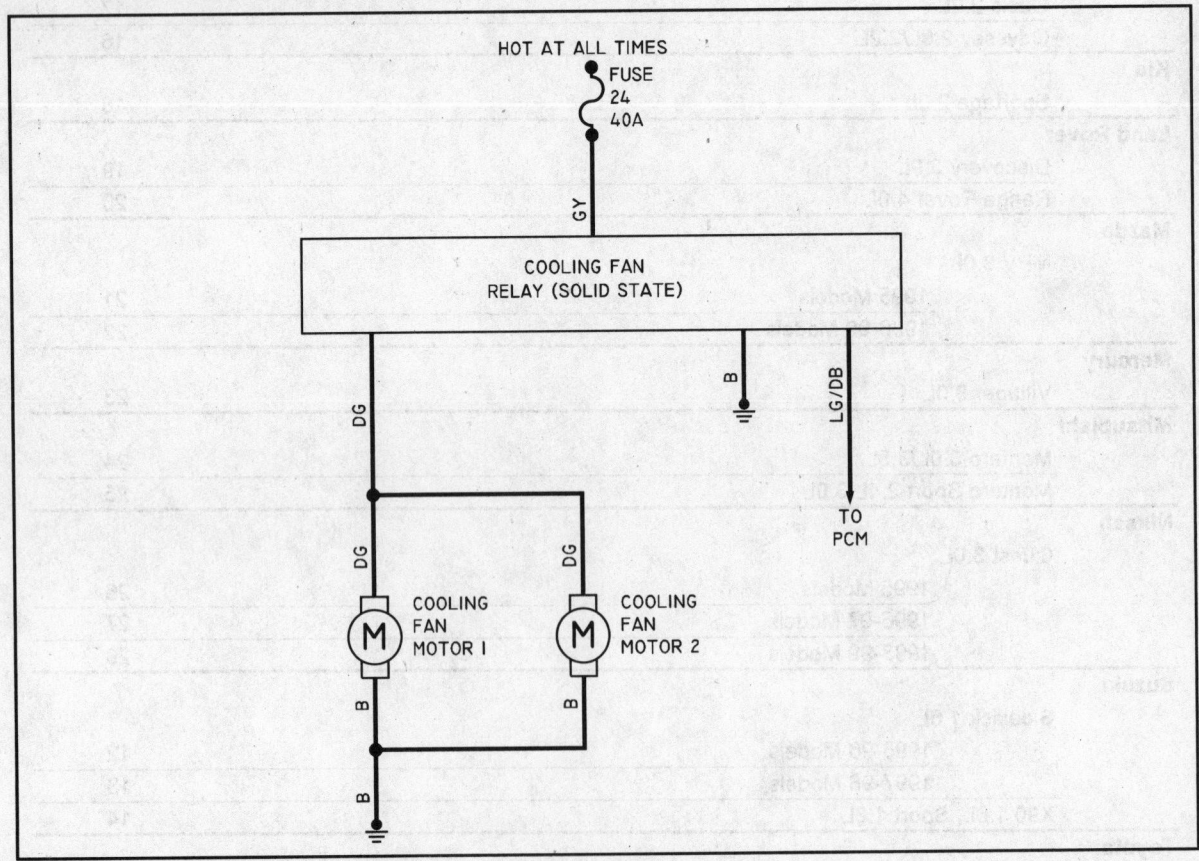

DIA. 2- 1996-99 Chrysler Town & Country/Caravan/Voyager 2.4L/3.0L/3.3L/3.8L

7924-1

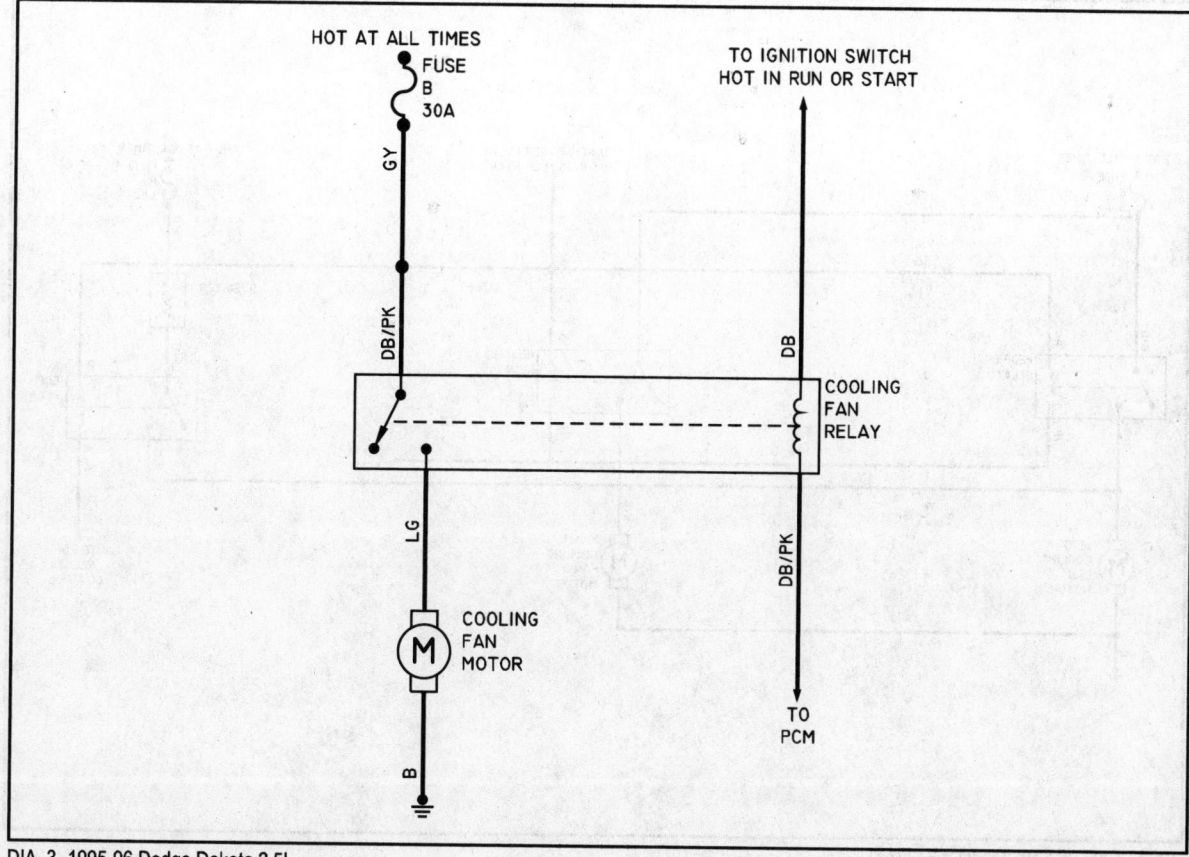

DIA. 3- 1995-96 Dodge Dakota 2.5L

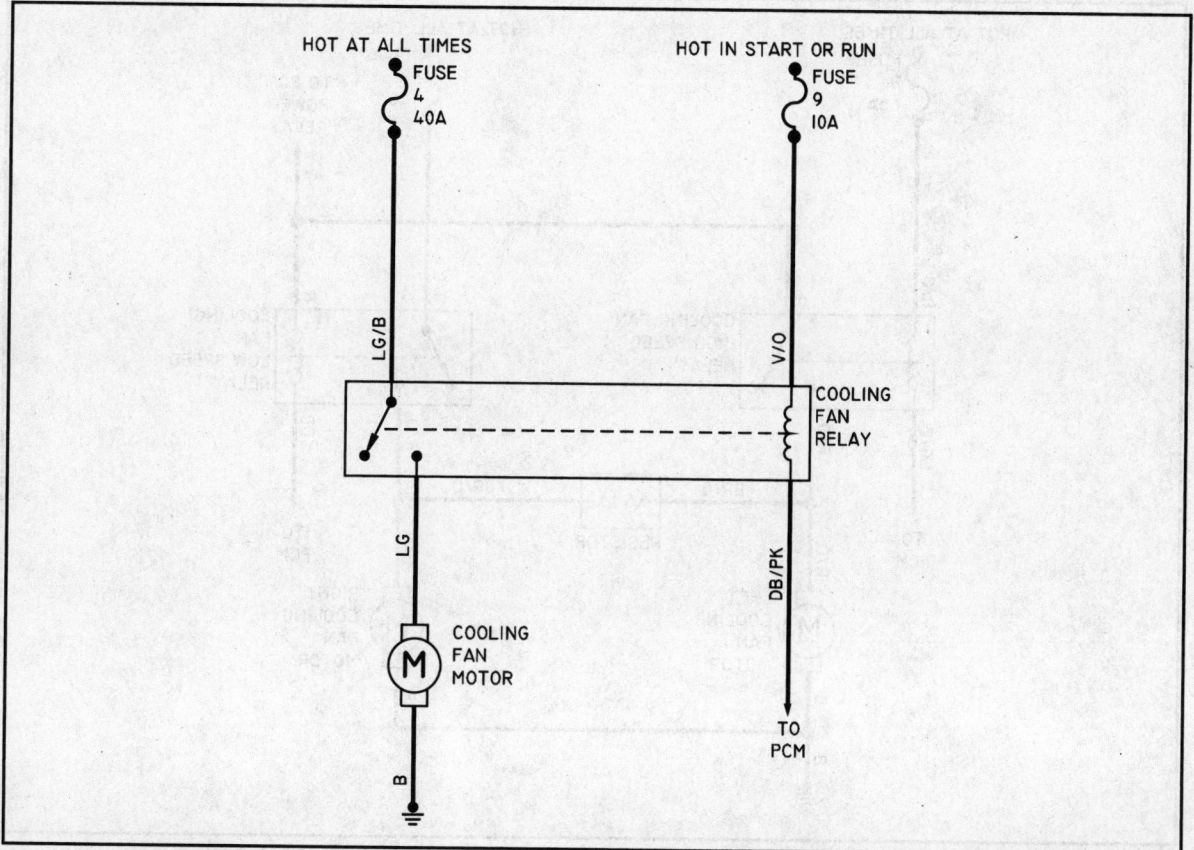

DIA. 4- 1997-99 Dodge Dakota 2.5L

7924-2

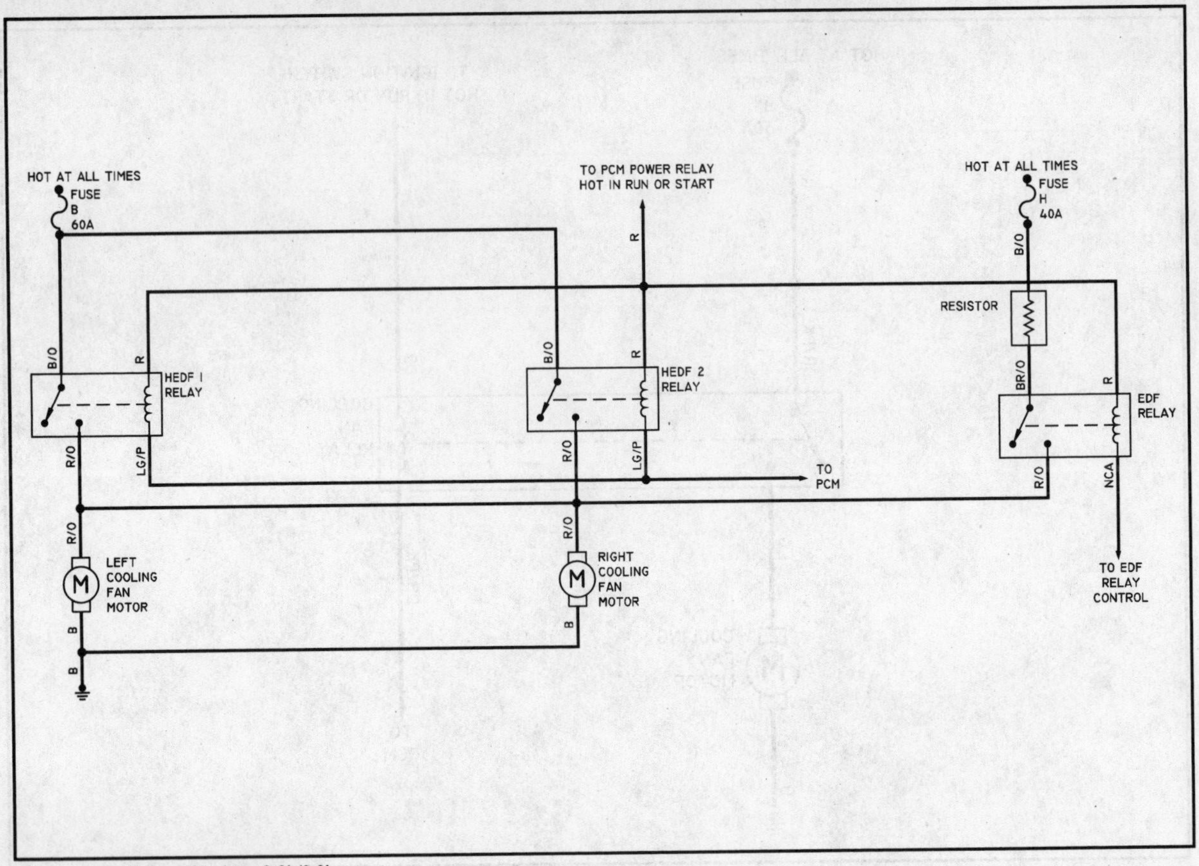

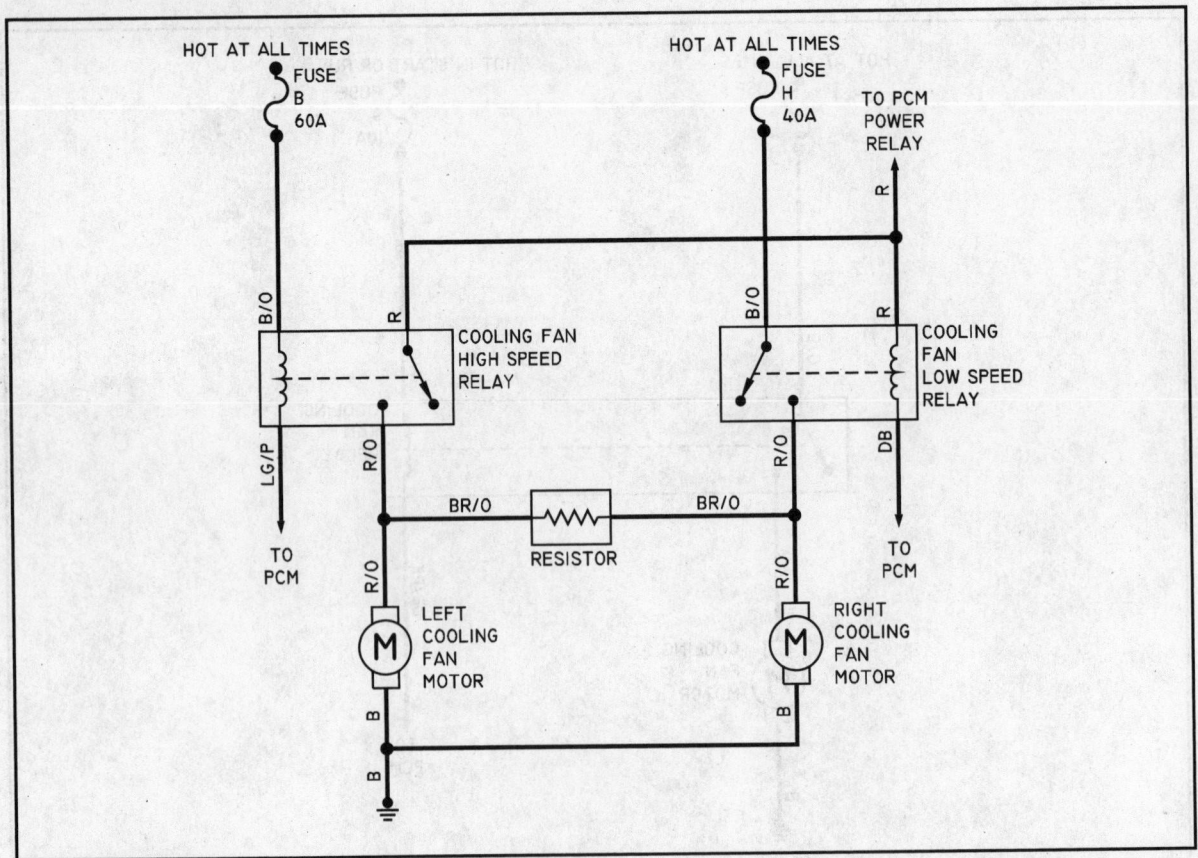

DIA. 5- 1995-96 Ford Windstar 3.0L/3.8L

DIA. 6- 1997-99 Ford Windstar 3.0L/3.8L

7924-3

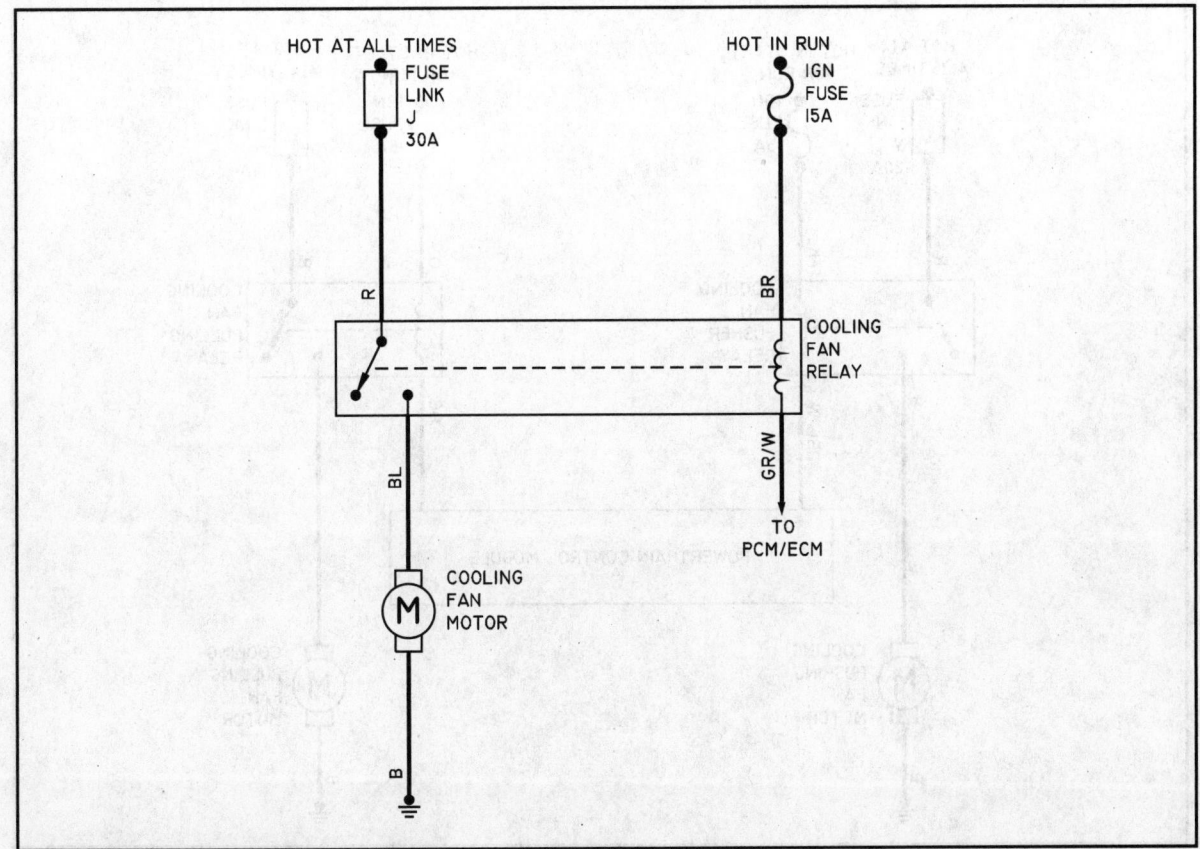

DIA. 7- 1995 GM Lumina APV/Silhouette/Trans Sport 3.1L

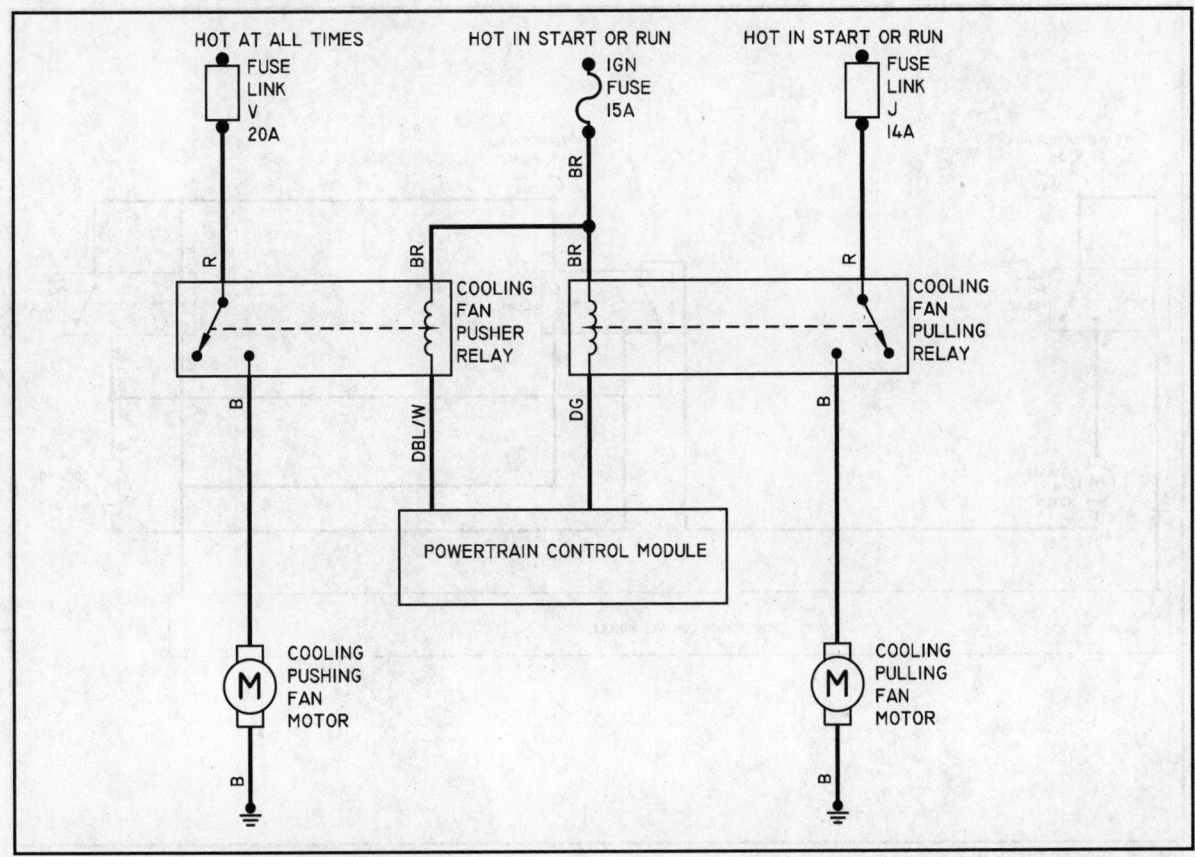

DIA. 8- 1995 GM Lumina APV/Silhouette/Trans Sport 3.8L

7924-4

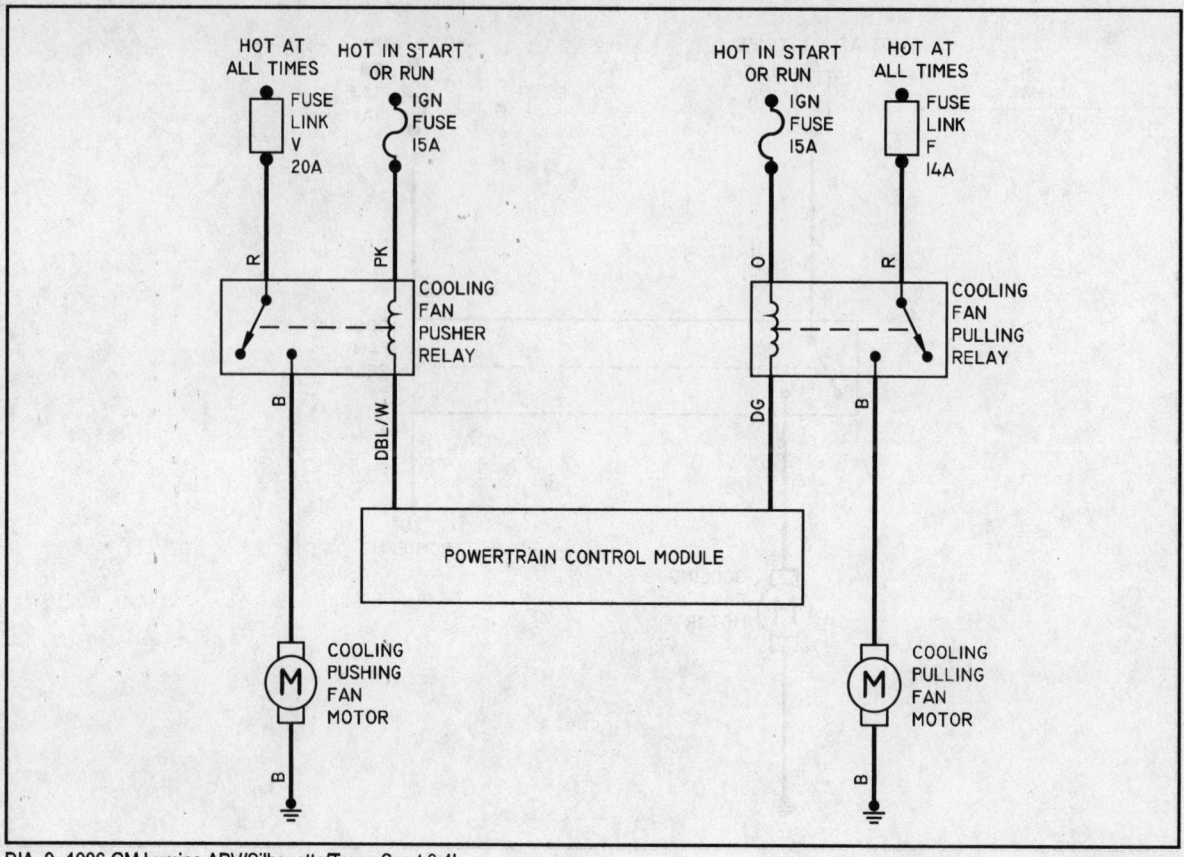

DIA. 9- 1996 GM Lumina APV/Silhouette/Trans Sport 3.4L

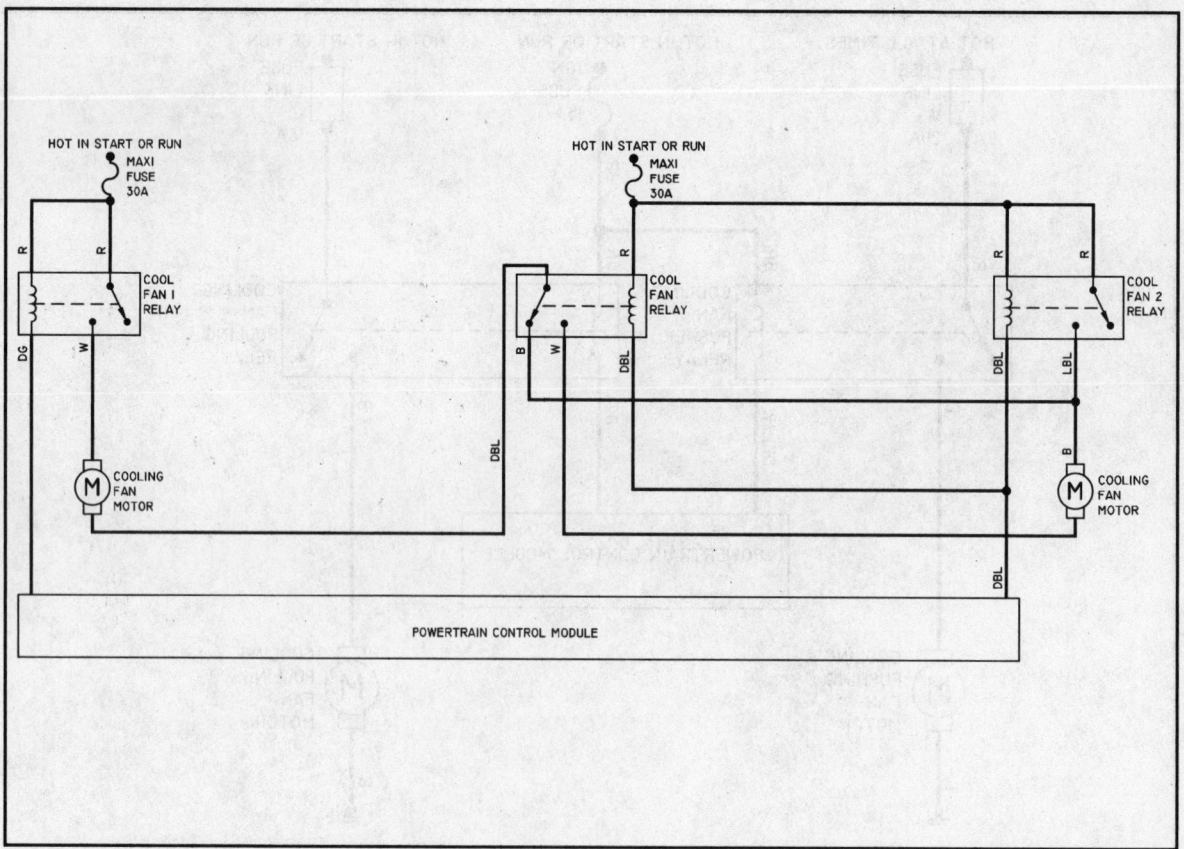

DIA. 10- 1997-99 GM Venture/Silhouette/Trans Sport 3.4L

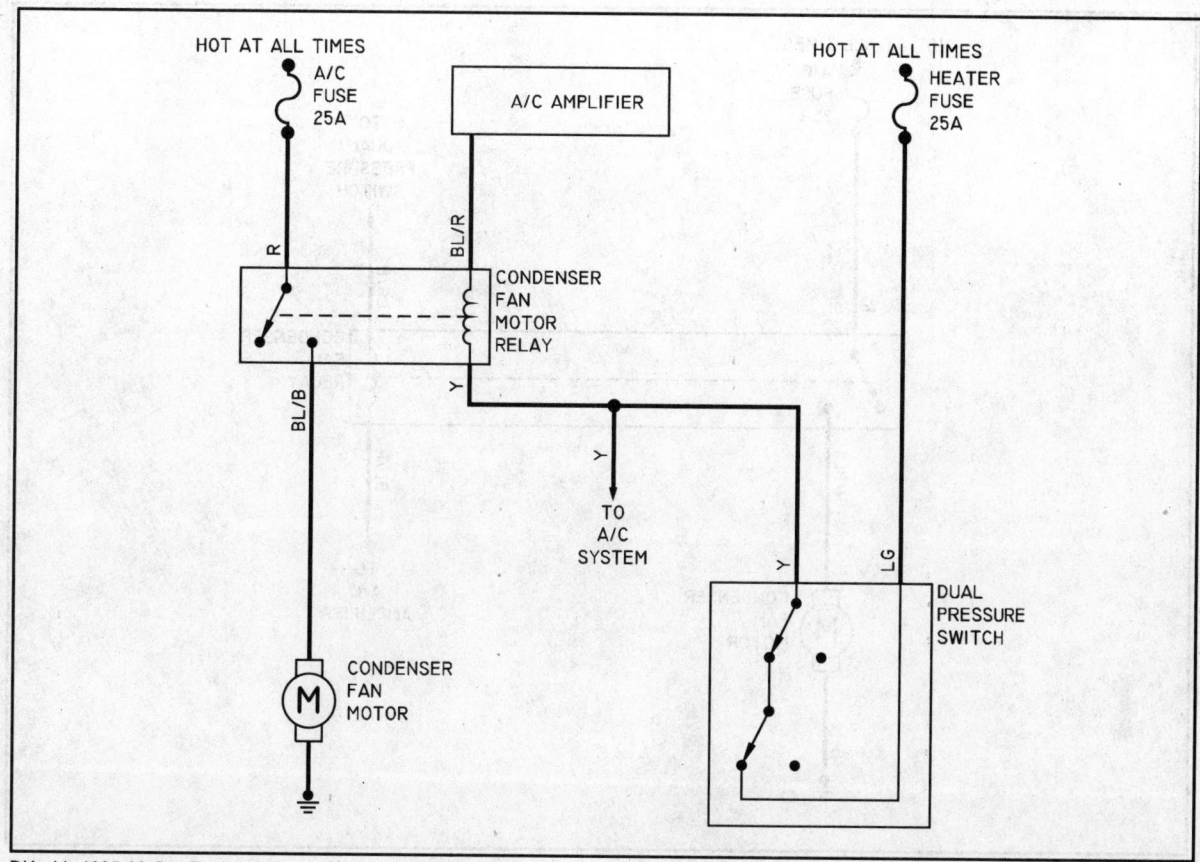

DIA. 11- 1995-98 Geo Tracker 1.6L

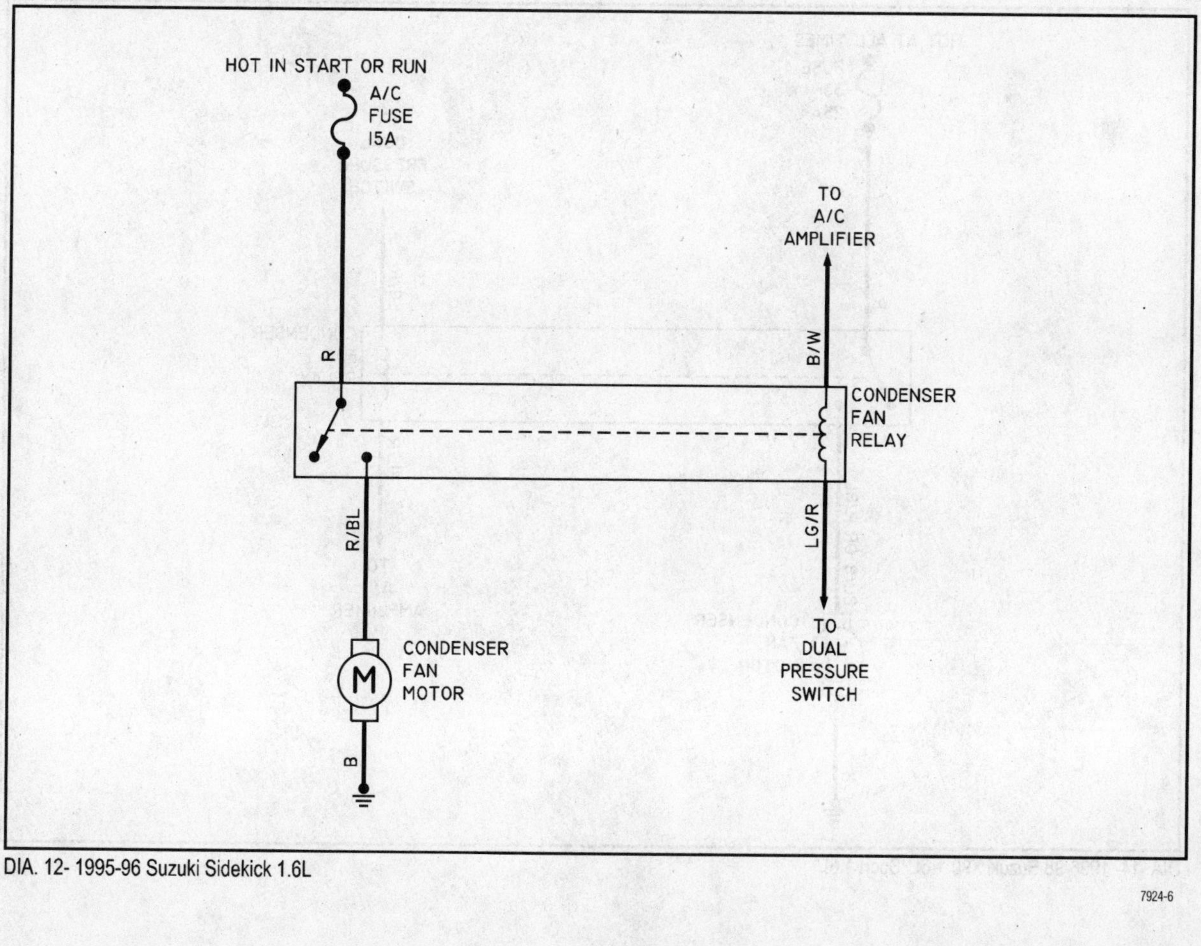

DIA. 12- 1995-96 Suzuki Sidekick 1.6L

7924-6

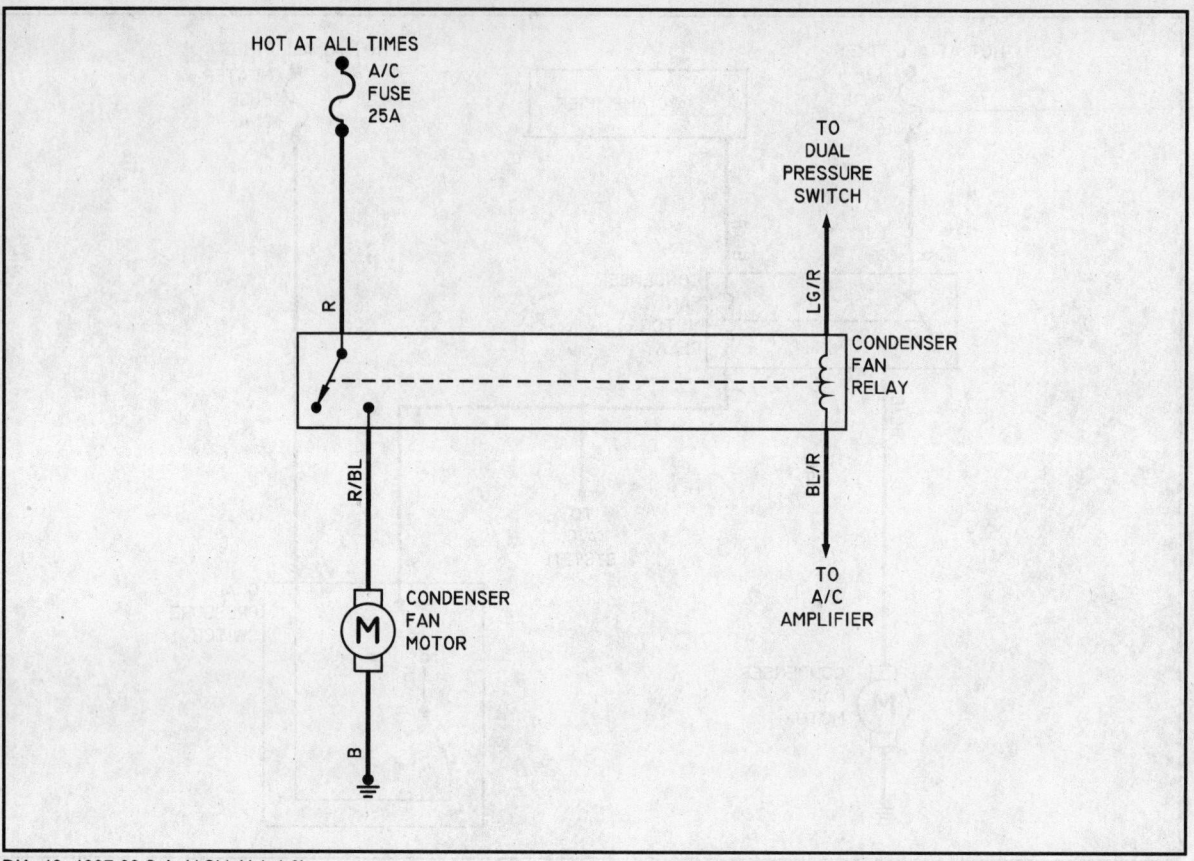

DIA. 13- 1997-98 Sukuki Sidekick 1.6L

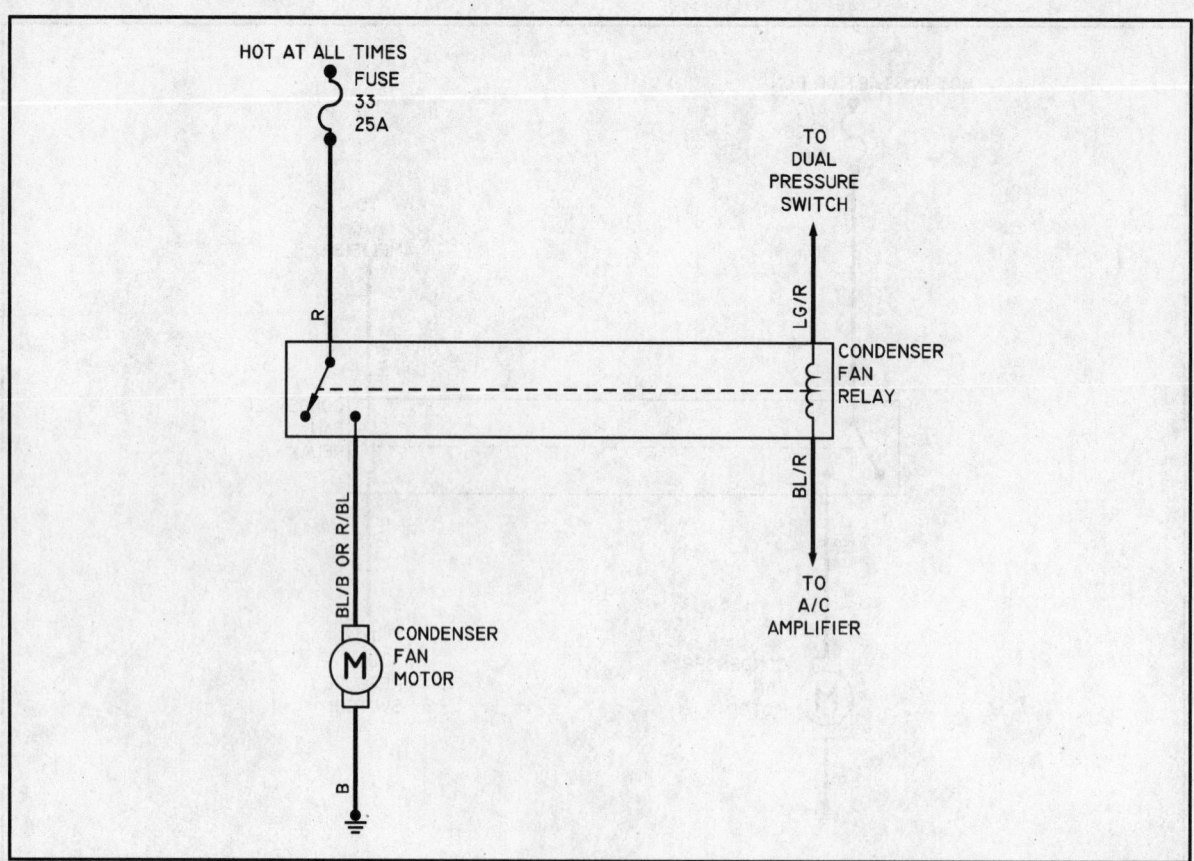

DIA. 14- 1996-98 Suzuki X90 1.6L, Sport 1.8L

STARTING AND CHARGING SYSTEMS

10

STARTING SYSTEM

General Information

The starting system includes the battery, starter motor, solenoid, ignition switch, and in some cases, a starter relay. An inhibitor (neutral safety) switch is included in the starting system circuit to prevent the vehicle from being started while in gear.

When the ignition key is turned to the **START** position, current flows and energizes the starter's solenoid coil. The energized coil becomes an electromagnet which pulls the plunger into the coil, the plunger closes a set of contacts which allow high current to reach the starter motor. On models where the solenoid is mounted on the starter, the plunger also serves to push the starter pinion to mesh with the teeth on the flywheel/flexplate.

To prevent damage to the starter motor when the engine starts, the pinion gear incorporates an over-running (one-way) clutch which is splined to the starter armature shaft. The rotation of the running engine may speed the rotation of the pinion but not the starter motor itself.

Some starting systems employ a starter relay in addition to the solenoid. This relay may be located under the instrument panel, in the kickpanel or in the fuse/relay center under the hood. This relay is used to reduce the amount of current the starting (ignition) switch must carry.

PRECAUTIONS

To prevent damage to the on-board computer, alternator and regulator, the following precautionary measures must be taken when working with the electrical system.

• Always disconnect the negative battery cable before servicing the starter motor. Battery voltage is always present at the large (**B**)

terminal on the solenoid. When removing the starter motor, be prepared to support its weight after the last bolt is removed because the starter motor is a fairly heavy component.

• Never operate the starter motor for more than 30 seconds at a time. Too much cranking will cause the starter motor to overheat, causing permanent damage. Allow the starter motor to cool for at least two minutes between starting attempts.

• Wear safety glasses when working on or near the battery.

• Don't wear a watch with a metal band when servicing the battery. Serious burns can result if the band completes the circuit between the positive battery terminal and ground.

TCCA1P02

Before servicing the electrical system always disconnect the negative battery cable to prevent system damage

• Be absolutely sure of the polarity of a booster battery before making connections. Connect the cables positive to positive, and negative to negative. Connect positive cables first and then make the last connection to ground on the body of the booster vehicle so that arcing cannot ignite hydrogen gas that may have accumulated near the battery. Even momentary connection of a booster battery with the polarity reversed will damage the alternator diodes.

• Disconnect both vehicle battery cables before attempting to charge a battery.

• Be cautious when using metal tools around a battery to avoid creating a short circuit between the terminals.

• When installing a battery, make sure that the positive and negative cables are not reversed.

• When jump-starting the car, be sure that like terminals are connected. This also applies to using a battery charger. Reversed polarity will burn out the alternator and regulator in a matter of seconds.

• Always disconnect the battery (negative cable first) when charging it.

System Testing

➡ **A good quality digital multimeter with at least 10 megohm/volt impedance should be used when testing modern automotive circuits. These meters can accurately detect very small amounts of voltage, current and resistance. This type of meter also has a high internal resistance that will not load the circuit being tested. Loading the circuit causes inaccurate readings, and may cause damage to sensitive computer circuits. Although we are not testing computer circuits in this section, accuracy is very important.**

WITH STARTER MOUNTED SOLENOID

1. Check the battery and clean the connections as follows:
   a. If the battery cells have removable caps, check the water level. Add distilled water if low. Load test the battery and charge if necessary. See Battery Testing in this section for the procedure.
   b. Remove the cables and clean them with a wire brush. Reconnect the cables.
2. Check the starter motor ground circuit with a voltage drop test as follows:

a. Set the meter to read DC voltage on the lowest possible scale.
   b. Connect the negative lead of your multimeter to the negative terminal of the battery.
   c. Connect the positive lead to the body of the starter. Make sure the starter mounting bolts are tight. The meter should read 0.2 volts or less. If the voltage reading is greater, remove and clean the negative battery connection on the engine block. The voltage reading should now be within specification: if not, replace the negative battery cable.
3. Check the motor feed circuit with a voltage drop test as follows:
   a. Disconnect the coil wire or the fuel injector harness to prevent the engine from starting.
   b. Connect the positive lead of your meter to the positive terminal of the battery.

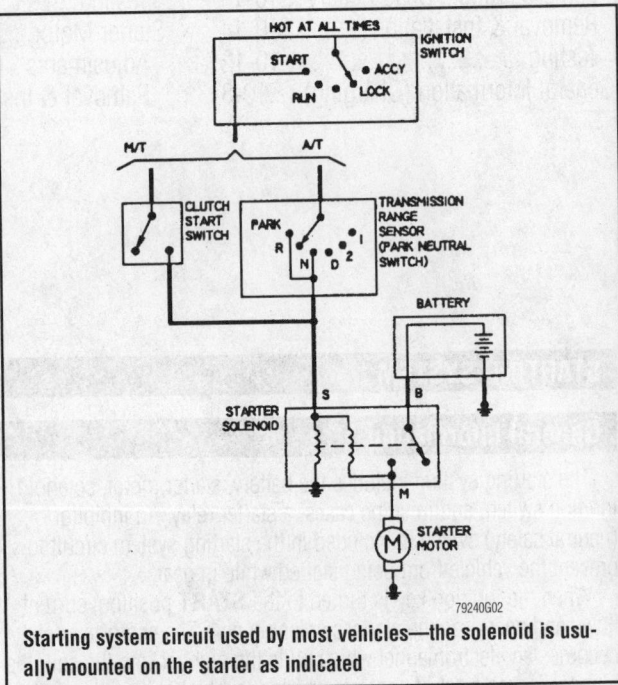

Starting system circuit used by most vehicles—the solenoid is usually mounted on the starter as indicated

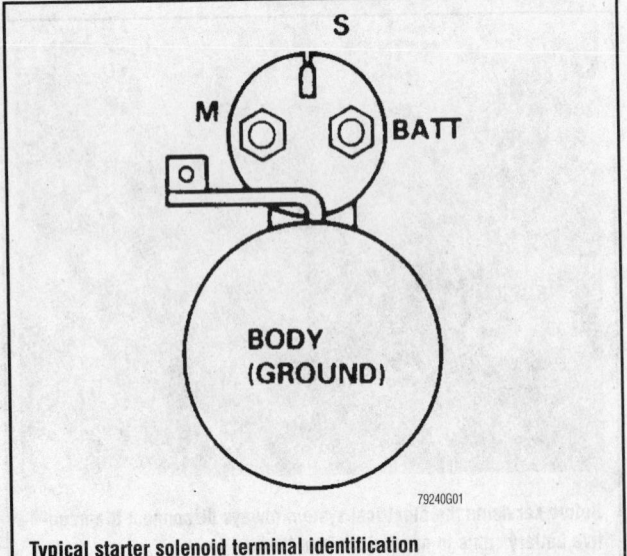

Typical starter solenoid terminal identification

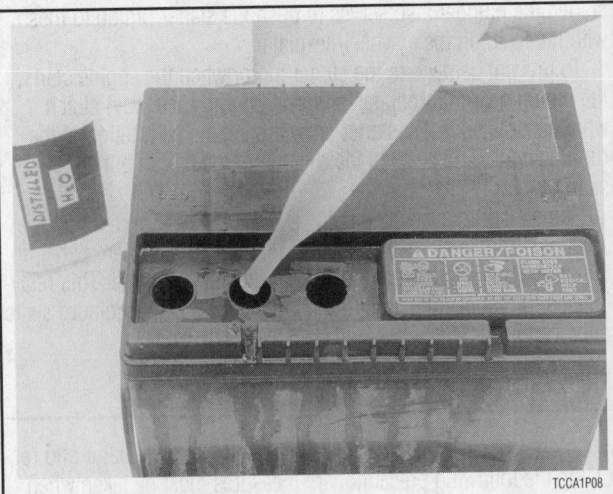

Before testing the system, be sure that the battery is in good shape, which includes ensuring the cells are full on serviceable batteries

Also, disconnect both battery cables (negative first) . . .

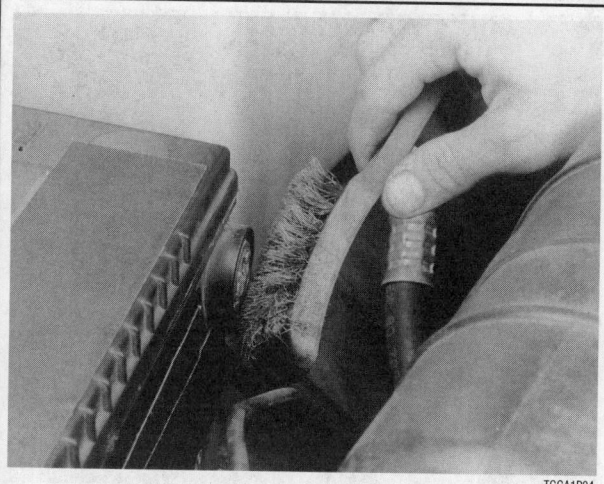

. . . clean the cable terminals of all dirt and corrosion with a wire brush . . .

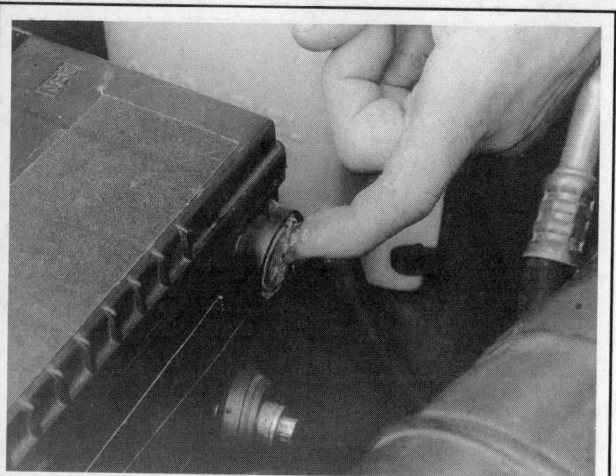

. . . and apply petroleum jelly or multi-purpose grease to the terminals before reattaching the cables

c. Connect the negative meter lead to the motor feed terminal. The motor feed terminal comes out of the body of the starter motor and connects to the solenoid.

d. Turn the ignition key to the **START** position. The meter should read 0.2 volts or less. If the voltage reading is greater, remove and clean the positive battery connection on the starter solenoid. The voltage reading should now be within specification, if not replace the positive battery cable.

e. Connect the coil wire or fuel injector harness.

4. Check for battery voltage at the **S** terminal on the starter solenoid as follows:

a. Disconnect the coil wire or the fuel injector harness to prevent the engine from starting.

b. Set the meter to read battery voltage. Move it to the next higher range if set on the 2 volt scale.

c. Connect the positive lead to the **S** terminal on the starter solenoid and the negative lead to a good ground.

d. Turn the ignition key to the **START** position and crank the engine. The meter should read battery voltage. If battery voltage is not present, check the inhibitor (neutral safety) switch, fuse(s) and wiring between the ignition switch and starter solenoid. If battery voltage is present at the **S** terminal on the solenoid and the starter does not operate, replace the starter and solenoid assembly.

e. Connect the coil wire or fuel injector harness.

WITH EXTERNAL SOLENOID

➡**Not all solenoids are mounted on the starter motor. Some models use a solenoid (relay) mounted on the inner fender or firewall. Both types of solenoids serve to make the connection between the battery and starter motor. Trace the wires for positive identification. The small wire comes from the ignition switch, one large cable from the battery and the other large cable to the starter. The terminals are S, B and M respectively.**

1. Check the battery and clean the connections as follows:

a. If the battery cells have removable caps, check the water level. Add distilled water if low. Load test the battery and charge if necessary. See Battery Testing in this section for the procedure.

❈❈ **CAUTION**

Alway remove the negative battery cable first, and install it last.

b. Remove the cables and clean them with a wire brush. Disconnect and clean the cables on the solenoid in the same manner. Reconnect the cables on the solenoid, then the battery.

2. Check the starter motor ground circuit with a voltage drop test as follows:

a. Set the meter to read DC voltage on the lowest possible scale.

b. Connect the negative lead of your multimeter to the negative terminal of the battery.

c. Connect the positive lead to the body of the starter. Make sure the starter mounting bolts are tight. The meter should read 0.2 volts or less. If the voltage reading is greater, remove and clean the negative battery connection on the engine block. The voltage reading should now be within specification; if not, replace the negative battery cable.

3. Check the motor feed circuit with a voltage drop test as follows:

a. Disconnect the coil wire or the fuel injector harness to prevent the engine from starting.

b. Connect the positive lead of your meter to the positive terminal of the battery.

c. Connect the negative meter lead to the motor feed terminal at the starter. This is the heavy cable on the starter. Turn the ignition key to the **START** position and crank the engine. The meter should read 0.2 volts or less. If the voltage reading is greater, remove and clean the positive battery connections on the starter and solenoid. The voltage reading should now be within specification; if not, replace the positive battery cable.

d. Connect the coil wire or fuel injector harness.

4. Check for battery voltage at the **S** terminal on the starter solenoid as follows:

a. Disconnect the coil wire or the fuel injector harness to prevent the engine from starting.

b. Set the meter to read battery voltage. Move it to next higher range, if previously set on the 2 volt scale.

c. Connect the positive lead to the **S** terminal on the starter solenoid and the negative lead to a good ground.

d. Turn the ignition key to the **START** position. The meter should read battery voltage. If battery voltage is not present, check the inhibitor (neutral safety) switch, fuse(s) and wiring between the ignition switch and starter solenoid. If battery voltage is present at the **S** and **B** terminals but not at the motor feed terminal, replace the solenoid. If battery voltage is present at all three terminals and the starter does not operate, replace the starter motor.

e. Connect the coil wire or fuel injector harness.

Starter Motor

REMOVAL & INSTALLATION

1. Disconnect the negative battery cable.

2. Remove all components necessary to gain access to the starter motor (such as exhaust pipes, air intake ducts, hoses, brackets and heat shields).

3. Disconnect the wiring from the starter. In some cases, the wiring may be more accessible after removing the mounting bolts and moving the starter.

. . . then pull the starter out of the transmission bell housing . . .

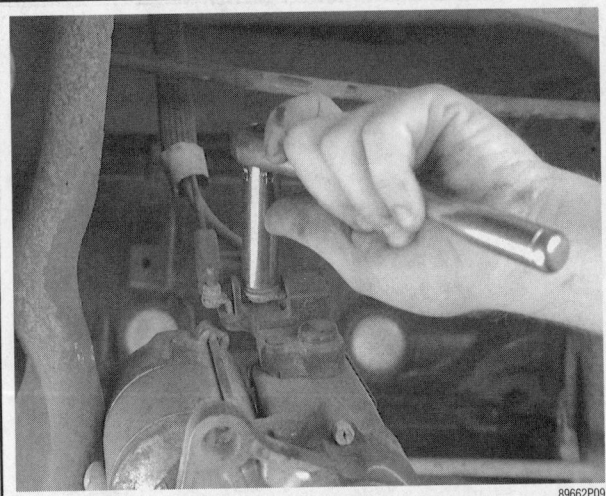

. . . and disconnect the starter motor wires, if not already done

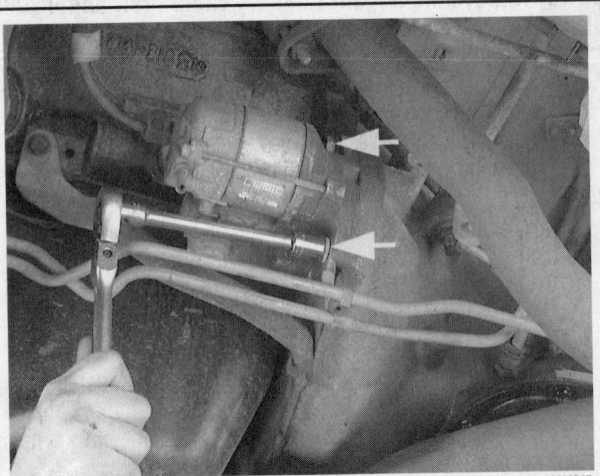

To remove a common starter, raise the vehicle if needed and loosen the starter mounting fasteners (arrows) . . .

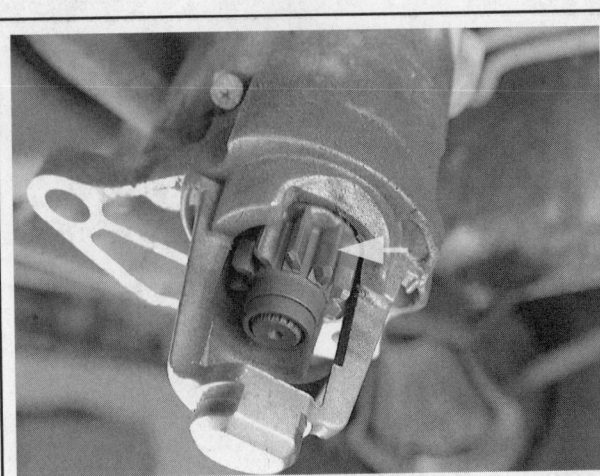

Before installing the starter motor, be sure to inspect the gear teeth (arrow) . . .

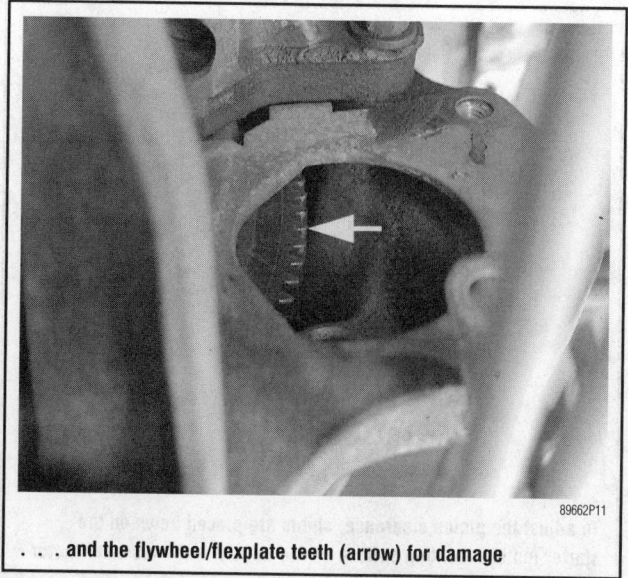

. . . and the flywheel/flexplate teeth (arrow) for damage

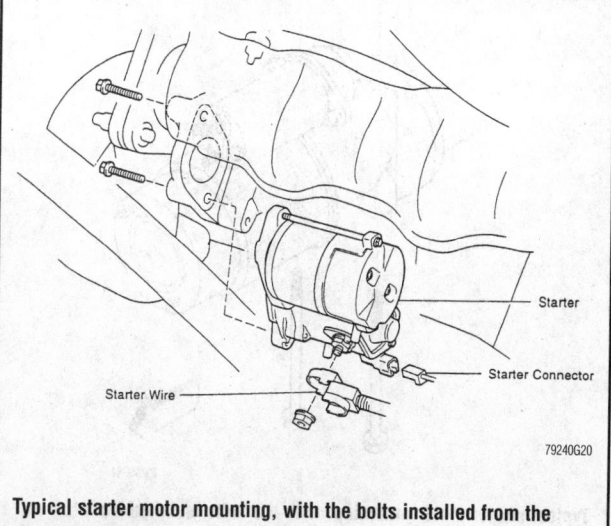

Typical starter motor mounting, with the bolts installed from the transmission side

4. Remove the starter mounting bolts, if not already done.
5. Remove the starter assembly from the vehicle. In some cases, the starter will have to be turned to a different angle to clear obstructions. Don't loose any shims that may fall out from between the starter and the mounting boss, they will need to be returned to their original position when installing the starter. The shims are used to adjust the clearance between the starter pinion and flywheel/flexplate teeth.

To install:

6. If necessary, measure and adjust the pinion-to-ring gear clearance.
7. Position the shim (if any) and the starter motor on the mounting boss. Tighten the mounting bolts securely.
8. Connect the wiring, if not already done.
9. Install any components that were removed to gain access to the starter.
10. Connect the negative battery cable.

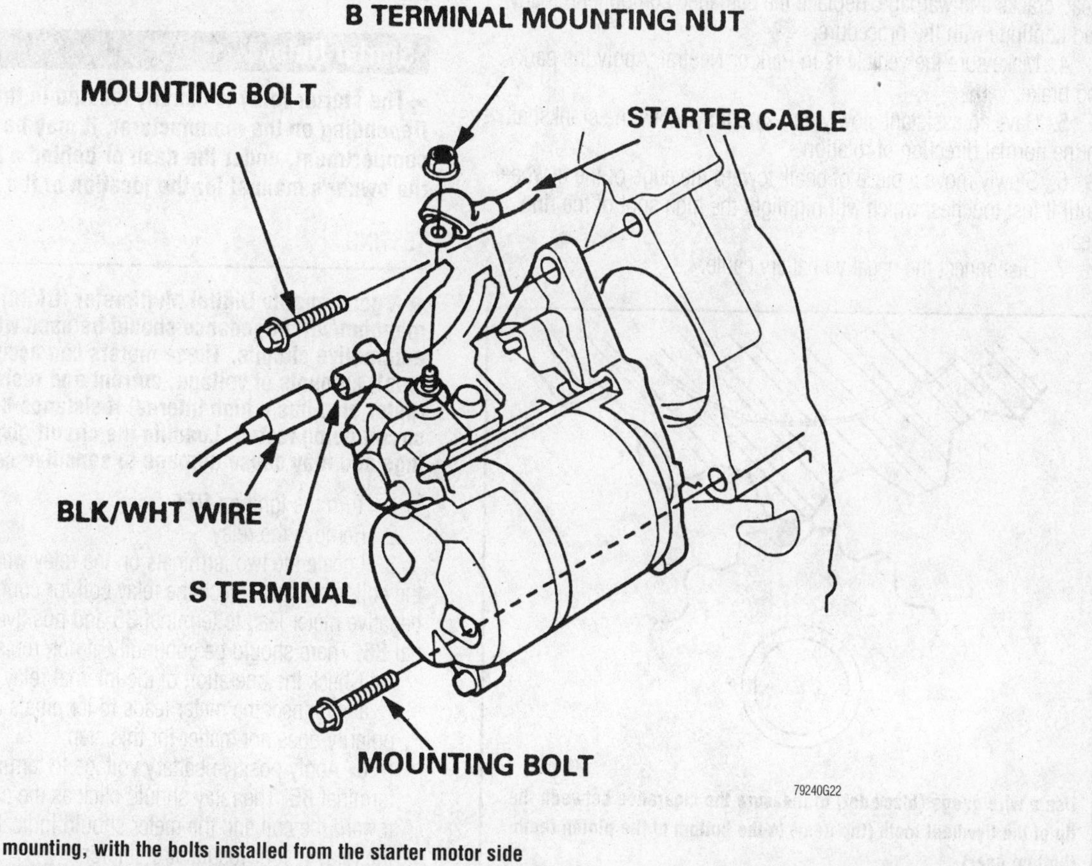

B TERMINAL MOUNTING NUT

MOUNTING BOLT

STARTER CABLE

BLK/WHT WIRE

S TERMINAL

MOUNTING BOLT

Typical starter motor mounting, with the bolts installed from the starter motor side

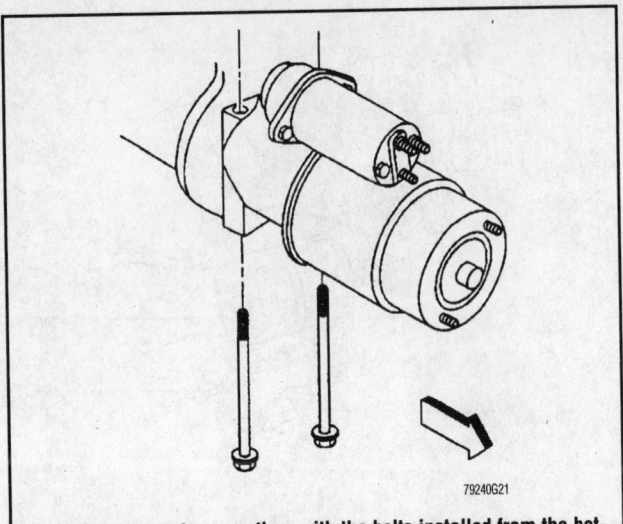

Typical starter motor mounting, with the bolts installed from the bottom of the starter motor

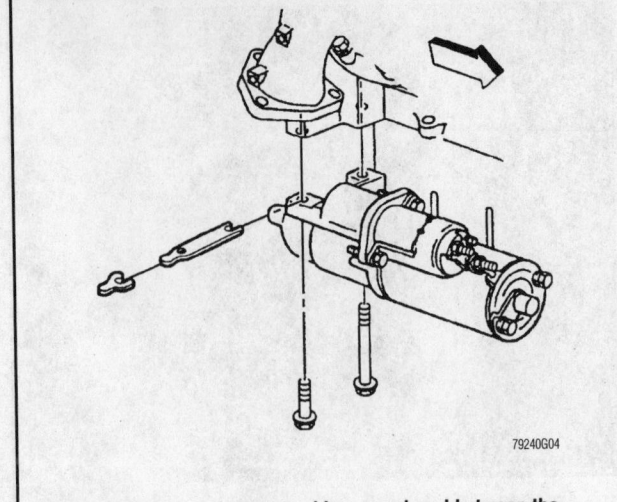

To adjust the pinion clearance, shims are placed between the starter motor mounting surface on the engine and the starter motor

ADJUSTMENTS

Starter Pinion Depth

➡This procedure is used to diagnose starter noise caused by incorrect clearance between the starter pinion and flywheel while the starter is engaged.

1. Raise and safely support the front of the vehicle securely.
2. Remove the flywheel cover.
3. Inspect the flywheel for chipped or missing teeth, abnormal wear, cracks and warpage. Replace the damaged component, if any, and continue with the procedure.
4. Make sure the vehicle is in Park or Neutral. Apply the parking brake.
5. Have an assistant slowly and smoothly rotate the crankshaft in the normal direction of rotation.
6. Slowly move a piece of chalk toward the edge of the flywheel until it just touches, which will highlight the high spot of the ring gear.
7. Disconnect the negative battery cable.

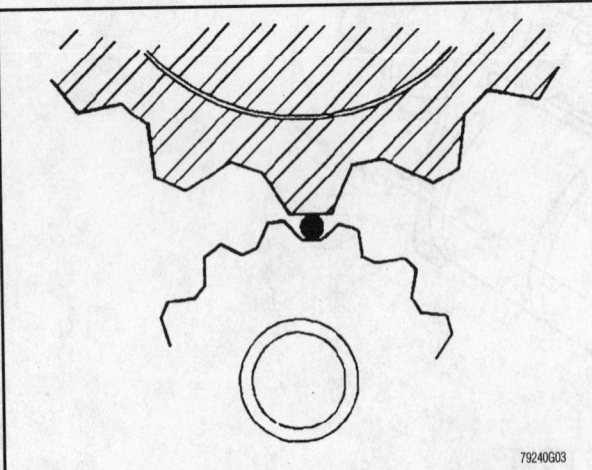

Use a wire gauge (black dot) to measure the clearance between the tip of the flywheel tooth (top gear) to the bottom of the pinion teeth (bottom gear)

8. Turn the high spot of the flywheel to the area of the starter drive pinion.
9. Using a wire gauge, measure the clearance between the tip of the ring gear tooth and the bottom of the pinion gear teeth. Clearance should generally be 0.02–0.06 in. (0.5–1.5mm).
10. Add or remove shims to adjust the clearance, if needed.
11. Install the flywheel cover.
12. Lower the vehicle to the floor.
13. Connect the negative battery cable.

Generally, add shims if the starter whines after the engine starts, and remove shims if the starter whines only during cranking.

Starter Relay

➡The starter relay is usually located in the fuse/relay panel. Depending on the manufacturer, it may be in the engine compartment, under the dash or behind a kickpanel. Refer to the owner's manual for the location of the fuse/relay box.

TESTING

➡A good quality Digital Multimeter (DMM) with at least 10 megohm/volt impedance should be used when testing modern automotive circuits. These meters can accurately detect very small amounts of voltage, current and resistance. This type of meter also has a high internal resistance that will not load the circuit being tested. Loading the circuit gives inaccurate readings and may cause damage to sensitive computer circuits.

1. Turn the ignition **OFF**.
2. Remove the relay.
3. Locate the two terminals on the relay which are connected to the coil windings. Check the relay coil for continuity. Connect the negative meter lead to terminal **85** and positive meter lead to terminal **86**. There should be continuity. If not, replace the relay.
4. Check the operation of the internal relay contacts, as follows:
   a. Connect the meter leads to terminals **30** and **87**. Meter polarity does not matter for this step.
   b. Apply positive battery voltage to terminal **86** and ground to terminal **85**. The relay should click as the contacts are drawn toward the coil and the meter should indicate continuity. Replace the relay if your results are different.

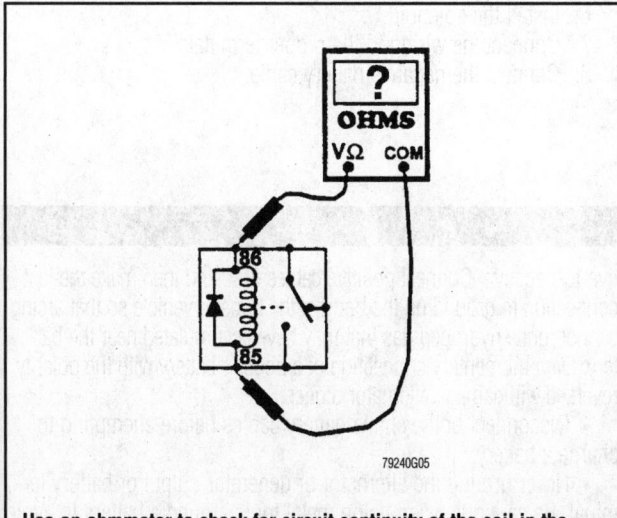

Use an ohmmeter to check for circuit continuity of the coil in the relay

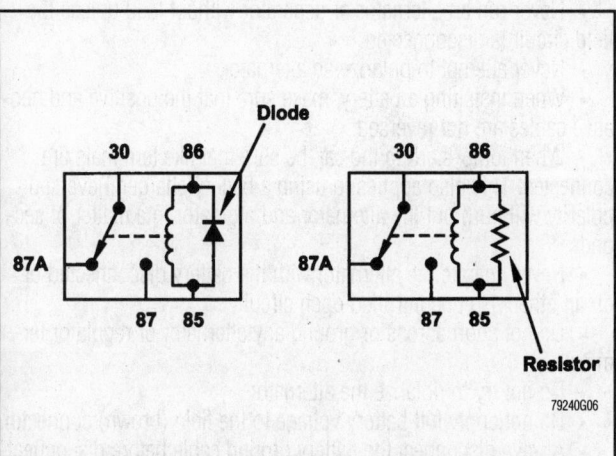

Terminal identification of the most common types of relays. Diodes and resistors in the relay prevent voltage spikes, induced when the current is removed from the coil, from damaging electronic components

Solenoid

TESTING

1. Disconnect the negative battery cable.
2. Remove the wire connections from the starter solenoid.
3. Using a self-powered test light or ohmmeter, check for continuity between the following:
• Solenoid **B** terminal and solenoid case or ground terminal—no continuity
  • **S** terminal and solenoid case or ground terminal—continuity
  • **S** terminal and **M** terminal—continuity
  • **M** terminal and solenoid case or ground terminal—continuity
4. If the actual results of the test are different than indicated, replace the starter solenoid.

REMOVAL & INSTALLATION

➡This procedure is for externally mounted starter solenoids only. For solenoids mounted on the starter, we recommend replacing the complete assembly.

1. Disconnect the negative battery cable.
2. Remove the wiring from the starter solenoid. Label the wires and the corresponding terminals if necessary for installation.

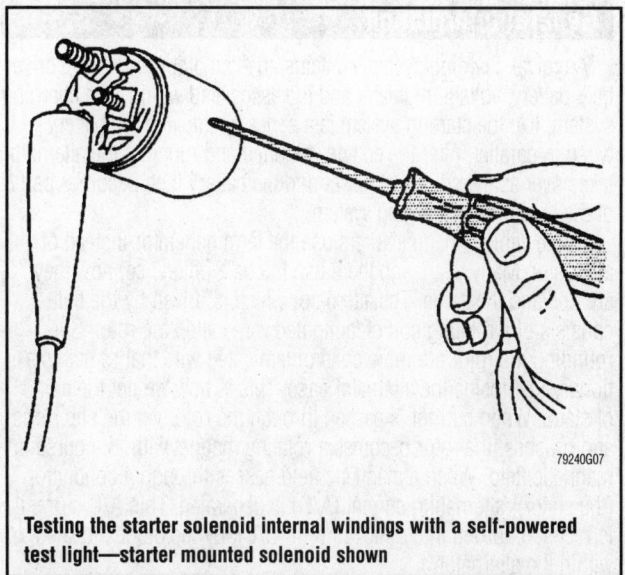

Testing the starter solenoid internal windings with a self-powered test light—starter mounted solenoid shown

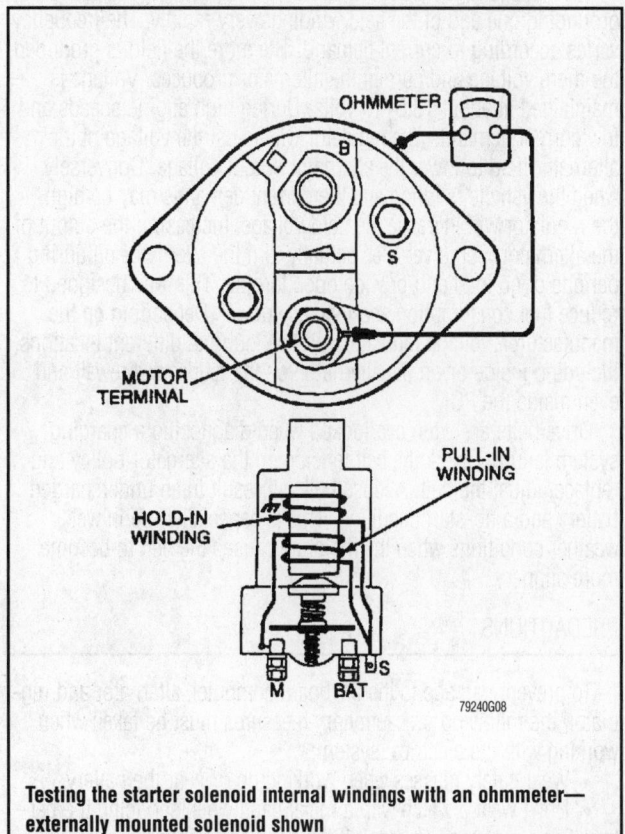

Testing the starter solenoid internal windings with an ohmmeter—externally mounted solenoid shown

3. Remove the fasteners securing the solenoid to the fender or firewall.

4. Remove the solenoid.

To install:

5. Clean the solenoid mounting and the solenoid to ensure good electrical contact.

6. Install the solenoid.

7. Connect the wiring to the proper terminals.

8. Connect the negative battery cable.

CHARGING SYSTEM

General Information

A typical charging system contains an alternator (generator), drive belt, battery, voltage regulator and the associated wiring. The charging system, like the starting system is a series circuit with the battery wired in parallel. After the engine is started and running, the alternator takes over as the source of power and the battery then becomes part of the load on the charging system.

Some vehicle manufacturers use the term generator instead of alternator. Many years ago there used to be a difference, now they are one and the same. The alternator which is driven by the belt, consists of a rotating coil of laminated wire called the rotor. Surrounding the rotor are more coils of laminated wire that remain stationary just inside the alternator case. This is how we get the name of stator. When current is passed through the rotor via the slip rings and brushes, the rotor becomes a rotating magnet with, of course, a magnetic field. When a magnetic field passes through a conductor (the stator), alternating current (A/C) is generated. This A/C current is rectified, turned into direct current (D/C), by the diodes located within the alternator.

The voltage regulator controls the alternator's field voltage by grounding one end of the field windings very rapidly. The frequency varies according to current demand. The more the field is grounded, the more voltage and current the alternator produces. Voltage is maintained at about 13.5–15 volts. During high engine speeds and low current demands, the regulator will adjust the voltage of the alternator field to lower the alternator output voltage. Conversely, when the vehicle is idling and the current demands may be high, the regulator will increase the field voltage, increasing the output of the alternator. Some vehicles actually turn the alternator off during periods of no load and/or wide open throttle. This was designed to reduce fuel consumption and increase power. Depending on the manufacturer, voltage regulators can be found in different locations, including inside or on the alternator, on the fender or firewall and even inside the PCM.

Drive belts are often overlooked when diagnosing a charging system failure. Check the belt tension on the alternator pulley and replace/adjust the belt. A loose belt will result in an undercharged battery and a no-start condition. This is especially true in wet weather conditions when the moisture causes the belt to become more slippery.

PRECAUTIONS

To prevent damage to the on-board computer, alternator and regulator, the following precautionary measures must be taken when working with the electrical system:

- Wear safety glasses when working on or near the battery.
- Don't wear a watch with a metal band when servicing the battery. Serious burns can result if the band completes the circuit between the positive battery terminal and ground.
- Be absolutely sure of the polarity of a booster battery before making connections. Connect the cables positive-to-positive, and negative-to-negative. Connect positive cables first, and then make the last connection to ground on the body of the booster vehicle so that arcing cannot ignite hydrogen gas that may have accumulated near the battery. Even momentary connection of a booster battery with the polarity reversed will damage alternator diodes.
- Disconnect both vehicle battery cables before attempting to charge a battery.
- Never ground the alternator or generator output or battery terminal. Be cautious when using metal tools around a battery to avoid creating a short circuit between the terminals.
- Never ground the field circuit between the alternator and regulator.
- Never run an alternator or generator without load unless the field circuit is disconnected.
- Never attempt to polarize an alternator.
- When installing a battery, make sure that the positive and negative cables are not reversed.
- When jump-starting the car, be sure that like terminals are connected. This also applies to using a battery charger. Reversed polarity will burn out the alternator and regulator in a matter of seconds.
- Never operate the alternator with the battery disconnected or on an otherwise uncontrolled open circuit.
- Do not short across or ground any alternator or regulator terminals.
- Do not try to polarize the alternator.
- Do not apply full battery voltage to the field (brown) connector.
- Always disconnect the battery ground cable before disconnecting the alternator lead.
- Always disconnect the battery (negative cable first) when charging it.
- Never subject the alternator to excessive heat or dampness. If you are steam cleaning the engine, cover the alternator.
- Never use arc-welding equipment on the car with the alternator connected.

SYSTEM TESTING

The charging system should be inspected if:

- A Diagnostic Trouble Code (DTC) is set relating to the charging system
- The charging system warning light is illuminated
- The voltmeter on the instrument panel indicates improper charging (either high or low) voltage
- The battery is overcharged (electrolyte level is low and/or boiling out)
- The battery is undercharged (insufficient power to crank the starter)

The starting point for all charging system problems begins with the inspection of the battery, related wiring and the alternator drive belt. The battery must be in good condition and fully charged before system testing. If a Diagnostic Trouble Code (DTC) is set, diagnose and repair the cause of the trouble code first.

If equipped, the charging system warning light will illuminate if the charging voltage is either too high or too low. The warning light should light when the key is turned to the **ON** position as a bulb check. When the alternator starts producing voltage due to the engine starting, the light should go out. A good sign of voltage that is too high are lights that burn out and/or burn very brightly. Overcharging can also cause damage to the battery and electronic circuits.

Alternator

TESTING

➡ **Before testing, make sure all connections and mounting bolts are clean and tight. Many charging system problems are related to loose and corroded terminals or bad grounds. Don't overlook the engine ground connection to the body, or the tension of the alternator drive belt.**

Voltage Drop Test

➡ **A good quality Digital Multimeter (DMM) with at least 10 megohm/volt impedance should be used when testing modern automotive circuits. These meters can accurately detect very small amounts of voltage, current and resistance. This type of meter also has a high internal resistance that will not load the circuit being tested. Loading the circuit gives inaccurate readings and may cause damage to sensitive computer circuits.**

1. Make sure the battery is in good condition and fully charged.
2. Perform a voltage drop test of the positive side of the circuit as follows:
   - Start the engine and allow it to reach normal operating temperature.
   - Turn the headlamps, heater blower motor and interior lights on.
   - Bring the engine to about 2,500 rpm and hold it there.
   - Connect the negative (-) voltmeter lead directly to the battery positive (+) terminal.
   - Touch the positive (+) voltmeter lead directly to the alternator **B+** output stud, not the nut. The meter should read no higher than about 0.5 volts. If it does, then there is higher than normal resistance between the positive side of the battery and the **B+** output at the alternator.
   - Move the positive (+) meter lead to the nut and compare the voltage reading with the previous measurement. If the voltage reading drops substantially, then there is resistance between the stud and the nut.

➡ **The theory is to keep moving closer to the battery terminal one connection at a time in order to find the area of high resistance (bad connection).**

3. Perform a voltage drop test of the negative side of the circuit as follows:
   a. Start the engine and allow it to reach normal operating temperature.
   b. Turn the headlamps, heater blower motor and interior lights ON.
   c. Bring the engine to about 2,500 rpm and hold it there.
   d. Connect the negative (-) voltmeter lead directly to the negative battery terminal.
   e. Touch the positive (+) voltmeter lead directly to the alternator case or ground connection. The meter should read no higher

than about 0.3 volts. If it does, then there is higher than normal resistance between the battery ground terminal and the alternator ground.
   f. Move the positive (+) meter lead to the alternator mounting bracket, if the voltage reading drops substantially then you know that there is a bad electrical connection between the alternator and the mounting bracket.

➡ **The theory is to keep moving closer to the battery terminal one connection at a time in order to find the area of high resistance (bad connection).**

Current Output Test

➡ **A good quality Digital Multimeter (DMM) with at least 10 megohm/volt impedance should be used when testing modern automotive circuits. These meters can accurately detect very small amounts of voltage, current and resistance. This type of meter also has a high internal resistance that will not load the circuit being tested. Loading the circuit gives inaccurate readings and may cause damage to sensitive computer circuits.**

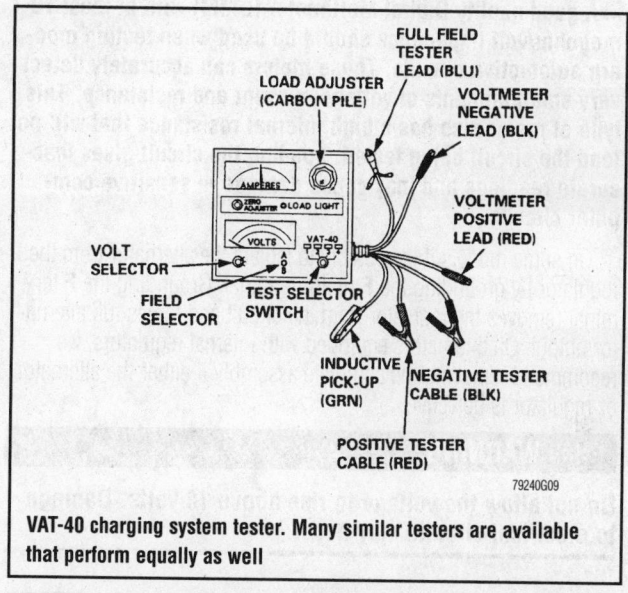

VAT-40 charging system tester. Many similar testers are available that perform equally as well

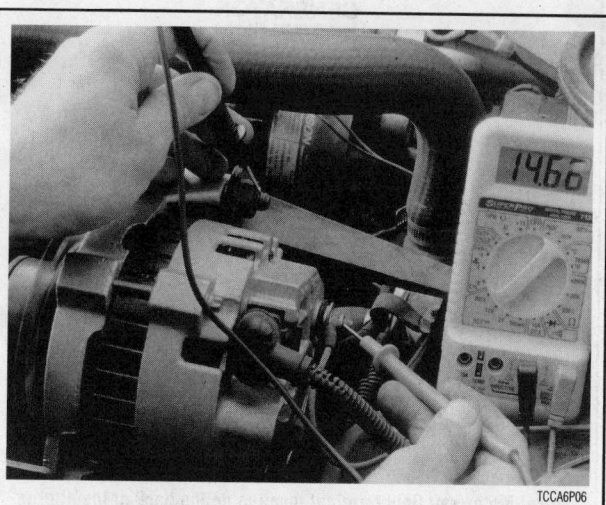

The output voltage of the alternator can be quickly measured by probing between the output terminal and a good ground

1. Perform a current output test as follows:

➡️ **The current output test requires the use of a volt/amp tester with battery load control and an inductive amperage pick-up. Follow the manufacturer's instructions on the use of the equipment.**

    a. Start the engine and allow it to reach normal operating temperature.

    b. Apply the parking brake and turn OFF all electrical accessories.

    c. Connect the tester to the battery terminals and cable according to the instructions.

    d. Bring the engine to about 2,500 rpm and hold it there.

    e. Apply a load to the charging system with the rheostat on the tester. Do not let the voltage drop below 12 volts.

    f. The alternator should deliver to within 10% of the rated output. If the amperage is not within 10% and all other components test good, replace the alternator.

Alternator Isolation Test

➡️ **A good quality Digital Multimeter (DMM) with at least 10 megohm/volt impedance should be used when testing modern automotive circuits. These meters can accurately detect very small amounts of voltage, current and resistance. This type of meter also has a high internal resistance that will not load the circuit being tested. Loading the circuit gives inaccurate readings and may cause damage to sensitive computer circuits.**

On some models it is possible to isolate the alternator from the regulator by grounding the **F** (field) terminal. Grounding the **F** terminal removes the regulator from the circuit and forces full alternator output. On alternators equipped with internal regulators, we recommend replacing the complete assembly if either the alternator or regulator is defective.

✳✳ WARNING

Do not allow the voltage to rise above 18 volts. Damage to electrical circuits may occur.

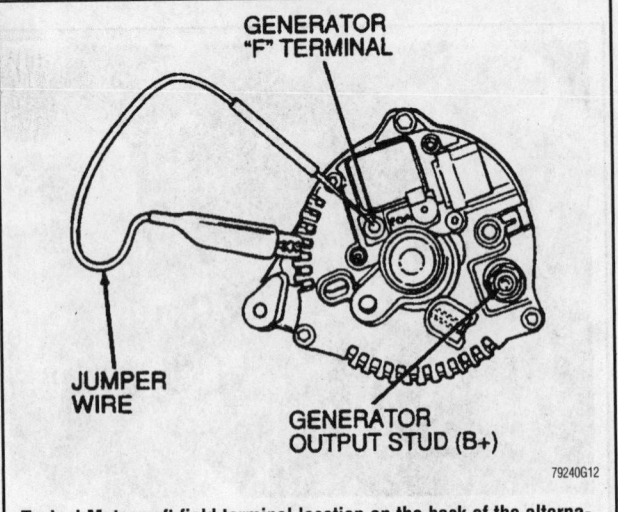

Typical Motorcraft field terminal location on the back of the alternator

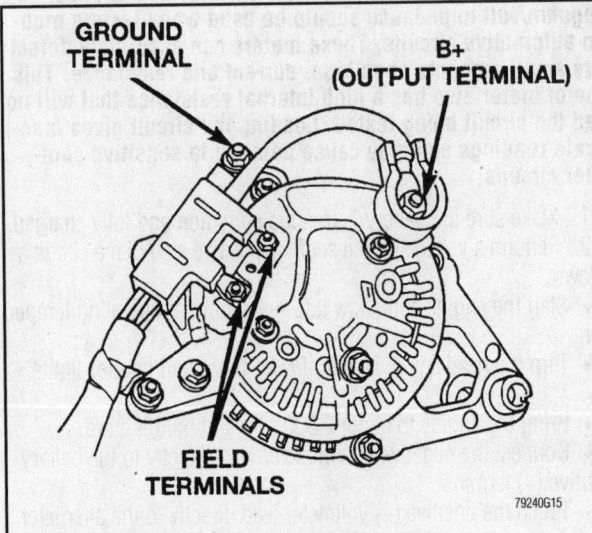

Typical Nippondenso field terminal location on the back of the alternator

Typical Mopar alternator terminal locations

1. Connect a voltmeter across the battery terminals so the voltage can be monitored.

2. Start the engine and allow it reach normal operating temperature.

3. Connect a jumper lead to a good ground.

4. Locate the field terminal (negative) on the back of the alternator.

5. Momentarily connect the grounded jumper to the field terminal. If the alternator is OK, the voltage will climb rapidly. Disconnect the jumper before the output reaches 18 volts. If the voltage does not rise, replace the alternator. If the voltage rises, then the regulator is bad.

➡️ **Chrysler models have two field terminals, one positive and one negative. The positive (+) terminal will have battery voltage present and the negative (-) terminal will have 3–5 volts less. Ground the negative (-) terminal when testing this type of alternator.**

REMOVAL & INSTALLATION

1. Disconnect the negative battery cable.
2. Remove the drive belt from the alternator pulley.

➡**In some cases, it may be easier to disconnect the wiring after the alternator has been removed. Be sure to support the alternator by hand while removing the wiring.**

3. Disconnect the wiring from the alternator.
4. Remove the alternator.

To install:

➡**If necessary, attach the wiring to the alternator before installation.**

5. Install the alternator and attach the wiring if not already done.
6. Install the drive belt on the alternator pulley. Adjust the belt if necessary.
7. Connect the negative battery cable.

When removing alternator fasteners, be sure to retain any washers, spacers or nuts for reassembly

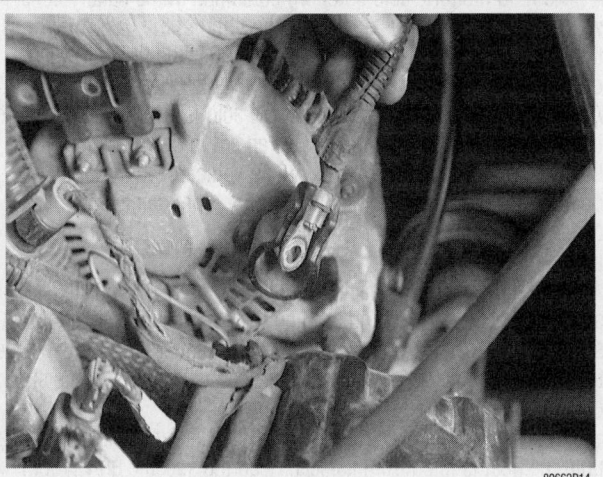

To remove a typical alternator, first detach the wiring terminals from it (if possible) . . .

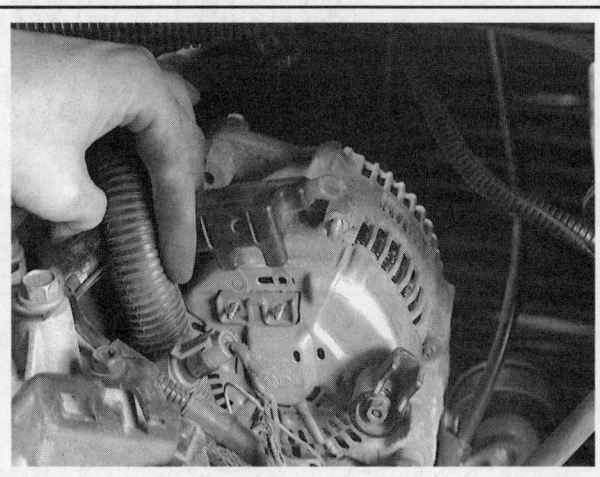

If not possible earlier, now disconnect any applicable wiring from the alternator

. . . then loosen the alternator mounting fasteners—this alternator uses mounting bolts (arrows)

Finally, carefully remove the alternator from the engine compartment—although not shown here some alternators must be dropped out the bottom of the vehicle

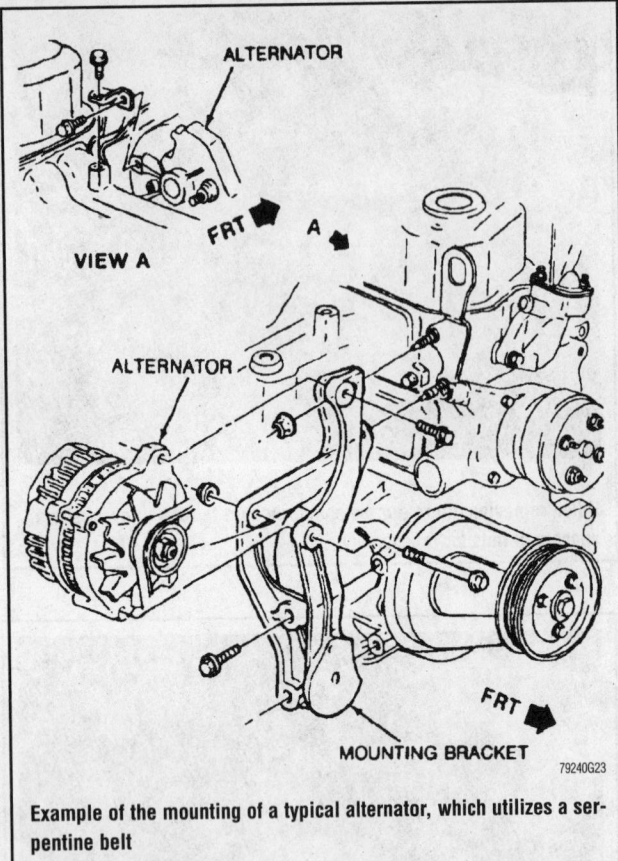

Example of the mounting of a typical alternator, which utilizes a serpentine belt

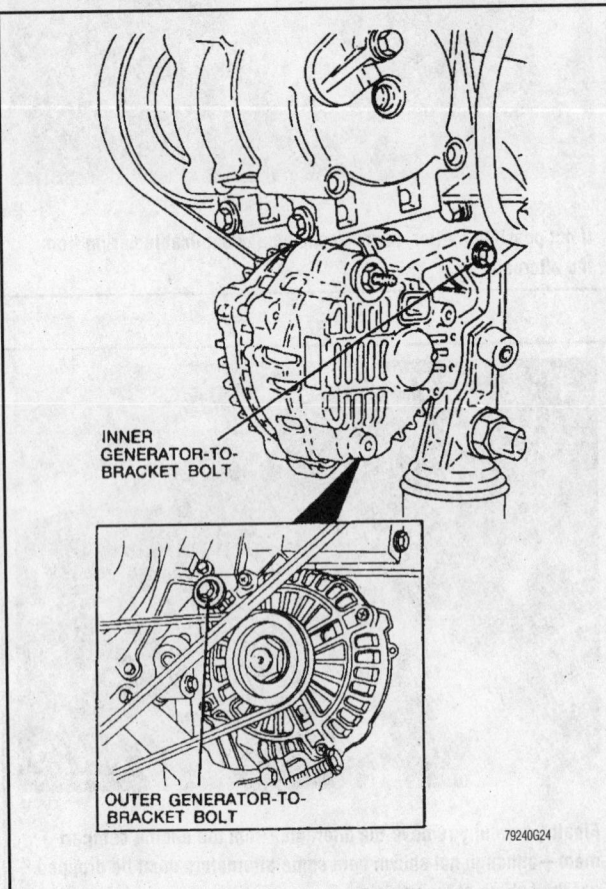

Example of the mounting for a common alternator that uses a V-belt

Battery

TESTING

> ### ☀☀ CAUTION
>
> **If the battery shows signs of freezing, cracking, leaking, loose posts or low electrolyte level, do not attempt to test, charge or jump start. Internal arcing may occur and cause the battery to explode. Always replace a battery that is physically damaged. If only the water level is low and the battery can be filled, add distilled water to the proper level. When charging, disconnect the battery cables, attach the connections to the battery first, then turn the charger ON. Never disconnect the battery cable(s) while the engine is running. Always wear safety glasses when servicing the battery.**

Specific Gravity Test

The fluid (sulfuric acid solution) contained in the battery cells will tell you many things about the condition of the battery. Because the cell plates must be kept submerged below the fluid level in order to operate, maintaining the fluid level is extremely important. And, because the specific gravity of the acid is an indication of electrical charge, testing the fluid can be an aid in determining if the battery must be replaced. A battery in a vehicle with a properly operating charging system should require little maintenance, but careful, periodic inspection should reveal problems before they leave you stranded.

At least once a year, check the specific gravity of the battery. It should be between 1.20 and 1.26 on the gravity scale. Most auto supply stores carry a variety of inexpensive battery testing hydrometers. These can be used on any non-sealed battery to test the specific gravity in each cell.

Draw some of the electrolyte from the battery into the hydrometer until the float is lifted from its seat. Read the specific gravity indicated by the position of the float. If the specific gravity is low in one or more cells, the battery should be slowly charged and checked again to see if the gravity has come up. Generally, if after charging,

If the battery cells are low on fluid, top them up with distilled water if possible

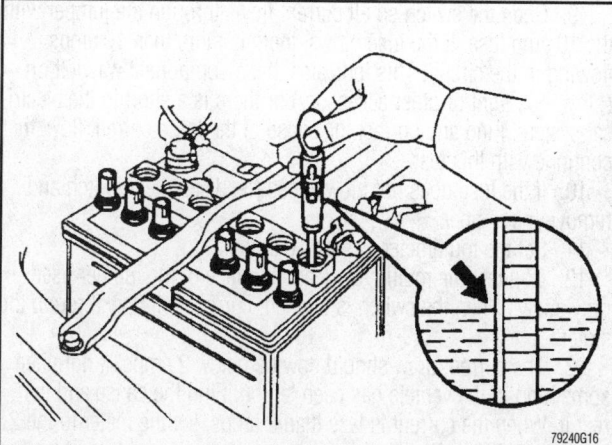

On serviceable batteries, draw some battery fluid into the hydrometer and read the specific gravity indicated by the float inside the tester

Open Circuit Voltage	
Open Circuit Volts	**Charge Percentage**
11.7 volts or less	0%
12.0 volts	25%
12.2 volts	50%
12.4 volts	75%
12.6 volts or more	100%

79240G10

Compare the actual voltage measured with these values to determine the percent of charge based on no load test results

the specific gravity between any two cells varies more than 50 points (0.50), replace the battery, as it can no longer produce sufficient voltage to guarantee proper operation.

No Load Voltage Test

➡A good quality Digital Multimeter (DMM) with at least 10 megohm/volt impedance should be used when testing modern automotive circuits. These meters can accurately detect very small amounts of voltage, current and resistance. This type of meter also has a high internal resistance that will not load the circuit being tested. Loading the circuit gives inaccurate readings and may cause damage to sensitive computer circuits.

1. Perform a no load voltage test to determine the state of charge by doing the following:
   a. If the battery has just been charged, remove the surface charge by turning on the headlamps for 15 seconds, then let the voltage stabilize for about 5 minutes before making any measurements.

   b. Disconnect the negative battery cable.
   c. Measure the battery voltage with a DMM.
   d. Compare the readings to the chart to determine the state of charge.

High Capacity Discharge Test

1. Perform a high capacity discharge test to determine the cranking capacity as follows:
   a. Fully charge the battery.
   b. Connect a VAT-40 or equivalent load tester to the battery.
   c. Apply a load equal to $FR1/2 of the Cold Cranking Amp (CCA) rating of the battery for 15 seconds. The CCA is usually found on the battery label, if not, apply a load equal to 200 amps.
   d. If the voltmeter reading falls below 9.6 volts at 70°F (21°C) or more, the battery should be replaced. The minimum battery voltage will be lower depending on the ambient temperature. Refer to the chart for testing in temperatures lower than 70°F (21°C).

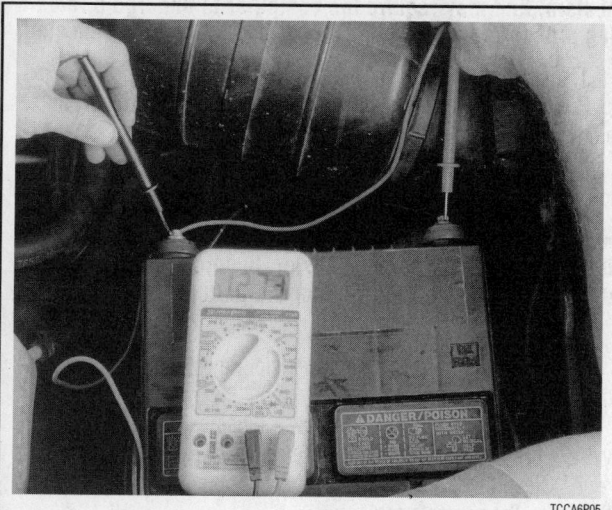

Use a high quality DMM to measure the battery voltage

Load Test Temperature		
Minimum Voltage	**Temperature**	
	°F	**°C**
9.6 volts	70° and above	21° and above
9.5 volts	60°	16°
9.4 volts	50°	10°
9.3 volts	40°	4°
9.1 volts	30°	-1°
8.9 volts	20°	-7°
8.7 volts	10°	-12°
8.5 volts	0°	-18°

79240G11

High capacity discharge test minimum voltage/temperature chart

Parisitic Draw Test

➡️A good quality Digital Multimeter (DMM) with at least 10 megohm/volt impedance should be used when testing modern automotive circuits. These meters can accurately detect very small amounts of voltage, current and resistance. This type of meter also has a high internal resistance that will not load the circuit being tested. Loading the circuit gives inaccurate readings and may cause damage to sensitive computer circuits.

This test measures the amount of current that the vehicle draws while it is parked and not in use. A small amount of current should be flowing for such things as the on-board computer memory, automatic climate control, clock, and radio station presets. If there is a short in the vehicle electrical system or something has been left on, the excess current draw will eventually drain the battery and cause a no-start condition.

1. Be sure all accessories are turned **OFF**. Disconnect the negative battery cable.

2. Install a battery quick-disconnect switch (such as GM Parasitic Draw Test Switch J 38758) between the negative cable and the negative battery terminal. A battery disconnect switch will work in most cases.

3. Road test the vehicle while activating all accessories including the radio and air conditioning. Then, turn all accessories **OFF**.

4. Turn the vehicle **OFF** and open the hood.

5. If equipped, disable the underhood light.

6. Allow approximately 20 minutes for the vehicle computer system(s) to power down.

7. Connect one end of a jumper with a 10 amp fuse to the side of the quick-disconnect switch closest to the negative battery terminal. Be sure the jumper is on the metal part of the switch.

8. Connect the remaining end of the jumper to the other side of the switch closest to the negative battery cable.

⁘ WARNING

Do not connect the multimeter to the circuit if more than 10 amps are flowing. Damage to the meter may occur.

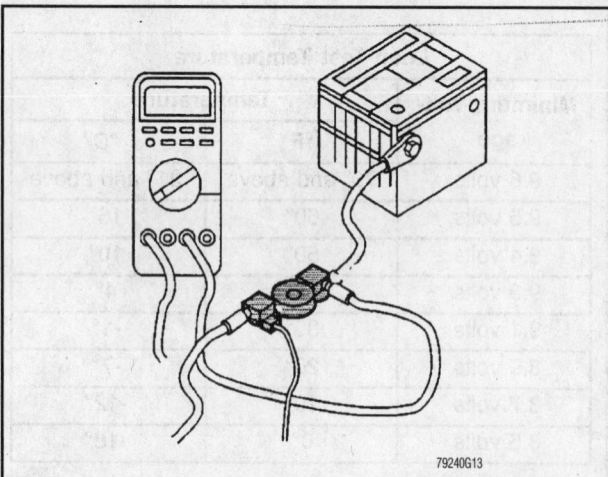

Before starting the parasitic draw test, install a battery disconnect switch between the negative battery cable and the battery terminal, as shown

9. Open the switch so all current flows through the jumper with the 10 amp fuse. If the fuse blows, there is more than 10 amps flowing in the circuit. This indicates that a component was left on (glove box light or other accessory) or there is a short in the electrical system. Find and correct the cause of the large current flow, then continue with this test.

10. If the fuse does not blow, close the disconnect switch and remove the jumper.

11. Set the multimeter to read 10 amps.

12. Connect the multimeter leads in place of the jumper used previously. When the switch is opened, current will flow through the meter.

13. The current draw should now be below 2 amps. If not, then something in the vehicle has been left on. Find the cause and correct it. When the current is less than 2 amps, set the meter to the 2 amp range. This will allow you to measure small amounts of current.

⁘ WARNING

Do not open the door of the vehicle. The interior lights coming on will blow the fuse of the meter while on the 2 amp range.

14. Normal current draw should be less than ¼ of the reserve capacity of the battery. If the reserve capacity is unknown, normal current draw should be somewhere in the range of 0.005–0.040 amps depending on the type and amount of equipment on the vehicle.

➡️The reserve capacity is the amount of time, in minutes, it takes for the battery voltage to fall below 10.5 volts at a discharge rate of 25 amps at 80°F (26.7°C). In most cases, this number can be found on the battery label.

15. If the current draw is higher than specified, pull fuses and/or disconnect components until the problem is found. Don't overlook the alternator connection.

REMOVAL & INSTALLATION

➡️Disconnecting the negative battery cable on some vehicles may interfere with the functioning of the on-board computer system, and may require the computer to undergo a relearning process once the negative battery cable is reconnected.

1. Turn the ignition key to the **OFF** position.

2. Disconnect the negative battery cable first. On some vehicles a cover or trim panel may have to be removed first.

3. Disconnect the positive battery cable.

4. Remove the battery hold-down.

➡️A battery strap or holding device can make removing or installing the battery much easier. In some cases it can be difficult to get your hands under the battery.

To install:

5. Position the battery in the vehicle. Pay attention to the location of the terminals.

6. Install the battery hold-down. A loose battery may cause a vehicle fire or severe damage to the electrical system.

7. Clean the terminals and connect the positive battery cable first, then the negative cable.

8. If equipped, install the cover or trim panel.

JUMP STARTING A DEAD BATTERY

Whenever a vehicle is jump started, precautions must be followed in order to prevent the possibility of personal injury. Remember that batteries contain a small amount of explosive hydrogen gas which is a by-product of battery charging. Sparks should always be avoided when working around batteries, especially when attaching jumper cables. To minimize the possibility of accidental sparks, follow the procedure carefully.

❄❄ CAUTION

NEVER hook the batteries up in a series circuit or the entire electrical system will go up in smoke, including the starter!

Vehicles equipped with a diesel engine may utilize two 12 volt batteries. If so, the batteries are connected in a parallel circuit (positive terminal-to-positive terminal, negative terminal-to-negative terminal). Hooking the batteries up in parallel circuit increases battery cranking power without increasing total battery voltage output. Output remains at 12 volts. On the other hand, hooking two 12 volt batteries up in a series circuit (positive terminal-to-negative terminal, positive terminal-to-negative terminal) increases total battery output to 24 volts (12 volts plus 12 volts).

Jump Starting Precautions

To avoid personal injury and/or vehicle damage, please read all of the following precautions prior to jump starting a discharged battery:

• NEVER hook the batteries up in a series circuit or the entire electrical system will go up in smoke, including the starter!
• Be sure that both batteries are of the same voltage. Vehicles covered by this manual and most vehicles on the road today utilize a 12 volt charging system.
• Be sure that both batteries are of the same polarity (have the same terminal, in most cases NEGATIVE grounded).
• Be sure that the vehicles are not touching, otherwise a short could occur.
• On serviceable batteries, be sure the vent cap holes are not obstructed.
• Do not smoke or allow sparks anywhere near the batteries.
• In cold weather, make sure the battery electrolyte is not frozen. This can occur more readily in a battery that has been in a state of discharge.
• Do not allow electrolyte to contact your skin or clothing.

Jump Starting Procedure

1. Make sure that the voltages of the two batteries are the same. Most batteries and charging systems are of the 12 volt variety.
2. Pull the vehicle with the good battery into a position so the jumper cables can reach the dead battery and that vehicle's engine compartment. Make sure that the vehicles DO NOT touch.

➡ Remote power terminals are usually provided on vehicles where the battery is located in the fender or other location that makes connecting jumper cables difficult. These power terminals are located in the engine compartment. If this is the situation, use the remote terminals instead of the terminals on the battery.

3. Place the transmissions/transaxles of both vehicles in Neutral, manual transmissions, or P (park), automatic transmissions, as applicable, then firmly set their parking brakes.

➡ If necessary for safety reasons, the hazard lights on both vehicles may be operated throughout the entire procedure without significantly increasing the difficulty of jumping the dead battery.

4. Turn all lights and accessories OFF on both vehicles. Be sure the ignition switches on both vehicles are turned to the **OFF** position.
5. Cover the battery cell caps with a rag, but do not cover the terminals.
6. Make sure the terminals on both batteries are clean and free of corrosion, otherwise proper electrical connection will be impeded. If necessary, clean the battery terminals before proceeding.
7. Identify the positive (+) and negative (-) terminals on both batteries.
8. Connect the first jumper cable to the positive (+) terminal of the dead battery, then attach the other end of that cable to the positive (+) terminal of the booster (good) battery.
9. Connect the clamp of the negative jumper cable to the negative (-) terminal on the good battery and the final cable clamp to an engine bolt head, alternator bracket or other solid, metallic point on the engine with the dead battery. Try to pick a ground on the engine that is positioned away from the battery in order to minimize the possibility of explosion due to the sparks created when the last connection is made. DO NOT connect this clamp to the negative terminal of the bad battery.

❄❄ WARNING

Be very careful to keep the jumper cables away from moving parts (cooling fan, belts, etc.) on both engines.

10. Ensure the cables are routed away from any moving parts, then start the donor vehicle's engine. Run the engine at moderate speed for several minutes to allow the dead battery a chance to receive some initial charge.
11. With the donor vehicle's engine still running slightly above idle, try to start the vehicle with the dead battery. Crank the engine for no more than 10 seconds at a time and let the starter cool for at least 20 seconds between tries. If the vehicle does not start in 3 tries, it is likely that something else is also wrong, or that the battery needs additional time to charge.
12. Once the vehicle is started, allow it to run at idle for a few seconds to make sure that it is operating properly.
13. Turn ON the headlights, heater blower and, if equipped, the rear defroster of both vehicles in order to reduce the severity of voltage spikes and subsequent risk of damage to the vehicles' electrical systems when the cables are disconnected. This step is especially important to any vehicle equipped with computer control modules.
14. Carefully remove the cables in the reverse order of connection. Start with the negative cable that is attached to the engine ground, then the negative cable on the donor battery. Disconnect the positive cable from the donor battery, and finally disconnect the positive cable from the formerly dead battery. Be careful when disconnecting the cables from the positive terminals not to allow the alligator clips to touch any metal on either vehicle or a short and sparks will occur.

Voltage Regulator

TESTING

➡**Most regulators are integral (built in) to the alternator or Powertrain Control Module (PCM). If the regulator is found to be defective on these models, the alternator or PCM should be replaced.**

For voltage regulator testing, refer to the Alternator Isolation test.

REMOVAL & INSTALLATION

➡**The following procedure is only for voltage regulators mounted on the back (outside) of the alternator or elsewhere in the engine compartment.**

1. Disconnect the negative battery cable.
2. If equipped, remove the exterior alternator cover to expose the regulator. Do not disassemble the alternator case that houses the rotor and stator.
3. If equipped, disengage the electrical connector from the regulator.
4. Remove the regulator mounting screws and remove the regulator.

To install:
5. Position the regulator in its original position and install the mounting screws.
6. Connect any wiring that was removed from the regulator.
7. If equipped, install the cover.
8. Connect the negative battery cable.

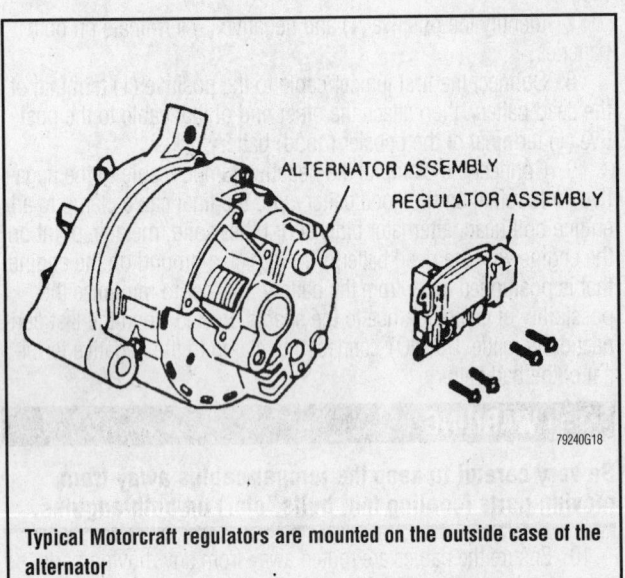

Typical Motorcraft regulators are mounted on the outside case of the alternator

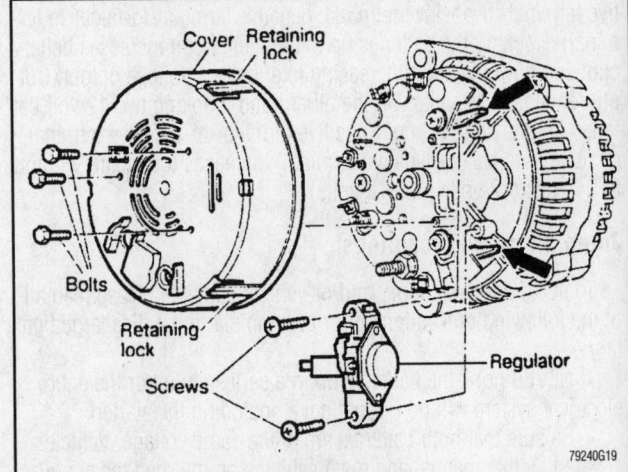

The regulator on a common Bosch alternator is mounted under a cover on the rear of the alternator

PISTON, PISTON RING & CONNECTING ROD POSITIONING

11

PISTON, PISTON RING AND CONNECTING ROD POSITIONING

When assembling the pistons, piston rings and connecting rods, and when installing these assemblies into the engine block, it is vitally important to ensure that these three components are properly positioned with respect to each other. Often times the engine block is designed so that if a connecting rod or piston is installed backwards, or in the wrong bank of cylinders, internal engine damage may occur once the engine is started. The piston ring end-gap spacing that is recommended by the engine manufacturer is often with the purpose of increased compression pressures during the engine break-in period. Failure to properly space the piston ring end-gaps may lead to increased oil consumption and extended break-in time. Therefore, always be sure to position the pistons, rings and connecting rods as shown in the accompanying illustrations.

✷✷ WARNING

Always be sure to matchmark the connecting rods and caps prior to disassembly so that they may be reassembled with their original counterparts. If the caps are not installed on their original connecting rods, the assemblies will most likely need machining to avoid bearing, connecting rod and/or crankshaft damage.

PISTON, PISTON RING AND CONNECTING ROD POSITIONING INDEX

MANUFACTURER ENGINE	DESCRIPTION	FIGURE
Acura		
All Engines	Piston Ring Positioning and Top Mark Locations	1
	Piston Ring End-Gap Spacing	2
	Piston-To-Engine Positioning	3
Chrysler		
All Engines	Bearing Cap and Connecting Rod Matchmarks	4
2.4L and 2.5L Engines	Piston Ring End-Gap Spacing	6
	Piston Ring Positioning	5
2.4L Engines	Piston-To-Engine Positioning Mark Locations	7
3.0L Engines	Piston-To-Engine Positioning Mark Locations	10
3.0L, 3.3L and 3.8L Engines	Piston and Connecting Rod Mark Locations	9
	Piston Ring End-Gap Spacing	8
	Piston-To-Engine Positioning Mark Locations	11
Dodge Truck		
Gasoline Engines	Piston Ring End-Gap Spacing	12
	Piston-To-Engine Positioning	13
Diesel Engines	Piston Ring Positioning	14
	Piston Ring End-Gap Spacing	15
	Oil Control Ring-To-Spacer End-Gap Spacing	16
	Piston and Connecting Rod Assembly Positioning	17
Ford		
All Engines	Piston Ring Positioning	18
	Piston Ring End-Gap Spacing	19
2.3L and 2.5L Engines	Piston and Connecting Rod Assembly Positioning	20
3.0L (VIN U) and 3.8L Engines	Piston and Connecting Rod Assembly Positioning	21
4.0L Engines	Piston and Connecting Rod Assembly Positioning	22
4.2L and 4.6L Engines	Piston and Connecting Rod Front Mark Locations	23
	Piston-To-Engine Orientation	24
4.9L Engines	Piston and Connecting Rod Assembly Positioning	25
5.0L, 5.4L, 5.8L and 6.8L Engines	Piston and Connecting Rod Assembly Positioning	26
7.3L Diesel Engines	Piston-To-Engine Orientation	27
7.5L Engines	Piston and Connecting Rod Assembly Positioning	28
	Piston-To-Engine Orientation	29
General Motors		
2.2L Engines	Piston Ring End-Gap Spacing	30
Except 2.2L Gasoline Engines	Piston Ring End-Gap Spacing	31
Diesel Engines	Piston Ring End-Gap Spacing	32
2.2L, 3.1L, 3.4L and 3.8L Engines	Piston/Connecting Rod-To-Engine Positioning	33
4.3L, 5.0L and 5.7L Engines	Piston and Connecting Rod Assembly Positioning	34
7.4L Engines	Piston and Connecting Rod Assembly Positioning	35
Geo/Chevrolet		
1.6L Engines	Piston Ring End-Gap Spacing	71
	Compression Ring Identification Mark Locations	72
	Piston and Connecting Rod Assembly Positioning	73
	Piston Installation	74
Honda		
All Engines	Piston Ring Positioning and Top Mark Locations	36

7924AC01

PISTON, PISTON RING AND CONNECTING ROD POSITIONING INDEX

MANUFACTURER ENGINE	DESCRIPTION	FIGURE
Honda (cont.)		
2.0L and 2.2L (F22B6) Engines	Piston Ring End-Gap Spacing	37
	Piston-To-Connecting Rod Assembly	38
	Compression Ring Identification	39
	Piston-To-Connecting Rod Assembly	40
2.2L (VIN W) Engines	Piston Ring End-Gap Spacing	41
2.6L and 3.2L Engines	Piston Ring Positioning	42
	Piston-To-Connecting Rod Assembly	43
	Piston Ring End-Gap Spacing	44
	Piston-To-Engine Positioning	45
Isuzu		
All Engines	Piston Ring Positioning and Top Mark Locations	36
2.0L Engines	Piston Ring End-Gap Spacing	37
	Piston-To-Connecting Rod Assembly	38
2.2L (F22B6) Engines	Piston Ring End-Gap Spacing	37
	Compression Ring Identification	39
	Piston-To-Connecting Rod Assembly	40
2.2L (VIN W) Engines	Piston Ring End-Gap Spacing	41
2.2L (VIN 4) Engines	Piston Ring End-Gap Spacing	30
	Piston/Connecting Rod-To-Engine Positioning	33
2.6L and 3.2L Engines	Piston Ring Positioning	42
	Piston-To-Connecting Rod Assembly	43
	Piston Ring End-Gap Spacing	44
	Piston-To-Engine Positioning	45
Jeep		
All Engines	Piston-To-Engine Positioning	48
4.0L Engines	Piston Ring End-Gap Spacing	46
Except 4.0L Engines	Piston Ring End-Gap Spacing	47
Kia		
2.0L Engines	Compression Ring Positioning Mark Locations	49
	Oil Control Ring Rail and Spacer Positioning	50
	Piston Ring End-Gap Spacing	51
Land Rover		
All Engines	Piston Ring Positioning	52
	Piston Ring End-Gap Spacing	53
	Connecting Rod Front Mark Location	54
Lexus		
4.5L Engines	Piston Ring End-Gap Spacing	55
	Piston/Connecting Rod-To-Engine Positioning	56
Mazda		
Except MPV Engines	Piston Ring Positioning	18
	Piston Ring End-Gap Spacing	19
2.3L and 2.5L Engines	Piston and Connecting Rod Assembly Positioning	20
3.0L (Except MPV) and 3.8L Engines	Piston and Connecting Rod Assembly Positioning	21
3.0L (MPV) Engines	Compression Ring Positioning	57
	Oil Control Ring Positioning	58
	Piston Ring End-Gap Spacing	59
4.0L Engines	Piston and Connecting Rod Assembly Positioning	22

7924AC02

PISTON, PISTON RING AND CONNECTING ROD POSITIONING INDEX

7924AC03

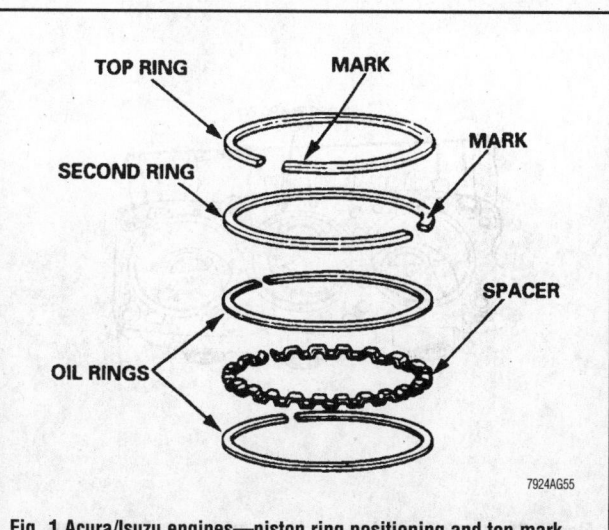

Fig. 1 Acura/Isuzu engines—piston ring positioning and top mark locations

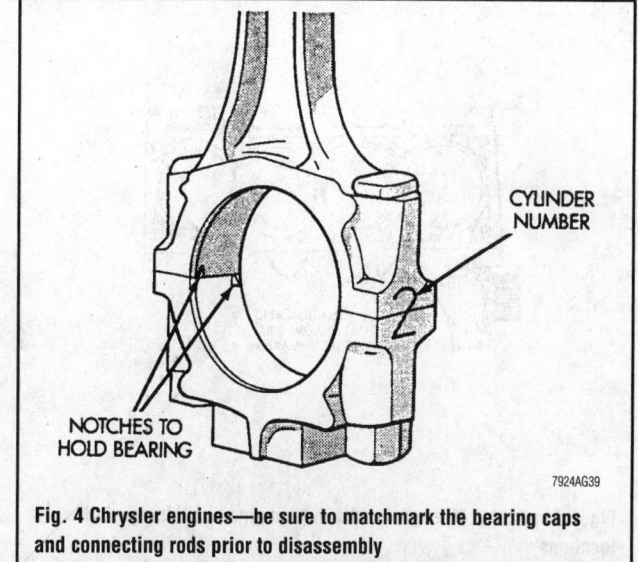

Fig. 4 Chrysler engines—be sure to matchmark the bearing caps and connecting rods prior to disassembly

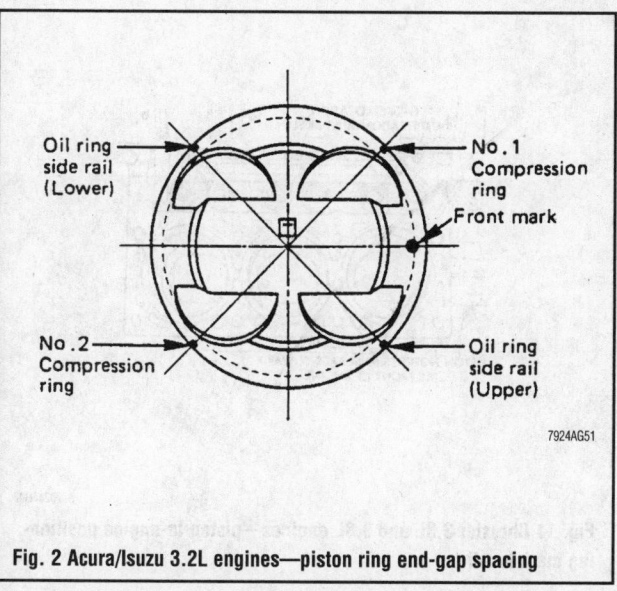

Fig. 2 Acura/Isuzu 3.2L engines—piston ring end-gap spacing

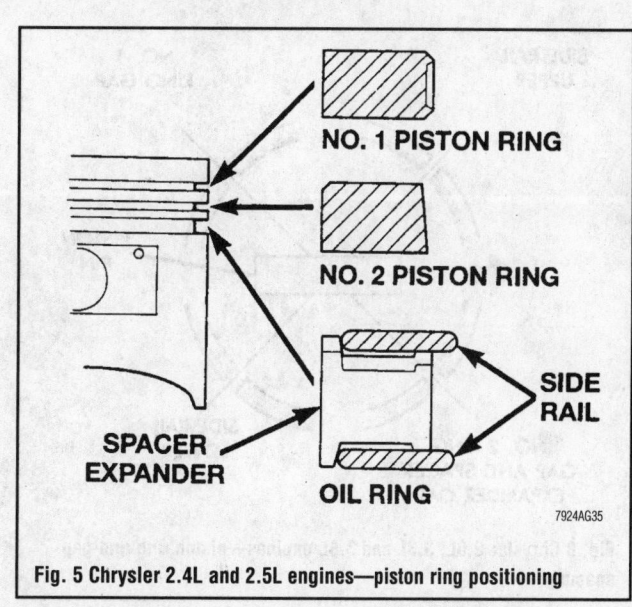

Fig. 5 Chrysler 2.4L and 2.5L engines—piston ring positioning

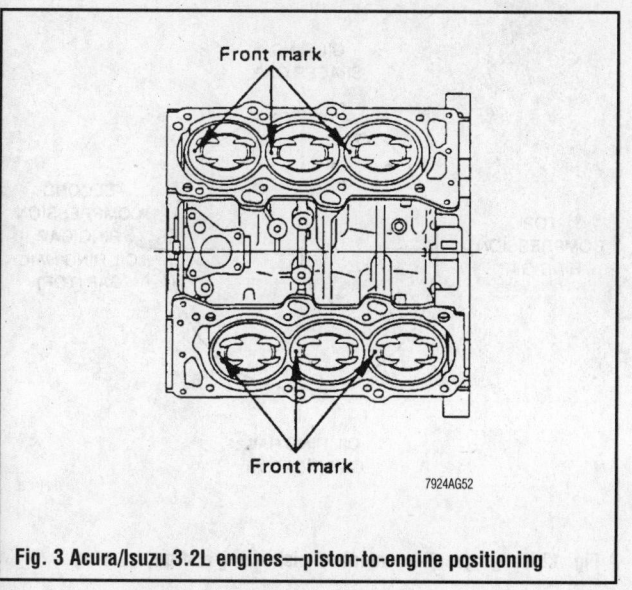

Fig. 3 Acura/Isuzu 3.2L engines—piston-to-engine positioning

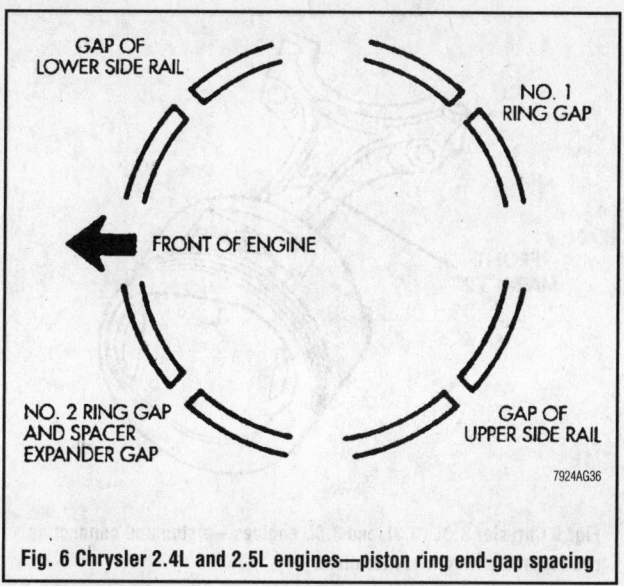

Fig. 6 Chrysler 2.4L and 2.5L engines—piston ring end-gap spacing

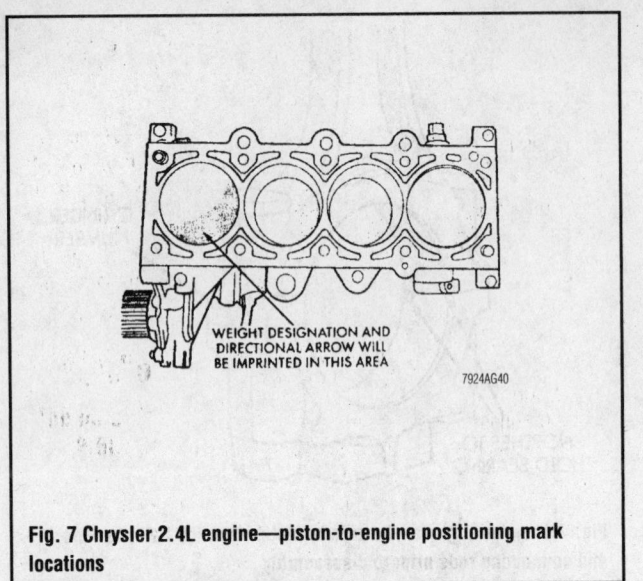

Fig. 7 Chrysler 2.4L engine—piston-to-engine positioning mark locations

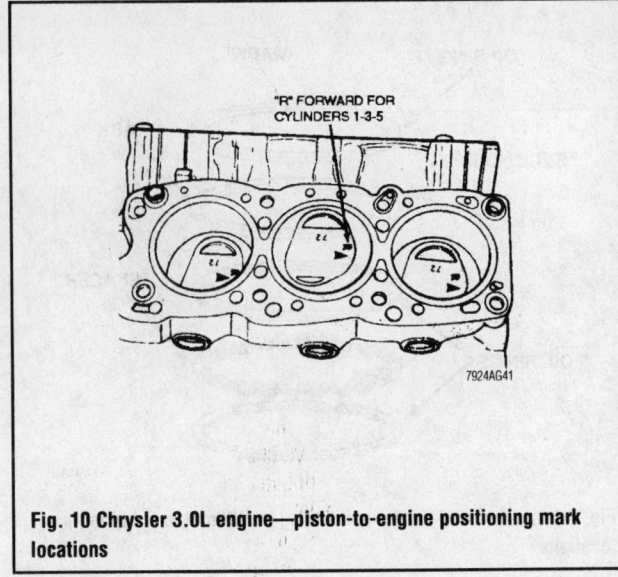

Fig. 10 Chrysler 3.0L engine—piston-to-engine positioning mark locations

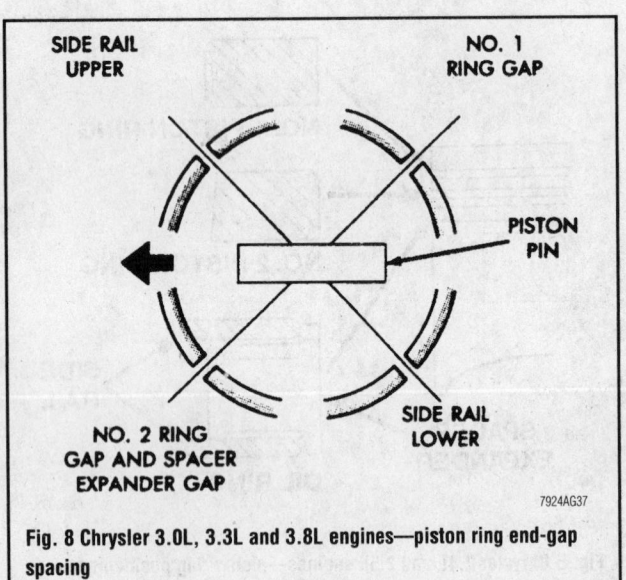

Fig. 8 Chrysler 3.0L, 3.3L and 3.8L engines—piston ring end-gap spacing

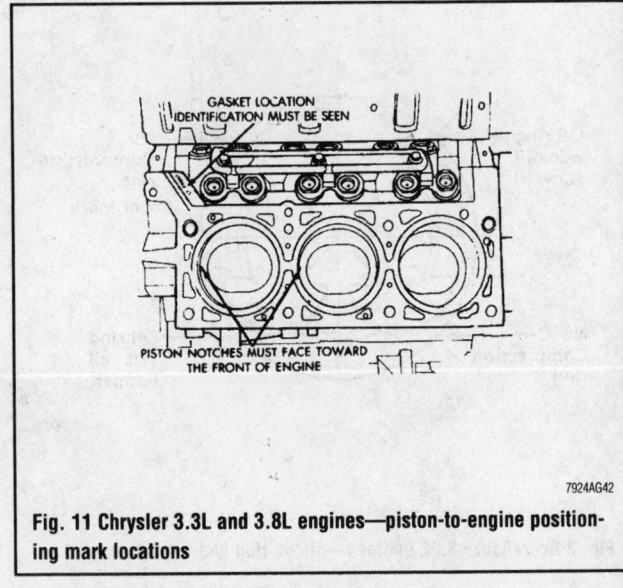

Fig. 11 Chrysler 3.3L and 3.8L engines—piston-to-engine positioning mark locations

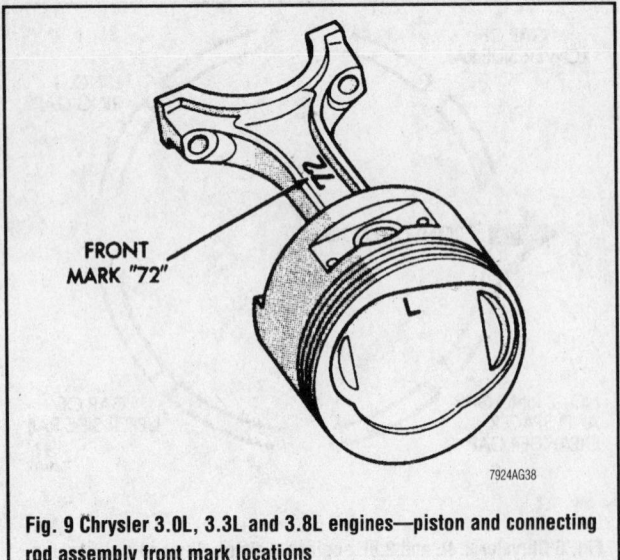

Fig. 9 Chrysler 3.0L, 3.3L and 3.8L engines—piston and connecting rod assembly front mark locations

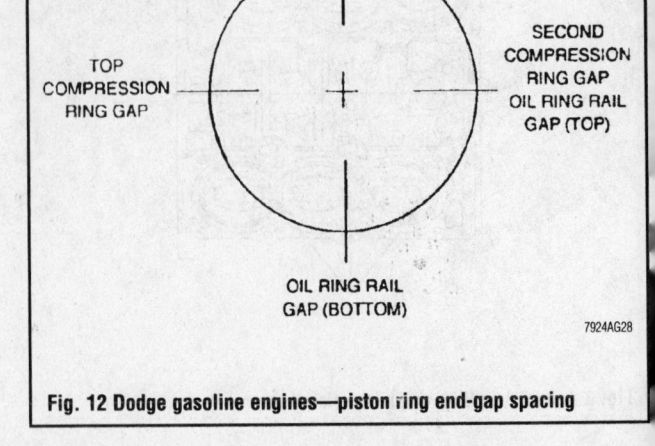

Fig. 12 Dodge gasoline engines—piston ring end-gap spacing

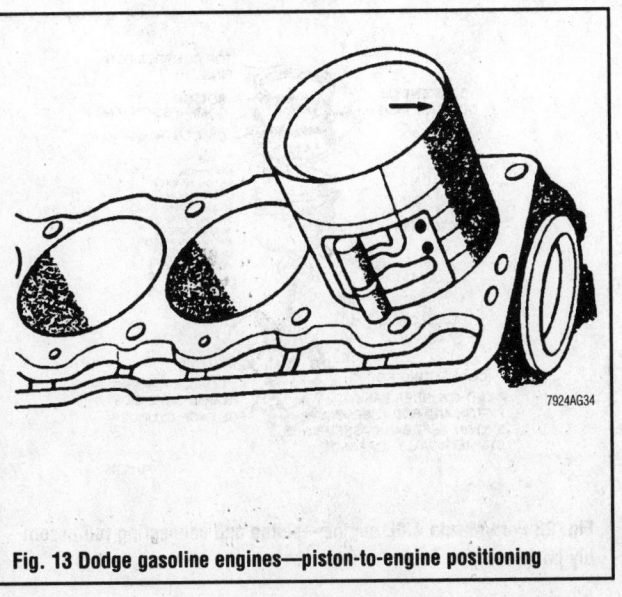

Fig. 13 Dodge gasoline engines—piston-to-engine positioning

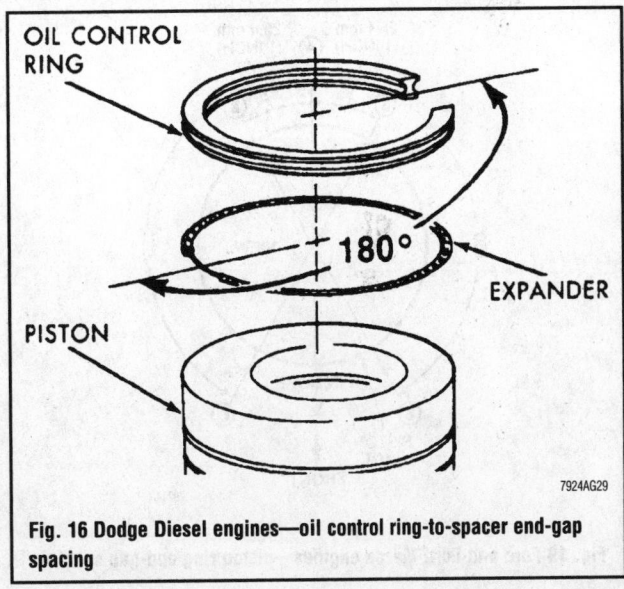

Fig. 16 Dodge Diesel engines—oil control ring-to-spacer end-gap spacing

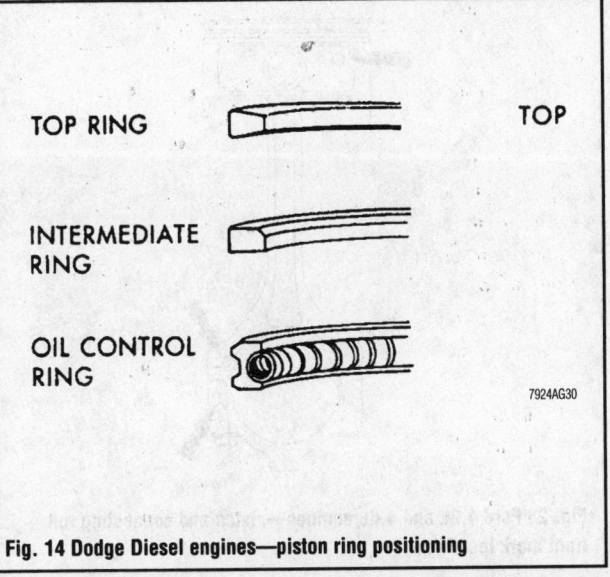

Fig. 14 Dodge Diesel engines—piston ring positioning

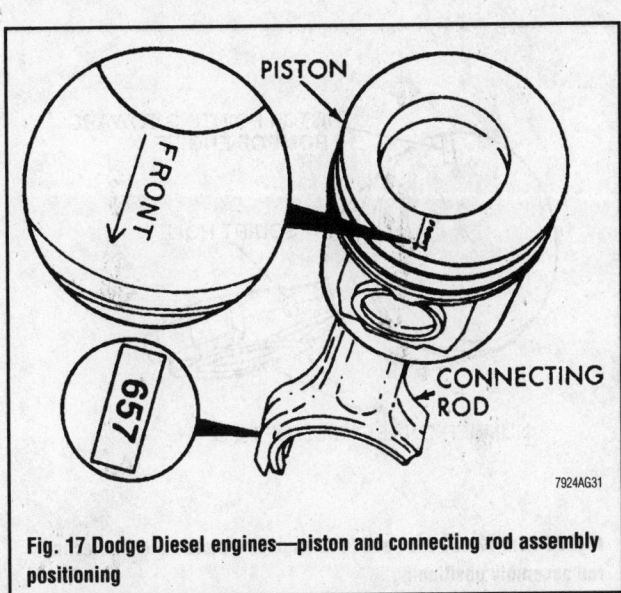

Fig. 17 Dodge Diesel engines—piston and connecting rod assembly positioning

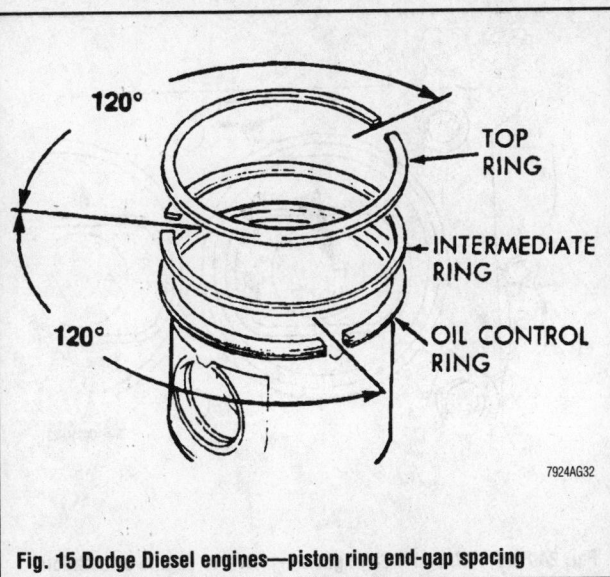

Fig. 15 Dodge Diesel engines—piston ring end-gap spacing

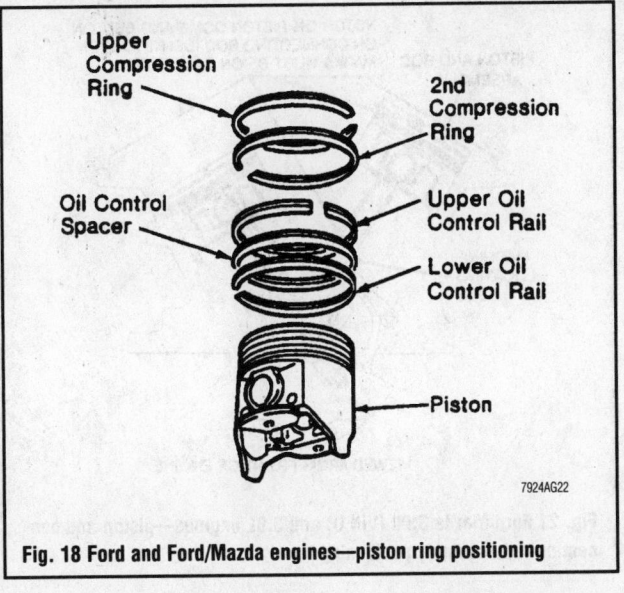

Fig. 18 Ford and Ford/Mazda engines—piston ring positioning

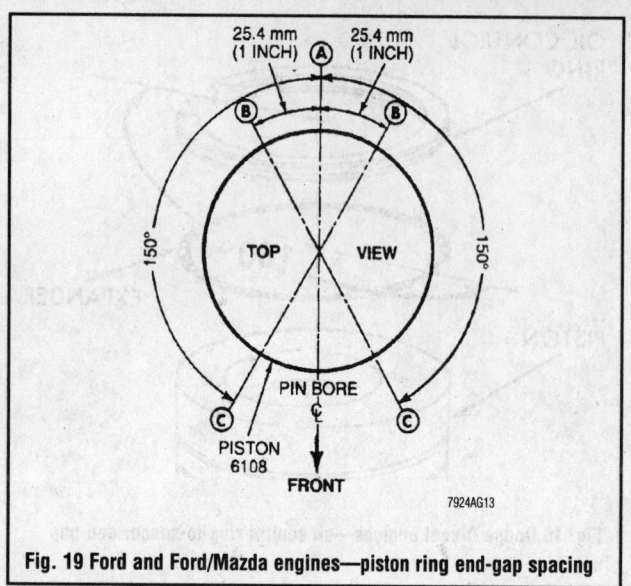

Fig. 19 Ford and Ford/Mazda engines—piston ring end-gap spacing

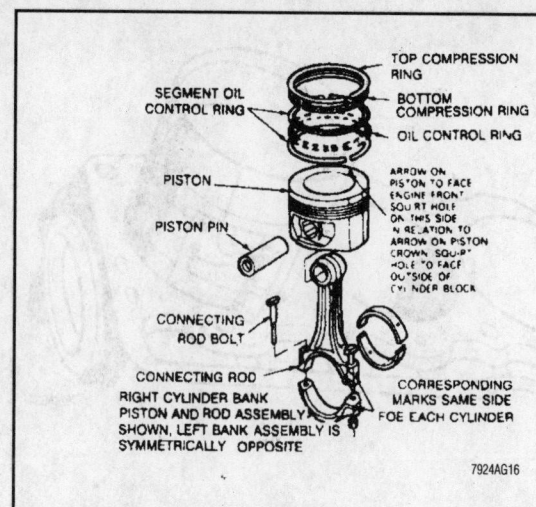

Fig. 22 Ford/Mazda 4.0L engine—piston and connecting rod assembly positioning

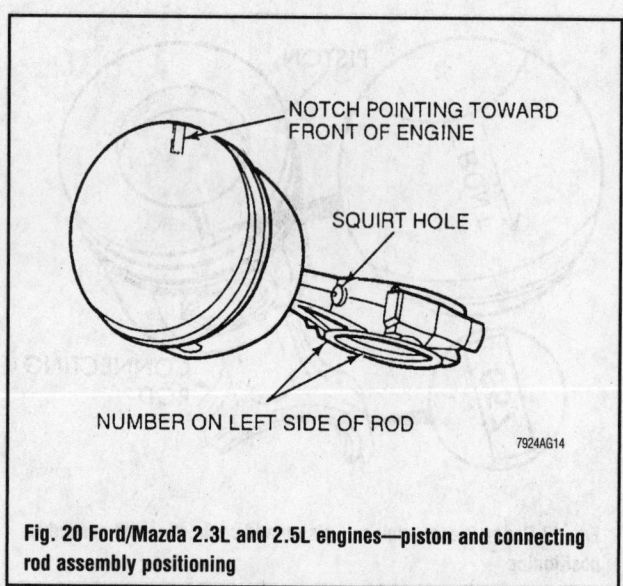

Fig. 20 Ford/Mazda 2.3L and 2.5L engines—piston and connecting rod assembly positioning

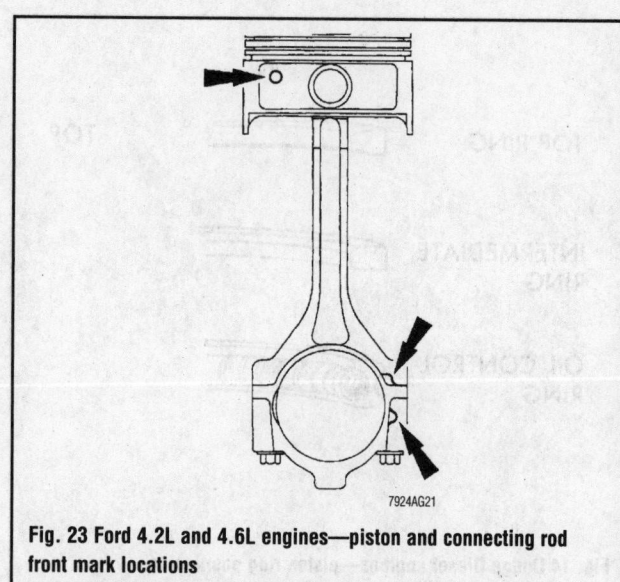

Fig. 23 Ford 4.2L and 4.6L engines—piston and connecting rod front mark locations

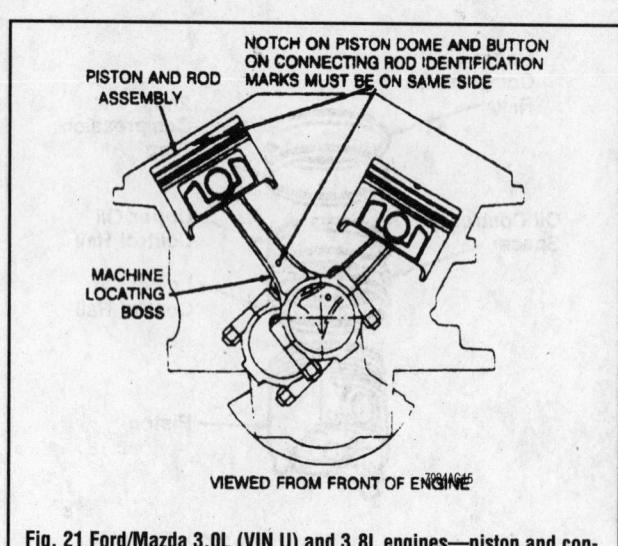

Fig. 21 Ford/Mazda 3.0L (VIN U) and 3.8L engines—piston and connecting rod assembly positioning

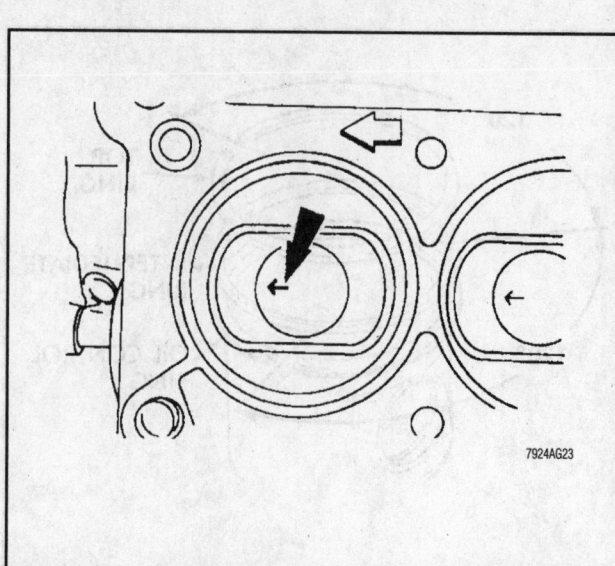

Fig. 24 Ford 4.2L and 4.6L engines—piston-to-engine orientation

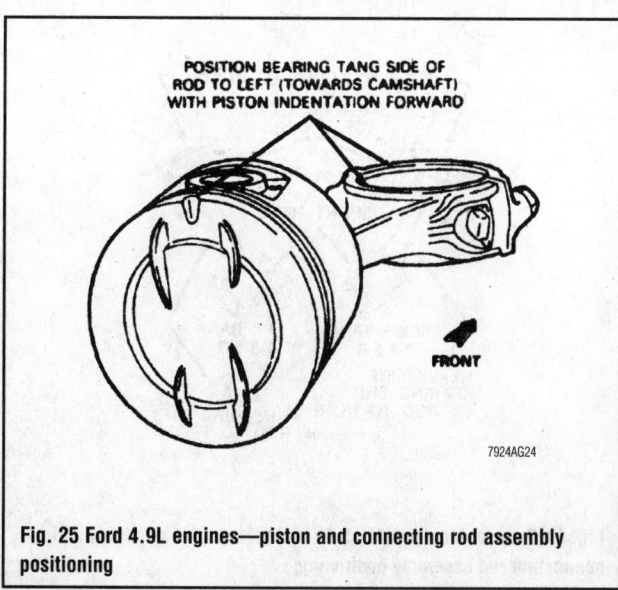

Fig. 25 Ford 4.9L engines—piston and connecting rod assembly positioning

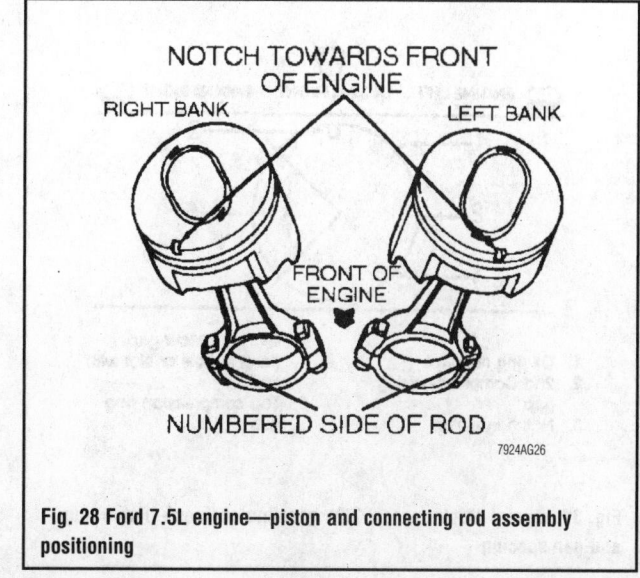

Fig. 28 Ford 7.5L engine—piston and connecting rod assembly positioning

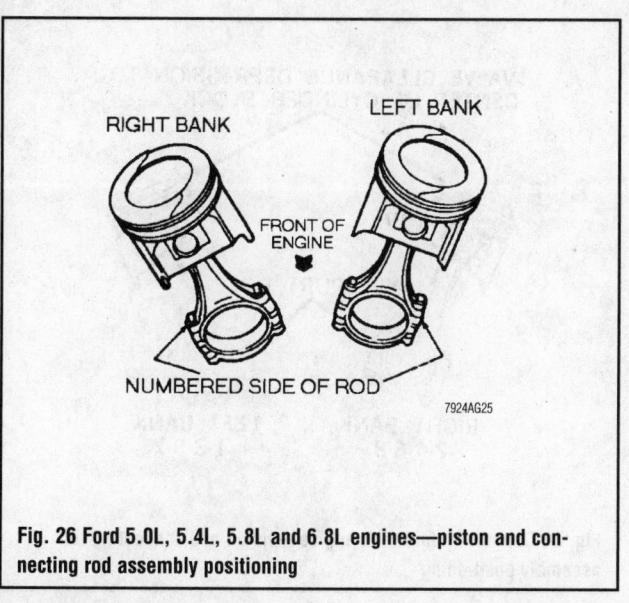

Fig. 26 Ford 5.0L, 5.4L, 5.8L and 6.8L engines—piston and connecting rod assembly positioning

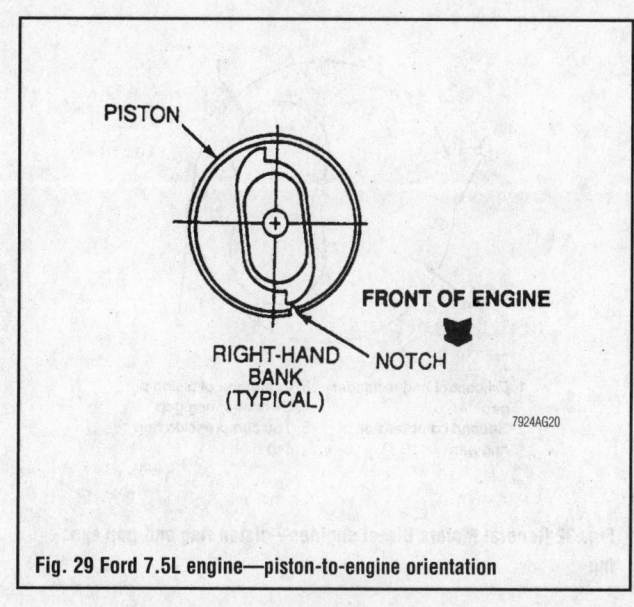

Fig. 29 Ford 7.5L engine—piston-to-engine orientation

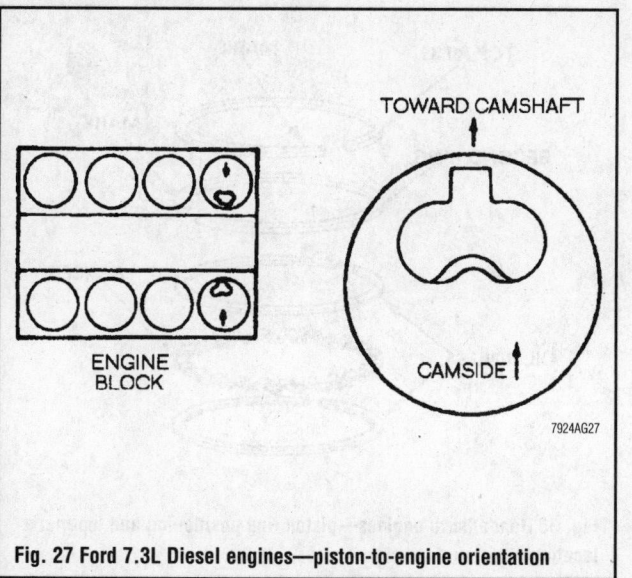

Fig. 27 Ford 7.3L Diesel engines—piston-to-engine orientation

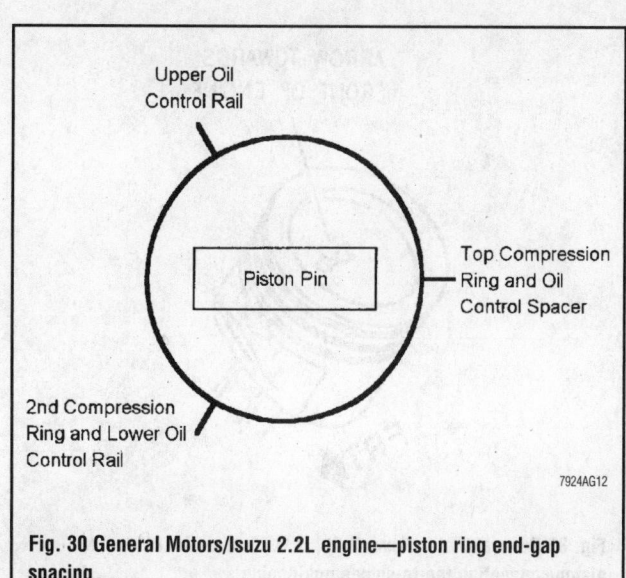

Fig. 30 General Motors/Isuzu 2.2L engine—piston ring end-gap spacing

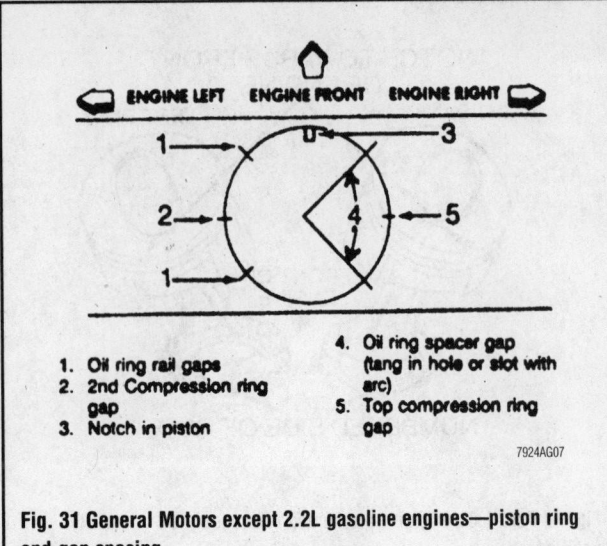

1. Oil ring rail gaps
2. 2nd Compression ring gap
3. Notch in piston
4. Oil ring spacer gap (tang in hole or slot with arc)
5. Top compression ring gap

7924AG07

Fig. 31 General Motors except 2.2L gasoline engines—piston ring end-gap spacing

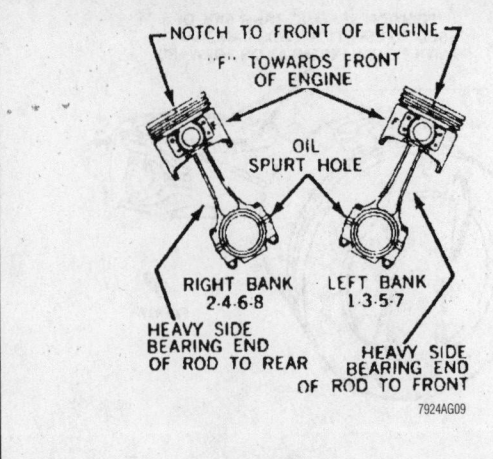

7924AG09

Fig. 34 General Motors 4.3L, 5.0L and 5.7L engines—piston and connecting rod assembly positioning

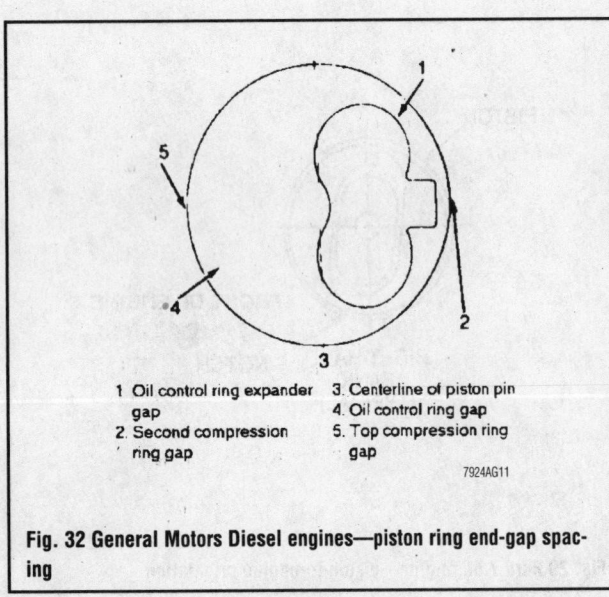

1 Oil control ring expander gap
2 Second compression ring gap
3 Centerline of piston pin
4 Oil control ring gap
5 Top compression ring gap

7924AG11

Fig. 32 General Motors Diesel engines—piston ring end-gap spacing

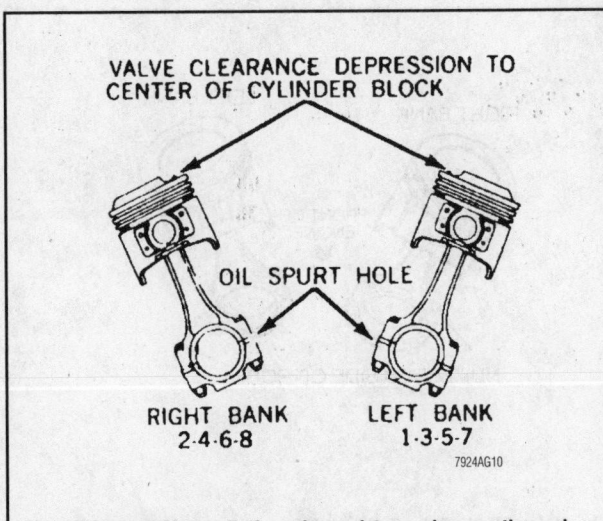

7924AG10

Fig. 35 General Motors 7.4L engine—piston and connecting rod assembly positioning

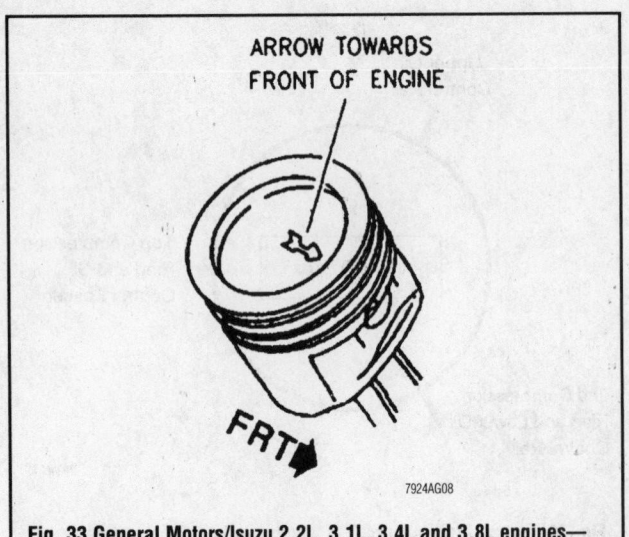

7924AG08

Fig. 33 General Motors/Isuzu 2.2L, 3.1L, 3.4L and 3.8L engines—piston/connecting rod-to-engine positioning

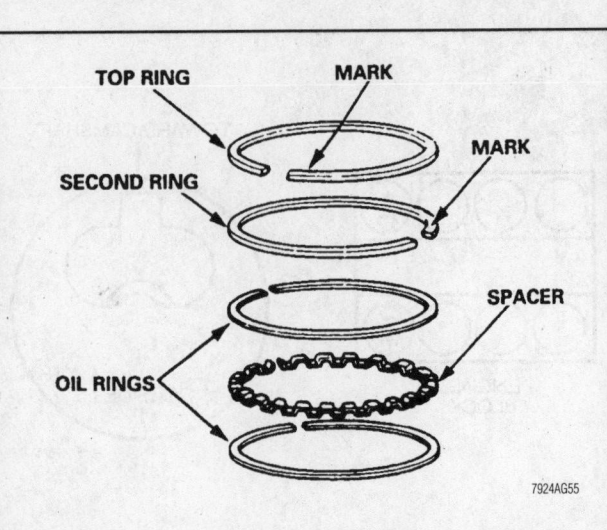

7924AG55

Fig. 36 Honda/Isuzu engines—piston ring positioning and top mark locations

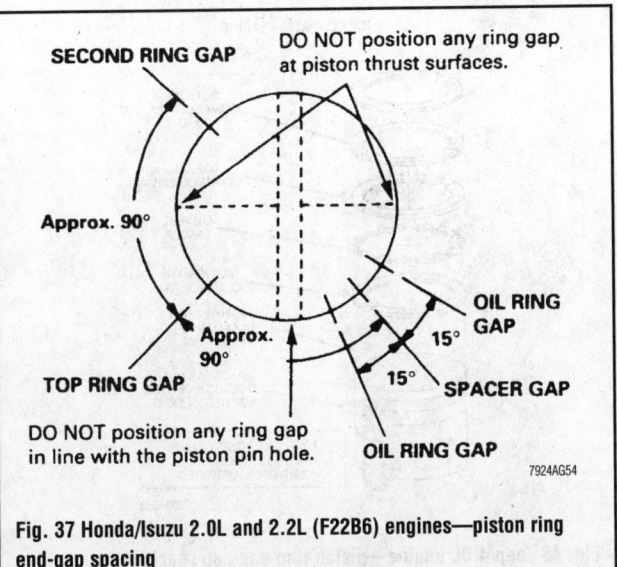

Fig. 37 Honda/Isuzu 2.0L and 2.2L (F22B6) engines—piston ring end-gap spacing

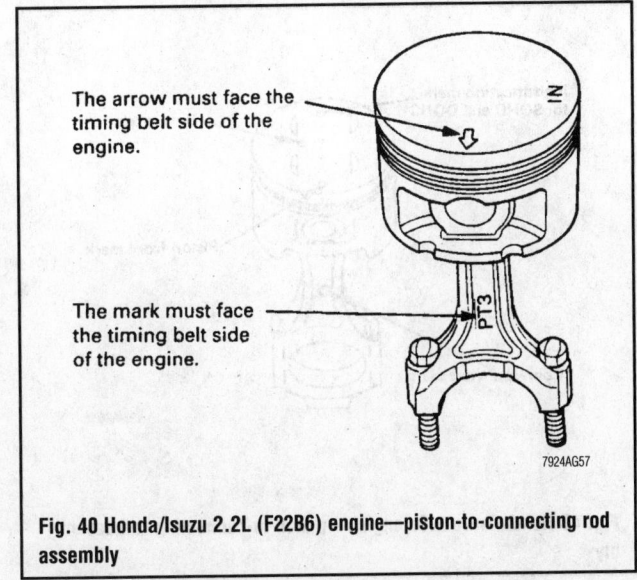

Fig. 40 Honda/Isuzu 2.2L (F22B6) engine—piston-to-connecting rod assembly

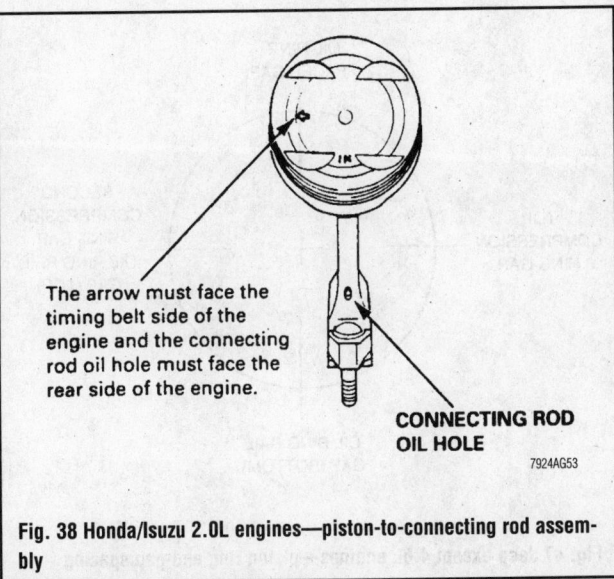

Fig. 38 Honda/Isuzu 2.0L engines—piston-to-connecting rod assembly

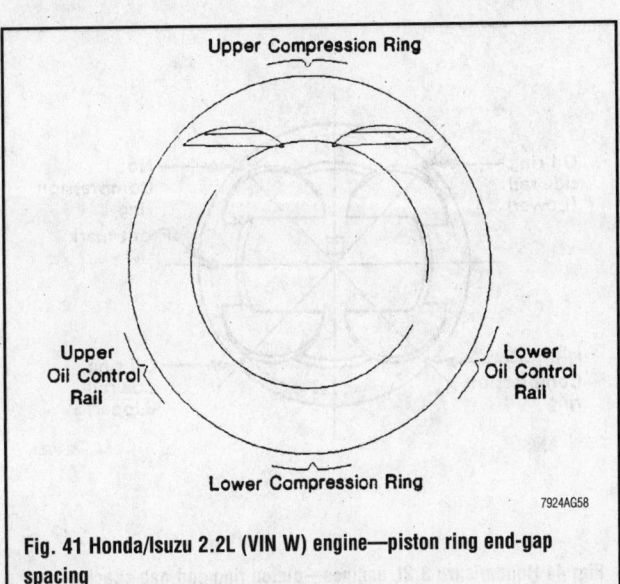

Fig. 41 Honda/Isuzu 2.2L (VIN W) engine—piston ring end-gap spacing

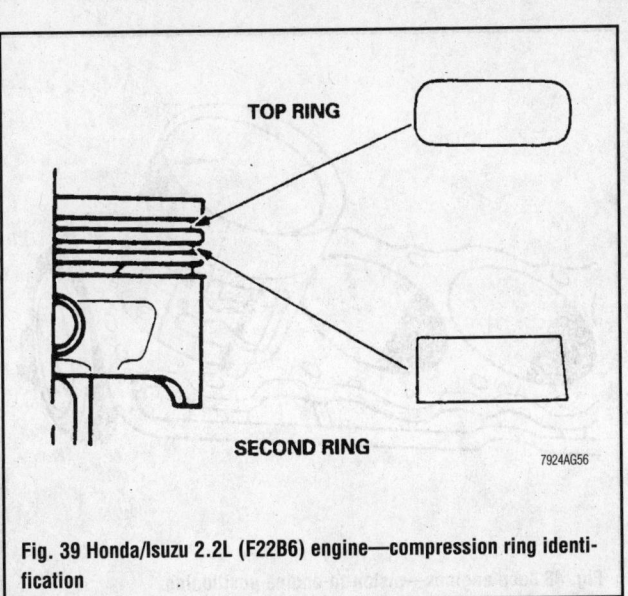

Fig. 39 Honda/Isuzu 2.2L (F22B6) engine—compression ring identification

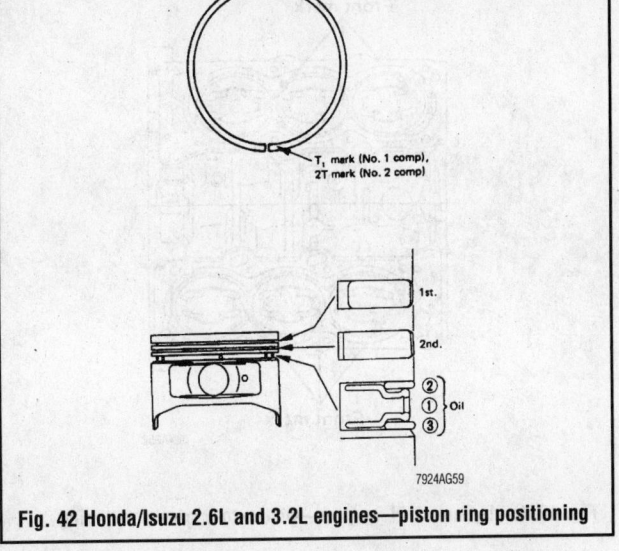

Fig. 42 Honda/Isuzu 2.6L and 3.2L engines—piston ring positioning

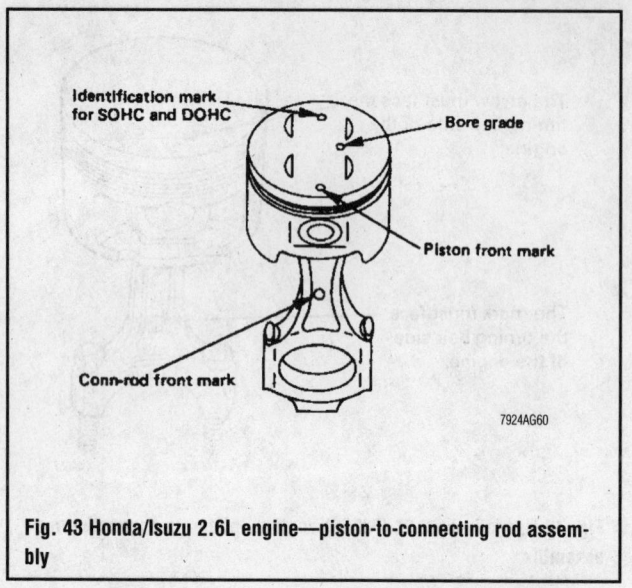

Fig. 43 Honda/Isuzu 2.6L engine—piston-to-connecting rod assembly

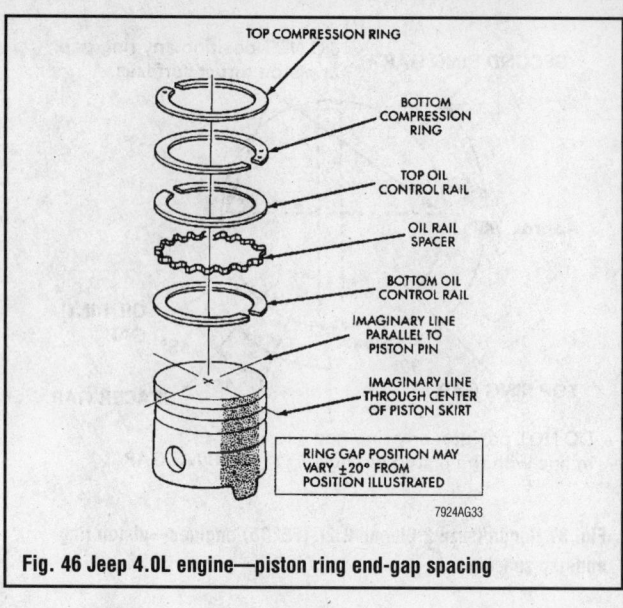

Fig. 46 Jeep 4.0L engine—piston ring end-gap spacing

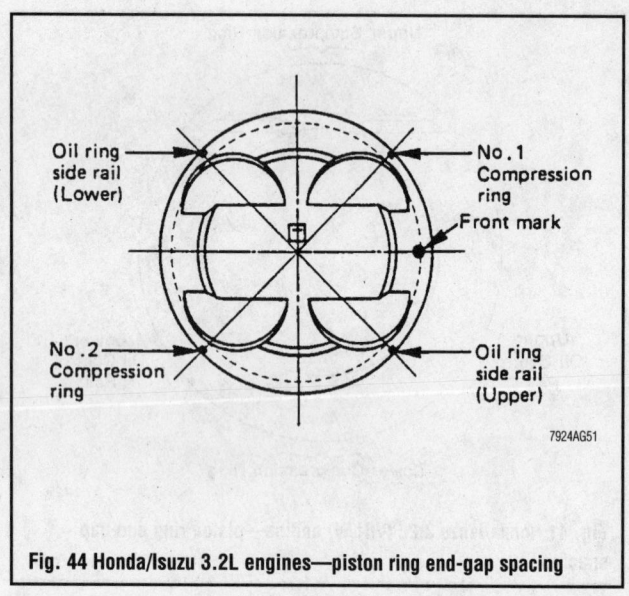

Fig. 44 Honda/Isuzu 3.2L engines—piston ring end-gap spacing

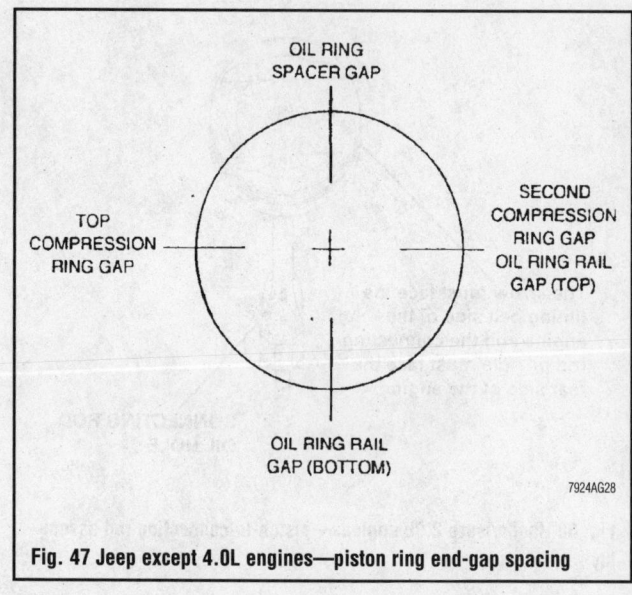

Fig. 47 Jeep except 4.0L engines—piston ring end-gap spacing

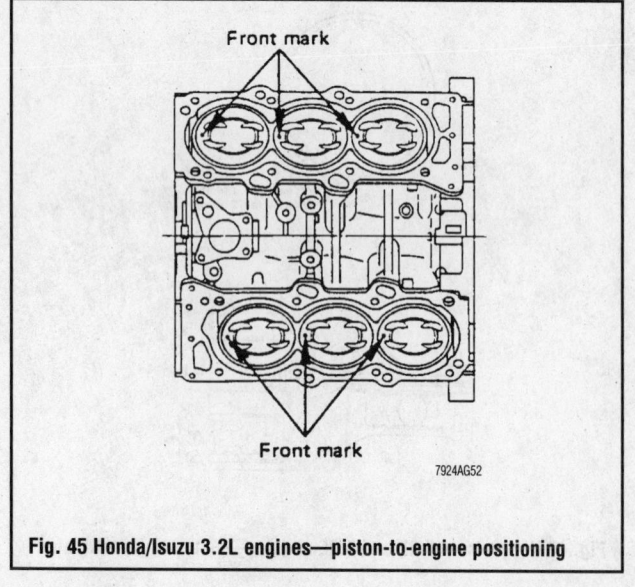

Fig. 45 Honda/Isuzu 3.2L engines—piston-to-engine positioning

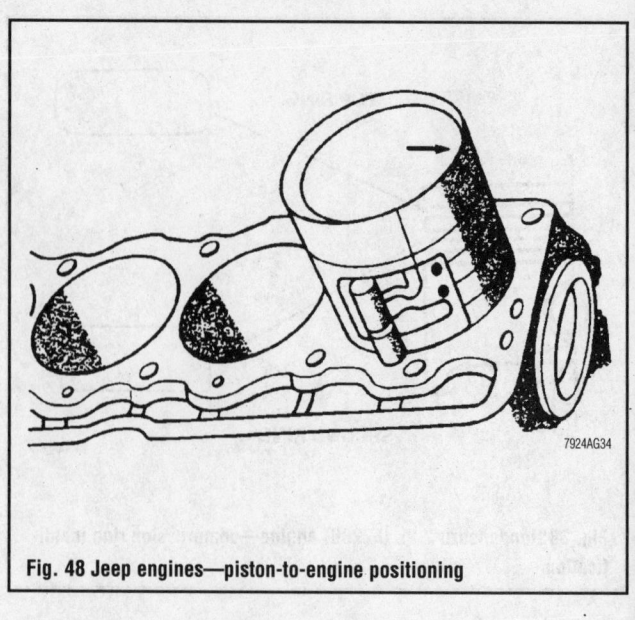

Fig. 48 Jeep engines—piston-to-engine positioning

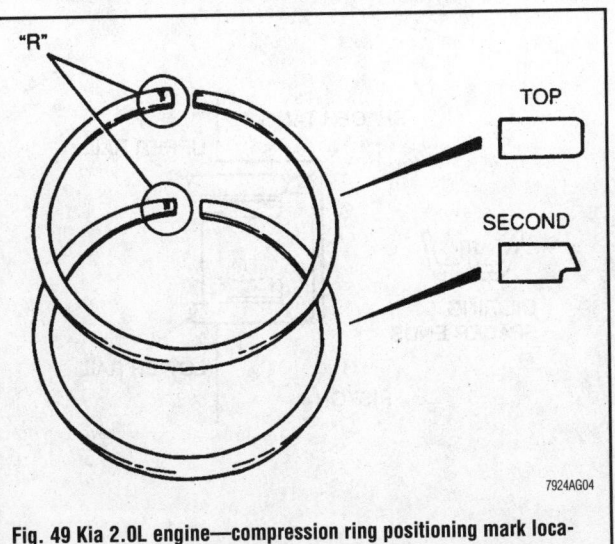

Fig. 49 Kia 2.0L engine—compression ring positioning mark locations

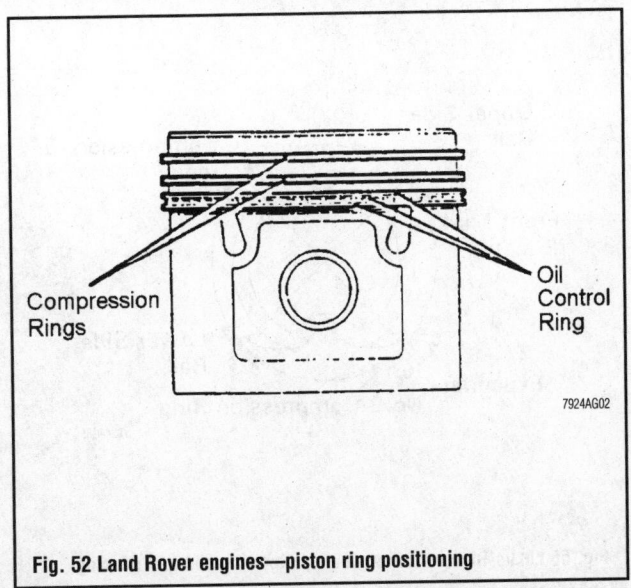

Fig. 52 Land Rover engines—piston ring positioning

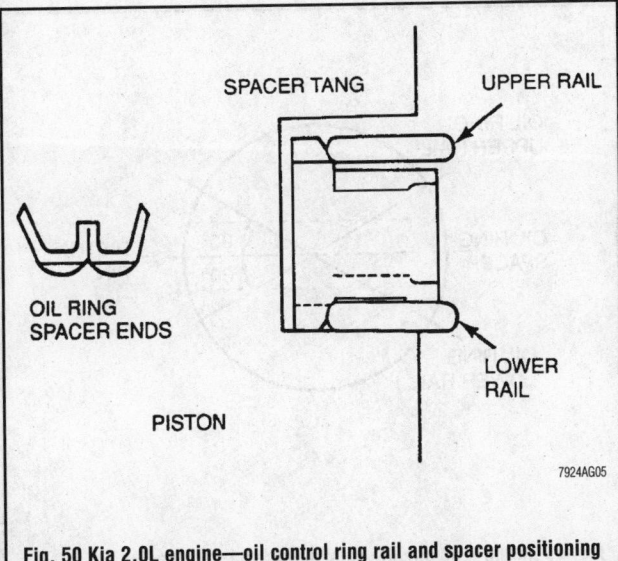

Fig. 50 Kia 2.0L engine—oil control ring rail and spacer positioning

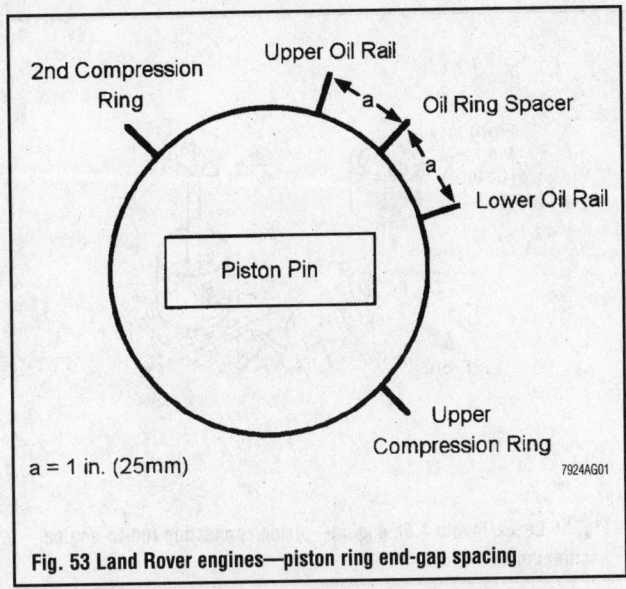

Fig. 53 Land Rover engines—piston ring end-gap spacing

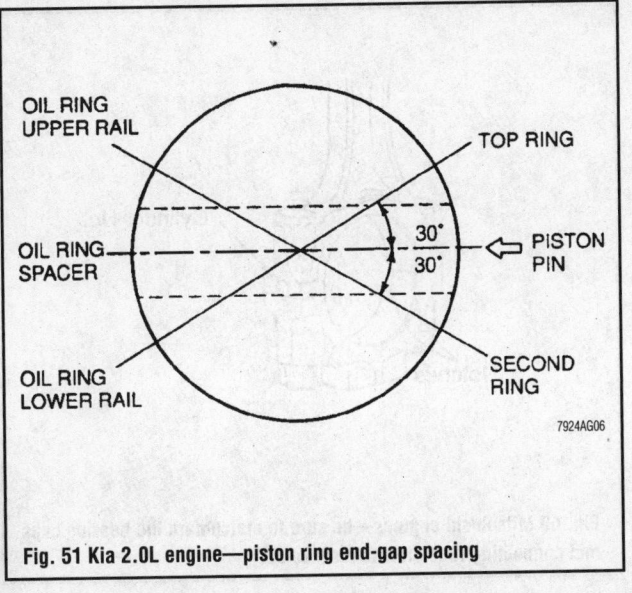

Fig. 51 Kia 2.0L engine—piston ring end-gap spacing

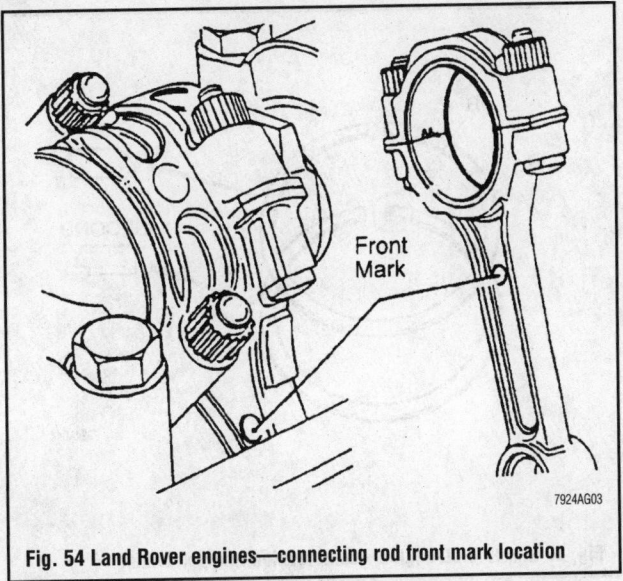

Fig. 54 Land Rover engines—connecting rod front mark location

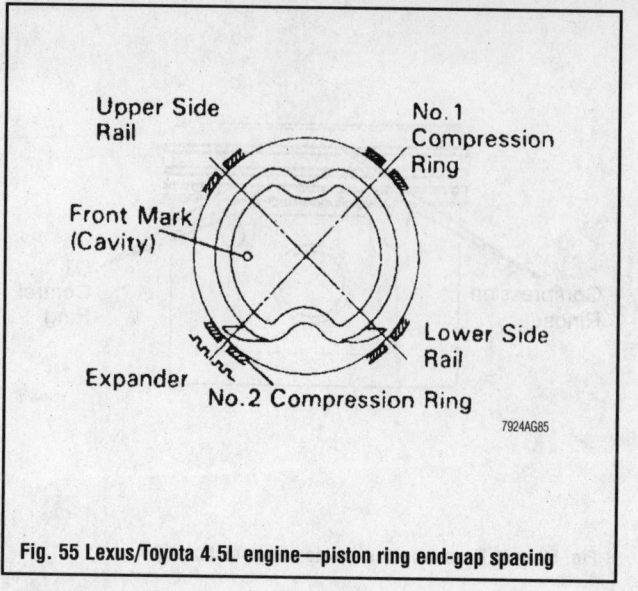

Fig. 55 Lexus/Toyota 4.5L engine—piston ring end-gap spacing

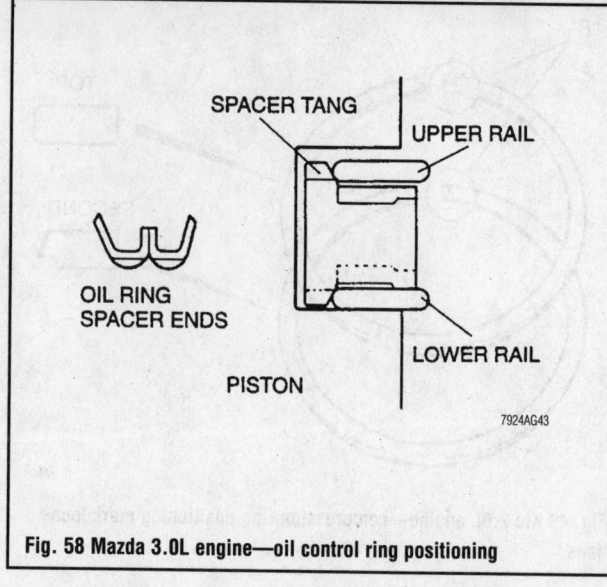

Fig. 58 Mazda 3.0L engine—oil control ring positioning

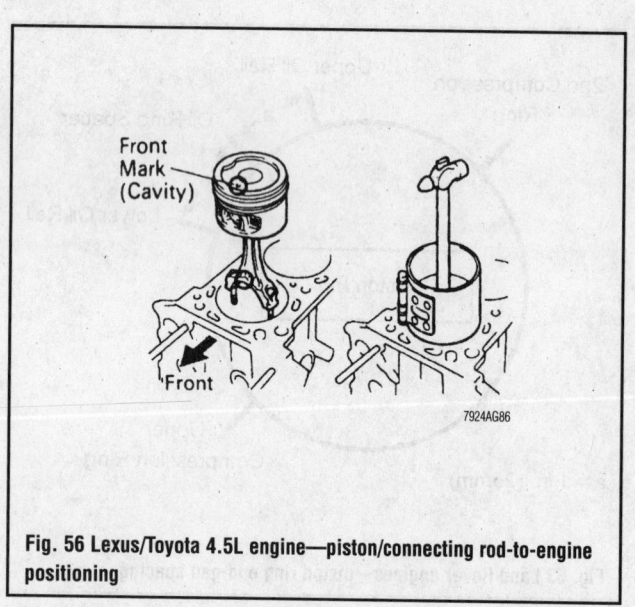

Fig. 56 Lexus/Toyota 4.5L engine—piston/connecting rod-to-engine positioning

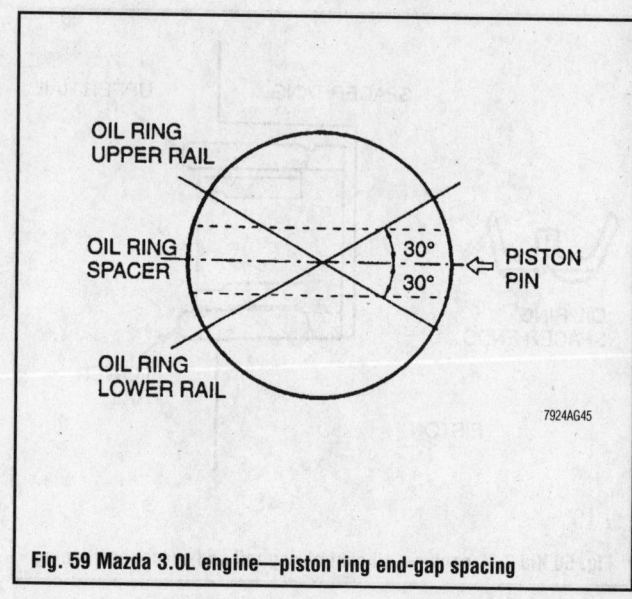

Fig. 59 Mazda 3.0L engine—piston ring end-gap spacing

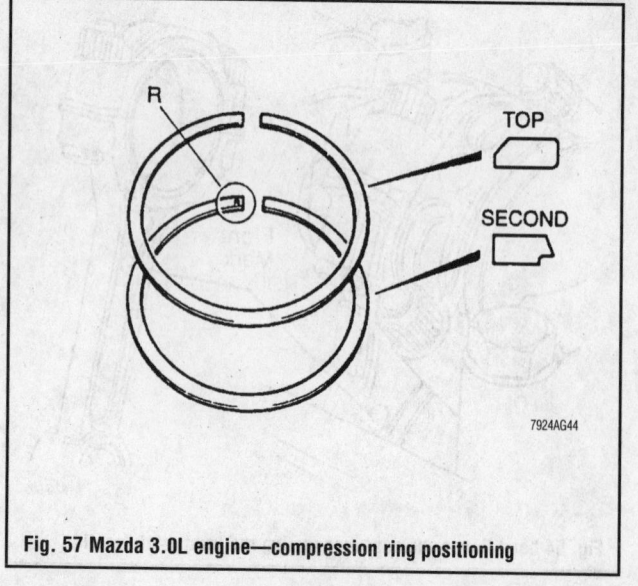

Fig. 57 Mazda 3.0L engine—compression ring positioning

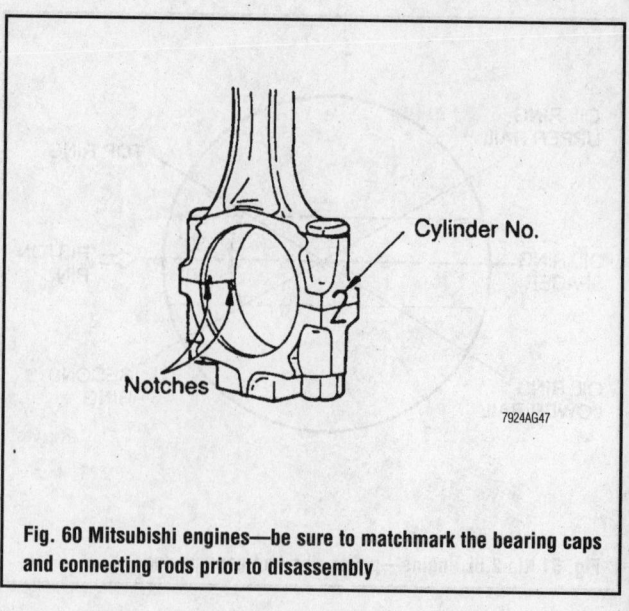

Fig. 60 Mitsubishi engines—be sure to matchmark the bearing caps and connecting rods prior to disassembly

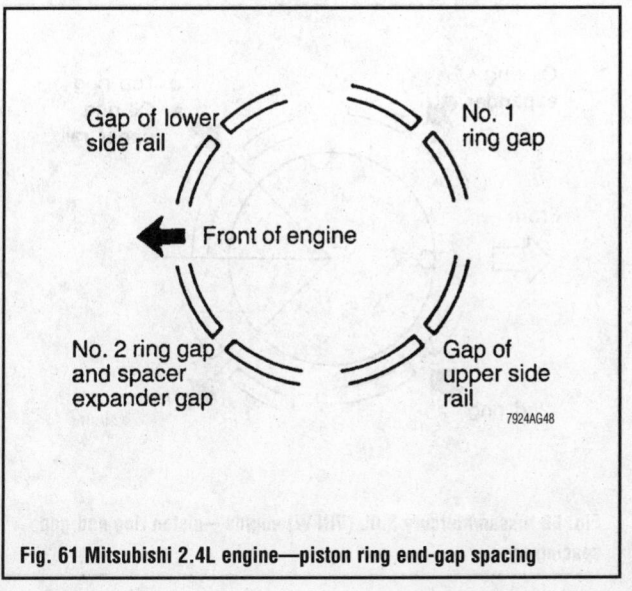

Fig. 61 Mitsubishi 2.4L engine—piston ring end-gap spacing

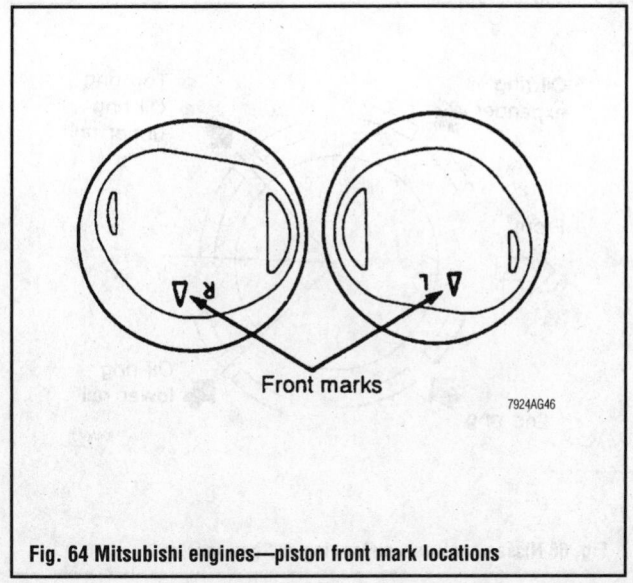

Fig. 64 Mitsubishi engines—piston front mark locations

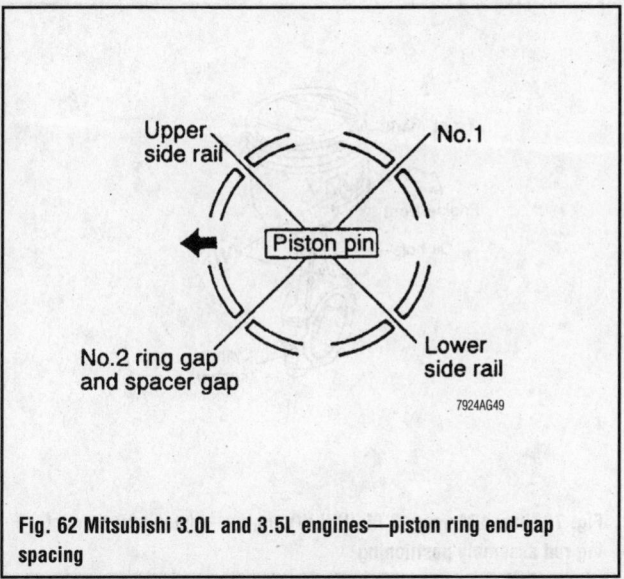

Fig. 62 Mitsubishi 3.0L and 3.5L engines—piston ring end-gap spacing

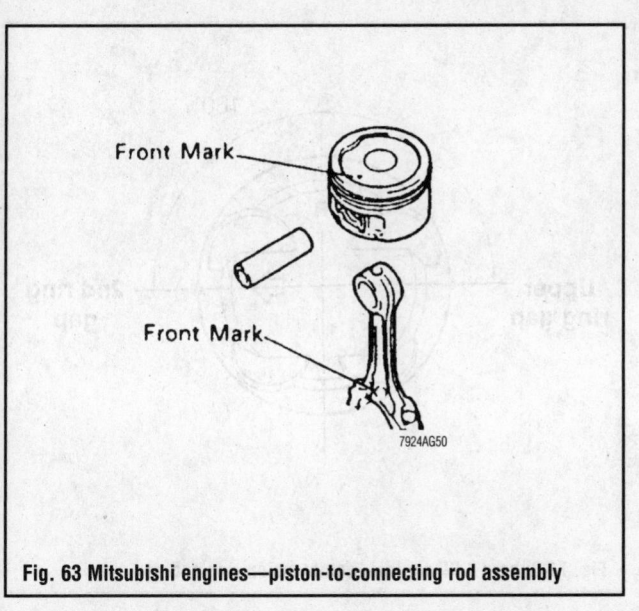

Fig. 63 Mitsubishi engines—piston-to-connecting rod assembly

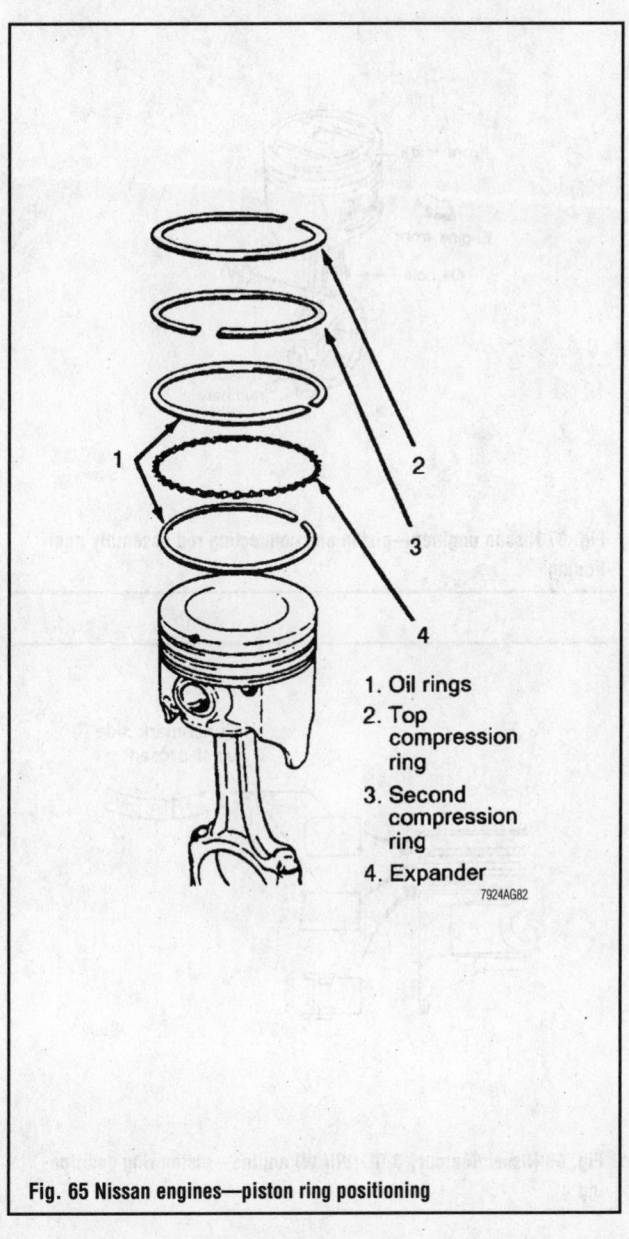

1. Oil rings
2. Top compression ring
3. Second compression ring
4. Expander

Fig. 65 Nissan engines—piston ring positioning

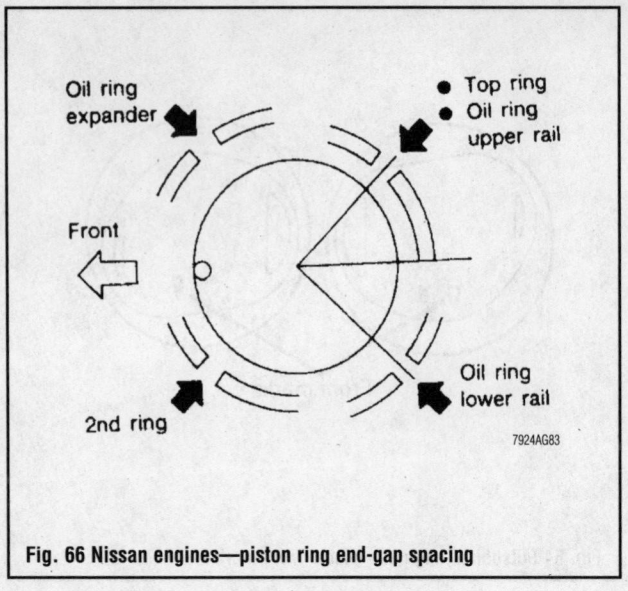

Fig. 66 Nissan engines—piston ring end-gap spacing

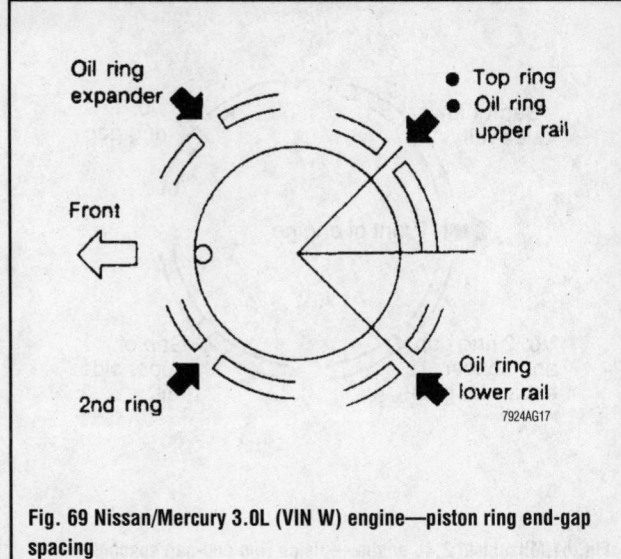

Fig. 69 Nissan/Mercury 3.0L (VIN W) engine—piston ring end-gap spacing

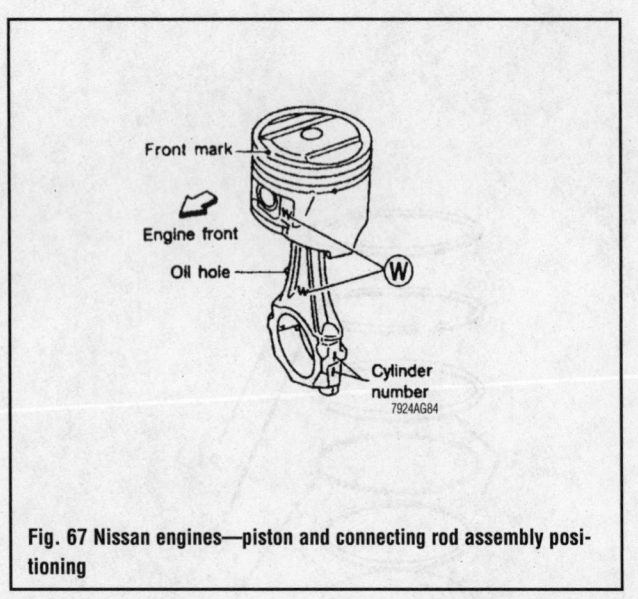

Fig. 67 Nissan engines—piston and connecting rod assembly positioning

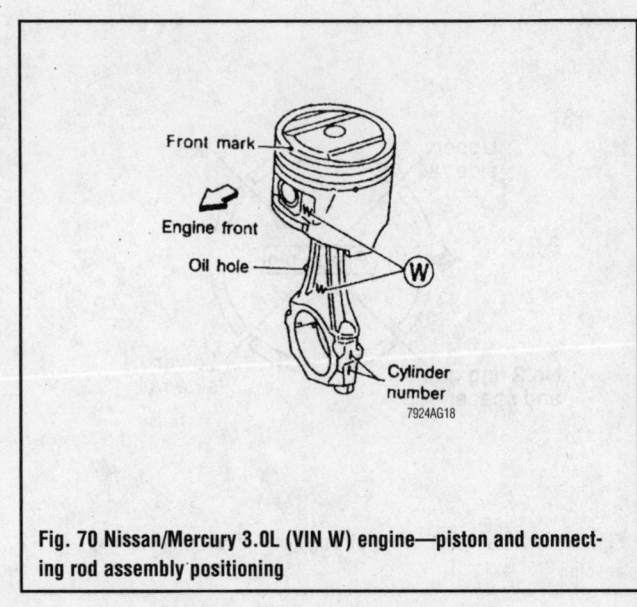

Fig. 70 Nissan/Mercury 3.0L (VIN W) engine—piston and connecting rod assembly positioning

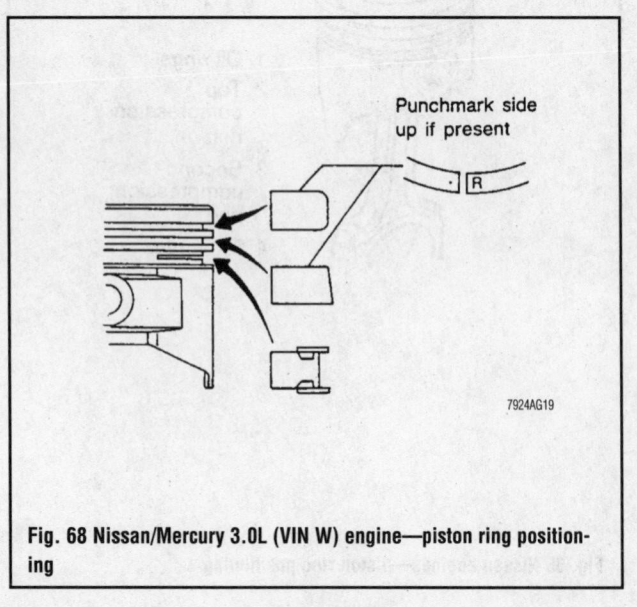

Fig. 68 Nissan/Mercury 3.0L (VIN W) engine—piston ring positioning

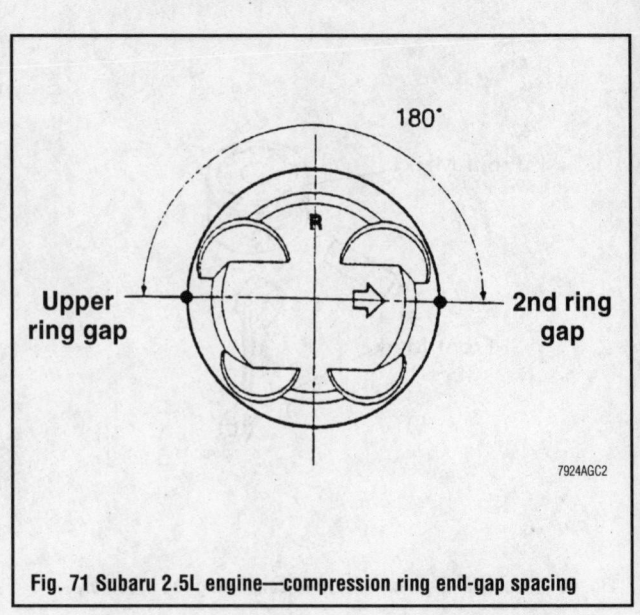

Fig. 71 Subaru 2.5L engine—compression ring end-gap spacing

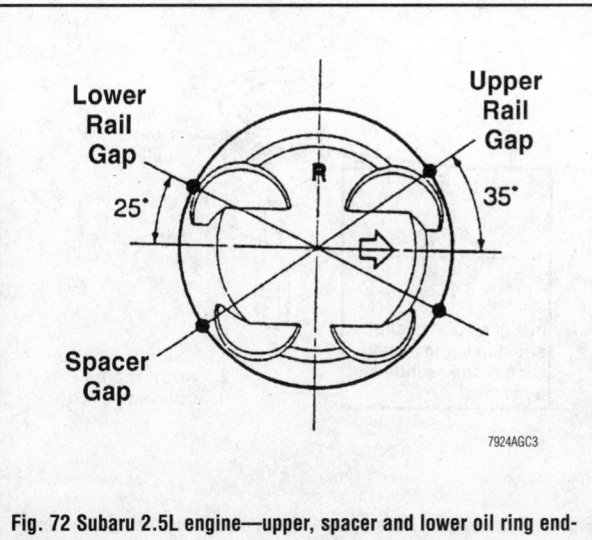

Fig. 72 Subaru 2.5L engine—upper, spacer and lower oil ring end-gap spacing

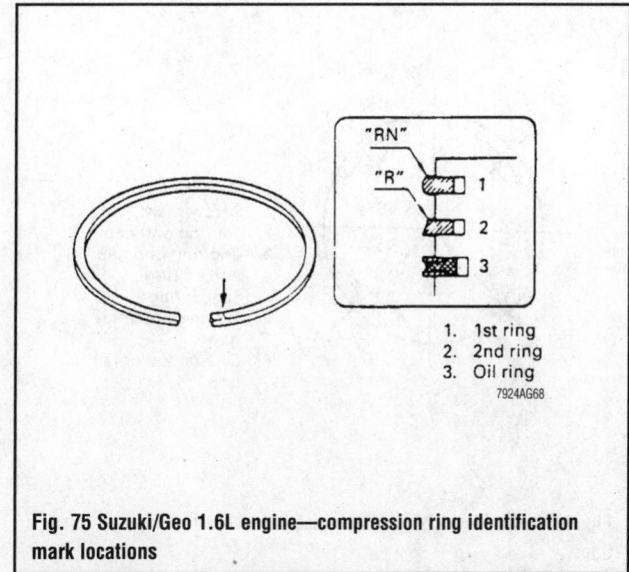

1. 1st ring
2. 2nd ring
3. Oil ring

Fig. 75 Suzuki/Geo 1.6L engine—compression ring identification mark locations

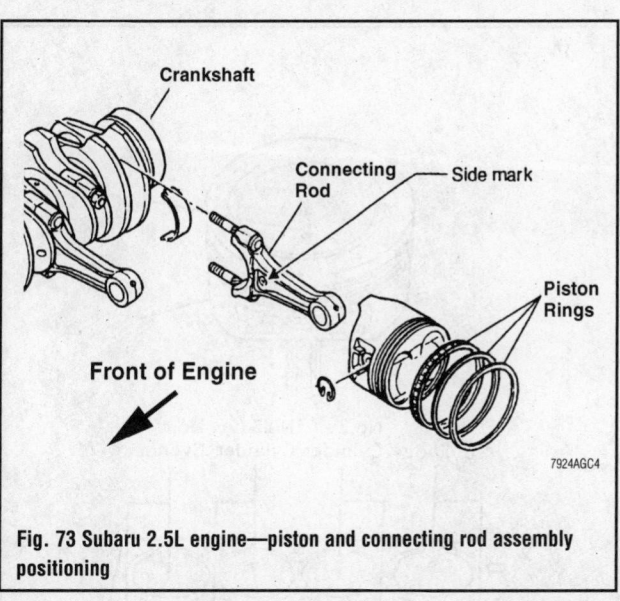

Fig. 73 Subaru 2.5L engine—piston and connecting rod assembly positioning

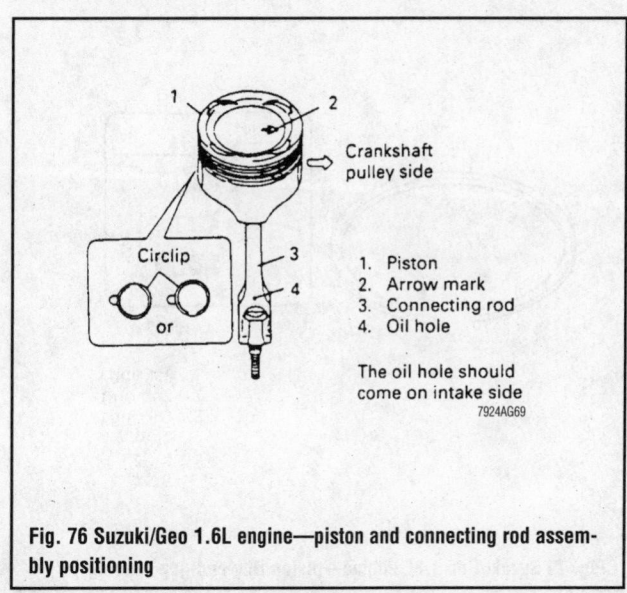

1. Piston
2. Arrow mark
3. Connecting rod
4. Oil hole

The oil hole should come on intake side

Fig. 76 Suzuki/Geo 1.6L engine—piston and connecting rod assembly positioning

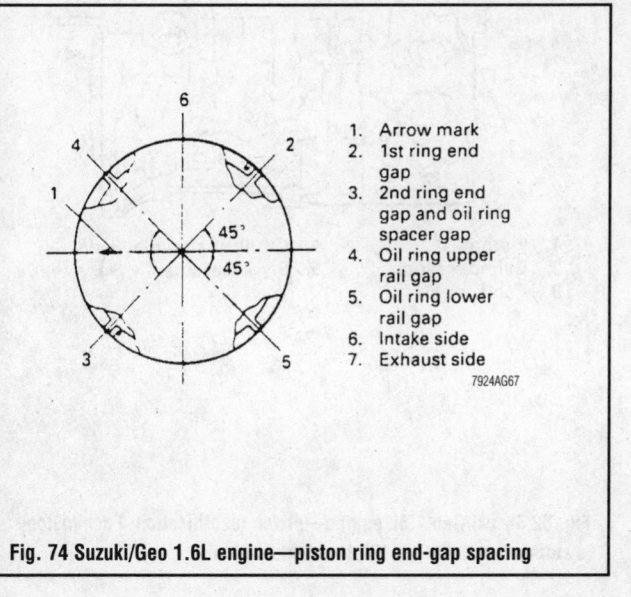

1. Arrow mark
2. 1st ring end gap
3. 2nd ring end gap and oil ring spacer gap
4. Oil ring upper rail gap
5. Oil ring lower rail gap
6. Intake side
7. Exhaust side

Fig. 74 Suzuki/Geo 1.6L engine—piston ring end-gap spacing

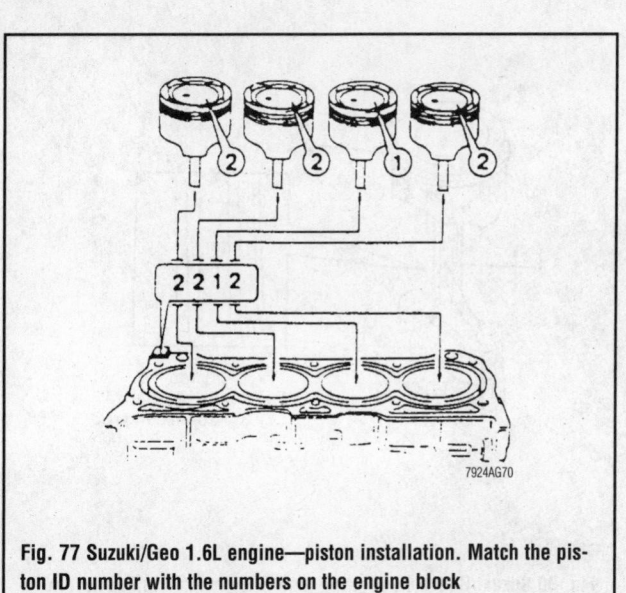

Fig. 77 Suzuki/Geo 1.6L engine—piston installation. Match the piston ID number with the numbers on the engine block

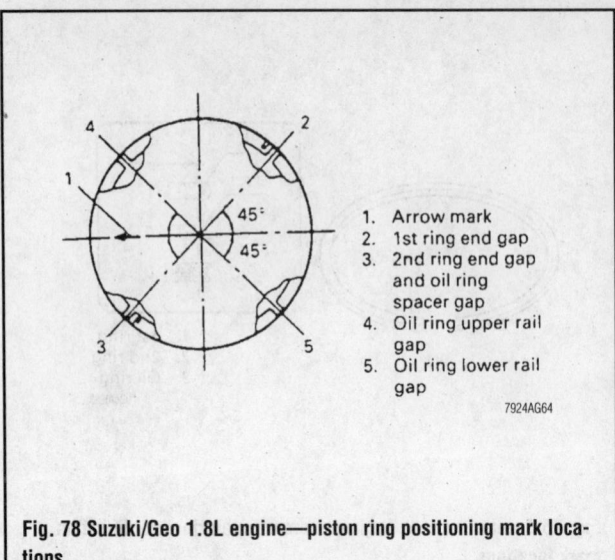

1. Arrow mark
2. 1st ring end gap
3. 2nd ring end gap and oil ring spacer gap
4. Oil ring upper rail gap
5. Oil ring lower rail gap

7924AG64

Fig. 78 Suzuki/Geo 1.8L engine—piston ring positioning mark locations

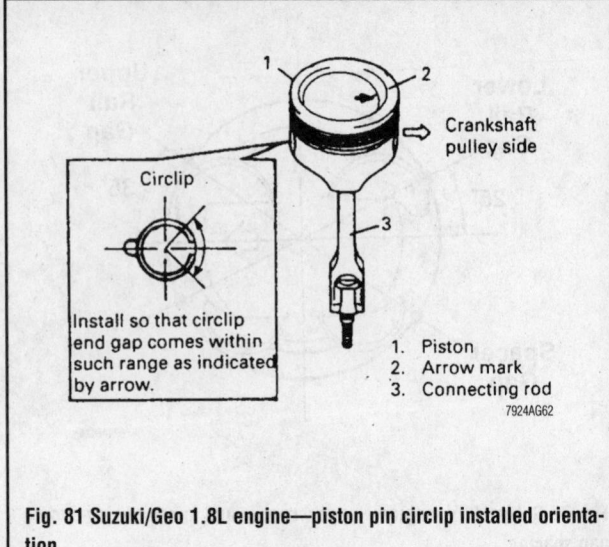

Circlip

Install so that circlip end gap comes within such range as indicated by arrow.

Crankshaft pulley side

1. Piston
2. Arrow mark
3. Connecting rod

7924AG62

Fig. 81 Suzuki/Geo 1.8L engine—piston pin circlip installed orientation

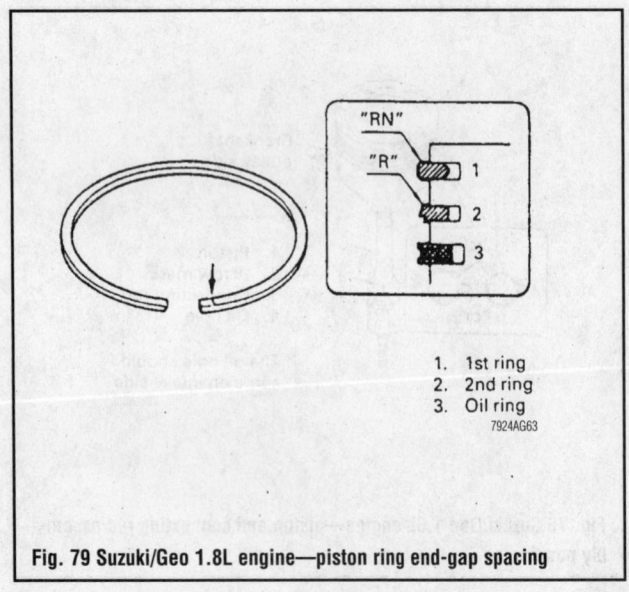

"RN"
"R"

1. 1st ring
2. 2nd ring
3. Oil ring

7924AG63

Fig. 79 Suzuki/Geo 1.8L engine—piston ring end-gap spacing

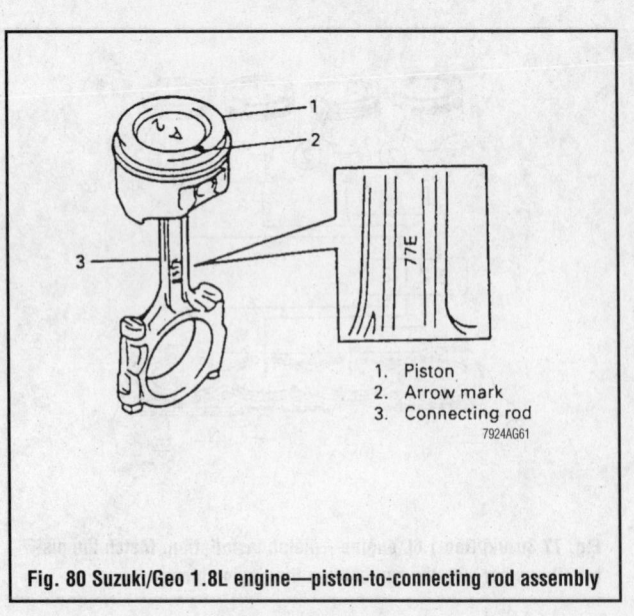

77E

1. Piston
2. Arrow mark
3. Connecting rod

7924AG61

Fig. 80 Suzuki/Geo 1.8L engine—piston-to-connecting rod assembly

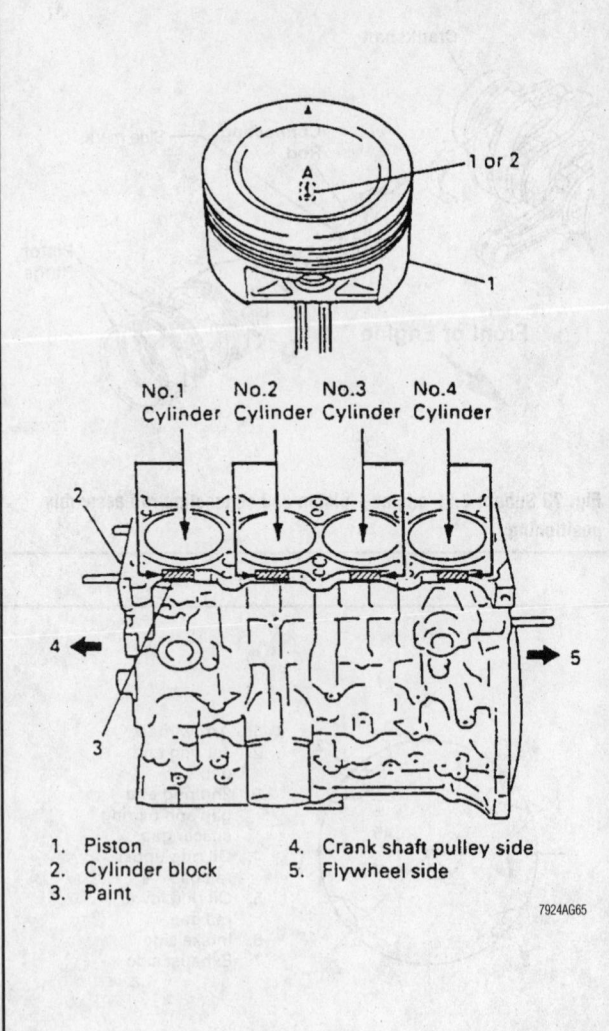

1 or 2

No.1 Cylinder   No.2 Cylinder   No.3 Cylinder   No.4 Cylinder

1. Piston
2. Cylinder block
3. Paint
4. Crank shaft pulley side
5. Flywheel side

7924AG65

Fig. 82 Suzuki/Geo 1.8L engine—piston identification. Each piston is stamped with "1" or "2" for identification

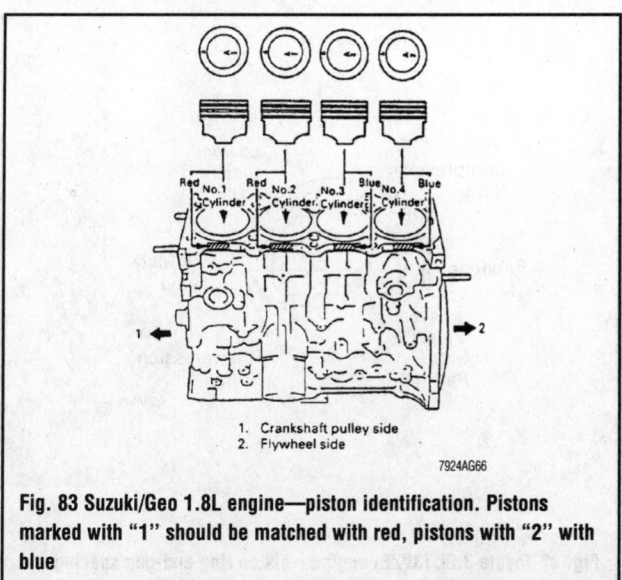

Fig. 83 Suzuki/Geo 1.8L engine—piston identification. Pistons marked with "1" should be matched with red, pistons with "2" with blue

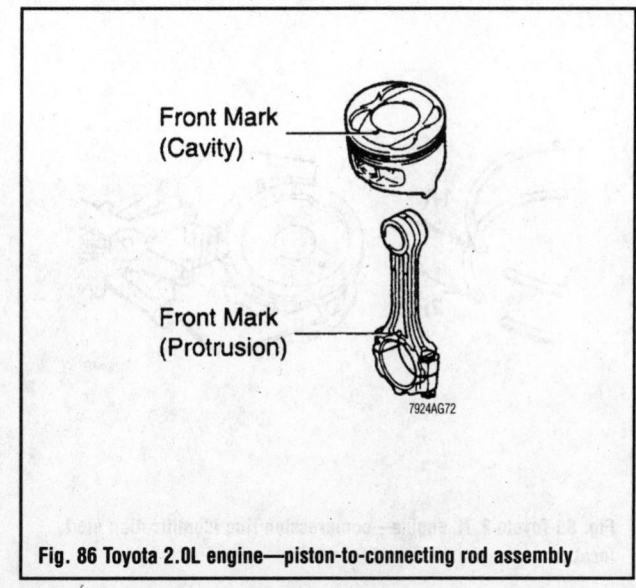

Fig. 86 Toyota 2.0L engine—piston-to-connecting rod assembly

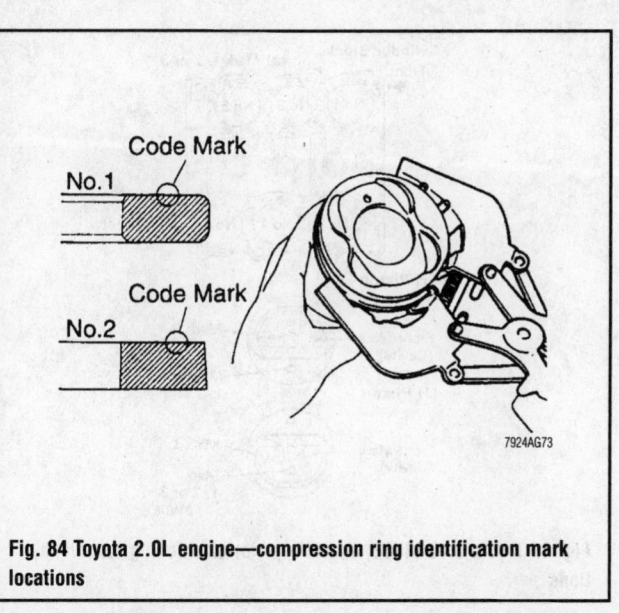

Fig. 84 Toyota 2.0L engine—compression ring identification mark locations

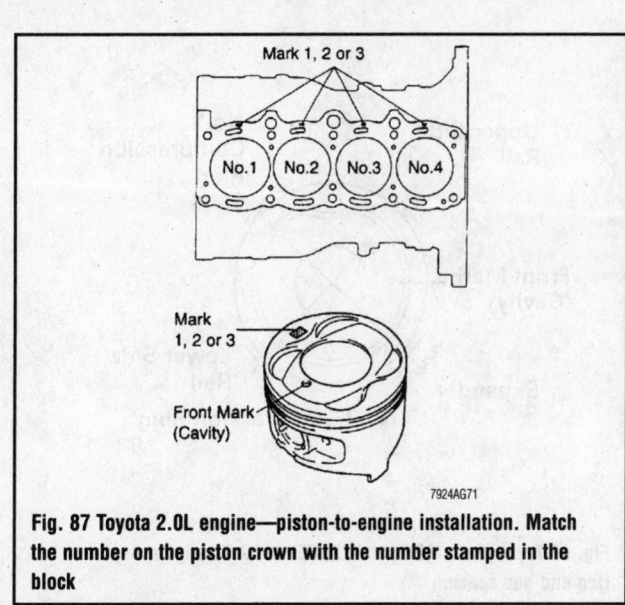

Fig. 87 Toyota 2.0L engine—piston-to-engine installation. Match the number on the piston crown with the number stamped in the block

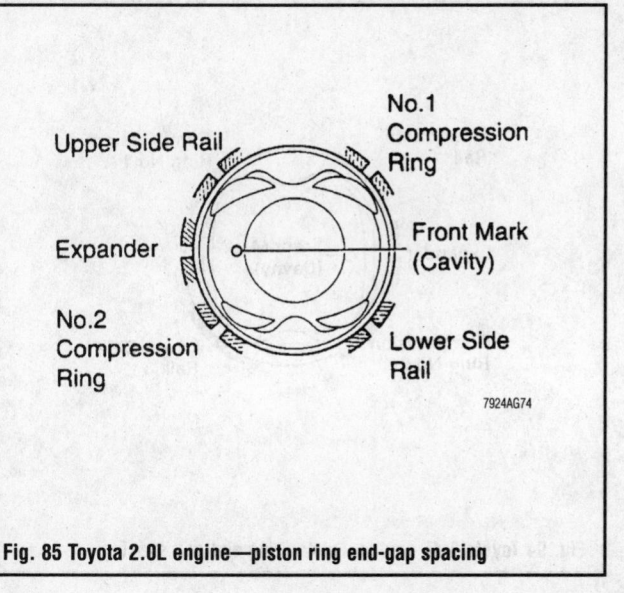

Fig. 85 Toyota 2.0L engine—piston ring end-gap spacing

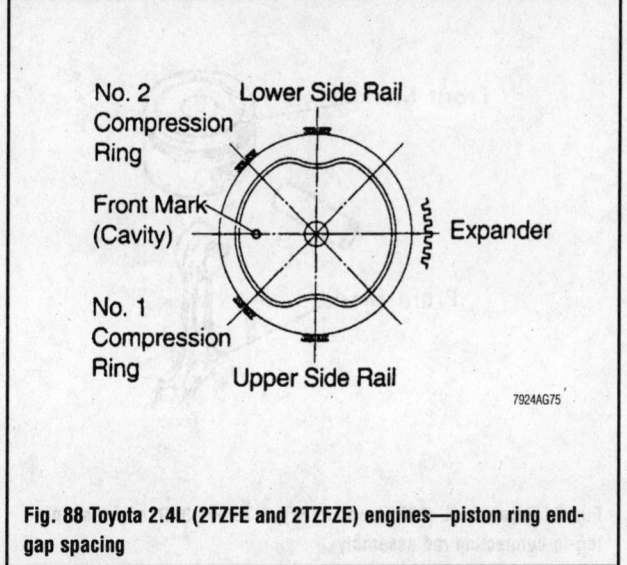

Fig. 88 Toyota 2.4L (2TZFE and 2TZFZE) engines—piston ring end-gap spacing

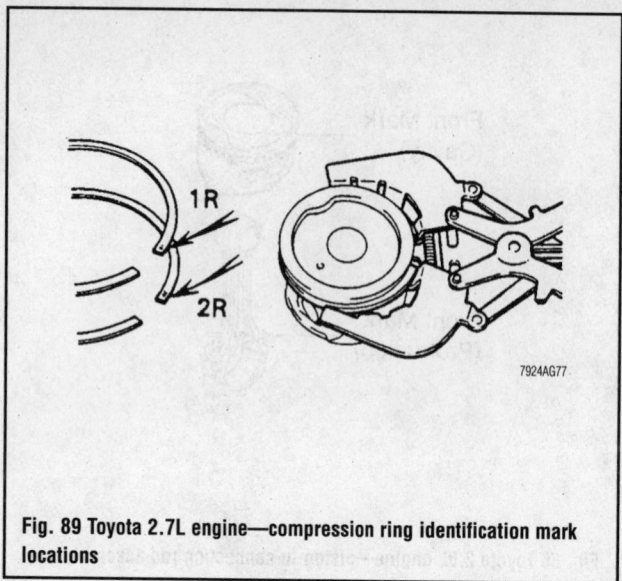

Fig. 89 Toyota 2.7L engine—compression ring identification mark locations

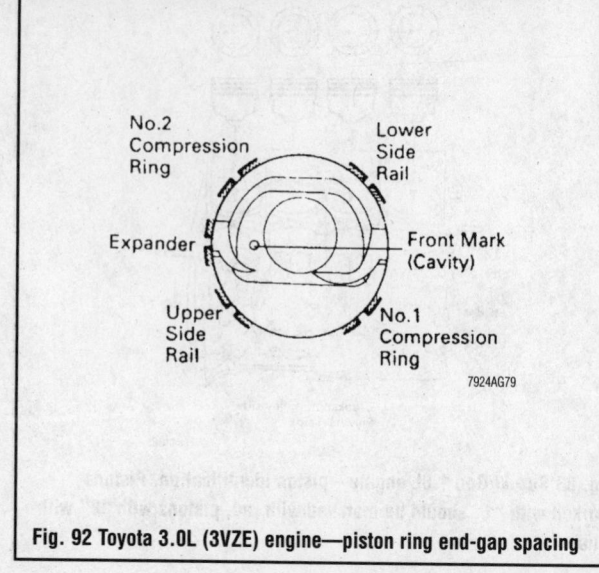

Fig. 92 Toyota 3.0L (3VZE) engine—piston ring end-gap spacing

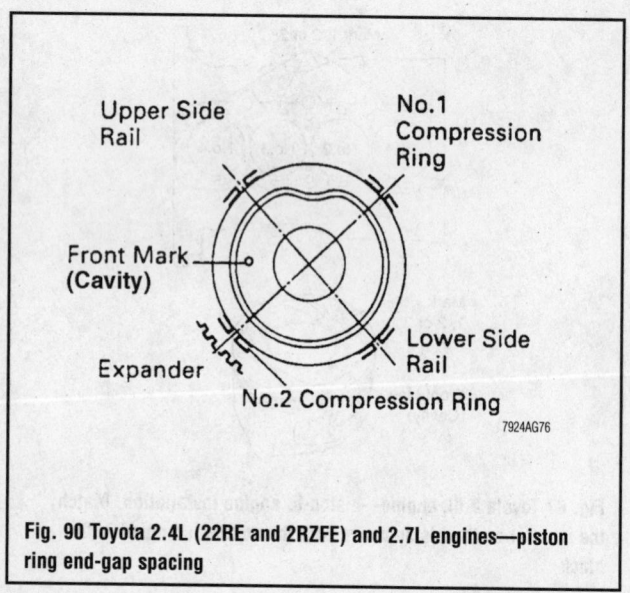

Fig. 90 Toyota 2.4L (22RE and 2RZFE) and 2.7L engines—piston ring end-gap spacing

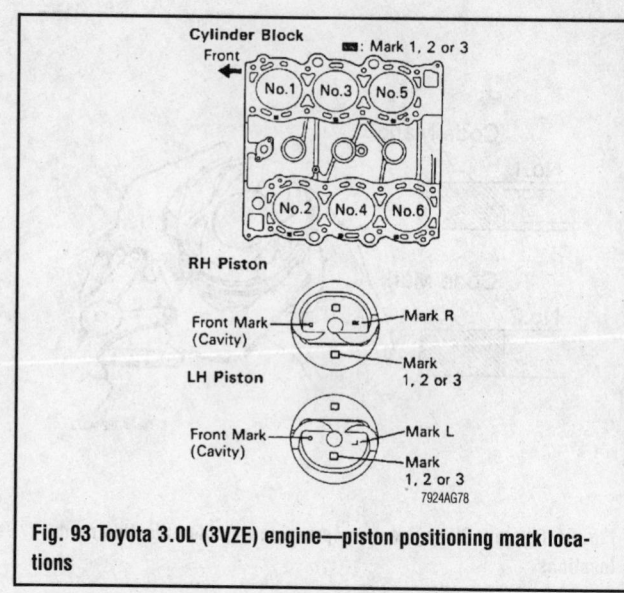

Fig. 93 Toyota 3.0L (3VZE) engine—piston positioning mark locations

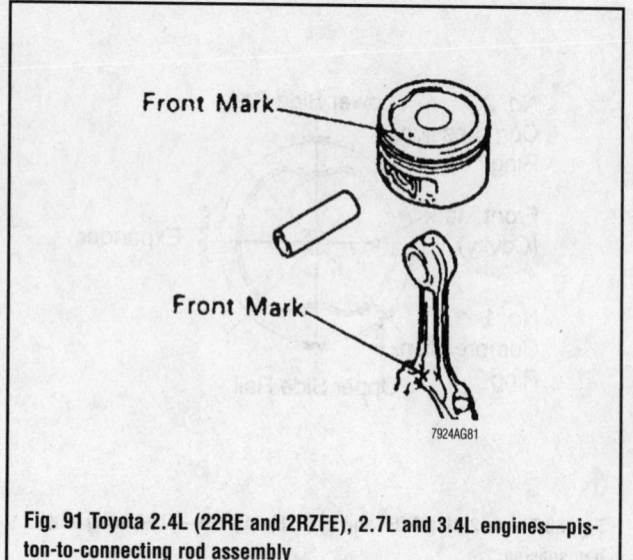

Fig. 91 Toyota 2.4L (22RE and 2RZFE), 2.7L and 3.4L engines—piston-to-connecting rod assembly

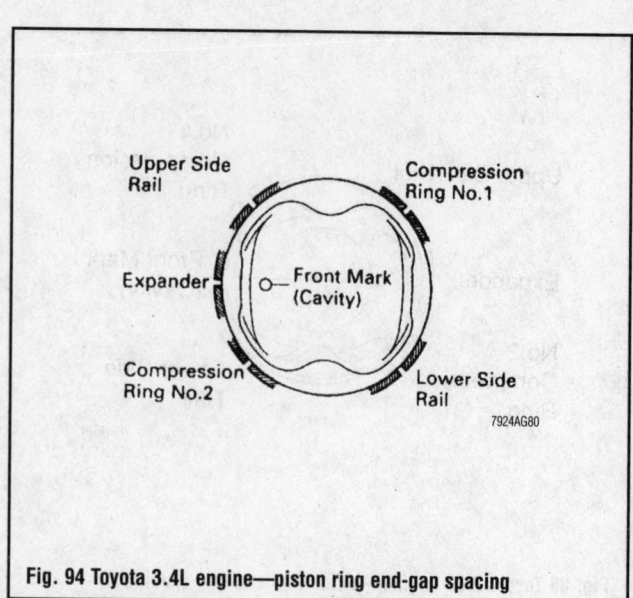

Fig. 94 Toyota 3.4L engine—piston ring end-gap spacing

ACURA and ISUZU

Acura-SLX • Isuzu-Trooper

12

ENGINE REPAIR

➡Disconnecting the negative battery cable on some vehicles may interfere with the functions of the on board computer systems and may require the computer to undergo a relearning process, once the negative battery cable is reconnected.

Ignition Timing

ADJUSTMENT

These engines are equipped with a distributorless ignition system. The ignition timing is controlled by the Powertrain Control Module (PCM) through the input of engine control system sensors.

Engine Assembly

REMOVAL & INSTALLATION

➡The transmission must be removed from the vehicle before the engine is removed.

1. Relieve the fuel system pressure.
2. Disconnect the negative and positive battery cables. Remove the battery. Reinstall the fuel pump relay.
3. Use a felt-tipped marker to match-mark the hood hinges to the hood. Remove the hood.
4. Remove the radiator skid plates.
5. Drain the coolant into a sealable container.
6. Remove the air cleaner duct and hose. Use a clean shop cloth to cover the air cleaner port to prevent dirt from entering the engine.
7. Label and detach the necessary hoses, electrical connectors, control cables, and control rods from the engine.
8. Label and disconnect the following items:
   - Air switch valve hose
   - Oxygen sensor harness
   - Vacuum switch valve hose
   - Thermal Vacuum Switching Valve (VSV) hose
   - Pressure regulator vacuum hose
   - Canister hose
   - Powertrain Control Module harness
   - Fuel inlet and return hoses

❋❋ CAUTION

The fuel injection system remains under pressure after the engine has been turned OFF. Properly relieve fuel pressure before disconnecting any fuel lines. Failure to do so may result in fire or personal injury.

9. Remove the grille from the deflector panel.
10. Disconnect the upper and lower radiator hoses and the reservoir tank hose.
11. Remove the fan shroud, fan blade assembly, and the radiator.
12. If equipped with air conditioning, remove the compressor from the engine and move it aside; do not disconnect the pressure hoses.
13. If equipped with a manual transmission, remove the gear shift lever by performing the following:
   a. Place the gear shift lever in neutral.
   b. Remove the front shift console.
   c. Pull the shift lever boot and grommet upward.
   d. Remove the shift lever cover bolts and the shift lever.
14. Remove the transfer shift lever by performing the following:
   a. Place the transfer shift lever in 2H.
   b. Pull the shift lever boot and dust cover upward.
   c. Remove the shift lever retaining bolts.
   d. Pull the shift lever from the transfer case.
15. Raise and safely support the vehicle. Remove the front wheels. Drain the oil from the engine.
16. Drain the transmission and transfer case fluid.
17. If equipped with an automatic transmission, perform the following:
   a. Remove the dipstick and its tube.
   b. Disconnect the shift select control link rod from the select lever.
   c. Disconnect the downshift cable and vehicle speed sensor cable from the transmission.
   d. Disconnect and plug the fluid coolant lines from the transmission.
18. Matchmark the front and rear driveshaft flanges. Unbolt and remove the front and rear driveshafts.
19. Remove the starter.
20. If equipped with a clutch slave cylinder, remove it from the transmission and move it aside.
21. Remove the exhaust pipe-to-exhaust manifold nuts, the exhaust pipe bracket-to-transmission bolts, the front exhaust pipe-to-second exhaust pipe bolts and the front exhaust pipe from the vehicle.
22. Attach engine hangers to the rear of the exhaust manifolds.

23. Connect a chain hoist to the engine hangers and support the engine.
24. Remove the transmission/transfer case assembly by performing the following:
   a. Place a transmission jack under the transmission for support.
   b. Remove the rear transmission mount nuts.
   c. Remove the rear mount crossmember support and then remove the crossmember and mount.
   d. Unbolt the transmission from the engine.
   e. Move the transmission assembly rearward.
   f. Carefully lower the transmission from the vehicle.
25. Raise the chain hoist to support the weight of the engine. Unbolt the engine mounts and separate them from the engine.
26. Verify that all vacuum lines, hoses, and wiring harnesses have been disconnected.
27. Slowly lift the engine out of the vehicle with the chain hoist. Position the front of the engine higher than the rear.
28. Place the engine on a work stand.
To install:

➡Note the position of the transmission dowel pins before installation. If the dowels are installed in the incorrect position, the transmission case may crack.

29. Using the hoist, slowly lower the engine into the vehicle; be sure to hold the front of the engine higher than the rear.
30. Install the engine mount nuts and bolts. Tighten the engine mount nuts to 30 ft. lbs. (41 Nm), and the bolts to 37 ft. lbs. (50 Nm).
31. Install the transmission/transfer assembly by performing the following:
   a. Raise the transmission into position.
   b. Move the transmission forward and engage it with the engine. Be sure the dowel pins engage with the holes in the engine block.
   c. Install the transmission mounting bolts. Tighten the upper six bolts to 56 ft. lbs. (76 Nm). Tighten the two bolts on the lower right to 35 ft. lbs. (48 Nm), and the bolt on the lower left (A/T only) to 20 ft. lbs. (27 Nm). Tighten the remaining bolts to 53 inch lbs. (6 Nm).
   d. Install the transmission mount and crossmember. Tighten the mount bolts to 37 ft. lbs. (50 Nm). Tighten the front crossmember bolts to 58 ft. lbs. (78 Nm). Tighten the third crossmember bolts to 37 ft. lbs. (50 Nm).
   e. Remove the transmission jack.

32. Remove the engine hoist and the engine hangers from the rear of the exhaust manifolds.

33. Install the front exhaust pipe, exhaust pipe-to-exhaust manifold nuts, the exhaust pipe bracket-to-transmission bolts, the front exhaust pipe-to-second exhaust pipe bolts. Tighten the manifold nuts to 49 ft. lbs. (67 Nm). Tighten the flange bolts to 32–37 ft. lbs. (43–50 Nm).

34. Install the clutch slave cylinder, if equipped. Tighten the slave cylinder bolts to 32 ft. lbs. (43 Nm).

35. Install the starter and tighten the mounting bolts to 30 ft. lbs. (40 Nm).

36. Install the front and rear driveshafts and tighten the flange bolts to 46 ft. lbs. (63 Nm).

37. If equipped with an automatic transmission, perform the following:

   a. Connect the fluid coolant lines to the transmission.

   b. Connect the downshift cable to the transmission.

   c. Connect the shift select control link rod to the select lever.

   d. Install the dipstick and tube.

38. Connect the backup light switch connector and the vehicle speed sensor cable to the transmission.

39. Install the front wheels and lower the vehicle.

40. Install the transfer shift lever by performing the following:

   a. Position the shift lever into the transfer case.

   b. Install the shift lever retaining bolts.

   c. Push the dust cover and the shift lever boot downward.

41. If equipped with a manual transmission, install the gear shift lever by performing the following:

   a. Install the shift lever and the shift lever cover bolts.

   b. Push the grommet and shift lever boot downward.

   c. Install the front console to the floor panel.

42. Install the power steering pump and bracket to the engine.

43. If equipped with air conditioning, install the compressor to the engine.

44. Install the radiator, the fan blade assembly, and the fan shroud.

45. Connect the upper and lower radiator hoses and the reservoir tank hose.

46. Install the grille to the deflector panel.

47. Connect the following items:
- Air switch valve hose
- Oxygen sensor harness
- Vacuum switch valve hose
- Thermal Vacuum Switching Valve (VSV) hose
- Pressure regulator vacuum hose
- Canister hose
- Powertrain Control Module harness
- Fuel hose(s)

48. Attach the necessary hoses, electrical connectors, control cables and control rods to the engine.

49. Install the air cleaner duct and hose.

50. Refill the engine, the transmission, and the transfer case.

51. Refill and bleed the cooling system. Install the radiator skid plate and tighten its bolts to 27 ft. lbs. (37 Nm).

52. Install the hood.

53. Verify that all vacuum lines and wiring harnesses have been reconnected properly.

54. Install the battery and connect the positive and negative cables.

55. Adjust the accessory belt tensions.

56. Start the engine and check for fluid and fuel leaks.

Water Pump

REMOVAL & INSTALLATION

1. Disconnect the negative battery cable.

2. Drain the engine coolant into a clean container.

3. Remove the upper radiator hose.

4. Remove the timing belt and the idler pulley.

➡**The timing belt must be replaced if it has been contaminated by oil or coolant.**

5. Unbolt and remove the water pump.

6. Remove the water pump gasket. Clean any gasket material or sealant residue from the water pump mating sealing surfaces.

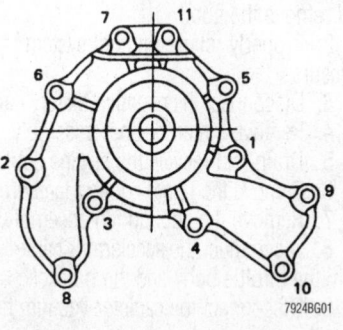

Water pump bolt tightening sequence

To install:

7. Install the water pump using a new gasket. Tighten the mounting bolts to 13 ft. lbs. (18 Nm) in a two-step crisscross sequence.

8. Install the idler pulley. Tighten the mounting bolt to 31 ft. lbs. (42 Nm).

9. Install and tension the timing belt.

10. Install the upper radiator hose.

11. Refill and bleed the cooling system.

12. Connect the negative battery cable. Start the engine and check for coolant leaks.

Cylinder Head

REMOVAL & INSTALLATION

3.2L Engine

1995 MODELS

❉❉ CAUTION

The fuel injection system remains under pressure even after the engine has been turned OFF. The fuel system pressure must be relieved before disconnecting any fuel lines. Failure to do so may result in fire and personal injury.

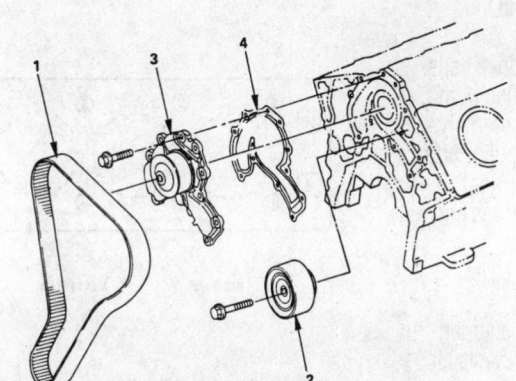

1. Timing belt
2. Idle pulley
3. Water pump assembly
4. Gasket

Exploded view of the water pump mounting

1. Matchmark the hinge to the hood and remove the hood.
2. Properly relieve the fuel system pressure.
3. Disconnect the negative battery cable.
4. Remove the air cleaner assembly.
5. Drain and recycle the engine coolant.
6. Remove the upper cooling fan shroud.
7. Remove the cooling fan assembly.
8. Disconnect the accelerator cable from the throttle body and the bracket.
9. Disconnect the canister vacuum hose from the vacuum pipe.
10. Disconnect the air vacuum hose from the common chamber.
11. Disconnect the vacuum booster hose from the common chamber.
12. Detach the MAP sensor, canister Vacuum Switching Valve (VSV), Exhaust Gas Recirculation (EGR) VSV, Intake Air Temperature (IAT) sensor and ground connectors.
13. Remove the spark plug wires from the cylinder head cover.
14. Remove the ignition control module assembly.
15. Remove the four bolts and the throttle body from the common chamber.
16. Disconnect the vacuum hoses from the throttle body.
17. Disconnect the positive crankcase ventilation hose from the common chamber.
18. Disconnect the fuel pressure control valve vacuum hose from the common chamber.
19. Disconnect the evaporative emission canister purge hose from the common chamber.
20. Remove the EGR valve assembly.
21. Remove the common chamber from the intake manifold.
22. Disconnect the fuel feed and return hoses from the fuel rail assembly.
23. Detach the connectors to the fuel injectors and the thermo sensor.
24. Remove the intake manifold.
25. Remove the engine coolant manifold by removing the heater hose and four mounting bolts.
26. Remove the accessory drive belts.
27. Remove the power steering pump.
28. Remove the fan pulley assembly.
29. Remove the crankshaft pulley and damper.
30. Remove the oil cooler hoses and bracket on the timing belt cover.
31. Remove the timing belt cover.
32. Align the timing marks.
33. Remove the timing belt auto-tensioner (pusher). The pusher prevents air from entering the oil chamber. Its rod must always be facing upward.

34. Remove the timing belt.
35. Remove the cylinder head cover.
36. Remove the power steering pump bracket.
37. Remove the front exhaust pipes from the exhaust manifolds.
38. Remove the dipstick tube bracket from the cylinder head.

✳✳ WARNING

To prevent warpage, the cylinder head and engine block must be at room temperature before removing the cylinder head.

39. Remove the cylinder head bolts in the reverse of the installation sequence, gradually and in two steps.
40. Remove the cylinder head.
   To install:
41. Install new camshaft seals and retaining plates onto the cylinder. Tighten the right camshaft seal retaining plate 6mm bolts to 65 inch lbs. (7.3 Nm). Tighten the left camshaft seal retaining plate 8mm bolts to 191 inch lbs. (22 Nm).
42. Thoroughly clean the cylinder head and engine block sealing surfaces.
43. Place a new head gasket on the engine block and carefully position the cylinder head on top of the new gasket.

➡**Do not reuse or apply oil to the cylinder head bolts.**

44. Install new cylinder head bolts and tighten them in sequence to 47 ft. lbs. (64 Nm) for the M11 bolts and 15 ft. lbs. (21 Nm) for the M8 bolts.
45. Install the dipstick tube bracket to the cylinder head.
46. Connect the front exhaust pipes to the exhaust manifolds. Tighten the exhaust bolts to 48 ft. lbs. (67 Nm).
47. Install the power steering pump bracket. Tighten the mounting bolts to 34 ft. lbs. (46 Nm).

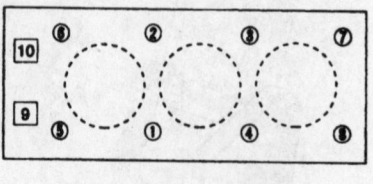

M8 Bolt   **M11 Bolt**

7924BG02

Cylinder head bolt tightening sequence— 1995 SOHC and DOHC engines

48. Install the cylinder head covers.
49. Install the timing belt and the auto-tensioner (pusher). Tighten the mounting bolt to 14 ft. lbs. (19 Nm).
50. Install the timing belt cover and oil cooler hoses and bracket.
51. Install the crankshaft pulley. Tighten the center bolt to 123 ft. lbs. (167 Nm).
52. Install the fan pulley assembly.
53. Install the power steering pump.
54. Install the accessory drive belts.
55. Install the engine coolant manifold and the heater hose.
56. Install the intake manifold. Tighten the nuts and bolts to 17 ft. lbs. (24 Nm) in a crisscross pattern.
57. Install the fuel injector connectors and the fuel hoses to the fuel rail.
58. Install the common chamber. Tighten the nuts and bolts to 17 ft. lbs. (24 Nm).
59. Install the EGR valve assembly. Tighten the mounting bolts on the valve side to 69 inch lbs. (8 Nm) and the bolts on the exhaust side to 21 ft. lbs. (28 Nm).
60. Connect the evaporative emission canister purge, the fuel pressure control valve vacuum hose, and the positive crankcase ventilation hoses.
61. Install the throttle body assembly. Tighten the mounting bolts to 14 ft. lbs. (19 Nm).
62. Connect the vacuum hoses to the throttle body.
63. Install the ignition control module and the spark plug wires.
64. Attach the MAP sensor, canister VSV, EGR VSV, IAT sensor and ground connectors.
65. Connect the vacuum booster hose.
66. Connect the air vacuum hose.
67. Connect the accelerator cable. Adjust the accelerator cable by pulling the cable housing while closing the throttle valve and tightening the adjusting nut and screw cap by hand temporarily. Now, loosen the adjusting nut by three turns and tighten the screw cap. Be sure the throttle valve reaches the screw stop when the throttle is closed.
68. Install the cooling fan assembly and the upper fan shroud.
69. Install the air cleaner assembly.
70. Connect the negative battery cable.
71. Refill and bleed the cooling system.
72. Refill and bleed the power steering pump if necessary.
73. Refill the engine with fresh oil.
74. Run the engine and check for leaks and proper compression.

1996–97 MODELS

➡**There was (only) an SOHC engine available for the 1996–97 vehicle years.**

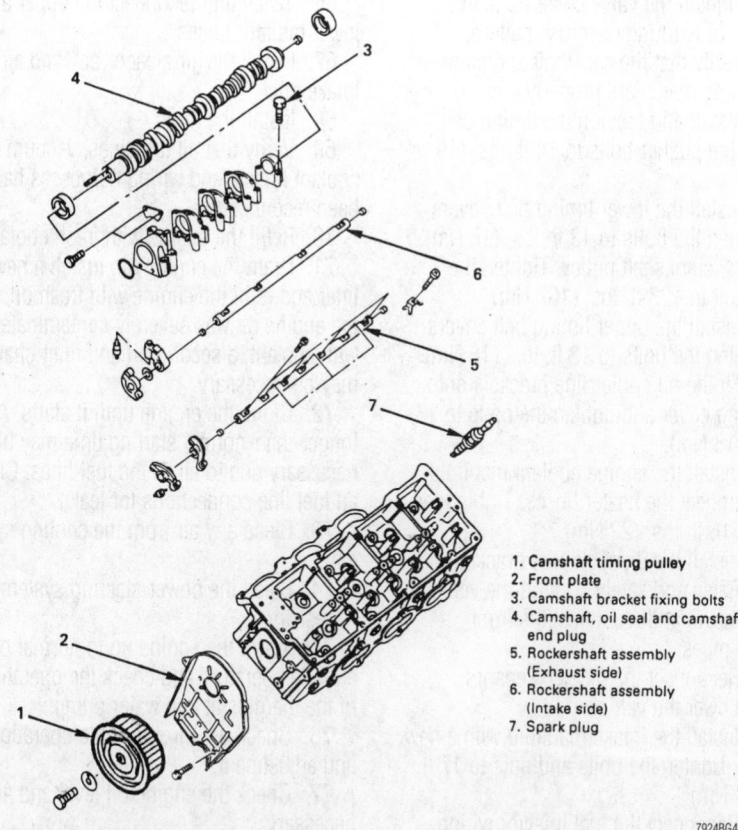

1. Camshaft timing pulley
2. Front plate
3. Camshaft bracket fixing bolts
4. Camshaft, oil seal and camshaft end plug
5. Rockershaft assembly (Exhaust side)
6. Rockershaft assembly (Intake side)
7. Spark plug

7924BG42

Exploded view of the cylinder head components—1996–97 3.2L engines

✳✳ WARNING

The cylinder head should be cool to the touch before it is removed. If the head bolts are loosened on a hot engine, the cylinder head may warp.

1. Properly relieve the fuel system pressure.

✳✳ CAUTION

The fuel injection system remains under pressure even after the engine has been turned OFF . The fuel system pressure must be relieved before disconnecting any fuel lines. Failure to do so may result in fire and personal injury.

2. Disconnect the negative battery cable and reinstall the fuel pump relay.

3. Raise and support the vehicle safely.

4. Disconnect the front exhaust pipes from the exhaust manifolds. If necessary, separate the front exhaust pipes from the crossover pipe. Label and disconnect the oxygen sensors.

5. Lower the vehicle.

6. Drain the coolant into a sealable container.

7. Use a felt-tipped marker to match-mark the hood hinge plates. Remove the hood.

8. Remove the air intake duct and the air cleaner box.

9. Disconnect and remove the upper and lower radiator hoses. Catch the coolant that runs out.

10. Loosen and remove the power steering pump, A/C compressor, and alternator drive belts.

11. Remove the cooling fan and its pulley assembly.

12. Unbolt the power steering pump mounting bracket. Move the pump and bracket out of the way without disconnecting the hydraulic lines.

13. Disconnect the throttle cable from the throttle body linkage.

14. Label and disconnect the following vacuum hoses from the intake manifold chamber:
- PCV hose
- EVAP canister vacuum hose
- Brake booster hose

15. Label and detach the following sensor connectors from the rear of the intake manifold chamber:
- Ignition control module connectors
- Linear EGR valve

- MAP sensor
- EVAP purge valve
- Throttle position sensor
- Idle air control valve
- Intake air temperature sensor

16. Disconnect the EGR valve supply tube and bracket.

17. First, remove the throttle body, and then remove the intake manifold chamber.

18. Carefully clean any dirt from the fuel rail and fuel fittings.

19. Disconnect the fuel feed and return lines from the front of the fuel rail. Clean up any spilled fuel.

20. Label and disconnect the fuel injector wiring harness.

21. Remove the fuel injectors and lower intake manifold as an assembly. If desired, the fuel rail and injectors may be removed separately as an assembly.

22. Remove the intake manifold gaskets. Be careful not to drop any pieces of the gaskets into the engine. Don't scratch or gouge the machined aluminum mating surfaces of the intake manifold and engine block.

23. Cover the intake openings with a sheet of plastic or clean shop towels to keep out dirt and foreign objects.

24. Label the ignition coil assemblies and disconnect them from the wiring harness. Remove the coil assemblies so they won't be damaged.

25. Unbolt the oil cooler line brackets from the timing belt covers.

26. Rotate the crankshaft to align the camshaft timing marks with the pointer dots on the back covers. When the timing marks are aligned, the No. 2 cylinder is at TDC/compression.

27. Remove the crankshaft pulley. Remove the lower timing belt cover.

28. Remove the pusher assembly (tensioner) from below the timing belt tensioner pulley. The pusher rod must always be facing upward to prevent oil leakage. Depress the pusher rod, and insert a wire pin into the hole to keep the pusher rod retracted.

29. Remove the timing belt.

✳✳ WARNING

If the timing belt is worn, damaged, or shows signs of oil or coolant contamination, it must be replaced.

30. Disconnect the heater hoses from the engine, then unbolt and remove the engine coolant manifold.

31. Unbolt the dipstick tube from the cylinder head.

32. Loosen the valve cover bolts in a crisscross sequence in the reverse of the

installation sequence. Remove the valve covers.

33. Loosen the cylinder head bolts in a two-step crisscross pattern working from the outer bolts to those at the center of the head. First, partially loosen the 11mm bolts, then partially loosen the 8mm bolts. Finally loosen all the bolts and remove them.

34. Remove the cylinder head. If it sticks, tap it with a wooden or plastic-faced mallet.

35. Remove the head gasket.

36. Inspect the cylinder head for cracking or warpage. Inspect the engine block for any signs of damage. Carefully clean the head gasket mating surfaces; don't scratch or gouge the machined aluminum surfaces.

37. Cover the engine block with a sheet of plastic or clean shop towels to keep any dirt and foreign objects out of the combustion chambers.

To install:

➡**Use new head bolts when installing the cylinder head. Do not apply oil to the head bolt threads.**

38. Be sure all mating surfaces are clean and free of oil, coolant, or gasket residue.

39. Install new cylinder head gaskets.

40. Install the cylinder head. Install the new head bolts and tighten them by hand only.

41. Follow these steps to tighten the cylinder head bolts to their final tighten specification:

a. Use a two-step crisscross pattern to tighten the 11mm bolts to 47 ft. lbs. (64 Nm). Start tightening with the center bolts, and work toward the outer bolts.

b. Tighten the 8mm bolts to 15 ft. lbs. (21 Nm). Start with the bolt closest to the exhaust side of the head and work toward the intake side.

42. Apply a ⅛ in. (3mm) bead of sealant to the joint where the camshaft holders meet the cylinder head. Install the valve cover with a new gasket before the sealant cures.

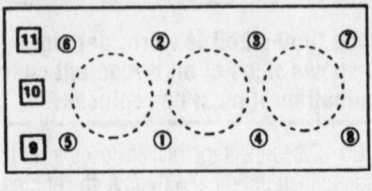

☐ M8 Bolt   ○ M11 Bolt

7924BG03

Cylinder head bolt tightening sequence— 1996–97 3.2L engines

43. Tighten the valve cover bolts to 72 inch lbs. (8 Nm) in crisscross pattern.

44. Verify that the camshaft and crankshaft timing marks are properly aligned.

45. Install and tension the timing belt. Tighten the pusher bolts to 14 ft. lbs. (19 Nm).

46. Install the lower timing belt covers and tighten the bolts to 13 ft. lbs. (18 Nm). Install the crankshaft pulley. Tighten the pulley bolt to 123 ft. lbs. (167 Nm).

47. Install the upper timing belt covers and tighten the bolts to 13 ft. lbs. (18 Nm).

48. Fit the oil cooler line brackets onto the timing cover and tighten the bolts to 13 ft. lbs. (18 Nm).

49. Install the engine coolant manifold and reconnect the heater hoses. Tighten the bolts to 16 ft. lbs. (22 Nm).

50. Install the dipstick tube bracket.

51. Raise and safely support the vehicle.

52. Install and reconnect the front exhaust pipes.

53. Reconnect the oxygen sensors.

54. Lower the vehicle.

55. Install the intake manifold with a new gasket . Tighten the bolts and nuts to 17 ft. lbs. (24 Nm).

56. Reconnect the fuel injector wiring harness. Reconnect the fuel feed and return lines.

57. Install and reconnect the ignition coil assemblies.

58. Install the intake manifold chamber and throttle body with new gaskets. Tighten the nuts and bolts to 17 ft. lbs. (24 Nm).

59. Reconnect the throttle cable to the throttle body linkage.

60. Reconnect the EGR valve supply tube and bracket.

61. Reconnect the following vacuum hoses to the intake manifold chamber:

• PCV hose
• EVAP canister vacuum hose
• Brake booster hose

62. Attach the following sensor connectors to the rear of the intake manifold chamber:

• Ignition control module connectors
• Linear EGR valve
• MAP sensor
• EVAP purge valve
• Throttle position sensor
• Idle air control valve
• Intake air temperature sensor

63. Install the power steering pump and mounting bracket.

64. Install the cooling fan and its pulley assembly.

65. Install and tension the alternator, A/C compressor, and power steering pump drive belts.

66. Install and reconnect the upper and lower radiator hoses.

67. Install the air cleaner box and air intake duct.

68. Install the hood.

69. Verify that all fuel lines, vacuum and coolant hoses, and wiring harnesses have been reconnected.

70. Refill the engine with fresh coolant.

71. Drain the engine oil. Install a new oil filter and refill the engine with fresh oil. If the engine oil was severely contaminated with coolant, a second oil and filter change may be necessary.

72. Crank the engine until it starts. A longer-than-normal starting time may be necessary due to air in the fuel lines. Check all fuel line connections for leaks.

73. Bleed any air from the cooling system.

74. Bleed the power steering system if necessary.

75. Warm the engine up to normal operating temperature and check the operation of the thermostat and water pump.

76. Check the throttle cable operation and adjustment.

77. Check the engine oil level and add if necessary.

3.5L Engine

1. Properly relieve the fuel system pressure.

2. Disconnect the negative battery cable.

> ❋❋ **CAUTION**
>
> **The fuel injection system remains under pressure even after the engine has been turned OFF. The fuel system pressure must be relieved before disconnecting any fuel lines. Failure to do so may result in fire and personal injury.**

3. Raise and support the vehicle safely.

4. Drain the coolant into a sealable container.

5. Use a felt-tipped marker to match-mark the hood hinge plates. Remove the hood.

6. Drain the engine oil into a suitable container.

> ❋❋ **CAUTION**
>
> **The EPA warns that prolonged contact with used engine oil may cause a number of skin disorders, including cancer! You should make every effort to minimize your exposure to used engine oil. Protective gloves should be worn when changing the oil. Wash**

your hands and any other exposed skin areas as soon as possible after exposure to used engine oil. Soap and water, or waterless hand cleaner should be used.

7. Remove the crankshaft pulley assembly using tool J-8614–01 or equivalent crankshaft holder. Hold the pulley, then remove the center bolt and pulley assembly.

8. Remove the timing belt, using the procedure in the timing belt unit repair section.

9. Label and disconnect all wiring harnesses and vacuum hoses related to cylinder head removal.

10. Remove the left and right valve covers.

11. Remove the common chamber.

12. Loosen and remove the eight cylinder head bolts, then remove the cylinder head.

13. Remove the head gasket.

14. Inspect the cylinder head for cracking or warpage. Inspect the engine block for any signs of damage. Carefully clean the head gasket mating surfaces, don't scratch or gouge the machined aluminum surfaces.

15. Cover the engine block with a sheet of plastic or clean shop towels to keep any dirt and foreign objects out of the combustion chambers.

To install:

16. Clean the mating surfaces of the cylinder head and engine block.

17. Place the head gasket on the engine block.

➡ **Do not reuse the head gaskets. There is a "R" mark for the right bank and a "L" for the left bank head gasket as shown.**

18. Install the cylinder head. Be sure the cylinder head is seated over the dowel pins on the engine block.

19. Tighten two bolts by hand to prevent the cylinder head from moving.

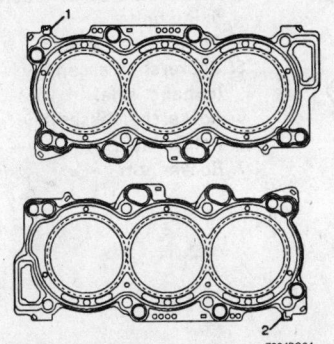

Right (1) and left (2) head gasket identification mark locations—3.5L engine

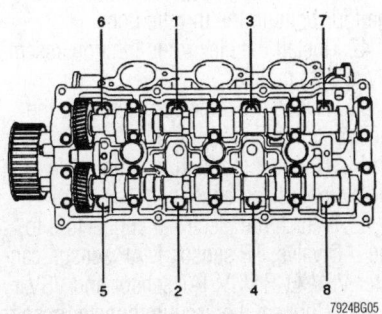

Cylinder head bolt tightening sequence—3.5L engine

➡ **Do not reuse cylinder head bolts. Do not apply any type of lubricant to the head bolts, it will affect the torque reading.**

20. Using Cylinder Head Bolt Wrench J-24239–01 or equivalent, tighten the head bolts in two steps in numerical order as illustrated. In the first step, tighten the head bolts to 21 ft. lbs. (29 Nm). In the second step, tighten the head bolts to 47 ft. lbs. (64 Nm).

21. Install the common chamber.

22. Install the left and right valve covers.

23. Connect all wiring harnesses and vacuum hoses.

24. Install the timing belt, using the procedure in the timing belt unit repair section

25. Install the crankshaft pulley and tighten the center bolt to 123 ft. lbs. (167 Nm).

26. Fill the engine with the correct amount and type of oil.

27. Refill the engine with fresh coolant.

28. Install the hood.

29. Warm the engine up to normal operating temperature and check the operation of the thermostat and water pump.

30. Check the throttle cable operation and adjustment.

31. Check the engine oil level and add if necessary.

Rocker Arm/Shafts

REMOVAL & INSTALLATION

3.2L Engine

1995 MODELS

1. Properly relieve the fuel system pressure.

❊ CAUTION

The fuel injection system remains under pressure even after the engine

has been turned OFF. The fuel system pressure must be relieved before disconnecting any fuel lines. Failure to do so may result in fire and personal injury.

2. Disconnect the negative battery cable.

3. Remove the air cleaner assembly.

4. Disconnect the accelerator pedal cable from the throttle body and cable brackets.

5. Disconnect the canister vacuum hose from the Vacuum Switching Valve (VSV).

6. Disconnect the vacuum booster hose from the common chamber duct.

7. Detach the electrical connectors from the Idle Air Control (IAC) valve, Throttle Position (TP) sensor, Manifold Absolute Pressure (MAP) sensor, canister VSV, EGR VSV, Intake Air Temperature (IAT) sensor and VSV.

8. Remove the high tension cable from the cylinder heads.

9. Disconnect the ignition module.

10. Remove the three bolts from the electronic ignition bracket and assembly.

11. Remove the four bolts from the throttle body and remove the throttle body.

12. Disconnect the canister VSV and the EGR VSV vacuum hoses from the throttle body.

13. Disconnect the fuel pressure control valve vacuum hose from the common chamber duct.

14. Disconnect the PCV hose from the common chamber duct.

15. Disconnect the evaporative emission canister purge hose from the common chamber duct.

16. Remove the four bolts from the EGR valve assembly common chamber duct and remove the exhaust manifold.

17. Remove the four bolts, four nuts and three manifold bracket fixing bolts from the common chamber duct.

18. Remove the ground cable fixing bolt from the rear of the common chamber duct.

19. Remove the six bolts and two nuts from the common chamber duct.

20. Remove the common chamber duct bracket fixing bolts from the rear of the common chamber duct.

21. Remove the timing belt.

22. Remove the cylinder head cover.

23. Remove the camshaft holders and camshaft.

24. Remove the rocker arm shaft bolts. Lift the rocker arm assembly from the cylinder head.

25. The hydraulic lifters are attached to the rocker arms. Remove them and inspect, bleed, or replace as necessary.

To install:

26. Install new camshaft seals and retaining plates onto the cylinder head. Tighten the right camshaft seal retaining plate 6mm bolts to 65 inch lbs. (7 Nm). Tighten the left camshaft seal retaining plate 8mm bolts to 191 inch lbs. (21 Nm).

27. Install the rocker arm assembly and tighten the bolts in sequence to 13 ft. lbs. (18 Nm).

28. Oil the camshaft bearing journals, camshaft lobes, and rocker arm contact areas.

29. Install the camshaft. Apply sealant to the contact edges of the camshaft holders. Tighten the camshaft 6mm holder bolts to 69 inch lbs. (8 Nm). Tighten the remaining bolts to 13 ft. lbs. (18 Nm).

30. Install the cylinder head covers and carefully tighten the bolts to 69 inch lbs. (8 Nm). Do not overtighten the head cover bolts: they crack very easily

31. Install the timing belt.

32. Install the common chamber duct bracket bolts to the rear of the common chamber duct.

33. Install the six bolts and two nuts to the common chamber duct.

34. Install the ground cable fixing bolt to the rear of the common chamber duct.

35. Install the four bolts, four nuts and three manifold bracket bolts to the common chamber duct.

36. Install the exhaust manifold and install the four bolts to the EGR valve assembly and common chamber duct.

37. Connect the evaporative emission canister purge hose to the common chamber duct.

38. Connect the PCV hose to the common chamber duct.

39. Connect the fuel pressure control valve vacuum hose to the common chamber duct.

40. Connect the canister VSV and the EGR VSV vacuum hose to the throttle body.

41. Install the throttle body and install the four bolts to the throttle body.

42. Install the electronic ignition assembly and its bracket.

43. Attach the three connectors to the electronic ignition module.

44. Connect the high tension cable to the cylinder head cover clips.

45. Attach the electrical connectors to the IAC valve, TP sensor, MAP sensor, canister VSV, EGR VSV, IAT sensor and VSV.

46. Connect the vacuum booster hose to the common chamber duct.

47. Connect the canister vacuum hose to the VSV.

48. Connect the accelerator pedal cable to the throttle body and cable brackets.

49. Verify that all vacuum hoses, lines, and wiring harnesses are reconnected.

50. Install the air cleaner assembly.

51. Connect the negative battery cable.

1996–97 MODELS

1. Properly relieve the fuel system pressure.

❋❋ CAUTION

The fuel injection system remains under pressure even after the engine has been turned OFF. The fuel system pressure must be relieved before disconnecting any fuel lines. Failure to do so may result in fire and personal injury.

2. Disconnect the negative battery cable.

3. Drain the coolant into a sealable container.

4. Support the hood as far open as possible.

5. Remove the air intake duct and the air cleaner box.

6. Disconnect and remove the upper and lower radiator hoses. Catch any coolant that runs out.

7. Loosen and remove the power steering pump, A/C compressor, and alternator drive belts.

8. Remove the cooling fan and its pulley assembly.

9. Unbolt the power steering pump mounting bracket. Move the pump and bracket out of the way without disconnecting the hydraulic lines.

10. Disconnect the throttle cable from the throttle body linkage.

11. Label and disconnect the following vacuum hoses from the intake manifold chamber:
   • PCV hose
   • EVAP canister vacuum hose

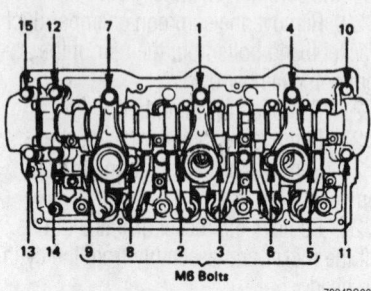

Camshaft holder bolt tightening sequence—1995 3.2L SOHC engine

7924BG06

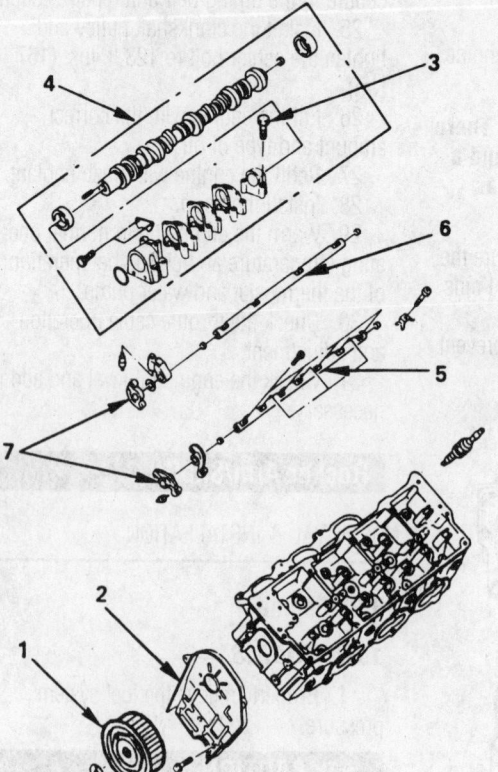

1. Camshaft timing pulley
2. Front plate
3. Camshaft bracket housing fixing bolts
4. Camshaft assembly
5. Rockershaft assembly (Exhaust side)
6. Rockershaft assembly (Intake side)
7. Rocker arm

Exploded view of the rocker arm/shaft assembly mounting—1996–97 models

7924BG43

- Brake booster hose

12. Label and detach the following sensor connectors from the rear of the intake manifold chamber:

- Ignition control module connectors
- Linear EGR valve
- MAP sensor
- EVAP purge valve
- Throttle position sensor
- Idle air control valve
- Intake air temperature sensor

13. Disconnect the EGR valve supply tube and bracket.

14. First, remove the throttle body, and then remove the intake manifold chamber.

15. Carefully clean any dirt from the fuel rail and fuel fittings.

16. Disconnect the fuel feed and return lines from the front of the fuel rail. Clean up any spilled fuel.

17. Remove the intake manifold gaskets. Be careful not to drop any pieces of the gaskets into the engine. Don't scratch or gouge the machined aluminum mating surfaces of the intake manifold and engine block.

18. Cover the intake openings with a sheet of plastic or clean shop towels to keep out dirt and foreign objects.

19. Label the ignition coil assemblies and disconnect them from the wiring harness. Remove the coil assemblies so they won't be damaged.

20. Unbolt the oil cooler line brackets from the timing belt covers.

21. Remove the timing belt using the recommended procedure in the timing belt unit repair section.

✷✷ WARNING

If the timing belt is worn, damaged, or shows signs of oil or coolant contamination, it must be replaced.

22. Loosen the valve cover bolts in a crisscross sequence. Remove the valve covers.

23. Remove the camshaft sprockets and back covers.

24. Loosen the camshaft holder bolts in a crisscross sequence to prevent warping.

25. Remove the camshaft and camshaft holders from the cylinder head.

26. Inspect the camshaft lobes and journals for signs of wear or damage.

27. Loosen the exhaust and intake rocker shaft bolts in a crisscross sequence to prevent warping.

28. Remove the intake and exhaust rocker shafts from the cylinder head.

29. If the rocker arms and shafts must be disassembled, label the parts so that they can be reassembled in the same positions.

30. If necessary, remove the hydraulic valve lash adjusters from the rocker arms.

31. Inspect the hydraulic lash adjusters for excess movement and replace if necessary. The hydraulic lash adjusters are designed to be self-bleeding, but new adjusters must be primed before installation, as follows:

   a. Use a small-diameter rod (0.08 in. or 2mm) to push in the adjuster's check ball.

   b. Submerge the adjuster in a tub of clean engine oil.

   c. Pump the plunger with your finger to fill the adjuster with oil and displace any air.

   d. Keep pumping the plunger until it's hard and no more air bubbles come out. Then, remove the rod to release the check ball.

To install:

32. Reassemble the rocker arm, shaft, and hydraulic lash adjuster components. Assemble the hydraulic lash adjusters to the rockers before removing them from the tub of oil. The intake rocker arms all face the same direction when installed.

33. Lubricate the rocker arms and shafts with clean engine oil.

34. Install the intake and exhaust rocker arms and shaft assemblies. Tighten the rocker

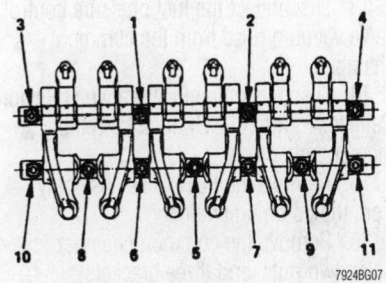

Rocker shaft bolt tightening pattern—1996–97 3.2L engines

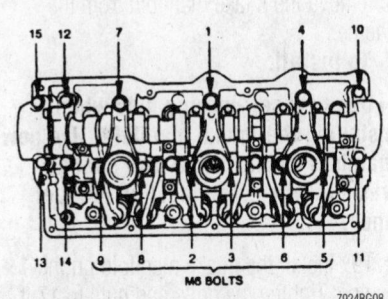

Camshaft mounting bolt tightening sequence—1996–97 3.2L engines

shaft holder bolts to 13 ft. lbs. (18 Nm), starting with the intake shaft and then moving to the exhaust shaft. Be sure the intake and exhaust rockers contact each other properly.

35. Be sure all mating surfaces are clean and free of oil, coolant, or gasket residue.

36. Lubricate the camshaft lobes and journals with clean engine oil.

37. Apply a bead of sealant to the front and rear camshaft holder mating surfaces on the cylinder head.

38. Install the camshaft and holder assembly onto the cylinder head before the sealant cures. Install the camshaft holder bolts, but don't tighten them yet.

39. Use a crisscross sequence to tighten the camshaft holder bolts. Tighten the 8mm bolts to 13 ft. lbs. (18 Nm). Tighten the 6mm bolts to 72 inch lbs. (8 Nm).

40. Use a seal driver to install a new camshaft seal.

41. Install the camshaft sprocket back covers and tighten their bolts to 12 ft. lbs. (17 Nm).

42. Install the camshaft sprockets so that the timing marks are aligned. Tighten the bolts to 46 ft. lbs. (64 Nm).

43. Apply a ⅛ in. (3mm) bead of sealant to the joint were the camshaft holders meet the cylinder head. Install the valve cover with a new gasket before the sealant cures.

44. Tighten the valve cover bolts to 6 ft. lbs. (8 Nm) in a crisscross pattern.

45. Verify that the camshaft and crankshaft timing marks are properly aligned.

46. Install and tension the timing belt. Tighten the pusher bolts to 14 ft. lbs. (19 Nm).

47. Install the lower timing belt covers and tighten the bolts to 13 ft. lbs. (18 Nm). Install the crankshaft pulley. Tighten the pulley bolt to 123 ft. lbs. (167 Nm).

48. Install the upper timing belt covers and tighten the bolts to 13 ft. lbs. (18 Nm).

49. Fit the oil cooler line brackets onto the timing cover and tighten the bolts to 13 ft. lbs. (18 Nm).

50. Reconnect the fuel feed and return lines.

51. Install and reconnect the ignition coil assemblies.

52. Install the intake manifold chamber and throttle body with new gaskets. Tighten the nuts and bolts to 17 ft. lbs. (24 Nm).

53. Reconnect the throttle cable to the throttle body linkage.

54. Reconnect the EGR valve supply tube and bracket.

55. Reconnect the following vacuum hoses to the intake manifold chamber:

- PCV hose
- EVAP canister vacuum hose
- Brake booster hose

56. Attach the following sensor connectors to the rear of the intake manifold chamber:

- Ignition control module connectors
- Linear EGR valve
- MAP sensor
- EVAP purge valve
- Throttle position sensor
- Idle air control valve
- Intake air temperature sensor

57. Install the power steering pump and mounting bracket.

58. Install the cooling fan and its pulley assembly.

59. Install and tension the alternator, A/C compressor, and power steering pump drive belts.

60. Install and reconnect the upper and lower radiator hoses.

61. Install the air cleaner box and air intake duct.

62. Verify that all fuel lines, vacuum and coolant hoses, and wiring harnesses have been reconnected.

63. Refill the engine with fresh coolant.

64. Crank the engine until it starts. A longer-than-normal starting time may be necessary due to air in the fuel lines. Check all fuel line connections for leaks.

65. Bleed any air from the cooling system.

66. Bleed the power steering system if necessary.

67. Check the throttle cable operation and adjustment.

68. Check the engine oil level and add if necessary.

3.5L Engine

The 3.5L engine does not utilize rocker arms nor shafts; it is designed with shims and cam followers, which sit directly between the camshaft lobes and the valve stems.

Intake Manifold

REMOVAL & INSTALLATION

1995 3.2L Engine

※※ CAUTION

The fuel injection system remains under pressure even after the engine has been turned OFF. The fuel system pressure must be relieved before disconnecting any fuel lines. Failure to do so may result in fire and personal injury.

1. Properly relieve the fuel system pressure.

2. Disconnect the negative battery cable.

3. Remove the air cleaner assembly.

4. Disconnect the accelerator pedal cable from the throttle body and bracket.

5. Disconnect the charcoal canister vacuum hose from the vacuum pipe.

6. Disconnect the air vacuum hose and the vacuum booster hose from the common chamber.

7. Detach the following electrical connectors:

- MAP sensor
- Charcoal canister Vacuum Switching Valve (VSV)
- Exhaust Gas Recirculation (EGR) VSV
- Intake Air Temperature (IAT) sensor
- Engine ground cable
- Fuel injector connectors
- Thermo sensor connector

8. Tag and disconnect the spark plug wires.

9. Remove the ignition module assembly with the spark plug wires attached.

10. Tag and disconnect the vacuum hoses from the throttle body.

11. Remove the four throttle body mounting bolts. Then, remove the throttle body.

12. Disconnect the PCV hose from the common chamber.

13. Disconnect the fuel pressure control valve vacuum hose from the common chamber.

14. Disconnect the Evaporative Emission Canister Purge (EECP) hose from the common chamber.

15. Remove the EGR valve assembly from the common chamber.

16. Remove the common chamber (six bolts, two nuts, and three brackets).

17. Disconnect the fuel feed and return hoses from the fuel rail. Remove the bracket mounting bolts from the cylinder head cover.

18. Remove the two bolts and four nuts to remove the intake manifold from the engine.

To install:

➡**Use new self-locking nuts when installing the intake manifold. Use new manifold gaskets. Use new sealing washers when reconnecting the fuel lines.**

19. Install the intake manifold on the engine. Tighten the bolts and nuts to 17 ft. lbs. (24 Nm). Tighten the bolts from the center towards the ends.

20. Attach the electrical connectors to the fuel injectors and the thermo sensor.

21. Connect the fuel return and feed hoses to the fuel rail.

22. Install the common chamber. Tighten the bolts and nuts to 17 ft. lbs. (24 Nm).

23. Install the EGR assembly. Tighten the bolts to 78 inch lbs. (9 Nm).

24. Connect the charcoal canister purge and fuel pressure regulator vacuum hoses to the common chamber.

25. Connect the PCV hose to the common chamber.

26. Install the throttle body assembly and connect the vacuum hoses to the throttle body. Tighten the throttle body mounting bolts to 16 ft. lbs. (22 Nm).

27. Install the ignition module assembly. Tighten the mounting bolts to 16 ft. lbs. (22 Nm).

28. Reconnect the spark plug wires.

29. Reattach the following electrical connectors:

- MAP sensor
- Charcoal canister
- EGR VSV
- IAT sensor
- Engine ground cable

30. Connect the air vacuum hose and the vacuum booster hose to the common chamber.

31. Connect the charcoal canister vacuum hose to the vacuum pipe.

32. Connect the accelerator cable to the throttle body and bracket. Adjust the cable so that the linkage rests against the stop when moved by hand.

33. Install the air cleaner assembly.

34. Verify that all electrical connectors and vacuum lines have been attached.

35. Connect the negative battery cable.

36. Turn the ignition key to the **ON** position for two seconds, then **OFF**. Turn the ignition **ON** again to pressurize the fuel system and check for leaks.

1996–97 3.2L and All 3.5L Engines

1. Properly relieve the fuel system pressure.

2. Disconnect the negative battery cable.

3. Remove the air cleaner and air intake duct.

4. Drain the engine coolant to a level below the upper radiator hose. Catch the coolant in a clean drain pan if it is to be reused.

5. Disconnect the throttle cable from the throttle body linkage.

6. Label and disconnect the following vacuum hoses from the intake manifold chamber:

- PCV hose
- EVAP canister vacuum hose
- Brake booster hose

7. Label and detach the following sensor connectors from the rear of the intake manifold chamber:

- Ignition control module connectors
- Linear EGR valve
- MAP sensor
- EVAP purge valve
- Throttle position sensor
- Idle air control valve
- Intake air temperature sensor

8. Disconnect the EGR valve supply tube and bracket.

9. Remove the throttle body.

※※ WARNING

Don't use solvent of any type when cleaning the gasket mating surfaces of the throttle body and intake manifold. Solvent may damage the machined surfaces of these components. Be careful not to scratch the mating surfaces.

10. Disconnect the MAP sensor tube, and then unbolt the MAP sensor bracket.

11. Unbolt the intake manifold chamber from its brackets which are located at its front and rear edges.

12. Loosen the six manifold mounting bolts and two nuts in a crisscross sequence.

13. Remove the bolts and nuts, and then lift the chamber off of the base of the intake manifold. Note the positions of the long and short bolts.

14. Cover the intake manifold with a sheet of plastic, or clean shop towels to keep out dirt and foreign objects.

15. Carefully clean any dirt from the fuel rail and fuel fittings.

※※ CAUTION

The fuel injection system remains under pressure even after the engine has been turned OFF. The fuel system pressure must be relieved before disconnecting any fuel lines. Failure to do so may result in fire and personal injury.

16. Disconnect the fuel feed and return lines from the front of the fuel rail. Clean up any spilled fuel.

17. Label and disconnect the fuel injector wiring harness.

18. Unbolt the fuel return line bracket from the front of the intake manifold.

19. Unbolt the fuel rail from the intake manifold.

20. Carefully lift the fuel rail and injectors off of the intake manifold as an assembly. Move the fuel rail out of the work area so it won't be damaged, then clean up any spilled fuel.

21. Remove the fuel rail spacer grommets from the sides of the intake manifold. Replace the spacer grommets if they are cracked or ripped.

22. Loosen the intake manifold nuts and bolts in a crisscross sequence working from the outer edges of the manifold toward the center.

23. Push the engine wiring harnesses aside and lift the intake manifold up and off of the engine block.

24. Remove the intake manifold gaskets. Be careful not to drop any pieces of the gaskets into the engine. Don't scratch or gouge the machined aluminum mating surfaces of the intake manifold and engine block.

25. Cover the intake openings with a sheet of plastic or clean shop towels to keep out dirt and foreign objects.

To install:

26. Remove the covers from the intake openings. Install new intake manifold gaskets.

27. Fit the intake manifold into position. Move the wiring harnesses back into position.

28. Install the intake manifold nuts and bolts. Tighten them in a two-step crisscross pattern to 15 ft. lbs. (20.5 Nm) working from the center of the manifold toward the outer edges.

29. Install the fuel rail spacer grommets.

30. Install the fuel rail assembly onto the intake manifold. Be sure all the fuel injectors are properly seated. Tighten the fuel rail bolts to 62 inch lbs. (7 Nm).

31. Reconnect the fuel feed and return lines. Install the return line bracket bolt.

32. Reconnect the fuel injector wiring harness.

33. Install a new intake manifold chamber gasket. Install the intake manifold chamber.

34. Install the six intake manifold chamber bolts and two nuts. Tighten the nuts and bolts to 15 ft. lbs. (20.5 Nm) in a crisscross sequence.

35. Install the intake manifold chamber bracket bolts.

36. Reconnect the MAP sensor tube and bracket.

37. Reconnect the EGR supply tube and bracket.

38. Install the throttle body with a new gasket. Tighten the bolts to 10 ft. lbs. (13.5 Nm) in a crisscross sequence.

39. Attach the following sensor connectors to the rear of the intake manifold chamber:

- Ignition control module connectors
- Linear EGR valve
- MAP sensor
- EVAP purge valve
- Throttle position sensor
- Idle air control valve
- Intake air temperature sensor

40. Reconnect the following vacuum hoses to the intake manifold chamber:

- PCV hose
- EVAP canister vacuum hose
- Brake booster hose

41. Reconnect the throttle cable to the throttle body linkage.

42. Install the air cleaner and air intake duct.

43. Reconnect the negative battery cable.

44. Refill and bleed the cooling system.

45. Crank the engine until it starts. Air trapped in the fuel lines may cause the engine to crank for a longer period of time than normal.

46. Check the fuel lines, fuel rail, and injectors for any signs of leakage.

47. Warm the engine up to normal operating temperature and check the operation of the throttle cable and linkage. Adjust if necessary.

48. Check the manifold and throttle body mating surfaces for vacuum leaks.

Exhaust Manifold

REMOVAL & INSTALLATION

➡**Allow the engine to cool completely before removing the exhaust manifolds.**

1. Disconnect the negative battery cable.

2. Raise and safely support the vehicle.

3. Remove the air duct.

4. If necessary, remove the EGR pipe mounting bolts from the exhaust manifold.

5. Raise and safely support the vehicle.

6. If necessary to gain extra working room, remove the transfer case skid plate.

7. Label and detach the oxygen sensor connectors.

8. Remove the two stud nuts and two bolts and nuts and separate the front exhaust pipes from the exhaust manifold. Be careful not to damage the oxygen sensors when working around the exhaust pipes.

9. Lower the vehicle.

10. Remove the engine hanger and the heat shield.

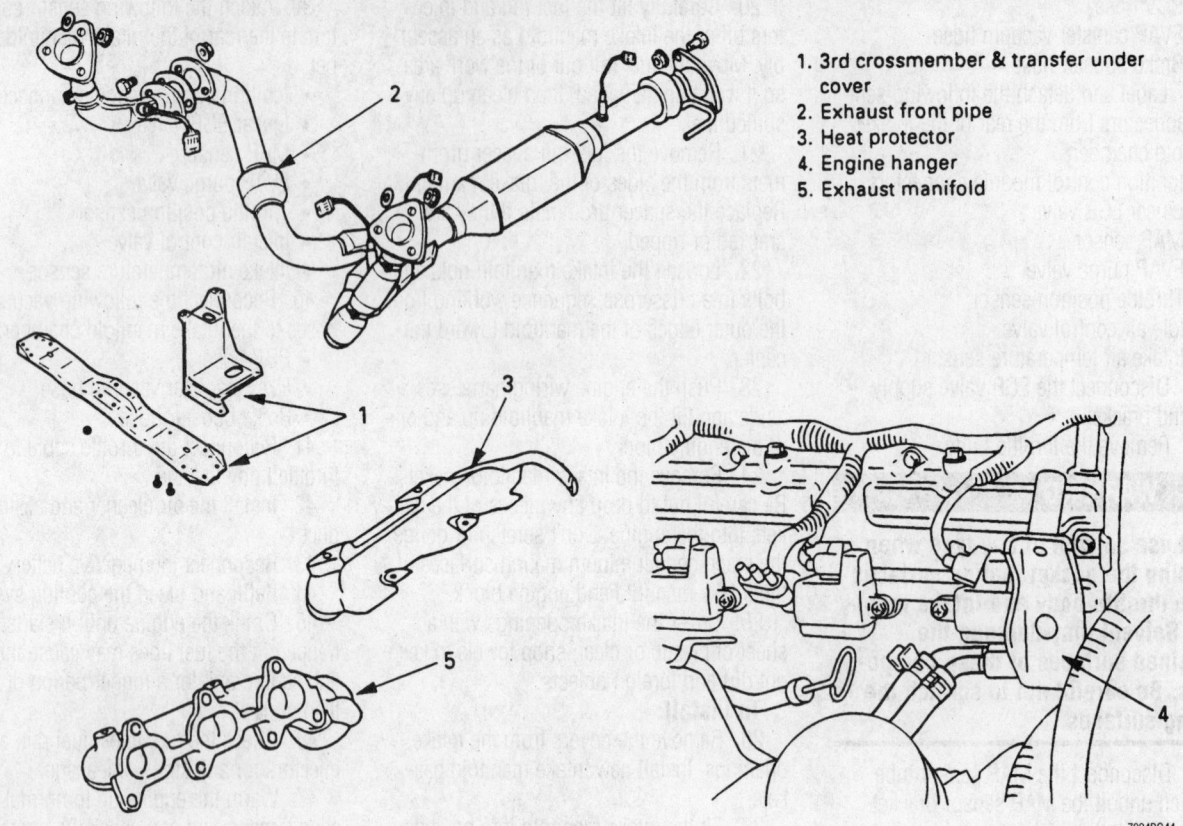

1. 3rd crossmember & transfer under cover
2. Exhaust front pipe
3. Heat protector
4. Engine hanger
5. Exhaust manifold

7924BG44

Identification of the right-hand exhaust manifold and related service components

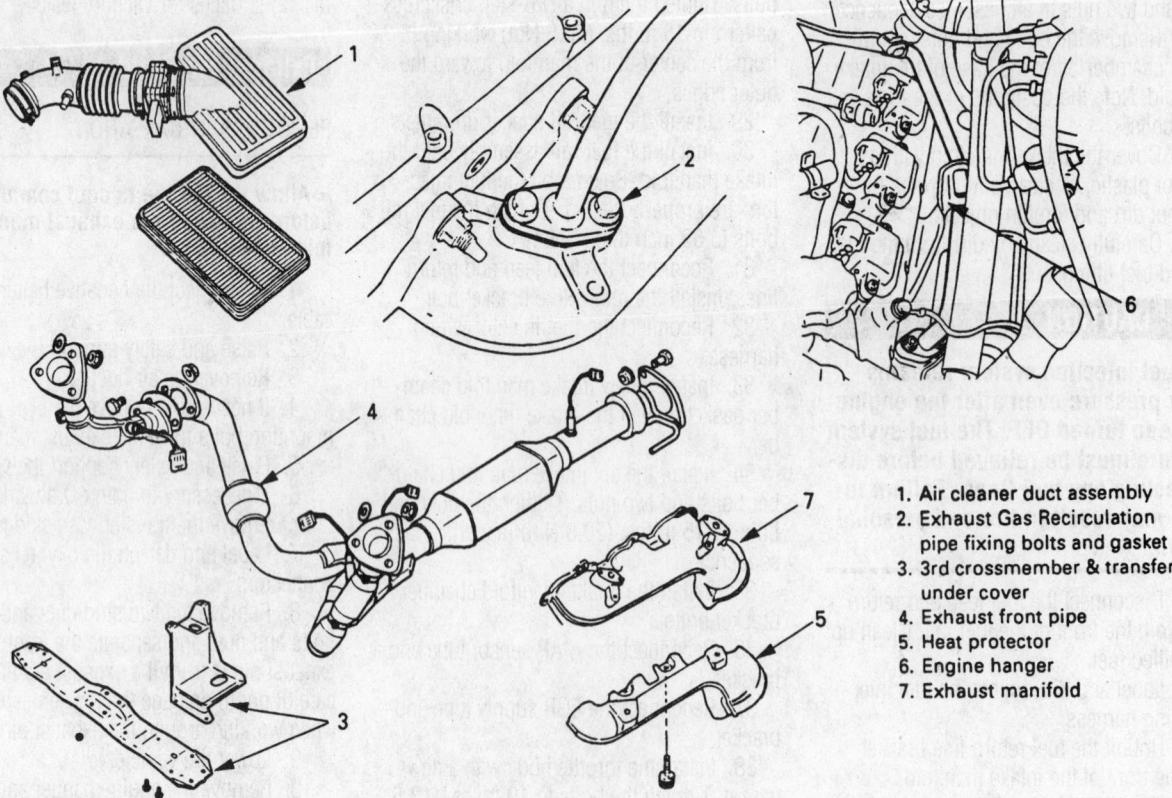

1. Air cleaner duct assembly
2. Exhaust Gas Recirculation pipe fixing bolts and gasket
3. 3rd crossmember & transfer under cover
4. Exhaust front pipe
5. Heat protector
6. Engine hanger
7. Exhaust manifold

7924BG45

Identification of the left-hand exhaust manifold and related service components

11. Remove the seven nuts and then remove the exhaust manifold from the cylinder head.

To install:

➡**Use new self-locking nuts and new gaskets when installing the exhaust manifolds.**

12. Install the exhaust manifold and gasket to the cylinder head using new nuts. Tighten the new nuts to 42 ft. lbs. (57 Nm) in a crisscross sequence.

13. Install the heat shield and the engine hanger.

14. Raise and safely support the vehicle.

15. Install the front exhaust pipes and reconnect the exhaust system. Tighten the exhaust fasteners to the following specifications:

- Stud nuts: 49 ft. lbs. (67 Nm)
- Flange nuts and bolts: 32–37 ft. lbs. (43–50 Nm)

16. Attach the oxygen sensor connectors.

17. Install the skid plate and tighten the bolts to 27 ft. lbs. (37 Nm).

18. Lower the vehicle.

19. Install the EGR pipe to the exhaust manifold. Tighten the mounting bolts to 21 ft. lbs. (28 Nm).

20. Install the air duct and connect the negative battery cable. Verify that all wires and vacuum lines have been reconnected.

21. Start the engine and check for exhaust leaks.

Front Crankshaft Oil Seal

REMOVAL & INSTALLATION

1. Disconnect the negative battery cable.

2. Loosen and remove the accessory drive belts.

3. Drain the coolant to a level below the upper radiator hose. Disconnect the upper radiator hose from the coolant inlet.

4. Remove the cooling fan and pulley assembly.

5. Remove the upper timing belt covers.

6. Rotate the crankshaft to align the timing marks.

7. Remove the crankshaft pulley and the lower timing belt cover.

8. Remove the timing belt.

9. Remove the crankshaft timing belt sprocket and key.

10. Use a seal puller and remove the crankshaft oil seal. Be careful not to damage the crankshaft or the oil pump sealing surface.

To install:

11. Apply engine oil to the lip of the seal and install the oil seal using Seal Installer tool No. J-39202 or an equivalent seal driver.

12. Install the crankshaft timing belt sprocket and key.

13. Install the timing belt.

14. Install the lower timing belt covers and tighten the bolts to 12 ft. lbs. (17 Nm).

15. Install the crankshaft pulley and tighten it to 123 ft. lbs. (167 Nm).

16. Verify that the timing belt has been installed correctly.

17. Install the upper timing belt covers.

18. Install the cooling fan and pulley assembly and tighten the bolts to 16 ft. lbs. (22 Nm).

19. Install and adjust the accessory drive belts.

20. Connect the upper radiator hose. Refill and bleed the cooling system.

21. Check the engine oil level and top up if necessary.

22. Reconnect the negative battery cable.

Camshaft

REMOVAL & INSTALLATION

3.2L Engine

1995 SOHC MODEL

➡**For timing belt removal and installation, please refer to the applicable procedure in the timing belt unit repair section.**

1. Disconnect the negative battery cable.

2. Remove the air cleaner assembly.

3. Remove the upper cooling fan shroud.

4. Remove the cooling fan assembly.

5. Disconnect the accelerator cable from the throttle body and remove it from the bracket.

6. Disconnect the canister vacuum hose from the vacuum pipe.

7. Disconnect the vacuum booster hose from the common chamber.

8. Detach the MAP sensor, canister Vacuum Switching Valve (VSV), Exhaust Gas Recirculation (EGR) VSV, Intake Air Temperature (IAT) sensor and ground connectors.

9. Remove the spark plug wires from the cylinder head cover.

10. Remove the ignition control module assembly.

11. Remove the four bolts and the throttle body from the common chamber.

12. Disconnect the vacuum hoses from the throttle body.

13. Disconnect the positive crankcase ventilation hose from the common chamber.

14. Disconnect the fuel pressure control valve vacuum hose from the common chamber.

15. Disconnect the evaporative emission canister purge hose from the common chamber.

16. Remove the EGR valve assembly.

17. Remove the common chamber from the intake manifold.

18. Disconnect the fuel feed and return hoses from the fuel rail assembly.

19. Remove the accessory drive belts.

20. Remove the power steering pump.

21. Remove the fan pulley assembly.

22. Remove the crankshaft pulley and damper.

23. Remove the timing belt cover.

24. Remove the timing belt auto-tensioner (pusher). The pusher prevents air from entering the oil chamber. Its rod must always be facing upward.

25. Align the timing marks and remove the timing belt. After the timing marks are aligned, the engine must not be disturbed.

26. Remove the cylinder head cover(s).

27. Remove the camshaft pulley.

28. Remove the camshaft front plate.

29. Remove the camshaft mounting bracket bolts and the camshaft.

To install:

30. Install new camshaft seals and retaining plates onto the cylinder. Tighten the right camshaft seal retaining plate 6mm bolts to 65 inch lbs. (7 Nm). Tighten the left camshaft seal retaining plate 8mm bolts to 191 inch lbs. (21 Nm).

31. Apply sealant to the mounting surfaces on the cylinder head where the front and rear camshaft mounting brackets attach to the cylinder head.

32. Install the camshaft and mounting brackets. Tighten the bolts in sequence to 69 inch lbs. (8 Nm) for the M6 bolts and 13 ft. lbs. (18 Nm) for the M8 bolts.

33. Install the front plate. Tighten the bolts to 12 ft. lbs. (17 Nm).

34. Install the camshaft pulley. Tighten the mounting bolts to 41 ft. lbs. (55 Nm).

35. Apply sealant to both sides of the front and rear camshaft mounting brackets and install the cylinder head cover(s). Tighten the bolts to 69 inch lbs. (8 Nm). Do not overtighten the cylinder head covers, they crack very easily.

36. Install the timing belt.

37. Install the timing belt auto-tensioner. Tighten the mounting bolts to 13 ft. lbs. (18 Nm).

38. Rotate the crankshaft by hand to verify that the timing belt is aligned properly and there is no piston-to-valve interference.

39. Install the timing belt cover.

40. Install the oil cooler hoses and bracket.

41. Install the crankshaft pulley assembly. Tighten the center bolt to 123 ft. lbs. (167 Nm).

42. Install the fan pulley assembly. Tighten the mounting bolts to 16 ft. lbs. (22 Nm).

43. Install the power steering pump.

44. Install the accessory drive belts.

45. Connect the fuel hoses to the fuel rail assembly.

46. Install the common chamber. Tighten the bolts and nuts to 17 ft. lbs. (24 Nm) in a crisscross pattern.

47. Install the EGR valve assembly. Tighten the mounting bolts on the valve side to 69 inch lbs. (8 Nm) and the bolts on the exhaust side to 21 ft. lbs. (28 Nm).

48. Connect the evaporative emission canister purge hose.

49. Connect the fuel pressure control valve vacuum hose.

50. Connect the positive crankcase ventilation hose.

51. Install the throttle body assembly. Tighten the mounting bolts to 14 ft. lbs. (19 Nm).

52. Connect the vacuum hoses to the throttle body.

53. Install the ignition control module and the spark plug wires.

54. Attach the MAP sensor, canister VSV, EGR VSV, IAT sensor and ground connectors.

55. Connect the vacuum booster hose.

56. Connect the air vacuum hose.

57. Connect the accelerator cable. Adjust the accelerator cable by pulling the cable housing while closing the throttle valve and tightening the adjusting nut and screw cap by hand temporarily. Now loosen the adjusting nut by three turns and then tighten the screw cap. Be sure the throttle valve reaches the screw stop when the throttle is closed.

58. Install the cooling fan assembly and the upper fan shroud.

59. Install the air cleaner assembly.

60. Connect the negative battery cable.

1995 DOHC MODEL

1. Properly relieve the fuel system pressure.

2. Disconnect the negative battery cable.

3. Remove the air cleaner assembly.

4. Remove the upper cooling fan shroud.

5. Remove the cooling fan assembly.

6. Disconnect the accelerator cable from the throttle body and remove it from the bracket.

7. Disconnect the canister vacuum hose from the vacuum pipe.

8. Disconnect the vacuum booster hose from the common chamber.

9. Detach the following connectors:
- MAP sensor
- Canister Vacuum Switching Valve (VSV)
- Exhaust Gas Recirculation (EGR) VSV
- Intake Air Temperature (IAT) sensor
- Ground cables

10. Disconnect and label the spark plug wires.

11. Remove the ignition control module assembly with the spark plug wire attached.

12. Remove the four bolts to separate the throttle body from the common chamber.

13. Disconnect the vacuum hoses from the throttle body.

14. Disconnect the Positive Crankcase Ventilation (PCV) hose from the common chamber.

15. Disconnect the fuel pressure control valve vacuum hose from the common chamber.

16. Disconnect the evaporative emission canister purge hose from the common chamber.

17. Remove the EGR valve assembly.

18. Remove the common chamber air duct.

19. Remove the common chamber from the intake manifold.

20. Disconnect the fuel feed and return hoses from the fuel rail assembly.

21. Remove the accessory drive belts.

22. Remove the power steering pump.

23. Remove the fan pulley assembly.

24. Remove the crankshaft pulley and damper.

25. Remove the oil cooler hoses and bracket.

26. Remove the timing belt cover.

27. Align the timing marks

28. Remove the timing belt auto-tensioner (pusher). The pusher prevents air from entering the oil chamber. Its rod must always be facing upward.

29. Remove the timing belt.

30. Remove the cylinder head covers.

31. Remove the camshaft pulley.

32. Remove the camshaft front plate.

33. Remove the camshaft mounting bracket bolts.

34. Remove the camshaft chain tensioner bolts.

35. Remove the camshafts with the timing chain and tensioner attached.

To install:

➡ **Install a retainer on the chain tensioner to prevent the plunger from moving. Remove the retainer after installation.**

36. Apply sealant to the cylinder head mounting surface of the front and rear camshaft mounting brackets.

37. Apply clean engine oil to the camshaft journals, lobes, and sprockets.

38. Install the camshaft assembly with the chain and tensioner. Take care not to install the wrong tensioner, the left and right tensioners are different and they are marked accordingly. Be sure the timing marks on the camshaft chain sprockets are aligned with the timing marks on the chain links. Tighten the camshaft holder mounting bolts to 87 inch lbs. (10 Nm).

39. Install the chain tensioner. Tighten the bolts to 14 ft. lbs. (19 Nm).

40. Install new camshaft seals and retaining plates onto the cylinder. Tighten the rear camshaft seal retaining plate 6mm bolts to 70 inch lbs. (8 Nm). Tighten the front camshaft seal retaining bolts to 65 inch lbs. (7 Nm).

41. Install the camshaft pulley. Hold the camshaft with an open end wrench to prevent it from turning and tighten the pulley bolts to 46 ft. lbs. (63 Nm).

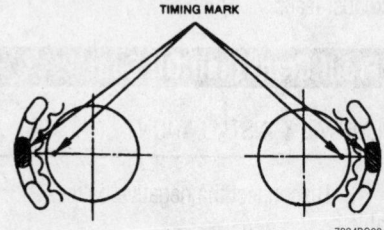

Camshaft and chain timing marks, properly aligned for installation—1995 DOHC engine

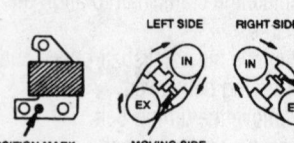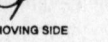

Chain tensioner installed position—1995 DOHC engine

42. Apply sealant to both sides of the front and rear camshaft mounting brackets and install the cylinder head cover(s). Tighten the bolts to 69 inch lbs. (8 Nm). Don't over-tighten the bolts, as the head covers may crack.

43. Install the timing belt.

44. Install the timing belt auto-tensioner. Tighten the mounting bolts to 14 ft. lbs. (19 Nm).

45. Install the timing belt cover.

46. Install the oil cooler hoses and bracket.

47. Install the crankshaft pulley assembly. Tighten the center bolt to 123 ft. lbs. (167 Nm).

48. Install the fan pulley assembly. Tighten the mounting bolts to 16 ft. lbs. (22 Nm).

49. Install the power steering pump.

50. Install the accessory drive belts.

51. Connect the fuel hoses to the fuel rail assembly.

52. Install the common chamber. Tighten the bolts and nuts to 17 ft. lbs. (24 Nm).

53. Install the common chamber air duct. Tighten the nuts and bolts to 17 ft. lbs. (24 Nm).

54. Connect the ground cable.

55. Install the EGR valve assembly. Tighten the mounting bolts on the valve side to 69 inch lbs. (8 Nm) and the bolts on the exhaust side to 21 ft. lbs. (28 Nm).

56. Connect the evaporative emission canister purge hose.

57. Connect the fuel pressure control valve vacuum hose.

58. Reconnect the fuel lines using new washers.

59. Connect the positive crankcase ventilation hose.

60. Install the throttle body assembly. Tighten the mounting bolts to 14 ft. lbs. (19 Nm).

61. Connect the vacuum hoses to the throttle body.

62. Install the ignition control module and the spark plug wires.

63. Attach the MAP sensor, canister VSV, EGR VSV, the IAT sensor connectors, and ground cables.

64. Connect the vacuum booster hose.

65. Connect the air vacuum hose.

66. Connect the accelerator cable. Adjust the accelerator cable by pulling the cable housing while closing the throttle valve and tightening the adjusting nut and screw cap by hand temporarily. Now loosen the adjusting nut by three turns and then tighten the screw cap. Be sure the throttle valve reaches the screw stop when the throttle is closed.

67. Install the cooling fan assembly and the upper fan shroud.

68. Install the air cleaner assembly.

69. Connect the negative battery cable.

70. Start the engine and check for fuel leaks.

1996–97 MODELS

✳✳ CAUTION

The fuel injection system remains under pressure even after the engine has been turned OFF. The fuel system pressure must be relieved before disconnecting any fuel lines. Failure to do so may result in fire and personal injury.

1. Properly relieve the fuel system pressure.

2. Disconnect the negative battery cable.

3. Drain the coolant into a sealable container.

4. Support the hood as far open as possible.

5. Remove the air intake duct and the air cleaner box.

6. Disconnect and remove the upper and lower radiator hoses. Catch any coolant that runs out.

7. Loosen and remove the power steering pump, A/C compressor, and alternator drive belts.

8. Remove the cooling fan and its pulley assembly.

9. Unbolt the power steering pump mounting bracket. Move the pump and bracket out of the way without disconnecting the hydraulic lines.

10. Disconnect the throttle cable from the throttle body linkage.

11. Label and disconnect all applicable vacuum hoses from the intake manifold chamber.

12. Label and detach all necessary sensor connectors from the rear of the intake manifold chamber.

13. Disconnect the EGR valve supply tube and bracket.

14. First, remove the throttle body, and then remove the intake manifold chamber.

15. Carefully clean any dirt from the fuel rail and fuel fittings.

16. Disconnect the fuel feed and return lines from the front of the fuel rail. Clean up any spilled fuel.

17. Remove the intake manifold gaskets. Be careful not to drop any pieces of the gaskets into the engine. Don't scratch or gouge the machined aluminum mating surfaces of the intake manifold and engine block.

18. Cover the intake openings with a sheet of plastic or clean shop towels to keep out dirt and foreign objects.

19. Label the ignition coil assemblies and disconnect them from the wiring harness. Remove the coil assemblies so they won't be damaged.

20. Unbolt the oil cooler line brackets from the timing belt covers.

21. Remove the upper timing belt covers.

22. Rotate the crankshaft to align the camshaft timing marks with the pointer dots on the back covers. When the timing marks are aligned, the No. 2 cylinder is at TDC/compression.

23. Remove the crankshaft pulley. Remove the lower timing belt cover.

24. Remove the pusher assembly (tensioner) from below the timing belt tensioner pulley. The pusher rod must always be facing upward to prevent oil leakage. Depress the pusher rod, and insert a wire pin into the hole to keep the rod retracted.

25. Remove the timing belt.

✳✳ WARNING

If the timing belt is worn, damaged, or shows signs of oil or coolant contamination it must be replaced.

26. Loosen the valve cover bolts in a crisscross sequence. Remove the valve covers.

27. Remove the camshaft sprockets and back covers.

28. Loosen the camshaft holder bolts in a crisscross sequence to prevent warping.

29. Remove the camshaft and camshaft holders from the cylinder head.

30. Inspect the camshaft lobes and journals for signs of wear or damage.

To install:

31. Be sure all mating surfaces are clean and free of oil, coolant, or gasket residue.

32. Lubricate the camshaft lobes and journals with clean engine oil.

33. Apply a bead of sealant to the front and rear camshaft holder mating surfaces on the cylinder head.

34. Install the camshaft and holder assembly onto the cylinder head before the sealant cures. Install the camshaft holder bolts, but don't tighten them yet.

35. Use a crisscross sequence to tighten the camshaft holder bolts. Tighten the 8mm bolts to 13 ft. lbs. (18 Nm). Tighten the 6mm bolts to 72 inch lbs. (8 Nm).

36. Use a seal driver to install a new camshaft seal.

37. Install the camshaft sprocket back covers and tighten their bolts to 12 ft. lbs. (17 Nm).

38. Install the camshaft sprockets so that the timing marks are aligned. Tighten the bolts to 46 ft. lbs. (64 Nm).

39. Apply a ⅛ in. (3mm) bead of sealant to the joint were the camshaft holders meet the cylinder head. Install the valve cover with a new gasket before the sealant cures.

40. Tighten the valve cover bolts to 6 ft. lbs. (8 Nm) in crisscross pattern.

41. Verify that the camshaft and crankshaft timing marks are properly aligned.

42. Install and tension the timing belt. Tighten the pusher bolts to 14 ft. lbs. (19 Nm).

43. Install the lower timing belt covers and tighten the bolts to 13 ft. lbs. (18 Nm). Install the crankshaft pulley. Tighten the pulley bolt to 123 ft. lbs. (167 Nm).

44. Install the upper timing belt covers and tighten the bolts to 13 ft. lbs. (18 Nm).

45. Fit the oil cooler line brackets onto the timing cover and tighten the bolts to 13 ft. lbs. (18 Nm).

46. Reconnect the fuel feed and return lines.

47. Install and reconnect the ignition coil assemblies.

48. Install the intake manifold chamber and throttle body with new gaskets. Tighten the nuts and bolts to 17 ft. lbs. (24 Nm).

49. Reconnect the throttle cable to the throttle body linkage.

50. Reconnect the EGR valve supply tube and bracket.

51. Reconnect all vacuum hoses to the intake manifold chamber.

52. Reattach all of the sensor connectors to the rear of the intake manifold chamber.

53. Install the power steering pump and mounting bracket.

54. Install the cooling fan and its pulley assembly.

55. Install and tension the alternator, A/C compressor, and power steering pump drive belts.

56. Install and reconnect the upper and lower radiator hoses.

57. Install the air cleaner box and air intake duct.

58. Verify that all fuel lines, vacuum and coolant hoses, and wiring harness have been reconnected.

59. Refill the engine with fresh coolant.

60. Crank the engine until it starts. A longer-than-normal starting time may be necessary due to air in the fuel lines. Check all fuel line connections for leaks.

61. Bleed any air from the cooling system.

62. Bleed the power steering system if necessary.

63. Check the throttle cable operation and adjustment.

64. Check the engine oil level and add if necessary.

3.5L Engine

1. Disconnect the negative battery cable.

2. Remove the crankshaft pulley assembly using tool J-8614–01 or equivalent crankshaft holder, hold the pulley then remove the center bolt and pulley assembly.

3. Remove the timing belt using the recommended procedure in the timing belt unit repair section.

4. Label, then disconnect all wiring harnesses and vacuum hoses related to valve cover removal.

5. Remove the left and right valve covers.

6. Remove the twenty bolts from the camshaft hold-down brackets, and remove the camshafts.

7. Remove the bolt holding the camshaft drive gear pulley.

8. Remove the three bolts from the camshaft drive gear retainer, then the camshaft drive gear assembly.

To install:

9. Install the camshaft drive gear assembly and tighten the three bolts to 89 inch lbs. (10 Nm).

10. Tighten the camshaft drive gear pulley bolt to 72 ft. lbs. (98 Nm).

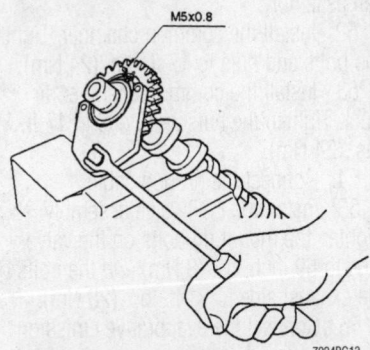

Aligning the sub gear with the Gear Spring Lever J-42686—3.5L engine

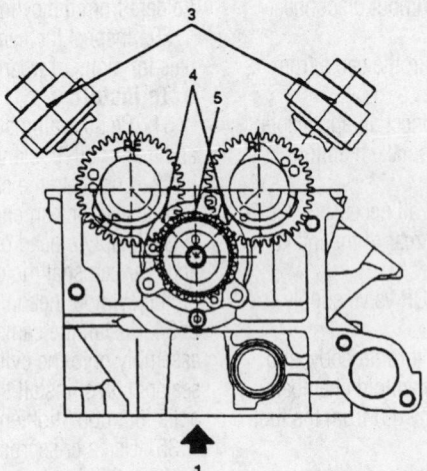

1	Right Bank	3	Alignment Mark on Camshaft Drive Gear
2	Left Bank	4	Alignment Mark on Camshaft
		5	Alignment Mark on Retainer

Camshaft alignment marks for the left and right cylinder heads—3.5L engine

7924BG11

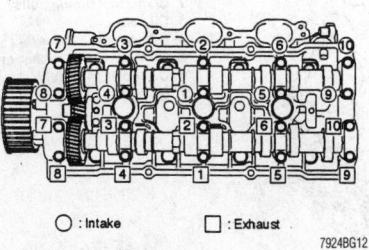

Camshaft retaining bracket tightening sequence—3.5L engine

○ : Intake   □ : Exhaust

7924BG12

11. Use the Gear Spring Lever J-42686 or equivalent to turn the sub gear until it aligns with the M5 bolt hole between the camshaft driven gear and the sub gear, then tighten the bolt to prevent the sub gear from moving.

12. Apply engine oil to the camshaft journal and bearing surface of the camshaft brackets.

13. Install and align the camshaft timing marks as shown.

14. Install the camshaft brackets and tighten the bolts in sequence to 89 inch lbs. (10 Nm).

15. The remaining steps of this procedure are the reverse of the installation, noting the crankshaft bolt should be tightened to 123 ft. lbs. (167 Nm).

Hydraulic Lash Adjusters

BLEEDING

➡ **The hydraulic lash adjuster is a precision component that relies on a clean operating environment. When handling the lash adjuster, make certain that no dirt or foreign particles are allowed to get inside the unit. Do not try to take the unit apart; it can't be rebuilt. The adjuster is filled with oil. Hold the adjuster in the upright position so that the oil does not spill out. Use only clean engine oil when bleeding the lifters.**

To bleed the hydraulic lash adjusters when disassembling the rocker arms or as part of an engine overhaul, perform the following:

1. Hold the adjuster upright in a container of clean engine oil.

2. Insert a 0.08 in. (2mm) diameter wire into the adjuster. Lightly press down on the steel check ball.

3. Hold the check ball in with the wire. Manually move the plunger up and down at

one second intervals until all air bubbles disappear.

4. Remove the wire from the adjuster and push down firmly on the plunger. If the plunger moves even slightly, repeat the previous steps until the plunger stops moving. If the plunger continues to move, the lifter must be replaced.

Oil Pan

REMOVAL & INSTALLATION

1. Disconnect the negative battery cable.
2. Raise and safely support the vehicle.
3. Drain the engine oil.
4. Remove the front wheels.
5. Remove the dipstick from the dipstick tube.
6. Remove the radiator skid plate and the radiator lower shroud.
7. Remove the suspension crossmember from below the oil pan.
8. Remove the flywheel dust cover.
9. Remove the outer tie rod end cotter pins and castle nuts. Use a ball joint separator tool to disconnect the tie rod ends from the steering knuckles.
10. Matchmark the Pitman arm to the steering shaft and use a puller, (tool No. J-29107 or equivalent) to remove the Pitman arm.
11. Unbolt the idler arm bracket from the frame. Then, remove the steering linkage from the vehicle as an assembly.
12. If equipped with four wheel drive, perform the following steps:

a. Support the axle assembly with a jack and safety stands.

b. Remove the mounting bolts and nuts from the axle assembly mounting brackets on both sides of the axle assembly.

c. Lower the axle assembly to gain access to the oil pan. Support the axle assembly so that the halfshafts aren't stressed.

13. Remove the oil pan mounting bolts. Use a sealer cutter to break the seal and separate the oil pan from the engine block.

To install:

14. Thoroughly clean and dry the sealing surface of the oil pan and engine block. Apply a continuous bead of sealant to the oil pan flange and install the oil pan to the engine block. Do not allow the sealant to cure before installation. Tighten the mounting bolts to 7.4 ft. lbs. (10 Nm) in a two-step crisscross sequence.

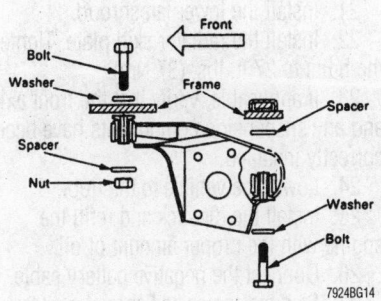

7924BG14

Exploded view of the axle bracket mounting bolt, spacer, and nut locations

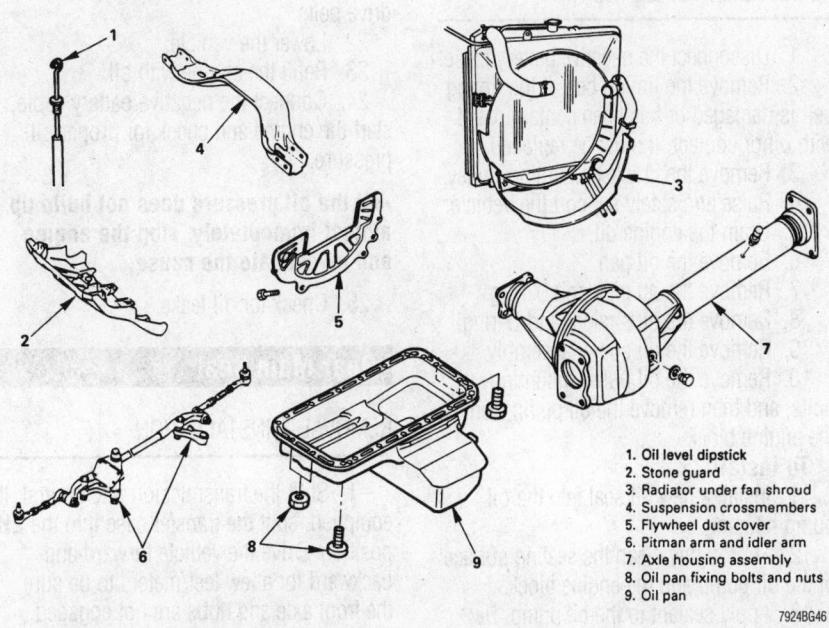

1. Oil level dipstick
2. Stone guard
3. Radiator under fan shroud
4. Suspension crossmembers
5. Flywheel dust cover
6. Pitman arm and idler arm
7. Axle housing assembly
8. Oil pan fixing bolts and nuts
9. Oil pan

7924BG46

Identification of the oil pan (9) and related service components

15. If equipped, install the axle housing assembly. Tighten the axle mounting bracket bolts and nut to 112 ft. lbs. (152 Nm). If the axle housing flange bolts at the mounting brackets were loosened or removed, tighten them to 61 ft. lbs. (82 Nm).

16. Install the idler arm. Tighten the mounting bolts to 33 ft. lbs. (45 Nm).

17. Align the matchmark and install the Pitman arm on the sector shaft. Tighten the nut to 160 ft. lbs. (216 Nm).

18. Install the flywheel dust cover.

19. Install the suspension crossmember. Tighten the mounting bolts to 58 ft. lb. (78 Nm).

20. Install the steering linkage, Pitman arm, and idler arm assembly. Tighten the castle nuts to the following specifications:

    a. Tie rod end castle nuts: 73 ft. lbs. (98 Nm).

    b. Idler arm bracket bolts: 33 ft. lbs. (44 Nm).

    c. Pitman arm nuts: 159 ft. lbs. (216 Nm).

21. Install the lower fan shroud.

22. Install the radiator skid plate. Tighten the bolts to 27 ft. lbs. (37 Nm).

23. If applicable, verify that the front axle and any suspension components have been correctly installed.

24. Lower the vehicle to the floor.

25. Install the dipstick and refill the engine with the proper amount of oil.

26. Connect the negative battery cable.

27. Start the engine and check for oil leaks.

Oil Pump

REMOVAL & INSTALLATION

1. Disconnect the negative battery cable.

2. Remove the timing belt. If the timing belt is damaged or has been contaminated with oil or coolant, it must be replaced.

3. Remove the crankshaft timing pulley.

4. Raise and safely support the vehicle.

5. Drain the engine oil.

6. Remove the oil pan.

7. Remove the oil pipe and O-ring.

8. Remove the oil strainer and O-ring.

9. Remove the oil cooler assembly.

10. Remove the oil pump mounting bolts, and then remove the oil pump from the engine block.

To install:

11. Install a new oil seal into the oil pump housing.

12. Thoroughly clean the sealing surface of the oil pump and the engine block.

13. Apply sealant to the oil pump. Be careful not to block the oil ports.

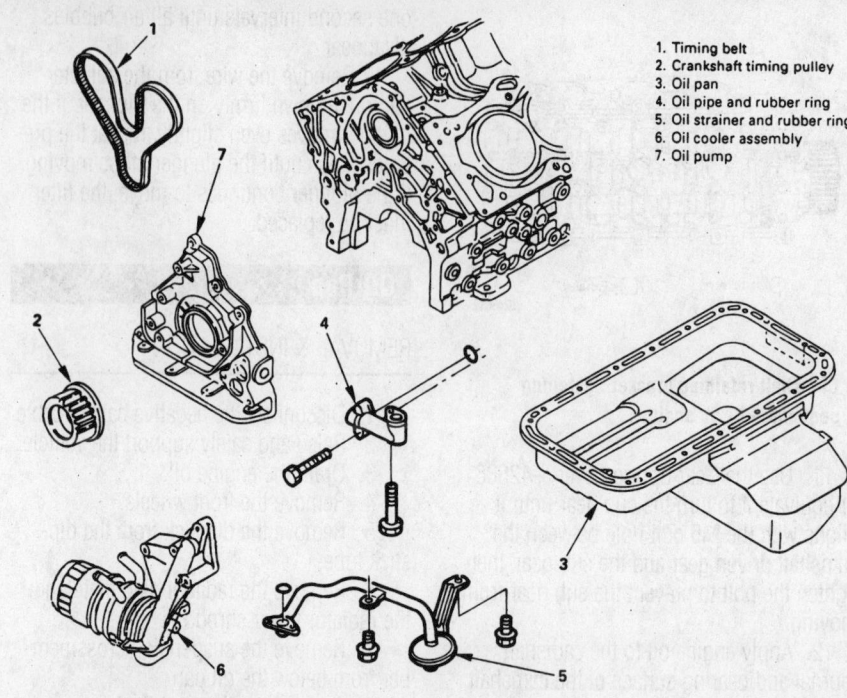

1. Timing belt
2. Crankshaft timing pulley
3. Oil pan
4. Oil pipe and rubber ring
5. Oil strainer and rubber ring
6. Oil cooler assembly
7. Oil pump

Identification of the oil pump (7) and related service components

14. Apply engine oil to the seal lip and install the oil pump on the engine block. Tighten the mounting bolts to 13 ft. lbs. (18 Nm). Take care not to drop the garter spring from the seal lid during installation.

15. Install the oil cooler assembly.

16. Install the oil pipe and O-ring.

17. Install the oil strainer and O-ring.

18. Install the oil pan.

19. Install the crankshaft timing pulley.

20. Install the timing belt.

21. Install the remaining accessories and drive belts.

22. Lower the vehicle.

23. Refill the engine with oil.

24. Connect the negative battery cable, start the engine and check for proper oil pressure.

➡**If the oil pressure does not build up almost immediately, stop the engine and investigate the cause.**

25. Check for oil leaks.

Rear Main Seal

REMOVAL & INSTALLATION

1. Shift the transmission into neutral. If equipped, shift the transfer case into the **2H** position. Drive the vehicle forward and backward for a few feet/meters to be sure the front axle and hubs are not engaged.

2. Disconnect the negative battery cable.

3. Remove the screws securing the shifter console. On manual transmission vehicles, remove both shift knobs. On automatic transmission vehicles, remove only the transfer case shift knob. Lift the console up and over the shift levers.

4. Unbolt and remove the transmission and transfer case shift levers from inside the vehicle.

5. Raise and safely support the vehicle.

6. Remove the transmission assembly.

➡**Note the positions of the transmission mounting dowels. Manual and automatic transmission use different dowel positions.**

7. If equipped with a manual transmission, remove the clutch assembly.

8. Loosen the flywheel bolts in a two-step crisscross sequence and remove the flywheel.

9. Using a seal puller, remove the rear main seal. Do not damage the crankshaft sealing surface.

To install:

10. Clean any oil from the crankshaft and flywheel mounting surfaces. Clean the bolt holes in the end of the crankshaft.

11. Apply engine oil to the oil seal lip and install the seal using Seal Installer J-39201 or equivalent.

12. Install the flywheel using new flywheel bolts. Tighten the bolts in a two-step crisscross pattern to 40 ft. lbs. (54 Nm).

7924BG47

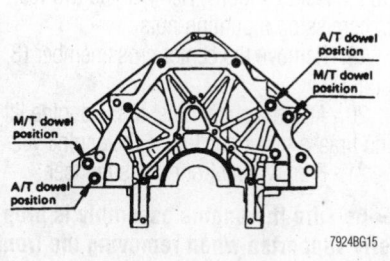

Automatic and manual transmission mounting dowel positions

✳✳ WARNING

Do not coat the flywheel bolts with oil or any type of thread-locking compound.

13. If removed, install the clutch assembly.

14. Verify that the transmission mounting dowels are installed correctly.

15. Install the transmission assembly.

16. Lower the vehicle and install the transmission and transaxle shift levers.

17. Install the console and shift knobs.

18. Connect the negative battery cable.

19. Check the oil and refill as necessary.

20. Road test the vehicle and check the transmission, clutch, and transfer case operation.

FUEL SYSTEM

Fuel System Service Precautions

Safety is the most important factor when performing not only fuel system maintenance but any type of maintenance. Failure to conduct maintenance and repairs in a safe manner may result in serious personal injury or death. Maintenance and testing of the vehicle's fuel system components can be accomplished safely and effectively by adhering to the following rules and guidelines:

• To avoid the possibility of fire and personal injury, always disconnect the negative battery cable unless the repair or test procedure requires that battery voltage be applied.

• Always relieve the fuel system pressure prior to disconnecting any fuel system component (injector, fuel rail, pressure regulator, etc.), fitting or fuel line connection.

Exercise extreme caution whenever relieving fuel system pressure to avoid exposing skin, face and eyes to fuel spray. Please be advised that fuel under pressure may penetrate the skin or any part of the body that it contacts.

• Always place a shop towel or cloth around the fitting or connection prior to loosening to absorb any excess fuel due to spillage. Ensure that all fuel spillage (should it occur) is quickly removed from engine surfaces. Ensure that all fuel soaked cloths or towels are deposited into a suitable waste container.

• Always keep a dry chemical (Class B) fire extinguisher near the work area.

• Do not allow fuel spray or fuel vapors to come into contact with a spark or open flame.

• Always use a backup wrench when loosening and tightening fuel line connection fittings. This will prevent unnecessary stress and torsion to fuel line piping. Always follow the proper tightening specifications.

• Always replace worn fuel fitting O-rings with new. Do not substitute fuel hose or equivalent, where fuel pipe is installed.

Fuel System Pressure

RELIEVING

1. Remove the fuel filler cap.

2. Remove the fuel pump relay from the underhood relay box.

3. Start the engine and let it run until it stalls, then crank the engine for an additional 30 seconds.

4. Turn the ignition switch to the **OFF** position and remove the key. Disconnect the negative battery cable and then install the fuel pump relay.

Fuel Filter

REMOVAL & INSTALLATION

✳✳ CAUTION

The fuel injection system remains under pressure even after the engine has been turned OFF. The fuel system pressure must be relieved before disconnecting any fuel lines. Failure to do so may result in fire and personal injury.

1. Properly relieve the fuel system pressure.

2. Raise and safely support the vehicle.

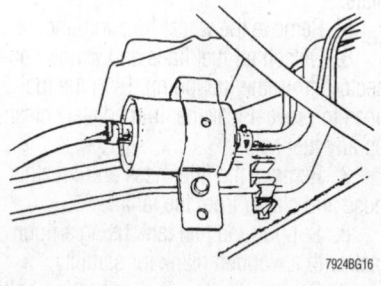

Fuel filter mounting location

3. Use block-off clamps to pinch the fuel lines shut. Disconnect the fuel lines from the fuel filter. Clean up any fuel spills.

4. Loosen the filter mounting bolt and remove the fuel filter.

To install:

5. Install the fuel filter with the label, arrow, or manufacturer's mark facing toward the front of the vehicle and tighten the bracket bolt.

6. Connect the fuel lines to the fuel filter.

7. Lower the vehicle to the floor and install the filler cap.

8. Reconnect the negative battery cable.

9. Start the engine and inspect the fuel filter connections for leaks.

Fuel Pump

REMOVAL & INSTALLATION

✳✳ CAUTION

The fuel injection system remains under pressure after the engine has been turned OFF. Properly relieve fuel pressure before disconnecting any fuel lines. Failure to do so may result in fire or personal injury.

1. Properly relieve the fuel pressure.

2. Raise and support the vehicle safely.

3. Drain the fuel into an approved sealable container. Install the drain plug with a new washer and tighten 8mm bolts to 14 ft. lbs. (20 Nm), or 14mm bolts to 22 ft. lbs. (29 Nm).

✳✳ CAUTION

Do not allow fuel spray or fuel vapors to come in contact with a spark or open flame. Keep a dry chemical fire extinguisher nearby. Never store fuel in an open container due to risk of fire or explosion.

4. Unbolt and remove the fuel tank skid plate.

5. Remove the wheel housing liner.

6. Detach all fuel lines and wiring connectors from the fuel pump. Plug the fuel lines to prevent leakage. Immediately clean up any fuel spills.

7. Remove the filler neck and breather hose and clamp from the tank.

8. Support the fuel tank using a floor jack with a wooden plank for stability.

9. Remove the fuel tank mounting bolts and lower the tank from the vehicle 3–4 in. (76–101mm). Disconnect the fuel filter line from the pump.

10. Lower the fuel tank from the vehicle.

11. Unbolt the fuel pump bracket plate. Lift the fuel pump assembly out of the tank and allow any fuel to drain into the tank.

12. Remove the fuel pump. If the pump is out of the tank for a long period of time, cover the tank opening to keep out moisture and foreign objects.

To install:

13. Install the fuel pump assembly into the tank with a new gasket. Evenly tighten the bracket plate bolts to prevent leakage.

14. Raise the tank and connect the hoses.

15. Connect the fuel lines to their ports on the pump.

16. Connect the breather hose and clamp.

17. Attach the wiring connectors.

18. Connect the fuel filler line and breather hose.

19. Raise the tank into position. Install the fuel tank mounting bolts and tighten them to 27–30 ft. lbs. (36–39 Nm). Lower the jack.

20. Install the fuel tank skid plate and tighten its bolts to 27 ft. lbs. (37 Nm).

21. Install the wheel arch liner.

22. Lower the vehicle safely.

23. Refill the fuel tank.

24. Install the fuel pump relay. Reconnect the negative battery cable.

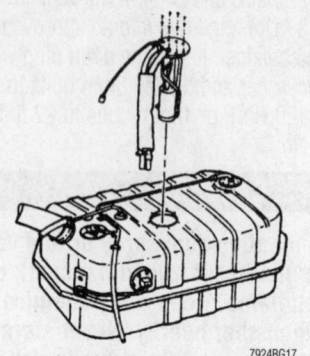

To remove the fuel pump, lift it up and out of the fuel tank

25. Turn the ignition to the **ON** position to pressurize the fuel system. Then, check for leaks and proper fuel system operation.

DRIVE TRAIN

Transmission Assembly

REMOVAL & INSTALLATION

Manual

➡**The transfer case is an integral part of the transmission housing. Although the two cases can be separated, the transfer case should be removed with the transmission.**

1. Shift the transmission into neutral. Shift the transfer case into **2H** and drive the vehicle forward and backward a few feet/meters to be sure the front axle is not engaged.

2. Use a felt-tipped marker to matchmark the hood to the hood hinges. Remove the hood.

3. Disconnect the negative battery cable.

4. Remove the shift knobs and the console and shift boots.

5. Remove the shift lever and the transfer case shift lever, by unbolting their mounting plates from the transmission case.

6. Raise and safely support the vehicle.

7. Drain the transmission fluid.

8. Remove the exhaust and transfer case skid plates.

9. Label and disconnect the oxygen sensor connectors from the transmission harness.

10. Remove the catalytic converter, left front, and center exhaust pipes.

11. Remove the harness heat protector.

12. Remove the slave cylinder heat protector. Unbolt and remove the slave cylinder. Don't disconnect the hydraulic line.

13. Remove the slave cylinder dust covers.

14. Matchmark the driveshafts at the their flanges. Unbolt and remove the driveshafts.

15. Label and detach the reverse light switch, 4WD indicator switch, and 1–2 and 3–4 indicator switch harness connectors.

16. Disconnect the speed sensor harness connector.

17. Remove the two harness clamps from the transmission case.

18. Using a transmission jack, raise the transmission slightly. Remove the two rear transmission mounting nuts.

19. Remove the center crossmember (8 bolts).

20. Attach a chain hoist to the engine lifting brackets, but don't raise the engine yet.

21. Remove the front crossmember.

➡**Be sure the engine assembly is properly supported when removing the front crossmember.**

22. Remove the three flywheel inspection cover bolts.

23. Use the Clutch Release Bearing Remover tool J-39207, or an equivalent, pry tool to release the bearing from the pressure plate. Push the release fork toward the rear of the vehicle. Insert the tool between the release bearing and the pressure plate collar. Move the lever toward the rear to separate.

24. Raise the engine slightly with a chain hoist and remove the bolts and nuts securing the transmission to the engine.

25. Carefully pull the transmission rearward. Lower the transmission from the vehicle.

To install:

➡**Be sure the transmission dowel pins are installed in the correct position. If the dowels are in the wrong hole, the**

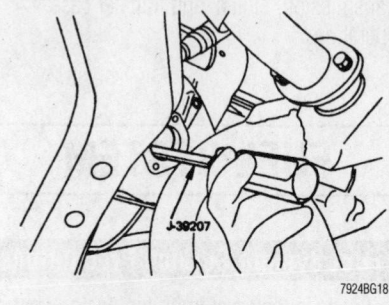

Insert the release bearing remover (J-39207 or equivalent) through the hole in the bell housing

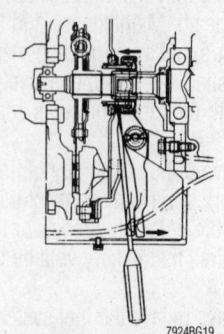

Pull the release bearing fork toward the transmission to release the bearing from the pressure plate

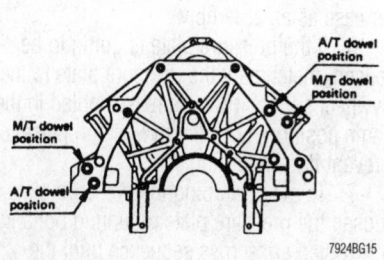

7924BG15

Transmission mounting dowel position

transmission case may crack when the engine-to-transmission bolts are tightened.

26. Apply a thin coating of molybdenum grease to the spline of the input shaft, slowly raise the transmission into position with the rear of the engine. Align the splines of the input shaft with the grooves of the clutch disc hub and install the transmission to the engine.

➡**It may be helpful the put the transmission in gear and rotate the driveshaft flange so the input shaft will turn and engage the grooves in the clutch disc hub.**

27. Install the transmission case bolts. Tighten the upper six bolts to 56 ft. lbs. (76 Nm). Tighten the two remaining large bolts to 56 ft. lbs. (76 Nm). Tighten the remaining three bolts to 52 inch lbs. (6 Nm).

28. Push the release bearing fork rearward with a force of 13–18 lbs. (59–78 N) to engage the release bearing with the pressure plate. A click sound will be heard when the bearing engages the pressure plate properly.

29. Install the flywheel inspection cover.

30. Install the front crossmember and the front driveshaft (if equipped). Tighten

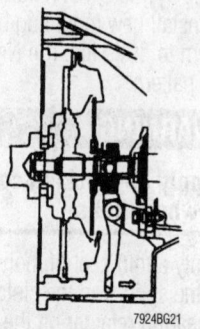

7924BG21

Push the release bearing fork toward the transmission to engage the release bearing with the pressure plate

the crossmember bolts to 58 ft. lbs. (78 Nm). Tighten the driveshaft flange bolts to 46 ft. lbs. (60 Nm).

31. Install the rear crossmember and mount. Tighten the crossmember mounting bolts to 37 ft. lbs. (50 Nm) and the transmission mounting nuts to 30 ft. lbs. (41 Nm).

32. Remove the transmission jack and the engine hoist.

33. Attach the transmission harness connectors and install the harness clamps.

34. Install the rear driveshaft and tighten the flange bolts to 46 ft. lbs. (60 Nm).

35. Apply grease to the dimple on the end of the release bearing fork and install the slave cylinder, heat protector, and dust covers. Tighten the slave cylinder bolts to 32 ft. lbs. (43 Nm). Tighten the dust cover bolts to 62 inch lbs. (6 Nm).

36. Install the exhaust pipes and heat protectors using new self-locking nuts. Tighten the exhaust manifold flange bolts to 49 ft. lbs. (67 Nm). Tighten the center pipe flange bolts to 32 ft. lbs. (43 Nm) and tighten the converter bolts to 25 ft. lbs. (34 Nm).

37. Attach the oxygen sensor connectors.

38. Refill the transmission with engine oil.

39. Install the skid plates and tighten their bolts to 27 ft. lbs. (37 Nm).

40. Lower the vehicle to the floor.

41. Install the shift levers. Tighten the shift lever mounting plate bolts to 15 ft. lbs. (20 Nm).

42. Install the console, shift boots, and shift knobs.

43. Align the matchmarks and install the hood.

44. Connect the negative battery cable.

45. Check the operation of the clutch and starter. Road test the vehicle.

Automatic

➡**The transfer case is an integral part of the transmission housing. Although the two cases can be separated, the transfer case should be removed with the transmission.**

1. Use a felt-tipped marker to matchmark the hood to the hood hinges. Remove the hood.

2. Disconnect the negative battery cable.

3. Shift the transmission into neutral, and the transfer case into the **2H** position.

4. Remove the air cleaner assembly.

5. Remove the transfer case shift knob. Remove the four console retaining screws.

6. Remove the center console assembly and detach the console switch wiring connectors.

7. Disconnect the shift lock cable and the shift control rod from the selector lever assembly.

8. Remove the transfer case control lever.

9. Raise and safely support the vehicle.

10. Remove the transmission and transfer case skid plates.

11. Remove the exhaust pipe protectors.

12. Drain the transmission fluid.

13. Label and detach the oxygen sensor connectors.

14. Remove the catalytic converter, the center and front exhaust pipes.

15. Matchmark the front (if equipped) and rear driveshafts to the differential and transfer case flanges.

16. If applicable, unbolt and remove the front driveshaft.

17. Unbolt and remove the rear driveshaft.

18. Detach the oxygen sensor connector from the transmission harness.

19. Disconnect the oil cooler lines from the transmission. Plug the lines to keep moisture out.

20. Remove the brackets securing the oil cooler lines to the stiffener.

21. Remove the front crossmember.

22. Remove the dipstick and tube. Disconnect the breather hoses from the tube.

23. Remove the five engine stiffener bracket bolts and the stiffener bracket.

24. Remove the heat protector.

25. Detach the transmission harness and the mode switch harness connector from the engine harness.

26. Disconnect the harness from the clamp bracket.

27. Disconnect the ground cable from the engine.

28. Remove the starter.

29. Remove the flexplate inspection cover.

30. Remove the three bolts securing the flexplate to the torque converter.

➡**Remove the radiator upper fan shroud and the cooling fan to access the crankshaft center bolt to turn the crankshaft.**

31. Place a suitable transmission jack under the transmission and transfer case unit for support.

32. Raise the transmission slightly and remove the eight bolts securing the rear mount and the transmission crossmember.

➡**Be sure the engine and transmission assembly is properly supported before removing the rear mount and third crossmember.**

33. Raise the engine slightly with a hoist and remove the transmission-to-engine bolts.

34. Separate the transmission from the engine and lower the transmission from the vehicle.

To install:

➡**Use new self-locking nuts when reconnecting the exhaust system. Replace any color-coded self-locking bolts when installing the frame cross-members.**

35. Install the transfer case onto the transmission case if the two were separated.

36. Install a new O-ring on the front pump shaft.

❈ WARNING

Be sure the transmission dowel pins are installed in the correct position. If the dowels are in the wrong hole, the transmission case may crack when the engine-to-transmission bolt are tightened.

37. Be sure the torque converter is fully seated in the front pump and slowly raise the front of the transmission into position until it is flush with the rear of the engine and install the transmission-to-engine bolts. Tighten the upper six mounting bolts to 56 ft. lbs. (76 Nm). Tighten the lower three mounting bolts to 35 ft. lbs. (48 Nm). Tighten the remaining two bolts to 53 inch lbs. (6 Nm).

38. Install the third crossmember. Tighten the bolts to 37 ft. lbs. (50 Nm).

39. Install the rear mount and lower the engine from the hoist. Tighten the nuts to 30 ft. lbs. (41 Nm).

40. Remove the transmission jack and the engine hoist.

41. Install the flexplate-to-torque converter bolts. Tighten the bolts in a crisscross pattern to 41 ft. lbs. (55 Nm).

42. Install the flexplate inspection cover.

43. Attach the transmission and mode switch harness connectors to the engine harness.

44. Connect the harness clamp to the clamp bracket and radiator clip.

45. Connect the ground cable to the engine.

46. Install the starter and tighten the mounting bolts to 30 ft. lbs. (40 Nm).

47. Install the heat protector.

48. Install the stiffener bracket. Tighten the bolts to 35 ft. lbs. (48 Nm).

49. Install the transmission dipstick and tube. Connect the breather hoses to the tube.

50. Install the front crossmember. Tighten the bolts to 58 ft. lbs. (79 Nm).

51. Connect the fluid cooling lines to the transmission. Tighten the line fittings to 40 ft. lbs. (54 Nm)

52. Install the front and rear driveshafts. Tighten the flange bolts to 46 ft. lbs. (64 Nm).

53. Install the exhaust pipes and the catalytic converter. Tighten the manifold nuts to 49 ft. lbs. (67 Nm). Tighten the center pipe and converter bolts to 32 ft. lbs. (43 Nm).

54. Reconnect the oxygen sensors.

55. Install the exhaust pipe protectors.

56. Install the skid plates and tighten their mounting bolts to 27 ft. lbs. (37 Nm).

57. Refill the transmission with DEXRON® III ATF.

58. Lower the vehicle.

59. Install the transfer case control lever.

60. Connect the shift control rod to the selector lever. Adjust the shift cable so that there is ⁵⁄₆₄ in. (2.0mm) of slack at the selector lever.

61. Connect the console switch connectors and the shift lock cable to the selector lever.

62. Install the center console assembly and the transfer case lever knob.

63. Install the upper fan shroud and the cooling fan.

64. Install the air cleaner assembly.

65. Align the matchmarks and install the hood.

66. Connect the negative battery cable.

67. Start the engine and shift through the gear range three times to circulate the ATF through the valve body.

68. Recheck the fluid level and top up if necessary.

69. Road test the vehicle.

Clutch Assembly

REMOVAL & INSTALLATION

1. Shift the transmission into neutral and the transfer case into **2H**. Drive the vehicle forward and backward for a few feet to be sure the front axle and hubs are disengaged.

2. Disconnect the negative battery cable.

3. Remove the shift knobs. Remove the console screws and lift the console and shift boots over the shift levers.

4. Raise and support the vehicle safely.

5. Remove the transmission and transfer case as an assembly.

6. If the pressure plate is going to be reused, matchmark the pressure plate to the flywheel so that it can be reassembled in the same position. Lock the flywheel in place to prevent it from turning.

7. To avoid warping the pressure plate, loosen the pressure plate mounting bolts in a two-step crisscross sequence until the spring tension is relieved.

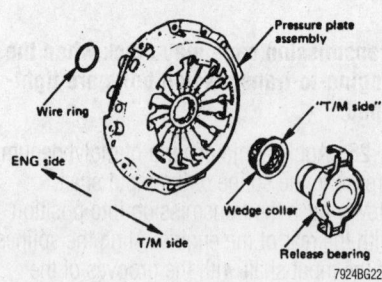

Exploded view of the pressure plate and release bearing components

8. Install a clutch alignment tool through the clutch disc and into the pilot bearing to support the pressure plate and cover assembly. Remove the pressure plate bolts and the clutch assembly.

9. Remove the release bearing from the input shaft and fork.

10. Remove the snap pin and release fork fulcrum pin. Remove the release fork.

11. Unbolt the release fork fulcrum bracket from the transmission case and inspect it for damage.

12. Inspect the flywheel for scoring, grooves and cracks, or discoloration from heat. Replace the flywheel if necessary.

13. Inspect the rear main oil seal for signs of leakage, and replace if necessary.

To install:

14. If the flywheel was removed or replaced, install new mounting bolts and tighten them to 40 ft. lbs. (54 Nm) in a crisscross pattern.

❈ WARNING

Do not apply any locking compound to the flywheel bolts.

15. Apply a thin coat of molybdenum grease to the splines in the clutch disc and the front bearing retainer on the transmission where the release fork and bearing slide.

16. Pack the recess in the release bearing with molybdenum grease and apply grease to the tabs where the clutch fork attaches to the release bearing.

17. Apply molybdenum grease the release fork pin and install it with a new locking cotter pin. Tighten the release-fork fulcrum bracket bolts to 28 ft. lbs. (38 Nm). Apply molybdenum grease to the contact surfaces of the release fork.

18. Install a new wedge collar and wire snapring into the pressure plate.

19. Using a clutch alignment tool, assemble the clutch disc and pressure plate into the flywheel. Be sure the clutch disc is facing the right direction.

20. If reusing the pressure plate, align the matchmarks and install the pressure plate mounting bolts. Tighten the bolts in a two-step crisscross pattern to 13 ft. lbs. (18 Nm). Remove the aligning tool.

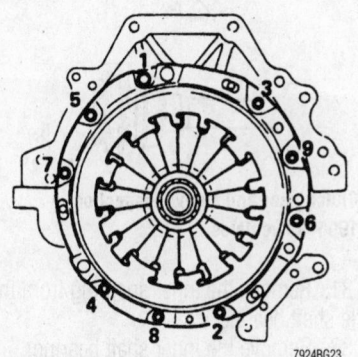

7924BG23

Pressure plate mounting bolt tightening sequence

21. Install the transmission and transfer case as an assembly.

22. Bleed the hydraulic clutch system.

23. Install the console and shift knobs.

24. Reconnect the negative battery cable.

25. Check the clutch pedal free-play and height.

26. Road test the vehicle and check for proper clutch action.

Hydraulic Clutch System

BLEEDING

1. Firmly set the parking brake.

2. Check the reservoir fluid level and refill as necessary.

3. Raise and safely support the front of the vehicle. Block the rear wheels.

4. Connect a vinyl tube to the bleeder screw. Submerge the other end of the tube in a container of brake fluid.

5. Have an assistant pump the clutch pedal slowly several times and hold it depressed.

6. Loosen the bleeder screw and allow the air bubbles and fluid to flow into the container. Tighten the bleeder screw.

7. Release the clutch pedal.

8. Add fluid to the reservoir if necessary. Don't let the reservoir run dry.

9. Continue to pump the pedal until the fluid draining from the tube is free of air bubbles.

10. Refill the reservoir to the full mark. Remove the vinyl tube. Tighten the bleeder screw and install its rubber cap.

Halfshaft

REMOVAL & INSTALLATION

➡The right axle shaft is an integral part of the right halfshaft assembly. Removal of the left axle shaft involves disassembling the shift-on-the-fly 4-wheel drive gearbox. The inboard joints of both halfshafts fit through the axle mounting brackets, which are bolted to the axle housing.

1. Shift the transfer case shift lever into **2H**. Drive the vehicle a few feet forward and reverse to verify that the front axle is disengaged.

2. Set the front wheels and steering wheel in the straight-ahead position. Lock the steering column in this position, and remove the key.

3. Raise and safely support the vehicle.

4. Remove the front wheels.

5. Remove the radiator skid plate.

6. Remove the transfer case skid plates.

7. Drain the oil from the differential.

8. Unbolt the calipers from their brackets. Support the calipers out of the way on wire hangers. Don't disconnect the brake hoses.

9. Remove the caliper mounting brackets from the steering knuckle.

10. Remove the automatic hub and brake rotor assemblies. Note the positions of the hub snaprings, shims, and lockwashers for reassembly.

11. If equipped with 4-wheel ABS, unbolt the front wheel sensor brackets from the steering knuckles. Move the sensors out of the work area. They don't need to be disconnected.

12. Remove the shift-on-the-fly 4-wheel drive actuator:

   a. Remove the skid plate from the shift-on-the-fly gear housing.

   b. Label and disconnect the Vacuum Switching Valve (VSV) hoses and **2P** connector from the gear housing.

   c. Unbolt the VSV assembly from the left axle tube and remove it so it won't be damaged. Don't disconnect the two vacuum hoses from the body of the VSV.

13. Use a ball joint separator tool to disconnect the upper and lower ball joints and tie rod ends, and then remove the steering knuckles.

14. Matchmark and disconnect the Pitman arm and idler arm. Remove the steering linkage as an assembly.

15. Unbolt and remove the suspension crossmember from its brackets at the lower control arms.

16. Matchmark the front driveshaft flanges to the differential flange and transfer case flange. Unbolt and remove the front driveshaft.

17. Support the front axle assembly with a floor jack and safety stands.

18. Remove the four bolts which secure the right halfshaft axle mounting bracket to the differential. Don't unbolt the left halfshaft from the axle.

19. Remove the mounting bracket bolts which secure the right and left axle mounting brackets to the vehicle's frame.

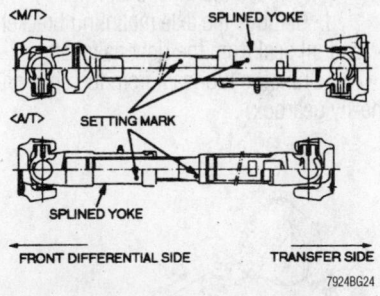

7924BG24

Driveshaft alignment mark locations— 1996–99 vehicles

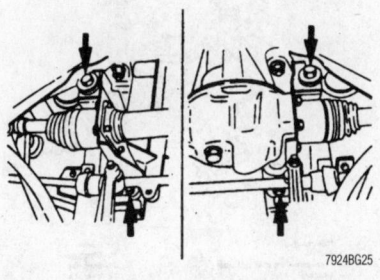

7924BG25

Axle assembly mounting bracket bolt locations—1996–99 vehicles

20. Separate the right halfshaft and axle mounting bracket assembly from the differential and allow it to rest on the lower control arm.

✳✳ WARNING

Be careful not to damage the CV-joints, boots, or splined shafts when removing the axle.

21. Follow these steps to remove the axle assembly:

a. Verify that the axle assembly is securely supported by the floor jack. Remove the safety stands.

b. First, slide the axle to the left to release the splined stub axle of the right halfshaft.

c. Next, lower the axle slightly and slide it to the right so that the left halfshaft clears the left lower control arm.

d. Finally, completely lower the axle from the vehicle.

22. Remove the right halfshaft from the vehicle. Follow these steps to remove the right axle shaft seal and bearing:

a. Remove the snapring from the splined shaft.

b. Remove the shaft bearing. Use a puller if necessary, but don't damage the shaft or splines.

c. Remove the inner snapring.

d. Remove the axle mounting bracket and oil seal from the right halfshaft.

23. Drain the lubricant from the shift-on-the-fly gearbox.

24. With the axle out of the vehicle, unbolt the left halfshaft from the axle case. Remove the halfshaft together with the left axle mounting bracket.

25. Loosen the shift-on-the-fly actuator mounting bolts in a crisscross pattern. Remove the actuator assembly from the gearbox.

26. Unbolt the shift-on-the-fly actuator gearbox from the axle tube flange.

27. Slowly pull the gearbox straight off of the axle tube. Be careful not to lose the sleeve and clutch gear if they fall out of the gearbox.

28. Remove the outer snapring from the left axle shaft. Draw the axle shaft out of the axle tube.

29. Remove the left axle shaft oil seal from the axle tube. Be careful not to damage the sealing surface.

30. Inspect the axle shaft, bearing, and clutch gear components for any damage:

a. Inspect the axle shaft splines for damage, and replace as necessary. Use a dial gauge and center blocks to inspect the shaft run-out. If run-out exceeds 0.02 in. (0.5mm), replace the shaft. Don't try to heat the axle shaft to correct excess run-out.

b. Insert the clutch gear into the axle shaft and inspect the motion and play of the inner bearing and needle bearing. If either bearing exhibits smoothness or play, they should be replaced.

c. Check the clutch sleeve for wear. First, coat the clutch gear with gear oil, and then slide the clutch sleeve back and forth over the gear to simulate operation. If the sleeve and gear exhibit smoothness or play, they must be replaced.

d. Check the clutch sleeve groove width. Groove width shouldn't exceed 0.28 in. (7.1mm).

e. Check the external diameter of the narrowest part of the clutch gear. The diameter shouldn't exceed 1.456 in. (36.98mm).

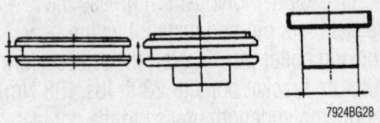

Several views of clutch sleeve inspection—1996–99 vehicles

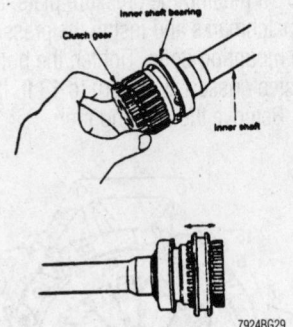

Clutch gear and sleeve inspection—1996–99 vehicles

31. Remove the inner snapring from the axle shaft bearing.

32. Remove the inner shaft bearing:

a. Install tool J-37452, or an equivalent bearing remover onto the axle shaft.

b. Place the axle shaft and the bearing remover into a press.

c. Press the bearing from the axle shaft. Don't damage the axle shaft.

33. Remove the needle bearing from the axle shaft clutch gear:

a. Support the axle shaft in a padded vise.

b. Install tool J-26941 into the bearing. Install tool J-2619–01, or an equiva-

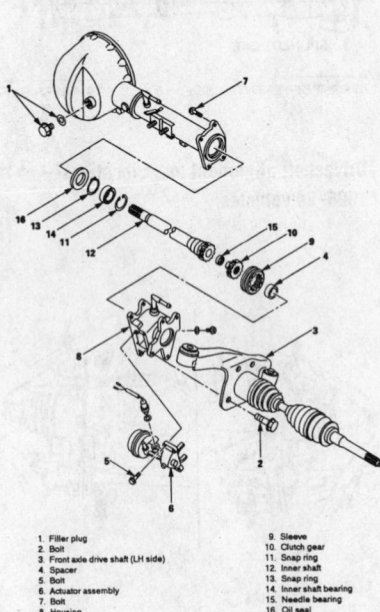

1. Filler plug
2. Bolt
3. Front axle drive shaft (LH side)
4. Spacer
5. Bolt
6. Actuator assembly
7. Bolt
8. Housing
9. Sleeve
10. Clutch gear
11. Snap ring
12. Inner shaft
13. Snap ring
14. Inner shaft bearing
15. Needle bearing
16. Oil seal

7924BG26

Exploded view of the left halfshaft, axle shaft, and gearbox—1996–99 vehicles

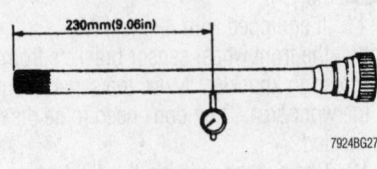

230mm(9.06in)

7924BG27

Use a dial indicator to inspect the axle shaft run-out—1996–99 vehicles

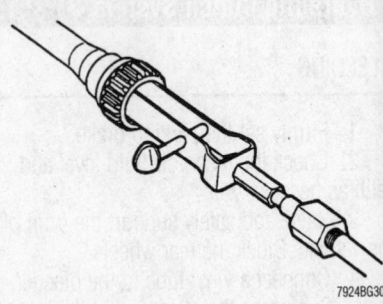

7924BG30

Use the needle bearing puller to draw the bearing off of the shaft—1996–99 vehicles

lent slide hammer onto the bearing removal tool.

    c. Work the slide hammer to gradually remove the needle bearing from the axle shaft.

34. Inspect the condition and operation of the shift-on-the-fly actuator:

    a. Disconnect the shift position switch.

    b. Use a vacuum pump to apply 15.7 in. Hg (400mm Hg) of vacuum to the port on the actuator's vacuum disc (port A). The shift fork should move toward the vacuum disc.

    c. Repeat substep B with vacuum applied to the port beside the shift position switch (port B). The shift fork should move away from the vacuum disc.

    d. If the actuator needs more than 15.7 in. Hg (400mm Hg) of vacuum to function, or if it functions incorrectly, it should be replaced.

    e. Inspect the shift fork for damage and wear. The internal distance between each end of the fork should be 2.52 in. (64.1mm). The width of the fork ends should be 0.26 in. (6.7mm) when measured at their narrowest point.

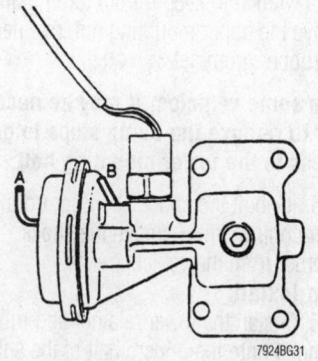

Shift-on-the-fly actuator port identification—1996–99 vehicles

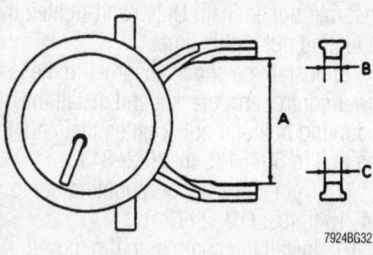

Actuator shift fork inspection points—1996–99 vehicles

To install:

→**Use new self-locking nuts and color-coded bolts when assembling the axle assembly mounts and suspension components. Suspension fasteners should be tightened to their final torque specifications when the vehicle is on the ground.**

35. Clean and dry all axle and shift-on-the-fly gearbox sealing surfaces.

36. Thoroughly lubricate a new axle shaft seal with clean gear oil. Use tool J-41693, or an equivalent seal driver to install it into the axle tube.

37. Use tool J-41694, or an equivalent bearing driver to install a new needle bearing.

38. Install a new inner snapring.

39. Use tool J-4169, or an equivalent press base to press a new bearing onto the axle shaft.

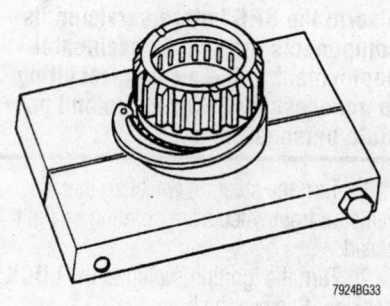

Axle shaft bearing installation press tool—1996–99 vehicles

40. Install the axle shaft into the axle tube. Don't damage the new oil seal. Install a new outer snapring.

41. Lubricate the clutch gear and clutch sleeve with SAE 75W-90 GL5-5 gear oil and install them.

42. Apply a 1/32 in. (1mm) wide bead of liquid gasket to the axle tube sealing surface. Install the shift-on-the-fly gearbox onto the axle tube before the sealant cures.

43. Tighten the shift-on-the-fly gearbox bolts to 85 ft. lbs. (116 Nm) in a 2-step crisscross pattern.

44. Install the shift position switch onto the actuator and tighten it to 29 ft. lbs. (39 Nm).

45. Install the actuator assembly:

    a. Apply a 1/32 in. (1mm) wide bead of liquid gasket to the actuator sealing surface. Don't allow the sealant to cure before installation.

    b. Align the shift fork arms with the groove of the clutch sleeve and install the actuator.

    c. Tighten the actuator mounting bolts to 10 ft. lbs. (13 Nm) in a crisscross sequence.

46. Install the spacer onto the left front halfshaft.

47. Install the halfshaft and axle mounting bracket assembly onto the axle. Tighten the bolts to 85 ft. lbs. (116 Nm) in a crisscross sequence.

48. After the sealant has fully cured, refill the shift-on-the-fly gearbox with SAE 75W-90 GL5—gear oil.

49. Assemble the right halfshaft and axle mounting bracket assembly:

    a. Install the axle mounting bracket onto the halfshaft.

    b. Lubricate and install a new oil seal.

    c. Install a new inner snapring.

    d. Install a new bearing.

    e. Install a new outer snapring.

50. Place the right halfshaft and axle mounting bracket assembly into position, and rest it on the right lower control arm.

51. Position the axle and left halfshaft assembly on a floor jack. Raise the axle into position.

52. Fit the right halfshaft and axle mounting bracket assembly into the differential. Be sure the splined stub axle shaft is fully seated. Be careful not to distort the oil seal when connecting the right mounting bracket to the differential.

53. Tighten the right halfshaft and axle mounting bracket assembly mounting bolts to 85 ft. lbs. (116 Nm) in a crisscross sequence.

54. Raise the axle assembly into its final position and support it with safety stands.

55. Install the mounting bracket bolts, nuts, and spacers. The washer fits under the bolt and the spacer is used with the nut. Tighten the mounting nuts and bolts to 112 ft. lbs. (152 Nm).

56. Install the steering knuckles and assemble any suspension components that were disconnected or removed.

57. Install the brake backing plates and the rotor and hub assemblies.

58. Install the caliper mounting brackets and the brake calipers.

59. Install the VSV onto the axle tube. Attach the vacuum hoses and the **2P** connector. Then, install the VSV skid plate.

60. Refill the differential with gear oil. Check the oil level in the gear case. Use new crush washers and tighten both drain plugs to 58 ft. lbs. (78 Nm).

61. Verify that all axle assembly mounting components have been installed.

62. Align the front driveshaft match-marks. Install the driveshaft flange bolts and tighten them to 46 ft. lbs. (63 Nm).

63. Install the suspension crossmember and tighten the bolts to 58 ft. lbs. (78 Nm).

64. Install the steering linkage assembly.

65. If equipped, reconnect the ABS front wheel sensors.

66. Install the radiator skid plate. Tighten the bolts to 58 ft. lbs. (78 Nm).

67. Install the transfer case skid plates and tighten their bolts to 27 ft. lbs. (37 Nm).

68. If necessary, bleed the brake system.

69. Install the front wheels.

70. Lower the vehicle.

71. Tighten the suspension bushing fasteners to their final torque specifications referring to the illustration.

72. Verify that the front axle and hubs engage and disengage properly.

73. If equipped with shift-on-the-fly 4-wheel drive, be sure the VSV and actuator function correctly.

74. Check and adjust the front wheel alignment and ride height.

75. Road test the vehicle.

STEERING AND SUSPENSION

Air Bag

※※ CAUTION

Some vehicles are equipped with an air bag system, also known as the Supplemental Restraint System (SRS). The system must be disabled before performing service on or around system components, steering column, instrument panel components, wiring and sensors. Failure to follow safety and disabling procedures could result in accidental air bag deployment, possible personal injury and unnecessary system repairs.

PRECAUTIONS

Several precautions must be observed when handling the inflator module to avoid accidental deployment and possible personal injury.

• Never carry the inflator module by the wires or connector on the underside of the module.

• When carrying a live inflator module, hold securely with both hands, and ensure that the bag and trim cover are pointed away from you.

• Place the inflator module on a bench or other surface with the bag and trim cover facing up.

• With the inflator module on the bench, never place anything on or close to the module which may be thrown in the event of an accidental deployment.

DISARMING

➡**This procedure is necessary for disabling and enabling the air bag system.**

※※ CAUTION

The Supplemental Restraint System (SRS) must be disarmed before any of its components are disconnected or the air bag is removed. Failing to disarm the SRS before servicing its components may cause accidental deployment of the air bag, resulting in unnecessary SRS repairs and possible personal injury.

1. Turn the steering wheel so that the vehicle's front wheels are pointing straight ahead.

2. Turn the ignition switch to the **LOCK** position. Remove the key.

3. Disconnect the negative and positive battery cables. Wait at least five minutes before working around the air bags.

➡**Removing the C-21, C-22, C-23, and C-24 SRS fuses (1995 Trooper) or C-21 and C-22 fuses (1996–99 Trooper and SLX) from the under-dash fuse block has the same effect as disconnecting the battery.**

4. For the driver's side air bag, detach the yellow 3-way SRS harness connector at the base of the steering column.

5. For the passenger's side air bag perform the following:

   a. Remove the glove box door and the passenger's air bag lower cover.

   b. Disconnect the yellow 4-way SRS harness connector from the passenger's side air bag.

   c. If the passenger's air bag module must be removed; carefully remove the lower mounting bracket and mounting nuts. Lift the air bag out of the dashboard and remove it from the vehicle.

After servicing is completed, enable the SRS.

6. For the driver's side air bag, attach the yellow 3-way SRS harness connector at the base of the steering column. Then, install the fuses if they were removed.

7. For the passenger's side air bag, install the lower mounting brackets, attach the yellow 4-way SRS connector, then the air bag lower cover and glove box door.

8. Reconnect the positive and negative battery cables.

9. Turn the ignition to the **ON** position, but don't start the engine. The AIR BAG warning light should turn ON and flash ON and OFF for seven seconds, and then turn OFF. This light sequence indicates that the SRS system is functioning normally. If the AIR BAG light doesn't come ON, or stays ON longer than seven seconds, the system must be diagnosed.

Shock Absorber

REMOVAL & INSTALLATION

Front

1. Raise and support the vehicle safely.

2. Remove the front wheels.

3. Support the lower control arm with a floor jack.

4. Hold the shaft of the shock absorber with a wrench to keep it from turning and remove the upper mounting nut, retainer, and rubber grommet.

➡**On some vehicles, it may be necessary to remove the bump stops to gain access to the lower mounting bolt.**

5. Unbolt the shock absorber from the lower control arm. Remove the shock absorber from the vehicle.

To install:

6. Install the lower retainer and rubber grommet onto the upper shaft of the shock absorber. Then, fully extend the shock and install the upper shaft into the mounting hole in the frame bracket.

7. Install the upper rubber grommet, retainer and attaching nut onto shock absorber upper shaft. Only hand-tighten the mounting nut at this time.

8. Install the shock absorber to the lower control arm bracket and install the mounting bolt and nut. Tighten the mounting bolt to 60–61 ft. lbs. (82–84 Nm).

9. Tighten the upper mounting nut to 14–15 ft. lbs. (19–20 Nm).

10. Install the bump stop if removed, and tighten the bolts to 30 ft. lbs. (41 Nm).

11. Install the front wheels and lower the vehicle.

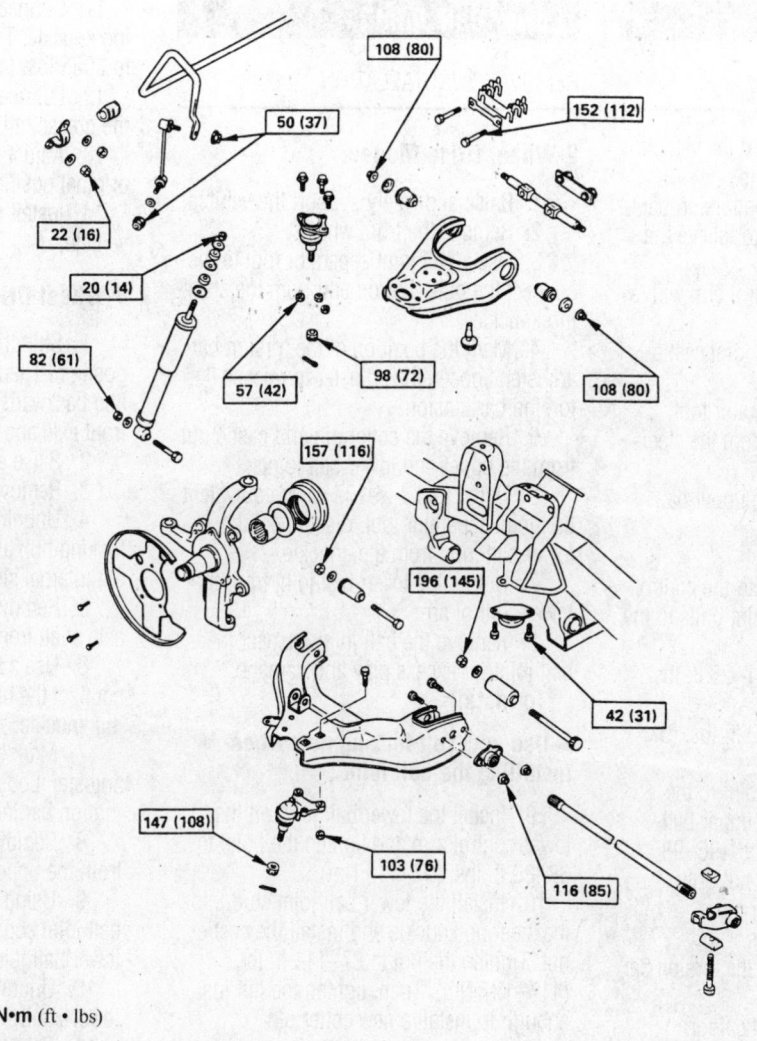

N•m (ft • lbs)

7924BG48

Exploded view of the front suspension, showing the tightening specifications

Rear

1. Raise and safely support the vehicle.
2. Support the rear axle with a floor jack and safety stands.
3. Remove the shock absorber lower mount nut, washers, and bushings.
4. Remove the shock absorber upper mount bolt and nut.
5. Remove the shock absorber.

To install:

6. Install the shock absorber. Install the upper mount nut and bolts, but only hand-tighten them at this time.
7. Fit the shock absorber, bushings, and washer to the lower mount. Only hand-tighten the nut at this time.
8. Remove the jack and stands from underneath the rear axle.
9. Lower the vehicle to the floor and tighten the upper mounting bolt to 70 ft. lbs. (95 Nm) and the lower mounting nut to 58 ft. lbs. (78 Nm).

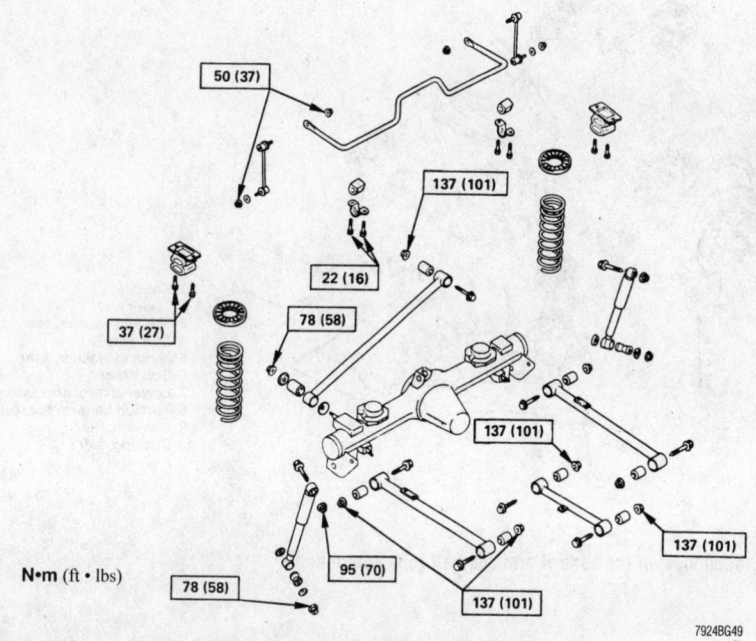

N•m (ft • lbs)

7924BG49

Exploded view of the rear suspension, showing the tightening specifications

Upper Ball Joint

REMOVAL & INSTALLATION

1. Raise and safely support the vehicle.
2. Remove the front wheels.
3. Mark the position of the torsion bar adjuster. Loosen the adjuster to relieve the torsion bar tension.
4. Support the lower control arm with a floor jack.
5. Remove the upper ball joint castle nut.
6. Using a ball joint separator tool, separate the upper ball joint from the steering knuckle.
7. Unbolt and remove the upper ball joint from the control arm.

To install:

8. Install the ball joint onto the control arm. Tighten the upper ball joint bolts to the following specification:
   - 4-bolt-style ball joint: 21–25 ft. lbs. (29–35 Nm)
   - 3-bolt style ball joint: 42 ft. lbs. (57 Nm)
9. Install the upper ball joint to the steering knuckle. Tighten the upper ball joint castle nut to 72–73 ft. lbs. (96–98 Nm). Then, tighten the castle nut only enough to install a new cotter pin.
10. Install the front wheels.
11. Adjust the tension on the torsion bar to its original position.
12. Lower the vehicle to the floor.

Lower Ball Joint

REMOVAL & INSTALLATION

2-Wheel Drive Models

1. Raise and safely support the vehicle.
2. Remove the front wheels.
3. Use a ball joint separator tool to disconnect the outer tie rod end from the steering knuckle.
4. Mark the position of the torsion bar adjuster. Loosen the adjuster to release the torsion bar tension.
5. Remove the cotter pin and castle nut from the upper and lower ball joints.
6. Using tool J-29107 or an equivalent ball joint separator tool, disconnect the lower ball joint from the knuckle.
7. Unbolt the lower ball joint from the lower control arm.
8. Remove the ball joint. Inspect the ball joint for excess play and damage.

To install:

➡ **Use new self-locking nuts when installing the ball joint.**

9. Install the lower ball joint on the lower control arm and tighten the bolts to 68–83 ft. lbs. (93–113 Nm).
10. Install the lower ball joint stud into the steering knuckle and install the castle nut. Tighten the nut to 87–111 ft. lbs. (117–137 Nm). Then, tighten the nut just enough to install a new cotter pin.

11. Connect the tie rod end to the steering knuckle. Tighten the tie rod securing nut to 80 ft. lbs. (109 Nm).
12. Lubricate the lower ball joint through the grease fitting.
13. Adjust the torsion bar tension to its original position.
14. Install the front wheels and lower the vehicle.

4-Wheel Drive Models

1. Shift the transfer case into the **2H** position. Then, drive the vehicle forward and backward for a few feet to be sure the front axle and hubs are disengaged.
2. Raise and safely support the vehicle.
3. Remove the front wheels.
4. Unbolt and remove the cap from the locking hub assembly. Do not lose any of the internal shims.
5. Remove the snapring to release the axle shaft from the hub.
6. Use a ball joint separator tool to disconnect the outer tie rod end from the steering knuckle.
7. Mark the position of the torsion bar adjuster. Loosen the adjuster to release the torsion bar tension.
8. Remove the cotter pin and castle nut from the upper and lower ball joints.
9. Using tool J-29107 or an equivalent ball joint separator tool, disconnect the lower ball joint from the knuckle.
10. Unbolt the lower ball joint from the lower control arm.
11. Remove the ball joint. Inspect the ball joint for excess play and damage.

To install:

➡ **Use new self-locking nuts when installing the ball joint.**

12. Install the lower ball joint on the lower control arm and tighten the bolts to 68–83 ft. lbs. (93–113 Nm).
13. Install the lower ball joint stud into the steering knuckle and install the castellated nut. Tighten the nut to 87–111 ft. lbs. (117–147 Nm). Then, tighten the nut just enough to align the cotter pin hole with a hole in the castellated nut. Install a new cotter pin.
14. Connect the tie rod end to the steering knuckle. Tighten the tie rod securing nut to 72–80 ft. lbs. (98–109 Nm). Then, tighten the nut only enough to install a new cotter pin.
15. Reinstall the shims and the snapring on the axle shaft. Repack the hub with heavy-duty multi-purpose grease.

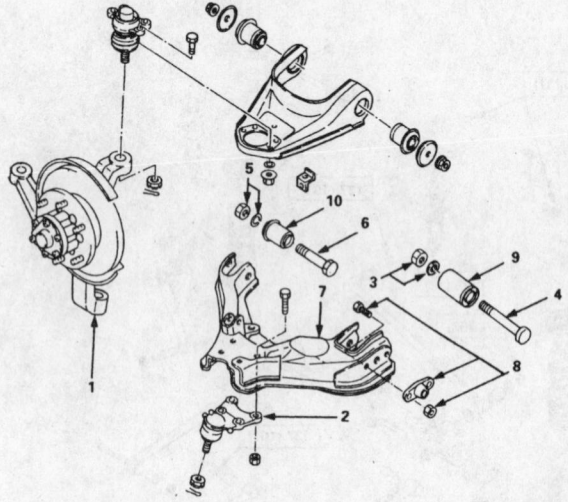

1. Knuckle
2. Lower end
3. Nut and washer, rear
4. Bolt, rear
5. Nut and washer, front
6. Bolt, front
7. Lower control arm assembly
8. Torsion bar arm bracket
9. Bushing, rear
10. Bushing, front

7924BG34

Exploded view of the control arm and ball joint components

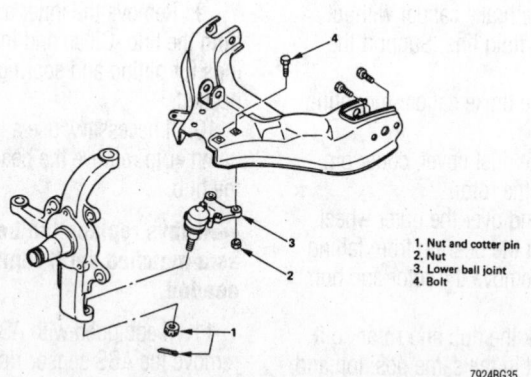

1. Nut and cotter pin
2. Nut
3. Lower ball joint
4. Bolt

7924BG35

Exploded view of the lower ball joint mounting and related components

16. Install the hub cover and tighten the bolts to 43 ft. lbs. (59 Nm).

17. Lubricate the lower ball joint through the grease fitting.

18. Adjust the torsion bar tension to its original position.

19. Install the front wheels and lower the vehicle.

20. Be sure the front hubs engage into four-wheel drive correctly.

21. Check the vehicle ride height and front wheel alignment and adjust as necessary.

Coil Spring

REMOVAL & INSTALLATION

Rear

1. Raise and safely support the vehicle under the frame.

2. Remove the rear wheels.

3. Place a floor jack under the rear axle housing and raise it slightly to compress the spring.

4. Unbolt the parking brake cable bracket from the trailing link.

5. Disconnect the stabilizer bar from its linkage.

6. Unbolt the shock absorber from the axle case mount.

✳✳✳ WARNING

Do not let the brake hose, parking brake cable or the breather hose extend to their full length.

7. Slowly lower the rear axle with the jack to release the coil spring tension. Remove the coil spring and the insulator.

To install:

8. Place the coil spring on the axle assembly and put the insulator on top of the spring.

9. Raise the axle assembly into position and connect the shock absorbers to the axle assembly. Tighten the shock absorber nut and bolt to 58 ft. lbs. (79 Nm).

10. Connect the stabilizer linkage to the bar. Tighten the nut to 37 ft. lbs. (50 Nm).

11. Install the parking brake bracket onto the trailing arm.

12. Install the rear wheels.

13. Lower the vehicle to the floor.

Wheel Bearings

ADJUSTMENT

Front

➡️**It is recommended that the wheel bearings be removed, cleaned, and repacked whenever an adjustment is necessary.**

1. Raise and safely support the vehicle.

2. Remove the front wheels.

3. If equipped, unbolt and remove the hub dust cap.

4. If equipped, remove the locking hub assembly.

5. Use a hub nut wrench to loosen the spindle nut.

6. Tighten the hub nut to 22 ft. lbs. (29 Nm) to seat the bearings and then fully loosen the nut. Use a spring scale connected to the wheel stud bolt at 90 degrees to measure bearing preload. Then, retighten the hub nut until the spring gauge measures a bearing preload of 4.4–5.5 lbs. (2.0–2.5 kg) for a new bearing and grease seal, or 2.6–4.0 lbs. (1.2–1.8 kg) for a used bearing and a new grease seal.

7. Install the lockwasher on the spindle nut. The larger diameter, tapered side of the lockwasher faces out, and the holes in the lockwasher must align with the holes in the spindle nut. If the holes don't align, reverse the lockwasher, or tighten the spindle just enough to bring them into alignment.

8. If equipped, install the locking hub assembly or the hub dust cap.

9. Install the front wheels.

10. Lower the vehicle.

Rear

The rear wheel and axle shaft bearings can't be adjusted. Failing or damaged rear wheel and axle shaft bearings produce a growling or grating noise that can be heard when the vehicle is coasting at low speeds. Bearing noise will remain constant no matter what type of road surface the vehicle is driven on.

To inspect the wheel and axle shaft bearings, first test drive the vehicle to confirm the noise. Then, inspect the chassis and suspension for damaged bushings or areas of metal-to-metal contact. Finally, inspect the bearings by raising the rear of the vehicle and spinning the wheels by hand. Rough wheel motion, or noise from the hub during rotation are signs of bearing failure.

REMOVAL & INSTALLATION

Front

➡️**For front wheel bearing replacement procedures, please refer to Front Brake Rotor and Hub, outlined in this section.**

Rear

1. Raise and safely support the vehicle.

2. Remove the rear wheels.

3. Remove the brake caliper and caliper brackets. Support the calipers up and out of the way with wire. Don't disconnect the brake hoses.

4. Remove the rear brake rotors. If equipped with ABS, unbolt the wheel sensor wire brackets from the axle tubes. Be careful not to damage the wheel sensors.

5. Disconnect the parking brake cable from the parking brake shoe assembly. Unbolt the brake cable bracket from the backing plate. Remove the parking brake shoe assembly.

6. Matchmark the bearing case to the axle tube flange. Mark the hubs so the axle shafts will not be confused.

7. Remove the bearing case mounting nuts from the axle tube flanges.

8. Remove the axle shafts from the axle.

9. Remove the snapring from the bearing case retainer.

10. Mount the axle shaft and bearing assembly into the Special Press Holder tool J-39211 or equivalent. Place the axle shaft assembly and the special holder into a hydraulic press.

11. Press the retainer from the axle shaft.

12. Remove the axle shaft bearing, bearing case, and brake rotor backing plate from the axle shaft.

13. Inspect the axle shaft and its splines for signs of wear, scoring, or other damage. Replace the retainer if it is damaged in the press.

To install:

14. Use tool J-39379 or a suitably-sized seal installer to install a new shaft seal into the bearing case.

15. Assemble the brake backing plate and bearing case onto the axle shaft.

16. Install a new bearing onto the axle shaft. Place the retainer into position on the axle shaft.

17. Mount the axle shaft and bearing assembly into the Special Press Base tool J-39212. Place the axle shaft assembly and the special base into a hydraulic press.

18. Press the retainer onto the axle shaft. Do not exert more pressing force than is necessary to fit the retainer and bearing assembly together.

19. Install a new snapring onto the bearing case retainer.

20. Install the axle shaft assembly into the axle tube. Be sure the splines engage into the differential carrier. Use the matchmarks as reference points.

21. Install the bearing case nuts with new lockwashers. Tighten the nuts to 54 ft. lbs. (74 Nm).

22. Install the parking brake shoes. Reconnect the parking brake cable to the shoes and install its backing plate bracket mounting bolts.

23. Install the ABS wheel sensor wire brackets onto the axle tubes.

24. Install the rear brake rotors.

25. Install the brake caliper brackets and calipers. Tighten the bracket bolts on 1995 vehicles to 109 ft. lbs. (148 Nm), or to 76 ft. lbs. (103 Nm) for 1996–99 vehicles. Tighten the caliper bolts to 32 ft. lbs. (44 Nm).

26. Install the rear wheels.

27. Bleed the brakes if necessary.

28. Check the differential oil level and refill as necessary.

29. Road test the vehicle and check for abnormal bearing and gear noises.

Front Brake Rotor and Hub

REMOVAL & INSTALLATION

1995 Models

2WD VEHICLES

1. Raise and safely support the vehicle. Remove the front wheels.

2. Remove the brake caliper without disconnecting the fluid line. Support the caliper aside.

3. Remove the brake caliper mounting bracket.

4. Remove the dust cover, cotter pin and locknut from the rotor.

5. Place a hand over the outer wheel bearing to prevent the bearing from falling on the floor and remove the rotor and hub from the spindle.

6. Matchmark the hub and rotor so it can be assembled in the same position and remove the bolts attaching the hub to the rotor, then pull off the rotor.

7. Thoroughly clean, inspect and repack the bearings with wheel bearing grease. Replace the bearings if needed. Use a brass drift to drive the bearing races out of the hub.

To install:

8. Use a suitably-sized bearing driver to drive new bearing races into the hub.

9. Install the rotor on the hub and tighten the bolts to 76 ft. lbs. (103 Nm).

10. Install the inner bearing and the grease seal in the hub if removed and position the hub and rotor assembly on the spindle.

11. Install the outer bearing and spindle nut in the hub. Adjust the wheel bearings.

12. Install the caliper mounting bracket and tighten the bolts to 115 ft. lbs. (156 Nm).

13. Install the caliper and tighten the mounting bolts to 20–27 ft. lbs. (28–38 Nm).

14. Install the front wheels and lower the vehicle.

15. Pump the brake pedal several times to test the system. Bleed the brakes if necessary.

4WD AUTOMATIC HUB VEHICLES

1. Move the transfer case shift lever into **2H** and move the vehicle forward and rearward about 3 ft. (1m).

2. Raise and safely support the vehicle. Remove the front wheels.

3. Remove the 4WD hub cap attaching bolts and the cap.

4. Remove the brake caliper and support the caliper on a wire; do not disconnect the brake hose. Remove the brake caliper support bracket from the steering knuckle.

5. Using snapring pliers, remove the snapring and shims from the hub assembly.

6. Remove the drive clutch assembly, the inner cam and lockwasher.

7. Using a hub nut wrench, loosen the hub nut.

8. Pull the hub from the spindle.

9. Remove the inner and outer bearings from the hub. Clean and inspect the bearings for pitting and scoring and replace if needed.

10. If necessary, use a brass drift and a hammer to remove the bearing races from the hub.

➡**Always replace the bearing and race as a matched set if replacement is needed.**

11. If equipped with ABS, unbolt and remove the ABS sensor ring.

12. If removing the disc from the hub, scribe matchmarks, remove the bolts attaching the hub to the rotor, then separate the disc from the hub.

To install:

13. Use a suitably-sized bearing driver to install new bearing races into the hub.

14. If the hub and rotor were separated, install the hub to the rotor and tighten the bolts to 76 ft. lbs. (103 Nm).

15. If equipped, install the ABS sensor ring and tighten the bolts to 13 ft. lbs. (18 Nm).

16. Pack the bearings with wheel bearing grease and install the inner bearing and a new grease seal in the hub. Position the hub and disc assembly on the spindle, place 1 to 1½ oz. (28–30 grams) of bearing grease in the hub and install the outer bearing and hub nut on the spindle.

17. When installing the hub nut, perform the following procedure:

   a. Tighten the hub nut to 22 ft. lbs. (30 Nm) and loosen the nut to seat the bearings.

   b. Use a spring gauge connected to the stud bolt at 90° to measure preload.

   c. Retighten the hub nut until the spring gauge measures a bearing preload of 4.4–5.5 lbs. (2.0–2.5 kg) for a new bearing and grease seal, or 2.6–4.0 lbs. (1.2–1.8 kg) for a new bearing and new oil seal, or 2.6–4.0 lbs. (11.8–17.7 N) for a used bearing and new oil seal.

18. Install the lockwasher with the larger diameter of the tapered bore facing out. If the bolt holes in the lockplate do not align with the corresponding holes in the nut, reverse the lockplate. If the bolt holes are still out of alignment, turn in the nut just enough to obtain alignment and install the lockscrew tightly so its head is lower than the surface of the washer.

19. Clean the flange surface of the hub, thread holes, the surface of the lockwasher and the splines of the axle shaft.

20. Install the inner cam by aligning the keyway of the inner cam with the groove of the knuckle. If the cam is difficult to install,

use the special tool or equivalent and a plastic hammer to lightly tap the inner cam into place.

21. Perform the following steps to select the proper shim:

a. Lower the vehicle and support the lower control arm with a block of wood and a floor jack to place the axle in the normal horizontal position.

b. Install the special adjusting tools onto the hub until it comes in contact with the lockwasher.

c. Pull the axle out as far as possible and using a feeler gauge, measure the clearance **t** between the hub and the snapring groove on the axle shaft.

d. If the clearance is larger than the snapring groove, selected shims must be installed so that clearance **t** is 0.000–0.039 in. (0.0–0.1mm). Shims come in thicknesses of, 0.2mm, 0.3mm, 0.5mm, and 1.0mm.

e. Remove the special tool J-36836 or equivalent.

22. Install the drive clutch assembly.

23. Install the shims by hand on the axle and use the following steps to install a new snapring:

a. Install special tool J-36835–2 or equivalent onto the axle.

b. Install the snapring on the tool.

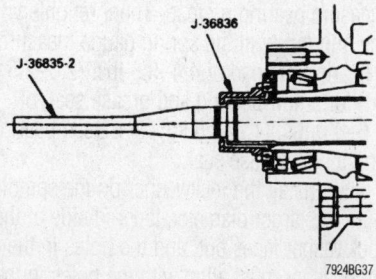

Use the special tools to measure shim thickness—1995 models with 4WD automatic hubs

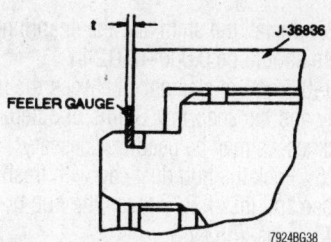

Measure the clearance "t" between the special tool and the snapring groove as shown—1995 models with 4WD automatic hubs

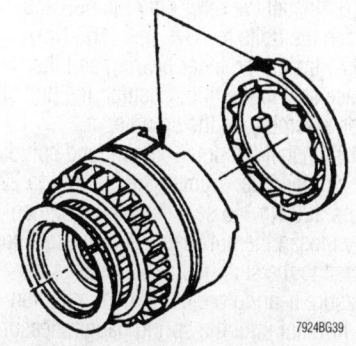

Drive clutch assembly installation—1995 models with 4WD automatic hubs

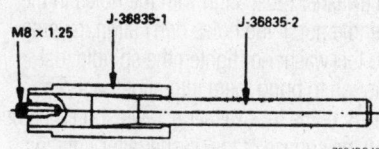

Special tools for snapring installation—1995 models with 4WD automatic hubs

c. Install tool driver J-36835–1 or equivalent.

d. Pull out the axle shaft by pulling tool J-36835–2 or equivalent. Install the snapring to the axle by pushing on tool J-36835–1.

e. Remove tool driver J-36835–2 or equivalent from the axle and check the fit of the snapring.

24. Install the housing assembly and cap. Tighten the cap bolts to 43.3 ft. lbs. (58.9 Nm).

25. Install the front wheels.

26. Lower the vehicle to the floor.

27. Pump the brake pedal several times to test the system. Bleed the brakes if necessary.

4WD MANUAL HUB VEHICLES

1. Shift the transfer case lever into "2H" and the manual hub into the "FREE" position.

2. Raise and safely support the vehicle.

3. Remove the front wheels.

4. Remove the brake caliper assembly and mounting bracket.

5. Loosen the 6 bolts and remove the housing assembly. Be careful not to lose the detent ball and spring.

6. Use snapring pliers and remove the snapring and shims.

7. Remove 6 bolts and remove the housing assembly from the hub.

8. Remove the lockwasher retaining screw and remove the lockwasher.

9. Use a special hub nut wrench to remove the hub nut.

10. Remove the hub and rotor assembly from the spindle.

11. To disassemble the clutch assembly, push the follower knob and turn the clutch assembly clockwise, then remove the clutch assembly from the knob. Remove the retaining spring from the clutch assembly by turning it counterclockwise.

12. Matchmark the hub and rotor. Place the rotor in a soft-jaw vise and remove the bolts attaching the rotor to the hub.

To install:

13. Align the matchmark and assemble the hub and rotor. Tighten the mounting bolts to 76 ft. lbs. (103 Nm).

14. Pack the clean bearings with wheel bearing grease and install the inner bearing and a new grease seal in the hub. Position the hub and disc assembly on the spindle, place 1 to 1½ oz. (28–32 grams) of bearing grease in the hub and install the outer bearing and hub nut on the spindle.

15. When installing the hub nut, perform the following procedure:

a. Tighten the hub nut to 22 ft. lbs. (29 Nm) and loosen the nut to seat the bearings.

b. Use a spring gauge connected to the stud bolt at 90°to measure preload.

c. Retighten the hub nut until the spring gauge measures a bearing preload of 4.4–5.5 lbs. (2.0–2.5 kg) a new bearing and oil seal or 2.6–4.0 lbs. (1.2–1.8 kg.) for a used bearing and new oil seal.

16. Install the lockwasher with the larger diameter of the tapered bore to the outer side of the vehicle. If the bolt holes in the lockplate do not align with the corresponding holes in the nut, reverse the lockplate. If the bolt holes are still out of alignment, turn in the nut just enough to obtain alignment and install the lockscrew tightly so its head is lower than the surface of the washer.

17. Apply grease to the spacer, ring, snapring and the splined part of the inner assembly.

18. Assemble the inner assembly and the clutch assembly.

19. Install the body assembly on the hub. Tighten the bolts to 103 inch lbs. (11.8 Nm).

20. Install the shim and a new snapring. The clearance between the free-wheeling hub body and the snapring should be 0.00–0.01 inch, shims are available in selective sizes.

21. Install the housing assembly. Tighten the bolts to 43 ft. lbs. (58 Nm).

22. Install the front wheels.

23. Lower the vehicle.

24. Pump the brake pedal several times to test the system. Bleed the brakes if necessary.

1996–99 Models

2WD VEHICLES

1. Raise and safely support the vehicle.

2. Remove the front wheels.

3. Unbolt and remove the hub dust cap.

4. Remove the brake caliper from its support bracket and support it with a wire hook. Don't disconnect the brake hose. Remove the brake caliper support bracket from the steering knuckle.

5. Remove the lockwasher.

6. Use a hub nut wrench to loosen and remove the spindle nut.

7. Pull the hub from the spindle.

8. Hold the outer wheel bearing with your hand to prevent it from falling out of the hub. Remove the hub and rotor assembly from the spindle.

9. Matchmark the hub and rotor so they can be assembled in the same positions. Place the hub and rotor assembly in a padded vice and then remove the six hub bolts to separate the rotor from the hub.

10. Thoroughly clean, inspect, and repack the bearings with wheel bearing grease. Replace the bearings if necessary.

11. Use a hammer and a brass drift to carefully tap the outer and inner bearing races and oil seals from the hub.

To install:

12. Clean the surfaces of the hub, lockwasher, and spindle.

13. Use bearing drivers to install new inner and outer bearing races and oil seals.

14. Pack the bearings and the hub cavity with wheel bearing grease.

15. If a new rotor is being installed, use brake cleaner or alcohol to remove any anti-rust coating.

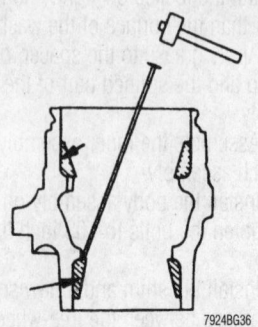

Use a brass drift and mallet to remove the inner or outer races as shown—1996–99 2WD vehicles

16. Install the rotor onto the hub and tighten the bolts to 76 ft. lbs. (103 Nm).

17. Install the inner bearing and the grease seal in the hub. Position the hub and rotor assembly on the spindle.

18. Install the outer bearing and spindle nut into the hub. Tighten the hub nut to 22 ft. lbs. (29 Nm) to seat the bearings, then fully loosen the nut. Use a spring scale connected to the stud bolt at 90 degrees to measure bearing preload. Then, retighten the hub nut until the spring gauge measures a bearing preload of 4.4–5.5 lbs. (2.0–2.5 kg) for a new bearing and grease seal, or 2.6–4.0 lbs. (1.2–1.8 kg) for a used bearing and a new grease seal.

19. Install the lockwasher on the spindle nut. The larger diameter, tapered side of the lockwasher faces out, and the holes in the lockwasher must align with the holes in the spindle nut. If the holes don't align, reverse the lockwasher or tighten the spindle just enough to bring them into alignment. The heads of the lockwasher screws should be below the surface of the washer after tightening.

20. Pack the hub dust cap with fresh grease and install it. Tighten the hub bolts to 43 ft. lbs. (59 Nm).

21. Install the brake caliper mount and the caliper.

22. Install the front wheels.

23. Lower the vehicle.

4WD VEHICLES

1. Move the transfer case shift lever into **2H** and move the vehicle forward and rearward about 3 ft. (1m) to be sure that the front axle isn't engaged.

2. Raise and safely support the vehicle.

3. Remove the front wheels.

4. Unbolt and remove the hub dust cap.

5. Remove the brake caliper and support it with a wire hook. Don't disconnect the brake hose. Remove the brake caliper support bracket from the steering knuckle.

6. Use snapring pliers to remove the snapring. Remove the shim.

7. Remove the hub flange assembly.

8. Remove the lockwasher and its screws.

9. Use a hub nut wrench to loosen and remove the spindle nut.

10. Pull the hub from the spindle.

11. Hold the outer wheel bearing with your hand to prevent it from falling out of the hub. Remove the hub and rotor assembly from the spindle.

12. Matchmark the hub and rotor so they can be assembled in the same positions. Place the hub and rotor assembly in a padded vice and then remove the six hub bolts to separate the rotor from the hub. If

equipped with ABS, unbolt the sensor ring from the rotor.

13. Thoroughly clean, inspect, and repack the bearings with wheel bearing grease. Replace the bearings if necessary.

14. Use a hammer and a brass drift to carefully tap the outer and inner bearing races and oil seals from the hub.

To install:

15. Clean the flange surface of the hub, the thread holes, the surface of the lockwasher, and the splines of the axle shaft.

16. Use bearing drivers to install new inner and outer bearing races and oil seals.

17. Pack the bearings and the hub cavity with wheel bearing grease.

18. If a new rotor is being installed, use brake cleaner or alcohol to remove any anti-rust coating.

19. Install the rotor onto the hub and tighten the bolts to 76 ft. lbs. (103 Nm).

20. If equipped, install the ABS sensor ring and tighten the bolts to 13 ft. lbs. (18 Nm).

21. Install the inner bearing and the grease seal in the hub. Position the hub and rotor assembly on the spindle.

22. Install the outer bearing and spindle nut into the hub. Tighten the hub nut to 22 ft. lbs. (29 Nm) to seat the bearings and then fully loosen the nut. Use a spring scale connected to the stud bolt at 90 degrees to measure bearing preload. Then, retighten the hub nut until the spring gauge measures a bearing preload of 4.4–5.5 lbs. (2.0–2.5 kg) for a new bearing and grease seal, or 2.6–4.0 lbs. (1.2–1.8 kg) for a used bearing and a new grease seal.

23. Install the lockwasher on the spindle nut. The larger diameter, tapered side of the lockwasher faces out, and the holes in the lockwasher must align with the holes in the spindle nut. If the holes don't align, reverse the lockwasher, or tighten the spindle just enough to bring them into alignment.

24. Apply sealant to the both mating surfaces of the hub flange assembly and install it.

25. Install the shim and the snapring. There should be 0.000–0.012 in. (0.0–0.3mm) of clearance between the hub body and the snapring. Shims of different thicknesses may be used if necessary.

26. Pack the hub dust cap with fresh grease and install it. Tighten the hub bolts to 43 ft. lbs. (59 Nm).

27. Install the front wheels.

28. Lower the vehicle.

29. Verify that the front axle engages and the shift-on-the-fly 4WD system works correctly.

CHRYSLER CORP.

13

Chrysler-Town & Country • **Dodge-**Caravan • **Plymouth-**Voyager

ENGINE REPAIR

➡**Disconnecting the negative battery cable on some vehicles may interfere with the functioning of the on-board computer system, and may require the computer to undergo a relearning process once the negative battery cable is reconnected.**

Distributor

➡**The following procedure only applies to the 2.5L and 3.0L engines. Only the 2.5L and 3.0L engines use a distributor, the other engines utilize a Distributorless Ignition System (DIS).**

REMOVAL

1. Disconnect the negative battery cable.
2. On the 2.5L engine, disconnect the distributor lead wires, and vacuum hose, if equipped. Then, remove the distributor splash shield.
3. On the 3.0L engine, detach the electrical connector from the distributor.
4. Remove the distributor cap.
5. Mark the position of the rotor in relation to the distributor housing and mark the position of the distributor housing in relation to the engine. It is a good idea to take the time to turn the crankshaft to TDC No. 1 cylinder, compression stroke (firing position). This aligns the distributor rotor with the No. 1 spark plug tower in the distributor cap. This is a good reference point and makes installation easier.
6. Remove the distributor hold-down bolt.
7. Carefully lift the distributor from the engine. The shaft will rotate slightly when the drive gear disengages as the distributor is removed.

INSTALLATION

Timing Not Disturbed

1. Lower the distributor into the engine, aligning the marks made during removal. Be sure the O-ring is properly seated on the distributor. Replace the O-ring with a new one if it is cracked or nicked.

➡**Be sure the distributor drive is engaged with the gear on the camshaft.**

2. Install the distributor cap.
3. Tighten the hold-down bolt.
4. On the 2.5L engine, install the distributor splash shield, if equipped. Then,

reattach the distributor pick-up lead connector and vacuum hose, if necessary.
5. On the 3.0L engine, reattach the electrical connector to the distributor.
6. Reconnect the negative battery cable.
7. Check, and if necessary, adjust the ignition timing.

Timing Disturbed

1. Remove the spark plug from No. 1 cylinder. Place a finger over the spark plug hole.
2. Rotate the crankshaft in the normal direction of rotation, until compression is felt at the spark plug hole (air will leak out of the plug hole past your finger).
3. On the 2.5L engine, continue rotating the crankshaft until the **O** mark on the flywheel is aligned with the pointer on the bell housing.
4. On the 3.0L engine, continue rotating the crankshaft until the No. 1 piston is at the top of the compression stroke.
5. Install the distributor so that the distributor housing is fully seated when the rotor points to the No. 1 spark plug terminal on the distributor cap.
6. Install the distributor cap.
7. Tighten the hold-down bolt. Reattach the electrical connector to the distributor.
8. Check, and if necessary, adjust the ignition timing.

Ignition Timing

ADJUSTMENT

2.4L, 3.3L, 3.8L and 1996–99 3.0L Engines

The basic ignition timing cannot be adjusted. The Powertrain Control Module (PCM) regulates the ignition timing.

1995 2.5L and 3.0L Engines

1. Apply the parking brake and/or block the wheels.
2. Place the gearshift in the **P** or **N** position. Be sure to turn all lights and accessories OFF.
3. If a magnetic timing light is being used, place the pick-up probe into the open receptacle next to the timing scale window. If a conventional timing light is being used, connect it to the No. 1 cylinder spark plug wire.

➡**Do not puncture boots or cables with test probes. This will damage them. Be sure proper adapters are always used.**

4. Start the engine and allow it to run until it reaches operating temperature.

5. Once the engine has reached operating temperature, detach the electrical connector for the engine coolant temperature sensor. This will turn ON the radiator cooling fan and the malfunction indicator lamp (Check Engine light).
6. Aim the timing light at the timing scale or read the magnetic timing unit. The timing is advanced if a flash occurs when the timing mark is ahead of the specified degree mark. Make the adjustment by turning the distributor housing in the direction of rotor rotation. The timing is retarded if flashing occurs when the timing mark is after the specified degree mark. The adjustment is made by turning the distributor housing against the direction of rotor rotation.
7. Refer to the Vehicle Emission Control Information (VECI) label, located under the hood, for the correct ignition timing specification.
8. If the timing is within 2 degrees of the timing specification:
   a. Turn **OFF** the engine.
   b. Remove the timing light or magnetic timing unit and tachometer.
   c. Reconnect the engine coolant temperature sensor and erase any diagnostic trouble codes using a DRB or equivalent scan tool.

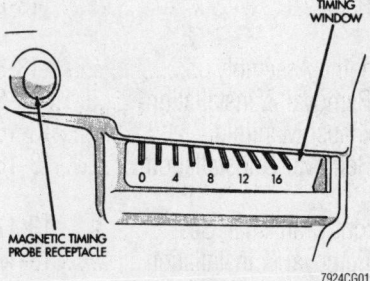

For the 1995 2.5L engine, the timing scale is located on the transaxle bell housing

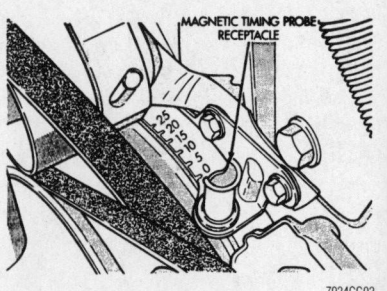

The timing scale is located on the harmonic balancer at the front of the engine—1995 3.0L engine

9. If the timing needs adjustment, proceed as follows:

a. Loosen the distributor hold-down bolt enough to rotate the distributor housing.

b. Turn the distributor housing to adjust the timing.

c. Tighten the distributor hold-down bolt and recheck the ignition timing.

d. Turn the engine **OFF**.

e. Remove the timing light or magnetic timing unit and tachometer.

f. Reconnect the engine coolant temperature sensor and erase any diagnostic trouble codes using a DRB or equivalent scan tool.

Engine Assembly

REMOVAL & INSTALLATION

✳✳ CAUTION

The fuel injection system remains under pressure, even after the engine has been turned OFF. The fuel system pressure must be relieved before disconnecting any fuel lines. Failure to do so may result in fire and/or personal injury.

1995 Models

1. Properly relieve the fuel system pressure.

2. Disconnect the battery cables from the battery, negative cable first.

3. Scribe the hood hinge outlines on the hood, and remove the hood.

4. Drain the cooling system. Remove the radiator hoses from the radiator and engine connections.

5. Remove the radiator and cooling fan assembly.

6. Remove the air cleaner assembly and related hoses.

7. Remove the air conditioner compressor from the engine and mounting brackets, leaving the refrigerant lines connected. Position the assembly aside and secure out of the way.

8. Remove the power steering pump from the engine with mounting brackets and hoses connected. Position the assembly aside and secure out of the way.

9. Label and detach all electrical connectors at the alternator, throttle body and engine.

10. Detach the fuel lines at the quick-disconnect fittings by pulling back on the fitting while pushing in on the plastic ring.

Disconnect the heater hose from the engine. Disconnect the accelerator cable.

11. Remove the alternator. Disconnect the clutch cable from the clutch lever, if equipped with a manual transaxle.

12. Remove the transaxle case lower cover.

13. If equipped with an automatic transaxle, matchmark the flexplate-to-torque converter position. Remove the bolts that mount the converter to the flexplate. Attach a small C-clamp to the front bottom of the converter housing to prevent the converter from falling out of the transaxle.

14. Disconnect the starter motor wiring and remove the starter motor.

15. Disconnect the exhaust pipe from the exhaust manifold.

16. Remove the right inner engine splash shield. Drain the engine oil and remove the oil filter. Disconnect the engine ground strap.

17. Attach a suitable hoist to the engine.

18. Support the transaxle. Apply slight upward pressure with the chain hoist and remove the throughbolt from the right (timing case cover) engine mount.

➡**If the complete engine mount is to be removed, mark the insulator position on the side rail to insure exact reinstallation location.**

19. Remove the transaxle-to-cylinder block mounting bolts.

20. Remove the front engine mount throughbolt. Remove the manual transaxle anti-roll strut, if equipped.

21. Remove the insulator throughbolt from the inside wheel house mount, or remove the insulator bracket-to-transaxle mounting bolt.

22. Raise the engine slowly with the hoist (transaxle supported). Separate the engine and transaxle and remove the engine.

To install:

23. With the hoist attached to the engine, lower the engine into the engine compartment.

24. Align the converter to the flexplate, if equipped, and the engine mounts. Install all mounting bolts loosely until all are in position, then tighten to 40 ft. lbs. (54 Nm).

25. Install the engine-to-transaxle mounting bolts. Tighten to 70 ft. lbs. (95 Nm) for the 2.5L and 3.0L engines or 75 ft. lbs. (102 Nm) for the 3.3L and 3.8L engines.

26. Remove the engine hoist and transaxle support.

27. Secure the engine ground strap.

28. Install the inner splash shield.

29. Install the starter assembly.

30. Install the exhaust system.

31. If equipped with a manual transaxle, install the transaxle case lower cover.

32. If equipped with an automatic transaxle, remove the C-clamp from the torque converter housing. Align the flexplate and torque converter with the marks made previously. Install the converter-to-flexplate mounting bolts and tighten to 55 ft. lbs. (75 Nm). Install the case lower cover.

33. If equipped with a manual transaxle, reconnect the clutch cable.

34. Install the power steering pump.

35. Install the air conditioning compressor.

36. Install the alternator.

37. Reconnect all wiring.

38. Install the radiator, cooling fan and shroud assembly.

39. Reconnect all cooling system hoses, accelerator cable and fuel lines.

40. Install the engine oil filter. Fill the crankcase to the proper oil level with clean engine oil.

41. Refill the cooling system.

42. Adjust linkages.

43. Install the air cleaner and hoses.

44. Install the hood.

45. Reconnect the battery cables, positive cable first. Ensure all electrical connections, cables, hoses, vacuum and fuel lines have been attached.

46. Start the engine and run until normal operating temperature is indicated. Check for leaks and check for proper engine operation.

1996–99 Models

1. Properly relieve the fuel system pressure.

2. Disconnect the battery cables from the battery, negative cable first.

3. Detach the fuel line-to-fuel rail quick-disconnect fittings by squeezing the retainer tabs together and pulling back on the fitting.

4. Scribe the hood hinge outlines on the hood and remove the hood.

5. Remove the wiper arm and blade assemblies.

6. Remove the cowl cover from the vehicle.

7. Detach the positive lock on the wiper unit electrical connector.

8. Detach the wiper unit electrical connector from the engine compartment wiring harness.

9. Disconnect the windshield washer hose from the hose coupling inside the wiper unit.

10. Remove the drain tubes from the bottom of the wiper unit.

11. Remove the sound absorbers from the ends of the wiper unit.

12. Remove the attaching nuts securing the wiper unit to the lower windshield fence.

13. Remove the attaching bolts securing the wiper unit to the dash panel.

14. Raise the wiper unit from the weld-studs on the lower windshield fence and remove the wiper unit from the vehicle.

15. Remove the air cleaner assembly and related hoses.

16. Remove the battery cover, battery, battery tray and integral vacuum reservoir from the vehicle.

17. If equipped, block off the rear heater to the rear heater unit.

18. Drain the cooling system. Remove the upper and lower radiator hoses from the radiator.

19. Disconnect and plug the heater hoses from the engine.

20. Remove the radiator and cooling fan assembly.

21. Disconnect the transaxle shift linkage.

22. Disconnect the throttle body linkage and all vacuum hoses to the throttle body.

23. Remove accessory drive belts.

24. Remove the air conditioner compressor from the engine and mounting brackets. Keep the A/C compressor hoses connected. Position the assembly aside and secure out of the way.

25. Detach the wiring harness connector to the alternator.

26. Remove the alternator.

27. Raise and safely support the vehicle. Remove the right and left halfshaft assemblies.

28. Disconnect the starter motor wiring and remove the starter motor.

29. Drain the engine oil and remove the oil filter.

30. Remove the right and left fender inner splash shields.

31. Disconnect the exhaust pipe from the exhaust manifold.

32. Remove the front motor mount and mount bracket as an assembly.

33. Remove the rear transaxle motor mount and bracket.

34. Remove the power steering pump and mounting bracket assembly from the engine.

35. Detach, label and remove the wiring harness and connectors from the front of the engine.

36. Remove bending braces.

37. Remove the transaxle case inspection cover.

38. Mark the flexplate-to-torque converter location.

39. Remove the bolts that mount the converter to the flexplate. Attach a small C-clamp to the front bottom of the converter housing to prevent the converter from falling out of the transaxle, if necessary.

40. Lower the vehicle to the ground.

41. Disconnect the engine ground straps.

42. Attach an engine lifting hoist to the engine.

43. Remove the right engine mount assembly and left transaxle mount through-bolt.

44. Using the engine hoist, raise the engine and transaxle assembly slowly out of the vehicle. Separate the engine and transaxle. Secure the engine on an engine stand.

To install:

45. Attach the transaxle to the engine assembly.

46. With the hoist attached to the engine/transaxle assembly, lower the engine/transaxle into the engine compartment.

47. Align the engine and transaxle motor mounts to their attaching points. Install the right engine and left transaxle mount bolts.

48. Remove the engine hoist.

49. Install bending braces.

50. Install the alternator.

51. Install and reattach the wiring harness connectors on the front of the engine.

52. Install the air conditioning compressor to the engine.

53. Install the power steering pump and mounting bracket to the engine.

54. Install accessory drive belts and adjust to the proper tension, if necessary.

55. Raise and safely support the vehicle.

56. Remove the C-clamp from the torque converter housing, if utilized.

57. Align the flexplate and torque converter with the marks made previously.

58. Install the converter-to-flexplate mounting bolts. Tighten them to 55 ft. lbs. (75 Nm).

59. Install the transaxle case lower inspection cover.

60. Install right and left halfshaft assemblies.

61. Install the engine and transaxle mount and bracket assemblies.

62. Install the exhaust system to the exhaust manifolds.

63. Install the left and right fender inner splash shields.

64. Install the starter assembly and reconnect the starter motor wiring.

65. Reconnect the automatic transaxle shift linkage.

66. Lower the vehicle.

67. Reattach the fuel line quick-disconnect fitting to the fuel rail.

68. Reconnect the cooling system heater hoses to the engine.

69. Unblock the heater hoses to the rear heater unit, if equipped.

70. Secure the engine ground straps.

71. Reattach the engine and throttle body vacuum connections and wiring harness connectors.

72. Reconnect the throttle body linkage.

73. Install the radiator, cooling fan and shroud assembly.

74. Reconnect the upper and lower radiator hoses.

75. Install the battery tray, battery and battery cover into the vehicle.

76. Install the air cleaner assembly and hoses.

77. Install a new engine oil filter. Fill the crankcase to the proper oil level with the correct type of clean engine oil.

78. Refill the cooling system to the proper level with a 50/50 mixture of clean ethylene glycol antifreeze and water.

79. Install the wiper unit into the vehicle engine compartment, making sure the wiper unit is installed properly over the weld-studs on the lower windshield fence. Install and tighten the attaching nuts to the weld-studs.

80. Install and tighten the attaching bolts securing the wiper unit to the dash panel.

81. Install the sound absorbers to each end of the wiper unit.

82. Install the drain tubes to the bottom of the wiper unit and reconnect the windshield washer hose to the hose coupling inside the wiper unit.

83. Reattach the wiper unit wiring connector to the engine wiring harness.

84. Place the cowl cover onto the vehicle. Reconnect the right side windshield washer hose to the right washer nozzle located on the underside of the cowl cover.

85. Engage the retainers that secure the cowl cover to the front fender edge.

86. Install and tighten the wing nuts that secure the front of the cowl cover to the wiper module.

87. Engage the quarter turn fasteners that secure the outer ends of the cowl cover to the wiper module.

88. Install and tighten the bolts that hold the lower area of the cowl cover to the wiper module.

89. Reconnect the positive battery cable, then the negative battery cable and verify that the wiper motor and wiper linkage are in the PARK position.

90. Install the wiper arm in the correct position over the wiper arm pivot. Align the wiper arm positions as follows:

a. Left arm should be no closer than 2.5 in. (65mm) from the lower edge of the windshield.

b. Right arm should be no closer than 1.5 in. (40mm) from the lower edge of the windshield.

91. Install the wiper arm-to-wiper arm pivot retaining nut and tighten to 26 ft. lbs. (35 Nm).

92. Push down the wiper arm cap cover and engage the clip that secures the outside end of the cover to the wiper arm.

93. Operate the windshield wipers and ensure they work and park properly.

94. Adjust linkages.

95. Install the hood.

96. Ensure all electrical connections, cables, hoses, vacuum and fuel lines have been attached.

97. Start the engine and run until normal operating temperature is indicated. Check for leaks. Road test the vehicle.

Water Pump

REMOVAL & INSTALLATION

2.4L Engine

1. Disconnect the negative battery cable.

➡This procedure requires removing the engine timing belt and the auto-tensioner. The factory specifies that the timing marks should always be aligned before removing the timing belt. Set the piston in the No. 1 cylinder to TDC on the compression stroke. This should align all timing marks on the crankshaft sprocket and both camshaft sprockets.

2. Raise and safely support the vehicle.

3. Remove the right inner splash shield.

4. Remove the accessory drive belts.

5. Place a drain pan under the radiator drain plug. Drain the cooling system.

6. Support the engine and remove the right motor mount.

7. Remove the power steering pump mounting bracket bolts and place the pump/bracket assembly off to one side. Do not disconnect the power steering fluid lines.

8. Remove the right engine mount bracket.

9. Remove the front timing belt upper and lower covers.

10. Loosen the timing belt tensioner bolts and remove the belt tensioner and timing belt.

✳✳ WARNING

With the timing belt removed, DO NOT rotate the camshaft or crankshaft or damage to the engine could occur.

11. Remove the camshaft sprockets. With the timing belt removed, remove both camshaft sprocket bolts. Do not allow the camshafts to turn when the camshaft sprockets are being removed.

12. Remove the rear timing belt cover to access the water pump.

13. Remove the water pump attaching bolts.

14. Remove the water pump.

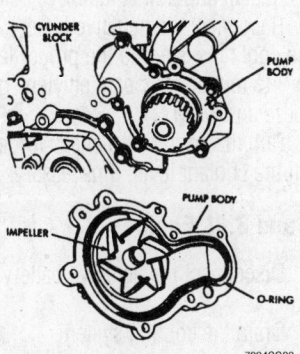

Water pump mounting location and pump component identification—2.4L engine

To install:

15. Thoroughly clean all parts. Replace the water pump if there are any cracks, signs of coolant leakage from the shaft seal, loose or rough turning bearing, damaged impeller or sprocket, or the sprocket flange is loose or damaged.

16. Clean all sealing surfaces. Install a new rubber O-ring into the water pump O-ring groove.

➡Be sure the O-ring is properly seated in the water pump groove before tightening the bolts. An improperly located O-ring may cause damage to the O-ring and cause a coolant leak.

17. Install the water pump to the engine and tighten the bolts to 105 inch lbs. (12 Nm).

18. Pressurize the cooling system to 15 psi (103 kPa) and check for leaks. If okay, release the pressure and continue the assembly process.

19. Install the rear timing belt cover.

20. Install the camshaft sprockets and tighten the attaching bolts to 75 ft. lbs. (101 Nm). To maintain timing mark alignment, DO NOT allow the camshafts to turn while the sprocket bolts are being tightened.

✳✳ WARNING

Do not attempt to compress the tensioner plunger with the tensioner assembly installed in the engine. This will cause damage to the tensioner and other related components. The tensioner MUST be compressed in a vise.

21. Install the timing belt tensioner and timing belt. Be sure to properly tension the timing belt.

22. Install the front upper and lower timing belt covers.

23. Install the right engine mount bracket and engine mount.

24. Install the crankshaft damper and tighten the center bolt to 105 ft. lbs. (142 Nm).

25. Install the right inner splash shield.

26. Lower the vehicle.

27. Install the power steering pump bracket and power steering pump. Tighten the bracket mounting bolts to 40 ft. lbs. (54 Nm).

28. Install and adjust the drive belts.

29. Refill the cooling system using a ⁵⁰/₅₀ mixture of water and ethylene glycol antifreeze. Bleed the cooling system.

30. Start the engine and check for proper operation.

31. Check and top off the cooling system, if necessary.

2.5L Engine

1. Disconnect the negative battery cable.

2. Drain the cooling system.

3. Remove the drive belts.

4. If equipped with air conditioning, remove the compressor from the bracket and position it aside. Do not disconnect the refrigerant lines.

5. Remove the alternator and mounting bracket and set it aside.

6. Raise and safely support the vehicle, if necessary. Remove the pulley from the water pump.

7. Disconnect the lower radiator hose and heater hose from the water pump.

8. Remove the water pump housing attaching bolts and remove the assembly from the vehicle. Discard the O-ring.

9. Remove the water pump from housing.

To install:

10. Using a new gasket or silicone sealer, install the water pump on the housing.

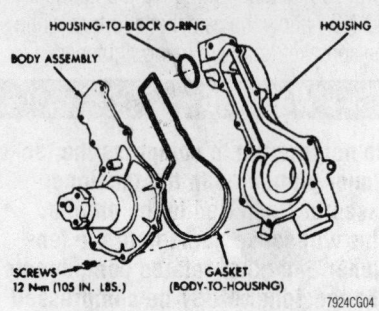

Exploded view of the water pump housing—2.5L engine

11. Install a new O-ring to the housing and install on the engine. Tighten the top three bolts to 21 ft. lbs. (30 Nm). Tighten the lower bolt to 50 ft. lbs. (68 Nm).

12. Install the water pump pulley. Tighten the water pump pulley bolts to 20 ft. lbs. (28 Nm). Connect the radiator hose and heater hose to the water pump.

13. Install the mounting bracket, alternator and A/C compressor, if removed. Install the drive belts and adjust the belts to the proper tension, if necessary.

14. Remove the hex head plug or vacuum switching valve on the top of the thermostat housing. Fill the radiator with coolant until the coolant comes out the plug hole. Install the plug or valve and continue to fill the radiator. Tighten the vent plug to 15 ft. lbs. (20 Nm).

15. Reconnect the negative battery cable, run the vehicle until the thermostat opens, and fill the overflow tank. Check for leaks.

16. Once the vehicle has cooled, recheck the coolant level.

3.0L Engine

1. Disconnect the negative battery cable.
2. Drain the cooling system.
3. Remove the drive belts. Remove the timing belt covers and the timing belt.
4. Remove the water pump mounting bolts.
5. Separate the water pump from the water inlet pipe and remove the water pump.
6. Inspect the water pump and replace as necessary.

To install:

7. Clean all gasket and O-ring surfaces on the water pump and water pipe inlet tube.
8. Wet a new O-ring with water and install it on the water inlet pipe.
9. Install a new gasket on the water pump.
10. Install the pump inlet opening over the water pipe and press until the pipe is completely inserted into the pump housing.

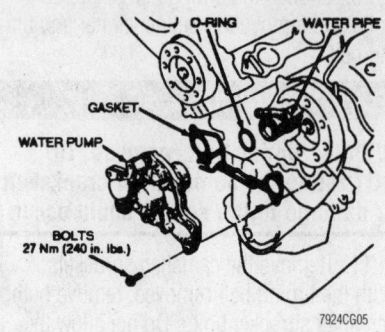

Exploded view of the water pump mounting—3.0L engine

11. Install the water pump-to-block mounting bolts and tighten to 20 ft. lbs. (27 Nm).

12. Install the timing belt and timing belt covers. Install and adjust the drive belts.

13. Reconnect the negative battery cable. Fill the cooling system to the proper level with a 50/50 mixture of clean, ethylene glycol antifreeze and water.

14. Run the engine and check for leaks. Top off the coolant level, if necessary.

3.3L and 3.8L Engines

1. Disconnect the negative battery cable.
2. Drain the cooling system.
3. Remove the serpentine belt.
4. Raise and safely support the vehicle. Remove the right front wheel and lower fender shield.
5. Remove the water pump pulley.
6. Remove the five mounting bolts and remove the pump from the engine.
7. Discard the O-ring. Clean the O-ring sealing surface and inspect the water pump for damage, cracks, seal leaks, and loose or rough turning bearings.

To install:

8. Install a new O-ring into the water pump groove. Install the pump onto the

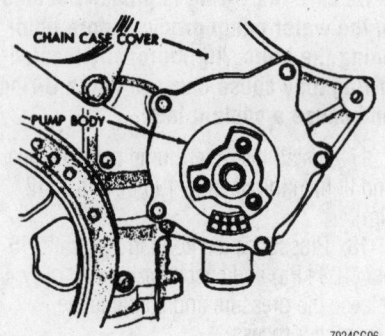

Water pump location—3.3L and 3.8L engines

engine. Tighten the mounting bolts to 108 inch lbs. (12 Nm).

9. Install the water pump pulley. Tighten the water pump pulley bolts to 20 ft. lbs. (28 Nm).

10. Install the fender shield and wheel. Tighten the wheel lug nuts, in sequence, to 95 ft. lbs. (129 Nm). Lower the vehicle.

11. Install the serpentine belt.

12. Refill the cooling system to the correct level with a 50/50 mixture of clean, ethylene glycol antifreeze and water. Bleed the cooling system.

13. Reconnect the negative battery cable, run the vehicle until the thermostat opens, fill the overflow tank and check for leaks.

14. Once the vehicle has cooled, recheck the coolant level.

Cylinder Head

REMOVAL & INSTALLATION

✳✳ CAUTION

The fuel injection system remains under pressure, even after the engine has been turned OFF. The fuel system pressure must be relieved before disconnecting any fuel lines. Failure to do so may result in fire and/or personal injury.

2.4L Engine

1. Properly relieve the fuel system pressure.
2. Disconnect the negative battery cable.
3. Place a large drain pan under the radiator drain plug. Open up the drain plug and drain the cooling system.
4. Remove the air cleaner assembly and disconnect all vacuum lines, electrical wiring and fuel lines from the throttle body.
5. Disconnect the throttle linkage.
6. Remove the accessory drive belts.
7. Disconnect the power brake vacuum hose from the intake manifold.
8. Raise and safely support the vehicle. Disconnect the exhaust pipe from the exhaust manifold.
9. Lower the vehicle as required to remove the power steering pump. Do not disconnect the fluid lines. Set the pump aside.
10. Label the spark plug wires for correct installation. Detach the coil pack wiring connector and remove the coil pack and spark plug wires from the engine.
11. Detach the cam sensor and fuel injector wiring connectors.

12. Remove the timing belt covers, timing belt and camshaft sprockets.

13. Remove the timing belt idler pulley and rear timing belt cover.

14. Remove the cylinder head cover mounting fasteners and cylinder head cover. Remove the ground strap.

15. Identify the camshafts, if they are to be reused, for later installation. The camshafts are not interchangeable. Remove the camshaft bearing caps and the camshafts.

16. Remove the camshaft followers. Any components that are to be reused must be installed in their original locations. Use care to identify and mark the positions of any removed valvetrain components so they may be reinstalled correctly.

17. Remove the intake and exhaust manifolds.

18. Remove the cylinder head bolts.

19. Remove the cylinder head from the vehicle, using care not to damage the aluminum gasket surfaces.

20. Remove all gasket material from the sealing surfaces of the cylinder head and engine block. Be careful not to gouge or scratch the surface of the aluminum head. The cylinder head should be checked for warpage using a good straightedge and feeler gauges. Place the straightedge on the bottom of the cylinder head from front to back, then try to insert the feeler gauge between the head and the straightedge in several places. If a feeler gauge larger than 0.004 in. (0.1mm) fits between the head and straightedge, take the cylinder head to a machine shop for service before installing it.

21. Inspect the camshaft bearing oil feed holes in the cylinder head for clogging. Inspect the camshaft bearing journals for wear or scoring. Check the cam surface for abnormal wear and damage. A visible worn groove in the roller path or on the cam lobes is cause for replacement. Valve service may be performed at this time.

To install:

22. Clean all parts well. Note that the cylinder head bolts are tightened using a new procedure. The cylinder head bolts should be checked carefully BEFORE reuse. If the threads are necked down the bolts should be replaced with new bolts. Necking can be checked by holding a steel scale or straightedge against the threads. If all the threads do not contact the scale, the bolt should be replaced. New cylinder head bolts are recommended for any engine rebuild, especially if known that the engine has been disassembled before.

23. Be sure both the top of the engine block and the bottom of the cylinder head are clean. Install a new gasket making sure all holes align with the openings in the engine block. Carefully set the cylinder head in place.

24. Apply clean engine oil to the threads and under the heads of the bolts. Install the bolts and tighten them in sequence using the following four substeps:

a. Tighten all bolts to 25 ft. lbs. (34 Nm).

b. Tighten all bolts to 50 ft. lbs. (68 Nm).

c. Tighten all bolts again to 50 ft. lbs. (68 Nm).

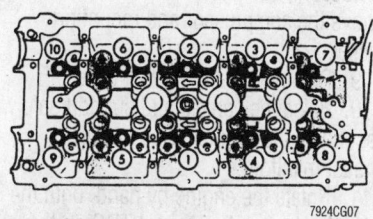

7924CG07

Cylinder head bolt tightening sequence— 2.4L engine

➡**Do not use a torque wrench for the fourth step.**

d. Tighten all bolts an additional ¼ turn.

25. The camshaft end-play should be checked using the following procedure:

a. Oil the camshaft journals and install the camshaft WITHOUT the cam follower assemblies. Install the rear cam caps and tighten them to 250 inch lbs. (28 Nm).

b. Carefully push the camshaft as far rearward as it will go.

c. Set up a dial indicator to bear against the front of the camshaft (the sprocket end). Zero the indicator.

d. Move the camshaft forward as far as it will go. Read the dial indicator. End-play specification is 0.002–0.010 in. (0.05–0.15mm).

26. When satisfied with the fit and condition of the camshafts, remove the camshafts for installation of the cam followers.

27. Lubricate the camshaft followers with clean engine oil. Install the cam followers in their original positions on the hydraulic adjuster and valve stem.

✳✳ WARNING

Be sure none of the pistons are at Top Dead Center (TDC) when installing the camshafts.

28. Lubricate the camshaft bearing journals and cam followers with clean engine oil and install the camshafts. Install right and left camshaft bearing caps No. 2 through No. 5 and right side No. 6. Tighten the M6 fasteners to 105 inch lbs. (12 Nm) in sequence.

29. Apply Mopar Gasket Maker or equivalent sealer to the No. 1 and No. 6 bearing caps. Install the bearing caps and tighten the M8 fasteners to 250 inch lbs. (28 Nm). The end caps must be installed before the seals may be installed.

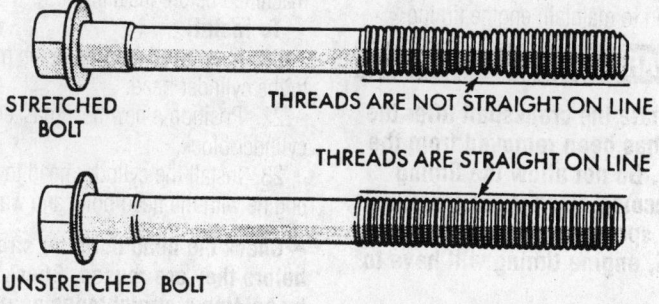

STRETCHED BOLT

THREADS ARE NOT STRAIGHT ON LINE

THREADS ARE STRAIGHT ON LINE

UNSTRETCHED BOLT

7924CG38

Examine the threads to determine if the old cylinder head bolts are stretched and need to be replaced—2.4L engine

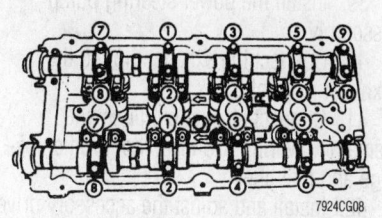

7924CG08

Camshaft bearing cap bolt tightening sequence—2.4L engine

30. Apply a light coating of clean engine oil to the lip of the new camshaft seal. Install the camshaft seal until it fits flush with the cylinder head.

➡ **Refer to the timing belt Unit Repair Section (URS) for information on the timing belts and sprockets.**

31. Install the camshaft sprockets, if removed. Install the rear timing belt cover and timing belt using care to be sure all timing marks are properly aligned. Install the timing belt cover.

✸✸ WARNING

Verify that all timing marks are correct. If the timing belt or sprockets are incorrectly installed, engine damage will occur. Take time to be sure all timing marks are correctly aligned.

32. Install the intake and exhaust manifolds.

33. Clean the cylinder head cover and cylinder head gasket rails (mating surfaces). Make certain the rails are flat.

34. Install new cylinder head cover gaskets. Use care; DO NOT allow oil or solvents to contact the timing belt as they can deteriorate the rubber and cause tooth skipping. Apply Mopar Silicone Rubber Adhesive Sealant, or equivalent, at the camshaft cap corners and at the top edge of the ½ round seal.

35. Install the cylinder head cover assembly to the head and tighten the fasteners in sequence using the following three substeps:

a. Tighten all cylinder head cover fasteners to 40 inch lbs. (4.5 Nm).

b. Tighten all fasteners to 80 inch lbs. (9 Nm).

c. Tighten all fasteners to 105 inch lbs. (12 Nm).

36. Install the ground strap.

37. Install the ignition coil pack and reconnect the spark plug wiring.

38. Connect the cam sensor and fuel injectors' wiring.

39. Install the power steering pump assembly.

40. Connect the exhaust pipe to the exhaust manifold.

41. Connect all vacuum lines and remaining wiring. Connect the throttle linkage and fuel lines.

42. Install and adjust the accessory drive belts.

43. Refill the cooling system. An oil and filter change is recommended since coolant can enter the oil system when a head is removed.

44. Connect the remaining air ducting. Connect the negative battery cable and test run the vehicle. Check for leaks and for proper operation.

2.5L Engine

1. Properly relieve the fuel system pressure.

2. Disconnect the negative battery cable and unbolt it from the head.

✸✸ CAUTION

Never open, service or drain the radiator or cooling system when hot; serious burns can occur from the steam and hot coolant.

3. Drain the cooling system. Remove the dipstick bracket nut from the thermostat housing and rotate the bracket from the stud.

4. Remove the air cleaner assembly. Remove the upper radiator hose and disconnect the heater hoses.

5. Remove the accessory drive belts.

6. Label and detach the vacuum lines, hoses and wiring connectors from the manifold, throttle body and cylinder head. Remove the air pump, if equipped.

7. Disconnect all linkage and the fuel lines from the throttle body. Disconnect the ground strap.

8. Remove the power steering pump and position aside, without disconnecting the hoses. If equipped with air conditioning, remove the compressor and mounting bracket.

9. Remove the timing belt cover.

10. Raise the vehicle and support safely. Disconnect the exhaust pipe from the exhaust manifold.

11. Rotate the engine by hand, until the piston in No. 1 cylinder is at TDC on the compression stroke (firing position). Be sure all the timing marks are aligned.

12. Lower the vehicle.

13. With the timing marks aligned, remove the camshaft sprocket. Suspend the camshaft sprocket and timing belt under light tension to maintain engine timing.

✸✸ WARNING

Do not rotate the crankshaft after the sprocket has been removed from the camshaft. Do not allow the timing belt to become disengaged from the camshaft sprocket. If the engine is disturbed, engine timing will have to be reset.

14. Label and disconnect the spark plug wires from the spark plugs. Detach the coil wiring connector and the coil wire from the coil.

15. Remove the valve cover. Remove the air/oil separation curtain, if equipped. Remove the cylinder head bolts and washers, starting from the ends of the cylinder head and working inward.

16. Remove the cylinder head from the engine.

17. If necessary, remove the intake and exhaust manifolds from the cylinder head.

✸✸ WARNING

Before disassembling or repairing any part of the cylinder head assembly, check for factory installed oversized components, indicated by green paint and the letters O/SJ. If equipped with oversize components, the tops of the bearing caps would be painted green and O/SJ would be stamped rearward of the oil gallery plug on the rear of the head. In addition, the barrel of the camshaft would be painted green and O/SJ stamped onto the rear end of the camshaft. Installing standard-sized parts in a head equipped with oversized parts, or oversized parts in a standard size head will cause severe engine damage.

18. Clean all gasket mating surfaces.

19. Remove all gasket material from the sealing surfaces of the cylinder head and engine block. Be careful not to gouge or scratch the surface of the head.

20. Inspect the cylinder head for warpage by using a good straightedge and feeler gauges. Place the straightedge longitudnally (lengthwise from the front to the rear of the cylinder head) and diagonally from corner-to-corner on the engine block-to-cylinder head surface, then attempt to insert the feeler gauge between the head and the straightedge in several places. The maximum allowable warpage is 0.004 in. (0.1mm). If more warpage is evident, have the cylinder head machined before installing it.

To install:

21. If removed, assemble the manifolds to the cylinder head.

22. Position a new head gasket on the cylinder block.

23. Install the cylinder head to the engine with the head bolts and washers.

➡ **Check the head bolts for stretching before they are reused. Check the bolts by holding a straightedge against the threads. If all threads do not contact the straightedge, replace the bolt.**

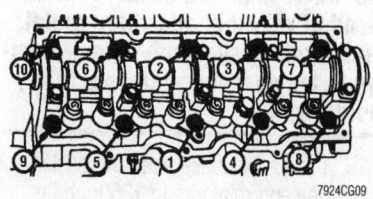

Cylinder head bolt tightening sequence—2.5L engine

24. Tighten the cylinder head bolts in sequence, as follows:

   a. Tighten the bolts, in sequence, to 45 ft. lbs. (61 Nm).

   b. Tighten the bolts, in sequence, to 65 ft. lbs. (88 Nm).

   c. Repeat substep b, making sure all bolts are tightened to 65 ft. lbs. (88 Nm).

   d. Turn each bolt, in sequence, an additional ¼ turn. Do not use a torque wrench for this step.

   e. Check the torque on each bolt after the ¼ turn. It should be over 90 ft. lbs. (122 Nm). If not, replace the bolt.

✳✳ WARNING

Head bolt diameter is 11mm. These bolts are identified with the No. 11 on the head of the bolt. 10mm bolts will thread into an 11mm bolt hole, but will strip the cylinder block bolt hole. Be sure the correct bolts are being used to avoid damaging the engine block.

25. Install the camshaft sprocket and tighten the bolt to 65 ft. lbs. (88 Nm). Be sure the timing marks are aligned.

26. Install the valve cover, curtain, timing belt and timing belt cover. Tighten the valve cover bolts to 105 inch lbs. (12 Nm).

27. Reconnect the wires to the spark plugs. Reattach the coil wire and coil wiring connector.

28. Raise and safely support the vehicle.

29. Reconnect the exhaust system to the exhaust manifold. Lower the vehicle.

30. If equipped, install the A/C compressor mounting bracket and compressor.

31. Install the power steering pump.

32. Install the air pump, if equipped.

33. Reconnect all linkages and the fuel lines to the throttle body. Install the ground strap.

34. Reattach all vacuum lines, hoses and wiring connectors in their proper locations, as labeled during the removal procedure.

35. Install the accessory drive belts and adjust to the proper tension.

36. Install the radiator hose and connect the heater hoses. Install the air cleaner assembly.

37. Install the dipstick bracket to the thermostat housing.

38. Refill the cooling system to the correct level with clean engine coolant. Reconnect the negative battery cable to the cylinder head and the negative terminal on the battery.

39. Start the engine and check for leaks and proper engine operation.

3.0L Engine

1. Properly relieve the fuel system pressure.

2. Disconnect the negative battery cable. Drain the cooling system.

3. Remove the accessory drive belts and the air conditioning compressor from its mount and support it aside. Remove the alternator and power steering pump from the brackets and move them aside.

4. Raise the vehicle and support safely. Remove the right front wheel and the right inner splash shield.

5. Remove the crankshaft pulleys and the torsional damper.

6. Lower the vehicle. Using a floor jack and a block of wood positioned under the oil pan, raise the engine slightly. Remove the engine mount bracket from the timing cover end of the engine.

7. Remove the timing belt covers.

8. Remove the timing belt as described in the unit repair section in the beginning of this manual.

9. Hold the camshaft sprocket using a suitable tool and remove the camshaft sprocket retaining bolt. Remove the sprocket and the inner timing belt cover (left bank) and/or alternator bracket (right bank).

10. Label and disconnect the spark plug wires from the spark plugs.

11. If removing the left (front) cylinder head, remove the distributor cap and spark plug wires. Mark the position of the rotor and distributor in relation to the cylinder head and remove the distributor. Remove the distributor drive adapter.

12. Remove the valve cover.

13. Install the auto lash adjuster retainers on the rocker arms. Remove the camshaft bearing cap-to-cylinder head bolts (do not remove the bolts from the assembly). Remove the rocker arms, rocker shafts and bearing caps as an assembly. Remove the camshaft from the cylinder head.

14. Remove the intake manifold assembly.

15. Remove the exhaust manifold and crossover pipe.

16. Remove the cylinder head bolts starting from the outside and working inward. Remove the cylinder head from the engine.

17. Clean the gasket mounting surfaces and check the head gasket surface for leaks, damage or warpage; the maximum warpage allowed is 0.008 in. (0.20mm).

To install:

18. Clean all parts well. Install the new cylinder head gasket over the dowels on the engine block.

19. Install the cylinder head on the engine block and tighten the 10mm Allen cylinder head bolts in sequence, using three even steps, to 80 ft. lbs. (108 Nm).

20. Install the intake and exhaust manifolds.

21. Lubricate the camshaft with clean engine oil and install on the cylinder head.

22. Apply silicone sealant to the cylinder head at the front and rear cam bearing cap contact areas. Install the rocker arm shaft

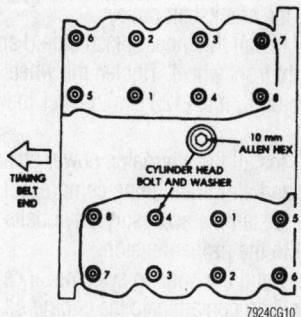

Tighten the cylinder head bolts clockwise starting from the middle as shown—3.0L engine

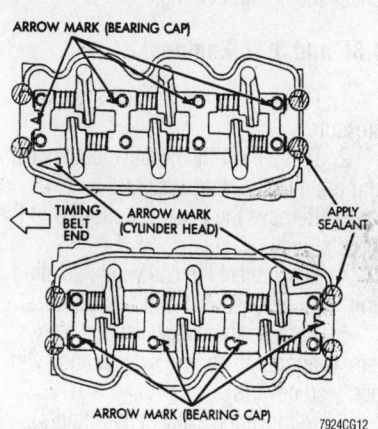

Apply sealant to the shaded areas on the cylinder head and be sure to tighten the bearing caps using the recommended sequence—3.0L engine

assembly and tighten the bearing cap bolts to 85 inch lbs. (10 Nm) in the following sequence: No. 3, No. 2, No. 1 and No. 4. Retighten to 180 inch lbs. (20 Nm) in the same sequence.

23. If removed, install the distributor drive adapter. Install a new camshaft seal.

24. Install the valve cover.

25. Install the inner timing belt cover (left bank) and/or alternator bracket (right bank).

26. Position the camshaft sprocket. Hold the sprocket using a suitable tool and install the camshaft sprocket bolt. Tighten the bolt to 70 ft. lbs. (95 Nm).

➡**Be sure the timing belt sprocket timing marks are aligned.**

27. If removed, install the distributor, aligning the marks made during removal. Install the distributor cap and spark plug wires.

28. Install the timing belt.

29. Install the timing belt covers and the engine support bracket.

30. Install the torsional damper and tighten the bolt to 112 ft. lbs. (151 Nm). Install the crankshaft pulleys.

31. Install the inner splash shield and the right front wheel. Tighten the wheel lug nuts to 95 ft. lbs. (129 Nm). Lower the vehicle.

32. Install the alternator, power steering pump and air conditioning compressor.

33. Install the accessory drive belts and adjust to the proper tension.

34. Refill the cooling system. Since coolant can contaminate the engine oil when a cylinder head is removed, an oil and filter change is recommended.

35. Reconnect the negative battery cable.

36. Start the engine and check for leaks. Check the ignition timing.

3.3L and 3.8L Engines

1. Properly relieve the fuel system pressure.

2. Disconnect the negative battery cable and drain the cooling system.

3. Remove the intake manifold with the throttle body.

4. Disconnect the coil wires, sending unit wire, heater hoses and bypass hose.

5. Remove the closed ventilation system, evaporation control system and cylinder head cover(s).

6. Remove the exhaust manifold(s).

7. Remove the rocker arm and shaft assemblies. Remove the pushrods and identify them to ensure installation in their original positions.

8. Loosen the cylinder head bolts in the reverse order of the torque sequence. Remove the cylinder head bolts and remove the cylinder head(s) from the block.

9. Clean all gasket mating surfaces.

To install:

10. Clean the gasket mounting surfaces and install a new head gasket to the block.

➡**The cylinder head bolts are tightened using the torque yield method. The bolts should be examined before they are reused. If the threads are stretched, the bolts should be replaced. Stretching can be checked by holding a straightedge against the threads. If all the threads do not contact the straightedge, the bolt should be replaced.**

11. Install the cylinder head to the block with the bolts.

12. Tighten cylinder head bolts Nos. 1 through 8 in sequence as follows:

a. Tighten each bolt to 45 ft. lbs. (61 Nm).

b. Repeat the sequence and tighten the bolts to 65 ft. lbs. (88 Nm).

c. Repeat the sequence again, making sure the bolts are at 65 ft. lbs. (88 Nm).

d. Finally, turn each bolt, in sequence, ¼ turn. Do not use a torque wrench for this step.

e. Check the torque of each bolt after the ¼ turn. The torque should be over 90 ft. lbs. (122 Nm). If not, replace the bolt.

13. Tighten head bolt No. 9 to 25 ft. lbs. (33 Nm) after the other eight bolts have been properly tightened.

14. Install the pushrods, rocker arms and shafts, and tighten the bolts to 250 inch lbs. (28 Nm).

15. Place a drop of silicone sealer onto each of the four manifold-to-cylinder head gasket corners.

Tighten the cylinder head bolts in the sequence shown—3.3L and 3.8L engines

The intake manifold gasket is composed of very thin and sharp metal. Handle this gasket with care or damage to the gasket or personal injury could result.

16. Install the intake manifold gasket and tighten the end retainers to 105 inch lbs. (12 Nm).

17. Install the intake manifold and tighten the bolts in sequence to 10 inch lbs. (1 Nm). Repeat the sequence, increasing the torque to 200 inch lbs. (22 Nm). Recheck each bolt, making sure it is at 200 inch lbs. (22 Nm) of torque. As the bolts are tightened, inspect the seals to ensure that they have not become dislodged.

18. Install the valve cover with a new gasket. Tighten the valve cover bolts to 105 inch lbs. (12 Nm). Install the exhaust manifold and crossover pipe. Tighten the bolts to 20 ft. lbs. (27 Nm) and the nuts to 15 ft. lbs. (20 Nm).

19. Install the closed ventilation system and the evaporation control system.

20. Reconnect the coil wires, sending unit wire, heater hoses and bypass hose.

21. Ensure all wiring connections, cables, hoses, vacuum and fuel lines have been reattached.

22. Refill the cooling system to the correct level. Coolant can contaminate the engine oil when performing cylinder head service. An oil and filter change is recommended.

23. Reconnect the negative battery cable. Start the engine and check for leaks.

Rocker Arms/Shafts

REMOVAL & INSTALLATION

3.0L Engine

The rocker arms and shafts are retained by the camshaft bearing journal caps. Four shafts are used, one for each intake and exhaust rocker arm assembly on each cylinder head. The hollow shafts provide a duct for lubricating oil from the cylinder head to the valve mechanisms. The rocker arms are lightweight die-cast with roller-type followers operating against the camshaft. The valve actuating end of the rocker arms are machined to retain hydraulic lash adjusters, eliminating valve lash adjustment.

1. Disconnect the negative battery cable. Disconnect and label the spark plug wires.

2. Remove the air cleaner assembly. Remove the accessory drive belts, and detach the vacuum connections.

3. Remove the valve cover.

4. Before removing the rocker arm shaft assembly, a function check can be made of the auto lash adjusters. Use the following procedure:

➡**The auto lash adjusters are precision units installed in machined openings in the valve actuating ends of the rocker arms. Do not disassemble the auto lash adjusters.**

a. Check the adjusters for free-play by inserting a small wire through the air bleed hole in the rocker arm.

b. VERY LIGHTLY push the auto adjuster check ball down.

c. While lightly holding the check ball down, move the rocker arm up and down to check for free-play. If there is no free-play, replace the adjuster.

5. Install Lash Adjuster Retainers MD998443 or equivalent, on the rocker arms.

6. Loosen all the camshaft bearing cap bolts. Do not remove the bolts from bearing caps. Remove the rocker arms, rocker shafts and bearing caps as an assembly.

7. Remove the bolts from the camshaft bearing caps and remove the rocker shafts

and arms. Keep all parts in order. Note the way the rocker shaft, rocker arms, bearing caps and springs are mounted. The rocker arm shaft on the intake side has a 3mm diameter oil passage hole from the cylinder head. The exhaust side does not have this oil passage.

8. Inspect the rocker arm mounting area and rocker for damage. Replace if worn or heavily damaged. Check oil passages for clogging and clean, if necessary.

To install:

9. Lubricate the rocker arms and shafts with clean engine oil prior to installation.

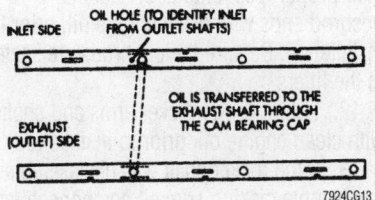

Be sure to install the rocker arm shaft with the oil hole on the intake side of the head—3.0L engine

10. Identify No. 1 bearing cap, (No. 1 and No. 4 caps are similar). Install the rocker shafts into the bearing cap with notches in proper position. Insert the attaching bolts to retain assembly.

11. Install the rocker arms, springs and bearing caps on shafts in numerical sequence.

12. Align the camshaft bearing caps with arrows (depending on the cylinder bank).

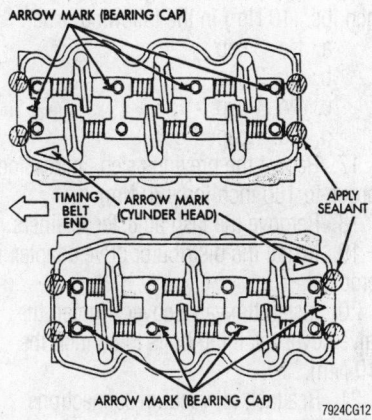

Apply silicone sealant to the cylinder head in the areas indicated—3.0L engine

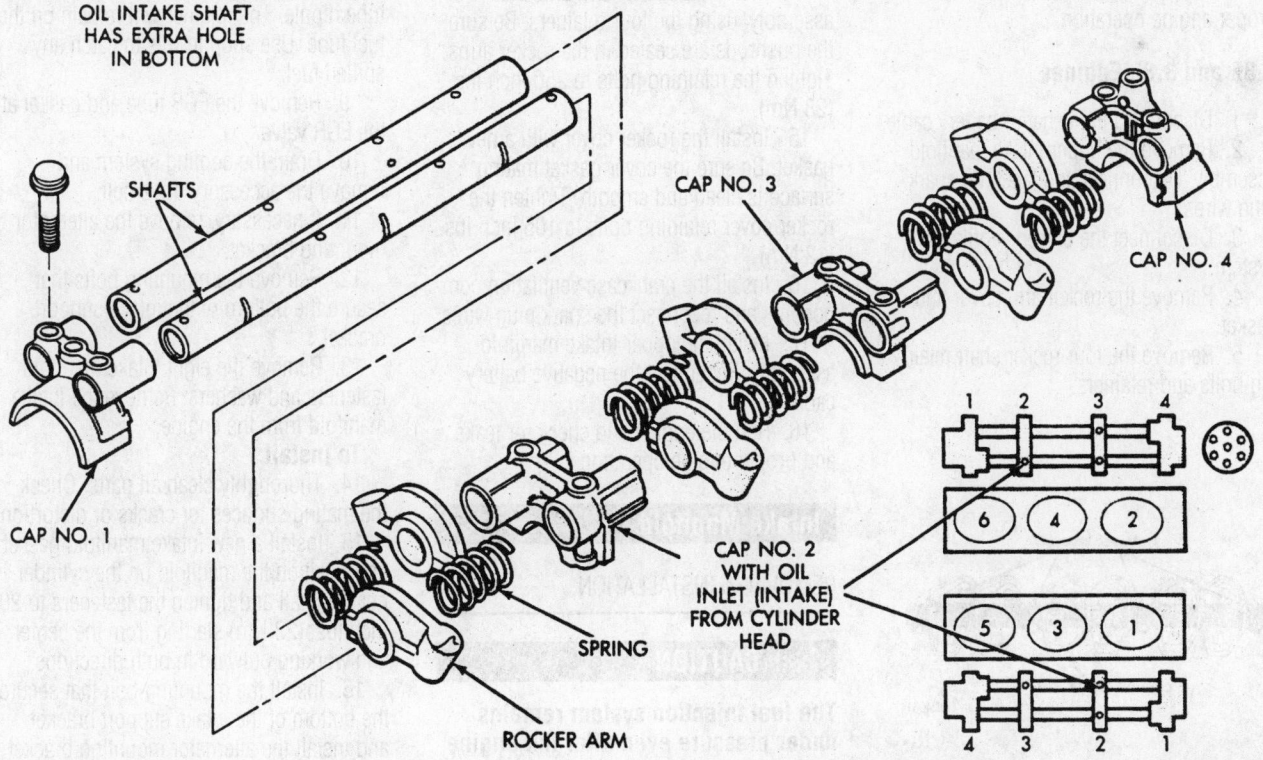

Exploded view of the rocker arms and shafts. Install cap No. 2 in the correct position on the oil hole—3.0L engine

13. Install the bolts in No. 4 cap to retain assembly.

14. Apply silicone rubber sealant at bearing cap ends.

15. Install the rocker arm shaft assembly.

➡ **Be sure the arrow mark on the bearing caps and the arrow mark on the cylinder heads are in the same direction. The direction of arrow marks on the front and rear assemblies are opposite to each other.**

16. Tighten the bearing caps bolts to 85 inch lbs. (10 Nm) in the following order:
   a. No. 3 cap
   b. No. 2 cap
   c. No. 1 cap
   d. No. 4 cap

17. Repeat the previous step, increasing torque to 180 inch lbs. (20 Nm).

18. Remove the lash adjuster retainers.

19. Install the distributor drive adapter, if removed.

20. Install the valve cover. Tighten the valve cover retaining bolts to 88 inch lbs. (10 Nm).

21. Reattach all vacuum connections. Install the accessory drive belts and adjust to the proper tension.

22. Install air cleaner assembly. Reconnect the spark plug wires.

23. Reconnect the negative battery cable. Run the engine and check for leaks and proper engine operation.

3.3L and 3.8L Engines

1. Disconnect the negative battery cable.

2. Remove the upper intake manifold assembly. Disconnect and label the spark plug wires.

3. Disconnect the closed ventilation system.

4. Remove the rocker arm cover and gasket.

5. Remove the four rocker shaft retaining bolts and retainers.

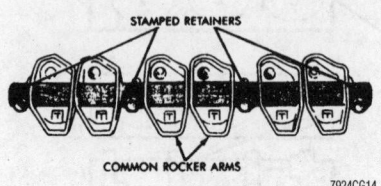

Rocker arms and shaft assembly—3.3L and 3.8L engines

6. Remove the rocker arms and shaft assembly.

7. If disassembling the rocker shaft, be sure to identify all components so they can be reinstalled in their original locations.

8. Inspect the rocker arms and shafts for wear and/or damage; replace components as necessary.

9. If necessary, remove the pushrods. Identify each pushrod as it is removed, so it can be reinstalled in its original location.

10. Inspect the pushrods for wear and/or damage. Roll each pushrod on a flat surface to check for a bent condition. Replace pushrods as necessary.

To install:

11. If removed, install the pushrods in their proper locations. Lubricate the pushrod ends with clean engine oil, prior to installation. Be sure the pushrods are seated in the lifters.

12. Lubricate the rocker arms and shafts with clean engine oil, prior to installation.

13. If the rocker shaft was disassembled, reassemble making sure all components are installed in their original locations.

➡ **The rocker arm shaft should be tightened slowly, starting with the center bolts. Allow 20 minutes, for tappet bleed down after installation, before engine operation.**

14. Install the rocker arm and shaft assembly, using the four retainers. Be sure the pushrods are seated in the rocker arms. Tighten the retaining bolts to 250 inch lbs. (28 Nm).

15. Install the rocker cover with a new gasket. Be sure the cover gasket mating surface is clean and smooth. Tighten the rocker cover retaining bolts to 105 inch lbs. (12 Nm).

16. Install the crankcase ventilation components and reconnect the spark plug wires.

17. Install the upper intake manifold assembly. Reconnect the negative battery cable.

18. Run the engine and check for leaks and proper engine operation.

Intake Manifold

REMOVAL & INSTALLATION

✳✳ CAUTION

The fuel injection system remains under pressure even after the engine has been turned OFF. The fuel system pressure must be relieved before disconnecting any fuel lines. Failure to do so may result in fire and/or personal injury.

2.4L Engine

The 2.4L (VIN B) DOHC (Dual Over Head Camshaft) engine intake manifold is a long branch design made of cast aluminum. It is attached to the cylinder head with eight fasteners.

1. Properly relieve the fuel system pressure.

2. Disconnect the negative battery cable.

3. Disconnect the air cleaner inlet hose from the throttle body.

4. Disconnect the throttle cable and speed control cable (if equipped) from the throttle lever and cable bracket. The cable(s) can be removed from the bracket by compressing the retaining tabs.

5. Detach the Idle Air Control (IAC) motor and Throttle Position Sensor (TPS) wiring connectors on the throttle body.

6. Disconnect the vacuum hoses from the intake plenum fittings.

7. Detach the electrical connectors from the Manifold Absolute Pressure (MAP) and Intake Air Temperature (IAT) sensors.

8. Detach the fuel line quick-disconnect fitting from the chassis fuel line tube by squeezing the retainer tabs together and pulling the fitting assembly from the fuel tube nipple. The retainer will remain on the fuel tube. Use shop towels to catch any spilled fuel.

9. Remove the EGR tube and gasket at the EGR valve.

10. Drain the cooling system and remove the accessory drive belt.

11. If necessary, remove the alternator mounting bracket.

12. Remove the mounting bolts that secure the bottom of the intake support bracket.

13. Remove the eight intake manifold fasteners and washers. Remove the intake manifold from the engine.

To install:

14. Thoroughly clean all parts. Check the mating surfaces for cracks or distortion.

15. Install a new intake manifold gasket and position the manifold on the cylinder head. Install and tighten the fasteners to 200 inch lbs. (23 Nm) starting from the center and working outward in both directions.

16. Install the mounting bolt that secures the bottom of the intake support bracket, and install the alternator mounting bracket (if removed).

17. Install the accessory drive belt and adjust to the proper tension.

18. Install the EGR gasket and tube to the EGR valve.

19. Inspect the quick-disconnect fittings for damage and repair as required. Lightly lube the fuel line tube with clean 30W engine oil. Reconnect the fuel hose quick-disconnect fitting to the chassis fuel tube. Push the quick-disconnect fitting onto the chassis fuel tube until it clicks into place. Check the connection by pulling on the fitting to insure it is locked in position.

20. Reconnect the MAP and IAT air temperature sensor wiring connectors.

21. Reconnect the vacuum hoses to the intake plenum fittings.

22. Reattach the IAC motor and TPS wiring connectors.

23. Install the throttle cables into the throttle cable bracket. Be sure to engage the retaining tabs.

24. Reconnect the throttle cable and speed control cable (if equipped) to the throttle lever.

25. Install the air cleaner inlet hose to the throttle body.

26. Refill the cooling system with a 50/50 mixture of clean, ethylene glycol antifreeze and water to the proper level.

27. Reconnect the negative battery cable.

28. Run the engine and check for leaks and proper operation.

2.5L Engine

➡The intake and exhaust manifolds share the same gasket. Both manifolds must be removed if either one of them requires service.

✳✳ CAUTION

The fuel injection system remains under pressure, even after the engine has been turned OFF. The fuel system pressure must be relieved before disconnecting any fuel lines. Failure to do so may result in fire and/or personal injury.

1. Properly relieve the fuel system pressure.

2. Disconnect the negative battery cable.

3. Drain the cooling system.

4. Remove the air cleaner. Label and disconnect all vacuum lines, electrical wiring and fuel lines from the throttle body and intake manifold.

5. Disconnect the throttle linkage.

6. Loosen the power steering pump and remove the power steering drive belt, if necessary.

7. Remove the power brake vacuum hose from the intake manifold.

8. Remove the EGR tube from the intake manifold. Remove the water hoses from the water crossover.

9. Raise and safely support the vehicle. Disconnect the exhaust pipe from the exhaust manifold.

10. Remove the power steering pump and set aside.

11. Remove the intake manifold support bracket, if equipped.

12. If equipped, remove the air injection tube bolts and the air injection tube assembly.

13. Remove the intake manifold bolts.

14. Lower the vehicle and remove the intake manifold.

15. Remove the exhaust manifold nuts and remove the exhaust manifold.

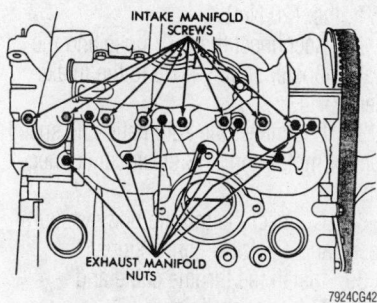

Intake and exhaust manifold fastener locations—2.5L engine

To install:

16. Clean the mating surfaces and install a new combination manifold gasket. Apply a light coat of gasket sealer to the manifold side of the new combination manifold gasket.

17. Install the exhaust manifold. Install the mounting nuts and tighten them to 17 ft. lbs. (23 Nm) starting from the middle and working outward.

18. Install the intake manifold.

19. Raise and safely support the vehicle. Install the mounting bolts and tighten them to 17 ft. lbs. (23 Nm) starting from the middle and working outward.

20. Connect the exhaust pipe to the exhaust manifold.

21. Install the air injection tube assembly, if equipped.

22. Install the EGR tube.

23. Install the intake support bracket.

24. Install the power steering pump and drive belt. Adjust the drive belt to the proper tension.

25. Lower the vehicle and install the water hoses to the water crossover.

26. Install the power brake vacuum hose to the intake manifold.

27. Connect the throttle linkage.

28. Install all vacuum lines, electrical wiring and fuel lines to the throttle body and intake manifold.

29. Install the air cleaner assembly.

30. Refill the cooling system to the proper level.

31. Reconnect the negative battery cable, run the engine and check the manifolds for leaks. Check the coolant level and top off, if necessary.

3.0L Engine

1. Properly relieve the fuel system pressure.

2. Disconnect the negative battery cable.

3. Drain the cooling system.

4. Remove the air cleaner-to-throttle body hose.

5. Remove the throttle cable and transaxle kickdown cable.

6. Remove the Automatic Idle Speed (AIS) motor and Throttle Position Sensor (TPS) electrical connectors from the throttle body. Detach the vacuum connections from throttle body.

7. Remove the PCV and brake booster hoses from the air intake plenum.

8. Remove the ignition coil from the intake plenum.

9. Remove the EGR tube from the intake plenum (if equipped).

10. Remove the electrical connection from the coolant temperature sensor.

11. Remove the vacuum connection from the fuel pressure regulator.

12. Remove the air intake connection from the air intake plenum. Remove the fuel hose-to-fuel rail connections.

13. Remove the air intake plenum-to-manifold bolts (8), and remove the air intake plenum and gasket.

✳✳ WARNING

Whenever the air intake plenum is removed, cover the intake manifold properly to keep objects from entering the cylinder head.

14. Label and disconnect the fuel injector wiring harness from the engine wiring harness.

15. Disconnect the vacuum hose from the fuel rail, then remove the pressure regulator attaching bolts and remove the pressure regulator from the rail.

16. Remove the fuel rail attaching bolts and remove the fuel rail.

17. Remove the radiator hose from the thermostat housing and the heater hose from the pipe.

18. Remove the intake manifold attaching nuts and washers, and remove the intake manifold.

19. Clean the gasket material from the cylinder head and manifold gasket surface. Check for cracks or damaged mounting surfaces.

To install:

20. Install a new gasket on the intake surface of the cylinder head and install the intake manifold.

21. Install the intake manifold washers and nuts. Tighten the intake manifold attaching nuts, in sequence, to 15 ft. lbs. (20 Nm).

22. Clean the injectors and lubricate the injector O-rings with a drop of clean engine oil.

23. Place the tip of each injector into their ports. Push the assembly into place until the injectors are seated in their ports.

24. Install the fuel rail attaching bolts and tighten to 115 inch lbs. (13 Nm).

25. On 1995 models, install the pressure regulator on the rail, then install the mounting bolts and tighten to 95 inch lbs. (11 Nm).

26. Install the fuel supply and return tube hold-down bolt and the vacuum crossover tube hold-down bolt. Tighten to 95 inch lbs. (11 Nm).

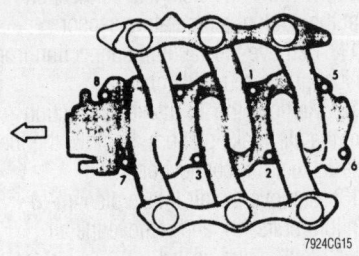

Tighten the intake manifold bolts in the correct sequence to prevent vacuum leaks—3.0L engine

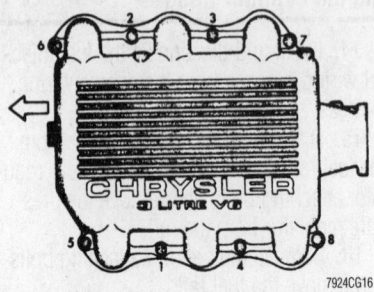

Tighten the intake plenum bolts in a clockwise spiral starting from No.1 as shown— 3.0L engine

27. On 1995 models, tighten the fuel pressure regulator hose clamps to 10 inch lbs. (1 Nm).

28. Connect the injector wiring harness to the engine wiring harness.

29. Remove the covering from the intake manifold.

30. Position the intake manifold gasket, beaded side up, on the intake manifold.

31. Place the air intake plenum in position. Install the attaching bolts and tighten, in sequence, to 115 inch lbs. (13 Nm).

32. Connect the fuel line to the fuel rail. Tighten the clamps to 10 inch lbs. (1 Nm).

33. Connect the vacuum hoses to the intake plenum.

34. Connect the electrical connection to the coolant temperature sensor.

35. Connect the EGR tube flange to the intake plenum (if equipped) and tighten to 15 ft. lbs. (20 Nm).

36. Reconnect the PCV hose and the brake booster supply hose to the intake plenum.

37. Reconnect the automatic idle speed control motor and TPS electrical connectors.

38. Connect the throttle body vacuum hoses and electrical connections.

39. Install the throttle cable and transaxle kickdown linkage.

40. Install the air inlet hose assembly.

41. Install the radiator and heater hose.

42. Refill the cooling system with a 50/50 mixture of clean, ethylene glycol antifreeze and water to the proper level.

43. Reconnect the negative battery cable. An engine oil and filter change is recommended.

44. Run the vehicle until it reaches operating temperature and check the cooling system, fuel system and engine for fuel, oil or coolant leaks.

3.3L and 3.8L Engines

1. Properly relieve the fuel system pressure.

2. Disconnect the negative battery cable.

3. On 1996–99 models, perform the following procedures:

   a. Remove the wiper arm and blade assemblies.

   b. Remove the cowl cover from the vehicle.

   c. Open the hood. Detach the positive lock on the wiper unit electrical connector.

   d. Detach the wiper unit electrical connector from the engine compartment wiring harness.

   e. Disconnect the windshield washer hose from the hose coupling inside the wiper unit.

   f. Remove the drain tubes from the bottom of the wiper unit.

   g. Remove the sound absorbers from the ends of the wiper unit.

   h. Remove the attaching nuts securing the wiper unit to the lower windshield fence.

   i. Remove the attaching bolts securing the wiper unit to the dash panel.

   j. Raise up the wiper unit from the from the weld-studs on the lower windshield fence and remove the wiper unit from the vehicle.

4. Drain the cooling system.

5. If equipped, remove the intake manifold cover.

6. Disconnect the air inlet resonator-to-throttle body hose assembly.

7. Disconnect the throttle cable and remove the wiring harness from the cable bracket.

8. Remove the Automatic Idle Speed (AIS) motor and Throttle Position Sensor (TPS) wiring connectors from the throttle body.

9. Remove the vacuum hose harness from the throttle body.

10. Remove the PCV and brake booster hoses from the air intake plenum.

11. Remove the EGR tube flange and the vacuum harness connectors from the intake plenum.

12. If equipped, detach the charge temperature sensor electrical connector. Remove the vacuum harness connectors from the intake plenum.

13. Remove the cylinder head-to-intake plenum strut.

14. Detach the MAP sensor wiring connector. Remove the engine mounted ground strap.

15. Remove the fuel hose quick-disconnect fittings from the fuel rail by pulling back on the fitting while pushing in the plastic ring. This may require the use of an open end wrench to push in the plastic ring. Be sure to plug the open fuel lines to prevent system contamination. Wrap a shop towel around the fuel hoses to absorb any fuel spill.

16. Remove the DIS coils and the alternator bracket to intake manifold bolt.

17. On 1996–99 models, remove the alternator wiring harness from the back of the upper intake manifold plenum.

18. Remove the attaching bolts and remove the upper intake manifold. Cover the intake manifold openings to prevent foreign material from entering the engine.

Tighten the upper intake manifold bolts in the sequence shown—1996–99 3.3L and 3.8L engines

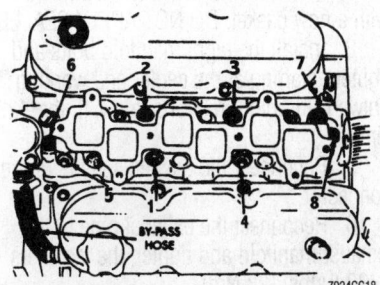

The lower intake manifold bolts must be tightened in sequence to prevent leaks—3.3L and 3.8L engines

19. On 1995 models, remove the vacuum harness connector from the fuel pressure regulator.

20. Remove the fuel tube retainer bracket bolt and fuel rail attaching bolts. Spread the retainer bracket to allow for clearance when removing the fuel tube.

21. On 1995 models, remove the fuel rail injector wiring clip from the alternator bracket.

22. Detach the cam sensor and coolant temperature sensor connectors.

23. Remove the injector wiring clip from the intake manifold water tube.

24. Remove the fuel rail. Be careful not to damage the fuel injector O-rings.

25. Remove the upper radiator hose, bypass hose and rear intake manifold hose.

26. Remove the intake manifold bolts and remove the manifold from the engine.

27. Remove the intake manifold seal retaining bolts and remove the manifold gasket.

28. Clean out any clogged end water passages and fuel runners.

To install:

29. Clean and dry all gasket mating surfaces.

30. Place a bead of approximately ¼ in. (6mm) diameter of silicone sealant onto each of the four manifold-to-cylinder head gasket corners.

☆☆ CAUTION

The intake manifold gasket is made of very thin material and could cause personal injury. Handle with care.

31. Carefully install the intake manifold gasket and tighten the end seal retainer bolts to 105 inch lbs. (12 Nm).

32. Install the intake manifold and eight retaining bolts and tighten to 10 inch lbs. (1 Nm). Then tighten the bolts, in sequence, to 200 inch lbs. (22 Nm).

33. When the bolts are tightened, inspect the seals to ensure that they have not become dislodged.

34. Lubricate the injector O-rings with clean oil to ease installation. Put the tip of each injector into its port and place the fuel rail in position. Install the fuel rail mounting bolts and tighten to 200 inch lbs. (22 Nm).

35. Install the fuel tube retaining bracket bolt. Tighten the bolt to 35 inch lbs. (4 Nm).

36. Connect the cam, coolant temperature and engine temperature sensors if equipped.

37. On 1995 models, perform the following:

a. Install the fuel injector harness wiring clip to the alternator bracket. Install the intake manifold water tube.

b. Install the vacuum harness to the pressure regulator.

38. Install the upper intake manifold with a new intake manifold gasket. Install the bolts only finger-tight. Install the alternator bracket to intake manifold bolt and the cylinder head to intake manifold strut bolts. Tighten the intake manifold mounting bolts to 250 inch lbs. (28 Nm) starting from the middle and working outward.

39. Tighten the alternator bracket and cylinder head-to-intake manifold strut bolts to 40 ft. lbs. (54 Nm).

40. On 1996–99 models, connect the alternator wiring harness to the alternator and to the rear of the intake manifold.

41. Reconnect the ground strap and MAP sensor connectors. On 1995 models, reconnect the oxygen sensor and charge temperature sensor connectors.

42. Reconnect the intake plenum vacuum harness.

43. Install a new gasket and connect the EGR tube to the intake manifold. Tighten the retaining fasteners to 200 inch lbs. (22 Nm).

44. Clip the wiring harness into the throttle cable bracket hole. Reconnect the Throttle Position Sensor (TPS) and Automatic Idle Speed (AIS) control motor wiring connectors.

45. Reconnect the throttle body vacuum harness and install the ignition coils. Tighten the ignition coil fasteners to 105 inch lbs. (12 Nm).

46. Lightly lubricate the ends of the fuel lines with clean, 30W engine oil. Reconnect the fuel hoses to the rail. Push the fittings in until they click in place. Pull back on the quick-disconnect fittings to ensure that the fuel lines are securely locked in place.

47. Reconnect the throttle cable.

48. Reconnect the fuel injector wiring harness.

49. On 1996–99 models, perform the following steps:

a. Install the air inlet resonator to throttle body hose assembly.

b. If equipped, install the intake manifold cover.

c. Install the wiper unit into the vehicle engine compartment. Be sure the wiper unit is installed properly over the weld-studs on the lower windshield fence. Install and tighten the attaching nuts to the weld-studs.

d. Install and tighten the attaching bolts securing the wiper unit to the dash panel.

e. Install the sound absorbers to each end of the wiper unit.

f. Install the drain tubes to the bottom of the wiper unit and reconnect the windshield washer hose to the hose coupling inside the wiper unit.

g. Reconnect the wiper unit wiring connector to the engine wiring harness.

h. Place the cowl cover onto the vehicle. Reconnect the right side windshield washer hose to the right washer nozzle located on the underside of the cowl cover.

i. Engage the retainers that secure the cowl cover to the front fender edge.

j. Install and tighten the wing nuts that secure the front of the cowl cover to the wiper module.

k. Engage the quarter turn fasteners that secure the outer ends of the cowl cover to the wiper module.

l. Install and tighten the bolts that hold the lower area of the cowl cover to the wiper module.

m. Reconnect the negative battery cable and verify that the wiper motor and wiper linkage are in the PARK position.

n. Install the wiper arm in correct position over the wiper arm pivot.

50. Install the wiper arm-to-wiper arm pivot retaining nut and tighten to 26 ft. lbs. (35 Nm).

51. Push down the wiper arm cap cover and engage the clip that secures the outside end of the cover to the wiper arm.

52. Reconnect the negative battery cable.

53. On 1996–99 models, operate the windshield wipers and ensure they work and park properly.

54. On 1996–99 models, refill the cooling system with a 50/50 mixture of clean, ethylene glycol antifreeze and water. Bleed the cooling system.

55. Start the engine. Check for leaks and proper engine operation.

Exhaust Manifold

REMOVAL & INSTALLATION

2.4L Engine

1. Disconnect the negative battery cable.
2. Raise and safely support the vehicle.
3. Disconnect the exhaust pipe from the exhaust manifold at the flex joint. Apply penetrating oil on the exhaust manifold-to-exhaust pipe flange bolts to aid in removal. It may be necessary to remove the entire exhaust system.
4. Detach the oxygen sensor wiring connector at the rear of the exhaust manifold.
5. Remove the eight manifold attaching bolts and remove the manifold from the cylinder head.

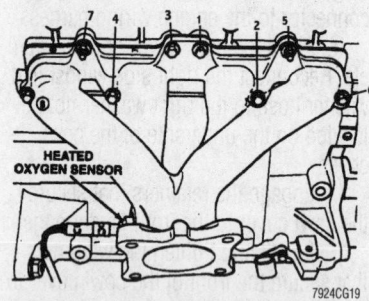

Tighten the exhaust manifold bolts in the sequence shown to prevent exhaust leaks—2.4L engine

To install:

6. Thoroughly clean all parts. Discard the gasket (if equipped) and clean all gasket surfaces of the manifold and cylinder head. Test the manifold gasket surface for flatness with a straightedge and feeler gauge. The surface must be flat within 0.006 inches per foot (0.15mm per 30cm) of manifold length. Inspect the manifold for cracks or distortion. Replace if necessary.

7. Install the manifold to the vehicle with a new gasket. DO NOT APPLY SEALER.

8. Install the eight manifold bolts and tighten, starting at the center and working outward in both directions, to 17 ft. lbs. (23 Nm).

9. Reconnect the oxygen sensor wiring connector.

10. Reconnect the exhaust pipe to the exhaust manifold and tighten the fasteners to 20 ft. lbs. (28 Nm).

11. Reconnect the negative battery cable. Start the engine and allow it to idle while inspecting the manifold for exhaust leaks.

2.5L Engine

➡The intake and exhaust manifolds share the same gasket. Both manifolds must be removed if either one of them requires service.

✳✳ CAUTION

The fuel injection system remains under pressure, even after the engine has been turned OFF. The fuel system pressure must be relieved before disconnecting any fuel lines. Failure to do so may result in fire and/or personal injury.

1. Properly relieve the fuel system pressure.
2. Disconnect the negative battery cable.
3. Drain the cooling system.
4. Remove the air cleaner. Label and disconnect all vacuum lines, electrical wiring and fuel lines from the throttle body and intake manifold.
5. Disconnect the throttle linkage.
6. Loosen the power steering pump and remove the power steering drive belt, if necessary.
7. Remove the power brake vacuum hose from the intake manifold.
8. Remove the EGR tube from the intake manifold. Remove the water hoses from the water crossover.
9. Raise and safely support the vehicle. Disconnect the exhaust pipe from the exhaust manifold.
10. Remove the power steering pump and set aside.
11. Remove the intake manifold support bracket, if equipped.
12. If equipped, remove the air injection tube bolts and the air injection tube assembly.
13. Remove the intake manifold bolts.
14. Lower the vehicle and remove the intake manifold.

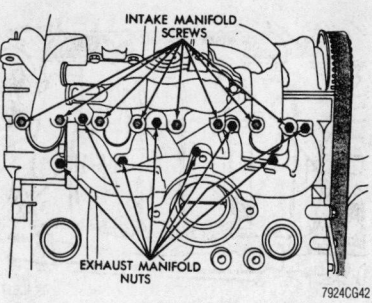

Intake and exhaust manifold fastener locations—2.5L engine

15. Remove the exhaust manifold nuts and remove the exhaust manifold.

To install:

16. Clean the mating surfaces and install a new combination manifold gasket. Apply a light coat of gasket sealer to the manifold side of the new combination manifold gasket.

17. Install the exhaust manifold. Install the mounting nuts and tighten them to 17 ft. lbs. (23 Nm) starting from the middle and working outward.

18. Install the intake manifold.

19. Raise and safely support the vehicle. Install the mounting bolts and tighten them to 17 ft. lbs. (23 Nm) starting from the middle and working outward.

20. Connect the exhaust pipe to the exhaust manifold.

21. Install the air injection tube assembly, if equipped.

22. Install the EGR tube.

23. Install the intake support bracket.

24. Install the power steering pump and drive belt. Adjust the drive belt to the proper tension.

25. Lower the vehicle and install the water hoses to the water crossover.

26. Install the power brake vacuum hose to the intake manifold.

27. Connect the throttle linkage.

28. Install all vacuum lines, electrical wiring and fuel lines to the throttle body and intake manifold.

29. Install the air cleaner assembly.

30. Refill the cooling system to the proper level.

31. Reconnect the negative battery cable, run the engine and check the manifolds for leaks. Check the coolant level and top off, if necessary.

3.0L Engine

1. Disconnect the negative battery cable. Raise and safely support the vehicle.

2. Disconnect the exhaust pipe from the rear exhaust manifold, at the flex joint.

3. On 1995 models, disconnect the EGR tube from the rear manifold, if equipped, and disconnect the oxygen sensor wire.

4. Remove the crossover pipe-to-exhaust manifold attaching bolts.

5. On 1996–99 models, remove the rear exhaust manifold heat shield.

6. Remove the rear exhaust manifold-to-cylinder head nuts and remove the exhaust manifold.

7. Lower the vehicle and remove the heat shield from the front exhaust manifold.

8. Remove the bolts fastening the crossover pipe to the front exhaust manifold. Remove the front exhaust manifold-to-cylinder head nuts and remove the exhaust manifold.

9. Clean the gasket mounting surfaces. Inspect the manifolds for cracks, flatness and/or damage.

To install:

➡**Install the gasket with the numbers 1–3–5 embossed on the top on the rear bank and those with the numbers 2–4–6 on the front (radiator side) bank.**

10. Raise and safely support the vehicle.

11. Install the new gasket and rear exhaust manifold to the cylinder head. Install and tighten the rear exhaust manifold-to-cylinder head nuts to 175 inch lbs. (20 Nm).

12. Attach the exhaust pipe to the exhaust manifold and tighten the shoulder bolts to 20 ft. lbs. (28 Nm)

13. Attach the crossover pipe to the exhaust manifold and tighten the bolts to 51 ft. lbs. (69 Nm).

14. On 1996–99 models, install the rear exhaust manifold heat shield and tighten the heat shield mounting fasteners to 130 inch lbs. (15 Nm).

15. Connect the EGR tube to the rear manifold, if removed, and reconnect the oxygen sensor.

16. Lower the vehicle.

17. Install the front exhaust manifold and attach the exhaust crossover pipe. Tighten the bolts to 51 ft. lbs. (69 Nm).

18. Install the front exhaust manifold heat shield and tighten the bolts to 130 inch lbs. (15 Nm).

19. Reconnect the negative battery cable. Start the engine and check for exhaust leaks.

3.3L and 3.8L Engines

1. Disconnect the negative battery cable.

2. On 1996–99 models, perform the following procedures:

a. Remove the accessory drive belt.

b. Remove the alternator.

3. Raise and safely support the vehicle.

4. Disconnect the exhaust pipe from the rear exhaust manifold at the flex joint.

5. Separate the EGR tube from the exhaust manifold and disconnect the oxygen sensor.

6. Remove the alternator/power steering support strut.

7. Remove the crossover pipe attaching bolts to the rear exhaust manifold.

8. On 1996–99 models, remove the rear exhaust manifold heat shield.

9. Remove the rear manifold-to-cylinder head nuts, then remove the rear manifold.

10. Lower the vehicle and remove the front exhaust manifold heat shield.

11. Remove the front exhaust manifold crossover pipe bolts.

12. Remove the front manifold-to-cylinder head nuts, then remove the front exhaust manifold.

13. Clean the mounting surfaces. Inspect the manifolds for cracks or other damage. With a straightedge and feeler gauge, check for flatness. Standard is 0.004 in. (0.1mm) with 0.008 in. (0.2mm) out-of-flatness the service limit. If distorted beyond these specifications, replace the manifold.

To install:

14. Raise and safely support the vehicle.

15. Install the new gasket and rear manifold. Tighten the manifold-to-cylinder head nuts to 17 ft. lbs. (23 Nm).

16. On 1996–99 models, perform the following procedures:

a. Install the rear exhaust manifold heat shield. Tighten the heat shield mounting fasteners to 17 ft. lbs. (23 Nm).

b. Install the alternator unit.

17. Connect the exhaust pipe to the exhaust manifold and tighten the bolt to 20 ft. lbs. (28 Nm).

18. Connect the crossover pipe to the rear manifold and tighten the bolts to 25 ft. lbs. (33 Nm).

19. Connect the oxygen sensor.

20. Connect the EGR tube to the exhaust manifold and install the alternator/power steering support strut.

21. Lower the vehicle.

22. Install the new gasket and front exhaust manifold to the cylinder head. Tighten the exhaust manifold-to-cylinder head nuts to 17 ft. lbs. (23 Nm).

23. Connect the crossover pipe to the front manifold.

24. Install the front manifold heat shield. Tighten the heat shield mounting fasteners to 200 inch lbs. (23 Nm).

25. On 1996–99 models, install the accessory drive belt.

26. Reconnect the negative battery cable.

27. Start the engine and check for exhaust leaks.

Front Crankshaft Seal

REMOVAL & INSTALLATION

2.4L Engine

The timing belt must be removed for this procedure. Use care that all timing marks are aligned after installation or the engine will become damaged.

1. Disconnect the negative battery cable.

2. Remove the accessory drive belts.

3. Raise and safely support the vehicle. Drain the engine oil.

4. Remove the crankshaft damper/pulley using a jaw puller tool.

5. Remove the timing belt cover and timing belt.

6. Remove the crankshaft timing belt sprocket using tool No. 6793 or equivalent.

✳✳ WARNING

Do not nick the seal surface of the crankshaft or the seal bore.

7. Remove the front crankshaft seal using tool No. 6771 or equivalent seal puller. Be careful not to damage the seal contact area of the crankshaft.

To install:

8. Apply a light coating of clean engine oil to the lip of the new oil seal. Install the new front crankshaft oil seal using Oil Seal Installer tool No. 6780–1 or equivalent. Install the new oil seal into the opening with the seal spring facing the inside of the engine. Be sure the oil seal is installed flush with the front cover.

9. Install the crankshaft timing belt sprocket using tool No. 6792 or equivalent.

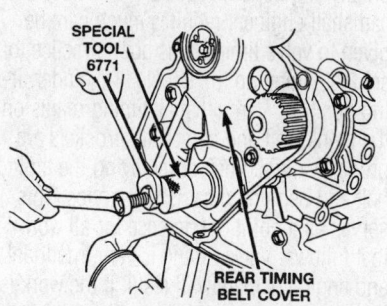

7924CG43

Use a seal puller such as No. 6771 to remove the front crankshaft seal—2.4L engine

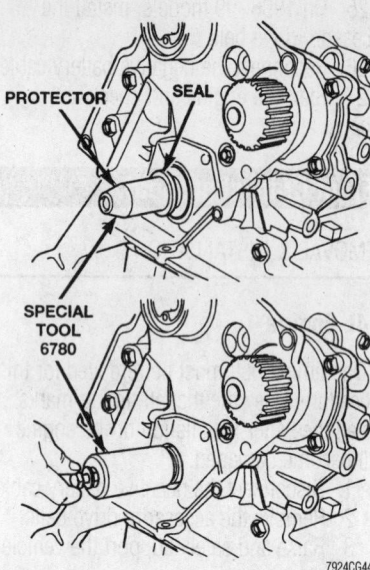

The front crankshaft seal can be installed using tool No. 6780 or equivalent—2.4L engine

➡**Be sure the word "FRONT" on the timing belt sprocket is facing you.**

10. Install the timing belt and timing belt cover.

11. Install the crankshaft damper/pulley onto the crankshaft. Use thrust bearing/washer and 12M-1.75 x 150mm bolt from special tool No. 6792. Install the crankshaft damper/pulley retaining bolt and tighten to 105 ft. lbs. (142 Nm).

12. Lower the vehicle.

13. Install the accessory drive belts. Adjust the belts to the proper tension.

14. Refill the engine with the correct amount of clean engine oil.

15. Reconnect the negative battery cable. Start the engine and check for leaks.

2.5L Engine

The timing belt must be removed to service the crankshaft front oil seal. Working on any engine, but especially overhead camshaft engines, requires much care be given to valve timing. It is good practice to set the engine up to TDC No. 1 cylinder firing position. Verify that all timing marks on the crankshaft and camshaft sprockets are properly aligned before removing the timing belt and beginning camshaft service. This serves as a point of reference for all work that follows. Valve timing is very important and engine damage will result if the work is incorrect.

1. Disconnect the negative battery cable.

2. Remove the accessory drive belts.

3. Raise and safely support the vehicle. Remove the right inner splash shield on front wheel drive models.

4. Loosen and remove the three water pump pulley mounting bolts and remove the pulley.

5. Remove the four crankshaft pulley retaining bolts and the crankshaft pulley.

6. Remove the nuts at upper the portion of the timing cover and the bolts from the lower portion, then remove both halves of the cover.

7. On front wheel drive models, separate the right engine mount. Loosen the timing belt tensioner and remove the timing belt.

8. Using the proper puller tool, remove the crankshaft sprocket.

9. Remove the oil seal using Seal Removal tool No. 6341, or equivalent seal puller.

To install:

10. Clean all parts well. Inspect the front of the crankshaft. The seal lip surface must be clean of any scores, dirt or varnish. It may be necessary to polish the surface with 400 grit sandpaper. Clean all debris well or the new seal will wear quickly.

11. Install a new oil seal using tool Nos. 6342 and 6343, or equivalent seal driver.

12. Install the crankshaft sprocket and the timing belt. Tighten the sprocket bolt to 85 ft. lbs. (115 Nm).

13. Reconnect the right side engine mount, if required.

14. Install the timing belt covers. Secure the upper section to the cylinder head with nuts and lower section to the cylinder block with bolts. Tighten the timing belt cover fasteners to 40 inch lbs. (4 Nm).

15. Install the crankshaft pulley and tighten the bolt to 20 ft. lbs. (27 Nm).

16. Install the water pump pulley. Tighten the bolts to 105 inch lbs. (12 Nm).

17. Install the right side inner fender splash shield, if required. Lower the vehicle.

18. Install the accessory drive belts.

19. Reconnect the negative battery cable.

20. Check the engine oil level; add oil as required.

21. Run the engine and check for leaks.

Camshaft and Valve Lifters

REMOVAL & INSTALLATION

※※ CAUTION

The fuel injection system remains under pressure, even after the engine has been turned OFF. The fuel system pressure must be relieved before disconnecting any fuel lines. Failure to do so may result in fire and/or personal injury.

2.4L Engine

This engine uses a Dual Over Head Camshaft (DOHC) 4-valves per cylinder crossflow aluminum cylinder head. The valves are actuated by roller cam followers which pivot on stationary hydraulic valve lash adjusters. Care must be taken to ensure all valve timing marks align after cylinder head and valvetrain service.

1. Properly relieve the fuel system pressure.

2. Disconnect the negative battery cable.

3. Label and disconnect the spark plug wires from the spark plugs.

4. Remove the ignition coil pack and spark plug wires.

5. Remove the cylinder head cover retaining fasteners and remove the cylinder head cover from the cylinder head. Discard the old cylinder head cover gasket.

6. Remove the ground strap.

7. Remove the timing belt covers, timing belt and camshaft sprockets.

8. Take note that the camshaft bearing caps are numbered for correct location dur-

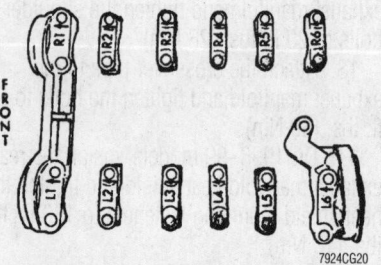

Lay the bearing caps out in their original positions so that they can be reinstalled correctly—2.4L engine

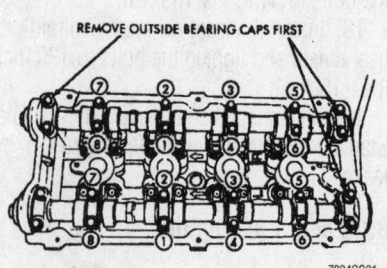

Loosen the bearing cap bolts in the sequence shown to prevent bending of the cam—2.4L engine

ing installation. Remove the outer bearing caps first.

9. Loosen, but do not remove, the camshaft bearing cap retaining fasteners in the correct sequence. Perform this step on one camshaft at a time.

10. Tag the camshafts for intake and exhaust, if they are to be reused, for later installation. The camshafts are not interchangeable. Remove the camshaft bearing caps and the camshafts.

11. Remove the camshaft followers. Any components that are to be reused must be installed in their original locations. Use care to identify and mark the positions of any removed valvetrain components so they may be reinstalled correctly.

12. Inspect the camshaft bearing oil feed holes in the cylinder head for clogging. Inspect the camshaft bearing journals for wear or scoring. Check the cam surface for abnormal wear and damage. A visible worn groove in the roller path or on the cam lobes is cause for replacement.

To install:

13. Thoroughly clean the camshafts and related parts.

14. The camshaft end-play should be checked using the following procedure:

   a. Oil the camshaft journals and install the camshaft **WITHOUT** the cam follower assemblies. Install the rear cam caps and tighten to 250 inch lbs. (28 Nm).

   b. Carefully push the camshaft as far rearward as it will go.

   c. Set up a dial indicator to bear against the front of the camshaft (the sprocket end). Zero the indicator.

   d. Move the camshaft forward as far as it will go. Read the dial indicator. End-play specification is 0.002–0.010 in. (0.05–0.15mm).

   e. If excessive end-play is present, inspect the cylinder head and camshaft for wear; replace if necessary.

15. If satisfied with the fit and condition of the camshafts, remove the camshafts for installation of the cam followers.

16. The hydraulic valve lash adjusters are inside the roller cam followers. Ensure they are clean, well-lubricated with clean engine oil and properly positioned. Install the cam followers in their original positions on the hydraulic lash adjuster and valve stem.

✳✳ WARNING

To avoid valve-to-piston contact, be sure NONE of the pistons are at Top Dead Center (TDC) when installing the camshafts.

17. Lubricate the camshaft bearing journals and cam followers with clean engine oil and install the camshafts. Install right and left camshaft bearing caps No. 2 through No. 5 and right side No. 6. Tighten the M6 fasteners to 105 inch lbs. (12 Nm) in correct sequence.

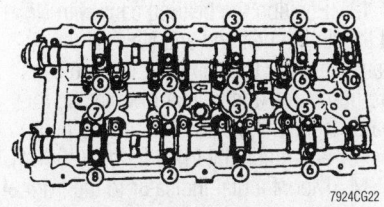

Tighten the camshaft bearing cap bolts in the correct sequence to insure proper operation—2.4L engine

18. Apply Mopar® Gasket Maker or equivalent sealer to the No. 1 and left side No. 6 bearing caps. Install the bearing caps and tighten the M8 fasteners to 250 inch lbs. (28 Nm). The end caps must be installed before the seals may be installed.

19. Install the camshaft end seals.

20. Install the camshaft sprockets, if removed. Install the timing belt using care to be sure all timing marks are properly aligned. Install the timing belt covers.

✳✳ WARNING

Verify that all timing marks are correct. If the timing belt or sprockets are incorrectly installed, engine damage will occur. Take time to be sure all timing marks are correctly aligned.

21. Clean the cylinder head cover and cylinder head gasket rails (mating surfaces). Make certain the rails are flat.

22. Install new cylinder head cover gaskets. Use care; DO NOT allow oil or solvents to contact the timing belt as they can deteriorate the rubber and cause tooth skipping. Apply Mopar Silicone Rubber Adhesive Sealant, or equivalent, at the camshaft cap corners and at the top edge of the ½ round seal.

➡**Inspect the spark plug well seals for cracking and/or swelling, and replace if necessary.**

23. Install the cylinder head cover assembly to the head and tighten the fasteners in sequence using the following three substeps:

   a. Tighten all cylinder head cover fasteners to 40 inch lbs. (4.5 Nm).

   b. Tighten all fasteners to 80 inch lbs. (9 Nm).

   c. Tighten all fasteners to 105 inch lbs. (12 Nm).

24. Install the ignition coil pack and connect the spark plug wiring to the correct spark plugs. Tighten the coil pack retaining fasteners to 105 inch lbs. (12 Nm).

25. Reconnect the ground strap.

26. Ensure all vacuum lines and remaining wiring have been reconnected.

➡**To remove any dirt particles that may have fallen in the engine, an oil and filter change is recommended.**

27. Connect the negative battery cable and test run the vehicle. Check for leaks and for proper operation.

2.5L Engine

The following procedure is performed with the engine in the vehicle.

➡**Removal of the camshaft requires removal of the camshaft sprocket. To maintain proper engine timing, the timing belt can be left indexed on the sprockets and suspended under light pressure. This will prevent the belt from coming off and will help maintain timing.**

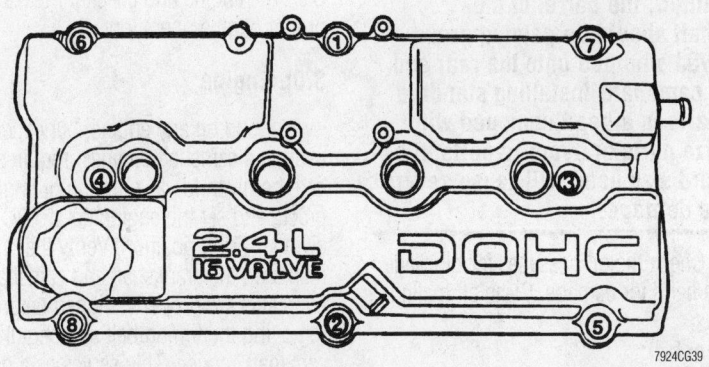

Cylinder head cover bolt tightening sequence—2.4L engine

1. Properly relieve the fuel system pressure.

2. Disconnect the negative battery cable.

➡It is good practice to turn the engine to TDC No. 1 cylinder, compression stroke (firing position) before beginning timing belt and camshaft service. This gives a point of reference to help verify timing mark alignment.

3. Remove the accessory drive belts. Detach and label any electrical connectors and/or vacuum hoses that will allow for easier removal of the camshaft.

4. Turn the crankshaft so the No. 1 piston is at TDC on the compression stroke. Remove the upper timing belt cover. Remove the air pump pulley, if equipped.

5. Remove the camshaft sprocket bolt and sprocket, and suspend tightly so the belt does not lose tension. If it does, the belt timing will have to be reset.

6. Remove the valve cover. Under the valve cover may be a large baffle or shroud that Chrysler calls an oil vapor curtain. Remove the curtain. Be careful not to lose the rubber bumpers on the top of the curtain.

7. If the rocker arms are being reused, mark them for installation identification since they must be reinstalled in same location from which they were removed. Loosen the camshaft bearing bolts, evenly and gradually.

8. Using a soft mallet, rap the rear of the camshaft a few times to break the bearing caps loose.

9. Remove the bolts, bearing caps and the camshaft with seals.

✳✳ WARNING

Before replacing the camshaft, identify factory installed oversize components. To do so, look for the tops of the bearing caps painted green and O/SJ stamped rearward of the oil gallery plug on the rear of the head. In addition, the barrel of the camshaft should be painted green and O/SJ stamped onto the rear end of the camshaft. Installing standard size parts in a head equipped with oversize parts or oversize parts in a standard size head will cause severe engine damage.

10. Check the oil passages for blockage and the parts for damage. Clean all mating surfaces.

To install:

11. Transfer the sprocket key to the new camshaft. New rocker arms and a new

camshaft sprocket bolt are normally included with the camshaft package. Install the rocker arms, lubricate the camshaft and position it on the cylinder head with end seals installed. NEVER install used rocker arms on a new camshaft. A new camshaft requires new rockers.

12. Apply RTV silicone gasket material to the No. 1 and No. 5 bearing caps. Install the bearing caps before the seals are installed.

13. Position the bearing caps with No. 1 at the timing belt end and No. 5 at the transaxle end. The camshaft bearing caps are numbered and have arrows facing forward. Tighten the camshaft bearing bolts evenly and gradually to 18 ft. lbs. (25 Nm).

14. Mount a dial indicator to the front of the engine and check the camshaft end-play. End-play must not exceed the 0.005–0.013 in. specification range.

15. Install the valve cover oil vapor curtain with the cutouts over the cam towers and contacting the cylinder head floor, then press the opposite distributor side into position below the cylinder head rail. Be sure the rubber bumpers are properly positioned on top of the curtain. Install the valve cover and a new gasket. Be sure the gasket mounting surface is clean of any oil or debris before installation. Tighten the mounting bolts to 105 inch lbs. (12 Nm).

16. Install the camshaft sprocket and the new bolt. Tighten the camshaft sprocket bolt to 65 ft. lbs. (89 Nm).

17. Install the timing belt. Verify that the valve timing is correct and that all timing marks are aligned.

18. Install the timing belt cover.

19. Install the air pump pulley, if equipped.

20. Install the accessory drive belts and adjust to the proper tension.

21. Reattach any remaining electrical connectors and/or vacuum hoses.

22. An oil and filter change are recommended after this procedure.

23. Reconnect the negative battery cable. Start the engine and check for leaks and proper engine operation.

3.0L Engine

Working on any engine, but especially overhead camshaft engines, requires much care be given to valve timing. It is good practice to set the engine up at TDC No. 1 cylinder firing position. Verify that all timing marks on the crankshaft and camshaft sprockets are properly aligned before removing the timing belt and beginning camshaft service. This serves as a point of reference for all work that follows. Valve

timing is very important and engine damage will result if the work is incorrect.

1. Disconnect the negative battery cable.

2. Remove the air cleaner assembly. Disconnect and label the spark plug wires.

3. Remove the accessory drive belts, if necessary.

4. Move away any wiring harnesses for easier access to the rear valve cover. Disconnect and label any vacuum hoses required for easier removal of the valve covers.

5. Remove the valve covers from the vehicle.

6. Remove the timing belt. Remove the camshaft timing belt sprockets.

7. Install auto lash adjuster retainers MD998443 or equivalent on the rocker arms.

8. If removing the right side (front) camshaft, remove the distributor adapter.

9. Remove the camshaft bearing caps, but do not remove the bolts from the caps.

10. Remove the rocker arms, rocker shafts and bearing caps, as an assembly.

➡Use care not to mix rocker arms and shafts. The oil holes are different in the shafts and each must be returned to its original location or the engine overhead will not be lubricated.

11. Remove the camshaft from the cylinder head.

12. Inspect the bearing journals on the camshaft, cylinder head and bearing caps.

To install:

13. Clean all parts well. Pay particular attention to the oil feed holes in the cylinder head, checking for clogging. If the camshafts are to be reused, check with a micrometer. Measure the cam height and replace if out of limit. Standard value is 1.624 in. (41.25mm) and wear limit is 1.604 in. (40.75mm).

14. Lubricate the camshaft journals and camshaft with clean engine oil and install the camshaft in the cylinder head.

15. Align the camshaft bearing caps with the arrow mark (depending on cylinder numbers) and in numerical order.

16. Apply sealer at the ends of the bearing caps and install the assembly.

17. Tighten the bearing cap bolts, in the following sequence: No. 3, No. 2, No. 1 and No. 4 to 85 inch lbs. (10 Nm).

18. Repeat the sequence, increasing torque to 180 inch lbs. (20 Nm).

19. Install the distributor adapter, if it was removed.

20. Remove the lash adjuster retainers. Install the camshaft timing belt sprockets and timing belt.

21. Install the valve covers and new gaskets. Tighten the valve cover retaining bolts to 88 inch lbs. (10 Nm).

22. Reconnect all vacuum hoses that were removed during the removal procedure. Be sure they are all reconnected to the correct vacuum lines.

23. Position the wiring harnesses back to their original locations. Install the accessory drive belts and adjust to the proper tension.

24. Reconnect the spark plug wires to the correct spark plugs.

25. Install the air cleaner assembly. Reconnect the negative battery cable.

26. Run the engine and check for leaks and proper engine operation.

3.3L and 3.8L Engines

1. Relieve the fuel system pressure. Disconnect the negative battery cable.

2. Remove the engine from the vehicle. Remove the intake manifold, cylinder heads, timing chain cover and timing chain from the engine.

3. Remove the rocker arm and shaft assemblies.

4. Label and remove the pushrod and lifters.

5. Remove the camshaft thrust plate.

6. Install a long bolt into the front of the camshaft to facilitate its removal. Remove the camshaft being careful not to damage the cam bearings with the cam lobes.

To install:

7. Install the camshaft to within 2 in. of its final installation position.

8. Install the camshaft thrust plate and two bolts. Tighten the bolts to 10 ft. lbs. (12 Nm).

9. Place both camshaft and crankshaft gears on the bench with the timing marks on the exact imaginary center line through both gear bores as they are installed on the engine. Place the timing chain around both sprockets.

10. Turn the crankshaft and camshaft so the keys line up with the key ways in the gears when the timing marks are in proper position.

11. Slide both gears over their respective shafts and use a straight edge to check timing mark alignment.

12. Measure camshaft end-play. If not within specifications, replace the thrust plate.

13. If the camshaft was not replaced, lubricate and install the lifters in their original locations. If the camshaft was replaced, new lifters must be used.

14. Install the pushrods and rocker shaft assemblies.

15. Install the timing chain cover, cylinder heads and intake manifold.

16. Install the engine in the vehicle.

17. When everything is bolted in place, change the engine oil and replace the oil filter.

➡ **If the camshaft or lifters have been replaced, add 1 pint of Mopar crankcase conditioner, or equivalent when replenishing the oil to aid in break in. This mixture should be left in the engine for a minimum of 500 miles and drained at the next normal oil change.**

18. Fill the radiator with coolant.

19. Connect the negative battery cable, set all adjustments to specifications and check for leaks.

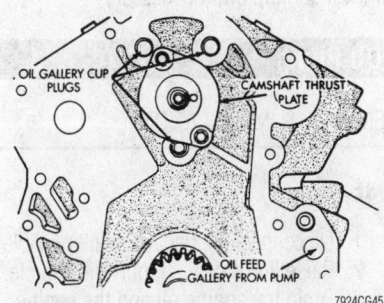

Remove the thrust plate and withdraw the camshaft from the engine—3.3L and 3.8L engines

Valve Lash

ADJUSTMENT

➡ **The 2.4, 2.5 and 3.0L engines utilize hydraulic auto lash adjusters. The auto lash adjusters automatically control valve lash. No adjustment is necessary.**

3.3L and 3.8L Engines

These engines are equipped with hydraulic lifters. The lifters are located in bores in the cylinder block. As the camshaft turns, a camshaft lobe forces the lifter upwards in its bore. The lifter drives a pushrod upwards and acts on a pivoting rocker arm, which in turn opens the valve. When the valve opening event is over and the base circle of the camshaft lobe contacts the hydraulic lifter, the valve spring pulls the valve closed. The function of the hydraulic valve lifter is to maintain zero valve lash

during the entire valve opening and closing process; any lash is instantaneously taken up by hydraulic action. The valve lash cannot be manually adjusted.

Oil Pan

REMOVAL & INSTALLATION

2.4L Engine

1. Disconnect the negative battery cable.

2. Raise and safely support the vehicle.

3. Drain the engine oil.

4. Remove the oil pan attaching bolts.

5. Remove the oil pan.

6. Thoroughly clean the inside of the oil pan. Clean the gasket mating surfaces of the oil pan and engine block.

To install:

7. Using a suitable gasket sealant, apply a ⅛ in. (3mm) bead at the oil pump-to-engine block parting line.

8. Install the new oil pan gasket.

9. Install the oil pan to the engine.

10. Tighten the oil pan attaching bolts to 105 inch lbs. (12 Nm).

11. Install the oil pan drain plug and gasket. Tighten the drain plug to 25 ft. lbs. (34 Nm).

12. Lower the vehicle.

13. Fill the engine with the correct type of clean engine oil to the proper level.

14. Reconnect the negative battery cable. Start the engine and check for leaks.

2.5L Engine

1. Disconnect the negative battery cable.

2. Raise and safely support the vehicle.

3. Drain the engine oil.

4. Remove the oil pan attaching bolts and remove the oil pan.

To install:

5. Thoroughly clean the inside of the oil pan. Clean the oil pan and engine block gasket mating surfaces.

6. Apply RTV sealant to the oil pan rail at the front seal retainer parting line.

7. Attach the oil pan side gaskets using RTV to hold the gasket in place.

8. Install the new oil pan seals and apply RTV sealant to the ends of the seals at the junction where the seals and gasket meets.

9. Install the oil pan and tighten the M8 bolts to 17 ft. lbs. (23 Nm), and the M6 bolts to 105 inch lbs. (12 Nm).

10. Install the oil pan drain plug, then tighten the drain plug to 20 ft. lbs. (27 Nm).

11. Lower the vehicle.

12. Refill the engine with the proper type and quantity of engine oil.

13. Connect the negative battery cable.

14. Start the engine and check for leaks. After the engine has been turned **OFF**, check the oil level and top off, if necessary.

3.0L Engine

1. Disconnect the negative battery cable.

2. Raise and safely support the vehicle.

3. Drain the engine oil.

4. Remove the oil pan attaching bolts and remove the oil pan.

To install:

5. Thoroughly clean the inside of the oil pan. Clean the oil pan and engine block gasket mating surfaces.

6. Apply RTV sealant to the oil pan.

7. Install the oil pan, then tighten the bolts in sequence to 50 inch lbs. (6 Nm).

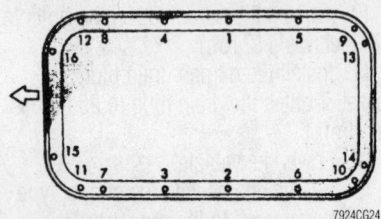

7924CG24

Tighten the oil pan bolts in the sequence shown for proper sealing—3.0L engine

8. Install the oil pan drain plug and tighten to 30 ft. lbs. (40 Nm).

9. Lower the vehicle.

10. Refill the engine with the proper type and quantity of oil.

11. Connect the negative battery cable.

12. Start the engine and check for leaks. After the engine has been turned **OFF**, check the oil level and top off, if necessary.

3.3L and 3.8L Engines

1. Disconnect the negative battery cable. Remove the engine oil dipstick.

2. Raise the vehicle and support safely.

3. Remove the torque converter bolt access cover, if equipped.

4. Drain the engine oil.

5. Remove the oil pan retaining bolts and remove the oil pan and gasket.

To install:

6. Thoroughly clean the inside of the oil pan. Thoroughly clean and dry all sealing surfaces, bolts and bolt holes.

7. Apply a ⅛ in. (3mm) bead of silicone sealer to the chain cover-to-block mating seam and the rear main seal retainer-to-block seam.

8. Install a new oil pan gasket and install the oil pan to the engine.

9. Install the retaining bolts and tighten to 105 inch lbs. (12 Nm).

10. Install the oil pan drain plug and tighten to 25 ft. lbs. (34 Nm).

11. Install the torque converter bolt access cover, if equipped. Lower the vehicle.

12. Install the oil dipstick. Refill the engine with the proper type and quantity of oil.

13. Reconnect the negative battery cable. Start the engine and check for leaks. After the engine has been turned **OFF**, check the oil level and top off, if necessary.

Oil Pump

REMOVAL & INSTALLATION

2.4L Engine

1. Disconnect the negative battery cable.

2. Raise and safely support the vehicle.

3. Drain the engine oil and the engine coolant into suitable containers.

4. Lower the vehicle.

5. Remove the accessory drive belts, as required.

6. Take up the weight of the engine with an engine support tool and remove the right engine mount and bracket. Be sure the engine is safely supported.

7. Remove the timing belt cover.

8. Loosen the timing belt tensioner bolts and remove the tensioner and the timing belt.

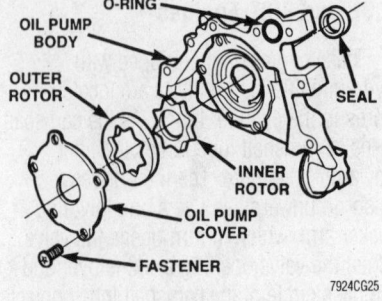

7924CG25

Exploded view of the oil pump—2.4L engine

9. Raise and safely support the vehicle.

10. Remove the oil pan assembly. Remove the oil pump pick-up tube and O-ring.

11. Using a suitable puller, remove the crankshaft damper from the front of the crankshaft.

12. Using a suitable puller, draw the crankshaft sprocket from the front of the crankshaft.

13. Loosen the oil pump bolts and remove. Take note of the location of each bolt for reassembly. Remove the oil pump from the face of the engine block. If necessary, tap lightly with a soft face mallet. Use care working with light alloy parts.

14. To remove the relief valve from the pump body, unscrew threaded plug, then pull out the spring and valve. Note the order of the parts' removal.

To install:

15. Clean all parts well for inspection. Remove the bolts holding the back cover to the pump body. Remove the pump rotors. The mating surface of the oil pump should be smooth. Replace the pump cover if scratched or grooved.

16. The pump should be checked for wear by carefully measuring the components.

➡**If oil pressure is low and the pump is within specifications, inspect for worn engine bearings or other reasons for oil pressure loss.**

17. Clean all oil pump parts in suitable solvent before assembly. Assemble the pump with new parts as required. **Install the inner rotor with the chamfer facing the cast iron oil pump cover (back of the pump).** Tighten the cover bolts to 105 inch lbs. (12 Nm).

18. Install the relief valve first, then the spring, gasket and cover cap into the pump body. Note that installing the spring first will seriously damage the engine. The relief valve goes in first. Tighten the cover cap to 40 ft. lbs. (55 Nm).

19. Prime the oil pump before installation by filling the rotor cavity with clean engine oil.

20. Insert a new oil ring seal in the oil pump counterbore on the pump body discharge passage. Apply Mopar Gasket Maker or equivalent anaerobic type gasket sealer, to the oil pump body flange. This material cures in the absence of air when squeezed between two flat machined metal surfaces. For this reason, the mating surfaces of both the pump body and the engine block must be spotlessly clean so all air will be

expelled when the parts are bolted together and tightened. Install the pump slowly onto the crankshaft aligning the oil pump rotor flats with the flats on the crankshaft until seated to the engine block. Tighten the fasteners to 20 ft. lbs. (28 Nm).

21. Install a new front oil seal. Install the seal with the spring side towards the inside of the engine. Tap the seal into place until flush with the cover.

22. Install the crankshaft sprocket. A special tool is used to draw the sprocket onto the end of the crankshaft. Use care if using a substitute tool.

23. Raise and safely support the vehicle.

24. Install the oil pump pick-up tube and O-ring. Tighten the oil pump pick-up tube mounting bolt to 20 ft. lbs. (28 Nm).

25. Thoroughly clean the oil pan and be sure the gasket rails are in good condition. Use Mopar Silicone Rubber Adhesive Sealant or equivalent sealer at the oil pump-to-engine block parting line. Use a new oil pan gasket, install the oil pan and tighten the 13 oil pan bolts to 105 inch lbs. (12 Nm).

26. Install a new oil filter.

27. Lower the vehicle.

28. Install the timing belt and covers using the recommended procedures. Use care to be sure all valve timing marks are aligned. This is most important. Failure to properly align the timing marks will result in severe engine damage.

29. Install the crankshaft damper. A special tool making use of a 12mm x 1.75 x 150mm bolt is used to draw the crankshaft damper onto the end of the crankshaft. Use care if using substitute tools. Tighten the center bolt to 105 ft. lbs. (142 Nm).

30. Install the accessory drive belts, as required. Adjust the accessory drive belts to the proper tension.

31. Install the engine mount and bracket, as required. Remove the engine support fixture from the vehicle.

32. Refill the engine with the correct type and amount of fresh, clean engine oil. Refill the cooling system with a 50/50 mixture of clean, ethylene glycol antifreeze and water.

33. Test run the vehicle to check for leaks. An oil pressure gauge should be installed to verify proper engine oil pressure.

2.5L Engine

1. Disconnect the negative battery cable.
2. Raise and safely support the vehicle.
3. Drain the oil and remove the engine oil pan.
4. Remove the oil pump pick-up tube from the oil pump. Discard the pick-up tube O-ring.

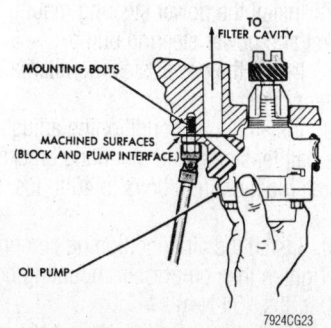

The oil pump is mounted to a machined surface on the bottom of the engine block under the oil pan—2.5L engine

5. Remove the pump mounting bolts.
6. Pull the pump down and out of the engine. Clean the pump-to-engine block mating surface.

To install:

7. Clean all parts well for inspection. Remove the bolts holding the pump cover to the pump body. Remove the pump rotors. The mating surface of the oil pump should be smooth. Replace the pump cover if scratched or grooved.

8. The pump should be checked for wear by carefully measuring the components.

➡**If oil pressure if low and the pump is within specifications, inspect for worn engine bearings or other reasons for oil pressure loss.**

9. Clean all oil pump parts in suitable solvent before assembly. Assemble the pump with new parts as required. **Install the outer rotor with the small chamfered edge facing the oil pump cover.** Tighten the cover bolts to 105 inch lbs. (12 Nm).

10. Install the relief valve first, then the spring, cup and cotter key into the pump body. Note that installing the spring first will seriously damage the engine. The relief valve goes in first.

11. Apply gasket sealer to the pump body-to-engine block mating surface.

12. Prime, by filling the pump with fresh oil. Check crankshaft/intermediate shaft timing and oil pump drive alignment. Adjust if necessary.

13. The slot in the oil pump shaft must be positioned parallel to the center line of the crankshaft when the crankshaft and intermediate shaft are properly timed.

14. Install the pump and rotate it back and forth slightly to ensure full surface contact of the pump and block.

15. While holding the pump in the fully seated position, install the pump mounting bolts to 17 ft. lbs. (23 Nm).

16. Install the oil pump pick-up tube and mounting bolt. Be sure to install a new O-ring on the pick-up tube. Tighten the mounting bolt to 20 ft. lbs. (28 Nm).

17. Install the engine oil pan.

18. Lower the vehicle and connect the negative battery cable

19. Refill the crankcase with the proper type and quantity of clean, fresh engine oil.

20. Start the engine and check for leaks. Check engine oil pressure.

3.0L Engine

The oil pump assembly is mounted at the front of the crankshaft. The oil pump housing also retains the front crankshaft oil seal. Since the timing belt must be removed to access the front cover, care must be taken. It is good practice to set the engine up to TDC No. 1 cylinder firing position. Verify that all timing marks on the crankshaft and camshaft sprockets are properly aligned before removing the timing belt. This serves as a point of reference for all work that follows. Valve timing is most important and engine damage will result if the work is incorrect.

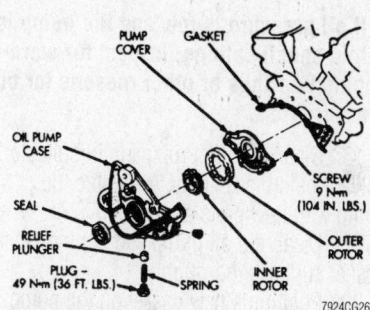

Exploded view of the oil pump assembly— 3.0L engine

1. Disconnect the negative battery cable.

2. Remove the accessory drive belts.

3. Remove the air conditioning compressor from the mounting bracket and lay it aside, if equipped. Remove the compressor mounting bracket and adjustable drive belt tensioner from the engine.

4. Remove the power steering pump/alternator belt tensioner mounting bolt and remove the tensioner.

5. Remove the power steering pump mounting bracket bolts, rear support locknut and set the power steering pump aside.

6. Raise and safely support the vehicle.

7. Drain the engine oil into a suitable container.

8. Remove the right inner fender inner shield.

9. Remove the crankshaft drive pulley bolt, drive pulley and torsional damper. Lower the vehicle.

10. Place a floor jack under the engine. Separate the engine mount insulator from the engine mount bracket.

11. Raise the engine slightly and remove the engine mount bracket.

12. Remove the timing belt cover and timing belt.

13. Remove the crankshaft sprocket.

14. Remove the five oil pump mounting bolts, then remove the oil pump assembly. Mark the mounting bolts for proper installation during reassembly.

15. Remove the front crankshaft oil seal from the oil pump cover.

To install:

16. Clean all parts well for inspection. Remove the bolts holding the rear cover to the oil pump body. Remove the pump rotors. The mating surface of the oil pump should be smooth. Replace the pump cover if scratched or grooved.

17. The pump should be checked for wear by carefully measuring the components.

➡If oil pressure is low and the pump is within specifications, inspect for worn engine bearings or other reasons for oil pressure loss.

18. Clean all oil pump parts in suitable solvent before assembly. Assemble the pump with new parts as required.

19. Clean the oil pump and engine block gasket surfaces thoroughly.

20. Position a new gasket on the pump assembly and install on the cylinder block. Be sure the correct length bolts are in their proper locations and tighten all bolts to 11 ft. lbs. (15 Nm).

21. Install a new front crankshaft oil seal into the oil pump using Seal Installer tool MD-998717 or an equivalent seal driver tool.

22. Install the crankshaft sprocket and timing belt. Recheck the engine timing marks.

23. Install the timing belt covers. Tighten the timing belt cover fasteners to 10 ft. lbs. (14 Nm).

24. Raise the engine slightly and install the engine mount bracket. Install the engine mount insulator into the engine mount bracket.

25. Install the torsional damper, crankshaft drive pulley and drive pulley bolt. Tighten to 112 ft. lbs. (151 Nm).

26. Install the right fender inner splash shield. Lower the vehicle.

27. Install the power steering mount bracket and power steering pump.

28. Install the power steering/alternator belt tensioner.

29. Install the air conditioning adjustable drive belt tensioner and mounting bracket. Tighten the mounting bolts to 40 ft. lbs. (54 Nm).

30. Install the air conditioning compressor. Tighten the compressor mounting bolts to 40 ft. lbs. (54 Nm).

31. Install the accessory drive belts and adjust to the proper tension.

32. Reconnect the negative battery cable.

33. Refill the crankcase with the correct amount of clean engine oil and start the engine. Check for leaks.

34. Check engine oil pressure.

3.3L and 3.8L Engines

1. Disconnect the negative battery cable. Remove the dipstick. Drain the cooling system.

2. Raise the vehicle and support safely. Support the engine and remove the right side engine mount.

3. Drain the oil and remove the oil pan.

4. Remove the oil pick-up tube. Remove the transaxle inspection cover, if necessary.

5. Remove the timing chain case cover.

6. Disassemble the oil pump as required.

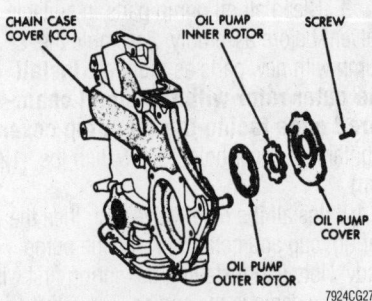

Exploded view of the oil pump assembly—3.3L and 3.8L engines

To install:

7. Clean all parts well for inspection. Remove the bolts holding the back cover to the pump body. Remove the pump rotors. The mating surface of the oil pump should be smooth. Replace the pump cover if scratched or grooved.

8. The pump should be checked for wear by carefully measuring the components.

➡If oil pressure is low and the pump is within specifications, inspect for worn engine bearings or other reasons for oil pressure loss.

9. Clean all oil pump parts in suitable solvent before assembly. Assemble the pump with new parts as required. **Install the inner rotor with the chamfer facing the cast iron oil pump cover (back of the pump).** Tighten the cover bolts to 105 inch lbs. (12 Nm).

10. Install the relief valve first, then the spring, then the cover cap into the pump body. Note that installing the spring first will seriously damage the engine. The relief valve goes in first.

11. Prime the oil pump by filling the rotor cavity with fresh oil and turning the rotors until oil comes out the pressure port. Repeat a few times until no air bubbles are present.

12. Install the chain case cover. Tighten the timing chain case cover bolts as follows:
   a. M8 x 1.25—20 ft. lbs. (27 Nm)
   b. M10 x 1.5—40 ft. lbs. (54 Nm)

13. Clean out the oil pick-up or replace as required. Replace the oil pick-up O-ring and install the pick-up to the pump. Tighten the pick-up tube retaining bolt to 250 inch lbs. (28 Nm).

14. Install the oil pan. Install the right side engine mount.

15. Lower the vehicle. Install the dipstick. Refill the engine with the proper amount of oil.

16. Refill the cooling system to the proper level.

17. Reconnect the negative battery cable and start the engine.

18. Check the area for leaks and check the oil pressure.

Rear Main Seal

REMOVAL & INSTALLATION

1. Disconnect the negative battery cable.
2. Remove the transaxle.
3. Remove the flexplate from the crankshaft.

✳✳ WARNING

Do not let the prytool damage the sealing surface of the crankshaft, or an oil leak may result.

4. Carefully pry the oil seal out of the bore with a flat-bladed prytool.
To install:

✳✳ WARNING

To prevent seal damage, remove any burrs or scratches on the edge of the crankshaft with 400 grit sandpaper before installing the seal.

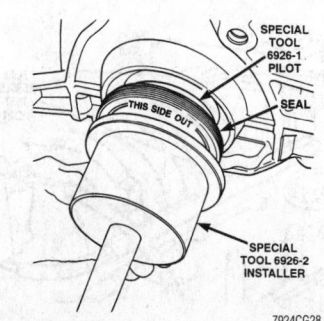

Rear main seal installation using the appropriate tools—2.4L engine shown, 2.5L is similar

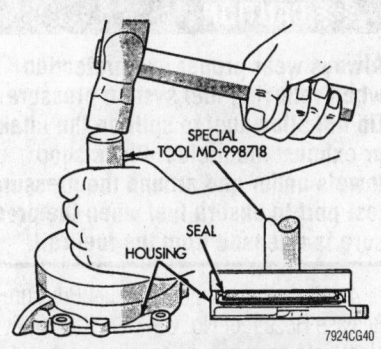

Rear main seal installation using the appropriate tools—3.0L engine

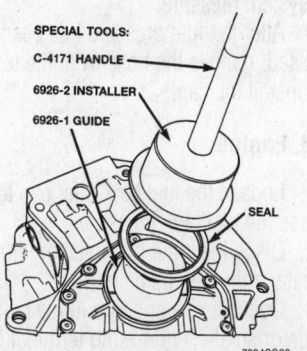

Install the rear main seal using the appropriate seal driver—3.3 and 3.8L engines

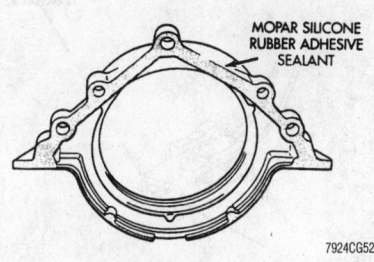

Apply silicone sealant to the seal retainer before installation—3.0l engine

5. Install the seal using the appropriate seal installation tool.

6. Install the flexplate and the transaxle assembly.

Timing Chain, Sprockets and Front Cover

REMOVAL & INSTALLATION

3.3L and 3.8L Engines

1. Turn the crankshaft and position the engine so the No. 1 piston is at TDC on the compression stroke (firing position).

2. Disconnect the negative battery cable. Drain the coolant.

3. Support the engine with a floor jack and remove the right engine mount.

4. Raise and safely support the vehicle. Drain the engine oil.

5. Remove the oil pan and oil pump pick-up tube. If necessary, remove the transaxle inspection cover.

6. Remove the right front wheel and inner fender splash shield.

7. Remove the accessory drive belt.

8. Remove the air conditioning compressor and set aside. Remove the A/C compressor mounting bracket.

9. Remove the crankshaft pulley bolt and remove the pulley using a suitable puller.

10. Remove the idler pulley from the engine bracket and remove the bracket.

11. Remove the cam sensor from the timing chain cover.

12. Unbolt and remove the cover from the engine. Be sure the oil pump inner rotor does not fall out. Remove the three O-rings from the coolant passages and the oil pump outlet.

13. Remove the camshaft sprocket attaching cup washer and remove the timing chain with both sprockets attached. Remove the timing chain snubber.

14. Remove the crankshaft sprocket using a suitable puller.

To install:

15. Assemble the timing chain and sprockets.

16. Turn the crankshaft and camshaft to line up with the keyway locations of the sprockets.

17. Slide both sprockets over their respective shafts and use a straightedge to confirm alignment.

18. Install the cup washer and camshaft bolt. Tighten the bolt to 40 ft. lbs. (54 Nm).

19. Check camshaft end-play. The specification with a new plate is 0.005–0.012 in.

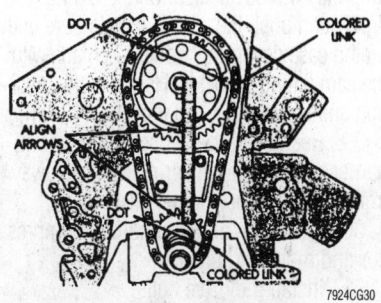

Align the timing marks as shown for timing chain installation—3.3L and 3.8L engines

(0.0127–0.3040mm) and 0.012 in. (0.310mm) maximum with a used plate. Replace the thrust plate if not within specifications.

20. Install the timing chain snubber.

21. Clean all parts well. Thoroughly clean and dry the gasket mating surfaces. Install new O-rings on the block.

22. Remove the crankshaft oil seal from the cover. The seal must be removed from the cover when installing to ensure proper oil pump engagement. The seal is installed from the front AFTER the timing chain front cover is completely installed as outlined below.

23. Using a new gasket, install the chain case cover on the engine. DO NOT adhere the new gasket to the cover. Be sure the lower edge of the gasket is flush to 0.020 in. (0.6mm) past the lower edge of the cover.

24. If necessary, rotate the crankshaft so the oil pump drive flats are vertical, and position the oil pump rotor so its mating flats are in the same position. Make certain that the oil pump is engaged onto the crankshaft before proceeding, or severe engine damage will result. Install the attaching bolts and tighten to 20 ft. lbs. (27 Nm).

25. Use tool C-4992 or an equivalent seal driver to install the crankshaft oil seal.

26. Install the crankshaft pulley using a bolt approximately 6 in. (150mm) long, a thrust bearing and a washer plate. Be sure the pulley bottoms out on the crankshaft seal diameter. Install the bolt and tighten to 40 ft. lbs. (54 Nm).

27. Install the engine bracket and tighten the bolts to 40 ft. lbs. (54 Nm). Install the idler pulley to the engine bracket.

28. To install the cam sensor, first clean off the old spacer from the sensor face completely. Inspect the O-ring for damage and replace if necessary. A new spacer must be attached to the cam sensor, prior to installation. If a new spacer is not used, engine per-

formance will be affected. Oil the O-ring lightly and push the sensor into its bore in the timing case cover until contact is made with the cam timing sprocket. Hold it in this position and tighten it to 108 inch lbs. (12 Nm).

29. Reconnect the sensor to the wiring harness. Position the wiring harness away from the accessory drive belt.

30. Install the air conditioning compressor and mounting bracket.

31. Install the drive belt.

32. Install the inner splash shield and the wheel. Tighten the wheel lug nuts, in a star pattern, to 95 ft. lbs. (129 Nm).

33. Install the oil pan with a new gasket. Tighten the oil pan bolts to 105 inch lbs. (12 Nm).

34. Install the motor mount.

35. Refill the cooling system to the correct level with a 50/50 mixture of clean, ethylene glycol antifreeze and water.

36. Fill the engine with the proper amount and type of clean engine oil.

37. Reconnect the negative battery cable. Road test the vehicle and check for leaks.

FUEL SYSTEM

Fuel System Service Precautions

Safety is the most important factor when performing not only fuel system maintenance but any type of maintenance. Failure to conduct maintenance and repairs in a safe manner may result in serious personal injury or death. Maintenance and testing of the vehicle's fuel system components can be accomplished safely and effectively by adhering to the following rules and guidelines.

• To avoid the possibility of fire and personal injury, always disconnect the negative battery cable unless the repair or test procedure requires that battery voltage be applied.

• Always relieve the fuel system pressure prior to disconnecting any fuel system component (injector, fuel rail, pressure regulator, etc.), fitting or fuel line connection. Exercise extreme caution whenever relieving fuel system pressure to avoid exposing skin, face and eyes to fuel spray. Please be advised that fuel under pressure may penetrate the skin or any part of the body that it contacts.

• Always place a shop towel or cloth around the fitting or connection prior to loosening to absorb any excess fuel due to spillage. Ensure that all fuel spillage

(should it occur) is quickly removed from engine surfaces. Ensure that all fuel soaked cloths or towels are deposited into a suitable waste container.

• Always keep a dry chemical (Class B) fire extinguisher near the work area.

• Do not allow fuel spray or fuel vapors to come into contact with a spark or open flame.

• Always use a back-up wrench when loosening and tightening fuel line connection fittings. This will prevent unnecessary stress and torsion to fuel line piping. Always follow the proper torque specifications.

• Always replace worn fuel fitting O-rings with new ones. Do not substitute fuel hose or equivalent, where fuel pipe is installed.

Fuel System Pressure

RELIEVING

❋❋ CAUTION

The fuel injection system remains under pressure even after the engine has been turned OFF. The fuel system pressure must be relieved before disconnecting any fuel lines. Failure to do may result in fire and/or personal injury.

2.4L, 3.3L and 3.8L Engines

➡**The following procedure requires Fuel Pressure Release Hose C-4799–1 or equivalent. Fuel Gauge C-4799-A contains a hose to direct fuel into an approved container.**

1. Disconnect the negative battery cable.
2. Remove the fuel tank filler cap to release the pressure in the fuel tank.
3. Remove the cap from the fuel pressure test port on the fuel rail.

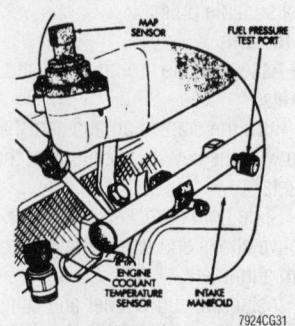

Connect the fuel pressure release hose to the test port to relieve the system pressure—2.4L engine

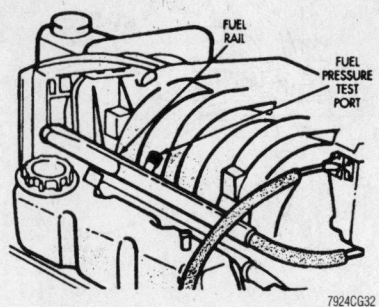

On 3.3L and 3.8L engines, the fuel pressure test port is usually located on the fuel rail

❋❋ CAUTION

Always wear proper eye protection when relieving fuel system pressure. Do not allow fuel to spill on the intake or exhaust manifolds. Place shop towels under and around the pressure test port to absorb fuel when the pressure is released from the fuel rail.

4. Place the open end of Fuel Pressure Release Hose tool No. C-4799–1, or equivalent, into an approved gasoline container. Place a shop towel under the test port.

5. Connect the other end of the hose onto the fuel pressure test port, to relieve the system pressure.

6. After the fuel pressure has been released, remove the hose from the test port and install the cap.

2.5L Engine

1. Loosen the fuel tank filler cap to release the fuel tank pressure.

2. Disconnect the fuel injector wiring harness from the engine harness.

3. Connect one end of a jumper wire (18 gauge or smaller) to ground terminal No. 1 of the injector harness. Connect the other end of the jumper wire to engine ground.

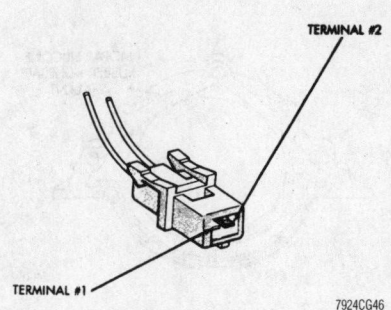

Fuel injector harness terminal identification—2.5L engine

4. Connect one end of a jumper wire (18 gauge or smaller) to the positive terminal No. 2 of the injector harness. Touch the other end of the jumper wire to the positive battery post (for no longer than five seconds), to release the fuel system pressure.

5. Remove the jumper wires and disconnect the negative battery cable, then continue with the service procedure.

3.0L Engine

1995 MODELS

1. Loosen the fuel tank filler cap to release the fuel tank pressure.

2. Disconnect the fuel injector wiring harness from the engine harness.

3. Connect one end of a jumper wire to the **A142** circuit terminal (dark green/orange colored wire) of the fuel injector harness connector (black connector located on fuel rail harness). Connect the other end of the jumper wire to a 12 volt power source.

4. Connect one end of a jumper wire to a good ground.

5. Momentarily ground one of the injectors by connecting the other end of the jumper wire to an injector terminal in the harness connector. Repeat the procedure for two or three injectors.

6. Remove the jumper wires and disconnect the negative battery cable, then continue with the service procedure.

➡Removal of the fuel pump relay from the PDC may have caused one or more **Diagnostic Trouble Codes (DTC's)** to be stored in the Powertrain Control Module (PCM) memory. Erasing any DTC's will require the use of a DRB scan tool or equivalent.

1996–99 MODELS

1. Remove the fuel pump relay from the Power Distribution Center (PDC) located on the left side in the engine compartment. Location of the relay can be verified by the label located on the underside of the PDC cover.

2. Start the engine and allow it to run until it stalls.

3. Continue to start the engine until it will no longer run.

4. Turn the ignition key to the **OFF** position.

✳✳ WARNING

Steps 1–4 must be performed to relieve the pressurized fuel from within the fuel rail. Do NOT use the following steps to relieve this fuel

pressure, as excessive fuel will be forced into a cylinder chamber.

5. Disconnect any fuel injector.

6. Connect one end of a jumper wire (18 gauge or smaller) to either injector terminal of the fuel injector harness connector (black connector located on fuel rail harness). Connect the other end of the jumper wire to the positive battery terminal.

7. Connect one end of a second jumper wire to the remaining (other) injector terminal.

✳✳ WARNING

Do NOT supply power to the injector for more than four seconds or permanent damage to the injector will result.

8. Momentarily touch the other end of the second jumper wire to the negative battery terminal for **no more than four seconds**.

9. Place a shop rag or towel under the fuel line at the quick-disconnect fitting to the fuel rail.

10. Disconnect the negative battery cable. Squeeze the quick-disconnect fitting retainer tabs together and pull the fuel tube/quick-disconnect fitting assembly off the fuel tube nipple. The retainer will remain on the fuel tube.

11. Install the fuel pump relay into the PDC.

Fuel Filter

REMOVAL & INSTALLATION

✳✳ CAUTION

The fuel injection system remains under pressure, even after the engine has been turned OFF. The fuel system pressure must be relieved before disconnecting any fuel lines. Failure to do so may result in fire and/or personal injury.

1995 Models

The Front Wheel Drive (FWD) and the All Wheel Drive (AWD) vans have different fuel delivery systems. The tanks, pumps and fuel lines are different. Both systems use quick-disconnect fittings in some locations. The fuel filter location is similar for both FWD and AWD vehicles.

1. Properly relieve the fuel system pressure.

2. Disconnect the negative battery cable.

3. Raise and safely support the vehicle.

4. On Front Wheel Drive models, locate the fuel filter in its mounting along the frame rail. Remove the filter retaining bolt and remove the filter from the rail.

5. On All Wheel Drive models, remove the converter support bracket and the exhaust pipe heat shield.

6. Loosen the hose clamps. Wrap a shop towel around the hoses to absorb excess fuel.

7. On All Wheel Drive models, remove the filter retaining bolt and remove the filter from the rail.

8. Disconnect the hoses from the filter and remove the filter from the vehicle. Discard the clamps.

To install:

9. Connect the hoses and new clamps to the filter.

10. Install the filter to the rail and tighten the retaining bolt to 75 inch lbs. (8 Nm).

11. Position and tighten the hose clamps.

12. On All Wheel Drive models, install the exhaust pipe heat shield and converter support bracket.

13. Lower the vehicle. Start the engine and check for leaks.

1996–99 Models

➡The fuel delivery system uses quick-disconnect fittings. The fuel filter mounts to the top of the fuel tank.

1. Properly relieve the fuel system pressure.

2. Disconnect the negative battery cable.

3. Raise and safely support the vehicle.

4. Locate the fuel filter in its mounting on top of the fuel tank. Detach the quick-disconnect fittings from the chassis fuel supply tube and fuel pump module by squeezing the quick-disconnect fitting retainer tabs together and pulling the fitting assembly away from the fuel line nipple. The retainer will remain on the fuel tube.

5. Remove the fuel filter mounting bolt and remove the fuel filter from the fuel tank.

To install:

6. Install the fuel filter on the top of the fuel tank and tighten the mounting bolt.

7. The fuel supply tube (to chassis fuel line), return tube (to pump module) and fuel supply (to fuel filter) tube are permanently attached to the fuel filter. The quick-disconnect fitting ends of the fuel supply and return tubes are of different sizes.

8. Apply a light coating of 30W engine oil to the nipples of the fuel filter.

9. Push the quick-disconnect fitting over the fuel line until the retainer seats and

clicks into place. Be sure the retainer tabs have locked into the case of the fitting.

10. Lower the vehicle. Start the engine and check for leaks.

Fuel Pump

REMOVAL & INSTALLATION

1995 Models

Although the Front Wheel Drive (FWD) van and All Wheel Drive (AWD) van have different fuel systems, both have fuel pump modules with an internal fuel reservoir, a fuel level sending unit and a fuel strainer mounted on the pump housing. Both systems use quick-disconnect fittings at the fuel tank.

✳✳ CAUTION

The fuel injection system remains under pressure, even after the engine has been turned OFF. The fuel system pressure must be relieved before disconnecting any fuel lines. Failure to do so may result in fire and/or personal injury.

FRONT WHEEL DRIVE

1. Properly relieve the fuel system pressure.
2. Disconnect the negative battery cable.
3. Raise the vehicle and support safely.

✳✳ CAUTION

Observe all applicable safety precautions when working around fuel. Do not allow fuel spray or fuel vapors to come in contact with a spark or open flame. Keep a dry chemical (Class B) fire extinguisher near the work area. Never drain or store fuel in an open container due to the possibility of fire or explosion.

4. Remove the fuel tank from the vehicle.
5. Clean all dirt and foreign material from the pump mounting area.
6. Using a hammer and **brass or plastic drift**, carefully tap the lockring counterclockwise to release the fuel pump.
7. Remove the fuel pump from the tank with the O-ring seal. Discard the O-ring seal.
8. Remove the fuel pump inlet strainer and O-ring from the fuel pump and discard.
9. Cover the fuel tank opening to prevent dirt from getting into the fuel tank.

To install:
10. Lubricate a new strainer O-ring with silicone grease or spray lube and install into the outlet of the strainer. The O-ring must sit evenly on the step of the filter outlet.
11. Push the new strainer onto the inlet of the fuel pump reservoir body, making sure the locking tabs on the reservoir body engage the tangs on the filter.
12. Wipe the seal area of the fuel tank clean. Install a new O-ring seal to the pump.
13. Install the fuel pump to the tank.
14. Install the lockring with a hammer and **brass or plastic drift**, tapping the ring clockwise to lock the pump in place.

✳✳ CAUTION

Over-tightening the pump lockring may result in a leak.

15. Install the fuel tank and lower the vehicle.
16. Reconnect the negative battery cable, start the engine and check for leaks.

ALL WHEEL DRIVE

1. Properly relieve the fuel system pressure.
2. Disconnect the negative battery cable.
3. Raise the vehicle and support safely.

✳✳ CAUTION

Observe all applicable safety precautions when working around fuel. Do not allow fuel spray or fuel vapors to come in contact with a spark or open flame. Keep a dry chemical (Class B) fire extinguisher near the work area. Never drain or store fuel in an open container due to the possibility of fire or explosion.

4. Remove the fuel tank from the vehicle.
5. Clean all dirt and foreign material from the pump mounting area.
6. Unclip the fuel vapor hose and fuel drain hose from the fuel tank.
7. While holding down on the fuel pump assembly, remove the band clamp.

➡ **The fuel pump assembly is spring-loaded and may rise up slightly when the band clamp is removed.**

8. Remove the fuel pump assembly from the tank and discard the rubber seal.
9. Bend the locking tabs on the fuel inlet strainer to clear the locking tangs on the fuel pump. Remove and discard the strainer.

To install:
10. Align the orientation tabs in the inlet strainer with the slot in the bottom of the fuel pump. Push the strainer onto the fuel pump inlet, making sure the locking tabs on the filter snap over the tangs on the pump module.
11. Clean the seal area of the tank and install a new seal on the pump.
12. Position the fuel pump assembly on the fuel tank, aligning the arrow on the edge of the pump between the two lines molded into the fuel tank.
13. Push the module down and install the band clamp. Tighten the clamp to 40 inch lbs. (5 Nm).
14. Install the fuel tank and lower the vehicle.
15. Reconnect the negative battery cable, start the engine and check for leaks.

1996–99 Models

The in-tank fuel pump module contains the fuel pump and pressure regulator which adjusts fuel system pressure to approximately 49 psi. (338kPa). Voltage to the fuel pump is supplied through the fuel pump relay.

The fuel pump is serviced as part of the fuel pump module. The fuel pump module is installed in the top of the fuel tank and contains the electric fuel pump, fuel pump reservoir, inlet strainer fuel gauge sending unit, fuel supply and return line connections and the pressure regulator. The inlet strainer, fuel pressure regulator and level sensor are the only serviceable items. If the fuel pump requires service, replace the fuel pump module, using the following procedure:

✳✳ CAUTION

The fuel injection system remains under pressure, even after the engine has been turned OFF. The fuel system pressure must be relieved before disconnecting any fuel lines. Failure to do so may result in fire and/or personal injury.

1. Remove the fuel filler cap and properly relieve the fuel system pressure.
2. Disconnect and isolate the negative battery cable.
3. Drain and remove the fuel tank.

✳✳ CAUTION

Observe all applicable safety precautions when working around fuel. Do not allow fuel spray or fuel vapors to come in contact with a spark or open

flame. Keep a dry chemical (Class B) fire extinguisher near the work area. Never drain or store fuel in an open container due to the possibility of fire or explosion.

4. Clean the top of the tank to remove any loose dirt.

5. Detach the fuel lines from the fuel pump module by squeezing the quick-disconnect fitting with your thumb and forefinger.

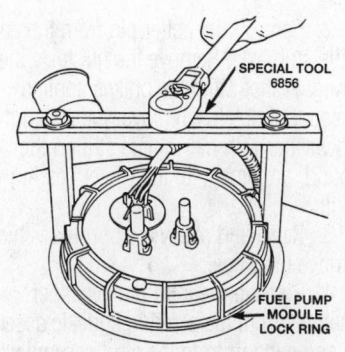

Using special tool No. 6856 or equivalent, remove the fuel pump module locknut

6. Detach the fuel pump module electrical connector from the top of the fuel pump module.

7. Using special tool No. 6856 or equivalent, remove the fuel pump locknut by turning it counterclockwise.

❋❋ CAUTION

The fuel reservoir of the fuel pump module does not empty out when the tank is drained. The fuel in the reservoir may spill out when the module is removed.

8. Remove the fuel pump and O-ring from the tank. Discard the O-ring.

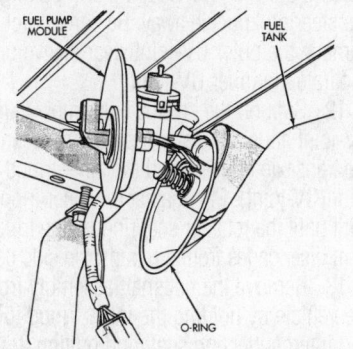

After the locknut is removed, pull the fuel pump module out of the tank—1996–99 models

To install:

9. Thoroughly clean all parts. Wipe the seal area of the tank clean. Place a new O-ring on the ledge between the tank threads and the pump module opening.

10. Position the fuel pump module in the tank. Be sure the alignment tab on the underside of the pump module flange sits in the corresponding notch in the fuel tank.

11. While holding the fuel pump module in place install the locking ring and tighten to 40 inch lbs. (5 Nm) using special tool No. 6856 or equivalent spanner-type tool.

12. Install the fuel tank assembly.

13. Connect fuel pump module electrical connector.

14. Reconnect the negative battery cable.

15. Fill the fuel tank with fuel. Install the fuel filler cap. Turn the ignition switch to the **ON** position to pressurize the system. Check the fuel system for leaks.

DRIVE TRAIN

➡**Refer to the Unit Repair Section for information on CV-joint boot replacement.**

Automatic Transaxle Assembly

REMOVAL & INSTALLATION

❋❋ WARNING

If the vehicle is going to be rolled on its own wheels while the transaxle is out of the vehicle, obtain two outer CV-joints to install to the hubs. If the vehicle is rolled without the proper torque applied to the front wheel bearings, the bearings will be destroyed.

1. Disconnect the negative battery cable. If equipped with the 3.0L engine, drain the coolant.

2. Use an engine support fixture to support the engine.

3. Remove the air cleaner assembly if preventing access to the upper bell housing bolts.

4. Disconnect the transaxle shift linkage at the manual valve lever.

5. Squeeze the grommet clips to disconnect the cable at the transaxle bracket.

6. Remove the transaxle oil dipstick tube.

7. Disconnect and plug the transaxle fluid cooler lines.

8. Remove the input and output speed sensors.

9. Remove the upper bell housing mounting bolts.

10. Raise and safely support the vehicle. Remove the front wheels.

11. Position a drain pan under the transaxle where the halfshafts enter the differential or extension housing. Remove the right and left halfshaft assemblies.

12. Drain the transaxle fluid.

13. Remove the torque converter dust shield (inspection cover), matchmark the torque converter to the flexplate and rotate the engine clockwise to remove the torque converter bolts.

14. Detach the wiring harness connections to the transaxle range switch and the Park/Neutral position switch.

15. Remove the front motor mount insulator and bracket.

16. If equipped with Distributorless (DIS) ignition system, remove the crankshaft position sensor from the bell housing.

17. Remove the starter motor mounting bolts and set the starter motor aside. Do not allow the starter motor to hang suspended from the battery cable.

18. Position a transmission jack under the transaxle.

19. With the transaxle mount firmly in position, remove the left transaxle mount.

20. Remove the lower bell housing bolts.

21. Pull the transaxle completely away from the engine and carefully lower it from the vehicle.

22. To prepare the vehicle for rolling, secure the engine with a suitable support or reinstall the front motor mount to the engine. Then, reinstall the ball joints to the steering knuckle and install the retaining bolt. Install the obtained outer CV-joints to the hubs, install the washers and tighten the axle nuts to 180 ft. lbs. (244 Nm). The vehicle may now be safely rolled.

To install:

23. Install the transaxle securely on the transmission jack. Rotate the converter so it will align with the positioning of the flexplate.

❋❋ WARNING

If the torque converter has been replaced, a torque converter clutch break-in procedure must be performed. This procedure will reset the transaxle control module break-in status. Failure to perform this procedure may cause transaxle shutter. To properly do this, a DRB or equivalent scan tool, is required to read or reset the break-in status.

24. Apply a coating of high temperature grease to the torque converter pilot hub.

25. Raise the transaxle into place and push it forward until the dowels engage and the bell housing is flush with the block.

26. Install the lower transaxle bell housing bolts.

27. Jack the transaxle up and install the left transaxle mount.

28. Install the starter to the transaxle. Tighten the starter motor mounting bolts to 40 ft. lbs. (54 Nm).

29. Remove the transaxle jack from under the vehicle.

30. If equipped with D.I.S. ignition system, clean off the old spacer on the crankshaft position sensor and install a new spacer. Install the crankshaft position sensor to the transaxle bell housing and push down until contact is made with the drive plate. Tighten the sensor retaining bolts to 105 inch lbs. (12 Nm).

31. Install the front engine mount insulator and bracket.

32. Reattach the wiring harness connectors to the Park/Neutral position switch and the transaxle range switch.

33. Align the torque converter to the flexplate mounting bolt holes. Install the torque converter bolts and tighten to 55 ft. lbs. (75 Nm). Install the torque converter inspection cover.

34. Install the right and left halfshaft assemblies. Install the ball joints to the steering knuckles. Tighten the axle nuts to 180 ft. lbs. (244 Nm) and install new cotter pins.

35. Install the front wheels and lug nuts. Tighten the lug nuts in a star pattern to 95 ft. lbs. (129 Nm).

36. Lower the vehicle.

37. Install the upper transaxle mounting bolts. Tighten the mounting bolts to 70 ft. lbs. (95 Nm).

38. Remove the engine support fixture.

39. Install the input and output speed sensors.

40. Reconnect the transaxle oil cooler lines.

41. Install the transaxle oil dipstick tube.

42. Attach the shift cable to the transaxle bracket.

43. Reconnect the transaxle shift linkage to the manual valve lever.

44. Install the air cleaner assembly. Fill the transaxle with the proper amount of clean, fresh MOPAR® ATF Plus 7176 automatic transmission fluid.

45. If equipped with the 3.0L engine, refill the cooling system to the correct level with a 50/50 mix of clean, ethylene glycol antifreeze and water. Bleed the cooling system.

46. Ensure all linkages, electrical connectors and fluid lines have been reattached.

47. Reconnect the negative battery cable and check the transaxle for proper operation. Perform the transaxle quick learn and torque converter clutch break-in procedures.

Power Transfer Unit

REMOVAL & INSTALLATION

1. Raise and safely support the vehicle.

2. Remove the front wheels.

3. Remove the right front halfshaft and plug the seal hole.

4. Matchmark the front driveshaft flange and remove the driveshaft from the Power Transfer Unit (PTU). Suspend the driveshaft with wire.

5. Remove the cradle plate.

6. Remove the PTU mounting bracket bolts at the rear of the unit.

7. Remove the right outboard support bracket near the right halfshaft.

8. Remove the mounting bolts and the PTU from the vehicle.

To install:

9. Position the PTU on the transaxle and install the mounting bolts. Tighten the bolts to 30 ft. lbs. (41 Nm).

10. Install the right outboard support bracket. Tighten the bolts to 37 ft. lbs. (50 Nm).

11. Install the rear mounting bracket. Tighten the bolts to 37 ft. lbs. (50 Nm).

12. Install the cradle plate. Tighten the bolts to 38 ft. lbs. (51 Nm).

13. Align the matchmark and install the driveshaft.

14. Install the right front halfshaft.

15. Install the front wheels and lower the vehicle to the floor.

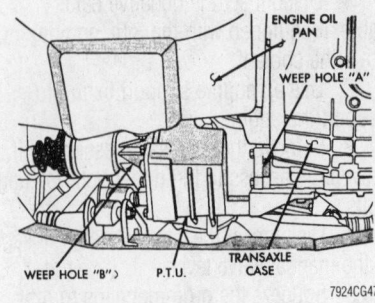

Power transfer unit and related components

Halfshaft

REMOVAL & INSTALLATION

Front

On vehicles with Anti-Lock Brakes (ABS), each outer CV-joint will be equipped with an ABS speed sensor tone wheel, which is utilized to determine vehicle speed for ABS brake operation.

1. Disconnect the negative battery cable.

2. Remove the cotter pin from the end of the stub axle. Remove the nut lock and spring washer. With the brakes applied, loosen, but do NOT remove, the axle nut and washer with the vehicle still on the ground, or damage to the wheel bearing will result.

3. Raise and safely support the vehicle. Remove the wheel.

4. Remove the front brake caliper assembly from the steering knuckle assembly and support from the strut assembly using a strong piece of wire.

5. Remove the front brake rotor from the hub/bearing assembly.

6. Remove the retaining nut and washer from the halfshaft stub axle.

7. Separate the outer tie rod from the steering knuckle.

8. If equipped, remove the ABS wheel speed sensor from the steering knuckle.

9. Remove the wheel stop from the steering knuckle, if equipped.

10. Remove the nut and bolt that clamps the steering knuckle to the ball joint stud. Using a prybar, pry the control arm down to release the ball stud from the steering knuckle. Be careful not to tear the ball joint grease seal when prying down from the steering knuckle.

11. Separate the outer CV-joint splined shaft from the hub and bearing assembly by holding the CV-joint housing and pulling the steering knuckle away. Be careful not to damage the outer CV-joint wear sleeve or separate the inner CV-joint.

12. Support the halfshaft assembly at the CV-joint housing. Install a prybar between the transaxle housing and the inner tripod joint (CV-joint). Pry against the inner tripod joint until the retainer snapring on the tripod joint disengages from the transaxle side gear.

13. Remove the halfshaft assembly from the vehicle by holding the inner tripod joint and interconnecting shaft and pulling it straight out of the transaxle side gear. Do

not allow the splines or the snapring of the shaft to drag across the sealing lip of the transaxle-to-tripod joint oil seal.

To install:

14. Thoroughly clean the tripod joint shaft splines and oil seal contact surface. Apply a light coating of clean transaxle lubricant to the oil seal sealing surface of the tripod joint.

15. Hold the halfshaft assembly by the tripod joint and interconnecting shaft and install the halfshaft assembly into the transaxle, being careful not to damage the oil seal. Be sure the inner joint clicks into place inside the differential. Check that the snapring is fully engaged by attempting to pull the halfshaft assembly out by hand. If it cannot be removed by hand, the snapring is engaged.

16. Thoroughly clean the steering knuckle and hub/bearing area of all debris and moisture, where the CV-joint will be installed into the steering knuckle. Also, thoroughly clean the bearing shield of the outer CV-joint.

17. Pull the front strut out and insert the splined outer CV-joint into the front hub.

18. Insert the ball joint stud into the steering knuckle clamp. Install a **new** steering knuckle-to-ball joint stud clamping nut and bolt. Be sure to use an exact replacement nut and bolt during installation. Tighten the bolt to 105 ft. lbs. (145 Nm).

19. Install the tie rod end into the steering knuckle. Tighten the tie rod end-to-steering knuckle nut to 45 ft. lbs. (61 Nm).

20. Install the disc brake rotor.

21. Install the brake caliper assembly onto the steering knuckle.

22. Install the axle washer and nut. Tighten the nut to about 45 ft. lbs. (61 Nm) temporarily.

23. If removed, install the ABS wheel speed sensor.

24. Install the wheel and lug nuts. Tighten the lug nuts in a star pattern, to 95 ft. lbs. (129 Nm).

25. Lower the vehicle. Do NOT roll the vehicle until the axle nut has been properly tightened or damage to the front wheel bearings will result.

26. With the vehicle's brakes applied, tighten the axle nut to 180 ft. lbs. (244 Nm). Install the spring washer, nut lock and a new cotter pin. Wrap the cotter pin prongs tightly around the axle nut lock.

27. Reconnect the negative battery cable. Road test the vehicle.

Rear

1. Disconnect the negative battery cable.
2. Raise and safely support the vehicle.

3. Remove the rear wheel.

4. Remove the cotter pin from the end of the halfshaft. Remove the nut lock, spring washer, axle nut and washer.

5. Remove the inner shaft retaining bolts. The halfshaft is spring loaded. Compress the inner halfshaft joint slightly and pull downward to clear the differential.

6. Remove the halfshaft.

To install:

7. Install the halfshaft.

8. Install and tighten the inner shaft retaining bolts to 45 ft. lbs. (61 Nm).

9. Install the washer and nut on the axle shaft. Tighten the nut to 180 ft. lbs. (244 Nm). Install the spring washer, nut lock and a new cotter pin.

10. Install the wheel and lug nuts. Tighten the lug nuts to 95 ft. lbs. (129 Nm).

11. Lower the vehicle and reconnect the negative battery cable.

STEERING AND SUSPENSION

Air Bag

✳✳ CAUTION

Some vehicles are equipped with an air bag system, also known as the Supplemental Inflatable Restraint (SIR) or Supplemental Restraint System (SRS). The system must be disabled before performing service on or around system components, steering column, instrument panel components, wiring and sensors. Failure to follow safety and disabling procedures could result in accidental air bag deployment, possible personal injury and unnecessary system repairs.

PRECAUTIONS

Several precautions must be observed when handling the inflator module to avoid accidental deployment and possible personal injury.

• Never carry the inflator module by the wires or connector on the underside of the module.

• When carrying a live inflator module, hold securely with both hands, and ensure that the bag and trim cover are pointed away from your body.

• Place the inflator module on a bench or other surface with the bag and trim cover facing up.

• With the inflator module on the bench, never place anything on or close to the module which may be thrown in the event of an accidental deployment.

DISARMING

✳✳ CAUTION

The Supplemental Inflatable Restraint (SIR) system must be disarmed before working around the air bag or SIR wiring. Failure to do so may cause accidental deployment of the air bag, resulting in unnecessary SIR system repairs and/or personal injury.

1. Disconnect and isolate the negative battery cable from the battery.

2. Allow the SIR system capacitor to discharge for at least two (2) minutes, before performing any removal procedures.

3. The air bag system is now disabled.

✳✳ CAUTION

When carrying a live air bag, be sure the bag and trim cover are pointed away from the body. In the unlikely event of an accidental deployment, the bag will then deploy with minimal chance of injury. When placing a live air bag on a bench or other surface, always face the bag and trim cover up, away from the surface. This will reduce the motion of the module if accidentally deployed.

Power Rack and Pinion Steering Gear

REMOVAL & INSTALLATION

1995 Models

FRONT WHEEL DRIVE

1. Disconnect the negative battery cable.

2. Loosen the wheel lug nuts slightly. Raise and safely support the front of the vehicle, not on the front crossmember.

3. Remove the front wheels. Remove the tie rod ends from the steering knuckles.

4. Disconnect the engine damper strut from the crossmember, if equipped.

5. If equipped, remove the air diverter valve from the left side of the crossmember.

6. Place a transmission jack, or floor jack with a wide lifting flange, under the front suspension K-crossmember. Support the crossmember, then remove the cross-member-to-frame attaching bolts. Slowly lower the crossmember until enough room is gained to disconnect the steering column from the steering gear assembly. Support the crossmember in this position.

7. Remove the splash and boot shields. Disconnect the power steering hoses. Place a drain pan under the power steering pump hoses to catch draining power steering fluid.

8. Remove the bolts that mount the steering gear assembly to the crossmember. Remove the assembly from the vehicle.

To install:

9. Install the steering gear rack assembly onto the front crossmember. Install the mounting bolts.

10. Using a transmission jack, raise the front crossmember and steering gear rack and pinion assembly up to the frame rails.

11. Line up the gear pinion with the column. Install the crossmember-to-frame mounting bolts and nut. The right rear crossmember stud is the alignment pilot for reinstallation.

12. Tighten all crossmember attaching bolts to 90 ft. lbs. (122 Nm). Steering gear rack mounting bolts are tightened to 50 ft. lbs. (68 Nm). To ensure proper alignment of the steering gear assembly, tighten the left front bolt first.

13. Reconnect the engine damper strut to the front crossmember, if equipped. Install the air diverter valve to the left side of the crossmember, if equipped.

14. Reconnect the power steering fluid hoses. Tighten the fluid pressure fittings to 23 ft. lbs. (31 Nm).

15. Install the outer tie rod ends to the steering knuckles. Tighten the tie rod end-to-steering knuckle nuts to 38 ft. lbs. (52 Nm). Install a new cotter pin.

16. Install the front wheels and lug nuts. Tighten the lug nuts, in a star sequence, to 95 ft. lbs. (129 Nm).

17. Lower the vehicle. Reconnect the negative battery cable.

18. Refill the power steering reservoir with the correct amount of clean power steering fluid and bleed the system.

19. Check the toe setting and adjust, if necessary. Road test the vehicle and check steering operation.

ALL WHEEL DRIVE

Before removing the steering gear on All Wheel Drive (AWD) vehicles, the steering column must be removed to provide clearance for steering rack removal.

1. Disconnect the negative battery cable.

2. Raise and safely support the vehicle. Remove the front wheels.

3. Remove the steering column assembly from the vehicle.

4. Remove the tie rod ends from the steering knuckle using a suitable puller.

5. Remove the two nuts and bolts that attach the bridge assembly to the crossmember. They can be reached through the access holes in the top of the bridge assembly.

6. Remove the bracket securing the power steering fluid hoses to the cross-member.

7. Remove the crossmember-to-frame rail attaching bolts and the nut from the locating stud. Use a jack to lower the crossmember so it is suspended from the lower control arms. It is not necessary to remove the cross-member completely from the vehicle.

8. Disconnect and plug the power steering lines from the steering gear. Place a drain pan under the power steering hoses to catch any spilling fluid.

9. Remove the four bolts that retain the steering gear to the bridge assembly.

➡**Note the position of each bolt as it is removed, there are different bolts for the left and right sides.**

10. Remove the lower steering column coupler from the steering gear. Drive the roll pin from the coupler using a punch. If this is not done, there will not be enough clearance for rack removal.

11. Remove the steering gear from the vehicle by pulling it out through the driver's side wheel well. Rotate the gear to clear the frame rail.

To install:

12. Install the steering gear in the vehicle. Work it in through the left wheel opening, rotating it as needed.

13. Install the steering column coupler; be sure to fully seat the roll pin.

14. Install the steering gear mounting bolts in their proper locations. Do not tighten them at this time.

15. Install the steering hose bracket and tighten to 17 ft. lbs. (23 Nm). Install the hoses to the correct fittings on the steering rack and tighten them to 25 ft. lbs. (34 Nm).

16. Raise the crossmember into position, install the bolts and tighten them as follows:

   a. Crossmember-to-frame rail stud nut—90 ft. lbs. (122 Nm)

   b. Crossmember-to-frame rail bolt and washer—90 ft. lbs. (122 Nm)

17. Install the bridge assembly onto the crossmember and tighten the mounting nut to 50 ft. lbs. (68 Nm).

18. Install the outer tie rod ends on the steering knuckle and tighten the nuts to 38 ft. lbs. (52 Nm). Be sure to install a new cotter pin.

19. Install the front wheels and lug nuts. Tighten the lug nuts, in a star sequence, to 95 ft. lbs. (129 Nm). Lower the vehicle.

20. Reconnect the negative battery cable. Refill the power steering fluid reservoir with

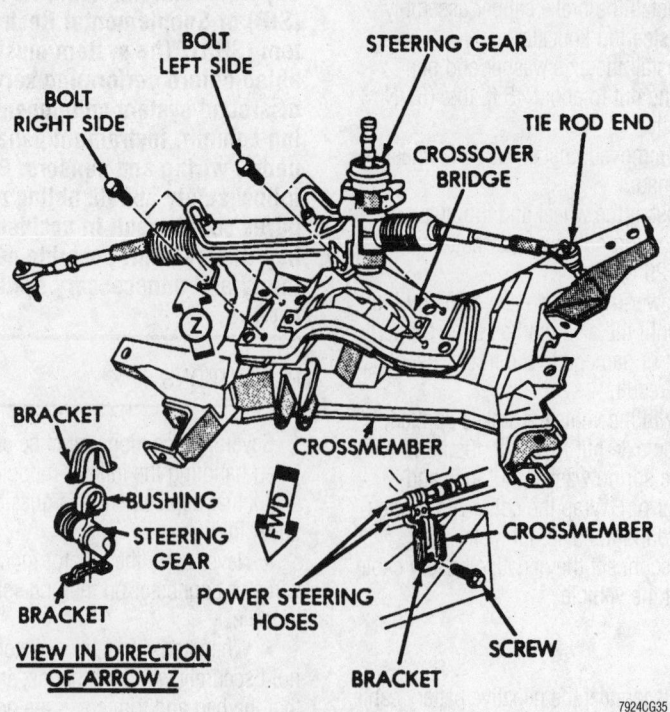

Exploded view of the power rack and pinion steering gear mounting—1995 AWD models

the correct amount of clean power steering fluid and bleed the system.

21. Check the toe setting and adjust, if necessary. Road test the vehicle and check steering operation.

1996–99 Models

1. With the ignition key in the **LOCK** position, turn the steering wheel to the left until the steering wheel locks itself in position.

2. Disconnect the negative battery cable.

3. Disconnect the steering column shaft coupler from the steering gear intermediate coupler.

4. Raise and safely support the vehicle.

5. Remove the front wheels.

6. Place a drain pan under the power steering fluid lines and disconnect the fluid hose from the metal tube portion of the power return line and allow the steering fluid to drain into the pan.

7. Remove the tie rod ends from the steering knuckles.

8. Remove the two bolts and loosen the third bolt, which mounts the ABS Hydraulic Control Unit (HCU) to the front suspension cradle. Rotate the HCU rearward to allow access to the cradle plate mounting nut and bolt just ahead of the HCU.

9. Remove the front suspension cradle plate from the front suspension cradle.

10. Remove the retaining bracket, which attaches the power steering fluid lines to the front suspension cradle.

11. Using an 18mm crowfoot, disconnect the power steering fluid pressure and return lines from the power steering gear.

12. Remove the three bolts and nuts mounting the rack and pinion steering gear to the suspension cradle.

13. Lower the steering gear from the suspension cradle enough to allow access to the steering column intermediate coupler roll pin. Remove the roll pin and separate the intermediate coupler from the steering gear shaft.

14. Remove the rack and pinion steering gear assembly from the front suspension cradle.

To install:

15. Install the rack and pinion steering gear assembly into the front suspension cradle, providing enough room to install the intermediate coupler.

16. Connect the steering gear shaft to the intermediate coupler and secure it with a roll pin.

17. Place the steering gear assembly in the correct position on the suspension cradle and install the three steering gear mounting bolts and nuts. Tighten the bolts and nuts to 100 ft. lbs. (136 Nm).

18. Connect the power steering fluid pressure and return lines to the correct fittings on the steering gear. Tighten the power steering fluid line tube fittings to 23 ft. lbs. lbs. (31 Nm).

19. Install the outer tie rod ends to the steering knuckles. Tighten the tie rod end-to-steering knuckle nuts to 40 ft. lbs. (54 Nm).

20. Install the front suspension cradle plate to the front suspension cradle. Tighten the 10 mounting bolts and nuts to 123 ft. lbs. (165 Nm).

21. Install the bracket mounting the power steering fluid lines to the suspension cradle. Be sure the protective heat shields cover the entire rubber hose hose-to-tube connection of both power steering fluid hoses.

22. Install the hose onto the metal tube portion of the power steering fluid return

line. Install a hose clamp on the return hose. Be sure the hose clamp is installed above the bead on the tube.

23. Install the front wheels and lug nuts. Tighten the lug nuts, in a star sequence, to 95 ft. lbs. (129 Nm).

24. Lower the vehicle just enough to access the interior of the vehicle.

25. Using the intermediate coupler, turn the front wheels to the left until the intermediate coupler shaft is correctly aligned with the steering column coupler. Connect the steering column shaft coupler to the steering gear intermediate coupler. Install the steering column coupler-to-intermediate shaft retaining pinch bolt. Tighten the pinch bolt nut to 21 ft. lbs. (28 Nm).

26. Reconnect the negative battery cable.

27. Refill power steering reservoir, with the correct amount of clean, fresh MOPAR® Power Steering Fluid or equivalent, and bleed the system.

28. Check the toe setting and adjust, if necessary. Road test the vehicle and check steering operation.

Strut

REMOVAL & INSTALLATION

Front

1. Raise and safely support the vehicle.

2. Remove the front wheel.

3. If equipped, remove the brake hose routing bracket and ABS speed sensor cable routing bracket from the strut damper brackets.

4. If equipped, remove the sway bar attaching link from the mounting bracket on the strut assembly. Hold the sway bar attaching link stud using a 6mm hex bit while removing the retaining nut with a box wrench.

5. Remove the steering knuckle-to-strut ssembly attaching bolts.

6. Remove the three nuts securing the strut assembly upper mount to the strut tower. Remove the strut assembly from the vehicle.

To install:

7. Position the strut assembly into the strut tower and loosely install the upper

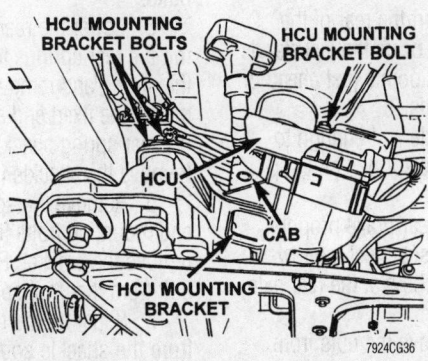

Remove the two HCU bracket mounting bolts and loosen the third to allow access to the cradle plate mounting nut—1996–99 models

washers and nuts. Tighten the three attaching upper mount nuts to 21 ft. lbs. (28 Nm).

8. Position the lower mount over the steering knuckle and loosely install the attaching bolts and nuts. If one of the attaching bolts is a cam bolt, the cam bolt must be installed in the lower slotted hole of the strut clevis bracket. Be sure that the attaching nuts face the front of the vehicle. Tighten the attaching nuts to 65 ft. lbs. (88 Nm) plus an additional ¼ turn.

9. If equipped, install the sway bar attaching link to the bracket on the strut assembly. Using a 6mm hex bit and crowfoot, tighten the sway bar link bracket attaching nut to 65 ft. lbs. (88 Nm).

10. If equipped, install the brake hose retaining bracket and ABS speed sensor cable routing bracket onto the strut assembly bracket. Tighten the mounting bolts to 10 ft. lbs. (13 Nm).

11. Install the front wheels and lug nuts. Tighten the lug nuts, in a star pattern sequence, to 95 ft. lbs. (129 Nm). Lower the vehicle and check the wheel alignment. Adjust, if necessary.

Shock Absorber

REMOVAL & INSTALLATION

Rear

1. Raise and safely support the vehicle.
2. Support the rear axle with a jack.
3. Remove the top and bottom shock absorber bolts.
4. Remove the shock absorbers.

To install:

5. Place the new shock in position and install the mounting bolts. For 1995 models, tighten the bolts to 80 ft. lbs. (108 Nm) for the lower bolts and 85 ft. lbs. (115 Nm) for the upper bolts. For 1996–99 models, tighten the shock absorber mounting bolts to 75 ft. lbs. (101 Nm).

6. Remove the jack supporting the rear axle and lower the vehicle to the ground.

Coil Spring

REMOVAL & INSTALLATION

Front

1. Remove the strut assembly from the vehicle.
2. Secure the strut assembly into a bench vise. Mount the strut assembly in the vertical position, clamping the strut assembly at the strut clevis bracket ONLY.

3. Using a coil spring compressor, compress the coil spring. It is required that the upper spring seat and second spring coil from the bottom be secured within the jaws of the spring compressor tool.

4. Using a wrench, hold the end of the strut shaft from rotating and remove the strut shaft nut.

5. Remove the upper strut mount from the strut assembly. Carefully remove the coil spring compressor tool from the coil spring.

6. Remove the upper spring seat and pivot bearing assembly from the coil spring.

7. Remove the coil spring from the strut. If the coil spring is being reused, mark the coil spring for correct reinstallation position.

To install:

8. With the strut assembly mounted firmly in the bench vise, install the coil spring on the strut assembly. Be sure the end of the coil spring's bottom coil aligns with the strut clevis bracket.

9. Install the upper spring seat onto the coil spring. Position the notch on the top of the spring seat in alignment with the strut clevis bracket.

10. Install the compressor tool onto the coil spring, and compress the spring.

11. Install the pivot bearing onto the top of the upper spring seat with the smaller diameter side of the bearing facing toward the spring seat.

12. Install the strut mount on the upper spring seat of the strut assembly. Loosely install the strut shaft nut.

13. Tighten the strut shaft nut to 70 ft. lbs. (94 Nm). This step is performed with the spring compressor tool still installed on the spring.

14. Check that the top of the coil spring is seated correctly against the upper spring seat. Carefully remove the spring compressor tool.

Rear

FRONT WHEEL DRIVE MODELS

1. Raise and support the rear of the vehicle on jackstands. Locate the jackstands under the frame contact points just ahead of the rear spring fixed ends.

2. Raise the rear axle just enough to relieve the weight on the springs and support on jackstands.

3. Disconnect the rear brake proportioning valve spring. Disconnect the lower ends of the shock absorbers at the rear axle bracket.

4. Loosen and remove the nuts from the U-bolts. Remove the washer and U-bolts.

5. Lower the rear axle assembly to permit the rear springs to hang free. Support the spring and remove the four bolts that mount the fixed end spring bracket. Remove the rear spring shackle nuts and plate. Remove the shackle from the spring.

6. Remove the spring. Remove the fixed end mounting bolts from the bracket and remove the bracket. Remove the front pivot bolt from the front spring hanger.

To install:

7. Install the spring on the rear shackle and hanger. Start the shackle nuts but do not completely tighten.

8. Assembly the front spring hanger on the spring. Raise the front of the spring and install the four mounting bolts. Tighten the mounting bolts to 45 ft. lbs. (61 Nm).

9. Raise the axle assembly and align the spring center bolts in correct position. Install the mounting U-bolts. Tighten the nuts to 60 ft. lbs. (82 Nm).

10. Install the rear shock absorber to the lower brackets.

11. Lower the vehicle to the ground so that the full weight is on the springs. Tighten the mounting components as follows: Front fixed end bolt, 95 ft. lbs. (130 Nm); Shackle nuts, 35 ft. lbs. (48 Nm); Shock absorber bolts, 50 ft. lbs. (68 Nm).

12. Raise and support the vehicle. Connect the brake valve spring and adjust the valve.

ALL WHEEL DRIVE MODELS

1. Raise and support the rear of the vehicle on jackstands. Locate the jackstands under the chassis, ahead of the springs.

2. Raise the rear axle just enough to relieve the weight on the springs and support on jackstands.

3. Disconnect the rear brake proportioning valve spring. Disconnect the lower ends of the shock absorbers at the rear axle bracket.

4. Loosen and remove the nuts from the U-bolts. Remove the washer and U-bolts.

5. Lower the rear axle assembly to permit the rear springs to hang free. Support the spring and remove the four bolts that mount the fixed end spring bracket. Remove the rear spring shackle nuts and plate. Remove the shackle from the spring.

6. Remove the spring. Remove the fixed end mounting bolts from the bracket and remove the bracket. Remove the front pivot bolt from the front spring hanger.

7. Separate the rear shackle plate from the shackle and pin assembly. Remove the shackle and pin assembly from the spring.

To install:

8. Assemble the shackle and pin assembly, bushing and shackle plate on rear of spring and spring hanger. Start the shackle and pin assembly through bolts, do not tighten.

9. Assemble the front spring hanger to the front of the spring eye and install pivot bolt and nut. Do not tighten.

➡ **The pivot bolt must installed inboard to prevent structural damage during spring installation.**

10. Raise the front of the spring into position and install the 4 hanger bolts, tighten them to 45 ft. lbs. (61 Nm). Connect the actuator assembly for the proportioning valve.

11. Raise the axle assembly into position, centered under the spring center bolt.

12. Install the U-bolts, nuts and washers. Tighten the U-bolt nuts to 65 ft. lbs. (88 Nm).

13. Install the shock absorbers and start the bolts.

14. Lower the vehicle to the ground, with the full weight of the vehicle on the wheels. Tighten all of the fasteners in the following sequence and to the listed values:

    a. Front pivot bolts—105 ft. lbs. (143 Nm)

    b. Shackle and pin assembly through-bolt nuts—35 ft. lbs. (48 Nm)

    c. Shackle and pin assembly retaining bolts—35 ft. lbs. (48 Nm)

    d. Shock absorber upper bolts—85 ft. lbs. (116 Nm)

    e. Shock absorber lower bolts—80 ft. lbs. (109 Nm)

15. Raise the vehicle and connect the rear brake proportioning valve spring.

Lower Ball Joint

REMOVAL & INSTALLATION

1995 Models

1. Raise and safely support the vehicle. Remove the front wheel.

2. Remove the lower control arm. Pry off the seal from the ball joint.

3. Position a receiving cup, or its equivalent to support the lower control arm.

4. Install a 1⅛ inch (28mm) deep socket over the stud and against the joint upper housing.

5. Press the ball joint assembly from the lower control arm.

To install:

6. Position the ball joint housing into the control arm cavity. Be sure the ball joint is not cocked in the control arm bore, or this will cause the ball joint to bind.

7. Position the assembly in a press.

8. Install the receiver cup special tool No. 6908–4 over the ball joint stud and down on the lower control arm assembly.

9. Carefully align the ball joint assembly. Using a press, apply pressure against the control arm assembly until the housing ledge of the ball joint assembly stops against the control arm cavity down flange.

10. To install a new seal, support the ball joint housing.

11. With a 1½ in. (38mm) socket, press the seal onto the joint housing with the seat against the control arm.

12. Install the control arm in the vehicle.

13. Install the front wheel and lug nuts. Tighten the lug nuts to 95 ft. lbs. (129 Nm).

14. Lower the vehicle. Check the wheel alignment.

1996–99 Models

➡ **Special Chrysler Tools 6758, 6908–4 and 6919 or equivalents, are required to remove and install the lower ball joint. An arbor press is also required to remove and install the ball joint.**

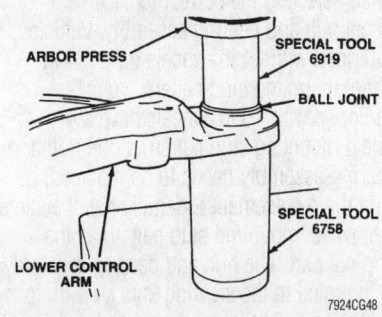

Using the special tools or equivalent size sockets, press the ball joint from the lower control arm—1996–99 models

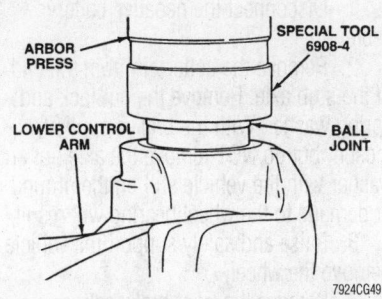

Using a press and the correct adapters, carefully press the new ball joint into the lower control arm—1996–99 models

1. Raise and safely support the vehicle. Remove the front wheel.

2. Remove the lower control arm. Using a flat blade tool, pry off the seal from the ball joint.

3. Position a receiving cup, special tool 6758 or its equivalent, to support the lower control arm.

4. Install remover, special tool 6919 or its equivalent, over the stud and against the joint upper housing.

5. Using an arbor press, press the joint assembly from the lower control arm.

To install:

6. Position the ball joint housing into the control arm cavity. Be sure the ball joint is not cocked in the control arm bore, or this will cause the ball joint to bind. The notch in the ball joint stud must face inward to the control arm to allow for clearance of the ball joint-to-steering knuckle clamp bolt.

7. Position the assembly in a press with special tool 6758 or its equivalent, supporting the control arm.

8. Position the ball joint installer special tool 6908–4 or its equivalent, on the bottom of the ball joint.

9. Carefully align the ball joint assembly. Using a press, apply pressure against the control arm assembly until the housing ledge of the ball joint assembly seats completely against the lower control arm surface and there is no gap between the lower control arm and ball joint. Be careful not to apply excessive force on the ball joint or the control arm.

10. Install a new seal boot as far as it will go on the ball joint by hand, first making sure the shield of the ball joint seal is facing outward from the end of the control arm.

11. Grease the ball joint using MOPAR® Multi-Mile grease, or equivalent. Do NOT over grease the ball joint or this will prevent the seal boot from being properly installed.

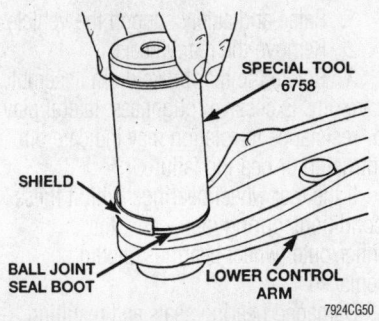

Install a new boot on the ball joint using tool No. 6758 or an equivalent socket—1996–99 models

12. To install the new seal, place special tool 6758 or equivalent, over the seal boot and align it squarely with the bottom edge of the seal boot. Apply hand pressure only, until the new seal boot is against the top surface of the control arm.

13. Install the control arm in the vehicle.

14. Install the front wheel and lug nuts. Tighten the lug nuts to 95 ft. lbs. (129 Nm).

15. Lower the vehicle. Check the wheel alignment.

16. Road test the vehicle.

Wheel Bearings

ADJUSTMENT

Front

The front wheel bearing is designed for the life of the vehicle and requires no type of adjustment or periodic maintenance. The bearing is a sealed unit with the wheel hub and can only be removed and/or replaced as an assembly.

If the wheel bearing is worn or damaged, the vehicle will produce vibration and noise. With the bearings loaded, the noise will generally change. Take the vehicle on a road test to determine the location of the worn/damaged bearing. Driving on a smooth, level road surface, accelerate the vehicle to a constant speed. Once the vehicle has reached a constant speed, swerve the vehicle back and forth from left to right. This will change the noise level by loading and unloading the wheel bearings. If the wheel bearing damage is slight, the noise will usually not be noticeable at speeds above 30 mph (48 km/h)

Rear

➡The following procedure is for the rear wheel bearings on the 1995 front wheel drive vehicles. All other bearings are serviced as a complete assembly and are not adjustable.

1. Raise and safely support the vehicle.
2. Remove the rear wheels.
3. Rotate the rear brake drum assembly carefully. Excessive roughness, lateral play or resistance to rotation may indicate dirt intrusion or bearing failure.

If the rear wheel bearings exhibit these conditions during inspection, the inner/outer wheel bearings should be replaced.

Damaged bearing seals and resulting excessive grease loss may also require bearing replacement. Moderate grease loss from the bearing is considered normal and

should not require replacement of the hub and bearing assembly.

If the wheel bearings are in good condition but the bearing grease is dirty, the wheel bearings may only need to be removed and cleaned with a good grease solvent. Then repack the wheel bearings using clean, fresh wheel bearing grease.

4. Adjust the wheel bearings as follows:

a. Remove the grease cap, cotter pin and nutlock. Loosen the adjusting nut.

b. Tighten the adjusting nut to 20–25 ft. lbs. (27–34 Nm) while rotating the brake drum.

c. Back off the adjusting nut ¼ turn (90 degrees).

d. Tighten the adjusting nut finger-tight.

e. Position the nutlock with one pair of slots in line with the cotter pin hole on the stub axle. Install a new cotter pin.

f. Clean and install the grease cap and wheel. Tighten the wheel lug nuts, in a star sequence, to 95 ft. lbs. (129 Nm).

5. Lower the vehicle.

REMOVAL & INSTALLATION

Front

The hub and wheel bearing unit is serviced as a complete assembly. Use care when selecting the correct replacement wheel hub and bearing assembly. Vehicles equipped with 14 in. wheels have a 4 in. wheel mounting stud pattern. Vehicles equipped with 15 in. wheels have a 4½ in. wheel mounting stud pattern. If the hub and bearing assembly needs to be replaced, be sure the replacement assembly has the same size wheel mounting stud pattern as the original part. The hub and bearing assembly is mounted to the steering knuckle with four mounting bolts that are removed from the rear of the steering knuckle. Replacement of the front drive hub and bearing assembly can be done without having to remove the steering knuckle from the vehicle.

1. Disconnect the negative battery cable.

2. Remove the cotter pin from the end of the stub axle. Remove the nut lock and spring washer. With the brakes applied, loosen, but do NOT remove the axle nut and washer with the vehicle still on the ground or damage to the wheel bearing will result.

3. Raise and safely support the vehicle. Remove the wheel.

4. Remove the front brake caliper assembly from the steering knuckle assembly and support it from the strut assembly using a strong piece of wire.

5. Remove the front brake rotor from the hub/bearing assembly.

6. Remove the retaining nut and washer from the halfshaft stub axle.

7. Separate the outer tie rod from the steering knuckle.

8. Remove the ABS wheel speed sensor from the steering knuckle.

9. Remove the wheel stop from the steering knuckle, if equipped.

10. Remove the nut and bolt that clamps the steering knuckle to the ball joint stud. Using a prybar, pry the control arm down to release the ball stud from the steering knuckle. Be careful not to tear the ball joint grease seal when prying down from the steering knuckle.

11. Separate the outer CV-joint splined shaft from the hub and bearing assembly by holding the CV-joint housing and pulling the steering knuckle away. Be careful not to damage the outer CV-joint wear sleeve or separate the inner CV-joint.

12. Support the halfshaft assembly at the CV-joint housing. Pull the knuckle assembly away from the halfshaft. Take care not to separate the halfshaft inner CV-Joint, or damage to the CV-joint will occur. Support the halfshaft.

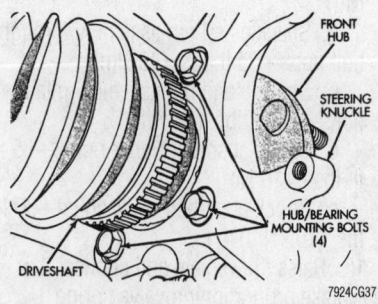

Four bolts are used to attach the hub assembly to the steering knuckle

✳✳ WARNING

The steering knuckle-to-strut assembly attaching bolts are of the serrated type and must NOT be turned during the removal procedure. Be sure to hold the bolts stationary in the steering knuckle while removing the nuts

13. Remove the two steering knuckle-to-strut clevis bracket mounting bolts, and remove the steering knuckle from the vehicle.

14. Remove the four hub and bearing assembly mounting bolts from behind the steering knuckle.

15. Remove the hub and bearing assembly from the steering knuckle.

To install:

16. Thoroughly clean the mating surfaces of the steering knuckle and the hub and bearing assembly of any foreign material or nicks so the surfaces are clean and smooth.

17. Install the new hub and bearing assembly, and tighten the mounting bolts in a crisscross pattern to 45 ft. lbs. (65 Nm). Be sure the hub and bearing assembly is seated squarely against the front steering knuckle.

➡**If equipped with eccentric strut assembly attaching bolts, it must be installed in the bottom (slotted) hole of the strut clevis bracket.**

18. Install the steering knuckle into the strut clevis bracket of the strut assembly. Install the strut-to-steering knuckle attaching bolts and tighten both attaching bolts to 65 ft. lbs. (90 Nm) plus an additional ¼ turn.

19. Thoroughly clean the steering knuckle and hub and bearing area of all debris and moisture, where the CV-joint will be installed into the steering knuckle. Also, thoroughly clean the bearing shield of the outer CV-joint.

20. Pull the front strut out and insert the splined outer CV-joint into the front hub.

21. Insert the ball joint stud into the steering knuckle clamp. Install a **new** steering knuckle-to-ball joint stud clamping nut and bolt. Be sure to use an exact replacement nut and bolt during installation. Tighten the bolt to 105 ft. lbs. (145 Nm).

22. Install the tie rod end into the steering knuckle. Tighten the tie rod end-to-steering knuckle nut to 45 ft. lbs. (61 Nm).

23. Install the disc brake rotor.

24. Install the brake caliper assembly onto the steering knuckle.

25. Install the axle washer and nut. Temporarily tighten the nut to about 45 ft. lbs. (61 Nm).

26. Install the ABS wheel speed sensor.

27. Install the wheel and lug nuts. Tighten the lug nuts, in a star pattern, to 95 ft. lbs. (129 Nm).

28. Lower the vehicle. Do NOT roll the vehicle until the axle nut has been properly tightened or damage to the front wheel bearings will result.

29. With the vehicle's brakes applied, tighten the axle nut to 180 ft. lbs. (244 Nm). Install the spring washer, nut lock and a new cotter pin. Wrap the cotter pin prongs tightly around the axle nut lock.

30. Reconnect the negative battery cable. Check the wheel alignment.

Rear

1995 FRONT WHEEL DRIVE VEHICLES

1. Raise and safely support the vehicle. Remove the rear wheel.

2. Remove the wheel grease cap, cotter pin, nutlock and bearing adjusting nut.

3. Remove the thrust washer and outer bearing.

4. Remove the drum from the spindle. Using a seal puller tool, remove the grease seal and inner bearing from the drum or hub. Discard the old grease seal.

5. Thoroughly clean the old lubricant from the bearings and hub cavity. Inspect the bearing rollers for pitting or other signs of wear that would indicate that replacement is necessary. Light discoloration is normal.

6. If the bearing cup requires replacement, remove the cup from the drum using a brass drift or suitable removal tool.

➡**Wheel bearings must be replaced as a set, both the cup and the bearing need to be replaced at the same time.**

To install:

7. Install the new bearing cup with a bearing cup installation tool, if necessary. Be sure the cup mounting area in the hub is clean and free from nicks and burrs that would keep the cup from completely seating in the hub.

8. Repack the bearings with high temperature multi-purpose EP grease and add a small amount of new grease to the hub cavity. Be sure to force the lubricant between all rollers in the bearing.

9. Install the new greased inner bearing into the grease coated hub and bearing cup. Install a new grease seal using an appropriate seal installation tool.

10. Install the drum on the spindle after coating the polished spindle surfaces with wheel bearing lubricant.

11. Install the outer wheel bearing, thrust washer and adjusting nut.

12. Tighten the adjusting nut to 20–25 ft. lbs. (27–34 Nm) while rotating the wheel.

13. Back off the adjusting nut ¼ turn (90 degrees).

14. Tighten the adjusting nut finger-tight.

15. Position the nutlock with one pair of slots in line with the cotter pin hole on the stub axle. Install a new cotter pin.

16. Clean and install the grease cap and wheel. Tighten the wheel lug nuts, in a star pattern, to 95 ft. lbs. (129 Nm).

17. Lower the vehicle.

1996–99 FRONT WHEEL DRIVE VEHICLES

1. Raise and safely support the vehicle.

2. Remove the rear wheel and brake drum.

3. Remove the ABS wheel speed sensor.

4. Remove the mounting bolts and the hub/bearing assembly.

To install:

5. Install the hub/bearing assembly on the axle. Tighten the four bolts in a crisscross pattern to 95 ft. lbs.(129 Nm).

6. Install the wheel speed sensor. Tighten the bolts to 105 inch lbs. (12 Nm).

7. Install the brake drum and wheel. Tighten the wheel nuts to 95 ft. lbs.(129 Nm).

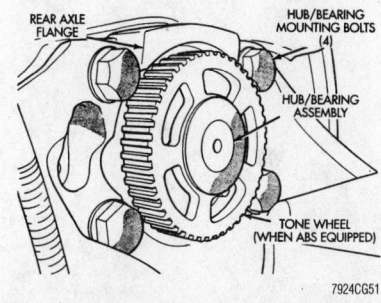

The rear hub/bearing assembly is attached to the axle by four bolts—1996–99 FWD models

ALL WHEEL DRIVE VEHICLES

The rear wheel bearings are serviced with the hub as a unit. The hub and bearing assembly bolts to the knuckle.

1. Raise and safely support the vehicle.

2. Remove the rear wheel. Remove the brake drum.

3. Remove the cotter pin, nut lock, spring washer and hub nut.

4. Remove the halfshaft flange retaining bolts and remove the halfshaft assembly.

5. Remove the hub assembly mounting bolts and remove the wheel bearing and hub assembly. Replace the grease seal.

To install:

6. Install the hub and bearing assembly. Tighten the bolts to 96 ft. lbs. (130 Nm) in a crisscross pattern.

➡**Thoroughly clean the seal and wear sleeve. Lubricate both before installation.**

7. Fully lubricate the seal and wear sleeve with a multi-purpose lubricant. Install the halfshaft assembly.

8. Install the washer and hub nut. With the brakes applied tighten the nut to 180 ft. lbs. (244 Nm).

9. Install the spring washer, nut lock and new cotter pin. Wrap the cotter pin prongs tightly around the nut lock.

10. Install the brake drum.

11. Install the rear wheel and lug nuts. Tighten the lug nuts, in a star pattern, to 95 ft. lbs. (129 Nm).

12. Lower the vehicle.

CHRYSLER CORP.

Dodge-Dakota • Durango • Ram Trucks • Ram Vans

GASOLINE ENGINE REPAIR

→Disconnecting the negative battery cable on some vehicles may interfere with the functioning of the on-board computer system, and may require the computer to undergo a relearning process once the negative battery cable is reconnected.

Distributor

REMOVAL

1. Remove the splash shield (if equipped).
2. If necessary, remove the air cleaner assembly and connecting tubes.
3. Unplug the pick-up lead wire connector(s) from the wiring harness.
4. Leaving the distributor wires connected, unfasten the clips or screws that retain the distributor cap and lift off the cap.

→If insufficient clearance is obtained by removing the cap with the wires still connected, all or some of the spark plug wires should be tagged and disconnected from the cap for better access.

5. Bump the engine around until the rotor is pointing at No. 1 cylinder firing position, with the timing marks on the front case and crank pulley are aligned. Disconnect the negative battery cable from the battery.
6. Mark the distributor body and the engine block to indicate the position of the distributor in the block. Mark the distributor body to indicate the rotor position. These marks are used as guides when installing the distributor.
7. Remove the distributor hold-down bolt and bracket. Carefully lift the distributor from the engine. The shaft may rotate slightly as the distributor is removed. Make a note of where the movement stops. That is where the rotor must point when the distributor is reinstalled into the block.

INSTALLATION

Engine Undisturbed

1. If the crankshaft has not been rotated while the distributor was removed from the engine, use the reference marks made before removal to correctly position the distributor in the block. The shaft may have to

be rotated slightly to engage the intermediate shaft gear (3.9L, 5.2L and 5.9L (OHV).
2. Install the distributor while holding the rotor in position, allowing it to move only enough to engage the slot in the drive gear.
3. Install pick-up coil leads and distributor cap. Ensure all high tension wires are firmly snapped in cap towers. Install distributor hold-down clamp screw and tighten to 200 inch lbs. (22.5 Nm).

Engine Disturbed

Perform this procedure if the crankshaft was rotated or otherwise disturbed (for example, during engine rebuilding) after the distributor was removed.

1. Rotate the crankshaft until the No. 1 cylinder is at Top Dead Center (TDC) of the compression stroke. The simplest way to do this is to remove the spark plug from the No. 1 cylinder and place your thumb over the hole. Slowly turn the engine by hand in the normal direction of rotation until compression is felt at the hole.
2. Ensure that the indicating mark on the crankshaft vibration damper is aligned to the 0 degree (TDC) mark on the timing chain cover.
3. Clean the distributor mounting at the engine and distributor base. Lightly oil the rubber O-ring seal on the distributor housing.
4. Hold the distributor over the mounting pad on the cylinder block so that the distributor body flange coincides with the mounting pad and the rotor points to the No. 1 cylinder firing position.
5. Install the distributor while holding the rotor in position, allowing it to move only enough to engage the slot in the drive gear.
6. Install pick-up coil leads and distributor cap. Ensure all high tension wires are firmly snapped in cap towers. Install distributor hold-down clamp screw and tighten to 200 inch lbs. (23 Nm).

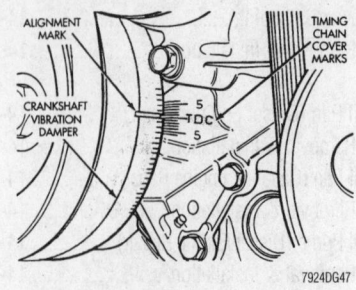

Rotate the crankshaft until the mark on the crankshaft puller and timing chain cover are aligned at Top Dead Center (TDC)

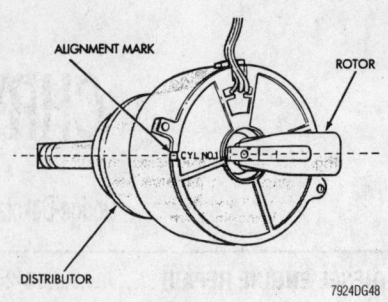

The alignment mark on the distributor is the No. 1 cylinder firing position

Ignition Timing

ADJUSTMENT

The ignition timing is automatically set by the Powertrain Control Module and is not adjustable.

Engine Assembly

❋❋ CAUTION

Observe all applicable safety precautions when working around fuel. Whenever servicing the fuel system, always work in a well ventilated area. Do not allow fuel spray or vapors to come in contact with a spark or open flame. Keep a dry chemical fire extinguisher near the work area. Always keep fuel in a container specifically designed for fuel storage; also, always properly seal fuel containers to avoid the possibility of fire or explosion.

REMOVAL & INSTALLATION

Ram Van

1. Mark hood hinge outlines on the hood and remove the hood from the vehicle.
2. Disconnect the battery terminals.
3. Remove the grille and support brace.
4. Remove the engine cover from inside the van.
5. Drain the coolant into a suitable container.

❋❋ CAUTION

When draining the coolant, keep in mind that cats and dogs are attracted by ethylene glycol antifreeze, and are quite likely to drink any that is left in an uncovered container or in puddles

on the ground. This will prove fatal in sufficient quantity. Always drain the coolant into a sealable container.

6. Remove the air cleaner.

✳✳ CAUTION

The EPA warns that prolonged contact with used engine oil may cause a number of skin disorders, including cancer! You should make every effort to minimize your exposure to used engine oil. Protective gloves should be worn when changing the oil. Wash your hands and any other exposed skin areas as soon as possible after exposure to used engine oil. Soap and water, or waterless hand cleaner should be used.

7. Depressurize the fuel injection system.
8. Have the A/C system properly discharged by an EPA certified technician.
9. If equipped, remove the transmission oil cooler.
10. Disconnect both radiator and heater hoses. Remove the radiator and set the fan shroud aside.
11. Remove the A/C condenser. Cap the open lines immediately with plastic wrap or duct tape to prevent moisture from entering the system.
12. Remove the power steering pump and air pump with the hoses attached and lay them aside.
13. Remove the washer bottle.
14. Disconnect and tag the vacuum lines.
15. Remove the A/C compressor and lay it aside with the hoses connected, if enough slack exists. If not, disconnect and plug the A/C hoses.
16. Disengage the throttle linkage and all electrical connection to the engine sensors. Tag all electrical connections before detaching them.
17. Remove the alternator, fan, pulley and drive belt(s).
18. If applicable, remove the distributor cap and wires.
19. Remove and plug the fuel line.
20. Remove the throttle body.
21. Remove the fuel rail assembly.
22. Remove the intake manifold.
23. Raise and safely support the vehicle.
24. Drain the engine oil and remove the oil filter.
25. Disconnect the exhaust pipe at the manifold.
26. Remove the starter.

27. Remove the bell housing bolts and inspection plate. Attach a C-clamp on the front bottom of the transmission torque converter housing to prevent the torque converter from coming out.
28. Remove the retaining bolts from the torque converter drive plate. Mark the converter and drive plate to aid in reassembly.
29. Remove the driveshaft and engine rear support.
30. Support the transmission with a suitable transmission jack.
31. Remove the transmission.
32. Remove the engine front mounts and insulators.
33. Using a boom hoist attached to the cylinder heads with the shortest hook-up possible, take up all tension and support the engine.
34. Carefully remove the engine from the front of the vehicle. Lift slowly and watch for any snagged or missed connections.

➡It may be necessary the raise the vehicle enough to keep the engine hoist horizontal.

35. Mount the engine on a suitable engine stand for further disassembly or service work.

To install:
36. Carefully guide the engine into the vehicle.
37. Install the engine front mounts and insulators.
38. Raise the vehicle.
39. Install the torque converter drive plate bolts.
40. Install the clutch or drive plate and the flywheel.
41. If equipped with an automatic transmission, install the transmission intact with the torque converter. Use the matchmarks to align the drive plate and torque converter mounting bolts.
42. Install the insulator on the bottom face of the transmission housing.
43. Install the driveshaft and engine rear support.
44. Install the starter.
45. Connect the exhaust pipe to the manifold.
46. Lower the vehicle.
47. If applicable, install the distributor cap and wires.
48. Reconnect the fuel line.
49. Install the alternator, fan, pulley and drive belt(s).
50. Detach the throttle linkage, heater and vacuum hoses and all electrical connections to the ignition, alternator, and all other electrical connections.

51. Install the A/C compressor with lines attached.
52. Install the power steering and air pumps.
53. Install the A/C condenser.
54. Install the radiator and fan shroud.
55. Install the crossover brace and grille.
56. Connect both radiator and heater hoses.
57. If equipped, install the transmission oil cooler.
58. Install the air cleaner.
59. Refill all fluids with the proper type and amount.
60. Evacuate, recharge and leak test the A/C system.

Except Ram Van

1. Disconnect the negative battery cable.
2. Remove the hood from the vehicle.
3. Drain the cooling system into a clean container. This coolant will be reused if not contaminated. If contaminated, have the coolant recycled.
4. On the Ram pick-up, remove the upper crossmember and top core support.
5. Remove the radiator, fan blade assembly and fan shroud.
6. Remove the air cleaner intake duct and if equipped, the upper air deflector.
7. Relieve the fuel system pressure using the recommended procedure.
8. Disconnect the accelerator cable, and, if equipped, the speed control actuator cable and transmission throttle valve cable.
9. Tag and disconnect all vacuum and heater hoses from the engine.
10. Remove the accessory drive belt(s).
11. On power steering pumps with the integral reservoir, it may be possible to remove the power steering pump and position it out of the way without disconnecting the hoses. Otherwise place a drain pan under the pump and disconnect the lines to the pump.
12. If equipped, remove and position the A/C compressor aside. Use wire or string to hold it out of the way. In some cases the refrigerant lines do not need to be disconnected. If the lines do need to be disconnected, cap them immediately to prevent moisture from entering the system.
13. Tag and disengage all wiring connections and if equipped, the engine block heater.
14. Disconnect the fuel lines from the fuel injection supply manifold, on some engines the upper intake manifold will need to be removed first. If the intake was removed, plug the open intake manifold holes with clean shop rags so that nothing (i.e. loose nuts or bolts) can fall down them.

15. Raise and safely support the vehicle.

16. Disconnect the exhaust system from the exhaust manifold.

17. Remove the starter motor.

18. On vehicles equipped with automatic transmissions, remove the transmission cooler hoses from the block mounted clip, if equipped. Also remove the transmission inspection cover, torque converter bolts and transmission-to-engine block bolts. Push the torque converter away from the flexplate toward the transmission so it does not come off with the engine.

19. On vehicles equipped with manual transmissions, remove the bell housing bolts and separate the bell housing from the engine assembly.

20. Support the front of the transmission with a jack. On automatic transmissions, place a block of wood on the jack to protect the oil pan.

21. Remove the right-hand and left-hand engine mount bolts .

22. Lower the vehicle.

❊❊ WARNING

Do not lift the engine by the intake manifold.

23. Install the engine lifting bracket or chain on the engine. Carefully examine the engine for anything that may interfere with removal and remove or disconnect as needed.

24. Slowly lift the engine out of the vehicle.

25. Mount the engine on an engine work stand for disassembly.

To install:

26. Carefully lower the engine into the vehicle and install the engine mount bolts. Tighten the bolt to 70 ft. lbs. (95 Nm).

27. Install the transmission assembly into the vehicle.

28. Install the starter motor.

29. Connect the exhaust system to the manifolds.

30. Connect the fuel lines and if removed, install the upper intake manifold.

31. Attach all wiring harness connections and if equipped, the engine block heater.

32. If removed, install the A/C compressor and power steering pump.

33. Connect all vacuum lines and hoses.

34. Connect the accelerator cable, and, if equipped, the speed control actuator cable and transmission throttle valve cable.

35. Install the radiator, fan blade assembly and fan shroud.

36. Install the engine air cleaner intake duct and if equipped, the upper air deflector.

37. Refill the cooling system.

38. Install the hood, then connect the negative battery cable.

39. Check fluid levels and add as needed.

40. Start the engine and check for leaks. Bleed the cooling system as needed.

41. The A/C system should be evacuated and recharged by an EPA-certified mechanic utilizing a refrigerant recovery/recycling machine.

Water Pump

REMOVAL & INSTALLATION

2.5L (SOHC) Engine

1. Disconnect the negative battery cable.

➡**When removing or installing the constant tension hose clamps from vehicles so equipped, use only the correct clamp tool, such as the Snap-On No. HPC-20, or equivalent.**

2. If the vehicle is equipped with air conditioning, remove the compressor from the bracket and position it to the side.

3. Raise the vehicle and support safely, if necessary and remove the alternator and bracket. Remove the pulley from the water pump.

4. Disconnect the lower radiator hose and heater hose from the water pump.

5. Remove the water pump housing attaching screws and remove the assembly from the vehicle. Discard the O-ring.

6. Remove the water pump from the housing.

To install:

7. Clean the mating surfaces prior to sealing the water pump.

➡**This component is subjected to constant high pressure from hot fluid and must be sealed correctly or it will leak.**

8. Using a new gasket or silicone sealer, install the water pump to the housing.

9. Install a new O-ring to the housing and install to the engine. Tighten the bolts to 21 ft. lbs. (30 Nm).

10. Install the water pump pulley. Connect the radiator hose and heater hose to the water pump.

11. Install all items removed to gain access to the water pump and adjust the belt(s).

12. Remove the hex-head plug on the top of the thermostat housing. Fill the radiator with coolant until the coolant comes out the plug hole. Install the plug or valve and continue to fill the radiator.

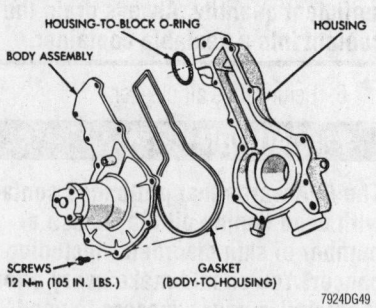

Exploded view of the water pump—2.5L (SOHC) engine

13. Connect the negative battery cable, run the vehicle until the thermostat opens, fill the radiator completely and check for leaks.

14. Once the vehicle has cooled, recheck the coolant level.

2.5L (OHV) Engine

➡**Be aware that on the 2.5L (OHV) engine, the impeller rotates in a counterclockwise direction. Check on the impeller for the letter R stamped on the blade. The use of a water pump from previous year engines will cause overheating.**

1. Disconnect the negative battery cable.

2. Drain the coolant into a suitable container.

3. Remove the drive belt.

4. Remove the power steering pump.

5. Remove the lower radiator hose and heater hose from the water pump connections.

➡**When removing or installing the constant tension hose clamps from vehicles so equipped, use only the correct clamp tool, such as the Snap-On No. HPC-20, or equivalent.**

6. Remove the four mounting bolts and remove the pump from the engine. Note that one of the mounting bolts is longer than the others.

To install:

7. Clean the mating surfaces of all dirt and gasket material.

8. If the pump is being replaced, remove the heater hose tube from the old pump, wrap the threads with Teflon® tape and install the pipe in the new pump.

9. Install the water pump and new gasket. Tighten the bolts to 22 ft. lbs. (30 Nm).

10. Install the water pump pulley.

11. Connect the lower radiator hose and heater hose to the water pump.

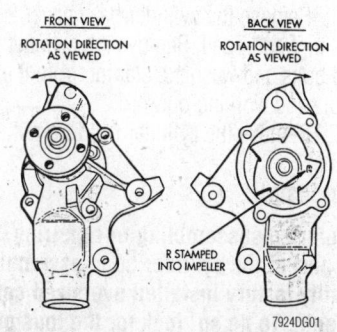

The R stamped on the impeller denotes a water pump which is designed to rotate counterclockwise—2.5L (OHV) engine

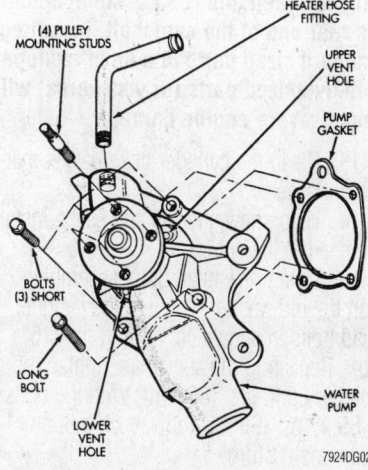

Water pump assembly—2.5L (OHV) engine

12. Install the power steering pump.
13. Install the drive belt.
14. Properly fill the cooling system.
15. Connect the negative battery cable.

3.9L, 5.2L and 5.9L Engines

1. Disconnect the negative battery cable.
2. Drain the coolant into a suitable container.

→**When removing or installing the constant tension hose clamps from vehicles so equipped, use only the correct clamp tool, such as the Snap-On No. HPC-20, or equivalent.**

3. Relax the tension on the tensioner pulley by rotating it clockwise. (Rotate it counterclockwise on 5.9L (HDC engine only). Remove the drive belt.
4. Disconnect the upper radiator hose from the radiator.
5. Disconnect the thermal clutch from the water pump shaft using Snap-On 36mm fan wrench SP346, or equivalent on the thermal clutch nut and a prybar between the water pump pulley bolts.

6. Remove the four fan shroud bolts. Ram vans have a 2-piece shroud. Remove the two attaching bolts from the middle of the shroud.
7. Remove the shroud and thermal clutch with the fan at the same time.

→**To prevent silicone fluid from draining into the drive bearing and ruining the lubricant, Chrysler Corporation recommends that you do not place the thermostatic fan drive unit with the shaft pointing downward.**

8. Remove the four bolts attaching the water pump pulley to the pump.
9. Disconnect the lower radiator hose and heater hoses from the water pump.
10. Remove the seven water pump bolts attaching the pump to the engine. Remove the pump.

To install:

→**This component is subjected to constant high pressure from hot fluid and must be sealed correctly or it will leak.**

11. Clean the mating surfaces prior to sealing the water pump.
12. Transfer the coolant return tube, with a new O-ring installed, to the replacement pump.
13. Install the water pump on the engine. Use a new gasket coated with sealer. Tighten the bolts to 30 ft. lbs. (41 Nm).
14. Reconnect all hoses to the water pump.
15. Install the water pump pulley. tighten the bolts to 20 ft. lbs. (27 Nm).
16. Install the drive belt.
17. Install the shroud and fan assembly. Tighten the thermal clutch nut 42 ft. lbs. (57 Nm).
18. Position the shroud and tighten the mounting bolts 50 inch lbs. (6 Nm).
19. Connect the upper radiator hose to the radiator.
20. Properly fill the cooling system.
21. Connect the negative battery cable.

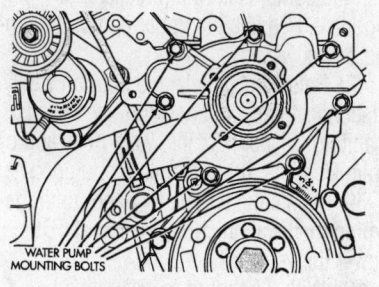

Water pump mounting bolt locations—3.9L, 5.2L and 5.9L engines, 8.0L engine is similar

8.0L Engine

1. Disconnect the negative battery cable.
2. Drain the coolant into a suitable container.
3. Remove the washer bottle from the fan shroud and disconnect the fan shroud but do not remove it from the radiator.
4. Remove the upper radiator hose from the radiator.

→**When removing or installing the constant tension hose clamps from vehicles so equipped, use only the correct clamp tool, such as the Snap-On No. HPC-20, or equivalent.**

5. Disconnect the thermal clutch from the water pump shaft using Snap-On 36mm fan wrench SP346, or equivalent on the thermal clutch nut and a prybar between the water pump pulley bolts.

→**To prevent silicone fluid from draining into the drive bearing and ruining the lubricant, Chrysler Corporation recommends that you do not place the thermostatic fan drive unit with the shaft pointing downward.**

6. Relax the tension on the tensioner pulley by rotating it counterclockwise. Remove the drive belt.
7. Remove the four water pump pulley-to-water pump hub bolts and remove the pulley from the vehicle.
8. Remove the lower radiator hose from the water pump.
9. Remove the heater hose at the water pump fitting.
10. Remove the seven water pump mounting bolts.
11. Loosen the clamp at the water pump end of the bypass hose. Slip the hose from the water pump while removing the pump from the vehicle. Do not remove the clamp from the bypass hose.
12. Discard the water pump-to-timing chain/case/cover O-ring seal.
13. Remove the heater hose fitting from the water pump if the pump replacement is necessary. Note the position (direction) of the fitting before removal. The fitting must be installed in the same position.

→**Do not disconnect any refrigerant lines from the compressor.**

To install:

❊❊ WARNING

This component is subjected to constant high pressure from hot fluid and must be sealed correctly or it will leak.

14. Clean the mating surfaces prior to sealing the water pump.

15. Install the water pump on the engine. Use a new gasket coated with sealer.

16. Install the water-pump-to-compressor front bracket mounting bolts. Tighten the bolts to 30 ft. lbs. (41 Nm).

17. If a new pump has been installed, then install the heater hose fitting to the pump. Tighten the fitting to 144 inch lbs. (16 Nm). After the fitting is tightened, position it as shown in the drawing. When positioning the fitting, do not back it off (rotate counterclockwise). Use a suitable Teflon® containing thread sealant. Refer to the directions on the package.

18. Clean the O-ring groove and install a new O-ring.

19. Apply a small amount of petroleum jelly to the O-ring to help it stay in place on the water pump.

20. Install the water pump to the engine as follows:

  a. Guide the pump fitting bypass hose as the hose is being installed.

  b. Install the water pump and tighten them to 30 ft. lbs. (40 Nm).

21. Position the bypass clamp on the hose.

22. Spin the water pump to be sure that the pump impeller does not rub against the timing chain case/cover.

23. Connect the radiator lower hose to the to the water pump.

24. Connect the heater hose and hose clamp to the heater hose fitting.

25. Install the water pump pulley. Tighten the bolts to 16 ft. lbs. (22 Nm). Place a prybar between the water pump pulley bolts to prevent the pulley from relaxing.

26. Install the serpentine belt.

27. Position the fan shroud assembly and fan blade/viscous fan drive assembly as a complete unit.

28. Install the fan shroud to the radiator. Tighten the bolts to 50 inch lbs. (6 Nm).

29. Install the fan blade and viscous fan drive to the water pump shaft.

30. Fill the cooling system, connect the negative battery cable and check for leaks.

Cylinder Head

REMOVAL & INSTALLATION

✱✱ CAUTION

Observe all applicable safety precautions when working around fuel. Whenever servicing the fuel system, always work in a well ventilated area. Do not allow fuel spray or vapors to come in contact with a spark or open flame. Keep a dry chemical fire extinguisher near the work area. Always keep fuel in a container specifically designed for fuel storage; also, always properly seal fuel containers to avoid the possibility of fire or explosion.

2.5L (SOHC) Engine

1. Relieve the fuel system pressure.

2. Disconnect the negative battery cable and unbolt it from the head.

3. Drain the cooling system. Remove the dipstick bracket nut from the thermostat housing.

4. Remove the air cleaner assembly. Remove the upper radiator hose and disconnect the heater hoses.

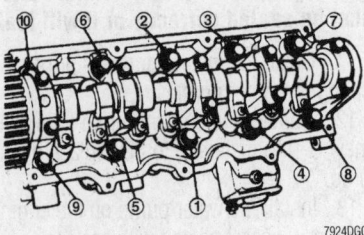

Cylinder head tightening sequence—2.5L (SOHC) engine

5. Detach and label the vacuum lines, hoses and wiring connectors from the manifold(s), throttle body and from the cylinder head. Remove the air pump, if equipped.

6. Disconnect the all linkages and the fuel line from the throttle body. Unbolt the cable bracket. Remove the ground strap attaching screw from the firewall.

7. If equipped with air conditioning, remove the upper compressor mounting bolts. The cylinder head can be removed with the compressor and bracket still mounted. Remove the upper part of the timing belt cover.

8. Raise the vehicle and support safely. Disconnect the converter from the exhaust manifold. Disconnect the water hose and oil drain from the turbocharger, if equipped.

9. Rotate the engine by hand, until the timing marks align (No. 1 piston at TDC). Lower the vehicle.

10. With the timing marks aligned, remove the camshaft sprocket. The camshaft sprocket can be suspended to keep the timing intact. Remove the spark plug wires from the spark plugs.

11. Remove the cylinder head cover and curtain, if equipped. Remove the cylinder head bolts and washers, starting from the middle and working outward.

12. Remove the cylinder head from the engine.

To install:

➡ Before disassembling or repairing any part of the cylinder head assembly, identify factory installed oversized components. To do so, look for the tops of the bearing caps painted green and O/SJ stamped rearward of the oil gallery plug on the rear of the head. In addition, the barrel of the camshaft is painted green and O/SJ is stamped onto the rear end of the camshaft. Installing standard sized parts in a head equipped with oversized parts (or visa versa) will cause severe engine damage.

13. Clean the cylinder head gasket mating surfaces.

14. Using new gaskets and seals, install the head to the engine.

15. Using new head bolts assembled with the old washers, tighten the cylinder head bolts in sequence, to 45 ft. lbs. (61 Nm). Repeating the sequence, tighten the bolts to 65 ft. lbs. (88 Nm). With the bolts at 65 ft. lbs. (88 Nm), turn each bolt an additional ¼ turn.

➡ Head bolt diameter is 11mm. These bolts are identified with the number 11 of the head of the bolt. The 10mm bolts used on some earlier vehicles will thread into an 11mm bolt hole, but will permanently damage the cylinder block. Ensure the correct bolts are being used when replacing old head bolts.

16. Install the timing belt.

17. Install or connect all items that were removed or disconnected during the removal procedure.

18. Refill the cooling system. Connect the negative battery cable. Start the engine and check for leaks.

2.5L (OHV) Engine

1. Disconnect the negative battery cable.

✱✱ CAUTION

Fuel injection systems remain under pressure, even after the engine has been turned OFF. The fuel system pressure must be relieved before disconnecting any fuel lines. Failure to do so may result in fire and/or personal injury.

2. Properly relieve the fuel system pressure, as required.

3. Drain the cooling system.

4. Disconnect the hoses at the thermostat housing.

5. Remove the air cleaner.

6. Disconnect the Crankcase Ventilation (CCV) vacuum hose and the fresh air inlet hose from the cylinder head cover.

7. Remove the cylinder head cover retaining bolts and remove the cylinder head cover. Use care. The plastic-like composite cylinder head cover used on many versions of this engine is easily damaged. Clean the cover with a suitable solvent and inspect the lower part of the cover. There may an indication PRY HERE showing a reinforced area where a putty knife can be inserted to help break the seal so the cover can be removed.

➡**On some late-model engines, the rocker arm cover (cylinder head cover) has a cured gasket attached to it. This gasket should not be removed. If sections of the gasket are missing or are compressed, replace the cover. However, minor damage such as small cracks, cuts or chips can be repaired with liquid gasket material.**

8. Remove the rocker arms and the pushrods. Any valvetrain parts that are to be reused must be returned to their original locations. Keep them in their original order for installation reference.

9. Remove the accessory drive belt(s).

10. If equipped with A/C, proceed as follows:

a. Remove the compressor and position it aside with the lines attached.

b. Remove the compressor bracket bolts from the cylinder head.

c. Loosen the throughbolt at the bottom of the bracket.

11. If equipped, remove the power steering pump and mounting bracket. Suspend the pump aside. Do not disconnect the pump hoses.

12. Disconnect the fuel lines.

13. Remove the intake and exhaust manifolds.

14. Remove the spark plugs and wires. Be sure to label each spark plug wire to ensure correct installation.

15. Disconnect the temperature sending unit wire.

16. Remove the ignition coil and bracket assembly.

17. Remove the cylinder head bolts in the reverse order of the installation torque sequence.

18. Lift the head from the engine and remove the head gasket. Prevent any foreign

material from entering the engine by stuffing clean, lint-free shop towels into the cylinder bores.

19. Thoroughly clean the gasket mating surfaces. Remove all traces of old gasket material. Remove all carbon deposits from the combustion chambers. Lay a straightedge across the head and check for flatness. Total deviation should not exceed 0.008 in. (0.20mm).

To install:

20. Fabricate two cylinder head alignment dowels from used head bolts. Using the longest bolts, trim the hex head off and cut slots for a large screwdriver into the top.

21. Install one dowel in cylinder head bolt hole No. 10 and the other dowel in bolt hole No. 8.

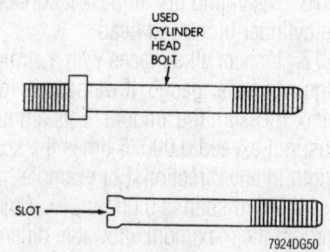

Fabricate two alignment dowels out of used cylinder head bolts—2.5L (OHV) engine

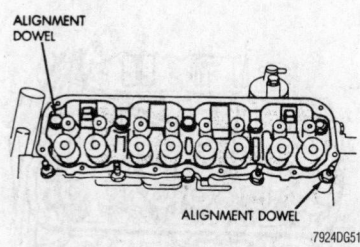

To properly position the cylinder head during assembly, temporarily install the alignment dowels in the corner bolt holes as shown—2.5L (OHV) engine

➡**Cylinder head bolts should be reused only once. Replace the head bolts which were previously used or are marked with paint. If head bolts are to be reused, mark each head bolt with paint for future reference. Head bolts should be installed using sealer. There were reports of head bolt breakage on this engine. New replacement head bolts are always recommended.**

22. Remove the shop towels from the cylinder bores and coat each cylinder bore with clean engine oil.

23. Install the new head gasket into position on the engine over the dowels. Be sure that the numbers on the gasket are facing up.

➡**Do not apply sealer as the cylinder head gaskets are of a composition type.**

24. Install the cylinder head into position on the engine over the gasket and dowels.

25. On cylinder head bolt No. 7 only, coat the threads with Loctite® PST sealant or equivalent.

26. Install all of the head bolts except for bolts No. 8 and 10.

27. Remove the dowels and install No. 8 and No. 10 head bolts.

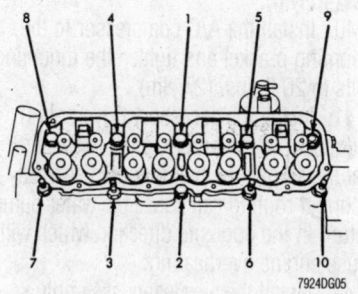

Tighten the cylinder head bolts in the sequence shown—2.5L (OHV) engine

28. Tighten the head bolts in the proper sequence to the following torque specifications:

a. Tighten all bolts to 22 ft. lbs. (30 Nm).

b. Tighten all bolts to 45 ft. lbs. (61 Nm).

c. Tighten bolts Nos. 1 through 6 to 110 ft. lbs. (150 Nm)

d. Tighten bolt No. 7 to 100 ft. lbs. (136 Nm).

e. Tighten bolts Nos. 8 through 10 to 110 ft. lbs. (150 Nm)

29. Check the head bolts in sequence to verify correct torque. If not already done, clean and dab a small coat of paint on the head of each bolt after tightening. This indicates that the bolts have now been tightened and if further service work is required in the future, indicates that the bolts should be replaced with new service replacement parts.

30. Install the ignition coil and bracket assembly.

31. Connect the temperature sending unit wire.

32. Install the spark plugs and tighten to 27 ft. lbs. (37 Nm).

33. Reconnect the spark plug wires. Be sure that the wires are connected to the correct spark plug.

34. Install the intake and exhaust manifolds.

35. Reconnect the fuel lines.

36. If equipped, install the power steering pump and bracket.

37. Install the pushrods and rocker arms in their original positions. Tighten the rocker arm bolts, alternately and evenly, to avoid damaging the bridge. Tighten the rocker arm bolts to 21 ft. lbs. (28 Nm).

38. Install the rocker arm cover. Tighten the rocker arm cover bolts to 115 inch lbs. (13 Nm).

39. Install the A/C compressor mounting bracket to the cylinder head and engine block. Tighten the mounting bolts to 30 ft. lbs. (40 Nm).

40. Install the A/C compressor to the mounting bracket and tighten the mounting bolts to 20 ft. lbs. (27 Nm).

41. Install the accessory drive belt(s) and adjust to the proper tension. Be sure that the serpentine belt is routed correctly. Incorrect routing can cause the water pump to turn in the opposite direction which will cause engine overheating.

42. Install the air cleaner assembly.

43. Reconnect the CCV vacuum hose and the fresh air inlet hose to the rocker arm cover.

44. Connect the hoses at the thermostat housing.

45. Fill the cooling system.

46. Change the engine oil and filter.

47. Connect the negative battery cable.

48. Adjust the automatic transmission throttle linkage and cable, if equipped.

49. Run the engine to normal operating temperature. Check for leaks and proper engine operation.

3.9L Engine

1. Relieve the fuel pressure.

2. Disconnect the negative battery cable.

3. Drain the cooling system.

4. Raise and safely support the vehicle.

5. Disconnect the exhaust pipe from the manifolds.

6. Remove the alternator if the right head is being removed and the air pump and negative battery cable if the left head is being removed.

7. Remove the air cleaner assembly. Unbolt the air conditioning compressor and lay it to the side, if equipped. Remove the distributor cap with all wires attached.

8. Disconnect all wires, hoses, linkages and cables from the throttle body. Disconnect and plug the fuel line.

9. Detach the ignition coil, coolant temperature sending unit wire and all other connectors along the wiring harness connected to items on the intake manifold.

10. Disconnect the heater hose, upper radiator hose and the lower bypass hose clamp.

11. Remove the cylinder head covers.

12. Remove the intake manifold assembly. Remove the exhaust manifolds.

13. Remove the rocker arms.

14. Remove the pushrods and identify them to ensure installation in their original locations.

15. Remove the head bolts and remove the cylinder head(s).

To install:

16. Clean and dry all gasket surfaces of the cylinder block and head.

17. Inspect all surfaces with a straightedge and feeler gauge. If warpage is indicated, measure the amount. This amount must not exceed 0.00075 times the span length in any direction. For example, if a 12 in. (305mm) span is 0.004 in. (0.10mm) warped, the maximum allowable difference is 12 x 0.00075 = 0.009 in. (305 x 0.00075 = 0.22875mm). In this case, the head is within limits. If the warpage exceeds the specified limits, either replace the head or lightly machine the head gasket surface.

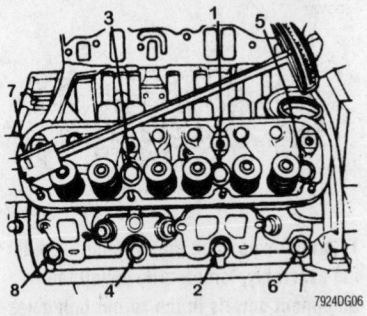

Cylinder head tightening sequence—3.9L engine

18. Using no sealer whatsoever, install the new head gasket(s) to the block. Clean, dry and lightly oil all head bolts threads. Install the heads and install the head bolts.

19. Tighten the head bolts in sequence to 50 ft. lbs. (68 Nm). Repeat the sequence retightening the bolts to a final torque of 105 ft. lbs. (143 Nm) and repeat the second step to ensure that all bolts are accurately tightened.

20. Assemble the rockers. Tighten the bolts evenly and gradually to 21 ft. lbs. (28 Nm).

21. Clean and dry the intake manifold contact surfaces. Coat the intake manifold side gaskets very lightly with sealer and install the gaskets to the heads. Cut-outs at the front of the gaskets differentiate the right and left sides.

22. Apply a thin uniform coat of quick dry cement to the front and rear intake manifold gaskets and mounting surfaces on the block and apply a thin bead of sealer to each of the four corners. Install the front and rear gaskets engaging the hole in the block and the tangs from the head gaskets. Apply a second thin bead of sealer above the gaskets in the four corners.

23. Carefully lower the intake manifold into position engaging the bypass hose; after it is satisfactorily in place, inspect the gasket to ensure they have not become dislodged.

24. Install the intake manifold bolts and tighten in sequence to 25 ft. lbs. (34 Nm). Repeat the sequence retightening the bolts to a final torque of 40 ft. lbs. (54 Nm) and repeat the second step to ensure all bolts are accurately tightened.

25. Install the exhaust manifold(s) and tighten the bolts to 20 ft. lbs. (27 Nm). Tighten the end nuts to 15 ft. lbs. (20 Nm).

26. Clean and dry the cylinder head cover mating surfaces, bolts and bolt holes. Install the cylinder head covers each with a new gasket.

27. Connect the heater hose, upper radiator hose and the lower bypass hose clamp.

28. Reattach the ignition coil, coolant temperature sending unit wire and all other connectors that were detached along the wiring harness.

29. Install the air conditioning compressor, if equipped. Install the distributor cap and all spark plug wires.

30. Install the alternator, battery ground and air pump, if they were removed.

31. Connect all wires, hoses, cables and the fuel line to the throttle body. Install the air cleaner assembly.

32. Raise the vehicle and safely support. Connect the exhaust pipe to the manifolds.

33. Fill the cooling system.

34. Connect the negative battery cable.

5.2L and 5.9L Engines

1. Drain the cooling system and disconnect the negative battery cable.

2. Remove the alternator, air cleaner, fuel line and any obstructing heat shields.

3. Disconnect the accelerator linkage.

4. Remove the vacuum lines from the throttle body. On vehicles with the 5.9L engine, remove the battery.

5. Remove the distributor cap and wires as an assembly.

6. Disconnect the coil wires, water temperature sending unit, heater hoses, and bypass hose. On the 5.9L engine, remove the distributor.

7. Remove the closed ventilation system, the evaporative control system, and the cylinder head covers.

8. Remove the intake manifold.

9. Remove the exhaust manifolds.

10. On 5.9L engines, tag the center bolts.

11. Remove the tappet chamber cover. Remove the rocker and shaft assemblies.

12. Remove the pushrods and keep them in order to ensure installation in their original locations. On the 5.9L engine, remove the water pump-to-head bolts.

13. Remove the head bolts from each cylinder head and remove the cylinder heads.

14. Clean all the gasket surfaces of the engine block and the cylinder heads. Install the spark plugs.

15. Inspect all surfaces with a straightedge and feeler gauge. If warpage is indicated, measure the amount. This amount must not exceed 0.00075 times the span length in any direction. For example, if a 12 in. (305mm) span is 0.004 in. (0.10mm) warped, the maximum allowable difference is 12 x 0.00075 = 0.009 in. (305 x 0.00075 = 0.22875mm). In this case, the head is within limits. If the warpage exceeds the specified limits, either replace the head or lightly machine the head gasket surface.

To install:

16. Coat new cylinder head gaskets with sealer, install the gaskets and install the cylinder heads.

17. Install the cylinder head bolts. Tighten the cylinder head bolts to 105 ft. lbs. (143 Nm) in the sequence indicated. Repeat this sequence to retighten all the cylinder head bolts to specifications.

18. Install the pushrods.

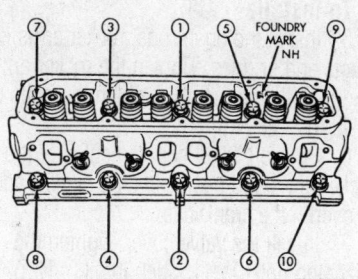

Tighten the cylinder head bolts in the sequence shown to ensure complete sealing and correct gasket crush—5.2L and 5.9L engines

19. On the 5.9L engine, install the water pump-to-head bolts.

✳✳ WARNING

To avoid valve to piston contact, ensure the piston in that cylinder is not at TDC while tightening the rocker arms.

20. Install the rocker arms. Tighten the bolts to 21 ft. lbs. (28 Nm).

21. Install the exhaust manifolds.

22. Install the intake manifold.

23. Install the PCV system, the evaporative control system and the cylinder head covers.

24. Connect the coil wires, water temperature sending unit, heater hoses, and bypass hose. On the 5.9L engine, install the distributor.

25. Install the distributor cap and wires as an assembly.

26. Install the vacuum lines to the throttle body. On vehicles with the 5.9L engine, install the battery.

27. Connect the accelerator linkage.

28. Install the alternator, air cleaner, and fuel line.

29. Fill the cooling system and connect the negative battery cable.

8.0L Engine

1. Disconnect the negative battery cable.

2. Drain the cooling system.

3. Remove the heat shields.

4. Remove the intake manifold-to-alternator bracket support rod. Remove the alternator.

5. Disconnect the accelerator linkage.

6. Remove the closed ventilation system, the evaporative control system, and the cylinder head covers.

7. Disconnect the evaporation control system.

8. Remove the air cleaner.

9. Relieve the fuel system pressure.

10. Disconnect the throttle linkage and the speed control and transmission kickdown cables, if equipped.

11. Remove the coil pack and bracket.

12. Disconnect the coil wires.

13. Disconnect the heat indicator sending unit wire.

14. Disconnect the heater hoses and bypass hose.

15. Remove the upper intake manifold and throttle body as an assembly.

16. Remove the cylinder head covers and (reusable) gaskets.

17. Remove the EGR tube. Discard the gasket for right side only.

18. Remove the lower intake manifold. Discard the flange side gaskets and the front and rear crossover gaskets.

19. Disconnect the exhaust pipe from the exhaust manifold.

20. Remove rocker arm assemblies and pushrods. Organize them so as to be able to install them in the exact same location.

21. Remove the head bolts from each cylinder head and remove the cylinder heads. Discard the gasket.

22. Remove the spark plugs.

23. Clean all the gasket surfaces of the engine block and the cylinder heads.

24. Inspect all surfaces with a straightedge. If warpage is indicated, measure the amount. The out-of-flatness specifications are 0.0004 in. per 1 in. (0.0007mm per 1mm) or 0.005 in. per 6 in. (0.127mm per 152mm) in any direction, or 0.010 in. (0.254mm) overall across the head. If exceeded, either replace the head, or machine the surface to skim it flat. The cylinder head surface finish should be 15–80 microinches (1.78–4.57 microns).

To install:

25. Install the gaskets and install the cylinder heads.

26. Install the cylinder head bolts. Tighten the cylinder head bolts in two steps:

    a. Tighten all cylinder head bolts in sequence to 43 ft. lbs. (58 Nm).

    b. Tighten all cylinder head bolts in sequence to 105 ft. lbs. (143 Nm).

✳✳ WARNING

To avoid valve-to-piston contact, ensure the piston in that cylinder is not at TDC while tightening the rocker arms.

27. Install the pushrods and rocker arm assemblies in their original positions. Tighten the bolts to 21 ft. lbs. (28 Nm).

28. Install the intake manifold gaskets. Be sure that the locator dowels are positioned in the head.

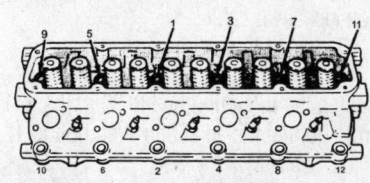

Cylinder head bolt tightening sequence— 8.0L engine

⁂ WARNING

Make absolutely sure the block sealing surface is free of oil.

29. Peel off the protective paper (blue—rear, and brown—front).

30. Align the slots in the end seals with the notches in the intake manifold.

31. Insert Mopar® Silicone Rubber Adhesive Sealant, or equivalent, into the four corner pockets. Fill—without overfilling—the pockets.

32. Install the lower intake manifold within NO MORE THAN three minutes of applying the sealant. Carefully lower the intake manifold into position on the cylinder block and heads. After the intake manifold is in place, inspect to ensure the seals and gaskets are in place.

33. Finger-start all bolts while alternating one side to the other.

34. Tighten the lower intake manifold bolts to 40 ft. lbs. (54 Nm).

35. Use a new gasket and position the upper intake manifold onto the lower intake manifold.

36. Tighten the upper intake manifold bolts to 16 ft. lbs. (22 Nm).

37. Install the exhaust pipe to the exhaust manifold. Tighten the bolts to 25 ft. lbs. (34 Nm).

38. Using a new gasket, position the EGR tube to the intake manifold and the exhaust manifold.

39. Tighten the nut to 25 ft. lbs. (34 Nm). Tighten the bolts to 15 ft. lbs. (20 Nm).

40. Install the heat shields and the washers. Ensure that the heat shields tabs hook over the exhaust gasket. Tighten the nuts to 11 ft. lbs. (15 Nm).

41. Install the spark plugs. Tighten to 30 ft. lbs. (41 Nm).

42. Install the coil packs and bracket. Tighten the bolts to 16 ft. lbs. (21 Nm). Connect the coil wires.

43. Connect the heat indicator sending unit wire.

44. Connect the heater hoses and bypass hose.

45. Connect the throttle linkage and (if equipped) the speed control and transmission kickdown cables.

46. Install the fuel line.

47. Install the alternator and drive belt. Tighten the mounting bolt to 30 ft. lbs. (41 Nm). Tighten the adjusting strap bolt to 17 ft. lbs. (23 Nm).

48. Install the cylinder head cover onto the gasket. Install the stud bolts and hex head bolts in their proper positions. Tighten the stud bolts to 12 ft. lbs. (16 Nm).

49. Install the closed crankcase ventilation system.

50. Connect the evaporation control system.

51. Install the air cleaner.

52. Fill the cooling system.

53. Connect the negative battery cable.

54. Start the vehicle and check for any leaks.

Rocker Arms/Shafts

REMOVAL & INSTALLATION

2.5L (SOHC) Engine

1. Disconnect the negative battery cable.
2. Remove the cylinder head cover.
3. Rotate the crankshaft until the low point of the desired cam lobe is contacting the rocker arm.
4. Using the special valve spring compressor tool, or equivalent, depress the valve spring (without dislodging the keeper) and slide the rocker arm out.

To install:

5. Depress the valve spring with the compressor tool and install in reverse order, turning the camshaft as necessary.
6. Install the cylinder head cover.
7. Connect the negative battery cable.

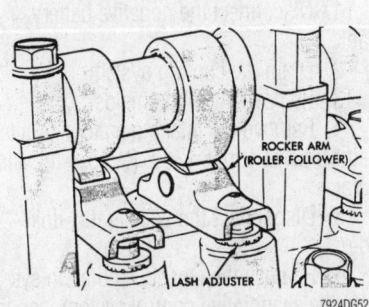

Close-up view of the rocker arm and lash adjuster—2.5L (SOHC) engine

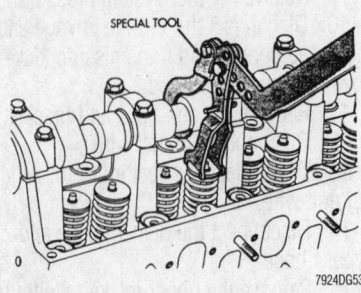

Compress the valve spring to remove the rocker arm assembly—2.5L (SOHC) engine

2.5L (OHV) Engine

1. Disconnect the negative battery cable.
2. Disconnect the Crankcase Ventilation (CCV) vacuum hose and the fresh air inlet hose from the cylinder head cover.
3. Remove the cylinder head cover retaining bolts and remove the cylinder head cover. Use care. The plastic-like composite cylinder head cover used on many versions of this engine is easily damaged. Clean the cover with a suitable solvent and inspect the lower part of the cover. There may an indication PRY HERE showing a reinforced area where a putty knife can be inserted to help break the seal so the cover can be removed.

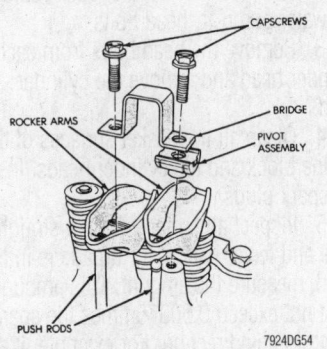

Exploded view of the rocker arm mounting assembly—2.5L (OHV) engine

➡On some late-model engines, the rocker arm cover (cylinder head cover) has a cured gasket attached to it. This gasket should not be removed. If sections of the gasket are missing or are compressed, replace the cover. However, minor damage such as small cracks, cuts or chips can be repaired with liquid gasket material.

4. Remove the rocker arms and the pushrods. Any valvetrain parts that are to be reused must be returned to their original locations. Keep them in their original order for installation reference.

To install:

5. Install the push rods, rocker arms, pivots and bridges. Tighten the rocker arm bridge bolts alternately. Tighten the bolts to 21 ft. lbs. (28 Nm).

6. Pour oil or oil supplement over the rocker arms and push rods. Be careful not to overfill the crankcase.

7. Install the valve cover. Tighten the attaching bolts to 115 inch lbs. (14 Nm).

8. Connect the Crankcase Ventilation (CCV) vacuum hose and the fresh air inlet hose to the cylinder head cover.

9. Connect the negative battery cable.

3.9L Engine

1. Disconnect the negative battery cable.
2. Remove the cylinder head cover and gasket.
3. Note the positioning of the oil notch and remove the rocker arms and pivots from the head.

To install:

4. Install the rocker arms in the same order they were removed. Tighten the bolts to 21 ft. lbs. (28 Nm).
5. Install the cylinder head cover. Tighten the attaching bolts 95 inch lbs. (11 Nm).
6. Connect the negative battery cable and check for leaks.

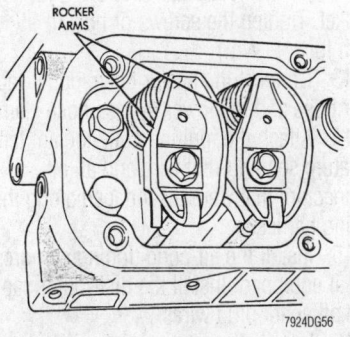

Pushrod and rocker arm positioning on the cylinder head—3.9L engine

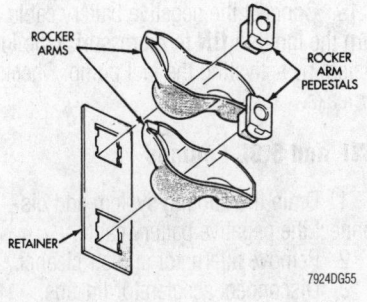

Exploded view of the rocker arm assembly—8.0L engine

5.2L, 5.9L and 8.0L Engines

1. Disconnect the closed ventilation system and evaporative control system (if so equipped) from the cylinder head cover.
2. Disconnect the spark plug wires.
3. Remove each valve cover and gasket.
4. Remove each rocker arm and pivot bolt. Keep everything in order for installation in the original position.

To install:

5. Rotate the crankshaft until the "V8" mark aligns with TDC mark on the timing chain cover.

✵✵ WARNING

To avoid valve to piston contact, ensure the piston in that cylinder is not at TDC while tightening the rocker arms.

6. Install the rocker arms and pivots. Tighten the rocker arm bolts to 21 ft. lbs. (28 Nm).

➡**Do not crank or rotate the engine for at least five minutes after the rocker arms are installed. This will give the lifters time to bleed down.**

7. Install the valve cover gasket.
8. Install the valve cover. Tighten the attaching bolts to 95 inch lbs. (11 Nm).
9. Connect the spark plug wires.

Intake Manifold

REMOVAL & INSTALLATION

2.5L (SOHC) Engine

➡**Both the intake and exhaust manifolds must be removed whenever either one is in need of repair, since they share one gasket.**

1. Disconnect the negative battery cable.
2. Drain the cooling system.
3. Depressurize the fuel system. Remove the air cleaner and disconnect all vacuum lines, electrical wiring and fuel lines from the throttle body and/or manifold.
4. Disconnect the throttle linkage.
5. Loosen the power steering pump and remove the drive belt. Remove the power steering and air pump support bracket.
6. Remove the power brake vacuum hose from the intake manifold.
7. On Canadian models, remove the coupling hose from the diverter valve to the exhaust manifold air injection tube assembly.
8. Remove the water hoses from the water crossover.
9. Raise and safely support the vehicle. Disconnect the exhaust pipe from the exhaust manifold.
10. Remove the power steering pump and set it aside.
11. Remove the intake manifold support bracket, if equipped.
12. Remove the EGR tube.
13. If equipped, remove the air injection tube bolts and the air injection tube assembly.
14. Remove the intake manifold bolts.

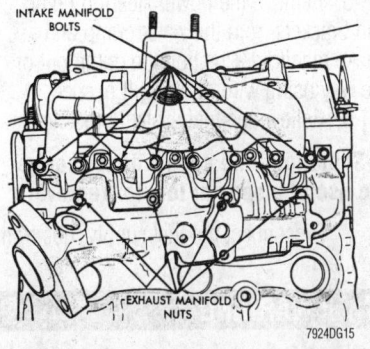

Intake and exhaust manifold attaching bolt locations—2.5L (SOHC) engine

15. Lower the vehicle and remove the intake manifold.
16. Remove the exhaust manifold nuts.
17. Remove the exhaust manifold.

To install:

18. Install a new combination manifold gasket lightly coated on the manifold side with Mopar® Gasket Sealer, or equivalent.
19. Set the gasket in place and install and tighten the nuts to 17 ft. lbs. (23 Nm). The tightening should begin in the center and progress outwards in both directions.
20. Repeat the tightening sequence until all nuts are at the specified torque.
21. Install the intake manifold strut. Tighten the bolt to 70 ft. lbs. (95 Nm). Tighten the nut to 40 ft. lbs. (54 Nm).
22. Install with a new gasket the EGR tube. Tighten to 17 ft. lbs. (23 Nm).
23. Install the exhaust pipe to the exhaust manifold, tightening to 20 ft. lbs. (27 Nm).
24. Connect the water hose.
25. Install the diverter valve assembly and the air injection tube to the exhaust manifold.
26. Install the power brake vacuum hose to the intake manifold.
27. Install the power steering and air pump support bracket.
28. Install the throttle linkage.
29. Install a new gasket and the throttle body onto the intake manifold. Tighten the bolts to 175 inch lbs. (20 Nm).
30. Install the air cleaner. Connect all the vacuum lines, electrical wiring and fuel lines to the throttle body.
31. Fill the cooling system and connect the negative battery cable.

2.5L (OHV) Engine

1. Disconnect the negative battery cable.
2. Remove the air cleaner assembly.
3. Drain the engine coolant.
4. Remove the belt from the power steering pump.

5. Remove the power steering pump and brackets from the water pump and intake manifold. Position the pump out of the way using wire or string if needed.

6. Relieve the fuel system pressure.

➡ **Some fuel line connections require the use of a special tool for removal.**

7. Disconnect the fuel supply tube from the fuel rail.

❋❋ WARNING

To prevent damage to the retainer, use finger pressure only when disconnecting the cruise control cable from the throttle body.

8. Disconnect the throttle cable and if equipped, the cruise control cable from the throttle body.

9. Disengage all electrical connectors and hoses from the intake manifold. Tag them for installation in their original positions.

10. Remove intake manifold bolts 2 through 5. Slightly loosen bolt No. 1 and nuts 6 and 7.

11. Remove the intake manifold and gasket.

To install:

12. Clean the intake manifold and cylinder head sealing surfaces.

13. Using a new gasket, install the intake manifold on the cylinder head. Using the proper sequence, tighten all bolts and nuts except No. 1 to 23 ft. lbs. (31 Nm) and tighten No. 1 bolt to 30 ft. lbs. (41Nm).

14. Install a new O-ring on the supply line and connect it to the fuel rail. Push them together until a "click" is heard.

15. Install the power steering pump and brackets.

16. Install the drive belt on the power steering pump.

17. Install all remaining components in the reverse order of removal.

18. Connect the negative battery cable.

19. Refill the cooling system with the proper amount and type of coolant.

3.9L Engine

1. Relieve fuel system pressure. Disconnect the negative battery cable from the battery and drain the cooling system.

2. Remove the air pump and bracket. Removal of the bracket will allow for easier installation of the left front corner of the intake manifold.

3. Remove the air cleaner assembly. Unbolt the air conditioning compressor and lay it to the side, if equipped. Remove the distributor cap with all wires attached.

4. Disconnect all wires, hoses, linkages and cables from the throttle body. Disconnect the fuel line.

5. Disengage the ignition coil, coolant temperature sending unit wire and all other connectors along the wiring harness attached to items on the intake manifold.

6. Disconnect the heater hose, upper radiator hose and the lower bypass hose clamp.

7. Unbolt the intake manifold from the heads and remove the intake manifold assembly. Disassemble the manifold as required and clean out the exhaust crossover passages.

To install:

8. Clean and dry the intake manifold contact surfaces. Coat the intake manifold side gaskets very lightly with sealer and install the gaskets to the heads. Cut-outs at the front of the gaskets differentiate the right and left sides.

9. Apply a thin uniform coat of quick dry cement to the front and rear intake manifold gaskets and mounting surfaces on the block and apply a thin bead of sealer to each of the four corners. Install the front and rear gaskets engaging the hole in the block and the tangs from the head gaskets. Apply a second thin bead of sealer above the gaskets in the four corners.

10. Carefully lower the intake manifold into position engaging the bypass hose; after it is satisfactorily in place, inspect the

gaskets to ensure they have not become dislodged.

11. Install the intake manifold bolts with the aspirator tube, air pump bracket and kickdown linkage bracket in place, if equipped. Tighten the bolts in four steps:

   a. Tighten bolt Nos. 1 and 2 to 72 inch lbs.(8 Nm)

   b. Tighten bolt Nos. 3 through 12, in sequence to 72 inch lbs. (8 Nm).

   c. Repeat the sequence retightening the bolts to 72 inch lbs. (8 Nm).

   d. Tighten all the bolts, in sequence, to a final torque of 144 inch lbs. (16 Nm).

12. Clean and dry the cylinder head cover mating surfaces, bolts and bolt holes. Install the cylinder head covers with a new gasket. Tighten the screws or nuts to 95 inch lbs. (11 Nm).

13. Connect the heater hose, upper radiator hose and the lower bypass hose clamp.

14. Attach the ignition coil, coolant temperature sending unit wire and all other connectors that were disengaged along the wiring harness.

15. Install the air conditioning compressor, if equipped. Install the distributor cap and all spark plug wires.

16. Install air pump.

17. Connect all wires, hoses, cables and the fuel line to the throttle body. Install the air cleaner assembly.

18. Fill the cooling system.

19. Connect the negative battery cable. Turn the ignition **ON** to repressurize the fuel system by activating the fuel pump. Check for leaks.

5.2L and 5.9L Engines

1. Drain the cooling system and disconnect the negative battery cable.

2. Remove alternator and air cleaner.

3. Disconnect accelerator linkage.

4. Depressurize the fuel system and disconnect the fuel line.

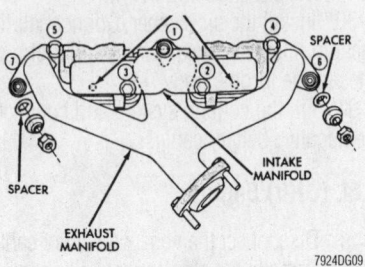

Tighten the intake manifold bolts and nuts in the correct sequence to prevent warpage—2.5L OHV engine

7924DG09

Tighten the intake manifold bolts in the proper sequence to prevent leakage—3.9L engine

7924DG10

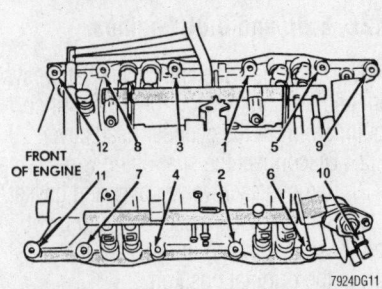

Tighten the intake manifold bolts using the numbered sequence as shown—5.2L and 5.9L engines

7924DG11

5. Remove the distributor cap and wires.

6. Disconnect coil wires, temperature sending unit wire, heater hoses and bypass hose.

7. Remove intake manifold and throttle body as an assembly.

To install:

8. Clean the gasket surfaces so they are clean and dry.

9. Position the intake manifold and install the mounting fasteners. If the fasteners were severely rusted when removed, use new fasteners.

10. Tighten the intake manifold bolts in four steps:

a. Tighten bolt Nos. 1 through 4 to 72 inch lbs.(8 Nm)

b. Tighten bolt Nos. 5 through 12, in sequence, to 72 inch lbs. (8 Nm).

c. Repeat the sequence retightening all the bolts to 72 inch lbs. (8 Nm).

d. Tighten all the bolts, in sequence, to a final torque of 144 inch lbs. (16 Nm).

11. Install the distributor cap and wires.

12. Connect coil wires, temperature sending unit wire, heater hoses and bypass hose.

13. Connect the fuel line.

14. Connect accelerator linkage.

15. Install the alternator and air cleaner.

16. Properly fill the cooling system.

17. Connect the negative battery cable.

8.0L Engines

1. Drain the cooling system and disconnect the negative battery cable.

2. Remove alternator and air cleaner.

3. Remove the serpentine belt.

4. Remove the alternator and its brace.

5. Remove the air conditioning compressor brace, then remove the compressor and set it aside without disconnecting the lines.

6. Remove the air cleaner cover and filter. Remove the filter. Discard the gasket.

7. Depressurize the fuel system and disconnect the fuel line.

8. Disconnect accelerator linkage, and if so equipped, the speed control and transmission kickdown cables.

9. Remove the coil assemblies. Disconnect the vacuum lines.

10. Disconnect the heater hoses and bypass hose.

11. Remove the closed crankcase ventilation and evaporation control systems.

12. Remove throttle body and lift it off the upper intake manifold. Discard the gasket.

13. Remove the front upper intake manifold bolts. Retain the three rear bolts in the up position with tape or rubber bands.

14. Lift the upper intake manifold out of the engine bay. Discard the gasket.

15. Remove the lower intake manifold bolts and remove the manifold. Discard the gasket.

To install:

16. Clean the manifolds and mounting surfaces with solvent and let dry or dry completely with compressed air.

17. Inspect the surfaces for cracks and for warpage using a straightedge.

18. With the locator dowels in place on the head, install the intake manifold side gaskets.

19. When sure that the block is oil free, peel off the paper (blue-rear, brown-front) and press firmly onto the block. Align the slots in the end seals with the notches in the intake manifold gaskets.

20. Into each of the four corner pockets, insert Mopar® Silicon Rubber Adhesive Sealant, or equivalent. Do NOT overfill.

21. The lower intake manifold must be installed within three minutes of the sealant having been applied. After in place, inspect to be sure all the gaskets and the seals are in their proper places. Finger-start all lower intake bolts.

22. Tighten the lower intake manifold bolts, in sequence, to 40 ft. lbs. (54 Nm).

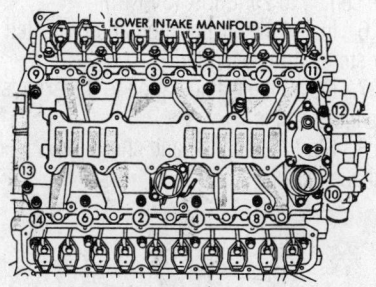

Lower intake manifold tightening sequence—8.0L engine

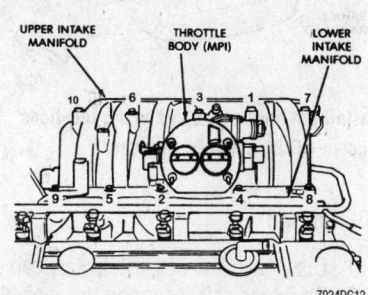

Upper intake manifold tightening sequence—8.0L engine

23. Install a new gasket and the upper intake manifold, then finger-tighten all bolts. Alternate from one side to the other.

24. Tighten the bolts to 16 ft. lbs. (22 Nm).

25. Install a new gasket and the throttle body onto the upper intake manifold. Tighten the bolts to 17 ft. lbs. (23 Nm).

26. Install the closed crankcase ventilation and evaporation control systems.

27. Connect the heater hoses and bypass hose.

28. Connect the vacuum lines.

29. Install the oil assemblies and the ignition cables.

30. Connect the accelerator linkage, and, if so equipped, the speed control and transmission kickdown cables.

31. Install the fuel lines.

32. Using a new gasket, install the air cleaner housing. Tighten the nuts to 96 inch lbs. (11 Nm). Install the air cleaner assembly and its cover.

33. Install the air conditioning compressor. Install the brace, tightening the bolts to 30 ft. lbs. (41 Nm).

34. Install the serpentine belt.

35. Fill the cooling system and connect the negative battery cable.

Exhaust Manifold

REMOVAL & INSTALLATION

2.5L (SOHC) Engine

Please refer to the intake manifold procedure for the 2.5L (SOHC) engine.

2.5L (OHV) Engine

1. Remove the intake manifold using the procedure in this section.

2. Raise and safely support the vehicle securely.

3. Remove the exhaust pipe from the exhaust manifold.

4. Safely lower the vehicle.

5. Remove fasteners Nos. 1, 6 and 7 and remove the exhaust manifold.

To install:

6. Clean the sealing surfaces of the cylinder head, intake and exhaust manifolds.

7. Install a new intake manifold gasket on the cylinder head alignment dowels.

8. Position the exhaust manifold on the cylinder head. Ensure the manifold is centered on the end studs and spacers. Tighten bolt No. 1 to 30 ft. lbs. (41 Nm). Do not install the nuts at this time.

9. Install the intake manifold and tighten bolts Nos. 2 through 5 to 23 ft. lbs. (31 Nm).

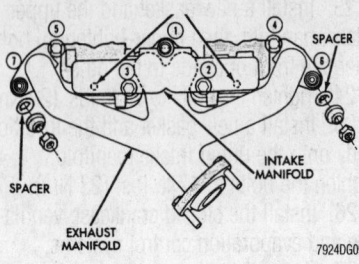

Tighten the intake and exhaust manifold bolts and nuts in the correct sequence to prevent vacuum and exhaust leaks—2.5L OHV engine

10. Install new spacers on the end studs and tighten the nuts to 23 ft. lbs. (31 Nm).

11. Tighten nuts Nos. 6 and 7 to 23 ft. lbs. (31 Nm).

12. Install any component that were previously removed.

13. Raise and safely support the vehicle securely.

14. Connect the exhaust pipe to the exhaust manifold. Tighten the bolts to 23 ft. lbs. (31Nm).

15. Safely lower the vehicle.

16. Connect the negative battery cable, start the engine and check for leaks.

3.9L Engine

1. Disconnect the negative battery cable.

2. Remove the hot air tube and heat shield, if necessary.

3. Raise the vehicle and safely support it.

4. Remove the exhaust pipe from the exhaust manifolds.

5. Lower the vehicle.

6. Take note of all conical washer locations and remove the bolts, nuts and washers attaching the manifold to the head.

7. Remove the manifold.

To install:

8. If either of the end studs came out with the nuts, install a new stud using sealer on the coarse threads.

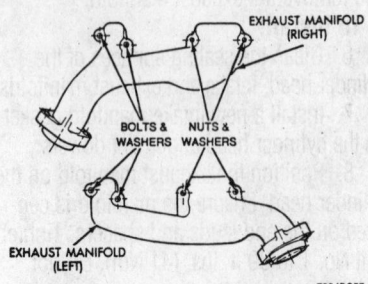

Be sure to install the studs and bolts in the correct locations, as indicated—3.9L engine

9. Position the manifold on the end studs. Install conical washers and nuts on the studs.

10. Install the remaining bolts and washers in their proper locations. Working outward from the center, tighten the bolts and nuts to 25 ft. lbs. (34 Nm).

11. Attach the exhaust pipe to the manifolds.

12. Connect the negative battery cable, start the engine, and check for exhaust leaks.

5.2L and 5.9L Engines

1. Disconnect the exhaust manifold at the flange where it mates to the exhaust pipe.

2. If the vehicle is equipped with air injection and/or a heated air stove, remove them.

3. Remove the exhaust manifold by removing the securing bolts and washers. To reach these bolts, it may be necessary to jack the engine slightly off its front mounts. When the exhaust manifold is removed, sometimes the securing studs will screw out with the nuts. If this occurs, the studs must be replaced with the aid of sealing compound on the coarse thread ends. If this is not done, water leaks may develop at the studs.

To install:

4. Clean the mounting surfaces so they are clean and dry.

5. Replace hardware, as necessary. Install a new gasket and the exhaust manifold. Tighten the nuts and bolts, starting in the center, to 25 ft. lbs. (34 Nm).

6. Install the air injection and/or heated air stove, if removed.

7. Install the exhaust pipe to the exhaust manifold flange.

8. Start the engine, and check for exhaust leaks.

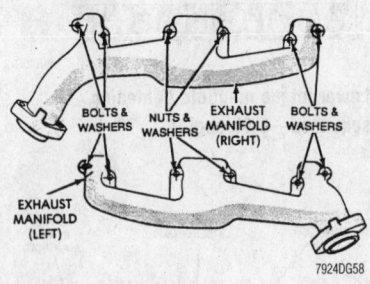

Install the bolts and studs in the locations shown—5.2L and 5.9L engines

8.0L Engine

1. Disconnect the negative battery cable.

2. Raise and safely support the vehicle.

3. Remove the hardware that attaches the exhaust pipe to the exhaust manifold.

4. Lower the vehicle.

5. Remove the exhaust heat shields.

6. To remove the right exhaust manifold:

   a. Remove the EGR tube. Discard the gasket.

   b. Remove the dipstick bracket from the manifold.

7. Unbolt and remove the exhaust manifold. Discard the gasket.

To install:

8. Clean the sealing surfaces so they are free of all old gasket material, oil and grease.

9. Install a new gasket and the exhaust manifold(s). Ensure to position the bolts and stud bolts in their proper positions. Tighten the stud bolts and bolts to 16 ft. lbs. (22 Nm).

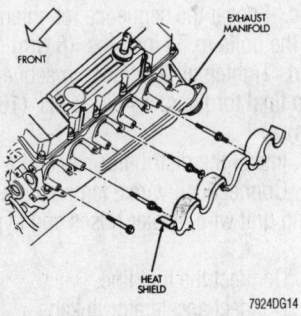

Install the stud bolts in the correct position so the heat shield can be attached—8.0L engine

10. For the right side manifold:

   a. Install a new gasket and the EGR tube. Tighten the tube assembly nut to 25 ft. lbs. (34 Nm). Tighten the two EGR nuts to 15 ft. lbs. (20 Nm).

   b. Install the dipstick bracket to the manifold.

11. Install the heat shield. Tighten the nuts on the stud bolts to 15 ft. lbs. (20 Nm).

12. Raise and safely support the vehicle.

13. Install the exhaust pipe. Tighten the bolts to 25 ft. lbs. (34 Nm).

14. Lower the vehicle and connect the negative battery cable.

Front Crankshaft Seal

REMOVAL & INSTALLATION

➡**This procedure only applies to engines which utilize timing belts.**

2.5L (SOHC) Engine

1. Remove the timing belt.

2. Remove the crankshaft sprocket using a sprocket or steering wheel puller. Ensure

not to damage the threads inside the crankshaft.

3. If necessary, hold the engine sprocket from turning with Special Tool No. C-4687 (or equivalent) and adapter while removing/installing the screw.

4. Remove the crankshaft seal using Special Tool No. 6341-A. Remove the intermediate and camshaft seals, if necessary, with Special Tool No. C-4679, or equivalent.

5. Shaft seal lip surface must be free of varnish, dirt and nicks. Polish with 1500 grit sandpaper, if necessary.

6. Install engine crankshaft seal into the retainer using Special Tool No. 6342 and 6343. Install the intermediate and camshaft seals, if removed, with Special tool C-4680. Install the seals until they fit flush.

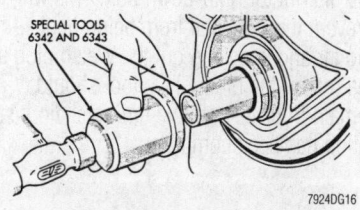

Install the seals until they fit flush with the cylinder block—2.5L (SOHC) engine

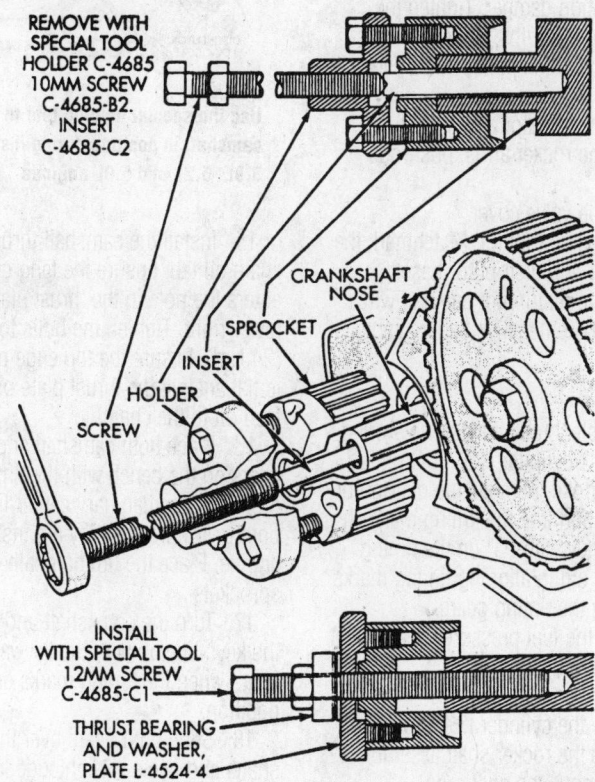

To prevent damage, use the proper tools when removing or installing the crankshaft sprocket— 2.5L (SOHC) engine

7. Using tool C-4685, 10mm screw C-4685-B2 and insert C-4685-C2, install the crankshaft sprocket, and if removed, install the intermediate shaft sprocket. Do not hammer on the sprocket, a crankshaft pulley installation tool should be used to install the sprocket on the crankshaft.

Camshaft and Valve Lifters

REMOVAL & INSTALLATION

2.5L (SOHC) Engine

1. Disconnect the negative battery cable.
2. Relieve the fuel pressure.
3. Turn the crankshaft so the No. 1 piston is at the TDC of the compression stroke. Remove the upper timing belt cover. Remove the air pump pulley, if equipped.
4. Remove the camshaft sprocket bolt and the sprocket and suspend tightly so the belt does not lose tension. If it does, the belt timing will have to be reset.
5. Remove the cylinder head cover.
6. If the rocker arms are being reused, mark them for installation identification and loosen the camshaft bearing bolts evenly and gradually.

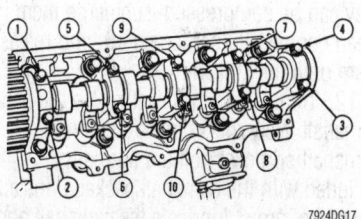

Loosen and remove the camshaft bearing cap bolts in the correct order—2.5L (SOHC) engine

7. Using a soft mallet, tap the rear of the camshaft a few times to break the bearing caps loose.
8. Remove the bolts, bearing caps and the camshaft with seals.
9. Remove the lash adjusters from their bores in the cylinder head.

➡ **Before replacing the camshaft, identify factory installed oversized components. To do so, look for the tops of the bearing caps to be painted green and O/SJ stamped rearward of the oil gallery plug on the rear of the head. In addition, the barrel of the camshaft is painted green and O/SJ is stamped onto the rear end of the camshaft. Installing standard sized parts in a head equipped with oversized parts— or vice versa—will cause severe engine damage. Also, take note of the color of the paint stripe on the rear camshaft seal. These stripes differentiate seal sizes. If a seal with a different color stripe is installed, a severe leak will develop if the seal is too small, or the cap will not be able to be fully installed if the seal is too big.**

10. Check the oil passages for blockages and the parts for wear and damage and replace parts, as required. Clean the gasket mounting surfaces.

To install:

11. Install the lash adjusters in their bores on the cylinder head. Be sure they are at least partially filled with oil. This is indi-

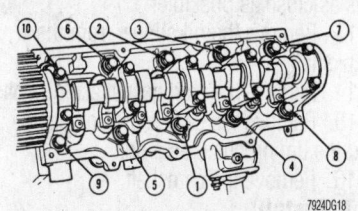

Camshaft bearing cap bolt tightening sequence—2.5L (SOHC) engine

cated by their inability to be compressed. If they can be compressed, submerge them in clean engine oil, then compress and release them until they fill with oil.

12. Transfer the sprocket key to the new camshaft. New rocker arms and a new camshaft sprocket bolt are normally included with the camshaft package. Install the rocker arms, lubricate the camshaft and install it with end seals already installed.

13. Place the bearing caps with No. 1 at the timing belt end and No. 5 at the transaxle end. The camshaft bearing caps are numbered and have arrows facing forward. Tighten the camshaft bearing bolts evenly and gradually to 18 ft. lbs. (24 Nm).

➡Apply RTV silicone gasket material to the No. 1 and 5 bearing caps. Install the bearing caps before the seals are installed.

14. Mount a dial indicator to the front of the engine and check the camshaft end-play. Play should not exceed 0.006 in. (0.15mm).

15. Install the camshaft sprocket and the new bolt. Install the air pump pulley, if equipped.

16. Install the cylinder head cover with a new gasket.

17. Connect the negative battery cable and check for leaks.

2.5L (OHV) Engine

1. Disconnect the negative battery cable.
2. Remove the belt(s) from the crankshaft pulley.
3. Remove the fan and shroud from the vehicle.
4. Remove the distributor cap and wires. Matchmark the distributor rotor and distributor base.
5. Remove the distributor.
6. Remove the valve cover.
7. Remove the rocker arms, pushrods and lifters.
8. Remove the crankshaft pulley.
9. Rotate the vibration damper to TDC.
10. Remove the vibration damper using the proper puller.
11. Remove the timing case cover.
12. Align the timing marks on the pulleys as close as possible.
13. Remove the oil slinger from the crankshaft.
14. Remove the cam gear attaching bolts.
15. Remove the cam and crank gears and the timing chain as a unit.
16. Remove the camshaft.

To install:

17. Coat the camshaft with assembly lube or an oil supplement.

18. Carefully install the camshaft. Be careful not to nick or damage the bearing surface or cam lobes.

19. Turn the chain tensioner to the unlocked position. Pull the tensioner block toward the lever to compress the spring. Hold the spring back and turn the lever to the lock position.

20. Install the cam and crank gear and the timing chain as a unit. Tighten the cam gear attaching bolts to 80 ft. lbs. (108 Nm).

21. Verify that the timing marks are properly aligned. The mark on the cam gear should be at 6 o'clock and the crank gear at 12 o'clock.

22. Turn the chain tensioner to the unlocked (down) position.

23. Position the oil slinger on the crankshaft.

24. Install a new oil seal in the timing case cover.

25. Apply a light coat of gasket sealant around the cover including the bottom surface where the cover joins the oil pan.

26. Position the new gasket and install the cover on the engine with the seal tool to properly align the cover. Tighten the ¼ inch cover bolts to 60 inch lbs. (7 Nm), the 5⁄16 inch bolts to 192 inch lbs. (22 Nm) and the oil pan-to-cover bolts to 84 inch lbs. (9.5 Nm). Remove the seal tool.

27. Install the keyway in the crankshaft. Install the vibration damper. Tighten the attaching bolt to 80 ft. lbs. (108 Nm).

28. Install the crank pulley and drive belt.

29. Install the radiator and shroud.

30. Install the rocker arms, pushrods and lifters.

31. Install the valve cover.

32. Install the distributor. Matchmark the distributor rotor and distributor base.

33. Install the distributor cap and wires.

34. Connect the negative battery cable.

3.9L Engine

1. If possible, crank the engine around so that the No. 1 cylinder is at TDC on the compression stroke. Remove the distributor cap to confirm and line the timing mark on the damper pulley with "O" on the timing scale. This will aid in aligning timing marks when installing the timing gears.

2. Relieve the fuel pressure. Disconnect the negative battery cable.

3. Drain the cooling system.

4. Remove the cylinder head cover(s).

5. Remove the rocker shaft assemblies. Identify and remove the pushrods.

6. Remove the intake manifold. Identify and remove all lifters.

7. Remove the distributor.

8. Lift out the oil pump and distributor driveshaft.

9. Remove the radiator, fan and all related parts.

10. Remove the fuel pump, if equipped. Remove the timing chain cover, timing chain and gears.

11. Note the location of the oil tab and remove the camshaft thrust plate.

12. Install a suitable long bolt into the front of the camshaft to facilitate removal. Remove the camshaft, being careful not to damage any of the cam bearings with the cam lobes.

To install:

13. Install the camshaft to within 2 in. (50mm) of its final installation position.

14. Install the Camshaft Blocking tool C-3509, or equivalent, and bolt it in place with the distributor hold-down bolt. This will prevent the camshaft from being pushed in too far and knocking out the Welsh plug at the rear of the block. This tool should remain in place until the timing chain installation has been completed.

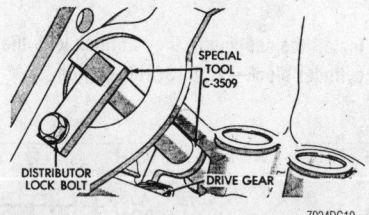

Use the special holding tool to keep the camshaft in position during installation— 3.9L, 5.2L and 5.9L engines

15. Install the camshaft thrust plate and chain oil tab. Ensure the tang of the oil tab enters the hole in the thrust plate at the lower right. Tighten the bolts to 18 ft. lbs. (24 Nm). Ensure the top edge of the toil tab is flat against the thrust plate or it will not feed oil to the chain.

16. Place both camshaft and crankshaft gears on the bench with the timing marks on the exact imaginary center link through both gear bores as they are installed on the engine. Place the timing chain around both sprockets.

17. Turn the crankshaft and camshaft so the keys line up with the keyways in the gears when the timing marks are in proper position.

18. Slide both gears over their respective shafts and use a straightedge to check timing mark alignment.

19. Install the fuel pump eccentric and cup washer, if equipped. Tighten the

camshaft gear retaining bolt to 35 ft. lbs. (47 Nm).

20. Remove the camshaft blocking tool, if it was installed.

21. Measure camshaft end-play. If the end-play is not 0.002–0.010 in. (0.051–0.254mm), replace the thrust plate.

22. Coat the oil pump and distributor driveshaft with oil. Install the shaft so that when the gear spirals into place and drops into the oil pump, the slot in the top of the gear is pointing directly to the left front intake manifold bolt hole.

23. If the camshaft was not replaced, lubricate and install the lifters in their original locations. If the camshaft was replaced, new lifters must be used.

24. Install the pushrods and rocker shaft assemblies.

25. Install the intake manifold, if it was removed. Install the cylinder head covers.

26. Install the distributor so the rotor points to the No. 1 spark plug wire position on the cap.

27. Install the timing chain cover and all related parts.

28. Install the fuel pump if equipped and radiator.

29. When everything is bolted in place, change the engine oil and replace the oil filter.

➡**If the camshaft or lifters have been replaced, add one pint of Mopar crankcase conditioner, or equivalent when replenishing the oil to aid in break in. This mixture should be left in the engine for a minimum of 500 miles (805 km) and drained at the next normal oil change.**

30. Fill the radiator with coolant.

31. Connect the negative battery cable, set all adjustments to specifications and check for leaks.

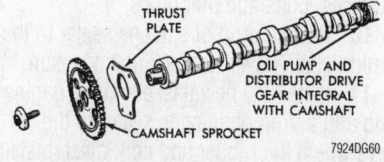

THRUST PLATE

OIL PUMP AND DISTRIBUTOR DRIVE GEAR INTEGRAL WITH CAMSHAFT

CAMSHAFT SPROCKET

7924DG60

If the camshaft end-play is not within specifications, the thrust plate must be replaced—3.9L, 5.2L, 5.9L and 8.0L engines

5.2L and 5.9L Engines

1. Drain the cooling system and disconnect the negative battery cable.

2. Remove the intake manifold, cylinder head covers, rocker arm assemblies, pushrods, and valve tappets, keeping them in order to insure the installation in their original locations.

3. Remove the timing gear cover, the camshaft and the crankshaft sprockets, and the timing chain.

4. Remove the distributor and lift out the oil pump and distributor driveshaft.

5. Remove the camshaft thrust plate.

6. Install a long bolt into the front of the camshaft and remove the camshaft, being careful not to damage the cam bearings with the cam lobes.

To install:

➡**Prior to installation, lubricate the camshaft lobes and bearings journals. It is recommended that one pint of Mopar® Crankcase Conditioner be added to the initial crankcase oil fill. Insert the camshaft into the engine block within 2 in. (51mm) of its final position in the block. Have an assistant support the camshaft with a suitable tool to prevent the camshaft from contacting the plug in the rear of the engine block. Position the suitable tool against the rear side of the cam gear and be careful not to damage the cam lobes.**

7. Install the camshaft thrust plate. If camshaft end-play exceeds 0.010 in. (0.25mm), install a new thrust plate. It should be 0.002–0.006 in (0.15mm) with the new plate.

8. Install the timing chain and sprockets, timing gear cover, and pulley.

9. Install the tappets, pushrods, rocker arms, and cylinder head covers. Install the fuel pump, if removed.

10. Install the distributor and oil pump driveshaft. If necessary, install a new bushing.

11. Install the distributor.

12. Install the remaining components which were removed, in reverse order of their removal.

13. Refill the cooling and reconnect the negative battery cable.

14. After starting the engine, adjust the ignition timing.

8.0L Engine

1. Drain the cooling system and disconnect the negative battery cable.

2. Remove the cylinder head covers, rocker arm assemblies, pushrods, and valve tappets, keeping them in order to insure the installation in their original locations.

➡**Keep all tappets in order for reassembly in the exact same position as when removed. The four corner tappets cannot be removed without removing the cylinder heads and gaskets. However they can be lifted and retained for camshaft removal.**

3. Remove the upper and lower intake manifold.

4. Remove the timing chain cover, timing chain, the timing chain thrust plate and the timing chain.

5. Remove the distributor and lift out the oil pump and distributor driveshaft.

6. Install a long bolt into the front of the camshaft and remove the camshaft, being careful not to damage the cam bearings with the cam lobes.

To install:

➡**Prior to installation, lubricate the camshaft lobes and bearings journals. It is recommended that one pint of Mopar® Crankcase Conditioner be added to the initial crankcase oil fill. Insert the camshaft into the engine block within 2 in. (51mm) of its final position in the block. Have an assistant support the camshaft with a suitable tool to prevent the camshaft from contacting the plug in the rear of the engine block. Position the suitable tool against the rear side of the cam gear and be careful not to damage the cam lobes.**

7. Install the camshaft thrust plate. If camshaft end-play exceeds 0.010 in. (0.254mm), install a new thrust plate. Normal tolerance for a new plate should be 0.051–0.152 in. (0.002–0.006mm).

8. Line up the key with the keyway in the sprocket and press on the crankshaft timing sprocket using special tool Nos. C-3688, C-3718 and MB990799, or equivalent, to seat the sprocket against the crankshaft shoulder.

9. Align the timing mark on the crankshaft sprocket with the centerline of the crankshaft and camshaft.

10. Put the timing chain on the camshaft sprocket.

11. Take the chain and camshaft sprockets and align the mark with the centerline of the crankshaft and camshaft install camshaft sprocket and chain to camshaft.

12. Install the camshaft bolt. Tighten the bolt to 55 ft. lbs. (75 Nm).

13. Install the timing chain cover.

14. Install the crankshaft pulley/damper using tool C7–3688, or equivalent.

15. Prime the oil pump by squirting oil in the oil filter mounting hole and filling

the J-trap of the front timing cover. Fill a new oil filter with fresh engine oil and install quickly just as the oil starts to come out.

16. Each tappet reused must be installed in the same position from which it was removed.

➡️**When the camshaft is replaced, all of the tappets must be replaced.**

17. Install the tappets and pushrods in their original location.

18. Install the rocker arms.

❋❋ WARNING

The cylinder head cover fasteners have a special plating and should not be substituted for another fastener.

19. Position the cylinder head cover onto the gasket. Install the stud bolts and hex head bolts in the proper positions. Tighten the stud bolts and the bolts to 144 inch lbs. (16 Nm).

➡️**The cylinder head cover gasket can be reused. For the left side, the number tab is at the front of the engine with the number up. For the right side the number tab is at the rear of the engine.**

20. Install the intake manifolds.

21. Refill the cooling and reconnect the negative battery cable.

Valve Lash

ADJUSTMENT

All gasoline engines covered in this section use hydraulic lifters and/or lash adjusters (SOHC engine). No maintenance or periodic adjustment is required.

Oil Pan

REMOVAL & INSTALLATION

2.5L (SOHC) Engine

1. Disconnect the negative battery cable. Remove the oil dipstick.

2. Disconnect the air pump relief valve upper hose.

3. Raise and safely support the vehicle.

4. Remove the clutch housing to engine strut and clutch inspection cover.

5. Remove the lower radiator hose support bracket.

6. Slightly loosen the right motor mount throughbolt just enough to relieve the tension.

7. Using the proper equipment, support the weight of the engine. Loosen the left motor mount throughbolt enough to clear the bracket.

8. Raise the left side of the engine about 2 in. (51mm).

9. Remove the oil pan retaining screws and remove the oil pan and gasket.

To install:

10. Thoroughly clean and dry all sealing surfaces, bolts and bolt holes.

11. Apply silicone sealer to the four end seal-to-block corners and install the end seals making sure the corners are not twisted.

12. Apply silicone to the four pan-to-block corners. Install a new pan gasket or apply a silicone sealer to the sealing surface of the pan and install to the engine making sure not to dislodge the end seals.

13. Install the pan retaining screws and tighten to 17 ft. lbs. (23 Nm).

14. Lower the engine. Tighten the motor mount throughbolts to 50 ft. lbs. (68 Nm).

15. Install the clutch inspection cover and housing to engine strut. Install the lower radiator hose support bracket. Lower the vehicle.

16. Install the dipstick and air pump hose. Fill the engine with the proper amount of oil.

17. Connect the negative battery cable and check for leaks.

2.5L (OHV) Engine

1. Disconnect the negative battery cable. Remove the oil dipstick.

2. Raise and safely support the vehicle.

3. Drain the engine oil.

4. Disconnect the front exhaust pipe from the manifold and the support bracket at the catalytic converter.

5. Remove the starter motor.

6. Remove the torque converter bolt access cover.

7. Position a jackstand under the vibration damper with a block of wood in between.

8. Remove the engine mount throughbolts.

9. Raise the jackstand until adequate clearance is obtained.

10. Remove the oil pan attaching bolts.

To install:

11. Clean the mating surfaces of dirt, oil and old gasket material.

12. Coat the pan with a gasket sealer to stick the gasket in place or fabricate alignment dowels from bolts to keep the gasket from shifting.

13. Install the attaching bolts. Tighten the ¼ inch bolts to 120 inch lbs. (14 Nm) and the 5⁄16 inch bolts to 156 inch lbs. (18 Nm).

14. Lower the jackstand enough to align the engine mount throughbolts. Tighten the throughbolts to 60 ft. lbs. (81 Nm).

15. Install the starter motor.

16. Install the torque converter bolt access cover.

17. Connect the front exhaust pipe to the manifold and the support bracket at the catalytic converter.

18. Lower the vehicle.

19. Fill the crankcase to the proper level with clean engine oil.

20. Connect the negative battery cable.

3.9L Engine

2-WHEEL DRIVE MODELS

1. Disconnect the negative battery cable. Remove the oil dipstick.

2. Disengage the distributor cap and remove it away from the firewall.

3. Raise the vehicle and support safely.

4. Drain the engine oil.

5. Remove the exhaust crossover.

6. Loosen the motor mount bolts. Using the proper equipment, raise the engine. When the engine is high enough, install replacement bolts (similar in size to the motor mount bolts), in the engine mount attaching points on the frame brackets.

7. Lower the engine so the bottom of the motor mounts rest on the two replacement bolts. Remove the torque converter inspection cover, if equipped.

8. Remove the oil pan retaining screws and remove the oil pan and gaskets.

To install:

9. Thoroughly clean and dry all sealing surfaces, bolts and bolt holes.

10. Place a drop of silicone sealer to the timing chain cover to block mating seam.

11. Install the new gaskets to the engine and add a drop of silicone sealer to the corners where the rubber and cork meet. Install the rubber seals to the pan.

12. Install the pan to the engine and tighten the retaining screws to 17 ft. lbs. (23 Nm). Install the torque converter inspection cover, if equipped.

13. Reinstall the engine to the mount and install the exhaust crossover. Lower the vehicle.

14. Install the distributor cap.

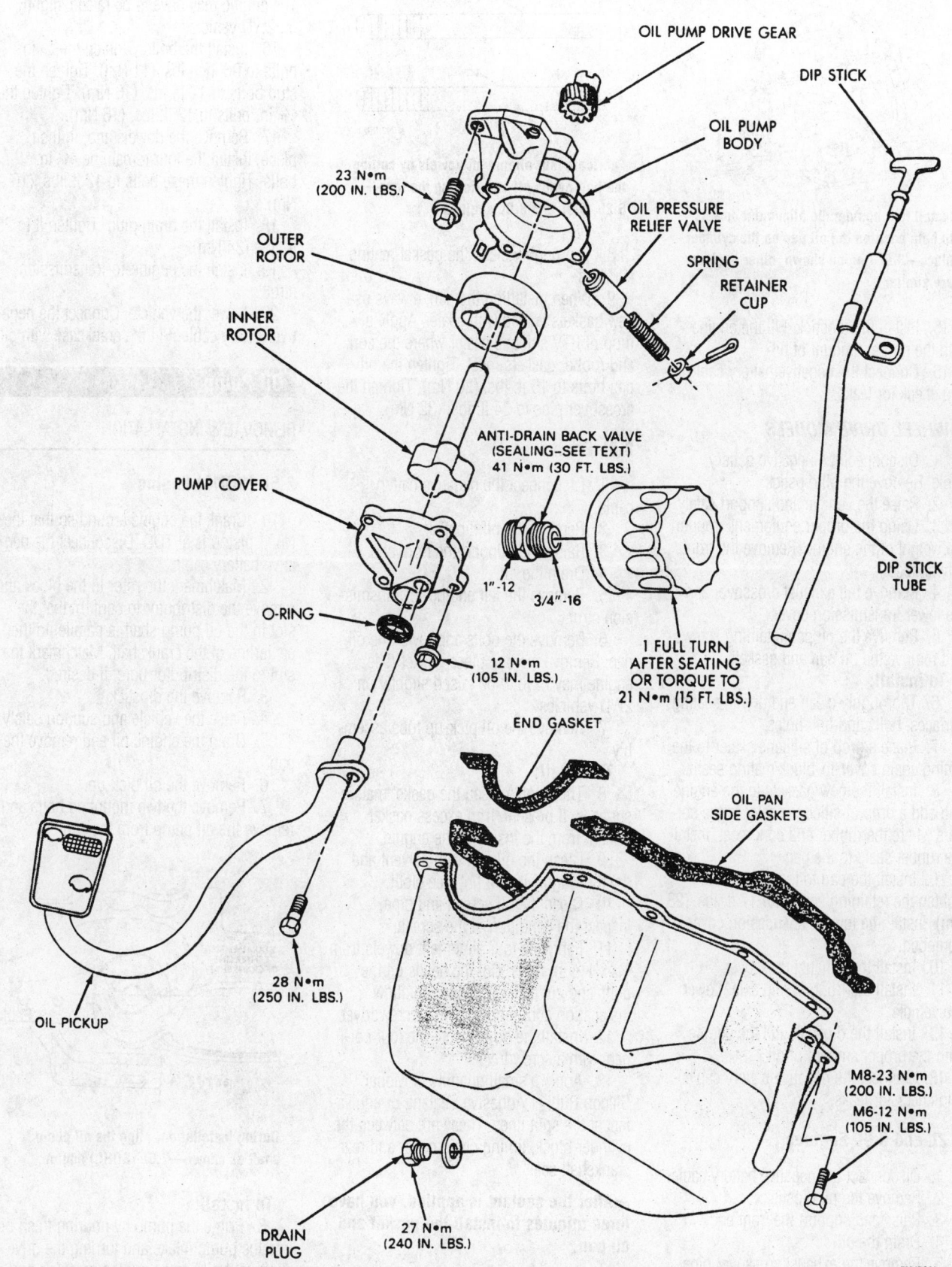

OIL PUMP DRIVE GEAR

DIP STICK

OIL PUMP BODY

23 N•m (200 IN. LBS.)

OIL PRESSURE RELIEF VALVE

SPRING

RETAINER CUP

OUTER ROTOR

INNER ROTOR

PUMP COVER

ANTI-DRAIN BACK VALVE (SEALING—SEE TEXT) 41 N•m (30 FT. LBS.)

DIP STICK TUBE

O-RING

1"-12

3/4"-16

1 FULL TURN AFTER SEATING OR TORQUE TO 21 N•m (15 FT. LBS.)

12 N•m (105 IN. LBS.)

END GASKET

OIL PAN SIDE GASKETS

OIL PICKUP

28 N•m (250 IN. LBS.)

M8-23 N•m (200 IN. LBS.)

M6-12 N•m (105 IN. LBS.)

DRAIN PLUG

27 N•m (240 IN. LBS.)

7924DG61

Exploded view of the engine lubrication system—2.5L (SOHC) engine

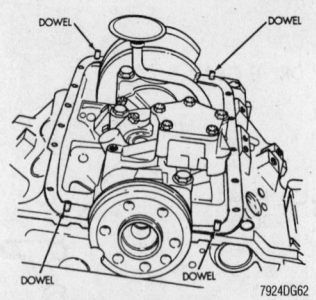

Install four homemade alignment dowels to help position the oil pan on the cylinder block—3.9L engine shown, other engines are similar

15. Install the dipstick. Fill the engine with the proper amount of oil.

16. Connect the negative battery cable and check for leaks.

4-WHEEL DRIVE MODELS

1. Disconnect the negative battery cable. Remove the oil dipstick.

2. Raise the vehicle and support safely.

3. Using the proper equipment, support the weight of the engine. Remove the front driving axle.

4. Remove the exhaust crossover and the lower transmission cover.

5. Remove the oil pan retaining screws and remove the oil pan and gaskets.

To install:

6. Thoroughly clean and dry all sealing surfaces, bolts and bolt holes.

7. Place a drop of silicone sealer to the timing chain cover-to-block mating seam.

8. Install the new gaskets to the engine and add a drop of silicone sealer to the corners where the rubber and cork meet. Install the rubber seals to the pan.

9. Install the pan to the engine and tighten the retaining screws to 17 ft. lbs. (23 Nm). Install the lower transmission cover, if equipped.

10. Install the exhaust crossover.

11. Install the front driving axle. Lower the vehicle.

12. Install the dipstick. Fill the engine with the proper amount of oil.

13. Connect the negative battery cable and check for leaks.

5.2L and 5.9L Engines

1. Disconnect the negative battery cable.

2. Remove the oil dipstick.

3. Raise and support the front end.

4. Drain the oil.

5. Remove the exhaust crossover pipe.

6. Remove the left engine-to-transmission strut.

7. Remove the bolts and lower the oil pan.

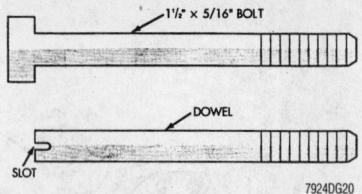

Fabricate four alignment dowels by cutting the bolt heads off and slotting the tops— 5.2L, 5.9L and 8.0L engines

8. Thoroughly clean the gasket mating surfaces.

9. When installing the pan, always use new gaskets coated with sealer. Apply a drop of RTV silicone sealer where the cork and rubber gaskets meet. Tighten the oil pan bolts to 15 ft. lbs. (20 Nm). Tighten the crossover pipe to 24 ft. lbs. (32 Nm).

8.0L Engine

1. Disconnect the negative battery cable.

2. Remove the oil dipstick.

3. Raise and support the front end.

4. Drain the oil.

5. Remove the left engine-to-transmission strut.

6. Remove the bolts and lower the oil pan. Remove the one piece gasket. The engine may have to be raised slightly on 2WD vehicles.

7. Remove the oil pick-up tube assembly.

To install:

8. Thoroughly clean the gasket mating surfaces. If present, trim excess gasket sealant from the inside of the engine.

9. Clean the oil pan with solvent and dry thoroughly with a lint-free cloth.

10. Clean the oil screen and pipe. Inspect the condition of the screen.

11. Fabricate four alignment dowels from 5/16 X 1½ in. bolts. Cut the heads off the bolts and cut a slot in the top to allow installation and removal with a screwdriver.

12. Install the dowels into the four corners with a screwdriver.

13. Apply a small quantity of Mopar® Silicon Rubber Adhesive Sealant, or equivalent at the split lines. These are between the cylinder block, timing chain cover and rear crankshaft seal.

➡**After the sealant is applied, you have three minutes to install the gasket and oil pan.**

14. Slide the one-piece gasket over the alignment dowels and position it on the block.

15. Position the oil pan over the gasket. The engine may have to be raised slightly on 2WD vehicles.

16. Install the bolts. Tighten the ½ in. bolts to 96 inch lbs. (11 Nm). Tighten the stud bolts to 12 ft. lbs. (16 Nm). Tighten the 5/16 in. bolts to 12 ft. lbs. (16 Nm).

17. Remove the dowels and, in their place, install the four remaining 5/16 in. bolts. Tighten these bolts to 12 ft. lbs. (16 Nm).

18. Install the drain plug. Tighten it to 25 ft. lbs. (34 Nm).

19. Install the engine-to-transmission strut.

20. Lower the vehicle. Connect the negative battery cable. Fill the crankcase with oil.

Oil Pump

REMOVAL & INSTALLATION

2.5L (SOHC) Engine

1. Crank the engine around so that the No. 1 piston is at TDC. Disconnect the negative battery cable.

2. Matchmark the rotor to the block and remove the distributor to confirm that the slot in the oil pump shaft is parallel to the centerline of the crankshaft. Matchmark the slot to the distributor bore, if desired.

3. Remove the dipstick.

4. Raise the vehicle and support safely.

5. Drain the engine oil and remove the pan.

6. Remove the oil pick-up.

7. Remove the two mounting bolts and remove the oil pump from the engine.

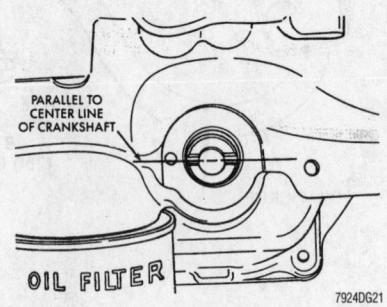

During installation, align the oil pump shaft as shown—2.5L (SOHC) engine

To install:

8. Prime the pump by pouring fresh oil into the pump intake and turning the driveshaft until oil comes out the pressure port. Repeat a few times until no air bubbles are present.

9. Apply sealer (Loctite® 515, or equivalent) to the pump body-to-block machined surface interface. Lubricate the oil pump and distributor driveshaft.

10. Align the slot so it will be in the same position as when it was removed. If it is not, the distributor will not be timed correctly. Install the pump fully and rotate back and forth to ensure proper positioning between the pump mounting surface and the machined surface of the block.

11. Install the mounting bolts finger-tight and lower the vehicle to confirm proper slot positioning. If the slot is not properly positioned, raise the vehicle and move the gear as required. If the slot is correct, hold the pump firmly against the block and tighten the mounting bolts to 17 ft. lbs. (23 Nm).

12. Clean out the oil pick-up or replace as required. Replace the oil pick-up O-ring and install the pick-up to the pump.

13. Install the oil pan using new gaskets. Lower the vehicle.

14. Install the distributor.

15. Install the dipstick. Fill the engine with the proper amount of oil.

16. Connect the negative battery cable, check the timing and the oil pressure.

2.5L (OHV) Engine

1. Remove the dipstick.
2. Raise and safely support the vehicle.
3. Drain the engine oil and remove the pan as described in this section.
4. Remove the two oil pump attaching bolts.
5. Remove the oil pump and strainer assembly from the engine.

➡ **To ensure a leakproof seal, a new oil inlet tube and strainer must be installed if the original oil inlet tube has been moved within the oil pump body.**

To install:

6. Prime the pump with clean engine oil.
7. Position the oil pump and new gasket. Install and tighten the two attaching bolts to 17 ft. lbs. (23 Nm).
8. Install the oil pan.
9. Lower the vehicle.
10. Fill the crankcase to the proper level with clean engine oil.

3.9L Engine

1. Disconnect the negative battery cable.
2. Raise the vehicle and support safely. Drain the oil and remove the oil pan.
3. Remove the screen.
4. Unbolt the oil pump from the rear main bearing cap and remove it from the vehicle.

To install:

5. Prime the pump by pouring fresh oil into the pump intake and turning the driveshaft until oil comes out the pressure port. Repeat a few times until no air bubbles are present. Install the oil pump with a rotating

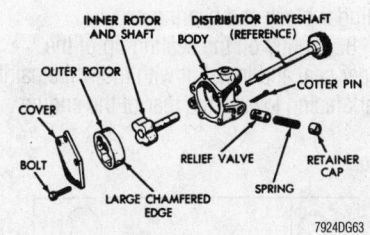

Exploded view of the oil pump assembly— 3.9L, 5.2L and 5.9L engines

motion to ensure proper pump driveshaft engagement.

6. Hold the pump flush against the main cap and finger-tighten the attaching bolts.
7. Tighten the bolts to 130 ft. lbs. (176 Nm).
8. Install the screen.
9. Install the oil pan with a new gasket.
10. Connect the negative battery cable and check the oil pressure.

5.2L and 5.9L Engines

It is necessary to remove the oil pan, and to remove the oil pump from the rear main bearing cap to service the oil pump.

1. Drain the engine oil and remove the oil pan.
2. Remove the oil pump mounting bolts and remove the oil pump from the rear main bearing cap.

To install:

3. Install the pump and tighten the cover bolts to 95 inch lbs. (11 Nm).
4. Prime the oil pump before installation by filling the rotor cavity with engine oil. Install the oil pump on the engine and tighten attaching bolts to 30 ft. lbs. (41 Nm).
5. Install the oil pan.
6. Fill the engine with the proper grade motor oil. Start the engine and check for leaks.

8.0L Engine

1. Remove the timing chain cover.
2. Remove the relief valve plug, gasket, spring and the valve. Discard the gasket.

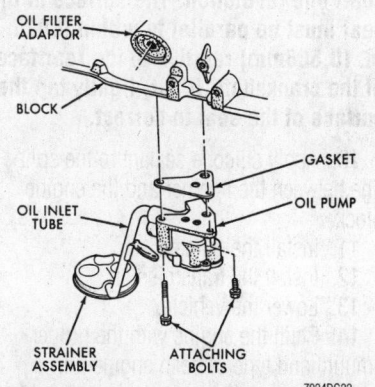

Exploded view of the oil pump assembly mounting—2.5L (OHV) engine

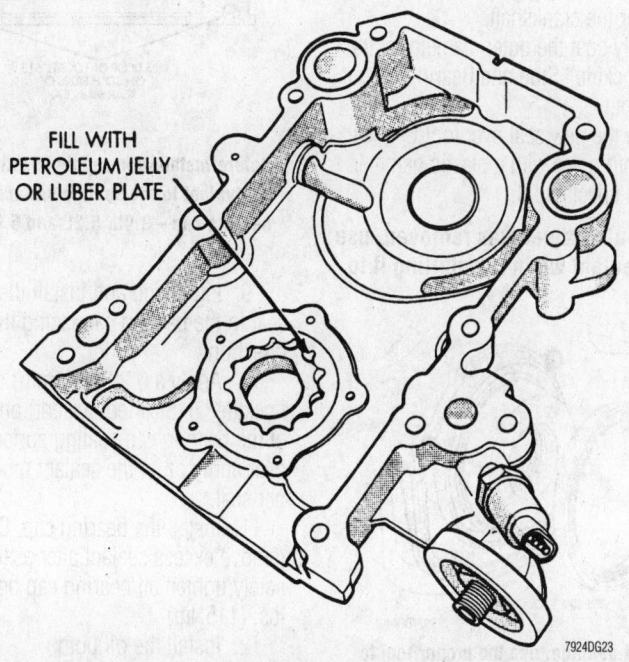

FILL WITH PETROLEUM JELLY OR LUBER PLATE

Lubricate the rotors as a means of priming the oil pump so it is not dry on start-up—8.0L (VIN W) engine

3. Remove the oil pump cover.

4. Remove the pump rotors.

To install:

5. Install the oil pump and lubricate the pump rotors with petroleum jelly or Lubriplate®.

6. Install the timing chain cover.

7. Position the oil pump cover onto the timing chain cover and tighten the cover bolts to 125 inch lbs. (14 Nm).

8. After the cover is installed, ensure that the inner ring can still move freely and does not bind in any way.

9. Install the timing chain cover. Squirt oil into the relief valve hole until oil runs out.

10. Using a new pressure relief valve gasket, install the relief valve plug, tightening it to 15 ft. lbs. (20 Nm).

11. Fill the oil filter with oil and install it on the engine.

Rear Main Seal

REMOVAL & INSTALLATION

2.5L Engines

1. Remove the transmission assembly.

2. Remove the flywheel or flexplate.

3. Carefully pry out the rear main oil seal. Be careful not to damage or scratch the crankshaft sealing surface or the seal bore.

To install:

4. For the SOHC engine place special tool C-4681 or equivalent over the crankshaft. For the OHV engine place special tool 6271-A over the crankshaft.

5. Lightly coat the outer diameter of the seal with Locktite® Stud and Bearing Mount or equivalent.

6. Place the new seal over tool C-4681 and tap it into place with a plastic hammer. Remove the special tool.

➡**If the seal retainer was removed, use silicone sealant when reinstalling it to**

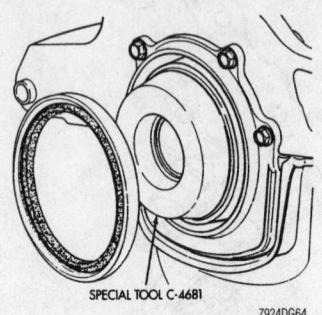

SPECIAL TOOL C-4681

7924DG64

To prevent damage, use the proper tool to guide the rear main seal over the crankshaft—2.5L (SOHC) engine

ensure no leaks. **Tighten the retainer bolts to 105 inch lbs. (12 Nm).**

7. Install the flexplate or flywheel.

8. Install the transmission.

3.9L, 5.2L and 5.9L Engines

1. Disconnect the negative battery cable.

2. Raise and safely support the vehicle securely.

3. Drain the engine oil and remove the oil pan.

4. Remove the oil pump from the rear main bearing cap.

5. Remove the rear main bearing cap.

➡**To ease removal and installation of the oil seal, loosen at least two of the bearing caps forward of the rear main cap.**

6. Carefully remove the upper half of the oil seal from the cylinder block. Remove the lower half from the bearing cap.

To install:

7. Clean the mating surface of the cylinder block and bearing cap.

8. Lightly oil the sealing lip of the upper seal and install it with the white paint mark facing toward the rear of the engine.

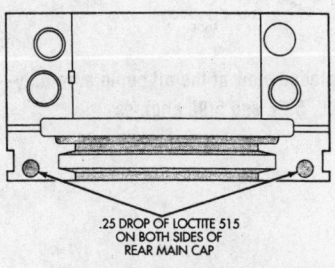

.25 DROP OF LOCTITE 515
ON BOTH SIDES OF
REAR MAIN CAP

7924DG24

Before installation, apply Locktite® 515 or equivalent to the sealing surface of the bearing cap—3.9L, 5.2L and 5.9L engines

9. Lightly oil and install the new lower seal in the bearing cap facing the same direction.

10. Apply a 0.20 in. (5mm) drop of Locktite® 515 (or equivalent) on both sides of the bearing cap sealing surface. Do not over apply or let the sealant touch the rubber seal.

11. Install the bearing cap. Do not remove excess sealant after assembly. Alternately tighten all bearing cap bolts to 85 ft. lbs. (115Nm).

12. Install the oil pump.

13. Install the oil pan.

14. Lower the vehicle.

15. Refill the engine with the proper amount and type of fresh engine oil.

16. Connect the negative battery cable, start the engine and check for leaks.

8.0L Engine

1. Disconnect the negative battery cable.

2. Drain the engine oil.

3. Raise and safely support the vehicle securely.

4. Remove the transmission assembly.

5. Remove the oil pan.

6. Remove the rear oil seal retainer and gasket. Remove the oil seal from the retainer.

To install:

7. Clean the oil seal retainer and retainer mounting surface on the engine block.

8. Lubricate the new seal lip and position it in the retainer.

9. Using special tool 6687 or equivalent, position the new gasket and retainer over the crankshaft. Install the tighten the bolts to 16 ft. lbs. (22Nm).

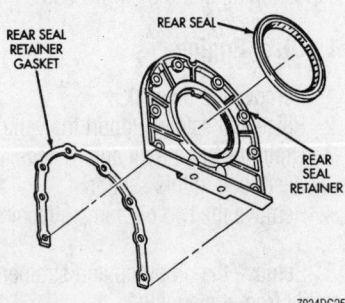

REAR SEAL
RETAINER
GASKET

REAR SEAL

REAR
SEAL
RETAINER

7924DG25

Exploded view of the rear oil seal and retainer—8.0L engine

➡**Mount a dial indicator on the rear of the crankshaft with the tip on the metal surface of the seal. Rotate the crankshaft one revolution. The surface of the seal must be parallel to within 0.020 in. (0.508mm) relative to the rear face of the crankshaft. If not, lightly tap the surface of the seal to correct.**

10. Apply silicone sealant to the split line between the retainer and the engine block.

11. Install the oil pan.

12. Install the transmission.

13. Lower the vehicle.

14. Refill the engine with the proper amount and type of fresh engine oil.

15. Connect the negative battery cable, start the engine and check for leaks.

Timing Chain, Sprockets and Front Cover

REMOVAL & INSTALLATION

2.5L (OHV) Engine

1. Disconnect the negative battery cable.
2. Remove the belt(s) from the crankshaft pulley.
3. Remove the fan and shroud from the vehicle.
4. Remove the crankshaft pulley.
5. Rotate the vibration damper to TDC.
6. Remove the vibration damper using the proper puller.
7. Remove the timing case cover.
8. Align the timing marks on the pulleys as close as possible.
9. Remove the oil slinger from the crankshaft.
10. Remove the cam gear attaching bolts.
11. Remove the cam and crank gear and the timing chain as a unit.

To install:

12. Turn the chain tensioner to the unlocked position. Pull the tensioner block toward the lever to compress the spring.

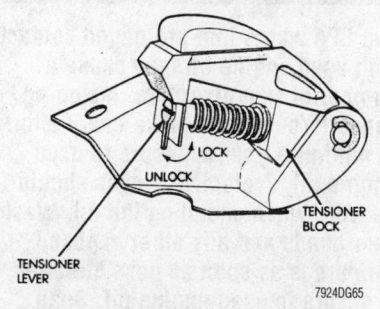

Close up view of the timing chain tensioner, showing the locking lever—2.5L (OHV) engine

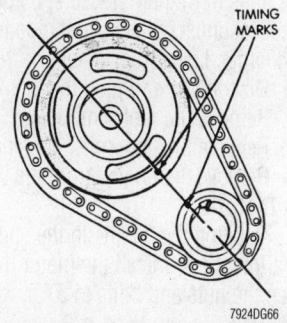

When installed properly, the timing marks should be aligned, as indicated—2.5L (OHV) engine

Hold the spring back and turn the lever to the lock position.

13. Install the cam and crank gear and the timing chain as a unit. Tighten the cam gear attaching bolts to 80 ft. lbs. (108 Nm).
14. Verify that the timing marks are properly aligned. The mark on the cam gear should be at 6 o'clock and the crank gear at 12 o'clock.
15. Turn the chain tensioner to the unlocked (down) position.
16. Position the oil slinger on the crankshaft.
17. Install a new oil seal in the timing case cover.
18. Apply a light coat of gasket sealant around the cover including the bottom surface where the cover joins the oil pan.
19. Position the new gasket and install the cover on the engine with the seal tool to properly align the cover. Tighten the ¼ inch cover bolts to 60 inch lbs. (7 Nm), the 5⁄16 inch bolts to 192 inch lbs. (22 Nm) and the oil pan-to-cover bolts to 84 inch lbs. (9.5 Nm). Remove the seal tool.
20. Install the keyway in the crankshaft. Install the vibration damper. Tighten the attaching bolt to 80 ft. lbs. (108 Nm).
21. Install the crank pulley and drive belt.
22. Install the radiator and shroud.
23. Connect the negative battery cable.

3.9L Engine

1. If possible, crank the engine around so that the No. 1 cylinder is at TDC on the compression stroke. Remove the distributor cap to confirm and line the timing mark on the damper pulley with "0" on the timing scale. This will aid in aligning timing marks when installing the timing gears.
2. Disconnect the negative battery cable.
3. Drain the cooling system.
4. Remove the radiator, fan and all related parts. Remove the water pump.
5. Remove the crankshaft pulley.
6. Remove the vibration damper using the proper puller.
7. Unbolt the chain cover from the block and remove, using caution to avoid damaging the oil pan gasket. Remove the fuel pump from the cover, if equipped.
8. Remove the camshaft gear retaining bolt and cup washer. Remove the timing chain and gears. If necessary, remove the chain and upper gear, then use a gear puller to remove the lower gears.

To install:

9. Place both camshaft and crankshaft gears on the bench with the timing marks on the exact imaginary center line through

both gear bores as they are installed on the engine.

10. Place the timing chain around both sprockets.
11. Turn the crankshaft and camshaft so the keys line up with the keyways in the gears when the timing marks are in proper position.
12. Slide both gears over their respective shafts and use a straightedge to check timing mark alignment.
13. Install the fuel pump eccentric and cup washer, if equipped. Tighten the camshaft gear retaining bolt to 35 ft. lbs. (47 Nm).
14. Clean and dry the mating surfaces of the timing chain cover and block. Apply a thin bead of sealer to the oil pan gasket.
15. Install a new cover gasket and install the cover. Tighten the bolts to 30 ft. lbs. (41 Nm).
16. Install the water pump with a new gasket, if it was removed.
17. Install the damper with tool C-3638, or equivalent, install the bolt and washer and tighten to specification. Apply a small amount of sealer to the bolts and install the crankshaft pulley.
18. Install the fuel pump using a new gasket, if equipped, and connect the fuel lines. Install the two oil pan bolts if they were removed.
19. Install the radiator, fan and all related parts.
20. Fill the cooling system.
21. Connect the negative battery cable, set all adjustments to specifications and check for leaks.

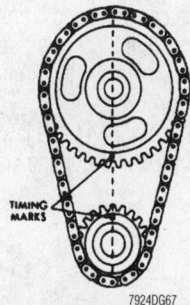

Be sure the timing alignment marks are inline, as shown—3.9L, 5.2L and 5.9L engines

5.2L and 5.9L Engines

1. Remove the front cover.
2. Remove the camshaft sprocket lockbolt, securing cup washer, and fuel pump eccentric. Remove the timing chain with both sprockets.

To install:

3. To begin the installation procedure, place the camshaft and crankshaft sprockets on a flat surface with the timing indicators on an imaginary centerline through both sprocket boxes. Place the timing chain around both sprockets. Be sure that the timing marks are in alignment.

✳✳ WARNING

When installing the timing chain, have an assistant support the camshaft with a suitable tool to prevent it from contacting the plug in the rear of the engine block. Remove the distributor and the oil pump/distributor drive gear. Position the suitable tool against the rear side of the cam gear and be careful not to damage the cam lobes.

4. Turn the crankshaft and camshaft to align them with the keyway location in the crankshaft sprocket and the keyway or dowel hole in the camshaft sprocket.

5. Lift the sprockets and timing chain while keeping the sprockets tight against the chain in the correct position. Slide both sprockets evenly onto their respective shafts.

6. Use a straightedge to measure the alignment of the sprocket timing marks. They must be perfectly aligned.

7. Install the fuel pump eccentric, cup washer, and camshaft sprocket lockbolt and tighten to 35 ft. lbs. (47 Nm). If camshaft end-play exceeds 0.010 in. (0.25mm), install a new thrust plate. It should be 0.002–0.006 in. (0.05–0.15mm) with the new plate.

8.0L Engine

1. Align the camshaft and crankshaft centerline. Remove the camshaft sprocket attaching bolt and remove the timing chain and camshaft sprockets.

2. Use puller No. 6444 and jaws No. 6920, or equivalents, to pull the crankshaft sprocket.

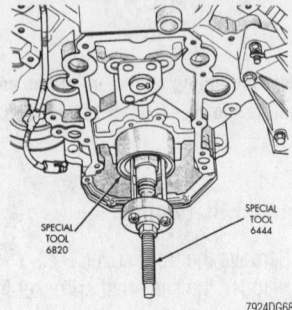

Use the proper puller tool to draw the sprocket off the end of the crankshaft—8.0L engine

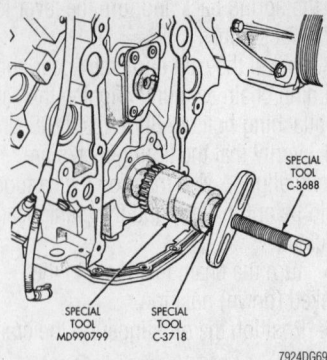

The proper tools also make installing the sprocket much easier—8.0L engine

To install:

3. Align the key in the crankshaft sprocket with the sprocket. Press on the crankshaft sprocket using tools No. C-3688, C-3718 and MB-990799, or their equivalents. Seat the sprocket against the crankshaft shoulder.

4. Turn the crankshaft to line up the timing mark with the crankshaft and camshaft centerline.

5. Put the timing chain on the camshaft sprocket.

6. Align the timing marks and install the chain and camshaft sprocket onto the crankshaft sprocket. Check to see that the timing marks are on the centerline of the crankshaft and camshaft centerline.

7. Install the camshaft bolt. Tighten the bolt to 45 ft. lbs. (61 Nm).

8. Check the camshaft end-play. If a new thrust plate was installed, the clearance should be 0.002–0.006 in. (0.051–0.152mm). If the used thrust plate is installed, the end-play clearance may be up to 0.010 in. (0.254mm).

9. If not within these limits, install a new thrust plate.

DIESEL ENGINE REPAIR

Engine Assembly

REMOVAL & INSTALLATION

✳✳ CAUTION

This engine has a dry weight of 880 lbs. (400 kg). Ensure the engine removal equipment is rated adequately, or serious personal injury could result.

1. Mark the hinge outlines on the hood and remove the hood from the vehicle.

2. Disconnect the negative cable from the battery and from the engine.

3. Drain the coolant.

4. Remove the upper crossmember and top core support.

5. Remove the radiator, shroud, belt, fan and all related parts.

6. Remove the intake and exhaust pipes from the turbocharger.

7. Have the A/C system discharged by an EPA-certified technician.

8. Detach the air conditioner connections. Cover the openings on the compressor.

9. Disengage the alternator and all other electrical connection to the engine.

10. Disconnect the accelerator linkage.

11. Disconnect the throttle linkage from the control lever, but do not remove the control lever from the injection pump.

12. Disconnect all engine-driven accessories.

13. Raise the vehicle and support safely. Remove the starter.

14. If equipped with an automatic transmission, remove the torque converter bolts and remove the lower bell housing bolts. If equipped with a manual transmission, remove the transmission.

15. Drain the oil from the engine.

✳✳ CAUTION

The EPA warns that prolonged contact with used engine oil may cause a number of skin disorders, including cancer! You should make every effort to minimize your exposure to used engine oil. Protective gloves should be worn when changing the oil. Wash your hands and any other exposed skin areas as soon as possible after exposure to used engine oil. Soap and water, or waterless hand cleaner should be used.

16. Disconnect the transmission oil cooler lines from their brackets, if equipped.

17. Disconnect the exhaust pipe from the turbocharger. Lower the vehicle.

18. Disconnect and plug the fuel lines.

19. Remove the motor mounts.

20. Remove the upper bell housing bolts.

21. Remove the engine from the vehicle.

To install:

22. Position the engine in the engine compartment and install the motor mounts. Tighten the nuts and bolts to 57 ft. lbs. (77 Nm).

23. Install the bell housing bolts and tighten the converter bolts, if equipped. Install the manual transmission, if equipped.

24. Install the starter.
25. Connect the exhaust pipe.
26. Connect the transmission oil cooler lines to their brackets, if equipped.
27. Connect the fuel lines.
28. Connect the power steering lines.
29. Connect all engine driven accessories.
30. Connect the accelerator linkage.
31. Connect the throttle linkage to the control lever.
32. Reattach the air conditioner connections.
33. Engage the alternator and all other electrical connections to the engine.
34. Install the intake and exhaust pipes to the turbocharger.
35. Install the fan and all related parts, shroud and radiator.
36. Fill the engine with the proper amount of Diesel engine oil.
37. Fill the radiator with coolant.
38. Connect the negative battery cable, set all adjustments to specifications and check for leaks.

Water Pump

REMOVAL & INSTALLATION

1. Disconnect the negative battery cable.

❊❊ CAUTION

Never open, service or drain the radiator or cooling system when hot; serious burns can occur from the steam and hot coolant. Always drain coolant into a sealable container. Coolant should be reused unless it is contaminated or is several years old.

2. Drain the coolant.
3. Use a ⅜ in. drive breaker bar to lift the belt tensioner and remove the belt.
4. Remove the two water pump retaining bolts and remove the pump from the engine.
5. Remove the O-ring from the pump groove.

To install:

6. Clean the O-ring groove and install a new O-ring.
7. Clean the pump mating surfaces and install the pump to the engine.
8. Tighten the mounting bolts to 18 ft. lbs. (24 Nm). Fill the radiator with coolant.
9. Install the drive belt.
10. Connect the negative battery cable, run the vehicle until the thermostat opens, fill the radiator completely and check for leaks.
11. Once the vehicle has cooled, recheck the coolant level.

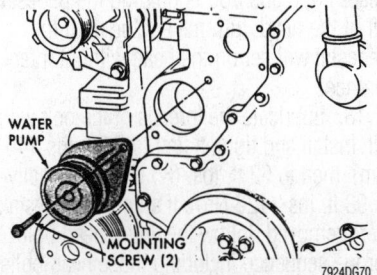

Exploded view of the water pump mounting—5.9L Diesel engine

Glow Plugs

REMOVAL & INSTALLATION

The 5.9L Diesel engine uses an intake manifold air heater instead of glow plugs to preheat the air for improved starting ability. The heater element is located within the intake manifold top cover. Refer to the intake manifold removal and installation procedure to service the intake manifold air heater.

Cylinder Head

REMOVAL & INSTALLATION

1. Disconnect the negative battery cable.
2. Drain the coolant.
3. Disconnect the radiator hose and heater hoses.
4. Remove the turbocharger and air crossover.
5. Remove the exhaust manifold.
6. Remove all fuel lines from the injection pump and injector nozzles. Remove the fuel filter.
7. Remove the cylinder head covers.
8. Remove the rocker arms and pushrods.

9. If the cylinder head is hot, gradually loosen the cylinder head bolts using the TIGHTENING sequence. If the engine is cold, then the loosening sequence for the head bolts is not important Remove the cylinder head.
10. Inspect the coolant passages. A large accumulation of rust or lime will require service to the block.
11. Inspect the surface of the head for flatness. The maximum variation is 0.0004 in. (0.010mm) within any 2 in. (50mm) diameter area or 0.012 in. (0.30mm) overall end-to-end or side-to-side.

To install:

12. Thoroughly clean and dry the mating surfaces of the head and block. Position the new head gasket on the dowels.
13. Install the head onto the dowels on the block.
14. Lubricate the pushrod sockets and install the pushrods and rocker arms.
15. Clean, dry and lightly lubricate the head bolts. Install and tighten in sequence first to 29 ft. lbs. (40 Nm), then to 62 ft. lbs. (85 Nm) and finally to 93 ft. lbs. (126 Nm).
16. Install the rocker arm pedestal bolts. Tighten to 18 ft. lbs. (85 Nm) and finally to 93 ft. lbs. (126 Nm).
17. Install the rocker arm pedestal bolts. Tighten to 18 ft. lbs. (24 Nm).
18. Adjust the valve clearance.
19. Install the cylinder head covers with new gaskets. Tighten the bolts to 18 ft. lbs. (24 Nm).
20. Install all fuel lines and the fuel filter.
21. Install the exhaust manifold.
22. Install the turbocharger and air crossover.
23. Connect the radiator hose and heater hoses.
24. Fill the radiator with coolant.
25. Connect the negative battery cable, set all adjustments to specifications and check for leaks.

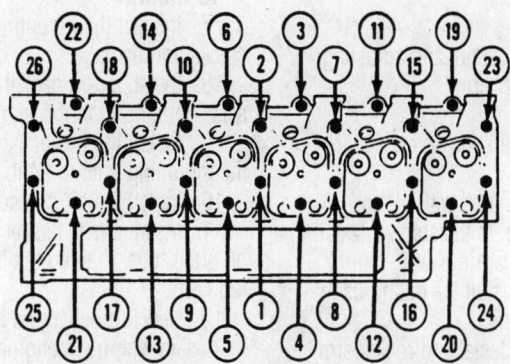

Cylinder head bolt tightening sequence—5.9L Diesel engine

Rocker Arms/Shafts

REMOVAL & INSTALLATION

1. Disconnect the negative battery cable.
2. Remove the cylinder head cover.
3. Loosen the adjusting screw locknuts. Loosen the screws until they stop.

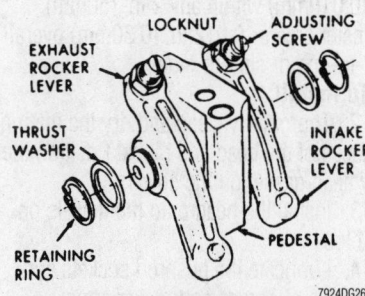

Rocker arms and related components— 5.9L Diesel engine

Rocker arm bolt tightening sequence— 5.9L Diesel engine

4. Remove the 8mm and 12mm bolt from the pedestal.
5. Remove the pedestal and rocker arm assembly. Remove the pushrods if necessary.
6. Remove the retaining ring and thrust washer.
7. Remove the rocker arm from the pedestal.

➡ **Do not disassemble the rocker shaft and pedestal; they must be replaced as an assembly.**

8. Remove the locknut and adjusting screw from the rocker arm.
To install:
9. Install the adjusting screw and locknut.
10. Lubricate the shaft with oil and install the rocker arm to the shaft. Install the thrust washer and snapring.
11. If removed, install the pushrods in their original locations.
12. Install the pedestal and rocker arm assembly to the head aligning the dowel in the pedestal with the dowel bore in the

head. If the pushrod is holding the pedestal off of the head, turn the engine until the pedestal will set on the head without interference.
13. Lubricate the threads of the bolt with oil. Install and tighten first to 29 ft. lbs. (40 Nm), then to 62 ft. lbs. (85 Nm) and finally to 93 ft. lbs. (126 Nm). If all of the pedestals were removed, follow the entire head bolt torque sequence including those head bolts that were not removed in this procedure.
14. Tighten the 8mm bolts to 18 ft. lbs. (24 Nm).
15. Adjust the valves.
16. Install the cylinder head cover with a new gasket. Tighten the bolts to 18 ft. lbs. (24 Nm).
17. Connect the negative battery cable.

Turbocharger

REMOVAL & INSTALLATION

1. Disconnect the negative battery cable.
2. Loosen the air crossover hose.

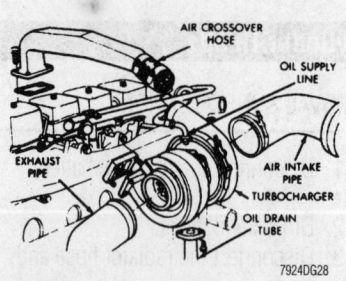

Exploded view of the turbocharger mounting—5.9L Diesel engine

3. Disconnect the intake hose and exhaust pipe.
4. Remove the oil drain tube bolts.
5. Remove the oil supply line from the turbocharger.
6. Remove the turbocharger mounting nuts and remove the turbocharger.
To install:
7. Inspect the mounting surface for cracks and damage.
8. Install a new gasket and apply anti-seize compound to the mounting studs.
9. Install the turbocharger and tighten the mounting nuts to 24 ft. lbs. (32 Nm).
10. Install the air crossover hose.
11. Install a new gasket and install the oil drain tube. Tighten the bolts to 18 ft. lbs. (24 Nm).
12. New turbochargers must be pre-lubricated with fresh engine oil before operation. To do so, pour about 2 or 3 oz. (65–100 ml) of oil into the supply fitting

and rotate the turbine wheel to circulate the oil.
13. Install the oil supply line to the turbocharger.
14. Connect the intake and exhaust piping.
15. Connect the negative battery cable and check for exhaust leaks.

Intake Manifold

REMOVAL & INSTALLATION

1. Disconnect the negative battery cable.
2. Remove the throttle control bracket and linkage.
3. Remove the high pressure fuel lines.
4. Disconnect the intake manifold heater.
5. Disconnect the fuel heater ground wire from the intake manifold.
6. Remove the air crossover tube and the intake manifold heater.
7. Remove the manifold cover and gasket. Clean the gasket sealing surface.
To install:
8. If removed, install the intake manifold heater and crossover tube.
9. Install the new gasket and cover. Some of the bolt holes are drilled through. Apply liquid Teflon® sealant to these bolts. Tighten all bolts to 18 ft. lbs. (24 Nm).
10. Assemble all intake piping and intake manifold heater with the throttle control bracket and linkage.
11. Connect the fuel heater ground wiring.
12. Install and bleed the high pressure fuel lines.
13. Connect the negative battery cable.

Exhaust Manifold

REMOVAL & INSTALLATION

1. Disconnect the negative battery cable.
2. Disconnect the air intake hose and exhaust pipe from the turbocharger.
3. Remove the turbocharger and gasket.
4. Remove the cab heater supply and return lines.
5. Remove the exhaust manifold and gasket. Clean the gasket sealing surface.
To install:
6. Install the new gasket and manifold. Starting from the center, tighten the exhaust manifold bolts to 32 ft. lbs. (43 Nm).
7. Install the cab heater supply and return lines.

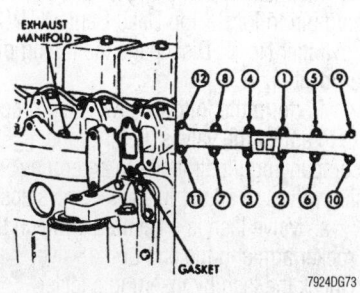

Tighten the exhaust manifold mounting bolts according to the sequence shown—5.9L Diesel engine

8. Install the turbocharger and gasket.

9. Connect the air intake and exhaust pipes.

10. Connect the negative battery cable, start the engine and check for exhaust leaks.

Front Crankshaft Seal

REMOVAL & INSTALLATION

1. Remove the drive belt.

2. Remove the vibration damper.

3. Drill two 1/8 in. (3mm) holes into the seal face, 180° apart.

4. Thread #10 sheet metal screws into the holes in the face of the seal.

5. Use a slide hammer tool to pull on the sheet metal screws, alternating from side-to-side until the seal is free.

To install:

6. Ensure the sealing surface on the crankshaft is completely free of all oil residue and debris to prevent seal leaks.

7. If the gear cover was replaced, use the alignment tool from the seal kit to ensure the cover is aligned with the crankshaft.

8. Apply a bead of Loctite® 277, or equivalent, to the outside diameter of the seal.

9. Install the pilot from the seal kit onto the crankshaft.

10. Install the seal onto the pilot and start it into the gear housing cover seal bore.

11. Remove the pilot.

12. Use the alignment/installation tool and a plastic hammer to install the seal to the correct depth.

13. Install the vibration damper, but DO NOT tighten the damper bolt until the belt is installed.

14. Install the drive belt.

15. Tighten the vibration damper bolts to 92 ft. lbs. (125 Nm). Use an engine barring tool to keep the engine from rotating during the tightening procedure.

Camshaft and Valve Lifters

REMOVAL & INSTALLATION

1. Disconnect the negative battery cable.

2. Remove the cylinder head cover.

3. Remove the rocker arm assemblies.

4. Remove the pushrods.

5. Remove the drive belt.

✳✳ CAUTION

Never open, service or drain the radiator or cooling system when hot; serious burns can occur from the steam and hot coolant. Always drain coolant into a sealable container. Coolant should be reused unless it is contaminated or is several years old.

6. Drain the cooling system. Remove the fan assembly, radiator and all related parts.

7. Remove the crankshaft pulley.

8. Remove the front gear cover.

9. Remove the fuel pump.

10. Insert the special dowels into the pushrod holes and onto the top of each lifter. When properly installed, the dowels can be used to hold the lifters up securely. Wrap rubber bands around the top of the dowels to prevent them from dropping down.

11. Rotate the crankshaft to align the crankshaft-to-camshaft timing marks.

12. Remove the bolts from the thrust plate.

13. Remove the camshaft and thrust plate.

14. Press the gear from the camshaft and remove the key.

15. To remove the lifters, install a trough into the camshaft opening to catch the lifters as the dowels are removed. If the lifters are to be reused, remove one lifter at a time and note it's positions.

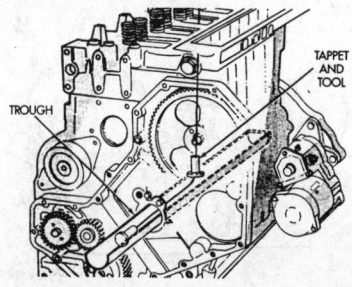

Use the special tools to remove or install the lifters as shown—5.9L Diesel engine

To install:

16. If the lifters have been removed, install the trough in the camshaft opening. Feed the lifter installation tool down the appropriate lifter bore.

17. Pull the trough out and retrieve the lifter installation tool. Attach the lifter to the tool, then slide the trough back in the cam bore while pulling the lifter into the lifter bore.

18. After the lifter is in the bore, turn the trough 1/2 turn so the round side of the trough holds the lifter in place.

19. Remove the lifter installation tool and reinstall the special dowel to hold the lifter in place while the remaining lifters are installed.

20. Install the key on the camshaft.

21. Heat the camshaft gear to 250°F (121°C) for 45 minutes. Lubricate the gear mount surface with Lubriplate 105®. Install the gear to the camshaft with the timing marks facing away from the shaft.

22. Lubricate the camshaft bores, lobes, journals and thrust washer with Lubriplate 105®.

✳✳ WARNING

Do not push the camshaft in too far or it may dislodge the plug in the rear of the camshaft bore, possibly creating a leak.

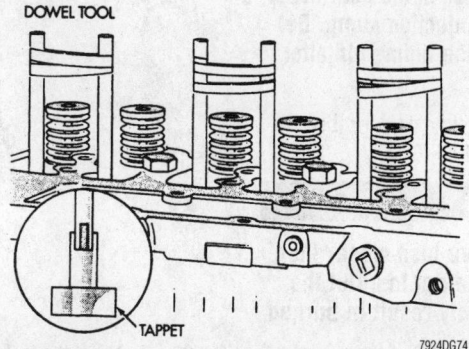

Use dowel tools and rubber bands to hold the lifters up in the bore while removing the camshaft—5.9L Diesel engine

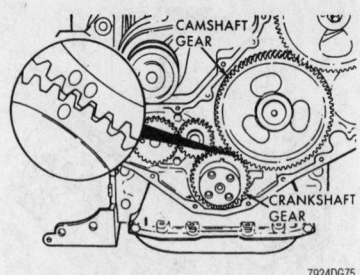

During assembly, be sure to align the marks on the camshaft and crankshaft gears as shown—5.9L Diesel engine

23. Install the camshaft and thrust washer so the timing marks on the camshaft and crankshaft gears align.

24. Install the thrust washer bolts and tighten to 18 ft. lbs. (24 Nm).

25. Check the end-play of the camshaft. The specification is 0.006–0.010 in. (0.152–0.254mm).

26. Check the backlash of the camshaft gear. The specification is 0.003–0.013 in. (0.080–0.330mm).

27. Install the pushrods.

28. Install the rocker pedestal and arm assemblies.

29. Install the front cover and crankshaft pulley.

30. Install the drive belt and fan assembly.

31. Install the fuel pump.

32. Adjust the valves.

33. Install the cylinder head covers.

34. Connect the negative battery cable and check for leaks.

Valve Lash

ADJUSTMENT

Valve adjustment is required on the Diesel engine every 36,000 miles (57,971 km).

➡The timing pin is used in this procedure to locate Top Dead Center (TDC). It is found at the back of the gear housing and below the injection pump. Be sure to disengage the timing pin after locating TDC.

1. The engine must be cold for this adjustment (below 140°F/60°C).

✳✳ WARNING

Do not set the valve lash closer than specified in an attempt to quiet the lifters. This will only result in burned valves.

2. Remove the valve cover.

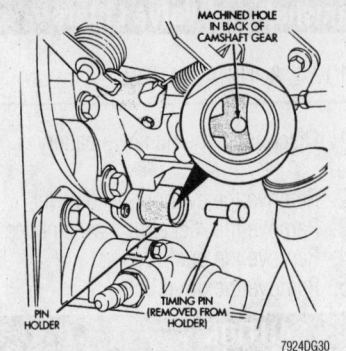

Use the timing pin to locate TDC for Diesel engines

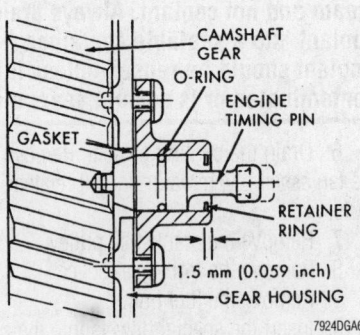

Cut away view of the timing pin entering the hole in the cam gear—Diesel engine

3. Manually turn the engine and use the timing pin to locate Top Dead Center (TDC) for cylinder No. 1. Disengage the timing pin after locating TDC.

4. Perform the following two-step procedure to adjust the valves. Refer to the accompanying illustrations to determine which valves to adjust in each of the steps:

a. Valve lash is measured between the rocker arm and the end of the valve. Check the lash by inserting a feeler gauge between the rocker arm and the valve. When the clearance is correct, there should be a slight drag on the feeler gauge. Adjust the clearance, if necessary. The clearance for the intake valves is 0.010 in. (0.254mm). The clearance for the exhaust valves is 0.20 in. (0.508mm). Adjust the lash by loosening the locknut on the rocker arm and turning the adjusting screw. After the adjustment is made, tighten the locknut to 18 ft. lbs. (24 Nm) and recheck the lash to be sure it did not change as the locknut was being tightened.

b. Double check that the timing pin is disengaged, then mark the pulley and turn the engine crankshaft 360° in the normal direction of rotation (clockwise). Check and adjust the clearance of the

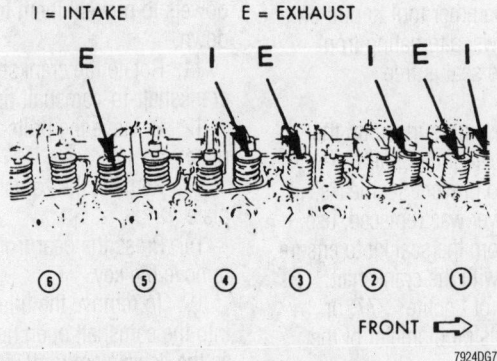

Step 1. Adjust the lash on the valves indicated—5.9L Diesel engine

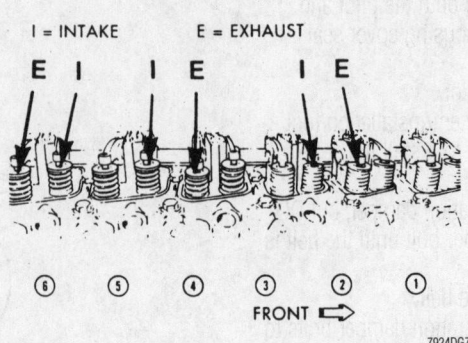

Step 2. Adjust the lash on the remaining valves not adjusted during Step 1—5.9L Diesel engine

indicated valves following the same specifications as in Step 4a.

5. Install the rocker cover with a new gasket, then attach the fuel line to the injector. Start the engine and check for leaks.

Oil Pan

REMOVAL & INSTALLATION

✷✷ CAUTION

The EPA warns that prolonged contact with used engine oil may cause a number of skin disorders, including cancer! You should make every effort to minimize your exposure to used engine oil. Protective gloves should be worn when changing the oil. Wash your hands and any other exposed skin areas as soon as possible after exposure to used engine oil. Soap and water, or waterless hand cleaner should be used.

1. Remove the engine and place it safely on an engine stand.
2. Drain the engine oil into a suitable container and dispose of it properly.
3. Remove the oil pan and gasket. Be sure to connect the support bracket.
4. If required, remove the suction tube and gasket.

To install:
5. Clean the sealing surface.
6. Install the suction tube and gasket. Tighten the bolts to 18 ft. lbs. (24 Nm).
7. Fill the joint between the pan rail/gear housing and pan rail/rear cover with sealant. Use Three Bond 1207-C, or equivalent.
8. Install the pan and gasket. Tighten the bolts to 18 ft. lbs. (24 Nm).
9. Install the drain plug with a new sealing washer. Tighten to 60 ft. lbs. (80 Nm).

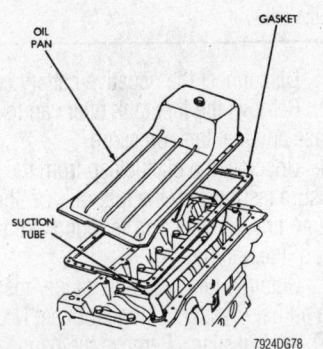

Exploded view of oil pan mounting—5.9L Diesel engine

10. Install the engine assembly into the vehicle.
11. Fill the engine with clean engine oil. Run the engine and check for leaks.
12. Stop the engine and let it set for five minutes. Check the oil level. Add oil, if necessary.

Oil Pump

REMOVAL & INSTALLATION

✷✷ CAUTION

The EPA warns that prolonged contact with used engine oil may cause a number of skin disorders, including cancer! You should make every effort to minimize your exposure to used engine oil. Protective gloves should be worn when changing the oil. Wash your hands and any other exposed skin areas as soon as possible after exposure to used engine oil. Soap and water, or waterless hand cleaner should be used.

1. Disconnect the negative battery cable.
2. Remove the drive belt.
3. Remove the radiator.
4. Remove the fan assembly.
5. Remove the oil fill tube and adapter.
6. Remove the crankshaft pulley.
7. Remove the front cover.
8. Remove the four pump mounting bolts and remove the pump from the block.

To install:
9. Prime the pump by pouring fresh oil into the pump intake and turning the driveshaft until oil comes out the pressure port. Repeat this a few times until no air bubbles are present.
10. Align the idler gear pin with the locating bore in the block and install the pump.
11. Tighten the mounting bolts in the proper sequence to 44 inch lbs. (5 Nm), then repeat the tightening sequence to 18 ft. lbs. (24 Nm).

➡**When the pump is correctly installed, the flange on the pump should not touch the block; the back plate on the pump seats against the bottom of the bore.**

12. Measure the backlash of the idler-to-pump drive gears. The specification is 0.003–0.013 in. (0.08–0.33mm).
13. Measure the backlash of the idler-to-crankshaft gears. The specification is 0.003–0.013 in. (0.08–0.33mm).

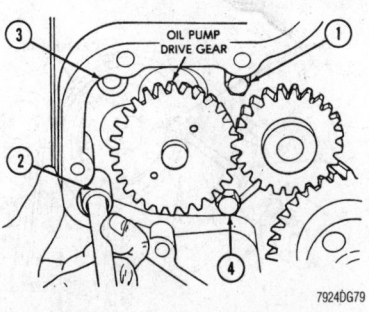

Be sure to tighten the oil pump mounting bolts in the sequence shown—5.9L Diesel engine

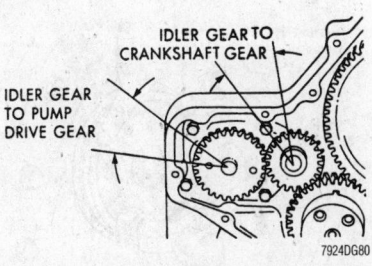

Replace the idler and oil pump gears if the camshaft gear backlash is not within specification—5.9L Diesel engine

14. Install the front cover and crankshaft pulley.
15. Install the oil fill tube and adapter.
16. Install the fan assembly, radiator and drive belt.
17. Connect the negative battery cable and check the oil pressure.

Rear Main Seal

REMOVAL & INSTALLATION

1. Remove the transmission.
2. Remove the pressure plate, clutch, and flywheel, if equipped.
3. Remove the rear seal housing and gasket.
4. Drill two ⅛ in. (3mm) holes into the seal face, 180° apart.
5. Thread #10 sheet metal screws into the holes in the face of the seal.
6. Use a slide hammer tool to pull on the sheet metal screws, alternating from side-to-side until the seal is free.

To install:
7. Clean the sealing surface of all dirt and oil.
8. Install the seal pilot, included with the replacement seal kit. punch the seal on the pilot and crankshaft.

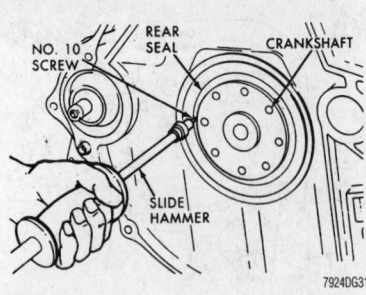

Carefully remove the rear seal with a sheet metal screw and slide hammer—5.9L Diesel

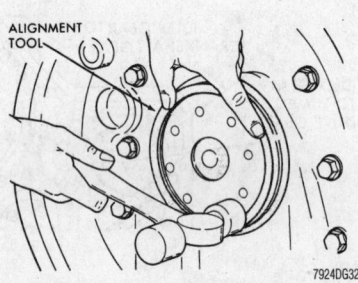

Place the alignment tool on the seal and tap the seal into place—5.9L Diesel

9. Install the seal by tapping the alignment/installation tool until the tool is flush with the block.

10. Install the rear seal housing and gasket. Tighten the housing bolts to 84 inch lbs. (9 Nm).

11. Install the pressure plate, clutch, and flywheel, if equipped.

12. Install the transmission.

Timing Gears

REMOVAL & INSTALLATION

1. Disconnect the negative battery cable.
2. Remove the fan drive assembly and the belt.
3. Remove the belt tensioner.
4. Remove the oil filler tube and adapter.
5. Remove the crankshaft vibration damper.
6. Remove the gear cover to gain access to the timing gears.
7. Simply unbolt the gears and slide them off their respective shafts.

To install:

8. Lubricate the gears with engine oil and install them on the camshaft and crankshaft. Be sure to align the timing marks.

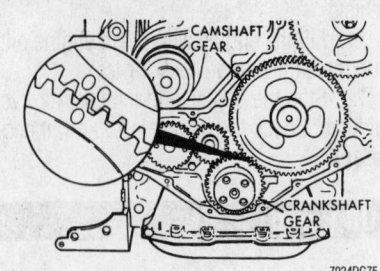

Timing gear alignment marks—5.9L Diesel engine

9. Install a new crankshaft seal in the gear cover.

10. Install the pilot tool from the seal kit on the crankshaft.

11. Install the gear cover with a new gasket. Tighten the bolts to 18 ft. lbs. (24 Nm), then remove the pilot tool.

12. Install the oil filler tube assembly. Tighten the mounting bolts to 32 ft. lbs. (43 Nm).

13. Install the belt tensioner. Tighten the mounting bolts to 32 ft. lbs. (43 Nm).

14. Raise the tensioner and install the belt.

15. Install the crankshaft vibration damper. Tighten the bolt to 92 ft. lbs. (125 Nm).

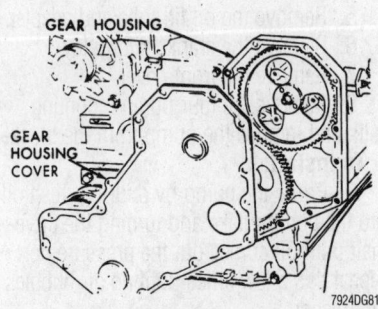

Remove the gear cover to replace the timing gears—5.9L Diesel engine

16. Install the fan drive assembly.
17. Connect the negative battery cable.

GASOLINE FUEL SYSTEM

Fuel System Service Precautions

Safety is the most important factor when performing not only fuel system maintenance but any type of maintenance. Failure to conduct maintenance and repairs in a safe manner may result in serious personal injury or death. Maintenance and testing of the vehicle's fuel system components can be accomplished safely and effectively by adhering to the following rules and guidelines.

• To avoid the possibility of fire and personal injury, always disconnect the negative battery cable unless the repair or test procedure requires that battery voltage be applied.

• Always relieve the fuel system pressure prior to detaching any fuel system component (injector, fuel rail, pressure regulator, etc.), fitting or fuel line connection. Exercise extreme caution whenever relieving fuel system pressure to avoid exposing skin, face and eyes to fuel spray. Please be advised that fuel under pressure may penetrate the skin or any part of the body that it contacts.

• Always place a shop towel or cloth around the fitting or connection prior to loosening to absorb any excess fuel due to spillage. Ensure that all fuel spillage (should it occur) is quickly removed from engine surfaces. Ensure that all fuel soaked cloths or towels are deposited into a suitable waste container.

• Always keep a dry chemical (Class B) fire extinguisher near the work area.

• Do not allow fuel spray or fuel vapors to come into contact with a spark or open flame.

• Always use a back-up wrench when loosening and tightening fuel line connection fittings. This will prevent unnecessary stress and torsion to fuel line piping. Always follow the proper torque specifications.

• Always replace worn fuel fitting O-rings with new. Do not substitute fuel hose or equivalent, where fuel pipe is installed.

Fuel System Pressure

RELIEVING

1. Disconnect the negative battery cable.
2. Remove the fuel tank filler cap to release any fuel tank pressure.
3. Unscrew the plastic cap from the pressure test port on the fuel rail. On the 8.0L engine, the test port is found at the front of the engine.
4. Obtain a fuel pressure gauge/hose from a Fuel Pressure Gauge tool set No. 5069, or equivalent. Remove the gauge, then place the gauge end of the hose into a suitable gasoline container.

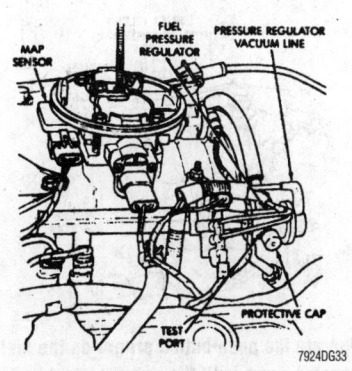

Fuel pressure test port—3.9L engine

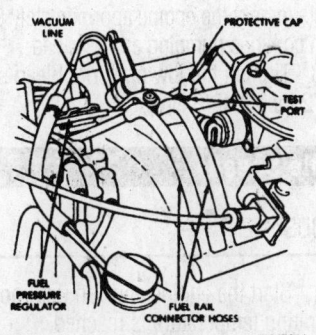

Fuel pressure test port—5.2L engine, 5.9L engine is similar

5. Place a shop towel under the test port.

6. Screw the other end of the hose onto the fuel pressure port to relieve the pressure.

7. When the pressure has been relieved, remove the hose and cap the port.

Fuel Filter

REMOVAL & INSTALLATION

❊❊ CAUTION

Observe all applicable safety precautions when working around fuel. Whenever servicing the fuel system, always work in a well ventilated area. Do not allow fuel spray or vapors to come in contact with a spark or open flame. Keep a dry chemical fire extinguisher near the work area. Always keep fuel in a container specifically designed for fuel storage; also, always properly seal fuel containers to avoid the possibility of fire or explosion.

The fuel filter is integrated into the fuel pressure regulator which is mounted in the fuel tank. This unit is not controlled by the PCM or engine vacuum. It is calibrated to deliver approximately 35–45 psi (241–310 kPa) of fuel pressure to the injectors. If the pressure exceeds the maximum of the specified range, an internal diaphragm closes to route fuel back into the fuel tank. This system eliminates the need for conventional return lines from the engine bay and accounts for the name of the "Returnless" fuel injection system employed in these vehicles.

➡**Fuel tank removal is required for this procedure. Also needed will be external snapring pliers and proper hose clamp pliers, such as No. C-4124 pliers (available through Plymouth/Dodge dealers), or equivalent.**

1. Properly relieve the fuel system pressure.
2. Drain the fuel tank and remove the tank.
3. Remove the fuel filter/regulator (which is pressed into a rubber grommet) by twisting and pulling it straight up.
4. Remove the snapring retaining the cover tube, then slide it down to reveal the clear plastic fuel tube and its retaining clamp.

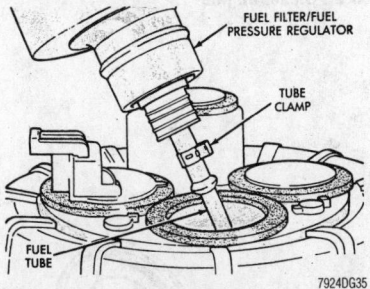

Pull and twist the filter/regulator to remove it from the top of the fuel pump module—5.9L Diesel engine

5. Gently cut off the old clamp without damaging the tube, then discard the clamp.
6. Carefully pull the tube off, then remove the filter/regulator from the fuel pump module.

To install:
7. Install a new clamp over the plastic fuel tube and attach it loosely to the filter/regulator. Rotate the unit in the line until it is pointed to the driver's side of the vehicle.

8. Tighten the clamp using hose clamp pliers.

➡**Do not use conventional side cutters to tighten the clamp.**

9. Slide the cover tube up to the bottom of the filter/regulator and install the snapring.
10. Carefully press the assembly back into the rubber grommet by hand. It should be pointed to the driver's side of the vehicle.
11. Install the fuel tank.

Fuel Pump

REMOVAL & INSTALLATION

❊❊ CAUTION

The fuel injection system is under a constant pressure. Before servicing any part of the fuel injection system, the system pressure must be released. Use a clean shop towel to catch any fuel spray and take precautions to avoid the risk of fire.

1. Perform the fuel pressure release procedure described previously in this section.

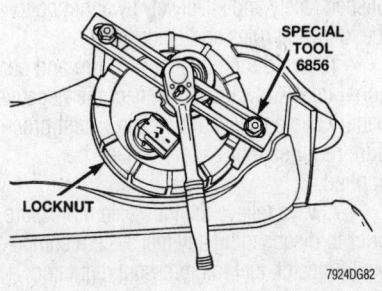

Remove the locknut ring to release the fuel pump module from the tank

2. Properly relieve the fuel system pressure and disconnect the negative battery cable.
3. Remove the fuel tank from the vehicle.
4. Note the direction of the fuel filter/fuel pressure regulator, the pressure relief/rollover valve and the pump electrical connector. These should all be pointed to the driver's side. Tag any lines for replacement, if necessary.

5. Remove the locknut threaded into the fuel tank. The fuel pump module will spring up when the locknut is removed.

6. Remove the module from the fuel tank.

To install:

7. Use a new gasket and position the fuel pump module in the opening. The fuel filter/fuel pressure regulator, the pressure relief/rollover valve and the fuel pump electrical connector should all be pointed to the driver's side of the vehicle when this unit is properly installed.

8. Position a new locknut over the top of the fuel pump module, then tighten the locknut.

9. Install the fuel tank.

DIESEL FUEL SYSTEM

Fuel System Service Precautions

Safety is the most important factor when performing not only fuel system maintenance but any type of maintenance. Failure to conduct maintenance and repairs in a safe manner may result in serious personal injury or death. Maintenance and testing of the vehicle's fuel system components can be accomplished safely and effectively by adhering to the following rules and guidelines.

• To avoid the possibility of fire and personal injury, always disconnect the negative battery cable unless the repair or test procedure requires that battery voltage be applied.

• Always relieve the fuel system pressure prior to disengaging any fuel system component (injector, fuel rail, pressure regulator, etc.), fitting or fuel line connection. Exercise extreme caution whenever relieving fuel system pressure to avoid exposing skin, face and eyes to fuel spray. Please be advised that fuel under pressure may penetrate the skin or any part of the body that it contacts.

• Always place a shop towel or cloth around the fitting or connection prior to loosening to absorb any excess fuel due to spillage. Ensure that all fuel spillage (should it occur) is quickly removed from engine surfaces. Ensure that all fuel soaked cloths or towels are deposited into a suitable waste container.

• Always keep a dry chemical (Class B) fire extinguisher near the work area.

• Do not allow fuel spray or fuel vapors to come into contact with a spark or open flame.

• Always use a back-up wrench when loosening and tightening fuel line connection fittings. This will prevent unnecessary stress and torsion to fuel line piping. Always follow the proper torque specifications.

• Always replace worn fuel fitting O-rings with new. Do not substitute fuel hose or equivalent, where fuel pipe is installed.

Fuel System

BLEEDING AIR

1. Loosen the low pressure bleed bolt.

2. Operate the rubber push-button primer on the fuel transfer pump. Do this until the fuel exiting the bleed screw is free of air. If the

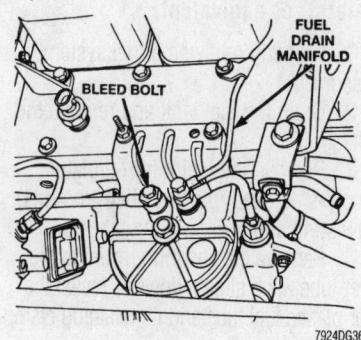

Location of the low pressure bleed bolt— 5.9L Diesel engine

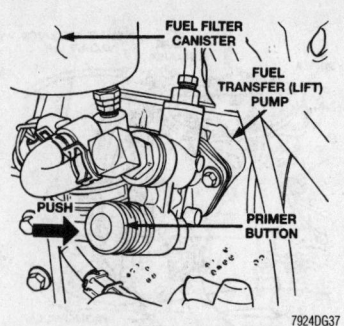

Operate the push-button primer on the fuel transfer pump until the escaping fuel is free of air

primer button feels as if it is not pumping, rotate (crank) the engine approximately 90°, then continue pumping as described.

3. Tighten the low pressure bleed screw to 72 inch lbs. (8 Nm).

Idle Speed

ADJUSTMENT

1. Start the engine and run until normal operating temperature is reached.

2. An optical tachometer must be used to read engine speed.

3. If equipped, turn the air conditioning ON.

4. Turn the idle speed screw until the desired idle speed is obtained. The specifi-

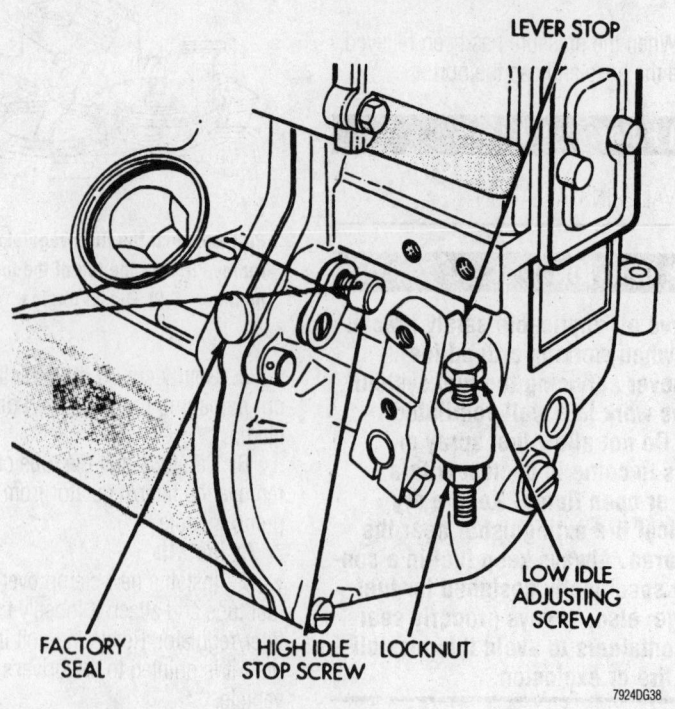

Idle speed adjusting screw location—5.9L Diesel engine

cation for a vehicle equipped with automatic transmission is 700 rpm. The specification for a vehicle equipped with manual transmission is 750 rpm.

Fuel Water Separator/Filter

DRAINING WATER

Filtration and separation of water from the fuel is important for trouble-free operation and long life of the fuel system. Regular maintenance, including draining moisture from the fuel/water separator filter is essential to keep water out of the fuel pump. To remove the collected water, simply unscrew the drain at the bottom of the Water In Filter (WIF) assembly located at the bottom of the filter separator.

REMOVAL & INSTALLATION

1. Disconnect the negative battery cable.
2. Disengage the WIF sensor connector.
3. Remove the separator filter assembly from the filter head with a standard oil filter wrench.
4. Remove the square cut O-ring from the filter mounting bushing.
5. Drain the fuel/water separator filter and remove the assembly from the fuel filter.

To install:

6. Install a new O-ring to the WIF assembly and install to the new separator filter.
7. Install a new square cut O-ring to the mounting bushing.
8. Fill the fuel/water separator filter with clean Diesel fuel.
9. Apply a light coat of oil to the sealing surface of the separator filter.
10. Install the assembly and tighten it ½ turn after the seal contacts the filter head.
11. Reattach the WIF sensor connector.
12. Connect the negative battery cable, start the engine and check for leaks.

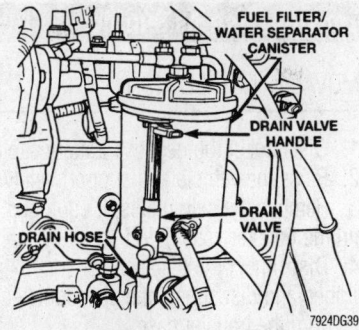

Fuel filter/water separator assembly and related components

Diesel Injection Pump

REMOVAL & INSTALLATION

➡**The Bosch VE lever is indexed to the shaft during pump calibration. Do not remove it from the pump during removal.**

1. Disconnect the negative battery cable.
2. Remove the throttle linkage and bracket.
3. Disconnect the fuel drain manifold.
4. Remove the injection pump supply line.
5. Remove the high pressure lines.
6. Disconnect the electrical wire to the fuel shut off valve.
7. Remove the fuel air control tube.
8. Remove the pump support bracket.
9. Remove the oil fill tube bracket and adapter from the front gear cover.
10. Place a shop towel in the gear cover opening in a position that will prevent the nut and washer from falling into the gear housing. Remove the gear retaining nut and washer.
11. Install the turning tool into the flywheel housing opening on the exhaust side of the engine. Place a ½ in. drive universal joint in the turning tool and attach enough extensions to the joint to make it convenient to turn the tool.
12. Using a ratchet to turn the barring tool, turn the engine until the keyway on the fuel pump shaft is pointing approximately in the six o'clock position.
13. Locate TDC for cylinder No. 1 by turning the engine slowly while pushing in on the TDC pin. Stop turning the engine as soon as the pin engages with the gear timing hole. Disengage the pin after locating TDC and remove the turning equipment.
14. Loosen the lockscrew, remove the special washer from the injection pump and wire it to the line above it so it will not get misplaced. Retighten the lockscrew to 22 ft. lbs. (30 Nm) to lock the driveshaft.
15. Using a suitable puller, pull the pump drive gear from the driveshaft.

➡**Be careful not to drop the drive gear key into the front cover when removing or installing the pump. If it does drop in, it must be removed before proceeding.**

16. Remove the three mounting nuts and remove the injection pump from the vehicle.
17. Remove the gasket and clean the mounting surface.

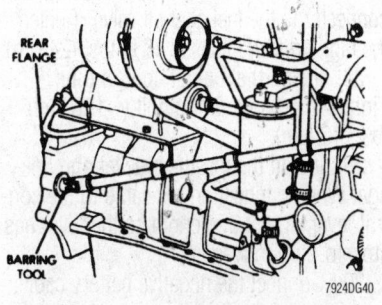

Use the barring tool to rotate the Diesel engine

To install:

18. Install a new gasket.

➡**The shaft of a new or reconditioned pump is locked so the key aligns with the drive gear keyway with cylinder No. 1 at TDC.**

19. Install the pump and finger-tighten the mounting nuts; the pump must be free to move in the slots.
20. Install the pump drive gear, washer and nut to the driveshaft. The pump will rotate slightly because of gear helix and clearance. This is acceptable providing the pump is free to move on the flange slots and the crankshaft does not move. Tighten the nut to 11–15 ft. lbs. (15–20 Nm). This is not the final torque; do not overtighten.
21. If installing the original pump, rotate the pump to align the original timing marks and tighten the mounting nuts to 18 ft. lbs. (24 Nm).
22. If installing a replacement pump, take up gear lash by rotating the pump counterclockwise toward the cylinder head, and tighten the mounting nuts to 18 ft. lbs. (24 Nm). Permanently mark the new injection pump flange to match the mark on the gear housing.
23. Loosen the lockscrew and install the special washer under the lockscrew; tighten to 13 ft. lbs. (18 Nm). Disengage the TDC pin.
24. Install the injection pump support bracket. Finger-tighten the bolts initially, then tighten them to 18 ft. lbs. (24 Nm) in the following sequence:
    a. Bracket to block bolts.
    b. Bracket to injection pump bolts.
    c. Throttle support bracket bolts.
25. Now perform the final tighten of the pump drive gear retaining nut to 48 ft. lbs. (65 Nm).
26. Install the oil filler tube assembly and clamp. Tighten the bolts to 32 ft. lbs. (43 Nm).

27. Install all fuel lines and the electrical connector to the fuel shut off valve. Tighten the high pressure lines to 18 ft. lbs. (24 Nm).

28. Install the fuel air control tube. Tighten the banjo fitting bolt to 108 inch lbs. (12 Nm).

29. Install the throttle bracket and linkage. When connecting the cable to the control lever, adjust the length so the lever has stop-to-stop movement.

30. Connect the negative battery cable.

✳ CAUTION

Do not place any part of the hand near the base of the high pressure line. A fuel leak from a high pressure fuel line has sufficient pressure to penetrate the skin and cause serious bodily harm. Do not bleed the lines if the engine is hot. Fuel spilling onto a hot exhaust manifold creates the danger of fire.

31. To bleed air from the system, run or crank the engine and carefully loosen the high pressure fitting from each injector one at a time. Retighten the fitting after the air has expelled before going on to the next injector fitting. The operation is complete when the engine runs smoothly. If the air cannot be removed, check the pump and supply line for suction leaks.

32. Adjust the idle speed if necessary.

DRIVE TRAIN

➡**For driveshaft, Universal joint (U-joint) and Constant Velocity joint (CV-joint) boot service, please refer to the Unit Repair Section (URS) in the front of this manual.**

Transmission Assembly

REMOVAL & INSTALLATION

1. Disconnect the negative battery cable.
2. On manual transmissions, remove the shifter boot and lever from inside the vehicle.
3. Raise and safely support the vehicle.
4. On 4-wheel drive vehicles, remove the front driveshaft and transfer case.
5. Disengage all cables, connectors and fluid lines that may interfere with transmission removal. Tag them if helpful for installation.
6. On automatic transmissions, unbolt the torque converter from the flexplate and disconnect the shift linkage.

7. Place a drain pan under the transmission and drain the fluid .
8. Position a transmission jack under the transmission and safety chain the case to the jack.
9. Matchmark and remove the driveshaft.
10. Remove the transmission rear mount.
11. Slightly raise the transmission and remove the crossmember.
12. On automatic transmissions, remove the transmission-to-engine block bolts. For manual transmissions, remove the bolts securing the transmission to the bell housing.

✳ WARNING

The torque converter will fall out of the transmission if it is tilted forward. Keep a hand on it while lowering the transmission out of the vehicle.

13. Roll the transmission rearward until the input shaft clears, lower the jack and remove the transmission.

To install:

14. Carefully raise the transmission to the engine or bell housing.
15. Roll the transmission forward and into position.
16. On automatic transmissions tighten the bolts to 65 ft. lbs. (87 Nm) for the Diesel; 50 ft. lbs. (67 Nm) for gasoline engines. On manual transmissions, tighten the bolts to 50 ft. lbs. (64 Nm).
17. Install the crossmember and tighten the bolts to 55 ft. lbs. (74 Nm).
18. Install the transmission rear insulator and lower retainer. Tighten the bolts to 60 ft. lbs. (81 Nm).
19. The rest of the installation is the reverse of removal.
20. Refill the transmission with the correct amount and type of fluid.

Clutch Assembly

REMOVAL & INSTALLATION

1. Disconnect the negative battery cable.
2. Raise the vehicle and support safely.
3. Remove the transmission and transfer case, if equipped.
4. Remove the inspection cover at the bottom of the bell housing.

✳ CAUTION

The pressure plate is heavy, do not drop it after removing the last bolt.

5. Remove the pressure plate bolts gradually. Rotate the engine if necessary for access with a flywheel turner.

6. Remove the pressure plate and disc by lowering it through the opening at the bottom of the housing.

To install:

7. Thoroughly clean all working surfaces of the flywheel and the pressure plate.
8. Inspect the flywheel for heat cracks and/or grooves. Have the flywheel resurfaced or replaced.
9. Sparingly apply anti-seize compound to the input shaft and clutch disc splines. Install a new release bearing.
10. Raise the pressure plate and disc into place, them use a suitable clutch aligning tool or spare input shaft to center the disc. Tighten all of the bolts finger-tight.
11. Gradually tighten the pressure plate bolts in a star pattern to 21 ft. lbs. (28 Nm).

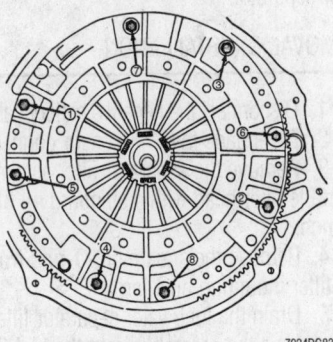

7924DG83

Tighten the pressure plate bolts in the sequence shown to ensure uniform seating

12. Install the transmission and transfer case, if equipped.
13. Install the inspection cover.
14. Connect the negative battery cable and check the clutch for proper operation.

Hydraulic Clutch System Bleeding

The system is self-bleeding. Press the clutch pedal repeatedly to release air from the fluid. The air will be vented from the reservoir.

Transfer Case Assembly

REMOVAL & INSTALLATION

1. Disconnect the negative battery cable.
2. Raise the vehicle and support safely.
3. Remove the skid plates, if equipped. Drain the transfer case fluid.
4. Disconnect the distance sensor, if equipped and disconnect the speedometer cable from the transfer case.
5. Matchmark and remove the drive-shafts.

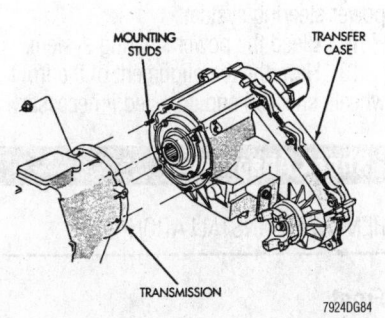

Typical transfer case mounting

6. Disconnect the Power Take-Off (PTO), if equipped.

7. Detach the linkage, electrical connectors and vacuum lines from the transfer case. Using a suitable jack, support the transfer case and remove the crossmember.

8. Unbolt the transfer case from the transmission and slide it backward to remove it from the vehicle.

9. The installation is the reverse of the removal procedure. Tighten the transfer case-to-transmission case nuts to 26 ft. lbs. (35 Nm). Fill the transfer case with Dexron® II automatic transmission fluid.

Halfshaft

REMOVAL & INSTALLATION

Dakota and Durango

1. Disconnect the negative battery cable.

2. Raise the vehicle and support safely.

3. Remove the tire and wheel assembly.

4. Remove the cotter pin from the end of the halfshaft. Remove the nut lock, spring washer, axle nut and washer.

5. Remove the ball joint retaining bolt and pry the control arm down to release the ball stud from the steering knuckle.

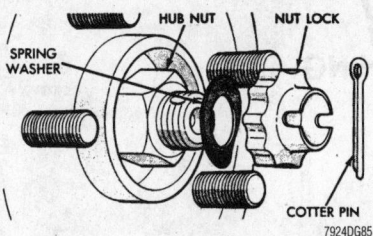

To separate the halfshaft from the hub, remove the cotter pin, nut lock and spring washer from the axle shaft

6. Position a drain pan under the transaxle where the halfshaft enters the differential or extension housing. Remove the halfshaft from the transaxle where the halfshaft enters the differential or extension housing. Remove the halfshaft from the transaxle or center bearing. Unbolt the center bearing from the block and remove the intermediate shaft from the transaxle, if equipped.

To install:

7. Install the halfshaft or intermediate shaft to the transaxle, being careful not to damage the side seals. Ensure the inner joint clicks into place inside the differential. Install the outer shaft to the center bearing if equipped. Install the outer shaft to the center bearing if equipped.

8. Pull the front strut out and insert the outer joint into the front hub.

9. Turn the ball joint stud, if necessary, to position the bolt retaining indent to the inside of the vehicle. Install the ball join stud into the steering knuckle. Install the retaining bolt and nut.

10. Install the axle nut washer and nut and tighten the nut to 180 ft. lbs. (244 Nm). Install the spring washer, nut lock and a new cotter pin.

11. Install the tire and wheel assembly.

STEERING AND SUSPENSION

Air Bag

❋❋ CAUTION

Some vehicles are equipped with an air bag system, also known as the Supplemental Inflatable Restraint (SIR) or Supplemental Restraint System (SRS). The system must be disabled before performing service on or around system components, steering column, instrument panel components, wiring and sensors. Failure to follow safety and disabling procedures could result in accidental air bag deployment, possible personal injury and unnecessary system repairs.

PRECAUTIONS

Several precautions must be observed when handling the inflator module to avoid accidental deployment and possible personal injury.

• Never carry the inflator module by the wires or connector on the underside of the module.

• When carrying a live inflator module, hold securely with both hands, and ensure that the bag and trim cover are pointed away.

• Place the inflator module on a bench or other surface with the bag and trim cover facing up.

• With the inflator module on the bench, never place anything on or close to the module which may be thrown in the event of an accidental deployment.

DISARMING

1. First read the system precautions.

2. Disconnect and isolate the negative battery cable.

3. If the air bag module is undeployed, wait two minutes for the system capacitor to discharge.

ARMING

Assuming the system components (air bag control module, sensors, air bag, etc.) are installed correctly and are in good working order, the system is armed whenever the battery positive and negative battery cables are connected.

If you have disarmed the air bag system for any reason, to rearm, ensure no one is in the vehicle (as an added safety measure), then connect the battery negative cable.

Recirculating Ball Power Steering Gear

REMOVAL & INSTALLATION

❋❋ WARNING

To avoid accidental air bag deployment and possible personal injury, always disconnect and isolate the negative battery cable. Allow two minutes to elapse before beginning any component removal.

1. Place the front wheels in straight-ahead position.

2. Disconnect the negative battery cable.

3. Remove the windshield washer solvent reservoir and the coolant overflow tank, if necessary.

4. Position a drain pan under the steering gear.

5. Disconnect and cap the fluid hoses from the power steering gear.

6. Disconnect the steering column shaft from the stub shaft.

7. Raise the vehicle and support safely.

8. On Dakota and Durango models, separate the Pitman arm from the center link. On Ram models, matchmark and remove the Pitman arm from the steering gear.

9. Remove the retaining bolts and remove the steering gear from the vehicle.

To install:

10. Position the steering gear at frame rail or reinforcement and install the bolts loosely. Connect the steering shaft to the stub shaft. Tighten the bolt to 33 ft. lbs. (45 Nm). Realign gear at frame and tighten the bolts to 100 ft. lbs. (136 Nm).

11. On Ram models, install the Pitman arm to steering shaft and tighten nut to 175 ft. lbs. (237 Nm). On Dakota and Durango models, install the Pitman arm to the center link. Tighten the nut to 65 ft. lbs. (88 Nm). Install replacement cotter pin(s).

12. Connect the fluid hoses to the steering gear. Tighten the fittings to 25 ft. lbs. (35 Nm)

13. Lower the vehicle.

14. Connect the negative battery cable.

2. Separate the tie rod ends from the knuckles using Puller C-3894-A, or equivalent.

3. Disconnect and cap the fluid hoses from the power steering gear.

4. Remove the lower coupler bolt and slide the coupler off the gear.

5. Remove the two steering gear-to-crossmember mounting bolts and remove the steering rack (gear) from the vehicle.

To install:

➡**Inspect the mounting bushings and replace them if worm or damaged.**

6. Install the steering gear on the crossmember. Tighten the bolts to 190 ft. lbs. (258 Nm).

7. Connect the shaft coupler onto the steering gear. Install the tighten the bolt to 36 ft. lbs. (49 Nm).

8. Clean the tie rod end studs and the tapered holes in the knuckles.

9. Connect the tie rod ends to the knuckles. Tighten the nuts to 65 ft. lbs. (88 Nm).

10. Connect the fluid hoses to the steering gear. Tighten the fittings to 25 ft. lbs. (35 Nm)

11. Safely lower the vehicle and refill the power steering system.

12. Bleed the power steering system.

13. Have the toe alignment of the front wheels checked and adjusted if necessary.

Shock Absorber

REMOVAL & INSTALLATION

Front

COIL SPRING SUSPENSION

1. Raise and support the vehicle with jackstands or floor jacks positioned at the extreme front ends of the frame rails.

2. Remove the wheel.

3. Remove the upper mounting bolt.

4. Remove the two lower mounting bolts.

5. Remove the shock absorber.

6. When installing the shock absorber, ensure the upper bushings are in the correct position. Replace any worn or cracked bushings. Tighten the top bolt to 25 ft. lbs. (34 Nm). Then, tighten the lower bolts to 15 ft. lbs. (20 Nm).

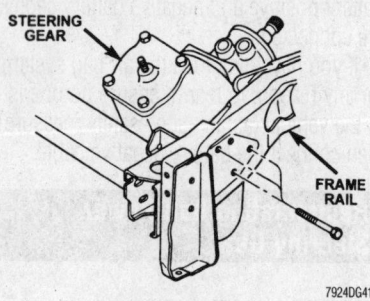

7924DG41

Typical recirculating ball power steering gear mounting

15. Refill the steering system with the correct amount of power steering fluid.

16. Start the engine and bleed the power steering system.

17. Road test the vehicle for proper operation.

Rack and Pinion Steering Gear

REMOVAL & INSTALLATION

1. Raise and safely support the front of the vehicle securely.

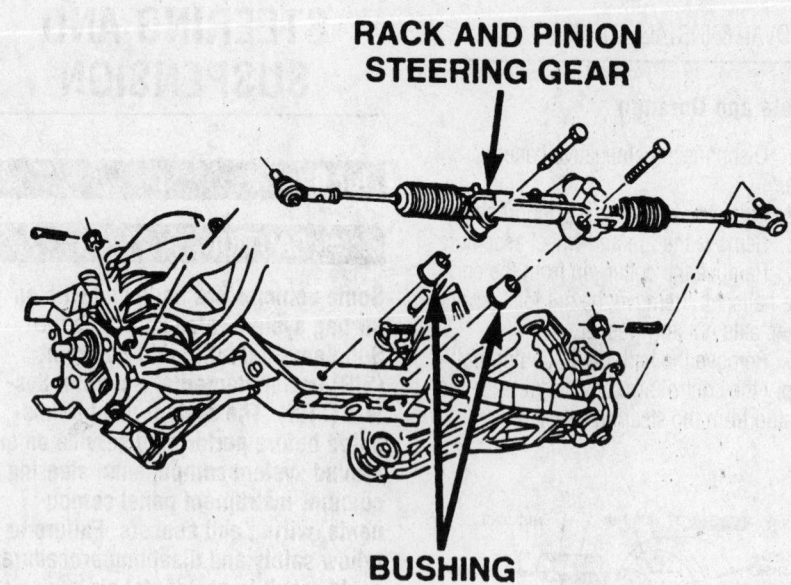

RACK AND PINION STEERING GEAR

BUSHING

7924DG42

Rack and pinion steering gear mounting used on the 2WD Dakota and Durango models

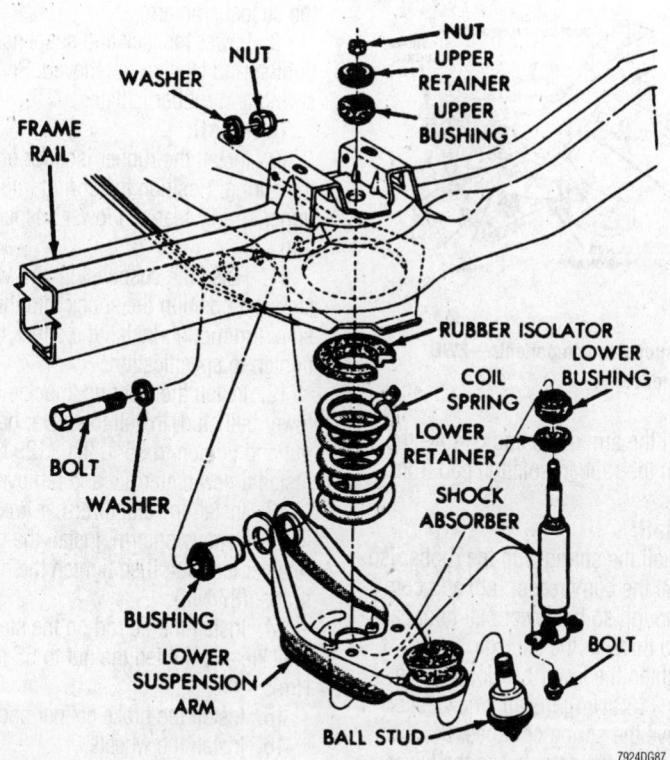

Front shock absorber mounting—Ram Van models

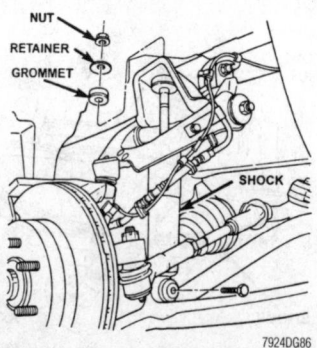

Front shock absorber mounting—4WD
Dakota and Durango models

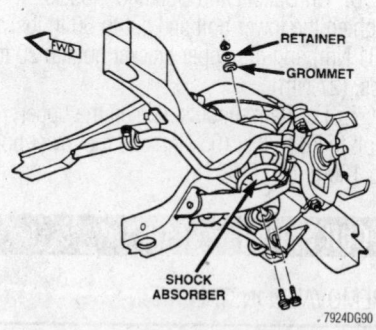

Front shock absorber mounting—2WD
Dakota and Durango models

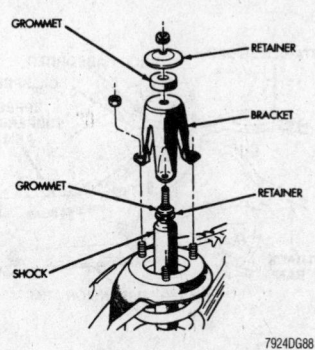

Upper shock absorber mounting—Ram
Truck models

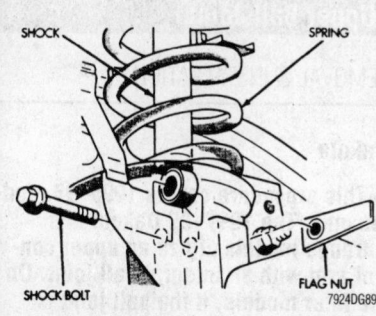

Lower shock absorber mounting—Ram
Truck models

LEAF SPRING SUSPENSION

1. Raise and safely support the vehicle.
2. Remove the wheel.
3. Remove the two upper shock absorber bracket-to-frame bolts.
4. Remove the lower bracket nut and remove the shock absorber.
5. If new shocks are being installed, remove the upper bracket from the old shock. Replace any worn or cracked bushings.
6. When installing the shock absorbers, tighten all fasteners to 50 ft. lbs. (68 Nm). Installation of the remaining components is the reverse of the removal.

TORSION BAR SUSPENSION

1. Remove the hardware from the shock absorber stud.
2. Raise and safely support the vehicle.
3. Remove the wheel.
4. Remove the lower bolt, then remove the shock absorber.

To install:
5. Install the lower retainer and grommet on the shock absorber stud. Insert the shock absorber through the frame hole. Install the lower bolt.
6. Tighten the bolt to 100 ft. lbs. (136 Nm.)

7. Install the upper grommet and retainer on the shock absorber stud. Install the bayonet nut and tighten to 30 ft. lbs. (41 Nm).

Rear

1. Raise and support the axle.
2. Remove the bolt and flag nut from the frame crossmember bracket.
3. Remove the bolt and nut from the axle bracket.
4. Remove the rear shock absorber from the vehicle.

To install:
5. Install the bolts through the brackets and shock.

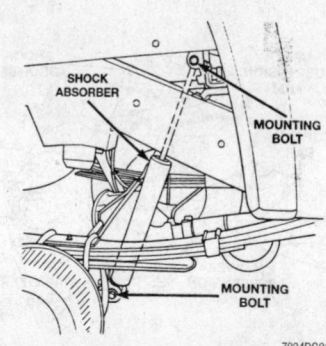

Rear shock absorber mounting—Dakota
and Durango models

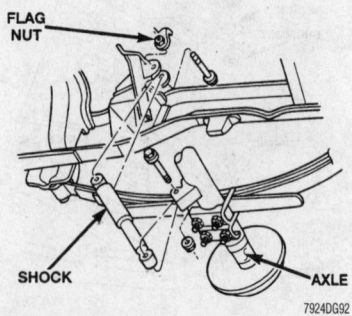

Rear shock absorber mounting—Ram Truck and Van models

6. On Dakota and Durango models, tighten the lower bolt and nut to 60 ft. lbs. (81 Nm) and the upper bracket nuts to 20 ft. lbs. (27 Nm).

7. For Ram models, tighten the upper bolt to 70 ft. lbs. (95 Nm) and the lower bolt to 100 ft. lbs. (136 Nm).

Coil Spring

REMOVAL & INSTALLATION

Dakota and Durango

1. Raise the vehicle and support safely.
2. Remove the shock absorber.
3. Disconnect the sway bar from the lower control arm, if equipped.
4. Install Spring Compressor tool DD-1278, or equivalent to the coil spring. Tighten the nut finger-tight, then loosen it ½ turn.
5. Remove the cotter pin and lower ball joint nut.
6. Release the lower ball joint taper using Ball Stud Loosening Tool C-3564-A, or equivalent.
7. Remove the tool and remove the ball stud from the control arm. Release the compressor tool from the coil spring.

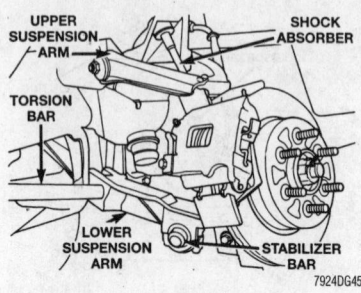

Front suspension components—4WD Dakota and Durango models

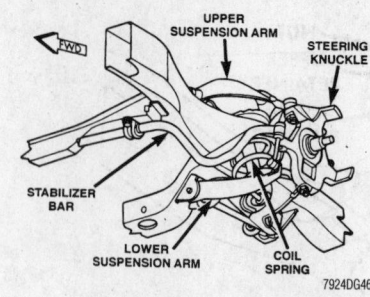

Front suspension components—2WD Dakota models

8. Pull the arm down and remove the spring with the rubber isolation pad from the vehicle.

To install:

9. Install the spring with the rubber isolator. Install the compressor tool and compress it enough so the lower ball joint can be inserted through the knuckle.

10. Tighten the lower ball joint nut to 135 ft. lbs. (183 Nm). Install a new cotter pin. Remove the spring compressor.

11. Connect the sway bar to the lower control arm, if equipped.

12. Install the shock absorber.

Except Dakota and Durango

1. Raise and safely support the vehicle.
2. Remove the wheel.
3. Remove the brake caliper and rotor.
4. Disconnect the tie rod from the steering knuckle.
5. Disconnect the stabilizer bar link from the lower arm.
6. Support the lower arm outboard end with a floor jack. Place the jack under the arm in front of the shock mount.
7. Remove the cotter pin and nut from the lower ball stud. Use remover tool C-4150A, or equivalent, to separate the ball stud.

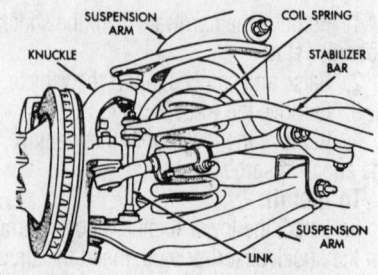

Independent front suspension components—2WD Ram Truck and Van models

8. Remove the lower shock bolt from the suspension arm.

9. Lower the jack and suspension arm until spring tension is relieved. Remove the spring and rubber isolator.

To install:

10. Install the rubber isolator on top of the spring. Position the spring into the upper spring seat and lower suspension arm.

11. Raise the suspension arm with the jack and position the shock into the suspension arm mount. Install the shock bolt and tighten to specification.

12. Install the steering knuckle on the lower ball stud. Install the lower ball stud nut and tighten to 95 ft. lbs. (129 Nm). Install a new cotter pin and remove the jack.

13. Install the stabilizer bar link on the lower suspension arm. Install the grommet, retainer and nut, then tighten the nut to 27 ft. lbs. (37 Nm).

14. Install the tie rod on the steering knuckle and tighten the nut to 65 ft. lbs. (88 Nm).

15. Install the brake caliper and rotor.
16. Install the wheels.
17. Remove the jackstands and lower the vehicle.
18. Align the front wheels.

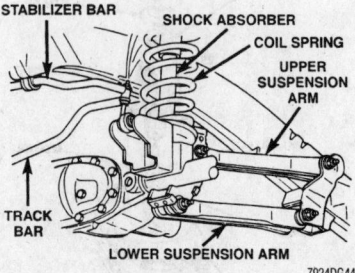

Link and coil front suspension components—4WD Ram Truck and Van models

Upper Ball Joint

REMOVAL & INSTALLATION

Dakota

➡This procedure covers 1995–96 models only. The 1997–99 Dakota and Durango models utilize an upper control arm with an integral ball joint. On the later models, if the ball joint is damaged or worn, the upper control arm must be replaced.

1. Raise and safely support the vehicle.

2. Position a support at the outer end of the lower control arm and lower the vehicle so that the support compresses the coil spring.

3. Remove the tire and wheel assembly.

4. Remove the cotter pin and stud nut.

5. Release the upper ball joint taper using Ball Stud Loosening tool C-3564-A, or equivalent.

6. Unthread the ball joint from the control arm with tool C-3561, or equivalent.

To install:

7. Tighten the ball joint itself to 125 ft. lbs. (169 Nm).

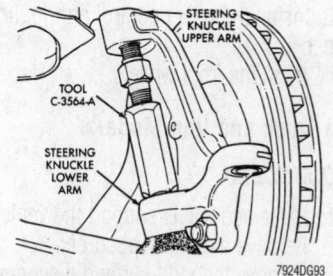

Separate the ball joint stud from the knuckle assembly using tool C-3564-A or equivalent—1995–96 2 and 4WD Dakota models

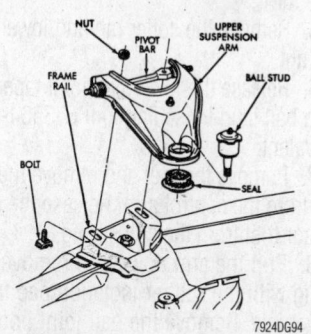

Upper suspension (control) arm and related components—1995–96 2 and 4WD Dakota models

8. Tighten the upper ball stud nut to 135 ft. lbs. (183 Nm) and install a new cotter pin.

9. The installation of the remaining components is the reverse of the removal procedure.

Except Dakota and Durango

2-WHEEL DRIVE

1. Raise and safely support the vehicle.

2. Position a support at the outer end of the lower control arm and lower the vehicle so that the support compresses the coil spring.

3. Remove the tire and wheel assembly.

4. Remove the cotter pin and the stud nut.

5. Release the upper ball joint taper using Ball Stud Loosening tool C-3564-A, or equivalent.

6. Unscrew the ball joint from the control arm with tool C-3561, or equivalent.

7. The installation is the reverse of the removal procedure. Tighten the ball joint itself to 125 ft. lbs. (169 Nm). Tighten the upper ball stud nut to 135 ft. lbs. (183 Nm).

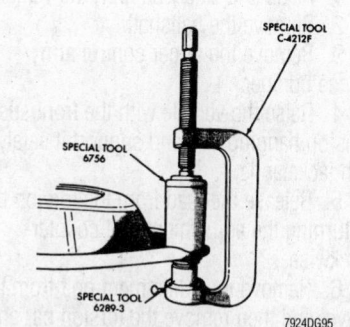

Upper ball joint removal—except Dakota models

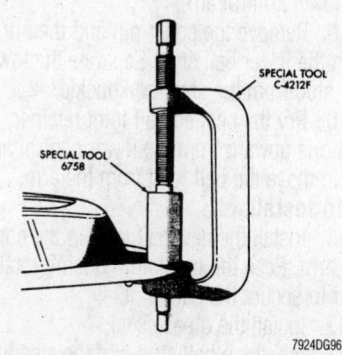

Upper ball joint installation—except Dakota models

4-WHEEL DRIVE

1. Raise and safely support the vehicle on jackstands.

2. Remove the front axle shaft.

3. Disconnect the tie rod end from the steering knuckle. On the left side, disconnect the drag link ball stud from the steering knuckle.

4. On the left side, remove the nuts and washers from the steering knuckle arm and remove the arm and spring, if equipped, from the knuckle.

5. If equipped with a Model 44 front axle, remove the ball joint nuts and discard

the lower nut. Use a brass drift and hammer to separate the steering knuckle from the axle tube yoke. Use tool C-4169 to remove the sleeve from the upper yoke arm.

6. Remove the snapring from the ball joint. Install the knuckle in a vise and use tools D-150-1, D-150-3 and C-4212-L to remove the ball joint from the knuckle.

7. If equipped with a Model 60 front axle, remove the bolts from the knuckle lower cap. Dislodge the cap from the steering knuckle and axle tube yoke. Remove the steering knuckle. Use tool D-192 to remove the upper socket pin from the axle tube upper arm bore. Remove the seal.

To install:

8. If equipped with a Model 44 front axle, use tools C-4212-L and C-4288 to force the upper ball joint into the steering knuckle. Install the snapring and install a new rubber boot. Thread the replacement sleeve into the upper yoke bore so that 2 threads are exposed at the top of the yoke. Position the knuckle on the axle tube yoke and install a new lower ball stud nut, then tighten to 80 ft. lbs. (108 Nm). Using the special socket, tighten the sleeve to 40 ft. lbs. (54 Nm). Install the upper ball stud nut and tighten to 100 ft. lbs. (136 Nm) and install a new cotter pin.

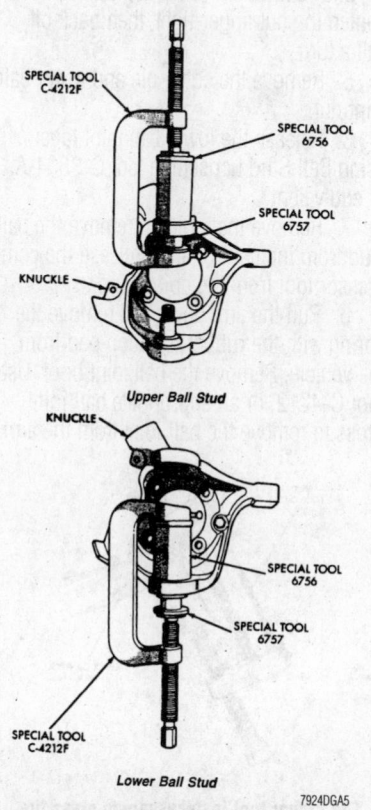

Typical ball joint removal using the special tools—248 FBI axle shown

9. If equipped with a Model 60 front axle, use tool D–192 to install the upper socket pin in the axle tube upper arm bore. Install a new seal. Tighten to 500–600 ft. lbs. (668–813 Nm). Position the knuckle over the socket pin. Fill the lower socket cavity with grease. Install the lower cap and tighten the bolts to 80 ft. lbs. (108 Nm).

10. On the left side, install the spring, if equipped and the steering knuckle arm to the steering knuckle.

11. Connect the tie rod to the end of the steering knuckle. On the left side, connect the drag link ball stud to the steering knuckle.

12. Install the front axle shaft and all related components.

Lower Ball Joint

REMOVAL & INSTALLATION

Dakota and Durango

2WD MODELS

1. Raise and safely support the vehicle.
2. Remove the shock absorber.
3. Disconnect the sway bar from the lower control arm, if equipped.
4. Install Spring Compressor tool DD-1278, or equivalent to the coil spring and tighten the nut finger-tight, then back off half a turn.
5. Remove the cotter pin and lower ball joint nut.
6. Release the lower ball joint taper using Ball Stud Loosening tool C-3564-A, or equivalent.
7. Remove the tool and remove the ball stud from the control arm. Release the compressor tool from the coil spring.
8. Pull the arm down and remove the spring with the rubber isolation pad from the vehicle. Remove the ball joint boot. Use tool C-4212, or an appropriate ball joint press to remove the ball joint from the arm.

To install:

9. Use the remover tool to press the ball joint into the arm. Install a new rubber boot. Install the spring with the rubber isolators. Install the compressor tool and compress it enough so the lower ball joint can be inserted through the knuckle.

10. Tighten the lower ball joint nut to 135 ft. lbs. (183 Nm). Install a new cotter pin. Remove the spring compressor.

11. Connect the sway bar to the lower control arm, if equipped.

12. Install the shock absorber.

4WD MODELS

1. Raise and safely support the vehicle.
2. Remove the halfshaft.
3. Remove the upper control arm jounce bumper.
4. Raise the vehicle with the front suspension hanging free and support it safely with jackstands.
5. Release the load from the torsion bar by turning the adjustment bolt counter-clockwise.
6. Remove the adjustment bolt from the swivel and then remove the torsion bar and the anchor together from the vehicle.
7. Remove the shock absorber lower attaching bolt.
8. Disconnect the stabilizer bar from the lower control arm.
9. Remove the cotter pin and the nut from the lower ball stud. Separate the lower ball stud from the steering knuckle.
10. Pry the peened ball joint retainer sections upward from the lower control arm and remove the ball joint from the arm.

To install:

11. Install the new ball joint in the control arm. Peen the ball joint housing retainer over to secure the ball joint.

12. Install the grease seal.

13. Insert the ball stud into the steering knuckle bore. Tighten the nut to 120 ft. lbs. (163 Nm) and install a new cotter pin.

14. Attach the stabilizer bar to the control arm and install the shock mount bolt.

15. Insert the torsion bar ends into the sockets.

16. Position the anchor and the bushing in the crossmember. Insert the adjustment bolt and thread it into the swivel.

17. Turn the adjustment bolt clockwise to apply a load on the bar.

18. Lower the vehicle.

19. Set the front suspension height. The difference between the distance from the surface that the tires are on to the lower control arm inner pivot and the distance that the tires are on to the outer end of the arm is 1–1½ in. (25–38mm).

20. Install the upper control arm jounce bumper.

21. Align the front wheels.

Ram Truck and Van Models

2WD MODELS

1. Raise and safely support the vehicle.
2. Remove the shock absorber.
3. Remove the strut bar and disconnect the sway bar from the lower control arm, if equipped.
4. Install spring compressor tool DD-1278, or equivalent to the coil spring and tighten the nut finger-tight, then back off half a turn.
5. Remove the cotter pin and lower ball joint nut.
6. Release the lower ball joint taper using ball stud loosening tool C-3564-A, or equivalent.
7. Remove the tool and remove the ball stud from the control arm. Release the compressor tool from the coil spring.
8. Pull the arm down and remove the spring with the rubber isolation pad from the vehicle. Remove the ball joint boot. Use tool C-4212, or an appropriate ball joint press to remove the ball joint from the arm.

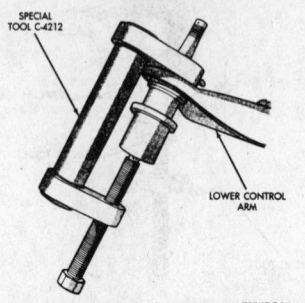

7924DGA4

The proper tool is necessary to press the old lower ball joint out of the lower control arm—2WD Dakota and Durango models

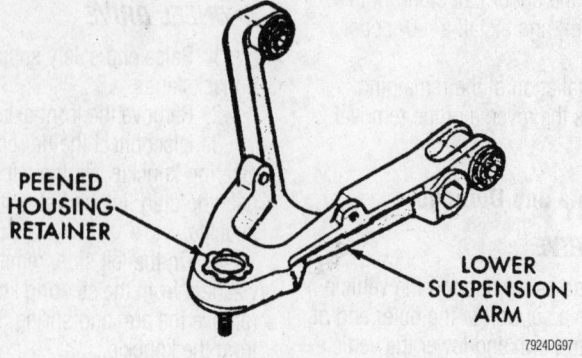

7924DG97

Lower control arm with ball joint assembly—4WD Dakota models

To install:

9. Use the remover tool to press the ball joint into the arm. Install a new rubber boot. Install the spring with the rubber isolators. Install the compressor tool and compress it enough so the lower ball joint can be inserted through the knuckle.

10. Tighten 1¹⁄₁₆ lower ball joint nuts to 135 ft. lbs. (183 Nm). Tighten ¾ nuts to 175 ft. lbs. (237 Nm). Install a new cotter pin. Remove the spring compressor.

11. Install the strut bar and connect the sway bar from the lower control arm, if equipped.

12. Install the shock absorber.

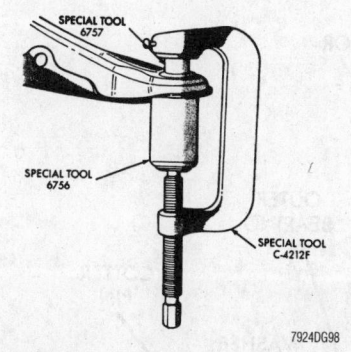

Lower ball joint removal—2WD Ram Truck and Van models

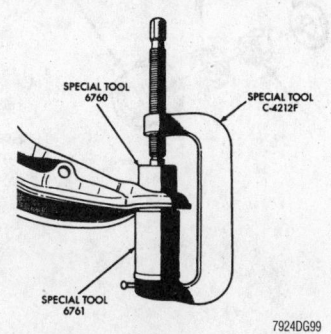

Lower ball joint installation—2WD Ram Truck and Van models

4WD MODELS

1. Raise and safely support the vehicle.

2. Remove the front axle shaft.

3. Disconnect the tie rod end from the steering knuckle. On the left side, disconnect the drag link ball stud from the steering knuckle.

4. On the left side, remove the nuts and washers from the steering knuckle arm and

remove the arm and spring, if equipped, from the knuckle.

5. If equipped with a Model 44 front axle, remove the ball joint nuts and discard the lower nut. Use a brass drift and hammer to separate the steering knuckle from the axle tube yoke.

6. Remove the snapring from the ball joint. Install the knuckle in a vise and use tools D-150–1, D-150–3 and C-4212-L to remove the ball joint from the knuckle.

7. If equipped with a Dana 60 front axle, use tools C-4212-L, C-4366–1 and C-4366–2 (or equivalents) to remove the lower ball joint.

To install:

8. If equipped with a Model 44 front axle, use tools C-4212-L and C-4288 to force the lower ball joint into the steering knuckle. Install the snapring and install a new rubber boot. Position the knuckle on the axle tube yoke and install a new lower ball stud nut. Tighten to 80 ft. lbs. (108 Nm). Install the upper ball stud nut and tighten to 100 ft. lbs. (136 Nm), then install a new cotter pin.

9. If equipped with a Model 60 front axle, use tools C-4212-L, C-4366–3 and C-4366–4 to install the seal and lower bearing cup into the axle tube yoke lower bore. Reposition the tools and install the lower bearing and seal into the bore. Position the knuckle over the socket pin. Fill the lower socket cavity with grease. Install the lower cap and tighten the bolts to 80 ft. lbs. (108 Nm).

10. On the left side, install the spring, if equipped and the steering knuckle arm to the steering knuckle.

11. Connect the tie rod to the end of the steering knuckle. On the left side, connect the drag link ball stud to the steering knuckle.

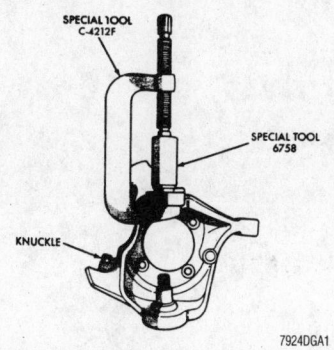

Lower ball joint installation—4WD Ram Truck and Van models

12. Install the front axle shaft and all related components.

Wheel Bearings

ADJUSTMENT

1997–99 Dakota and Durango

The 1997–99 Dakota and Durango models utilize a hub/bearing assembly which is not adjustable.

Except 1997–99 Dakota and Durango

1. Tighten the wheel bearing nut to 30–40 ft. lbs. (41–54 Nm) while turning the rotor.

2. Loosen the wheel bearing adjusting nut completely.

3. Tighten the nut finger-tight.

4. Check the wheel bearing end-play. The specification is 0.001–0.003 in. (0.025–0.076mm).

5. Install the nut lock and cotter pin.

REMOVAL & INSTALLATION

1997–99 Dakota and Durango

1. Raise and safely support the vehicle.

2. Remove the front wheel.

3. Remove the brake caliper and rotor.

4. Remove the spindle nut, then the hub/bearing assembly.

To install:

5. Position the hub/bearing assembly on the spindle.

6. Install the spindle nut. Tighten the nut to 185 ft. lbs. (251 Nm).

7. Install the rotor and caliper.

8. Install the front wheel.

9. Lower the vehicle to the floor.

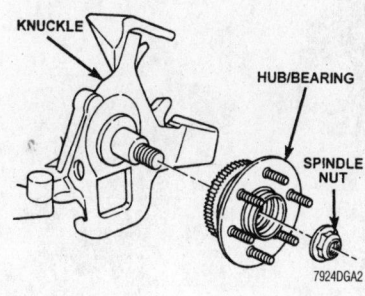

Hub/bearing assembly—1997–99 Dakota and Durango

Except 1997–99 Dakota and Durango

1. Raise and safely support the vehicle.
2. Remove the tire and wheel assembly.
3. Remove the caliper and disc brake pads.
4. Remove the dust cap.
5. Remove the cotter pin, castellated nut lock, wheel bearing nut and washer from the spindle.
6. Remove the outer wheel bearing.
7. Remove the rotor with the inner wheel bearing from the spindle. Remove the grease seal.

To install:

8. Lubricate and install the inner wheel bearing. Install a new grease seal.
9. Install the rotor to the spindle.
10. Lubricate and install the outer wheel bearing, washer and nut. When the bearing preload is properly set, install the nut lock and a new cotter pin.
11. Install the grease cap.
12. Install the brake pads and caliper.
13. Install the wheel.

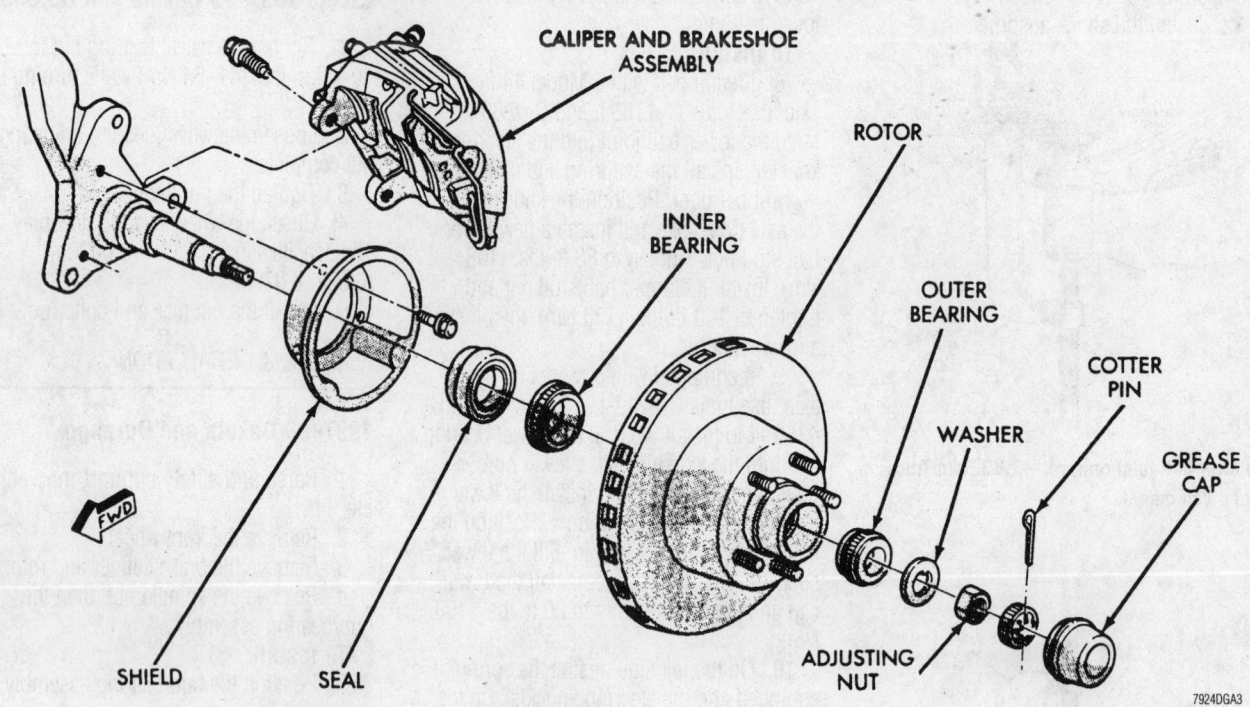

Exploded view of the front rotor, caliper and bearing mounting—except 1997–99 Dakota and Durango

FORD MOTOR CO. and MAZDA

15

Ford-Aerostar • Explorer • Mountaineer • Ranger • **Mazda**-B Series Pick-Ups

ENGINE REPAIR

Distributor

REMOVAL

➡**Except for the cap, adapter, rotor, Hall effect stator, TFI module (if applicable) and O-ring, no other distributor assembly parts are replaceable. There is no calibration required with the universal distributor. The distributor assembly can be identified by the part number information printed on a decal attached to the side of the distributor base.**

1. Disconnect the negative battery cable.

2. Set the No. 1 cylinder at TDC on the compression stroke with the timing marks aligned. To set the engine to TDC on No. 1 cylinder compression stroke, perform the following:

  a. Remove the rocker arm cover over the valves of cylinder No. 1.

  b. Rotate the crankshaft until the timing mark aligns with the TDC mark. While rotating the crankshaft watch the valves of cylinder No. 1. Since the crankshaft timing mark can align with the TDC mark on both the exhaust stroke and compression stroke, it is imperative that both the intake and exhaust valves are closed when cylinder No. 1 reaches TDC. If this is the case, cylinder No. 1 is at TDC on the compression stroke. If the exhaust valve is still not completely closed, or if the intake valve has opened some what, cylinder No. 1 is on the exhaust stroke and must be rotated another 360°.

3. Disengage the primary wiring connector from the distributor.

4. Mark the position of the No. 1 spark plug wire tower on the distributor base for future reference before removing the distributor cap.

5. Use a screwdriver to remove the distributor cap and adapter. Position the distributor cap and any attached ignition wires to the side.

6. Place a mark on the distributor housing indicating the direction the rotor is pointing.

7. Scribe a mark on the distributor body and the engine block to indicate the position of the distributor in the engine.

8. Remove the rotor.

9. Remove the distributor hold-down bolt and clamp. Some engines may be

equipped with a security type hold-down bolt, requiring a Torx® bit of the proper size to remove it.

10. Remove the distributor by lifting it straight up.

INSTALLATION

Timing Not Disturbed

1. Make sure that the No. 1 piston is on TDC of its compression stroke.

2. If equipped, check that the O-ring is installed and in good condition on the distributor body.

3. Rotate the distributor shaft so that the rotor tip is pointing toward the mark previously made on the distributor base (No. 1 spark plug tower). Continue rotating slightly so that the leading edge of the vane is centered in the vane switch stator assembly.

4. Rotate the distributor in the engine block to align the leading edge of the vane and the vane switch and verify that the rotor is pointing at the No. 1 cap terminal.

➡**If the vane and vane switch stator cannot be aligned by rotating the distributor in the engine block, pull the distributor out of the block enough to disengage the distributor, then rotate the distributor to engage a different distributor gear tooth.**

5. Install the distributor hold-down bolt and clamp, but do not tighten yet.

6. Connect the distributor TFI and primary wiring harnesses.

7. Install the distributor rotor and tighten the attaching screws to 24–36 inch lbs. (3–4 Nm).

8. Install the distributor cap and tighten the attaching screws to 18–23 inch lbs. (2–3 Nm). Check that the ignition wires are securely attached to the cap towers.

9. Install the ignition wires to the spark plugs, making sure they are in the correct firing order and tight on the spark plugs.

10. Check and adjust the initial timing to specifications with a timing light. Refer to the underhood emission control sticker for initial timing specifications. No attempt should be made to alter the timing from factory specifications.

11. Once the initial timing is set, tighten the distributor hold-down bolt to 17–25 ft. lbs. (23–34 Nm). Recheck the timing, then remove the timing light.

Timing Disturbed

1. If the engine was rotated while the distributor was removed, once again set the

No. 1 piston at TDC on the compression stroke with the timing marks aligned for correct initial timing. To set the engine to TDC on No. 1 cylinder compression stroke, perform the following:

  a. Remove the rocker arm cover over the valves of cylinder No. 1.

  b. Rotate the crankshaft until the timing mark aligns with the TDC mark. While rotating the crankshaft watch the valves of cylinder No. 1. Since the crankshaft timing mark can align with the TDC mark on both the exhaust stroke and compression stroke, it is imperative that both the intake and exhaust valves are closed when cylinder No. 1 reaches TDC. If this is the case, cylinder No. 1 is at TDC on the compression stroke. If the exhaust valve is still not completely closed, or if the intake valve has opened some what, cylinder No. 1 is on the exhaust stroke and must be rotated another 360°.

2. Rotate the distributor shaft so that the rotor tip is pointing toward the mark previously made on the distributor base (No. 1 spark plug tower). Continue rotating slightly so that the leading edge of the vane is centered in the vane switch stator assembly.

3. Rotate the distributor in the engine block to align the leading edge of the vane and the vane switch and verify that the rotor is pointing at No. 1 cap terminal.

➡**If the vane and vane switch stator cannot be aligned by rotating the distributor in the engine block, pull the distributor out of the block enough to disengage the distributor and rotate the distributor to engage a different distributor gear tooth.**

4. Install the distributor hold-down bolt and clamp, but do not tighten yet.

5. Connect the distributor TFI and primary wiring harnesses.

6. Install the distributor rotor and tighten the attaching screws to 24–36 inch lbs. (3–4 Nm).

7. Install the distributor cap and tighten the attaching screws to 18–23 inch lbs. (2–3 Nm). Check that the ignition wires are securely attached to the cap towers.

8. Install the ignition wires to the spark plugs, making sure they are in the correct firing order and tight on the spark plugs.

9. Check and adjust the initial timing to specifications with a timing light. Refer to the underhood emission control sticker for initial timing specifications. No attempt should be made to alter the timing from factory specifications.

10. Once the initial timing is set, tighten the distributor hold-down bolt to 17–25 ft. lbs. (23–34 Nm). Recheck the timing, then remove the timing light.

Ignition Timing

ADJUSTMENT

The only time in which timing should be set or checked is when the cylinder head and/or engine block have been disassembled or when the distributor has been removed from the engine. To do so, perform the following:

1. Disengage the one-wire SPOUT (spark out) connector or remove the shorting bar from the double wire SPOUT connector.

➡ **Adjust the timing to the specification on the Vehicle Emission Control Information label (VECI) located in the engine compartment. If this procedure is different from the procedure on the label, use the one on the label.**

2. Connect an inductive timing light to the engine using the manufacturers instructions.

3. Start the engine and allow it to warm up to normal operating temperature.

4. Loosen the distributor hold-down bolt and carefully rotate the distributor until the timing marks are as specified on the VECI.

5. Tighten the hold-down bolt to 14–21 ft. lbs. (19–28 Nm).

6. Install the shorting bar to the double wire SPOUT connector or engage the single wire SPOUT connector.

7. Turn **OFF** the engine and remove the timing light.

Engine Assembly

REMOVAL & INSTALLATION

Aerostar

➡ **The engine and subframe assembly on the Aerostar is removed from the bottom of the vehicle. An assortment of jacks and stands are needed to perform this procedure safely. The vehicle must be raised about 3–4ft. off the ground for clearance to remove the engine assembly.**

1. Disconnect the negative battery cable, then the positive cable. Relieve the fuel system pressure. Drain the cooling system.

2. Drain the engine oil.

☀☀ CAUTION

The EPA warns that prolonged contact with used engine oil may cause a number of skin disorders, including cancer! You should make every effort to minimize your exposure to used engine oil.

3. Disconnect the upper and lower radiator hoses and remove the radiator.

4. Remove the air cleaner hose assembly.

5. Disengage all electrical connectors, fuel lines, throttle linkage and hoses from the engine. Label the connections as needed to aid installation.

6. Remove the accessory drive belts.

7. If equipped with A/C, remove the compressor without disconnecting the lines and position it out of the way with wire.

8. From inside the vehicle, remove the engine cover.

9. Raise and safely support the vehicle.

10. Remove the transmission.

11. Disconnect and remove the exhaust pipe and catalytic converter.

12. Remove the front wheels.

13. Remove the engine block ground straps.

14. Disconnect the stabilizer bar from the lower control arms. Discard the bar nuts.

15. Behind the spindles, disconnect and plug the brake lines at the bracket on the frame.

16. Position a jack under the lower control arm and raise the arm until tension is applied to the coil spring. Install safety chains around the lower control arm and spring seat. Remove the bolt and nut retaining the spindle to the upper control arm ball joint. Slowly lower the jack to disconnect the spindle from the ball joint. Install safety chains around the lower control arm and spring seat.

17. Position drive train removal lift 109–00002 or equivalent, under the crossmember and engine assembly.

18. Slowly lower the vehicle until the crossmember rests on the removal lift. Place wood blocks under the front crossmember and rear of the engine block to keep the engine and crossmember assembly level. Install safety chains around the crossmember and lift.

19. With the engine and crossmember securely supported on the lift, remove the nuts that retain the engine crossmember assembly to the frame on each side of the vehicle.

20. Carefully lower the engine assembly out of the vehicle, making sure the A/C compressor, wiring and hoses do not interfere. When the assembly is clear, roll the lift away from the vehicle.

To install:

21. Position the engine assembly on the crossmember. Install the retaining nuts and tighten to 71–94 ft. lbs. (96–127 Nm).

22. Roll the removal lift under the vehicle. Align the lift, engine and crossmember assembly so the mounting bolts on each side of the frame are in alignment with the holes in the crossmember.

23. Slowly lower the vehicle so the bolts are piloted in the crossmember holes. Raise the lift or lower the vehicle so the crossmember is against the frame. Install the nuts retaining the crossmember to the frame and tighten to 145–195 ft. lbs. (196–264 Nm). Raise the vehicle and remove the lift.

24. Install the transmission assembly.

25. Remove the safety chains from around the lower control arms and spring seat. Install a jack under the lower control arms. Slowly raise the control arm until the coil spring is under tension. Continue to raise the arm until the spindle is connected to the upper arm ball joint. Install a new nut and bolt and tighten to 27–37 ft. lbs. (37–50 Nm).

26. Connect the stabilizer bar and new bar nuts.

27. Connect the front brake lines to the caliper hoses at the frame brackets. Install the wheel and tire assemblies.

28. Install the driveshaft, aligning the marks that were made during removal.

29. Install new gaskets on the exhaust manifold and catalytic converter. Install the exhaust pipe and catalytic converter. Tighten the converter-to-muffler nuts and bolts to 18–26 ft. lbs. (25–35 Nm). Tighten the exhaust pipe-to-exhaust manifold nuts to 25–34 ft. lbs. (34–46 Nm).

30. Connect all electrical connectors, fuel lines, throttle linkage, speedometer cable and hoses to the engine.

31. If equipped with A/C, install the compressor using the attaching bolts.

32. Install the drive belts. Place the injector harness behind the belt tension idler arm and tighten the idler arm.

33. Connect the throttle linkage to the ball stud located on the throttle body. Connect the shroud covering the throttle body.

34. Install the fan and fan shroud.

35. Connect the positive and negative battery cables and install the air cleaner and duct assembly.

36. Bleed the brakes. Fill and bleed the cooling system. Check all fluid levels.

37. Run the engine and check for leaks and proper operation. Check the front end alignment.

Except Aerostar

1. If equipped with air conditioning, have the system discharged into a recovery station by an EPA certified technician and remove the A/C compressor from the engine.
2. Disconnect the negative battery cable.
3. Remove the hood.
4. Remove the air intake tube and the accessory drive belt.

✳✳ CAUTION

Never open, service or drain the radiator or cooling system when hot; serious burns can occur from the steam and hot coolant. Coolant should be reused unless it is contaminated or is several years old.

5. Drain the cooling system.
6. Remove the cooling fan, shroud, radiator and all cooling system hoses.
7. Label and detach all engine wiring and vacuum hoses which will interfere with engine removal. Position the wire harness out of the way.
8. Unbolt the power steering pump from the engine and position it out of the way. The fluid lines do not have to be disconnected.
9. Detach the accelerator and transmission control cables from the throttle body, and the control cables' mounting bracket from the engine.

✳✳ CAUTION

Observe all applicable safety precautions when working around fuel. Always keep fuel in a container specifically designed for fuel storage; also, always properly seal fuel containers to avoid the possibility of fire or explosion.

10. Relieve the fuel system pressure, then disconnect the fuel supply and return lines from the engine.
11. Remove any remaining mounting brackets and/or drive belt tensioners.
12. Raise and safely support the vehicle.

✳✳ CAUTION

The EPA warns that prolonged contact with used engine oil may cause a number of skin disorders, including cancer! You should make every effort to minimize your exposure to used engine oil.

13. Drain the engine oil and remove the oil filter.
14. Detach the exhaust system from the exhaust manifolds. Use wire to position the pipe out of the way.
15. Remove the starter motor and starter motor wiring from the engine.
16. Label and detach any undervehicle engine wiring, which will interfere with engine removal.
17. On vehicles equipped with automatic transmissions, matchmark the position of the torque converter to the flywheel. Remove the bolts.
18. Remove all of the engine-to-transmission bolts.

➡**All Ford engines use a plate between the engine and the transmission. Some models may have a smaller, removable, flywheel/flexplate cover.**

19. If equipped, remove the transmission oil cooler line retainers-to-engine bolts.
20. Remove the front engine support insulator-to-crossmember retaining fasteners.
21. If equipped, remove the engine damper mounting bracket from the engine. The bracket may use two Torx® bolts for the lower mounting points.
22. Partially lower the vehicle.
23. Support the transmission with a floor jack.
24. Using a engine crane or hoist, lift the engine out of the vehicle. Be sure to lift the engine slowly and check often that nothing (such as wires, hoses, etc.) will cause the engine to hang up on the vehicle.
25. At this point, the engine can be installed on an engine stand.

To install:

➡**Lightly oil all bolts and stud threads, except those specifying special sealant, prior to installation.**

26. Using the hoist or engine crane, slowly and carefully position the engine in the vehicle. Make sure the exhaust manifolds are properly aligned with the exhaust pipes.
27. Align the engine to the transmission and install two engine-to-transmission bolts.

➡**Seat the left-hand side, front engine support insulator locating pin prior to the right-hand side, front engine support insulator.**

28. Lower the engine onto the front engine support insulators.
29. Detach the engine crane or hoist from the engine.

30. Remove the floor jack from beneath the transmission fluid pan.
31. Tighten the two installed engine-to-transmission bolts, then raise and securely support the vehicle.
32. Install and tighten the remaining engine-to-transmission bolts.
33. The remainder of installation is the reverse of the removal procedure. Be sure to tighten the fasteners to:
   • Transmission to engine: 28–38 ft. lbs. (38–51 Nm)
   • Rear insulator to transmission: 60–80 ft. lbs. (82–108 Nm)
   • Rear insulator to crossmember: 71–94 ft. lbs. (97–127 Nm)

✳✳ WARNING

Do NOT start the engine without first filling it with the proper type and amount of clean engine oil, and installing a new oil filter. Otherwise, severe engine damage will result.

34. Fill the crankcase with the proper type and quantity of engine oil. If necessary, adjust the transmission and/or throttle linkage.
35. Install the air intake duct assembly.
36. Connect the negative battery cable, then fill and bleed the cooling system.
37. Bring the engine to normal operating temperature, then check for leaks.
38. Stop the engine and check all fluid levels.
39. Install the hood, aligning the marks that were made during removal.
40. If equipped, have the A/C system properly leak-tested, evacuated and charged by a MVAC-trained, EPA-certified, automotive technician.

Water Pump

REMOVAL & INSTALLATION

✳✳ CAUTION

Never open, service or drain the radiator or cooling system when hot; serious burns can occur from the steam and hot coolant.

2.3L Engine

1. Disconnect the negative battery cable.
2. Drain the cooling system.
3. Remove the two bolts that retain the fan shroud and position the shroud back over the fan.

4. Remove the four bolts that retain the cooling fan. Remove the fan and shroud.

5. Loosen and remove the accessory drive belt. Earlier models may have two drive belts, remove them both.

6. Remove the water pump pulley and, if necessary, the vent hose to the emissions canister.

7. Remove the heater hose at the water pump.

8. Remove the timing belt cover. Remove the lower radiator hose from the water pump.

9. Remove the water pump mounting bolts and the water pump. Clean all gasket mounting surfaces.

10. Install the water pump in the reverse order of removal. Coat the threads of the mounting bolts with sealer before installation.

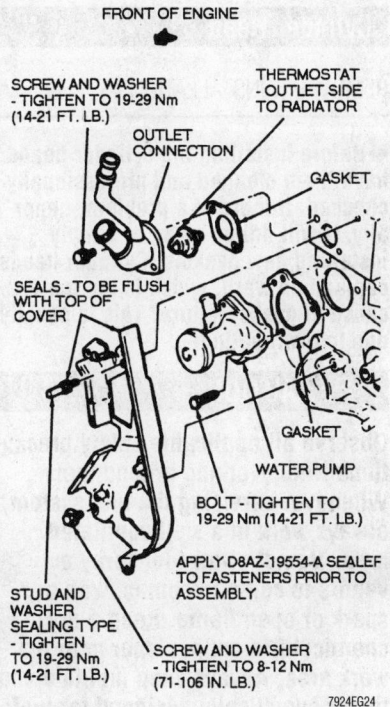

FRONT OF ENGINE

SCREW AND WASHER
- TIGHTEN TO 19-29 Nm
(14-21 FT. LB.)

THERMOSTAT
- OUTLET SIDE
TO RADIATOR

OUTLET
CONNECTION

GASKET

SEALS - TO BE FLUSH
WITH TOP OF
COVER

GASKET

WATER PUMP

BOLT - TIGHTEN TO
19-29 Nm (14-21 FT. LB.)

APPLY D8AZ-19554-A SEALER
TO FASTENERS PRIOR TO
ASSEMBLY

STUD AND
WASHER
SEALING TYPE
- TIGHTEN
TO 19-29 Nm
(14-21 FT. LB.)

SCREW AND WASHER
- TIGHTEN TO 8-12 Nm
(71-106 IN. LB.)

7924EG24

Exploded view of the water pump and coolant outlet—2.3L engine

3.0L Engine

1. Disconnect the negative battery cable.

2. Drain the cooling system.

3. Remove the engine air cleaner outlet tube.

✳✳ WARNING

The following procedures for removing the fan clutch gives the factory recommended loosening and tightening directions for the fan hub nut. However, it has been our experience that certain aftermarket parts manufacturers have changed this to enable use of universal fit parts. We recommend trying the factory direction first, then, if the nut doesn't seem to be moving, reverse the direction. Placing too much load on the water pump snout will break it.

4. Remove the engine fan and radiator shield.

5. Loosen, but do not remove at this time, the four water pump pulley bolts.

6. Remove the accessory drive belts.

7. Remove the four water pump pulley retaining screws, then remove the pulley itself.

8. Disconnect the engine wiring harness from the alternator.

9. Remove the oil fill tube retaining nut at the alternator stud, then lift the tube from the stud.

10. Remove the alternator adjusting arm and throttle body brace.

11. Using a Torx® 50 driver, remove the engine drive belt tensioner assembly.

12. If equipped with an auxiliary heater, remove the screw retaining the auxiliary heater tube bracket at the power steering pump support bracket.

13. Remove the lower radiator hose.

14. Disconnect the heater hose at the water pump.

15. For vehicles equipped with power steering, remove the 5 screws retaining the power steering pump support bracket to the engine, then secure the power steering pump and bracket assembly near the battery tray. Do not disconnect the power steering hoses from the pump.

16. Remove the water pump attaching bolts. Note their location for reinstallation.

17. Remove the pump from the engine and discard the old gasket.

To install:

➡**Lightly oil all bolts and stud threads before installation except those retaining special sealant.**

18. Thoroughly clean the pump and engine mating surfaces.

19. Using an adhesive type sealer (Trim Adhesive D7AZ-19B508-B or equivalent), position a new gasket on the timing cover.

20. Position the water pump onto the engine, then install the retaining bolts. When all the bolts are started, tighten them to 84 inch lbs. (9 Nm).

21. Install the lower hose and connect the radiator hose.

22. If applicable, install the power steering pump support bracket to the engine. Tighten the fasteners to 30–40 ft. lbs. (40–54 Nm).

23. Install the water pump pulley, then hand tighten the 4 bolts.

24. Reinstall the engine accessory drive belt tensioner, then tighten the bolts to 27–33 ft. lbs. (35–45 Nm).

25. Install the heater water return hose at the water pump fitting, then tighten the hose clamp.

26. If equipped with an auxiliary heater, reinstall the auxiliary heater tube bracket at the power steering pump support bracket. Tighten the bolts to 6–8 ft. lbs. (8–12 Nm).

27. Install the alternator adjusting arm and brace. Tighten the retaining bolts to 30–40 ft. lbs. (40–54 Nm).

28. Attach the engine harness wiring to the alternator.

29. Install the oil fill tube bracket over the stud at the alternator, then tighten the retaining nuts to 32–37 ft. lbs. (42–50 Nm).

30. Install the accessory drive belts.

31. Tighten the 4 pulley bolts to 19 ft. lbs.

32. Install the fan/clutch assembly and the fan shroud.

33. Connect the negative battery cable.

34. Fill and bleed the cooling system.

35. Run the engine and check for leaks.

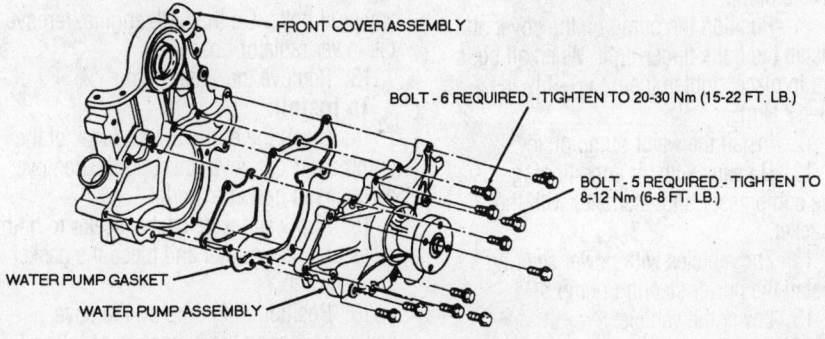

FRONT COVER ASSEMBLY

BOLT - 6 REQUIRED - TIGHTEN TO 20-30 Nm (15-22 FT. LB.)

BOLT - 5 REQUIRED - TIGHTEN TO
8-12 Nm (6-8 FT. LB.)

WATER PUMP GASKET

WATER PUMP ASSEMBLY

7924EG25

Exploded view of the water pump and front cover—3.0L engine

4.0L OHV Engine

1. Raise and safely support the vehicle so that access to the engine can be gained from both the top and the bottom of the engine compartment.
2. Drain the cooling system.
3. Remove the lower radiator hose and heater return hose from the water pump.
4. Remove the fan and fan clutch assembly.

✳✳ WARNING

The following procedures for removing the fan clutch gives the factory recommended loosening and tightening directions for the fan hub nut. However, it has been our experience that certain aftermarket parts manufacturers have changed this to enable use of universal fit parts. We recommend trying the factory direction first, then, if the nut doesn't seem to be moving, reverse the direction. Placing too much load on the water pump snout will break it.

5. Loosen the alternator mounting bolts and remove the belt. On vans with air conditioning, remove the compressor and bracket without disconnecting the A/C lines. Support the compressor from the vehicle frame rail with strong cord or wire.
6. If equipped with power steering, remove the power steering pump. Set it aside as with the A/C compressor.
7. Remove the water pump pulley.
8. Remove the water pump attaching bolts, then remove the water pump.
To install:
9. Clean the mounting surfaces of the pump and front cover thoroughly. Remove all traces of gasket material.
10. Apply adhesive gasket sealer to both sides of a new gasket and place the gasket on the pump.
11. Position the pump on the cover and install the bolts finger-tight. When all bolts are in place, tighten them to 6–9 ft. lbs. (9–12 Nm).
12. Install the water pump pulley.
13. On vans with air conditioning, install the compressor and alternator with the bracket.
14. For vehicles with power steering, install the power steering pump.
15. Lower the vehicle.
16. Install and adjust the accessory drive belt.

17. Connect the coolant hoses to the water pump, then tighten the clamps.
18. Install the fan and clutch assembly.

➡**The fan/clutch retaining nut is tightened counterclockwise.**

19. Fill and bleed the cooling system. Start the engine and check for leaks.

4.0L SOHC and 5.0L Engines

1. Disconnect the negative battery cable.
2. Drain the cooling system.

✳✳ WARNING

The following procedures for re-moving the fan clutch gives the factory recommended loosening and tightening directions for the fan hub nut. However, it has been our experience that certain aftermarket parts manufacturers have changed this to enable use of universal fit parts. We recommend trying the factory direction first, then, if the nut doesn't seem to be moving, reverse the direction. Placing too much load on the water pump snout will break it.

3. Remove the fan and fan clutch assembly.
4. Remove the radiator if required for clearance.
5. Loosen the water pump pulley attaching bolts.
6. Remove the accessory drive belt and the idler pulley.
7. Slide the bypass hose clamp back, away from the pump.
8. Disconnect the heater hose at the pump.
9. On the 4.0L engine, remove the lower radiator hose.
10. Remove the water pump pulley.
11. On the 5.0L engine remove the water bypass hose and engine harness bracket.
12. Remove all of the water pump attaching bolts. Pay attention to the locations of any stud bolts. On the 5.0L engine, remove the lower radiator hose.
13. Remove the water pump.
To install:
14. Clean the mounting surfaces of the pump and front cover thoroughly. Remove all traces of gasket material.
15. Apply adhesive gasket sealer to both sides of a new gasket and place the gasket on the pump.
16. Position the pump on the cover, while connecting the bypass hose to the pump, and install the bolts finger–tight. On

the 4.0L SOHC engine, tighten the bolts to 72–108 inch lbs. (6–9 ft. lbs. or 8.5–12 Nm). On the 5.0L engine, tighten the bolts to 15–21 ft. lbs. (20–28 Nm).
17. Position the bypass hose clamp back to its original position.
18. Install the water pump pulley and its attaching bolts. Snug the bolts.
19. Connect the lower radiator and heater hoses to the water pump.
20. Install the belt idler pulley.
21. Lift the accessory drive belt tensioner and install the belt.
22. Securely tighten the water pump pulley attaching bolts.
23. If removed, install the radiator.
24. Install the engine fan/clutch assembly.
25. Refill and bleed the cooling system.
26. Connect the negative battery cable, start the engine and check for leaks.

Cylinder Head

REMOVAL & INSTALLATION

➡**Before installing the cylinder heads, have them cleaned and professionally checked. If there is a problem, generally, it will not go away by simply installing new gaskets. Cylinder heads can and do warp, which is the major cause of gasket failure. This is usually due to overheating.**

✳✳ CAUTION

Observe all applicable safety precautions when working around fuel. Whenever servicing the fuel system, always work in a well ventilated area. Do not allow fuel spray or vapors to come in contact with a spark or open flame. Keep a dry chemical fire extinguisher near the work area. Always keep fuel in a container specifically designed for fuel storage; also, always properly seal fuel containers to avoid the possibility of fire or explosion.

2.3L and 2.5L Engines

1. Disconnect the negative battery cable.
2. Drain the cooling system.
3. Remove the intake manifold.
4. Disconnect the upper radiator hose at both ends and remove from the vehicle.
5. Remove the timing belt cover bolts and remove the cover. For power steering-equipped vehicles, unbolt the power steering pump bracket and position it off to the side.

6. Loosen the timing belt idler retaining bolts. Position the idler in the unloaded position and tighten the retaining bolts.

7. Remove the timing belt and exhaust manifold.

8. Remove the timing belt idler, two bracket bolts and spring from the cylinder head.

9. Disconnect the oil sending unit lead wire.

10. Remove the cylinder head retaining bolts.

11. Remove the cylinder head.

12. Clean the cylinder head, intake manifold and exhaust manifold gasket surfaces.

13. Blow oil out of the cylinder head bolt block hoses.

14. Clean the valve cover gasket surface on the head and clean the deck of the block.

15. Check the cylinder head for flatness.

To install:

16. Position a new head gasket on the block.

17. Clean the rocker arm cover (cam cover).

18. Install the valve cover gasket to the valve cover.

19. Position the cylinder head to the block.

20. Install cylinder head retaining bolts and tighten to specifications.

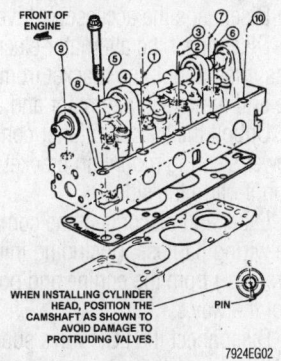

Cylinder head bolt tightening sequence—2.3L and 2.5L engines

21. Install the remaining components in the reverse order of removal.

22. Install the heater hose-to-valve cover retaining screws.

23. Change the engine oil and filter.

24. Fill and bleed the cooling system.

25. Install the air cleaner.

26. Connect the negative battery cable.

27. Start the engine and check for leaks.

3.0L Engine

➡**On Aerostar, Ford suggests that the cylinder head(s) is removable while the** engine is in the vehicle although this can be quite difficult. In some cases, cylinder head removal will be easier with the engine removed.

1. Have the air conditioning system discharged by an EPA certified, professional mechanic utilizing a refrigerant recovery/recycling machine.

2. Drain the cooling system (engine cold) into a clean container and save the coolant for reuse.

3. Remove the power steering pump and bracket assembly. DO NOT disconnect the hoses. Tie the assembly out of the way.

4. Disconnect the negative battery cable.

➡**Regardless of which cylinder head is being removed, the No.3 cylinder intake valve pushrod must be removed to allow removal of the intake manifold.**

5. Remove the intake manifold.

6. Remove the exhaust manifold(s).

7. Remove the rocker arm covers.

8. On the Aerostar with A/C, disconnect the liquid line at the A/C condenser and the suction hose from the accumulator. Remove the bolt securing the air conditioning hose assembly to the compressor housing, then remove the hose assembly from the vehicle. Plug or cap all openings in the air conditioning system to keep dirt, foreign material and excess moisture out of the system.

9. Remove the exhaust inlet pipe and exhaust manifolds.

10. Loosen the rocker arm fulcrum attaching bolts enough to allow the rocker arm to lifted off the pushrod and rotated to one side. Remove the pushrods, keeping them in order so they may be installed in their original locations.

11. Loosen the cylinder head attaching bolts in reverse of the tightening sequence, then remove the bolts and lift off the cylinder head(s). Remove and discard the old cylinder head gasket(s).

To install:

12. Clean the cylinder heads, intake manifold, valve rocker arm cover and cylinder block gasket surfaces of all traces of old gasket material and/or sealer. Clean the carbon in the combustion chambers with a wire cup brush mounted in a drill.

13. Lightly oil all bolt and stud bolt threads except those specifying special sealant. Position the new head gasket(s) on the cylinder block, using the dowels for alignment. The dowels should be replaced if damaged.

14. Set the cylinder head onto the engine block.

15. Install NEW cylinder head retaining bolts hand-tight.

16. Tighten the retaining bolts, in sequence, to 60 ft. lbs. (80 Nm). Then back off all bolts a minimum of one full rotation (360°).

17. Retighten the head bolts in sequence, in two stages: first step to 33–41 ft. lbs. (45–55 Nm), second step to 63–73 ft. lbs. (85–99 Nm).

➡**When the cylinder head retaining bolts have been tightened using this procedure, it is not necessary to retighten the bolts after extended engine operation. However, the bolts can be checked for tightness if desired.**

18. Install the intake manifold and remaining components in the reverse order of removal.

19. If needed, recharge the A/C system and check for leaks.

20. Fill the cooling system.

➡**Change the engine oil. Coolant contamination of the engine oil often occurs during cylinder head removal.**

21. Connect the negative battery cable.

22. Start the engine and check for leaks. Adjust the throttle linkage and speed control, as necessary.

23. Recharge the A/C system using the recommended procedure, this means using a recovery/recycling machine.

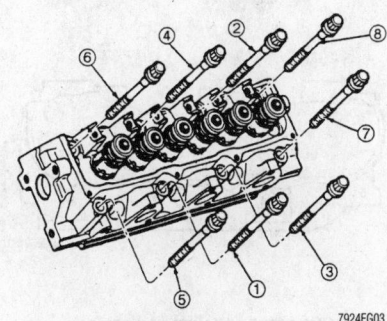

Tighten the cylinder head bolts in the correct order for proper sealing—all 3.0L engines

4.0L OHV Engine

➡**New cylinder head bolts must be used when installing the cylinder head on the 4.0L engine.**

1. Relieve the fuel system pressure.

2. Disconnect the negative battery cable.

3. Drain the cooling system (engine cold) into a clean container and save the coolant for reuse.

4. Remove the upper and lower intake manifolds.

5. Remove the accessory drive belt.

6. If the left cylinder head is being removed, perform the following:

　a. Remove the air conditioning compressor. Remove the screw securing the A/C manifold tube to the upper intake manifold. Set the compressor aside without disconnecting the A/C refrigerant lines.

　b. Remove the power steering pump and bracket assembly. DO NOT disconnect the hoses. Tie the assembly out of the way.

7. If the right head is being removed, perform the following:

　a. Remove the alternator and bracket.

　b. Remove the ignition coil and bracket.

8. Remove the spark plugs, exhaust manifolds(s), rocker arm covers and rocker arm assemblies.

9. Remove the pushrods, keeping them in order so they may be installed in their original locations.

10. Loosen the cylinder head attaching bolts in reverse of the tightening sequence, then remove the bolts and discard them. They cannot be reused.

11. Lift the cylinder head(s) off of the engine block. Do not pry between the cylinder head and engine block.

12. Remove and discard the old cylinder head gasket(s).

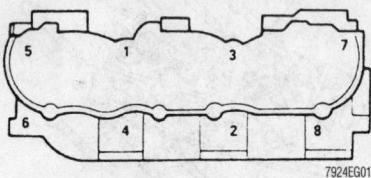

7924EG01

To ensure proper sealing and correct cylinder head installation, tighten the retaining bolts in the sequence shown— 4.0L OHV engine

To install:

13. Clean the cylinder heads, intake manifold, valve rocker arm cover and cylinder block gasket surfaces of all traces of old gasket material and/or sealer. Clean the carbon in the combustion chambers with a wire cup brush mounted in a drill.

➡**The 4.0L engine should always be assembled using new cylinder head bolts.**

14. Lightly oil all bolt and stud bolt threads except those specifying special sealant. Replace the alignment dowels if they are damaged. Position the new head gasket(s) on the cylinder block, using the dowels for alignment. The installation arrows on the head gaskets must be pointing to the front of the engine. If the alignment dowels are damaged, replace them.

15. Position the cylinder head(s) on the block.

16. Install and tighten the cylinder head bolts in the sequence shown, in the following steps:

- Step 1—22–26 ft. lbs. (30–35 Nm)
- Step 2—52–56 ft. lbs. (70–75 Nm)
- Step 3—tighten bolts an additional 90°

17. Install the lower intake manifold.

18. Install the valve lifters, if removed.

19. Dip each pushrod in heavy engine oil, then install the pushrods in their original locations.

20. Install the rocker shaft assemblies.

21. Apply another bead of RTV sealer at the 4 corners where the intake manifold and heads meet.

22. Install the rocker arm covers.

23. Install the upper intake manifold. Tighten upper intake manifold nuts to 15–18 ft. lbs. (20–33 Nm).

24. If applicable, install the A/C compressor on the bracket. Install the screw retaining the A/C manifold tube to the upper intake manifold. Tighten the compressor bolts to 15–21 ft. lbs. (20–29 Nm).

25. Install the exhaust manifold(s).

26. Install the spark plugs and wires.

27. If the left head was removed, install the power steering pump, compressor and drive belt.

28. Install the A/C compressor with bracket onto the engine.

29. If the right head was removed, install the ignition coil and bracket, alternator and bracket, and the drive belt.

30. Install the air cleaner.

31. Fill the cooling system.

➡**Change the engine oil and filter. Coolant contamination of the engine oil occurs during cylinder head removal.**

32. Connect the negative battery cable.

➡**When the battery has been disconnected and reconnected, some abnormal drive symptoms may occur while the Powertrain Control Module (PCM) relearns its adaptive strategy. The vehicle may need to be driven about 10 miles (16 km) or more to relearn the strategy.**

33. Start the engine and check for leaks. Adjust the throttle linkage and speed control, as necessary.

4.0L SOHC Engine

➡**If only one cylinder head is to be removed, only follow the procedures which apply. The following tools, or their equivalents are absolutely necessary to properly perform this procedure:**

- Cam Chain Tensioner tool T97T-6K254-A
- Cam Gear Removal tool T97T-6256-F
- Cam Gear Torque adapter T97T-6256-G
- Camshaft Gear Positioning/Holding tool T97T-6256-B
- Camshaft Gear Positioning/Holding tool adapter T97T-6256-A
- Camshaft Holding tool T97T-6256-C
- Crankshaft Holding tool T97T-6303-A
- Camshaft Holding tool adapter T97T-6256-D

1. Disconnect the negative battery cable.

2. Drain the cooling system.

3. Remove the upper and lower intake manifolds.

4. Remove the engine fan/clutch assembly and the shroud.

5. Remove the top radiator hose and connecting tube.

6. Disengage the accessory drive belt.

7. Disconnect the alternator electrical harness, then unbolt the bracket from the engine and remove it, alternator and all.

8. Unbolt the air conditioner compressor/power steering mounting bracket and position it off to the side.

9. Label and disconnect the complete engine wiring harness, (including injectors, sensors, etc.) from the engine and position it out of the way.

10. Disconnect the EGR valve supply tube, then remove the valve.

11. Remove the heater hose bracket from the engine and disconnect the transmission dipstick tube.

12. Disconnect both heater hoses from the engine.

13. Remove the thermostat housing bypass hose.

14. Remove the thermostat housing assembly from the engine.

15. Label and disconnect the spark plug wires from the plugs.

16. Remove the fuel line upper bracket bolts.

17. Remove the fuel injection supply manifold.

18. Remove both valve covers.

19. Remove the six fuel injectors.
20. Remove the crankcase vent separator spring steel clip, then remove the separator.
21. Remove the engine oil dipstick and tube.
22. Raise and safely support the vehicle.
23. Unbolt both exhaust pipe inlets from the exhaust manifolds.
24. Lower the vehicle.
25. Remove the LH hydraulic chain tensioner and the camshaft sprocket bolt.
26. Remove the LH cassette (chain guide) retaining bolt.
27. Remove the eight 12mm and two 8mm bolts, in sequence, from the LH cylinder head.
28. Remove the cylinder head and discard the old gasket.

※※ WARNING

The RH exhaust manifold must be removed before the RH cylinder head to avoid breaking the camshaft cassette.

29. Remove the RH exhaust manifold.
30. Remove the RH hydraulic chain tensioner. Access to the tensioner is through the RH fender well.

※※ WARNING

The RH camshaft sprocket bolt uses left-hand threads.

31. Use the cam gear torque adapter tool to remove the RH camshaft sprocket bolt.
32. Remove the RH cassette retaining bolt.

※※ WARNING

You must remove the camshaft sprocket from the chain and cassette to gain clearance to remove the cylinder head and avoid breaking the cassette.

→Secure the chain to the cassette with a rubber band to aid removal and prevent the chain from falling into the cylinder block.

33. Remove the RH camshaft sprocket from the cassette.
34. Remove the eight 12mm and two 8mm bolts, in sequence, from the RH cylinder head.
35. Remove the cylinder head and discard the old gasket.

To install:
36. Thoroughly clean all gasket mating surfaces. Remove all traces of old gasket material, oil, grease or dirt.

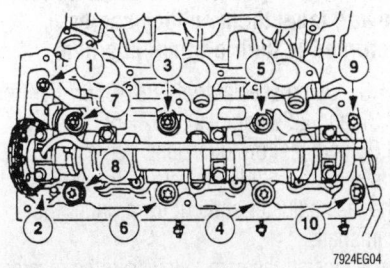

The correct cylinder head bolt loosening and tightening sequence must be used to prevent warpage and for proper sealing—4.0L SOHC engine

37. Insure that the rubber band is holding the RH chain to the cassette.
38. Install the RH cylinder head gasket.
39. Position the RH cylinder head to the engine block and install the attaching bolts. Tighten the bolts in three stages: first tighten to 26 ft. lbs. (35 Nm), second, rotate 90°, third, rotate an additional 90°.
40. Install a rubber band to the LH cassette to hold the chain and sprocket in place.
41. Install the LH cylinder head gasket.
42. Position the LH cylinder head to the engine block and install the attaching bolts. Tighten the bolts in three stages: first, tighten to 26 ft. lbs. (35 Nm); second, tighten an additional 90 degrees; third, rotate another additional 90 degrees.
43. Install all components that were removed in the reverse order of removal.
44. Change the engine oil and filter.
45. Refill and bleed the cooling system.
46. Connect the negative battery cable, then start and run the engine. Check for leaks.

5.0L Engine

→According to the manufacturer, you must use new cylinder head bolts when installing the head(s).

1. Disconnect the negative battery cable.
2. Drain the cooling system.
3. Remove one or both valve covers.
4. Remove the upper and lower intake manifolds.
5. Remove the accessory drive belt.
6. If the LH cylinder head is being removed, proceed as follows:
   a. Remove the A/C compressor/power steering mounting bracket and position it out of the way.
   b. Remove the engine oil dipstick and tube.
7. If the RH cylinder head is being removed, proceed as follows:

   a. Disconnect the alternator electrical harness.
   b. Unbolt the alternator mounting bracket and remove it from the engine.
8. Remove the exhaust manifolds from the engine.
9. Loosen the rocker arm bolts so that the rockers can be rotated to the side.
10. Remove the pushrods, keeping them in order so that they can be installed to their original locations.
11. Remove and discard the cylinder head bolts. Lift the cylinder head from the engine and discard the old gasket.

To install:
12. Thoroughly clean all of the gasket mating surfaces.
13. Position a new cylinder head gasket to the engine block, then install the cylinder head.

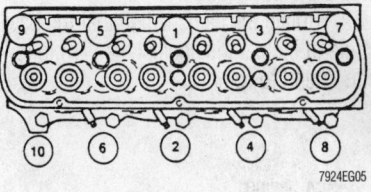

Use the proper torque sequence to insure good cylinder head sealing—5.0L engine

14. Install new cylinder head bolts and tighten in three steps:
   a. Tighten all bolts in sequence to 25–37 ft. lbs. (34–47 Nm).
   b. Tighten all bolts in sequence to 45–55 ft. lbs. (61–75 Nm).
   c. Tighten all bolts an additional 85–95° (¼ turn).
15. Ensure that the oil passages within the pushrods are clear and install them in the engine.

→Lubricate the ends of the pushrods with SAE 50W engine oil.

16. Thoroughly coat the rocker arms with SAE 50W engine oil and rotate them back to their proper positions. Tighten the attaching bolts to 18–25 ft. lbs. (24–34 Nm).
17. Install the exhaust manifolds.
18. If the RH cylinder head was removed, proceed as follows:
   a. Attach the alternator mounting bracket and to the engine.
   b. Connect the alternator electrical harness.

19. If the LH cylinder head was removed, proceed as follows:

a. Install the engine oil dipstick and tube.

b. Install the A/C compressor/power steering mounting bracket to the engine.

20. Install the lower and upper intake manifolds.

21. Install the one or both valve covers.

22. Refill and bleed the cooling system.

23. Connect the negative battery cable, then start and run the engine. Check for leaks.

Rocker Arms/Shafts

REMOVAL & INSTALLATION

2.3L , 2.5L and 4.0L SOHC Engines

➡**A special tool is required to compress the valve spring.**

1. Remove the valve cover and associated parts as required.

2. Rotate the camshaft so that the base circle of the cam is against the cam follower you intend to remove.

➡**If removing more than one cam follower, label them so they can be returned to their original position.**

3. Using special tool T88T-6565-BH (for 2.3L and 2.5L engines), T97T-6565-A (for 4.0L engines) or equivalent, depress the valve spring. Slide the cam follower over the lash adjuster and out from under the camshaft.

4. Install the cam follower in the reverse order of removal. Lubricate the followers with SAE 50W engine oil meeting Ford specification WSE-M2C908-A1 and API SJ or better, prior to installation.

3.0L and 5.0L Engines

1. Remove the rocker arm covers.

2. Remove the single retaining bolt at each rocker arm.

3. The rocker arm and pushrod may then be removed from the engine. Keep all rocker arms and pushrods in order so they may be installed in their original locations.

4. Installation is the reverse of removal. Lubricate the rocker arm assemblies with SAE 50W engine oil. Insure that the fulcrums are properly seated into the cylinder

head (3.0L engines) or the fulcrum guide (5.0L engines). Tighten the rocker arm fulcrum bolts to 19–28 ft. lbs. (26–38 Nm).

4.0L OHV Engine

1. Remove the rocker arm covers.

2. Remove the rocker arm shaft stand attaching bolts by loosening the bolts two turns at a time, in sequence (from the end of the shaft to the middle of the shaft).

3. Lift off the rocker arm and shaft assembly. If equipped, remove the oil baffle.

To install:

4. If equipped, loosen the valve lash adjusting screws a few turns. Apply engine oil to the assembly to provide initial lubrication.

5. If equipped, install the oil baffle.

6. Install the rocker arm shaft assembly to the cylinder head and guide adjusting screws on to the pushrods.

7. Install and tighten the rocker arm stand attaching bolts to 46–52 ft. lbs. (62–70 Nm), two turns at a time, in sequence (from middle of shaft to the end of the shaft).

8. Adjust the valve lash to the cold specified setting.

9. Install the rocker arm covers.

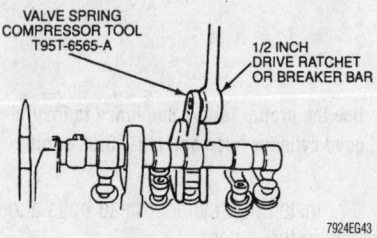

To remove the cam follower (rocker arm), use the special tool to depress the valve spring, then remove the cam follower— 2.3L SOHC engine

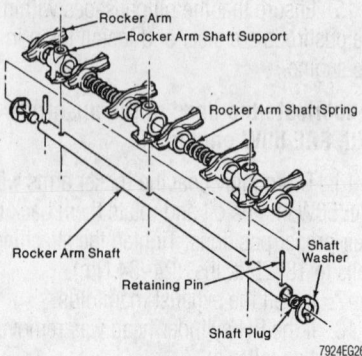

Rocker arm and shaft assembly—4.0L SOHC engine

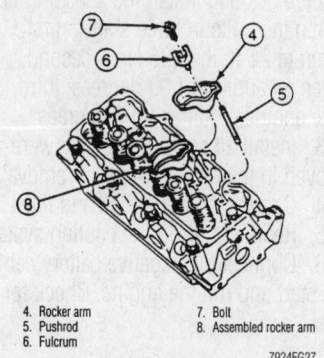

4. Rocker arm
5. Pushrod
6. Fulcrum
7. Bolt
8. Assembled rocker arm

Exploded view of the rocker arm assembly—3.0L engine

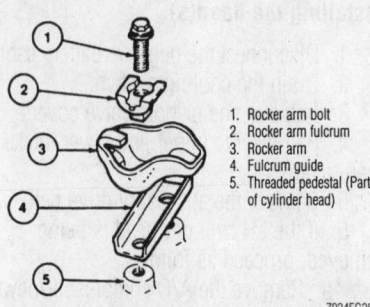

1. Rocker arm bolt
2. Rocker arm fulcrum
3. Rocker arm
4. Fulcrum guide
5. Threaded pedestal (Part of cylinder head)

Exploded view of the rocker arm assembly. Notice the fulcrum guide between the pedestals used for extra stability—5.0L engine

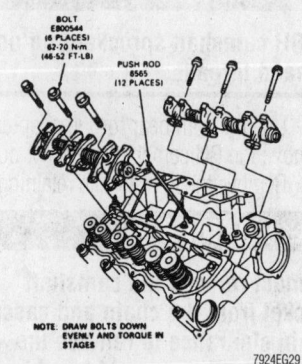

Rocker arm and shaft assemblies—4.0L OHV engine

Intake Manifold

➡**Although Ford suggests that this component is removable while the engine is installed in the vehicle, depending on the particular options with which your Aerostar is equipped, working clearance may be extremely tight and this procedure may be much easier to perform with the engine removed. Before commencing, read through this procedure and make certain enough clearance, or working** room, exists with the engine in the vehicle; if there is not enough space, the engine should be removed.

REMOVAL & INSTALLATION

✳✳ CAUTION

Observe all applicable safety precautions when working around fuel. Whenever servicing the fuel system, always work in a well ventilated area. Do not allow fuel spray or vapors to come in contact with a spark or open flame. Keep a dry chemical fire extinguisher near the work area. Always keep fuel in a container specifically designed for fuel storage; also, always properly seal fuel containers to avoid the possibility of fire or explosion.

2.3L and 2.5L Engines

The engines covered by this manual utilize an upper and lower intake manifold assembly. If necessary, only the upper intake manifold may be removed by following the intake manifold procedure up to that point. Obviously, installation would also begin at the upper intake steps.

✳✳ WARNING

Anytime the upper or lower intake manifold has been removed, cover all openings with a rag or a sheet of plastic to prevent dirt and debris from falling into the engine.

The intake manifold is a two-piece (upper and lower) aluminum casting. Runner lengths are tuned to optimize engine torque and power output. The manifold provides mounting flanges for the air throttle body assembly, fuel supply manifold, accelerator control bracket and the EGR valve and supply tube. A vacuum fitting is installed to provide vacuum to various engine accessories. Pockets for the fuel injectors are machined to prevent both air and fuel leakage. The following procedure is for the removal of the intake manifold with the fuel charging assembly attached.

1. Make sure the ignition is **OFF**, then drain the coolant from the radiator (engine cold).

✳✳ CAUTION

Never open, service or drain the radiator or cooling system when hot; serious burns can occur from the steam and hot coolant.

2. Disconnect the negative battery cable and secure it out of the way.

3. Remove the fuel filler cap to vent tank pressure. Release the pressure from the fuel system at the fuel pressure relief valve using EFI pressure gauge T80L–9974–A or equivalent. The fuel pressure relief valve is located on the fuel line in the upper right-hand corner of the engine compartment. Remove the valve cap to gain access to the valve.

4. Label and unplug any electrical connectors related to the intake manifold assemblies being removed.

5. Tag and disconnect the vacuum lines at the upper intake manifold vacuum tree, at the EGR valve and at the fuel pressure regulator and canister purge line as necessary.

6. Remove the throttle linkage shield and disconnect the throttle linkage and speed control cable (if equipped). Unbolt the accelerator cable from the bracket and position the cable out of the way.

7. Disconnect the air intake hose, air bypass hose and crankcase vent hose.

8. Disconnect the PCV hose from the fitting on the underside of the upper intake manifold.

9. Loosen the clamp on the coolant bypass line at the lower intake manifold and disconnect the hose.

10. Disconnect the EGR tube from the EGR valve by removing the flange nut.

11. Remove the upper intake manifold retaining nuts. Remove the upper intake manifold and throttle body assembly.

➡**If you only need to remove the upper intake manifold, stop at this point. Otherwise, continue with the procedure to also remove the lower intake manifold.**

12. Disengage the push-connect fitting at the fuel supply manifold and fuel return lines. Disconnect the fuel return line from the fuel supply manifold.

13. Remove the engine oil dipstick bracket retaining bolt.

14. Unplug the electrical connectors from all four fuel injectors and move the harness aside.

15. Remove the four bottom retaining bolts from the lower manifold. The front two bolts also secure an engine lifting bracket. Once the bolts are removed, remove the lower intake manifold.

16. Clean and inspect the mounting faces of the lower intake manifold and cylinder head. Both surfaces must be clean and flat. If the intake manifold upper or lower section is being replaced, it will be necessary to transfer components from the old to the new part.

To install:

17. Clean and oil the manifold bolt threads. Install a new lower manifold gasket.

18. Position the lower manifold assembly to the head and install the engine lifting bracket. Install the four top manifold retaining bolts finger-tight. Install the four remaining manifold bolts and tighten to 15–22 ft. lbs. (20–30 Nm) following the sequence illustrated.

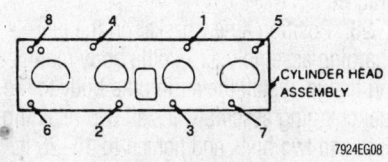

Lower intake manifold torque sequence—2.3L engine

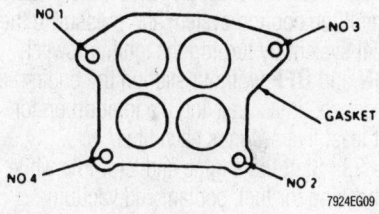

Upper intake manifold torque sequence—2.3L engine

19. Engage the four electrical connectors to the injectors.

20. Install the engine oil dipstick, then connect the fuel return and supply lines to the fuel supply manifold.

➡**The following procedures are for installing the upper intake manifold.**

21. Make sure the gasket surfaces of the upper and lower intake manifolds are clean. Place a gasket on the lower intake manifold assembly, then place the upper intake manifold in position.

22. Install the retaining bolts and tighten in sequence to 15–22 ft. lbs. (20–30 Nm).

23. Connect the EGR tube to the EGR valve and tighten it to 18 ft. lbs. (25 Nm).

24. Connect the coolant bypass line and tighten the clamp. Connect the PCV system

hose to the fitting on the underside of the upper intake manifold.

25. If removed, install the vacuum tee on the upper intake manifold. Use Teflon® tape on the threads and tighten to 12–18 ft. lbs. (16–24 Nm). Reconnect the vacuum lines to the tee, the EGR valve and the fuel pressure regulator and canister purge line as necessary.

26. Hold the accelerator cable bracket in position on the upper intake manifold and install the retaining bolt. Tighten the bolt to 10–15 ft. lbs. (14–20 Nm).

27. Install the accelerator cable to the bracket.

28. Position a new gasket on the fuel charging assembly air throttle body mounting flange. Install the air throttle body to the fuel charging assembly. Install two retaining nuts and two bolts and tighten to 15–25 ft. lbs. (20–30 Nm).

29. Connect the accelerator and speed control cable (if equipped), then install the throttle linkage shield.

30. Reconnect any electrical harness plugs which were removed.

31. Connect the air intake hose, air bypass hose and crankcase ventilation hose.

32. Reconnect the negative battery cable. Refill the cooling system and pressurize the fuel system by turning the ignition switch **ON** and **OFF** (without starting the engine) at least six times. Leaving the ignition on for at least five seconds each time.

33. Start the engine and let it idle while checking for fuel, coolant and vacuum leaks. Correct as necessary. Road test the vehicle for proper operation.

3.0L Engine

➡ **The upper intake manifold is integral with the throttle body.**

1. Drain the cooling system with the engine cold.

✳✳ CAUTION

Never open, service or drain the radiator or cooling system when hot; serious burns can occur from the steam and hot coolant.

2. Disconnect the negative battery cable.

3. Disengage the wiring harness connectors from the engine components.

4. Depressurize the fuel system and remove the throttle body.

5. Disconnect the fuel return and supply lines.

6. Disconnect the fuel injection wiring retaining standoffs from the inboard rocker arm cover studs. Carefully remove the fuel injector wiring harness from each the fuel injectors.

7. Disconnect the upper radiator hose.

8. Disconnect the water outlet heater hose.

9. If equipped with distributor ignition, disconnect the distributor cap with the spark plug wires attached. Matchmark and remove the distributor assembly.

10. If equipped with distributorless ignition, Label and remove the spark plug wires from the coil pack, then remove the coil pack.

11. Remove the spark plug from cylinder No. 1, then bring the No. 1 piston to top dead center on the compression stroke.

12. Remove the rocker arm covers.

13. Loosen the No. 3 intake valve rocker arm bolt and rotate the rocker arm off of the pushrod and away from the top of the valve stem. Remove the pushrod from the engine.

14. Remove the intake manifold attaching bolts and studs using a Torx® head socket.

15. Lift the intake manifold off the engine. Use a plastic mallet to tap lightly around the intake manifold to break it loose, if necessary. Do not pry between the manifold and cylinder head with any sharp instrument. The manifold can be removed with the fuel rails and injectors in place.

To install:

➡ **When cleaning the engine surfaces, lay a clean cloth or shop rag in the lifter valley to catch any gasket material. After scraping, carefully lift the cloth from the lifter valley preventing any particles from entering oil drain holes or cylinder head.**

16. Remove the manifold side gaskets and end seals and discard. If the manifold is being replaced, transfer the fuel injector and fuel rail components to the new manifold on a clean workbench. Clean all gasket mating surfaces.

17. Lightly oil all attaching bolts and stud threads. The intake manifold, cylinder head and cylinder block mating surfaces should be clean and free of old silicone rubber sealer or old gasket material. Use a suitable solvent to clean these areas.

18. If installing a new intake manifold, transfer the fuel injectors, fuel injection supply manifold, ECT sensor, water thermostat, thermostat housing, hot water heater elbow connection, and the coolant temperature sending unit to the new manifold.

19. Apply silicone rubber sealer (D6AZ–19562–A or equivalent) to the intersection of the cylinder block assembly and head assembly at the four corners.

➡ **When using silicone rubber sealer, assembly must occur within 15 minutes after sealer application. After this time, the sealer may start to set and its sealing effectiveness may be reduced. In high temperature/humidity conditions, the RTV will start to skin over in about 5 minutes.**

20. Install the front intake manifold seal and rear intake manifold seal, then secure them with retaining features.

21. Position the intake manifold gaskets in place and insert the locking tabs over the tabs on the cylinder head gaskets.

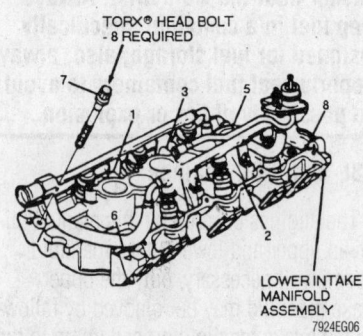

Make certain to tighten the intake manifold attaching bolts in the prescribed order—3.0L engine

22. Carefully lower the intake manifold into position on the cylinder block and cylinder heads to prevent smearing the silicone sealer and causing gasket leaks.

23. Install the retaining bolts and tighten in two stages, in the sequence illustrated, first to 11 ft. lbs. (15 Nm) and then to 18 ft. lbs. (24 Nm).

24. Install the camshaft position sensor.

➡ **The rocker arm fulcrum must be fully seated into the cylinder head and the pushrod must be fully seated in the rocker arm and lifter sockets prior to tightening.**

25. Apply Engine Assembly Lubricant D9AZ-19579-D or the equivalent to cylinder No. 3 intake valve pushrod and rocker arm. Install the pushrod. Move the rocker arm into position with the pushrod, then snug the rocker arm bolt. Rotate the crankshaft to position the camshaft lobe straight down and away from the valve lifter. Tighten the retaining bolt to 8 ft. lbs. (11 Nm) to seat the rocker arm fulcrum into the cylinder head. Final-tighten the bolt to 19–28 ft. lbs. (26–38 Nm) in any position.

26. Install the rocker arm covers.

27. Reconnect the fuel lines.

28. Install the air intake throttle body.

29. If equipped with distributor ignition, install the distributor assembly, using the matchmarks made earlier to insure correct alignment. Install the distributor cap and spark plug wires.

30. If equipped with distributorless ignition, install the coil pack then install the spark plug wires to their original locations.

31. Reconnect the wiring harness to the engine electrical components.

32. Reconnect the negative battery cable and refill the cooling system.

33. Run the engine and check for coolant, fuel, oil and vacuum leaks.

34. If equipped with distributor ignition, verify base initial timing. Check and adjust engine idle as necessary.

4.0L OHV Engine

The intake manifold is a 4-piece assembly, consisting of the upper intake manifold, the throttle body, the fuel supply manifold, and the lower intake manifold.

1. Disconnect the battery ground cable.
2. Remove the weather shield.
3. Remove the air cleaner intake duct.
4. Disconnect the throttle cable and bracket.
5. Tag and unplug all vacuum lines connected to the manifold.
6. Tag and disconnect all electrical wires attached to the upper manifold assembly.
7. Relieve the fuel system pressure.
8. Remove the throttle body.
9. Tag and remove the spark plug wires.
10. Remove the ignition coil and bracket.
11. Remove the 6 attaching nuts and lift off the upper manifold.

➡**If you only need to remove the upper intake manifold, stop at this point. Otherwise, continue with the procedure to also remove the lower intake manifold.**

12. Disconnect the fuel supply and return lines from the injector fuel rail.

13. Label and disconnect all of the electrical connections on the lower manifold assembly.

14. Remove the valve covers.

15. Remove the lower intake manifold bolts. Tap the manifold lightly with a plastic mallet and remove it.

16. Clean all surfaces of old gasket material.

To install:

17. Apply Silicone Rubber D6AZ-19562-BA or equivalent to the block and cylinder head mating surfaces where the parts of the gaskets come together. Install the intake manifold gasket and again apply sealer to the same four corner locations.

➡**This material will set within 15 minutes, so work quickly!**

18. Position the lower manifold onto the engine block, then install the nuts hand-tight. Tighten the nuts, in 3 stages, in the sequence shown, to 18 ft. lbs. (25 Nm).

   a. 6 ft. lbs. (8 Nm).
   b. 11 ft. lbs. (15 Nm).
   c. 16 ft. lbs. (22 Nm).

19. Install the valve covers using new gaskets.

➡**The following procedures are for installing the upper intake manifold.**

20. Position a new gasket and install the upper manifold. Tighten the nuts to 18 ft. lbs.

21. Install the ignition coil.
22. Connect the fuel and return lines.
23. Install the throttle body.
24. Connect all wires.
25. Connect all vacuum lines.
26. Connect the throttle linkage.
27. Install the weather shield.
28. Install the air cleaner and duct.
29. Fill and bleed the cooling system.
30. Connect the battery ground.
31. Run the engine and check for leaks.

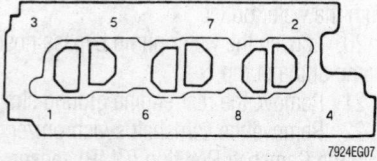

To insure good gasket crush, tighten the intake manifold retaining bolts in the correct order—4.0L engine

4.0L SOHC Engine

1. Disconnect the negative battery cable.
2. Remove the air cleaner-to-intake tube.
3. Remove the accelerator splash shield.
4. Disconnect the accelerator and, if equipped with cruise control, speed control cables from the throttle control cam.
5. Remove the accelerator cable retaining bracket from the upper intake manifold.
6. Label and disengage all vacuum and electrical connections on the intake manifold.

7. Remove the eight upper intake manifold attaching bolts.

8. Lift up on the manifold and remove both fuel Vapor Management Valve (VMV) hoses.

9. Remove the upper intake manifold.

➡**If you only need to remove the upper intake manifold, stop at this point. Otherwise, continue with the procedure to also remove the lower intake manifold.**

10. Remove the 12 lower manifold attaching bolts.

11. Remove the lower intake manifold from the engine.

To install:

➡**Ford does not specify a sequence for either upper or lower intake manifolds, but it is recommended that you start tightening in the middle and work your way out to the ends. Repeat the tightening sequence several times until the bolts will no longer turn at the specified torque.**

12. Install the lower intake manifold to the engine.

13. Install the attaching bolts. Tighten the bolts to 8.8–10.3 ft. lbs. (12–14 Nm).

➡**The following procedures are for installing the upper intake manifold.**

14. Position the upper manifold on the lower manifold.

15. Attach both VMV hoses to the manifold, then install the upper manifold attaching bolts. Tighten the bolts to 53–62 inch lbs. (6–7 Nm). See the note at the beginning of the installation procedure.

16. Attach any vacuum and electrical connections that were removed.

17. Connect the accelerator cable bracket to the intake and the cable (or cables if equipped with cruise control) to the throttle cam.

18. Install the accelerator splash shield.

19. Install the air cleaner-to-intake supply tube.

20. Connect the negative battery cable. Run the engine until normal operating temperature is reached and check for vacuum leaks.

5.0L Engine

➡**To remove the lower intake manifold on the 5.0L engine, the camshaft synchronizer, which houses the Camshaft Position (CMP) sensor, must be removed. There are special tools necessary for this procedure. If the tools** are not available and you remove the CMP, the fuel injectors will not remain timed to the engine and the vehicle will not run.

1. Disconnect the negative battery cable.
2. Drain the cooling system.

✲✲ CAUTION

Never open, service or drain the radiator or cooling system when hot; serious burns can occur from the steam and hot coolant.

3. Remove the air cleaner outlet tube.
4. Disconnect the crankcase ventilation hose from the throttle body.
5. Remove the throttle body shield, then disengage the cables from the throttle shaft.
6. Label and disconnect any electrical components or vacuum lines attached to the upper intake manifold.
7. Remove the accelerator cable brackets from the manifold and position aside.
8. If equipped, disconnect the EGR spacer (or throttle body) coolant hoses.
9. Label and remove the spark plug wires from the coil packs.
10. Disconnect the ignition coil wiring harness plug.
11. Unbolt and remove the coil packs from the engine.
12. Remove the intake cover plate to gain access to the two long upper intake manifold attaching bolts.
13. Remove any wires or loom holders from the upper manifold.
14. Loosen and remove the upper intake manifold attaching bolts in the sequence given.
15. Remove the upper manifold and its gasket.

➡**If you only need to remove the upper intake manifold, stop at this point. Otherwise, continue with the procedure to also remove the lower intake manifold.**

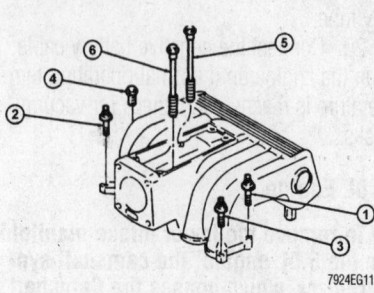

Upper intake manifold bolt loosening sequence—5.0L engine

✲✲ CAUTION

Fuel lines on fuel injected vehicles will remain pressurized after the engine is shut OFF. Fuel pressure must be relieved before servicing the fuel system.

16. Properly relieve the fuel system pressure then disconnect the fuel supply and return lines from the fuel supply manifold.

✲✲ CAUTION

Observe all applicable safety precautions when working around fuel. Whenever servicing the fuel system, always work in a well ventilated area. Do not allow fuel spray or vapors to come in contact with a spark or open flame. Keep a dry chemical fire extinguisher near the work area. Always keep fuel in a container specifically designed for fuel storage; also, always properly seal fuel containers to avoid the possibility of fire or explosion.

17. Disconnect the heater hoses from the hot water tube. Also remove the tube from the lower manifold.
18. Label and disconnect any electrical components, sensors or vacuum lines connected to the lower manifold.
19. Disconnect the upper radiator hose from the water outlet.
20. Loosen the water pump bypass hose clamp at the pump.
21. Remove the rear engine ground strap.
22. Remove the camshaft synchronizer with the Camshaft Position (CMP) sensor installed.
23. Loosen and remove the lower manifold attaching bolts in the sequence given.
24. Remove the lower manifold and its gasket from the engine.

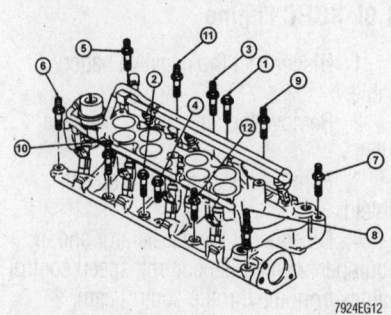

To prevent manifold warpage, loosen the lower intake manifold using the proper sequence—5.0L engine

To install:

25. Ensure that all of the gasket mating surfaces are clean and free of grease, oil or dirt. Also ensure that the EGR passages in the manifolds and heads are clear.
26. Apply a ¹⁄₁₆ in. (1.6mm) bead of silicone sealer to the points where the cylinder block rails meet the cylinder heads.
27. Position new seals on the cylinder block and new gaskets on the cylinder heads with the gaskets interlocked with the seal tabs. Make sure the holes in the gaskets are aligned with the holes in the cylinder heads.
28. Apply a ¹⁄₁₆ in. (1.6mm) bead of sealer to the outer end of each intake manifold seal for the full width of the seal. Make sure the silicone sealer will not fall into the engine and possibly block oil passages.
29. Using guide pins to ease installation, carefully lower the intake manifold into position on the cylinder block and cylinder heads. Also, ensure that the water pump bypass hose is installed at the same time.

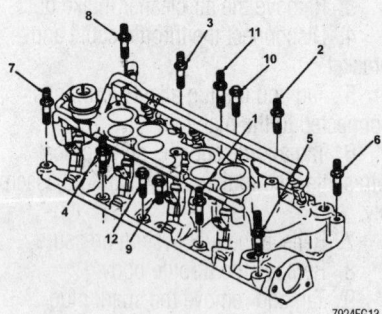

Be sure to tighten the lower intake manifold bolts in the proper order—5.0L engine

30. Install the lower manifold attaching bolts and tighten, in the sequence shown, to specification in two steps as follows:
   a. Tighten all bolts to 5–10 ft. lbs. (6–14 Nm).
   b. Tighten all bolts to 23–25 ft. lbs. (31–34 Nm).
31. Install the camshaft synchronizer.
32. Install the rear engine ground strap.
33. Tighten the water pump bypass hose clamp at the water pump.
34. Install the upper radiator hose to the water outlet.
35. Connect any electrical components or vacuum lines which were removed from the lower manifold.
36. Install the hot water tube to the engine, and connect the heater hose to the tube.
37. Connect the fuel lines, then temporarily connect the negative battery cable.

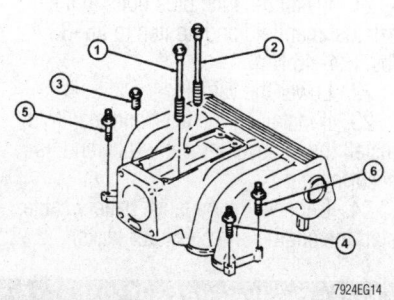

Tighten the upper intake manifold bolts in the order shown to ensure proper sealing—5.0L engine

7924EG14

Cycle the ignition to pressurize the fuel system and check for leaks. Turn the ignition key back and forth (from **ON** to **OFF**) at least 6 times, leaving the key **ON** for 5 seconds each time. If no leaks are found, disconnect the negative battery cable and continue the installation.

➡**The following procedures are for installing the upper intake manifold.**

38. Install the upper intake manifold using a new gasket. Tighten the attaching bolts in the sequence shown to 12–18 ft. lbs. (16–25 Nm).

39. Install the intake cover plate.

40. Install the ignition coils to the engine and connect the harness plug and spark plug wires to them.

41. If equipped, install the EGR spacer (throttle body) coolant hoses.

42. Install the accelerator cable brackets and cables to the intake and throttle shaft.

43. Connect any electrical components or vacuum lines removed from the upper manifold.

44. Install the throttle body shield.

45. Install the air cleaner outlet tube to the throttle body.

46. Refill and bleed the cooling system.

47. Connect the negative battery cable then start and run the engine until it reaches normal operating temperatures. Check for leaks.

Exhaust Manifold

REMOVAL & INSTALLATION

✳✳ CAUTION

Allow the engine to cool before attempting to remove the manifolds. Serious injury can result from contact with hot exhaust manifolds.

2.3L and 2.5L Engines

1. Disconnect the negative battery cable.
2. Remove the air cleaner outlet tube.
3. Remove the EGR transducer lines at the tube. Loosen and remove the EGR valve-to-exhaust manifold tube.
4. Disconnect the exhaust pipe from the exhaust manifold.
5. Remove the two nuts securing the lifting bracket/transducer mount, and remove the bracket.
6. Remove the exhaust manifold mounting bolts/nuts and remove the manifold.
7. Install the exhaust manifold in the reverse of the removal. Tighten the manifold in sequence in two steps: first to 5–7 ft. lbs. (7–10 Nm), then to 16–23 ft. lbs. (22–31 Nm).

3.0L Engine

1. Disconnect the negative battery cable.

➡**Spray retaining fasteners with rust penetrant prior to loosening in order to prevent removal of the stud when the nut is loosened.**

2. Raise and safely support the vehicle.
3. Loosen and remove the muffler inlet pipe retaining nuts from the exhaust manifolds.
4. Lower the vehicle.
5. To remove the right-hand side manifold, perform the following:
   a. Remove the spark plugs.
   b. Remove the exhaust manifold retaining bolts and studs.
6. To remove the left-hand side manifold, perform the following:
   a. Remove the two hoses from the EGR valve to the exhaust manifold tube.
   b. Loosen the EGR tube flare nut at the EGR valve.
   c. Remove the EGR tube bolt at the exhaust manifold.
   d. Remove the engine oil level indicator dipstick tube retaining nut. Rotate or remove the tube out of the way.
   e. Remove the spark plugs.
   f. Remove the exhaust manifold retaining bolts and studs.
7. Remove the exhaust manifolds.
   To install:
8. Clean the mating surfaces of the exhaust manifolds and cylinder heads of carbon deposits.
9. Position the right or left exhaust manifold in place on the cylinder head, then hand-tighten the retaining bolts and studs. Tighten to 15–22 ft. lbs. (20–30 Nm).

10. Install the oil level indicator tube to the appropriate exhaust stud, and install the retaining nut if the exhaust manifold was removed. Tighten the nut to 12–15 ft. lbs. (16–20 Nm). Apply ESE-M4G217-A or equivalent sealer to the tube prior to installation, if the tube was removed.

11. Install the spark plugs. Tighten the plugs to 8 ft. lbs. (11 Nm).

12. Install the EGR tube flare nut into the EGR valve finger-tight.

13. Install the EGR valve-to-exhaust manifold tube, at the exhaust manifold, then tighten to 15–22 ft. lbs. (20–30 Nm).

14. Tighten the flare nut to 26–48 ft. lbs. (35–65 Nm).

15. Connect the two hoses onto the ports in the EGR tube.

16. Connect the exhaust inlet pipe to the left exhaust manifold and right exhaust manifold. Tighten the retaining nuts to 30 ft. lbs. (41 Nm). Tighten both nuts in equal amounts to correctly seat the inlet pipe flange.

➡**When the battery has been disconnected and reconnected, some abnormal drive symptoms may occur while the Powertrain Control Module (PCM) relearns its adaptive strategy. The vehicle may need to be driven about 10 miles (16 km) or more to relearn the strategy.**

17. Start the engine and check for oil and exhaust leaks.

4.0L OHV Engine

1. Disconnect the negative battery cable. Remove the dipstick tube bracket.
2. Raise and safely support the vehicle.
3. Remove the exhaust pipe-to-manifold bolts.
4. Lower the vehicle.
5. If removing the left-hand manifold, disconnect the power steering pump hoses.
6. If removing the right-hand manifold, remove the hot air intake shroud which is bolted around the manifold.
7. Unbolt and remove the manifold.
8. Clean and lightly oil all fastener threads.
9. Installation is the reverse of removal. Replace all gaskets if so equipped. Tighten the manifold bolts to 19 ft. lbs. (26 Nm); the exhaust pipe nuts to 20 ft. lbs. (27 Nm). Tighten both exhaust pipe retaining nuts in equal amounts to correctly seat inlet pipe flange.
10. If installing the left-hand manifold, reconnect the power steering pump hoses, then fill and bleed the system.

11. Connect the negative battery cable and run the engine to check for leaks.

4.0L SOHC Engine

➡**When installing the exhaust manifold, always use a new gasket and attaching nuts.**

1. Disconnect the negative battery cable.

2. Raise and safely support the vehicle.

3. Remove the exhaust inlet pipe-to-manifold attaching bolts.

4. Lower the vehicle.

5. For removing the left-hand manifold, proceed as follows:

   a. Disconnect the hoses from the Differential Pressure Feedback EGR (DPFE) transducer, which is mounted to the valve cover.

   b. Disconnect the EGR tube from the manifold and the valve and remove the tube.

6. Remove the six manifold-to-engine attaching nuts.

7. Remove the manifold and the gasket. Discard the old gasket.

   To install:

8. Clean the gasket mating surfaces.

9. Position the exhaust manifold and new gasket to the cylinder head and install new attaching nuts. Tighten the nuts to 15–18 ft. lbs. (20–25 Nm).

10. For installing the left-hand manifold, proceed as follows:

    a. Position and install the EGR tube to the manifold and the valve.

    b. Connect the two hoses to the DPFE transducer.

11. Raise and safely support the vehicle.

12. Install the exhaust inlet pipe-to-manifold attaching bolts, and tighten them to 25–32 ft. lbs. (34–46 Nm).

13. Lower the vehicle and reconnect the negative battery cable.

14. Start the engine and check for leaks.

5.0L Engine

➡**When installing the manifolds, always use new gaskets.**

1. Disconnect the negative battery cable.

2. Remove the accessory drive belt.

3. For the right-hand manifold, proceed as follows:

   a. Drain the cooling system.

   b. Disconnect the alternator wires.

   c. Remove the drive belt tensioner.

   d. Disconnect the heater hot water tube from the water pump.

   e. Remove the alternator bracket bolts and then remove the assembly.

4. For the left-hand manifold, proceed as follows:

   a. Remove A/C compressor bolts and position it off to the side.

   b. Remove the oil level indicator tube nut and tube.

5. Raise and safely support the vehicle.

6. Remove the exhaust inlet pipe-to-manifold bolts.

7. Lower the vehicle to a suitable height to access the manifold-to-engine attaching bolts.

➡**Access to the manifold-to-engine bolts is through the wheel well opening.**

8. Remove the pushpins securing the fender apron, and remove it for access to the manifold attaching bolts.

9. Label and disconnect the spark plug wires from the side of the engine on which the manifold is to be removed.

10. Remove the exhaust manifold attaching bolts.

11. Lower the vehicle and remove the manifold.

12. Discard the old gasket.

   To install:

13. Clean the gasket mating surfaces. Inspect the manifold for cracks and damaged gasket mating surfaces. Replace if damaged.

14. Position the exhaust manifold and gasket to the cylinder head and install all of the bolts finger-tight.

15. For installing the left-hand manifold, proceed as follows:

    a. Position the A/C compressor to its bracket and install the attaching bolts. Tighten the bolts to 16–21 ft. lbs. (21–29 Nm).

    b. Install the drive belt.

16. For installing the right-hand manifold, proceed as follows:

    a. Install the generator bracket to the cylinder head.

    b. Connect the heater hot water tube to the water pump.

    c. Install the drive belt tensioner. Tighten the bolts to 15–22 ft. lbs. (20–30 Nm).

    d. Connect the alternator wires.

    e. Install the accessory drive belt.

    f. Refill the cooling system.

17. Raise and safely support the vehicle to allow access to the exhaust manifold-to-engine attaching bolts.

18. Tighten the exhaust manifold bolts to 26–35 ft. lbs. (35–44 Nm).

19. Connect the spark plug wires to the plugs.

20. Install the fender apron and its pushpins.

21. Install the inlet pipe bolts to the exhaust manifold and tighten to 26–33 ft. lbs. (34–46 Nm).

22. Lower the vehicle.

23. If installing the left-hand manifold, install the oil level indicator tube and its securing nut.

24. Connect the negative battery cable, start the engine and check for leaks.

Front Crankshaft Seal

REMOVAL & INSTALLATION

➡**The 3.0L, 4.0L and 5.0L engines use timing chains; only engines using timing belts will be covered here.**

2.3L and 2.5L Engines

1. Disconnect the negative battery cable.

2. Remove the accessory drive belts.

3. Remove the timing belt cover.

4. Align the crankshaft and camshaft timing marks and remove the timing belt.

5. Remove the crankshaft pulley center bolt and slide the pulley off of the crankshaft.

6. Remove the key from the crankshaft.

✱✱ WARNING

Do not damage the crankshaft sealing surface while removing the oil seal.

7. Position the Crankshaft Seal Remover tool T74P-6700-B or equivalent on the crankshaft and into the oil seal. Remove the seal and clean the seal journal.

   To install:

8. Apply clean engine oil to the rubber lip of the new seal to aid installation.

9. Using Cam Bearing Adapter Tube T72C-6250, or equivalent, and crankshaft center bolt, carefully install the new oil seal until flush with the engine.

10. Install the key and crankshaft pulley, washer and bolt. Tighten the bolt to 92–121 ft. lbs. (125–165 Nm).

11. Install the timing belt and cover.

12. Install the drive belts and connect the negative battery cable.

Camshaft and Valve Lifters

➡**Although Ford suggests that this component is removable while the engine is installed in the vehicle, depending on the particular options with which your Aerostar is equipped, working clearance may be extremely** tight and this procedure may be much easier to perform with the engine removed. Before commencing, read through this procedure and make certain enough clearance, or working room, exists with the engine in the vehicle; if there is not enough space, the engine should be removed.

REMOVAL & INSTALLATION

2.3L and 2.5L Engines

➡The following procedure covers camshaft removal and installation with the cylinder head on or off the engine. If the cylinder head has been removed, follow Steps 7–9, then skip to Step 12.

1. Drain the cooling system. Remove the air cleaner assembly and disconnect the negative battery cable.

✳✳ CAUTION

Never open, service or drain the radiator or cooling system when hot; serious burns can occur from the steam and hot coolant.

2. Remove the spark plug wires from the plugs, disconnect the retainer from the valve cover and position the wires out of the way. Disconnect all vacuum lines as necessary.
3. Remove all drive belts. Remove the alternator mounting bracket-to-cylinder head mounting bolts, position the bracket and alternator out of the way.
4. Disconnect and remove the upper radiator hose. Disconnect the radiator shroud.
5. Remove the fan blades, water pump pulley and fan shroud. Remove the timing belt and valve covers.
6. Align engine timing marks at TDC for No. 1 cylinder. Remove the timing belt.
7. Remove the rocker arms (camshaft followers).
8. Remove the camshaft drive gear and belt guide using a suitable puller. Remove the front oil seal with Front Seal Replacer T74P-6150-A or equivalent.
9. Remove the camshaft retainer located on the rear mounting stand by removing the two bolts.
10. Raise and safely support the vehicle. Remove the front motor mount bolts. Disconnect the lower radiator hose from the radiator. If equipped, disconnect and plug the automatic transmission cooler lines.
11. Position a piece of wood on a floor jack and raise the engine carefully as far as it will go. Place blocks of wood between the

engine mounts and crossmember pedestals.
12. Remove the camshaft by carefully withdrawing it toward the front of the engine. Caution should be used to prevent damage to cam bearings, lobes and journals.
13. Check the camshaft journals and lobes for wear. Inspect the cam bearings, if worn (unless the proper bearing installing tool is on hand), the cylinder head must be removed for new bearings to be installed by a machine shop.
14. Camshaft installation is in the reverse order of service removal procedure. Coat the camshaft with a heavy SF (or better) grade oil before sliding it into the cylinder head. Install a new front seal. Apply a coat of sealer or Teflon® tape to the cam drive gear bolt before installation. After any procedure requiring removal of the rocker arms, each lash adjuster must be fully collapsed after assembly, then released using Valve Spring Compressor T95T-6565-A or equivalent. This must be done before the camshaft is turned.
15. Refill the cooling system. Start the engine and check for leaks. Roadtest the vehicle for proper operation.

3.0L Engine

1. Have the A/C system discharged by a qualified, professional mechanic utilizing a recovery/recycling machine.
2. Disconnect the negative battery cable.
3. Remove the air cleaner hoses.
4. Remove the fan and spacer, and shroud.
5. Drain the cooling system. Remove the radiator.

✳✳ CAUTION

Never open, service or drain the radiator or cooling system when hot; serious burns can occur from the steam and hot coolant.

6. Rotate the crankshaft so that No. 1 piston is at TDC on the compression stroke.
7. Remove the A/C system condenser.
8. Relieve the fuel system pressure.

✳✳ CAUTION

Observe all applicable safety precautions when working around fuel. Whenever servicing the fuel system, always work in a well ventilated area. Do not allow fuel spray or vapors to come in contact with a spark or open flame. Keep a dry chemical fire extinguisher near the

work area. Always keep fuel in a container specifically designed for fuel storage; also, always properly seal fuel containers to avoid the possibility of fire or explosion.

9. Remove the fuel lines at the fuel supply manifold.
10. Tag and disconnect all vacuum hoses in the way.
11. Tag and disconnect all wires in the way.
12. Remove the engine front cover and water pump.
13. Remove the alternator.
14. Remove the power steering pump and secure it out of the way. DO NOT disconnect the hoses!
15. Remove the air conditioning compressor and secure it out of the way. DO NOT disconnect the hoses!
16. Remove the throttle body.
17. Remove the fuel injection harness.
18. Drain the engine oil into a suitable container and dispose of it properly.

✳✳ CAUTION

The EPA warns that prolonged contact with used engine oil may cause a number of skin disorders, including cancer! You should make every effort to minimize your exposure to used engine oil.

19. Turn the engine by hand to 0 BTDC of the power stroke on No. 1 cylinder.
20. Disconnect the spark plug wires from the plugs.
21. If equipped, remove the distributor cap with the spark plug wires as an assembly.
22. If equipped, matchmark the rotor, distributor body and engine. Disconnect the distributor wiring harness and remove the distributor.
23. Remove the rocker arm covers.
24. Remove the intake manifold as previously described.
25. Loosen the rocker arm bolts enough to pivot the rocker arms out of the way and remove the pushrods. Identify them for installation. They must be installed in their original positions!
26. Remove the lifters. Identify them for installation. They must be installed in their original positions!
27. Remove the crankshaft pulley/damper.
28. Remove the starter.
29. Remove the oil pan as previously described.
30. If not already performed, turn the engine by hand until the timing marks align

at TDC of the power stroke on the No.1 cylinder.

31. Check the camshaft end-play. If excessive, you'll have to replace the thrust plate.

32. Remove the camshaft gear attaching bolt and washer, then slide the gear off the camshaft.

33. Remove the camshaft thrust plate.

34. Carefully slide the camshaft out of the engine block, using caution to avoid any damage to the camshaft bearings.

To install:

35. Oil the camshaft journals and cam lobes with heavy SJ engine oil (50W). Install the spacer ring with the chamfered side toward the camshaft, then insert the camshaft key.

36. Install the camshaft in the block, using caution to avoid any damage to the camshaft bearings.

37. Install the thrust plate. Tighten the attaching screws to 84 inch lbs.

38. Rotate the camshaft and crankshaft as necessary to align the timing marks. Install the camshaft gear and chain. Tighten the attaching bolt to 46 ft. lbs. (62 Nm).

39. Coat the tappets with 50W engine oil and place them in their original locations.

40. Apply 50W engine oil to both ends of the pushrods. Install the pushrods in their original locations.

41. Pivot the rocker arms into position. Tighten the fulcrum bolts to 8 ft. lbs. (11 Nm).

42. Rotate the engine until both timing marks are at the tops of their sprockets and aligned. Tighten the following fulcrum bolts to 18 ft. lbs. (24 Nm):
- No.1 intake
- No.2 exhaust
- No.4 intake
- No.5 exhaust

43. Rotate the engine until the camshaft timing mark is at the bottom of the sprocket and the crankshaft timing mark is at the top of the sprocket, and both are aligned. Tighten the following fulcrum bolts to 18 ft. lbs. (24 Nm):
- No.1 exhaust
- No.2 intake
- No.3 intake and exhaust
- No.4 exhaust
- No.5 intake
- No.6 intake and exhaust

44. Now, tighten all the bolts to 24 ft. lbs. (33 Nm).

45. Turn the engine by hand to 0 BTDC of the power stroke on No. 1 cylinder.

46. Install the engine front cover and water pump assembly.

47. Install the oil pan.

48. Install the crankshaft damper/pulley and tighten the retaining bolt to 107 ft. lbs. (145 Nm).

49. Install the intake manifold and tighten the mounting bolts to the specifications and in the sequence described under Intake Manifold removal and installation.

50. Install all components that were removed.

51. Refill the cooling system.

52. Replace the oil filter and refill the crankcase with the specified amount of engine oil.

53. Reconnect the battery ground cable.

54. Start the engine and check the ignition timing and idle speed. Adjust if necessary. Run the engine at fast idle and check for coolant, fuel, vacuum or oil leaks.

➡When the battery has been disconnected and reconnected, some abnormal drive symptoms may occur while the Powertrain Control Module (PCM) relearns its adaptive strategy. The vehicle may need to be driven about 10 miles (16 km) or more to relearn the strategy.

4.0L OHV Engine

➡It is necessary to replace the oil pan gasket when removing and installing the engine front cover. It will also be necessary to remove the transmission to properly reseal the oil pan.

1. If the vehicle is equipped with an Air Conditioning (A/C) system, discharge the system utilizing a refrigerant recovery/recycling machine OR do not disconnect the A/C refrigerant lines from the condenser or from the compressor.

2. Disconnect the negative battery cable.

3. Drain the engine oil into a suitable container and dispose of it properly.

4. Drain the cooling system.

✳✳ CAUTION

Never open, service or drain the radiator or cooling system when hot; serious burns can occur from the steam and hot coolant.

5. Remove the radiator.

6. Remove the A/C condenser from the engine. If the A/C was not discharged, remove the A/C compressor from its mounting bracket and secure it aside with the refrigerant lines still attached.

7. Remove the fan and spacer, and shroud.

8. Remove the air cleaner hoses.

9. Tag and remove the spark plug wires.

10. Remove the ignition coil and bracket.

11. Remove the crankshaft pulley/damper.

12. Remove the clamp, bolt and oil pump drive from the rear of the block.

13. Remove the alternator.

14. Relieve the fuel system pressure.

✳✳ CAUTION

Observe all applicable safety precautions when working around fuel. Whenever servicing the fuel system, always work in a well ventilated area. Do not allow fuel spray or vapors to come in contact with a spark or open flame. Keep a dry chemical fire extinguisher near the work area. Always keep fuel in a container specifically designed for fuel storage; also, always properly seal fuel containers to avoid the possibility of fire or explosion.

15. Remove the fuel lines at the fuel supply manifold.

16. Remove the upper and lower intake manifolds.

17. Remove the rocker arm covers.

18. Remove the rocker arm shaft assemblies.

19. Remove the pushrods. Identify them for installation. They must be installed in their original positions!

20. Remove the tappets. Identify them for installation. They must be installed in their original positions!

21. Remove the oil pan, engine front cover and water pump.

22. Turn the engine by hand until the timing marks align at TDC of the power stroke on No.1 piston.

23. Place the timing chain tensioner in the retracted position and install the retaining clip.

24. Check the camshaft end-play. If excessive, you'll have to replace the thrust plate.

25. Remove the camshaft gear attaching bolt and washer, then slide the gear off the camshaft.

26. Remove the camshaft thrust plate.

27. Carefully slide the camshaft out of the engine block, using caution to avoid any damage to the camshaft bearings.

To install:

28. Lubricate the camshaft using a good assembly lubricant.

29. Install the camshaft in the block, using caution to avoid any damage to the camshaft bearings.

30. Install the thrust plate. Make sure that it covers the main oil gallery. Tighten the attaching screws to 7–10 ft. lbs. (9–13 Nm).

31. Rotate the camshaft and crankshaft, as necessary, to align the timing marks. Install the camshaft gear and chain. Tighten the attaching bolt to 44–50 ft. lbs. (60–68 Nm).

32. Remove the clip from the chain tensioner.

33. Install the engine front cover and water pump assembly.

34. Install the crankshaft damper/pulley and tighten the retaining bolt to 107 ft. lbs. (146 Nm).

35. Install the oil pan. It is important to adhere to the procedures given in this section.

36. Coat the tappets with 50W engine oil and place them in their original locations.

37. Apply 50W engine oil to both ends of the pushrods. Install the pushrods in their original locations.

38. Install the remaining components in the reverse of the removal.

39. Replace the oil filter and refill the crankcase with the specified amount of engine oil.

40. Reconnect the negative battery cable.

➡**When the battery has been disconnected and reconnected, some abnormal drive symptoms may occur while the Powertrain Control Module (PCM) relearns its adaptive strategy. The vehicle may need to be driven about 10 miles (16 km) or more to relearn the strategy.**

41. Start the engine and check the ignition timing and idle speed; adjust if necessary. Run the engine at fast idle and check for coolant, fuel, vacuum or oil leaks.

4.0L SOHC Engine

1. Disconnect the negative battery cable for safety.

2. Remove the valve cover.

3. Remove the hydraulic camshaft tensioner.

➡**The RH camshaft sprocket bolt uses left-hand threads.**

4. For the RH camshaft use the Cam Gear Torque Adapter tool T97T-6256-F, or its equivalent, to remove the camshaft sprocket bolt.

5. For the LH camshaft, remove the sprocket bolt.

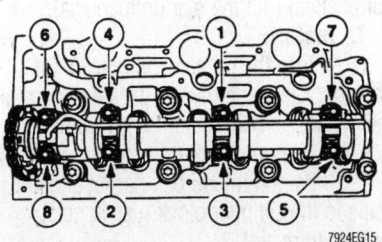

7924EG15

Use the proper sequence to prevent damage to the camshaft both when installing and removing the bearing caps—4.0L SOHC engine

➡**When removing the followers, label them so that they may be returned to their original positions.**

6. Using the Valve Spring Compressor tool ST1330-A, or its equivalent, remove the camshaft roller followers.

7. Remove the camshaft bearing cap bolts and the oil rail.

8. Remove the camshaft.

To install:

9. Lubricate all of the moving parts with SAE 50W engine oil.

10. Install the camshaft onto the cylinder head.

11. Position the oil rail and install the bearing caps and bolts. Tighten the bolts in two steps.
   a. Step 1—53.5 inch lbs. (6 Nm)
   b. Step 2—11–12.5 ft. lbs. (15–17 Nm)

12. Install the camshaft followers in the same manner as removal.

13. Install the camshaft sprocket bolt. Do not tighten the bolt.

14. Install the Camshaft Chain Tensioner T97T-6K254-A or equivalent in the hole that the hydraulic chain tensioner was in.

15. Turn the crankshaft one revolution clockwise until No. 1 piston is TDC.

16. Install Crankshaft Holding tool T97T-6303-A or equivalent on the crankshaft to keep it from turning.

17. Position the timing slot on the rear of the camshaft to fit Camshaft Holding tool T97T-6256-C or equivalent and install the holding tool on the rear of the head.

18. Install the Camshaft Gear Holding tool T97T-6256-B and Camshaft Gear Holding tool tool T97T-6256-A or equivalents on the front of the cylinder head to securely hold the camshaft gear.

19. Tighten the camshaft sprocket bolt to 63 ft. lbs. (85 Nm).

20. Remove the Camshaft Chain Tensioner tool and install the hydraulic chain tensioner, tighten the tensioner to 35–39 ft. lbs. (47–53 Nm).

21. Remove the special tools from the engine.

22. Install the valve cover.

23. Connect the negative battery cable.

24. Start the engine and check for leaks.

5.0L Engine

1. Disconnect the negative battery cable.

2. Remove the timing chain cover.

3. Remove the camshaft sprocket and chain assembly.

4. Remove the upper and lower intake manifolds.

5. Remove both valve covers.

6. Loosen the rocker arm bolts and rotate the rocker arms to the side.

7. Remove the pushrods in sequence so that they may be installed to their original positions.

8. Remove all of the lifters, also keeping them in order.

9. Remove the camshaft thrust plate bolts and the plate.

10. Withdraw the camshaft from the engine, taking care not to damage the bearings, lobes or journals.

To install:

11. Apply SAE 50W engine oil to the camshaft lobes and journals.

12. Carefully install the camshaft into position in the cylinder block.

13. Apply SAE 50W engine oil to the camshaft thrust plate.

14. Position the thrust plate with the groove toward the block and install the retaining bolts. Tighten to 9–12 ft. lbs. (13–16 Nm).

15. Apply SAE 50W engine oil to the valve tappets and install them. If reusing the old lifters, place them in their original positions.

16. Install the pushrods to their original positions.

17. Reposition the rocker arms and tighten them as outlined earlier in this Section.

18. Install the valve covers, the lower and the upper intake manifolds.

19. Install the camshaft sprocket and chain assembly. Ensure that the timing marks on the cam and crankshaft sprockets are aligned.

20. Install the timing chain cover.

21. Connect the negative battery cable. Start the engine and check for leaks.

Oil Pan

REMOVAL & INSTALLATION

> ※ **CAUTION**
>
> **The EPA warns that prolonged contact with used engine oil may cause a number of skin disorders, including cancer! You should make every effort to minimize your exposure to used engine oil.**

2.3L and 2.5L Engines

1. Disconnect the negative battery cable.
2. Remove the air cleaner assembly. Remove the oil dipstick. Remove the engine mount retaining nuts.
3. Remove the oil cooler lines at the radiator, if so equipped. Remove the two bolts retaining the fan shroud to the radiator and remove shroud.
4. On automatic transmissions, remove the radiator retaining bolts. Position the radiator upward and secure to the hood.
5. Raise and safely support the vehicle.

> ※ **CAUTION**
>
> **The EPA warns that prolonged contact with used engine oil may cause a number of skin disorders, including cancer! You should make every effort to minimize your exposure to used engine oil.**

6. Drain the oil from crankcase.
7. Remove the cable from the starter, then remove the starter.
8. Disconnect the exhaust manifold tube to the inlet pipe bracket at the Thermactor check valve.
9. Remove the transmission mount retaining nuts to the crossmember.
10. Remove the bellcrank from the converter housing (automatic only).
11. Remove the oil cooler lines from retainer at the block (automatic only).
12. Remove the front crossmember (automatic only).
13. Disconnect the right front lower shock absorber mount (manual only).
14. Position the jack under the engine, raise and block with a piece of wood approximately 2½ in. high. Remove the jack.
15. Position the jack under the transmission and raise it slightly (automatic only).
16. Remove the oil pan retaining bolts, lower the pan to the chassis. Remove the oil pump drive and pick-up tube assembly.

17. Remove the oil pan (out the front on automatics) (out the rear on manuals).

To install:

18. Clean the oil pan and inspect for damage. Clean the oil pan gasket surface at the cylinder block. Clean the oil pump exterior and oil pump pick-up tube screen.
19. Position the oil pan gasket and end seals to the cylinder block (use contact cement to retain).
20. Position the oil pan to the crossmember.
21. Install the oil pump and pick-up tube assembly. Install the oil pan to cylinder block with retaining bolts.
22. Lower the jack under the transmission (automatic only).
23. Position the jack under the engine, raise it slightly, and remove the wood spacer block.
24. Replace the oil filter.
25. Connect the exhaust manifold tube to the inlet pipe bracket at the Thermactor check valve.
26. Install the transmission mount to the crossmember.
27. Install the oil cooler lines to the retainer at the block (automatic only).
28. Install the bellcrank to the converter housing (automatic only).
29. Install the right front lower shock absorber mount (manual only). Install the front crossmember (automatic only).
30. Install the starter and connect the cable. Lower the vehicle.
31. Install the engine mount bolts.
32. Locate the radiator to the supports and install the (2) retaining bracket bolts (automatic only). Install the fan shroud on the radiator.
33. Connect the oil cooler lines to the radiator (automatic only).
34. Install the air cleaner assembly.
35. Install the oil dipstick. Fill the crankcase with oil.
36. Start the engine and check for leaks.

3.0L Engine

1. Disconnect the negative battery cable.
2. Remove the oil level dipstick.
3. Remove the fan shroud. Leave the fan shroud over the fan assembly.
4. Remove the motor mount nuts from the frame.

> ※ **WARNING**
>
> **On models equipped with distributor ignition, failure to remove the distributor will damage or break it when the engine is lifted.**

5. If equipped, mark and remove the distributor assembly from the engine.
6. Raise and support the vehicle safely. Remove the oil level sensor wire.

> ※ **CAUTION**
>
> **The EPA warns that prolonged contact with used engine oil may cause a number of skin disorders, including cancer! You should make every effort to minimize your exposure to used engine oil.**

7. Remove the starter motor from the engine.
8. Remove the transmission inspection cover.
9. Remove the right hand axle I-beam. The brake caliper must be removed and secured out of the way.
10. Remove the oil pan attaching bolts, using a suitable lifting device, raise the engine about 2 in. (5cm).
11. Remove the oil pan from the engine block.

➡ **The oil pan fits tightly between the transmission spacer plate and oil pump pick-up tube. Use care when removing the oil pan from the engine.**

12. Clean all gasket surfaces on the engine and oil pan. Remove all traces of old gasket and/or sealer.

To install:

13. Apply a ⅛ (4mm) bead of RTV sealer to the junctions of the rear main bearing cap and block, and the front cover and block. The sealer sets in 15 minutes, so work quickly!
14. Apply adhesive to the gasket surfaces and install the oil pan gasket.
15. Install the oil pan on the engine block.
16. Tighten the pan bolts EVENLY to 9 ft. lbs. (12 Nm) working from the center to the end position on the oil pan.
17. Install the low oil level sensor connector. Lower the engine assembly to the original position.
18. Install the right hand axle I-beam. Install the brake caliper.
19. Install the transmission inspection cover. Install the starter motor.
20. Lower the vehicle and install the fan shroud.
21. Install the motor mount retaining nuts. If removed, install the distributor assembly.
22. Replace the oil level dipstick. Connect the battery ground. Fill the crankcase with the correct amount of new engine oil. Start the engine and check for leaks.

23. If equipped with distributor ignition, check the base ignition timing.

4.0L OHV Engine

➡**Review the complete service procedure before starting this repair.**

1. Disconnect the negative battery cable. Remove the complete engine assembly from the vehicle. Refer to the necessary service procedures in this section.

2. Mount the engine on a suitable engine stand with the oil pan facing up.

3. Remove the oil pan attaching bolts (note location of 2 spacers) and remove the pan from the engine block.

4. Remove the oil pan gasket and crankshaft rear main bearing cap wedge seal.

5. Clean all gasket surfaces on the engine and oil pan. Remove all traces of old gasket and/or sealer.

To install:

6. Install a new crankshaft rear main bearing cap wedge seal. The seal should fit snugly into the sides of the rear main bearing cap.

7. Position the oil pan gasket to the engine block and place the oil pan in correct position on the 4 locating studs.

8. Tighten the oil pan retaining bolts EVENLY to 5–7 ft. lbs.

9. The transmission bolts to the engine and oil pan. There are 2 spacers on the rear of the oil pan to allow proper mating of the transmission and oil pan. If these spacers were lost, or the oil pan was replaced, you must determine the proper spacers to install. To do this:

   a. With the oil pan installed, place a straightedge across the machined mating surface of the rear of the block, extending over the oil pan-to-transmission mounting surface.

   b. Using a feeler gauge, measure the gap between the oil pan mounting pad and the straightedge.

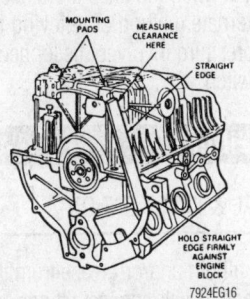

The correct spacer must be used to extend the mounting surface of the oil pan so it is flush with the mounting surface of the engine block—4.0L OHV engine

c. Repeat the procedure for the other side.

   d. Select the spacers as follows:
   • Gap = 0.011–0.020 in.—spacer = 0.010 in.
   • Gap = 0.021–0.029 in.—spacer = 0.020 in.
   • Gap = 0.030–0.039 in.—spacer = 0.030 in.

➡**Failure to use the correct spacers will result in damage to the oil pan and oil leakage.**

10. Install the selected spacers to the mounting pads on the rear of the oil pan before bolting the engine and transmission together. Install the engine assembly in the vehicle.

11. Connect the negative battery cable. Start the engine and check for leaks.

4.0L SOHC Engine

➡**The 4.0L SOHC engine does not use an oil pan in the conventional sense. There is a separate access panel that unbolts from what would be considered the oil pan (which is now known as the ladder frame).**

1. Disconnect the negative battery cable.
2. Raise and safely support the vehicle.

✳✳ CAUTION

The EPA warns that prolonged contact with used engine oil may cause a number of skin disorders, including cancer! You should make every effort to minimize your exposure to used engine oil.

3. Drain the engine oil.
4. Remove the ten oil pan bolts and remove the pan.
5. Installation is the reverse of removal. Clean the gasket mating surfaces and install a new gasket. Tighten the pan bolts to 6–7.4 ft. lbs. (8–10 Nm). Refill the engine with clean oil.

5.0L Engine

➡**The oil pan cannot be removed with the engine in the vehicle.**

1. Remove the engine.

✳✳ CAUTION

The EPA warns that prolonged contact with used engine oil may cause a number of skin disorders, including cancer! You should make every effort to minimize your exposure to used engine oil.

2. Drain the engine oil.
3. Remove the oil pan bolts.
4. Remove the oil pan and gasket.
5. Installation is the reverse of removal. Use RTV sealer at the four corners of the block and pan. Tighten the four oil pan end attaching bolts to 12–18 ft. lbs. (16–25 Nm) and the 18 oil pan bolts to 110–144 inch lbs. (13–16 Nm).

Oil Pump

REMOVAL & INSTALLATION

Except 4.0L SOHC and 5.0L Engines

➡**The oil pumps are not serviceable. If defective, they must be replaced.**

1. Remove the oil pan assembly.
2. Remove the oil pick-up and tube assembly from the pump.
3. Remove the oil pump retainer bolts and remove the oil pump.

To install:

4. Prime the oil pump with clean engine oil by filling either the inlet or outlet port with clean engine oil. Rotate the pump shaft to distribute the oil within the pump body.

5. Install the pump and tighten the mounting bolts to:
   • 14–21 ft. lbs. (19–29 Nm) on 2.3L engines
   • 8–12 ft. lbs. (11–16 Nm) on 2.5L engines
   • 30–40 ft. lbs. (41–54 Nm) on 3.0L engines
   • 13–15 ft. lbs. (18–20 Nm) on 4.0L engines

✳✳ WARNING

Do not force the oil pump if it does not seat readily. The oil pump driveshaft may be misaligned with the distributor or shaft assembly. If the pump is tightened down with the driveshaft misaligned, damage to the pump could occur. To align, rotate the intermediate driveshaft into a new position.

6. Install the oil pick-up and tube assembly to the pump. If there is a gasket between the pump and the pick-up, use a new gasket when installing.
7. Install the oil pan.

4.0L SOHC and 5.0L Engines

➡**The oil pump cannot be removed with the engine in the vehicle.**

1. Remove the engine.
2. Remove the oil pan.

3. Unbolt the oil pick-up tube.

4. On the 4.0L engine, perform the following:

a. Remove the eight ladder frame bolts which were under the oil pan.

b. Remove the two rear outer ladder frame bolts.

c. Remove the seven LH and the eight RH ladder frame bolts.

d. Lift the ladder frame from the engine.

5. Remove the two oil pump attaching bolts and the pump.

To install:

6. On the 4.0L engine, do the following:

a. Position the ladder frame on the engine.

b. Install the eight RH and seven LH ladder frame bolts.

c. Install the two rear outer and the eight frame bolts under the pan.

7. Installation is the reverse of removal. Submerge the pump in clean engine oil to prime it. Tighten the pump attaching bolts to 13–15 ft. lbs. (17–21 Nm) for the 4.0L engine and 23–31 ft. lbs. (30–43 Nm) for the 5.0L engine.

Rear Main Seal

REMOVAL & INSTALLATION

If the crankshaft rear oil seal replacement is the only operation being performed, it can be done in the vehicle as detailed in the following procedure. If the oil seal is being replaced in conjunction with a rear main bearing replacement, the engine must be removed from the vehicle and installed on a work stand.

1. Remove the starter.

2. Remove the transmission from the vehicle.

3. On vehicles with a manual transmission, remove the pressure plate and clutch disc.

4. Remove the flywheel attaching bolts and remove the flywheel and engine rear cover plate.

5. Use an awl to punch two holes in the crankshaft rear oil seal. Punch the holes on opposite sides of the crankshaft and just above the bearing cap to cylinder block split line. Install a sheet metal screw in each hole. Use two small pry bars and pry against both screws at the same time to remove the crankshaft rear oil seal. It may be necessary to place small blocks of wood against the cylinder block to provide a fulcrum point for the pry bars. Use caution throughout this procedure to avoid scratching or otherwise damaging the crankshaft oil seal surface.

To install:

6. Clean the oil seal recess in the cylinder block and main bearing cap.

7. Clean, inspect and polish the rear oil seal rubbing surface on the crankshaft. Coat a new oil seal and the crankshaft with a light film of engine oil. Start the seal in the recess with the seal lip facing forward and install it with a seal driver. Keep the tool, T82L-6701-A (4-cyl. engines) or T72C-6165 (6-cyl. engine) straight with the centerline of the crankshaft and install the seal until the tool contacts the cylinder block surface. Remove the tool and inspect the seal to be sure it was not damaged during installation.

8. On 8-cylinder engines, coat the new oil seal and crankshaft with a light film of clean engine oil. Start the seal in the recess with the seal lip facing forward and install it with Rear Crank Seal Replacer T95P-6701-BH and Spacer T96T-6701-B (or equivalents).

9. Install the engine rear cover plate. Position the flywheel on the crankshaft flange. Coat the threads of the flywheel attaching bolts with oil-resistant sealer and install the bolts. Tighten the bolts in sequence across from each other to 75–85 ft. lbs. (102–115 Nm).

10. On vehicles with a manual transmission, install the clutch disc and the pressure plate.

11. Install the transmission.

FUEL SYSTEM

Fuel System Service Precautions

Safety is the most important factor when performing not only fuel system maintenance, but any type of maintenance. Failure to conduct maintenance and repairs in a safe manner may result in serious personal injury or death. Work on a vehicle's fuel system components can be accomplished safely and effectively by adhering to the following rules and guidelines.

• To avoid the possibility of fire and personal injury, always disconnect the negative battery cable unless the repair or test procedure requires that battery voltage be applied.

• Always relieve the fuel system pressure prior to disconnecting any fuel system component (injector, fuel rail, pressure regulator, etc.) fitting or fuel line connection. Exercise extreme caution whenever relieving fuel system pressure to avoid exposing skin, face and eyes to fuel spray. Please be advised that fuel under pressure may penetrate the skin or any part of the body that it contacts.

• Always place a shop towel or cloth around the fitting or connection prior to loosening to absorb any excess fuel due to spillage. Ensure that all fuel spillage is quickly remove from engine surfaces. Ensure that all fuel-soaked cloths or towels are deposited into a flame-proof waste container with a lid.

• Always keep a dry chemical (Class B) fire extinguisher near the work area.

• Do not allow fuel spray or fuel vapors to come into contact with a light bulb, spark or open flame.

• Always use a second wrench when loosening or tightening fuel line connection fittings. This will prevent unnecessary stress and torsion to fuel piping. Always follow the proper torque specifications.

• Always replace worn fuel fitting O-rings with new ones. Do not substitute fuel hose where rigid pipe is installed.

Fuel System Pressure

RELIEVING

All SFI fuel injected engines are equipped with a pressure relief valve located on the fuel supply manifold. Remove the fuel tank cap and attach fuel pressure gauge T80L-9974-B, or equivalent, to the valve to release the fuel pressure. Be sure to drain the fuel into a suitable container and to avoid gasoline spillage. If a pressure gauge is not available, disconnect the vacuum hose from the fuel pressure regulator and attach a hand-held vacuum pump. Apply about 25 in. Hg (84 kPa) of vacuum to the regulator to vent the fuel system pressure into the fuel tank through the fuel return hose. Note that this procedure will remove the fuel pressure from the lines, but not the fuel. Take precautions to avoid the risk of fire and use clean rags to soak up any spilled fuel when the lines are disconnected.

An alternate method of relieving the fuel system pressure involves disconnecting the inertia switch.

Fuel Filter

REMOVAL & INSTALLATION

Clean all dirt and/or grease from the fuel filter fittings. Quick-connect fittings are used on all models. These fittings must be disconnected using the proper procedure or the fittings may be damaged. The fuel filter uses a "hairpin" clip retainer. Spread the two hairpin clip legs about ⅛ in. (3mm) each to disengage it from the fitting, then pull the clip outward. Use finger pressure only; do not use any tools. Push the quick-connect fittings onto the filter ends. Ford recommends that the retaining clips be replaced whenever removed. The fuel tubes used on these fuel systems are manufactured in ⅝ in. and ⅜ in. diameters. Each fuel tube takes a different size hairpin clip, so keep this in mind when purchasing new clips. A click will be heard when the hairpin clip snaps into its proper position. Pull on the lines with moderate pressure to ensure proper connection. Start the engine and check for fuel leaks. If the inertia switch (reset switch) was disconnected to relieve the fuel system pressure, cycle the ignition switch from the **OFF** to **ON** position several times to re-charge the fuel system before attempting to start the engine.

➡The inline reservoir-type fuel filter should last the life of the vehicle under normal driving conditions. If the filter does need to be replaced, proceed as follows:

✳✳ CAUTION

If the fuel filter is being serviced with the rear of the vehicle higher than the front, or if the tank is pressurized, fuel leakage or siphoning from the tank fuel lines could occur. To prevent this condition, maintain the vehicle front end at or above the level of the rear of vehicle. Also, relieve tank pressure by loosening the fuel fill cap. The cap should be tightened after the pressure is relieved.

1. Shut the engine **OFF**. Depressurize the fuel system as follows:

    a. Disengage the electrical connector from the inertia switch.

    b. Start the engine and let it run until it stalls. Crank the engine for about 15–30 seconds more to relieve residual pressure.

2. Raise and safely support the vehicle.

3. Detach the fuel lines from both ends of the fuel filter by disengaging both push connect fittings. Install new retainer clips in each quick-connect fitting.

4. Note which way the **flow** direction arrow points on the old filter.

5. Remove the filter from the bracket by loosening the filter retaining clamp enough to allow the filter to pass through.

To install:

➡The flow direction arrow should be positioned as installed in the bracket to ensure proper flow of fuel through the replacement filter.

6. Install the filter in the bracket, ensuring proper direction of flow as noted by arrow. Tighten clamp to 15–25 in. lbs. (0.7–2.5 Nm).

7. Install the quick-connect fittings at both ends of the filter.

8. Lower the vehicle.

9. Start the engine and check for leaks.

Fuel Pump

REMOVAL & INSTALLATION

➡**To gain access to the fuel pump, it is necessary to remove the fuel tank.**

1. Depressurize the fuel system and remove the fuel tank from the vehicle.

2. Remove any dirt that has accumulated around the fuel pump attaching flange, to prevent it from entering the tank during service.

3. Turn the fuel pump locking ring counterclockwise using a locking ring removal tool, then remove the locking ring.

4. Remove the fuel pump and bracket assembly.

5. Remove the seal gasket and discard it.

To install:

6. Put a light coating of heavy grease on a new seal ring to hold it in place during assembly. Install it in fuel tank ring groove.

7. Insert the fuel pump assembly into the fuel tank, then secure it in place with the locking ring. Tighten the ring until secure.

8. Install the tank in the vehicle.

9. Install a minimum of 10 gallons of fuel and check for leaks.

10. Install a pressure gauge on the throttle body valve and turn the ignition **ON** for 3 seconds. Turn the key **OFF**, then repeat the key cycle five to ten times until the pressure gauge shows at least 30 psi (207 kPa). Reinspect for leaks at the fittings.

11. Remove the pressure gauge. Start the engine and check for fuel leaks.

DRIVE TRAIN

Transmission Assembly

REMOVAL & INSTALLATION

Manual Transmission

1. Disconnect the negative battery cable.

2. Remove the gearshift lever assembly from the control housing.

3. Cover the opening in the control housing with a cloth to prevent dirt from falling into the unit.

4. Raise and safely support the vehicle.

5. On 2WD vehicles, matchmark the driveshaft to the rear axle flange. Position a drain pan under the rear of the transmission. Remove the driveshaft-to-rear axle flange fasteners and pull the driveshaft rearward to disengage it from the transmission.

6. Disconnect the clutch hydraulic line at the clutch housing. Plug the lines.

7. Disconnect the speedometer from the transfer case/extension housing.

8. Disengage the starter motor, back-up lamp and, if equipped, neutral sensing switch harness connectors.

9. Place a wood block on a service jack and position the jack under the engine oil pan.

10. On 4WD vehicles, remove the transfer case from the vehicle.

11. Remove the starter motor.

12. Position a transmission jack under the transmission.

13. Remove the transmission-to-engine retaining bolts and washers.

14. Remove the nuts and bolts attaching the transmission mount and damper to the crossmember.

15. Remove the nuts and bolts attaching the crossmember to the frame side rails, then remove the crossmember.

16. Lower the engine jack slightly to angle the transmission assembly. Work the clutch housing off the locating dowels and slide the clutch housing and the transmission rearward until the input shaft clears the clutch disc.

17. Lower the transmission jack and remove the transmission from the vehicle.

To install:

18. Check that the mating surfaces of the clutch housing, engine rear and dowel holes are free of burrs, dirt and paint.

19. Place the transmission on the transmission jack. Position the transmission under the vehicle, then raise it into position. Align the input shaft splines with the clutch disc splines and work the transmission forward onto the locating dowels.

20. Install the transmission-to-engine retaining bolts and washers. Tighten the retaining bolts to 30–41 ft. lbs. (40–55 Nm).

21. Install the starter motor. Tighten the attaching nuts.

22. Raise the engine and install the rear crossmember and mount. Tighten to the following specifications

• Mount to transmission bolts—64–81 ft. lbs. (87–110 Nm)

• Crossmember to frame bolts—63–87 ft. lbs. (85–118 Nm)

• Transmission mount to crossmember nuts—73–97 ft. lbs. (98–132 Nm)

23. On 4WD vehicles, install the transfer case.

24. Install the driveshaft(s).

25. Connect the starter motor, back-up lamp and, if equipped, neutral sensing switch connectors.

26. Connect the hydraulic clutch line and bleed the system.

27. Install the speedometer cable.

28. Check and adjust the fluid level.

29. Lower the vehicle.

30. Install the gearshift lever assembly. Install the boot cover and bolts.

31. Reconnect the negative battery cable.

32. Check for proper shifting and operation of the transmission.

Automatic Transmission

1. Disconnect the negative battery cable.

2. Raise and safely support the vehicle.

3. Position a drain pan under the transmission pan and drain the transmission fluid.

4. Remove the converter access cover from the lower right side of the converter housing on the 3.0L engine. Remove the cover from the bottom of the engine oil pan on the 2.3L and 2.5L engines. Remove the access cover and adapter plate bolts from the lower left side of the converter housing on all other applications.

5. Remove the flywheel-to-converter attaching nuts. Use a socket and breaker bar on the crankshaft pulley attaching bolt. Rotate the pulley clockwise as viewed from the front to gain access to each of the nuts.

➡**On belt-driven overhead cam engines, never rotate the pulley in a counterclockwise direction as viewed from the front.**

6. Remove the speedometer cable and/or vehicle speed sensor from the transfer case (4WD) or extension housing (2WD).

7. Scribe a mark indexing the driveshaft(s) to the axle flange(s). Remove the driveshaft(s).

8. On 4WD vehicles, remove the transfer case.

9. Disconnect the shift rod or cable at the transmission manual lever and retainer bracket.

10. Disconnect the downshift cable from the downshift lever. Depress the tab on the

retainer and remove the kickdown cable from the bracket.

11. Disconnect all of the transmission wire harness plugs.

12. Remove the starter mounting bolts and the ground cable. Remove the starter.

13. If equipped, remove the vacuum line from the transmission vacuum modulator.

14. Remove the filler tube from the transmission.

15. Position a transmission jack under the transmission and raise it slightly.

16. Remove the engine rear support-to-crossmember bolts.

17. Remove the crossmember-to-frame side support attaching nuts and bolts. Remove the crossmember.

18. Remove the converter housing-to-engine bolts.

19. Slightly lower the jack to gain access to the oil cooler lines. Disconnect the oil cooler lines at the transmission. Plug all openings to keep dirt and contamination out.

20. Move the transmission to the rear so it disengages from the dowel pins and the converter is disengaged from the flywheel. Lower the transmission from the vehicle.

21. If necessary, remove the torque converter from the transmission.

➡**If the transmission is to be removed for a period of time, support the engine with a safety stand and wood block.**

To install:

22. Install the converter on the transmission.

✳✳ WARNING

Before installing an automatic transmission, always check that the torque converter is fully seated into the transmission. Typically, the converter has notches or tangs on the hub which must engage the transmission fluid pump. If they are not engaged in the pump, the transmission will not mate to the engine properly, as the converter will be holding it away. Severe damage to the pump, converter or transmission casting can occur if the transmission-to-engine bolts are tightened to force the transmission to mate to the engine.

Proper installation of the converter requires full engagement of the converter hub in the pump gear. To accomplish this, the converter must be pushed and at the same time rotated through what feels like 2 notches or bumps. When fully installed, rotation of the converter will usually result

in a clicking noise heard, caused by the converter surface touching the housing to case bolts.

This should not be a concern, but an indication of proper converter installation since, when the converter is attached to the engine flywheel, it will be pulled slightly forward, away from the bolt heads. Besides the clicking sound, the converter should rotate freely with no binding.

For reference, a properly installed converter will have a distance from the converter pilot nose from face-to-converter housing outer face of 13/32–9/16 in. (10.5–14.5mm).

23. With the converter properly installed, position the transmission on the jack.

24. Rotate the converter so that the drive studs are in alignment with the holes in the flywheel.

25. Move the converter and transmission assembly forward into position, being careful not to damage the flywheel and converter pilot. The converter housing is piloted into position by the dowels in the rear of the engine block.

➡**During this move, to avoid damage, do not allow the transmission to get into a nose down position as this will cause the converter to move forward and disengage from the pump gear.**

26. Install the converter housing to engine attaching bolts and tighten to 30–41 ft. lbs. (40–55 Nm). The 2 longer bolts are located at the dowel holes.

27. Remove the jack supporting the engine.

28. The rest of the installation procedure is the reverse of removal. Before starting the engine, add three quarts of transmission fluid (six quarts if the converter was drained) through the dipstick tube. Allow the engine to warm up. Apply the service brakes and run the gear selector through the gears. Check the fluid level using the dipstick, adding fluid if needed. Do not over-fill the transmission.

Clutch

REMOVAL & INSTALLATION

✳✳ CAUTION

The clutch disc may contain asbestos, which has been determined to be a cancer causing agent. Never clean clutch surfaces with compressed air! Avoid inhaling any dust from any clutch surface! When

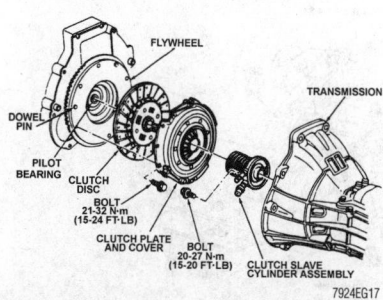

Clutch disc, pressure plate and bearing assembly for 2.3L, 2.5L, 3.0L and 4.0L engines—the 5.0L engine is similar

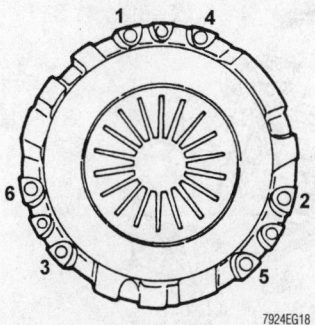

Tighten the bolts gradually in the correct sequence to avoid warping the pressure plate

cleaning clutch surface, use a commercially available brake cleaning fluid.

1. Disconnect the negative battery cable.
2. Disconnect the clutch hydraulic system master cylinder from the clutch pedal and remove.
3. Raise and safely support the vehicle.
4. Remove the starter.
5. Disconnect the hydraulic coupling at the transmission.

➡ **Clean the area around the hose and slave cylinder to prevent fluid contamination.**

6. Remove the transmission from the vehicle.
7. Mark the assembled position of the pressure plate in relation to the flywheel, to aid during reassemble.

✳✳ CAUTION

The pressure plate is heavy. Hold the assembly securely and remove the uppermost bolt last so the assembly won't swing to the side.

8. Loosen the pressure plate and cover attaching bolts evenly until the pressure

plate springs are expanded, and remove the bolts.

9. Remove the pressure plate and cover assembly and the clutch disc from the flywheel. Remove the pilot bearing only for replacement.

To install:

10. Position the clutch disc on the flywheel so that the Clutch Alignment Shaft tool T74P-7137-K (or equivalent) can enter the clutch pilot bearing and align the disc.
11. When reinstalling the original pressure plate and cover assembly, align the assembly and flywheel according to the marks made during the removal operations. Position the pressure plate and cover assembly on the flywheel, align the pressure plate and disc, and install the retaining bolts that fasten the assembly to the flywheel. Tighten the bolts to 15–25 ft. lbs. (21–35 Nm) in the proper sequence. Remove the clutch disc pilot tool.
12. Install the transmission into the vehicle.
13. Connect the coupling by pushing the male coupling into the slave cylinder.
14. Connect the hydraulic clutch master cylinder pushrod to the clutch pedal.
15. Bleed the hydraulic clutch system.

ADJUSTMENT

Because the clutch is hydraulically driven, there is no clutch cable or linkage installed, and therefore no adjustments are needed.

In the event the clutch pedal develops a squeak or uneven feel when depressing, spray the pedal bushing assembly with a suitable penetrating oil and work the pedal back-and-forth.

Hydraulic Clutch System

BLEEDING

The following procedure is recommended for bleeding the clutch hydraulic system installed on the vehicle. It is recommended that the original clutch tube, with quick-connect fitting be replaced when servicing the hydraulic system, because air can be trapped in the quick-connect fitting and prevent complete bleeding of the system. The replacement tube does not include a quick-connect fitting.

1. Clean the dirt and grease from the dust cap.
2. Remove the cap and diaphragm and fill the reservoir to the top with approved brake fluid C6AZ-19542-AA or BA, (ESA-M6C25-A) or equivalent.

➡ **To keep brake fluid from entering the clutch housing, route a suitable rubber tube of appropriate inside diameter from the bleed screw to a container.**

3. Loosen the bleed screw, located in the slave cylinder body, next to the inlet connection. Fluid will now begin to move from the master cylinder down the tube to the slave cylinder.

➡ **The reservoir must be kept full at all time during the bleeding operation, to ensure no additional air enters the system.**

4. Observe the bleed screw outlet. When the slave cylinder is full, a steady stream of fluid will flow from the outlet port. Tighten the bleed screw.
5. Depress the clutch pedal to the floor and hold for 1–2 seconds. Release the pedal as rapidly as possible. The pedal must be released completely. Pause for 1–2 seconds. Repeat 10 times.
6. Check the fluid level in the reservoir. The fluid should be level with the step when the diaphragm is removed.
7. Hold the pedal to the floor, slightly open the bleed screw to allow any additional air to escape. Close the bleed screw, then release the pedal.
8. Check the fluid in the reservoir. The hydraulic system should now be fully bled, and should actuate the clutch.
9. Check the vehicle by starting, pushing the clutch pedal to the floor and selecting reverse gear. There should be no grating of gears. If there is, and the hydraulic system still contains air; repeat the bleeding procedure.

Transfer Case Assembly

REMOVAL & INSTALLATION

✳✳ CAUTION

The catalytic converter is located beside the transfer case. Be careful when working around the catalytic converter because of the extremely high temperatures generated by the converter.

1. Disconnect the negative battery cable.
2. Raise and safely support the vehicle.
3. If so equipped, remove the skid plate from the frame.
4. Remove the damper from the transfer case, if so equipped.
5. On electronic-shift models, disengage the wire connector from rear of the

transfer case. Be sure to squeeze the locking tabs, then pull the connectors apart.

6. Disconnect the front driveshaft from the axle input yoke.

7. If equipped, loosen the clamp retaining the front driveshaft boot to the transfer case, and pull the driveshaft and front boot assembly out of the transfer case front output shaft.

8. Disconnect the rear driveshaft from the transfer case output shaft yoke.

9. If equipped, disconnect the speedometer driven gear from the transfer case rear cover.

10. If equipped, disconnect the electrical plug from the Vehicle Speed Sensor (VSS).

11. Disconnect the vent hose from the mounting bracket.

12. On manual-shift models, perform the following:

  a. Remove the shift lever retaining nut and remove the lever.

  b. Remove the bolts that retains the shifter to the extension housing. Note the size and location of the bolts to aid during installation. Remove the lever assembly and bushing.

13. If equipped, remove the heat shield from the transfer case.

14. Support the transfer case with a transmission jack.

15. Remove the 5 bolts (6 bolts on the AWD transfer case) retaining the transfer case to the transmission and the extension housing.

16. Slide the transfer case rearward off the transmission output shaft and lower the transfer case from the vehicle. Remove the gasket from between the transfer case and extension housing.

To install:

17. Install the heat shield onto the transfer case, if equipped, and place a new gasket between the transfer case and adapter.

18. Raise the transfer case with a suitable transmission jack (or equivalent), high

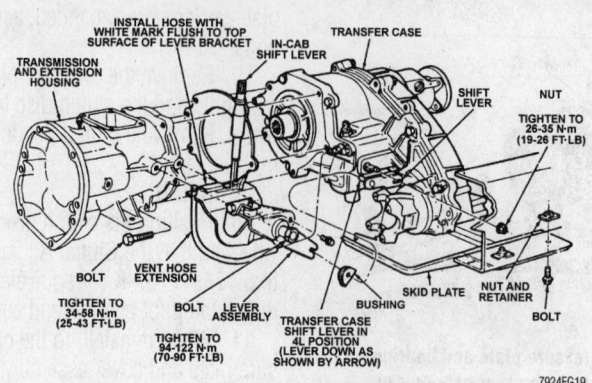

Exploded view of the 13-54 mechanical shift transfer case-to-transmission mounting

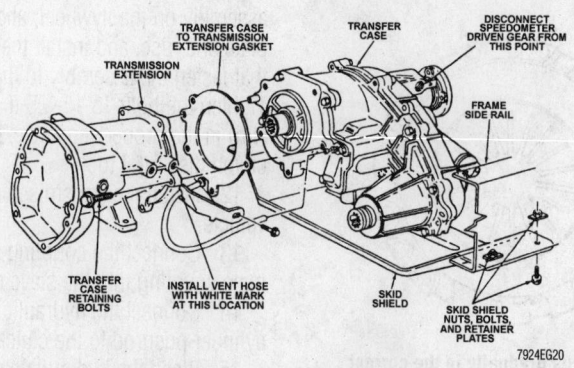

Exploded view of the 13-54 electronic shift transfer case-to-transmission mounting—44-05 model is similar

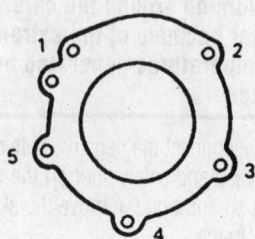

Transfer case-to-extension bolt torque sequence

enough so that the transmission output shaft aligns with the splined transfer case input shaft.

19. Slide the transfer case forward on to the transmission output shaft and onto the dowel pin. Install the transfer case retaining bolts and tighten them to 25–35 ft. lbs. (34–47 Nm).

20. The remainder of the installation procedure is the reverse of removal. Check the fluid level and, if necessary, top off with Dexron/Mercon automatic transmission fluid to achieve the proper level.

ADJUSTMENTS

Manual Shift Models

The following procedure should be used, if a partial or incomplete engagement of the transfer case shift lever detent is experienced, or if the control assembly requires removal.

1. Disconnect the negative battery cable.

2. Raise the shift boot to expose the top surface of the cam plates.

3. Loosen the one large and one small bolt, approximately one turn. Move the transfer case shift lever to the **4L** position (lever down).

4. Move the cam plate rearward until the bottom chamfered corner of the neutral lug

just contacts the forward right edge of the shift lever.

5. Hold the cam plate in this position and tighten the larger bolt first to 70–90 ft. lbs. (95–122 Nm) and tighten the smaller bolt to 31–42 ft. lbs. (42–57 Nm).

6. Move the transfer case shift lever in the vehicle, check for positive engagement in all positions. There should be a clearance between the shift lever and the cam plate in the **2H** front and **4H** rear (clearance not to exceed 0.13in./3.3mm) and **4L** shift positions.

7. Install the shift boot assembly.

8. Reconnect the negative battery cable.

Except Manual Shift Models

Both of the electronic shift and the AWD model transfer cases do not require any linkage adjustments, nor are any possible.

Locking Hubs

REMOVAL & INSTALLATION

Manual Type

1. Loosen the front wheel lug nuts.

2. Raise and safely support the front of the vehicle.

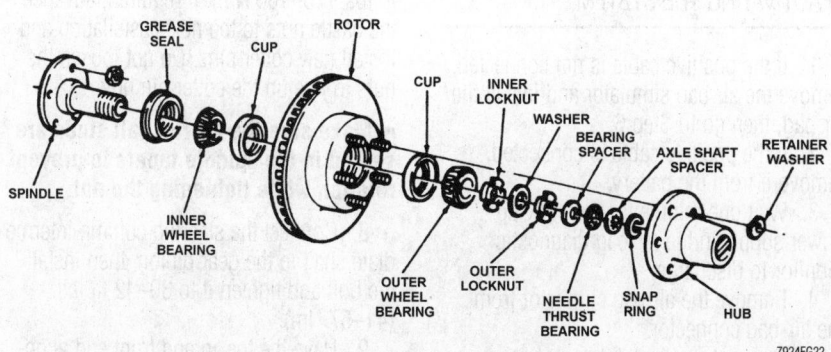

Exploded view of the manual locking hub assembly and related components

3. Remove the lug nuts and wheel/tire assembly.

4. If equipped, remove the lug nut retainer washers from the wheel studs.

➡**Some gentle tapping with a soft faced hammer may help to loosen the locking hub if it seems stuck.**

5. Remove the manual locking hub assembly from the rotor by pulling straight outward.

6. Inspect the O-ring seal on the back side of the hub assembly and, if damaged, replace it.

7. Installation is the reverse of the removal procedure. Ensure that the rotor mounting face is flat and free of burrs, dirt or grease, especially where the O-ring seal makes contact.

Automatic Type

MOUNTAINEER AND EXPLORER

The Mountaineer and Explorer models use a locking mechanism mounted in the differential. This system is called a vacuum disconnect axle lock.

EXCEPT MOUNTAINEER AND EXPLORER

1. Raise and safely support the vehicle.
2. Remove the wheel/tire assembly.

3. If equipped, remove the lug nut retainer washers from the wheel studs.

➡**Some gentle tapping with a soft faced hammer may help to loosen the locking hub cover if it seems stuck.**

4. Remove the automatic locking hub cover assembly from the rotor by pulling straight outward.

5. Inspect the O-ring seal on the back side of the hub assembly and, if damaged, replace it.

6. Remove the snap-ring from the end of the splined axle shaft.

7. Remove the axle shaft spacer(s).

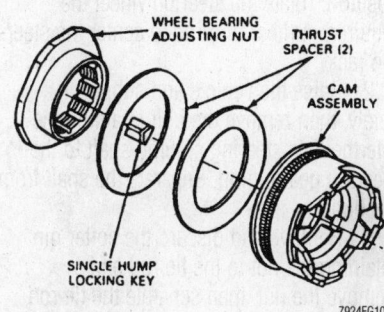

Exploded view of the locking cam assembly—4WD except Mountaineer and Explorer models

✲✲ WARNING

Do not pry on the locking cam or thrust spacers during removal. Prying may damage the cam or spacers.

8. Pull the locking cam assembly and the two thrust spacers (behind cam assembly) from the wheel bearing adjusting nut.

9. Installation is the reverse of the removal procedure. Make sure to install the two thrust spacers first. Also, when pushing or pressing the locking cam into position, ensure that the key in the cam assembly is aligned with the keyway of the front spindle.

✲✲ WARNING

Extreme care must be taken when aligning the locking cam key with the keyway on the front spindle to prevent damage to the fixed cam.

10. Install the outer hub cover. Ensure that the rotor mounting face is flat and free of burrs, dirt or grease, especially where the O-ring seal makes contact.

STEERING AND SUSPENSION

Air Bag

✲✲ CAUTION

Some models covered by this manual may be equipped with a Supplemental Restraint System (SRS), or Supplemental Inflatable Restraint (SIR) which uses an air bag. Whenever working near any of the SRS/SIR components, such as the impact sensors, the air bag module, steering column and instrument panel, disable the SRS/SIR.

PRECAUTIONS

• Always wear safety glasses when servicing an air bag vehicle, and when handling an air bag.

• Never attempt to service the steering wheel or steering column on an air bag-equipped vehicle without first properly disarming the air bag system. The air bag system should be properly disarmed whenever ANY service procedure in this manual indicates that you should do so.

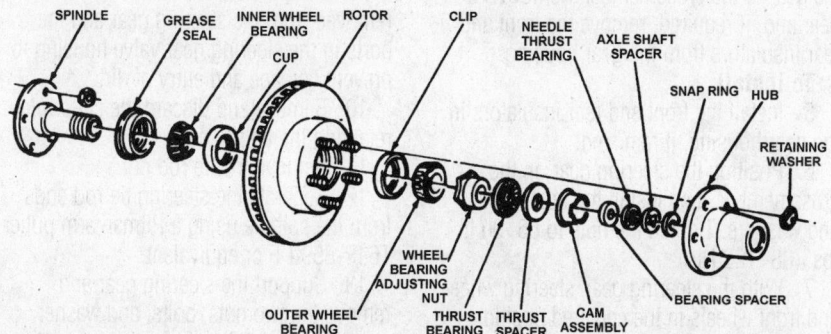

Exploded view of the automatic locking hubs and related components—4WD except Mountaineer and Explorer models

• When carrying a live air bag module, always make sure the bag and trim cover are pointed away from your body. In the unlikely event of an accidental deployment, the bag will then deploy with minimal chance of injury.

• When placing a live air bag on a bench or other surface, always face the bag and trim cover up, away from the surface. This will reduce the motion of the air bag if it is accidentally deployed.

• If you should come in contact with a deployed air bag, be advised that the air bag surface may contain deposits of sodium hydroxide, which is a product of the gas combustion and is irritating to the skin. Always wear gloves and safety glasses when handling a deployed air bag, and wash your hands with mild soap and water afterwards.

DISARMING THE SYSTEM

1. Disconnect the positive battery cable.
2. Wait one minute for the back-up power supply in the air bag diagnostic monitor to discharge.

➡**If the ignition must be turned ON or the engine started, the air bag should be removed and replaced with the air bag simulator such as Rotunda 105-R0011. The back-up power supply allows air bag deployment if the battery or battery cables are damaged in an accident before the crash sensors close.**

3. Remove the nut and washer assemblies retaining the driver air bag module to the steering wheel.
4. Unplug the driver air bag module connector and attach an air bag simulator such as Rotunda 105-R0011.
5. If needed, reconnect the positive battery cable.

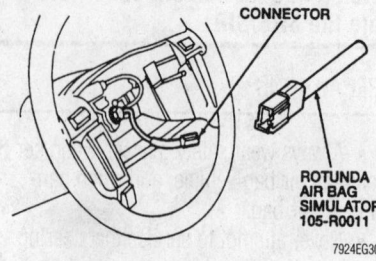

CONNECTOR

ROTUNDA AIR BAG SIMULATOR 105-R0011

7924EG30

Install the air bag simulator in place of the air bag if the ignition must be turned ON or the engine started

REACTIVATING THE SYSTEM

1. If the positive cable is not connected, remove the air bag simulator and install the air bag, then go to Step 6.
2. If the positive cable is connected, remove it from the battery.
3. Wait one minute for the back-up power supply in the air-bag diagnostic monitor to discharge.
4. Remove the air bag simulator from the air-bag connector.
5. Connect and install the air bag assembly.
6. Connect the positive battery cable. Verify that the air bag light illuminates for approximately six seconds when the ignition switch first is turned **ON**.

Manual Rack and Pinion Steering Gear

REMOVAL & INSTALLATION

1. Rotate the steering wheel from lock-to-lock (entire gear travel) and record the number of steering wheel rotations. Divide the number of steering wheel rotations by two to get the required number of turns to place the steering wheel in the centered (straight ahead) position. From one lock position, rotate the steering wheel the required number of turns to center the steering rack.
2. Raise the vehicle and support it safely, then remove the bolt retaining the intermediate steering column shaft to the steering gear pinion. Separate the shaft from the pinion.
3. Remove and discard the cotter pin retaining the nut to the tie rod ends. Remove the nut, then separate the tie rod ends from the spindle arms using remover tool T64P-3590-F, or equivalent.
4. Support the steering gear and remove the two nuts, bolts and washers retaining the gear to the crossmember. Remove the gear and, if required, remove the front and rear insulators from the gear housing.

To install:
5. Install the front and rear insulators in the gear housing, if removed.
6. Position the steering gear on the crossmember, then install the nuts, bolts and washers. Tighten the nuts to 65–90 ft. lbs. (88–122 Nm).
7. With the steering gear, steering wheel and front wheels in the centered position, attach the tie rod ends to the spindle arms. Install the nuts and tighten them to 52–73

ft. lbs. (70–100 Nm). If required, advance the castle nuts to the next castellation and install new cotter pins. Do not loosen the nuts to line up the cotter pin hole.

➡**Make sure the tie rod ball studs are seated in the spindle tapers to prevent rotation while tightening the nut.**

8. Connect the steering column intermediate shaft to the gear pinion, then install the bolt and tighten it to 30–42 ft. lbs. (41–57 Nm).
9. Have the toe-in and front end alignment checked at a qualified service shop.

Power Rack and Pinion Steering Gear

REMOVAL & INSTALLATION

Aerostar

REAR WHEEL DRIVE MODELS

1. Turn the ignition key to the **OFF** position.
2. On vehicles equipped with automatic transmissions, position the transmission selector in PARK and set the hand brake.
3. For vehicles equipped with manual transmissions, put the gear shift lever in Reverse and set the hand brake.
4. Raise and safely support the front of the vehicle.
5. Remove the front wheels.
6. Remove the lower intermediate steering column shaft retaining bolt from the steering gear.
7. Disconnect the shaft from the steering gear.
8. Unscrew the fitting for the power steering pressure and return lines at the rack and pinion steering gear valve housing.

➡**Do not remove the right and left transfer tube fitting.**

9. Plug the ends of the fluid lines removed from the steering gear and the ports in the steering gear valve housing to prevent damage and entry of dirt.
10. Remove and discard the cotter pin retaining the tie rod end.
11. Remove the tie rod nut.
12. Separate the steering tie rod ends from the spindle using a Pitman arm puller T64P-3590-F or equivalent.
13. Support the steering gear and remove the two nuts, bolts, and washer assemblies retaining the steering gear to the vehicle crossmember.

14. Remove the steering gear from the vehicle. If required, remove the front and rear insulators from the gear housing.

To install:

15. If removed, install the insulators into the steering gear housing.

➡**The larger end of the inner sleeve faces the rear of the vehicle and contacts the crossmember.**

16. Push the insulators in until there is no space between the lip on the insulator and edge of the steering gear housing.

17. Position the steering gear on the crossmember. Install the nuts, bolts and washers retaining the gear to the crossmember. Tighten the nuts to 80–105 ft. lbs. (108–142 Nm).

18. Unplug the power steering fluid lines and steering gear valve housing.

19. If required, replace the TFE seal on the power steering pressure and return line quick connect fitting. Install a new seal as follows:

    a. Unscrew the tube nut, then replace the plastic seal washer.

    b. To facilitate assembly of the new TFE seal, a tapered shaft may be required to stretch the washer so that it may be slipped over the tube nut threads. Recommended tools are D90P-3517-A2 and D90P-3517-A3 or their equivalents.

20. Connect the pressure and return lines to the appropriate ports on the steering gear valve housing. Tighten the fittings to 20–25 ft. lbs. (27–34 Nm).

➡**The fittings' design allows the hoses to swivel when properly tightened. Do not attempt to eliminate looseness by over-tightening, since this can cause damage to the fittings.**

21. With the steering gear, steering wheel and front wheels in the straight ahead position, attach the tie rod ends to the spindle arms. Tighten the tie rod nuts to 52–73 ft. lbs. (70–100 Nm).

➡**Make sure that the tie rod end ball studs are seated in the tapered spindle holes to prevent rotation while tightening the nut.**

22. If necessary, advance the tie rod nuts to the next slot, then install a new cotter pin.

23. Position the steering column lower intermediate shaft over the steering gear input shaft spline and dust seal. Replace the pinch bolt and tighten it to 30–42 ft. lbs. (41–56 Nm).

➡**Verify that no rotation from the straight ahead position has occurred.**

24. Install the front wheels.

25. Lower the vehicle.

26. Refill the power steering pump reservoir with the proper fluid.

27. Purge air from the power steering system. Verify the absence of any unusual power steering noise.

28. Have the front end aligned by a qualified automotive mechanic.

29. Make sure that the power steering system operates correctly and is not leaking.

30. Check and adjust the fluid level in the power steering pump reservoir.

ALL WHEEL DRIVE MODELS

1. Start the engine, then rotate the steering wheel from lock-to-lock (entire gear travel) and record the number of rotations. Divide the number of steering wheel rotations by two to get the required number of turns to place the steering wheel in the centered (straight ahead) position. From one lock position, rotate the steering wheel the required number of turns to center the steering rack.

➡**Verify that the front wheels and steering wheel are in the straight ahead position.**

2. Stop the engine, then disconnect the negative battery cable.

3. On automatic transmissions, put the transmission selector in PARK and set the hand brake.

4. On manual transmissions, put the gear shift lever in Reverse and set the hand brake.

5. Remove the front hub caps.

6. Remove the hub nut and washer.

7. Raise and safely support the vehicle.

8. Remove the front wheels.

9. Remove the bolt retaining the lower intermediate steering column shaft to the steering gear, then disconnect the shaft from the gear.

10. Disconnect the pressure and return lines from the steering gear valve housing. Plug the lines and ports in the steering gear valve housing to prevent the entry of dirt into the system.

11. Remove both steering knuckles.

12. Remove both lower control arms.

13. Remove the five nuts from the forward edge of the crossmember lower plate assembly.

14. Remove the nut from the driver side rear edge of the crossmember lower plate assembly.

15. Remove the two bolts from the center and passenger side rear edge of the crossmember lower plate assembly.

16. Remove the lower plate.

17. While supporting the steering gear, remove the two bolts and spacers retaining the steering gear to the crossmember.

18. Remove the power steering rack and pinion gear from the vehicle.

To install:

19. If removed, install the insulators into the steering gear housing.

➡**The larger end of the inner sleeve faces the rear of the vehicle and contacts the crossmember.**

20. Push the insulators in until there is no space between the lip on the insulator and edge of the steering gear housing.

21. Position the steering gear on the crossmember. Install the nuts, bolts and washers retaining the gear to the crossmember. Tighten the nuts to 61–82 ft. lbs. (83–111 Nm).

22. Install the crossmember lower plate by inserting the studs on the plate through the front edge of the crossmember.

23. Install the two bolts in the center and passenger side of the crossmember lower plate assembly. Tighten the nuts to 35–47 ft. lbs. (47–64 Nm).

24. Install the nut on the stud located at the driver side rear edge of the crossmember lower plate assembly. Tighten it to 22–30 ft. lbs. (30–41 Nm).

25. Install the five nuts on the studs at the forward edge of the crossmember lower plate assembly. Tighten these nuts also to 22–30 ft. lbs. (30–41 Nm).

26. Install both front lower control arms.

27. Install both steering knuckles, as described earlier in this section.

➡**Make sure that the steering gear and front wheels are in the straight ahead position before attaching the tie rod ends to the steering knuckles. Make sure that the tie rod end ball studs are seated in the tapered spindle holes to prevent rotation while tightening the nut.**

28. Unplug the power steering fluid lines and steering gear valve housing.

29. If required, replace the TFE seal on the power steering pressure and return line Quick Connect fitting. Install a new seal as follows:

    a. Unscrew the tube nut, then replace the plastic seal washer.

    b. To facilitate assembly of the new TFE seal, a tapered shaft may be required to stretch the washer so that it may be slipped over the tube nut threads. Recommended tools are D90P-3517-A2 and D90P-3517-A3 or their equivalents.

30. Connect the pressure and return lines to the appropriate ports on the steering gear valve housing. Tighten the fittings to 20–25 ft. lbs. (27–34 Nm).

➡ **The fittings' design allows the hoses to swivel when properly tightened. Do not attempt to eliminate looseness by over-tightening, since this can cause damage to the fittings.**

31. Position the steering column lower intermediate shaft over the steering gear input shaft spline and dust seal. Replace the pinch bolt and tighten it to 30–42 ft. lbs. (41–56 Nm).

➡ **Verify that no rotation from the straight ahead position has occurred.**

32. Install the front wheels.
33. Install the hub nuts and washers.
34. Install the front hub caps.
35. Lower the vehicle.
36. Refill the power steering pump reservoir with the proper fluid.
37. Purge air from the power steering system. Verify the absence of any unusual power steering noise.
38. Have the front end aligned by a qualified automotive mechanic.
39. Make sure that the power steering system operates correctly and is not leaking.
40. Check and adjust the fluid level in the power steering pump reservoir.

Except Aerostar

✳✳ WARNING

If equipped, always turn off the Automatic Ride Control (ARC) service switch before lifting the vehicle off of the ground. Failure to do so could damage the ARC system components.

1. Raise and safely support the front of the vehicle, block the rear wheels and apply the parking brake.
2. Start the engine then rotate the steering wheel from lock-to-lock and record the number of rotations.
3. Divide the number of rotations by two. This gives the number of rotations to achieve true center of the steering. Turn the wheel in one direction to the full lock.
4. Turn the wheel in the opposite direction the number of turns equal to true steering (lock-to-lock number divided by two).

✳✳ WARNING

Do not rotate the steering wheel when the shaft is disconnected from the steering gear as damage to the clock spring could occur.

5. Remove the bolt retaining the lower steering column shaft to the steering gear input shaft and disconnect the two.
6. Remove the stabilizer bar.
7. Unscrew the quick-connect fittings for the power steering pressure and return hoses at the steering gear housing.
8. Plug the ends of the lines and the fitting in the rack to avoid dirt contamination.
9. Remove the two nuts securing the power steering cooler and remove the cooler.
10. Remove the outer tie rod ends.
11. Remove the two nuts, bolts and washer assemblies retaining the steering gear housing to the front crossmember.

To install:
12. Position the steering gear to the front crossmember and install the nuts, bolts and washer assemblies. Tighten to 94–127 ft. lbs. (128–172 Nm).
13. Install the power steering cooler and two retaining bolts.
14. Connect the power steering lines to the steering gear housing and tighten the fittings to 20–25 ft. lbs. (27–34 Nm).
15. Install the outer tie rod ends.
16. Ensure that the steering shaft or gear input shaft has not been rotated, then connect the two.
17. Install the intermediate shaft-to-steering input shaft retaining (pinch) bolt and tighten to 30–42 ft. lbs. (41–56 Nm).
18. Lower the vehicle and refill the power steering pump reservoir.
19. Bleed the air from the power steering system.
20. Ensure that there are no leaks and the fluid is maintained at the proper level.
21. Have the alignment checked and adjusted by a professional repair shop.

Recirculating Ball Manual Steering Gear

REMOVAL & INSTALLATION

1. Raise and safely support the vehicle. Disengage the flex coupling shield from the steering gear input shaft shield and slide it up the intermediate shaft.
2. Remove the bolt that retains the flex coupling to the steering gear.
3. Remove the steering gear input shaft shield.
4. Remove the nut and washer that secures the Pitman arm to the sector shaft. Remove the Pitman arm using a Pitman

Arm Puller tool T64P-3590-F or equivalent. Do not hammer on the end of the puller as this can damage the steering gear.
5. Remove the bolts and washers that attach the steering gear to the side rail. Remove the gear.

To install:
6. Rotate the gear input shaft (wormshaft) from stop-to-stop, counting the total number of turns. Then turn back exactly half-way, placing the gear on center.
7. Slide the steering gear input shaft shield on the steering gear input shaft.
8. Position the flex coupling on the steering gear input shaft. Ensure that the flat on the gear input shaft is facing straight up and aligns with the flat on the flex coupling. Install the steering gear to side rail with bolts and washers. Tighten the bolts to 66 ft. lbs. (89 Nm).
9. Place the Pitman arm on the sector shaft and install the attaching washer and nut. Align the 2 blocked teeth on the Pitman arm with 4 missing teeth on the steering gear sector shaft. Tighten the nut to 170–228 ft. lbs. (230–310 Nm).
10. Install the flex coupling to steering gear input shaft attaching bolt and tighten to 50–62 ft. lbs. (68–84 Nm).
11. Snap the flex coupling shield to the steering gear input shield.
12. Check the system to ensure equal turns from center to each lock position.

Recirculating Ball Power Steering Gear

REMOVAL & INSTALLATION

1. Disconnect the pressure and return lines from the steering gear. Plug the lines and the ports in the gear to prevent entry of dirt.
2. Remove the upper and lower steering gear shaft U-joint shield from the flex coupling. Remove the bolts that secure the flex coupling to the steering gear and to the column steering shaft assembly.
3. Raise the vehicle and remove the Pitman arm attaching nut and washer.
4. Remove the Pitman arm from the sector shaft using tool T64P-3590-F. Remove the tool from the Pitman arm. Do not damage the seals.
5. Support the steering gear, and remove the steering gear attaching bolts.
6. Work the steering gear free of the flex coupling. Remove the steering gear from the vehicle.

To install:
7. Install the lower U-joint shield onto the steering gear lugs. Slide the upper U-

joint shield into place on the steering shaft assembly.

8. Slide the flex coupling into place on the steering shaft assembly. Turn the steering wheel so that the spokes are in the horizontal position. Center the steering gear input shaft.

9. Slide the steering gear input shaft into the flex coupling and into place on the frame side rail. Install the attaching bolts and tighten to 50–62 ft. lbs. (68–84 Nm). Tighten the flex coupling bolt 30–40 ft. lbs. (41–54 Nm).

10. Be sure the wheels are in the straight-ahead position, then install the Pitman arm on the sector shaft. Install the Pitman arm attaching washer and nut. Tighten nut to 170–228 ft. lbs. (230–310 Nm).

11. Connect and tighten the pressure and the return lines to the steering gear.

12. Disconnect the coil wire. Fill the reservoir. Turn on the ignition and turn the steering wheel from left-to-right to distribute the fluid.

13. Recheck fluid level and add fluid, if necessary. Connect the coil wire, start the engine and turn the steering wheel from side-to-side. Inspect for fluid leaks.

Shock Absorber

REMOVAL & INSTALLATION

➡**Low pressure gas shocks are charged with Nitrogen gas. Do not attempt to open, puncture or apply heat to them. Prior to installing a new shock absorber, hold it upright and extend it fully. Invert it and fully compress and extend it at least 3 times. This will bleed trapped air.**

Mountaineer, Explorer, 1998–99 Ranger and B Series Models

1. Raise the front of the vehicle and place jackstands under the lower control arms. Ensure that the lower shock attaching nuts do not become obstructed by the jackstands.

2. Remove the upper shock-to-frame attaching nut, washer and insulator assembly.

3. Remove the two lower shock-to-control arm attaching nuts.

4. Slightly compress the shock absorber by hand and remove it from the vehicle.

To install:

5. Position the lower washer and insulator on the shock absorber rod and position the shock absorber to the upper frame bracket mount.

6. Position the upper insulator and washer on the shock absorber rod and install the attaching nut loosely.

7. Position the lower shock absorber mounting studs into the control arm and install the attaching nuts loosely.

8. Tighten the lower shock attaching nuts to 15–21 ft. lbs. (21–29 Nm), and the upper shock attaching bolts to 30–40 ft. lbs. (40–55 Nm).

1995–97 Ranger and B Series Models

1. Raise the vehicle, as required to provide additional access and remove the nut attaching the shock absorber to the lower mounting stud on the radius arm.

2. Slide the lower shock absorber end off of the stud.

3. Remove the nut, washer and insulator from the upper shock absorber mount at the frame bracket and remove the shock absorber.

➡**A second wrench may be needed to hold the shock absorber from turning while removing the upper attaching nut.**

To install:

4. Position the washer and insulator on the shock absorber rod and position the shock absorber to the upper frame bracket mount.

5. Position the insulator and washer on the shock absorber rod and install the attaching nut loosely.

6. Position the shock absorber to the lower mounting stud and install the attaching nut loosely.

7. Tighten the lower shock attaching bolts to 39–53 ft. lbs. (53–72 Nm), and the upper shock attaching bolts to 25–34 ft. lbs. (34–46 Nm).

Coil Spring

REMOVAL & INSTALLATION

Aerostar

FRONT

1. Place the steering wheel and front wheels in the centered (straight ahead) position.

➡**Whenever the steering linkage is disconnected, the steering system must be centered prior to beginning any work.**

2. Raise the vehicle and support it safely beneath the frame at the jacking pads.

3. Remove the tire and wheel assemblies.

4. Disconnect the stabilizer bar link bolt from the lower control arm.

5. Remove the two bolts attaching the shock absorber to the lower arm assembly.

6. Remove the upper nut and washer retaining the shock absorber and remove the shock absorber from the vehicle.

7. Using spring compressor tool D78P-5310-A or equivalent, install one plate with the pivot ball seat facing downward into the coils of the spring. Rotate the plate so that it is flush with the upper surface of the lower arm.

8. Install the other plate with the pivot ball seat facing upward into the coils of the spring, so that the nut rests in the upper plate.

9. Insert the compression rod into the opening in the lower arm, through the upper and lower plate and upper ball nut. Insert the securing pin through the upper ball nut and compression rod. This pin can only be inserted one way into the upper ball nut because of a stepped hole design.

10. With the upper ball nut secured, turn the upper plate so that it walks up the coil until it contacts the upper spring seat, then back it off ½ turn.

11. Install the lower ball nut and thrust washer on the compression rod, then screw on the forcing nut. Tighten the forcing nut until the spring is compressed enough so that it is free in its seat.

12. Loosen the two lower arm pivot bolts. Remove the cotter pin and loosen, but do not remove the nut attaching the lower ball joint to the spindle. Using Pitman arm puller T64P-3590-F, or equivalent, loosen the lower ball joint.

13. Remove the puller tool.

14. Support the lower control arm with a hydraulic jack, then remove the ball joint

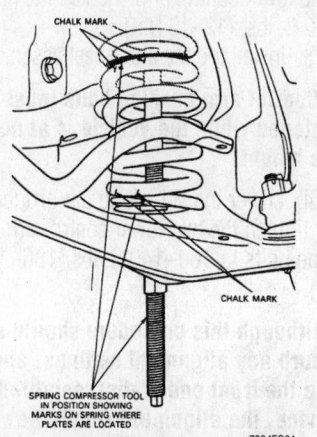

Mark the position of the spring compressor on the spring so the new spring can be mounted in the same position—Aerostar front coil spring shown

nut. Slowly lower the control arm and remove the coil spring.

> ※※ **CAUTION**
>
> **Handle the coil spring with care. A compressed coil spring has enough stored energy to be dangerous if suddenly released. Mount the spring securely in a vise and slowly loosen the spring compressor if the spring is being replaced.**

15. If the coil spring is being replaced, measure the compressed length of the old spring and mark the position of the compressor plates on the old spring with chalk. Remove the spring compressor from the old spring carefully.

To install:

16. Install the spring compressor on the new spring, placing the compressor plates in the same position as marked on the old spring. Make sure the upper ball nut securing pin is installed properly, then compress the new spring to the compressed length of the old spring.

17. Position the coil spring assembly into the lower control arm.

18. Place a hydraulic jack under the lower control arm and slowly raise it into position. Reconnect the ball joint and install the nut. Tighten the ball joint castle nut to 80–120 ft. lbs. (108–163 Nm) and install a new cotter pin. The nut may be tightened slightly to align the cotter pin hole, but not loosened.

19. Slowly release the spring compressor and remove it from the coil spring.

20. Reconnect the steering center link to the Pitman arm.

21. Install the shock absorber.

22. Reconnect the stabilizer bar link bolt to the lower control arm. Tighten the nuts to 9–12 ft. lbs. (12–16 Nm).

23. Install the tire and wheel assembly.

➡ **Control arm bushing bolts must be tightened while the vehicle is at normal ride height.**

24. Lower the vehicle to the ground and tighten the two lower control arm-to-frame bolts to 100–140 ft. lbs.(136–190 Nm).

➡**Although this procedure should not disturb any alignment settings, anytime the front end is disassembled for service, the alignment should be checked.**

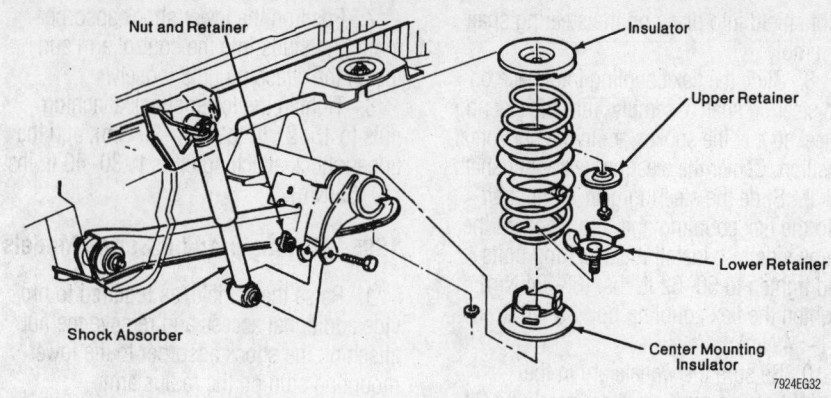

Exploded view of the rear spring assembly—Aerostar

REAR

1. Raise the vehicle and support it safely with jackstands placed beneath the frame rear lift points or under the rear bumper support brackets.

2. Support the rear axle assembly by placing a hydraulic floor jack under the differential housing.

3. Remove the nut and bolt retaining the shock absorber to the axle mount on the lower control arm. Disconnect the shock absorber from the axle bracket.

4. Carefully lower the rear axle until the coil springs are no longer under compression.

5. Remove the nut securing the lower retainer and spring to the control arm.

6. Remove the bolt securing the upper retainer and spring to the frame.

7. Remove the spring and retainers, then remove the upper and lower insulators.

To install:

8. Before installing the spring, first make sure the axle is in the lowered (spring unloaded) position. Place the lower insulator on the control arm and the upper insulator at the top of the spring.

9. Install the coil spring in position between the control arm and vehicle frame. The small diameter, tapered coils (white colored) must face upward.

10. Install the upper retainer and bolt, then tighten the bolt to 30–40 ft. lbs. (40–55 Nm).

11. Install the lower retainer and nut, then tighten the nut to 41–65 ft. lbs. (55–88 Nm).

12. Raise the axle to the normal ride position with the hydraulic floor jack.

13. Position the shock absorber in the axle bracket, then install the bolt so the head is positioned outboard of the bracket.

Install the nut and tighten it to 41–65 ft. lbs. (55–88 Nm).

14. Remove the jackstands and lower the vehicle.

1995–97 Ranger and B Series Models

2-WHEEL DRIVE VEHICLES

1. Raise the front of the vehicle and place jackstands under the frame and a jack under the axle.

> ※※ **WARNING**
>
> **The axle must not be permitted to hang by the brake hose. If the length of the brake hoses is not sufficient to provide adequate clearance for removal and installation of the spring, the disc brake caliper must be removed from the spindle. A Strut Spring Compressor, T81P-5310-A or equivalent may be used to compress the spring sufficiently, so that the caliper does not have to be removed. After removal, the caliper must be placed on the frame or otherwise supported to prevent suspending the caliper from the brake hose. These precautions are absolutely necessary to prevent serious damage to the tube portion of the brake hose!**

2. Disconnect the shock absorber at the lower shock stud. Remove the nut securing the lower retainer to spring seat. Remove the lower retainer.

3. Lower the axle as far as it will go without stretching the brake hose and tube assembly. The axle should now be unsupported without hanging by the brake hose. If not, then either remove the caliper or use

Strut Spring Compressor tool, T81P-5310-A or equivalent. Remove the spring.

4. If there is a lot of slack in the brake hose assembly, a pry bar can be used to lift the spring over the bolt that passes through the lower spring seat.

5. Rotate the spring so the built-in retainer on the upper spring seat is cleared.

6. Remove the spring from the vehicle.

To install:

7. If removed, install the bolt in the axle arm and install the nut all the way down. Install the spring lower seat and insulator.

8. With the axle in the lowest position, install the top of the spring in the upper seat. Rotate the spring into position.

9. Lift the lower end of the spring over the bolt.

10. Raise the axle slowly until the spring is seated in the lower spring seat. Install the lower retainer and nut.

11. Connect the shock absorber to the lower shock stud.

12. Remove the jack and jackstands and lower vehicle.

4-WHEEL DRIVE VEHICLES

1. Raise the vehicle and install jackstands under the frame. Position a jack beneath the spring under the axle. Raise the jack and compress the spring.

2. Remove the nut retaining the shock absorber to the radius arm. Slide the shock out from the stud.

3. Remove the nut that retains the spring to the axle and radius arm. Remove the retainer.

4. Slowly lower the axle until all spring tension is released and adequate clearance exists to remove the spring from its mounting.

5. Remove the spring by rotating the upper coil out of the tabs in the upper spring seat. Remove the spacer and the seat.

✳✳ WARNING

The axle must be supported on the jack throughout spring removal and installation, and must not be permitted to hang by the brake hose. If the length of the brake hose is not sufficient to provide adequate clearance for removal and installation of the spring, the disc brake caliper must be removed from the spindle. After removal, the caliper must be placed on the frame or otherwise supported to prevent suspending the caliper from the brake line hose. These precautions are absolutely necessary to prevent serious damage to the tube portion of the caliper hose assembly!

6. If required, remove the stud from the axle assembly.

To install:

7. If removed, install the stud on the axle and tighten it 190–230 ft. lbs. (258–313 Nm). Install the lower seat and spacer over the stud.

8. Place the spring in position and slowly raise the front axle. Ensure springs are positioned correctly in the upper spring seats.

9. Position the spring lower retainer over the stud and lower seat and tighten the attaching nut to 70–100 ft. lbs. (95–136 Nm).

10. Position the shock absorber to the lower stud and install the attaching nut. Tighten the nut to 41–63 ft. lbs. (56–85 Nm). Lower the vehicle.

Upper Ball Joint

REMOVAL & INSTALLATION

Aerostar

➡**Ford Motor Company recommends replacement of the upper control arm and ball joint as an assembly, rather than replacement of the ball joint alone. However, aftermarket ball joints are available. The following procedure is for replacement of the ball joint only.**

1. Raise the van and support it safely with jackstands placed under the frame lifting pads. Allow the front suspension to hang unsupported.

2. Remove the front wheels.

3. Place a hydraulic floor jack under the lower control arm and raise the jack until it just contacts the arm.

4. Drill a ⅛ in. (3mm) hole completely through each ball joint attaching rivet.

5. Use a chisel to cut the head off of each rivet, then drive them from the upper control arm with a suitable small drift or blunt punch.

6. Raise the lower control arm about 6 in. (15cm) with the hydraulic jack.

7. Remove the pinch nut and bolt holding the ball joint stud from the spindle.

8. Using a suitable tool, loosen the ball joint stud from the spindle, then remove the ball joint from the upper arm.

To install:

9. Clean all metal burrs from the upper arm and install a new ball joint, using the service part nuts and bolts to attach the ball joint to the upper arm. Do not attempt to rivet the ball joint again once it has been removed.

10. Attach the ball joint stud to the spindle, then install the pinch bolt and nut and tighten to 27–37 ft. lbs. (36–50 Nm).

11. Install the tire and wheel assemblies.

12. Remove the hydraulic jack and lower the van. Have the front end alignment checked.

1995–97 Ranger and B Series Models

➡**The ball joints are arranged such that, if the upper ball joint is to be removed, the lower ball joint must be removed first. Conversely, the upper ball joint must be installed first, before the lower ball joint. Failure to install the upper ball joint before the lower, will result in a lack of clearance for the installation tool.**

1. Remove the steering knuckle.

2. Place the knuckle in a vise and remove the snapring from the bottom ball joint socket, if so equipped.

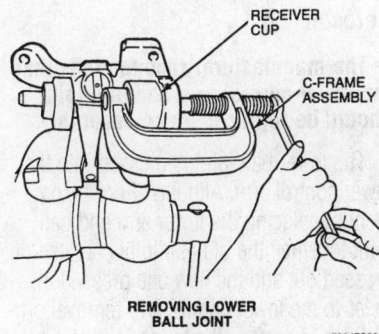

Press the ball joint out of the knuckle using the special tools—2WD Ranger shown

3. Assemble the C-frame, T74P-4635-C, forcing screw, D79T-3010-AE and ball joint remover T83T-3050-A (or equivalent) on the lower ball joint.

4. Turn the forcing screw clockwise until the lower ball joint is removed from the steering knuckle.

5. Repeat Steps 2–4 for the upper ball joint.

➡**Always remove the lower ball joint first**

To install:

6. Clean the steering knuckle bore and insert the lower ball joint in the knuckle as straight as possible. The lower ball joint doesn't have a cotter pin hole in the stud.

7. Assemble the C-frame, T74P-4635-C, forcing screw, D790T-3010-AE, ball joint installer, T83T-3050-A and receiver cup T80T-3010-A3 (or equivalent tools), to install the upper ball joint.

8. Turn the forcing screw clockwise until the upper ball joint is firmly seated.

➡If the ball joint cannot be installed to the proper depth, realignment of the receiver cup and ball joint installer will be necessary.

9. Repeat Steps 6–8 for the lower ball joint. Install the snapring on the lower ball joint.

10. Install the steering knuckle.

Mountaineer, Explorer, 1998–99 Ranger and B Series Models

The ball joints on the Mountaineer and Explorer are integral with the control arm. If the ball joint is defective, the entire control arm must be replaced.

Lower Ball Joint

REMOVAL & INSTALLATION

Aerostar

➡The manufacturer recommends that the lower control arm and ball joint should be replaced as an assembly.

The lower ball joint is pressed into the lower control arm. Although Ford recommends replacing the lower arm and ball joint together, the old ball joint can also be pressed out and the new one pressed in. Refer to the lower control arm removal procedures to remove the lower control arm. Once the lower control arm is removed, the ball joint may be pressed out.

1995–97 Ranger and B Series Models

Refer to the procedure for the upper ball joint outlined earlier in this section.

Mountaineer, Explorer, 1998–99 Ranger and B Series Models

The ball joints on the Mountaineer and Explorer are integral with the control arm. If the ball joint is defective, the entire control arm must be replaced.

Wheel Bearings

ADJUSTMENT

Explorer, Mountaineer and AWD Aerostar

The wheel bearings on the Explorer, Mountaineer and all wheel drive Aerostar are not adjustable. If they become loose or make noise, they must be replaced.

Ranger, B Series Models and 2WD Aerostar

2-WHEEL DRIVE VEHICLES

1. Raise and support the vehicle safely. Remove the wheel cover. Remove the grease cap from the hub.

2. Wipe the excess grease from the end of the spindle. Remove the cotter pin and retainer. Discard the cotter pin.

3. Loosen the adjusting nut 3 turns.

✸✸ WARNING

Obtain running clearance between the disc brake rotor surface and shoe linings by rocking the entire wheel assembly in and out several times in order to push the caliper and brake pads away from the rotor. An alternate method to obtain proper running clearance is to tap lightly on the caliper housing. Be sure not to tap on any other area that may damage the disc brake rotor or the brake lining surfaces. Do not pry on the phenolic caliper piston. The running clearance must be maintained throughout the adjustment procedure. If proper clearance cannot be maintained, the caliper must be removed from its mounting.

4. While rotating the wheel assembly, tighten the adjusting nut to 17–25 ft. lbs. in order to seat the bearings. Loosen the adjusting nut a half turn. Retighten the adjusting nut 18–20 inch lbs.

5. Place the retainer on the adjusting nut. The castellations on the retainer must be in alignment with the cotter pin holes in the spindle. Once this is accomplished install a new cotter pin and bend the ends to insure its being locked in place.

6. Check for proper wheel rotation. If correct, install the grease cap and wheel

cover. If rotation is noisy or rough, recheck your work and correct as required.

7. Lower the vehicle and tighten the lug nuts to 100 ft. lbs., (136 Nm) if the wheel was removed. Before driving the vehicle, pump the brake pedal several times to restore normal brake pedal travel.

✸✸ CAUTION

If the wheel was removed, retighten the wheel lug nuts to specification after about 500 miles (804km) of driving. Failure to do this could result in the wheel coming off while the vehicle is in motion causing loss of vehicle control or collision.

4-WHEEL DRIVE WITH MANUAL HUBS

1. Raise and safely support the vehicle.

2. Remove the wheel and tire assembly.

3. Remove the retainer washers from the lug nut studs and remove the manual locking hub assembly from the spindle.

4. Remove the snapring and spacer from the end of the spindle shaft.

5. Remove the outer wheel bearing locknut from the spindle using 4 prong spindle nut spanner wrench, T86T-1197-A or equivalent. Make sure the tabs on the tool engage the slots in the locknut.

6. Remove the locknut washer from the spindle.

7. Loosen the inner wheel bearing locknut using 4 prong spindle nut spanner wrench, tool T86T-1197-A or equivalent. Make sure that the tabs on the tool engage the slots in the locknut and that the slot in the tool is over the pin on the locknut.

8. Tighten the inner locknut to 35 ft. lbs. (47 Nm) to seat the bearings.

9. Spin the rotor and back off the inner locknut ¼ turn. Install the lockwasher on the spindle. Retighten the inner locknut to

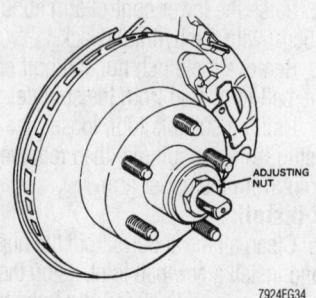

Loosen the adjusting nut three turns, then rock the entire wheel assembly in-and-out to spread the brake pads before attempting to adjust the bearing—2wd vehicles

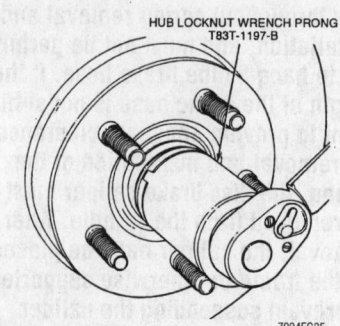

Obtain the special socket to properly adjust the wheel bearing—manual locking hub shown

16 inch lbs. (1.8 Nm). It may be necessary to turn the inner locknut slightly so that the pin on the locknut aligns with the closest hole in the lockwasher.

10. Install the outer wheel bearing locknut using 4 prong spindle nut spanner wrench, tool T86T-1197-A or equivalent. Tighten locknut to 150 ft. lbs. (203 Nm).

11. Install the axle shaft spacer.

12. Clip the snapring onto the end of the spindle.

13. Install the manual hub assembly over the spindle. Install the retainer washers.

14. Install the wheel and tire assembly. Install and tighten lug nuts to specification.

15. Check the end-play of the wheel and tire assembly on the spindle. End-play should be 0.001–0.003 in. (0.025–0.076mm) and the maximum torque to rotate the hub should be 25 inch lbs. (2.8 Nm).

4-WHEEL DRIVE WITH AUTOMATIC HUBS

1. Raise and safely support the vehicle.
2. Remove the wheel and tire assembly.
3. Remove the retainer washers from the lug nut studs and remove the automatic locking hub assembly from the spindle.
4. Remove the snapring and spacer from the end of the spindle shaft.
5. Pull the locking cam assembly and the two plastic spacers off of the wheel bearing adjusting nut.
6. Use a magnet and remove the locking key from under the adjusting nut. If required, rotate the adjusting nut slightly to relieve pressure against the locking key.

✷✷ WARNING

To prevent damage to the adjusting nut and spindle threads on vehicles equipped with automatic hubs, look into the spindle keyway under the adjusting nut and remove the separate locking key before removing the adjusting nut.

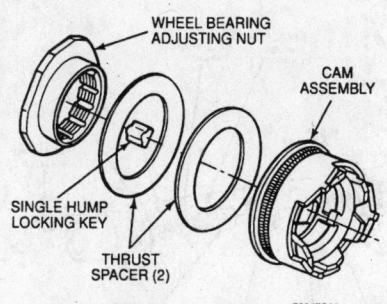

WHEEL BEARING ADJUSTING NUT

CAM ASSEMBLY

SINGLE HUMP LOCKING KEY

THRUST SPACER (2)

7924EG36

Exploded view of the wheel bearing adjusting nut and related components— automatic locking hub shown

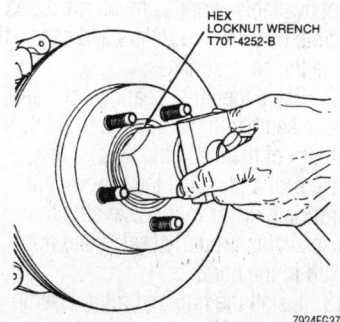

HEX LOCKNUT WRENCH T70T-4252-B

7924EG37

An oversize socket is needed to properly adjust the wheel bearing—automatic locking hub shown

7. Loosen the wheel bearing locknut using a 2-⅜ inch (60.3mm) hex socket, such as Hex Locknut Wrench T70T-4252-B (or equivalent).

8. Tighten the inner locknut to 35 ft. lbs. (47 Nm) to seat the bearings.

9. Spin the rotor and back off the inner locknut ¼ turn (90°). Retighten the locknut to 16 inch lbs. (1.8 Nm).

10. Align the closest lug in the bearing adjusting nut with the center of the spindle keyway slot. Advance the nut to the next if required.

11. Install the separate locking key in the spindle keyway under the adjusting nut.

✷✷ CAUTION

Extreme care must be taken when aligning the adjusting nut with the center of the spindle keyway slot to prevent damage to the separate locking key. The wheel and tire assembly may come off while the vehicle is in motion if the key is damaged.

12. Install the two plastic thrust spacers and push or press the cam assembly onto the adjusting nut by lining up the keyway in the cam assembly with the separate locking key.

✷✷ WARNING

Do not damage the locking key when installing the cam assembly.

13. Install the axle shaft spacer.

14. Clip the snapring onto the end of the spindle.

15. Install the manual hub assembly over the spindle. Install the retainer washers.

16. Install the wheel and tire assembly. Install and tighten the lug nuts to specification.

17. Check the end-play of the wheel and tire assembly on the spindle. End-play should be 0.001–0.003 in. (0.025–

0.076mm) and the maximum torque to rotate the hub should be 25 inch lbs. (2.8 Nm).

REMOVAL & INSTALLATION

Aerostar

2-WHEEL VEHICLES

If wheel bearing adjustment will not eliminate looseness or rough and noisy operation, the hub and bearings should be cleaned, inspected and repacked with lithium base wheel bearing grease. If the bearing cups or the cone and roller assemblies are worn or damaged, they must be replaced as follows:

➡**Sodium based grease is not compatible with lithium based grease and the two should not be mixed. Do not lubricate the front and/or rear wheel bearings without first identifying the type of grease being used. Use of incompatible wheel bearing lubricant could result in premature lubricant breakdown and subsequent bearing damage.**

1. Raise and safely support the vehicle.

2. Remove the brake caliper from the spindle, then wire it to the underbody. Do not let the caliper hang by the brake hose.

3. Remove the grease cap from the hub. Remove the cotter pin, castellated retainer, adjusting nut and flat washer from the spindle. Remove the outer bearing assembly.

4. Pull the hub and rotor assembly off the spindle.

5. Place the hub and rotor on a clean workbench, with the back side facing up, and remove the grease seal using a suitable seal remover or small prybar. Discard the grease seal.

6. Remove the inner bearing assembly from the hub.

7. Clean the inner and outer bearing races with solvent. Inspect them for scratches, pits, scoring, excessive wear and

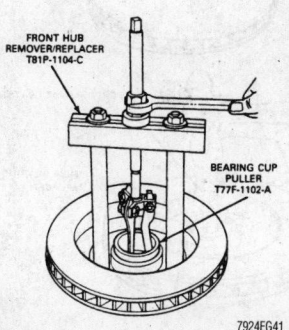

FRONT HUB REMOVER/REPLACER T81P-1104-C

BEARING CUP PULLER T77F-1102-A

7924EG41

Use an internal three jaw puller to remove the races (cups) from the hub assembly— 2WD models

Examine the bearings and races (cups) for damage and excessive wear

other damage. If the cups are worn or damaged, remove them with a bearing cup puller (T77F–1102–A or equivalent) as illustrated.

8. Wipe all old lubricant from the spindle and the inside of the hub with a clean rag. Cover the spindle and brush all loose dirt and dust from the dust shield. Remove the cover cloth carefully to prevent dirt from falling on the spindle.

To install:

9. If the inner or outer bearing cups were removed, install replacement cups using a suitable driver tool (T80T–4000–W or equivalent) and bearing cup replacer. Make sure the cups are seated properly in the hub and not cocked in the bore.

10. Thoroughly clean all old grease from the surrounding surfaces.

11. Pack the bearing and cone assemblies with suitable wheel bearing grease using a bearing packer tool. If a packer tool

is not available, work as much grease as possible between the rollers and cages, then grease the cone surfaces.

12. Place the inner bearing cone and roller assembly in the inner cup. Apply a light film of grease to the lip of a new grease seal and install the seal with an appropriate driver tool. Make sure the grease seal is properly seated and not cocked in the bore.

13. Install the hub and rotor assembly on the spindle. Keep the hub centered on the spindle to prevent damage to the retainer and the spindle threads.

14. Install the outer bearing cone and roller assembly (after being fully greased) and the flat washer on the spindle, then install the adjusting nut finger-tight. Adjust the wheel bearing.

15. Install the caliper onto the spindle.

16. Install the front wheel, then lower the van and tighten the lug nuts to 85–115 ft. lbs. (115–155 Nm). Install the wheel cover.

17. Repeat steps 1–16 for the other front wheel if needed.

18. Before moving the vehicle, pump the brake pedal several times to restore normal brake travel.

ALL-WHEEL DRIVE VEHICLES

➡The front wheel bearing/hub assembly is non-serviceable and non-adjustable. The assembly has to be replaced as a complete unit.

1. Raise and safely support the vehicle.
2. Remove the front wheels.
3. Remove the brake caliper and rotor.
4. Remove the three bolts retaining the hub/bearing assembly to the spindle, then remove the hub/bearing assembly from the spindle.

To install:

5. Install the hub/bearing assembly to the steering knuckle, then tighten the three retaining bolts to 65 ft. lbs. (88 Nm).

6. Install the brake rotor and caliper.

7. Install the front wheels.
8. Lower the vehicle.

Mountaineer and Explorer

✸✸ WARNING

If equipped, always turn off the Automatic Ride Control (ARC) service switch before lifting the vehicle off of the ground. Failure to do so could damage the ARC system components.

1. Loosen the wheel lug nuts then raise and safely support the front of the vehicle.

2. Remove the wheels.

3. Remove the front disc brake caliper, bracket and rotor. Also remove the rotor splash shield.

4. If equipped, unbolt the front wheel ABS sensor and wire harness from the steering knuckle.

5. Remove the front wheel hub nut and washer.

✸✸ WARNING

Never reuse the wheel hub nut and washer. This nut is a torque prevailing design and cannot be reused.

➡The hub shaft is a slip fit into the wheel hub and bearing; a press is not normally required.

6. Ensure that the wheel hub shaft can be pushed inwards. If not, assemble a press to the front wheel studs and press the wheel hub shaft inwards slightly to break it loose.

7. Remove the three wheel hub/bearing to steering knuckle retaining bolts. Remove the hub and bearing assembly.

To install:

8. Install the ABS sensor to the wheel hub then position the hub to the front axle shaft and steering knuckle.

9. Install the three retaining bolts and tighten them to 70–80 ft. lbs. (95–108 Nm).

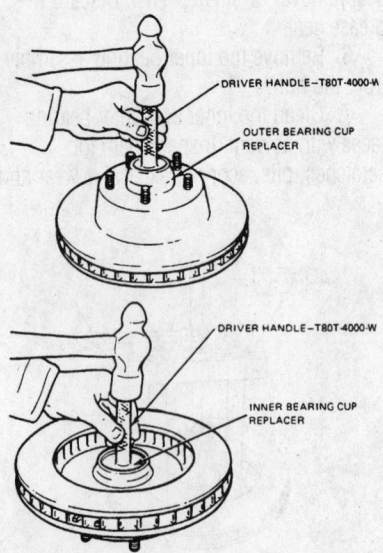

Install the races using the proper size driver, then apply wheel bearing grease to the surface of the race—2WD models

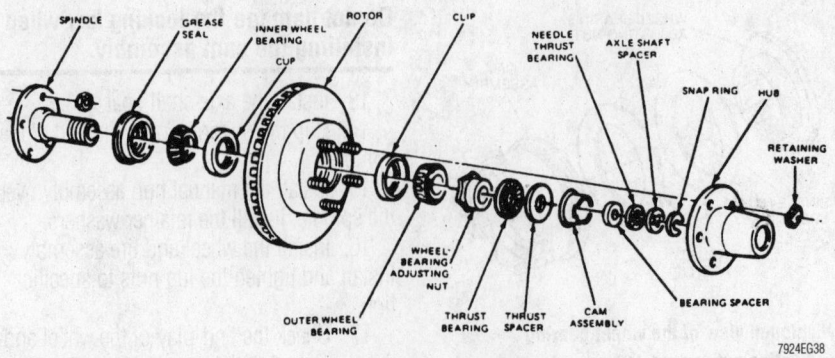

Exploded view of the wheel bearing and automatic locking hub assembly

10. Install the hub washer and nut and tighten to 157–213 ft. lbs. (212–288 Nm).

11. Install the ABS sensor retaining bolt.

12. Install the front brake rotor shield, rotor, bracket and caliper.

13. Install the wheel and snug the lug nuts.

14. Lower the vehicle and tighten the lug nuts to 100 ft. lbs. (135 Nm).

Ranger and B Series Models

WITH MANUAL HUBS

1. Raise and safely support the vehicle.

2. Remove the wheel and tire assembly.

3. Remove the retainer washers from the lug nut studs and remove the manual locking hub assembly from the spindle.

4. Remove the snapring and spacer from the end of the spindle shaft.

5. Remove the outer wheel bearing locknut from the spindle using 4 prong spindle nut spanner wrench, T86T-1197-A or equivalent. Make sure the tabs on the tool engage the slots in the locknut.

6. Remove the locknut washer from the spindle.

7. Remove the inner wheel bearing locknut from the spindle using 4 prong spindle nut spanner wrench, T86T-1197-A or equivalent. Make sure the tabs on the tool engage the slots in the locknut.

8. Remove the outer bearing cone and roller assembly from the hub. Remove the hub and rotor from the spindle.

9. Using seal removal tool 1175-AC or equivalent remove and discard the grease seal. Remove the inner bearing cone and roller assembly from the hub.

10. Clean the inner and outer bearing assemblies in solvent. Inspect the bearings and the cones for wear and damage. Replace defective parts, as required.

11. If the cups are worn or damaged, remove them with front hub remover tool T81P-1104-C and tool T77F-1102-A or equivalent.

12. Wipe the old grease from the spindle. Check the spindle for excessive wear or damage. Replace defective parts, as required.

To install:

13. If the inner and outer cups were removed, use bearing driver handle tool (T80-4000-W or equivalent) and replace the cups. Be sure to seat the cups properly in the hub.

14. Use a bearing packer tool and properly repack the wheel bearings with the proper grade and type grease. If a bearing packer is not available work as much of the grease as possible between the rollers and cages. Also, grease the cone surfaces.

15. Position the inner bearing cone and roller assembly in the inner cup. A light film of grease should be included between the lips of the new grease seal.

16. Install the grease seal by driving in place with Hub Seal Replacer tool T83T-1175-B and Driver Handle T80T-4000-W, or their equivalents.

17. Install the hub and rotor assembly onto the spindle. Keep the hub centered on the spindle to prevent damage to the spindle and the retainer.

18. Install the outer bearing cone and roller assembly

19. Carefully install the rotor onto the spindle. Install the outer wheel bearing in the rotor.

20. Install the inner adjusting nut with the pin facing out. Tighten the inner adjusting nut to 35 ft. lbs. (47 Nm) to seat the bearings.

21. Follow the appropriate wheel bearing adjustment procedures.

WITH AUTOMATIC LOCKING HUBS

1. Raise and safely support the vehicle.

2. Remove the wheel and tire assembly.

3. Remove the retainer washers from the lug nut studs and remove the automatic locking hub assembly from the spindle.

4. Remove the snapring and spacer from the end of the spindle shaft.

5. Pull the locking cam assembly and the two plastic spacers off of the wheel bearing adjusting nut.

6. Use a magnet and remove the locking key from under the adjusting nut. If required, rotate the adjusting nut slightly to relieve pressure against the locking key.

✳✳ WARNING

To prevent damage to the adjusting nut and spindle threads on vehicles equipped with automatic hubs, look into the spindle keyway under the adjusting nut and remove the separate locking key before removing the adjusting nut.

7. Remove the wheel bearing locknut using a 2-⅜ inch (60.3mm) hex socket, such as Hex Locknut Wrench T70T-4252-B, or their equivalents.

8. Remove the outer bearing cone and roller assembly from the hub. Remove the hub and rotor from the spindle.

9. Using seal removal tool 1175-AC (or equivalent) remove and discard the grease seal. Remove the inner bearing cone and roller assembly from the hub.

10. Clean the inner and outer bearing assemblies in solvent. Inspect the bearings and the cones for wear and damage. Replace defective parts, as required.

11. If the cups are worn or damaged, remove them with front hub remover tool T81P-1104-C and tool T77F-1102-A, or their equivalents.

12. Wipe the old grease from the spindle. Check the spindle for excessive wear or damage. Replace defective parts, as required.

To install:

13. If the inner and outer cups were removed, use bearing driver handle tool T80-4000-W (or equivalent) and replace the cups. Be sure to seat the cups properly in the hub.

14. Use a bearing packer tool and properly repack the wheel bearings with the proper grade and type of grease. If a bearing packer is not available, work as much of the grease as possible between the rollers and cages. Also, grease the cone surfaces.

15. Position the inner bearing cone and roller assembly in the inner cup. A light film of grease should be included between the lips of the new grease seal.

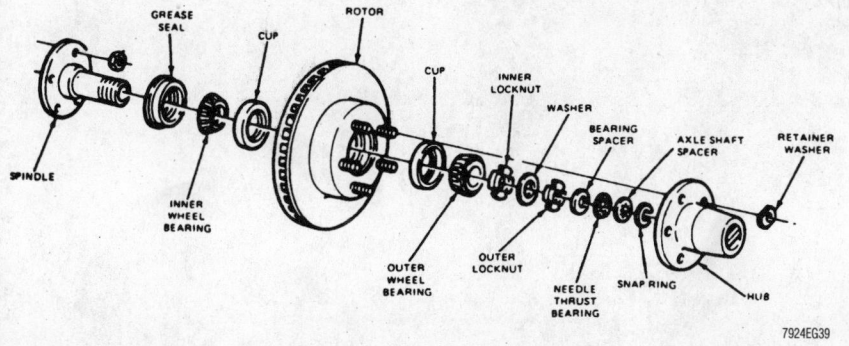

Exploded view of the wheel bearing and manual locking hub assembly

16. Install the grease seal by driving in place with Hub Seal Replacer tool T83T-1175-B and Driver Handle T80T-4000-W, or their equivalents.

17. Install the hub and rotor assembly onto the spindle. Keep the hub centered on the spindle to prevent damage to the spindle and the retainer.

18. Install the outer bearing cone and roller assembly.

19. Carefully install the rotor onto the spindle. Install the outer wheel bearing in the rotor.

20. Install the adjusting nut and tighten 35 ft. lbs. (47 Nm) to seat the bearings.

21. Follow the appropriate wheel bearing adjustment procedures.

FORD MOTOR CO.

16

Ford-Bronco • E-Series (Vans) • Expedition • F-Series (Pick-Ups) • **Lincoln**-Navigator

GASOLINE ENGINE REPAIR

➡ **Disconnecting the negative battery cable on some vehicles may interfere with the functions of the on board computer systems and may require the computer to undergo a relearning process, once the negative battery cable is reconnected.**

Distributor

REMOVAL

1. Disengage the primary wiring connector from the distributor.
2. Mark the position of the cap's No. 1 terminal on the distributor base.
3. Unclip and remove the cap. Remove the adapter.
4. Remove the rotor.
5. Remove the TFI connector.
6. Matchmark the distributor base and engine for installation reference.
7. Remove the hold-down bolt and lift out the distributor.

INSTALLATION

Timing Not Disturbed

1. Visually inspect the distributor. The O-ring should fit tightly onto the housing and be free of cuts. The drive gear should be free of nicks, cracks or excessive wear. The distributor shaft should rotate freely, without any binding.
2. Lubricate the distributor gear teeth with a coating of engine assembly lubricant, such as Ford D9AZ-19579-D, or with fresh motor oil meeting Ford specification ESR-M99C80-A.
3. Align the locating boss and fully seat the distributor rotor on the distributor shaft, if removed.
4. Rotate the distributor shaft so that the distributor rotor blade points toward the marked position on the distributor base adapter.
5. Install the distributor assembly into the engine block with a slight side-to-side twist.

➡ **If the vane and vane switch assembly cannot be kept on the leading edge after installation, remove the distributor from the cylinder block by pulling upward enough for the distributor gear to disengage the distributor gear from the camshaft gear. Rotate the distributor rotor enough so that the gear will align on the next tooth of the camshaft gear.**

6. Install the distributor hold-down clamp and bolt; leave it snug.
7. On V8 engines, position the adapter base in place, then install the attaching bolts.
8. Attach the electrical connector to the distributor.
9. Install the distributor cap. On 4.9L (VIN Y and Z) engines, tighten the distributor cap hold-down screws to 18–23 inch lbs. (2.0–2.6 Nm). On V8 engines, secure the distributor cap using the spring clips. If the spark plug wires were removed from the distributor cap, install them in their proper position, as marked during the removal procedure.
10. Connect the negative battery cable. Check the initial timing according to the proper procedure.
11. Adjust the timing, as necessary, then tighten the distributor hold-down bolt to 17–25 ft. lbs. (23–34 Nm).

Timing Disturbed

1. Disconnect the No. 1 spark plug wire and remove the No. 1 spark plug.
2. Place a finger over the spark plug

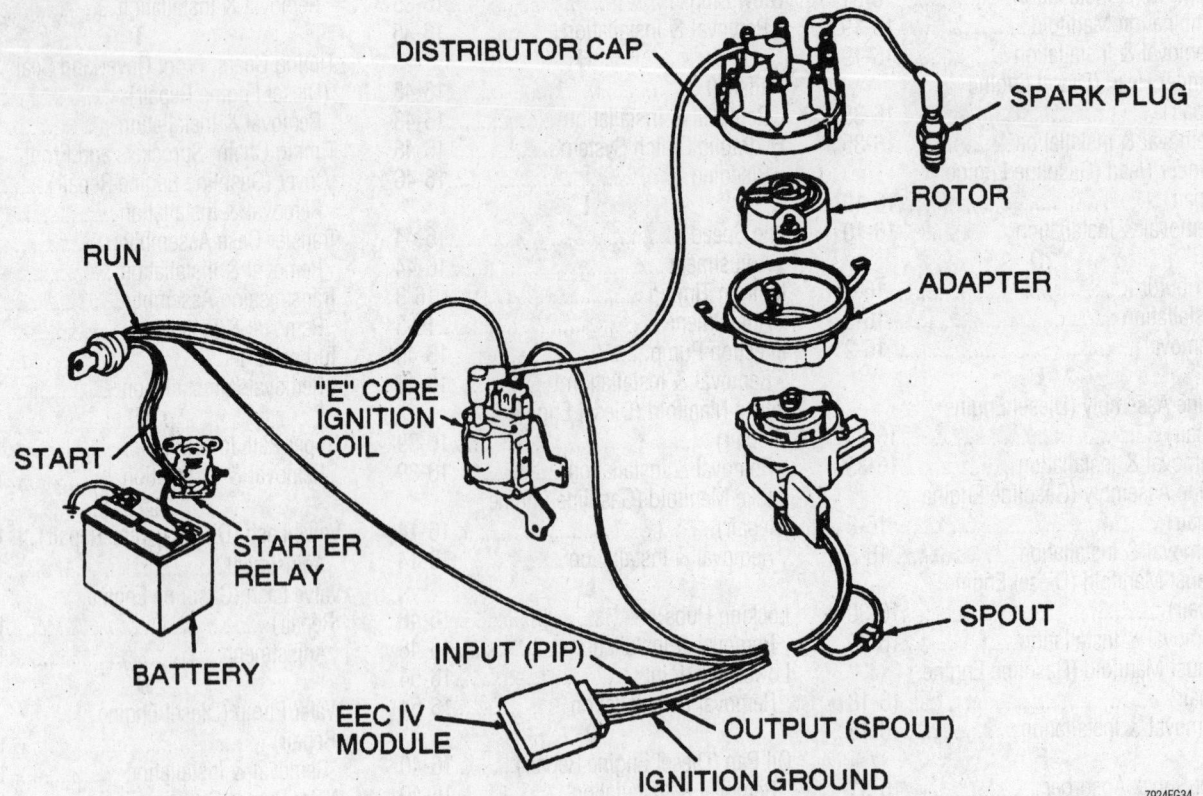

Exploded view of the Thick Film integrated (TFI) ignition system with universal distributor—4.9L (VIN Y and Z), 5.0L (VIN N), 5.8L (VIN H and R) and 7.5L (VIN G) engines

hole and crank the engine slowly until compression is felt.

3. Align the TDC mark on the crankshaft pulley with the pointer on the timing cover. This places the No. 1 cylinder at TDC on the compression stroke.

4. Turn the distributor shaft until the rotor points to the No. 1 spark plug tower on the cap.

5. Install the distributor assembly into the engine block with a slight side-to-side twist.

➡**If the vane and vane switch assembly cannot be kept on the leading edge after installation, remove the distributor from the cylinder block by pulling upward enough for the distributor gear to disengage the distributor gear from the camshaft gear. Rotate the distributor rotor enough so that the gear will align on the next tooth of the camshaft gear.**

6. Install the distributor hold-down clamp and bolt; leave it snug.

7. On V8 engines, position the adapter base in place, then install the attaching bolts.

8. Attach the electrical connector to the distributor.

9. Install the distributor cap. On 4.9L (VIN Y and Z) engines, tighten the distributor cap hold-down screws to 18–23 inch lbs. (2.0–2.6 Nm). On V8 engines, secure the distributor cap using the spring clips. If the spark plug wires were removed from the distributor cap, install them in their proper position, as marked during the removal procedure.

10. Connect the negative battery cable. Check the initial timing according to the proper procedure.

11. Adjust the timing, as necessary, then tighten the distributor hold-down bolt to 17–25 ft. lbs. (23–34 Nm).

Ignition Timing

ADJUSTMENT

➡**Always refer to the Vehicle Emission Control Information (VECI) label to verify the timing adjustment procedure and ignition specifics, which may have changed during the manufacture year.**

Distributorless Ignition System

Base timing for 5.4L and 6.8L distributorless ignition engines is set at the factory at 10 degrees BTDC and is not adjustable.

Distributor Ignition System

1. Place automatic transmissions in **P** or manual transmissions in Neutral. The air

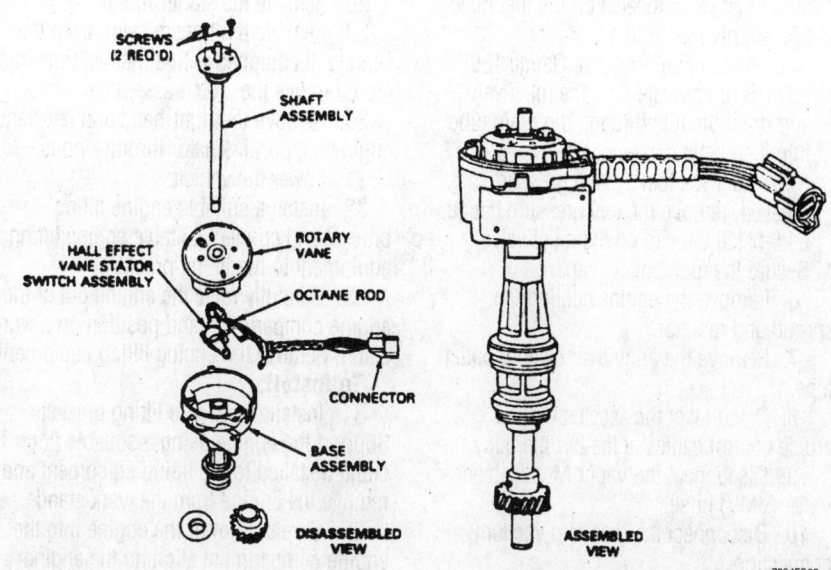

Exploded view of the closed bowl distributor used on the 5.0L, 5.8L, and 7.5L engines—4.9L engines similar

conditioning and heater controls should be in the OFF position.

2. Connect a suitable inductive timing light and a tachometer according to the manufacturer's instructions.

3. Disengage the single wire inline spout connector or remove the shorting bar from the double wire spout connector.

4. Start the engine and bring it up to normal operating temperature.

➡**To set timing correctly, a remote starter should not be used. Use the ignition key only to start the vehicle. Disconnecting the start wire at the starter relay will cause the TFI module to revert to "start mode timing" after the vehicle is started. Reconnecting the start wire after the vehicle is running will not correct the timing.**

5. With the engine running at the timing rpm specified, check the initial timing by aiming the timing light at the timing marks and pointer. Refer to the underhood Vehicle Emission Control Information (VECI) label for specific specifications.

6. If the marks do not align, shut the engine **OFF** and loosen the distributor hold-down clamp bolt. Start the engine, and while watching the timing marks with the timing light, turn the distributor until the marks are correctly aligned. Shut the engine **OFF**, then tighten the distributor hold-down clamp bolt to 17–25 ft. lbs. (23–34 Nm).

7. Reattach the single wire inline spout connector or reinstall the shorting bar on

the double wire spout connector. Check the timing advance to verify the distributor is advancing beyond the initial setting.

8. Remove the timing light and tachometer.

Engine Assembly

REMOVAL & INSTALLATION

4.2L Engine

❊❊ CAUTION

Fuel injection systems remain under pressure, even after the engine has been turned OFF. The fuel system pressure must be relieved before disconnecting any fuel lines. Failure to do so may result in fire and/or personal injury.

1. Disconnect both battery cables, negative cable first.

2. Mark the position of the hood on the hinges and remove the hood.

3. Drain the engine cooling system into a suitable container.

4. Recover the refrigerant from the A/C system using approved recovery equipment.

5. Relieve the fuel system pressure as follows:

   a. Remove the fuel tank fill cap to relieve the pressure in the fuel tank.

   b. Remove the cap from the fuel pressure relief valve located on the fuel injection supply manifold.

c. Attach Fuel Pressure Gauge T80L-9974-B or equivalent, to the relief valve and drain the fuel through the drain tube into a suitable container.

d. After the fuel system pressure is relieved, remove the fuel pressure gauge and install the cap on the relief valve. Secure the fuel tank fill cap.

6. Remove the engine cooling fan, shroud and radiator.

7. Remove the engine air cleaner outlet tube.

8. Disconnect the accelerator and cruise control cables at the throttle body.

9. Disconnect the Vapor Management Valve (VMV) hose.

10. Disconnect the manifold vacuum connection.

11. Disconnect the Intake Manifold Runner Control (IMRC) vacuum connectors, fuel pressure regulator vacuum connector, IMRC solenoid vacuum connector and vacuum reservoir connector.

12. Disconnect the EGR valve vacuum connector.

13. Remove the three power steering reservoir retaining bolts and position aside.

14. If equipped with air conditioning, remove the A/C compressor manifold bolt and disconnect, then position the A/C lines aside.

15. Remove the four power steering pump retaining bolts and position the pump aside.

16. Disconnect the alternator electrical harness connectors. Remove the positive battery cable nut and disconnect the battery cable.

17. Disconnect the electrical harness connectors to the fuel injectors.

18. Tag and disconnect the ignition wires at the spark plugs.

19. Disconnect both heater hoses.

20. Disconnect the brake booster vacuum hose.

21. Disconnect the EGR Differential Pressure Feedback (DPFE) transducer hose.

22. Remove the breather tube from the cylinder head cover.

23. Remove the upper intake manifold.

24. Disconnect and plug the fuel supply and return lines and remove the fuel injection supply manifold.

25. Raise and safely support the vehicle.

26. If equipped, disconnect the block heater cable.

27. Disconnect the exhaust system from the exhaust manifolds and support with wire hung from the crossmember.

28. Remove the starter motor.

29. Remove the transmission from the vehicle. If equipped with a manual transmission, remove the clutch assembly.

30. Remove the right-hand and left-hand engine support insulator through-bolts.

31. Lower the vehicle.

32. Install a suitable engine lifting bracket and connect suitable engine lifting equipment to the lifting brackets.

33. Carefully raise the engine out of the engine compartment and position on a work stand. Remove the engine lifting equipment.

To install:

34. Install the engine lifting brackets. Support the engine using a suitable floor crane installed to the lifting equipment and remove the engine from the work stand.

35. Carefully lower the engine into the engine compartment aligning the engine support insulators.

36. Remove the engine lifting equipment and brackets.

37. Raise and safely support the vehicle.

38. Install the left-hand and right-hand engine support insulator through-bolts and tighten them to 51–67 ft. lbs. (68–92 Nm).

39. If equipped with a manual transmission, install the clutch assembly.

40. Install the transmission.

41. Install the starter motor.

42. Connect the exhaust pipes to the exhaust manifolds and tighten to 30 ft. lbs. (41 Nm).

43. If equipped with a block heater, connect the heater cable.

44. Lower the vehicle.

45. Connect the alternator electrical harness connectors and install the positive battery cable to the retaining stud. Tighten the retaining nut to 96 inch lbs. (11 Nm).

46. Install the fuel injectors and the fuel injection supply manifold.

47. Install a new upper intake gasket and install the upper intake manifold.

48. Install the electrical harness connectors to the fuel injectors.

49. Place the power steering pump in position and install four retaining bolts. Tighten the bolts to 17–20 ft. lbs. (22–28 Nm).

➡Ensure that the air conditioning manifold O-rings are in place.

50. Install the A/C manifold to the compressor and install the retaining bolt. Tighten the bolt to 14–18 ft. lbs. (18–24 Nm).

51. Install the power steering reservoir and three retaining bolts. Tighten the bolts to 107 inch lbs. (12 Nm).

52. Connect the VMV hose.

53. Connect the IMRC vacuum connectors, fuel pressure regulator vacuum connector, IMRC solenoid vacuum connector and vacuum reservoir connector.

54. Connect the EGR valve vacuum connector.

55. Connect the brake booster hose.

56. Connect the two EGR DPFE transducer hoses.

57. Connect the manifold vacuum connection.

58. Place the accelerator cable and speed control cable in position, if equipped. Tighten the speed control cable retaining bolt to 72 inch lbs. (8 Nm) and accelerator cable retaining bolt to 25 inch lbs. (3 Nm).

59. Install the radiator, cooling fan and shroud.

60. Install the engine air cleaner outlet tube.

61. If needed, fill the crankcase with the proper type and quantity of engine oil.

62. Fill and bleed the engine cooling system.

63. Connect both battery cables, negative cable last.

64. Start the engine and allow to reach normal operating temperature while checking for leaks.

65. Check all fluid levels.

66. Properly evacuate and recharge the A/C system.

67. Install the hood, aligning the marks that were made during removal.

68. Road test the vehicle and check the engine and transmission for proper operation.

4.6L, 5.4L and 6.8L Engines

⁑⁑ CAUTION

Fuel injection systems remain under pressure, even after the engine has been turned OFF. The fuel system pressure must be relieved before disconnecting any fuel lines. Failure to do so may result in fire and/or personal injury.

1. Disconnect both battery cables, negative cable first.

2. Mark the position of the hood on the hinges and remove the hood.

3. Drain the engine cooling system into a suitable container.

4. Have the A/C system discharged by an EPA-certified technician.

5. Relieve the fuel system pressure as follows:

a. Remove the fuel tank fill cap to relieve the pressure in the fuel tank.

b. Remove the cap from the fuel pressure relief valve located on the fuel injection supply manifold.

c. Attach Fuel Pressure Gauge T80L-9974-B or equivalent, to the relief valve and drain the fuel through the drain tube into a suitable container.

d. After the fuel system pressure is relieved, remove the fuel pressure gauge and install the cap on the relief valve. Secure the fuel tank fill cap.

6. Remove the engine cooling fan, shroud and radiator.

7. Remove the accessory drive belt.

8. Remove the engine air cleaner outlet tube.

9. Remove the intake manifold assembly.

10. Remove the bulkhead connector cover and disconnect the bulkhead connector.

11. Remove the three power steering reservoir bracket retaining bolts and move the reservoir aside.

12. Disconnect two DPFE transducer hoses.

13. Disconnect the upper and lower EGR valve to exhaust manifold tube fittings and remove the tube.

14. Slide the heater water hose clamp back and remove the hose.

15. Remove the ignition coil, radio capacitor and Camshaft Position (CMP) sensor electrical harness connectors.

16. Remove both ignition coils and mounting bracket bolts and remove the coil and bracket assemblies.

17. Raise and safely support the vehicle.

18. Remove the starter motor.

19. Remove the three lower radiator air deflector screws. Remove the five clips and remove the air deflector.

20. Disconnect the A/C compressor electrical harness connector.

21. Remove the A/C manifold-to-compressor bolt and remove the manifold and tube assembly.

22. Remove the three A/C compressor retaining bolts and remove the A/C compressor.

23. Remove the fluid cooler hoses from the block mounted clip.

24. On vehicles with automatic transmissions, remove the inspection cover, torque converter bolts and transmission-to-engine retaining bolts.

25. On vehicles with manual transmission, remove the transmission and the clutch assembly.

26. Remove the upper and lower power steering pump bolts and move the power steering pump aside.

27. Disconnect the exhaust system from the exhaust manifolds and support with wire hung from the crossmember.

28. Remove the right-hand and left-hand engine support insulator (mount) through-bolts.

29. Lower the vehicle.

30. Install a suitable engine lifting bracket and connect suitable engine lifting equipment to the lifting brackets.

31. Carefully raise the engine out of the engine compartment and place on a work stand. Remove the engine lifting equipment.

To install:

32. Install the engine lifting brackets. Support the engine using a suitable floor crane installed to the lifting equipment and remove the engine from the work stand.

33. Carefully lower the engine into the engine compartment. Start the converter pilot into the flywheel and align the paint marks on the flywheel and torque converter. Be sure the studs on the torque converter align with the holes in the flywheel.

34. Fully engage the engine to the transmission and lower onto the engine support insulators.

35. Remove the engine lifting equipment and brackets.

36. Raise and safely support the vehicle.

37. If equipped with a manual transmission, install the clutch and transmission assemblies.

38. Install the six engine-to-transmission retaining bolts and tighten to 30–44 ft. lbs. (40–60 Nm).

39. Install the engine support insulator through-bolts and tighten to 15–22 ft. lbs. (20–30 Nm).

40. Install the four torque converter retaining nuts and tighten to 22–25 ft. lbs. (20–30 Nm).

41. Install the transmission housing cover to the cylinder block.

42. Connect the exhaust pipes to the exhaust manifolds and tighten to 30 ft. lbs. (41 Nm).

43. Place the power steering pump in position on the cylinder block and install four retaining nuts. Tighten to 15–20 ft. lbs. (20–30 Nm).

44. Install the starter motor.

45. Install the transmission fluid cooler hoses into the cylinder block mounted clip.

46. Place the A/C compressor in position and install the three retaining bolts. Tighten the bolts to 15–22 ft. lbs. (20–30 Nm).

47. Place the A/C manifold and tube assembly on the compressor and install the retaining bolt. Tighten the bolt to 14–18 ft. lbs. (18–24 Nm).

48. Connect the A/C compressor clutch electrical harness connector.

49. Install the lower radiator air deflector.

50. Install the ignition coil and bracket assemblies. Tighten the retaining nuts to 15–23 ft. lbs. (20–30 Nm).

51. Connect the ignition coil, radio capacitor and CMP sensor electrical harness connectors.

52. Install the rear heater water hose and compress and slide the clamp in position.

53. Place the EGR valve to exhaust manifold tube and tighten the upper and lower fittings to 26–33 ft. lbs. (35–45 Nm).

54. Connect the DPFE transducer hoses.

55. Place the power steering pump reservoir in position and install the three retaining bolts. Tighten the bolts to 71–107 inch lbs. (8–12 Nm).

56. Connect the engine bulkhead connector and install the retaining bolt. Tighten the bolt to 36–50 inch lbs. (4–6 Nm). Install the cover.

57. Install the intake manifold assembly.

58. Install the accessory drive belt.

59. Install the radiator, cooling fan and shroud.

60. Install the engine air cleaner outlet tube.

61. If needed, fill the crankcase with the proper type and quantity of engine oil.

62. Fill and bleed the engine cooling system.

63. Connect both battery cables, negative cable last.

64. Start the engine and allow to reach normal operating temperature while checking for leaks.

65. Check all fluid levels.

66. Properly evacuate and recharge the A/C system.

67. Install the hood, aligning the marks that were made during removal.

68. Road test the vehicle and check the engine and transmission for proper operation.

4.9L Engine

E-SERIES

1. Drain the cooling system and the crankcase.

2. Remove the engine cover.

3. Remove the throttle body inlet tubes.

4. Disconnect the positive battery cable.

5. Have the A/C system discharged by

an EPA-certified technician using recovery equipment.

6. Disconnect the refrigerant lines at the compressor. Cap all openings at once.

7. Remove the compressor.

8. Disconnect the refrigerant lines at the condenser. Cap all openings at once.

9. Remove the condenser.

10. Disconnect the heater hose from the water pump and coolant outlet housing.

11. Disconnect the flexible fuel line from the fuel pump.

12. Remove the radiator.

13. Remove the bumper.

14. Remove the grille.

15. Remove the gravel deflector.

16. Remove the fan, water pump pulley, and fan belt.

17. Disconnect the accelerator cable.

18. Disconnect the brake booster vacuum hose at the intake manifold.

19. On vehicles with automatic transmission, disconnect the transmission kickdown rod at the bellcrank assembly.

20. Disconnect the exhaust pipe from the exhaust manifold.

21. Disconnect the Electronic Engine Control (EEC) harness from all the sensors.

22. Disconnect the body ground strap and the battery ground cable from the engine.

23. Disconnect the engine wiring harness at the ignition coil, the coolant temperature sending unit, and the oil pressure sensing unit. Position the wiring harness out of the way.

24. Remove the alternator mounting bolts and position the alternator out of the way.

25. Remove the power steering pump from the mounting brackets and move it to one side, leaving the lines attached.

26. Raise and safely support the vehicle.

27. Remove the starter.

28. Remove the automatic transmission filler tube bracket, if so equipped.

29. Remove the rear engine plate upper right bolt.

30. On manual transmission equipped vehicles:

    a. Remove the flywheel housing lower attaching bolts.

    b. Disconnect the clutch return spring.

31. On automatic transmission equipped vehicles:

    a. Remove the converter housing access cover assembly.

    b. Remove the flywheel-to-converter attaching nuts.

    c. Secure the converter in the hous-

ing so it does not fall out when the engine is removed.

    d. Remove the transmission oil cooler lines from the retaining clip at the engine.

    e. Remove the lower converter housing-to-engine attaching bolts.

32. Remove the nut from each of the two front engine mounts.

33. Lower the vehicle and position a jack under the transmission and support it.

34. Remove the remaining bell housing-to-engine attaching bolts.

35. Attach an engine lifting device and raise the engine slightly and carefully pull it from the transmission. Lift the engine out of the vehicle.

To install:

36. Place a new gasket on the muffler inlet pipe.

37. Carefully lower the engine into the vehicle. Be sure that the dowels in the engine block engage the holes in the bell housing.

38. On vehicles with manual transmissions, turn the crankshaft until the input shaft splines mesh with the clutch disc splines.

39. On vehicles with automatic transmissions, engage the converter pilot into the crankshaft.

40. Install the bell housing upper attaching bolts. Tighten the bolts to 50 ft. lbs. (68 Nm).

41. Remove the jack supporting the transmission.

42. Remove the lifting device.

43. Install the engine mount nuts and tighten them to 70 ft. lbs. (95 Nm).

44. Install the automatic transmission coil cooler lines bracket, if equipped.

45. Install the remaining bell housing attaching bolts. Tighten them to 50 ft. lbs. (68 Nm).

46. Connect the clutch return spring, if so equipped.

47. Install the starter and connect the starter cable.

48. If removed, install the fluid filler tube bracket.

49. On vehicles with automatic transmissions, install the transmission oil cooler lines in the bracket at the cylinder block.

50. Connect the exhaust pipe to the exhaust manifold. Tighten the nuts to 25–35 ft. lbs. (34–48 Nm).

51. Connect the engine ground strap and negative battery cable.

52. On vehicles with an automatic transmission, connect the kickdown rod to the bellcrank assembly on the intake manifold.

53. Connect the accelerator linkage.

54. Connect the brake booster vacuum line to the intake manifold.

55. Connect the coil primary wire, oil pressure and coolant temperature sending unit wires, fuel line, heater hoses, and the battery positive cable.

56. Connect the EEC sensors.

57. Install the alternator on its mounting bracket.

58. Install the power steering pump on its bracket.

59. Install the water pump pulley, spacer, fan, and fan belt. Adjust the belt tension.

60. Install the air conditioning compressor. Connect the refrigerant lines.

61. Install the gravel deflector.

62. Install the grille.

63. Install the bumper.

64. Install the radiator.

65. Install the condenser and connect the refrigerant lines.

66. Evacuate, recharge and leak test the A/C system.

67. Connect the upper and lower radiator hoses to the radiator and engine.

68. If removed, connect the automatic transmission oil cooler lines.

69. Install the engine cover.

70. Fill the cooling system.

71. Fill the crankcase.

72. Start the engine and check for leaks.

73. Bleed the cooling system.

74. Adjust the clutch pedal free-play or the automatic transmission control linkage.

75. Install the air cleaner.

EXCEPT E-SERIES

1. Drain the cooling system and the crankcase.

2. Remove the hood.

3. Remove the throttle body inlet tubes.

4. Disconnect the positive battery cable.

5. Have the A/C system discharged by an EPA-certified technician.

6. Disconnect the refrigerant lines at the compressor. Cap all openings at once.

7. Remove the compressor.

8. Disconnect the refrigerant lines at the condenser. Cap all openings at once.

9. Remove the condenser.

10. Disconnect the heater hose from the water pump and coolant outlet housing.

11. Disconnect the flexible fuel line from the fuel pump.

12. Remove the radiator.

13. Remove the fan, water pump pulley, and fan belt.

14. Disconnect the accelerator cable.

15. Disconnect the brake booster vacuum hose at the intake manifold.

16. On vehicles with automatic transmission, disconnect the transmission kickdown rod at the bellcrank assembly.

17. Disconnect the exhaust pipe from the exhaust manifold.

18. Disconnect the Electronic Engine Control (EEC) harness from all the sensors.

19. Disconnect the body ground strap and the battery ground cable from the engine.

20. Disconnect the engine wiring harness at the ignition coil, the coolant temperature sending unit, and the oil pressure sensing unit. Position the wiring harness out of the way.

21. Remove the alternator mounting bolts and position the alternator out of the way.

22. Remove the power steering pump from the mounting brackets and move it to one side, leaving the lines attached.

23. Raise and safely support the vehicle on jackstands.

24. Remove the starter.

25. Remove the automatic transmission filler tube bracket, if so equipped.

26. Remove the rear engine plate upper right bolt.

27. On manual transmission equipped vehicles:

   a. Remove the flywheel housing lower attaching bolts.

   b. Disconnect the clutch return spring.

28. On automatic transmission equipped vehicles:

   a. Remove the converter housing access cover assembly.

   b. Remove the flywheel-to-converter attaching nuts.

   c. Secure the converter in the housing.

   d. Remove the transmission oil cooler lines from the retaining clip at the engine.

   e. Remove the lower converter housing-to-engine attaching bolts.

29. Remove the nut from each of the two front engine mounts.

30. Lower the vehicle and position a jack under the transmission and support it.

31. Remove the remaining bell housing-to-engine attaching bolts.

32. Attach an engine lifting device and raise the engine slightly and carefully pull it from the transmission. Lift the engine out of the vehicle.

To install:

33. Remove the engine mount brackets from the frame. Attach them to the engine mounts, making the nuts just tight enough to hold the brackets securely to the mounts.

34. Place a new gasket on the muffler inlet pipe.

35. Carefully lower the engine into the vehicle. Be sure that the dowels in the engine block engage the holes in the bell housing and the mount bracket holes align with the frame holes.

36. On manual transmission equipped vehicles, turn the crankshaft until the input shaft splines mesh with the clutch disc splines.

37. On automatic transmission equipped vehicles, be sure the converter pilot fits into the crankshaft.

38. Install the bolts securing the mount brackets to the frame.

39. Install the bell housing upper attaching bolts. Tighten the bolts to 50 ft. lbs. (68 Nm).

40. Remove the jack supporting the transmission.

41. Remove the lifting device.

42. Install the engine mount nuts and tighten them to 70 ft. lbs. (95 Nm). Tighten the bracket-to-frame bolts to 70 ft. lbs. (95 Nm).

43. Install the automatic transmission coil cooler lines bracket, if equipped.

44. Install the remaining bell housing attaching bolts. Tighten them to 50 ft. lbs. (68 Nm).

45. If equipped with an automatic transmission, install the torque converter-to-flywheel bolts.

46. Connect the clutch return spring, if so equipped.

47. Install the starter and connect the starter cable.

48. Attach the automatic transmission fluid filler tube bracket, if so equipped.

49. On vehicles with automatic transmissions, install the transmission oil cooler lines in the bracket at the cylinder block.

50. Connect the exhaust pipe to the exhaust manifold. Tighten the nuts to 25–35 ft. lbs. (34–48 Nm).

51. Connect the engine ground strap and negative battery cable.

52. On a vehicle with an automatic transmission, connect the kickdown rod to the bellcrank assembly on the intake manifold.

53. Connect the accelerator linkage.

54. Connect the brake booster vacuum line to the intake manifold.

55. Connect the coil primary wire, oil pressure and coolant temperature sending unit wires, fuel line, heater hoses, and the battery positive cable.

56. Connect the EEC sensors.

57. Install the alternator on its mounting bracket.

58. Install the power steering pump on its bracket.

59. Install the water pump pulley, spacer, fan, and fan belt. Adjust the belt tension.

60. Install the air conditioning compressor. Connect the refrigerant lines.

61. Install the radiator.

62. Install the condenser and connect the refrigerant lines.

63. Charge the refrigerant system.

64. Connect the upper and lower radiator hoses to the radiator and engine.

65. Connect the automatic transmission oil cooler lines, if so equipped.

66. Install and adjust the hood.

67. Fill the cooling system.

68. Fill the crankcase.

69. Start the engine and check for leaks.

70. Bleed the cooling system.

71. Adjust the clutch pedal free-play or the automatic transmission control linkage.

72. Install the air cleaner.

5.0L and 5.8L Engines

E-SERIES

1. Remove the engine cover.

2. Drain the cooling system and crankcase.

3. Disconnect the battery and alternator cables.

4. Remove the air intake hoses, PCV tube and carbon canister hose.

5. Disconnect the upper and lower radiator hoses.

6. Have the A/C system discharged by an EPA-certified technician.

7. Disconnect the refrigerant lines at the compressor. Cap all openings immediately.

8. If so equipped, disconnect the automatic transmission oil cooler lines.

9. Remove the fan shroud and lay it over the fan.

10. Remove the radiator and fan, shroud, fan, spacer, pulley and belt.

11. Remove the grille.

12. Remove the gravel deflector.

13. Remove the bumper.

14. Remove the upper grille support bracket.

15. Remove the hood lock support.

16. Remove the alternator pivot and adjusting bolts. Remove the alternator.

17. Disconnect the oil pressure sending unit lead from the sending unit.

18. Disconnect the fuel supply and return lines at the fuel injector rails.

19. Disconnect the accelerator linkage and speed control linkage at the throttle body.

20. Disconnect the automatic transmission kick-down rod and remove the return spring, if so equipped.

21. Disconnect the power brake booster vacuum hose.

22. Disconnect the throttle bracket from the upper intake manifold and swing it out of the way with the cables still attached.

23. Disconnect the heater hoses from the water pump and intake manifold or tee.

24. Disconnect the temperature sending unit wire from the sending unit.

25. Remove the upper bell housing-to-engine attaching bolts.

26. Remove the wiring harness from the left rocker arm cover and position the wires out of the way.

27. Disconnect the ground strap from the cylinder block.

28. Disconnect the air conditioning compressor clutch wire.

29. Raise and safely support the front of the vehicle and disconnect the starter cable from the starter.

30. Remove the starter.

31. Disconnect the exhaust pipe from the exhaust manifolds.

32. Disconnect the engine mounts from the brackets on the frame.

33. On vehicles with automatic transmissions, remove the converter inspection plate and remove the torque converter-to-flywheel attaching bolts.

34. Remove the remaining bell housing-to-engine attaching bolts.

35. Lower the vehicle and support the transmission with a jack.

36. Install an engine lifting device.

37. Raise the engine slightly and carefully pull it out of the transmission. Lift the engine out of the engine compartment.

To install:

38. Lower the engine carefully into the transmission. Be sure that the dowels in the engine block engage the holes in the bell housing through the rear cover plate. If equipped with a manual transmission, turn the crankshaft with the transmission in gear until the input shaft splines mesh with the clutch disc splines.

39. Install the lower bell housing-to-engine attaching bolts. Tighten the bolts to 50 ft. lbs. (68 Nm).

40. Install the engine mount nuts and washers. Tighten the nuts to 80 ft. lbs. (109 Nm).

41. Remove the engine lifting device.

42. Remove the transmission support jack.

43. On vehicles with automatic transmissions, install the torque converter-to-fly-

wheel attaching bolts. Tighten the bolts to 30 ft. lbs. (41 Nm).

44. Install the converter inspection plate. Tighten the bolts to 60 inch lbs. (7 Nm).

45. Connect the exhaust pipe to the exhaust manifolds. Tighten the exhaust pipe-to-exhaust manifold nuts to 25–35 ft. lbs. (34–48 Nm).

46. Install the starter. Tighten the mounting bolts to 20 ft. lbs. (27 Nm).

47. Connect the starter cable to the starter.

48. Lower the vehicle.

49. Install the upper bell housing-to-engine attaching bolts. Tighten the bolts to 50 ft. lbs. (68 Nm).

50. Connect the wiring harness at the left rocker arm cover.

51. Connect the ground strap to the cylinder block.

52. Connect the air conditioning compressor clutch wire.

53. Connect the heater hoses at the water pump and intake manifold or tee.

54. Connect the temperature sending unit wire at the sending unit.

55. Connect the accelerator linkage and speed control linkage at the throttle body.

56. Connect the automatic transmission kick-down rod and install the return spring, if so equipped.

57. Connect the power brake booster vacuum hose.

58. Connect the throttle bracket to the upper intake manifold.

59. Connect the fuel lines.

60. Connect the oil pressure sending unit lead to the sending unit.

61. Install the alternator.

62. Connect the refrigerant lines to the compressor.

63. Install the hood lock support.

64. Install the upper grille support bracket.

65. Install the bumper.

66. Install the gravel deflector.

67. Install the grille.

68. Install the radiator, shroud, fan and spacer, pulley and belt.

69. Connect the upper and lower radiator hoses,, if so equipped, the automatic transmission oil cooler lines.

70. Install the air intake hoses, PCV tube and carbon canister hose.

71. Connect the battery and alternator cables.

72. Fill the cooling system and crankcase.

73. Evacuate, recharge and leak test the A/C system.

74. Install the engine cover.

EXCEPT E-SERIES

1. Remove the hood.

2. Drain the cooling system and crankcase.

3. Disconnect the battery and alternator cables.

4. On carbureted engines, remove the air cleaner and intake duct assembly, plus the crankcase ventilation hose.

5. On fuel injected engines, remove the air intake hoses, PCV tube and carbon canister hose.

6. Disconnect the upper and lower radiator hoses.

7. Have the A/C system discharged by an EPA-certified technician.

8. Disconnect the refrigerant lines at the compressor. Cap all openings immediately.

9. If so equipped, disconnect the automatic transmission oil cooler lines.

10. Remove the fan shroud and lay it over the fan.

11. Remove the radiator and fan, shroud fan, spacer, pulley and belt.

12. Remove the alternator pivot and adjusting bolts. Remove the alternator.

13. Disconnect the oil pressure sending unit lead from the sending unit.

14. Disconnect the fuel tank-to-pump fuel line at the fuel pump and plug the line.

15. Disconnect the chassis fuel line at the fuel rails.

16. Disconnect the accelerator linkage and speed control linkage at the carburetor or throttle body.

17. Disconnect the automatic transmission kick-down rod and remove the return spring, if so equipped.

18. Disconnect the power brake booster vacuum hose.

19. Disconnect the throttle bracket from the upper intake manifold and swing it out of the way with the cables still attached.

20. Disconnect the heater hoses from the water pump and intake manifold or tee.

21. Disconnect the temperature sending unit wire from the sending unit.

22. Remove the upper bell housing-to-engine attaching bolts.

23. Remove the wiring harness from the left rocker arm cover and position the wires out of the way.

24. Disconnect the ground strap from the cylinder block.

25. Disconnect the air conditioning compressor clutch wire.

26. Raise and safely support the front of the vehicle and disconnect the starter cable from the starter.

27. Remove the starter.

28. Disconnect the exhaust pipe from the exhaust manifolds.

29. Disconnect the engine mounts from the brackets on the frame.

30. On vehicles with automatic transmissions, remove the converter inspection plate and remove the torque converter-to-flywheel attaching bolts.

31. Remove the remaining bell housing-to-engine attaching bolts.

32. Lower the vehicle and support the transmission with a jack.

33. Install an engine lifting device.

34. Raise the engine slightly and carefully pull it out of the transmission. Lift the engine out of the engine compartment.

To install:

35. Remove the engine mount brackets from the frame and attach them to the engine mounts. Tighten the mount-to-bracket nuts just enough to hold them securely.

36. Lower the engine carefully into the transmission. Be sure that the dowel in the engine block engage the holes in the bell housing through the rear cover plate. If the engine hangs up after the transmission input shaft enters the clutch disc (manual transmission only), turn the crankshaft with the transmission in gear until the input shaft splines mesh with the clutch disc splines.

37. Install the engine mount nuts and washers. Tighten the nuts to 80 ft. lbs. Tighten the bracket-to-frame bolts to 70 ft. lbs. (95 Nm).

38. Remove the engine lifting device.

39. Install the lower bell housing-to-engine attaching bolts. Tighten the bolts to 50 ft. lbs. (68 Nm).

40. Remove the transmission support jack.

41. On vehicles with automatic transmissions, install the torque converter-to-flywheel attaching bolts. Tighten the bolts to 30 ft. lbs. (41 Nm).

42. Install the converter inspection plate. Tighten the bolts to 60 inch lbs. (7 Nm).

43. Connect the exhaust pipe to the exhaust manifolds. Tighten the exhaust pipe-to-exhaust manifold nuts to 25–35 ft. lbs. (34–48 Nm).

44. Install the starter. Tighten the mounting bolts to 20 ft. lbs. (27 Nm).

45. Connect the starter cable to the starter.

46. Lower the vehicle.

47. Install the upper bell housing-to-engine attaching bolts. Tighten the bolts to 50 ft. lbs. (68 Nm).

48. Connect the wiring harness at the left rocker arm cover.

49. Connect the ground strap to the cylinder block.

50. Connect the air conditioning compressor clutch wire.

51. Connect the heater hoses at the water pump and intake manifold or tee.

52. Connect the temperature sending unit wire at the sending unit.

53. Connect the accelerator linkage and speed control linkage at the carburetor or throttle body.

54. Connect the automatic transmission kick-down rod and install the return spring, if so equipped.

55. Connect the power brake booster vacuum hose.

56. Connect the throttle bracket to the upper intake manifold.

57. Connect the fuel tank-to-pump fuel line at the fuel pump. Connect the chassis fuel line at the fuel rails.

58. Connect the oil pressure sending unit lead to the sending unit.

59. Install the alternator.

60. Connect the refrigerant lines to the compressor.

61. Install the radiator and fan, shroud, fan, spacer, pulley and belt.

62. Connect the upper and lower radiator hoses,, if so equipped, the automatic transmission oil cooler lines.

63. On carbureted engines, install the air cleaner and intake duct assembly, plus the crankcase ventilation hose.

64. On fuel injected engines, install the air intake hoses, PCV tube and carbon canister hose.

65. Connect the battery and alternator cables.

66. Fill the cooling system and crankcase.

67. Charge the air conditioning system.

68. Install the hood.

Water Pump

REMOVAL & INSTALLATION

1. Disconnect the negative battery cable.

2. Remove the radiator, fan blade assembly and fan shroud.

3. Remove the accessory drive belt.

4. Remove the water pump pulley.

5. If equipped, disconnect the heater hose from the water pump.

6. Remove the water pump bolts and nuts. Note the locations of the bolts if different lengths.

7. Remove the water pump stud bolt, the water pump and the water pump hous-

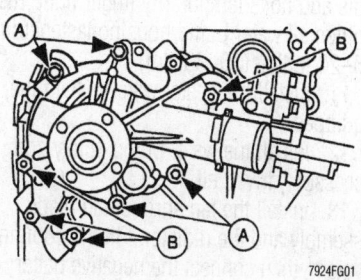

When removing the water pump, note the locations of the mounting bolts (A) and nuts (B)—4.2L engine

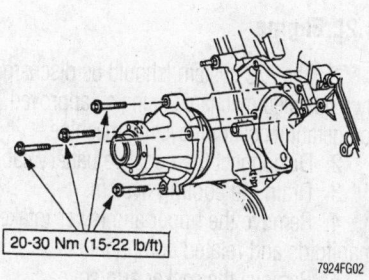

20-30 Nm (15-22 lb/ft)

Exploded view of the water pump mounting—4.6L, 5.4 and 6.8L engines

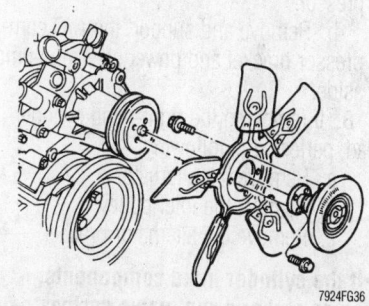

Exploded view of the cooling fan which is mounted on the water pump—7.5L engine

ing gasket. Discard the water pump housing gasket.

To install:

8. Before water pump installation, be sure to aptly clean the water pump mounting surfaces of all dirt, grime and old gasket material.

➡**All water pump housing bolts, nuts and studs are tightened to 15–22 ft. lbs. (20–30 Nm).**

9. Install the water pump onto the engine with a new gasket. Install the water pump stud bolt temporarily finger-tight.

10. Install the water pump mounting

nuts and bolts temporarily finger-tight, then tighten all water pump housing fasteners to 15–22 ft. lbs. (20–30 Nm).

11. Install the heater water outlet tube, if equipped.

12. Install the water pump pulley and accessory drive belt.

13. Install the fan shroud, fan blade assembly and the radiator. Fill the cooling system, then connect the negative battery cable.

Cylinder Head

REMOVAL & INSTALLATION

4.2L Engine

1. The A/C system, should be discharged by a qualified mechanic using an approved refrigerant recovery/recycling machine.

2. Disconnect the negative battery cable.

3. Drain the cooling system.

4. Remove the upper and lower intake manifolds and related components.

5. Remove the rocker arm covers.

6. Remove the exhaust manifold.

7. If removing the left-hand cylinder head, perform the following:

    a. Position the power steering pump reservoir aside and remove the A/C compressor.

    b. Remove and support the A/C compressor bracket and power steering pump aside.

8. If removing the right-hand cylinder head, perform the following:

    a. Remove the alternator.

    b. Remove the idler pulley.

    c. Remove the alternator bracket.

➡ **If the cylinder head components, such as rocker arms, valve springs, etc., are to be reinstalled, they must be installed in the same position. Mark the components for original location.**

9. Remove the six rocker arms by removing the retaining bolts.

10. Pull the pushrods out of the engine. Once again, be sure to label or mark the components removed for reinstallation in their original location.

11. Remove and discard the eight cylinder head mounting bolts. New bolts are a must for installation.

12. Lift the cylinder head off of the engine block. Remove the cylinder head gasket and discard.

To install:

13. Clean and inspect the cylinder head for flatness.

14. Install a new cylinder head gasket on the cylinder block with the small hole to the front of the engine, then install the cylinder head.

✳✳ WARNING

Always use new cylinder head bolts for installation.

15. Lubricate the cylinder head bolts with clean engine oil prior to installation.

➡ **Be sure to tighten the cylinder head bolts in three (3) steps.**

16. Install the new cylinder head bolts. Tighten the cylinder head bolts in the sequence shown and in three steps to the following values:

- Step 1—14 ft. lbs. (20 Nm)
- Step 2—29 ft. lbs. (40 Nm)
- Step 3—36 ft. lbs. (50 Nm)

✳✳ WARNING

Do not loosen all of the cylinder head bolts at one time. Each cylinder head bolt must be loosened and the final tightening performed prior to loosening the next bolt in the sequence.

17. In the same sequence as used previously, loosen the cylinder head bolt three turns, then tighten the cylinder head bolt to the specific value according to its length. The short bolts (A) should be tightened to

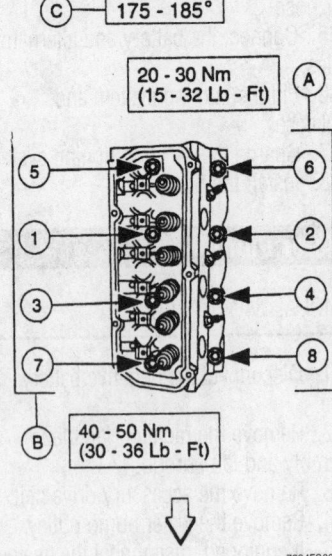

7924FG03

Final-tighten the cylinder head bolts one at a time in the sequence shown here, then tighten an additional 175–185 degrees—4.2L engines

15–32 ft. lbs. (20–30 Nm) and the long bolts (B) to 30–36 ft. lbs. (40–50 Nm). Finally, tighten each cylinder head bolt, in sequence, an additional 175–185 degrees (C). It would be very helpful to utilize a degree socket wrench for this last step.

18. Lubricate the pushrods with clean engine oil prior to installation, then install them into their original positions.

19. Install the rocker arms. Tighten the rocker arm mounting bolts to 23–29 ft. lbs. (30–40 Nm).

20. If the valvetrain components were replaced with new components, inspect the valve clearance.

21. If installing the right-hand cylinder head, perform the following:

    a. Position the alternator bracket in place, then install the two long bolts to 31–39 ft. lbs. (41–54 Nm). Install the short bolt and tighten to 18–22 ft. lbs. (24–31 Nm).

    b. Install the idler pulley. Tighten the center retaining bolt to 35–46 ft. lbs. (47–63 Nm).

    c. Install the alternator.

22. If installing the left-hand cylinder head, complete the following steps:

    a. Position the A/C compressor bracket and power steering pump in place, then start the A/C compressor bracket bolt. Install the three compressor bracket bolts to 30–40 ft. lbs. (40–55 Nm). Then, install the two compressor bracket nuts to 16–21 ft. lbs. (21–29 Nm).

    b. Install the A/C compressor.

    c. Install the power steering pump reservoir. Tighten the hold-down bolts to 80–107 inch lbs. (9–12 Nm).

23. Install the exhaust manifold.

24. Install the two rocker arm covers. Inspect the rocker arm cover gaskets for damage prior to installation; replace them if necessary.

25. Install the lower intake manifold and related components.

26. Install the upper intake manifold and related components.

27. Fill the cooling system and connect the negative battery cable.

28. Start the engine and check for any fuel, coolant and vacuum leaks.

4.6L, 5.4L and 6.8L Engines

➡ **To correctly tighten the cylinder head bolts an angle torque wrench is needed.**

1. The A/C system, should be discharged by a qualified mechanic utilizing

the appropriate refrigerant recovery/recycling machine.

2. Disconnect the negative battery cable.

3. Remove the cylinder head covers.

4. Remove the intake manifold.

5. Remove the timing chains from the engine.

6. Remove the exhaust manifolds.

7. Remove the two heater hose retaining bolts, then compress and slide the hose clamp back to remove the heater water hose.

8. Remove the cylinder head bolts, then lift the cylinder head from the engine block. Discard the cylinder head gasket and clean the engine block surface.

To install:

❋❋ WARNING

Cylinder head bolts must be replaced with new ones. They are torque-to-yield designed and cannot be reused.

9. Turn the crankshaft to position the keyway at the 12 o'clock position.

10. Clean and inspect the cylinder head for damage or warpage. Install the cylinder head gasket over the dowel pins. Then, install the cylinder head onto the engine block. Loosely install NEW cylinder head bolts.

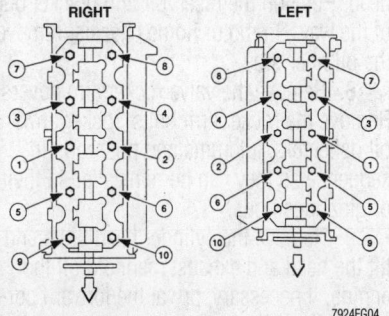

Tighten the cylinder head bolts in three steps using the sequence shown—4.6L and 5.4L engines

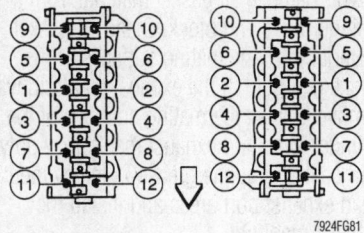

Be sure to tighten the cylinder head bolts in three steps using the sequence shown—6.8L engine

➡**Be sure to tighten the head bolts in three steps.**

11. Tighten the cylinder head bolts in the sequence shown in three steps, as follows:

a. Step 1—27–31 ft. lbs. (37–43 Nm).

b. Step 2—tighten an additional 85–95 degrees.

c. Step 3—tighten another 85–95 degrees.

12. Install the heater water hose and slide the hose clamp back into position. Install the two heater water hose bolts.

13. Install the exhaust manifolds.

14. Install the timing chains.

15. Install the intake manifold.

16. Install the cylinder head covers.

17. Connect the negative battery cable, then start the engine and check for leaks.

18. If the vehicle is equipped with A/C, have the system evacuated and recharged by a qualified mechanic utilizing the appropriate refrigerant recovery/recycling machine.

4.9L Engine

1. Drain the cooling system. Remove the hood.

2. Remove the throttle body inlet tubes.

3. Remove the air conditioning compressor.

4. Remove the condenser.

5. Disconnect the battery ground cable.

6. Disconnect the heater hoses from the water pump and coolant outlet housing.

7. Disconnect the fuel line at the fuel pump.

8. Remove the radiator.

9. Remove the engine fan and fan drive, the water pump pulley and the drive belt.

10. Disconnect the accelerator cable and retracting spring.

11. Disconnect the power brake hose at the manifold.

12. Disconnect the transmission kickdown rod on vehicles with automatic transmission.

13. Disconnect the muffler inlet pipe at the exhaust manifold. Pull the muffler inlet pipe down. Remove the gasket.

14. Disconnect the EEC harness from all the sensors.

15. Tag and disconnect all remaining wiring from the head and related components.

16. Remove the alternator, leaving the wires connected and position it out of the way.

17. Remove the air pump and bracket.

18. Remove the power steering pump and position it out of the way with the hoses still connected.

19. If the vehicle is equipped with an air compressor, bleed the two pressure lines and remove the compressor and bracket.

20. Remove the rocker arm cover.

21. Loosen the rocker arm bolts so they can be pivoted out of the way. Remove the pushrods in sequence so they can be identified and reinstalled in their original positions.

22. Disconnect the spark plug wires at the spark plugs.

23. Remove the cylinder head bolts and remove the cylinder head. Do not pry between the cylinder head and the block as the gasket surfaces maybe damaged.

To install:

24. Clean the head and block gasket surfaces. If the cylinder head was removed for a gasket change, check the flatness of the cylinder head and block.

25. Position the gasket on the cylinder block.

26. Install a new gasket on the flange of the muffler inlet pipe.

27. Lift the cylinder head above the cylinder block and lower it into position using two head bolts installed through the head as guides.

28. Coat the threads of the Nos. 1 and 6 bolts for the right side of the cylinder head with a small mount of water-resistant sealer. Oil the threads of the remaining bolts. Install, but do not tighten, two bolts at the opposite ends of the head to hold the head and gasket in position.

29. The cylinder head bolts are tightened in three progressive steps. Tighten them (in the proper sequence):

a. First tighten the bolts to 50–55 ft. lbs. (68–75 Nm).

b. Then, tighten the cylinder head bolts to 60–65 ft. lbs. (82–88 Nm).

c. Finally, tighten the head bolts to 70–85 ft. lbs. (95–116 Nm).

30. Apply Lubriplate® to both ends of the pushrods and install them in their original positions.

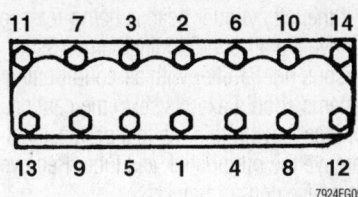

Tighten the cylinder head bolts using three steps in the proper sequence—4.9L engine

31. Apply Lubriplate® to both the fulcrum and seat and position the rocker arms on the valves and pushrods.

32. Adjust the valves.

33. Install the rocker arm cover.

34. Install the air compressor and bracket.

35. Install the power steering pump.

36. Install the air pump and bracket.

37. Install the alternator.

38. Connect all wiring at the head and related components.

39. Connect the EEC harness to all the sensors.

40. Connect the muffler inlet pipe at the exhaust manifold.

41. Connect the transmission kickdown rod on vehicles with automatic transmission.

42. Connect the power brake hose at the manifold.

43. Connect the accelerator cable and retracting spring.

44. Install the water pump pulley, the engine fan and fan drive, and the drive belt.

45. Install the radiator.

46. Attach the fuel line at the fuel pump.

47. Connect the heater hoses at the water pump and coolant outlet housing.

48. Connect the battery ground cable.

49. Install the condenser.

50. Install the air conditioning compressor.

51. Install the throttle body inlet tubes.

52. Fill and bleed the cooling system.

53. Install the hood.

5.0L and 5.8L Engines

1. Drain the cooling system.

2. Remove the intake manifold and throttle body.

3. Remove the rocker arm cover(s).

4. If the right cylinder head is to be removed, lift the tensioner and remove the drive belt. Loosen the alternator adjusting arm bolt and remove the alternator mounting bracket bolt and spacer. Swing the alternator down and out of the way. Remove the air cleaner inlet duct.

If the left cylinder head is being removed, remove the air conditioning compressor. Persons not familiar with air conditioning systems should exercise extreme caution, perhaps leaving this job to a professional. Remove the oil dipstick and tube. Remove the cruise control bracket.

5. Disconnect the exhaust manifold(s) from the muffler inlet pipe(s).

6. Loosen the rocker arm stud nuts so the rocker arms can be rotated to the side. Remove the pushrods and identify them so

they can be reinstalled in their original positions.

7. Disconnect the Thermactor air supply hoses at the check valves. Cover the check valve openings.

8. Remove the cylinder head bolts and lift the cylinder head from the block. Remove the discard the gasket.

To install:

9. Clean the cylinder head, intake manifold, the valve cover and the head gasket surfaces.

10. A specially treated composition head gasket is used. Do not apply sealer to a composition gasket. Position the new gasket over the locating dowels on the cylinder block. Then, position the cylinder head on the block and install the attaching bolts.

11. The cylinder head bolts are tightened in progressive steps. Tighten all the bolts in the proper sequence to:

5.0L (VIN N) Engine
- Step 1—55–65 ft. lbs. (75–89 Nm)
- Step 2—66–72 ft. lbs. (90–98 Nm)

5.8L (VIN H and R) Engine
- Step 1—85 ft. lbs.(115 Nm)
- Step 2—95 ft. lbs. (129 Nm)
- Step 3—105–112 ft. lbs. (143–152 Nm)

12. Clean the pushrods. Blow out the oil passage in the rods with compressed air. Check the pushrods for straightness by rolling them on a piece of glass. Never try to straighten a pushrod; always replace it.

13. Apply Lubriplate® to the ends of the pushrods and install them in their original positions.

14. Apply Lubriplate® to the rocker arms and their fulcrum seats and install the rocker arms. Adjust the valves.

15. Position a new gasket(s) on the muffler inlet pipe(s) as necessary. Connect the exhaust manifold(s) at the muffler inlet pipe(s).

16. If the right cylinder head was removed, install the alternator, and air

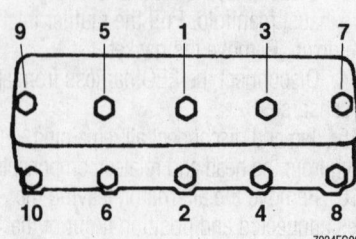

**Cylinder head bolt torque sequence—
5.0L, 5.8L and 7.5L engines**

cleaner duct. Install the drive belt. If the left cylinder head was removed, install the compressor. Install the dipstick and cruise control bracket.

17. Clean the valve rocker arm cover and the cylinder head gasket surfaces. Place the new gaskets in the covers, making sure the tabs of the gasket engage the notches provided in the cover. Evacuate, charge and leak test the air conditioning system.

18. Install the intake manifold and related parts. Install the Thermactor hoses.

19. Fill and bleed the cooling system.

7.5L Engine

1. Drain the cooling system.

2. Remove the upper and lower intake manifolds..

3. Disconnect the exhaust pipe from the exhaust manifold.

4. Loosen the air conditioning compressor drive belt, if equipped.

5. Loosen the alternator attaching bolts and remove the bolt attaching the alternator bracket to the right cylinder head.

6. Disconnect the air conditioning compressor from the engine and move it aside, out of the way. Do not discharge the air conditioning system.

7. Remove the bolts securing the power steering reservoir bracket to the left cylinder head. Position the reservoir and bracket out of the way. On motor home chassis, remove the oil filler tube.

8. Remove the valve rocker arm covers. Remove the rocker arm bolts, rocker arms, oil deflectors, fulcrums and pushrods in sequence so they can be reinstalled in their original positions.

9. Remove the cylinder head bolts and lift the head and exhaust manifold off the engine. If necessary, pry at the forward corners of the cylinder head against the casting bosses provided on the cylinder block. Do not damage the gasket mating surfaces of the cylinder head and block by prying against them.

To install:

10. Remove all gasket material from the cylinder head and block. Clean all gasket material from the mating surfaces of the intake manifold. If the exhaust manifold was removed, clean the mating surfaces of the cylinder head and exhaust manifold. Apply a thin coat of graphite grease to the cylinder head exhaust port areas and install the exhaust manifold.

11. Position the two long cylinder head bolts in the two rear lower bolt holes of the left cylinder head. Place a long cylinder head bolt in the rear lower bolt hole of the

right cylinder head. Use rubber bands to keep the bolts in position until the cylinder heads are installed on the cylinder block.

12. Position new cylinder head gaskets on the cylinder block dowels. Do not apply sealer to the gaskets, heads, or block.

13. Place the cylinder heads on the block, guiding the exhaust manifold studs into the exhaust pipe connections. Install the remaining cylinder head bolts. The longer bolts go in the lower row of holes.

14. Tighten all the cylinder head attaching bolts in the proper sequence in three stages: 80–90 ft. lbs. (109–122 Nm), 100–110 ft. lbs. (136–149 Nm), and finally to 130–140 ft. lbs. (176–190 Nm). When this procedure is used, it is not necessary to retighten the heads after extended use.

15. Be sure the oil holes in the pushrods are open and install the pushrods in their original positions. Place a dab of Lubriplate® to the ends of the pushrods before installing them.

16. Lubricate and install the valve rockers. Be sure the pushrods remain seated in their lifters.

17. Connect the exhaust pipes to the exhaust manifolds.

18. Install the upper and lower intake manifolds.

19. Install the air conditioning compressor.

20. Install the power steering reservoir.

21. Apply oil-resistant sealer to one side of the new valve cover gaskets and lay the cemented side in place in the valve cover. Install the covers.

22. Install the alternator and adjust the drive belt.

23. Adjust the air conditioning compressor drive belt tension.

24. On motor home chassis, install the oil filler tube.

25. Fill and bleed the cooling system.

26. Start the engine and check for leaks.

Rocker Arms

REMOVAL & INSTALLATION

4.2L Engine

➡**If removing more than one rocker arm, mark the components for proper location.**

1. Disconnect the negative battery cable.
2. Remove the lower intake manifold.
3. Remove the rocker arm cover.
4. Remove the rocker arm hold-down bolt, then remove the rocker arm from the cylinder head.

To install:

5. Position the rocker arms in place, then install the hold-down bolts. Tighten the bolts to 23–29 ft. lbs. (30–40 Nm).

6. Install the rocker arm cover and the lower intake manifold.

7. Connect the negative battery cable.

4.6L and 5.4L Engines

1. Disconnect the negative battery cable.
2. Remove the camshaft covers.
3. Position the cylinder being serviced at the bottom of it's travel.

➡**Two different valve spring compressor tools are used for this procedure. Valve Spring Compressor (T91P-6565-A) is used on the exhaust camshaft and Valve Spring Compressor (T93P-6565-A) is used on the intake camshaft.**

4. Compress the valve spring and remove the rocker arm.

To install:

5. Position the cylinder being serviced at the bottom of it's travel.

6. Apply clean engine oil to the rocker arm, valve stem tip and tappet bore.

➡**Valve tappet should have no more than 1/16 inch (1.5mm) of travel before installing the rocker arm.**

7. Compress the valve spring using the correct tool and install the rocker arm.

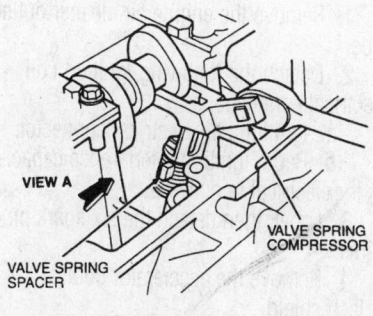

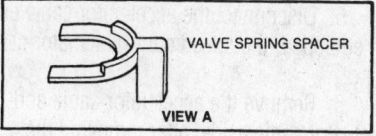

Using the proper tool, compress the valve spring and remove the rocker arm—4.6L and 5.4L engines

4.9L Engine

1. Disconnect the inlet hose at the crankcase filler cap.

2. Remove the throttle body inlet tubes.

3. Disconnect the accelerator cable at the throttle body. Remove the cable retracting spring. Remove the accelerator cable bracket from the upper intake manifold and position the cable and bracket out of the way.

4. Remove the fuel line from the fuel rail. Be careful not to kink the line.

5. Remove the upper intake manifold and throttle body assembly.

6. Remove the ignition coil and wires.

7. Remove the rocker arm cover.

8. Remove the spark plug wires.

9. Remove the distributor cap.

10. Remove the pushrod cover (engine side cover).

11. Remove the rocker arm bolts, then lift the rocker arms off of the cylinder head; keep the rocker arms in order for installation.

To install:

12. Engage the rocker arms with the pushrods and tighten the rocker arm bolts enough to hold the pushrods in place.

13. Adjust the valve clearance.

14. Install the pushrod cover (engine side cover).

15. Install the distributor cap.

16. Install the spark plug wires.

17. Install the rocker arm cover.

18. Install the ignition coil and wires.

19. Install the upper intake manifold and throttle body assembly.

20. Install the fuel line at the fuel rail.

21. Install the accelerator cable bracket on the upper intake manifold. Install the cable retracting spring. Attach the accelerator cable at the throttle body.

22. Install the throttle body inlet tubes.

23. Connect the inlet hose at the crankcase filler cap.

5.0L and 5.8L Engines

1. Remove the intake manifold.

2. Disconnect the Thermactor air supply hose at the pump.

3. Remove the rocker arm covers.

4. Loosen the rocker arm fulcrum bolts, fulcrum seats and rocker arms; keep all parts in order for installation.

To install:

5. Apply multipurpose grease to the valve stem tips, the fulcrum seats and sockets.

6. Install the fulcrum guides, rocker arms, seats and bolts. Tighten the bolts to 18–25 ft. lbs. (25–34 Nm).

7. Install the rocker arm covers.

8. Connect the Thermactor air supply hose at the pump.

9. Install the intake manifold.

6.8L Engine

1. Disconnect the negative battery cable.

2. Remove the camshaft covers.

3. Position the base circle of the camshaft lobe on the rocker arm to be serviced. Also, be sure the piston is not at the top of it's travel near the valve.

4. Compress the valve spring using Valve Spring Compressor 303–381(TJ91P-6565-A) and remove the rocker arm.

To install:

5. Position the base circle of the camshaft lobe over the place where the rocker arm is to be installed.

6. Apply clean engine oil to the rocker arm, valve stem tip and tappet bore.

7. Compress the valve using the special tool and install the rocker arm.

8. Install the rocker arm covers and the remaining components.

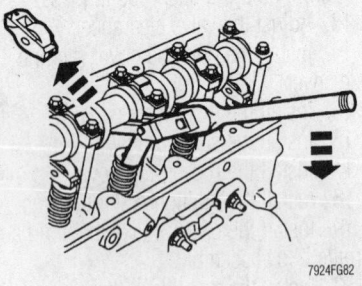

Compress the valve spring and remove the rocker arm—6.8L engine

7.5L Engine

1. Remove the intake manifold.

2. Remove the rocker arm covers.

3. Loosen the rocker arm fulcrum bolts, fulcrum, oil deflector, seat and rocker arms; keep everything in order for installation.

To install:

4. Coat each end of each pushrod with multipurpose grease.

5. Coat the top of the valve stems, the rocker arms and the fulcrum seats with multipurpose grease.

6. Rotate the crankshaft by hand until No. 1 piston is at TDC of compression. The firing order marks on the damper will be aligned at TDC with the timing pointer.

7. Install the rocker arms, seats, deflectors and bolts on the following valves:

- No. 1 intake and exhaust
- No. 3 intake
- No. 8 exhaust
- No. 7 intake
- No. 5 exhaust
- No. 8 intake
- No. 4 exhaust

8. Engage the rocker arms with the pushrods and tighten the rocker arm fulcrum bolts to 18–25 ft. lbs. (25–34 Nm).

9. Rotate the crankshaft one full turn—360 degrees—and realign the TDC mark and pointer. Install the parts and tighten the bolts on the following valves:

- No. 2 intake and exhaust
- No. 4 intake
- No. 3 exhaust
- No. 5 intake
- No. 6 intake and exhaust
- No. 7 exhaust

10. Install the rocker arm covers.

11. Install the intake manifold.

12. Check the valve clearance as described under Hydraulic Valve Clearance, below.

Intake Manifold

REMOVAL & INSTALLATION

➡When the battery is disconnected and reconnected, some abnormal drive symptoms may occur while the vehicle relearns its adaptive strategy. The vehicle may need to be driven 10 miles (16 km) or more to relearn the strategy.

4.2L Engine

1. Remove the engine air cleaner outlet tube.

2. Detach the following ignition coil electrical connections:

   a. Ignition coil electrical connector.

   b. Radio ignition interference capacitor electrical connector.

3. Label, then detach the six spark plug wires.

4. Remove the accelerator control splash shield.

5. Disconnect the accelerator cable end, if equipped, the speed control actuator cable end.

6. Remove the accelerator cable and actuator cable aside, after removing the hold-down bolts.

7. Disconnect the Vapor Management Valve (VMV) hose.

8. Disconnect the brake booster vacuum hose.

9. Disengage the manifold vacuum connection.

10. Remove the PCV valve from the rocker arm cover.

11. Position the Engine Vacuum Regulator (EVR) bracket aside.

12. Detach the throttle position sensor and idle air control valve electrical connectors.

13. Remove the breather from the rocker arm cover.

14. Remove the 12 upper intake manifold retaining bolts, then lift the manifold off of the engine. Discard the intake manifold upper gasket.

15. Detach the six fuel injector electrical connectors.

16. Detach the engine coolant temperature sensor and the water temperature indicator sending unit electrical connectors.

17. Disconnect the EGR valve vacuum hose.

18. Remove the EGR valve tube upper fitting.

19. Remove the radiator hose from the lower intake manifold.

20. Position the Intake Manifold Runner Control (IMRC) actuator brackets aside.

21. Disconnect the fuel pressure regulator vacuum line.

22. Disconnect the fuel lines.

23. Disconnect the water pump bypass hose.

➡Remove the lower intake manifold with the fuel injection supply manifold and fuel injectors as one unit.

24. Remove the six long bolts and the eight short bolts, then lift the lower intake manifold off of the engine.

25. Remove and discard the lower intake manifold sealing components.

To install:

26. Clean all components of dirt, grease and old gasket material.

27. Install the lower intake manifold front and rear end seals as follows:

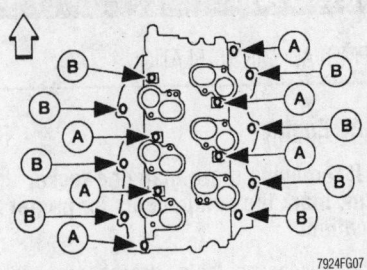

Make certain to install the long bolts (A) and the short bolts (B) in the correct lower intake manifold holes—4.2L engine

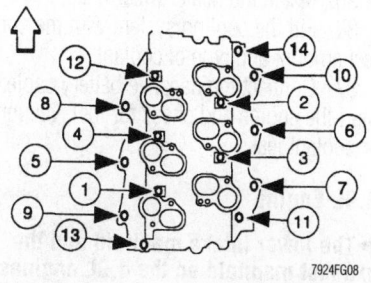

Tighten the lower manifold bolts in the sequence shown—4.2L engine

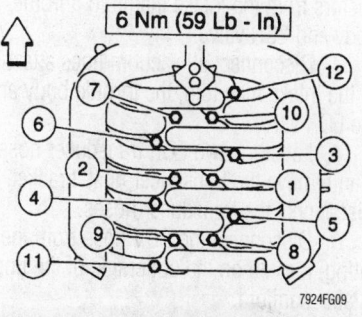

Tighten the upper intake manifold mounting bolts in the sequence shown—4.2L engine

a. Apply a bead of sealant (Silicone Gasket and Sealant F6AZ-19562-A or equivalent) to the intake manifold front and rear end seal mounting points.

b. Install the lower intake manifold front and rear end seals.

28. Install new lower intake manifold gaskets onto the cylinder heads.

➡**The lower intake manifold must be installed within 15 minutes of applying sealant.**

29. Apply a bead of the same sealant to the end of the lower intake manifold end seals, where they stop on the cylinder head surface. Position the intake manifold onto the engine block and cylinder heads.

30. Install the lower intake manifold mounting bolts in the correct positions. Refer to the illustration for the correct placement of the long (A) and the short (B) mounting bolts.

➡**Be sure to tighten the intake manifold bolts in two steps.**

31. Tighten the lower intake manifold mounting bolts in the sequence shown, first to 44 inch lbs. (5 Nm), then to 71–101 inch lbs. (8.0–11.5 Nm).

32. Connect the water bypass hose.
33. Attach the fuel lines.
34. Connect the fuel pressure regulator vacuum line.
35. Install the IMRC actuators. Tighten the bolts on the brackets to 71–102 inch lbs. (8.0–11.5 Nm).
36. Install the upper radiator hose to the lower intake manifold.
37. Connect the EGR valve vacuum hose.
38. Install the EGR tube upper fitting to 25–34 ft. lbs. (37–47 Nm).
39. Attach the engine coolant temperature sensor and water temperature indicator sending unit electrical connectors.
40. Install the six fuel injector electrical connectors.
41. Install a new intake manifold upper gasket.

➡**Be sure to tighten the upper intake manifold bolts in the sequence shown.**

42. Position the upper intake manifold onto the lower intake manifold, then tighten the upper intake manifold bolts in the sequence shown to 59 inch lbs. (6 Nm). Tighten the upper intake manifold bolts to 6–8 ft. lbs. (8.0–11.5 Nm) in the sequence shown.
43. Install the breather into the rocker arm cover.
44. Attach the throttle position sensor and idle air control valve electrical connectors.
45. Install the Engine Vacuum Regulator (EVR) bracket.
46. Attach the manifold vacuum connection.
47. Install the PCV valve.
48. Connect the brake booster vacuum hose.
49. Connect the Vapor Management Valve (VMV) hose.
50. Install the accelerator and speed actuator cables.
51. Install the accelerator control splash shield.
52. Install the spark plug wires.
53. Attach the ignition coil and the radio ignition interference capacitor electrical connectors.
54. Install the engine air cleaner outlet tube.

4.6L, 5.4L and 6.8L Engine

1. Disconnect the negative battery cable.
2. Relieve the fuel system pressure.
3. Drain the cooling system.
4. Disconnect the upper radiator hose from the intake manifold.
5. Remove the engine air cleaner outlet tube.
6. Disconnect the accelerator cable from the bracket and the throttle body cam.
7. If equipped, remove the speed control actuator cable from the throttle body.
8. Disconnect all vacuum hoses, fuel lines and electrical wires from the throttle body and intake manifold.
9. Remove the brake booster vacuum hose bracket.
10. Remove the EGR valve-to-exhaust manifold tube.
11. Detach the fuel injector electrical connectors.
12. On the 6.8L engine, remove the radio interference capacitors from the left side of the intake manifold.
13. Label, then remove the spark plug wires, if necessary.
14. Remove the accessory drive belt.
15. Remove the alternator.
16. Remove the power steering oil reservoir bracket and set aside.
17. Disconnect the heater water hose from the intake manifold.
18. Remove the intake manifold bolts.
19. Lift the intake manifold off of the engine, then detach the Intake Manifold Tuning Valve (IMTV) electrical connector. Remove and discard the upper intake manifold gaskets.
20. Remove the upper-to-lower intake manifold bolts, then separate the upper intake manifold from the lower intake manifold. Discard the old gasket.

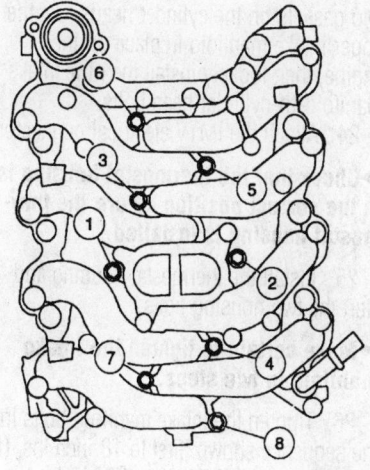

Tighten the lower-to-upper intake manifold bolts in two steps following the sequence shown—4.6L engine shown, 5.4L engine similar

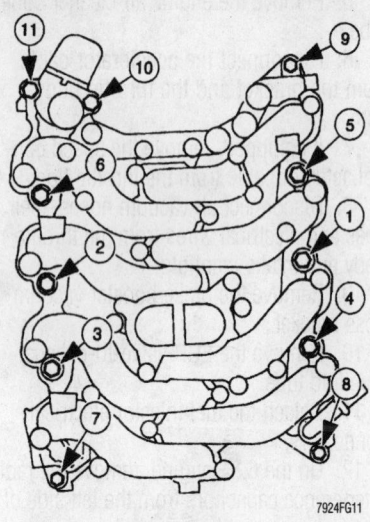

Tighten the upper intake manifold-to-cylinder head mounting bolts in the sequence shown—4.6L engine shown, 5.4L engine similar

To install:

21. Position the lower intake manifold gasket and the upper intake manifold onto the lower intake manifold, then loosely install the upper-to-lower intake manifold bolts.

➡Be sure to tighten the lower-to-upper manifold bolts in two steps.

22. Tighten the eight lower-to-upper intake manifold bolts in two steps following the tightening sequence shown. The first step should be to 18 inch lbs. (2 Nm) and the second step to 6–8 ft. lbs. (8–12 Nm).

23. Position the two upper intake manifold gaskets on the cylinder heads. Set the upper intake manifold in place on the engine, then loosely install the nine intake manifold-to-cylinder head bolts.

24. Attach the IMTV electrical connector.

➡Check that the thermostat housing is in the correct position before the thermostat housing is installed.

25. Install the thermostat housing and start the two housing bolts.

➡Make certain to tighten the intake manifold in two steps.

26. Tighten the intake manifold bolts in the sequence shown, first to 18 inch lbs. (2 Nm), then to 15–22 ft. lbs. (20–30 Nm).

27. Install the heater water hose.

28. Position the power steering bracket and install the power steering pump bracket bolts to 71–107 inch lbs. (8–12 Nm).

29. Attach all electrical connections, fuel

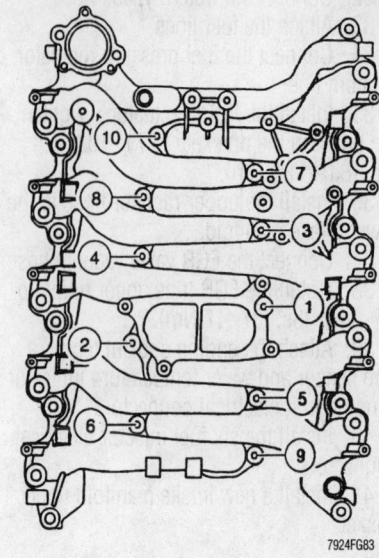

Tighten the upper-to-lower intake manifold bolts in the sequence shown—6.8L engine

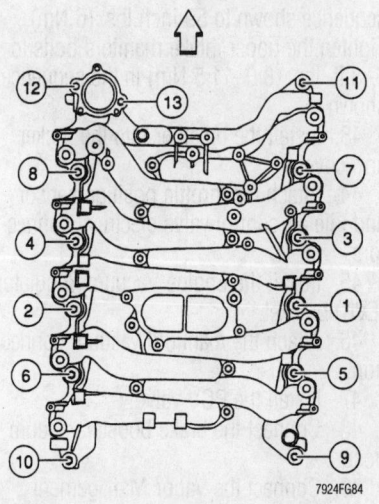

Tighten the upper intake manifold-to-cylinder head bolts using the sequence shown—6.8L engine

lines, vacuum tubes and coolant hoses to the intake manifold, fuel injectors and throttle body assembly.

30. Install the alternator and the accessory drive belt.

31. Install the spark plug wires.

32. Install the EGR valve-to-exhaust manifold tube. The tube fittings should be tightened to 26–33 ft. lbs. (35–45 Nm).

33. Install the speed actuator cable, if equipped, and the accelerator cable to the throttle body.

34. Install the engine air cleaner outlet tube.

35. Install the heater water hose.

36. Fill the cooling system with the correct amount and type of coolant.

37. Connect the negative battery cable. Start the engine and check for fuel, vacuum or coolant leaks.

4.9L Engine

➡The lower intake manifold and the exhaust manifold on the 4.9L engines is considered a combination manifold. Only the upper intake manifold is covered here.

1. Disconnect the negative battery cable.

2. Label, then detach all electrical connectors from the intake manifold, throttle body and EGR valve.

3. Disconnect all vacuum lines attached to the intake manifold, the throttle body and the EGR valve.

4. Label the two EGR transducer hoses, then remove the transducer hose bracket. Disconnect the transducer hoses.

5. Disconnect the PCV hose from the fitting, located on the underside of the upper intake manifold.

✳✳ WARNING

When disconnecting the throttle cable from the ball stud, use a prytool or similar tool close to the ball stud to pry it off. Removing it by hand may damage the cable.

6. Remove the accelerator control splash shield, then disconnect the accelerator cable and speed actuator cable from the throttle body. Position the cables away from the engine.

7. Disconnect the air cleaner outlet tube from the throttle body.

8. Remove the EGR valve-to-exhaust manifold tube.

9. Remove the bolt and washer attaching the intake manifold support to the upper intake manifold.

10. Remove the seven studs that retain the upper intake manifold.

11. Remove the upper intake manifold and throttle body from the lower intake manifold.

To install:

12. Position a new intake manifold upper gasket on the lower intake manifold, using the lower manifold dowels to position the intake manifold upper gasket.

13. Position the upper intake manifold onto the lower intake manifold. Install the seven studs to attach the upper manifold to the lower manifold hand-tight.

14. Tighten the seven upper manifold-to-lower manifold studs to 12–18 ft. lbs. (16–24 Nm).

15. Install the upper manifold support and tighten the mounting bolt to 22–32 ft. lbs. (30–43 Nm).

16. Install the EGR valve-to-exhaust manifold tube. The tube is routed between lower intake manifold runners No. 5 and No. 6. Tighten both of the fittings to 25–35 ft. lbs. (34–47 Nm).

17. Attach the PCV hose to the fitting, located on the valve cover under the upper intake manifold.

18. Attach the accelerator cable, transmission kickdown cable, and speed control actuator cable. Install the accelerator control splash shield.

19. Connect the air cleaner outlet tube to the throttle body.

20. Install the EGR transducer and bracket. Tighten the bolts to 12–18 ft. lbs. (17–24 Nm). Connect the EGR transducer hoses.

21. Reattach all of the vacuum lines removed from the throttle body, the intake manifold and the EGR valve assemblies. Also install all of the electrical connections to the same components.

22. Connect the negative battery cable.

5.0L, 5.8L and 7.5L Engines

➡ **Relieve the fuel system pressure before starting any work that involves disconnecting fuel system lines.**

UPPER INTAKE MANIFOLD

1. Remove the air cleaner. Disengage the electrical connectors at the air bypass valve, throttle position sensor and EGR position sensor.

2. Disconnect the throttle linkage at the throttle ball and the AOD transmission linkage from the throttle body. Remove the bolts that secure the bracket to the intake and position the bracket and cables out of the way.

3. Disengage the upper manifold vacuum fitting connections by removing all the vacuum lines at the vacuum tree (label lines for position identification). Remove the vacuum lines to the EGR valve and fuel pressure regulator.

4. Disengage the PCV system by disconnecting the hose from the fitting at the rear of the upper manifold.

5. Remove the two canister purge lines from the fittings at the throttle body.

6. Disconnect the EGR tube from the EGR valve by loosening the flange nut.

7. Remove the bolt from the upper

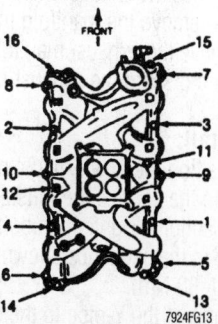

Tighten the intake manifold bolts according to the sequence shown—7.5L engine

intake support bracket to upper manifold. Remove the upper manifold retaining bolts and remove the upper intake manifold and throttle body as an assembly.

8. Clean and inspect all mounting surfaces of the upper and lower intake manifolds.

To install:

9. Position a new mounting gasket on the lower intake manifold.

10. Install the upper intake manifold and throttle body as an assembly. Install the upper manifold retaining bolts and install the bolt at the upper intake support bracket. Mounting bolts are tightened to 12–18 ft. lbs. (16–25 Nm).

11. Connect the EGR tube at the EGR valve.

12. Install the two canister purge lines at the fittings at the throttle body.

13. Connect the PCV system hose at the fitting at the rear of the upper manifold.

14. Connect the upper manifold vacuum lines at the vacuum tree. Install the vacuum lines at the EGR valve and fuel pressure regulator.

15. Install the throttle bracket on the intake manifold. Attach the throttle linkage at the throttle ball and the AOD transmission linkage at the throttle body.

16. Attach the electrical connectors at the air bypass valve, throttle position sensor and EGR position sensor.

17. Install the air cleaner.

LOWER INTAKE MANIFOLD

1. Upper manifold and throttle body must be removed first.

2. Drain the cooling system.

3. Remove the distributor assembly, cap and wires.

4. Disengage the electrical connectors at the engine, coolant temperature sensor and sending unit, at the air charge temperature sensor and at the knock sensor.

5. Disconnect the injector wiring har-

ness from the main harness assembly. Remove the ground wire from the intake manifold stud. The ground wire must be installed at the same position it was removed from.

6. Disconnect the fuel supply and return lines from the fuel rails.

7. Remove the upper radiator hose from the thermostat housing. Remove the bypass hose. Remove the heater outlet hose at the intake manifold.

8. Remove the air cleaner mounting bracket. Remove the intake manifold mounting bolts and studs. Pay attention to the location of the bolts and studs for reinstallation. Remove the lower intake manifold assembly.

To install:

9. Clean and inspect the mounting surfaces of the heads and manifold.

10. Apply a 1/16 in. (1.5mm) bead of RTV sealer to the ends of the manifold seal (the junction point of the seals and gaskets). Install the end seals and intake gaskets on the cylinder heads. The gaskets must interlock with the seal tabs.

11. Install locator bolts at opposite ends of each head and carefully lower the intake manifold into position. Install and tighten the mounting bolts and studs to 23–25 ft. lbs. (31–34 Nm).

12. Install the lower intake manifold assembly. Install the intake manifold mounting bolts and studs. Pay attention to the location of the bolts. Install the air cleaner mounting bracket.

13. Install the heater outlet hose at the intake manifold.

14. Install the bypass hose.

15. Install the upper radiator hose.

16. Connect the fuel supply and return lines at the fuel rails.

17. Connect the injector wiring harness from the main harness assembly. Install the ground wire from the intake manifold stud.

18. Attach the electrical connectors at

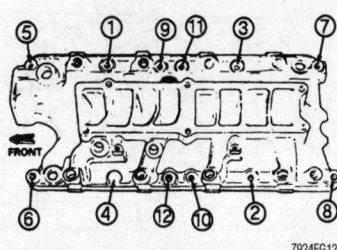

Lower intake manifold bolt tightening sequence—5.0L and 5.8L engines

the engine, coolant temperature sensor and sending unit, at the air charge temperature sensor and at the knock sensor.

19. Install the distributor assembly, cap and wires.

20. Fill the cooling system.

Exhaust Manifold

REMOVAL & INSTALLATION

4.2L Engine

1. Disconnect the negative battery cable.

2. For the right-hand manifold, remove the EGR valve-to-exhaust manifold tube.

3. For the left-hand manifold, remove the oil level indicator tube bracket nut, then remove the oil level indicator tube. Remove and discard the oil level indicator tube O-ring.

4. Raise and safely support the front of the vehicle.

5. Detach the heated oxygen sensor electrical connector.

6. Remove the two catalytic converter-to-exhaust manifold nuts, then disconnect the Y-pipe from the left-hand exhaust manifold.

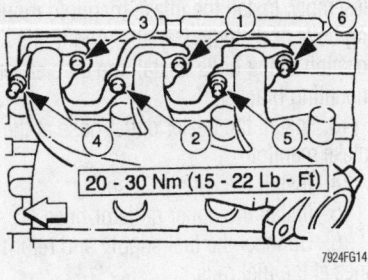

Tighten the left-hand exhaust manifold bolts in the order shown—4.2L engine

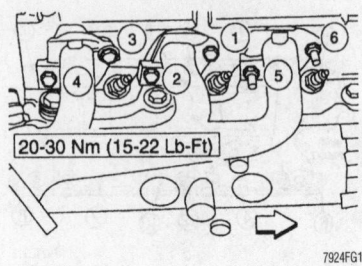

Tighten the right-hand exhaust manifold bolts in the order shown—4.2L engine

7. Remove the exhaust manifold stud bolts, then remove the manifold mounting bolts. Remove the exhaust manifold. Remove and discard the exhaust manifold gasket.

To install:

8. Position the new exhaust manifold gasket onto the engine, then install the exhaust manifold. Tighten the bolts and stud bolts in the sequence shown to 15–22 ft. lbs. (20–30 Nm).

9. Connect the Y-pipe to the exhaust manifold, then install and tighten the catalytic converter nuts to 25–34 ft. lbs. (34–46 Nm).

10. Attach the oxygen sensor connector, then lower the vehicle.

11. For the left-hand exhaust manifold, install a new oil level indicator tube O-ring onto the tube. Insert the tube into the engine block and tighten the bracket retaining nut to 15–22 ft. lbs. (20–30 Nm).

12. For the right-hand exhaust manifold, Install the EGR valve-to-exhaust manifold tube. Tighten the upper and lower fittings to 25–34 ft. lbs. (34–47 Nm).

13. Connect the negative battery cable.

4.6L, 5.4 and 6.8L Engines

1. Raise and safely support the vehicle.

2. Remove the front fender splash shield.

3. For the left-hand exhaust manifold, remove the EGR valve-to-exhaust manifold tube and if equipped, the Differential Pressure Feedback Exhaust (DPFE) gas recirculation transducer hoses.

4. On the 4.6L and 5.4L engines, remove the catalytic converter-to-exhaust manifold bolts. On the 6.8L engine, remove the front exhaust pipe from the manifold.

5. Remove the exhaust manifold mounting nuts, then remove the exhaust manifold itself.

6. Remove and discard the old gasket.

7. Clean and inspect the exhaust manifold for damage.

To install:

8. Position a new gasket and the exhaust manifold onto the engine block. Install the mounting nuts and tighten, in the sequence shown, to 13–16 ft. lbs. (18–22 Nm) on the 4.6L and 5.4L engines. Tighten the nuts to 17–20 ft. lbs. (23–27 Nm) on the 6.8L engine.

9. On the 6.8L engine, tighten the exhaust manifold-to-front pipe fasteners to 27–34 ft. lbs. (34–46 Nm).

10. On the 4.6L and 5.4L engines, attach the catalytic converter to the exhaust manifold, then install the catalytic converter-to-

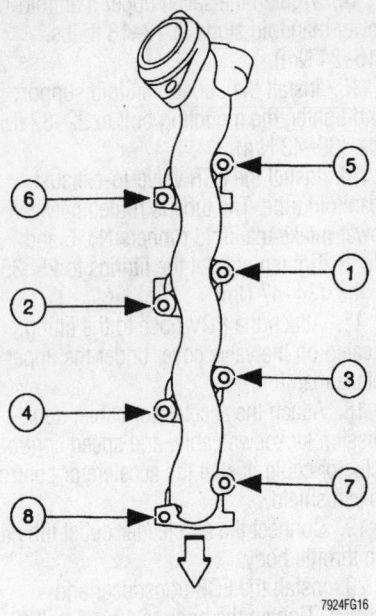

Tighten the exhaust manifold bolts in the sequence shown—4.6L engine shown, 5.4L engine similar

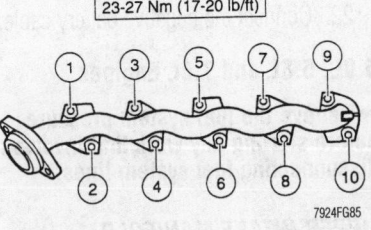

23-27 Nm (17-20 lb/ft)

Tighten the exhaust manifold bolts in the sequence shown—right side of 6.8L engine shown

exhaust manifold bolts and tighten to 25–34 ft. lbs. (34–46 Nm).

11. For the left-hand exhaust manifold, install the DPFE transducer hoses if equipped, and the EGR valve-to-exhaust manifold tube. Tighten the upper and lower fittings to 26–33 ft. lbs. (35–45 Nm).

12. Install the front fender splash shield.

13. Lower the vehicle to the ground.

5.0L, 5.8L and 7.5L Engines

1. On the 5.0L (VIN N) engine, remove the dipstick bracket.

2. Disconnect the exhaust pipe or catalytic converter from the exhaust manifold. Remove and discard the doughnut gasket.

3. Remove the exhaust manifold attaching screws and remove the manifold from the cylinder head.

To install:

4. Apply a light coat of graphite grease to the mating surface of the manifold. Install and tighten the attaching bolts, starting from the center and working to both ends alternately. Tighten to the proper specifications.

5. Install the exhaust pipe or catalytic converter to the exhaust manifold using a new doughnut gasket.

6. If necessary, install the dipstick bracket.

Combination Manifold

REMOVAL & INSTALLATION

4.9L Engine

➡The lower intake and exhaust manifolds on these engines are known as combination manifolds and are serviced as a unit.

1. Remove the air inlet hose at the crankcase filter cap.

2. Remove the throttle body inlet hoses.

3. Disconnect the accelerator cable at the throttle body.

4. Remove the cable retracting spring.

5. Remove the cable bracket from the upper intake manifold.

6. Disconnect the fuel inlet line at the fuel rail. Don't kink the line!

7. Remove the upper intake and throttle body as an assembly.

8. Tag and disconnect all vacuum lines attached to the parts in question.

9. Disconnect the inlet pipe from the exhaust manifold.

10. Disconnect the power brake vacuum line, if equipped.

11. Remove the bolts and nuts attaching the manifolds to the cylinder head. Lift the manifold assemblies from the engine. Remove and discard the gaskets.

12. To separate the manifold, remove the nuts joining the intake and exhaust manifolds.

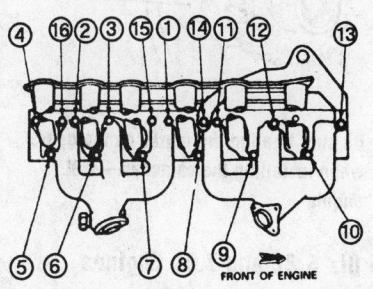

Combination manifold bolt torque sequence—4.9L engine

To install:

13. Clean the mating surfaces of the cylinder head and the manifolds.

14. If the intake and exhaust manifolds have been separated, coat the mating surfaces lightly with graphite grease and place the exhaust manifold over the studs on the intake manifold. Install the lockwashers and nuts. Tighten them finger-tight.

15. Install a new intake manifold gasket.

16. Coat the mating surfaces lightly with graphite grease. Place the manifold assemblies in position against the cylinder head. Be sure the gaskets have not become dislodged. Install the attaching nuts and bolts in the proper sequence to 26 ft. lbs. (35 Nm). If the intake and exhaust manifolds were separated, tighten the nuts joining them.

17. Position a new gasket on the muffler inlet pipe and connect the inlet pipe to the exhaust manifold.

18. Connect the crankcase vent hose to the intake manifold inlet tube and position the hose clamp.

19. Connect the power brake vacuum line, if equipped.

20. Connect the inlet pipe at the exhaust manifold.

21. Attach all vacuum lines.

22. Install the upper intake and throttle body as an assembly.

23. Connect the fuel inlet line at the fuel rail.

24. Install the accelerator cable bracket at the upper intake manifold.

25. Install the cable retracting spring.

26. Attach the accelerator cable at the throttle body.

27. Install the throttle body inlet hoses.

28. Install the air inlet hose at the crankcase filter cap.

Camshaft and Valve Lifters

REMOVAL & INSTALLATION

4.2L Engine

1. Disconnect the negative battery cable.

2. Remove the lower intake manifold.

3. Remove the rocker arm cover.

4. Remove the rocker arm hold-down bolt, then remove the rocker arm from the cylinder head.

5. Remove the pushrods.

6. Remove the valve lifters by pulling them up out of their bores.

7. Remove the timing chain and sprockets.

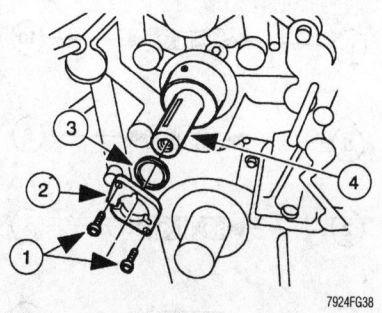

Exploded view of the camshaft retaining hardware—4.2L engine

8. Remove the camshaft key from the end of the camshaft, then slide the engine dynamic balance shaft drive gear off of the camshaft.

9. Remove the two camshaft thrust plate retaining bolts (1), then remove the thrust plate (2). Remove the camshaft spacer (3), then slide the camshaft (4) out of the front of the engine block. Be cautious not to gouge or scratch the camshaft bearing journals.

To install:

10. Lubricate the camshaft with engine oil prior to installation.

11. Carefully slide the camshaft into the camshaft bore. Do not scratch the bearing surfaces.

12. Install the camshaft thrust plat with the spacer. Tighten the thrust plate mounting bolts to 6–10 ft. lbs. (8–14 Nm).

13. Slide the engine dynamic balance shaft drive gear onto the camshaft. Install the camshaft key to the camshaft groove.

14. Install the timing chain and sprockets.

15. Install the valve lifters, pushrods, intake manifolds and rocker arm covers.

4.6L and 5.4L Engines

1. Remove the cylinder head covers from the engine.

2. Remove the timing chain.

✻✻ CAUTION

At no time, when the timing chains are removed and the cylinder heads are installed may the crankshaft or camshaft be rotated. Severe piston and valve damage will occur.

3. On the 6.8L engine, remove the six bolts securing the balance shaft to the cylinder head and remove the shaft.

4. Remove the camshaft roller lifters.

5. On VIN W engines, remove the tim-

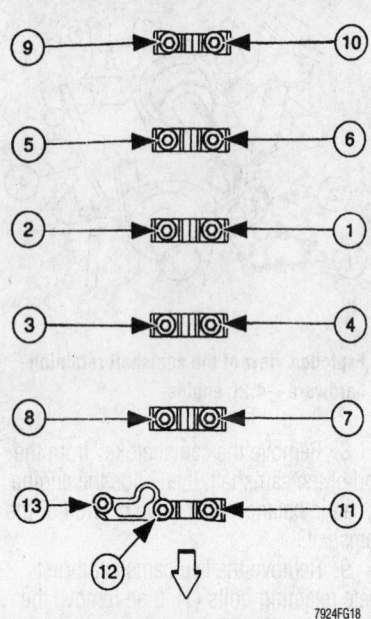

7924FG18

Tighten the bearing caps in the sequence shown—4.6L and 5.4L engines

8-12 Nm (71-106 lb/in)

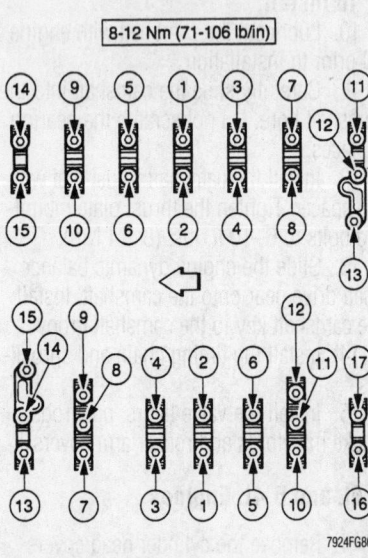

7924FG86

Camshaft bearing cap bolt tightening sequence—6.8L engine

ing chain camshaft gear by removing the gear retaining bolt.

➡**Keep the bearing caps in order so they can be installed in the same position.**

6. Remove the camshaft bearing cap bolts, then lift the camshaft bearing caps off of the cylinder head.

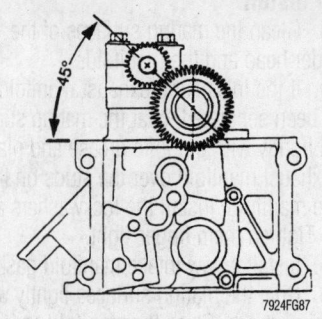

7924FG87

Be sure to align the balance shaft timing mark with the mark on the camshaft gear—6.8L engine

8-12 Nm (71-106 lb/in)

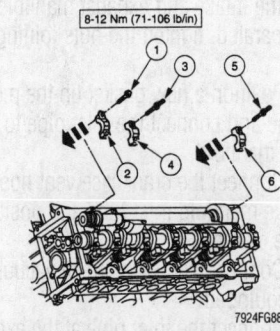

7924FG88

Tighten the balance shaft bearing cap bolts in the sequence shown—6.8L engine

7. Lift the camshaft from the cylinder head.

8. Remove the rocker arms and pull the lash adjusters out of their bores. Keep all the parts in order. They must be installed in their original positions.

To install:

9. Install the lash adjusters and rocker arms in their original positions.

10. Lubricate the camshaft journals and bearing caps with Super Premium SAE 5W30 engine oil which meets Ford specifications WSS-M2C153-G. On the 6.8L engine, lubricate the balance shaft journals and bearing caps with the same lubricant.

11. Lower the camshaft onto the camshaft bearing journals.

12. Install the camshaft bearing caps, then loosely install the bearing cap bolts.

13. Tighten the camshaft bearing cap mounting bolts, in the sequence shown for the particular engine, to 71–107 inch lbs. (8–12 Nm).

14. On the 6.8L engine, align the timing marks and position the balance shaft on the journals, then install the bearing caps. Tighten the bolts in sequence to 71–106 in lbs. (8–12 Nm).

15. On VIN W engines, install the

camshaft timing chain gear by tightening the retaining bolt to 81–95 ft. lbs. (110–130 Nm).

16. Install the valve lifters.

17. Install the timing chain and sprockets, if applicable.

18. Install the cylinder head covers.

4.9L Engine

1. Remove the grille, radiator, air conditioner condenser, and timing cover.

2. Remove the distributor, fuel pump, oil pan and oil pump.

3. Align the timing marks. Unbolt the camshaft thrust plate, working through the holes in the camshaft gear.

4. Loosen the rocker arms, remove the pushrods, take off the side cover and remove the valve lifters with a magnet.

5. Remove the camshaft very carefully to prevent nicking the bearings.

To install:

6. Oil the camshaft bearing journals and use Lubriplate® or something similar on the lobes. Install the camshaft, gear, and thrust plate, aligning the gear marks. Tighten down the thrust plate. Be sure the camshaft end-play is not excessive.

7. Install the rocker arms, pushrods and valve lifters.

8. Install the camshaft timing gear.

9. Install the oil pump, oil pan and fuel pump.

10. Install the distributor. The rotor should be at the firing position for No. 1 cylinder, with the timing gear marks aligned.

11. Install the front engine cover, A/C condenser, the radiator and the grille.

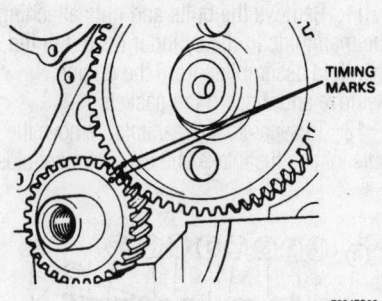

7924FG39

Be sure to align the marks on the gears when installing the camshaft—4.9L engine

5.0L, 5.8L and 7.5L Engines

1. Remove the intake manifold and valley pan, if equipped.

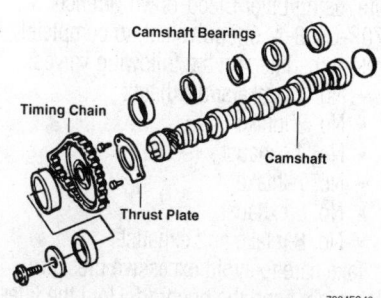

Camshaft and related components—5.0L, 5.8L and 7.5L engines

2. Remove the rocker covers, and loosen the rockers on their pivots and remove the pushrods. The pushrods must be reinstalled in their original positions.

3. Remove the valve lifters in sequence with a magnet. They must be replaced in their original positions.

4. Remove the timing gear cover, timing chain and sprockets.

5. In addition to the radiator and air conditioning condenser, if equipped, it may be necessary to remove the front grille assembly and the hook lock assembly to gain the necessary clearance to code the camshaft out of the front of the engine.

➡A camshaft removal tool, Ford part no. T65L-6250-A and adapter 14-0314 are needed to remove the Diesel camshaft.

To install:

6. Coat the camshaft liberally with clean engine oil before installing it. Slide the camshaft into the engine very carefully so as not to scratch the bearing bores with the camshaft lobes. Install the camshaft thrust plate and tighten the attaching screws to 9–12 ft. lbs. (12–16 Nm). Measure the camshaft end-play. If the end-play is more than 0.009 in. (0.228mm), replace the thrust plate. Assemble the remaining components in the reverse order of removal.

7. Install the radiator, front grille, hood lock assembly and air conditioning condenser, if removed.

8. Install the timing chain and front cover.

9. Install the valve lifters. They must be replaced in their original positions.

10. Install the pushrods, the rocker arms and the rocker arm covers.

11. Install the intake manifold and valley pan, if equipped.

Valve Lash

ADJUSTMENT

4.2L, 4.6L, 5.4L and 6.8L Engines

The 4.2L, 4.6L and 5.4L engines do not require valve lash adjusting, because they utilize hydraulic lash components in their valve actuation systems. The 4.2L engine uses hydraulic valve lifters, whereas the 4.6L and 5.4L engines utilize hydraulic lash adjusters, all of which automatically adjust the valve lash. No valve lash adjustment is necessary.

4.9L Engine

1. Rotate the crankshaft by hand so that the No. 1 piston is at TDC of the compression stroke. Make a chalk mark on the damper at that point, then make two more chalk marks about 120 degrees apart, dividing the damper into three equal parts.

2. With No. 1 at TDC, tighten the rocker arm bolts on No. 1 cylinder intake and exhaust to 17–23 ft. lbs. (23–31 Nm). Then, slowly apply pressure, using Lifter Bleed-down Wrench T70P-6513-A, or equivalent, to completely bottom the lifter. Take care to avoid excessive pressure that might bend the pushrod. Hold the lifter in this position and check the clearance between the rocker

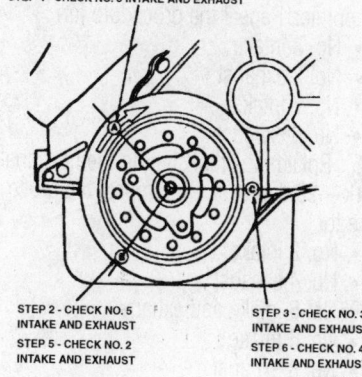

STEP 1 - SET NO. 1 PISTON ON T.D.C. AT END OF COMPRESSION STROKE ADJUST NO. 1 (INTAKE AND EXHAUST)
STEP 4 - CHECK NO. 6 INTAKE AND EXHAUST

STEP 2 - CHECK NO. 5 INTAKE AND EXHAUST
STEP 5 - CHECK NO. 2 INTAKE AND EXHAUST

STEP 3 - CHECK NO. 3 INTAKE AND EXHAUST
STEP 6 - CHECK NO. 4 INTAKE AND EXHAUST

7924FG19

Valve clearance adjustment positions on the crankshaft damper—4.9L engine

arm and the valve stem tip. Allowable clearance is 0.10–0.20 in. (2.5–5.0mm) with a desired clearance of 0.125–0.175 in. (3.0–4.5mm)

3. If the clearance is less than specified, install a shorter pushrod. If the clearance is greater than specified, install a longer pushrod.

4. Rotate the crankshaft clockwise—viewed from the front—until the next chalk mark is aligned with the timing pointer. Repeat the procedure for No. 5 intake and exhaust.

5. Rotate the crankshaft to the next chalk mark and repeat the procedure for No. 3 intake and exhaust.

6. Repeat the rotation/checking procedure for the remaining valves in firing order, that is: 6–2–4.

5.0L Engine

1. Rotate the crankshaft by hand so No. 1 piston is at TDC of the compression stroke. Make a chalk mark on the damper at that point, then, make two more chalk marks about 90 degrees apart in a clockwise direction.

2. With No. 1 at TDC, slowly apply pressure, using Lifter Bleed-down Wrench T70P-6513-A, or equivalent, to completely bottom the lifter, on the following valves:

- No. 1 intake and exhaust
- No. 7 intake
- No. 5 exhaust
- No. 8 intake
- No. 4 exhaust

Take care to avoid excessive pressure that might bend the pushrod. Hold the lifter in this position and check the clearance between the rocker arm and the valve stem tip. Allowable clearance is 0.071–0.193 in. (1.8–4.9mm) with a desired clearance of 0.096–0.165 in. (2.4–4.2mm).

3. If the clearance is less than specified, install a shorter pushrod. If the clearance is greater than specified, install a longer pushrod.

4. Rotate the crankshaft clockwise—viewed from the front—180 degrees, until the next chalk mark is aligned with the timing pointer. Repeat the procedure for:

- No. 5 intake
- No. 2 exhaust
- No. 4 intake
- No. 6 exhaust

5. Rotate the crankshaft to the next chalk

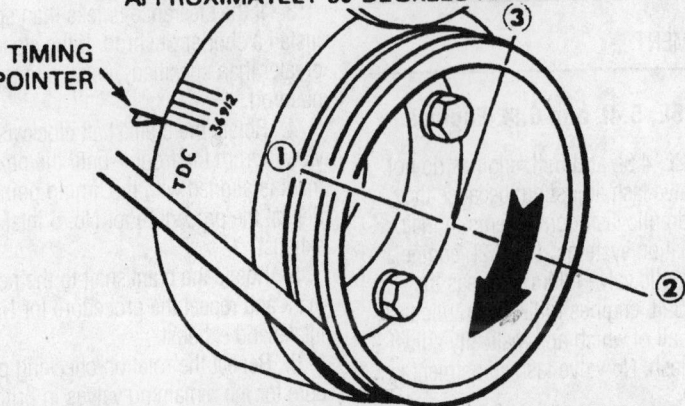

WITH NO. 1 AT TDC AT END OF COMPRESSION
STROKE MAKE A CHALK MARK AT POINTS 2 AND 3
APPROXIMATELY 90 DEGREES APART.

TIMING POINTER

TDC 34912

POSITION 1 — NO. 1 AT TDC AT END OF COMPRESSION
STROKE.

POSITION 2 — ROTATE THE CRANKSHAFT 180 DEGREES
(1/2 REVOLUTION) CLOCKWISE FROM
POSITION 1.

POSITION 3 — ROTATE THE CRANKSHAFT 270 DEGREES
(3/4 REVOLUTION) CLOCKWISE FROM
POSITION 2.

7924FG20

Valve clearance adjustment positions on the crankshaft damper/pulley—5.0L, 5.8L and 7.5L engines

mark—90 degrees—and repeat the procedure for:
- No. 2 intake
- No. 7 exhaust
- No. 3 intake and exhaust
- No. 6 intake
- No. 8 exhaust

5.8L Engine

1. Rotate the crankshaft by hand so No. 1 piston is at TDC of the compression stroke. Make a chalk mark on the damper at that point, then, make two more chalk marks about 90 degrees apart in a clockwise direction.

2. With No. 1 at TDC, slowly apply pressure, using Lifter Bleed-down Wrench T70P-6513-A, or equivalent, to completely bottom the lifter, on the following valves:
- No. 1 intake and exhaust
- No. 4 intake
- No. 3 exhaust
- No. 8 intake
- No. 7 exhaust

Take care to avoid excessive pressure that might bend the pushrod. Hold the lifter in this position and check the clearance between the rocker arm and the valve stem tip. Allowable clearance is 0.098–0.198 in.

(2.5–5.0mm) with a desired clearance of 0.123–0.173 in. (3.1–4.4mm).

3. If the clearance is less than specified, install a shorter pushrod. If the clearance is greater than specified, install a longer pushrod.

4. Rotate the crankshaft clockwise—viewed from the front—180 degrees, until the next chalk mark is aligned with the timing pointer. Repeat the procedure for:
- No. 3 intake
- No. 2 exhaust
- No. 7 intake
- No. 6 exhaust

5. Rotate the crankshaft to the next chalk mark—90 degrees—and repeat the procedure for:
- No. 2 intake
- No. 4 exhaust
- No. 5 intake and exhaust
- No. 6 intake
- No. 8 exhaust

7.5L Engine

1. Rotate the crankshaft by hand so No. 1 piston is at TDC of the compression stroke. Make a chalk mark on the damper at that point.

2. With No. 1 at TDC, slowly apply pres-

sure, using Lifter Bleed-down Wrench T70P-6513-A, or equivalent, to completely bottom the lifter, on the following valves:
- No. 1 intake and exhaust
- No. 3 intake
- No. 4 exhaust
- No. 7 intake
- No. 5 exhaust
- No. 8 intake and exhaust

Take care to avoid excessive pressure that might bend the pushrod. Hold the lifter in this position and check the clearance between the rocker arm and the valve stem tip. Allowable clearance is 0.075–0.175 in. (1.9–4.4mm) with a desired clearance of 0.100–0.150 in. (2.5–3.8mm).

3. If the clearance is less than specified, install a shorter pushrod. If the clearance is greater than specified, install a longer pushrod.

4. Rotate the crankshaft clockwise—viewed from the front—360 degrees, until the chalk mark is once again aligned with the timing pointer. Repeat the procedure for:
- No. 2 intake and exhaust
- No. 4 intake
- No. 3 exhaust
- No. 5 intake
- No. 7 exhaust
- No. 6 intake and exhaust

Oil Pan

REMOVAL & INSTALLATION

4.2L Engine

1. Raise and safely support the front of the vehicle.

2. Remove the oil pan plug and drain the engine oil.

3. Remove the two front wheel driveshafts and joints, if so equipped.

4. Remove the front differential from the vehicle.

5. Remove the front differential support.

6. Remove the three oil pan-to-transmission bolts.

7. Remove the 15 oil pan-to-cylinder block mounting bolts, then lower the oil pan.

To install:

➡ **If the oil pan is not installed within 15 minutes, remove the sealer and reapply.**

8. Temporarily install two locator dowels in two of the oil pan-to-engine block corner mounting bolt holes.

9. Clean and apply sealant to the rear main bearing cap, the oil pan mating sur-

face, the front cover mounting area, the front cover-to-engine block joints, then install the oil pan rear seal. Make certain to use Silicone Gasket and Sealant F6AZ-19562-A or equivalent for the sealant.

10. Position the oil pan, then install 13 of the oil pan mounting bolts loosely.

11. Remove the two locator dowels and install the remaining two oil pan-to-engine block bolts.

12. Starting from the rear and alternating from the right to left, tighten the 15 oil pan mounting bolts to 36–44 inch lbs. (4–5 Nm), then retighten the bolts to 80–106 inch lbs. (9–12 Nm).

13. Install the oil pan-to-transmission bolts to 28–38 ft. lbs. (38–51 Nm).

14. Install the oil pan drain plug to 16–22 ft. lbs. (22–30 Nm).

15. Install the front differential and front differential support.

16. Install the front driveshafts and joints.

17. Lower the vehicle to the ground.

18. Fill the engine with the correct type and amount of engine oil.

4.6L and 5.4L Engines

1. Raise and safely support the vehicle.

2. Remove the front axle housing from the vehicle.

3. Drain the engine oil.

4. Remove the 16 oil pan-to-engine block bolts.

5. Remove the oil pan and old oil pan gasket.

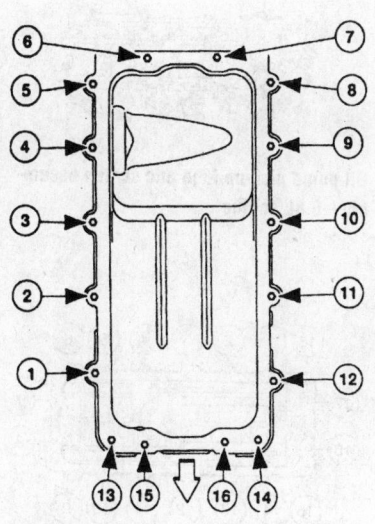

7924FG21

Tighten the oil pan-to-engine block bolts in three steps following the sequence shown—4.6L and 5.4L engines

To install:

6. Clean the oil pan and engine block mating surfaces of oil and old gasket material.

7. Install the new oil pan gasket and the oil pan, then install the 16 oil pan-to-engine block bolts loosely.

➡**Be sure to tighten the oil pan bolts in three steps.**

8. Tighten the oil pan-to-engine bolts in the sequence shown, in the following three steps:

    a. Step 1—18 inch lbs. (2 Nm).
    b. Step 2—15 ft. lbs. (20 Nm).
    c. Step 3—tighten an additional 60 degrees (one hex head flat of a bolt).

9. Install the oil drain plug.

10. Install the front axle housing.

11. Lower the vehicle.

12. Fill the engine with the correct amount and type of engine oil.

4.9L Engine

1. Drain the crankcase.

2. Drain the cooling system.

3. Remove the upper intake manifold and throttle body.

4. Remove the starter.

5. Remove the engine front support insulator-to-support bracket nuts and washers on both supports.

6. Raise the front of the engine with a transmission jack and wood block and place 1 in. (25mm) thick wood blocks between the front support insulators and support brackets.

7. Lower the engine and remove the transmission jack.

8. Remove the oil pan attaching bolts and lower the pan to the crossmember.

9. Remove the two oil pump inlet tube and screw assembly bolts and drop the assembly in the pan.

10. Remove the oil pan.

11. Remove the oil pump inlet tube attaching bolts.

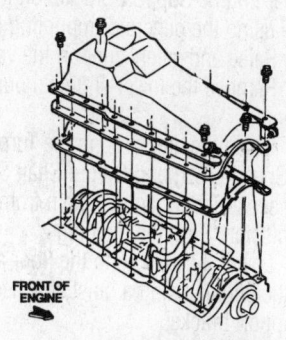

FRONT OF ENGINE

7924FG41

Exploded view of the oil pan mounting— 4.9L engine

12. Remove the inlet tube and screen assembly from the oil pump and leave it in the bottom of the oil pan.

13. Remove the oil pan gaskets.

14. Remove the inlet tube and screen from the oil pan.

To install:

15. Clean the gasket surfaces of the oil pump, oil pan and cylinder block. Remove the rear main bearing cap-to-oil pan seal and cylinder front cover-to-oil pan seal. Clean the seal grooves.

16. Apply oil-resistant sealer in the cavities between the bearing cap and cylinder block. Install a new seal in the rear main bearing cap and apply a bead of oil-resistant sealer to the tapered ends of the seal.

17. Install new side gaskets on the oil pan with oil-resistant sealer. Position a new oil pan-to-cylinder front cover seal on the oil pan.

18. Clean the inlet tube and screen assembly and place it in the oil pan.

19. Position the oil pan under the engine. Install the inlet tube and screen assembly on the oil pump with a new gasket. Tighten the screws to 5–7 ft. lbs. (6.8–9.5 Nm). Position the oil pan against the cylinder block and install the attaching bolts. Tighten the bolts in sequence to 10–12 ft. lbs. (14–16 Nm).

20. Raise the engine with a transmission jack and remove the wood blocks from the engine front supports. Lower the engine until the front support insulators are positioned on the support brackets. Install the washers and nuts on the insulator studs and tighten the nuts.

21. Install the starter and connect the starter cable.

22. Install the manifold and throttle body.

23. Fill the crankcase and cooling system.

24. Start the engine and check for coolant and oil leaks.

5.0L and 5.8L Engines

1. Drain the cooling system.

2. Remove the bolts attaching the fan shroud to the radiator and position the shroud over the fan.

3. Remove the upper intake manifold and throttle body.

4. Remove the nuts and lockwashers attaching the engine support insulators to the chassis bracket.

5. If equipped with an automatic transmission, disconnect the oil cooler line at the left side of the radiator.

6. Remove the exhaust system.

7. Raise the engine and place wood blocks under the engine supports.

8. Drain the crankcase.

9. Support the transmission with a floor jack and remove the transmission crossmember.

10. Remove the oil pan attaching bolts and lower the oil pan onto the crossmember.

11. Remove the two bolts attaching the oil pump pick-up tube to the oil pump. Remove the nut attaching the oil pump pick-up tube to the No. 3 main bearing cap stud. Lower the pick-up tube and screen into the oil pan.

12. Remove the oil pan from the vehicle.

To install:

13. Clean the oil pan, inlet tube and gasket surfaces. Inspect the gasket sealing surface for damages and distortion due to over tightening of the bolts. Repair and straighten as required.

14. Position a new oil pan gasket and seal to the cylinder block.

15. Position the oil pick-up tube and screen to the oil pump, and install the lower attaching bolt and gasket loosely. Install the nut on the No. 3 main bearing cap stud.

16. Place the oil pan on the crossmember. Install the upper pick-up tube bolt. Tighten the pick-up tube bolts.

17. Position the oil pan to the cylinder block and install the attaching bolts. Tighten to 10–12 ft. lbs. (14–16 Nm).

18. Install the transmission crossmember.

19. Raise the engine and remove the blocks under the engine supports. Bolt the engine to the supports.

20. Install the exhaust system.

21. If equipped with an automatic transmission, connect the oil cooler line at the left side of the radiator.

22. Install the nuts and lockwashers attaching the engine support insulators to the chassis bracket.

23. Install the upper intake manifold and throttle body.

24. Install the fan shroud.

25. Fill the crankcase.

26. Fill and bleed the cooling system.

6.8L Engine

1. Disconnect the negative battery cable.

2. Relieve the fuel system pressure.

3. Drain the cooling system.

4. Disconnect the upper radiator hose from the intake manifold.

5. Remove the engine air cleaner outlet tube.

6. Disconnect the accelerator cable from the bracket and the throttle body cam.

7. If equipped, remove the speed control actuator cable from the throttle body.

8. Disconnect all vacuum hoses, fuel lines and electrical wires from the throttle body and intake manifold.

9. Remove the brake booster vacuum hose bracket.

10. Remove the EGR valve-to-exhaust manifold tube and disconnect the vacuum line.

11. Remove the connector and vacuum line from the Engine Vacuum Regulator (EVR) solenoid.

12. Remove the four bolts and the throttle body adapter.

13. Remove the upper fan shroud mounting screws and position the shroud toward the engine.

14. Remove the generator and install the Modular Engine Support Bracket on the engine using the generator mounting holes.

15. Raise and safely support the vehicle.

16. Remove the lower engine mount-to-frame nuts.

17. Clean the area around the Turbine Shaft Speed (TSS) and Output Shaft Speed (OSS) sensors and remove them from the transmission. Plug the openings.

18. Lower the vehicle to the floor and raise the engine using a hoist attached to the support bracket.

19. Install an engine support fixture with a J hook to keep the engine raised, then remove the hoist.

20. Raise and safely support the vehicle.

21. Drain the engine oil and remove the bypass filter.

22. Remove the dual converter Y-pipe and the flywheel inspection cover.

23. Remove the driveshaft and the two transmission mounting nuts.

24. Raise the transmission assembly with a transmission jack. Be sure to support the transmission along the rails of the pan to avoid damage.

25. Remove the oil pan mounting bolts and partially lower the pan.

26. Remove the oil pump pick-up tube and screen assembly and allow it to drop into the pan.

27. Remove the oil pan towards the rear of the vehicle.

To install:

✳✳ WARNING

To prevent possible oil leaks, use only a plastic scraper to clean the oil pan mounting surface.

28. Clean the oil pan-to-engine mounting surface.

29. Place the oil pump pick-up tube and screen assembly in the oil pan, then position the pan and gasket near the engine.

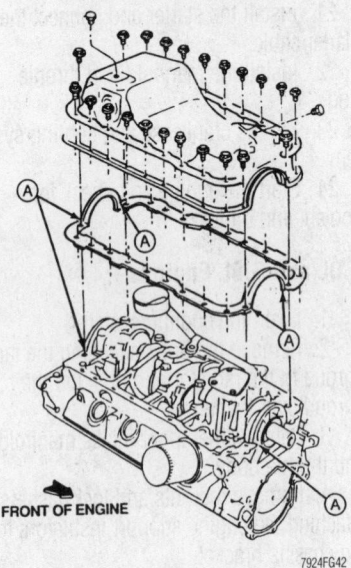

FRONT OF ENGINE

7924FG42

Exploded view of the oil pan mounting. Apply RTV sealant to the areas marked "A"—5.0L and 5.8L engines

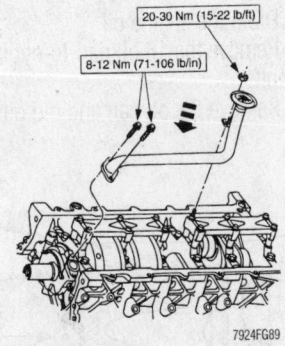

7924FG89

Oil pump pick-up tube and screen assembly—6.8L engine

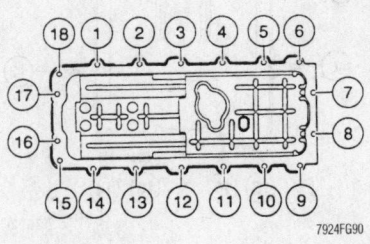

7924FG90

To prevent leaks, tighten the oil pan bolts in the order shown—6.8L engine

30. Install the oil pump pick-up tube and screen assembly. Tighten the nut to 15–22 ft. lbs. (20–30 Nm) and the two bolts to 71–106 in lbs. (8–12 Nm).

31. Apply a bead of silicone sealant to the areas where the front cover and rear bearing cap meet the engine block.

32. Install the oil pan. Tighten the bolts in sequence using three steps as follows:
- Step one—18 inch lbs. (2 Nm)
- Step two—15 ft. lbs. (20 Nm)
- Step three—tighten an additional 60°

33. Lower the transmission and install the two mounting nuts. Tighten the nuts to 60–80 ft. lbs. (81–108 Nm).

34. Install the driveshaft and the TSS and OSS sensors.

35. Install the flywheel cover and dual converter Y-pipe.

36. Install the oil bypass filter.

37. Lower the vehicle and remove the engine support fixture.

38. Install the engine mounting nuts. Tighten the nuts to 66 ft. lbs. (90 Nm).

39. Remove the modular engine support bracket and install the generator.

40. Install the fan shroud.

41. Use a new gasket and install the throttle body adapter. Tighten the bolts in two steps first to 71–88 in lbs. (8–10 Nm), then tighten them an additional 85–95°.

42. Connect the EVR solenoid harness and vacuum line.

43. Attach the vacuum line to the EGR valve.

44. Install the EGR valve-to-exhaust manifold tube. Tighten the fittings to 55 ft. lbs. (41 Nm).

45. Install the EGR transducer.

46. Install all remaining components in the reverse of the removal.

✳✳ WARNING

Operating the engine without the proper amount and type of engine oil will result in severe engine damage.

47. Fill the engine with Super Premium SAE 5W30 Motor Oil meeting ford specification WSS-M2C153-G to the proper level.

48. Fill and bleed the cooling system.

7.5L Engine

1. Remove the hood.
2. Disconnect the battery ground cable.
3. Drain the cooling system.
4. Remove the air intake tube and air cleaner assembly.

5. Disconnect the throttle linkage at the throttle body.

6. Disconnect the power brake vacuum line at the manifold.

7. Disconnect the fuel lines at the fuel rail.

8. Disconnect the air tubes at the throttle body.

9. Remove the radiator.

10. Remove the power steering pump and position it out of the way without disconnecting the lines.

11. Remove the oil dipstick tube. On motor home chassis, remove the oil filler tube.

12. Remove the front engine mount through-bolts.

13. Position the air conditioner refrigerant hoses so they are clear of the firewall. If necessary, discharge the system and remove the compressor.

14. Remove the upper intake manifold and throttle body.

15. Drain the crankcase. Remove the oil filter.

16. Disconnect the exhaust pipe at the manifolds.

17. Disconnect the transmission linkage at the transmission.

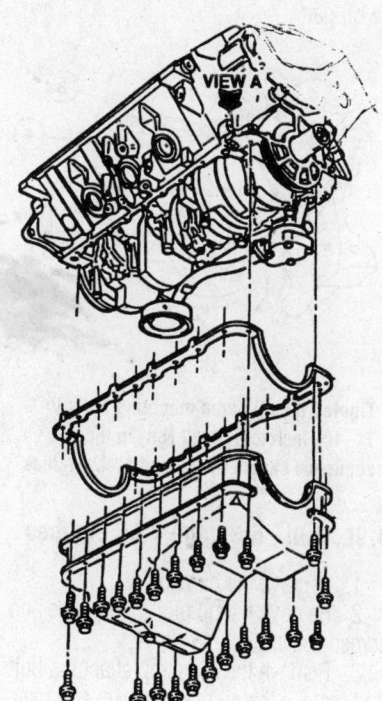

7924FG43

Exploded view of the oil pan mounting—7.5L engine

18. Remove the driveshaft(s).

19. Remove the transmission fill tube.

20. Raise the engine with a jack placed under the crankshaft damper and a block of wood to act as a cushion. Raise the engine until the transmission contacts the underside of the floor. Place wood blocks under the engine supports. The engine **must** remain centralized at a point at least 4 in. (102mm) above the mounts, to remove the oil pan!

21. Remove the oil pan attaching screws and lower the oil pan onto the crossmember. Remove the two bolts attaching the oil pump pick-up tube to the oil pump. Lower the assembly from the oil pump. Leave it on the bottom of the oil pan. Remove the oil pan and gaskets. Remove the inlet tube and screen from the oil pan.

To install:

22. Clean the gasket surfaces of the oil pan and cylinder block.

23. Apply a coating of gasket adhesive on the block mating surface and stick the one-piece silicone gasket on the block.

24. Clean the inlet tube and screen assembly and place on the pump.

25. Position the oil pan against the cylinder block and install the retaining bolts. Tighten all bolts to 10 ft. lbs. (14 Nm).

26. Lower the engine and bolt it in place.

27. Install the transmission fill tube.

28. Install the driveshaft(s).

29. Attach the transmission linkage at the transmission.

30. Connect the exhaust pipe at the manifolds.

31. Install the oil filter.

32. Install the upper intake manifold and throttle body.

33. Install the compressor or reposition the hoses.

34. Install the oil dipstick tube. On motor home chassis, install the oil filler tube.

35. Install the power steering pump.

36. Install the radiator.

37. Connect the air tubes at the throttle body.

38. Connect the fuel lines at the fuel rail.

39. Connect the power brake vacuum line at the manifold.

40. Connect the throttle linkage at the throttle body.

41. Install the air intake tube and air cleaner assembly.

42. Fill and bleed the cooling system.

43. Fill the crankcase.

44. Connect the battery ground cable.

45. Install the hood.

Oil Pump

REMOVAL & INSTALLATION

4.2L Engine

1. Raise and support the front of the vehicle.
2. Drain the engine oil and dispose.
3. Remove the oil filter.
4. Remove the six oil pump bolts, then remove the oil pump drive gear, the oil pump driven gear, the oil pump O-ring and the oil pump itself. Discard the used oil pump O-ring.
5. Inspect the oil pump components for damage or excessive wear.
6. Check the oil pump face for warpage with a flat edge ruler. The face cannot exhibit more than 0.00157 in. (0.04mm) of distortion.
7. Remove the plug over the oil pressure relief; valve.
8. Remove the oil pressure relief valve ball and spring, then clean the parts.

To install:

➡**Lubricate the parts with clean engine oil before assembly.**

9. Assemble the oil pressure relief valve ball and spring with a new plug.
10. Install the oil pump, along with a new O-ring, the oil pump driven gear and the drive gear. Install and tighten the six oil pump mounting bolts to the torque value specifications indicated in the illustration.
11. Apply a film of clean engine oil to the rubber O-ring on the new filter, then install the filter onto the filter mount.
12. Install the oil pan drain plug.
13. Lower the vehicle.
14. Fill the engine with the correct amount and type of new engine oil.
15. Start the engine and make certain that the oil light on the instrument panel

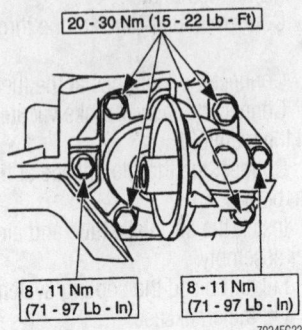

Tighten the oil pump mounting bolts to the specifications shown—4.2L engines

extinguishes within 6–8 seconds after the engine starts.

4.6L and 5.4L Engines

1. Remove the timing chain.
2. Remove the oil pan.
3. Remove the three oil pump screen and cover bolts, then remove the screen and cover.
4. Remove the oil pump screen and cover spacer.
5. Remove the four oil pump mounting bolts, then remove the oil pump from the engine.

To install:

6. Clean and inspect the mating surfaces.
7. Install the oil pump and loosely install the four oil pump mounting bolts. Tighten the four oil pump bolts in the sequence shown to 71–106 inch lbs. (8–12 Nm).
8. Install the oil pump screen and cover spacer to 15–22 ft. lbs. (20–30 Nm).
9. Install the oil pump screen and cover, then install the three oil pump screen and cover bolts. Tighten the bolts near the oil pick-up screen to 15–22 ft. lbs. (20–30 Nm) and the bolts at the opposite end of the pick-up to 70–106 inch lbs. (8–12 Nm).
10. Install the timing chains, then install the oil pan.

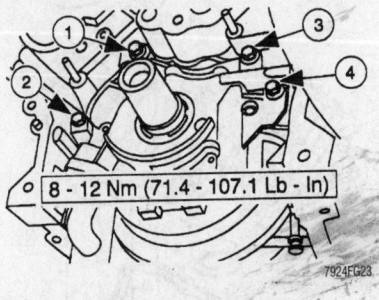

Tighten the oil pump mounting bolts to 71–106 inch lbs. (8–12 Nm) in the sequence shown—4.6L and 5.4L engines

4.9L, 5.0L, 5.8L and 7.5L Engines

1. Remove the oil pan.
2. Remove the oil pump inlet tube and screen assembly.
3. Remove the oil pump attaching bolts and remove the oil pump gasket and intermediate driveshaft.

To install:

4. Before installing the oil pump, prime it by filling the inlet and outlet port with oil and rotating the shaft of the pump to distribute it.

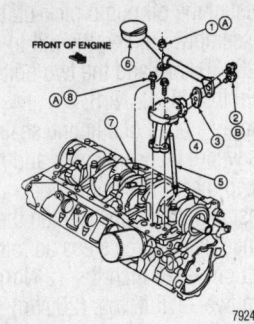

7924FG44

Exploded view of the oil pump mounting—5.0L engine shown, 5.8L and 7.5L engines are similar

5. Position the intermediate driveshaft into the distributor socket.
6. Position the new gasket on the pump body and insert the intermediate driveshaft into the pump body.
7. Install the pump and intermediate driveshaft as an assembly. Do not force the pump if it does not seal readily. The driveshaft may be misaligned with the distributor shaft. To align it, rotate the intermediate driveshaft into a new position.
8. Install the oil pump attaching bolts and tighten them to 12–15 ft. lbs. (16–20 Nm) on the 4.9L engine and to 20–25 ft. lbs. (27–34 Nm) on the 5.0L, 5.8L and 7.5L engines.

6.8L Engine

1. Disconnect the negative battery cable.
2. Remove the engine front cover and crankshaft sprocket.
3. Remove the oil pan.
4. Remove the three oil pump mounting bolts, then remove the oil pump from the engine.

To install:

5. Clean and inspect the mating surfaces.
6. Install the oil pump and loosely install the oil pump mounting bolts. Tighten

7924FG91

Be sure to tighten the oil pump mounting bolts in the sequence shown—6.8L engine

the bolts in the sequence shown to 71–106 inch lbs. (8–12 Nm).

7. Install the oil pan.

8. Install the crankshaft sprocket and timing chains.

9. Install the front cover.

Rear Main Seal

REMOVAL & INSTALLATION

7.5L Engine

1. Raise and safely support the vehicle.
2. Remove the oil pan.
3. Loosen all the crankshaft main bearing cap bolts and lower the crankshaft no more than 1/32 in. (0.7938mm).

✳✳ CAUTION

Be careful that the crankshaft sealing surfaces are not damaged in this process

4. Remove the rear main bearing cap and remove the seal. On the cylinder block half of the seal, use a seal removal tool, or install a small metal screw in one end of the seal and pull on the screw to remove the seal.

To install:

5. Clean the seal groove in the crankshaft main bearing cap and the block using a brush and a solvent such as Metal Surface Cleaner F4AZ-19A536-RA or its equivalent.

6. Clean the areas where the sealer is to be applied later and dry the area thoroughly so that no solvent contacts the rear main seal.

7. Dip the seal halves in engine oil.

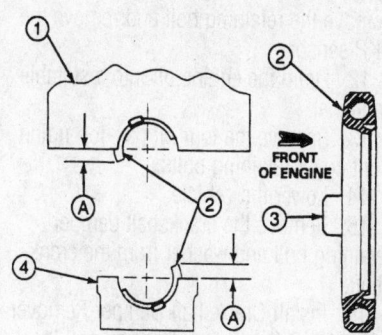

1 Cylinder Block
2 Crankshaft Rear Oil Seal
3 Tab
4 Crankshaft Rear Main Bearing Cap
5 9.53mm (3/8 Inch)

7924FG45

Rear main seal and related components— 7.5L engines

✳✳ CAUTION

Be sure no rubber has been removed from the outside diameter of the seal by the bottom edge of the groove. Do not allow oil to get on the sealer.

8. Install the upper half of the seal (cylinder block side) into its groove with the undercut side of the seal towards the front of the engine (with the tab side of the seal towards the rear face of the block), by rotating it on the seal journal until approximately 3/8 in. (9.525mm) protrudes below the parting surface.

9. Tighten all the crankshaft main bearing cap bolts, EXCEPT THE REAR MAIN BEARING, to 95–105 ft. lbs. (129–142 Nm).

10. Install the lower half of the seal in the rear crankshaft main bearing cap with the undercut side of the seal towards the front of the engine (with the tab side of the seal towards the rear face of the block), Allow the seal to protrude 3/8 in. (9.525mm) above the parting surface to mate with the upper half of the seal.

11. Apply a 1/16 in. (1.588mm) bead of Gasket Maker E2AZ-19562-B or its equivalent to the rear oil seal area of the block starting from the forward face of the return groove and overlaying the end of the wire seal retainer.

➡**Do not allow the sealer to contact the inside diameter of the seal.**

12. Install the rear main cap and tighten the bolts to 95–105 ft. lbs. (129–142 Nm).

13. Install the oil pan.

Except 7.5L Engine

If the crankshaft rear oil seal replacement is the only operation being performed, it can be done in the vehicle as detailed in the following procedure. If the oil seal is being replaced in conjunction with a rear main bearing replacement, the engine must be removed from the vehicle and installed on a work stand.

1. Remove the transmission from the vehicle.

2. Remove the flywheel/flexplate. If equipped, remove the crankshaft oil slinger from the crankshaft.

3. Use an awl to punch two holes in the crankshaft rear oil seal. Punch the holes on opposite sides of the crankshaft and just above the bearing cap-to-cylinder block split line.

4. Install a sheet metal screw in each hole. Use two small prybars to pry against both screws at the same time to remove the

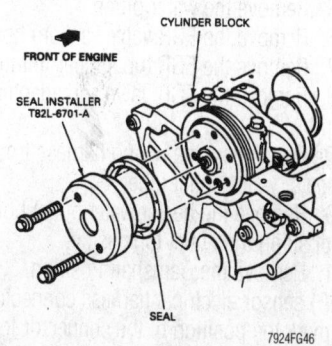

Rear main oil seal installation—5.0L and 5.8L engines

crankshaft rear oil seal. It may be necessary to place small blocks of wood against the cylinder block to provide a fulcrum point for the prybars. Use caution throughout this procedure to avoid scratching or otherwise damaging the crankshaft oil seal surface.

5. Clean the oil seal recess in the cylinder block and main bearing cap.

To install:

6. Clean, inspect and polish the rear oil seal rubbing surface on the crankshaft.

7. Coat the new oil seal and the crankshaft with a light film of engine oil.

8. Start the seal in the recess with the seal lip facing forward and install it with a seal driver. Keep the tool straight with the centerline of the crankshaft and install the seal until the tool contacts the cylinder block surface. Remove the tool and inspect the seal to be sure it was not damaged during installation.

9. If equipped, install the crankshaft oil slinger.

10. Position the flywheel on the crankshaft flange. Coat the threads of the flywheel attaching bolts with Locktite® and install the bolts. Tighten the bolts in sequence across from each other to 75–85 ft. lbs. (102–115 Nm).

11. Install the transmission, following the recommended procedure.

Timing Chain, Sprockets and Front Cover

REMOVAL & INSTALLATION

4.2L Engine

1. Disconnect the negative battery cable.

2. Remove the accessory drive belt.

3. Drain the coolant into a suitable clean container.

4. Remove the radiator, fan blade assembly and fan shroud.

5. Remove the water pump.

6. Remove the EGR valve vacuum hose.

7. Remove the EGR tube upper fitting.

8. Remove the EGR valve and adapter assembly.

9. Disconnect the wiring harness from the heater water outlet tube.

10. Remove the heater water outlet bolt and position the outlet tube aside.

11. Remove the Camshaft Position (CMP) sensor electrical harness connector and mark the position of the connector for proper installation.

12. Rotate the crankshaft until the TDC timing mark lines up with the timing mark.

13. Remove the two bolts retaining the CMP and remove the CMP from the camshaft synchronizer.

14. Remove the camshaft synchronizer adjustment bolt and remove the camshaft synchronizer.

➡**The oil pump drive shaft may come out with the camshaft synchronizer.**

15. Raise and safely support the vehicle.

16. Remove the oil pan drain plug and drain the engine oil in a suitable container.

17. Remove the crankshaft pulley and damper using appropriate tools.

18. Remove the engine oil pan.

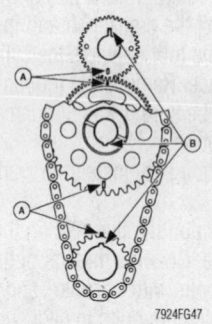

7924FG47

After removing the CMP drive gear, be sure the timing marks (A) and keyways (B) are aligned—4.2L engine

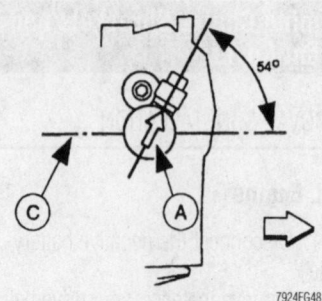

7924FG48

Install the camshaft synchronizer (A) in the orientation shown (white arrow points to front of engine)—4.2L engines

19. Remove the two engine front cover stud bolts, front cover bolt and cap screw.

20. Slide the engine front cover and gasket off the dowels and discard the gasket.

21. Remove the CMP sensor drive gear bolt and remove the drive gear.

22. Be sure the timing marks and keyways align.

23. Compress and install a retaining pin to hold the timing chain tensioner.

24. Slide both sprockets and timing chain forward and remove as an assembly.

25. Remove the three bolts retaining the timing chain tensioner and remove the tensioner.

26. Check the timing chain and sprockets for excessive wear. Replace if necessary.

To install:

27. Before installation, clean and inspect all parts. Clean the gasket material from the engine oil pan, cylinder block and engine front cover.

28. Place the timing chain tensioner in position and install the three retaining bolts. Tighten the bolts to 6–10 ft. lbs. (8–14 Nm).

29. Verify that the balance shaft timing gears are in correct alignment.

30. Slide both sprockets and the timing chain onto the camshaft and crankshaft with the timing marks aligned. Install the CMP sensor drive gear and bolt. Tighten the bolt to 30–36 ft. lbs. (40–50 Nm).

31. Remove the retaining pin.

32. Inspect the engine front cover seal for wear or damage and replace if necessary.

33. Install the engine front cover gasket and front cover onto the guide studs.

34. Install the two engine front cover stud bolts, front cover bolt and cap screw. Tighten the bolts to 15–22 ft. lbs. (20–30 Nm).

35. Install the engine oil pan.

36. Install the crankshaft damper and pulley.

❊❊ WARNING

A Synchro Positioning Tool must be used prior to installation. Failure to using this procedure will result in the fuel system being out of time, possibly causing engine damage.

37. Install Synchro Positioning Tool T89P-12200-A or equivalent, on the camshaft synchronizer by rotating the tool until it engages the notch in the housing.

38. Install the camshaft synchronizer housing assembly so that the arrow on the tool is 54 degrees from the centerline of the engine.

39. Install the adjustment bolt and tighten to 15–22 ft. lbs. (21–30 Nm).

40. Remove the tool and position the CMP sensor. Install the 2 bolts and 40–70 inch lbs. (5–8 Nm) and install the CMP electrical harness connector.

41. Install the water pump.

42. Install the fan shroud, fan blade assembly and radiator.

43. Install the accessory drive belt.

44. Fill the crankcase with the correct amount and type of engine oil.

45. Fill and bleed the engine cooling system.

46. Connect the negative battery cable.

47. Start the engine and check for coolant and oil leaks.

48. Road test the vehicle and check for proper engine operation.

4.6L Engine

1. Disconnect the negative battery cable.

2. Remove the radiator, fan blade and fan shroud assembly.

3. Remove the accessory drive belt.

4. Remove the water pump pulley.

5. Disconnect the electrical harness connectors from both ignition coils.

6. Remove both ignition coils with their brackets attached.

7. Remove the left-hand and right-hand cylinder head covers.

8. Raise and safely support the vehicle.

9. Remove the two upper power steering pump retaining bolts.

10. Remove the two lower power steering pump retaining bolts and move the pump aside.

11. Disconnect the Crankshaft Position (CKP) sensor electrical harness connector. Remove the retaining bolt and remove the CKP sensor.

12. Drain the engine oil into a suitable container.

13. Remove the four oil pan-to-engine front cover retaining bolts.

14. Lower the vehicle.

15. Remove the crankshaft damper retaining bolt and washer from the crankshaft.

16. Install Crankshaft Damper Remover T58P-6316-D or equivalent and pull the damper from the crankshaft.

17. Remove the Camshaft Position (CMP) sensor retaining bolt and remove the CMP sensor.

18. Remove the idler pulley bolt and remove the pulley.

19. Remove the three belt tensioner retaining bolts and remove the tensioner.

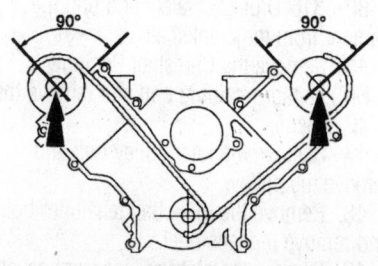

When removing the timing chains, rotate the crankshaft so that the camshaft keyways are positioned as shown—4.6L engines

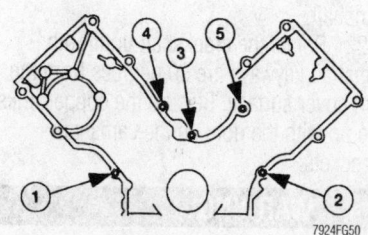

Tighten the first five front cover fasteners in the sequence shown—4.6L engine

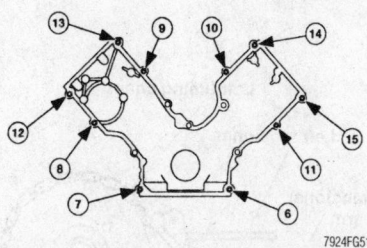

Continue tightening the remaining fasteners in the sequence shown here—4.6L engine

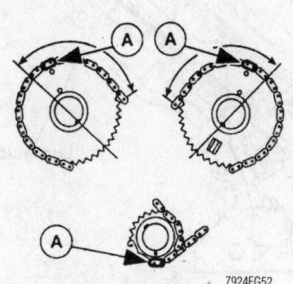

When installing the timing chains, make certain that the copper colored links (A) are aligned with the timing marks—4.6L engine

20. Remove the eight engine front cover retaining bolts and the seven nuts. Swing the top of the cover out off the dowel pins and remove the cover.

21. Remove the sensor ring from the crankshaft.

22. Use Camshaft Positioning Tool T91P-6256-A and Camshaft Positioning Adapters T92P-6256-A or equivalents, to position the camshaft.

23. Rotate the crankshaft until both camshaft keyways are 90 degrees from the cam cover surface. Be sure the copper links line up with the dots on the camshaft sprockets.

✳✳ WARNING

At no time, when the timing chains are removed and the cylinder heads are installed may the crankshaft or the camshaft be rotated. Severe piston and valve damage will occur.

24. Remove the two left-hand and right-hand tensioner bolts and remove the timing chain tensioners.

25. Slide the left-hand and right-hand tensioner guides off the dowel pins.

26. Remove the right-hand timing chain from the camshaft sprocket.

27. Remove the left-hand timing chain from the camshaft sprocket.

28. Remove the left-hand and right-hand timing chain guide bolts and remove the timing chain guides.

29. If necessary, remove the camshaft gear bolt and remove the camshaft gear.

To install:

30. Examine the timing chains, looking for the copper links. If the copper links are not visible, lay the chain on a flat surface and pull the chain taught until the opposite sides of the chain contact one another. Mark the links at each end of the chain and use these marks in place of the copper links.

➡**If the engine jumped time, damage has been done to valves and possibly pistons and/or connecting rods. Any damage must be corrected before installing the timing chains.**

31. If removed, install the camshaft gears and tighten the retaining bolt to 81–95 ft. lbs. (110–130 Nm).

32. Install the left-hand and right-hand timing chain guides and retaining bolts. Tighten the retaining bolts to 71–106 inch lbs. (8–12 Nm).

33. If removed, install the left-hand crankshaft sprocket with the tapered part of the sprocket facing away from the engine block.

➡**The crankshaft sprockets are identical. They may only be installed one way, with the tapered part of the sprockets facing each other. Ensure that the keyway and timing marks on the crankshaft sprockets are aligned.**

34. Install the left-hand timing chain on the camshaft and crankshaft sprockets. Be sure the copper links of the timing chain line up with the timing marks on both sprockets.

35. If removed, install the right-hand crankshaft sprocket with the tapered part of the sprocket facing the left-hand crankshaft sprocket.

36. Install the right-hand timing chain on the camshaft and crankshaft sprockets. Be sure the copper links of the timing chain line up with the timing marks on both sprockets.

37. It is necessary to bleed the timing chain tensioners before installation. Proceed as follows:

   a. Place the timing chain tensioner in a soft-jawed vise.

   b. Using a small pick or similar tool, hold the ratchet lock mechanism away from the ratchet stem and slowly compress the tensioner plunger by rotating the vise handle.

✳✳ WARNING

The tensioner must be compressed slowly or damage to the internal seals will result.

   c. Once the tensioner plunger bottoms in the tensioner bore, continue to hold the ratchet lock mechanism and push down on the ratchet stem until flush with the tensioner face.

   d. While holding the ratchet stem flush to the tensioner face, release the ratchet lock mechanism and install a paper clip or similar tool in the tensioner body to lock the tensioner in the collapsed position.

   e. The paper clip must not be removed until the timing chain, tensioner, tensioner arm and timing chain guide are completely installed on the engine.

38. Install the left-hand and right-hand timing chain tensioner guides on the dowel pins.

39. Place the left-hand and right-hand timing chain tensioners in position and install the retaining bolts. Tighten the bolts to 15–22 ft. lbs. (20–30 Nm).

40. Remove the retaining pins from the timing chain tensioners.

41. Remove Camshaft Positioning Tool

T91P-6256-A and Camshaft Positioning Adapters T92P-6256-A or equivalent, from the camshaft.

42. Install the crankshaft sensor ring on the crankshaft.

43. Apply a bead of silicone sealer along the cylinder head-to-cylinder block and the oil pan-to-cylinder block sealing surfaces.

44. Install the engine front cover carefully onto the dowel pins.

45. Tighten the engine front cover bolts, in sequence, to 15–22 ft. lbs. (20–30 Nm).

46. Place the idler pulley in position and install the retaining bolt. Tighten the bolt to 15–22 ft. lbs. (20–30 Nm).

47. Install the drive belt tensioner and the three retaining bolts. Tighten the bolts to 15–22 ft. lbs. (20–30 Nm).

48. Install the CMP sensor and the retaining bolt. Tighten the bolt to 106 inch lbs. (12 Nm).

49. Place the damper on the crankshaft. Ensure the crankshaft key and keyway are aligned.

50. Using Crankshaft Damper Replacer T74P-6316-B or equivalent, install the crankshaft damper.

51. Raise and safely support the vehicle.

52. Install the four oil pan-to-engine front cover bolts and tighten, in sequence, in 2 steps:

 a. Tighten the bolts to 15 ft. lbs. (20 Nm).

 b. Rotate the bolts an additional 60 degrees.

53. Install the CKP sensor and the retaining bolt. Tighten the bolt to 106 inch lbs. (12 Nm). Connect the CKP sensor electrical harness connector.

54. Install the power steering pump and the two upper and two lower retaining bolts. Tighten the bolts to 15–20 ft. lbs. (20–30 Nm).

55. Lower the vehicle.

56. Place the ignition coil and brackets on the engine front cover and install the bracket bolts. Tighten the bolts to 15–22 ft. lbs. (20–30 Nm).

57. Connect the ignition coil and capacitor electrical harness connectors.

58. Connect the CMP electrical harness connector.

59. Install the water pump pulley and tighten the bolts to 15–22 ft. lbs. (20–30 Nm).

60. Install the radiator, fan blade and fan shroud assembly.

61. Install the accessory drive belt.

62. Connect the negative battery cable.

63. Start the engine and check for leaks.

64. Road test the vehicle and check for proper engine operation.

5.4L and 6.8L Engines

1. Disconnect the negative battery cable.

2. Remove the radiator, fan blade and fan shroud assembly.

3. Remove the accessory drive belt.

4. Remove the water pump pulley.

5. Disconnect the electrical harness connectors from both ignition coils.

6. Remove both ignition coils with their brackets attached.

7. Raise and safely support the vehicle.

8. Remove the two upper power steering pump retaining bolts.

9. Remove the two lower power steering pump retaining bolts and move the pump aside.

10. Disconnect the Crankshaft Position (CKP) sensor electrical harness connector. Remove the retaining bolt and remove the CKP sensor.

11. Drain the engine oil into a suitable container.

12. Remove the four oil pan-to-engine front cover retaining bolts.

13. Lower the vehicle.

14. Remove the crankshaft damper retaining bolt and washer from the crankshaft.

15. Install Crankshaft Damper Remover

T58P-6316-D or equivalent and pull the damper from the crankshaft.

16. Remove the Camshaft Position (CMP) sensor retaining bolt and remove the CMP sensor.

17. Remove the idler pulley bolt and remove the pulley.

18. Remove the three belt tensioner bolts and remove the tensioner.

19. Remove the eight engine front cover retaining bolts and the seven nuts. Swing the top of the cover out off the dowel pins and remove the cover.

20. Remove the sensor ring from the crankshaft.

21. Use Camshaft Positioning Tool T96T-6256-A or equivalent, to position the camshaft.

22. Rotate the crankshaft until both camshaft keyways are 90 degrees from the cam cover surface. Be sure the copper links line up with the dots on the camshaft sprockets.

✳✳ WARNING

At no time, when the timing chains are removed and the cylinder heads are installed may the crankshaft or the camshaft be rotated. Severe piston and valve damage will occur.

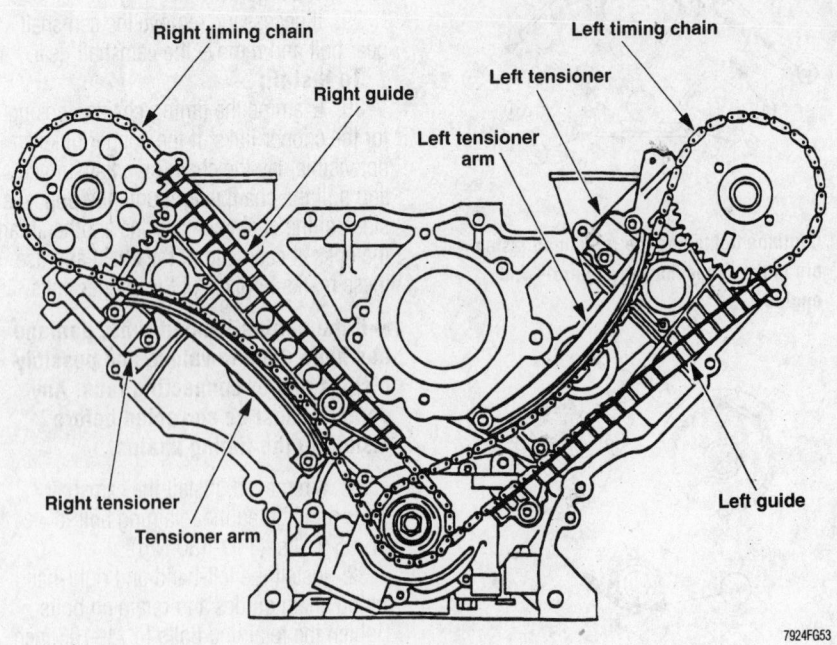

Timing chains and related components—5.4L and 6.8L engines

7924FG53

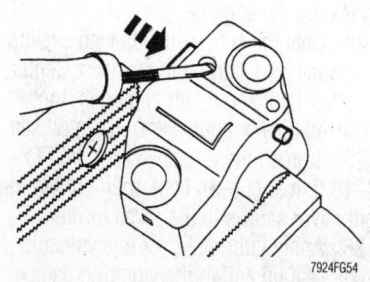

7924FG54

Compress the tensioner while holding the ratchet mechanism with a suitable tool—5.4L and 6.8L engines

23. Install Camshaft Holding Tool T96T-6256-B or equivalent, on the camshaft.

24. Remove 2 left-hand and right-hand tensioner bolts and remove the timing chain tensioners.

25. Slide the left-hand and right-hand tensioner guides off the dowel pins.

26. Remove the right-hand timing chain from the camshaft sprocket.

27. Remove the left-hand timing chain from the camshaft sprocket.

28. Remove the left-hand and right-hand timing chain guide bolts and remove the timing chain guides.

To install:

29. Examine the timing chains, looking for the copper links. If the copper links are not visible, lay the chain on a flat surface and pull the chain taught until the opposite sides of the chain contact one another. Mark the links at each end of the chain and use these marks in place of the copper links.

➡**If the engine jumped time, damage has been done to valves and possibly pistons and/or connecting rods. Any damage must be corrected before installing the timing chains.**

30. Install the left-hand and right-hand timing chain guides and retaining bolts.

7924FG92

If the copper links are not visible, mark one link on one end of the chain and two links on the opposite end of the chain—5.4L and 6.8L engines

Tighten the retaining bolts to 71–106 inch lbs. (8–12 Nm).

31. If removed, install the left-hand crankshaft sprocket with the tapered part of the sprocket facing away from the engine block.

➡**The crankshaft sprockets are identical. They may only be installed one way, with the tapered part of the sprockets facing each other. Ensure that the keyway and timing marks on the crankshaft sprockets are aligned.**

32. Install the left-hand timing chain on the camshaft and crankshaft sprockets. Be sure the one copper link aligns with the mark on the crankshaft sprocket and the two copper links align with the mark on the camshaft sprocket.

33. Install the right-hand crankshaft sprocket with the tapered part of the sprocket facing the left-hand crankshaft sprocket.

34. Install the right-hand timing chain on the camshaft and crankshaft sprockets. Be sure the copper links of the timing chain line up with the timing marks on both sprockets.

35. It is necessary to bleed the timing chain tensioners before installation. Proceed as follows:

a. Place the timing chain tensioner in a soft-jawed vise.

b. Using a small pick or similar tool, hold the ratchet lock mechanism away from the ratchet stem and slowly compress the tensioner plunger by rotating the vise handle.

✴✴ WARNING

The tensioner must be compressed slowly or damage to the internal seals will result.

c. Once the tensioner plunger bottoms in the tensioner bore, continue to hold the ratchet lock mechanism and push down on the ratchet stem until flush with the tensioner face.

d. While holding the ratchet stem flush to the tensioner face, release the ratchet lock mechanism and install a paper clip or similar tool in the tensioner body to lock the tensioner in the collapsed position.

e. The paper clip must not be removed until the timing chain, tensioner, tensioner arm and timing chain guide are completely installed on the engine.

36. Install the left-hand and right-hand timing chain tensioner guides on the dowel pins.

37. Place the left-hand and right-hand timing chain tensioners in position and install the retaining bolts. Tighten the bolts to 15–22 ft. lbs. (20–30 Nm).

38. Remove the retaining pins from the timing chain tensioners.

39. Remove Cam Holding Tool T96T-6256-B or equivalent, from the camshaft.

40. Install the crankshaft sensor ring on the crankshaft.

41. Apply silicone gasket along the cylinder head-to-cylinder block and engine oil pan-to-cylinder block sealing surfaces.

42. Install the engine front cover carefully onto the dowel pins.

43. Tighten the engine front cover bolts, in sequence, in the following manner:

a. Tighten bolts 1 through 5 to 15–22 ft. lbs. (20–30 Nm).

b. Tighten bolts 6 through 15 to 29–40 ft. lbs. (40–55 Nm).

44. Place the idler pulley in position and install the retaining bolt. Tighten the bolt to 15–22 ft. lbs. (20–30 Nm).

45. Place the drive belt tensioner in position and install 3 retaining bolts. Tighten the bolts to 15–22 ft. lbs. (20–30 Nm).

46. Place the CMP sensor in position and install the retaining bolt. Tighten the bolt to 106 inch lbs. (12 Nm).

47. Install the damper on the crankshaft. Ensure the crankshaft key and keyway are aligned.

48. Using Crankshaft Damper Replacer T74P-6316-B or equivalent, install the crankshaft damper.

49. Raise and safely support the vehicle.

50. Install 4 front oil pan-to-engine front cover retaining bolts and tighten in sequence in 2 steps:

a. Tighten the bolts to 15 ft. lbs. (20 Nm).

b. Rotate the bolts an additional 60 degrees.

51. Place the CKP sensor in position and install the retaining bolt. Tighten the bolt to 106 inch lbs. (12 Nm).

52. Connect the CKP sensor electrical harness connector.

53. Install the power steering pump with 2 upper and 2 lower retaining bolts. Tighten the bolts to 15–20 ft. lbs. (20–30 Nm).

54. Lower the vehicle.

55. Place the ignition coil and brackets on the engine front cover and install the bracket bolts. Tighten the bolts to 15–22 ft. lbs. (20–30 Nm).

56. Connect the ignition coil and capacitor electrical harness connectors.

57. Connect the CMP electrical harness connector.

58. Install the water pump pulley and tighten the bolts to 15–22 ft. lbs. (20–30 Nm).

59. Install the radiator.

60. Connect the negative battery cable.

61. Start the engine and check for leaks.

62. Road test the vehicle and check for proper engine operation.

5.0L and 5.8L Engines

1. Drain the cooling system and the crankcase.

2. Disconnect the upper and lower radiator hoses and the transmission oil cooler lines and remove the radiator.

3. Disconnect the heater hose from the water pump. Slide the water pump bypass hose clamp toward the water pump.

4. Loosen the alternator pivot bolt and the bolt which secures the alternator adjusting arm to the water pump. Position the alternator out of the way.

5. Remove the power steering pump and air conditioning compressor from their mounting brackets, if so equipped.

6. Remove the fan, spacer, pulley and drive belts.

7. Remove the crankshaft pulley from the crankshaft damper. Remove the damper attaching bolt and washer and remove the damper with a puller.

8. If equipped with a manual fuel pump, disconnect the fuel pump outlet line at the fuel pump. Disconnect the vacuum inlet and outlet lines from the fuel pump. Remove the fuel pump attaching bolts and lay the pump to one side with the fuel inlet line still attached.

9. If necessary, remove the oil level dipstick and the bolt holding the dipstick tube to the exhaust manifold.

10. Remove the oil pan-to-cylinder front cover attaching bolts. Use a sharp, thin cutting blade to cut the oil pan gasket flush with the cylinder block. Remove the front cover and water pump as an assembly.

11. Discard the front cover gasket. If necessary, properly support the cover and carefully drive the oil seal out towards the front of the cover.

12. Rotate the crankshaft counterclockwise to take up the slack on the left side of the chain.

13. Establish a reference point on the cylinder block and measure from this point to the chain.

14. Rotate the crankshaft in the opposite direction to take up the slack on the right side of the chain.

15. Force the left side of the chain out with your fingers and measure the distance between the reference point and the chain. The timing chain deflection is the difference between the two measurements. If the deflection exceeds ½ in. (13mm), replace the timing chain and sprockets.

16. Turn the crankshaft until the timing marks on the sprockets are aligned vertically.

17. Remove the camshaft sprocket retaining screw, if equipped, remove the fuel pump eccentric and washers.

18. Alternately slide both of the sprockets and timing chain off the crankshaft and camshaft until free of the engine.

To install:

19. Clean the front cover mating surfaces of all gasket material and/or sealer. If the front cover seal is being replaced, support the cover to prevent damage and drive out the seal. Coat the new seal with heavy SJ engine oil and install it in the cover, making sure it is not cocked.

20. Position the timing chain on the sprockets so that the timing marks on the sprockets are aligned vertically. Alternately slide the sprockets and chain onto the crankshaft and camshaft sprockets.

21. If equipped with a manual fuel pump, install the fuel pump eccentric washers and attaching bolt on the camshaft sprocket. Tighten to 40–45 ft. lbs.

22. Cut the new oil pan gasket as needed for a correct fit. Apply sealing compound to the oil pan and fit the gasket into place.

23. Apply sealing compound to the gasket surfaces on the cylinder block and back side of the front cover. Position the gasket

TIMING MARKS

7924FG55

Align the marks when installing the timing chain—5.0L, 5.8L and 7.5L engines

onto the cylinder block and fit the front cover onto the engine.

24. Coat the screw threads with sealing compound and start all the screws. Center the cover by inserting an alignment tool (T74P-6019-a or equivalent) in the oil seal.

25. Tighten the oil pan screws first to 12–18 ft. lbs. (17–24 Nm), then tighten the front cover screws to the same torque.

26. Apply Lubriplate® or equivalent to the oil seal lip and to the vibration damper to prevent damage to the seal. Coat the front of the crankshaft with engine oil for damper installation.

27. Line up the damper keyway with the key on the crankshaft and push the damper onto the crankshaft. Install the bolt and washer and tighten to 80 ft. lbs. Install the crankshaft pulley.

28. Install the fan, spacer, pulley and drive belts.

29. Install the bolts holding the fan shroud to the radiator, if so equipped.

30. Install the power steering pump and air conditioning compressor.

31. Install the alternator and adjust the belt tension.

32. Connect the heater hose at the water pump.

33. Install the radiator.

34. Connect the upper and lower radiator hoses, and transmission oil cooler lines.

35. Fill the cooling system and the crankcase.

7.5L Engine

1. Drain the cooling system and crankcase.

2. Remove the radiator shroud and fan.

3. Disconnect the upper and lower radiator hoses and the automatic transmission oil cooler lines from the radiator.

4. Remove the radiator upper support and remove the radiator.

5. Loosen the alternator attaching bolts and air conditioning compressor idler pulley and remove the drive belts with the water pump pulley. Remove the bolts attaching the compressor support to the water pump and remove the bracket (support), if equipped.

6. Remove the crankshaft pulley from the vibration damper. Remove the bolt and washer attaching the crankshaft damper and remove the damper with a puller. Remove the woodruff key from the crankshaft.

7. Loosen the bypass hose at the water pump and disconnect the heater return tube at the water pump.

8. Disconnect and plug the fuel inlet

and outlet lines at the fuel pump and remove the fuel pump.

9. Remove the bolts attaching the front cover to the cylinder block. Cut the oil pan seal flush with the cylinder block face with a thin knife blade prior to separating the cover from the cylinder block. Remove the cover and water pump as an assembly. Discard the front cover gasket and oil pan seal.

10. Rotate the crankshaft counterclockwise to take up the slack on the left side of the chain.

11. Establish a reference point on the cylinder block and measure from this point to the chain.

12. Rotate the crankshaft in the opposite direction to take up the slack on the right side of the chain.

13. Force the left side of the chain out and measure the distance between the reference point and the chain. The timing chain deflection is the difference between the two measurements. If the deflection exceeds ½ in. (13mm), replace the timing chain and sprockets.

14. Turn the crankshaft until the timing marks on the sprockets are aligned vertically.

15. Remove the camshaft sprocket retaining screw and remove the fuel pump eccentric and washers.

16. Alternately slide both of the sprockets and timing chain off the crankshaft and camshaft until free of the engine.

To install:

17. Position the timing chain on the sprockets so the timing marks on the sprockets are aligned vertically. Alternately slide the sprockets and chain onto the crankshaft and camshaft sprockets.

18. Install the fuel pump eccentric washers and attaching bolt on the camshaft sprocket. Tighten to 40–45 ft. lbs. (54–61 Nm).

19. Transfer the water pump if a new cover is going to be installed. Clean all of the gasket sealing surfaces on both the front cover and the cylinder block.

20. Coat the gasket surface of the oil pan with sealer. Cut and position the required sections of a new seal on the oil pan. Apply sealer to the corners.

21. Drive out the old front cover oil seal with a pin punch. Clean out the seal recess in the cover. coat a new seal with Lubriplate® or equivalent grease. Install the seal, making sure the seal spring remains in the proper position. A front cover seal tool,

Ford part No. T72J-117 or equivalent, makes installation easier.

22. Coat the gasket surfaces of the cylinder block and cover with sealer and position the new gasket on the block.

23. Position the front cover on the cylinder block. Use care not to damage the seal and gasket or misplace them.

24. Coat the front cover attaching screws with sealer and install them.

➡**It may be necessary to force the front cover downward to compress the oil pan seal in order to install the front cover attaching bolts. Use a prybar or drift to engage the cover screw holes through the cover and pry downward.**

25. Install the fuel pump.

26. Connect the fuel inlet and outlet lines at the fuel pump.

27. Tighten the bypass hose at the water pump.

28. Connect the heater return tube at the water pump.

29. Install the woodruff key from the crankshaft.

30. Install the damper.

31. Install the crankshaft pulley on the vibration damper.

32. Install the compressor support on the water pump and install the bracket (support), if equipped.

33. Install the drive belts with the water pump pulley.

34. Install the radiator and upper support.

35. Connect the upper and lower radiator hoses and the automatic transmission oil cooler lines.

36. Install the radiator shroud and fan.

37. Fill the cooling system and crankcase.

Tighten the fasteners to the following specifications:

• Front cover bolts: 15–20 ft. lbs. (20–27 Nm)
• Water pump attaching screws: 12–15 ft. lbs. (16–20 Nm)
• Crankshaft damper: 70–90 ft. lbs. (95–122 Nm)
• Crankshaft pulley: 35–50 ft. lbs. (47–68 Nm)
• Fuel pump: 19–27 ft. lbs. (26–37 Nm)
• Oil pan bolts: 9–11 ft. lbs. (12.2–15.0 Nm) for the 5⁄16 in. (7.9mm) screws and to 7–9 ft. lbs. (9.5–12.2 Nm) for the ¼ in. (6.3mm) screws
• Alternator pivot bolt: 45–57 ft. lbs. (61–77 Nm)

DIESEL ENGINE REPAIR

Engine Assembly

REMOVAL & INSTALLATION

1. Remove the hood.

2. Drain the coolant.

3. Remove the air cleaner and intake duct assembly and cover the air intake opening with a clean rag to keep out the dirt.

4. Remove the upper grille support bracket and upper air conditioning condenser mounting bracket.

5. If equipped with air conditioning, the system MUST be discharged to remove the condenser.

6. Remove the radiator fan shroud halves.

7. Remove the fan and clutch assembly.

8. Detach the radiator hoses and the transmission cooler lines, if equipped.

9. Remove the condenser. Cap all openings at once!

10. Remove the radiator.

11. Remove the power steering pump and position it out of the way.

12. Disconnect the fuel supply line heater and alternator wires at the alternator.

13. Disconnect the oil pressure sending unit wire at the sending unit, remove the sender from the firewall and lay it on the engine.

14. Disconnect the accelerator cable and the speed control cable, if equipped, from the injection pump. Remove the cable bracket with the cables attached, from the intake manifold and position it out of the way.

15. Disconnect the transmission kickdown rod from the injection pump, if equipped.

16. Disengage the main wiring harness connector from the right side of the engine and the ground strap from the rear of the engine.

17. Remove the fuel return hose from the left rear of the engine.

18. Remove the two upper transmission-to-engine attaching bolts.

19. Disconnect the heater hoses.

20. Disconnect the water temperature sender wire.

21. Disconnect the overheat light switch wire and position the wire out of the way.

22. Raise the vehicle and support it securely.

23. Disconnect the battery ground cables from the front of the engine and the cables from the starter.

24. Remove the fuel inlet line and plug the fuel line at the fuel pump.

25. Detach the exhaust pipe at the exhaust manifold.

26. Disconnect the engine insulators from the No. 1 crossmember.

27. Remove the flywheel inspection plate and the four converter-to-flywheel attaching nuts, if equipped with automatic transmission.

28. Lower the vehicle.

29. Supporting the transmission on a jack.

30. Remove the four lower transmission attaching bolts.

31. Attach an engine lifting sling and remove the engine from the vehicle.

To install:

32. Lower the engine into vehicle.

33. Align the converter to the flexplate and the engine dowels to the transmission.

34. Install the engine mount bolts and tighten them to 80 ft. lbs. (109 Nm).

35. Remove the engine lifting sling.

36. Install the four lower transmission attaching bolts. Tighten the bolts to 65 ft. lbs. (88 Nm).

37. Remove transmission jack.

38. Raise and support the front end.

39. If equipped with an automatic transmission, install the four converter-to-flywheel attaching nuts. Tighten the nuts to 34 ft. lbs. (47 Nm).

40. Install the flywheel inspection plate. Tighten the bolts to 60–90 inch lbs. (6.7–10.0 Nm).

41. Attach the exhaust pipe at the exhaust manifold.

42. Connect the fuel inlet line.

43. Attach the battery ground cables to the front of the engine.

44. Connect the starter cables at the starter.

45. Lower the vehicle.

46. Connect the overheat light switch wire.

47. Attach the water temperature sender wire.

48. Connect the heater hoses.

49. Install the two upper transmission-to-engine attaching bolts. Tighten the bolts to 65 ft. lbs. (88 Nm).

50. Connect the fuel return hose at the left rear of the engine.

51. Attach the main wiring harness connector at the right side of the engine and the ground strap from the rear of the engine.

52. Connect the transmission kickdown rod at the injection pump, if equipped.

53. Connect the accelerator cable and the speed control cable, if equipped, at the injection pump.

54. Install the cable bracket with the cables attached, to the intake manifold.

55. Install the oil pressure sending unit.

56. Connect the oil pressure sending unit wire at the sending unit.

57. Connect the fuel supply line heater and alternator wires at the alternator.

58. Install the power steering pump.

59. Install the radiator.

60. Install the condenser.

61. Connect the radiator hoses and the transmission cooler lines, if equipped.

62. Install the fan and clutch assembly.

63. Install the radiator fan shroud halves.

64. If equipped with air conditioning, charge the system.

65. Install the upper grille support bracket and upper air conditioning condenser mounting bracket.

66. Install the air cleaner and intake duct assembly.

67. Fill the cooling system.

68. Install the hood.

Water Pump

REMOVAL & INSTALLATION

1. Disconnect both battery ground cables.

2. Drain the cooling system.

3. Remove the radiator shroud halves.

4. Remove the fan clutch and fan.

➡**The fan clutch bolts are left-hand thread. Remove them by turning clockwise.**

5. Remove the power steering pump belt.

6. Remove the air conditioning compressor belt.

7. Remove the vacuum pump drive belt.

8. Remove the alternator drive belt.

9. Remove the water pump pulley.

10. Disconnect the heater hose at the water pump.

11. If installing a new pump, remove the heater hose fitting from the old pump.

12. Remove the alternator adjusting arm and bracket.

13. Unbolt the air conditioning compres-sor and position it out of the way; do not disconnect the refrigerant lines!

14. Remove the air conditioning compressor brackets.

15. Unbolt the power steering pump and bracket and position it out of the way; do not disconnect the power steering fluid lines.

16. Remove the bolts attaching the water pump to the front cover and lift off the pump.

To install:

17. Thoroughly clean the mating surfaces of the pump and front cover.

18. Obtain two dowel pins, anything that will fit into two mounting bolt holes in the front cover, when installing the water pump.

19. Using a new gasket, position the water pump over the dowel pins and into place on the front cover.

20. Install the attaching bolts. The two top center and two bottom center bolts must be coated with RTV silicone sealant prior to installation. Also, the four bolts marked No. 1 are a different length than the other bolts. Tighten the bolts to 14 ft. lbs. (19 Nm).

21. Install the water pump pulley.

22. Wrap the heater hose fitting threads with Teflon® tape and screw the fitting into the water pump. Tighten it to 18 ft. lbs. (25 Nm).

23. Connect the heater hose to the pump.

24. Install the power steering pump and bracket. Install the belt.

25. Install the air conditioning compressor bracket.

26. Install the air conditioning compressor. Install the belt.

27. Install the alternator adjusting arm and install the belt.

28. Install the vacuum pump drive belt.

29. Adjust all the drive belts.

30. Install the fan and clutch. Remember that the bolts are left-hand thread. Turn

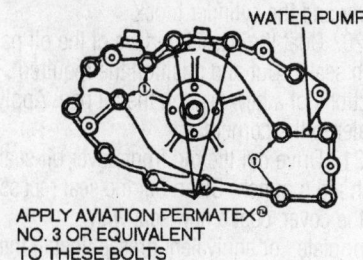

WATER PUMP

APPLY AVIATION PERMATEX®
NO. 3 OR EQUIVALENT
TO THESE BOLTS

① THESE BOLTS 2 3/4 IN. LONG
ALL OTHERS ARE 1 1/2 IN. LONG

7924FG56

Apply aviation grade RTV sealant to the bolts indicated—7.3L Diesel engine

them counterclockwise to tighten them. Tighten them to 45 ft. lbs. (61 Nm).

31. Install the fan shroud halves.
32. Fill and bleed the cooling system.
33. Connect the battery ground cables.
34. Start the engine and check for leaks.

Glow Plugs

REMOVAL & INSTALLATION

✳✳ CAUTION

The red-striped wiring harness carries 115V direct current. Severe electrical shock may be received. DO NOT pierce.

1. Disconnect the negative battery cable.
2. Remove the rocker arm cover.
3. Disconnect the glow plug electrical leads using a pair of pliers.
4. Remove the glow plugs by unscrewing them from the cylinder head with a 10mm socket and wrench.
5. Inspect the tips of the plugs for any evidence of distortion or missing tip ends; replace them if necessary.

To install:

6. Install the glow plug into the cylinder head. Tighten the glow plugs to 14 ft. lbs. (19 Nm).
7. Attach the glow plug electrical connector. Be sure that the glow plug wiring is routed to avoid moving components in the engine bay.
8. Install the rocker arm cover.

➡ **When the battery is disengaged and reconnected, some abnormal drive symptoms may occur while the vehicle relearns its adaptive strategy. The vehicle may need to be driven 10 miles (16 km) or more to relearn this strategy.**

9. Connect the negative battery cable.

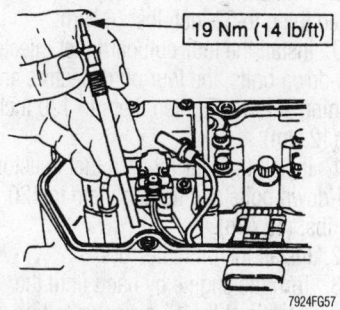

Tighten the glow plugs to 14 ft. lbs. (19 Nm) and attach the connector—7.3L engine

Cylinder Head

REMOVAL & INSTALLATION

F-Series

1. Remove the hood.
2. Drain the coolant.
3. Remove the air cleaner and intake duct assembly and cover the air intake opening with a clean rag to keep out the dirt.
4. Remove the upper grille support bracket and upper air conditioning condenser mounting bracket.
5. If equipped with air conditioning, the system MUST be discharged to remove the condenser.
6. Remove the radiator fan shroud halves.
7. Remove the fan and clutch assembly.
8. Detach the radiator hoses and the transmission cooler lines, if equipped.
9. Remove the condenser. Cap all openings at once!
10. Remove the radiator.
11. Remove the power steering pump and position it out of the way.
12. Disconnect the fuel supply line heater and alternator wires at the alternator.
13. Disconnect the oil pressure sending unit wire at the sending unit, remove the sender from the firewall and lay it on the engine.
14. Disconnect the accelerator cable and the speed control cable, if equipped, from the injection pump. Remove the cable bracket, with the cables attached, from the intake manifold and position it out of the way.
15. Disconnect the transmission kickdown rod from the injection pump, if equipped.
16. Disengage the main wiring harness connector from the right side of the engine and the ground strap from the rear of the engine.
17. Remove the fuel return hose from the left rear of the engine.
18. Remove the two upper transmission-to-engine attaching bolts.
19. Disconnect the heater hoses.
20. Disconnect the water temperature sender wire.
21. Disconnect the overheat light switch wire and position the wire out of the way.
22. Raise the vehicle and support it securely.
23. Disconnect the battery ground cables from the front of the engine and the starter cables from the starter.

24. Remove the fuel inlet line and plug the fuel line at the fuel pump.
25. Detach the exhaust pipe at the exhaust manifold.
26. Disconnect the engine insulators from the No. 1 crossmember.
27. Remove the flywheel inspection plate and the four converter-to-flywheel attaching nuts, if equipped with an automatic transmission.
28. Lower the vehicle.
29. Supporting the transmission on a jack.
30. Remove the four lower transmission attaching bolts.
31. Attach an engine lifting sling and remove the engine from the vehicle.

To install:

32. Lower the engine into the vehicle.
33. Align the converter to the flexplate and the engine dowels to the transmission.
34. Install the engine mount bolts and tighten them to 80 ft. lbs. (109 Nm).
35. Remove the engine lifting sling.
36. Install the four lower transmission attaching bolts. Tighten the bolts to 65 ft. lbs. (88 Nm).
37. Remove transmission jack.
38. Raise and support the front end.
39. If equipped with automatic transmission, install the four converter-to-flywheel attaching nuts. Tighten the nuts to 34 ft. lbs. (47 Nm).
40. Install the flywheel inspection plate. Tighten the bolts to 60–90 inch lbs. (6.7–10.0 Nm).
41. Attach the exhaust pipe at the exhaust manifold.
42. Connect the fuel inlet line.
43. Connect the battery ground cables to the front of the engine.
44. Attach the starter cables at the starter.
45. Lower the vehicle.
46. Connect the overheat light switch wire.
47. Connect the water temperature sender wire.
48. Connect the heater hoses.
49. Install the two upper transmission-to-engine attaching bolts. Tighten the bolts to 65 ft. lbs. (88 Nm).
50. Connect the fuel return hose at the left rear of the engine.
51. Attach the main wiring harness connector at the right side of the engine and the ground strap from the rear of the engine.
52. Connect the transmission kickdown rod at the injection pump, if equipped.
53. Connect the accelerator cable and the speed control cable, if equipped, at the injection pump.

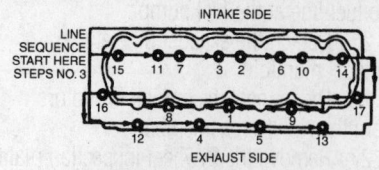

STEP 1. TIGHTEN BOLTS TO (65 FT-LB) IN NUMBERED
SEQUENCE SHOWN ABOVE
STEP 2. TIGHTEN BOLTS TO (85 FT-LB) IN NUMBERED
SEQUENCE SHOWN ABOVE
STEP 3. TIGHTEN BOLTS TO (100 FT-LB) IN LINE
SEQUENCE SHOWN ABOVE
STEP 4. REPEAT STEP NO. 3

7924FG24

**Tighten the cylinder head bolts according
to the sequence shown—Diesel engines**

54. Install the cable bracket with the cables attached, to the intake manifold.

55. Install the oil pressure sending unit.

56. Connect the oil pressure sending unit wire at the sending unit.

57. Connect the fuel supply line heater and alternator wires at the alternator.

58. Install the power steering pump.

59. Install the radiator.

60. Install the condenser.

61. Connect the radiator hoses and the transmission cooler lines, if equipped.

62. Install the fan and clutch assembly.

63. Install the radiator fan shroud halves.

64. If equipped with air conditioning, charge the system.

65. Install the upper grille support bracket and upper air conditioning condenser mounting bracket.

66. Install the air cleaner and intake duct assembly.

67. Fill the cooling system.

68. Install the hood.

E-Series

RIGHT CYLINDER HEAD

1. Remove the engine from the van.

2. Disengage the vacuum hose from the right intake manifold.

3. Disengage both electrical harness connectors from the valve cover.

4. Remove the valve cover and gasket.

5. Remove the intake manifold covers.

6. Disengage the fuel injector electrical injectors.

7. Disengage and cap the fuel lines from the heads.

8. Loosen the banjo bolt from the fuel supply line at the fuel pump.

9. Remove the fuel supply assembly.

10. Disconnect the heater hose.

11. Loosen the dipstick tube bracket retainer at the right exhaust manifold.

12. Remove the rocker arms and pushrods.

13. Loosen the four inboard fuel injector hold-down bolts.

14. Remove the four outboard fuel injector hold-down bolts, retaining screws and four oil deflectors.

✳✳ WARNING

Remove the oil drain plugs prior to removing the injectors or oil could enter the combustion chamber which could result in hydrostatic lock and severe engine damage.

15. Remove the oil rail drain plugs.

16. Remove the fuel injectors using Injector Remover No. T94T-9000-AH1, or equivalent. Position the tool's fulcrum beneath the fuel injector hold-down plate and over the edge of the cylinder head. Install the remover screw in the threaded hole of the fuel injector plate (see illustration). Tighten the screw to lift out the injector from its bore. Place the injector in a suitable protective sleeve such as Rotunda Injector Protective Sleeve, No. 014–00933–2, and set the injector in a suitable holding rack.

➡️**During removal the injector tab (located above the fuel injector) must be bent completely flat and flush with the cowl and heat shield.**

17. Use a suitable vacuum tool, such as Rotunda Vacuum Pump, No. 021–00037, or equivalent to remove the oil and fuel left over in the injector bores.

18. Remove the four glow plugs.

19. Disconnect the high pressure oil pump supply line from the cylinder head.

20. Remove the grille opening reinforcement, headlamp assembly, radiator, oil reservoir and fuel filter assembly.

21. Remove the exhaust back pressure line.

22. Loosen the glow plug relay bracket retainers and disengage the ground wire.

23. Disconnect the fuel return line at the front of the cylinder head.

24. Loosen the cylinder head bolts.

25. Carefully lift the cylinder head out of the engine compartment and remove the head gaskets.

To install:

➡️**To prepare a good seat for the fuel injector O-rings, use a suitable injector sleeve brush to clean any debris from the bore.**

26. Carefully clean the cylinder block and head mating surfaces.

27. Position the cylinder head gasket on the engine block and carefully lower the cylinder head in place.

28. Install the cylinder head bolt and tighten in three steps using the sequence shown in the illustration.

➡️**Lubricate the threads and the mating surfaces of the bolt heads and washers with engine oil.**

29. Connect the fuel return line to the cylinder head.

30. Install the glow plug relay bracket and ground wire, then tighten the retainers.

31. Install the exhaust back pressure line.

32. Install the high pressure fuel supply line and tighten the fitting to 19 ft. lbs. (26 Nm).

33. Connect the heater hose to the cylinder head.

34. Connect the manifold hoses.

35. Connect the fuel supply lines to the rear of the cylinder head.

36. Install the banjo bolt through the fuel line and into the pump. Tighten the pump to 40 ft. lbs. (55 Nm).

37. Coat the glow plugs with anti-seize compound and install them. Tighten the glow plugs to 14 ft. lbs. (19 Nm).

38. Install the fuel injectors using special tools as follows:

   a. Lubricate the injector O-rings with clean engine oil. Using new copper washers, carefully push the injectors square into the bore using hand pressure only to seat the O-rings.

   b. Position the open end of Injector replacer, No. T94T-9000-AH2, or equivalent between the fuel injector body and injector hold-down plate, while positioning the opposite end of the tool over the edge of the cylinder head.

   c. Align the hole in the tool with the threaded hole in the cylinder head and install the bolt from the tool kit. Tighten the bolt to fully seat the injector, then remove the bolt and tool.

39. Install the oil rail drain plugs and tighten them to 53 inch lbs. (6 Nm).

40. Install the four outboard fuel injector hold-down bolts, the four oil deflectors and retaining screws. Tighten them to 120 inch lbs. (12 Nm).

41. Install the four inboard fuel injector hold-down bolts and tighten them to 120 inch lbs. (12 Nm).

42. Install the oil deflectors.

43. Turn the engine by hand until the timing mark is at the 11 o'clock position as viewed from the front.

44. Dip the pushrod ends in clean

engine oil and install the pushrods with the copper colored ends toward the rocker arms, making sure the pushrods are fully seated in the tappet pushrod seats.

45. Install the rocker arms and posts in their original positions. Apply multipurpose grease to the valve stem tips. Install the rocker arm posts, bolts and tighten to 27 ft. lbs. (37 Nm).

46. Install the valve cover gasket.

47. Connect the wiring to the fuel injectors and glow plugs.

48. Install the valve cover, tightening the bolts to 97 inch lbs. (11 Nm).

49. Connect both electrical harness connectors to the valve cover.

50. Connect the vacuum hose to the right intake valve manifold cover.

51. Install the engine in the van.

LEFT CYLINDER HEAD

1. Remove the engine from the van.
2. Remove the wiring harness bracket.
3. Disengage the electrical connections from the valve cover gasket, then remove the valve cover.
4. Disengage the electrical connections from the fuel injectors and glow plugs.
5. Remove the valve cover gasket.
6. Remove the rocker arms and pushrods, KEEPING EVERYTHING IN ORDER.
7. Remove the four inboard fuel injector hold-down bolts.

✳✳ WARNING

Remove the oil drain plugs prior to removing the injectors or oil could enter the combustion chamber which could result in hydrostatic lock and severe engine damage.

8. Remove the oil rail drain plugs.

✳✳ CAUTION

Be sure to retrieve the fuel injector copper washer, located at the tip of the injector during removal.

9. Remove the four outboard fuel injector hold-down bolts, retaining screws and four oil deflectors.

10. Remove the fuel injectors using Injector Remover No. T94T-9000-AH1, or equivalent. Position the tool's fulcrum beneath the fuel injector hold-down plate and over the edge of the cylinder head. Install the remover screw in the threaded hole of the fuel injector plate (see illustration). Tighten the screw to lift out the injec-

tor from its bore. Place the injector in a suitable protective sleeve such as Rotunda Injector Protective Sleeve, No. 014–00933–2, and set the injector in a suitable holding rack.

11. Use a suitable vacuum tool, such as Rotunda Vacuum Pump, No. 021–00037, or equivalent to remove the oil and fuel left over in the injector bores.

12. Remove the four glow plugs.

13. Disconnect the fuel supply lines from the rear of the cylinder head.

14. Remove the banjo bolt from the fuel line at the pump.

15. Disengage the oil line from the high pressure oil pump.

16. Disengage the electrical connection from the injection control pressure sensor.

17. Remove the high pressure oil supply line from the left cylinder head.

18. Loosen the fuel line nut from the intake manifold stud, then disconnect the fuel return line from the left cylinder head.

19. Loosen the fuel return line block screws at the front of the left cylinder head.

20. Remove the fuel line retaining clamp from the intake manifold cover.

21. Loosen the cylinder head bolts.

22. Remove the oil reservoir and fuel filter.

23. Remove the cylinder head and gasket.

To install:

➡**To prepare a good seat for the fuel injector O-rings, use a suitable injector sleeve brush to clean any debris from the bore.**

24. Carefully clean the cylinder block and head mating surfaces.

25. Position the cylinder head gasket on the engine block and carefully lower the cylinder head in place.

26. Install the cylinder head bolt and tighten in three steps using the sequence shown in the illustration.

27. Install the fuel line retaining clamp to the intake manifold cover.

28. Tighten the fuel return line block screws at the front of the left cylinder head.

29. Connect the fuel return line to the left cylinder head.

30. Tighten the fuel line nut from the intake manifold stud.

31. Install the high pressure oil supply line to the left cylinder head.

32. Engage the electrical connection to the injection control pressure sensor.

33. Engage the oil line to the high pressure oil pump.

34. Install the manifold hoses.

35. Install the fuel supply line.

36. Install the banjo bolt through the fuel line into the pump. Tighten the bolt to 40 ft. lbs. (55 Nm).

37. Connect the fuel supply lines at the rear of the cylinder heads.

38. Coat the glow plugs with anti-seize compound and install them. Tighten the glow plugs to 14 ft. lbs. (19 Nm).

39. Install the fuel injectors using special tools as follows:

   a. Lubricate the injector O-rings with clean engine oil. Using new copper washers, carefully push the injectors square into the bore using hand pressure only to seat the O-rings.

   b. Position the open end of Injector replacer, No. T94T-9000-AH2, or equivalent between the fuel injector body and injector hold-down plate, while positioning the opposite end of the tool over the edge of the cylinder head.

   c. Align the hole in the tool with the threaded hole in the cylinder head and install the bolt from the tool kit. Tighten the bolt to fully seat the injector, then remove the bolt and tool.

40. Install the four outboard fuel injector hold-down bolts, the four oil deflectors and retaining screws. Tighten them to 120 inch lbs. (12 Nm).

41. Install the oil deflectors and tighten the bolts to 120 inch lbs. (12 Nm).

42. Install the oil rail drain plugs and tighten them to 53 inch lbs. (6 Nm).

43. Install the four inboard fuel injector hold-down bolts and tighten them to 120 inch lbs. (12 Nm).

44. Turn the engine over by hand until the timing mark is at the 11 o'clock position as viewed from the front.

45. Dip the pushrod ends in clean engine oil and install the pushrods with the copper colored ends toward the rocker arms, making sure the pushrods are fully seated in the tappet pushrod seats.

46. Install the rocker arms and posts in their original positions. Apply multipurpose grease to the valve stem tips. Install the rocker arm posts, bolts and tighten to 27 ft. lbs. (37 Nm).

47. Install the valve cover gasket.

48. Connect the wiring to the fuel injectors and glow plugs.

49. Install the valve cover, tightening the bolts to 97 inch lbs. (11 Nm).

50. Connect both electrical harness connectors to the valve cover.

51. Install the engine in the van.

Rocker Arms

REMOVAL & INSTALLATION

1. Disconnect the ground cables from both batteries.

2. Remove the valve cover attaching screws and remove both valve cover.

3. Remove the valve rocker arm post mounting bolts. Remove the rocker arms and posts in order and mark them with tape so they can be installed in their original positions.

4. If the cylinder heads are to be removed, then the pushrods can now be removed. Make a holder for the pushrods out of a piece of wood or cardboard and remove the pushrods in order. It is very important that the pushrods be reinstalled in their original order. The pushrods can remain in position if no further disassembly is required.

To install:

5. If the pushrods were removed, install them in their original locations. be sure they are fully seated in the tappet seats.

➡**The copper colored end of the pushrod goes toward the rocker arm.**

6. Apply a polyethylene grease to the valve stem tips. Install the rocker arms and posts in their original positions.

7. Turn the engine over by hand until the valve timing mark is at the 11:00 o'clock

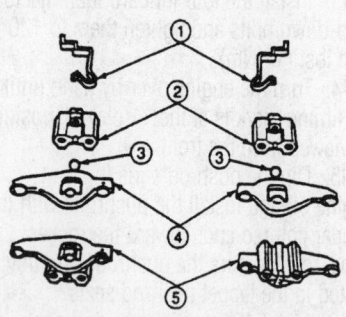

Item	Description
1	Snap Retaining Clip
2	Rocker Arm Pedestal
3	Rocker Arm Ball
4	Rocker Arm
5	Rocker Arm Assembly

7924FG58

Exploded view of the rocker arm assembly—Diesel engines

position, as viewed from the front of the engine. Install all of the rocker arm post attaching bolts and tighten to 20 ft. lbs. (27 Nm).

8. Install new valve cover gaskets and install the valve cover. Install the battery cables, start the engine and check for leaks.

Turbocharger

REMOVAL & INSTALLATION

1. Disconnect the negative battery cable.

2. Remove the two air intake tube assembly bolts, clamps at the turbocharger, crankcase breather assembly, engine air cleaner and air intake tube and hoses.

3. Remove the exhaust outlet clamp from the turbocharger.

4. Raise and safely support the vehicle.

5. Remove the engine charge exhaust pipe bolt from the transmission, if so equipped.

6. Remove the bolts and nuts from the catalytic converter-to-engine charge exhaust pipe, if so equipped.

7. Loosen two bolts retaining the turbocharger exhaust inlet pipe to the left exhaust manifold.

8. For automatic transmissions, remove the bolts retaining the left turbocharger exhaust inlet pipe to the turbocharger exhaust inlet adapter.

9. Loosen the two bolts retaining the turbocharger exhaust inlet pipe to the right exhaust manifold.

10. Remove the lower bolt retaining the right turbocharger exhaust inlet pipe to the turbocharger exhaust inlet adapter.

11. Lower the vehicle.

12. For automatic transmissions, remove the upper bolts retaining the right and left turbocharger exhaust inlet pipes to the turbocharger exhaust inlet adapter.

13. Remove the right engine lift hook and bolt.

14. Loosen the air inlet hose clamp at the turbocharger. Disconnect the hose and lay aside.

15. For automatic transmissions, loosen the four intake manifold hose clamps, and one clamp retaining the compressor manifold to the turbocharger. Remove the compressor manifold.

16. Remove the four bolts retaining the turbocharger pedestal assembly to the cylinder block.

17. Remove the turbocharger assembly and detach all electrical connectors from it.

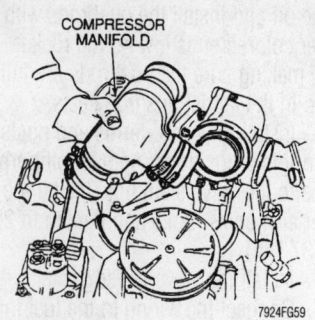

COMPRESSOR MANIFOLD

7924FG59

After loosening the clamps, the compressor manifold can be removed—Diesel engines

➡**If the turbocharger is not being removed for service, install the Fuel/Oil Turbo Protector Cap Set T94T-9395-AH or equivalent.**

18. Remove the oil gallery O-rings.

To install:

19. Install new oil gallery O-rings.

20. Attach the turbocharger electrical connectors and install the turbocharger assembly.

21. Install the four bolts retaining the turbocharger pedestal assembly to the engine block. Tighten the bolts to 18 ft. lbs. (25 Nm).

22. Loosely install the four bolts retaining the right and left turbocharger exhaust inlet pipes to the turbocharger exhaust inlet adapter.

23. Install the compressor manifold, intake manifold hoses and clamps. Be sure the compressor outlet seal is in position.

24. Install the right engine lift hook and bolt.

25. Raise and safely support the front of the vehicle.

26. Tighten the two right and left lower bolts (retaining the turbocharger exhaust inlet pipes to the turbocharger exhaust inlet adapter) to 36 ft. lbs. (49 Nm).

27. Tighten the four right and left bolts and nuts (retaining the turbocharger exhaust inlet pipes to the exhaust manifolds) to 36 ft. lbs. (49 Nm).

28. Install the catalytic converter to the engine charge exhaust pipe bolts and nuts.

29. Lower the vehicle.

30. Tighten the two right and left upper bolts (retaining the turbocharger exhaust inlet pipes to the turbocharger exhaust inlet adapter) to 36 ft. lbs. (49 Nm).

31. Install the exhaust outlet clamp to the turbocharger.

32. Install the air intake tube and hose assembly.

33. Connect the negative battery cable.

Intake Manifold

REMOVAL & INSTALLATION

1. Open the hood and remove both battery ground cables.

2. Remove the air cleaner and install clean rags into the air intake of the intake manifold. It is important that no dirt or foreign objects get into the intake.

3. Remove the injection pump.

4. Remove the fuel return hose from No. 7 and No. 8 rear nozzles and remove the return hose to the fuel tank.

5. Label the positions of the wires and remove the engine wiring harness from the engine.

➡ **The engine harness ground cables must be removed from the back of the left cylinder head.**

6. Remove the bolts attaching the intake manifold to the cylinder heads and remove the manifold.

7. Remove the CDR tube grommet from the valley pan.

8. Remove the bolts attaching the valley pan strap to the front of the engine block and remove the strap.

9. Remove the valley pan drain plug and remove the valley pan.

To install:

10. Apply a ⅛ in. (3mm) bead of RTV sealer to each end of the cylinder block.

➡ **The RTV sealer should be applied immediately prior to the valley pan installation.**

11. Install the valley pan drain plug, CDR tube and new grommet into the valley pan.

12. Install a new O-ring and new back-up ring on the CDR valve.

13. Install the valley pan strap on the front of the valley pan.

14. Install the intake manifold and tighten the bolts to 24 ft. lbs. (33 Nm) using the sequence.

15. Reconnect the engine wiring harness and the engine ground wire located to the rear of the left cylinder head.

16. Install the injection pump.

17. Install the No. 7 and No. 8 fuel return hoses and the fuel tank return hose.

18. Remove the rag from the intake manifold and replace the air cleaner. Reconnect the battery ground cables to both batteries.

➡ **If necessary, purge the nozzle high pressure lines of air by loosening the connector one half to one turn and cranking the engine until solid stream of fuel, devoid of any bubbles, flows from the connection.**

19. Run the engine and check for oil and fuel leaks.

※※ CAUTION

Keep eyes and hands away from the nozzle spray. Fuel spraying from the nozzle under high pressure can penetrate the skin.

20. Check and adjust the injection pump timing.

Exhaust Manifold

REMOVAL & INSTALLATION

1. Disconnect the ground cables from both batteries.

2. Raise the vehicle and safely support it.

3. Disconnect the muffler inlet pipe from the exhaust manifolds.

4. If removing the right manifold, lower the vehicle. When removing the left manifold, raise the vehicle and remove the manifold from underneath. Bend the tabs on the manifold attaching bolts, then remove the bolts and manifold.

To install:

5. Before installing, clean all mounting surfaces on the cylinder heads and the manifold. Apply an anti-seize compound on the manifold both threads and install the left manifold, using a new gasket and new locking tabs.

6. Tighten the bolts to 45 ft. lbs. (61 Nm) and bend the tabs over the flats on the bolt heads to prevent the bolts from loosening.

7. Raise the vehicle to install the right manifold. Install the right manifold Steps 5 and 6.

8. Connect the inlet pipes to the manifold and tighten. Lower the vehicle, connect the batteries and run the engine to check for exhaust leaks.

Camshaft and Valve Lifters

REMOVAL & INSTALLATION

➡ **Ford recommends removing the Diesel engine from the vehicle for camshaft removal.**

1. Remove the intake manifold and valley pan, if equipped.

2. Remove the rocker covers, and either remove the rocker arm shafts or loosen the rockers on their pivots and remove the pushrods. The pushrods must be reinstalled in their original positions.

3. Remove the valve lifters in sequence with a magnet. They must be replaced in their original positions.

4. Remove the timing gear cover, timing gear and sprockets.

➡ **A camshaft removal tool, Ford part No. T65L-6250-A and Adapter 14-0314, or equivalents, are needed to remove the Diesel camshaft.**

To install:

5. Liberally coat the camshaft with oil before installing it. Slide the camshaft into the engine very carefully so as not to scratch the bearing bores with the camshaft lobes. Install the camshaft thrust plate and tighten the attaching screws to 9–12 ft. lbs. (12–16 Nm). Measure the camshaft end-play. If the end-play is more than 0.009 in. (0.228mm), replace the thrust plate. Assemble the remaining components in the reverse order of removal.

6. Install the timing gear and front cover.

7. Install the valve lifters. They must be replaced in their original positions.

8. Install the pushrods, the rocker arms and the rocker arm covers.

9. Install the intake manifold and valley pan, if equipped.

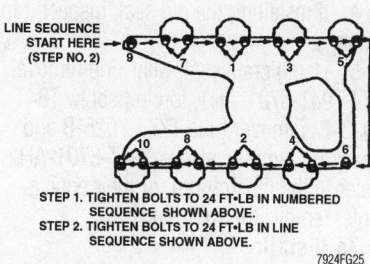

STEP 1. TIGHTEN BOLTS TO 24 FT•LB IN NUMBERED SEQUENCE SHOWN ABOVE.
STEP 2. TIGHTEN BOLTS TO 24 FT•LB IN LINE SEQUENCE SHOWN ABOVE.

7924FG25

Intake manifold bolt torque sequence— Diesel engines

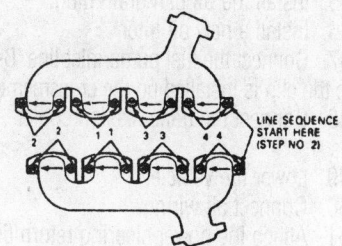

STEP1. TIGHTEN BOLTS TO 35 FT.LB. IN NUMBERED SEQUENCE SHOWN ABOVE
STEP2. TIGHTEN BOLTS TO 35 FT.LB. IN LINE SEQUENCE SHOWN ABOVE

7924FG26

Tighten the manifold bolts in the proper sequence—Diesel engines

Valve Lash

ADJUSTMENT

Valve lash on the 7.3L Diesel engine is not adjustable. The hydraulic tappets automatically adjust the valve lash for the proper clearance.

Oil Pan

REMOVAL & INSTALLATION

1. Disconnect both battery ground cables.
2. Remove the oil dipstick.
3. Remove the transmission oil dipstick.
4. Remove the air cleaner and cover the intake opening.
5. Remove the fan and fan clutch.

➡**The fan uses left-hand threads. Remove them by turning them clockwise.**

6. Drain the cooling system.
7. Disconnect the lower radiator hose.
8. Disconnect the power steering return hose and plug the line and pump.
9. Disconnect the alternator wiring harness.
10. Disengage the fuel line heater connector from the alternator.
11. Raise and support the front end.
12. If equipped with automatic transmission, disconnect the transmission cooler lines at the radiator and plug them.
13. Disconnect and plug the fuel pump inlet line.
14. Drain the crankcase and remove the oil filter.
15. Remove the oil filler tube.
16. Disconnect the exhaust pipes at the manifolds.
17. Disconnect the muffler inlet pipe from the muffler and remove the pipe.
18. Remove the upper inlet mounting stud from the right exhaust manifold.
19. Unbolt the engine from the No. 1 crossmember.
20. Lower the vehicle.
21. Install lifting brackets on the front of the engine.
22. Raise the engine until the transmission contacts the body.
23. Install wood blocks—2 ¾ in. (70mm) on the left side, 2 in. (50mm) on the right side—between the engine insulators and crossmember.
24. Lower the engine onto the blocks.

25. Raise and support the front end.
26. Remove the flywheel inspection plate.
27. Position fuel pump inlet line No. 1 rearward of the crossmember and position the oil cooler lines out of the way.
28. Remove the oil pan bolts.
29. Lower the oil pan.

➡**The oil pan is sealed to the crankcase with RTV silicone sealant in place of a gasket. It may be necessary to separate the pan from the crankcase with a utility knife. Also, the crankshaft may have to be turned to allow the pan to clear the crankshaft throws.**

30. Clean the pan and crankcase mating surfaces thoroughly.
   To install:
31. Apply a ⅛ in. (3mm) bead of RTV silicone sealant to the pan mating surfaces, and a ¼ in. (6mm) bead on the front and rear covers and in the corners; you have 15 minutes within which to install the pan!
32. Install the locating dowels into position.
33. Position the pan on the engine and install the pan bolts loosely.
34. Remove the dowels.
35. Tighten the pan bolts to 7 ft. lbs. (9.5 Nm) for ¼ in.-20 bolts; 14 ft. lbs. (19 Nm) for 5/16 in.-18 bolts; 24 ft. lbs. (33 Nm) for ⅜ in.-16 bolts.
36. Install the flywheel inspection cover.
37. Lower the truss.
38. Raise the engine and remove the wood blocks.
39. Lower the engine onto the crossmember and remove the lifting brackets.
40. Raise and support the front end.
41. Tighten the engine-to-crossmember nuts to 70 ft. lbs. (95 Nm).
42. Install the upper inlet pipe mounting stud.
43. Install the inlet pipe, using a new gasket.
44. Install the transmission oil filler tube, using a new gasket.
45. Install the oil pan drain plug.
46. Install a new oil filter.
47. Connect the fuel pump inlet line. Be sure the clip is installed on the crossmember.
48. Connect the transmission cooler lines.
49. Lower the vehicle.
50. Connect all wiring.
51. Attach the power steering return line.
52. Connect the lower radiator hose.
53. Install the fan and fan clutch.

➡**The fan uses left-hand threads. Install them by turning them counterclockwise.**

54. Remove the cover and install the air cleaner.
55. Install the dipstick.
56. Fill the crankcase.
57. Fill and bleed the cooling system.
58. Fill the power steering reservoir.
59. Connect the batteries.
60. Run the engine and check for leaks.

Oil Pump

REMOVAL & INSTALLATION

1. Remove the oil pan.
2. Remove the oil pick-up tube from the pump.
3. Unbolt and remove the oil pump.
   To install:
4. Assemble the pick-up tube and pump. Use a new gasket.
5. Install the oil pump and tighten the bolts to 14 ft. lbs. (19 Nm).

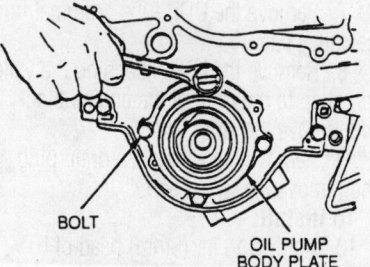

BOLT

OIL PUMP
BODY PLATE

7924FG62

The oil pump is mounted on the cylinder block with four bolts—Diesel engine

Rear Main Seal

REMOVAL & INSTALLATION

1. Remove the flywheel.
2. Loosen the crankshaft rear oil seal bolts and remove the seal.
3. Clean the seal mating surfaces.
4. If installing the old seal, inspect it for damage.
5. Using crankshaft wear ring removal tool T94T-6701-AH1, forcing screw T84T-7025-B, remover tube T77J-7025-B and wear ring remover sleeve T94T-6701-AH2 (refer to the illustration), or their equivalents, remove the wear ring.
   To install:
6. Apply silicone sealant D6AZ-19562-BA, or equivalent, to the seal retaining ring and the seal retaining bolts.

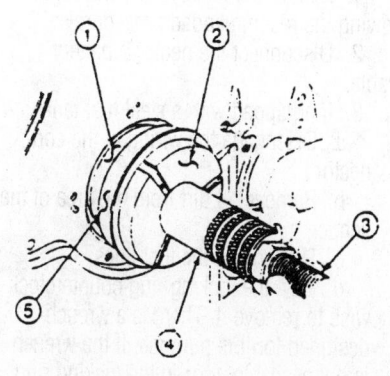

Item	Description
1	Crankshaft Wear Ring
2	Crankshaft Rear Wear Ring Remover
3	Forcing Screw
4	Remover Tube
5	Crankshaft Rear Wear Ring Remover Sleeve

7924FG61

Assemble the seal and wear ring removal tools, then remove the wear ring—diesel engine

7. Using seal replacers T94T-6701-AH3 and T94T-AH4, driver sleeve T79T-6316-A4 (part of T79T-6316-A) and guide pins T94P-7000-P or their equivalents, install the wear ring and oil seal.

8. Install the seal retaining bolts and tighten them to specifications.

9. Remove the installation tools and install the flywheel.

Timing Gears, Front Cover and Seal

REMOVAL & INSTALLATION

➡ **The crankshaft gear sprocket is not serviced separately from the crankshaft. Do not try to remove the sprocket or you will damage the crankshaft.**

Remove the camshaft sprocket as follows:

1. Remove the camshaft.
2. Use a press to remove the sprocket from the camshaft.
3. Remove the thrust plate and sprocket key.
4. Inspect the camshaft and related parts for wear and damage.

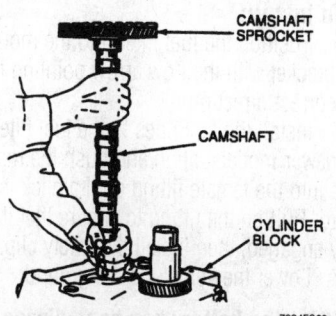

The camshaft gear is removed with the camshaft,, then pressed off—1995–96 Diesel engines

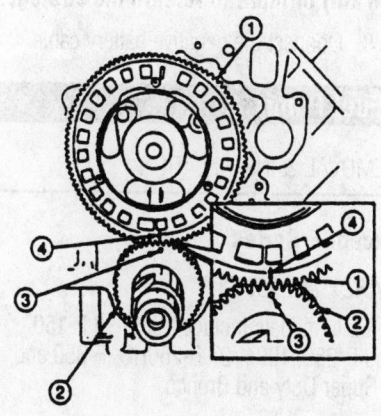

Item	Description
1	Camshaft Sprocket
2	Crankshaft Sprocket
3	Crankshaft Sprocket Timing Mark
4	Camshaft Sprocket Timing Mark

7924FG64

Be sure the timing marks are aligned as illustrated—Diesel engines

To install:

5. Clean the nose of the camshaft and install the thrust plate.
6. Place the key in the keyway on the camshaft.
7. Heat the sprocket in an oven to 500° F (260° C).
8. Remove the sprocket from the oven, align the sprocket keyway with the camshaft key and install the sprocket on the camshaft until it is fully seated. Allow the camshaft assembly to cool before installation
9. Install the camshaft in the engine and align the timing marks on the gears.

GASOLINE FUEL SYSTEM

Fuel System Service Precautions

Safety is the most important factor when performing not only fuel system maintenance but any type of maintenance. Failure to conduct maintenance and repairs in a safe manner may result in serious personal injury or death. Maintenance and testing of the vehicle's fuel system components can be accomplished safely and effectively by adhering to the following rules and guidelines.

• To avoid the possibility of fire and personal injury, always disconnect the negative battery cable unless the repair or test procedure requires that battery voltage be applied.

• Always relieve the fuel system pressure prior to disconnecting any fuel system component (injector, fuel rail, pressure regulator, etc.), fitting or fuel line connection. Exercise extreme caution whenever relieving fuel system pressure to avoid exposing skin, face and eyes to fuel spray. Please be advised that fuel under pressure may penetrate the skin or any part of the body that it contacts.

• Always place a shop towel or cloth around the fitting or connection prior to loosening to absorb any excess fuel due to spillage. Ensure that all fuel spillage (should it occur) is quickly removed from engine surfaces. Ensure that all fuel soaked cloths or towels are deposited into a suitable waste container.

• Always keep a dry chemical (Class B) fire extinguisher near the work area.

• Do not allow fuel spray or fuel vapors to come into contact with a spark or open flame.

• Always use a back-up wrench when loosening and tightening fuel line connection fittings. This will prevent unnecessary stress and torsion to fuel line piping. Always follow the proper torque specifications.

• Always replace worn fuel fitting O-rings with new. Do not substitute fuel hose or equivalent where fuel pipe is installed.

Fuel System Pressure

RELIEVING

➡ **A fuel pressure gauge, such as Ford Tool T80L-9974-B, is needed to correctly perform this procedure.**

1. Disconnect the negative battery cable and remove the fuel filler cap.

2. Remove the cap from the pressure relief valve on the fuel supply manifold. Install Pressure Gauge T80L-9974-B or equivalent, to the pressure relief valve.

3. Direct the gauge drain hose into a suitable container and depress the pressure relief button.

4. Remove the gauge and replace the cap on the pressure relief valve.

➡ **As an alternate method on models except 1997–99 F-150 and Expedition, disconnect the inertia switch and crank the engine for 15–20 seconds until the pressure is relieved.**

Fuel Filter

REMOVAL & INSTALLATION

➡ **On newer vehicles, especially the 1997–1999 F-150 and Expedition vehicles, a fuel line disconnect tool such as Ford tool T93T-9550-AH is needed for this procedure.**

1. Disconnect the negative battery cable and relieve the fuel system pressure.

2. Raise and support the vehicle safely.

3. Disconnect the fuel lines from the fuel filter. Have a drain pan handy to catch any residual fuel once the lines are separated. On newer models, disconnect the fuel lines from the filter as follows:

   a. Disconnect the safety clip from the male hose.

   b. Install and push the fuel line disconnect tool into the female fitting.

   c. Separate the male and female fittings.

   d. Inspect the fuel lines for any damage after the fuel is finished draining.

4. Remove the fuel filter from the bracket and the retainer, if equipped. Note the direction of the flow arrow so the replacement filter can be installed correctly.

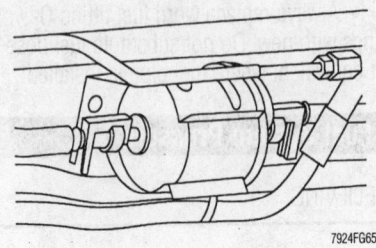

7924FG65

Typical fuel filter mounting along an under vehicle frame rail

To install:

5. Position the fuel filter into the mounting bracket with the flow arrow pointing in the correct direction.

6. Install the fuel lines to the fuel filter. On newer models, align and push the male tube into the female fitting until a click is heard. Pull on the fitting to ensure that it is fully engaged, then install the safety clip.

7. Lower the vehicle to the ground.

➡ **When the battery has been disconnected and reconnected, some abnormal drive symptoms may occur while the Powertrain Control Module (PCM) relearns its adaptive strategy. The vehicle may need to be driven 10 miles (16 km) or more to relearn the strategy.**

8. Connect the negative battery cable.

Fuel Pump

REMOVAL & INSTALLATION

Except E-Series

EARLY MODELS

Early models include 1995–96 F-150 and F-250, 1994–97 F-250HD, F-350 and F-Super Duty and Bronco.

1. Relieve the fuel system pressure following the recommended procedure.

2. Disconnect the negative battery cable.

3. If equipped with a steel fuel tank:

   a. Disengage the wiring at the connector.

   b. Remove all dirt from the area of the sender.

   c. Disconnect the fuel lines.

   d. Turn the locking ring counterclockwise to remove it. There is a wrench designed for this purpose. If the wrench is not available, loosen the locking ring by placing a wood dowel against the tabs on the locking ring and carefully hammering it loose. Never use a metal drift!

   e. Lift out the fuel pump and sending unit. Discard the gasket.

✳✳ CAUTION

Use of a metal drift may result in sparks which could cause an explosion!

4. If equipped with a plastic fuel tank:

   a. Disengage the wiring at the connector.

   b. Remove all dirt from the area of the sender.

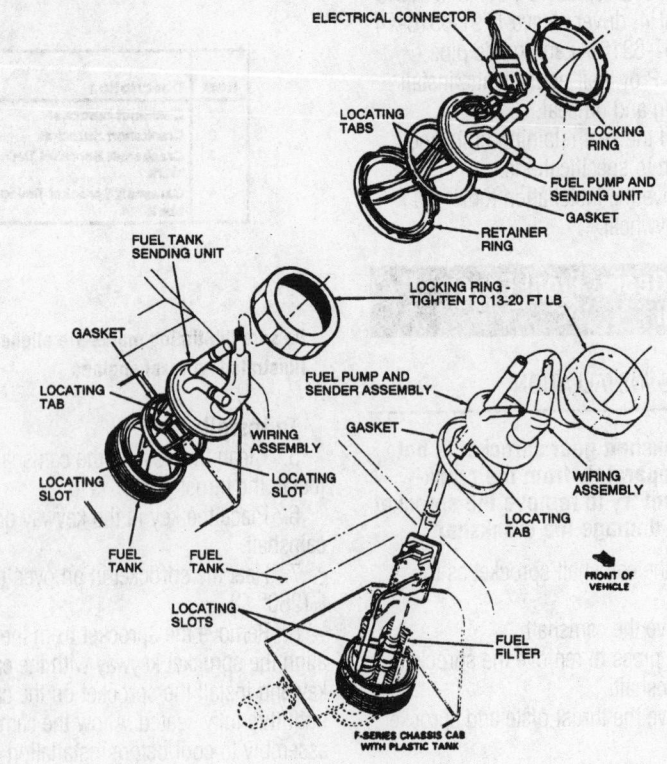

7924FG66

Exploded view of the in-tank fuel pump assembly—except late model F-series and Expedition vehicles

c. Disconnect the fuel lines.

d. Turn the locking ring counterclockwise to remove it. A band-type oil filter wrench is ideal for this purpose. Lift out the fuel pump and sending unit. Discard the gasket.

To install:

5. Place a new gasket in position in the groove in the tank.

6. Place the sending unit/fuel pump assembly in the tank, indexing the tabs with the slots in the tank. Be sure the gasket stays in place.

7. Hold the assembly in place and position the locking ring.

8. On steel tanks, turn the locking ring clockwise until the stop is against the retainer ring tab.

9. On plastic tanks, turn the retaining ring clockwise until hand-tight, then tighten it to 40–55 ft. lbs. (54–75 Nm).

10. Be sure the gasket is still in place.

11. Connect the fuel lines and wiring.

12. Install the tank.

LATE MODELS

Late models include 1997–99 F-150, F-250 and Expedition and 1998–99 F-250HD, F-350, F-450 and Navigator

1. Disconnect the negative battery cable.

2. Remove the fuel tank:

a. Relieve the fuel pressure.

b. Raise and support the vehicle.

c. Remove the fuel tank skid plate bolts and lower the skid plate.

d. Drain the fuel tank.

e. Remove the hose clamp on the fuel filler pipe support and disconnect the fuel tank filler pipe hose from the tank.

f. Disconnect the fuel tank filler pipe vent hose from the tank.

g. Disconnect the fuel lines from the fuel pump.

h. Disengage the front fuel tank connections.

i. Release the rear EVAP hose clamp and disconnect the hose.

j. Disengage the electrical connector from the fuel pump.

k. Support the fuel tank with a jack.

l. Remove the fuel tank support strap bolts and remove the fuel tank straps.

m. Remove the fuel tank.

3. Remove the fuel pump bolts.

4. Remove the fuel pump.

To install:

5. Install the fuel tank.

6. Tighten the fuel tank bolts to 66–91 inch lbs. (7.6–10.4 Nm).

7. Install the fuel tank.

a. Tighten the fuel tank strap bolts to 22–30 ft. lbs. (29.7–40.7 Nm).

b. Tighten the skid plate bolts to 9.3–12.7 ft. lbs. (12.7–17.3 Nm).

8. Connect the negative battery.

E-Series

1. Release the fuel system pressure.

2. Disconnect the negative battery cable.

3. Drain the fuel tank.

4. Raise and support the vehicle.

5. Disconnect the fuel tank filler pipe vent hose and fuel tank filler pipe from the tank.

6. Support the fuel tank with a jack.

7. Remove the two fuel tank support strap nuts and remove the two fuel tank support straps.

8. Lower the fuel tank to allow access to the electrical connections.

9. Disengage the fuel tank connections.

10. Disengage the fuel and electrical connections from the fuel pump.

11. Remove the fuel tank.

12. Remove the fuel tank screws/nuts, fuel pump and sender.

To install:

13. Install the fuel sender and fuel pump into the fuel tank. Tighten the screws/nuts.

14. Raise the fuel tank.

15. Attach the fuel and electrical connections to the fuel pump.

16. Attach the fuel tank connections.

17. Raise the fuel tank into place.

18. Install the fuel tank support straps and tighten the nuts to 13–17 ft. lbs. (17–23 Nm).

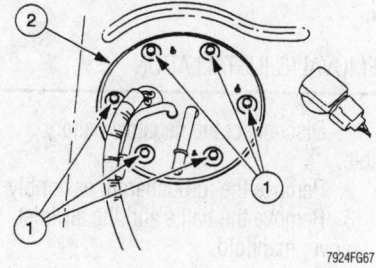

Remove the mounting bolts (1), then lift the fuel pump assembly (2) out of the tank—E-Series

7924FG67

19. Connect the fuel tank filler pipe vent hose and the fuel tank filler pipe to the tank.

20. Connect the negative battery cable.

DIESEL FUEL SYSTEM

Fuel System Service Precautions

Safety is the most important factor when performing not only fuel system maintenance but any type of maintenance. Failure to conduct maintenance and repairs in a safe manner may result in serious personal injury or death. Maintenance and testing of the vehicle's fuel system components can be accomplished safely and effectively by adhering to the following rules and guidelines.

• To avoid the possibility of fire and personal injury, always disconnect the negative battery cable unless the repair or test procedure requires that battery voltage be applied.

• Always relieve the fuel system pressure prior to disconnecting any fuel system component (injector, fuel rail, pressure regulator, etc.), fitting or fuel line connection. Exercise extreme caution whenever relieving fuel system pressure to avoid exposing skin, face and eyes to fuel spray. Please be advised that fuel under pressure may penetrate the skin or any part of the body that it contacts.

• Always place a shop towel or cloth around the fitting or connection prior to loosening to absorb any excess fuel due to spillage. Ensure that all fuel spillage (should it occur) is quickly removed from engine surfaces. Ensure that all fuel soaked cloths or towels are deposited into a suitable waste container.

• Always keep a dry chemical (Class B) fire extinguisher near the work area.

• Do not allow fuel spray or fuel vapors to come into contact with a spark or open flame.

• Always use a back-up wrench when loosening and tightening fuel line connection fittings. This will prevent unnecessary stress and torsion to fuel line piping. Always follow the proper torque specifications.

• Always replace worn fuel fitting O-rings with new. Do not substitute fuel hose or equivalent where fuel pipe is installed.

Fuel System Pressure

RELIEVING

✳✳ CAUTION

Before removing the fuel tank filler cap, turn the fuel tank filler cap ¼ to ¾ turn counterclockwise and wait for the tank pressure to be relieved. Personal injury may result if the fuel tank filler cap is removed without the pressure fully relieved.

1. Remove the fuel tank filler cap to relieve any pressure in the fuel tank.
2. When servicing the fuel lines, loosen the fuel fitting to allow any residual fuel line pressure to be relieved.

Idle Speed

ADJUSTMENT

1. Place the transmission in Neutral (manual transmissions) or **P** (automatic transmissions).
2. Bring the engine up to normal operating temperature.

➡**Idle speed is measured with the manual transmission in Neutral or the automatic transmission in D.**

3. Ensure that the curb idle adjusting screw is against the stop. If not, correct the vehicle linkage.
4. Check curb idle speed, using either Rotunda 055–00108 or an equivalent tachometer. Curb idle speed is specified on the Vehicle Emissions Control Information (VECI) decal on the underside of the vehicle's hood. Adjust the idle speed to specification using the idle speed adjusting screw.
5. Place the transmission in Neutral (manual) or **P**. Rev the engine momentarily,

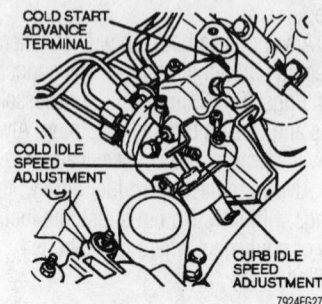

Raise or lower the curb idle speed by turning the curb idle speed adjusting screw—Diesel engines

then place the transmission in the specified gear and recheck the idle speed. Adjust again if necessary.
6. Remove the tachometer and close the hood.

Fuel Filter/Water Separator

DRAINING WATER

➡**Drain water from the water separator manual drain valve whenever the warning light comes ON or every 5000 miles (8000km). The "Water in Fuel" light will glow when approximately 3.5 oz. (103.5ml) of water accumulates in the separator.**

The Diesel engines are equipped with a fuel/water separator in the fuel supply line. A "Water in Fuel" indicator light is provided on the instrument panel to alert the driver. The light should glow when the ignition switch is in the **START** position to indicate proper light and water sensor function. If the light glows continuously while the engine is running, the water must be drained from the separator as soon as possible to prevent damage to the fuel injection system.

1. Shut the engine **OFF**. Failure to shut the engine **OFF** before draining the separator will cause air to enter the system.
2. Unscrew the vent on the top center of the separator unit 2½ –3 turns.
3. Unscrew the drain screw on the bottom of the separator 1½ –2 turns and drain the water into an appropriate container.
4. After the water is completely drained, close the water drain finger-tight.
5. Tighten the vent until snug, then turn it an additional ¼ turn.
6. Start the engine and check the "Water in Fuel" indicator light; it should not be lit. If it is lit and continues to stay so, there is a problem somewhere else in the fuel system.

REMOVAL & INSTALLATION

1. Disconnect the negative battery cable.
2. Remove the turbocharger assembly.
3. Remove the baffle and the air inlet crossover manifold.
4. Place a suitable container under the drain hose and open the filter drain.
5. Remove the two capscrews securing the fuel filter base to the crankcase.
6. Disconnect the water drain hose from the filter.
7. Disconnect the fuel outlet hose, located between the fuel and filter housing,

and the fuel return hose from the fuel pressure regulator valve.
8. Disconnect the two fuel supply hoses that connect the regulator block to the cylinder head fuel rails.
9. Loosen the clamp at the fuel pump end of the hose which connects the fuel filter to the inlet of high pressure stage at the fuel pump.
10. Disengage the wiring harness from the right side of the filter housing.
11. Disengage the electrical connections from the Water In Fuel (WIF) sensor and the fuel heater.
12. Remove the fuel filter.
13. Use a prybar to remove the fuel filter cap and the filter element will come out with the cap.
14. Depress the element locking tabs and remove the element from the cap.
To install:
15. Clean the mating surfaces and install the filter element onto the cap, making sure the tabs engage.
16. Install the filter gap and press down firmly, but gently, to engage it.
17. Engage the wiring connections to the fuel heater and WIF sensor.
18. Engage the wiring harness to the filter housing.
19. Tighten the clamp at the fuel pump

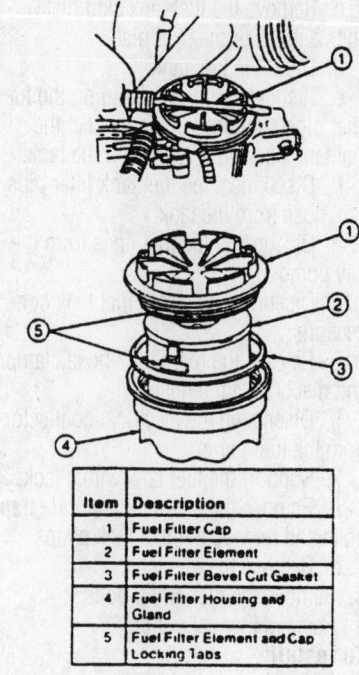

Item	Description
1	Fuel Filter Cap
2	Fuel Filter Element
3	Fuel Filter Bevel Cut Gasket
4	Fuel Filter Housing and Gland
5	Fuel Filter Element and Cap Locking Tabs

Use a prytool to remove the fuel filter cap to gain access to the filter element-Diesel fuel systems

end of the hose which connects the fuel filter to the inlet of high pressure stage at the fuel pump.

20. Engage the two fuel supply hoses that connect the regulator block to the cylinder head fuel rails.

21. Connect the fuel outlet hose, located between the fuel and filter housing, and the fuel return hose from the fuel pressure regulator valve.

22. Connect the water drain hose to the filter.

23. Install the two capscrews securing the fuel filter base to the crankcase.

24. Install the air inlet crossover manifold and baffle.

25. Install the turbocharger assembly.

26. Connect the negative battery cable.

Injection Pump

REMOVAL & INSTALLATION

1. If equipped, remove the turbocharger assembly.

2. Remove the fuel line banjo bolt at the pump.

3. Remove the fuel line fittings at the rear of the cylinder heads.

4. Remove the fuel lines assembly.

5. Loosen the three hose clamps at the injection pump fittings.

6. Disconnect the water drain hose at the fuel filter.

7. Disconnect the filter and position it forward.

8. Remove the injection pump retaining bolts, then lift the pump out of the crankcase bore.

9. Remove the injection pump tappet from the crankcase bore.

To install:

10. Rotate the engine so the injection pump eccentric is on the base circle.

11. Install the injection pump tappet in the base of the injection pump.

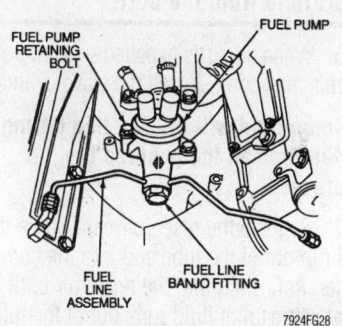

Injection pump assembly components— Diesel engines

12. Replace the O-ring on the injection pump base.

13. Install the injection pump and tighten the bolts to 19–27 ft. lbs. (26–37 Nm).

14. Install the fuel filter and connect the water drain hose.

15. Connect the three fuel hoses at the front of the injection pump.

16. Tighten the fuel line clamps and install the fuel filter retaining bolts.

17. Install the fuel line assembly and new seal rings at the rear of the pump.

18. Loosely install the fuel line fittings at the rear of the cylinder heads.

19. Install the fuel line banjo fitting at the pump. Tighten the bolt to 18 ft. lbs. (24 Nm).

20. Tighten the fuel line fittings.

21. Install the turbocharger assembly.

DRIVE TRAIN

Transmission Assembly

REMOVAL & INSTALLATION

1. Disconnect the negative battery cable.

2. On manual transmissions, remove the shifter boot and lever from inside the vehicle.

3. Raise and safely support the vehicle.

4. On 4-wheel drive vehicles, remove the front driveshaft and transfer case.

5. Disengage all cables, connectors and fluid lines that may interfere with transmission removal. Tag them if helpful for installation.

6. On automatic transmissions, unbolt the torque converter from the flexplate and disconnect the shift linkage.

7. Place a drain pan under the transmission and drain the fluid.

8. Position a transmission jack under the transmission and safety-chain the case to the jack.

9. Matchmark and remove the driveshaft.

10. Remove the transmission rear mount.

11. Slightly raise the transmission and remove the crossmember.

12. On automatic transmissions, remove the transmission-to-engine block bolts. For manual transmissions, remove the bolts securing the transmission to the bell housing.

✳✳ CAUTION

The torque converter will fall out of the transmission if it is tilted forward. Keep a hand on it while lowering the transmission out of the vehicle.

13. Roll the transmission rearward until the input shaft clears, lower the jack and remove the transmission.

To install:

14. Carefully raise the transmission to the engine or bell housing.

15. Roll the transmission forward and into position.

16. On automatic transmissions tighten the bolts to 65 ft. lbs. (87 Nm) for the Diesel or to 50 ft. lbs. (67 Nm) for gasoline engines. On manual transmissions, tighten the bolts to 50 ft. lbs. (64 Nm).

17. Install the crossmember and tighten the bolts to 55 ft. lbs. (74 Nm).

18. Install the transmission rear insulator and lower retainer. Tighten the bolts to 60 ft. lbs. (81 Nm).

19. The rest of the installation is the reverse of removal.

20. Refill all transmissions with the correct amount of Motorcraft MERCON® automatic transmission fluid.

Clutch

REMOVAL & INSTALLATION

1. Raise and safely support the vehicle.

2. On vehicles with the externally mounted slave cylinder, remove the clutch slave cylinder. On vehicles with an internally mounted slave cylinder, disengage the quick-disconnect coupling with a spring coupling tool such as T88T-70522-A.

3. Remove the transmission.

4. On gasoline engine models, except the 7.5L engine, remove the starter. Remove the flywheel housing attaching bolts and remove the housing. On Diesel engine models and the 7.5L gasoline engine, remove the cover, then remove the release lever and bearing from the clutch housing. To remove the release lever:

   a. Remove the dust boot.

   b. Push the release lever forward to compress the slave cylinder.

   c. Remove the slave cylinder by prying on the steel clip to free the tangs while pulling the cylinder clear.

   d. Remove the release lever by pulling it outward.

5. Mark the pressure plate and cover

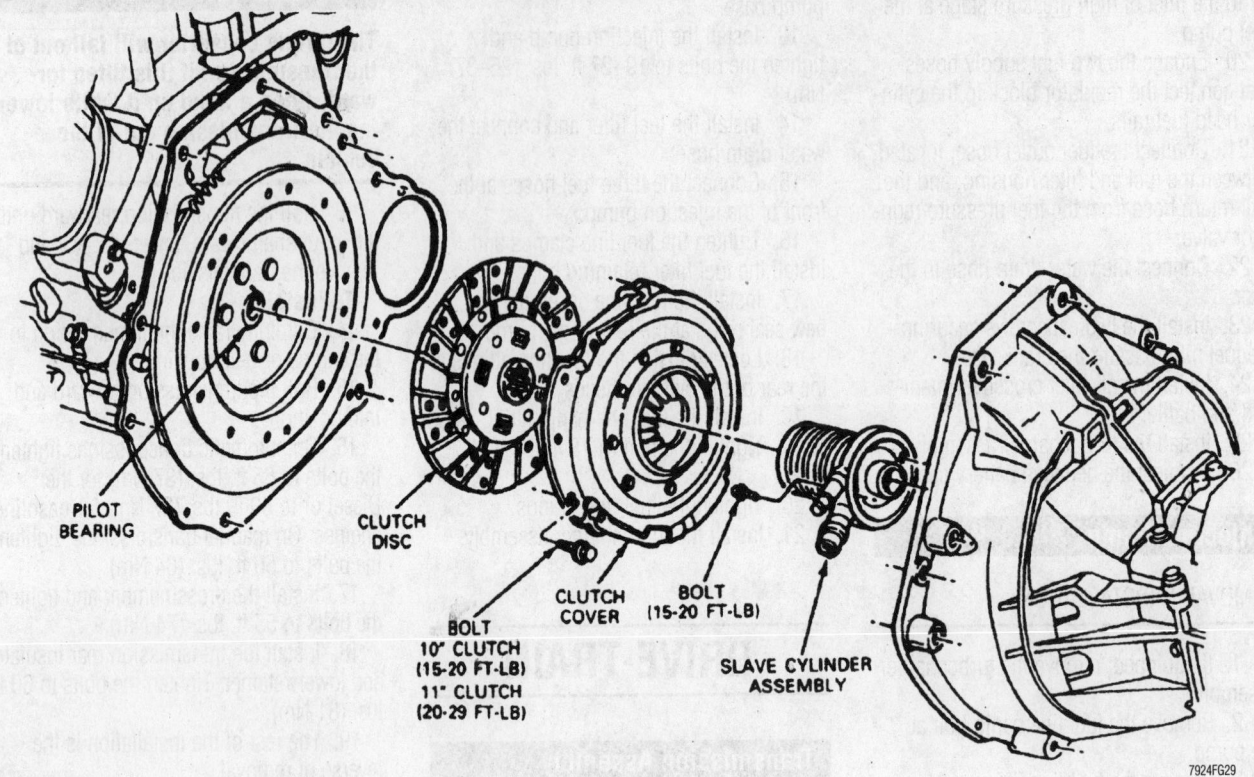

Typical clutch assembly with internal slave cylinder

assembly and the flywheel so that they can be reinstalled in the same relative position.

6. Loosen the pressure plate and cover attaching bolts evenly in a staggered sequence a turn at time until the pressure plate springs are relieved of their tension. Remove the attaching bolts.

7. Remove the pressure plate and cover assembly and the clutch disc from the flywheel.

8. Inspect the flywheel for wear, damage and flatness.

To install:

9. Position the clutch disc on the flywheel so that an aligning tool or spare transmission mainshaft can enter the clutch pilot bearing and align the disc.

10. When reinstalling the original pressure plate and cover assembly, align the assembly and flywheel according to the marks made during removal. Position the pressure plate and cover assembly on the flywheel, align the pressure plate and disc, and install the retaining bolts. Tighten the bolts in an alternating sequence a few turns at a time until the proper torque is reached:

- 10 in. clutch: 15–20 ft. lbs. (20–27 Nm)
- 11 in. clutch: 20–29 ft. lbs. (27–39 Nm)

11. Remove the tool used to align the clutch disc.

12. With the clutch fully released, apply a light coat of grease on the sides of the driving lugs.

13. Position the clutch release bearing and the bearing hub on the release lever. Install the release lever on the fulcrum in the flywheel housing. Apply a light coating of grease to the release lever fingers and the fulcrum. Fill the groove of the release bearing hub with grease.

14. If the flywheel housing has been removed, position it against the rear engine cover plate and install the attaching bolts and tighten them to 40–50 ft. lbs. (54–68 Nm).

15. Install the starter motor, if removed.

16. Install the transmission.

17. Install the slave cylinder and bleed the system.

Hydraulic Clutch System

BLEEDING

Externally Mounted Slave Cylinder

1. Clean the reservoir cap and the slave cylinder connection.

2. Remove the slave cylinder from the housing.

3. Using a ³⁄₃₂ in. punch, drive out the pin that holds the tube in place.

4. Remove the tube from the slave cylinder and place the end of the tube in a container.

5. Hold the slave cylinder so the connector port is at the highest point, by tipping it about 30 degrees from horizontal. Fill the cylinder with DOT 3 brake fluid through the port. It may be necessary to rock the cylinder or slightly depress the pushrod to expel all the air.

✷✷ CAUTION

Pushing too hard on the pushrod will spurt fluid from the port!

6. When all air is expelled—no more bubble are seen—install the slave cylinder.

➡**Some fluid will be expelled during installation as the pushrod is depressed.**

7. Remove the reservoir cap. Some fluid will run out of the tube end into the container. Pour fluid into the reservoir until a steady stream of fluid runs out of the tube and the reservoir is filled. Quickly install the diaphragm and cap. The flow should stop.

8. Connect the tube and install the pin. Check the fluid level.

9. Check the clutch operation.

Internally Mounted Slave Cylinder

EXCEPT 1997–99 F-150 AND EXPEDITION MODELS

➡With the quick-disconnect coupling, no air should enter the system when the coupling is disengaged. However, if air should somehow enter the system, it must be bled.

1. Remove the reservoir cap and diaphragm. Fill the reservoir with DOT 3 brake fluid.

2. Connect a piece of rubber tubing to the slave cylinder bleed screw. Place the other end in a container.

3. Loosen the bleed screw. Gravity will force fluid from the master cylinder to flow down to the slave cylinder, forcing air out of the bleed screw. When a steady stream—no bubbles—flows out, the system is bled. Close the bleed screw.

➡Check periodically to be sure the master cylinder reservoir doesn't run dry.

4. Add fluid to fill the master cylinder reservoir.

5. Fully depress the clutch pedal. Release it as quickly as possible. Pause for two seconds. Repeat this procedure 10 times.

6. Check the fluid level. Refill it if necessary. It should be kept full.

7. Repeat Steps 5 and 6 five more times.

8. Install the diaphragm and cap.

9. Have an assistant hold the pedal to the floor while you crack the bleed screw—not too far—just far enough to expel any trapped air. Close the bleed screw, then release the pedal.

10. Check, and if necessary, fill the reservoir.

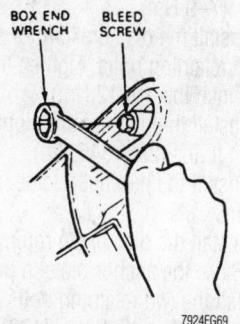

Bleed screw location for internally mounted slave cylinders

1997–99 F-150 AND EXPEDITION MODELS

➡Be sure to keep the clutch master cylinder reservoir full of brake fluid during the bleeding process to prevent air from entering the clutch master cylinder.

1. Raise and safely support the front of the vehicle.

➡It is necessary to have the assistance of a helper to bleed this system.

2. Fill the clutch system reservoir with Ford High Performance DOT 3 Brake Fluid C6AZ-19542-AA or equivalent.

3. Have your assistant depress the clutch pedal rapidly for 5–10 strokes.

4. Wait 1–3 minutes.

5. Repeat Steps 3 and 4 three more times.

6. Loosen the bleeder screw on the transmission for the slave cylinder.

7. Have the helper fully depress the clutch pedal and hold it down.

8. Tighten the bleeder screw.

9. The helper should now release the clutch pedal.

10. Apply pressure to the clutch pedal. If the clutch pedal travels more than 6–7 in. (15.3–17.7 cm), repeat the bleeding process.

Transfer Case Assembly

REMOVAL & INSTALLATION

❊❊ CAUTION

The catalytic converter is located beside the transfer case. Due to the extreme high temperatures generated by the converter, be careful when removing the transfer case or personal injury may result.

1. Raise and support the vehicle safely.

2. Drain the fluid from the transfer case.

3. Disengage the 4WD indicator switch wire connector at the transfer case.

4. Remove the skid plate from the frame, if equipped.

5. Matchmark and disconnect the front driveshaft from the front output yoke.

6. Matchmark and disconnect the rear driveshaft from the rear output shaft yoke.

7. Disconnect the speedometer driven gear from the transfer case rear bearing retainer.

8. Remove the retaining rings and shift rod from the transfer case shift lever.

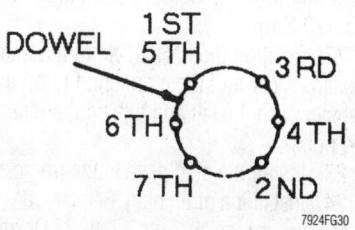

Transfer case-to-adapter bolt torque sequence—Borg-Warner model 13–45

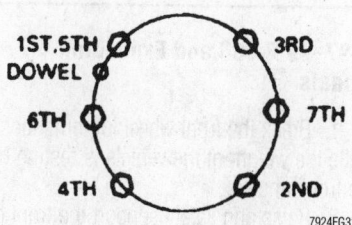

Transfer case-to-adapter bolt torque sequence—Borg-Warner 13–56 electronic and manual shift transfer case used in Bronco, F-Series and E-Series

9. Disconnect the vent hose from the transfer case.

10. Remove the heat shield from the frame.

11. Support the transfer case with a transmission jack.

12. Remove the bolts retaining the transfer case to the transmission adapter.

13. Lower the transfer case from the vehicle.

To install:

14. When installing place a new gasket between the transfer case and the adapter.

15. Raise the transfer case with the transmission jack so the transmission output shaft aligns with the splined transfer case input shaft. Install the bolts retaining the transfer case to the adapter.

16. Remove the transmission jack from the transfer case.

17. Connect the rear driveshaft to the rear output shaft yoke. Tighten the bolts to 15 ft. lbs. (20 Nm).

18. Install the shift lever to the transfer case and install the retaining nut.

19. Connect the speedometer driven gear to the transfer case.

20. Attach the 4WD indicator switch wire connector at the transfer case.

21. Connect the front driveshaft to the

front output yoke. Tighten the bolts to 15 ft. lbs. (20 Nm).

22. Position the heat shield to the frame crossmember and the mounting lug on the transfer case. Install and tighten the retaining bolts.

23. Install the skid plate to the frame.

24. Install the drain plug. Remove the filler plug and install 6 pts. (2.8L) of Dexron® II type transmission fluid or equivalent.

25. Lower the vehicle.

Halfshaft

REMOVAL & INSTALLATION

1997–99 F-150 and Expedition Models

1. Break the front wheel lug nuts loose while the weight of the vehicle is resting on the front wheels.

2. Raise and safely support the front of the vehicle.

3. Remove the front wheels.

4. Remove the front hub cotter pin, retainer and nut.

5. Using a floor hydraulic jack, support the lower suspension arm.

6. Remove the upper ball joint cotter pin and castle nut.

7. Use a Pitman arm puller, such as Ford Tool T64P-3590-F, to separate the front wheel knuckle from the front suspension upper arm.

8. Lower the lower suspension arm and steering knuckle slightly to facilitate easier halfshaft removal.

9. Remove the two disc caliper mounting bolts, then lift the front disc caliper off of the front disc brake caliper anchor plate and position aside. Do not allow the caliper to hang by the brake hose; suspend it from the vehicle's frame with strong cord or wire.

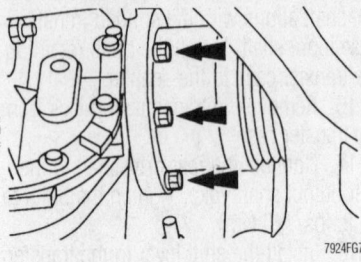

Halfshaft-to-differential mounting bolts (three of the six bolts shown)—1997–99 F-150 and Expedition models

10. Remove the six front halfshaft-to-differential bolts.

✳✳ WARNING

Use care to avoid damaging the hub seal when removing the front halfshaft.

11. Remove the inboard end of the halfshaft from the differential case or extension axle case. Separate the front halfshaft and joints from the hub, then remove the halfshaft and joints from the vehicle.

To install:

12. Slide the halfshaft outboard end into the hub, making sure that the splines engage.

13. Situate the inboard end of the halfshaft against the front differential flange and install the six halfshaft-to-differential bolts. Tighten the halfshaft bolts to 51–67 ft. lbs. (68–92 Nm).

14. Install the front disc brake caliper onto the rotor and anchor plate, then install and tighten the two caliper mounting bolts to 21–26 ft. lbs. (28–36 Nm).

15. Lift the lower suspension arm and steering knuckle up until the upper ball joint stud is inserted into the steering knuckle. Install the upper ball joint castle nut and tighten to 57–76 ft. lbs. (77–104 Nm). Install a new cotter pin.

16. Install the hub nut onto the halfshaft and tighten the hub nut to 188–254 ft. lbs. (255–345 Nm).

17. Install the hub nut retainer and a new cotter pin.

18. Install the front wheels and tighten the lug nuts in a star-shaped sequence to 83–112 ft. lbs. (113–153 Nm).

19. Lower the vehicle to the ground.

Locking Hubs

REMOVAL & INSTALLATION

Manual

1. Raise and safely support the vehicle.

2. Remove the wheel and tire assembly.

3. Remove the disc brake caliper.

4. Remove the disc brake pads and anti-rattle clips.

5. Remove two anchor plate retaining bolts and remove the anchor plate.

6. Remove the disc brake rotor.

7. Remove the three disc brake rotor shield retaining bolts and the shield.

8. If equipped with 4-wheel ABS,

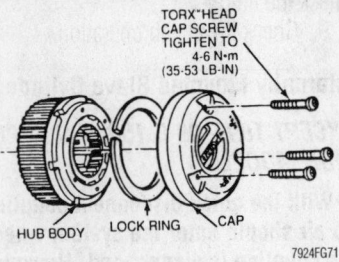

Exploded view of the manual locking hub assembly—1995–96 F series and Bronco

remove the speed sensor retaining bolt and move the speed sensor and harness aside.

9. Remove the hub nut cotter pin, retainer and the hub nut. Discard the cotter pin.

10. Remove the three hub assembly retaining bolts from the inside of the steering knuckle.

11. Push the CV-joint inward and remove the hub assembly.

➡If necessary, use a suitable puller to separate the hub assembly from the CV-joint. Use care not to over-extend the CV-joint and boot when removing the hub assembly.

12. If required, remove the grease seal from the steering knuckle.

To install:

13. If removed, install a new grease seal with Seal Installer T96T-1175-A, Drawbar T77F-1176-A and Cup Replacer T80T-4000-P, or equivalents.

14. Install the hub assembly to the steering knuckle and secure with the three retaining bolts. Tighten the bolts to 110–145 ft. lbs. (149–201 Nm).

15. Align the splines, then push the CV-joint into the hub assembly.

16. If equipped with 4-wheel ABS, install the speed sensor and secure with one retaining bolt. Tighten the bolt to 60–84 inch lbs. (7–9 Nm).

17. Install the disc brake rotor shield and the three retaining bolts. Tighten the bolts to 80–107 inch lbs. (9–12 Nm).

18. Install the hub nut and tighten to 188–254 ft. lbs. 255–345 Nm).

19. Install the hub nut retainer and a new cotter pin.

20. Install the disc brake rotor.

21. Place the anchor plate in position and install the two retaining bolts. Tighten the bolts to 125–168 ft. lbs. (170–230 Nm

22. Install the brake pad anti-rattle clips and install the disc brake pads.

23. Install the disc brake caliper.

24. Install the wheel and tire assembly. Tighten the lug nuts to 83–112 ft. lbs. (113–153 Nm).

25. Lower the vehicle.

26. Pump the brake pedal several times to position the brake pads prior to moving the vehicle.

27. Road test the vehicle and check for proper operation.

Automatic

1. Raise and safely support the vehicle.
2. Remove the tire.
3. Remove the three screws and separate the cap from the body.
4. Remove the lockring seated in the groove of the hub assembly.
5. Remove the body assembly from the brake rotor/hub.
6. Remove the snapring from the groove in the stub-shaft.
7. Remove the three thrust washers from the stub-shaft.
8. Pull the cam assembly to remove it.

To install:

9. Align the fixed cam retaining key on the cam assembly with the keyway on the spindle. Firmly push the cam assembly on the wheel retaining nut.

10. Install the metal, plastic, then the splined washers on the stub-shaft.

11. Install the snapring in the groove of the stub-shaft. It may be necessary to push the stub-shaft outward from the back of the knuckle assembly.

✳✳ WARNING

Do not pack the hub assembly with grease. Too much grease will damage the hub assembly.

12. Rotate the moving cam assembly to the one o'clock position in relation to the fixed cam retaining key. Use any one of the three stops.

13. Install the body assembly onto the hub by lining up the three legs with the three pockets in the cam assembly. Be sure the assembly is in far enough to see the groove in the hub.

14. Install the large lockring in the groove on the hub. Ensure the lockring is seated completely.

15. Install the cap using the three screws. Tighten the screws to 35–53 inch lbs. (4–6 Nm).

16. Install the tire and lower the vehicle to the floor.

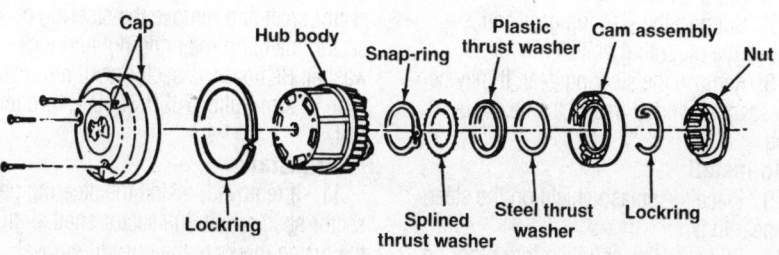

Exploded view of the typical automatic locking hub assembly

7924FG93

STEERING AND SUSPENSION

Air Bag

✳✳ CAUTION

Some vehicles are equipped with an air bag system, also known as the Supplemental Inflatable Restraint (SIR) or Supplemental Restraint System (SRS). The system must be disabled before performing service on or around system components, steering column, instrument panel components, wiring and sensors. Failure to follow safety and disabling procedures could result in accidental air bag deployment, possible personal injury and unnecessary system repairs.

PRECAUTIONS

Several precautions must be observed when handling the inflator module to avoid accidental deployment and possible personal injury:

• Never carry the inflator module by the wires or connector on the underside of the module.

• When carrying a live inflator module, hold securely with both hands and ensure that the bag and trim cover are pointed away.

• Place the inflator module on a bench or other surface with the bag and trim cover facing up.

• With the inflator module on the bench, never place anything on or close to the module which may be thrown in the event of an accidental deployment.

DISARMING

✳✳ CAUTION

The Supplemental Inflatable Restraint (SIR) system must be disarmed before performing service around SIR system components or SIR system wiring. Failure to do so may cause accidental deployment of the air bag, resulting in unnecessary SIR system repairs and/or personal injury.

For the Air Bag system on the Ford full-sized vehicles, the positive battery cable must be disconnected for a minimum of one minute before beginning any air bag work to de-energize the back-up power supply. It is a good idea to disengage both the positive and negative battery cables to ensure that the Air Bag system is definitely discharged.

Steering Gear

REMOVAL & INSTALLATION

Except E-Series

EARLY MODELS

➡Early models include 1995–1996 F-150 and F-250; 1995–97 F-250HD, F-350, F-Super Duty and Bronco.

1. Raise and support the front end on jackstands.

2. Place the wheels in the straight-ahead position.

3. Place a drain pan under the gear and disconnect the pressure and return lines. Cap the openings.

4. Remove the splash shield from the flex coupling.

5. Disconnect the flex coupling at the gear.

6. Matchmark and remove the Pitman arm from the sector shaft.

7. Support the steering gear and remove the mounting bolts.

8. Remove the steering gear. It may be necessary to work it free of the flex coupling.

To install:

9. Place the splash shield on the steering gear lugs.

10. Slide the flex coupling into place on the steering shaft. Be sure the steering wheel spokes are still horizontal.

11. Center the steering gear input shaft with the indexing flat facing downward.

12. Slide the steering gear input shaft into the flex coupling and into place on the frame side rail. Install the flex coupling bolt and tighten it to 30 ft. lbs. (41 Nm).

13. Install the gear mounting bolts and tighten them to 65 ft. lbs. (88 Nm).

14. Be sure that the wheels are still straight ahead and install the Pitman arm. Tighten the nut to 230 ft. lbs. (312 Nm).

15. Connect the pressure, then, the return lines. Tighten the pressure line to 25 ft. lbs. (34 Nm).

16. Snap the flex coupling shield into place.

17. Fill the steering reservoir.

18. Run the engine and turn the steering wheel lock-to-lock several times to expel air. Check for leaks.

LATE MODELS

➡**Late models include 1997–99 F-150, F-250 and Expedition and 1998–99 F-250HD, F-350, F-450 and Navigator.**

1. Raise and safely support the vehicle.

2. If equipped, remove the retaining bolts securing the skid plate and remove the skid plate.

3. Remove the three screws and five push clips retaining the lower radiator air deflector and remove.

4. Remove the Pitman arm (steering sector shaft arm) cotter pin and castellated nut from the drag link.

5. Using Pitman Arm Puller T65P-3590-F or equivalent, separate the Pitman arm from the drag link.

6. Remove the dust cover from the steering shaft valve housing.

7. Remove the intermediate shaft pinch bolt and slide the shaft off the steering gear input shaft.

8. Disconnect the power steering pressure hoses at the steering gear.

9. Remove the steering gear-to-frame rail retaining bolts and remove the steering gear.

10. If replacing or servicing the steering gear, match mark the sector shaft arm to the sector shaft and remove the steering gear sector shaft arm retaining nut and lockwasher. Remove the sector shaft arm using Pitman Arm Puller T65P-3590-F, or equivalent.

To install:

11. If removed, install the steering gear sector shaft arm to the sector shaft aligning the match marks made during removal. Install the retaining nut and lockwasher and tighten to 170–228 ft. lbs. (234–316 Nm).

12. Mount the steering gear in position. Install the three retaining bolts and tighten them to 50–68 ft. lbs. (68–92 Nm).

13. Connect the power steering pressure hoses to the steering gear using new seals, if necessary.

14. Slide the intermediate shaft on the steering gear input shaft and install the shaft pinch bolt. Tighten the pinch bolt to 30–42 ft. lbs. (41–57 Nm).

15. Install the dust cover over the steering shaft valve housing.

16. Connect the Pitman arm to the drag link. Install the castellated nut and tighten to 57–76 ft. lbs. (77–104 Nm). Install a new cotter pin.

17. Install the radiator air deflector and secure with the retaining screws and push clips.

18. Install the skid plate and secure it with the retaining bolts.

19. Lower the vehicle.

20. Fill and bleed the power steering system.

21. Road test the vehicle and check the steering system for proper operation.

E-Series

1. Raise and support the front end on jackstands.

2. Place the wheels in the straight-ahead position.

3. Place a drain pan under the gear and disconnect the pressure and return lines. Cap the openings.

4. Remove the splash shield from the flex coupling.

5. Disconnect the flex coupling at the gear.

6. Matchmark and remove the Pitman arm from the sector shaft.

7. Support the steering gear and remove the mounting bolts.

8. Remove the steering gear. It may be necessary to work it free of the flex coupling.

To install:

9. Place the splash shield on the steering gear lugs.

10. Slide the flex coupling into place on the steering shaft. Be sure the steering wheel spokes are still horizontal.

11. Center the steering gear input shaft with the indexing flat facing downward.

12. Slide the steering gear input shaft into the flex coupling and into place on the frame side rail. Install the flex coupling bolt and tighten it to 30 ft. lbs. (41 Nm).

13. Install the gear mounting bolts and tighten them to 65 ft. lbs. (88 Nm).

14. Be sure that the wheels are still straight ahead and install the Pitman arm. Tighten the nut to 230 ft. lbs. (312 Nm).

15. Connect the pressure, then, the return lines. Tighten the pressure line to 25 ft. lbs. (34 Nm).

16. Snap the flex coupling shield into place.

17. Fill the steering reservoir.

18. Run the engine and turn the steering wheel lock-to-lock several times to expel air. Check for leaks.

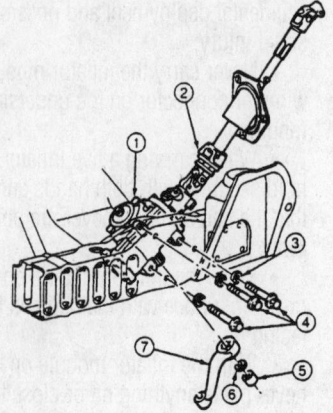

Item	Description
1	Steering Gear
2	Lower Steering Column Shaft
3	Washer
4	Bolt 73-90 N-m (54-66 Ft-Lb)
5	Nut 230-310 N-m (170-228 Ft-Lb)
6	Washer
7	Steering Gear Sector Shaft Arm

7924FG72

The steering gear is mounted on the left frame rail as shown—E-Series

Front Shock Absorber

REMOVAL & INSTALLATION

Except E-Series

EARLY MODELS

➡**Early models include 1995–1996 F-150 and F-250; 1995–97 F-250HD, F-350 and F-Super Duty and Bronco.**

1. Raise the vehicle and secure on support stands.

2. Remove the self-locking nut, steel washer, and rubber bushings at the upper end of the shock absorber.

3. Remove the bolt and nut at the lower end and remove the shock absorber.

To install:

4. When installing a new shock absorber, use new rubber bushings. Position the shock absorber on the mounting brackets with the stud end at the top. Install the upper bushing, steel washer and self-locking nut at the upper end, and the bolt and nut at the lower end.

5. Tighten the upper mounting studs to 18–22 ft. lbs. and the lower mounting nuts to 40–60 ft. lbs.

LATE MODELS

➡**Late models include 1997–99 F-150, F-250 and Expedition; 1998–99 F-250HD, F-350, F-450 and Navigator.**

1. If equipped with 2-wheel drive, perform the following:

   a. Hold the shock absorber stem and remove the nut from the top of the shock.

   b. Raise and safely support the vehicle.

   c. Remove the two lower retaining nuts and remove the shock absorber.

2. If equipped with 4-wheel drive, perform the following:

   a. Hold the shock absorber stem and remove the nut, washer and bushing from the top of the shock absorber stud.

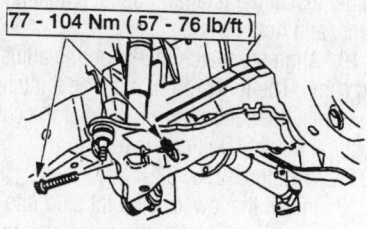

77 - 104 Nm (57 - 76 lb/ft)

7924FG73

Lower shock absorber mounting—late models with torsion bar suspension

b. Raise and support the vehicle.

c. Remove the lower retaining nut and bolt.

d. Remove the shock absorber from the vehicle.

To install:

3. On 4-wheel drive models, perform the following:

   a. Install the washer and bushing to the top stem of the shock absorber.

   b. Place the shock absorber up through the coil spring.

   c. Install the two lower retaining nuts. Tighten the nuts to 19–25 ft. lbs. (26–34 Nm).

   d. Lower the vehicle.

   e. Install the bushing, washer and retaining nut to the top of the shock absorber stud. Tighten the nut to 34–46 ft. lbs. (47–63 Nm).

4. On 4-wheel drive models, perform the following:

   a. Install the washer and bushing to the top stem of the shock absorber.

   b. Place the shock absorber up through the coil spring.

   c. Install the lower retaining nut and bolt. Tighten to 57–76 ft. lbs. (77–104 Nm).

   d. Lower the vehicle.

   e. Install the shock absorber upper bushing, washer and retaining nut. Tighten the nut to 22–29 ft. lbs. (30–40 Nm).

5. Road test the vehicle and check for proper operation.

E-Series

1. Raise the vehicle and secure on support stands.

2. Remove the self-locking nut, steel washer, and rubber bushings at the upper end of the shock absorber.

3. Remove the bolt and nut at the lower end and remove the shock absorber.

To install:

4. When installing a new shock absorber, use new rubber bushings. Position the shock absorber on the mounting brackets with the stud end at the top. Install the upper bushing, steel washer and self-locking nut at the upper end, and the bolt and nut at the lower end.

5. Tighten the upper mounting studs to 18–22 ft. lbs. and the lower mounting nuts to 40–60 ft. lbs.

Rear Shock Absorber

1. Raise the vehicle and secure on support stands.

2. Remove the self-locking nut, steel washer, and rubber bushings at the upper end of the shock absorber.

3. Remove the bolt and nut at the lower end and remove the shock absorber. If needed, raise the rear axle assembly slightly with a jack.

To install:

4. When installing a new shock absorber, use new rubber bushings. Position the shock absorber on the mounting brackets with the stud end at the top. Install the upper bushing, steel washer and self-locking nut at the upper end, and the bolt and nut at the lower end.

5. Tighten the upper mounting studs to 18–22 ft. lbs. and the lower mounting nuts to 40–60 ft. lbs.

Coil Spring

REMOVAL & INSTALLATION

Early Models

➡**Early models include 1995–1996 F-150 and F-250; 1995–97 F-250HD, F-350 and F-Super Duty and Bronco.**

1. Raise and safely support the vehicle. Place a jack under the lower control arm.

2. Remove the wheels.

3. Disconnect the shock absorber from the lower bracket.

4. Remove one bolt and nut and remove the rebound bracket.

5. Remove the two spring upper retainer attaching bolts from the top of the spring upper seat and remove the retainer.

6. Remove the nut attaching the spring lower retainer to the lower seat and axle and remove the retainer.

7. Place a safety chain through the spring to prevent it from suddenly coming loose. Slowly lower the control arm and remove the spring.

To install:

8. Place the spring in position and raise the lower control arm with a jack.

9. Position the spring lower retainer over the stud and lower seat, and install the two attaching bolts.

10. Position the upper retainer over the spring coil and against the spring upper seat, and install the two attaching bolts.

11. Tighten the upper retaining bolts to 13–18 ft. lbs.; the lower retainer attaching nuts to 70–100 ft. lbs. (95–136 Nm).

12. Connect the shock absorber to the lower bracket. Tighten the bolt and nut to

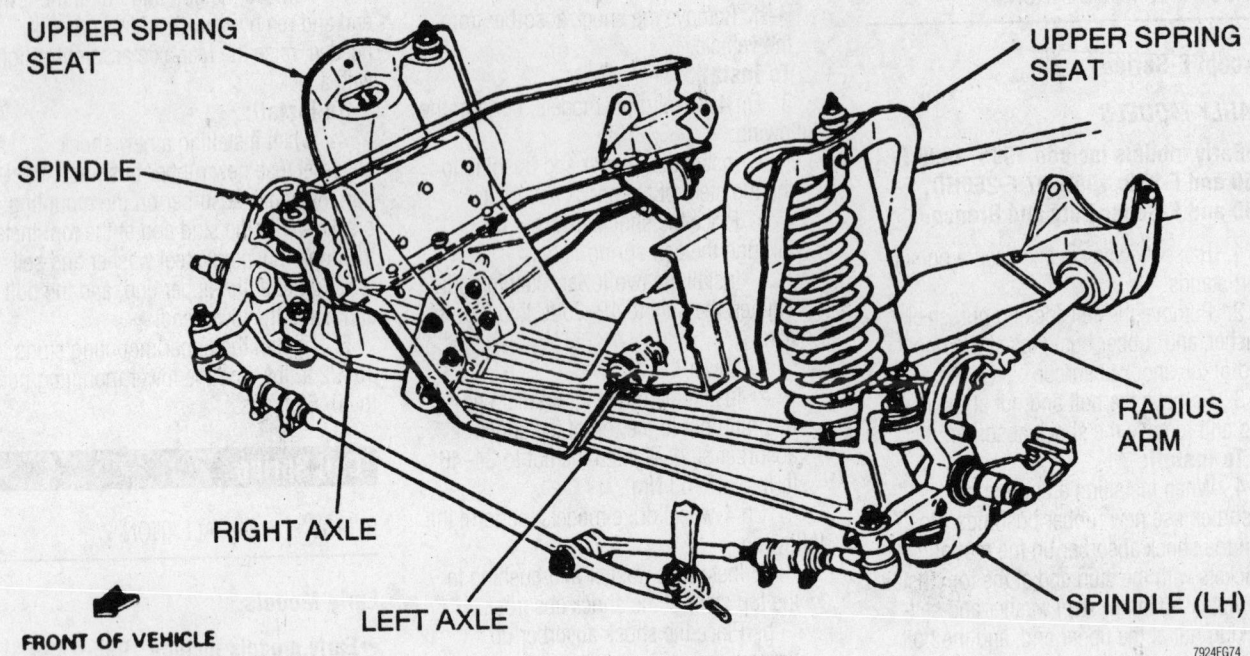

UPPER SPRING SEAT

UPPER SPRING SEAT

SPINDLE

RADIUS ARM

RIGHT AXLE

SPINDLE (LH)

FRONT OF VEHICLE

LEFT AXLE

7924FG74

Front suspension assembly—1995–96 F-Series

40–60 ft. lbs. (54–81 Nm). Install the rebound bracket.

13. Remove the jack and safety stands.

Late Models

➡**Late models include 1997–99 F-150, F-250 and Expedition and 1998–99 F-250HD, F-350, F-450 and Navigator. Also, this procedure applies to 2-wheel drive vehicles only. In order to remove the coil spring, the lower control arm must also be removed.**

1. Raise and safely support the vehicle.

2. Remove the wheel and tire assembly.

3. Remove the disc brake caliper and support aside with wire to prevent damage to the brake hose.

4. Remove the two retaining bolts and the disc brake adapter.

5. Remove the brake rotor.

6. Remove the three retaining bolts and the brake rotor splash shield.

7. Remove the shock absorber.

8. Remove the bracket supporting the brake hose.

9. Remove the sway bar link retaining nut and bushing from the lower control arm. Separate the sway bar link from the lower control arm.

10. Install Coil Spring Compressor D78P-5310A or equivalent and compress the coil spring enough to relieve the tension of the spring between the upper and lower control arms.

11. Remove the cotter pin and castellated nut from the lower ball joint. Using Pitman Arm Puller T64P-3590-F or equivalent, separate the lower ball joint from the wheel spindle.

12. Matchmark the lower control arm alignment cams for installation reference.

13. Remove the lower control arm retaining nuts and bolts.

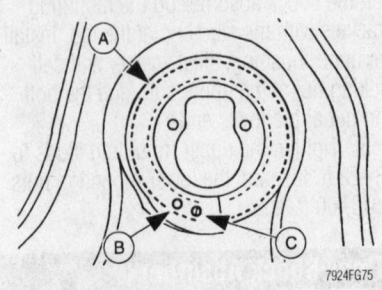

7924FG75

Be sure the coil spring is mounted correctly in the lower control arm

14. Remove the lower control arm and the compressed coil spring as an assembly.

15. Loosen the coil spring compressor and remove the coil spring from the lower control arm.

16. Inspect the coil spring and replace as needed.

To install:

17. Place the coil spring correctly in the saddle of the lower control arm. Install the coil spring compressor and compress the coil spring. The end of the coil spring **A**, must cover the hole designated **B** and be visible in the second hole designated as **C**, for proper installation.

18. Place the lower control arm to the frame. Install the retaining bolts, adjusting cams, and nuts.

19. Align the match marks on the adjusting cams. The forward nut must be tighten first while the control arm is held at the curb position height. Tighten the nuts to 197–241 ft. lbs. (270–330 Nm).

20. Install the lower ball joint stud into the wheel spindle. Install the castellated nut and tighten to 83–113 ft. lbs. (113–153 Nm). Install a new cotter pin.

21. Connect the sway bar link to the lower control arm. Install the bushing and

retaining nut. Tighten to 15–21 ft. lbs. (21–29 Nm).

22. Remove the coil spring compressor.

23. Install the shock absorber. Tighten the lower bolts to 22 ft. lbs. (32 Nm) and the top nut to 45 ft. lbs. (61 Nm).

24. Connect the brake hose bracket.

25. Install the brake rotor splash shield. Install the three retaining bolts and tighten them to 90–107 inch lbs. (10–14 Nm).

26. Install the disc brake rotor and caliper assemblies.

27. Install the wheel and tire assembly.

28. Lower the vehicle.

29. Pump the brake pedal several times to position the brake pads prior to moving the vehicle.

30. Check the alignment and adjust if out of specification.

31. Road test the vehicle and check for proper operation.

Upper Ball Joint

REMOVAL & INSTALLATION

Early Models

➡ **Early models include 1995–96 F-150 and F-250, 1994–97 F-250HD, F-350 and F-Super Duty, Bronco and E-Series**

⁂ WARNING

Do not use a forked ball joint removal tool to separate the ball joints as this will damage the seal and the ball joint socket. Do not use heat to aid in removal or installation of any suspension parts or components.

1. Raise and safely support the vehicle.
2. Remove the wheel and tire assembly.
3. Remove the front brake caliper assembly and support it from the body way with wire. Do not disconnect the brake line or allow the caliper to hang on the brake line.
4. Remove the brake rotor/hub assembly.
5. Remove the wheel spindle or steering knuckle assembly with the ball joints attached.
6. Place the spindle/knuckle assembly in a suitable vise.
7. Remove the lower ball joint. This is required to access the upper ball joint.
8. Remove the snapring from the upper ball joint.
9. Install U-Joint Remover T74P-4635-C and Receiver Cup D81T-3010-A or equivalents, on the upper ball joint. Turn the

forcing screw clockwise until the upper ball joint is pressed out of the spindle/knuckle assembly.

To install:

10. Place the new upper ball joint to the spindle/knuckle assembly and install U-Joint Remover T74P-4635-C, Receiver Cup D81T-3010-A5 and Installation Cup D81T-3010-A1, or equivalents. Turn the forcing screw clockwise until the upper ball joint is seated.

11. Install the snapring onto the upper ball joint.

12. Install the lower ball joint.

13. Remove the spindle/knuckle assembly from the vise and install to the axle arm.

14. Install the brake rotor/hub assembly.

15. Install the front brake caliper assembly.

16. Install the wheel and tire assembly. Tighten the lug nuts to 100 ft. lbs. (135 Nm).

17. Lower the vehicle.

18. Pump the brake pedal several times to position the brake pads before moving the vehicle.

19. Check the alignment and adjust if not within specification.

20. Road test the vehicle and check for proper operation.

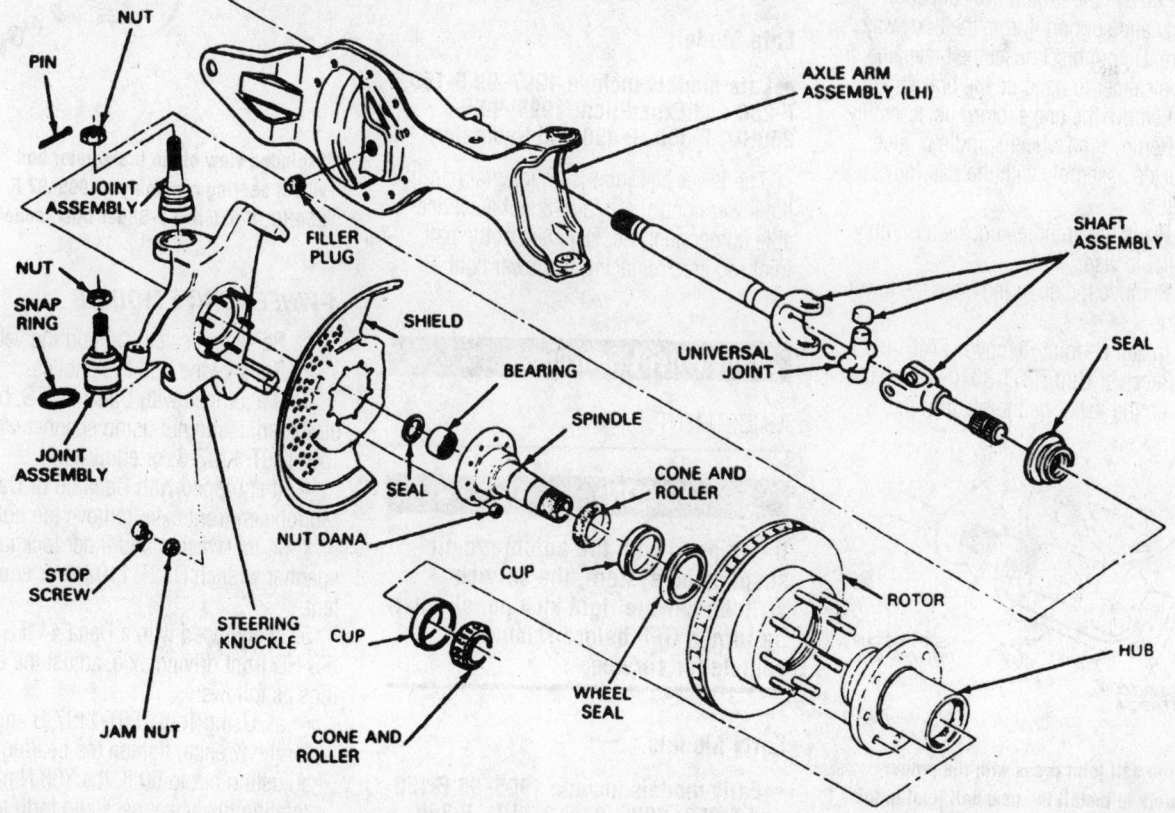

Dana models 44 and 50 axle shaft and joint assemblies—early models

7924FG76

Late Models

→Late models include 1997–99 F-150, F-250 and Expedition, 1998–99 F-250HD, F-350, F-450 and Navigator.

The upper ball joint is an integral part of the upper control arm and is not a serviceable component. Replacement of the ball joint requires replacing the upper control arm assembly.

Lower Ball Joint

REMOVAL & INSTALLATION

Early Models

→Early models include 1995–96 F-150 and F-250, 1995–97 F-250HD, F-350 and F-Super Duty, Bronco and E-Series.

※※ WARNING

Do not use a forked ball joint removal tool to separate the ball joints as this will damage the seal and the ball joint socket. Do not use heat to aid in removal or installation of any suspension parts or components.

1. Raise and safely support the vehicle.
2. Remove the wheel and tire assembly.
3. Remove the front brake caliper assembly and support it from the body way with wire. Do not disconnect the brake line or allow the caliper to hang on the brake line.
4. Remove the brake rotor/hub assembly.
5. Remove the wheel spindle or steering knuckle assembly with the ball joints attached.
6. Place the spindle/knuckle assembly in a suitable vise.
7. Remove the snapring from the lower ball joint.
8. Install U-Joint Remover T74P-4635-C and Receiver Cup D81T-3010-A or equivalents, on the lower ball joint. Turn the

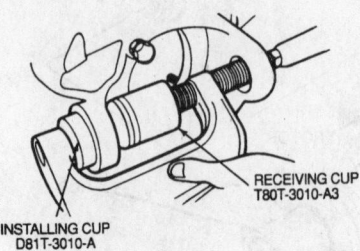

RECEIVING CUP
T80T-3010-A3

INSTALLING CUP
D81T-3010-A

7924FG77

Use the ball joint press with the proper adapters to install the new ball joint in the control arm—E-Series and early non-E-Series models

forcing screw clockwise until the lower ball joint is pressed out of the spindle/knuckle assembly.

To install:

9. Inspect the upper ball joint for wear. If replacing the upper ball joint, do so before installing the lower ball joint.
10. Place the new lower ball joint to the spindle/knuckle assembly and install U-Joint Remover T74P-4635-C, Receiver Cup D81T-3010-A5 and Installation Cup D81T-3010-A1, or equivalents. Turn the forcing screw clockwise until the lower ball joint is seated.
11. Install the snapring onto the lower ball joint.
12. Remove the spindle/knuckle assembly from the vise and install to the axle arm.
13. Install the brake rotor/hub assembly.
14. Install the front brake caliper assembly.
15. Install the wheel and tire assembly. Tighten the lug nuts to 100 ft. lbs. (135 Nm).
16. Lower the vehicle.
17. Pump the brake pedal several times to position the brake pads before moving the vehicle.
18. Check the alignment and adjust if not within specification.
19. Road test the vehicle and check for proper operation.

Late Models

→Late models include 1997–99 F-150, F-250 and Expedition, 1998–99 F-250HD, F-350, F-450 and Navigator.

The lower ball joint is an integral part of the lower control arm and is not a serviceable component. Replacement of the ball joint requires replacing the lower control arm.

Front Wheel Bearings

ADJUSTMENT

※※ CAUTION

If equipped with the automatic air suspension system, the service switch near the right kick panel must be turned OFF before raising the vehicle for service.

Early Models

→Early models include 1995–96 F-150 and F-250, 1995–97 F-250HD, F-350 and F-Super Duty, Bronco and E-Series.

2-WHEEL DRIVE MODELS

1. Raise and support the front end.
2. Remove the wheel cover.
3. Remove the grease cap from the hub. Then, remove the cotter pin and nut lock. Back off the adjusting nut.
4. Adjust the wheel bearings by tightening the adjusting nut to 17–25 ft. lbs. (23–34 Nm) with the wheel rotating to seat the bearing. Then, back off the adjusting nut ½ turn. Retighten the adjusting nut to 10–15 inch lbs. (1–1.7 Nm). Install the locknut so that the castellations are aligned with the cotter pin hole. Install the cotter pin. Bend the ends of the cotter pin around the castellations of the locknut to prevent interference with the radio static collector in the grease cap. Install the grease cap.
5. Install the wheels.
6. Check the wheel for proper rotation, then install the grease cap. If the wheel still does not rotate properly, inspect and clean or replace the wheel bearings and cups.
7. Install the wheel cover.

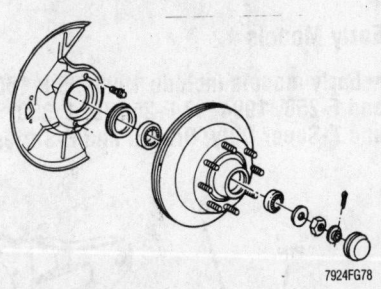

7924FG78

Exploded view of the brake rotor and wheel bearing assembly—1995–97 F-250HD, F-350 and F-Super Duty models

4-WHEEL DRIVE MODELS

1. Raise and safely support the vehicle.
2. Remove the hub assemblies.
3. If equipped with Dana 44 IFS, back off the adjusting nut using spanner wrench, Tool T59T-1197-B, or equivalent.
4. If equipped with Dana 50 or Dana 60 Monobeam front axle, remove the outer locknut, lockwasher and inner locknut with spanner wrench D85T-1197-A or equivalent.
5. If equipped with a Dana 44 IFS of 44 IFS HD front driving axle, adjust the bearings as follows:
   a. Using Tool T59T-1197-B and a torque wrench, tighten the bearing adjusting nut to 50 ft. lbs. (68 Nm), while rotating the wheel back and forth to seat the bearings.

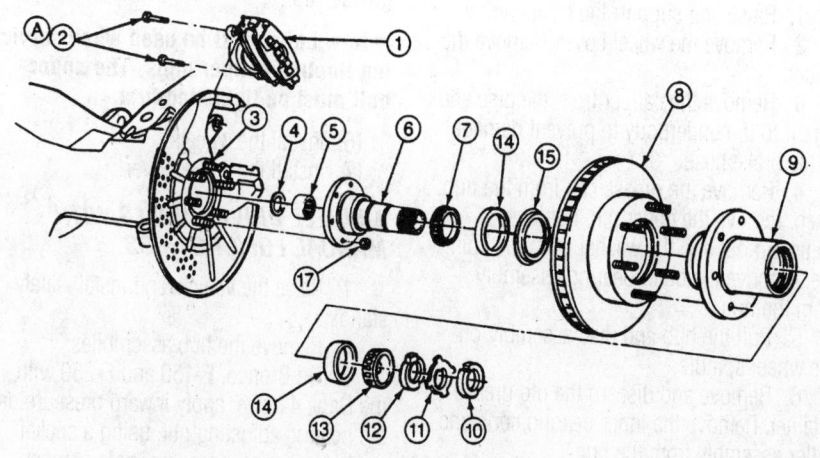

1 Disc Brake Caliper	9 Hub
2 Bolt, M8-1.25 x 65	10 Locknut
3 Front Disc Brake Rotor	11 Retainer
Shield	12 Locknut
4 Seal	13 Differential Bearing
5 Bearing	14 Differential Bearing Cup
6 Front Wheel Spindle	15 Wheel Seal
7 Differential Bearing	17 Dana Nut
8 Front Disc Brake Hub and	A Tighten to 30-36 N·m
Rotor	(22-26 Lb-Ft)

7924FG79

Exploded view of the brake rotor and wheel bearing assembly—1995–97 F-150 and Bronco
models with automatic locking hubs

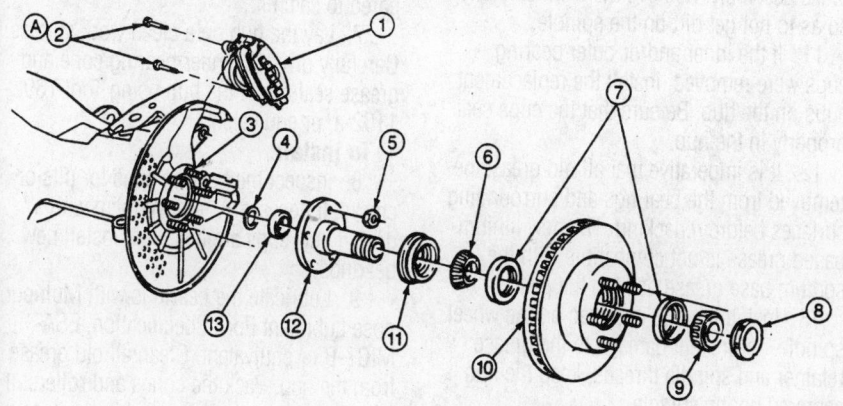

1 Disc Brake Caliper	8 Adjusting Nut
2 Bolt, M8-1.25 x 65	9 Differential Bearing
3 Front Disc Brake Rotor	10 Front Disc Brake Hub and
Shield	Rotor
4 Spindle Seal	11 Grease Seal
5 Nut	12 Front Wheel Spindle
6 Differential Bearing	13 Spindle Needle Bearing
7 Bearing Cups	A Tighten to 30-36 N·m
	(22-26 Lb-Ft)

7924FG80

Exploded view of the brake rotor and wheel bearing assembly—1995–97 F-150 and Bronco
models with manual locking hubs

b. Back off the adjusting nut approximately 60°.

c. Retighten the adjusting nut to 15 ft. lbs. (20 Nm). Remove the tool and the torque wrench.

d. Inspect the end-play of the wheel on the spindle. It should be 0.000–0.006 in. (0.00–0.15mm). If excess end-play is present, retighten the bearings.

6. If equipped with a Dana 50 IFS and Dana 60 Monobeam front driving axle, adjust bearings as follows:

a. With the outer locknut and lock-washer removed, tighten the inner lock-nut to 50 ft. lbs. (68 Nm) while rotating the wheel back and forth to seat the bearings.

b. Back off the adjusting nut and retighten to 31–39 ft. lbs. (42–53 Nm).

c. While rotating the hub, back off the locknut 135–150°.

d. Install the lockwasher so the key is positioned in the spindle groove. Rotate the inner locknut so the pin is aligned into the nearest lockwasher hole.

e. Install the outer locknut and tighten to 160–205 ft. lbs. (218–279 Nm) using the spanner wrench.

f. Inspect the end-play of the wheel on the spindle. It should be 0–0.006 in. (0–0.15mm). If excess end-play is present, retighten the bearings.

7. Assemble the hub parts.

8. Install the caliper.

9. Safely lower the vehicle.

Late Models

➡Late models include 1997–99 F-150, F-250 and Expedition; 1998–99 F-250HD, F-350, F-450 and Navigator.

➡On 4-wheel drive vehicles, the front wheel bearings are not adjustable.

1. Raise and safely support the vehicle.
2. Support the front end.
3. Remove the wheel cover, if equipped.
4. Remove the grease cap.

➡Check the wheel bearings for sufficient grease.

5. Remove the cotter pin and retaining washer. Back off the spindle nut. Discard the cotter pin.

6. Adjust the wheel bearings as follows:

a. Tighten the spindle nut to 17–24 ft. lbs. (23–34 Nm) while rotating the wheel and tire assembly to seat the wheel bearings.

b. Back off the spindle nut no less than ½ turn.

c. Tighten the spindle nut to 17 inch lbs. (2 Nm).

7. Install the retaining washer so the castellations are aligned with the cotter pin hole. Install a new cotter pin.

8. Check the wheel and tire assembly for proper rotation, then install the grease cap. If the wheel still does not rotate properly, inspect and clean or replace the wheel bearings and cups.

9. Install the wheel cover, if equipped.

10. Lower the vehicle.

11. Road test the vehicle and check for proper operation.

REMOVAL & INSTALLATION

Before handling the bearings, there are a few things that you should remember to do and not to do.

Remember to DO the following:

• Remove all outside dirt from the housing before exposing the bearing.

• Treat a used bearing as gently as you would a new one.

• Work with clean tools in clean surroundings.

• Use clean, dry canvas gloves, or at least clean, dry hands.

• Clean solvents and flushing fluids are a must.

• Use clean paper when laying out the bearings to dry.

• Protect disassembled bearings from rust and dirt. Cover them up.

• Use clean rags to wipe bearings.

• Keep the bearings in oil-proof paper when they are to be stored or are not in use.

• Clean the inside of the housing before replacing the bearing.

Do NOT do the following:

• Don't work in dirty surroundings.

• Don't use dirty, chipped or damaged tools.

• Try not to work on wooden work benches or use wooden mallets.

• Don't handle bearings with dirty or moist hands.

• Do not use gasoline for cleaning; use a safe solvent.

• Spin-dry bearings with compressed air. They may come apart and cause personal injury.

• Avoid using cotton waste or dirty cloths to wipe bearings.

• Scratch or nick bearing surfaces.

• Do not allow the bearing to come in contact with dirt or rust at any time.

Early Models

➡**Early models include 1995–96 F-150 and F-250, 1995–97 F-250HD, F-350 and F-Super Duty, Bronco and E-Series.**

2-WHEEL DRIVE

1. Raise and support the front end.

2. Remove the wheel cover. Remove the wheel.

3. Remove the caliper from the disc and wire it to the underbody to prevent damage to the brake hose.

4. Remove the grease cap from the hub. Then, remove the cotter pin, nut lock, adjusting nut and flat washer from the spindle. Remove the outer bearing assembly from the hub.

5. Pull the hub and disc assembly off the wheel spindle.

6. Remove and discard the old grease retainer. Remove the inner bearing cone and roller assembly from the hub.

7. Clean all grease from the inner and outer bearing cups with solvent. Inspect the cups for pits, scratches, or excessive wear. If the cups are damaged, remove them with a drift.

8. Clean the inner and outer cone and roller assemblies with solvent and shake them dry. If the cone and roller assemblies show excessive wear or damage, replace them with the bearing cups as a unit.

9. Clean the spindle and the inside of the hub with solvent to thoroughly remove all old grease.

10. Covering the spindle with a clean cloth, brush all loose dirt and dust from the brake assembly. Remove the cloth carefully so as to not get dirt on the spindle.

11. If the inner and/or outer bearing cups were removed, install the replacement cups on the hub. Be sure that the cups seat properly in the hub.

12. It is imperative that all old grease be removed from the bearings and surrounding surfaces before repacking. The new lithium-based grease is not compatible with the sodium base grease used in the past.

13. Install the hub and disc on the wheel spindle. To prevent damage to the grease retainer and spindle threads, keep the hub centered on the spindle.

14. Install the outer bearing cone and roller assembly and the flat washer on the spindle. Install the adjusting nut.

15. Adjust the wheel bearings by tightening the adjusting nut to 17–25 ft. lbs. (23–34 Nm) with the wheel rotating to seat the bearing. Then, back off the adjusting nut ½ turn. Retighten the adjusting nut to 10–15 inch lbs. (1–1.7 Nm). Install the locknut so that the castellations are aligned with the cotter pin hole. Install the cotter pin. Bend the ends of the cotter pin around the castellations of the locknut to prevent interference with the radio static

collector in the grease cap. Install the grease cap.

➡**New bolts must be used when servicing floating caliper units. The upper bolt must be tightened first.**

16. Install the wheels.

17. Install the wheel cover.

4-WHEEL DRIVE MODELS WITH MANUAL LOCKING HUBS

1. Raise the vehicle and install safety stands.

2. Remove the hub assemblies.

3. On Bronco, F-150 and F-250 with the Dana 44 axle: apply inward pressure on the bearing adjusting nut, using a socket made for that purpose, available at most auto parts stores, to disengage the adjusting nut locking splines, while turning it counterclockwise to remove it.

4. On F-250 HD (Dana 50 axle) and F-350, use the hub nut tool to unscrew the outer locking nut. Then, remove the lock ring from the bearing adjusting nut. This can be done with your finger tips or a screwdriver. Use the locknut socket to remove the bearing adjusting nut.

5. Remove the caliper and suspend it out of the way.

6. Slide the hub and disc assembly off of the spindle. The outer wheel bearing will slide out as the hub is removed, so be prepared to catch it.

7. Lay the hub on a clean work surface. Carefully drive the inner bearing cone and grease seal out of the hub using Tool T69L-1102-a, or equivalent.

To install:

8. Inspect the bearing cups for pits or cracks. If necessary, remove them with a drift. If new cups are installed, install new bearings.

9. Lubricate the bearings with Multipurpose Lubricant Ford Specification, ESA-MIC7-B or equivalent. Clean all old grease from the hub. Pack the cones and rollers. If a bearing packer is not available, work as much lubricant as possible between the rollers and the cages.

10. Drive new cups into place with a driver, making sure that they are fully seated.

11. Position the inner bearing cone and roller in the inner cup and install the grease retainer.

12. Carefully position the hub and disc assembly on the spindle.

13. Install the outer bearing cone and roller, and the adjusting nut.

14. On Bronco, F-150 and F-250 LD with the Dana 44 axle:

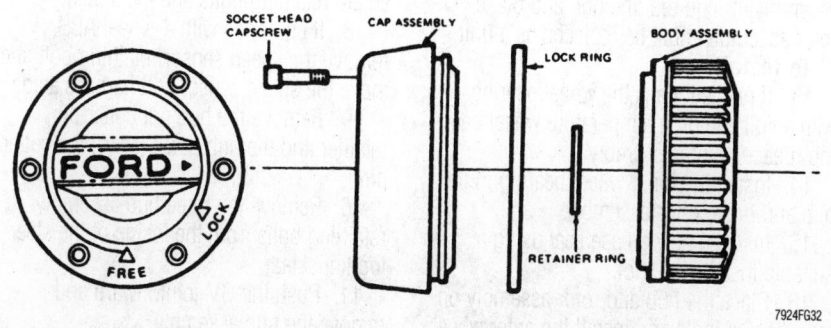

Manual locking hub assembly—1995-96 Bronco and F-Series

a. Be sure the metal stamping on the adjusting nut faces inboard and the inner diameter key on the nut enters the spindle keyway.

b. Apply inward pressure on the hub nut wrench and tighten the adjusting nut to 70 ft. lbs. (95 Nm) while rotating the hub back and forth to seat the bearings.

c. Apply inward pressure on the wrench and back off the nut about 90°, then retighten the nut to 15—20 ft. lbs. (20—27 Nm).

d. Remove the wrench. End-play of the hub/rotor assembly should be 0 (zero) and the torque required to rotate the hub assembly should not exceed 20 inch lbs. (2.2 Nm).

15. On the F-250 HD (Dana 50 axle) and F-350:

➡ **The adjusting nut has a small dowel on one side. This dowel faces outward to engage the locking ring.**

a. Using the hub nut socket and a torque wrench, tighten the bearing adjusting nut to 50 ft. lbs. (68 Nm), while rotating the wheel back and forth to seat the bearings.

b. Back off the adjusting nut approximately 90°.

c. Install the lock ring by turning the nut to the nearest hole and inserting the dowel pin.

✳✳ CAUTION

The dowel pin must seat in a lock ring hole for proper bearing adjustment and wheel retention!

d. Install the outer locknut and tighten to 160—205 ft. lbs. (218—279 Nm). Final end-play of the wheel on the spindle should be 0—0.004 in. (0—0.15mm).

16. Assemble the hub parts.

17. Install the caliper.

18. Safety lower the vehicle.

4-WHEEL DRIVE MODELS WITH AUTOMATIC LOCKING HUBS

1. Raise and safely support the vehicle.

2. Remove the hub assemblies.

3. Using a socket made for that purpose, available at most auto parts stores, use the hub nut tool to unscrew the outer locking nut.

4. Remove the lockring from the bearing adjusting nut. This can be done with your finger tips or a screwdriver.

5. Use the locknut socket to remove the bearing adjusting nut.

6. Remove the caliper and suspend it out of the way.

7. Slide the hub and disc assembly off of the spindle. The outer wheel bearing will slide out as the hub is removed, so be prepared to catch it.

8. Lay the hub on a clean work surface. Carefully drive the inner bearing cone and grease seal out of the hub using Tool T69L-1102-A, or equivalent.

To install:

9. Inspect the bearing cups for pits or cracks. If necessary, remove them with a drift. If new cups are installed, install new bearings.

10. Lubricate the bearings with Multipurpose Lubricant Ford Specification, ESA-MIC7-B or equivalent. Clean all old grease from the hub. Pack the cones and rollers. If a bearing packer is not available, work as much lubricant as possible between the rollers and the cages.

11. Drive new cups into place with a driver, making sure that they are fully seated.

12. Position the inner bearing cone and roller in the inner cup and install the grease retainer.

13. Carefully position the hub and disc assembly on the spindle.

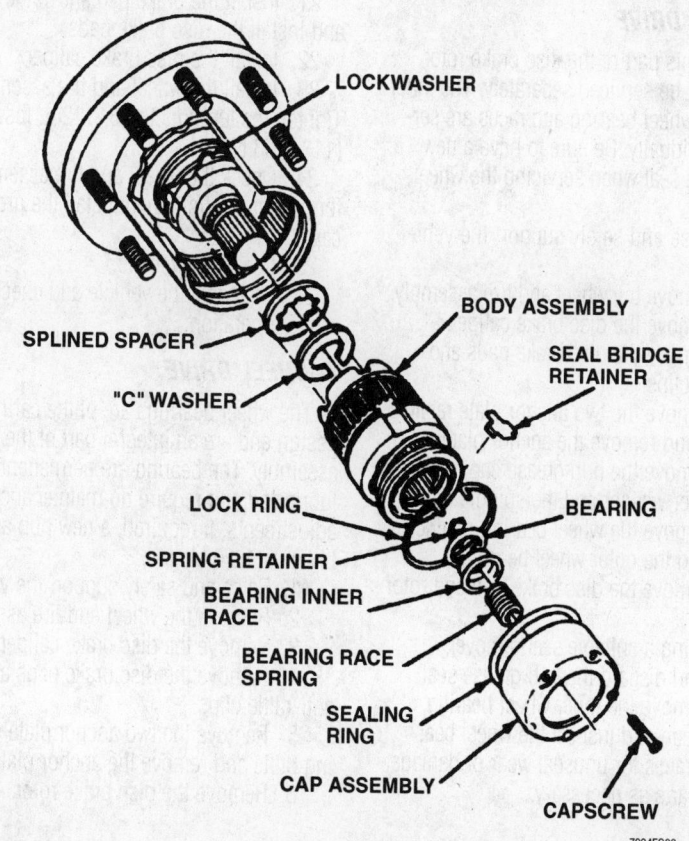

Exploded view of the automatic locking hub assembly—1995-96 Bronco and F-Series

14. Install the outer bearing cone and roller, and the adjusting nut.

➡**The adjusting nut has a small dowel on one side. This dowel faces outward to engage the locking ring.**

15. Using the hub nut socket and a torque wrench, tighten the bearing adjusting nut to 50 ft. lbs. (68 Nm), while rotating the wheel back and forth to seat the bearings.

16. Back off the adjusting nut approximately 90°.

17. Install the lockring by turning the nut to the nearest hole and inserting the dowel pin.

➡**The dowel pin must seat in a lockring hole for proper bearing adjustment and wheel retention.**

18. Install the outer locknut and tighten to 160–205 ft. lbs. (218–279 Nm). Final end-play of the wheel on the spindle should be 0–0.004 in. (0–0.15mm).

19. Assemble the hub parts.

20. Install the caliper.

21. Lower the vehicle.

Late Models

➡**Late models include 1997–99 F-150, F-250 and Expedition and 1998–99 F-250HD, F-350, F-450 and Navigator.**

2-WHEEL DRIVE

The hub is part of the disc brake rotor and cannot be serviced separately. The inner and outer wheel bearing and races are serviced individually. Be sure to have a new hub grease seal when servicing the wheel bearings.

1. Raise and safely support the vehicle.

2. Remove the wheel and tire assembly.

3. Remove the disc brake caliper.

4. Remove the disc brake pads and anti-rattle clips.

5. Remove the two anchor plate retaining bolts and remove the anchor plate.

6. Remove the hub grease cap, cotter pin, retainer washer and the spindle nut.

7. Remove the wheel bearing retainer washer and the outer wheel bearing.

8. Remove the disc brake hub and rotor assembly.

9. Using a suitable seal remover, remove and discard the hub grease seal.

10. Remove the inner wheel bearing.

11. Clean and inspect the wheel bearings and races for unusual wear or damage. Replace parts as necessary.

12. Inspect the hub and brake rotor assembly. If required, the hub and brake rotor assembly must be replaced as a unit.

To install:

13. If needed, pack the wheel bearing with a suitable high temperature wheel bearing grease before assembly.

14. Install the inner wheel bearing in the hub and brake rotor assembly.

15. Install a new grease seal using a suitable installation tool.

16. Place the hub and rotor assembly on the wheel spindle and install the outer wheel bearing.

17. Install the retainer washer and the spindle nut.

18. Adjust the wheel bearings as follows:

   a. Tighten the spindle nut to 17–24 ft. lbs. (23–34 Nm) while rotating the wheel and tire assembly to seat the wheel bearings.

   b. Back off the spindle nut no less than ½ turn.

   c. Tighten the spindle nut to 17 inch lbs. (2 Nm).

19. Install the retaining washer so the castellations are aligned with the cotter pin hole. Install a new cotter pin.

20. Replace the anchor plate and install two retaining bolts. Tighten the bolts to 125–168 ft. lbs. (170–230 Nm).

21. Install the brake pad anti-rattle clips and install the disc brake pads.

22. Install the disc brake caliper.

23. Install the wheel and tire assembly. Tighten the lug nuts to 83–112 ft. lbs. (113–153 Nm).

24. Check the wheel and tire assembly for proper rotation, then install the grease cap.

25. Lower the vehicle.

26. Road test the vehicle and check for proper operation.

4-WHEEL DRIVE

The wheel bearings are of the cartridge design and are an integral part of the hub assembly. The bearing are permanently lubricated and require no maintenance or adjustments. If required, a new hub assembly must be installed.

1. Raise and safely support the vehicle.

2. Remove the wheel and tire assembly.

3. Remove the disc brake caliper.

4. Remove the disc brake pads and anti-rattle clips.

5. Remove the two anchor plate retaining bolts and remove the anchor plate.

6. Remove the disc brake rotor.

7. Remove the three disc brake rotor shield retaining bolts and the shield.

8. If equipped with 4-wheel ABS, remove the speed sensor retaining bolt and move the speed sensor and harness aside.

9. Remove the hub nut cotter pin, retainer and the hub nut. Discard the cotter pin.

10. Remove the three hub assembly retaining bolts from the inside of the steering knuckle.

11. Push the CV-joint inward and remove the hub assembly.

➡**If necessary, use a suitable puller to separate the hub assembly from the CV-joint. Use care not to over extend the CV-joint and boot when removing the hub assembly.**

12. If required, remove the grease seal from the steering knuckle.

To install:

13. If removed, install a new grease seal with Seal Installer T96T-1175-A, Drawbar T77F-1176-A and Cup Replacer T80T-4000-P, or equivalents.

14. Install the hub assembly to the steering knuckle and secure with the three retaining bolts. Tighten the bolts to 110–145 ft. lbs. (149–201 Nm).

15. Align the splines, then push the CV-joint into the hub assembly.

16. If equipped with 4-wheel ABS, install the speed sensor and secure with the retaining bolt. Tighten the bolt to 60–84 inch lbs. (7–9 Nm).

17. Install the disc brake rotor shield and the three retaining bolts. Tighten the bolts to 80–107 inch lbs. (9–12 Nm).

18. Install the hub nut and tighten to 188–254 ft. lbs. 255–345 Nm).

19. Install the hub nut retainer and a new cotter pin.

20. Install the disc brake rotor.

21. Place the anchor plate in position and install the two retaining bolts. Tighten the bolts to 125–168 ft. lbs. (170–230 Nm).

22. Install the brake pad anti-rattle clips and install the disc brake pads.

23. Install the disc brake caliper.

24. Install the wheel and tire assembly. Tighten the lug nuts to 83–112 ft. lbs. (113–153 Nm).

25. Lower the vehicle.

26. Pump the brake pedal several times to position the brake pads prior to moving the vehicle.

27. Road test the vehicle and check for proper operation.

ENGINE REPAIR

Engine Assembly

REMOVAL & INSTALLATION

In the process of removing the engine, you will come across a number of steps which call for the removal of a separate component or system, such as "disconnect the exhaust system" or "remove the intake manifold." In most instances, a detailed removal procedure can be found elsewhere in this section.

It is virtually impossible to list each individual wire and hose which must be disconnected, simply because so many different model and engine combinations have been manufactured. Careful observation and common sense are the best possible approaches to any repair procedure.

Removal and installation of the engine can be made easier if you follow these basic points:

• If you have to drain any of the fluids, use a suitable container.

• Always tag any wires or hoses and, if possible, the components they came from before disconnecting them.

• Because there are so many bolts and fasteners involved, store and label the retainers from components separately in muffin pans, jars or coffee cans. This will prevent confusion during installation.

• After unbolting the transmission or transaxle, always make sure it is properly supported.

• If it is necessary to disconnect the air conditioning system, have this service performed by a qualified technician using a recovery/recycling station. If the system does not have to be disconnected, unbolt the compressor and set it aside.

• When unbolting the engine mounts, always make sure the engine is properly supported. When removing the engine, make sure that any lifting devices are properly attached to the engine. It is recommended that if your engine is supplied with lifting hooks, your lifting apparatus be attached to them.

• Lift the engine from its compartment slowly, checking that no hoses, wires or other components are still connected.

• After the engine is clear of the compartment, place it on an engine stand or workbench.

• After the engine has been removed, you can perform a partial or full teardown of the engine using the procedures outlined in this manual.

1. Disconnect the negative battery cable.
2. Drain the engine coolant.
3. Have the A/C system discharged by an EPA-certified technician into a refrigerant recovery station.
4. Remove the cowl top vent panel.
5. Disconnect the wiring from the alternator.
6. Loosen the air cleaner outlet tube clamp at the throttle body and remove the engine air cleaner assembly.
7. Disconnect the upper and lower radiator hoses from the engine.
8. Disconnect the heater water hoses and secure to body.
9. Disconnect and plug the A/C discharge and suction hoses, then secure them to the engine. Cap the open lines to prevent moisture from entering the system.
10. Disconnect accelerator cable and speed control cable from the throttle body lever.
11. Remove the accelerator cable bracket from the throttle body.
12. Disconnect the fuel supply and return lines from the fuel injection supply manifold.
13. Label and disconnect all engine wiring harnesses from the engine and secure to the body.
14. Label and disconnect all vacuum hoses from the engine.
15. Remove the gear shift cable from the transaxle.

➡Damage to the steering column air bag wiring can result if the steering wheel is allowed to rotate freely. The wire is wound like a watch spring and can be overtightened and break if the steering wheel is rotated too far in either direction.

16. Lock the steering wheel with the wheels in the straight ahead position by turning the ignition to the **OFF** position.
17. Raise and support the vehicle safely.
18. Remove the front wheels.
19. Drain the engine oil.
20. Disconnect and plug the transaxle cooler lines at the transaxle. Secure the lines to the radiator.
21. Disconnect the heated oxygen sensor wiring harness.

➡Do not allow the flex connector of the duel converter Y-pipe to hang unsupported or damage to the flex joint will result.

22. Disconnect the dual converter Y-pipe and support it from the body.

➡The routing of the battery ground cable to the cylinder block is critical. It should go between the transaxle and the bracket. Take note during disassembly.

23. Disconnect the starter motor wires and secure out of the way.
24. Remove the starter motor.
25. Remove the engine rear plate and torque converter-to-flywheel nuts.
26. Disconnect and plug the power steering cooler lines.
27. Remove the upper bolt from the sway bar links.
28. Remove the dust boot from the steering rack pinion support by gently spreading the integral tension ring and pushing upward.
29. Remove the steering coupling pinch bolt from the steering column intermediate shaft at the steering gear.
30. Remove the intermediate shaft from the steering gear.
31. Remove the front stabilizer bar links.
32. Separate the front suspension lower arms from the knuckles at the ball joint.
33. Separate the tie rod ends from the knuckle.
34. Remove the front axle wheel hub retainers from the halfshaft ends.
35. Remove the halfshafts from the front wheel knuckle.
36. Support the front subframe, engine and transaxle assembly.
37. Remove the four retaining bolts and lower the engine transaxle and front subframe from the vehicle.
38. Disconnect the power steering pressure hose from the power steering pump.
39. Attach an engine hoist and lift the engine slightly.
40. Remove the engine support insulators.
41. Lift the engine and transaxle assembly from the front subframe.
42. Lower the engine and transaxle.
43. Support the transaxle on a level stationary surface and separate the engine from the transaxle.

To install:

44. If removed, install the transaxle on the engine.
45. Install the engine on the subframe.
46. Install the engine support insulators.
47. Connect the power steering hose to the pump.
48. Carefully raise the engine/transaxle assembly into the vehicle.
49. Install the four subframe bolts.
50. The remainder of the installation is the reverse of removal.

51. Please note the following torque specifications:

- Engine-to-transaxle—30–44 ft. lbs. (40–60 Nm)
- Torque converter nuts—20–34 ft. lbs. (27–46 Nm)
- Subframe-to-body bolts—57–76 ft. lbs. (77–103 Nm)
- Steering coupling pinch bolt—25–34 ft. lbs. (34–46 Nm)
- Dual converter Y-pipe-to-exhaust manifold—25–34 ft. lbs. (34–46 Nm)
- Flex pipe retaining bolts—25–34 ft. lbs. (34–46 Nm)
- Accelerator cable bracket retaining bolts—71–106 inch lbs. (8–12 Nm)
- Front engine support insulator-to-subframe (3.8L engine)—50–68 ft. lbs. (68–92 Nm)
- Front engine support insulator-to-subframe (3.0L engine)—65–87 ft. lbs. (88–119 Nm)
- Transmission insulator-to-subframe—65–87 ft. lbs. (88–119 Nm)
- Rear engine and transaxle support insulator-to-subframe—56–75 ft. lbs. (76–103 Nm)

52. Refill the engine, transaxle and cooling system with the correct amount of the appropriate fluids before starting the engine.

Water Pump

REMOVAL & INSTALLATION

3.0L Engine

1. Disconnect the negative battery cable.
2. Drain the engine coolant.
3. Loosen the four water pump pulley retaining bolts while the accessory drive belts are still tight.
4. Rotate the automatic tensioner down and to the left, then remove the accessory drive belt.

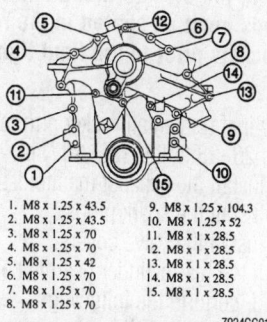

1. M8 x 1.25 x 43.5	9. M8 x 1.25 x 104.3
2. M8 x 1.25 x 43.5	10. M8 x 1.25 x 52
3. M8 x 1.25 x 70	11. M8 x 1 x 28.5
4. M8 x 1.25 x 70	12. M8 x 1 x 28.5
5. M8 x 1.25 x 42	13. M8 x 1 x 28.5
6. M8 x 1.25 x 70	14. M8 x 1 x 28.5
7. M8 x 1.25 x 70	15. M8 x 1 x 28.5
8. M8 x 1.25 x 70	

7924GG01

Water pump bolts come in different sizes, make sure the bolts go back into the correct holes—3.0L engine

5. Remove the two nuts and bolt retaining the drive belt automatic tensioner to the engine, then remove the tensioner.
6. Disconnect and remove the lower radiator and heater hose from the water pump.
7. Remove the eleven water pump-to-engine retaining bolts, then lift the water pump and pulley up and out of the vehicle.

➡ **Bolts 1, 2, 3 and 10 are for the front cover and are not removed for this procedure.**

8. Remove the water pump pulley retaining bolts, then remove the pulley from the water pump.

To install:

❄❄ WARNING

Be careful not to gouge the aluminum surfaces when scraping the old gasket material from the mating surfaces of the water pump and front cover.

9. Clean the gasket surfaces on the water pump and front cover. Lightly oil all bolt and stud threads, except those requiring special sealant.
10. Position a new water pump housing gasket on the water pump sealing surface using gasket sealant to hold the gasket in place.
11. With the water pump pulley and retaining bolts loosely installed on the water pump, align the water pump-to-engine front cover, then install the retaining bolts.
12. Tighten the bolts to the following specifications:

  a. Bolt numbers 4, 5, 6, 7, 8 and 9—22 ft. lbs. (20–30 Nm).
  b. Bolt numbers 11–15—71–106 inch lbs. (8–12 Nm).

13. Hand-tighten the water pump pulley retaining bolts.
14. Install the automatic belt tensioner assembly. Tighten the two retaining nuts and bolt to 35 ft. lbs. (47 Nm).
15. Install the alternator and power steering belts. Final tighten the water pump pulley retaining bolts to 15–22 ft. lbs. (22–30 Nm).
16. Position the hose clamps between the alignment marks on both ends of the hose, then slide the hose on the connection. Tighten the hose clamps to 20–30 inch lbs. (2.2–3.4 Nm).
17. Fill and bleed the cooling system.
18. Connect the negative battery cable.
19. Start the engine and check for leaks.

3.8L Engine

1. Disconnect the negative battery cable.
2. Drain the engine coolant.
3. Loosen the drive belt tensioner, then remove the drive belts.
4. Remove the lower radiator hose.
5. Remove the lower nut on both front engine supports.
6. Remove the alternator.
7. Position a drain pan under the power steering pump.
8. Disconnect power steering pressure line from the pump using a Fuel Line Disconnect tool (T90T-9550-S), or equivalent.
9. Remove the power steering reservoir filler cap.

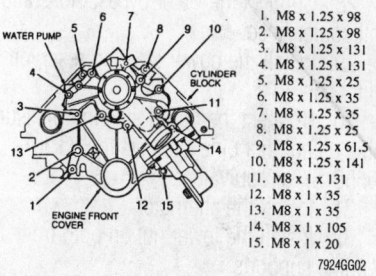

1. M8 x 1.25 x 98	
2. M8 x 1.25 x 98	
3. M8 x 1.25 x 131	
4. M8 x 1.25 x 131	
5. M8 x 1.25 x 25	
6. M8 x 1.25 x 35	
7. M8 x 1.25 x 35	
8. M8 x 1.25 x 25	
9. M8 x 1.25 x 61.5	
10. M8 x 1.25 x 141	
11. M8 x 1 x 131	
12. M8 x 1 x 35	
13. M8 x 1 x 35	
14. M8 x 1 x 105	
15. M8 x 1 x 20	

7924GG02

Because of their varying lengths, be sure to install the water pump bolts in the correct bolt holes—3.8L engine

10. Disconnect the water bypass hose and oil cooler hose from the heater water outlet tube.
11. Remove the retaining bolt and disconnect the heater water outlet tube from the water pump.
12. Remove the A/C bracket brace.
13. Raise the engine approximately 2 inches (51mm) to provide necessary clearance for water pump removal.
14. Remove the water pump pulley.
15. Remove the drive belt tensioner form the power steering pump brace.
16. Remove the power steering pump brace and place the pump and brace aside in the engine compartment.
17. Remove the water pump.

To install:

18. Clean all gasket mating surfaces thoroughly.

❄❄ WARNING

Be careful not to gouge the aluminum surfaces when scraping the old gasket material from the mating surfaces of the water pump and front cover.

19. Coat the threads of the No. 1 engine front cover stud with Teflon® pipe sealant, or equivalent.

20. Position a new water pump housing gasket on the water pump sealing surface using gasket sealant to hold the gasket in place.

21. Install the water pump and tighten the bolts to 15–22 ft. lbs. (20–30 Nm) and the nuts to 71–106 inch lbs. (8–12 Nm).

22. Install the power steering pump brace.

23. Install the drive belt tensioner.

24. Install the water pump pulley.

25. Lower the engine.

26. Install the A/C bracket brace.

27. Connect the heater water outlet tube.

28. Connect the water bypass hose and oil cooler hose.

29. Install the power steering reservoir filler cap.

30. Connect the power steering pressure line using Fuel Line Connect tool (T90T-9550-S), or equivalent.

31. Install the alternator.

32. Install the lower nut on both front engine supports.

33. Install the lower radiator hose.

34. Install the drive belts.

35. Fill and bleed the cooling system.

36. Connect the negative battery cable.

Cylinder Head

REMOVAL & INSTALLATION

3.0L Engine

1. Rotate the crankshaft to 0° TDC on the compression stroke.

2. Disconnect the negative battery cable.

3. Drain and recycle the engine coolant.

4. Remove the cowl top vent panel.

5. Remove the air cleaner outlet tube from the throttle body.

6. Label and disconnect all necessary vacuum lines.

7. Disconnect the EGR backpressure transducer from the EGR valve.

8. Loosen the lower EGR valve-to-exhaust manifold tube nut and rotate the EGR valve tube away from the valve.

9. Label and disconnect the all necessary engine wiring.

10. Properly relieve the fuel system pressure.

11. Remove the fuel line safety clips and disconnect the fuel lines.

➡The fuel injectors and fuel injection supply manifold may be removed with the lower intake manifold as an assembly.

12. Remove the ignition wires and ignition coil pack.

13. Disconnect the upper radiator and heater hoses.

14. Remove the camshaft position sensor.

15. If the front cylinder head is being removed, perform the following:

    a. Disconnect the alternator electrical harness.

    b. Rotate the tensioner clockwise and remove the accessory drive belt.

    c. Remove the automatic belt tensioner assembly.

    d. Remove the alternator.

    e. Remove the power steering mounting bracket retaining bolts. Leave the hoses connected and place the pump aside in a position to prevent fluid from leaking out.

    f. Remove the engine oil dipstick tube from the exhaust manifold.

16. If the rear cylinder head is being removed, perform the following:

    a. Remove the alternator belt tensioner bracket.

    b. Remove the heater supply tube retaining brackets from the exhaust manifold.

    c. Remove the vehicle speed sensor cable retaining bolt.

    d. Remove the EGR vacuum regulator sensor and bracket.

17. Remove the valve covers.

➡Pushrods must be installed in their original positions. Note pushrod location during removal.

18. Loosen the rocker arm seat retaining bolts and remove the pushrods.

19. Remove the lower intake manifold.

20. Remove the spark plugs.

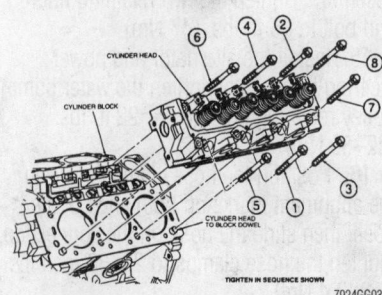

TIGHTEN IN SEQUENCE SHOWN 7924GG03

Cylinder head bolts must be tightened in the correct sequence to ensure proper cylinder sealing—3.0L engine

21. Remove the exhaust manifolds.

22. Remove the cylinder head bolts.

23. Lift the cylinder head from the engine block and discard the gaskets.

To install:

24. The cylinder head should be cleaned and inspected prior to installation.

25. Lightly oil all bolt and stud bolt threads before installation.

26. Clean all gasket mating surfaces thoroughly.

27. Position new head gaskets on the cylinder block, noting the UP position mark on the gasket face and using the dowels in the engine block for alignment. If the dowels are damaged, they must be replaced.

✳✳ WARNING

Always use new cylinder head bolts when installing the cylinder head or damage to the engine may occur.

28. Position the cylinder head on the cylinder block. Tighten the cylinder head bolts in 2 steps following the proper torque sequence. The first step is 37 ft. lbs. (50 Nm) and the second step is 68 ft. lbs. (92 Nm).

➡When the cylinder head attaching bolts have been tightened using this procedure, it is not necessary to retighten the bolts after extended engine operation. The bolts can be rechecked for tightness if desired.

29. Install the intake manifold.

30. Engage all engine wiring harnesses previously disconnected.

31. Dip each pushrod end in engine assembly lubricant. Install the pushrods in their original positions.

32. Lubricate all rocker arm components with engine assembly lubricant.

33. Install the rocker arms, seats and retaining bolts. Tighten the bolts to 62–132 inch lbs. (7–15 Nm).

➡The rocker arm seats must be fully seated in the cylinder head and the pushrods must be seated in the rocker arm sockets prior to the final tightening.

34. Final tighten all rocker arm retaining bolts to 20–28 ft. lbs. (26–38 Nm).

35. Install the exhaust manifolds.

36. Install the spark plugs.

37. Install the valve covers.

38. If the rear cylinder head is being installed, perform the following:

    a. Install the EGR vacuum regulator sensor and bracket.

    b. Install the vehicle speed sensor cable retaining bolt.

c. Install the heater supply tube retaining brackets from the exhaust manifold.

d. Install the alternator belt tensioner bracket.

e. Install the engine oil dipstick tube from the exhaust manifold.

39. If the front cylinder head is being installed, perform the following:

a. Install the power steering mounting bracket retaining bolts. Leave the hoses connected and place the pump aside in a position to prevent fluid from leaking out.

b. Install the alternator.

c. Install the automatic belt tensioner assembly.

d. Rotate the tensioner clockwise and install the accessory drive belt.

e. Connect the alternator electrical harness.

40. Install the camshaft position sensor.

41. Connect the upper radiator and heater hoses.

42. Install the ignition coil pack and spark plug wires.

43. Connect the fuel lines and install the fuel line safety clips.

44. Loosen the lower EGR valve-to-exhaust manifold tube nut and rotate the EGR valve tube away from the valve.

45. Connect the EGR backpressure transducer from the EGR valve.

46. Connect all necessary vacuum lines.

47. Install the air cleaner outlet tube to the throttle body.

48. Install the cowl top vent panel.

49. Fill and bleed the cooling system.

➡**Engine coolant is corrosive to engine bearing material. Replace the engine oil after removal of any coolant carrying component to help prevent potential bearing damage.**

50. Change the engine oil and filter

51. Connect the negative battery cable.

52. Start the engine and check for leaks.

3.8L Engine

1. Disconnect the negative battery cable.

2. Drain the engine coolant.

3. Remove the air cleaner assembly.

4. Remove the cowl top vent panel.

5. Rotate the tensioner clockwise and remove the accessory drive belt.

6. If the front cylinder head is being removed, perform the following:

a. Remove the oil filler cap.

b. Remove the A/C compressor mounting bracket and set the A/C compressor aside with the refrigerant lines still connected.

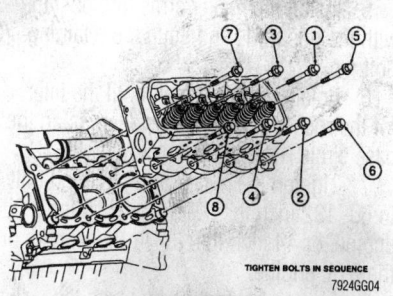

TIGHTEN BOLTS IN SEQUENCE

7924GG04

During installation, tighten the cylinder head bolts in the proper order—3.8L engine

c. Remove the power steering pump and bracket. Leave the power steering hoses connected and place the pump aside in the engine compartment.

d. Remove the alternator and alternator mounting bracket.

7. If the rear cylinder head is being removed, perform the following:

a. Remove the accessory drive belt tensioner.

b. Remove the PCV valve

c. Remove the power steering line bracket.

d. Remove the tensioner bracket.

e. Remove the coil pack assembly.

8. Remove the upper intake manifold.

9. Remove the valve cover.

10. Properly relieve the fuel system pressure.

11. Remove the fuel charging assembly.

12. Remove the lower intake manifold.

13. Remove the exhaust manifolds.

➡**Pushrods must be installed in their original positions. Note pushrod location during removal.**

14. Loosen the rocker arm seat retaining bolts and remove the pushrods.

15. Remove the cylinder head bolts.

16. Lift the cylinder head from the engine block and discard the gaskets.

To install:

17. The cylinder head should be cleaned and inspected prior to installation.

18. Lightly oil all bolt and stud bolt threads before installation.

19. Clean all gasket mating surfaces thoroughly.

20. Position new head gaskets on the cylinder block, noting the UP position mark on the gasket face, using the dowels in the engine block for alignment. If the dowels are damaged, they must be replaced.

Always use new cylinder head bolts when installing cylinder head or damage to the engine may occur.

21. Position the cylinder head on the cylinder block.

22. Lubricate the cylinder head bolts with 50 weight oil and install. Tighten the cylinder head bolts in 3 steps following the proper torque sequence. The first step is 15 ft. lbs. (20 Nm), the second step is 29 ft. lbs. (40 Nm) and the third step is 37 ft. lbs. (50 Nm).

➡**Do not loosen all of the cylinder head bolts at once. Only work on one bolt at a time or damage to the engine may occur.**

23. In sequence, loosen each cylinder head bolt 2–3 turns and retighten in 2 steps.

a. On long bolts, the first step is 29–37 ft. lbs. (40–50 Nm). The second step is tighten an additional 175–185 degrees.

b. On short bolts, the first step is 15–22 ft. lbs. (20–30 Nm). The second step is tighten an additional 175–185 degrees.

24. Dip each pushrod end in engine assembly lubricant. Install the pushrods in their original positions.

25. Lubricate all rocker arm components with engine assembly lubricant.

26. For the rocker arm being installed, rotate the engine clockwise until the valve tappet rests on the heel (base circle) of the camshaft lobe.

27. Install the rocker arms, seats and bolts and tighten to 44 inch lbs. (5 Nm).

28. Perform the previous two steps for each rocker arm.

29. After all rocker arms have been installed, final tighten all bolts to 22–29 ft. lbs. (30–40 Nm).

30. Install the exhaust manifolds.

31. Install the lower intake manifold.

32. Install the fuel injection charging assembly.

33. Install the valve covers with new gaskets. Tighten the bolts to 71–97 inch lbs. (8–11 Nm).

34. Install the upper intake manifold.

35. Install the spark plugs and ignition wires.

36. If the front cylinder head is being installed, perform the following:

a. Install the alternator and alternator mounting bracket.

b. Install the power steering pump and bracket.

c. Install the A/C compressor mounting bracket.

d. Install the oil filler cap.

37. If the rear cylinder head is being installed, perform the following:

a. Install the coil pack assembly.

b. Install the tensioner bracket.

c. Install the power steering line bracket.

d. Install the PCV valve.

e. Install the accessory drive belt tensioner.

f. Rotate the tensioner clockwise and install the accessory drive belt.

38. Install the cowl top vent panel.

39. Install the air cleaner assembly.

40. Fill and bleed the cooling system.

41. Connect the negative battery cable.

Rocker Arms

REMOVAL & INSTALLATION

1. Remove the valve cover.

2. Remove the rocker arm retaining bolt.

➡Rocker the arms should be installed in their original location during assembly.

3. Remove the rocker arms. If more than one rocker arm is to be removed, identify each rocker arm location.

To install:

4. Lubricate the pushrods and rocker arms with Engine Assembly Lubricant (D9AZ-19579-D), or equivalent. Lubricate the retaining bolts with engine oil.

➡Prior to final tightening, the rocker arm seats must be fully seated into the cylinder head. The pushrods must be fully seated in the rocker arm and valve tappet sockets.

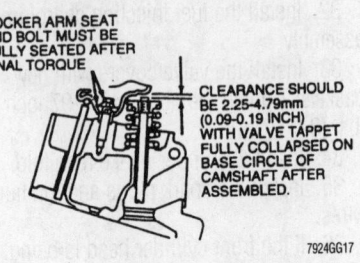

ROCKER ARM SEAT AND BOLT MUST BE FULLY SEATED AFTER FINAL TORQUE

CLEARANCE SHOULD BE 2.25-4.79mm (0.09-0.19 INCH) WITH VALVE TAPPET FULLY COLLAPSED ON BASE CIRCLE OF CAMSHAFT AFTER ASSEMBLED.

7924GG17

When the lifter is fully collapsed and on the base circle of the cam, check for proper clearance between the tip of the valve and the rocker arm—3.8L engine

5. Install the rocker arms into position with the pushrods and snug the retaining bolt.

6. Rotate the crankshaft until the lifter for the rocker arm being installed, is on the base circle (heel) of the cam lobe.

7. Tighten the rocker arm retaining bolt to 60–132 inch lbs. (7–15 Nm) on the 3.0L engine, or 44 inch lbs. (5 Nm) maximum on the 3.8L engine.

8. Finally, tighten the bolt with the camshaft in any position to 20–28 ft. lbs. (26–38 Nm).

9. Install the valve cover.

Intake Manifold

REMOVAL & INSTALLATION

3.0L Engine

1. Disconnect the negative battery cable.

2. Drain the engine coolant.

3. Remove the crankcase ventilation tube from the valve cover.

4. Remove the air cleaner inlet and outlet tube.

5. Properly relieve the fuel system pressure.

6. Remove the fuel line safety clips and disconnect the fuel lines.

7. Label and disconnect the vacuum lines.

8. Label and disconnect all electrical wiring attached to the intake manifold.

9. Label and disconnect all control cables.

10. Disconnect the radiator and heater hoses.

11. Remove the alternator brace.

12. Remove the EGR tube and EGR valve.

13. Remove the fuel charging assembly.

14. Label and remove the ignition wires. Remove the ignition coil pack.

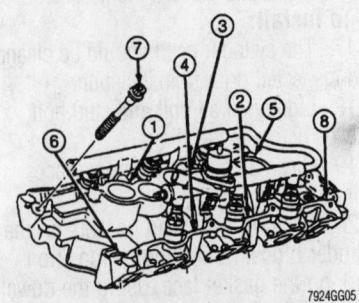

7924GG05

To prevent vacuum leaks and to ensure proper sealing, be sure to tighten the intake manifold bolts in the sequence shown—3.0L engine

15. Remove the camshaft position sensor.

16. Remove the valve covers.

17. Loosen cylinder No. 3 intake valve rocker arm seat retaining bolt. Rotate the rocker arm off its pushrod and away from the top of the valve stem. Remove the pushrod.

➡The lower intake manifold may be removed with the fuel injection supply manifold and fuel injectors in place as an assembly.

18. Remove the lower intake manifold retaining bolts using a Torx® head socket. Use a soft-faced mallet to tap the manifold upward if it is hard to remove.

19. Remove the intake manifold.

To install:

20. Throughly clean all gasket mating surfaces on the intake manifold and cylinder head.

❊❊❊ WARNING

When cleaning the cylinder head gasket surfaces, lay a clean cloth in the valve tappet area to prevent any particles from entering the oil drain-back area.

21. Apply a 0.25 in. (5–6mm) bead of silicone rubber sealant to the intersection of the cylinder block and cylinder head at the four corners of the intake manifold.

22. Install the intake manifold gaskets, aligning the intake gasket locking tabs to provisions on the head. Install the front and rear intake manifold end seals and secure with retainers.

23. Install the intake manifold.

24. Install intake bolts Nos. 1, 2, 3, and 4. Tighten by hand.

25. Install the remaining intake bolts and tighten all bolts in sequence to 15–22 ft. lbs. (20–30 Nm). Tighten again in sequence to 20–23 ft. lbs. (26–32 Nm).

26. Install the No. 3 intake valve pushrod and rocker arm.

27. Install the valve covers.

28. Lubricate the camshaft position sensor and install.

29. Install the coil pack and ignition wires.

30. Install the fuel charging assembly, and tighten the bolts to 71–106 inch lbs. (8–12 Nm).

31. Install the EGR tube and EGR valve.

32. Install the alternator brace.

33. Connect the radiator and heater hoses.

34. Connect and adjust the control cables.

35. Connect all electrical wiring

36. Connect the vacuum lines.
37. Connect the fuel lines and install the fuel line safety clips.
38. Install the air cleaner inlet and outlet tube.
39. Install the crankcase ventilation tube.
40. Fill and bleed the engine cooling system.
41. Connect the negative battery cable.
42. Start the engine and check for leaks.

3.8L Engine

UPPER MANIFOLD

1. Remove the air cleaner outlet tube.
2. Disconnect the accelerator cable and speed control actuator cable at the throttle body.
3. Remove the accelerator cable bracket and position it aside.
4. Label and disconnect the vacuum lines.
5. Label and disconnect the necessary electrical harnesses.
6. Disconnect the crankcase ventilation tube from the PCV valve.
7. Remove the throttle body.
8. Remove the idle air control valve.
9. Remove the intake manifold retaining bolts, noting their positions.

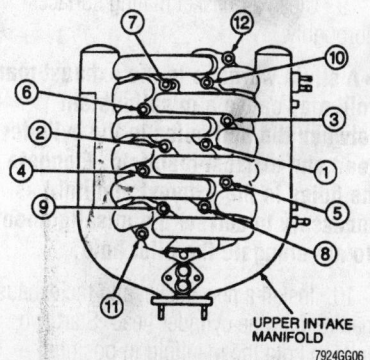

UPPER INTAKE MANIFOLD

7924GG06

Upper intake manifold bolt tightening sequence—3.8L engine

➡️**Keep the intake manifold bolts in order, so they can be installed in their original positions.**

10. Remove the upper intake manifold.
To install:
11. Inspect the intake gasket to ensure the seals are completely installed in the manifold groove and the seals show no signs of damage.
12. Install the intake manifold and tighten the bolts to 71–106 inch lbs. (8–12 Nm) in the sequence shown.
13. Install the idle air control valve.
14. Install the throttle body.

15. Connect the crankcase ventilation tube to the PCV valve.
16. Connect the electrical harnesses.
17. Connect the vacuum lines.
18. Install the accelerator cable bracket and tighten the bolts to 71–106 inch lbs. (8–12 Nm).
19. Connect the accelerator cable and speed control actuator cable at the throttle body.
20. Install the air cleaner outlet tube.

LOWER MANIFOLD

1. Drain the engine coolant.
2. Remove the upper intake manifold.
3. Disconnect the water bypass hose from the heater water outlet tube.
4. Remove the bypass hose from the lower intake manifold.
5. Properly relieve the fuel system pressure.
6. Label and disconnect the electrical wiring harnesses.
7. Remove the fuel injectors and fuel charging assembly.
8. Disconnect the vacuum motor and bracket assemblies.
9. Disconnect the valve assembly and linkage from the Intake Manifold Runner

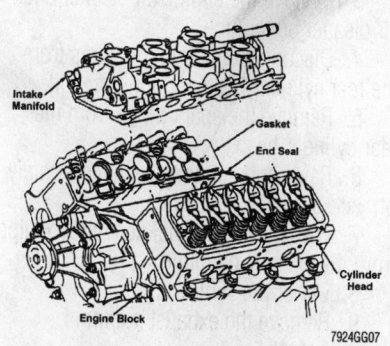

Intake Manifold

Gasket

End Seal

Cylinder Head

Engine Block

7924GG07

Lower intake manifold and related components—3.8L engine

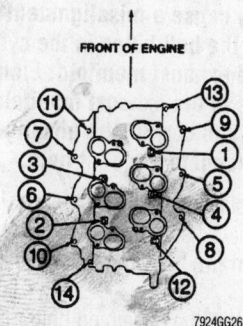

FRONT OF ENGINE

7924GG26

Tighten the lower intake manifold bolts in the sequence shown to prevent possible vacuum leaks—3.8L engine

Control (IMRC) lever and bushing by using a prytool.
10. Remove the tube retaining bolts.
11. Remove the old bushing from the lever.
12. Remove the EGR valve and adapter.
13. Remove the lower intake manifold retaining bolts.

➡️**The lower intake manifold is sealed at each corner with sealer. To break the seal it may be necessary to pry on the front of the intake manifold with a pry-bar. If it is necessary, use care to prevent damage to the machined surfaces.**

14. Remove the lower intake manifold.
To install:
15. Thoroughly clean all gasket mating surfaces.

➡️**When using silicone rubber sealer, assembly must occur within 15 minutes after sealer application. After this time, the sealer may start to set up and its sealing effectiveness may be reduced.**

16. Install new bushings into the IMRC levers.
17. Apply a 0.125 in. (3mm) bead of silicone rubber sealer at each corner where the cylinder head joins the engine block.
18. Install the front and rear intake manifold seals.
19. Using new gaskets, carefully install the lower intake manifold into position on the cylinder block.
20. Apply pipe sealant to the intake bolts and install in their original locations. Tighten in sequence to 71–106 inch lbs. (8–12 Nm).
21. Install the EGR valve and adapter.
22. Install the IMRC vacuum motors, and tighten the retaining bolts to 71–106 inch lbs. (8–12 Nm).
23. Install the fuel injectors and charging assembly. Tighten retaining bolts to 71–97 inch lbs. (8–11 Nm).
24. Install the water bypass tube to the lower intake manifold. Tighten the retaining bolts to 71–97 inch lbs. (8–11 Nm).
25. Connect the water bypass tube hose to the outlet tube and tighten the hose clamp securely.
26. Connect the electrical wiring harnesses.
27. Connect the vacuum lines to the IMRC motors.
28. Connect the upper radiator hose to water hose connection and tighten the hose clamp securely.
29. Install the upper intake manifold.
30. Fill and bleed the cooling system.
31. Start the engine and check for leaks.

Exhaust Manifold

REMOVAL & INSTALLATION

→Spray the exhaust system fasteners with penetrating lubricant before removing them to help prevent broken studs and bolts. The use of a 6-point socket is highly recommended when removing exhaust system fasteners.

❊❊ CAUTION

To prevent serious burns, allow the exhaust manifold to cool down before attempting to remove it.

3.0L Engine

REAR MANIFOLD

1. Disconnect the negative battery cable.
2. Remove the cowl vent panel.
3. Label and disconnect the EGR backpressure transducer hoses.
4. Remove the EGR valve tube from the exhaust manifold.

→**Use a backup wrench to prevent damaging the tube.**

5. Raise and support the vehicle safely.
6. Disconnect the dual converter Y-pipe from the exhaust manifold.
7. Lower the vehicle.
8. Remove the exhaust manifold.

To install:

9. Clean all mating surfaces thoroughly.
10. Position the exhaust manifold and tighten the bolts to 15–22 ft. lbs. (20–30 Nm).
11. Raise and support the vehicle safely.
12. Connect the dual converter Y-pipe and tighten the bolts to 25–34 ft. lbs. (34–47 Nm).
13. Lower the vehicle.
14. Install the EGR valve tube and tighten the fitting to 26–48 ft. lbs. (35–65 Nm).
15. Connect the EGR backpressure transducer hoses.
16. Install the cowl vent panel.
17. Connect the negative battery cable.
18. Start the engine and check for exhaust leaks.

FRONT MANIFOLD

1. Disconnect the negative battery cable.
2. Remove the oil level indicator tube support bracket and retaining nut.
3. Remove the oil level dipstick and oil level indicator tube.
4. Raise and support the vehicle safely.
5. Disconnect the dual converter Y-pipe from the exhaust manifold.
6. Lower the vehicle.
7. Remove the exhaust manifold.

To install:

8. Clean all mating surfaces thoroughly.
9. Position exhaust manifold and tighten the bolts to 15–22 ft. lbs. (20–30 Nm).
10. Raise and support the vehicle safely.
11. Connect the dual converter Y-pipe and tighten the bolts to 25–34 ft. lbs. (34–47 Nm).
12. Lower the vehicle.
13. Install the oil level dipstick and oil level indicator tube. Tighten the nut to 11–14 ft. lbs. (15–20 Nm).
14. Connect the negative battery cable.
15. Start the engine and check for exhaust leaks.

3.8L Engine

REAR MANIFOLD

1. Disconnect the negative battery cable.
2. Remove the cowl vent panel.
3. Remove the engine air cleaner and air cleaner outlet tube.
4. Disconnect the ignition wires from the rear cylinder head and ignition coil.
5. Remove the spark plugs from the rear cylinder head.
6. Raise and support the vehicle safely on jackstands.
7. Disconnect the dual converter Y-pipe from the exhaust manifold.
8. Lower the vehicle.
9. Remove the exhaust manifold.

To install:

10. Clean all gasket mating surfaces thoroughly.

→**A slight warpage in the exhaust manifold may cause a misalignment between the bolt holes in the cylinder head and exhaust manifold. Elongate the holes in the exhaust manifold as necessary to correct the misalignment. Do not elongate the pilot hole.**

11. Install a new exhaust manifold gasket and the exhaust manifold on the cylinder head. Start two bolts to hold the manifold in position.
12. Install the remaining bolts. Tighten the bolts beginning from the center port and working outward to 15–22 ft. lbs. (20–30 Nm).

13. Raise and support the vehicle safely.
14. Connect the dual converter Y-pipe and tighten the bolts to 25–34 ft. lbs. (34–47 Nm).
15. Lower the vehicle.
16. Install the spark plugs in the rear cylinder head.
17. Connect the ignition wires.
18. Install the engine air cleaner and air cleaner outlet tube.
19. Install the cowl vent panel.
20. Connect the negative battery cable.
21. Start the engine and check for exhaust leaks.

FRONT MANIFOLD

1. Disconnect the negative battery cable.
2. Remove the oil level indicator tube.
3. Label and disconnect the ignition wires from the front cylinder head.
4. Disconnect the EGR-to-exhaust manifold tube.
5. Raise and support the vehicle safely on jackstands.
6. Disconnect the dual converter Y-pipe from the exhaust manifold.
7. Lower the vehicle.
8. Remove the exhaust manifold.

To install:

9. Clean all gasket mating surfaces thoroughly.

→**A slight warpage in the exhaust manifold may cause a misalignment between the bolt holes in the cylinder head and exhaust manifold. Elongate the holes in the exhaust manifold as necessary to correct the misalignment. Do not elongate the pilot hole.**

10. Install a new gasket and the exhaust manifold on the cylinder head. Start two bolts to hold the manifold in position.
11. Install the remaining bolts. Tighten the bolts, starting from the center port and working outward, to 15–22 ft. lbs. (20–30 Nm).
12. Raise and support the vehicle safely on jackstands.
13. Connect the dual converter Y-pipe and tighten the bolts to 25–34 ft. lbs. (34–47 Nm).
14. Lower the vehicle.
15. Connect the EGR to the exhaust manifold tube.
16. Connect the ignition wires.
17. Install the oil level indicator tube.
18. Connect the negative battery cable.
19. Start the engine and check for exhaust leaks.

Camshaft and Lifters

REMOVAL & INSTALLATION

3.0L Engine

1. Remove the engine from the vehicle.
2. Rotate the crankshaft until the No. 1 piston is at the TDC on its compression stroke and the timing marks are aligned.
3. Remove the throttle body.
4. Label and remove the ignition wires.
5. Remove the camshaft position sensor.
6. Remove the ignition coil.
7. Remove the valve covers.
8. Loosen the cylinder No. 3 intake valve rocker arm seat retaining bolt and rotate the rocker arm off of the pushrod and away from the top of the valve stem, then remove the pushrod.
9. Remove the alternator and mounting brackets.
10. Remove the drive belt tensioner and drive belt.
11. Remove the intake manifold.
12. Loosen the rocker arm fulcrum nuts and position the rocker arms to the side for easy access to the pushrods.
13. Remove the pushrods and label so they may be installed in their original positions.
14. Remove the tappet guide plate from the valve tappets by lifting straight up.
15. Using a suitable magnet or tappet removal tool, remove the hydraulic tappets. Keep them in order so they can be installed in their original positions.
16. Remove the crankshaft pulley and damper.
17. Remove the oil pan assembly.
18. Remove the engine front cover assembly.
19. Align the timing marks on the camshaft and crankshaft sprockets.
20. Check the camshaft end-play as follows:

a. Push the camshaft toward the rear of the engine and install a dial indicator, so the indicator point is on the camshaft sprocket attaching screw.

b. Zero the dial indicator. Position a small prybar, or equivalent, between the camshaft sprocket and block.

c. Pull the camshaft forward and release it. Camshaft end-play should be 0.007 in. (0.17mm) or less.

d. If the camshaft end-play is not within specification, replace the thrust plate.

21. Remove the timing chain and sprockets.

22. Remove the camshaft thrust plate.
23. Carefully remove the camshaft by pulling it toward the front of the engine. Remove it slowly to avoid damaging the bearings, journals and lobes.

To install:

24. Clean and inspect all parts before installation.
25. Lubricate the camshaft lobes and journals with Molylube® or heavy engine oil.
26. Carefully install the camshaft.

➡**If a new camshaft is being installed, recheck camshaft end-play.**

27. Lubricate the engine thrust plate with Engine Assembly Lubricant D9AZ-19579-D or equivalent, then install the thrust plate. Tighten the retaining bolts to 7 ft. lbs. (10 Nm).
28. Install the timing chain and sprockets.

➡**Check the camshaft sprocket bolt for blockage of the drilled oil passages prior to installation, and clean if necessary.**

29. Install the engine front cover.
30. Install the crankshaft damper and pulley.
31. Lubricate and install the hydraulic tappets in their original bores.
32. Align the valve tappet flats and install the tappet guide plate with the word "UP" facing you.
33. Install the intake manifold assembly.
34. Lubricate and install the pushrods and rocker arms.
35. Install the oil pan.
36. Install the valve covers.
37. Install the alternator and brackets.
38. Install the drive belt tensioner and the drive belt.
39. Install the throttle body.
40. Connect the ignition wires.
41. Install the engine assembly into the vehicle.

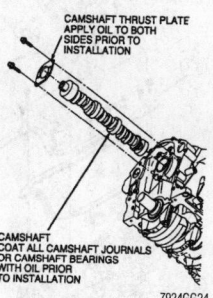

CAMSHAFT THRUST PLATE
APPLY OIL TO BOTH SIDES PRIOR TO INSTALLATION

CAMSHAFT
COAT ALL CAMSHAFT JOURNALS OR CAMSHAFT BEARINGS WITH OIL PRIOR TO INSTALLATION

7924GG24

The thrust plate, which holds the camshaft in position, comes in different thicknesses to allow for camshaft free-play adjustment—3.0L and 3.8L engines

3.8L Engine

1. Remove the engine from the vehicle.
2. Rotate the crankshaft until the No. 1 piston is at the TDC on its compression stroke and the timing marks are aligned.
3. Remove the valve covers.
4. Remove the intake manifolds.
5. Loosen the rocker arm bolts and position the rocker arms to the side for easy access to the pushrods.
6. Label and remove the pushrods so they may be installed in their original positions.
7. Remove the tappet guide plate from the valve tappets by lifting straight up.
8. Using a suitable magnet or tappet removal tool, remove the hydraulic tappets and keep them in order so they can be installed in their original positions.
9. Remove the crankshaft pulley and damper.
10. Remove the oil pan assembly.
11. Remove the engine front cover assembly.
12. Align the timing marks on the camshaft and crankshaft sprockets.
13. Check the camshaft end-play as follows:

a. Push the camshaft toward the rear of the engine and install a dial indicator, so the indicator point is on the camshaft sprocket attaching screw.

b. Zero the dial indicator. Position a small prybar or equivalent, between the camshaft sprocket or gear and block.

c. Pull the camshaft forward and release it. Camshaft end-play should be 0.001–0.006 in. (0.025–0.15mm).

d. If the camshaft end-play is not within specification, replace the thrust plate upon reassembly.

14. Remove the timing chain and sprockets.
15. Remove the camshaft thrust plate.
16. Carefully remove the camshaft by pulling it toward the front of the engine. Remove it slowly to avoid damaging the bearings, journals and lobes.

To install:

17. Clean and inspect all parts before installation.
18. Lubricate the camshaft lobes and journals with Molylube® or heavy engine oil.
19. Carefully install the camshaft.

➡**If a new camshaft is being installed, recheck camshaft end-play.**

20. Lubricate the engine thrust plate with Engine Assembly Lubricant D9AZ-19579-D or equivalent, then install the thrust plate.

Tighten the retaining bolts to 71–124 inch lbs. (8–14 Nm).

21. Install the timing chain and sprockets.

➡ **Check the camshaft sprocket bolt for blockage of the drilled oil passages prior to installation, and clean if necessary.**

22. Install the engine front cover.
23. Install the crankshaft damper and pulley.
24. Lubricate and install the hydraulic tappets into their original bores.
25. Align the valve tappet flats and install the tappet guide plate with the word "UP" facing you.
26. Install the intake manifold assembly.
27. Lubricate and install the pushrods and rocker arms.
28. Install the oil pan.
29. Install the valve covers.
30. Install the engine assembly into the vehicle.

Oil Pan

❊ CAUTION

The EPA warns that prolonged contact with used engine oil may cause a number of skin disorders, including cancer! You should make every effort to minimize your exposure to used engine oil. Protective gloves should be worn when changing the oil. Wash your hands and any other exposed skin areas as soon as possible after exposure to used engine oil. Soap and water, or waterless hand cleaner, should be used.

REMOVAL & INSTALLATION

3.0L Engine

1. Disconnect the negative battery cable.
2. Remove the oil level dipstick.
3. Raise and safely support the vehicle.
4. Remove the retainer clip at the low oil level sensor. Disconnect the wiring harness from the sensor.
5. Drain the crankcase.
6. Disconnect the wiring harness from the oxygen sensors.
7. Remove the dual converter Y-pipe.
8. Remove the starter motor.

9. Remove the lower engine/flywheel dust cover from the torque converter housing.
10. Remove the oil pan bolts, then slowly remove the oil pan, making sure the internal pan baffle does not snag the oil pump screen cover and tube.
11. Remove the oil pan gasket.

To install:

12. Clean the gasket mating surfaces thoroughly.

➡ **When using a silicone sealer, the assembly process should occur within 15 minutes after the sealer has been applied. After this time, the sealer may start to set-up and its sealing effectiveness may be affected.**

13. Apply a 0.25 in. (6mm) thick bead of silicone sealer to the junction of the rear main bearing cap and cylinder block junction of the front cover assembly and cylinder block.
14. Position the oil pan gasket to the oil pan with sealing bends against the oil pan surface and secure with gasket adhesive.
15. Position the oil pan on the engine block and install the oil pan attaching bolts. Tighten the bolts to 8–10 ft. lbs. (10–14 Nm).
16. Back off all of the bolts and retighten them.
17. Install the lower engine/flywheel dust cover to the torque converter housing.
18. Install the starter motor.
19. Install the dual converter Y-pipe.
20. Connect the wiring harness to the oxygen sensors.
21. Connect the low oil level sensor wiring harness and install the retainer clip.
22. Lower the vehicle.
23. Install the oil level dipstick.
24. Connect the negative battery cable.
25. Fill the engine with oil.
26. Start the engine and check for leaks.

3.8L Engine

1. Disconnect the negative battery cable.
2. Raise and support the vehicle safely on jackstands.
3. Drain the engine oil.
4. Remove the oil filter.
5. Remove the dual converter Y-pipe assembly.
6. Remove the starter motor.
7. Remove the engine rear plate/converter housing cover.
8. Remove the retaining bolts and remove the oil pan.

To install:

9. Clean the gasket mating surfaces thoroughly.
10. Trial fit the oil pan to the cylinder block. Ensure that enough clearance has been provided to allow the oil pan to be installed without sealant being scraped off when pan is positioned under the engine.
11. Apply a bead of silicone sealer to the oil pan flange. Also apply a bead of sealer to the front cover/cylinder block joint and fill the grooves on both sides of the rear main seal cap.

➡ **When using silicone rubber sealer, assembly must occur within 15 minutes after sealer application. After this time, the sealer may start to harden and its sealing effectiveness may be reduced.**

12. Install the oil pan and secure to the block with the attaching screws. Tighten the screws to 80–106 inch lbs. (9–12 Nm).
13. Install a new oil filter.
14. Install the engine rear plate/converter housing cover.
15. Install the starter motor.
16. Install the Y-pipe converter assembly.
17. Lower the vehicle.

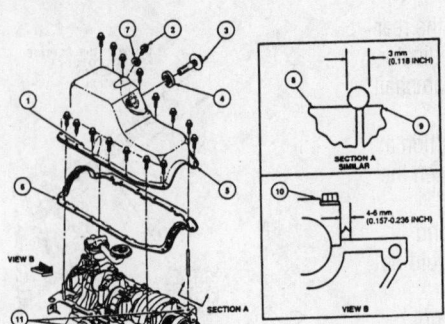

1. Bolt
2. Oil pan drain plug
3. Low oil level sensor
4. Low oil level sensor washer
5. Oil pan
6. Oil pan gasket
7. Drain plug gasket
8. Cylinder block
9. Engine front cover
10. Rear main bearing cap
11. Silicone gasket and sealant

7924GG18

Oil pan and related components. Apply silicone gasket sealant in the places shown—3.0L engine

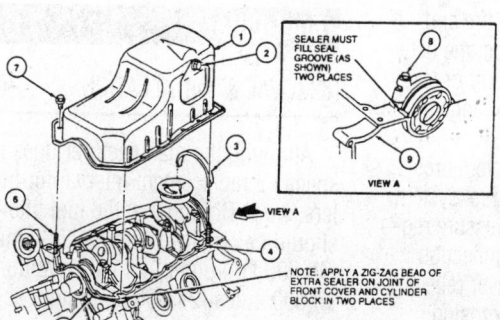

1. Oil pan
2. Oil pan drain plug
3. End seal
4. Silicone gasket and sealant
5. Engine front cover
6. Guide pin
7. Bolt
8. Rear bearing cap
9. Cylinder block

SEALER MUST FILL SEAL GROOVE (AS SHOWN) TWO PLACES

VIEW A

NOTE. APPLY A ZIG-ZAG BEAD OF EXTRA SEALER ON JOINT OF FRONT COVER AND CYLINDER BLOCK IN TWO PLACES

7924GG19

Exploded view of the oil pan and related components. Apply silicone gasket sealant in the places shown—3.8L engine

18. Fill the engine with the proper type and amount of clean oil.
19. Connect the negative battery cable.
20. Start the engine and check for leaks.

Oil Pump

REMOVAL & INSTALLATION

3.0L Engine

1. Disconnect the negative battery cable.
2. Remove the oil pan.
3. Remove the oil pump attaching bolts. Lift the oil pump from the engine.
4. If replacing the oil pump, remove the oil pump intermediate shaft.

To install:

5. Prime the oil pump by filling either the inlet or the outlet port with engine oil. Rotate the pump shaft to distribute the oil within the oil pump body cavity.
6. Insert the oil pump intermediate shaft assembly into the hex drive hole in the oil pump assembly until the retainer "clicks" into place.
7. Place the oil pump in the proper position with a new gasket and install the retaining bolt.
8. Tighten the oil pump retaining bolt to 30–40 ft. lbs. (40–55 Nm).
9. Install the oil pan.
10. Fill the engine with clean oil.
11. Connect the negative battery cable.

➡**Check for proper engine oil pressure immediately after starting the engine. If engine oil pressure is not within specification a few seconds after starting the engine, stop the engine and determine the reason for the low oil pressure condition. Running an engine with low oil pressure may result in serious engine damage.**

12. Start the engine and check for leaks.

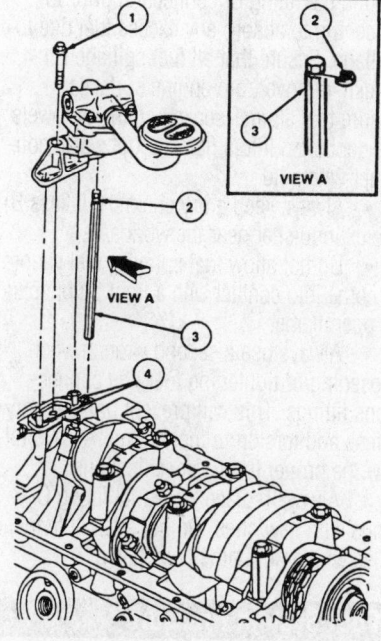

VIEW A

VIEW A

1	Bolt
2	Oil Pump Intermediate Shaft Retaining Ring
3	Oil Pump Intermediate Shaft
4	Dowel

7924GG20

Remove the oil pan to gain access to the oil pump assembly on the 3.0L engine

3.8L Engine

➡**The oil pump, oil pressure relief valve and drive intermediate shaft are contained in the front cover assembly.**

1. Disconnect the negative battery cable.
2. If necessary for access, remove the oil filter.
3. Remove the oil pump and filter body-to-engine front cover retaining bolts, then remove the oil pump and filter body from the engine front cover.

4. Inspect the oil pump body seal, oil pump and filter body, and engine front cover for distortion. Replace damaged components as necessary.

To install:

5. Position the oil pump and filter body on the engine front cover, then install the retaining bolts.
6. Tighten the four large engine front cover retaining bolts to 17–23 ft. lbs. (23–32 Nm), then tighten the remaining retaining bolts to 71–97 ft. lbs. (8–11 Nm).
7. If removed, install the oil filter.

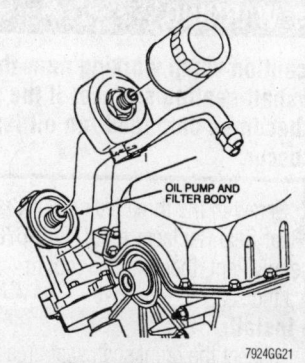

OIL PUMP AND FILTER BODY

7924GG21

The oil pump assembly is mounted to the side of the 3.8L engine

8. Connect the negative battery cable.

➡**Check for proper engine oil pressure immediately after starting the engine. If engine oil pressure is not within specification a few seconds after starting the engine, stop the engine and determine the reason for the low oil pressure condition. Running an engine with low oil pressure may result in serious engine damage.**

9. Start engine and check for leaks.

Rear Main Seal

REMOVAL & INSTALLATION

1. Disconnect the negative battery cable.
2. Raise and support the vehicle safely on jackstands.
3. Remove the transaxle.
4. Remove the flywheel and the rear cover plate, if necessary.
5. Using a sharp awl, punch one hole into the crankshaft rear oil seal metal surface between the seal lip and the cylinder block.

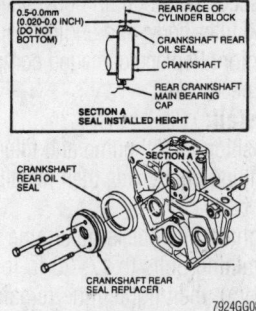

The rear main seal must be installed with the proper tools to avoid damaging the seal or crankshaft

✳✳ WARNING

Use caution when working near the crankshaft sealing surface. If the surface becomes damaged, an oil leak may occur.

6. Screw in the threaded end of Crankshaft Rear Seal Replacer tool (T88L-6701-A) or equivalent, then use the tool to remove the seal.

To install:

7. Inspect the crankshaft seal area for any damage which may cause the seal to leak. If damage is evident, service or replace the crankshaft as necessary.

8. Coat the crankshaft seal area and the seal lip with engine oil.

9. Using a Rear Crankshaft Seal Replacer tool T88P-6701-A, or equivalent, install the seal. Tighten the bolts of the seal installer tool evenly so the seal is straight and seats without misalignment.

10. Install the flywheel.

11. Install the rear cover plate, if necessary.

12. Install the transaxle, lower the vehicle and connect the battery.

FUEL SYSTEM

Fuel System Service Precautions

Safety is the most important factor when performing not only fuel system maintenance, but any type of maintenance. Failure to conduct maintenance and repairs in a safe manner may result in serious personal injury or death. Work on a vehicle's fuel system components can be accomplished safely and effectively by adhering to the following rules and guidelines.

• To avoid the possibility of fire and personal injury, always disconnect the negative battery cable unless the repair or test procedure requires that battery voltage by applied.

• Always relieve the fuel system pressure prior to disconnecting any fuel system component (injector, fuel rail, pressure regulator, etc.) fitting or fuel line connection. Exercise extreme caution whenever relieving fuel system pressure to avoid exposing skin, face and eyes to fuel spray. Please be advised that fuel under pressure may penetrate the skin or any part of the body that it contacts.

• Always place a shop towel or cloth around the fitting or connection prior to loosening to absorb any excess fuel due to spillage. Ensure that all fuel spillage is quickly remove from engine surfaces. Ensure that all fuel-soaked cloths or towels are deposited into a flame-proof waste container with a lid.

• Always keep a dry chemical (Class B) fire extinguisher near the work area.

• Do not allow fuel spray or fuel vapors to come into contact with a light bulb, spark or open flame.

• Always use a second wrench when loosening or tightening fuel line connections fittings. This will prevent unnecessary stress and torsion to fuel piping. Always follow the proper torque specifications.

• Always replace worn fuel fitting O-rings with new ones. Do not substitute fuel hose where rigid pipe is installed.

Fuel System Pressure

RELIEVING

All Sequential Electronic Fuel Injection (SEFI) engines are equipped with a pressure relief valve located on the fuel supply manifold. Remove the fuel tank cap and attach fuel pressure gauge T80L-9974-B, or equivalent, to the valve to release the fuel pressure. Be sure to drain the fuel into a suitable container and to avoid gasoline spillage. If a pressure gauge is not available, disconnect the vacuum hose from the fuel pressure regulator and attach a hand-held vacuum pump. Apply about 25 in. Hg (84 kPa) of vacuum to the regulator to vent the fuel system pressure into the fuel tank through the fuel return hose. Note that this procedure will remove the fuel pressure from the lines, but not the fuel. Take precautions to avoid the risk of fire and use clean rags to soak up any spilled fuel when the lines are disconnected.

Fuel Filter

REMOVAL & INSTALLATION

Although the manufacturer does not specify a replacement interval for fuel filters, we at Chilton feel the fuel filter should be replaced every 30,000 miles (48,000 km) under normal conditions or 15,000 miles (24,000 km) under severe conditions.

1. Relieve the fuel system pressure.

2. Raise and support the vehicle safely on jackstands.

3. Place a rag under the fuel filter to catch any residual fuel that may leak out when the filter is removed.

4. Remove the push-connect fittings at both ends of the fuel filter.

5. Install retainer clips in each fitting.

6. Note the flow arrow direction for installation reference.

7. Remove the fuel filter by pulling it from the bracket.

To install:

8. Install the fuel filter in its bracket, ensuring proper direction of flow as noted earlier.

9. Install push-connect fittings at both ends of the fuel filter.

10. Start the engine and check the filter connections for leaks by running the tip of your finger around each connection.

11. Turn the engine off and lower the vehicle.

Fuel Pump

REMOVAL & INSTALLATION

➡**To gain access to the fuel pump, it is necessary to remove the fuel tank.**

1. Depressurize the fuel system and remove the fuel tank from the vehicle.

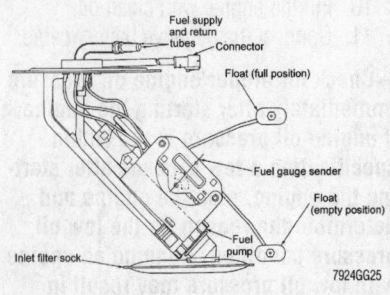

In-tank electric fuel pump and related components

2. Remove any dirt that has accumulated around the fuel pump module attaching flange to prevent it from entering the tank during service.

3. Turn the fuel pump module locking ring counterclockwise using a locking ring removal tool or a brass drift, and remove the locking ring.

4. Remove the fuel pump module.

5. Remove the seal gasket and discard it.

To install:

6. Put a light coating of grease on a new seal ring to hold it in place during assembly. Install it in the fuel tank ring groove.

7. Insert the fuel pump module into the fuel tank, then secure it in place with the locking ring. Tighten the ring until secure.

8. Install the tank in the vehicle.

9. Install a minimum of 10 gallons of fuel and check for leaks.

10. Install a pressure gauge on the throttle body valve and turn the ignition **ON** for 3 seconds. Turn the key **OFF**, then repeat the key cycle five to ten times until the pressure gauge shows at least 30 psi. (207 kPa).

11. Check for fuel leaks at the fittings.

12. Remove the pressure gauge.

13. Start the engine and check for fuel leaks.

DRIVE TRAIN

Transaxle Assembly

REMOVAL & INSTALLATION

1. Disconnect the negative, then the positive battery cables.

2. Remove the battery and battery tray.

3. Remove the hood and cowl vent.

4. Remove the air cleaner assembly.

5. Label and disconnect all transaxle electrical harnesses.

6. Disconnect the transaxle shift cable from the lever by unsnapping the shift cable end from the lever ball stud.

7. Disconnect and plug the transaxle fluid cooler lines.

➡**Leave the two lower engine-to-transaxle bolts in place to hold the transaxle secure against the engine block until a suitable jack can be placed under the transaxle to support it during removal.**

8. Remove the upper transaxle-to-engine bolts.

9. Disconnect the engine electrical harness bracket.

10. Disconnect the battery cable bracket.

➡**Install Engine Lifting Eyes (D94P-6001-A) or equivalent, to support the engine during transaxle removal.**

11. Install an Engine Support Kit (014-00792) or equivalent, and suitably support the engine.

12. Raise and support the vehicle safely on jackstands.

13. Loosen the drain pan bolts. Drain and recycle the transaxle fluid.

14. Remove the front wheels.

15. Remove the halfshafts.

16. Remove the bolts retaining the rear engine support to the transaxle.

17. Remove the front subframe.

18. Disconnect the speedometer cable from the vehicle speed sensor.

19. Remove the starter.

20. Remove the transaxle housing cover.

21. Remove the four flexplate-to-converter nuts.

22. Support the transaxle with a suitable jack and remove the remaining transaxle-to-engine bolts.

23. Remove the engine bracket-to-transaxle bolts.

24. Separate the transaxle from the engine block by carefully moving the transaxle rearward until enough clearance exists to remove the transaxle from the engine compartment.

25. Slowly lower the transaxle from the engine compartment.

To install:

26. Place the transaxle on a suitable jack and position it in place.

27. Install the engine bracket-to-transaxle bolts, and tighten the bolts to 39–53 ft. lbs. (53–72 Nm).

28. Install the lower transaxle-to-engine bolts, and tighten the bolts to 39–53 ft. lbs. (53–72 Nm).

29. Install the flexplate-to-torque converter bolts, and tighten them to 20–34 ft. lbs. (27–46 Nm).

30. Install the transaxle housing cover, and tighten the bolts to 80–106 inch lbs. (9–12 Nm).

31. Install the starter motor, and connect the electrical harness.

32. Connect the speedometer cable.

33. Install and align the front subframe.

34. Install the four bolts retaining the rear engine support, and tighten them to 39–53 ft. lbs. (53–72 Nm).

35. Install both halfshafts.

36. Install the front wheels and lower the vehicle.

37. Remove the engine support kit.

38. Connect the transaxle electrical harnesses.

39. Install the upper transaxle-to-engine bolts and tighten to 39–53 ft. lbs. (53–72 Nm).

40. Connect the fluid cooler-to-transaxle lines.

41. Connect the transaxle shift cable to the manual lever ball stud.

42. Install the air cleaner assembly.

43. Install the cowl vent and hood.

44. Install the battery tray and battery.

45. Fill the transaxle with proper amount of Mercon® fluid.

46. Connect the positive, then the negative battery cable.

Halfshafts

REMOVAL & INSTALLATION

➡**Do not begin this removal procedure unless a new wheel hub retainer nut, a new retainer circlip and a new lower ball joint-to-front wheel knuckle retaining bolt and nut are available. Once removed, these parts must not be reused during assembly. Their torque holding ability, or retention capability, is diminished during removal.**

1. Raise and support the vehicle safely.

2. Remove the front wheels.

3. Insert a steel rod into the brake rotor to prevent the rotor from turning, and loosen the axle wheel hub nut. Discard the nut.

4. Remove the ball joint-to-front wheel knuckle retaining nut. Drive the bolt out of the front wheel knuckle using a punch and hammer.

5. Remove the front brake anti-lock sensor and position it out of the way.

6. Separate the ball joint from the front wheel knuckle using a prybar. Position the end of the prybar outside of the bushing pocket to avoid damage to the bushing.

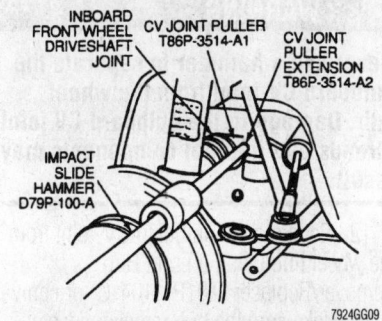

Remove the halfshaft from the transaxle using a CV-joint puller, extension and impact slide hammer

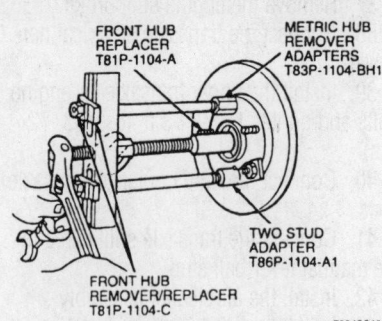

FRONT HUB REPLACER T81P-1104-A

METRIC HUB REMOVER ADAPTERS T83P-1104-BH1

TWO STUD ADAPTER T86P-1104-A1

FRONT HUB REMOVER/REPLACER T81P-1104-C

7924GG10

The front hub adapter must be used to remove the hub without damage

❋❋ WARNING

Use care to prevent damage to the CV-joint boot.

7. Remove the stabilizer bar link at the front stabilizer bar.

➡**Make sure the CV-joint puller does not contact the transaxle shaft speed sensor. Damage to the sensor will result.**

8. Install a CV-Joint Puller (T86P-3514-A1) or equivalent between the inboard CV-joint and the transaxle case.

9. Install a CV-Joint Extension (T86P-3514-A2) or equivalent into the puller and hand-tighten.

10. Using a slide hammer, remove the driveshaft from the transaxle.

❋❋ WARNING

Do not allow the halfshaft to hang unsupported. Damage to the CV-joint may result. Do not wrap wire around the joint boot. Damage to the boot may result.

11. Support the end of the halfshaft assembly by suspending it from the chassis using a length of wire.

❋❋ WARNING

Never use a hammer to separate the outboard CV-joint from the wheel hub. Damage to the outboard CV joint threads and internal components may result.

12. Separate the outboard CV-joint from the wheel hub using a Front Hub Remover/Replacer (T81P-1104-C) or equivalent. Make sure the hub remover adapter is fully threaded onto the hub stud.

❋❋ WARNING

Do not move vehicle without the outboard CV-joint properly installed as damage to the bearing may occur.

13. Remove the halfshaft assembly from the vehicle.

To install:

❋❋ WARNING

Do not reuse the retainer circlip. A new circlip must be installed each time the inboard CV-joint stub shaft is installed into the transaxle differential.

14. Install a new retainer circlip on the inboard CV-joint stub shaft by starting one end in the groove and working the retainer circlip over the inboard shaft housing end and into the groove. The will avoid over-expanding the circlip.

➡**A non-metallic mallet may be used to aid in seating the retainer circlip into the differential side gear groove. If a mallet is necessary, tap only on the outboard CV-joint shaft.**

15. Carefully align the splines of the inboard CV-joint stub shaft housing with the splines in the differential. Exerting some force, push the inboard CV-joint stub shaft housing into the differential until the retainer circlip is felt to seat in the differential side gear. Use care to prevent damage to the inboard CV-joint stub shaft and transaxle seal.

16. Carefully align the splines of the outboard CV-joint with the splines in the wheel hub and push the shaft into the wheel hub as far as possible.

17. Temporarily fasten the front disc brake rotor to the wheel hub with washers and two lug nuts. Insert a steel rod into the front disc brake rotor and rotate clockwise to contact the front wheel knuckle to prevent the front disc brake rotor from turning when the nut is tightened.

➡**A new front axle wheel hub retaining nut must be installed.**

18. Manually thread the front axle wheel hub retaining nut onto the outboard CV-joint stub shaft housing as far as possible.

➡**A new bolt and nut must be used to connect the front suspension arm to the knuckle.**

19. Connect the front suspension lower arm to the front wheel knuckle. Tighten the nut and bolt to 40–55 ft. lbs. (54–74 Nm).

20. Install the front brake anti-lock sensor.

21. Connect the front stabilizer bark link and tighten to 35–45 ft. lbs. (47–65 Nm).

➡**Do not use power or impact tools to tighten the hub nut.**

22. Tighten front axle wheel hub retaining nut to 157–212 ft. lbs. 213–287 Nm).

23. Install the front wheels and lower the vehicle.

24. Fill the transaxle to the proper level with Mercon®automatic transmission fluid.

STEERING AND SUSPENSION

Air Bag (Supplemental Restraint) System

The Supplemental Restraint System (SRS) is designed to work in conjunction with the standard three-point safety belts to reduce injury in a head-on collision.

❋❋ CAUTION

The SRS can actually cause physical injury or death if the safety belts are not used, or if the manufacturer's warnings are not followed. The manufacturer's warnings can be found in your owner's manual, or, in some cases, on your sun visors.

The SRS is comprised of the following components:
- Driver's side air bag module
- Passenger's side air bag module
- Right-hand and left-hand primary crash front air bag sensors
- Air bag diagnostic monitor computer
- Electrical wiring

The SRS primary crash front air bag sensors are hard-wired to the air bag modules and determine when the air bags are deployed. During a frontal collision, the sensors quickly inflate the two air bags to reduce injury by cushioning the driver and front passenger from striking the dashboard, windshield, steering wheel and any other hard surfaces. The air bag inflates so quickly (in a fraction of a second) that in most cases it is fully inflated before you actually start to move during an automotive collision.

Since the SRS is a complicated and essentially important system, its components are constantly being tested by a diagnostic monitor computer, which illuminates the air bag indicator light on the instrument cluster for approximately 6 seconds when the ignition switch is turned to the **RUN** position when the SRS is functioning properly. After being illuminated for the 6 seconds, the indicator light should then turn off.

If the air bag light does not illuminate at all, stays on continuously, or flashes at any time, a problem has been detected by the diagnostic monitor computer.

✳✳ CAUTION

If at any time the air bag light indicates that the computer has noted a problem, have your vehicle's SRS serviced immediately by a qualified automotive technician. A faulty SRS can cause severe physical injury or death.

SERVICE PRECAUTIONS

Whenever working around, or on, the air bag supplemental restraint system, ALWAYS adhere to the following warnings and cautions.

• Always wear safety glasses when servicing an air bag vehicle and when handling an air bag module.

• Carry a live air bag module with the bag and trim cover facing away from your body, so that an accidental deployment of the air bag will have a small chance of personal injury.

• Place an air bag module on a table or other flat surface with the bag and trim cover pointing up.

• Wear gloves, a dust mask and safety glasses whenever handling a deployed air bag module. The air bag surface may contain traces of sodium hydroxide, a byproduct of the gas that inflates the air bag and which can cause skin irritation.

• Ensure to wash your hands with mild soap and water after handling a deployed air bag.

• All air bag modules with discolored or damaged cover trim must be replaced, not repainted.

• All component replacement and wiring service must be made with the nega-tive and positive battery cables disconnected from the battery for a minimum of one minute prior to attempting service or replacement.

• NEVER probe the air bag electrical terminals. Doing so could result in air bag deployment, which can cause serious physical injury.

• If the vehicle is involved in a fender-bender which results in a damaged front bumper or grille, have the air bag sensors inspected by a qualified automotive technician to ensure that they were not damaged.

• If at any time, the air bag light indicates that the computer has noted a problem, have your vehicle's SRS serviced immediately by a qualified automotive technician. A faulty SRS can cause severe physical injury or death.

DISARMING THE SYSTEM

✳✳ WARNING

If you are disarming the system with the intent of testing the system, do not! The SRS is a sensitive, complex system and should only be tested or serviced by a qualified automotive technician. Also, specific tools are needed for SRS testing.

1. Disconnect the negative battery cable from the battery.
2. Disconnect the positive battery cable from the battery.3. Wait one minute. This time is required for the back-up power supply in the air bag diagnostic monitor to completely drain. The system is now disarmed.

ARMING THE SYSTEM

1. Connect the positive battery cable.
2. Connect the negative battery cable.
3. Stand outside the vehicle and carefully turn the ignition to the **RUN** position. Be sure that no part of your body is in front of the air bag module on the steering wheel, to prevent injury in case of an accidental air bag deployment.
4. Ensure the air bag indicator light turns off after approximately 6 seconds. If the light does not illuminate at all, does not turn off, or starts to flash, have the system tested by a qualified automotive technician. If the light does turn off after 6 seconds and does not flash, the SRS is working properly.

Rack and Pinion Steering Gear

REMOVAL & INSTALLATION

1. Loosen the lug nuts on the front wheels.
2. Raise and safely support the vehicle securely.
3. Remove the front wheels.
4. Remove the tie rod end cotter pins and castle nuts.
5. Disconnect the tie rod ends from the knuckles.
6. Remove the front stabilizer bar.
7. Position the dash opening weather seal for the steering column out of the way.
8. Remove the pinch bolt retaining the steering column intermediate shaft coupling.
9. Remove the steering gear retaining nuts/bolts.
10. Remove the rear subframe bolts.

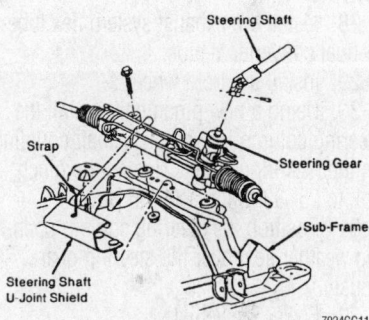

Exploded view of the rack and pinion steering gear mounting on the front subframe of the vehicle

➡Use wire or string to support exhaust components unless you are removing them completely.

11. Support the exhaust system flex tube and remove the flex tube-to-dual converter Y-pipe attachment.
12. Lower the vehicle slightly until the rear subframe separates from the body approximately four inches (10cm).
13. Remove the heat shield band and fold the heat shield down.
14. Rotate the rack and pinion assembly to clear the bolts from the front subframe, and pull toward the driver's side of the vehicle.
15. Place a drain pan under the vehicle and disconnect the power steering lines.
16. Remove the rack and pinion assembly through the driver's side of the vehicle.

To install:

17. Install new Teflon® O-rings on the power steering line fittings.

18. Place the rack and pinion retaining bolts in the gear housing.

19. Install the rack and pinion assembly through the driver's side of the vehicle.

20. Install the power steering lines on the rack and pinion assembly.

21. Position the rack and pinion assembly on the subframe.

22. Install the strap on the heat shield.

23. Install the tie rod ends to the front wheel knuckles. Tighten the castle nuts and install the cotter pins.

24. Install the stabilizer bar.

25. Install the rack and pinion assembly retaining bolts, and tighten them to 85–99 ft. lbs. (115–135 Nm).

26. Raise the vehicle until the subframe contacts the body.

27. Install the rear subframe retaining bolts, and tighten them to 83–112 ft. lbs. (113–153 Nm).

28. Install the exhaust system flex tube-to-dual converter Y-pipe.

29. Install the front wheels.

30. Using a new pinch bolt, install the steering column intermediate shaft coupling on the rack input shaft. Tighten the pinch bolt to 25–33 ft. lbs. (34–46 Nm).

31. Position the steering column opening weather seal over the steering gear housing.

32. Lower the vehicle.

33. Fill the power steering oil reservoir.

34 Start the vehicle and check for leaks.

35 Check for proper wheel alignment and steering wheel position.

MacPherson Struts

REMOVAL & INSTALLATION

➡ **Do not begin this procedure unless a new front axle wheel hub nut, a new lower arm ball joint pinch bolt and nut, and a new tie rod and knuckle nut and cotter pin are available. Once removed, these parts must not be reused during assembly. Their torque holding ability, or retention capability, is diminished during removal.**

1. Turn the ignition switch **OFF** and place the steering column in the unlocked position.

2. Loosen but do not remove the three front strut-to-tower nuts.

❉❉ WARNING

Do not raise the vehicle by the lower arms.

3. Remove the front wheel.

4. Without disconnecting the brake hose, remove the disc brake caliper and suspend it out of the way using a piece of wire.

➡ **The hydraulic brake system will need to be bled if the brake hose is removed from the caliper.**

5. Remove the front disc brake rotor.

6. Remove the front axle wheel hub nut.

7. Using a Front Wheel Hub Remover (T81P-1104-C) or equivalent, remove the front wheel bearing and front wheel knuckle as an assembly.

8. Using a Tie Rod End Remover (TOOL-3290-D) or equivalent, disconnect the tie rod end from the front wheel knuckle.

❉❉ WARNING

Use extreme care to not damage the link ball joint boot seal.

9. Remove the stabilizer bar link from the lower arm.

10. Remove and discard the front suspension lower arm-to-front wheel knuckle pinch bolt and nut.

11. Using a prybar, slightly spread the front wheel knuckle-to-lower arm pinch joint, and remove the lower arm from the front wheel knuckle.

12. Remove the speed sensor bracket and speed sensor from the front wheel knuckle.

13. Remove the strut-to-front wheel knuckle pinch bolt.

14. Using a prybar, slightly spread the front wheel knuckle-to-front strut pinch joint as required for removal.

15. Remove the front wheel knuckle and wheel hub assembly from the front strut.

16. Remove the three front strut mounting bracket-to-strut tower nuts and remove the front strut assembly from the vehicle.

To install:

17. Install the front strut assembly and tighten the strut-to-strut tower mounting bolts hand-tight.

18. Install the front wheel knuckle and wheel hub assembly.

19. Install the new strut-to-wheel knuckle pinch bolt and tighten to 85–97 ft. lbs. (115–132 Nm).

20. Install the halfshaft into the wheel hub.

21. Install the lower arm ensuring that the ball stud groove is properly posi-

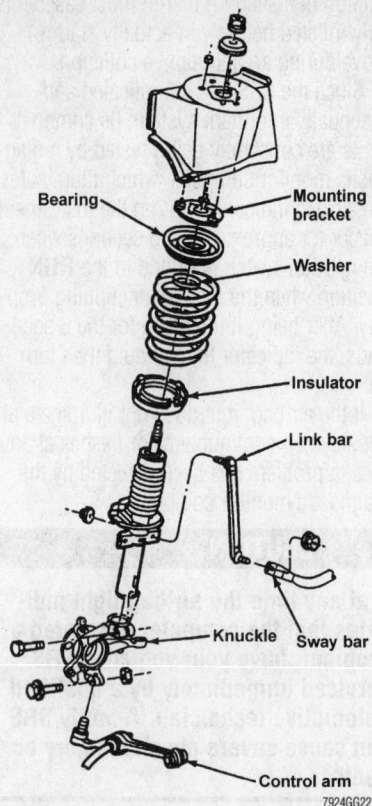

Exploded view of the MacPherson strut and related components

tioned. Use extreme care not to damage the ball joint seal. Install a new pinch bolt and nut. Tighten to 46–52 ft. lbs. (62–71 Nm).

22. Install the front wheel knuckle making sure the front stabilizer bar link is properly positioned. Use care not to damage the ball joint seal.

➡ **Top left-hand (Top LH) and top right-hand (Top RH) are molded into the stabilizer bar link for correct assembly reference.**

23. Install a new stabilizer link nut and tighten to 66–74 ft. lbs. (90–100 Nm).

24. Install the tie rod to the front wheel knuckle. Tighten the new tie rod castellated nut to 66–74 ft. lbs. (90–100 Nm).

25. Install the front brake anti-lock sensor and bracket on the knuckle.

26. Install the front disc brake rotor.

27. Install the front wheel.

28. Tighten the three strut mounting bracket-to-strut tower bolts to 25–30 ft. lbs. (35–40 Nm).

29. Lower the vehicle and tighten the front axle wheel hub nut to 170–202 ft. lbs. (230–275 Nm).

30. Check the wheel alignment.

Shock Absorbers

REMOVAL & INSTALLATION

1. Loosen the lug nuts on the rear wheels.
2. Raise and safely support the vehicle.
3. Remove the rear wheels.
4. Position a jack under the rear axle assembly and raise slightly to put the suspension at normal ride height.
5. Remove the lower shock absorber bolt/nut and disconnect the shock from the rear axle.
6. Lower the rear axle slightly to help aid removal of the upper shock absorber bolt/nut.
7. Remove the shock absorber.

To install:

8. Attach the shock absorber to the upper mounting bracket and install a new retaining bolt/nut.
9. Slowly raise the rear axle assembly with a jack, and guide the lower shock absorber into the bracket on the rear axle assembly. Install a new retaining bolt/nut.
10. Raise the rear suspension to normal ride height and tighten the shock absorber retaining bolts to 50–68 ft. lbs. (68–92 Nm).
11. Install the wheels.
12. Lower the vehicle.

Coil Springs

REMOVAL & INSTALLATION

1. Raise and safely support the vehicle.
2. Remove the rear wheels.

➡The rear axle will need to be supported when the shock absorbers are removed.

3. Position an adjustable stand or jack under the rear axle.
4. Remove the shock absorber-to-rear axle nut and disconnect the shock from the rear axle.
5. Slowly lower the rear axle assembly until the rear spring can be removed.
6. Remove the rear spring.

To install:

7. Position the rear spring insulator on the rear axle assembly and press the insulator downward into place. Verify rear spring insulator is properly seated into correct position.
8. Slowly raise the rear axle assembly with a jack, and guide the upper rear spring insulator onto the upper spring seat on the underbody.

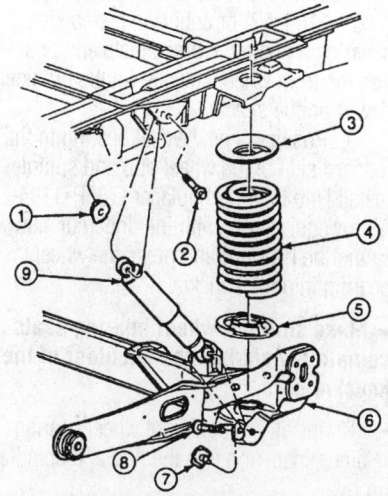

1	J-Nut
2	Bolt M12-1.75 x 66 Hex Flanged Head
3	Rear Spring Insulator (Upper)
4	Rear Spring
5	Rear Spring Insulator (Lower)
6	Axle Assembly
7	Nut
8	Bolt
9	Shock Absorber

7924GG12

Exploded view of the rear coil spring and shock absorber mounting between the axle and chassis

9. Position the shock absorber on the lower rear axle assembly and install new nuts and bolts. Tighten to 50–68 ft. lbs. (68–92 Nm).
10. Install the wheels.
11. Lower the vehicle.

Lower Ball Joint

REMOVAL & INSTALLATION

The lower ball joint and seal are an integral part of the lower control arm assembly and can not be replaced separately. If the lower ball joint or seal is found to be defective, the lower control arm must be replaced as an assembly.

Hub and Wheel Bearing

ADJUSTMENT

Front

The front wheel bearings on the Ford Windstar are not adjustable. If the bearings become loose or make noise they must be replaced as an assembly.

Rear

1. Loosen the lug nuts on the rear wheel(s).
2. Block the front wheels, then raise and safely support the rear of the vehicle securely on jackstands.
3. Remove the rear wheel(s).
4. Remove the hub grease cap.
5. Remove the cotter pin.
6. Tighten nut to 18–23 ft. lbs. (24–31 Nm) while rotating the hub to set the end play. Back off the nut and retighten to 18 inch lbs. (2 Nm).
7. Install a new cotter pin.
8. Install the hub grease cap.
9. Install the brake drum or disc.
10. Install the wheels
11. Lower the vehicle.

REMOVAL & INSTALLATION

Front

1. Remove the front wheel knuckle.

➡Make sure the shaft protector is centered, clears the bearing ID and rests on the end-face of the wheel hub journal.

2. On a work bench, install a 2 jaw puller and Shaft Protector (D80L-625-1) or

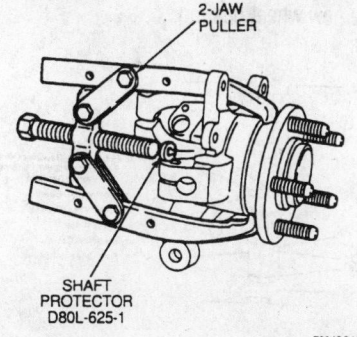

7924GG13

Use a 2-jaw puller to separate the knuckle

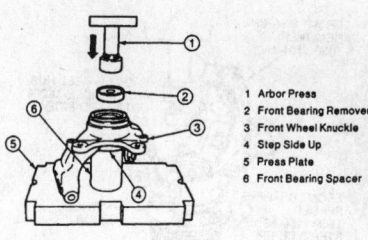

1	Arbor Press
2	Front Bearing Remover
3	Front Wheel Knuckle
4	Step Side Up
5	Press Plate
6	Front Bearing Spacer

7924GG14

Remove the wheel bearing from the knuckle using a hydraulic press and the proper adapters

equivalent, with the jaws of the puller on the knuckle bosses.

3. Separate the front wheel knuckle from the hub.

4. Remove and discard the wheel bearing retainer snap-ring.

5. Using a hydraulic press, place a Front Bearing Spacer (T86P-1104-A2) or equivalent, step side up on a press plate and position the knuckle (outboard side up) on the spacer.

6. Install Front Wheel Bearing Remover (T83P-1104-AH2) or equivalent, centered on the front wheel bearing outer race, and press the front wheel bearing out of the knuckle.

To install:

➡**If the wheel bearing journal is scored or damaged, replace the wheel hub. Do not attempt to service it. The front wheel bearings are of a cartridge design. Wheel bearings are pregreased, sealed and require no scheduled maintenance. The front wheel bearings are preset and cannot be adjusted. If a front wheel bearing is disassembled for any reason, it must be replaced as an assembly. No individual components are available.**

7. Throughly clean the wheel hub and bearing journal to ensure correct seating of the new wheel bearing.

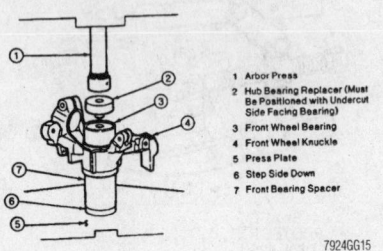

1 Arbor Press
2 Hub Bearing Replacer (Must Be Positioned with Undercut Side Facing Bearing)
3 Front Wheel Bearing
4 Front Wheel Knuckle
5 Press Plate
6 Step Side Down
7 Front Bearing Spacer

7924GG15

Installing the wheel bearing into the knuckle using a hydraulic press and the proper adapters

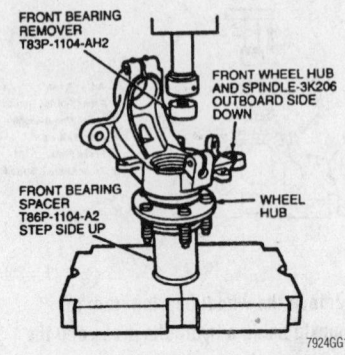

FRONT BEARING REMOVER T83P-1104-AH2

FRONT WHEEL HUB AND SPINDLE-3K206 OUTBOARD SIDE DOWN

FRONT BEARING SPACER T86P-1104-A2 STEP SIDE UP

WHEEL HUB

7924GG16

Press the knuckle and hub together using a hydraulic press, as shown

8. Place the Front Bearing Spacer (T86P-1104-A2) or equivalent, step side down on a hydraulic press plate and position the front wheel knuckle (outboard side down) on the spacer.

9. Position a new wheel bearing in the inboard side of the wheel hub and spindle. Install Hub Bearing Replacer (T86P-1104-A3) or equivalent, (with the undercut side facing the bearing) and press the wheel bearing into the knuckle.

➡**Make sure the wheel bearing seats completely against the shoulder of the knuckle bore.**

10. Install the new front wheel bearing retainer snap-ring into the hub and spindle groove.

11. Place Front Bearing Spacer (T86P-1104-A2) or equivalent, on the arbor press plate and position the wheel hub on the front bearing spacer with the lug bolts facing downward.

12. Position the wheel hub and knuckle (outboard side down) on the press. Place Front Bearing Remover (T83P-1104-AH2) or equivalent, flat side down, centered on the inner race of the front wheel bearing and press down until the bearing is fully seated.

13. Ensure the wheel hub rotates freely in the knuckle after installation.

14. Remove the front halfshaft from the vehicle.

15. Install the halfshaft and knuckle.

Rear

➡**Sodium-based grease is not compatible with lithium-based grease. Do not lubricate the wheel bearings without first thoroughly cleaning all old grease from the bearing. Use of incompatible bearing lubricants could result in premature lubricant breakdown.**

1. Loosen the lug nuts on the rear wheel(s).

2. Raise and safely support the vehicle.

3. Remove the rear wheel(s).

4. Remove the brake drum or brake disc.

5. Remove the hub grease cap.

6. Remove the cotter pin retainer, adjusting nut and flatwasher from the rear wheel spindle. Discard the cotter pin.

7. Remove the outer bearing and cone assembly.

8. Remove the rear hub from the rear wheel spindle.

9. Using Seal Remover (TOOL-1175-AC) or equivalent, remove and discard the oil seal.

10. Remove the inner bearing cone and roller assembly.

11. Clean the inner and outer bearing

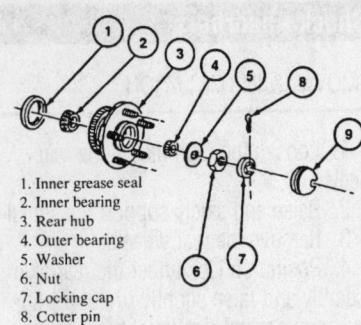

1. Inner grease seal
2. Inner bearing
3. Rear hub
4. Outer bearing
5. Washer
6. Nut
7. Locking cap
8. Cotter pin
9. Grease cap

7924GG23

Exploded view of the rear wheel bearing and hub assembly

cups with solvent. Inspect the bearing cups for scratches, pits, excessive wear and other damage. If the bearing cups are worn or damaged, remove them using a Bearing Cup Puller (T77F-1102A) or equivalent.

To install:

12. Throughly clean all old grease from the surrounding surfaces. If a new hub assembly is being installed, remove the protective coating using degreaser.

13. If the inner or outer bearing cups were removed, install replacement cups using Bearing Cup Replacer (T73-1202-A) or equivalent. Seat the cups properly in the hub.

➡**If a bearing packer is not available, work as much grease as possible between the rollers and cages. Grease the cone surfaces.**

14. Using a bearing packer, pack the bearing cone and roller assemblies with a premium bearing grease.

15. Place the inner bearing cone and roller assembly in the inner cup. A light film of grease should be included between the lips of the new grease retainer.

16. Install the retainer with Hub Seal Replacer (T83T-1175-B) or equivalent. Be sure retainer is properly seated.

➡**Keep the hub centered on the spindle to prevent damage to the retainer and spindle threads.**

17. Install the hub assembly on the spindle.

18. Install the outer bearing cone and roller assembly on the spindle.

19. Install the flat washer and nut. Tighten the nut to 18–23 ft. lbs. (24–31 Nm) while rotating the hub to set the endplay. Back off the nut and retighten it to 18 inch lbs. (2 Nm).

20. Insall a new cotter pin.

21. Install the hub grease cap.

22. Install the brake drum or disc.

23. Install the rear wheels

24. Lower the vehicle.

GEO/CHEVROLET and SUZUKI

Geo/Chevrolet-Tracker • Suzuki-Sidekick • Sidekick Sport • X-90

ENGINE REPAIR

➡ **Disconnecting the negative battery cable on some vehicles may interfere with the functioning of the on-board computer system, and may require the computer to undergo a relearning process once the negative battery cable is reconnected.**

Distributor

REMOVAL

1. Disconnect the negative battery terminal at the battery.
2. Label the distributor cap terminal towers to correspond with their applicable cylinders. For example, trace the No. 1 cylinder spark plug wire to the distributor cap and number that cap terminal tower with a "1".
3. Label and disconnect all wires and vacuum hoses from the distributor.

➡ **Do not bend or twist the spark plug wires, otherwise internal plug wire damage may result. Grip the wire boot when removing or installing the wires.**

4. Matchmark the No. 1 cylinder terminal tower to the distributor housing, then remove the distributor cap.
5. Remove the distributor cap.
6. Rotate the crankshaft clockwise until the distributor rotor points to the No. 1 cylinder mark on the distributor housing.
7. Matchmark the distributor housing position on the engine.
8. Remove the distributor flange bolt, and remove the distributor by carefully sliding it up and out of the engine.

➡ **Do not crank the engine with the distributor removed.**

INSTALLATION

Timing Not Disturbed

1. Insert the distributor in the engine, ensuring that matchmarks on the distributor housing, engine and rotor are aligned.
2. Tighten the flange bolt until secure, then install the distributor cap.
3. Reattach all of the wiring and vacuum hoses to the distributor.
4. Connect the negative battery cable, then inspect and adjust ignition timing as necessary.

Timing Disturbed

1. Rotate the crankshaft in a clockwise position until the Top Dead Center (TDC) timing mark on the crankshaft pulley is aligned with the TDC timing matchmark on the engine timing mark tab.

➡ **After aligning the 2 timing marks, remove the cylinder head cover to visually ensure that neither rocker arm is riding on the peak of its camshaft lobe at the No. 1 cylinder. If one (or both) of the arms is (are) found to be riding on the camshaft lobe peaks, turn the crankshaft another 360 degrees until the same 2 marks are realigned.**

2. Turn the distributor rotor so that it points to the No. 1 cylinder terminal tower mark on the distributor housing.
3. Connect all vacuum and electrical wires to the distributor and cap.
4. Connect the negative battery cable and set the timing to specification.

Ignition Timing

ADJUSTMENT

➡ **Since manufacturing specifications often change during the model-year, refer to the Vehicle Emission Control Information (VECI) label, located in the engine compartment, for your engine's specific ignition timing specification, prior to adjusting the ignition timing. If the specification indicated on your vehicle's VECI label differs from that given in this procedure, use the VECI label specification.**

1. Start the engine and allow it to warm up to normal operating temperature.
2. Be sure that all of the electrical loads, except for the ignition switch, are OFF. If equipped, ensure that the air conditioning is OFF.

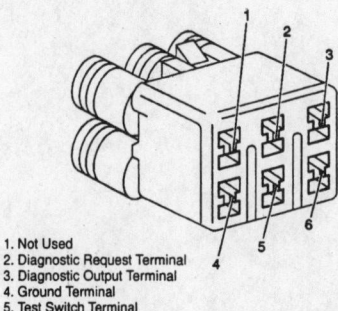

1. Not Used
2. Diagnostic Request Terminal
3. Diagnostic Output Terminal
4. Ground Terminal
5. Test Switch Terminal
6. Duty Check Terminal

7924HG01

Duty Check Data Link Connector terminal identification for ignition timing

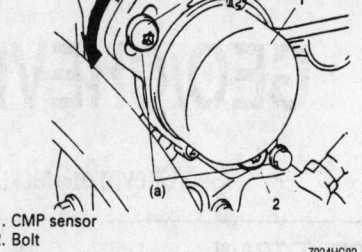

1. CMP sensor
2. Bolt

7924HG82

On 1.8L engines, the Camshaft Position (CMP) sensor must be rotated to adjust the ignition timing

3. Apply the parking brake and place the gearshift in Neutral (manual transmissions) or Park (automatic transmissions).
4. Check the idle speed to ensure that it is within specifications. If it is not, adjust the idle speed prior to adjusting ignition timing.
5. Turn the ignition switch **OFF**.
6. Install the inductive timing light according to the manufacturer's instructions. The timing light input lead should always be attached to the No. 1 cylinder spark plug wire.
7. Remove the cover from the Duty Check Data Link Connector (DC-DLC), located beside the right-hand head light assembly (1995–96 models) or next to the battery (1997–99 models), and, using a jumper wire, connect DC-DLC connector cavities **4** and **5**. Grounding cavity **5** to cavity **4** will fix the ignition timing (prevent the timing from moving).
8. Start the engine and allow it to idle.

✳✳ CAUTION

When using the timing light, be sure to keep yourself and all tools away from moving and hot engine components. Do not wear loose clothing when working around a running engine, otherwise personal injury or death may result.

9. Use the timing light to read the ignition timing by pointing the light at the timing marks (located on the crankshaft pulley and timing belt cover tab) and pulling the light's trigger.

➡ **If the timing fluctuates or changes, cavity 5 is not properly grounded to cavity 4.**

10. If the ignition timing, with terminal **5** properly grounded, is not 4–6 degrees BTDC at 750–850 rpm engine speed, adjust the timing.

11. To adjust the ignition timing, loosen the distributor flange bolts (1.6L engines) or the Camshaft Position (CMP) sensor bolts, and rotate it until the proper ignition timing is within specifications. Turn it counter-clockwise to advance, and clockwise to retard the ignition timing. Tighten the distributor flange bolts securely, or tighten the CMP bolts to 133 inch lbs. (15 Nm).

12. After adjusting the ignition timing and tightening the distributor flange bolts, recheck the ignition timing to ensure that it did not change while tightening the distributor flange bolts.

13. Detach the jumper wire from the DC-DLC.

➡**With this jumper wire removed, the ignition timing may fluctuate, which is normal.**

14. Observe the ignition timing marks with the timing light while increasing engine speed. The ignition timing should advance as the engine speed increases. If the engine speed does not advance, inspect the TP sensor, the test switch terminal circuit, the engine start signal circuit and PCM.

❊❊ WARNING

Driving the vehicle with the DC-DLC terminals grounded will result in catalytic converter damage; be sure to remove the jumper wire prior to driving the vehicle.

15. Shut the engine **OFF**, then remove the inductive timing light from the vehicle. Install the DC-DLC cover.

Engine Assembly

REMOVAL & INSTALLATION

➡**Only a MVAC-trained, EPA-certified automotive technician should service the A/C system or its components.**

The engine is separated from the transmission in the vehicle and removed by itself.

1. Properly relieve fuel system pressure.
2. Disconnect the negative, then the positive battery cables.
3. Remove the hood.

❊❊ CAUTION

The EPA warns that prolonged contact with used engine oil may cause a number of skin disorders, including cancer! You should make every effort to minimize your exposure to used

engine oil. Protective gloves should be worn when changing the oil. Wash your hands and any other exposed skin areas as soon as possible after exposure to used engine oil. Soap and water, or waterless hand cleaner should be used.

4. Drain the engine cooling system and the engine oil.

5. Remove the cooling fan, cooling fan clutch, fan shroud and radiator from the vehicle for added clearance.

6. Disconnect the accelerator and, if equipped, transmission kick-down cable from the carburetor or throttle body.

7. If equipped, remove the strut tower reinforcement bar.

8. Remove the air inlet hose and upper case from the throttle body.

9. On models equipped with automatic transmissions, remove the engine oil level gauge and transmission fluid level gauge guide.

10. Label and detach all wiring attached to the engine and related components (which will inhibit engine removal), then position the wiring harnesses aside. Be sure to disconnect all wiring, even wiring under the vehicle.

11. Label and disconnect all vacuum and cooling system hoses which will interfere with engine removal.

12. Apply the parking brake, block the rear wheels, then raise and safely support the vehicle securely .

13. If equipped, remove the front differential housing from the vehicle.

14. Remove the No. 1 exhaust pipe from the exhaust manifold and the No. 2 exhaust pipe.

15. For 1.6L engines, detach the clutch cable from the clutch release lever, then remove the clutch cable mount from the engine block.

16. If applicable, remove the automatic transmission fluid line clamps from the right-hand side transmission support brace, then remove the support brace.

17. Remove the lower transmission inspection cover.

18. For models equipped with automatic transmissions, have an assistant hold the center crankshaft pulley bolt with a large breaker bar while you remove the torque converter-to-flywheel bolts through the lower transmission inspection cover access hole. If an assistant is not available, the crankshaft can be kept from rotating by installing a tool designed for this purpose, or by holding it steady with a large prytool on the flywheel teeth.

19. Lower the vehicle.

20. Remove the starter motor, then support the front edge of the transmission on a jackstand. Use a piece of wood between the jackstand and the transmission to avoid damaging the transmission housing.

❊❊ WARNING

Do not raise or support automatic transmissions from beneath the fluid pan; damage to the fluid pan may result.

21. If equipped, remove the power steering pump and/or A/C compressor and brackets, leaving the hoses attached. Suspend the power steering pump and/or A/C compressor with strong cord or wire from the side frame rail so that it does not leak.

22. Remove the transmission-to-engine block attaching bolts and nuts.

23. Attach the engine lifting hoist to the engine by means of a hoisting chain. The engine comes from the factory equipped with hoisting eye hooks (one on each side of the engine); use these hooks when removing the engine.

24. From above the vehicle, remove the left- and right-hand engine-to-mount attaching fasteners.

25. Separate the engine from the transmission by pulling the engine forward. On models with manual transmissions, the engine must be pulled forward enough so that the transmission input shaft clears the clutch disc and pressure plate assembly.

26. Slowly and carefully lift the engine and transmission up and out of the engine compartment. When lifting the engine, double check that all wires, hoses and cables have been disconnected and will not hinder engine and transmission removal.

27. If necessary, install the engine on an engine work stand.

28. Installation is the reverse of the removal procedure. During installation, be sure to keep the following points in mind:

❊❊ WARNING

After lowering the engine into the engine compartment, slide the engine back until it is completely mated with the transmission. Failure to properly seat the engine on the transmission can lead to component damage when the attaching fasteners are tightened.

• When installing the engine into the engine compartment, do not remove the hoist support from the assembly until all of the mount fasteners are fully tightened.

• For 1.6L engines with 8-valve cylinder heads, tighten the transmission-to-engine bolts to 62 ft. lbs. (85 Nm), the engine mount nuts to 29–37 ft. lbs. (40–50 Nm), the engine mount (chassis side) bracket bolts to 37–43 ft. lbs. (50–60 Nm), the engine mount (engine side) bracket bolts to 36–43 ft. lbs. (50–60 Nm), the torque converter bolts to 37–43 ft. lbs. (50–60 Nm), and the No. 1 exhaust pipe nuts to 29–43 ft. lbs. (40–60 Nm).

• For 1.6L engines with 16-valve cylinder heads, tighten the engine-to-transmission nuts and bolts to 51–72 ft. lbs. (70–100 Nm), the torque converter bolts to 47 ft. lbs. (65 Nm), the exhaust pipe nuts and bolts to 37 ft. lbs. (50 Nm), the transmission support brace bolts to 37 ft. lbs. (50 Nm), and the engine mount bolts to 29–43 ft. lbs. (40–60 Nm).

• For 1.8L engines, tighten the transmission-to-engine bolts to 58 ft. lbs. (80 Nm), the engine-to-mount fasteners to 36 ft. lbs. (50 Nm), the torque converter-to-flywheel bolts to 47 ft. lbs. (65 Nm), the transmission support brace bolts to 36 ft. lbs. (50 Nm), and the No. 1 exhaust pipe fasteners to 36 ft. lbs. (50 Nm).

• When installing the power steering pump bracket, install the bolt which points toward the rear of the vehicle first, then the other two bolts.

• Be sure to adjust the accelerator cable play and the clutch cable or kickdown cable play.

• If necessary, the emission control label on the underside of the vehicle's hood can be referred to for proper vacuum hose routing.

• Before starting the engine, BE SURE to fill the engine with the proper amount and type of oil and coolant, and the transmission with the proper type of clean lubricant.

• Be sure to properly tension the water pump drive belt.

• Before starting the engine, double-check the connection and routing of all wires, hoses and cables one last time to prevent avoidable component damage.

• After the engine is started, check for oil, coolant and fuel leaks; repair them, if necessary. Also, listen for any strange engine or transmission sounds, which may indicate unseen internal damage.

Water Pump

REMOVAL & INSTALLATION

1.6L Engines

1. Remove the timing belt cover, timing belt, tensioner, plate and spring.

2. Remove the water pump mounting bolts.

3. Remove the one (1.6L MFI engines) or two (1.6L TFI engines) small rubber seals from between the water pump and the oil pump, and the water pump and the cylinder head.

4. If necessary for clearance, remove the oil level dipstick tube retaining bolt from the engine block and the alternator adjusting brace.

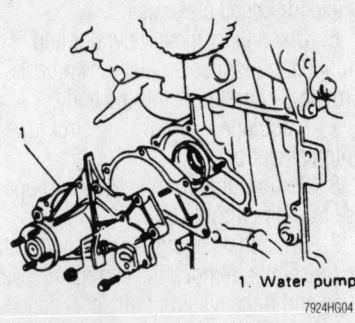

1. Water pump

7924HG04

Exploded view of the water pump mounting—1.6L engines

✳✳ WARNING

Do NOT use a prybar between the water pump housing and the engine block to separate the two components; this can cause scratches and/or gouges, which can prevent proper sealing.

5. Pull the water pump off of the engine block. If the water pump is difficult to remove from the engine block, use a soft-faced mallet to tap the water pump housing until it loosens.

➡**Do not disassemble the water pump; if the water pump is damaged or defective, the entire unit is replaced.**

6. Thoroughly clean the water pump gasket mating surfaces of old gasket material and corrosion.

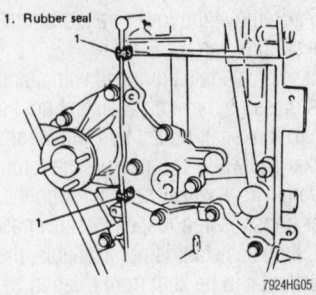

1. Rubber seal

7924HG05

During water pump installation, be sure to install two new rubber seals as shown—1.6L engines

To install:

7. Along with a new gasket, install the water pump on the engine block. Tighten the water pump mounting bolts to 88–115 inch lbs. (10–13 Nm).

8. On TFI 1.6L engines, install two new rubber seals: one between the water pump and oil pump, and the other between the water pump and the cylinder head. The MFI 1.6L engines only use one rubber seal, located between the water and oil pumps.

9. If removed, install the alternator adjusting brace and the oil level dipstick retaining bolt.

10. Install the timing belt, tensioner, plate, spring and cover.

1.8L Engine

1. Disconnect the negative battery cable.
2. Drain the engine cooling system.
3. Disconnect the upper radiator hose from the thermostat housing.
4. Remove the heater outlet pipe bolt.
5. Remove the alternator belt.

➡**When removing the water pump, do not misplace the dowel pin.**

6. Remove the four water pump mounting bolts, then remove the water pump from the engine. Discard the old water pump mounting bolts.

7. Remove the water pump O-ring and discard it.

To install:

8. Install a new O-ring on the water pump, and ensure that the dowel pins are still mounted in the water pump prior to installation.

9. Position the water pump on the engine and install NEW mounting bolts. Tighten the bolts to 221 inch lbs. (25 Nm). Failure to use four new bolts when installing the water pump may lead to coolant leakage.

10. Install the heater outlet pipe bolt.

11. Install the alternator drive belt.

12. Reattach the upper radiator hose to the thermostat housing.

13. Fill the cooling system.

14. Connect the negative battery cable.

Cylinder Head

REMOVAL & INSTALLATION

1.6L 8-Valve Engine

➡**The manufacturer recommends removing the cylinder head with the distributor, exhaust manifold and intake manifold installed. If desired, these items can be removed from the** cylinder head before the head is removed.

1. Relieve fuel system pressure.
2. Disconnect the negative battery cable.

✳✳ CAUTION

Never open, service or drain the radiator or cooling system when hot; serious burns can occur from the steam and hot coolant.

3. Drain the engine cooling system.
4. Remove the air intake case and throttle body.
5. Label and detach all cooling system and vacuum hoses from the intake manifold, throttle body and cylinder head.
6. Label and disengage all electrical wires from the distributor, intake manifold, cylinder head, throttle body and oxygen sensor. Detach the wiring harness from any retaining clamps on the cylinder head.
7. Detach the fuel lines from the throttle body and pressure regulator.
8. Remove the timing belt.
9. Detach the exhaust pipe from the exhaust manifold by removing the attaching nuts.

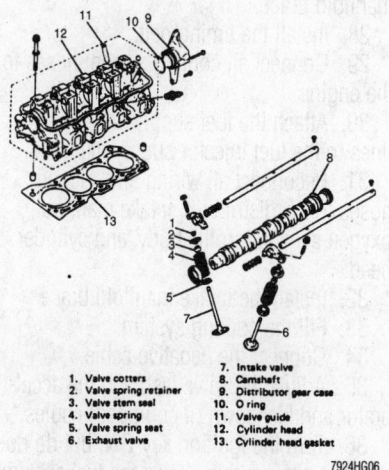

1. Valve cotters	7. Intake valve
2. Valve spring retainer	8. Camshaft
3. Valve stem seal	9. Distributor gear case
4. Valve spring	10. O ring
5. Valve spring seat	11. Valve guide
6. Exhaust valve	12. Cylinder head
	13. Cylinder head gasket

7924HG06

Exploded view of an 8-valve cylinder head—1.6L engines

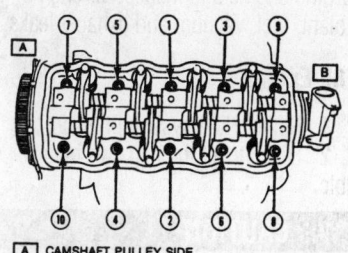

A CAMSHAFT PULLEY SIDE
B DISTRIBUTOR SIDE

7924HG07

1.6L 8-valve cylinder head bolt tightening sequence—loosen the bolts in the reverse order

➡**Only a MVAC-trained, EPA-certified automotive technician should service the A/C system or its components.**

10. If applicable, remove the air conditioning compressor adjusting brace from the cylinder head. Do NOT disconnect any of the A/C refrigerant lines, unless the system is properly discharged.
11. Remove the rocker arm cover, then loosen all of the valve lash locknuts and adjusting screws until all pressure is relieved from the camshaft.
12. Loosen and remove the 10 cylinder head mounting bolts.

✳✳ WARNING

Do not attempt to loosen the cylinder head from the engine block by striking it with a hammer or mallet; the cylinder head is positioned on the engine block by locating pins. If the cylinder head is struck with a hammer or mallet, the cylinder head may be damaged, or the locating pins may shear off in the block.

13. Lift the cylinder head up and off of the engine block. If the cylinder head is difficult to separate from the engine block, lift the cylinder head with a prytool positioned under one of the bolt bosses located on the side of the cylinder head. Do not strike the side of the cylinder head with a mallet or hammer because the head is positioned on the engine block with locating pins.
14. If desired, the intake manifold, exhaust manifold, distributor and distributor case may now be removed from the cylinder head.
15. Insert clean shop rags in the cylinder bores to prevent dirt and other contaminants from falling into the cylinder bores.
16. Remove the old gasket from the engine block. Clean all gasket mating surfaces on the engine block and cylinder head, including the intake manifold, exhaust manifold, and throttle body.
17. Inspect the locating pins and engine block gasket surface for damage. If the locating pins are damaged, new ones must be installed. Engine block damage will require engine disassembly and must be repaired, preferably by a qualified automotive machine shop.
18. Clean and inspect the cylinder head for damage.

To install:

19. Ensure that the locating pins are properly installed in the engine block.

20. Position the new cylinder head gasket on the engine block so that it is retained by the locating pins. The gasket should be situated so that the word TOP is facing upward (away from the engine block) and toward the front (timing belt end) of the engine.
21. Remove the shop rags from the cylinder bores.
22. If necessary, install the intake manifold, exhaust manifold, distributor case and distributor on the cylinder head. These components can also be installed after the cylinder head is mated to the engine block.
23. Gently set the cylinder head on the engine block ensuring that the intake and exhaust sides of the cylinder head are facing the proper directions.

➡**Although not required by the manufacturer, it is always a good idea to use new cylinder head bolts.**

24. Lightly lubricate the threads of the cylinder head bolts, then install them finger-tight.
25. Using the sequence shown in the accompanying illustration, tighten the cylinder head bolts to 27 ft. lbs. (37 Nm), then to 40 ft. lbs. (54 Nm), and finally to 52 ft. lbs. (71 Nm).
26. If equipped, install the air conditioning compressor adjusting brace onto the cylinder head.
27. Install the timing belt inside cover, camshaft sprocket, timing belt, outer cover and all other related items, such as the crankshaft pulley, water pump pulley, water pump drive belt, cooling fan and shroud.
28. Reattach all fuel lines, vacuum hoses, cooling system hoses and wiring to their respective components.
29. Reattach the exhaust pipe to the exhaust manifold.
30. Connect the accelerator and, if equipped, the kickdown cables to the throttle body. Adjust cable end-play as described in Section 5 (accelerator cable) and Section 7 (kickdown cable).
31. Fill the engine cooling system with the proper amount and type of coolant.
32. Adjust the valve lash for all valves.
33. Install the rocker arm cover.
34. Install the air intake case.
35. Connect the negative battery cable.
36. Start the engine and check for coolant, fuel, vacuum and exhaust leaks.

1.6L 16-Valve Engine

1. Relieve fuel system pressure.
2. Disconnect the negative cable.

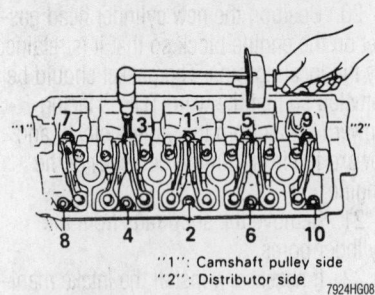

1.6L 16-valve cylinder head bolt tightening sequence—loosen the bolts in the reverse order

※※ CAUTION

Never open, service or drain the radiator or cooling system when hot; serious burns can occur from the steam and hot coolant.

3. Drain the cooling system.

4. Remove the intake manifold brace from the engine.

5. Label and detach all wiring and vacuum hoses from the distributor, intake manifold, oxygen sensor, throttle body, and cylinder head.

6. Detach the fuel supply and return lines from the fuel injector supply manifold.

7. Remove the rocker arm cover, then loosen all of the valve lash locknuts and adjusting screws.

8. Disconnect all cooling system hoses from the engine.

9. Remove the timing belt from the engine.

10. Separate the exhaust pipe from the exhaust manifold and remove the exhaust manifold brace.

※※ WARNING

Failure to properly loosen the cylinder head bolts may result in cylinder head warpage.

11. Loosen the cylinder head mounting bolts using Suzuki Tools 09900–00415 (A) and 09900–00411 (B) (or their equivalents) in the order shown in the accompanying illustration, then remove the bolts from the cylinder head.

12. Ensure that all wires, hoses and cables, which would restrict cylinder head removal, have been detached.

※※ WARNING

Do not attempt to loosen the cylinder head from the engine block by striking it with a hammer or mallet; the cylinder head is positioned on the

engine block by locating pins. If the cylinder head is struck with a hammer or mallet, the cylinder head may be damaged, or the locating pins may shear off in the block.

13. Lift the cylinder head up and off of the engine block. If the cylinder head is difficult to separate from the engine block, lift the cylinder head with a prytool positioned under one of the bolt bosses located on the side of the cylinder head. Do not strike the side of the cylinder head with a mallet or hammer because the head is positioned on the engine block with locating pins.

14. If desired, the intake manifold, exhaust manifold, distributor and distributor case may now be removed from the cylinder head.

15. Insert clean shop rags in the cylinder bores to prevent dirt and other contaminants from falling into the cylinder bores.

16. Remove the old gasket from the engine block. Clean all gasket mating surfaces on the engine block and cylinder head, including the intake manifold, exhaust manifold, and throttle body.

17. Inspect the locating pins and engine block gasket surface for damage. If the locating pins are damaged, new ones must be installed. Engine block damage will require engine disassembly and must be repaired, preferably by a qualified automotive machine shop.

18. Clean and inspect the cylinder head for damage.

To install:

19. Ensure that the locating pins are properly installed in the engine block.

20. Position the new cylinder head gasket on the engine block so that it is retained by the locating pins. The gasket should be situated so that the word TOP is facing upward (away from the engine block) and toward the front (timing belt end) of the engine.

21. Ensure the oil jet (venturi plug) is installed in the cylinder head, and that it is

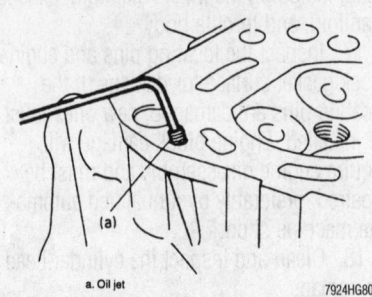

Prior to installing the cylinder head and gasket, ensure that the oil jet (venturi plug) is tightened to 35–53 inch lbs. (4–6 Nm)

not clogged. If it is not installed, tighten it to 35–53 inch lbs. (4–6 Nm).

22. Remove the shop rags from the cylinder bores.

23. If necessary, install the intake manifold, exhaust manifold, distributor case and distributor on the cylinder head. These components can also be installed after the cylinder head is mated to the engine block.

24. Gently set the cylinder head on the engine block ensuring that the intake and exhaust sides of the cylinder head are facing the proper directions.

→**Although not required by the manufacturer, it is always a good idea to use new cylinder head bolts.**

25. Lightly lubricate the threads of the cylinder head bolts, then install them finger-tight.

26. Using the sequence shown in the accompanying illustration and using Suzuki Tools 09900–00415 (A) and 09900–00411 (B) (or their equivalents), tighten all of the cylinder head bolts to 25 ft. lbs. (35 Nm), then to 40 ft. lbs. (55 Nm), and finally to 47.5–50.5 ft. lbs. (65–70 Nm).

27. Reattach the exhaust pipe to the exhaust manifold, and install the exhaust manifold brace.

28. Install the timing belt.

29. Connect all cooling system hoses to the engine.

30. Attach the fuel supply and return lines to the fuel injector supply manifold.

31. Reconnect all wiring and vacuum hoses to the distributor, intake manifold, oxygen sensor, throttle body, and cylinder head.

32. Install the intake manifold brace.

33. Fill the cooling system.

34. Connect the negative cable.

35. Adjust the valve lash and the accelerator and kickdown (if equipped) cables.

36. Turn the ignition key **ON**, but do not start the engine; this allows the fuel system to pressurize. Check for fuel system leaks.

37. After checking for fuel system leaks, start the engine and inspect for engine coolant, fuel, vacuum and exhaust leaks.

1.8L Engine

1. Relieve fuel system pressure.

2. Disconnect the negative battery cable.

※※ CAUTION

Never open, service or drain the radiator or cooling system when hot; serious burns can occur from the steam and hot coolant.

1. Crankshaft pulley side
2. Flywheel side
3. Bolt (M6)

7924HG09

1.8L engine cylinder head bolt loosening sequence—be sure to tighten the bolts in the reverse order

3. Drain the engine cooling system.
4. Remove the camshafts and valve lash adjusters.
5. Remove the intake manifold brace from the engine.
6. Label and detach all wiring and vacuum hoses, which will interfere with cylinder head removal, from the intake manifold, oxygen sensor, throttle body, and cylinder head.
7. Detach the fuel supply and return lines from the fuel injector supply manifold.
8. Detach the water pipe from the intake manifold, and disconnect any other cooling system hoses from the engine.
9. Separate the exhaust pipe from the exhaust manifold and remove the exhaust manifold brace.

❊❊ WARNING

Failure to properly loosen the cylinder head bolts may result in cylinder head warpage.

10. Loosen the cylinder head mounting bolts in the order shown in the accompanying illustration, then remove the bolts from the cylinder head.

➡**Be sure to loosen the small M6 bolt located on the side of the cylinder head, near the crankshaft pulley end.**

11. Ensure that all wires, hoses and cables, which would restrict cylinder head removal, have been detached.

❊❊ WARNING

Do not attempt to loosen the cylinder head from the engine block by striking it with a hammer or mallet; the cylinder head is positioned on the engine block by locating pins. If the cylinder head is struck with a hammer or mallet, the cylinder head may be damaged, or the locating pins may shear off in the block.

12. Lift the cylinder head up and off of the engine block. If the cylinder head is dif-

ficult to separate from the engine block, lift the cylinder head with a prytool positioned under one of the bolt bosses located on the side of the cylinder head. Do not strike the side of the cylinder head with a mallet or hammer because the head is held in position on the engine block with locating pins.
13. If desired, the intake manifold, exhaust manifold, and any other items may now be removed from the cylinder head.
14. Insert clean shop rags in the cylinder bores to prevent dirt and other contaminants from falling into the cylinder bores.
15. Remove the old gasket from the engine block. Clean all gasket mating surfaces on the engine block and cylinder head, including the intake manifold and exhaust manifold.
16. Inspect the locating pins and engine block gasket surface for damage. If the locating pins are damaged, new ones must be installed. Engine block damage will require engine disassembly and must be repaired, preferably by a qualified automotive machine shop.
17. Clean and inspect the cylinder head for damage.

To install:
18. Ensure that the locating pins are properly installed in the engine block.
19. Position the new cylinder head gasket on the engine block a shown in the accompanying illustration.
20. Remove the shop rags from the cylinder bores.
21. If necessary, install the intake manifold, exhaust manifold, and any other components on the cylinder head. These components can also be installed after the cylinder head is mated to the engine block.
22. Gently set the cylinder head on the engine block ensuring that the intake and exhaust sides of the cylinder head are facing the proper directions.

➡**Although not required by the manufacturer, it is always a good idea to use new cylinder head bolts.**

23. Lightly lubricate the threads of the cylinder head bolts, then install them finger-tight.

➡**Be sure to follow the cylinder head bolt tightening sequence exactly, otherwise improperly cylinder head sealing will occur.**

24. Using the sequence shown in the accompanying illustration, tighten the cylinder head bolts in the following steps:
   a. Tighten all cylinder head bolts, in sequence, to 38.5 ft. lbs. (53 Nm).

   b. Tighten the head bolts, in sequence, to 61 ft. lbs. (84 Nm).
   c. Loosen all of the head bolts until they can be rotated by hand.
   d. Retighten the head bolts, in sequence, to 27 ft. lbs. (37 Nm).
   e. Tighten the cylinder head bolts, in sequence, to 76 ft. lbs. (105 Nm).

➡**Tighten the small M6 bolt ONLY after the other bolts are fully tightened.**

   f. Tighten the small M6 bolt to 97 inch lbs. (11 Nm).
25. Ensure that the crankshaft sprocket key is aligned with the timing mark on the engine block. If it is not aligned, rotate the crankshaft until it is.
26. Attach the exhaust pipe to the exhaust manifold, and install the exhaust manifold brace.
27. Reattach the water pipe to the intake manifold, and connect any other cooling system hoses to the engine.
28. Reattach the fuel supply and return lines to the fuel injector supply manifold.
29. Attach all applicable wiring and vacuum hoses to the intake manifold, oxygen sensor, throttle body, and cylinder head.
30. Install the intake manifold brace from the engine.
31. Install the camshafts, valve lash adjusters and all other related components.
32. Fill the cooling system.
33. Connect the negative cable.
34. Adjust the accelerator and kickdown (if equipped) cables.
35. Turn the ignition key **ON**, but do not start the engine; this allows the fuel system to pressurize. Check for fuel system leaks.
36. After checking for fuel system leaks, start the engine and inspect for engine coolant, fuel, vacuum and exhaust leaks.

Rocker Arms/Shafts

REMOVAL & INSTALLATION

1.6L 8-Valve Engine

➡**Only a MVAC-trained, EPA-certified automotive technician should service the A/C system or its components.**

1. Disconnect the negative battery cable and drain the engine cooling system.
2. It may be necessary to remove several components from the front of the vehicle for enough clearance to remove the rocker arm shafts. If this added clearance is necessary for rocker arm shaft removal, perform the following:

✳✳ CAUTION

Some models covered by this manual may be equipped with a Supplemental Restraint System (SRS). Whenever working near any of the SRS components, such as the impact sensors, the air bag module, steering column and instrument panel, refer to the SRS precautions. Failure to heed all precautions may result in accidental air bag deployment, which could easily result in severe personal injury or death. Also, never attempt any electrical diagnosis or service to the SRS components and wiring; this work should only be performed by a qualified automotive technician.

a. If your vehicle is equipped with air conditioning, the manufacturer states that the compressor must be disconnected from the A/C hoses.

b. Remove the hood.

c. Remove the front grille.

d. Remove the hood lock from the front upper member, then remove the front upper member.

e. Remove the radiator cooling fan, shroud and radiator.

3. Loosen the water pump pulley mounting bolts, then remove the accessory drive belts and water pump pulley.

4. If equipped, remove the air intake case from the throttle body.

5. Remove the rocker arm cover.

6. Loosen all of the valve lash locknuts and adjusting screws so that there is no pressure exerted on the camshaft lobes.

7. Remove the timing belt cover, timing belt, camshaft sprocket and inside timing belt cover from the engine.

8. Matchmark the rocker arms so that they can be reinstalled in their original positions.

9. Loosen the 10 rocker arm shaft securing screws, then slowly slide one of the rocker arm shafts out of the front of the cylinder head. While withdrawing the rocker arm shaft from the cylinder head, remove the rocker arms and springs. Set the rocker arms and springs aside in order.

10. Remove the other rocker arm shaft in the same manner.

11. Clean the rocker arms, shafts, springs and rocker arm shaft bores in the cylinder head.

To install:

12. Inspect the tip of the adjusting screw for excessive wear; if wear is evident, replace the adjusting screw.

13. Inspect the face of the rocker arm cam-riding surface for wear; if excessive wear is evident, replace the rocker arm.

14. Inspect the rocker arm-to-shaft clearance by using a micrometer to measure the outside diameter of the rocker arm shaft where the rocker arms ride on it. Use a bore gauge to measure the inside diameter of the rocker arm bore. Subtract the rocker arm bore inside diameter from the rocker arm shaft outside diameter to calculate the rocker arm-to-shaft clearance. Compare your findings with the values presented in the engine rebuilding specification charts at the end of this section. If your findings are not within the values in the charts, replace the rocker arm, shaft, or both.

15. Position the rocker arm shaft in two wooden blocks with V cutouts in them (refer to the accompanying illustration). Using a dial indicator, measure the amount of runout at the center of the rocker arm shaft. If your measurement is not within the values presented in the engine rebuilding specifications chart.

16. Apply clean engine oil to the rocker arms and shafts.

➡The two rocker arm shafts are not identical. To distinguish between the two, inspect the stepped ends of the rocker arm shafts. The intake rocker arm shaft stepped end is 0.55 in. (14mm) wide and the exhaust side rocker arm shaft stepped end is 0.59 in. (15mm) wide. Refer to the accompanying illustration.

17. Insert the intake side rocker arm shaft into the cylinder head bore so that the stepped end is toward the timing belt end of the engine, then slowly slide the rocker arm shaft into position, while installing the rocker arms and springs onto the shaft. Ensure that the rocker arms and springs are installed in their original positions.

18. Install the exhaust side rocker arm shaft in the cylinder head so that the stepped end faces the distributor end of the engine. While slowly sliding the rocker arm shaft into position, install the rocker arms and springs onto the shaft. Ensure that the rocker arms and springs are installed in their original positions.

19. Tighten the rocker arm shaft retaining screws to 80–106 inch lbs. (9–12 Nm). Do not attempt to adjust the valve lash at this point.

20. Install the timing belt inside cover, camshaft sprocket, timing belt and outer cover.

21. Adjust the valve lash as described in Section 1.

22. Install the rocker arm cover.

23. If equipped, install the air intake case onto the throttle body.

24. Install the water pump pulley and tighten the mounting bolts until snug. Install the accessory drive belts and tighten the water pump pulley bolts fully.

25. If the front components were removed from the vehicle for added shaft removal clearance, perform the following:

a. Install the radiator, shroud and cooling fan.

b. Install the hood lock and the front upper member.

c. Install the front grille.

d. Install the hood.

26. Connect the negative battery cable and fill the engine cooling system.

1.6L 16-Valve Engine

1. Disconnect the negative battery cable.

2. Remove the camshaft from the cylinder head.

3. Matchmark the positions of the intake rocker arms so that they can be reinstalled in their original positions.

4. Remove the rocker arm shaft plug and timing belt inside cover from the cylinder head.

✳✳ WARNING

Be sure not to bend the rocker arm clip during removal.

5. Remove each intake rocker arm with its clip from the rocker arm shaft.

6. Matchmark the positions of the exhaust rocker arms so that they can be reinstalled in their original positions.

7. Remove the rocker arm shaft bolts, then slide the rocker arm shaft toward the rear of the engine until the O-ring is exposed. Remove the O-ring from the rocker arm shaft groove.

8. Remove the exhaust rocker arms by sliding the rocker arm shaft out of the front of the cylinder head.

9. Position the rocker arms and springs on a clean work surface in order so that they can be reinstalled in their original positions.

10. Clean the rocker arms, springs and rocker arm shaft.

To install:

11. Inspect the tip of the adjusting screw for excessive wear; if wear is evident, replace the adjusting screw.

12. Inspect the face of the rocker arm cam-riding surface for wear; if excessive wear is evident, replace the rocker arm.

13. Inspect the exhaust rocker arm-to-shaft clearance as follows:

a. Use a micrometer to measure the outside diameter of the rocker arm shaft where the rocker arms ride on it. Use a bore gauge to measure the inside diameter of the rocker arm bore. Subtract the rocker arm bore inside diameter from the rocker arm shaft outside diameter to calculate the rocker arm-to-shaft clearance. Compare your findings with the values presented in the engine rebuilding specification charts at the end of this section. If your findings are not within the values in the charts, replace the rocker arm, shaft, or both.

14. Position the rocker arm shaft in two wooden blocks with V cutouts in them (refer to the accompanying illustration). Using a dial indicator, measure the amount of run-out at the center of the rocker arm shaft. If your measurement is not within the values presented in the engine rebuilding specifications chart.

15. Apply clean engine oil to the rocker arms and shafts.

➡ **Ensure that the rocker arm is installed in the cylinder head so that the O-ring groove is toward the distributor end of the engine.**

16. Insert the rocker arm shaft into the cylinder head bore so that the O-ring groove end is toward the distributor end of the engine, while installing the exhaust rocker arms and springs onto the shaft. Ensure that the rocker arms and springs are installed in their original positions.

17. Push the rocker arm out of the back side of the cylinder head until the O-ring groove is exposed. Install a new O-ring into the groove, then slide the rocker arm shaft back into the cylinder head so that both ends are flush with the head. Rotate the shaft so that the flat surface on its front end faces down and is parallel with the cylinder head-to-engine block gasket surface.

18. Install and tighten the rocker arm shaft retaining screws to 97 inch lbs. (11 Nm). Do not attempt to adjust the valve lash at this point.

19. Fill a small amount of clean engine oil into the rocker arm pivot holding part of the rocker arm shaft, then install the intake rocker arms so that their clips are properly engaged on the shaft.

20. Install the camshaft and all other items removed earlier.

21. Adjust the valve lash, then install the rocker arm cover.

1.8L Engine

The 1.8L DOHC engine does not utilize rocker arms, nor rocker arm shafts.

Intake Manifold

REMOVAL & INSTALLATION

1.6L TFI Engine

1. Remove the throttle body from the intake manifold.
2. Disconnect the PCV hose from the rocker arm cover.
3. Detach the pressure sensor hose from the gas filter.
4. Disconnect the brake booster and automatic transmission (if equipped) vacuum hoses from the intake manifold.
5. Detach the Vacuum Switching Valve (VSV) hose for the throttle opener and EVAP canister from the intake manifold.
6. Detach the upper radiator hose from the thermostat housing, and the heater inlet and water bypass hoses from the intake manifold.
7. Disconnect the EGR valve hoses from the EGR valve.
8. Detach and label all wiring and wiring connectors from components mounted on the intake manifold.

※※ WARNING

Never use a prybar between the intake manifold and cylinder head mating surface to attempt to separate the two components; damage to the intake manifold or cylinder head may occur, necessitating component replacement.

9. Remove the intake manifold mounting fasteners, then remove the intake manifold from the cylinder head. It may be necessary to tap the manifold with a soft-faced mallet to free it from the cylinder head.
10. Insert clean shop rags in the intake holes in the cylinder head to prevent accidentally dropping anything, which would require cylinder head removal, into the cylinders.
11. At this time, the PCV valve, EGR valve, gas filter, thermostat, sensors, switch and gauge may be removed from the intake manifold.
12. Clean the intake manifold-to-cylinder head gasket mating surfaces thoroughly.

To install:

13. If a new manifold is being installed, transfer the PCV valve, EGR valve, gas filter, thermostat, sensors, switch and gauge to the new manifold.
14. Remove the rags from the cylinder head intake holes.

15. Position a new intake manifold gasket on the cylinder head, then install the intake manifold. Tighten the intake manifold-to-cylinder head bolts to 159–248 inch lbs. (18–28 Nm). Be sure to tighten the center bolts first, then work your way out to both ends of the manifold.
16. Reattach all wiring to the intake manifold and related components.
17. Reconnect the upper radiator, the bypass, the heater inlet, the pressure sensor, the VSV, the brake booster, the EGR valve, the automatic transmission, and the PCV hoses.
18. Install the throttle body.

1.6L MFI Engine

1. Relieve the fuel system pressure.
2. Disconnect the negative battery cable.

※※ CAUTION

Never open, service or drain the radiator or cooling system when hot; serious burns can occur from the steam and hot coolant.

3. Drain the engine cooling system.
4. Remove the air intake pipe.
5. Detach the accelerator cable and automatic transmission kickdown cable (if equipped) from the throttle body.
6. Detach the vacuum hose from the PCV valve.
7. Disengage all wiring from the intake manifold and throttle body which will interfere with removal.
8. Disconnect all vacuum and cooling system hoses from the throttle body and intake manifold which will restrict removal.

※※ WARNING

Be sure to plug the fuel lines to prevent dirt or other contaminants, which may cause future fuel system damage, from entering the fuel system.

9. Disconnect and plug the fuel feed line from the junction near the firewall. Be sure to use a second wrench to hold the junction steady while loosening the fuel line flare nut. Also, detach the fuel return line.
10. Remove the alternator adjusting arm brace.
11. Remove the intake manifold brace, the No. 1 brace and the No. 2 brace with the EGR pressure transducer.
12. Detach the cooling system bypass hose from the intake manifold.

❋❋ WARNING

Never use a prybar between the intake manifold and cylinder head mating surface to attempt to separate the two components; damage to the intake manifold or cylinder head may occur, necessitating component replacement.

13. Remove the intake manifold-to-cylinder head mounting fasteners, then separate the intake manifold and throttle body assembly from the engine. It may be necessary to tap the manifold with a soft-faced mallet to free it from the cylinder head. At this point the throttle body and intake surge tank can be separated from the intake manifold by removing the attaching fasteners. Be sure to discard any old gaskets used between these components.

14. Insert clean shop rags in the intake holes in the cylinder head to prevent accidentally dropping anything (such as dirt, nuts, bolts, etc.), which would require cylinder head removal, into the cylinders.

15. Remove and discard the old intake manifold-to-cylinder head gasket.

16. Clean all gasket mating surfaces thoroughly.

To install:

17. If necessary, assemble the intake surge tank and throttle body onto the intake manifold, making sure to use new gaskets and to tighten the attaching fasteners to 203 inch lbs. (23 Nm).

18. Remove the rags from the cylinder head intake holes.

19. Position a new intake manifold gasket on the cylinder head studs, then install the intake manifold and throttle body assembly onto the cylinder head. Be sure to install the wiring clamps on their original studs.

20. Starting in the middle of the intake manifold and working outward toward the ends, tighten the manifold mounting fasteners to 203 inch lbs. (23 Nm).

21. Reattach the cooling system bypass hose to the intake manifold.

22. Install the intake manifold brace, the No. 1 brace and the No. 2 brace with the EGR pressure transducer. Tighten the brace fasteners to 36 ft. lbs. (50 Nm).

23. Install the alternator adjusting arm brace. Tighten the brace bolts to 36 ft. lbs. (50 Nm).

24. Attach the fuel feed line to the junction near the firewall. Be sure to use a second wrench to hold the junction steady while tightening the fuel line flare nut to 32.5 ft. lbs. (45 Nm). Also, reattach the fuel return line.

25. Connect all vacuum and cooling system hoses to the throttle body and intake manifold.

26. Engage all wiring to the intake manifold and throttle body.

27. Connect the PCV vacuum hose to the valve.

28. Reattach the accelerator cable and automatic transmission kickdown cable (if equipped) to the throttle body. Be sure to adjust the accelerator and kickdown (if equipped) cables play.

29. Install the air intake pipe.

30. Fill the engine cooling system.

31. Connect the negative battery cable.

32. Turn the ignition key **ON**, but do not start the engine; this allows the fuel system to pressurize. Check for fuel system leaks.

33. After checking for fuel system leaks, start the engine and inspect for engine coolant leaks.

1.8L Engine

1. Remove the throttle body from the intake manifold.

2. Remove the front and rear intake manifold braces.

3. Detach the water pipe from the intake manifold.

❋❋ WARNING

Never use a prybar between the intake manifold and cylinder head mating surface to attempt to separate the two components; damage to the intake manifold or cylinder head may occur, necessitating component replacement.

4. Remove the intake manifold mounting fasteners, then separate the manifold from the cylinder head. It may be necessary to tap the manifold with a soft-faced mallet to free it from the cylinder head.

5. Insert clean shop rags in the intake holes in the cylinder head to prevent accidentally dropping anything (such as dirt, nuts, bolts, etc.), which would require cylinder head removal, into the cylinders.

6. Remove and discard the old intake manifold gasket. Thoroughly clean all gasket mating surfaces of all dirt and old gasket material.

7. If a new manifold is being installed, remove the EVAP solenoid purge valve, Exhaust Gas Recirculation (EGR) valve, Idle Air Control (IAC) valve, EGR pipe and vacuum pipe from the old intake manifold.

To install:

8. Clean all gasket mating surfaces thoroughly.

9. If necessary, transfer the EVAP solenoid purge valve, Exhaust Gas Recirculation (EGR) valve, Idle Air Control (IAC) valve, EGR pipe and vacuum pipe onto the new intake manifold.

10. Remove the rags from the cylinder head intake holes.

11. Along with a new gasket, install the intake manifold onto the cylinder head. Starting in the middle of the intake manifold and working outward toward the ends, tighten the manifold mounting fasteners to 203 inch lbs. (23 Nm).

12. Connect the water pipe to the intake manifold.

Install the front and rear intake manifold braces. Tighten the front brace bolts to 36.5 ft. lbs. (50 Nm), and the rear manifold brace bolts to 221 inch lbs. (25 Nm).

13. Install the throttle body.

Exhaust Manifold

REMOVAL & INSTALLATION

1. For 1.8L engines, remove the strut tower bar.

2. If added working room is needed, remove the air cleaner outlet tube and the air inlet pipe.

3. Disengage the oxygen sensor wiring harness connector. Remove the air intake case support brace from the engine.

4. Remove the exhaust manifold shield retaining bolts, then separate the shield(s) from the manifold.

5. If equipped, remove the exhaust manifold brace from the engine block and manifold.

6. Detach the exhaust pipe from the manifold by removing the mounting nuts.

7. Remove the exhaust manifold retaining bolts and nuts, then pull the exhaust manifold off of the cylinder head.

8. Remove and discard the old exhaust manifold gasket.

9. Thoroughly clean the manifold-to-cylinder head gasket mating surfaces of all dirt, carbon and old gasket material.

To install:

10. Install a new exhaust manifold-to-cylinder head gasket on the cylinder head studs, then install the exhaust manifold. Tighten the exhaust manifold mounting fasteners to 159–248 inch lbs. (18–28 Nm). Be sure to tighten the center bolts first, then work your way out to both ends of the manifold.

11. Install the air intake case support brace, and tighten the mounting bolts to 159–248 inch lbs. (18–28 Nm).

12. Reattach the exhaust pipe to the exhaust manifold, using a new gasket, then tighten the attaching nuts to 29–43 ft. lbs. (40–60 Nm).

13. If equipped, install the exhaust manifold brace, and tighten the brace-to-engine block bolt to 36.5–43 ft. lbs. (50–60 Nm) and the brace-to-exhaust manifold nut to 29–43 ft. lbs. (40–60 Nm).

14. Position the cover on the exhaust manifold and install the retaining bolts. Tighten the retaining bolts securely.

15. Reattach the oxygen sensor lead wire to the wiring harness. Be sure to properly retain the wire with the wiring clamp.

16. If removed, install the air cleaner outlet tube and the air inlet pipe.

17. For 1.8L engines, install the strut tower bar. Tighten the mounting nuts and bolts to 66 ft. lbs. (90 Nm).

Front Crankshaft Seal

REMOVAL & INSTALLATION

1.6L Engines

➡This procedure only applies to 1.6L engines; the 1.8L engine oil pump does not use an oil seal.

1. Remove the timing belt, crankshaft sprocket and belt guide.

2. Using a small prytool, carefully pry the old oil seal out of the oil pump housing bore. Take care not to scratch the oil pump housing bore, otherwise oil leakage may occur. Wrapping a piece of tape around the end of the prytool may help reduce the change of scoring the bore by covering any sharp corners on the tool.

3. Clean the oil pump housing bore and the crankshaft.

4. If the oil seal is being removed because of oil leakage, inspect the crankshaft surface where the oil seal contacts it. If there is a wear groove on the crankshaft, a new oil seal will probably not cure the oil leak. Before removing the crankshaft and replacing it, attempt to repair it by installing a metal sleeve, designed just for this problem, over the crankshaft end. The crankshaft sleeve will present a new, flat surface for the new oil seal to seal against. Crankshaft sleeves are usually available from automotive parts stores.

5. Apply a thin coat of clean engine oil to the oil seal lip, then slide the oil seal over the crankshaft and into position against the oil pump housing. Use a deep socket, which is the same diameter as the oil seal, and a hammer to drive the oil seal into the

housing. Drive the new seal into the housing only until the outer seal edge is flush with the oil pump housing surface.

6. Install the timing belt guide, crankshaft sprocket, timing belt, and all other related components.

1.8L Engine

1. If extra clearance is necessary for crankshaft drive belt pulley removal is necessary, remove the cooling fan, pulley and accessory drive belt.

2. Remove the alternator drive belt.

3. Using a long breaker bar and socket, remove the crankshaft pulley bolt. To secure the crankshaft from rotating, install a pulley holding tool (such as Suzuki Tool 09917–68221). Use M8, P1.25 bolts with a strength rating of at least 7T to attach the holding tool to the pulley.

4. Once the bolt is removed, install a steering wheel puller on the crankshaft drive belt pulley, then draw the pulley off of the end of the crankshaft.

❄❄ WARNING

Take care not to damage the oil seal bore during removal, otherwise oil leakage may occur.

5. Using a small prytool, carefully pry the old oil seal out of the cover bore. Take care not to scratch the oil pump housing bore, otherwise oil leakage may occur. Wrapping a piece of tape around the end of the prytool may help reduce the change of scoring the bore by covering any sharp corners on the tool.

6. Clean the oil seal bore and the crankshaft.

7. If the oil seal is being removed because of oil leakage, inspect the crankshaft surface where the oil seal contacts it. If there is a wear groove on the crankshaft, a new oil seal will probably not cure the oil leak. Before removing the crankshaft and replacing it, explore the possibility of repairing it by installing a metal sleeve, designed just for this problem, over the crankshaft end. The crankshaft sleeve will present a new, flat surface, with which the new oil seal can make contact. Crankshaft sleeves are usually available from automotive parts stores.

To install:

8. Apply a thin coat of clean engine oil to the oil seal lip, then slide the oil seal over the crankshaft and into position against the timing chain cover. Use a deep socket, which is the same diameter as the oil seal, and a hammer to drive the oil seal into the housing. Drive the new seal into the hous-

ing only until the outer seal edge is flush with the cover surface.

➡**The new oil seal must be installed so that the side of the seal where the seal spring is visible is installed inward (toward the engine block).**

9. Slide the crankshaft pulley onto the end of the crankshaft and install the center bolt. Install the crankshaft holding tool, then tighten the center bolt to 109 ft. lbs. (150 Nm). The center bolt will draw the pulley onto the crankshaft when tightened to the specified torque value.

10. Install the cooling fan belt, pulley and fan.

11. Install the alternator drive belt.

Camshaft and Valve Lifters

REMOVAL & INSTALLATION

1.6L 8-Valve Engine

➡**During this procedure, identify all components removed from the engine so that they may be reinstalled in their original positions. If discarding the old components so that new components can be installed, identifying the old items is not necessary.**

1. Remove the cylinder head.

2. Remove the intake and exhaust manifolds, the rocker arms and shafts, and the camshaft sprocket from the cylinder head.

3. Remove the distributor from the distributor case.

4. Remove the distributor case mounting fasteners, then separate the case from the cylinder head.

5. Carefully slide the camshaft out of the rear (distributor) end of the cylinder head.

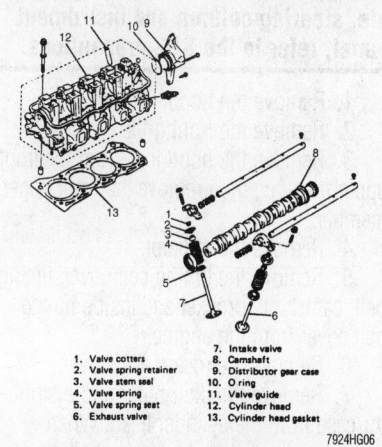

1. Valve cotters	7. Intake valve
2. Valve spring retainer	8. Camshaft
3. Valve stem seal	9. Distributor gear case
4. Valve spring	10. O ring
5. Valve spring seat	11. Valve guide
6. Exhaust valve	12. Cylinder head
	13. Cylinder head gasket

7924HG06

Exploded view of camshaft and valve train mounting—1.6L 8-valve cyclinder head

6. Using a seal removal tool or small prytool, carefully pry the old camshaft oil seal out of the front of the cylinder head bore. Take care not to scratch the camshaft oil seal bore, otherwise oil leakage may occur. If using a small prytool, wrapping a piece of tape around the end of the tool may help reduce the change of scoring the bore by covering any sharp corners on the tool.

7. Clean the cylinder head, camshaft, rocker arms and shafts and all gasket mating surfaces of all dirt, corrosion and oil.

To install:

8. Apply a thin coat of clean engine oil to the oil seal lip, then drive the new seal into the housing only until the outer seal edge is flush with the cover surface.

➡**The new oil seal must be installed so that the side of the seal where the seal spring is visible is installed inward (toward the engine block).**

9. Apply clean engine oil to the camshaft, then carefully slide it into the rear end of the cylinder head.

10. Install the distributor case and tighten the mounting fasteners until secure.

11. Install the rocker arm shafts and rocker arms, the inside timing belt cover, the timing belt sprocket, the intake and exhaust manifolds, and the distributor.

12. Install the cylinder head on the engine block. Install all related items.

1.6L 16-Valve Engine

✳✳ CAUTION

Some models covered by this manual may be equipped with a Supplemental Restraint System (SRS), which uses an air bag. Whenever working near any of the SRS components, such as the impact sensors, the air bag module, steering column and instrument panel, refer to the SRS precautions.

1. Remove the hood.
2. Remove the front grille.
3. Remove the hood lock from the front upper member, then remove the front upper member.
4. Remove the radiator.
5. Remove the timing belt cover, timing belt, camshaft sprocket and inside timing belt cover from the engine.
6. Remove the rocker arm cover.
7. Remove the distributor and distributor case from the cylinder head. When removing the distributor case, position a small catch pan or rag beneath it to catch any oil which may leak during removal.

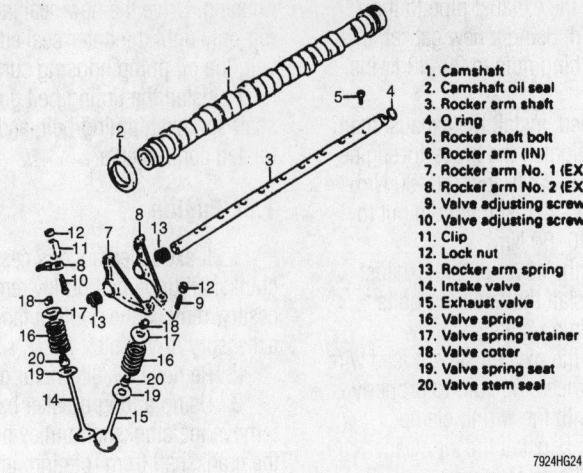

1. Camshaft
2. Camshaft oil seal
3. Rocker arm shaft
4. O ring
5. Rocker shaft bolt
6. Rocker arm (IN)
7. Rocker arm No. 1 (EX)
8. Rocker arm No. 2 (EX)
9. Valve adjusting screw
10. Valve adjusting screw
11. Clip
12. Lock nut
13. Rocker arm spring
14. Intake valve
15. Exhaust valve
16. Valve spring
17. Valve spring retainer
18. Valve cotter
19. Valve spring seat
20. Valve stem seal

Exploded view of the cylinder head valve train—1.6L 16-valve engine

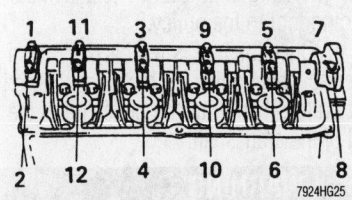

1.6L 16-valve camshaft housing bolt tightening sequence—loosen the bolts in the reverse of the order shown

8. Loosen all of the valve lash locknuts and adjusting screws so that there is no pressure exerted on the camshaft lobes.

9. Loosen the camshaft bearing cap bolts in the sequence shown in the accompanying illustration, then remove the caps from the cylinder head. Discard the old camshaft oil seal.

10. Lift the camshaft up and off of the cylinder head.

11. Clean the camshaft and bearing caps of all dirt and oil.

To install:

12. Apply clean engine oil to the camshaft journals and cam lobes, then set the camshaft on the cylinder head.

13. Apply clean engine oil to the camshaft bearing surfaces. Apply silicone sealant to the mating surface of bearing cap No. 6. The camshaft bearing caps are marked with their position numbers (no. 1 is the foremost cap, and No. 6 is the rearmost bearing cap). Install bearing cap No. 1 first, then install the remaining bearing caps. Oil the camshaft bearing cap bolt threads, then tighten all of them first to 24 inch lbs. (2.7 Nm), then to 48 inch lbs. (5.4 Nm), then to 72 inch lbs. (8.2 Nm), and finally to 97 inch lbs. 911 Nm) in the sequence shown in the accompanying illustration.

14. Apply a thin coat of clean engine oil to the oil seal lip, then drive the new seal

into the housing only until the outer seal edge is flush with the cover surface.

➡**The new oil seal must be installed so that the open side of the seal is installed inward (toward the cylinder head).**

15. Apply silicone sealant on the distributor case, as shown in the accompanying illustration, then install the distributor case on the cylinder head. Tighten the mounting fasteners to 97 inch lbs. (11 Nm). Install the distributor to the distributor case.

16. Install the timing belt cover, timing belt, camshaft sprocket and inside timing belt cover on the engine.

17. Install the remaining components in the reverse order of removal.

18. Fill the cooling system.

1.8L Engine

✳✳ CAUTION

The EPA warns that prolonged contact with used engine oil may cause a number of skin disorders, including cancer!

1. Drain the engine oil.
2. Remove the oil pan and oil pump pick-up.
3. Remove the rocker arm cover, the timing chain cover, the outer timing chain, and the Camshaft Position (CMP) sensor.
4. After removing the outer timing chain, rotate the crankshaft 90 degrees clockwise so that interference between the valves and pistons is avoided.
5. Loosen the camshaft retaining housing bolts in sequence, then remove them.
6. Lift the camshaft housings from the cylinder head.

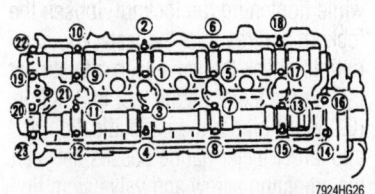

1.8L engine camshaft housing bolt tightening sequence—be sure to loosen the bolts in the reverse of the order shown

7. Lift the camshafts up and off of the cylinder heads. Set the camshaft aside so that they will not be confused, and can be reinstalled in their original positions. Marking them with an I (intake) and an E (exhaust) with a grease pencil, is a good idea to prevent mixing them up.

❊❊❊ WARNING

Never disassemble the hydraulic valve lash adjusters. Also, do not apply force to the body of the adjusters, otherwise the oil in the adjuster's high pressure chamber will leak.

8. Remove the hydraulic lash adjusters from the cylinder head. Set the adjusters in a container filled with clean engine oil; they should be kept submerged in clean engine oil until ready to reinstall them. If a container of clean engine oil is not available, only set the lash adjusters with their bucket bodies facing down. NEVER set the lash adjusters on their sides or with their bucket bodies facing up.

9. Clean the camshaft journals on the cylinder head, the camshaft housings (caps) and the camshafts of all dirt, grime and oil.

To install:

10. Fill the oil passages of the cylinder head with clean engine oil by pouring the oil through the oil holes. Ensure that the oil comes out from the oil holes in the sliding part of the valve lash adjuster bore.

11. Apply clean engine oil to the sides of the lash adjusters, then slide them into their bores in the cylinder head.

12. Rotate the crankshaft 1¾ revolutions (630 degrees) so that the crankshaft key is aligned with the mark on the engine block and the timing mark (dot) on the idler gear is positioned up.

13. Oil the camshaft journals and lobes, and lubricate the camshaft seats in the cylinder head, then set the camshafts on the cylinder head. Rotate the camshafts so that the locating pins on their front ends are positioned as shown.

➡ Install the camshafts so that the end with the CMP sensor groove is oriented toward the exhaust side.

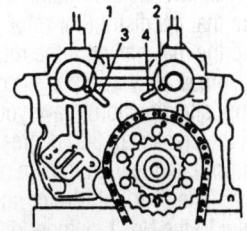

1. Knock pin of intake camshaft
2. Knock pin of exhaust camshaft
3. Match mark of intake camshaft
4. Match mark of exhaust camshaft

Position the camshafts as indicated prior to installing the timing chains—1.8L engine

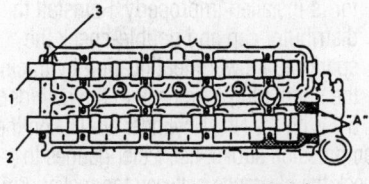

1. Intake camshaft
2. Exhaust camshaft
3. Pin installation position

Apply sealant to the areas marked "A"— 1.8L engine

14. Install the camshaft housing pins in the cylinder head, then apply silicone sealant to the areas marked A in the accompanying illustration.

➡ Each camshaft housing (bearing cap) is equipped with identification as to whether it is an intake or exhaust camshaft housing, what position it should take from the timing chain end of the engine, and an arrow (should point to the timing chain end of the engine).

15. Lubricate the camshaft housing-to-camshaft surfaces with clean engine oil, then position the housings on the cylinder head.

16. Lubricate the housing mounting bolts with clean engine oil, then install them finger-tight.

17. Using an inch-pound torque wrench and the proper sequence, tighten all of the housing mounting bolts first to 24 inch lbs. (2.7 Nm), then to 48 inch lbs. (5.4 Nm), then to 72 inch lbs. (8.2 Nm), and finally to 97 inch lbs. (11 Nm).

18. Install the remaining components in the reverse order of removal.

19. Refill the cooling system with coolant, the front differential with gear oil and the engine with clean engine oil.

❊❊❊ WARNING

Do not turn the camshafts or start the engine for approximately ½ hour after installing the valve lash adjusters and camshafts; it takes time for the valves to settle in place. Operating the engine before this time period may result n interference between the valves and pistons, which can easily cause expensive and time-consuming internal engine damage.

20. After a half hour period, start the engine. Check the ignition timing and adjust it as necessary.

21. If the valve lash adjusters make a tapping sound when the engine started, air may be trapped in the adjuster(s). In such a case, run the engine at 2000 rpm for approximately ½ hour, after which the air should be bled from the adjuster(s). If the tapping sound continues, the adjuster is most likely defective. To check the adjuster, perform the following:

a. Remove the rocker arm cover.

b. With you thumb, press down on each of the adjusters with a force of less than 44 lbs. (20 kg) when the valve adjuster is resting against the camshaft lobe base circle. If clearance exists between the adjuster and the camshaft, that adjuster is defective.

c. Perform this inspection for all 16 adjusters. Replace any that are found to be defective.

d. Install the rocker arm cover.

Valve Lash

ADJUSTMENT

➡ This procedure can be performed with the engine cold (overnight cold) or hot (normal operating temperature), but it is not a good idea to perform it when the engine is warm. Either allow the engine to cool completely or operate the engine until normal operating temperature has been reached. If you decide to perform the adjustment with the engine hot and it takes too long to complete the procedure, you may need to stop in the middle to warm the engine up again. Therefore, it is better to allow the engine to cool completely before adjusting the valve lash.

➡ Through-out this procedure the cylinders are referred to by their number. The number of each cylinder is dependent upon their location; the front-most cylinder is No. 1 and the numbering proceeds for each cylinder toward the rear-most cylinder, which is No. 4. Therefore, the cylinders are numbered, from front-to-back, 1 through 4.

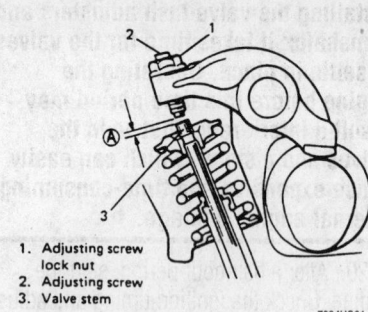

1. Adjusting screw lock nut
2. Adjusting screw
3. Valve stem

7924HG81

Valve lash is the clearance (A) between the bottom surface of the adjusting screw (2) and the valve stem tip (3)

1.6L Engines

1. Remove the cylinder head cover.

2. Using a large wrench, or a socket (typically 17mm) and large ratchet on the crankshaft pulley center nut, turn the crankshaft clockwise (viewing the crankshaft from the front of the engine) until the Top Dead Center (TDC) line on the crankshaft pulley is aligned with the 0 mark on the timing mark tab attached to the timing belt cover.

3. Locate the spark plug wire tower on the distributor cap that corresponds to the No. 1 cylinder spark plug. Matchmark the position of the No. 1 distributor tower with the engine block or other engine component.

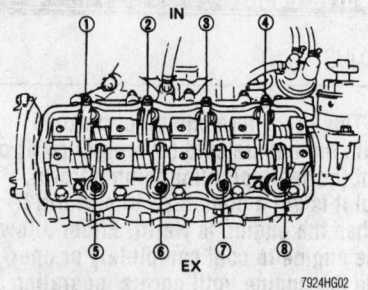

Valve identification for 8-valve cylinder heads

7924HG02

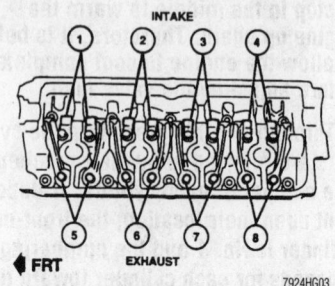

Valve pair identification for 16-valve cylinder heads—adjust both valves in each pair at the same time

7924HG03

4. Remove the distributor cap and ensure that the distributor rotor points toward the matchmark. If the rotor points 180 degrees away from the matchmark (in the opposite direction), have your assistant rotate the crankshaft 360 degrees (one full revolution) and realign the timing marks.

  a. The distributor rotor should now point to the No. 1 cylinder distributor cap tower matchmark. The engine is now positioned with the No. 1 cylinder at Top Dead Center (TDC) on the compression stroke.

  b. If the rotor still does not point to the matchmark, you may have match-marked the wrong tower, or the distributor is installed improperly. Reinstall the distributor cap and double-check the spark plug tower identification by tracing the spark plug wire back to the cylinder.

5. With the No. 1 cylinder at TDC on the compression stroke, use feeler gauges to check the clearance between the rocker arm adjusting screws and the tips of the valve stems for valves, or valve pairs 1, 2, 5 and 7.

6. If the valve clearance is not within the specifications shown in the tune-up specifications chart, adjust the clearance as follows:

  a. Loosen the adjusting screw locknut of the valve needing adjustment.

  b. Pass the appropriately-sized feeler gauge between the adjusting screw and the valve stem tip. Tighten or loosen the adjusting screw until a slight drag can be felt on the feeler gauge.

➡**Valve lash adjusting tools (such as Suzuki Tool 09917–18210) can be purchased to help make adjusting the valve lash easier, but are by no means necessary for this procedure.**

  c. Once this drag is felt, tighten the locknut to 133–168 inch lbs. (15–19 Nm) while holding the adjusting screw to prevent it from turning. This may seem harder than it sounds and may take you several attempts to get the knack for it.

❊❊ WARNING

Although it is important to properly adjust the valve lash, it is better to have too loose of a valve lash clearance than too tight of a clearance. Too tight of a valve lash clearance may cause the valves to burn or prematurely wear, necessitating expensive engine repairs.

  d. Once the locknut is tightened, double-check the valve lash of the valve you just adjusted. If the valve lash changed

while tightening the locknut, loosen the locknut and readjust the lash. You can double-check the valve lash adjustment by using a feeler gauge that is 0.002 in. (0.05mm) bigger than specification. If this larger feeler gauge passes between the adjusting screw and valve stem tip with less than moderate effort, the adjustment is still too loose.

➡**A set of stepped feeler gauges can also be used to ensure that the clearance is not too great.**

7. Once you have the lash properly adjusted on valves (or valve pairs) 1, 2, 5 and 7. Have your assistant rotate the crankshaft another full revolution (360 degrees) until the timing marks are once again aligned. At this point, valves (or valve pairs) 3, 4, 6 and 8 can be adjusted by repeating substeps 7a through 7d for each of them.

8. Once all of the valves have been properly adjusted, install the cylinder head cover and distributor cap.

1.8L Engine

The 1.8L engine utilizes automatic hydraulic lash adjusters to maintain proper valve lash at all times. The valve lash for this engine is not manually adjustable. Therefore, periodic valve lash inspection and adjustment is not necessary or possible.

Oil Pan

REMOVAL & INSTALLATION

1.6L Engines

1. Raise and safely support the vehicle.

2. On 4-wheel drive models, remove the front differential from the vehicle.

3. On MFI engines, remove the Crankshaft Position (CKP) sensor.

❊❊ CAUTION

The EPA warns that prolonged contact with used engine oil may cause a number of skin disorders, including cancer! You should make every effort to minimize your exposure to used engine oil.

4. Drain the engine oil.

5. On MFI engines equipped with automatic transmissions, remove the left-hand side transmission brace.

6. Remove the clutch or torque converter inspection cover.

Do not use a prytool between the oil pan and engine block mating surfaces to separate the two components, otherwise the gasket surface may be damaged.

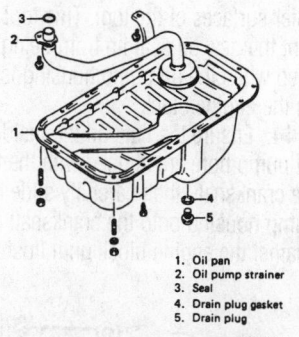

1. Oil pan
2. Oil pump strainer
3. Seal
4. Drain plug gasket
5. Drain plug

7924HG11

Exploded view of common oil pan and pump pick-up mounting—1.6L engines

7. Remove the oil pan mounting bolts and lower the oil pan from the engine. Remove the oil pick-up tube retaining bolt(s), then remove the tube from the engine.

➡**If removing the oil pan with the oil pick-up still attached to the oil pump is difficult, lower the oil pan as far as possible, detach the pick-up attaching bolts, then remove the pan with the pick-up from the vehicle.**

8. Thoroughly clean the engine block gasket surface, the oil pan and the pick-up tube of all oil, grime and old silicone sealant. Remove the old seal from the pick-up tube mounting hole.

To install:

9. Apply a continuous bead of silicone sealant (such as Suzuki Sealant 99000–31150) to the oil pan gasket surface.

10. Install a new pick-up seal into the tube mounting hole.

11. Raise the oil pan and oil pump pick-up into position. Install the pick-up tube, and tighten the retaining bolts to 80–106 inch lbs. (9–12 Nm).

12. Position the oil pan against the engine block and install all of the mounting bolts finger-tight.

13. Tighten the oil pan mounting bolts to 80–106 inch lbs. (9–12 Nm) by starting at the center of the oil pan and working your way toward both ends.

14. Install the oil pan drain plug, along with a new washer. Tighten the drain plug to 22–28 ft. lbs. (30–40 Nm).

15. Install the lower clutch or torque converter inspection cover.

16. If applicable, install the left-hand side transmission brace, and tighten the mounting bolts to 36 ft. lbs. (50 Nm).

17. If applicable, install the front differential assembly, then refill it.

18. If equipped, install the CKP sensor.

19. Lower the vehicle.

Wait at least 30 minutes after tightening the oil pan mounting bolts before adding oil to the engine. This time period will allow the silicone sealant to set properly.

20. Fill the engine with engine oil.

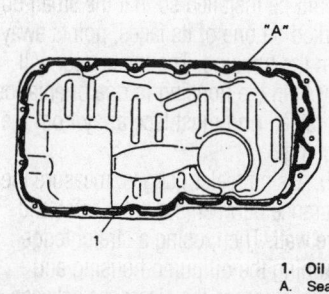

1. Oil pan
A. Sealant

7924HG79

Before installing the oil pan, apply a continuous bead of silicone sealant to the oil pan mating flange—all engines

1.8L Engine

1. Raise and safely support the vehicle.
2. Remove the engine oil dipstick and tube from the engine.
3. Remove the front differential, along with the tie rods, center link and idler arm.

The EPA warns that prolonged contact with used engine oil may cause a number of skin disorders, including cancer! You should make every effort to minimize your exposure to used engine oil.

4. Drain the engine oil.
5. Remove the clutch or torque converter inspection cover.

Do not use a prytool between the oil pan and engine block mating surfaces to separate the two components, otherwise the gasket surface may be damaged.

6. Remove the oil pan mounting bolts and lower the oil pan from the engine; tem-porarily support it on the vehicle cross-member. Using an open end wrench inserted between the lowered oil pan and the engine block, remove the oil pick-up tube retaining bolt(s). Separate the pick-up tube from the oil pump, then remove the tube and oil pan from the vehicle together.

7. Thoroughly clean the engine block gasket surface, the oil pan and the pick-up tube of all oil, grime and old silicone sealant. Remove the old seal from the pick-up tube mounting hole.

To install:

8. Apply a continuous bead of silicone sealant (such as Suzuki Sealant 99000–31150) to the oil pan mating surface.

9. Install a new pick-up seal into the tube mounting hole.

10. Raise the oil pan and oil pump pick-up into position. Install the pick-up tube, and tighten the retaining bolts to 80–106 inch lbs. (9–12 Nm).

11. Position the oil pan against the engine block and install all of the mounting bolts finger-tight.

12. Tighten the oil pan mounting bolts to 80–106 inch lbs. (9–12 Nm) by starting at the center of the oil pan and working your way toward both ends.

13. Install the oil pan drain plug, along with a new washer. Tighten the drain plug to 22–28 ft. lbs. (30–40 Nm).

14. Install the lower clutch or torque converter inspection cover and all other applicable components.

Wait at least 30 minutes after tightening the oil pan mounting bolts before adding oil to the engine. This time period will allow the silicone sealant to set properly.

15. Fill the engine with oil.
16. Start the engine and check for oil leaks.

Oil Pump

REMOVAL & INSTALLATION

1.6L Engines

1. Remove the timing belt and tensioner.
2. Remove the alternator and mounting bracket.

➡**Only MVAC-trained, EPA-certified automotive technicians should service the A/C system or its components.**

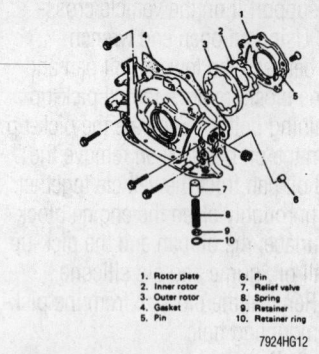

1. Rotor plate
2. Inner rotor
3. Outer rotor
4. Gasket
5. Pin
6. Pin
7. Relief valve
8. Spring
9. Retainer
10. Retainer ring

7924HG12

Exploded view of the oil pump housing—1.6L engines

3. If equipped, remove the air conditioning compressor bracket bolts and support the compressor out of the way. Do NOT disconnect any of the A/C refrigerant lines.

4. Raise and safely support the vehicle .

✳✳ CAUTION

The EPA warns that prolonged contact with used engine oil may cause a number of skin disorders, including cancer! You should make every effort to minimize your exposure to used engine oil.

5. Drain the engine and front differential oil.

6. Remove the lower transmission (clutch or torque converter) inspection cover.

7. Remove the crankshaft timing belt sprocket by loosening the center bolt, while preventing the crankshaft from rotating. To hold the crankshaft from turning, you can use Suzuki Tool 09927–56010 (or equivalent), or a large prybar inserted in the transmission housing slot and the flywheel teeth.

8. Remove the oil pan and oil pump pick-up.

9. Loosen the seven mounting bolts, and remove the oil pump housing. If the housing is difficult to separate from the engine block, use a small prytool against one of the mounting bolt bosses—do NOT slide the prybar between the oil pump housing and engine block mating surfaces.

10. If the oil pump is to be inspected, disassemble and inspect it as follows:

a. Remove the oil level dipstick tube bracket bolt, then pull the dipstick and tube out of the oil pump housing.

b. Remove the oil pump rotor plate by loosening the five retaining screws. The rotor plate retaining screws can be difficult to remove, be sure not to strip the Philips grooves in the heads of the screws, otherwise the screw will need to be drilled out of the housing.

c. Remove the outer and inner rotors from the housing.

d. Without scratching the oil seal bore surface, use a small punch to drive the oil seal, or a seal removal tool to pry the oil seal, from the housing.

e. Remove the oil pressure relief retaining snap-ring from the housing bore, then remove the retainer cap, spring and piston from the housing. Ensure that the piston bore is thoroughly cleaned when cleaning the housing.

f. Clean the oil pump housing, rotors and pressure regulating components until free of all dirt and oil.

g. Install the inner and outer rotors in the oil housing bore. The outer rotor should be installed so that the small dot, marked on one of its faces, points away from the housing. The inner rotor will only fit in the housing in one orientation: the raised ring must face away from the housing.

h. Using feeler gauges, measure the clearance between the outer rotor and bore wall. Then, using a straightedge spanning the oil pump housing and rotors, measure the clearance between the edge of the straightedge and oil pump rotors. Compare your findings with the engine rebuilding specification charts at the end of this section.

i. If the measurements were not within the specified ranges, or one or more oil pump components shows evidence of excessive wear or damage, the oil pump components should be replaced.

To install:

11. If the oil pump was disassembled for inspection, assemble it as follows:

a. Using an oil seal installer, or an aptly-sized socket, drive the new oil seal into the seal bore. Ensure that the seal is driven in straight and is fully seated in the housing.

b. Apply a thin coat of clean engine oil to the oil pump rotors, the oil seal lip, the inside surfaces of the oil pump housing (including the regulator bore), and to the oil pressure piston (relief valve).

c. Install the relief valve, the spring, the retainer cap and a new snap-ring in the housing bore.

d. Install the inner and outer pump rotors in the housing.

e. Position the rotor plate on the housing, then install and tighten the five retaining screws securely. After tightening the screws, ensure that the oil pump rotors spin smoothly by hand.

f. Install a new oil level dipstick tube seal in the housing, then insert the end

of the tube in the housing. Install the tube bracket bolt and tighten it securely.

12. Ensure the two oil pump locating pins are installed in the engine block, then install a new oil pump housing-to-engine block gasket.

13. Install Suzuki Tool 09926–18210, or its equivalent, onto the crankshaft. Apply a thin coating of clean engine oil onto the outer surfaces of the tool. This tool will prevent the new oil seal lip from being damaged when the oil pump housing is installed on the engine block.

14. Ensure the flats on the inside of the oil pump bore are aligned with the flats on the crankshaft, then carefully slide the oil pump housing onto the crankshaft and against the engine block until flush. Install

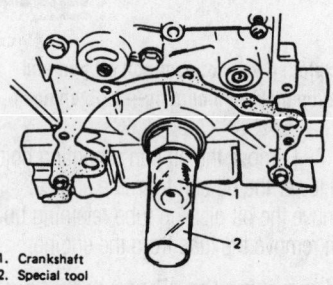

1. Crankshaft
2. Special tool
(Oil seal guide (Vinyl resin) 09926-18210)

7924HG78

Use the special tool when installing the oil pump so that the new oil seal will not be damaged

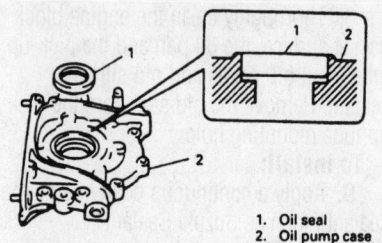

1. Oil seal
2. Oil pump case

7924HG13

Install the new oil pump seal until flush with the outer pump case—1.6L engines

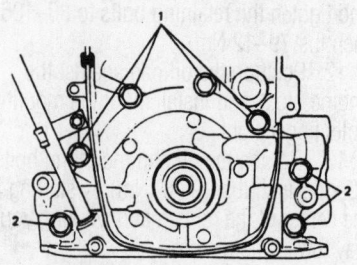

1. No. 1 bolts (short)
2. No. 2 bolts (long)

7924HG14

Oil pump housing short (1) and long bolt (2) locations—1.6L engines

the oil pump housing mounting bolts and tighten them to 97 inch lbs. (11 Nm). When installing the oil pump mounting bolts, be sure to install the shorter oil pump bolts as indicated in the accompanying illustration. The three other bolts are slightly longer and should be installed in the lower holes.

15. Install a new rubber seal between the oil pump and the water pump.

16. Inspect the lower oil pump edge to ensure that the gasket is not protruding past the surface. If the gasket protrudes from the oil pump, cut it flush with a sharp knife.

17. Install the remaining applicable components.

➡**Be sure to adjust the valve lash on all valves, then refill the engine with the proper type and amount of engine oil.**

18. After filling the engine with engine oil, start the vehicle and inspect the oil pressure as follows:

a. Remove the oil pressure switch from the engine block. The switch is located next to the exhaust manifold.

b. Install an oil pressure gauge in the oil pressure switch threaded hole. Ensure that your oil pressure gauge hose is equipped with the same thread pitch and size as the oil pressure switch, otherwise damage to the oil pressure switch hole will result.

c. Start the engine and allow it to warm up to normal operating temperature.

d. Place the transmission in Neutral (manual models) or Park (automatic models), apply the parking brake and block the drive wheels.

e. Raise the engine speed to 3000 rpm and read the value indicated by the oil pressure gauge. The oil pressure should match the values presented in the engine rebuilding specifications charts at the end of this section. If the oil pressure is not as indicated, there is a defect in the engine lubrication system.

f. After inspecting the oil pressure, stop the engine and remove the oil pressure gauge.

g. Install the oil pressure switch, making sure to wrap its threads with Teflon® sealing tape. Tighten the switch to 124 inch lbs. (14 Nm). Cut off any exposed Teflon® tape.

19. Start the engine and inspect the oil pan and pressure switch for leaks.

1.8L Engine

1. Remove the oil pan and oil pump pick-up.

2. Remove the oil pump sprocket cover mounting bolts, then separate the cover from the engine.

➡**When separating the pump from the engine, do not loose the small locating dowels; they will be needed upon installation.**

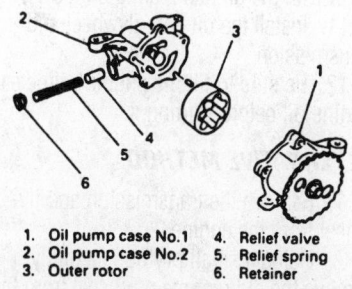

1. Oil pump case No.1
2. Oil pump case No.2
3. Outer rotor
4. Relief valve
5. Relief spring
6. Retainer

7924HG15

Exploded view of oil pump and sprocket cover mounting—1.8L engine

3. Remove the oil pump mounting bolts, and remove the oil pump from the lower crankcase. If the oil pump is difficult to separate from the engine, tap it lightly with a soft-faced mallet until it works loose.

⁕⁕ WARNING

Do NOT remove the sprocket from the oil pump, otherwise damage to the oil pump center shaft and abnormal pump operation may be the result.

4. Clean the oil pump-to-engine mating surface thoroughly.

5. If the oil pump is to be inspected, disassemble and inspect it as follows:

a. Separate the two oil pump case halves by loosening the five attaching screws.

b. Remove the outer and inner rotors from the housing.

c. Remove the oil pressure relief retainer from the housing, then remove the spring and piston (relief valve) from the housing. Ensure that the piston bore is thoroughly cleaned when cleaning the housing.

d. Clean the oil pump housing, rotors and pressure relief components until free of all dirt and oil.

e. Install the inner and outer rotors in the oil housing bore.

f. Using feeler gauges, measure the clearance between the outer rotor and bore wall. Then, using a straightedge spanning the oil pump housing and rotors, measure the clearance between the edge of the straightedge and oil

pump rotors. Compare your findings with the engine rebuilding specification charts at the end of this section.

g. If the measurements were not within the specified ranges, or one or more oil pump components shows evidence of excessive wear or damage, the oil pump should be replaced.

To install:

6. If the oil pump was disassembled for inspection, assemble it as follows:

a. Apply a thin coat of clean engine oil to the oil pump rotors, the inside surfaces of the oil pump housing (including the relief bore), and to the oil pressure piston (relief valve).

b. Install the relief valve, the spring, and the retainer in the housing bore.

c. Install the inner and outer pump rotors in the housing.

d. Assemble the two oil pump halves, then install and tighten the five retaining screws to 106 inch lbs. (12 Nm). After tightening the screws, ensure that the oil pump gear spins smoothly by hand.

⁕⁕ WARNING

When installing the oil pump, do not allow the locating dowels to fall out of position.

7. Ensure the locating dowels are installed in the pump, then position the oil pump sprocket in the drive chain and the pump housing against the lower crankcase. Install and tighten the mounting bolts to 177 inch lbs. (20 Nm).

8. Install the oil pump sprocket cover and tighten the mounting bolts to 97 inch lbs. (11 Nm).

9. Install the oil pump pick-up and oil pan.

10. Refill the engine with oil.

11. After filling the engine with engine oil, start the vehicle and inspect the oil pressure as follows:

a. Remove the oil pressure switch from the engine block. The switch is located on the left-hand side of the engine, near the flywheel.

b. Install an oil pressure gauge in the oil pressure switch threaded hole. Ensure that your oil pressure gauge hose is equipped with the same threads as the oil pressure switch, otherwise damage to the oil pressure switch hole will result.

c. Start the engine and allow it to warm up to normal operating temperature.

d. Place the transmission in Neutral (manual models) or Park (automatic models), apply the parking brake and block the drive wheels.

e. Raise the engine speed to 4000 rpm and read the value indicated by the oil pressure gauge. The oil pressure should be 55–66 psi (390–470 kPa). If the oil pressure is not as indicated, there is a defect in the engine lubrication system.

f. After inspecting the oil pressure, stop the engine and remove the oil pressure gauge.

g. Install the oil pressure switch, making sure to wrap its threads with Teflon® sealing tape. Tighten the switch to 124 inch lbs. (14 Nm). Cut off any exposed Teflon® tape.

h. Start the engine and inspect for leaks.

Rear Main Seal

REMOVAL & INSTALLATION

1.6L Engines

RECOMMENDED METHOD

It is recommended to replace the oil seal by dropping the oil pan and removing the entire rear main seal housing. It is possible to remove the oil seal without doing this, but it is often difficult because the oil seal is extremely stiff and can easily cock in the bore. If you wish to try to replace the seal without removing the oil pan, refer to the alternative method.

1. Remove the transmission, the flywheel and the oil pan from the vehicle.

2. Loosen the five rear main seal housing mounting bolts, then separate the housing from the back of the engine block. Discard the old gasket.

3. Position the housing on wooden blocks so that the outer side is facing down and the seal is not being supported. Use a blunt drift and hammer to drive the old seal out of the housing.

4. Clean the rear main seal housing and engine block mating surfaces of all dirt, grime, oil and old gasket material.

To install:

5. Position the rear main seal housing with the outer surface facing up, and on a flat, clean piece of wood.

6. Position the new oil seal in the seal housing bore and using a seal driver, or a large flat block of wood, and a hammer seat the seal in the housing until the outer edge of the seal is flush with the outer surface of the housing.

7. Apply a thin coat of clean engine oil to the oil seal lip.

8. Along with a new gasket, position the oil seal housing against the engine block, and install the mounting bolts finger-tight.

9. Tighten the housing mounting bolts to 80–106 inch lbs. (9–12 Nm).

10. After the oil seal housing mounting bolts are tightened, the gasket may protrude out from the bottom of the housing. Use a sharp knife or razor blade to trim the gasket flush with the oil pan mating surface.

11. Install the oil pan, flywheel and transmission.

12. Be sure to fill the engine with clean engine oil before starting it.

ALTERNATIVE METHOD

1. Remove the transmission and flywheel from the engine.

2. Using a small prytool, carefully remove the old rear main oil seal from the oil seal housing. Be careful not to scratch or gouge the oil seal bore.

To install:

3. Apply a thin coat of clean engine oil to the new oil seal lip.

4. Position the new oil seal in the seal housing bore and using a seal driver and a hammer seat the seal in the housing until the outer edge of the seal is flush with the outer surface of the housing.

5. Install the flywheel and transmission.

1.8L Engine

1. Remove the transmission and flywheel from the engine.

2. Using a small prytool, carefully remove the old rear main oil seal from the oil seal housing. Be careful not to scratch or gouge the oil seal bore.

To install:

3. Apply a thin coat of clean engine oil to the new oil seal lip.

4. Using Suzuki Tools 09911–97710 and 09911–97810, or their equivalents (essentially drive adapters), install the new oil seal in the seal housing bore. Seat the seal in the housing until the outer edge of the seal is flush with the outer surface of the housing.

5. Install the flywheel and transmission.

Timing Chain, Sprockets, Front Cover and Seal

REMOVAL & INSTALLATION

1.8L Engine

✳✳ CAUTION

The EPA warns that prolonged contact with used engine oil may cause a
number of skin disorders, including cancer! You should make every effort to minimize your exposure to used engine oil.

1. Drain the engine oil.

✳✳ CAUTION

Never open, service or drain the radiator or cooling system when hot; serious burns can occur from the steam and hot coolant.

2. Drain the engine coolant.

3. Remove the oil pan and oil pump pick-up.

4. Remove the rocker arm cover.

5. Remove the water bypass pipe and bypass hose No. 2.

6. Remove the cooling fan, pulley and accessory drive belt.

7. Remove the alternator drive belt.

8. Remove the water pump pulley.

9. Remove the alternator drive belt tensioner and idler pulley.

10. Detach the upper radiator hose from the thermostat housing.

11. If equipped, without disconnecting the A/C hoses, separate the A/C compressor from the compressor mounting bracket. Position the compressor aside and secure it with strong cord or wire.

12. If equipped, remove the A/C compressor mounting bracket from the engine.

13. Using a long breaker bar and socket, remove the crankshaft pulley bolt. To secure the crankshaft from rotating, install a pulley holding tool (such as Suzuki Tool 09917–68221). Use M8, P1.25 bolts with a strength rating of at least 7T to attach the holding tool to the pulley.

14. Once the bolt is removed, install a steering wheel puller on the crankshaft drive belt pulley, then draw the pulley off of the end of the crankshaft.

15. Remove the timing chain cover retaining fasteners, then pull the cover off of the engine and cylinder head.

16. Clean the gasket mating surfaces on the timing chain cover, engine block, and cylinder head of all oil, dirt and old sealant material.

✳✳ WARNING

Take care not to damage the oil seal bore during removal, otherwise oil leakage may occur.

17. Remove the crankshaft oil seal from the cover by using a seal prytool, or by driving it out of the cover with a drift and hammer.

→All of the descriptions in the following step are as if you are looking at the front of the engine.

18. Rotate the crankshaft clockwise until either the timing chain gear-to-engine alignment marks are positioned as shown, or until the crankshaft gear, idler gear, camshaft gear and timing chain colored links are positioned as follows (both methods will position the engine properly):

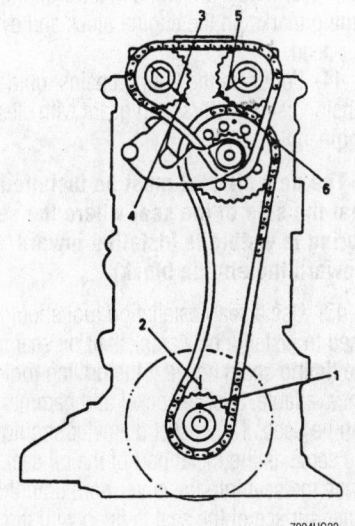

7924HG22

Before removing the timing chains ensure that all gear-to-engine alignment marks are positioned, as shown

- Right-hand camshaft gear matchmark and colored link—between 2 and 3 o'clock position
- Left-hand camshaft gear matchmark and colored link—11 o'clock position
- Idler gear inner timing chain matchmark and colored link—12 o'clock position
- Idler gear outer timing chain matchmark and colored link—6 o'clock position
- Crankshaft gear matchmark and colored link—6 o'clock position

→Ensuring the crankshaft, idler gear and camshafts are correctly positioned is vital for timing chain installation.

19. Remove the outer timing chain tension adjuster mounting bolts, then remove the adjuster from the cylinder head by rotating the intake camshaft counterclockwise a little while depressing the adjuster contact pad.

20. Hold the intake camshaft steady by using an open end wrench on the hexagonal section of the camshaft, then loosen the camshaft gear retaining bolt. Loosen the exhaust camshaft gear retaining bolt in the same manner.

21. Remove the camshaft gears from the ends of the camshafts, then lift the gears and chain up and off of the engine.

※※ WARNING

Do not rotate the crankshaft or camshafts once the outer timing chain is removed from the engine, otherwise damage to the pistons and/or valves may occur.

22. Remove the inner timing chain guide by loosening the mounting fasteners and separating it from the engine block.

23. Remove the inner timing chain tension adjuster and contact arm (tensioner) from the engine block.

24. Remove the timing chain idler pulley from the cylinder head, drop the idler gear and inner timing chain down to disengage it from the crankshaft gear, then remove the inner timing chain and idler pulley from the engine.

25. Slide the crankshaft timing chain gear off of the end of the crankshaft.

26. Clean all components of all dirt and oil.

To install:

27. Inspect the following timing chain components for wear and/or damage. Replace any items found to be worn, defective or damaged:

- Inner and outer timing chain guide contact surfaces
- Inner and outer timing chain tensioner contact surfaces
- Crankshaft and camshaft timing chain gear teeth and bushings
- Idler sprocket teeth and bushing
- Inner and outer timing chains

28. Ensure that the latch and tooth sur-

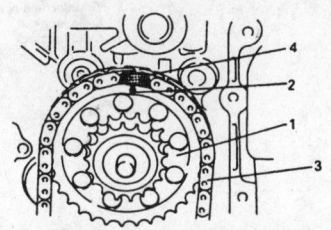

1. Idler sprocket
2. Match mark on idler sprocket
3. 1st timing chain
4. Dark blue plate

7924HG16

Install the inner timing chain so that the dark blue chain link is aligned with the matchmark on the idler gear . . .

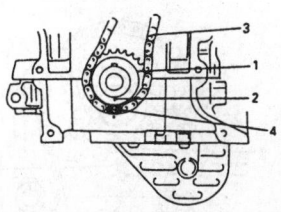

1. Crankshaft timing sprocket
2. Match mark
3. 1st timing chain
4. Yellow plate

7924HG17

. . . and the yellow link is aligned with the matchmark on the crankshaft gear

faces of the inner timing chain tensioner are free from damage and that the latch functions properly.

29. Ensure that the crankshaft timing sprocket key is aligned with the timing mark on the engine block. If the engine was assembled so that the crankshaft is not properly positioned, remove the hydraulic valve lash adjusters so that the crankshaft may be rotated to the proper position.

30. Slide the crankshaft timing gear onto the end of the crankshaft so that the key in the crankshaft is aligned with the slot in the gear.

31. Apply clean engine oil to the idler gear bushing, then drape the inner timing chain on the idler gear so that the dark blue chain piece is aligned with the matchmark on the idler gear. Then, position the timing chain around and under the crankshaft gear

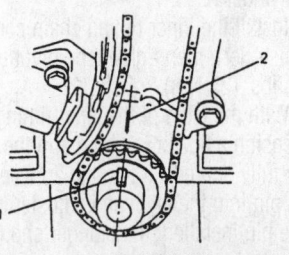

1. Crank timing sprocket key
2. Timing mark

7924HG18

To install the outer chain, ensure that the crankshaft gear is still aligned with the mark on the engine block

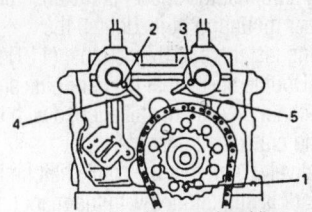

1. Arrow mark on idler sprocket
2. Knock pin of intake camshaft
3. Knock pin of exhaust camsaft
4. Timing mark of intake side
5. Timing mark of exhaust side

7924HG19

Then, position the camshafts so that they are oriented as shown

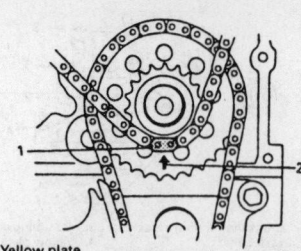

1. Yellow plate
2. Match mark of 2nd timing chain (Arrow mark)

7924HG20

Install the outer chain on the idler gear so that the yellow link is aligned with the arrow mark

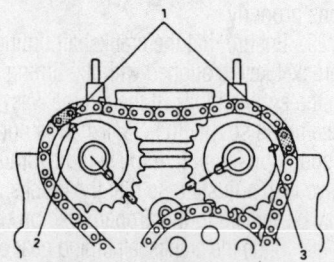

1. Dark blue
2. Arrow mark on intake camshaft timing sprocket
3. Arrow mark on exhaust camshaft timing sprocket

7924HG21

The outer timing chain must be installed on the camshaft gears as indicated

so that the yellow chain piece is aligned with the matchmark on the crankshaft gear. Install the idler sprocket and sprocket shaft in the cylinder head.

32. Install the inner timing chain contact arm (tensioner), then tighten the pivot bolt to 19 ft. lbs. (25 Nm).

33. With the latch of the inner timing chain tension adjuster returned and the plunger fully depressed into the adjuster, insert a pin into the latch and adjuster body. With the pin installed, the plunger should not come out of the adjuster body.

34. Install the inner timing chain tension adjuster onto the engine block, then tighten the mounting bolts to 97 inch lbs. (11 Nm). Pull the pin out of the adjuster.

35. Install the inner timing chain guide, making sure that the spacer is installed on the upper mounting bolt. Tighten the mounting fasteners to 97 inch lbs. (11 Nm).

36. Double check that the dark blue and yellow chain pieces are still aligned with the idler and crankshaft gear matchmarks (respectively), and that the crankshaft timing gear mark is still aligned with the mark on the engine block..

37. Ensure that the locating dowels of the intake and exhaust camshafts are aligned with the timing marks on the cylinder head, as shown in the accompanying illustration.

38. Hold both of the camshaft gears and have an assistant drape the outer timing chain over the gears so that the blue chain pieces are aligned with the matchmarks on the gears. It may be necessary to turn the chain around if the blue marks are too far apart. Position the gears and chain so that the chain is engaged on the underside of the idler gear, and so that the yellow chain link is aligned with the mark (arrow) on the idler sprocket. Once the chain is properly positioned with respect to all three gears, install the intake and exhaust camshaft gears on their respective camshafts.

➡**The camshaft sprockets do not have a specific direction of orientation; either side of both gears may be positioned so that it faces away from the camshaft.**

39. Install the camshaft gear retaining bolts finger-tight, then secure the camshafts (one at a time) with an open end wrench. Tighten both gear bolts to 44 ft. lbs. (60 Nm).

40. With the plunger of the outer timing chain tension adjuster fully depressed into the adjuster, insert a push-pin into the latch and adjuster body. With the push-pin

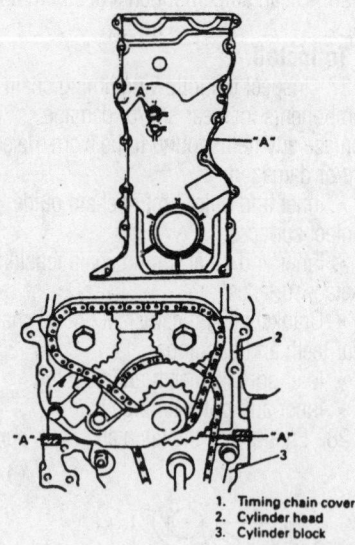

1. Timing chain cover
2. Cylinder head
3. Cylinder block

7924HG10

Prior to installing the timing chain cover on the engine block and cylinder head, apply silicone sealant to the cover as indicated (areas marked A)

installed, the plunger should not come out of the adjuster body.

41. Install the outer timing chain tension adjuster, along with a new gasket, onto the cylinder head. Tighten the mounting bolts to 97 inch lbs. (11 Nm) and the idler gear nut to 33 ft. lbs. (45 Nm).

42. Pull the push-pin out of the adjuster body to tension the outer timing chain.

43. Rotate the crankshaft clockwise two full revolutions, and realign the crankshaft gear timing mark with the mark on the engine block. Ensure that the camshaft and idler gear marks are still aligned with the timing marks on the engine block and cylinder head.

44. Lubricate the timing chains, guides, tensioners, adjusters, and gears with clean engine oil.

➡**The new oil seal must be installed so that the side of the seal where the seal spring is visible is installed inward (toward the engine block).**

45. Use a seal installation tool should be used to install a new crankshaft oil seal in the timing chain cover. If the driving tool is not available, a large socket and hammer can be used. The socket diameter should be the same as the metal part of the oil seal. Drive the seal into the cover bore until the outer surface of the seal is flush with the timing chain cover.

46. Apply silicone sealant, such as Suzuki sealant 99000–31150, to the timing chain cover on the cover-to-engine mating surface. Refer to the accompanying illustration for sealant application.

47. Ensure that the timing chain cover locating dowel is installed in the engine block. Apply a thin coat of clean engine oil to the crankshaft oil seal lip, then install the cover. Tighten the cover retaining bolts to 97 inch lbs. (11 Nm).

48. Slide the crankshaft pulley onto the end of the crankshaft and install the center bolt. Install the crankshaft holding tool, then tighten the center bolt to 109 ft. lbs. (150 Nm). The center bolt will draw the pulley onto the crankshaft when tightened to the specified torque value.

49. Reinstall all applicable components in the reverse order of removal.

50. Refill the cooling system, front differential and the engine.

❊❊ WARNING

Operating the engine without the proper amount and type of engine oil can result in severe engine damage.

FUEL SYSTEM

Fuel System Service Precautions

Safety is the most important factor when performing not only fuel system maintenance but any type of maintenance. Failure to conduct maintenance and repairs in a safe manner may result in serious personal injury or death. Maintenance and testing of the vehicle's fuel system components can be accomplished safely and effectively by adhering to the following rules and guidelines.

• To avoid the possibility of fire and personal injury, always disconnect the negative battery cable unless the repair or test procedure requires that battery voltage be applied.

• Always relieve the fuel system pressure prior to disconnecting any fuel system component (injector, fuel rail, pressure regulator, etc.), fitting or fuel line connection. Exercise extreme caution whenever relieving fuel system pressure to avoid exposing skin, face and eyes to fuel spray. Please be advised that fuel under pressure may penetrate the skin or any part of the body that it contacts.

• Always place a shop towel or cloth around the fitting or connection prior to loosening to absorb any excess fuel due to spillage. Ensure that all fuel spillage (should it occur) is quickly removed from engine surfaces. Ensure that all fuel soaked cloths or towels are deposited into a suitable waste container.

• Always keep a dry chemical (Class B) fire extinguisher near the work area.

• Do not allow fuel spray or fuel vapors to come into contact with a spark or open flame.

• Always use a backup wrench when loosening and tightening fuel line connection fittings. This will prevent unnecessary stress and torsion to fuel line piping. Always follow the proper torque specifications.

• Always replace worn fuel fitting O-rings with new. Do not substitute fuel hose or equivalent, where fuel pipe is installed.

Fuel System Pressure

RELIEVING

✴✴ CAUTION

This procedure can NOT be performed when the engine is hot. If done so, it may cause an adverse effect to the catalytic converter, or create a dangerous, explosive condition.

1. Place the transmission gearshift lever in Neutral (manual transmissions) or Park (automatic transmissions), apply the parking brake, and block the drive wheels.

2. Detach the wiring harness connector from the fuel pump relay, located under the left-hand side of the instrument panel near the ECM.

3. Remove the fuel tank filler cap to release fuel vapor pressure in the fuel tank. Reinstall the filler cap.

4. Start the engine and run it until it stops from lack of fuel. Crank the engine 2 or 3 times for a three second period. The fuel lines should now be depressurized.

5. After servicing the component(s), reattach the wiring harness connector to the fuel pump relay.

Fuel Filter

REMOVAL & INSTALLATION

✴✴ CAUTION

This procedure must not be done when the engine is hot. Allow the engine to cool down completely prior to performing this procedure. Eye goggles and gloves should be worn to prevent direct contact with gasoline, which is toxic.

➡**New fuel filter gaskets (which should. be included along with a new filter) are necessary for this procedure.**

The fuel filter is located in front of the fuel tank on the right-hand underside of the vehicle's chassis on carbureted models.

➡**The fuel filter is not designed to be disassembled. The entire assembly is replaced when necessary.**

1. If necessary, allow the engine to cool down until it is completely cold.

2. On fuel-injected vehicles, relieve fuel system pressure, as described in Section 5.

3. Disconnect the negative battery cable.

4. Remove the fuel tank filler cap to release any built-up fuel tank pressure, then reinstall the fuel tank filler cap.

5. Raise and safely support the vehicle .

6. Along with a small drain pan and plenty of clean shop rags, crawl under the right-side of the vehicle between the front and rear wheels. Position the drain pan under the fuel filter to catch any fuel in the lines. If necessary, use the rags to soak up any spilled gasoline.

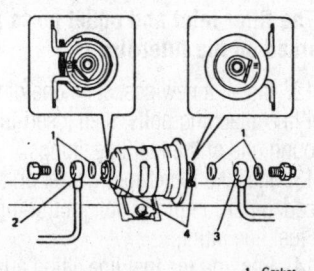

1. Gasket
2. Outlet pipe
3. Inlet pipe
4. Recess

7924HG34

Exploded view of the fuel line mounting on the fuel filter

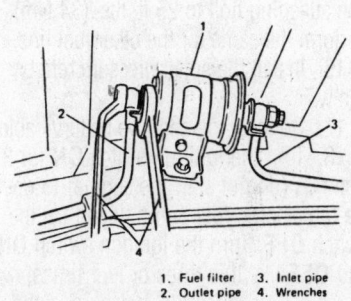

1. Fuel filter
2. Outlet pipe
3. Inlet pipe
4. Wrenches

7924HG33

Whenever loosening or tightening the fuel lines at the filter, always use a backup wrench on the fittings to prevent damage

7. Remove the small pressure relief screw from the end of the filter (to release any residual fuel pressure in the lines). Position an aptly-sized open end wrench on the fuel filter flange and a box end wench on the fuel line fitting attaching bolt. Use the open end wrench to hold the filter steady while loosening the line bolt with the box end wrench. Disconnect the other fuel line from the filter in the same manner. Once the lines are detached from the filter, pour any residual fuel into the catch pan.

➡**Note the position of the fuel filter inlet and outlet nipples before removing it from the vehicle.**

8. Remove the lower clamp bolt, then loosen the clamp adjusting bolt until the filter can be removed from the mounting clamp.

9. Pull the fuel filter out of the mounting clamp.

To install:

10. Slide the new filter into the mounting clamp. Tighten the adjusting bolt until snug, then turn the filter in the clamp until the inlet and outlet nipples are positioned as before removal, then tighten the adjusting bolt securely.

11. Install the lower clamp bolt and tighten it securely.

➡ **The filter inlet and outlet ports are marked on the filter itself.**

12. Install a new gasket on one of the fuel line attaching bolts, then insert the bolt through one of the fuel line fittings.

13. Install another new gasket on the threaded portion of the bolt protruding from the fuel line fitting.

14. Position the fuel line fitting against the fuel filter so that the fuel line is positioned between the two arms of the line bracket, and thread the attaching bolt by hand.

15. Using the open end wrench and a torque wrench and socket, tighten the fuel line attaching bolt to 25 ft. lbs. (34 Nm). Perform the same for the other fuel line.

16. Install the small pressure relief screw.

17. Connect the negative battery cable.

18. Turn the ignition switch **ON** for 3 seconds (do not start the engine) to operate the electric fuel pump, then turn the switch **OFF**. Turn the ignition switch **ON** and **OFF** like this three or four times, which will produce fuel pressure in the lines. Inspect the fuel filter and lines for gasoline leaks.

19. Lower the vehicle.

Fuel Pump

REMOVAL & INSTALLATION

➡ **This is a general procedure and slight differences may be applicable for your vehicle. Change the procedure as necessary.**

1. Release the fuel system pressure.
2. Disconnect the negative battery cable.
3. Remove the fuel tank filler cap, then reinstall it.
4. Raise and safely support the vehicle securely .

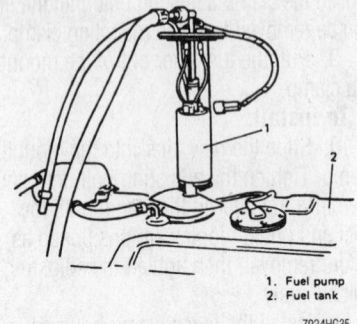

1. Fuel pump
2. Fuel tank

7924HG35

The fuel pump is mounted in the top of the fuel tank

5. Detach and label the fuel level gauge sending unit and fuel pump wires.

6. Due to the lack of a drain plug, drain the fuel from the tank through the filler tube with a pump designed for this purpose.

7. Remove the upper filler hose protector, then detach the breather hose from the filler neck and the vapor hose from the separator.

8. Remove the lower filler hose protector, then disconnect the filler hose from the fuel tank.

9. Detach the fuel filter inlet pipe from the filter.

10. Detach the fuel line and pipe clamps from the chassis, then remove the fuel filter inlet pipe from the clamp.

11. Position a floor jack beneath the fuel tank, then remove the fuel tank mounting fasteners. Slowly lower the fuel tank slightly until the fuel supply and return lines can be disconnected from the fuel tank, then lower the tank completely from the vehicle.

12. Loosen the fuel pump retaining bolts, then lift the fuel pump up and out of the fuel tank.

13. Remove and discard the old fuel pump gasket.

To install:

14. Position the fuel pump, along with a new gasket, in the fuel tank, ensuring that the fuel pump retaining bolt holes are aligned.

15. Install and tighten the fuel pump retaining bolts securely and evenly.

16. Position the fuel tank on the floor jack, then slowly raise it up into position beneath the vehicle. Reattach the fuel supply and return lines to the fuel tank, then raise the fuel tank fully. Install the mounting fasteners and tighten them to 36.5 ft. lbs. (50 Nm).

17. Install the fuel filter inlet pipe to the clamp, then reattach the fuel line and pipe clamps to the chassis.

18. Reattach the fuel filter inlet pipe to the filter, ensuring that you use new gaskets. Tighten the fuel filter union bolt to 22–28 ft. lbs. (30–40 Nm).

19. Reattach the filler hose to the fuel tank, then install the lower filler hose protector.

20. Reconnect the breather hose to the filler neck, and the vapor hose to the separator. Install the upper filler hose protector.

21. Reattach the fuel level gauge sending unit and fuel pump wires.

22. Lower the vehicle.

23. Connect the negative battery cable.

24. Turn the ignition switch **ON** for 3 seconds (do not start the engine) to operate the electric fuel pump, then turn the switch **OFF**. Turn the ignition switch **ON** and **OFF** like this three or four times, which will pro-

duce fuel pressure in the lines. Inspect the fuel tank and associated lines for fuel leaks.

DRIVE TRAIN

Transmission Assembly

REMOVAL & INSTALLATION

On 4WD models, the transmission and transfer case are removed as an assembly, since the transfer case is bolted directly to the rear of the transmission and takes the place of an extension housing.

1. Disconnect the negative battery cable for safety.
2. From inside the passenger compartment, remove the transmission shift handle.
3. On 4WD vehicles, remove the transfer case shift handle from inside the passenger compartment.

➡ **Although removal of the transfer case shift handle is recommended, it is not always necessary. If you would like to leave the handle in place, remove the center console for access, then remove the shift knob (which usually means unthreading the setscrew to free the knob), boot and boot cover. But, be very careful not to damage the shift lever when lowering the transmission/transfer case assembly from the vehicle. If the lever is left in place, it may be necessary to shift the transfer case lever to different positions while lowering or raising the assembly.**

4. For 3-speed automatic models, perform the following:

   a. Free the transmission breather hose from the clamp at the rear of the cylinder head.

   b. Bend back the rubber-coated metal clamp usually found at the rear of the intake manifold to free up the wiring harness. Then, disconnect the harness couplers.

   c. Disconnect the kick-down cable at the throttle body.

   d. Remove the vacuum modulator hose at the intake manifold.

5. For 4-speed automatic models, perform the following:

   a. Remove the battery, dipstick and oil filler tube.

   b. Disconnect the throttle cable from the throttle cam and bracket.

   c. Tag and disconnect the wiring harness couplers.

6. For manual models, perform the following:

    a. Free the transmission breather hose from the clamp at the rear of the cylinder head.

    b. Bend back the rubber-coated metal clamp at the rear of the intake manifold to free up the wiring harness. Then, disconnect the harness coupler.

7. Tag and disconnect the lead wires from the starter motor which is mounted to the transmission bell-housing through an engine mounted flange. This step can be avoided, if the starter motor is supported to hold its position relative to the engine and the engine mounted flange which sits between the engine and transmission bell-housing. You'll have to use your judgment here, but if you try this, be sure that the starter is properly supported (so as not to stress or damage any of the wiring). This can be done with a combination of a large block of wood below the starter motor, leaving the upper starter mounting bolt in place (after removing the nut) and using a few creative wire ties or bungee cords.

➡**If you decide to remove the starter, you will need an ignition wrench or a creative combination of ¼ in. drive tools to reach the wiring retainers.**

8. Remove the starter mounting bolts and remove it completely from the vehicle, or remove the lower mounting bolt and remove the nut from the transmission side of the upper mounting bolt, then secure the starter in position so the wiring will not be damaged.

9. Remove the 2 upper transmission-to-engine mounting bolts. Unfortunately, this is another tight spot. You will either need a large breaker bar with a very short socket or a large combination wrench with a slight offset to really get at the bolt on the driver's side of the vehicle.

➡**On some automatic models (including all 4-speed transmissions) the right side upper transmission-to-engine bolt is longer. This may make the bolt somewhat more difficult to remove, and be sure not to mix it up with the shorter bolt on the opposite side come installation time.**

10. Unbolt the fan shroud (usually 4 bolts) and hang it from the front of the engine. This will allow the engine to pivot slightly on the motor mounts once the transmission is removed, without jeopardizing the cooling fan.

11. Raise and support the vehicle safely at a height which will be convenient to work from both above and below the vehicle.

➡**Later in this procedure, the exhaust center pipe must be removed for clearance. If you do not have air tools (which make exhaust fastener removal MUCH easier) take a moment now to spray the exhaust center pipe fasteners with penetrating oil to help loosen them. Spray both the 3 nuts and studs at the exhaust manifold and the 2 through-bolts at the rear of the converter.**

12. If equipped, remove the front skid plate for better access.

13. On 2WD vehicles, either drain the oil from the transmission case or have a transmission case plug handy for the extension housing (some aftermarket tool companies like Lisle® make plastic transmission plugs for just this purpose. If a plug is not available, a large plastic bag can be stretched across the extension housing and secured with a rubber band. This second method will catch some fluid, but if the transmission is left with the rear downward for any length of time you will wind up with smelly gear oil on the garage floor or in the driveway.

14. On 4WD vehicles, you do not have to drain the transmission, but it is probably smartest to drain the transfer case. If you are really adamant about not draining either, you've got 2 options. Either buy 2 transmission plugs that will fit where the front and rear driveshaft slip-yokes go or buy 1 plug and leave the front driveshaft in position, just unbolted at the front differential. Both have the potential to be a pain and to be really messy, but it's your call.

15. Matchmark and remove the rear driveshaft from the vehicle.

16. On 4WD vehicles, matchmark and remove the front driveshaft between the transfer case and the front differential.

17. For automatic models, perform the following:

    a. Disconnect the gear select cable from the transmission by removing the locknut from the end of the cable, and the E-ring from the bracket. Remove the two bracket bolts and the bracket.

    b. On automatic models, remove the exhaust center pipe to provide the necessary clearance for transmission removal. This pipe runs from the exhaust manifold to the flange at the front of the muffler pipe. On most late-model vehicles it is a 2 piece unit, one from the manifold to a flange at the front of the converter, and the second which contains the converter and bolts to the front of the muffler pipe. This 2 piece unit can usually be removed as an assembly, which saves you the trouble of breaking one gasket surface and one set of bolts free.

➡**On automatic models which utilize a 2 piece center pipe assembly, it may be possible to only remove the converter portion. If this is attempted, take great care not to damage the down pipe which remains attached to the bottom of the exhaust manifold. But, remember, nothing is gained if you later decide to remove the down pipe, since you now have one more exhaust joint to seal during installation than you would have if you had removed the 2 piece center pipe as an assembly.**

    c. For automatic transmission-equipped models, loosen the clamps, then disconnect and plug the transmission oil cooler hoses from the cooler pipes.

    d. Remove the torque converter housing lower plate and disconnect and plug the oil cooler lines.

    e. Hold the flywheel in place with Special tool 09927–56010, or an equivalent flywheel holding tool and remove the three torque converter mounting bolts at the flywheel.

➡**There are many different types of flywheel tools available. The most convenient (but more rare and expensive) are the types that bolt in place leaving your other hands free. But others are available which are essentially prytools that can be used to hold the teeth of the flywheel to keep if from turning. You can usually get away with using a large prybar, but if this is attempted be VERY CAREFUL not to damage the flywheel teeth.**

18. For manual models, perform the following:

    a. Disconnect the clutch cable from the throw-out arm, then remove the clutch housing lower plate.

    b. Unbolt and remove the center exhaust pipe, as follows: first, use a long extension (or a few shorter ones) and a deep socket to loosen and remove the 3 center pipe-to-exhaust manifold stud nuts. Second, loosen the 2 spring loaded center pipe through-bolts at the rear of the converter. Finally, hold the center pipe and remove the 2 converter bracket-to-transmission bolts (located on the side of the transmission for 2WD vehicles or underneath the transmission on 4WD vehicles.

➡**On manual models, center pipe removal is another item you may be able to get away without, but ONLY on 2WD vehicles. Although the exhaust pipe itself does not really interfere with** transmission removal (it does make it a little more awkward), the problem occurs with the 4WD converter-to-transmission bracket. If you do not remove the bracket, you CANNOT remove the transmission. And, the seemingly easier solution of removing the bracket from the converter is usually impossible because the fasteners which attach the bracket to the top of the converter are so badly cooked and rusted from the extreme heat that they break off on the first attempt (TRUST US ON THIS ONE).

c. Remove the 2 nuts from the studs attaching the bottom of the transmission bell-housing to the back of the engine.

19. Disconnect the speedometer cable from the transmission (2WD) or the transfer case (4WD), as applicable. On some models there is a ground wire bolted to the case by the speedometer cable retainer, be sure to remove the ground wire before lowering the transmission.

20. For automatic models, perform the following steps:

a. If equipped, remove the left-side transmission case stiffener bracket.

b. If equipped, unbolt the right-side transmission case stiffener bracket from the transmission. On certain models this right-side bracket is attached to the engine using 3 bolts, if this is so on your vehicle remove the rear 2, but only loosen the front-most bracket-to-engine retaining bolt.

c. Remove the transmission-to-engine retaining nuts.

21. Position a transmission jack to take the weight off of the rear transmission mount (crossmember).

22. Unbolt the rear transmission mount from the chassis at either side and from the transmission at the center, then remove the mount from the vehicle.

➡The 2 bolts on the passenger side of the transmission mount are locked in place using small metal tabs. This is done to help assure they cannot loosen, since the mount simply hangs from the bolts on that side and there is no ledge for it to sit on should they come out. Be sure to carefully bend these tabs out of the way before trying to loosen those 2 bolts.

23. Place a wooden block 8 in. (200mm) tall by 4–6 in. (100–150mm) wide by 1.8 in. (45mm) thick on its side below the distributor cap, between the cylinder head distributor housing and the firewall. Lower the transmission jack slightly, to preload the wood. This wood will keep the engine from pivoting any

further and possibly causing damage to the distributor or to the motor mounts.

✳✳ WARNING

On 3-speed automatic models where the transmission dipstick tube has not been removed, take care not to damage the tube during removal. Also, check for a tube guide hook on the engine. If necessary, remove the oil filler tube bolt and set the guide hook free.

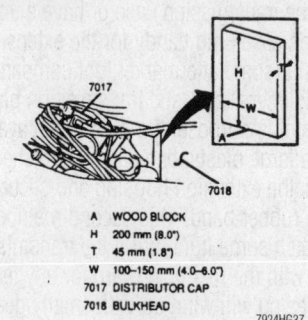

A	WOOD BLOCK
H	200 mm (8.0")
T	45 mm (1.8")
W	100–150 mm (4.0–6.0")
7017	DISTRIBUTOR CAP
7018	BULKHEAD

7924HG37

A wooden block should be positioned between the engine and firewall to prevent the possibility of damage, if the engine should pivot back while the transmission is being removed

24. For manual models so equipped, remove the 2 bolts from the right and or left transmission-to-engine reinforcement braces.

25. Carefully pull the transmission (and transfer case assembly on 4WD vehicles) toward the rear of the vehicle until the input shaft pulls clear of the clutch and pressure plate assembly (manual models or until the torque converter is clear of the flywheel, and until the transmission case/bell housing pulls off of the lower engine-to-transmission studs/bolts. Although one person can do this, if you don't have a transmission jack we REALLY recommend that you get a friend to help you with this step. Lower the transmission from the vehicle.

➡For automatic models, do your best to keep the transmission level while it is being lowered to minimize the chance of fluid draining from the unit and to keep the torque converter from falling out. The torque converter is a heavy and relatively expensive component, it would be wise to secure the converter to the transmission housing. This can be done by bolting metal tabs (which can be made from metal stock) to the housing in a position where a portion of the tab protrudes over the

converter. This can also be done using large wire ties running from the bolt holes on either side of the transmission housing, through holes on the converter.

To install:
26. Clean and inspect the transmission components located in the bell-housing. Check the throw-out bearing, fork and pivots, along with the input shaft for wear. Now is a very good time to check or replace the clutch and throw-out bearing.

27. Apply a light coating of high temperature lithium grease to the input shaft pilot and splines. Be careful not to apply too much grease which could contaminate the clutch and pressure plate assembly.

28. Apply a light coating of high temperature lithium grease to the thrust surfaces of the throw-out bearing.

➡There should be 1 or 2 metal bushings which press into the transmission housing at the lower bell-housing bolt holes or they may be left on the lower engine-to-transmission bolts/studs. If used on your application, make sure they are in position before you crawl under with the transmission assembly.

29. Carefully raise the transmission assembly into position using the transmission jack and/or a friend. With the transmission raised to the proper height, carefully slide the assembly forward, inserting the input shaft through the clutch and pressure plate, until the splines mesh (manual models), or until the torque converter is mated to the flywheel (automatic models).

✳✳ WARNING

DO NOT force the input shaft into the clutch since damage may occur to the input shaft or the clutch and pressure plate. If the splines do not easily mesh, pull the transmission back enough to carefully insert your CLEAN hand and turn the shaft slightly. Another method (when the friend comes in handy again) is to temporarily insert the driveshaft into the back of the transmission and rotate it to turn the input shaft as the transmission is carefully pushed toward the engine. Of course this second trick only works when the transmission is in gear.

30. Once the transmission is in place, install the transmission-to-engine bolts and nuts finger-tight.

31. Lift the transmission jack slightly to

pivot the engine forward and remove the wooden block.

32. Install the engine rear mounting member, then tighten member retaining bolts 29–43 ft. lbs. (40–60 Nm).

33. Remove the transmission jack and install the left and/or right transmission-to-engine reinforcement bracket bolts, as applicable. Tighten the reinforcement bracket bolts to 44–51 ft. lbs. (60–70 Nm).

➡ **Some of the Geo factory manuals list a specification for the reinforcement bracket bolts of 51–72 ft. lbs. (70–100 Nm), but on 2 of our teardown vehicles the bolts began to strip before that specification was reached. Some of the Suzuki technical manuals don't mention these brackets, while others use a lower bolt torque. We recommend that the lower torque be used in order to prevent possible damage to the transmission case or the engine block. If you are at all unsure, use a thread locking compound such as Loctite® to be certain that these bolts won't loosen in service.**

34. On automatic models, align the bolt holes in the flywheel and the torque converter, then install the flywheel-to-converter bolts and tighten gradually (using multiple passes) to 44–51 ft. lbs. (60–70 Nm).

35. Position the center exhaust pipe using a new pipe-to-manifold gasket. Torque the mounting bolts, spring-loaded bolts and stud nuts all to 29–43 ft. lbs. (40–60 Nm).

36. Tighten the engine-to-transmission bolts and nuts to 51–72 ft. lbs. (70–100 Nm).

37. The balance of the installation procedure is the reverse of removal. If the transmission and/or transfer case was drained, or even if it wasn't but some fluid leaked during the procedure, be sure to check and fill the transmission when you are finished.

38. If the clutch was replaced, or if the adjuster was disturbed when removing the cable, check and adjust the clutch before attempting to drive the vehicle.

39. When you are finished, double check all wiring connections, wiring clamps, breather hoses, etc. to make sure everything is back the way you found it.

40. For automatic models, check and adjust the select cable (which should not have changed unless the adjusting nut was disturbed, or unless other mechanical components vary, such as a different transmission was installed). Check and adjust the throttle cable, as necessary.

41. Connect the negative battery cable. Start the engine; check for any leaks and proper clutch operation.

ADJUSTMENTS

Throttle (Kickdown) Cable— Automatic Transmissions

The throttle or kickdown cable can also be known as the Throttle Valve (TV) cable. The purpose of the cable is to signal the transmission that a downshift should occur when the accelerator is pushed all the way to the floor. To check and adjust the cable:

1. Check the accelerator cable end-play

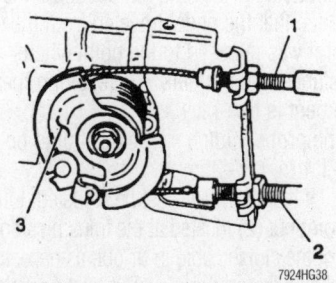

7924HG38

Before checking the throttle cable (2) adjustment at the throttle cam (3), make sure the accelerator pedal cable (1) end-play is within specification

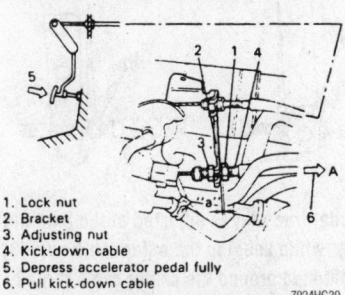

1. Lock nut
2. Bracket
3. Adjusting nut
4. Kick-down cable
5. Depress accelerator pedal fully
6. Pull kick-down cable

7924HG39

When adjusting the throttle cable on 3-speed models, make sure that clearance "a" is 0.0–0.039 in. (0–1mm) with the pedal fully depressed and the cable housing pulled tight away from the throttle body (direction A)

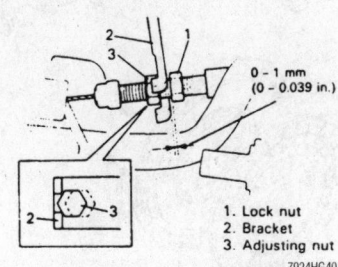

0 - 1 mm
(0 - 0.039 in.)

1. Lock nut
2. Bracket
3. Adjusting nut

7924HG40

While adjusting the cable on 3-speed models, make sure the adjusting nut is fit into the adjusting bracket (as shown with the dotted lines) while holding the specified clearance, then tighten the locknut nut against the bracket

and make sure it is within specification. There should be 0.4–0.6 in. (10-15mm) of end-play. If not, loosen the cable locknut located at the bracket on the throttle body and turn the adjusting nut until the proper play is achieved. Hold the adjusting nut and tighten the locknut when you are finished.

➡ **The accelerator cable is the upper cable, while the transmission throttle (kickdown) cable is the lower of the 2 cables on the throttle cam and bracket. If you are in doubt, follow the cables back, one will go through the fire-wall, while the other will continue down to the automatic transmission.**

2. Make sure the ignition switch is in the **LOCK** position and have an assistant fully depress and hold the accelerator pedal. If an assistant is not available, a large brick should suffice, but make it is heavy enough that the pedal does not move during adjustment.

3. For 3-speed models:

a. Loosen the locknut and adjusting nut so that both are loose and not in contact with the TV cable bracket on the throttle body.

b. Pull the throttle cable casing AWAY from the throttle body until tight and no cable deflection exists, then with the cable held in this position, tighten the cable locknut to within 0.0–0.039 in. (0–1mm) of the cable bracket.

➡ **Make sure that the cable adjusting nut is not in contact with the bracket at this point.**

c. Release the accelerator pedal while keeping the cable locknut-to-cable bracket clearance at 0.039 in. (1mm).

d. Tighten the cable adjusting nut until it engages and fits into the cable bracket.

e. Now, with the adjusting nut positioned flush with the cable bracket, fully tighten the cable locknut.

4. For 4-speed models:

a. Measure the distance between the tip end of the cable adjustment mark and the end of the boot. The distance should be 0.031–0.059 in. (0.8–1.5mm). If not, loosen the locknut and turn the adjusting nut until this measurement is achieved.

b. Snug the throttle cable locknut with the throttle cable pulled tight.

c. Double-check the measurement between the tip end of the cable and the adjustment mark to make sure it did not change while tightening the locknut.

5. Operate the vehicle and verify that the transmission kickdown shift is occurring properly.

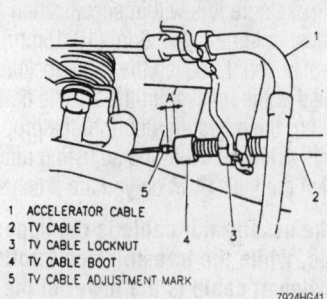

1 ACCELERATOR CABLE
2 TV CABLE
3 TV CABLE LOCKNUT
4 TV CABLE BOOT
5 TV CABLE ADJUSTMENT MARK

7924HG41

Throttle cable adjustment on the 4-speed is a relatively simple matter of obtaining a 0.031–0.059 in. (0.8–1.5mm) measurement between the cable tip end adjustment mark and the end of the boot.

Shifter Select Cable—Automatic Transmissions

The shifter select cable attaches the manual gear selector in the passenger compartment to the shift shaft in the side of the transmission housing. It should not require periodic attention, but may need adjustment if other mechanical components at either end of the cable are replaced. Adjustment is simply a case of removing excessive play from the transmission end of the cable using the adjusting nut and locknut on the threaded portion of the cable which passes through the shift shaft. For more details, please refer to procedures for the Neutral Safety Switch information.

Clutch

※※ CAUTION

The clutch driven disc may contain asbestos, which has been determined to be a cancer causing agent. Never clean clutch surfaces with compressed air! Avoid inhaling any dust from any clutch surface! When cleaning clutch surfaces, use a commercially available brake cleaning fluid.

ADJUSTMENTS

Clutch Pedal Height

The clutch pedal height should not be a periodic adjustment, but may be checked from time-to-time. More importantly, it should be checked after components of the mechanical clutch system have been replaced.

The proper clutch pedal height should be 0.2 in. (5mm) above the brake pedal height

If adjustment is necessary, loosen the locknut and turn the adjusting bolt until the

appropriate height is reached. Once set, keep the bolt from turning and tighten the locknut to secure the adjustment.

Clutch Pedal Free-Play

Unlike pedal height, which should not change because of wear, the clutch pedal free-play will tend to change as the clutch disc wears. The pedal free-play should therefore be checked from time-to-time in order to assure proper clutch operation.

To check pedal free-play depress the clutch pedal, but stop and hold the moment that clutch resistance is felt. Measure the distance that the pedal traveled from the point it was released to the point where resistance was first felt. The resulting measurement is free-play.

The proper clutch free-play should be 0.6–1.1 in. (15–25mm).

1. If the free-play must be adjusted, turn the joint nut (2) located at the transmission end of the clutch cable in or out, as necessary to achieve the proper play. But, make sure that

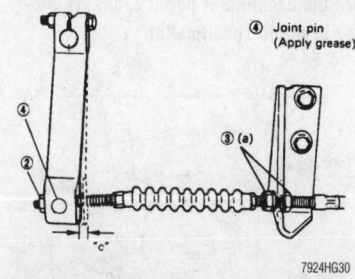

④ Joint pin (Apply grease)

7924HG30

Pedal free-play is adjusted at the joint nut (2), while keeping the outer cable nuts (3) tightened around the center cable thread portion

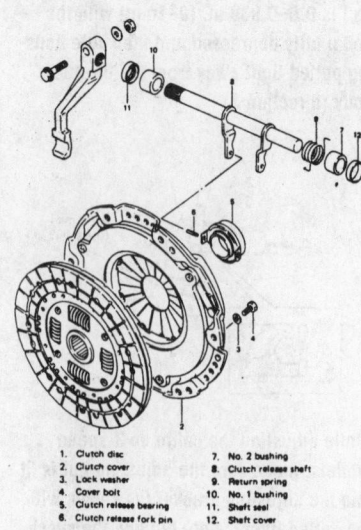

1. Clutch disc
2. Clutch cover
3. Lock washer
4. Cover bolt
5. Clutch release bearing
6. Clutch release fork pin
7. No. 2 bushing
8. Clutch release shaft
9. Return spring
10. No. 1 bushing
11. Shaft seal
12. Shaft cover

7924HG29

Exploded view of the typical clutch assembly

the outer cable nuts (3) are tightened around the center of the outer cable thread portion. Once the correct pedal free-play is obtained, check the free-play on the release arm itself (c), it should be 0.02–0.06 in. (0.5–1.5mm).

2. On all models, once you are finished, check pedal adjustment and clutch function with the engine running.

REMOVAL & INSTALLATION

1. Remove the transmission from the vehicle.
2. Hold the flywheel from turning using a large prytool or a special flywheel holding tool which locks the flywheel teeth. With the flywheel held stationary, loosen and remove the pressure plate mounting bolts.

※※ WARNING

In case you plan on using the pressure plate again, be careful to loosen the bolts gradually, one turn at a time, in a star pattern. This gradual release of the pressure plate spring tension helps to prevent possible warping and damage to the plate and spring assembly.

3. Remove the pressure plate and the clutch disc from the flywheel.
4. Inspect the flywheel for wear and/or scoring and machine or replace, as necessary. Check the flywheel for excessive runout using a dial indicator; if runout exceeds 0.0078 in. (0.2mm) the flywheel should be machined or replaced.
5. Inspect the clutch disc and pressure plate for wear or damage. Check the depth of the rivet heads on the clutch disc. Standard depth is 0.06 in. (1.6mm) for other models. The wear limit is 0.02 in. (0.5mm) for all models. Any clutch disc showing wear equal to or greater than the wear limit should be replaced.
6. Check the clutch disc, pressure plate and flywheel for light burnt or glazed surfaces. A glazed component will have a glass-like surface. If found, break the glass-like finish of light glazing using 120-200 grit sandpaper.
7. Clean and inspect the clutch release bearing and the input shaft. Many of the late-model trucks covered by this manual utilize a sealed release bearing. DO NOT submerge a sealed bearing in or soak it with solvent.

➡The clutch release bearing is a relatively inexpensive part, that performs a very important function. ALWAYS replace the bearing when a new disc and pressure plate is installed. For that matter, it s not a bad idea to replace it even when other components are being reused.

8. Make sure that the pressure plate retaining bolts and the bolt holes in the flywheel are clean and free of grease or oil to assure proper fastening during installation.

To install:

9. Apply a high temperature lithium grease to the thrust surfaces on the throw-out bearing and input shaft. Be careful not to apply too much grease which could come off and contaminate the surfaces of the clutch disc.

10. Make sure the clutch disc, pressure plate and flywheel surfaces are clean and free of all grease, oil or other possible contaminants.

11. Using a clutch alignment arbor (preferably the exact one with the proper sized pilot and proper splines), position and hold the clutch disc against the flywheel.

12. Install the pressure plate over the disc, flywheel and alignment arbor assembly, then carefully finger-tighten the bolts.

13. Check to make sure that the clutch disc is properly aligned to the pressure plate and flywheel. To do this, first look down the length of the alignment arbor and make sure it seems perfectly straight. Next, grasp the end of the arbor and withdraw it gently (note the pressure plate bolts must have been tightened sufficiently to hold the clutch plate at this stage or it will fall out of alignment the second that the arbor is removed). If the clutch disc is properly aligned, the arbor will slide smoothly out from the flywheel pilot and the clutch splines. The arbor should also insert back into both without any hesitation.

14. Once you are satisfied that the disc is properly aligned, tighten the pressure plate bolts and evenly using multiple passes of a star pattern until all are tightened to 159–248 inch lbs. (18–28 Nm).

15. Remove the clutch disc alignment tool.

16. Install the transmission.

17. Check and adjust the clutch, as necessary.

Hydraulic Clutch System

BLEEDING

contact with a painted surface; it will damage the paint.

The clutch system bleeder valve is located on the slave cylinder.

1. Raise and safely support the vehicle.

2. Fill the master cylinder reservoir to the MAX line with clean brake fluid and keep it at least half full throughout the bleeding procedure.

3. From beneath the vehicle, remove the bleeder plug cap, then attach a clear vinyl tube to the slave cylinder bleeder plug. Insert the open end of the hose into a container.

4. Have an assistant depress the clutch pedal, and while your helper holds the pedal in the depressed position, loosen the bleeder plug one-third to one-half of a turn (or until brake fluid starts to exit the bleeder valve).

5. When the fluid pressure is almost gone, retighten the bleeder plug, THEN have your assistant release the clutch pedal. It is very important that the pedal stay depressed while the bleeder valve is open, because air will be sucked into the clutch system if the pedal is released while the valve is still open.

6. If the fluid is level in the master cylinder is low, fill it with clean DOT 3 fluid.

7. Repeat Steps 3 through 5 until all air bubbles are gone from the hydraulic fluid, which is emitted from the bleeder valve.

8. If equipped, install the bleeder plug cap.

9. After completing the bleeding procedure, have your assistant apply fluid pressure to the pipe line (by depressing the clutch pedal) while you check for fluid leaks.

10. Fill the clutch master cylinder fluid reservoir to the specified full level.

11. Check clutch pedal for a spongy feeling; if any sponginess exists, repeat the entire procedure.

Transfer Case Assembly

REMOVAL & INSTALLATION

On these models the transfer case is bolted to the transmission and takes the place of the 2WD transmission's extension housing. Therefore, transfer case is removed or installed as part of the transmission assembly.

Halfshaft

REMOVAL & INSTALLATION

1. Raise and safely support the vehicle.

2. If equipped, remove the front skid plate.

3. Remove the front wheel(s).

4. If equipped, remove the locking hub.

5. Remove the snapring from the end of the halfshaft and remove the spindle washer.

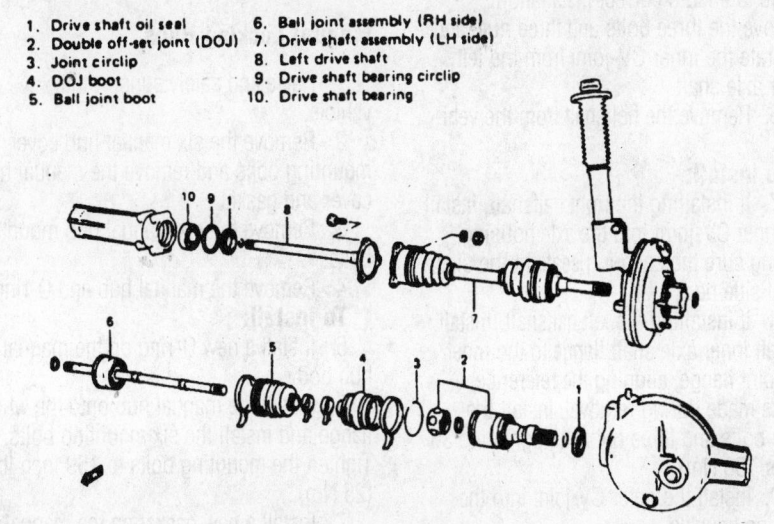

1. Drive shaft oil seal	6. Ball joint assembly (RH side)
2. Double off-set joint (DOJ)	7. Drive shaft assembly (LH side)
3. Joint circlip	8. Left drive shaft
4. DOJ boot	9. Drive shaft bearing circlip
5. Ball joint boot	10. Drive shaft bearing

7924HG31

Exploded view of the left- and right-hand halfshaft assemblies

6. Remove the sway bar nut from the lower control arm.

7. Separate the tie rod end from the steering knuckle.

8. Remove the brake caliper from the knuckle and suspend with wire, without disconnecting the brake line. Do not let the caliper hang from the brake hose.

✳✳ CAUTION

The coil spring is under extreme pressure. Make sure the control arm is firmly supported with a hydraulic jack before removing the lower ball joint nut. After the lower ball joint nut has been removed, lower the hydraulic jack slowly to relieve coil spring pressure. If this precaution is not observed, serious bodily injury may result.

9. Support the lower control arm with a hydraulic jack.

10. Remove the cotter pin and nut attaching the ball joint to the lower control arm.

11. Remove the nuts and bolts connecting the strut to the steering knuckle. Separate the steering knuckle from the strut and lower control arm.

12. Slowly lower the hydraulic jack until coil spring pressure is relieved.

13. Remove the outer CV-joint from the steering knuckle.

14. If removing the right side halfshaft, place tool J 37780 or equivalent, between the front axle housing and the inner CV-joint. Gently tap the inner CV-joint away and out of the front axle housing.

15. If removing the left side halfshaft, scribe a reference mark on the left inner axle shaft flange and the inner CV-joint flange to ensure correct installation. Remove the three bolts and three nuts and separate the inner CV-joint from the left inner axle shaft.

16. Remove the halfshaft from the vehicle.

To install:

17. If installing the right halfshaft, install the inner CV-joint into the axle housing, making sure the snapring seats in the differential side gear.

18. If installing the left halfshaft, install the left inner axle shaft flange to the inner CV-joint flange, aligning the reference marks made during removal. Install the three bolts and three nuts and tighten to 36 ft. lbs. (50 Nm).

19. Install the outer CV-joint into the steering knuckle.

20. Support the lower control arm with the hydraulic jack.

21. Attach the steering knuckle and lower ball joint to the lower control arm. Tighten the strut bolts and nuts to 65 ft. lbs. (90 Nm). Tighten the ball joint nut to 42 ft. lbs. (58 Nm) and install a new cotter pin.

22. Remove the hydraulic jack from the lower control arm.

23. Install the brake caliper to the knuckle.

24. Install the tie rod end to the steering knuckle and tighten the nut to 30 ft. lbs. (40 Nm). Install a new cotter pin.

25. Install the spindle washer and snapring to the end of the halfshaft.

26. If equipped, install the locking hub and/or skid plate.

27. Install the front wheel(s).

Locking Hubs

REMOVAL & INSTALLATION

Automatic Locking Hubs

1. Raise and safely support the vehicle .

2. Unscrew the automatic hub cover and remove the cover and O-ring.

3. Remove the hub assembly mounting bolts and remove the hub assembly.

To install:

4. Install a new O-ring on the hub assembly.

5. Install the hub assembly onto the wheel flange. Make sure the tab on the hub fits into the notch on the spindle.

6. Install the six mounting bolts and tighten to 221 inch lbs. (25 Nm).

7. Install a new O-ring on the hub cover.

8. Install the hub cover.

Manual Locking Hubs

1. Raise and safely support the vehicle.

2. Remove the six manual hub cover mounting bolts and remove the manual hub cover and gasket.

3. Remove the six manual hub mounting bolts.

4. Remove the manual hub and O-ring.

To install:

5. Install a new O-ring on the manual hub body.

6. Install the manual hub onto the wheel flange and install the six mounting bolts. Tighten the mounting bolts to 159 inch lbs. (25 Nm).

7. Install a new gasket on the manual hub cover.

8. Install the hub cover on the hub. The

lever must be in the **FREE** position with the clutch pulled out toward the cover.

9. Install the six mounting bolts. Tighten the bolts to 106 inch lbs. (12 Nm).

STEERING AND SUSPENSION

Air Bag

✳✳ CAUTION

Some vehicles are equipped with an air bag system, also known as the Supplemental Inflatable Restraint (SIR) or Supplemental Restraint System (SRS). The system must be disarmed before performing service on, or around, system components, the steering column, instrument panel components, wiring and sensors. Failure to follow the safety precautions and the disarming procedure could result in accidental air bag deployment, possible personal injury and unnecessary system repairs.

PRECAUTIONS

Several precautions must be observed when handling the inflator module to avoid accidental deployment and possible personal injury.

• Never carry the inflator module by the wires or connector on the underside of the module.

• When carrying a live inflator module, hold securely with both hands, and ensure that the bag and trim cover are pointed away.

• Place the inflator module on a bench or other surface with the bag and trim cover facing up.

• With the inflator module on the bench, never place anything on or close to the module which may be thrown in the event of an accidental deployment.

• Never use air bag component parts from another vehicle.

• If the vehicle will be exposed to temperatures above 200°F (93°C), remove the air bag module from the steering wheel.

• If there is a chance of electrical shock to any of the air bag components, remove the air bag module before servicing the vehicle.

DISARMING

1. Turn the steering wheel so the front wheels are in the straight ahead position.
2. Turn the ignition switch to the **LOCK** position and remove the key.
3. Remove the AIR BAG fuse from the air bag fuse box.
4. Remove the steering wheel side cap and disconnect the yellow connector inside the inflator module housing.
5. Remove the glove box, by disengaging both glove box stoppers from each side, then disconnect the yellow passenger air bag inflator module connector.

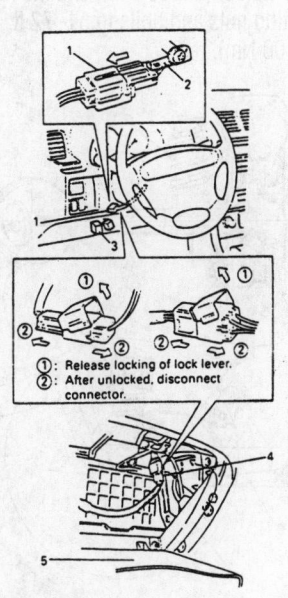

1. Yellow connector of driver air bag (inflator) module
2. Connector stay
3. Air bag fuse box
4. Yellow connector of passenger air bag (inflator) module
5. Glove box

7924HG36

To disable the air bag system, the air bag fuse needs to be removed and the yellow air bag module connector needs to be disengaged

Recirculating Ball Steering Gear

REMOVAL & INSTALLATION

Manual Steering

1. Raise and support the vehicle safely.
2. Remove the lower skid plate.
3. Disconnect the steering lower shaft-to-steering gear input shaft attaching bolt.
4. Using a tie rod end separator (such as Suzuki Special Tool J29107), detach the center link end from the Pitman arm.
5. Remove the 3 steering gear box mounting bolts, noting the position of the longer bolt.

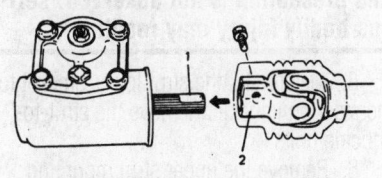

1. Flat part
2. Shaft joint

7924HG72

Remove the lower steering shaft mounting bolt and disconnect the shaft from the gear box

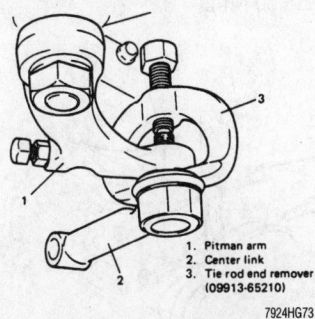

1. Pitman arm
2. Center link
3. Tie rod end remover (09913-65210)

7924HG73

Remove the center link-to-Pitman arm attaching nut and, using a tie rod end remover, press out the link from the arm

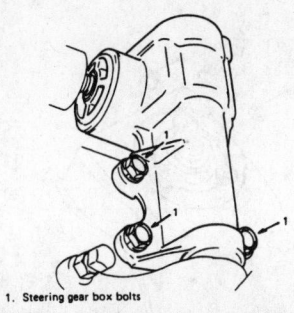

1. Steering gear box bolts

7924HG74

Remove the three steering gear box attaching bolts and remove the box from the vehicle

6. Disconnect the steering lower shaft joint and remove the steering gear.
7. If replacing the steering gear, remove the nut connecting the Pitman arm to the steering gear. Use a Pitman arm puller to remove the Pitman arm from the steering gear.
   To install:
8. Connect the Pitman arm to the steering gear, then install the nut. Tighten the nut to 101–129 ft. lbs. (140–180 Nm).
9. Install the steering gear box by connecting to the lower shaft joint.

➡**Align the flat part of the steering gear worm shaft with the bolt hole of the lower shaft joint.**

10. Install the steering gear box mounting bolts, then tighten them to 51–72 ft. lbs. (70–100 Nm).
11. Attach the center link to the Pitman arm, then tighten the nut to 22–50 ft. lbs. (30–70 Nm).
12. Connect the lower steering shaft mounting bolts, then tighten them to 15–22 ft. lbs. (20–30 Nm).
13. Install the lower skid plate.
14. Lower the vehicle.

Power Steering

1. Remove the coolant reservoir tank from the radiator.

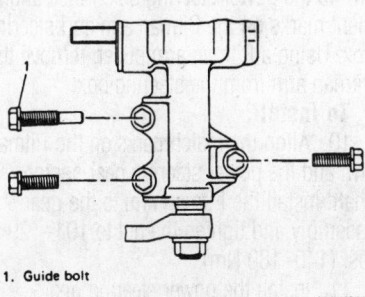

1. Guide bolt

7924HG75

When installing the steering gear box, note the position of the longer guide bolt

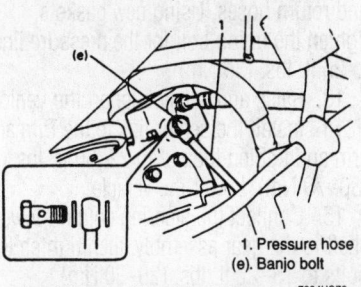

1. Pressure hose
(e). Banjo bolt

7924HG76

On power steering models, disconnect the fluid lines from the gear box—be sure to install new washers on the lines during installation

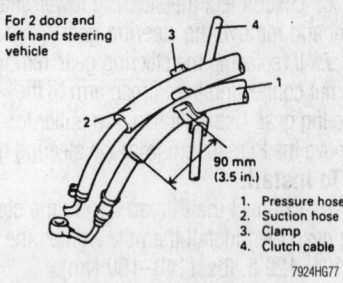

For 2 door and left hand steering vehicle

90 mm (3.5 in.)

1. Pressure hose
2. Suction hose
3. Clamp
4. Clutch cable

7924HG77

During installation, be sure to route and position the power steering fluid lines as shown

2. Disconnect the steering column lower shaft from the gear box by removing the bolt.

3. Raise and safely support the vehicle.

4. Remove the center link nut and lock-washer holding the Pitman arm to the center link. Using a Pitman arm puller, disconnect the center link from the Pitman arm.

5. Lower the vehicle and place a fluid catch pan under the power steering gear box. Remove the pressure hose from the power steering gear assembly and plug the line.

6. Disconnect the return hose and plug the line.

7. Remove the three power steering gear mounting bolts.

8. Remove the power steering gear.

9. Remove the nut holding the Pitman arm to the power steering box. Place alignment marks on the Pitman arm and steering box. Using a Pitman arm puller, remove the Pitman arm from the steering box.

To install:

10. Align the matchmarks on the Pitman arm and the power steering gear sector shaft. Install the Pitman arm to the gear assembly and tighten the nut to 101–129 ft. lbs. (140–180 Nm).

11. Install the power steering gear assembly on the vehicle, then tighten the mounting bolts to 59–73 ft. lbs. (80–100 Nm).

12. Connect the power steering pressure and return hoses. Using new gaskets, tighten the union bolt for the pressure line to 26 ft. lbs. (35 Nm).

13. Raise and safely support the vehicle.

14. Install the center link to the Pitman arm and tighten the nut to 22–50 ft. lbs. (30–70 Nm). Lower the vehicle.

15. Connect the steering column lower shaft to the gear assembly, then tighten the bolts to 15–22 ft. lbs. (20–30 Nm).

16. Install the coolant reservoir tank to the radiator.

17. Refill and bleed the power steering pump.

Strut

REMOVAL & INSTALLATION

1. Raise and safely support the vehicle. Allow the front suspension to hang free.

2. Remove the front wheel.

3. Disconnect the E-clip mounting the brake hose and remove the brake hose from the strut bracket.

4. Support the control arm with a stand or floor jack.

✳✳ CAUTION

The coil spring is under extreme pressure. Make sure control arm is firmly supported with a hydraulic jack before continuing with procedure. If this precaution is not observed, serious bodily injury may result.

5. Matchmark the strut lower bracket to the steering knuckle. Remove the strut-to-knuckle bolts.

6. Remove the upper strut mounting nuts. Hold the strut to prevent it from falling. Remove the strut assembly from the vehicle.

To install:

7. Install the strut assembly and tighten the upper strut mounting nuts to 15–22 ft. lbs. (20–30 Nm).

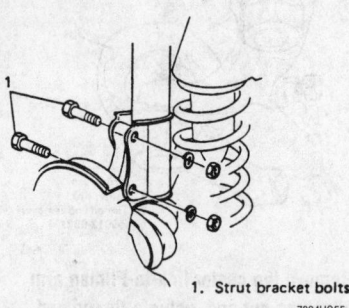

1. Strut bracket bolts

7924HG55

The lower end of the strut is secured to the knuckle with two large bolts and nuts

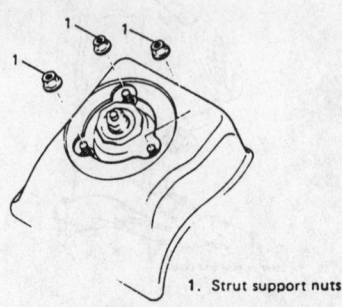

1. Strut support nuts

7924HG56

The upper end of the strut is mounted to the body with three nuts

8. Install the strut-to-knuckle bolts, ensuring that the matchmarks are aligned, then tighten the nuts and bolts to 58–75 ft. lbs. (80–100 Nm).

9. Remove the stand or jack.

10. Connect the brake hose to the strut bracket using the E-clip.

11. Install the front wheels and lower the vehicle.

12. Check the vehicle's alignment.

Shock Absorber

REMOVAL & INSTALLATION

1. Raise and safely support the vehicle .

2. Support the rear axle housing with a hydraulic jack or stand.

3. Remove the shock absorber upper locknut and retaining nut.

4. Remove the lower shock absorber from the axle housing by removing the mounting nut and bolt.

5. Remove the rear shock absorber.

To install:

6. Install the rear shock absorber.

7. Install the lower mounting nut and bolt. The lower bolt should head should point in toward the center of the vehicle.

8. Install the upper retaining nut and locknut.

9. Tighten the upper mounting nuts to 16–25 ft. lbs. (22–35 Nm), and the lower mounting nuts and bolts to 51–72 ft. lbs. (70–100 Nm).

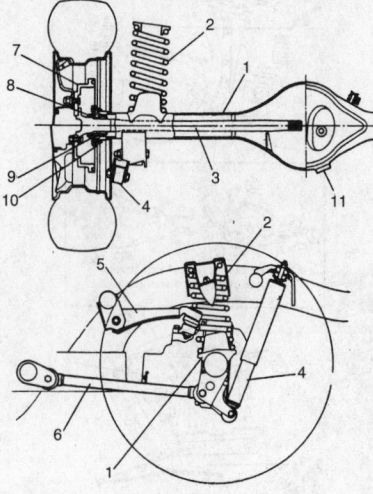

1. Rear axle housing
2. Coil spring
3. Axle shaft
4. Shock absorber
5. Upper arm
6. Trailing rod
7. Brake drum
8. Wheel bearing retainer
9. Rear wheel bearing
10. Brake back plate
11. Oil drain plug

7924HG32

Rear suspension component identification

10. Remove the jack or stand from the rear axle assembly.

11. Lower the vehicle.

Coil Spring

REMOVAL & INSTALLATION

Front

1. Raise and safely support the vehicle securely .

2. Remove the front wheel. Unbolt the brake caliper and suspend it out of the way.

3. Remove the engine skid plate, if equipped.

4. Support the lower control arm with a hydraulic jack or stand.

5. Remove the 3 nuts and bolts from the control arm, or remove the castellated nut from the ball joint, and separate the control arm from the steering knuckle.

6. Disconnect the stabilizer bar from the control arm.

※※ CAUTION

The coil spring is under pressure. Make sure control arm is firmly supported with a hydraulic jack before continuing with procedure. If this precaution is not observed, serious bodily injury may result.

7. Remove the lower strut mounting bracket bolts and disconnect the strut bracket from the steering knuckle.

8. Lower the control arm enough to remove the steering knuckle assembly.

9. Lower the jack until all tension is removed from the coil spring. Remove the coil spring from the vehicle.

To install:

10. Install the coil spring onto the control arm and slowly raise the jack.

➡**The bottom of the spring has a larger diameter than the top. Make sure that the spring is installed correctly.**

11. Install the strut-to-knuckle mounting nuts and bolts and tighten to specifications.

12. Connect the stabilizer link to the control arm. Tighten the nut to the value specified in the torque chart located at the end of this section.

13. If removed, install the 3 nuts and bolts connecting the control arm and ball joint. Tighten the nuts to 51–75 ft. lbs. (70–100 Nm).

14. If removed, install and tighten the ball joint castle nut to 33–50 ft. lbs. (45–70 Nm). Insert a new cotter pin through the castle nut and ball joint stud holes, then bend the cotter pin ends over. If none of the castle nut grooves are aligned with the hole in the ball joint stud hole, continue tightening the castle nut until one of the grooves is aligned with the stud hole, then install the cotter pin.

15. Install the engine skid plate. Tighten the bolts to 40 ft. lbs. (54 Nm).

16. Install the brake caliper and the front wheels, lower the vehicle.

Rear

1. Raise and safely support the vehicle .

2. Remove the rear wheels.

3. Support the rear axle housing with a hydraulic jack or stand.

4. Remove the shock absorber lower mounting nut and bolt. Remove the bolts and nuts from both shock absorbers.

5. Disconnect the parking brake cables from the hangers on the trailing arms.

6. Lower the rear axle housing so the coil spring can be removed.

※※ WARNING

Take care to avoid stretching the brake hose!

7. Remove the coil spring from the vehicle.

To install:

8. Install the coil spring to the spring seat and raise the axle housing. Make sure the spring is seated correctly. The end of the coil spring should be seated in the stepped part of the spring seat.

9. Install the lower shock absorber mounting bolts and nuts but do not tighten.

10. Connect the parking brake cable hangers and install the rear wheels.

11. Lower the vehicle and tighten the lower shock absorber nuts.

Lower Ball Joint

REMOVAL & INSTALLATION

1. Raise and safely support the vehicle.

2. Remove the front wheel.

3. Position a hydraulic jack beneath the lower control arm, then raise the jack until it supports the lower control arm.

4. On 4-wheel drive models, remove the wheel hub.

5. Detach the tie rod end from the steering knuckle arm.

6. Secure the bottom of the coil spring to the lower control arm with a chain. This will prevent the spring from accidentally dislodging form the control arm and possibly causing personal injury.

7. Detach the lower ball joint from the steering knuckle.

8. While slowly raising the floor jack, pull outward on the spindle/knuckle assembly to detach it from the ball joint stud. Support the spindle/knuckle assembly aside.

9. Lower the control arm to relieve the spring pressure and unbolt the ball joint from the control arm.

To install:

10. Position the new ball joint on the control arm. Install the bolts and tighten the nuts to 51–75 ft. lbs. (70–100 Nm).

➡**During installation ensure that the coil spring is properly positioned on its lower and upper seats.**

11. Raise the jack enough to allow the spindle/knuckle assembly to clear the ball joint stud.

12. While slowly lowering the floor jack, position the spindle/knuckle assembly onto the lower ball joint stud.

13. Install and tighten the ball joint castle nut to 33–50 ft. lbs. (45–70 Nm). Insert a new cotter pin through the castle nut and ball joint stud holes, then bend the cotter pin ends over. If none of the castle nut grooves are aligned with the hole in the ball joint stud hole, continue tightening the castle nut until one of the grooves is aligned with the stud hole, then install the cotter pin.

14. Remove the chain securing the coil spring to the lower control arm.

15. Attach the tie rod end to the steering knuckle arm, then tighten the tie rod end castle nut to 22–39 ft. lbs. (30–55 Nm). Insert a new cotter pin through the castle nut and tie rod end stud holes, then bend the cotter pin ends over. If none of the castle nut grooves are aligned with the hole in the tie rod end stud hole, continue tightening the castle nut until one of the grooves is aligned with the stud hole, then install the cotter pin.

16. On 4-wheel drive models, install the hub.

17. Lower the floor jack and remove it from beneath the lower control arm.

18. Install the front wheel.

19. Lower the vehicle.

Wheel Bearings

ADJUSTMENT

Front

The front wheel bearings are a cartridge type design and cannot be adjusted. To check for a loose wheel bearing, proceed as follows:

1. Raise and safely support the vehicle.
2. Remove the front wheel.
3. Compress the brake caliper piston to free the caliper assembly.
4. Using a suitable dial indicator, measure the thrust play.
5. Push and pull the brake rotor by hand. If rotor movement exceeds 0.002 in. (0.05mm), replace the wheel bearings.
6. Install the wheel and lower the vehicle.
7. Apply the brakes several times before moving the vehicle, to seat the caliper piston.

Rear

The rear wheel bearings are not adjustable.

REMOVAL & INSTALLATION

Front

1. Raise and safely support the vehicle .
2. Remove the wheels.
3. If equipped with 4WD and automatic locking hubs, perform the following:
   a. Unscrew the automatic hub cover and remove the cover and O-ring.
   b. Remove the hub assembly mounting bolts and remove the hub assembly.
4. If equipped with 4WD and manual locking hubs, perform the following:
   a. Remove the six manual hub cover mounting bolts and remove he manual hub cover and gasket.
   b. Remove the six manual hub mounting bolts.
   c. Remove the manual hub and O-ring.
5. If equipped with 2WD, remove the hub cap from the hub.
6. Remove the caliper mounting bracket and position the caliper out of the way. Support the caliper.
7. Remove the brake rotor from the wheel hub. Remove the front wheel bearing lockplate screws, lockplate and washer.
8. Remove the wheel bearing locknut and washer and remove the wheel hub complete with bearings and seals.

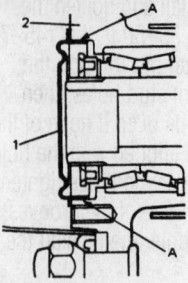

1. Hub cap
2. Wheel hub
A: Apply water tight sealant (99000-31090)

7924HG62

Cross-sectional view of the 2WD front wheel hub and bearing components

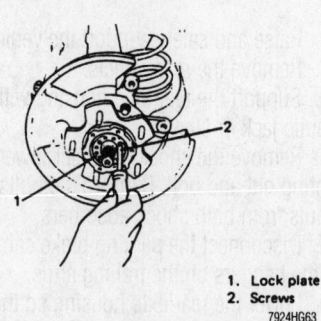

1. Lock plate
2. Screws

7924HG63

To remove the wheel hub and bearings, first remove the bearing lockplate which is secured by four screws

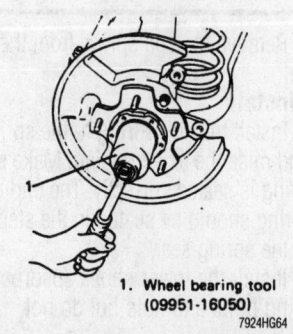

1. Wheel bearing tool (09951-16050)

7924HG64

Remove the wheel bearing locknut and washer using the special tool shown

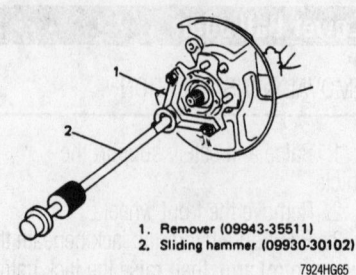

1. Remover (09943-35511)
2. Sliding hammer (09930-30102)

7924HG65

A slide hammer may be necessary to remove the assembly from the spindle

➡If wheel hub can not be removed by hand, use a sliding hammer and the proper adapter (such as Suzuki Special Tools J37781 with a J2619–01 for the Tracker or 09943–35511 with 09930–30102 for the Sidekick, Sidekick Sport and X-90 models)

9. If equipped with ABS brakes, remove the sensor rotor from the wheel hub.

➡Pull out the sensor rotor from the wheel hub gradually and evenly. Pulling it out partially may cause it to deform.

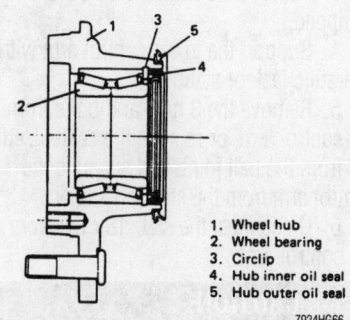

1. Wheel hub
2. Wheel bearing
3. Circlip
4. Hub inner oil seal
5. Hub outer oil seal

7924HG66

Cross-sectional view of the front wheel hub and bearing assembly

7924HG67

To remove the bearings from the hub, remove the snapring and rear oil seal . . .

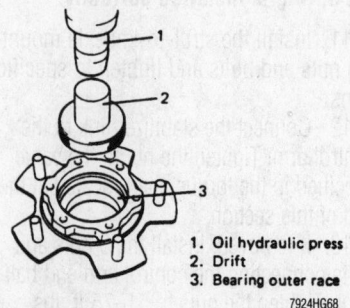

1. Oil hydraulic press
2. Drift
3. Bearing outer race

7924HG68

. . . then press the bearing assembly out of the wheel hub

10. Remove the inner bearing grease seal.

11. Remove the snapring and remove the inner bearing.

12. Clean and inspect the hub and bearing seats.

To install:

13. Drive the inner race into the hub, then install the wheel bearing snapring. Drive the two new hub oil seals into the hub. Apply lithium grease to the lip portion of the oil seal.

14. If equipped with ABS brakes, install the sensor rotor.

15. Install the hub on the spindle and install the outer bearing, locknut, washer, and lockplate. Tighten the locknut to 123–180 ft. lbs. (170–250 Nm).

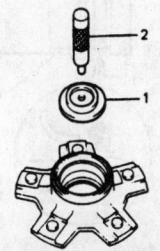

1. Bearing & oil seal installer (09944-68210)
2. Installer handle (09924-74510)

7924HG69

Press the new bearing into the wheel hub until it contacts the stepped surface of the hub

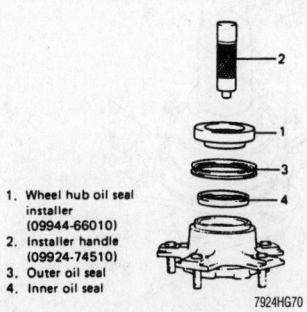

1. Wheel hub oil seal installer (09944-66010)
2. Installer handle (09924-74510)
3. Outer oil seal
4. Inner oil seal

7924HG70

Drive in the two wheel hub oil seals—if the tool shown is not available, you may need to install the seals one at a time

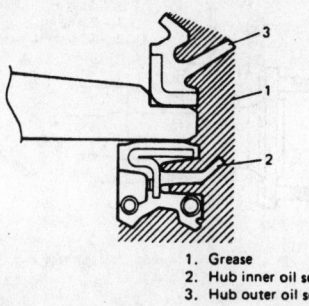

1. Grease
2. Hub inner oil seal
3. Hub outer oil seal

7924HG71

Apply lithium grease to the wheel hub seals as shown

16. Install the four lockplate mounting screws. Tighten the screws to 8.8–17.7 inch lbs. (1–2 Nm).

17. Install the brake rotor.

18. Install the caliper and tighten the bolts to 61 ft. lbs. (85 Nm).

19. If equipped with 2WD, install the locking hub cap.

20. If equipped with 4WD and automatic locking hubs, perform the following:

   a. Install a new O-ring on the hub assembly.

   b. Install the hub assembly onto the wheel flange. Make sure the tab on the hub fits into the notch on the spindle.

   c. Install the six mounting bolts and tighten to 18 ft. lbs. (25 Nm).

   d. Install a new O-ring on the hub cover.

   e. Install the hub cover.

21. If equipped with 4WD and manual locking hubs, complete the following substeps:

   a. Install a new O-ring on the manual hub body.

   b. Install the manual hub onto the wheel flange and install the six mounting bolts. Tighten the mounting bolts to 18 ft. lbs. (25 Nm).

   c. Install a new gasket on the manual hub cover.

   d. Install the hub cover on the hub. The lever must be in the **FREE** position with the clutch pulled out toward the cover.

   e. Install the six mounting bolts. Tighten the bolts to 71–106 inch lbs. (8–12 Nm).

22. Install the wheels and lower the vehicle.

Rear

1. Raise and safely support the vehicle.

2. Remove the rear wheels and remove the rear brake drums from the vehicle.

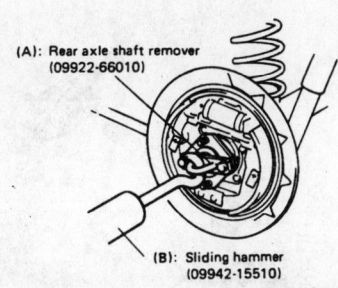

(A): Rear axle shaft remover (09922-66010)

(B): Sliding hammer (09942-15510)

7924HG42

To remove the rear wheel bearings, pull the axle shaft out of the rear housing—it may be necessary to use a slide hammer and adapter, as shown

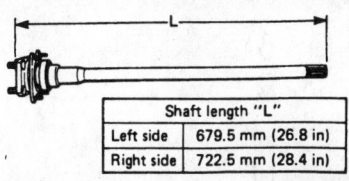

Shaft length "L"	
Left side	679.5 mm (26.8 in)
Right side	722.5 mm (28.4 in)

7924HG43

If removing both of the rear axle shafts from the housing, be sure to mark the shafts so that they can be reinstalled in their original positions—if you mix the shafts up, measure their length

3. Drain the gear oil from the rear axle housing.

4. Remove the rear brake return springs.

➡**If both axles are being removed mark the axles left and right. The axles are different lengths and must be installed in the correct position.**

5. Remove the rear wheel bearing retainer nuts from the rear axle housing.

6. Using an axle puller remove the axle shaft from the housing.

✳✳ WARNING

Do not remove the backing plate with the axle; this may cause damage to the inner seal.

7. If the axle, axle bearing or seal is being replaced, support the axle in a vise with additional support under the shaft next to the bearing.

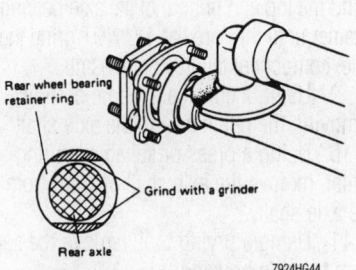

Rear wheel bearing retainer ring

Grind with a grinder

Rear axle

7924HG44

Use an angle grinder to shave the rear axle shaft bearing retainer ring as shown (shaded areas) . . .

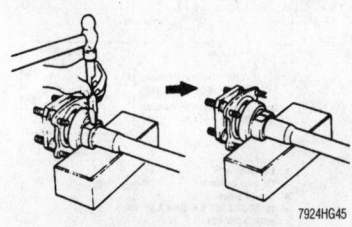

7924HG45

. . . then use a chisel and hammer to break the retainer ring in half

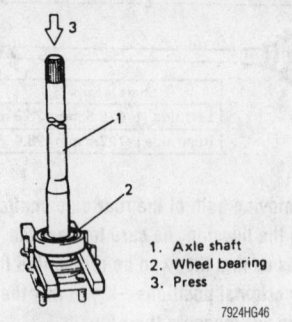

1. Axle shaft
2. Wheel bearing
3. Press

7924HG46

Remove the axle shaft bearing by pressing it off, as shown

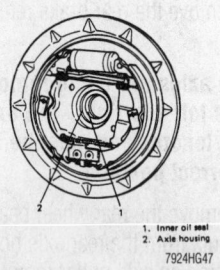

1. Inner oil seal
2. Axle housing

7924HG47

Once the axle shaft is removed from the housing, the inner oil seal should be pried out of the housing—whenever the axle shaft is removed, the oil seal should be replaced with a new one

✳✳ CAUTION

Eye protection must be worn during rear axle bearing and seal removal. Failure to do so could cause injury.

8. With the axle shaft supported properly, grind the top and bottom of the axle bearing retainer until they are flat. DO NOT grind the axle, component failure could result.

9. Using a chisel and hammer, finish removing the retainer from the axle shaft.

10. Using a press or suitable bearing puller, remove the axle shaft bearing from the axle shaft.

11. Using a prying tool, remove the seal from the axle housing.

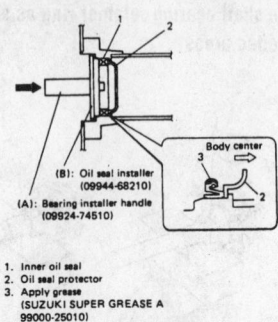

(B): Oil seal installer (09944-68210)
(A): Bearing installer handle (09924-74510)

Body center

1. Inner oil seal
2. Oil seal protector
3. Apply grease (SUZUKI SUPER GREASE A 99000-25010)

7924HG48

Use a seal driver to press the new inner oil seal into the housing, as shown

To install:

12. Using a seal driver, install the new seal with the lip facing the housing to the same depth as the old seal.

13. Apply grease to the axle shaft inner oil seal lip.

14. Install the new bearing and the retainer on the axle shaft using a suitable press.

15. Apply a bead of sealant on the outer face of the bearing retainer.

16. Install the axle shaft into the rear axle housing and replace the rear wheel bearing retaining nuts, tighten to 159–248 inch lbs. (18–28 Nm). When sliding the axle shaft into the housing, be careful not to damage the inner oil seal.

17. Install the rear brake return springs.

18. Replace the rear brake drums and replace the rear tires on the vehicle.

19. Refill the rear axle housing with the proper gear oil and safely lower the vehicle.

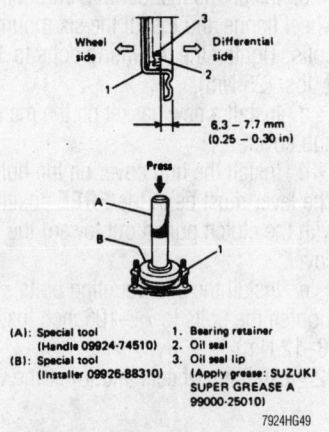

Wheel side   Differential side

6.3 – 7.7 mm
(0.25 – 0.30 in)

Press

A
B

(A): Special tool (Handle 09924-74510)
(B): Special tool (Installer 09926-88310)

1. Bearing retainer
2. Oil seal
3. Oil seal lip (Apply grease: SUZUKI SUPER GREASE A 99000-25010)

7924HG49

Use an appropriately-sized seal driver to press the new oil seal into the bearing retainer—be sure to apply a small amount of grease to its outside circumference before installation

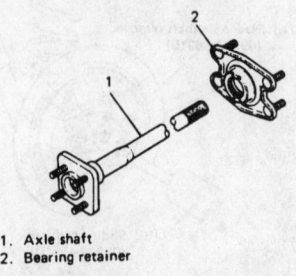

1. Axle shaft
2. Bearing retainer

7924HG50

Slide the bearing retainer onto the axle shaft so that the retainer studs point away from the axle flange

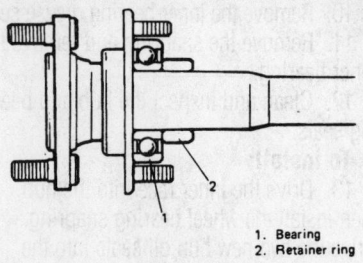

1. Bearing
2. Retainer ring

7924HG51

Press the axle bearing and bearing retainer ring onto the axle shaft, as shown—be sure not to damage the outside face of the retainer ring

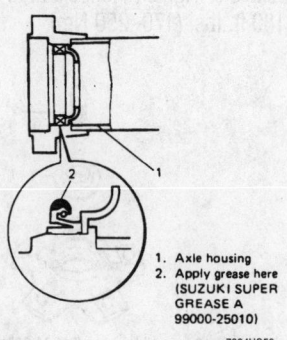

1. Axle housing
2. Apply grease here (SUZUKI SUPER GREASE A 99000-25010)

7924HG52

Apply grease to the axle shaft inner oil seal lip, as shown

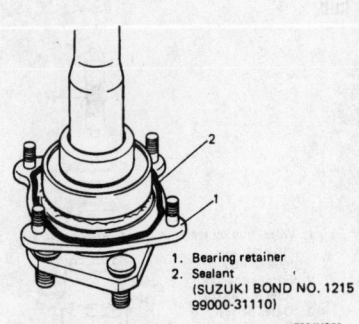

1. Bearing retainer
2. Sealant (SUZUKI BOND NO. 1215 99000-31110)

7924HG53

Apply a bead of sealant to the bearing retainer . . .

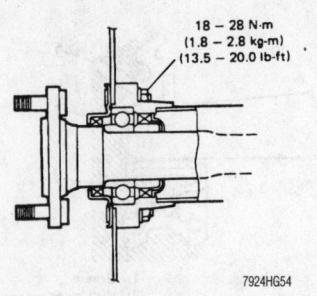

18 – 28 N·m
(1.8 – 2.8 kg-m)
(13.5 – 20.0 lb-ft)

7924HG54

. . . then slide the axle shaft into the housing—tighten the bearing retainer stud nuts to 159–248 inch lbs. (18–28 Nm)

ENGINE REPAIR

➡Disconnecting the negative battery cable on some vehicles may interfere with the functions of the on board computer systems and may require the computer to undergo a relearning process, once the negative battery cable is reconnected.

❋❋ WARNING

NEVER disconnect the negative battery cable with the ignition ON. Removing power from the computer control module with the ignition ON may destroy the module.

Distributor

Most distributor equipped vehicles covered by this manual utilize a commonly recognizable distributor ignition system. On Board Diagnostics II (OBD II) equipped engines utilize a distributor-like ignition system in which the commonly recognized distributor is replaced by a High Voltage Switch (HVS). The HVS is mounted in a fixed position so the Powertrain Control Module (PCM) or Vehicle Control Module (VCM) may control ignition timing using input from a camshaft position sensor (usually integral to the HVS). Because of this, proper rotor alignment when removing and installing the HVS is even more critical than with the conventional distributor systems.

REMOVAL

1. Disconnect the negative battery cable.
2. Remove all necessary components in order to gain access to the distributor assembly. Remove the air cleaner assembly for access.
3. Tag and disengage the distributor electrical connectors. If equipped, disconnect the vacuum line.
4. Remove the distributor cap. If necessary, tag and disconnect the spark plug wires.
5. Matchmark the rotor and the distributor body. Matchmark the distributor assembly and the engine block.

➡Although it is only necessary to matchmark the distributor position for ease of installation, it may be advisable to first position the engine at TDC. If the distributor is being removed for

further engine repair which may involve rotating the crankshaft (such as to align timing marks), setting the engine to TDC at this time will keep the distributor matchmarks valid during installation.

6. Remove the distributor hold-down bolt. Carefully remove the distributor.

➡As the distributor is removed from the engine, the rotor will turn counterclockwise. Observe and mark the finish position of the rotor. When reinstalling, position the rotor at the last mark and set the distributor into the engine. As the distributor drops into place, the rotor should turn to its original position, providing the engine crankshaft has not been rotated with the distributor out.

INSTALLATION

Timing Not Disturbed

➡To ensure correct ignition timing if the engine has not been disturbed, the distributor must be installed with the rotor in the same position as when removed.

1. Align the rotor to the last mark made and install the distributor in the engine.

➡On vehicles not equipped with the HVS system, if the distributor shaft cannot drop into the engine, remove the distributor and use a small prytool through the mounting hole to rotate the oil pump driveshaft until it can align with the distributor gear.

2. As the distributor is fully seated, the rotor should turn and end up at the first mark made. Ensure the distributor and oil pump rod are fully engaged.

➡When the distributor is fully seated on the engine, be sure the rotor is properly aligned with the first timing mark. If equipped with the HVS system, if the rotor does not align with the mark, the gear teeth of the HVS and camshaft have meshed one or more teeth out of time and the HVS procedure for timing disturbed should be used to assure proper installation.

3. Reconnect the distributor cap and wires. If applicable, attach the vacuum line to the distributor.
4. Tighten the distributor hold-down bolt, then install the air cleaner or duct, if removed for access.

➡If a check engine light is illuminated after HVS installation (Astro/Safari) and a Diagnostic Trouble Code (DTC) 1345 is found, the HVS has been installed incorrectly. Proceed to the timing disturbed installation (HVS) procedure for the 1995–99 Astro/Safari Van.

5. Check and adjust the ignition timing.

Timing Disturbed

EXCEPT HVS IGNITION SYSTEMS

1. Set the No. 1 piston to TDC on the compression stroke: Remove the No. 1 spark plug. Place a finger over the spark plug hole and rotate the engine in the normal direction of rotation slowly, until compression is felt.
2. Align the timing mark on the crankshaft pulley to the **0** on the engine timing indicator by slowly rotating the engine in the same direction. The No. 1 piston is now at TDC on the compression stroke.

➡An alternate method may be used to assure the engine is at TDC if the valve cover is removed. Watch the rocker arms for the No. 1 cylinder as the engine is turned. If the valves move as the crankshaft timing marker approaches the scale, the No. 1 cylinder is on its exhaust stroke. If the valves remain closed as the timing mark approaches the scale, then the No. 1 cylinder is approaching TDC of the compression stroke.

3. Install the distributor to the engine so the rotor is pointing to the No. 1 spark plug tower on the distributor cap once the distributor is fully seated in the engine.
4. Install the distributor cap, spark plug, wiring and connectors. If applicable, attach the vacuum line to the distributor.
5. If removed for access, install the air duct and/or air cleaner assembly, as applicable.
6. Check and adjust ignition timing.

HVS IGNITION SYSTEMS

1. Set the engine to TDC: Remove the No. 1 spark plug. Place a finger over the spark plug hole and rotate the engine in the normal direction of rotation slowly, until compression is felt.
2. Align the timing mark on the crankshaft pulley to the **0** on the engine timing indicator by rotating the engine in the same direction slowly. The engine is now set on No. 1 TDC.

➡An alternate method may be used to assure the engine is at TDC if the valve cover is removed. **Watch the rocker arms for the No. 1 cylinder as the engine is turned. If the valves move as the crankshaft timing marker approaches the scale, the No. 1 cylinder is on its exhaust stroke. If the valves remain closed as the timing mark approaches the scale, then the No. 1 cylinder is approaching TDC of the compression stroke.**

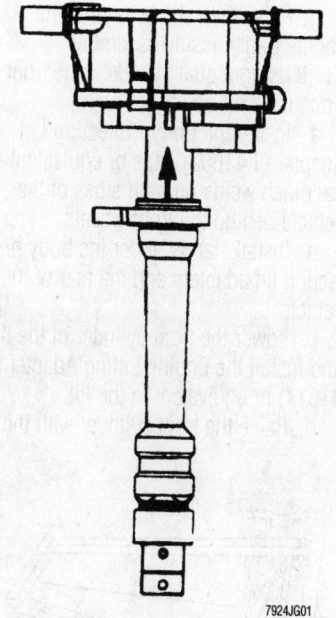

Proper HVS alignment for installation— models with the HVS ignition system

3. Remove the HVS cap screws and cap to expose the rotor.

4. Align the pre-drilled indent hole in the HVS driven gear with the arrow cast into the upper portion of the shaft housing. The rotor should point to the cap hold-down mount nearest the flat side of the housing.

5. Using a long prytool, align the oil pump driveshaft in the engine to the mating drive tab in the HVS.

6. Guide the HVS into place, making sure the locating slot in the HVS base fits over the dowel pin in the intake manifold.

7. Once the HVS is FULLY SEATED, the rotor tip should be aligned with the pointer cast into the HVS base. this pointer will have a "6" cast into it, indicating the HVS component is designed for use in a 6-cylinder

engine. If the rotor tip does not align within a few degrees of the pointer, the gear mesh between the HVS and camshaft is likely off by a tooth or more. If so, repeat the procedure again to achieve proper alignment.

8. Install the cap and mounting screws.

9. Install the HVS mounting clamp and tighten to 20 ft. lbs. (27 Nm).

10. Engage the 3-wire camshaft position sensor connector to the base of the HVS assembly.

11. Connect the spark plug and coil leads to the HVS cap. If a check engine light is illuminated and a Diagnostic Trouble Code (DTC) 1345 is found, either the HVS has been installed incorrectly or an incorrect HVS assembly has been installed.

Ignition Timing

The timing may be adjusted only on those that are equipped with a conventional distributor ignition. If equipped with the distributorless Electronic Ignition (EI) system, the control module sets timing and makes all necessary spark changes. On these systems, the crankshaft position sensor is mounted in a fixed position, therefore not allowing for adjustment. The HVS system found on some 1995–99 vehicles is also not adjustable.

➡**The 4.3L (VIN W) engine is equipped with a new distributor-like ignition component called the High Voltage Switch (HVS). On the HVS ignition system the Powertrain Control Module (PCM) utilizes a camshaft position sensor to determine spark timing, dwell and firing of the ignition coil. Because of this, positioning of the HVS (and ignition timing) is fixed and non-adjustable. Do not attempt to adjust the ignition timing by rotating the HVS or cross/misfiring may occur.**

Connect the timing light and tachometer to the engine according to the tool manufacturer's instructions. Be sure the timing light is connected to the No. 1 spark plug wire. A digital tachometer must be used.

ADJUSTMENT

1. Locate and clean the timing marks on the crankshaft pulley and the front of the timing case cover.

2. Use chalk or white paint to color the mark on the scale that will indicate the correct timing, when aligned with the mark on the pulley or the pointer.

3. Attach a tachometer and a timing light to the engine.

4. On early model vehicles that are not equipped with Electronic Spark Control (ESC), disconnect and plug the vacuum lines to the distributor.

5. If equipped with ESC, the electronic spark timing must be disabled or bypassed to prevent the control module from advancing timing while attempting to set it. This would obviously lead to an incorrect base timing setting. There are two possible methods of disabling the EST system, depending on the type of engine:

• On all other engines using the EST distributor, disengage the timing connector wire. Refer to the Vehicle Emission Control Information (VECI) label for details on the particular engine. Most vehicles are equipped with a single wire timing bypass connector. The bypass wire is normally a tan wire with a black stripe, that breaks out of a taped section just below the heater case in the passenger compartment.

6. Start the engine, then check and adjust the idle speed, as necessary.

7. Loosen the distributor lockbolt slightly to permit the distributor to be turned.

8. With the timing light aimed at the pulley and the marks on the engine, turn the distributor in the direction of rotor rotation to retard the spark or in the opposite direction of rotor rotation to advance the spark. Align the marks on the pulley and the engine with the flashes of the timing light.

9. Turn the engine **OFF**, tighten the distributor hold-down bolt and recheck the timing.

Engine Assembly

REMOVAL & INSTALLATION

➡**The engine assembly on the Astro and Safari is removed from the bottom of the vehicle. A special engine lifting table is necessary to perform the following procedure.**

Astro and Safari

1. Disconnect the negative and positive battery cables.

2. Remove the battery.

3. Remove the engine cover.

4. Remove the air cleaner assembly.

5. Disconnect the throttle cable and if equipped, the cruise control cable from the throttle body.

6. Have the A/C system discharged by an EPA-certified technician.

7. Disconnect the A/C lines at the condenser and accumulator.

✷✷ CAUTION

Never open, service or drain the radiator or cooling system when hot; serious burns can occur from the steam and hot coolant. Always drain coolant into a sealable container. Coolant should be reused unless it is contaminated or is several years old.

8. Drain the coolant and remove the radiator.

9. Remove the power steering reservoir and drain the fluid.

10. Disconnect the lines from the hydroboost unit.

11. Remove the master cylinder from the hydroboost unit and secure it to the oil fill tube.

12. Disconnect the steering shaft from the steering gear.

13. Disconnect the heater hoses and vacuum lines from the engine. Tag them if needed for installation.

14. Detach the fuse box and wiring harness from the bulkhead connector and lay the harness on the engine.

➡ The engine/transmission assembly is removed from the bottom of the vehicle. Raise the vehicle so the rear of the vehicle is slightly higher than the front. When the frame bolts are removed, the body will be lifted away from the engine/transmission assembly.

15. Raise and safely support the vehicle on a twin post or side post lift.

✷✷ CAUTION

The EPA warns that prolonged contact with used engine oil may cause a number of skin disorders, including cancer! You should make every effort to minimize your exposure to used engine oil. Protective gloves should be worn when changing the oil. Wash your hands and any other exposed skin areas as soon as possible after exposure to used engine oil. Soap and water, or waterless hand cleaner should be used.

16. Drain the engine oil.

17. Matchmark and remove the driveshaft.

18. Remove the starter and starter opening cover.

19. Remove the torque converter bolts through the starter opening.

20. Disconnect the shift linkage from the transmission.

21. Remove the exhaust pipe from the rear of the catalytic converter.

22. Remove the parking brake bracket from the frame.

23. Disconnect the rear brake line from the Brake Pressure Modulator Valve (BPMV).

24. Remove the front bumper and the power steering cooler from the front air deflector.

25. Disengage the Supplemental Inflatable Restraint (SIR) connector.

26. Remove the splash shields from the wheel openings.

27. If equipped, disconnect the rear A/C lines at the rear crossmember. Leave the lines attached to the engine assembly.

✷✷ CAUTION

Observe all applicable safety precautions when working around fuel. Whenever servicing the fuel system, always work in a well ventilated area. Do not allow fuel spray or vapors to come in contact with a spark or open flame. Keep a dry chemical fire extinguisher near the work area. Always keep fuel in a container specifically designed for fuel storage; also, always properly seal fuel containers to avoid the possibility of fire or explosion.

28. Disconnect the fuel lines at the filter, then pull the lines through the crossmember and lay them on the transmission.

29. Disengage the fuel tank electrical connector.

30. On all wheel drive models, disconnect the transfer case vent tube.

31. Be sure all lines and connections are free between the engine/transmission assembly and the body.

32. If using a twin post lift (side lift) do the following:

a. Lower the vehicle to the floor

b. Install the Body Protection Lift Adapter (J 41602) pads or equivalent to the pinch welds on both sides of the vehicle behind the front wheels.

c. Position the front lifting arms of the lift under the body protection adapters.

d. Be sure the rear of the vehicle will be slightly higher than the front when the lift is raised.

e. Raise the lift about halfway up.

f. Place suitable stands under the frame attached to the engine/transmission assembly and remove the frame mounting bolts.

g. Raise the vehicle to clear the engine/transmission assembly.

33. If using a dual cylinder (one front and one rear) lift do the following:

a. Install the Body Protection Lift Adapter (J 41602) pads or equivalent to the pinch welds on both sides of the vehicle behind the front wheels.

b. Install stands under the body protection lift adapters and the rear of the vehicle.

c. Lower the front cylinder of the lift and install the Engine Lifting Adapter (J 41617) or equivalent to the lift.

d. Raise the front cylinder with the

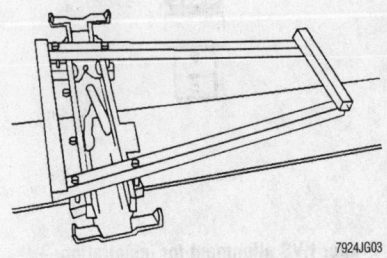

Install the engine lifting adapter to the front cylinder of the twin cylinder lift if applicable—Astro and Safari models

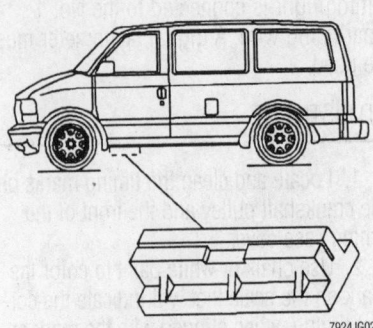

Attach the body protection pads to the pinch welds on both sides before raising the vehicle—Astro and Safari models

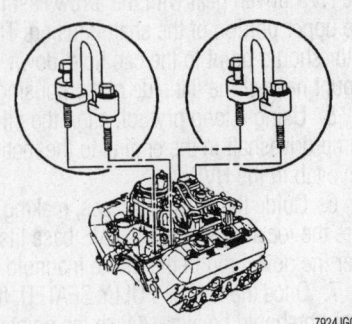

Install the engine lifting brackets at the right rear and left front of the intake manifold—Astro and Safari models

adapter attached until it touches the engine/transmission assembly.

e. Remove the frame mounting bolts and lower the engine/transmission assembly from the vehicle.

34. Remove the two right rear and two left front intake manifold bolts. Install engine lifting brackets (J 41427) or equivalent on the intake manifold to provide lifting points for an engine hoist. Install the hoist and raise the engine/transmission assembly.

35. Remove the transmission from the engine and remove the engine from the frame.

To install:

36. Lower the engine onto the frame and connect the transmission to the engine.

37. Tighten the engine mount through-bolts to 74 ft. lbs. (100 Nm).

38. Position the assembly on suitable stands or on the engine lifting adapter depending on which type of lift you used.

39. Remove the engine lifting brackets from the intake manifold and replace the bolts.

40. Position the engine/transmission assembly in the vehicle and install the frame bolts. Tighten the frame bolts in the following order:

a. Right center bolt to 114 ft. lbs. (155Nm)

b. Left center bolt to 114 ft. lbs. (155Nm)

c. Right front bolt to 66 ft. lbs. (90Nm)

d. Left rear bolt to 66 ft. lbs. (90Nm)

e. Left front bolt to 66 ft. lbs. (90Nm)

f. Right rear bolt to 66 ft. lbs. (90Nm)

41. Remove the stands or the engine lifting adapter. If the engine lifting adapter was used, raise the vehicle and remove the stands.

42. Remove the body protection adapter from the pinchwelds.

43. Install the splash shields in the wheel openings.

44. Connect the steering shaft to the steering gear.

45. Install the power steering cooler.

46. Connect the lines to the hydroboost unit.

47. Connect the hose to the power steering reservoir.

48. Connect the wiring harness to the bulkhead and fuse box.

49. Attach the heater hoses and the SIR connector.

50. Connect the throttle cable and if equipped, the cruise control cable.

51. Install the radiator and air cleaner assembly.

52. Install the master cylinder to the booster.

53. Connect the rear brake line to the BMPV.

54. Install the parking brake bracket to the frame.

55. Connect the fuel lines and if equipped, the A/C lines at the rear cross-member.

56. On all wheel drive models, connect the transfer case vent tube.

57. Install the front bumper and transmission shift linkage.

58. Install the torque converter bolts, starter and driveshaft.

59. Install the exhaust pipe and lower the vehicle.

60. Refill the power steering, engine crankcase, brake system, cooling system and transmission with the proper type and amount of fluid.

61. Have the A/C system charged by an EPA-certified technician.

62. Install the engine cover and battery.

63. Bleed the brake system.

64. Start the engine and check for leaks.

Except Astro and Safari

2.2L ENGINE

> **❊❊❊ CAUTION**
>
> **The manufacturer recommends the discharge and recovery of the A/C system R-134a refrigerant for this procedure. Do not attempt this without the proper equipment. R-134a should not be mixed with R-12 refrigerant.**

1. Disconnect the negative battery cable and properly relieve the fuel system pressure.

2. Disconnect the vacuum reservoir and/or the underhood light from the hood (as equipped).

3. Disconnect the windshield washer line from the hood, then remove the outer cowl vent grilles.

4. Matchmark and remove the hood.

5. Raise and support the vehicle safely. It will be most convenient if the vehicle can be supported so underhood access is still possible. Otherwise, the vehicle will have to be raised and lowered multiple times during the procedure for the necessary access.

6. Properly recover the R-134a refrigerant from the A/C system.

7. Drain the engine cooling system and the engine oil into separate drain pans.

8. Disconnect the oxygen sensor and/or wiring.

9. Disconnect the exhaust at the mani-

folds and loosen the hanger at the catalytic converter.

10. Remove the pencil braces from the engine to the transmission.

11. Remove the inspection cover.

12. Tag and remove the wiring from the starter, then remove the starter.

13. Remove the engine mount through-bolts, then remove the bell housing bolts.

14. Remove the battery from the vehicle, then disconnect the battery ground (negative cable) from the engine.

15. Remove the air cleaner assembly and duct work.

16. Remove the upper fan shroud, then remove the engine cooling fan.

17. Remove the multi-ribbed serpentine drive belt, then remove the water pump assembly.

18. Disconnect the upper radiator hose, then remove the A/C compressor, if equipped, and position aside.

19. Disconnect the lower radiator hose, then remove the radiator and lower fan shroud.

20. Disconnect the power steering hoses from the pump, then cap the openings to prevent system contamination or excessive fluid loss.

21. Detach the heater hose from the intake manifold, then disconnect the ground straps at the rear of the engine.

22. Tag, disconnect and remove the wiring harness and vacuum lines from the engine.

23. Detach the throttle cable, then disconnect the fuel lines from the fuel rail and engine.

24. Support the transmission.

25. Install a suitable lifting device and carefully lift the engine. Pause several times while lifting the engine to be sure no wires or hoses have become snagged.

To install:

26. Carefully lower the engine into the vehicle, then remove the support and engine lifting device.

27. Connect the fuel lines and install the bracket(s).

28. Connect the throttle cable.

29. Attach the vacuum lines and wiring harness connectors as noted during removal. Engage the wiring harness clips and connect the ground straps to the rear of the engine.

30. Attach the heater hose, then uncap and connect the power steering hoses.

31. Install the lower shroud, then install the radiator.

32. Reposition and secure the A/C compressor to the engine.

33. Install the upper radiator hose, then install the water pump.

34. Install the serpentine drive belt, then install the fan assembly.

35. Install the upper radiator shroud, then install the air cleaner and ducts.

36. Connect the battery ground strap to the engine block.

37. Install the bell housing bolts.

38. Install the engine mount through-bolts.

39. Install the starter motor.

40. Install the flywheel cover.

41. Install the pencil braces.

42. Install the catalytic converter pipe assembly and hangers.

43. Carefully lower the vehicle.

44. Align the marks made during removal and install the hood.

45. Install the outer cowl vent grilles, then connect the vacuum reservoir and/or the underhood light to the hood (as equipped).

46. Connect the windshield washer hoses.

47. Check all powertrain fluid levels and add, as necessary. Be sure to properly fill the engine crankcase with clean engine oil.

48. Connect the negative battery cable and properly fill the engine cooling system.

49. Start and run the engine, then check for leaks.

4.3L ENGINES

1. Disconnect the negative battery cable and properly relieve the fuel system pressure.

2. Disconnect the vacuum reservoir and/or the underhood light from the hood (as equipped), then remove the outer cowl vent grilles.

3. Matchmark and remove the hood.

4. Raise and support the front of the vehicle safely. It will be most convenient if the vehicle can be supported so underhood access is still possible. Otherwise, the vehicle will have to be raised and lowered multiple times during the procedure for the necessary access.

5. Drain the engine cooling system and the engine oil into separate drain pans.

6. Disconnect the oxygen sensor and/or wiring.

7. Disconnect the exhaust at the manifolds and loosen the hanger at the catalytic converter. This is necessary to remove the rear catalytic converter cushion mounts for removal of the exhaust assembly.

8. If equipped, remove the skid plate.

9. Remove the pencil braces from the engine to the transmission.

10. If equipped, remove the slave cylinder and position aside.

11. Disconnect the line clamp at the bell housing.

12. Tag and remove the wiring from the starter, remove the flywheel cover and remove the starter.

13. Remove the oil filter.

14. Remove the engine mount through-bolts and remove all of the bell housing bolts, except the upper left.

15. Disconnect the battery ground (negative) cable from the engine.

16. On 4WD vehicles, remove the front drive axle bolts and roll the axle downward.

17. Remove the air cleaner assembly and duct work.

18. Remove the upper radiator shroud, then remove the fan assembly.

19. Remove the multi-ribbed serpentine drive belt, then remove the water pump pulley.

20. Disconnect the upper radiator hose, then remove the A/C compressor, if equipped, and position aside with the lines intact.

21. Disconnect the lower radiator hose, then detach the oil cooler and overflow lines from the radiator. Plug the cooler line openings to prevent system contamination or excessive fluid loss.

22. Remove the radiator from the vehicle, then remove the lower radiator shroud.

23. Disconnect the power steering hoses from the steering gear, then cap the openings to prevent system contamination or excessive fluid loss.

24. Disconnect the heater hoses from the intake manifold and the water pump.

25. Tag, disconnect and remove the wiring harness and vacuum lines from the engine

26. Disconnect the throttle cables, then remove the distributor cap.

27. Remove the remaining bolt from the bell housing.

28. Disconnect the fuel lines and remove the bracket.

29. Remove the ground strap(s) from the rear of the cylinder head.

30. On 4WD vehicles, loosen the front body mount bolts.

31. Support the transmission.

32. Install a suitable lifting device and carefully lift the engine. Pause several times while lifting the engine to be sure no wires or hoses have become snagged.

To install:

33. Carefully lower the engine into the vehicle.

34. On 4WD vehicles, tighten the front body mount bolts.

35. Install the ground strap(s) to the rear of the cylinder head.

36. Connect the fuel lines and install the bracket.

37. Install the upper left bell housing bolt.

38. Install the distributor cap and wires.

39. Connect the throttle cables.

40. Attach the vacuum lines and wiring harness connectors as noted during removal.

41. Connect the heater hoses, then uncap and attach the power steering hoses.

42. Install the lower shroud, then install the radiator.

43. Uncap and attach the oil cooler lines to the radiator, then connect the overflow hose.

44. Connect the lower radiator hose, then if equipped, reposition and secure the A/C compressor to the engine.

45. Install the upper radiator hose, then install the water pump pulley.

46. Install the serpentine drive belt, then install the fan assembly.

47. Install the upper radiator shroud, then install the air cleaner and ducts.

48. For 4WD vehicles, roll the front axle up into position, then install and tighten the retaining bolts.

49. Connect the battery ground strap to the engine block.

50. Install the remaining bell housing bolts.

51. Install the engine mount through-bolts and tighten to 49 ft. lbs. (66 Nm).

52. Install a new oil filter, then install the starter motor.

53. Install the flywheel cover.

54. If equipped, reposition and secure the clutch slave cylinder.

55. Install the pencil brace and the skid plate, as equipped.

56. Install the catalytic converter Y-pipe assembly and hangers.

57. Carefully lower the vehicle.

58. Align the marks made during removal and install the hood.

59. Install the outer cowl vent grilles, then connect the vacuum reservoir and/or the underhood light to the hood (as equipped).

60. Check all powertrain fluid levels and add, as necessary. Be sure to properly fill the engine crankcase with clean engine oil.

61. Connect the negative battery cable and properly fill the engine cooling system.

62. Start and run the engine, then check for leaks.

Water Pump

REMOVAL & INSTALLATION

Astro and Safari

1. Disconnect the negative battery cable, then drain the engine cooling system.

2. For 1996–99 vehicles, remove the Mass Air Flow (MAF) sensor clamp and the air cleaner housing.

3. Remove the upper fan shroud.

4. For 1996–99 vehicles, remove the drive belt and the fan and clutch assembly.

5. Remove the water pump pulley.

6. Loosen the clamps and disconnect any remaining hoses from the water pump, as applicable.

7. Remove the water pump retaining bolts. Remove the water pump assembly from the engine.

➡ **On some engines, the pump retaining bolts will vary in size and thread. Be sure to note the positioning of all**

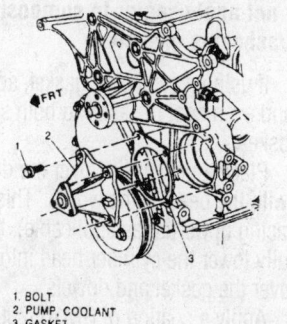

1. BOLT
2. PUMP, COOLANT
3. GASKET

7924JG05

Exploded view of the water pump mounting—2.2L engine

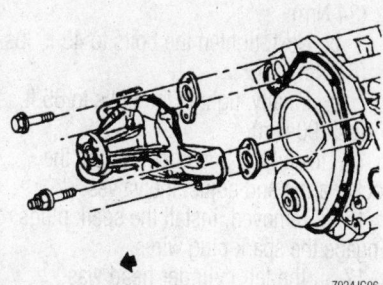

7924JG06

Exploded view of the water pump assembly mounting—4.3L engine

bolts during removal to assure proper installation.

8. Clean gasket mounting surface.

To install:

9. Install water pump, and tighten to 33 ft. lbs. (45 Nm).

10. Install coolant hoses, and replace clamps.

11. Install the pulley, and clutch assembly as needed.

12. Replace drive belt and fan shroud.

13. Install MAF sensor, if needed

14. Connect the negative battery cable.

15. Refill cooling system

Except Astro and Safari

1. Disconnect the negative battery cable, then drain the engine cooling system.

2. Relieve the belt tension, then remove the accessory drive belts or the serpentine drive belt, as applicable.

3. Remove the upper fan shroud, then remove the fan or fan and clutch assembly, as applicable.

4. Remove the water pump pulley.

5. Loosen the clamp and disconnect the coolant hose(s) from the water pump.

➡ **For the hoses on some engines, removal may be easier if the hose is left attached until the pump is free from the block. Once the pump is removed from the engine, the pump may be pulled (giving a better grip and greater leverage) from the tight hose connection.**

6. Remove the retainers, then remove the water pump from the engine. Note the positions of all retainers as some engines will utilize different length fasteners in different locations and/or bolts and studs in different locations.

To install:

7. Using a gasket scraper, carefully clean the gasket mounting surfaces.

➡ **The water pumps on some of the earlier engines covered may have been installed using sealer only, no gasket, at the factory. If a gasket is supplied with the replacement part, it should be used. Otherwise, a ⅛ in. (3mm) bead of RTV sealer should be used around the sealing surface of the pump.**

8. Apply GM 1052080 or equivalent sealant to the threads of the water pump retainers. Install the water pump to the engine using a new gasket, then thread the retainers in order to hold it in position.

9. Tighten the water pump retainers to specification:

a. For 2.2L gasoline engines, tighten the water pump-to-engine retainers to 18 ft. lbs. (23 Nm).

b. For the 4.3L engine, tighten the bolts and studs to 30 ft. lbs. (41 Nm).

10. Connect the coolant hose(s) and secure using the retaining clamp(s).

11. Install the water pump pulley, then install the fan or fan and clutch assembly.

12. If equipped with a serpentine drive belt, position the belt over the pulleys, then carefully allow the tensioner back into contact with the belt.

13. If equipped with V-belts, install the accessory drive belts and adjust the tension.

14. Install the upper fan shroud, then connect the negative battery cable.

15. Properly refill the engine cooling system, then run the engine and check for leaks.

Cylinder Head

REMOVAL & INSTALLATION

2.2L Engine

1. Properly relieve the fuel system pressure, then disconnect the negative battery cable.

2. Drain the engine cooling system, then disconnect the air duct from the air inlet.

3. Disconnect the upper radiator hose, then remove the upper fan shroud.

4. Remove the radiator assembly, then remove the lower fan shroud.

5. Remove the fan assembly, then remove the serpentine drive belt.

6. Remove the water pump pulley.

7. Disconnect the heater hose from the intake manifold and the thermostat housing, then remove the thermostat housing.

8. Remove the alternator support brace and disengage the alternator wiring.

9. If equipped, remove the A/C compressor with brackets, then position them aside. Do not disconnect the refrigerant lines, but be careful not to kink and damage them.

10. Disconnect and reposition the accessory bracket along with the alternator and power steering pump still attached. Be careful not to damage the steering pump lines.

11. Disconnect the throttle cable and cable support linkage, then detach the heater hose from the water pump.

12. Remove the oil fill tube, then disconnect the exhaust pipe and the oxygen sensor.

13. Remove the exhaust manifold bolts, then remove the manifold.

14. Tag and disconnect both the electrical wiring and the vacuum hoses from the upper intake manifold.

15. Remove the upper intake manifold, then tag and disconnect the wiring from the lower intake manifold.

16. Disconnect the fuel lines, then tag and detach the spark plug wires.

17. Remove the lower intake manifold from the engine.

18. Remove the rocker arm cover from the cylinder head.

19. Remove the rocker arms and pushrods.

20. Disconnect the engine lift bracket from the rear of the engine.

21. Remove the cylinder head bolts and studs, then carefully lift the cylinder head from the engine.

To install:

22. Carefully clean and inspect the gasket mounting surfaces.

➡**The gasket surfaces on both the head and block must be clean of any foreign matter and free of nicks or heavy scratches. The cylinder bolt threads in the block and thread on the bolts must be cleaned (dirt will affect the bolt torque).**

23. Place a new gasket over the dowel pins (do not use any sealer on the gasket), then position the cylinder head over the gasket and dowels.

24. Apply a coating of GM 1052080 or equivalent sealer to the cylinder head bolt threads. Install the cylinder head bolts within 15 minutes of sealer application, then tighten them in the proper sequence first to a torque of 46 ft. lbs. (63 Nm) for long bolts or to 43 ft. lbs. (58 Nm) for short bolts, and then tighten all bolts an additional 90 degree turn using a torque angle meter.

25. Install the engine lift bracket.

26. Install the rocker arms and pushrods.

27. Install the rocker arm cover.

28. Install the lower intake manifold.

29. Connect the spark plug wires and the fuel lines.

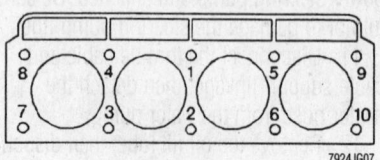

Tighten the cylinder head bolts in the proper sequence to ensure good cylinder sealing—2.2L engine

30. Engage the wiring to the lower intake manifold.

31. Install the upper intake manifold.

32. Connect the vacuum hoses and electrical wiring to the upper intake, as tagged during removal.

33. Install the oil fill tube assembly.

34. Install the exhaust manifold, then connect the exhaust pipe and oxygen sensor.

35. Connect the heater hose to the water pump, then attach the throttle cable support and throttle cable.

36. Install the accessory support bracket and components.

37. If equipped, reposition and secure the A/C compressor.

38. Install the power steering support brace and the alternator support brace. Engage the alternator wiring.

39. Install the thermostat housing, then connect the heater hose to the housing.

40. Install the water pump pulley and the serpentine drive belt.

41. Install the fan assembly, then install the radiator and the lower fan shroud.

42. Install the upper fan shroud, then connect the upper radiator hose.

43. Connect the air inlet duct work, then attach the negative battery cable.

44. Properly refill the engine cooling system and check for leaks.

4.3L Engine

✳✳ CAUTION

Relieve the pressure on the fuel system before disconnecting any fuel line connection.

1. Properly relieve the fuel system pressure, then disconnect the negative battery cable.

2. Drain the engine cooling system.

3. Remove the rocker arm cover.

4. Remove the intake manifold.

5. Remove the exhaust manifold.

6. If removing the right cylinder head, remove or detach the following:
- Electrical connector at the sensor.
- Dipstick tube at the cylinder head bracket.
- Air conditioning compressor (position it aside with the refrigerant lines attached), if equipped.
- A/C compressor, if equipped/belt tensioner bracket.

7. If removing the left cylinder head, remove or detach the following:
- Alternator (position it aside).
- Left side engine accessory bracket with power steering pump (position the

pump aside with the lines attached) and brackets, if equipped.

8. Tag and disconnect the wiring from the spark plugs. If necessary, remove the spark plugs from the cylinder head.

9. Loosen the rocker arms and remove the pushrods.

➡**If valvetrain components, such as the rocker arms or pushrods, are to be reused, they must be tagged or arranged to insure installation in their original locations.**

10. Remove the cylinder head bolts by loosening them in the reverse of the torque sequence, then carefully remove the cylinder head.

To install:

✳✳ WARNING

The gasket surfaces on both the head and block must be clean of any foreign matter and free of nicks or heavy scratches. The cylinder bolt threads in the block and thread on the bolts must be cleaned (dirt will affect the bolt torque).

11. Carefully clean and inspect the gasket mounting surfaces.

➡**Do not apply sealer to composition steel/asbestos gaskets.**

12. If using a steel only gasket, apply a thin and even coat of sealer to both sides of the gaskets.

13. Place a new gasket over the dowel pins with the bead or the words "This Side Up" facing upwards (as applicable), then carefully lower the cylinder head into position over the gasket and dowels.

14. Apply a coating of GM1052080 or equivalent sealer to the threads of the cylinder head bolts, then thread the bolts into position until finger-tight. Using the proper torque sequence, tighten the bolts in three steps:

    a. First, tighten the bolts to 25 ft. lbs. (34 Nm).

    b. Next, tighten the bolts to 45 ft. lbs. (61 Nm).

    c. Finally, tighten the bolts to 65 ft. lbs. (90 Nm).

15. Install the pushrods, secure the rocker arms and adjust the valves.

16. If removed, install the spark plugs. Engage the spark plug wires.

17. If the left cylinder head was removed, reposition and secure the engine accessory bracket with the power steering pump and brackets, as equipped. Install the alternator.

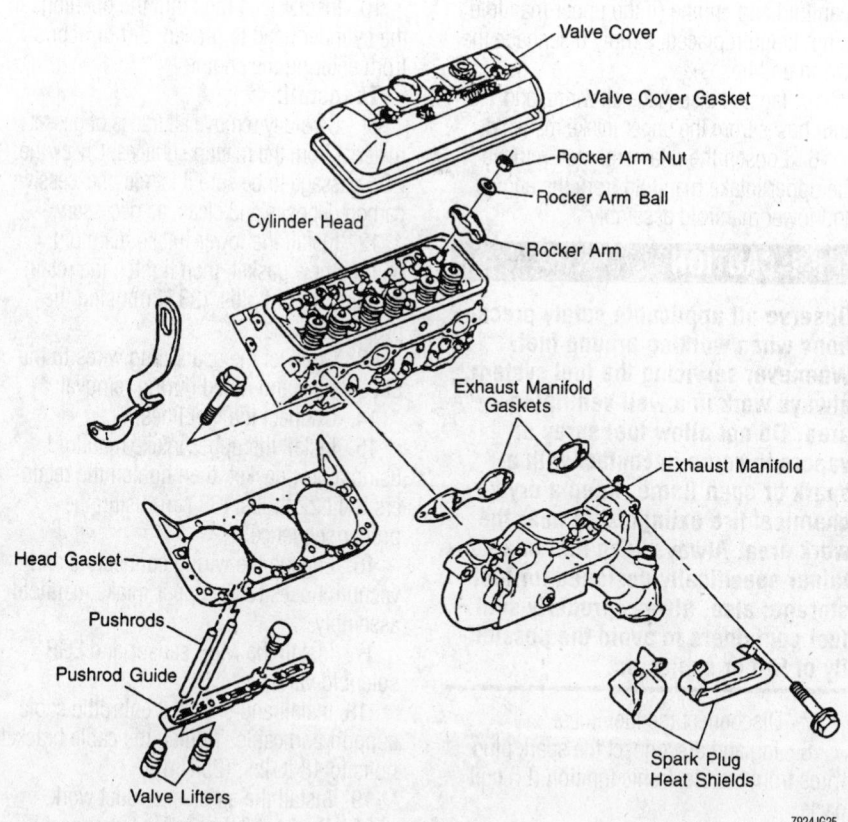

Cylinder head and related components—4.3L engine

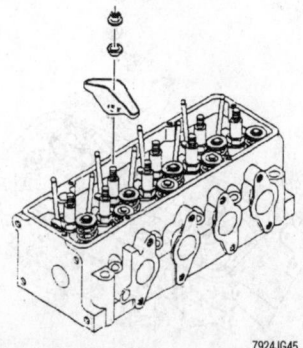

Exploded view of the rocker arm assembly—2.2L engine

18. If the right cylinder head was removed, install the A/C compressor, if equipped, and A/C compressor/belt tensioner bracket, then install the dipstick tube bracket and engage the sensor electrical connector.

19. Install the exhaust manifold.

20. Install the intake manifold.

21. Install the rocker arm cover.

22. Connect the negative battery cable, then properly refill the engine cooling system.

23. Run the engine to check for leaks, then check and/or adjust the ignition timing.

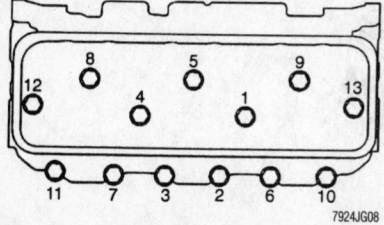

Cylinder head bolt torque sequence—4.3L engine

Rocker Arms

REMOVAL & INSTALLATION

2.2L Engine

1. Remove the rocker arm cover from the cylinder head.

2. Remove the rocker arm retaining nut, then remove the arm and ball. If necessary, withdraw the pushrod form the cylinder head.

➡Valvetrain components which are to be reused must be installed in their original positions. If removed, be sure to tag or arrange all rocker arms and pushrods to assure proper installation.

To install:

3. Inspect the rocker arms, balls and pushrods for damage or wear and replace, as necessary:

• Check the rocker arms, balls and their mating surfaces. Be sure the surfaces are smooth and free from scoring or other damage.

• Check the rocker arm areas that contact the valve stems and the sockets that contact the pushrods, be sure these areas

are smooth and free of both damage and wear.

• Be sure the pushrods are not bent which can be determined by rolling them on a flat surface. Check the ends of the pushrods for scoring or roughness

• Inspect the rocker arm bolts for thread damage. Check the rocker arm bolts in the shoulder area for contact damage with the rocker arm.

4. If removed, install the pushrods making sure they are seated within the lifters.

5. If installing new rocker arms and balls, coat the friction surfaces using Dri-Slide Molykote® or equivalent pre-lube.

✳✳ WARNING

When tightening the rocker arm retainers, be sure the lifter for that valve is resting on the base circle of the camshaft not on the lobe, otherwise the valvetrain can be damaged. Do not overtighten the retainers.

6. Install the rocker arms and ball, then tighten the retaining nuts to 22 ft. lbs. (30 Nm).

➡Valve lash is not adjustable on the 2.2L engine.

7. Install the rocker arm cover, then start and run the engine to check for leaks.

4.3L Engines

1. Remove the rocker arm cover(s) from the cylinder head.

2. Remove the rocker arm nut, the rocker arm and the ball washer.

➡If only the pushrod is to be removed, loosen the rocker arm nut, swing the rocker arm to the side and remove the pushrod.

3. Withdraw the pushrod from the cylinder head.

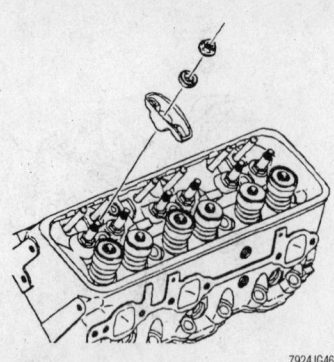

Exploded view of the rocker arm assembly—4.3L engine

To install:

4. Inspect and replace components if worn or damaged.

5. Coat the bearing surfaces of the rocker arms and the rocker arm ball washers with Molykote® or equivalent pre-lube.

6. Install the pushrods making sure they seat properly in the lifter.

7. Install the rocker arms, ball washers and the nuts.

➡The 4.3L engines are equipped with screw-in type rocker arm studs with positive stop shoulders.

8. Tighten the rocker arm adjusting nuts against the stop shoulders to 18 ft. lbs. (24 Nm). No further adjustment is necessary or possible.

9. If possible, properly adjust the valve lash.

10. Install the rocker arm cover(s) to the cylinder head.

11. Start and run the engine, then check for leaks and for proper ignition timing adjustment.

Intake Manifold

REMOVAL & INSTALLATION

2.2L Engine

The 2.2L engine was introduced in 1994 and utilizes a Multi-Port Fuel Injection (MFI) system. The intake manifold is an assembly of separate components, an upper and a lower manifold.

1. Properly release the fuel system pressure (if the lower manifold assembly is being removed) and disconnect the negative battery cable.

2. Remove the air cleaner duct work.

3. Disconnect the throttle cable support and cable from the manifold.

4. Remove the MAP sensor and the EGR solenoid valve from the upper intake

manifold and engine (if the upper manifold is not being replaced, simply disengage the wiring and hoses).

5. Tag and disengage all wiring and vacuum hoses from the upper intake manifold.

6. Loosen the retainers, then remove the upper intake manifold from the engine and lower manifold assembly.

✳✳ CAUTION

Observe all applicable safety precautions when working around fuel. Whenever servicing the fuel system, always work in a well ventilated area. Do not allow fuel spray or vapors to come in contact with a spark or open flame. Keep a dry chemical fire extinguisher near the work area. Always keep fuel in a container specifically designed for fuel storage; also, always properly seal fuel containers to avoid the possibility of fire or explosion.

7. Disconnect the fuel lines.

8. Tag and disconnect the spark plug wires from the Electronic Ignition (EI) coil pack.

9. Remove the lower intake manifold retaining nuts, then remove the lower intake manifold and gasket.

10. Insert clean rags into the openings in the cylinder head to prevent dirt or debris from entering the engine.

To install:

11. Carefully remove all traces of gasket material from the mating surfaces. Check the EGR passage to be sure it is free of excessive carbon deposits and clean, as necessary.

12. Install the lower intake manifold using a new gasket, then tighten the retaining nuts to 24 ft. lbs. (33 Nm) using the proper sequence.

13. Connect the spark plug wires to the EI coil pack and noted during removal.

14. Connect the fuel lines.

15. Install the upper intake manifold using a new gasket, then tighten the retainers to fit 22 ft. lbs. (33 Nm) using the proper sequence.

16. Engage the wiring connectors and vacuum hoses to the upper intake manifold assembly.

17. Install the MAP sensor and EGR solenoid valve.

18. Install and secure the throttle cable support and cable. Tighten the cable bracket bolts to 18 ft. lbs. (25 Nm).

19. Install the air cleaner duct work.

20. Connect the negative battery cable, then start and run the engine to check for leaks.

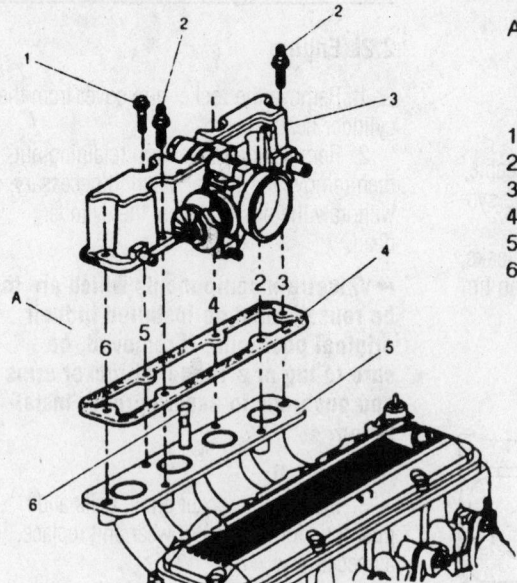

A Upper intake manifold assembly tightening sequence

1 Bolt
2 Stud
3 Upper intake manifold assembly
4 Gasket
5 Lower intake manifold
6 EGR valve injector

Exploded view of the upper intake manifold mounting showing the torque sequence—2.2L engine

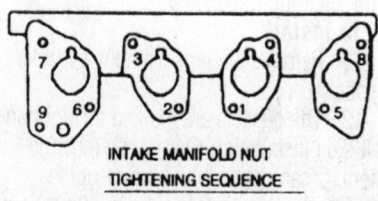

**INTAKE MANIFOLD NUT
TIGHTENING SEQUENCE**

7924JG10

**Lower intake manifold bolt tightening
sequence—2.2L engine**

4.3L (VIN W and X) Engines

➡**If only the upper intake manifold is
being removed, the fuel system pressure does not need to be released.
ALWAYS release the pressure before
disconnecting any fuel lines.**

1. Remove the plastic cover, then properly relieve the fuel system pressure and
disconnect the negative battery cable.

2. Drain the engine cooling system, then
remove the air cleaner and air inlet duct.

3. For 1995 vehicles, disengage the
wiring harness from the necessary upper
intake components including:
- Throttle Position (TP) sensor
- Idle Air Control (IAC) motor
- Manifold Absolute Pressure (MAP)
sensor
- Intake Manifold Tuning Valve
(IMTV)

4. For 1996–99 vehicles, disengage the
wiring harness connectors and brackets and
move aside.

5. Disengage the throttle linkage from
the upper intake manifold, then remove the
ignition coil.

6. Disconnect the fuel lines and bracket
from the rear of the lower intake manifold.
Detach the brake booster vacuum hose at
the upper intake manifold.

7. Disconnect the PCV hose at the rear
of the upper intake manifold, then tag and
disengage the vacuum hoses from both the
front and rear of the upper intake.

8. For 1996–99 vehicles, remove the
purge solenoid and bracket.

9. Remove the upper intake manifold
bolts and studs, making sure to note or
mark the location of all studs to assure
proper installation. Remove the upper intake
manifold from the engine.

10. Disengage the distributor or HVS
wiring (as equipped), then matchmark the
distributor or HVS and remove the assembly
from the engine.

11. Disconnect the upper radiator hose

at the thermostat housing and the heater
hose at the lower intake manifold.

12. Remove any necessary wiring harnesses and brackets.

13. For 1996–99 vehicles, remove the
automatic transmission dipstick tube.
Remove the EGR tube, clamp and tube.

14. Remove the pencil brace (A/C
compressor bracket-to-lower intake manifold).

15. For 1996–99 vehicles, remove the
alternator bracket bolts next to the thermostat housing.

16. For 1995 vehicles, disengage the
wiring harness connectors from the necessary lower intake components including:
- Fuel injector
- Exhaust Gas Recirculation (EGR)
valve
- Engine Coolant Temperature (ECT)
sensor

17. Remove the lower intake manifold
retaining bolts, then remove the manifold
from the engine.

18. Insert clean rags into the openings
in the cylinder head to prevent dirt and
debris from entering the engine.

19. Using a gasket scraper, carefully
clean the gasket mounting surfaces. Be sure
to inspect the manifold for warpage and/or
cracks; if necessary, replace it.

To install:

20. Remove the rags from the cylinder
heads.

21. Position the gaskets to the cylinder
head with the port blocking plates to the rear
and the "This Side Up" stamps facing upward,
then apply a 3/16 in. (5mm) bead of RTV sealant
to the front and rear of the engine block at the
block-to-manifold mating surface. Extend the
bead 1/2 in. (13mm) up each cylinder head to
seal and retain the gaskets.

22. Install the lower intake manifold taking care not to disturb the gaskets, then
tighten the manifold retainers to 35 ft. lbs.
(48 Nm) using the proper torque sequence
for vehicles through 1995.

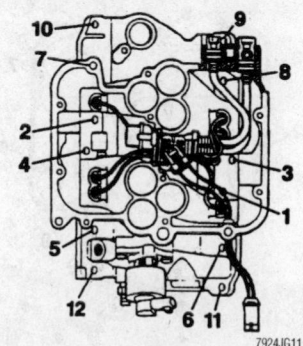

7924JG11

**Lower intake manifold torque sequence—
1995 4.3L (VIN W) engine**

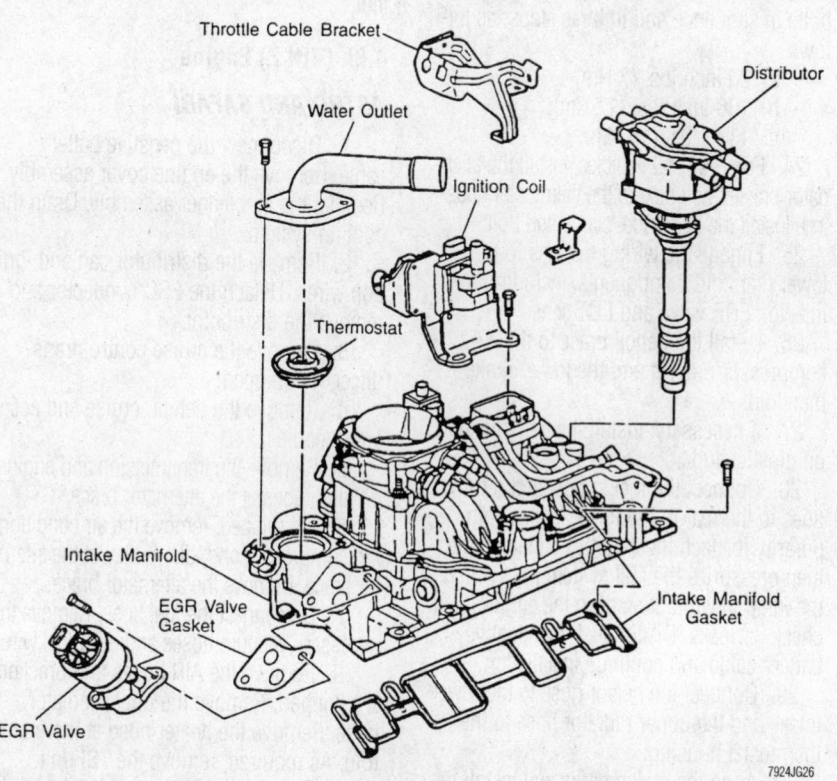

7924JG26

Intake manifold and related components—4.3L engine

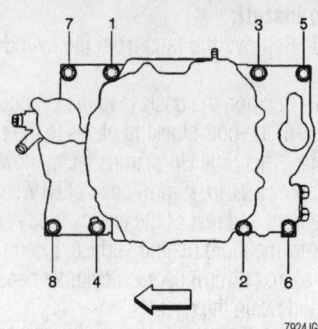

Lower intake manifold tightening sequence—1996–99 4.3L (VIN W and X) engines

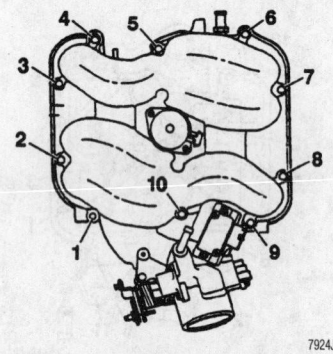

Upper intake manifold torque sequence— 1995 4.3L (VIN W) engine

23. For 1996–99 vehicles, tighten the bolts in sequence and in three steps, as follows:

   a. 26 inch lbs. (3 Nm).
   b. 106 inch lbs. (12 Nm).
   c. 11 ft. lbs. (15 Nm).

24. For 1996–99 vehicles, install the alternator bracket bolt next to the thermostat housing. Install the EGR tube, clamp and bolt.

25. Engage the wiring harness to the lower manifold components, including the injector, EGR valve and ECT sensor.

26. Install the pencil brace to the A/C compressor bracket and the lower intake manifold.

27. If necessary, install the transmission oil dipstick tube.

28. Connect the fuel supply and return lines to the rear of the lower intake. Temporarily reattach the negative battery cable, then pressurize the fuel system (by cycling the ignition without starting the engine) and check for leaks. Disconnect the negative battery cable and continue installation.

29. Connect the heater hose to the lower intake and the upper radiator hose to the thermostat housing.

30. Align the matchmarks and install the distributor assembly, then engage the wiring.

31. Connect the vacuum hoses to the upper and lower intake manifold.

32. Position a new upper intake manifold gasket on the engine, making sure the green sealing lines are facing upward.

33. Install the upper intake manifold being careful not to pinch the fuel injector wires between the manifolds.

34. Install the manifold retainers, making sure the studs are properly positioned, then tighten them using the proper sequence to 124 inch lbs. (14 Nm) for vehicles through 1995. For 1996–99 vehicles, tighten the bolts to 88 inch lbs. (10 Nm).

35. For 1996–99 vehicles, install the purge solenoid and bracket. Connect the brake booster vacuum hose at the upper intake manifold.

36. If removed, connect the PCV hose to the rear of the upper intake manifold and the vacuum hoses to both the front and rear of the manifold assembly.

37. Connect the throttle linkage to the upper intake, then install the ignition coil.

38. If necessary, engage the necessary wiring to the upper intake components including the TP sensor, IAC motor, MAP sensor and the IMTV.

39. Install the plastic cover, the air cleaner and air inlet duct.

40. Connect the negative battery cable, then properly refill the engine cooling system.

4.3L (VIN Z) Engine

ASTRO AND SAFARI

1. Disconnect the negative battery cable. Remove the engine cover assembly. Remove the air cleaner assembly. Drain the cooling system.

2. Remove the distributor cap and ignition wires. Detach the ESC connector and remove the distributor.

3. Remove the cruise control transducer, if equipped.

4. Remove the detent, cruise and accelerator cables.

5. Remove the transmission and engine oil filler tubes at the alternator brace.

6. If equipped, remove the air conditioning compressor and idler pulley at the alternator brace. Remove the alternator brace.

7. Disconnect the fuel lines. Remove the necessary vacuum hoses and electrical wires.

8. Remove the AIR hoses and brackets, if equipped. Remove the upper radiator hose. Remove the heater hose at the manifold. As required, remove the TBI unit.

9. Remove the intake manifold retaining bolts. Remove the intake manifold from the engine.

10. Install rags into the openings in the cylinder heads to prevent dirt and debris from entering the engine.

To install:

11. Remove the rags from the cylinder heads.

12. The gaskets are marked for right and left side installation. Do not interchange them. Clean the sealing surface of the engine block and apply a 5/16 in. (8mm) bead of silicone sealer to each ridge.

13. Install the new gaskets onto the heads. The gaskets will have to be cut slightly to fit past the center pushrods. Do not cut any more material than necessary. Hold the gaskets in place by extending the ridge bead of sealer 1/4 in. (6mm) onto the gasket ends. (When the intake manifold is installed the area between the ridges and the manifold should be completely sealed.)

14. Install the intake manifold and tighten the bolts in sequence to 35 ft. lbs. (47 Nm) except position No. 9 which is tightened to 41 ft. lbs. (56 Nm).

15. Install the AIR hoses and brackets, if equipped. Install the upper radiator hose. Install the heater hose at the manifold. Install the TBI unit.

16. Connect the fuel lines. Install the necessary vacuum hoses and electrical wires.

17. If equipped, install the air conditioning compressor and idler pulley at the alternator brace. Install the alternator brace.

18. Install the transmission and engine oil filler tubes at the alternator brace.

19. Install the detent, cruise and accelerator cables.

20. Install the cruise control transducer, if equipped.

21. Install the distributor, cap and ignition wires. Attach the ESC connector to the distributor.

22. Install the air cleaner assembly. Fill the cooling system with the proper type and quantity of antifreeze.

23. Connect the negative battery cable. Install the engine cover assembly.

EXCEPT ASTRO AND SAFARI

1. Disconnect the negative battery cable and properly relieve the fuel system pressure.

2. Drain the engine cooling system.

3. Remove the air cleaner and heat stove tube.

4. Remove the two braces at the rear of the serpentine drive belt tensioner.

5. Disconnect the upper radiator hose.

6. Remove the emissions relays along with the bracket, then disconnect the wiring harness from the retaining clips and posi-

tion aside. Disconnect the ground cable from the intake manifold stud.

7. Remove the power brake vacuum pipe, then disconnect the heater hose pipe at the manifold and fuel lines at the TBI unit.

8. Remove the ignition coil, then disengage the electrical connectors at the sensors on the manifold.

9. Matchmark and remove the distributor from the engine. For details, please refer to the procedure earlier in this section.

➡**For ease of installation, do not crank the engine with the distributor removed.**

10. Tag and disengage the wires and hoses from the TBI unit.

11. Detach the EGR hose, then disconnect the throttle, TVS and cruise control cables (as equipped).

12. Remove the intake manifold retaining studs and/or bolts, then remove the manifold and gaskets.

13. Install rags into the openings in the cylinder heads to prevent dirt and debris from entering the engine.

To install:

14. Using a gasket scraper, carefully clean the gasket mounting surfaces. Be sure to inspect the manifold for warpage and/or cracks; if necessary, replace it.

15. Remove the rags from the cylinder heads.

16. Position the gaskets to the cylinder head with the port blocking plates to the rear, then apply a ³⁄₁₆ in. (5mm) bead of RTV sealant to the front and rear of the engine block at the block-to-manifold mating surface. Extend the bead ½ in. (13mm) up each cylinder head to seal and retain the gaskets.

17. Install the intake manifold taking care not to disturb the gaskets, then tighten the manifold retainers to 35 ft. lbs. (48 Nm) using the proper torque sequence.

18. Engage the TVS, cruise control and/or throttle cables, as equipped.

19. Connect the EGR hose then engage the wires and hoses at the TBI unit as noted during removal

20. Align and install the distributor assembly.

21. Install the ignition coil, then connect the fuel pipes.

22. Connect the heater hose pipe and the power brake vacuum pipe.

23. Attach the ground cable to the intake manifold stud, then position and secure the wiring harness using the clips.

24. Install the emissions relays along with their bracket, then connect the upper radiator hose.

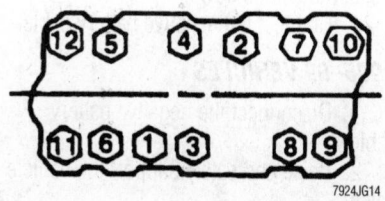

Intake manifold torque sequence—1995 4.3L (VIN Z) engine

25. Install the brace at the rear of the drive belt tensioner, then install the air cleaner and heat stove tube.

26. Connect the negative battery cable, then properly refill the engine cooling system.

27. Run the engine and check for leaks.

Exhaust Manifold

REMOVAL & INSTALLATION

2.2L Engine

1. Disconnect the negative battery cable, then remove the air cleaner and duct work.

2. Either disengage the wiring or remove the oxygen sensor from the manifold. If the manifold or sensor is to be replaced, remove the sensor.

➡**For oxygen sensor service, refer to the Unit Repair Section in the beginning of this manual.**

3. Remove the oil fill tube assembly.

4. Remove the Power Steering (P/S) brace.

5. Remove the heater hose brace, and the A/C brace.

6. Loosen and remove the exhaust manifold retaining nuts, first disconnect the pipe from the manifold, then remove the manifold from the engine.

To install:

7. Carefully clean the threads of the exhaust manifold retainers, then remove all remaining traces of gasket from the mating surfaces.

8. Install the manifold to the engine using a new gasket, then tighten the manifold nuts to 115 inch lbs. (13 Nm). Connect the pipe to the manifold and tighten the retainers.

9. Install the A/C, heater hose, and the P/S braces.

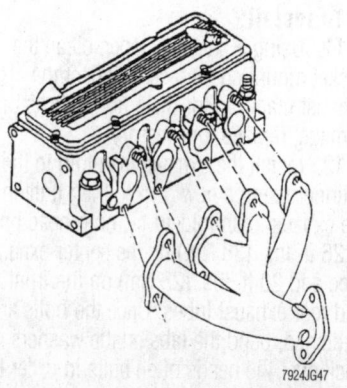

Exhaust manifold mounting—2.2L engine

10. Install the oil fill tube assembly.

11. Either install the oxygen sensor or engage the wiring, as applicable. If the sensor or manifold was replaced, tighten the oxygen sensor to 30 ft. lbs. (41 Nm).

12. Install the air cleaner and duct work.

13. Connect the negative battery cable.

4.3L Engines

1995 VEHICLES

1. Disconnect the negative battery cable.

2. On the Astro and Safari, remove the engine cover.

3. Raise and support the vehicle safely.

4. Disconnect the exhaust pipe from the exhaust manifold.

5. Lower the vehicle for underhood access.

6. Tag and disconnect the spark plug wires from the plugs and from the retaining clips.

7. If removing the left side manifold:

   a. Remove the air cleaner with heat stove pipe and cold air intake pipe.

   b. Remove the power steering/alternator rear bracket.

   c. Check for sufficient clearance between the manifold and the intermediate steering shaft. On some models it will be necessary to disconnect the intermediate shaft from the steering gear in order to reposition the shaft for clearance.

8. If necessary when removing the right side manifold, unbolt the A/C compressor and bracket, then position the assembly aside. Do not disconnect the lines or allow them to become kinked or otherwise damaged.

9. If necessary for the right side manifold, remove the spark plugs, dipstick tube and wiring.

10. Unbend the lock tangs then remove the exhaust manifold retaining bolts, washers and tab washers. Remove the exhaust manifold, then remove and discard the old gaskets.

To install:

11. Using a gasket scraper, clean the gasket mounting surfaces. Inspect the exhaust manifold for distortion, cracks or damage; replace if necessary.

12. Install the exhaust manifold to the cylinder using a new gasket, then tighten the exhaust manifold-to-cylinder head bolts to 26 ft. lbs. (36 Nm) on the center exhaust tube and 20 ft. lbs. (28 Nm) on the front and rear exhaust tubes. Once the bolts are tightened, bend the tabs on the washers back over the heads of all bolts in order to lock them in position.

13. If removed on the right side, install the spark plugs, dipstick tube and wiring.

14. If unbolted, reposition and secure the A/C compressor and bracket assembly.

15. If the left manifold was removed:

    a. If unbolted, reconnect the intermediate shaft to the steering gear.

    b. Install the power steering/alternator rear bracket.

    c. Install the air cleaner along with the heat stove pipe and cold air intake pipe.

16. Connect the spark plug wires to the retainer clips and to the plugs as noted during removal.

17. Raise and support the vehicle safely.

18. Connect the exhaust pipe to the manifold.

19. Lower the vehicle.

20. On the Astro and Safari models, install the engine cover.

21. Connect the negative battery cable.

1996–99 VEHICLES

1. Disconnect the negative battery cable.

2. Raise and safely support the vehicle safely.

3. Disconnect the exhaust pipe from the exhaust manifold.

4. Remove the front tires to gain access to the rear manifold bolts through the front wheel well opening.

5. Remove spark plug wires from spark plugs.

6. If removing the left side manifold, remove the EGR inlet pipe.

7. Carefully lower the vehicle.

8. Remove the exhaust manifold bolts, washers and tab washers.

9. Remove the heat shields, then remove the exhaust manifold from the vehicle.

10. Using a gasket scraper, clean the mounting surfaces. Inspect the exhaust manifold for distortion, cracks or damage; replace if necessary.

To install:

11. Position the exhaust manifold and the heat shields.

12. Install the manifold bolts, washers and tab washers. Tighten the bolts to 22 ft. lbs. (30 Nm), then bend the tab washers over the heads of all the bolts.

13. If installing the left side manifold, connect the EGR inlet pipe.

14. Raise and safely support the vehicle safely.

15. Install the rear manifold bolts through the front wheel opening. Install the front tires.

16. Connect the front exhaust pipe to the manifold.

17. Install spark plug wires.

18. Carefully lower the vehicle, then connect the negative battery cable.

Camshaft and Valve Lifters

REMOVAL & INSTALLATION

2.2L Engine

1. Properly relieve the fuel system pressure, then disconnect the negative battery cable.

2. Drain the engine cooling system and the engine oil.

3. Remove the radiator.

4. Remove the rocker arm cover.

5. Remove the cylinder head.

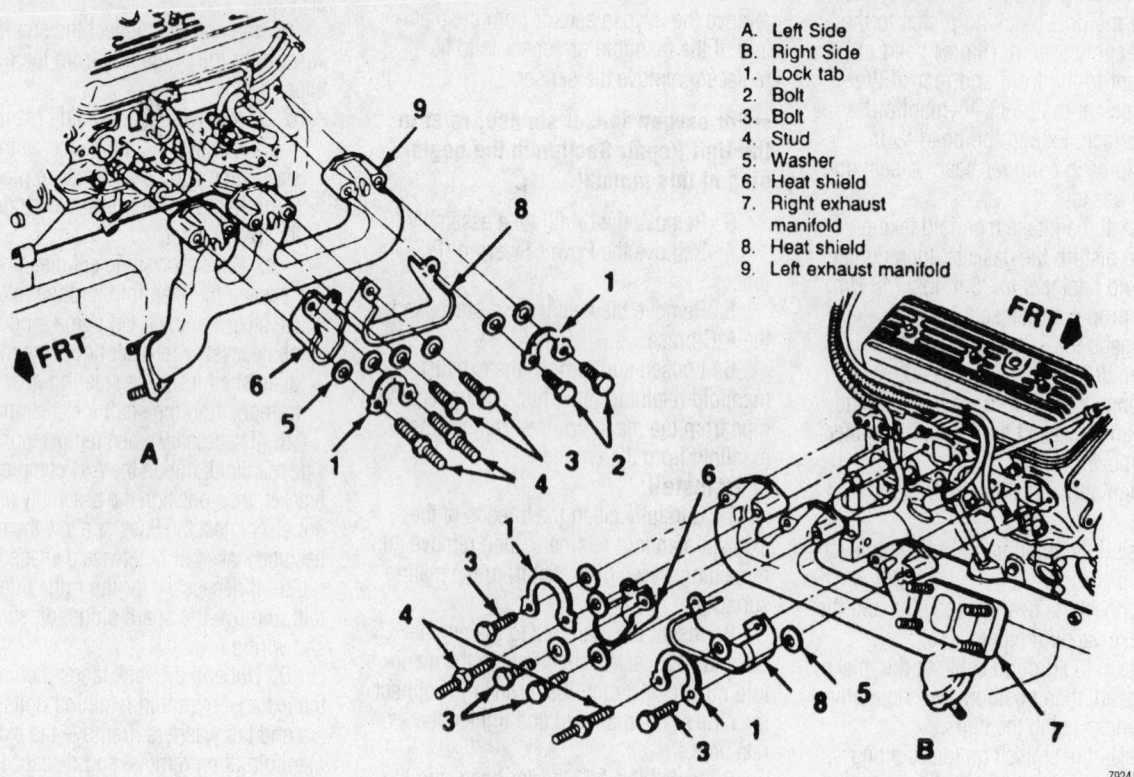

A. Left Side
B. Right Side
1. Lock tab
2. Bolt
3. Bolt
4. Stud
5. Washer
6. Heat shield
7. Right exhaust manifold
8. Heat shield
9. Left exhaust manifold

Exhaust manifold mounting—1995 4.3L engine

7924JG48

6. Remove the anti-rotation bracket bolts and brackets, then remove the valve lifters.

7. Remove the oil pump drive retaining bolt, then remove the drive by lifting and twisting.

8. Remove the crankshaft pulley and hub.

9. Remove the serpentine drive belt idler pulley.

10. Remove the timing cover from the engine.

11. Remove the timing chain and camshaft sprocket.

12. Remove the camshaft thrust plate retaining bolts, then remove the plate from the block.

13. Pull the camshaft straight out of the engine, turning slightly as it is withdrawn and taking care not to damage the bearings.

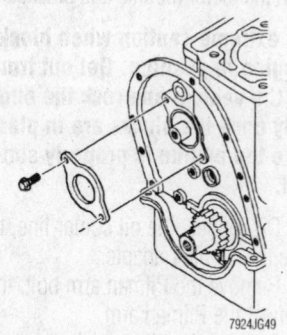

Remove the camshaft thrust plate, then withdraw the camshaft from the engine— 2.2L engine

To install:

14. Inspect the camshaft, journals and lobes for wear and replace, if necessary.

15. If removed, use the camshaft bearing tool to install a new set of bearings.

16. Coat the camshaft lobes and journals with a high viscosity oil with zinc such as GM 12345501 or equivalent.

17. Carefully insert the camshaft in the engine, turning it slightly from side-to-side as it is inserted.

18. Install the thrust plate and tighten the retaining bolts to 106 inch lbs. (12 Nm).

19. Install the timing chain and camshaft sprocket.

20. Install the timing cover to the engine.

21. Install the serpentine drive belt idler pulley.

22. Install the crankshaft pulley and hub.

23. Install the oil pump drive by inserting while twisting, then install the retaining bolt and tighten to 18 ft. lbs. (25 Nm).

24. Install the valve lifters and the anti-rotation brackets.

25. Install the cylinder head and tighten to 46 ft. lbs. (62 Nm) plus 90° turn.

26. Install the rocker arm cover.

27. Install the radiator.

28. Connect the negative battery cable and properly refill the engine cooling system.

4.3L Engines

1. Properly relieve the fuel system pressure, then disconnect the negative battery cable.

2. Drain the engine cooling system.

3. Remove the radiator.

4. Properly recover the refrigerant from the A/C system.

5. Remove the A/C condenser.

6. Remove the rocker arm covers from the engine.

7. Remove the intake manifold assembly.

8. Remove the rocker arms, pushrods and lifters.

9. Remove the crankshaft pulley and hub.

10. Remove the engine front (timing) cover.

11. Align the timing marks on the crankshaft and camshaft sprockets.

12. Remove the camshaft sprocket and timing chain.

13. If equipped, remove the balance shaft drive gear.

14. Remove the camshaft thrust plate.

15. Install the sprocket bolts or longer bolts of the same thread into the end of the camshaft as a handle, then remove the camshaft front the front of the engine while turning slightly from side to side, as necessary. Take care not to damage the camshaft bearings when removing the camshaft.

To install:

16. Lubricate the camshaft journals with clean engine oil or a suitable pre-lube, then install the camshaft into the block being extremely careful not to contact the bearings with the cam lobes.

17. Install the camshaft thrust plate.

18. If equipped, install the balance shaft drive gear.

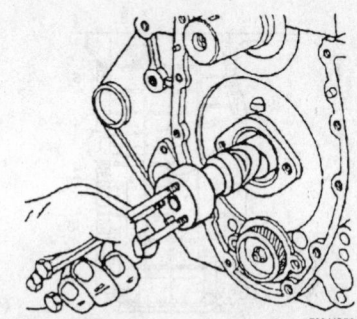

Thread three long bolts into the camshaft to use as a handle, then withdraw the it from the engine

19. Install the timing chain and camshaft sprocket.

20. Install the engine front (timing) cover.

21. Install the crankshaft pulley and hub.

22. Install the valve lifters, then install the pushrods and rocker arms. Properly adjust the valve clearance.

23. Install the intake manifold assembly.

24. Install the rocker arm covers to the engine.

25. Install the radiator to the vehicle.

26. Connect the negative battery cable and properly refill the engine cooling system.

Valve Lash

ADJUSTMENT

2.2L Engine

Because the rocker arm fasteners are secured and tightened, valve lash is not adjustable on the 2.2L engine. If a valvetrain problem is suspected, check that the rocker arm nuts are tightened to 22 ft. lbs. (30 Nm). Be very careful not to overtighten the rocker arm nuts. ONLY tighten the nuts when the hydraulic lifter is resting on the base circle of the camshaft and not when it is held upward on the lobe. When valve lash falls out of specification (valve tap is heard), replace the rocker arm, pushrod and hydraulic lifter on the offending cylinder.

4.3L Engines

The 4.3L engines are equipped with screw-in type rocker arm studs with positive stop shoulders. Because the shoulders allow the rocker arms to be tightened into proper position, no adjustments are necessary or possible. If a valvetrain problem is suspected, check that the rocker arm nuts are tightened to 18 ft. lbs. (24 Nm). When valve lash falls out of specification (valve tap is heard), replace the rocker arm, pushrod and hydraulic lifter on the offending cylinder.

Oil Pan

REMOVAL & INSTALLATION

2.2L Engine

1. Remove the engine assembly.

2. If equipped, remove the clutch pressure plate and disc from the engine.

3. Remove the flywheel.

4. Remove the oil pan nuts and bolts, then remove the pan from the bottom of the block.

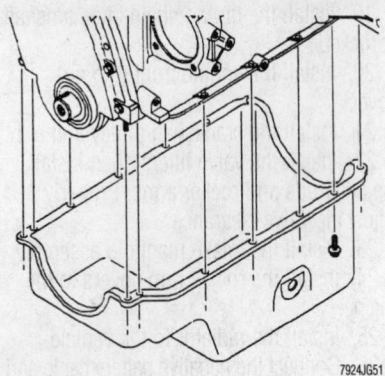

Oil pan mounting—2.2L engine

To install:

5. Carefully clean the gasket mating surfaces of any remaining old gasket or sealer material.

6. Position a new gasket and seal onto the oil pan. Use a thin bead of sealant at either side of the seal.

7. Install the oil pan and tighten the retaining bolts to 89 inch. lbs. (10 Nm).

8. Install the flywheel.

9. If equipped, install the pressure plate and disc.

10. Install the engine to the vehicle.

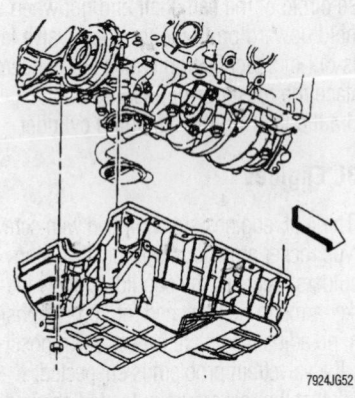

Oil pan mounting—4.3L engines

4.3L Engines

2WD MODELS

1. Remove the engine.

2. Remove the oil pan retainers (nuts, studs and/or bolts) and rail reinforcements, if equipped. Remove the oil pan from the block.

3. For the 1996–99 engines, remove the rubber bell housing plugs and gasket.

To install:

4. Using a gasket scraper, clean the gasket mounting surfaces. Be sure all sealing surfaces are clean and free of oil.

➡On 1996 and later models, the alignments between the rear of the oil pan

and the rear of the block is critical. The oil pan must be flush or slightly forward of the rear of the block to allow for proper alignment with the transmission housing. Use a feeler gauge to measure the clearance between the three oil pan-to-transmission contact points. If the clearance exceeds 0.011 in. (0.3mm) at any of the three points, realign the oil pan.

5. Apply GM 1052080 or equivalent sealant to the oil pan rail where it contacts the timing cover-to-block joint (front) and the crankshaft rear seal retainer-to-block joint (rear). Continue the bead of sealant about 1 in. (25mm) in both directions from each of the four corners.

6. Using a new gasket, install the rubber bell housing plugs (if equipped), the oil pan, reinforcements, if equipped, and retainers. Tighten the retainers to specification:

• 1995 engines:
• Bolts: 100 inch lbs. (11 Nm)
• Nuts at corners: 17 ft. lbs. (23 Nm)
• 1996–99 engines:
• Bolts and studs (in sequence): 18 ft. lbs. (25 Nm).

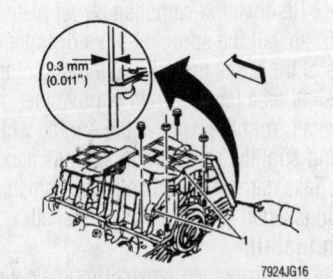

If the clearance between the three oil pan-to-transmission contact points exceeds 0.011 in. (0.3mm) at any of the three points, realign the oil pan—1996–99 4.3L engine

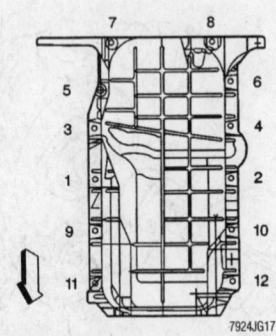

Tighten the bolts in sequence to prevent warping the sealing surface of the oil pan—1996–99 4.3L vehicles

7. Install the engine into the vehicle. Refill the crankcase with fresh oil. Start the engine, establish normal operating temperatures and check for leaks.

4WD MODELS

1. Disconnect the negative battery cable.

2. Remove the dipstick.

3. Raise and support the vehicle safely.

4. Remove the drive belt splash shield, the front axle shield, and the transfer case shield.

5. Remove the front skid plate and drain the engine crankcase oil, then remove the flywheel cover.

6. Remove the left and right motor mount throughbolts.

7. Raise the engine using a suitable lifting device and block in position. This may be accomplished using large wooden blocks between the motor mounts and brackets.

➡Use extreme caution when blocking the engine in position. Get out from under the vehicle and rock the engine slightly once the blocks are in place to be sure the engine is properly supported.

8. Disconnect the oil cooler line, then remove the oil filter adapter.

9. Remove the Pitman arm bolt, then disconnect the Pitman arm.

10. Remove the idler arm bolts, then disconnect the idler arm.

11. Remove the front differential throughbolts, then disconnect or remove the front driveshaft, if necessary.

12. Roll the differential assembly forward for clearance.

13. Remove the starter motor retaining bolts, then lower the starter and either remove it from the vehicle or suspend it out of the way using mechanic's wire.

14. Remove the oil pan bolts, nuts and reinforcements, then lower the oil pan and gasket.

To install:

15. Using a gasket scraper, clean the gasket mounting surfaces. Be sure all sealing surfaces are clean and free of oil.

➡On 1996 and later models, the alignments between the rear of the oil pan and the rear of the block is critical. The oil pan must be flush or slightly forward of the rear of the block to allow for proper alignment with the transmission housing. Use a feeler gauge to measure the clearance between the three oil pan-to-transmission contact points. If the clearance exceeds 0.011 in. (0.3mm) at any of the three points, realign the oil pan.

16. Apply GM 1052080 or equivalent sealant to the oil pan rail where it contacts the timing cover-to-block joint (front) and the crankshaft rear seal retainer-to-block joint (rear). Continue the bead of sealant about 1 in. (25mm) in both directions from each of the four corners.
17. Using a new gasket, install the oil pan, reinforcements and retainers. On 1995 models, tighten the bolts to 100 inch lbs. (11 Nm) and the nuts at the corners to 17 ft. lbs. (23 Nm). On 1996–99 models, tighten the retainers, in sequence, to 18 ft. lbs. (25 Nm).
18. Install the starter motor and secure using the mounting bolts.
19. Roll the differential back into position, then install the front driveshaft. Install the front differential throughbolts.
20. Connect the idler arm and secure using the retaining bolts, then connect the Pitman arm and secure using the bolts.
21. Install the transfer case shield.
22. Install the flywheel cover, then install the front skid plate.
23. Install the front axle shield, then install the drive belt splash shield.
24. Carefully lower the vehicle.
25. Install the dipstick, then properly refill the engine crankcase.
26. Connect the negative battery cable.
27. Start the engine, establish normal operating temperatures and check for leaks.

Oil Pump

REMOVAL & INSTALLATION

1. Remove the oil pan.
2. Remove the oil pump attaching bolt, and if equipped, the pick-up tube nut/bolt. Then remove the pump along with the pick-up tube and shaft, as necessary.
3. If necessary for the 2.2L engine, remove the extension shaft and retainer (being careful not the crack the retainer) from the pump.

To install:

4. For the 2.2L engine, if the extension shaft was removed, heat the extension shaft retainer in hot water, then install the shaft and retainer to the oil pump. Be sure the retainer does not crack during installation.
5. Ensure that the pump pick-up tube is tight in the pump body. If the tube should come loose, oil pressure will be lost and oil starvation will occur. If the pick-up tube is loose it should be replaced.

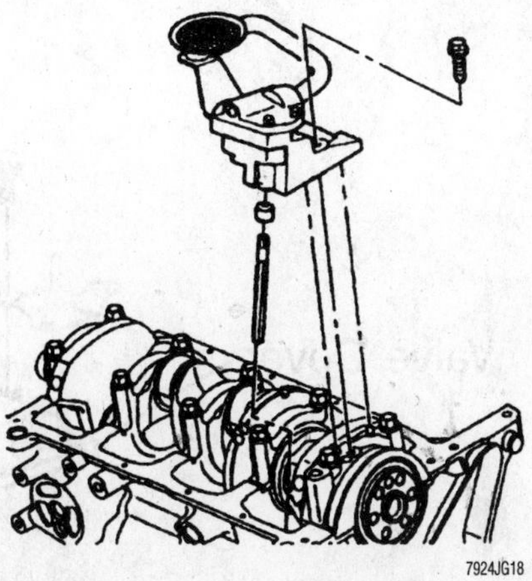

Exploded view of the oil pump mounting—2.2L engine

6. If the pump has been disassembled and is being replaced or for any reason oil has been removed, it must be primed. It can either be filled with oil before installing the cover plate and oil kept within the pump during handling or the entire pump cavity can be filled with petroleum jelly.

➡If the pump is not primed, the engine could be damaged before it receives adequate lubrication when the engine is started.

7. Install the pump aligning the pump shaft with the distributor drive gear as necessary. Tighten oil pump/pick-up tube retainer(s) to specification:
   - 2.2L engine: 32 ft. lbs. (44 Nm)
   - 4.3L engines: 65 ft. lbs. (90 Nm)

➡If the oil pump does not build up oil pressure almost immediately, remove the pan and check for a loose oil pump-to-pick-up tube attachment. If necessary dismantle the pump and pack the pump cavity with petroleum jelly.

8. Install the oil pan and refill the engine crankcase. Disable the ignition system; crank engine for approximately 10 seconds to aid in priming the oil pump and reducing the risk of engine damage.

WARNING

Running the engine without measurable oil pressure will cause extensive damage.

Rear Main Oil Seal

REMOVAL & INSTALLATION

2.2L Engines

Please note that the transmission assembly (and transfer case, if equipped) must be removed to perform this procedure.

1. Disconnect the negative battery cable.
2. Remove the transmission assembly (and transfer case, if equipped) using the recommended procedure.
3. Remove the flexplate on vehicles with automatic transmissions.
4. On vehicles equipped with a manual transmission, remove the clutch assembly and flywheel.
5. Remove the crankshaft seal using a suitable prying tool. Pry the seal out moving the tool around the seal as required until it is removed.

➡Care must be taken not to damage the crankshaft seal surface with the prying tool.

To install:

6. Lubricate the new seal with engine oil.
7. Install the seal using special tool J-34686 or equivalent.
8. Slide the seal over the mandrel until the dust lip bottoms squarely against the tool collar.
9. Align the dowel pin of the tool with the dowel pin hole in the crankshaft and attach the tool to crankshaft.

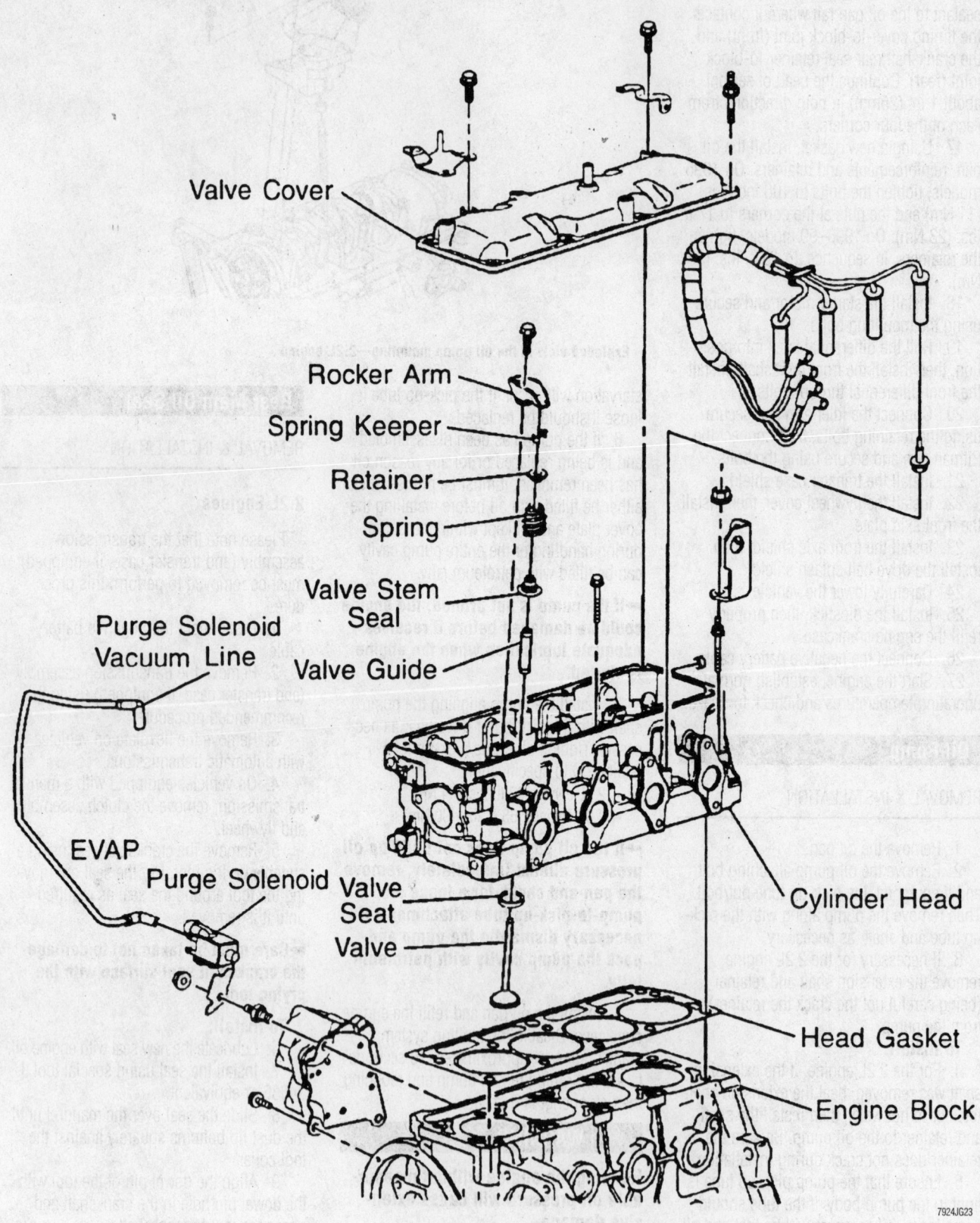

Valve Cover

Rocker Arm

Spring Keeper

Retainer

Spring

Valve Stem
Seal

Valve Guide

Purge Solenoid
Vacuum Line

EVAP

Purge Solenoid

Valve
Seat

Valve

Cylinder Head

Head Gasket

Engine Block

7924JG23

Cylinder head and related components—2.2L engine

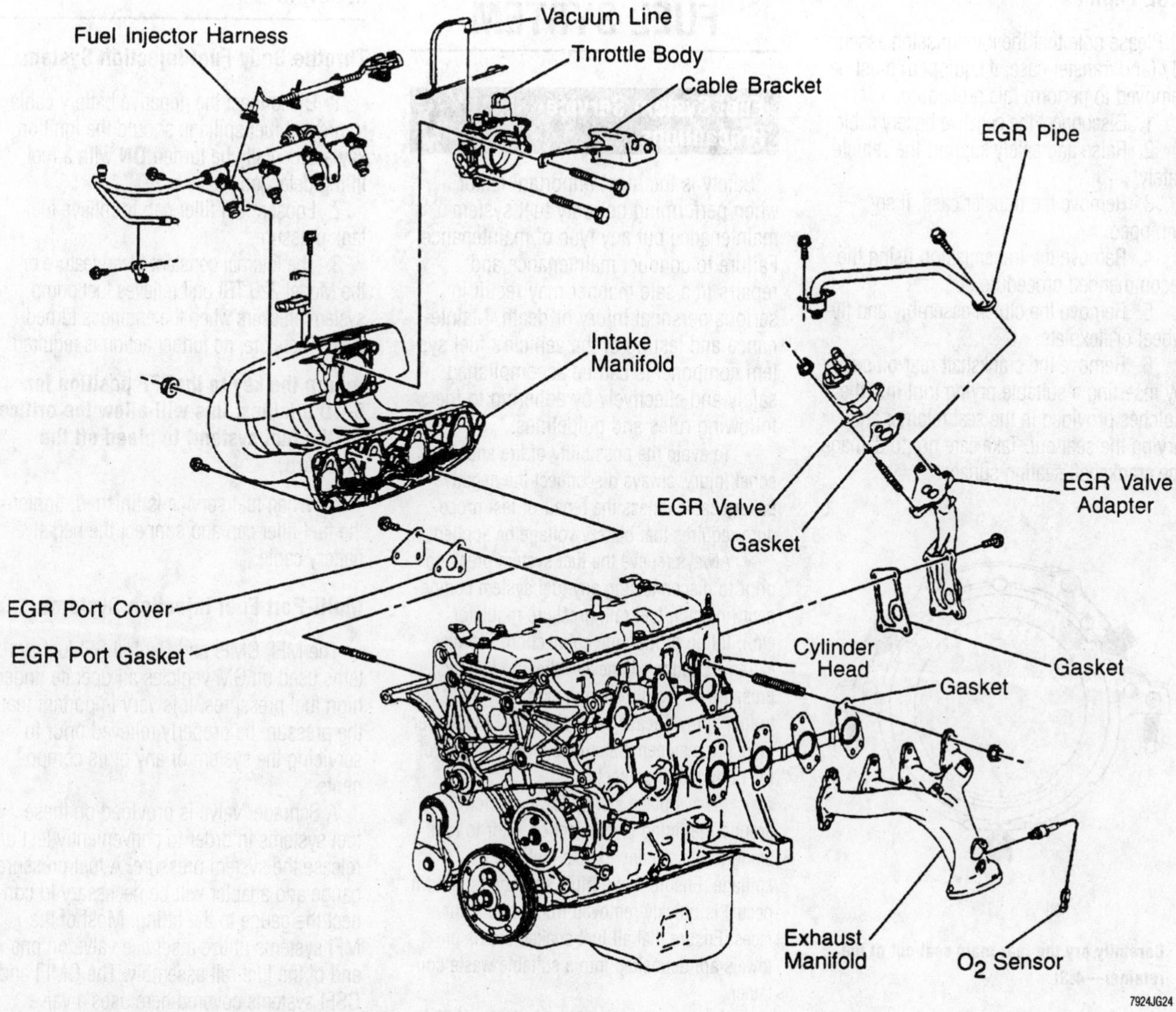

Fuel Injector Harness

Vacuum Line

Throttle Body

Cable Bracket

EGR Pipe

Intake Manifold

EGR Valve Adapter

EGR Valve

Gasket

EGR Port Cover

EGR Port Gasket

Cylinder Head

Gasket

Gasket

Exhaust Manifold

O₂ Sensor

7924JG24

Exploded view of the engine intake and exhaust components—2.2L engine

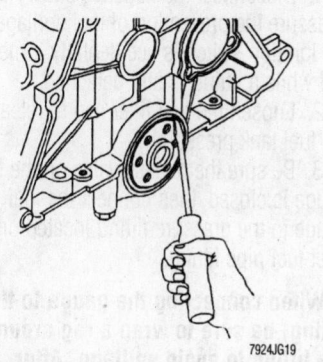

7924JG19

Carefully pry the rear main oil seal out of it's bore—2.2L engine

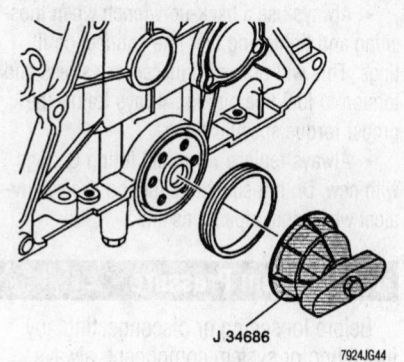

J 34686

7924JG44

Rear main oil seal installation using tool J-34686—2.2L engine

10. Tighten the "T" handle of the tool to push the seal into the bore. Continue until the tool collar is flush against the block.

11. Loosen the "T" handle completely. Remove the attaching screws and tool. Check to be sure the seal is seated squarely in the bore.

12. Install the flywheel and clutch assembly or the flexplate.

13. Install the transmission assembly using the recommended procedure.

14. Connect the negative battery cable.

15. Start the engine and check for leaks.

4.3L Engines

Please note that the transmission assembly (and transfer case, if equipped) must be removed to perform this procedure.

1. Disconnect the negative battery cable.
2. Raise and safely support the vehicle safely.
3. Remove the transfer case, if so equipped.
4. Remove the transmission using the recommended procedure.
5. Remove the clutch assembly and flywheel or flexplate.
6. Remove the crankshaft rear oil seal by inserting a suitable prying tool into the notches provided in the seal retainer and prying the seal out. Take care not to damage the crankshaft sealing surface.

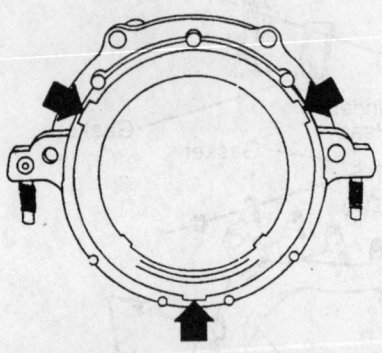

7924JG20

Carefully pry the rear main seal out of the retainer—4.3L

To install:

7. Inspect the crankshaft for grit, rust or burrs and correct as necessary.
8. Clean the running surface of the crankshaft with a non-abrasive cleaner.
9. Lubricate the inner and outer diameter of the new seal. Using a suitable seal installer, install the crankshaft rear oil seal.
10. Install the flywheel and clutch or flexplate.
11. Install the transmission using the recommended procedure.
12. Install the transfer case, if so equipped.
13. Lower the vehicle.
14. Start the engine and verify no oil leaks.

FUEL SYSTEM

Fuel System Service Precautions

Safety is the most important factor when performing not only fuel system maintenance but any type of maintenance. Failure to conduct maintenance and repairs in a safe manner may result in serious personal injury or death. Maintenance and testing of the vehicle's fuel system components can be accomplished safely and effectively by adhering to the following rules and guidelines.

• To avoid the possibility of fire and personal injury, always disconnect the negative battery cable unless the repair or test procedure requires that battery voltage be applied.

• Always relieve the fuel system pressure prior to disconnecting any fuel system component (injector, fuel rail, pressure regulator, etc.), fitting or fuel line connection. Exercise extreme caution whenever relieving fuel system pressure to avoid exposing skin, face and eyes to fuel spray. Please be advised that fuel under pressure may penetrate the skin or any part of the body that it contacts.

• Always place a shop towel or cloth around the fitting or connection prior to loosening to absorb any excess fuel due to spillage. Ensure that all fuel spillage (should it occur) is quickly removed from engine surfaces. Ensure that all fuel soaked cloths or towels are deposited into a suitable waste container.

• Always keep a dry chemical (Class B) fire extinguisher near the work area.

• Do not allow fuel spray or fuel vapors to come into contact with a spark or open flame.

• Always use a back-up wrench when loosening and tightening fuel line connection fittings. This will prevent unnecessary stress and torsion to fuel line piping. Always follow the proper torque specifications.

• Always replace worn fuel fitting O-rings with new. Do not substitute fuel hose or equivalent where fuel pipe is installed.

Fuel System Pressure

Before loosening or disconnecting any fuel fitting or system component, always relieve the fuel system pressure in order to help prevent the danger of fire or injury.

RELIEVING

Throttle Body Fuel Injection System

1. Disconnect the negative battery cable to prevent fuel spillage should the ignition key accidentally be turned **ON** with a fuel fitting detached.
2. Loosen fuel filler cap to relieve fuel tank pressure.
3. The internal constant bleed feature of the Model 220 TBI unit relieves fuel pump system pressure when the engine is turned **OFF**. Therefore, no further action is required.

➡**Turn the key to the OFF position for 5–10 minutes; this will allow the orifice (in the fuel system) to bleed off the pressure.**

4. When fuel service is finished, tighten the fuel filler cap and connect the negative battery cable.

Multi-Port Fuel Injection Systems

The MFI, CMFI and Central SFI fuel systems used on GM vehicles all operate under high fuel pressures. It is very important that the pressure be properly relieved prior to servicing the system or any of its components.

A Schrader valve is provided on these fuel systems in order to conveniently test or release the system pressure. A fuel pressure gauge and adapter will be necessary to connect the gauge to the fitting. Most of the MFI systems utilize a service valve on one end of the fuel rail assembly. The CMFI and CSFI systems covered here uses a valve located on the inlet pipe fitting, immediately before it enters the CMFI/CSFI assembly (towards the rear of the engine).

1. Disconnect the negative battery cable to assure the prevention of fuel spillage if the ignition switch is accidentally turned **ON** while a fitting is still detached.
2. Loosen the fuel filler cap to release the fuel tank pressure.
3. Be sure the release valve on the fuel gauge is closed, then connect the fuel gauge to the pressure fitting located on the inlet fuel pipe fitting.

➡**When connecting the gauge to the fitting, be sure to wrap a rag around the fitting to avoid spillage. After repairs, place the rag in an approved container.**

4. Install the bleed hose portion of the fuel gauge assembly into an approved container, then open the gauge release valve and bleed the fuel pressure from the system.

5. When the gauge is removed, be sure to open the bleed valve and drain all fuel from the gauge assembly.

6. When fuel service is finished, tighten the fuel filler cap and connect the negative battery cable.

Fuel Filter

REMOVAL & INSTALLATION

✳✳ CAUTION

Observe all applicable safety precautions when working around fuel. Whenever servicing the fuel system, always work in a well ventilated area. Do not allow fuel spray or vapors to come in contact with a spark or open flame. Keep a dry chemical fire extinguisher near the work area. Always keep fuel in a container specifically designed for fuel storage; also, always properly seal fuel containers to avoid the possibility of fire or explosion.

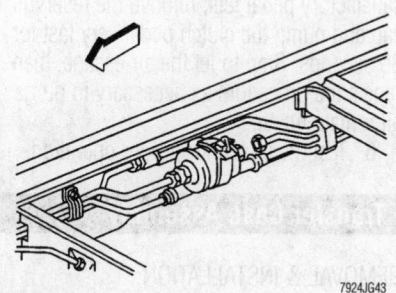

Typical fuel filter location along frame rail

The fuel filter is normally located along the frame rail of the vehicle. On some vehicles however, it may have been relocated to the engine compartment. When in doubt, trace a fuel line from the engine backwards or from the tank forward in order to locate the filter.

1. Properly relieve the fuel system pressure.

2. Remove fuel filler cap.

3. Disconnect the negative battery cable.

4. Raise and support the vehicle safely.

5. Disengage the fuel line connections from the filter.

6. Remove the bolt from the filter mounting clamp, then remove the clamp and filter assembly. Separate the filter from the clamp.

To install:

7. Install the fuel filter on frame rail.

8. Connect the fuel lines.

9. Lower the vehicle.

10. Connect the negative battery cable.

11. Install the fuel filler cap run engine and check for leaks.

➡**The filter has an arrow (fuel flow direction) on the side of the case, be sure to install it correctly in the system, the with arrow facing away from the fuel tank.**

Fuel Pump

REMOVAL & INSTALLATION

✳✳ CAUTION

Observe all applicable safety precautions when working around fuel. Whenever servicing the fuel system, always work in a well ventilated area. Do not allow fuel spray or vapors to come in contact with a spark or open flame. Keep a dry chemical fire extinguisher near the work area. Always keep fuel in a container specifically designed for fuel storage; also, always properly seal fuel containers to avoid the possibility of fire or explosion.

1. Properly relieve the fuel system pressure.

2. Disconnect the negative battery cable.

3. Drain the fuel tank.

4. Remove the filler neck from the tank.

5. Support the fuel tank.

6. Remove the shield from tank, and remove tank straps.

7. Remove the fuel lines and vapor hose from pump

8. Detach the electrical connection from fuel pump

9. Remove the fuel tank

10. Using a suitable spanner wrench, turn the fuel pump/sending unit assembly locking ring (located on top of the fuel tank) counterclockwise, then carefully lift the assembly from the tank and remove the pump from the fuel lever sending device.

To install:

11. Install the fuel pump in tank with new seal around opening.

12. Raise tank and connect fuel lines and vapor hose.

13. Install the tank to the frame and tighten the fasteners to 33 ft. lbs. (45 Nm).

14. Install the shield.

15. Install the fuel filler neck and clamp.

16. Refill the tank.

17. Connect the negative battery cable.

18. Run the engine and check for leaks

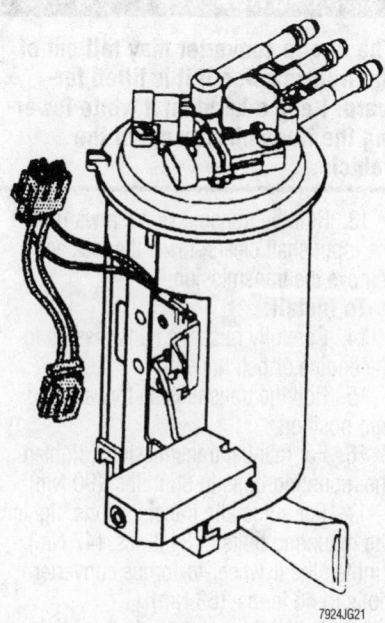

View of the in-tank fuel pump assembly

DRIVE TRAIN

Transmission Assembly

REMOVAL & INSTALLATION

1. Disconnect the negative battery cable.

2. On manual transmissions, remove the shifter boot and lever from inside the vehicle.

3. Raise and safely support the vehicle safely.

4. On 4-wheel drive vehicles, remove the front driveshaft and transfer case.

5. Detach all cables, connectors and fluid lines that may interfere with transmission removal. Tag them if helpful for installation.

6. On automatic transmissions, unbolt

the torque converter from the flexplate and disconnect the shift linkage.

7. Place a drain pan under the transmission and drain the fluid .

8. Position a transmission jack under the transmission and safety-chain the case to the jack.

9. Matchmark and remove the driveshaft.

10. Remove the transmission rear mount.

11. Slightly raise the transmission and remove the crossmember.

12. On automatic transmissions, remove the transmission-to-engine block bolts. For manual transmissions, remove the bolts securing the transmission to the bell housing.

✳✳ CAUTION

The torque converter may fall out of the transmission if it is tilted forward. Keep a hand on it while lowering the transmission out of the vehicle.

13. Roll the transmission rearward until the input shaft clears, lower the jack and remove the transmission.

To install:

14. Carefully raise the transmission to the engine or bell housing.

15. Roll the transmission forward and into position.

16. For manual transmissions, tighten the mounting bolts to 66 ft. lbs. (90 Nm).

17. For automatic transmissions, tighten the mounting bolts to 34 ft. lbs. (47 Nm). Tighten the flywheel-to-torque converter bolts to 46 ft. lbs. (63 Nm).

18. Install the crossmember and tighten the bolts to 55 ft. lbs. (74 Nm).

19. Install the transmission rear insulator and lower retainer. Tighten the bolts to 35 ft. lbs. (47 Nm).

20. The rest of the installation is the reverse of removal.

21. Refill all transmissions with MERCON® automatic transmission fluid to the correct level.

Clutch

REMOVAL & INSTALLATION

1. Disconnect the negative battery cable.

2. Remove the transmission.

3. Install a clutch alignment tool or a used transmission input shaft to support the clutch.

4. If the clutch assembly is going to be reused, mark the flywheel, clutch cover and a pressure plate lug for alignment when installing.

5. Remove the clutch cover bolts and washers.

6. Remove the clutch cover assembly and the clutch plate. Remove the clutch alignment tool.

7. Clean all parts and inspect for damage.

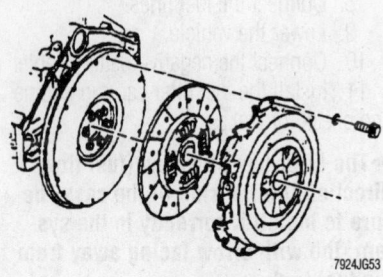

Exploded view of the clutch disc and related components

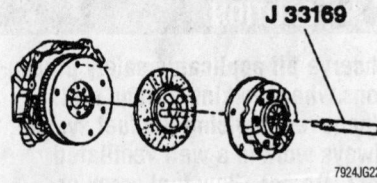

Use the clutch alignment tool to center and support the clutch disc during installation

To install:

8. Install a clutch alignment tool, or equivalent, to support the clutch.

9. Align the marks made during removal or, if new, align the lightest part of the clutch cover, identified by a yellow dot, with the heaviest part identified by an "X".

10. Install the washers and bolts securing the clutch plate and clutch cover assembly to the flywheel.

11. Tighten each screw one turn at a time to avoid warping the clutch cover. Tighten the clutch cover and plate-to-flywheel bolts to 33 ft. lbs. (45 Nm) for 2.2L engines, or to 29 ft. lbs. (40 Nm) for 4.3L engines.

12. Remove the clutch alignment tool, then install the transmission in the vehicle.

13. Connect the negative battery cable.

Hydraulic Clutch System

Bleeding air from the hydraulic clutch system is necessary whenever any part of

the system has been disconnected or the fluid level (in the reservoir) has been allowed to fall so low, that air has been drawn into the master cylinder.

BLEEDING

1. Fill master cylinder reservoir with new brake fluid conforming to DOT 3 specifications.

✳✳ CAUTION

Always use new fluid from a sealed container. Never, under any circumstances, use fluid which has been bled from a system to fill the reservoir as it may be aerated, have too much moisture content and possibly be contaminated.

2. Have an assistant fully depress and hold the clutch pedal, then open the bleeder screw.

3. Close the bleeder screw and have your assistant release the clutch pedal.

4. Repeat the procedure until all of the air is evacuated from the system. Check and refill master cylinder reservoir as required to prevent air from being drawn through the master cylinder.

➡**Never release a depressed clutch pedal with the bleeder screw open or air will be drawn into the system.**

5. If the previous steps do not result in satisfactory pedal feel, remove the reservoir cap and pump the clutch pedal very fast for 30 seconds. Stop to let the air escape, then repeat the procedure as necessary to purge all remaining air.

6. Test the clutch for proper operation.

Transfer Case Assembly

REMOVAL & INSTALLATION

Astro and Safari

1. Disconnect the negative battery cable. If necessary, shift the transfer case into the **4HI** position to ease linkage removal.

2. Raise and support the vehicle safely.

3. If equipped, remove the skid plate.

4. Drain the fluid from the transfer case.

5. Remove the front and rear driveshafts.

6. Disconnect the breather hose.

7. Detach the electrical connections.

8. Support the transfer case with a suitable jack.

9. Remove the adapter-to-transfer case bolts.

10. Remove the nuts and washers.

11. Remove the transfer case, then remove the transfer case-to-adapter gasket.

To install:

12. Position a new transfer case-to-adapter gasket. Use sealer to hold the gasket in place.

13. Support the transfer case with a suitable jack.

14. Fasten the transfer case adapter to the transfer case.

15. Install the transfer case retaining bolts and tighten the bolts to 38 ft. lbs. (52 Nm).

16. Remove the jack from the transfer case.

17. Install the washers and nuts. Tighten the nuts to 26 ft. lbs. (35 Nm).

18. Install the support brace, washers and bolts. Tighten the bolts to 74 ft. lbs. (100 Nm).

19. Attach the electrical connectors.

20. Connect the breather hose.

21. Install the front and rear driveshafts.

22. Fill the transfer case with the proper lubricant.

23. Lower the vehicle, then connect the negative battery cable.

Except Astro and Safari

1. Disconnect the negative battery cable.

2. Shift the transfer case into the **4HI** range.

3. Raise and support the vehicle safely.

4. If equipped, remove the skid plate/transfer case shield bolts, then remove the skid plate/transfer case from under the transmission/transfer case assembly.

5. Remove the plug and drain the transfer case fluid.

6. Matchmark and remove the front and rear driveshafts from the transfer case.

7. Tag and detach the vacuum lines and/or the electrical connectors, as equipped.

8. If applicable, disconnect the transfer case shift rod/cable from the case.

9. If applicable, remove the support brace-to-transfer case bolts.

10. Support the transfer case, then remove the transfer case-to-transmission retaining bolts

11. Slide the transfer case rearward and off the transmission output shaft, then carefully lower it.

12. Remove all traces of old gasket material from the mating surfaces.

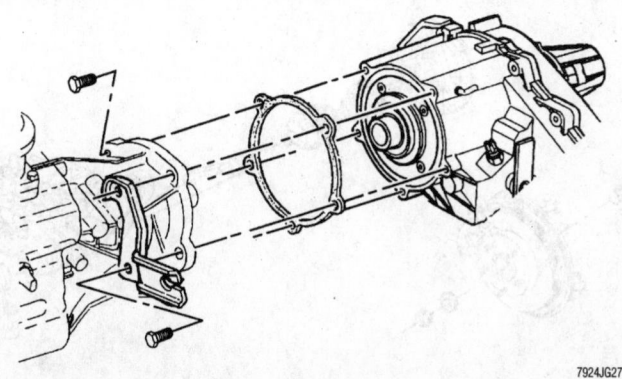

Typical transfer case-to-manual transmission mounting

7924JG27

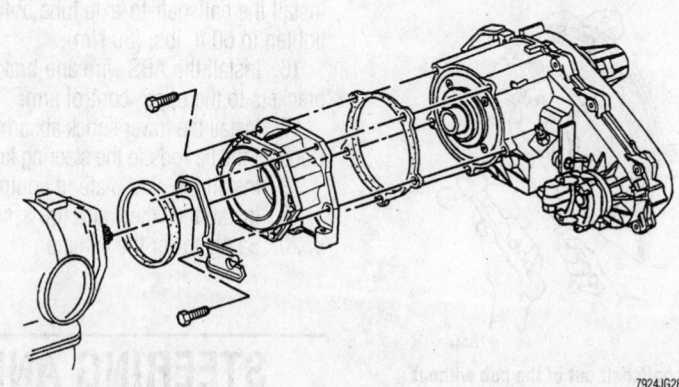

Typical transfer case-to-automatic transmission mounting

7924JG28

To install:

13. Carefully raise the transfer case into position behind the transmission. Position a new gasket, using sealer to hold it in position, then slide the transfer case onto the transmission output shaft.

14. Install the transfer case-to-transmission retaining bolts, then tighten to 24 ft. lbs. (33 Nm) for 1995 models or to 41 ft. lbs. (55 Nm) for 1996–99 models.

15. If equipped, install the support brace bolts and tighten to 35–37 ft. lbs. (47–50 Nm).

16. Remove the support from the transfer case.

17. If equipped, connect the shift rod to the case.

18. Engage the vacuum lines and/or electrical connections, as necessary.

19. Align and install the front and rear driveshafts.

20. Be sure the vehicle is level, then properly refill the transfer case through the filler plug.

21. If equipped, install the skid plate.

22. Carefully lower the vehicle.

23. Connect the negative battery cable.

Halfshaft

REMOVAL & INSTALLATION

1. Unlock the steering column so the steering linkage is free to move.

2. Disconnect the negative battery cable.

3. Raise and support the vehicle safely. Remove the front tire and wheel assemblies.

4. Place a drift through the caliper into the edge of the rotor to keep the rotor from turning when the nut is removed. Remove the cotter pin, retainer, nut and washer.

5. Remove the brake line support bracket and ABS wire bracket from the upper control arm.

6. Matchmark the axle tube-to-flange location. Loosen but do not remove the axle tube-to-flange bolts. Using the proper tools, remove the tie rods at the steering knuckles.

7. Remove the lower shock absorber retaining bolts and move the shock absorbers aside.

8. Separate the upper ball joint from the steering knuckle. Suspend the steering knuckle from the frame using a piece of wire.

9. If equipped, remove the skid plate. Remove the halfshaft-to-axle tube bolts.

10. Move the inner part of the halfshaft

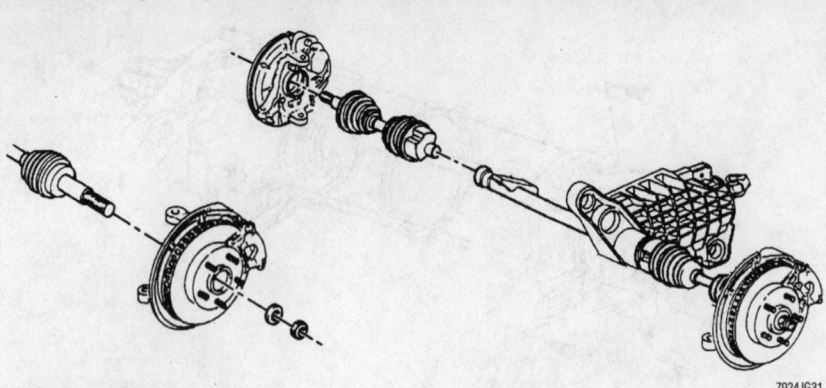

Halfshafts and related components

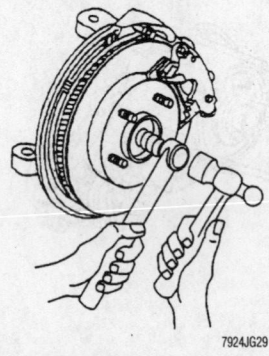

Tap the halfshaft out of the hub without damaging the threads

forward. Support it away from the frame. Using a suitable tool, remove the shaft from the hub and bearing assembly.

11. Remove the halfshaft from the differential using a block of wood and a hammer.

To install:

12. Install the halfshaft into to differential, then push it into the hub.

13. Connect the steering knuckle to the upper ball joint.

14. Install the shaft retaining nut and washer. Tighten the halfshaft retaining nut to 160–200 ft. lbs. (220–270 Nm).

15. Install the cotter pin in the nut.

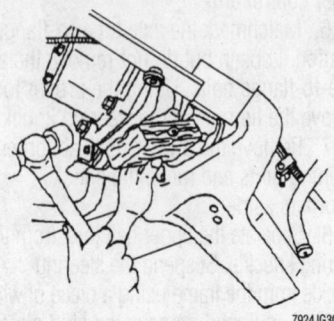

Using a block of wood and a mallet, disengage the halfshaft from the differential assembly

Install the halfshaft-to-axle tube bolts and tighten to 60 ft. lbs. (80 Nm).

16. Install the ABS wire and brake hose brackets to the upper control arm.

17. Install the lower shock absorber bolts. Connect the tie rods to the steering knuckle.

18. Install the skid plate, if equipped.

19. Install the wheel and tire assemblies.

20. Safely lower the vehicle.

STEERING AND SUSPENSION

Air Bag

✳✳ CAUTION

Some vehicles are equipped with an air bag system, also known as the Supplemental Inflatable Restraint (SIR) system. The system must be disabled before performing service on or around system components, steering column, instrument panel components, wiring and sensors. Failure to follow safety and disabling procedures could result in accidental air bag deployment, possible personal injury and unnecessary system repairs.

PRECAUTIONS

Several precautions must be observed when handling the inflator module to avoid accidental deployment and possible personal injury.

• Never carry the inflator module by the wires or connector on the underside of the module.

• When carrying a live inflator module, hold securely with both hands, and ensure that the bag and trim cover are pointed away.

• Place the inflator module on a bench or other surface with the bag and trim cover facing up.

• With the inflator module on the bench, never place anything on or close to the module which may be thrown in the event of an accidental deployment.

DISARMING

✳✳ CAUTION

The Supplemental Inflatable Restraint (SIR) system must be disarmed before performing many in-vehicle service procedures. Failure to do so may cause accidental deployment of the air bag, resulting in unnecessary SIR system repairs and/or personal injury.

1. Turn the steering wheel so the vehicle's wheels are pointing straight ahead.

2. Turn the ignition switch to the **LOCK** position and remove the key.

3. Disconnect the negative battery cable.

4. Remove the AIR BAG or SIR fuse from the instrument panel fuse block, as applicable.

5. Remove the steering column filler panel or left-hand sound insulator, as applicable for access to the SIR wiring harness.

6. Remove the Connector Position Assurance (CPA) device, then disengage the yellow 2-way connector at the base of the steering column.

➡ **With the fuse removed, the AIR BAG or SIR light will illuminate if the ignition switch is turned ON at any time. This is normal and does not indicate a problem when the system is disarmed.**

7. If equipped with passenger side air bags, remove the right-hand sound insulator, then detach the CPA and yellow 2-way connector from the passenger inflator module pigtail.

To enable:

8. Be sure the ignition is in the **LOCK** position.

9. If equipped with passenger side air bags, attach the yellow 2-way connector and CPA to the passenger inflator module pigtail, then install the right-hand sound insulator.

10. Attach the yellow 2-way connector and CPA at the base of the steering column. After installing the CPA, clip the connector to the flange on the steering column support.

11. Install the steering column filler or left-hand insulator, as applicable.

12. Install the AIR BAG or SIR fuse into the instrument panel fuse block.

13. Disconnect the negative battery cable.

14. Turn the ignition switch to the **RUN** position and verify that the AIR BAG warning lamp flashes 7–9 times and then extinguishes. If it does not go out, it indicates a fault in the air bag system.

Manual Recirculating Ball Steering Gear

REMOVAL & INSTALLATION

➡ **The following procedure requires the use of the GM Pitman Arm Remover tool No. J-6632 or equivalent.**

1. Disconnect the negative battery cable from the battery.

2. Raise and safely support the front frame of the vehicle. Position the wheel in the straight ahead direction.

3. Remove the intermediate shaft-to-steering gear pinch bolt.

4. Remove the Pitman arm-to-Pitman shaft nut, mark the relationship the arm to the shaft. Using a heavy duty 2-jaw puller, separate the Pitman arm from the Pitman shaft.

➡ **When separating the Pitman arm from the shaft, DO NOT use a hammer or apply heat to the arm.**

5. Remove the steering gear-to-frame bolts and the gear from the vehicle.

➡ **When installing the steering gear, be sure that the intermediate shaft bottoms on the worm shaft, so that the pinch bolt passes through the undercut on the worm shaft. Check and/or adjust the alignment of the Pitman arm-to-Pitman shaft.**

To install:

6. Line up the steering gear and install the three steering gear-to-frame mounting bolts and tighten to 60 ft. lbs. (81 Nm).

7. Install the Pitman arm on the Pitman arm shaft and tighten the attaching bolt to 185 ft. lbs. (252 Nm).

8. Install the intermediate shaft-to-steering gear bolt and tighten to 30 ft. lbs. (40 Nm).

9. Lower the vehicle, connect the negative battery cable and road test.

Power Recirculating Ball Steering Gear

REMOVAL & INSTALLATION

The power steering gear is has a recirculating ball system that acts as a rolling thread between the worn shaft and the rack piston. The worm shaft is supported by a preloaded thrust bearing and two conical thrust races at the lower end and a bearing assembly in the adjuster plug at the upper end. If the steering system becomes damaged and looses hydraulic pressure, the vehicle can be controlled manually. Steering gear adjustments are not required as part of periodic maintenance. Service this vehicle with GM Hydraulic Power Steering Fluid GM P/N 1052884 or equivalent.

1. Raise and safely support the vehicle safely.

2. Turn the wheels to the straight ahead position.

3. Mark the position of the Pitman arm in relation to the Pitman arm shaft.

4. Remove the Pitman arm using the proper puller. Do not use heat or a hammer on the arm.

5. Position a fluid catch pan under the power steering gear.

6. At the power steering gear, disconnect and plug the hoses. Catch any excess fluid in a drain pan.

➡ **Plug the hoses and the openings of the power steering pump to keep dirt out of the system.**

7. Remove the intermediate shaft-to-steering gear bolt. Matchmark the intermediate shaft-to-power steering gear and separate the shaft from the gear.

8. Remove the power steering gear-to-frame bolts, washers and the steering gear from the vehicle.

To install:

9. Install the steering gear into the vehicle and tighten the power steering gear-to-frame bolts to 55 ft. lbs. (75 Nm).

10. Align the intermediate shaft and the steering gear on the marks previously made and tighten the intermediate shaft-to-power steering gear bolt to 30 ft. lbs. (39 Nm).

11. Install the Pitman arm, making sure to line up the reference marks made at removal. Tighten the Pitman arm-to-sector shaft nut to 185 ft. lbs. (250 Nm).

12. Connect the hoses to the power steering gear, refill the power steering reservoir and bleed the power steering system.

13. Lower the vehicle and road test.

Shock Absorbers

REMOVAL & INSTALLATION

Astro and Safari

1. Raise and safely support the vehicle. Support the lower control arm (front) or axle assembly (rear)

2. Remove the tire and wheel assembly. If removing the front shock absorber, remove the inner wheel well splash shield.

3. Remove the lower nut, washer and bolt.

4. Remove the upper nut, washer and bolt. Compress the front shock absorber to make removal easier.

5. Remove the shock absorber from the vehicle.

6. Installation is the reverse of the removal procedure. Tighten the lower shock absorber mounting nut to 62 ft. lbs. (84 Nm) and the upper bolts to 18 ft. lbs. (25 Nm).

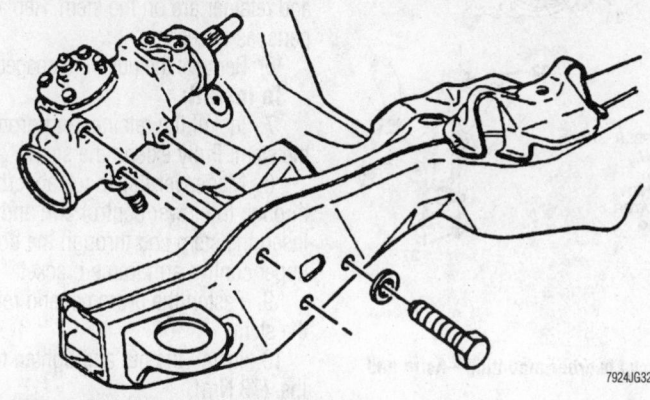

Exploded view of a typical steering gear mounting

7924JG32

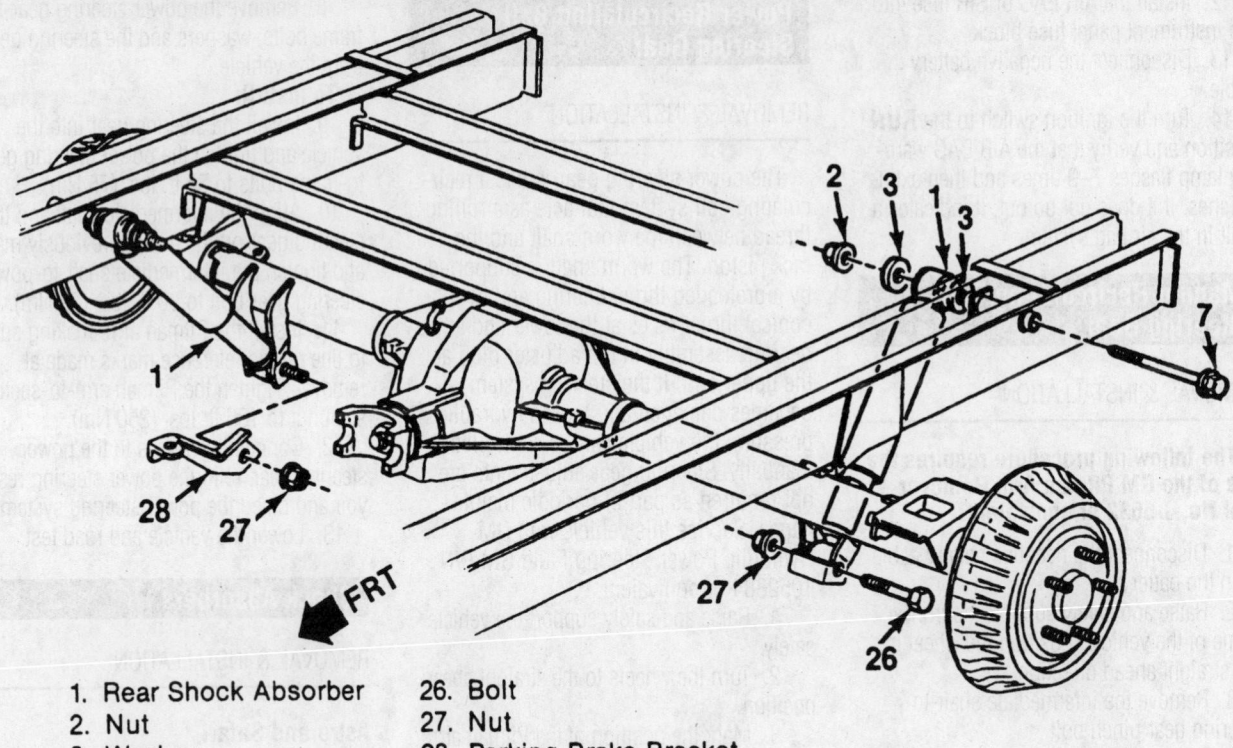

1. Rear Shock Absorber
2. Nut
3. Washer
7. Bolt

26. Bolt
27. Nut
28. Parking Brake Bracket

7924JG36

Rear shock absorber mounting—Astro and Safari

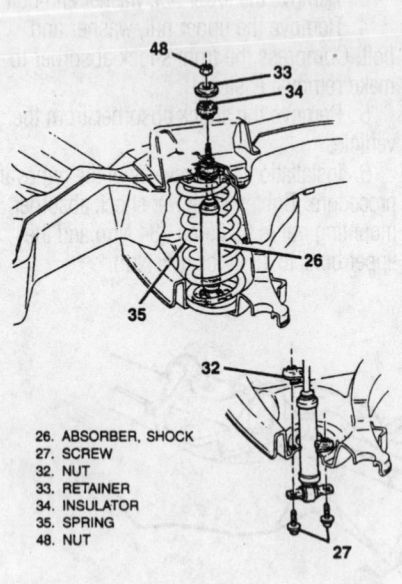

26. ABSORBER, SHOCK
27. SCREW
32. NUT
33. RETAINER
34. INSULATOR
35. SPRING
48. NUT

7924JG37

Front shock absorber mounting—Astro and Safari

Except Astro and Safari

FRONT—2WD MODELS

1. Raise and safely support the vehicle.
2. Remove the wheel and wire assembly.
3. Remove the nut. Hold the shock absorber stem with a wrench while backing the nut off.
4. Unfasten the retainer and remove the grommet.
5. Remove the bolts. Pull the shock absorber out from below. Lower grommet and retainer are on the stem. Replace the parts, as necessary.
6. Remove the nuts, if damaged or worn.

To install:

7. Install the retainer and grommet on the stem. Fully extend the stem.
8. Maneuver the shock absorber up through the lower control arm and spring. Insert the stem end through the hold in the upper control arm frame bracket.
9. Fasten the grommet and retainer to the stem.
10. Install the nut and tighten to 54 ft. lbs. (73 Nm).
11. If needed, install new nuts.

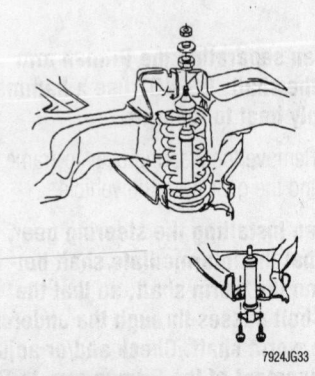

7924JG33

Front shock absorber mounting—2WD except Astro and Safari

12. Install the bolt through the pivot holes to the lower control arm holes. Tighten to 54 ft. lbs. (73 Nm).
13. Install the wheel and tire assembly, then carefully lower the vehicle.

FRONT—4WD MODELS

1. Raise and safely support the vehicle safely.
2. Remove the wheel and tire assembly.
3. Remove the lower nut and bolt, then collapse the shock absorber.

4. Unfasten the upper nut and bolt, then remove the shock absorber from the vehicle.

To install:

5. Position the shock absorber to the bracket.

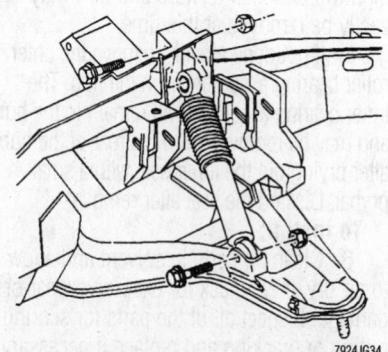

Front shock absorber mounting—4WD except Astro and Safari

6. Install the bolts, the install the nuts. Tighten the nuts to 54 ft. lbs. (73 Nm).

7. Install the tire and wheel assembly, then carefully lower the vehicle.

REAR

1. Raise and support the vehicle safely.

2. Properly support the rear axle assembly.

3. Remove the frame bracket nuts and bolts from the shock absorber.

4. Remove the anchor plate nut and washer from the shock absorber.

5. Remove the shock absorber from the vehicle.

To install:

6. Position the shock absorber to the lower anchor plate. Do not attach the washer or nut.

7. Install the shock absorber-to-frame bracket with the bolts and nuts.

8. Install the lower anchor plate washer and nut and tighten to 62 ft. lbs. (84 Nm).

9. Lower the vehicle.

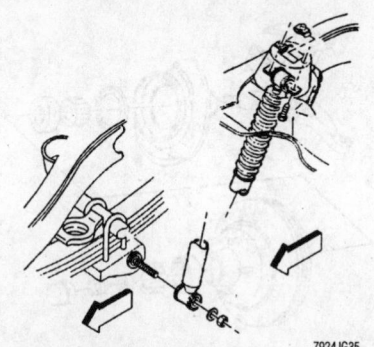

Rear shock absorber mounting—except Astro and Safari models

Coil Springs

REMOVAL & INSTALLATION

1. Raise and support the vehicle safely. Remove the wheel and tire assembly. Remove the shock absorber lower retaining bolts.

2. Push the shock absorber through the control arm and into the spring.

3. With the vehicle supported so the control arms hang free, install tool J-23028 or equivalent, onto a support and into the lower control arm bushings.

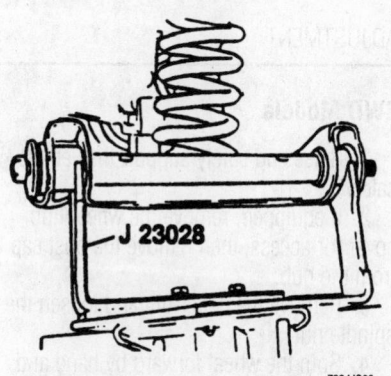

Secure tool J 23028 or equivalent to a jack, then raise the jack to remove the tension on the lower control arm bolts—Astro and Safari models

A. ALIGNMENT OF SPRING END,
   INSULATOR EDGE
   AND DRAIN HOLE
35. SPRING
36. INSULATOR, UPPER
37. INSULATOR, LOWER

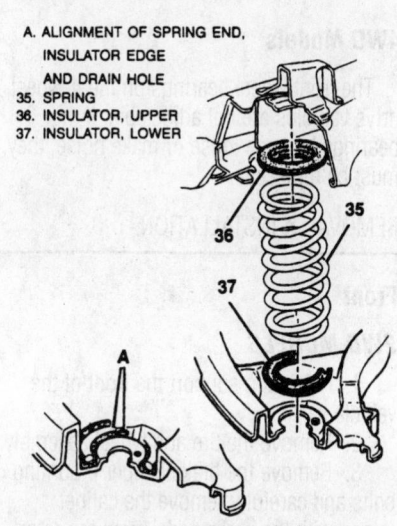

Exploded view of the coil spring mounting—Astro and Safari

Remove the stabilizer bar from the control arm.

4. Remove the stabilizer to lower control arm attachment. Raise and remove the tension on the lower control arm bolts.

5. Install a safety chain around the spring and through the lower control arm. Remove the lower control arm rear pivot bolt, than remove the other pivot bolt.

6. Lower and allow the lower control arm to hang free. Remove the spring assembly.

7. Installation is the reverse of the removal procedure. When positioning the spring in the lower control arm, be sure the spring insulator is in the proper position before lifting the control arm in place.

Upper Ball Joint

REMOVAL & INSTALLATION

1. Raise and support the vehicle safely. Properly support the lower control arm.

➡**The control arm must be supported so the spring and the control arm remain intact.**

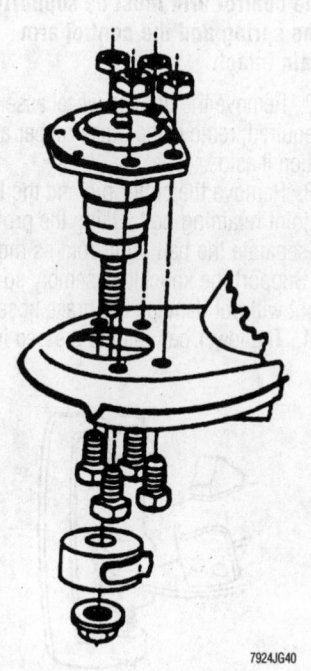

The replacement ball joint comes with nuts and bolts for installation

2. Remove the tire and wheel assembly. As required, remove the brake caliper and position it aside.

3. Remove the cotter pin and the upper ball joint retaining bolt. Using the proper tool separate the upper joint from its mounting. Support the knuckle assembly so its weight will not damage the brake hose.

4. Remove the rivets from the ball joint assembly, using a drill with an ⅛ inch and then ½ inch bit. Remove the ball joint from the upper control arm.

To install:

5. Install ball joint to control arm and bolt in place tighten to 17 ft. lbs. or (23 Nm)

6. Install ball joint to knuckle.

7. Lower nut to 79 ft. lbs. (108 Nm).

8. Upper nut to 61 ft. lbs. (83 Nm).

9. Install cotter pins

10. Install wheels and tires.

11. Lower the vehicle.

12. Align the front end.

Lower Ball Joint

REMOVAL & INSTALLATION

1. Raise and support the vehicle safely. Properly support the lower control arm.

➡**The control arm must be supported so the spring and the control arm remain intact.**

2. Remove the tire and wheel assembly. As required, remove the brake caliper and position it aside.

3. Remove the cotter pin and the lower ball joint retaining bolt. Using the proper tool separate the ball joint from its mounting. Support the knuckle assembly so its weight will not damage the brake hose.

4. The lower ball joint is pressed into

the control arm. Remove ball joint by using a good ball joint press that resembles a C-clamp.

To install:

5. Install ball joint to control arm and press into place.

6. Install control arm to steering knuckle and tighten to 79 ft. lbs. (108 Nm).

7. Install new cotter pin and install brake caliper.

8. Install tires and wheels.

9. Lower the vehicle.

10. Align front end.

Front Wheel Bearings

ADJUSTMENT

2WD Models

1. Raise and safely support the vehicle safely.

2. If equipped, remove the wheel/hub cover for access, then remove the dust cap from the hub.

3. Remove the cotter pin and loosen the spindle nut.

4. Spin the wheel forward by hand and tighten the nut to 12 ft. lbs. (16 Nm) in order to fully seat the bearings and remove any burrs from the threads.

5. Back off the nut until it is just loose, then finger-tighten the nut.

6. Loosen the nut ¼–½ turn until either hole in the spindle lines up with a slot in the nut, then install a new cotter pin. This may appear to be too loose, but it is the proper adjustment.

7. Proper adjustment creates 0.001–0.005 in. (0.025–0.127mm) end-play.

4WD Models

The front wheel bearings on the 4-wheel drive vehicles are not adjustable. If the bearings become loose or make noise, they must be replaced.

REMOVAL & INSTALLATION

Front

2WD MODELS

1. Raise and support the front of the vehicle safely.

2. Remove the tire and wheel assembly.

3. Remove the brake caliper mounting bolts and carefully remove the caliper (along with the brake pads) from the rotor. Do not disconnect the brake line; instead wire the caliper out of the way with the line still attached.

4. Carefully pry out the grease cap, then remove the cotter pin, spindle nut, and washer. Remove the hub, being careful not to drop the outer wheel bearings. As the hub is pulled forward, the outer wheel bearings will often fall forward and they may easily be removed at this time.

5. If not done already, remove the outer roller bearing assembly from the hub. The inner bearing assembly will remain in the hub and may be removed from the rear of the hub after prying out the inner seal with a small prybar. Discard the seal after removal.

To install:

6. Clean all parts in solvent and allow to air dry, then check for excessive wear or damage. Inspect all of the parts for scoring, pitting or cracking and replace if necessary.

➡**DO NOT remove the bearing races from the hub, unless they show signs of damage.**

7. If it is necessary to remove the wheel bearing races, use the GM Front Bearing Race Removal tool No. J-29117 or equivalent, to drive the races from the hub/disc assembly. A hammer and brass drift may also be used to drive the races from the hub, but the race removal tool is quicker.

8. If the bearing races were removed, position the replacement races in the freezer for a few minutes and then install them to the hub:

    a. Lightly lubricate the inside of the hub/disc assembly using wheel bearing grease.

    b. Using the GM Seal Installation tools No. J-8092 and J-8850 or equivalent, drive the inner bearing race into the hub/disc assembly until it seats. Be sure the race is properly seated against the hub shoulder and is not cocked.

➡**When installing the bearing races, be sure to support the hub/disc assem-**

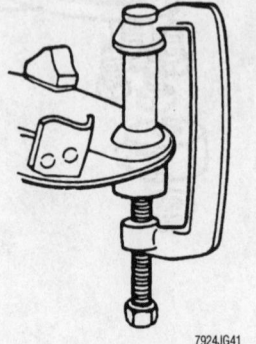

Press the lower ball joint out of the control arm with a portable ball joint press

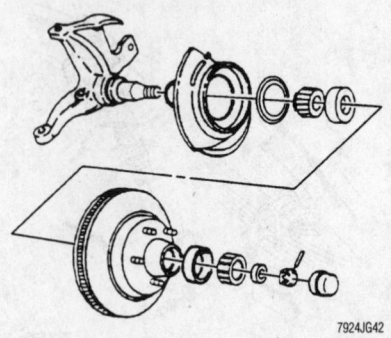

Wheel bearings, races and related components—2WD vehicles

bly with GM tool No. J-9746–02 or equivalent.

c. Using the GM Seal Installation tools No. J-8092 and J-8457 or equivalent, drive the outer race into the hub/disc assembly until it seats.

9. Using a suitable high melting point wheel bearing grease, lubricate the bearings, races and spindle; be sure to place a gob of grease (inside the hub/disc assembly) between the races to provide an ample supply of lubricant.

➡To lubricate each bearing, place a gob of grease in the palm of the hand, then scoop the bearing through the grease until it is well lubricated.

10. Place the inner bearing in the hub, then apply a thin coating of grease to the sealing lip and install a new inner seal, making sure the seal flange faces the bearing cup.

➡Although a seal installation tool is preferable, a section of pipe with a smooth edge or a suitably sized socket may be used to drive the seal into position. Be sure the seal is flush with the outer surface of the hub assembly.

11. Carefully install the wheel hub over the spindle.

12. Using your hands, firmly press the outer bearing into the hub.

13. Loosely install the spindle washer and nut, but do not install the cotter pin or dust cap at this time.

14. Install the brake caliper.

15. Install the tire and wheel assembly.

16. Properly adjust the wheel bearings, then install a new cotter pin and the dust cap.

17. Install the wheel/hub cover, then remove the supports and carefully lower the vehicle.

4WD Models

1. Raise and support the front of the vehicle safely.

2. Install Torsion Bar Unloading tool J 36202 on the torsion bar adjusting bolt and remove the bolt. To aid during installation, count the number of turns required to remove the bolt.

3. Remove the tire and wheel assembly.

4. Install an axle shaft boot seal protector to the Tri-pot axle joint.

5. At the wheel hub, remove the cotter pin and retainer, then loosen and remove the castle nut and the thrust washer. In order to hold the hub from turning when loosening the nut, insert a drift through the caliper and into the rotor vanes.

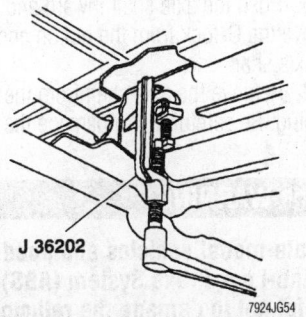

Use Torsion Bar Unloading tool J 36202 to remove the adjusting bolt and unload the torsion bar

6. Remove the brake caliper and support it aside using wire or a coat hanger. Be sure the brake line is not stretched or damaged.

7. Remove the brake disc from the wheel hub.

8. Remove the bolts retaining the hub/bearing assembly to the knuckle, then carefully pull the assembly from the splined end of the halfshaft. If available, use J-28733-A, or an equivalent spindle remover to prevent damage to the shaft or hub/bearing assembly.

➡When removed, lay the hub and bearing assembly on the hub bolt (outboard) side in order to prevent damage or contamination of the bearing seal.

9. Remove the splash shield.

10. Remove the cotter pin and castle nut from the tie rod end, then separate the end from the knuckle using a suitable steering linkage puller.

11. Remove the cotter pins from the ball joints, then loosen the stud nuts.

12. Use the Ball Joint Separator tool J-36607 or equivalent to loosen the ball joints in the steering knuckle.

13. Remove the ball joint nuts, then separate the ball joints from the knuckle and remove the knuckle from the vehicle.

14. Remove the spacer and the seal from the steering knuckle.

15. Clean and inspect the parts for nicks, scores and/or damage, then replace them as necessary.

To install:

16. Install a new seal into the steering knuckle, using a Knuckle Seal Installation tool such as J-28574 or equivalent.

17. Install the spacer, then position the knuckle and insert the upper and lower ball joints.

18. Install the upper and lower ball joint stud nuts and tighten to specification, then install new cotter pins.

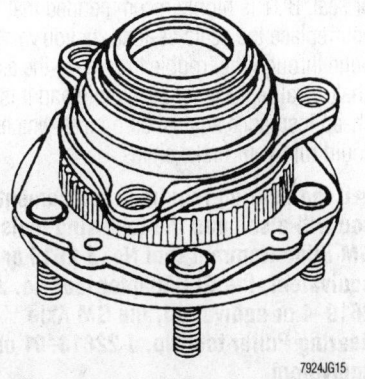

Hub and bearing assembly—4WD vehicles

a. Tighten upper nut to 61 ft. lbs. (83 Nm).

b. Tighten lower nut to 79 ft. lbs. (108 Nm).

19. Align the splash shield to the knuckle, then install the hub and bearing assembly, aligning the threaded holes. Install the retaining bolts and tighten to 77 ft. lbs. (105 Nm).

20. Install the tie rod end to the steering knuckle, then secure using the retaining nut and a new cotter pin.

21. Install the brake disc.

22. Reposition and secure the brake caliper.

23. Install the washer and retaining nut to the end of the halfshaft. Insert a brass drift to keep the rotor and hub from turning, then tighten the shaft nut to 180 ft. lbs. (245 Nm).

24. Install the retainer and a new cotter pin, but DO NOT back off specification in order to insert the cotter pin.

25. Remove the torsion bar unloader tool and the drive axle boot protector.

26. Install the tire and wheel assembly.

27. Carefully lower the vehicle.

28. Check and/or adjust the vehicle trim height, as necessary.

Rear

A new pinion shaft lockbolt should be installed whenever either of the axle shafts are removed. You should probably purchase this and 2 new seals if you are planning on removing or replacing any components covered by this procedure.

➡Axle shaft seal removal and installation uses the following special tools: the GM Axle Shaft Seal Installer tool No. J-33782 or equivalent and the Axle Shaft Bearing Installer tool No. J-34974 or equivalent.

The axle shaft and seal may be removed and replaced without disturbing the bearing or seal, BUT is highly recommended that you replace the seals as-long-as you've gone through the trouble to remove the axle shaft. Seal replacement is simple and it is cheap insurance against an oil leak which could ruin your brake shoes.

➡**If the bearing requires replacement, you will also need the following tools: GM Slide Hammer tool No. J-2619 or equivalent, the GM Adapter tool No. J-2619–4 or equivalent, the GM Axle Bearing Puller tool No. J-22813–01 or equivalent.**

1. Raise and support the rear of the vehicle safely using jackstands.

2. Remove the rear wheel assemblies, then remove the brake drums.

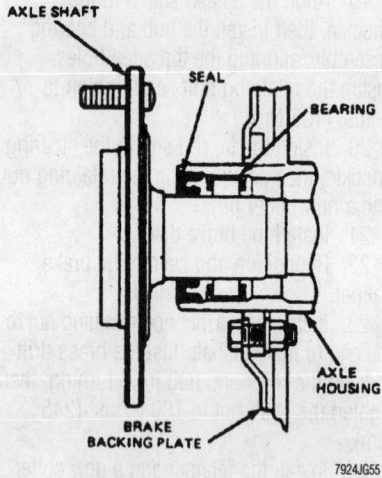

Cross-sectional view of the rear axle, bearing and seal assembly

> ✳ **CAUTION**
>
> **Brake shoes may contain asbestos, which has been determined to be a cancer causing agent. Never clean the brake surfaces with compressed air! Avoid inhaling any dust from any brake surface! When cleaning brake surfaces, use a commercially available brake cleaning fluid.**

3. Using a wire brush, clean the dirt/rust from around the rear axle cover.

4. Place a catch pan under the differential, then remove the drain plug (if equipped) or rear axle cover and drain the fluid (discard the old fluid).

5. At the differential, remove the rear pinion shaft lockbolt and the pinion shaft.

6. Push the axle shaft inward and remove the C-lock from the button end of the axle shaft.

7. Remove the axle shaft from the axle housing. Be careful not to damage the oil seal.

> ✳ **WARNING**
>
> **On late-model vehicles equipped with an Anti-Lock Brake System (ABS) be careful not to damage the reluctor ring on the axle shaft or the speed sensor bolted to the backing plate, immediately adjacent to the shaft.**

8. Using a putty knife, clean the gasket mounting surfaces.

➡**It is recommended, when the axle shaft is removed, to replace the oil seal.**

9. To replace the oil seal use a medium prybar or, better yet, an inexpensive seal removal tool, to pry the oil seal from the end of the rear axle housing. DO NOT damage the housing oil seal surface. And again, on late-model ABS equipped vehicles, STAY CLEAR OF THE SPEED SENSOR.

10. If replacing the wheel bearing, perform the following procedures:

a. Using the GM Slide Hammer tool No. J-2619 or equivalent, the GM Adapter tool No. J-2619–4 or equivalent and the GM Axle Bearing Puller tool No. J-22813–01 or equivalent, install the tool assembly so that the tangs engage the outer race of the bearing.

b. Using the action of the slide hammer, pull the wheel bearing from the axle housing.

To install:

11. If the wheel bearing was removed:

a. Using solvent, thoroughly clean the wheel bearing, then blow dry with compressed air. Inspect the wheel bearing for excessive wear or damage, then replace it (if necessary).

b. With a new or the reused bearing, thoroughly coat the bearing with gear lubricant.

c. Using the Axle Shaft Bearing Installer tool No. J-34974 or equivalent, drive the bearing into the axle housing until it bottoms against the seat. Be sure the bearing installer does not contact and damage the speed sensor on ABS equipped vehicles.

12. If the axle shaft seal was removed:

a. Clean and inspect the axle tube housing.

b. Using the GM Axle Shaft Seal Installer tool No. J-33782 or an equivalent driver, seat the new seal into the housing until it is flush with the axle tube. Be sure the seal installer does not contact and damage the speed sensor on ABS equipped vehicles.

c. Using gear oil, lubricate the new seal lips.

13. Slide the axle shaft into the rear axle housing and engage the splines of the axle shaft with the splines of the rear axle side gear, then install the C-lock retainer on the axle shaft button end.

> ✳ **WARNING**
>
> **BE CAREFUL not to damage the wheel bearing seal with the splines on the axle shaft. And, do we even have to mention the SPEED SENSOR again on ABS equipped vehicles!**

14. After the C-lock is installed, pull the axle shaft outward to seat the C-lock retainer in the counterbore of the side gears.

15. Install the pinion shaft through the case and the pinions, then install a NEW pinion shaft lockbolt. Torque the new lockbolt to 25 ft. lbs. (34 Nm) for 1985–93 vehicles or to 27 ft. lbs. (36 Nm) for 1994–96 vehicles.

16. Use a new rear axle cover gasket and install the housing cover.

17. Install the brake drums, followed by the tire and wheel assemblies.

18. Properly refill the housing. For details, please refer to the information in Section 1 of this manual. REMEMBER that the vehicle must be completely level, meaning that if the rear is still raised and supported, the front should also be raised.

19. Remove the jackstands and carefully lower the vehicle.

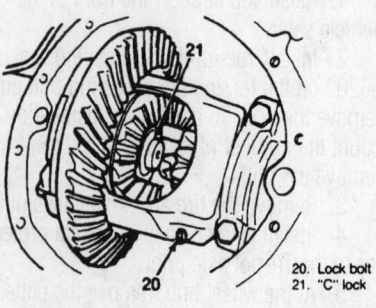

20. Lock bolt	
21. "C" lock	

Pinion shaft lockbolt and axle C-lock locations, inside the differential

GENERAL MOTORS

Cadillac-Escalade • **Chevrolet**-C/K Pick-Ups • Express • G/P Vans • Suburban • Tahoe
GMC-C/K Pick-Ups • Denali • G/P Vans • Savana • Sierra • Yukon

20

GASOLINE ENGINE REPAIR

➡Disconnecting the negative battery cable on some vehicles may interfere with the functioning of the on-board computer system, and may require the computer to undergo a relearning process once the negative battery cable is reconnected.

Distributor

REMOVAL

1. Disconnect the negative battery cable.
2. Tag and remove the spark plug wires and the coil leads from the distributor.
3. Disengage the electrical connector at the base of the distributor.
4. Loosen the distributor cap fasteners and remove the cap.
5. Using a marker, matchmark the rotor-to-housing and housing-to-engine block positions so that they can be matched during installation.
6. Loosen and remove the distributor hold-down bolt.
7. Remove the distributor from the engine.

INSTALLATION

Timing Not Disturbed

1. Install the distributor in the engine making sure that the matchmarks are properly aligned.
2. Install and tighten the hold-down bolt.
3. Install the distributor cap and engage the electrical connector at the base of the distributor.
4. Install the spark plug wires and coil leads.
5. Connect the negative battery cable.

Timing Disturbed

1. Remove the No. 1 cylinder spark plug. Turn the engine using a socket wrench on the large bolt on the front of the crankshaft pulley. Place a finger near the No. 1 spark plug hole and turn the crankshaft until the piston reaches Top Dead Center (TDC). As the engine approaches TDC, you will feel air being expelled through the No. 1 cylinder spark plug hole. The timing mark on the crankshaft pulley should now be aligned with the **0** mark on the timing scale. If the position is not being met, turn the engine

another full turn (360 degrees). Once the engine's position is correct, install the spark plug.

➡Before installation, position the rotor so it points to the No. 2 terminal on the cap. As the distributor is lowered into the engine, the rotor will rotate clockwise and stop at the No. 1 terminal. This is the desired position.

2. Turn the rotor so that it will point to the No. 1 terminal of the distributor cap when it is fully seated in the engine.
3. Install the distributor in the engine. It may be necessary to turn the rotor a little in either direction, in order to engage the gears.

➡If the distributor will not seat completely in the engine, remove the distributor and align the groove on the top of the oil pump drive shaft with a long screwdriver to match the tab on the bottom of the distributor shaft. Reinstall the distributor.

4. Tap the starter a few times to ensure that the oil pump shaft is mated to the distributor shaft.
5. Bring the engine to TDC again and check that the rotor is pointed toward the No. 1 terminal of the cap. If the marks are all aligned, install and tighten the hold-down bolt.
6. Install the cap and fasten the mounting screws.
7. Engage the electrical connections and the spark plug wires.

Ignition Timing

ADJUSTMENT

Always refer to the Vehicle Emissions Control Information label in the engine compartment for base ignition timing specification and adjustment procedures.

Engine Assembly

REMOVAL & INSTALLATION

It is virtually impossible to list each individual wire and hose which must be disconnected, simply because so many different model and engine combinations have been manufactured. Careful observation and common sense are the best possible approaches to any repair procedure.

Removal and installation of the engine can be made easier if you follow these basic points:

• If you have to drain any of the fluids, use a suitable container.
• Always tag any wires or hoses, if possible, the components they came from before disconnecting them.
• Because there are so many bolts and fasteners involved, store and label the retainers from components separately in muffin pans, jars or coffee cans. This will prevent confusion during installation.
• After unbolting the transmission, always be sure it is properly supported.
• If it is necessary to disconnect the air conditioning system, have this service performed by a qualified technician using a recovery/recycling station. If the system does not have to be disconnected, unbolt the compressor and set it aside.
• When unbolting the engine mounts, always be sure the engine is properly supported. When removing the engine, be sure that any lifting devices are properly attached to the engine. It is recommended that if your engine is supplied with lifting hooks, your lifting apparatus be attached to them.
• Lift the engine from its compartment slowly, checking that no hoses, wires or other components are still connected.
• After the engine is clear of the compartment, place it on an engine stand or workbench.

1. Disconnect the negative battery cable, then the positive battery cable, at the battery.
2. Matchmark the remove the hood.

❋❋ CAUTION

Never open, service or drain the radiator or cooling system when hot; serious burns can occur from the steam and hot coolant.

3. Drain the cooling system and remove the radiator.
4. Remove the air cleaner assembly.
5. Remove the radiator coolant reservoir bottle.
6. On the van perform the following:
   a. Remove the upper radiator support.
   b. Remove the grille and the lower grille valance.
   c. If necessary, remove the front bumper.
7. Have the A/C system discharged by an EPA certified technician.
8. Remove the air conditioning condenser from in front of the radiator.
9. If equipped, with an automatic transmission, remove the fluid cooler lines from the radiator.
10. Disconnect the accelerator and cruise control linkages.

11. Relieve the fuel system pressure, then disconnect the fuel hoses at the engine.

12. On turbocharged engines, remove the turbocharger assembly.

13. Disconnect the engine wiring harness from the firewall connection.

14. Tag and disconnect all vacuum lines.

15. Remove the power steering pump. Position the pump out of the way with wire, it's not necessary to disconnect the hoses.

16. Disconnect the heater hoses at the engine.

17. On some models it may be necessary to remove the thermostat housing.

18. Remove the oil filler and automatic transmission tubes.

19. Raise and support the vehicle safely.

20. Remove the cruise control servo, servo bracket and transducer.

21. Drain the engine oil.

22. Disconnect the exhaust pipes at the manifolds.

23. Remove the driveshaft and plug the end of the transmission.

24. Disconnect the transmission shift linkage and the speedometer cable.

25. Remove the fuel line from the fuel tank and at the fuel pump.

26. Remove the transmission mounting bolts.

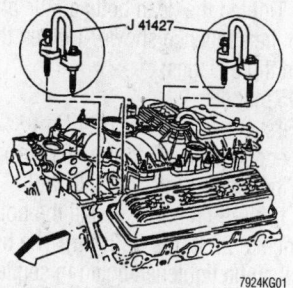

7924KG01

To remove the engine, universal lift brackets should be installed in place of the proper intake manifold bolts to prevent damage to the engine

27. Lower the vehicle, support the transmission and engine.

28. Install the engine lifting hooks J-41427 as follows:

• Disconnect the spark plug wires and remove the distributor cap.

• Remove the two right rear lower intake manifold retainers and install lifting hook J-41427 (the one marked "right"). Tighten the bolts to 11 ft. lbs. (15 Nm).

• Remove the air conditioning compressor and the accessory drive bracket.

• Disconnect the EGR tube and the two left lower bolts from the intake manifold.

• Install the lifting hook J-41427 (the one marked "left") and tighten the bolts to 11 ft. lbs. (15 Nm).

29. Remove the engine mount bracket-to-frame bolts.

30. Remove the engine mount through-bolts.

31. Raise the engine slightly and remove the engine mounts. Support the engine with wood between the oil pan and the crossmember.

32. Remove the manual transmission and clutch as follows:

a. Remove the clutch housing rear bolts.

b. Remove the bolts attaching the clutch housing to the engine and remove the transmission and clutch as a unit.

➡**Support the transmission as the last bolt is being removed to prevent damaging the clutch.**

c. Remove the starter and clutch housing rear cover.

d. Loosen the clutch mounting bolts a little at a time to prevent distorting the disc until spring pressure is released. Remove all of the bolts, the clutch disc and the pressure plate.

33. Remove the automatic transmission as follows:

a. Lower the engine and support it on blocks.

b. Remove the starter and converter housing cover.

c. Remove the flywheel/flexplate-to-converter attaching bolts.

d. Support the transmission on blocks.

e. Disconnect the detent cable on the Turbo Hydra-Matic.

f. Remove the transmission-to-engine mounting bolts.

34. Attach an engine crane to the engine.

35. Carefully lift the engine out of the vehicle. Be sure all hoses, cables and wires have been disconnected.

To install:

36. Carefully lower the engine into the vehicle and install the engine mounts. Tighten the bolts to specification.

37. Install the manual transmission and clutch as follows:

a. Install the clutch disc and the pressure plate. Tighten the clutch mounting bolts a little at a time to prevent distorting the disc.

b. Install the starter and clutch housing rear cover.

c. Install the bolts attaching the clutch housing to the engine and install the transmission and clutch as a unit. Tighten the bolts to specification.

d. Install the clutch housing rear bolts.

38. Install the automatic transmission as follows:

a. Position the transmission.

b. Install the transmission-to-engine mounting bolts.

c. Connect the throttle linkage and detent cable.

d. Install the flywheel/flexplate-to-converter attaching bolts. Tighten the bolts to specification.

e. Install the starter and converter housing cover.

39. Install the engine mount through-bolts. Tighten the bolts to specification.

40. Install the engine mount bracket-to-frame bolts. Tighten the bolts to specification.

41. Install the clutch cross-shaft.

42. Install the transmission mounting bolts. Tighten the bolts to specification.

43. Connect the transmission shift linkage and the speedometer cable.

44. Install the driveshaft.

45. Remove the lifting hooks and install the compressor, EGR valve tube and intake manifold retaining bolts.

46. Install the condenser.

47. If removed, install the hood latch support.

48. Install the lower fan shroud and filler panel.

49. Install the transmission dipstick tube and the accelerator cable at the tube.

50. Install the coolant hose at the intake manifold and the PCV valve.

51. Install the distributor cap.

52. Install the cruise control servo, servo bracket and transducer.

53. Install the oil filler pipe and automatic transmission filler pipe.

54. Install the engine dipstick tube.

55. If removed, install the thermostat housing.

56. Connect the heater hoses at the engine.

57. Connect the engine wiring harness to the firewall connection.

58. Install the radiator, shroud and radiator support bracket.

59. Connect the fuel lines.

60. Connect the accelerator and cruise control linkages.

61. Install the windshield wiper jar and bracket.

62. Install the air conditioning condenser.

63. Install the air conditioning vacuum reservoir.

64. Have the air conditioning system charged by a qualified technician using a recovery/recycling station.

65. If the vehicle is equipped with an automatic transmission, install the fluid cooler lines at the radiator.

66. Install the radiator coolant reservoir bottle.

67. If removed, install the grille and the lower grille valance.

68. Install the air cleaner assembly

69. If equipped, install the engine cover.

70. Install the hood.

71. Refill the engine, cooling system and if equipped, the automatic transmission with the correct type and amount of fluid.

72. Connect the battery cables.

Water Pump

REMOVAL & INSTALLATION

4.3L, 5.0L, 5.7L and 7.4L Engines

1. Disconnect the negative battery cable.

2. Drain the radiator. Remove the fan shroud.

3. Remove the drive belt(s).

4. Remove the alternator and other accessories, if necessary.

5. Remove the fan, fan clutch and pulley.

6. Remove any accessory brackets that might interfere with water pump removal.

7. Disconnect the lower radiator hose from the water pump inlet and the heater hose from the nipple on the pump. On the 7.4L engine, remove the bypass hose.

8. Remove the bolts, then pull the water pump assembly away from the timing cover.

To install:

9. Clean all old gasket material from the timing chain cover.

10. Install the pump assembly with a new gasket. Tighten the bolts to 30 ft. lbs. (41 Nm).

11. Connect the hose between the water pump inlet and the nipple on the pump. Connect the heater hose and the bypass hose (7.4L only).

12. Install the fan, fan clutch and pulley.

13. Install and adjust the alternator and other accessories, if necessary.

14. Install the drive belt (s). Install the upper radiator shroud

15. Fill the cooling system. Connect the battery.

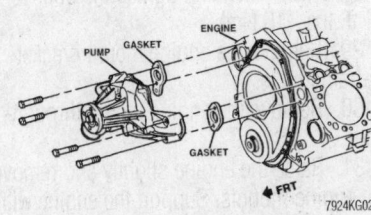

Exploded view of the water pump mounting—4.3L engine

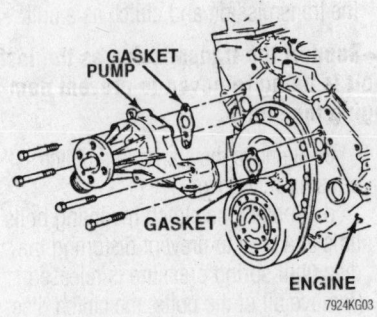

Exploded view of the water pump mounting—5.0L and 5.7L engines

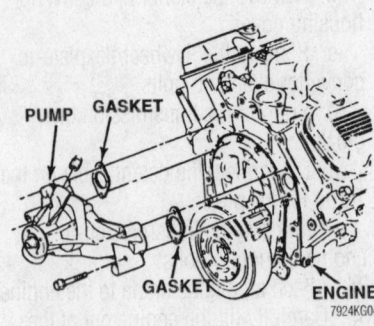

Exploded view of the water pump mounting—7.4L engine

Cylinder Head

REMOVAL & INSTALLATION

4.3L Engine

1. Disconnect the negative battery cable.

2. Remove the engine cover.

3. Drain the coolant.

4. Remove the intake manifold.

5. Remove the exhaust manifold.

6. If applicable, remove the air pipe at the rear of the right cylinder head.

7. Remove the alternator mounting bolt at the right cylinder head, if necessary, the alternator.

8. Remove the power steering pump and brackets from the left cylinder head, and lay them aside.

9. Remove the air conditioner compressor, and lay it aside. Remove the spark plug wires at their brackets, the ground strap from the right side and the coolant sensor wire from the left head.

10. Remove the cylinder cover.

11. Remove the spark plugs.

12. Remove the pushrods.

13. Remove the cylinder head bolts in the reverse order of the tightening sequence.

14. Remove the cylinder head and gasket.

To install:

15. Clean all gasket mating surfaces, install a new gasket and reinstall the cylinder head. Install the cylinder heads using new gaskets. Be sure the gasket has the word **HEAD** up.

➡ **Coat a steel gasket on both sides with sealer. If a composition gasket is used, do not use sealer.**

16. Clean the cylinder head bolts, apply sealer to the threads, and install them hand-tight.

17. Tighten the head bolts a little at a time in the sequence shown. Tighten the bolts in three stages:

1995 models:
- First pass: 25 ft. lbs. (34 Nm)
- Second pass: 45 ft. lbs. (61 Nm)
- Final pass: 65 ft. lbs. (90 Nm)

On 1996–99 models, install the bolts in sequence to 22 ft. lbs. (30 Nm). The bolts must, then be tightened again in sequence in the following order:
- Short length bolt: (11, 7, 3, 2, 6, 10) 55 degrees
- Medium length bolt: (12, 13) 65 degrees

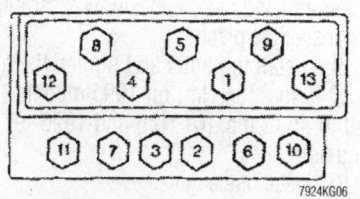

7924KG06

Tighten the cylinder head bolts in the correct sequence for proper gasket crush and to ensure cylinder sealing—4.3L engine

- Long length bolts: (1, 4, 8, 5, 9) 75 degrees

18. Install the pushrods.
19. If necessary, adjust the rocker arms.
20. Install the spark plugs.
21. Install the rocker arm cover.
22. Install the air conditioner compressor.
23. Install the power steering pump and brackets.
24. Install the alternator or the alternator mounting bolt at the cylinder head.
25. If removed, install the air pipe at the rear of the head.
26. Install the exhaust manifold.
27. Install the intake manifold.
28. Fill the engine with coolant.
29. Connect the negative battery cable.
30. Install the engine cover.

5.0L and 5.7L Engines

1. Disconnect the negative battery cable and drain the coolant.
2. Remove the engine cover.
3. If applicable, remove the coolant recovery reservoir.
4. Remove the intake manifold.
5. Remove the exhaust manifolds and position them out of the way.
6. Remove the ground strap at the rear of the right AIR pipe, If equipped.
7. If the van is equipped with air conditioning, remove the air conditioning compressor and the forward mounting bracket and lay the compressor aside. Do not disconnect any of the refrigerant lines.
8. Remove the EGR inlet tube.
9. On the right side cylinder head, disconnect the fuel pipe and move it out of the way. Remove the spark plug wires and disconnect the wiring harness bracket.
10. Remove the nut and stud attaching the main accessory bracket to the cylinder head. You may have to loosen the remaining bolts and studs in order to remove the head.

11. Tag and disconnect the coolant sensor wire. Remove the spark plug wire bracket.
12. Remove the cylinder head covers. Remove the spark plugs.
13. Back off the rocker arm nuts and pivot the rocker arms out of the way so that the pushrods can be removed. Identify the pushrods so that they can be installed in their original positions.
14. Remove the cylinder head bolts in the reverse order of the tightening sequence, then remove the heads.

To install:

15. Inspect the cylinder head and block mating surfaces. Clean all old gasket material.
16. Install the cylinder heads using new gaskets. Install the gaskets with the word **HEAD** up.

➡ **Coat a steel gasket on both sides with sealer. If a composition gasket is used, do not use sealer.**

17. Clean the bolts, apply sealer to the threads, and install them hand-tight.
18. Tighten the cylinder head bolts a little at a time, in the sequence shown. Tighten the bolts in three stages:
19. Install the cylinder head bolts in sequence to 22 ft. lbs. (30 Nm). The bolts must, then be tightened again in sequence in the following order:
   - Short length bolt: (3, 4, 7, 8, 11, 12, 15, 16) 55 degrees
   - Medium length bolt: (14, 17) 65 degrees
   - Long length bolts: (1, 2, 5, 6, 9, 10, 13) 75 degrees
20. Install the pushrods so that they are in their original positions. Swing the rocker arms into position and tighten the bolts.
21. Install the cylinder head covers. Install the spark plugs.
22. Connect the coolant sensor wire. Install the spark plug wire bracket.
23. Install main accessory bracket to the cylinder head.
24. Install the EGR vent tube.
25. Connect the fuel pipe. Install the spark plug wires and Connect the wiring harness bracket.
26. Install the air conditioning compressor and the forward mounting bracket.
27. Connect the ground strap to the rear of the right AIR pipe.
28. Install the exhaust manifolds.
29. Install the intake manifold.
30. If removed, install the coolant recovery reservoir.
31. Connect the negative battery cable and fill the engine with coolant.
32. Install the engine cover.

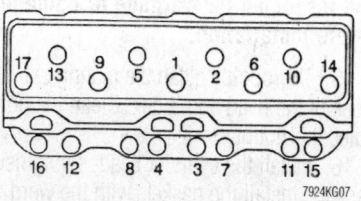

7924KG07

Cylinder head bolt tightening sequence—5.0L and 5.7L engines

7.4L Engines

1. Disconnect the negative battery cable and drain the cooling system.
2. Remove the engine cover.
3. Remove the intake manifold.
4. Remove the exhaust manifolds.
5. Remove the alternator and bracket.
6. Remove the AIR pump, if equipped.
7. If the vehicle is equipped with air conditioning, remove the air conditioning compressor and the forward mounting bracket and lay the compressor aside. Do not disconnect any of the refrigerant lines.
8. Remove the rocker arm cover.
9. Remove the spark plugs.

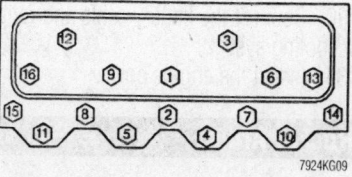

7924KG09

Cylinder head bolt tightening sequence—7.4L engines

10. Remove the AIR pipes at the rear of the head, if equipped.
11. Disconnect the ground strap at the rear of the head.
12. Disconnect the temperature sensor wire.
13. Back off the rocker arm nuts and pivot the rocker arms out of the way so that the pushrods can be removed. Identify the pushrods so that they can be installed in their original positions.
14. Remove the cylinder head bolts and remove the heads.

To install:

➡ **The cylinder head should be cleaned and inspected for warpage or damage before installation.**

15. Thoroughly clean the mating surfaces of the head and block. Clean the bolt holes thoroughly.

16. Install the cylinder heads using new gaskets. Install the gaskets with the word **HEAD** up.

➡ **Coat a steel gasket on both sides with sealer. If a composition gasket is used, do not use sealer.**

17. Clean the bolts, apply sealer to the threads, and install them hand-tight.

18. Tighten the head bolts a little at a time in the sequence in three stages, first to 30 ft. lbs. (40 Nm), then to 60 ft. lbs. (80 Nm) and finally to 85 ft. lbs. (115 Nm).

19. Install the intake and exhaust manifolds.

20. Install the pushrods.

21. Install the rocker arms.

22. Connect the temperature sensor wire.

23. Connect the ground strap at the rear of the head.

24. Install the AIR pipes at the rear of the head.

25. Install the spark plugs.

26. Install the rocker arm cover.

27. Install the air conditioning compressor and the forward mounting bracket.

28. Install the AIR pump.

29. Install the alternator.

30. Connect the battery cable and refill the cooling system.

31. Install the engine cover.

Rocker Arms

REMOVAL & INSTALLATION

4.3L, 5.0L and 5.7L Engines

1. Remove the engine cover.
2. Remove the cylinder head cover.
3. Remove the rocker arm nut. If you are only replacing the pushrod, back the nut off until you can swing the rocker out of the way.
4. Remove the rocker arms and balls as a unit.

➡ **Always remove each set of rocker arms (one set per cylinder) as a unit.**

5. Lift out the pushrods and pushrod guides.

To install:

6. Install the pushrods and their guides. Be sure that they seat properly in each lifter.

7. Position a set of rocker arms (for one cylinder) in the proper location.

➡ **Install the rocker arms for each cylinder only when the lifters are off the cam lobe and both valves are closed.**

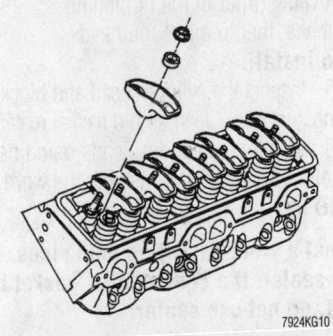

Exploded view to the rocker arm and related components—4.3L, 5.0L and 5.7L engines

8. Coat the replacement rocker arm with Molykote® or its equivalent, and the rocker arm and pivot with SAE 90 gear oil, and install the pivots.

9. Install the nuts and tighten alternately as detailed in the valve lash adjustment procedure later in this section.

10. Install the engine cover.

7.4L Engines

1. Remove the engine cover.
2. Remove the cylinder head cover.
3. Remove the rocker arm bolt. If you are only replacing the pushrod, back the nut off until you can swing the rocker out of the way.
4. Remove the rocker arms and balls as a unit.

➡ **Always remove each set of rocker arms (one set per cylinder) as a unit.**

5. Lift out the pushrods and pushrod guides.

To install:

6. Install the pushrods and their guides. Be sure that they seat properly in each lifter.

7. Position a set of rocker arms (for one cylinder) in the proper location.

➡ **Install the rocker arms for each cylinder only when the lifters are off the cam lobe and both valves are closed.**

8. Coat the replacement rocker arm with Molykote® or its equivalent, and the rocker arm and pivot with SAE 90 gear oil, and install the pivots.

9. Install the bolts and tighten them to 40 ft. lbs. (54 Nm) on 1995 models and to 45 ft. lbs. (61 Nm) and 1996–97 models.

10. Install the engine cover.

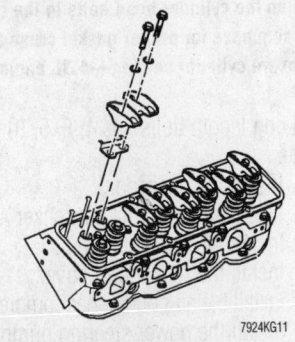

Exploded view of the rocker arms and related components—7.4L engines

Intake Manifold

REMOVAL & INSTALLATION

4.3L Engine

1995 MODELS

1. Disconnect the negative battery cable.
2. Remove the air cleaner assembly.
3. Remove the distributor.

❋❋ CAUTION

Never open, service or drain the radiator or cooling system when hot; serious burns can occur from the steam and hot coolant.

4. Disconnect the accelerator and if equipped, the cruise control cable from the throttle body.

5. Remove the A/C compressor rear bracket from the manifold.

6. Remove the alternator bracket from the manifold.

7. Remove the idler pulley.

❋❋ CAUTION

Observe all applicable safety precautions when working around fuel. Whenever servicing the fuel system,

always work in a well ventilated area. Do not allow fuel spray or vapors to come in contact with a spark or open flame. Keep a dry chemical fire extinguisher near the work area. Always keep fuel in a container specifically designed for fuel storage; also, always properly seal fuel containers to avoid the possibility of fire or explosion.

8. Relieve the fuel system pressure, then disconnect the fuel lines, electrical connectors and vacuum lines from the throttle body and manifold.

9. Remove the heater pipe.

10. Remove the bracket with sensors and the wiring harness from the right side and position them out of the way.

11. Remove the intake manifold mounting bolts and the intake manifold.

To install:

12. Clean all gasket mating surfaces.

13. Position the new gaskets on the cylinder heads, then apply a ³⁄₁₆ inch (5mm) bead of RTV to the front and rear sealing surfaces on the engine block. Extend the bead ½ inch (13mm) up each cylinder head to retain the gasket.

14. Carefully lower the intake manifold onto the engine.

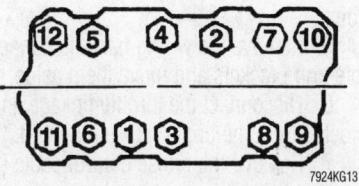

Intake manifold bolt tightening sequence—1995 4.3L engine

15. Install the intake manifold bolts, then tighten the bolts to 35 ft. lbs. (47 Nm) using the sequence shown.

16. Install the sensor bracket and wiring harness.

17. Install the heater pipe.

18. Connect the fuel lines, vacuum lines and electrical connectors to the throttle body.

19. Install the idler pulley, alternator and A/C compressor brackets.

20. Connect the accelerator cable and if equipped, the cruise control cable to the throttle body.

21. Install the distributor and air cleaner assembly.

22. Connect the negative battery cable.

1996–99 MODELS

➡The 1996–99 4.3L engines utilize a two piece intake manifold. The following procedure outlines the removal and installation for both manifolds.

1. Disconnect the negative battery cable.

2. Remove the air intake duct.

3. Unplug the wiring harness connectors from the manifold.

4. Remove the wiring harness brackets and position them out of the way.

5. Remove the throttle linkage and bracket from the upper manifold and position them out of the way.

6. If equipped, disconnect the cruise control cable.

> ✳✳ **CAUTION**

Observe all applicable safety precautions when working around fuel. Whenever servicing the fuel system, always work in a well ventilated area. Do not allow fuel spray or vapors to come in contact with a spark or open flame. Keep a dry chemical fire extinguisher near the work area. Always keep fuel in a container specifically designed for fuel storage; also, always properly seal fuel containers to avoid the possibility of fire or explosion.

7. Relieve the fuel system pressure, then disconnect the fuel lines at the rear of the lower intake manifold.

8. Remove the brake booster vacuum hose from the upper intake manifold.

9. Remove the ignition coil and bracket.

10. Remove the purge solenoid and bracket.

11. Mark the location of all of the upper intake manifold studs for correct assembly, then remove the studs and intake manifold attaching bolts.

12. Remove the upper intake manifold.

13. Mark the locations of the distributor housing and rotor for correct assembly, then remove the distributor.

14. Disconnect the upper radiator hose from the thermostat housing.

15. Disconnect the heater hoses and the bypass hose from the lower intake manifold.

16. Remove the Exhaust Gas Recirculation (EGR) valve.

17. If equipped, remove the transmission dipstick tube.

18. Remove the Positive Crankcase Ventilation (PCV) valve and hoses.

19. Without disconnecting the lines, remove the A/C compressor and bracket. Position the compressor out of the way.

20. If needed, remove the alternator bracket and bolt next to the thermostat housing.

21. Remove the lower intake manifold mounting bolts and the lower manifold.

To install:

22. Clean all gasket mating surfaces thoroughly.

23. Position the new gaskets on the cylinder heads with the port blocking plates at the rear and the words **THIS SIDE UP** facing up.

24. Apply a ³⁄₁₆ inch (5mm) bead of RTV to the front and rear sealing surfaces on the engine block. Extend the bead ½ inch (13mm) up each cylinder head to retain the gasket.

25. Carefully position the lower intake manifold onto the engine.

26. Apply GM 1052080 or equivalent sealer to the lower intake manifold bolts.

27. Install and tighten the bolts using three steps in the sequence shown, fist to 2 ft. lbs. (3 Nm), then to 9 ft. lbs. (12 Nm) and finally to 11 ft. lbs. (15 Nm).

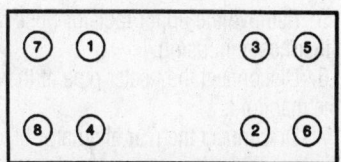

INTAKE SEQUENCE

Tighten the lower intake manifold bolts in the sequence shown to ensure good sealing—1996–99 4.3L engines

28. If removed, install the alternator bracket and bolts near the thermostat housing.

29. Install the A/C compressor.

30. Install the PCV valve and hose.

31. If equipped, install the transmission dipstick tube.

32. Install the EGR valve.

33. Connect the upper radiator and bypass hose to the thermostat housing.

34. Install the distributor.

35. Position the upper intake manifold gasket on the lower manifold.

⁕⁕ WARNING

Be careful not to pinch the injector tubes between the upper and lower manifolds.

36. Install the upper intake manifold, be sure the studs are returned to their original positions. Install the two corner studs first to help align the two halves, Tighten the bolts and studs to 88 inch lbs. (10 Nm).

37. Install the purge control bracket and valve.

38. Install the ignition coil.

39. Connect the brake booster vacuum.

40. Install the fuel lines.

41. Connect the accelerator cable and if equipped, the cruise control cable.

42. Install the wiring harness brackets and make all connections.

43. Install the air intake duct.

44. Connect the negative battery cable.

45. Refill and bleed the cooling system.

46. Pressurize the fuel system and check for leaks.

5.0L, 5.7L and 7.4L Engines

1995 MODELS

1. Disconnect the negative battery cable. Drain the cooling system.

2. Remove the engine cover.

3. Remove the air cleaner assembly.

4. If necessary, remove the coolant recovery reservoir.

5. Remove the upper radiator hose from the thermostat housing.

6. Disconnect the heater pipe at the rear of the manifold.

7. Disconnect the rear alternator brace at the manifold.

8. Disengage all electrical connections and vacuum lines from the manifold. Remove the EGR valve if necessary.

➡ **Mark the relationship of the distributor and rotor for proper reassembly**

9. Remove the distributor.

10. Disconnect the fuel line at the intake manifold.

11. Remove the accelerator and cruise control linkage.

12. Remove the air conditioner compressor rear bracket.

13. Remove the brake booster vacuum pipe, then disconnect the coil wires.

14. Remove the emission control sensors and their bracket from the right side.

15. Remove the fuel line bracket at the rear of the manifold and position the fuel lines out of the way.

16. Remove the bracket behind the idler pulley.

17. Remove the TBI unit if necessary.

➡ **Mark the location of the intake manifold studs for proper reassembly**

18. Remove the intake manifold bolts. Remove the manifold and the gaskets. Remember to reinstall the O-ring between the intake manifold and timing chain cover during assembly, if so equipped.

To install:

➡ **Before installing the intake manifold, be sure that the gasket surfaces are thoroughly clean.**

19. Use plastic gasket retainers to prevent the manifold gasket from slipping out of place, if so equipped. Place a ³⁄₁₆ in. (5mm) bead of RTV type silicone sealer on the front and rear ridges of the cylinder block-to-manifold mating surfaces. Extend the bead ½ in. (13mm) up each cylinder head to seal and retain the manifold side gaskets.

20. Install the manifold and the gaskets. Remember to reinstall the O-ring between the intake manifold and timing chain cover, if so equipped.

21. Install the intake manifold bolts and tighten in the proper sequence. Tighten the bolts to 35 ft. lbs. (48Nm) on the 5.0L and 5.7L engines or 30 ft. lbs. (40 Nm) on the 7.4L engines.

22. Install the TBI unit if removed.

23. Install the bracket behind the idler pulley.

24. Install the fuel line bracket at the rear of the manifold.

25. Install the emission control sensors and their bracket on the right side.

26. Install the brake booster vacuum pipe, then connect the coil wires.

27. Install the air conditioner compressor rear bracket.

28. Install the accelerator and cruise control linkage.

29. Install the fuel line and bracket.

30. Install the distributor.

31. Engage all electrical connections and vacuum lines at the manifold. Install the EGR valve.

32. Connect the rear alternator brace at the manifold.

33. Connect the heater pipe to the rear of the manifold.

34. Install the upper radiator hose.

35. If removed, install the coolant recovery reservoir.

36. Install the air cleaner assembly.

37. Connect the negative battery cable. Fill the cooling system.

1996–99 MODELS

1. Disconnect the negative battery cable.

2. Remove the engine cover.

3. Remove the air cleaner intake duct.

4. Remove the coolant recovery reservoir.

5. Remove the wiring harness connectors and brackets and move them aside.

6. Disconnect the throttle linkage and bracket from the upper intake manifold.

7. Remove the cruise control cable (if equipped).

8. Remove the fuel lines and the bracket from the rear of the intake manifold.

9. Remove the PCV valve and hose.

10. Remove the ignition coil and bracket.

11. Remove the purge solenoid and bracket.

➡ **Note the location of the manifold bolts and studs before removal for reassembly in their original positions.**

12. Remove the intake manifold bolts and studs.

➡ **Do not disassemble the CSFI unit.**

13. Remove the upper intake manifold.

14. Clean the old gasket residue from both mating surfaces.

15. Remove the distributor.

16. Disconnect the upper radiator hose from the thermostat housing.

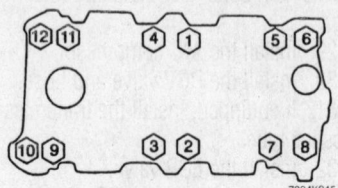

Tighten the intake manifold bolts according to the sequence shown—1995 5.0L and 5.7L engines

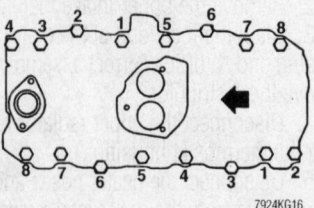

Tighten the intake manifold bolts in the sequence shown—1995 7.4L engines

17. Disconnect the heater hose from the lower intake manifold.

18. Remove the coolant bypass hose.

19. Remove the EGR valve.

20. Disconnect the fuel pressure and return lines from the lower intake manifold.

21. Disconnect the wiring harnesses and brackets from the lower manifold.

22. Remove the left side valve cover.

23. Remove the transmission oil level indicator and tube, if equipped.

24. Remove the EGR tube, clamp and bolt.

25. Remove the PCV valve and hose.

26. Remove the A/C compressor and bracket, but do NOT disconnect the lines. Move the compressor out of the way. Take care not to kink the A/C lines.

27. Loosen the compressor mounting bracket and slide it forward, but do NOT remove it.

28. Remove the power brake vacuum tube.

29. Remove the lower intake manifold bolts.

30. Remove the lower intake manifold.

To install:

31. Clean all gasket surfaces completely.

32. Install the intake manifold gaskets with the port blocking plates facing the rear. Factory gaskets should have the words "This Side Up" visible.

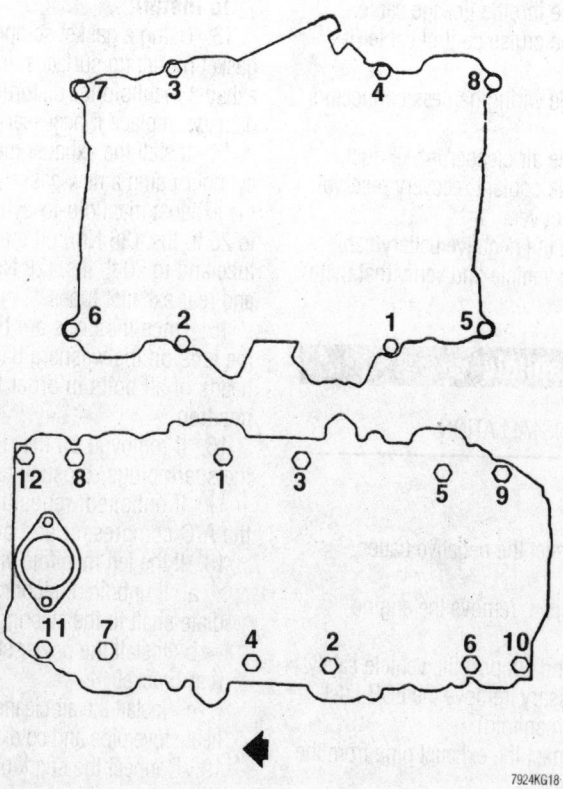

Tighten the upper and lower intake manifolds according to the sequence shown—1996–99 7.4L engines

33. Apply gasket sealer to the front and rear sealing surfaces of the engine block. Extend the sealer approximately ½ -inch (13mm) onto the heads.

34. Install the lower intake manifold.

35. Apply sealer to the lower intake manifold bolts prior to installation.

36. On the 5.0L and 5.7L engines, install the bolts and tighten in sequence and in three steps as follows:

   a. First step to 71 inch lbs. (8 Nm).

   b. Second step to 106 inch lbs. (12 Nm).

   c. Final step to 11 ft. lbs. (15 Nm).

37. On the 7.4L engine, tighten the bolts to 30 ft. lbs. (40 Nm) in the sequence shown.

38. Install the power brake vacuum tube.

39. Install the PCV valve and hose.

40. Install the EGR tube, clamp and bolt.

41. Install the transmission oil level indicator and tube, if equipped.

42. Install the left side valve cover.

43. Connect the wiring harnesses and brackets to the lower manifold.

44. Connect the fuel pressure and return lines to the lower intake manifold.

45. Install the EGR valve.

46. Install the coolant bypass hose.

47. Connect the heater hose to the lower intake manifold.

48. Connect the upper radiator hose to the thermostat housing.

49. Install the A/C compressor and bracket.

50. Install the distributor.

51. Install the upper intake manifold gasket.

52. Install the upper intake manifold.

✳✳ WARNING

When installing the upper intake manifold be careful not to pinch the injector wires between the upper and lower intake manifolds.

53. Install the upper intake manifold mounting bolts and studs in the same locations as prior to removal.

54. Tighten the bolts and studs in a crisscross pattern in two steps, first to 44 inch lbs. (5 Nm), then to 83 inch lbs. (10 Nm).

55. Install the purge solenoid and bracket.

56. Install the PCV hose.

57. Install the fuel lines and the bracket at the rear of the intake manifold.

58. Install the ignition coil and bracket.

59. Install the throttle linkage and bracket to the upper intake manifold.

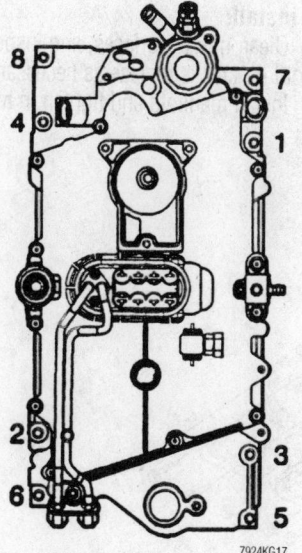

Tighten the lower intake manifold bolts in the sequence shown—1996–99 5.0L and 5.7L engines

60. Install the throttle linkage cable.
61. Install the cruise control cable (if equipped).
62. Install the wiring harness connectors and brackets.
63. Install the air cleaner intake duct.
64. Install the coolant recovery reservoir and the engine cover.
65. Connect the negative battery cable.
66. Start the vehicle and verify that there are no leaks.

Exhaust Manifold

REMOVAL & INSTALLATION

4.3L Engines

1. Disconnect the negative battery cable.
2. If equipped, remove the engine cover.
3. Raise and support the vehicle safely.
4. If necessary, remove the EGR inlet pipe (left side manifold).
5. Disconnect the exhaust pipe from the exhaust manifold.
6. Lower the vehicle for underhood access.
7. Tag and disconnect the spark plug wires from the plugs and from the retaining clips.
8. Remove the heat shields.
9. If removing the left side manifold:
   a. If needed, remove the power steering/alternator rear bracket.
   b. Check for sufficient clearance between the manifold and the intermediate steering shaft. On some models it will be necessary to disconnect the intermediate shaft from the steering gear in order to reposition the shaft for clearance.
10. If necessary when removing the right side manifold, unbolt the A/C compressor and bracket, then position the assembly aside. Do not disconnect the lines or allow them to become kinked or otherwise damaged.
11. If necessary for the right side manifold, remove the spark plugs, dipstick tube and wiring.
12. Unbend the lock tangs, then remove the exhaust manifold retaining bolts, washers and tab washers. Remove the exhaust manifold, then remove and discard the old gaskets.

To install:
13. Using a gasket scraper, clean the gasket mounting surfaces. Inspect the exhaust manifold for distortion, cracks or damage; replace if necessary.
14. Install the exhaust manifold to the cylinder using a new gasket, then tighten the exhaust manifold-to-cylinder head bolts to 26 ft. lbs. (36 Nm) on the center exhaust tube and to 20 ft. lbs. (28 Nm) on the front and rear exhaust tubes.
15. Once the bolts are tightened, bend the tabs on the washers back over the heads of all bolts in order to lock them in position.
16. If removed on the right side, install the spark plugs, dipstick tube and wiring.
17. If unbolted, reposition and secure the A/C compressor and bracket assembly.
18. If the left manifold was removed:
   a. If unbolted, reconnect the intermediate shaft to the steering gear.
   b. Install the power steering/alternator rear bracket.
   c. Install the air cleaner along with the heat stove pipe and cold air intake pipe.
19. Connect the spark plug wires to the retainer clips and to the plugs as noted during removal.
20. Raise and support the vehicle safely.
21. Connect the exhaust pipe to the manifold.
22. Lower the vehicle.
23. On Van models, install the engine cover.
24. Connect the negative battery cable.

5.0L and 5.7L Engines

1. Disconnect the negative battery cable.

2. If equipped, remove the engine cover.
3. If needed, remove the air cleaner.
4. Raise and support the vehicle safely.
5. Disconnect the exhaust pipe at the manifold, then lower the vehicle for underhood access.
6. If removing a manifold with the oxygen sensor mounted to it, disengage the sensor wiring.
7. If necessary, disconnect the AIR hose at the check valve and remove the EGR inlet pipe.
8. Disconnect the heat stove pipe and remove the dipstick tube bracket, if working on the right side of the engine.
9. If removing the left side manifold, remove the power steering pump rear bracket at the manifold.
10. If necessary, loosen the alternator and remove the lower bracket. Then, if needed, remove the air conditioner compressor rear bracket and the diverter valve and bracket.

➡**On models with air conditioning, it may be necessary to remove the compressor, and tie it out of the way. Do not disconnect the compressor lines.**

11. Remove the manifold bolts and remove the manifold(s). Some models have lock tabs on the front and rear manifold bolts which must be removed before removing the bolts.

To install:
12. Clean gasket surfaces, and inspect manifold for cracks replace as necessary.
13. Install manifold and tighten in two steps.

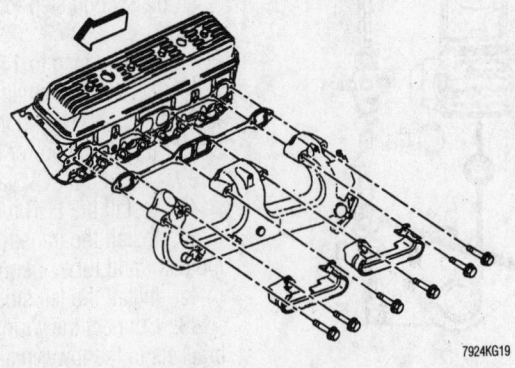

Exploded view of the left exhaust manifold, the right side is similar—5.0L and 5.7L engines

7924KG19

- First tighten to 15 ft. lbs. (20 Nm)
- Next tighten to 22 ft. lbs. (30 Nm)

14. If removed, install alternator, A/C compressor, diverter, and P/S brackets.

15. Install dipstick tube on right side.

16. Install the EGR inlet pipe

17. Install oxygen sensor connector.

18. Connect exhaust pipes

19. Connect the negative battery cable.

7.4L Engines

1. Disconnect the negative battery cable.

2. On Van models, remove the engine cover.

3. If removing the right side manifold, remove the heat stove pipe and the dipstick tube.

4. Remove the EGR inlet pipe.

5. If removing a manifold with the oxygen sensor mounted to it, disengage the sensor wiring.

6. If applicable, disconnect the AIR hose at the check valve.

7. Remove the spark plugs and wires.

8. Remove the exhaust manifold bolts and the spark plug heat shields. Leave the front nut (left manifold) or rear nut (right manifold) in place for support.

9. Raise and support the vehicle safely for access.

10. If equipped, remove the heat shield bolts from the engine mount and bell housing, then remove the shield.

11. Disconnect the exhaust pipe at the manifold.

➡It may be necessary to raise the engine slightly to gain sufficient clearance for removal of the manifold on some van models.

12. Remove the exhaust manifold.

To install:

13. Clean the mating surfaces and the retainer threads.

14. Install the manifold, spark plug heat shields and nuts. Tighten the nuts to 22 ft. lbs. (30 Nm). starting from the center bolts and working towards the outside.

15. Install the exhaust pipe.

16. Install spark plugs and wires.

17. If removed, installed AIR hose at check valve.

18. Install oxygen sensor connector.

19. Install EGR pipe and dipstick tube.

20. Connect the negative battery cable.

21. Run engine and check for leaks.

Camshaft and Valve Lifters

REMOVAL & INSTALLATION

4.3L Engines

✳✳ CAUTION

Observe all applicable safety precautions when working around fuel. Whenever servicing the fuel system, always work in a well ventilated area. Do not allow fuel spray or vapors to come in contact with a spark or open flame. Keep a dry chemical fire extinguisher near the work area. Always keep fuel in a container specifically designed for fuel storage; also, always properly seal fuel containers to avoid the possibility of fire or explosion.

1. Properly relieve the fuel system pressure, then disconnect the negative battery cable.

✳✳ CAUTION

Never open, service or drain the radiator or cooling system when hot; serious burns can occur from the steam and hot coolant.

2. Drain the engine cooling system.

3. Remove the radiator.

4. Remove the cooling fan.

5. Remove the water pump.

6. Remove the rocker arm covers from the engine.

7. Remove the intake manifold assembly.

8. Remove the rocker arms, pushrods and lifters.

9. Remove the crankshaft pulley and hub.

10. Remove the engine front (timing) cover.

11. Align the timing marks on the crankshaft and camshaft sprockets.

12. Remove the camshaft sprocket and timing chain.

13. If equipped, remove the balance shaft drive gear.

14. Remove the camshaft thrust plate.

15. Install the sprocket bolts or longer bolts of the same thread into the end of the camshaft as a handle, then remove the camshaft front the front of the engine while turning slightly from side to side, as necessary. Take care not to damage the camshaft bearings when removing the camshaft.

To install:

16. Lubricate the camshaft journals with clean engine oil or a suitable pre-lube, then install the camshaft into the block being extremely careful not to contact the bearings with the cam lobes.

17. Install the camshaft thrust plate.

18. If equipped, install the balance shaft drive gear.

19. Install the timing chain and camshaft sprocket.

20. Install the engine front (timing) cover.

21. Install the crankshaft pulley and hub.

22. Install the valve lifters, then install the pushrods and rocker arms. Properly adjust the valve clearance.

23. Install the intake manifold assembly.

24. Install the rocker arm covers to the engine.

25. Install the radiator to the vehicle.

26. Connect the negative battery cable and properly refill the engine cooling system.

5.0L and 5.7L Engines

✳✳ CAUTION

Observe all applicable safety precautions when working around fuel. Whenever servicing the fuel system, always work in a well ventilated area. Do not allow fuel spray or vapors to come in contact with a spark or open flame. Keep a dry chemical fire extinguisher near the work area. Always keep fuel in a container specifically designed for fuel storage; also, always properly seal fuel containers to avoid the possibility of fire or explosion.

1. Disconnect the negative battery cable, drain the cooling system and properly relieve the fuel system pressure.

2. On van models, remove the engine cover.

3. Remove the air cleaner.

4. On van models, remove the grille and center support.

5. Remove the air conditioning condenser and swing the condenser forward from its mounting, if equipped.

6. Remove the fan, the shroud and the radiator.

7. Remove the valve covers.

8. Remove the water pump assembly.

9. Align the timing marks and remove the torsional damper.

10. Remove the timing chain cover.

11. Disconnect the electrical and vacuum connections at the intake manifold.

12. Mark the distributor rotor-to-housing location. Remove the distributor assembly.

13. Remove the intake manifold, pushrods and hydraulic lifters.

14. Remove the camshaft sprocket bolts, camshaft sprocket and timing chain. Tap the sprocket on its lower edge to loosen it.

15. Remove the crankshaft sprocket, as required.

16. As required, remove the front engine mount through-bolts and raise the engine to gain sufficient clearance for camshaft removal.

17. Install two or three ⁵⁄₁₆ –18 bolts 4–5 in. long into the camshaft threaded holes; carefully pull the camshaft from the block.

18. Inspect the shaft for signs of excessive wear or damage.

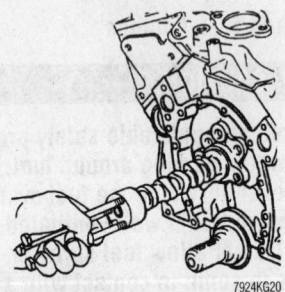

7924KG20

Install two or three long bolts into the camshaft to use as a handle for easy removal or installation—7.4L engine shown, other engines similar

To install:

19. Liberally coat camshaft and bearing with heavy engine oil or engine assembly lubricant and insert the cam into the engine.

20. Lower the engine and install the engine mount through-bolts.

21. Align the timing marks on the camshaft and crankshaft gears.

22. Install the camshaft sprocket and chain and tighten the bolts to specification.

23. Install the hydraulic lifters and pushrods and adjust the valves.

24. Install the distributor assembly.

25. Install the timing chain cover.

26. Install the torsional damper.

27. Install the water pump.

28. Install the valve covers.

29. Install the fan, the shroud and radiator.

30. Install the air conditioning condenser, if equipped.

31. On van models, install the grille and center support.

32. Install the air cleaner.

33. On van models, install the engine cover.

34. Connect the battery cable and fill the cooling system.

7.4L Engines

> **✳✳ CAUTION**
>
> **Observe all applicable safety precautions when working around fuel. Whenever servicing the fuel system, always work in a well ventilated area. Do not allow fuel spray or vapors to come in contact with a spark or open flame. Keep a dry chemical fire extinguisher near the work area. Always keep fuel in a container specifically designed for fuel storage; also, always properly seal fuel containers to avoid the possibility of fire or explosion.**

1. Disconnect the negative battery cable. Properly relieve the fuel system pressure.

2. On van models, remove the engine cover. Remove the air cleaner assembly.

3. Remove the grille and center support section, as required.

4. Properly discharge the air conditioning system. Remove the air conditioning compressor, condenser and auxiliary fan, if equipped.

5. Drain the cooling system.

6. Remove the fan, the shroud and radiator.

7. Remove the alternator belt, remove the alternator assembly, as required.

8. Remove the valve covers.

9. Disconnect the hoses from the water pump.

10. Remove the water pump.

11. Align the timing marks at TDC. Remove the harmonic balancer and pulley.

12. Remove the engine front cover.

13. Mark the distributor rotor-to-housing location. Remove the distributor assembly.

14. Remove the intake manifold assembly.

15. Remove the lifters, pushrods, and rocker arms.

16. Rotate the camshaft so the timing marks align.

17. Remove the camshaft sprocket bolts. Remove the camshaft sprocket and timing.

18. Remove the engine mount through-bolts. Raise and support the engine to aid in camshaft removal, as required.

19. Install two or three ⁵⁄₁₆ –18 bolts in the holes in the front of the camshaft and carefully pull the camshaft from the block.

To install:

20. Liberally coat camshaft and bearing with heavy engine oil or engine assembly lubricant and insert the cam into the engine.

21. Align the timing marks on the camshaft sprocket and crankshaft gears.

22. Install the camshaft sprocket and chain and tighten the bolts to specification. Lower the engine and install the engine mount bolts.

23. Install the lifters and pushrods and adjust the valves.

24. Install the intake manifold.

25. Install the distributor using the locating marks made during removal.

26. Install the engine front cover.

27. Install the harmonic balancer and pulley.

28. Install the water pump.

29. Connect the hoses at the water pump.

30. Install the valve covers.

31. Install the alternator.

32. Install the fan shroud and radiator.

33. Fill the cooling system with the proper type and quantity of antifreeze.

34. Install the air conditioning condenser and compressor.

35. Install the grille and center support.

36. Install the air cleaner assembly.

37. Connect the battery.

Valve Lash

ADJUSTMENT

All engines use hydraulic lifters, which require no periodic adjustment.

Oil Pan

REMOVAL & INSTALLATION

4.3L Engines

1. Disconnect the negative battery cable. Raise and support the vehicle safely. Drain the engine oil into a suitable container.

2. Remove the exhaust crossover pipe.

3. Remove the torque converter cover, if equipped with automatic transmission.

4. Remove the cooler lines from guides and the oil filter adapter.

5. Remove the strut rods at the flywheel/flexplate cover, if equipped.

6. Remove the strut rod at the front engine mounts, if equipped.

7. Remove the starter assembly.

8. Remove front drive axle tube nuts, and lower axle bushing bolts

9. Remove the oil pan bolts, nuts and reinforcements.

10. Remove the oil pan and gaskets.

To install:

11. Thoroughly clean all gasket surfaces and install a new gasket, using only a small amount of sealer at the front and rear corners of the oil pan.

12. Install the oil pan and new gaskets.

13. Install the oil pan bolts, nuts and reinforcements. Tighten bolts to 18 ft. lbs. (25 Nm).

14. Install the front drive axle tube nuts and lower axle bushing bolts.

15. Install the starter assembly.

16. Install the strut rod brackets at the front engine mounts.

17. Install the strut rods at the flywheel/flexplate cover.

18. Install cooler lines into guides and oil filter adapter with new filter

19. Install the torque converter cover, if equipped with automatic transmission.

20. Install the exhaust crossover pipe. Lower the vehicle.

21. Connect the negative battery cable.

22. Fill the engine with the proper quantity and type of oil.

5.0L and 5.7L Engines

1. Disconnect the negative battery cable. Raise and safely support the vehicle, and drain the engine oil.

2. Remove under body protectors shields

3. Remove transmission and engine oil lines from guides.

4. If needed, remove the front driveshaft.

5. If needed, remove the front drive axles.

6. Remove the exhaust crossover pipe.

7. Remove the flywheel/flexplate or torque converter cover.

8. Remove the oil filter and adapter.

9. Remove the strut rods at the front engine mounting, if used.

10. Remove the oil pan bolts, nuts and reinforcements.

11. Remove the oil pan and gaskets.

To install:

12. Thoroughly clean all gasket surfaces and install a new gasket, using a small amount of sealer at the front and rear corners of the oil pan.

13. Install the oil pan and new gaskets.

14. Install the oil pan bolts, nuts and reinforcements. Tighten bolts to 18 ft. lbs. (25 Nm).

15. Install the strut rods at the front engine mounting.

16. Install oil filter and adapter.

17. Install the torque converter or flywheel/flexplate cover.

18. Install the exhaust crossover pipe.

19. If equipped, install the front drive axles and driveshaft.

20. Install transmission and engine oil lines to guides.

21. Install under body protectors.

22. Connect the negative battery cable.

23. Fill the crankcase with the proper type and quantity of oil.

7.4L Engines

➡**Removal of the transmission may be necessary on van vehicles.**

1. Disconnect the negative battery cable.

2. Remove the fan shroud.

3. Remove the air cleaner.

4. Remove the distributor cap.

5. Raise and support the vehicle safely.

6. If needed remove underbody protectors.

7. If needed remove front driveshaft and front drive axles. Use the procedure outlined in the driveline section.

8. Drain the engine oil. Remove the starter assembly, if equipped with manual transmission.

9. Remove the torque converter or clutch housing cover.

10. Remove the oil filter and adapter.

11. Remove the oil pressure line from the side of the block.

12. Support the engine.

13. Remove the engine mount through-bolts.

14. Raise the engine just enough to remove the pan.

15. Remove the oil pan bolts, the oil pan and discard the gaskets.

To install:

16. Clean all sealing surfaces.

17. Apply RTV gasket material to the front and rear corners of the gaskets.

18. Coat the gaskets with adhesive sealer and position them on the block.

19. Install the rear pan seal in the pan with the seal ends mating with the gaskets.

20. Install the front seal on the bottom of the front cover, pressing the locating tabs into the holes in the cover.

21. Install the oil pan.

22. Install the pan bolts, clips and reinforcements. Tighten the pan bolts to 18 ft. lbs. (25 Nm).

23. Lower the engine onto the mounts.

24. Install the engine mount through-bolts.

25. Install the oil pressure line.

26. Install the oil filter. Install the starter assembly, if removed.

27. Install the torque converter or clutch housing cover.

28. If removed, install front drive axles and driveshaft.

29. If removed, install underbody protectors.

30. Install the distributor cap.

31. Install the air cleaner.

32. Install the fan shroud.

33. Connect the battery.

34. Fill the crankcase with the proper type and quantity of oil.

Oil Pump

REMOVAL & INSTALLATION

1. Remove the oil pan.

2. Remove the oil pump attaching bolt, if equipped, the pick-up tube nut/bolt, then remove the pump along with the pick-up tube and shaft, as necessary.

3. Clean all sealing surfaces

To install:

4. Ensure that the pump pick-up tube is tight in the pump body. If the tube should come loose, oil pressure will be lost and oil starvation will occur. If the pick-up tube is loose it should be replaced.

5. If the pump has been disassembled and is being replaced or for any reason oil has been removed, it must be primed. It can either be filled with oil before installing the cover plate and oil kept within the pump during handling or the entire pump cavity can be filled with petroleum jelly.

➡**If the pump is not primed, the engine could be damaged before it receives adequate lubrication when the engine is started.**

6. Install the pump aligning the pump shaft with the oil pump drive gear as necessary. Tighten oil pump/pick-up tube retainer(s) to 65 ft. lbs. (90 Nm) on all 4.3L and all V8 Engines.

7. Install the oil pan and refill the engine crankcase. Disable the ignition system; crank engine for approximately 10 seconds to aid in priming the oil pump and reducing the risk of engine damage.

➡**If the oil pump does not build up oil pressure almost immediately, remove the pan and check for a loose oil pump-to-pick-up tube attachment. If necessary dismantle the pump and pack the pump cavity with petroleum jelly. Running the engine without measurable oil pressure will cause extensive damage.**

Rear Main Seal

REMOVAL & INSTALLATION

Please note that the entire transmission assembly and flywheel/flexplate must be removed to perform this procedure.

1. Disconnect the negative battery cable.
2. Raise and safely support the vehicle.
3. Remove the transfer case, if so equipped.
4. Remove the transmission assembly using the recommended procedure.
5. Remove the clutch assembly and flywheel if equipped with manual transmission.
6. Remove the flexplate if equipped with automatic transmission.
7. Remove the crankshaft rear main oil seal by inserting a suitable prying tool and prying the seal out. Take care not to damage the crankshaft sealing surface.

To install:

8. Clean the oil seal bore in the block thoroughly before installation of the new seal.
9. Inspect the crankshaft for grit, rust or burrs and correct as necessary. Also inspect the portion of the crankshaft where the oil seal makes contact, for wear due to the rubbing action of the oil seal.
10. Clean the seal running surface of the crankshaft with a non-abrasive cleaner.
11. Lubricate the inner diameter of the new seal and the outer diameter of the crankshaft with engine oil. Using a J 38841 installation tool, install the rear crankshaft oil seal until the tool bottoms against the block and crankshaft rear main bearing cap.
12. If equipped with manual transmission, inspect the clutch assembly. If the rear crankshaft oil seal was leaking badly, the clutch plate may be contaminated with oil and require replacement. Service and install the flywheel and clutch as required.
13. If equipped with automatic transmission, inspect the flexplate for cracks. Check the condition of the starter motor teeth on the flexplate and replace as required.
14. Install the transmission assembly using the recommended procedure.
15. Install the transfer case, if so equipped.
16. Lower the vehicle.
17. Connect the negative battery cable.
18. Start the engine and verify no oil leaks.

Timing Chain, Sprockets, Front Cover and Seal

It is recommended by the manufacturer that the front cover oil seal be replaced whenever the cover is removed.

REMOVAL & INSTALLATION

1. Disconnect the negative battery cable.
2. Drain the cooling system. Remove the fan shroud assembly.
3. Remove the belts, pulleys and water pump assembly.
4. Remove the crankshaft pulley and damper.
5. Remove the oil pan-to-front cover bolts.
6. Remove the water pump.

➡**Anytime the composite front cover is removed, it must be replaced with a new one. Reusing the front cover may result in oil leaks.**

7. Remove the screws holding the timing chain cover to the block.
8. Pull off the cover and gaskets.
9. Use a suitable tool to pry the old seal out of the front face of the cover.
10. Rotate the crankshaft until the timing marks on the camshaft and crankshaft sprockets are in proper alignment.
11. Remove the camshaft sprocket-to-camshaft nut and/or bolts, then remove the camshaft sprocket (along with the timing chain). If the sprocket is difficult to remove, use a plastic mallet to bump the sprocket from the camshaft.

➡**The camshaft sprocket (located by a dowel) is lightly pressed onto the camshaft and should come off easily. The chain comes off with the camshaft sprocket.**

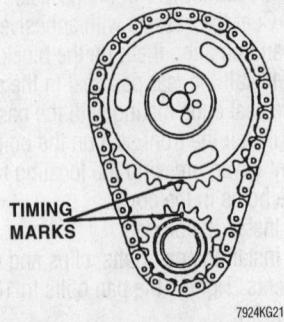

TIMING MARKS

7924KG21

Timing mark alignment for timing chain removal and installation—4.3L, 5.0L, 5.7L, and 7.4L engines

12. If necessary use J-5825-A or equivalent crankshaft sprocket removal tool to free the timing sprocket from the crankshaft.

To install:

13. Inspect the timing chain and the timing sprockets for wear or damage, replace the damaged parts as necessary.
14. Using a gasket scraper, clean the gasket mounting surfaces of all remaining traces of old gasket. Using solvent, clean the oil and grease from the gasket mounting surfaces.
15. If removed, use J-5590 or equivalent crankshaft sprocket installation tool and a hammer to drive the crankshaft sprocket onto the crankshaft, without disturbing the position of the engine.

➡**During installation, coat the thrust surfaces lightly with Molykote® or equivalent pre-lube.**

16. Position the timing chain over the camshaft sprocket. Arrange the camshaft sprocket in such a way that the timing marks will align between the shaft centers and the camshaft locating dowel will enter the dowel hole in the cam sprocket.
17. Position the chain under the crankshaft sprocket, then place the cam sprocket, with the chain still mounted over it, in position on the front of the camshaft. Install and tighten the camshaft sprocket-to-camshaft retainers.
18. With the timing chain installed, turn the crankshaft two complete revolutions, then check to make certain that the timing marks are in correct alignment between the shaft centers.
19. Using seal driver J-22102 or equivalent, install the new seal so the open end is toward the inside of the cover.

➡**Coat the lip of the new seal with oil prior to installation.**

20. Install a new front pan seal, cutting the tabs off.
21. Coat a new cover gasket with adhesive sealer and position it on the block.
22. Apply a ⅛ in. (3mm) bead of RTV gasket material to the front cover. Install the cover carefully onto the locating dowels.
23. Tighten the attaching screws.
24. If removed, install the oil pan.
25. Tighten the cover-to-pan bolts.
26. Install the torsional damper.
27. Install the water pump assembly.
28. Connect the negative battery cable.
29. Fill the cooling system with the proper type and quantity of antifreeze.

DIESEL ENGINE REPAIR

Engine Assembly

REMOVAL & INSTALLATION

1. Disconnect the negative battery cable, then the positive battery cable, at the battery.

2. If equipped, remove the engine cover.

✳✳ CAUTION

Never open, service or drain the radiator or cooling system when hot; serious burns can occur from the steam and hot coolant.

3. Drain the cooling system.

4. Remove the air cleaner.

5. Remove the radiator coolant reservoir bottle.

6. Remove the upper radiator support.

7. Remove the grille and the lower grille valance.

8. If necessary, remove the front bumper.

➡**Only a MVAC-trained, EPA-certified, automotive technician should service the A/C system or its components.**

9. Discharge the air conditioning system and remove the air conditioning vacuum reservoir.

10. Remove the air conditioning condenser from in front of the radiator.

11. If the van is equipped with an automatic transmission, remove the fluid cooler lines from the radiator.

12. Disconnect the radiator hoses at the radiator.

13. Loosen the radiator support bracket and remove the radiator and the shroud.

14. Disconnect the accelerator and cruise control linkages.

15. Disconnect all hoses and wires at the fuel unit.

16. Remove the fuel supply unit and cap the lines.

17. On 1996–99 models, remove the intake manifold.

18. On turbocharged engines, remove the turbocharger assembly.

19. Remove the lower intake manifold if equipped.

20. Remove the exhaust manifolds if necessary.

21. Disconnect the engine wiring harness from the firewall connection.

22. Remove the power steering pump. It's not necessary to disconnect the hoses; just lay the pump aside.

23. Disconnect the heater hoses at the engine.

24. On some models it may be necessary to remove the thermostat housing.

25. Remove the oil filler and automatic transmission tubes.

26. Raise and safely support the vehicle.

27. If equipped, remove the cruise control servo, servo bracket and transducer.

28. Drain the engine oil.

29. Disconnect the exhaust pipes at the manifolds.

30. Remove the driveshaft and plug the end of the transmission.

31. Disconnect the transmission shift linkage and the speedometer cable.

32. Remove the fuel line from the fuel tank and at the fuel pump.

33. Remove the transmission mounting bolts.

34. Lower the vehicle, support the transmission and engine.

35. On 1996–99 models, install lifting hooks J-41427 as follows:

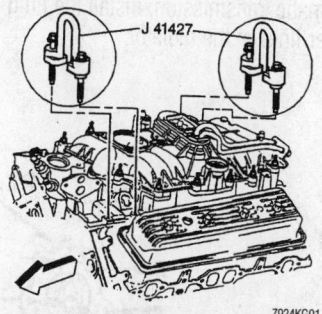

7924KG01

For engine removal and installation, universal lift brackets should be installed in place of the proper intake manifold bolts

• Remove the two right rear lower intake manifold retainers and install lifting hook J-41427 (the one marked "right"). Tighten the bolts to 11 ft. lbs. (15 Nm).

• Remove the air conditioning compressor and the accessory drive bracket.

• If equipped, disconnect the EGR tube and the two left lower bolts from the intake manifold.

• Install the lifting hook J-41427 (the one marked "left") and tighten the bolts to 11 ft. lbs. (15 Nm).

36. Remove the engine mount bracket-to-frame bolts.

37. Remove the engine mount through-bolts.

38. Raise the engine slightly and remove the engine mounts. Support the engine with wood between the oil pan and the crossmember.

39. Remove the manual transmission and clutch as follows:

a. Remove the clutch housing rear bolts.

b. Remove the bolts attaching the clutch housing to the engine and remove the transmission and clutch as a unit.

➡**Support the transmission as the last bolt is being removed to prevent damaging the clutch.**

c. Remove the starter and clutch housing rear cover.

d. Loosen the clutch mounting bolts a little at a time to prevent distorting the disc until spring pressure is released. Remove all of the bolts, the clutch disc and the pressure plate.

40. Remove the automatic transmission as follows:

a. Lower the engine and support it on blocks.

b. Remove the starter and converter housing cover.

c. Remove the flexplate-to-converter attaching bolts.

d. Support the transmission on blocks.

e. Disconnect the detent cable on the Turbo Hydra-Matic.

f. Remove the transmission-to-engine mounting bolts.

41. Attach an engine crane to the engine.

a. Remove the blocks from the engine only and glide the engine away from the transmission.

To install:

42. Raise the engine slightly and install the engine mounts. Tighten the bolts to specification.

43. Install the manual transmission and clutch as follows:

a. Install the clutch disc and the pressure plate. Tighten the clutch mounting bolts a little at a time to prevent distorting the disc.

b. Install the starter and clutch housing rear cover.

c. Install the bolts attaching the clutch housing to the engine and install the transmission and clutch as a unit. Tighten the bolts to specification.

d. Install the clutch housing rear bolts.

44. Install the automatic transmission as follows:

a. Position the transmission.

b. Install the transmission-to-engine mounting bolts.

c. Connect the throttle linkage and detent cable.

d. Install the flexplate-to-converter attaching bolts. Tighten the bolts to specification.

e. Install the starter and converter housing cover.

45. Install the engine mount through-bolts. Tighten the bolts to specification.

46. Install the engine mount bracket-to-frame bolts. Tighten the bolts to specification.

47. Install the clutch cross-shaft.

48. Install the transmission mounting bolts. Tighten the bolts to specification.

49. Connect the transmission shift linkage and the speedometer cable.

50. Install the driveshaft.

51. Remove the lifting hooks and install the compressor, EGR valve tube and intake manifold retaining bolts.

52. Install the condenser.

53. Install the hood latch support.

54. Install the lower fan shroud and filler panel.

55. Install the transmission dipstick tube and the accelerator cable at the tube.

56. Install the coolant hose at the intake manifold and the PCV valve.

57. Install the distributor cap.

58. Install the cruise control servo, servo bracket and transducer.

59. Install the oil filler pipe and automatic transmission filler pipe.

60. Install the engine dipstick tube.

61. If removed, install the thermostat housing.

62. Connect the heater hoses at the engine.

63. Connect the engine wiring harness to the firewall connection.

64. Install the radiator and the shroud.

65. Install the radiator support bracket.

66. Install the exhaust manifolds if removed.

67. Install the lower intake manifold if removed.

68. On 1996–99 models, install the intake manifold.

69. Install the fuel supply unit.

70. Connect the lines to the fuel supply unit.

71. Connect the accelerator and cruise control linkages.

72. Install the windshield wiper jar and bracket.

73. Install the air conditioning condenser.

74. If equipped, install the air conditioning vacuum reservoir.

75. Have the air conditioning system charged by a qualified technician using a recovery/recycling station.

76. If the vehicle is equipped with an automatic transmission, install the fluid cooler lines at the radiator.

77. Install the radiator coolant reservoir bottle.

78. Connect the radiator hoses at the radiator.

79. Install the upper radiator support the grille and the lower grille valance.

80. Install the air cleaner.

81. Install the air stove pipe.

82. Install the engine cover.

83. Fill the cooling system.

84. Connect the battery cables.

Water Pump

REMOVAL & INSTALLATION

1. Disconnect the negative battery cables.

2. Remove the fan and fan shroud.

3. Drain the engine coolant into a suitable container.

4. If necessary, remove the air conditioning hose bracket and/or the oil filler tube, as required.

5. Remove the engine accessory drive belt(s).

6. Raise and support the vehicle safely.

7. Remove the vacuum pump mounting bracket nuts, then remove the bolt holding the pump and alternator. Remove the vacuum pump and bracket.

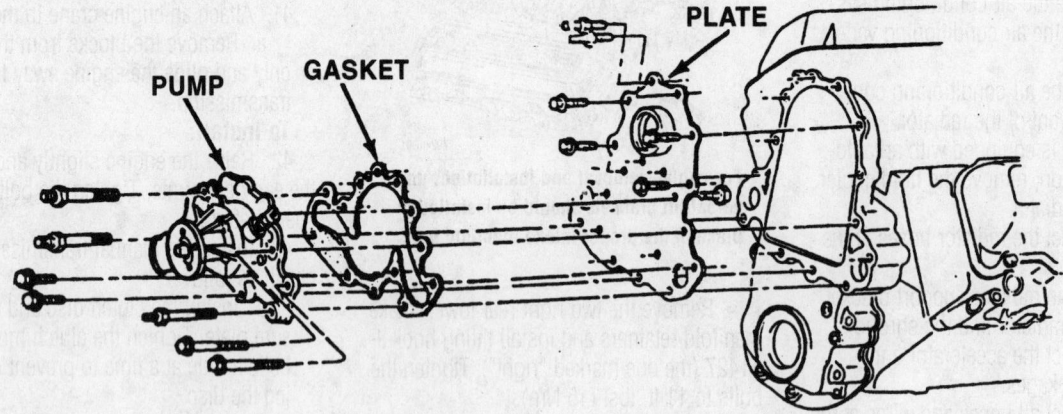

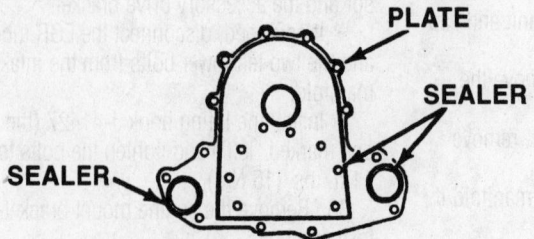

Exploded view of the water pump assembly and related components—6.5L Diesel engines

7924KG05

8. Remove the power steering pump and bracket, then support the assembly aside.

9. Lower the vehicle, then disconnect the coolant hoses from the pump.

10. Remove the water pump plate retaining bolts, then remove the pump and plate assembly from the engine.

11. Remove the bolt on the rear of the water pump plate, then separate the pump and gasket from the plate.

To install:

12. Install the water pump and a new gasket to the plate. Tighten the retaining bolt (at the rear of the plate) to 17 ft. lbs. (23 Nm).

13. Be sure the block mating surface and the plate flanges are free of oil. Apply an anaerobic sealer GM part 1052357 or equivalent.

➡**The sealer must be wet to the touch when the bolts are tightened.**

14. Attach the water pump and plate assembly, then install and tighten the retainers.

15. Connect the coolant hoses to the pump assembly.

16. Raise and support the vehicle safely, then reposition and secure the power steering pump and bracket.

17. Install the vacuum pump and bracket, along with the bolt holding the pump and alternator.

18. Lower the vehicle, then install the fan and pulley.

19. Install the engine accessory drive belt(s).

20. If removed, install the oil filler tube and/or air conditioning hose bracket nuts.

21. Install the fan shroud.

22. Connect the batteries.

23. Fill the radiator with the proper type and quantity of antifreeze.

Glow Plugs

REMOVAL & INSTALLATION

1. Disconnect the negative battery cables.

2. Disconnect the glow plug lead wires, then remove the plugs. You'll need a ⅜ in. deep-well socket.

3. Raise the vehicle and support it with safety stands. Remove the right front tire.

4. Remove the inner splash shield from the fender well.

5. Remove the lead wire from the plug at the No. 2 cylinder. Remove the lead wires from plugs in the Nos. 4 and 6 cylinders at the harness connectors.

6. Remove the heat shroud for the plug in the No. 4 cylinder. Remove the heat shroud for the plug in the No. 6 cylinder. Slide the shrouds back just far enough to allow access so you can unplug the wires.

7. Remove the plugs in cylinders No. 2, 4 and 6.

8. Reach up under the vehicle and disconnect the lead wire at No. 8. Remove the glow plug. You may find that removing the exhaust down pipe make this a bit easier when working on Nos. 6 and 8.

To install:

9. Install glow plugs and tighten to 16 ft. lbs. (22 Nm).

10. Install heat shrouds and electrical connection.

11. If removed, install exhaust pipe.

12. If removed, install splash shields.

13. Lower the vehicle.

14. Disconnect the negative battery cable.

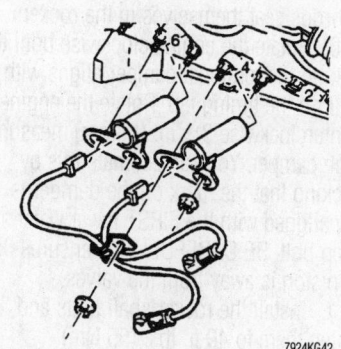

Exploded view of the heat shrouds and glow plug wiring—Diesel engines

Cylinder Head

REMOVAL & INSTALLATION

1. Disconnect the negative battery cable, relieve the fuel system pressure and drain the coolant system.

2. If equipped, have the A/C system discharged by an EPA certified technician.

3. Remove the intake manifold. Remove the fan upper shroud. Remove the compressor assembly, if equipped.

4. If equipped, remove the turbocharger.

5. Raise and support the vehicle safely.

6. Remove the exhaust manifold. Lower the vehicle.

7. Remove the valve cover, rocker arm assemblies and pushrods. Mark all components so they may be returned to their original location.

8. Remove the air cleaner resonator and bracket.

9. Remove the transmission and oil dipstick tube; remove the oil fill tube from the coolant crossover pipe.

10. Remove the heater, radiator and bypass hoses.

11. Remove the alternator upper bracket and alternator.

12. Remove power steering pump.

13. Remove vacuum pump.

14. Remove the fuel bleeder valve at the coolant crossover pipe.

15. Remove the fuel return crossover line clamp bolts from both cylinder heads.

16. Disconnect the wire connector from the sensor in the coolant crossover pipe.

17. Remove electrical connection and brackets from cylinder head.

18. Remove the coolant crossover pipe/thermostat assembly.

19. Remove the head bolts and the cylinder heads.

To install:

20. Clean the mating surfaces of the heads and block thoroughly.

21. Install a new head gaskets on the engine block. Do not coat the gaskets with any sealer on either engine. The gaskets have a special coating that eliminates the need for sealer. Install the rear head bolt, Install the cylinder heads onto the block.

22. Clean the head bolts thoroughly. Coat the threads of the head bolts with sealing compound GM part 1052080 or equivalent, before installation. Tighten the head bolts to 20 ft. lbs. (25 Nm), in the proper sequence, next tighten all bolts to 50 ft. lbs. (65 Nm) in the proper sequence, and finally tighten all bolts an additional 90 degree (¼ turn).

23. Install the coolant crossover pipe and thermostat.

24. Install the fuel valve and alternator assembly.

25. Connect the bypass hose.

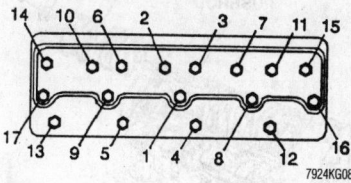

Tighten the cylinder head bolts according to the sequence shown for proper cylinder sealing—6.5L Diesel engines

26. Connect the upper radiator hose.

27. Connect the heater hoses at the head.

28. Install the transmission and oil dipstick tube.

29. Install the air cleaner resonator and bracket.

30. Install the pushrods, hardened ends facing up.

31. Install the rocker arm assemblies.

32. Adjust the valves.

33. Install the valve cover. Install the alternator assembly.

34. Raise and support the vehicle safely.

35. Install the exhaust manifolds and tighten to 22 ft. lbs. (30 Nm) Lower the vehicle.

36. Install the upper fan shroud.

37. Install the intake manifold.

38. If equipped, install the turbocharger, vacuum pump, and A/C compressor.

39. Install engine electrical connection.

40. Fill the cooling system with the proper type and quantity of antifreeze. Evacuate and recharge the air conditioning system.

41. Connect the negative battery cable.

Rocker Arms/Shaft

REMOVAL & INSTALLATION

1. Remove the engine cover.

➡**Rotate the engine until the mark on the crankshaft balancer is at two o'clock. Rotate the crankshaft counterclockwise 3½ in. (88mm) aligning the crankshaft balancer mark with the first lower water pump bolt, about 12:30. This will ensure that no valves are close to a piston crown.**

2. Remove the cylinder head cover.

3. The rocker assemblies are mounted on two short rocker shafts per cylinder head, with each shaft operating four rockers. Remove the two bolts which secure each rocker shaft assembly, and remove the shaft.

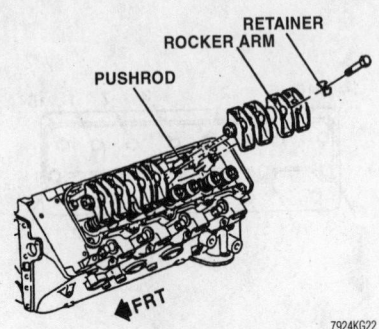

Rocker shaft assembly and related components—Diesel engines

7924KG22

Mark the shafts so they can be installed in their original locations.

4. Remove the pushrods. The pushrods MUST be installed in the original direction! A paint stripe usually identifies the upper end of each rod, but if you can't see it, be sure to mark each rod yourself.

5. Insert a small prybar into the end of the rocker shaft bore and break off the end of the nylon retainers. Pull off the retainers with pliers, then slide off the rockers.

To install:

6. Be sure first that the rocker arms and springs go back on the shafts in the exact order in which they were removed. Its a good idea to coat them with engine oil.

7. Center the rockers on the corresponding holes in the shaft and install new plastic retainers using a ½ in. (13mm) drift.

8. Install the pushrods with their marked ends up.

9. Install the rocker shaft assemblies and be sure that the ball ends of the pushrods seat themselves in the rockers.

10. Rotate the engine clockwise until the mark on the torsional damper aligns with the **0** on the timing tab. Rotate the engine counterclockwise 3½ in. (88mm) measured at the damper. You can estimate this by checking that the mark on the damper is now aligned with the FIRST lower water pump bolt. BE CAREFUL! This ensures that the piston is away from the valves.

11. Install the rocker shaft bolts and tighten them to 40 ft. lbs. (55 Nm).

12. Install the cylinder head cover.

13. Install the engine cover.

Turbocharger

REMOVAL & INSTALLATION

1. Disconnect the negative battery cable.

2. Remove the air inlet duct.

3. Disconnect the oil feed line from the top of the turbocharger.

4. Remove the screw retaining the Crankcase Depression Regulator (CDR) valve vent bracket.

5. Remove the CDR valve and vent tube.

6. Loosen the fasteners retaining the air cleaner assembly to the wheel well and remove the air cleaner assembly.

7. Loosen the heat shield retainers and remove the shield.

8. Remove the right front tire assembly and the splash shield.

9. Loosen the exhaust pipe-to-turbocharger exhaust outlet elbow V-band clamp.

10. Remove the oil drain tube-to-turbocharger center bearing bolts.

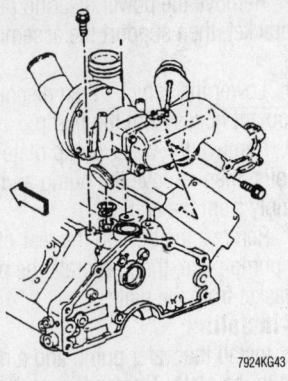

7924KG43

Turbocharger mounting—Diesel engines

11. Remove the exhaust manifold-to-turbocharger nuts.

12. Remove the turbocharger.

To install:

➡**Use anti-seize compound on all threaded fasteners connected to the turbocharger**

13. Engage the turbocharger to the exhaust manifold and tighten the nuts to 37 ft. lbs. (50 Nm).

14. Install a new oil drain tube flange gasket and the oil drain tube. Tighten the bolts to 19 ft. lbs. (26 Nm).

➡**Use 0.03–0.07 fl. oz. (1–2ml) of engine oil to feed the oil feed hole at the top of the turbocharger and hand rotate the compressor wheel/shaft. This will pre-lube the shaft bearings**

15. Connect the oil feed line and tighten the connection to 13 ft. lbs. (17 Nm).

16. Engage the exhaust pipe to the turbocharger exhaust elbow V-band clamp and tighten the clamp to 71 inch. lbs. (8 Nm).

17. Disengage the injection pump fuel shutdown solenoid connector and crank the engine for no more 15 seconds to prime the oil system. Do not let the engine start.

18. Install the right front splash shield and wheel.

19. Install the heat shield. Apply Loctite or equivalent to the bolts and install the bolts and tighten to 56 inch. lbs. (6 Nm).

20. Install the air cleaner assembly and the rubber connector to the intake duct and the turbocharger compressor outlet.

21. Install the CDR valve, tube and bracket and tighten the screw.

22. Install the air intake duct.

23. Install the CDR vent tubes into the air intake duct.

➡**Operate the engine at idle for at least three minutes after installing the turbocharger**

Intake Manifold

REMOVAL & INSTALLATION

1995–97 Models

1. Disconnect both negative battery cables.
2. Drain the cooling system and properly relieve the fuel system pressure.
3. For van models, remove the engine cover.
4. Remove the air cleaner assembly.
5. Remove the EPR/EGR valve bracket from the intake manifold.
6. Remove the CDR valve.
7. Remove the crankcase ventilator hose and EGR.
8. Remove the air conditioning rear bracket, if equipped.
9. Remove the fuel line bracket and ground strap.
10. Remove the fuel filter bracket at the intake manifold.
11. Remove the intake manifold bolts. The injection line clips are retained by these bolts.
12. Remove the intake manifold.

➡ If the engine is to be further serviced with the manifold removed, install protective covers over the intake ports.

To install:

13. Clean the manifold gasket surfaces on the cylinder heads and install new gaskets before installing the manifold.

➡ The gaskets have an opening for the EGR valve on light duty installations. An insert covers this opening on heavy duty installations.

14. Install the intake manifold.
15. Install the intake manifold bolts and fuel injection line clips.
16. Install the fuel filter bracket at the intake manifold.
17. Install the fuel line bracket and ground strap.
18. Install the air conditioning rear bracket, if equipped.
19. Install the crankcase ventilator hose and EGR.
20. Install the CDR valve.
21. Install the EPR/EGR valve bracket from the intake manifold.
22. Install the air cleaner assembly.
23. For van models, install the engine cover.
24. Fill the cooling system with the proper type and quantity of antifreeze.

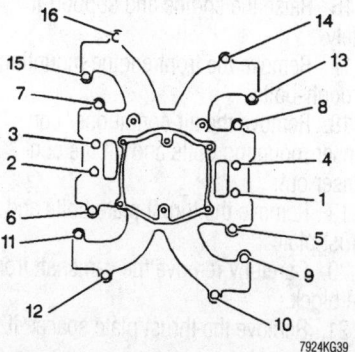

Intake manifold installation and bolt tightening sequence—1995–97 models

25. Connect both negative battery cables. Inspect for leaks.

1998–99 Models

1. Disconnect the negative battery cable.
2. Remove the air cleaner assembly.
3. Remove fuel lines, and electrical connections.
4. Remove the engine and transmission oil level tubes.
5. Recover A/C system, and reposition A/C lines.

❄❄❄ WARNING

Do not remove the center intake and side intakes as an assembly. Damage to the center intake and turbocharger may occur.

6. Remove the center intake assembly, and glow plug relay.
7. Remove the side intake bolts and fuel retaining clips.
8. Remove the side intakes.
9. Clean all gaskets surface.

To install:

10. Install the side intakes and new gaskets tighten the bolts to 31 ft. lbs. (42 Nm).

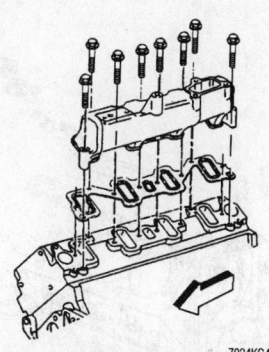

Exploded view of the side intake manifold mounting—1998–99 models

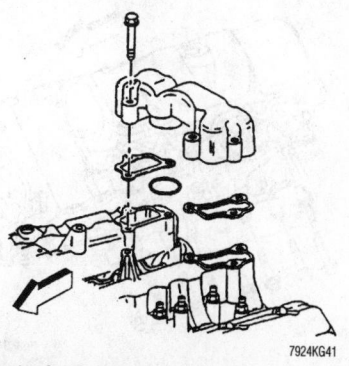

Center intake manifold mounting— 1998–99 models

11. Install the fuel lines retaining clips and electrical connection.
12. Install the center intake with new gaskets. Tighten the bolts to 17 ft. lbs. (23 Nm).
13. Install the engine oil and transmission oil level tubes.
14. Install the glow plug relay.
15. Install the A/C lines and recharge system
16. Install air cleaner assembly.
17. Connect the negative battery cable.

Exhaust Manifold

REMOVAL & INSTALLATION

1. Disconnect the batteries.
2. Raise and support the vehicle safely.
3. Disconnect the exhaust pipe from the manifold flange and lower the vehicle.
4. Remove the engine oil and transmission oil fill tubes.
5. Remove the engine cover and disconnect the glow plug wires.
6. Remove the glow plugs. Remove the turbocharger assembly, as required.
7. Remove the air conditioner compressor rear bracket, as required.

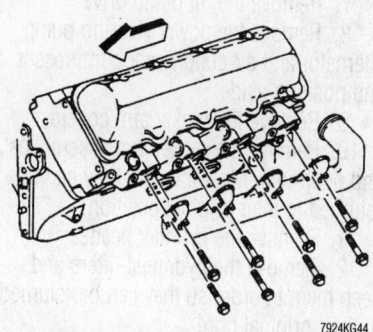

Exploded view of the left exhaust manifold mounting—Diesel engines

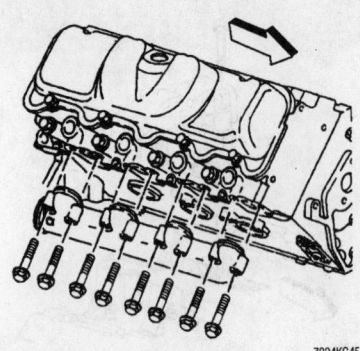

Exploded view of the right exhaust manifold mounting—Diesel engines

8. Remove the manifold bolts and remove the manifold.

To install:

9. Install exhaust manifold and tighten bolts to 26 ft. lbs. (35 Nm).

10. Install exhaust pipe.

11. Install glow plugs and electrical connection.

12. Install engine and transmission oil fill tubes.

13. Install A/C compressor bracket.

14. Connect the negative battery cable.

Camshaft and Valve Lifters

REMOVAL & INSTALLATION

Except Van Models

1. Disconnect the battery cables and relieve the fuel system pressure.

2. Drain the cooling system.

3. Have the A/C system discharged by an EPA certified technician.

4. Remove the radiator, condenser, shroud and fan assembly.

5. Remove the grille and parking light assembly.

6. Remove the hood latch and brace assembly.

7. Remove the oil pump drive.

8. Remove the power steering pump, alternator and air conditioner compressor and position aside.

9. Remove the rocker arm covers.

10. Remove the rocker arm assemblies and pushrods. Mark them so they can be returned to their original position.

11. Remove the cylinder heads.

12. Remove the hydraulic lifters and keep them in order so they can be returned to their original bore.

13. Remove the front cover.

14. Remove the timing chain and camshaft sprocket.

15. Remove the injector pump.

16. Raise the engine and support it safely.

17. Remove the front engine mounting through-bolts.

18. Remove the air conditioner condenser mounting bolts and lift the condenser out.

19. Remove the thrust plate bolts and thrust plate.

20. Carefully remove the camshaft from the block.

21. Remove the thrust plate spacer, if necessary.

To install:

22. Install the spacer with the ID chamfer toward the camshaft.

➡**It is recommended that the engine oil, oil filter and hydraulic lifters be replaced when installing a new camshaft.**

23. Coat the camshaft lobes with Molykote® or equivalent.

24. Lubricate the camshaft journals with engine oil.

25. Insert the camshaft carefully into the block, install the thrust plate and bolts. Tighten to 17 ft. lbs. (23 Nm).

26. Lower the engine and install the engine mount through-bolts.

27. Align the timing marks and install the timing chain and sprockets.

28. Install the air conditioner condenser, if equipped.

29. Install the injector pump.

30. Install the front cover.

31. Install cylinder head.

32. Install the hydraulic lifters in the same bore as they were removed.

33. Install the rocker arm assemblies and pushrods in their original locations.

34. Install the rocker arm covers.

35. Install the power steering pump, alternator and air conditioner compressor.

36. Install the oil pump drive.

37. Install the hood latch and brace.

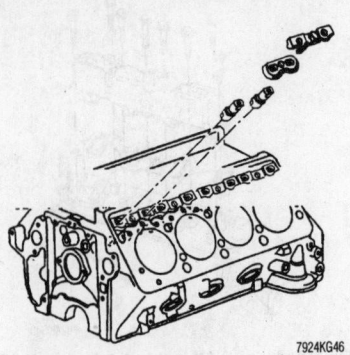

Exploded view of the lifter, guide plate and clamp—Diesel engines

38. Install the grille and parking light assembly.

39. Install the radiator, the shroud and fan assembly.

40. Fill the cooling system with the proper type and quantity of antifreeze.

41. Connect the negative battery cables.

Van Models

1. Disconnect the battery cables and relieve the fuel system pressure.

2. Remove the headlight bezels.

3. Remove the grille, bumper and lower valance panel.

4. Remove the hood latch.

5. Remove the coolant recovery bottle.

6. Remove the upper tie bar.

7. Remove the air conditioner compressor.

❊ CAUTION

Never open, service or drain the radiator or cooling system when hot; serious burns can occur from the steam and hot coolant.

8. Drain the cooling system and remove the radiator and fan.

9. Remove the oil pump drive.

10. Remove the cylinder heads to gain clearance for lifter removal.

11. Remove the alternator lower bracket.

12. Remove the water pump.

13. Remove the torsional damper.

14. Remove the front cover.

15. Remove the injection pump.

16. Remove the rocker arm covers.

17. Remove the rocker arm assemblies and pushrods. Mark them so they can be returned to their original position.

18. Remove the hydraulic lifters and keep them in order so they can be returned to their original bore.

19. Remove the timing chain and camshaft sprocket.

20. Remove the thrust plate bolts and thrust plate.

21. Carefully remove the camshaft from the block.

22. Remove the thrust plate spacer, if necessary.

To install:

23. Install the spacer with the ID chamfer toward the camshaft.

➡**It is recommended that the engine oil, oil filter and hydraulic lifters be replaced when installing a new camshaft.**

24. Coat the camshaft lobes with Molykote® or equivalent.

25. Lubricate the camshaft journals with engine oil.

26. Insert the camshaft carefully into the block, install the thrust plate and bolts and tighten to 17 ft. lbs.

27. Align the timing marks and install the timing chain and sprockets.

28. Install the hydraulic lifters in the same bore as they were removed.

29. Install the rocker arm assemblies and pushrods in their original locations.

30. Install the rocker arm covers.

31. Install the fuel pump.

32. Install the front cover.

33. Install the torsional damper and water pump.

34. Install the alternator lower bracket.

35. Install the cylinder heads.

36. Install the oil pump drive.

37. Install the radiator and fan.

38. Install the air conditioner compressor.

39. Install the upper tie bar.

40. Install the coolant recovery bottle.

41. Install the hood latch.

42. Install the grille, bumper and lower valance panel.

43. Install the headlight bezels.

44. Install the battery cables.

45. Fill the cooling system.

46. Evacuate and charge the air conditioner system.

Valve Lash

ADJUSTMENT

All engines use hydraulic lifters, which require no periodic adjustment.

Oil Pan

REMOVAL & INSTALLATION

Except Van Models

1. Disconnect the battery cables.

2. Raise and safely support the vehicle.

3. Drain the engine oil. Remove the oil dipstick.

4. Remove the flywheel/flexplate cover.

5. Remove oil cooler line guides

6. If needed, remove front driveshaft. Use the procedure outlined in the driveline section.

7. If needed, remove front axle. Use the procedure outlined in the driveline section.

8. Disconnect the exhaust pipes from the manifolds.

9. Remove the front engine mount through-bolts and raise the engine.

10. Remove the oil pan bolts and remove the oil pan.

11. Remove the oil pan rear seal.

To install:

12. Clean all sealing surfaces.

13. Apply a ³⁄₁₆ in. (5mm) bead of RTV sealant to the oil pan sealing surface, inboard of the bolt holes. The sealant must be wet to the touch when the oil pan is to be installed.

14. Install the oil pan rear seal.

15. Install the oil pan to the engine and install the retaining bolts. Tighten all except the rear two bolts to 84 inch lbs. (9.4 Nm). Tighten the rear the bolts to 17 ft. lbs. (23 Nm).

16. Lower the engine.

17. Install the engine mounting through-bolt and nut.

18. If removed, install front axles and front driveshaft.

19. Install oil cooler lines in guides.

20. Install the oil dipstick. Install the exhaust pipes to the manifolds.

21. Install the flywheel/flexplate cover and lower the vehicle.

22. Refill with the proper grade and quantity of oil.

23. Connect the negative battery cable.

Van Models

1. Disconnect the battery cables.

2. Remove the engine cover.

3. Remove the engine oil dipstick.

4. Remove the engine oil dipstick tube at the rocker cover.

5. Raise the vehicle and support it safely.

6. Remove the transmission flywheel/flexplate cover.

7. Drain the engine oil.

8. Disconnect the oil cooler lines at the block.

9. Remove the starter. Remove the battery cables, transmission cooler lines and attaching clamps from the oil pan.

10. Remove the oil pan bolts and remove the oil pan and oil pan rear seal.

To install:

11. Clean all sealing surfaces

12. Apply a ³⁄₁₆ in. (5mm) bead of RTV sealant to the oil pan sealing surface, inboard of the bolt holes. The sealant must be wet to the touch when the oil pan is to be installed.

13. Install the oil pan rear seal.

14. Install the oil pan to the engine and install the retaining bolts.

15. Install the starter. Install the transmission, battery cables and attaching clamps to the oil pan.

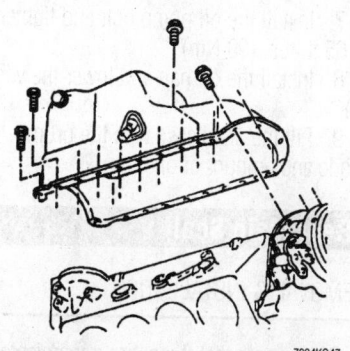

Exploded view of the oil pan mounting— Diesel engines

16. Install the engine oil cooler lines.

17. Install the transmission flywheel/flexplate cover.

18. Lower the vehicle.

19. Install the engine oil dipstick tube at the rocker cover.

20. Install the engine oil dipstick.

21. Install the engine cover.

22. Refill with the proper grade and quantity of oil.

23. Install the battery cables.

Oil Pump

REMOVAL & INSTALLATION

1. Raise the support the vehicle safely.

2. Drain the oil and remove the oil pan.

3. Remove the oil pump to crankshaft rear main bearing attaching bolt.

4. Remove the oil pump and hex drive.

To install:

5. Inspect the oil pan pick up tube and screen for damage and the hex drive for cracks.

6. Install the oil pump and extension shaft to the engine. Align the extension shaft hex with the drive hex, the oil pump should push easily into place.

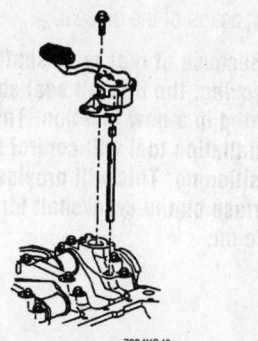

Exploded view of the oil pump mounting— Diesel engines

7. Install the oil pump bolt and tighten to 65 ft. lbs. (90 Nm).

8. Install the oil pan and lower the vehicle.

9. Fill the crankcase with the proper grade and amount of oil.

Rear Main Seal

REMOVAL & INSTALLATION

Please note that the entire transmission assembly must be removed before performing this procedure. Before a new seal is installed, the Crankcase Depression Regulator (CDR) and crankcase ventilation system should be cleaned and inspected. In addition, use care removing the flywheel. Some models use a heavy, dual mass flywheel which must be handled with care.

1. Disconnect the negative battery cables.

2. Raise and safely support the vehicle.

3. Remove the transfer case, if so equipped.

4. Remove the transmission assembly using the recommended procedure.

5. Remove the clutch assembly and flywheel if equipped with manual transmission. Use care removing the flywheel. Some models use a heavy, dual mass flywheel which must be handled with care.

6. Remove the flexplate if equipped with automatic transmission.

7. Remove the crankshaft rear main oil seal by inserting a suitable prying tool and prying the seal out. Take care not to damage the crankshaft sealing surface.

To install:

8. Clean the oil seal bore in the block thoroughly before installation of the new seal.

9. Inspect the crankshaft for grit, rust or burrs and correct as necessary. Also inspect the portion of the crankshaft where the oil seal makes contact, for wear due to the rubbing action of the oil seal.

➡**Because of rear crankshaft wear or grooving, the new oil seal should be seated in a new location. The J 39084 installation tool will control the seal positioning. This will provide a new surface on the crankshaft for the seal to ride on.**

10. Clean the running surface of the crankshaft with a non-abrasive cleaner.

11. Lubricate the inner diameter of the new seal and the outer diameter of the crankshaft with engine oil. Using a J 39084 installation tool, install the rear main oil seal until the tool bottoms against the block and crankshaft rear main bearing cap.

12. If equipped with manual transmission and if the rear crankshaft oil seal was leaking badly, the clutch plate may be contaminated with oil and require replacement. Service the clutch as required. Install the flywheel. Use threadlocking compound on the flywheel retainer bolts and tighten to 45 ft. lbs. (60 Nm). New replacement flywheel bolts are recommended.

13. Install the transmission assembly using the recommended procedure.

14. Install the transfer case, if so equipped.

15. Lower the vehicle.

16. Connect the negative battery cables.

17. Start the engine and verify no oil leaks.

Timing Chain, Sprockets, Front Cover and Seal

REMOVAL & INSTALLATION

1. Disconnect both negative battery cables. Drain the cooling system.

2. Remove the water pump and pulleys.

3. Rotate the crankshaft to align the marks on the torsional damper with the **0** mark on the timing tab.

4. Scribe a mark aligning the injection pump flange and the front cover, if not already marked.

➡**The outer ring (weight) of the torsional damper is bonded to the hub with rubber. The damper must be removed with a puller which acts on the inner hub only. Pulling on the outer portion of the damper will break the rubber bond or destroy the tuning of the unit.**

5. Remove the crankshaft pulley and torsional damper.

6. Remove the front cover-to-oil pan bolts (4).

7. Remove the two fuel return line clips.

8. Remove the injection pump gear.

9. Remove the injection pump retaining nuts from the front cover.

10. Remove the crankshaft sensor.

11. Remove the baffle. Remove the remaining cover bolts and remove the front cover.

12. Remove the injection pump gear.

13. Align the camshaft timing gear marks and remove the bolt and washer attaching the camshaft gear.

14. Remove the camshaft sprocket with the timing chain. Remove the crankshaft sprocket.

To install:

15. Install the cam sprocket, timing chain and crankshaft sprocket as a unit, aligning the timing marks on the sprockets.

16. Rotate the crankshaft to align the injection pump and camshaft gears. Install the injection pump gear.

17. If the front cover oil seal is to be replaced, it can now be pried out of the cover with a suitable prying tool. Press the new seal into the cover evenly.

18. Clean both sealing surfaces until all traces of old sealer are gone. Apply a 3/32 in. (2mm) bead of GM sealant 1052357 or equivalent to the sealing surface. Apply a 3/16 in. (5mm) bead of RTV type sealer to the bottom portion of the front cover which attaches to the oil pan. Install the front cover.

19. Install the baffle.

20. Install the injection pump, making sure the scribe marks on the pump and front cover are aligned. Tighten the nuts to 31 ft. lbs. (42 Nm).

21. Install the injection pump driven gear, making sure the marks on the cam gear and pump are aligned. Tighten the injection pump gear bolts to 17 ft. lbs. (23 Nm).

➡**Verify that there is a minimum clearance of 0.040 in. (1.0mm) between the injection pump gear and baffle or noise may be result.**

22. Install the fuel line clips, the front cover-to-oil bolts, and the torsional damper and crankshaft pulley. Tighten the pan and the damper bolt to specification.

23. Install the water pump and pulley assembly.

24. Fill the cooling system with the proper type and quantity of antifreeze.

25. Connect the negative battery cables. Inspect engine for leaks.

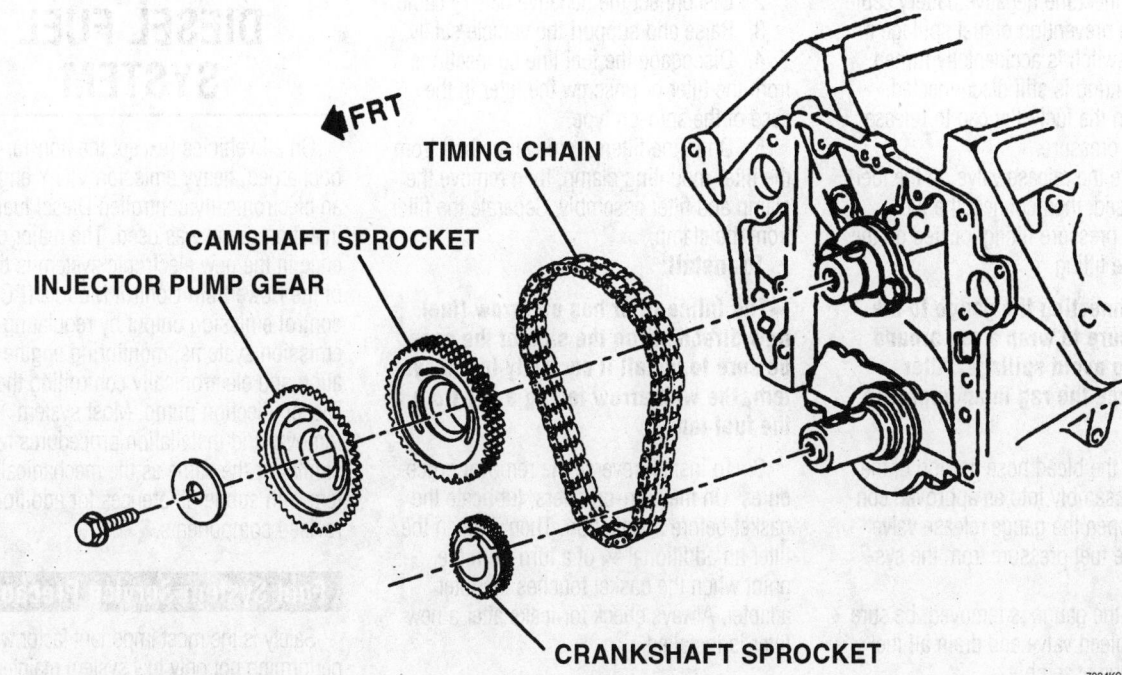

◀FRT

TIMING CHAIN

CAMSHAFT SPROCKET

INJECTOR PUMP GEAR

CRANKSHAFT SPROCKET

7924KG23

Timing chain and related components—6.5L Diesel engines

GASOLINE FUEL SYSTEM

Fuel System Service Precautions

Safety is the most important factor when performing not only fuel system maintenance but any type of maintenance. Failure to conduct maintenance and repairs in a safe manner may result in serious personal injury or death. Maintenance and testing of the vehicle's fuel system components can be accomplished safely and effectively by adhering to the following rules and guidelines.

• To avoid the possibility of fire and personal injury, always disconnect the negative battery cable unless the repair or test procedure requires that battery voltage be applied.

• Always relieve the fuel system pressure prior to disconnecting any fuel system component (injector, fuel rail, pressure regulator, etc.), fitting or fuel line connection. Exercise extreme caution whenever relieving fuel system pressure to avoid exposing skin, face and eyes to fuel spray. Please be advised that fuel under pressure may penetrate the skin or any part of the body that it contacts.

• Always place a shop towel or cloth around the fitting or connection prior to loosening to absorb any excess fuel due to

spillage. Ensure that all fuel spillage (should it occur) is quickly removed from engine surfaces. Ensure that all fuel soaked cloths or towels are deposited into a suitable waste container.

• Always keep a dry chemical (Class B) fire extinguisher near the work area.

• Do not allow fuel spray or fuel vapors to come into contact with a spark or open flame.

• Always use a back-up wrench when loosening and tightening fuel line connection fittings. This will prevent unnecessary stress and torsion to fuel line piping. Always follow the proper torque specifications.

• Always replace worn fuel fitting O-rings with new. Do not substitute fuel hose or equivalent, where fuel pipe is installed.

Fuel System Pressure

RELIEVING

Before loosening or disconnecting any fuel fitting or system component, always relieve the fuel system pressure in order to help prevent the danger of fire or injury.

Throttle Body Fuel Injection Systems

GM vehicles with TBI engines utilize an automatic pressure bleed down feature. But, some fuel pressure related steps should still be taken to assure safer working conditions.

1. Disconnect the negative battery cable to prevent fuel spillage should the ignition key accidentally be turned **ON** with a fuel fitting disconnected.

2. Loosen fuel filler cap to relieve fuel tank pressure.

3. The internal constant bleed feature of the Model 220 TBI unit relieves fuel pump system pressure when the engine is turned **OFF**. Therefore, no further action is required.

➡**Allow the engine to set for 5–10 minutes; this will allow the orifice (in the fuel system) to bleed off the pressure.**

4. When fuel service is finished, tighten the fuel filler cap and connect the negative battery cable.

Multi-Port Fuel Injection Systems

The MFI and CMFI fuel systems used on GM vehicles all operate under high fuel pressures. It is very important that the pressure be properly relieved prior to servicing the system or any of its components.

A Schrader valve is provided on these fuel systems in order to conveniently test or release the system pressure. A fuel pressure gauge and adapter will be necessary to connect the gauge to the fitting. Most of the MFI systems utilize a service valve on one end of the fuel rail assembly. The CMFI system covered here uses a valve located on the inlet pipe fitting, immediately before it enters the CMFI assembly (towards the rear of the engine).

1. Disconnect the negative battery cable to assure the prevention of fuel spillage if the ignition switch is accidentally turned **ON** while a fitting is still disconnected.

2. Loosen the fuel filler cap to release the fuel tank pressure.

3. Be sure the release valve on the fuel gauge is closed, then connect the fuel gauge to the pressure fitting located on the inlet fuel pipe fitting.

➡**When connecting the gauge to the fitting, be sure to wrap a rag around the fitting to avoid spillage. After repairs, place the rag in an approved container.**

4. Install the bleed hose portion of the fuel gauge assembly into an approved container, then open the gauge release valve and bleed the fuel pressure from the system.

5. When the gauge is removed, be sure to open the bleed valve and drain all fuel from the gauge assembly.

6. When fuel service is finished, tighten the fuel filler cap and connect the negative battery cable.

Fuel Filter

REMOVAL & INSTALLATION

The fuel filter is normally located along the frame rail of the vehicle. On some vehicles however, it may have been relocated to the engine compartment. When in doubt, trace a fuel line from the engine backwards or from the tank forward in order to locate the filter.

Some vehicles utilize a spin-on fuel filter located on the frame rail. This filter can be turned counterclockwise after the fuel pressure is relieved.

1. Properly relieve the fuel system pressure.

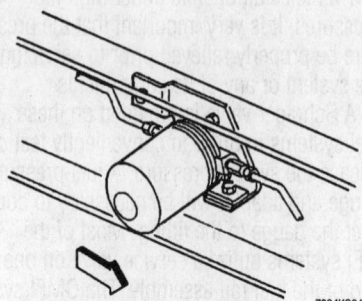

The spin-on fuel filter is serviced in the same manner as a spin-on oil filter— 5.0L, 5.7L and 7.4L engines

2. Disconnect the negative battery cable.

3. Raise and support the vehicle safely.

4. Disengage the fuel line connections from the filter or unscrew the filter in the case of the spin-on type.

5. On inline filters, remove the bolt from the filter mounting clamp, then remove the clamp and filter assembly. Separate the filter from the clamp.

To install:

➡**The inline filter has an arrow (fuel flow direction) on the side of the case, be sure to install it correctly in the system, the with arrow facing away from the fuel tank.**

6. To install, reverse the removal procedures. On the spin-on filters, lubricate the gasket before installation. Then, tighten the filter an additional ¾ of a turn from the point when the gasket touches the filter adapter. Always check for leaks after a new filter is installed.

Fuel Pump

REMOVAL & INSTALLATION

1. Properly relieve the fuel system pressure.

2. Disconnect the negative battery cable.

3. Drain and remove the fuel tank from the vehicle

4. Using a suitable spanner wrench, turn the fuel pump/sending unit assembly locking ring (located on top of the fuel tank) counterclockwise, then carefully lift the assembly from the tank and remove the pump from the fuel lever sending device.

To install:

5. Install fuel pump and secure with locking ring.

6. Secure fuel tank to vehicle and refill.

7. Connect the negative battery cable.

8. Run engine and check for fuel leaks.

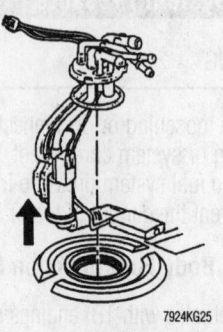

Lift the fuel pump assembly out of the tank after removing the locking ring—4.3L, 5.0L, 5.7L and 7.4L engines

DIESEL FUEL SYSTEM

On all vehicles (except the non-turbocharged, heavy emission VIN Y engine), an electronically controlled Diesel fuel injection pump was used. The major difference in the new electronic system is the use of the Powertrain Control Module (PCM) to control emission output by regulating the emission systems, monitoring engine operation and electronically controlling the Diesel injection pump. Most system removal and installation procedures remain similar or the same as the mechanical system with subtle differences for additional or revised components.

Fuel System Service Precautions

Safety is the most important factor when performing not only fuel system maintenance but any type of maintenance. Failure to conduct maintenance and repairs in a safe manner may result in serious personal injury or death. Maintenance and testing of the vehicle's fuel system components can be accomplished safely and effectively by adhering to the following rules and guidelines.

• To avoid the possibility of fire and personal injury, always disconnect the negative battery cable unless the repair or test procedure requires that battery voltage be applied.

• Always relieve the fuel system pressure prior to disconnecting any fuel system component (injector, fuel rail, pressure regulator, etc.), fitting or fuel line connection. Exercise extreme caution whenever relieving fuel system pressure to avoid exposing skin, face and eyes to fuel spray. Please be advised that fuel under pressure may penetrate the skin or any part of the body that it contacts.

• Always place a shop towel or cloth around the fitting or connection prior to loosening to absorb any excess fuel due to spillage. Ensure that all fuel spillage (should it occur) is quickly removed from engine surfaces. Ensure that all fuel soaked cloths or towels are deposited into a suitable waste container.

• Always keep a dry chemical (Class B) fire extinguisher near the work area.

• Do not allow fuel spray or fuel vapors to come into contact with a spark or open flame.

• Always use a back-up wrench when loosening and tightening fuel line connection fittings. This will prevent unnecessary stress and torsion to fuel line piping. Always follow the proper torque specifications.

• Always replace worn fuel fitting O-rings with new. Do not substitute fuel hose or equivalent, where fuel pipe is installed.

Fuel System Pressure

RELIEVING

Fuel system pressure can be released by wrapping a fuel fitting in a heavy shop towel and cracking loose the fitting. NEVER perform this with any source of ignition nearby!

Fuel System Air

BLEEDING

1. Open the air bleed valve on the fuel manager/filter.
2. Connect a hose to the air bleed valve and place the other of the hose in a suitable container.

✳✳ CAUTION

The Diesel/water mixture is flammable and may be hot. to avoid personal injury or property damage, do not allow the Diesel/water mixture to come in contact with skin, open flame or a hot engine. Do not overfill the container holding the fuel mixture as heat from a warm engine or any another heat source may cause the fuel to expand and leak from the container which may lead to a fire.

3. Remove the F/SOL fuse from the fuse panel.
4. Crank the engine in short intervals of 10-to-15 seconds until clear fuel is observed at the air bleed hose (wait for one minute between cranking intervals).
5. Remove the hose and close the air bleed valve.
6. Install the F/SOL fuse and start the vehicle. Allow the vehicle to run at idle for 5 minutes.
7. Check for fuel leaks and clear any Diagnostic Trouble Code's (DTC's).

Idle Speed

ADJUSTMENT

Idle speed and injection timing is controlled by the Powertrain Control Module (PCM), there is no provision for adjustment.

Fuel Filter

REMOVAL & INSTALLATION

1. Turn the ignition **OFF**. Remove the fuel tank cap to release any pressure or vacuum in the tank.

➡**It is not necessary to drain all the fuel from the header in order to change the element since the fuel will remain in the header's cavity.**

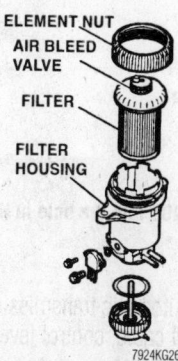

ELEMENT NUT
AIR BLEED VALVE
FILTER
FILTER HOUSING

7924KG26

Exploded view of the fuel filter assembly—6.5L Diesel engines

2. Open the air bleed valve to relieve residual pressure.
3. Remove the element nut, turning it by hand to the left. If necessary, a strap wrench may be used to loosen the nut.
4. Remove the element by lifting straight up and out of the header assembly.
To install:
5. Be sure the mating surface between the element assembly and the header assembly is clean.
6. Install the new element by aligning the widest key slot located under the element assembly cap with the widest key in the header assembly.
7. Carefully push the element downward until the mating surfaces make contact.
8. Install the element nut and tighten securely by hand.
9. If not already done, open the air bleed valve on top of the fuel manager/filter assembly, then connect a length of hose placing the other end in a suitable container.

➡**Be extremely cautious when handling Diesel fuel. Do not expose the fuel to sparks or open flames. Also, be cautious as the fuel coming out of the drain hose could be hot.**

10. Disconnect the fuel injection pump shutdown solenoid wire.

11. Crank the engine for 10–15 seconds, then wait one minute for the starter motor to cool. Repeat until clear fuel is observed coming from the air bleed.
12. Close the air bleed valve, reconnect the injection pump solenoid wire and replace the fuel tank cap.
13. Start the engine, allow it to idle for 5 minutes and check the fuel manager/filter assembly for leaks.

Diesel Injection Pump

All vehicles are equipped with an electronically controlled pump. The electronic pump is driven by gears and rotates at the same speed as the camshaft. An electronic stepper motor used to control injection timing and a fuel solenoid driver used to control the fuel injection solenoid on the electronic model.

REMOVAL & INSTALLATION

1. Disconnect both battery negative cables.
2. Remove the intake manifold.
3. Relieve the fuel system pressure.
4. Remove the fuel injection and inlet lines.
5. Tag and disconnect the necessary wires, cables and hoses at the injection pump.
6. Disconnect the fuel return line at the top of the injection pump.
7. If necessary, disconnect the fuel feed line at the injection pump.
8. Remove the oil filler tube grommet.

➡**Do not engage the starter in order to rotate the engine with the injection pump removed. The pump driven gear could jam in the front housing resulting in a sheared crankshaft or camshaft gear key and possible valvetrain damage.**

9. Scribe or paint a matchmark on the front cover and the injection pump flange.
10. Rotate the crankshaft by hand and remove the injection pump driven gear bolts, accessing the bolts through the oil filler neck hole.
11. Remove the injection pump-to-front cover attaching nuts. Remove the pump. Be sure to cap all open lines and nozzles in order to prevent system contamination and damage.
To install:
12. Align the locating pin on the pump hub with the slot in the injection pump driven gear (the SLOT not the hole in the gear). At the same time, align the timing marks.

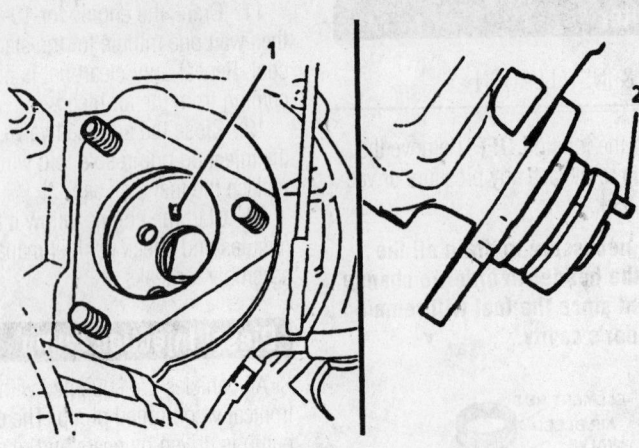

1   SLOT IN DRIVEN GEAR

2   PUMP HUB

7924KG27

Align the pin on the pump hub with the slot in the driven gear, NOT into the hole in the gear—Diesel engines

13. Attach the injection pump to the front cover, checking the timing marks before tightening the nuts to 30 ft. lbs. (40 Nm).

14. Install the driven gear-to-injection pump bolts, tightening the bolts to 20 ft. lbs. (25 Nm).

15. Install the grommet and oil fill tube.

16. If applicable, install the air conditioning bracket.

17. Connect the fuel feed line and tighten to 20 ft. lbs. (25 Nm).

18. If removed, connect the fuel return line to the pump.

19. Connect all wires, cables and hoses previously removed.

20. Connect the injector lines and install the intake manifold.

21. Connect the negative battery cables. Start the engine and check for leaks.

DRIVE TRAIN

➡**Refer to the Unit Repair Section for information on driveshaft and Universal Joint (U-joint) service.**

Transmission Assembly

REMOVAL & INSTALLATION

1. Disconnect the negative battery cable.

2. On manual transmissions, remove the shifter boot and lever.

3. Raise and safely support the vehicle, then drain the transmission.

4. On automatic transmissions, disconnect the shift cable, control lever and remove the bracket.

5. Remove the exhaust pipes and parking brake cables.

6. Matchmark and remove the driveshaft.

7. If equipped, remove the transfer case.

8. Remove the transmission to engine braces (vehicles equipped with Diesel engines have only one brace).

9. Disconnect the wiring harness at the transmission.

10. Support the transmission with a transmission jack.

11. Remove the nut securing the transmission mount to the cross member.

12. Position a transmission jack or equivalent, under the transmission for support.

13. Remove the crossmember. Visually inspect to see if other equipment, brackets or lines, must be removed to permit removal of transmission.

➡**Mark position of crossmember when removing to prevent incorrect installation. The tapered surface should face the rear.**

14. On automatic transmissions perform the following:

    a. Remove the torque converter inspection cover.

    b. Mark the alignment of the torque converter to the flexplate.

    c. Remove the torque converter to flexplate bolts. Remove the dipstick tube and seal from the transmission. Plug the opening to avoid contamination.

    d. Disconnect both transmission lines at the transmission and plug them to avoid contamination and leakage.

    e. Position a J 21366 converter holding strap onto the transmission/torque converter to keep the torque converter from sliding off of the transmission turbine shaft.

15. On manual transmissions, except the New Venture Gear 3500 (NV 3500), remove the top two transmission to housing bolts and insert two guide pins.

➡**The clutch housing on the New Venture Gear 3500 is integral to the transmission as is the converter housing on the automatic transmissions.**

16. On the New Venture Gear 3500, remove the bolts securing the clutch housing to the engine.

➡**The use of guide pins will not only support the transmission but will prevent damage to the clutch disc. Guide pins can be made by taking two bolts, the same as those just removed only longer, and cutting off the heads. Make an adjustment slot. Be sure to support the clutch release bearing and support assembly during removal of the transmission.**

17. Remove the remaining bolts and slide transmission straight back from engine. Use care to keep the transmission drive gear straight in line with clutch disc hub.

18. Remove/disconnect all vehicle wiring, clips, tubes and brackets etc., which would interfere with the removal of the transmission.

➡**Ensure that the engine is supported with a jack stand before detaching the transmission from the engine.**

19. Remove the studs and/or bolts securing the transmission to the engine.

20. Pull the transmission straight back from the engine.

21. Carefully lower the transmission using the transmission jack.

To install:

22. Check the area behind the torque converter for leaks. Replace the front seal, if required.

23. It is good practice to examine the area around the rear crankshaft seal, checking for leaks. If necessary, remove the flywheel/flexplate and replace the seal.

24. Inspect the flywheel/flexplate ring gear teeth. If damaged, replace the flywheel/flexplate.

25. Since the driveshaft has been removed, inspect the universal joints for wear and service, as required.

26. On manual transmissions, perform the following steps:

a. Place the transmission in high gear. Lightly coat the input shaft splines with high temperature grease.

b. Raise the transmission into position.

c. On transmissions with a separate clutch housing, install the guide pins in the top two bolt holes if they have been removed.

d. Roll the transmission forward and engage the clutch splines. Keep pushing the transmission forward until it mates with the engine.

e. On transmissions with a separate clutch housing, remove the guide pins and install the bolts, tighten the bolts to 23 ft. lbs. (31 Nm).

f. On the NV 3500, install the transmission-to-engine bolts. Tighten the bolts to 35 ft. lbs. (47 Nm).

27. On automatic transmissions, perform the following steps:

a. With the J 21366 or equivalent, torque converter holding strap in place, raise the transmission into position with a transmission jack.

b. Remove the torque converter holding strap and slide the transmission into place. Slide the transmission straight onto the locating pins while lining up the marks on the flywheel/flexplate and the torque converter. Be sure the transmission is fully seated against the rear of the engine block and the locating pins are completely engaged.

❊❊ WARNING

DO NOT attempt to draw the transmission to the block with the mounting bolts. If the transmission is not properly seated, the bolts will break the transmission case.

➡ **The torque converter must be flush with the flexplate and rotate freely by hand.**

c. When satisfied that the transmission is properly seated, install and tighten the transmission-to-engine bolts and/or studs to 34 ft. lbs. (47 Nm).

28. On automatic transmissions perform the following steps:

a. Install the dipstick tube and seal.

b. Check the alignment marks on the torque converter and flexplate to be sure that they are properly aligned. Install the torque converter bolts. Finger-tighten the bolts to ensure proper converter seating. When the converter is properly seated, tighten the bolts to 46 ft. lbs. (63 Nm).

c. Install the torque converter cover. Tighten the retaining bolts to 24 ft. lbs. (33 Nm) on the 4.3L engines or 89 inch lbs. (10 Nm) on the V8 engines.

29. Install all vehicle wiring, clips, tubes and brackets etc., which were removed during the removal of the transmission. All components must be positioned as they were when removed. Use care to be sure the transmission cooling lines will not rub the frame or related components, causing a leak later. Connect both transmission lines to the transmission.

a. Connect the shifter cable.

30. Install the starter.

31. Install the exhaust components previously removed.

32. Install the transmission crossmember. Tighten the bolts securing the crossmember to the frame to 56 ft. lbs. (77 Nm).

33. Install the transmission mount on the transmission and tighten the bolts to 35 ft. lbs. (47 Nm).

34. Install the nut and washer securing the transmission mount to the crossmember and tighten the nut to 38 ft. lbs. (52 Nm).

35. Install the transmission to engine brace(s). Tighten the bolts to 41 ft. lbs. (55 Nm) for gasoline engines and to 51 ft. lbs. (70 Nm) for Diesel engines.

36. If equipped, install the transfer case, if equipped.

37. Remove the transmission jack and engine support stands.

38. Install the driveshaft.

39. Lower the vehicle and fill the transmission with the proper grade and quantity of fluid.

40. On manual transmissions, install the shifter lever and boot.

41. Connect the negative battery cable.

42. Road test the vehicle and test for proper operation. Check for leaks.

Clutch

ADJUSTMENTS

The hydraulic clutch system requires no periodic adjustment.

REMOVAL & INSTALLATION

1. Disconnect the negative battery cable.
2. Disconnect the slave cylinder.
3. Remove the transmission assembly.

4. Install the clutch removal tool and support the clutch assembly.

➡ **Before removing the clutch from the flywheel, matchmark the flywheel, clutch cover and one pressure plate lug, so these parts may be assembled in their same relative positions and retain the factory balance.**

5. Loosen the clutch plate retaining bolts slowly and evenly one at a time until all pressure is released from the pressure plate assembly.

6. Remove the clutch, pressure plate and removal tool. Check the flywheel for damage, repair or replace, as required.

7. Check the clutch assembly and flywheel for signs of wear, scoring, overheating, etc. If the clutch plate, flywheel or pressure plate is oil-soaked, inspect the engine rear main seal and the transmission input shaft seal and correct leakage as required. Replace any damaged parts.

To install:

8. Assemble the pressure plate and disc assembly, as required.

➡ **The manufacturer recommends that new pressure plate bolts and washers be used.**

9. Turn the flywheel until the previously applied mark is at the bottom.

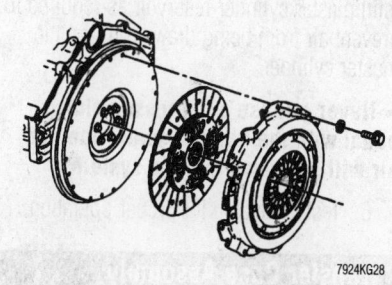

Exploded view of the typical clutch assembly

10. Install the clutch disc, pressure plate and cover, using a suitable clutch aligning tool.

11. Turn the clutch until the matchmark on the clutch cover aligns with the mark on the flywheel.

12. Install the attaching bolts and tighten in a crossing pattern until the spring pressure is taken up tighten bolts to 29 ft. lbs. (34 Nm).

13. Remove the aligning tool.

14. Apply high temperature grease to transmission input shaft.

15. Install the transmission and tighten bell housing bolts to 35 ft. lbs. (47 Nm).

16. Connect the negative battery cable.

17. Bleed the hydraulic system.

Hydraulic Clutch System

BLEEDING

Bleeding air from the hydraulic clutch system is necessary whenever any part of the system has been disconnected or the fluid level (in the reservoir) has been allowed to fall so low, that air has been drawn into the master cylinder.

1. Fill master cylinder reservoir with new brake fluid conforming to DOT 3 specifications.

❊❊ CAUTION

Never, under any circumstances, use fluid which has been bled from a system to fill the reservoir as it may be aerated, have too much moisture content and possibly be contaminated.

2. Raise and support vehicle safely.

3. Have an assistant fully depress and hold the clutch pedal, then open the bleeder screw.

4. Close the bleeder screw and have your assistant release the clutch pedal.

5. Repeat the procedure until all of the air is evacuated from the system. Check and refill master cylinder reservoir as required to prevent air from being drawn through the master cylinder.

➡**Never release a depressed clutch pedal with the bleeder screw open or air will be drawn into the system.**

6. Test the clutch for proper operation.

Transfer Case Assembly

REMOVAL & INSTALLATION

1. Disconnect the negative battery cable.

2. Raise and safely support the vehicle.

3. Remove the skid plate, if equipped.

4. Remove the drain plug and allow the transfer case to drain before proceeding.

5. Loosen the vent hose clamp at the transfer case and disconnect the hose.

6. Disconnect the front driveshaft at the transfer case and support it aside.

7. Disconnect the rear driveshaft and support it aside.

8. Disconnect the electrical connections at the transfer case.

9. Disconnect the transfer case shift linkage.

10. Support the transfer case with a transmission jack. Remove the transmission to transfer case bolts and spring washers.

11. Remove the transfer case assembly and gasket.

12. Carefully lower the transfer case.

To install:

13. Carefully raise the transfer case into position.

14. Install a new gasket to the transmission using gasket sealer to hold it in place.

15. Install the transfer case onto the transmission or transmission adapter. Tighten the bolts to 33 ft. lbs. (45 Nm).

16. Connect the electrical harness connectors to the transfer case connections.

17. Attach the transfer case shift linkage and make the proper adjustments.

18. Reconnect the front and rear driveshafts.

19. Refill the transfer case with DEXRON® IIE automatic transmission fluid.

20. Install the skid plate, if equipped.

21. Lower the vehicle and connect the negative battery cable.

22. Test drive for proper operation.

Halfshaft

REMOVAL & INSTALLATION

1. Raise and support the vehicle safely.

2. Remove the front wheel and tire assembly.

3. If equipped, remove the skid plate, as required.

4. Remove the drive axle hub nut and washer.

5. Remove the brake line and wheel speed sensor support bracket from the upper control arm to allow extra travel of the control arm.

6. Remove the left outer tie rod attaching nut and cotter pin. Separate the tie rod from the steering knuckle.

7. Position the tie rod aside and push steering linkage to the opposite side of the vehicle.

8. Remove the lower shock attaching nut and bolt; position the shock aside.

9. Remove the left stabilizer bar bracket and bushing at the frame. Remove the stabilizer bar bolt, spacer and bushings at the lower control arm.

10. Lower the vehicle, taking pressure off the upper control arm by placing a support below the lower control arm between the spring seat and the ball joint.

11. Remove the upper ball joint cotter pin and loosen (do not remove) the upper ball joint attaching nut. Separate the ball joint stud from the steering knuckle. Remove the attaching nut.

➡**Cover the shock mounting bracket and lower ball joint stud with a towel to prevent the axle boot from tearing during removal and installation.**

12. Separate the axle shaft from the hub and rotor using tool J-28733 or equivalent.

13. Remove the axle shaft inner flange bolts. Remove the shaft.

To install:

14. Lubricate the axle and hub splines with an approved high temperature wheel bearing grease. Position the shaft in the hub and install the inboard CV-joint-to-flange bolts.

15. Install the upper ball joint to steering knuckle and tighten the stud nut to 61 ft. lbs. (83 Nm). Install a cotter pin through the upper ball joint stud and nut. Lubricate the ball joint as required.

16. Install the left stabilizer bar bracket and bushing at the frame. Install the stabilizer bar bolt, spacer and bushings at the lower control arm.

17. Position the lower shock in the mount bracket and install the attaching nut and bolt.

18. Connect the left tie rod end at the steering knuckle. Tighten the nut to 35 ft. lbs. (47 Nm). Install a cotter pin through the tie rod stud and nut.

19. Connect the brake line bracket to the control arm, ensuring the line and/or hose is not twisted or kinked.

20. Install the skid plate, as required.

21. Install the axle hub washer and nut. Insert a drift through the rotor vanes to keep the axle from turning and tighten the hub nut to 180 ft. lbs. (245 Nm) and the inboard CV-joint flange bolts to 60 ft. lbs. (80 Nm).

22. Remove the drift, install the wheel and tire assembly.

23. Safely lower the vehicle to the floor.

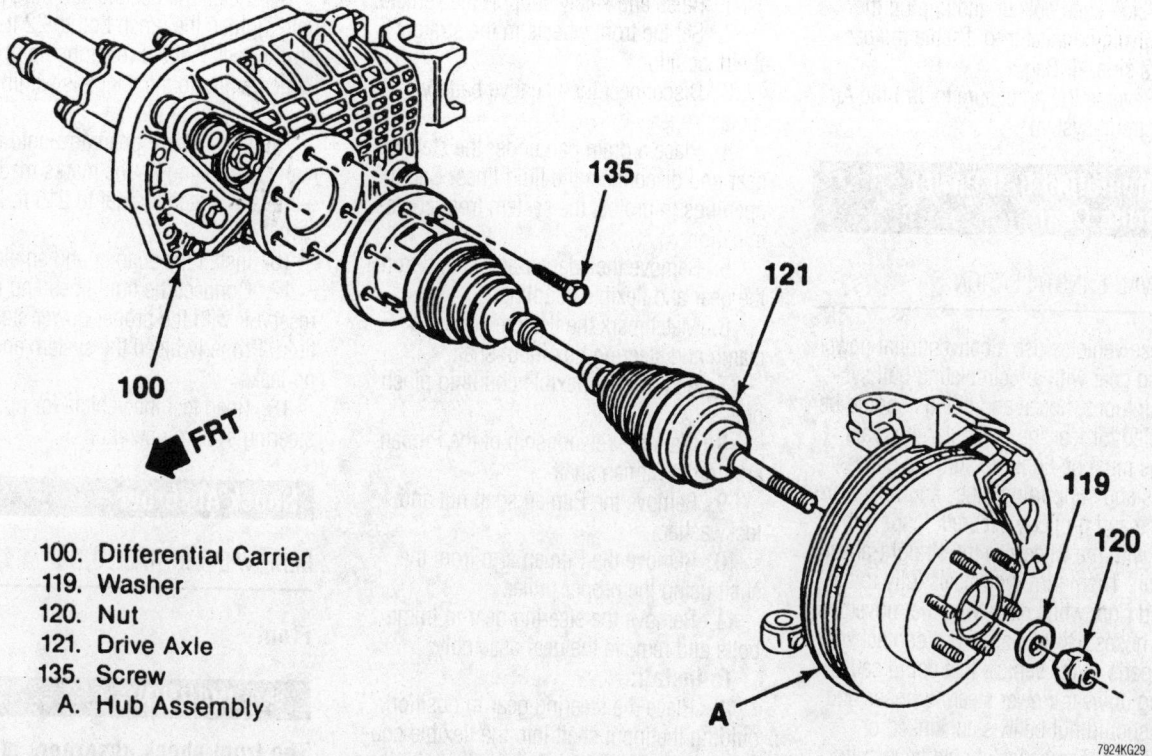

100. Differential Carrier
119. Washer
120. Nut
121. Drive Axle
135. Screw
A. Hub Assembly

The halfshaft is mounted to the flange on the differential (100) and through the hub assembly (A)—4-wheel drive models

STEERING AND SUSPENSION

Air Bag

✳✳ CAUTION

Some vehicles are equipped with an air bag system, also known as the Supplemental Inflatable Restraint (SIR) or Supplemental Restraint System (SRS). The system must be disabled before performing service on or around system components, steering column, instrument panel components, wiring and sensors. Failure to follow safety and disabling procedures could result in accidental air bag deployment, possible personal injury and unnecessary system repairs.

PRECAUTIONS

Several precautions must be observed when handling the inflator module to avoid accidental deployment and possible personal injury.

• Never carry the inflator module by the wires or connector on the underside of the module.
• When carrying a live inflator module, hold securely with both hands, and ensure that the bag and trim cover are pointed away.
• Place the inflator module on a bench or other surface with the bag and trim cover facing up.
• With the inflator module on the bench, never place anything on or close to the module which may be thrown in the event of an accidental deployment.

DISARMING

1. Turn the front wheels to the straight ahead position.
2. Turn the ignition switch to the **LOCK** position and remove the key.

➡ If the key is in the RUN position when the Air Bag fuse is removed or open (blown), the Air Bag warning lamp in the dash will light up. This is normal operation, not a sign of a malfunction.

3. Remove the Air Bag fuse from the fuse panel.
4. Remove the drivers side knee bolster and unplug the yellow 2-pin connector at the base of the steering column to disarm the driver's side Air Bag. Remove the pas-

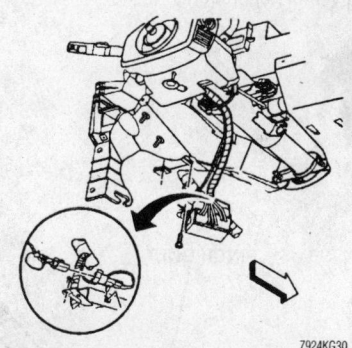

Typical air bag connector location—driver's side

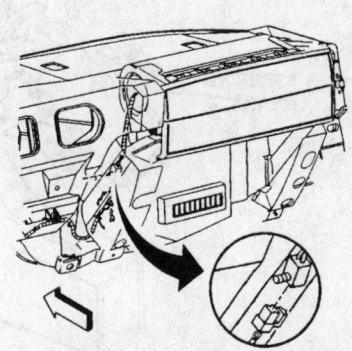

Typical air bag connector location—passenger's side

senger side knee bolster and unplug the yellow 2-pin connector to disable the passenger's side Air Bag.

5. Reverse the procedure to arm the Air Bag restraint system.

Recirculating Ball Power Steering Gear

REMOVAL & INSTALLATION

These vehicles use a conventional power steering gear with a recirculating ball system. All tubes, hoses and fittings should be inspected for leakage at regular intervals. Fittings must be tight. Be sure the clips, clamps and supporting tubes and hoses are in place and properly secured. Inspect the hoses with the wheels in the straight-ahead position. Then, turn the wheels fully to the left and right while observing the movement of the hoses. Correct any hose contact with other parts of the vehicle that could cause chafing or wear. Power steering hoses and pipes should not be twisted, kinked or tightly bent. The hoses should have sufficient natural curvature in the routing to absorb movement and hose shortening during vehicle operation.

1. Raise and safely support the vehicle.
2. Set the front wheels in the straight ahead position.
3. Disconnect the negative battery cable.
4. Place a drain pan under the steering gear and disconnect the fluid lines. Cap the openings to protect the system from contamination.
5. Remove the adapter and shield from the gear and flexible coupling.
6. Matchmark the flexible coupling clamp and steering box input shaft.
7. Remove the flexible coupling pinch bolt.
8. Mark the relationship of the Pitman arm to the Pitman shaft.
9. Remove the Pitman shaft nut and lockwasher.
10. Remove the Pitman arm from the shaft using the proper puller.
11. Remove the steering gear to frame bolts and remove the gear assembly.

To install:

12. Place the steering gear in position, guiding the input shaft into the flexible coupling. Align the flat in the coupling with the flat on the input shaft.
13. Install the steering gear-to-frame bolts and tighten to 100 ft. lbs. (135 Nm).

14. Install the flexible coupling pinch bolt. Tighten the pinch bolt to 22 ft. lbs. (30 Nm). Check that the relationship of the flexible coupling to the flange is within ¼ to ¾ in. (6–19mm) of the flat.
15. Install the Pitman arm onto the Pitman shaft, lining up the marks made at removal. Tighten the nut to 215 ft. lbs. (285 Nm).
16. Install the adapter and shield.
17. Connect the fluid lines and refill the reservoir with the proper power steering fluid. Properly bleed the system and verify no leaks
18. Road test the vehicle for proper steering system operation.

Shock Absorber

REMOVAL & INSTALLATION

Front

> ☀ **WARNING**
>
> **The front shock absorbers on these vehicle are multifunctional. They not only aid in a smooth ride but serve as the suspension stop when the suspension is fully extended. When replacing front shocks, a shock of equivalent length and strength must be used. Use of a shock that does not comply mat result in suspension over travel and component failure.**

1. Raise and support the front of the vehicle safely under the lower control arms.
2. Remove the tire and wheel assembly.
3. Remove the upper and lower shock absorber retaining fastener(s).

➡**Vehicles equipped with quad shocks have a spacer between them.**

4. Remove the shock absorber.

To install:

5. Install the shock absorber.
6. On 2-wheel drive vehicles, tighten the upper bolt to 12 ft. lbs. (16 Nm) and the lower bolts to 24 ft. lbs. (33 Nm).
7. On 4-wheel drive vehicles, tighten the nuts to 66 ft. lbs. (90 Nm). Be sure the bolts are inserted in the proper direction. The upper bolt head should be forward; the bottom bolt head should be rearward.
8. Safely lower the vehicle.

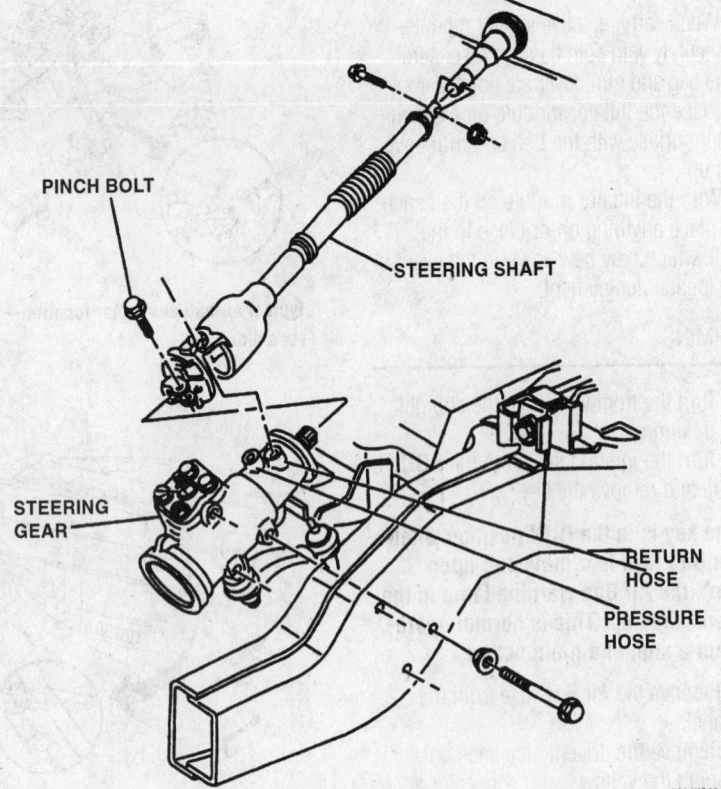

PINCH BOLT

STEERING SHAFT

STEERING GEAR

RETURN HOSE

PRESSURE HOSE

7924KG32

Three long bolts attach the power steering gear to the driver's side frame rail

Rear

✳✳ WARNING

Original equipment shock absorbers serve additionally as suspension drop cutoffs. Replacement shock absorbers must have a built in suspension cutoff feature and must not be longer than original shocks when they are fully extended or serious vehicle or component damage could result.

1. Raise and safely support the rear of the vehicle. The vehicle's weight should rest on correctly placed safety stands located under the frame. Chock the front wheels to prevent vehicle movement.
2. Support the rear axle with a floor jack.
3. If the vehicle is equipped with air lift type shocks, bleed the air from the lines and disconnect the line from the shock absorber.
4. Disconnect the shock absorber at the top by removing the two mounting bolts and nuts from the frame bracket.
5. Remove the nut, washers and bolt from the bottom mount.
6. Remove the shock from the vehicle.
   To install:
7. Install the shock onto the vehicle.
8. Tighten the upper mounting nuts.
9. Safely return the vehicle to the ground. Check that no parts such as exhaust components bind on the shock absorbers.

Coil Springs

REMOVAL & INSTALLATION

Van Models

1. Raise and support the vehicle safely under the frame rails. The control arms should hang freely.
2. Remove the wheel.
3. Disconnect the shock absorber at the lower end and move it aside.
4. Disconnect the stabilizer bar from the lower control arm.
5. Support the lower control arm and install a spring compressor on the spring or chain the spring to the control arm as a safety precaution.

➡ **If equipped with an air cylinder inside the spring, remove the valve core from the cylinder and expel the air by compressing the cylinder with a pry-**

bar. With the cylinder compressed, replace the valve core so the cylinder will stay in the compressed position. Push the cylinder as far as possible towards the top of the spring.

6. Raise to remove the tension from the lower control arm the bolts securing the control arm.

➡ **The cross-shaft and lower control arm keeps the coil spring compressed. Use care when lowering the assembly.**

7. Slowly lower the control arm until the spring can be removed. Be sure all compression is relieved from the spring.
8. If the coil spring was chained, remove the chain and spring. If a compressor was used, remove the spring and slowly release the compressor.
9. Remove the air cylinder, if equipped.
   To install:
10. Install the air cylinder so the protector plate is towards the upper control arm. The Schrader valve should protrude through the hole in the lower control arm.
11. Install the chain and spring or compress the spring and install the assembly.
12. Slowly raise the control arm.
13. Install the bolts securing the control arm tighten the nuts to 115 ft. lbs. (155 Nm).
14. Connect the stabilizer bar to the lower control arm. Tighten the nuts to 24 ft. lbs.
15. Connect the shock absorber at the lower end. Tighten the bolt to specification.
16. Remove the support.
17. If equipped with air cylinders, inflate the cylinder to 60 psi.
18. Install the wheel.
19. Lower the vehicle. Once the weight of the vehicle is on the wheels, reduce the air cylinder pressure to 50 psi.
20. Check the front end alignment.

Except Van Models

1. Raise and support the vehicle safely. Allow the control arms to hang free. Remove the tire and wheel assembly. Remove the shock absorber assembly.
2. Install tool J-23028 under the lower control arm and a jack. Install a safety chain around the spring and through the lower control arm.
3. Remove the stabilizer shaft from the lower control arm. Raise and remove the tension on the lower control arm bolts.

4. Remove the lower control arm rear bolt, than remove the other retaining bolt.
5. Lower allow the lower control arm to hang free. Remove the spring assembly.
   To install:
6. Install the chain and spring. If you used spring compressors, install the spring and compressors.
   a. Be sure the insulator is in place and the tape is towards the bottom of the spring.
   b. Position the gripper notch on the top coil is in the frame bracket.
   c. Be sure one drain hole in the lower arm is covered by the bottom coil and the other is open.
7. Slowly raise the lower control arm. Guide the control arm into place with a pry-bar.
8. Install the pivot shaft bolts, front one first. The bolts must be installed with the heads towards the front of the vehicle. Remove the safety chain or spring compressors.

➡ **Do not tighten the bolts yet. The bolts must be tightened with the vehicle at its proper ride height.**

9. Remove the jack.
10. Connect the stabilizer bar to the lower control arm. Tighten the nuts to specification.
11. Install the shock absorber.
12. Install the wheel.
13. Lower the vehicle. Once the weight of the vehicle is on the wheels, bounce the vehicle two or three times by pushing down on the front bumper a couple of inches. When the vehicle settles, tighten the front nut first,, then the rear nut to 101 ft. lbs. (137 Nm).

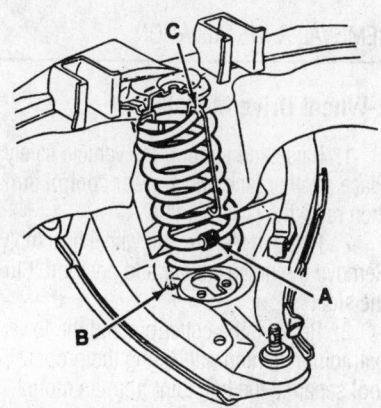

7924KG33

Position the coil spring so the bottom end of the spring covers only one drain hole—the other hole must remain open

Upper Ball Joint

REMOVAL & INSTALLATION

1. Raise and support the vehicle safely.
2. Remove the wheel.
3. Unbolt the brake hose bracket from the control arm.
4. Using a ⅛ in. drill bit, drill a pilot hole through each ball joint rivet.
5. Drill out the rivets with a ½ in. drill bit. Punch out any remaining rivet material.
6. Remove the cotter pin and nut from the ball stud.
7. Support the lower control arm.
8. Using a ball joint separator, separate the stud from the knuckle.

To install:

9. Position the new ball joint on control arm.

➡**Service replacement ball joints come with nuts and bolts to replace the rivets.**

10. Install the bolts and nuts. Tighten the nuts to 17 ft. lbs. (23 Nm) for 15 and 25-Series, 52 ft. lbs. (70 Nm) for 35-Series.

➡**The bolts are inserted from the bottom.**

11. Start the ball stud into the knuckle. Ensure it is squarely seated. Install the ball stud nut and pull the ball stud into the knuckle with the nut. Tighten the nut after the vehicle wheel are on the ground and the suspension is loaded.
12. Install the wheel.
13. Lower the vehicle. Once the weight of the vehicle is on the wheels tighten the nut to 84 ft. lbs. (115 Nm).

Lower Ball Joint

REMOVAL & INSTALLATION

2-Wheel Drive Models

1. Raise and support the vehicle safely. Place another jack under lower control arm, then raise the jack slightly.
2. Remove the tire and wheel assembly. Remove the brake caliper and position it to the side.
3. Remove the cotter pin and the lower ball joint retaining nut. Using the proper tool separate the ball joint from its mounting. Support the knuckle assembly so its weight will not damage the brake hose.
4. Press the ball joint out of the lower control arm, using tool J-9519–30-D or equivalent.

To install:

5. Start the new ball joint into the control arm. Position the bleed vent in the rubber boot facing inward.
6. Press the ball joint into the control arm until fully seated.
7. Lower the upper arm and insert the lower ball joint stud into the steering knuckle.
8. Install the brake caliper, if removed.
9. Install the ball stud nut and tighten to 90 ft. lbs. (122 Nm) plus the additional tighten necessary to align the cotter pin hole. Do not exceed 130 ft. lbs. (175 Nm) or back the nut off to align the holes with the pin.
10. Install a new lube fitting and lubricate the new joint.
11. Install the tire and wheel.
12. Lower the vehicle.

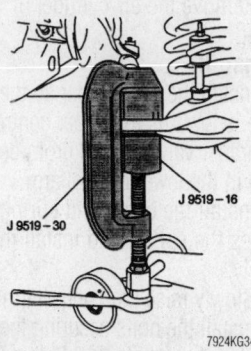

Installing the lower ball joint into the lower control arm—2-wheel drive

4-Wheel Drive Models

1. Raise and support the vehicle safely.
2. Remove the wheel.
3. Remove the splash shield from the knuckle.
4. Disconnect the inner tie rod end from the relay rod using a ball joint separator.
5. Remove the hub nut and washer. Insert a long drift or dowel through the vanes in the brake rotor to hold the rotor in place.
6. Remove the axle shaft inner flange bolts.
7. Using a puller, force the outer end of the axle shaft out of the hub. Remove the shaft.
8. Remove the cotter pin and nut from the ball stud.
9. Support the lower control arm.
10. Matchmark the both torsion bar adjustment bolt positions.
11. Using tool J-36202 or equivalent, increase the tension on the adjusting arm.
12. Remove the adjustment bolt and retaining plate.

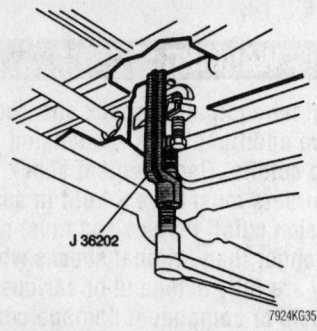

A special tool is available for removing or installing the torsion bar adjusting bolt—4-wheel drive

13. Move the tool aside.
14. Slide the torsion bars forward.
15. Using a screw-type forcing tool, separate the ball joint from the knuckle.
16. Remove the lower control arm.
17. Press lower ball joint out of control arm with tool J-9519-E or equivalent ball joint press.

To install:

18. Position the new ball joint on control arm.
19. Press the ball joint into with tool J-9519-E or equivalent.
20. Install the lower control arm.
21. Using tool J-36202 or equivalent, increase tension on both torsion bars.
22. Install the adjustment retainer plate and bolt on both torsion bars.
23. Set the adjustment bolt to the marked position.
24. Release the tension on the torsion bar until the load is take up by the adjustment bolt.
25. Remove the tool.
26. Position the shaft in the hub and install the washer and hub nut. Leave the drift in the rotor vanes and tighten the hub nut to 175 ft. lbs. (238 Nm).
27. Install the flange bolts. Tighten them to 59 ft. lbs. (80 Nm). Remove the drift.
28. Connect the inner tie rod end at the steering relay rod. Tighten the nut to 35 ft. lbs. (48 Nm).
29. Install the splash shield.
30. Install the wheel.
31. Lower the vehicle. Once the weight of the vehicle is on the wheels:
   a. Lift the front bumper about 1½ in. (38mm) and let it drop.
   b. Repeat this procedure 2–3 more times.
   c. Draw a line on the side of the lower control arm from the centerline of the control arm pivot shaft, dead level to the outer end of the control arm.

d. Measure the distance between the lowest corner of the steering knuckle and the line on the control arm. Record the figure.

e. Push down about 1½ in. (38mm) on the front bumper and let it return. Repeat the procedure 2–3 more times.

f. Re-measure the distance at the control arm.

g. Determine the average of the two measurements. This is the "Z" height

measurement. The "Z" height should be as specified in the chart.

h. If the figure is correct, tighten the control arm pivot nuts to 94 ft. lbs. (128 Nm).

i. If the figure is not correct, tighten the pivot bolts to 94 ft. lbs. (128 Nm) and have the front end alignment corrected.

Wheel Bearings

ADJUSTMENT

➡**Only the front wheel bearings on the 2-wheel drive vehicles are adjustable.**

1. Raise and safely support the vehicle.
2. Remove the dust cap, cotter pin.
3. Loosen the spindle nut.

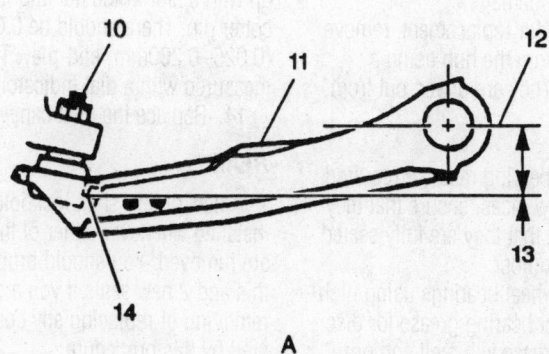

A

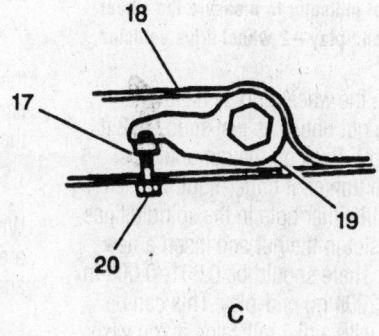

C

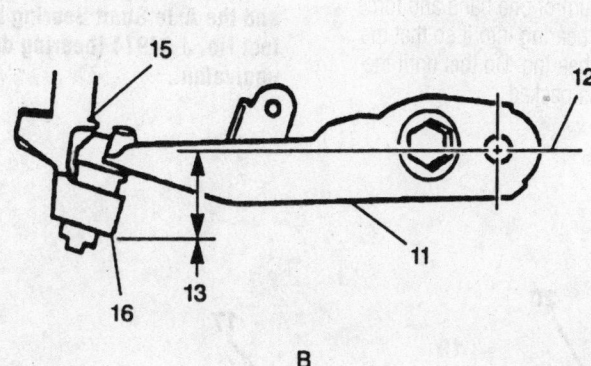

B

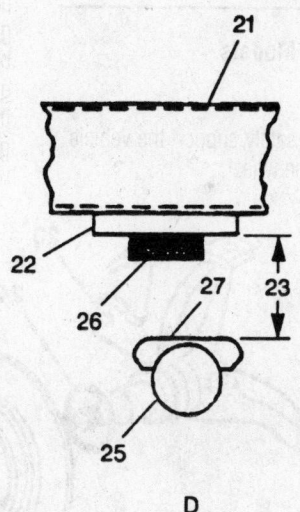

D

A. "C" MODEL
B. "K" MODEL
C. "K" MODEL TORSION BAR ADJUSTER
D. "CK" MODEL REAR SUSPENSION
10. LOWER BALL JOINT
11. LOWER CONTROL ARM
12. PIVOT BOLT CENTER LINE
13. "Z" HEIGHT
   C 1,2,3   95.0 ± 6.0mm
   K 1,2   157.0 ± 6.0mm
   K 3   145.0 ± 6.0mm
14. LOWER BALL JOINT EXTRUSION

15. STEERING KNUCKLE
16. STEERING KNUCKLE LOWER CORNER
17. NUT
18. TORSION BAR SUPPORT ASM.
19. TORSION BAR ADJUSTMENT ARM
20. BOLT – ONE TURN EQUALS 6mm HEIGHT CHANGE
21. FRAME
22. BOTTOM SURFACE OF JOUNCE BRACKET
23. "D" HEIGHT
25. REAR AXLE
26. JOUNCE BUMPER
27. AXLE JOUNCE PAD

7924KG36

Use these specifications and diagrams to determine if the **vehicle ride height is correct**

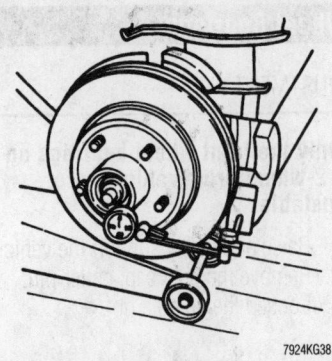

Use a dial indicator to measure the wheel bearing end-play—2-wheel drive vehicles

4. Spin the wheel hub by hand and tighten the nut until it is just snug—12 ft. lbs. (16 Nm). Back off the nut until it is loose, then tighten it finger-tight. Loosen the nut until either hole in the spindle lines up with a slot in the nut and insert a new cotter pin. There should be 0.001–0.008 in. (0.025–0.200mm) end-play. This can be measured with a dial indicator, if you wish.

5. Replace the dust cap, wheel and tire.

REMOVAL & INSTALLATION

2-Wheel Drive Models

FRONT

1. Raise and safely support the vehicle.
2. Remove the wheel.

3. Dismount the caliper and wire it out of the way.

4. Pry out the grease cap, remove the cotter pin, spindle nut, and washer, then remove the hub. Do not drop the wheel bearings.

5. Remove the outer roller bearing assembly from the hub. The inner bearing assembly will remain in the hub and may be removed after prying out the inner seal. Discard the seal.

6. Clean all parts in a non-flammable solvent and let them air dry. Never spin-dry a bearing with compressed air! Check for excessive wear and damage.

7. If necessary for replacement, remove the bearing races from the hub using a hammer and drift. They are driven out from the inside out.

To install:

8. Install new bearing races, if required. When installing new races, ensure that they are not cocked and that they are fully seated against the hub shoulder.

9. Pack both wheel bearings using high melting point wheel bearing grease for disc brakes. Ordinary grease will melt and ooze out ruining the pads. Bearings should be packed using a cone-type wheel bearing greaser tool. If one is not available they may be packed by hand. Place a healthy glob of grease in the palm of one hand and force the edge of the bearing into it so that the grease fills the bearing. Do this until the whole bearing is packed.

10. Place the inner bearing in the hub and install a new inner seal, making sure that the seal flange faces the bearing race.

11. Carefully install the wheel hub over the spindle.

12. Using your hands, firmly press the outer bearing into the hub. Install the spindle washer and nut.

13. Spin the wheel hub by hand and tighten the nut until it is just snug—12 ft. lbs. (16 Nm). Back off the nut until it is loose, then tighten it finger-tight. Loosen the nut until either hole in the spindle lines up with a slot in the nut and insert a new cotter pin. There should be 0.001–0.008 in. (0.025–0.200mm) end-play. This can be measured with a dial indicator, if you wish.

14. Replace the dust cap, wheel and tire.

REAR

A new pinion shaft lockbolt should be installed whenever either of the axle shafts are removed. You should probably purchase this and 2 new seals if you are planning on removing or replacing any components covered by this procedure.

➡**Axle shaft seal removal and installation uses the following special tools: the GM Axle Shaft Seal Installer tool No. J-33782 (seal driver) or equivalent and the Axle Shaft Bearing Installer tool No. J-34974 (bearing driver) or equivalent.**

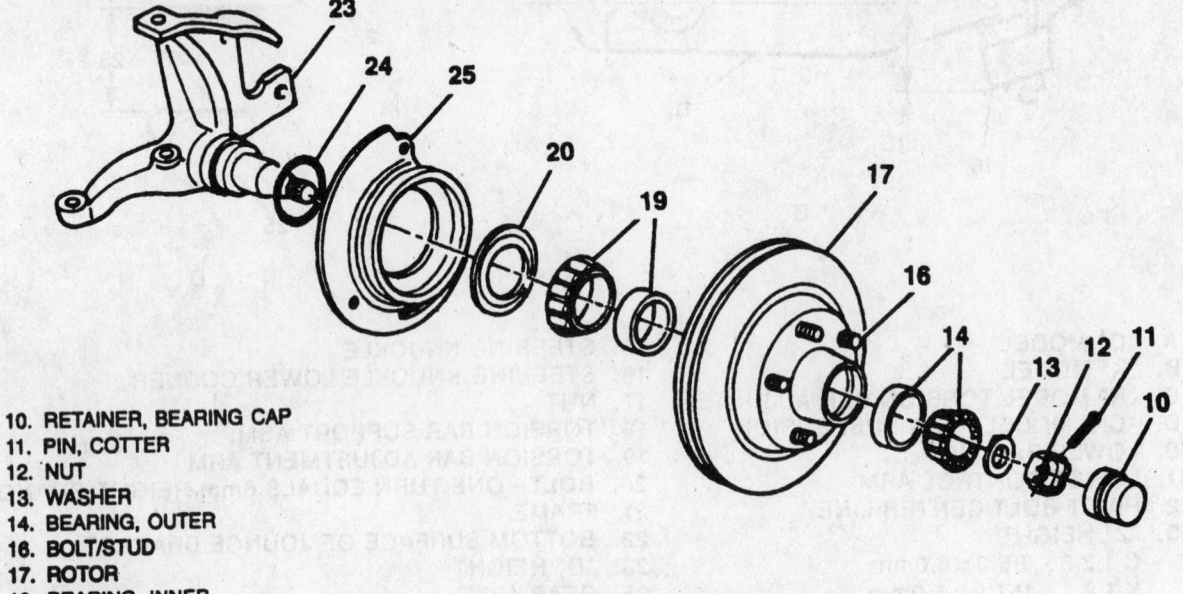

10. RETAINER, BEARING CAP
11. PIN, COTTER
12. NUT
13. WASHER
14. BEARING, OUTER
16. BOLT/STUD
17. ROTOR
19. BEARING, INNER
20. SEAL
23. KNUCKLE
24. GASKET
25. SHIELD

Exploded view of the front wheel bearing and related components—2-wheel drive models

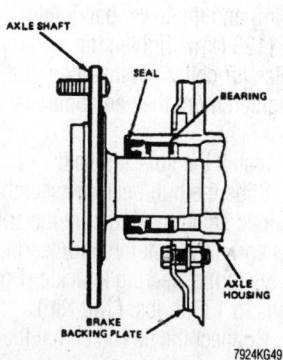

Cutaway view of the rear axle shaft and bearing assembly

The axle shaft and seal may be removed and replaced without disturbing the bearing or seal, BUT is highly recommended that you replace the seals as-long-as you've gone through the trouble to remove the axle shaft. Seal replacement is simple and it is cheap insurance against an oil leak which could ruin your brake shoes.

➡️**If the bearing requires replacement, you will also need a slide hammer with adapters to reach behind the bearing in the axle tube and pull it out.**

1. Raise and support the rear of the vehicle safely using jackstands.
2. Remove the rear wheel assemblies, then remove the brake drums.

✳✳ CAUTION

Brake shoes may contain asbestos, which has been determined to be a cancer causing agent. Never clean the brake surfaces with compressed air! Avoid inhaling any dust from any brake surface! When cleaning brake surfaces, use a commercially available brake cleaning fluid.

3. Using a wire brush, clean the dirt/rust from around the rear axle cover.

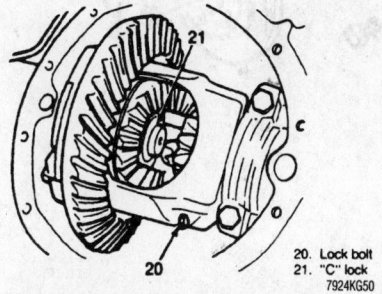

20. Lock bolt
21. "C" lock
7924KG50

Remove the lockbolt and pinion shaft, then push in the axle shaft and remove the C-lock

4. Place a catch pan under the differential, then remove the drain plug (if equipped) or rear axle cover and drain the fluid (discard the old fluid).
5. At the differential, remove the rear pinion shaft lockbolt and the pinion shaft.
6. Push the axle shaft inward and remove the C-lock from the button end of the axle shaft.
7. Remove the axle shaft from the axle housing. Be careful not to damage the oil seal.

✳✳ WARNING

On vehicles equipped with an Anti-Lock Brake System (ABS) be careful not to damage the reluctor ring on the axle shaft or the speed sensor bolted to the backing plate, immediately adjacent to the shaft.

8. Using a putty knife, clean the gasket mounting surfaces.

➡️**It is recommended, when the axle shaft is removed, to replace the oil seal.**

9. To replace the oil seal use a medium prybar or, better yet, an inexpensive seal removal tool, to pry the oil seal from the end of the rear axle housing. DO NOT damage the housing oil seal surface. And again, on late-model ABS equipped vehicles, STAY CLEAR OF THE SPEED SENSOR.

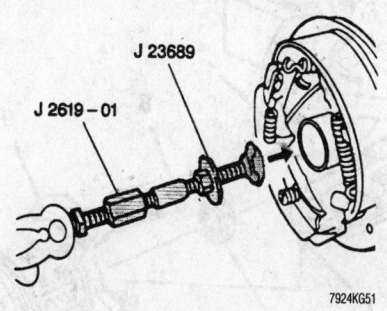

Using a slide hammer and adapters, remove the axle bearing and seal

10. Using the slide hammer and adapter, pull the bearing out of the axle tube.
To install:
11. If the wheel bearing was removed:
   a. Using solvent, thoroughly clean the wheel bearing, then blow dry with compressed air. Inspect the wheel bearing for excessive wear or damage, then replace it (if necessary).
   b. With a new or the reused bearing, thoroughly coat the bearing with gear lubricant.
   c. Using the Axle Shaft Bearing Installer tool No. J-34974 or equivalent,

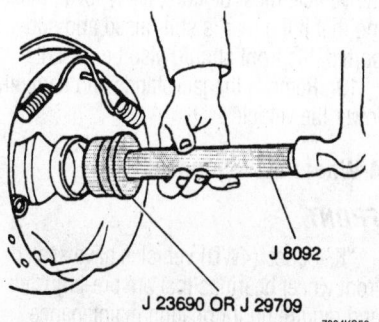

Axle and seal installation using a bearing driver

drive the bearing into the axle housing until it bottoms against the seat. Be sure the bearing installer does not contact and damage the speed sensor on ABS equipped vehicles.
12. If the axle shaft seal was removed:
   a. Clean and inspect the axle tube housing.
   b. Using the GM Axle Shaft Seal Installer tool No. J-33782 or an equivalent driver, seat the new seal into the housing until it is flush with the axle tube. Be sure the seal installer does not contact and damage the speed sensor on ABS equipped vehicles.
   c. Using gear oil, lubricate the new seal lips.
13. Slide the axle shaft into the rear axle housing and engage the splines of the axle shaft with the splines of the rear axle side gear, then install the C-lock retainer on the axle shaft button end.

✳✳ WARNING

BE CAREFUL not to damage the wheel bearing seal with the splines on the axle shaft., do we even have to mention the SPEED SENSOR again on ABS equipped vehicles!

14. After the C-lock is installed, pull the axle shaft outward to seat the C-lock retainer in the counterbore of the side gears.
15. Install the pinion shaft through the case and the pinions, then install a NEW pinion shaft lockbolt. Torque the new lockbolt to 25 ft. lbs. (34 Nm) for 1985–93 vehicles or to 27 ft. lbs. (36 Nm) for 1994–96 vehicles.
16. Use a new rear axle cover gasket and install the housing cover.
17. Install the brake drums, followed by the tire and wheel assemblies.
18. Properly refill the housing. For details, please refer to the information in Section 1 of this manual. REMEMBER that

the vehicle must be completely level, meaning that if the rear is still raised and supported, the front should also be raised.

19. Remove the jackstands and carefully lower the vehicle.

4-Wheel Drive Models

FRONT

"K" Series (4WD) vehicles have sealed front wheel bearings that are pre-adjusted and require no lubrication maintenance. Darkened areas on the bearing assembly are cause by a heat treatment process and do not require bearing replacement.

1. Raise and safely support the vehicle.
2. Remove the wheel.
3. Wrap shop towels around the CV-Joint boots to protect them from damage during this procedure.
4. Remove the brake caliper and support it out of the way on a wire. Do not let the caliper hang by the brake hose.
5. Remove the brake rotor.
6. Remove the cotter pin, retainer, castle nut, and thrust washer from the axle shaft.
7. Disconnect the tie rod end from the knuckle.

8. Remove the hub/bearing assembly retaining bolts.
9. Using a J-28733-B hub/bearing puller tool or equivalent, press the hub/bearing assembly from the splined shaft.

✷✷ WARNING

After removal, lay the hub and bearing assembly on the outboard side. This will prevent damage and/or contamination of the bearing seal.

10. Remove the splash shield.
11. Support the lower control arm with a jack stand.
12. Disconnect the upper and lower ball joints from the steering knuckle.
13. Separate the ball joints from the steering knuckle.
14. Remove the seal from the steering knuckle.

To install:

15. Using a J 36605 seal installer, install a new seal in the steering knuckle.
16. Position the steering knuckle on the ball joints and install the retaining nuts. Tighten the upper ball joint nut to 74 ft. lbs.

(100 Nm) and the lower ball joint nut to 94 ft. lbs. (128 Nm). Tighten the nuts to align the holes for cotter pin insertion, but do NOT tighten more than an additional ⅛ - turn.

17. Install the splash shield.
18. Slide the hub/bearing assembly over the splined shaft, making sure the splines line up correctly. Bolt the hub/bearing assembly to the steering knuckle. Tighten the bolts to 133 ft. lbs. (180 Nm).
19. Connect the tie rod end at the steering knuckle.
20. Install the thrust washer and axle nut, tighten the nut to 165 ft. lbs. (225 Nm). Install the retainer and cotter pin.
21. Install the rotor and caliper.
22. Remove the shop towels from the CV-Joint boot.
23. Tighten the wheels to specification and lower the vehicle.
24. Check and adjust the front end alignment and road test the vehicle.

REAR

Please refer to the procedure for the 2-wheel drive vehicles for servicing the rear axle shaft bearings.

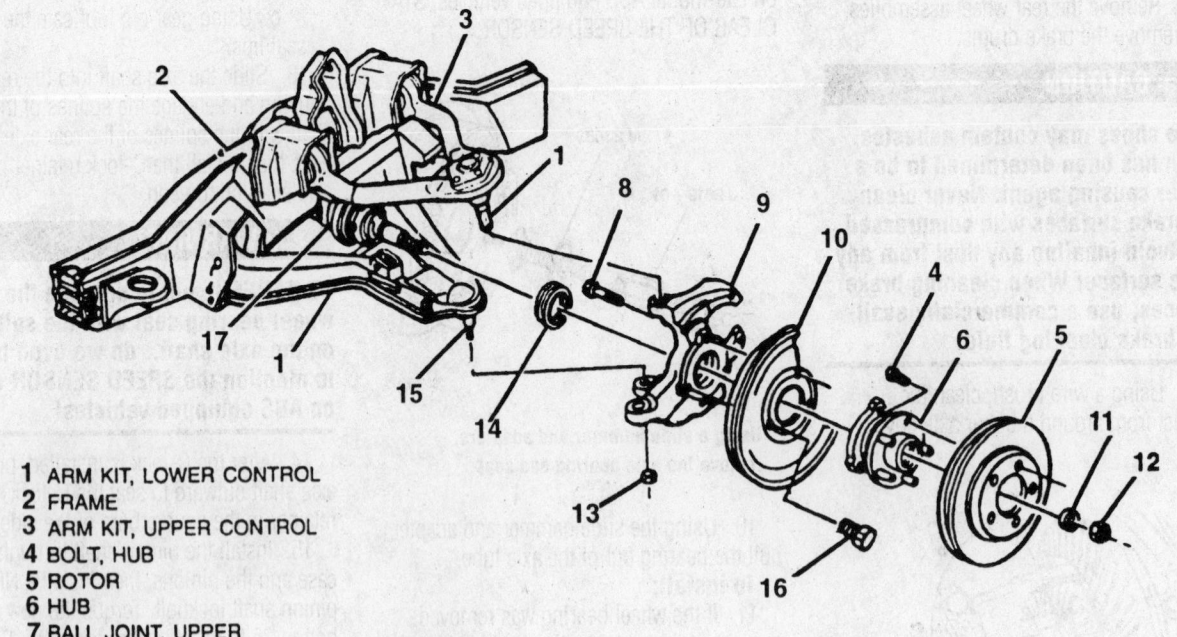

1 ARM KIT, LOWER CONTROL
2 FRAME
3 ARM KIT, UPPER CONTROL
4 BOLT, HUB
5 ROTOR
6 HUB
7 BALL JOINT, UPPER
8 BOLT
9 KNUCKLE, STEERING
10 SHIELD
11 WASHER
12 NUT
13 NUT
14 SEAL
15 BALL JOINT, LOWER
16 BOLT
17 JOINT KIT, FRONT AXLE

7924KG37

Exploded view of the front hub and knuckle assembly—4-wheel drive

GENERAL MOTORS

Chevrolet-Lumina APV • Venture • Oldsmobile-Silhouette
Pontiac-Trans Sport • Montana

21

ENGINE REPAIR

➡Disconnecting the negative battery cable on some vehicles may interfere with the functions of the on board computer systems and may require the computer to undergo a relearning process, once the negative battery cable is reconnected.

❊❊ WARNING

NEVER disconnect the negative battery cable with the ignition ON. Removing power from the computer control module with the ignition ON may destroy the module.

Distributor

The 3.1L engine is equipped with a distributor ignition system. This distributor is comprised of an internal magnetic pick-up assembly which contains a permanent magnet, a pole piece with internal teeth and a pick-up coil.

The 3.4L and 3.8L engines utilize a Distributorless Ignition System (DIS). There is no distributor to remove and no provision for adjustment.

REMOVAL

1. Disconnect the negative battery cable.
2. Remove all necessary components in order to gain access to the distributor assembly. On most V-type engines it will be necessary to remove the air cleaner assembly for access.
3. Tag and disengage the distributor electrical connectors. If equipped, disconnect the vacuum line.
4. Remove the distributor cap. If necessary, tag and disconnect the spark plug wires.
5. Matchmark the rotor and the distributor body. Matchmark the distributor assembly and the engine block.

➡Although it is only necessary to matchmark the distributor position for ease of installation, it may be advisable to first position the engine at TDC. If the distributor is being removed for further engine repair which may involve rotating the crankshaft (such as to align timing marks), setting the engine to TDC at this time will keep the distributor matchmarks valid during installation.

6. Remove the distributor hold-down bolt. Carefully remove the distributor.

➡As the distributor is removed from the engine, the rotor will turn counterclockwise. Observe and mark the finish position of the rotor. When reinstalling, position the rotor at the last mark and set the distributor into the engine. As the distributor drops into place, the rotor should turn to its original position, providing the engine crankshaft has not been rotated with the distributor out.

INSTALLATION

Timing Not Disturbed

➡To ensure correct ignition timing if the engine has not been disturbed, the distributor must be installed with the rotor in the same position as when removed.

1. Align the rotor to the last mark made and install the distributor in the engine.

➡On vehicles not equipped with the High Voltage Spark (HVS) system, if the distributor shaft cannot drop into the engine, remove the distributor and use a small prytool through the mounting hole to rotate the oil pump driveshaft until it can align with the distributor gear.

2. As the distributor is fully seated, the rotor should turn and end up at the first mark made. Ensure the distributor and oil pump rod are fully engaged.

➡When the distributor is fully seated on the engine, be sure the rotor is properly aligned with the first timing mark. If equipped with the HVS system, if the rotor does not align with the mark, the gear teeth of the HVS and camshaft have meshed one or more teeth out of time and the HVS procedure for timing disturbed should be used to assure proper installation.

3. Reconnect the distributor cap and wires. If applicable, connect the vacuum line to the distributor.
4. Tighten the distributor hold-down bolt, then install the air cleaner or duct, if removed for access.
5. Check and adjust the ignition timing.

Timing Disturbed

1. Set the engine to TDC: Remove the No. 1 spark plug. Place a finger over the spark plug hole and rotate the engine in the normal direction of rotation slowly, until compression is felt.
2. Align the timing mark on the crankshaft pulley to the **0** on the engine timing

indicator by rotating the engine in the same direction slowly. The engine is now set on No. 1 TDC.

➡An alternate method may be used to assure the engine is at TDC if the valve cover is removed. Watch the rocker arms for the No. 1 cylinder as the engine is turned. If the valves move as the crankshaft timing marker approaches the scale, the No. 1 cylinder is on its exhaust stroke. If the valves remain closed as the timing mark approaches the scale, then the No. 1 cylinder is approaching TDC of the compression stroke.

3. Install the distributor to the engine so the rotor is pointing to the No. 1 spark plug tower on the distributor cap once the distributor is fully seated in the engine.
4. Install the distributor cap, spark plug, wiring and connectors. If applicable, connect the vacuum line to the distributor.
5. If removed for access, install the air duct and/or air cleaner assembly, as applicable.
6. Check and adjust ignition timing.

Ignition Timing

The ignition timing may be adjusted only on those engines equipped with a conventional distributor ignition. If equipped with the distributorless Electronic Ignition (EI) system, the control module sets timing and makes all necessary spark changes. On these systems, the crankshaft position sensor is mounted in a fixed position, therefore not allowing for adjustment.

Connect the timing light and tachometer to the engine according to the tool manufacturers' instructions. Be sure the timing light is connected to the No. 1 spark plug wire.

ADJUSTMENT

1. Locate and clean the timing marks on the crankshaft pulley and the front of the timing case cover.
2. Use chalk or white paint to color the mark on the scale that will indicate the correct timing, when aligned with the mark on the pulley or the pointer.
3. Put the ignition system in the bypass mode by locating and grounding the bypass wire coming from the distributor.

➡Most vehicles are equipped with a single wire timing bypass connector. The bypass wire is normally a tan wire with a black stripe, that breaks out of the wiring harness conduit adjacent to the distributor, but it may break out of a taped section just below the heater case in the passenger compartment.

4. Connect a timing light with inductive spark pick-up according the manufacturers instructions.

5. Start the engine, then aim the timing light at the marks on the harmonic balancer and note the ignition timing.

6. If the timing needs adjustment, loosen the distributor lockbolt and turn the distributor slightly until the timing is adjusted to the specification on the underhood emissions label on the right strut tower.

7. Turn the engine **OFF**, tighten the distributor hold-down bolt and recheck the timing.

8. Remove the ground wire from the bypass connector.

Engine Assembly

REMOVAL & INSTALLATION

1. Disconnect the negative battery cable.
2. If equipped, remove the fuel injector sight shield.
3. Disconnect the air flow tube from the air cleaner.

✳✳ CAUTION

Never open, service or drain the radiator or cooling system when hot; serious burns can occur from the steam and hot coolant.

4. Drain the cooling system and remove the radiator hoses.
5. Remove engine mount struts.

✳✳ CAUTION

Observe all applicable safety precautions when working around fuel. Whenever servicing the fuel system, always work in a well ventilated area. Do not allow fuel spray or vapors to come in contact with a spark or open flame. Keep a dry chemical fire extinguisher near the work area. Always keep fuel in a container specifically designed for fuel storage; also, always properly seal fuel containers to avoid the possibility of fire or explosion.

6. Relieve the fuel system pressure and disconnect the fuel lines from the engine.

7. Disconnect the throttle and Throttle Valve (TV) cables.

8. Unplug the engine wiring harness connectors.

9. Tag and disconnect the vacuum lines from the engine, including the brake booster hose.

10. Disconnect the shifter cable from the transaxle.

11. Raise and safely support the vehicle.

12. Drain the engine oil.

13. Remove the wiring harness grounds from the engine.

14. Without disconnecting the lines, remove the air conditioning compressor from the bracket and support it out of the way.

15. Disconnect the exhaust pipe at the manifold.

16. Remove the front wheel and tire assemblies.

17. Disconnect the halfshafts from the transaxle and wire them to the strut assemblies.

18. Disconnect the steering shaft pinch bolt from intermediate shaft and remove the shaft from the rack and pinion steering gear.

19. Disconnect the fluid cooler lines at the transaxle.

20. Support the engine and sub-frame with a suitable jack.

21. Remove the sub-frame bolts and lower the engine/transaxle and subframe from the vehicle.

22. Separate the engine from the transaxle.

To install:

23. Raise the engine/transaxle assembly into position and install the subframe bolts. On 1995–96 models, tighten the bolts to 103 ft. lbs. (140 Nm). On 1997–99 models, tighten the bolts to 133 ft. lbs. (180 Nm).

24. Connect the exhaust pipe at the rear manifold. Install the starter.

25. Install the intermediate shaft to the steering gear, then install the pinch to secure the steering column to the intermediate shaft.

26. Connect the fluid cooler lines at the transaxle.

27. Install the halfshafts to the transaxle and assemble the related suspension.

28. Install the front wheels.

29. Connect the exhaust pipe to the manifold.

30. Install the A/C compressor.

31. Reattach the wiring harness grounds and connectors.

32. Connect the throttle and TV cables to the throttle body.

33. Connect the fuel lines. For quick-connect fittings, use the following procedure:

✳✳ CAUTION

To reduce the risk of fire and personal injury, before connecting a quick-connect fitting, apply a few drops of clean engine oil to the male pipe end. This will ensure proper reconnection and prevent a possible fuel leak. During normal operation, the O-rings located in the female connector will swell and prevent proper reconnection if not lubricated.

a. Apply a few drops of clean engine oil to the male pipe end.

b. Push both sides of the fitting together to cause the retaining tabs/fingers to snap into place.

c. Once installed, pull on both sides of the fitting to be sure the connection is secure.

34. Install the engine mount struts.

35. Install the radiator hoses.

36. If equipped, install the fuel injector sight shield.

37. Make a visual inspection of the engine compartment to ensure all connections have been made.

38. Install the intake air duct.

39. Connect the negative battery cable.

40. Refill the engine, transmission and radiator with the correct type and amount of fluid.

41. Start the engine and check for leaks.

Water Pump

REMOVAL & INSTALLATION

All Engines

1. Disconnect the negative battery cable, then drain the engine cooling system.

2. On the 3.1L engine, disconnect the heater hose.

3. If needed, remove the serpentine drive belt shield.

4. For the 3.8L and 3.4L engines, loosen but do not remove the water pump pulley bolts.

5. Remove the serpentine drive belt. On the 3.1L engine, a ⅜ in. drive breaker bar may be used to pivot the belt tensioner.

6. Remove the water pump pulley.

7. Loosen the clamps and disconnect any remaining hoses from the water pump, as applicable.

8. Remove the water pump retaining bolts. Remove the water pump assembly from the engine.

➡**On some engines, then pump retaining bolts will vary in size and thread. Be sure to note the positioning of all bolts during removal to assure proper installation.**

To install:

9. Clean all sealing surfaces

10. Install a new gasket and the water pump.

11. Tighten all water pump retainers as follows:

- 3.1L and 3.4L engines—89 inch. lbs. (10 Nm)
- 3.8L engine—short bolts to 11 ft. lbs. (15 Nm) +80°
- 3.8L engine—long bolts to 22 ft. lbs. (30 Nm)

12. Install hoses and new hose clamps as needed.

13. Install water pump pulley and loosely install retaining bolts.

14. Install serpentine belt and tighten pulley bolts to 18 ft. lbs. (25 Nm).

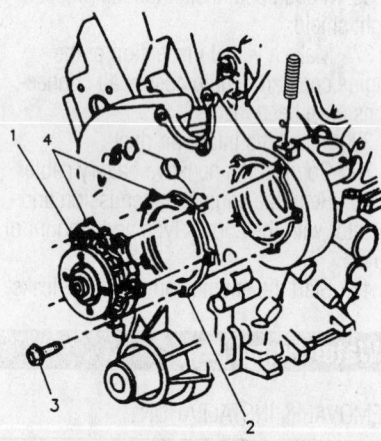

1 WATER PUMP
2 GASKET
3 10 N•m (89 LB. IN.)
4 LOCATOR – MUST BE VERTICAL

7924LG01

Water pump assembly mounting—3.1L and 3.4L engines

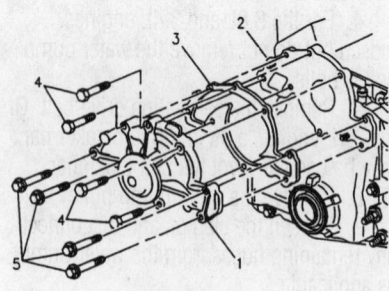

1 COOLANT PUMP
2 ENGINE FRONT COVER
3 GASKET
4 15 N•m (11 LB. FT.) +80°
5 30 N•m (22 LB. FT.)

7924LG02

Water pump assembly mounting—3.8L engine

15. If removed, install serpentine belt shield.

16. Refill cooling system.

17. Connect the negative battery cable.

Cylinder Head

REMOVAL & INSTALLATION

3.1L Engine

⁎⁎ CAUTION

Observe all applicable safety precautions when working around fuel. Whenever servicing the fuel system, always work in a well ventilated area. Do not allow fuel spray or vapors to come in contact with a spark or open flame. Keep a dry chemical fire extinguisher near the work area. Always keep fuel in a container specifically designed for fuel storage; also, always properly seal fuel containers to avoid the possibility of fire or explosion.

1. Relieve the fuel system pressure, then disconnect the negative battery cable.

⁎⁎ CAUTION

Never open, service or drain the radiator or cooling system when hot; serious burns can occur from the steam and hot coolant.

2. Drain the engine cooling system.

3. Remove the valve cover(s).

4. Remove the intake manifold.

5. Raise and support the vehicle safely, then disconnect the exhaust crossover pipe.

6. If necessary, disconnect the dipstick tube attachment and/or the alternator bracket.

7. Lower the vehicle.

8. Loosen the rocker arms and remove the pushrods. Keep all valvetrain components in order for reinstallation.

9. Remove the cylinder head retaining bolts. Remove the cylinder head from the engine along with the exhaust manifold.

To install:

10. Ensure that the cylinder bolt threads in the block and threads on the bolts are cleaned, as dirt will affect bolt torque.

11. Position the new gasket over the dowel pins with THIS SIDE UP showing, then install the cylinder head (along with the exhaust manifold).

12. Coat the threads of the cylinder head bolts with Sealing Compound 1052080 or equivalent.

13. Install the cylinder head retaining bolts. Tighten the cylinder head bolts gradually in the proper sequence to 33 ft. lbs. (45 Nm). Then turn each bolt an additional 90 degrees.

14. Position the intake gasket, then install the pushrods in their original locations. Be sure the lower ends of the pushrods are properly positioned in the lifter seats.

15. Properly adjust the valve lash.

16. Install the intake manifold assembly.

17. Install the valve cover(s).

18. Raise and support the vehicle safely. If removed, install the dipstick tube and/or the alternator bracket.

19. Connect the exhaust crossover pipe. Lower the vehicle.

20. Connect the negative battery cable and properly refill the engine cooling system.

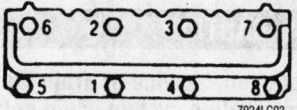

7924LG03

Tighten the cylinder head bolts in the sequence shown to ensure good cylinder sealing—3.1L and 3.4L engines

3.4L Engine

The 3.4L engine uses aluminum cylinder heads. Use care when working with light alloy parts. Valve guides are pressed in. Roller rocker arms are located on a pedestal in a slot in the cylinder head and are retained on individual threaded bolts.

The cylinder heads are retained by torque-to-yield bolts. A torque angle meter is required for proper torque at assembly. New replacement head bolts are recommended.

Before removing the cylinder head(s) from the engine and before disassembling the valve mechanism, perform a compression test and note the results. During disassembly, be sure that the valvetrain components are kept together and identified so that they can be installed in their original locations.

✳✳✳ CAUTION

Fuel injection systems remain under pressure, even after the engine has been turned OFF. The fuel system pressure must be relieved before disconnecting any fuel lines. Failure to do so may result in fire and/or personal injury.

LEFT (FRONT) SIDE

1. Evacuate the air conditioning system refrigerant and recover, using approved refrigerant recycling equipment.
2. Relieve the fuel system pressure using the recommended procedure.
3. Drain the cooling system.
4. Disconnect the negative battery cable.
5. Remove the upper and lower intake manifold assembly.
6. Remove the exhaust crossover pipe.
7. Disconnect the spark plug wires.
8. Remove the rocker arm cover.

➡**Any valvetrain components that are to be reused must be returned to their original locations. Keep the parts organized and in order.**

9. Removed the rocker arms and pushrods.
10. Remove the engine mount strut.
11. Remove the oil level indicator (dipstick) retainer bolts and remove the dipstick and tube.
12. Remove the A/C compressor using the following procedure.
   a. Remove the top compressor bolts.
   b. Raise and safely support the vehicle.
   c. Remove the bottom A/C compressor bolts.
   d. Disconnect the lines from the back of the A/C compressor.
   e. Remove the A/C compressor from the vehicle.
13. Remove the lower A/C compressor bracket bolts.
14. Lower the vehicle.
15. Remove the top A/C compressor bracket bolts and remove the bracket.
16. Remove the cylinder head bolts.
17. Remove the cylinder head and gasket.

To install:

18. Clean all parts well. Clean all gasket surfaces. Carefully remove all varnish soot and carbon to the bare metal. DO NOT use a motorized wire brush on any gasket surface since the soft aluminum will be damaged. If necessary, the head can be disassembled for thorough inspection and reconditioning.

19. Inspect the cylinder head for cracks. Do not attempt to weld the cylinder head. If cracked, replace it. Check the cylinder head deck, intake and exhaust manifold mating surfaces for flatness. These surfaces may be reconditioned by milling. If the surfaces are warped more than 0.005 in. (0.127mm), the surface should be milled. If more than 0.010 in. (0.251mm) of metal must be removed from the head, the head should be replaced.

20. Clean the cylinder head bolts and the bolt holes. Check the head bolts for damaged threads or stretching. New replacement head bolts are recommended.

21. Inspect the new head gasket. It should be marked which side is "UP". Place new cylinder head gasket on the block, over the dowel pins.

22. To avoid damage, install the spark plugs after the cylinder head has been installed on the engine block assembly. Install the cylinder head onto block.

23. Coat the cylinder head bolt threads with GM 1052080 sealer or equivalent.

24. Install the cylinder head bolts, and tighten in sequence. Tighten bolts 33 ft. lbs. (45 Nm) and then an additional 90 degrees (¼-turn).

25. Install the A/C compressor bracket and upper bolts.
26. Raise and safely support the vehicle.
27. Install the lower bracket bolts.
28. Install the A/C compressor.
29. Connect the A/C lines to the back of the compressor.
30. Install the lower A/C compressor bolts.
31. Lower the vehicle.
32. Install the top compressor bolts.
33. Install the oil level indicator.
34. Install the pushrods and rocker arms.
35. Tighten the rocker arm bolts to 89 inch lbs. (10 Nm) plus an additional 30 degrees. Do not overtighten or the threads in the aluminum head may be damaged.
36. Install the rocker arm covers.
37. Install the exhaust crossover pipe.
38. Install the spark plug wires.
39. Install the lower intake manifold and tighten the bolts to 115 inch lbs. (13 Nm).
40. Install the upper intake manifold and tighten in sequence to 18 ft. lbs. (25 Nm).
41. Install engine mount strut.
42. Refill the coolant system.
43. Connect the negative battery cable.
44. Since dirt, debris and coolant can enter the crankcase through the oil drain-back holes when a head is removed, an oil and filter change is recommended.

45. Start the vehicle and verify no leaks, abnormal noises and correct engine operation.
46. When satisfied with the repair, charge the A/C system.

RIGHT (REAR) SIDE

✳✳✳ CAUTION

Fuel injection systems remain under pressure, even after the engine has been turned OFF. The fuel system pressure must be relieved before disconnecting any fuel lines. Failure to do so may result in fire and/or personal injury.

1. Relieve the fuel system pressure using the recommended procedure.
2. Disconnect the negative battery cable.
3. Drain the coolant system.
4. Remove the intake manifold assembly using the recommended procedure.
5. Disconnect the ignition coil connection.
6. Remove the alternator.
7. Remove the exhaust crossover pipe.
8. Remove the oxygen sensor.
9. Raise and safely support the vehicle.
10. Remove the exhaust pipe from the manifold.
11. Lower the vehicle.
12. Remove the rocker arm cover.

➡**Any valvetrain components that are to be reused must be returned to their original locations. Keep the parts in organized and in order.**

13. Loosen the rocker arms until able to remove the pushrods using the recommended rocker arm removal procedure. Remove the pushrods.
14. Remove the cylinder head bolts and remove the cylinder head.

To install:

15. Clean all parts well. Clean all gasket surfaces. Carefully remove all varnish soot and carbon to the bare metal. DO NOT use a motorized wire brush on any gasket surface since the soft aluminum will be damaged. If necessary, the head can be disassembled for thorough inspection and reconditioning.

16. Inspect the cylinder head for cracks. Do not attempt to weld the cylinder head. If cracked, replace it. Check the cylinder head deck, intake and exhaust manifold mating surfaces for flatness. These surfaces may be reconditioned by milling. If the surfaces are "out of flat" by more than 0.005 inch, the surface should be milled. If more than 0.010

inch of metal must be removed from the head, the head should be replaced.

17. Clean the cylinder head bolts and the bolt holes. Check the head bolts for damaged threads or stretching. New replacement head bolts are recommended.

18. Inspect the new head gasket. It should be marked which side is "UP". Place new cylinder head gasket on the block, over the dowel pins.

19. To avoid damage, install the spark plugs after the cylinder head has been installed on the engine block assembly. Install the cylinder head onto block.

20. Coat the cylinder head bolt threads with GM 1052080 sealer or equivalent.

21. Install the cylinder head bolts, and tighten in sequence. Tighten bolts 33 ft. lbs. (45 Nm) and then an additional 90 degrees (¼-turn).

22. Install the pushrods and rocker arms.

23. Tighten the rocker arm bolts to 89 inch lbs. (10 Nm) plus an additional 30 degrees. Do not overtighten or the threads in the aluminum head may be damaged.

24. Install the rocker arm cover.

25. Raise and safely support the vehicle.

26. Install the exhaust pipe to the manifold.

27. Lower the vehicle.

28. Install the oxygen sensor.

29. Install the exhaust crossover pipe.

30. Install the alternator.

31. Attach the coil connections.

32. Install the lower intake manifold and tighten to 115 inch lbs. (10 Nm).

33. Install the upper intake manifold and tighten in sequence to 18 ft. lbs. (24 Nm).

34. Refill the coolant system.

35. Connect the negative battery cable.

36. Since dirt, debris and coolant can enter the crankcase through the oil drainback holes when a head is removed, an oil and filter change is recommended.

37. Start the vehicle and verify no leaks, abnormal noises and correct engine operation.

3.8L Engine

1. Relieve the fuel system pressure, then disconnect the negative battery cable.

2. Drain the engine cooling system.

3. Remove the intake manifold.

4. Remove the exhaust manifold(s).

5. Remove the valve cover(s).

6. Disconnect the electronic ignition and spark plug wires.

7. Remove the alternator bracket and one air conditioner bracket bolt. Remove the power steering pump.

8. As necessary, remove the belt tensioner assembly and/or the fuel pipe heat shield.

9. Remove the rocker arms, pushrods and guide plates. Keep them in order for reinstallation.

10. Remove the cylinder head retaining bolts. Remove the cylinder head from the engine.

To install:

11. Ensure that the cylinder bolt threads in the block are cleaned, as dirt will affect bolt torque.

➡This engine uses special torque-to-yield head bolts. The procedure must be followed carefully and new bolts must be used whenever the head is removed.

12. Position the new gasket with the arrow pointing to the front of the engine.

13. Install the cylinder head onto the engine.

14. Coat the underside of the bolt heads with Sealing Compound 1052080 or equivalent. Coat the threads of the bolts with a suitable thread locking compound.

15. Install the cylinder head onto the engine. Install the cylinder head retaining bolts and tighten as follows:

a. Tighten the cylinder head bolts gradually in the proper sequence to 35 ft. lbs. (47 Nm).

b. Then turn each bolt an additional 130 degree turn in sequence.

c. Finally turn the four center bolts an additional 30 degrees.

16. Install the rocker arms, pushrods and guide plates in the same position from which they were removed. Apply a suitable thread locking compound to the rocker arm pedestal bolts, then install and tighten to 11 ft. lbs. (15 Nm) plus 90 degrees.

17. Install the intake manifold and the valve cover(s).

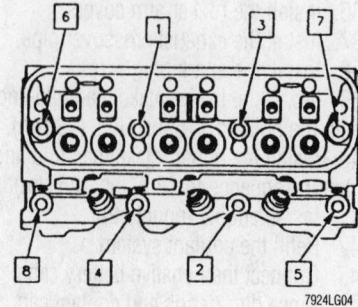

Cylinder head bolt torque sequence—3.8L engine

18. Raise and support the vehicle safely. Connect the exhaust manifold(s) to the exhaust pipe(s). Lower the vehicle.

19. Install the air conditioner bracket bolt. Install the alternator and bracket.

20. Install the ignition coil and spark plug wires.

21. Install the belt tensioner, then install power steering pump assembly.

22. Install the fuel pump heat shield.

23. Connect the negative battery cable, then properly refill the engine cooling system.

Rocker Arms

REMOVAL & INSTALLATION

➡Valvetrain components which are to be reused must be installed in their original positions. If removed, be sure to tag or arrange all rocker arms and pushrods to assure proper installation.

3.1L Engine

1. Remove the rocker arm cover(s) from the cylinder head.

2. Remove the rocker arm nut, the rocker arm and the ball washer.

➡If only the pushrod is to be removed, loosen the rocker arm nut, swing the rocker arm to the side and remove the pushrod.

3. Withdraw the pushrod from the cylinder head.

To install:

4. Inspect the components, and replace them if worn or damaged.

5. Coat the bearing surfaces of the rocker arms and the rocker arm ball washers with Molykote® or equivalent pre-lube.

6. Install the pushrods making sure they seat properly in the lifter.

7. Install the rocker arms, ball washers and the nuts, then tighten the rocker arm nuts until there is little or no valve lash.

➡Each valve must be adjusted when the lifter is sitting on the base circle of the camshaft, not the raised section of the lobe.

8. Properly adjust the valve lash.

9. Install the rocker arm cover.

10. Start and run the engine, then check for leaks. Check and adjust the timing, as necessary.

3.4L and 3.8L Engines

1. Remove the rocker arm cover from the engine.

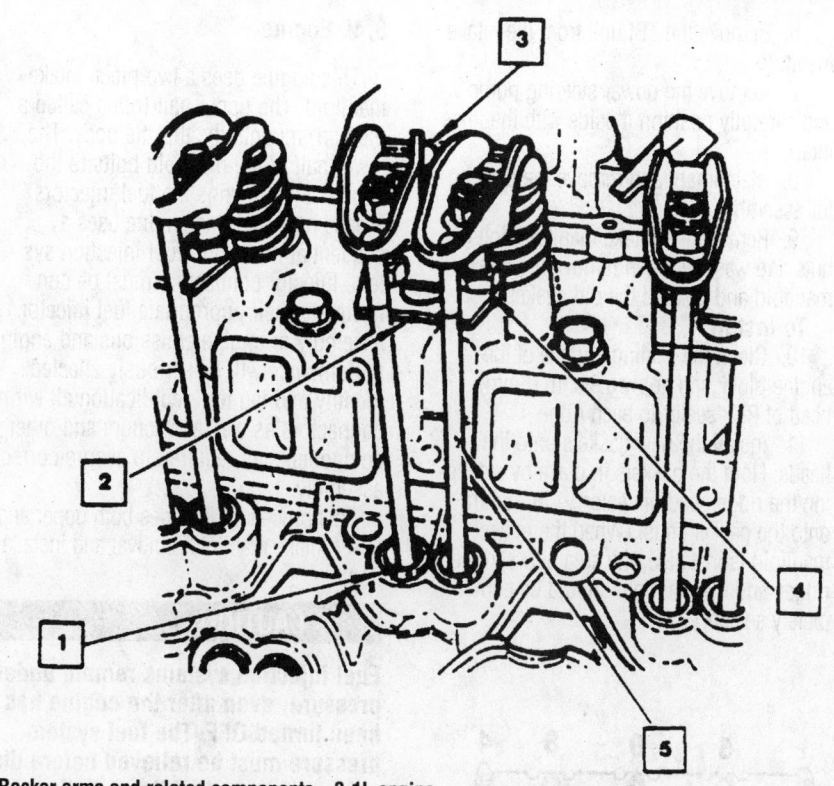

LIFTERS 1

PUSH ROD GUIDE 2

ROCKER ARM 3

ROCKER ARM STUDS 4

PUSH RODS 5

Rocker arms and related components—3.1L engine

7924LG26

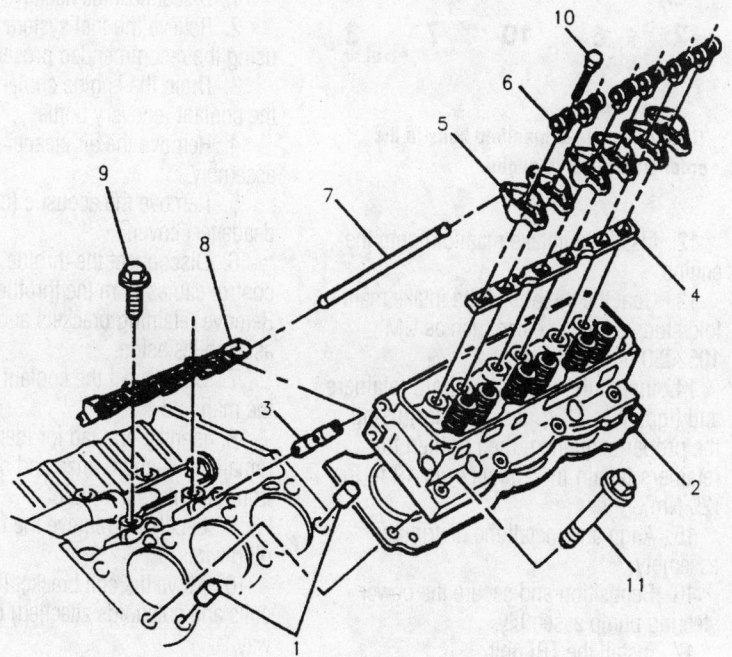

1 DOWEL PIN
2 HEAD GASKET
3 VALVE LIFTER
4 PUSHROD GUIDE
5 ROCKER ARM
6 ROCKER ARM BEARING
7 PUSHROD
8 LIFTER GUIDE RETAINER
9 BOLT
10 BOLT
11 HEAD BOLT

7924LG27

Exploded view of the rocker arms and related components—3.8L engine

2. Remove the rocker arm pedestal retaining bolts.

3. Remove the pedestal and rocker arm assembly.

To install:

4. Inspect and replace components if worn or damaged. Clean all old thread locking material from the pedestal bolts.

5. Install the rocker arms and pedestals. Apply a suitable thread locking compound to the rocker arm pedestal bolts, install and tighten to the following specifications:

- 3.4L engine—89 inch lbs. (10 Nm) plus 30 degrees (1/12 turn)
- 3.8L engine—11 ft. lbs. (15 Nm) plus 90 degrees (1/4 turn)

6. Install the rocker arm covers.

Intake Manifold

REMOVAL & INSTALLATION

3.1L Engine

> ❊❊❊ **CAUTION**
>
> Observe all applicable safety precautions when working around fuel. Whenever servicing the fuel system, always work in a well ventilated area. Do not allow fuel spray or vapors to come in contact with a spark or open flame. Keep a dry chemical fire extinguisher near the work area. Always keep fuel in a container specifically designed for fuel storage; also, always properly seal fuel containers to avoid the possibility of fire or explosion.

1. Properly relieve the fuel system pressure, then disconnect the negative battery cable.

> ❊❊❊ **CAUTION**
>
> Never open, service or drain the radiator or cooling system when hot; serious burns can occur from the steam and hot coolant.

2. Drain the engine cooling system and remove the air cleaner.

3. Tag and disconnect the necessary wiring and vacuum hoses in order to remove the valve covers.

4. Disconnect the fuel lines from the TBI unit and reposition for access.

5. Remove the valve covers. It will be necessary to remove or reposition the alternator (with brackets) and disconnect some coolant hoses for valve cover removal.

6. Remove the TBI unit from the intake manifold.

7. Remove the power steering pump and carefully position it aside with the lines intact.

8. Matchmark and remove the distributor assembly.

9. Remove the intake manifold bolts, nuts and washers, then remove the intake manifold and discard the old gasket.

To install:

10. Clean the sealing surface of the engine block and apply a 3/16 in. (5mm) bead of RTV sealer to each ridge.

11. Install the new gaskets onto the heads. Hold the gaskets in place by extending the ridge bead of sealer 1/4 in. (6mm) onto the gasket ends. (When the intake manifold is installed, the area between the ridges and the manifold should be completely sealed.)

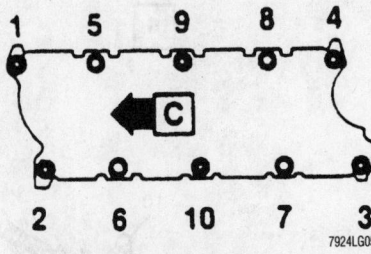

Tighten the intake manifold bolts in the order shown—3.1L engine

12. Install the intake manifold onto the engine.

13. Coat the threads of the intake manifold studs using a sealer such as GM 1052080 or equivalent.

14. Install the intake manifold retainers and tighten to 13 ft. lbs. (18 Nm) using the proper sequence, then tighten the retainers (again in sequence) to 19 ft. lbs. (26 Nm).

15. Align and install the distributor assembly.

16. Reposition and secure the power steering pump assembly.

17. Install the TBI unit.

18. Install the valve covers. Reposition and secure the alternator (with brackets) and connect the coolant hoses.

19. Connect the fuel lines to the TBI unit.

20. Connect the wiring and vacuum hoses as tagged during removal.

21. Install the air cleaner.

22. Connect the negative battery cable, then properly refill the engine cooling system.

3.4L Engine

This engine uses a two piece intake manifold. The upper half (often called a plenum) mounts the throttle body. The lower half of the manifold bolts to the engine and contains the fuel injectors. Please note that this engine uses a sequential multi-port fuel injection system. Injector connectors must be connected to their appropriate fuel injector assembly or engine emissions and engine performance will be seriously affected. Identify and tag for identification all wiring connectors as well as vacuum and other components as required to assure correct assembly.

This procedure includes both upper and lower intake manifold removal and installation.

> ❊❊❊ **CAUTION**
>
> Fuel injection systems remain under pressure, even after the engine has been turned OFF. The fuel system pressure must be relieved before disconnecting any fuel lines. Failure to do so may result in fire and/or personal injury.

1. Disconnect the negative battery cable.

2. Relieve the fuel system pressure using the recommended procedure.

3. Drain the engine coolant. Remove the coolant recovery bottle.

4. Remove the air cleaner and duct assembly.

5. Remove the acoustic (engine sound deadener) cover.

6. Disconnect the throttle and cruise control cables from the throttle body. Remove retaining brackets and set cable assemblies aside.

7. Disconnect the coolant hoses from the manifold.

8. Identify and tag for identification any remaining vacuum lines and disconnect from the intake manifold.

9. Label and remove the front spark plug wires.

10. Move the coil bracket (leaving the coils and solenoids attached) out of the way.

11. Disengage the electrical connectors from the ignition coil assembly.

12. Remove the manifold air pressure sensor.

13. Remove the brake booster hose.

14. Remove the EGR valve assembly.

15. Remove the thermostat bypass pipe nut from the upper intake manifold.

16. Remove the upper intake manifold

studs and bolts, then remove the upper intake manifold and gaskets.

17. If the lower intake manifold needs to be removed, remove the fuel injector rail bolts and remove the fuel injector rail assembly.

18. Remove the heater inlet pipe assembly, upper radiator hose and tie straps retaining the heater outlet pipe and ignition wiring assembly. Disconnect the heater pipe from the heater core to the coolant pump.

19. Remove the power steering pump bolts and pump.

20. Remove the rocker arm covers.

21. Remove the rocker arms and pushrods.

➡**Valvetrain components which are to be reused must be installed in their original positions. If removed, be sure to tag or arrange all rocker arms and pushrods to assure proper installation.**

22. Remove the lower intake manifold retaining bolts and remove the lower intake manifold and gasket.

To install:

23. Clean all parts well. Use care in cleaning old gasket material from the machined aluminum surfaces on the plenum and manifold as sharp tools may damage sealing surfaces.

24. Clean the mating surfaces to the intake manifold and engine block. Remove any loose pieces of RTV sealer.

25. Install the lower intake manifold to the engine block. Apply sealant GM 12345739 or equivalent at the engine block to manifold mating surface. The bead should be ⅛ in. (3mm) wide and ³⁄₁₆ in. (5mm) thick.

26. Install the lower intake manifold retaining bolts. Apply sealant GM 12345382 or equivalent to the threads of the bolts. Tighten bolts in sequence to 115 inch lbs. (13 Nm).

27. Install the valve rocker covers.

28. Connect the heater pipe from the heater core to the coolant pump. Install new tie straps around the heater outlet pipe and ignition harness assembly. Connect the upper radiator hose to the engine and the heater inlet pipe to the manifold assembly.

29. Install the power steering pump and pulley.

30. Remove the injector O-ring seals from both the spray tip ends and the fuel rail end of each injector. Discard the seals. With the spray tip end O-ring removed, the O-ring back-up piece may slip off of the injector. Be sure to retain the O-ring back-up for reuse. Be sure that the O-ring back-up piece is in place on the spray tip end of the injector before installing a new O-ring. Lubricate new

injector O-ring seals with clean engine oil and install on the injector assembly.

31. Install the fuel rail assembly to the intake manifold. Tilt the rail assembly to install the injectors. Install the fuel rail attaching bolts and tighten to 89 inch lbs. (10 Nm).

32. Attach the injector electrical connectors.

33. Install new O-rings on the fuel lines and install the fuel feed and return pipes. Tighten the fuel rail nuts to 13 ft. lbs. (17 Nm). Use a back-up wrench on the fittings to prevent them from turning.

34. Using new gaskets, install the intake manifold plenum. Be sure to route the MAP sensor electrical connector to the outside of the plenum gasket. Tighten the bolts to 18 ft. lbs. (25 Nm).

35. Install the serpentine drive belt. Install the coolant recovery tank.

36. Install the MAP sensor, braces to the alternator, ignition coil front bolts and the EGR-to-plenum bolts. Connect the vacuum lines as noted during removal.

37. If the throttle body was removed from the upper intake manifold, inspect the throttle body before installation. Throttle body bore and valve deposits may be cleaned using carburetor cleaner and a parts cleaning brush. DO NOT use a cleaner that contains methylethylketone (MEK), an extremely strong solvent and not necessary for this type of deposit. The TP sensor and IAC valve should NOT come into contact with solvents or cleaners as they may be damaged. Verify that the gasket surfaces are clean, and, using a new flange gasket, install the throttle body. Tighten the fasteners to 18 ft. lbs. (25 Nm).

38. Connect the throttle and cruise control cables.

39. Attach the IAC valve and TP sensor electrical connectors. Connect the air inlet duct. Check that the accelerator pedal is free by depressing the pedal to the floor and releasing.

40. Attach all remaining electrical connections and vacuum lines. Be sure the alternator braces are secure.

41. Refill the cooling system. GM recommends adding two engine coolant sealant pellets GM 3634621 or equivalent. Starting with the 1996 model year, these vehicles were filled at the factory with a new type of antifreeze/coolant called GM Goodwrench DEX-COOL®. When adding coolant, it is important that you use GM Goodwrench DEX-COOL® (orange-colored, silicate-free) coolant. A 50/50 mixture of DEX-COOL® and clean water will provide all the recommended protection. **DO NOT use**

DEX-COOL® in pre-1996 vehicles. DO NOT mix DEX-COOL® with any other type of antifreeze.

42. Since coolant can get into the engine's oil system when the intake manifold is removed, change the engine oil and filter.

43. Connect the negative battery cable.

44. Turn the key to the **ON** position several times to pressurize the fuel system and check for fuel leaks.

45. After the engine is running, bleed the cooling system.

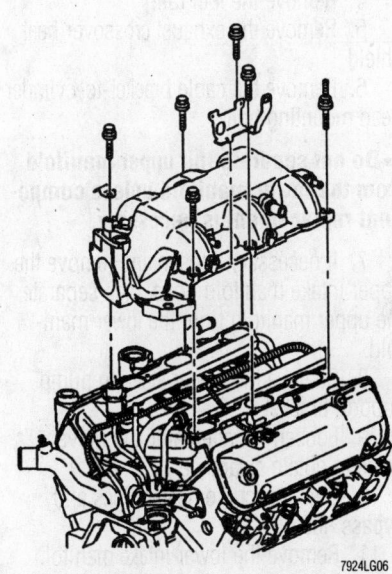

Upper intake manifold assembly—3.4L engine

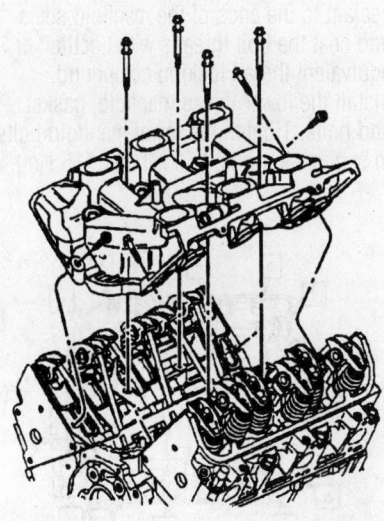

Lower intake manifold assembly—3.4L engine

3.8L Engine

The 3.8L engine utilizes a 2-piece intake manifold assembly. The entire assembly may be removed without separating the upper half from the lower half.

1. Properly relieve the fuel system pressure, then disconnect the negative battery cable.

2. Drain the engine cooling system and disconnect the air intake duct.

3. Tag and disconnect the spark plug wires on the right side of the engine, then position the wires aside.

4. Remove the fuel rail.

5. Remove the exhaust crossover heat shield.

6. Remove the cable bracket-to-cylinder head mounting bolt.

➡**Do not separate the upper manifold from the lower manifold unless component replacement is necessary.**

7. If necessary, loosen and remove the upper intake manifold bolts, then separate the upper manifold from the lower manifold.

8. Remove the power steering pump support bracket.

9. Loosen the alternator and move aside to obtain clearance.

10. Disconnect the heater pipes and bypass hose.

11. Remove the lower intake manifold bolts, then remove the manifold or manifold assembly (as applicable) from the engine.

To install:

12. Thoroughly clean all manifold mating surfaces, bolts and bolt holes. Apply sealant to the ends of the manifold seals and coat the bolt threads with Loctite® or equivalent thread locking compound. Install the lower intake manifold, gasket and bolts. Tighten the lower manifold bolts in sequence, twice, to 11 ft. lbs. (15 Nm).

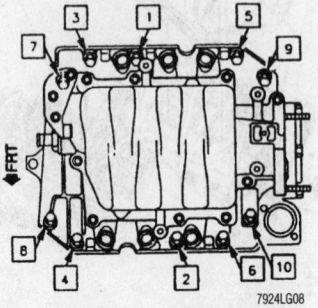

Tighten the intake manifold bolts according to the sequence shown in order to ensure proper sealing—3.8L engine

---

※ WARNING

The manifold surfaces used on this engine should not be scraped or wire brushed in order to clean the gaskets. The surfaces could be easily damaged if this is ignored. Instead, use a commercially available solvent to clean the mating surfaces.

13. Connect the heater pipes and bypass hose.

14. Reposition and secure the alternator.

15. Install the power steering pump support bracket.

16. If removed, prepare the upper intake manifold mating surface for installation. Apply a 1/16 in. (1.5mm) bead of Loctite Instant Gasket Eliminator GM 1052942 or equivalent to the mating surface on the lower manifold. Be sure to circle all bolt holes. Install the upper manifold assembly, then apply thread locking compound to the retainers. Install the upper intake manifold retainers and tighten to 11 ft. lbs. (15 Nm).

17. Install the cable bracket-to-cylinder head mounting bolt.

18. Install the exhaust heat shield.

19. Install the fuel rail.

20. Connect the spark plug wires on the right side of the engine.

21. Connect the air intake duct.

22. Connect the negative battery cable, then properly refill the engine cooling system.

Exhaust Manifold

REMOVAL & INSTALLATION

3.1L Engine

1. Disconnect the negative battery cable.

2. To remove the left (front) exhaust manifold:

a. Remove the serpentine belt and the air conditioning compressor. Position the compressor aside with the lines intact.

b. Remove the engine strut and bracket.

c. Disconnect the crossover pipe.

d. Remove the exhaust manifold attaching bolts and remove the manifold.

3. To remove the right (rear) exhaust manifold:

a. Disconnect the oxygen sensor wire.

b. Remove the crossover pipe.

c. Raise and support the vehicle safely.

---

d. Disconnect the exhaust pipe.

e. Support the rear center of the frame.

f. Remove the rear frame mount bolts.

g. Lower the frame 8–10 in. (20–25cm) for access.

h. Remove the exhaust manifold bolts and remove the assembly.

To install:

4. Clean the exhaust manifold mounting surfaces.

5. Install the manifold on the cylinder head. Tighten the mounting bolts to 25 ft. lbs. (34 Nm).

6. The rest of the installation is the reverse of the removal.

3.4L Engine

The exhaust manifolds are conventional iron castings. Left and right manifolds are connected by a crossover pipe. Use care with the exhaust manifold-to-cylinder head fasteners. The cylinder heads are aluminum.

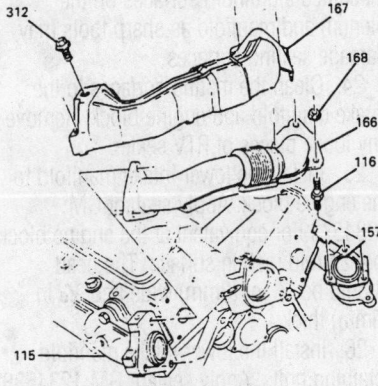

115	MANIFOLD, LEFT HAND EXHAUST
116	STUD, EXHAUST CROSSOVER
157	MANIFOLD, RIGHT HAND EXHAUST
166	CROSSOVER PIPE, EXHAUST
167	SHIELD, EXHAUST CROSSOVER UPPER HEAT
168	NUT, EXHAUST CROSSOVER
312	BOLT/SCREW, EXHAUST CROSSOVER UPPER HEAT SHIELD

7924LG28

Exploded view of the exhaust crossover and heat shield mounting—3.4L engine

LEFT (FRONT) SIDE

1. Disconnect the negative battery cable.

2. Drain the cooling system.

3. Remove the air cleaner.

4. Remove the crossover heat shield.

5. Remove the crossover pipe nuts and pipe.

6. Remove the engine strut bolts and strut.

7. Remove upper radiator hose.

8. Remove transaxle vacuum modulator pipe.

9. Remove the thermostat bypass pipe.

10. Remove the serpentine drive belt.

11. Remove the A/C compressor (leaving the hoses attached) and lay it aside.

12. Remove the engine strut and air conditioning bracket.

13. Remove the exhaust manifold heat shield.

14. Remove the exhaust manifold retaining bolts and remove the exhaust manifold.

To install:

15. Clean the mating surfaces, install the exhaust manifold and tighten the retaining bolts to 12 ft. lbs. (16 Nm).

16. Install the exhaust manifold heat shield.

17. Install the engine strut and air conditioning bracket.

18. Install the air conditioning compressor.

19. Install the serpentine belt.

20. Install thermostat bypass pipe.

21. Install transaxle vacuum modulator pipe.

22. Install upper radiator hose.

23. Install the engine strut bolts and strut.

24. Install the crossover pipe.

25. Install the crossover pipe heat shield.

26. Install the air cleaner.

27. Refill cooling system.

28. Connect the negative battery cable.

29. Start the vehicle and check for leaks.

RIGHT (REAR) SIDE

1. Disconnect the negative battery cable.

2. Drain cooling system.

3. Remove the upper radiator

4. Disconnect transaxle vacuum modulator line.

5. Remove the crossover heat shield.

6. Disconnect the crossover pipe at the manifold.

7. Remove the EGR tube.

8. Raise and safely support the vehicle.

9. Remove the oxygen sensor.

10. Disconnect the exhaust pipe.

11. Remove the transaxle fill tube.

12. Remove the heat shield bolts and heat shield.

13. Remove the exhaust manifold nuts and remove the exhaust manifold.

To install:

14. Clean the mating surfaces, install the exhaust manifold and tighten the retaining bolts to 12 ft. lbs. (16 Nm).

15. Install the heat shield.

16. Install the transaxle fill tube.

17. Install the exhaust pipe.

18. Connect the oxygen sensor wire.

19. Lower the vehicle.

20. Install the EGR tube.

21. Install the crossover pipe.

22. Install the crossover heat shield.

23. Connect transaxle vacuum line.

24. Install upper radiator hose.

25. Refill cooling system.

26. Connect the negative battery cable.

27. Start the engine and check for leaks.

3.8L Engine

1. Disconnect the negative battery cable.

2. To remove the left (front) exhaust manifold:

a. Remove the crossover pipe.

b. Tag and disconnect the spark plug wires from the plugs.

c. Remove the exhaust manifold bolts/studs and the oil dipstick tube.

d. Remove the exhaust manifold assembly.

3. To remove the right (rear) exhaust manifolds:

a. Tag and disconnect the spark plug wires.

b. Remove the throttle cable bracket.

c. Remove the crossover pipe heat shield.

d. Remove the transaxle dipstick and tube assembly.

e. Disconnect the oxygen sensor wire.

f. Remove the fasteners connecting the crossover pipe to the manifold.

g. Remove the plastic vacuum tank mounted on the cowl.

h. Raise and support the vehicle safely.

i. Remove the catalytic converter heat shield and hanger.

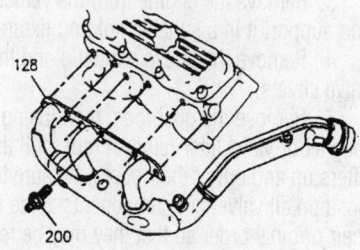

128 LEFT (FRONT) EXHAUST MANIFOLD
200 STUD

7924LG31

Left exhaust manifold and crossover pipe mounting—3.8L engine

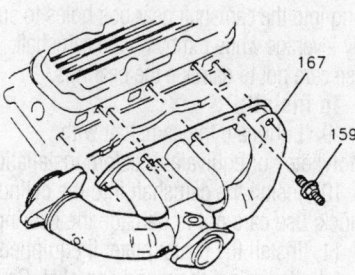

159 STUD
167 RIGHT (REAR) EXHAUST MANIFOLD

7924LG32

Right (rear) exhaust manifold mounting—3.8L engine

![Left exhaust manifold mounting 3.4L engine diagram]

111 SCREW, LH EXHAUST MANIFOLD HEAT SHIELD
112 SHIELD LH EXHAUST MANIFOLD
113 NUT, LH EXHAUST MANIFOLD
114 STUD, LH EXHAUST MANIFOLD
115 MANIFOLD, LH EXHAUST
117 GASKET, LH EXHAUST MANIFOLD
121 HEAD, LH CYLINDER

7924LG29

Left exhaust manifold mounting—3.4L engine

![Exploded view of right exhaust manifold mounting 3.4L engine diagram]

121 HEAD, CYLINDER
155 STUD, EXHAUST MANIFOLD
156 NUT, EXHAUST MANIFOLD
157 MANIFOLD RIGHT EXHAUST
160 GASKET, RIGHT EXHAUST MANIFOLD

7924LG30

Exploded view of the right exhaust manifold mounting—3.4L engine

j. Remove the front exhaust pipe-to-manifold attaching nuts.

k. Remove the front exhaust pipe from the manifold. Lower the vehicle.

l. Remove the engine lift bracket and remove the manifold attaching nuts.

m. Remove the exhaust manifold assembly.

4. Installation is the reverse of the removal procedure. Tighten the stud nuts to 38 ft. lbs. (32 Nm).

Camshaft and Valve Lifters

REMOVAL & INSTALLATION

➡**When removing valvetrain components, be sure to keep them in order for reassembly purposes. Note that new lifters are a must when replacing a camshaft. Used or worn lifters installed on a new camshaft will rapidly fail the camshaft.**

3.1L and 3.8L Engines

Please note that the factory recommends that the engine assembly be removed from the vehicle to remove the camshaft from this engine.

1. Disconnect the negative battery cable.

2. Drain the cooling system.

3. Remove the engine from the vehicle and support it in a suitable holding fixture.

4. Remove the intake manifold and the valve covers.

5. Remove the pushrods, then, using a magnet or valve lifter removal tool, pull the lifters up and out of their bores. Be sure to position all valvetrain components aside in their original order so that they may be reinserted into their proper bores.

6. Remove the front cover assembly.

7. Remove the timing chain and sprocket. Remove the thrust plate, if equipped.

8. Remove the camshaft from the block. Insert three bolts approximately 3 in. (76mm) long into the camshaft gear bolt holes to supply leverage while removing the camshaft. Use care not to damage the bearings.

To install:

9. Lubricate the camshaft with Molykote® or equivalent, before installation.

10. Install the camshaft into the cylinder block, use care not to damage the bearings.

11. Install the thrust plate, if equipped. Install the timing chain and sprocket. Be sure the timing marks align correctly.

12. Install the front cover assembly.

13. Lubricate the bottom of each valve lifter with Molykote® or equivalent, then slide them into their respective bores. Install

the pushrods and other valvetrain components.

14. Install the intake manifold assembly and the valve cover.

15. Install the engine into the vehicle.

16. Fill the cooling system to the correct level and connect the negative battery cable.

3.4L Engine

Please note that the factory recommends that the engine assembly be removed from the vehicle to remove the camshaft from this engine.

1. Disconnect the negative battery cable.

2. Drain the cooling system.

3. Remove the engine from the vehicle using the proper procedure outlined in this section. This a lengthy process involving the removal of the engine/transaxle/subframe assembly. Safely secure the engine in a suitable holding fixture.

4. Remove the rocker arm covers.

5. Remove the oil pump drive gear hold down clamp bolt and clamp.

6. Remove the oil pump driven gear assembly.

7. Remove the intake manifold.

8. Remove the rocker arms and components. Keep the parts in a rack so they may be reinstalled in the same location. Remove the pushrods and the valve lifters. Valve lifters and pushrods should be kept in order so they may be reinstalled in their original position. Some engines may have standard and oversize (O.S.) valve lifters. Where O.S. lifters are used, the cylinder block should be marked with a daub of white paint and "0.25 in. (6mm) O.S." stamped on the lifter boss.

9. Remove the dampener retaining bolt. It will probably be necessary to lock the crankshaft by freezing the flywheel with a prybar or other suitable locking device before attempting to remove the crank dampener bolt. Use a puller to draw the dampener off the crankshaft. Use care not to loose the key on the crankshaft.

10. Remove the front cover assembly. Note that the oil pan will need to be removed, or at least loosened.

11. Be sure that the timing marks on the cam and crank sprockets are aligned. Use the alignment marks on the dampener stamping or cast alignment marks on the cylinder block. Piston No. 1 should still be at top dead center of the compression stroke from the distributor removal procedure. Carefully check the camshaft sprocket timing marks. With piston No. 1 at top dead center of the compression

stroke, the camshaft sprocket timing mark should be at the 12 o'clock position, with the marks on the camshaft sprocket and crankshaft sprocket aligned. Confirm the position of the timing marks before disassembly so replacement parts may be assembled in the same relationship. Unbolt the camshaft sprocket and remove the timing chain from the crankshaft sprocket. If the camshaft sprocket does not come off easily, a light blow on the lower edge of the sprocket with a plastic mallet should dislodge the sprocket. A puller may be required to draw off the crankshaft sprocket, if required.

12. Remove the camshaft thrust plate screws and remove the thrust plate.

13. Remove the camshaft from the block by pulling it out. The camshaft is supported by four journals. All camshaft journals are the same diameter and care must be exercised in removing the camshaft to avoid damage to the bearings. A good aid in removing or installing a camshaft is using extra long bolts in the camshaft sprocket bolt holes to act as a handle.

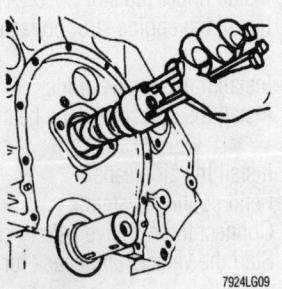

7924LG09

Using three long bolts threaded into the camshaft allow better control for removal—all engines

To install:

14. Inspect the camshaft for scratches, pitting and/or wear on the bearing and lobe surfaces. Note the timing marks on the sprockets (cam and crankshaft). It is absolutely necessary that the camshaft sprocket timing mark and the crankshaft timing mark be properly aligned. These marks are often indistinct and hard to see. If new sprockets are to be installed, or as an aid in reinstalling the original sprockets, it may be helpful to place a small dot of light-color paint on each timing mark. This will make aligning the marks easier under the low-light conditions found in many work areas. If the camshaft timing marks are misaligned, the engine will run poorly or may even suffer engine damage from being out-of-time.

15. Be sure all parts are clean. The

camshaft and its related components are precision made and dirt or other foreign material can easily damage new parts. Work as cleanly as possible. This means tools, parts, hands, cleaning cloths and work area. Liberally coat the camshaft with special camshaft lubricant (GM usually recommends Molykote® or equivalent) before installation. This is important. Camshaft lobes are lubricated by throw-off oil from the crankshaft and connecting rods. If not enough oil is thrown onto the cam lobes, the cam could be damaged or become worn out in a matter of minutes. An oil supplement and/or camshaft lubricant (often, but not always supplied with a new camshaft) is recommended.

16. Install the camshaft into the cylinder block, using care not to damage the bearings. It will probably be necessary to turn the camshaft, especially as it nears the rear bearing.

17. Install the camshaft thrust plate and tighten the screws to 89 inch lbs. (10 Nm).

18. Install the crankshaft sprocket, if previously removed.

19. Lubricate the cam sprocket thrust surface with camshaft lube. Hold the cam sprocket with the chain hanging down and slip the chain around the crankshaft sprocket, aligning the marks on the camshaft and crankshaft sprockets. Align the dowel in the camshaft end with the dowel hole in the cam sprocket. Draw the cam sprocket onto the camshaft using the mounting bolts. Tighten the bolts to 21 ft. lbs. (28 Nm). Lubricate the timing chain with engine oil. Recheck the timing marks to be sure they are correctly aligned.

20. Inspect the front cover sealing surfaces both on the cover and on the engine block. Clean with a suitable degreaser. Apply sealer to the bottom sealing surface of the front cover. Install the front cover assembly. If the oil pan gasket is damaged, the oil pan should be removed, thoroughly cleaned and reinstalled using a new gasket.

21. Coat the front cover seal contact area (on the dampener) with engine oil. Apply sealant to the key and keyway. Place the dampener in position over the key in the crankshaft and pull the dampener into position. Install the dampener retaining bolt. It will probably be necessary to lock the crankshaft by freezing the flywheel with a prybar or other suitable locking device before attempting to tighten the crank dampener bolt. Tighten to 76 ft. lbs. (103 Nm).

22. Clean sealing material from the sealing surfaces of the intake manifold and front

and rear ridges of the engine block. Clean the sealing surfaces with a degreaser. Apply a 3⁄16-inch (5mm) bead of RTV sealer on each ridge. Install the intake manifold gaskets and lay the intake manifold assembly in place. Liberally coat the lifters with camshaft lube and install. If the original lifters are being reinstalled on the original camshaft, use care that each lifter is returned to the location from which it was removed. Installing used lifters on a new camshaft will quickly fail the new camshaft. In nearly all cases, a new replacement camshaft should be installed with new lifters, as a set.

23. Install the pushrods, making sure the lower ends of the pushrods are centered in the lifter seats. Coat the bearing surfaces of the rockers and pivot balls with camshaft lube and install.

24. Tighten the rocker arms to 89 inch lbs. (10 Nm).

25. Install the rocker arm covers.

26. Install the engine into the vehicle/transaxle/subframe assembly using care to align the supports and engine mounts properly.

27. Change the engine oil and filter. Always check that the proper quantity of oil is in the crankcase. An oil supplement may be used to help the camshaft through its break-in period. Fill the cooling system to the correct level but leave off the radiator cap in case coolant needs to be added as the engine warms up. With the ignition **OFF** or disconnected, crank the engine several times. Listen for any unusual noises or evidence that any parts are binding.

28. Connect the negative battery cable. Start the engine, listening for any unusual noises and checking for leaks. Run the engine at about 1000 rpm until the engine is at operating temperature. Listen for improperly adjusted valves or any unusual noises. Check for oil, fuel, coolant and vacuum leaks. With the engine at operating temperature, set

the ignition timing to the underhood tune-up and emission label specification

29. Road test vehicle.

➡Whenever the vehicle sub-frame is removed or lowered, the wheel alignment should be checked.

Valve Lash

ADJUSTMENT

3.1L Engine

➡These engines utilize hydraulic valve lifters which means that a valve adjustment is not a regular maintenance item. The valves must only be adjusted if the rockers arms have been disturbed for any reason such as cylinder head, camshaft, pushrod or lifter removal.

1. Remove the air cleaner and the rocker arm cover(s).

2. Rotate the crankshaft until the mark on the crankshaft pulley aligns with the **0** mark on the timing plate. Be sure the No. 1 cylinder is positioned on the compression stroke. The No. 1 piston is on it's compression stroke when both the intake and exhaust valves remain closed as the crankshaft damper mark approaches the timing scale.

➡Another method to tell when the piston is coming up on the compression stroke is by removing the spark plug and placing a finger over the hole in order to feel air being forced out of the spark plug hole. Stop turning the crankshaft when pressure is felt and the TDC timing mark on the crankshaft pulley is directly aligned with the timing mark pointer or the 0 mark on the scale.

3. When the engine is on the No. 1 firing position, adjust the following valves:

- Intake—1, 5 and 6
- Exhaust—1, 2 and 3

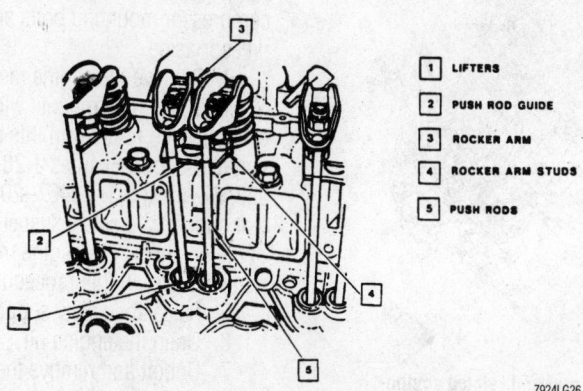

1	LIFTERS
2	PUSH ROD GUIDE
3	ROCKER ARM
4	ROCKER ARM STUDS
5	PUSH RODS

7924LG26

Valve rocker arm and related components—3.1L engine

4. To adjust the valves, back out the adjusting nut until lash can be felt at the pushrod, then turn the nut until all of the lash is removed.

➡️**To determine is all of the lash is removed, turn the pushrod between two fingers until the movement is removed.**

5. When all of the lash has been removed, turn the adjusting an additional 1½ turns; this will center the lifter plunger.

6. Crank the engine one complete revolution until the timing tab (**0** degree mark) and the crankshaft pulley mark are again in alignment. Now the engine is in the No. 4 firing position. Adjust the following valves:
- Intake—2, 3 and 4
- Exhaust—4, 5 and 6

7. Install the rocker arm cover(s).

8. Start and run the engine, then check and adjust the timing, as necessary.

3.4L and 3.8L Engines

Because the rocker arm fasteners are secured and tightened, valve lash is not adjustable on the 3.4L or 3.8L engines. If a valvetrain problem is suspected, check that the rocker arm pedestals bolts are tightened to specification. During initial installation the bolts are coated with a suitable thread locking compound. If they are sufficiently loosened to cause valvetrain noise, they should be removed and thoroughly cleaned. Apply a suitable thread locking compound to the rocker arm pedestal bolts, install and tighten to the following specifications:
- On 3.4L engine, tighten to 89 inch lbs. (10 Nm) plus 30 degrees
- On 3.8L engine, tighten to 11 ft. lbs. (15 Nm) plus 90 degrees

When valve lash falls out of specification (valve tap is heard) and tightening the bolts does not solve the problem, replace the rocker arm, pushrod and hydraulic lifter on the offending cylinder.

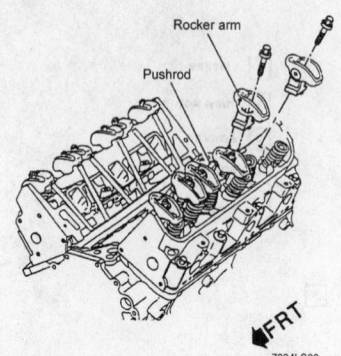

Valve rocker arm and related components—3.4L engine

Rocker arm
Pushrod
FRT
7924LG33

Oil Pan

REMOVAL & INSTALLATION

3.1L Engine

1. Disconnect the negative battery cable. Remove the accessory drive belt.
2. Raise and safely support the vehicle.
3. Remove the crankshaft damper and pulley.
4. Drain the engine oil. Remove the flywheel shields.
5. Remove the starter. Support the engine.
6. Remove the engine mounting bolts.
7. Raise the engine slightly.
8. Remove the oil pan bolts and the oil pan.

To install:

9. Install a new oil pan gasket and install the oil pan. Tighten M8 oil pan bolts to 19 ft. lbs. (25 Nm) and the M6 oil pan bolts to 85 inch lbs. (10 Nm).
10. Lower the engine and install the engine mounting bolts.
11. Install the starter and flywheel shields.
12. Install the crankshaft damper and pulley.
13. Lower the vehicle and install the accessory drive belt.
14. Refill the crankcase and connect the negative battery cable.

3.4L Engine

Use care when servicing the oil pan on the 3.4L engine. The engine main bearing caps are drilled and tapped for structural oil pan side bolts. Do not overlook the side bolts when attempting to remove the oil pan.

1. Disconnect the negative battery cable.
2. On 1997–99 models remove the A/C compressor mounting bolts and set compressor aside.
3. Remove the engine mount struts.
4. The engine mounts will have to be disconnected to perform this service. Install Engine Support Fixtures J-28467-A, J28467–90 and J-28467–200 or equivalent supports to safely suspend the weight of the engine so the engine mounts can be removed later in this procedure.
5. Raise and safely support the vehicle.
6. Drain the engine oil.
7. Unbolt and remove the oil filter drip shield.

8. Disconnect the exhaust pipe from the manifold.
9. Remove the engine mount-to-frame nuts.
10. Remove the transaxle mount-to-frame nuts.
11. On 1996 models perform the following steps:
   a. Support the vehicle under the subframe at the front and rear.
   b. Loosen the rear frame bolts, but do not remove.
   c. Remove the front frame bolts and lower the front of the frame to gain clearance to oil pan.
12. On 1997–99 models raise the engine with the engine support fixture to gain clearance to oil pan.
13. Remove the engine mount.
14. Remove the engine mount bracket.
15. Remove the starter motor.
16. Remove the transaxle brace.
17. Disengage the oil level wiring harness connector at the oil pan.
18. Remove the oil pan side bolts.
19. Remove the oil pan bottom bolts.
20. Remove the oil pan.

To install:

21. Clean all parts well. Clean all gasket sealing surfaces on the oil pan flanges, the oil pan rail and the front cover. Clean the threaded holes in the main bearing cap and all threaded holes.

➡️**Apply a small amount of sealer on either side of the rear main bearing cap, where the seal surface on the cap meets the cylinder block.**

22. Install the oil pan and gasket.
23. Install the oil pan bolts and tighten the bottom bolts to 18 ft. lbs. (25 Nm) and tighten the side bolts to 37 ft. lbs. (50 Nm). Tool J 39505 or equivalent is required to properly tighten the structural side bolts.
24. Install the oil level wiring harness connector.
25. Install the transaxle brace.
26. Install the starter.
27. Install the engine mount bracket.
28. Install the engine mount.
29. On 1996 models perform the following steps:
   a. Raise the frame to proper position and install new bolts.
   b. Tighten the side-to-crossmember bolts to 40 ft. lbs. (54 Nm) and tighten the left-hand frame insulator bolt to 103 ft. lbs. (140 Nm).
   c. Remove the vehicle supports.
30. On 1997–99 models lower engine and engage engine mounts to the subframe.

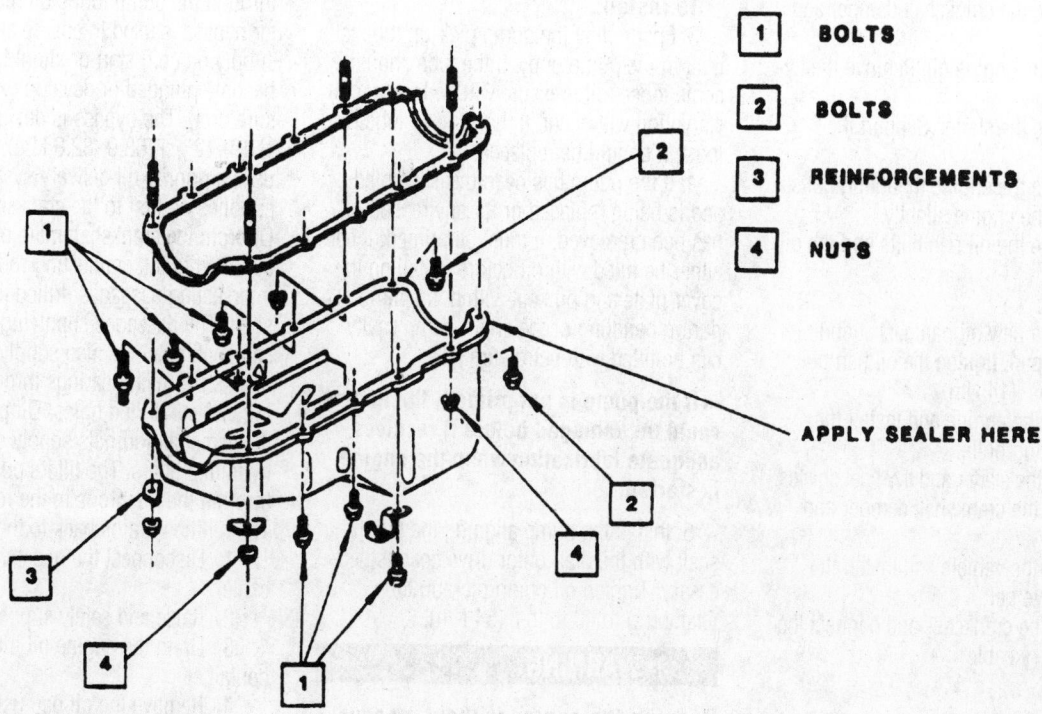

1	BOLTS
2	BOLTS
3	REINFORCEMENTS
4	NUTS

APPLY SEALER HERE

7924LG34

Exploded view of the oil pan mounting—3.1L engine

31. Install the exhaust pipe to the manifold.

32. Install the transaxle mount nuts and tighten to 32 ft. lbs. (44 Nm).

33. Install the engine mount nuts and tighten to 32 ft. lbs. (44 Nm).

34. Install the oil filter drip shield.

35. Lower the vehicle.

36. Remove the engine support fixtures.

37. Fill the crankcase with engine oil. A filter change is recommended.

38. On 1997–99 models install the A/C compressor.

39. Connect the negative battery cable.

40. Start the engine and verify no leaks.

➡Whenever the vehicle sub-frame is removed or lowered, the wheel alignment should be checked.

3.8L Engine

1. Disconnect the negative battery cable. Remove the accessory drive belt.

2. Raise and safely support the vehicle.

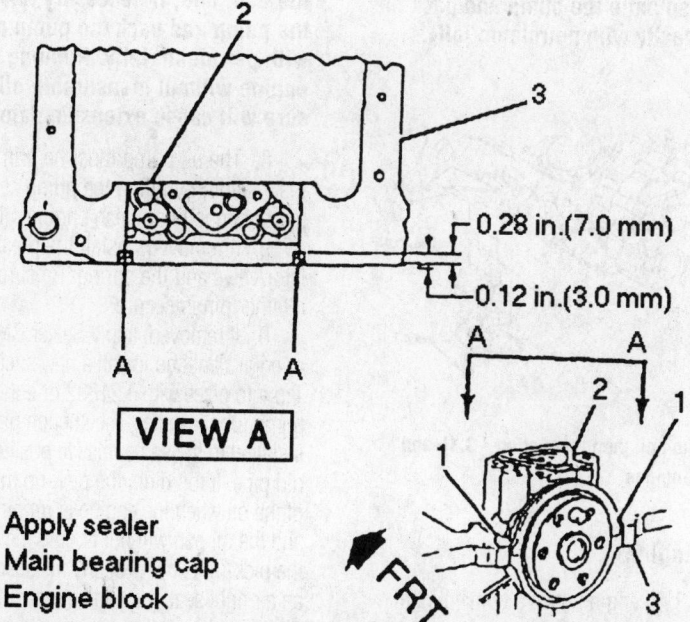

0.28 in.(7.0 mm)

0.12 in.(3.0 mm)

VIEW A

1. Apply sealer
2. Main bearing cap
3. Engine block

7924LG10

Prior to installation, apply sealer to the rear main bearing cap as shown—3.4L engine

1. Oil pan
2. Oil pan side bolt
3. Oil pan retaining bolt
4. Engine block

7924LG35

Exploded view of the oil pan mounting—3.4L engine

3. Remove the crankshaft damper and pulley.

4. Drain the engine oil. Remove the flywheel shields.

5. Remove the starter. Support the engine.

6. Remove the engine mounting bolts.

7. Raise the engine slightly.

8. Remove the oil pan bolts and the oil pan.

To install:

9. Install a new oil pan gasket and install the oil pan. tighten the oil pan bolts to 124 inch lbs. (14 Nm).

10. Lower the engine and install the engine mounting bolts.

11. Install the starter and flywheel shields.

12. Install the crankshaft damper and pulley.

13. Lower the vehicle and install the accessory drive belt.

14. Refill the crankcase and connect the negative battery cable.

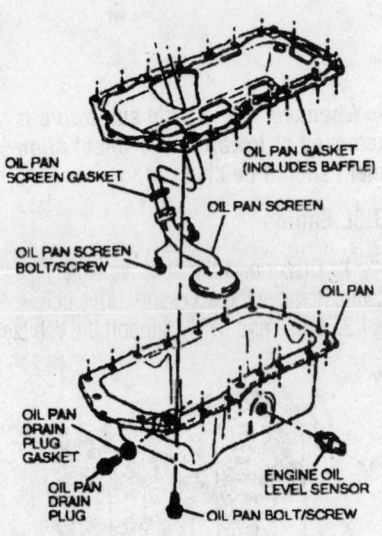

OIL PAN SCREEN GASKET
OIL PAN GASKET (INCLUDES BAFFLE)
OIL PAN SCREEN
OIL PAN SCREEN BOLT/SCREW
OIL PAN
OIL PAN DRAIN PLUG GASKET
OIL PAN DRAIN PLUG
ENGINE OIL LEVEL SENSOR
OIL PAN BOLT/SCREW

7924LG36

Exploded view of the oil pan mounting—3.8L engine

Oil Pump

REMOVAL & INSTALLATION

3.1L Engine

1. Remove the oil pan.

2. Remove the oil pump attaching bolt and, if equipped, the pick-up tube nut/bolt, then remove the pump along with the pick-up tube and shaft, as necessary.

To install:

3. Ensure that the pump pick-up tube is tight in the pump body. If the tube should come loose, oil pressure will be lost and oil starvation will occur. If the pick-up tube is loose it should be replaced.

4. If the pump has been disassembled and is being replaced or for any reason oil has been removed, it must be primed. It can either be filled with oil before installing the cover plate and oil kept within the pump during handling or the entire pump cavity can be filled with petroleum jelly.

➡ **If the pump is not primed, the engine could be damaged before it receives adequate lubrication when the engine is started.**

5. Install the pump aligning the pump shaft with the distributor drive gear as necessary. Tighten oil pump/pick-up tube retainer(s) to 40 ft. lbs. (54 Nm).

✳✳ WARNING

Running the engine without measurable oil pressure will cause extensive damage.

6. Install the oil pan and refill the engine crankcase. Disable the ignition system; crank engine for approximately 10 seconds to aid in priming the oil pump and reducing the risk of engine damage.

➡ **If the oil pump does not build up oil pressure almost immediately, remove the pan and check for a loose oil pump-to-pick-up tube attachment. If necessary dismantle the pump and pack the pump cavity with petroleum jelly.**

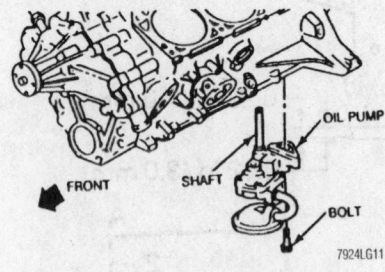

OIL PUMP
SHAFT
BOLT
FRONT

7924LG11

Common oil pump mounting—3.1L and 3.4L engines

3.4L Engine

The 3.4L engine uses a conventional gear type oil pump. Oil is drawn up through the pick-up screen and tube and passed

through the pump to the oil filter. An oil filter bypass is used to ensure adequate oil supply on cold start or should the filter become plugged or develop excessive pressure drop. The bypass is designed to open at 10–12 psi (68.9–82.6 kPa). The engine uses a priority oil delivery system which supplies oil first to the crankshaft journals. Oil from the crankshaft main bearings is supplied to the connecting rod bearings by intersecting passages drilled in the crankshaft. The passages supplying oil to the camshaft bearings also supply oil to the crankshaft main bearings through intersecting vertical drilled holes. Oil passages from the camshaft journals supply oil to the hydraulic lifters. The lifters pump up through the pushrods to the rocker arms. The oil then drains back to the oil pan.

1. Disconnect the negative battery cable.

2. Raise and safely support the vehicle.

3. Drain the engine oil into a suitable container.

4. Remove the oil pan using the recommended procedure outlined in this section. Use care; there are "hidden" bolts that go through the side of the oil pan which must not be overlooked.

5. Remove the oil pump bolt.

6. Remove the oil pump and oil pump driveshaft extension.

To install:

7. Clean all parts well. The oil pan may contain sludge which should be removed.

➡ **If the oil pump does not build up oil pressure almost immediately, remove the pan, and, if necessary, dismantle the pump and pack the pump cavity with petroleum jelly. Running the engine without measurable oil pressure will cause extensive damage.**

8. The oil pump must be primed before installation, by filling the pump cavity with petroleum jelly or clean engine oil.

9. If removed, install the pressure regulator valve and the spring. Be sure the retainer pin is secure.

10. If removed, apply Sealer GM 1050026 or equivalent and install a new suction pipe. Tap into place with J 21882 or equivalent oil pump tube installer. The suction pipe must be installed in the same relative position as the old pipe. If too high, the pick-up may be out of the oil when the engine is running. Too low and the oil pan will not fit onto the engine. If the pick-up is not properly installed and with an air tight seal, the engine may not develop oil pressure and the engine will be severely damaged.

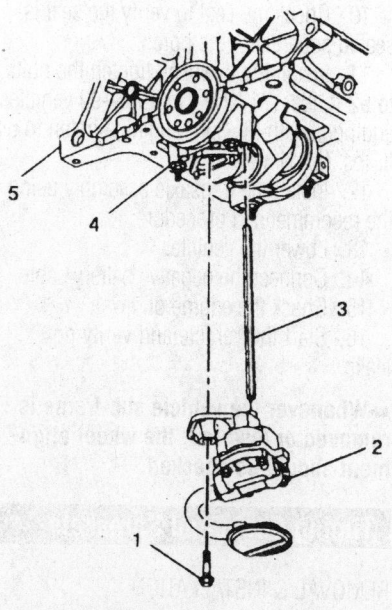

1. Oil pump bolt
2. Oil pump
3. Oil pump drive rod
4. Main bearing cap
5. Engine block

7924LG37

**Exploded view of the oil pump mounting—
3.4L engine**

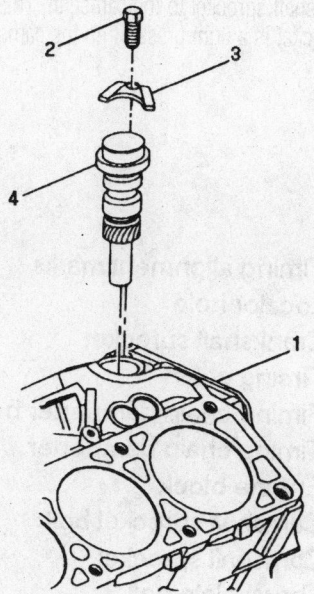

1. Engine block
2. Oil pump drive clamp bolt
3. Oil pump drive clamp
4. Oil pump drive

7924LG38

**Oil pump drive gear assembly mounting—
3.4L engine**

11. Engage the drive shaft into the drive gear and install the rear bearing cap bolt. Tighten the bolt to 30 ft. lbs. (41 Nm).

12. Install the oil splash shield. Install the oil pan using the recommended procedure. Use care when installing the side bolts that thread into the sides of the main bearing caps. Connect the oil level sensor.

13. Lower the vehicle.

14. Fill the crankcase with new engine oil. A new oil filter is recommended.

15. Connect the negative battery cable.

16. Start the engine and verify oil pressure and no leaks.

3.8L Engine

1. Disconnect the negative battery cable.

2. Remove the front timing cover assembly.

3. Remove the oil pump cover attaching screws and remove the pump gears.

4. Remove the oil filter drip shield.

5. Remove the oil filter.

6. Remove the four bolts securing the adapter to the front cover.

7. Remove the adapter, gasket, the oil pressure valve and spring.

To install:

8. Clean all part in solvent and remove the old gaskets.

9. Check all parts for scoring, cracks or excessive wear. Check pressure regulator spring for loss of tension and replace as necessary.

10. Install the oil filter adapter, a new gasket, the oil pressure valve and spring.

11. Install the four adapter to front cover screws and tighten to 22 ft. lbs. (30 Nm).

12. Install the oil filter.

13. Install the oil filter drip shield.

14. Install the pump gears into the cover assembly and pack with petroleum jelly. Install the oil pump cover attaching screws and tighten to 97 inch lbs. (11 Nm).

15. Install the front timing cover assembly.

16. Connect the negative battery cable.

➡**If the oil pump pick-up screen is thought to be possibly clogged or dirty the oil pan should be removed and the screen removed and cleaned with solvent.**

Rear Main Seal

REMOVAL & INSTALLATION

The entire transaxle assembly must be removed to perform this service. This requires special tooling to support the engine assembly while the transaxle and sub-frame are lowered from under the vehicle.

1. Disconnect the negative battery cable.

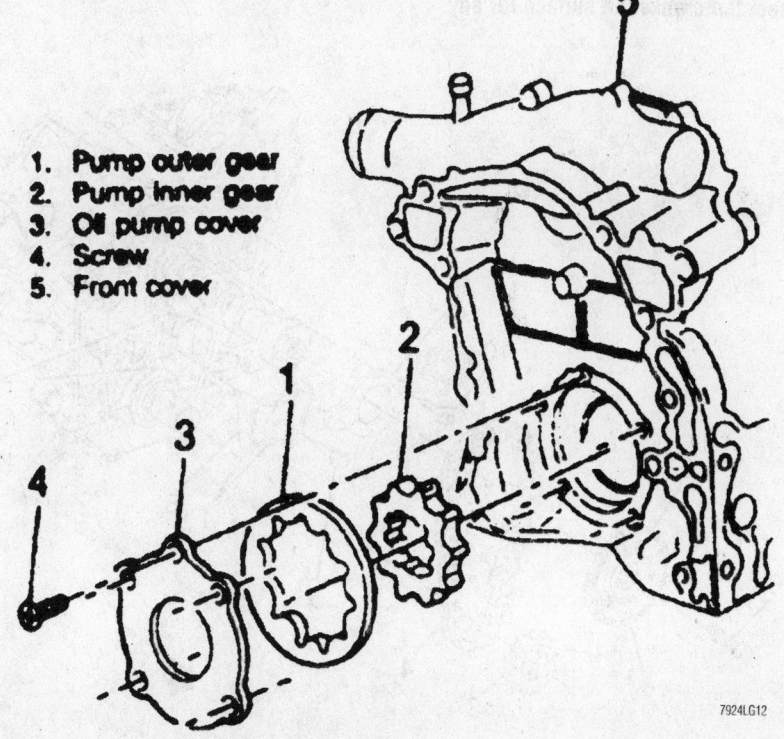

1. Pump outer gear
2. Pump inner gear
3. Oil pump cover
4. Screw
5. Front cover

7924LG12

Exploded view of the oil pump gears and housing—3.8L engine

2. Remove the transaxle assembly using the recommended procedure.

3. Remove the flywheel.

4. Examine the oil seal. The replacement seal should be installed to the same depth as the original. Checking this now saves time later.

5. Remove the old seal. Insert a suitable prytool through the dust lip at an angle.

6. Pry out the seal by moving handle of the tool towards the end of the crankshaft. Repeat this procedure until the seal is removed.

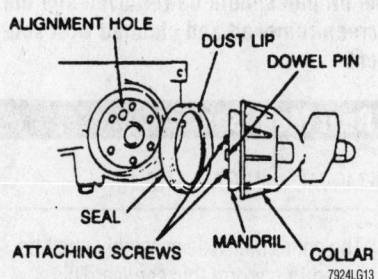

Rear main oil seal installation

➡ **When removing the seal, use care so that no damage occurs to the crankshaft. Once the seal is removed, inspect the crankshaft surface for any** nicks or burrs. **Repair or replace crankshaft as necessary.**

To install:

7. Clean all parts well. Install the new seal using tool J-34686 or equivalent for the 3.1L and 3.4L engines and J-38196 or equivalent for the 3.8L engine.

8. Apply a light coat of oil to inside diameter of the new seal and install over the tool's mandrel. Slide the seal onto the mandrel until the dust lip (back of the seal) bottoms squarely against collar of the tool.

9. Align the dowel pin of the tool (or its equivalent, if available) with the dowel pin hole in crankshaft and attach the tool to crankshaft. The factory recommended tool has a center screw which, when turned, forces the seal into place. Turn the tool by hand or tighten the center screw to 53 inch lbs. (6 Nm). On 1996–99 vehicles, with 3.4L engines, tighten to 44 inch lbs. (5 Nm). Turn the T handle of the tool so that collar pushes seal into the bore, turn handle until the collar is tight against the case. This will insure that the seal is seated properly.

➡ **Some aftermarket seals and engine overhaul gasket kits which include the rear crankshaft oil seal may come with a similar but more simple plastic installation tool. Whatever tool is used, the goal is to seat the seal squarely in its bore without damage to the seal or the crankshaft.**

10. Check the seal to verify the seal is seated squarely in the bore.

11. Install the flywheel, tighten the bolts to 52 ft. lbs. (71 Nm). On 1996–99 vehicles equipped with the 3.4L engine, tighten to 61 ft. lbs. (83 Nm).

12. Install the transaxle assembly using the recommended procedure.

13. Lower the vehicle.

14. Connect the negative battery cable.

15. Check the engine oil level.

16. Start the vehicle and verify no leaks.

➡ **Whenever the vehicle sub-frame is removed or lowered, the wheel alignment should be checked.**

Timing Chain and Sprockets

REMOVAL & INSTALLATION

3.1L Engine

1. Disconnect the negative battery cable. Drain the engine cooling system.

2. Remove the right front tire and wheel assembly.

3. Remove the front cover assembly. Ensure the marks on the crankshaft and camshaft gears are aligned using the marks on the damper stamping or cast alignment marks on the cylinder head and case.

4. Remove the bolts that hold the camshaft sprocket to the camshaft. This sprocket is a light press fit on the camshaft.

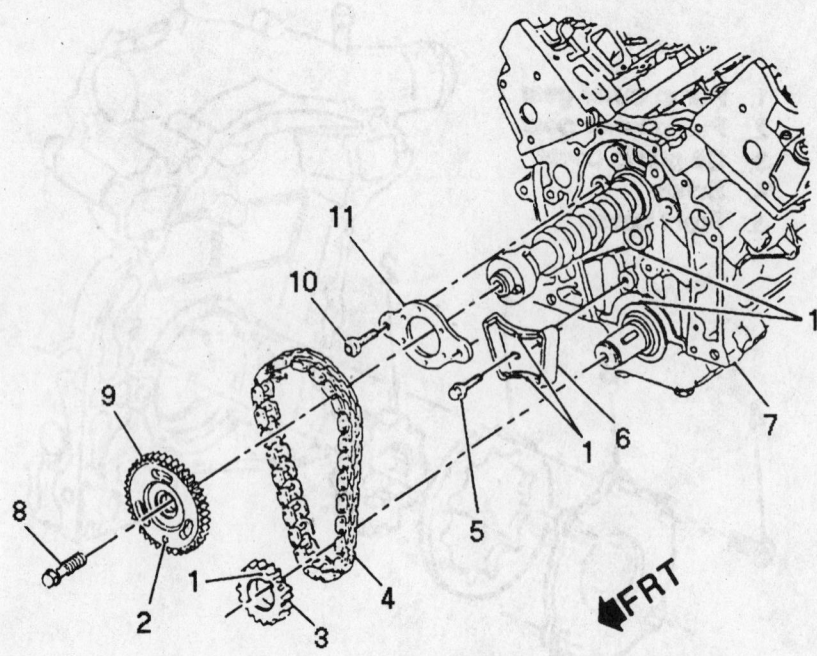

1. Timing alignment marks
2. Locator hole
3. Crankshaft sprocket
4. Timing chain
5. Timing chain dampener bolt
6. Timing chain dampener
7. Engine block
8. Camshaft sprocket bolt
9. Camshaft sprocket
10. Thrust plate bolt
11. Thrust plate

Exploded view of the timing chain assembly—3.4L engine shown, 3.1L engine is similar

5. Remove the timing chain. Using a suitable puller, remove the crankshaft sprocket, as required.

To install:

6. Install the crankshaft sprocket.

7. Lubricate the camshaft thrust plate surface with Molykote® or equivalent. Install the chain onto camshaft sprocket.

8. Holding the sprocket vertically with the chain hanging down, align the marks on the camshaft and crankshaft sprockets and install the assembly onto the camshaft.

9. Install the camshaft to gear attaching bolts and tighten to 18 ft. lbs. (24 Nm). After the sprockets are in place, turn the engine two full revolutions to make certain the timing marks are in correct alignment between the shaft centers.

10. Lubricate the chain with engine oil and install the front cover.

11. Connect the negative battery cable, then properly refill the engine cooling system.

3.4L Engine

The camshaft drive uses a conventional timing chain and sprockets. The front cover (timing chain cover) houses the front crankshaft oil seal and also mounts the water pump.

1. Disconnect the negative battery cable.

2. Raise and safely support the vehicle.

3. Drain the engine oil into a suitable container.

4. Remove the oil pan.

5. Lower the vehicle.

6. Drain the coolant system into a suitable container.

7. Remove the serpentine drive belt.

8. Remove the alternator and brackets.

9. Remove the power steering pump.

10. Remove the serpentine belt tensioner.

11. Remove the coolant bypass adapter.

12. Remove the coolant hose from the water pump.

13. Remove the water pump pulley.

14. Remove the crankshaft balancer using the following steps:

   a. Raise and safely support the vehicle.

   b. Remove the right front tire and wheel.

   c. Remove the inner fender splash shield.

   d. Remove the balancer center bolt. An assistant may be required to keep the flywheel from turning.

   e. The inertia weight section of the crankshaft balancer is assembled to the hub with a rubber sleeve. Use a puller to draw the balancer from the front of the crankshaft. If improperly removed, the inertia weight section of the balancer may shift, destroying the tuning of the balancer.

15. Remove the crankshaft position sensor.

16. Remove the front cover bolts. There are different size bolts so use care to note the location of each bolt.

17. Remove the front cover as outlined to this section.

18. Rotate the crankshaft until the timing marks on the crankshaft sprocket and camshaft sprocket locator hole are aligned to the marks on the timing chain dampener. This is the No. 1 piston at Top Dead Center (TDC) of the compression stroke.

19. Remove the camshaft sprocket bolt.

20. Remove the camshaft sprocket and timing chain.

21. Remove the crankshaft sprocket using J-5825-A or equivalent gear puller.

To install:

22. Clean all parts well. The sealing surfaces of the engine block and the front cover must be clean of old sealer and oil. Use a suitable degreaser solvent. The front seal can be replaced. Pry out the old seal. Drive in the replacement seal until it is flush with the front cover seal bore.

23. Coat all parts with a suitable lubricant. GM recommends their Engine Oil Supplement 1052367. Apply to all parts and especially the thrust face surface of the camshaft sprocket.

24. Install the crankshaft sprocket using J-38612 or equivalent. This tool threads into the front of the crankshaft to draw the sprocket onto the crankshaft. Use care if using substitutes. Install the sprocket until it is fully seated on the flange of the crankshaft nose.

25. Install the camshaft sprocket and timing chain.

26. Install the chain damper to the block and tighten to 15 ft. lbs. (21 Nm).

27. Align the crankshaft mark to the timing mark on the bottom of the chain damper.

28. Hold camshaft sprocket with the chain hanging down and drape the chain over the crankshaft sprocket.

29. Align the timing mark on the camshaft gear (center line of locator hole) with the timing mark on top of the chain damper.

30. Install the camshaft sprocket bolt and tighten 103 ft. lbs. (140 Nm).

31. Lubricate the timing chain with engine oil.

32. Clean all gasket sealing surfaces.

33. Apply sealer GM 1052080 or equivalent sealer to both sides of the lower edges of the front cover gasket where the gasket contacts the oil pan gasket.

34. Install new front cover gasket.

35. Install the front cover and bolts.

36. Tighten the large bolts to 35 ft. lbs. (47 Nm) and tighten the small bolts to 15 ft. lbs. (21 Nm).

37. Install the crankshaft position sensor.

38. Install the crankshaft balancer and tighten the bolt to 76 ft. lbs. (103 Nm).

39. Install the inner fender splash shield and the right front tire and wheel.

40. Install the water pump pulley.

41. Install the coolant hose to the water pump.

42. Install the coolant bypass adapter.

43. Install the belt tensioner.

44. Install the power steering pump.

45. Install the alternator brackets.

46. Install the alternator.

47. Install the serpentine drive belt.

48. Refill the cooling system.

49. Install the oil pan.

50. Refill the crankcase with new engine oil. A filter change is recommended.

51. Connect the negative battery cable.

52. Start the engine and verify no leaks.

53. Road test the vehicle.

3.8L Engine

1. Disconnect the negative battery cable, then drain the engine cooling system.

2. Align the marks on the crankshaft damper and timing cover with the engine positioned so that the No. 1 cylinder is at TDC/compression stroke. Remove the front cover assembly.

3. Ensure the marks on the crankshaft and camshaft gears are aligned.

4. Remove the timing chain damper and camshaft sprocket.

5. Remove the timing chain. Using a suitable puller, remove the crankshaft sprocket, as required.

To install:

6. Install the crankshaft sprocket.

7. Install the chain onto camshaft sprocket.

8. Holding the sprocket vertically with the chain hanging down, align the marks on the camshaft and crankshaft sprockets and install the assembly onto the camshaft.

9. Install the camshaft to gear attaching bolt and tighten to 74 ft. lbs. (100 Nm) and then an additional 105 degree turn. Install the damper and tighten to 16 ft. lbs. (22 Nm).

10. After the sprockets are in place, turn

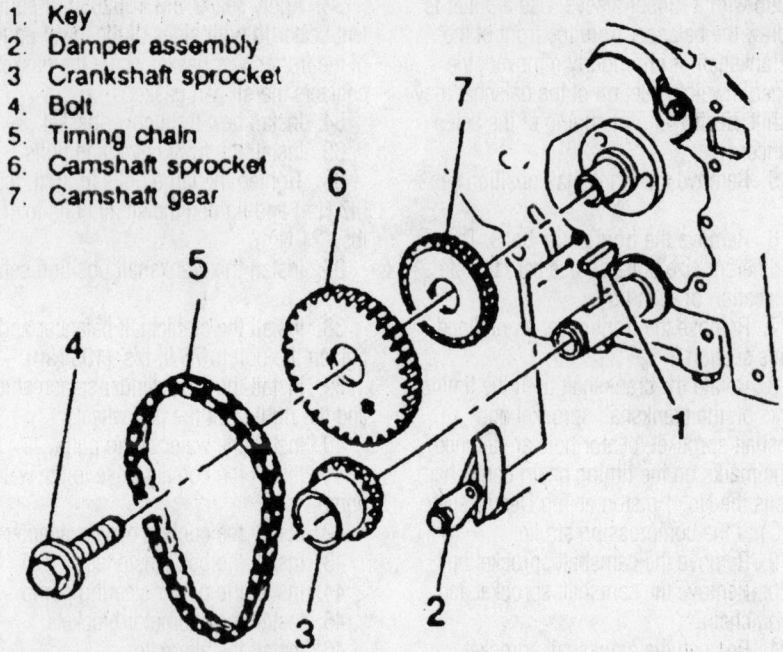

1. Key
2. Damper assembly
3. Crankshaft sprocket
4. Bolt
5. Timing chain
6. Camshaft sprocket
7. Camshaft gear

Exploded view of the timing chain assembly—3.8L engine

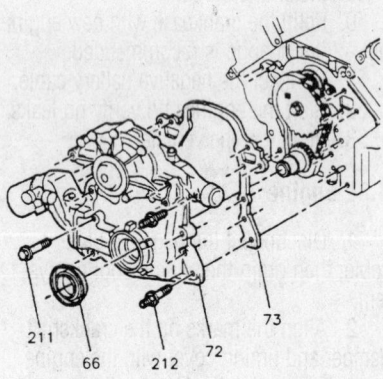

66 OIL SEAL
72 COVER
73 GASKET
211 BOLT
212 STUD

Front cover assemby removal—3.8L engine

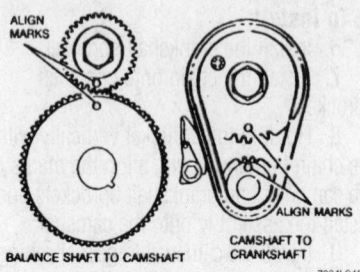

Be sure that the timing marks are properly aligned (facing each other) after the chain is installed—3.8L engine

the engine two full revolutions to make certain the timing marks are in correct alignment between the shaft centers.

11. Lubricate the chain with engine oil and install the front cover.

12. Connect the negative battery cable, then properly refill the engine cooling system.

FUEL SYSTEM

Fuel System Service Precaution

Safety is the most important factor when performing not only fuel system maintenance but any type of maintenance. Failure to conduct maintenance and repairs in a safe manner may result in serious personal injury or death. Maintenance and testing of the vehicle's fuel system components can be accomplished safely and effectively by adhering to the following rules and guidelines.

• To avoid the possibility of fire and personal injury, always disconnect the negative battery cable unless the repair or test procedure requires that battery voltage be applied.

• Always relieve the fuel system pressure prior to disconnecting any fuel system component (injector, fuel rail, pressure regulator, etc.), fitting or fuel line connection. Exercise extreme caution whenever relieving fuel system pressure to avoid exposing skin, face and eyes to fuel spray. Please be advised that fuel

under pressure may penetrate the skin or any part of the body that it contacts.

• Always place a shop towel or cloth around the fitting or connection prior to loosening to absorb any excess fuel due to spillage. Ensure that all fuel spillage (should it occur) is quickly removed from engine surfaces. Ensure that all fuel soaked cloths or towels are deposited into a suitable waste container.

• Always keep a dry chemical (Class B) fire extinguisher near the work area.

• Do not allow fuel spray or fuel vapors to come into contact with a spark or open flame.

• Always use a back-up wrench when loosening and tightening fuel line connection fittings. This will prevent unnecessary stress and torsion to fuel line piping. Always follow the proper torque specifications.

• Always replace worn fuel fitting O-rings with new ones. Do not substitute fuel hose or equivalent, where fuel pipe is installed.

Fuel System Pressure

RELIEVING

Before loosening or disconnecting any fuel fitting or system component, always relieve the fuel system pressure in order to help prevent the danger of fire or injury.

3.1L Engine

Unlike most TBI engines, the TBI system used in the 3.1L engine utilizes an automatic pressure bleed down feature. But, some fuel pressure related steps should still be taken to assure safer working conditions.

1. Disconnect the negative battery cable to prevent fuel spillage should the ignition key accidentally be turned **ON** with a fuel fitting disengaged.

2. Loosen fuel filler cap to relieve fuel tank pressure.

3. The internal constant bleed feature of the Model 220 TBI unit relieves fuel pump system pressure when the engine is turned **OFF**. Therefore, no further action is required.

➡**Allow the engine to remain OFF for 5–10 minutes; this will allow the orifice (in the fuel system) to bleed off the pressure.**

4. When fuel service is finished, tighten the fuel filler cap and connect the negative battery cable.

3.4L and 3.8L Engines

The MFI fuel systems used on GM vehicles all operate under high fuel pressures. It is very important that the pressure be properly relieved prior to servicing the system or any of its components.

A Schrader valve is provided on these fuel systems in order to conveniently test or release the system pressure. A fuel pressure gauge and adapter will be necessary to connect the gauge to the fitting. Most of the MFI systems utilize a service valve on one end of the fuel rail assembly.

1. Disengage the negative battery cable to assure the prevention of fuel spillage if the ignition switch is accidentally turned **ON** while a fitting is still disconnected.
2. Loosen the fuel filler cap to release the fuel tank pressure.
3. Be sure the release valve on the fuel gauge is closed, then connect the fuel gauge to the pressure fitting located on the inlet fuel pipe fitting.

➡**When connecting the gauge to the fitting, be sure to wrap a rag around the fitting to avoid spillage. After repairs, place the rag in an approved container.**

4. Install the bleed hose portion of the fuel gauge assembly into an approved container, then open the gauge release valve and bleed the fuel pressure from the system.
5. When the gauge is removed, be sure to open the bleed valve and drain all fuel from the gauge assembly.
6. When fuel service is finished, tighten the fuel filler cap and connect the negative battery cable.

Fuel Filter

REMOVAL & INSTALLATION

1. Disconnect the negative battery cable.
2. Relieve fuel system pressure.
3. Remove the fuel line connections from the filter. If equipped with quick connect fittings, use tool J-37088A or equivalent to separate the lines.

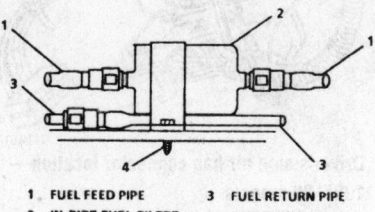

1	FUEL FEED PIPE	3	FUEL RETURN PIPE
2	IN-PIPE FUEL FILTER	4	SCREWS (2)

7924LG16

Fuel filter component identification—the fuel filter is located on the frame rail near the tank

➡**Before disengaging quick-connect fittings, always twist each side of the connection ¼ turn (in opposite directions) in order to loosen any dirt, then use compressed air (while wearing safety glasses) to blow the dirt free of the fittings.**

4. Remove the filter mounting clamp bolt and remove the filter.
5. Installation is the reverse of the removal procedure. Before installing any quick-connect fittings, apply a few drops of clean engine oil to the male connector to assure proper seal and prevent a possible leak.

Electric Fuel Pump

REMOVAL & INSTALLATION

1. Properly relieve the fuel system pressure.
2. Disconnect the negative battery cable.
3. Drain and remove the fuel tank from the vehicle
4. Clean dirt off area around fuel pump sender to prevent dirt from falling into tank when pump is removed.
5. Using a suitable spanner wrench, turn the fuel pump/sending unit assembly locking ring (located on top of the fuel tank) counterclockwise, then carefully lift the assembly from the tank and remove the pump from the fuel lever sending device.

To install:

6. The fuel pump/sending unit assembly O-ring should be replaced whenever the tank is removed.
7. Install fuel pump and new O-ring into fuel tank and secure with locking ring.
8. Install fuel tank to vehicle and refill.
9. Connect the negative battery cable.
10. Start engine and check for fuel leaks.

DRIVE TRAIN

Transaxle Assembly

REMOVAL & INSTALLATION

The automatic transaxle can be removed only as an assembly with the engine and subframe. See the procedures under Engine Removal and Installation.

Halfshaft

REMOVAL & INSTALLATION

1. Raise and safely support the vehicle.
2. Remove the tire and wheel assemblies.

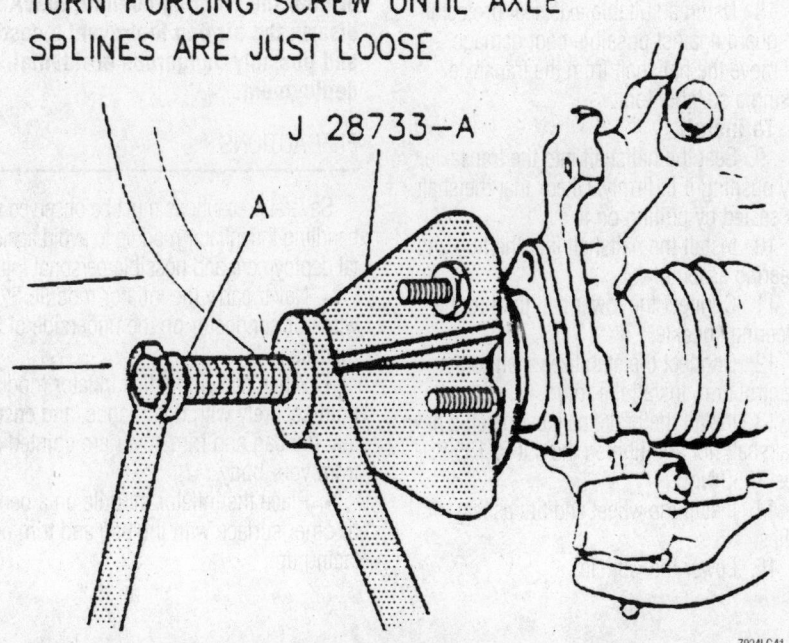

A TURN FORCING SCREW UNTIL AXLE SPLINES ARE JUST LOOSE

J 28733-A

7924LG41

Use the special tools if needed to push the halfshaft throught the hub assembly

1 SHAFT ASSEMBLY, FRONT WHEEL DRIVE
2 KNUCKLE AND HUB ASSEMBLY, STEERING
3 NUT, FRONT WHEEL DRIVE SHAFT
4 SHAFT ASSEMBLY, FRONT WHEEL DRIVE OUTER

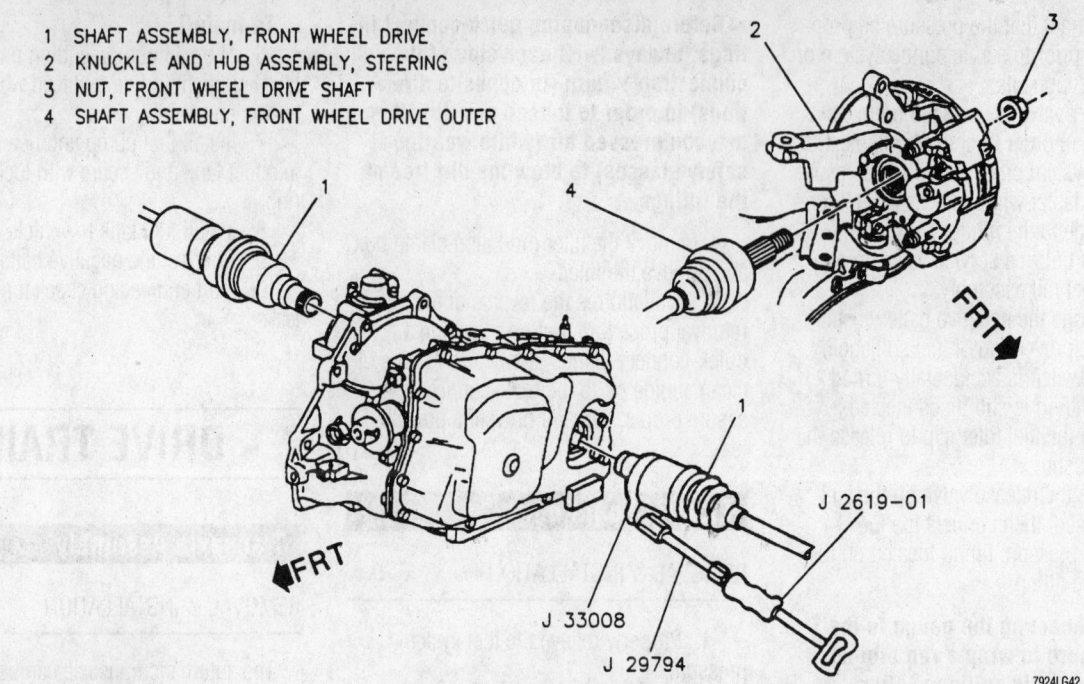

Halfshaft removal or installation

3. Remove the halfshaft retaining nut and washer.

4. Remove the brake caliper from the rotor and support it aside.

5. Remove the brake rotor from the hub.

6. Disengage the stabilizer shaft from the control arm and disconnect the ball joint from the steering knuckle.

7. Remove the halfshaft from the hub and bearing assembly.

8. Using a suitable axle seal protector to guard against possible boot damage. Remove the halfshaft from the transaxle using a suitable tool.

To install:

9. Seat the halfshaft into the transaxle, by pushing it in firmly. Check that the shaft is seated by pulling on it.

10. Install the halfshaft into the hub and bearing assembly.

11. Connect the lower ball joint to the steering knuckle.

12. Connect the stabilizer shaft to the control arm. Install the rotor.

13. Install the brake caliper. Install a new halfshaft nut and tighten the nut to 185 ft. lbs. (250 Nm).

14. Install the wheel and tire assemblies.

15. Lower the vehicle.

STEERING AND SUSPENSION

Air Bag

➡All models are equipped with a Supplemental Inflatable Restraint (SIR) system. Before attempting any work on or near the steering column, ALWAYS disarm the air bag to prevent a costly and possibly dangerous accidental deployment.

PRECAUTIONS

Several precautions must be observed when handling the inflator module to avoid accidental deployment and possible personal injury.

• Never carry the inflator module by the wires or connector on the underside of the module.

• When carrying a live inflator module, hold securely with both hands, and ensure that the bag and trim cover are pointed away from your body.

• Place the inflator module on a bench or other surface with the bag and trim cover facing up.

• With the inflator module on the bench, never place anything on or close to the module which may be thrown in the event of an accidental deployment.

DISARMING

1. Turn the wheels to the straight-ahead position, then turn the ignition switch to **LOCK.**

2. Remove the instrument panel lower extension for access to the fuse block.

3. Remove the "AIR BAG" or "SIR" fuse from the block, as applicable.

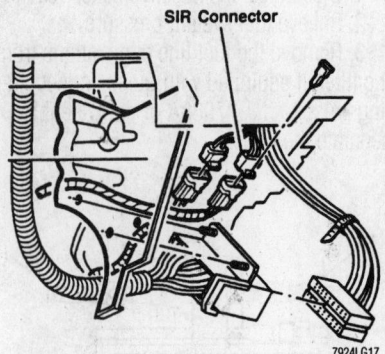

Driver's side air bag connector location—1997–99 models

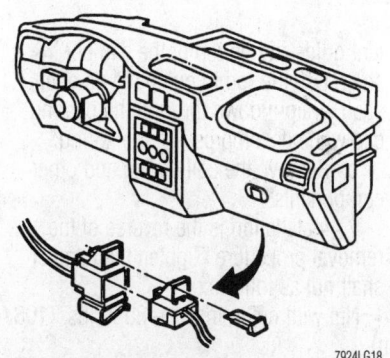

Passenger's side air bag connector location—1997–99 models

7924LG18

4. Remove the steering column filler panel or left-hand sound insulator, as applicable, for access to the SIR wiring harness.

5. Remove the Connector Position Assurance (CPA) device, then disengage the yellow 2-way connector at the base of the steering column.

➡**With the fuse removed, the AIR BAG or SIR light will illuminate if the ignition switch is turned ON at any time. This is normal and does not indicate a problem when the system is disarmed.**

To enable:

6. Be sure the ignition is in the **LOCK** position.

7. Engage the yellow SIR connector, then secure using the CPA device.

8. Install the steering column filler or sound insulator panel, as applicable.

9. Install the SIR system fuse to the fuse block.

10. Turn the ignition switch to the **ON** position and verify that the AIR BAG indicator light flashes seven times, then extinguishes. If it does not go out, troubleshoot the SIR system fault.

11. Install the instrument panel lower extension.

Power Rack and Pinion Steering Gear

REMOVAL & INSTALLATION

1. Disconnect the negative battery cable.

2. Remove the air cleaner assembly.

3. Remove the dust boot from the steering gear.

4. Remove the intermediate shaft lower pinch bolt and disconnect the intermediate shaft from the lower stub shaft.

5. Remove the fluid line retaining clips at the pump and disconnect the lines.

6. Raise and safely support the vehicle.

7. Remove the wheel and tire assemblies. Disconnect the tie rod ends at the steering knuckle.

8. Remove the remaining brackets and clips at the crossmember. Support the body

safely with the appropriate equipment, to allow lowering of the subframe.

9. Support the subframe

10. Remove the rear subframe mounting bolts and carefully lower the rear of the subframe approximately 5 in. (128mm).

11. If equipped, remove the heat shield.

12. Remove the rack and pinion mounting bolts and remove the rack through the left wheel opening.

To install:

13. Install the rack and pinion through the left wheel opening.

14. Install the rack and pinion mounting nuts, tighten to 59 ft. lbs. (80 Nm).

15. If equipped, install the heat shield.

16. Raise the subframe assembly and install the rear mounting bolts.

17. Remove any supports and install the brackets and clips to the crossmember.

18. Install the wheel and tire assemblies. Lower the vehicle.

19. Connect the fluid lines at the pump and tighten to 20 ft. lbs. (27 Nm).

20. Install the line retaining clips. Connect the intermediate shaft to the stub shaft.

21. Install the dust boot over the steering gear.

22. Install the air cleaner assembly and connect the negative battery cable.

23. Fill and bleed the steering system.

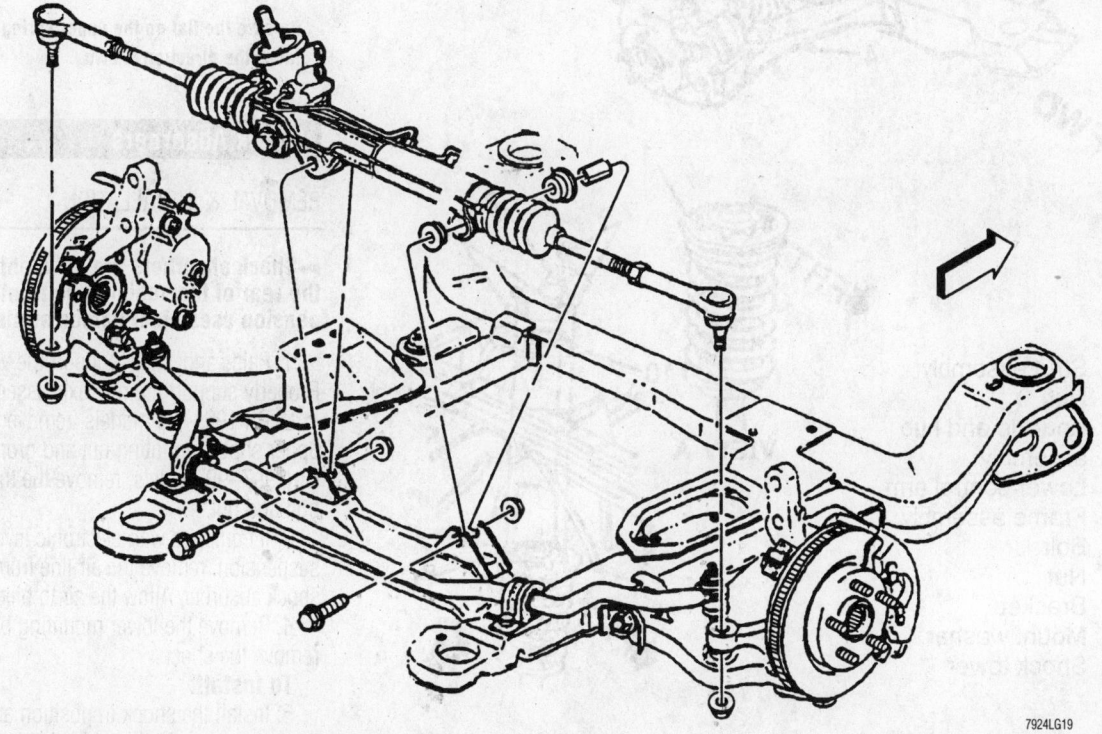

The rack and pinion steering gear is bolted to the rear of the subframe, as shown

7924LG19

Strut

REMOVAL & INSTALLATION

❋❋ CAUTION

Do not remove the top center nut from the strut assembly. This nut should only be removed when the strut assembly is out of the vehicle, mounted in a holding fixture and the coil spring is in a compressed position using the proper strut coil spring compressor.

1. Remove the three nuts that retain the top of the strut assembly to the body.
2. Raise and safely support the vehicle.
3. Remove the wheel and tire assembly.

Remove the brake line bracket from the strut mount.

4. Remove the lower strut mounting bolts after marking the installed position of the strut on the steering knuckle. If the strut is not installed to the steering knuckle in the original position, the camber angle of the front wheel will change and wheel alignment will be affected.

5. Remove the strut assembly from the vehicle and place the strut in a suitable holding fixture.

6. Disassemble the strut as follows:
   a. With the strut coil spring in a compressed position approximately ½ its normal length, remove the nut from the top of the strut.
   b. Place tool J-34013–27 or equivalent guide rod on top of the damper shaft. Use the rod to guide the damper shaft straight down through the bearing cap while decompressing the spring.
   c. Remove the coil spring and other components.

7. Installation is the reverse of the removal procedure. Tighten the damper shaft nut as follows:
   • Nut with nylon insert—80 ft. lbs. (108 Nm)
   • Nut without nylon insert—65 ft. lbs. (85 Nm)

➡**Ensure the spring seat flat faces 10 degrees forward of the centerline of the strut assembly spindle.**

8. Tighten the strut lower bolts to 140 ft. lbs. (190 Nm) and the upper mounting nuts to 18 ft. lbs. (25 Nm).
9. Install the strut assembly in the vehicle.

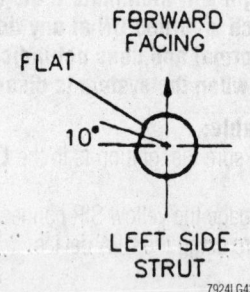

Be sure the flat on the upper spring seat faces the direction shown

Shock Absorbers

REMOVAL & INSTALLATION

➡**Shock absorbers are used only on the rear of the vehicle; the front suspension uses MacPherson struts.**

1. Raise and safely support the vehicle. Properly support the rear axle assembly.
2. On 1995–96 models, remove the upper shock mounting nut and grommet. On 1997–99 models, remove the through-bolt and nut.
3. If equipped with electronic level control suspension, remove the air line from the shock absorber. Allow the air to bleed off.
4. Remove the lower mounting bolt and remove the shock.

To install:

5. Install the shock in position and install the lower mounting bolt. On 1995–96 models, tighten the bolt to 50 ft. lbs. (68 Nm).

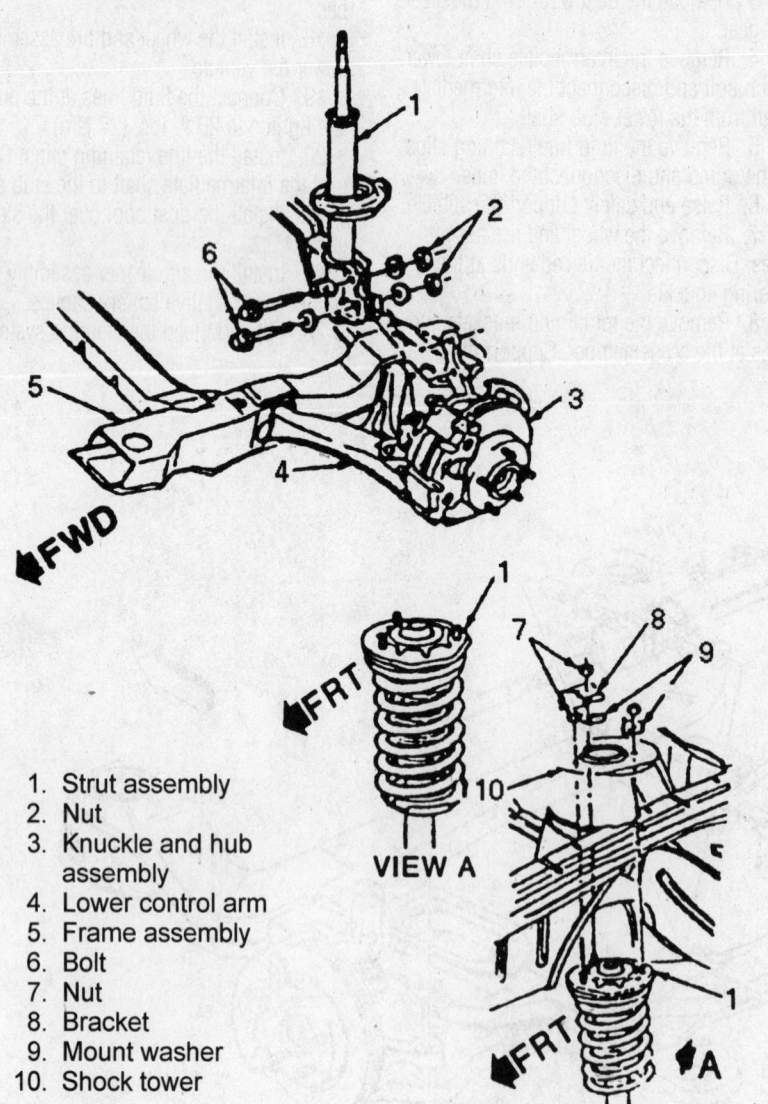

1. Strut assembly
2. Nut
3. Knuckle and hub assembly
4. Lower control arm
5. Frame assembly
6. Bolt
7. Nut
8. Bracket
9. Mount washer
10. Shock tower

MacPherson strut assembly/disassembly

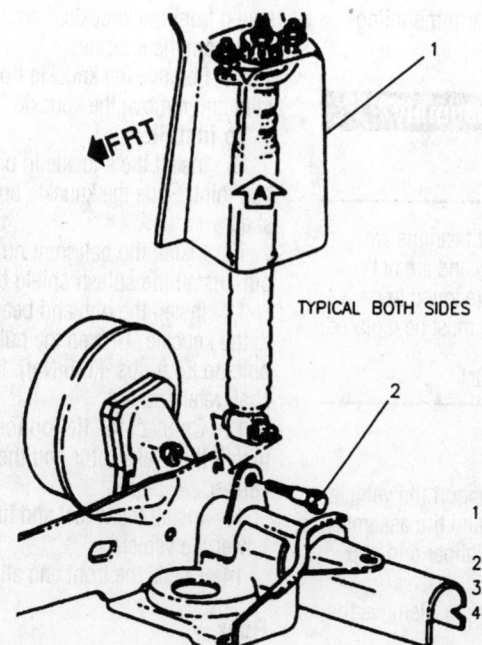

TYPICAL BOTH SIDES

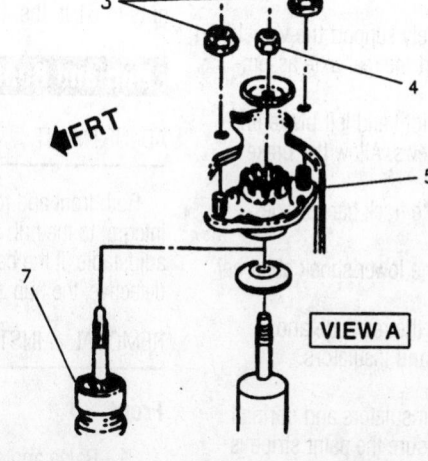

VIEW A

1	BODY SHOCK TOWER	5	UPPER SHOCK ABSORBER MOUNT
	MOUNTING BRACKET	6	AIR LIFT SHOCK ABSORBER
2	68 N•m (50 LB. FT.)	7	STANDARD SHOCK ABSORBER
3	26 N•m (19 LB. FT.)		
4	28 N•m (21 LB. FT.)		

7924LG24

Rear shock absorber mounting—1995–96 models

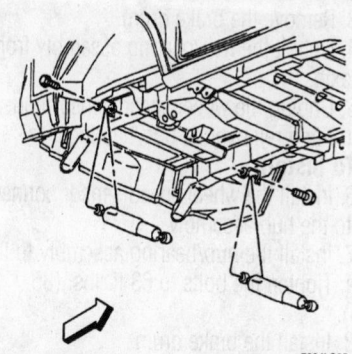

7924LG25

**Rear shock absorber upper mounting—
1997–99 models**

On 1997–99 models, tighten the bolt to 63 ft. lbs. (85 Nm).

6. If equipped, connect the air line to the shock.

7. Lower the vehicle and install the upper shock mounting. On 1995–96 models, tighten the nut to 21 ft. lbs. (28 Nm). On 1997–99 models, tighten the throughbolt to 63 ft. lbs. (85 Nm).

Coil Springs

REMOVAL & INSTALLATION

Front

The service procedure for the front coil springs is covered under MacPherson Strut removal and installation.

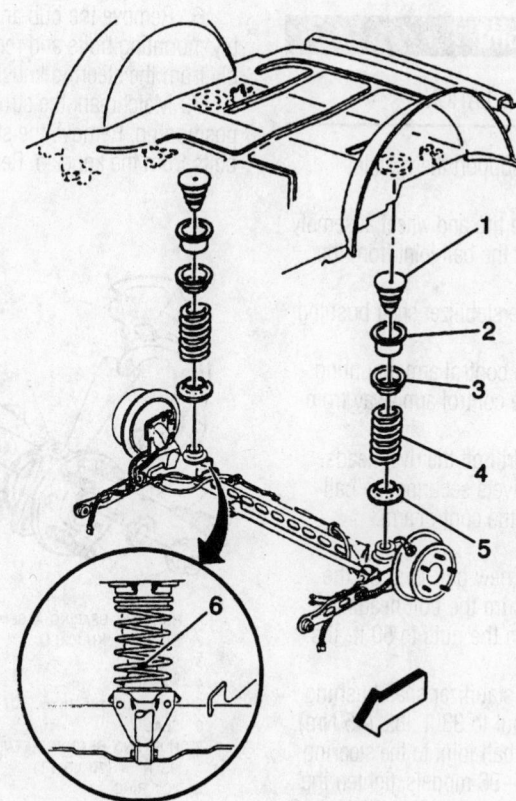

Legend

(1) Rear Suspension Jounce Bumper
(2) Rear Suspension Jounce Bumper Retainer
(3) Rear Suspension Insulator
(4) Rear Spring
(5) Rear Spring Insulator
(6) Paint Stripe

7924LG21

Exploded view of the coil spring assembly—1997–99 models, 1995–96 models similar

Rear

1. Raise and safely support the vehicle.
2. Safely support the rear axle assembly.
3. Remove the right and left brake line-to-axle attaching screws. Allow the brake lines to hang freely.
4. Disconnect the track bar-to-axle attaching bolt.
5. Disconnect the lower shock absorber mounting bolts.
6. Slowly lower the rear axle and remove the springs and insulators.

To install:

7. Position the insulators and springs on the rear axle. Be sure the paint stripe is facing rearward.
8. Raise the axle assembly and connect the tie rod.
9. Install both shock absorbers.
10. Install the brake line brackets.
11. Remove the axle supports and lower the vehicle to the floor.

Lower Ball Joint

REMOVAL & INSTALLATION

1. Raise and support the vehicle safely.
2. Remove the tire and wheel assembly.
3. Disconnect the ball joint from the steering knuckle.
4. Remove the stabilizer shaft bushing assembly nut.
5. Loosen the control arm mounting nuts and move the control arm away from the knuckle.
6. Carefully drill off the rivet heads, then remove the rivets securing the ball joint assembly to the control arm.

To install:

7. Install the new ball joint on the control arm. Be sure the bolt heads are facing up. Tighten the nuts to 50 ft. lbs. (68 Nm).
8. Install the stabilizer shaft bushing nut. Tighten the nut to 33 ft. lbs. (45 Nm).
9. Install the ball joint to the steering knuckle. On 1995–96 models, tighten the pinch bolt and nut to 56 ft. lbs. (76 Nm). On 1997–99 models, tighten the nut to 40 ft. lbs. (54 Nm).
10. Install the wheels and lower the vehicle to the floor.

11. Tighten the control arm mounting nuts to 61 ft. lbs. (83 Nm).

Hub and Bearing Assembly

ADJUSTMENT

Both front and rear wheel bearings are integral to the hub assembly and are not adjustable. If the bearings are found to be defective, the hub assemlby must be replaced.

REMOVAL & INSTALLATION

Front

1. Raise and safely support the vehicle.
2. Remove the wheel and tire assembly.
3. Remove the brake caliper and support it aside.
4. Remove the brake rotor. Remove the halfshaft retaining nut.
5. Disconnect the tie rod end from the knuckle.
6. Remove the hub and bearing assembly mounting bolts and remove the assembly from the steering knuckle.
7. Matchmark the strut-to-knuckle positioning. Remove the strut mounting bolts from the knuckle. Remove the splash

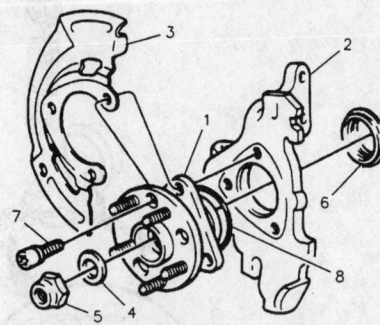

1	HUB AND BEARING ASSEMBLY
2	STEERING KNUCKLE
3	SHIELD
4	WASHER
5	HUB NUT 145 N·m (107 LB FT)
6	SEAL
7	HUB AND BEARING RETAINING BOLT (55 TORX) 95 N·m (70 LB FT)
8	"O" RING

7924LG22

Exploded view of the front hub and bearing assembly mounting

shield from the knuckle. Remove the ball joint from the knuckle.

8. Remove the knuckle from the ball joint and remove the knuckle.

To install:

9. Install the knuckle in position on the ball joint. Slide the knuckle onto the half-shaft.
10. Install the ball joint nut and cotter pin. Install the splash shield to the knuckle.
11. Install the hub and bearing assembly to the knuckle. Tighten the hub mounting bolts to 86 ft. lbs. (116 Nm). Install the half-shaft retaining nut.
12. Connect the tie rod to the knuckle. Install the brake rotor and the brake caliper.
13. Install the wheel and tire assembly. Lower the vehicle.
14. Check the front end alignment.

Rear

1. Raise and safely support the vehicle securely.
2. Remove the wheel.
3. Remove the brake drum.
4. Unbolt the hub/bearing assembly from the axle.
5. Unplug the wheel speed sensor connector and remove the hub/bearing.

To install:

6. Install the wheel speed sensor connector to the hub/assembly.
7. Install the hub/bearing assembly to the axle. Tighten the bolts to 63 ft. lbs. (85 Nm).
8. Install the brake drum.
9. Install the wheel and lower the vehicle to the floor.

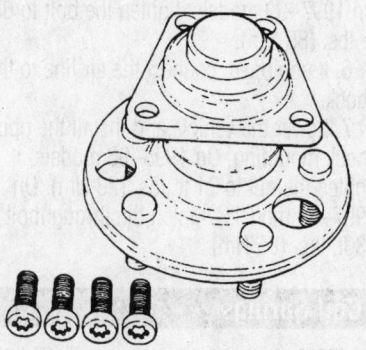

7924LG23

The rear wheel hub is mounted with four Torx® head bolts

HONDA and ISUZU

Honda-CR-V • Odyssey • Isuzu-Oasis

22

ENGINE REPAIR

➡Disconnecting the negative battery cable on some vehicles may interfere with the functions of the on-board computer systems and may require the computer to undergo a relearning process, once the negative battery cable is reconnected.

Distributor

REMOVAL & INSTALLATION

➡The radio may contain a coded theft protection circuit. Always make note of your code before disconnecting the battery.

1. Disconnect the negative battery cable.

2. Unclamp the cruise control cable from the valve cover and carefully move it out of the way.

3. The intake air duct may be removed for more access to the distributor.

4. Set the No. 1 cylinder at Top Dead Center (TDC) for the compression stroke. Once the engine is in this position, it must not be disturbed.

5. Label and disconnect the ignition wires.

6. Uncouple the connectors and remove them from their clips on the side of the distributor.

7. Remove the three distributor mounting bolts.

8. Remove the distributor from the cylinder head.

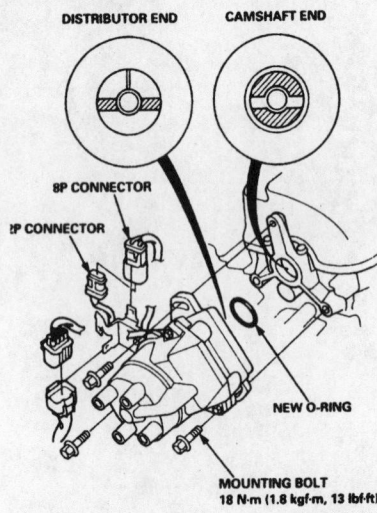

Exploded view of the distributor mounting—except CR-V models

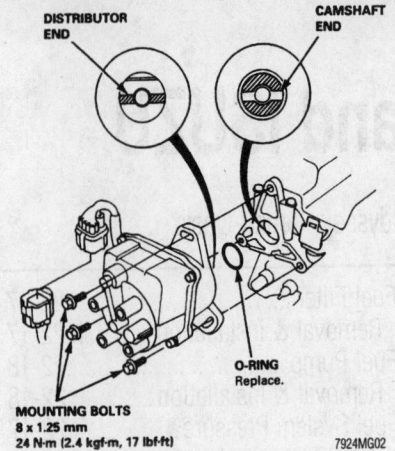

Exploded view of the distributor mounting—CR-V model

To install:

➡If the camshaft or crankshaft has rotated during assembly, rotate the crankshaft to bring the engine to TDC/compression for the No. 1 cylinder. Be sure the UP mark on the camshaft is facing up, and that the crankshaft TDC mark aligns with the pointer on the lower timing cover.

9. Coat a new O-ring with clean engine oil and install it onto the distributor shaft.

10. Install the distributor into the cylinder head. The lugs on the distributor shaft fit into the groove on the end of the camshaft.

11. Install the three mounting bolts, only hand-tighten them at this time.

12. Couple the connectors and install them onto their clips.

13. Connect the ignition wires.

14. Reconnect the negative battery cable.

15. Check and adjust the ignition timing. To adjust the ignition timing: first, loosen the distributor mounting bolts. Then, turn the distributor housing clockwise to retard the timing, or counterclockwise to advance the timing.

16. Tighten the mounting bolts to 13 ft. lbs. (18 Nm).

17. Install the cruise control cable back into its clamp on the valve cover.

18. Install the intake air duct if it was removed.

19. Enter the radio security code.

Ignition Timing

ADJUSTMENT

1. Start the engine and run it at 3000 rpm with the transaxle in the **N** or **P** posi-

tion. When the radiator fan comes on (normal operating temperature), let the engine run at idle.

2. Remove the connector holder underneath the left edge of the glove box. Remove the 2-P service check connector from the holder.

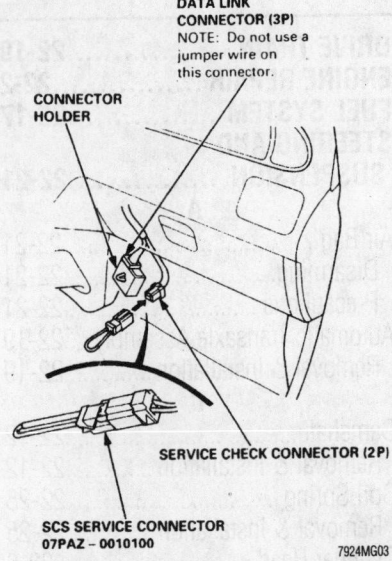

Service check connector and special connector locations

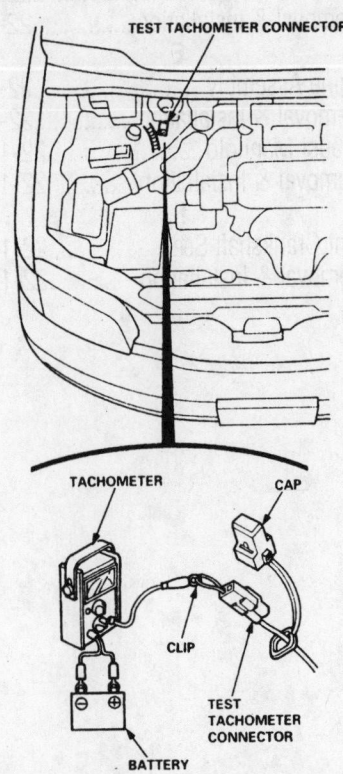

Test tachometer connector location—attach a tachometer to the test connector as shown

3. Attach a SCS service connector, 07PAZ-0010100 or equivalent, to the service check connector.

4. Check the idle speed:

a. Connect a test tachometer to the test tachometer lead located on the right side of the firewall near the shock tower.

b. The idle speed should be 650–750 rpm with the transaxle in **N** or **P**, and all electrical accessories OFF.

5. Connect a timing light to the No. 1 ignition wire.

6. The timing should be 13°–17° BTDC (red timing mark on crankshaft pulley) at 650–750 rpm.

7. If the timing is out of range, adjust it. First, loosen the distributor mounting bolts. Next, turn the distributor housing counter-clockwise to advance the timing, or clockwise to retard the timing. Then, tighten the distributor mounting bolts to 13 ft. lbs. (18 Nm).

8. Recheck the ignition timing.

9. Remove the service connector from the check connector. Install the check connector back into its holder. Tuck the connector holder back under the glove box.

Engine Assembly

REMOVAL & INSTALLATION

Except CR-V

➡The radio may contain a coded theft protection circuit. Always make note of your code number before disconnecting the battery.

The engine and transaxle are removed from the vehicle as a unit. A hydraulic lift is helpful for this procedure since the front subframe must be removed for the engine/transaxle assembly to be lowered from the vehicle.

1. Disconnect negative and positive battery cables.

2. Raise the hood and mark the positions of the hinge plates with a felt-tipped marker. Remove the hood and move it out of the work area to avoid scratching the paint.

3. Remove the battery. Disconnect the ground cable from the battery tray. Unbolt and remove the battery tray and its support bracket.

4. Loosen the throttle cable and cruise control cable locknuts. Remove both cables from their brackets and slip them out of their linkages.

5. Remove the intake air duct.

6. Drain the engine oil, transaxle fluid, and engine coolant into separate, sealable containers.

7. Disconnect the battery cable from the fuse/relay box and ABS relay box.

8. Uncouple the three engine harness connectors located on the right side of the engine compartment.

9. Disconnect the brake booster vacuum hose and EVAP canister hose from the intake manifold plenum.

10. Relieve the fuel pressure by loosening the fuel rail service bolt one turn.

❊❊ CAUTION

The fuel injection system remains under pressure after the engine has been turned OFF. Properly relieve fuel pressure before disconnecting any fuel lines. Failure to do so may result in fire or personal injury.

11. Disconnect the fuel line and the return hose from the fuel rail. Clean up any spilled fuel.

12. Disconnect the three vacuum hoses from the left side of the intake manifold. Unclamp the power steering hose.

13. Uncouple the three engine harness connectors located on the left side of the engine compartment. Unplug the fuel injector resistor connector.

14. Loosen the power steering pump adjusting bolt and mount bolts. Slip the belt off its pulleys.

15. Unbolt and remove the power steering pump, but do not disconnect its hydraulic lines. Wire the pump away from the engine.

16. Disconnect the alternator. Loosen and remove the alternator belt.

17. Unbolt and remove the alternator and its mounting bracket.

18. Disconnect and remove the upper and lower radiator hoses.

19. Remove the radiator with both cooling fans attached to it.

20. Disconnect the heater hoses from the coolant pipe under the intake manifold.

21. Disconnect the two cooler lines from the front of the transaxle case. Plug the inlet lines on the transaxle case to prevent moisture contamination.

22. Unbolt the A/C compressor from the side of the engine block. Don't disconnect the A/C lines. Wire the compressor to the radiator support so it's out of the work area.

23. Attach a chain hoist to the lifting hooks on either side of the engine block.

24. Raise and safely support the vehicle. Take up the slack of the chain hoist.

25. Disconnect the heated oxygen sensor wiring harness. Leave the sensor installed in the exhaust pipe.

26. Separate the exhaust pipe from the catalytic converter and the exhaust manifold and remove it.

27. Matchmark the subframe center beam to the rear beam. Remove the two bolts, but don't remove the center beam at this point.

28. Remove the shift cable cover. Disconnect the shift cable from the control shaft and suspend it out of the way with wire.

29. Remove the splash shield.

30. Remove the front wheels.

31. Remove the damper fork flange bolts from the lower control arms. Unbolt the damper fork from the strut and remove it from the vehicle.

32. Use a ball joint removal tool to separate the lower control arms from the steering knuckles.

33. Use a prytool to detach the left and right halfshafts from the intermediate shaft and transaxle case. Tap the splined shafts out of the hubs using a plastic mallet. Move the halfshafts out of the way and wire them up. Tie plastic bags over the halfshaft ends to protect the boots and splined shafts.

34. Unbolt and remove the front engine mount bracket.

❊❊ WARNING

The next step involves the removal of the subframe front beam. Be sure the vehicle is securely supported. Take up any slack in the chain hoist to support the weight of the engine and transaxle assembly.

35. Support the subframe front beam with a transmission jack and a sturdy wood plank.

36. Unbolt the front beam with the center beam, front engine mount, radius rods, and lower control arms attached to it. Lower the front beam assembly from the vehicle.

37. Remove the through-bolt to separate the rear mount from its bracket.

38. Unbolt and remove the side engine mount.

39. Unbolt and remove the transaxle side mount.

40. Verify that all hoses, wires, and vacuum lines have been disconnected from the engine.

41. Slowly lower the engine/transaxle assembly from the vehicle.

42. Move the engine out from under the vehicle and mount it securely on an engine stand.

43. Support the front of the vehicle with jackstands and block the rear wheels, as the front suspension has been disassembled. If the vehicle must be moved with the engine

out, install the front beam and all the suspension components.

To install:

➡ Use new self-locking nuts and color-coded self-locking bolts when installing the engine mounts, subframe components, and suspension components.

44. Raise and safely support the vehicle.

45. Move the engine/transaxle assembly into position under the vehicle and attach a chain hoist to the engine lifting hooks.

46. Carefully lift the engine/transaxle assembly into position in the vehicle.

47. Install the side engine mount. Use a 6mm punch or similarly-sized tool to steady the mount. Install the nut and bolts. Tighten the bolt to 47 ft. lbs. (64 Nm), and remove the 6mm punch. Only hand-tighten the nut and bolt on the engine side of the mount.

48. Install the transaxle side mount. Use a 6mm punch or similarly-sized tool to steady the mount. Install and only hand-tighten the nuts. Install the bolt and tighten it to 47 ft. lbs. (64 Nm).

49. Connect the rear mount bracket and tighten its bolt to 47 ft. lbs. (64 Nm).

50. Install the front beam assembly. Install and only hand-tighten the new color-coded bolts.

51. Align the center beam and rear beam matchmarks. Install the two bolts and tighten them to 37 ft. lbs. (50 Nm).

52. Tighten the front beam bolts to 47 ft. lbs. (64 Nm).

53. Connect the lower control arms and stabilizer bar links. Be sure that all the stabilizer link spacers and washers are properly positioned. Only hand-tighten the fasteners at this time.

54. Install the front engine mount bracket and only hand-tighten the three bolts. Tighten the mount through-bolt to 47 ft. lbs. (64 Nm).

55. Tighten the side engine mount nut and bolt to 47 ft. lbs. (64 Nm) each.

56. Tighten each of the three transaxle side mount nuts to 28 ft. lbs. (38 Nm).

57. Tighten each of the three front engine mount bracket bolts to 28 ft. lbs. (38 Nm).

58. Install new set rings onto the inboard splined shaft of halfshaft. Install the halfshafts, making sure that each snaps securely into place.

59. Reconnect the lower control arm to the steering knuckle ball joint. Tighten the castle nut to 36–43 ft. lbs. (49–59 Nm). Then, tighten the nut only enough to install a new cotter pin.

60. Connect the shift cable linkage with a new lockwasher and tighten the bolt to 88 inch lbs. (10 Nm). Install the shift cable cover.

61. Install the exhaust pipe using new gaskets. Tighten the manifold nuts to 40 ft. lbs. (54 Nm). Tighten the converter flange nuts to 16 ft. lbs. (22 Nm). Tighten the bracket nuts to 13 ft. lbs. (18 Nm). Reconnect the oxygen sensor.

62. Move the A/C compressor back into position and tighten the bolts to 16 ft. lbs. (22 Nm).

63. Reconnect the transaxle cooler line hoses.

64. Install the front wheels and splash shield.

65. Lower the vehicle and remove the chain hoist.

66. With the vehicle on the ground, tighten the damper fork bolt to 47 ft. lbs. (64 Nm). Tighten the pinch bolt to 32 ft. lbs. (44 Nm). Tighten the control arm flange bolt to 40 ft. lbs. (50 Nm). Tighten the stabilizer bar linkage nut to 14 ft. lbs. (19 Nm).

67. Install the radiator and its upper brackets. Reconnect the cooling and condenser fan motors.

68. Install the upper and lower radiator hoses. Connect the heater hoses to the coolant pipe.

69. Install the alternator bracket and tighten the three bracket bolts to 36 ft. lbs. (49 Nm).

70. Install and connect the alternator. Install the A/C compressor belt and the alternator belt. Tighten the alternator mounting bolt to 33 ft. lbs. (44 Nm). Adjust the tension of both belts.

71. Move the power steering pump into place and install its mounting nuts. Install the power steering belt. Tighten the mounting nuts to 16 ft. lbs. (22 Nm). Adjust the belt tension.

TRANSMISSION MOUNT:

REAR MOUNT:

FRONT MOUNT:

Torque Specifications:
A: 10 x 1.25 mm
59 N·m (6.0 kgf·m, 43 lbf·ft)
B: 12 x 1.25 mm
93 N·m (9.5 kgf·m, 69 lbf·ft)
C: 10 x 1.25 mm
39 N·m (3.9 kgf·m, 28 lbf·ft)
D: 12 x 1.25 mm
54 N·m (5.5 kgf·m, 40 lbf·ft)
Replace.

7924MG09

Engine mount locations and fastener tightening specifications—except CR-V models

72. Couple the three engine wiring harness connectors on the left side of the engine compartment. Reconnect the fuel injector resistor.

73. Reconnect the vacuum hoses at the intake manifold plenum. Connect the power steering pump clamp.

74. Reconnect the fuel line to the fuel rail using new sealing washers. Tighten the fitting to 16 ft. lbs. (22 Nm), and the service bolt to 106 inch lbs. (12 Nm).

75. Reconnect the EVAP and brake booster hoses to the intake manifold plenum.

76. Couple the three engine wiring harness connectors on the right side of the engine compartment.

77. Refill the engine with fresh oil.

78. Refill the transaxle with fresh Automatic Transmission Fluid (ATF).

79. Refill the radiator with fresh coolant and bleed the cooling system.

80. Reconnect the battery cable to the fuse/relay box and the ABS fuse/relay box.

81. Reconnect the throttle and cruise control cables to their linkages. Replace the cables if they are kinked. Adjust each cable's deflection.

82. Install the intake air duct.

83. Install the battery tray and its support bracket. Connect the ground cable to the bracket.

84. Install the battery and connect the positive and negative cables.

85. Fit the hood into position and loosely install the hinge bolts. Align the hinges with their matchmarks and tighten the bolts to 88 inch lbs. (10 Nm). Close the hood and check its alignment with the fenders, bumper, and windshield. Be sure the windshield washer fluid tube is connected.

86. Check the shift cable adjustment.

87. Start the engine and allow it to reach normal operating temperature.

88. Check all fluid levels and top off if necessary. Check for signs of fluid or fuel leaks.

89. Check the operation of the heater and air conditioner.

90. Check and adjust the front wheel alignment.

91. Enter the radio security code.

92. Road test the vehicle and check the transaxle shift points and the operation of all engine-driven accessories, such as power steering and air conditioning.

CR-V

➡ **The radio may contain a coded theft protection circuit. Always make note of your code number before disconnecting the battery.**

The engine and transaxle are removed from the vehicle as a unit.

1. Disconnect the negative and positive battery cables.

2. Raise the hood and mark the positions of the hinge plates with a felt-tipped marker. Remove the hood and move it out of the work area to avoid scratching the paint.

3. Remove the battery. Disconnect the ground cable from the battery tray. Unbolt and remove the battery tray and its support bracket.

4. Relieve the fuel system pressure. Drain the cooling system.

5. Drain the engine oil and the automatic transmission fluid.

✳✳ CAUTION

The EPA warns that prolonged contact with used engine oil may cause a number of skin disorders, including cancer! You should make every effort to minimize your exposure to used engine oil. Protective gloves should be worn when changing the oil. Wash your hands and any other exposed skin areas as soon as possible after exposure to used engine oil. Soap and water, or waterless hand cleaner should be used.

6. Disconnect the upper and lower radiator hoses.

7. Remove the air cleaner hose assembly.

8. Remove the engine fan and shroud.

9. Disengage all electrical connectors, fuel lines, throttle linkage and hoses from the engine. Label the connections as needed to aid installation.

10. Remove the accessory drive belts.

11. Disconnect the transmission cooler lines from the radiator. Remove the radiator.

12. If equipped with A/C, remove the compressor and position it out of the way with wire.

13. Raise and safely support the vehicle.

14. Disconnect and remove the front exhaust pipe.

15. Remove the shift cable. Be careful not to bend or kink the cable when removing it.

16. Remove the front wheels.

17. Remove the halfshafts.

18. Make matchmarks on the driveshaft prior to removal for proper alignment when reinstalling.

19. Remove the driveshaft.

20. Attach a hoist to the engine and take up the engine weight. Remove the left and right front engine mounts. Remove the rear

mount bracket bolt. Note that some engine mount pieces have arrows on them for proper assembly. Double check that all cables, hoses, wiring harnesses, etc., are disconnected from the engine. Lift the engine slowly from the engine compartment.

To install:

➡ **Install the mounting nuts/bolts in the sequence presented. Failure to follow the sequence may cause excessive noise and vibration, and reduce bushing life.**

21. Slowly lower the engine into the engine compartment.

22. Install the rear mount bracket mounting bolt, but do not tighten it yet.

23. Install the mount and bracket. Tighten the bolts on the frame side to 47 ft. lbs. (64 Nm).

24. Install the upper bracket and tighten the nuts according to the accompanying illustration.

25. Tighten the nuts/bolts on the transmission mount in sequence as shown.

26. Tighten the rear mount bracket mounting bolt to 43 ft. lbs. (59 Nm).

27. Install and tighten the right and left front mounts/brackets according to the accompanying illustration.

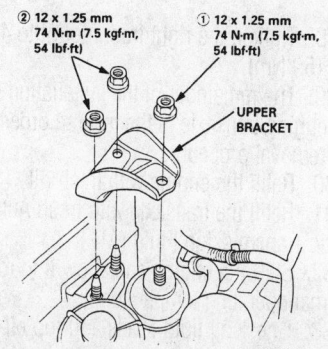

Upper bracket tightening sequence—CR-V models

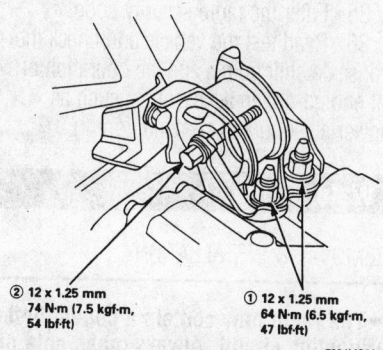

Transition mount fastener tightening sequence and values—CR-V models

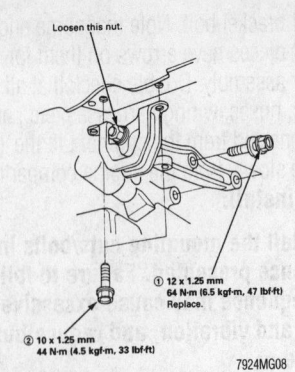

Right front mount/bracket tightening sequence—CR-V models

① 12 x 1.25 mm
64 N-m (6.5 kgf-m, 47 lbf-ft)
Replace.

② 10 x 1.25 mm
44 N-m (4.5 kgf-m, 33 lbf-ft)

Loosen this nut.

7924MG08

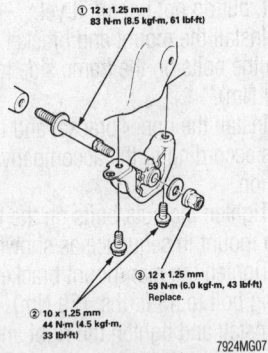

Left front mount/bracket tightening sequence—CR-V models

① 12 x 1.25 mm
83 N-m (8.5 kgf-m, 61 lbf-ft)

① 12 x 1.25 mm
59 N-m (6.0 kgf-m, 43 lbf-ft)
Replace.

② 10 x 1.25 mm
44 N-m (4.5 kgf-m, 33 lbf-ft)

7924MG07

28. Tighten the right front mount to 43 ft. lbs. (59 Nm).

29. The remainder of the installation of the engine/transaxle is the reverse order of the removal procedure.

30. Refill the engine with fresh oil.

31. Refill the transaxle with fresh Automatic Transmission Fluid (ATF).

32. Start the engine and allow it to reach normal operating temperature.

33. Check all fluid levels and top off if necessary. Check for signs of fluid or fuel leaks.

34. Check the operation of the heater and air conditioner.

35. Enter the radio security code.

36. Road test the vehicle and check the transaxle shift points and the operation of all engine-driven accessories, such as power steering and air conditioning.

Water Pump

REMOVAL & INSTALLATION

➡ **The radio may contain a coded theft protection circuit. Always make note of your code number before disconnecting the battery.**

1. Disconnect the negative battery cable.

2. Drain the coolant from the radiator.

3. Remove the timing belt. Refer to the Timing Belt unit repair section.

4. Remove the five bolts (6x1.0mm), that attach the pump to the cylinder block.

5. Remove the water pump.

To install:

6. Inspect and clean the O-ring mating surface on cylinder block.

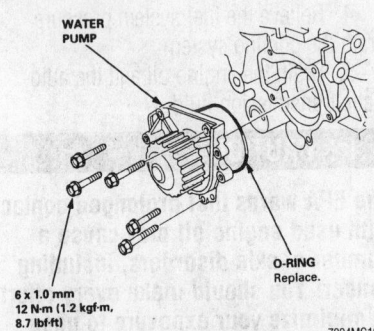

WATER PUMP

6 x 1.0 mm
12 N-m (1.2 kgf-m, 8.7 lbf-ft)

O-RING
Replace.

7924MG10

Exploded view of the water pump mounting—be sure to replace the O-ring during assembly

7. Install a new O-ring on the water pump.

8. Install the water pump onto the cylinder block with the five 6x1.0mm bolts and tighten the bolts to 106 inch lbs. (12 Nm).

9. Install the timing belt.

10. Open the cooling system bleed bolt. It is located on the thermostat housing.

11. Refill the radiator with a coolant mixture containing 50–60% antifreeze. Use only antifreeze formulated to prevent the corrosion of aluminum parts. Fill the radiator until the coolant draining from the bleed bolt is free of air bubbles. Then, tighten the bleed bolt to 88 inch lbs. (10 Nm).

12. Install the radiator cap. Reconnect the negative battery cable.

13. Run the engine until it is at normal operating temperature. Turn the heater ON. Check for coolant leaks. Be sure the cooling fan turns ON.

14. Recheck the coolant level and add more if necessary.

15. Enter the radio security code.

Cylinder Head

REMOVAL & INSTALLATION

Except CR-V

➡ **The radio may contain a coded theft protection circuit. Always make note of your code number before disconnecting the battery.**

1. Disconnect the negative and positive battery cables.

2. Remove the valve cover and the upper timing belt cover.

3. Turn the crankshaft to align the TDC marks and set cylinder No. 1 to TDC/compression. The white mark on the crankshaft pulley should align with the pointer on the timing belt cover.

4. Drain the engine coolant into a sealable container.

5. Drain the engine oil.

6. Disconnect the throttle cable and throttle control cable from the throttle body. If equipped with cruise control, remove the cruise control cable.

➡ **Be careful not to bend the cable when removing it. Do not use pliers to remove the cable from the linkage. Always replace a kinked cable with a new one.**

7. Remove the intake air duct.

8. Disconnect and label the breather hose, Positive Crankcase Ventilation (PCV) hose, and evaporative emissions (EVAP) control canister hose.

9. Relieve the fuel system pressure by loosening the service bolt on the fuel rail.

✳✳ CAUTION

The fuel injection system remains under pressure after the engine has been turned OFF. Properly relieve fuel pressure before disconnecting any fuel lines. Failure to do so may result in fire or personal injury.

10. Disconnect the fuel feed and return hoses from the fuel rail.

11. Disconnect the vacuum hoses located near the fuel feed and return hoses.

12. Remove the brake booster vacuum hose from the intake manifold. Label and remove the other vacuum hoses from the intake manifold.

13. Remove the clamp holding the power steering hose to the strut tower.

14. Remove the wiring harness clamp and the ground cable from the intake manifold.

15. Remove the connector and the terminal from the alternator. Then, remove the engine wiring harness from the valve cover.

16. Remove the mounting bolts and drive belt from the power steering pump. Pull the pump away from the mounting bracket without disconnecting the hoses. Support the pump out of the way.

17. Loosen the adjusting and mounting bolts for the alternator and remove the drive belt.

18. Unclamp the engine wiring harness and bypass hose from the lower side of the intake manifold.

19. Disconnect and label the following engine wiring harness connectors:
- Fuel injector connectors
- Intake Air Temperature (IAT) sensor connector
- Idle Air Control (IAC) valve connector
- Throttle Position Sensor (TPS) connector
- Manifold Absolute Pressure (MAP) sensor connector
- Heated Oxygen Sensor (HO2S) connector
- Engine Coolant Temperature (ECT) sensor connector
- ECT switch connector
- ECT gauge sending unit connector
- Exhaust Gas Recirculation (EGR) valve lift sensor
- CKP/TDC/CYP sensor connector
- Ignition coil connector

20. Label and detach the electrical connectors and ignition wires from the distributor.

21. Mark the position of the distributor and remove it from the cylinder head. Disconnect the ignition coil wire from the distributor.

22. Remove the upper radiator hose and the heater inlet hose from the cylinder head.

23. Remove the lower radiator hose from the thermostat housing.

24. Remove the coolant bypass hoses.

25. Use a jack with a cushioned pad to support the weight of the engine. Remove the through-bolt from the side engine mount and remove the mount.

26. Remove the cylinder head cover. Replace the rubber seals if they're damaged or deteriorated.

27. Remove the timing belt covers and the timing belt.

28. Remove the camshaft sprocket and the back cover. Do not lose the sprocket key.

29. Raise and safely support the vehicle.

30. Remove the splash shield.

31. Disconnect the exhaust pipe from the exhaust manifold.

32. Lower the vehicle.

33. Remove the exhaust manifold and the exhaust manifold heat insulator.

34. Remove the thermostat housing mounting bolts. Remove the thermostat housing from the intake manifold and the connecting pipe by pulling and twisting the housing. Discard the O-rings.

35. Remove the fuel rail and fuel injectors.

36. Remove the intake manifold bracket bolts.

37. Remove the intake manifold chamber with the throttle body attached.

38. Remove the intake manifold.

39. Remove the cylinder head bolts in the proper crisscross sequence starting at the outer edges and working inward. Then, remove the cylinder head.

➡To prevent warpage, loosen the bolts in reverse of the tightening sequence ⅓ turn at a time. Repeat the sequence until all bolts are loosened.

To install:

40. Be sure all cylinder head and block gasket surfaces are clean. Check the cylinder head for warpage. If warpage is less than 0.002 in. (0.05mm), cylinder head resurfacing is not required. The maximum resurface limit is 0.008 in. (0.2mm) based on a cylinder head total height of 3.94 in. (100mm).

41. Install a new head gasket.

42. Be sure the No. 1 cylinder is at TDC/compression.

43. Clean the oil control orifice and install a new O-ring. Replace the oil control orifice if necessary.

44. Install the dowel pins to the engine block.

45. Install the bolts that secure the intake manifold to its bracket, but do not tighten them.

46. Position the camshaft so that the **UP** mark is facing upward.

47. Install the cylinder head and be sure it is properly seated onto its dowel pins.

48. On 1995–97 engines, apply clean engine oil to the threads of the cylinder head bolts and to the underside of their heads. Install all of the head bolts. Following a crisscross pattern, tighten the bolts sequentially in three steps:
- Step 1: 29 ft. lbs. (39 Nm)
- Step 2: 51 ft. lbs. (69 Nm)
- Step 3: 72.3 ft. lbs. (98.1 Nm)

49. On 1998–99 engines, apply clean engine oil to the threads of the cylinder head bolts and to the underside of their heads. Install all of the head bolts. Tighten the cylinder head bolts following this procedure:
- Step 1: 22 ft. lbs. (29 Nm)
- Step 2: Mark the bolt head and cylinder head as illustrated
- Step 3: Tighten the head bolts in sequence 90°
- Step 4: Tighten the head bolts in sequence another 90°
- Step 5: If using new head bolts, tighten the head bolts in sequence a final 90°

50. Install the intake manifold, manifold chamber and throttle body.

51. Install new O-rings, cushion rings, and seal rings onto the fuel injectors, fuel rail, and intake orifice. Install the fuel rail to the intake manifold as an assembly with the fuel injectors.

52. Install the exhaust manifold.

53. Lower the vehicle.

54. Install the timing belt back cover to the cylinder head. Tighten the cover bolts to 106 inch lbs. (12 Nm).

55. Install the key into the camshaft groove, then install the camshaft sprocket. Tighten the sprocket bolt to 27 ft. lbs. (37 Nm).

56. Be sure the camshaft sprocket and the crankshaft pulleys are aligned to TDC for the compression stroke. The camshaft sprocket **UP** mark should face up. The camshaft keyway should also face up.

57. Install the timing belt and the balancer shaft belt.

58. Install the lower timing belt cover and tighten the bolts to 106 inch lbs. (12 Nm).

59. Install a new seal around the adjusting nut. Do not loosen the adjusting nut.

60. Install the crankshaft pulley. Coat the threads and seating face of the pulley bolt with engine oil. Install and tighten the bolt to 181 ft. lbs. (245 Nm).

61. Install the side engine mount. Tighten the bolt and nut attaching the mount to the engine to 40 ft. lbs. (55 Nm). Tighten the through-bolt and nut to 47 ft. lbs. (65 Nm). Remove the jack from under the engine.

62. Adjust the valves.

63. Install the upper timing belt cover. Tighten the bolt on the intake side of the head to 106 inch lbs. (12 Nm) and tighten the bolt on the exhaust side of the head to 88 inch lbs. (10 Nm).

64. Tighten the crankshaft pulley bolt to 181 ft. lbs. (245 Nm) if it broke loose while adjusting the valves.

65. Install the splash shield.

66. Thoroughly clean the valve cover gasket mating surfaces . Install the valve cover gasket to the groove of the cylinder head cover. Be sure the gasket is seated securely in the corners of the recesses.

67. Apply sealant to the four corners of the recesses of the valve cover gasket. Do not install the parts if 5 minutes or more have elapsed since applying sealant. After assembly, wait at least 20 minutes before filling the engine with oil.

68. Clean the valve cover contact surface with a shop towel. Install the valve cover. Tighten the valve cover capnuts in a clockwise sequence to 88 inch lbs. (10 Nm).

69. Install a new O-ring to the coolant connecting pipe, and to the thermostat housing. Install the housing to the coolant pipe and the intake manifold. Tighten the mounting bolts to 16 ft. lbs. (22 Nm).

70. Install the coolant bypass hoses.

71. Connect the lower radiator hose to the thermostat housing.

72. Connect the upper radiator hose and the heater inlet hose to the cylinder head.

73. Install the distributor to the cylinder head. Only hand-tighten the mounting bolts at this time.

74. Connect the ignition wires to the spark plugs. Attach the distributor electrical connectors. Install the ignition coil wire to the distributor.

75. Attach the following engine wiring harness connectors:
- Fuel injector connectors
- Intake Air Temperature (IAT) sensor connector
- Idle Air Control (IAC) valve connector
- Throttle Position (TP) sensor connector
- Manifold Absolute Pressure (MAP) sensor connector
- Heated Oxygen Sensor (HO2S) connector
- Engine Coolant Temperature (ECT) sensor connector
- ECT switch connector
- ECT gauge sending unit connector
- Exhaust Gas Recirculation (EGR) valve lift sensor
- CKP/TDC/CYP sensor connector
- Ignition coil connector

76. Install the engine wiring harness and bypass hose to the lower side of the intake manifold.

77. Install and adjust the alternator drive belt.

78. Install the power steering pump to the power steering pump mounting bracket.

79. Install and adjust the power steering belt.

80. Install the alternator wiring harness to the valve cover. Connect the terminal to the alternator.

81. Connect the ground cable and the wiring harness clamp to the intake manifold.

82. Connect the power steering hose clamp to the engine block.

83. Connect the brake booster vacuum hose to the intake manifold. Connect the other vacuum hoses to the intake manifold.

84. Connect the vacuum hoses located near the fuel feed and return hoses.

85. Connect the fuel return hose and the fuel feed hose to the fuel rail. Install new washers to the fuel feed hose connection. Tighten the fuel feed hose banjo bolt to 16 ft. lbs. (22 Nm). Tighten the service bolt to 106 inch lbs. (12 Nm).

86. Install the breather hose, PCV hose, and the EVAP control canister hose.

87. Install the air intake duct.

88. Connect and adjust the throttle cable. Connect and adjust the throttle control cable and the cruise control cable, if equipped.

89. Drain the engine oil into a sealable container. Install the drain plug with a new crush washer.

90. Refill the engine with clean oil.

91. Fill and bleed the air from the cooling system.

92. Connect the positive and the negative battery cables. Enter the radio security code.

93. Start the engine, checking carefully for any coolant, fuel, oil, or air leaks.

94. Check the ignition timing and adjust it if necessary, then tighten the distributor bolts to 13 ft. lbs. (18 Nm).

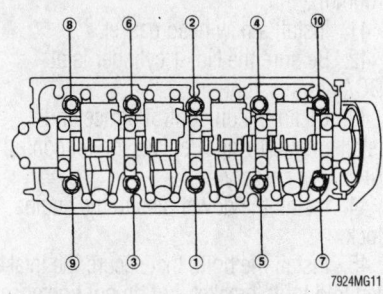

Cylinder head bolt tightening sequence—all models

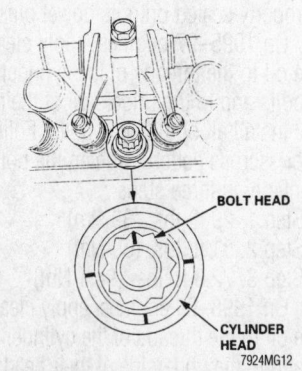

Cylinder head bolt tightening marks—1998–99 models

CR-V

✳✳ WARNING

To avoid damaging the cylinder head, allow the coolant temperature to drop below 100°F (38°C) before removing the head bolts.

1. Disconnect the negative battery cable.

2. Drain and recycle the engine coolant.

3. Remove the air cleaner and the air intake duct.

4. Remove the accessory drive belts.

5. Remove the power steering pump and its bracket.

6. Properly relieve the fuel system pressure.

7. Disengage all electrical connectors, fuel lines, throttle linkage and hoses from the engine associated with cylinder head removal. Label the connections as needed to aid installation.

8. Remove the spark plug caps and distributor from the cylinder head.

9. Use a jack with a cushioned pad to support the weight of the engine.

10. Remove the valve cover.

11. Remove the timing belt, camshaft pulley and back cover.

12. Remove the exhaust header pipe from the manifold.

13. Remove the intake manifold mounting bolts and water bypass hose.

14. Remove the camshafts.

15. Remove the cylinder head bolts. To prevent warpage of the cylinder head, loosen the bolts ⅓ of a turn in sequence. Remove the cylinder head.

To install:

16. Clean the cylinder head and block surface.

17. Clean the oil control orifice and install on the block. Be sure the dowel pins are properly located in the block.

18. Place the new head gasket on the engine block.

19. Apply clean engine oil to the bolt threads and under the bolt heads.

20. Carefully mate the cylinder head to the engine block.

21. Tighten the cylinder head in two steps following the illustrated sequence. First step, tighten the bolts to 22 ft. lbs. (29 Nm); in the final step, tighten the bolts to 63 ft. lbs. (85 Nm).

22. Tighten the intake manifold bolts to 17 ft. lbs. (24 Nm).

23. Install the exhaust header pipe to the exhaust manifold.

24. Install the camshafts.

25. Install the back cover.

26. Set the camshafts at TDC for No. 1 piston, align the holes in the camshafts with the holes in the No. 1 camshaft holder and insert a 5mm pin punch into the holes.

27. Install the keys into the keyways on the camshafts and install the pulleys, then tighten the retaining bolts to 27 ft. lbs. (37 Nm).

28. Install the timing belt.

29. Adjust the valve clearance.

30. Install the valve cover and tighten the nuts to 86 inch lbs. (9.8 Nm).

31. Verify that all tubes, hoses and connectors are installed properly.

Rocker Arms/Shafts

REMOVAL & INSTALLATION

➡ **The original radio contains a coded anti-theft circuit. Obtain the security code number before disconnecting the battery.**

1. Disconnect the negative battery cable.

2. Remove the valve cover and the upper timing belt cover.

3. Set the No. 1 cylinder to TDC for the compression stroke. Verify that the TDC marks are correctly aligned. Once the engine is set in this position, it must not be disturbed.

4. Remove the distributor as an assembly.

5. Loosen the valve adjusting screws.

6. Cover the timing belt with a clean shop towel to protect it from engine oil. If the belt is contaminated with oil, it must be replaced.

7. Loosen, but DO NOT remove the camshaft holder bolts from their holes in the shaft assemblies. Unscrew the bolts two turns at a time in a crisscross pattern to prevent damaging the valves. camshaft, or rocker arm assembly.

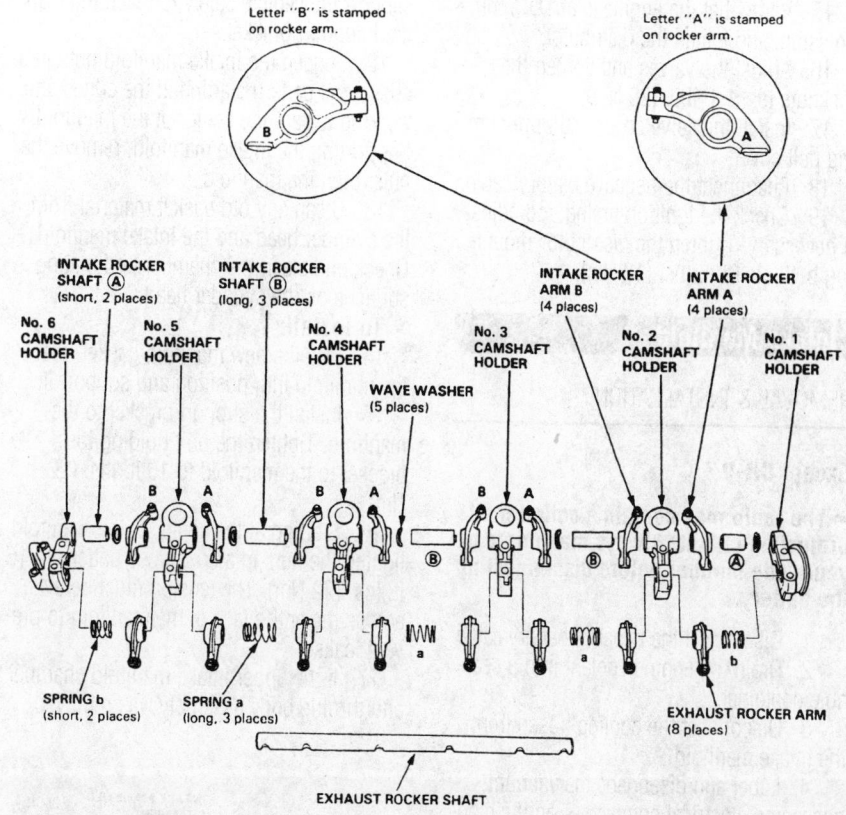

Exploded view of the rocker arm and shaft components

7924MG45

➡ **The rocker arms and shafts are an assembly; they must be removed from the engine as a unit. To prevent warpage, always follow the tightening sequence carefully when removing or installing the rocker shaft assembly.**

8. Remove the rocker arm and shaft assemblies. Do not remove the camshaft holder bolts from the camshaft holder. The bolts keep the camshaft bearing caps, springs, and rocker arms in place on the shafts.

9. If the rocker arms or shafts are to be replaced, identify the parts as they are removed from the shafts to ensure reinstallation in the original location.

To install:

10. Verify that the engine is set to TDC/compression for the No. 1 cylinder. The camshaft keyway faces up when the engine is at TDC/compression.

11. Lubricate the camshaft journals and lobes with clean engine oil. Install a new camshaft seal if necessary.

12. If necessary, assemble the rocker arms, shafts, and camshaft bearing caps.

13. Apply sealant to the mating surfaces of the first and last camshaft bearing caps. Do not allow the sealant to cure before the rocker arm assembly is installed.

14. Set the rocker arm assembly in place and loosely install the bolts. Tighten each bolt two turns at a time in the proper sequence to ensure that the rockers do not bind on the valves. Tighten the 8mm rocker arm bolts to 16 ft. lbs. (22 Nm). Tighten the 6mm bolts to 8.7 ft. lbs. (12 Nm).

Specified torque:
8 mm bolts: 22 N·m (2.2 kgf·m, 16 lbf·ft)
6 mm bolts: 12 N·m (1.2 kgf·m, 8.7 lbf·ft)

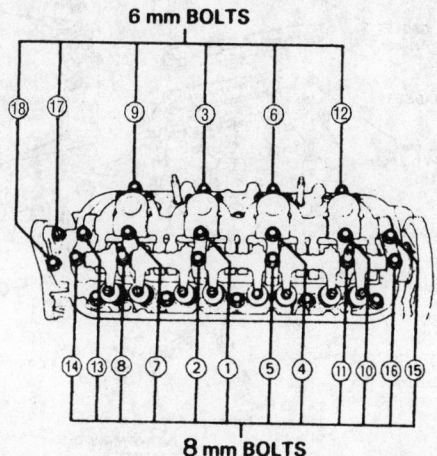

7924MG44

Camshaft holder bolt tightening sequence

15. Verify that the engine is at TDC/compression, and install the distributor.

16. Adjust the valves and tighten the locknuts to 14 ft. lbs. (20 Nm).

17. Install the valve cover and upper timing belt cover.

18. Reconnect the negative battery cable.

19. Check the ignition timing and adjust if necessary. Tighten the distributor mounting bolts to 13 ft. lbs. (18 Nm).

Intake Manifold

REMOVAL & INSTALLATION

Except CR-V

➡ **The radio may contain a coded theft protection circuit. Always make note of your code number before disconnecting the battery.**

1. Disconnect the negative battery cable.

2. Drain the engine coolant into a sealable container.

3. Disconnect the cooling hoses from the intake manifold.

4. Label and disengage the vacuum hoses and electrical connectors on the manifold and throttle body. Unplug the connector from the Exhaust Gas Recirculation (EGR) valve. Position the wiring harnesses out of the way.

5. Disconnect the throttle cable from the throttle body.

6. Disconnect the cruise control cable from the throttle linkage. Unbolt the cable clamp from the valve cover and move the cable out of the way.

7. Relieve the fuel pressure.

✷✷ CAUTION

The fuel injection system remains under pressure after the engine has been turned OFF. Properly relieve fuel pressure before disconnecting any fuel lines. Failure to do so may result in fire or personal injury.

8. Remove the fuel rail and fuel injectors.

9. Remove the thermostat housing mounting bolts. Gently pull and twist the thermostat housing to remove it from the intake manifold and the coolant connecting pipe. Discard the O-rings.

10. It may be necessary to remove the upper intake manifold chamber and throttle body assembly in order to access the nuts securing the manifold to the head.

11. Unbolt and remove the intake manifold support bracket. If necessary, raise and support the vehicle safely to reach the manifold support bracket.

12. Loosen the intake manifold nuts in a crisscross pattern starting at the edges and working toward the center of the manifold. Supporting the intake manifold, remove the nuts, then the manifold.

13. Clean any old gasket material from the cylinder head and the intake manifold. Check and clean the chamber and mating surfaces on the cylinder head.

To install:

14. Install a new manifold gasket. Place the manifold into position and support it.

15. Install the support bracket to the manifold. Tighten the bolt holding the bracket to the manifold to 16 ft. lbs. (22 Nm).

16. Starting at the center of the manifold, tighten the nuts in a crisscross pattern to 16 ft. lbs. (22 Nm). The tension must be even across the entire face of the manifold to prevent leaks.

17. If the upper intake manifold chamber and throttle body assembly was removed, install it with a new gasket. Tighten the nuts and bolts to 16 ft. lbs. (22 Nm).

18. Install new O-rings onto the coolant connecting pipe and the thermostat housing. Install the housing to the coolant pipe and the intake manifold. Tighten the mounting bolts to 16 ft. lbs. (22 Nm).

19. Connect and adjust the throttle cable.

20. Install the fuel rail and injector assembly. Reconnect the fuel lines using new sealing washers.

21. Properly position the wiring harnesses and engage the electrical connectors.

22. Connect the cruise control cable and place it back into its clamp on the valve cover. Adjust the cruise control cable so there is 0.18–0.22 in. (4.5–5.5mm) of free-play at the linkage.

23. Connect the vacuum hoses.

24. Fill and bleed the air from the cooling system.

25. Connect the negative battery cable and enter the radio security code.

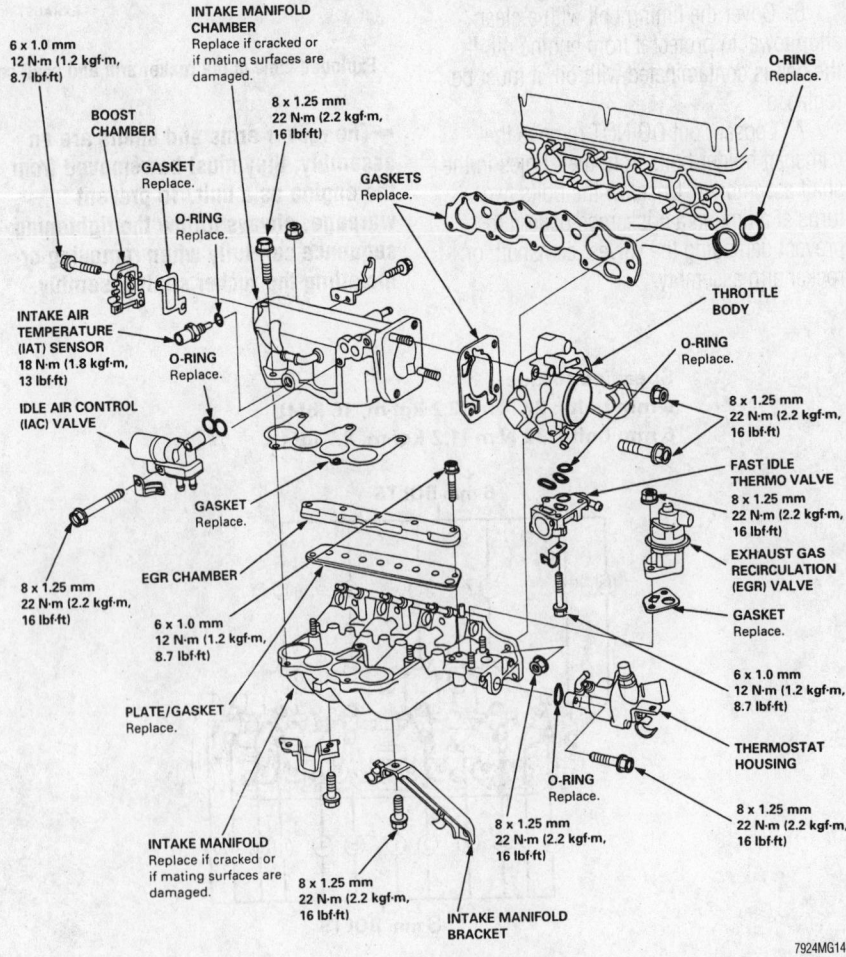

Exploded view of the intake manifold, showing associated components—except CR-V models

7924MG14

26. Start the engine and check carefully for any fuel, coolant, or vacuum leaks. Check the manifold gasket areas carefully for any vacuum leaks.

CR-V

➡The radio may contain a coded theft protection circuit. Always make note of the code number before disconnecting the battery.

1. Disconnect the negative battery cable.

2. Drain and recycle the engine coolant.

3. Properly relieve the fuel system pressure.

4. If equipped, disconnect the cruise control cable from the throttle linkage.

5. Detach all electrical connectors associated with the throttle body and intake manifold.

6. Remove the intake air hose, throttle cable, and coolant hoses from the throttle body.

7. Remove the two bolts and two nuts mounting the throttle body to the intake manifold.

8. Remove the two intake manifold support brackets.

9. Unbolt the intake manifold from the cylinder head. Inspect the manifold for cracks or damage to any of the gasket surfaces. Replace if a fault is found.

To install:

➡Use new O-rings and gaskets when reassembling.

10. Clean any old gasket material from the cylinder head, intake manifold, and throttle body.

11. Install a new manifold gasket. Place the manifold into position and support it.

12. Install the support brackets to the manifold. Tighten the bolt holding the brackets to the manifold to 17 ft. lbs. (24 Nm).

13. Starting at the center of the manifold, tighten the nuts in a crisscross pattern to 17 ft. lbs. (24 Nm). The tension must be even across the entire face of the manifold to prevent leaks.

14. Place the throttle body into position and tighten the nuts/bolts to 16 ft. lbs. (22 Nm).

15. The completion of installation is the reverse of the removal procedure.

16. Refill the engine with coolant.

17. Connect the negative battery cable.

Exhaust Manifold

REMOVAL & INSTALLATION

➡The radio may contain a coded theft protection circuit. Always make note of the code number before disconnecting the battery.

1. Disconnect the negative battery cable.

2. Safely raise and support the vehicle.

3. Remove the nuts attaching the front exhaust pipe to the exhaust manifold. Separate the pipe from the manifold and discard the gasket. A long extension may be helpful for reaching the nuts.

4. Remove the exhaust manifold heat shield.

5. Remove the exhaust manifold bracket bolts and remove the bracket.

6. Loosen the exhaust manifold nuts in a crisscross pattern starting at the edges of the manifold and working toward its center. Remove the nuts.

7. Remove the manifold and discard the gaskets. Clean the manifold and cylinder head mating surfaces.

To install:

8. Install a new exhaust manifold gasket. Place the manifold into position and support it. Install the nuts snugly onto the studs.

9. Install the support bracket below the manifold. Tighten the bracket mounting bolts to 33 ft. lbs. (44 Nm).

10. Starting with the inner or center nuts, tighten the nuts in a crisscross pattern to 23 ft. lbs. (31 Nm). The tension must be even across the entire face of the manifold to prevent leaks.

11. Install the exhaust manifold heat shield and tighten its bolts to 16 ft. lbs. (22 Nm).

12. Connect the front exhaust pipe, using new gaskets and self-locking nuts. Tighten the exhaust pipe attaching nuts to 40 ft. lbs. (55 Nm).

13. Connect the negative battery cable and enter the radio security code.

14. Start the engine and check for exhaust leaks.

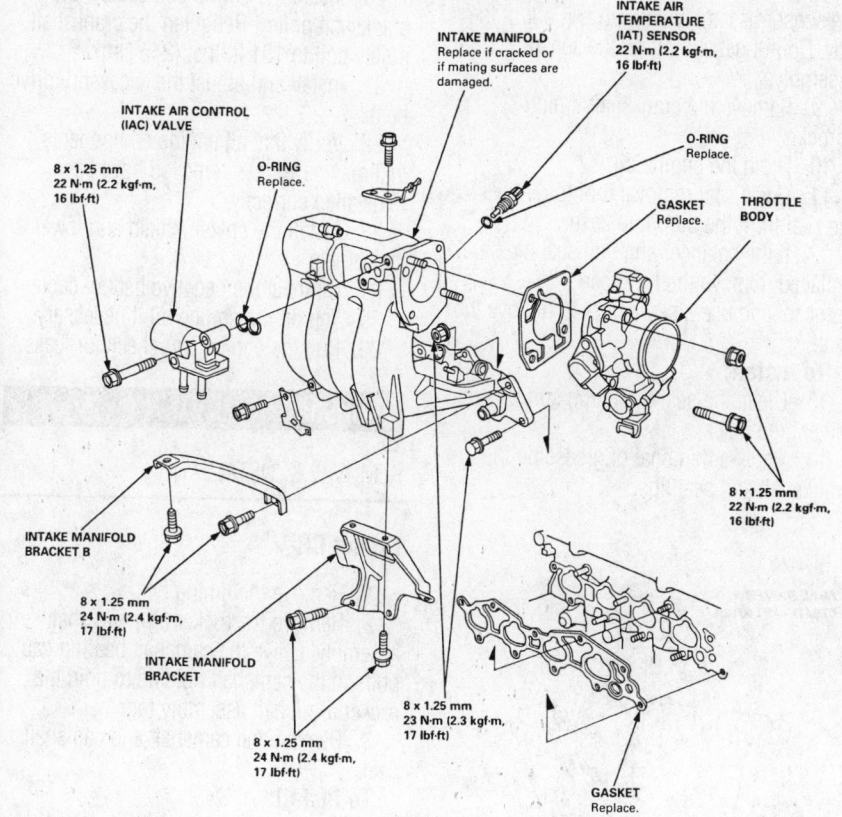

INTAKE AIR CONTROL (IAC) VALVE

8 x 1.25 mm 22 N·m (2.2 kgf·m, 16 lbf·ft)

O-RING Replace.

INTAKE MANIFOLD Replace if cracked or if mating surfaces are damaged.

INTAKE AIR TEMPERATURE (IAT) SENSOR 22 N·m (2.2 kgf·m, 16 lbf·ft)

O-RING Replace.

GASKET Replace.

THROTTLE BODY

INTAKE MANIFOLD BRACKET B

8 x 1.25 mm 24 N·m (2.4 kgf·m, 17 lbf·ft)

INTAKE MANIFOLD BRACKET

8 x 1.25 mm 24 N·m (2.4 kgf·m, 17 lbf·ft)

8 x 1.25 mm 23 N·m (2.3 kgf·m, 17 lbf·ft)

8 x 1.25 mm 22 N·m (2.2 kgf·m, 16 lbf·ft)

GASKET Replace.

7924MG13

Exploded view of the intake manifold—CR-V models

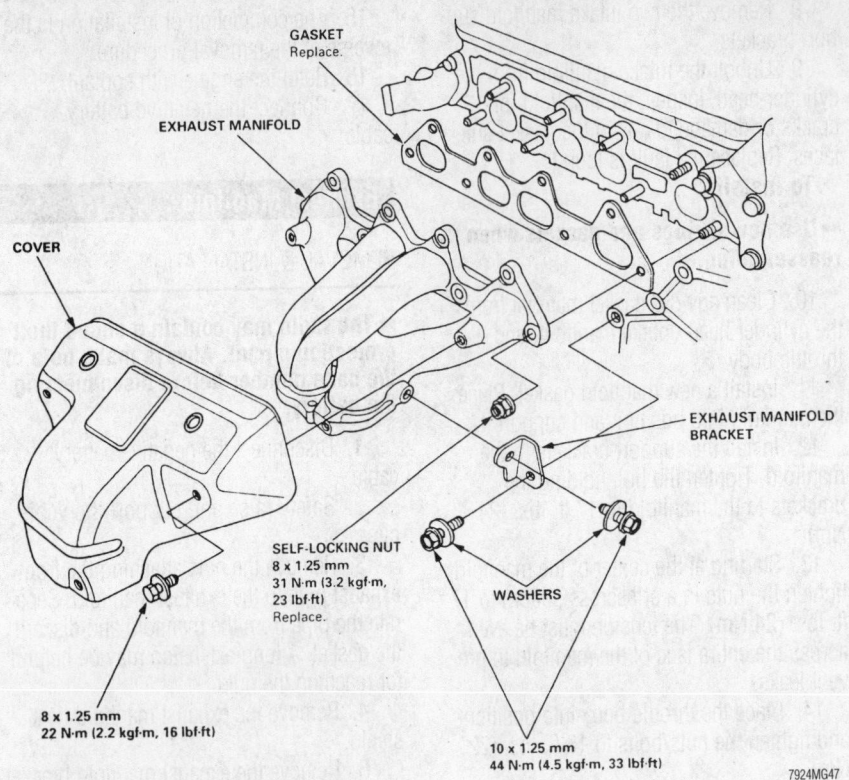

GASKET
Replace.

EXHAUST MANIFOLD

COVER

EXHAUST MANIFOLD
BRACKET

SELF-LOCKING NUT
8 x 1.25 mm
31 N·m (3.2 kgf·m,
23 lbf·ft)
Replace.

WASHERS

8 x 1.25 mm
22 N·m (2.2 kgf·m, 16 lbf·ft)

10 x 1.25 mm
44 N·m (4.5 kgf·m, 33 lbf·ft)

7924MG47

Exploded view of the exhaust manifold

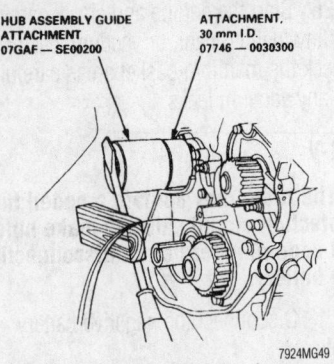

HUB ASSEMBLY GUIDE
ATTACHMENT
07GAF — SE00200

ATTACHMENT,
30 mm I.D.
07746 — 0030300

7924MG49

Drive the balancer shaft oil seal in until it bottoms on the oil pump

Front Crankshaft Seal

REMOVAL & INSTALLATION

➡**The original radio may contain a coded anti-theft circuit. Obtain the security code number before disconnecting the battery cables.**

1. Disconnect the negative battery cable.
2. Raise and safely support the vehicle.
3. Remove the splash shield.
4. Remove the engine accessory drive belts.
5. Set the engine at TDC for the No. 1 piston on the compression stroke. The crankshaft pulley mark must be aligned with the white mark on the lower timing cover. Once in this position, the engine must not be disturbed.
6. Remove the upper timing belt cover and crankshaft pulley. Remove the lower timing belt cover.

➡**Mark the direction of the timing belt's rotation if it is to be reinstalled. If there is any doubt about the condition of the timing belt, or if it has been contaminated by oil, it must be replaced.**

7. Remove the timing belt.
8. If equipped with a TDC sensor mounted on the oil pump housing, unbolt the sensor assembly and move it out of the way. Do not get any oil on the sensor assembly.
9. Remove the crankshaft timing sprocket.
10. Drain the engine oil.
11. Use a seal removal tool to remove the seal from the oil pump case.
12. If the balancer shaft oil seal must be replaced, remove the balancer shaft sprocket and use a seal puller to remove the seal.

To install:

13. Clean the seal mounting surfaces on the engine block.
14. Apply a thin coat of grease on the crankshaft and seal lips.

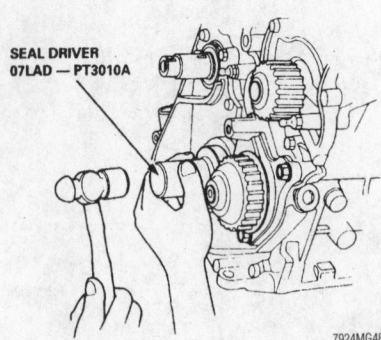

SEAL DRIVER
07LAD — PT3010A

7924MG48

Lubricate the lip of the oil seal to prevent premature wear

15. Install the seal with the part number facing out. Use a seal driver to seat the seal against the oil pump. Clean any excess grease off the crankshaft and be sure the seal lip is not distorted.
16. Install the TDC sensor assembly into position. Tighten the sensor mounting bolts to 106 inch lbs. (12 Nm).
17. Install the crankshaft timing sprocket.
18. Refill the engine with oil.
19. Verify that the engine is at TDC for the No. 1 cylinder on the compression stroke. Install and tension the timing belt.
20. Install the timing belt covers and crankshaft pulley. Retighten the crankshaft pulley bolt to 181 ft. lbs. (245 Nm).
21. Install and adjust the accessory drive belts.
22. Verify that all engine components that may have been removed have been reinstalled correctly.
23. Install the splash shield and lower the vehicle.
24. Connect the negative battery cable.
25. Top off the engine oil if necessary.
26. Run the engine and check for leaks.

Camshaft

REMOVAL & INSTALLATION

Except CR-V

1. Remove the timing belt.
2. Remove the rocker arm and shaft assembly. Leave the camshaft bearing cap bolts in the camshaft holders to hold the rocker arm/shaft assembly together.
3. Remove the camshaft and camshaft seal.

To install:

➡**Use new O-rings, seals, and gaskets when installing the camshaft.**

4. Clean and inspect the camshaft bearing caps in the cylinder head.

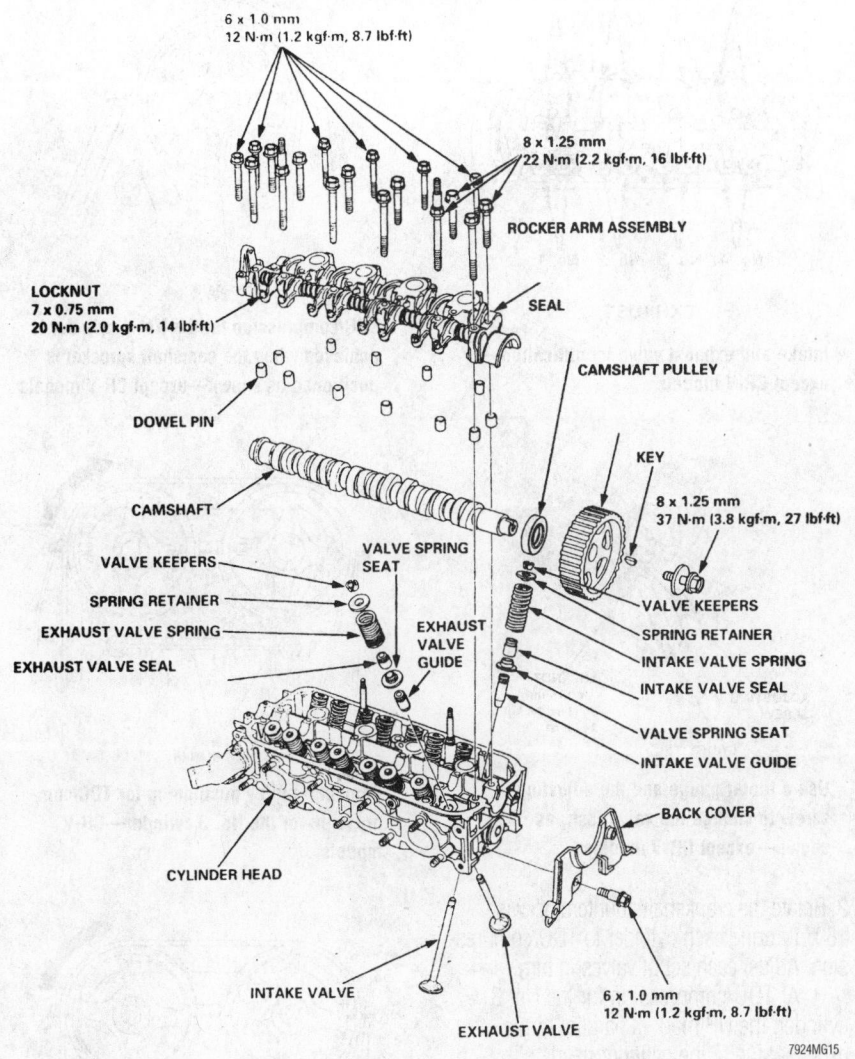

Exploded view of the camshaft and valve components—except CR-V models

5. Lubricate the lobes and journals of the camshaft prior to installation. Install the camshaft with the keyway facing up so that the engine remains at TDC/compression for the No. 1 piston.

6. Lubricate a new camshaft seal with engine oil. Use Camshaft Installer Shaft 07NAF-PT0020A, Installer Cap 07NAF-PT0010A, and Seal Guide 07NAG-PT0010A, or their equivalents to install the camshaft seal.

7. Install the rocker shaft assembly.

8. Install the timing belt.

CR-V

1. Disconnect the negative battery cable.

2. Label and disconnect the spark plug wires.

3. Remove the valve cover.

4. Remove the timing belt.

5. Loosen the rocker arm locknuts and adjusting screws.

6. Remove the camshaft bearing caps, then the camshafts.

7. Remove the rocker arms, noting their positions, and inspect for wear.

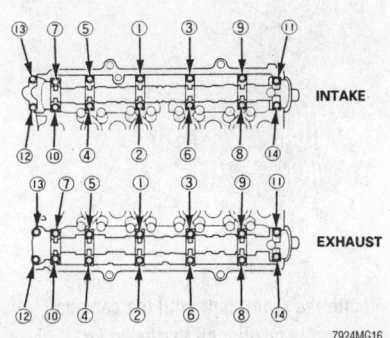

Camshaft bearing tightening sequence— CR-V models

To install:

➡**Use new O-rings, seals, and gaskets when installing the camshaft.**

8. Clean and inspect the camshaft bearing caps in the cylinder head.

9. Lubricate the lobes and journals of the camshaft prior to installation. Install the camshaft and the bearing caps.

10. Tighten the camshaft bearing caps to 86 inch lbs. (9.8 Nm), in sequence.

11. Set the camshafts at TDC for No. 1 piston, align the holes in the camshafts with the holes in the No. 1 camshaft holder and insert a 5mm pin punch into the holes.

12. Install the keys into the keyways on the camshafts and install the pulleys, then tighten the retaining bolts to 27 ft. lbs. (37 Nm).

13. Install the timing belt.

14. Adjust the valve clearance.

15. Install the valve cover and tighten the nuts to 86 inch lbs. (9.8 Nm).

16. Connect the spark plug wires.

17. Connect the negative battery cable.

Valve Lash

ADJUSTMENT

➡**The radio may contain a coded anti-theft circuit. Obtain the security code before disconnecting the battery.**

1. Disconnect the negative battery cable.

2. The valves should be adjusted when the engine is cold. If the engine has been run, allow it to cool to below 100°F (38°C) before beginning adjustments.

3. Remove the cylinder head cover and the upper timing belt cover.

4. Rotate the crankshaft to align the white TDC mark on the crankshaft pulley with the pointer on the cover for the No. 1 cylinder compression stroke. Be sure the **UP** mark on the camshaft sprocket(s) is up

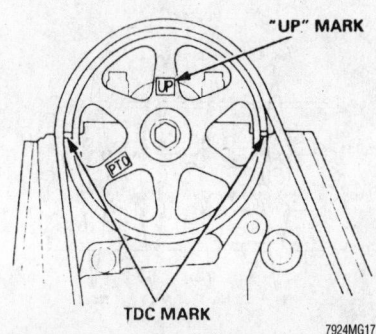

Camshaft timing belt sprocket positioning for TDC/compression for the No. 1 cylinder—except CR-V models

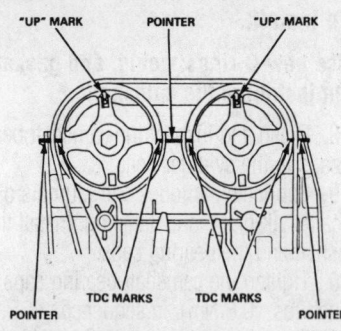

Camshaft pulley positioning for TDC/compression for the No. 1 cylinder—CR-V models

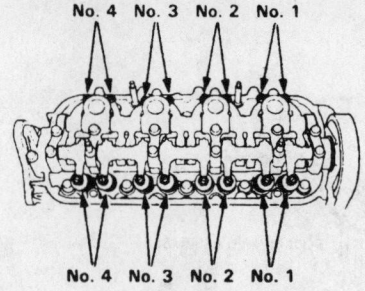

EXHAUST

Intake and exhaust valve identification—except CR-V models

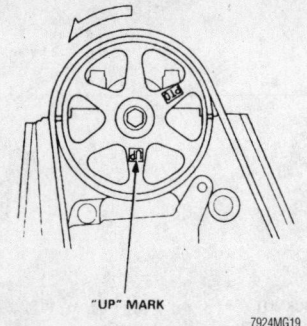

TDC/compression for the No. 4 cylinder is achieved when the camshaft sprocket is positioned as shown—except CR-V models

and the TDC marks align with the edge of the cylinder head.

5. Except CR-V models, perform the following:

a. Hold the No. 1 cylinder rocker arm against the camshaft and use a feeler gauge to check the clearance at the valve stem; intake valve clearance should be 0.010 in. (0.26mm), exhaust valve clearance should be 0.012 in. (0.30mm). The service limit for both intake and exhaust valves is plus or minus 0.0008 in. (0.02mm).

b. Loosen the locknut and turn the adjusting screw to adjust the clearance.

c. Tighten the locknut and recheck the clearance.

6. On CR-V models, perform the following:

a. Adjust the valves for the No. 1 cylinder to the following specifications:
• Intake valves: 0.003–0.005 in. (0.08–0.12mm)
• Exhaust valves: 0.006–0.008 in. (0.16–0.20mm)

b. Loosen the locknut and turn the adjusting screw to adjust the clearance.

c. Tighten the locknut and recheck the clearance.

7. The adjustment order is cylinder No. 1, cylinder No. 3, cylinder No. 4, cylinder No.

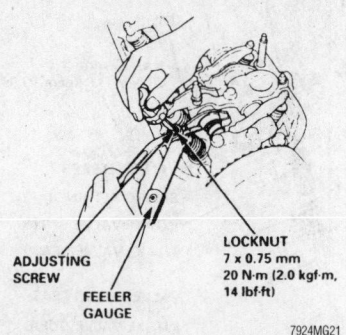

ADJUSTING SCREW **FEELER GAUGE** **LOCKNUT 7 x 0.75 mm 20 N·m (2.0 kgf·m, 14 lbf·ft)**

Use a feeler gauge and the adjusting screw to change the valve lash, as shown—except CR-V models

2. Rotate the crankshaft counterclockwise 180° to bring each cylinder to TDC/compression. Adjust each set of valves in turn.
• At TDC/compression for the No. 3 cylinder, the UP mark is parallel to the exhaust side of the cylinder head.
• At TDC/compression for the No. 4 cylinder, the UP mark is pointed straight down, and the TDC marks align with the edge of the cylinder head.
• At TDC/compression for the No. 2 cylinder, the UP mark is parallel to the intake side of the cylinder head.

8. After adjusting the valves, retighten the crankshaft pulley bolt to 130 ft. lbs.

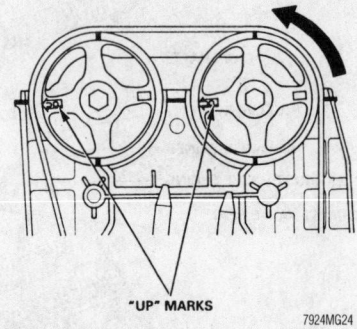

Camshaft pulley positioning for TDC/compression for the No. 3 cylinder—CR-V models

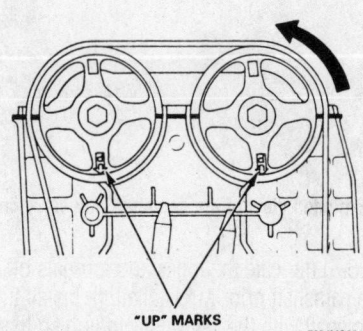

Camshaft pulley positioning for TDC/compression for the No. 4 cylinder—CR-V models

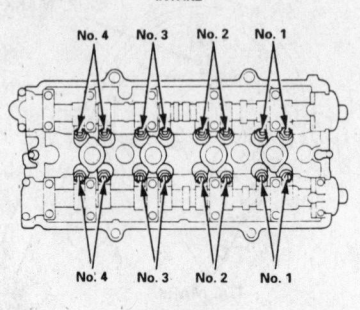

INTAKE

EXHAUST

Intake and exhaust valve identification—CR-V models

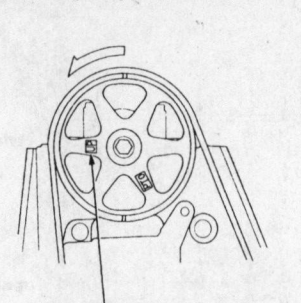

Rotate the crankshaft until the camshaft sprocket is positioned as shown for TDC/compression for the No. 3 cylinder—except CR-V models

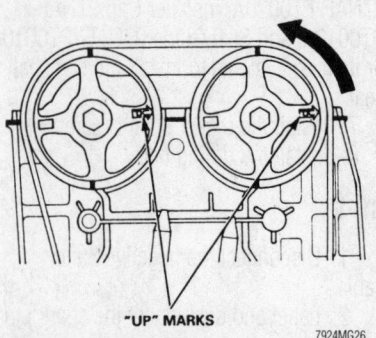

Camshaft pulley positioning for TDC/compression for the No. 2 cylinder—CR-V models

(177 Nm) for CR-V models, or to 181 ft. lbs. (245 Nm) for all other models.

9. Install the cylinder head and timing belt covers.

10. Reconnect the negative battery cable. Enter the radio security code.

Oil Pan

REMOVAL & INSTALLATION

Except CR-V

➡The radio may contain a coded theft protection circuit. Always make note of the code number before disconnecting the battery.

1. Disconnect the negative battery cable.

2. Raise and safely support the vehicle.

3. Drain the engine oil into a sealable container.

4. Install the drain bolt with a new crush washer, tighten the bolt to 33 ft. lbs. (44 Nm).

5. Remove the front wheels and the splash shield.

6. Remove the subframe center beam.

7. Disconnect the oxygen sensor electrical wiring harness.

8. Remove the nuts attaching the exhaust pipe to the exhaust manifold and the mid pipe. Remove the exhaust pipe and discard the gaskets.

9. Remove the torque converter cover.

10. Loosen the oil pan nuts and bolts in a crisscross pattern. Remove the oil pan. If necessary, use a seal cutter, or a mallet to tap the corners of the oil pan. Do not pry on the pan to get it loose.

11. Clean the oil pan mounting surface of old gasket material and engine oil.

12. Inspect the oil screen and pick-up tube for blockage, residue, or build-up. Replace the oil screen and pick-up tube if necessary.

To install:

13. Apply sealant where the oil pump and rear oil seal housing attach to the engine block. Work quickly so that the sealant doesn't set before the oil pan is installed.

14. Apply sealant to the corners of the curved section of the oil pan gasket. Install the oil pan and gasket to the engine block.

15. Install the oil pan nuts and bolts. Evenly finger-tighten the nuts and bolts.

16. Tighten the nuts and bolts in a three-step, crisscross pattern to 123 inch lbs. (14 Nm). Do not over-tighten the bolts, this can distort the gasket and cause oil leakage.

17. Install the torque converter cover and tighten the bolts to 106 inch lbs. (12 Nm).

18. Install the exhaust pipe with new gaskets and new locknuts. Tighten the nuts attaching the exhaust pipe to the exhaust manifold to 40 ft. lbs. (54 Nm), tighten the nuts attaching the exhaust pipe to the middle pipe to 16 ft. lbs. (22 Nm). Install the nuts to the exhaust pipe support bracket and tighten the nuts to 13 ft. lbs. (18 Nm).

19. Connect the oxygen sensor electrical wiring harness.

20. Install the center beam, tighten the mounting bolts to 37 ft. lbs. (50 Nm).

21. Install the splash shield.

22. Lower the vehicle. Ensure the sealant has cured and fill the engine with oil.

23. Connect the negative battery cable and enter the radio security code.

24. Start the engine and check for oil leaks.

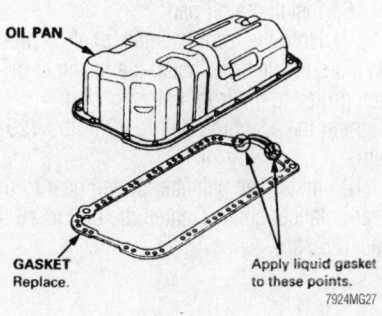

Oil pan gasket installation—all models

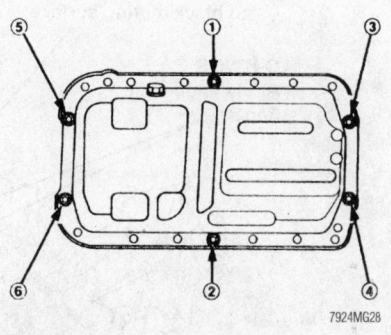

Oil pan fastener tightening sequence—all models

CR-V

➡The radio may contain a coded theft protection circuit. Always make note of the code number before disconnecting the battery.

1. Disconnect the negative battery cable.

2. Raise and safely support the vehicle.

3. Drain the engine oil into a sealable container.

4. Install the drain bolt with a new crush washer, tighten the bolt to 33 ft. lbs. (44 Nm).

5. Loosen the oil pan nuts and bolts in a crisscross pattern. Remove the oil pan. If necessary, use a seal cutter, or a mallet to tap the corners of the oil pan. Do not pry on the pan to get it loose.

6. Clean the oil pan mounting surface of old gasket material and engine oil.

7. Inspect the oil screen and pick-up tube for blockage, residue, or build-up. Replace the oil screen and pick-up tube if necessary.

To install:

8. Apply sealant where the oil pump and rear oil seal housing attach to the engine block. Work quickly so that the sealant doesn't set before the oil pan is installed.

9. Install the oil pan gasket and oil pan.

10. Tighten the nuts and bolts at six points as shown.

11. Tighten the nuts and bolts in a three-step, crisscross pattern to 123 inch lbs. (14 Nm). Do not over-tighten the bolts, this can distort the gasket and cause oil leakage.

12. Lower the vehicle. Ensure the sealant has cured and fill the engine with oil.

13. Connect the negative battery cable and enter the radio security code.

14. Start the engine and check for oil leaks.

Oil Pump

REMOVAL & INSTALLATION

Except CR-V

1. Remove the timing belt, balancer belt and tensioners.

2. If equipped with a TDC sensor mounted on the oil pump housing, unbolt the sensor assembly and move it out of the way.

3. Remove the timing belt drive sprocket from the crankshaft.

4. Insert a pin punch or holder tool into the maintenance hole in the front balancer shaft (located behind the balancer sprocket). Hold the shaft steady with the tool and remove the balancer sprocket. Ensure the tool or bolt used is strong enough to resist bending when torque is applied to the sprocket nut.

➡Front refers to the side of the engine facing the vehicle's radiator. Rear refers to the side of the engine facing the vehicle's firewall.

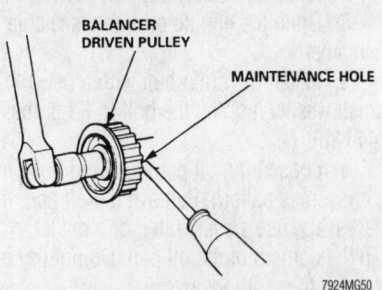

Use a prytool through the maintenance hole to hold the balance shaft while removing the pulley retaining bolt—except CR-V

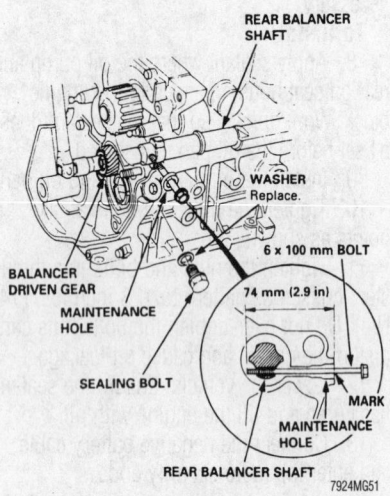

Cut away view of the service bolt installation—except CR-V

5. Align the rear timing balancer sprocket using a 6 x 100mm bolt, rod, or pin punch. Mark the bolt or rod at a point 2.9 in. (74mm) from its end.

6. Remove the 12mm sealing bolt from the maintenance hole on the right side of the block below the water pump. Insert the bolt/rod into the hole until the 2.9 in. (74mm) mark you made on it is aligned with the face of the hole. This bolt/rod will act as a pin to hold the shaft in place.

7. Remove the balancer gear case and the dowel pins. Discard the O-ring.

8. Unbolt and remove the balancer driven gear. Leave the holder tool in the maintenance hole.

9. Remove the oil pan and the oil screen. Discard the screen gasket. Replace the oil screen if it shows signs of blockage.

10. Remove the oil pump mounting bolts and remove the oil pump assembly. Remove the dowel pins from the engine and clean the oil pump mating surfaces of old gasket material, oil, and sludge. Discard the O-rings.

To install:

11. Install new crankshaft and balancer shaft seals into the oil pump housing using an appropriately-sized seal driver.

12. Install the two dowel pins and new O-rings to the cylinder block.

13. Be sure that the mating surfaces are clean and dry. Apply liquid gasket evenly in a narrow bead, centered on the mating surface. Once the sealant is applied, do not wait longer than 20 minutes to install the parts; the sealant will become ineffective. After final assembly, wait at least 30 minutes before adding oil to the engine to give the sealant time to set. To prevent leakage of oil, apply a suitable thread sealer to the inner threads of the bolt holes.

14. Install the oil pump to the engine block. Tighten the mounting bolts to 106 inch lbs. (12 Nm).

15. Install the oil screen. Tighten the screen mounting bolts and nuts to 106 inch lbs. (12 Nm).

16. Install the oil pan.

17. Hold the front balancer shaft in place with a suitable tool. Install the balancer driven pulley to the front balancer shaft. Tighten the attaching bolt to 22 ft. lbs. (29 Nm).

18. Install the balancer driven gear to the rear balancer shaft. Tighten the bolt to 18 ft. lbs. (25 Nm).

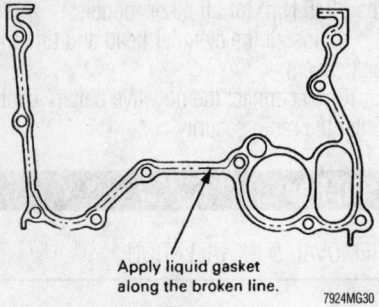

Apply the oil pump housing gasket sealer as indicated—except CR-V

19. Before installing the balancer driven gear and the gear case, apply molybdenum disulfide (lithium) grease to the thrust surfaces of the balancer gears.

20. Align the groove on the pulley edge to the pointer on the balancer gear case.

21. Install the balancer gear case to the engine with a new O-ring. Install the mounting bolts and nut. The rear balancer shaft should be held in place with a 6 x 100mm bolt/rod. Tighten the mounting bolts and nut to 18 ft. lbs. (25 Nm).

22. Check the alignment of the pointer on the balancer pulley to the pointer on the oil pump.

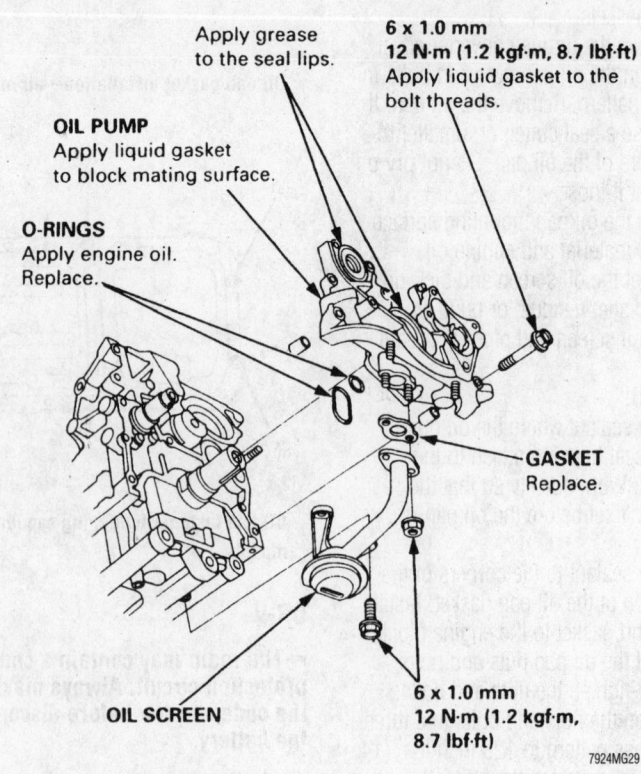

Exploded view of the oil pump and screen—except CR-V

23. Remove the 6 x 100mm holder bolt/rod from the maintenance hole. Install the sealing bolt with a new crush washer. Tighten it to 22 ft. lbs. (29 Nm).

24. Install the timing belt and tensioners.

CR-V

1. Disconnect the negative battery cable.

2. Drain the engine oil into a sealable container.

3. Raise and safely support the vehicle.

4. Set the engine to No. 1 cylinder TDC.

5. Remove the valve cover and the timing belt middle cover.

6. Remove the accessory drive belts.

7. Remove the crankshaft pulley and the lower timing cover.

8. Remove the timing belt and drive pulley.

9. Remove the oil pan and the oil screen. Discard the screen gasket. Replace the oil screen if it shows signs of blockage.

10. Remove the oil pump mounting bolts and remove the oil pump assembly. Remove the dowel pins from the engine and clean the oil pump mating surfaces of old gasket material, oil, and sludge. Discard the O-rings.

To install:

11. Install a new crankshaft seal into the oil pump housing using an appropriately-sized seal driver.

12. Install the two dowel pins and new O-ring to the cylinder block.

13. Be sure that the mating surfaces are clean and dry. Apply liquid gasket evenly, in a narrow bead centered on the mating surface. Once the sealant is applied, do not wait longer than 20 minutes to install the parts; the sealant will become ineffective. After final assembly, wait at least 30 minutes before adding oil to the engine to give the sealant time to set. To prevent leakage of oil, apply a suitable thread sealer to the inner threads of the bolt holes.

14. Apply engine oil to the lip of the oil pump seal.

15. Install the oil pump and oil screen, tightening the mounting bolts/nuts to 7.2 ft. lbs. (9.8 Nm).

16. Install the oil pan.

17. The completion of installation is the reverse of the removal procedure.

18. Lower the vehicle.

19. Fill the engine with clean engine oil.

20. Connect the negative battery cable.

FUEL SYSTEM

Fuel System Service Precautions

Safety is the most important factor when performing not only fuel system maintenance, but any type of maintenance. Failure to conduct maintenance and repairs in a safe manner may result in serious personal injury or death. Maintenance and testing of the vehicle's fuel system components can be accomplished safely and effectively by adhering to the following rules and guidelines:

• To avoid the possibility of fire and personal injury, always disconnect the negative battery cable unless the repair or test procedure requires that battery voltage be applied.

• Always relieve the fuel system pressure prior to disconnecting any fuel system component (injector, fuel rail, pressure regulator, etc.), fitting or fuel line connection. Exercise extreme caution whenever relieving fuel system pressure to avoid exposing skin, face and eyes to fuel spray. Please be advised that fuel under pressure may penetrate the skin or any part of the body that it contacts.

• Always place a shop towel or cloth around the fitting or connection prior to loosening to absorb any excess fuel due to spillage. Ensure that all fuel spillage (should it occur) is quickly removed from engine surfaces. Ensure that all fuel soaked cloths or towels are deposited into a suitable waste container.

• Always keep a dry chemical (Class B) fire extinguisher near the work area.

• Do not allow fuel spray or fuel vapors to come into contact with a spark or open flame.

• Always use a backup wrench when loosening and tightening fuel line connection fittings. This will prevent unnecessary stress and torsion to fuel line piping. Always follow the proper torque specifications.

• Always replace worn fuel fitting O-rings with new. Do not substitute fuel hose or equivalent, where fuel pipe is installed.

Fuel System Pressure

RELIEVING

✳✳ CAUTION

Do not allow fuel spray or fuel vapors to come in contact with a spark or open flame. Keep a dry chemical fire

extinguisher nearby. Never store fuel in an open container due to risk of fire or explosion.

1. Ensure the ignition is **OFF**.

➡**Always install new sealing washers whenever fuel system service bolts and banjo bolts are loosened or removed. Replace any stripped or damaged banjo bolts.**

2. Disconnect the negative battery cable.

3. Remove the fuel filler cap.

4. Hold the fuel rail inlet banjo bolt with a flare nut wrench. Hold the service bolt with a box end wrench.

5. Place a shop towel over the fitting to absorb leakage.

6. Slowly loosen the service bolt one turn.

The fuel system pressure will be relieved through the service bolt. To repressureize the system after service is complete, continue with the procedure.

7. After service has been completed, always install a new sealing washer on the service bolt.

8. Tighten the service bolt to 77 inch lbs. (12 Nm). If the entire fitting has been removed, tighten the service bolt to 16 ft. lbs. (22 Nm).

9. Reconnect the negative battery cable. Install the fuel filler cap.

10. Turn the ignition to the **ON** position, but don't start the engine. Then, turn the ignition **OFF**. Repeat this step two or three times to pressurize the fuel system.

11. Check the fuel fittings for any signs of leaks.

Fuel Filter

REMOVAL & INSTALLATION

✳✳ CAUTION

The fuel injection system remains under pressure after the engine has been turned OFF. Properly relieve fuel pressure before disconnecting any fuel lines. Failure to do so may result in fire or personal injury.

➡**The radio may contain a coded theft protection circuit. Always make note of the code number before disconnecting the battery.**

1. Disconnect the negative battery cable.

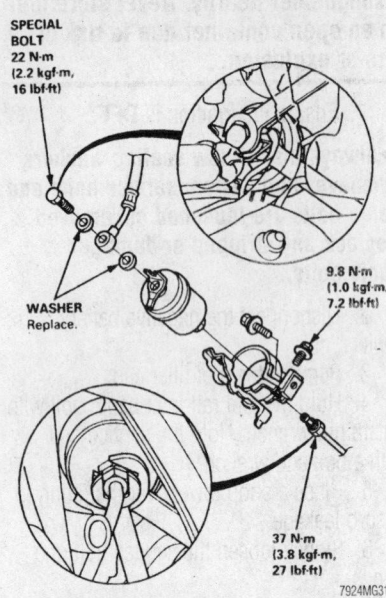

SPECIAL
BOLT
22 N·m
(2.2 kgf·m,
16 lbf·ft)

WASHER
Replace.

9.8 N·m
(1.0 kgf·m,
7.2 lbf·ft)

37 N·m
(3.8 kgf·m,
27 lbf·ft)

7924MG31

Exploded view of the fuel line connections' mounting to the filter

2. Place a shop towel under the fuel filter to absorb leakage.

3. Relieve the fuel system pressure.

4. Remove the service bolt and disconnect the fuel line from the filter. Due to the restricted location of the fuel filter, a flare nut wrench and socket may be needed to loosen the fuel filter fittings.

5. Use flare nut wrenches to loosen the fuel inlet line from the bottom of the filter.

6. Unbolt the fuel filter clamp from the vehicle's firewall. Remove the fuel filter.

To install:

➡**Always use new sealing washers when installing the fuel filter and reconnecting fuel lines. Replace any stripped banjo bolts.**

7. Clean the fuel line fittings before installing the fuel filter.

8. Install the fuel filter and bracket. Connect the fuel inlet line to the filter and carefully tighten it to 27 ft. lbs. (37 Nm).

9. Connect the fuel line to the top of the filter with new sealing washers. Carefully tighten the service bolt to 16 ft. lbs. (22 Nm).

10. Install new sealing washers onto the fuel rail fitting. Then, tighten the fitting to 16 ft. lbs. (22 Nm).

11. Install the fuel filler cap.

12. Reconnect the negative battery cable.

13. Turn the ignition **ON**, but don't start the engine. Then, turn the ignition **OFF**. Repeat this step two or three times to pressurize the fuel system.

14. Check the fuel filter and fuel rail fittings for leakage.

15. Enter the radio security code.

Fuel Pump

REMOVAL & INSTALLATION

Except CR-V

➡**The radio may contain a coded theft protection circuit. Always make note of the code number before disconnecting the battery.**

1. Remove the fuel tank as follows:

a. Disconnect the negative battery cable.

b. Relieve the fuel system pressure by loosening the fuel filler cap and the fuel rail service bolt.

✱✱ CAUTION

The fuel injection system remains under pressure after the engine has been turned OFF. Properly relieve fuel pressure before disconnecting any fuel lines. Failure to do so may result in fire or personal injury.

c. Raise and safely support the vehicle.

d. Remove the fuel tank drain plug. Drain the fuel into an approved gasoline storage container. Install the drain plug with a new crush washer.

✱✱ CAUTION

Do not allow fuel spray or fuel vapors to come in contact with a spark or open flame. Keep a dry chemical fire extinguisher nearby. Never store fuel in an open container due to risk of fire or explosion.

e. Remove the fuel hose protector.

f. Disconnect the fuel filler hoses.

g. Disconnect the fuel hose and the quick-connect fitting. Hold the connector with one hand, press the retainer tabs down and in with your other hand, then pull the connectors apart. Plug the fuel lines after removal to keep dirt out.

➡**Don't use tools to separate the connectors. If the connection is tight, gently wiggle the connector in and out to free it from retainer. Don't remove the retainer from the fuel pipe.**

h. Use a transmission jack and a broad piece of wood under the tank.

Adjust the position of the jack as necessary to evenly support the fuel tank.

i. Remove the bolts attaching the tank straps, then remove the straps.

j. Lower the tank to disengage the fuel pump and fuel gauge sender connectors. If the tank is stuck to the vehicle's undercoating, gently pry it loose with a piece of wood.

k. Remove the fuel tank.

2. Disconnect the 2-P connector from the fuel pump.

3. Unbolt the fuel pump from the fuel tank.

4. Lift the fuel pump up and allow any fuel in it to drain into the tank. Then, remove the fuel pump.

To install:

5. Fit the fuel pump into the tank.

6. Install the mounting nuts and tighten them to 53 inch lbs. (6 Nm).

7. Reconnect the 2-P connector.

8. Install the fuel tank as follows:

a. Position the fuel tank under the vehicle and raise it enough to engage the fuel pump and fuel gauge sender connectors.

b. Position the tank in the vehicle and install the tank straps. Tighten the bolts attaching the straps to 28 ft. lbs. (38 Nm).

c. Remove the transmission jack from under the fuel tank.

d. Clean the quick-connect fuel line fittings and apply a light coat of oil to the

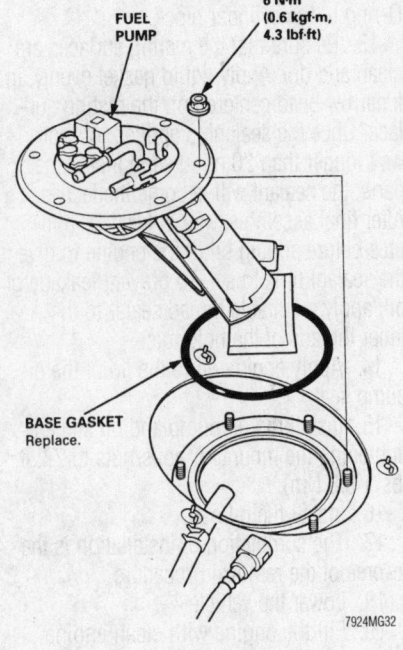

FUEL
PUMP

6 N·m
(0.6 kgf·m,
4.3 lbf·ft)

BASE GASKET
Replace.

7924MG32

Always use a new base gasket when servicing the fuel pump—all models

contact areas. Install a new retainer on the fuel pump connector.

   e. Connect the fuel filler hoses and install new clamps if necessary.

   f. Install the fuel hose protector.

   g. Ensure the drain plug has been installed. Tighten the drain plug to 36 ft. lbs. (49 Nm).

9. Install a new sealing washer on the fuel rail service bolt. Tighten the bolt to 106 inch lbs. (12 Nm).

10. Lower the vehicle.

11. Refill the fuel tank.

12. Connect the negative battery cable and enter the radio security code.

13. Switch the ignition **ON** but don't start the engine. The fuel pump should run for approximately two seconds, building pressure within the lines. Switch the ignition **OFF**, then **ON** two or three more times to build full system pressure. Check for fuel leaks.

CR-V

1. Disconnect the negative battery cable.

2. Properly relieve the fuel system pressure.

3. Access the base frame cover located under the left rear seat cushion.

4. Disconnect the fuel lines from the fuel pump.

5. Detach the electrical connector from the fuel pump.

6. Remove the fuel pump mounting bolts and withdrawal the pump from the tank.

To install:

7. Replace the base gasket and fit the fuel pump into the tank.

8. Tighten the mounting bolts to 52 inch lbs. (6 Nm).

9. Attach the electrical connector to the pump.

10. Connect the fuel lines to the fuel pump.

11. Install the cover and restore the seat cushion.

12. Connect the negative battery cable and enter the radio security code.

13. Switch the ignition **ON** but don't start the engine. The fuel pump should run for approximately two seconds, building pressure within the lines. Switch the ignition **OFF**, then **ON** two or three more times to build full system pressure. Check for fuel leaks.

DRIVE TRAIN

Automatic Transaxle Assembly

REMOVAL & INSTALLATION

➡**The radio may contain a code anti-theft circuit. Always obtain the security code number before disconnecting the battery cables.**

1. Shift the transaxle into **N** (neutral).

2. Disconnect the negative and positive battery cables and remove the battery.

3. Remove the air intake hose, air cleaner case, and battery tray. Disconnect the ground cable and the cable bracket from the battery tray.

4. Disconnect the throttle cable from the throttle control lever.

5. Detach the transaxle ground cable and the mainshaft speed sensor connectors. Detach the shift and lock-up solenoid valve connectors.

6. Disconnect the starter cables from the starter motor and unbolt the cable bracket from the transaxle case.

7. Disconnect the Vehicle Speed Sensor (VSS) wiring harness. The VSS is located on the rear of the transaxle case near the cooler line inlet.

8. Detach the gear position switch and counter shaft speed sensor connectors.

9. Loosen the four upper transaxle case bolts, but leave them threaded into the engine block. Move the lower radiator hose slightly upward and toward the engine block if more clearance is needed for a socket and extension.

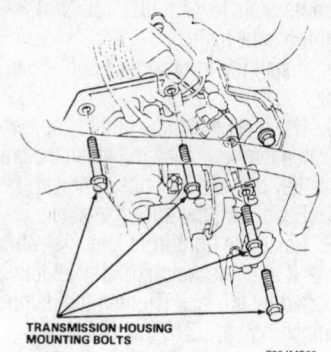

7924MG33

Upper transaxle case bolt locations—all models

10. Loosen, but do not remove, the three front engine mount bracket bolts.

11. Raise and safely support the vehicle. Remove the front wheels.

12. Remove the splash shield.

13. Drain the transaxle fluid and reinstall the drain plug with a new crush washer.

14. Disconnect the transaxle cooler hoses from the joint pipes. Plug the hoses to keep out dirt and moisture.

15. Remove the subframe center beam.

16. Remove the cotter pins and lower arm ball joint nuts, then separate the ball joints from the lower arms using a suitable ball joint tool.

17. Remove the right damper pinch bolt. Separate the damper fork from the strut.

18. Unbolt the right radius rod from the right lower control arm and remove it from the front subframe beam.

19. Using a suitable tool, carefully pry the right and left halfshafts out of the differential. Pull on the inboard CV-joints and remove the right and left halfshafts. Tie plastic bags over the halfshaft ends to prevent damage to the CV boots and splines. Don't let the left halfshaft hang by its own weight; use a piece of wire to suspend it out of the way.

20. Turn the right driveshaft toward the front of the vehicle so that it is resting on the lower control arm. Use a piece of wire to support the halfshaft.

21. Remove the intermediate shaft by unbolting its mounting bracket from the engine block.

22. Remove the torque converter cover and shift cable holder.

23. Remove the shift control cable by removing the lockbolt. Remove the shift cable lever from the control shaft. Don't disconnect the control lever from the shift cable. Wire the shift cable out of the work area and be careful not to kink it.

24. On CR-V models, remove the drive-shaft bolts from the transfer unit.

25. Remove the eight drive plate bolts one at a time while rotating the crankshaft pulley.

26. Place a suitable jack under the transaxle and raise the jack just enough to take weight off of the mounts.

27. Remove the transaxle mount from the transaxle case. Don't remove the bracket.

28. Remove the upper and lower transaxle case bolts. Unbolt the rear engine mount bracket from the transaxle case. The

rear engine mount bracket through-bolt may have to be loosened first.

29. Pull the transaxle away from the engine until it clears the dowel pins. Lower the transaxle from the vehicle.

To install:

➡**Use new self-locking nuts when assembling the front suspension components. Install new set rings onto the halfshaft inboard joint splines. Replace any color-coded self-locking bolts.**

30. Flush the transaxle cooler lines:

a. Use a pressurized flusher (Honda or Kent-Moore part No. J38405-A or equivalent). Use only Honda flushing fluid (Honda part No. J35944–20); other fluids will damage the system.

b. Fill the flusher with 21 ounces (600 grams) of fluid. Pressurize the flusher to 80–120 psi (551–827 kPa), following the procedure on the fluid container and flusher.

c. Clamp the discharge hose of the flusher to the cooler return line. Clamp the drain hose to the cooler inlet line and route it into a bucket or drain tank. Ensure the drain hose is securely clamped to the drain container.

d. Connect the flusher to air and water lines. Use hot water if its available.

e. Open the flusher water valve and flush the cooler for ten seconds.

f. Depress the flusher trigger to mix flushing fluid with the water. Flush for two minutes, turning the air valve on and off for five seconds every 15–20 seconds. The maximum air pressure for the flushing procedure is 120 psi (827 kPa).

g. After finishing one flushing cycle, reverse the hoses and flush in the opposite direction.

h. Dry the cooler lines with compressed air so that NO moisture is left in the cooler system.

31. Ensure the two 14mm dowel pins are installed into the torque converter housing.

32. Install the torque converter onto the transaxle mainshaft with a new hub O-ring. Install the starter motor onto the transaxle case and tighten its mounting bolts to 33 ft. lbs. (44 Nm).

33. Raise the transaxle into position and install the transaxle housing mounting bolts. Evenly tighten the bolts to 47 ft. lbs. (65 Nm).

34. Connect the rear engine mount bracket to the transaxle case and evenly tighten the three new self-locking bolts to 40 ft. lbs. (54 Nm). Tighten the rear mount through-bolt to 47 ft. lbs. (64 Nm) if it was loosened.

35. Install the intake manifold bracket and tighten the bolts to 16 ft. lbs. (22 Nm).

36. Install the transaxle mount and hand-tighten the through-bolt. Tighten the three nuts to 28 ft. lbs. (38 Nm). Tighten the through-bolt to 47 ft. lbs. (65 Nm).

37. Tighten the three front engine mount bracket bolts to 28 ft. lbs. (38 Nm).

38. Remove the transmission jack.

39. Attach the torque converter to the drive plate and hand-tighten the bolts. Tighten the eight bolts in two steps in a crisscross pattern: first to 53 inch lbs. (6 Nm), and finally to 106 inch lbs. (12 Nm). Check for free rotation after tightening the last bolt.

40. Install the shift control cable and control cable holder. Tighten the shift cable lockbolt to 123 inch lbs. (14 Nm). Tighten the control cable holder bolts to 13 ft. lbs. (18 Nm).

41. Install the torque converter cover and tighten the bolts to 106 inch lbs. (12 Nm).

42. Install the intermediate shaft and tighten the mounting bolts to 28 ft. lbs. (38 Nm).

43. On CR-V models, install the driveshaft to the transfer unit.

44. Install the radius rod and damper fork.

45. Install a new set ring on the end of each halfshaft.

46. Turn the right steering knuckle fully outward and slide the axle into the differential until the set ring snaps into the differential side gear. Repeat the procedure on the left side. Ensure the halfshafts are fully seated in the differential and intermediate shaft.

47. Install the damper fork bolts and ball joint nuts to the lower arms. Tighten the ball joint nut to 40 ft. lbs. (55 Nm) and install a new cotter pin.

48. Install the subframe center beam and tighten its bolts to 37 ft. lbs. (50 Nm). Install the splash shield.

49. Install the front wheels and lower the vehicle.

50. Use a pulley holder tool in conjunction with a torque wrench to tighten the crankshaft pulley bolt to 181 ft. lbs. (245 Nm).

51. Reconnect the speed sensor.

52. Raise the right front knuckle with a floor jack until the weight of the vehicle is supported by the jack. Tighten the damper fork pinch bolt to 32 ft. lbs. (44 Nm). Tighten the radius rod bolts to 76 ft. lbs. (103 Nm), and the radius rod nut to 32 ft. lbs. (44 Nm). Hold the damper fork bolt with a wrench and tighten the nut to 40 ft. lbs. (55 Nm).

53. Connect the cables to the starter. Place the radiator hose back into its bracket.

54. Reconnect the throttle control cable.

55. Attach the lock-up control solenoid valve and shift control solenoid valve connectors.

56. Attach the mainshaft and countershaft speed sensor connectors and the transaxle ground cable.

57. Connect the transaxle cooler hoses to the joint pipes.

58. Install the battery base, air cleaner case and air intake hose. Reconnect the ground cable and the cable bracket to the battery base.

59. Install the battery. Connect the positive and negative battery cables.

60. Refill the transaxle with ATF. Use only Honda Premium ATF or an equivalent DEXRON®II or III ATF. Connect the battery cables.

61. Leave the flusher drain hose attached to the cooler return line.

62. With the transaxle in **P** (park), run the engine for 30 seconds, or until approximately 1 qt. (0.9 L) of fluid is discharged. This completes the cooler flushing process.

63. Remove the drain hose and reconnect the cooler return line.

64. Refill the transaxle fluid to the proper level.

65. Start the engine, set the parking brake and shift the transaxle through all gears three times. Check for proper shift cable and throttle cable adjustment.

66. Let the engine reach operating temperature with the transaxle in **P** or **N** (neutral). Then, shut the engine **OFF** and check the fluid level.

67. Road test the vehicle. Check for proper shifting.

68. After road testing the vehicle, loosen the front engine mount bracket bolts, then retighten them to 28 ft. lbs. (39 Nm).

69. Check and adjust the vehicle's front wheel alignment. Tighten the front wheel nuts to 80 ft. lbs. (110 Nm).

70. Enter the radio security code.

Transfer Assembly

REMOVAL & INSTALLATION

1. Raise and safely support the front of the vehicle.

2. Drain the Automatic Transmission Fluid (ATF) and remove the guard bar/splash shield.

3. Disconnect the header pipe from the vehicle.

4. Remove the shift cable. Do not bend the shift cable excessively.

5. Matchmark the driveshaft and transfer assembly flanges, then separate.

6. Clean the area around the transfer assembly to keep dirt and debris form entering the transaxle.

7. Remove the mounting bolts for the rear stiffener and the transfer assembly.

➡**Do not allow dust or other foreign material to enter the transmission while servicing the transfer unit.**

To install:

8. Install a new O-ring on the transfer assembly.

9. Install the transfer assembly and tighten the mounting bolts to 33 ft. lbs. (44 Nm).

10. Install and tighten the rear stiffener mounting bolts to 17 ft. lbs. (24 Nm).

11. Line up the marks on the driveshaft and tighten the bolts to 24 ft. lbs. (32 Nm).

12. Install the shift cable and cover.

13. Attach the exhaust header pipe and install the splash shield/guard bar.

14. Fill the transmission with ATF, then start the engine and check the level.

Halfshaft

REMOVAL & INSTALLATION

1. Loosen the front spindle nut.
2. Raise and safely support the vehicle.
3. Remove the front wheels and the spindle nut.

➡**If the halfshaft to be removed is installed into the intermediate shaft, the transaxle fluid does not need to be drained.**

4. Drain the transaxle fluid and install the drain plug with a new crush washer.

5. Remove the damper fork nut and damper pinch bolt.

6. Remove the damper fork.

7. Remove the cotter pin and castle nut from the lower arm ball joint. Install a hex nut flush onto the ball joint stud to prevent the ball joint tool from damaging the stud threads.

8. Using a ball joint tool, separate the lower arm from the knuckle.

9. Pull the knuckle outward. Remove the halfshaft outboard joint from the hub by tapping it with a plastic hammer.

10. Carefully pry the inner CV-joint away from the transaxle case to force the halfshaft set ring out of the groove.

11. Pull on the inboard CV-joint and remove the halfshaft from the differential case or intermediate shaft.

➡**Do not pull on the halfshaft as the CV-joint may come apart. Use care when prying out the assembly and pull it straight to avoid damaging the differential oil seal or intermediate shaft oil or dust seals.**

To install:

12. Replace the differential oil seal or intermediate shaft seal if either were damaged during removal.

13. Install new set rings on the ends of the halfshafts.

14. Install the halfshafts and be sure the set ring locks in the differential gear groove and the halfshaft bottoms in the differential or intermediate shaft.

15. Install the outboard joint into the hub. Be sure the splines mesh together and the joint is fully seated into the hub.

16. Fit the ball joint stud into the lower control arm. Install the damper fork into position. Tighten the upper damper pinch bolt to 32 ft. lbs. (44 Nm) and the fork nut to 47 ft. lbs. (65 Nm).

17. Tighten the ball joint castle nut to 40 ft. lbs. (55 Nm); then, tighten the nut just enough to install a new cotter pin.

18. Install the front wheels. Install a new spindle nut, but don't tighten it yet.

19. Lower the vehicle.

20. Tighten the spindle nut to 181 ft. lbs. (245 Nm) and stake its tab. Tighten the wheel nuts to 80 ft. lbs. (110 Nm).

21. Refill the transaxle with ATF. Use only Honda Premium ATF or an equivalent DEXRON®II or III ATF.

22. Warm the engine up, check the transaxle fluid level, and road test the vehicle.

STEERING AND SUSPENSION

Air Bag

✳✳ CAUTION

Some vehicles are equipped with an air bag system, also known as the Supplemental Inflatable Restraint (SIR) or Supplemental Restraint System (SRS). The system must be disabled before performing service on or around system components, steering column, instrument panel components, wiring and sensors. Failure to follow safety and disabling procedures could result in accidental air bag deployment, possible personal injury and unnecessary system repairs.

PRECAUTIONS

Please take note of the following precautions whenever working on or near air bag system components:

• When carrying a live air bag module, point the bag and trim cushion away from your body. When placing a live air bag on a bench or other surface, always face the bag and trim cushion up, away from the surface. Following these precautions will reduce the chance of injury if the air bag is accidentally deployed.

• Use only a digital multimeter when checking any part of the air bag system. The multimeter's output must be 0.01A (10mA) or less when it is switched to its smallest ohmmeter range value.

• Do not bump, strike, or drop any SRS component. Store SRS components away from moisture, oil, grease, and extreme heat and humidity.

• Do not cut, damage, or attempt to alter the SRS wiring harness or its yellow insulation.

• Do not install SRS components which have been recovered from wrecked or dismantled vehicles.

• Always disconnect both battery cables when working around SRS components or wiring.

• Always disable the air bag when working under the dashboard.

• Always check the alignment of the air bag cable reel during steering-related service procedures.

DISARMING

✳✳ CAUTION

The Supplemental Restraint System (SRS) must be disarmed before any of its components are disconnected or the air bag is removed. Failing to disarm the SRS before servicing its components may cause accidental deployment of the air bag, resulting in unnecessary SRS repairs and possible personal injury.

Driver's Air Bag

1995 VEHICLES

➡**The radio may contain a coded theft protection circuit. Always make note of the code number before disconnecting the battery.**

1. Disconnect the negative and positive battery cables.

2. Always wait at least three minutes after disconnecting the battery before working around the air bag.

3. Remove the steering wheel lower access cover.

4. Remove the clip securing the air bag module/cable reel connection to the steering column.

5. Remove the red shorting connector from its holder on the access cover.

6. Uncouple the air bag and cable reel connection. Immediately install the red shorting connector onto the air bag module connector.

7. After service has been completed, detach the shorting connector from the air bag module connector. Immediately couple the air bag and cable reel connectors.

8. Install the clip securing the air bag/cable reel connection to the steering column.

9. Place the red shorting connector back into its holder on the access cover. Install the access cover.

10. Reconnect the positive and negative battery cables.

11. Turn the ignition switch to the **ON** position, but don't start the engine. The SRS indicator light should turn ON for six seconds and then turn OFF. If the SRS indicator light doesn't come ON, or stays ON longer than six seconds, the system fault must be diagnosed.

12. Enter the radio security code.

1996–99 VEHICLES

➡The radio may contain a coded theft protection circuit. Always make note of the code number before disconnecting the battery.

1. Disconnect the negative and positive battery cables.

2. Always wait at least three minutes after disconnecting the battery before working around the air bag.

3. Remove the steering wheel lower access cover.

4. Remove the clip securing the air bag module/cable reel plug to the steering column.

➡Two types of air bag connections are used: spring-loaded ones, and ones requiring a shorting connector upon uncoupling.

5. Uncouple the air bag and cable reel connection. Immediately install the red shorting connector onto the air bag module plug.

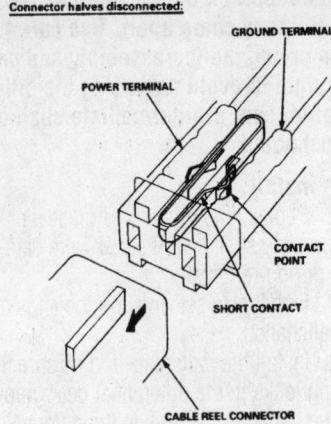

Connector halves disconnected:

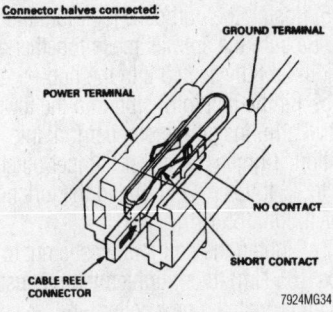

Connector halves connected:

7924MG34

Cut away view of a spring-loaded connector—1996–99 vehicles

➡Spring-loaded air bag connectors contain a spring-contact self-disabling contact. A shorting connector doesn't need to be installed on the driver's air bag.

6. If the vehicle is equipped with spring-loaded connectors:

   a. Hold the connector body, not the wiring.

   b. Pull the spring-loaded locking sleeve toward its stop while holding the opposite half of the connector.

   c. After releasing the locking sleeve, uncouple the connectors.

7. After servicing has been completed, couple the air bag and cable reel connectors. For spring-loaded connectors, press the sleeve side of the connector into the pawl side until the sleeve locks the connectors together.

8. Install the clip securing the air bag/cable reel connection to the steering column.

9. Install the access cover.

10. Reconnect the positive and negative battery cables.

11. Turn the ignition switch to the **ON** position, but don't start the engine. The SRS indicator light should turn ON for six seconds and then turn OFF. If the SRS indicator light doesn't come ON, or stays ON longer

than six seconds, the system fault must be diagnosed.

12. Enter the radio security code.

Passenger's Air Bag

1995 VEHICLES

➡The radio may contain a coded theft protection circuit. Always make note of the code number before disconnecting the battery.

1. Disconnect the negative and positive battery cables.

2. Always wait at least three minutes after disconnecting the battery before working around the air bag.

3. Remove the glove box door and frame. Remove any lower mounting brackets that may cover the air bag connection.

4. Remove the shorting connector from its holder. On some vehicles, the shorting connector is permanently attached to the passenger's air bag frame; this type of shorting connector stays in place.

5. Uncouple the air bag module connector from the yellow SRS main wiring harness. Immediately attach the air bag wiring plug to the red shorting connector.

6. After servicing has been completed, detach the shorting connector from the air bag module connector. Immediately couple the air bag and cable reel connectors.

7. Install the clip securing the air bag/SRS harness connection to the passenger's air bag frame.

8. Place the red shorting connector back into its holder.

9. Install any lower mounting brackets that may have been removed. Install the glove box frame and glove box door.

10. Reconnect the positive and negative battery cables.

11. Turn the ignition switch to the **ON** position, but don't start the engine. The SRS indicator light should turn ON for six seconds and then turn OFF. If the SRS indicator light doesn't come ON, or stays ON longer than six seconds, the system fault must be diagnosed.

12. Enter the radio security code.

1996–99 VEHICLES

➡The radio may contain a coded theft protection circuit. Always make note of the code number before disconnecting the battery.

1. Disconnect the negative and positive battery cables.

2. Always wait at least three minutes after disconnecting the battery before working around the air bag.

3. Remove the dashboard storage compartment.

4. Uncouple the air bag module connector from the yellow SRS main wiring harness. Immediately attach the air bag connector to the red shorting plug. Two types of air bag connections are used: spring-loaded ones, and ones requiring a shorting connector upon uncoupling.

➡**Spring-loaded air bag connectors contain a spring-contact self-disabling contact.**

5. If the vehicle is equipped with spring-loaded connectors:

   a. Hold the connector body, not the wiring.

   b. Pull the spring-loaded locking sleeve toward its stop while holding the opposite half of the connector.

   c. After releasing the locking sleeve, uncouple the connectors.

6. After servicing has been completed, detach the shorting connector from the air bag module. Immediately couple the air bag and SRS harness plugs. For spring-loaded connectors, press the sleeve side of the connector into the pawl side until the sleeve locks the connectors together.

7. Install the clip securing the air bag/SRS harness connection to the passenger's air bag frame.

8. Install the dashboard storage compartment.

9. Reconnect the positive and negative battery cables.

10. Turn the ignition switch to the **ON** position, but don't start the engine. The SRS indicator light should turn ON for six seconds and then turn OFF. If the SRS indicator light doesn't come ON, or stays ON longer than six seconds, the system fault must be diagnosed.

11. Enter the radio security code.

Power Rack and Pinion Steering Gear

REMOVAL & INSTALLATION

➡**The original radio may contain a coded anti-theft circuit. Obtain the security code number before disconnecting the battery cable.**

1. Drain the fluid from the power steering system:

   a. Lift the power steering reservoir off of its mount and disconnect the inlet hose.

   b. Insert a length of tubing into the inlet hose and route the tubing into a drain container.

   c. With the engine running at idle, turn the steering wheel lock-to-lock several times until fluid stops running out of the hose. Then, immediately shut the engine **OFF**.

2. Position the front wheels straight ahead. Lock the steering column with the ignition key. Reconnect the reservoir inlet hose.

3. Disconnect the negative and positive battery cables. Wait at least three minutes before working around the airbags.

4. Raise and safely support the vehicle.

5. Remove the steering joint cover and remove the lower steering joint bolts. Disconnect the steering joint by sliding it up toward the steering column.

6. Remove the front wheels.

7. Remove the tie-rod ends, cotter pins and castle nuts. Install a 10mm nut onto the end of the ball joint stud so the threads won't be damaged by the ball joint remover. Using a ball joint removal tool, disconnect the tie rod ends from the steering knuckles.

8. Remove the self-locking nuts and separate the catalytic converter and the joint pipe from exhaust pipe and the front muffler. Remove the catalytic converter. Be careful not to damage the oxygen sensors; disconnect their electrical leads if necessary.

9. Unbolt the fluid return line clamp from the top of the rear subframe beam.

10. Use a flare nut wrench to disconnect the two hydraulic lines from the rack valve body. Plug the lines to keep dirt and moisture out. Carefully move the disconnected lines to the rear of the rack assembly so that they are not damaged when the rack is removed.

11. Push on the left side inner tie rod to position the rack all the way to the right.

12. Remove the rack stiffener plate, then remove the right steering rack mounting bolts.

13. Pull the steering rack down to release it from the pinion shaft.

14. Drop the steering rack far enough to permit the end of the pinion shaft to come out of the hole in the frame channel.

15. Slide the steering rack to the right until the left tie rod clears the subframe, then drop it down and out of the vehicle to the left.

To install:

➡**Use new gaskets and self-locking nuts when installing the catalytic converter.**

16. Before installing the rack and pinion, slide the rack's ends all the way to the right. Install the pinion shaft grommet. The lug on the pinion shaft grommet aligns with the slot on the top of the valve body.

✳✳ WARNING

Use only genuine Honda Type-V power steering fluid. Any other type or brand of fluid will damage the power steering pump.

17. Install the steering rack into position. Install the pinion shaft grommet and insert the pinion through the hole in the firewall.

18. Install the rack mounting bolts. Tighten the bracket bolts to 28 ft. lbs. (39 Nm). Tighten the stiffener plate mounting bolts to 32 ft. lbs. (43 Nm).

19. Reconnect the two hydraulic lines to the rack valve body. Carefully tighten the 14mm inlet fitting to 27 ft. lbs. (37 Nm) and the 16mm outlet fitting to 21 ft. lbs. (28 Nm).

20. Install the catalytic converter using new gaskets and self-locking nuts. Tighten the self-locking nuts to 16 ft. lbs. (22 Nm). Reconnect the oxygen sensors if they were disconnected.

21. Center the rack ends within their steering strokes.

22. Install the tie rod ends onto the rack ends. Connect the tie rod ends to the steering knuckles and install the castle nuts. Install the front wheels.

23. Verify that the rack is centered within its strokes.

24. Center the SRS cable reel as follows:

   a. Turn the steering wheel clockwise until it stops.

   b. Turn the steering wheel counterclockwise until the yellow gear tooth lines up with the alignment mark on the lower column cover.

25. Line up the bolt hole in the steering joint with the groove in the pinion shaft. Slip the joint onto the pinion shaft. Pull the joint up and down to be sure the splines are fully seated. Tighten the joint bolts to 22 ft. lbs. (30 Nm).

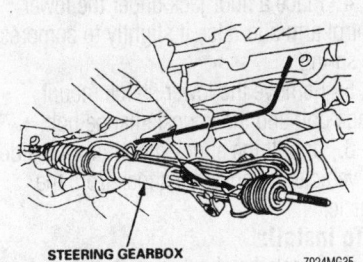

STEERING GEARBOX   7924MG35

Remove and install the rack and pinion gear by moving it as shown

➡**Connect the steering joint and pinion shaft with the cable reel and steering rack centered. Verify that the lower joint bolt is securely seated in the pinion shaft groove. If the steering wheel and rack are not centered, reposition the serrations at the lower end of the steering joint.**

26. Install the steering joint cover.
27. Tighten the tie rod end castle nuts to 29–35 ft. lbs. (40–48 Nm). Then, tighten them only enough to install new cotter pins.
28. Lower the vehicle. Reconnect the negative battery cable.
29. Be sure the reservoir inlet line has been reconnected. Fill the reservoir to the upper line with Honda power steering fluid. Run the engine at idle and turn the steering wheel lock-to-lock several times to bleed any air from the system and fill the rack valve body with fluid. Recheck the fluid level and add more if necessary.
30. Check the power steering system for leaks.
31. Check the front wheel alignment and steering wheel spoke angle. Make adjustments by turning the left and right tie rod ends equally.
32. Road test the vehicle.
33. Enter the radio security code.

Strut

REMOVAL & INSTALLATION

Front

1. Raise and safely support the vehicle.
2. Remove the front wheels.
3. Remove the brake hose clamp bolts from the strut.
4. Remove the damper fork bolts and remove the damper fork.
5. Remove the three strut mounting nuts. Remove the strut from the vehicle.

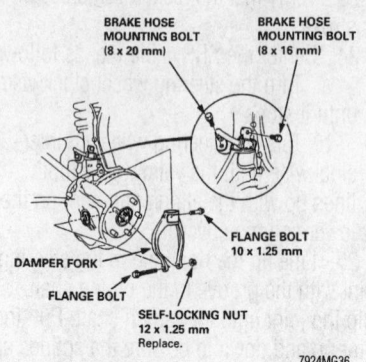

BRAKE HOSE MOUNTING BOLT (8 x 20 mm)
BRAKE HOSE MOUNTING BOLT (8 x 16 mm)
FLANGE BOLT 10 x 1.25 mm
DAMPER FORK
FLANGE BOLT
SELF-LOCKING NUT 12 x 1.25 mm Replace.

7924MG36

Identification of some of the front suspension components

6. Inspect the strut mounts for wear and damage. Replace any damaged or worn parts.

To install:

➡**Use new self-locking bolts when installing the struts and assembling the damper forks.**

7. Install the strut into the vehicle. Hand-tighten the mounting nuts.
8. Install the strut into the damper fork. The alignment mark on the strut tube fits into the groove on the damper fork.
9. Install the pinch bolt and damper fork bolt. Only hand-tighten these bolts.
10. Install the front wheels and lower the vehicle.
11. With all four of the vehicle's wheels on the ground, tighten the damper fork nut to 47 ft. lbs. (65 Nm) while holding the damper fork bolt. Tighten the damper fork pinch bolt to 32 ft. lbs. (44 Nm). Tighten the strut mounting nuts to 28 ft. lbs. (39 Nm).
12. Tighten the wheel nuts to 80 ft. lbs. (110 Nm).
13. Check and adjust the vehicle's front end alignment.

Shock

REMOVAL & INSTALLATION

Rear

➡**After removing and installing any rear suspension component, the Load Sensing Proportioning Valve (LSPV) spring length must be checked and adjusted. This step is important; the LSPV determines the fluid pressure for the rear brakes.**

1. Remove the cup holder from the top of the right rear interior trim. Remove the cup holder, storage tray, and jack from the left rear interior trim.
2. Raise and support the vehicle safely.
3. Remove the rear wheels.
4. Place a floor jack under the lower control arm and raise it slightly to compress the spring.
5. Remove the lower shock mount flange bolt and the knuckle flange bolt.
6. Unbolt the shock mount from inside the vehicle. Remove the shock from the vehicle.

To install:

7. Check the shock mount and bushings and replace any that are damaged. Assemble the mount, bushing, and stopper on the shock.

8. Install the shock into the vehicle. Install new self-locking upper mounting nuts.
9. Raise the lower control arm with a floor jack. Be sure the coil spring is properly seated.
10. Install the shock absorber and knuckle flange bolts.
11. Raise the jack enough to take up the weight of the vehicle. Tighten both of the flange bolts to 76 ft. lbs. (103 Nm).
12. Lower the floor jack. Install the rear wheels. Lower the vehicle.
13. Tighten the shock mount nuts to 28 ft. lbs. (39 Nm), and the shock piston nut to 22 ft. lbs. (29 Nm). Install the jack, tool tray, and cup holders.
14. Tighten the wheel nuts to 80 ft. lbs. (110 Nm).
15. Check the adjustment of the load sensing proportioning valve spring as follows:
    a. Be sure the vehicle is not loaded with cargo. Release the parking brake.
    b. Note the level of fuel in the tank and compare it with the chart to determine the degree of adjustment needed.
    c. Insert a metal pin 5.0–5.3mm in diameter into the 5mm diameter hole in the LSPV arm.
    d. Use a caliper to measure the distance between the 5mm pin and the 8mm adjusting bolt thread. This is the length of the LSPV spring.
    e. If the measurement is out of specification, loosen the 8mm adjusting bolt and adjust the spring length to specifica-

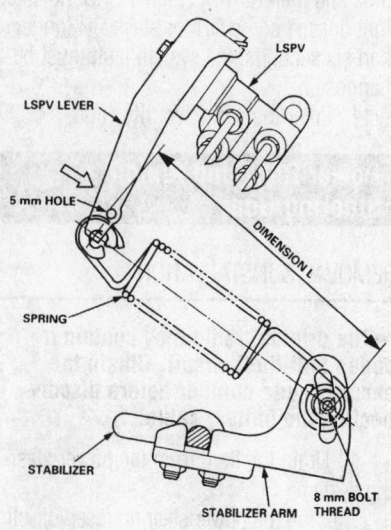

LSPV
LSPV LEVER
5 mm HOLE
DIMENSION L
SPRING
STABILIZER
STABILIZER ARM
8 mm BOLT THREAD

7924MG40

Load sensing proportioning valve spring length and adjusting bolt

Example:

Type: LX, 7-passenger
Fuel level: 1/2

The table shows that dimension L of fully-fueled 7-passenger LX 7 is 132 mm (5.197 in). However, because the fuel level is 1/2, dimension L should be compensated by – 2 mm.
Therefore, it should be adjusted at 130 mm (5.118 in).

U.S.A. MODEL:

mm (in)

Type	Dimension L	Fuel Level			
		3/4	1/2	1/4	0
LX 6-pass.	131 (5.157)	– 1 (– 0.039)	– 2 (– 0.079)	– 3 (– 0.012)	– 4 (– 0.157)
LX 7-pass.	132 (5.197)	↑	↑	↑	↑
EX	131 (5.157)	↑	↑	↑	↑

CANADA MODEL:

mm (in)

Passenger	Dimension L	Fuel Level			
		3/4	1/2	1/4	0
6-pass.	131 (5.157)	– 1 (– 0.039)	– 2 (– 0.079)	– 3 (– 0.012)	– 4 (– 0.157)
7-pass.	132 (5.197)	↑	↑	↑	↑

NOTE: If the vehicle is equipped with a trailer hitch, add 3 mm to dimension L before compensating for the fuel level.

7924MG41

Load sensing proportioning valve specifications chart

tion according to the values on the chart. Tighten the adjusting nut to 106 inch lbs. (12 Nm).
16. Check and adjust the rear wheel alignment.
17. Test drive the vehicle and check for proper brake system operation.

Coil Spring

REMOVAL & INSTALLATION

Front

1. Remove the strut from the vehicle.
2. Place the strut in a vice and install a spring compressor onto the coil spring. Follow the spring compressor manufacturer's instructions.
3. Compress the spring and remove the self-locking nut from the top of the strut. Disassemble the strut mounts and remove the coil spring.
To install:
➡**Use new self-locking nuts when assembling and installing the struts.**

4. Install the spring compressor onto the coil spring. Set the spring onto the strut cartridge. The flat part of the coil spring is its top.

5. Assemble the strut mount and its washer onto the strut. Tighten the self-locking nut to 22 ft. lbs. (29 Nm). Remove the spring compressor.
6. Install the strut into the vehicle.

Rear

➡**After removing and installing any rear suspension component, the Load Sensing Proportioning Valve (LSPV) spring length must be checked and adjusted. This step is important; the LSPV determines the fluid pressure for the rear brakes.**

1. Raise and safely support the vehicle.
2. Remove the rear wheels.
3. Place a floor jack under the lower control arm spring perch and raise it slightly.
4. Remove the shock absorber flange bolt and the knuckle flange bolt from the lower control arm.
5. Slowly lower the floor jack to release the tension on the coil spring.
6. Remove the coil spring and the upper and lower spring seats.
To install:
7. Replace the upper and lower spring seats if they are distorted or have disintegrated.
8. Install the spring seats into position.
9. Install the coil spring. Align the ends of the coil with the notches on the spring seats.
10. Raise the floor jack under the lower control arm to compress the spring.
11. Install the shock absorber and knuckle flange bolts. Hand-tighten them only at this point.
12. Lower the jack and move it under the knuckle. Raise the jack under the knuckle until it is supporting the weight of the vehicle. Tighten each of the two flange bolts to 76 ft. lbs. (103 Nm).
13. Lower the floor jack. Install the rear wheels. Lower the vehicle.
14. Tighten the wheel nuts to 80 ft. lbs. (110 Nm).
15. Check the adjustment of the load sensing proportioning valve spring:
   a. Be sure the vehicle is not loaded with cargo. Release the parking brake.
   b. Note the level of fuel in the tank and compare it with the chart to determine the degree of adjustment needed.
   c. Insert a metal pin 5.0mm in diameter into the 5mm diameter hole in the LSPV arm.
   d. Use a caliper to measure the distance between the 5mm pin and the 8mm

adjusting bolt thread. This is the length of the LSPV spring.
   e. If the measurement is out of specification, loosen the 8mm adjusting bolt and adjust the spring length to specification according to the values on the chart. Tighten the adjusting nut to 106 inch lbs. (12 Nm).
16. Check and adjust the rear wheel alignment.
17. Test drive the vehicle and check for proper brake system operation.

Upper Ball Joint

REMOVAL & INSTALLATION

The upper ball joints cannot be replaced separately. If the ball joints become worn or damaged, the whole upper arm must be replaced.

Upper Control Arm

REMOVAL & INSTALLATION

1. Raise and safely support the vehicle.
2. Remove the front wheels. Support the lower control arm assembly with a floor jack.
3. Remove the damper fork bolt and damper fork pinch bolt. Remove the damper fork.
4. Separate the upper ball joint from the steering knuckle using a ball joint separator tool.
5. Unbolt the brake hose clips from the strut tube.
6. Remove the three strut mounting nuts. Remove the strut from the vehicle.
7. Remove the self-locking nuts from the upper arm anchor bolts. Remove the upper arm from the vehicle.

➡**Do not disassemble the upper arm. If the ball joint or bushings are faulty, or the upper arm is damaged, the entire upper arm must be replaced.**

To install:
➡**Use new self-locking nuts when installing the upper arm and strut.**

8. Install the upper control arm assembly into the strut tower.
9. Install the strut into the vehicle. Connect the damper fork bolt and pinch bolt.
10. Connect the upper ball joint. Connect the brake hose clips to the strut tube.
11. Install the front wheels and lower the vehicle.
12. With all four of the vehicle's wheels on the ground, tighten the upper control

arm nuts to 47 ft. lbs. (65 Nm). Tighten the strut mounting nuts to 28 ft. lbs. (39 Nm). Tighten the damper fork pinch bolt to 32 ft. lbs. (44 Nm) and the damper fork bolt to 47 ft. lbs. (65 Nm). Tighten the castle nut to 32 ft. lbs. (44 Nm), then tighten it only enough to install a new cotter pin.

13. Tighten the wheel nuts to 80 ft. lbs. (110 Nm).

14. Check and adjust the vehicle's front end alignment.

Lower Ball Joint

REMOVAL & INSTALLATION

➡This procedure is performed after the removal of the steering knuckle and requires the use of special tools or their equivalent: Ball Joint Remover/Installation tools. Additionally, a large vise will be required. At installation, Clip Guide tool No. 07974-SA50700 or 07GAG-SD40700 will be required.

1. Remove the steering knuckle assembly from the vehicle.

2. Remove the brake disc and hub assembly from the knuckle.

3. Pry off the snapring and remove the ball joint boot.

4. Pry the snapring out of the groove in the ball joint.

5. Install the ball joint removal tool with the large end facing out. Install and tighten the ball joint castle nut to hold the tool in position.

6. Position the removal base tool on the ball joint and set the assembly in a large vise. Press the ball joint out of the steering knuckle.

To install:

7. Position the new ball joint into the hole of the steering knuckle.

8. Install the ball joint installer tool with the small end facing out.

9. Position the installation base tool on the ball joint and set the assembly in a large vise. Press the ball joint into the steering knuckle.

10. Seat the snapring in the groove of the ball joint.

11. Pack the interior of the ball joint boot with grease.

12. Adjust the boot clip guide tool with its adjusting bolt until the end of the tool aligns with the groove on the boot. Slide the clip over the tool and into position on the ball joint boot.

13. Install the hub assembly and brake disc onto the knuckle.

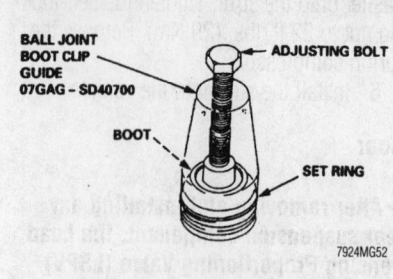

Be sure to adjust the ball joint boot clip before installing it into the steering knuckle

14. Install the steering knuckle assembly into the vehicle.

15. Check and adjust the vehicle's front wheel alignment.

Wheel Bearings

ADJUSTMENT

1. Raise and support the vehicle safely.

2. Remove the front and/or rear wheels.

3. Install the lug nuts and tighten them to 80 ft. lbs. (110 Nm).

4. Use a dial gauge to measure front

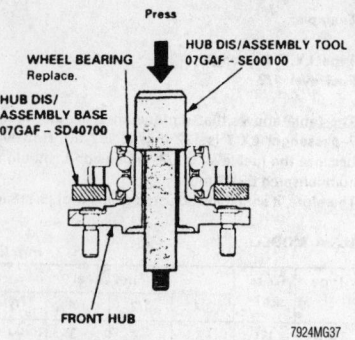

Removing the hub from the wheel bearing using the disassembly tools

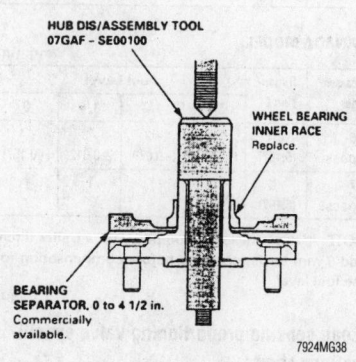

Pressing out the wheel bearing inner race

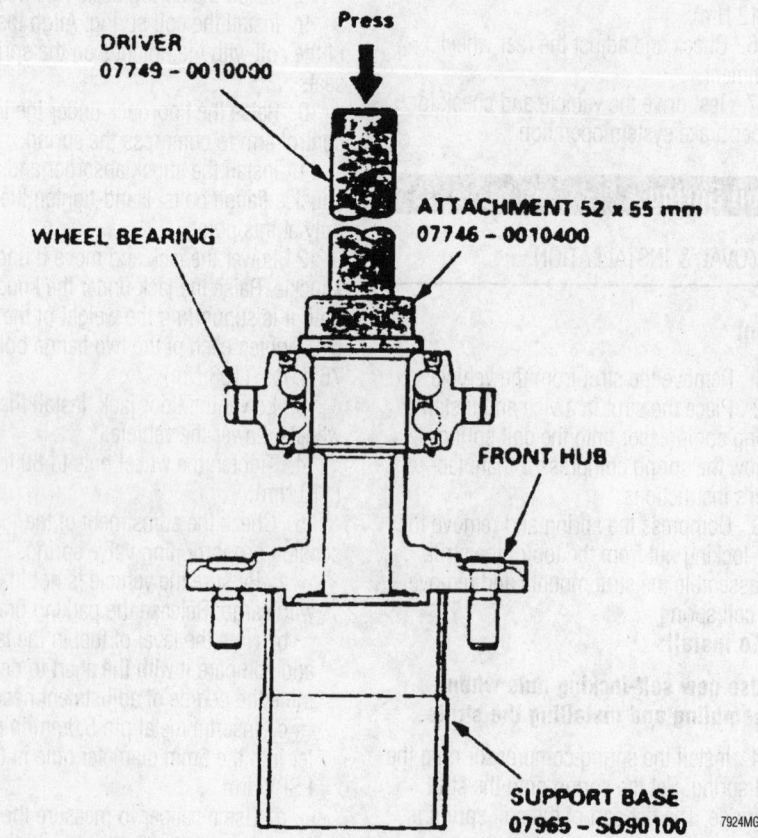

Utilizing the Hub Support Base and Driving Attachment tools to install the new wheel bearing

bearing end-play at the hub flange, or to measure rear bearing end-play at the center of the hub's grease cap.

5. Pull the rotor assembly in and out to measure the bearing play. Compare the dial gauge readings.

6. The standard bearing end-play for both front and rear wheels is 0.000–0.002 in. (0.00–0.05mm). If the end-play measurement exceeds the standard, the wheel bearings must be replaced. The wheel bearings cannot be adjusted.

REMOVAL & INSTALLATION

Front

➡**Once the wheel bearing is removed, it must be replaced. A hydraulic press and bearing drivers are required to remove and install the wheel bearing. The following Honda tools or their equivalents are needed: Hub Assembly tool 07GAF-SE0100, Hub Bases 07965-SD90100, Bearing Driver 07749–0010000, 52 x 55mm Driving Attachment 07746–0010400.**

1. Pry the spindle nut stake away from the spindle, then loosen the nut. Do not tighten or loosen a spindle nut unless the vehicle is sitting on all four wheels. The torque is high enough to cause the vehicle to fall even when properly supported.

2. Raise and safely support the vehicle.

3. Remove the wheel and the spindle nut.

4. Unbolt the ABS wheel sensor and its cable from the knuckle. Don't disconnect the cable, wire it out of the way.

5. Remove the brake disc. If the disc if seized, evenly screw two 8mm bolts into the threaded holes to pop it loose.

6. Remove the caliper mounting bolts and the caliper. Support the caliper out of the way with a length of wire. Do not let the caliper hang from the brake hose.

7. Remove the cotter pin from the tie rod castle nut, then remove the nut. Separate the tie rod ball joint using a ball joint remover, then lift the tie rod out of the knuckle.

8. Remove the cotter pin and loosen the lower arm ball joint nut half the length of the joint threads. The nut will keep the arm from flying off of the joint.

9. Separate the ball joint and lower arm using a ball joint puller with the pawls applied to the lower arm. Avoid damaging the ball joint boot. If necessary, apply penetrating lubricant to loosen the ball joint.

10. Pull the knuckle outward to separate

it from the halfshaft outboard joint. If necessary, use a soft-faced mallet to drive the knuckle off the axle shaft.

11. Remove the cotter pin and the upper ball joint nut. Separate the upper ball joint and remove the knuckle assembly.

12. Remove the four bolts and remove the hub unit from the knuckle.

13. Remove the splash guard from the knuckle.

14. Position the hub in a hydraulic press. Press the hub out of the wheel bearing. The inner bearing race may stay on the hub.

15. Remove the outboard bearing inner race from the hub using a bearing puller.

To install:

16. Clean the knuckle and hub thoroughly.

17. Position the hub in a hydraulic press. Press a new wheel bearing onto the hub using a press driver of the correct diameter. Be sure the press tool contacts only the inner bearing race.

18. Install the splash shield.

19. Install the hub/bearing assembly onto the knuckle and tighten the bolts to 33 ft. lbs. (45 Nm).

20. Install the knuckle/hub assembly on the vehicle. Be sure the hub is fully seated onto the axle shaft. Tighten the upper ball joint nut and the tie rod nut to 32 ft. lbs. (44 Nm) and install new cotter pins. Tighten the lower ball joint nut to 40 ft. lbs. (55 Nm) and install a new cotter pin.

21. Install the brake disc and caliper. Tighten the brake caliper bolts to 80 ft. lbs. (110 Nm). Tighten the brake disc retaining screws to 88 inch lbs. (10 Nm).

22. Install the front wheels and lower the vehicle.

23. With all four wheels resting on the ground, install a new spindle nut and tighten it to 180 ft. lbs. (245 Nm). After tightening, use a drift to stake the spindle nut shoulder against the spindle.

24. Tighten the wheel nuts to 80 ft. lbs. (110 Nm).

25. Check and adjust the vehicle's front wheel alignment.

Rear

EXCEPT CR-V

1. Loosen the hub spindle nut.

2. Raise and safely support the vehicle and remove the rear wheels.

3. Engage the parking brake.

4. Remove the caliper bracket mounting bolts. Use a piece of wire to hang the caliper out of the way.

5. Remove the two 6mm brake rotor retaining screws. If the rotor has seized onto the hub, screw two 8mm bolts into the threaded holes to pop the rotor loose.

6. Release the parking brake. Remove the brake rotor.

7. Remove the spindle nut and washer. Remove the hub unit.

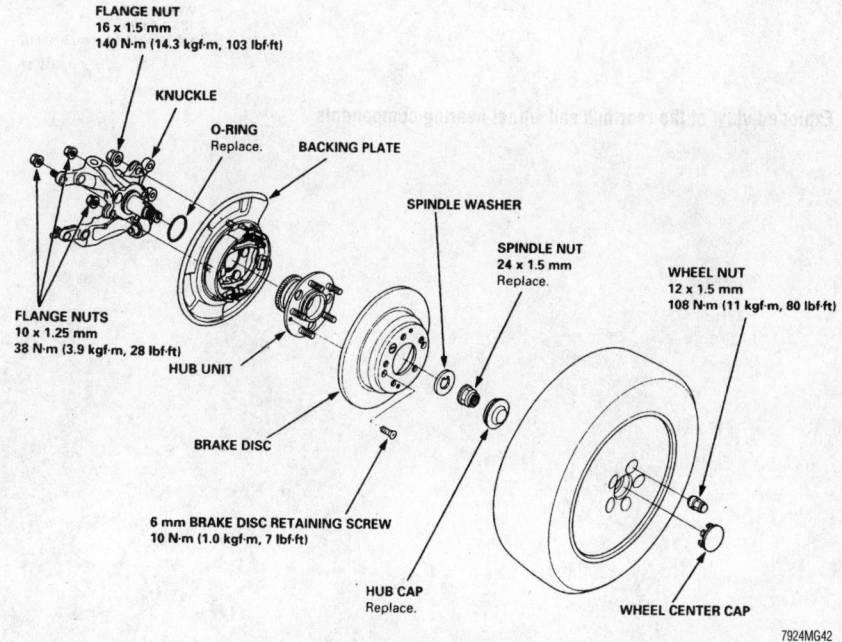

Exploded view of the rear hub and wheel bearing components

To install:

8. Clean the hub unit in solvent. Inspect the hub unit and wheel bearing for damage.

➡ **If the wheel bearing is faulty, the entire hub unit must be replaced.**

9. Clean excess brake dust and grease from the backing plate and brake rotor.

10. Install the hub unit, spindle washer, and spindle nut. Only hand-tighten the spindle nut.

11. Install the brake rotor and tighten the retaining screws to 88 inch lbs. (10 Nm).

12. Install the caliper and tighten the bracket mounting bolts to 28 ft. lbs. (39 Nm).

13. Install the rear wheels and lower the vehicle.

14. Tighten the spindle nut to 181 ft. lbs. (245 Nm).

15. Tighten the wheel nuts to 80 ft. lbs. (110 Nm).

16. Road test the vehicle and check the operation of the brakes.

CR-V

1. Raise and safely support the rear of the vehicle and remove the wheels.

2. Remove the spindle nut.

3. Remove the brake drum, shoes and parking brake cable.

4. Disconnect the brake line from the wheel cylinder.

5. Remove the three flange bolts from the trailing arms.

6. Remove the backing plate/hub bearing assembly from the trailing arm using a two jaw puller.

7. Press the wheel bearing inner race from the hub using the illustrated tools.

8. Remove the four flange bolts from the backing plate and separate the two, then remove the O-ring.

To install:

9. Install a new O-ring on the hub bearing unit.

10. Attach the hub to the backing plate and tighten the bolts to 47 ft. lbs. (64 Nm).

11. Install the bearing unit to the rear hub as shown.

12. Bolt the backing plate to the trailing arm and tighten to 76 ft. lbs. (103 Nm).

13. Connect the brake line to the wheel cylinder.

14. The completion of installation is the reverse order of the removal procedure, noting the following items:

• Apply some motor oil to the seating surface of the nut and tighten to 134 ft. lbs. (181 Nm).

• Using a drift, stake the spindle nut shoulder to the halfshaft.

• Bleed the brakes.

• Adjust the parking brake.

• Check the rear wheel alignment.

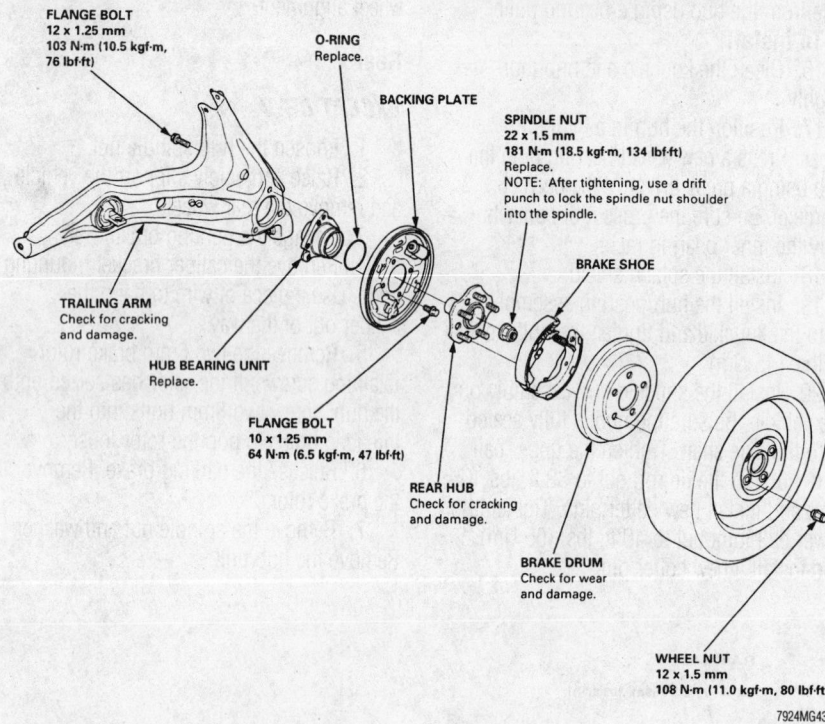

FLANGE BOLT
12 x 1.25 mm
103 N·m (10.5 kgf·m, 76 lbf·ft)

O-RING
Replace.

BACKING PLATE

SPINDLE NUT
22 x 1.5 mm
181 N·m (18.5 kgf·m, 134 lbf·ft)
Replace.
NOTE: After tightening, use a drift punch to lock the spindle nut shoulder into the spindle.

BRAKE SHOE

TRAILING ARM
Check for cracking and damage.

HUB BEARING UNIT
Replace.

FLANGE BOLT
10 x 1.25 mm
64 N·m (6.5 kgf·m, 47 lbf·ft)

REAR HUB
Check for cracking and damage.

BRAKE DRUM
Check for wear and damage.

WHEEL NUT
12 x 1.5 mm
108 N·m (11.0 kgf·m, 80 lbf·ft)

7924MG43

Exploded view of the rear hub and wheel bearing components

HONDA and ISUZU

23

Honda-Passport • **Isuzu**-Rodeo • Amigo

ENGINE REPAIR

➡ **Disconnecting the negative battery cable on some vehicles may interfere with the functions of the on-board computer systems and may require the computer to undergo a relearning process, once the negative battery cable is reconnected.**

Distributor

REMOVAL

1. Rotate the engine and bring No. 1 cylinder to top dead center of its compression stroke.

➡ **To bring the engine to TDC of the No. 1 compression stroke, remove the spark plug for the No. 1 cylinder. With the engine cool, turn the crankshaft over until compression is forced out of the spark plug hole. Watch the crankshaft damper while feeling for compression. When compression is felt, align the mark on the crankshaft damper with the 0° mark on the timing cover.**

2. Remove the air cleaner assembly, if needed.

3. Disconnect the negative battery cable. Detach and tag all the electrical connectors along with the spark plug wires from the distributor.

➡ **Disconnecting the negative battery cable on some vehicles may interfere with the functions of the on-board computer systems and may require the computer to undergo a relearning process once the negative battery cable is reconnected.**

4. Disconnect the vacuum hoses from the vacuum advance, if so equipped.

5. Make an alignment mark on the base of the distributor and the cylinder head or engine block. Also mark the position of the rotor to the distributor housing.

6. Remove the distributor hold-down bolt and bracket, then lift out the distributor assembly. Remove the distributor housing seal and discard it. Lubricate the new seal with clean engine oil before installation.

INSTALLATION

Timing Not Disturbed

1. Install the distributor assembly. Be sure to align the scribe marks made during disassembly. Lightly tighten the distributor mounting bolt.

2. Connect the vacuum hoses to the vacuum controller.

3. Attach all the electrical connectors along with the spark plug wires to the distributor.

4. Install the air cleaner, if removed.

5. Connect the negative battery cable. Start the engine, set the timing and check the idle speed. Tighten the distributor mounting bolts to 14 ft. lbs. (19 Nm).

Timing Disturbed

1. Remove the No. 1 spark plug.

2. Rotate the crankshaft in the direction of rotation until compression is felt at the spark plug hole.

3. Continue rotating the engine in the same direction until the timing marks line up and when No. 1 cylinder is at TDC.

4. Align the rotor with the No. 1 tower on the distributor cap and install the distributor.

5. Reconnect the distributor wiring, ignition wires, and vacuum hoses.

6. Install the air cleaner, if removed. Connect the negative battery cable.

7. Check and/or adjust the ignition timing when finished.

8. Tighten the distributor mounting bolts to 14 ft. lbs. (19 Nm).

Ignition Timing

ADJUSTMENT

2.2L and 3.2L Engines

Vehicles with these engine are equipped with a distributorless ignition system. The ignition timing is controlled by the PCM through the input of engine control system sensors. The ignition timing cannot be adjusted.

2.6L Engine

1. Connect a timing light to the No. 1 spark plug wire.

2. Set the parking brake and block the wheels.

3. Start the engine and allow it to warm up.

4. Be sure the air conditioner is OFF.

5. Disconnect and plug the evaporative emission canister purge line.

6. Disconnect and plug the exhaust gas recirculation vacuum lines.

7. While the engine idles, point the timing light at the notched line on the crankshaft pulley.

8. The ignition timing should be 12° BTDC at 900 rpm.

9. If adjustment is needed, loosen the distributor mounting bolts and turn the distributor counterclockwise to advance the timing or clockwise to retard the timing.

10. Tighten the distributor mounting bolts and recheck the timing and the idle.

➡ **When tightening the distributor mounting bolt, be sure that the distributor body does not rotate together with the mounting bolt.**

11. After everything has been rechecked, reconnect the vacuum lines and remove the timing light.

Engine Assembly

REMOVAL & INSTALLATION

2.2L Engines

1. Disconnect the negative and positive battery cable, then remove the battery.

2. Use a felt-tipped marker to match-mark the hood hinge plates. Remove the hood.

3. Drain and recycle the engine coolant.

4. Properly relieve the fuel system pressure.

5. Disconnect the throttle cable.

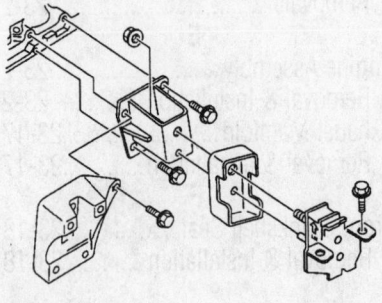

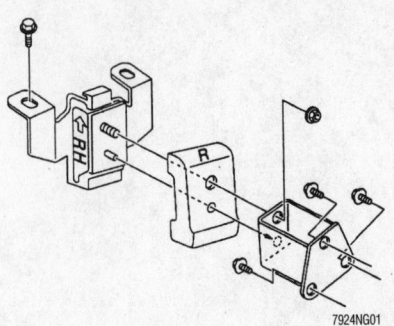

7924NG01

Left and right motor mount nut and bolt locations—2.2L engine

6. Remove the air cleaner and duct from the vehicle.

7. Label and detach all electrical connectors and vacuum lines form the engine assembly.

8. Disconnect the engine to body ground cables.

9. Remove the clutch cover from the front of transmission.

10. Disconnect the clutch hydraulics.

11. Remove the exhaust header pipe from the vehicle.

12. Remove the transmission crossmember.

13. Disconnect and plug the two fuel pipes from the right side of the transmission.

14. Remove the driveshaft from the vehicle.

15. Remove the engine-to-transmission bolts, then remove the transmission.

16. Unbolt the power steering pump from its bracket. Move the pump out of the way with the hydraulic line connected.

17. Remove the heater and radiator hoses, then the radiator.

18. Unbolt the engine mounts.

19. Raise the engine slightly. Verify that all vacuum lines and wiring harnesses have been disconnected so that the engine removal is not obstructed.

20. Raise the chain hoist to lift the engine out of the vehicle. If necessary, keep the front of the engine higher than the rear to clear the bulkhead.

To install:

➡**When assembling the engine and transmission, be sure the dowels are mounted in the specified positions on the engine block.**

21. Position the engine in the engine compartment and tighten the mounting bolts to the frame to 30 ft. lbs. (41 Nm).

22. Install the heater and radiator hoses, then the radiator.

23. Install the power steering assembly.

24. Install the transmission assembly, referring to the transmission repair section.

25. Attach the driveshaft to the transmission.

26. Connect the two fuel pipes to the right side of the transmission.

27. Install the clutch cover and hydraulic line.

28. Install and tighten the transmission crossmember mounting nuts/bolts to 36 ft. lbs. (50 Nm).

29. Install the exhaust header pipe.

30. Connect the engine to body ground cables.

31. Engage all electrical connectors and vacuum lines form the engine assembly.

32. Install the air cleaner and duct to the vehicle.

33. Connect the throttle cable.

34. Align the matchmarks and install the hood.

35. Verify that all vacuum lines, cooling hoses, control cables, and electrical connections have been routed and attached properly.

36. Install the battery. Reconnect the positive and negative battery cables.

37. Crank the engine until it starts to remove any air from the fuel lines. Carefully check all fuel lines and fittings for signs of leakage.

38. Warm it up to normal operating temperature. Check the adjustment and operation of the throttle cable.

39. Shut the engine off. Recheck the drive belt tensions.

40. Check all fluid levels and refill as necessary.

41. Test drive the vehicle.

2.6L Engine

➡**The transmission and transfer case assembly (if equipped) should be completely removed from the vehicle before the engine is removed. If you chose to leave the transmission in the vehicle after separating it from the engine, it must be securely supported. The transfer case shouldn't be separated from the transmission.**

1. Relieve the fuel pressure.

2. Disconnect the negative and positive battery cables. Remove the battery.

3. Use a felt-tipped marker to matchmark the hinge plates. Remove the hood.

4. Remove the radiator skid plate. Drain the coolant from the radiator and engine block.

5. Remove the air cleaner box and the intake air duct. Cover the throttle body port with a shop towel to prevent dirt from entering the engine.

6. Disconnect the throttle cable from the throttle body linkage.

7. Label and disconnect the following:
- Air switch valve hose
- Oxygen sensor harness
- Power booster vacuum hose
- Alternator wiring harness
- Fuel pressure regulator vacuum hose
- Canister hose
- Engine wiring harness connectors located on the right wheel well
- Inlet and return fuel lines
- Starter motor cables
- Engine ground cables
- Oil pressure switch connectors

8. Remove the radiator grille from the deflector panel.

9. Remove the radiator.

10. If equipped with A/C, remove the compressor from the engine and move it aside. Do not disconnect the A/C lines.

11. If equipped with a manual transmission, remove the gear shift lever by performing the following procedure:

a. Place the gear shift lever in neutral.

b. Remove the front console.

c. Pull the shift lever boot and grommet upward.

d. Remove the shift lever cover bolts and the shift lever.

12. If equipped with four-wheel drive, remove the transfer case shift lever by performing the following steps:

a. Place the transfer shift lever in **2H**.

b. Pull the shift lever boot and dust cover upward.

c. Remove the shift lever retaining bolts.

d. Pull the shift lever from the transfer case.

13. Raise and safely support the vehicle. Remove the front wheels.

14. Disconnect the back-up light switch and the vehicle speed sensor cable from the transmission.

15. Remove the transmission and transfer case skid plates.

16. Drain the oil from the engine, transmission and transfer case, if equipped.

17. If equipped with an automatic transmission, perform the following steps:

a. Remove the dipstick and the tube.

b. Disconnect the shift select control link rod from the select lever.

c. Disconnect the downshift cable from the transmission.

d. Disconnect and plug the fluid coolant lines from the transmission.

18. Remove the driveshaft(s).

19. Remove the starter.

20. If equipped with a clutch slave cylinder, remove it from the transmission and move it aside.

21. Unbolt the front exhaust pipe flanges and separate the front pipe from the exhaust system.

22. Attach engine hangers to the engine and connect a chain hoist to the engine hangers and support the engine.

23. Remove the transmission and transfer case assembly (if equipped).

24. Support the weight of the engine with the chain hoist. Unbolt the engine mounts.

25. Using the hoist, slowly lift the engine from the vehicle; be sure to hold the front of the engine higher than the rear.

26. Place the engine on a workstand.

To install:

27. Using the hoist, slowly lower the engine into the vehicle. Be sure to hold the front of the engine higher than the rear.

28. Install the engine mount nuts and bolts. Tighten the engine mount and transmission mount bolts to 30 ft. lbs. (41 Nm). Tighten the engine mount nuts to 62 ft. lbs. (83 Nm). Tighten the transmission mount nuts to 30 ft. lbs. (41 Nm).

29. Install the transmission and/or transfer assembly.

30. Remove the engine hoist and the engine hangers from the engine.

31. Install the front exhaust pipe and reconnect the exhaust system with new self-locking nuts.

32. If equipped with a clutch slave cylinder, install it onto the transmission.

33. Install the starter.

34. Install the driveshafts.

35. If equipped with an automatic transmission, perform the following steps:

a. Connect the fluid coolant lines to the transmission.

b. Connect the downshift cable to the transmission.

c. Connect the shift select control link rod to the select lever.

d. Install the oil level gauge and the tube.

36. Connect the back-up light switch and the vehicle speed sensor cable to the transmission.

37. Install the front wheels and lower the vehicle.

38. Install the transfer case shift lever by performing the following procedure:

a. Position the shift lever into the transfer case.

b. Install the shift lever retaining bolts.

c. Push the dust cover and the shift lever boot downward.

39. Install the gear shift lever by performing the following steps:

a. Install the shift lever and the shift lever cover bolts.

b. Push the grommet and shift lever boot downward.

c. Install the front console.

40. Install the power steering pump and bracket to the engine.

41. If equipped with A/C, install the compressor to the engine.

42. Install the radiator and grille.

43. Reconnect the following:

- Air switch valve hose
- Oxygen sensor harness
- Power booster vacuum hose
- Alternator wiring harness

- Fuel pressure regulator vacuum hose
- Canister hose
- Engine wiring harness connectors located on the right wheel well
- Inlet and return fuel lines
- Starter motor cables
- Engine ground cables
- Oil pressure switch connectors

44. Reconnect the throttle cable.

45. Install the air cleaner duct and hose.

46. Refill the engine with SAE 10W-30, the transmission with Dexron® III, and the transfer case with SAE 10W-30.

47. Refill and bleed the cooling system.

48. Refill and bleed the power steering system.

49. Install the radiator skid plate and tighten the bolts to 27 ft. lbs. (37 Nm).

50. Align the matchmarks and install the hood.

51. Install the battery. Connect the positive and negative battery cables.

52. Start the engine, check for fuel, coolant, and oil leaks.

53. After the engine has warmed up, check the throttle cable deflection and operation. Check the belt tension and adjust if necessary.

3.2L Engines

✳✳ WARNING

The transmission and transfer case assembly may be completely removed from the vehicle before the engine is removed. If you chose to leave the transmission and transfer case assembly in the vehicle, it must be securely supported.

1. Shift the transmission into the **N** or neutral position. If equipped with four-wheel drive, shift the transfer case into the **2H** position and verify that the front axle and hubs are not engaged. Set the parking brake and securely block the rear wheels while the vehicle is on the ground.

2. Relieve the fuel pressure.

3. Disconnect the negative and positive battery cables. Remove the battery.

4. Use a felt-tipped marker to matchmark the hood hinge plates. Remove the hood.

5. If equipped with a manual transmission, remove the gear shift lever as follows:

a. Verify that the transmission is in neutral.

b. Remove retaining screws from the front console.

c. Remove the shift knob.

d. Pull the shift lever boot and grommet upward.

e. Remove the shift lever cover bolts and the shift lever.

6. If equipped with an automatic transmission, perform the following steps:

a. Verify that the transmission is in **N**.

b. Remove the retaining screws from the front console.

c. Disconnect the shift lock cable.

d. Label and uncouple the wiring connectors.

e. After the vehicle has been raised and supported, disconnect the shift control rod from the selector lever linkage.

7. If equipped with four-wheel drive, remove the transfer case shift lever as follows:

a. Verify that the transfer case is in **2H**.

b. Pull the shift lever boot and dust cover upward.

c. Remove the shift lever retaining bolts.

d. Pull the shift lever from the transfer case.

8. If equipped, remove the radiator skid plate.

9. Drain the engine coolant into a container.

10. Remove the air cleaner and the intake air duct. Use a clean shop cloth to plug the throttle body port to prevent dirt from entering the engine.

11. If necessary, remove the vehicle's grille to prevent it from being damaged.

12. Remove the radiator.

13. Remove the drive belts.

14. If equipped with A/C, unbolt the compressor and move it out of the work area. Don't disconnect the A/C lines.

15. Unbolt the power steering pump from its bracket. Move the pump out of the way with the hydraulic line connected.

16. Remove the starter.

17. Disconnect the throttle cable from the throttle body linkage.

18. Label and disconnect the following vacuum hoses from the intake manifold chamber:

- PCV hose
- EVAP canister vacuum hose
- Brake booster hose

19. Label and detach the following sensor connectors from the rear of the intake manifold chamber:

- Ignition control module connectors
- Linear EGR valve
- MAP sensor
- EVAP purge valve
- Throttle position sensor
- Idle air control valve
- Intake air temperature sensor

20. Disconnect the EGR valve supply tube and bracket.

21. Disconnect the MAP sensor tube, then unbolt the MAP sensor bracket.

22. If necessary, the intake manifold chamber may be removed to avoid damage. If removed, cover the intake ports to keep dirt or foreign objects out.

23. If necessary, the ignition coil assembly may be removed to avoid damaged. Label the spark plug wires to avoid confusion.

24. If necessary, the cruise control actuator and cable brackets may be unbolted to move the actuator and cable out of the work area.

25. Raise and safely support the vehicle. Remove the front wheels.

26. Drain the oil from the engine, transmission and transfer case.

27. Detach the back-up light switch connector and the speed sensor connector from the transmission.

28. If equipped with an automatic transmission, perform the following steps:

   a. Remove the dipstick and the tube.

   b. Disconnect the shift select control link rod from the select lever.

   c. Disconnect the downshift cable from the transmission.

   d. Disconnect and plug the fluid coolant lines from the transmission.

29. Unbolt and remove the rear driveshaft. Unbolt the center bearing and lower the driveshaft from the vehicle.

30. If equipped, remove the front driveshaft splined yoke flange-to-transfer case bolts and separate the front driveshaft from the transfer case; do not allow the splined flange to fall away from the driveshaft.

31. If equipped with a clutch slave cylinder, remove it from the transmission and move it aside.

32. Label and detach the oxygen sensor connectors.

33. Unbolt the front exhaust pipe flanges from the exhaust manifolds and catalytic converters. Separate the exhaust system from the engine, and move it out of the work area. If necessary, the front part of the exhaust system may be removed from the vehicle.

34. Attach an engine lifting chain to the engine hangers. The engine hangers are located on the right and left sides of the engine below the valve covers.

35. Be sure the engine is safely supported.

➡ **Be sure that no engine components will be damaged by the lifting chain.**

36. Remove the transmission/transfer case assembly.

37. Unbolt the engine mounts.

38. Raise the engine slightly. Verify that all vacuum lines and electrical connectors have been detached so that the engine removal is not obstructed.

39. Raise the chain hoist to lift the engine out of the vehicle. If necessary, keep the front of the engine higher than the rear to clear the bulkhead.

40. Secure the engine to a workstand.

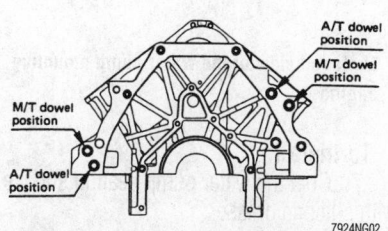

7924NG02

Dowel pin locations for automatic and manual transmissions used with 3.2L engines

To install:

41. Using the chain hoist, slowly lower the engine into the vehicle. Be sure to hold the front of the engine higher than the rear.

42. Install the engine mount nuts and bolts. Tighten the engine mount bolts to 30 ft. lbs. (41 Nm). Tighten the engine mount nuts to 37 ft. lbs. (50 Nm).

❋❋ WARNING

Be sure that the transmission mounting dowels are in the correct locations for the type of transmission (M/T or A/T). Incorrect dowel positioning can crack the transmission case.

43. Install the transmission/transfer case assembly.

44. Install the front exhaust system components. Use new self-locking nuts where necessary.

45. Reattach the oxygen sensor connectors.

46. If equipped with a clutch slave cylinder, install it onto the transmission.

47. Install the front driveshaft splined yoke flange-to-transfer case bolts. Tighten the flange bolts to 46 ft. lbs. (63 Nm).

48. Install the driveshaft into the transmission and the driveshaft flange nuts to 46 ft. lbs. (63 Nm).

49. If equipped with an automatic transmission, perform the following steps:

   a. Connect the fluid coolant lines to the transmission.

   b. Connect the downshift cable to the transmission.

   c. Connect the shift select control link rod to the select lever.

   d. Install the dipstick and the tube.

50. Attach the back-up light switch connector and the vehicle speed sensor connector to the transmission.

51. Install the front wheels and lower the vehicle.

52. Remove the engine lifting chain.

53. Install the power steering pump and bracket to the engine.

54. Install the A/C compressor to the engine.

55. If removed, install the intake manifold chamber.

56. If removed, install the ignition coil assembly and reconnect the spark plug wires.

57. Reconnect the EGR valve supply tube and bracket.

58. Reconnect the MAP sensor tube, then install the MAP sensor bracket.

59. Reconnect the following vacuum hoses to the intake manifold chamber:
- PCV hose
- EVAP canister vacuum hose
- Brake booster hose

60. Reattach the following sensor connectors to the rear of the intake manifold chamber:
- Ignition control module connectors
- Linear EGR valve
- MAP sensor
- EVAP purge valve
- Throttle position sensor
- Idle air control valve
- Intake air temperature sensor

61. Reconnect the throttle cable from the throttle body linkage.

62. Install the starter.

63. Install and adjust the drive belts.

64. If removed, install the cruise control actuator and cable brackets.

65. Install the radiator.

66. Install the radiator skid plate and tighten the bolts to 27 ft. lbs. (37 Nm).

67. Reconnect the heater hoses.

68. Install the air cleaner duct and hose.

69. If equipped with a manual transmission, install the gear shift lever:

   a. Install the shift lever and its mounting cover bolts.

   b. Install the shift lever boot and grommet.

   c. Install the front console.

   d. Install the shift knob.

   e. Verify that the transmission is in neutral.

70. If equipped with an automatic transmission:

   a. Reconnect the shift control rod from the selector lever linkage.

b. Reattach the wiring connectors.

c. Reconnect the shift lock cable.

d. Install the front console.

e. Verify that the transmission is in **N**.

71. If equipped with four-wheel drive, install the transfer case shift lever:

a. Install the shift lever onto the transfer case.

b. Install the shift lever retaining bolts.

c. Install the shift lever dust cover and boot.

d. Verify that the transfer case is in **2H**.

72. Refill the engine with SAE 10W-30, the transmission with Dexron® III, and the transfer case with SAE 10W-30.

73. Refill and bleed the cooling system.

74. Refill and bleed the power steering system if the hydraulic lines were opened.

75. Align the matchmarks and install the hood.

76. Verify that all vacuum lines, cooling hoses, control cables, and electrical connections have been routed and attached properly.

77. Install the battery. Reconnect the positive and negative battery cables.

78. Crank the engine until it starts to remove any air from the fuel lines. Carefully check all fuel lines and fittings for signs of leakage.

79. Warm it up to normal operating temperature. Check the adjustment and operation of the throttle cable.

80. Shut the engine **OFF**. Recheck the drive belt tensions.

81. Check all fluid levels and refill as necessary.

82. Test drive the vehicle.

Water Pump

REMOVAL & INSTALLATION

2.2L Engine

➡Be sure to note the position of the mounting lug on the water pump. Failure to position the water pump correctly will cause difficulty in adjusting the timing belt and may cause overheating.

1. Disconnect the negative battery cable.

2. Drain and recycle the engine coolant.

3. Remove the radiator hose on the inlet side of the water pump.

4. Remove the timing belt, refer to the timing belt unit repair section.

5. Remove the water pump mounting bolts, then the pump.

6. Clean the water pump mounting surface.

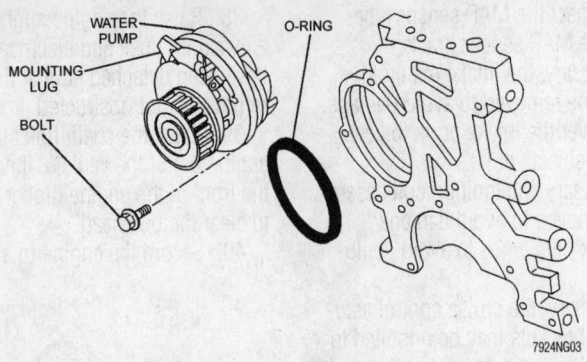

Exploded view of the water pump mounting, showing the location of the mounting lug—2.2L engine

To install:

7. Coat the water pump sealing surface with silicone grease.

8. Install the water pump and O-ring and tighten the bolts to 18 ft. lbs. (25 Nm).

9. Install the timing belt.

10. Connect the radiator hose.

11. Fill and bleed the cooling system.

12. Connect the negative battery cable.

2.6L Engine

1. Disconnect the negative battery cable.

2. Drain the coolant from the radiator into a sealable container.

3. Disconnect the radiator hoses from the radiator.

4. Remove the air duct assembly.

5. Remove the lower fan guide clips and the bottom lock, then remove the lower fan shroud.

6. Remove the upper fan shroud bolts and remove the shroud.

7. Remove the nuts attaching the fan to the water pump, then remove the fan.

8. If equipped with power steering, remove the drive belt.

9. If equipped with A/C, loosen the A/C idler pulley nuts, then remove the mounting bolts and idler pulley. Remove the A/C compressor belt.

10. Remove the alternator belt.

11. Remove the pulley from the water pump.

12. Rotate the crankshaft to align the crankshaft pulley timing marks.

13. Remove the starter and install flywheel holder (part No. J-38674) or equivalent.

14. Remove the crankshaft pulley bolt and pulley.

15. Remove the upper and lower timing belt covers.

16. Remove the four bolts and one nut from the water pump and remove the pump from the engine.

To install:

17. Clean the water pump mounting surface.

18. Install the water pump with a new gasket. Tighten the mounting bolts to 14 ft. lbs. (19 Nm), and the nut to 20 ft. lbs. (25 Nm).

19. Install the timing belt lower and upper covers. Tighten the timing belt cover bolts to 4 ft. lbs. (6 Nm).

20. Install the crankshaft pulley, tighten the bolt to 90 ft. lbs. (122 Nm).

21. Install the starter motor. Tighten the mounting bolts to 30 ft. lbs. (40 Nm).

22. Install the water pump pulley.

23. Install the alternator bracket and belt, do not tension the belt at this time.

24. If equipped with A/C, install the and idler pulley, then adjust the belt tension.

25. If equipped with power steering, install the bracket and belt, then adjust the drive belt.

26. Install the fan pulley to the water pump, and adjust the alternator belt tension. Tighten the fan attaching nuts to 20 ft. lbs. (27 Nm). Install the cooling fan.

27. Install the upper and lower fan shroud.

28. Install the air duct assembly.

29. Connect the radiator hoses.

30. Fill and bleed the cooling system.

31. Connect the negative battery cable.

3.2L Engines

1. Disconnect the negative battery cable.

2. Drain the engine coolant into a sealable container.

3. Remove the upper radiator hose.

4. Remove the timing belt and idler pulley. The timing belt must be replaced if it has been contaminated by oil or coolant.

5. Unbolt and remove the water pump. Clean any gasket material or sealant residue from the water pump mating sealing surfaces.

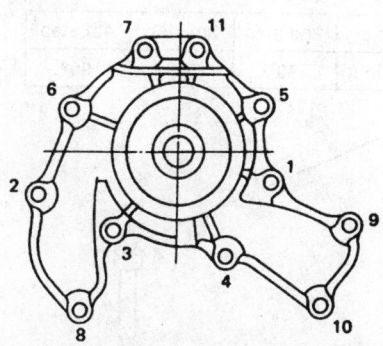

Water pump bolt tightening sequence—3.2L engines

To install:

6. Install the water pump using a new gasket. Tighten the mounting bolts to 13 ft. lbs. (18 Nm) in a two-step crisscross sequence.

7. Install the idler pulley. Tighten the mounting bolt to 31 ft. lbs. (42 Nm).

8. Install and tension the timing belt.

9. Install the upper radiator hose.

10. Refill and bleed the cooling system.

11. Connect the negative battery cable. Start the engine and check for coolant leaks.

Cylinder Head

REMOVAL & INSTALLATION

2.2L Engine

1. Disconnect the negative battery cable.

2. Remove the air intake assembly and ducting.

3. Drain and recycle the engine coolant.

4. Remove the upper radiator hose.

5. Label and disconnect any vacuum hoses associated with the intake manifold or cylinder head.

6. Disconnect the exhaust header pipe from the manifold.

7. Remove the serpentine drive belt.

8. Remove the alternator and its brackets.

9. Disconnect the crankshaft angle and knock sensors.

10. Remove the coolant hoses from the cylinder head and throttle body.

11. Properly relieve the fuel system pressure and disconnect the fuel lines from the fuel rail.

12. Remove the intake manifold supports.

13. Remove the four crankshaft pulley mounting bolts and pulley.

14. Label and disconnect the ignition cables.

15. Disconnect and remove the camshaft angle sensor.

16. Remove the timing belt.

17. Remove the two idler pulleys and the rear timing belt cover.

18. Remove the exhaust side camshaft.

19. Using tool J-42623 or equivalent, loosen and remove the ten cylinder head bolts.

To install:

20. Clean the mating surfaces of the cylinder head and engine block.

21. Place the head gasket on the engine block.

22. Install the cylinder head and tighten the bolts (following the illustrated sequence) in the following four steps:

   a. Step 1—18 ft. lbs. (25 Nm).

   b. Step 2—tighten all bolts an additional 90° (¼ turn).

   c. Step 3—an additional 90° (¼ turn).

   d. Step 4—an additional 90° (¼ turn).

23. Install the exhaust camshaft and tighten the bearing caps to 5.9 ft. lbs. (8 Nm) as shown.

24. Install the camshaft angle sensor.

25. Install and tighten the rear timing belt cover M6 mounting bolts to 4.4 ft. lbs. (6 Nm) and the M8 mounting bolts to 5.8 ft. lbs. (8 Nm).

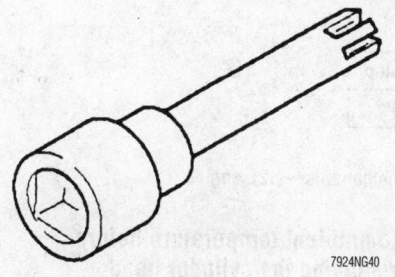

Cylinder head bolt removal tool J-42623

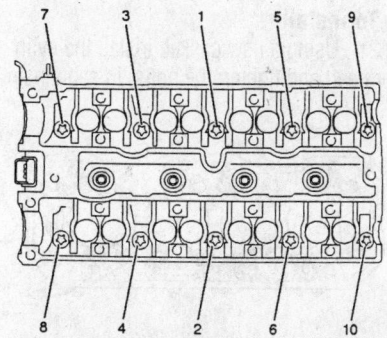

Cylinder head mounting bolt tightening sequence—2.2L engine

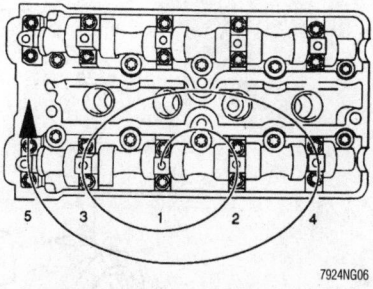

Camshaft bearing cap bolt tightening sequence—2.2L engine

26. Install the idler pulleys.

27. Install the timing belt.

28. Tighten the valve cover to 5.9 ft. lbs. (8 Nm).

29. Connect the ignition cables.

30. Install the crankshaft pulley and tighten the bolts to 14 ft. lbs. (20 Nm).

31. Connect the camshaft sensor and knock sensor.

32. Install the intake manifold supports.

33. Attach the fuel line to the fuel rail and ensure the ground connections are clean.

34. Connect the water hoses that were removed.

35. Connect the wiring harness to the crankshaft sensor.

36. Install the alternator and tighten the long bolt to 25 ft. lbs. (35 Nm) and the short bolt to 14 ft. lbs. (20 Nm).

37. Install the exhaust header pipe.

38. Install the upper radiator hose and the serpentine belt.

39. Install the air intake assembly and ducting.

40. Connect the negative battery cable.

41. Fill engine coolant to the recommended level.

2.6L Engine

1. Relieve the fuel system pressure.

2. Disconnect the negative battery cable. Drain the cooling system.

3. Remove the drive belts.

4. Remove the air pump switching valve.

5. Remove the air pump hoses from the manifold and air pump.

6. Remove the MAP sensor hose, charcoal canister hose, vacuum booster and Vacuum Switching Valve (VSV) hoses.

7. Rotate the engine to position the No. 1 cylinder on TDC.

8. Remove the distributor.

9. Remove the exhaust manifold-to-exhaust pipe bolts.

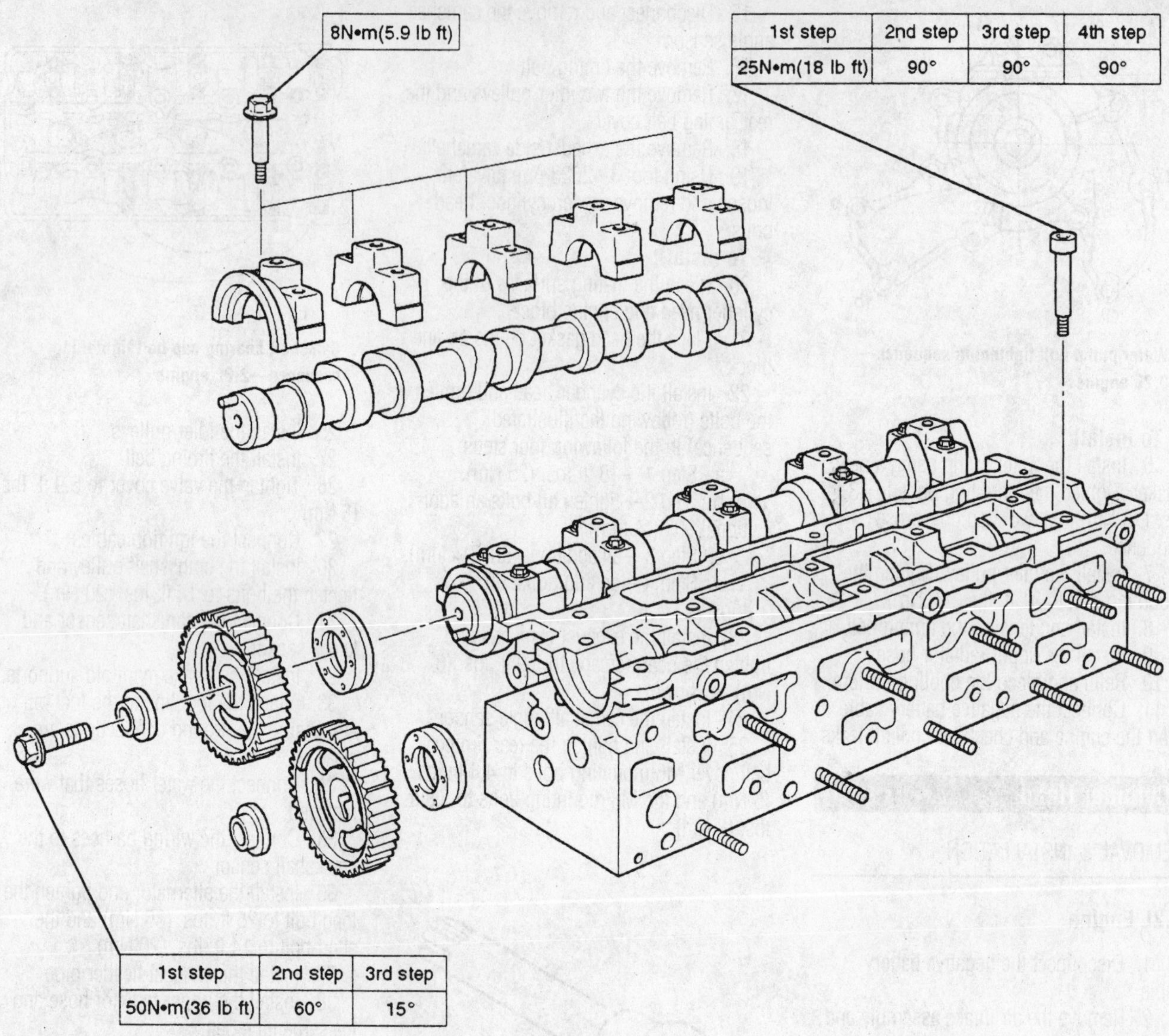

8N•m(5.9 lb ft)

1st step	2nd step	3rd step	4th step
25N•m(18 lb ft)	90°	90°	90°

1st step	2nd step	3rd step
50N•m(36 lb ft)	60°	15°

7924NG41

Exploded view of the cylinder head and camshaft components—2.2L engine

10. Label and detach the connectors and vacuum hoses which may be in the way.

11. Remove the linkage to the carburetor or throttle body.

12. Remove the coolant hoses and cooling fan. It is not necessary to remove the radiator.

13. Remove the crankshaft pulley.

14. Remove the timing belt.

15. Remove the camshaft pulley and camshaft boss.

16. Remove the timing belt guide plate and the cylinder head front plate.

17. Remove the rocker arm cover and gasket.

✳✳ WARNING

Cylinder head warpage can result if the cylinder head is removed from a hot engine. Allow the engine to cool to ambient temperature before removing the cylinder head.

18. Remove the cylinder head-to-engine bolts slowly and in sequence.

19. Remove the cylinder head and gasket.

20. Clean the gasket mounting surfaces.

To install:

21. Using a new gasket, install the cylinder head and tighten the bolts, in sequence to 58 ft. lbs. (79 Nm) in the 1st step, and to 65–79 ft. lbs. (88–107 Nm) in the final step.

22. Install the cylinder head front plate.

23. Install the camshaft pulley.

24. Align the camshaft pulley mark with the mark on the front plate. Be sure the keyway on the crankshaft if facing upward, aimed at the pointer on the engine block.

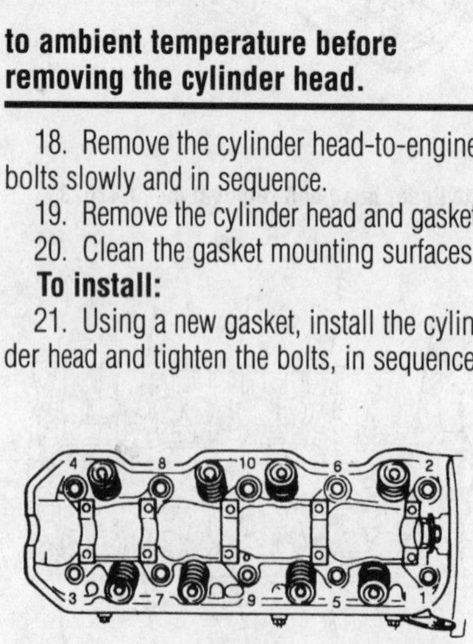

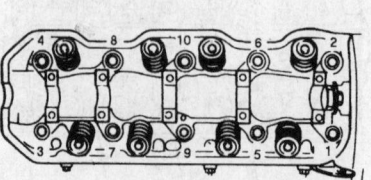

Cylinder head bolt loosening sequence— 2.6L engine

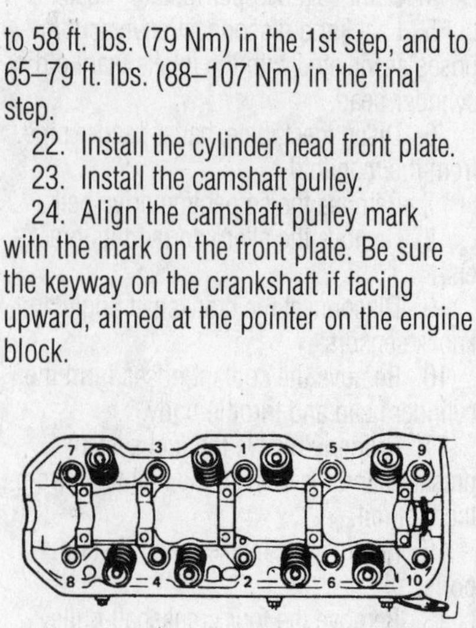

7924NG07

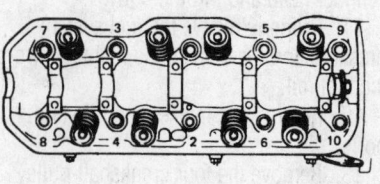

Cylinder head bolt torque sequence—2.6L engine

7924NG08

25. Install the timing belt. Install the timing belt covers, using a new gasket.

26. Install the crankshaft pulley.

27. Install the cooling fan and coolant hoses. If removed, reinstall the radiator.

28. Install the linkage to the carburetor or throttle body.

29. Reconnect the harnesses and vacuum hoses.

30. Install the exhaust manifold-to-exhaust pipe bolts.

31. Install the distributor.

32. Install the drive belts.

33. Adjust the valve lash to specification.

34. Using a new gasket, install the rocker arm cover. Tighten the bolts evenly in a crisscross pattern.

35. Reconnect the negative battery cable. Refill the cooling system.

36. Start the engine and check for leaks.

3.2L Engines

1995 MODELS

1. Relieve the fuel system pressure and disconnect the negative battery cable.

2. Remove the air cleaner assembly.

3. Remove the upper cooling fan shroud and the cooling fan assembly.

4. Disconnect the accelerator cable from the throttle body and the bracket.

5. Disconnect the canister vacuum hose from the vacuum pipe.

6. Disconnect the air vacuum hose from the common chamber.

7. Disconnect the vacuum booster hose from the common chamber.

8. Detach the Manifold Absolute Pressure (MAP) sensor, canister Vacuum Switching Valve (VSV), Exhaust Gas Recirculation (EGR) VSV, Intake Air Temperature (IAT) sensor and ground connectors.

9. Remove the spark plug wires from the cylinder head cover.

10. Remove the ignition control module assembly.

11. Remove the four bolts and the throttle body from the common chamber.

12. Disconnect the vacuum hoses from the throttle body.

13. Disconnect the positive crankcase ventilation hose from the common chamber.

14. Disconnect the fuel pressure control valve vacuum hose from the common chamber.

15. Disconnect the evaporative emission canister purge hose from the common chamber.

16. Remove the EGR valve assembly.

17. Remove the common chamber from the intake manifold.

18. Disconnect the fuel feed and return hoses from the fuel rail assembly.

19. Disconnect the harnesses to the fuel injectors and the thermo sensor.

20. Remove the intake manifold.

21. Remove the engine coolant manifold by removing the heater hose and mounting bolts.

22. Remove the drive belts.

23. Remove the power steering pump.

24. Remove the fan pulley assembly.

25. Remove the crankshaft pulley and damper.

26. Remove the oil cooler hoses and bracket on the timing belt cover.

27. Remove the timing belt.

28. Remove the cylinder head cover.

29. Remove the power steering pump bracket.

30. Remove the front exhaust pipes from the exhaust manifolds.

31. Remove the dipstick tube bracket from the cylinder head.

✳✳ WARNING

The cylinder head and engine block must be at room temperature before removing the cylinder head.

32. Remove the cylinder head bolts in the illustrated sequence, gradually and in two steps.

33. Remove the cylinder head.

To install:

34. Install new camshaft seals and retaining plates onto the cylinder. Tighten the right camshaft seal retaining plate bolts to 65 inch lbs. (7 Nm). Tighten the left camshaft seal retaining plate bolts to 191 inch lbs. (22 Nm).

35. Thoroughly clean the cylinder head and engine block sealing surfaces.

36. Place a new head gasket on the engine block and carefully position the cylinder head on top of the new gasket.

➡**Do not reuse or apply oil to the cylinder head bolts.**

37. Install new cylinder head bolts and tighten them in sequence to 47 ft. lbs. (64 Nm) for the M11 bolts and 15 ft. lbs. (21 Nm) for the M8 bolts.

38. Install the dipstick tube bracket to the cylinder head.

39. Connect the front exhaust pipes to the exhaust manifolds. Tighten the exhaust bolts to 48 ft. lbs. (67 Nm).

40. Install the power steering pump bracket. Tighten the mounting bolts to 34 ft. lbs. (46 Nm).

41. Install the cylinder head covers.

42. Install the timing belt.

43. Install the timing belt cover and oil cooler hoses and bracket.

44. Install the crankshaft pulley. Tighten the center bolt to 123 ft. lbs. (167 Nm).

45. Install the fan pulley assembly.

46. Install the power steering pump.

47. Install the drive belts.

48. Install the engine coolant manifold and the heater hose.

49. Install the intake manifold.

50. Install the fuel injector harnesses and the fuel hoses to the fuel rail.

51. Install the common chamber. Tighten the nuts and bolts to 17 ft. lbs. (24 Nm).

52. Install the EGR valve assembly. Tighten the mounting bolts on the valve side to 69 inch lbs. (8 Nm) and the bolts on the exhaust side to 21 ft. lbs. (28 Nm).

53. Connect the evaporative emission canister purge hose.

54. Connect the fuel pressure control valve vacuum hose.

55. Connect the positive crankcase ventilation hose.

56. Install the throttle body assembly. Tighten the mounting bolts to 14 ft. lbs. (19 Nm).

57. Connect the vacuum hoses to the throttle body.

58. Install the ignition control module and the spark plug wires.

59. Attach the MAP sensor, canister VSV, EGR VSV, IAT sensor and ground connectors.

60. Connect the vacuum booster hose.

61. Connect the air vacuum hose.

62. Connect the accelerator cable. Adjust the accelerator cable by pulling the cable housing while closing the throttle valve and tightening the adjusting nut and screw cap by hand temporarily. Now loosen the adjusting nut by three turns, then tightening the screw cap. Be sure the throttle valve reaches the screw stop when the throttle is closed.

63. Install the cooling fan assembly and the upper fan shroud.

64. Install the air cleaner assembly.

65. Connect the negative battery cable.

66. Refill and bleed the cooling system.

67. Refill and bleed the power steering pump if necessary.

68. Refill the engine with fresh oil.

69. Run the engine and check for leaks and proper compression.

1996–97 MODELS

✳✳ WARNING

The cylinder head should be cool to the touch before it is removed. If the head bolts are loosened on a hot engine, the cylinder head may warp.

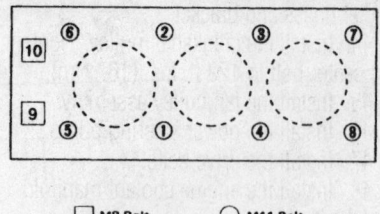

☐ M8 Bolt   ○ M11 Bolt

7924NG09

**Cylinder head bolt loosening sequence—
1996–97 3.2L engine**

1. Relieve the fuel system pressure.

☀ CAUTION

The fuel injection system remains under pressure even after the engine has been turned OFF. The fuel system pressure must be relieved before disconnecting any fuel lines. Failure to do so may result in fire and personal injury.

2. Disconnect the negative battery cable and reinstall the fuel pump relay.

3. Raise and support the vehicle safely.

4. Disconnect the front exhaust pipes from the exhaust manifolds. If necessary, separate the front exhaust pipes from the crossover pipe. Label and disconnect the oxygen sensors.

5. Lower the vehicle.

6. Drain the coolant into a sealable container. Disconnect and remove the upper and lower radiator hoses.

7. Remove the air intake duct and the air cleaner box.

8. Loosen and remove the drive belts.

9. Remove the cooling fan and pulley assembly.

10. Unbolt the power steering pump mounting bracket. Move the pump and bracket out of the way without disconnecting the hydraulic lines.

11. Disconnect the throttle cable from the throttle body linkage.

12. Label and disconnect the following vacuum hoses from the intake manifold chamber:
- PCV hose
- EVAP canister vacuum hose
- Brake booster hose

13. Label and detach the following sensor connectors from the rear of the intake manifold chamber:
- Ignition control module connectors
- Linear EGR valve
- MAP sensor
- EVAP purge valve
- Throttle position sensor

- Idle air control valve
- Intake air temperature sensor

14. Disconnect the EGR valve supply tube and bracket.

15. Remove the throttle body, then the intake manifold chamber.

16. Carefully clean any dirt from the fuel rail and fuel fittings.

17. Disconnect the fuel feed and return lines from the front of the fuel rail. Clean up any spilled fuel.

18. Label and disconnect the fuel injector wiring harness.

19. Remove the fuel injectors and lower intake manifold as an assembly. If desired, the fuel rail and injectors may be removed separately as an assembly. Remove the intake manifold gaskets.

20. Cover the intake openings with a sheet of plastic or clean shop towels to keep out dirt and foreign objects.

21. Label the ignition coil assemblies, then disconnect and remove.

22. Unbolt the oil cooler line brackets from the timing belt covers.

23. Remove the timing belt.

☀ WARNING

If the timing belt is worn, damaged, or shows signs of oil or coolant contamination, it must be replaced.

24. Disconnect the heater hoses from the engine, then unbolt and remove the engine coolant manifold.

25. Unbolt the dipstick tube from the cylinder head.

26. Loosen the valve cover bolts in a crisscross sequence. Remove the valve covers.

27. Loosen the cylinder head bolts in a two-step crisscross pattern, working from the outer bolts to those at the center of the head, as illustrated. First, partially loosen the 11mm bolts, then partially loosen the 8mm bolts. Finally, loosen all the bolts, then remove them.

28. Remove the cylinder head. If it sticks, tap it with a wooden or plastic-faced mallet.

29. Remove the head gasket.

30. Inspect the cylinder head for cracking or warpage. Inspect the engine block for any signs of damage. Carefully clean the head gasket mating surfaces, don't scratch or gouge the machined aluminum surfaces.

To install:

➡ **Use new head bolts when installing the cylinder head. Do not apply oil to the head bolt threads.**

31. Be sure all mating surfaces are clean and free of oil, coolant, or gasket residue.

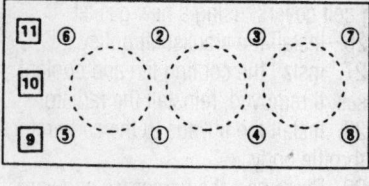

☐ M8 Bolt   ○ M11 Bolt

7924NG10

**Cylinder head bolt tightening sequence—
1996–97 3.2L engine**

32. Install new cylinder head gaskets.

33. Install the cylinder head. Install the new head bolts and tighten them by hand only.

34. Follow these steps to tighten the cylinder head bolts to their final tighten specification:

   a. Use a two-step crisscross pattern to tighten the 11mm bolts to 47 ft. lbs. (64 Nm). Start tightening with the center bolts, and work toward the outer bolts.

   b. Tighten the 8mm bolts to 15 ft. lbs. (21 Nm). Start with the bolt closest to the exhaust side of the head and work toward the intake side.

35. Apply a 0.7–0.8 in. (2–3mm) bead of sealant to the joint where the camshaft holders meet the cylinder head. Install the valve cover with a new gasket before the sealant cures.

36. Tighten the valve cover bolts to 6 ft. lbs. (8 Nm) in a crisscross pattern.

37. Verify that the camshaft and crankshaft timing marks are properly aligned.

38. Install the timing belt..

39. Fit the oil cooler line brackets onto the timing cover and tighten the bolts to 13 ft. lbs. (18 Nm).

40. Install the engine coolant manifold and reconnect the heater hoses. Tighten the bolts to 16 ft. lbs. (22 Nm).

41. Install the dipstick tube bracket.

42. Raise and safely support the vehicle. Install and reconnect the front exhaust pipes. Reconnect the oxygen sensors. Lower the vehicle.

43. Install the intake manifold with a new gasket. Tighten the bolts and nuts to 17 ft. lbs. (23 Nm).

44. Reconnect the fuel injector wiring harness. Reconnect the fuel feed and return lines.

45. Install and reconnect the ignition coil assemblies.

46. Install the intake manifold chamber and throttle body with new gaskets. Tighten the nuts and bolts to 17 ft. lbs. (23 Nm).

47. Reconnect the throttle cable to the throttle body linkage.

48. Reconnect the EGR valve supply tube and bracket.

49. Reconnect the following vacuum hoses to the intake manifold chamber:
- PCV hose
- EVAP canister vacuum hose
- Brake booster hose

50. Reattach the following sensor connectors to the rear of the intake manifold chamber:
- Ignition control module connectors
- Linear EGR valve
- MAP sensor
- EVAP purge valve
- Throttle position sensor
- Idle air control valve
- Intake air temperature sensor

51. Install the power steering pump and mounting bracket.

52. Install the cooling fan and pulley assembly.

53. Install and tension the drive belts.

54. Connect the upper and lower radiator hoses.

55. Install the air cleaner box and air intake duct.

56. Install the hood.

57. Verify that all fuel lines, vacuum and coolant hoses, and wiring harness have been reconnected.

58. Refill the engine with fresh coolant.

59. Install a new oil filter and refill the engine with fresh oil.

60. Crank the engine until it starts. A longer than normal starting time may be necessary due to air in the fuel lines. Check all fuel line connections for leaks.

61. Bleed any air from the cooling system.

62. Bleed the power steering system if necessary.

63. Warm the engine up to normal operating temperature and check the operation of the thermostat and water pump.

64. Check the throttle cable operation and adjustment.

65. Check the engine oil level and add if necessary.

1998–99 MODELS

1. Properly relieve the fuel system pressure.

2. Disconnect the negative battery cable.

✳✳ CAUTION

The fuel injection system remains under pressure even after the engine has been turned OFF. The fuel system pressure must be relieved before disconnecting any fuel lines. Failure to do so may result in fire and personal injury.

3. Raise and support the vehicle safely.

4. Drain the coolant into a sealable container.

5. Use a felt-tipped marker to matchmark the hood hinge plates. Remove the hood.

6. Drain the engine oil into a suitable container.

7. To remove the crankshaft pulley assembly using tool J-8614–01 or equivalent crankshaft holder, hold the pulley, then remove the center bolt and pulley assembly.

8. Remove the timing belt, using the procedure in the timing belt unit repair section.

9. Label, then disconnect all wiring harnesses and vacuum hoses related to cylinder head removal.

10. Remove the left and right valve covers.

11. Remove the common chamber.

12. Loosen and remove the eight cylinder head bolts, then remove the cylinder head.

13. Remove the head gasket.

14. Inspect the cylinder head for cracking or warpage. Inspect the engine block for any signs of damage. Carefully clean the head gasket mating surfaces, don't scratch or gouge the machined aluminum surfaces.

15. Cover the engine block with a sheet of plastic or clean shop towels to keep any dirt and foreign objects out of the combustion chambers.

To install:

16. Clean the mating surfaces of the cylinder head and engine block.

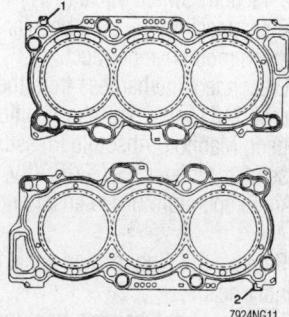

Right (1) and left (2) head gasket identification mark locations—1998–99 3.2L engine

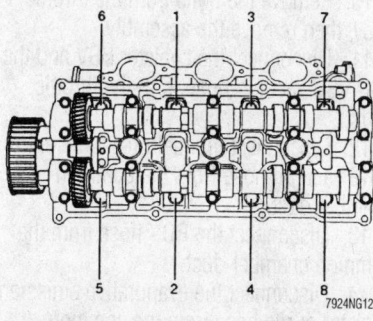

Cylinder head mounting bolt tightening sequence—1998–99 3.2L engine

17. Place the head gasket on the engine block.

➡**Do not reuse the head gaskets. There is an "R" mark for the right bank and an "L" for the left bank head gasket as shown.**

18. Install the cylinder head. Be sure the cylinder head is seated over the dowel pins on the engine block.

19. Tighten two bolts to prevent the cylinder head from moving, by hand.

➡**Do not reuse cylinder head bolts. Do not apply any type of lubricant to the head bolts, it will affect the torque reading.**

20. Using Cylinder Head Bolt Wrench J-24239–01 or equivalent, tighten the head bolts in two steps in numerical order as illustrated. In the first step, tighten the head bolts to 21 ft. lbs. (29 Nm). In the second step, tighten the head bolts to 47 ft. lbs. (64 Nm).

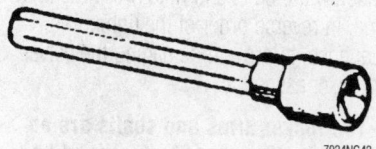

Cylinder head bolt wrench J-24239–01

21. Install the common chamber.

22. Install the left and right valve covers.

23. Connect all wiring harnesses and vacuum hoses.

24. Install the timing belt, using the procedure in the timing belt unit repair section

25. Install the crankshaft pulley and tighten the center bolt to 123 ft. lbs. (167 Nm).

26. Fill the engine with the correct amount and type.

27. Refill the engine with fresh coolant.

28. Install the hood.

29. Warm the engine up to normal operating temperature and check the operation of the thermostat and water pump.

30. Check the throttle cable operation and adjustment.

31. Check the engine oil level and add if necessary.

Rocker Arms/Shafts

REMOVAL & INSTALLATION

2.6L Engine

1. Disconnect the negative battery cable.
2. Remove the drive belts.

3. Remove the cooling fan and the water pump pulley.

4. Unbolt and remove the power steering pump. Unbolt the hydraulic line brackets from the upper timing cover. Move the pump out without disconnecting the hydraulic lines.

5. Remove the valve cover and the upper timing cover.

6. Rotate the crankshaft to set the engine at TDC/compression for the No. 1 cylinder. The arrow mark on camshaft sprocket aligns with mark on the rear timing cover.

7. Remove the lower timing belt cover.

8. Remove the crankshaft pulley.

9. Verify that the engine is set at TDC/compression for the No. 1 cylinder. The notch on the crankshaft sprocket aligns with the pointer on the oil seal retainer.

10. Remove the timing belt.

11. Loosen the valve adjusting screws.

12. Unfasten the camshaft holder bolts and nuts, **but do not remove them.** Unscrew the bolts and nuts two turns at a time, in reverse order of the tightening sequence, to prevent damaging the valves or rocker assembly.

➡**The rocker arms and shafts are an assembly; they must be removed from the engine as a unit. Always follow the tightening sequence carefully when removing or installing the rocker shaft assembly.**

13. Remove the rocker arm/shaft assemblies, with the bolts still in place. The bolts keep the camshaft bearing caps, springs, and rocker arms in place on the shafts.

14. If the rocker arms or shafts are to be replaced, identify the parts as they are removed from the shafts to ensure reinstallation in the original location. The longer of the two rocker shafts is for the exhaust side of the cylinder head.

To install:

15. Lubricate the camshaft journals and lobes with clean engine oil.

16. Apply sealant to the contact surface of the No. 1 camshaft holder. Set the rocker arm assembly in place and loosely thread

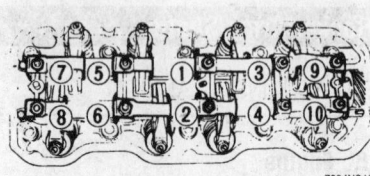

Camshaft holder mounting bolt tightening sequence—2.6L engine

the camshaft holder bolts and nuts. The punch mark on each rocker shaft faces the front of the engine.

17. Tighten the rocker arm bolts and nuts to the following specifications in a two-step crisscross pattern:

   a. Tighten the rocker arm bolts to 6 ft. lbs. (8 Nm).

   b. Tighten rocker arm nuts to 16 ft. lbs. (22 Nm).

18. Install the timing belt.

19. Install the lower timing cover and the crankshaft pulley.

20. Adjust the valves and tighten the locknuts to 9 ft. lbs. (12 Nm).

21. Install the timing belt covers.

22. Install the water pump pulley, tightening the nuts to 20 ft. lbs. (27 Nm). Install the cooling fan.

23. Install the engine drive belts.

24. Change the engine oil and filter.

25. Reconnect the negative battery cable.

26. Start the engine. Check and adjust the ignition timing.

27. Check for oil leaks.

1995 3.2L Engine

1. Disconnect negative battery cable.

2. Remove the air cleaner assembly.

3. Disconnect the accelerator pedal cable from the throttle body and cable brackets.

4. Disconnect the canister vacuum hose from the Vacuum Switch Valve (VSV).

5. Disconnect the vacuum booster hose from the common chamber duct.

6. Disconnect the harness from the Idle Air Control (IAC) valve, Throttle Position (TP) sensor, Manifold Absolute Pressure (MAP) sensor, canister VSV, EGR VSV, Intake Air Temperature (IAT) sensor and VSV.

7. Remove the high tension cable from the cylinder heads.

8. Disconnect the harness from the ignition module.

9. Remove the three bolts from the electronic ignition bracket and assembly.

10. Remove the bolts from the throttle body, then remove the assembly.

11. Disconnect the canister VSV and the EGR VSV vacuum hose from the throttle body.

12. Disconnect the fuel pressure control valve vacuum hose from the common chamber duct.

13. Disconnect the PCV hose from the common chamber duct.

14. Disconnect the evaporative emission canister purge hose from the common chamber duct.

15. Remove the four bolts from the EGR valve assembly common chamber duct and remove the exhaust manifold.

16. Remove the four bolts, four nuts and three manifold bracket fixing bolts from the common chamber duct.

17. Remove the ground cable fixing bolt from the rear of the common chamber duct.

18. Remove the six bolts and two nuts from the common chamber duct.

19. Remove the common chamber duct bracket fixing bolts from the rear of the common chamber duct.

20. Remove the timing belt.

21. Remove the cylinder head cover.

22. Remove the camshaft holders and camshaft.

23. Remove the rocker arm shaft bolts. Lift the rocker arm assembly from the cylinder head.

24. The hydraulic lifters are attached to the rocker arms. Remove them and inspect, bleed, or replace as necessary.

To install:

25. Install new camshaft seals and retaining plates onto the cylinder head. Tighten the right camshaft seal retaining plate bolts to 65 inch lbs. (7 Nm). Tighten the left camshaft seal retaining plate bolts to 191 inch lbs. (21 Nm).

26. Install the rocker arm assembly and tighten the bolts in sequence to 13 ft. lbs. (18 Nm).

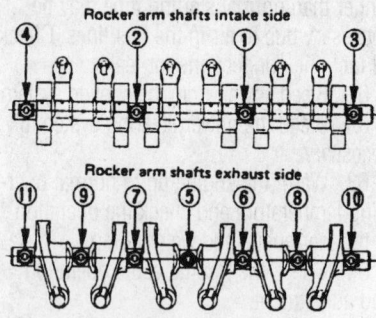

Rocker arm shaft bolt tightening sequence—1995 3.2L engine

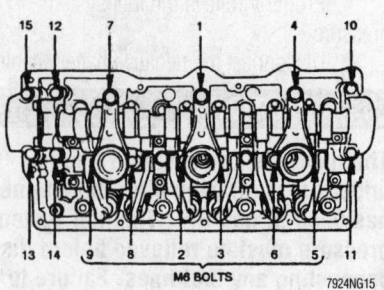

Camshaft holder bolt tightening sequence—1995 3.2L engine

27. Oil the camshaft bearing journals, camshaft lobes, and rocker arm contact areas.

28. Install the camshaft. Apply sealant to the contact edges of the camshaft holders. Tighten the camshaft holder bolts to 69 inch lbs. (8 Nm). Tighten the remaining bolts to 13 ft. lbs. (18 Nm).

29. Install the cylinder head covers and carefully tighten the bolts to 69 inch lbs. (8 Nm). Do not overtighten the head cover bolts; they crack very easily.

30. Install the timing belt.

31. Install the common chamber duct bracket bolts to the rear of the common chamber duct.

32. Install the six bolts and two nuts to the common chamber duct.

33. Install the ground cable fixing bolt to the rear of the common chamber duct.

34. Install the four bolts, four nuts and three manifold bracket bolts to the common chamber duct.

35. Install the exhaust manifold and install the four bolts to the EGR valve assembly and common chamber duct.

36. Connect the evaporative emission canister purge hose to the common chamber duct.

37. Connect the PCV hose to the common chamber duct.

38. Connect the fuel pressure control valve vacuum hose to the common chamber duct.

39. Connect the canister VSV and the EGR VSV vacuum hose to the throttle body.

40. Install the remaining components.

41. Verify that all vacuum hoses, lines, and wiring harnesses are reconnected.

42. Install the air cleaner assembly.

43. Connect the negative battery cable.

1996–97 3.2L Engine

1. Relieve the fuel system pressure.

2. Disconnect the negative battery cable and reinstall the fuel pump relay.

3. Drain the coolant into a sealable container.

4. Remove the air intake duct and the air cleaner box.

5. Disconnect and remove the upper and lower radiator hoses.

6. Remove the drive belts.

7. Remove the cooling fan and pulley assembly.

8. Unbolt the power steering pump mounting bracket. Move the pump and bracket out of the way without disconnecting the hydraulic lines.

9. Disconnect the throttle cable from the throttle body linkage.

10. Label and disconnect the following vacuum hoses from the intake manifold chamber:
- PCV hose
- EVAP canister vacuum hose
- Brake booster hose

11. Label and detach the following sensor connectors from the rear of the intake manifold chamber:
- Ignition control module connectors
- Linear EGR valve
- MAP sensor
- EVAP purge valve
- Throttle position sensor
- Idle air control valve
- Intake air temperature sensor

12. Disconnect the EGR valve supply tube and bracket.

13. Remove the throttle body, then remove the intake manifold chamber.

14. Carefully clean any dirt from the fuel rail and fuel fittings.

✳✳ CAUTION

The fuel injection system remains under pressure even after the engine has been turned OFF. The fuel system pressure must be relieved before disconnecting any fuel lines. Failure to do so may result in fire and personal injury.

15. Disconnect the fuel feed and return lines from the front of the fuel rail. Clean up any spilled fuel.

16. Remove the intake manifold gaskets.

17. Label and disconnect the ignition coil from the wiring harness. Remove the coil assemblies.

18. Unbolt the oil cooler line brackets from the timing belt covers.

19. Remove the timing belt.

✳✳ WARNING

If the timing belt is worn, damaged, or shows signs of oil or coolant contamination, it must be replaced.

20. Loosen the valve cover bolts in a crisscross sequence. Remove the valve covers.

21. Remove the camshaft sprockets and back covers.

22. Loosen the camshaft holder bolts in a crisscross sequence to prevent warping.

23. Remove the camshaft and camshaft holders from the cylinder head.

24. Inspect the camshaft lobes and journals for signs of wear or damage.

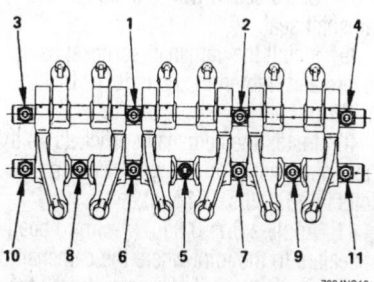

Rocker shaft bolt tightening pattern—1996–97 3.2L engine

25. Loosen the exhaust and intake rocker shaft bolts in a crisscross sequence to prevent warping.

26. Remove the intake and exhaust rocker shafts from the cylinder head.

27. If the rocker arms and shafts must be disassembled, label the parts and wave washers so that they can be reassembled in the same positions.

28. If necessary, remove the hydraulic valve lash adjusters from the rocker arms.

29. Inspect the hydraulic lash adjusters for excess movement and replace if necessary. The hydraulic lash adjusters are designed to be self-bleeding, but new adjusters must be primed before installation.

To install:

30. Reassemble the rocker arm, shaft, and hydraulic lash adjuster components. Assemble the hydraulic lash adjusters to the rockers before removing them from the tub of oil. The intake rocker arms all face the same direction when installed.

31. Lubricate the rocker arms and shafts with clean engine oil.

32. Install the intake and exhaust rocker arms and shaft assemblies. Tighten the shaft holder bolts to 13 ft. lbs. (18 Nm), starting with the intake shaft, then moving to the exhaust shaft. Be sure the intake and exhaust rockers contact each other properly.

33. Be sure all mating surfaces are clean and free of oil, coolant, or gasket residue.

34. Lubricate the camshaft lobes and journals with clean engine oil.

35. Apply a bead of sealant to the front and rear camshaft holder mating surfaces on the cylinder head.

36. Install the camshaft and holder assembly onto the cylinder head before the sealant cures. Install the camshaft holder bolts, but don't tighten them yet.

37. Use a crisscross sequence to tighten the camshaft holder bolts. Tighten the 8mm bolts to 13 ft. lbs. (18 Nm). Tighten the 6mm bolts to 6 ft. lbs. (8 Nm).

38. Use a seal driver to install a new camshaft seal.

39. Install the camshaft sprocket back covers and tighten the bolts to 12 ft. lbs. (17 Nm).

40. Install the camshaft sprockets so that the timing marks are aligned. Tighten the bolts to 46 ft. lbs. (64 Nm).

41. Apply a 0.7–0.8 in. (2–3mm) bead of sealant to the joint where the camshaft holders meet the cylinder head. Install the valve cover with a new gasket before the sealant cures.

42. Tighten the valve cover bolts to 6 ft. lbs. (8 Nm) in crisscross pattern.

43. Verify that the camshaft and crankshaft timing marks are properly aligned.

44. Install the timing belt.

45. Fit the oil cooler line brackets onto the timing cover and tighten the bolts to 13 ft. lbs. (18 Nm).

46. Reconnect the fuel feed and return lines.

47. Install and reconnect the ignition coil assemblies.

48. Install the intake manifold chamber and throttle body with new gaskets. Tighten the nuts and bolts to 17 ft. lbs. (24 Nm).

49. Reconnect the throttle cable to the throttle body linkage.

50. Reconnect the EGR valve supply tube and bracket.

51. Reconnect the following vacuum hoses to the intake manifold chamber:
- PCV hose
- EVAP canister vacuum hose
- Brake booster hose

52. Reattach the following sensor connectors to the rear of the intake manifold chamber:
- Ignition control module
- Linear EGR valve
- MAP sensor
- EVAP purge valve
- Throttle position sensor
- Idle air control valve
- Intake air temperature sensor

53. Install the power steering pump and mounting bracket.

54. Install the cooling fan and pulley assembly.

55. Install the drive belts.

56. Install and reconnect the upper and lower radiator hoses.

57. Install the air cleaner box and air intake duct.

58. Verify that all fuel lines, vacuum and coolant hoses, and wiring harness have been reconnected.

59. Refill the engine with fresh coolant.

60. Crank the engine until it starts. A longer than normal starting time may be necessary due to air in the fuel lines. Check all fuel line connections for leaks.

61. Bleed any air from the cooling system.

62. Bleed the power steering system if necessary.

63. Check the throttle cable operation and adjustment.

64. Check the engine oil level and add if necessary.

1998–99 3.2L and 2.2L Engines

The 1998–99 models do not use rocker arms/shafts, the camshaft directly actuates the valve.

Intake Manifold

REMOVAL & INSTALLATION

2.2L Engine

1. Disconnect the negative battery cable.

2. Properly relieve the fuel system pressure.

3. Drain and recycle the engine coolant.

4. Remove the air intake duct assembly from the vehicle.

5. Label and disconnect any vacuum hoses attached to the intake manifold and throttle body.

6. Disconnect the fuel lines from the fuel injector rail.

7. Label and disconnect the engine wiring harness from the Throttle Position (TP) sensor, Idle Air Control (IAC) valve and fuel injectors.

8. Remove the accessory drive belt and alternator brackets.

9. Disconnect the cooling hoses from the intake manifold and throttle body.

10. Remove the intake manifold support bracket.

11. Remove the ignition coil mounting bracket.

12. Remove the intake manifold mounting nuts/bolts from the cylinder head, then remove the manifold

To install:

13. Clean any old gasket material from the cylinder head and intake manifold mating surfaces.

14. Install the intake manifold using a new gasket and tighten the bolts to 16 ft. lbs. (22 Nm).

15. Install the ignition coil mounting bracket.

16. Install the intake manifold support bracket and tighten the bolts to 16 ft. lbs. (22 Nm).

17. Attach the cooling hoses to the intake manifold and throttle body.

18. Install the alternator brackets, then the accessory drive belt.

19. Attach the engine wiring harness to the Throttle Position (TP) sensor, Idle Air Control (IAC) valve and fuel injectors.

20. Connect the fuel lines to the fuel injector rail.

21. Detach the hoses that were disconnected during removal.

22. Install the air intake duct assembly to the vehicle.

23. Refill and bleed the cooling system.

24. Connect the negative battery cable.

25. Start the vehicle and check for leaks.

2.6L Engine

1. Relieve the fuel pressure.

2. Disconnect the negative battery cable.

3. Drain the engine coolant.

4. Disconnect the throttle cable from the throttle body linkage.

5. Disconnect and remove the air intake duct.

6. Disconnect the vacuum hose from the EGR valve. If equipped, detach the EGR temperature sensor connector.

7. Disconnect the EGR fuel pressure control rubber hose.

8. Use a flare nut wrench to disconnect the EGR pipe fitting from the intake manifold.

✷✷ CAUTION

The fuel injection system remains under pressure after the engine has been turned OFF. Properly relieve fuel pressure before disconnecting any fuel lines. Failure to do so may result in fire or personal injury.

9. Disconnect the fuel feed hose from the fuel rail.

10. Disconnect the coolant hoses from the intake manifold.

11. Disconnect the air regulator hose and harness from the lower rear of the intake manifold chamber.

12. Disconnect the throttle position sensor harness and the coolant hoses from the throttle body.

13. Loosen the throttle body mounting bolts, and remove the throttle body from the intake manifold chamber.

14. Loosen the intake manifold chamber mounting nuts and bolts, then remove the intake manifold chamber from the lower part of the manifold assembly.

15. Remove the fuel injector rail attaching bolts.

16. Label and disconnect the fuel injector harnesses.

17. Carefully lift the fuel rail and injectors from the intake manifold as an assembly.

18. Loosen the intake manifold mounting bolts and nuts in a crisscross sequence.

19. Remove the intake manifold from the engine. Clean any old gasket material from the cylinder head and intake manifold mating surfaces.

To install:

20. Install a new intake manifold gasket onto the cylinder head. Next, position the intake manifold onto the cylinder head mounting studs.

21. Install the intake manifold attaching bolts and nuts, Tighten the bolts and nuts in a two-step crisscross pattern beginning in the center and working outward. The final torque specification is 16 ft. lbs. (22 Nm).

22. Lubricate new O-rings with a small amount of clean engine oil, then install them onto the fuel injectors. Next, install the fuel injectors into the fuel rail, if they were removed. Install the fuel injectors and fuel rail assembly onto the intake manifold. Tighten the fuel rail attaching bolts to 14 ft. lbs. (19 Nm).

23. Reattach the fuel injector wiring harness connectors.

24. Install the intake manifold chamber to the lower part of the manifold using a new gasket. Tighten the bolts and nuts to 20 ft. lbs. (27 Nm) in a two-step crisscross pattern.

25. Install a new throttle body gasket, then install the throttle body. Tighten the mounting bolts to 14 ft. lbs. (19 Nm).

26. Reattach the throttle position sensor connector. Reattach the coolant hoses to the throttle body.

27. Connect the EGR pipe to the intake manifold, tighten the flange nut to 33 ft. lbs. (45 Nm).

28. Connect the EGR fuel pressure control rubber hose. If equipped, reattach the EGR temperature sensor connector.

29. Connect the vacuum hose to the EGR valve.

30. Connect the air regulator hose and harness.

31. Connect the coolant hoses to the intake manifold.

32. Connect the fuel feed hose to the fuel rail using new sealing washers.

33. Install the air intake duct and reconnect the vacuum hose.

34. Connect the throttle cable to the throttle body linkage and adjust as necessary.

35. Refill and bleed the cooling system.

36. Connect the negative battery cable.

37. Turn the ignition switch **ON** and check for fuel leaks at the fuel rail.

38. Check the manifold coolant hoses for leaks. Check the intake manifold mating surfaces for leaks.

3.2L Engine

1995 MODELS

✳✳ CAUTION

The fuel injection system remains under pressure even after the engine has been turned OFF. The fuel system pressure must be relieved before disconnecting any fuel lines. Failure to do so may result in fire and personal injury.

1. Relieve the fuel pressure.

2. Disconnect the negative battery cable.

3. Remove the air cleaner assembly.

4. Disconnect the accelerator pedal cable from the throttle body and bracket.

5. Disconnect the charcoal canister vacuum hose from the vacuum pipe.

6. Disconnect the air vacuum hose and the vacuum booster hose from the common chamber.

7. Detach the following electrical connectors:

- MAP sensor
- Charcoal canister Vacuum Switching Valve (VSV)
- Exhaust gas recirculation VSV
- Intake air temperature sensor
- Engine ground cable
- Fuel injector connectors
- Thermo sensor connector

8. Disconnect the spark plug wires.

9. Remove the ignition module assembly with the spark plug wires attached.

10. Disconnect the vacuum hoses from the throttle body.

11. Remove the four throttle body mounting bolts. Then, remove the throttle body.

12. Disconnect the PCV hose from the common chamber.

13. Disconnect the fuel pressure control valve vacuum hose from the common chamber.

14. Disconnect the evaporative emission canister purge hose from the common chamber.

15. Remove the EGR valve assembly from the common chamber.

16. Remove the common chamber (six bolts, two nuts, and three brackets).

17. Disconnect the fuel feed and return hoses from the fuel rail. Remove the bracket

mounting bolts from the cylinder head cover.

18. Remove the two bolts and four nuts to remove the intake manifold from the engine.

To install:

➡Use new self-locking nuts when installing the intake manifold. Use new manifold gaskets. Use new sealing washers when reconnecting the fuel lines.

19. Install the intake manifold on the engine. Tighten the bolts and nuts to 17 ft. lbs. (23 Nm). Tighten the bolts from the center toward the ends.

20. Attach the electrical connectors to the fuel injectors and the thermo sensor.

21. Connect the fuel return and feed hoses to the fuel rail.

22. Install the common chamber. Tighten the bolts and nuts to 17 ft. lbs. (23 Nm).

23. Install the EGR assembly. Tighten the bolts to 78 inch lbs. (9 Nm).

24. Connect the charcoal canister purge and fuel pressure regulator vacuum hoses to the common chamber.

25. Connect the PCV hose to the common chamber.

26. Install the throttle body assembly and connect the vacuum hoses to the throttle body. Tighten the throttle body mounting bolts to 16 ft. lbs. (22 Nm).

27. Install the ignition module assembly. Tighten the mounting bolts to 16 ft. lbs. (22 Nm).

28. Reconnect the spark plug wires.

29. Reattach the following electrical connectors:

- MAP sensor
- Charcoal canister Vacuum Switching Valve (VSV)
- Exhaust gas recirculation VSV
- Intake air temperature sensor
- Engine ground cable

30. Connect the air vacuum hose and the vacuum booster hose to the common chamber.

31. Connect the charcoal canister vacuum hose to the vacuum pipe.

32. Connect the accelerator cable to the throttle body and bracket. Adjust the cable so that the linkage rests against the stop when moved by hand.

33. Install the air cleaner assembly.

34. Verify that all electrical connectors and vacuum lines have been reattached.

35. Connect the negative battery cable.

36. Turn the ignition key to the **ON** position for two seconds, then **OFF**. Turn the ignition **ON** again to pressurize the fuel system and check for leaks.

1996–97 MODELS

1. Relieve the fuel system pressure.
2. Disconnect the negative battery cable.
3. Remove the air cleaner and intake duct.
4. Drain the engine coolant to a level below the upper radiator hose.
5. Disconnect the throttle cable from the throttle body linkage.
6. Label and disconnect the following vacuum hoses from the intake manifold chamber:
   - PCV hose
   - EVAP canister vacuum hose
   - Brake booster hose
7. Label and detach the following sensor connectors from the rear of the intake manifold chamber:
   - Ignition control module
   - Linear EGR valve
   - MAP sensor
   - EVAP purge valve
   - Throttle position sensor
   - Idle air control valve
   - Intake air temperature sensor
8. Disconnect the EGR valve supply tube and bracket.
9. Remove the throttle body.

✴✴ WARNING

Don't use solvent of any type when cleaning the gasket mating surfaces of the throttle body and intake manifold. Solvent may damage the machined surfaces of these components. Be careful not to scratch the mating surfaces.

10. Disconnect the MAP sensor tube, then unbolt the MAP sensor bracket.
11. Unbolt the intake manifold chamber from the brackets which are located at its front and rear edges.
12. Loosen the manifold mounting bolts and two nuts in a crisscross sequence.
13. Remove the bolts and nuts, then lift the chamber off of the base of the intake manifold. Note the positions of the long and short bolts.
14. Cover the intake manifold with a sheet of plastic, or clean shop towels to keep out dirt and foreign objects.
15. Carefully clean any dirt from the fuel rail and fuel fittings.

✴✴ CAUTION

The fuel injection system remains under pressure even after the engine has been turned OFF. The fuel system pressure must be relieved before dis-

connecting any fuel lines. Failure to do so may result in fire and personal injury.

16. Disconnect the fuel feed and return lines from the front of the fuel rail. Clean up any spilled fuel.
17. Label and disconnect the fuel injector wiring harness.
18. Unbolt the fuel return line bracket from the front of the intake manifold.
19. Unbolt the fuel rail from the intake manifold.
20. Carefully lift the fuel rail and injectors off of the intake manifold as an assembly. Move the fuel rail out of the way so it won't be damaged.
21. Remove the fuel rail spacer grommets from the sides of the intake manifold. Replace the spacer grommets if they are cracked or ripped.
22. Loosen the intake manifold nuts and bolts in a crisscross sequence working from the outer edges of the manifold toward the center.
23. Push the engine wiring harnesses aside and lift the intake manifold up and off of the engine block.
24. Remove the intake manifold gaskets. Be careful not to drop any pieces of the gaskets into the engine.
25. Cover the intake openings with a sheet of plastic or clean shop towels to keep out dirt and foreign objects.

To install:

26. Remove the covers from the intake openings. Install new intake manifold gaskets.
27. Fit the intake manifold into position. Move the wiring harness back into position.
28. Install the intake manifold nuts and bolts. Tighten them in a two-step crisscross pattern to 15 ft. lbs. (20 Nm) working from the center of the manifold toward the outer edges.
29. Install the fuel rail spacer grommets.
30. Install the fuel rail assembly onto the intake manifold. Be sure all the fuel injectors are properly seated. Tighten the fuel rail bolts to 5 ft. lbs. (7 Nm).
31. Reconnect the fuel feed and return lines. Install the return line bracket bolt.
32. Reconnect the fuel injector wiring harness.
33. Install a new intake manifold chamber gasket. Install the intake manifold chamber.
34. Install the six intake manifold chamber bolts and two nuts. Tighten the nuts and bolts to 15 ft. lbs. (20 Nm) in a crisscross sequence.
35. Install the intake manifold chamber bracket bolts.

36. Reconnect the MAP sensor tube and bracket.
37. Reconnect the EGR supply tube and bracket.
38. Install the throttle body with a new gasket. Tighten the bolts to 10 ft. lbs. (14 Nm) in a crisscross sequence.
39. Reattach the following sensor connectors to the rear of the intake manifold chamber:
   - Ignition control module connectors
   - Linear EGR valve
   - MAP sensor
   - EVAP purge valve
   - Throttle position sensor
   - Idle air control valve
   - Intake air temperature sensor
40. Reconnect the following vacuum hoses to the intake manifold chamber:
   - PCV hose
   - EVAP canister vacuum hose
   - Brake booster hose
41. Reconnect the throttle cable to the throttle body linkage.
42. Install the air cleaner and air intake duct.
43. Reconnect the negative battery cable.
44. Refill and bleed the cooling system.
45. Crank the engine until it starts.

➡ **Air trapped in the fuel lines may cause the engine to crank for a longer period of time than normal.**

46. Check the fuel lines, fuel rail, and injectors for any signs of leakage.
47. Warm the engine up to normal operating temperature and check the operation of the throttle cable and linkage. Adjust if necessary.
48. Check the manifold and throttle body mating surfaces for vacuum leaks.

1998–99 MODELS

On this engine the intake manifold is referred to as the common chamber.

1. Disconnect the negative battery cable.
2. Properly relieve the fuel system pressure.
3. Remove the air cleaner and intake duct.
4. Drain the engine coolant to a level below the upper radiator hose.
5. Disconnect the throttle cable from the throttle body linkage.
6. Label and disconnect all vacuum hoses from the common chamber.
7. Label and disconnect all electrical sensors from the common chamber.
8. Disconnect the EGR valve supply tube and bracket.
9. Remove the water hoses and mounting bolts from the throttle body.

10. Remove the throttle body.

11. Remove the two bolts mounting the fuel rail at the rear of the common chamber.

12. Loosen and remove the common chamber mounting nuts/bolts.

13. Cover the intake openings with a sheet of plastic or clean shop towels to keep out dirt and foreign objects.

To install:

14. Install the common chamber and tighten the nuts/bolts to 18 ft. lbs. (25 Nm).

15. Install the fuel rail bracket and tighten the bolts to 89 inch lbs. (10 Nm).

16. Install the throttle body and tighten the mounting bolts to 18 ft. lbs. (25 Nm).

17. Connect the water hoses to the throttle body.

18. Connect the EGR valve supply tube and bracket.

19. Reconnect all electrical sensors and vacuum hoses from the common chamber.

20. Connect the throttle cable to the throttle body linkage.

21. Fill the engine with coolant.

22. Install the air cleaner and intake duct assembly.

23. Connect the negative battery cable.

24. Crank the engine until it starts.

➡**Air trapped in the fuel lines may cause the engine to crank for a longer period of time than normal.**

25. Check the fuel lines, fuel rail, and injectors for any signs of leakage.

26. Warm the engine up to normal operating temperature and check the operation of the throttle cable and linkage. Adjust if necessary.

27. Check the manifold and throttle body mating surfaces for vacuum leaks.

Exhaust Manifold

REMOVAL & INSTALLATION

2.2L Engine

✳✳ CAUTION

To avoid injury from burns, allow the engine to cool to ambient temperature before removing the exhaust manifold.

1. Disconnect the negative battery cable.

2. Disconnect the intake air duct assembly.

3. Remove the exhaust heat shield.

4. Remove the exhaust header pipe.

5. Remove the 10 exhaust manifold mounting bolts, then remove the manifold.

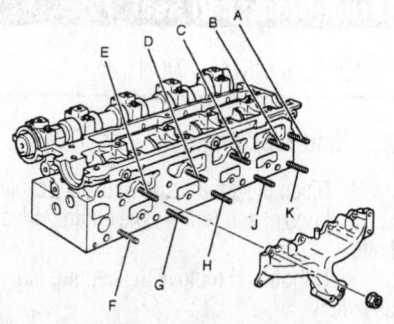

- Tightening sequence:
  Step1: J G H B D C J G B D
  Step2: A B C D E F G H J K
  Step3: A B C D E F G H J K
- Tightening torque:
  Step1: 14 N·m (10 lb ft)
  Step2: 20 N·m (14 lb ft)
  Step3: 20 N·m (14 lb ft)

7924NG17

Exhaust manifold mounting stud identification—2.2L engine

To install:

6. Install the exhaust manifold and tighten the mounting nuts following the sequence shown in the accompanying illustration.

7. Install the exhaust header pipe.

8. Install the exhaust manifold heat shield.

9. Connect the intake air assembly to the throttle body.

10. Connect the negative battery cable.

11. Start the engine and check the exhaust for leaks.

2.6L Engine

✳✳ CAUTION

To avoid injury from burns, allow the engine to cool to ambient temperature before removing the exhaust manifold.

1. Disconnect the negative battery cable.

2. Remove the intake air duct.

3. Disconnect the hoses from the air pump.

4. Remove the air pump mounting bolts. Slip the drive belt off the pulley and remove the air pump.

5. Remove the manifold heat shield.

6. Remove the EGR pipe clamp bolt from the rear of the cylinder head.

7. Raise and safely support the vehicle.

8. Disconnect the EGR pipe from the exhaust manifold.

➡**The dipstick and tube may be removed for extra access to the oxygen sensor and EGR pipe.**

9. Disconnect the front exhaust pipe from the exhaust manifold.

10. Uncouple the oxygen sensor harness.

11. Loosen the exhaust manifold nuts in a crisscross pattern.

12. Remove the exhaust manifold from the cylinder head. If it sticks, tap it with a soft-faced mallet.

To install:

➡**Install the new exhaust manifold gasket with the stamped mark facing outward.**

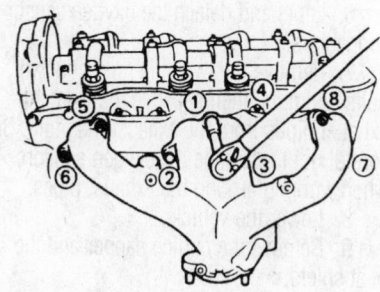

7924NG18

Exhaust manifold bolt tightening sequence—2.6L engine

13. Inspect the exhaust manifold mating surfaces for warpage or other damage. Replace if necessary. The warpage limit is 0.016 in. (0.4mm).

14. Using a new gasket, install the exhaust manifold. Tighten the manifold nuts in a crisscross pattern starting in the center and working outward to 33 ft. lbs. (44 Nm).

15. Install the tube and the dipstick if they were removed.

16. Connect the exhaust pipe to the exhaust manifold. Tighten the nuts to 49 ft. lbs. (67 Nm).

17. If the oxygen sensor was removed, coat its threads with a small amount of anti-seize compound. Don't get any anti-seize on the sensor's tip. Install the sensor and tighten its fitting to 31 ft. lbs. (42 Nm).

18. Connect the harness to the oxygen sensor.

19. Install the EGR pipe to the intake and exhaust manifolds and tighten the bolts to 17 ft. lbs. (24 Nm).

20. Lower the vehicle.

21. Install the manifold heat shield.

22. Install the EGR pipe clamp bolt to the rear of the cylinder head.

23. Install the air pump and drive belt.

24. Connect the hoses to the air pump.

25. Install the intake air duct.

26. Connect the negative battery cable.

3.2L Engines

➡ **To avoid injury from burns, allow the engine to cool completely before removing the exhaust manifolds.**

1. Disconnect the negative battery cable.

2. If removing the left manifold, remove the air duct.

3. If removing the left manifold, remove the EGR pipe mounting bolts from the exhaust manifold.

4. Raise and safely support the vehicle.

5. If necessary to gain extra working room, remove the transfer case skid plate.

6. Label and detach the oxygen sensor connectors.

7. Remove the two stud nuts and two bolts and nuts, then separate the front exhaust pipes from the exhaust manifold. Be careful not to damage the oxygen sensors when working around the exhaust pipes.

8. Lower the vehicle.

9. Remove the engine hanger and the heat shield.

10. Remove the seven nuts, then remove the exhaust manifold from the cylinder head.

To install:

➡ **Use new self-locking nuts and new gaskets when installing the exhaust manifolds.**

11. Install the exhaust manifold and gasket to the cylinder head using new nuts. Tighten the new nuts to 42 ft. lbs. (57 Nm) in a crisscross sequence.

12. Install the heat shield and the engine hanger.

13. Raise and safely support the vehicle.

14. Install the front exhaust pipes and reconnect the exhaust system. Tighten the exhaust fasteners to the following specifications:

• Stud nuts: 49 ft. lbs. (67 Nm)
• Flange nuts and bolts: 32–37 ft. lbs. (43–50 Nm).

15. Reattach the oxygen sensor connectors.

16. Install the skid plate and tighten the bolts to 27 ft. lbs. (37 Nm).

17. Lower the vehicle.

18. If installing the left manifold, install the EGR pipe to the exhaust manifold. Tighten the mounting bolts to 21 ft. lbs. (28 Nm).

19. If installing the left manifold, install the air duct and connect the negative battery cable. Verify that all wires and vacuum lines have been reattached.

20. Start the engine and check for exhaust leaks.

Front Crankshaft Seal

REMOVAL & INSTALLATION

2.2L and 2.6L Engines

1. Disconnect the negative battery cable.

2. Loosen and remove the engine drive belts.

3. Unbolt and remove the cooling fan assembly.

4. If equipped with A/C, remove the belt tensioner.

5. Remove the water pump pulley.

6. Remove the power steering pump from the mount. Don't disconnect the hydraulic lines.

7. Remove the upper timing belt cover.

8. Rotate the crankshaft to set the engine at TDC/compression for the No. 1 cylinder. The mark on the camshaft sprocket aligns with the mark on the rear timing cover. The crankshaft sprocket keyway aligns with the mark on the oil seal retainer cover.

9. Remove the crankshaft pulley.

10. Remove the lower timing belt cover.

11. Disconnect and remove the starter motor if a flywheel holder tool is to be installed.

12. Remove the timing belt or timing chain.

13. Use tool No. J-22888 or an equivalent gear puller to remove the oil pump sprocket.

14. If necessary, use a prytool to remove the oil pump seal.

15. Using the appropriate seal remover, pull the crankshaft oil seal from the oil seal retainer. Be careful not to damage the crankshaft sealing surface.

To install:

16. Lubricate a new crankshaft oil seal with clean engine oil and tap it into the retainer with an oil seal installation tool. Install a new oil pump oil seal if necessary.

17. Install the crankshaft sprocket so that its keyway and alignment mark point to the alignment mark on the oil seal retainer.

18. Install the oil pump pulley. Apply a small amount of thread-locking compound to the sprocket nut threads. Tighten the nut to 56 ft. lbs. (76 Nm).

19. Verify that all the timing marks are aligned and install the timing belt or chain.

20. Install the lower timing cover and tighten the bolts to 4.4 ft. lbs. (6 Nm).

21. Install the crankshaft pulley and tighten the bolt to 87 ft. lbs. (118 Nm).

22. Install the starter motor.

23. Install the upper timing cover and tighten the bolts to 4.4 ft. lbs. (6 Nm).

24. Install the power steering pump.

25. Install the cooling fan and tighten the bolts to 20 ft. lbs. (26 Nm).

26. Install the A/C belt tensioner.

27. Install the accessory drive belts.

28. Reconnect the negative battery cable.

29. Start the engine and check for leaks. Add oil if necessary.

30. Check and adjust the ignition timing as necessary.

3.2L Engines

1. Disconnect the negative battery cable.

2. Remove the drive belts.

3. Drain the coolant to a level below the upper radiator hose. Disconnect the upper radiator hose from the coolant inlet.

4. Remove the cooling fan and pulley assembly.

5. Remove the upper timing cover.

6. Rotate the crankshaft to align the timing marks.

7. Remove the crankshaft pulley and the lower timing belt cover.

8. Remove the timing belt.

9. Remove the crankshaft timing sprocket and key.

10. Use a seal puller and remove the crankshaft oil seal. Be careful not to damage the crankshaft or the oil pump sealing surface.

To install:

11. Apply engine oil to the lip of the seal and install the oil seal using seal installer J-39202 or an equivalent seal driver.

12. Install the crankshaft timing sprocket and key.

13. Install the timing belt.

14. Install the lower timing belt covers and tighten the bolts to 12 ft. lbs. (17 Nm).

15. Install the crankshaft pulley and tighten it to 123 ft. lbs. (167 Nm).

16. Install the upper timing covers.

17. Install the cooling fan and pulley assembly and tighten the bolts to 16 ft. lbs. (22 Nm).

18. Install the accessory drive belts.

19. Connect the upper radiator hose. Refill and bleed the cooling system.

20. Check the engine oil level and top up if necessary.

21. Reconnect the negative battery cable.

Camshaft

REMOVAL & INSTALLATION

2.2L Engine

1. Disconnect the negative battery cable.

2. Remove the valve cover and the timing belt cover.

3. Set the engine to Top Dead Center (TDC) for cylinder No. 1.

4. Loosen the timing belt tensioner and remove the timing belt.

5. Utilizing an adjustable wrench to hold the hexagonal part of the camshaft, remove the camshaft pulley bolt.

6. Remove the intake and exhaust camshaft pulleys.

7. Remove the twenty camshaft bearing cap mounting bolts, then the caps.

8. Carefully remove the camshafts from the engine and remove the oil seals.

9. If necessary, the Hydraulic Lash Adjusters (HLA's) may now be removed from the cylinder head.

10. Cover the cylinder head to prevent foreign materials from entering.

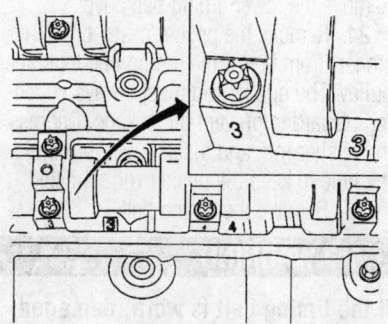

Camshaft bearing cap identification locations—2.2L engine

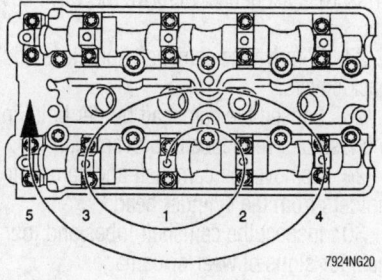

Camshaft bearing cap tightening sequence—2.2L engine

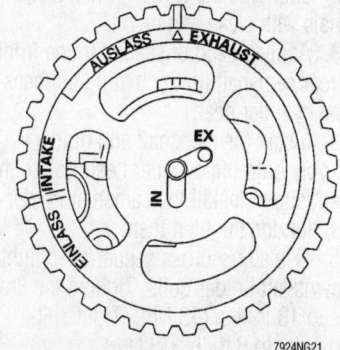

Guide pin location for the exhaust cam gear—2.2L engine

To install:

11. Clean the camshaft bearing journals and the top of the HLA's.

12. Apply clean engine oil to the camshaft bearing journals and the top of the HLA's.

13. Install the camshafts.

14. Install the camshaft bearing caps according to the cap number. Bracket numbers are as follows:
- Exhaust: 1–5 from the front
- Intake: 6–10 from the front

15. Tighten the bearing caps to 5.9 ft. lbs. (8 Nm).

16. Install the camshaft oil seal using the appropriate seal driver.

➡ **The camshaft drive gears are interchangeable.**

17. Install the cam drive gear and align the timing mark to be centered on the camshaft bearing cap lug.

18. Insert the guide pin in the appropriate slot for the designated camshaft.

19. Tighten the drive gear bolt to 36 ft. lbs. (50 Nm).

20. Install the timing belt and valve cover.

21. Connect the negative battery cable.

2.6L Engine

1. Disconnect the negative battery cable.

2. Loosen and remove the engine accessory drive belts.

3. Remove the cooling fan assembly.

4. Remove the water pump pulley.

5. Label and disconnect the ignition wires.

6. Remove the upper timing belt cover.

7. Rotate the crankshaft to position the No. 1 cylinder on the TDC of its compression stroke. The mark on the camshaft sprocket aligns with the pointer on the back timing cover. Do not disturb the engine once it is in this position.

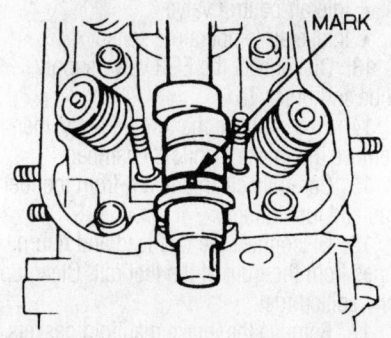

Camshaft installation positioning mark location—2.6L engine

➡**Remove the starter motor if flywheel holder tool No. J-38674 is to be installed. The use of alternate types of flywheel/crankshaft pulley holder tools may not require removal of the starter motor.**

8. Remove the valve cover.

9. Remove the crankshaft pulley and the lower timing belt cover.

10. Remove the timing belt. Mark its direction of rotation if it is to be reinstalled.

11. Unbolt and remove the camshaft pulley.

12. Loosen the camshaft holder bolts in a two-step crisscross sequence to prevent warpage. Remove the bolts and lift the rocker arm assembly off of the cylinder head. Label the positions of the rocker arms and shafts if the rockers and shafts are to be disassembled.

13. Lift the camshaft from the cylinder head. Remove the camshaft oil seals.

14. Inspect the camshaft lobes and journals for signs of wear. Replace parts as necessary.

15. Cover the cylinder head with a sheet of plastic or clean shop towels to keep out dust and foreign objects.

To install:

16. Lubricate the camshaft and its journals with clean engine oil and position it onto the cylinder head. The marks on the camshaft must face upwards when installed to ensure that the engine remains at TDC/compression for the No. 1 cylinder.

17. Assemble the rocker arm and shaft assemblies and camshaft holders.

18. Apply silicone gasket sealer to the front side of the contact surface of the No. 1 camshaft holder prior to installation. Don't allow the sealant to cure before installing the holder.

19. Install the rocker arm and shaft assembly. Install all the retaining bolts, but only hand-tighten them at this point.

20. Starting in the center of the cylinder head and working outward, use a two-step crisscross pattern to tighten the holder bolts to 6 ft. lbs. (8 Nm) and the holder nuts to 16 ft. lbs. (22 Nm).

21. Lubricate a new camshaft oil seal when clean engine oil. Using the appropriate seal installer, drive the seal into position.

22. Install the back timing cover.

23. Install the camshaft sprocket and its key. Tighten the bolt to 43 ft. lbs. (59 Nm).

24. Verify that the timing marks are aligned and that the engine is at TDC/compression for the No. 1 cylinder. Install the timing belt and tension the timing belt.

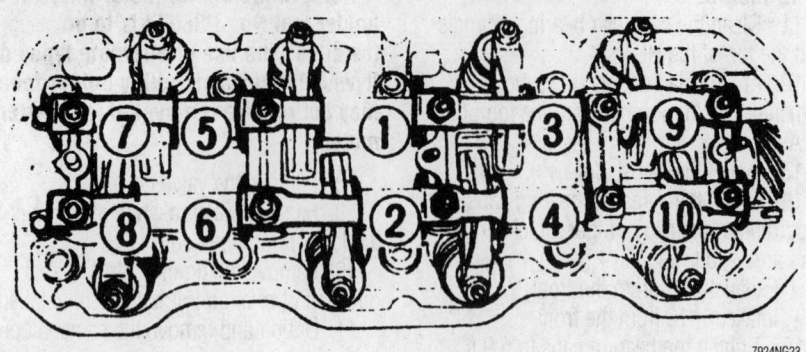

Camshaft holder bolt tightening sequence—2.6L engine

25. Install the lower timing belt cover. Install the crankshaft pulley and tighten the bolt to 87 ft. lbs. (118 Nm).

26. Adjust the valve lash. Tighten the adjusting screw locknuts to 9 ft. lbs. (13 Nm).

27. Rotate the engine manually two times to check that there is no piston-to-valve interference.

28. Apply silicone sealant to the front and rear valve cover mating surfaces of the cylinder head. Immediately install a new gasket and the valve cover. Tighten the nuts to 7 ft. lbs. (10 Nm).

29. Install the upper timing belt cover.

30. Install the starter motor if it was removed. Tighten the mounting bolts to 30 ft. lbs. (41 Nm).

31. Install the water pump pulley and tighten its bolts to 6 ft. lbs. (8 Nm). Install the cooling fan.

32. Install the engine accessory drive belts and adjust their tensions.

33. Change the engine oil and filter.

34. Verify that all electrical connectors and vacuum lines have been reattached.

35. Reconnect the negative battery cable.

36. Start the engine. Check and adjust the ignition timing.

3.2L (VIN V) Engine

> ✳✳ **CAUTION**
>
> **The fuel injection system remains under pressure even after the engine has been turned OFF. The fuel system pressure must be relieved before disconnecting any fuel lines. Failure to do so may result in fire and personal injury.**

1. Relieve the fuel system pressure.

2. Disconnect the negative battery cable and reinstall the fuel pump relay.

3. Drain the coolant into a sealable container.

4. Support the hood as far open as possible.

5. Remove the air intake duct and the air cleaner box.

6. Disconnect and remove the upper and lower radiator hoses. Catch any coolant that runs out.

7. Loosen and remove the power steering pump, A/C compressor, and alternator drive belts.

8. Remove the cooling fan and its pulley assembly.

9. Unbolt the power steering pump mounting bracket. Move the pump and bracket out of the way without disconnecting the hydraulic lines.

10. Disconnect the throttle cable from the throttle body linkage.

11. Label and disconnect the following vacuum hoses from the intake manifold chamber:
- PCV hose
- EVAP canister vacuum hose
- Brake booster hose

12. Label and detach the following sensor connectors from the rear of the intake manifold chamber:
- Ignition control module connectors
- Linear EGR valve
- MAP sensor
- EVAP purge valve
- Throttle position sensor
- Idle air control valve
- Intake air temperature sensor

13. Disconnect the EGR valve supply tube and bracket.

14. First, remove the throttle body, then remove the intake manifold chamber.

15. Carefully clean any dirt from the fuel rail and fuel fittings.

16. Disconnect the fuel feed and return lines from the front of the fuel rail. Clean up any spilled fuel.

17. Remove the intake manifold gaskets. Be careful not to drop any pieces of the gaskets into the engine. Don't scratch or gouge the machined aluminum mating surfaces of the intake manifold and engine block.

18. Cover the intake openings with a sheet of plastic or clean shop towels to keep out dirt and foreign objects.

19. Label the ignition coil assemblies and disconnect them from the wiring harness. Remove the coil assemblies so they won't be damaged.

20. Unbolt the oil cooler line brackets from the timing belt covers.

21. Remove the upper timing belt covers.

22. Rotate the crankshaft to align the camshaft timing marks with the pointer dots on the back covers. When the timing marks are aligned, the No. 2 cylinder is at TDC/compression.

23. Remove the crankshaft pulley. Remove the lower timing belt cover.

24. Remove the pusher assembly (tensioner) from below the timing belt tensioner pulley. The pusher rod must always be facing upward to prevent oil leakage. Depress the pusher rod, and insert a wire pin into the hole to keep the pusher rod retracted.

25. Remove the timing belt.

> ✳✳ **WARNING**
>
> **If the timing belt is worn, damaged, or shows signs of oil or coolant contamination it must be replaced.**

26. Loosen the valve cover bolts in a crisscross sequence. Remove the valve covers.

27. Remove the camshaft sprockets and back covers.

28. Loosen the camshaft holder bolts in a crisscross sequence to prevent warping.

29. Remove the camshaft and camshaft holders from the cylinder head.

30. Inspect the camshaft lobes and journals for signs of wear or damage.

To install:

31. Be sure all mating surfaces are clean and free of oil, coolant, or gasket residue.

32. Lubricate the camshaft lobes and journals with clean engine oil.

33. Apply a bead of sealant to the front and rear camshaft holder mating surfaces on the cylinder head.

34. Install the camshaft and holder assembly onto the cylinder head before the sealant cures. Install the camshaft holder bolts, but don't tighten them yet.

35. Use a crisscross sequence to tighten the camshaft holder bolts. Tighten the 8mm bolts to 13 ft. lbs. (18 Nm). Tighten the 6mm bolts to 6 ft. lbs. (8 Nm).

36. Use a seal driver to install a new camshaft seal.

37. Install the camshaft sprocket back covers and tighten their bolts to 12 ft. lbs. (16 Nm).

38. Install the camshaft sprockets so that the timing marks are aligned. Tighten the bolts to 46 ft. lbs. (64 Nm).

39. Apply a 1/8 inch (3mm) bead of sealant to the joint were the camshaft holders meet the cylinder head. Install the valve cover with a new gasket before the sealant cures.

40. Tighten the valve cover bolts to 6 ft. lbs. (8 Nm) in a crisscross pattern.

41. Verify that the camshaft and crankshaft timing marks are properly aligned.

42. Install and tension the timing belt. Tighten the pusher bolts to 14 ft. lbs. (19 Nm).

43. Install the lower timing belt covers and tighten the bolts to 13 ft. lbs. (18 Nm). Install the crankshaft pulley. Tighten the pulley bolt to 123 ft. lbs. (167 Nm).

44. Install the upper timing belt covers and tighten the bolts to 13 ft. lbs. (18 Nm).

45. Fit the oil cooler line brackets onto the timing cover and tighten the bolts to 13 ft. lbs. (18 Nm).

46. Reconnect the fuel feed and return lines.

47. Install and reconnect the ignition coil assemblies.

48. Install the intake manifold chamber and throttle body with new gaskets. Tighten the nuts and bolts to 17 ft. lbs. (24 Nm).

49. Reconnect the throttle cable to the throttle body linkage.

50. Reconnect the EGR valve supply tube and bracket.

51. Reconnect the following vacuum to the intake manifold chamber:
- PCV hose
- EVAP canister vacuum hose
- Brake booster hose

52. Reattach the following sensor connectors to the rear of the intake manifold chamber:
- Ignition control module connectors
- Linear EGR valve
- MAP sensor
- EVAP purge valve
- Throttle position sensor
- Idle air control valve
- Intake air temperature sensor

53. Install the power steering pump and mounting bracket.

54. Install the cooling fan and its pulley assembly.

55. Install and tension the alternator, A/C compressor, and power steering pump drive belts.

56. Install and reconnect the upper and lower radiator hoses.

57. Install the air cleaner box and air intake duct.

58. Verify that all fuel lines, vacuum and coolant hoses, and wiring harness have been reconnected.

59. Refill the engine with fresh coolant.

60. Crank the engine until it starts. A longer than normal starting time may be necessary due to air in the fuel lines. Check all fuel line connections for leaks.

61. Bleed any air from the cooling system.

62. Bleed the power steering system if necessary.

63. Check the throttle cable operation and adjustment.

64. Check the engine oil level and add if necessary.

3.2L (VIN W) Engine

1. Disconnect the negative battery cable.

2. Remove the crankshaft pulley assembly using tool J-8614-01 or equivalent crankshaft holder, hold the pulley, then remove the center bolt and pulley assembly.

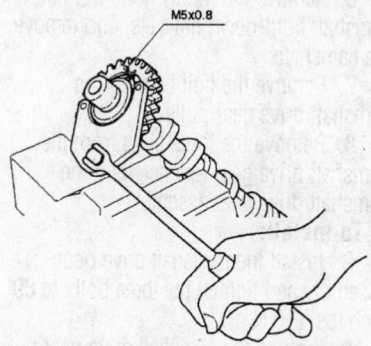

Aligning the sub gear with the Gear Spring Lever J-42686—3.2L (VIN W) engine

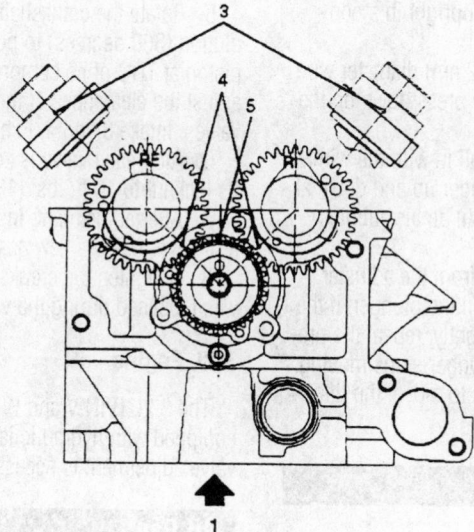

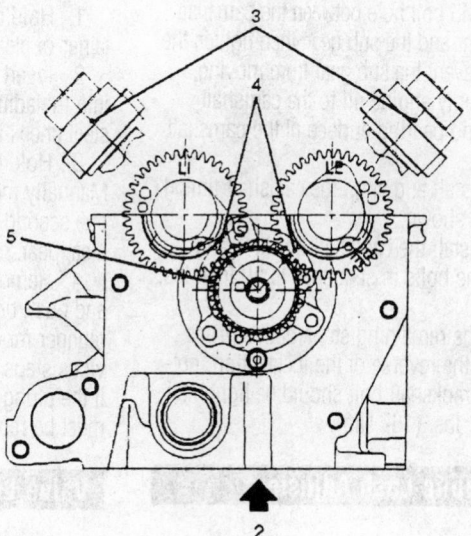

Legend
(1) Right Bank
(2) Left Bank
(3) Alignment Mark on Camshaft Drive Gear
(4) Alignment Mark on Camshaft
(5) Alignment Mark on Retainer

Camshaft alignment marks for the left and right cylinder heads—3.2L (VIN W) engine

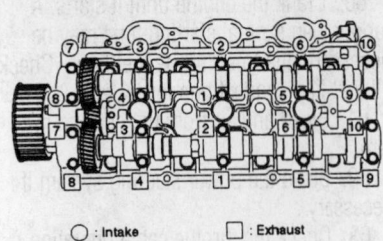

○ : Intake □ : Exhaust

7924NG47

Camshaft retaining bracket tightening sequence—3.2L (VIN W) engine

3. Remove the timing belt using the recommended procedure in the timing belt unit repair section.

4. Label, then disconnect all wiring harnesses and vacuum hoses related to valve cover removal.

5. Remove the left and right valve covers.

6. Remove the twenty bolts from the camshaft hold-down brackets, and remove the camshafts.

7. Remove the bolt holding the camshaft drive gear pulley.

8. Remove the three bolts from the camshaft drive gear retainer, then the camshaft drive gear assembly.

To install:

9. Install the camshaft drive gear assembly and tighten the three bolts to 89 inch lbs. (10 Nm).

10. Tighten the camshaft drive gear pulley bolt to 72 ft. lbs. (98 Nm).

11. Use the Gear Spring Lever J-42686 or equivalent to turn the sub gear until it aligns with the M5 bolt hole between the camshaft driven gear and the sub gear, then tighten the bolt to prevent the sub gear from moving.

12. Apply engine oil to the camshaft journal and bearing surface of the camshaft brackets.

13. Install and align the camshaft timing marks as shown.

14. Install the camshaft brackets and tighten the bolts in sequence to 89 inch lbs. (10 Nm).

15. The remaining steps of this procedure are the reverse of the installation, noting the crankshaft bolt should be tightened to 123 ft. lbs. (167 Nm).

Hydraulic Lash Adjusters

BLEEDING

➡The hydraulic lash adjuster is a precision component that relies on a clean operating environment. When handling the lash adjuster, make certain that no dirt or foreign particles are allowed to get inside the unit. Do not try to take the unit apart; it can't be rebuilt. The adjuster is filled with oil. Hold the adjuster in the upright position so that the oil does not spill out. Use only clean engine oil when bleeding the lifters.

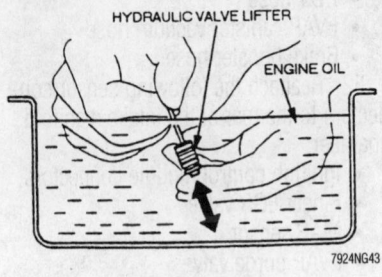

Priming the valve lash adjuster

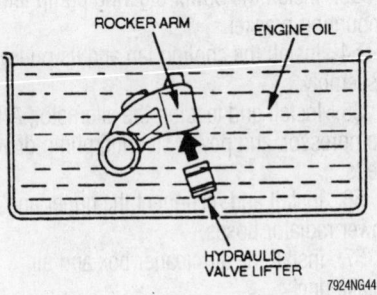

7924NG44

Installing the valve lash adjuster into the rocker arm

To bleed the hydraulic lash adjusters when disassembling the rocker arms or as part of an engine overhaul, perform the following:

1. Hold the adjuster upright in a container of clean engine oil.

2. Insert a 0.08 in. (2mm) diameter wire into the adjuster. Lightly press down on the steel check ball.

3. Hold the check ball in with the wire. Manually move the plunger up and down at one second intervals until all air bubbles disappear.

4. Remove the wire from the adjuster and push down firmly on the plunger. If the plunger moves even slightly, repeat the previous steps until the plunger stops moving. If the plunger continues to move, the lifter must be replaced.

Valve Lash

ADJUSTMENT

2.2L Engine

The 2.2L engine is equipped with hydraulic lash adjusters. No valve adjustment is necessary.

2.6L Engine

➡The valves are adjusted with the engine COLD. It is best to allow an engine to sit overnight before beginning a valve adjustment. While all valve adjustments must be made as accurately as possible, it is better to have the valve adjustment slightly loose rather than slightly tight. A burned valve may result from overly tight valve adjustments.

1. Remove the rocker arm cover and discard the gasket.

2. Rotate the crankshaft pulley until the No. 1 piston is at TDC of the compression stroke.

➡To be sure the piston is on the correct stroke, remove the spark plug and place your finger over the hole. Feel for air being forced out of the spark plug hole. Both valves on No. 1 cylinder will be closed. Stop turning the crankshaft when the TDC timing mark on the crankshaft pulley is directly aligned with the timing mark pointer.

3. With the No. 1 piston at TDC of the compression stroke, adjust the clearances of the following valves: Intake 1 and 2, Exhaust 1 and 3.

4. Adjust the clearance by loosening the locknut and turning the adjusting screw. Retightening the locknut when the proper thickness feeler gauge passes between the rocker arm and valve stem with a slight drag. Clearance is 0.006 in. (0.15mm) for intake, and 0.010 in. (0.25mm) for exhaust.

5. Rotate the crankshaft 1 complete revolution (360 degrees) to position the No. 4 piston at TDC of its compression stroke and adjust the clearances of the following valves: Intake 3 and 4, Exhaust 2 and 4.

6. After each valve is adjusted, tighten its locknut to 10 ft. lbs. (13 Nm).

7. After adjustment, install the rocker arm cover using a new gasket and sealant.

8. Retighten the crankshaft pulley if it was loosened during the valve adjustment.

3.2L Engine

The 3.2L (VIN V and W) engines are equipped with hydraulic lash adjusters. No valve adjustment is necessary.

Oil Pan

REMOVAL & INSTALLATION

2.2L and 2.6L Engines

1. Disconnect the negative battery cable
2. Drain the engine oil.

3. Attach a chain hoist to the engine lifting hooks.

4. Raise and safely support the vehicle.

5. Remove the front wheels.

6. Remove the dipstick from the dipstick tube.

7. Remove the radiator skid plate, if equipped and the radiator lower shroud.

8. Remove the suspension crossmember.

9. Remove the flywheel dust cover.

10. Matchmark the Pitman arm to the steering shaft. Use a puller to remove the Pitman arm.

11. Unbolt the idler arm assembly from the frame.

12. If equipped with 4WD, support the axle assembly with a jack and remove the mounting bolts on both sides of the axle assembly. Lower the axle assembly to gain access to the oil pan. After lowering the axle, support it securely.

➡ **The lower section of the oil pan may be separated and removed from the upper section of the oil pan.**

13. Raise the chain hoist to take the engine's weight off the mounts.

14. Unbolt the engine mounts from the brackets on either side of the oil pan.

15. Remove the oil pan mounting bolts and bolt retainers. Use a sealer cutter to break the seal and remove the oil pan from the engine block.

To install:

16. Thoroughly clean and dry the sealing surface of the oil pan and engine block. Apply beads of sealant to the front and rear oil seal retainer surfaces. Install the oil pan to the engine block within five minutes of sealer application. Install the bolt retainers and all the mounting bolts. Tighten the mounting bolts in sequence to 13 ft. lbs. (18 Nm).

17. Connect the engine mounts to the brackets.

18. Lower the hoist.

19. On 4WD models, raise the axle housing assembly into position. Tighten the axle-to-frame mounting bolts and nuts to 112 ft. lbs. (152 Nm).

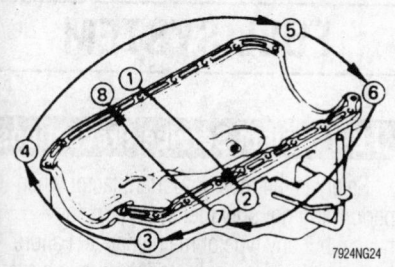

Oil pan bolt tightening sequence—2.2L and 2.6L engines

20. Install the idler arm bracket. Tighten the mounting bolts to 33 ft. lbs. (45 Nm).

21. Align the matchmarks and install the Pitman arm on the selector shaft. Tighten the nut to 160 ft. lbs. (216 Nm).

22. Install the remaining components.

23. Refill the engine with the proper amount of oil.

24. Connect the negative battery cable.

25. Start the engine and check for leaks.

26. Check and adjust the front wheel alignment and the steering wheel spoke angle.

3.2L Engines

1. Disconnect the negative battery cable.

2. Drain the engine oil.

3. Raise and safely support the vehicle.

4. Remove the front wheels.

5. Remove the dipstick from the dipstick tube.

6. Remove the radiator skid plate and the radiator lower shroud.

7. Remove the suspension crossmember from below the oil pan.

8. Remove the flywheel dust cover.

9. Disconnect the tie rod ends from the steering knuckles.

10. Matchmark the Pitman arm to the steering shaft and use a puller (tool No. J-29107 or equivalent) to remove the Pitman arm.

11. Unbolt the idler arm bracket from the frame. Then, remove the steering linkage from the vehicle as an assembly.

12. If equipped with 4WD, perform the following steps:

 a. Support the axle assembly securely with a jack.

 b. Remove the mounting bolts and nuts from the axle assembly mounting brackets on both sides of the axle assembly.

 c. Lower the axle assembly to gain access to the oil pan. Support the axle assembly so that the halfshafts aren't stressed.

13. Remove the oil pan mounting bolts. Use a sealer cutter to break the seal and separate the oil pan from the engine block.

To install:

14. Thoroughly clean and dry the sealing surface of the oil pan and engine block. Apply a continuous bead of sealant to the oil pan flange and install the oil pan to the engine block. Do not allow the sealant to cure before installation. Tighten the mounting bolts to 74 ft. lbs. (10 Nm) in a two-step crisscross sequence.

15. Install the axle housing assembly. Tighten the axle mounting bracket bolts and nut to 112 ft. lbs. (152 Nm). If the axle

housing flange bolts at the mounting brackets were loosened or removed, tighten them to 61 ft. lbs. (82 Nm).

16. Install the idler arm. Tighten the mounting bolts to 33 ft. lbs. (45 Nm).

17. Align the matchmark and install the Pitman arm on the sector shaft. Tighten the nut to 160 ft. lbs. (216 Nm).

18. Install the flywheel dust cover.

19. Install the suspension crossmember. Tighten the mounting bolts to 58 ft. lb. (78 Nm).

20. Install the steering linkage, Pitman arm, and idler arm assembly. Tighten the fasteners to the following specifications:

 • Tie rod end castle nuts: 73 ft. lbs. (98 Nm)

 • Idler arm bracket bolts: 33 ft. lbs. (44 Nm)

 • Pitman arm nuts: 159 ft. lbs. (216 Nm).

21. Install the lower fan shroud.

22. Install the radiator skid plate. Tighten the bolts to 27 ft. lbs. (37 Nm).

23. Verify that the front axle and any suspension components have been correctly installed.

24. Lower the vehicle to the floor.

25. Install the dipstick and refill the engine with the proper amount of oil.

26. Connect the negative battery cable.

27. Start the engine and check for oil leaks.

Oil Pump

REMOVAL & INSTALLATION

2.2L and 2.6L Engines

The oil pump is attached to the front, lower right side of the engine and is driven by the timing belt.

1. Disconnect the negative battery cable.

2. Drain the engine oil.

3. Loosen and remove the engine accessory drive belts.

4. Unbolt and remove the cooling fan assembly.

5. If equipped with A/C, remove the belt tensioner.

6. Remove the water pump pulley.

7. Remove the power steering pump from its mount. Don't disconnect the hydraulic lines.

8. Disconnect and remove the starter motor if a flywheel holding tool is to be used.

9. Remove the upper timing belt cover.

10. Rotate the crankshaft to set the engine at TDC/compression for the No. 1 cylinder. The mark on the camshaft sprocket will align with the mark on the rear timing cover.

11. Remove the crankshaft pulley.

12. Remove the lower timing belt cover. Verify that the engine is set at TDC/compression for the No. 1 cylinder; the pointer on the crankshaft sprocket aligns with the pointer on the oil seal retainer.

13. Loosen the timing belt tensioner and relax the tension, then remove the timing belt from the crankshaft sprocket.

14. Use tool No. J-22888 or an equivalent puller to remove the oil pump sprocket.

15. Unbolt the oil pump and remove it from the engine.

16. Remove the O-ring seal from the oil pump housing.

To install:

✷✷ WARNING

The timing belt must be replaced if it is damaged, or has come in contact with oil or coolant.

17. Inspect the oil pump and its rotors for signs of scoring and damage. Replace the pump or any damaged parts.

18. Lubricate and install a new O-ring seal.

19. Install the oil pump and tighten the bolts to 14 ft. lbs. (19 Nm).

20. Align the timing marks and install the oil pump sprocket. Apply a small amount of thread-locking compound to the nut threads and tighten it to 56 ft. lbs. (76 Nm).

21. Verify that the engine is at TDC/compression for the No. 1 cylinder.

22. Install the timing belt.

23. If a flywheel holder tool was used, remove it.

24. Apply the tensioner pulley spring pressure to the timing belt.

➡**Remove the crankshaft holder before rotating the crankshaft to tension the timing belt.**

25. Rotate the crankshaft counterclockwise for two complete revolutions and realign the timing marks.

26. Loosen the tensioner pulley bolt to allow the spring to adjust the correct tension. Tighten the tensioner pulley bolt to 14 ft. lbs. (19 Nm).

27. Install the lower timing cover and tighten the bolts to 53 inch lbs. (6 Nm).

28. Install the crankshaft pulley and tighten the bolt to 87 ft. lbs. (118 Nm).

29. Install the upper timing cover.

30. Install the starter motor and tighten the mounting bolts to 30 ft. lbs. (40 Nm).

31. Install the power steering pump. If the hydraulic lines were disconnected, refill and bleed the power steering system.

32. Install the cooling fan and tighten the bolts to 20 ft. lbs. (26 Nm).

33. Install and adjust the accessory drive belts.

34. Refill the engine with fresh oil.

35. Reconnect the negative battery cable.

36. Start the engine and check for leaks.

37. Check the engine's oil pressure.

3.2L Engines

1. Disconnect the negative battery cable.

2. Remove the timing belt. If the timing belt is damaged or has been contaminated with oil or coolant, it must be replaced.

3. Remove the crankshaft timing pulley.

4. Raise and safely support the vehicle.

5. Drain the engine oil.

6. Remove the oil pan.

7. Remove the oil pipe and O-ring.

8. Remove the oil strainer and O-ring.

9. Remove the oil cooler assembly.

10. Remove the oil pump mounting bolts, then remove the oil pump from the engine block.

To install:

11. Install a new oil seal into the oil pump housing.

12. Thoroughly clean the sealing surface of the oil pump and the engine block.

13. Apply sealant to the oil pump. Be careful not to block the oil ports.

14. Apply engine oil to the seal lip and install the oil pump on the engine block. Tighten the mounting bolts to 13 ft. lbs. (18 Nm). Take care not to drop the garter spring from the seal lid during installation.

15. Install the oil cooler assembly.

16. Install the oil pipe and O-ring.

17. Install the oil strainer and O-ring.

18. Install the oil pan.

19. Install the crankshaft timing pulley.

20. Install the timing belt.

21. Install the remaining accessories and drive belts.

22. Lower the vehicle.

23. Refill the engine with oil.

24. Connect the negative battery cable, start the engine and check for proper oil pressure.

➡**If the oil pressure does not build up almost immediately, stop the engine and investigate the cause.**

25. Check for oil leaks.

Rear Main Seal

REMOVAL & INSTALLATION

1. Disconnect the negative battery cable. Raise and safely support the vehicle.

2. Drain the engine oil and remove the oil pan.

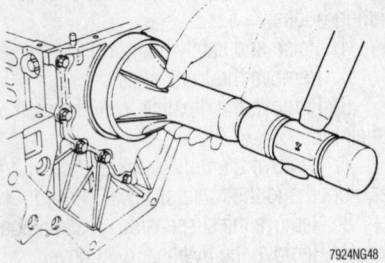

Installing a one-piece rear crankshaft oil seal

✷✷ CAUTION

The EPA warns that prolonged contact with used engine oil may cause a number of skin disorders, including cancer! You should make every effort to minimize your exposure to used engine oil. Protective gloves should be worn when changing the oil. Wash your hands and any other exposed skin areas as soon as possible after exposure to used engine oil. Soap and water, or waterless hand cleaner should be used.

3. If equipped with an automatic transmission, remove the transmission. If equipped with a manual transmission, remove the transmission and clutch assembly.

4. Remove the starter without disconnecting the wires and secure it aside.

5. Remove the flywheel-to-crankshaft bolts and the flywheel.

6. Carefully, remove the oil seal, using a small prybar; work the tool around the diameter of the seal until the seal begins to lift out. Use care not to damage the seat and area around the seal.

7. Fill the space between the seal lips with grease and lubricate the seal lips with clean engine oil. Install the new oil seal.

8. Install the flywheel, transmission and starter motor.

9. Connect the battery cable, start the engine and check for leaks.

FUEL SYSTEM

Fuel System Service Precautions

Safety is the most important factor when performing not only fuel system maintenance, but any type of maintenance. Failure to conduct maintenance and repairs in a safe manner may result in serious personal injury or death. Maintenance and testing of the vehicle's fuel system components can be accomplished safely and effectively by adhering to the following rules and guidelines:

• To avoid the possibility of fire and personal injury, always disconnect the negative battery cable unless the repair or test procedure requires that battery voltage be applied.

• Always relieve the fuel system pressure prior to detaching any fuel system component (injector, fuel rail, pressure regulator, etc.), fitting or fuel line connection. Exercise extreme caution whenever relieving fuel system pressure to avoid exposing skin, face and eyes to fuel spray. Please be advised that fuel under pressure may penetrate the skin or any part of the body that it contacts.

• Always place a shop towel or cloth around the fitting or connection prior to loosening to absorb any excess fuel due to spillage. Ensure that all fuel spillage (should it occur) is quickly removed from engine surfaces. Ensure that all fuel soaked cloths or towels are deposited into a suitable waste container.

• Always keep a dry chemical (Class B) fire extinguisher near the work area.

• Do not allow fuel spray or fuel vapors to come into contact with a spark or open flame.

• Always use a back-up wrench when loosening and tightening fuel line connection fittings. This will prevent unnecessary stress and torsion to fuel line piping. Always follow the proper tighten specifications.

• Always replace worn fuel fitting O-rings with new. Do not substitute fuel hose or equivalent, where fuel pipe is installed.

Fuel System Pressure

RELIEVING

1. Remove the fuel filler cap.
2. Remove the fuel pump relay from the underhood relay box.
3. Start the engine and let it run until it stalls, then crank the engine for an additional 30 seconds.
4. Turn the ignition switch to the **OFF** position and remove the key. Disconnect the negative battery cable, then install the fuel pump relay.

Fuel Filter

REMOVAL & INSTALLATION

❋❋ CAUTION

The fuel injection system remains under pressure even after the engine has been turned OFF. The fuel system

pressure must be relieved before disconnecting any fuel lines. Failure to do so may result in fire and personal injury.

1. Properly relieve the fuel system pressure.
2. Raise and safely support the vehicle.
3. Use block-off clamps to pinch the fuel lines shut. Disconnect the fuel lines from the fuel filter. Clean up any fuel spills.
4. Loosen the filter mounting bolt and remove the fuel filter.

To install:
5. Install the fuel filter with the label, arrow, or manufacturer's mark facing toward the front of the vehicle and tighten the bracket bolt.
6. Connect the fuel lines to the fuel filter.
7. Lower the vehicle to the floor and install the filler cap.
8. Reconnect the negative battery cable.
9. Start the engine and inspect the fuel filter connections for leaks.

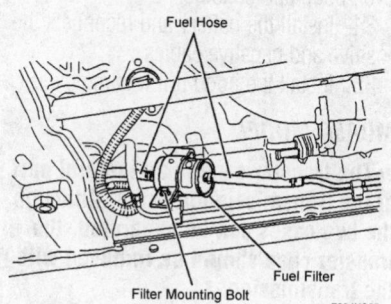

Fuel filter mounting location under the vehicle

Fuel Pump

REMOVAL & INSTALLATION

❋❋ CAUTION

The fuel injection system remains under pressure even after the engine has been turned OFF. The fuel system pressure must be relieved before disconnecting any fuel lines. Failure to do so may result in fire and personal injury.

1. Disconnect the negative battery cable.
2. Relieve the fuel system pressure.
3. Raise and safely support the vehicle.
4. Drain the fuel into an approved sealable container. Install the drain plug with a

new washer and tighten it to 14 ft. lbs. (19 Nm) for 8mm bolts, or 22 ft. lbs. (30 Nm) for 14mm bolts.

❋❋ CAUTION

Observe all applicable safety precautions when working around fuel. Do not allow fuel spray or vapor to come in contact with a spark or open flame. Keep a class B dry chemical fire extinguisher near your work area. Never drain or store fuel in an open container due to the risk of fire or explosion.

5. If equipped, remove the fuel tank undercover.
6. Disconnect the feed, return, and vapor hoses.
7. Detach all wiring harness connectors.
8. Disconnect the filler neck.
9. Position a suitable jack under the tank, remove the retainers and lower the tank far enough to disconnect any remaining wiring or hoses.
10. Remove the fuel tank from the vehicle.

To install:
11. Position a floor jack under the tank, then raise it far enough to reconnect the wiring and hoses. Install the retainers. Tighten the mounting bolts to 27 ft. lbs. (37 Nm).
12. Connect the filler neck.
13. Attach all harness connectors.
14. Connect the feed, return, and vapor hoses.
15. Refill the tank. Install the fuel tank undercover.
16. Connect the negative battery cable and check for leaks.

DRIVE TRAIN

Transmission Assembly

REMOVAL & INSTALLATION

Manual Transmissions

MUA5 AND MSG5C MODELS

1. Disconnect the negative and positive battery cables.
2. Remove the battery.
3. Support the hood as far open as possible. If you choose to remove the hood,

first matchmark the hood hinge plates with a felt-tipped marker.

4. Remove the console and shift boot. Unbolt the shift lever from the transmission case and remove it. Cover the quadrant box hole to prevent contaminants from entering the transmission.

5. Raise and support the vehicle safely.

6. Drain the transmission oil. Install the drain plug with a new washer.

7. Remove the transfer case skid plates, if equipped.

8. If equipped with a two-piece driveshaft, remove the center bearing retainer bolts.

9. If 4WD equipped, mark the driveshafts to the differential flanges and remove the front and rear driveshafts.

10. Matchmark the driveshaft flanges to the differential and transmission flanges. Remove the driveshaft.

11. Remove the starter.

12. Disconnect the speedometer cable.

13. For MUA5 transmissions, unbolt the slave cylinder from the side of the transmission case. Don't disconnect the hydraulic line.

14. For MSG5C transmissions, disconnect the clutch cable from the release fork.

15. Remove the exhaust pipe bracket from the transmission case. Disconnect the front exhaust pipe from the exhaust manifold and the second exhaust pipe.

16. Support the engine with a lifting chain or jack. Support the transmission with a jack.

17. Remove the rear housing mount from the transmission. Remove the mount bracket from the third crossmember.

18. Remove the quadrant box from the transmission, if equipped.

➡**The frame crossmember may interfere with transmission removal. An assistant will be helpful for shifting the transmission back and away from the engine.**

19. Position a jack under the transmission and remove the engine-to-transmission bolts. Move the transmission as far to the rear of the vehicle as possible, then lower the clutch housing end of the transmission toward the jack.

To install:

20. Raise the transmission into position.

21. Using a transmission jack, position the transmission-to-engine and tighten the retaining bolts to 28 ft. lbs. (38 Nm). Install the quadrant box.

22. Install the mount bracket and tighten the bolt to 27 ft. lbs. (37 Nm).

23. Install the frame-to-rear housing mount bolts at the number three crossmember, and tighten to 62 ft. lbs. (83 Nm).

24. Tighten the engine mount nuts to 30 ft. lbs. (41 Nm).

25. Install the exhaust pipe and bracket.

26. If equipped with 4WD, install the front and rear driveshafts in the marked locations. Tighten the flange bolts to 43 ft. lbs. (63 Nm).

27. Align the matchmarks and install the driveshaft. Tighten the center bearing bolts to 46 ft. lbs. (65 Nm). Position the retaining bolts with the head facing rearward and tighten to 22 ft. lbs. (30 Nm).

28. Install the speedometer cable and starter motor. Tighten the starter motor bolts to 30 ft. lbs. (41 Nm).

29. For MUA5 transmissions, install the slave cylinder. If necessary, refill and bleed the clutch hydraulic system.

30. For MSG5C transmissions, connect the clutch cable to the release fork.

31. Install the transfer case skid plates, if equipped.

32. Refill the transmission with the proper type of oil.

33. Lower the vehicle and install the shift lever, boot, and console.

34. Install the battery and reconnect the positive and negative cables.

35. Install the hood if it was removed.

MUA5C MODEL

➡**The transfer case is an integral part of the transmission housing. Although the two cases can be separated, the transfer case should be removed with the transmission.**

1. Shift the transmission into neutral. Shift the transfer case into **2H** and drive the vehicle forward and backward a few feet/meters to be sure the front axle is not engaged.

2. Use a felt-tipped marker to matchmark the hood to the hood hinges. Remove the hood.

3. Disconnect the negative battery cable.

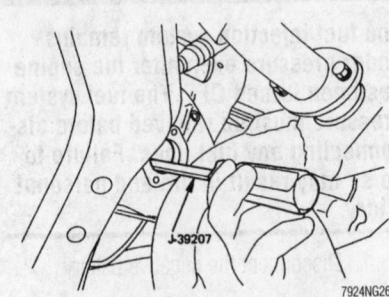

Insert the release bearing remover (J-39207) through the bell housing—MUA5C manual transmission

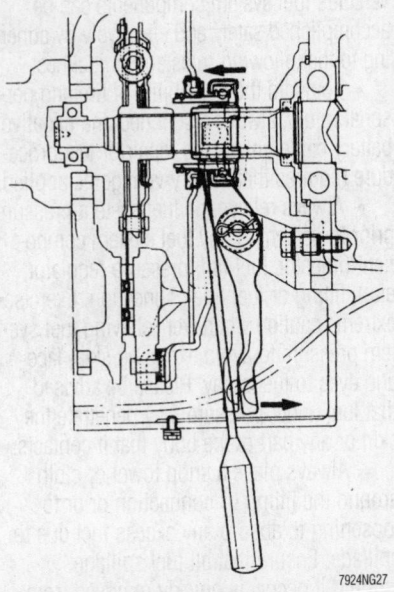

Push the release bearing fork toward the transmission to release the bearing from the pressure plate—MUA5C manual transmission

4. Remove the shift knobs and the console and shift boots.

5. Remove the shift lever and the transfer case shift lever (if equipped), by unbolting their mounting plates from the transmission case.

6. Raise and safely support the vehicle.

7. Drain the oil from the transmission and transfer case.

8. Remove the exhaust and transfer case skid plates.

9. Detach the oxygen sensor connector from the transmission harness.

10. Unbolt the exhaust flanges. Separate and remove the catalytic converter, left front, and center exhaust pipes.

11. Remove the harness heat protector.

12. Remove the slave cylinder heat protector. Unbolt and remove the slave cylinder. Don't disconnect the hydraulic line.

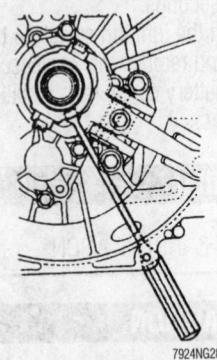

Insert the tool between the wedge collar and the release bearing

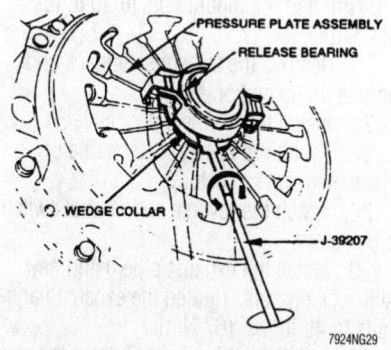

Turn the remover to separate the release bearing

13. Remove the slave cylinder dust covers.

14. Matchmark the driveshafts at the flanges and remove them.

15. Detach the reverse light switch, 4WD indicator switch, 1–2 and 3–4 indicator switch harness connectors.

16. Detach the speed sensor harness connector.

17. Remove the two harness clamps from the transmission case.

18. Attach a chain hoist to the engine.

19. Using a transmission jack, raise the transmission slightly. Remove the two rear transmission mounting nuts.

20. Remove the center crossmember (eight bolts).

21. On 4WD vehicles, remove the front crossmember and the front driveshaft.

➡️**Be sure the engine assembly is properly supported when removing the front crossmember.**

22. Remove the three flywheel inspection cover bolts.

23. Use Clutch Release Bearing Remover J-39207 or an equivalent prytool to release the bearing from the pressure plate. Push the release fork toward the rear of the vehicle. Insert the tool between the release bearing and the pressure plate collar. Move the lever to the rear to pry.

24. Raise the engine slightly with a chain hoist and remove the bolts and nuts securing the transmission to the engine.

25. Carefully pull the transmission rearward. Lower the transmission from the vehicle.

To install:

26. Apply a thin coating of molybdenum grease to the splines of the input shaft, then slowly raise the transmission into position against the rear of the engine. Align the splines of the input shaft with the grooves of the clutch disc hub and install the transmission to the engine.

➡️**It may be helpful the put the transmission in gear and rotate the driveshaft flange so the input shaft will turn and engage the grooves in the clutch disc hub.**

27. Install the transmission case bolts. Tighten the upper six bolts to 56 ft. lbs. (76 Nm). Tighten the two remaining large bolts to 56 ft. lbs. (76 Nm). Tighten the remaining three bolts to 4.4 ft. lbs. (6 Nm).

28. Push the release bearing fork rearward with a force of 13–18 ft. lbs. (18–24 Nm) to engage the release bearing with the pressure plate. A click sound will be heard when the bearing engages the pressure plate properly.

29. Install the flywheel inspection cover.

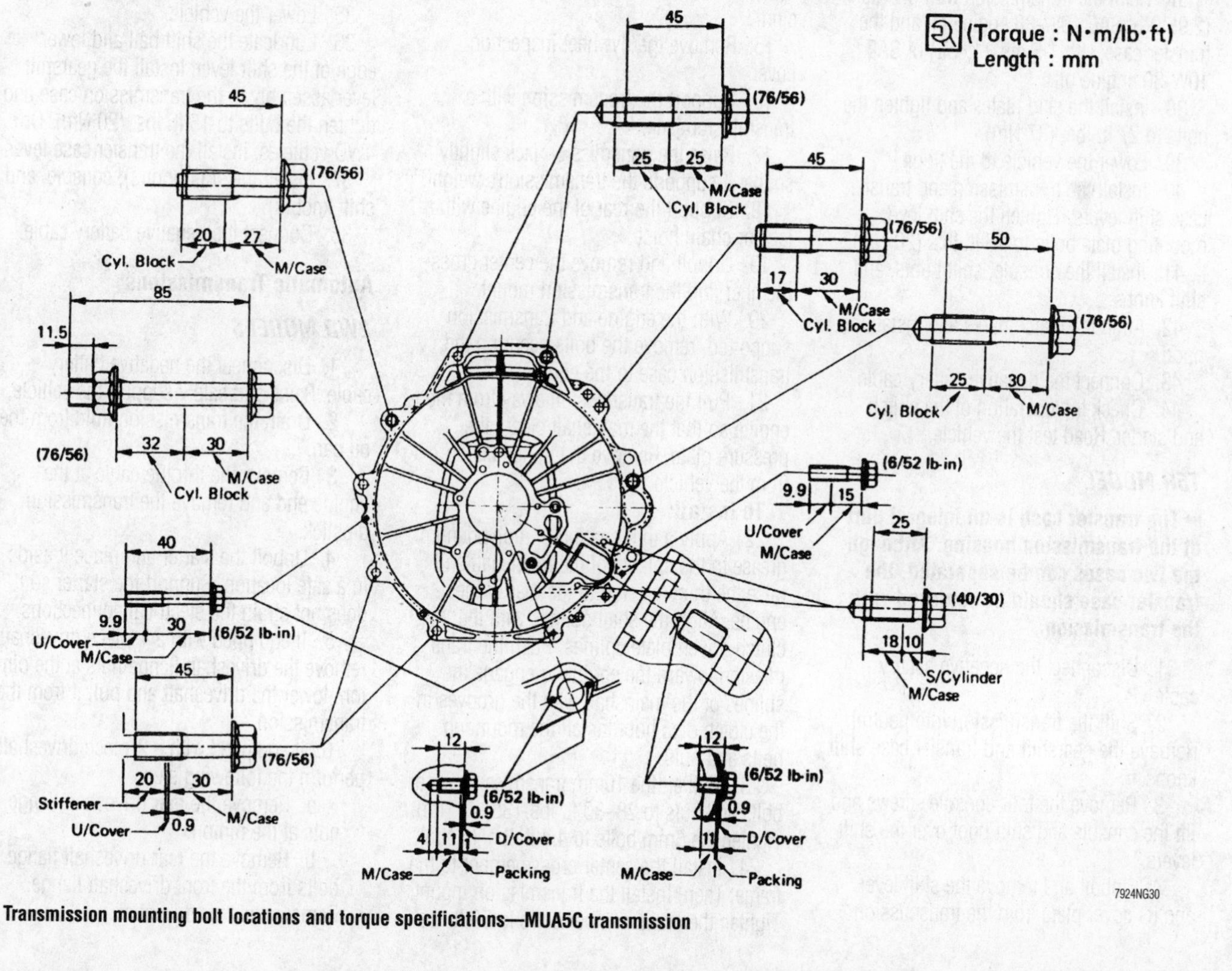

(Torque : N·m/lb·ft)
Length : mm

Transmission mounting bolt locations and torque specifications—MUA5C transmission

30. On 4WD vehicles, install the front crossmember and the front driveshaft. Tighten the crossmember bolts to 58 ft. lbs. (78 Nm). Tighten the driveshaft flange bolts to 46 ft. lbs. (62 Nm).

31. Install the center crossmember and transmission mount. Tighten the center crossmember mounting bolts to 37 ft. lbs. (50 Nm) and the transmission mounting nuts to 30 ft. lbs. (41 Nm).

32. Remove the transmission jack and the engine hoist.

33. Connect the transmission wiring harness and install the harness clamps.

34. Install the rear driveshaft and tighten the flange bolts to 46 ft. lbs. (62 Nm).

35. Apply grease to the dimple on the end of the release bearing fork and install the slave cylinder, heat protector, and dust covers. Tighten the slave cylinder bolts to 32 ft. lbs. (43 Nm). Tighten the dust cover bolts to 4.4 ft. lbs. (6 Nm).

36. Install the exhaust pipes and heat protectors using new self-locking nuts. Tighten the exhaust manifold flange bolts to 49 ft. lbs. (67 Nm). Tighten the center pipe flange bolts to 32 ft. lbs. (43 Nm).

37. Refill the transmission with 3.1 qts. (2.95L) of SAE 10W-30 engine oil and the transfer case with 1.5 qts. (1.45L) of SAE 10W-30 engine oil.

38. Install the skid plates and tighten the bolts to 27 ft. lbs. (37 Nm).

39. Lower the vehicle to the floor.

40. Install the transmission and transfer case shift levers. Tighten the shift lever mounting plate bolts to 15 ft. lbs. (20 Nm).

41. Install the console, shift boots, and shift knobs.

42. Align the matchmarks and install the hood.

43. Connect the negative battery cable.

44. Check the operation of the clutch and starter. Road test the vehicle.

T5R MODEL

➡The transfer case is an integral part of the transmission housing. Although the two cases can be separated, the transfer case should be removed with the transmission.

1. Disconnect the negative battery cable.

2. Shift the transmission into neutral. Remove the gearshift and transfer case shift knobs.

3. Remove the four console screws and lift the console and shift boot over the shift levers.

4. Unbolt and remove the shift lever and its cover plate from the transmission

case. If equipped, remove the transfer case shift lever.

5. Raise and safely support the vehicle.

6. Drain the transmission oil.

7. Disconnect and remove the starter.

8. Unbolt the slave cylinder from the transmission case. Don't disconnect the hydraulic line.

9. Matchmark the driveshaft U-joints to the transmission and differential flanges.

10. If the vehicle is equipped with 4WD, matchmark the front driveshaft U-joints to the differential and transfer case.

➡On 4WD vehicles, the transfer case skid plates and front exhaust pipe must be removed before removing the front drive shaft.

11. Remove the driveshaft and the center bearing.

12. Disconnect the front exhaust pipe from the manifold and the catalytic converter. It is not necessary to remove the exhaust pipe from the chassis.

13. Detach the reverse and neutral switch connectors from the transmission.

14. Detach the speedometer cable or speed sensor connector from the transmission.

15. Remove the flywheel inspection cover.

16. Support the transmission with a transmission jack.

17. Raise the transmission jack slightly so that it supports the transmission's weight.

18. Support the rear of the engine with a jack or chain hoist.

19. Unbolt and remove the center crossmember and the transmission mount.

20. With the engine and transmission supported, remove the bolts securing the transmission case to the engine.

21. Pull the transmission away from the engine so that the mainshaft clears the pressure plate. Remove the transmission from the vehicle.

To install:

22. Apply a thin coating of molybdenum grease to the splines of the mainshaft and raise the transmission to the rear of the engine. Align the shaft splines with the clutch driven plate splines. Push the transmission toward the engine to engage the splines of the mainshaft with the grooves in the clutch disc hub. Install the mounting bolts and nuts.

23. Tighten the 10mm transmission case bolts and nuts to 28–30 ft. lbs. (38–41 Nm). Tighten the 6mm bolts to 4.4 ft. lbs. (6 Nm).

24. Install the center crossmember to the frame. Then, install the transmission mount. Tighten the crossmember bolts to 56 ft. lbs.

(76 Nm) and the mount nuts to 30 ft. lbs. (41 Nm).

25. Remove the transmission jack and engine lifting equipment.

26. Install the flywheel dust cover.

27. Attach the speedometer cable or speed sensor connector.

28. Attach the reverse and neutral switch connectors.

29. Install the exhaust pipe using new self-locking nuts. Tighten the exhaust flange nuts to 49 ft. lbs. (67 Nm).

30. Install the driveshafts. Tighten the center bearing bolts to 45 ft. lbs. (60 Nm). Tighten the flange bolts to 46 ft. lbs. (63 Nm).

31. Install the skid plates, if equipped. Tighten the skid plate bolts to 27 ft. lbs. (37 Nm).

32. Coat the tip of the slave cylinder with molybdenum grease. Install the slave cylinder assembly and tighten the bolts to 37 ft. lbs. (50 Nm).

33. Install the starter and tighten the bolts to 30 ft. lbs. (41 Nm).

34. Refill the transmission with 2.4 qts. (2.25L) of Dexron® IIE ATF and the transfer case with 1.5 qts. (1.45L) of SAE 10W-30 engine oil.

35. Lower the vehicle.

36. Lubricate the shift ball and lower edge of the shift lever. Install the gearshift lever assembly to the transmission case and tighten the bolts to 15 ft. lbs. (20 Nm). On 4WD vehicles, install the transfer case lever.

37. Install the shift boot(s), console, and shift knob(s).

38. Connect the negative battery cable.

Automatic Transmissions

2WD MODELS

1. Disconnect the negative battery cable. Raise and safely support the vehicle.

2. Drain the transmission fluid from the oil pan.

3. Remove the throttle cable at the engine end and remove the transmission dipstick.

4. Unbolt the starter and place it aside in a safe location. Support the starter so it does not strain the electrical connections.

5. If equipped with a 1-piece driveshaft, remove the driveshaft flange nuts at the pinion, lower the driveshaft and pull it from the transmission.

6. If equipped with a 2-piece driveshaft, perform the following steps:

   a. Remove the rear driveshaft flange nuts at the pinion.

   b. Remove the rear driveshaft flange bolts from the front driveshaft flange. Remove the rear driveshaft.

c. Remove the center bearing bolts from the chassis, move the front driveshaft rearward and from the transmission.

7. Disconnect the shift lever at the shifter end.

8. Disconnect the speedometer cable.

9. Disconnect the oil cooler lines and place the cooler bypass line close to the transmission case to prevent damage during transmission removal.

10. Remove the torque converter bolts from the flexplate through the starter hole.

11. Using a transmission jack, place it under the transmission and raise it slightly.

12. Remove the rear mount nuts from the transmission.

13. Remove the rear mount nuts/bolts from the crossmember. Remove the mount.

14. Remove the transmission-to-engine bolts.

15. Move the transmission back and lower the transmission out of the vehicle.

To install:

➡**Installation of the transmission will require an assistant.**

16. Raise the transmission into position.

17. Raise the rear of the transmission and move it into position on the crossmember.

18. Move the transmission forward and engage it with the engine.

❇❇ WARNING

Be sure the transmission dowel pins are installed in the correct position. If the dowels are in the wrong hole, the transmission case may crack.

19. Install the engine-to-transmission bolts and tighten to 47 ft. lbs. (64 Nm).

➡**Be sure the torque converter is seated properly. The transmission should mount flush to the engine block; if the transmission does not mount flush, check the torque converter. Do not use the mounting bolts to pull the transmission flush to the engine or torque converter/transmission damage may occur.**

20. Install the mount and the rear mount nuts/bolts to the crossmember.

21. Install the rear mount nuts to the transmission. Tighten to 30 ft. lbs. (41 Nm).

22. Install the torque converter bolts to the flexplate through the starter hole, and tighten to 22 ft. lbs. (30 Nm).

23. Connect the oil cooler lines, speedometer cable, and shift linkage.

24. If equipped with a 2-piece driveshaft, perform the following steps:

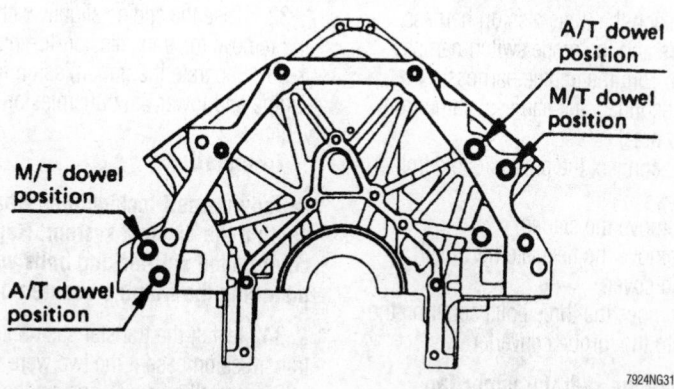

Transmission mounting dowel positions

a. Install the front driveshaft into the transmission, and the center bearing bolts to the chassis.

b. Install the rear driveshaft, and the rear driveshaft flange bolts to the front driveshaft flange.

c. Install the rear driveshaft flange nuts to the pinion.

25. If equipped with a 1-piece driveshaft, install the driveshaft into the transmission and the driveshaft flange nuts to the pinion.

26. Install the starter.

27. Connect the throttle (downshift) cable to the transmission.

28. Install the oil level dipstick and the tube.

29. Refill the transmission with Dexron® IIE ATF.

30. Connect the negative battery cable.

31. Start the engine and check for leaks.

4WD MODELS

➡**The transfer case is an integral part of the transmission housing. Although the two cases can be separated, the transfer case should be removed with the transmission.**

1. Use a felt-tipped marker to matchmark the hood to the hood hinges. Remove the hood.

2. Shift the transmission into the **N** position, and the transfer case into the **2H** position.

3. Disconnect the negative battery cable.

4. Remove the air cleaner assembly.

5. Remove the transfer case shift knob. Remove the four console retaining screws.

6. Remove the center console assembly and detach the console switch wiring connectors.

7. Disconnect the shift lock cable and the shift control rod from the selector lever assembly.

8. Unbolt and remove the transfer case control lever.

9. Raise and safely support the vehicle.

10. Remove the transmission and transfer case skid plates.

11. Remove the exhaust pipe protectors.

12. Detach the oxygen sensor connectors from the transmission harness.

13. Remove the nuts and bolts, then separate the catalytic converters, center, and front exhaust pipes from the exhaust manifold and tailpipe. Remove the exhaust components from the vehicle, or move them out of the way.

14. Matchmark the front and rear driveshafts to the differential and transfer case flanges.

15. Unbolt and remove the front driveshaft.

16. Unbolt the front part of the rear driveshaft from the U-joint behind the driveshaft center bearing. Unbolt the rear driveshaft from the transfer case. Unbolt the center bearing from the frame, and remove the driveshaft.

17. Disconnect the oil cooler lines from the transmission. Plug the lines to prevent fluid loss and contamination.

18. Remove the brackets securing the oil cooler lines to the engine stiffener.

19. Remove the front suspension crossmember.

20. Remove the dipstick and tube. Disconnect the breather hoses from the tube.

21. Place a transmission jack under the transmission and transfer case unit for support.

22. Raise the transmission slightly and remove the eight bolts securing the rear mount and the third crossmember.

➡**Be sure the engine and transmission assembly is properly supported before removing the rear mount and third crossmember.**

23. Remove the five engine stiffener bolts and the stiffener.

24. Remove the heat protector.

25. Detach the transmission harness connectors and the mode switch harness connector from the engine harness.

26. Disconnect the harness clamp from the clamp bracket.

27. Disconnect the ground cable from the engine.

28. Remove the starter.

29. Remove the flexplate (flywheel) inspection cover.

30. Remove the three bolts securing the flexplate to the torque converter.

➡**Remove the radiator upper fan shroud and the cooling fan to access the crankshaft center bolt to turn the crankshaft.**

31. Attach a chain hoist to the engine lifting points.

32. Raise the engine slightly with a hoist and remove the transmission-to-engine bolts.

33. Separate the transmission from the engine and lower the transmission from the vehicle.

To install:

➡**Use new self-locking nuts when reconnecting the exhaust system. Replace any color-coded self-locking bolts when installing the frame crossmembers.**

34. Install the transfer case onto the transmission case if the two were separated.

35. Install a new O-ring on the front pump shaft.

✲✲ WARNING

Be sure the transmission dowel pins are installed in the correct position.

If the dowels are in the wrong hole, the transmission case may crack.

36. Be sure the torque converter is fully seated in the front pump and slowly raise the front of the transmission into position until it is flush with the rear of the engine and install the transmission-to-engine bolts. Tighten the upper six mounting bolts to 56 ft. lbs. (76 Nm). Tighten the lower three mounting bolts to 35 ft. lbs. (47 Nm). Tighten the remaining two bolts to 52 ft. lbs. (70 Nm).

➡**If the transmission does not seat flush against the engine block before the bolts are installed, check to see that the torque converter is seated properly in the transmission, do not use the bolts to draw the transmission to the engine block.**

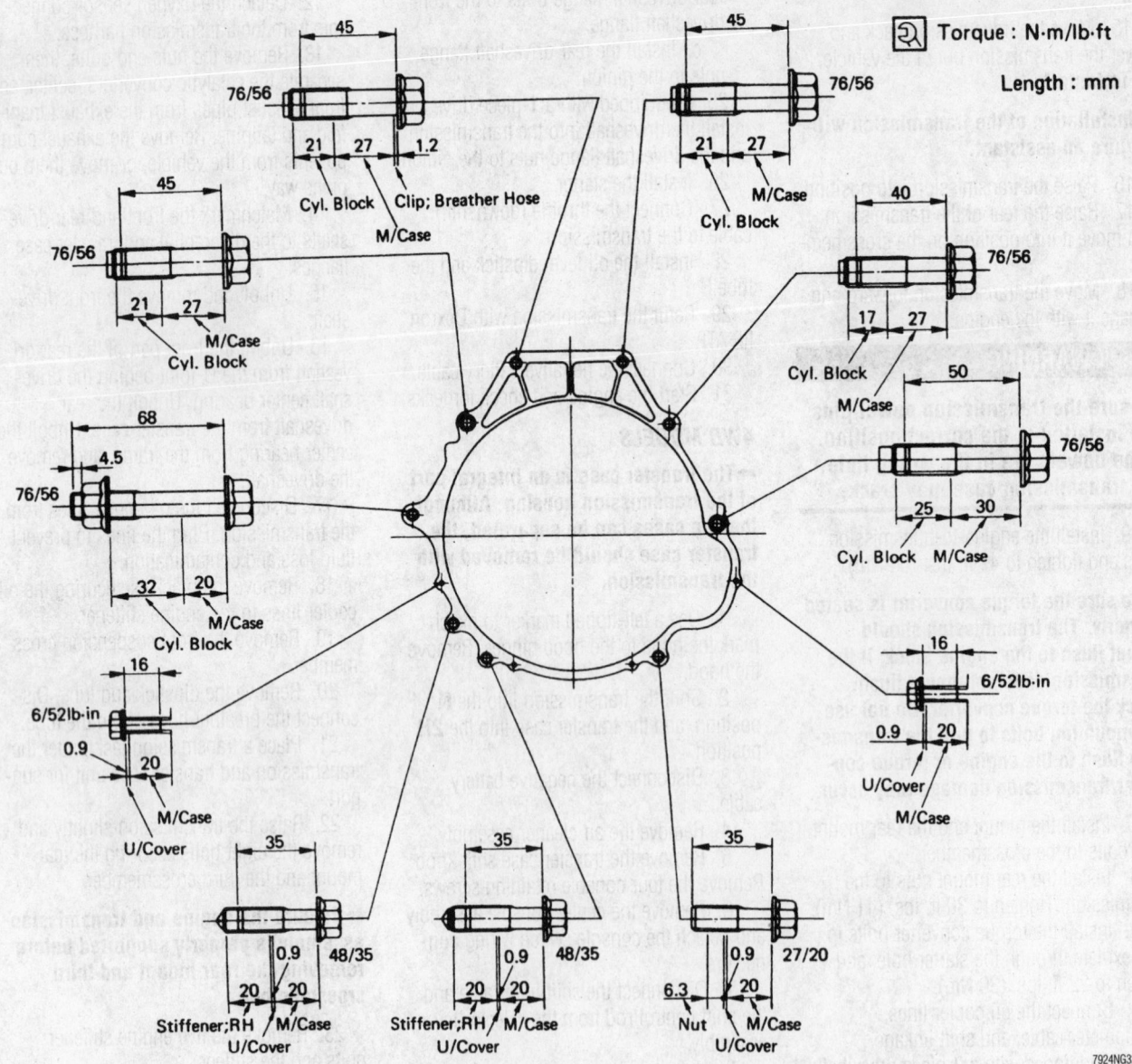

Automatic transmission mounting bolt locations and torque specifications

7924NG32

37. Install the third crossmember. Tighten the bolts to 37 ft. lbs. (50 Nm).

38. Install the rear mount and lower the engine from the hoist. Tighten the nuts to 30 ft. lbs. (41 Nm).

39. Remove the transmission jack and the engine hoist.

40. Install the flexplate-to-torque converter bolts. Tighten the bolts in a crisscross pattern to 41 ft. lbs. (55 Nm).

41. Install the flexplate inspection cover.

42. Connect the transmission and mode switch harness to the engine harness.

43. Connect the harness clamp to the clamp bracket and radiator clip.

44. Connect the ground cable to the engine.

45. Install the starter and tighten the mounting bolts to 30 ft. lbs. (41 Nm).

46. Install the heat protector.

47. Install the stiffener. Tighten the bolts to 35 ft. lbs. (47 Nm).

48. Install the transmission dipstick tube and dipstick. Connect the breather hoses to the tube.

49. Install the front suspension crossmember. Tighten the bolts to 58 ft. lbs. (79 Nm).

50. Connect the transmission fluid cooling lines to the transmission.

51. Install the front and rear driveshafts. Tighten the flange bolts to 46 ft. lbs. (62 Nm).

52. Install the exhaust pipes and the catalytic converter. Tighten the manifold nuts to 49 ft. lbs. (67 Nm). Tighten the center pipe and converter bolts to 32 ft. lbs. (43 Nm).

53. Reconnect the oxygen sensors.

54. Install the exhaust pipe protectors.

55. Install the skid plates and tighten their mounting bolts to 27 ft. lbs. (37 Nm).

56. Lower the vehicle.

57. Install the transfer case control lever.

58. Connect the shift control rod to the selector lever. Adjust the shift cable so that there is 0.059–0.098 in. (1.5–2.5mm) of slack at the selector lever.

59. Attach the console switch connectors and the shift lock cable to the selector lever.

60. Install the center console assembly and the transfer case lever knob.

61. Install the upper fan shroud and the cooling fan.

62. Install the air cleaner assembly.

63. Align the matchmarks and install the hood.

64. Refill the transmission with Dexron® II ATF.

65. Connect the negative battery cable.

66. Start the engine and shift through the gear range three times to circulate the ATF through the valve body.

67. Recheck the fluid level and top up if necessary.

68. Road test the vehicle.

Clutch

ADJUSTMENTS

Clutch Pedal

1. Locate the clutch switch (with cruise control) or clutch pedal stop bolt (without cruise control) at the top of the clutch pedal under the dash.

2. If equipped, disconnect the clutch switch.

3. Loosen the clutch switch or stop bolt as equipped. Loosen the clutch master cylinder pushrod yoke nut.

4. Adjust the clutch master cylinder pushrod to obtain a clutch pedal height: 7.28–7.68 in. (18.5–19.5cm) for the 2.6L engine, or 7.64–8.03 in. (19.4–20.4cm) for the 3.2L (VIN V) engine or for the 3.2L (VIN W) engine 7.01–7.40 in. (17.8–18.8 cm).

5. After adjusting the pedal height, tighten the pushrod yoke nut to 12 ft. lbs. (16 Nm). Screw in the clutch switch (cruise control) until the plunger is fully depressed, then unscrew the switch ½ turn and tighten the locknut. There should be 0.020–0.059 in. (0.5–1.5mm) of clearance. Tighten the pedal stop bolt (no cruise control) so it just touches the pedal, then tighten the locknut.

6. After the pedal height has been completely adjusted, check the clutch pedal free-play and adjust if necessary. Free-play should fall within the range of 0.20–0.59 in. (5–15mm).

7. Reconnect the clutch switch.

REMOVAL & INSTALLATION

1. Disconnect the negative battery cable.

2. Raise and support the vehicle safely.

3. On 2WD vehicles, remove the transmission. On 4WD models, remove the transmission and transfer case as an assembly.

4. If the pressure plate is going to be reused, matchmark the pressure plate to the flywheel so the pressure plate can be reassembled in the same position. Lock the flywheel in place to prevent it from turning.

➡It is recommended that the clutch disc, pressure plate and release bearing be replaced with new parts when the clutch assembly requires service.

5. Loosen the pressure plate mounting bolts, one turn at a time in an crisscross

sequence. Loosening in a sequence will relieve the spring tension to avoid distorting or bending the pressure plate.

6. Install a clutch alignment tool through the clutch disc into the pilot bearing to support the pressure plate and cover assembly. Remove the bolts and clutch assembly.

7. Remove the release bearing from the input shaft and fork.

8. Inspect the flywheel for scoring, grooves and cracks or discoloration from heat. Replace or resurface the flywheel as needed.

➡The clutch disc, pressure plate, and release bearing should be replaced with new parts when the clutch assembly requires service.

To install:

9. If the flywheel was removed or replaced, install it with new mounting bolts. Apply Loctite® to the threads, then tighten the bolts in a two-step crisscross pattern to 40–43 ft. lbs. (54–58 Nm). After installation, clean any excess Loctite® from the flywheel.

10. Apply a thin coat of lubricant to the splines in the clutch disc and the front bearing retainer on the transmission where the release bearing slides.

11. Pack the recess in the release bearing with grease and apply grease to the groove where the clutch fork attaches to the release bearing. Install the release bearing on the input shaft and clutch fork.

12. Using a clutch alignment tool, assemble the clutch disc and pressure plate onto the flywheel. Be sure the clutch disc is facing the right direction.

13. Align the matchmarks, if reusing the pressure plate and install the pressure plate mounting bolts and tighten to 13 ft. lbs. (18 Nm) using a star pattern tightening sequence. Remove the aligning tool.

14. On 2WD vehicles, install the transmission. On 4WD models, install the transmission and transfer case as an assembly.

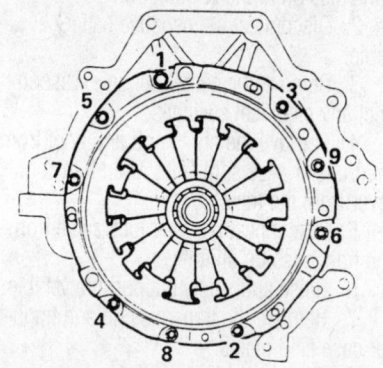

Pressure plate tightening sequence

7924NG49

15. Adjust the clutch cable linkage, if equipped. Bleed the clutch hydraulic system, if equipped.

16. Reconnect the negative battery cable.

Hydraulic Clutch System

BLEEDING

1. Firmly set the parking brake.
2. Check the reservoir fluid level and refill as necessary.
3. Vehicles with V6 engines use a damper cylinder for the hydraulic clutch system. Bleed the damper cylinder before bleeding the slave cylinder.

➡**The bleeding procedure is the same for both the slave cylinder and the damper cylinder.**

4. Connect a vinyl tube to the bleeder screw. Submerge the other end of the tube in a container of brake fluid.
5. Pump the clutch pedal slowly several times and hold it depressed.
6. Loosen the bleeder screw and allow the air bubbles and fluid to flow into the container. Tighten the bleeder screw.
7. Release the clutch pedal.
8. Add fluid to the reservoir if necessary. Don't let the reservoir run dry.
9. Continue to pump the pedal until the fluid draining from the tube is clear of air bubbles.
10. Refill the reservoir to the full mark. Remove the vinyl tube. Tighten the bleeder screw and install its rubber cap.

Transfer Case Assembly

REMOVAL & INSTALLATION

1. Shift the transfer case into the **2H** position. Drive the vehicle forward and backward for a few feet to be sure the front axle and hubs are disengaged. Shift the transmission to the **N** position.
2. Disconnect the negative battery cable.
3. Remove the center console. Disconnect any electrical switches.
4. Remove the shift knob and boot from the transfer case shift lever. Unbolt the shift lever from the transfer case.
5. Disconnect the shift lock cable from the transmission shift lever.
6. Raise and safely support the vehicle.
7. Remove the transmission and transfer case skid plates.
8. Disconnect the oxygen sensors from the front exhaust pipe.

9. Unbolt the front exhaust pipe from the exhaust manifolds and the catalytic converter. Remove the exhaust pipe and converter assembly.
10. Matchmark the front and rear driveshaft U-joints to the differential and transfer case flanges.
11. Unbolt and remove the front driveshaft.
12. Unbolt the rear driveshaft from the rear differential and transfer case flanges.
13. Unbolt the center bearing and remove the rear driveshaft.
14. Drain the transfer case oil.
15. Disconnect the transmission shift linkage from the shift lever rod.
16. Label and disconnect the two wiring harnesses from the transfer case.
17. Support the transfer case with a transmission jack.
18. Remove the bolts securing the transfer case to the transmission.
19. Separate the transfer case from the transmission output shaft. Lower the transfer case from the vehicle.

To install:

➡**Use new self-locking nuts when installing the exhaust pipe and converter.**

20. Apply a thin coating of molybdenum grease to transfer case input shaft splines.
21. Raise the transfer case to the level of the transmission and align the output and input shaft splines.
22. Install the transfer case-to-transmission case bolts and tighten them to 30–34 ft. lbs. (41–46 Nm).
23. Fill the transfer case with SAE 10W-30 engine oil.
24. Remove the transmission jack.
25. Connect the two wiring harnesses. Connect the shift lever rod to the shift linkage.
26. Align the matchmarks and install the front and rear driveshafts. Tighten the center bearing bolts to 45 ft. lbs. (61 Nm). Tighten the flange bolts to 46 ft. lbs. (62 Nm).
27. Install the exhaust pipe and catalytic converter. Tighten the nuts to 49 ft. lbs. (67 Nm), and the bolts to 32 ft. lbs. (43 Nm).
28. Install the skid plates and tighten their bolts to 27 ft. lbs. (37 Nm).
29. Lower the vehicle.
30. Install the transfer case shift lever.
31. Install the console, shift boot, and shift knob.
32. Reconnect the negative battery cable.
33. Check to be sure that the transmission, transfer case, and front axle engage correctly.

Halfshaft

REMOVAL & INSTALLATION

➡**The front axle must be lowered from the vehicle if the inboard CV-joint cases must be removed.**

1. Shift the transfer case to the **2H** position and verify that the front axle is not engaged.
2. Raise and safely support the vehicle.
3. Remove the front wheels.
4. Remove the boot bands from the inboard joint boot. Push the boot away from the inboard joint case to expose the circlip. Remove the circlip from the inboard joint case.
5. Separate the halfshaft from the inboard joint case. Remove the halfshaft from the vehicle. The inboard joint case remains attached to the axle.

To install:

6. Fit the inboard joint boot over the edge of the case and into position. Gently squeeze the boot while lifting up the inner edge to release any trapped air. The joint's extended length from the end of the case to the outer edge of the boot should be 6.5 in. (16.5cm).
7. Install new boot bands and clench them so that the crimp faces the front of the vehicle.
8. Assemble any suspension components that were disconnected to remove the halfshaft.
9. Install the front wheels.
10. Lower the vehicle. Tighten any suspension fasteners to their final torque specifications.

Automatic Locking Hubs

REMOVAL & INSTALLATION

1995 Models

1. Move the transfer case shift lever into **2H** and move the vehicle forward and rearward about 3 ft. to be sure the transfer case is in Neutral.
2. Raise and safely support the vehicle.
3. Remove the front wheel(s).
4. Remove the locking hub cap-to-housing bolts and the cap.
5. Remove the brake caliper-to-steering knuckle bolts and support the caliper on a wire; do not allow the caliper to hang on the brake hose. Do not disconnect the brake hose.
6. Remove the brake caliper support bracket from the steering knuckle.

1. Bolt
2. Hub cap
3. Housing assembly
4. Snap ring and shims
5. Drive clutch assembly
6. Inner cam
7. Lock washer
8. Hub nut
9. Hub and disc assembly
10. Outer bearing and outer race
11. Oil seal
12. Inner bearing and outer race
13. Bolt
14. Wheel pin

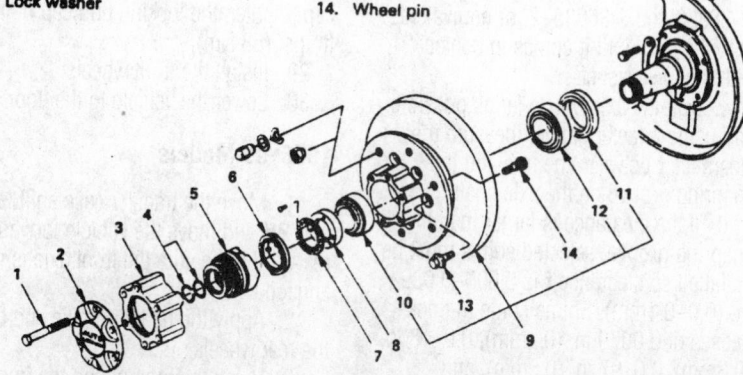

Exploded view of the hubs on all 1995 4WD models equipped with automatic locking hubs

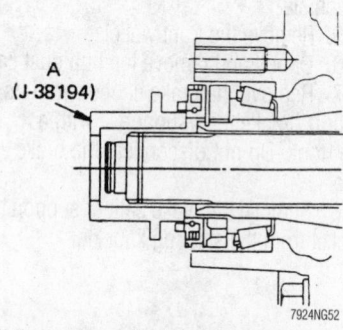

If the inner cam is difficult to install, use the special tool (J-38194 or equivalent) and a hammer to tap it into position

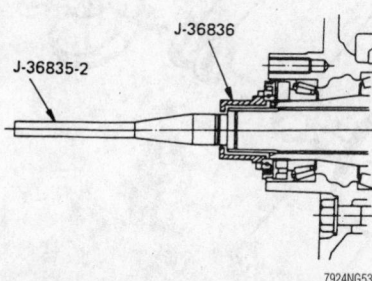

During shim selection, install special tool J-36836 (or equivalent) onto the axle shaft with tool J-36835-2 until it comes into contact with the lockwasher

7. Using snapring pliers, remove the snapring and shims.

8. Remove the drive clutch assembly, the inner cam and lockwasher.

9. Using a hub nut wrench, loosen the hub nut.

10. Pull the hub from the spindle.

11. Remove the outer bearing from the hub.

12. Using a prytool, pry the inner bearing lip seal out of the inboard side of the hub, then remove the inner bearing assembly with your fingers.

13. Wash all parts in a cleaning solvent and dry with compressed air. Do NOT allow the bearings to spin while drying them with the compressed air.

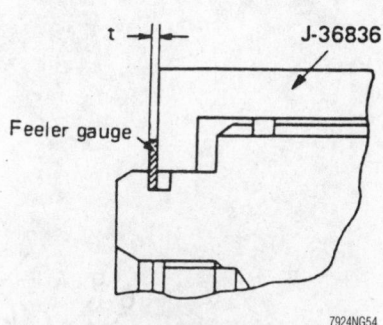

Using feeler gauges, measure the clearance (t) between the tool and the snapring groove of the axle shaft

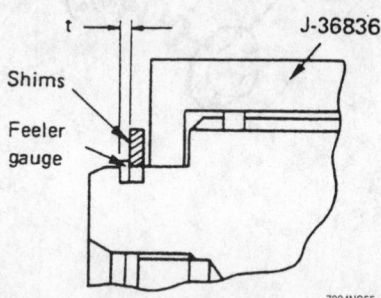

If clearance (t) is larger than the snapring groove, shims must be installed so clearance (t) is 0.0000–0.0039 in. (0.0–0.1mm)

➡Always replace the bearing and race as a matched set, if replacement bearings are needed.

14. Check the bearings for pitting or scoring. Also check for smooth rotation and lack of noise as follows:

   a. Once the bearings and bearing races, located in the hub assembly, have been cleaned of all old grease and build up, oil the bearings with regular, clean engine oil.

   b. Position each bearing, 1 at a time, in its respective bearing race. Provide slight inward pressure and turn the bearings.

   c. While pressing in and rotating the bearings, note how they turn in the bearing races. The bearings should turn smoothly and evenly.

   d. If the bearings turn irregularly or make noises while rotating, they must be replaced with new ones. Any time the bearings are replaced, the bearing races should also be replaced.

➡Special installation tools are necessary to assemble new bearing races into the hub assembly.

15. If necessary, remove the bearing races as follows:

   a. Place the hub assembly on a piece of clean cardboard with the outboard side facing down.

   b. Using a long, brass drift pin or punch, tap the outer bearing race out of the hub assembly. It will be necessary to work the punch around the bearing in a circular motion, otherwise the bearing race can become excessive cocked in the hub bore; this makes removal almost impossible. If this happens, simply tap on the opposite side of the bearing race until it has leveled out in the bore.

   c. After the outer bearing race is removed, flip the hub over and perform the same with the inner bearing race.

16. If removing the disc from the hub, scribe matchmarks, remove the disc-to-hub bolts and separate the disc from the hub.

To install:

17. Clean the flange surface of the hub, thread holes, the surface of the lockwasher and the splines of the axle shaft.

18. If the rotor was removed from the hub unit, position the rotor on the hub so the matchmarks align. Install the hub-to-rotor bolts and tighten in a crisscross fashion to 68–83 ft. lbs. (92–113 Nm).

19. If new wheel bearings are to be installed, and the old bearing races were removed, install the new inner and outer bearing races as follows:

a. Position the hub unit so the outer side faces down on a clean work surface (cardboard or wood surface to prevent damage to the hub case). Position the new inner bearing race in the hub bore and drive it in place with the Bearing Race Installer tool J-36829 and Installer Grip J-8092, or their equivalents. Tap the race into the hub housing until it is seated completely in the bore.

b. Turn the hub unit over so the inner side faces down and position the new outer bearing race in the hub bore. Drive it into the hub housing with Bearing Race Installer tool J-29015 and Grip J-8092, or their equivalents, as with the inner bearing race.

20. Pack the bearings with wheel bearing grease, then install the inner bearing and a new grease seal in the hub. Use the Oil Seal Installer tool J-36830 and Grip J-8092, or their equivalents.

21. Carefully position the hub and disc assembly on the spindle. Place 1 to 1½ oz. of bearing grease in the hub and install the outer bearing and hub nut on the spindle.

22. When installing the hub nut, perform the following procedures:

a. Install the nut on the front spindle threads by hand. Be sure the side of the hub nut with chamfers in the machine screw holes faces outward; the machine screws must sit flush and if the hub nut is installed wrong, they cannot do so.

b. Tighten the hub nut to 22 ft. lbs. (29 Nm), then completely loosen the nut; this seats the bearings on the spindle.

c. Use a spring gauge connected at 90 degrees to the uppermost wheel stud to measure preload.

d. Tighten the hub nut until the spring gauge measures a bearing preload of 4.4–5.5 lbs. (2.0–2.5 kg) for a new bearing and a new oil seal, or 2.6–4.0 lbs. (1.2–1.8 kg) for a used bearing and a new oil seal.

23. Install the lockwasher with the larger diameter of the tapered bore to the outer side of the vehicle. If the bolt holes in the lock plate do not align with the corresponding holes in the nut, reverse the lock plate. If the bolt holes are still out of alignment, turn in the nut just enough to obtain alignment and install the lockscrew tightly, so its head is lower than the surface of the washer.

24. Install the inner cam by aligning the keyway of the inner cam with the groove of the knuckle. If the cam is difficult to install, use tool No. J-38194 or equivalent and a plastic hammer to lightly tap the inner cam into place.

25. Select the proper shim:

a. Lower the vehicle and support the lower control arm with a block of wood and a floor jack to place the axle in the normal horizontal position.

b. Install the special adjusting tools (J-36836 and J-36835–2, or equivalents) onto the hub until it comes in contact with the lockwasher.

c. Pull the axle out as far as possible and using a feeler gauge, measure the clearance **t** between the hub and the snapring groove on the axle shaft.

d. If the clearance is larger than the snapring groove, selected shims must be installed so clearance **t** is 0.000–0.039 in. (0.0–0.1mm). Shims come in thicknesses of 0.0079 in. (0.2mm), 0.0118 in. (0.3mm), 0.0197 in. (0.5mm), and 0.0393 in. (1.0mm).

e. Remove the adjusting tools.

26. Install the drive clutch assembly.

27. Install the shims selected above by hand on the axle and use the following steps to install a new snapring.

a. Install special tool J-36835–2 or equivalent onto the axle.

b. Install the snapring on the tool.

c. Install tool driver J-36835–1 or equivalent.

d. Pull out the axle shaft by pulling tool J-36835–2 or equivalent. Install the snapring to the axle by pushing on tool J-36835–1.

e. Remove tool driver J-36835–2 or equivalent from the axle and check the fit of the snapring.

28. Install the housing assembly and cap. Tighten the locking hub cap bolts to 43 ft. lbs. (58 Nm).

29. Install the front wheels.

30. Lower the vehicle to the floor.

1996–99 Models

1. Move the transfer case shift lever into **2H** and move the vehicle forward and rearward to be sure the front axle is not engaged.

2. Apply the parking brake and block the rear wheels.

3. Break the lug nuts on the front wheels loose—do not completely loosen the lug nuts at this time.

4. Raise and safely support the front of the vehicle.

5. Remove the front wheels.

6. Unbolt and remove the hub dust cap.

7. Remove the brake caliper from its support bracket and support it with a wire hook. Do not disconnect the brake hose.

8. Remove the brake caliper support bracket from the steering knuckle.

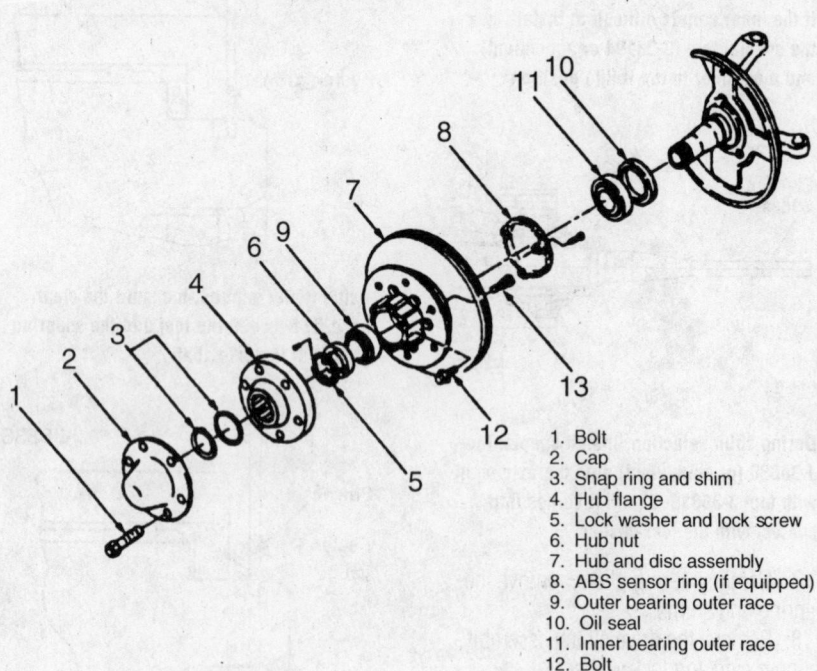

1. Bolt
2. Cap
3. Snap ring and shim
4. Hub flange
5. Lock washer and lock screw
6. Hub nut
7. Hub and disc assembly
8. ABS sensor ring (if equipped)
9. Outer bearing outer race
10. Oil seal
11. Inner bearing outer race
12. Bolt
13. Wheel pin

7924NG56

Exploded view of the front automatic locking hub assembly—1996–99 Models

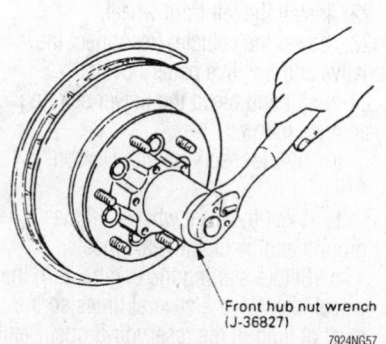

Front hub nut wrench
(J-36827)

7924NG57

Use a hub nut wrench, such as J-36827, to loosen and remove the hub nut

9. Use snapring pliers to remove the snapring from the end of the axle shaft. Remove the shim.

10. Slide the hub flange assembly off the axle shaft.

11. Remove the lockwasher retaining screws, then pull the lockwasher out of the hub assembly.

12. Use a hub nut wrench (J-36827 or equivalent) to loosen and remove the hub nut.

13. Carefully pull the hub from the spindle; do not drag the hub assembly bearings across the spindle threads. Hold the outer wheel bearing with your thumb during removal to prevent it from falling out of the hub.

14. Matchmark the hub and rotor so they can be assembled in the same positions. Place the hub and rotor assembly in a padded vice, then remove the 6 hub-to-rotor bolts to separate the rotor from the hub. If equipped with ABS, unbolt the sensor ring from the inside of the rotor.

15. Thoroughly clean, inspect, and repack the bearings with wheel bearing grease. Replace the bearings if they show signs of damage (such as: nicks, burrs, bent or dented cage, flat spots on the rollers, etc.).

16. If the bearings are to be replaced, or if the bearing races shows signs of damage, use a hammer and a brass drift to carefully tap the outer and inner bearing races and oil seals from the hub.

To install:

17. Clean the flange surface of the hub, thread holes, the surface of the lockwasher, and the splines of the axle shaft.

18. Use bearing drivers (inner bearing race, J-36829; outer bearing race, J-36828 or their equivalents) to install new inner and outer bearing races.

19. Pack the bearings and the hub cavity with high-temperature wheel bearing grease.

20. Install the rotor onto the hub and tighten the bolts to 76 ft. lbs. (103 Nm).

21. If equipped, install the ABS sensor ring and tighten the bolts to 13 ft. lbs. (18 Nm).

22. Install the inner bearing and the grease seal in the hub (use special tool J-36830 or equivalent). Position the hub and rotor assembly on the spindle.

23. Install the outer bearing and spindle nut into the hub. Tighten the hub nut to 22 ft. lbs. (29 Nm) to seat the bearings and, then fully loosen the nut. Use a spring scale connected to the stud bolt at a 90 degree angle to measure bearing preload. Then, retighten the hub nut until the spring gauge measures a bearing preload of 4.4–5.5 lbs. (19.6–24.5 N) for a new bearing and grease seal, or 2.6–4.0 lbs. (11.8–17.7 N) for a used bearing and a new grease seal.

24. Install the lockwasher on the spindle nut. The larger diameter, tapered side of the lockwasher faces out, and the holes in the lockwasher must align with the holes in the spindle nut. If the holes don't align, reverse the washer or tighten the spindle just enough to bring them into alignment.

25. Apply sealant to the both mating surfaces of the hub flange assembly and install it.

26. Install the shim and the snapring. There should be 0.000–0.012 in. (0.0–0.3mm) of clearance between the hub body and the snapring. Shims of different thicknesses may be used, if necessary.

27. Pack the hub dust cap with fresh grease and install it. Tighten the hub bolts to 43 ft. lbs. (59 Nm).

28. Install the front wheels.

29. Lower the vehicle.

30. Verify that the front axle engages and the Shift-on-the-Fly 4WD system works correctly.

STEERING AND SUSPENSION

Air Bag

✸✸ CAUTION

Some vehicles are equipped with an air bag system, also known as the Supplemental Inflatable Restraint (SIR) or Supplemental Restraint System (SRS). The system must be disabled before performing service on or around system components, steering column, instrument panel components, wiring and sensors. Failure to follow safety and disabling proce-

dures could result in accidental air bag deployment, possible personal injury and unnecessary system repairs.

PRECAUTIONS

Several precautions must be observed when handling the inflator module to avoid accidental deployment and possible personal injury.

• Never carry the inflator module by the wires or connector on the underside of the module.

• When carrying a live inflator module, hold securely with both hands, and ensure that the bag and trim cover are pointed away from you.

• Place the inflator module on a bench or other surface with the bag and trim cover facing up.

• With the inflator module on the bench, never place anything on or close to the module which may be thrown in the event of an accidental deployment.

DISARMING

Driver's Side

✸✸ CAUTION

The Air Bag system must be disarmed before any of its components are disconnected or removed. Failing to disarm the system before servicing components may cause accidental deployment of the Air Bag, resulting in repairs and possible personal injury.

1. Turn the steering wheel so that the vehicle's front wheels are pointing straight ahead.

2. Turn the ignition switch to the **LOCK** position. Remove the key.

3. Disconnect the negative and positive battery cables. Wait at least five minutes before working around the air bags.

➡**Removing the SRS-1 and SRS-2 fuses from the under-dash fuse block has the same effect as disconnecting the battery.**

4. Detach the yellow 3-way Air Bag harness connector at the base of the steering column.

After servicing is completed, enable the Air Bag as follows:

5. Reconnect the yellow 3-way Air Bag harness connector at the base of the steering column. Then, install the fuses if they were removed.

6. Reconnect the positive and negative battery cables.

7. Turn the ignition to the **ON** position, but don't start the engine. The AIR BAG warning light should turn ON and flash ON and OFF for seven seconds, then turn OFF. This light sequence indicates that the Air Bag system is functioning normally. If the AIR BAG light doesn't come ON, or stays ON longer than seven seconds, the system must be diagnosed.

Passenger's Side

1. Turn the steering wheel so that the vehicle's front wheels are pointing straight ahead.

2. Turn the ignition switch to the **LOCK** position. Remove the key.

3. Disconnect the negative and positive battery cables. Wait at least five minutes before working around the air bags.

➡**Removing the SRS-1 and SRS-2 fuses from the underdash fuse block has the same effect as disconnecting the battery.**

4. Remove the glove box door and the passenger's air bag lower cover.

5. Detach the yellow 4-way Air Bag harness connector from the passenger's side air bag.

6. If the passenger's air bag module must be removed, carefully remove the lower mounting bracket and mounting nuts. Lift the air bag out of the dashboard and remove it from the vehicle.

7. After servicing is completed, enable the Air Bag.

8. Reconnect the yellow 4-way Air Bag harness connector.

9. Reconnect the positive and negative battery cables.

10. Turn the ignition to the **ON** position, but don't start the engine. The AIR BAG warning light should turn ON and flash ON and OFF for seven seconds, then turn OFF. This light sequence indicates that the Air Bag system is functioning normally. If the AIR BAG light doesn't come ON, or stays ON longer than seven seconds, the system must be diagnosed.

Power Steering Gear

REMOVAL & INSTALLATION

1. Set the front wheels and the steering wheel in the straight ahead position. Lock the steering column in this position and remove the key.

2. Disconnect the negative and positive battery cables. Wait at least five minutes before working around the Air Bags, if equipped.

✳✳ WARNING

If the vehicle is equipped with a driver's Air Bag, be careful not to violently jar or strike the steering column, otherwise air bag deployment may occur.

3. Drain the power steering fluid from the pump reservoir.

4. Raise and safely support the vehicle.

5. Remove the left front wheel.

6. Remove the radiator skid plate, if equipped.

7. Remove the lower fan shroud.

8. Disconnect the stabilizer bar linkages and loosen the stabilizer bar bracket bolts.

9. Matchmark the Pitman arm to the Pitman shaft. Remove the nut and washer from the Pitman arm.

10. Use a puller tool to draw the Pitman arm from the Pitman shaft.

11. Use a flare nut wrench to disconnect the power steering lines from the steering gear. Plug the lines to prevent fluid loss and contamination.

12. Matchmark the steering column universal joint at the gearbox. Remove the clamp bolt from the universal joint.

13. Push the stabilizer aside and loosen the steering gear mounting bolts.

14. Remove the steering gear from the vehicle.

To install:

15. Support the steering gear in position. Install the mounting bolts and tighten them to 33 ft. lbs. (45 Nm).

16. Align the matchmark and connect the steering column universal joint to the steering gear.

17. Install the universal joint clamp bolt and tighten it to 18 ft. lbs. (25 Nm).

18. Align the matchmark and install the Pitman arm onto the Pitman shaft. Be sure the notched teeth are aligned. Install a new washer and tighten the Pitman arm nut to 145–174 ft. lbs. (196–236 Nm).

19. Connect the hydraulic lines to the steering gear. Carefully tighten the line fittings to 33 ft. lbs. (45 Nm).

20. Connect the stabilizer bar to the control arms. Tighten the linkage bolts to 8 ft. lbs. (10 Nm), and tighten the mounting bracket bolts to 21 ft. lbs. (28 Nm). Tighten the linkage nuts to 37 ft. lbs. (50 Nm).

21. Install the skid plate, and tighten its bolts to 27 ft. lbs. (37 Nm).

22. Install the left front wheel.

23. Lower the vehicle. Reconnect the negative and positive battery cable.

24. Refill and bleed the power steering system, as follows:

   a. Fill the reservoir with Dexron® II ATF.

   b. Raise the front wheels off the ground and block the rear wheels.

   c. Before starting the engine, turn the wheel lock-to-lock several times so the level of fluid in the reservoir drops. Refill the reservoir as needed to bring the fluid to the specified level.

   d. Start the engine and turn the wheel lock-to-lock 3 or 4 more times until any buzzing sound disappears and the wheel turns smoothly. Do not hold the wheel in the lock position for more than 5 seconds.

   e. Straighten the wheels and turn **OFF** the engine. If the fluid level in the reservoir increases, repeat the bleeding procedure. If the fluid level stays the same, bleeding is completed.

25. Verify that the steering wheel spokes are centered when the front wheels are in the straight ahead position. Check and adjust the front wheel alignment.

Shock Absorber

REMOVAL & INSTALLATION

Front

1. Raise and support the vehicle safely.

2. Remove the front wheels.

3. Support the lower control arm with a floor jack.

4. Hold the shaft of the shock absorber with a wrench to keep it from turning and remove the upper mounting nut, retainer, and rubber grommet.

➡**On some vehicles, it may be necessary to remove the bump stops to gain access to the lower mounting bolt.**

5. Unbolt the shock absorber from the lower control arm. Remove the shock absorber from the vehicle.

To install:

6. Install the lower retainer and rubber grommet onto the upper shaft of the shock absorber. Then, fully extend the shock and install the upper shaft into the mounting hole in the frame bracket.

7. Install the upper rubber grommet, retainer and attaching nut onto the shock absorber upper shaft. Only hand-tighten the mounting nut at this time.

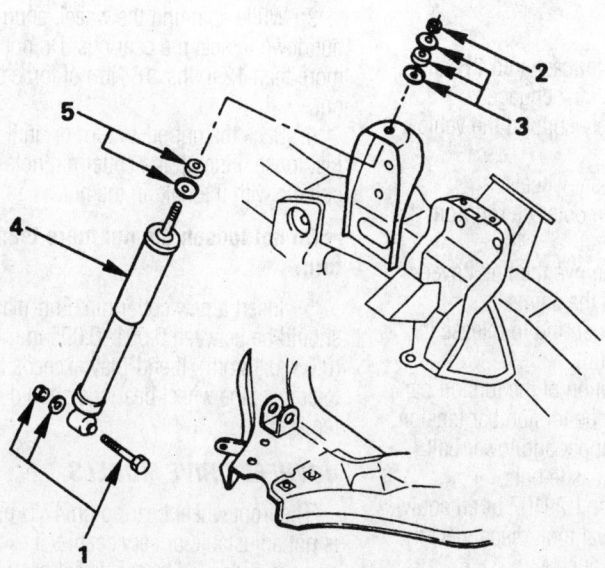

1. Bolt, nut and washer
2. Nut
3. Rubber bushing and washer
4. Shock absober
5. Rubber bushing and washer

Exploded view of the front shock absorber mounting

8. Install the shock absorber to the lower control arm bracket and install the mounting bolt and nut. Tighten the mounting bolt to 60–61 ft. lbs. (82–84 Nm).

9. Tighten the upper mounting nut to 14–15 ft. lbs. (19–20 Nm).

10. Install the bump stop if removed, and tighten the bolts to 30 ft. lbs. (41 Nm).

11. Install the front wheels and lower the vehicle.

Rear

1. Raise and safely support the vehicle.
2. Remove the shock absorber-to-lower mount nut, washers, and bushings.
3. Remove the shock absorber-to-chassis nut, washers, and bushings.

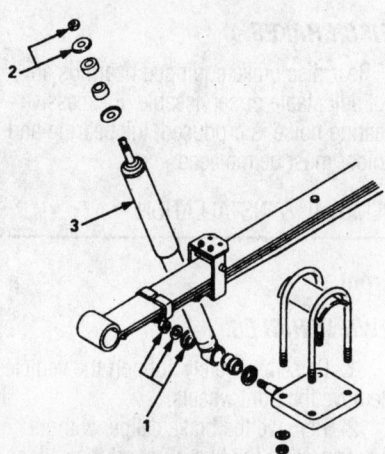

1 Lower nut and washer
2 Upper nut and washer
3 Shock absorber

Exploded view of the rear shock absorber mounting

4. Remove the shock absorber.

To install:

5. Install the shock absorber to the upper mount. Do not tighten the nut until the vehicle is on the floor.

6. Install the shock absorber on the lower mount. Do not tighten the nut until the vehicle is on the floor.

7. Lower the vehicle to the floor and tighten the mounting nuts to 25–32 ft. lbs. (34–43 Nm.)

Upper Ball Joint

REMOVAL & INSTALLATION

1. Raise and safely support the vehicle.
2. Remove the front wheels.
3. Mark the position of the torsion bar adjuster. Loosen the adjuster to relieve the torsion bar tension.
4. Support the lower control arm with a floor jack.
5. Remove the upper ball joint castle nut.
6. Using a ball joint separator tool, separate the upper ball joint from the steering knuckle.
7. Remove the upper ball joint from the control arm.

To install:

8. Install the ball joint onto the control arm. Tighten the upper ball joint bolts to 42 ft. lbs. (57 Nm).

9. Install the upper ball joint to the steering knuckle. Tighten the upper ball joint castle nut to 72–73 ft. lbs. (96–98 Nm). Then, tighten the castle nut only enough to install a new cotter pin.

10. Install the front wheels.

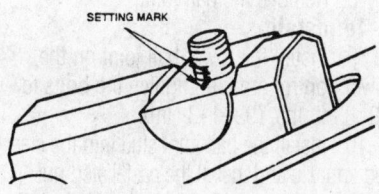

Matchmarking the height adjustment bolt and collar

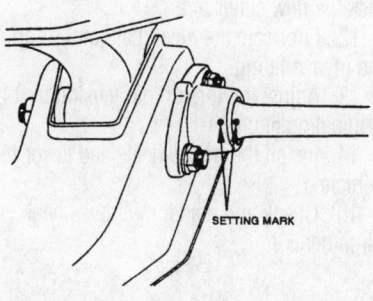

Matchmarking the torsion bar to the lower control arm

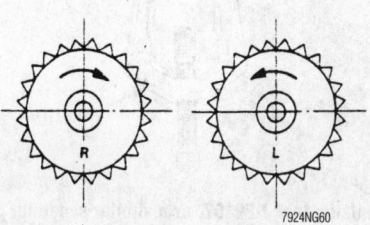

Left and right torsion bar identification

11. Adjust the tension on the torsion bar to the original position.

12. Lower the vehicle to the floor.

Lower Ball Joint

REMOVAL & INSTALLATION

2WD Models

1. Raise and safely support the vehicle under the frame.

2. Remove the front wheels.

3. Disconnect the outer tie rod from the steering knuckle.

4. Mark the position of the torsion bar adjuster and release the torsion bar tension.

5. Remove the lower ball joint cotter pin and castellated nut.

6. Using J-29107 or equivalent, disconnect the ball joint from the knuckle.

7. Remove the lower ball joint mounting bolts from the lower control arm.

8. Remove the ball joint.

To install:

9. Install the lower ball joint on the lower control arm and tighten the bolts to 68–83 ft. lbs. (93–113 Nm).

10. Install the ball joint stud into the steering knuckle and install the castellated nut. Tighten the nut to 87–101 ft. lbs. (117–137 Nm), with just enough additional tighten to align the cotter pin hole with a castellation on the nut. Install a new cotter pin.

11. Connect the tie rod end to the steering knuckle. Tighten the castle nut to 72 ft. lbs. (98 Nm), then tighten only enough to install a new cotter pin.

12. Lubricate the lower ball joint through the grease fitting.

13. Adjust the torsion bar tension to its original position.

14. Install the front wheels and lower the vehicle.

15. Check and adjust the front wheel alignment.

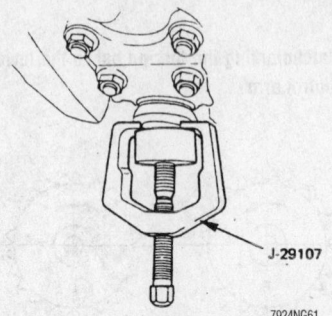

J-29107

7924NG61

Using tool J-29107, or a similar separator tool, to separate the lower ball joint from the steering knuckle

4WD Models

1. Shift the transfer case into **2H** and be sure the front axle isn't engaged.

2. Raise and safely support the vehicle under the frame.

3. Remove the front wheels.

4. Disconnect the outer tie rod from the steering knuckle.

5. Unbolt and remove the hub cover. Note the positions of the shims.

6. Remove the snapring to release the axle shaft from the hub.

7. Mark the position of the torsion bar adjuster and release the torsion bar tension.

8. Remove the upper and lower ball joint cotter pins and castle nuts.

9. Using tool No. J-29107 or an equivalent ball joint removal tool, disconnect the ball joints from the knuckle.

10. Remove the knuckle/hub assembly from the axle shaft.

11. Remove the lower ball joint mounting bolts from the lower control arm.

12. Remove the ball joint.

To install:

13. Install the lower ball joint on the lower control arm and tighten the bolts to 69–83 ft. lbs. (93–113 Nm).

14. Install the ball joint studs into the steering knuckle and install the castellated nut. Tighten the lower nut to 87–101 ft. lbs. (117–137 Nm), tighten the upper nut to 72 ft. lbs. (98 Nm). Then, tighten the nuts just enough to align the cotter pin hole with a castellation on the nut. Install a new cotter pin.

15. Connect the tie rod end to the steering knuckle. Tighten the castle nut to 72 ft. lbs. (98 Nm), then tighten only enough to install a new cotter pin.

16. Reinstall the shims and the snapring on the axle shaft.

17. Lubricate the lower ball joint through the grease fitting.

18. Adjust the torsion bar tension to the original position.

19. Install the front wheels and lower the vehicle.

20. Check and adjust the front wheel alignment.

Wheel Bearings

ADJUSTMENT

Front

2-WHEEL DRIVE MODELS

1. Raise and safely support the vehicle.

2. Remove the hub dust cover and spindle cotter pin. Loosen the nut.

3. While spinning the wheel, snug the nut down to seat the bearings. Do not exert more than 12 ft. lbs. 16 Nm) of force on the nut.

4. Back the nut off ¼ turn or until it is just loose. Line up the cotter pin hole in the spindle with the hole in the nut.

➡ **Do not loosen the nut more than ½ turn.**

5. Insert a new cotter pin. End-play should be between 0.001–0.005 in. (0.03–0.13mm). If end-play exceeds this tolerance, the wheel bearings should be replaced.

4-WHEEL DRIVE MODELS

The front wheel bearing on 4WD models is not adjustable or serviceable. If excessive noise is produced by the wheel bearing, the assembly must be replaced.

Rear

DRUM BRAKES

1. Raise and safely support the vehicle.

2. Remove the hub dust cover and spindle cotter pin. Loosen the nut.

3. While spinning the wheel, snug the nut down to seat the bearings. Do not exert over 12 ft. lbs. (16 Nm) of force on the nut.

4. Back the nut off ¼ turn or until it is just loose. Line up the cotter pin hole in the spindle with the hole in the nut.

➡ **Do not loosen the nut more than ½ turn.**

5. Insert a new cotter pin. End-play should be between 0.001–0.005 in. (0.03–0.13mm). If play exceeds this amount, the wheel bearings should be replaced.

DISC BRAKES

Rear disc brake equipped bearings are not adjustable or serviceable. If excessive bearing noise is produced, the bearing and holder must be replaced.

REMOVAL & INSTALLATION

Front

2WD VEHICLES

1. Raise and safely support the vehicle. Remove the front wheels.

2. Remove the brake caliper without disconnecting the line. Support the caliper aside.

3. Remove the brake caliper mounting bracket.

4. Remove the dust cover, cotter pin and locknut from the rotor.

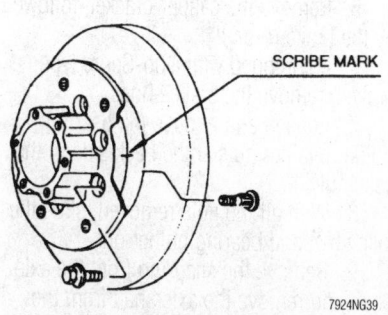

Matchmark the hub and rotor for reassembly

5. Place a hand over the outer wheel bearing to prevent the bearing from falling on the floor and remove the rotor and hub from the spindle.

6. Matchmark the hub and rotor so it can be assembled in the same position, then remove the bolts attaching the hub to the rotor. Pull off the rotor.

7. Thoroughly clean, inspect and repack the bearings with wheel bearing grease. Replace the bearings if needed. Use a brass drift to drive the bearing races out of the hub.

To install:

8. Use a suitably-sized bearing driver to drive new bearing races into the hub.

9. Install the rotor on the hub and tighten the bolts to 76 ft. lbs. (103 Nm).

10. Install the inner bearing and the grease seal in the hub, if removed, and position the hub and rotor assembly on the spindle.

11. Install the outer bearing and spindle nut in the hub. Tighten the hub nut to 22 ft. lbs. (29 Nm) to seat the bearings, then loosen the nut. Adjust the wheel bearing.

12. Install the caliper mounting bracket and tighten the bolts to 115 ft. lbs. (156 Nm).

13. Install the brake caliper.

14. Install the front wheels and lower the vehicle.

15. Pump the brake pedal several times to test the system. Bleed the brakes if necessary.

4WD VEHICLES

1. Shift the transfer case shift lever into **2H** and move the vehicle forward and rearward about 3 ft. (0.91m).

2. Raise and safely support the vehicle. Remove the front wheels.

3. Remove the 4WD hub cap attaching bolts and the cap.

4. Remove the brake caliper and support the caliper on a wire; do not disconnect the brake hose. Remove the caliper support bracket from the steering knuckle.

5. Using snapring pliers, remove the snapring and shims from the hub assembly.

6. Remove the drive clutch assembly, inner cam and lockwasher.

7. Using a hub nut wrench, loosen the hub nut and pull the hub from the spindle.

8. Remove the inner and outer bearings from the hub. Clean and inspect the bearings for pitting and scoring and replace if needed.

9. If necessary, use a brass drift to remove the bearing races from the hub.

➡**Always replace the bearing and race as a matched set if replacement is needed.**

10. If equipped with ABS, unbolt and remove the ABS sensor ring.

11. If removing the disc from the hub, scribe matchmarks, then remove the bolts attaching the hub to the rotor. Separate the disc from the hub.

To install:

12. Use a suitably-sized bearing driver to install new bearing races into the hub.

13. If the hub and rotor were separated, install the hub to the rotor and tighten the bolts to 76 ft. lbs. (103 Nm).

14. If equipped, install the ABS sensor ring and tighten the bolts to 13 ft. lbs. (18 Nm).

15. Pack the bearings with wheel bearing grease and install the inner bearing and a new grease seal in the hub. Position the hub and disc assembly on the spindle, place 1 to 1½ oz. (28–42 grams) of bearing grease in the hub and install the outer bearing and hub nut on the spindle.

16. When installing the hub nut, perform the following steps:

   a. Tighten the hub nut to 22 ft. lbs. (30 Nm) and loosen the nut to seat the bearings.

   b. Use a spring gauge connected to the stud bolt at 90° to measure preload.

   c. Retighten the hub nut until the spring gauge measures a bearing preload of 4.4–5.5 lbs. (2.0–2.5 kg) for a new bearing and new oil seal, or 2.6–4.0 lbs. (1.2–1.8 kg) for a used bearing and new oil seal.

17. Install the lockwasher with the larger diameter of the tapered bore facing out. If the bolt holes in the lock plate do not align with the corresponding holes in the nut, reverse the lock plate. If the bolt holes are still out of alignment, turn in the nut just enough to obtain alignment and install the lockscrew tightly so the head is lower than the surface of the washer.

18. Clean the flange surface of the hub, thread holes, surface of the lockwasher and the splines of the axle shaft.

19. Install the inner cam by aligning the keyway of the inner cam with the groove of the knuckle. If the cam is difficult to install, use the special tool or equivalent and a plastic hammer to lightly tap the inner cam into place.

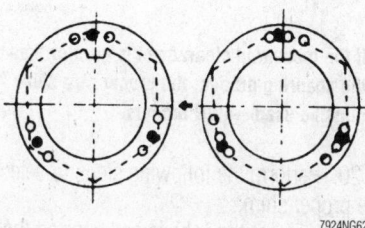

Lockwasher alignment—4WD models

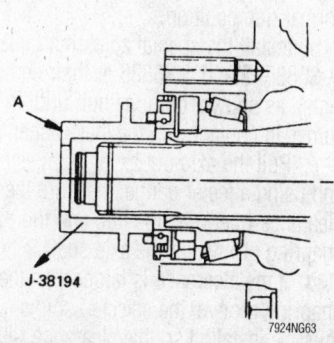

Special tool J-38194 or equivalent should be used to install the inner cam—4WD models

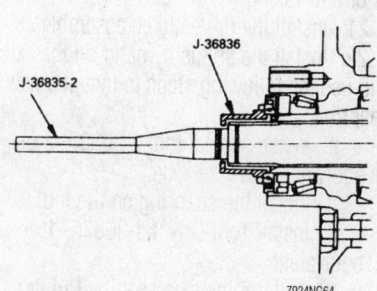

Install the special tool to measure the shim thickness—4WD models

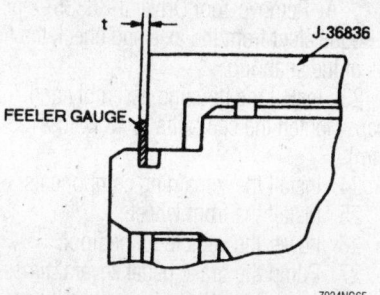

Measure clearance t between the tool and the snapring groove—4WD models

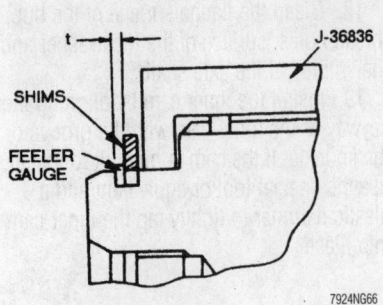

7924NG66

If the measured clearance t is greater than the snapring groove, the proper size shim must be used—4WD models

20. Perform the following steps to select the proper shim:

a. Lower the vehicle and support the lower control arm with a block of wood and a floor jack to place the axle in the normal ride position.

b. Install the special adjusting tools (J-36835–2 and J-36836 or their equivalents), as shown, onto the hub until it comes in contact with the lockwasher.

c. Pull the axle out as far as possible, and using a feeler gauge, measure the clearance **t** between the hub and the snapring groove on the axle shaft.

d. If the clearance is larger than the snapring groove, the selected shims must be installed so that clearance **t** is 0–0.039 in. (0–0.1mm). Shims come in thickness of 0.2, 0.3, 0.5, and 1.0mm.

e. Remove the special tool J-36836 or equivalent.

21. Install the drive clutch assembly.

22. Install the shims by hand on the axle and use the following steps to install a new snapring:

a. Install special tool J-36835–2 or equivalent onto the axle.

b. Install the snapring on the tool.

c. Install Tool Driver J-36835–1 or equivalent.

d. Pull out the axle shaft by Pulling Tool J-36835–2 or equivalent. Install the snapring to the axle by pushing on tool J-36835–1.

e. Remove Tool Driver J-36835–2 or equivalent from the axle and check the fit of the snapring.

23. Install the housing assembly and cap. Tighten the cap bolts to 43 ft. lbs. (58 Nm).

24. Install the remaining components.

25. Install the front wheels.

26. Lower the vehicle to the floor.

27. Pump the brake pedal several times to test the system. Bleed the brakes if necessary.

Rear

DRUM BRAKE MODELS

✳✳ CAUTION

Some brake shoes may contain asbestos, which has been determined to be a cancer causing agent. Never clean the brake surfaces with compressed air! Avoid inhaling any dust from any brake surface! When cleaning brake surfaces, use a commercially available brake cleaning fluid.

1. Raise and safely support the vehicle. Remove the wheel.

2. Remove the dust cap, cotter pin, castle nut, thrust washer and outside wheel bearing. Pull the drum assembly from the spindle. Place the drum on a clean work surface.

3. Using a suitable seal puller, pry out the inner seal. With the seal removed, lift out the inner bearing. Remove the inner races. It may be necessary to use a hammer and a brass drift to drive the bearing races from the hub.

4. Clean all parts in kerosene or equivalent, DO NOT use gasoline. After cleaning, dry the parts with compressed air and check parts for excessive wear and replace damaged parts. Do NOT allow the bearings to spin while blowing dry.

To install:

5. Apply a coating of grease to the inside of the hub. Install the bearing races into hub, using a hammer and a brass drift. Drive the races in until they seat against the shoulder of the hub.

6. Pack the bearings with grease and install the inner bearing in the hub. Install a new grease seal, being careful not to damage the seal.

7. Install the drum assembly onto the spindle. Install the outer bearing, thrust washer and castle nut. Tighten the nut until the wheel does not turn freely.

8. Back off the nut until the wheel turns freely and install the cotter pin. Install dust cap.

9. Install the wheel and lower the vehicle.

DISC BRAKE MODELS

1. Raise and safely support the vehicle.

2. Remove the wheel and tire assembly.

3. Disconnect the parking brake cable from the caliper or if equipped the brake shoe.

4. Remove the caliper and suspend it with wire. The flexible hose does not need to be disconnected.

5. Remove the caliper bracket, followed by the brake rotor.

6. If equipped with Duo-Servo type brakes, remove the brake shoes.

7. Loosen and remove the bearing holder retainer nuts at the back side of the assembly.

8. With all the nuts removed, slide the axle shaft and bearing holder out.

9. Remove the snapring from the axle shaft and remove the axle shaft from the bearing housing.

10. Remove the bearing seal from the housing. Press out the bearing and discard.

To install:

11. Apply grease around the outside of the bearing assembly, and press into the bearing housing.

12. Install a new seal to the bearing housing.

13. Install the axle shaft into the bearing housing, and secure in place with the snapring.

14. Install the axle shaft and bearing housing to the rear axle, and secure in place with the retainer nuts and bolts. Tighten the nuts to 54 ft. lbs. (74 Nm).

15. If equipped with Duo-Servo brakes, install the brake shoes to the backing plate.

16. Install the brake rotor and caliper.

17. Install the parking brake cable to the brake assembly.

18. Install the wheel and tire assembly, then lower the vehicle.

PACKING

Proper and regular greasing of the wheel bearing can prolong the life of the bearing.

Clean the wheel bearings thoroughly with solvent and check their condition before installation.

✳✳ WARNING

Do not blow the bearing dry with compressed air as this would allow the bearing to turn without lubrication.

The are several different types of inexpensive wheel bearing packing tools available at most automotive parts stores. Although they all work well. they really do not work any better than the old-fashion method of hand greasing.

To hand grease a bearing place a sizable amount of lubricant in the palm of one hand. Using your other hand, work the bearing into the lubricant so that the grease is pushed through the rollers and out the other side. Keep rotating the bearing while continuing to push the lubricant through it.

JEEP

Cherokee • Grand Cherokee • Wrangler

ENGINE REPAIR

✷✷ WARNING

NEVER disconnect the negative battery cable with the ignition ON. Removing power from the computer control module with the ignition ON may destroy the module. Also, disconnecting the negative battery cable on some vehicles may interfere with the functions of the on board computer systems and may require the computer to undergo a relearning process, once the negative battery cable is reconnected.

Distributor

REMOVAL

2.5L and 4.0L Engines

1. Disconnect the negative battery cable.
2. Unfasten the distributor cap retaining screws. Remove the distributor cap with the coil and spark plug wires attached and position them aside.

3. Disconnect the distributor primary wiring connector.
4. Remove the No. 1 spark plug.
5. Hold a finger over the spark plug hole and rotate the engine until compression pressure is felt. Slowly continue to rotate the engine until the timing index on the vibration damper pulley aligns the Top Dead Center (TDC) mark (0 degree) on the timing degree scale.

➡ **Always rotate the engine in the direction of normal rotation. Do not turn the engine backward to align the timing marks.**

6. Scribe a mark on the distributor housing inline with the tip of the rotor.
7. Note the position of the rotor and distributor housing in relation to the surrounding engine components as reference points for installing the distributor.
8. Remove the distributor hold-down bolt and clamp.
9. Lift the distributor straight up and out of the engine.

5.2L and 5.9L Engines

1. Disconnect the negative battery cable.
2. Remove the air cleaner tube at the throttle body.

3. Label and remove the high tension wires from the distributor cap.
4. Remove the distributor cap. Note the position of the rotor and distributor. Scribe a mark on the base of the distributor and the engine as an installation reference.
5. Turn the engine clockwise, using a socket on the end of the crankshaft damper bolt, until the rotor is pointing to the No. 1 spark plug wire post and the timing mark on the damper aligns with the **0** on the timing scale; No. 1 cylinder is at TDC on the compression stroke.

➡ **The timing mark is on the edge of the vibration damper, closest to the front engine cover.**

6. Unplug the camshaft position sensor wiring connector.
7. Remove the rotor.
8. Remove the bolt for the distributor hold-down clamp.
9. Remove the distributor from the engine.

INSTALLATION

2.5L and 4.0L Engines

1. If the crankshaft has been rotated after distributor removal, cylinder No. 1 must be returned to TDC of the compression stroke. Follow Steps 4 and 5 of the removal procedure if necessary.
2. Check the position of the oil pump slot. On 2.5L engines, it should be slightly before the 10 o'clock position. On 4.0L engines, it should be slightly before the 11 o'clock position. If necessary, rotate the oil pump gear to position using a screwdriver.

➡ **Factory replacement distributors are equipped with a plastic alignment pin installed. If this pin is in place, the next step may be skipped.**

3. Remove the camshaft position sensor from the distributor. Four different alignment

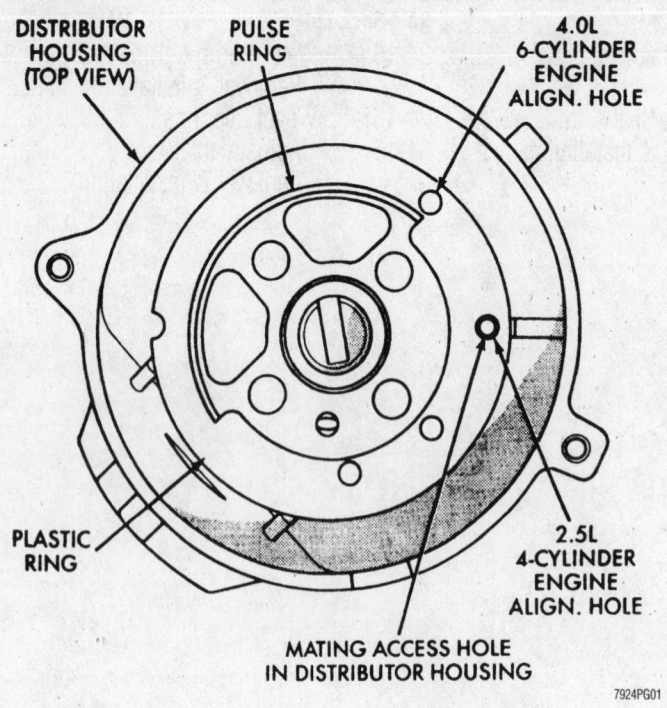

DISTRIBUTOR HOUSING (TOP VIEW)

PULSE RING

4.0L 6-CYLINDER ENGINE ALIGN. HOLE

PLASTIC RING

2.5L 4-CYLINDER ENGINE ALIGN. HOLE

MATING ACCESS HOLE IN DISTRIBUTOR HOUSING

7924PG01

Distributor pin alignment holes—2.5L and 4.0L engines

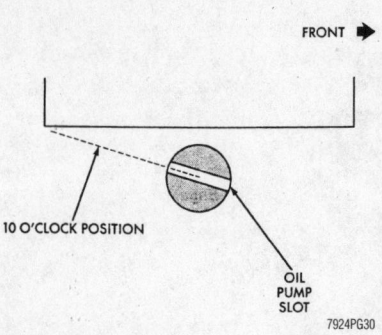

FRONT ➡

10 O'CLOCK POSITION

OIL PUMP SLOT

7924PG30

Slot in the oil pump gear at 10 o'clock position—2.5L engine

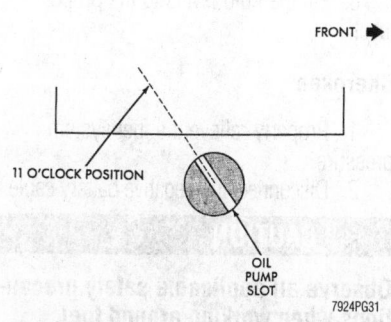

Slot in the oil pump gear at 11 o'clock position—4.0L engine

holes are provided on the plastic ring. Rotate the distributor shaft, then insert a ³⁄₁₆ in. pin punch through the proper alignment hole and mating access hole in the distributor. This prevents the shaft from rotating.

4. Clean the distributor mounting area of the cylinder block.

5. Install a new distributor mounting gasket.

➡**There is a fork on the distributor housing where the housing seats against the engine block. The slot in the fork aligns with the distributor hold-down bolt hole in the engine block. The distributor is properly installed when the rotor is in the correct position. Because of the fork in the distributor housing, initial ignition timing is not adjustable (the distributor cannot be rotated).**

6. Position the distributor shaft in the cylinder block, while holding the centerline of the base slot in the 1 o'clock position.

7. When the distributor is fully seated in the block, the centerline of the base slot should be aligned to the clamp bolt mounting hole on the engine. On the 2.5L engines, the rotor should point slightly past the 3 o'clock position. On 4.0L engines it should be pointed at the 5 o'clock position.

➡**It may be necessary to move the rotor and shaft (slightly) to engage the distributor shaft with the slot in the oil pump shaft. the same may have to be done to engage the distributor gear with the camshaft gear.**

5.2L and 5.9L Engines

1. If the crankshaft has been rotated after distributor removal, cylinder No. 1 must be returned to TDC of the compression stroke. Use the following sub-steps if necessary:

a. Remove the No. 1 spark plug.

b. Hold a finger over the spark plug hole and rotate the engine until compression pressure is felt. Slowly continue to rotate the engine until the timing index on the vibration damper pulley aligns the Top Dead Center (TDC) mark (0 degree) on the timing degree scale.

2. Clean the top of the cylinder block.

3. Lubricate the oil seal on the distributor with engine oil.

4. Install the rotor.

5. Position the distributor into the block and engage the tongue of the distributor shaft with the slot in the distributor oil pump drive gear. Position the rotor to the No. 1 spark plug terminal position.

6. Install the hold-down clamp and loosely install the bolt.

7. Rotate the distributor housing until the rotor is aligned to the **cylinder No. 1** alignment mark on the camshaft position sensor.

8. Tighten the hold-down clamp bolt to 17 ft. lbs. (22 Nm).

9. Connect the camshaft position sensor wiring harness to the main engine harness.

10. Install the distributor cap and tighten the screws.

11. Ensure the spark plug wires are firmly connected to their terminals.

12. Connect the negative battery cable. Start the engine and check for proper operation.

Ignition Timing

ADJUSTMENT

Base ignition timing is not adjustable. The distributor does not have a built-in centrifugal or vacuum assisted advance. Base ignition timing and timing advance are controlled by the Powertrain Control Module (PCM) which monitors inputs from various sensors to determine and adjust correct ignition timing. On the 5.2L and 5.9L engines, proper distributor position can be checked as follows:

1. Connect the DRB or equivalent scan tool to the data link connector in the engine compartment.

2. Gain access to the SET SYNC screen on the scan tool.

✳✳ WARNING

The engine will be running. Be careful not to stand inline with the fan blades or belt. Do not wear loose clothing.

3. Follow the directions on the scan tool screen and start the engine. With the engine running, the words IN RANGE should appear on the screen along with 0°.

4. If a plus or minus is displayed and/or the degree displayed is not zero, loosen the distributor hold-down bolt and rotate the distributor until IN RANGE appears. Continue to rotate until achieving as close to **0** as possible. After adjustment, tighten the clamp bolt.

➡**The degree scale on the SET SYNC screen is referring to fuel synchronization only. It is not referring to ignition timing. Do not attempt to adjust ignition timing using this method. Rotating the distributor will have no effect on ignition timing.**

Engine Assembly

REMOVAL & INSTALLATION

Wrangler

1. Place a protective cloth on the windshield frame. Raise the hood and rest it on the frame.

2. Properly relieve the fuel system pressure.

3. Disconnect the battery cables and remove the battery.

✳✳ CAUTION

Observe all applicable safety precautions when working around fuel. Whenever servicing the fuel system, always work in a well ventilated area. Do not allow fuel spray or vapors to come in contact with a spark or open flame. Keep a dry chemical fire extinguisher near the work area. Always keep fuel in a container specifically designed for fuel storage; also, always properly seal fuel containers to avoid the possibility of fire or explosion.

➡**Label all electrical connectors and vacuum lines prior to disconnecting them, so they can be reinstalled in their proper locations.**

4. Drain the cooling system.

5. Disconnect the wires from the alternator.

6. Disconnect the ignition coil and distributor wire connections.

7. Disconnect the oil pressure sending unit connector.

8. Disconnect the wires from the starter.

9. Disconnect the fuel injection wires.

10. Disconnect the fuel lines from the fuel rails.

11. Remove the fuel line bracket from the intake manifold.

12. Disconnect the engine ground strap.

13. Remove the air cleaner assembly.

14. Disconnect the canister purge hose from the vapor canister "T" connector.

15. Disconnect the idle speed actuator wire connector.

16. Disconnect the throttle cable and remove it from the bracket.

17. Disconnect the throttle rod from the bellcrank.

18. If equipped, disconnect the cruise control cable.

19. Disconnect the oxygen sensor electrical connector.

20. Disconnect the upper and lower hoses from the radiator.

21. Disconnect the coolant hoses from the rear of the intake manifold and thermostat housing.

22. Disconnect the heater hoses.

23. Remove the fan shroud screws.

24. Remove the radiator and fan shroud.

25. Remove the engine cooling fan.

26. Remove the engine cooling fan and install a 5/16 x 1/2 inch cap screw through the fan pulley into the water pump flange. This will maintain the pulley and water pump in alignment when the crankshaft is rotated.

27. If equipped, disconnect the check valve from the power brake booster.

28. If equipped with power steering, perform the following:

a. Disconnect the steering hoses from the fittings at the steering gear.

b. Drain the pump reservoir.

c. Cap all fittings once removed.

29. Raise and support the vehicle safely.

30. Remove the oil filter.

31. Remove the starter.

32. Remove the flywheel access cover.

33. Remove the engine support cushion-to-bracket through-bolts.

34. Disconnect the exhaust pipe from the manifold.

35. Remove the upper flywheel housing bolts and loosen the bottom bolts.

36. Remove the engine shock damper bracket from the sill.

37. Lower the vehicle.

38. Attach a lifting device to the engine.

39. Place a support under the bell housing.

40. Remove the remaining flywheel housing bolts.

41. Lift the engine from the vehicle.

42. Install the oil filter to keep foreign material out of the engine.

To install:

43. Remove the oil filter.

44. Lower the engine into the vehicle. To ease installation, remove the engine support cushions to aid in engine-to-transmission alignment.

45. Insert the transmission shaft into the clutch spline.

46. Align the flywheel housing with the engine.

47. Install and tighten the flywheel housing bolts finger-tight.

48. If removed, install the engine support cushions.

49. Lower the engine into place and remove the lifting device.

50. Raise and support the vehicle safely.

51. Attach the engine shock damper bracket to the sill.

52. Attach the exhaust pipe to the manifold and tighten the nuts to 23 ft. lbs. (31 Nm).

53. Install the flywheel access cover.

54. Install the remaining flywheel housing bolts and tighten them to 28 ft. lbs. (38 Nm).

55. Install the starter.

56. Install a new oil filter.

57. Lower the vehicle.

58. Connect the coolant lines and tighten the clamps.

59. If equipped with power steering:

a. Connect the hoses to the steering gear and tighten the nut to 38 ft. lbs. (52 Nm).

b. Fill the pump reservoir with fluid.

60. Remove the alignment cap screw and install the fan assembly.

61. Install the accessory drive belt.

62. Install the radiator and shroud.

63. Connect the radiator hoses.

64. Connect the oxygen sensor electrical connector.

65. Connect the throttle valve rod and retainer. Connect the throttle cable and install the rod and spring.

66. If equipped, connect the speed control cable.

67. Install the vacuum hose and check valve to the brake booster.

68. Connect the electrical connections disconnected during removal.

69. Connect the fuel lines to the fuel rail.

70. Install the fuel line bracket to the intake manifold.

71. Install the air cleaner.

72. Install the battery and connect the cables.

73. Fill the engine to the proper level with fresh oil.

74. Fill the cooling system.

75. Start the engine and check for leaks.

76. Fill the fluid levels to the proper level.

Cherokee

1. Properly relieve the fuel system pressure.

2. Disconnect the negative battery cable.

✳✳ CAUTION

Observe all applicable safety precautions when working around fuel. Whenever servicing the fuel system, always work in a well ventilated area. Do not allow fuel spray or vapors to come in contact with a spark or open flame. Keep a dry chemical fire extinguisher near the work area. Always keep fuel in a container specifically designed for fuel storage; also, always properly seal fuel containers to avoid the possibility of fire or explosion.

3. If equipped with A/C, properly discharge the system using a recovery/recycling machine.

4. Matchmark the hood and hinges and remove the hood.

✳✳ CAUTION

Never open, service or drain the radiator or cooling system when hot; serious burns can occur from the steam and hot coolant.

5. Drain the cooling system.

➡**Label all electrical connectors and vacuum lines prior to disconnecting them, so they can be reinstalled in their proper locations.**

6. Remove the upper, lower and coolant recovery cooling system hoses.

7. Remove the fan shroud.

8. If equipped with an automatic transmission, disconnect the fluid cooler lines.

9. Remove the radiator and, if equipped, the A/C condenser.

10. Remove the engine cooling fan and install a 5/16 x 1/2 inch cap screw through the fan pulley into the water pump flange. This will maintain the pulley and water pump in alignment when the crankshaft is rotated.

11. Disconnect the heater hoses.

12. Disconnect the throttle linkages, speed control cable, if equipped, and throttle valve rod.

13. Disconnect the oxygen sensor electrical connector.

14. Disconnect the fuel injection harness connectors.

15. Disconnect the quick-connection fuel lines at the fuel rail and return line.

16. Remove the fuel line bracket from the intake manifold.

17. Remove the air cleaner assembly.

18. If equipped with A/C, remove the service valves and cap the compressor ports.

19. Remove the power brake vacuum check valve from the booster, if equipped.

20. If equipped with power steering, perform the following:

   a. Disconnect the steering hoses from the fittings at the steering gear.

   b. Drain the pump reservoir.

   c. Cap all fittings once removed.

21. Disconnect the coolant hoses from the rear of the intake manifold.

22. Identify, tag and disconnect all necessary wires and vacuum lines.

23. Raise and support the vehicle safely.

24. Remove the oil filter.

25. Remove the starter.

26. Disconnect the exhaust pipe from the manifold.

27. Remove the flywheel/converter housing access cover.

28. If equipped with an automatic transmission, matchmark the converter to the driveplate and remove the bolts.

29. Remove the upper flywheel/converter housing bolts and loosen the bottoms bolts.

30. Remove the engine mount-to-engine compartment bracket bolts.

31. Remove the engine shock damper bracket from the sill.

32. Lower the vehicle.

33. Attach a lifting device to the engine.

34. Raise the engine slightly off the front supports.

35. Place a support stand under the transmission housing.

36. Remove the remaining flywheel bolts.

37. Lift the engine out of the vehicle.

38. Install the oil filter to keep foreign material out of the engine.

To install:

39. Remove the oil filter.

40. Lower the engine into the vehicle. To ease installation, remove the engine mounts to aid in engine-to-transmission alignment.

41. If equipped with a manual transmission, perform the following:

   a. Insert the transmission shaft into the clutch spline.

   b. Align the flywheel housing with the engine.

   c. Install and tighten the flywheel housing bolts finger-tight.

42. If equipped with an automatic transmission, perform the following:

   a. Align the torque converter housing with the engine.

   b. Loosely install the converter housing lower bolts and install the next higher nut and bolt on each side.

   c. Tighten all four bolts finger-tight.

43. If removed, install the engine mounts.

44. Lower the engine into place and remove the lifting device.

45. Raise and support the vehicle safely.

46. If equipped with an automatic transmission, perform the following:

   a. Align the torque converter to the driveplate.

   b. Install the bolts and tighten them to 40 ft. lbs. (54 Nm).

   c. Install the access cover.

   d. Install the exhaust pipe support.

47. Install the remaining converter/flywheel bolts finger-tight.

48. Install the starter.

49. Tighten the engine support cushion bolts/nuts.

50. Tighten the loose converter/flywheel bolts to 28 ft. lbs. (38 Nm).

51. Install a new oil filter.

52. Connect the exhaust pipe to the manifold.

53. Lower the vehicle.

54. Connect the coolant hoses and tighten the clamps.

55. If equipped with power steering, perform the following:

   a. Unplug the lines and connect them to the steering gear. Tighten the fittings to 38 ft. lbs. (52 Nm).

   b. Fill the pump reservoir with fluid.

56. Remove the alignment cap screw and install the fan.

57. Install the radiator, condenser, if equipped and fan shroud.

58. Connect the radiator hoses.

59. If equipped with an automatic transmission, connect the cooling lines.

60. Connect the oxygen sensor electrical connector.

61. Connect the throttle valve rod and retainer. Connect the throttle cable and install the rod and spring.

62. If equipped, connect the cruise control cable.

63. Connect the fuel lines to the throttle body.

64. Connect all vacuum lines and electrical connectors disconnected during removal.

65. If equipped with A/C, connect the service valves to the compressor ports.

66. Install the air cleaner.

67. Install the hood.

68. Connect the battery cables.

69. Fill the cooling system.

70. Start the engine and check for leaks.

71. If equipped, recharge the A/C system.

72. Check and top off fluid levels.

Grand Cherokee

4.0L ENGINES

1. Matchmark the hood to the hinges and remove the hood.

2. Properly relieve the fuel system pressure.

3. Remove the battery.

✳✳ CAUTION

Observe all applicable safety precautions when working around fuel. Whenever servicing the fuel system, always work in a well ventilated area. Do not allow fuel spray or vapors to come in contact with a spark or open flame. Keep a dry chemical fire extinguisher near the work area. Always keep fuel in a container specifically designed for fuel storage; also, always properly seal fuel containers to avoid the possibility of fire or explosion.

4. Drain the cooling system.

5. Remove the air cleaner and tube.

6. Remove the radiator.

7. Remove the heater hoses.

8. Label and disconnect the necessary vacuum lines.

9. Remove the distributor cap and wiring.

10. Disconnect the accelerator linkage.

11. Remove the air duct from the throttle body.

12. Label and disconnect the Manifold Absolute Pressure (MAP) sensor, Idle Air Control (IAC) motor and Throttle Position Sensor (TPS) electrical connectors from the throttle body.

13. Disconnect the vacuum line from the throttle body.

14. Disconnect (unsnap) the control cables from the throttle body (lever) arm.

15. Remove the throttle body from the intake manifold. Discard the gasket.

16. Disconnect the oil pressure electrical connector.

17. If equipped, properly discharge the A/C system.

18. Disconnect the A/C lines from the compressor.

19. If equipped with power steering, disconnect the lines from the pump.

20. Remove the starter.

21. Remove the alternator.

22. Raise and support the vehicle safely.

23. Disconnect the fuel line connections coming from the fuel rail.

24. Disconnect the exhaust pipe from the manifold.

25. Support the transmission with a jack.

26. Remove the bell housing bolts and inspection plate.

27. If equipped with an automatic transmission, attach a C-clamp to the bottom of the torque converter housing to prevent the torque converter from coming out. Matchmark the torque converter to the driveplate and remove the bolts. Disconnect the engine from the torque converter driveplate.

28. Install a suitable lifting device to the engine.

❊❊ WARNING

Do not lift the engine by the intake manifold.

29. Remove the front engine mount through-bolts.

30. Lower the vehicle.

31. Remove the engine from the vehicle and mount on a suitable workstand.

To install:

32. Remove the engine from the workstand and position it in the engine compartment.

33. Raise and support the vehicle.

34. On automatic transmissions, position the torque converter and driveplate. Tighten the bolts to 23 ft. lbs. (31 Nm).

35. Install the front engine mount through-bolts.

36. Install the bell housing bolts and tighten them to 30 ft. lbs. (41 Nm).

37. Remove the C-clamp and install the inspection plate. Remove the stand from the transmission.

38. Connect the exhaust pipe to the manifold.

39. Connect the fuel rail lines.

40. Lower the vehicle.

41. Install the starter.

42. Install the alternator.

43. If equipped, install the power steering hoses.

44. If equipped, connect the A/C hoses.

45. Connect the accelerator linkage.

46. Connect the starter wires.

47. Connect the oil pressure electrical connector.

48. Install the distributor cap and wires.

49. Connect the vacuum lines.

50. Install the radiator, radiator hoses and heater hoses.

51. Install the fan shroud into position.

52. Install the air cleaner.

53. Install the battery.

54. Fill the cooling system.

55. Start the engine and check for leaks.

56. If equipped, recharge the A/C system.

57. Install the hood.

58. Road test the vehicle.

5.2L AND 5.9L ENGINES

1. Matchmark the hood to the hinges and remove the hood.

2. Properly relieve the fuel system pressure.

3. Remove the battery.

❊❊ CAUTION

Never open, service or drain the radiator or cooling system when hot; serious burns can occur from the steam and hot coolant.

4. Drain the cooling system.

5. Remove the air cleaner and tube.

6. Remove the radiator.

7. Remove the heater hoses.

8. Label and disconnect the necessary vacuum lines.

9. Remove the distributor cap and wiring.

10. Disconnect the accelerator linkage.

11. Remove the air duct from the throttle body.

12. Label and disconnect the Manifold Absolute Pressure (MAP) sensor, Idle Air Control (IAC) motor and Throttle Position Sensor (TPS) electrical connectors from the throttle body.

13. Disconnect the vacuum line from the throttle body.

14. Disconnect (unsnap) the control cables from the throttle body (lever) arm.

15. Remove the throttle body from the intake manifold. Discard the gasket.

16. Disconnect the oil pressure electrical connector.

17. If equipped, properly discharge the A/C system.

18. Disconnect the A/C lines from the compressor.

19. If equipped with power steering, disconnect the lines from the pump.

20. Remove the starter.

21. Remove the alternator.

22. Raise and support the vehicle safely.

23. Disconnect the fuel line connections coming from the fuel rail.

24. Disconnect the exhaust system from the engine.

25. Support the transmission with a stand.

26. Remove the bell housing bolts and inspection plate.

27. Attach a C-clamp to the bottom of the torque converter housing to prevent the torque converter from coming out.

28. Matchmark the torque converter to the driveplate and remove the bolts.

29. Disconnect the engine from the torque converter driveplate.

30. Install a suitable lifting device to the engine.

❊❊ WARNING

Do not lift the engine by the intake manifold.

31. Remove the front engine mount through-bolts.

32. Lower the vehicle.

33. Remove the engine from the vehicle and mount on a suitable workstand.

To install:

34. Remove the engine from the workstand and position it in the engine compartment.

35. Raise and support the vehicle.

36. Position the torque converter and driveplate. Tighten the bolts to 271 inch lbs. (31 Nm).

37. Install the front engine mount through-bolts.

38. Install the bell housing bolts and tighten them to 30 ft. lbs. (41 Nm).

39. Remove the C-clamp and install the inspection plate. Remove the stand from the transmission.

40. Connect the exhaust system.

41. Connect the fuel rail lines.

42. Lower the vehicle.

43. Install the starter.

44. Install the alternator.

45. If equipped, install the power steering hoses.

46. If equipped, connect the A/C hoses.

47. Connect the accelerator linkage.

48. Connect the starter wires.

49. Connect the oil pressure electrical connector.

50. Install the distributor cap and wires.

51. Connect the vacuum lines.

52. Install the radiator, radiator hoses and heater hoses.

53. Install the fan shroud into position.

54. Install the air cleaner.

55. Install the battery.

56. Fill the cooling system.

❊❊ WARNING

Operating the engine without the proper amount and type of engine oil will result in severe engine damage.

57. Fill the engine to the full level with oil.

58. Start the engine and check for leaks.

59. If equipped, recharge the A/C system.

60. Install the hood.

61. Road test the vehicle.

Water Pump

REMOVAL & INSTALLATION

2.5L and 4.0L Engines

➡ **Some vehicles use a serpentine drive belt and have a reverse rotating water pump coupled with a viscous fan drive assembly. The components are identified by the words REVERSE stamped on the cover of the viscous drive and on the inner side of the fan. The word REV is also cast into the body of the water pump.**

1. Disconnect the negative battery cable.

2. Drain the cooling system.

3. Disconnect the hoses at the pump.

4. Remove the drive belts.

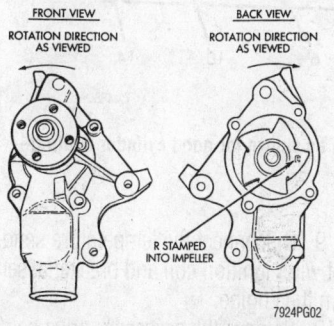

Water pumps that have an R stamped into the impeller are for use with serpentine drive belts only—2.5L and 4.0L engines

5. Remove the power steering pump bracket.

6. Remove the fan and shroud.

7. If equipped, remove the idler pulley to gain clearance for pump removal.

8. Unbolt and remove the pump.

To install:

9. Clean the mating surfaces thoroughly.

10. Using a new gasket, install the pump and tighten the bolts to 22 ft. lbs. (30 Nm).

11. If removed, install the idler pulley.

12. Reconnect the hoses at the pump and install accessory drive belt.

13. Install the power steering pump bracket. Install the fan and shroud.

14. Adjust the belt tension and fill the cooling system to the correct level.

15. Operate the engine with the heater control valve in the **HEAT** position until the thermostat opens to purge air from the system. Check coolant level and fill as required.

5.2L and 5.9L Engines

1. Disconnect the negative battery cable.

✳✳ CAUTION

Never open, service or drain the radiator or cooling system when hot; serious burns can occur from the steam and hot coolant.

2. Open the radiator valve and drain the cooling system.

3. Remove the cooling fan and shroud as an assembly.

4. Remove the accessory drive belt.

5. Remove the water pump pulley from the hub.

6. Disconnect the hoses from the water pump.

7. Loosen the heater hose coolant return tube mounting bolt and nut and remove the tube. Discard the O-ring.

8. Remove the water pump mounting bolts.

9. Loosen the clamp at the water pump end of the bypass hose. Slip the bypass hose from the water pump while removing the pump from the engine. Discard the gasket.

To install:

10. Clean all gasket mating surfaces.

11. Guide the water pump and new gasket into position while connecting the bypass hose to the pump. Tighten the water pump bolts to 30 ft. lbs. (40 Nm).

12. Install the bypass hose clamp.

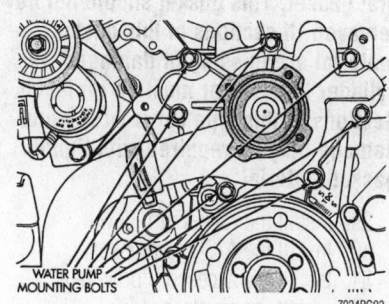

Water pump mounting bolt locations— 5.2L and 5.9L engines

13. Spin the water pump to ensure the pump impeller does not rub against the timing chain cover.

14. Coat a new O-ring with coolant and install it to the heater hose coolant return tube.

15. Install the coolant return tube to the engine. Ensure the slot in the tube bracket is bottomed to the mounting bolt. This will properly position the return tube.

16. Connect the radiator hose to the water pump.

17. Connect the heater hose and clamp to the return tube.

18. Install the water pump pulley and tighten the bolts to 20 ft. lbs. (27 Nm).

19. Install the accessory drive belt.

20. Install the cooling fan and shroud.

21. Fill the cooling system.

22. Connect the negative battery cable.

23. Start the engine and check for leaks.

Cylinder Head

REMOVAL & INSTALLATION

2.5L Engine

✳✳ CAUTION

Observe all applicable safety precautions when working around fuel. Whenever servicing the fuel system, always work in a well ventilated area. Do not allow fuel spray or vapors to come in contact with a spark or open flame. Keep a dry chemical fire extinguisher near the work area. Always keep fuel in a container specifically designed for fuel storage; also, always properly seal fuel containers to avoid the possibility of fire or explosion.

1. Properly relieve the fuel system pressure.

2. Disconnect the negative battery cable.

3. Drain the cooling system.

4. Disconnect the hoses at the thermostat housing.

5. Remove the air cleaner.

6. Remove the rocker arm cover.

7. Remove the rocker arms and the pushrods. Keep them in their original order for installation.

8. Remove the power steering pump bracket. Suspend the pump aside.

9. Remove the intake and exhaust manifolds.

10. If equipped with A/C, perform the following:

a. Remove the compressor and position it aside with the lines attached.

b. Remove the compressor bracket bolts from the cylinder head.

c. Loosen the through-bolt at the bottom of the bracket.

11. Disconnect the fuel lines and vacuum advance hose.

12. Remove the intake and exhaust manifolds.

13. Remove the spark plugs and wires.

14. Remove the ignition coil and bracket assembly.

15. Disconnect the temperature sending unit wire.

16. Remove the cylinder head bolts in the reverse order of the installation torque sequence.

17. Lift the head off the engine and remove the head gasket.

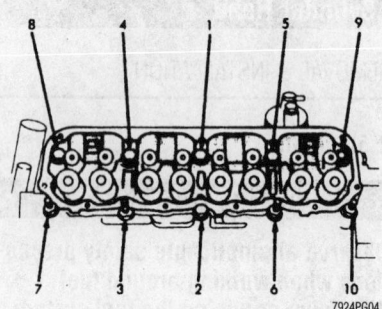

Be sure to tighten the cylinder head bolts in the correct sequence—2.5L engine

18. Thoroughly clean the gasket mating surfaces. Remove all traces of old gasket material. Remove all carbon deposits from the combustion chambers. Lay a straight-edge across the head and check for flatness. Total deviation should not exceed 0.008 in. (0.20mm).

To install:

19. Install a new head gasket.

➡**Do not apply sealer as the cylinder head gaskets are of a composition type.**

20. Fabricate two cylinder head alignment dowels from used head bolts. Using the longest bolts, trim the hex head off and cut slots into the top.

➡**Cylinder head bolts should be reused only once. Replace the head bolts which were previously used or are marked with paint. If head bolts are to be reused, mark each head bolt with paint for future reference.**

21. Coat the threads of bolt No. 7 with Loctite® PST sealant or equivalent.

22. Install the cylinder head and tighten the head bolts in the proper sequence to the following torque specifications:

a. Tighten all bolts to 22 ft. lbs. (30 Nm).

b. Tighten all bolts to 45 ft. lbs. (61 Nm).

c. Tighten bolts 1 through 6 to 110 ft. lbs. (150 Nm)

d. Tighten bolt 7 to 100 ft. lbs. (136 Nm).

e. Tighten bolts 8 through 10 to 110 ft. lbs. (150 Nm)

23. Connect the temperature sending unit wire.

24. Install the ignition coil, spark plugs and wires.

25. Install the intake and exhaust manifolds.

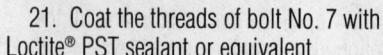
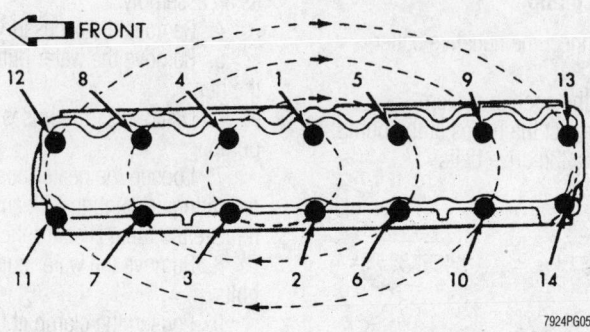

Tighten the cylinder head bolts in a clockwise fashion as shown for good cylinder sealing—4.0L engine

26. Install the fuel lines and vacuum advance hose.

27. If equipped, install the power steering pump and bracket.

28. Install the pushrods and rocker arms in their original positions.

➡**Some vehicles are equipped with a cylinder head cover containing an integral gasket. This gasket should not be removed. If sections of this gasket material are missing or damaged, the cylinder head cover must be replaced. Sections of this type cover with minor damage may be repaired with liquid gasket material.**

29. Install the rocker arm cover.

30. Install the A/C compressor.

31. Install the accessory drive belt.

32. Install the air cleaner.

33. Connect the hoses at the thermostat housing.

34. Fill the cooling system.

35. Connect the negative battery cable.

36. Run the engine to normal operating temperature and check for leaks.

4.0L Engines

1. Properly relieve the fuel system pressure.

2. Disconnect the negative battery cable.

3. Drain the cooling system.

4. Disconnect the hoses at the thermostat housing.

5. Remove the cylinder head cover.

6. Remove the pushrods, bridges, pivots and rocker arms.

➡**The valvetrain components must be replaced in their original positions.**

7. Remove the intake and exhaust manifolds from the cylinder head.

8. Disconnect the spark plug wires and remove the spark plugs.

9. Disconnect the temperature sending unit wire, ignition coil and bracket assembly from the engine.

10. Remove the accessory drive belt(s).

11. Unbolt and set aside the power steering pump and bracket. Do not disconnect the hoses.

12. Remove the intake and exhaust manifold assembly.

13. If equipped with A/C, perform the following:

a. Remove the compressor and position it aside with the lines attached.

b. Remove the compressor bracket bolts from the cylinder head.

c. Loosen the through-bolt at the bottom of the bracket.

14. Remove the alternator.

15. Remove the cylinder head bolts, the cylinder head and gasket from the block.

➡**Bolt No. 14 cannot be removed until the head is moved forward. Pull the bolt out as far as it will go and suspend in place by wrapping with tape.**

16. Discard the gasket. Thoroughly clean the head and block mating surfaces. Check them for warpage with a straight-edge. Deviation should not exceed 0.002 in. (0.05mm) in a 6 in. (152mm) span.

To install:

➡**Cylinder head bolts should be reused only once. Replace the head bolts which were previously used or are marked with paint. Head bolts should be installed using sealer.**

17. Coat a new head gasket with suitable sealing compound and place it on the block. Most replacement gaskets will have the word **TOP** stamped on them. Be sure the word **TOP** is facing up to ensure correct oil return hole alignment.

18. Install the cylinder head and bolts. The threads of bolt No. 11 must be coated with Loctite® 592 sealant before installation. Tighten the bolts in three steps, using the correct sequence:

a. Tighten all bolts to 22 ft. lbs. (30 Nm).

b. Tighten all bolts to 45 ft. lbs. (61 Nm).

c. Retorque all bolts to 45 ft. lbs. (61 Nm).

d. Tighten bolts 1 through 10 to 110 ft. lbs. (150 Nm).

e. Tighten bolt 11 to 100 ft. lbs. (136 Nm).

f. Tighten bolts 12 through 14 to 110 ft. lbs. (149 Nm)

19. Install the ignition coil.

20. Install the air conditioning compressor.

21. Install the alternator.

22. Install the intake and exhaust manifolds.

23. Install the power steering pump and bracket.

24. Install the accessory drive belt(s).

25. Connect the temperature sending unit wire, ignition coil and bracket.

26. Install the spark plugs and wires.

➡**Some vehicles are equipped with a cylinder head cover containing an integral gasket. This gasket should not be removed. If sections of this gasket material are missing or damaged, the cylinder head cover must be replaced. Sections of this type cover with minor damage may be repaired with liquid gasket material.**

27. Install the pushrods, rocker arm assembly, gasket and cylinder head cover.

28. Connect the hoses at the thermostat housing.

29. Fill the cooling system.

30. Connect the negative battery cable.

31. Run the engine to normal operating temperature and check for leaks.

5.2L and 5.9L Engines

1. Properly relieve the fuel system pressure.

2. Disconnect the negative battery cable.

✴✴ CAUTION

Observe all applicable safety precautions when working around fuel. Whenever servicing the fuel system, always work in a well ventilated area. Do not allow fuel spray or vapors to come in contact with a spark or open flame. Keep a dry chemical fire extinguisher near the work area. Always keep fuel in a container specifically designed for fuel storage; also, always properly seal fuel containers to avoid the possibility of fire or explosion.

3. Drain the cooling system.

4. Remove the alternator.

5. Disconnect the PCV valve.

6. Disconnect the EVAP fuel lines.

7. Remove the air cleaner and disconnect the fuel lines.

8. Disconnect the accelerator linkage, speed control cable, if equipped, and transmission kickdown cables.

9. Remove the return spring.

10. Remove the distributor cap and wires.

11. Disconnect the coil wires.

12. Disconnect the heat indicator sending unit wire.

13. Disconnect the heater and bypass hoses.

14. Remove the cylinder head covers, discard the gaskets.

15. Remove the intake manifold and throttle body.

16. Remove the exhaust manifolds.

17. Remove the rocker arm assemblies and pushrods.

➡**Identify the rocker arms and pushrods for installation purposes.**

18. Remove the spark plugs.

19. Remove the cylinder head bolts.

20. Remove the heads and discard the gaskets.

21. Thoroughly clean the gasket mating surfaces.

To install:

22. Apply Perfect Sealer No. 5 or equivalent, to the inner corners of the new head gaskets.

23. Position the head gaskets onto the block.

24. Position the cylinder heads onto the cylinder block.

25. Install the cylinder head bolts as follows:

a. Tighten all bolts, in sequence, to 50 ft. lbs. (68 Nm).

b. Tighten all bolts, in sequence, to 105 ft. lbs. (143 Nm).

c. Repeat Step b to ensure the torque is correct.

26. Install the pushrods and rocker arms to their original positions.

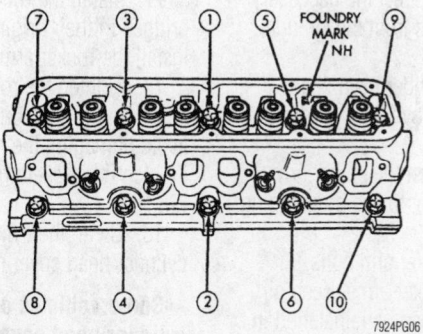

In order for the cylinder head gasket to seal properly, the bolts must be tightened in the sequence shown—5.2L and 5.9L engines

27. Install the intake and exhaust manifolds.

28. Install the spark plugs.

29. Install the coil wires.

30. Connect the heat indicating sending unit wire.

31. Connect the heater and bypass hoses.

32. Install the distributor cap and wires.

33. Hook up the return spring.

34. Connect the accelerator linkage, transmission kickdown cables, and if equipped, the cruise control cable.

35. Install the fuel lines.

36. Install the alternator.
37. Install the intake manifold-to-alternator bracket support rod.

➡**Some vehicles are equipped with a cylinder head cover containing an integral gasket. This gasket should not be removed. If sections of this gasket material are missing or damaged, the cylinder head cover must be replaced. Sections of this type cover with minor damage may be repaired with liquid gasket material.**

38. Place new cylinder head cover gaskets into position and install the cylinder head covers. Tighten the bolts to 96 inch lbs. (11 Nm).
39. Install the PCV valve.
40. Connect the EVAP lines.
41. Install the air cleaner.
42. Fill the cooling system.
43. Connect the negative battery cable. Start the engine and check for leaks.

Rocker Arms

REMOVAL & INSTALLATION

2.5L and 4.0L Engines

1. Disconnect the negative battery cable.
2. Label and disconnect the necessary hoses and vacuum lines from the cylinder head cover.
3. Remove the cylinder head cover retaining bolts and remove the cylinder head cover.
4. Alternately loosen the rocker arm bolts, one turn at a time, to avoid damaging the rocker arm bridge.
5. Remove the rocker arm bolts, bridges, pivots and rocker arms. Keep all parts in order so they can be reinstalled in their original locations.
6. Remove the pushrods. Keep them in order so they can be reinstalled in their original locations.

To install:

7. Clean all parts in solvent and allow to dry. If available, blow out the oil passages in the rocker arms and pushrods with compressed air.
8. Inspect all parts for wear or damage and replace as necessary.
9. Lubricate the pushrod tips and rocker arm bearing surfaces with clean engine oil.
10. Install the pushrods in their original locations.

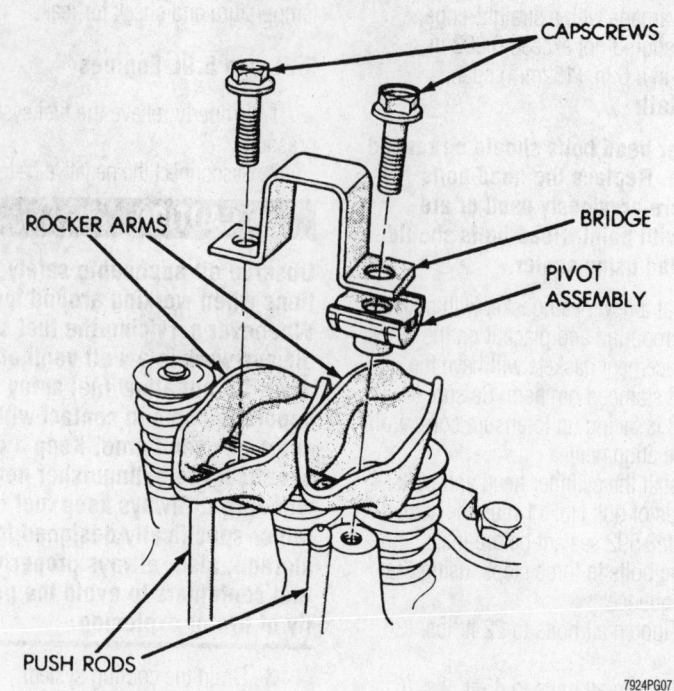

Exploded view of the rocker arm assembly—2.5L and 4.0L engines

11. Install the rocker arms, pivots and bridges in their original locations. Loosely install the rocker arm bolts.
12. Tighten the rocker arm bolts alternately and evenly, one turn at a time, to avoid damaging the rocker arm bridges. Tighten the rocker arm bolts to 21 ft. lbs. (28 Nm).
13. Clean the cylinder head cover and cylinder head cover mating surfaces.

➡**Some vehicles are equipped with a cylinder head cover containing an integral gasket. This gasket should not be removed. If sections of this gasket material are missing or damaged, the cylinder head cover must be replaced. Sections of this type cover with minor damage may be repaired with liquid gasket material.**

14. If applicable, install the cylinder head cover using a new gasket or sealant, as required.
15. Connect the hoses and vacuum lines.
16. Connect the negative battery cable.

5.2L and 5.9L Engines

1. Disconnect the negative battery cable.
2. Label and disconnect the necessary hoses from the cylinder head cover.
3. If removing the left cylinder head cover, remove the coolant tube bracket.
4. Remove the spark plug wires from the holders and disconnect them from the spark plugs.
5. Remove the cylinder head cover and gasket. The steel backed silicon gasket can be used again if not damaged.
6. Remove the rocker arm bolts and remove the rocker arm pivots and rocker arms. Keep the rocker arm assemblies in order so they can be reinstalled in their original locations.
7. Remove the pushrods, keeping them in order so they can be reinstalled in their original locations.

To install:

8. Rotate the crankshaft until the **V8** mark lines up with the TDC mark on the timing chain cover (located 147° After Top Dead Center (ATDC) from the No. 1 firing mark).

Do not rotate or crank the engine during or immediately after rocker arm installation. Allow about five minutes for the hydraulic lifters to bleed down.

9. Install the pushrods in their original locations. Be sure they are seated in the lifters.

10. Lubricate the pushrod tips, rocker arm bearing surfaces and rocker arm pivots with clean engine oil.

11. Install the rocker arm and pivot assemblies in their original locations. Tighten the bolts to 21 ft. lbs. (28 Nm).

12. Install the cylinder head cover and gasket. On the left cover, install the coolant tube bracket. Tighten the cylinder head cover retaining bolts to 96 inch lbs. (11 Nm).

13. Connect the spark plug wires to the spark plugs and install the wires in the holders.

14. Connect all hoses that were disconnected during the removal procedure.

15. Connect the negative battery cable.

Intake Manifold

REMOVAL & INSTALLATION

2.5L Engines

1. Properly relieve the fuel system pressure.
2. Disconnect the negative battery cable.

❊❊ CAUTION

Observe all applicable safety precautions when working around fuel. Whenever servicing the fuel system, always work in a well ventilated area. Do not allow fuel spray or vapors to come in contact with a spark or open flame. Keep a dry chemical fire extinguisher near the work area. Always keep fuel in a container specifically designed for fuel storage; also, always properly seal fuel containers to avoid the possibility of fire or explosion.

3. Drain the cooling system.
4. Remove the air cleaner.
5. Remove the power steering pump from its bracket. Position it aside.
6. Disconnect the fuel lines from the fuel rail.

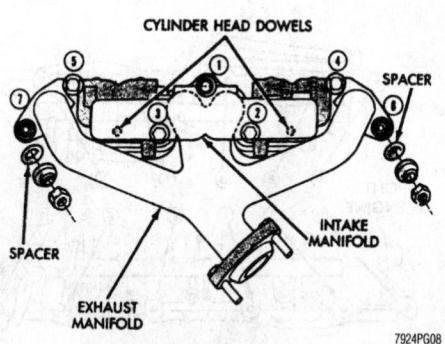

Intake and exhaust manifold bolt identification—2.5L engines

7. Disconnect the accelerator cable from the throttle body and hold-down bracket.

❊❊ WARNING

When disconnecting the cruise control connector at the throttle body, do not pry the connector off with tools. Use finger pressure only. Prying the connector off could break it.

8. Label and unplug the electrical and vacuum connections from the intake manifold.

9. Remove the molded vacuum harness.

10. Remove bolts 2 through 5 securing the intake manifold to the cylinder head. Slightly loosen bolt No. 1 and nuts 6 and 7.

11. Remove the intake manifold and gaskets.

To install:

12. Clean the gasket mating surfaces.

13. Install a new gasket over the locating dowels.

14. Position the manifold in place and finger-tighten the mounting bolts.

15. Tighten the fasteners in sequence as follows:

   a. Tighten fastener No. 1 to 30 ft. lbs. (41 Nm).

   b. Tighten fasteners No. 2 through No. 7 to 23 ft. lbs. (31 Nm).

16. Installation of the remaining components is the reverse of removal.

5.2L and 5.9L Engines

1. Properly relieve the fuel system pressure.
2. Disconnect the negative battery cable.
3. Drain the cooling system.
4. Remove the air cleaner.
5. Remove the alternator.
6. Remove the fuel lines and fuel rail.

7. Disconnect the accelerator linkage, if equipped, the cruise control and transmission kickdown cables.

8. Remove the return spring.

9. Remove the distributor cap and wires.

10. Disconnect the coil wires.

11. Disconnect the heat indicator sending unit wire.

12. Disconnect the heater and bypass hoses.

13. Disconnect the PCV and EVAP lines.

14. If equipped with A/C, remove the compressor and position it aside with the lines attached.

15. Remove the support bracket from the mounting bracket and intake manifold.

16. Remove the intake manifold and discard the gaskets.

17. Remove the throttle body and discard the gasket.

18. Turn the intake manifold upside down and support it. Remove the bolts and lift the plenum pan off the manifold. Discard the gasket.

19. Clean all gasket mating surfaces. Clean the intake manifold with solvent and blow dry with compressed air. The plenum pan rail must be clean, dry and free of all foreign material.

Tighten the plenum pan bolts in sequence to prevent possible leakage—5.2L and 5.9L engines

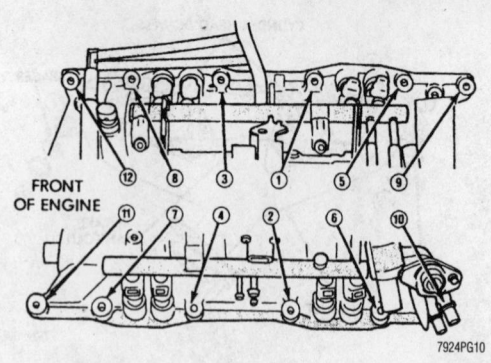

7924PG10

Be sure to tighten the intake manifold bolts in the sequence shown—5.2L and 5.9L engines

To install:

20. Place a new plenum pan gasket onto the seal rail of the intake manifold.

21. Position the pan over the gasket and align the holes. Hand-tighten the bolts.

22. Tighten the bolts as follows:

    a. Tighten all bolts, in sequence, to 24 inch lbs. (2.7 Nm).

    b. Tighten all bolts, in sequence, to 48 inch lbs. (5.4 Nm).

    c. Tighten all bolts, in sequence, to 84 inch lbs. (9.5 Nm).

    d. Repeat Step c to ensure proper torque.

23. Using a new gasket, install the throttle body onto the intake manifold. Tighten the bolts to 17 ft. lbs. (23 Nm).

24. Place the four plastic locator dowels into the holes in the block.

25. Apply Mopar rubber adhesive sealant or equivalent, to the four corner joints.

➡**An excessive amount of sealant is not required to ensure a leak proof seal, however, and excessive amount of sealant may reduce the effectiveness of the flange gasket. The sealant should be slightly higher than the crossover gaskets (approximately 0.2 inch/5mm).**

26. Install the front and rear crossover gaskets onto the dowels.

27. Install the flange gaskets. Ensure the vertical port alignment tab is resting on the deck face of the block. Also, the horizontal mating alignment tabs must be in position with the mating cylinder head gasket tabs. The words MANIFOLD SIDE should be visible on the center of each flange gasket.

28. Carefully lower the intake manifold into place. Use the alignment dowels in the crossover gaskets to position the manifold. Once in place ensure the gaskets are still in position.

29. Tighten the manifold bolts in sequence to the following specifications:

    a. Bolts 1–4—72 inch lbs. (8 Nm) in 12 inch lbs. (1.4 Nm) intervals

    b. Bolts 5–12—72 inch lbs. (8 Nm)

    c. Repeat Steps a and b to ensure proper torque.

    d. All bolts—12 ft. lbs. (16 Nm)

    e. Repeat Step d to ensure proper torque.

30. Connect the PCV and EVAP lines.

31. Install the coil wires.

32. Connect the heat indicator sending unit wire.

33. Connect the heater and bypass hoses.

34. Install the distributor cap and wires.

35. Hook up the return spring.

36. Connect the accelerator linkage, if equipped, cruise control and transmission kickdown cables.

37. Install the fuel lines and fuel rail.

38. Install the support bracket.

39. Install the alternator and drive belt.

40. If equipped with A/C, install the compressor.

41. Install the air cleaner.

42. Fill the cooling system.

43. Connect the negative battery cable.

44. Start the engine and check for leaks.

Exhaust Manifold

REMOVAL & INSTALLATION

2.5L Engine

1. Disconnect the negative battery cable.

2. Remove the intake manifold.

3. Disconnect the exhaust pipe at the manifold.

4. Remove the fasteners and exhaust manifold.

To install:

5. Clean the intake manifold and cylinder head mating surfaces.

6. Using a new intake manifold gasket, position the intake and exhaust manifolds on the cylinder head and place spacers over the end studs to center the exhaust manifold. Install the end stud nuts and washer clamps but do not tighten.

7. Install washer clamp and bolt at position 1 and tighten to 30 ft. lbs. (41 Nm).

8. Install bolts and washers at positions 2–5 and tighten to 23 ft. lbs. (31 Nm).

9. Tighten end stud nuts (positions 6 and 7) to 23 ft. lbs. (31 Nm).

10. Install all components removed from the intake manifold.

11. Connect the exhaust pipe and tighten the bolts to 23 ft. lbs. (31 Nm).

5.2L and 5.9L Engines

1. Disconnect the negative battery cable.

2. Remove the exhaust manifold heat shields.

3. Remove the spark plug wire loom and cables from the mounting stud at the rear of the valve cover and position the cables at the top of the valve cover.

4. Label and disconnect the two hoses from the EGR valve.

5. Disconnect the electrical connector and hoses from the EGR transducer.

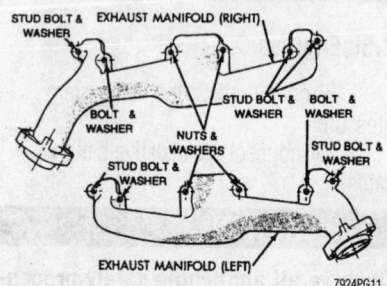

7924PG11

Be sure to install the fasteners in their original locations—5.2L and 5.9L engines

6. Remove the EGR valve and discard the gasket.

7. Disconnect the oil pressure sending unit electrical connector.

8. Using oil pressure sending unit remover C-4597 or equivalent, remove the sending unit.

9. Loosen the EGR mounting nut from the intake manifold.

10. Remove the mounting bolts and EGR tube. Discard the gasket.

11. Raise and safely support the vehicle.

12. Disconnect the exhaust pipes from the manifolds.

13. Lower the vehicle.

14. Remove the fasteners and exhaust manifold.

To install:

➡**If the manifold mounting studs came out with the fasteners, replace the studs.**

15. Position the manifold and install the conical washers on the studs.

16. Install new bolt and washer assemblies into the remaining holes. Working from the center outward, tighten the fasteners to 20 ft. lbs. (27 Nm).

17. Raise and support the vehicle safely.

18. Connect the exhaust pipes to the manifolds and tighten the fasteners to 23 ft. lbs. (31 Nm).

19. Lower the vehicle.

20. Clean the EGR and tube gasket mating surfaces.

21. Install a new gasket onto the exhaust manifold ends of the EGR tube and install the tube. Tighten the tube nut to the intake manifold and tighten the tube-to-exhaust manifold bolts to 17 ft. lbs. (23 Nm).

22. Coat the threads of the oil pressure sending unit with sealer taking care not to apply sealant to the opening. Install the sending unit and tighten it to 11 ft. lbs. (14 Nm) and connect the electrical connector.

23. Install the EGR valve and new gasket to the intake manifold and tighten the bolts to 17 ft. lbs. (23 Nm).

24. Position the EGR transducer and connect the vacuum lines and electrical connector.

25. Position the spark plug cables and loom into place and connect the cables.

26. Install the exhaust heat shields and tighten the bolts to 20 ft. lbs. (27 Nm).

27. Connect the negative battery cable.

Combination Manifold

REMOVAL & INSTALLATION

4.0L Engine

➡**The intake and exhaust manifold are mounted externally on the left side of the engine and are attached to the cylinder head. They are removed as a unit.**

1. Disconnect the negative battery cable.

2. Remove the air cleaner assembly.

3. Disconnect the accelerator cable, cruise control cable, if equipped and transmission line pressure cable.

4. Disconnect all electrical connectors on the intake manifold.

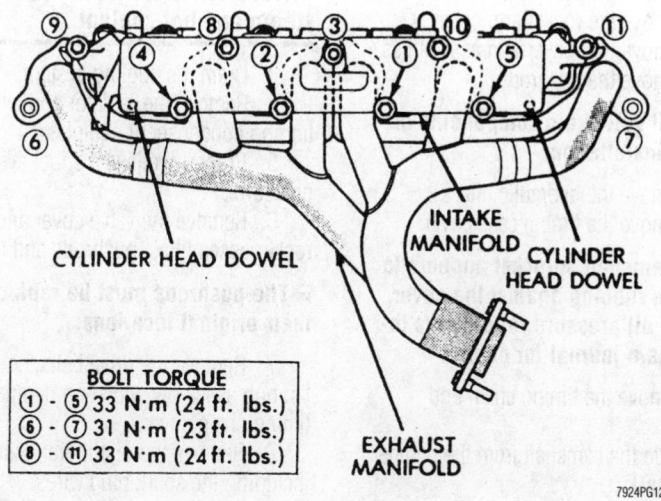

CYLINDER HEAD DOWEL

INTAKE MANIFOLD CYLINDER HEAD DOWEL

EXHAUST MANIFOLD

BOLT TORQUE	
① - ⑤	33 N·m (24 ft. lbs.)
⑥ - ⑦	31 N·m (23 ft. lbs.)
⑧ - ⑪	33 N·m (24 ft. lbs.)

7924PG12

Tighten the combination manifold bolts in the correct sequence for proper sealing—4.0L engines

5. Disconnect and remove the fuel supply and return lines from the fuel rail assembly.

6. Remove the fuel rail and injectors.

7. Loosen the accessory drive belts.

8. Remove the power steering pump.

9. Disconnect the exhaust pipe from the manifold and discard the seal.

10. Remove the combination manifold.

To install:

11. Clean the gasket mating surfaces thoroughly. Install a new gasket over the alignment dowels and position the exhaust manifold to the cylinder head. Install bolt No. 3 finger-tight.

12. Install the intake manifold and the remaining bolts and washers.

13. Tighten bolts, in sequence, to the following torque specifications:

   a. Bolts 1–5—24 ft. lbs. (33 Nm)

   b. Bolts 6 and 7—23 ft. lbs. (31 Nm)

   c. Bolts 8–11—24 ft. lbs. (33 Nm)

14. Install the fuel rail and injectors.

15. Install the power steering pump and belt.

16. Using new O-rings, install the fuel supply and return lines.

17. Connect all electrical connectors, vacuum connectors, throttle cable, cruise

control cable and transmission lines pressure cable.

18. Install the air cleaner assembly.

19. Using a new seal, connect the exhaust pipe to the manifold and tighten the bolts to 23 ft. lbs. (31 Nm).

20. Connect the negative battery cable.

21. Start the engine and check for leaks.

Camshaft and Valve Lifters

REMOVAL & INSTALLATION

2.5L Engine

1. Disconnect the negative battery cable.

2. If equipped with air conditioning, have the system discharged by an EPA certified technician.

✳✳ CAUTION

Never open, service or drain the radiator or cooling system when hot; serious burns can occur from the steam and hot coolant.

3. Drain the cooling system.

4. Remove the radiator and air conditioning condenser, if equipped.

5. Matchmark the distributor housing and engine for installation.

6. Matchmark the rotor position by marking it on the distributor body.

7. Remove the distributor and wires.

8. Remove the rocker arm cover.

9. Remove the rocker arm assemblies.

10. Remove the pushrods.

➡**Keep all valvetrain components in order for installation.**

11. Remove the hydraulic lifters.

12. Remove the timing case cover.

➡**If the camshaft sprocket appears to have been rubbing against the cover, check the oil pressure relief holes in the rear cam journal for debris.**

13. Remove the timing chain and sprockets.

14. Slide the camshaft from the engine.

To install:

15. Inspect the camshaft for wear and damage.

16. Lubricate the camshaft with an engine oil supplement.

17. Install the camshaft, taking care not to damage the cam bearings.

18. Install the timing chain and sprockets. Ensure that the timing marks are correctly positioned.

19. Install the timing case cover.

20. Lubricate and install the hydraulic lifters, pushrods and rocker arm assemblies.

21. Install the cylinder head cover, distributor and ignition wires.

22. Install the distributor. The distributor rotor should align with the position of the No. 1 spark plug terminal on the distributor cap when the distributor shaft is down in place.

➡**It may be necessary to rotate the oil pump shaft with a long flat-blade screwdriver to engage the oil pump drive tang.**

23. Install all other components in reverse order of removal.

24. Start the engine and allow it to reach operating temperature. Check the ignition timing.

4.0L Engine

1. Disconnect the negative battery cable.

2. If equipped with air conditioning, have the system discharged by an EPA certified technician.

✳✳ CAUTION

Never open, service or drain the radiator or cooling system when hot; serious burns can occur from the steam and hot coolant.

3. Drain the cooling system.

4. Remove the radiator and air conditioning condenser, if equipped.

5. Properly relieve the fuel system pressure.

6. Remove the valve cover and gasket, rocker assemblies, pushrods and lifters.

➡**The pushrods must be replaced in their original locations.**

7. Remove the drive belts, cooling fan, fan hub assembly, vibration damper and timing chain cover.

8. Remove the distributor assembly, including the spark plug wires.

9. Remove the cylinder head.

10. Remove the valve lifters. Keep them in order for installation.

11. Rotate the crankshaft until the timing mark of the crankshaft sprocket is adjacent to and on a center line with the timing mark of the camshaft sprocket.

12. Remove the crankshaft sprocket, camshaft sprocket and timing chain as an assembly.

13. Remove the front bumper or grille as required and carefully slide out the camshaft.

To install:

14. Lubricate the camshaft with an engine oil supplement.

15. Slide the camshaft into the block carefully to avoid damage to the bearings.

16. Install the crankshaft sprocket, camshaft sprocket and the timing chain as an assembly.

17. Be sure the timing mark of the crankshaft sprocket is adjacent to and on a centerline with, the timing mark of the camshaft sprocket. Tighten the camshaft sprocket bolt to 80 ft. lbs. (109 Nm).

18. Lubricate the tension spring, thrust pin and pin bore in the preload bolt. Install the assembly on the preload bolt. Install the timing cover. Install the vibration damper.

19. Install the valve lifters, cylinder head, pushrods, rocker assemblies, valve cover and gasket.

20. Install the distributor assembly. The rotor should be aligned with the No. 1 cylinder spark plug terminal on the cap

---

when the distributor is fully seated on the cylinder block.

21. Install the condenser and receiver/drier.

22. Install the radiator and fill the cooling system.

23. Connect the negative battery cable.

24. Start the engine and allow it to reach operating temperature.

25. Check the ignition timing.

5.2L and 5.9L Engines

1. Disconnect the negative battery cable.

2. Remove the engine from the vehicle and mount on a suitable workstand.

3. Remove the intake manifold.

4. Remove the valve covers.

5. Remove the timing case cover and timing chain.

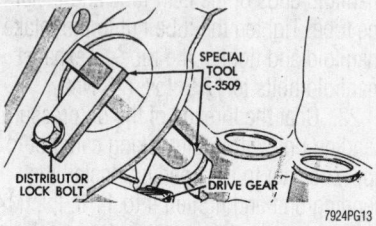

Install the camshaft holding tool to prevent the camshaft from being pushed in too far and knocking out the Welsh plug on the back of the engine block—5.2L and 5.9L engines

6. Remove the rocker arms.

7. Remove the pushrods and valve lifters.

8. Remove the distributor and lift out the oil pump and distributor driveshaft.

9. Remove the camshaft thrust plate and note the location of the oil tab.

10. Install a long bolt into the camshaft to facilitate the removal of the camshaft.

11. Remove the camshaft, being careful not to damage the cam bearings with the cam lobes.

To install:

➡**To prevent premature camshaft damage, always install new lifters with a new camshaft.**

12. Lubricate the camshaft lobes and bearings with engine oil and insert the camshaft to within 2 inches (51mm) of its final position in the cylinder block.

13. Install camshaft gear installer tool C-3509 or equivalent, with the tool secured in the distributor drive gear position. Hold the tool in position with a distributor lockplate bolt.

➡ **The tool will restrict the camshaft from being pushed in too far and prevent knocking out the plug in the rear of the cylinder block. The tool should remain installed until the camshaft and crankshaft sprockets with the chain are installed.**

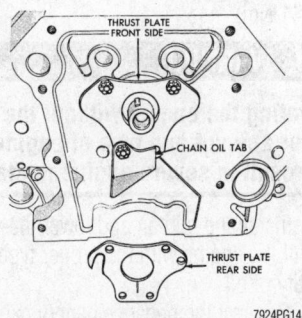

7924PG14

Be sure the tang on the oil tab enters the lower right hole on the thrust plate—5.2L and 5.9L engines

14. Install the camshaft thrust plate and chain oil tab. Ensure the tang enters the lower right hole in the thrust plate and tighten the bolts to 18 ft. lbs. (24 Nm). The top edge of the tab should be flat against the thrust plate in order to catch oil for chain lubrication.
15. Install the timing chain and sprockets.
16. Remove installer tool C-3509.
17. Ensure the camshaft end-play is 0.002–0.010 inch (0.051–0.76mm); if not, install a new thrust plate.
18. Install the driveshaft and distributor.
19. Lubricate (with new engine oil) and install the lifters and pushrods.
20. Install the rocker arms.
21. Install the valve covers with new gaskets.
22. Install the intake manifold.
23. Install the engine.
24. Connect the negative battery cable.

Valve Lash

ADJUSTMENT

These engines are equipped with hydraulic valve lifters. No valve clearance adjustments are possible or necessary.

Oil Pan

REMOVAL & INSTALLATION

2.5L and 4.0L Engines

1. Disconnect the negative battery cable.
2. Raise and support the vehicle safely.

❋❋ **CAUTION**

The EPA warns that prolonged contact with used engine oil may cause a number of skin disorders, including cancer! You should make every effort to minimize your exposure to used engine oil. Protective gloves should be worn when changing the oil. Wash your hands and any other exposed skin areas as soon as possible after exposure to used engine oil. Soap and water, or waterless hand cleaner should be used.

3. Drain the engine oil.
4. Disconnect the exhaust pipe at the manifold.
5. Disconnect the exhaust hanger at the catalytic converter and lower the pipe.
6. Remove the starter.
7. Remove the bell housing access cover.
8. If equipped, disconnect the oil level sensor.
9. Position a jackstand directly under the vibration damper. Place a piece of wood between the jack and the vibration damper.
10. Remove the engine mount through-bolts. Using the jack, raise the engine until there is enough room to remove the oil pan.
11. Remove the oil pan bolts and remove the oil pan.
12. Clean all sealant and old gasket material from the oil pan and cylinder block mating surfaces. Thoroughly clean the oil pan.

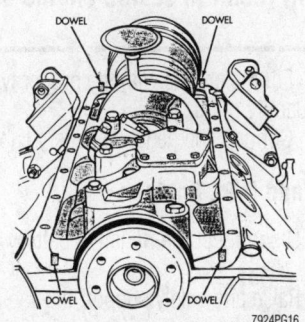

7924PG16

Temporarily install the alignment dowels in the four bolt holes—2.5L and 4.0L engines

To install:

13. Fabricate four alignment dowels from 1 ½ in. x ¼ in. bolts. Cut the heads off the bolts and cut a slot into the top of the dowel to allow installation/removal with a screwdriver.
14. Install two dowels in the timing case cover and the other two in the cylinder block. Slide the one-piece gasket over the dowels and onto the block and timing case cover.
15. Position the oil pan over the dowels and onto the gasket. Install the ¼ in. pan bolts and tighten to 120 inch lbs. (14 Nm), except on 1997–98 Wrangler. On 1997–98 Wrangler, tighten the bolts to 84 inch lbs. (10 Nm). Install the ⁵⁄₁₆ in. pan bolts and tighten to 156 inch lbs. (18 Nm), except on 1997–98 Wrangler. On 1997–98 Wrangler, tighten the bolts to 132 inch lbs. (15 Nm).
16. Remove the dowels and install the remaining ¼ in. pan bolts.
17. Lower the engine and install the engine mount through-bolts and nuts. Lower the jack and remove the piece of wood.
18. Connect the oil level sensor if equipped.
19. Install the bell housing access cover.
20. Install the starter.
21. Connect the exhaust pipe to the hanger and the exhaust manifold.
22. Install the oil pan drain plug and tighten to 25 ft. lbs. (34 Nm).
23. Lower the vehicle.
24. Fill the crankcase to the proper level with fresh engine oil.
25. Connect the negative battery cable. Start the engine and check for leaks.

5.2L and 5.9L Engines

1. Disconnect the negative battery cable.
2. Raise and safely support the vehicle.

❋❋ **CAUTION**

The EPA warns that prolonged contact with used engine oil may cause a number of skin disorders, including cancer! You should make every effort to minimize your exposure to used engine oil. Protective gloves should be worn when changing the oil. Wash your hands and any other exposed skin areas as soon as possible after exposure to used engine oil. Soap and water, or waterless hand cleaner should be used.

3. Remove the oil pan drain plug and drain the engine oil.

4. Remove the oil filter.

5. Remove the starter.

6. If equipped, disconnect the oil level sensor.

7. Position the oil cooler lines out of the way.

8. Disconnect the oxygen sensor and remove the exhaust pipe.

9. Remove the oil pan bolts and carefully slide the oil pan to the rear. If equipped, be careful not to damage the oil level sensor.

10. Clean all sealant and old gasket material from the oil pan and cylinder block mating surfaces. Thoroughly clean the oil pan.

To install:

11. Fabricate four alignment dowels from 1 ½ x 5⁄16 in. bolts. Cut the heads off the bolts and cut a slot in the dowel to allow installation/removal with a screwdriver.

12. Install the dowels in the cylinder block. Apply a small amount of silicone sealant in the corner of the cap and cylinder block.

13. Slide the one-piece gasket over the dowels and onto the block. Position the oil pan over the dowels and onto the gasket. If equipped, be careful not to damage the oil level sensor.

14. Install the oil pan bolts and tighten to 215 inch lbs. (24 Nm). Remove the dowels and install the remaining bolts. Tighten to 215 inch lbs. (24 Nm).

15. Install the drain plug and tighten to 25 ft. lbs. (34 Nm).

16. Install the exhaust pipe and connect the oxygen sensor.

17. Install the oil filter. If equipped, connect the oil level sensor.

18. Install the starter. Move the oil cooler lines back into position.

19. Lower the vehicle and connect the negative battery cable.

✳✳ WARNING

Operating the engine without the proper amount and type of engine oil will result in severe engine damage.

20. Fill the engine with the proper type and quantity of oil. Start the engine and check for leaks.

Oil Pump

REMOVAL & INSTALLATION

2.5L and 4.0L Engines

1. Disconnect the negative battery cable. Raise and safely support the vehicle.

2. Drain the engine oil and remove the oil pan.

3. Unbolt and remove the pump assembly from the block. Discard the gasket.

✳✳ WARNING

If the oil pump is not to be serviced, do not disturb the position of the oil inlet tube and strainer assembly in the pump body. If the tube is moved within the pump body, a replacement tube and strainer assembly must be installed to assure an airtight seal.

To install:

4. If a new pump is being installed, prime the pump by submerging the strainer in clean engine oil and turning the pump gears until oil emerges from the pump feed hole.

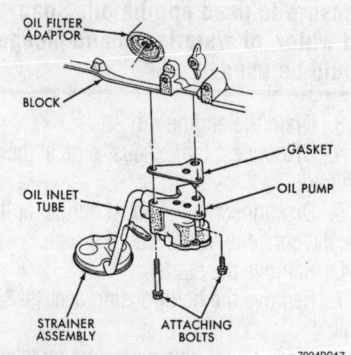

OIL FILTER ADAPTOR

BLOCK

GASKET

OIL INLET TUBE

OIL PUMP

STRAINER ASSEMBLY

ATTACHING BOLTS

7924PG17

Exploded view of the oil pump assembly— 2.5L and 4.0L engine

5. Using a new gasket, install the pump on the cylinder block. Tighten the mounting bolts to 17 ft. lbs. (23 Nm).

6. Install the oil pan and lower the vehicle.

✳✳ WARNING

Operating the engine without the proper amount and type of engine oil will result in severe engine damage.

7. Fill the engine with the proper type and quantity of oil.

8. Connect the negative battery cable.

5.2L and 5.9L Engines

1. Disconnect the negative battery cable.

2. Raise and safely support the vehicle.

3. Drain the engine oil and remove the oil pan.

4. Unbolt and remove the pump assembly from the rear main bearing cap.

5. If a new pump is being installed, prime the pump by submerging the pick-up in clean engine oil and turning the pump gears until oil emerges from the pump feed hole.

6. Install the oil pump. During installation, slowly rotate the pump body to ensure driveshaft-to-pump rotor shaft engagement.

7. Hold the oil pump base flush against the mating surface of the rear main bearing cap and finger-tighten the pump mounting bolts. Tighten the mounting bolts to 30 ft. lbs. (41 Nm).

✳✳ WARNING

Operating the engine without the proper amount and type of engine oil will result in severe engine damage.

8. Install the oil pan and lower the vehicle. Fill the engine with the proper type and quantity of oil.

9. Connect the negative battery cable.

Rear Main Seal

REMOVAL & INSTALLATION

2.5L Engine

1. Raise and safely support the vehicle.

2. Remove the transmission.

3. Remove the flywheel or torque converter drive plate. Discard the mounting bolts.

4. Pry out the seal from around the crankshaft flange. Take care not to damage the seal housing.

To install:

5. Coat the lip of the new seal with clean engine oil.

6. Carefully position the seal into place. Using an oil seal driver tool, install the seal flush with the cylinder block.

➡ The felt lip must be located inside the flywheel mounting surface. If the lip is not positioned correctly the flywheel could tear the seal.

7. Using new bolts, install the flywheel or converter drive plate and tighten the bolts to 50 ft. lbs. (68 Nm), plus an additional 60 degree turn.

8. Install the transmission.

9. Lower the vehicle.

10. Check the engine oil level; add oil as necessary.

11. Run the engine and check the seal area for oil leaks.

4.0L Engines

1. Raise and safely support the vehicle.
2. Remove the transmission.
3. Remove the flywheel or converter drive plate.
4. Remove the oil pan.
5. Remove the rear main bearing cap (No. 7).
6. Push the upper seal out of the groove. Be sure the crankshaft and seal groove are not damaged.
7. Remove the lower seal from the bearing cap.

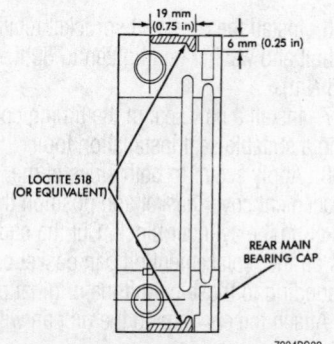

Apply sealant to the rear main bearing cap as shown—4.0L engine

To install:

8. Wipe the seal surface area of the crankshaft until it is clean, then apply a thin coat of clean engine oil.
9. Coat the lip of a new seal with clean engine oil.
10. Carefully position the upper seal into the groove in the cylinder block. The lip of the seal must face the front of the engine.
11. Place the lower half of the seal into the bearing cap (No. 7).
12. Coat the outer curved surface of the lower seal with soap and the lip of the seal with clean engine oil.
13. Position the lower seal into the bearing cap recess and seat it firmly. Be sure it is flush with the cylinder block pan rail.
14. Apply Loctite® 518 or equivalent, on the rear bearing cap. The bead should be 0.125 in. (3mm) thick. Do not apply Loctite® 518 or equivalent, to the seal lip.
15. Install the rear main bearing cap. Do not strike the cap more than twice for proper engagement.
16. Tighten the cap bolts to 80 ft. lbs. (108 Nm).

17. Install the oil pan.
18. Install the flywheel or converter drive plate and tighten the bolts to 105 ft. lbs. (143 Nm).
19. Lower the vehicle.
20. Check the engine oil level; add oil as necessary.
21. Run the engine and check for oil leaks at the seal.

5.2L and 5.9L Engines

The rear main bearing seal is a two piece, viton seal. The seals should be replaced as a pair.

1. Raise and safely support the vehicle.
2. Remove the oil pan.
3. Remove the oil pump.
4. Remove the rear main bearing cap. Remove and discard the old lower oil seal.
5. Carefully remove and discard the old upper oil seal.

To install:

6. Before installing the new oil seal, clean the cylinder block mating surfaces. Inspect for burrs at the oil hole on the cylinder block mating surface to rear cap.

7. To ease installation, loosen at least the two main bearing caps forward of the rear cap.
8. Lightly lubricate the new seals with engine oil.
9. Rotate the seal into the cylinder block, being careful not to shave or cut the outer surface of the seal.
10. Install the new lower seal into the bearing cap.

➡**If using factory seals, install them with the yellow (or white) paint facing towards the rear of the engine.**

11. Apply 0.20 inch (5mm) of Loctite® 518 or equivalent, on each side of the rear main bearing cap. Assemble the bearing cap immediately after applying the sealant. Align the bearing cap by using the cap slot, alignment dowels and cap bolts.

✳✳ WARNING

Do not over apply the sealant or allow it to contact the seal. Do not remove excess material after assembly or strike the rear cap more than two times for proper alignment.

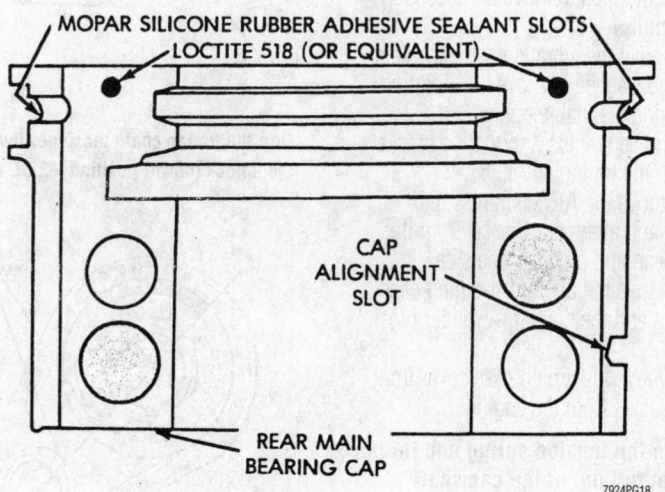

Apply sealant to the slot shown on the rear main bearing cap—5.2L and 5.9L engines

12. Install the rear main bearing cap with clean and oiled cap bolts. Alternately tighten all cap bolts to 85 ft. lbs. (115 Nm).

13. Install the oil pump.

14. Apply silicone sealant at the bearing cap-to-block joint to provide cap-to-block and oil pan sealing. Apply enough sealant until a small amount is squeezed out. Withdraw the nozzle and wipe the excess sealant off the oil pan seal groove.

15. Immediately install the oil pan.

16. Lower the vehicle.

❄❄ WARNING

Operating the engine without the proper amount and type of engine oil will result in severe engine damage.

17. Fill the engine to the proper level with engine oil. Start the engine and check for leaks.

Timing Chain, Sprockets, Tensioner and Front Cover

REMOVAL & INSTALLATION

2.5L and 4.0L Engines

1. Disconnect the negative battery cable.

2. Remove the drive belt(s), fan and fan shroud. If equipped, remove the accessory drive belt pulley.

3. Remove the vibration damper retaining bolt and washer. Remove the vibration damper using a suitable puller.

4. Remove the accessory drive brackets attached to the timing cover.

5. Remove the A/C compressor, if equipped, and alternator bracket from the cylinder head and move to one side.

6. Remove the oil pan-to-timing case cover bolts and the cover-to-cylinder block bolts.

7. Remove the timing case cover front seal and gasket from the engine.

➡ **Be sure the tension spring and thrust pin do not fall out of the camshaft sprocket retaining preload bolt.**

8. Cut off the oil pan side gasket end tabs and oil pan front seal tabs flush with the front face of the cylinder block. Remove the gasket tabs.

9. Clean the timing case cover, oil pan and cylinder block gasket surfaces.

10. Remove the crankshaft seal oil seal from the cover by prying it out with a suitable tool.

11. Rotate the crankshaft until the **0** timing mark on the crankshaft sprocket is closest to and on a center line with the timing mark on the camshaft sprocket.

12. Remove the oil slinger from the crankshaft.

13. Remove the camshaft retaining bolt and remove the sprockets and chain as an assembly. If the timing chain tensioner is to be replaced, the oil pan must be removed.

To install:

14. Turn the timing chain tensioner lever to the unlock (down) position. Pull the tensioner block toward the tensioner lever to compress the spring. Hold the block and turn the tensioner lever to the lock (up) position.

15. Install the sprockets and timing chain. Ensure the timing marks on the sprockets are properly aligned.

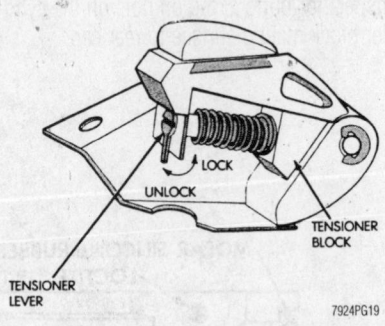

Turn the timing chain tensioner lever to the unlock (down) position—2.5L engines

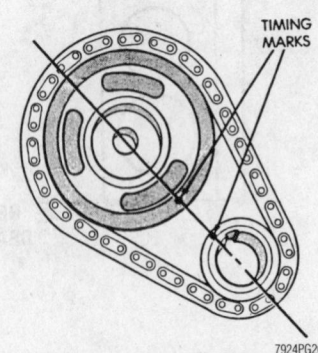

Ensure that the timing marks are facing each other—2.5L and 4.0L engines

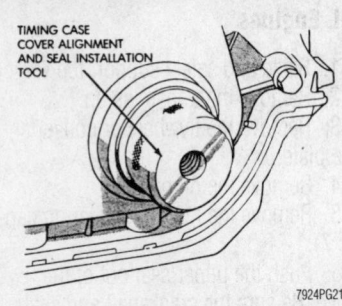

Install the cover alignment/seal installation tool on the crankshaft to ensure that the timing chain cover is centered—2.5L and 4.0L engines

16. Install the camshaft sprocket retaining bolt and washer and tighten to 80 ft. lbs. (108 Nm).

17. Install a new seal in the timing cover using a suitable seal installation tool.

18. Apply sealer to both sides of the replacement cover gasket and position the gasket on the cylinder block. Cut the end tabs off the replacement oil pan gasket corresponding to those cut off the original gasket. Attach the end tabs to the oil pan with sealer.

19. Coat the front cover seal end tab recesses generously with sealer and position the seal on the timing cover.

20. Apply engine oil to the seal/oil pan contact surface, then position the cover on the cylinder block.

➡ **Be sure the tension spring and thrust pin are in place in the preload bolt, before installing the cover.**

21. Insert timing case cover alignment tool 6139 or equivalent, in the crankshaft opening. Install the cover bolts and tighten the bolts as follows:

   a. ¼ in. cover-to-block bolts to 60 inch lbs. (7 Nm).

   b. 5/16 in. cover-to-block bolts to 192 inch lbs. (22 Nm).

   c. ¼ in. oil pan-to-cover bolts to 84 inch lbs. (10 Nm) on 4.0L engines or 120 inch lbs. (14 Nm) on 2.5L engines.

   d. 5/16 in. oil pan-to-cover bolts to 132 inch lbs. (15 Nm) on 4.0L engines or 84 inch lbs. (10 Nm) on 2.5L engines.

22. Remove the cover alignment tool.

23. Apply a light film of oil to the vibration damper hub seal contact surface. Install the vibration damper using a suitable installation tool.

24. Install and tighten the crankshaft vibration damper bolt to 80 ft. lbs. (108 Nm).

25. If equipped, install the crankshaft pulley and tighten the bolts to 20 ft. lbs. (27 Nm).

26. Install the accessory brackets.

27. Install the fan and fan shroud.

28. Install the drive belt(s) and adjust to the proper tension.

29. Connect the negative battery cable.

30. Start the engine and check for leaks.

5.2L and 5.9L Engines

1. Disconnect the negative battery cable.

✳✳ CAUTION

Observe all applicable safety precautions when working around fuel. Whenever servicing the fuel system, always work in a well ventilated area. Do not allow fuel spray or vapors to come in contact with a spark or open flame. Keep a dry chemical fire extinguisher near the work area. Always keep fuel in a container specifically designed for fuel storage; also, always properly seal fuel containers to avoid the possibility of fire or explosion.

2. Properly relieve the fuel system pressure.

3. Drain the cooling system.

4. Remove the serpentine belt.

5. Remove the cooling fan shroud and position it on the engine.

6. Remove the water pump.

7. Remove the power steering pump.

8. Remove the vibration damper using puller C-3688 or equivalent.

9. Disconnect the fuel lines.

10. Loosen the oil pan bolts and remove the front bolt at each side.

11. Remove the timing chain cover bolts. Remove the chain cover and gasket using extreme caution to avoid damaging the oil pan gasket.

12. Place a scale next to the timing chain so any movement of the chain can be measured.

13. Place a torque wrench and socket over the camshaft sprocket attaching bolt. Apply 30 ft. lbs. (41 Nm) with the cylinder heads installed or 15 ft. lbs. (20 Nm) with the cylinder heads removed.

➡**With the torque applied the crankshaft sprocket should not be permitted to move, but it may be necessary to block the crankshaft to prevent rotation.**

14. Hold the scale with the dimension reading even with the edge of a chain link. Apply 30 ft. lbs. (41 Nm) with the cylinder heads installed or 15 ft. lbs. (20 Nm) with the cylinder heads removed, in the reverse direction. Note the amount of chain movement.

15. Install a new timing chain if the movement exceeds ⅛ inch (3.175mm).

16. Remove the camshaft sprocket retaining bolt.

17. Remove the timing chain and sprockets.

To install:

18. Position the camshaft and crankshaft sprockets on a bench with the timing marks facing each other.

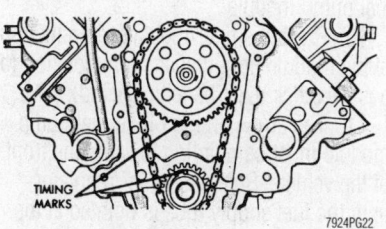

7924PG22

When the chain is installed, the timing marks should be facing each other—5.2L and 5.9L engines

19. Position the timing chain onto the sprockets.

20. Turn the crankshaft and camshaft to align with the keyway location in the crankshaft and camshaft sprockets.

21. Keeping tension on the chain, slide the sprocket and chain assembly onto the engine.

22. Ensure the timing marks are still aligned by using a straight edge.

23. Install the camshaft bolt and tighten it to 50 ft. lbs. (68 Nm).

24. Install a new timing chain cover gasket to the chain cover. Apply a small amount of Mopar® silicone rubber adhesive sealant or equivalent, at the joint where the chain cover and oil pan gasket meet.

25. Install the timing chain cover taking care not to damage to oil pan. Tighten the

timing chain cover bolts to 30 ft. lbs. (41 Nm) and oil pan bolts to 215 inch lbs. (24 Nm).

26. Install the vibration damper. Tighten the crankshaft bolt to 135 ft. lbs. (183 Nm) and pulley bolt to 17 ft. lbs. (23 Nm).

27. Connect the fuel lines.

28. Install the water pump.

29. Install the power steering pump.

30. Install the serpentine belt.

31. Install the cooling fan shroud.

32. Fill the cooling system.

33. Connect the negative battery cable.

FUEL SYSTEM

Fuel System Service Precautions

Safety is the most important factor when performing not only fuel system maintenance but any type of maintenance. Failure to conduct maintenance and repairs in a safe manner may result in serious personal injury or death. Maintenance and testing of the vehicle's fuel system components can be accomplished safely and effectively by adhering to the following rules and guidelines.

• To avoid the possibility of fire and personal injury, always disconnect the negative battery cable unless the repair or test procedure requires that battery voltage be applied.

• Always relieve the fuel system pressure prior to disconnecting any fuel system component (injector, fuel rail, pressure regulator, etc.), fitting or fuel line connection. Exercise extreme caution whenever relieving fuel system pressure to avoid exposing skin, face and eyes to fuel spray. Please be advised that fuel under pressure may penetrate the skin or any part of the body that it contacts.

• Always place a shop towel or cloth around the fitting or connection prior to loosening to absorb any excess fuel due to spillage. Ensure that all fuel spillage (should it occur) is quickly removed from engine surfaces. Ensure that all fuel soaked cloths or towels are deposited into a suitable waste container.

• Always keep a dry chemical (Class B) fire extinguisher near the work area.

• Do not allow fuel spray or fuel vapors to come into contact with a spark or open flame.

• Always use a back-up wrench when loosening and tightening fuel line connection fittings. This will prevent unnecessary stress and torsion to fuel line piping. Always follow the proper torque specifications.

• Always replace worn fuel fitting O-rings with new. Do not substitute fuel hose or equivalent where fuel pipe is installed.

Fuel System Pressure

RELIEVING

✳✳ CAUTION

The fuel system is under constant pressure, even with the engine OFF. Fuel pressure must be released before servicing any fuel supply or fuel return system component. Do not allow fuel to spill onto the engine intake or exhaust manifolds. Place shop towels under and around the any fittings to be disconnected. This will absorb any fuel spilled when residual pressure is released from the fuel system.

1. Remove the fuel pump relay.
2. Start the engine and allow it to run until it stalls.
3. Attempt restarting the engine until it no longer runs.
4. Turn the ignition key to the **OFF** position.

Fuel Filter

REMOVAL & INSTALLATION

1995–96 Models

1. Relieve the fuel system pressure.
2. Disconnect the battery ground cable.
3. Raise and support the rear of the vehicle safely.
4. Remove the filter shield, if equipped.
5. Disconnect the fuel lines from the filter.
6. Remove the filter strap bolt and remove the filter.

To install:

➡**The filter is marked for installation. IN goes towards the fuel tank; OUT towards the engine.**

7. Place the new filter on the frame rail and tighten the strap bolt.
8. Connect the fuel lines to the filter.
9. Install the filter shield, if applicable.
10. Connect the negative battery cable.

1997–99 Models

1. Relieve the fuel system pressure.
2. Disconnect the battery ground cable.
3. Remove the fuel tank.
4. Clean around the filter/regulator.
5. Remove and discard the retaining clamp from the filter/regulator.
6. Pry the filter/regulator from the top of the fuel pump module with two small prytools.
7. Discard the gasket below the filter/regulator. Discard the O-rings on the bottom of the filter/regulator. If the smallest of the two O-rings cannot be found on the bottom of the filter/regulator, it may be lodged in the fuel inlet passage.

To install:

8. Install new O-rings on the filter/regulator. Apply a small amount of clean engine oil to the O-rings.
9. Install a new gasket to the top of the fuel pump module.
10. Press the filter/regulator into the top of the module until it snaps into position (a positive click must be felt or heard).
11. The arrow on top of the fuel pump module must be pointing towards the front of the vehicle. Rotate the filter/regulator until the fuel supply tube is pointed at the 10 o'clock position.

12. Install a new retainer clamp to the top of the filter/regulator.
13. Install the fuel tank.
14. Connect the negative battery cable.

Fuel Pump

REMOVAL & INSTALLATION

➡**The fuel pump is not serviceable. If the fuel pump is found to be defective, the entire fuel pump module assembly must be replaced.**

1. Relieve the fuel system pressure.
2. Disconnect the battery ground cable.

✳✳ CAUTION

Observe all applicable safety precautions when working around fuel. Whenever servicing the fuel system, always work in a well ventilated area. Do not allow fuel spray or vapors to come in contact with a spark or open flame. Keep a dry chemical fire extinguisher near the work area. Always keep fuel in a container specifically designed for fuel storage; also, always properly seal fuel containers to avoid the possibility of fire or explosion.

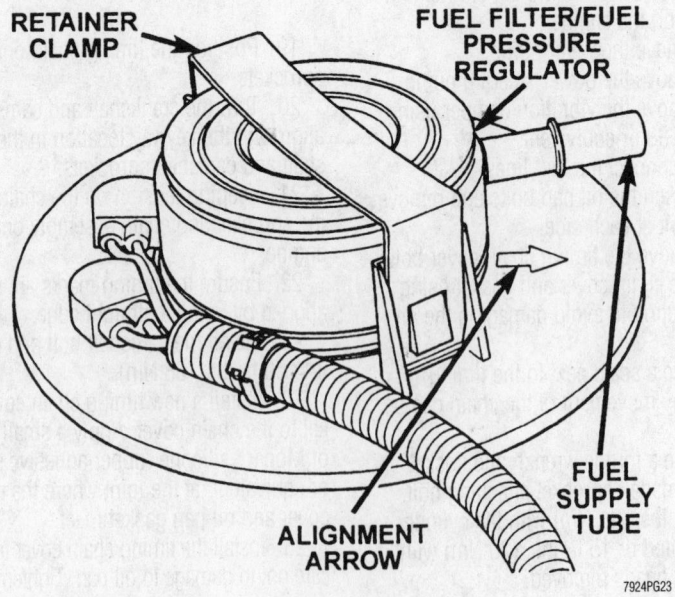

The fuel filter/pressure regulator is located on top of the fuel pump module—1997–99 models

RETAINER CLAMP

FUEL FILTER/FUEL PRESSURE REGULATOR

ALIGNMENT ARROW

FUEL SUPPLY TUBE

7924PG23

3. Remove the fuel tank.

4. Install special tool 6856 or equivalent to the fuel pump module locknut and remove the locknut. The fuel pump module will spring up when the locknut is removed.

5. Remove the module from the tank.

To install:

6. Using a new gasket, position the fuel pump module into the opening in the fuel tank.

7. Position the locknut over the module. Rotate the module until the arrow is pointed toward the front of the vehicle.

8. Install special tool 6856 or equivalent to the locknut. Tighten the locknut to 25 ft. lbs. (34 Nm).

9. Rotate the fuel filter/pressure regulator until the fitting is pointed to the 10 o'clock position.

10. Install the fuel tank.

11. Connect the negative battery cable.

DRIVE TRAIN

Transmission Assembly

REMOVAL & INSTALLATION

1. Disconnect the negative battery cable.

2. On manual transmissions, remove the shifter boot and lever from inside the vehicle.

3. Raise and safely support the vehicle.

4. On 4-wheel drive vehicles, remove the front driveshaft and transfer case.

5. Disconnect all cables, connectors and fluid lines that may interfere with transmission removal. Tag them if helpful for installation.

6. On automatic transmissions, unbolt the torque converter from the flexplate and disconnect the shift linkage.

7. Place a drain pan under the transmission and drain the fluid.

8. Position a transmission jack under the transmission and safety-chain the case to the jack.

9. Matchmark and remove the driveshaft(s).

10. Remove the transmission rear mount.

11. Slightly raise the transmission and remove the crossmember.

12. On automatic transmissions, remove the transmission-to-engine block bolts. For manual transmissions, remove the bolts securing the transmission to the bell housing.

The torque converter will fall out of the transmission if it is tilted forward. Keep a hand on it while lowering the transmission out of the vehicle.

13. Roll the transmission rearward until the input shaft clears, lower the jack and remove the transmission.

To install:

14. Carefully raise the transmission to the engine or bell housing.

15. Roll the transmission forward and into position.

16. On automatic transmissions tighten the bolts to 65 ft. lbs. (87 Nm). On manual transmissions, tighten the bolts to 50 ft. lbs. (64 Nm).

17. Install the crossmember and tighten the bolts to 55 ft. lbs. (74 Nm).

18. Install the transmission rear insulator and lower retainer. Tighten the bolts to 60 ft. lbs. (81 Nm).

19. The rest of the installation is the reverse of removal.

20. Refill the transmission with the correct amount and type of fluid. Use 75W–90 Grade GL-3 in the manual transmissions and Mopar® ATF Plus 3 automatic transmission fluid in the automatic transmissions.

Clutch

REMOVAL & INSTALLATION

1. Raise and safely support the vehicle.

2. Remove the transmission or transmission/transfer case assembly.

3. Matchmark the pressure plate and flywheel. Loosen the pressure plate bolts, a little at a time, in rotation, to avoid warpage.

4. Remove the pressure plate and clutch disc.

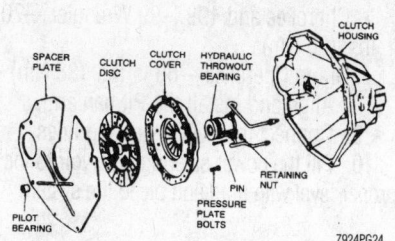

Exploded view of the clutch components

5. Inspect the flywheel for scoring, cracks, warpage or other wear; resurface or replace as necessary.

6. Inspect the pilot bearing for excessive wear or damage and replace as necessary.

To install:

7. If removed, install the pilot bearing after lubricating with grease. Seat the bearing in the crankshaft with a clutch alignment tool.

8. Install the clutch alignment tool in the pilot bearing.

9. Install the clutch disc on the tool.

10. Install the pressure plate and tighten the bolts finger-tight. The pressure plate bolts must be tightened a little at a time, in rotation, to avoid warpage. Tighten the pressure plate bolts as follows:

- 2.5L engine—23 ft. lbs. (31 Nm)
- 4.0L engine—40 ft. lbs. (54 Nm)
- 5.2L and 5.9L engines— ⁵⁄₁₆ inch bolts 17 ft. lbs. (23 Nm) and ³⁄₈ inch bolts 30 ft. lbs. (41 Nm)

11. Install the transmission or transmission/transfer case assembly and lower the vehicle.

Hydraulic Clutch System

BLEEDING

➡ **The replacement reservoir, master cylinder, slave cylinder and fluid line assembly is pre-filled with fluid from the factory. Bleeding is not necessary or possible on these models.**

Transfer Case Assembly

REMOVAL & INSTALLATION

All Models

1. Shift the transfer case into **N**.

2. Raise and support the vehicle safely.

3. Drain the lubricant.

4. Matchmark and remove the front and rear driveshafts.

5. Support the transmission with a jack.

6. Remove the rear crossmember.

7. Disconnect the speedometer cable or speed sensor connector.

8. Disconnect the linkage.

9. Disconnect the vent and vacuum hoses and the indicator wire.

10. Support the transfer case with a transmission jack.

11. Remove the transfer case-to-transmission bolts.

12. Pull the case rearward to disengage it and lower it from the vehicle.

To install:

13. Raise the transfer case into position. Make certain the case and transmission are mated without binding, before tightening the attaching bolts. Tighten the bolts to 26 ft. lbs. (35 Nm).

14. Connect the shift linkage at the case.

15. Connect the vacuum hoses.

16. Install the driveshafts. New strap bolts should be used. Tighten the nuts to 14 ft. lbs. (19 Nm), tighten the flange nuts to 35 ft. lbs. (48 Nm).

17. Install the rear crossmember. Tighten the bolts to 30 ft. lbs. (41 Nm).

18. Remove the transmission floor jack.

19. Connect the speedometer cable or speed sensor connector and vent hose.

20. Connect the shift lever link at the operating lever.

21. Refill the transfer case to the correct level.

22. Lower the vehicle.

STEERING AND SUSPENSION

Air Bag

✳✳ CAUTION

Some vehicles are equipped with an air bag system, also known as the Supplemental Inflatable Restraint (SIR) or Supplemental Restraint System (SRS). The system must be disabled before performing service on or around system components, steering column, instrument panel components, wiring and sensors. Failure to follow safety and disabling procedures could result in accidental air bag deployment, possible personal injury and unnecessary system repairs.

PRECAUTIONS

Several precautions must be observed when handling the inflator module to avoid accidental deployment and possible personal injury.

- Never carry the inflator module by the wires or connector on the underside of the module.
- When carrying a live inflator module, hold securely with both hands, and ensure that the bag and trim cover are pointed away.
- Place the inflator module on a bench or other surface with the bag and trim cover facing up.
- With the inflator module on the bench, never place anything on or close to the module which may be thrown in the event of an accidental deployment.

DISARMING

1. Disconnect and isolate the negative battery cable. Wait two minutes for the system capacitor to discharge before performing any service.

2. To arm the system, connect the negative battery cable.

Recirculating Ball Power Steering Gear

REMOVAL & INSTALLATION

All Models

1. Place the wheels in the straight ahead position.

2. Disconnect and cap the power steering lines from the steering gear.

3. Remove the column coupler shaft from the gear.

4. Matchmark the Pitman arm to the gear and remove the Pitman arm.

5. Remove the steering gear retaining bolts and remove the gear.

To install:

6. Align the column coupler shaft to the steering gear.

7. Position the steering gear and bracket on the frame rail and install the bolts. Tighten the bolts to:

- 1995–96 Wrangler—78 ft. lbs. (105 Nm)
- Cherokee and 1997–99 Wrangler—70 ft. lbs. (96 Nm)
- Grand Cherokee—65 ft. lbs. (88 Nm)

8. Align and install the Pitman arm.

9. Connect the power steering lines.

10. Fill the power steering reservoir to the proper level with fluid and bleed the system.

Shock Absorber

REMOVAL & INSTALLATION

Front

1. Remove the locknuts and washers from the upper stud.

2. Raise and support the vehicle safely.

3. If necessary for access, remove the wheels.

4. Remove the lower attaching nuts and bolts.

5. Pull the shock absorber eyes and rubber bushings from the mounting pins.

To install:

➡Before installing new shocks, they should be purged of air. To do this, hold the shock upright and fully extend it, then invert and compress it. Do this several times.

6. Position the shock on the vehicle and install the mounting hardware.

7. Tighten the upper mounting nut to 15 ft. lbs. (20 Nm) and the lower end bolts to 20 ft. lbs. (27 Nm).

8. Install the wheels and lower the vehicle.

Rear

➡Before installing new shocks, they should be purged of air. To do this, hold the shock upright and fully extend it, then invert and compress it. Do this several times.

1. Raise and safely support the rear of the vehicle.

2. Using a suitable jack, relieve the axle weight from the springs.

3. On the Grand Cherokee, remove the upper locknut and washer from the frame bracket stud. On the Wrangler and Cherokee, remove the upper mounting bolts.

4. Remove the lower bolt, nut and washers from the axle shaft tube bracket.

5. Remove the shock absorber.

To install:

6. Place the top end of the shock absorber in position. Install the mounting fasteners. On the Grand Cherokee, tighten the nut to 52 ft. lbs. (70 Nm). On the Wrangler and Cherokee, tighten the bolts to 23 ft. lbs. (31 Nm).

7. Position the shock absorber lower eye in the axle shaft tube bracket and install the bolt, washers and nut. On the Grand Cherokee, tighten the nut to 68 ft. lbs. (92 Nm). On the Cherokee, tighten the nut to 46 ft. lbs. (62 Nm) and on the Wrangler to 74 ft. lbs. (100 Nm).

8. Remove the supports and lower the vehicle.

Coil Spring

REMOVAL & INSTALLATION

Front

CHEROKEE

✳✳ CAUTION

Coil springs are under a great deal of tension when installed in the vehicle. Serious injury or death may result from being hit by an expanding spring. A piece of chain fastened to the frame and wrapped around the coil spring will keep the spring from flying out if it should slip before it is fully expanded.

1. Raise and support the vehicle safely.
2. Support the axle with a floor jack.
3. Remove the wheels.
4. On 4WD vehicles, matchmark and disconnect the front driveshaft from the axle.
5. Disconnect the lower control arm at the axle.
6. Disconnect the stabilizer bar links and the shock absorbers at the axle.
7. Disconnect the track bar at the sill bracket.
8. Disconnect the tie rod at the Pitman arm.
9. Lower the axle until tension is removed from the spring, then loosen the spring retainer and remove the spring.

   To install:

10. Position the replacement spring on the retainer, tighten the spring retainer bracket screw and lift the axle into position.
11. Connect the lower control arm to the axle. Tighten the control arm-to-axle bolt to 133 ft. lbs. (180 Nm).
12. Remove the support jack.
13. Connect the stabilizer bar links and the shock absorbers at the axle. Tighten the shock absorber-to-axle bolt to 14 ft. lbs. (19 Nm) and the stabilizer bar-to-axle bolt to 70 ft. lbs. (95 Nm).
14. Connect the track bar at the sill bracket. Tighten the track bar-to-frame rail bolt to 35 ft. lbs. (47 Nm).

15. Connect the tie rod at the Pitman arm. Tighten the center link-to-Pitman arm bolt to 35 ft. lbs. (47 Nm).

➡**New strap bolts must be used each time the driveshaft is disconnected.**

16. On 4WD vehicles, connect the front driveshaft. Tighten U-joint-to-axle bolt to 14 ft. lbs. (19 Nm).
17. Install the wheels and lower the vehicle.

GRAND CHEROKEE AND 1997–99 WRANGLER

1. Raise and safely support the vehicle, allowing the front axle to hang.
2. Support the axle with a jack.
3. Paint or scribe alignment marks on the cam adjusters and axle bracket for installation reference.
4. Matchmark and disconnect the front driveshaft from the axle.
5. Disconnect the lower suspension arm nut, cam and cam bolt from the axle.
6. Disconnect the stabilizer bar links and shock absorbers at the axle.
7. Disconnect the track bar at the frame rail bracket.
8. Disconnect the drag link at the Pitman arm.

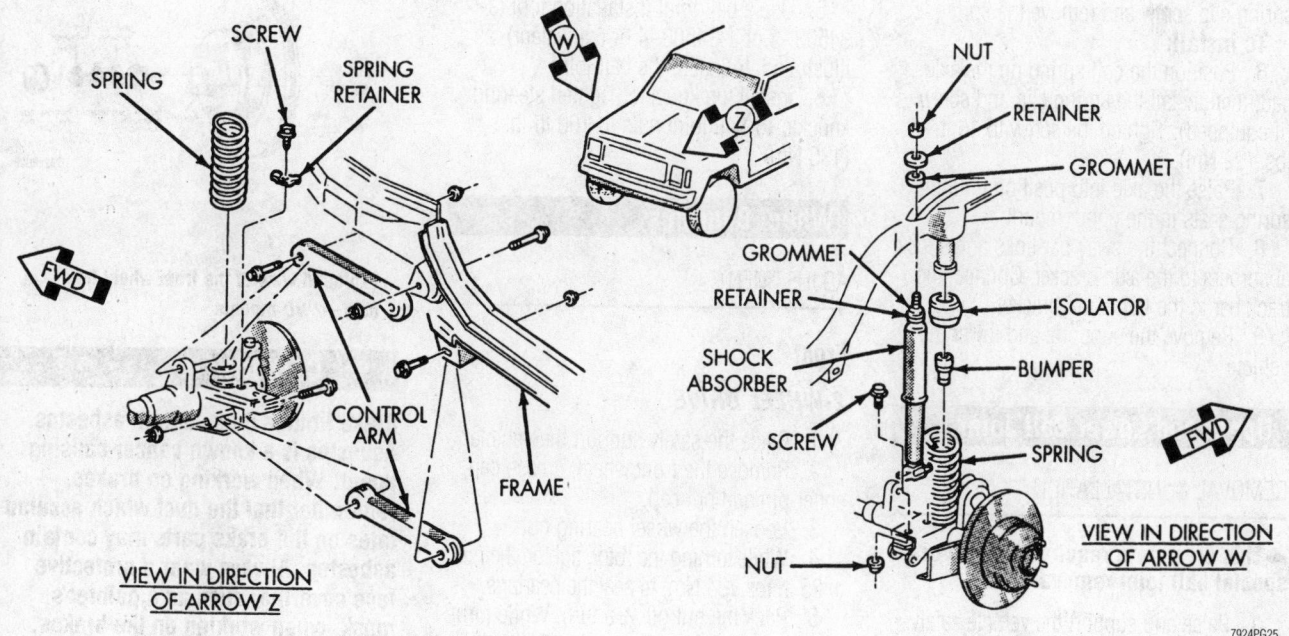

Exploded view of the front suspension—1995–98 Cherokee-Grand Cherokee is similar

7924PG25

9. Lower the axle until spring is free from the upper mount, then remove the coil spring clip screw and remove the spring.

10. If necessary, remove the jounce bumper from the upper spring mount.

To install:

11. If removed, install the jounce bumper and tighten the bolts to 31 ft. lbs. (42 Nm).

12. Position the replacement spring on axle pad. Install the spring clip and tighten the screw to 16 ft. lbs. (21 Nm).

13. Raise the axle into position until the spring seats in the upper mount.

14. Connect the stabilizer bar links and shock absorbers to the axle bracket. Connect the track bar to the frame rail bracket.

15. Install the lower suspension arm to the axle.

16. Connect the driveshaft to the yoke.

17. Lower the vehicle.

Rear

GRAND CHEROKEE AND 1997–99 WRANGLER

1. Raise and safely support the vehicle.

2. Support the axle with a suitable jack.

3. Disconnect the sway bar links and shock absorbers from the axle bracket.

4. Disconnect the track bar from the frame rail bracket.

5. Lower the axle until the spring is free from the upper mount seat. Remove the coil spring clip screw and remove the spring.

To install:

6. Position the coil spring on the axle pad, then install the spring clip and screw (if equipped). Tighten the screw to 16 ft. lbs. (22 Nm).

7. Raise the axle into position until the spring seats in the upper mount.

8. Connect the sway bar links and shock absorbers to the axle bracket. Connect the track bar to the frame rail bracket.

9. Remove the supports and lower the vehicle.

Upper and Lower Ball Joint

REMOVAL & INSTALLATION

➡**This procedure requires the use of a special ball joint removal tool.**

1. Raise and support the vehicle safely.

2. Remove the wheel. Remove the steering knuckle.

3. Position a ball joint removal tool (J-34503–1 and 34503–3 or equivalent), as illustrated, to remove the ball joint.

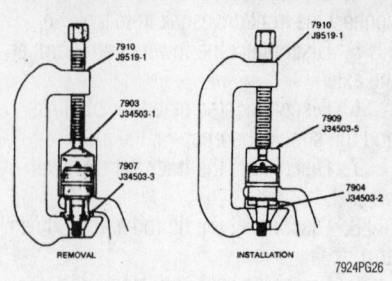

Upper ball joint removal and installation

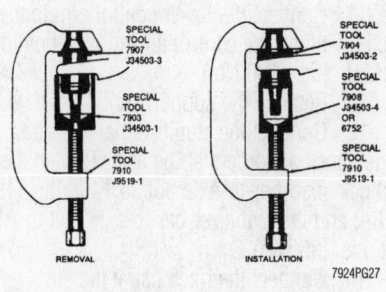

Lower ball joint removal and installation

4. Tighten the clamp screw to remove the joint.

To install:

5. Use a ball joint installation tool (J-34503–5 or J-34503–4 or equivalent), as illustrated, to install the ball joint.

6. Install the knuckle. Tighten steering knuckle-to-ball joint nuts to 100 ft. lbs. (135 Nm).

Wheel Bearings

ADJUSTMENT

Front

2-WHEEL DRIVE

1. Raise the safely support the vehicle.

2. Remove the front wheel, grease cap, cotter pin and nut cap.

3. Loosen the wheel bearing nut.

4. While turning the rotor, tighten the nut to 25 ft. lbs. (34 Nm) to seat the bearings.

5. Back the nut off ½ a turn. While turning the rotor, tighten the nut to 19 inch lbs. (2 Nm).

6. Install the nut cap and a new cotter pin. Install the grease cap.

7. Install the wheel and lower the vehicle.

4-WHEEL DRIVE

The front wheel bearing are not adjustable. If the bearings make noise or become loose, they must be replaced.

Rear

The rear axle bearing are not adjustable. If the bearings make noise or become loose, they must be replaced.

REMOVAL & INSTALLATION

Front

➡**Sodium-based grease is not compatible with lithium-based grease. Read the package labels and be careful not to mix the two types. If there is any doubt as to the type of grease used, completely clean the old grease from the bearing and hub before replacing.**

2WD MODELS

1. Raise and support the vehicle safely.

2. Remove the wheels.

3. Remove the caliper without disconnecting the brake line. Suspend it out of the way using a piece of wire to prevent damage.

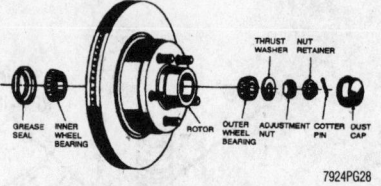

Exploded view of the front wheel bearings—2WD models

✷✷ CAUTION

Brake linings may contain asbestos. Asbestos is a known cancer-causing agent. When working on brakes, remember that the dust which accumulates on the brake parts may contain asbestos. Always wear a protective face covering, such as a painter's mask, when working on the brakes. NEVER blow the dust from the brakes or drum! There are solvents made for the purpose of cleaning brake parts. Use them!

4. Remove the grease cap, cotter pin, nut cap, nut, and washer from the spindle. Discard the cotter pin.

5. Slowly remove the hub and rotor. Catch the outer bearing as it falls.

6. Carefully drive out the inner bearing and seal from the hub, using a wood block.

7. Inspect the bearing races for excessive wear, pitting or grooves. If they are cracked or grooved, or if pitting and excess wear is present, drive them out with a drift or punch.

8. Check the bearing for excess wear, pitting or cracks, or excess looseness.

➡ **If it is necessary to replace either the bearing or the race, replace both. Never replace just a bearing or a race. These parts wear in a mating pattern. If just one is replaced, premature failure of the new part will result.**

To install:

9. On vehicles with drum brakes, cover the spindle with a cloth and thoroughly brush all dirt from the brakes.

❊❊ CAUTION

Brake linings may contain asbestos. Asbestos is a known cancer-causing agent. When working on brakes, remember that the dust which accumulates on the brake parts and/or in the drum may contain asbestos. Always

wear a protective face covering, such as a painter's mask, when working on the brakes. NEVER blow the dust from the brakes or drum! There are solvents made for the purpose of cleaning brake parts. Use them!

10. Remove the cloth and thoroughly clean the spindle and the inside of the hub.

11. Pack the inside of the hub with high temperature wheel bearing grease. Add grease to the hub until it is flush with the inside diameter of the bearing cup.

12. Pack the bearings with the same grease. A needle-shaped wheel bearing packer is best for this operation. If one is not available, place a large amount of grease in the palm of your hand and slide the edge of the bearing cage through the grease to pick up as much as possible, then work the grease in until it squeezes through the bearing.

13. If a new race is being installed, very carefully drive it into position until it bottoms all around, using a brass drift. Be careful to avoid scratching the surface.

14. Place the inner bearing in the race and install a new grease seal.

15. Clean the rotor contact surface if necessary.

16. Position the hub and rotor on the spindle and install the outer bearing.

17. Install the washer and nut.

18. While turning the rotor, tighten the nut to 25 ft. lbs. (34 Nm) to seat the bearings.

19. Back the nut off ½ a turn. While turning the rotor, tighten the nut to 19 inch lbs. (2 Nm).

20. Install the nut cap and a new cotter pin. Install the grease cap.

21. Install the caliper.

22. Install the wheels.

4WD MODELS

1. Raise and support the vehicle safely.

2. Remove the wheels.

3. Remove, but do not disconnect, the caliper. Suspend it out of the way.

❊❊ CAUTION

Brake linings may contain asbestos. Asbestos is a known cancer-causing agent. When working on brakes, remember that the dust which accumulates on the brake parts and/or in the drum may contain asbestos. Always wear a protective face covering, such as a painter's mask, when working on the brakes. NEVER blow the dust from the brakes or drum! There are solvents made for the purpose of cleaning brake parts. Use them!

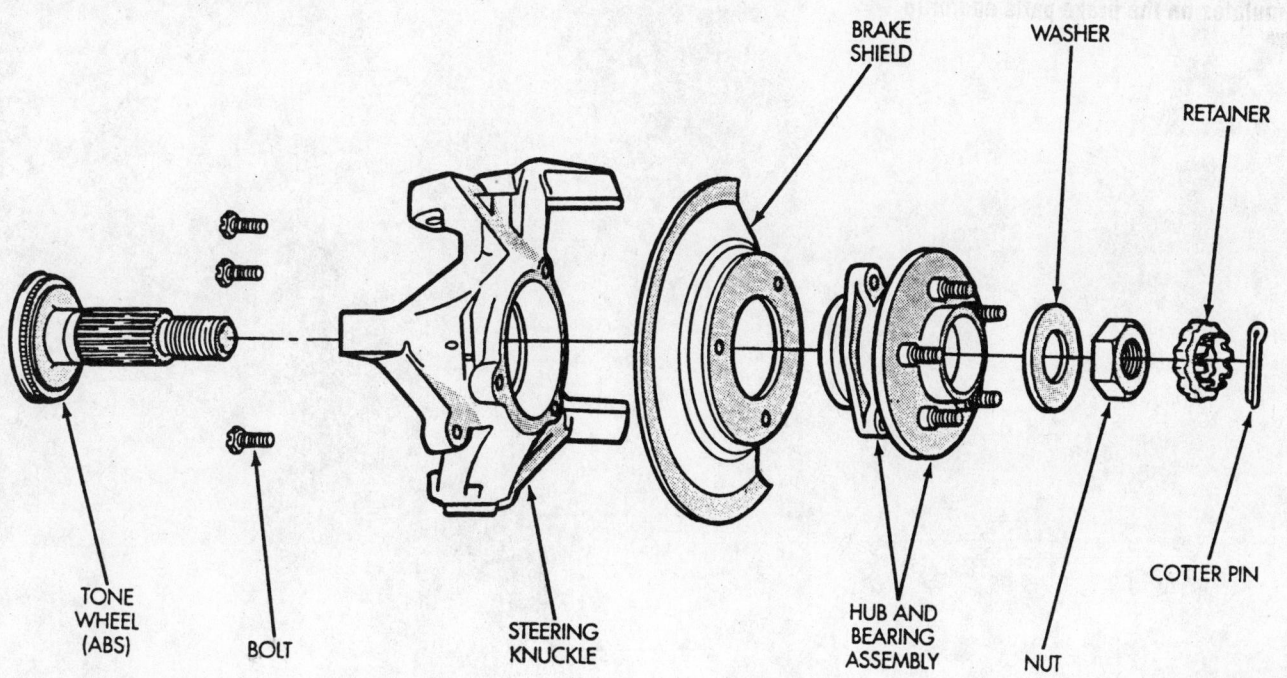

Exploded view of the hub assembly—4WD models

TONE WHEEL (ABS) · BOLT · STEERING KNUCKLE · BRAKE SHIELD · HUB AND BEARING ASSEMBLY · WASHER · NUT · RETAINER · COTTER PIN

7924PG29

4. Remove the rotor.

5. If equipped, remove the ABS wheel speed sensor.

6. Remove the cotter pin, nut retainer, axle nut and washer.

7. Remove the three hub-to-steering knuckle attaching bolts.

8. Remove the hub/bearing carrier and the dust shield.

To assemble and install:

9. Install the dust shield and hub/bearing assembly on the axle shaft.

10. Coat the carrier bolt threads with Loctite®. Install them and tighten to 75 ft. lbs. (102 Nm).

11. Install the washer and axle shaft nut. Tighten the nut to 175 ft. lbs. (237 Nm).

12. Install the nut retainer and cotter pin. NEVER back off the nut to install the cotter pin! ALWAYS advance it!

13. Install the rotor and caliper.

14. Install the wheel.

Rear

1. Remove the wheel and tire and brake drum/disc assembly.

✳✳ CAUTION

Brake linings may contain asbestos. Asbestos is a known cancer-causing agent. When working on brakes, remember that the dust which accumulates on the brake parts and/or in the drum may contain asbestos. Always wear a protective face covering, such as a painter's mask, when working on the brakes. NEVER blow the dust from the brakes or drum! There are solvents made for the purpose of cleaning brake parts. Use them!

2. Remove the nuts that attach the outer seal retainer (and brake backing plate) to the axle shaft tube. Discard the nuts.

Remove the axle shaft from the housing with an axle puller attached to a slide hammer.

3. Discard the inner axle seal. Position the axle shaft in a vise.

4. Remove the retaining ring by drilling a ¼ in. (6mm) hole about ¾ of the way through the ring, then using a cold chisel over the hole, split the ring.

5. Remove the bearing with an arbor press, discard the seal and remove the retainer plate.

To install:

6. Clean and then apply a thin coating of wheel bearing lubricant to the bearing and seal contact surfaces. Apply wheel bearing lubricant to the lips of the replacement inner and outer seals.

7. Install the inner seal with the open end of the seal facing inward. Ensure it is completely seated.

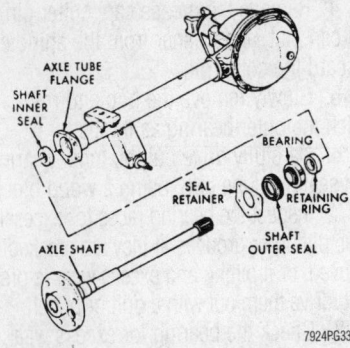

Exploded view of the typical axle and bearing assembly

➡It is helpful to cool the axle shaft before pressing on the bearing and retainer. Leave the bearing and retainer at room temperature.

8. Install the retainer plate and the outer seal on the shaft. Ensure the open end of the seal faces toward the axle shaft bearing.

9. Pack the replacement bearing with wheel bearing lubricant and position on the axle shaft. Press into place.

10. Press a replacement bearing retainer on the axle shaft against the bearing.

11. Install the axle into the axle tube. Position and align the seal retainer and brake support plate and install replacement attaching nuts. Tighten nuts to 32 ft. lbs. (43 Nm).

12. Install brake assembly wheels and tires. Lower vehicle.

KIA

Sportage

ENGINE REPAIR

➡️**Disconnecting the negative battery cable on some vehicles may interfere with the functions of the on-board computer systems and may require the computer to undergo a relearning process, once the negative battery cable is reconnected.**

Ignition Timing

ADJUSTMENT

The 2.0L engine in the Sportage is equipped with a distributorless ignition system. The ignition timing is controlled by the Powertrain Control Module (PCM) through the input of engine control system sensors. The ignition timing is set at 4 degrees BTDC for vehicles equipped with manual or automatic transmissions. The ignition timing cannot be adjusted.

Engine Assembly

REMOVAL & INSTALLATION

1. Disconnect the battery cables.
2. Properly relieve the fuel system pressure.
3. Disconnect the windshield washer hose from the hood.
4. With the aid of an assistant, remove the four hood attaching bolts and remove the hood.
5. Remove the two air duct mounting bolts from the top of the radiator. Loosen the clamp at the air intake housing and remove the duct.
6. Disconnect the accelerator cable by pulling the throttle back and rotating the cable until it aligns with the slot in the pulley.
7. Disconnect the transmission control cable.
8. Remove the resonance chamber mounting bolt, chamber bolt and air silencer.
9. Remove the IAC air hose, breather hose and vacuum line from the air intake tube.
10. Disconnect the MAF sensor connector.
11. Loosen the air inlet hose clamp from the MAF sensor.
12. Remove the three bolts from the air intake tube to the throttle body.
13. Remove the air intake hose and tube as an assembly.

14. Remove the radiator cap.
15. Drain the coolant into a suitable container.
16. Drain the engine oil.
17. Remove the upper radiator hose.
18. Remove the four clutch fan nuts.
19. Remove the five cooling fan shroud bolts.
20. Remove the fan and shroud at the same time.
21. Loosen the adjusting and mounting bolt and remove the alternator drive belt.
22. Remove the fan pulley.
23. Disconnect the electrical terminal connectors from the alternator.
24. Remove the EGR solenoid valve connector on the intake manifold in front of the dynamic chamber.
25. Remove both heater hoses from the pipes.
26. Disconnect the engine-to-body ground wire from the intake manifold and the harness bracket.
27. Remove the brake booster vacuum hose from the dynamic chamber.
28. Disconnect the fuel lines and fuel pressure regulator from the rear of the dynamic chamber.
29. Disconnect the vacuum hose from the bottom of the EGR valve.
30. Remove the purge solenoid valve vacuum hose from dynamic chamber.
31. Remove the two vacuum hoses from the top of the charcoal canister.
32. Slide the charcoal canister up and out of the bracket.
33. Disconnect the lower radiator hose.
34. If equipped with an automatic transmission, place a clean drain pan under the radiator. Disconnect the ATF cooling lines from the radiator.
35. Remove the four radiator attaching bolts at the corners of the radiator. Disconnect the overflow tube from the radiator.
36. Remove the radiator.
37. Raise and safely support the vehicle.
38. Remove the four attaching bolts and remove the lower splash panel.
39. Loosen the A/C idler pulley locknut and adjusting bolt. Remove the drive belt.
40. Remove the two A/C idler pulley bracket bolts and remove the pulley assembly.
41. Remove the four A/C compressor mounting bolts. Position the A/C compressor out of the way.
42. Loosen the power steering pump lock and mounting bolts. Remove the power steering drive belt.
43. Remove the three intake manifold support bracket bolts and the bracket.
44. Disconnect the starter wiring harness.

45. Remove the starter.
46. Disconnect the front exhaust pipe from the exhaust manifold.
47. Remove the bracket bolt from the front exhaust pipe.
48. Remove the two exhaust-to-clutch (manual transmission) or converter (automatic transmission) housing bolts and the bracket.
49. Remove the clutch housing (manual transmission) or converter (automatic transmission) housing bolts.
50. On vehicles with automatic transmission, remove the six drive plate-to-torque converter bolts.
51. Lower the vehicle.
52. Support the transmission from underneath the vehicle.
53. Connect the engine hoist to the engine assembly.
54. Remove the three left side engine mounting bolts.
55. Remove the three right side engine mounting bolts.
56. Lift the engine up and forward slightly to provide access to three electrical connectors on the rear of the cylinder head.
57. Disconnect the three electrical connectors from the Camshaft Position (CMP) sensor, coil and condenser on the rear of the cylinder head.
58. Slowly remove the engine from the vehicle.

To install:
59. Lower the engine enough to connect the three electrical connectors to the CMP sensor, coil and condenser on the rear of the cylinder head.
60. Position the engine to the transmission. Install the transmission bolts and tighten the bolts. Torque according to bolt size:
   - 14mm bolts to 80 ft. lbs. (108 Nm)
   - 10mm bolts to 28 ft. lbs. (38 Nm)
   - 6mm bolts to 60 inch lbs. (7 Nm)
61. Install the three right and three left side engine mounting bolts. Tighten to 27–38 ft. lbs. (37–52 Nm).
62. Disconnect the engine hoist from the engine assembly.
63. Raise and safely support the vehicle.
64. On vehicles with automatic transmission, install the six drive plate-to-torque converter bolts.
65. Connect the front exhaust pipe to the exhaust manifold. Tighten the flange bolts to 24 ft. lbs. (31 Nm).
66. Install the bracket bolt to the front exhaust pipe. Tighten to 20 ft. lbs. (27 Nm).
67. Install the starter.
68. Connect the starter wiring harness.

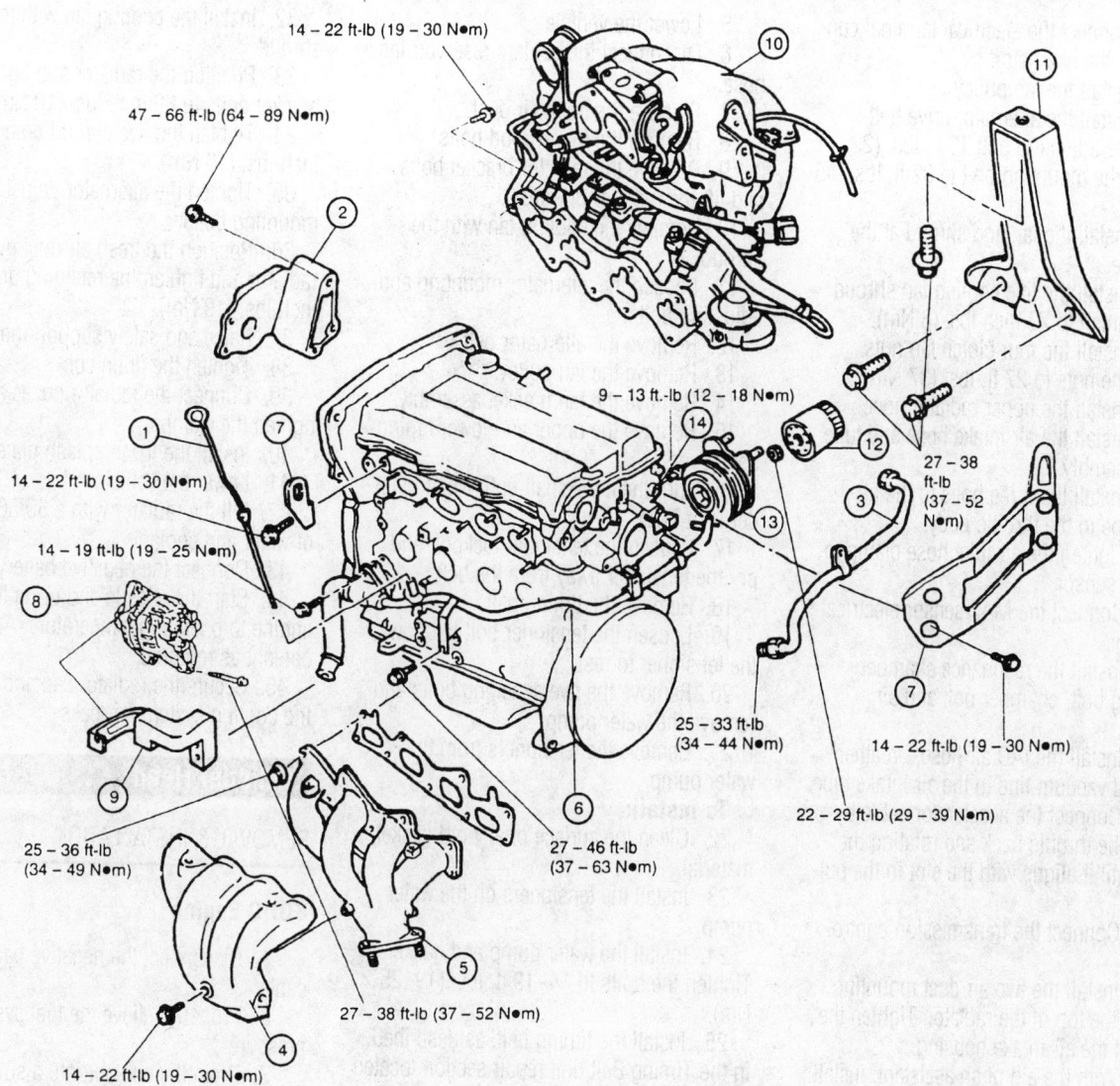

14 – 22 ft-lb (19 – 30 N•m)

47 – 66 ft-lb (64 – 89 N•m)

14 – 22 ft-lb (19 – 30 N•m)

14 – 19 ft-lb (19 – 25 N•m)

9 – 13 ft.-lb (12 – 18 N•m)

27 – 38 ft-lb (37 – 52 N•m)

25 – 33 ft-lb (34 – 44 N•m)

14 – 22 ft-lb (19 – 30 N•m)

22 – 29 ft-lb (29 – 39 N•m)

25 – 36 ft-lb (34 – 49 N•m)

27 – 46 ft-lb (37 – 63 N•m)

27 – 38 ft-lb (37 – 52 N•m)

14 – 22 ft-lb (19 – 30 N•m)

1. Oil Level Gauge
2. Thermo-Modulated Fan Bracket
3. EGR Pipe
4. Exhaust Manifold Heat Shield
5. Exhaust Manifold
6. Coolant Inlet Pipe and Bypass Pipe
7. Engine Hanger
8. Generator
9. Generator Strap and Bracket
10. Intake Manifold Assembly
11. Intake Manifold Support Bracket
12. Oil Filter
13. Oil Cooler
14. Oil Pressure Switch

7924QG36

Exploded view of some peripheral engine component mountings—DOHC engine

69. Lower the vehicle.

70. Install the three intake manifold support bracket bolts and the bracket.

71. Install the power steering pump lock and mounting bolts. Install the power steering drive belt. Tighten the bolts to 30 ft. lbs. (42 Nm).

72. Position the A/C compressor and install the A/C compressor mounting bolts. Tighten to 18 ft. lbs. (24 Nm).

73. Install the A/C belt pulley assembly and drive belt. Install the two A/C idler pulley bracket bolts and tighten to 24 ft. lbs. (32 Nm).

74. Install the four attaching bolts attaching the lower splash panel.

75. Lower the vehicle.

76. Install the radiator.

77. Connect the ATF cooling lines to the radiator.

78. Connect the lower radiator hose.

79. Slide the charcoal canister in the bracket.

80. Install the two vacuum hoses to the top of the charcoal canister.

81. Connect the engine-to-body ground wire to the intake manifold and the harness bracket.

82. Install the brake booster vacuum hose to the dynamic chamber.

83. Connect the fuel lines and fuel pressure regulator to the rear of the dynamic chamber.

84. Connect the vacuum hose to the bottom of the EGR valve.

85. Install the purge solenoid valve vacuum hose to dynamic chamber.

86. Install the EGR solenoid valve connector on the intake manifold in front of the dynamic chamber.

87. Install both heater hoses to the pipes.

88. Connect the electrical terminal connectors to the alternator.

89. Install the fan pulley.

90. Install the alternator drive belt. Tighten the adjusting bolt 16 ft. lbs. (22 Nm) and the mounting bolt to 32 ft. lbs. (45 Nm).

91. Install the fan and shroud at the same time.

92. Install the five cooling fan shroud bolts. Tighten to 72 inch lbs. (8 Nm).

93. Install the four clutch fan nuts. Tighten the nuts to 27 ft. lbs. (37 Nm).

94. Install the upper radiator hose.

95. Install the air intake hose and tube as an assembly.

96. Install the three bolts to the air intake tube to the throttle body.

97. Tighten the air inlet hose clamp to the MAF sensor.

98. Connect the MAF sensor electrical connector.

99. Install the resonance chamber mounting bolt, chamber bolt and air silencer.

100. Install the IAC air hose, breather hose and vacuum line to the air intake tube.

101. Connect the accelerator cable by pulling the throttle back and rotating the cable until it aligns with the slot in the pulley.

102. Connect the transmission control cable.

103. Install the two air duct mounting bolts to the top of the radiator. Tighten the clamp at the air intake housing.

104. With the aid of an assistant, install the hood and the four hood attaching bolts.

105. Connect the windshield washer hose to the hood.

106. Fill the engine with fresh engine oil.

107. Connect the battery cables.

108. Fill the cooling system.

109. Start the vehicle. Check for:
- Fuel leaks
- Coolant leaks
- Vacuum leaks

110. Road test the vehicle to check engine performance.

Water Pump

REMOVAL & INSTALLATION

1. Disconnect the negative battery cable.

2. Raise and safely support the vehicle.

3. Remove the lower splash plate.

4. Remove the radiator drain cock and drain the engine coolant into a suitable container.

5. Lower the vehicle.

6. Disconnect the coolant reservoir tank hose.

7. Remove the fresh air duct.

8. Remove the five shroud bolts.

9. Remove the radiator bracket bolts and lift the radiator upward.

10. Remove the cooling fan with the shroud.

11. Loosen the alternator mounting and adjusting bolts.

12. Remove the alternator belt.

13. Remove the fan pulley.

14. Remove the fan bracket assembly.

15. Remove the upper and lower timing belt covers.

16. Turn the crankshaft until No. 1 cylinder is at TDC.

17. Loosen the tensioner lockbolt and pry the tensioner away from the belt.

18. Remove the timing belt.

19. Loosen the tensioner bolt to allow the tensioner to rest.

20. Remove the five attaching bolts and remove the water pump.

21. Remove the tensioners from the water pump.

To install:

22. Clean the surface of any old gasket material.

23. Install the tensioners on the water pump.

24. Install the water pump and gasket. Tighten the bolts to 14–19 ft. lbs. (19–25 Nm).

25. Install the timing belt, as described in the Timing Belt unit repair section located in the front of this manual.

26. Loosen the tensioner lockbolt and allow the tensioner to rest against the belt.

27. Tighten the tensioner lockbolt 32 ft. lbs. (43 Nm).

28. Install the upper and lower timing belt covers.

29. Install the fan bracket assembly.

30. Install the fan pulley.

31. Install the alternator drive belt.

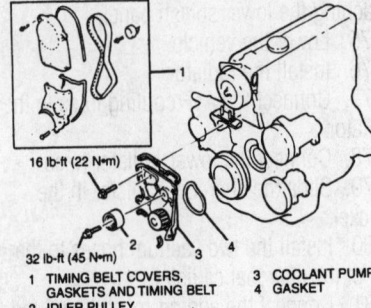

16 lb-ft (22 N•m)

32 lb-ft (45 N•m)

1 TIMING BELT COVERS, GASKETS AND TIMING BELT
2 IDLER PULLEY
3 COOLANT PUMP
4 GASKET

79240G01

Exploded view of the water pump mounting

32. Install the cooling fan with the shroud.

33. Position the radiator and tighten the bracket bolts to 89 inch lbs. (10 Nm).

34. Tighten the five shroud bolts to 89 inch lbs. (10 Nm).

35. Tighten the alternator adjusting and mounting bolts.

36. Position the fresh air duct over the radiator and tighten the retaining bolt 89 inch lbs. (10 Nm).

37. Raise and safely support the vehicle.

38. Tighten the drain cock.

39. Connect the radiator hoses and tighten the clamps.

40. Install the lower splash plate.

41. Lower the vehicle.

42. Fill the radiator with a 50/50 mixture of water and coolant.

43. Connect the negative battery cable.

44. Start the vehicle and bring the engine to operating temperature. Add coolant as required.

45. Secure the radiator cap and check the cooling system for leaks.

Cylinder Head

REMOVAL & INSTALLATION

SOHC Engine

1. Disconnect the negative battery cable.

2. Properly relieve the fuel system pressure.

3. Drain the coolant into a suitable container.

4. Remove the alternator, power steering and A/C drive belts.

5. Remove the fresh air duct from the top of the radiator.

6. Remove the upper radiator hose.

7. Remove the four attaching nuts to the clutch fan.

8. Remove the five fan shroud bolts. Remove the fan and shroud as an assembly.

9. Remove the five attaching bolts to the crankshaft pulley.

10. Remove the damper pulley.

11. Remove the four upper timing belt cover bolts and remove the cover.

12. Remove the two lower timing belt cover bolts and remove the cover.

13. Align the timing marks.

✳✳ WARNING

When aligning the timing marks, do not turn the timing gear counterclockwise. Damage to the engine will occur.

14. Loosen the tensioner bolt. Pry the tensioner away from the belt. Tighten tensioner bolt to relieve the pressure against the timing belt.

15. Remove the timing belt.

16. Remove the three bolts attaching the accelerator cable and air intake hose from the valve cover.

17. Remove the air, ISC, breather, and vacuum hoses from the air intake hose.

18. Disconnect the electrical terminal from the mass air flow sensor.

19. Remove the air hose attaching bolt from the air cleaner housing cover.

20. Loosen the air intake hose clamps at the throttle body and air cleaner housing cover.

21. Remove the air intake hose.

22. Remove the PCV valve from the valve cover.

23. Remove the ISC air hose from the idle speed actuator.

24. Remove the air hose from behind the intake manifold.

25. Remove the two screws securing the ignition wire guides to the valve cover.

26. Remove the oxygen sensor wire loom from the holder at the rear of the valve cover.

27. Remove the two bolts attaching the vacuum pipe to the valve cover and remove the pipe.

28. Remove the seven valve cover bolts.

29. Remove the valve cover and discard the old gasket.

30. Disconnect the two electrical connectors from the ignition coil.

31. Disconnect the fuel injector wiring harness at the rear of the cylinder head.

32. Disconnect the oxygen sensor electrical terminal.

33. Disconnect the electrical terminals from the throttle position sensor, oil pressure switch and idle speed actuator.

34. Remove the heater hose from the elbow.

35. Remove the brake booster vacuum hose from the surge tank.

36. Remove the cruise control vacuum hose from the surge tank.

37. Disconnect the wiring harness from the two clips on the rear of the cylinder head.

38. Remove the purge control solenoid valve vacuum hose from the intake manifold.

39. Disconnect the electrical terminal from the EGR vacuum solenoid on the intake manifold.

40. Disconnect the fuel hose from the fuel connector assembly at the rear of the cylinder head.

41. Remove the one bottom bolt from the ISC mounting bracket to remove the ground bracket.

42. Remove the four intake manifold bracket bolts and the bracket.

43. Remove the three starter attaching bolts and move the starter aside.

44. Remove the water bypass hose from the water bypass pipe.

45. Remove the two water hoses from the oil cooler.

46. Remove the two protective cover bolts from the exhaust manifold.

47. Remove the three exhaust manifold flange nuts.

48. Remove the front exhaust bracket bolt and lower the exhaust pipe.

49. Remove the 10 cylinder head bolts in the order shown in the accompanying illustration.

50. Remove the cylinder head, intake and exhaust manifolds as an assembly.

51. Discard the old head gasket.

52. If necessary, the intake and exhaust manifolds may now be removed from the cylinder head.

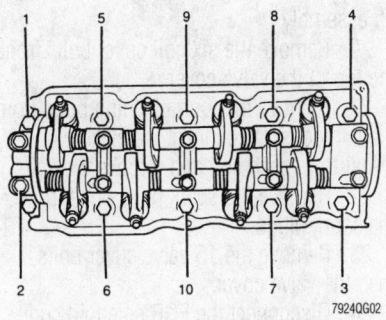

Cylinder head bolt removal order—SOHC engine

To install:

53. If necessary, install the exhaust and intake manifolds onto the cylinder head.

54. Position the new head gasket on the engine block.

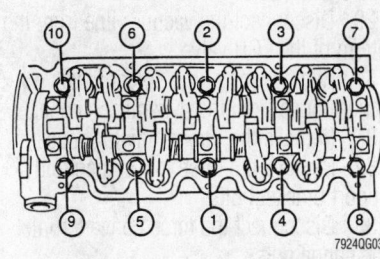

Cylinder head tightening sequence—SOHC engine

55. Install the cylinder head with the intake and exhaust manifolds attached.

56. Install the cylinder head bolts and washers.

57. Tighten the cylinder head bolts, in the proper sequence, to 62 ft. lbs. (84 Nm).

58. Connect the front exhaust bracket bolt to the lower the exhaust pipe.

59. Install the three exhaust manifold flange nuts.

60. Install the two protective cover bolts to the exhaust manifold.

61. Attach the water bypass hose to the water bypass pipe.

62. Install the two water hoses to the oil cooler.

63. Install the three starter attaching bolts to secure the starter.

64. Install the four intake manifold bracket bolts and the bracket.

65. Install the one bottom bolt to the ISC mounting bracket to secure the ground bracket.

66. Install the brake booster vacuum hose to the surge tank.

67. Install the cruise control vacuum hose to the surge tank.

68. Connect the wiring harness to the two clips on the rear of the cylinder head.

69. Install the purge control solenoid valve vacuum hose to the intake manifold.

70. Connect the electrical terminal to the EGR vacuum solenoid on the intake manifold.

71. Connect the fuel hose to the fuel connector assembly at the rear of the cylinder head.

72. Install the one bottom bolt to the ISC mounting bracket to install the ground bracket.

73. Connect the two electrical connectors to the ignition coil.

74. Connect the fuel injector wiring harness at the rear of the cylinder head.

75. Connect the oxygen sensor electrical terminal.

76. Connect the electrical terminals to the throttle position sensor, oil pressure switch and idle speed actuator.

77. Install the heater hose to the elbow.

78. Install the valve cover.

79. Install the seven valve cover bolts.

80. Install the air hose to behind the intake manifold.

81. Install the two screws securing the ignition wire guides to the valve cover.

82. Install the oxygen sensor wire loom to the holder at the rear of the valve cover.

83. Install the two bolts attaching the vacuum pipe to the valve cover and install the pipe.

84. Install the PCV valve to the valve cover.

85. Install the ISC air hose to the idle speed actuator.

86. Install the air intake hose.

87. Position the air intake hose clamps at the throttle body and air cleaner housing cover.

88. Install the air hose attaching bolt to the air cleaner housing cover.

89. Install the air, ISC, breather, and vacuum hoses to the air intake hose.

90. Connect the electrical terminal to the mass air flow sensor.

91. Install the three bolts attaching the accelerator cable and air intake hose to the valve cover.

92. Install the timing belt.

93. Allow the tensioner to press against the timing belt. Tighten tensioner bolt 33 ft. lbs. (45 Nm).

94. Check the timing belt deflection between the crankshaft and camshaft pulleys. The deflection should not exceed 0.43–0.51 in. (11–13mm) at 22 lbs. (10kg).

➡**If the deflection exceeds specification, replace the tensioner spring.**

95. Install the lower timing belt cover and the two cover bolts.

96. Install the upper timing belt cover. Install the four cover bolts.

97. Install the five attaching bolts to secure the crankshaft pulley.

98. Install the fan and shroud as an assembly.

99. Install the four attaching nuts to clutch fan.

100. Install the five fan shroud bolts.

101. Install the upper radiator hose.

102. Install the fresh air duct to the top of the radiator.

103. Install the alternator, power steering and A/C drive belts.

104. Properly fill the cooling system.

105. Connect the negative battery cable.

106. Start the engine and check for leaks.

107. Road test the vehicle.

DOHC Engine

1. Disconnect the negative battery cable.

2. Properly relieve the fuel system pressure.

3. Drain the coolant into a suitable container.

4. Remove the alternator drive belt.

5. Remove the fresh air duct from the top of the radiator.

6. Remove the upper radiator hose.

7. Remove the four attaching nuts to the clutch fan.

8. Remove the five fan shroud bolts. Remove the fan and shroud as an assembly.

9. Remove the four splash guard mounting bolts and the splash guard.

10. Loosen the lockbolts and loosen the A/C drive belt.

11. Loosen the power steering lock and mounting bolt. Remove the power steering belt.

12. Remove the timing belt, as described in the Timing Belt unit repair section in the front of this manual.

13. Remove the resonance chamber attaching bolt and chamber.

14. Remove the two accelerator cable bracket bolts from the valve cover.

15. Remove the two air intake tube attaching bolts.

16. Disconnect the accelerator cable by pulling the throttle back all the way until the slot aligns with the hole in the accelerator cable.

17. Loosen the clamp attaching the air tube to the MAF sensor.

18. Disconnect and mark the hoses from the air tube assembly.

19. Remove the three air tube attaching bolts. Remove the air tube and air hose as an assembly.

20. Remove the six coil cover bolts from the top of the valve cover.

21. Remove the four attaching bolts from the two coils. Disconnect the electrical terminal from each coil and remove the coils.

22. Remove the spark plug wires from the spark plugs.

23. Remove the 15 valve cover bolts, then the valve cover.

24. Disconnect the EGR solenoid connector in front of the dynamic chamber.

25. Remove the brake booster vacuum hose from the dynamic chamber.

26. Remove the fuel line from the pressure regulator and the return line located on the rear of dynamic chamber.

27. Disconnect the cruise control vacuum hose from the intake manifold.

28. Remove the engine-to-body ground strap from the intake manifold and the harness bracket below it.

29. Disconnect the vacuum line from the bottom of the EGR valve.

30. Remove the purge solenoid valve vacuum hose from the dynamic chamber.

31. Remove the upper radiator hose.

32. Remove the three intake manifold support bracket bolts.

33. Disconnect the three converter inlet pipe flange nuts.

34. Remove the 10 head bolts and move the cylinder head, with the intake and exhaust manifolds attached, enough to disconnect the wiring connectors from the rear of the cylinder head. Remove the cylinder head.

To install:

35. Place the new head gasket on the engine block.

36. Install the cylinder head with the intake and exhaust manifolds attached.

37. Connect the three wiring connectors at the rear of the cylinder head.

38. Install the 10 head bolts. Tighten the bolts in three equal steps, in proper sequence, to 59–64 ft. lbs. (80–87 Nm).

39. Connect the three converter inlet pipe flange nuts and tighten to 24 ft. lbs. (33 Nm).

40. Install the three intake manifold support bracket and bolts.

41. Install the upper radiator hose.

42. Install the purge solenoid valve vacuum hose to the dynamic chamber.

43. Install the brake booster vacuum hose to the dynamic chamber.

44. Install the fuel line to the pressure regulator and the return line located on the rear of dynamic chamber.

45. Connect the cruise control vacuum hose to the intake manifold.

46. Install the engine-to-body ground strap to the intake manifold and the harness bracket below it.

47. Connect the vacuum line to the bottom of the EGR valve.

48. Connect the EGR solenoid connector in front of the dynamic chamber.

49. Install the 15 valve cover bolts with the valve cover in place. Tighten the bolts 35–52 inch lbs. (4–6 Nm).

50. Install the spark plug wires to the spark plugs.

51. Connect the electrical terminal to each coil and Install the coils.

52. Install the six coil cover bolts to the top of the valve cover.

53. Connect the accelerator cable by pulling the throttle back all the way until the slot aligns with the hole in the accelerator cable.

54. Tighten the clamp attaching the air tube to the MAF sensor.

55. Connect the hoses to the air tube assembly.

56. Install the three air tube attaching bolts. Install the air tube and air hose as an assembly.

57. Install the resonance chamber attaching bolt and chamber.

58. Install the two accelerator cable bracket bolts to the valve cover.

59. Install the two air intake tube attaching bolts.

60. Install the timing belt and covers.

61. Install and adjust the A/C and power steering drive belts.

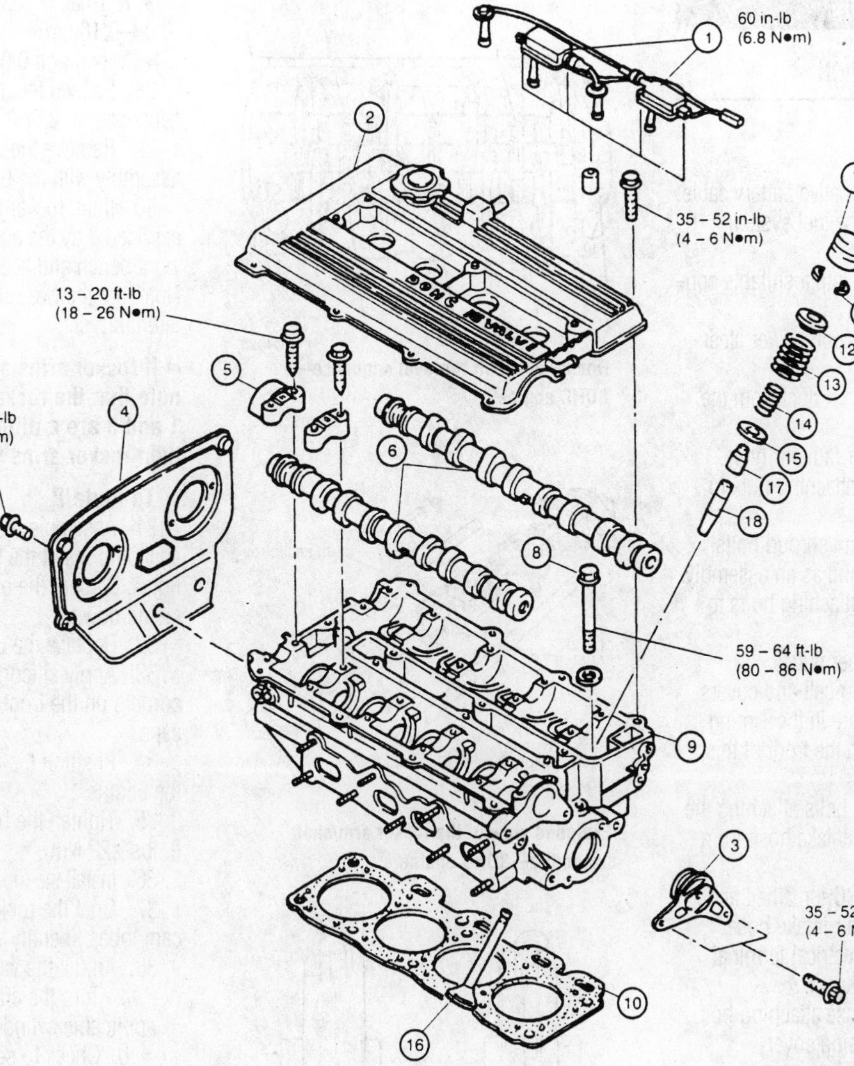

60 in-lb
(6.8 N•m)
35 – 52 in-lb
(4 – 6 N•m)
13 – 20 ft-lb
(18 – 26 N•m)
72 – 108 in-lb
(8 – 12 N•m)
59 – 64 ft-lb
(80 – 86 N•m)
35 – 52 in-lb
(4 – 6 N•m)
1
2
3
4
5
6
7
8
9
10
11
12
13
14
15
16
17
18

1. Ignition Coils and High Tension Leads
2. Cylinder Head Cover
3. Camshaft Position Sensor
4. Seal Plate
5. Camshaft Caps
6. Camshafts
7. Hydraulic Lash Adjuster
8. Cylinder Head Bolt
9. Cylinder Head
10. Cylinder Head Gasket
11. Valve Locks
12. Upper Spring Seat
13. Outer Valve Spring
14. Inner Valve Spring
15. Lower Spring Seat
16. Valve
17. Valve Stem Seal
18. Valve Guide

7924QG37

Exploded view of the cylinder head assembly—DOHC engine

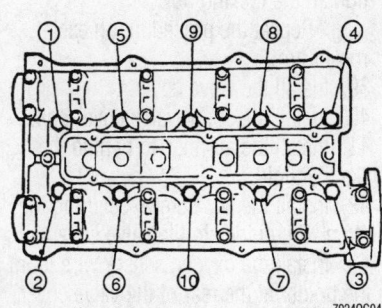

1
5
9
8
4
2
6
10
7
3
7924QG04

Cylinder head bolt removal order—DOHC engine

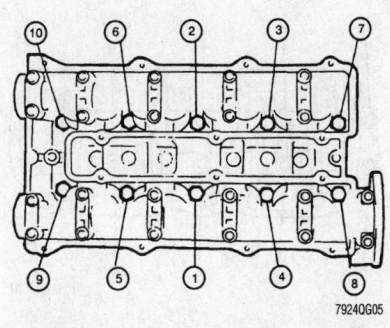

10
6
2
3
7
9
5
1
4
8
7924QG05

Cylinder head tightening sequence—DOHC engine

62. Install the four splash guard mounting bolts attaching the splash guard.
63. Install and tighten the alternator belt.
64. Install the upper radiator hose.
65. Install the fan and shroud as an assembly.
66. Install the four attaching nuts to the clutch fan.
67. Install the five fan shroud bolts.
68. Install the fresh air duct to the top of the radiator.
69. Properly fill the cooling system.
70. Connect the negative battery cable.
71. Start the engine and check for leaks.
72. Road test the vehicle.

Rocker Arms/Shafts

REMOVAL & INSTALLATION

SOHC Engine

1. Disconnect the negative battery cable.

2. Properly relieve the fuel system pressure.

3. Drain the coolant into a suitable container.

4. Remove the alternator, power steering and A/C drive belts.

5. Remove the fresh air duct from the top of the radiator.

6. Remove the upper radiator hose.

7. Remove the four attaching nuts to the clutch fan.

8. Remove the five fan shroud bolts. Remove the fan and shroud as an assembly.

9. Remove the five attaching bolts to the crankshaft pulley.

10. Remove the damper pulley.

11. Remove the timing belt and covers, according to the procedure in the Timing Belt unit repair section in the front of this manual.

12. Remove the three bolts attaching the accelerator cable and air intake hose from the valve cover.

13. Remove the air, ISC, breather, and vacuum hoses from the air intake hose.

14. Disconnect the electrical terminal from the mass air flow sensor.

15. Remove the air hose attaching bolt from the air cleaner housing cover.

16. Loosen the air intake hose clamps at the throttle body and air cleaner housing cover.

17. Remove the air intake hose.

18. Remove the PCV valve from the valve cover.

19. Remove the ISC air hose from the idle speed actuator.

20. Remove the air hose from behind the intake manifold.

21. Remove the two screws securing the ignition wire guides to the valve cover.

22. Remove the oxygen sensor wire loom from the holder at the rear of the valve cover.

23. Remove the two bolts attaching the vacuum pipe to the valve cover and remove the pipe.

24. Remove the seven valve cover bolts.

25. Remove the valve cover and discard the old gasket.

26. Remove the top bolt on the front housing.

27. Before removing the rocker arm assembly, measure the camshaft end-play. The end-play specifications are:

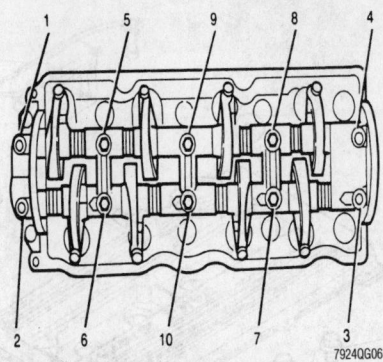

Rocker arm bolt removal sequence—SOHC engine

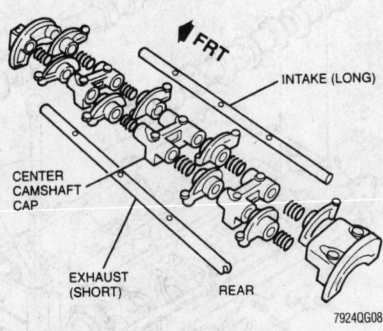

Exploded view of the rocker arm/shaft assembly—SOHC engine

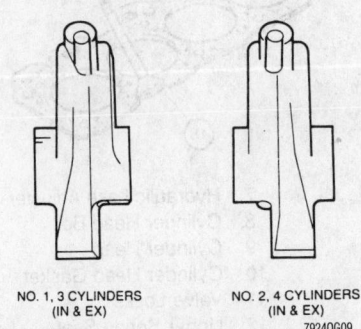

Note that there are different rocker arms for even and odd number cylinders—SOHC engine

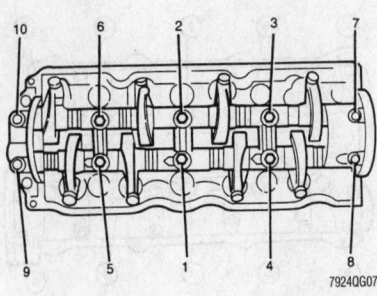

Rocker arm bolt tightening sequence—SOHC engine

- Normal: 0.003–0.006 in. (0.08–0.16mm)
- Maximum: 0.008 in. (0.20mm)

28. Loosen the 10 attaching bolts, in sequence, in 2 or 3 steps.

29. Remove the rocker arm and shaft assembly with the bolts in place.

30. If the rocker arm and shaft are to be serviced, Lay the assembly out on a clean work bench and keep the parts in order. Note that the two rocker arm shafts are different sizes.

➡**If rocker arms are to be replaced, note that the rocker arms for cylinders 1 and 3 are a different configuration than rocker arms for cylinders 2 and 4.**

To install:

31. Assemble the rocker arms on the shaft and insert the bolts in there original holes. Be sure the oil holes in the shafts are facing down.

32. Replace the oil seal on the rear cap.

33. Apply silicone sealant to the four corners on the front and rear rocker arm caps.

34. Position the rocker arm assembly on the engine.

35. Tighten the bolts, in sequence, to 16 ft. lbs. (22 Nm).

36. Install the top bolt on the front housing.

37. Coat the rocker arm assembly and cam lobes liberally with clean engine oil.

38. Adjust the valve lash, as follows:

   a. Turn the crankshaft so that the applicable cylinder is at TDC.

   b. Check to see if the rocker arms are loose.

   c. Loosen the locknut.

   d. Place the feeler gauge between the adjuster bolt and the top of the valve.

➡**Use a feeler gauge 0.012 inches on both the intake and exhaust valves.**

   e. Slowly turn the adjusting bolt until the feeler gauge can be removed and inserted with slight resistance.

   f. Hold the adjusting bolt still and tighten the locking nut.

   g. Repeat the procedure on each rocker arm.

39. Install the valve cover.

40. Install the seven valve cover bolts.

41. Install the air hose to behind the intake manifold.

42. Install the two screws securing the ignition wire guides to the valve cover.

43. Install the oxygen sensor wire loom to the holder at the rear of the valve cover.

44. Install the two bolts attaching the vacuum pipe to the valve cover and Install the pipe.

45. Install the PCV valve to the valve cover.

46. Install the ISC air hose to the idle speed actuator.

47. Install the air intake hose.

48. Position the air intake hose clamps at the throttle body and air cleaner housing cover.

49. Install the air hose attaching bolt to the air cleaner housing cover.

50. Install the air, ISC, breather, and vacuum hoses to the air intake hose.

51. Connect the electrical terminal to the mass air flow sensor.

52. Install the three bolts attaching the accelerator cable and air intake hose to the valve cover.

53. Install the timing belt.

54. Allow the tensioner to press against the timing belt. Tighten tensioner bolt 33 ft. lbs. (45 Nm).

55. Check the timing belt deflection between the crankshaft and camshaft pulleys. The deflection should not exceed 0.43–0.51 in. (11–13mm) at 22 lbs. (10kg).

➡**If the deflection exceeds specification, replace the tensioner spring.**

56. Install the lower timing belt cover and the two cover bolts.

57. Install the upper timing belt cover. Install the four cover bolts.

58. Install the five attaching bolts to secure the crankshaft pulley.

59. Install the fan and shroud as an assembly.

60. Install the four attaching nuts to the clutch fan.

61. Install the five fan shroud bolts.

62. Install the upper radiator hose.

63. Install the fresh air duct to the top of the radiator.

64. Install the alternator, power steering and A/C drive belts.

65. Properly fill the cooling system.

66. Connect the negative battery cable.

67. Start the engine and check for leaks.

68. Road test the vehicle.

Intake Manifold

REMOVAL & INSTALLATION

SOHC Engine

UPPER INTAKE MANIFOLD (SURGE TANK)

1. Disconnect the negative battery cable.

2. Properly relieve the fuel system pressure.

3. Drain the coolant into a suitable container.

4. Remove the three bolts attaching the accelerator cable and air intake hose from the valve cover.

5. Remove the air, ISC, breather, and vacuum hoses from the air intake hose.

6. Disconnect the electrical terminal from the mass air flow sensor.

7. Remove the air hose attaching bolt from the air cleaner housing cover.

8. Loosen the air intake hose clamps at the throttle body and air cleaner housing cover.

9. Remove the air intake hose.

10. Remove the PCV valve from the intake manifold.

11. Remove the ISC air hose from the idle speed actuator.

12. Remove the air hose from behind the intake manifold.

13. Disconnect the fuel injector wiring harness at the rear of the cylinder head.

14. Disconnect the oxygen sensor electrical terminal.

15. Disconnect the electrical terminals from the throttle position sensor, oil pressure switch and idle speed actuator.

16. Remove the heater hose from the elbow.

17. Remove the brake booster vacuum hose from the upper intake manifold.

18. Remove the cruise control vacuum hose from the upper intake manifold.

19. Disconnect the wiring harness from the two clips on the rear of the cylinder head.

20. Remove the purge control solenoid valve vacuum hose from the intake manifold.

21. Disconnect the electrical terminal from the EGR vacuum solenoid on the intake manifold.

22. Disconnect the fuel hose from the fuel connector assembly at the rear of the cylinder head.

23. Remove the one bottom bolt from the ISC mounting bracket to remove the ground bracket.

24. Remove the four intake manifold bracket bolts and the bracket.

25. Remove the two water hoses from under the throttle body.

26. Remove the three bolts and three nuts from the intake manifold.

27. Remove the upper manifold assembly.

To install:

28. Position the upper intake manifold and gasket. Tighten the three nuts and three bolts to 16 ft. lbs. (22 Nm).

29. Connect the two water hoses under the throttle body.

30. Install the four intake manifold bracket bolts and the bracket.

31. Install the one bottom bolt to the ISC mounting bracket to secure the ground bracket.

32. Install the brake booster vacuum hose to the upper intake manifold.

33. Install the cruise control vacuum hose to the upper intake manifold.

34. Connect the wiring harness to the two clips on the rear of the cylinder head.

35. Install the purge control solenoid valve vacuum hose to the intake manifold.

36. Connect the electrical terminal to the EGR vacuum solenoid on the intake manifold.

37. Connect the fuel hose to the fuel connector assembly at the rear of the cylinder head.

38. Install the one bottom bolt to the ISC mounting bracket to install the ground bracket.

39. Connect the two electrical connectors to the ignition coil.

40. Connect the fuel injector wiring harness at the rear of the cylinder head.

41. Connect the oxygen sensor electrical terminal.

42. Connect the electrical terminals to the throttle position sensor, oil pressure switch and idle speed actuator.

43. Install the heater hose to the elbow.

44. Install the PCV valve to the valve cover.

45. Install the ISC air hose to the idle speed actuator.

46. Install the air intake hose.

47. Position the air intake hose clamps at the throttle body and air cleaner housing cover.

48. Install the air hose attaching bolt to the air cleaner housing cover.

49. Install the air, ISC, breather, and vacuum hoses to the air intake hose.

50. Connect the electrical terminal to the mass air flow sensor.

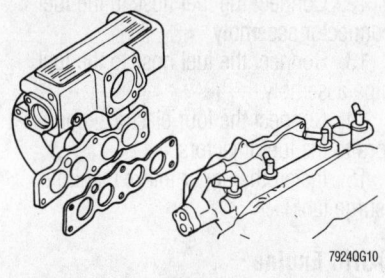

79240G10

Exploded view of the upper intake manifold (Surge Tank) mounting—SOHC engine

51. Install the three bolts attaching the accelerator cable and air intake hose to the valve cover.
52. Properly fill the cooling system.
53. Connect the negative battery cable.
54. Start the engine and check for leaks.
55. Road test the vehicle.

LOWER INTAKE MANIFOLD

1. Remove the upper intake manifold (surge tank).
2. Disconnect the four electrical connectors from the fuel injectors.
3. Remove the fuel hose from the fuel regulator.
4. Remove the fuel hose from the fuel connector assembly.
5. Remove the fuel hose the fuel pipe assembly.
6. Remove the oil filter.
7. Remove the four bolts and two nuts from the intake manifold.
8. Remove the bypass pipe, intake manifold and old gasket.

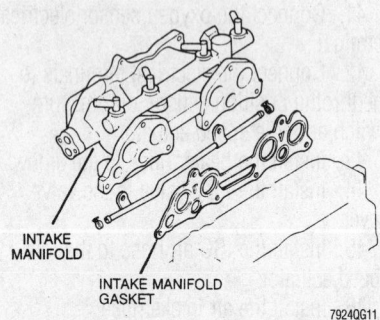

INTAKE MANIFOLD

INTAKE MANIFOLD GASKET

79240G11

Exploded view of the lower intake manifold and gasket mounting—SOHC engine

To install:

9. Position the new gasket, intake manifold and bypass pipe.
10. Install the four bolts and two nuts to the intake manifold. Tighten to 16 ft. lbs. (22 Nm).
11. Connect the fuel hose to the fuel regulator.
12. Connect the fuel hose to the fuel connector assembly.
13. Connect the fuel hose to the fuel pipe assembly.
14. Connect the four electrical connectors to the fuel injectors.
15. Install the upper intake manifold (surge tank).

DOHC Engine

1. Disconnect the negative battery cable.

2. Properly relieve the fuel system pressure.
3. Drain the coolant into a suitable container.
4. Remove the two accelerator cable bracket bolts from the valve cover.
5. Remove the two air intake tube attaching bolts.
6. Disconnect the accelerator cable by pulling the throttle back all the way until the slot aligns with the hole in the accelerator cable.
7. Loosen the clamp attaching the air tube to the MAF sensor.
8. Disconnect and mark the hoses from the air tube assembly.
9. Remove the three air tube attaching bolts. Remove the air tube and air hose as an assembly.
10. Disconnect the EGR solenoid connector in front of the dynamic chamber.
11. Remove the brake booster vacuum hose from the dynamic chamber.
12. Remove the fuel line from the pressure regulator and the return line located on the rear of dynamic chamber.
13. Disconnect the cruise control vacuum hose from the intake manifold.
14. Remove the engine-to-body ground strap from the intake manifold and the harness bracket below it.
15. Disconnect the vacuum line from the bottom of the EGR valve.
16. Remove the purge solenoid valve vacuum hose from the dynamic chamber.
17. Remove the three intake manifold support bracket and bolts.
18. Disconnect the four electrical connectors to the fuel injectors.
19. Remove the oil filter.
20. Remove the four bolts and four nuts from the intake manifold.
21. Remove the bypass pipe from the heater hose.
22. Remove the intake manifold and old gasket.

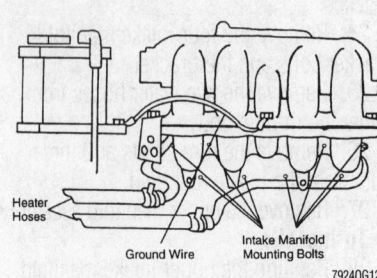

Heater Hoses

Ground Wire

Intake Manifold Mounting Bolts

79240G12

Intake manifold mounting bolt locations. Be sure to connect the ground cable— DOHC engine

To install:

23. Position the intake manifold with a new gasket to the cylinder head.
24. Connect the heater hose to the bypass pipe.
25. Install the four bolts and four nuts attaching the intake manifold to the cylinder head. Tighten the nuts and bolts 14–22 ft. lbs. (19–30 Nm).
26. Install the oil filter.
27. Install the three intake manifold support bracket and bolts. Tighten the bolts to 27–38 ft. lbs. (37–52 Nm).
28. Install the purge solenoid valve vacuum hose to the dynamic chamber.
29. Install the brake booster vacuum hose to the dynamic chamber.
30. Install the fuel line to the pressure regulator and the return line located on the rear of dynamic chamber.
31. Connect the cruise control vacuum hose to the intake manifold.
32. Install the engine-to-body ground strap to the intake manifold and the harness bracket below it.
33. Connect the vacuum line to the bottom of the EGR valve.
34. Connect the EGR solenoid connector in front of the dynamic chamber.
35. Connect the accelerator cable by pulling the throttle back all the way until the slot aligns with the hole in the accelerator cable.
36. Tighten the clamp attaching the air tube to the MAF sensor.
37. Connect the hoses to the air tube assembly.
38. Install the three air tube attaching bolts. Install the air tube and air hose as an assembly.
39. Install the resonance chamber attaching bolt and chamber.
40. Install the two accelerator cable bracket bolts to the valve cover.
41. Install the two air intake tube attaching bolts.
42. Properly fill the cooling system.
43. Connect the negative battery cable.
44. Start the engine and check for leaks.
45. Road test the vehicle.

Exhaust Manifold

REMOVAL & INSTALLATION

SOHC Engine

1. Disconnect the negative battery cable.
2. Remove the two air intake hose mounting bolts.

3. Loosen the clamp located at the throttle body end of the air intake hose.

4. Remove the air hose, the ISC hose, the breather hose and vacuum line from the air intake hose.

5. Disconnect the Mass Air Flow (MAF) electrical connector.

6. Loosen the clamp on the air cleaner housing cover.

7. Remove the bolt from the air intake hose bracket.

8. Remove the air intake hose.

9. Disconnect the wire holder for the oxygen sensor wire.

10. Disconnect the electrical connector from the oxygen sensor.

11. Remove the oxygen sensor wire from the holder.

12. Remove the three insulator attaching bolts and remove the insulator.

13. Raise and safely support the vehicle.

14. Remove the two bolts attaching the protective cover to the exhaust flange.

15. Remove the three flange nuts.

16. Remove the bolt attaching the front exhaust pipe the support bracket.

17. Lower the vehicle.

18. Remove the six attaching nuts from the exhaust manifold.

19. Remove the exhaust manifold and discard the old gasket.

To install:

20. Position the manifold and new gasket. Tighten the attaching nuts to 31 ft. lbs.

21. Raise and safely support the vehicle.

22. Install a new flange gasket. Tighten the three flange bolts to 24 ft. lbs. (33 Nm).

23. Install the two bolts attaching the protective cover to the exhaust flange. Tighten the bolts to 18 ft. lbs. (25 Nm)

24. Connect the front pipe to the support bracket.

25. Lower the vehicle.

26. Install the insulator. Tighten the three bolts to 18 ft. lbs. (25 Nm)

27. Place the Oxygen sensor wire through the wire loom and connect the wire to the harness.

28. Install the air intake hose.

29. Connect the clamp located at the throttle body end of the air intake hose.

30. Connect the air hose, the ISC hose, the breather hose and vacuum line to the air intake hose.

31. Connect the MAF electrical connector.

32. Tighten the clamp on the air cleaner housing cover.

33. Tighten the bolts to the air intake hose bracket.

34. Connect the negative battery cable.

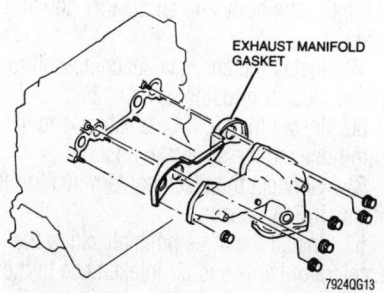

Exhaust manifold and gasket mounting— SOHC engine

DOHC Engine

1. Disconnect the negative battery cable.

2. Remove the two air intake hose mounting bolts.

3. Loosen the clamp located at the throttle body end of the air intake hose.

4. Remove the air hose, the ISC hose, the breather hose and vacuum line from the air intake hose.

5. Disconnect the Mass Air Flow (MAF) electrical connector.

6. Remove the six heat shield attaching bolts and remove the shield.

7. Raise and safely support the vehicle.

8. Remove the three flange nuts.

9. Remove the nine manifold bolts.

10. Remove the exhaust manifold and discard the old gasket.

To install:

11. Position the manifold and new gasket. Tighten the attaching bolts to 31 ft. lbs. (42 Nm).

12. Raise and safely support the vehicle.

13. Install a new flange gasket. Tighten the three flange bolts to 24 ft. lbs. (33 Nm).

14. Lower the vehicle.

15. Install the heat shield. Tighten the attaching bolts to 18 ft. lbs. (25 Nm)

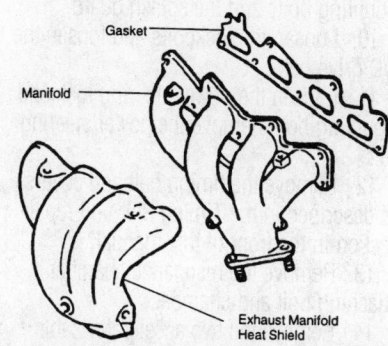

Exploded view of the exhaust manifold assembly—DOHC engine

16. Connect the two air intake hose mounting bolts.

17. Tighten the clamp located at the throttle body end of the air intake hose.

18. Connect the air hose, the ISC hose, the breather hose and vacuum line to the air intake hose.

19. Connect the MAF electrical connector.

20. Connect the negative battery cable.

Front Crankshaft Seal

REMOVAL & INSTALLATION

The front crankshaft seal is pressed into the external oil pump housing. Refer to oil pump removal and installation.

Camshaft

REMOVAL & INSTALLATION

SOHC Engine

1. Disconnect the negative battery cable.

2. Properly relieve the fuel system pressure.

3. Drain the coolant into a suitable container.

4. Remove the alternator, power steering and A/C drive belts.

5. Remove the fresh air duct from the top of the radiator.

6. Remove the upper radiator hose.

7. Remove the four attaching nuts to the clutch fan.

8. Remove the five fan shroud bolts. Remove the fan and shroud as an assembly.

9. Remove the five attaching bolts to the crankshaft pulley.

10. Remove the damper pulley.

11. Remove the timing belt and covers, as described in the Timing Belt unit repair section in the front of this manual.

12. Remove the three bolts attaching the accelerator cable and air intake hose from the valve cover.

13. Remove the air, ISC, breather, and vacuum hoses from the air intake hose.

14. Disconnect the electrical terminal from the mass air flow sensor.

15. Remove the air hose attaching bolt from the air cleaner housing cover.

16. Loosen the air intake hose clamps at the throttle body and air cleaner housing cover.

17. Remove the air intake hose.

18. Remove the PCV valve from the valve cover.

19. Remove the ISC air hose from the idle speed actuator.

20. Remove the air hose from behind the intake manifold.

21. Remove the two screws securing the ignition wire guides to the valve cover.

22. Remove the oxygen sensor wire loom from the holder at the rear of the valve cover.

23. Remove the two bolts attaching the vacuum pipe to the valve cover and remove the pipe.

24. Remove the seven valve cover bolts.

25. Remove the valve cover and discard the old gasket.

26. Remove the two nuts and two bolts from the front housing. Pull the housing forward to remove.

27. Before removing the rocker arm assembly, measure the camshaft end-play. The end-play specifications are:
- Normal: 0.003–0.006 in. (0.08–0.16mm)
- Maximum: 0.008 in. (0.20mm)

28. Loosen the 10 attaching bolts, in sequence, in 2 or 3 steps.

29. Remove the rocker arm and shaft assembly with the bolts in place.

30. Remove the camshaft.

To install:

31. Coat the camshaft with assembly lube and place it on the cylinder head.

32. Replace the oil seal on the rear rocker arm cap.

33. Apply silicone sealant to the four corners on the front and rear rocker arm caps.

34. Position the rocker arm assembly on the engine.

35. Tighten the bolts, in sequence, to 16 ft. lbs. (22 Nm).

36. Install the front housing. Tighten the nuts and bolts 16 ft. lbs. (22 Nm).

37. Coat the rocker arm assembly and cam lobes liberally with clean engine oil.

38. Adjust the valve lash.

39. Install the valve cover.

40. Install the seven valve cover bolts.

41. Install the air hose to behind the intake manifold.

42. Install the two screws securing the ignition wire guides to the valve cover.

43. Install the oxygen sensor wire loom to the holder at the rear of the valve cover.

44. Install the two bolts attaching the vacuum pipe to the valve cover and Install the pipe.

45. Install the PCV valve to the valve cover.

46. Install the ISC air hose to the idle speed actuator.

47. Install the air intake hose.

48. Position the air intake hose clamps at the throttle body and air cleaner housing cover.

49. Install the air hose attaching bolt to the air cleaner housing cover.

50. Install the air, ISC, breather, and vacuum hoses to the air intake hose.

51. Connect the electrical terminal to the mass air flow sensor.

52. Install the three bolts attaching the accelerator cable and air intake hose to the valve cover.

53. Install the timing belt and covers.

54. Install the five attaching bolts to secure the crankshaft pulley.

55. Install the fan and shroud as an assembly.

56. Install the four attaching nuts to the clutch fan.

57. Install the five fan shroud bolts.

58. Install the upper radiator hose.

59. Install the fresh air duct to the top of the radiator.

60. Install the alternator, power steering and A/C drive belts.

61. Properly fill the cooling system.

62. Connect the negative battery cable.

63. Start the engine and check for leaks.

64. Road test the vehicle.

DOHC Engine

1. Disconnect the negative battery cable.

2. Properly relieve the fuel system pressure.

3. Drain the coolant into a suitable container.

4. Remove the alternator drive belt.

5. Remove the fresh air duct from the top of the radiator.

6. Remove the upper radiator hose.

7. Remove the four attaching nuts to the clutch fan.

8. Remove the five fan shroud bolts. Remove the fan and shroud as an assembly.

9. Remove the four splash guard mounting bolts and the splash guard.

10. Loosen the lockbolts and loosen the A/C drive belt.

11. Loosen the power steering lock and mounting bolt. Remove the power steering belt.

12. Remove the timing belt and covers, as described in the Timing Belt unit repair section in the front of this manual.

13. Remove the resonance chamber attaching bolt and chamber.

14. Remove the two accelerator cable bracket bolts from the valve cover.

15. Remove the two air intake tube attaching bolts.

16. Disconnect the accelerator cable by pulling the throttle back all the way until the slot aligns with the hole in the accelerator cable.

17. Loosen the clamp attaching the air tube to the MAF sensor.

18. Disconnect and mark the hoses from the air tube assembly.

19. Remove the three air tube attaching bolts. Remove the air tube and air hose as an assembly.

20. Remove the six coil cover bolts from the top of the valve cover.

21. Remove the four attaching bolts from the two coils. Disconnect the electrical terminal from each coil and remove the coils.

22. Remove the spark plug wires from the spark plugs.

23. Remove the 15 valve cover bolts, then the valve cover.

24. Remove the camshaft pulley attaching bolts. Use a driver placed through one of the holes in the pulley to prevent it from moving when the attaching bolt is removed. Remove and mark the pulleys.

25. Remove the front cover attaching bolts and pull the front cover forward.

26. Remove the camshaft cap bolts in sequence.

27. Remove the caps and place them in order so they can be installed on the same journals.

28. Remove the camshafts.

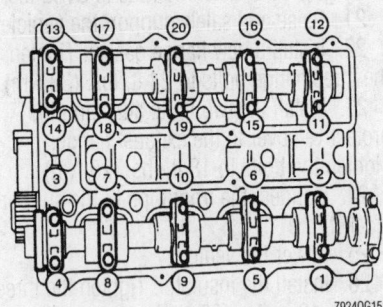

Camshaft cap bolt removal sequence— DOHC engine

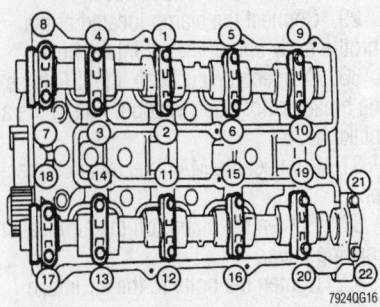

Camshaft journal bolt tightening sequence—DOHC engine

To install:

29. Install new oil seals on the ends of the camshafts.

30. Coat the camshafts with assembly lube and install them in the head with the dowel pins facing straight up.

➡**The exhaust cam has a steel dowel pin at the rear of the cam for the camshaft position sensor.**

31. Install the journal caps. Tighten the caps, in sequence, to 13–20 ft. lbs. (18–26 Nm).

32. Install the front cover on the alignment dowels. Tighten the attaching bolts to 72–108 inch lbs. (8–12 Nm).

33. Install the camshaft pulleys. Tighten the bolts to 35–48 ft. lbs. (47–65 Nm).

34. Install the 15 valve cover bolts with the valve cover in place. Tighten the bolts 35–52 inch lbs. (4–6 Nm).

35. Install the spark plug wires to the spark plugs.

36. Connect the electrical terminal to each coil and Install the coils.

37. Install the six coil cover bolts to the top of the valve cover.

38. Connect the accelerator cable by pulling the throttle back all the way until the slot aligns with the hole in the accelerator cable.

39. Tighten the clamp attaching the air tube to the MAF sensor.

40. Connect the hoses to the air tube assembly.

41. Install the three air tube attaching bolts. Install the air tube and air hose as an assembly.

42. Install the resonance chamber attaching bolt and chamber.

43. Install the two accelerator cable bracket bolts to the valve cover.

44. Install the two air intake tube attaching bolts.

45. Install the timing belt and covers.

46. Install and adjust the A/C and power steering drive belts.

47. Install the four splash guard mounting bolts attaching the splash guard.

48. Install and tighten the alternator belt.

49. Install the upper radiator hose.

50. Install the fan and shroud as an assembly.

51. Install the four attaching nuts to the clutch fan.

52. Install the five fan shroud bolts.

53. Install the fresh air duct to the top of the radiator.

54. Properly fill the cooling system.

55. Connect the negative battery cable.

56. Start the engine and check for leaks.

57. Road test the vehicle.

Valve Lash

ADJUSTMENT

SOHC Engine

1. Start the engine and bring the engine to normal operating temperature.

2. Shut the engine **OFF**.

3. Remove the valve cover.

4. Adjust both valves of each cylinder as follows:

   a. Turn the crankshaft so that the first cylinder is positioned at Top Dead Center (TDC) of the compression stroke (both the intake and exhaust valves of the cylinder will be fully closed).

   b. Check to see if the rocker arms are loose; if they are not, the engine is not at TDC/compression for that cylinder.

   c. Loosen the rocker arm adjuster locknut.

   d. Insert a 0.012 in. thick feeler gauge between the adjuster bolt and the top of the intake valve.

   e. Slowly turn the adjusting bolt until the feeler gauge can be removed and inserted with slight resistance.

   f. Hold the adjusting bolt still and tighten the locking nut.

   g. Adjust the lash on the exhaust valve in the same manner.

5. Repeat the procedure for each cylinder.

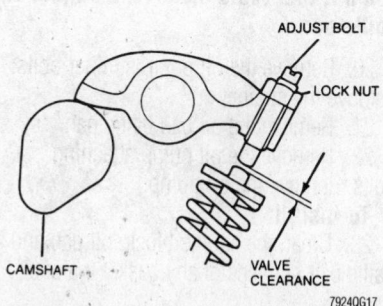

Valve lash is the clearance between the rocker arm adjust bolt and the valve stem tips

DOHC Engine

The DOHC engine uses Hydraulic Lash Adjusters (HLA's), which automatically maintain the proper amount of valve lash. Therefore, the DOHC engine does not need manual valve lash adjustment.

Oil Pan

REMOVAL & INSTALLATION

2WD Models

1. Raise and safely support the vehicle.

2. Remove the splash shield.

3. Drain the engine oil from the oil pan.

4. Remove the two transmission undercover bolts. Remove the cover.

5. Remove the 19 oil pan attaching bolts.

6. Remove the oil pan.

➡**If the oil pan is stuck to the engine block, strike the pan with a rubber mallet to break the seal. Do not pry the pan off, this could distort the oil pan or baffle pan. Also, if the oil pan will not clear the oil pump strainer, remove the engine mount bolts and raise the engine for more clearance.**

7. Remove the oil pump strainer bolts. Remove the strainer.

8. Remove the oil pan baffle pan.

To install:

9. Clean the engine block, oil pan and baffle pan surfaces of any gasket material.

10. Apply a continuous bead of Locktite Ultra Black 598® or equivalent silicone sealant around the baffle pan and position it on the engine.

11. Install the oil strainer with a new gasket to the oil pump. Tighten the two attaching bolts to 72–108 inch lbs. (8–12 Nm).

12. Apply a continuous bead of Locktite Ultra Black 598® or equivalent silicone sealant around the oil pan and position it on the engine. Tighten the attaching bolts to 72–108 inch lbs. (8–12 Nm).

13. Install the transmission cover.

14. Lower the engine on the mounts, if disconnected.

15. Install the splash shield.

16. Lower the vehicle.

17. Fill the crankcase with 4.4 qts. (4.16L) of clean oil.

4WD Models

1. Raise and safely support the vehicle.

2. Remove the splash shield.

3. Drain the engine oil from the oil pan.

4. Place a support stand under the front axle assembly.

5. Remove the three axle attaching bolts and lower the axle enough to allow the oil pan to clear during removal.

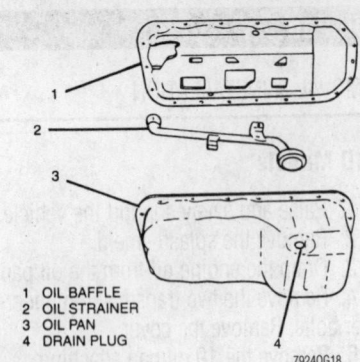

1 OIL BAFFLE
2 OIL STRAINER
3 OIL PAN
4 DRAIN PLUG

79240G18

Exploded view of the oil pan assembly mounting

6. Remove the two transmission under-cover bolts. Remove the cover.

7. Remove the 19 oil pan attaching bolts.

8. Remove the oil pan.

➡**If the oil pan is stuck to the engine block, strike the pan with a rubber mallet to break the seal. Do not pry the pan off, this could distort the oil pan or baffle pan.**

9. Remove the oil pump strainer bolts. Remove the strainer.

10. Remove the oil pan baffle pan.

To install:

11. Clean the engine block, oil pan and baffle pan surfaces of any gasket material.

12. Apply a continuous bead of Locktite Ultra Black 598® or equivalent silicone sealant around the baffle pan and position it on the engine.

13. Install the oil strainer with a new gasket to the oil pump. Tighten the two attaching bolts to 72–108 inch lbs. (8–12 Nm).

14. Apply a continuous bead of Locktite Ultra Black 598® or equivalent silicone sealant around the oil pan and position it on the engine. Tighten the attaching bolts to 72–108 inch lbs. (8–12 Nm).

15. Reposition the front axle and tighten the axle housing to frame bolt to 63 ft. lbs. (85 Nm). Tighten the differential and axle to frame bolts to 48 ft. lbs. (65 Nm).

16. Install the transmission cover.

17. Install the splash shield.

18. Lower the vehicle.

19. Fill the crankcase with 4.4 qts. (4.16L) of clean oil.

Oil Pump

REMOVAL & INSTALLATION

➡**The oil pump is externally-mounted, but still requires the removal of the oil pan to disconnect the oil pump strainer.**

SOHC Engine

1. Disconnect the negative battery cable.

2. Properly relieve the fuel system pressure.

3. Remove the alternator, power steering and A/C drive belts.

4. Remove the fresh air duct from the top of the radiator.

5. Remove the upper radiator hose.

6. Remove the four attaching nuts to the clutch fan.

7. Remove the five fan shroud bolts. Remove the fan and shroud as an assembly.

8. Remove the six attaching bolts to the crankshaft pulley.

9. Remove the damper pulley.

10. Remove the timing belt and covers, as described in the Timing Belt unit repair section in the front of the manual.

11. Remove the lower timing belt pulley.

12. Raise and safely support the vehicle.

13. Remove the splash shield.

14. Drain the engine oil.

15. Place a support stand under the front axle assembly.

16. Remove the three axle attaching bolts and lower the axle enough to allow the oil pan to clear during removal.

17. Remove the two transmission under-cover bolts.

18. Remove the 19 oil pan attaching bolts.

19. Remove the oil pan.

➡**If the oil pan is stuck to the engine block, strike the pan with a rubber mallet to break the seal. Do not pry the pan off, this could distort the oil pan or baffle pan.**

20. Remove the oil pump strainer bolts. Remove the strainer.

21. Remove the oil pan baffle pan.

22. Remove the oil pump attaching bolts. Remove the oil pump.

To install:

23. Clean the engine block, oil pan and baffle pan surfaces of any gasket material.

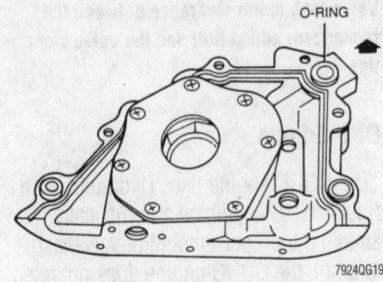

O-RING

79240G19

Be sure the oil pump O-ring is in the proper location prior to installation

24. Apply a continuous bead of Locktite Ultra Black 598® or equivalent silicone sealant around the oil pump and install a new O-ring. Position the pump on the engine.

➡**Do not allow sealant to get in the oil passages when applying sealant to the contact surface.**

25. Install the oil pump bolts and tighten the inside bolts 14–19 ft. lbs. (19–25 Nm) and the outer bolts 27–38 ft. lbs. (37–52 Nm).

26. Apply a continuous bead of Locktite Ultra Black 598® or equivalent silicone sealant around the baffle pan and position it on the engine.

27. Install the oil strainer with a new gasket to the oil pump. Tighten the attaching bolts to 72–108 inch lbs. (8–12 Nm).

28. Apply a continuous bead of Locktite Ultra Black 598® or equivalent silicone sealant around the oil pan and position it on the engine. Tighten the attaching bolts to 72–108 inch lbs. (8–12 Nm).

29. Reposition the front axle and tighten the axle housing to frame bolt to 63 ft. lbs. (85 Nm). Tighten the differential and axle to frame bolts to 48 ft. lbs. (65 Nm).

30. Install the splash shield.

31. Lower the vehicle.

32. Install the crankshaft pulley and belt guide. Tighten the attaching bolt to 132 inch lbs. (15 Nm).

33. Install the timing belt and covers.

34. Install the six attaching bolts to secure the crankshaft damper pulley. Tighten the bolts to 120 ft. lbs. (162 Nm).

35. Install the fan and shroud as an assembly.

36. Install the four attaching nuts to the clutch fan.

37. Install the five fan shroud bolts.

38. Install the upper radiator hose.

39. Install the fresh air duct to the top of the radiator.

40. Install the alternator, power steering and A/C drive belts.

41. Connect the negative battery cable.

42. Properly fill the cooling system.

43. Fill the crankcase with 4.4 qts. (4.16L) of clean engine oil.

44. Start the engine and check for leaks.

45. Road test the vehicle.

DOHC Engine

1. Disconnect the negative battery cable.

2. Properly relieve the fuel system pressure.

3. Remove the alternator drive belt.

4. Remove the fresh air duct from the top of the radiator.

5. Remove the upper radiator hose.

6. Remove the four attaching nuts to the clutch fan.

7. Remove the five fan shroud bolts. Remove the fan and shroud as an assembly.

8. Remove the four splash guard mounting bolts and the splash guard.

9. Loosen the lockbolts and loosen the A/C drive belt.

10. Loosen the power steering lock and mounting bolt. Remove the power steering belt.

11. Remove the timing belt and covers, as described in the Timing Belt unit repair section in the front of this manual.

12. Remove the lower timing belt pulley and locking bolt.

13. Raise and safely support the vehicle.

14. Remove the splash shield.

15. Drain the engine oil.

16. Place a support stand under the front axle assembly.

17. Remove the three axle attaching bolts and lower the axle enough to allow the oil pan to clear during removal.

18. Remove the two transmission undercover bolts.

19. Remove the 19 oil pan attaching bolts.

20. Remove the oil pan.

➡ **If the oil pan is stuck to the engine block, strike the pan with a rubber mallet to break the seal. Do not pry the pan off, this could distort the oil pan or baffle pan.**

21. Remove the oil pump strainer bolts. Remove the strainer.

22. Remove the oil pan baffle pan.

23. Remove the oil pump attaching bolts. Remove the oil pump.

To install:

24. Clean the engine block, oil pan and baffle pan surfaces of any gasket material.

25. Apply a continuous bead of Locktite Ultra Black 598® or equivalent silicone sealant around the oil pump and install a new O-ring. Position the pump on the engine.

➡ **Do not allow sealant to get in the oil passages when applying sealant to the contact surface.**

26. Install the oil pump bolts and tighten the inside bolts 14–19 ft. lbs. (19–25 Nm) and the outer bolts 27–38 ft. lbs. (37–52 Nm).

27. Apply a continuous bead of Locktite Ultra Black 598® or equivalent silicone sealant around the baffle pan and position it on the engine.

28. Install the oil strainer with a new gasket to the oil pump. Tighten the attach-ing bolts to 72–108 inch lbs. (8–12 Nm).

29. Apply a continuous bead of Locktite Ultra Black 598® or equivalent silicone sealant around the oil pan and position it on the engine. Tighten the attaching bolts to 72–108 inch lbs. (8–12 Nm).

30. Reposition the front axle and tighten the axle housing to frame bolt to 63 ft. lbs. (85 Nm). Tighten the differential and axle to frame bolts to 48 ft. lbs. (65 Nm).

31. Install the splash shield.

32. Lower the vehicle.

33. Install the lower timing belt pulley and locking bolt. Tighten the bolt to 120 ft. lbs. (162 Nm).

34. Install the timing belt and covers.

35. Install and adjust the A/C and power steering drive belts.

36. Install the four splash guard mounting bolts attaching the splash guard.

37. Install and tighten the alternator belt.

38. Install the upper radiator hose.

39. Install the fan and shroud as an assembly.

40. Install the four attaching nuts to the clutch fan.

41. Install the five fan shroud bolts.

42. Install the fresh air duct to the top of the radiator.

43. Properly fill the cooling system.

44. Fill the crankcase with 4.4 qts. (4.16L) of clean engine oil.

45. Connect the negative battery cable.

46. Start the engine and check for leaks.

47. Road test the vehicle.

Rear Main Seal

REMOVAL & INSTALLATION

1. Remove the transmission assembly from the vehicle.

2. Remove the clutch and flywheel assembly on manual transmission equipped

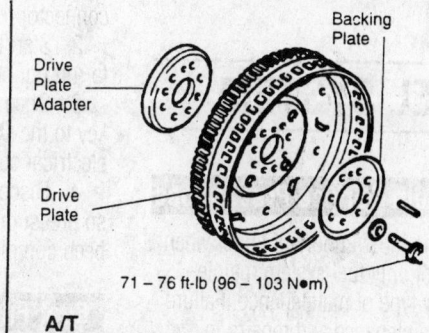

Drive Plate Adapter

Backing Plate

Drive Plate

71 – 76 ft-lb (96 – 103 N•m)

A/T

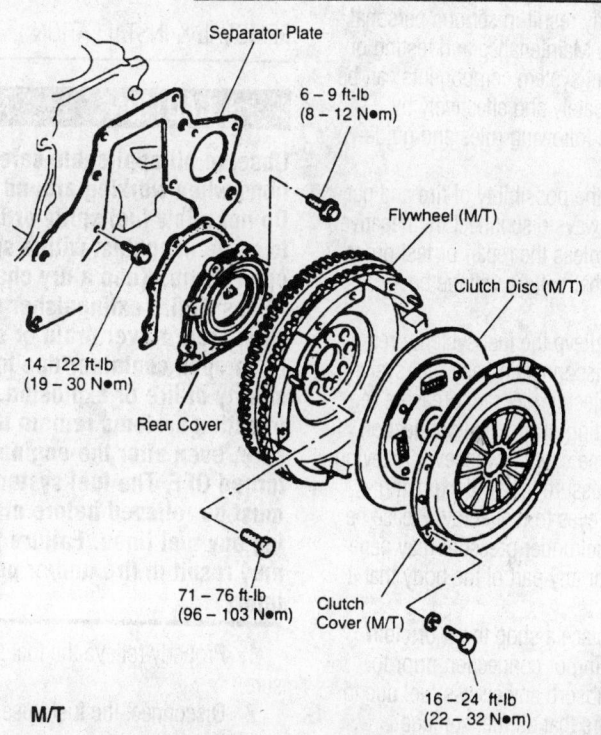

Separator Plate

6 – 9 ft-lb (8 – 12 N•m)

Flywheel (M/T)

Clutch Disc (M/T)

14 – 22 ft-lb (19 – 30 N•m)

Rear Cover

71 – 76 ft-lb (96 – 103 N•m)

Clutch Cover (M/T)

16 – 24 ft-lb (22 – 32 N•m)

M/T

79240G38

Exploded view of the rear main seal and related components

vehicles or the flexplate on vehicles with automatic transmissions.

3. Remove the six bolts attaching the rear cover to the engine block. Remove the rear cover.

4. Press the old seal out of the cover.

To install:

5. Coat the new seal with clean oil and press the seal into the cover.

6. Install the rear cover and new gasket to the engine block. Tighten the bolts to 69–104 inch lbs. (8–12 Nm).

7. Cut away the portion of gasket that projects out from the rear cover toward the oil pan.

➡**Do not scratch the rear cover assembly when cutting the gasket.**

8. Install the clutch and flywheel assembly on manual transmission equipped vehicles or the flexplate on vehicles with automatic transmissions.

9. Install the engine assembly.

FUEL SYSTEM

Fuel System Service Precautions

Safety is the most important factor when performing not only fuel system maintenance but any type of maintenance. Failure to conduct maintenance and repairs in a safe manner may result in serious personal injury or death. Maintenance and testing of the vehicle's fuel system components can be accomplished safely and effectively by adhering to the following rules and guidelines.

• To avoid the possibility of fire and personal injury, always disconnect the negative battery cable unless the repair or test procedure requires that battery voltage be applied.

• Always relieve the fuel system pressure prior to disconnecting any fuel system component (injector, fuel rail, pressure regulator, etc.), fitting or fuel line connection. Exercise extreme caution whenever relieving fuel system pressure to avoid exposing skin, face and eyes to fuel spray. Please be advised that fuel under pressure may penetrate the skin or any part of the body that it contacts.

• Always place a shop towel or cloth around the fitting or connection prior to loosening to absorb any excess fuel due to spillage. Ensure that all fuel spillage (should it occur) is quickly removed from

engine surfaces. Ensure that all fuel soaked cloths or towels are deposited into a suitable waste container.

• Always keep a dry chemical (Class B) fire extinguisher near the work area.

• Do not allow fuel spray or fuel vapors to come into contact with a spark or open flame.

• Always use a back-up wrench when loosening and tightening fuel line connection fittings. This will prevent unnecessary stress and torsion to fuel line piping. Always follow the proper torque specifications.

• Always replace worn fuel fitting O-rings with new. Do not substitute fuel hose or equivalent where fuel pipe is installed.

Fuel System Pressure

RELIEVING

1. Disconnect the fuel pump harness connector located behind the rear seat.

2. Start the engine and allow the engine to run out of fuel.

3. Once the engine has stalled, turn the key to the **OFF** position and connect the electrical connector.

4. Disconnect the negative battery cable so pressure cannot build up until work has been completed.

Fuel Filter

REMOVAL & INSTALLATION

✳ CAUTION

Observe all applicable safety precautions when working around gasoline. Do not allow fuel spray or fuel vapors to come in contact with a spark or open flame. Keep a dry chemical (Class B) fire extinguisher near the work area. Never drain or store fuel in an open container due to the possibility of fire or explosion. Fuel injection systems remain under pressure, even after the engine has been turned OFF. The fuel system pressure must be relieved before disconnecting any fuel lines. Failure to do so may result in fire and/or personal injury.

1. Properly relieve the fuel system pressure.

2. Disconnect the fuel hoses from the fuel filter.

3. Loosen the bracket bolt and remove the fuel filter from the bracket.

To install:

4. Place the fuel filter in the bracket and tighten the bolt.

5. Install the fuel hoses on the filter connection fully.

6. Tighten the clamps.

7. Start the engine and check for fuel leaks.

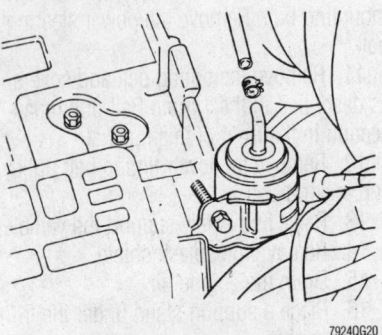

Fuel filter underhood mounting location

Fuel Pump

REMOVAL & INSTALLATION

✳ CAUTION

Observe all applicable safety precautions when working around gasoline. Do not allow fuel spray or fuel vapors to come in contact with a spark or open flame. Keep a dry chemical (Class B) fire extinguisher near the work area. Never drain or store fuel in an open container due to the possibility of fire or explosion. Fuel injection systems remain under pressure, even after the engine has been turned OFF. The fuel system pressure must be relieved before disconnecting any fuel lines. Failure to do so may result in fire and/or personal injury.

1. Disconnect the negative battery cable.

2. Properly relieve the fuel system pressure.

3. Open the vehicle widows for ventilation.

4. From inside the rear compartment of the vehicle tilt the rear seat forward.

5. Pull the carpet back to uncover the fuel tank access panel on the floor.

6. Disconnect the wiring terminals at the access cover.

7. Remove the screw securing the ground wire lug.

8. Remove the four access panel screws and lift up the panel.

9. Remove the two clamps from the fuel lines and disconnect the hoses from the fuel tank.

10. Remove the eight retaining screws from the fuel pump assembly.

11. Carefully remove the fuel pump assembly from the tank. Tilting the fuel pump assembly during removal may be required for clearance.

12. Wrap the fuel pump assembly in a rag before removing the assembly from the vehicle.

13. Cover or seal the fuel tank until installing the fuel pump assembly.

To install:

➡**The fuel pump is part of the assembly and is replaced as a complete unit.**

14. Remove any protective covers from the ports on the replacement fuel pump assembly.

15. Carefully guide the fuel pump assembly into the tank. Tilting the fuel pump assembly during installation may be required for clearance.

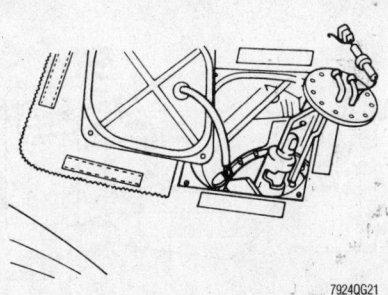

7924QG21

Removing the fuel pump through the access panel

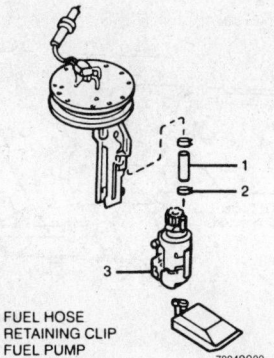

1 FUEL HOSE
2 RETAINING CLIP
3 FUEL PUMP

7924QG22

Exploded view of the fuel pump assembly

16. Install the eight retaining screws in the fuel pump assembly.

17. Install the fuel hoses and tighten the clamps.

18. Connect the wiring terminals at the access cover.

19. Reattach the screw securing the ground wire lug.

20. Cover the fuel tank access panel with the carpet.

21. Reposition the seat.

22. Connect the negative battery cable.

23. Start the vehicle and let the engine run to check the fuel pump operation.

DRIVE TRAIN

Transmission Assembly

REMOVAL & INSTALLATION

Manual

➡**The removal of the manual transmission is virtually the same for 4WD and 2WD vehicles.**

1. Disconnect the negative battery cable.

2. From the inside of the vehicle, remove the rear portion of the center console.

  a. Remove the two screws at the rear and the three screws in the front of the console.

  b. Raise the emergency brake handle.

  c. Unscrew the shifter knobs.

3. Lift the console up and place it out of the way.

4. Remove the five nuts and one bolt attaching the boot cover from the floorboard.

5. Remove the floor shifter lever by:

  a. Removing the four shift lever bolts attaching the lever to the transmission.

  b. Tilt the shift lever back to access the linkage.

  c. Remove the retaining clip from the linkage pin and remove the pin.

6. Remove the 4WD transfer lever by:

  a. Removing the five shift lever bolts attaching the lever to the transfer case.

  b. Tilt the shift lever back to access the linkage.

  c. Remove the retaining clip from the vehicle speed sensor cable.

7. Raise and safely support the vehicle.

When supporting the vehicle, know that when the transmission is removed the center of gravity will shift to the front of the vehicle. Place support stands in the front of the vehicle to prevent tipping when the transmission is removed.

8. Drain the fluid from the transmission and transfer case.

9. Matchmark the driveshafts. Remove the four attaching bolts from each end of the driveshafts and remove the driveshafts.

10. Disconnect the back-up light switch wiring connector.

11. Disconnect the 4WD indicator switch connector.

12. Disconnect the vehicle speed sensor connector.

13. Remove the crankshaft position sensor attaching bolt and remove the sensor from the transmission housing.

14. Remove the two clutch release cylinder bolts and move the cylinder aside.

15. Remove the three attaching bolts from the exhaust bracket.

16. Remove the five lower front transmission housing bolts.

17. Remove the transfer case side mount nuts and bolts.

18. Support the transmission and transfer case with a transmission jack.

19. Remove the four attaching bolts from each side of the crossmember.

20. Remove the two bolts attaching the transmission mount to the crossmember.

21. Remove the three starter bolts.

22. Lower the transmission enough to access the upper clutch housing bolts. Remove the bolts.

23. Remove the transmission from the vehicle.

To install:

24. Raise the transmission and push forward enough to engage the housing to the dowel pins.

25. Install a clutch housing bolt in to secure the transmission.

26. Route the vehicle speed sensor wiring harness over the top of the transmission.

27. Install the clutch housing bolts. Tighten the:

- Six 14mm bolts to 80 ft. lbs. (108 Nm).
- Five 10mm bolts to 29 ft. lbs. (39 Nm).
- One 6mm bolt to 5 ft. lbs. (7 Nm).

28. Install the three starter bolts. Tighten to 29 ft. lbs. (39 Nm).

29. Raise the transmission and install the crossmember. Tighten the eight attaching bolts to 32 ft. lbs. (44 Nm).

30. Install the two rear transmission mount bolts. Tighten to 80 ft. lbs. (108 Nm).

31. Install the transfer case side mount nuts and bolts.

32. Remove the transmission jack.

33. Install the three attaching bolts to the exhaust bracket.

34. Position the clutch release cylinder and tighten the attaching bolts 29 ft. lbs. (39 Nm).

35. Install the crankshaft position sensor in the clutch housing. Tighten the bolt 5 ft. lbs. (7 Nm).

36. Connect the back-up light switch wiring connector.

37. Connect the 4WD indicator switch connector.

38. Connect the vehicle speed sensor connector.

39. Install the drive shafts with the matchmarks aligned. Tighten the attaching bolts 27 ft. lbs. (36 Nm).

40. Add 1.3 qts. (1.25L) of API GL-4 or GL-5 75W-90 weight gear oil in the transmission.

41. Lower the vehicle.

42. Connect the shift lever linkage. Mount the shift lever on the transmission and tighten the attaching bolts to 18 ft. lbs. (24 Nm).

43. Connect the transfer case shift lever linkage. Mount the lever on the transmission and tighten the attaching bolts to 18 ft. lbs. (24 Nm).

44. Clip the vehicle speed sensor harness in position on top of the transmission.

45. Install the center console and shift lever knobs.

46. Connect the negative battery cable.

Automatic

1. Disconnect the negative battery cable.

2. Disconnect the automatic transmission control cable from the throttle body.

3. Remove the air silencer.

4. Provide automatic transmission control cable slack by gently pulling the cable to the left until the cable pin has rotated sufficiently to line the automatic transmission control cable up with the slot in the rear of the throttle body bellcrank.

5. Slide the automatic transmission control cable and cable pin to the rear and out of the bellcrank.

6. Loosen the locknut and remove the throttle kickdown cable from the mounting bracket.

7. Position the cable so that it can be slid out through the bottom of the engine compartment.

8. Remove the two rear console mounting screws. Slide the console forward to clear the parking brake handle and set aside.

9. Remove the three mounting screws from the front console. Untie the shift boot draw strings and open the boot to allow clearance over the shift lever handle.

10. Loosen the transfer case shift lever locknut and remove the lever knob.

11. Pull the console up to access the Power/Economy switch wiring connector. Unplug the connector and remove the console.

12. Shift the transfer lever to the 4L position.

13. Remove the four cover plate bolts and remove the plate.

14. Remove the five retaining bolts from the transfer case and lift the shifter lever assembly straight out.

➡ **Cover the hole in the transfer case with cardboard or plastic secured by tape to keep dirt and foreign material out of the transfer case.**

15. Shift the transmission shift selector lever to the **P** (Park) position. Remove the four shift selector lever nuts.

16. Raise and safely support the vehicle.

17. Drain the fluid from the transmission.

18. Remove the split pin from the shift selector lever. Disconnect the shift rod and washers.

19. Lower the vehicle.

20. Disconnect the four wiring connectors under the shift selector lever.

➡ **There is a fifth wire connection (Park Position). This wire is hard-wired, do not disconnect it.**

21. Remove the shift selector lever in the following steps, do not exert any force to remove the shift selector lever.

   a. Slide the shift selector lever assembly to the right as far as possible. (2–3 in./5.1–7.6cm).

   b. Tilt the shift selector lever assembly to the left.

   c. Lift the assembly up slightly and push forward (still tilted).

   d. Slide the shift selector lever assembly to the right while pulling up until the lever rod pin clears.

   e. Remove the shift selector lever and set it aside.

22. Disconnect the upper dipstick tube from the lower tube.

23. Raise and safely support the vehicle.

⁂ CAUTION

When supporting the vehicle, know that when the transmission is removed the center of gravity will shift to the front of the vehicle. Place support stands in the front of the vehicle to prevent tipping when the transmission is removed.

24. Remove the undercover splash shield.

25. Unplug the following connectors:

• Input/Turbine speed sensor from the top rear of the transmission.

• Shift solenoids from the lower left side of the transmission.

• Vehicle speed sensor from the center of the transfer case.

26. Matchmark the front and rear driveshafts. Remove the attaching bolts from both flanges and remove the driveshafts.

27. Position a drain under the oil cooler lines before loosening the clamps.

28. Remove the two oil cooling tube clamps and disconnect the lines.

29. Disconnect the oil cooler pipes at the transmission.

30. Allow the fluid to fully drain from the cooling lines.

31. Remove the three starter attaching bolts.

32. Remove the two transfer case mounting bolts located at the right center of the transfer case.

33. Remove the four transfer case nuts from the crossmember.

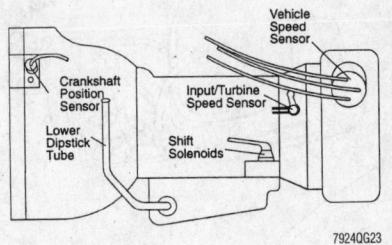

Automatic transmission wiring connections

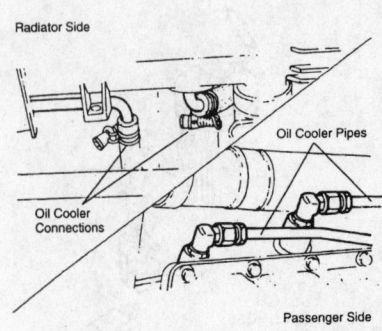

Oil cooler connections—models equipped with automatic transmissions

34. Place the transmission jack under the transmission. Secure the transmission to the jack.

35. Remove the eight crossmember bolts and remove the crossmember.

36. Remove the transmission mounting bolts from the transmission.

➡**When removing the transmission bolts, the following directions will be helpful:**

• When removing the two lower right-hand side bolts the gusset will drop free.

• When removing the left-hand side exhaust bracket bolt, a 21 inch extension, universal joint and 14mm socket are required. Access to the bolt is gained from behind and above the lower portion of the transmission dipstick tube. Loosen the lower exhaust bracket mounting bolt and slip the bracket out of the way after the top two mounting bolts have been removed.

• When removing the left-hand side upper mounting bolts, 40 inches of extension, a universal joint and a 17mm socket are required. Access to the bolt is gained from between the selector rod and transmission housing at the rear of the transmission oil pan with the transmission in the hanging position, supported with the transmission jack. Do not allow the transmission to hang free.

• When removing the right-hand side upper mounting bolts, a 20 inch extension, a universal joint and a 17mm socket are required. Access to the bolt is gained from below the transmission with the transmission in the hanging position, supported with the transmission jack. Do not allow the transmission to hang free.

37. After removing the transmission bolts, raise the transmission until it is level.

38. Remove the four front splash guard mounting bolts and the guard.

39. Remove the left-hand side lower gusset to allow access to the torque converter inspection cover bolt.

40. Remove the torque converter inspection cover bolts and cover.

41. Slide the transmission back about 2–3 in. (5.1–7.6cm) to access the six torque converter-to-drive plate bolts. Remove the bolts.

42. Lower the transmission only enough to disconnect the three electrical connectors on the top of the transmission.

43. Remove the crankshaft position sensor attaching bolt and the sensor.

44. Disconnect the 4WD, 4WD LOW indicator switches and the Vehicle Speed Sensor (VSS) from the transfer case.

45. Lower the transmission, be sure all wiring is clear and disconnected. Be sure the throttle cables come out without binding or attaching to anything.

To install:

46. Lift the transmission into position to attach the sensor and indicator switch wiring. Be sure the throttle cables are guided into the engine compartment without binding or attaching to anything.

47. Connect the 4WD, 4WD LOW indicator switches and the VSS to the transfer case.

48. Install the crankshaft position sensor.

49. Raise the transmission and position it to the engine. Install the upper housing bolts. Tighten the:

• 10mm bolts to 38–60 ft. lbs. (57–81 Nm).

• 12mm bolts to 51–65 ft. lbs. (69–88 Nm).

50. Install the exhaust hanger and bracket. Tighten the bolts to 38–60 ft. lbs. (57–81 Nm).

51. Install the crossmember. Tighten the eight crossmember bolts to 23–34 ft. lbs. (31–46 Nm).

52. Install the two transfer case mounting bolts located at the right center of the transfer case, and tighten to 23–34 ft. lbs. (31–46 Nm).

53. Install the four transfer case nuts to the crossmember. Tighten to 23–34 ft. lbs. (31–46 Nm).

54. Install the six torque converter-to-drive plate bolts. Tighten the bolts to 12–20 ft. lbs. (16–27 Nm).

55. Install the torque converter inspection cover bolts and cover. Tighten the bolts to 41–62 inch lbs. (5–7 Nm).

56. Install the four front splash guard mounting bolts and the guard. Tighten the bolts to 41–62 inch lbs. (5–7 Nm).

57. Install the three starter attaching bolts. Tighten the bolts to 27–40 ft. lbs. (37–54 Nm).

58. Install the left-hand side lower gusset. Tighten the attaching bolts to 38–60 ft. lbs. (57–81 Nm).

59. Install the right-hand side lower gusset in 3 steps.

a. Install the bottom mounting bolts, but do not tighten.

b. Install the top bolt to the intake manifold support bracket and manifold. Tighten to 27–40 ft. lbs. (37–54 Nm).

c. Secure the manifold intake bracket by tightening the two attaching bolts to 38–60 ft. lbs. (57–81 Nm).

60. Connect the oil cooler pipes at the transmission. Tighten the lines to 42–62 inch lbs. (5–7 Nm).

61. Install the two oil cooling tube clamps to the lines.

62. Install the driveshafts with the matchmarks aligned. Tighten the attaching bolts to the differential flanges to 20–22 ft. lbs. (27–30 Nm) and the transfer case flange bolts to 36–43 ft. lbs. (49–59 Nm).

63. Install the undercover splash shield. Tighten the bolts to 42–62 inch lbs. (5–7 Nm).

64. Connect the upper dipstick tube to the lower tube.

65. Lower the vehicle.

66. Provide automatic transmission control cable slack by gently pulling the cable to the left until the cable pin has rotated sufficiently to line the automatic transmission control cable up with the slot in the rear of the throttle body bellcrank.

67. Slide the automatic transmission control cable and cable pin into the bellcrank.

68. Tighten the locknut.

69. Install the throttle kickdown cable to the mounting bracket.

70. Connect the automatic transmission control cable to the throttle body.

71. Install the air silencer.

72. Connect the shifter lever assembly to the transfer case. Tighten the retaining bolts to 72–102 inch lbs. (22–28 Nm)

73. Connect the four wiring connectors under the shift selector lever.

74. Install the shift selector lever, do not exert any force when installing the shift selector lever.

75. Install the four shift selector lever nuts. Tighten to 72–102 inch lbs. (22–28 Nm).

76. Raise and safely support the vehicle.

77. Connect the shift rod and washers to the selector lever. Install the split pin to the shift selector lever.

78. Lower the vehicle.

79. Connect the Power/Economy switch wiring connector. Install the console.

80. Install the three mounting screws to the front console. Tie the shift boot draw strings.

81. Connect the negative battery cable.

82. Fill the transmission with 7 qts. (6.6L) Dexron® II or III fluid.

83. Start the vehicle. Check for leaks.

84. Road test the vehicle to check for proper operation.

Clutch

ADJUSTMENT

Pedal Height

1. Pull back the carpet to measure the distance from the firewall to the top of the pedal. The standard height is 9.84 in. (25cm).

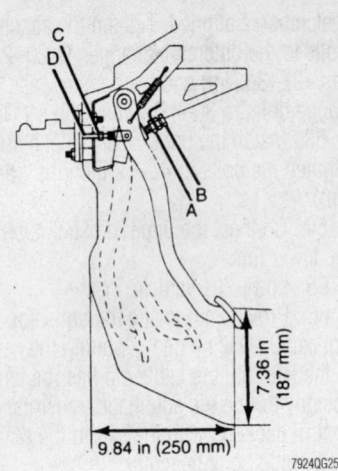

Clutch pedal height and free-play adjustment points

2. If adjustment is required, loosen the locknut and turn the stopper bolt.

3. After adjustment is made tighten the locknut to 12 ft. lbs. (16 Nm).

Free-Play

1. Depress the clutch pedal gently by hand and measure the amount of free-play (distance the pedal travels before resistance is felt). The proper amount of free-play is 0.5 in. (12.7mm). If the free-play is not within the proper specifications, continue with the procedure.

2. Measure from the floor pan to the middle point of the clutch pedal when the pedal is in the fully released position. The proper clutch pedal height is 7.36 in. (187mm).

3. If the pedal height is incorrect, loosen locknut (A) and turn the pedal adjusting bolt (B) until the proper height is achieved, then retighten the locknut.

4. Remeasure the free-play. If it is still out of specification, loosen the clutch pushrod locknut (C) and turn the pushrod (D) until the proper free-play is achieved. Tighten locknut (C) securely.

REMOVAL & INSTALLATION

1. Remove the transmission from the vehicle.

2. Remove the six pressure plate bolts. Remove the clutch plate and disc.

To install:

3. Install the clutch disc using a centering tool.

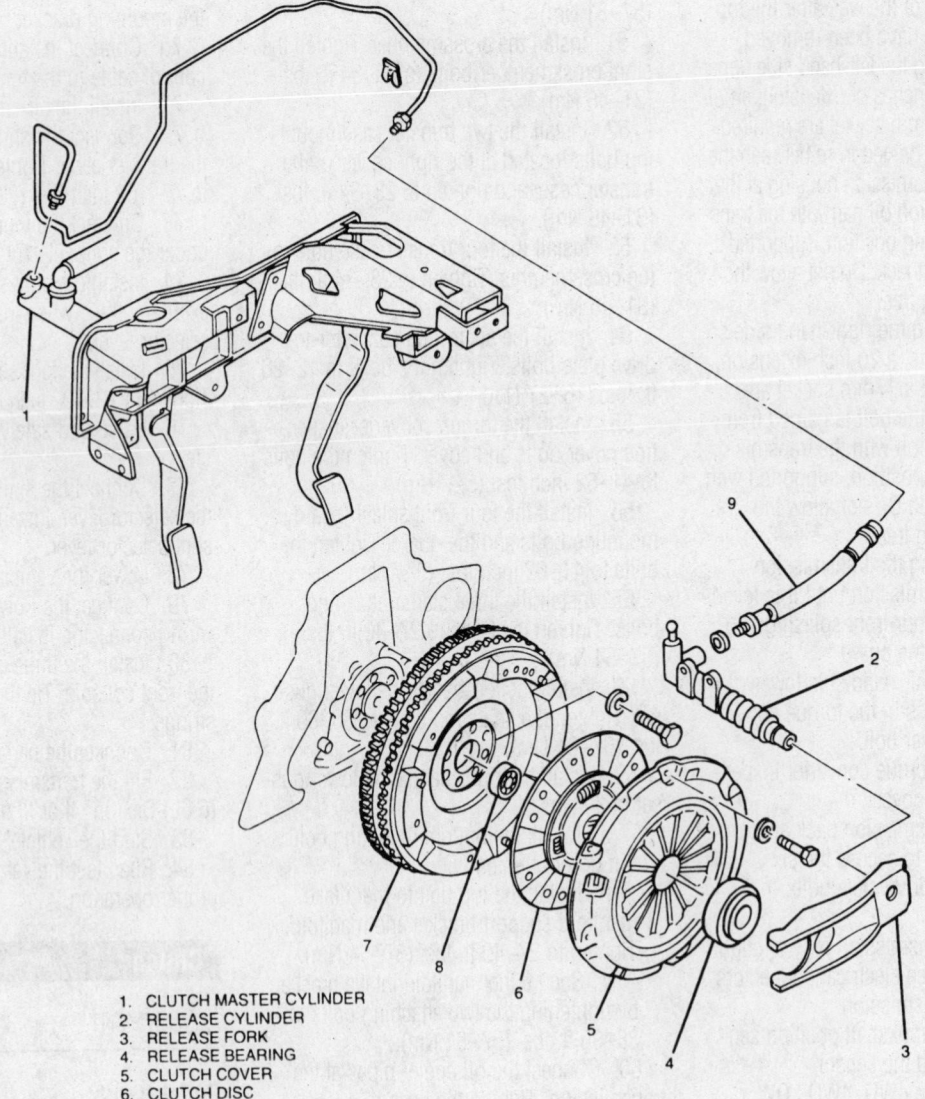

1. CLUTCH MASTER CYLINDER
2. RELEASE CYLINDER
3. RELEASE FORK
4. RELEASE BEARING
5. CLUTCH COVER
6. CLUTCH DISC
7. FLYWHEEL
8. PILOT BEARING
9. FLEXIBLE HOSE

Exploded view of the clutch assembly

4. Position the clutch cover and tighten the bolts to 73 ft. lbs. (99 Nm).

5. Remove the centering tool.

6. Check the release bearing condition. Lubricate or replace as necessary.

7. Install the transmission.

Hydraulic Clutch System

BLEEDING

1. With an assistant in the vehicle, raise and safely support the vehicle.

2. Have your assistant pump the clutch pedal three times and hold the pedal to the floor.

3. Open the bleeder valve on the clutch slave cylinder until the air is purged from the cylinder.

4. Tighten the bleeder valve.

5. Have your assistant release the clutch pedal.

6. Fill the clutch master cylinder if below minimum.

· 7. Repeat Steps 2 through 6 until no air exits from the bleeder valve.

8. Lower the vehicle.

9. Fill the clutch master cylinder fluid reservoir.

Transfer Case Assembly

REMOVAL & INSTALLATION

1. Disconnect the negative battery cable.

2. Remove the two rear console mounting screws. Slide the console forward to clear the parking brake handle and set aside.

3. Remove the three mounting screws from the front console. Untie the shift boot draw strings and open the boot to allow clearance over the shift lever handle.

4. Loosen the transfer case shift lever locknut and remove the lever knob.

5. Pull the console up to access the Power/Economy switch wiring connector. Unplug the connector and remove the console.

6. Shift the transfer lever to the 4L position.

7. Remove the four cover plate bolts and remove the plate.

8. Remove the five retaining bolts from the transfer case and lift the shifter lever assembly straight out.

9. Raise and safely support the vehicle.

10. Drain the fluid from the transfer case.

11. Support the transmission with a post jack.

12. Place the transmission jack under the transfer case.

13. Matchmark the driveshafts at the flanges. Remove the bolts from the flanges and remove the driveshafts.

14. Remove the eight crossmember bolts and remove the crossmember.

15. Disconnect the 4WD light switch connector.

16. Disconnect the speed censor connector.

17. Remove the two transfer case mounting bolts located at the right center of the transfer case.

18. Remove the four transfer case nuts from the crossmember.

19. Remove the 10 bolts attaching the transfer case to the transmission.

20. Separate the transfer case from the transmission by striking the transfer case with a plastic mallet at the seal area.

21. Lower the transfer case from the vehicle.

To install:

22. Raise the transfer case in position with a new gasket in place.

23. Tighten the 10 bolts attaching the transfer case to the transmission to 32 ft. lbs. (44 Nm).

24. Install the crossmember. Tighten the eight crossmember bolts to 32 ft. lbs. (44 Nm).

25. Install the two transfer case mounting bolts located at the right center of the transfer case. 38 ft. lbs. (52 Nm).

26. Install the four transfer case nuts from the crossmember. Tighten the nuts to 15 ft. lbs. (20 Nm).

27. Align the matchmarks on the driveshafts to the flanges. Tighten the bolts at the flanges to 27 ft. lbs. (36 Nm).

28. Remove the transmission and post jacks.

29. Attach the 4WD light switch connector.

30. Attach the speed censor connector.

31. Add 1.4 qts. (1.3L) of Dexron® II, or equivalent fluid, until level with the fill hole of the transfer case.

32. Lower the vehicle.

33. Install the shifter lever assembly. Tighten the five retaining bolts to the transfer case to 20 ft. lbs. (27 Nm).

34. Install the cover plate. Tighten the bolts to 15 ft. lbs. (20 Nm).

35. Attach the Power/Economy switch wiring connector.

36. Install the front console.

37. Install the lever knobs.

38. Install three mounting screws on the front console. Tie the shift boot draw strings.

39. Slide the console over the parking brake handle. Install the two rear console mounting screws.

40. Connect the negative battery cable.

41. Road the vehicle to assure proper operation.

Halfshaft

REMOVAL & INSTALLATION

1. Raise and safely support the vehicle.

2. Remove the tire and wheel assembly.

3. Remove the six free wheel hub bolts. Remove the hub body.

4. Remove the snapring and spacer from the hub.

5. Carefully remove the fixed cam assembly.

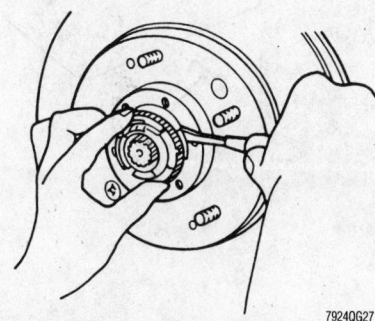

79240G27

Removing the 4WD fixed cam assembly

6. Remove the two attaching bolts and remove the caliper from the rotor.

7. Remove the upper control arm link lockbolt, spring washer and nut.

8. Disconnect the tie rod end from the steering knuckle.

9. Loosen the drop link lower locknut.

10. Loosen the four upper drop link locknuts.

11. Spread open the drop link fork with a rubber mallet.

12. Matchmark the halfshaft and differential.

13. Carefully pry the halfshaft from the differential and remove the halfshaft from the vehicle.

To install:

14. Install the halfshaft with the matchmarks aligned.

15. Tighten the upper and lower drop link nuts to 36 ft. lbs. (49 Nm).

16. Connect the tie rod end to the steering knuckle. tighten the locknut to 27 ft. lbs. (36 Nm) and install a new cotter pin.

17. Install the upper control arm link lockbolt, spring washer and nut. Tighten the nut and bolt to 36 ft. lbs. (49 Nm).

18. Assemble the fixed cam assembly.

19. Install the snapring and spacer in the hub.

20. Install the wheel hub and the six free wheel hub bolts. Tighten the bolts in two passes, 14 ft. lbs. (17 Nm), then to 23 ft. lbs. (31 Nm).

21. Install the tire and wheel assembly. Tighten the lug nuts to 73 ft. lbs. (99 Nm).

22. Lower the vehicle.

Locking Hubs

REMOVAL & INSTALLATION

1. Raise and safely support the vehicle.

2. Remove the tire and wheel assembly.

3. Remove the six free wheel hub bolts. Remove the hub body.

4. Remove the snapring and spacer from the hub.

5. Carefully remove the fixed cam assembly.

To install:

6. Install the fixed cam assembly.

7. Install the snapring and spacer in the hub.

8. Install the wheel hub and the six free wheel hub bolts. Tighten the bolts in two passes, 14 ft. lbs. (17 Nm), then to 23 ft. lbs. (31 Nm).

9. Install the tire and wheel assembly. Tighten the lug nuts to 73 ft. lbs. (99 Nm).

10. Lower the vehicle.

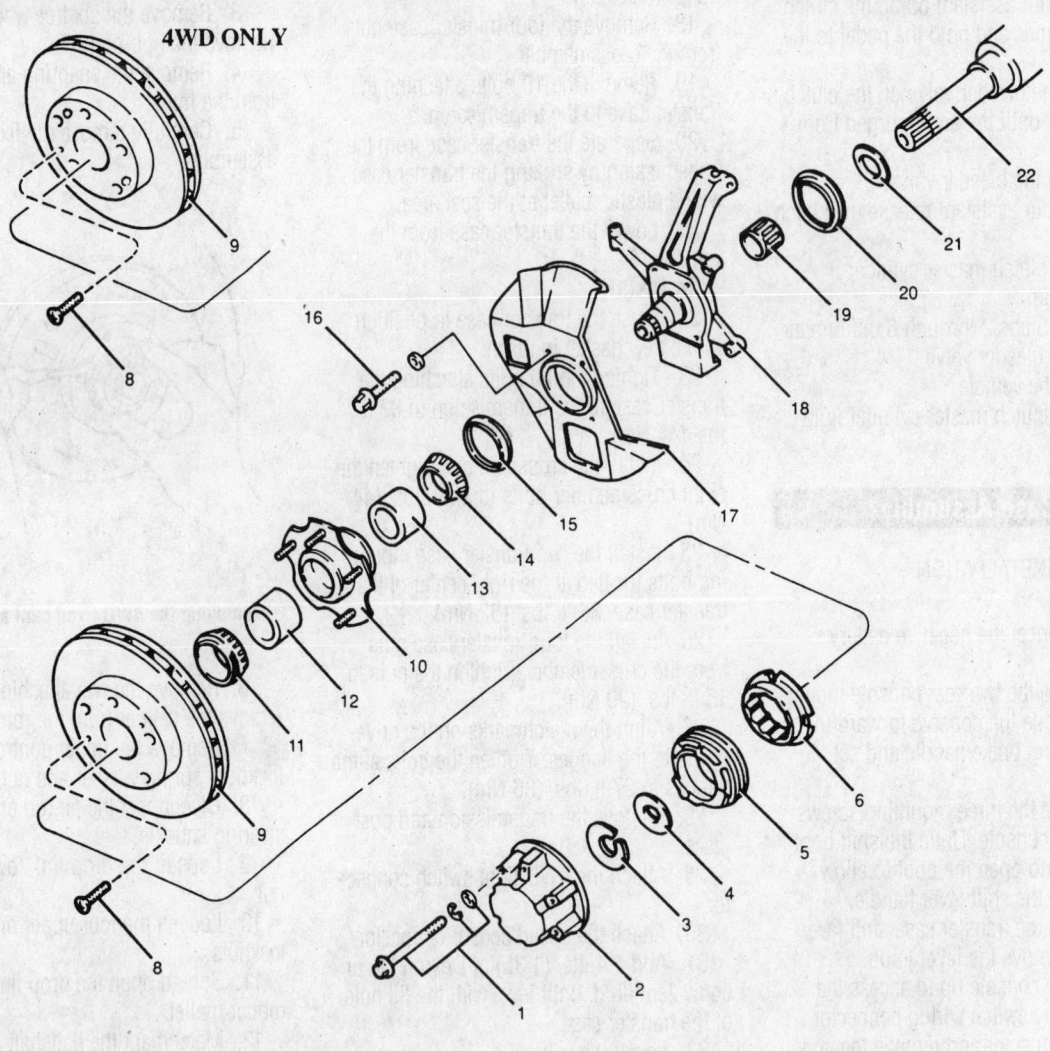

4WD ONLY

1 BOLT/WASHER	9 ROTOR	16 BOLT & SPRING WASHER
2 FREE WHEEL HUB BODY	10 WHEEL HUB	17 DUST COVER
3 SNAP RING	11 INNER BEARING INNER RACE	18 KNUCKLE
4 SPACER	12 INNER BEARING OUTER RACE	19 NEEDLE BEARING
5 FIXED CAM ASSEMBLY	13 OUTER BEARING OUTER RACE	20 OIL SEAL
6 LOCK NUT	14 OUTER BEARING INNER RACE	21 SPACER
8 SCREW	15 OIL SEAL	22 DRIVE SHAFT (LH)

7924QG28

Exploded view of the 4WD locking hub assembly

STEERING AND SUSPENSION

Air Bag

※※ CAUTION

Some vehicles are equipped with an air bag system, also known as the Supplemental Inflatable Restraint (SIR) or Supplemental Restraint System (SRS). The system must be disabled before performing service on or around system components, steering column, instrument panel components, wiring and sensors. Failure to follow safety and disabling procedures could result in accidental air bag deployment, possible personal injury and unnecessary system repairs.

PRECAUTIONS

Several precautions must be observed when handling the inflator module to avoid accidental deployment and possible personal injury.

• Never carry the inflator module by the wires or connector on the underside of the module.

• When carrying a live inflator module, hold securely with both hands, and ensure that the bag and trim cover are pointed away from you.

• Place the inflator module on a bench or other surface with the bag and trim cover facing up.

• With the inflator module on the bench, never place anything on or close to the module which may be thrown in the event of an accidental deployment.

DISARMING

1. Turn the ignition switch to the **LOCK** position.
2. Disconnect the negative battery cable.
3. Wait 10 minutes for the back-up power to discharge.

ARMING

Assuming the system components (air bag control module, sensors, air bag, etc.) are installed correctly and are in good working order, the system is armed whenever the battery positive and negative battery cables are connected.

If you have disarmed the air bag system for any reason, to rearm, be sure no one is in the vehicle (as an added safety measure), then connect the battery negative cable.

Power Steering Gear

ADJUSTMENT

1. Place the steering gear in a vise with protective jaws.
2. Place a torque wrench on the Pitman arm end of the shaft.
3. Loosen the locknut on the adjusting bolt.
4. Slowly turn the adjusting bolt to until the breakaway torque is 65 ft. lbs. (88 Nm).
5. Hold the adjusting bolt in position and tighten the locknut to 25 ft. lbs. (34 Nm).

REMOVAL & INSTALLATION

1. Center the steering wheel.
2. Disconnect the two power steering fluid lines from the steering gear.
3. Matchmark the steering column intermediate shaft and the steering gear shaft.
4. Remove the intermediate shaft set bolt. Compress the intermediate shaft to disengage intermediate shaft and the steering gear shaft.
5. Raise and safely support the vehicle.
6. Remove the left front tire.
7. Remove the Pitman arm-to-center-link attaching nut.
8. Separate the Pitman arm from the centerlink with a ball joint puller.
9. Remove the four steering gear-to-frame bolts.
10. Remove the steering gear.
To install:
11. Position the steering gear to the frame. Install the attaching bolts and tighten to 159 ft. lbs. (215 Nm).
12. Connect the Pitman arm to the centerlink. Tighten the attaching bolt to 36 ft. lbs. (49 Nm) and install a new cotter pin.
13. Install the left front tire.
14. Lower the vehicle.
15. Align the matchmarks on the steering column intermediate shaft and the steering gear shaft.
16. Install the intermediate shaft set bolt. tighten to 25 ft. lbs. (34 Nm).
17. Connect the two power steering fluid lines to the steering gear.
18. Fill the power steering fluid reservoir to the proper level.

19. Start the vehicle. Turn the steering wheel stop-to-stop to check steering operations.
20. Road test the vehicle to check that the steering wheel is straight.

Strut

REMOVAL & INSTALLATION

Front

1. From inside the engine compartment remove the three strut mounting plate nuts.
2. Raise and safely support the vehicle.
3. Remove the front wheels.

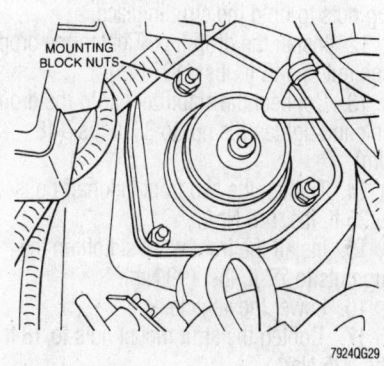

MOUNTING BLOCK NUTS

79240G29

Strut mounting plate nut locations

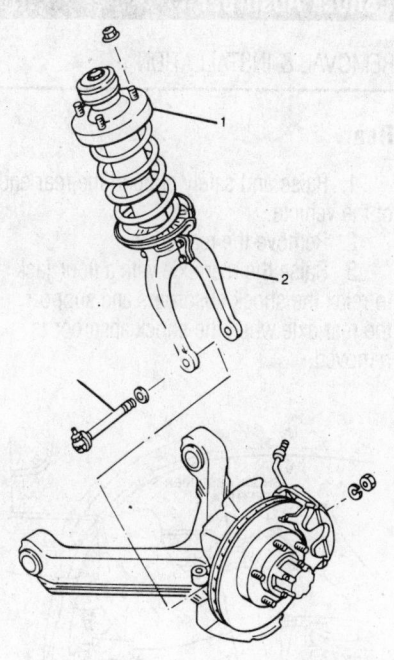

1 FRONT SHOCK ABSORBER & COIL SPRING ASSEMBLY
2 FRONT FORK
3 DROP LINK

79240G30

Exploded view of the front strut assembly

4. Disconnect the stabilizer bar from the front of the drop link.

5. Remove the drop link nut attaching the strut fork to the lower control arm.

6. Remove the two front fork half bolts. Remove the drop link.

7. Remove the strut assembly from the vehicle.

8. Place the strut in a spring compressor tool.

9. Compress the spring and remove the strut mount and coil spring.

To install:

10. Install the coil spring and strut mount.

11. Position the strut assembly on the vehicle. place the strut mount studs through the holes in the strut tower. Start the mounting nuts to hold the strut in place.

12. Install the drop link. Tighten the drop link nut to 145 ft. lbs. (197 Nm).

13. Connect the stabilizer bar to the drop link and tighten the nut to 36 ft. lbs. (48 Nm).

14. Tighten the two front fork half bolts to 36 ft. lbs. (48 Nm).

15. Install the front wheels. tighten the lug nuts to 77 ft. lbs. (99 Nm).

16. Lower the vehicle.

17. Tighten the strut mount nuts to 18 ft. lbs. (25 Nm).

18. Check the alignment.

Shock Absorber

REMOVAL & INSTALLATION

Rear

1. Raise and safely support the rear end of the vehicle.

2. Remove the rear wheels.

3. Raise the rear axle with a floor jack to relax the shock absorbers and support the rear axle when the shock absorber is removed.

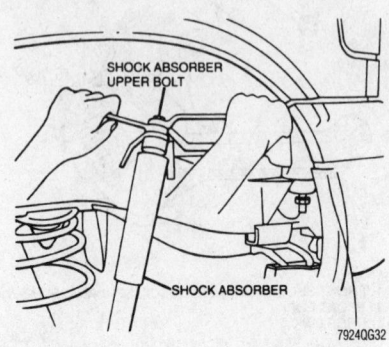

Upper shock absorber mounting nut

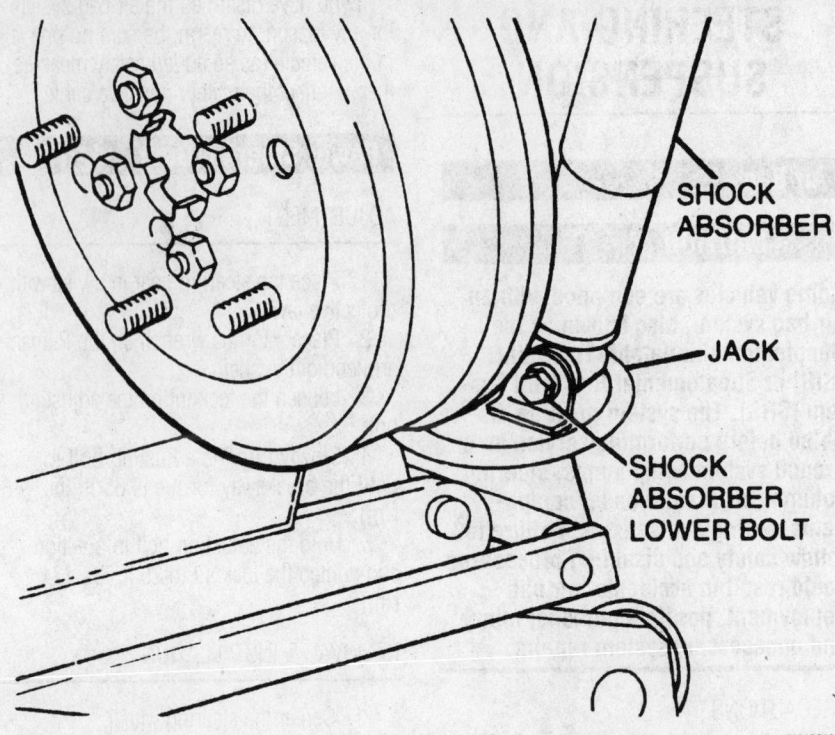

Lower shock absorber mounting bolt

4. Remove the upper safety and mounting nuts, washer and rubber cushion.

5. Remove the lower mounting bolt and remove the shock absorber.

To install:

6. Install the bottom washer and rubber cushion on the top of the shock absorber. Position the shock absorber on the vehicle.

7. Install the lower bolt.

8. Install the rubber cushion, washer and nut. Tighten to 53 ft. lbs. (72 Nm). Install the safety nut.

9. Tighten the lower bolt to 62 ft. lbs. (84 Nm).

10. Install the rear wheels. Tighten the lugs to 77 ft. lbs. (99 Nm).

11. Lower the vehicle.

Coil Spring

REMOVAL & INSTALLATION

Front

The replacement of the springs is covered in strut removal and installation.

Rear

1. Raise and safely support the rear end of the vehicle.

2. Remove the rear wheels.

3. Raise the rear axle with a floor jack to relax the shock absorbers and support the rear axle when the shock absorber is removed.

→**For easier installation, complete one side at a time.**

4. Remove the lower mounting bolt, then the shock absorber.

5. Slowly lower the floor jack until the coil spring is fully expanded. Remove the spring.

6. Inspect the upper and lower rubber spring seats and jounce stop for wear or damage, replace if necessary.

To install:

7. Position the spring in the upper and lower saddles.

8. Raise the floor jack and connect the lower shock absorber bolt. Tighten the bolt to 62 ft. lbs. (84 Nm).

9. Install the rear wheels. Tighten the lugs to 77 ft. lbs. (99 Nm).

10. Lower the vehicle.

Upper Ball Joint

REMOVAL & INSTALLATION

The upper ball joint is an integral part of the upper control arm. If the ball joint is worn, replacement of the upper control arm is necessary.

Lower Ball Joint

REMOVAL & INSTALLATION

1. Raise and safely support the vehicle.
2. Remove the front wheel assembly.
3. Remove and discard the cotter pin.
4. Remove the lower ball joint nut.
5. Separate the lower ball joint from the spindle with a puller tool.
6. Pry down on the spindle to separate it from the lower ball joint.
7. Remove the four lower ball joint attaching bolts. Remove the lower ball joint.

To install:

8. Position the lower ball joint and install the attaching nuts and bolts. Tighten to 36 ft. lbs. (48 Nm).
9. Pry down and guide the spindle onto the lower ball joint. Tighten the nut 87 ft. lbs. (118 Nm) and install a new cotter pin.
10. Install the front wheel assembly.
11. Lower the vehicle.

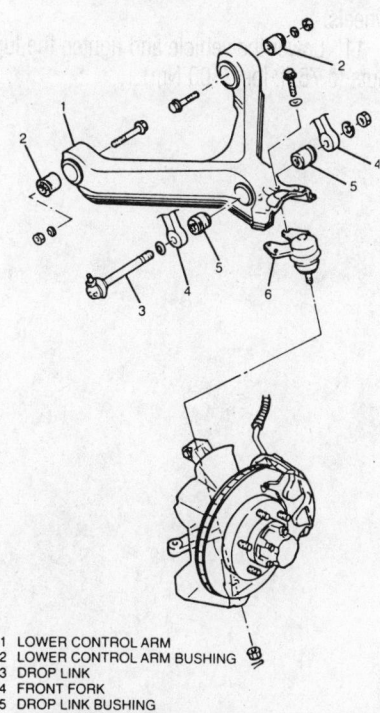

1 LOWER CONTROL ARM
2 LOWER CONTROL ARM BUSHING
3 DROP LINK
4 FRONT FORK
5 DROP LINK BUSHING
6 LOWER CONTROL ARM BALL JOINT

79240G34

Exploded view of the lower control arm and ball joint assembly

Upper Control Arm

REMOVAL & INSTALLATION

1. Raise and safely support the vehicle.
2. Remove the front wheel assembly.
3. Remove the through-bolt securing the upper ball joint to the steering knuckle.

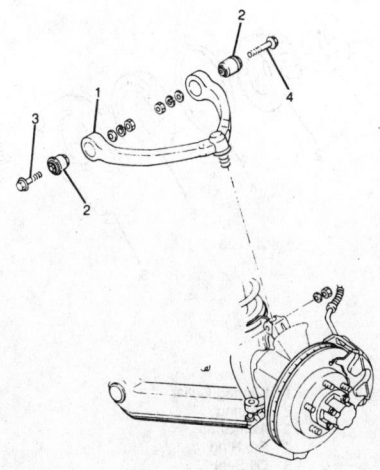

1 UPPER CONTROL ARM
2 UPPER CONTROL ARM BUSHING
3 FRONT SPINDLE
4 REAR SPINDLE

79240G31

Exploded view of the upper control arm assembly

➡**Note the matchmark setting on the upper control arm mounting bolts before removal.**

4. Remove the two upper control arm mounting bolts. Remove the upper control arm from the vehicle.

To install:

5. Position the upper control arms in the frame mounting. Install but do not tighten the bolts.
6. Position the ball joint in the spindle. Install the through-bolt and tighten the nut and bolt to 36 ft. lbs. (48 Nm).

➡**Be sure the slot in the ball joint aligns with the through-bolt during installation.**

7. Align the upper control arm bolts to the previous settings. Tighten the nut and bolts to 62 ft. lbs. (108 Nm).
8. Install the wheels. Tighten the lugs 77 ft. lbs. (99 Nm).
9. Lower the vehicle.
10. Check the alignment.

Wheel Bearings

ADJUSTMENT

Front

1. Raise and safely support the vehicle.
2. Remove the wheels.
3. Remove the caliper and hang it out of the way.

4. Remove the brake rotor.
5. Attach a dial indicator to the axle hub and measure the bearing play.
6. If the play exceeds .004 inch (.10mm), check and adjust locknut torque. The bolts should be tightened to 23 ft. lbs. (31 Nm).
7. Lower the vehicle.

Rear

The rear wheel bearings are not adjustable.

REMOVAL & INSTALLATION

Front

1. Raise and safely support the vehicle.
2. Remove the six bolts attaching the free wheel hub body.
3. Remove the snapring and spacer.
4. Carefully pry out the fixed cam assembly.
5. Remove the two caliper bolts and place the caliper out of the way.
6. Remove the rotor retaining screws and remove the rotor.
7. Remove the locknut.
8. Remove the hub assembly.
9. Remove the dust cover.
10. Remove the rear oil seal and discard.
11. Remove the inner and outer bearings. Using a press or a drift, remove the bearing races.

To install:

12. Pack the new bearings with grease.
13. From the rear of the hub, install the inner bearing and race. Install a new oil seal.
14. From the front of the hub, install the outer bearing and race. Secure the dust cover with four screws.
15. Apply grease to the new bearings and the lip of the oil seal.
16. Install the hub assembly in the steering knuckle.
17. Screw the locknut against the hub assembly until there is 10 inch lbs. (1.3 Nm) of preload on the hub.
18. Install the brake rotor and retaining screws.
19. Attach a runout gauge to check the rotor runout. The runout should not exceed 0.004 inch (0.10mm)
20. Install the brake caliper.
21. Install the fixed cam assembly with the cam key aligned with the locknut groove. Install the spacer and snapring.
22. Apply a light coat of sealant on the free wheel hub body. Install the body on the hub. Tighten the bolts in 2 passes. Tighten

the bolts on the first pass to 14 ft. lbs. (17 Nm). Tighten the bolts on the second pass to 23 ft. lbs. (31 Nm).

23. Lower the vehicle.

Rear

1. Raise and safely support the vehicle.
2. Remove the rear wheels, then the brake drums.
3. Remove the four nuts holding the oil seal retainer flange.

➡ **The axle shafts are different from side-to-side, mark the to ensure they are returned to the proper side.**

4. Using a slide hammer, remove the axle shaft assembly.
5. Using a hydraulic press, remove the bearing collar and bearing from the axle.
6. Remove the oil seal from the differential.

To install:

7. Using the appropriate seal driver, install the new axle seal into the differential.

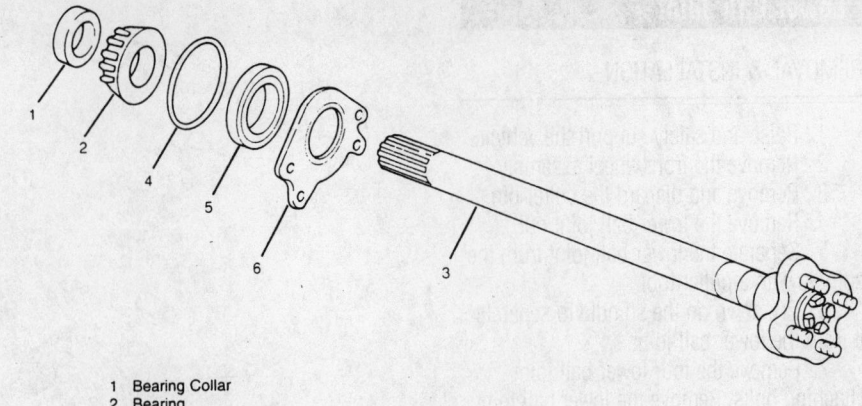

1 Bearing Collar
2 Bearing
3 Axle Shaft
4 Rib Ring
5 Oil Seal
6 Oil Seal Retainer

79240G35

Rear axle bearing component identification

➡ **The left-hand axle is 25.5 inches (647mm) long, and the right-hand axle is 27.4 inches (697mm) long.**

8. Using a hydraulic press, install the new wheel bearing and retainer collar to the axle shaft.

9. Insert the axle shaft into the carrier, and tighten the four mounting nuts to 75 ft. lbs. (100 Nm).
10. Install the brake drum and rear wheels.
11. Lower the vehicle and tighten the lug nuts to 75 ft. lbs. (100 Nm).

LAND ROVER

Defender 90 • Discovery • Range Rover

26

ENGINE REPAIR

➡️**Disconnecting the negative battery cable on some vehicles may interfere with the functions of the on board computer systems and may require the computer to undergo a relearning process, once the negative battery cable is reconnected.**

Distributor

REMOVAL

3.9L Engine

1. Disconnect the battery ground cable.
2. If necessary, remove the air cleaner assembly and connecting tubes.
3. Unplug the vacuum lines.
4. Unplug the pick-up lead wire connector(s) from the wiring harness.
5. Leaving the distributor wires connected, unfasten the clips that retain the distributor cap and lift off the cap.

➡️**If insufficient clearance is obtained by removing the cap with the wires still connected, all or some of the spark plug wires should be tagged and disconnected from the cap for better access.**

6. Bump the engine around until the rotor is pointing at No. 1 cylinder firing position, with the timing marks on the front case and crank pulley are aligned.
7. Mark the distributor body and the front case to indicate the position of the distributor in the block. Mark the distributor body to indicate the rotor position. These marks are used as guides when installing the distributor.
8. Remove the distributor hold-down bolt and bracket. Carefully lift the distributor from the engine. The shaft may rotate slightly as the distributor is removed. Make a second mark to note where the movement stops. That is where the rotor must point when the distributor is reinstalled into the block.

INSTALLATION

Timing Not Disturbed

1. If the crankshaft has not been rotated while the distributor was removed from the engine, use the reference marks made during removal to correctly position the distrib-

utor in the block. The shaft may have to be rotated slightly prior to installation.
2. Install the distributor while holding the rotor in position, allowing it to move only enough to engage the drive gear.
3. Install pick-up coil leads and distributor cap. Be sure all high tension wires are firmly snapped in cap towers. Install distributor hold-down clamp screw and tighten to 200 inch lbs. (22.5 Nm).

Timing Disturbed

Perform this procedure if the crankshaft was rotated or otherwise disturbed (for example, during engine rebuilding) after the distributor was removed.

1. Rotate the crankshaft until the No. 1 cylinder is at Top Dead Center (TDC) of the compression stroke. The simplest way to do this is to remove the spark plug from the No. 1 cylinder and place your thumb over the hole. Slowly turn the engine by hand in the normal direction of rotation until compression is felt at the hole.
2. Ensure that the indicating mark on the crankshaft vibration damper is aligned to the 0 degree (TDC) mark on the timing chain cover.
3. Clean the distributor mounting at the engine and distributor base. Lightly oil the rubber O-ring seal on the distributor housing.
4. Hold the distributor over the mounting pad on the cylinder block so that the distributor body flange coincides with the mounting pad and the rotor points to the No. 1 cylinder firing position.
5. Install the distributor while holding the rotor in position, allowing it to move only enough to engage the slot in the drive gear.
6. Install pick-up coil leads and distributor cap. Be sure all high tension wires are firmly snapped in cap towers. Install distributor hold-down clamp screw and tighten to 200 inch lbs. (22.5 Nm).

Ignition Timing

ADJUSTMENT

➡️**Except for the 3.9L engine, the timing is not adjustable. It is controlled by the PCM. Timing for the 3.9L engine should not need adjustment other than following distributor installation. Always refer to the Vehicle Emission Information (VECI) Label in the engine compartment to verify the timing adjustment procedure.**

3.9L Engine

1. Place automatic transmissions in **P** or manual transmissions in Neutral. The air conditioning and heater controls should be in the **OFF** position.
2. Connect a suitable inductive timing light and a tachometer according to the manufacturer's instructions.
3. Disconnect the vacuum lines and plug them.
4. Start the engine and bring it up to normal operating temperature. Don't exceed 3,000 rpm during the warm-up.
5. With the engine running at 750 rpm or less, check the initial timing by aiming the timing light at the timing marks and pointer.
6. If the marks do not align, shut the engine **OFF** and loosen the distributor hold-down clamp bolt. Start the engine, while watching the timing marks with the timing light, turn the distributor until the marks are correctly aligned. Shut the engine **OFF**, then tighten the distributor hold-down clamp bolt.
7. Reconnect the vacuum lines.
8. Remove the timing light and tachometer.

Engine Assembly

REMOVAL & INSTALLATION

1. Remove the hood.
2. Depressurize the fuel system.
3. Drain the cooling system and crankcase.
4. Disconnect the battery and alternator cables.
5. Remove the air intake hoses, PCV tube and carbon canister hose.
6. Disconnect the upper and lower radiator hoses.
7. Dismount the A/C compressor and position it out of the way. DO NOT discharge the system.
8. Disconnect the oil cooler lines.
9. Remove the fan shroud and lay it over the fan.
10. Remove the radiator, fan and shroud.
11. Remove the alternator.
12. Remove the plenum chamber and upper intake manifold.
13. Remove the fuel rail/injectors assembly.
14. Dismount the power steering pump and position it out of the way. DO NOT disconnect the lines.

15. Disconnect the ground strap from the cylinder block.

16. Remove the starter.

17. Disconnect the exhaust pipe from the exhaust manifolds.

18. Tag and disconnect all remaining hoses and wires attached to the engine.

19. Remove the 2 clamps securing the transmission cooler lines to the block.

20. Remove the upper bell housing-to-engine attaching bolts.

21. Disconnect the engine mounts from the brackets on the frame.

22. Remove the remaining bell housing-to-engine attaching bolts.

23. Lower the vehicle and support the transmission with a jack.

24. Install an engine lifting device.

25. Raise the engine slightly and carefully pull it out of the transmission. Lift the engine out of the engine compartment.

To install:

26. Remove the engine mount brackets from the frame and attach them to the engine mounts. Tighten the mount-to-bracket nuts just enough to hold them securely.

27. Lower the engine carefully into the transmission. Be sure the dowel in the engine block engage the holes in the bell housing through the rear cover plate. If the engine hangs up after the transmission input shaft enters the clutch disc (manual transmission only), turn the crankshaft with the transmission in gear until the input shaft splines mesh with the clutch disc splines.

28. Install the engine mount nuts and washers.

29. Remove the engine lifting device.

30. Install the lower bell housing-to-engine attaching bolts. Tighten the bolts to 30 ft. lbs. (40 Nm)

31. Remove the transmission support jack.

32. Connect the exhaust pipe to the exhaust manifolds. Tighten the exhaust pipe-to-exhaust manifold nuts to 25–35 ft. lbs. (32–45 Nm).

33. Install the starter.

34. Connect the starter cable to the starter.

35. Lower the vehicle.

36. Install the upper bell housing-to-engine attaching bolts. Tighten the bolts to 30 ft. lbs. (40 Nm)

37. Connect the wiring harness at the left rocker arm cover.

38. Connect the ground strap to the cylinder block.

39. Connect the throttle bracket to the upper intake manifold.

40. Install the upper intake manifold and plenum chamber.

41. Connect the fuel lines.

42. Install the alternator, power steering pump and compressor.

43. Connect all remaining hoses, wires and lines.

44. Install the radiator, fan and shroud.

45. Connect the upper and lower radiator hoses, and oil cooler lines.

46. Install the air intake hoses, PCV tube and carbon canister hose.

47. Connect the battery and alternator cables.

48. Fill the cooling system and crankcase.

49. Install the hood.

Water Pump

REMOVAL & INSTALLATION

3.9L and 4.0L Engines

1. Disconnect the negative battery cable and drain the cooling system. Remove the lower hose.

2. Loosen the nut that attaches the fan clutch to the water pump. The nut has left-hand thread and must be turned clockwise to remove. Remove the fan and clutch assembly.

3. Remove the accessory drive belt(s).

4. Remove the alternator adjustment bracket.

5. Remove the water pump attaching bolts, noting their positions for reinstallation.

6. Remove the water pump. Clean all gasket mating surfaces.

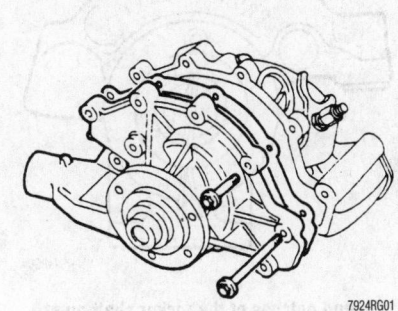

Exploded view of the engine water pump mounting

To install:

7. Position the water pump and new gasket in place.

8. Apply sealer to the long bolts. Tighten all the bolts to 21 ft. lbs. (28 Nm).

9. Attach the lower radiator to the water pump.

10. Install the alternator bracket.

11. Install and adjust the drive belts.

12. Install the fan, fan clutch and shroud. Tighten the fan clutch nut to 30–100 ft. lbs. (40–135 Nm).

13. Connect the negative battery cable. Fill and bleed the cooling system.

Cylinder Head

REMOVAL & INSTALLATION

3.9L Engine

1. Disconnect the negative battery cable and relieve the fuel system pressure. Drain the cooling system.

2. Remove the distributor cap, if equipped. Mark the position of the rotor in relation to the distributor housing and the position of the distributor housing in relation to the intake manifold. Remove the distributor hold-down bolt and clamp and remove the distributor. Tag and disconnect the spark plug wires from the spark plugs and remove the distributor cap and wires assembly.

3. Remove the intake manifold.

4. Remove the exhaust manifold.

5. Remove the water pump.

6. Remove the spark plugs.

7. Remove the rocker arm covers.

8. Remove the rocker arm shaft bolts and lift off the rocker arm shaft assemblies.

9. Remove the pushrods, marking them so they can be reinstalled in their original positions.

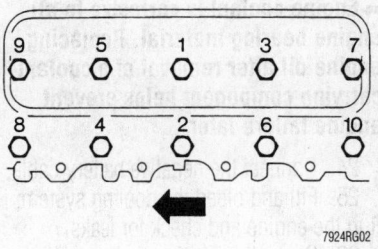

Cylinder head torque sequence—3.9L engine

10. Remove the 14 cylinder head attaching bolts, in reverse of the torque sequence, and lift off the cylinder head(s).

11. Clean all gasket mating surfaces. Check the cylinder head for flatness using a straight edge and a feeler gauge. The cylinder head must not be warped more than 0.003 in. (0.076mm) in any 6 in. (152mm) or more than 0.006 in. (0.15mm) overall.

To install:

12. Position new head gasket(s) on the cylinder block, with the word TOP up, using the dowels for alignment.

13. Install the cylinder head(s) on the block. Oil the threads of new cylinder head bolts and hand-tighten.

14. Tighten the cylinder head bolts, in sequence, and in progressive steps, to a final torque of 44 ft. lbs. (60 Nm) for the outer row, and 66 ft. lbs. (90 Nm) for the center and inner rows.

15. Install the intake manifolds.

16. Install the distributor, if equipped, aligning the marks that were made during removal. Install the hold-down bolt and clamp.

17. Dip each pushrod in heavy engine oil and install them in their original positions.

18. Position the rocker arms over the valves and pushrods and tighten to 28 ft. lbs. (38 Nm).

➡**If the original valvetrain components are being installed, a valve clearance check is not required. If a component has been replaced, perform a valve clearance check.**

19. Install the exhaust manifolds and the spark plugs.

20. Install the rocker arm covers. Install the dipstick tube.

21. Install the fuel injector/rail assembly.

22. Install the distributor cap and connect the spark plug wires to the spark plugs.

23. Change the engine oil and filter.

➡**Engine coolant is corrosive to all engine bearing material. Replacing engine oil after removal of a coolant carrying component helps prevent engine failure later.**

24. Connect the negative battery cable.

25. Fill and bleed the cooling system. Run the engine and check for leaks.

26. Check the ignition timing and idle speed and adjust, if necessary.

Rocker Arms/Shafts

REMOVAL & INSTALLATION

1. Disconnect the negative battery cable and relieve the fuel system pressure. Drain the cooling system.

2. Remove the rocker arm covers.

3. Remove the 4 rocker arm shaft bolts and lift off the rocker arm shaft assemblies.

4. Remove the cotter pin from either end of the shaft and slide the components from the shaft, keeping them in order for reassembly.

5. Inspect the parts for wear or damage and replace any suspect parts. Discard any weak springs.

To install:

6. Reassemble the shaft and components. Note the position of the oil feed holes. Use new cotter pin(s).

7. Position the rocker shaft assembly on the head. Be sure that the shaft is installed with the notches at each end on the up side.

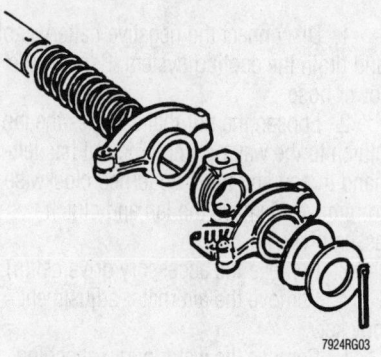

Exploded view of the rocker arm shaft components—3.9L engine

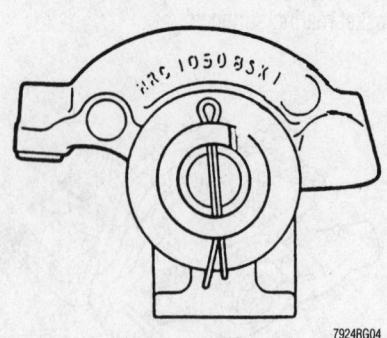

The end notches of the rocker shaft must face up—3.9L engine

Be sure that each rocker ball stud engages its respective pushrod.

8. Install the bolts and tighten them, gradually, to 28 ft. lbs. (38 Nm), starting with the 2 inner, then the 2 outer bolts.

Intake Manifold

REMOVAL & INSTALLATION

1. Disconnect the negative battery cable and relieve the fuel system pressure. Drain the cooling system.

2. Remove the distributor cap, if equipped. Mark the position of the rotor in relation to the distributor housing and the position of the distributor housing in relation to the intake manifold. Remove the distributor hold-down bolt and clamp and remove the distributor. Tag and disconnect the spark plug wires from the spark plugs and remove the distributor cap and wires assembly.

3. Remove the fuel injector/rail assembly.

4. Remove the 12 intake manifold bolts. Remove the intake manifold.

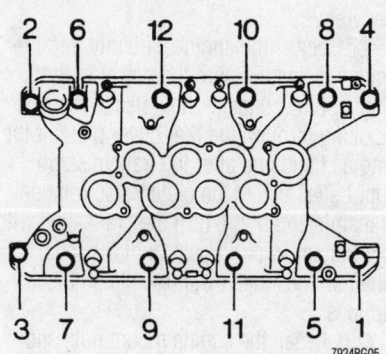

Intake manifold bolt removal sequence

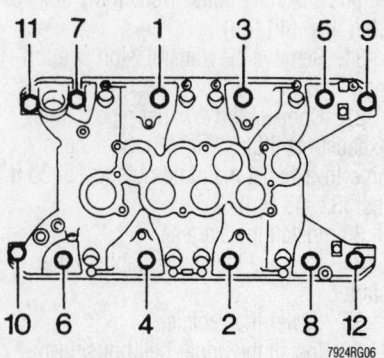

Intake manifold bolt tightening sequence

5. Remove the gasket retainer bolts, retainers, gasket and end seals.

To install:

6. Clean the gasket sealing surfaces on the lower intake manifold, cylinder heads and cylinder block.

7. Apply a bead of silicone sealant to the 4 corners of the end seals where the cylinder heads and cylinder block meet. Install the front and rear end seals.

8. Install the lower intake manifold gaskets to the cylinder heads.

➡**The intake manifold must be installed within 15 minutes of applying the silicone sealant.**

9. Install the gasket retainers, but finger-tighten the bolts at this time.

10. Apply a second bead of silicone sealant to the corners of the front and rear end seals and carefully position the lower intake manifold onto the engine. Visually inspect the gaskets and end seals for misalignment and correct as needed.

11. Install the intake manifold retaining bolts in their correct positions and tighten in a sequence, from the center to the ends, in 2 steps. First tighten to 15 ft. lbs. (20 Nm), then to 30 ft. lbs. (40 Nm).

12. Tighten the gasket retainer bolts to 13 ft. lbs. (18 Nm).

13. The remainder of installation is the reverse of removal.

Exhaust Manifold

REMOVAL & INSTALLATION

3.9L Engine

1. Disconnect the negative battery cable.
2. Raise and safely support the vehicle.
3. Remove the exhaust manifold-to-exhaust pipe retaining nuts.

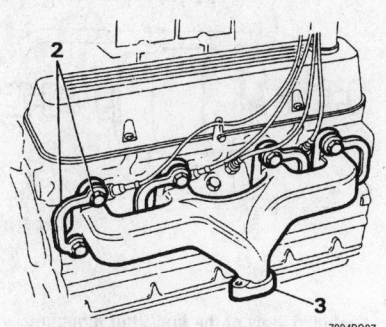

View of one of the exhaust manifolds (3), showing the mounting bolt locktabs (2)—3.9L engine

4. Lower the vehicle.
5. Remove the 8 manifold bolts. Remove the exhaust manifold and discard the gasket.

To install:

6. Clean all mating surfaces.
7. Place the exhaust manifold in position on the cylinder head using a new gasket. Install the exhaust manifold bolts and tighten to 15 ft. lbs. (20 Nm).
8. Install the exhaust pipe to the exhaust manifold and tighten the exhaust pipe retaining nuts to 25–34 ft. lbs. (34–47 Nm).
9. Lower the vehicle.
10. Connect negative battery cable.
11. Start the engine and check for exhaust leaks.
12. Road test the vehicle and check for proper engine operation.

Camshaft

REMOVAL & INSTALLATION

1. Remove the timing cover.
2. Remove the timing chain and sprockets.
3. Carefully, pull the camshaft from the block.

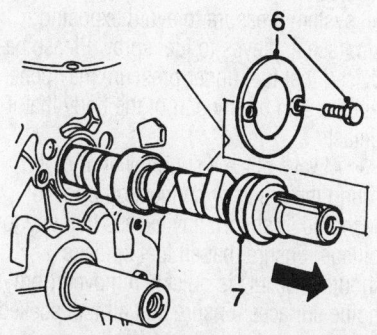

Exploded view of the camshaft (7) and thrust plate (6) mounting—3.9L engine

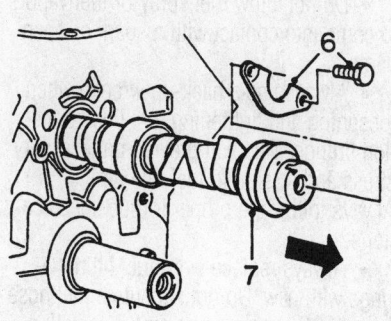

Exploded view of the camshaft (7) and thrust plate (6) mounting—except 3.9L engine

To install:

4. Coat the camshaft with assembly lube and carefully slide it into the block.
5. Install the timing chain, sprockets and cover.

Oil Pan

REMOVAL & INSTALLATION

※※ CAUTION

The EPA warns that prolonged contact with used engine oil may cause a number of skin disorders, including cancer! You should make every effort to minimize your exposure to used engine oil. Protective gloves should be worn when changing the oil. Wash your hands and any other exposed skin areas as soon as possible after exposure to used engine oil. Soap and water, or waterless hand cleaner should be used.

1. Raise and safely support the vehicle.
2. Drain the engine oil, then refit the drain plug with a new sealing washer and tighten to 30 ft. lbs. (40 Nm).
3. If equipped, disconnect the oil level sensor.
4. Remove the bolt securing the dipstick tube to the rocker cover.
5. Remove the bolts securing the oil pan.
6. Remove the oil pan, and clean/degrease the gasket mating surfaces of old sealant and gasket material.

To install:

7. Apply RTV Hylosil White sealant to the oil pan.
8. Install the oil pan and tighten the bolts to 15 ft. lbs. (18 Nm) in sequence as shown.

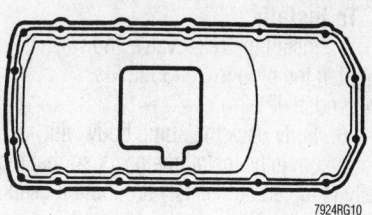

Oil pan sealant application for formed-in-place gaskets

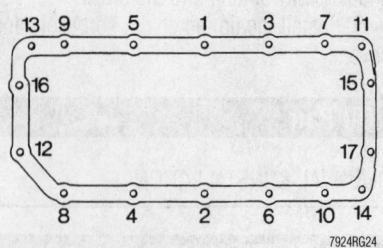

7924RG24

Oil pan bolt tightening sequence

9. Install the dipstick tube to the rocker cover.

10. If equipped, connect the oil level sensor.

✳✳ WARNING

Operating the engine without the proper amount and type of engine oil will result in severe engine damage.

11. Fill the engine with oil, start the engine and check for leaks.

Oil Pump

REMOVAL & INSTALLATION

3.9L Engine

1. Raise and support the front end on jackstands.

2. Drain the oil.

3. Remove the oil filter.

4. Remove the cover bolts and cover. It may be necessary to rap the cover with a soft mallet.

5. Remove the gears, relief valve and spring.

6. Check all parts for wear or damage. Place the gears in the cover and put a straightedge across them. Clearance between the straightedge and cover should be at least 0.002 in. (0.05mm).

To install:

7. Install the relief valve and washer. Tighten the plug to 30–35 ft. lbs. (40&ndsh;47 Nm).

8. Fully pack the pump body with petroleum jelly. Install the gears so that the jelly is forced completely around the gears. Leave no cavities.

9. Place a new gasket on the pump,

position the cover and tighten the bolts to 10 ft. lbs. (13 Nm).

10. Install a new filter and fill the sump.

FUEL SYSTEM

Fuel System Service Precautions

Safety is the most important factor when performing not only fuel system maintenance but any type of maintenance. Failure to conduct maintenance and repairs in a safe manner may result in serious personal injury or death. Maintenance and testing of the vehicle's fuel system components can be accomplished safely and effectively by adhering to the following rules and guidelines.

• To avoid the possibility of fire and personal injury, always disconnect the negative battery cable unless the repair or test procedure requires that battery voltage be applied.

• Always relieve the fuel system pressure prior to disconnecting any fuel system component (injector, fuel rail, pressure regulator, etc.), fitting or fuel line connection. Exercise extreme caution whenever relieving fuel system pressure to avoid exposing skin, face and eyes to fuel spray. Please be advised that fuel under pressure may penetrate the skin or any part of the body that it contacts.

• Always place a shop towel or cloth around the fitting or connection prior to loosening to absorb any excess fuel due to spillage. Ensure that all fuel spillage (should it occur) is quickly removed from engine surfaces. Ensure that all fuel soaked cloths or towels are deposited into a suitable waste container.

• Always keep a dry chemical (Class B) fire extinguisher near the work area.

• Do not allow fuel spray or fuel vapors to come into contact with a spark or open flame.

• Always use a back-up wrench when loosening and tightening fuel line connection fittings. This will prevent unnecessary stress and torsion to fuel line piping. Always follow the proper torque specifications.

• Always replace worn fuel fitting O-rings with new. Do not substitute fuel hose or equivalent where fuel pipe is installed.

Fuel System Pressure

RELIEVING

To relieve the high pressure in the fuel system, simply remove the fuel pump relay module and crank the engine. Usually the engine won't start, just the cranking process will deplete the pressure. If the engine does start, it will run for only a few seconds. Remember, there will still be some fuel in any line that is opened, so always exercise extreme caution when replacing parts or opening lines!

Fuel Filter

REMOVAL & INSTALLATION

1. Disconnect the negative battery cable and relieve the fuel system pressure.

2. Raise and support the vehicle safely.

3. Remove the bracket cover over the filter, if equipped.

➡**In the following step, fuel will be spilled. Take all necessary precautions to avoid personal injury.**

4. Place a pan under the filter. Using 2 wrenches, disconnect the fuel lines from the filter.

5. Remove the fuel filter from the bracket and retainer, if equipped. Note the direction of the flow arrow so the replacement filter can be installed correctly.

To install:

6. Install the fuel filter into the bracket making sure the flow direction is correct. Tighten the clamp to 15–25 inch lbs. (2–3 Nm).

7. Connect the fuel lines using back-up

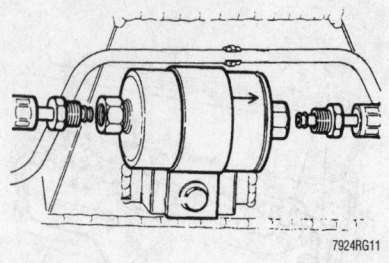

7924RG11

Exploded view of the fuel filter mounting

wrenches. Tighten the fittings to 22 ft. lbs. (30 Nm).

8. Lower the vehicle Start the engine and check for leaks.

Electric Fuel Pump

REMOVAL & INSTALLATION

1. Relieve the fuel system pressure. Disconnect the negative battery cable.

2. Raise and safely support the vehicle.

3. Remove the fuel tank from the vehicle.

4. Remove any dirt that has accumulated around the fuel pump flange so it will not enter the fuel tank during removal and installation.

5. Disconnect the hoses from the pump.

6. Using Locking Ring Tool (LRT) 19–001 or equivalent tool designed for retaining ring removal, or, using a hammer and brass drift, loosen and unscrew the retaining ring from the pump flange.

7. Remove the fuel pump and discard the seal ring. Separate the fuel pump from the sending unit, if required.

To install:

8. Clean the fuel pump mounting flange and tank mounting surface.

9. Apply a light coating of sealer on a new seal ring.

10. Install the fuel pump on the sending unit, if removed. Install the fuel pump assembly in the tank and secure it with the 5 screws, making sure the outlet pipe is facing forward.

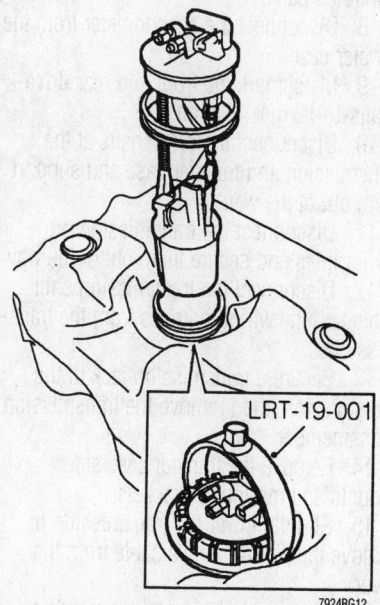

Fuel pump mounting—Defender 90

11. Install the fuel tank in the vehicle.

12. Lower the vehicle and fill the fuel tank with at least 10 gallons of fuel. Connect the negative battery cable. Turn the ignition key to **RUN** for 3 seconds repeatedly, 5–10 times, to pressurize the system. Check for leaks.

13. Start the engine and check for leaks.

DRIVE TRAIN

Transmission Assembly

REMOVAL & INSTALLATION

Manual

➡**The transmission and transfer case are removed as a unit.**

1. Raise and support the vehicle safely. Disconnect the battery ground cable. Prop the clutch pedal in the full up position with a block of wood.

❋❋ CAUTION

The clutch driven disc may contain asbestos, which has been determined to be a cancer causing agent. Never clean clutch surfaces with compressed air! Avoid inhaling any dust from any clutch surface! When cleaning clutch surfaces, use a commercially available brake cleaning fluid.

2. Remove the gear lever and transfer case knobs.

3. Remove the shift lever cover.

4. Remove the nut and washer and disconnect the shift lever from the gear case lever.

5. Place the transfer case lever in the HIGH position.

6. Remove the fan shroud.

7. Disconnect the speedometer cable at the transmission.

8. Disconnect the transmission breather pipes and starter harness at the engine.

9. Remove the air filter inlet hose.

10. Unbolt the crossmember from the frame rails and remove it.

11. Drain the transmission and transfer case (3 plugs) and put the plugs back in.

12. Remove the intermediate exhaust pipe and muffler.

13. Matchmark the front and rear driveshafts-to-flanges relation.

14. Disconnect the driveshafts at the transmission and transfer case and support them out of the way.

15. Remove the parking brake cable.

16. Remove the slave cylinder from the bell housing.

17. Using the accompanying illustration, fabricate a cradle and attach it to a transmission jack. To achieve a safe balance for the transmission, it is essential that point A is situated over the center of the jack or hoist ram. Drill holes at B to secure the fixture to the hoist. Secure the transmission to the lifting bracket at C using the lower bolts that retain the transfer case rear cover.

18. Bolt the transfer case to the cradle. Be sure that the tube in the center of the cradle goes over the extension housing drain plug.

19. Take up the weight of the transmission with the a transmission jack.

20. Remove the 3 nuts and bolts securing the transfer case brackets to the chassis. Remove the brackets.

21. Lower the unit enough to allow the transfer case lever to clear the tunnel.

22. Disconnect the 4wd indicator connector.

23. Disconnect the back-up light switch wiring.

24. Place a jackstand, cushioned with a 2x4 piece of lumber, under the oil pan.

25. Remove the 11 bell housing nuts and roll the unit back, away from the engine.

❋❋ WARNING

Do not depress the clutch pedal with the transmission removed.

26. Place the unit on a workbench. Remove the 4 transfer case-to-transmission bolts. Remove the 2 cross-shaft lever pivot bolts. Disconnect the lock lever link and pivot arm. Remove the high/low link and breather pipes.

To install:

27. Assemble the transfer case to the transmission. Tighten the bolts to 33 ft. lbs. (45 Nm).

28. Install the unit in the cradle.

29. Clean all mating surfaces thoroughly.

30. Temporarily install the gear lever and place the transmission in any gear. Run a bead of RTV sealant around the mating edge of the bell housing. Raise the unit into position, engage the input shaft and clutch disc, slide the unit forward, against the

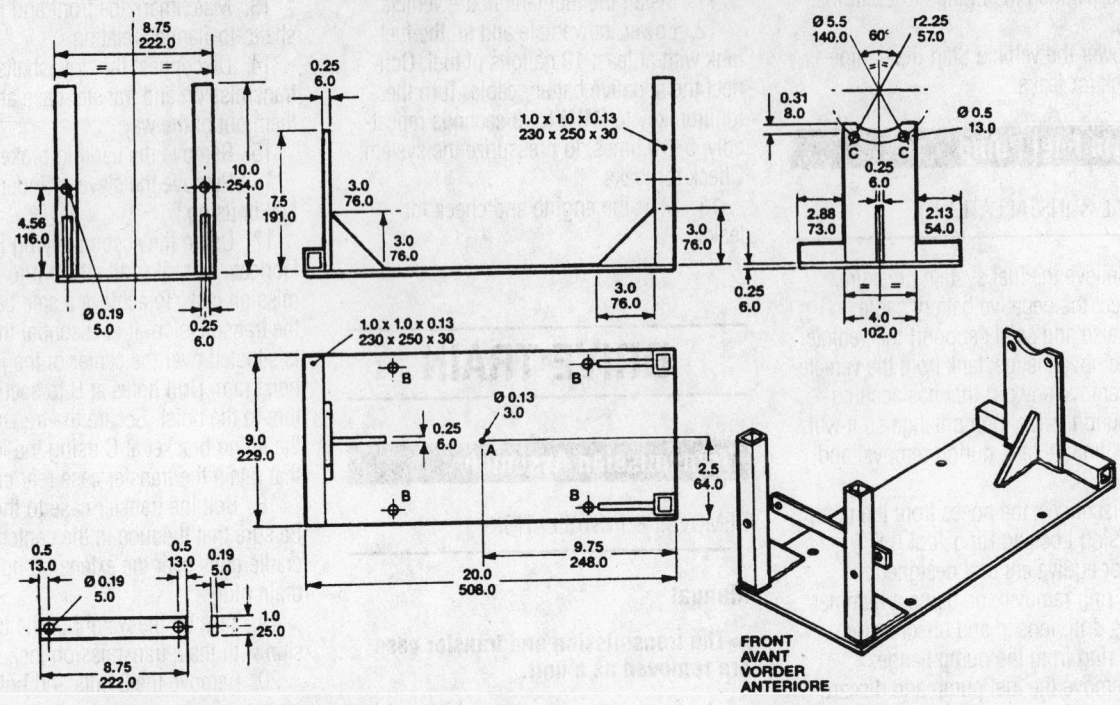

FRONT
AVANT
VORDER
ANTERIORE

7924RG13

MATERIAL AND WELDING SPECIFICATION

Steel Plate	BS 1449 (Grade 4 or 14)
Tube	BS 4848 (Part 2)
Arc Welding	BS 5135

Fabricate the cradle using these dimensions

engine making sure no wires, hose or cable are trapped. Install the 11 nuts, noting that the top right nut holds the speedometer cable clip. Tighten the nuts to 40 ft. lbs. (54 Nm).

31. Connect all wiring to the unit.

32. Raise the transmission just enough to install the transfer case mounting brackets. Install the brackets, but just snug down the bolts at this time. Install the rubber isolator nuts and lower the unit onto the mounts. Tighten the mount bolts to 66 ft. lbs. (90 Nm).

33. Remove the jack from under the engine and detach the cradle from the unit. Replace the 2 transfer case bolts, coating them with Loctite® 290. Tighten the bolts to 33 ft. lbs. (45 Nm). Note that the left bolt holds the speedometer cable clip.

34. Install the slave cylinder. Tighten the bolts to 16–21 ft. lbs. (22–28 Nm).

35. Install the parking brake cable.

36. Connect the speedometer cable at the transmission.

37. Connect the driveshafts.

38. Install the exhaust system.

39. Install the crossmember on the frame rails. Tighten the bolts to 80 ft. lbs. (104 Nm).

40. Install the shift control on the extension housing and transmission case.

41. Connect the shift rods at the shift levers.

42. Fill the units.

Automatic

1. Disconnect the negative battery cable.

2. Remove the fan shroud from the radiator.

3. Disconnect the transmission breather pipes from the right cylinder head at the rear and the dipstick.

4. Disconnect the kickdown cable from the throttle linkage.

5. Remove the shift boot knob and boot from the center console.

6. Drain the fluid from the transmission and the transfer case.

7. Remove the exhaust system from the manifolds back.

8. Disconnect the speedometer from the transfer case.

9. Matchmark the front and rear drive-shafts-to-flanges relation.

10. Disconnect the driveshafts at the transmission and transfer case and support them out of the way.

11. Disconnect the transmission oil cooler lines and secure them out of the way.

12. Disconnect the transmission shift cable and the wiring harness from the transmission.

13. Secure a transmission jack to the transmission, then remove the transmission crossmember.

14. Remove the transfer case side mounts and mounting brackets.

15. Slightly lower the transmission to remove the parking brake cable from the lever.

16. Remove the driveplate inspection cover and matchmark the torque converter to the driveplate.

17. Remove the torque converter bolts and the fill tube from the transmission.

18. Remove the bell housing to engine bolts.

19. Pull the transmission rearward, slightly, secure the torque converter to the transmission.

20. Remove the transmission from the vehicle.

To install:

21. Install the transmission to the vehicle.

22. Install the bell housing to engine bolts and tighten to 31 ft. lbs. (42 Nm).

23. Align the matchmarks for the torque converter to the driveplate.

24. Apply Loctite® to the torque converter bolts and tighten to 29 ft. lbs. (39 Nm).

25. Fit the fill tube to the transmission.

26. Install the driveplate inspection cover.

27. With the transmission slightly lowered, connect the parking brake cable to the lever.

28. Install the transfer case side mounts and tighten the bolts to 33 ft. lbs. (45 Nm).

29. Install the transmission crossmember, then remove the transmission jack from the transmission.

30. Connect the transmission shift cable and the wiring harness to the transmission.

31. Connect the transmission oil cooler lines and secure them out of the way.

32. Connect the driveshafts at the transmission and transfer case and support them out of the way.

33. Connect the speedometer to the transfer case.

34. Install the exhaust system.

35. Fill the transmission and the transfer case with ATF Dexron® II.

36. Install the shift boot and knob.

37. Connect and adjust the kickdown cable to the throttle linkage.

38. Connect the transmission breather pipes to the right cylinder head at the rear.

39. Install the fan shroud to the radiator.

40. Connect the negative battery cable.

Clutch Assembly

REMOVAL & INSTALLATION

✱✱ CAUTION

The clutch driven disc may contain asbestos, which has been determined to be a cancer causing agent. Never clean clutch surfaces with compressed air! Avoid inhaling any dust from any clutch surface! When cleaning clutch surfaces, use a commercially available brake cleaning fluid.

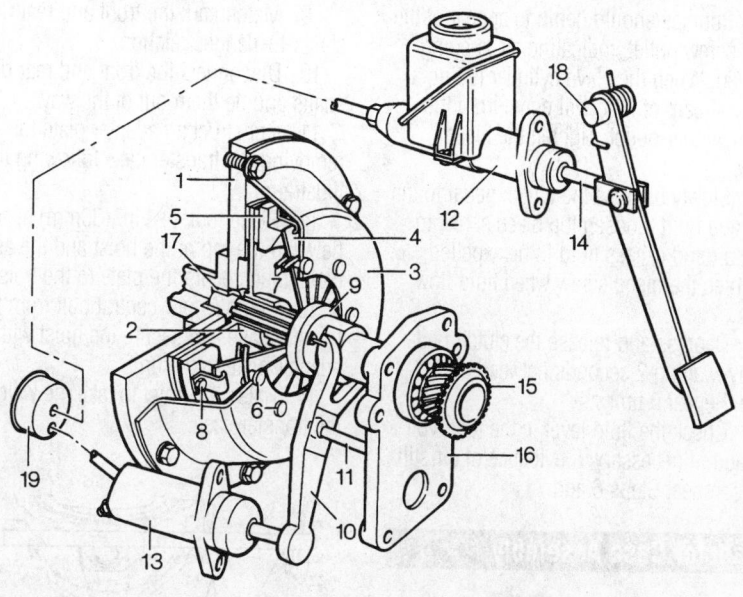

1. Crankshaft and flywheel
2. Friction plate
3. Clutch cover
4. Diaphragm spring
5. Pressure plate
6. Fulcrum posts (9) for diaphragm spring
7. Bearing rings (2) for diaphragm spring
8. Retraction links and bolts (3) for pressure plate
9. Release bearing
10. Release lever
11. Release lever pivot post
12. Master cylinder
13. Slave cylinder
14. Master cylinder pedal pushrod
15. Primary shaft and taper bearing (in gearbox)
16. Gearbox front cover
17. Primary shaft flywheel bush
18. Pedal pivot and return spring
19. Hydraulic damper (Diesel only)

7924RG14

Cut away view of the clutch assembly

1. Disconnect the negative battery cable.

2. Remove the transmission.

3. Mark the position of the pressure plate on the flywheel so if the pressure plate is reused, it can be reinstalled in the same position.

4. Loosen the pressure plate attaching bolts evenly until the diaphragm spring is expanded. Remove the bolts, pressure plate and clutch disc.

5. Inspect the flywheel for wear, scoring and cracks. Machine or replace, as necessary. Inspect the clutch pilot bearing for wear and free movement. If replacement is necessary, remove using puller tool T58L-101-b or equivalent.

6. Inspect the clutch release bearing for wear and free movement; replace as necessary. Remove the release bearing by twisting it until resistance is felt. Turning further will allow the preload spring to push the bearing assembly off the slave cylinder.

To install:

7. If the pilot bearing was removed, a new one must be installed. Install the pilot bearing with the seal facing the transmission so the adapter is not cocked.

8. Position the clutch disc on the flywheel so alignment tool can enter the pilot bearing and align the disc.

9. Install the pressure plate. If the original pressure plate is being reused, align the marks that were made during the removal procedure. Install the attaching bolts and tighten, in a star-shaped sequence, to 21 ft. lbs. (28 Nm), then remove the alignment tool.

10. Install the transmission.

11. Connect the negative battery cable.

Hydraulic Clutch System

BLEEDING

1. Clean all dirt and grease from around the reservoir cap.

2. Remove the cap and fill the reservoir with DOT 3 heavy duty brake fluid.

3. Raise and safely support the vehicle, as necessary. Connect a length of clear tubing to the slave cylinder bleeder screw and immerse the other end in a jar half full of clean, fresh brake fluid. Loosen the bleed screw, located in the slave cylinder body, next to the inlet connection.

4. Fluid will now begin to flow from the master cylinder, down the tube and into the slave cylinder.

➡**Keep the reservoir full at all times to be sure no additional air is drawn into the system.**

5. Bubbles should begin to appear at the bleed screw outlet, indicating air is being expelled. When the slave cylinder is full, a steady stream of fluid will come from the slave cylinder outlet. Tighten the bleed screw.

6. Slowly depress the clutch pedal to the floor and hold. Loosen the bleed screw to allow air and excess fluid to be expelled. Retighten the bleed screw when fluid flow stops.

7. Depress and release the clutch pedal slowly, waiting 2 seconds between each cycle. Repeat 5 times.

8. Check the fluid level in the reservoir and add, if necessary. If evidence of air still exists, repeat Steps 6 and 7.

Transfer Case Assembly

REMOVAL & INSTALLATION

1. Disconnect the negative battery cable.

2. Remove the radiator fan shroud.

3. Remove the transfer case shift knob and boot.

4. Raise and safely support the vehicle.

5. Drain the transfer case and refit the plug.

6. Detach the heat shield from the front exhaust pipe and remove the catalytic converter assembly.

7. Carefully remove the crossmember from under the transfer case.

8. Remove the speedometer cable and tie it out of the way.

9. Matchmark the front and rear driveshafts-to-flanges relation.

10. Disconnect the front and rear driveshafts and tie them out of the way.

11. Construct an adapter plate for removing the transfer case following the illustration.

12. Place four 1³⁄₁₆ in. (30mm) spacers between the top of the hoist and the adapter plate, then secure the plate to the hoist.

13. Remove four central bolt from the transfer case and secure the hoist with the adapter plate to the unit.

14. Raise the hoist to take the weight off the transfer case.

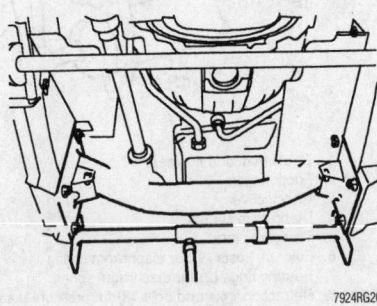

7924RG20

Be sure to support the transmission before removing the crossmember bolts

15. Remove the left and right transfer case rubber mounts.

16. Slowly lower the hoist until the park brake drum clears the passenger footwell. Be sure the engine does not crush any components.

17. Loosen the park brake adjustment nut and remove the park brake drum assembly from the rear output flange.

18. Label and unplug all sensors and switches from the transfer case.

19. Remove the transfer case breather banjo bolt and position it aside.

20. Disconnect the differential lock engaging rod.

21. Place the transfer case in low range.

22. Remove the range selector rod lower nut, then the rod from the yoke.

23. Support the transmission with a wooden block.

24. Remove the upper and lower transfer case mounting bolts.

25. Fit guide studs 18G 1425 to the transmission and move the transfer case rearward to remove.

To install:

26. Be sure the mating surface of the transmission and transfer case is free from dirt and debris.

27. Raise the transfer case until it is

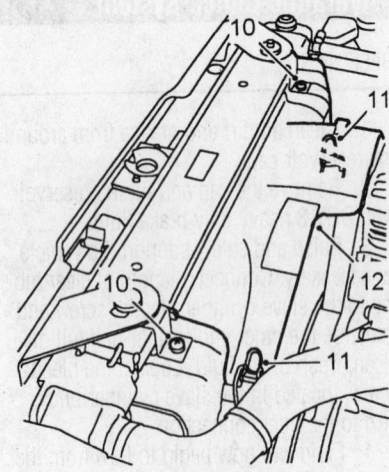

7924RG19

Remove the fan shroud to prevent the fan from damaging it

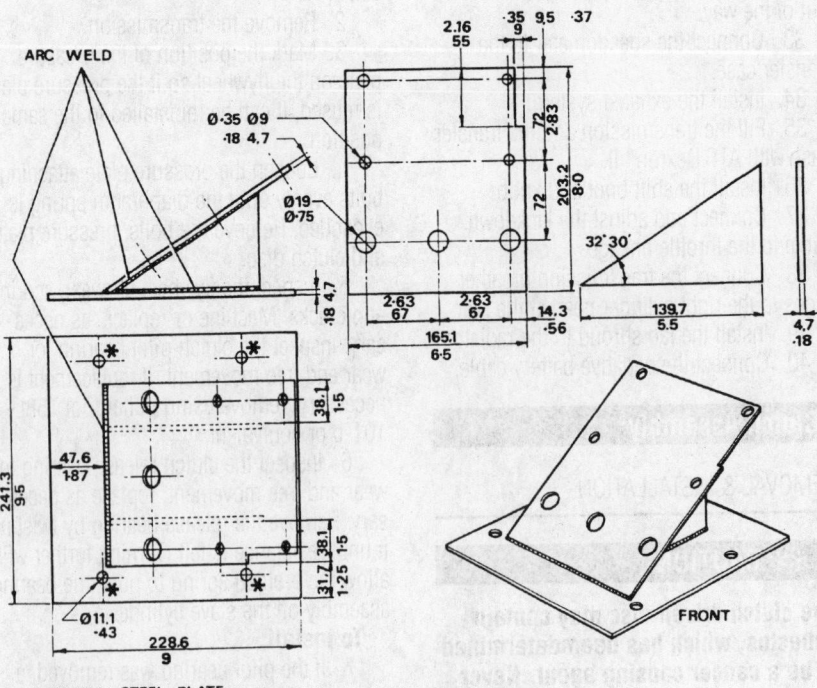

ARC WELD

MATERIAL: STEEL PLATE

✳ = TO BE DRILLED TO FIT TRANSMISSION JACK BEING USED

7924RG22

Dimensions for the adapter plate

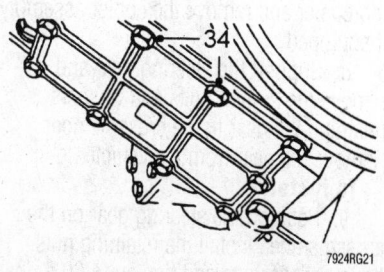

Location of the four central bolts—mount the adapter plate after removing these bolts

7924RG21

located over the guide studs 18G 1425, then slide it forward on to the transmission.

28. Remove the guide studs and bolt the transfer case to the transmission.

29. Complete the installation procedure is the reverse of the removal procedure, noting the following items.

30. After removing the adapter plate, clean the four bottom cover bolts and coat them with Loctite.® 290, then tighten to 19 ft. lbs. (25 Nm).

31. Fill the transfer case with 90W oil.

32. Check and adjust the park brake cable.

Halfshaft

REMOVAL & INSTALLATION

1. Remove the front hub assembly.
2. Drain the knuckle housing and reinstall the drain plug.
3. Remove the bolts securing the stub axle to the knuckle housing.
4. Remove the mud shield.
5. Remove the stub axle from the housing.
6. Pull the axle shaft from the differential case.

1. Brake caliper.
2. Mud shield.
3. Stub axle.
4. Joint washer.
5. Oil seal.
6. Bearing.
7. Brake disc shield.
8. Constant velocity joint.
9. Circlip.
10. Bush.
11. Inner driveshaft.
12. Top swivel pin and jump hose bracket.
13. Shim.
14. Swivel pin housing.
15. Joint washer.
16. Lower swivel pin.
17. Damper and shield bracket.
18. Thrust washer, ABS.
19. Bush and housing, ABS.
20. Swivel pin bearing housing.
21. Lower swivel pin bearing (and upper bearing non ABS).
22. Oil seal.
23. Oil seal retaining plate.
24. Joint washer.
25. Oil seal, ABS.
26. Sensor bush, ABS.
27. Thrust bearing, ABS.

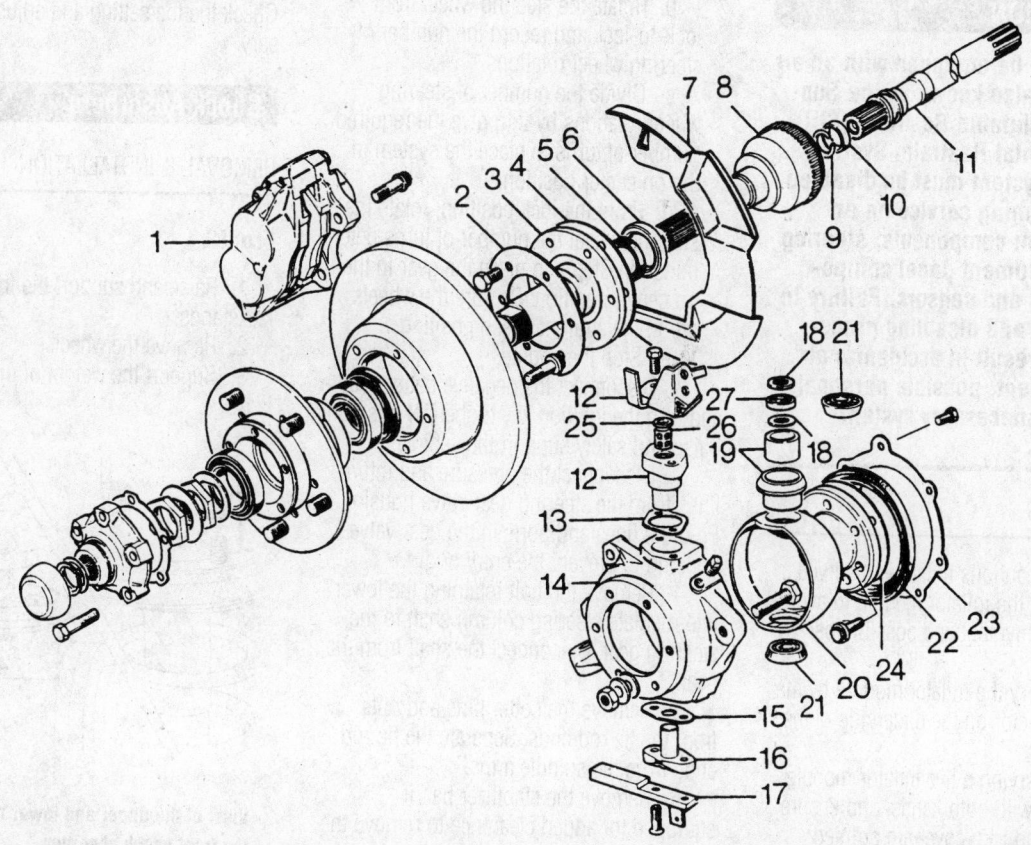

Exploded view of the front hub and swivel joint

7924RG15

To install:

7. Carefully slide the axle shaft into the differential housing. Avoid damaging the seals. Turn the shaft slightly to engage the splines and slide the shaft in until seated.

8. Place a new gasket on the knuckle housing and install the stub axle with the keyway at the 12 o'clock position.

➡**Be sure that the CV-joint bearing journal fully engages the bronze bushing in the rear of the stub axle before installing the stub axle bolts. Bushing damage will occur if this is not done.**

9. Install the mud shield and the 6 bolts. Tighten the bolts to 44–52 ft. lbs. (60–70 Nm).

10. Install the hub assembly. Fill the knuckle housing with 80W-90 gear oil, until the oil runs out the filler hole.

STEERING AND SUSPENSION

Air Bag

✳✳ CAUTION

Vehicles may be equipped with an air bag system, also known as the Supplemental Inflatable Restraint (SIR) or Supplemental Restraint System (SRS). The system must be disabled before performing service on or around system components, steering column, instrument panel components, wiring and sensors. Failure to follow safety and disabling procedures could result in accidental air bag deployment, possible personal injury and unnecessary system repairs.

PRECAUTIONS

Several precautions must be observed when handling the inflator module to avoid accidental deployment and possible personal injury.

• Never carry the inflator module by the wires or connector on the underside of the module.

• When carrying a live inflator module, hold securely with both hands, and ensure that the bag and trim cover are pointed away.

• Place the inflator module on a bench or other surface with the bag and trim cover facing up.

• With the inflator module on the bench, never place anything on or close to the module which may be thrown in the event of an accidental deployment.

DISARMING

1. Remove the key from the ignition.
2. Disconnect the negative battery cable, first, then the positive.
3. Wait 20 minutes for the back-up power to discharge.
4. After performing the required service, rearm the SRS by reconnecting the battery.
5. Start the vehicle and the SRS service light should go OFF after 8 seconds on 1995–96 models and 5 seconds on 1997–99 models.

Power Rack and Pinion Steering Gear

REMOVAL & INSTALLATION

1. Place the steering system in the on center position as follows:
   a. Start the engine.
   b. Rotate the steering wheel from lock-to-lock and record the number of steering wheel rotations.
   c. Divide the number of steering wheel rotations by 2 to give the required number of turns to place the system in the on center position.
   d. From the lock position, rotate the steering wheel the number of turns determined in Step c to place the gear in the on center position. Be sure the wheels are in the straight ahead position.
   e. Stop the engine.

2. Disconnect the negative battery cable and turn the ignition key to the **ON** position. Raise and safely support the vehicle.

3. Disconnect the pressure and return lines from the steering gear valve housing. Plug the lines and ports in the gear valve housing to prevent the entry of dirt.

4. Remove the bolt retaining the lower intermediate steering column shaft to the steering gear. Disconnect the shaft from the gear.

5. Remove the cotter pins and nuts from the tie rod ends. Separate the tie rod ends from the spindle arms.

6. Remove the stabilizer bar, if equipped for added clearance to remove the assembly.

7. Remove the nuts for the power steering cooler and remove the cooler assembly, if equipped.

8. Support the steering gear and remove the 2 nuts, bolts and washers retaining the gear to the crossmember. Remove the gear from the vehicle.

To install:

9. Position the steering gear on the crossmember. Install the retaining nuts, bolts and washers and tighten to 60 ft. lbs. (81 Nm).

10. Install the power steering cooler.

11. Connect the pressure and return lines to the gear valve housing ports. Tighten the fittings to specification according to their size. Tighten 16mm fittings to 15 ft. lbs. (20 Nm) and 14mm fittings to 11 ft. lbs. (15 Nm).

12. With the steering gear, steering wheel and front wheels in the on center position, attach the tie rod ends to the spindle arms.

13. Connect the steering column lower intermediate shaft to the gear. Install the bolt and tighten to 30–42 ft. lbs. (41–56 Nm).

14. Install the stabilizer bar, if equipped.

15. Lower the vehicle and turn the ignition key to the **OFF** position. Connect the negative battery cable.

16. Fill and bleed the steering system. Check the toe setting and adjust, if necessary.

Shock Absorbers

REMOVAL & INSTALLATION

Front

1. Raise and support the front end on jackstands.
2. Remove the wheel.
3. Support the weight of the axle with a jack.

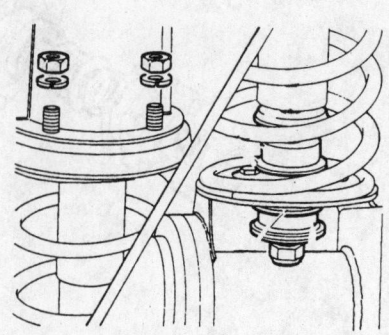

7924RG16

View of the upper and lower mountings for the front shock absorber

4. Remove the lower shock absorber nut and bushings.

5. Remove the 4 upper shock absorber bracket nuts and remove the shock absorber and bracket assembly. Remove the nut and separate the bracket from the shock absorber.

6. Installation is the reverse of removal.

Rear

1. Raise and safely support the vehicle.

2. Place a jack under the rear axle and raise slightly to take the load off the shock absorbers.

3. Remove the shock absorber lower attaching nut pull the lower end free of the mounting bracket on the axle housing.

4. Remove the upper attaching nut and remove the shock absorber.

To install:

5. Install the shock absorber onto the upper bracket and screw on the attaching nut. Don't tighten it yet.

6. Swing the shock absorber down and position in the lower mounting bracket. Attach the lower mounting nut.

7. Tighten the lower attaching nut to 55 ft. lbs. (75 Nm). Tighten the upper attaching nut to 60 ft. lbs. (82 Nm).

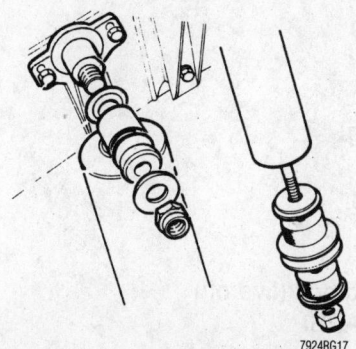

7924RG17

Exploded view of the upper and lower mountings for the rear shock absorber

Coil Springs

REMOVAL & INSTALLATION

Front

1. Raise and safely support the vehicle.
2. Remove the shock absorber.
3. Slowly lower the axle to relieve the spring tension. DO NOT STRETCH THE BRAKE HOSE! Remove the spring.

To install:

4. Install the spring in the upper seat.

5. Raise the axle until the spring is seated in the lower spring seat.
6. Install the shock absorber.

Rear

1. Raise and safely support the vehicle. Remove the wheels. Place jackstands on the frame rear lift points.

2. Remove the nut and bolt retaining the shock absorber to the axle mount on the lower control arm. Disconnect the shock absorber from the axle bracket.

3. Remove the spring retainer plate.

4. Lower the rear axle until the coil springs are no longer under compression. DO NOT STRETCH THE BRAKE LINE!

To install:

5. Be sure the axle is in the spring unloaded position.

6. Install the coil spring, top end first and locate in the lower perch using a twisting motion.

7. Raise the axle to the normal ride position and install the shock absorber. Install the wheels. Lower the vehicle.

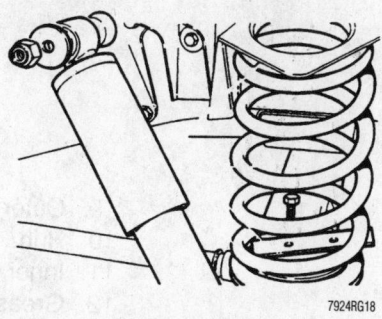

7924RG18

View of the rear spring mounting

Wheel Bearings

ADJUSTMENT

The front and rear wheel bearings are not adjustable.

REMOVAL & INSTALLATION

Front

1. Remove the front hub. Remove the outer bearing.

2. Use a seal removal tool and remove and discard the grease retainer. Remove the inner bearing.

3. Using a hammer and drift, drive out the bearing races.

To install:

4. Drive in new races until completely seated.

5. Using your hand or a bearing packer, pack the bearing with a suitable long life bearing grease. Grease the cone surfaces.

6. Place the inner bearing cone and roller assembly in the inner race. A light film of grease should be included between the lips of the new grease seal. Install the new seal using an appropriate driver tool.

7. Install the hub and outer bearing. Adjust the bearings.

Rear

1. Raise and safely support the vehicle.

2. Remove the rear wheels.

3. Release the brake hose clip, then remove the caliper and position it aside taking care not to kink the brake hose.

4. Remove the five axle retaining bolts, then withdraw the axle shaft.

5. Remove the joint washer.

6. Straighten the lockwasher tabs, then remove the locknut and lockwasher.

7. Remove the hub adjusting nut and spacing washer.

8. Remove the hub and brake rotor as an assembly.

9. Remove the outer bearing.

10. Remove the five nyloc nuts, then the ABS tone ring.

11. Matchmark the hub to the rotor for reassembly.

12. Remove the five bolts and separate the hub from the brake rotor.

13. Remove the grease seal with the appropriate seal puller, then the inner bearing.

14. Remove the inner and outer bearing races.

15. Clean the hub of any old grease and dry.

To install:

16. Install the inner and outer races.

17. Pack the inner bearing with grease and install it to the hub.

18. Using a seal driver, install the inner grease seal, and lubricate the seal lips to prevent premature wear.

19. Install the brake rotor to the hub, align the matchmarks, and apply Loctite® 270 or equivalent to the mounting bolts and tighten to 54 ft. lbs. (73 Nm).

20. Install the ABS tone ring and tighten the new nyloc nuts to 80 inch lbs. (9 Nm).

21. Pack the outer bearing with grease and install it to the hub.

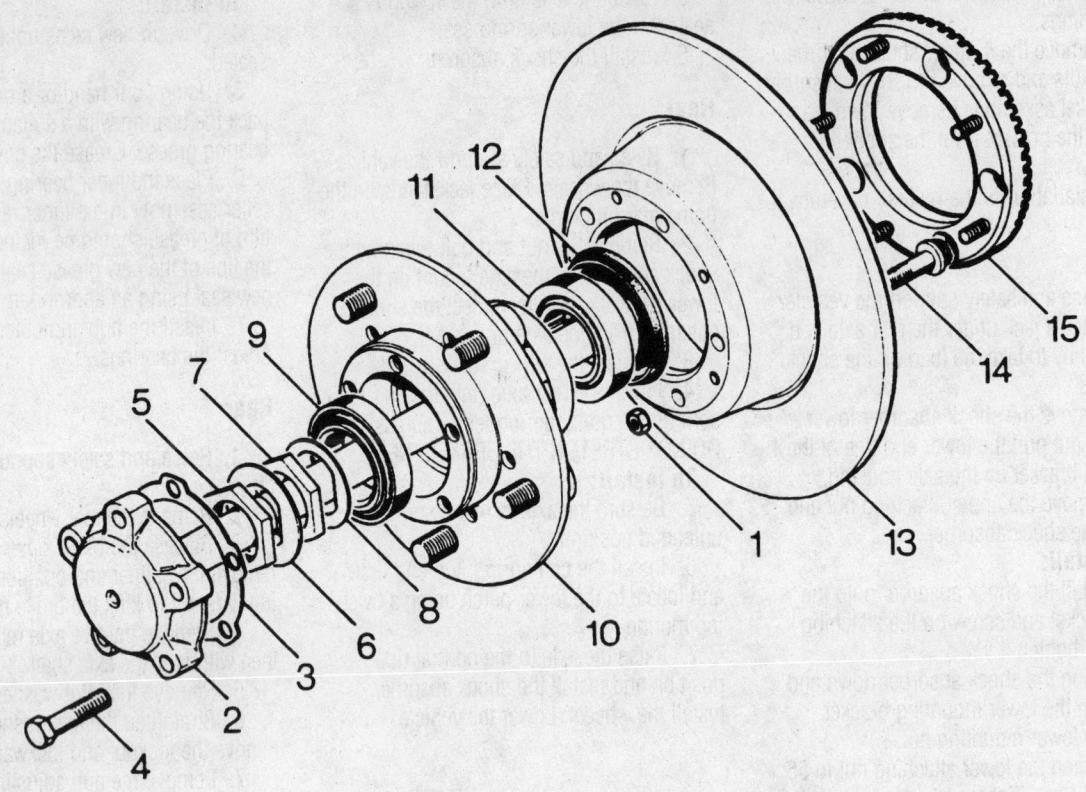

1. Sensor ring retaining nut ABS.
2. Axle shaft.
3. Axle shaft joint washer.
4. Axle shaft retaining bolt (five off).
5. Lock nut.
6. Lock washer.
7. Hub adjusting nut.
8. Spacing washer.
9. Outer bearing.
10. Hub.
11. Inner bearing.
12. Grease seal.
13. Brake disc.
14. Disc retaining bolt (five off).
15. Sensor ring ABS.

Exploded view of the rear hub components

22. Retract the ABS sensor slightly.
23. Install the hub and brake rotor assembly.
24. Install the spacing washer and nut. Tighten the nut to 45 ft. lbs. (61 Nm), loosen ½ turn (90°), then tighten to 35 inch lbs. (4 Nm).
25. Install a new lockwasher.

26. Install the locknut and tighten to 45 ft. lbs. (61 Nm).
27. Fold the tab of the lockwasher over the locknut.
28. Using a new joint washer install the axle shaft to the hub, and tighten the five bolts to 48 ft. lbs. (65 Nm).
29. Install the brake caliper and tighten

the mounting bolt to 60 ft. lbs. (82 Nm), and install the brake hose clip.
30. Reinstall the ABS sensor.
31. Install the wheels and tighten the nuts to 93 ft. lbs. (126 Nm), and operate the foot brake to seat the brake pads.
32. Check the fluid level in the axle.
33. Lower the vehicle and test drive.

ENGINE REPAIR

➡ **Disconnecting the negative battery cable on some vehicles may interfere with the functions of the on board computer systems and may require the computer to undergo a relearning process, once the negative battery cable is reconnected.**

✳✳ WARNING

NEVER disconnect the negative battery cable with the ignition ON or the engine running. Removing power from the computer control module with the ignition ON may destroy the module.

Distributor

REMOVAL

1. Disconnect the negative battery cable.
2. Detach the distributor connectors.
3. Remove the distributor cap without disconnecting the secondary leads. Position aside.
4. Matchmark the rotor with the distributor housing and housing with the cylinder block.
5. Remove the distributor hold-down bolt and pull the distributor from the cylinder block.

INSTALLATION

Engine Not Disturbed

1. Install a new O-ring to the distributor and lubricate with engine oil.
2. Insert the distributor into the cylinder block, while aligning the matchmarks made during removal. Install the distributor hold-down bolt.
3. Install the distributor cap. Attach the distributor connector.
4. Connect the negative battery cable. Start the engine and allow normal operating temperature to be reached.
5. Check and if necessary, adjust the ignition timing.

Engine Disturbed

1. Install a new O-ring to the distributor and lubricate it with engine oil.

2. Remove the No. 1 cylinder spark plug.
   a. Place a finger or compression gauge over the spark plug hole.
   b. Turn the crankshaft until compression starts to build up. Continue turning the crankshaft until the crankshaft pulley groove align with the timing mark "0" of the timing chain.
3. If necessary, remove the valve cover.
   a. Check that the timing marks with one and two dots are in straight line on the cylinder head surface.
   b. If not, turn the crankshaft one revolution (360 degrees) and align the crankshaft pulley groove with the timing mark "0" of the timing chain.
4. Align the groove of the distributor housing with the protrusion on the driven gear.
5. Insert the distributor into the cylinder block. Install the distributor hold-down bolt.
6. Install the distributor cap. Attach the distributor connector.
7. Connect the negative battery cable. Start the engine and allow normal operating temperature to be reached.
8. Check and if necessary, adjust the ignition timing.

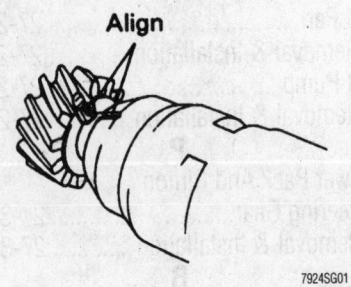

Align the groove on the distributor housing with the protrusion on the driven gear—4.5L (1FZ-FE) engine

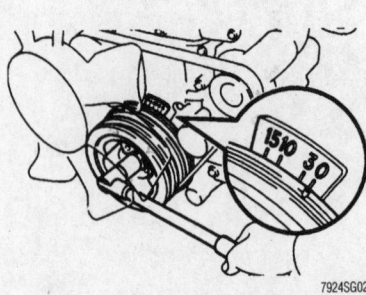

Align the groove on the crankshaft pulley with the timing mark "0" on the timing chain cover—4.5L (1FZ-FE) engine

Ignition Timing

ADJUSTMENT

4.5L (1FZ-FE) Engine

1. Warm up the engine to normal operating temperature.
2. Connect a Toyota or Lexus hand-held tester or the OBD II scan tool. Remove the fuse cover on the instrument panel. Connect the tool to the DLC3.
3. Connect the timing light to the engine.
4. Check the idle speed by racing the engine speed to 2,500 rpm for approximately 90 seconds and letting it return to idle. Idle speed should be 600–700 rpm.
5. Inspect and adjust ignition timing.
   a. Using tool No. SST 09843–18020 (jumper wire), or equivalent, connect terminals TE1 and E1 of the DLC1.
   b. Using a timing light, check the ignition timing. It should be 3° BTDC at idle.
6. Adjustment can be made by loosening the hold-down bolt and turning the distributor. Tighten the hold-down bolt and recheck the timing.
7. Remove the SST from the DLC1.
8. Disconnect the timing light from the engine.
9. Remove the tester or scan tool.

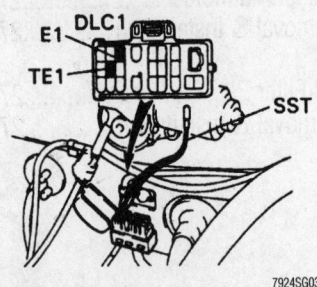

Connect a jumper wire such as service tool No. SST 9843–18020, or equivalent, to terminals TE1 and E1 of Data Link Connector 1 (DLC1)—4.5L (1FZ-FE) engine

4.7L (2UZ-FE) Engine

The ignition timing on the 4.7L engine is controlled by the engine control module and is not adjustable, however it can be checked using the following procedure.

1. Remove the battery cover and the air intake duct.

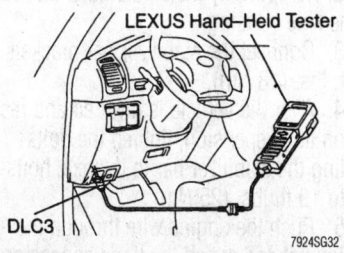

LEXUS Hand–Held Tester

DLC3

7924SG32

Connect the scan tool to Data Link Connector 3 (DLC3) at the lower left side of the instrument panel—4.5L (1FZ-FE) engine

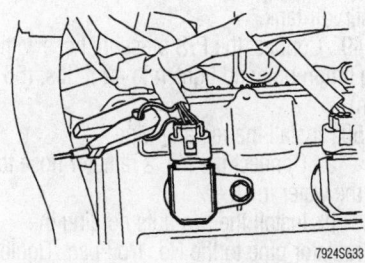

7924SG33

Be sure to connect the inductive pick-up of the timing light to the correct wire of the ignition coil connector—4.5L (1FZ-FE) engine

2. Remove the V-bank (engine) cover.

3. Start the engine and allow it to reach normal operating temperature.

4. Plug the scan tool into the Data Link Connector 3 (DLC3) at the lower left corner of the instrument panel.

5. Connect the timing light to the battery and to the wire shown on the No. 1 ignition coil connector.

6. Run the engine at about 2500 rpm for about 90 seconds, then return it to idle and check the idle speed. Correct idle speed should be 650–750 rpm, if not, inspect the air intake system.

7. Using tool No. SST 09843–18020 (jumper wire), or equivalent, connect terminals TE1 and E1 of the DLC1.

8. Using a timing light, check the ignition timing. It should be 5–15° BTDC at idle with the transmission in neutral.

9. Remove the jumper wire from the data link connector and turn the engine **OFF**.

10. Remove the timing light.

11. Unplug the scan tool.

12. Install the engine cover, intake air duct and battery cover.

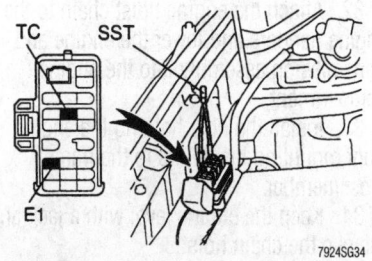

TC   SST

E1

7924SG34

Connect terminals TE1 and E1 of the DLC1 with a jumper wire such as tool No. SST 09843–18020—4.5L (1FZ-FE) engine

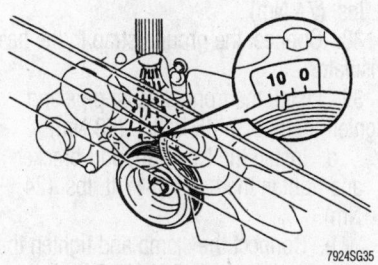

7924SG35

Aim the timing light at the crankshaft pulley to inspect the ignition timing—4.5L (1FZ-FE) engine

Engine Assembly

REMOVAL & INSTALLATION

4.5L (1FZ-FE) Engine

1. Properly relieve the fuel system pressure.

2. Disconnect the battery cables and remove the battery and the battery tray.

✸✸ CAUTION

Wait at least 90 seconds from the time the ignition switch is turned to the LOCK position and the negative battery cable is disconnected before starting work to avoid accidental air bag deployment.

3. Raise and safely support the vehicle.

✸✸ CAUTION

Never open, service or drain the radiator or cooling system when hot; serious burns can occur from the steam and hot coolant.

4. Drain the engine coolant, transmission oil and the engine oil.

5. Remove the hood.

6. Remove the radiator grille and remove the radiator.

7. Disconnect the oil cooler hose from the oil cooler pipe.

8. Remove the air cleaner hose, cap and the air cleaner case.

9. Disconnect the cruise control actuator cable from the throttle body.

10. Disconnect the accelerator cable from the throttle body.

11. Disconnect the heater hoses.

12. Disconnect the engine wiring harness and the heater valve from the cowl panel.

13. Disconnect the brake booster vacuum hose.

✸✸ CAUTION

Fuel injection systems remain under pressure after the engine has been turned OFF. Properly relieve fuel pressure before disconnecting any fuel lines. Failure to do so may result in fire or personal injury.

14. Disconnect the EVAP and fuel hoses.

15. Detach the following wires and connectors:

   a. Two heated oxygen sensor connectors.

   b. DCL1 clamp.

   c. Two oil pressure gauge connectors.

   d. Alternator wire and connector.

   e. Connector on the intake manifold from the fender apron.

   f. High tension cord from the ignition coil.

   g. Ground strap from the No. 1 engine hanger.

   h. Ground strap from the air intake chamber.

   i. Starter wire.

   j. Ground cable from the cylinder block.

16. Loosen the idler pulley nut and adjusting bolt and remove the A/C drive belt.

17. Disconnect the A/C compressor and remove the bracket.

18. Remove the radiator pipe.

   a. Remove the two nuts holding the radiator pipe to the No. 1 oil pan.

   b. Disconnect the No. 2 radiator hose from the water inlet and remove the radiator pipe.

19. Remove the union bolt and the two gaskets and disconnect the pressure hose from the P/S pump.

20. Disconnect the return hose from the P/S reservoir tank.

21. Disconnect the engine wire from the cabin.

   a. Remove the glove compartment door.

   b. Remove the screw and speaker panel.

   c. Disconnect the A/C amplifier.

   d. Detach the connector from the Powertrain Control Module and the cowl wire.

   e. Pull out the engine wire from the cabin.

22. Remove the stabilizer bar.

23. Put matchmarks on the flanges and remove the front and rear driveshafts.

24. Remove the transfer shift lever.

   a. Remove the nut and the transmission control rod.

   b. Remove the transfer shift lever knob.

   c. Lift up the console slightly in order to detach the connector.

   d. Remove the shifter console.

   e. Remove the center console box.

   f. Detach the connectors and remove the transfer shift lever boot and the transmission shift lever assembly.

   g. Pull out the pin and disconnect the shift rod.

   h. Remove the hose clamp and the transfer shift lever.

25. Remove the front exhaust pipe.

   a. Detach the heated oxygen sensor connector.

   b. Remove the nuts and bolts holding the exhaust to the rear catalytic converter.

   c. Loosen the clamp bolt and disconnect the clamp from the No. 1 support bracket.

   d. Remove the No. 1 support bracket.

   e. Remove the front exhaust pipe.

26. Disconnect the ground strap from the heat insulator.

27. Place a jack under the transmission. Put a block of wood between the jack and the transmission oil pan to prevent damage to the pan.

28. Remove the frame crossmember.

29. Attach the engine hoist chain to the two engine hangers.

30. Remove the nuts holding the engine front mounting insulators to the frame.

31. Lift the engine with the transmission out of the vehicle slowly and carefully. Be sure that the engine is clear of all wiring and hoses.

To install:

32. Attach the engine hoist chain to the engine hangers and lower the engine and transmission assembly into the engine compartment.

33. Install the nuts holding the engine front mounting insulators to the frame crossmember.

34. Keep the engine level with a jack and remove the chain hoist.

35. Install the frame crossmember and tighten the bolts to 45 ft. lbs. (61 Nm).

36. Tighten the nuts holding the crossmember to the engine rear mounting insulator to 54 ft. lbs. (74 Nm).

37. Tighten the nuts holding the engine front mounting insulators to the frame to 54 ft. lbs. (74 Nm).

38. Connect the ground strap to the heat insulator.

39. Install the front exhaust pipe and tighten the nuts to 46 ft. lbs. (63 Nm).

   a. Install the No. 1 support bracket and tighten the bolts to 17 ft. lbs. (24 Nm).

   b. Connect the clamp and tighten the clamp bolt to 14 ft. lbs. (19.5 Nm).

   c. Connect the front exhaust pipe to the rear catalytic converter and tighten the bolts to 34 ft. lbs. (46 Nm).

40. Install the transfer shift lever and hose clamp. Tighten the bolts to 13 ft. lbs. (18 Nm).

   a. Connect the shift rod and install the pin.

   b. Install the transfer shift lever boot and transmission shift lever assembly and tighten the bolts to 4 ft. lbs. (5.4 Nm).

   c. Connect the connectors to the transmission shift lever assembly.

   d. Install the center console box.

   e. Connect the pattern select switch connector.

   f. Install the shifter console. Install the transfer shift lever knob.

   g. Shift the shift lever to N position.

   h. Fully turn the control shaft lever back and return two notches. It is now in the neutral position.

   i. Connect the transmission control rod and tighten the nut to 9 ft. lbs. (13 Nm).

41. Install the front and rear driveshafts.

   a. At the differential side, align the matchmarks on the flanges and tighten the front shaft to 54 ft. lbs. (74 Nm) and the rear shaft to 65 ft. lbs. (88 Nm).

   b. At the transfer side, align the matchmarks on the flanges and tighten the front shaft to 54 ft. lbs. (74 Nm) and the rear shaft to 65 ft. lbs. (88 Nm).

42. Temporarily install the stabilizer bar to the axle housing.

43. Connect the stabilizer bar brackets to 13 ft. lbs. (18 Nm).

44. After the vehicle is lowered and resting on its suspension, tighten the bolts holding the stabilizer bar to the axle housing to 19 ft. lbs. (25 Nm).

45. Push the engine wire through the cowl panel and attach the three connectors to the Powertrain Control Module and the two connectors to the cowl wire.

46. Connect the A/C amplifier with its screw.

47. Install the speaker panel and the glove compartment door.

48. Connect the return hose to the P/S reservoir tank.

49. Connect the P/S pressure hose with the union bolt and tighten to 42 ft. lbs. (56 Nm).

50. Install the radiator pipe.

   a. Connect the No. 2 radiator hose to the water inlet.

   b. Install the two nuts holding the radiator pipe to the No. 1 oil pan. Tighten the nuts to 15 ft. lbs. (21 Nm).

51. Install the A/C bracket and tighten the bolts to 27 ft. lbs. (37 Nm).

52. Install the A/C compressor and tighten the bolts to 18 ft. lbs. (25 Nm).

53. Install and adjust the A/C drive belt.

54. Connect the following wires and connectors:

   a. Two heated oxygen sensor connectors.

   b. DCL1 clamp.

   c. Two oil pressure gauge connectors.

   d. Alternator wire and connector.

   e. Connector on the intake manifold from the fender apron.

   f. High tension cord from the ignition coil.

   g. Ground strap from the No. 1 engine hanger.

   h. Ground strap from the air intake chamber.

   i. Starter wire.

   j. Ground cable from the cylinder block.

55. Connect the fuel inlet hose to the fuel filter and tighten the union bolt to 22 ft. lbs. (39 Nm). Connect the fuel return hose.

56. Connect the EVAP hose and the brake booster vacuum hose.

57. Connect the heater valve and the engine wire to the cowl panel. Connect the engine wire and the ground strap.

58. Connect the heater hoses.

59. Connect the accelerator cable to the throttle body.

60. Connect the cruise control actuator cable to the throttle body.

61. Install the air cleaner case, the hose and the cap.

62. Connect the oil cooler hose to the oil cooler pipe.

63. Install the radiator.

64. Install the radiator grille.

65. Install the battery tray and the battery. Connect the battery cables.

66. Fill with engine with oil and fill the transmission with the proper amount and type of fluid.

67. Fill the radiator with engine coolant.

68. Start the engine and check for leaks.

69. Check the automatic transmission fluid level.

70. Check the ignition timing.

71. Install the hood and test drive the vehicle.

72. Recheck the engine coolant and oil levels.

4.7L (2UZ-FE) Engine

1. Matchmark the hood hinges to the hood, then remove the hood.

2. Remove the engine undercovers.

✳✳ CAUTION

Never open, service or drain the radiator or cooling system when hot; serious burns can occur from the steam and hot coolant. Also, when draining engine coolant, keep in mind that cats and dogs are attracted to ethylene glycol antifreeze and could drink any that is left in an uncovered container or in puddles on the ground. This will prove fatal in sufficient quantities. Always drain coolant into a sealable container. Coolant should be reused unless it is contaminated or is several years old.

3. Drain the engine coolant into a suitable container.

✳✳ CAUTION

The EPA warns that prolonged contact with used engine oil may cause a number of skin disorders, including cancer! You should make every effort to minimize your exposure to used engine oil. Protective gloves should be worn when changing the oil. Wash your hands and any other exposed skin areas as soon as possible after exposure to used engine oil. Soap and water, or waterless hand cleaner should be used.

4. Drain the engine oil.

5. Remove the engine V-bank cover.

6. Disconnect the negative battery cable first, then the positive cable. Remove the battery.

7. Remove the air cleaner assembly.

8. Remove the radiator coolant reservoir.

9. Disconnect the upper radiator hose from the front water bypass joint.

10. Remove the A/C discharge tube from the bracket. Then, remove the bracket.

11. Remove the three bolts attaching the fan shroud to the radiator and remove the shroud.

12. Remove the lower radiator hose.

13. Disconnect the transmission cooler lines from the radiator.

14. Remove the radiator.

15. Remove the generator drive belt, fan, fluid coupling and fan pulley.

16. Remove the glove box door, then the No. 2 lower panel. Detach the three electrical connectors from the ECM. Remove the ECM from the bracket and detach the three wiring harness connectors.

17. Detach/remove the following hoses, wires, clamps, cables and connectors:

✳✳ CAUTION

Observe all applicable safety precautions when working around fuel. Whenever servicing the fuel system, always work in a well ventilated area. Do not allow fuel spray or vapors to come in contact with a spark or open flame. Keep a dry chemical fire extinguisher near the work area. Always keep fuel in a container specifically designed for fuel storage; also, always properly seal fuel containers to avoid the possibility of fire or explosion.

a. Accelerator cable

b. Power steering air control valve vacuum hoses and clamp from the timing belt cover.

c. Generator wire and connector

d. Power steering air control valve vacuum hose from the upper intake manifold.

e. Heater hoses

f. Engine wiring harness, clamp and grommet from the cowl panel

g. Ground strap connector

h. Fuel supply and return hoses with clamps

i. Hoses from the charcoal canister

j. Engine wiring harness from clamp on the right fender

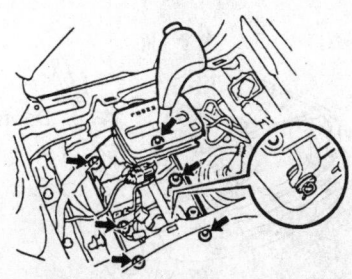

7924SG36

The transmission shift lever is attached with six mounting bolts. The linkage is held to the lever with a clip—4.7L (2UZ-FE) engine

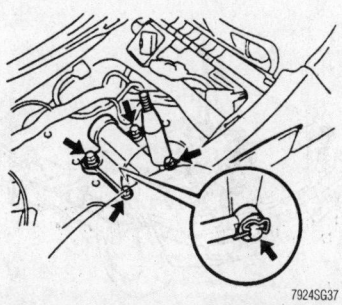

7924SG37

The transfer case shift lever is mounted with four bolts. Don't forget to remove the clip for the linkage—4.7L (2UZ-FE) engine

k. Battery cables and related retainers

18. Remove the transmission and transfer case shift levers after removing the upper console.

19. Remove the front exhaust pipes.

20. Matchmark and remove the driveshafts.

21. Remove the stabilizer bar.

22. Remove the A/C compressor from the engine without disconnecting the lines and suspend it out of the way with wire or string.

23. Remove the power steering pump from the engine without disconnecting the lines and suspend it out of the way with wire or string.

24. Attach a chain hoist to the engine and remove the bolts attaching the engine mounts to the frame brackets.

25. Remove the transfer case protector.

26. Remove the center crossmember.

27. Carefully raise the engine/transmission assembly up while tilting the transmission end of the assembly downward and remove it from the vehicle. Be sure the assembly is free of all wiring, hoses and cables.

28. Remove the transmission assembly from the engine.

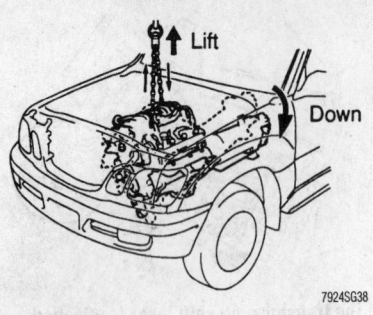

Tilt the transmission end of the assembly down to remove it from the vehicle—4.7L (2UZ-FE) engine

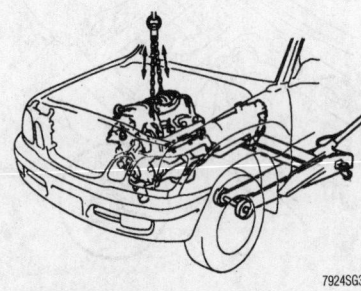

Raise the transmission with a jack to keep the assembly level while installing the engine mounts—4.7L (2UZ-FE) engine

To install:

29. Install the transmission on the engine.

➡**Support the transmission with a jack to keep the assembly level.**

30. Install the engine/transmission assembly into the vehicle and install the engine mount-to-frame bolts. Tighten the fasteners to 22 ft. lbs. (30 Nm).

31. Install the center crossmember. Tighten the nuts to 55 ft. lbs. (74 Nm) and the bolts to 37 ft. lbs. (50 Nm).

32. Install the power steering pump to the engine. Tighten the bolts to 13 ft. lbs. (17 Nm).

33. Install the A/C compressor to the engine. Tighten the mounting bolts to 36 ft. lbs. (49 Nm).

34. Install the stabilizer bar.

35. Align the matchmark and install the driveshafts.

36. Install the front exhaust pipes.

37. Install the transmission and transfer case shift levers. Then, replace the console.

38. Install all hoses, connectors, grommets, clamps and cables.

39. Install the engine wiring harness into the cabin and connect it to the ECM.

Then, replace the No. 2 trim panel and glove box door.

40. Install the fan pulley, fan and fluid coupling. Install the generator drive belt.

41. Install the radiator and fan shroud and coolant hoses.

42. Connect the transmission fluid cooler lines to the radiator.

43. Install the bracket and the A/C discharge hose.

44. Install the radiator coolant reservoir.

45. Install the air cleaner assembly.

46. Install the battery. Remember to connect the negative battery cable last.

47. Refill the cooling system with the proper amount of coolant.

⁂ WARNING

Operating the engine without the proper amount and type of engine oil will result in severe engine damage.

48. Refill the engine with the proper type and amount of engine oil.

49. Start the check all hoses and connections for leaks. Repair any leaks and turn the engine **OFF**.

50. Install the engine V-bank cover and the undercovers.

51. Install the hood using the matchmarks for proper alignment.

52. Road test the vehicle and recheck the fluid levels.

Water Pump

REMOVAL & INSTALLATION

4.5L (1FZ-FE) Engine

1. Disconnect the negative battery cable.

2. Drain the engine coolant.

3. Disconnect the No. 3 water bypass and radiator inlet hoses.

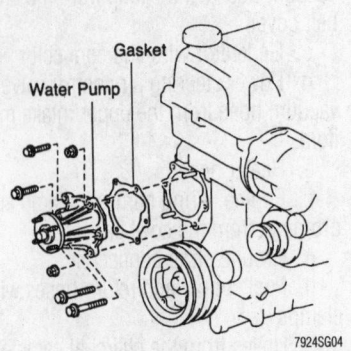

Water pump and related components— 4.5L (1FZ-FE) engine

4. Remove the drive belts, fan assembly and the fan shroud.

5. Disconnect the oil cooler hose from the clamp on the fan shroud. Remove the bolts holding the fan shroud to the radiator.

6. Remove the four bolts, two nuts, water pump and the gasket.

To install:

7. Install the water pump using a new gasket. Tighten the fasteners to 15 ft. lbs. (21 Nm).

8. Install the water pump pulley, fan shroud and the drive belts.

   a. Place the fan with the fluid coupling, water pump pulley and the fan shroud in position.

   b. Temporarily install the fan pulley mounting nuts.

   c. Install the fan shroud and tighten the bolts to 4.9 ft. lbs. (5.4 Nm).

   d. Connect the oil cooler hose to the clamp on the fan shroud.

9. Connect the No. 3 water bypass and radiator hoses. Fill the cooling system.

10. Connect the negative battery cable, start the engine and check for leaks.

11. Recheck the coolant level.

4.7L (2UZ-FE) Engine

1. Disconnect the negative battery cable.

2. Drain the coolant.

3. Remove the timing belt.

4. Remove the No. 2 idler pulley.

5. Disconnect the water bypass hose from the water inlet housing.

6. Remove the two bolts attaching the water inlet housing to the water pump.

7. Disconnect the water bypass pipe, then remove the water inlet housing from the water pump.

8. Remove the five bolts, two stud bolts and nut attaching the water pump to the engine and remove the water pump.

9. Remove the O-ring from the water bypass pipe.

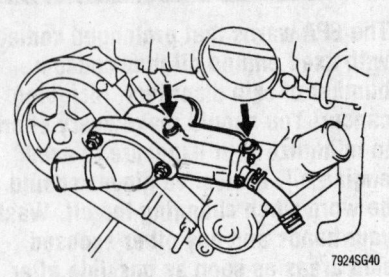

Water inlet housing attaching bolts—4.7L (2UZ-FE) engine

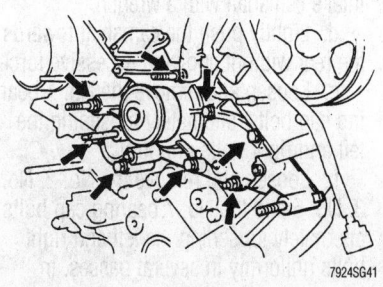

Water pump mounting bolts, stud bolts and nut locations—4.7L (2UZ-FE) engine

To install:

10. Apply soapy water a new O-ring and install it on the water bypass pipe.

11. Install the bypass pipe in the water pump and install the water pump using a new gasket. Tighten the bolts to 15 ft. lbs. (21 Nm) and the remaining fasteners to 13 ft. lbs. (18 Nm). Be sure to tighten them evenly in several passes.

12. Clean all old silicone material off of the water inlet housing.

13. Apply soapy water to a new O-ring and install it on the water inlet housing.

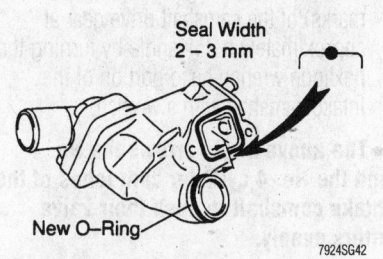

Install a new O-ring and apply the specified sealant or equivalent to the water inlet housing—4.7L (2UZ-FE) engine

14. Apply a bead of Seal Packing (RTV silicone) No. 08826–00100 or equivalent in the sealing groove on the water inlet housing.

15. Install the water inlet housing on the water pump. Alternately tighten the two bolts to 13 ft. lbs. (18 Nm).

16. Install the No. 2 idler pulley.

17. Install the timing belt.

18. Refill the engine with coolant.

19. Start the engine and check for leaks.

20. Allow the engine to cool down, then recheck the coolant level.

Cylinder Head

REMOVAL & INSTALLATION

4.5L (1FZ-FE) Engine

1. Properly relieve the fuel system pressure.

2. Disconnect the battery cables and remove the battery and the battery tray.

❊❊ CAUTION

Never open, service or drain the radiator or cooling system when hot; serious burns can occur from the steam and hot coolant. Also, when draining engine coolant, keep in mind that cats and dogs are attracted to ethylene glycol antifreeze and could drink any that is left in an uncovered container or in puddles on the ground. This will prove fatal in sufficient quantities. Always drain coolant into a sealable container. Coolant should be reused unless it is contaminated or is several years old.

3. Drain the engine coolant.

4. Remove the air cleaner hose and cap.

5. Disconnect the cruise control actuator cable from the throttle body.

6. Disconnect the accelerator cable from the throttle body.

7. Detach the engine ground strap from the No. 1 engine hanger and the ground strap from the air intake chamber.

8. Detach the connector on the intake manifold from the left fender apron.

9. Disconnect the brake booster vacuum hose.

10. Disconnect the EVAP hose and disconnect the fuel return hose.

❊❊ CAUTION

Fuel injection systems remain under pressure after the engine has been turned OFF. Properly relieve fuel pressure before disconnecting any fuel lines. Failure to do so may result in fire or personal injury.

11. Disconnect the heater hoses.

12. Disconnect the engine wire and heater valve from the cowl panel.

13. Remove the No. 2 and the No. 3 cylinder head covers

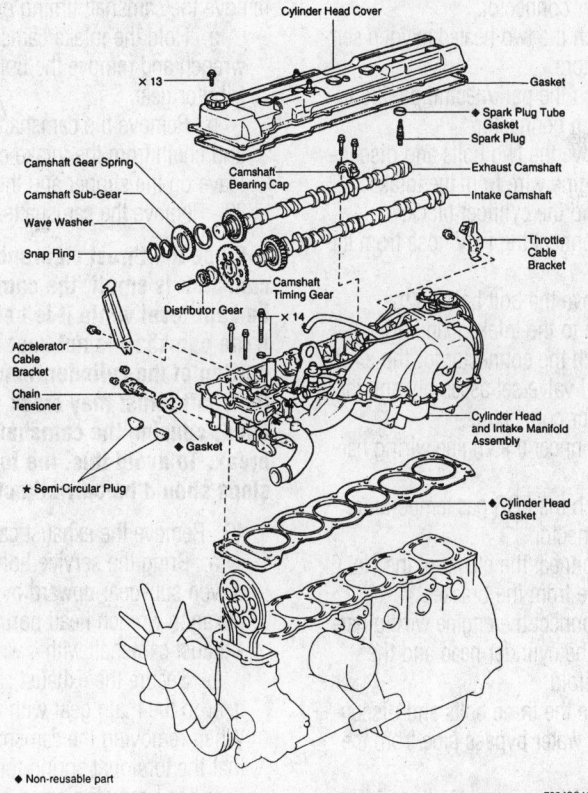

Exploded view of the cylinder head and related components—4.5L (1FZ-FE) engine

14. Remove the distributor.

15. Disconnect the P/S reservoir tank.

16. Disconnect the radiator inlet hose and the No. 3 water bypass hose.

17. Remove the alternator.

18. Remove the throttle body.

19. Remove the oil dipsticks and guides for the engine and transmission. Pull out the dipstick together with the dipstick guide and remove the O-ring from the dipstick guide.

20. Remove the intake manifold stay.

21. Disconnect the fuel inlet hose from the fuel filter.

22. Detach the following connectors:

   a. ECT sender gauge connector, the ECT cut switch connector and the ECT sensor connector.

   b. Knock sensor connector.

   c. Crankshaft position sensor connector.

23. Remove the bolt and disconnect the engine wiring harness from the cylinder block.

24. Detach the following:

   a. Oil level sensor connector.

   b. Detach the heated oxygen sensor connector.

   c. Two connectors from the transmission.

   d. Starter connector.

   e. Detach the two heated oxygen sensor connectors.

   f. Detach the park/neutral position (PNP) switch connector.

   g. Remove the two bolts and disconnect the engine wire from the intake manifold and the cylinder block.

   h. Disconnect the PCV hose from the PCV valve.

   i. Remove the bolt holding the engine wire to the intake manifold.

   j. Detach the connector for the emission control valve set assembly and the three injector connectors.

   k. Disconnect the engine wiring harness clamp.

   l. Detach the EGR gas temperature sensor connector.

   m. Disconnect the clamp of the No. 6 injector wire from the bracket.

   n. Disconnect the engine wiring harness from the cylinder head and the intake manifold.

25. Remove the three bolts and disconnect the No. 2 water bypass pipe from the cylinder head.

26. Remove the nuts and bolts holding the front exhaust pipe to the rear catalytic converter.

27. Disconnect the front exhaust pipe and remove the gasket.

28. Remove the clamp from the No. 1 support bracket and remove the bracket.

29. Remove the front exhaust pipe and the gaskets.

30. Remove the No. 1 and No. 2 exhaust manifolds. Remove the No. 1 and No. 2 heat insulators.

31. Remove the ground cable, heater pipe and gasket.

32. Remove the water bypass outlet and the pipe. Remove the three O-rings from the water bypass outlet and the pipe.

33. Remove the cylinder head cover.

34. Remove the semi-circular plug from the cylinder head.

35. Remove the spark plugs.

36. Set the No. 1 cylinder to TDC of the compression stroke.

   a. Turn the crankshaft pulley and align its groove with the **0** mark on the timing chain cover.

   b. Check that the timing marks (one and two dots) of the camshaft drive and driven gears are in straight line on the cylinder head surface. If not, turn the crankshaft one revolution (360°) and align the marks as above.

37. Remove the chain tensioner.

38. Place matchmarks on the camshaft timing gear and the timing chain and remove the camshaft timing gear.

   a. Hold the intake camshaft with a wrench and remove the bolt and the distributor gear.

   b. Remove the camshaft timing gear and chain from the intake camshaft and leave on the slipper and the damper.

39. Remove the camshafts.

➡**Since the thrust clearance of the camshaft is small, the camshaft must be kept level while it is being removed. If the camshaft is not kept level, the portion of the cylinder head receiving the shaft thrust may crack or be damaged, causing the camshaft to seize or break. To avoid this, the following steps should be carried out.**

40. Remove the exhaust camshaft.

   a. Bring the service bolt hole of the driven sub-gear upward by turning the hexagon wrench head portion of the exhaust camshaft with a wrench.

   b. Secure the exhaust camshaft sub-gear to the main gear with a service bolt. When removing the camshaft, be sure that the torsional spring force of the sub-gear has been eliminated by the above operation.

   c. Set the timing mark (two dot marks) of the camshaft driven gear at approximately 35° angle by turning the hexagon wrench head portion of the intake camshaft with a wrench.

   d. Lightly push the camshaft towards the rear without applying excessive force.

   e. Loosen and remove the No. 1 bearing cap bolts, alternately loosening the left and right bolts uniformly.

   f. Loosen and remove the No. 2, No. 3, No. 5 and the No. 7 bearing cap bolts, alternately loosening the left and right bolts uniformly in several passes, in sequence.

➡**Do not remove the No. 4 and No. 6 bearing cap bolts at this stage.**

   g. Remove the four bearing caps.

   h. Alternately and uniformly loosen and remove the No. 4 and the No. 6 bearing cap bolts.

   i. If the camshaft is not being lifted out straight and level, retighten the four No. 4 and No. 6 bearing cap bolts. Then, reverse the preceding order of Substeps g through e and repeat Substeps c to h once again.

   j. Remove the two bearing caps and exhaust camshaft. Do not pry on or attempt to force the camshaft with a tool or any other object.

41. Remove the intake camshaft.

   a. Set the timing mark (two dot marks) of the camshaft drive gear at approximately a 25° angle by turning the hexagon wrench head portion of the intake camshaft with a wrench.

➡**The above angle arrows the No. 1 and the No. 4 cylinder cam lobes of the intake camshaft to push their valve lifters evenly.**

   b. Lightly push the intake camshaft towards the front without applying excessive force.

   c. Loosen and remove the No. 1 bearing cap bolts, alternately loosening the left and the right bolts uniformly.

   d. Loosen and remove the No. 3, No. 4, No. 6 and the No. 7 bearing cap bolts, alternately loosening the left and right bolts uniformly in several passes in sequence.

➡**Do not remove the No. 2 and No. 5 bearing cap bolts at this stage.**

   e. Remove the four bearing caps.

   f. Alternately and uniformly loosen and remove the No. 2 and the No. 5 bearing cap bolts.

   g. If the camshaft is not being lifted out straight and level, retighten the four No. 2 and No. 5 bearing cap bolts. Then, reverse the preceding order of Substeps

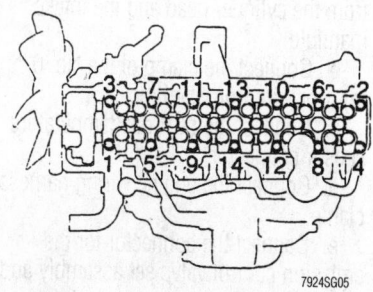

Loosen the cylinder head bolts according to the sequence shown to prevent warping the head—4.5L (1FZ-FE) engine

e through c and repeat Substeps a to f once again.

h. Remove the two bearing caps and the exhaust camshaft.

42. Remove the cylinder head and the intake manifold assembly.

a. Remove the two bolts in front of the head before the other head bolts are removed.

b. Loosen and remove the 14 cylinder head bolts in sequence using several passes.

✳✳ CAUTION

Cylinder head warpage or cracking could result from removing bolts in incorrect order.

c. Lift the cylinder head from the dowels on the cylinder block and place the cylinder head on wooden blocks on the bench.

d. If the cylinder head is difficult to lift off, pry between the cylinder head and the cylinder block with a flat prying tool.

43. Remove the alternator bracket.

44. Remove the two nuts, the water outlet and the gasket.

45. Loosen the union nut and remove the EGR pipe and gasket.

46. Remove the heater inlet pipe and hose.

47. Remove the air intake chamber and the intake manifold assembly.

a. Disconnect the vacuum hoses from the TVV.

b. Remove the 10 bolts, the two nuts and the intake manifold and gasket.

48. Remove the No. 1 water bypass hose.

49. Remove the No. 1 and the No. 2 engine hangers.

50. Remove the two engine wire clamp brackets.

51. Remove the accelerator cable bracket and the throttle cable bracket.

52. Remove the valve lifters and shims. Arrange the valve lifters and shims in correct order for reinstallation.

To install:

53. Install the valve lifters and shims. Check to be sure that the valve lifter rotates smoothly by hand.

54. Install the accelerator cable bracket and the throttle cable bracket.

55. Install the engine wire clamp brackets.

56. Install the No. 1 and No. 2 engine hangers and tighten the bolts to 30 ft. lbs. (41 Nm).

57. Install the air intake chamber and intake manifold assembly.

a. Place a new gasket so that the rear mark is toward the rear side.

b. Tighten the intake manifold bolts to 15 ft. lbs. (21 Nm).

c. Connect the vacuum hoses to the TVV.

58. Install the heater hose to the cylinder head and connect the pipe to the intake manifold. Tighten the bolts to 15 ft. lbs. (21 Nm).

59. Temporarily install the union nut to the EGR valve. Install the EGR pipe to the cylinder head. Tighten the bolts to 15 ft. lbs. (21 Nm). Tighten the union nut to 58 ft. lbs. (78 Nm).

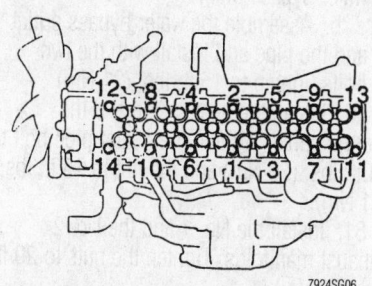

Tighten the cylinder head bolts in the correct order to ensure proper cylinder sealing and to prevent leaks—4.5L (1FZ-FE) engine

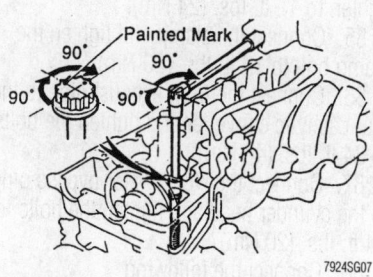

Mark the front of the bolt head with paint, then tighten the bolts an additional 90 degrees, in sequence, twice—4.5L (1FZ-FE) engine

60. Install a new gasket and the water outlet. Tighten the nuts to 15 ft. lbs. (21 Nm).

61. Install the alternator bracket and tighten the bolts to 32 ft. lbs. (43 Nm).

62. Install the cylinder head and the intake manifold assembly.

a. Apply seal packing on the end of the engine block by the timing belt.

b. Install a new cylinder head gasket on the cylinder block.

c. Install the cylinder head.

63. Lubricate the cylinder head bolts lightly with clean engine oil, then install them.

64. Tighten the cylinder head bolts progressively in sequence to 29 ft. lbs. 39 Nm).

65. Mark the front of the cylinder head bolt head with paint.

66. Retighten the cylinder head bolts by 90° in numerical order.

67. Retighten the cylinder head bolts an additional 90° so that the painted mark is now facing to the rear.

✳✳ WARNING

Do not combine Substeps d and e, the above steps must be followed exactly and in order to prevent cylinder head damage or pre-mature gasket failure.

a. Install and tighten the two mounting bolts to 15 ft. lbs. (21 Nm).

68. Install the camshafts.

➡**Since the thrust clearance of the camshaft is small, the camshaft must be kept level while it is being installed. If the camshaft is not kept level, the portion of the cylinder head receiving the shaft thrust may crack or be damaged, causing the camshaft to seize or break. To avoid this, the following steps should be carried out.**

a. Apply engine oil to the thrust portion of the intake camshaft.

b. Lightly place the intake camshaft on top of the cylinder head so that the No. 1 and the No. 4 cylinder cam lobes face downward.

c. Lightly push the camshaft towards the front without applying excessive force. Place the No. 2 and the No. 5 bearing caps in their proper location.

d. Temporarily tighten these bearing cap bolts uniformly and alternately in several passes until the bearing caps are snug with the cylinder head.

e. Place the No. 3, No. 4, No. 6 and the No. 7 bearing caps in their proper location. Temporarily tighten these bearing cap bolts, alternately tightening the left and right bolts uniformly.

f. Place the No. 1 bearing cap in its proper location. Check that there is no gap between the cylinder head and the contact surface of the bearing cap.

g. Uniformly tighten the 14 bearing cap bolts in several passes to 12 ft. lbs. (16 Nm).

69. Install the exhaust camshaft.

a. Set the timing mark (two dot marks) of the camshaft drive gear at approximately 35° angle by turning the hexagon wrench head portion of the intake camshaft with a wrench.

b. Apply engine oil to the thrust portion of the exhaust camshaft. Engage the exhaust camshaft gear to the intake camshaft hear by matching the timing marks (two dot marks) on each gear.

c. Roll down the exhaust camshaft onto the bearing journals while engaging the gears with each other. Lightly push the intake camshaft towards the front without applying excessive force.

d. Install the No. 4 and the No. 6 bearing caps in their proper location. Temporarily tighten these bearing cap bolts, alternately tightening the left and right bolts uniformly.

e. Place the No. 2, No. 3, No. 5 and the No. 7 bearing caps in their proper location. Temporarily tighten these bearing cap bolts, alternately tightening the left and right bolts uniformly.

f. Tighten the 14 bearing cap bolts in several passes to 12 ft. lbs. (16 Nm).

g. Bring the service bolt installed in the driven sub-gear upward by turning the hexagon wrench head portion of the camshaft with a wrench. Remove the service bolt.

h. Check that the intake and the exhaust camshafts turn smoothly.

70. Set the No. 1 cylinder to TDC of the compression stroke. Turn the crankshaft pulley and align its groove with the timing mark **0** of the timing chain cover. Turn the camshaft so that the timing marks with one and two dots will be in straight line on the cylinder head surface.

71. Install the camshaft timing gear.

a. Check that the matchmarks on the camshaft timing gear and the timing chain are aligned. Place the gear over the straight pin of the intake camshaft.

b. Align the straight pin of the distributor gear with the straight pin groove of the intake camshaft gear.

c. Hold the intake camshaft with a wrench, install and tighten the bolt to 54 ft. lbs. (74 Nm).

72. Install the chain tensioner. Push the tensioner by hand until it touches the head installation surface, then install and tighten the two nuts to 15 ft. lbs. (21 Nm).

73. Check the valve timing.

a. Turn the crankshaft pulley and align its groove with the timing mark **0** of the timing chain cover. Always turn the crankshaft clockwise.

b. Check that the timing marks (one and two dots) of the camshaft drive and driven gears are in straight line on the cylinder head surface. If not, turn the crankshaft one revolution (360°) and align the marks.

74. Check valve clearance and adjust if necessary.

75. Install the spark plugs.

76. Install the semi-circular plug to the cylinder head.

77. Be sure that the No. 1 cylinder is in TDC of the compression stroke.

78. Install the cylinder head cover.

79. Install the water bypass outlet and the pipe.

a. Install and new O-ring to the water bypass outlet. Install new O-rings to the water bypass pipe.

b. Assemble the water bypass outlet and the pipe and install with the two bolts tighten to 15 ft. lbs. (21 Nm).

80. Install the heater pipe and the ground cable. Tighten the heater pipe bolt to 14 ft. lbs. (20 Nm) and the nut to 15 ft. lbs. (21 Nm).

81. Install the No. 1 and the No. 2 exhaust manifolds. Tighten the nuts to 29 ft. lbs. (39 Nm).

82. Install the No. 1 insulator and No. 2 heat insulator and tighten the bolts to 14 ft. lbs. (20 Nm).

83. Install the front exhaust pipe. Tighten the nuts to 46 ft. lbs. (63 Nm).

84. Install the No. 1 support bracket and tighten to 17 ft. lbs. (24 Nm).

85. Connect the clamp and tighten the clamp bolt to 14 ft. lbs. (20 Nm).

86. Connect the front exhaust pipe to the rear catalytic converter and tighten the bolts to 34 ft. lbs. (46 Nm).

87. Connect the No. 2 water bypass pipe to the cylinder head and tighten the bolts to 14 ft. lbs. (20 Nm).

88. Connect the following:

a. Connect the engine wiring harness from the cylinder head and the intake manifold.

b. Connect the clamp of the No. 6 injector wire from the bracket.

c. Connect the EGR gas temperature sensor connector.

d. Connect the engine wiring harness clamp.

e. Connect the connector for the emission control valve set assembly and the three injector connectors.

f. Install the bolt holding the engine wire to the intake manifold.

g. Connect the PCV hose from the PCV valve.

h. Install the two bolts and connect the engine wire from the intake manifold and the cylinder block.

i. Connect the Park/Neutral Position (PNP) switch connector.

j. Connect the two heated oxygen sensor connectors.

k. Starter connector.

l. Two connectors from the transmission.

m. Oil level sensor connector.

89. Install the bolt and connect the engine wiring harness from the cylinder block.

90. Connect the following:

a. ECT sender gauge connector, the ECT cut switch connector and the ECT sensor connector.

b. Knock sensor connector.

c. Crankshaft position sensor connector.

91. Connect the fuel inlet hose to the fuel filter with the union bolt. Tighten to 22 ft. lbs. (29 Nm).

92. Install the intake manifold stay and tighten the bolts to 26 ft. lbs. (36 Nm).

93. Install the oil dipsticks and the guides for the engine and the transmission. Tighten the oil dipstick guide bolts to 14 ft. lbs. (20 Nm).

94. Install the throttle body.

95. Install the alternator and the drive belts.

96. Connect the No. 3 water bypass hose.

97. Connect the radiator inlet hose.

98. Connect the P/S reservoir tank and tighten the bolts to 14 ft. lbs. (20 Nm).

99. Install the distributor.

100. Install the No. 2 and the No. 3 cylinder head covers.

101. Connect the heater valve and the engine wiring harness to the cowl panel.

102. Connect the heater hoses.
103. Connect the fuel return hose, the EVAP hose and the brake booster vacuum hose.
104. Connect the connector on the intake manifold to the left fender apron.
105. Connect the ground strap to the No. 1 engine hanger and the air intake chamber.
106. Connect the throttle cable to the throttle body. Adjust the throttle cable.
107. Connect the accelerator cable to the throttle body.
108. Connect the cruise control actuator cable to the throttle body.
109. Install the air cleaner hose and cap.
110. Install the battery tray, the battery and connect the cables.
111. Refill the engine coolant.
112. Start the engine and check for leaks.
113. Make necessary engine adjustments.
114. Road test the vehicle and recheck the engine coolant level.

4.7L (2UZ-FE) Engine

1. Disconnect the negative battery cable.

❈❈ CAUTION

Never open, service or drain the radiator or cooling system when hot; serious burns can occur from the steam and hot coolant. Also, when draining engine coolant, keep in mind that cats and dogs are attracted to ethylene glycol antifreeze and could drink any that is left in an uncovered container or in puddles on the ground. This will prove fatal in sufficient quantities. Always drain coolant into a sealable container. Coolant should be reused unless it is contaminated or is several years old.

2. Drain the engine coolant.
3. Remove the timing belt from the camshaft pulleys, then remove the pulleys.
4. Remove the camshaft position sensor.
5. Remove the power steering pump and position it to the side with a piece of wire.
6. Remove the front exhaust pipes from the manifolds.
7. If equipped with an automatic transmission, remove the dipstick and tube from the transmission.

8. Remove the ignition coils and the timing belt rear covers.
9. Remove the V-bank cover.

❈❈ CAUTION

Observe all applicable safety precautions when working around fuel. Whenever servicing the fuel system, always work in a well ventilated area. Do not allow fuel spray or vapors to come in contact with a spark or open flame. Keep a dry chemical fire extinguisher near the work area. Always keep fuel in a container specifically designed for fuel storage; also, always properly seal fuel containers to avoid the possibility of fire or explosion.

10. Disconnect the fuel supply hose.
11. Remove the intake manifold.
12. Remove the water inlet housing assembly.
13. Remove the front and rear water bypass joints.
14. Remove the engine hangers and cylinder head covers.

➡ **If necessary, remove the semi-circular plugs and the camshaft housing plugs.**

15. Turn the crankshaft pulley so the timing mark on the pulley is in line with the centers of the No. 2 idler pulley bolt and the crankshaft pulley bolt.
16. Remove the camshafts.

❈❈ WARNING

Be careful not to damage the camshaft timing tube.

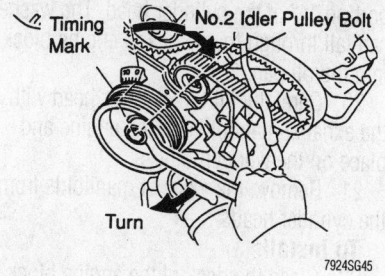

Be sure to align the timing mark on the crankshaft pulley as shown when removing or installing the camshafts—4.7L (2UZ-FE) engine

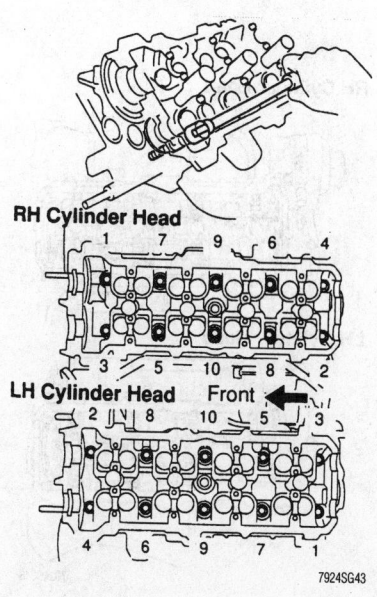

To avoid damaging the cylinder head, loosen the bolts in the sequence shown—4.7L (2UZ-FE) engine

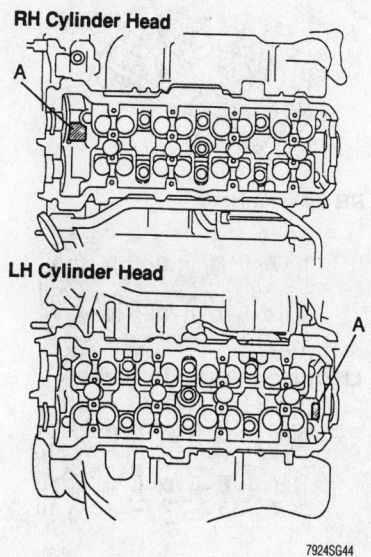

The washer will fall into the oil pan if dropped into section "A" of the cylinder head—4.7L (2UZ-FE) engine

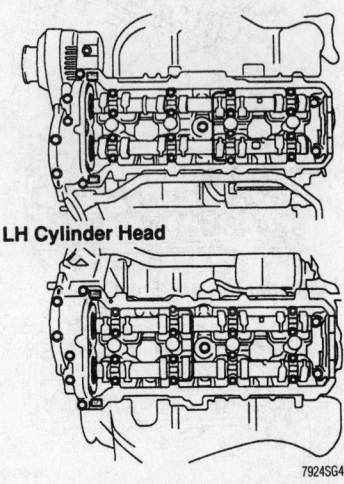

RH Cylinder Head

LH Cylinder Head

7924SG48

RH Cylinder Head

2UR

LH Cylinder Head

2UL

7924SG47

The cylinder head gaskets are marked right and left, be sure to use the correct one—4.7L (2UZ-FE) engine

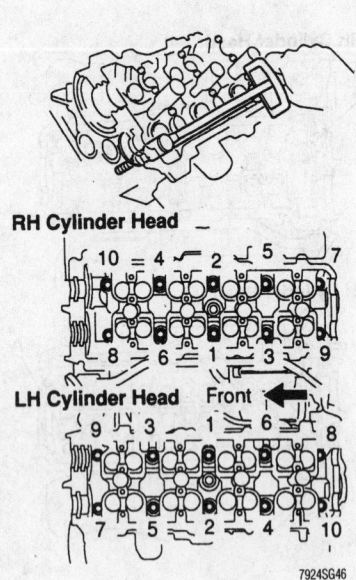

RH Cylinder Head

LH Cylinder Head   Front

7924SG46

Be sure to tighten the cylinder head bolts in the sequence shown—4.7L (2UZ-FE) engine

Be sure to install the correct gaskets on the right and left sides of the engine—4.7L (2UZ-FE) engine

17. Remove the oil seal from the intake camshaft.

18. Remove the spark plugs.

✵✵ WARNING

Cylinder head warping or cracking may result if the bolts are not removed gradually and in the correct order.

19. Loosen the cylinder head bolts evenly, in several passes, using the correct sequence. Do not drop the washers into portion "A" of the cylinder head. The washer can fall through the head and engine block into the oil pan.

20. Carefully lift the cylinder head with the exhaust manifolds off the engine and place on the wooden blocks.

21. Remove the exhaust manifolds from the cylinder heads.

To install:

22. Clean the deck of the engine block and place new gaskets on the engine. The rear of the gaskets are marked "2UR" for the right side and "2UL" for the left side. Be sure to install the correct gasket on the proper side of the engine.

23. Carefully position the cylinder heads on the engine. Lightly oil the threads and under the heads of each bolt.

24. Tighten the cylinder head bolts first to 24 ft. lbs. (32 Nm) in the correct sequence. Be sure to use several passes and tighten the bolts evenly.

25. Mark the front of each bolt with paint, tighten each bolt in sequence 90°.

26. Tighten each bolt in sequence an additional 90° so that the paint mark now faces toward the rear of the engine.

27. Install the spark plugs.

28. If not already done, turn the crankshaft pulley so the timing mark on the pulley is inline with the centers of the No. 2 idler pulley bolt and the crankshaft pulley bolt.

✵✵ WARNING

The valves may come into contact with the pistons when installing the camshafts if the crankshaft is not at the proper position.

29. Install the camshafts using the recommended procedure.

30. Check and adjust the valve clearance if necessary.

31. Install the camshaft housing plugs.

32. Apply sealant to either side of the camshaft sprockets on both cylinder heads.

33. Using new gaskets, install the cylinder head covers. Tighten the bolts to 53 inch lbs. (6 Nm).

34. Install the rear water bypass joint using new gaskets. Tighten the nuts to 13 ft. lbs. (18 Nm).

35. Install the front water bypass joint using new gaskets. Tighten the nuts to 13 ft. lbs. (18 Nm).

36. Install the water inlet assembly.

37. Install the intake manifold assembly and related components. Tighten the six bolts and four nuts to 13 ft. lbs. (18 Nm).

38. Connect the fuel supply hose.

39. Install the timing belt rear covers.

40. Install the V-bank cover.

41. Install the ignition coils.

42. If removed, install the transmission dipstick and tube.

43. Connect the front exhaust pipes to the manifolds.

44. Install the power steering pump.

45. Install the camshaft position sensor.

46. Install the camshaft timing pulleys.

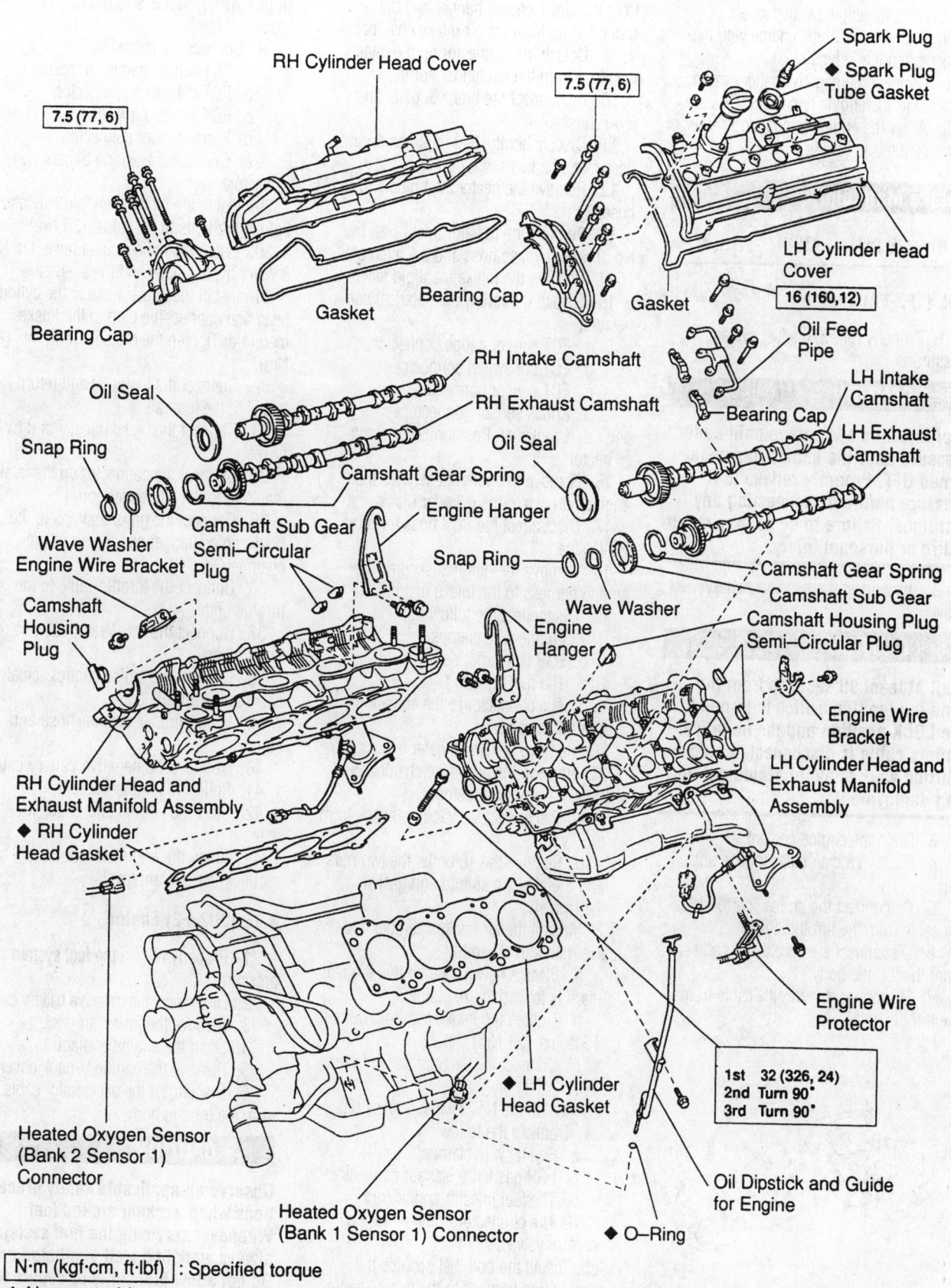

7.5 (77, 6)

RH Cylinder Head Cover

7.5 (77, 6)

Spark Plug

◆ Spark Plug Tube Gasket

7.5 (77, 6)

Bearing Cap

Gasket

Bearing Cap

Gasket

LH Cylinder Head Cover

16 (160, 12)

Oil Feed Pipe

Bearing Cap

Oil Seal

RH Intake Camshaft

RH Exhaust Camshaft

LH Intake Camshaft

LH Exhaust Camshaft

Snap Ring

Oil Seal

Camshaft Gear Spring

Camshaft Sub Gear

Semi–Circular Plug

Engine Hanger

Snap Ring

Wave Washer

Camshaft Gear Spring

Camshaft Sub Gear

Wave Washer Engine Wire Bracket

Camshaft Housing Plug

Engine Hanger

Camshaft Housing Plug Semi–Circular Plug

Engine Wire Bracket

RH Cylinder Head and Exhaust Manifold Assembly

◆ RH Cylinder Head Gasket

LH Cylinder Head and Exhaust Manifold Assembly

Engine Wire Protector

◆ LH Cylinder Head Gasket

Heated Oxygen Sensor (Bank 2 Sensor 1) Connector

Heated Oxygen Sensor (Bank 1 Sensor 1) Connector

◆ O–Ring

1st	32 (326, 24)
2nd	Turn 90°
3rd	Turn 90°

Oil Dipstick and Guide for Engine

N·m (kgf·cm, ft·lbf) : Specified torque

◆ Non–reusable part

Exploded view illustrating the cylinder head removal procedure—4.7L (2UZ-FE) engine

7924SG49

47. Install the timing belt.
48. Refill the engine with coolant.
49. Drain and refill the engine with the correct amount and type of oil.
50. Connect the negative battery cable.
51. Start the engine and check for leaks.
52. Allow the engine to cool and recheck the coolant level.

Intake Manifold

REMOVAL & INSTALLATION

4.5L (1FZ-FE) Engine

1. Properly relieve the fuel system pressure.

❄❄ CAUTION

Fuel injection systems remain under pressure after the engine has been turned OFF. Properly relieve fuel pressure before disconnecting any fuel lines. Failure to do so may result in fire or personal injury.

2. Disconnect the negative battery cable.

❄❄ CAUTION

Wait at least 90 seconds from the time the ignition switch is turned to the LOCK position and the negative battery cable is disconnected before starting work to avoid accidental air bag deployment.

3. Drain the engine coolant.
4. Remove the air cleaner hose and cap.
5. Disconnect the cruise control actuator cable from the throttle body.
6. Disconnect the accelerator cable from the throttle body.
7. Disconnect the throttle cable from the throttle body.

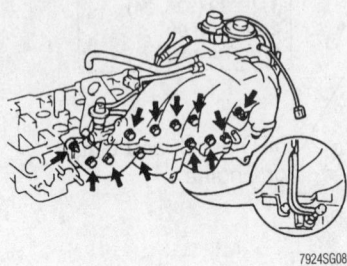

Intake chamber and intake manifold mounting bolts—4.5L (1FZ-FE) engine

7924SG08

8. Disconnect the engine ground strap from the No. 1 engine hanger and the ground strap from the air intake chamber.
9. Detach the connector on the intake manifold from the left fender apron.
10. Disconnect the brake booster and EVAP hoses.
11. Disconnect the fuel inlet and return lines from the fuel rail.
12. Remove the heater inlet pipe and hose.
13. Remove the radiator inlet hose, the No. 3 water bypass hose and the alternator.
14. Remove the intake manifold stay.
15. Detach the following electrical connectors:
    a. ECT sender gauge connector.
    b. ECT cut switch connector.
    c. ECT sensor connector.
    d. Knock sensor connector.
    e. Crankshaft Position Sensor connector.
16. Remove the bolt that secures the engine harness to the cylinder block.
17. Disconnect the PCV hose from the PCV valve.
18. Remove the bolt that secures the engine harness to the intake manifold.
19. Disconnect the following:
    a. Engine wire clamps.
    b. EGR gas temp. sensor
    c. The fuel injectors
    d. The connector to the emission control valve set.
20. Remove the No. 2 water bypass pipe.
21. Remove the air intake chamber and the intake manifold assembly.
    a. Disconnect the vacuum hoses from the TVV.
    b. Remove the 10 bolts, the two nuts and the intake manifold and gasket.

To install:

22. Install the air intake chamber and intake manifold assembly.
    a. Place a new gasket so that the rear mark is toward the rear side.
    b. Tighten the intake manifold bolts to 15 ft. lbs. (21 Nm).
    c. Connect the vacuum hoses to the TVV.
23. Install the No. 2 water bypass pipe.
24. Connect the following:
    a. Engine wire clamps.
    b. EGR gas temp. sensor connector.
    c. The fuel injector connectors.
    d. The connector to the emission control valve set.
25. Install the bolt that secures the engine wiring harness to the intake manifold.
26. Connect the PCV hose to the PCV valve.

27. Install the bolt that secures the engine wiring harness to the cylinder block.
28. Connect the following:
    a. ECT sender gauge connector.
    b. ECT cut switch connector.
    c. ECT sensor connector.
    d. Knock sensor connector.
    e. Crankshaft Position Sensor connector.
29. Install the intake manifold stay and tighten the bolts to 26 ft. lbs. (36 Nm).
30. Install the radiator inlet hose, the No. 3 water bypass hose and the alternator.
31. Install the heater hose to the cylinder head and connect the pipe to the intake manifold. Tighten the bolts to 15 ft. lbs. (21 Nm).
32. Connect the fuel inlet and return hoses to the fuel rail.
33. Connect the brake booster and EVAP hoses.
34. Connect the connector on the intake manifold to the left fender apron.
35. Connect the ground straps to the No.1 engine hanger and the air intake chamber.
36. Connect the throttle cable to the throttle body.
37. Connect the accelerator cable to the throttle body.
38. Connect the cruise control actuator cable to the throttle body.
39. Install the air cleaner hose and cap.
40. Connect the negative battery cable.
41. Refill the engine coolant.
42. Start the engine and check for leaks.
43. Allow the engine to cool and recheck the coolant level.

4.7L (2UZ-FE) Engine

1. Properly relieve the fuel system pressure.
2. Disconnect the negative battery cable.
3. Remove the intake air duct.
4. Drain the engine coolant.
5. Remove the engine V-bank cover.
6. Disconnect the accelerator cable from the throttle body.

❄❄ CAUTION

Observe all applicable safety precautions when working around fuel. Whenever servicing the fuel system, always work in a well ventilated area. Do not allow fuel spray or vapors to come in contact with a spark or open flame. Keep a dry chemical fire extinguisher near the work area. Always

keep fuel in a container specifically designed for fuel storage; also, always properly seal fuel containers to avoid the possibility of fire or explosion.

7. Disconnect the fuel supply hose.

8. Detach the following connectors:

- Throttle position sensor connector
- Accelerator position sensor connector
- Throttle motor connector
- Vacuum switching valve connector for EVAP
- Fuel injector connectors
- Engine coolant temperature sensor connector
- Oxygen sensor connectors

9. Disconnect the fuel pressure regulator vacuum hose from vacuum pipe.

10. Disconnect the hose to the PCV valve on the left cylinder head.

11. Disconnect the EVAP hoses from the vacuum switching valve and intake air connector on the intake manifold

12. Remove the coolant hoses from the throttle body.

13. Remove the fuel supply hose brackets from the right side of the manifold.

14. Remove the engine wiring harness protector from the rear water bypass joint and right cylinder head.

15. Remove the guide for the automatic transmission from the rear of the left cylinder head.

16. Remove the ground cables from the rear of the cylinder heads.

17. Remove the engine wiring harness protector from the intake manifold.

18. Remove the wiring harness from the engine hanger and bracket.

19. Remove the EVAP pipe.

20. Remove the right rear and left front V-bank cover brackets.

21. Remove the six bolts, four nuts, then remove the intake manifold assembly.

To install:

22. Position two new intake manifold gaskets on the cylinder heads with the white paint mark facing upward. Be sure to align the port holes on the gaskets with the ports on the head.

23. Place the intake manifold on the heads and install the fasteners. tighten the bolts and nuts evenly to 13 ft. lbs. (18 Nm).

24. Install the EVAP pipe on the manifold.

25. Install the V-bank cover brackets.

26. Install the wiring harness protector.

27. Install the engine wiring harness in it' original position.

28. Install the harness protector on the rear of the rear water bypass joint and right cylinder head.

29. Install the fuel supply brackets and hose on the right side of the manifold.

30. Connect the throttle body coolant hoses.

31. Connect the EVAP and PCV hoses.

32. Connect the fuel pressure regulator vacuum hose.

33. Attach all electrical connectors.

34. Connect the fuel supply hose.

35. Connect the accelerator cable to the throttle body.

36. Install the V-bank cover.

37. Install the intake air duct.

38. Refill the cooling system.

39. Connect the negative battery cable.

40. Start the engine and check for leaks.

41. Allow the engine to cool, then recheck the coolant level and add coolant as needed.

Exhaust Manifold

REMOVAL & INSTALLATION

4.5L (1FZ-FE) Engine

1. Disconnect the negative battery cable.

2. Raise and safely support the vehicle.

3. Working from under the vehicle, detach the heated oxygen sensor connector.

4. Remove the nuts and bolts holding the front exhaust pipe to the rear catalytic converter.

5. Loosen the pipe clamp bolt.

6. Remove the two bolts and the pipe bracket.

Exploded view of the intake manifold removal procedure—4.7L (2UZ-FE) engine

7922SG50

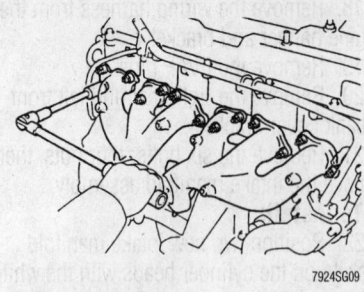

7924SG09

Exhaust manifold mounting bolts—4.5L (1FZ-FE) engine

7. Remove the four nuts and disconnect the front exhaust pipe. Remove the gasket.

8. Lower the vehicle and remove the six bolts and the exhaust manifold heat insulators.

9. Remove the 13 nuts, the No. 1 and the No. 2 exhaust manifolds and the gaskets.

To install:

10. Install the new gaskets and the No. 1 and the No. 2 exhaust manifolds. Uniformly tighten the nuts in several passes. Tighten the nuts to 29 ft. lbs. (39 Nm).

11. Install the exhaust manifold heat insulators with the bolts and tighten to 14 ft. lbs. (19 Nm).

12. Connect the exhaust pipe to the exhaust manifold with a new gasket and tighten the four new nuts to 46 ft. lbs. (63 Nm).

13. Install the No. 1 support bracket and tighten the bolts to 17 ft. lbs. (24 Nm).

14. Connect the clamp and tighten the clamp bolt to 14 ft. lbs. (19.5 Nm).

15. Connect the front exhaust pipe to the rear catalytic converter with a new gasket and tighten to 34 ft. lbs. (46 Nm).

16. Connect the heated oxygen sensor connector.

17. Connect the negative battery cable.

18. Start the engine and be sure that there are no exhaust leaks.

4.7L (2UZ-FE) Engine

The manufacturer recommends removing the cylinder head with the exhaust manifold attached, then removing the manifold from the cylinder head, however it may be possible to remove the exhaust manifold with the engine in the vehicle.

1. Disconnect the front exhaust pipe from the exhaust manifold.

2. Remove the cylinder head.

3. Remove the heat shield from the manifold.

4. Remove the eight nuts and the exhaust manifold with gasket from the cylinder head.

To install:

5. Position a new gasket on the cylinder head with the white paint mark facing the manifold.

6. Install the manifold on the head and tighten the nuts evenly to 32 ft. lbs. (44 Nm).

7. Install the heat shields on the exhaust manifolds.

8. Install the cylinder head on the engine.

9. Use new gaskets and connect the front exhaust pipes to the exhaust manifolds. Tighten the nuts to 30 ft. lbs. (40 Nm).

Camshaft

REMOVAL & INSTALLATION

4.5L (1FZ-FE) Engine

1. Disconnect the battery cables and remove the battery and the battery tray.

2. Drain the engine coolant.

3. Remove the air cleaner hose and cap.

4. Disconnect the cruise control actuator cable from the throttle body.

5. Disconnect the accelerator cable from the throttle body.

6. Disconnect the throttle cable from the throttle body.

7. Disconnect the engine ground strap from the No. 1 engine hanger and the ground strap from the air intake chamber.

8. Disconnect the brake booster vacuum hose.

9. Disconnect the heater hoses.

10. Disconnect the engine wire and heater valve from the cowl panel.

11. Remove the No. 2 and the No. 3 cylinder head covers

12. Remove the distributor.

13. Disconnect the P/S reservoir tank.

14. Disconnect the radiator inlet hose and the No. 3 water bypass hose.

15. Remove the alternator.

16. Remove the throttle body.

17. Detach the following connectors:

a. ECT sender gauge connector, the ECT cut switch connector and the ECT sensor connector.

b. Knock sensor connector.

c. Crankshaft position sensor connector.

18. Remove the bolt and disconnect the engine wiring harness from the cylinder block.

19. Disconnect the following:

a. Oil level sensor

b. Two connectors from the transmission.

c. Starter connector.

d. Detach the two heated oxygen sensor connectors.

e. Detach the park/neutral position (PNP) switch connector.

f. Remove the two bolts and disconnect the engine wire from the intake manifold and the cylinder block.

g. Disconnect the PCV hose from the PCV valve.

h. Remove the bolt holding the engine wire to the intake manifold.

i. Detach the connector for the emission control valve set assembly and the three injector connectors.

j. Disconnect the engine wiring harness clamp.

k. Detach the EGR gas temperature sensor connector.

l. Disconnect the clamp of the No. 6 injector wire from the bracket.

m. Disconnect the engine wiring harness from the cylinder head and the intake manifold.

20. Remove the cylinder head cover.

21. Remove the semi-circular plug from the cylinder head.

22. Remove the spark plugs.

23. Set the No. 1 cylinder to TDC of the compression stroke.

a. Turn the crankshaft pulley and align its groove with the **0** mark on the timing chain cover.

b. Check that the timing marks (one and two dots) of the camshaft drive and driven gears are in straight line on the cylinder head surface. If not, turn the crankshaft one revolution (360°) and align the marks.

24. Remove the chain tensioner.

25. Place matchmarks on the camshaft timing gear and the timing chain and remove the camshaft timing gear.

a. Hold the intake camshaft with a wrench and remove the bolt and the distributor gear.

b. Remove the camshaft timing gear and chain from the intake camshaft and leave on the slipper and the damper.

26. Remove the camshafts.

➡**Since the thrust clearance of the camshaft is small, the camshaft must be kept level while it is being removed. If the camshaft is not kept level, the portion of the cylinder head receiving the shaft thrust may crack or be dam-**

aged, causing the camshaft to seize or break. To avoid this, the following steps should be carried out.

27. Remove the exhaust camshaft.
    a. Bring the service bolt hole of the driven sub-gear upward by turning the hexagon wrench head portion of the exhaust camshaft with a wrench.
    b. Secure the exhaust camshaft sub-gear to the main gear with a service bolt. When removing the camshaft, be sure that the torsional spring force of the sub-gear has been eliminated by the above operation.
    c. Set the timing mark (two dot marks) of the camshaft driven gear at approximately 35° angle by turning the hexagon wrench head portion of the intake camshaft with a wrench.
    d. Lightly push the camshaft towards the rear without applying excessive force.
    e. Loosen and remove the No. 1 bearing cap bolts, alternately loosening the left and right bolts uniformly.
    f. Loosen and remove the No. 2, No. 3, No. 5 and the No. 7 bearing cap bolts, alternately loosening the left and right bolts uniformly in several passes, in sequence.

➡ **Do not remove the No. 4 and No. 6 bearing cap bolts at this stage.**

    g. Remove the four bearing caps.
    h. Alternately and uniformly loosen and remove the No. 4 and the No. 6 bearing cap bolts.
    i. If the camshaft is not being lifted out straight and level, retighten the four No. 4 and No. 6 bearing cap bolts. Then, reverse the order of the above Substeps from g to e and repeat Substeps from c to h once again.
    j. Remove the two bearing caps and exhaust camshaft. Do not pry on or attempt to force the camshaft with a tool or any other object.
28. Remove the intake camshaft.
    a. Set the timing mark (two dot marks) of the camshaft drive gear at approximately a 25° angle by turning the hexagon wrench head portion of the intake camshaft with a wrench.

➡ **The above angle ensures that the No. 1 and the No. 4 cylinder cam lobes of the intake camshaft push their valve lifters evenly.**

    b. Lightly push the intake camshaft towards the front without applying excessive force.
    c. Loosen and remove the No. 1 bear-

ing cap bolts, alternately loosening the left and the right bolts uniformly.
    d. Loosen and remove the No. 3, No. 4, No. 6 and the No. 7 bearing cap bolts, alternately loosening the left and right bolts uniformly in several passes in sequence.

➡ **Do not remove the No. 2 and No. 5 bearing cap bolts at this stage.**

    e. Remove the four bearing caps.
    f. Alternately and uniformly loosen and remove the No. 2 and the No. 5 bearing cap bolts.
    g. If the camshaft is not being lifted out straight and level, retighten the four No. 2 and No. 5 bearing cap bolts. Then, reverse the order of the above Substeps from e to c and repeat Substeps from a to f once again.
    h. Remove the two bearing caps and the exhaust camshaft.
29. Remove the valve lifters and shims. Arrange the valve lifters and shims in correct order for reinstallation.
    To install:
30. Install the valve lifters and shims. Check to be sure that the valve lifter rotates smoothly by hand.
31. Install the camshafts.

➡ **Since the thrust clearance of the camshaft is small, the camshaft must be kept level while it is being installed. If the camshaft is not kept level, the portion of the cylinder head receiving the shaft thrust may crack or be damaged, causing the camshaft to seize or break. To avoid this, the following steps should be carried out.**

    a. Apply engine oil to the thrust portion of the intake camshaft.
    b. Lightly place the intake camshaft on top of the cylinder head so that the No. 1 and the No. 4 cylinder cam lobes face downward.
    c. Lightly push the camshaft towards the front without applying excessive force. Place the No. 2 and the No. 5 bearing caps in their proper location.
    d. Temporarily tighten these bearing cap bolts uniformly and alternately in several passes until the bearing caps are snug with the cylinder head.
    e. Place the No. 3, No. 4, No. 6 and the No. 7 bearing caps in their proper location. Temporarily tighten these bearing cap bolts, alternately tightening the left and right bolts uniformly.
    f. Place the No. 1 bearing cap in its proper location. Check that there is no gap between the cylinder head and the contact surface of the bearing cap.

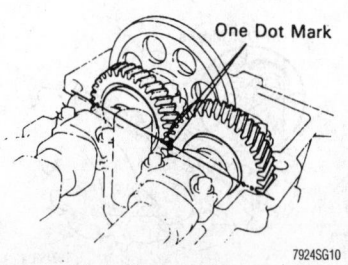

Check that the timing marks (one and two dots) of the camshaft drive and driven gears are in straight line on the cylinder head surface —4.5L (1FZ-FE) engine

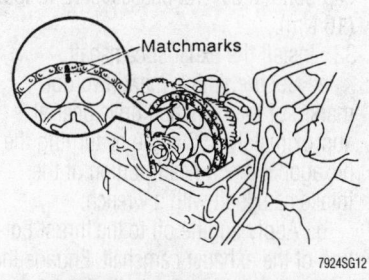

Matchmark the chain to the camshaft gear —4.5L (1FZ-FE) engine

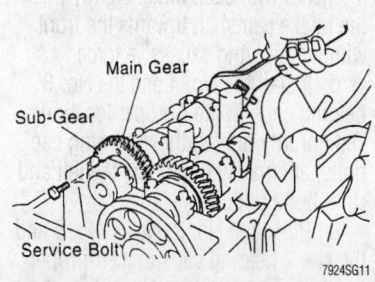

Secure the exhaust camshaft sub-gear to the main gear with a service bolt —4.5L (1FZ-FE) engine

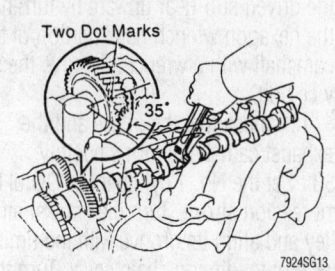

Set the timing mark (two dot marks) of the camshaft driven gear at approximately 35° angle —4.5L (1FZ-FE) engine

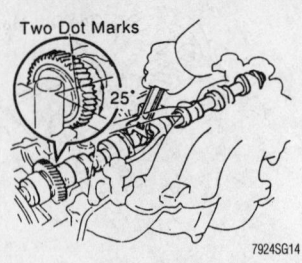

Two Dot Marks

25°

7924SG14

Set the timing mark (two dot marks) of the camshaft driven gear at approximately 25° angle —4.5L (1FZ-FE) engine

g. Uniformly tighten the 14 bearing cap bolts in several passes to 12 ft. lbs. (16 Nm).

32. Install the exhaust camshaft.

a. Set the timing mark (two dot marks) of the camshaft drive gear at approximately 35° angle by turning the hexagon wrench head portion of the intake camshaft with a wrench.

b. Apply engine oil to the thrust portion of the exhaust camshaft. Engage the exhaust camshaft gear to the intake camshaft hear by matching the timing marks (two dot marks) on each gear.

c. Roll down the exhaust camshaft onto the bearing journals while engaging the gears with each other. Lightly push the intake camshaft towards the front without applying excessive force.

d. Install the No. 4 and the No. 6 bearing caps in their proper location. Temporarily tighten these bearing cap bolts, alternately tightening the left and right bolts uniformly.

e. Place the No. 2, No. 3, No. 5 and the No. 7 bearing caps in their proper location. Temporarily tighten these bearing cap bolts, alternately tightening the left and right bolts uniformly.

f. Tighten the 14 bearing cap bolts in several passes to 12 ft. lbs. (16 Nm).

g. Bring the service bolt installed in the driven sub-gear upward by turning the hexagon wrench head portion of the camshaft with a wrench. Remove the service bolt.

h. Check that the intake and the exhaust camshafts turn smoothly.

33. Set the No. 1 cylinder to TDC of the compression stroke. Turn the crankshaft pulley and align its groove with the timing mark **O** of the timing chain cover. Turn the camshaft so that the timing marks with one

and two dots will be in straight line on the cylinder head surface.

34. Install the camshaft timing gear.

a. Check that the matchmarks on the camshaft timing gear and the timing chain are aligned. Place the gear over the straight pin of the intake camshaft.

b. Align the straight pin of the distributor gear with the straight pin groove of the intake camshaft gear.

c. Hold the intake camshaft with a wrench, install and tighten the bolt to 54 ft. lbs. (74 Nm).

35. Install the chain tensioner. Push the tensioner by hand until it touches the head installation surface, then install and tighten the two nuts to 15 ft. lbs. (21 Nm).

36. Check the valve timing.

a. Turn the crankshaft pulley and align its groove with the timing mark **O** of the timing chain cover. Always turn the crankshaft clockwise.

b. Check that the timing marks (one and two dots) of the camshaft drive and driven gears are in straight line on the cylinder head surface. If not, turn the crankshaft one revolution (360°) and align the marks.

37. Check valve clearance and adjust if necessary.

38. Install the spark plugs.

39. Install the semi-circular plug to the cylinder head.

40. Be sure that the No. 1 cylinder is in TDC of the compression stroke.

41. Install the cylinder head cover.

42. Connect the following:

a. Connect the engine wiring harness from the cylinder head and the intake manifold.

b. Connect the clamp of the No. 6 injector wire from the bracket.

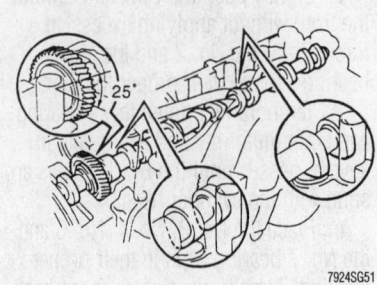

25°

7924SG51

Intake camshaft with No. 1 and No. 4 cam lobes down—4.5L (1FZ-FE) engine

c. Connect the EGR gas temperature sensor connector.

d. Connect the engine wiring harness clamp.

e. Connect the connector for the emission control valve set assembly and the three injector connectors.

f. Install the bolt holding the engine wire to the intake manifold

g. Connect the PCV hose from the PCV valve.

h. Install the two bolts and connect the engine wire from the intake manifold and the cylinder block.

i. Connect the park/neutral position (PNP) switch connector.

j. Connect the two heated oxygen sensor connectors.

k. Starter connector.

l. Two connectors from the transmission.

m. Oil level sensor connector.

43. Install the bolt and connect the engine wiring harness from the cylinder block.

44. Connect the following:

a. ECT sender gauge connector, the ECT cut switch connector and the ECT sensor connector.

b. Knock sensor connector.

c. Crankshaft position sensor connector.

45. Install the throttle body.

46. Install the alternator and the drive belts.

47. Connect the No. 3 water bypass hose.

48. Connect the radiator inlet hose.

49. Connect the P/S reservoir tank and tighten the bolts to 14 ft. lbs. (20 Nm).

50. Install the distributor.

51. Install the No. 2 and the No. 3 cylinder head covers.

52. Connect the heater valve and the engine wiring harness to the cowl panel.

53. Connect the heater hoses.

54. Connect the ground strap to the No. 1 engine hanger and the air intake chamber.

55. Connect the brake booster vacuum hose.

56. Connect the throttle cable to the throttle body. Adjust the throttle cable.

57. Connect the accelerator cable to the throttle body.

58. Connect the cruise control actuator cable to the throttle body.

59. Install the air cleaner hose and cap.

60. Install the battery tray, the battery and connect the cables.

61. Refill the engine coolant.
62. Start the engine and check for leaks.
63. Make necessary engine adjustments.
64. Road test the vehicle and recheck the engine coolant level.

4.7L (2UZ-FE) Engine

1. Disconnect the negative battery cable.
2. Remove the timing belt from the camshaft pulleys.
3. Remove the camshaft position sensor.
4. Remove the timing belt rear covers.
5. Remove the cylinder head covers.
6. Remove the V-bank cover.
7. Turn the crankshaft pulley so that the timing mark is aligned with the centers of the No. 2 idler pulley and the crankshaft pulley bolts.

➡The camshaft bearing caps must be returned to their original positions during installation of the camshafts. Keep them in order during removal.

8. To remove the right side camshafts:
   a. Bring the service bolt hole of the sub-gear up by turning the hexagon shaped portion of the camshaft with a wrench.
   b. Using a 6M X 1.0mm bolt, about 16–20mm long, secure the sub-gear to the main gear.
   c. Turn the timing mark (one dot) on the main gear to approximately a 10° angle by turning the hexagon shaped portion of the camshaft with a wrench.
   d. Using the sequence shown, gradually loosen the bearing caps bolts.
   e. Remove the oil feed pipe, bearing caps, oil control valve and camshafts.
9. To remove the left side camshafts:
   a. Bring the service bolt hole of the sub-gear up by turning the hexagon shaped portion of the camshaft with a wrench.

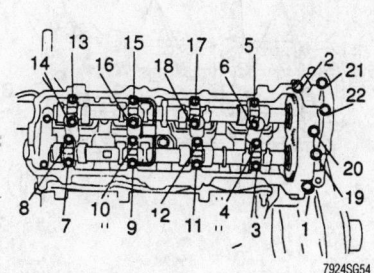

Be sure that the timing mark on the crankshaft pulley is facing the No. 2 idler pulley before removing or installing the camshafts—4.7L (2UZ-FE) engine

   b. Using a 6M X 1.0mm bolt, about 16–20mm long, secure the sub-gear to the main gear.
   c. Align the timing marks (two dot marks) by turning the hexagon shaped portion of the camshaft with a wrench.
   d. Using the sequence shown, gradually loosen the bearing caps bolts.
   e. Remove the oil feed pipe, bearing caps and camshafts.
10. To disassemble the exhaust camshafts:

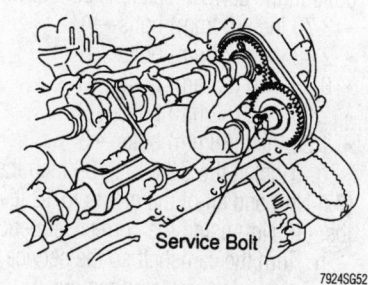

Turn the right side camshafts so the service bolt hole is up and install the bolt—4.7L (2UZ-FE) engine

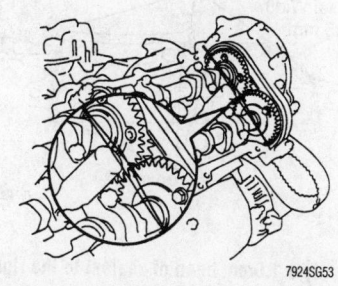

Turn the right camshafts timing mark (one dot mark) to a 10° angle—4.7L (2UZ-FE) engine

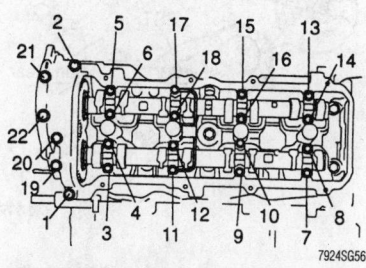

Be sure to loosen the bearing cap bolts in the sequence shown on the right cylinder head—4.7L (2UZ-FE) engine

   a. Clamp the exhaust camshaft in a vise by the hexagon shaped portion of the shaft.
   b. Turn the sub-gear clockwise with a suitable spanner wrench and remove the service bolt.
   c. Remove the snap-ring, wave washer, sub-gear and gear spring.

To install:

11. To assembly the exhaust camshafts:
    a. Place the gear spring on the main gear.
    b. Install the sub-gear while attaching

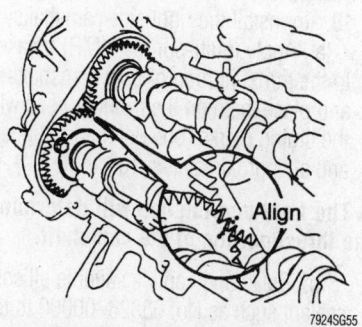

Align the timing marks (two dot marks) on the left camshafts before removing the bearing cap bolts—4.7L (2UZ-FE) engine

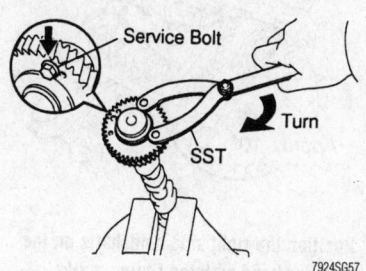

Loosen the bearing cap bolts on the left cylinder head in the sequence shown to prevent cam and head damage—4.7L (2UZ-FE) engine

Turn the sub-gear clockwise and remove the service bolt on the exhaust camshaft—4.7L (2UZ-FE) engine

the pins on the gears to the ends of the spring.

c. Install the snap-ring.

d. Clamp the exhaust camshaft in a vise by the hexagon shaped portion of the shaft.

e. Using the spanner wrench, turn the sub-gear counter clockwise and install the service bolt.

f. Align the teeth of both gears and tighten the service bolt.

12. Be sure the timing mark on the crankshaft pulley is still facing the No. 2 idler pulley. If not, turn the crankshaft to the correct position.

13. To install the right side camshafts:

a. Apply multi-purpose (MP) grease to the thrust portion of both camshafts and place them on the cylinder head with the timing marks (one dot mark) aligned and at approximately a 10° angle.

➡**The front bearing cap will determine the thrust portion of the camshaft.**

b. Clean and apply a suitable silicone sealant such as No. 08826–00080 to the front bearing cap as shown. Be sure not

to apply sealant to the bearing cap grooves.

c. Immediately install the front bearing cap along with the remaining caps.

➡**Be sure to align the directional marks on the bearing caps with the marks at the front and rear of the cylinder head.**

d. Push in a new camshaft oil seal.

e. Lightly oil the threads and under the heads of the bolts indicated by D and E. Do not lubricate the A, B and C bolts.

f. Install the oil feed pipe and the bearing cap bolts. Be sure to install the bolts in the correct locations as follows:

• 3.70 inch (94mm) bolts—A
• 2.83 inch (72mm) bolts—B
• 0.98 inch (25mm) bolts—C
• 2.05 inch (52mm) bolts—D
• 1.50 inch (38mm) bolts—E

g. Tighten the C bolts to 69 inch lbs. (7.5 Nm) and all of the others to 12 ft. lbs. (16 Nm) using the correct sequence.

h. Turn the camshaft so the service bolt is facing upward and remove it.

14. To install the left side camshafts:

a. Apply multi-purpose (MP) grease to the thrust portion of both camshafts and place them on the cylinder head with the timing marks (two dots) aligned.

➡**The front bearing cap will determine the thrust portion of the camshaft.**

b. Clean and apply silicone sealant such as No. 08826–00080 to the front bearing cap as shown. Be sure not to apply sealant to the bearing cap grooves.

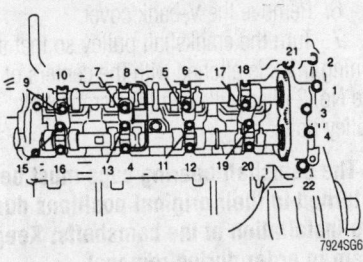

Right cylinder head camshaft bearing cap bolt tightening sequence—4.7L (2UZ-FE) engine

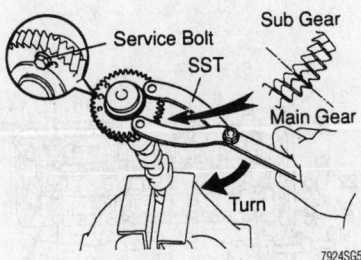

Align the teeth of both exhaust camshaft gears before tightening the service bolt— 4.7L (2UZ-FE) engine

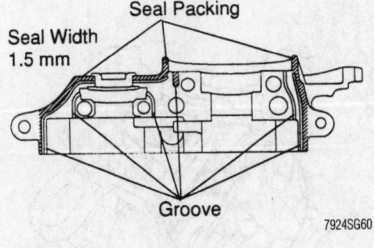

Apply a 1.5mm bead of sealant to the right front bearing cap and install it within five minutes—4.7L (2UZ-FE) engine

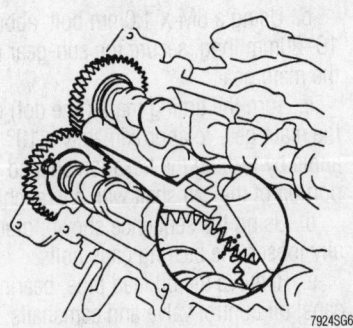

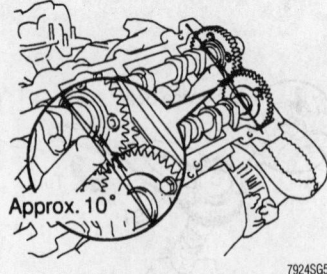

Position the right side camshafts on the cylinder head with the timing marks aligned and at a 10° angle to the head— 4.7L (2UZ-FE) engine

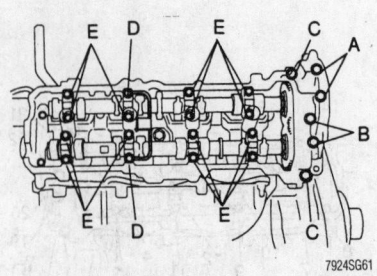

Be sure to install the right side camshaft bearing cap bolts in the correct locations—4.7L (2UZ-FE) engine

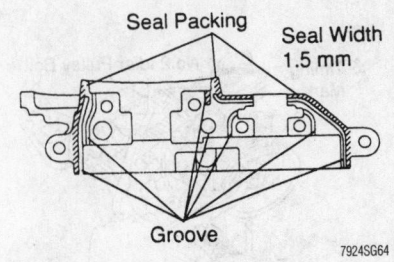

Apply a 1.5mm bead of sealant to the left front camshaft bearing cap as shown— 4.7L (2UZ-FE) engine

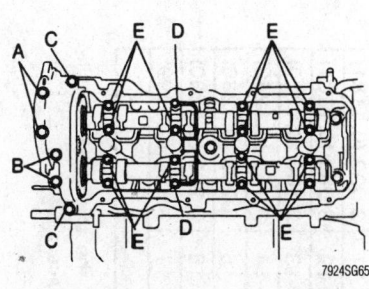

Left camshaft bearing cap bolt locations—4.7L (2UZ-FE) engine

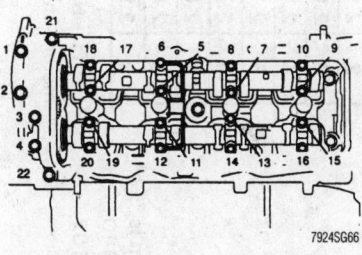

Tighten the left camshaft bearing cap bolts in the sequence shown to prevent damage to the cylinder head and camshaft—4.7L (2UZ-FE) engine

c. Immediately install the front bearing cap along with the remaining caps.

➡**Be sure to align the directional marks on the bearing caps with the marks at the front and rear of the cylinder head.**

d. Push in a new camshaft oil seal.

e. Lightly oil the threads and under the heads of the bolts indicated by D and E. Do not lubricate the A, B and C bolts.

f. Install the oil feed pipe and the bearing cap bolts. Be sure to install the bolts in the correct locations as follows:
- 3.70 inch (94mm) bolts—A
- 2.83 inch (72mm) bolts—B
- 0.98 inch (25mm) bolts—C
- 2.05 inch (52mm) bolts—D
- 1.50 inch (38mm) bolts—E

g. Tighten the C bolts to 69 inch lbs. (7.5 Nm) and all of the others to 12 ft. lbs. (16 Nm) using the correct sequence.

h. Turn the camshaft so the service bolt is facing upward and remove it.

15. Check and adjust the valve clearance and adjust if necessary.

16. Clean and apply new sealant to the center grooves on the semi-circular plugs and install them.

17. Install the cylinder head covers, tighten the bolts in several passes to 53 inch lbs. (6.0 Nm).

18. The remaining installation procedure is the reverse of removal.

Valve Lash

ADJUSTMENT

4.5L (1FZ-FE) Engine

1. Disconnect the negative battery cable.

2. Drain the engine coolant.

3. Remove the PCV hoses.

4. Remove the air cleaner cap, MAF meter and the resonator.

5. Detach the following connectors:
   a. ECT sensor connector.
   b. Oil pressure sensor connector.
   c. If detached, the A/C compressor connector.

6. Remove the throttle body.

7. Disconnect the engine wire and the heater valve from the cowl panel.

8. Disconnect the spark plug wires.

9. Remove the cylinder head cover.

10. Set the No. 1 cylinder to TDC of the compression stroke.

   a. Turn the crankshaft pulley clockwise and align its groove with the **0** mark on the timing chain cover.

   b. Check that the timing marks (one and two dots) of the camshaft drive and driven gears are in a straight line on the cylinder head surface. If not, turn the

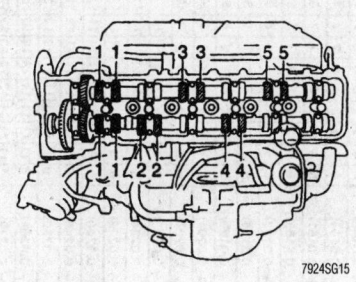

With the crankshaft in the first position, measure the clearance on the valve shown—4.5L (1FZ-FE) engine

crankshaft one revolution (360°) and align the marks.

11. Inspect the valve clearance.

   a. Measure the clearance between the valve lifter and the camshaft. Measure the first, second and fourth intake and the first, third and fifth exhaust valves.

   b. Turn the crankshaft pulley one revolution (360°) and align the marks as above. Measure the third, fifth and sixth intake and the second, fourth and sixth exhaust valves.

12. Valve clearance cold should be:
   - Intake: 0.006–0.010 in. (0.15–0.25mm)
   - Exhaust: 0.010–0.014 in. (0.25–0.35mm)

13. Adjust the valve clearance by using adjusting shims.

   a. Turn the equipment driveshaft so that the camshaft lobe for the valve to be adjusted faces up.

   b. Using Valve Lifter Press (tool No. 09248–05510), or equivalent, press down the valve lifter and place Valve Lifter Stopper (tool No. 09248–06020) or equivalent,

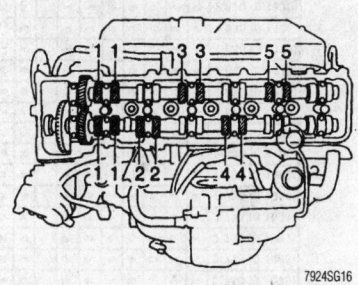

With the crankshaft in the second position, measure the clearance on the valve shown—4.5L (1FZ-FE) engine

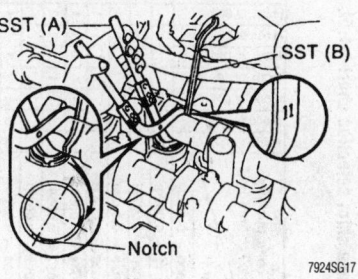

Use the special tool to remove or install the adjusting shim—4.5L (1FZ-FE) engine

Intake valve clearance shim selection chart—4.5L (1FZ-FE) engine

Measured clearance mm (in.) / Installed shim thickness mm (in.)	2.500 (0.0984)	2.520 (0.0992)	2.540 (0.1000)	2.550 (0.1004)	2.560 (0.1008)	2.580 (0.1016)	2.600 (0.1024)	2.620 (0.1031)	2.640 (0.1039)	2.650 (0.1043)	2.660 (0.1047)	2.680 (0.1055)	2.700 (0.1063)	2.710 (0.1067)	2.720 (0.1071)	2.730 (0.1075)	2.740 (0.1079)	2.750 (0.1083)	2.760 (0.1087)	2.770 (0.1091)	2.780 (0.1094)	2.790 (0.1098)	2.800 (0.1102)	2.810 (0.1106)	2.820 (0.1110)	2.830 (0.1114)	2.840 (0.1118)	2.850 (0.1122)	2.860 (0.1126)	2.870 (0.1130)	2.880 (0.1134)	2.890 (0.1138)	2.900 (0.1142)	2.910 (0.1146)	2.920 (0.1150)	2.930 (0.1154)	2.940 (0.1157)	2.950 (0.1161)	2.960 (0.1166)	2.970 (0.1169)	2.980 (0.1173)	2.990 (0.1177)	3.000 (0.1181)	3.010 (0.1185)	3.020 (0.1189)	3.030 (0.1193)	3.040 (0.1197)	3.050 (0.1201)	3.060 (0.1205)	3.070 (0.1209)	3.080 (0.1213)	3.090 (0.1217)	3.100 (0.1220)	3.120 (0.1228)	3.140 (0.1236)	3.150 (0.1240)	3.160 (0.1244)	3.180 (0.1252)	3.200 (0.1260)	3.220 (0.1268)	3.240 (0.1278)	3.250 (0.1280)	3.260 (0.1283)	3.280 (0.1291)	3.300 (0.1299)
0.000 - 0.030 (0.0000 - 0.0012)										1	1	1	1	2	2	2	2	2	3	3	3	3	3	4	4	4	4	4	5	5	5	5	5	6	6	6	6	6	7	7	7	7	7	8	8	8	8	8	9	9	9	9	9	10	10	10	11	11	11	12	12	12	13	13	13
0.031 - 0.050 (0.0012 - 0.0020)								1	1	1	1	1	2	2	2	2	3	3	3	3	3	4	4	4	4	4	5	5	5	5	5	6	6	6	6	6	7	7	7	7	7	8	8	8	8	8	9	9	9	9	9	10	10	10	11	11	11	11	12	12	13	13	13	13	14
0.051 - 0.070 (0.0020 - 0.0028)							1	1	1	1	1	2	2	2	3	3	3	3	3	4	4	4	4	4	5	5	5	5	5	6	6	C	6	6	7	7	7	7	7	8	8	8	8	8	9	9	9	9	9	10	10	10	10	11	11	11	11	12	12	13	13	13	13	14	14
0.071 - 0.090 (0.0028 - 0.0035)						1	1	1	1	2	2	2	3	3	3	3	3	4	4	4	4	4	5	5	5	5	5	6	6	6	6	6	7	7	7	7	7	8	8	8	8	8	9	9	9	9	9	10	10	10	10	10	11	11	11	12	12	12	13	13	13	14	14	14	15
0.091 - 0.110 (0.0036 - 0.0043)					1	1	1	1	2	2	2	3	3	3	3	4	4	4	4	4	5	5	5	5	5	6	6	6	6	6	7	7	7	7	7	8	8	8	8	8	9	9	9	9	9	10	10	10	10	10	11	11	11	11	12	12	12	13	13	13	14	14	14	15	15
0.111 - 0.130 (0.0044 - 0.0051)			1	1	1	1	1	2	2	2	3	3	3	4	4	4	4	4	5	5	5	5	5	6	6	6	6	6	7	7	7	7	7	8	8	8	8	8	9	9	9	9	9	10	10	10	10	10	11	11	11	11	11	12	12	12	13	13	13	14	14	14	15	15	15
0.131 - 0.149 (0.0052 - 0.0059)		1	1	1	1	1	2	2	3	3	3	3	4	4	4	4	5	5	5	5	5	6	6	6	6	6	7	7	7	7	7	8	8	8	8	8	9	9	9	9	9	10	10	10	10	10	11	11	11	11	11	12	12	12	13	13	13	13	14	14	15	15	15	15	16
0.150 - 0.250 (0.0059 - 0.0098)																																																																	
0.251 - 0.270 (0.0099 - 0.0106)	2	3	3	3	3	4	4	5	5	5	5	6	6	6	7	7	7	7	7	8	8	8	8	8	9	9	9	9	9	10	10	10	10	10	11	11	11	11	11	12	12	12	12	12	13	13	13	13	13	14	14	14	14	15	15	15	15	16	16	17	17	17	17	17	
0.271 - 0.290 (0.0107 - 0.0114)	3	3	3	4	4	4	5	5	5	6	6	6	7	7	7	7	7	8	8	8	8	8	9	9	9	9	9	10	10	10	10	10	11	11	11	11	11	12	12	12	12	12	13	13	13	13	13	14	14	14	14	14	15	15	15	16	16	16	17	17	17	17	17		
0.291 - 0.310 (0.0115 - 0.0122)	3	3	4	4	4	5	5	5	6	6	6	7	7	7	7	8	8	8	8	8	9	9	9	9	9	10	10	10	10	10	11	11	11	11	11	12	12	12	12	12	13	13	13	13	13	14	14	14	14	14	15	15	15	15	16	16	16	17	17	17	17				
0.311 - 0.330 (0.0122 - 0.0130)	3	4	4	4	5	5	5	6	6	6	7	7	7	8	8	8	8	8	8	9	9	9	9	10	10	10	10	10	11	11	11	11	11	12	12	12	12	12	13	13	13	13	13	14	14	14	14	14	15	15	15	15	16	16	16	16	17	17	17	17					
0.331 - 0.350 (0.0130 - 0.0138)	4	4	5	5	5	5	6	6	7	7	7	7	8	8	8	8	9	9	9	9	9	10	10	10	10	10	11	11	11	11	11	12	12	12	12	12	13	13	13	13	13	14	14	14	14	14	15	15	15	15	15	16	16	16	16	17	17	17	17						
0.351 - 0.370 (0.0138 - 0.0146)	4	5	5	5	5	6	6	7	7	7	7	8	8	8	9	9	9	9	9	10	10	10	10	10	11	11	11	11	11	12	12	12	12	12	13	13	13	13	13	14	14	14	14	14	15	15	15	15	15	16	16	16	16	17	17	17	17	17							
0.371 - 0.390 (0.0146 - 0.0154)	5	5	5	6	6	6	7	7	7	8	8	8	9	9	9	9	9	10	10	10	10	10	11	11	11	11	11	12	12	12	12	12	13	13	13	13	13	14	14	14	14	14	15	15	15	15	15	16	16	16	16	16	17	17	17	17	17								
0.391 - 0.410 (0.0154 - 0.0161)	5	5	6	6	6	7	7	7	8	8	8	9	9	9	9	10	10	10	10	10	11	11	11	11	11	12	12	12	12	12	13	13	13	13	13	14	14	14	14	14	15	15	15	15	15	16	16	16	16	16	17	17	17	17	17										
0.411 - 0.430 (0.0162 - 0.0169)	5	6	6	6	7	7	7	8	8	8	9	9	9	10	10	10	10	10	11	11	11	11	11	12	12	12	12	12	13	13	13	13	13	14	14	14	14	14	15	15	15	15	15	16	16	16	16	16	17	17	17	17	17	17											
0.431 - 0.450 (0.0170 - 0.0177)	6	6	7	7	7	7	8	8	9	9	9	9	10	10	10	10	11	11	11	11	11	12	12	12	12	12	13	13	13	13	13	14	14	14	14	14	15	15	15	15	15	16	16	16	16	16	17	17	17	17	17	17	17												
0.451 - 0.470 (0.0178 - 0.0185)	6	7	7	7	7	8	8	9	9	9	9	10	10	10	11	11	11	11	11	12	12	12	12	12	13	13	13	13	13	14	14	14	14	14	15	15	15	15	15	16	16	16	16	16	17	17	17	17	17	17	17														
0.471 - 0.490 (0.0185 - 0.0193)	7	7	7	8	8	8	9	9	9	10	10	10	11	11	11	11	11	12	12	12	12	12	13	13	13	13	13	14	14	14	14	14	15	15	15	15	15	16	16	16	16	16	17	17	17	17	17	17	17																
0.491 - 0.510 (0.0193 - 0.0201)	7	7	8	8	8	9	9	9	10	10	10	11	11	11	11	12	12	12	12	12	13	13	13	13	13	14	14	14	14	14	15	15	15	15	15	16	16	16	16	16	17	17	17	17	17	17	17																		
0.511 - 0.530 (0.0201 - 0.0209)	8	8	8	8	9	9	9	10	10	10	11	11	11	12	12	12	12	12	13	13	13	13	13	14	14	14	14	14	15	15	15	15	15	16	16	16	16	16	17	17	17	17	17	17	17																				
0.531 - 0.550 (0.0209 - 0.0217)	8	8	9	9	9	9	10	10	11	11	11	11	12	12	12	12	13	13	13	13	13	14	14	14	14	14	15	15	15	15	15	16	16	16	16	16	17	17	17	17	17	17	17																						
0.551 - 0.570 (0.0217 - 0.0224)	8	9	9	9	9	10	10	11	11	11	11	12	12	12	13	13	13	13	13	14	14	14	14	14	16	15	15	15	15	16	16	16	16	16	17	17	17	17	17	17	17																								
0.571 - 0.590 (0.0225 - 0.0232)	9	9	9	10	10	10	11	11	11	12	12	12	13	13	13	13	13	14	14	14	14	14	15	15	15	15	15	16	16	16	16	16	17	17	17	17	17	17	17																										
0.591 - 0.610 (0.0233 - 0.0240)	9	9	10	10	10	11	11	11	12	12	12	13	13	13	13	14	14	14	14	14	15	15	15	15	15	16	16	16	16	16	17	17	17	17	17	17	17																												
0.611 - 0.630 (0.0241 - 0.0248)	10	10	10	10	11	11	11	12	12	12	13	13	13	14	14	14	14	14	15	15	15	15	15	16	16	16	16	16	17	17	17	17	17	17	17																														
0.631 - 0.650 (0.0248 - 0.0256)	10	10	11	11	11	11	12	12	13	13	13	13	14	14	14	14	15	15	15	15	15	16	16	16	16	16	17	17	17	17	17	17	17																																
0.651 - 0.670 (0.0256 - 0.0264)	10	11	11	11	11	12	12	13	13	13	13	14	14	14	15	15	15	15	15	16	16	16	16	16	17	17	17	17	17	17	17																																		
0.671 - 0.690 (0.0264 - 0.0272)	11	11	11	11	12	12	13	13	13	14	14	14	15	15	15	15	15	16	16	16	16	16	17	17	17	17	17	17	17																																				
0.691 - 0.710 (0.0272 - 0.0280)	11	11	12	12	12	13	13	13	14	14	14	15	15	15	15	16	16	16	16	16	17	17	17	17	17	17	17																																						
0.711 - 0.730 (0.0280 - 0.0287)	12	12	12	12	13	13	13	14	14	14	15	15	15	16	16	16	16	16	17	17	17	17	17	17	17																																								
0.731 - 0.750 (0.0288 - 0.0295)	12	12	13	13	13	13	14	14	15	15	15	15	16	16	16	16	17	17	17	17	17	17	17																																										
0.751 - 0.770 (0.0296 - 0.0303)	12	13	13	13	13	14	14	15	15	15	15	16	16	16	16	17	17	17	17	17	17																																												
0.771 - 0.790 (0.0304 - 0.0311)	13	13	13	13	14	14	15	15	15	16	16	16	17	17	17	17	17	17	17																																																
0.791 - 0.810 (0.0311 - 0.0319)	13	13	14	14	14	15	15	15	16	16	16	17	17	17	17	17	17																																																		
0.811 - 0.830 (0.0319 - 0.0327)	14	14	14	14	15	15	15	16	16	16	17	17	17	17	17																																																				
0.831 - 0.850 (0.0327 - 0.0335)	14	14	15	15	15	15	16	16	17	17	17	17	17																																																						
0.851 - 0.870 (0.0335 - 0.0343)	14	15	15	15	15	16	16	17	17	17	17	17																																																							
0.871 - 0.890 (0.0343 - 0.0350)	15	15	15	15	16	16	17	17	17	17	17																																																								
0.891 - 0.910 (0.0351 - 0.0358)	15	15	16	16	16	17	17	17	17																																																												
0.911 - 0.930 (0.0359 - 0.0366)	16	16	16	16	17	17	17	17																																																													
0.931 - 0.950 (0.0367 - 0.0374)	16	16	17	17	17	17	17																																																															
0.951 - 0.970 (0.0374 - 0.0382)	16	17	17	17	17	17																																																															
0.971 - 0.990 (0.0382 - 0.0390)	17	17	17	17																																																																	
0.991 - 1.010 (0.0390 - 0.0398)	17	17	17																																																																		
1.011 - 1.030 (0.0398 - 0.0406)	17	17																																																																			
1.031 - 1.050 (0.0406 - 0.0413)	17																																																																					

Intake valve clearance (Cold):
0.15 – 0.25 mm (0.006 – 0.010 in.)

EXAMPLE: The 2.800 mm (0.1102 in.) shim is installed, and the measured clearance is 0.440 mm (0.0173 in.). Replace the 2.800 mm (0.1102 in.) shim with a No. 12 shim.

New shim thickness mm (in.)

Shim No.	Thickness	Shim No.	Thickness
1	2.500 (0.0984)	10	2.950 (0.1161)
2	2.550 (0.1004)	11	3.000 (0.1181)
3	2.600 (0.1024)	12	3.050 (0.1201)
4	2.650 (0.1043)	13	3.100 (0.1220)
5	2.700 (0.1063)	14	3.150 (0.1240)
6	2.750 (0.1083)	15	3.200 (0.1260)
7	2.800 (0.1102)	16	3.250 (0.1280)
8	2.850 (0.1122)	17	3.300 (0.1299)
9	2.900 (0.1142)		

HINT: New shims have the thickness in millimeters imprinted on the face.

301612

7924SG67

Exhaust valve clearance shim selection chart—4.5L (1FZ-FE) engine

New shim thickness　　mm (in.)

Shim No.	Thickness	Shim No.	Thickness
1	2.500 (0.0984)	10	2.950 (0.1161)
2	2.550 (0.1004)	11	3.000 (0.1181)
3	2.600 (0.1024)	12	3.050 (0.1201)
4	2.650 (0.1043)	13	3.100 (0.1220)
5	2.700 (0.1063)	14	3.150 (0.1240)
6	2.750 (0.1083)	15	3.200 (0.1260)
7	2.800 (0.1102)	16	3.250 (0.1280)
8	2.850 (0.1122)	17	3.300 (0.1299)
9	2.900 (0.1142)		

HINT: New shims have the thickness in milli-meters imprinted on the face.

Exhaust valve clearance (Cold):
0.25 – 0.35 mm (0.010 – 0.014 in.)

EXAMPLE: The 2.800 mm (0.1102 in.) shim is installed, and the measured clearance is 0.440 mm (0.0173 in.). Replace the 2.800 mm (0.1102 in.) shim with a No. 10 shim.

7924SG68

between the camshaft and the valve lifter. Remove the Valve Lifter Press.

c. Remove the adjusting shim with a small flat prying tool and a magnetic finger.

d. Determine the replacement adjusting shim size according to the following formula, or use the adjusting shim charts.

e. Using a micrometer, measure the thickness of the removed shim. Calculate the thickness of a new shim so that the valve clearance comes within the specified value.

- T = Thickness of the removed shim
- A = Measured valve clearance
- N = Thickness of the new shim

f. Intake: $N = T + (A—0.008$ in. (0.20mm))

g. Exhaust: $N = T + (A—0.012$ in. (0.30mm))

h. Install a new adjusting shim. Place it on the valve lifter. Using Valve Lifter Press (tool No. 09248–05510), or equivalent, press down the valve lifter and remove the Valve Lifter Stopper (tool No. 09248–06020)

i. Recheck the valve clearance.

14. Reinstall the cylinder head cover.

15. Reconnect the engine wire and clamps.

16. Connect the following:

a. ECT sensor connector.

b. Oil pressure sensor connector.

c. If detached, the A/C compressor connector.

17. Install the throttle body.

18. Install the spark plug wires.

19. Install the PCV hoses.

20. Install the air cleaner cap, MAF meter and the resonator.

21. Refill with engine coolant.

22. Check the ignition timing.

4.7L (2UZ-FE) Engine

➡**The procedure should be done while the engine is cold.**

1. Drain the engine coolant.

2. Remove the battery clamp cover.

3. Remove the V-bank cover.

4. Remove the air cleaner and duct assembly.

5. Remove the No. 3 timing belt covers.

6. Remove the ignition coils.

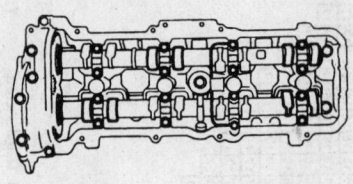

RH Cylinder Head

LH Cylinder Head   Front ⬅

Measure the clearance of the valves indicated while the crankshaft is in the 1st position—4.5L (1FZ-FE) engine

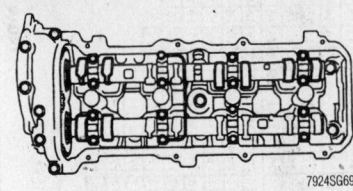

RH Cylinder Head

LH Cylinder Head   Front ⬅

Turn the crankshaft one revolution and measure the clearance of the valves shown-this is the 2nd position—4.5L (1FZ-FE) engine

7. Remove the cylinder head covers.

8. Turn the crankshaft so the No. 1 piston is at TDC on compression (1st position). Be sure the camshaft timing marks align with the marks on the rear covers.

9. Measure the clearance with a feeler gauge of the valves indicated and record the measurement of any valve out of specification.

- Intake—0.006–0.010 inch (0.15–0.25mm)
- Exhaust—0.010–0.014 inch (0.25–0.35mm)

10. Turn the crankshaft one revolution and measure the clearance of the valves indicated (2nd position). Record the measurement of any valves out of specification.

➡**The camshafts must be removed to allow access to the adjusting shims.**

11. Remove the timing belt and camshafts.

12. Remove the valve lifter and adjusting shim.

13. Using a micrometer, measure the thickness of the removed shim. Calculate the thickness of a new shim so that the valve clearance comes within the specified value.

14. Determine the size of the replacement shim by using the following formula, or the adjusting shim charts.

- T = Thickness of the removed shim
- A = Measured valve clearance
- N = Thickness of the new shim

a. Intake: $N = T + (A—0.008$ in. (0.20mm))

b. Exhaust: $N = T + (A—0.012$ in. (0.30mm))

15. Install the lifter with the new shim.

16. Install the camshafts and timing belt.

17. Recheck the valve clearance.

18. Install the cylinder head covers.

19. Install the ignition coils and the No. 3 timing belt covers.

20. Install the air cleaner and duct assembly.

21. Refill the cooling system.

22. Start the engine and check for leaks.

23. Install the V-bank cover.

24. Recheck the coolant level.

25. Install the battery clamp cover.

Intake valve clearance shim selection chart—4.7L (2UZ-FE) engine

Intake valve clearance shim selection chart with installed shim thickness (columns, mm/in. from 2.000 (0.0787) to 2.800 (0.1102)) versus measured clearance (rows, mm/in. from 0.000–0.030 (0.0000–0.0012) to 1.031–1.050 (0.0406–0.0413)).

New shim thickness

Shim No.	Thickness mm (in.)	Shim No.	Thickness mm (in.)	Shim No.	Thickness mm (in.)
00	2.000 (0.0787)	28	2.280 (0.0898)	56	2.560 (0.1008)
02	2.020 (0.0795)	30	2.300 (0.0906)	58	2.580 (0.1016)
04	2.040 (0.0803)	32	2.320 (0.0913)	60	2.600 (0.1024)
06	2.060 (0.0811)	34	2.340 (0.0921)	62	2.620 (0.1031)
08	2.080 (0.0819)	36	2.360 (0.0929)	64	2.640 (0.1039)
10	2.100 (0.0827)	38	2.380 (0.0937)	66	2.660 (0.1047)
12	2.120 (0.0835)	40	2.400 (0.0945)	68	2.680 (0.1055)
14	2.140 (0.0843)	42	2.420 (0.0953)	70	2.700 (0.1063)
16	2.160 (0.0850)	44	2.440 (0.0961)	72	2.720 (0.1071)
18	2.180 (0.0858)	46	2.460 (0.0969)	74	2.740 (0.1079)
20	2.200 (0.0866)	48	2.480 (0.0976)	76	2.760 (0.1087)
22	2.220 (0.0874)	50	2.500 (0.0984)	78	2.780 (0.1094)
24	2.240 (0.0882)	52	2.520 (0.0992)	80	2.800 (0.1102)
26	2.260 (0.0890)	54	2.540 (0.1000)		

Intake valve clearance (Cold):
0.15 – 0.25 mm (0.006 – 0.010 in.)

EXAMPLE:
The 2.300 mm (0.0906 in.) shim is installed, and the measured clearance is 0.440 mm (0.0173 in.). Replace the 2.300 mm (0.0906 in.) shim with a No. 54 shim.

7924SG71

Exhaust valve clearance shim selection chart—4.7L (2UZ-FE) engine

Installed shim thickness mm (in.) — columns (left to right): 2.000 (0.0787), 2.020 (0.0795), 2.040 (0.0803), 2.060 (0.0811), 2.080 (0.0819), 2.100 (0.0827), 2.120 (0.0835), 2.140 (0.0843), 2.160 (0.0850), 2.180 (0.0858), 2.200 (0.0866), 2.210 (0.0870), 2.220 (0.0874), 2.230 (0.0878), 2.240 (0.0882), 2.250 (0.0886), 2.260 (0.0890), 2.270 (0.0894), 2.280 (0.0898), 2.290 (0.0902), 2.300 (0.0906), 2.310 (0.0909), 2.320 (0.0913), 2.330 (0.0917), 2.340 (0.0921), 2.350 (0.0925), 2.360 (0.0929), 2.370 (0.0933), 2.380 (0.0937), 2.390 (0.0941), 2.400 (0.0945), 2.410 (0.0949), 2.420 (0.0953), 2.430 (0.0957), 2.440 (0.0961), 2.450 (0.0965), 2.460 (0.0969), 2.470 (0.0972), 2.480 (0.0976), 2.490 (0.0980), 2.500 (0.0984), 2.510 (0.0988), 2.520 (0.0992), 2.530 (0.0996), 2.540 (0.1000), 2.550 (0.1004), 2.560 (0.1008), 2.570 (0.1012), 2.580 (0.1016), 2.590 (0.1020), 2.600 (0.1024), 2.620 (0.1031), 2.640 (0.1039), 2.660 (0.1047), 2.680 (0.1055), 2.700 (0.1063), 2.720 (0.1071), 2.740 (0.1079), 2.760 (0.1087), 2.780 (0.1094), 2.800 (0.1102)

Measured clearance mm (in.) — rows:

Measured clearance mm (in.)
0.000–0.030 (0.0000–0.0012)
0.031–0.050 (0.0012–0.0020)
0.051–0.070 (0.0020–0.0028)
0.071–0.090 (0.0028–0.0035)
0.091–0.110 (0.0036–0.0043)
0.111–0.130 (0.0044–0.0051)
0.131–0.150 (0.0052–0.0059)
0.151–0.170 (0.0059–0.0067)
0.171–0.190 (0.0067–0.0075)
0.191–0.210 (0.0075–0.0083)
0.211–0.230 (0.0083–0.0091)
0.231–0.249 (0.0091–0.0098)
0.250–0.350 (0.0098–0.0138)
0.351–0.370 (0.0138–0.0146)
0.371–0.390 (0.0146–0.0154)
0.391–0.410 (0.0154–0.0161)
0.411–0.430 (0.0162–0.0169)
0.431–0.450 (0.0170–0.0177)
0.451–0.470 (0.0178–0.0185)
0.471–0.490 (0.0185–0.0193)
0.491–0.510 (0.0193–0.0201)
0.511–0.530 (0.0201–0.0209)
0.531–0.550 (0.0209–0.0217)
0.551–0.570 (0.0217–0.0224)
0.571–0.590 (0.0225–0.0232)
0.591–0.610 (0.0233–0.0240)
0.611–0.630 (0.0241–0.0248)
0.631–0.650 (0.0248–0.0256)
0.651–0.670 (0.0256–0.0264)
0.671–0.690 (0.0264–0.0272)
0.691–0.710 (0.0272–0.0280)
0.711–0.730 (0.0280–0.0287)
0.731–0.750 (0.0288–0.0295)
0.751–0.770 (0.0296–0.0303)
0.771–0.790 (0.0304–0.0311)
0.791–0.810 (0.0311–0.0319)
0.811–0.830 (0.0319–0.0327)
0.831–0.850 (0.0327–0.0335)
0.851–0.870 (0.0335–0.0343)
0.871–0.890 (0.0343–0.0350)
0.891–0.910 (0.0351–0.0358)
0.911–0.930 (0.0359–0.0366)
0.931–0.950 (0.0367–0.0374)
0.951–0.970 (0.0374–0.0382)
0.971–0.990 (0.0382–0.0390)
0.991–1.010 (0.0390–0.0398)
1.011–1.030 (0.0398–0.0406)
1.031–1.050 (0.0406–0.0413)
1.051–1.070 (0.0414–0.0421)
1.071–1.090 (0.0422–0.0429)
1.091–1.110 (0.0430–0.0437)
1.111–1.130 (0.0437–0.0445)
1.131–1.150 (0.0445–0.0453)

New shim thickness   mm (in.)

Shim No.	Thickness	Shim No.	Thickness	Shim No.	Thickness
00	2.000 (0.0787)	28	2.280 (0.0898)	56	2.560 (0.1008)
02	2.020 (0.0795)	30	2.300 (0.0906)	58	2.580 (0.1016)
04	2.040 (0.0803)	32	2.320 (0.0913)	60	2.600 (0.1024)
06	2.060 (0.0811)	34	2.340 (0.0921)	62	2.620 (0.1031)
08	2.080 (0.0819)	36	2.360 (0.0929)	64	2.640 (0.1039)
10	2.100 (0.0827)	38	2.380 (0.0937)	66	2.660 (0.1047)
12	2.120 (0.0835)	40	2.400 (0.0945)	68	2.680 (0.1055)
14	2.140 (0.0843)	42	2.420 (0.0953)	70	2.700 (0.1063)
16	2.160 (0.0850)	44	2.440 (0.0961)	72	2.720 (0.1071)
18	2.180 (0.0858)	46	2.460 (0.0969)	74	2.740 (0.1079)
20	2.200 (0.0866)	48	2.480 (0.0976)	76	2.760 (0.1087)
22	2.220 (0.0874)	50	2.500 (0.0984)	78	2.780 (0.1094)
24	2.240 (0.0882)	52	2.520 (0.0992)	80	2.800 (0.1102)
26	2.260 (0.0890)	54	2.540 (0.1000)		

Exhaust valve clearance (Cold):
0.25 – 0.35 mm (0.010 – 0.014 in.)

EXAMPLE:

The 2.300 mm (0.0906 in.) shim is installed, and the measured clearance is 0.440 mm (0.0173 in.). Replace the 2.300 mm (0.0906 in.) shim with a No. 44 shim.

7924SG72

Oil Pan

REMOVAL & INSTALLATION

4.5L (1FZ-FE) Engine

1. Disconnect the negative battery cable.

❋❋ CAUTION

Wait at least 90 seconds from the time the ignition switch is turned to the LOCK position and the negative battery cable is disconnected before starting work to avoid accidental air bag deployment.

2. Raise and safely support the vehicle.
3. Drain the engine oil.
4. Remove the engine undercover.
5. Disconnect the oil cooler pipe bracket from the No. 1 oil pan.
6. Remove the oil level sensor.
7. Remove the bolts holding the No. 1 oil pan to the transmission housing.
8. Remove the No. 2 oil pan.
9. Remove the No. 1 oil pan.

To install:

10. Install the No. 1 oil pan.
   a. Apply RTV sealant to the No. 1 oil pan.
   b. Install the oil pan and tighten the 14mm bolts to 32 ft. lbs. (44 Nm) and the 12mm bolts to 14 ft. lbs. (20 Nm).
11. Apply RTV sealant to the No. 2 oil pan and tighten the bolt to 5.7 ft. lbs. (7.8 Nm) and the nuts to 6.5 ft. lbs. (8.8 Nm).
12. Install the bolts holding the No. 1 oil pan to the transmission housing and tighten to 53 ft. lbs. (72 Nm).

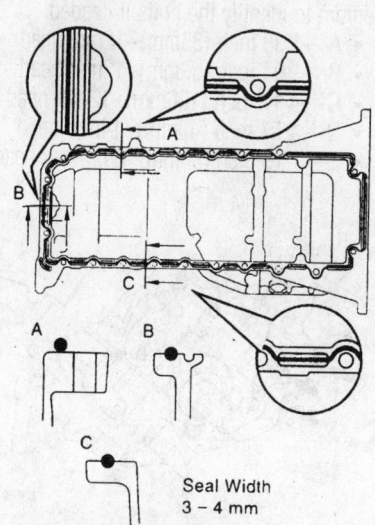

Apply RTV sealant as shown to the No. 1 oil pan—4.5L (1FZ-FE) engine

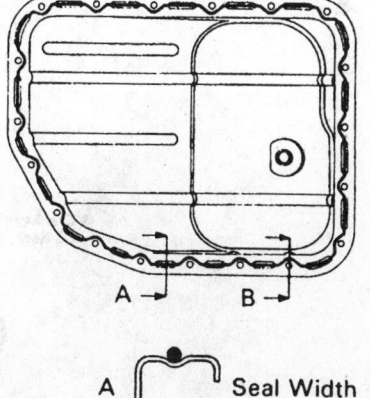

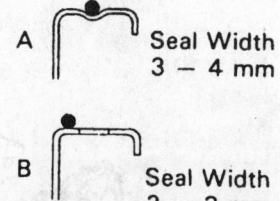

A — Seal Width 3 – 4 mm

B — Seal Width 2 – 3 mm

7924SG20

Apply RTV sealant as shown to the No. 2 oil pan—4.5L (1FZ-FE) engine

13. Install the oil level sensor and tighten the bolts to 4.9 ft. lbs. (5.4 Nm).
14. Connect the oil cooler pipe bracket to the No. 1 oil pan.
15. Install the engine undercover.
16. Fill with engine oil.
17. Connect the negative battery cable.
18. Start the engine and check for leaks.

4.7L (2UZ-FE) Engine

1. Remove the engine from the vehicle and install it on a stand.
2. Remove the oil dipstick and tube.
3. Remove the 20 bolts attaching the No. 2 (lower) oil pan to the No. 1 (upper) oil pan.
4. Insert the blade of seal breaker tool between the oil pans to break the seal, then remove the No. 2 oil pan.
5. Remove the bolts and nuts attaching the No. 1 oil pan to the engine and carefully pry at the four corners to remove the pan.

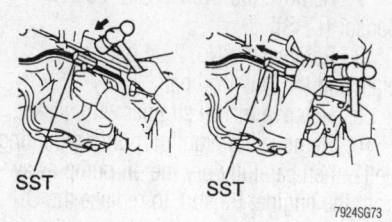

SST   SST

7924SG73

Use the seal breaker to separate the No. 2 oil pan from the No. 1 oil pan—4.7L (2UZ-FE) engine

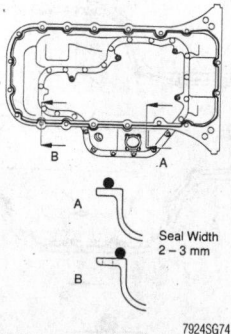

7924SG74

Apply a thin bead of sealant to the No. 1 oil pan as shown—4.7L (2UZ-FE) engine

To install:

6. Remove all of the old sealant from the sealing surfaces.
7. Clean the sealing surfaces of the engine block and No. 1 oil pan with solvent that will not leave a residue.
8. Apply a 0.08–0.12 inch (2–3mm) bead of sealant to the No. 1 oil pan as shown and position the pan on the engine. Temporarily install the fasteners. The bolts are different lengths, use the diagram to identify the correct bolt positions if needed.
   • A—0.79 inch (20mm)–10mm head
   • B—0.98 inch (25mm)–12mm head
   • C—2.36 inch (60mm)–12mm head
   • D—1.38 inch (35mm)–10mm head

➡**The parts with sealant applied must be installed within five minutes of application or the sealant must be reapplied.**

9. Be sure the clearance between the end of the pan and the end of the block is 0.008 inch (0.2mm) or less or the pan will be stretched when the fasteners are tightened. Tighten the 10mm heads to 66 inch lbs. (7.5 Nm) and the 12mm heads to 21 ft. lbs. (28 Nm).
10. Clean the sealing surfaces of the No. 1 and No. 2 oil pans with solvent that will not leave a residue.

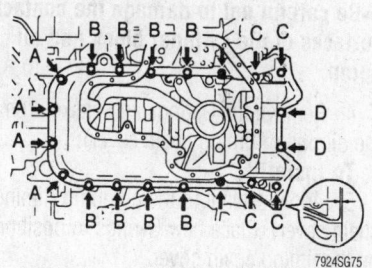

7924SG75

Be sure the rear ends of the No. 1 oil pan and engine block are as specified and the bolts are in the correct locations—4.7L (2UZ-FE) engine

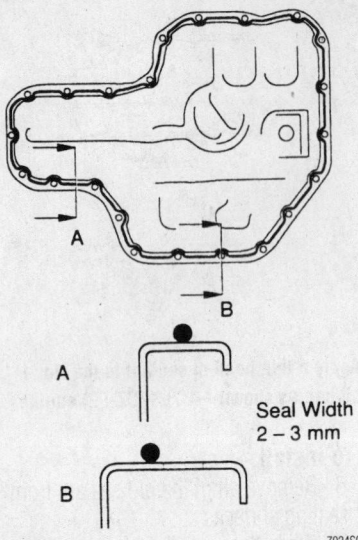

Seal Width
2 – 3 mm

Apply a 0.08–0.12 inch (2–3mm) bead of sealant to the No. 2 oil pan as shown— 4.7L (2UZ-FE) engine

11. Apply a 0.08–0.12 inch (2–3mm) bead of No. 08826–00080 or equivalent sealant to the No. 2 oil pan as shown.

12. Install the No. 2 oil pan. Tighten the fasteners to 66 inch lbs. (7.5 Nm).

13. Install the dipstick tube and dipstick.

14. Install the engine in the vehicle.

Oil Pump

REMOVAL & INSTALLATION

4.5L (1FZ-FE) Engine

1. Disconnect the negative battery cable.
2. Remove the timing chain cover.

➡**The oil pump is incorporated in the timing chain cover.**

3. Remove the oil pump by prying the portions between the cylinder block and oil pump.

➡**Be careful not to damage the contact surfaces of the cylinder block and oil pump.**

4. Remove the O-rings and gasket from the oil pump (timing chain cover).

To install:

5. Apply sealant to the oil pump (timing chain cover). Place new O-rings in position on the timing chain cover.

6. Engage the gear of the oil pump drive rotor with the gear of the oil pump drive gear and slide the oil pump.

7. Install the oil pump and drive belt adjusting bar. Install each bolt in their

proper location. Tighten each bolt to 15 ft. lbs. (21 Nm).

8. Complete installation of the timing chain cover.

9. Reconnect the negative battery cable. Start the engine and check for leaks.

4.7L (2UZ-FE) Engine

1. Remove the engine and place it on a stand to allow access to the oil pan.
2. Remove the timing belt.
3. Remove the idler pulleys.
4. Remove the crankshaft timing pulley.
5. Remove the dipstick and tube assembly.
6. Remove the oil filter, oil cooler and bracket assembly.
7. Remove the Crankshaft Position Sensor (CPS).
8. Remove the No. 2 oil pan, baffle plate and the No. 1 oil pan.
9. Remove the oil strainer with gasket.
10. Remove the eight oil pump mounting bolts, then carefully pry the oil pump away from the engine. Be sure to replace the O-ring with a new one.

To install:

11. Remove the old sealant and clean the sealing surfaces of the engine block and oil pump with a greaseless solvent.

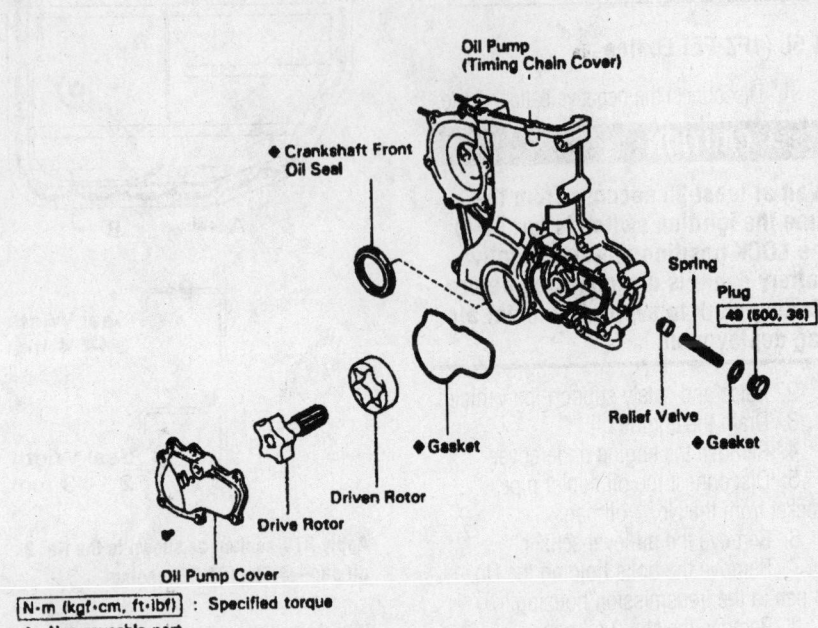

Oil pump (timing chain cover) components, exploded view—4.5L (1FZ-FE) engine

12. Install a new O-ring on the engine block.

13. Apply a 0.08–0.12 inch (2–3mm) bead of sealant to the oil pump sealing surface, then install the oil pump. Tighten the 12mm and 6mm hex head bolts to 11 ft. lbs. (15.5 Nm) and the 14mm head bolts to 22 ft. lbs. (30.5 Nm). The bolts are different sizes, use the following list along with the diagram to identify the bolts if needed.

- A—1.38 inch (35mm)–12mm head
- B—1.97 inch (50mm)–12mm head
- C—4.17 inch (106mm)–12mm head
- D—1.57 inch (40mm)–14mm head
- E—1.18 inch (30mm)–6mm hex head

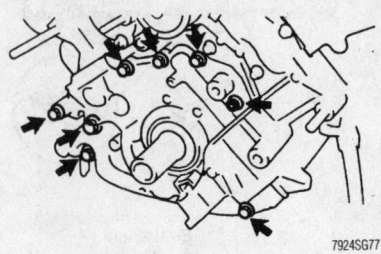

Oil pump mounting bolts—4.7L (2UZ-FE) engine

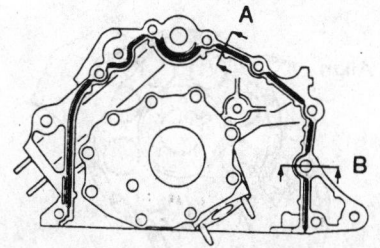

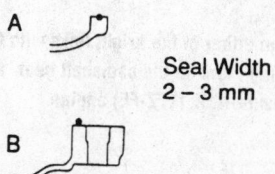

Seal Width 2 – 3 mm

A

B

7924SG78

Apply a thin bead of sealant to the oil pump mounting surface as shown—4.7L (2UZ-FE) engine

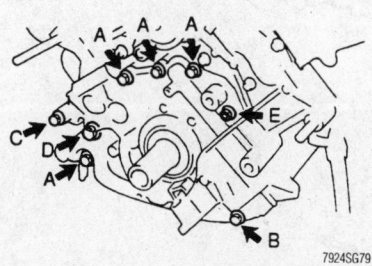

7924SG79

Be sure to install the oil pump bolts in their correct location—4.7L (2UZ-FE) engine

Rear Main Seal

REMOVAL & INSTALLATION

4.5L (1FZ-FE) Engine

➡There are two methods to replace the oil seal. Either method requires that the engine/transmission assembly be removed from the vehicle. They are as follows:

OIL SEAL RETAINER INSTALLED ON CYLINDER BLOCK

1. Cut off the rubber lip portion of the seal with a sharp knife.
2. Using a suitable tool, pry out the oil seal. Be careful not to damage the crankshaft.
3. Apply MP grease to the new oil seal lip.
4. Using a suitable seal installer and a mallet, tap in the new oil seal until its surface is flush with the rear oil seal retainer edge.

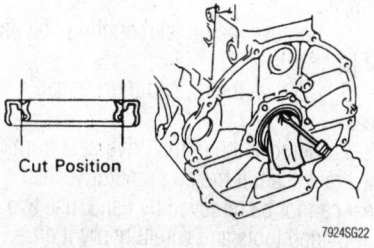

Cut Position

7924SG22

Carefully pry out the old seal without damaging the crankshaft—4.5L (1FZ-FE) engine shown

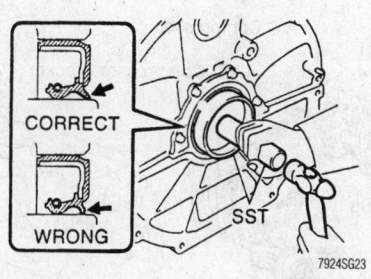

CORRECT

WRONG

SST

7924SG23

Tap the new seal until it is flush with the surface of the retainer—4.5L (1FZ-FE) engine

OIL SEAL RETAINER REMOVED FROM CYLINDER BLOCK

1. Remove the oil seal retainer mounting bolts and rear oil seal retainer from the cylinder block.
2. Carefully tap the oil seal out the oil seal retainer.
3. Apply MP grease to the new oil seal lip.
4. Using a suitable seal installer and a mallet, tap in the new oil seal until its surface is flush with the rear oil seal retainer edge.
5. Apply sealant to the oil seal retainer. Install the rear oil seal retainer to the cylinder block. Tighten the mounting bolts to 15 ft. lbs. (21 Nm).

4.7L (2UZ-FE) Engine

It is recommended that the oil seal be removed together with the seal retainer. However it may be possible to remove the seal by punching a small hole in the metal portion of the seal and using a slide hammer with a sheet metal screw to pull the seal.

1. Remove the transmission assembly.
2. Remove the flexplate.
3. Remove the bolts attaching the oil seal retainer to the engine block, then carefully pry the retainer from the engine.

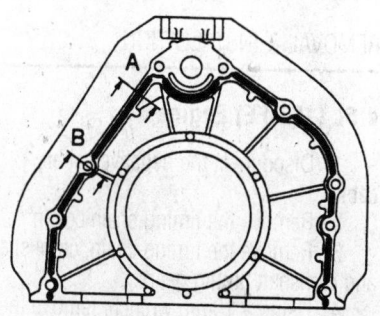

A

B

Seal Width 2 – 3 mm

A

B

7924SG79B

Apply a thin bead of sealant to the seal retainer and install it within five minutes—4.7L (2UZ-FE) engine shown

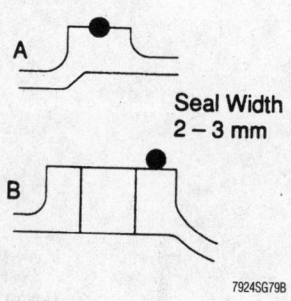

New O–Ring

7924SG80

Rear main oil seal O-ring and mounting bolt locations—4.7L (2UZ-FE) engine shown

4. Remove the O-ring and drive the oil seal out of the retainer.

To install:

5. Remove all of the old sealant from the retainer and engine block. Clean the sealing surfaces with a greaseless solvent.
6. Install a new seal and O-ring into the retainer.
7. Apply a 0.08–0.12 inch (2–3mm) bead of sealant to the sealing surface of the seal retainer as shown. Install the retainer and tighten the bolts to 71 inch lbs. (8.0 Nm).
8. Install the flexplate and the transmission assembly.

Timing Chain, Guide, Tensioner and Sprocket

REMOVAL & INSTALLATION

4.5L (1FZ-FE) Engine

1. Disconnect the negative battery cable.
2. Remove the timing chain cover.
3. Remove the timing chain, crankshaft and camshaft timing gears.
4. Using a 10mm wrench, remove the chain tensioner slipper and vibration damper.
5. Remove the oil jet mounting bolt and oil jet.
6. Remove the oil pump driveshaft gear.
7. Remove the pump drive gear from the crankshaft. If the oil pump driveshaft gear cannot be removed by hand, use two flat-bladed tools and carefully pry it off.

To install:

8. Turn the crankshaft until the set key is facing downward. Install the pump drive gear.

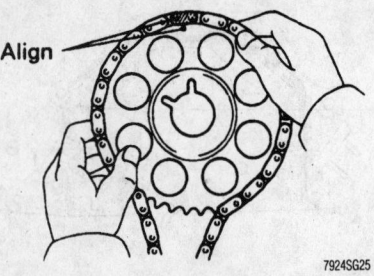

Align
7924SG25

Align either of the bright links with the timing mark on the camshaft gear, as shown—4.5L (1FZ-FE) engine

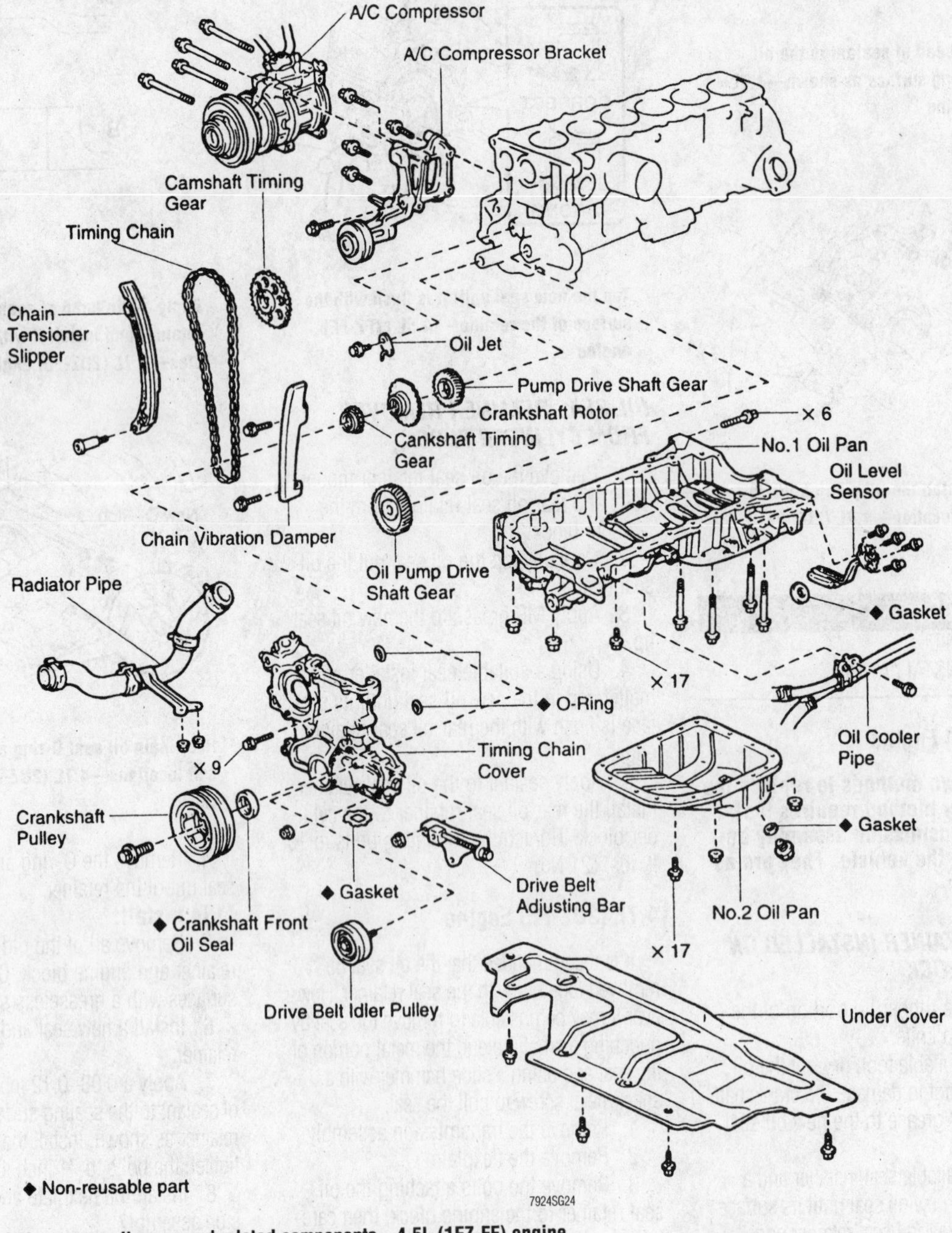

A/C Compressor
A/C Compressor Bracket
Camshaft Timing Gear
Timing Chain
Chain Tensioner Slipper
Oil Jet
Pump Drive Shaft Gear
Crankshaft Rotor
Cankshaft Timing Gear
× 6
No.1 Oil Pan
Oil Level Sensor
Chain Vibration Damper
Oil Pump Drive Shaft Gear
Radiator Pipe
◆ Gasket
× 17
◆ O-Ring
Timing Chain Cover
× 9
Oil Cooler Pipe
Crankshaft Pulley
◆ Gasket
◆ Gasket
Drive Belt Adjusting Bar
No.2 Oil Pan
◆ Crankshaft Front Oil Seal
× 17
Drive Belt Idler Pulley
Under Cover
◆ Non-reusable part
7924SG24

Timing chain, timing cover, oil pump and related components—4.5L (1FZ-FE) engine

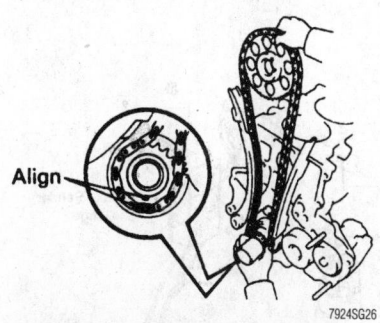

Align either of the bright links with the timing mark on the crankshaft timing gear, as shown—4.5L (1FZ-FE) engine

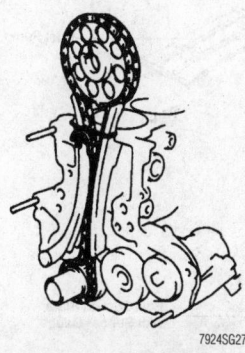

Tie the timing chain with a cord as shown—4.5L (1FZ-FE) engine

9. Lubricate the shaft portion of the oil pump driveshaft gear and install the gear.

10. Inspect the oil jet. Replace, if necessary. Install the oil jet and tighten the mounting bolt to 14 ft. lbs. (20 Nm).

11. Install the chain tensioner slipper and vibration damper. Tighten the damper bolts to 14 ft. lbs. (20 Nm) and the slipper bolt to 51 inch lbs. (69 Nm). Check that the tensioner slipper moves freely.

12. Install the crankshaft timing gear.

  a. Install the timing chain on the camshaft timing gear with the bright link aligned with the timing mark on the camshaft timing gear.

  b. Install the timing chain on the crankshaft timing gear with the other bright link aligned with the timing mark on the crankshaft timing gear.

  c. Wrap the timing chain with a piece of cord and be sure it doesn't come loose.

13. Install the timing chain cover.

14. Install the drive belt idler pulley.

15. Install the crankshaft timing gear.

FUEL SYSTEM

Fuel System Service Precautions

Safety is the most important factor when performing not only fuel system maintenance but any type of maintenance. Failure to conduct maintenance and repairs in a safe manner may result in serious personal injury or death. Maintenance and testing of the vehicle's fuel system components can be accomplished safely and effectively by adhering to the following rules and guidelines.

• To avoid the possibility of fire and personal injury, always disconnect the negative battery cable unless the repair or test procedure requires that battery voltage be applied.

• Always relieve the fuel system pressure prior to disconnecting any fuel system component (injector, fuel rail, pressure regulator, etc.), fitting or fuel line connection. Exercise extreme caution whenever relieving fuel system pressure to avoid exposing skin, face and eyes to fuel spray. Please be advised that fuel under pressure may penetrate the skin or any part of the body that it contacts.

• Always place a shop towel or cloth around the fitting or connection prior to loosening to absorb any excess fuel due to spillage. Ensure that all fuel spillage (should it occur) is quickly removed from engine surfaces. Ensure that all fuel soaked cloths or towels are deposited into a suitable waste container.

• Always keep a dry chemical (Class B) fire extinguisher near the work area.

• Do not allow fuel spray or fuel vapors to come into contact with a spark or open flame.

• Always use a back-up wrench when loosening and tightening fuel line connection fittings. This will prevent unnecessary stress and torsion to fuel line piping. Always follow the proper torque specifications.

• Always replace worn fuel fitting O-rings with new. Do not substitute fuel hose or equivalent, where fuel pipe is installed.

Fuel System Pressure

RELIEVING

✳✳ CAUTION

Fuel injection systems remain under pressure after the engine has been turned OFF. Properly relieve fuel pressure before disconnecting any fuel lines. Failure to do so may result in fire or personal injury.

1. Detach the fuel pump connector near the fuel tank.

2. Start the engine and allow it to run until it stalls. Crank the engine for a few seconds to relieve additional fuel pressure.

3. Turn the ignition switch to the **OFF** position.

4. Place a catch-pan under the joint to be disconnected. A large quantity of fuel may be released when the joint is opened.

5. Wear eye or full face protection.

6. Place a shop towel over the area and slowly loosen the joint using a wrench of the correct size. Use a back-up wrench if needed.

7. Allow the fuel left in the line to bleed off slowly before fully disconnecting the joint.

8. Plug the opened lines immediately to prevent fuel spillage or the entry of dirt.

9. Dispose of the released fuel properly.

10. Turn the ignition to the **ON** position to energize the fuel pump, then check for leaks and repair as needed.

Fuel Filter

REMOVAL & INSTALLATION

1. Disconnect the negative battery cable.

✳✳ CAUTION

Fuel injection systems remain under pressure after the engine has been turned OFF. Properly relieve fuel pressure before disconnecting any fuel lines. Failure to do so may result in fire or personal injury.

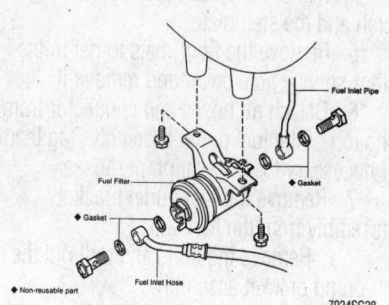

Always use new gaskets when replacing the fuel filter

2. Relieve the fuel system pressure.

3. Disconnect the negative battery cable.

4. Disconnect and plug the inlet and outlet lines from the filter. Always use a back-up wrench on the filter to keep it from turning.

5. Remove the fuel filter retaining bolts and remove the filter.

To install:

6. Install the fuel filter.

7. Use new washers and tighten the bolts to 21 ft. lbs. (29 Nm) on the 1FZ-FE engine or tighten the flare nuts to 28 ft. lbs. (38 Nm) on the 2UZ-FE engine.

8. Connect the negative battery cable.

9. Start the engine and check for leaks.

Fuel Pump

REMOVAL & INSTALLATION

1. Relieve the fuel system pressure.

❊❊ CAUTION

Fuel injection systems remain under pressure after the engine has been turned OFF. Properly relieve fuel pressure before disconnecting any fuel lines. Failure to do so may result in fire or personal injury.

2. Disconnect the negative battery cable.

❊❊ CAUTION

Wait at least 90 seconds from the time the ignition switch is turned to the LOCK position and the negative battery cable is disconnected before starting work to avoid accidental air bag deployment.

3. Remove the second seats.

4. Remove the scuff plate, the side garnish and the step plate.

5. Remove the floor mats to get to the floor service hole cover and remove it.

6. Detach all hoses and connector from the top of the fuel pump assembly. Tag them if necessary for installation purposes.

7. Remove the fuel pump bracket assembly from the fuel tank.

   a. Remove the bolts and pull out the pump bracket assembly.

   b. Remove the gasket from the pump bracket.

8. Remove the fuel pump from the fuel pump bracket.

   a. Pull off the lower side of the fuel pump from the pump bracket.

   b. Detach the fuel pump connector.

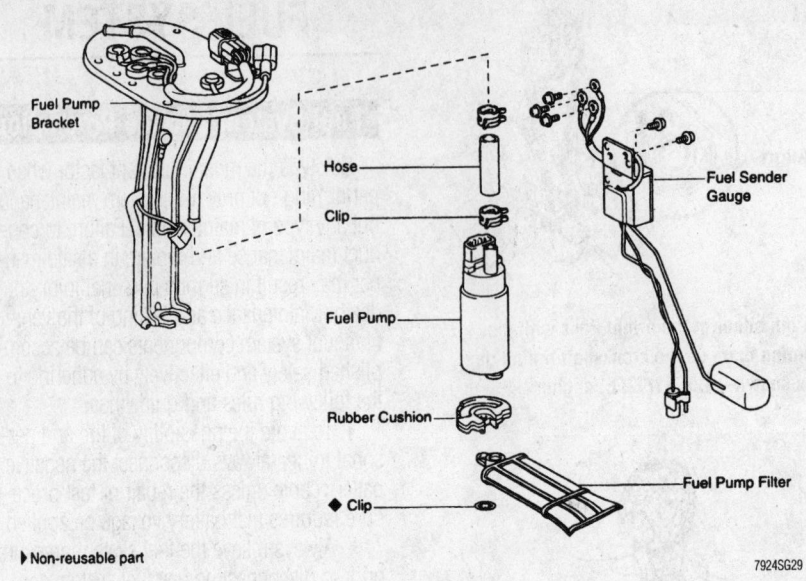

▶ Non-reusable part

7924SG29

Fuel pump and related components—4.5L (1FZ-FE) engine

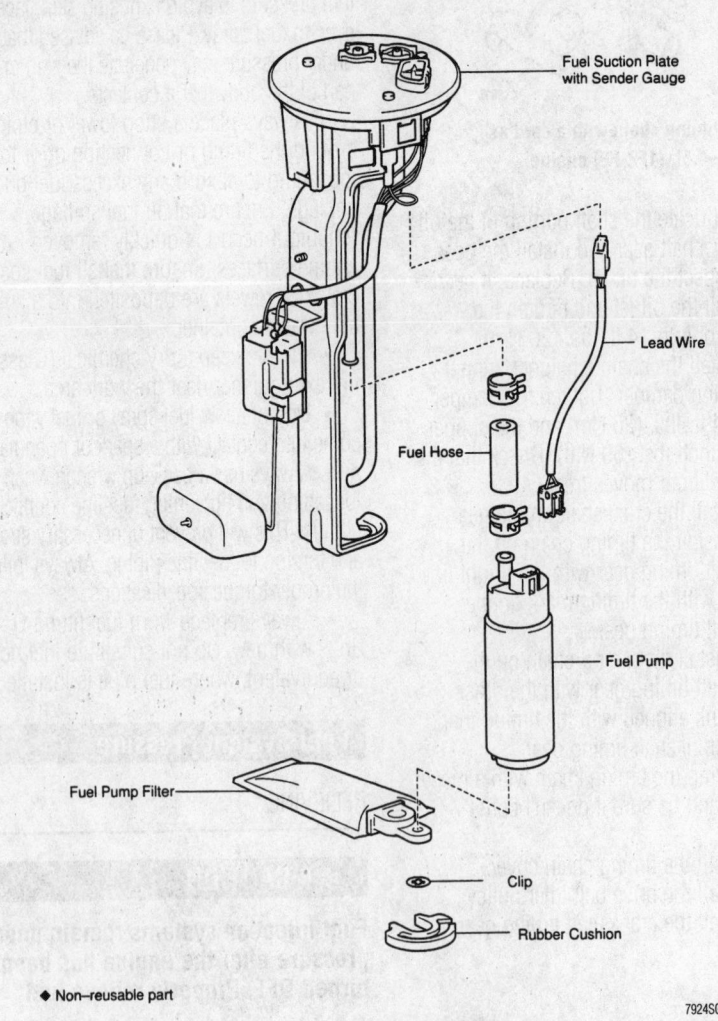

◆ Non–reusable part

7924SG81

Fuel pump and related components—4.7L (2UZ-FE) engine

c. Disconnect the fuel hose from the fuel pump and remove the fuel pump.

d. Remove the rubber cushion from the fuel pump.

9. Remove the fuel pump filter from the fuel pump by removing the clip.

To install:

10. Install the fuel pump filter to the fuel pump with a new clip.

11. Install the rubber cushion, the fuel hose and the fuel pump connector to the fuel pump. Install the fuel pump on the bracket.

12. Install a new gasket to the pump bracket and install the bracket in the fuel tank. Tighten the bolts to 2.8 ft. lbs. (3.9 Nm).

13. Attach the hoses and connectors.

14. Install the service hole cover.

15. Install the floor mats.

16. Install the step plate, the side garnish and the scuff plate.

17. Install the second seats and tighten the bolts to 29 ft. lbs. (39 Nm).

18. Connect the negative battery cable.

19. Start the engine and check for leaks.

DRIVE TRAIN

➡**Refer to the Unit Repair Section for information on driveshaft service.**

Transmission Assembly

REMOVAL & INSTALLATION

1995–97 Models

1. Disconnect the battery cables and remove the battery and the battery tray.

✳✳ CAUTION

To avoid accidental air bag deployment, disconnect the negative battery cable and wait at least 90 seconds before servicing the vehicle.

2. Loosen the fan shroud of the cooling fan to avoid damage to the fan.

3. Disconnect the throttle cable.

4. Raise and support the vehicle.

5. Drain the transmission and transfer case fluid.

6. Remove the upper starter mounting bolt.

7. Remove the transmission select lever and the transfer shift lever.

a. Remove the clip, washer and the wave washer and disconnect the link.

b. Remove the nut and washer and disconnect the link.

c. Remove the transfer shift lever knob.

d. Remove the console and the transfer shift lever boot.

e. Remove the center console box and disconnect the three connectors.

f. Remove the transmission shift lever assembly and the transfer shift lever.

8. Disconnect the No. 1 and No. 2 vehicle speed sensors, the park/neutral switch, the solenoid connector and the A/T fluid temperature sensor.

9. Detach the connectors and hoses from the transfer.

10. Remove the front and rear propeller (drive) shaft.

11. Remove the oil lever gauge, the upper side mounting bolt and the filler pipe.

12. Loosen the two oil cooler pipe union nuts.

13. Remove the four stabilizer bar bracket mounting bolts.

14. Remove the engine undercover.

15. Remove the torque converter clutch mounting bolts.

a. Remove the converter hole plug.

b. Turn the crankshaft to gain access to each bolt.

c. Hold the crankshaft pulley nut with a wrench and remove the bolts.

16. Remove the front exhaust pipe assembly.

a. Loosen the clamp bolt and disconnect the clamp from the No. 1 support bracket.

b. Remove the No. 1 support bracket.

17. Disconnect the wiring and remove the starter.

18. Place a jack under the transmission and remove the crossmember.

19. Lower the rear end of the transmission.

20. Separate the wiring harness from the transmission and the transfer case.

21. Remove the oil cooler pipe mounting bolts from the torque converter clutch housing and disconnect the two oil cooler pipes from the elbows.

22. Remove the ten bolts and remove the transmission.

23. Remove the transfer case from the transmission.

To install:

24. Install the transfer case to the transmission.

25. Install the transmission and tighten the bolts to 53 ft. lbs. (72 Nm).

26. Connect the two oil cooler pipes and install the oil cooler pipe mounting bolts to the torque converter clutch housing.

27. Install the crossmember and tighten the bolts to 45 ft. lbs. (61 Nm) and the nuts to 54 ft. lbs. (74 Nm).

28. Install the starter and tighten the bolt to 29 ft. lbs. (39 Nm). Attach the terminal and the connector.

29. Install the No. 1 support bracket and tighten the bolts to 29 ft. lbs. (39 Nm).

30. Connect the clamp from the No. 1 support bracket and tighten the bolt to 14 ft. lbs. (19 Nm).

31. Install the front exhaust pipe assembly and tighten the bolts to 29 ft. lbs. (39 Nm).

32. Install the torque converter clutch mounting bolts and tighten to 40 ft. lbs. (55 Nm).

➡**First install the gray colored bolt, then the five other bolts.**

33. Install the engine undercover. Tighten the bolts to 21 ft. lbs. (28 Nm).

34. Install the stabilizer bar bracket mounting and tighten the bolts to 13 ft. lbs. (18 Nm).

35. Tighten the oil cooler pipe union nuts.

36. Install a new O-ring and install the filler pipe. Install the upper side mounting bolt and the level gauge.

37. Install the front and rear propeller (drive) shaft.

38. Connect the connectors and hoses to the transfer case.

39. Connect the A/T fluid temperature sensor, the solenoid connector, the park/neutral switch and the No. 1 and No. 2 vehicle speed sensors.

40. Install the transmission select lever and transfer shift lever.

a. Install the transmission shift lever assembly and the transfer shift lever. Tighten the bolts for the shift lever to 13 ft. lbs. (18 Nm).

b. Install the center console box and attach the three connectors.

c. Install the console and the transfer shift lever boot.

d. Install the transfer shift lever knob.

e. Install the nut and washer and connect the link.

f. Install the clip, washer and the wave washer and connect the link.

41. Install the upper side starter mounting bolt and tighten to 29 ft. lbs. (39 Nm).

42. Lower the vehicle and fill the transfer case and transmission with the proper fluid.

43. Connect the throttle cable.

44. Tighten the fan shroud of the cooling fan.

45. Install the battery tray and the battery. Connect the battery cables.

46. Test drive the vehicle and check the shifting operation.

1998–99 Models

1. Disconnect the negative cable first, then the positive cable. Remove the battery.

2. Remove the air cleaner assembly, drive belt, fan assembly, shroud and reservoir.

3. Remove the upper mounting bolt on the oil dipstick tube.

4. Remove the knob, console panel and the boot from the transfer case shift lever.

5. Raise and safely support the vehicle.

6. Remove the engine undercovers.

7. Disconnect the front exhaust pipes from the manifolds.

8. Place a drain pan under the rear of the transmission. Matchmark, then remove the front and rear driveshafts.

9. Remove the oil dipstick tube.

10. Disconnect the shift linkage from the transmission and transfer case.

11. Detach the following connectors:
- No. 1 and No. 2 vehicle speed sensors
- O/D direct clutch speed sensor
- Solenoid
- ATF temperature sensor
- P/N position switch
- Center differential lock indicator switch
  - Motor actuator
  - L4 position switch
  - Neutral position switch

12. Remove the inspection cover, then remove the six bolts attaching the torque converter to the flexplate.

13. Disconnect the fluid cooler lines from the transmission.

14. Support the transmission assembly with a jack, then remove the crossmember. Be sure to place a block of wood between the jack and transmission.

✳✳ CAUTION

The torque converter will fall out of the transmission if the transmission is tilted forward.

15. Carefully lower the transmission to gain access to the mounting bolts. Remove the mounting bolts and slowly lower the transmission from the vehicle.

To install:

16. If removed, install the torque converter into the transmission. Slowly rotate the converter while gently pushing it towards the transmission until it is fully seated. When the torque converter is fully

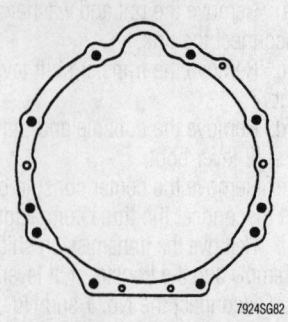

Transmission mounting bolt locations—1998–99 models

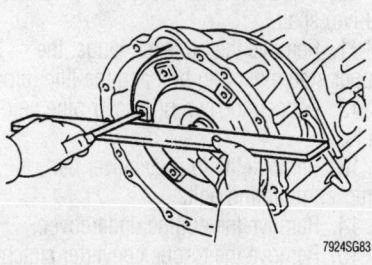

Measure the distance shown and compare it to the specification to be sure the torque converter is fully seated in the transmission—1998–99 models

seated, the distance from the mounting surface of the transmission to the mounting boss for the flexplate mounting bolts should be 0.673 inch (17.1mm) or more.

17. Raise and position the transmission assembly on the rear of the engine. Install the mounting bolts and tighten them to 53 ft. lbs. (72 Nm).

18. Install the crossmember. Tighten the bolts to 37 ft. lbs. (50 Nm). Tighten the nuts to 54 ft. lbs. (74 Nm).

19. Connect the fluid cooler lines to the transmission, tighten the union nuts to 24 ft. lbs. (32 Nm).

20. Install the six torque converter-to-flexplate bolts. Install the green bolt first, then the remaining bolts. Turn the crankshaft pulley bolt to gain access to the bolt holes. Tighten the bolts to 35 ft. lbs. (48 Nm).

21. Reattach the connectors.

22. Connect the linkage to the transmission and transfer case. Tighten the nut for the transmission control rod to 9 ft. lbs. (13 Nm).

23. Using a new O-ring, install the dipstick tube in the transmission.

24. Align the matchmarks and install the driveshafts in their original positions.

25. Use new gaskets and connect the exhaust pipes to the manifolds.

26. Install the under covers and lower the vehicle.

27. Install the boot, console and knob on the transfer case shift lever.

28. Install the coolant reservoir, fan assembly and shroud.

29. Install the battery and the air cleaner assembly.

30. Refill the transmission with the correct amount of fluid. First check the level, then add fluid as needed. Be sure not to overfill the transmission.

31. Check for leaks and road test the vehicle.

Transfer Case

REMOVAL & INSTALLATION

1. Disconnect the negative battery cable.

2. Raise and safely support the vehicle.

3. Remove the transfer case protector.

4. Drain the oil from the transfer case.

5. Matchmark and remove the driveshafts.

6. Disconnect the linkage from the transfer case.

7. Disconnect the ground cable.

8. Support the transmission assembly with a jack, then remove the crossmember. Be sure to place a block of wood between the jack and transmission.

9. Disconnect the breather hose.

10. Detach the No. 1 speed sensor, differential lock indicator switch, neutral position switch and actuator connectors.

11. Raise the transfer case slightly with a jack, then remove the mounting bolts and remove the transfer case assembly.

To install:

12. Apply MP grease to the transfer case adapter oil seal.

13. Raise the transfer case into position and install the mounting bolts. Tighten the bolts to 51 ft. lbs. (69 Nm). Remove the jack.

14. Reattach the connectors and the breather hose.

15. Install the crossmember and remove the jack and wooden block. Tighten the bolts to 37 ft. lbs. (50 Nm). Tighten the nuts to 54 ft. lbs. (74 Nm).

16. Connect the ground cable and shift linkage.

17. Align the matchmarks and install the driveshafts in their original positions.

18. Refill the transfer case with 1.4 Qts. (1.3L) of SAE 75–90W GL-4 or GL-5 gear oil. The oil should be near the bottom of the filler hole when full.

19. Install the transfer case protector and lower the vehicle to the floor.

20. Road test the vehicle.

Halfshaft

REMOVAL & INSTALLATION

1. Safely raise and support the vehicle.

2. Remove the brake caliper and position it out of the way with wire or string. Do not let it hang by the hydraulic hose.

3. Remove the grease cap from the hub assembly, then remove the snap ring from the end of the halfshaft.

4. Remove the ABS wheel speed sensor and wiring harness from the knuckle assembly.

5. Remove the two bolts and the steering arm from the knuckle assembly.

6. Remove the cotter pin and nut from the lower ball joint. Disconnect the ball joint stud from the knuckle assembly using a suitable two jaw puller.

7. Support the lower control arm with a jack and disconnect the upper ball joint from the knuckle assembly. Remove the knuckle/hub assembly while pushing the halfshaft through.

8. Using a hammer and brass drift, remove the halfshaft from the front differential.

To install:

9. Install a new snapring on the end of the halfshaft that goes into the differential. The opening should be facing down. Apply MP grease to the lip of the seal and install the halfshaft in the differential. Confirm proper installation by pulling on the halfshaft to be sure it is engaged in the differential.

10. Insert the outer end of the halfshaft through the knuckle/hub assembly and connect the upper ball joint to the knuckle. Tighten the nut to 81 ft. lbs. (110 Nm) and install a new cotter pin.

11. Install the lower ball joint stud on the knuckle. Tighten the nut to 118 ft. lbs. (159 Nm) and install a new cotter pin.

12. Apply thread locking compound to the bolt threads and install the steering arm on the knuckle. Tighten the bolts to 108 ft. lbs. (147 Nm).

13. Install the ABS wheel speed sensor and wiring harness.

14. Install the snapring, and grease cap.

15. Install the brake caliper assembly. Tighten the mounting bolts to 90 ft. lbs. (123 Nm).

16. Install the front wheel. Tighten the nuts to 97 ft. lbs. (131 Nm).

17. Lower the vehicle to the floor.

STEERING AND SUSPENSION

Air Bag

✳✳ CAUTION

These vehicles are equipped with an air bag system, also known as the Supplemental Inflatable Restraint (SIR) or Supplemental Restraint System (SRS). The system must be disabled before performing service on or around system components, steering column, instrument panel components, wiring and sensors. Failure to follow safety and disabling procedures could result in accidental air bag deployment, possible personal injury and unnecessary system repairs.

PRECAUTIONS

Several precautions must be observed when handling the inflator module to avoid accidental deployment and possible personal injury.

• Never carry the inflator module by the wires or connector on the underside of the module.

• When carrying a live inflator module, hold securely with both hands and ensure that the bag and trim cover are pointed away.

• Place the inflator module on a bench or other surface with the bag and trim cover facing up.

• With the inflator module on the bench, never place anything on or close to the module which may be thrown in the event of an accidental deployment.

DISARMING

To avoid personal injury when working on vehicles equipped with an air bag, the negative battery cable must be disconnected and at least 90 seconds must elapse before working on the system. Failure to do so may result in deployment of the air bag.

Worm and Sector Power Steering Gear

REMOVAL & INSTALLATION

1. Place the front wheels in the straight ahead position and turn the key to the **LOCK** position.

2. Disconnect the negative battery cable.

3. Raise and safely support the vehicle.

4. Remove the left front wheel.

5. Remove the link joint protector.

6. Matchmark the Pitman arm to the cross shaft, then remove the Pitman arm from the steering gear using a suitable puller.

7. Place a drain pan under the steering gear and disconnect the pressure and return lines.

8. Matchmark the flexible joint to the shafts. Loosen the top bolt and remove the lower bolt, then remove the joint from the steering gear input shaft.

9. Remove the four mounting bolts and remove the steering gear from the vehicle.

To install:

10. Installation is the reverse of the removal procedure.

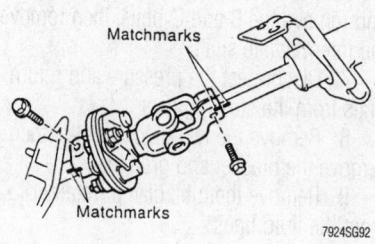

Be sure to matchmark the flex joint before removing it from the steering gear— 1995–97 models

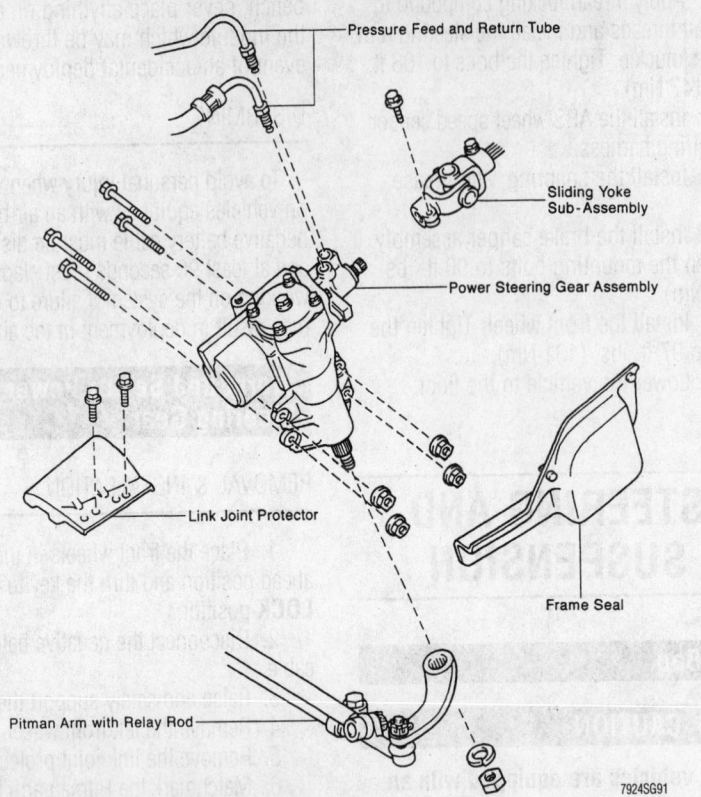

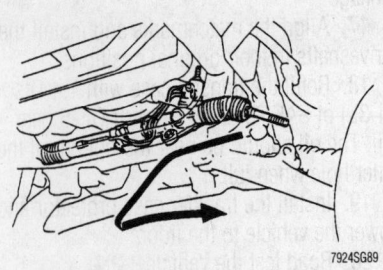

Exploded view of the worm and sector power steering gear mounting—1995–97 models

Power Rack And Pinion Steering Gear

REMOVAL & INSTALLATION

1. Disconnect the negative battery cable.
2. Raise and safely support the vehicle.
3. Remove the engine under covers.
4. Disconnect the outer tie rod ends.
5. Remove the engine oil filter hoses from the filter assembly and remove the assembly.
6. Matchmark the intermediate shaft to the control valve shaft. Loosen the A bolt and remove the B and C bolts, then remove the intermediate shaft.
7. Disconnect the pressure and return lines from the steering gear.
8. Remove the two bracket bolts, then remove the bracket and grommet.
9. Remove the tube clamp (retainer) from the fluid lines.
10. Matchmark the inner tie rod ends and nuts, then remove them.
11. Remove the two mounting bolts, slide the assembly to the right, then downward and out of the vehicle.

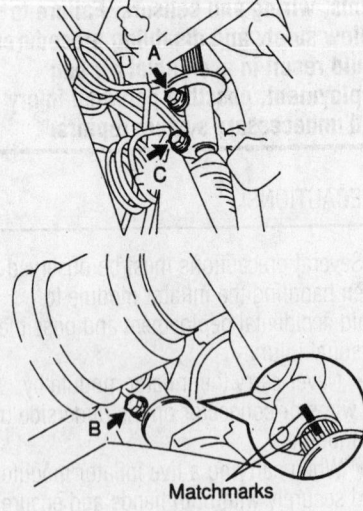

Loosen the A bolt, then remove bolts B and C to remove the intermediate shaft—1998–99 models

Power rack and pinion steering gear removal—1998–99 models

To install:

12. Position the steering gear in the vehicle and install two new mounting bolts. Tighten the bolts to 74 ft. lbs. (100 Nm).
13. Install the tie rods to the steering rack assembly using the matchmarks made during removal. Tighten the nuts to 41 ft. lbs. (55 Nm).
14. Install the tube clamp on the fluid lines. Tighten the bolt to 13 ft. lbs. (18 Nm).
15. Install the housing grommet, bracket and two new bolts. Tighten the bolts to 74 ft. lbs. (100 Nm).
16. Connect the return line. Tighten the fitting to 26 ft. lbs. (36 Nm).
17. Connect the pressure line using a new gasket. Tighten the fitting to 36 ft. lbs. (49 Nm).
18. Install the intermediate shaft assembly. Tighten the bolts to 25 ft. lbs. (34 Nm).
19. Install the oil filter assembly. Tighten the two bolts and nut to 13 ft. lbs. (18 Nm).
20. Connect the tie rod ends to the knuckles. Tighten the nut to 90 ft. lbs. (122 Nm).
21. Install the under covers and the front wheels.
22. Connect the negative battery cable.
23. Refill the power steering fluid reservoir and turn the steering wheel from lock to lock several times with the engine off.
24. Lower the vehicle to the floor and start the engine. Allow the engine to run for several minutes, then turn the steering wheel from lock to lock several times holding the wheel in the lock position for 2–3 seconds each time.
25. Turn the engine **OFF** and recheck the fluid level. Add fluid as needed.

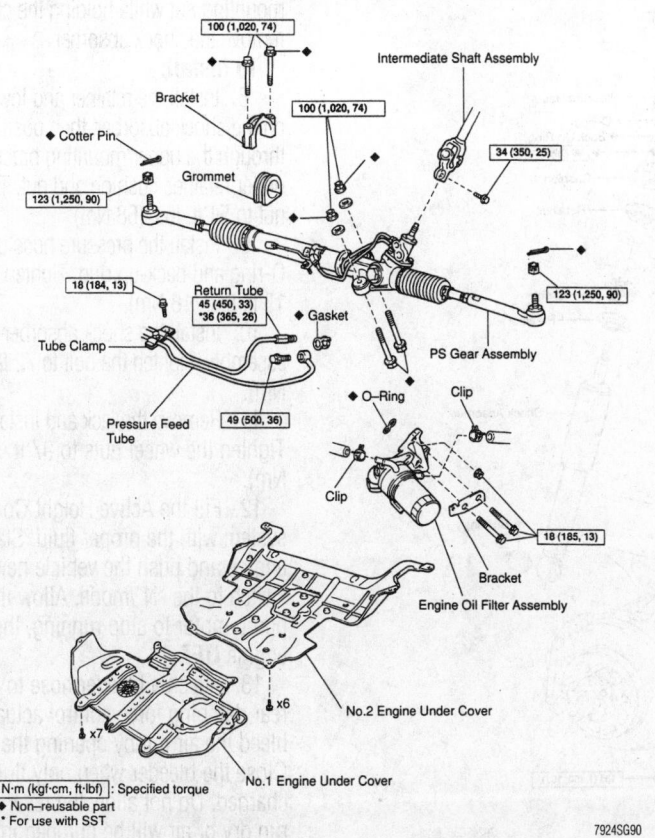

Exploded view of the rack and pinion steering gear mounting—1998–99 models

---

Shock Absorber

REMOVAL & INSTALLATION

1995–97 Models

FRONT

1. Disconnect the negative battery cable.

> ※ **CAUTION**
>
> **To avoid accidental air bag deployment, disconnect the negative battery cable and wait at least 90 seconds before servicing the vehicle.**

2. Raise the vehicle and support the body.
3. Remove the front wheels.
4. Remove the front shock absorber.
   a. Supporting the axle housing, hold the piston rod and remove the upper mounting nut.
   b. Hold the shock absorber and remove the lower mounting nut, shock absorber, cushions and the retainers.

To install:

5. Install the retainers, cushions, shock absorber and the lower mounting nut. Tighten the nut to 51 ft. lbs. (69 Nm).
6. Install the upper mounting nut for the shock absorber and tighten to 51 ft. lbs. (69 Nm).
7. Install the wheels.
8. Lower the vehicle and connect the battery cable.
9. Test drive the vehicle.

REAR

1. Raise and safely support the frame with stands.
2. Support the axle housing with a floor jack.
3. Remove the wheels.
4. Lower the floor jack to take the tension off of the spring.
5. Disconnect the shock absorber from the frame and spring seat. Remove the shock absorber.
6. Installation is the reverse of the removal procedure. Tighten the shock absorber-to-frame mounting bolt to 47 ft. lbs. (64 Nm). Tighten the shock absorber-to-spring seat bolt to 27 ft. lbs. (37 Nm).

1998–99 Models

FRONT

1. Raise and safely support the vehicle.
2. Attach a hose to the bleeder valve on the damping force control actuator. Open the bleeder and drain the fluid. Tighten the bleeder valve to 73 inch lbs. (8.3 Nm) after the pressure is relieved and the fluid has drained.
3. Remove the front wheel.
4. Remove the splash shield.
5. Remove the nut and bolt that attaches the lower end of the shock absorber to the suspension.
6. Remove the two bolts securing the pressure hose to the top of the shock absorber.
7. Remove the top shock absorber mounting nut while holding the cushion, then remove the shock absorber.

To install:

8. Install the retainer and lower cushion on the shock absorber, then position the shock through the upper mounting bracket. Install the upper retainer, cushion and nut. Tighten the nut to 51 ft. lbs. (68 Nm).
9. Install the pressure hose using a new O-ring and back-up ring. Tighten the bolts to 13 ft. lbs. (18 Nm).
10. Install the shock absorber on the lower control arm and install the nut and bolt. Tighten the nut to 101 ft. lbs. (135 Nm).
11. Install the splash shield and front wheel. Tighten the wheel nuts to 97 ft. lbs. (131 Nm).

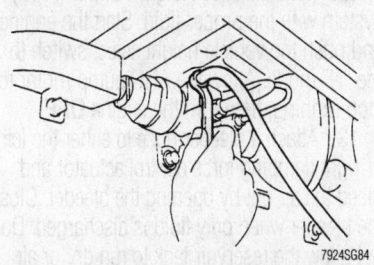

Be sure to attach a hose to the bleeder valve on the control actuator while bleeding the air out of the system—1998–99 models

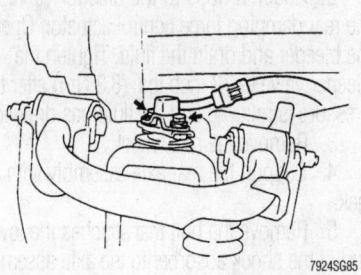

Remove the two bolts and detach the fluid hose from the shock absorber—1998–99 models

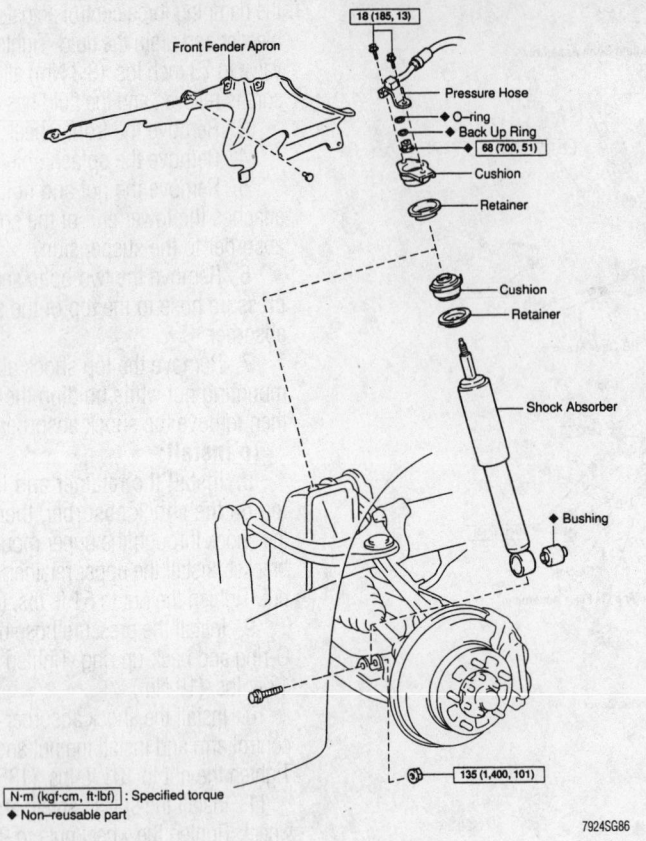

Exploded view of the front shock absorber mounting—1998–99 models

mounting nut while holding the cushion, then remove the shock absorber.

To install:

8. Install the retainer and lower cushion on the shock absorber, then position the shock through the upper mounting bracket. Install the upper retainer, cushion and nut. Tighten the nut to 51 ft. lbs. (68 Nm).

9. Install the pressure hose using a new O-ring and back-up ring. Tighten the bolts to 13 ft. lbs. (18 Nm).

10. Install the shock absorber on the axle assembly. Tighten the bolt to 72 ft. lbs. (98 Nm).

11. Remove the jack and install the wheel. Tighten the wheel nuts to 97 ft. lbs. (131 Nm).

12. Fill the Active Height Control (AHC) system with the proper fluid. Start the engine and push the vehicle height select switch to the "N" mode. Allow the AHC pump motor to stop running, then turn the engine **OFF**.

13. Attach a bleeder hose to either the rear damping force control actuator and bleed the air out by opening the bleeder. Close the bleeder when only fluid is discharged. Do not allow the reservoir tank to run dry or air will be pumped into the shock absorber.

12. Fill the Active Height Control (AHC) system with the proper fluid. Start the engine and push the vehicle height select switch to the "N" mode. Allow the AHC pump motor to stop running, then turn the engine **OFF**.

13. Attach a bleeder hose to either the left or right damping force control actuator and bleed the air out by opening the bleeder. Close the bleeder when only fluid is discharged. Do not allow the reservoir tank to run dry or air will be pumped into the shock absorber.

14. Repeat the procedure for the actuator on the opposite side.

REAR

1. Raise and safely support the vehicle.

2. Attach a hose to the bleeder valve on the rear damping force control actuator. Open the bleeder and drain the fluid. Tighten the bleeder valve to 73 inch lbs. (8.3 Nm) after the pressure is relieved and the fluid has drained.

3. Remove the rear wheel.

4. Support the rear axle assembly with a jack.

5. Remove the bolt that attaches the lower end of the shock absorber to the axle assembly.

6. Remove the two bolts securing the pressure hose to the top of the shock absorber.

7. Remove the top shock absorber

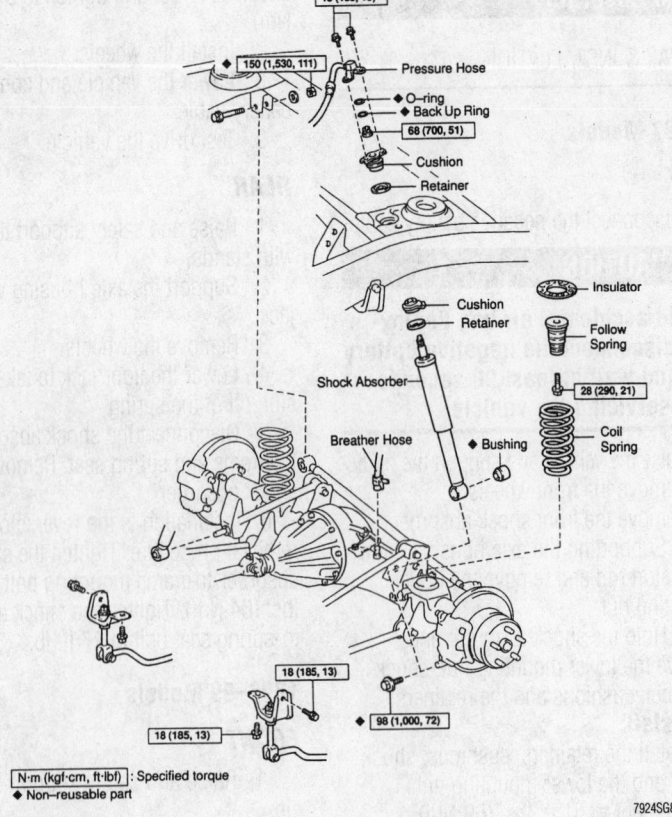

Exploded view of the rear shock absorber mounting—1998–99 models

Coil Spring

REMOVAL & INSTALLATION

1995–97 Models

FRONT

1. Disconnect the negative battery cable.

✳✳ CAUTION

To avoid accidental air bag deployment, disconnect the negative battery cable and wait at least 90 seconds before servicing the vehicle.

2. Raise the vehicle and support the body.
3. Remove the front wheels.
4. Remove the front shock absorber.
5. Disconnect the stabilizer bar from the axle housing.

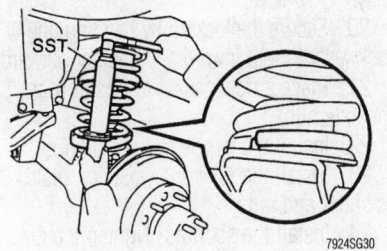

Compress the coil spring, them remove it from the vehicle—1995–97 models

6. Remove the coil spring.
   a. Using a spring compressor, compress the spring and remove it.
7. Remove the follow spring.

To install:

8. Install the follow spring. Tighten the nuts to 82 inch lbs. (9.2 Nm).
9. With the coil spring compressed, align the end with the lower seat and install the coil spring.
10. Remove the spring compressor.
11. Connect the stabilizer bar to the axle housing.
12. Install the retainers, cushions, shock absorber and the lower mounting nut. Tighten the nut to 51 ft. lbs. (69 Nm).
13. Install the upper mounting nut for the shock absorber and tighten to 51 ft. lbs. (69 Nm).
14. Install the wheels.

15. Lower the vehicle and connect the battery cable.
16. Test drive the vehicle.

REAR

1. Disconnect the negative battery cable.

✳✳ CAUTION

Wait at least 90 seconds from the time the ignition switch is turned to the LOCK position and the negative battery cable is disconnected before starting any work to avoid accidental air bag deployment.

2. Raise the vehicle and support the body. Hold the rear axle housing with a jack.
3. Remove the rear wheels.
4. Remove the rear shock absorber.
   a. Remove the bolt holding the shock absorber from the rear axle housing and disconnect the shock absorber.
   b. Remove the two upper bolts and the shock absorber.
5. Disconnect the stabilizer bar brackets from the rear axle housing.
6. Disconnect the lateral control rod from the rear axle housing.
7. Remove the coil spring. Begin to slowly lower the axle housing. While lowering the rear axle housing, remove the coil spring and the insulator.
8. Remove the follow spring from the frame.

To install:

9. Install the follow spring. Tighten the nuts to 21 ft. lbs. (28 Nm).
10. Install the coil spring with a jack under the axle housing. Slowly raise the jack.
11. Connect the lateral control rod to the rear axle housing.
12. Connect the stabilizer bar to the axle housing.
13. Install the shock absorber and the upper bolt. Tighten the nut to 37 ft. lbs. (50 Nm).
14. Install the shock absorber to the rear axle housing and tighten to 47 ft. lbs. (64 Nm).
15. Install the wheels.
16. Lower the vehicle and connect the battery cable.

1998–99 Models

The 1998–99 models utilize a front torsion bar. The following procedure is for the rear coil spring.

1. Raise and safely support the vehicle.
2. Position a jack under the rear axle assembly.

3. Remove the shock absorber.
4. Remove both right and left stabilizer bar brackets from the frame.
5. Disconnect the lateral control rod from the frame.
6. Disconnect the breather hose from the axle assembly.
7. Carefully lower the axle assembly and remove the coil spring and insulators. Be sure not to break the parking break cable with the weight of the axle.

To install:

8. Position the insulator and spring on the axle assembly.
9. Slowly raise the axle assembly into the normal position. Be sure the insulators are in place.
10. Attach the lateral control rod to the frame. Tighten the nut to 111 ft. lbs. (150 Nm).
11. Install the stabilizer bar bracket to the frame. Tighten the mounting bolts to 13 ft. lbs. (18 Nm).
12. Install the shock absorber and the wheel.
13. Lower the vehicle and bleed the air from the Active Height Control system using the procedure from shock absorber removal and installation.

Upper Ball Joint

The upper ball joint is an integral part of the upper control arm. If the ball joint is worn, the control arm must be replaced.

Lower Ball Joint

The lower ball joint is an integral part of the lower control arm. If the ball joint is worn, the control arm must be replaced.

Wheel Bearing

ADJUSTMENT

1. Raise the safely support the vehicle.
2. Remove the front wheel.
3. Remove the grease cap from the flange.
4. Remove the snapring from the axle.
5. Remove the six nuts and washers, then using a brass drift and hammer, tap on the studs to loosen the cone washers, then remove them.
6. Remove the flange and gasket.
7. Using a prytool, release the lockwasher, then remove the locknut and washer.
8. Adjust the preload:

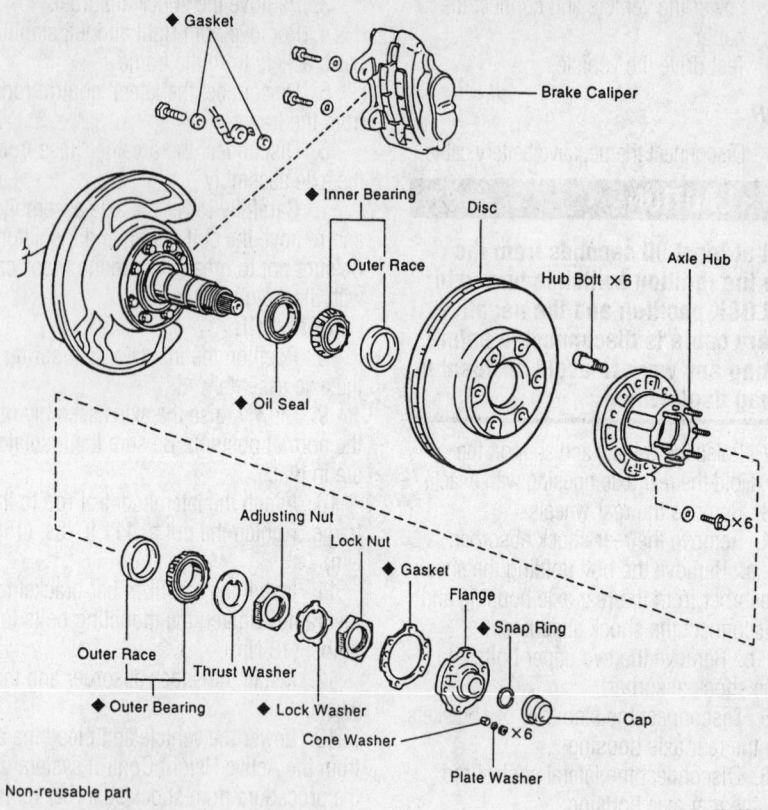

◆ Non-reusable part

7924SG31

Exploded view of the front hub and related components

a. Tighten the bearing adjusting nut to 43 ft. lbs. (59 Nm).

b. Turn the hub right and left two or three times and retighten.

c. Loosen the nut until it can be turned by hand.

d. Retighten to 4 ft. lbs. (5.4 Nm)

e. Check the bearing preload with a spring scale. The preload should be 6.4–12.6 ft. lbs. (28–56 Nm).

9. Install a new lockwasher and nut. Tighten to 47 ft. lbs. (64 Nm)

10. Check that there is no bearing end-play.

11. Secure the locknut by bending one lockwasher tooth inward and another outward.

12. Place a new gasket in position on the axle hub.

13. Install the flange, six cone washers, plate washers and nuts. Tighten the nuts to 26 ft. lbs. (35 Nm).

14. Install the bolt and tighten the bolt to 13 ft. lbs. (18 Nm).

15. Install the grease cap.

REMOVAL & INSTALLATION

1. Raise and safely support the vehicle.
2. Remove the front wheel.

3. Remove the grease cap from the flange.

4. Remove the snapring from the axle.

5. Remove the six nuts and washers, then using a brass drift and hammer, tap on the studs to loosen the cone washers, then remove them.

6. Remove the flange and gasket.

7. Using a prytool, release the lock-washer, then remove the locknut and washer.

8. Remove the adjusting nut and thrust washer.

9. Remove the hub and disc assembly with the outer bearing.

10. Using a seal puller, remove the grease seal, then remove the inner bearing.

11. Clean the bearings in solvent and dry them with compressed air. Replace them if there are signs of damage or abnormal wear.

※※ CAUTION

Do not spin the bearing with the compressed air while drying them. The wheel bearing may come apart causing serious injury.

12. If replacing bearing outer race, drive

out the outer bearing race using a brass bar and hammer.

To install:

13. If removed, drive in a new bearing outer race using a suitable installer.

14. Pack the bearings with MP grease. Coat the inside of the hub and cap with MP grease.

15. Install the inner bearing and oil seal. Coat the oil seal with MP grease.

16. Install the hub on the spindle. Install the outer bearing and claw washer.

17. Adjust the preload:

a. Tighten the bearing adjusting nut to 43 ft. lbs. (59 Nm).

b. Turn the hub right and left two or three times and retighten.

c. Loosen the nut until it can be turned by hand.

d. Retighten to 4 ft. lbs. (5.4 Nm)

e. Check the bearing preload with a spring scale. The preload should be 6.4–12.6 ft. lbs. (28–56 Nm).

18. Install the lockwasher and nut. Tighten to 47 ft. lbs. (64 Nm)

19. Check that there is no bearing end-play.

20. Secure the locknut by bending one lockwasher tooth inward and another outward.

21. Place a new gasket in position on the axle hub.

22. Install the flange to the axle hub.

23. Install the six cone washers, plate washers and nuts.

24. Install the six nuts. Tighten the nuts to 26 ft. lbs. (35 Nm).

25. Install a bolt in the center of the axle shaft, then pull out the shaft and install the snapring. Remove the bolt and install the grease cap.

26. Install the brake caliper support bracket to the steering knuckle and tighten each bolt to 90 ft. lbs. (123 Nm).

27. Clean the threads of the bolts and steering knuckle.

28. Apply sealant to the bolt threads and connect the knuckle arm with the brake line bracket to the steering knuckle. Tighten the bolts to 135 ft. lbs. (183 Nm).

29. If equipped with ABS brakes, connect the ABS speed sensor to the steering knuckle.

30. Install the front wheel and lower the vehicle.

31. Check the ABS speed sensor signal.

MAZDA

MPV

ENGINE REPAIR

→**Disconnecting the negative battery cable on some vehicles may interfere with the functions of the on board computer systems and may require the computer to undergo a relearning process, once the negative battery cable is reconnected.**

Distributor

REMOVAL

1. Disconnect the negative battery cable.
2. Label and remove the spark plug wires.
3. Turn the crankshaft so the No. 1 cylinder is at TDC of compression.
4. Disconnect the electrical connector from the distributor.
5. Mark the position of the rotor in relation to the distributor housing and the position of the distributor housing on the cylinder head.
6. Remove the distributor hold-down bolt(s), then remove the distributor.
7. Check the distributor O-ring for cuts or other damage and replace, if necessary.

INSTALLATION

Timing Not Disturbed

1. Lubricate the distributor O-ring with clean engine oil.
2. Align the matchmarks on the distributor shaft and the distributor drive gear.
3. Install the distributor with the hold-down bolt, aligning the marks that were made during removal. Snug the distributor hold-down bolt.
4. Connect the electrical connector.

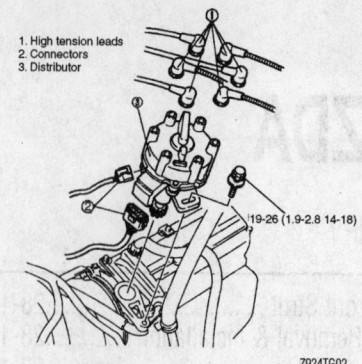

1. High tension leads
2. Connectors
3. Distributor

[19-26 (1.9-2.8 14-18)

Distributor assembly—1996–99

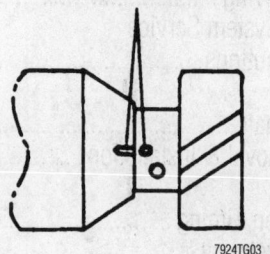

MATCHING MARKS

Distributor drive matchmarks—1996–99

5. Connect the spark plug wires.
6. Connect the negative battery cable. Start the engine and adjust the ignition timing. Tighten the distributor hold-down bolt to 14–18 ft. lbs. (19–25 Nm) after the timing has been set.

Timing Disturbed

1. Disconnect the spark plug wire from the No. 1 cylinder spark plug and remove the spark plug. After making sure the engine is cool enough to touch, place a finger over the spark plug hole.
2. Turn the crankshaft in the normal

direction of rotation until compression is felt at the spark plug hole.
3. Align the mark on the crankshaft pulley with the TDC mark on the timing belt cover.
4. Lubricate the distributor O-ring with clean engine oil.
5. Turn the distributor shaft until the rotor points to the No. 1 spark plug tower on the distributor cap and install the distributor. Install the distributor hold-down bolt(s) and align the distributor housing with the mark made on the cylinder head during removal. Snug the bolt(s).
6. Connect the electrical connector.
7. Connect the spark plug wires.
8. Install the spark plug in the No. 1 cylinder and connect the spark plug wire.
9. Connect the negative battery cable. Start the engine and adjust the ignition timing. Tighten the distributor hold-down bolt to 14–18 ft. lbs. (19–25 Nm) after the timing has been set.

Ignition Timing

ADJUSTMENT

1995 Models

1. Apply the parking brake and place the gearshift lever in **P**.
2. Start the engine and bring to normal operating temperature. Be sure all electrical loads and the A/C switch are **OFF**.
3. Connect a timing light and connect a tachometer to the **IG** terminal of the diagnostic connector.
4. Connect a jumper wire between the **TEN** and **GND** terminals of the diagnostic connector.
5. Check the idle speed and adjust, if necessary. The idle speed should be 780–820 rpm.
6. Aim the timing light at the timing

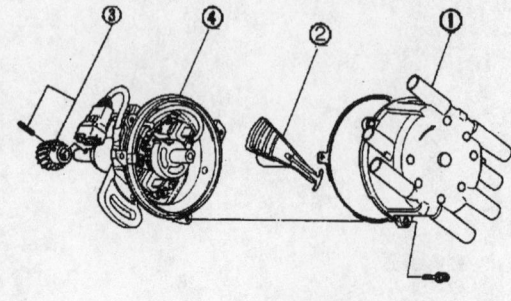

1. Cap
2. Distributor rotor
3. Coupling set
4. Distributor set

Distributor assembly (exploded view)—1995 3.0L engines

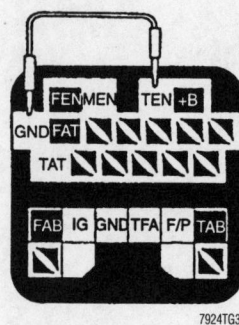

FEN MEN TEN +B
GND FAT
TAT
FAB IG GND TFA F/P TAB

Connect the two terminals with a jumper wire, as shown—1995 models

scale on the timing belt cover and the timing mark on the crankshaft pulley. The timing should be 4–6 degrees BTDC.

7. If the timing is not as specified, loosen the distributor mounting bolt and turn the distributor to adjust. After adjustment, tighten the bolt to 19 ft. lbs. (26 Nm) and recheck the timing.

8. Remove the jumper wire and all test equipment.

1996–99 Models

➡ **To perform this procedure System Selector tool 49 B019 9A0 or equivalent will be needed.**

1. Apply the parking brake and place the gearshift lever in **P**.

2. Start the engine and bring to normal operating temperature. Be sure all electrical loads and the A/C switch are **OFF**.

3. Connect a timing light to the No. 1 cylinder high-tension lead.

4. Connect the SST 49 B019 9A0 (system selector) to the data link connector.

5. Set the switch A to position No. 1 and set the test switch to SELF TEST. Verify that the idle speed is within specifications. If not, adjust it. Specification is 500–900 rpm.

6. Verify that the timing mark (yellow) is within the specification. Specification is BTDC 10–12°.

7. If not as specified, loosen the distributor lockbolts and turn the distributor to make the adjustment.

8. Tighten the distributor lockbolts to 14–18 ft. lbs. (19–25 Nm).

9. Disconnect the SST (system selector) and verify that the timing mark (yellow) is within specification. Specification is ATDC 2-bTDC 34°.

10. Stop the engine and disconnect the timing light.

Engine Assembly

REMOVAL & INSTALLATION

✳✳ CAUTION

The fuel injection system remains under pressure after the engine has been turned OFF. Properly relieve fuel pressure before disconnecting any fuel lines. Failure to do so may result in fire or personal injury. Do not allow fuel spray or fuel vapors to come in contact with a spark or open flame. Keep a dry chemical fire extinguisher nearby. Never store fuel in an open container due to risk of fire or explosion.

1. Relieve the fuel system pressure.

2. Disconnect the battery cables and remove the battery.

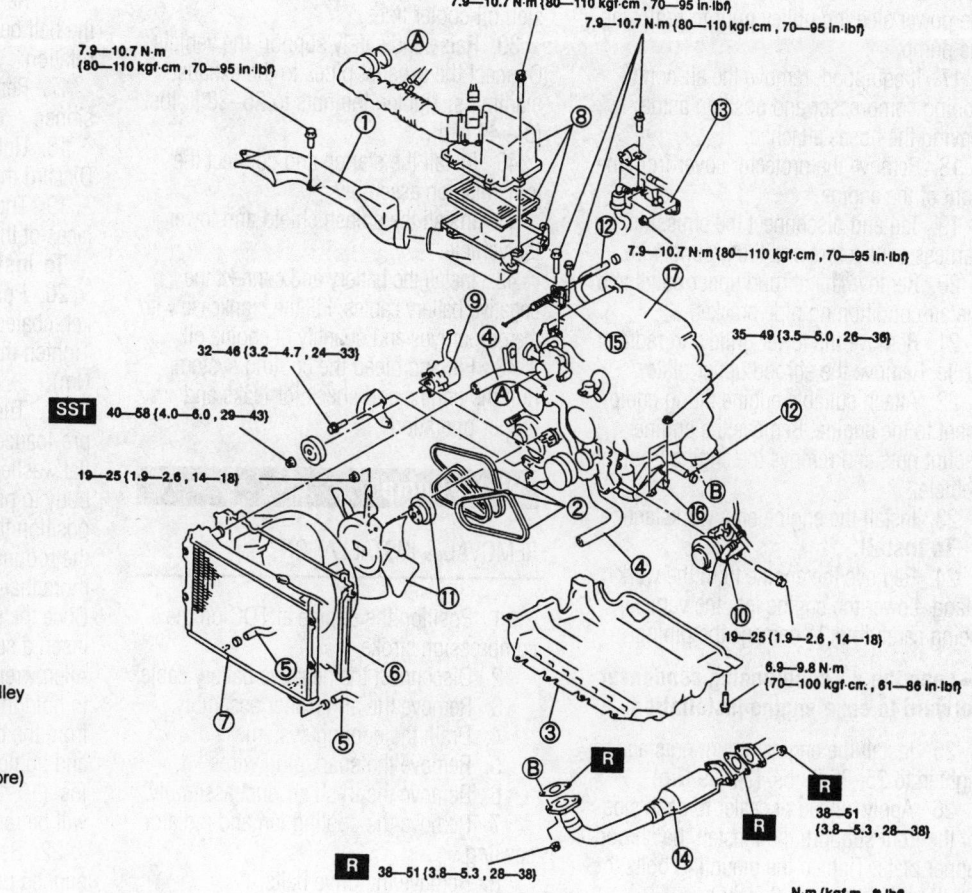

1	Fresh-air duct
2	Drive belt
3	Splash shield
4	Upper and lower radiator hose
5	Oil cooler hose
6	Radiator cowling
7	Radiator
8	Air cleaner and air duct
9	P/S oil pump
10	A/C compressor
11	Cooling fan and water pump pulley
12	Heater hose
13	Solenoid and bracket
14	Three way catalytic converter (pre)
15	Accelerator cable
16	Engine mount
17	Engine

Exploded view and tightening specifications for engine installation—3.0L engines

3. Raise and safely support the vehicle. Drain the engine oil and coolant.

4. Remove the splash shield.

5. Remove the starter, and disconnect the transmission from the engine.

6. Disconnect the exhaust pipes from the exhaust manifolds and lower the vehicle.

7. Remove the fresh air duct and the radiator hoses.

8. Disconnect the transmission oil cooler lines from the radiator, if equipped.

9. Remove the radiator, fan shroud and cooling fan.

10. Remove the accessory drive belts.

11. Tag and disconnect the volume air flow sensor connector. Remove the air cleaner and the volume air flow sensor.

12. Disconnect the brake vacuum hose, heater hoses and fuel lines.

13. Disconnect the vacuum hose to the vacuum actuator and the canister hose.

14. Disconnect the accelerator cable.

15. Remove the alternator.

16. If equipped, remove the power steering pump and position aside, leaving the hoses attached. It is necessary to remove the power steering pulley prior to removing the pump.

17. If equipped, remove the air conditioning compressor and position aside, leaving the hoses attached.

18. Remove the protector cover from the front of the engine.

19. Tag and disconnect the emission harness connectors and the ground wire.

20. Remove the shroud upper panel and the air conditioning pipe bracket.

21. Remove the lower grille and radiator grille. Remove the shroud upper plate.

22. Attach suitable engine lifting equipment to the engine. Remove the engine mount nuts and remove the engine from the vehicle.

23. Install the engine on a workstand.

To install:

24. Remove the engine from the workstand. Lower the engine into the vehicle, being careful not to damage the piping.

➡**Lean the air conditioning condenser forward to ease engine installation.**

25. Install the engine mount nuts and tighten to 25–36 ft. lbs. (34–49 Nm).

26. Apply a bead of sealer to each side of the front support, then install the shroud upper plate. Tighten the mounting bolts to 61–87 inch lbs. (6.9–9.8 Nm).

27. Install the radiator grille and lower grille.

28. Install the air conditioning compressor, if equipped. Tighten the mounting bolts to 13–20 ft. lbs. (18–26 Nm). Install the air conditioner pipe bracket and tighten the mounting nuts to 61–87 inch lbs. (6.9–9.8 Nm).

29. Install the power steering pump, if equipped. Tighten the mounting bolts to 23–34 ft. lbs. (31–46 Nm). Install the pump pulley and tighten the nut to 29–43 ft. lbs. (39–59 Nm).

30. Install the shroud upper panel and tighten the bolts to 69–95 inch lbs. (7.8–11.0 Nm).

31. Install the alternator.

32. Connect the accelerator cable. Install the air cleaner and volume airflow sensor.

33. Connect all electrical connectors and vacuum hoses.

34. Connect the brake vacuum hose, heater hoses and fuel lines.

35. Connect the vacuum hose to the vacuum actuator and the canister hose.

36. Install the accessory drive belts and the cooling fan. Install the fan shroud and the radiator. Adjust the drive belt tension.

37. Install the radiator hoses and fresh air duct.

38. If equipped, connect the transmission oil cooler lines.

39. Raise and safely support the vehicle. Connect the exhaust pipes to the exhaust manifolds. Tighten the nuts to 25–36 ft. lbs. (34–49 Nm).

40. Install the starter, and connect the transmission assembly.

41. Install the splash shield and lower the vehicle.

42. Install the battery and connect the negative battery cables. Fill the crankcase with the proper type and quantity of engine oil.

43. Fill and bleed the cooling system. Run the engine and check for leaks and proper operation.

Water Pump

REMOVAL & INSTALLATION

1. Position the engine at TDC on the compression stroke.

2. Disconnect the negative battery cable.

3. Remove the air cleaner assembly.

4. Drain the cooling system.

5. Remove the spark plug wires.

6. Remove the fresh air duct assembly.

7. Remove the cooling fan and radiator cowling.

8. Remove the drive belts.

9. Remove the air conditioning compressor idler pulley. Remove the compressor, and position it to the side without disconnecting the hose lines.

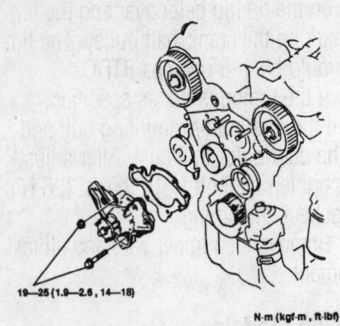

19–25 (1.9—2.6 , 14—18)

N·m (kgf·m , ft·lbf)

7924TG05

Exploded view of the water pump assembly

10. Remove the crankshaft pulley and baffle plate.

11. Remove the coolant bypass hose.

12. Remove the upper radiator hose.

13. Remove the timing belt cover retaining bolts. Remove the timing belt covers and gasket.

14. Turn the crankshaft to align the mating marks of the pulleys.

15. Remove the upper idler pulley.

16. Remove the timing belt. If reusing the belt be sure to mark the direction of rotation.

17. Remove the timing belt auto tensioner.

18. Unbolt and remove the water pump. Discard the gasket.

19. Thoroughly clean the mating surfaces of the pump and engine.

To install:

20. Position the pump and a new gasket, coated with sealer, on the engine. Tighten the bolts to 14–18 ft. lbs. (19–25 Nm).

21. The automatic tensioner must be pre-loaded. To load the tensioner, place a flat washer on the bottom of the tensioner body to prevent damage to the body and position the unit on an arbor press. Press the rod into the tensioner body. Do not use more than 2000 lbs. (8900N) of pressure. Once the rod is fully inserted into the body, insert a suitable L-shaped pin or a small Allen wrench through the body and the rod to hold the rod in place. Remove the unit from the press and install onto the block and tighten the mounting bolt to 14–19 ft. lbs. (19–26 Nm) Leave the pin in place, it will be removed later.

22. Be sure all the timing marks are aligned properly. With the upper idler pulley removed, hang the timing belt on each pulley in the proper order. Install the upper idler pulley and tighten the mounting bolt to 27–38 ft. lbs. (37–51 Nm). Rotate the

crankshaft twice in the normal direction of rotation to align all the timing marks.

23. Be sure all the marks are aligned correctly. If not, repeat the previous step.

24. Remove the pin from the auto tensioner. Again turn the crankshaft twice in the normal direction of rotation and be sure all the timing marks are aligned properly.

25. Check the timing belt deflection by applying 22 lbs. of force (98N). If the deflection is not 0.20–0.28 inch (5–7mm), repeat the adjustment procedure.

➡**Excessive belt deflection is caused by auto tensioner failure or an excessively stretched timing belt.**

26. Install the timing belt covers and new gasket.

27. Install the timing belt cover retaining bolts.

28. Install the upper radiator hose.

29. Install the coolant bypass hose.

30. Install the crankshaft pulley and baffle plate.

31. Install the compressor.

32. Install the air conditioning compressor idler pulley.

33. Install and adjust the drive belts.

34. Install the cooling fan and radiator cowling.

35. Install the fresh air duct assembly.

36. Install the spark plug wires.

37. Fill the cooling system.

38. Install the air cleaner assembly.

39. Connect the negative battery cable.

Cylinder Head

REMOVAL & INSTALLATION

✻✻ CAUTION

Fuel injection systems remain under pressure after the engine has been turned OFF. Properly relieve fuel pressure before disconnecting any fuel lines. Failure to do so may result in fire or personal injury.

1. Position the engine at TDC on the compression stroke.

2. Properly relieve the fuel system pressure.

✻✻ CAUTION

Do not allow fuel spray or fuel vapors to come in contact with a spark or open flame. Keep a dry chemical fire extinguisher nearby. Never store fuel in an open container due to risk of fire or explosion.

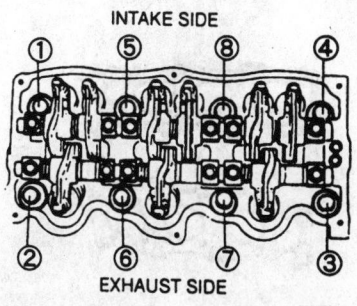

Cylinder head removal sequence

3. Disconnect the negative battery cable.

4. Remove the air cleaner assembly.

5. Disconnect the accelerator cable.

6. Drain the cooling system.

7. Remove the spark plug wires.

8. Remove the fresh air duct assembly.

9. Remove the cooling fan and radiator cowling.

10. Remove the drive belts.

11. Remove the air conditioning compressor idler pulley. If necessary, remove the compressor and position it to the side.

12. Remove the crankshaft pulley and baffle plate.

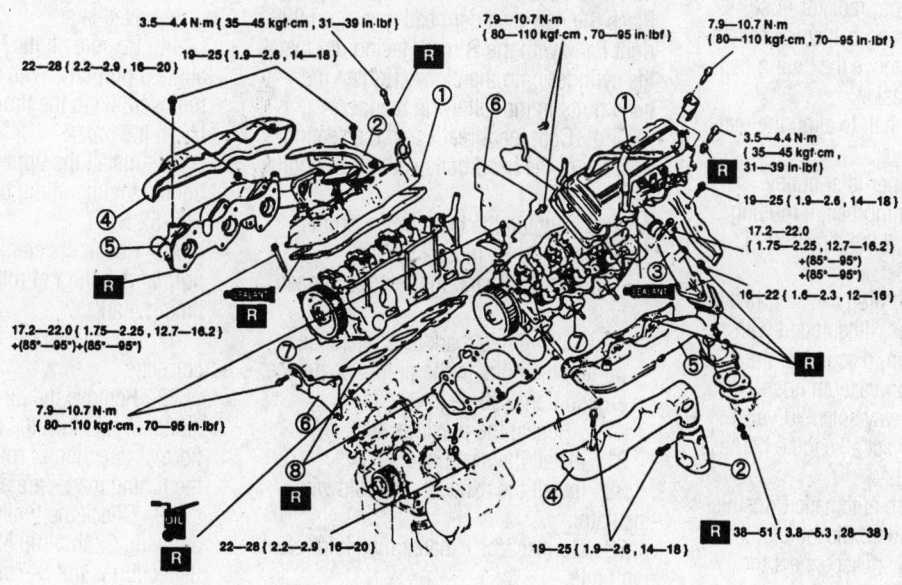

1 Cylinder head cover
  ☞ Installation Note
2 Center exhaust pipe insulator
3 Center exhaust pipe
4 Exhaust manifold insulator
5 Exhaust manifold

6 Seal plate
7 Cylinder head
8 Cylinder head gasket

N·m { kgf·m , ft·lbf }

R = replace

Exploded view of the cylinder head components

7924TG06

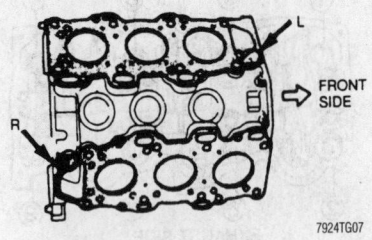

7924TG07

Head gasket positioning—3.0L engine

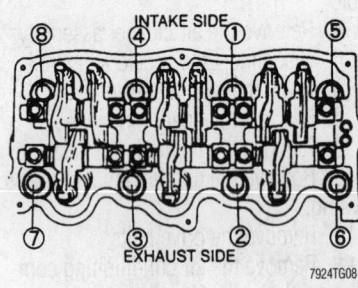

7924TG08

Cylinder head bolt tightening sequence— 3.0L engine

13. Remove the coolant bypass hose.

14. Remove the upper radiator hose.

15. Remove the timing belt cover assembly retaining bolts. Remove the timing belt cover assembly and gasket.

16. Turn the crankshaft to align the mating marks of the pulleys.

17. Remove the upper idler pulley.

18. Remove the timing belt. If reusing the belt be sure to mark the direction of rotation.

19. Disconnect and plug canister, brake vacuum and fuel hoses. If equipped with automatic transmission, disconnect the automatic transmission vacuum hose.

20. Remove the 3-way solenoid valve assembly and disconnect all engine harness connector and grounds.

21. If equipped with automatic transmission, remove the dipstick. Disconnect the required vacuum hoses. Disconnect the accelerator linkage.

22. Remove the distributor and the EGR pipe.

23. Remove the 6 extension manifolds. Remove the O-rings from the extension manifolds and replace with new ones. Remove the intake manifold by loosening the retaining bolts in the proper sequence.

24. Remove the cylinder head cover, gasket and seal washers.

25. Remove the center exhaust pipe insulator and pipe. Disconnect the exhaust manifold retaining bolts. Remove the exhaust manifold with insulator.

26. Remove the seal plate.

27. Remove the cylinder head retaining bolts in the proper sequence in 2 or 3 stages. Remove the cylinder head from the vehicle.

28. Thoroughly clean the cylinder head and cylinder block contact surfaces to remove any dirt or oil. Check the cylinder head for warpage and cracks. The maximum allowable warpage is 0.10mm. Inspect the cylinder head bolts for damaged threads and be sure they are free from grease and dirt. After the bolts are cleaned, measure the length of each bolt and replace out of specifications bolts as required.

    a. Length: Intake—4.25 in. (108mm); Exhaust—5.43 in. (138mm)

    b. Maximum: Intake—4.29 in. (109mm); Exhaust—5.47 in. (139mm)

29. Check the oil control plug projection at the cylinder block. Projection should be 0.02–0.03 in. (0.53–0.57mm). If correct, apply clean engine oil to a new O-ring and position it on the control plug.

To install:

30. Place the new cylinder head gasket on the left bank with the **L** mark facing up. Place the new cylinder head gasket on the right bank with the **R** mark facing up. Install the cylinder onto the block. Tighten the head bolts in the following manner:

    a. Coat the threads and the seating faces of the head bolts with clean engine oil.

    b. Tighten the bolts in the proper sequence to 14 ft. lbs.(19 Nm).

    c. Paint a mark on the head of each bolt.

    d. Using this mark as a reference, tighten the bolts in the proper sequence an additional 90 degrees.

    e. Repeat the previous step.

31. Install the seal plate.

32. Install the exhaust manifold with insulator.

33. Connect the exhaust manifold retaining bolts.

34. Install the center exhaust pipe insulator and pipe.

35. Install the cylinder head cover, gasket and seal washers.

36. Install the intake manifold by loosening the retaining bolts in the proper sequence.

37. Install the O-rings from the extension manifolds.

38. Install the 6 extension manifolds.

39. Install the distributor and the EGR pipe.

40. If equipped with automatic transmission, install the dipstick. Connect the required vacuum hoses. Connect the accelerator linkage.

41. Install the 3-way solenoid valve assembly and connect all engine harness connector and grounds.

42. Connect the canister, brake vacuum and fuel hoses. If equipped with automatic transmission, connect the automatic transmission vacuum hose.

43. To install the timing belt, first the automatic tensioner must be loaded. To load the tensioner:

    a. Place a flat washer on the bottom of the tensioner body to prevent damage to the body and position the unit on an arbor press.

    b. Press the rod into the tensioner body. Do not use more than 2000 lbs. (8900N) of pressure.

    c. Once the rod is fully inserted into the body, insert a suitable L-shaped pin or a small Allen wrench through the body and the rod to hold the rod in place.

    d. Remove the unit from the press and install onto the block and tighten the mounting bolt to 14–19 ft. lbs. (19–26 Nm).

    e. Leave the pin in place, it will be removed later.

44. Be sure all the timing marks are aligned properly. With the upper idler pulley removed, hang the timing belt on each pulley in the order.

45. Install the upper idler pulley and tighten the mounting bolt to 27–38 ft. lbs. (37–52 Nm).

46. Rotate the crankshaft twice in the normal direction of rotation to align all the timing marks.

47. Be sure all the marks are aligned correctly.

48. Remove the pin from the auto tensioner. Again turn the crankshaft twice in the normal direction of rotation and be sure all the timing marks are aligned properly.

49. Check the timing belt deflection by applying 22 lbs. (98 N) of force. If the deflection is not 5–7mm, repeat the adjustment procedure.

➡**Excessive belt deflection is caused by auto tensioner failure or an excessively stretched timing belt.**

50. Install the upper idler pulley.

51. Install the timing belt cover assembly and new gasket.

52. Install the upper radiator hose.

53. Install the coolant bypass hose.

54. Install the crankshaft pulley and baffle plate.

55. Install the A/C compressor.

56. Install the air conditioning compressor idler pulley.

57. Install the accessory drive belts.

58. Install the cooling fan and radiator cowling.

59. Install the fresh air duct assembly.

60. Install the spark plug wires.

61. Connect and adjust the accelerator cable.

62. Fill the cooling system.

63. Install the air cleaner assembly.

64. Connect the negative battery cable.

Rocker Arms/Shafts

REMOVAL & INSTALLATION

1. Disconnect the negative battery cable.

2. If removing the driver's side rocker arm/shaft assembly, proceed as follows:

   a. Remove the air inlet tube.

   b. Tag and disconnect the necessary electrical connectors and vacuum hoses from the throttle body and intake air pipe.

   c. Disconnect the throttle cable.

   d. Remove the throttle body and intake air pipe.

3. Remove the rocker arm cover.

4. Loosen the rocker arm and shaft assembly mounting bolts in sequence, in 2–3 steps. Remove the assembly with the bolts.

5. If necessary, disassemble the rocker arm/shaft assembly, noting the position of each component to ease reassembly.

6. Remove the valve lifter (Hydraulic Lash Adjuster—HLA) and inspect it. Replace the HLA as necessary.

7. Check for wear or damage to the contact surfaces of the shafts and rocker arms; replace as necessary.

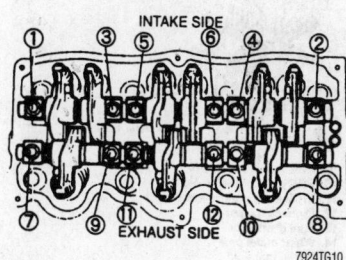

Rocker arm/shaft retaining bolt removal sequence

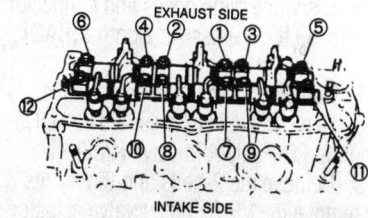

Rocker arm/shaft retaining bolt tightening sequence

8. Measure the rocker arm inner diameter, it should be 0.7480–0.7493 in. (19.000–19.033mm). Measure the rocker arm shaft diameter, it should be 0.7464–0.7472 in. (18.959–18.980mm).

9. Subtract the shaft diameter from the rocker arm diameter to get the oil clearance. The oil clearance should be 0.0008–0.0029 in. (0.020–0.074mm) and should not exceed 0.004 in. (0.10mm). Replace parts, as necessary, if the oil clearance is not within specification.

To install:

10. To install the HLA, pour engine oil into the rocker arm oil reservoir. Apply engine oil to the HLA, and carefully install the HLA into the rocker arm.

11. Apply clean engine oil to the rocker arm shafts and rocker arms and assemble the rocker arm/shaft assembly. The intake side shaft has twice as many oil holes as the exhaust side shaft.

12. Apply clean engine oil to the camshaft journals and valve stem tips.

13. Install the rocker arm/shaft assembly and tighten the mounting bolts, in sequence, in 2–3 steps to a maximum torque of 14–19 ft. lbs. (19–26 Nm).

➡ **Be careful that the rocker arm shaft spring does not get caught between the shaft and mounting boss during installation.**

14. Coat a new gasket with silicone sealant and install on the rocker arm cover. Install the rocker arm cover with new seal washers and tighten the bolts to 30–39 inch lbs. (3.4–4.4 Nm).

15. Install the intake air pipe, throttle body and air intake tube, if removed. Connect the throttle cable and the necessary electrical connectors and vacuum hoses.

16. Connect the negative battery cable, start the engine and check for leaks and proper operation.

Intake Manifold

REMOVAL & INSTALLATION

1995 Models

✳✳ CAUTION

Fuel injection systems remain under pressure after the engine has been turned OFF. Properly relieve fuel pressure before disconnecting any fuel lines. Failure to do so may result in fire or personal injury. Do not allow fuel spray or fuel vapors to come in contact with a spark or open flame. Keep a dry chemical fire extinguisher nearby. Never store fuel in an open container due to risk of fire or explosion.

1. Relieve the fuel system pressure, and disconnect the negative battery cable. Drain the cooling system.

2. Remove the air intake tube from the throttle body. Disconnect the accelerator cable.

3. Disconnect the throttle sensor connector and the coolant hoses. Remove the throttle body.

4. Tag and disconnect the vacuum hoses. Remove the bypass air control valve and the intake air pipe.

5. Remove the extension manifolds. Remove the dynamic chamber (upper intake plenum) with the shutter valve actuator.

6. Remove the fuel supply manifold and the injectors. Disconnect the coolant hoses.

7. Loosen the lower intake manifold nuts, in sequence, in 2 steps, then remove the lower intake manifold.

To install:

8. Clean all gasket mating surfaces.

9. Position new lower intake manifold gaskets, and install the lower intake manifold.

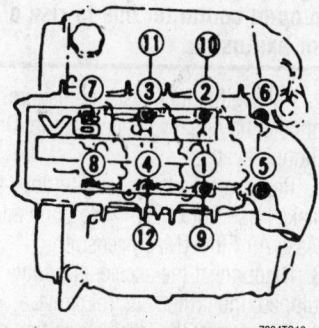

Intake manifold torque sequence tightening—1995 3.0L engine

10. Install the intake manifold washers with the white paint mark upward. Install the nuts and tighten, in sequence, in 2 steps to a maximum torque of 14–19 ft. lbs. (19–26 Nm).

11. Install the injectors and the fuel supply manifold. Tighten the attaching bolts to 14–19 ft. lbs. (19–26 Nm).

12. Connect the coolant hoses.

13. Install a new O-ring on the lower intake manifold and install the upper intake plenum. Apply clean engine oil to new O-rings and install on the extension manifolds. Position new gaskets and install the extension manifolds. Tighten the attaching nuts to 14–19 ft. lbs. (19–26 Nm).

14. Position a new gasket and install the intake air pipe. Install the bypass air control valve. Tighten the attaching bolts/nuts to 14–19 ft. lbs. (19–26 Nm).

15. Position a new gasket and install the throttle body. Tighten the attaching nuts to 14–19 ft. lbs. (19–26 Nm).

16. Connect the coolant and vacuum hoses. Connect the throttle sensor connector and accelerator cable.

17. Adjust the accelerator cable deflection to 0.039–0.118 inch (1–3mm).

18. Connect the air intake tube and the negative battery cable.

19. Fill and bleed the cooling system. Run the engine and check for leaks and proper operation.

1996–99 Models

✳✳ CAUTION

Fuel injection systems remain under pressure after the engine has been turned OFF. Properly relieve fuel pressure before disconnecting any fuel lines. Failure to do so may result in fire or personal injury. Do not allow fuel spray or fuel vapors to come in contact with a spark or open flame. Keep a dry chemical fire extinguisher nearby. Never store fuel in an open container due to risk of fire or explosion.

1. Relieve the fuel system pressure, and disconnect the negative battery cable. Drain the cooling system.

2. Remove the clamps and remove the air intake hose from the throttle body and the Mass Air Flow (MAF) sensor.

3. Disconnect the accelerator cable and if equipped, the cruise control cable.

4. Disconnect the throttle position sensor connector.

5. Remove the 2 bolts and the 2 nuts

and remove the throttle body unit and gasket.

6. Disconnect the hoses and connector and remove the Bypass Air Control (BAC) valve.

7. Remove the VRIS solenoid valve.

8. Remove the PRC solenoid valve No. 1 and the PRC solenoid valve No. 2.

9. Remove the 2 bolts and the 2 nuts and remove the VRIS shutter valve actuator and the gasket.

10. Remove the mounting bolts and the dynamic chamber.

11. Disconnect the fuel delivery pipe from the fuel distributor. Disconnect the fuel injector connectors, remove the mounting nuts and remove the fuel distributor with the injectors.

12. Remove the water hoses from the water outlet and remove the water outlet.

13. Disconnect any water hoses and vacuum hoses from the intake manifold. Remove the mounting nuts, washers, and remove the intake manifold and gaskets.

To install:

➡**Face the bead of the intake manifold gasket toward the intake manifold.**

14. Install the intake manifold with 2 new gaskets, the mounting washers and nuts.

➡**Be sure to install the manifold washers with the white paint marks facing up. Tighten the mounting nuts to 14–18 ft. lbs. (19–25 Nm).**

15. Reinstall any water hoses and vacuum hoses to the intake manifold.

16. Install the water outlet to the intake manifold with a new gasket. Tighten the mounting nuts to 14–18 ft. lbs. (19–25 Nm). Reinstall the water hoses with the clamps.

17. Reinstall the fuel delivery pipe with the injectors and tighten the mounting nuts to 14–18 ft. lbs. (19–25 Nm). Reconnect the fuel delivery pipe and reconnect the fuel injector connectors.

➡**Face the bead of the dynamic chamber gasket toward the dynamic chamber.**

18. Install the dynamic chamber with a new gasket. Tighten the mounting bolts to 70–95.4 inch lbs. (7.9–10.7 Nm).

19. Install the VRIS shutter valve actuator with a new gasket and the 2 nuts and 2 bolts. Tighten the bolts and nuts to 70–95.4 inch lbs. (7.9–10.7 Nm).

20. Install the PRC solenoid valve No. 2 and the PRC solenoid valve No. 1.

21. Reinstall the VRIS solenoid and connect the connectors.

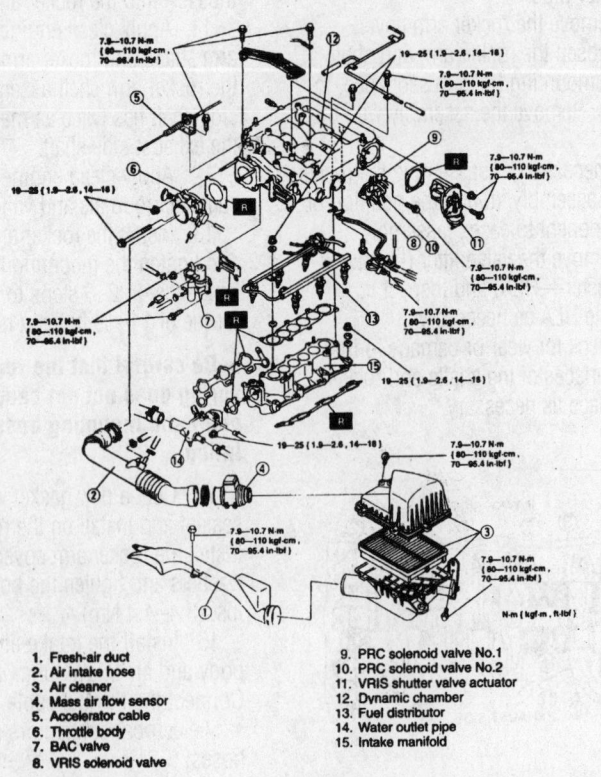

1. Fresh-air duct
2. Air intake hose
3. Air cleaner
4. Mass air flow sensor
5. Accelerator cable
6. Throttle body
7. BAC valve
8. VRIS solenoid valve
9. PRC solenoid valve No.1
10. PRC solenoid valve No.2
11. VRIS shutter valve actuator
12. Dynamic chamber
13. Fuel distributor
14. Water outlet pipe
15. Intake manifold

Exploded view of the intake air system—3.0L engine 1996–99

Intake manifold tightening sequence and mounting washers—3.0L engine 1996–99

7924TG14

22. Install the BAC valve with a new gasket. Tighten the 2 mounting bolts and the 2 nuts to 70–95.4 inch lbs. (7.9–10.7 Nm). Reconnect the hoses and connector.

23. Using a new gasket, install the throttle body assembly with the 2 bolts and the 2 nuts. Tighten the bolts and nuts to 14–16 ft. lbs. (19–22 Nm).

24. Reconnect the throttle position sensor connector.

25. Reconnect the accelerator cable and if equipped, the cruise control cable.

26. Install the air intake hose to the throttle body assembly and the MAF sensor with the 2 clamps.

27. Refill the cooling system, reconnect the negative battery cable, start the engine and operate until normal temperature, check for coolant leaks, and proper operation.

Exhaust Manifold

REMOVAL & INSTALLATION

1. Disconnect the negative battery cable. Wait at least 90 seconds before performing any work.

2. Raise and safely support the vehicle.

3. If necessary remove the engine splash shield.

4. Disconnect the oxygen sensor connector.

5. Remove the nuts attaching the 3 way pre-catalytic converter to the main catalytic converter and the LH exhaust manifold. Remove the 3 way pre-catalytic converter and the gaskets.

6. Remove the bolts and the LH and the RH exhaust manifold insulators.

7. Remove the attaching bolts and nuts and the center exhaust pipe insulators.

8. Remove the bolts securing the center exhaust manifold to the LH and RH exhaust manifolds. Remove the center exhaust manifold and the gaskets.

9. Remove the mounting nuts, the LH and the RH exhaust manifolds, and the gaskets.

To install:

10. Before installation be sure all mating surfaces are clean of any gasket material.

11. Install the RH and the LH exhaust manifolds with new gaskets and tighten the mounting nuts to 16–20 ft. lbs. (22–36 Nm).

12. Install the center exhaust manifold with 2 new gaskets to the RH and LH exhaust manifolds. Tighten the bolts to 14–18 ft. lbs. (19–24 Nm).

13. Reinstall the RH and the LH exhaust manifold insulators with the attaching bolts.

Tighten the bolts to 14–18 ft. lbs.

14. Install the exhaust insulator to the center exhaust pipe with the bolts and nuts. Tighten the bolts and nuts to 14–18 ft. lbs. (19–24 Nm).

15. Install the 3 way pre-catalytic converter using new gaskets to the LH exhaust manifold and the main catalytic converter. Tighten the nuts to 28–38 ft. lbs. (38–52 Nm).

16. Reconnect the oxygen sensor connector.

17. If necessary, reinstall the engine splash shield.

18. Safely lower the vehicle and reconnect the negative battery cable.

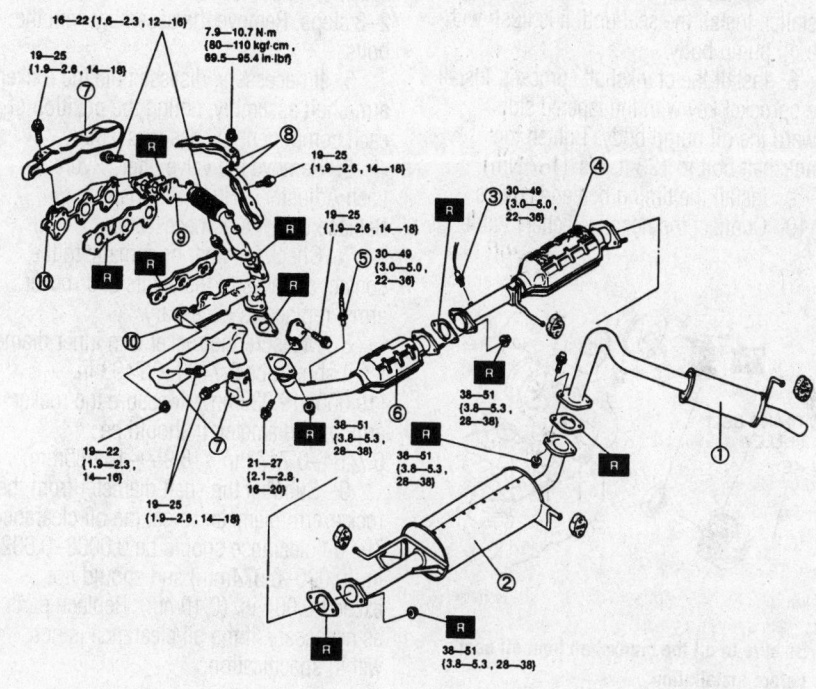

N·m (kgf·m, ft·lbf)

1	After-silencer	6	Three way catalytic converter (pre)
2	Main silencer	7	Exhaust manifold insulator
3	Heated oxygen sensor (rear)	8	Insulator
4	Three way catalytic converter (Main)	9	Center exhaust pipe
5	Heated oxygen sensor (front)	10	Exhaust manifold

Exploded view of the exhaust system

7924TG15

Front Crankshaft Seal

REMOVAL & INSTALLATION

1. Disconnect the negative battery cable.

2. Remove the timing belt covers and the timing belt.

3. If not removed during the timing belt removal procedure, remove the crankshaft sprocket bolt.

4. Remove the crankshaft sprocket, using a suitable puller.

5. Cut the seal lip with a razor knife. Protect the crankshaft with a shop towel and pry the seal from the engine.

To install:

6. Lubricate the seal lip with clean engine oil and push the seal slightly in by hand.

7. Tap the seal in evenly using a seal installer. Install the seal until it is flush with the oil pump body.

8. Install the crankshaft sprocket. Install the sprocket key with the tapered side toward the oil pump body. Tighten the crankshaft bolt to 123 ft. lbs. (167 Nm).

9. Install the timing belt and covers.

10. Connect the negative battery cable.

Be sure to oil the crankshaft front oil seal before installation

Valve Lifters

BLEEDING

➡The manufacturer does not recommend that the Hydraulic Lash Adjusters (HLA) be bled. Removing an HLA from its rocker arm will release its oil. If the HLA's are removed from the rocker arms, new ones should be installed using new O-rings.

1. Before installing new HLA's, fill the rocker arm oil reservoir with fresh engine oil.

2. Apply fresh engine oil to the new HLA and its O-ring.

3. Install the new HLA into the rocker arm, taking care not to damage or distort its O-ring.

REMOVAL & INSTALLATION

1. Disconnect the negative battery cable.

2. If removing the driver's side rocker arm/shaft assembly, proceed as follows:

   a. Remove the air inlet tube.

   b. Tag and disconnect the necessary electrical connectors and vacuum hoses from the throttle body and intake air pipe.

   c. Disconnect the throttle cable.

   d. Remove the throttle body and intake air pipe.

3. Remove the rocker arm cover.

4. Loosen the rocker arm and shaft assembly mounting bolts in sequence, in 2–3 steps. Remove the assembly with the bolts.

5. If necessary, disassemble the rocker arm/shaft assembly, noting the position of each component to ease reassembly.

6. Remove the valve lifter (Hydraulic Lash Adjuster—HLA) and inspect it. Replace the HLA as necessary.

7. Check for wear or damage to the contact surfaces of the shafts and rocker arms; replace as necessary.

8. Measure the rocker arm inner diameter, it should be 0.7480–0.7493 in. (19.000–19.033mm). Measure the rocker arm shaft diameter, it should be 0.7464–0.7472 in. (18.959–18.980mm).

9. Subtract the shaft diameter from the rocker arm diameter to get the oil clearance. The oil clearance should be 0.0008–0.0029 in. (0.020–0.074mm) and should not exceed 0.004 in. (0.10mm). Replace parts, as necessary, if the oil clearance is not within specification.

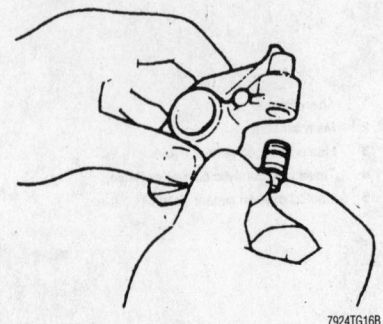

Valve lifter (Hydraulic Lash Adjuster—HLA)—3.0L engine

To install:

10. To install the HLA, pour engine oil into the rocker arm oil reservoir. Apply engine oil to the HLA, and carefully install the HLA into the rocker arm.

11. Apply clean engine oil to the rocker arm shafts and rocker arms and assemble the rocker arm/shaft assembly. The intake side shaft has twice as many oil holes as the exhaust side shaft.

12. Apply clean engine oil to the camshaft journals and valve stem tips.

13. Install the rocker arm/shaft assembly following the recommended procedure.

➡Be careful that the rocker arm shaft spring does not get caught between the shaft and mounting boss during installation.

14. Coat a new gasket with silicone sealant and install on the rocker arm cover. Install the rocker arm cover with new seal washers and tighten the bolts to 30–39 inch lbs. (3.4–4.4 Nm).

15. Install the intake air pipe, throttle body and air intake tube, if removed. Connect the throttle cable and the necessary electrical connectors and vacuum hoses.

16. Connect the negative battery cable, start the engine and check for leaks and proper operation.

Camshaft

REMOVAL & INSTALLATION

✲✲ CAUTION

Fuel injection systems remain under pressure after the engine has been turned OFF. Properly relieve fuel pressure before disconnecting any fuel lines. Failure to do so may result in fire or personal injury. Do not allow fuel spray or fuel vapors to come in contact with a spark or open flame. Keep a dry chemical fire extinguisher nearby. Never store fuel in an open container due to risk of fire or explosion.

1. Disconnect the negative battery cable.

2. Drain the cooling system.

3. Relieve the fuel system pressure.

4. Remove the PCV valve and blind cover.

5. Remove the cylinder head.

6. If removing the driver's side camshaft, remove the distributor and the distributor spacer.

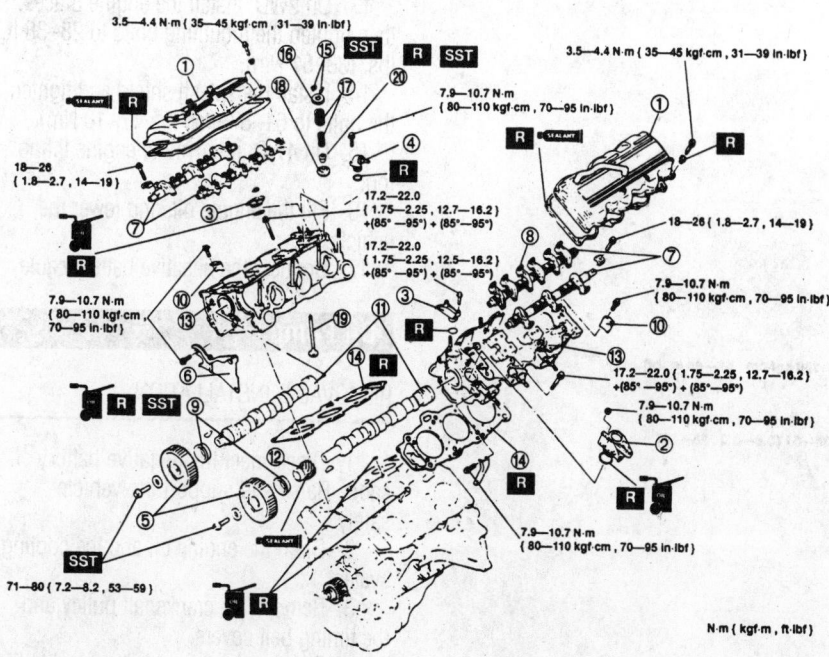

Exploded view of cylinder head components

1	Cylinder head cover
2	Distributor spacer
3	Blind cover
4	PCV valve
5	Camshaft pulley
6	Seal plate
7	Rocker arm and rocker arm shaft
8	HLA
9	Camshaft oil seal
10	Thrust plate
11	Camshaft
12	Distributor drive gear
13	Cylinder head
14	Cylinder head gasket
15	Valve keeper
16	Upper valve spring seat
17	Outer and inner valve spring
18	Lower valve spring seat
19	Valve
20	Valve seal

7924TG17

7. Remove the camshaft sprocket bolt and sprocket.

8. Remove the seal plate. Pry out the camshaft seal, being careful not to damage the seal housing.

9. Remove the rocker arm/shaft assembly.

10. Remove the thrust plate bolts and remove the thrust plate. Slide the camshaft out of the cylinder head. If removing the driver's side camshaft, remove the distributor drive gear.

11. Inspect the camshaft for wear and/or damage and replace if necessary.

To install:

➡**If installing the driver's side camshaft, remove all old sealer from the distributor drive gear and apply sealer to the gear face, then seat the gear fully on the camshaft.**

12. Apply clean engine oil to the camshaft journals, lobes and bearings. Install the camshaft and the thrust plate.

Tighten the thrust plate to 69–95 inch lbs. (7.8–11.0 Nm).

13. Apply clean engine oil to a new camshaft seal lip and press the seal into the cylinder head, using a seal installer.

14. Install the rocker arm/shaft assembly and tighten the bolts, in sequence, in 2–3 steps to 14–19 ft. lbs. (19–26 Nm). Be sure the rocker arm shaft spring does not get caught between the shaft and mounting boss during installation.

15. Install the seal plates and tighten the bolts to 69–95 inch lbs. (7.8–11.0 Nm).

16. Align and install the camshaft sprocket. Tighten the bolt to 52–59 ft. lbs. (71–80 Nm).

17. If installing the driver's side camshaft, apply clean engine oil to a new O-ring and install on the distributor spacer. Install the spacer and tighten the nuts to 69–95 inch lbs. (7.8–11.0 Nm). Install the distributor.

18. Install the cylinder head.

19. Install the blind cover and PCV valve.

20. Connect the negative battery cable.

21. Fill and bleed the cooling system.

22. Run the engine and check for leaks and proper operation. Check the idle speed and ignition timing.

Valve Lash

ADJUSTMENT

The engines covered by this manual are equipped with Hydraulic Lash Adjusters (HLA's). Valve clearance adjustments are not necessary or possible.

Oil Pan

REMOVAL & INSTALLATION

1. Disconnect the negative battery cable.

2. Raise and safely support the vehicle.

3. On 4WD, attach a suitable engine lifting tool to hold the engine.

4. Drain the oil.

5. Remove the splash shield.

6. On 2WD, remove the 2 engine braces (gusset plates).

7. On 4WD, perform the following procedures:

a. Remove the fresh-air duct and the cooling fan cowling.

b. Remove the driver's side engine mount.

c. Remove the transmission lower mount.

d. Remove the oil cooler hose and pipe.

e. Remove the stabilizer brackets.

8. Unbolt and remove the oil pan.

To install:

9. Clean the oil pan and engine block gasket surfaces thoroughly.

10. Install a new pan gasket.

11. Install the oil pan and tighten the bolts to 61–87 inch lbs. (7–10 Nm).

12. On 4WD, perform the following procedures:

a. Install the stabilizer brackets, then tighten the bolts to 14–18 ft. lbs. (19–25 Nm).

b. Install the oil cooler hose and pipe.

c. Install the transmission lower mount. Tighten the mounting bolt to 32–44 ft. lbs. (44–60 Nm), and the nut to 24–33 ft. lbs. (32–46 Nm).

d. Install the engine mount. Tighten the mounting bolts to 26–36 ft. lbs. (35–49 Nm).

e. Install the cooling fan cowling and the fresh-air duct.

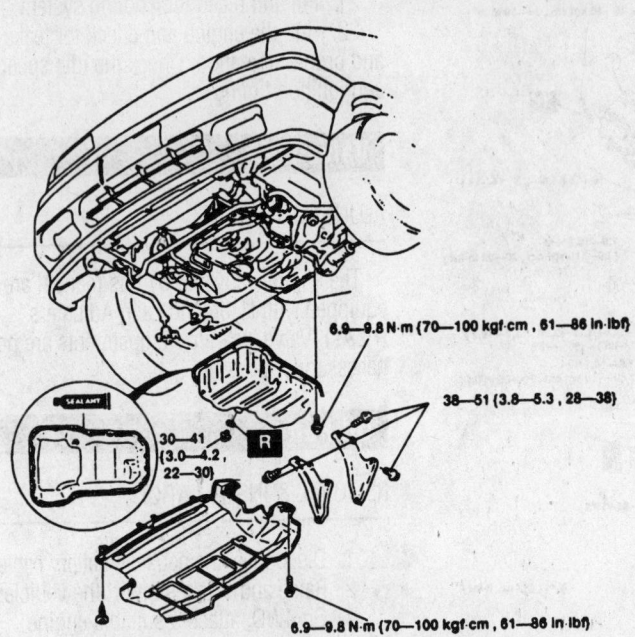

6.9—9.8 N·m (70—100 kgf·cm , 61—86 in·lbf)

38—51 (3.8—5.3 , 28—38)

30—41 {3.0—4.2 , 22—30}

R

6.9—9.8 N·m (70—100 kgf·cm , 61—86 in·lbf)

N·m {kgf·m , ft·lbf}

7924TG18

Exploded view of oil pan mounting—2WD vehicles

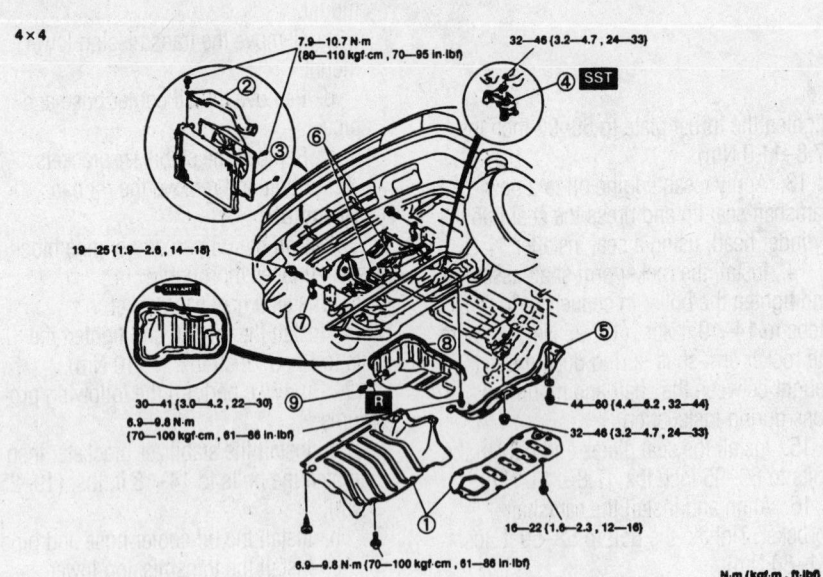

4 × 4

7.9—10.7 N·m
(80—110 kgf·cm , 70—95 in·lbf)

32—46 {3.2—4.7 , 24—33}

④ SST

19—25 {1.9—2.6 , 14—18}

30—41 {3.0—4.2 , 22—30}
6.9—9.8 N·m
(70—100 kgf·cm , 61—86 in·lbf)

R

44—60 {4.4—6.2 , 32—44}

32—46 {3.2—4.7 , 24—33}

16—22 {1.6—2.3 , 12—16}

6.9—9.8 N·m (70—100 kgf·cm , 61—86 in·lbf)

N·m {kgf·m , ft·lbf}

7924TG19

1 Splash shield
2 Fresh-air duct
3 Fan cowling
4 Engine mount
5 Transmission lower mount
6 Oil cooler hose and pipe
7 Stabilizer bracket
8 Oil pan
9 Drain plug

Exploded view of oil pan mounting—4WD vehicles

13. On 2WD, install the engine braces, then tighten the mounting bolts to 28–38 ft. lbs. (38–51 Nm).

14. Install the splash shield and tighten the bolts to 61–87 inch lbs. (7–10 Nm).

15. On 4WD, remove the engine lifting tool.

16. Fill the engine oil, and lower the vehicle.

17. Connect the negative battery cable.

Oil Pump

REMOVAL & INSTALLATION

1. Disconnect the negative battery cable. Raise and support the vehicle safely.

2. Drain the engine oil and the cooling system.

3. Remove the crankshaft pulley and the timing belt covers.

4. Remove the timing belt, crankshaft sprocket and key.

5. Remove the thermostat and gasket.

6. Remove the oil pan, oil strainer and O-ring.

7. Unbolt and remove the oil pump and gasket.

To install:

8. Press in a new oil seal and coat the seal lip with clean engine oil. Use a new gasket, O-ring and sealant as required. Tighten the oil pump retaining bolts to 14–18 ft. lbs. (19–25 Nm).

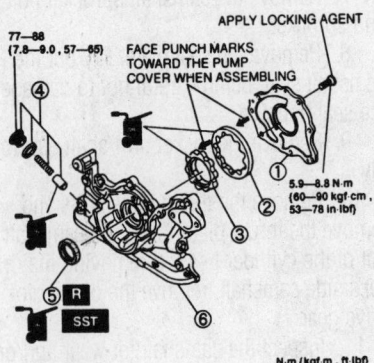

77—88
(7.8—9.0 , 57—65)

APPLY LOCKING AGENT

FACE PUNCH MARKS
TOWARD THE PUMP
COVER WHEN ASSEMBLING

5.9—8.8 N·m
(60—90 kgf·cm ,
53—78 in·lbf)

R
SST

N·m {kgf·m , ft·lbf}

7924TG20

1 Pump cover
2 Outer rotor
3 Inner rotor
4 Pressure relief valve
5 Oil seal
6 Oil pump body

Exploded view of the oil pump and related components—3.0L engine

9. Install the oil pan and tighten the pan bolts 5–8 ft. lbs. (8–11 Nm).

10. Install the crankshaft sprocket and key.

11. Install the timing belt and covers. Install the crankshaft pulley and tighten the pulley bolt to 116–123 ft. lbs. (157–167 Nm).

12. Install the thermostat and gasket. Tighten the thermostat housing bolts 14–18 ft. lbs. (19–25 Nm).

13. Fill the crankcase to the recommended level with fresh oil. Fill the cooling system.

14. Crank the engine to prime the oil pump.

15. Start the engine and check for leaks.

Rear Main Oil Seal

REMOVAL & INSTALLATION

1. Disconnect the negative battery cable.

2. Raise and support the vehicle safely.

3. Remove the transmission.

4. If equipped with a manual transmission, remove the pressure plate, the clutch disc and the flywheel. If equipped with an automatic transmission, remove the drive plate from the crankshaft.

5. Using a razor blade, carefully cut the oil seal lip.

6. Using a small prying tool protected with a rag, remove the oil seal.

To install:

7. Apply clean engine oil to the new seal.

8. Push the new oil seal slightly into place by hand.

9. Using a seal driver and a hammer, tap the oil seal in evenly.

10. Install the driveplate.

11. Install the transmission.

12. Lower the vehicle.

13. Connect the negative battery cable.

FUEL SYSTEM

Fuel System Service Precautions

Safety is the most important factor when performing not only fuel system maintenance but any type of maintenance. Failure to conduct maintenance and repairs in a safe manner may result in serious personal injury or death. Maintenance and testing of

the vehicle's fuel system components can be accomplished safely and effectively by adhering to the following rules and guidelines.

• To avoid the possibility of fire and personal injury, always disconnect the negative battery cable unless the repair or test procedure requires that battery voltage be applied.

• Always relieve the fuel system pressure prior to disconnecting any fuel system component (injector, fuel rail, pressure regulator, etc.), fitting or fuel line connection. Exercise extreme caution whenever relieving fuel system pressure to avoid exposing skin, face and eyes to fuel spray. Please be advised that fuel under pressure may penetrate the skin or any part of the body that it contacts.

• Always place a shop towel or cloth around the fitting or connection prior to loosening to absorb any excess fuel due to spillage. Ensure that all fuel spillage (should it occur) is quickly removed from engine surfaces. Ensure that all fuel soaked cloths or towels are deposited into a suitable waste container.

• Always keep a dry chemical (Class B) fire extinguisher near the work area.

• Do not allow fuel spray or fuel vapors to come into contact with a spark or open flame.

• Always use a back-up wrench when loosening and tightening fuel line connection fittings. This will prevent unnecessary stress and torsion to fuel line piping. Always follow the proper tighten specifications.

• Always replace worn fuel fitting O-rings with new. Do not substitute fuel hose or equivalent, where fuel pipe is installed.

Fuel System Pressure

RELIEVING

✳✳ CAUTION

The fuel injection system remains under pressure after the engine has been turned OFF. Properly relieve fuel pressure before disconnecting any fuel lines. Failure to do so may result in fire or personal injury. Do not allow fuel spray or fuel vapors to come in contact with a spark or open flame. Keep a dry chemical fire extinguisher nearby. Never store fuel in an open container due to risk of fire or explosion.

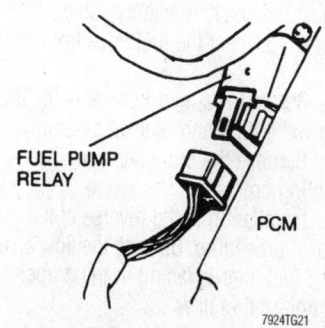

FUEL PUMP
RELAY

PCM

7924TG21

Fuel pump connector

1. Start the engine.

2. Disconnect the fuel pump relay connector, located at the ECM.

3. After the engine stalls, turn the ignition switch **OFF**.

4. Connect the fuel pump relay connector.

Fuel Filter

REMOVAL & INSTALLATION

✳✳ CAUTION

Fuel injection systems remain under pressure after the engine has been turned OFF. Properly relieve fuel pressure before disconnecting any fuel lines. Failure to do so may result in fire or personal injury. Do not allow fuel spray or fuel vapors to come in contact with a spark or open flame. Keep a dry chemical fire extinguisher nearby. Never store fuel in an open container due to risk of fire or explosion.

➡The fuel filter is located in the engine compartment, next to the pulsation damper.

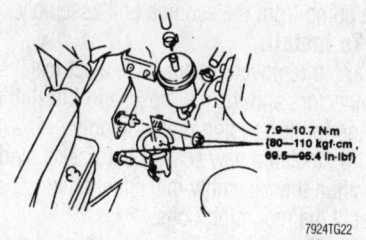

7.9–10.7 N·m
(80–110 kgf·cm,
69.5–95.4 in·lbf)

7924TG22

Fuel filter mounting

1. Relieve the fuel system pressure. Disconnect the negative battery cable.

2. Disconnect the fuel lines from the filter.

3. Remove the filter bracket bolts, then remove the filter and bracket assembly.

4. Remove the fuel filter from the mounting bracket.

5. Installation is the reverse of the removal procedure. Be sure the flow arrow on the fuel filter is facing in the proper direction of fuel flow.

Fuel Pump

REMOVAL & INSTALLATION

✳✳ CAUTION

Fuel injection systems remain under pressure after the engine has been turned OFF. Properly relieve fuel pressure before disconnecting any fuel lines. Failure to do so may result in fire or personal injury. Do not allow fuel spray or fuel vapors to come in contact with a spark or open flame. Keep a dry chemical fire extinguisher nearby. Never store fuel in an open container due to risk of fire or explosion.

1. Relieve the fuel system pressure, and disconnect the negative battery cable.

2. Remove the rear seat, and lift up the rear floormat. Remove the fuel pump cover.

3. Disconnect the sending unit/fuel pump assembly electrical connector and the fuel lines.

4. Remove any dirt that has accumulated around the sending unit/fuel pump assembly so it will not enter the fuel tank during removal and installation.

5. Remove the attaching screws and remove the sending unit/fuel pump assembly.

6. If necessary, disconnect the electrical connectors and the fuel hose, and remove the pump from the sending unit assembly.

To install:

7. If removed, connect the electrical connectors and the fuel hose, and install the pump to the sending unit assembly.

8. Install a new seal rubber gasket, and position the assembly in the fuel tank. Install the mounting bolts.

9. Connect the sending unit/fuel pump assembly electrical connector and the fuel lines.

10. Install the fuel pump cover, and replace the floormat.

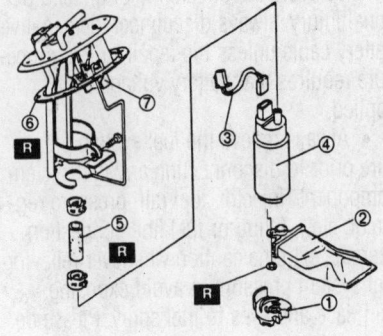

1	Rubber mount
2	Fuel filter (low-pressure side)
3	Fuel pump connector
4	Fuel pump
	☞ Assembly Note
5	Fuel hose
	☞ Assembly Note
6	Seal rubber
7	Fuel tank gauge sender unit

7924TG23

Exploded view of the fuel pump assembly

11. Install the rear seat.

12. Connect the negative battery cable.

DRIVE TRAIN

Automatic Transmission Assembly

REMOVAL & INSTALLATION

2WD Models

1. Disconnect the negative battery cable.

2. Raise and support the vehicle.

3. Drain the transmission fluid but do not remove the pan.

4. Disconnect the speedometer cable.

5. Label for identification and location, then disconnect all electrical wiring at the transmission.

6. Remove the exhaust pipe and heat shield.

7. Matchmark and disconnect the driveshaft.

8. Remove the selector cable from the

transmission shift lever and the cable bracket and remove the shift selector cable.

9. Remove the filler tube.

10. Remove the access cover from the lower front of the converter housing.

11. Matchmark the drive plate (flywheel) and torque converter for reassembly. Remove the 4 bolts holding the torque converter to the drive plate.

12. Remove the starter.

13. Remove the exhaust pipe bracket.

14. Remove the bolts connecting the crossmember to the transmission.

15. Disconnect and plug the cooler lines from the radiator at the transmission.

16. Remove the gusset plates.

17. Support the transmission with a jack. Remove the crossmember-to-frame bolts, and remove the crossmember.

18. Be sure the transmission is securely supported. Secure it to the jack with a safety chain, if necessary.

19. Remove the converter housing-to-engine bolts.

20. With a prybar, exert light pressure between the converter and the drive plate to prevent the converter from disengaging from the transmission as it is removed.

21. Lower the transmission and converter as an assembly. Be careful not to let the converter fall out.

To install:

22. Place the transmission on the jack. Be sure the converter is properly installed.

23. Raise the transmission into place. Install the converter housing-to-engine bolts, and tighten in 2 stages to 28–38 ft. lbs. (38–51 Nm).

24. Install the gusset plates. Tighten the bolts to 28–38 ft. lbs. (38–51 Nm).

25. Install the exhaust pipe bracket, tighten the bolts to 14–18 ft. lbs. (19–25 Nm).

26. Connect the oil cooler pipes.

27. Install the transmission crossmember mounting bolts. Tighten to 32–44 ft. lbs. (44–60 Nm).

28. Loosely and evenly tighten the torque converter mounting bolts. Tighten to 27–39 ft. lbs. (37–53 Nm).

29. Install the access cover. Remove the jack.

30. Install the starter.

31. Install the fluid filler tube with a new O-ring.

32. Connect the electrical connectors, and replace the wires in the clip. Install the vacuum hose.

33. Reconnect the speedometer cable.

34. Reconnect the shift selector cable to the transmission gear selector lever. Secure

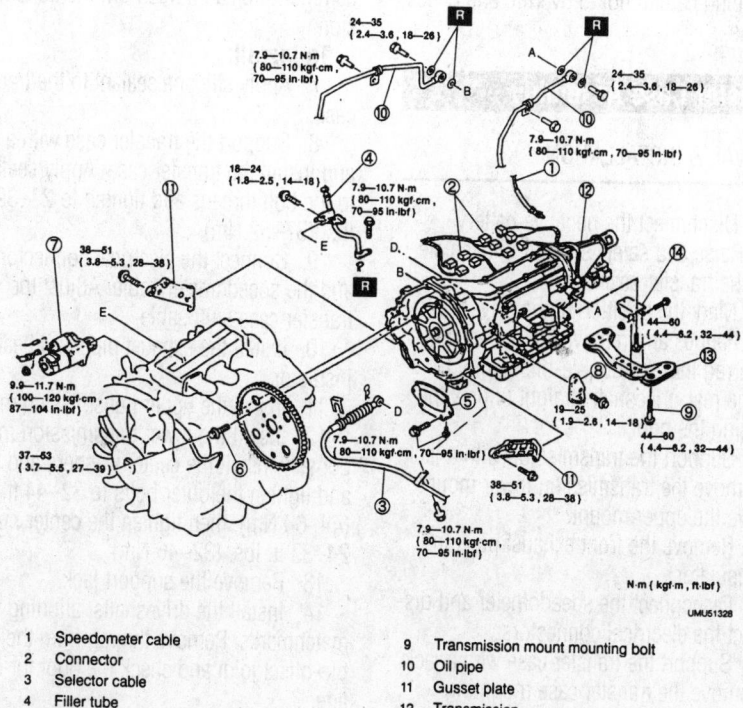

8. Remove the selector cable from the transmission shift lever and the cable bracket and remove the shift selector cable.

9. Remove the filler tube.

10. Remove the access cover from the lower front of the converter housing.

11. Matchmark the drive plate (flywheel) and torque converter for reassembly. Remove the 4 bolts holding the torque converter to the drive plate.

12. Remove the starter.

13. Remove the exhaust pipe bracket.

14. Remove the bolts connecting the crossmember to the transmission.

15. Disconnect and plug the cooler lines from the radiator at the transmission.

16. Remove the gusset plates.

17. Support the transmission with a jack. Remove the crossmember-to-frame bolts, and remove the crossmember.

18. Be sure the transmission is securely supported. Secure it to the jack with a safety chain, if necessary.

19. Support the transfer case with a jack and remove the transfer case mounting bolts. Remove the transfer case. It may be necessary to tap the case with a plastic hammer to break it free from the transmission.

1 Speedometer cable	9 Transmission mount mounting bolt
2 Connector	10 Oil pipe
3 Selector cable	11 Gusset plate
4 Filler tube	12 Transmission
5 Undercover	13 Transmission lower mount
6 Torque converter mounting bolt	14 Transmission upper mount
7 Starter	
8 Exhaust pipe bracket	

Exploded view of the transmission mounting—2WD vehicles

the cable end with the washer and the hitch pin. Reinstall the selector cable to the cable bracket with the clip.

35. Install the heat insulator and exhaust pipe.

36. Insert the driveshaft into the transmission. Install the center bearing support. Bolt the driveshaft to the rear of the axle flange.

37. Install a new pan gasket and the fluid pan, if this has not already been done.

38. Lower the vehicle. Connect the negative battery cable. Fill the transmission through the dipstick tube with the specified fluid, being careful not to overfill, and check for leaks.

4WD Models

1. Disconnect the negative battery cable.

2. Raise and support the vehicle.

3. Drain the transmission fluid but do not remove the pan.

4. Disconnect the speedometer cable.

5. Label for identification and location, then disconnect all electrical wiring at the transmission.

6. Remove the exhaust pipe and heat shield.

7. Matchmark and disconnect the front driveshaft and the rear driveshaft.

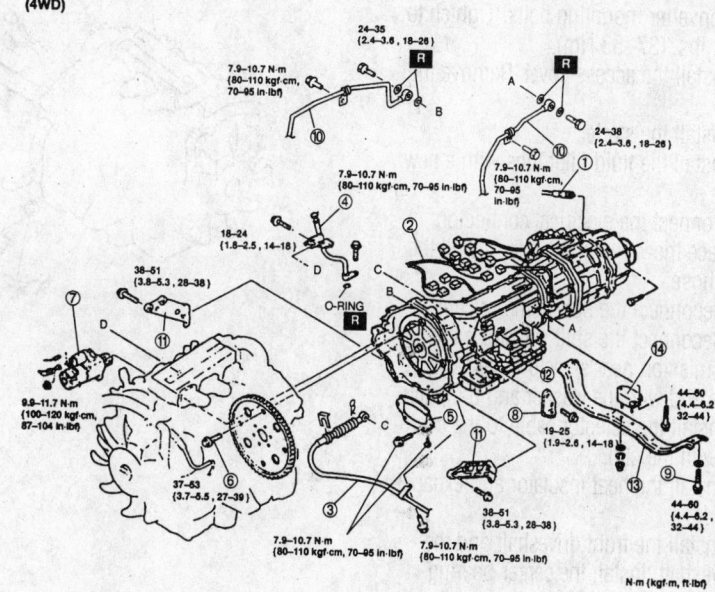

1 Speedometer cable	9 Transmission mount mounting bolt
2 Connector	10 Oil pipe
3 Selector cable	11 Gusset plate
4 Filler tube	12 Transmission
5 Undercover	13 Transmission lower mount
6 Torque converter mounting bolt	14 Transmission upper mount
7 Starter	
8 Exhaust pipe bracket	

Exploded view of the transmission mounting—4WD vehicles

20. Remove the converter housing-to-engine bolts.

21. With a prybar, exert light pressure between the converter and the drive plate to prevent the converter from disengaging from the transmission as it is removed.

22. Lower the transmission and converter as an assembly. Be careful not to let the converter fall out.

To install:

23. Place the transmission on the jack. Be sure the converter is properly installed.

24. Raise the transmission into place. Install the converter housing-to-engine bolts, and tighten in 2 stages to 28–38 ft. lbs. (38–51 Nm).

25. Apply silicone sealant to the transfer case.

26. Support the transfer case with a jack and install the transfer case. Apply sealant to the bolt threads and tighten to 27–39 ft. lbs. (37–52 Nm).

27. Install the gusset plates. Tighten the bolts to 28–38 ft. lbs. (38–51 Nm).

28. Install the exhaust pipe bracket, tighten the bolts to 14–18 ft. lbs. (19–25 Nm).

29. Connect the oil cooler pipes.

30. Install the transmission crossmember mounting bolts. Tighten to 32–44 ft. lbs. (44–60 Nm).

31. Loosely and evenly tighten the torque converter mounting bolts. Tighten to 27–39 ft. lbs. (37–53 Nm).

32. Install the access cover. Remove the jack.

33. Install the starter.

34. Install the fluid filler tube with a new O-ring.

35. Connect the electrical connectors, and replace the wires in the clip. Install the vacuum hose.

36. Reconnect the speedometer cable.

37. Reconnect the shift selector cable to the transmission gear selector lever. Secure the cable end with the washer and the hitch pin. Reinstall the selector cable to the cable bracket with the clip.

38. Install the heat insulator and exhaust pipe.

39. Install the front driveshaft and the rear driveshaft. Install the center bearing support.

40. Install a new pan gasket and the fluid pan, if this has not already been done.

41. Lower the vehicle. Connect the negative battery cable. Fill the transmission through the dipstick tube with the specified fluid, being careful not to overfill, and check for leaks.

Transfer Case Assembly

REMOVAL & INSTALLATION

1. Disconnect the negative battery cable. Raise and safely support the vehicle. Drain the transfer case.

2. Mark the position of the driveshafts on the flanges and remove the driveshafts. Push a rag into the double-offset joint to hold the rear driveshaft straight to prevent damaging the boot.

3. Support the transmission with a jack and remove the transmission lower mount. Remove the upper mount.

4. Remove the front exhaust pipe and heat insulator.

5. Disconnect the speedometer and disconnect the electrical connectors.

6. Support the transfer case with a jack and remove the transfer case mounting bolts. Remove the transfer case. It may be necessary to tap the case with a plastic hammer to break it free from the transmission.

To install:

7. Apply silicone sealant to the transfer case.

8. Support the transfer case with a jack and install the transfer case. Apply sealant to the bolt threads and tighten to 27–39 ft. lbs. (37–52 Nm).

9. Connect the electrical connectors and the speedometer cable. Adjust the transfer case shift cable.

10. Install the exhaust pipe and heat insulator.

11. Install the upper transmission mount.

12. Install the lower transmission mount. Loosely install the center washers and nuts and tighten the outer bolts to 32–44 ft. lbs. (44–60 Nm), then tighten the center nuts to 24–33 ft. lbs. (32–46 Nm).

13. Remove the support jack.

14. Install the driveshafts, aligning the matchmarks. Remove the rag from the double-offset joint and check the boot for damage.

15. Fill the transfer case with the proper type and quantity of fluid.

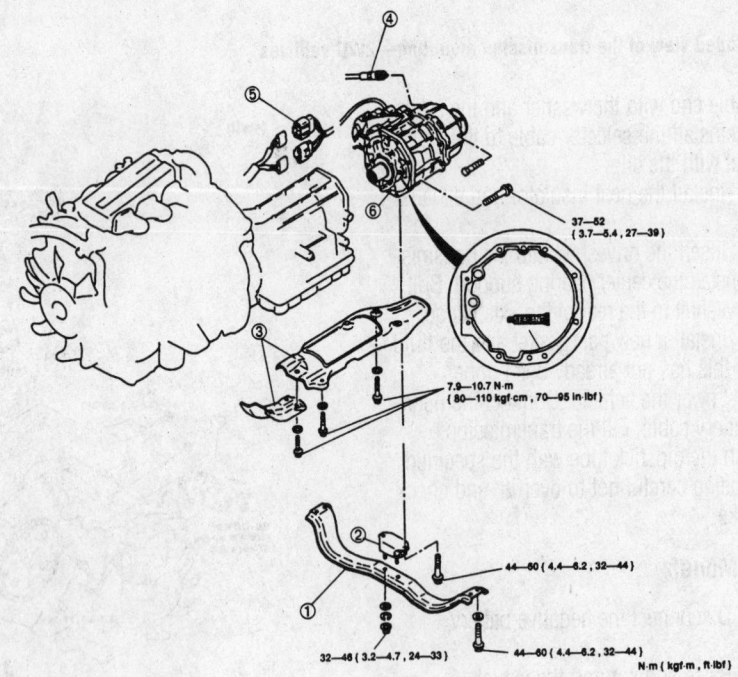

1	Transmission lower mount	4	Speedometer cable
2	Transmission upper mount	5	Connectors
3	Heat insulator	6	Transfer case

7924TG26

Exploded view of the transfer case and related components

16. Lower the vehicle and connect the negative battery cable. Check the transfer case for leaks and proper operation.

Halfshaft

REMOVAL & INSTALLATION

1. Raise and safely support the vehicle. Remove the wheel and tire assembly.
2. Drain the differential gear oil.
3. Remove and discard the halfshaft locknut.
4. Disconnect the tie rod end from the knuckle.
5. Remove the caliper and brake rotor from the knuckle. Support the caliper aside with rope or mechanics wire; do not let it hang by the brake hose.
6. Remove the nut and bolts and remove the lower ball joint. Remove the bolts and nuts and remove the knuckle/hub assembly from the strut.

➡**If the halfshaft is stuck to the hub, install a used locknut so it is flush with the end of the shaft, then tap the nut with a soft mallet.**

7. Remove the splash shield.

8. Using a prybar, pry out the halfshaft from the differential and remove the halfshaft from the vehicle. Be careful not to damage the dust cover or oil seal.

To install:

9. Install a new clip on the halfshaft. Coat the differential seal with clean transmission fluid.
10. Install the halfshaft in the differential, being careful not to damage the seal. After installation, attempt to pull the halfshaft outward to be sure it does not come out.
11. Install the knuckle/hub assembly to the strut.
12. Install the lower ball joint.
13. Install the brake rotor and caliper assembly.
14. Connect the tie rod end to the knuckle.
15. Install a new locknut and tighten to 174–231 ft. lbs. (236–313 Nm). After tightening, stake the locknut using a blunt chisel.
16. Install the splash shield, tighten the mounting bolts to 12–16 ft. lbs. (16–22 Nm).
17. Install the wheel and tire assembly and lower the vehicle.
18. Fill the differential with gear oil.
19. Check the front end alignment.

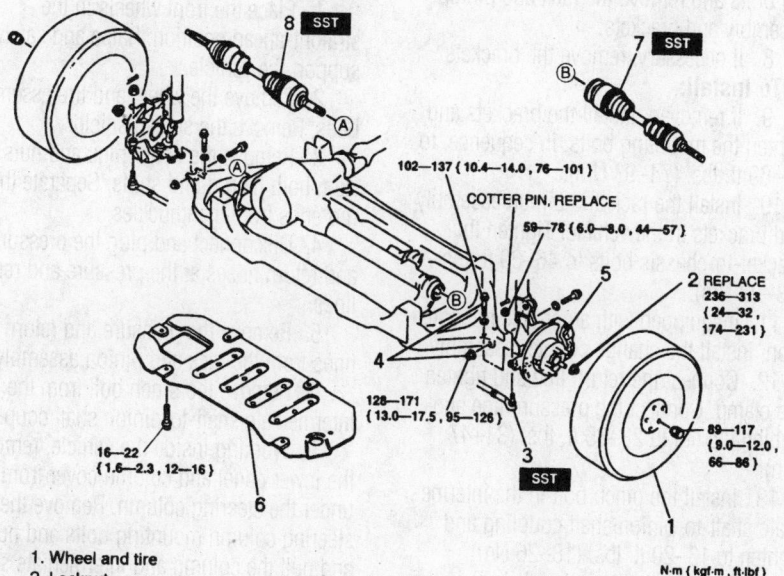

102–137 {10.4–14.0 , 76–101}

COTTER PIN, REPLACE

59–78 {6.0–8.0, 44–57}

2 REPLACE
236–313
{24–32,
174–231}

128–171
{13.0–17.5, 95–126}

16–22
{1.6–2.3, 12–16}

89–117
{9.0–12.0,
66–86}

N·m (kgf·m , ft·lbf)

1. Wheel and tire
2. Locknut
3. Tie rod end
4. Ball joint bolt and nut
5. Front axle
6. Engine undercover
7. Left drive shaft
8. Right drive shaft

7924TG27

Exploded view of the halfshaft and related components

STEERING AND SUSPENSION

Air Bag

✳✳ CAUTION

All vehicles are equipped with an air bag system, also known as the Supplemental Inflatable Restraint (SIR) or Supplemental Restraint System (SRS). The system must be disabled before performing service on or around system components, steering column, instrument panel components, wiring and sensors. Failure to follow safety and disabling procedures could result in accidental air bag deployment, possible personal injury and unnecessary system repairs.

PRECAUTIONS

Several precautions must be observed when handling the inflator module to avoid accidental deployment and possible personal injury.

• Never carry the inflator module by the wires or connector on the underside of the module.

• When carrying a live inflator module, hold securely with both hands, and ensure that the bag and trim cover are pointed away.

• Place the inflator module on a bench or other surface with the bag and trim cover facing up.

• With the inflator module on the bench, never place anything on or close to the module which may be thrown in the event of an accidental deployment.

DISARMING

Driver's Side Air Bag

1. Turn the ignition switch to **LOCK**.
2. Disconnect the negative battery cable and wait for more than 1 minute to allow the back-up power supply to deplete its stored power.
3. Remove the lower column cover and disconnect the orange and blue clock spring connectors.
4. After servicing, connect the negative battery cable. Turn the ignition switch to **ON**. Verify that the air bag system warning

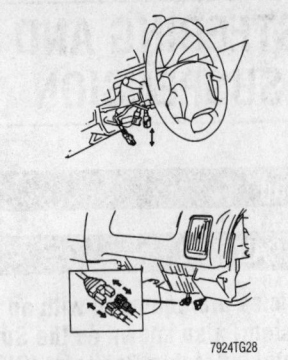

Air bag connector—Driver and passenger sides shown

7924TG28

light illuminates for 4–6 seconds, then goes OFF.

Passenger's Side Air Bag

1. Turn the ignition switch to **LOCK**.
2. Disconnect the negative battery cable and wait for more than 1 minute to allow the back-up power supply to deplete its stored power.
3. Remove the glove compartment cover.
4. Disconnect the orange and blue passenger-side air bag module connector.
5. After servicing, connect the negative battery cable. Turn the ignition switch to **ON**. Verify that the air bag system warning light illuminates for 4–6 seconds, then goes OFF.

Power Rack and Pinion Steering Gear

REMOVAL & INSTALLATION

2WD Models

1. Place the front wheels in the straight-ahead position. Raise and safely support the vehicle.
2. Remove the wheel and tire assemblies. Remove the splash shield.
3. Remove the cotter pins and nuts from both tie rod end studs. Separate the tie rod ends from the knuckles.
4. Remove the pinch bolt from the intermediate shaft-to-pinion shaft coupling.
5. Disconnect and plug the pressure line from the rack and pinion assembly. Loosen the clamp and disconnect the return line from the rack and pinion assembly. Plug the line.
6. If equipped with automatic transmission, remove the change counter assembly to remove the protector plate mounting bolt.

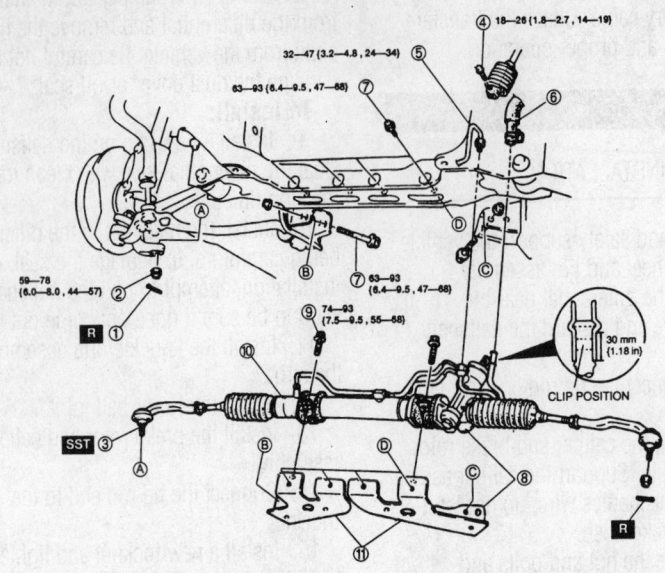

1	Cotter pin	7	Steering bracket mounting bolt
2	Nut	8	Steering gear, linkage, and steering bracket
3	Tie-rod end ball joint	9	Mounting bracket bolt
4	Fixing bolt (intermediate shaft/pinion shaft)	10	Steering gear and linkage
5	Pressure pipe	11	Steering brackets
6	Return hose		

7924TG29

Exploded view of the rack and pinion mounting—2WD vehicles

7. Remove the steering bracket mounting bolts and remove the rack and pinion assembly and brackets.
8. If necessary, remove the brackets.

To install:

9. If removed, install the brackets and tighten the mounting bolts, in sequence, to 54–69 ft. lbs. (74–93 Nm).
10. Install the rack and pinion assembly and brackets in the vehicle. Tighten the bracket-to-chassis bolts to 46–69 ft. lbs. (63–93 Nm).
11. If equipped with automatic transmission, install the change counter assembly.
12. Connect the return line and tighten the clamp. Connect the pressure line and tighten the nut to 23–35 ft. lbs. (31–47 Nm).
13. Install the pinch bolt in the intermediate shaft-to-pinion shaft coupling and tighten to 13–20 ft. lbs. (18–26 Nm).
14. Position the tie rod end studs in the knuckles and install the nuts. Tighten the nuts to 43–58 ft. lbs. (59–78 Nm) and install new cotter pins.
15. Install the splash shield and the wheel and tire assemblies. Lower the vehicle and bleed the power steering system.

4WD Models

1. Place the front wheels in the straight-ahead position. Raise and safely support the vehicle.
2. Remove the wheel and tire assemblies. Remove the splash shield.
3. Remove the cotter pins and nuts from both tie rod end studs. Separate the tie rod ends from the knuckles.
4. Disconnect and plug the pressure and return hoses at the pressure and return lines.
5. Remove the pressure and return lines from the rack and pinion assembly.
6. Remove the pinch bolt from the intermediate shaft-to-pinion shaft coupling.
7. Working inside the vehicle, remove the lower panel and column cover from under the steering column. Remove the steering column mounting bolts and nuts and pull the column and intermediate shaft rearward to separate the intermediate shaft from the pinion shaft.
8. Mark the position of the front driveshaft on the axle flange and remove the front driveshaft.
9. Remove the rack and pinion assembly

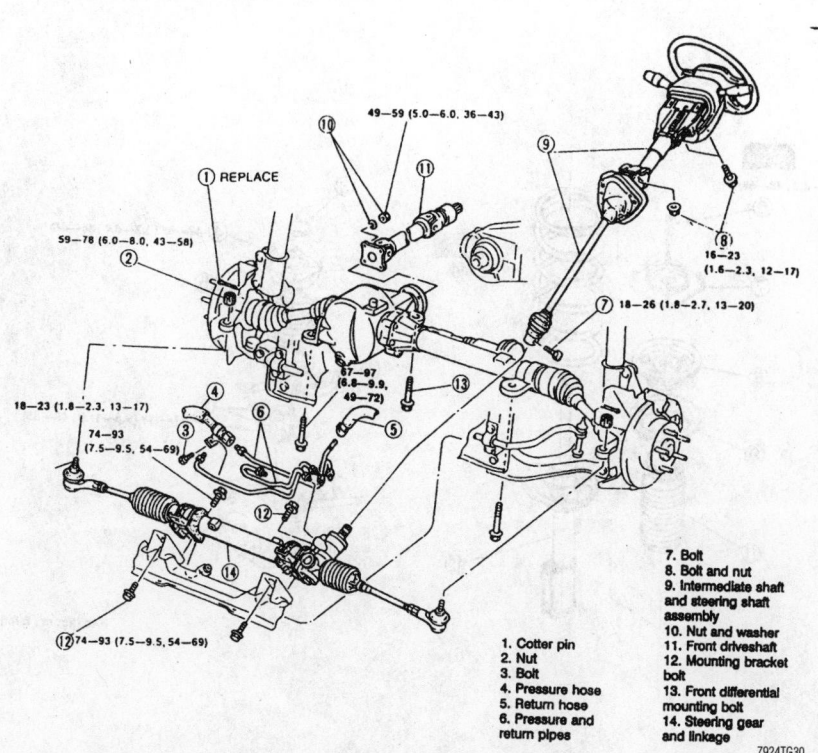

49—59 (5.0—6.0, 36—43)

① REPLACE

59—78 (6.0—8.0, 43—58)

18—23 (1.8—2.3, 13—17)

74—93 (7.5—9.5, 54—69)

⑫ 74—93 (7.5—9.5, 54—69)

87—97 (6.8—9.9, 49—72)

16—23 (1.6—2.3, 12—17)

18—26 (1.8—2.7, 13—20)

1. Cotter pin
2. Nut
3. Bolt
4. Pressure hose
5. Return hose
6. Pressure and return pipes
7. Bolt
8. Bolt and nut
9. Intermediate shaft and steering shaft assembly
10. Nut and washer
11. Front driveshaft
12. Mounting bracket bolt
13. Front differential mounting bolt
14. Steering gear and linkage

7924TG30

Rack and pinion mounting, exploded view—4WD vehicles

mounting bracket bolts and the front differential/joint shaft assembly mounting bolts.

10. Slide the differential/joint shaft assembly rearward. Slide the rack and pinion assembly rearward and turn it 90 degrees, then remove it from the left side of the vehicle.

To install:

11. Install the rack and pinion assembly from the left side of the vehicle, turn it 90 degrees and move it forward into position. Install the mounting bolts and tighten, in sequence, to 54–69 ft. lbs. (74–93 Nm).

12. Move the differential/joint shaft assembly forward, install the mounting bolts and tighten to 49–72 ft. lbs. (67–97 Nm).

13. Install the driveshaft, aligning the marks made during removal.

14. Working inside the vehicle, move the steering column and intermediate shaft forward to engage the intermediate shaft with the pinion shaft. Install and tighten the steering column nuts and bolts to 12–17 ft.

lbs. (16–23 Nm). Install the lower panel and column cover.

15. Install the pinch bolt in the intermediate shaft-to-pinion shaft coupling and tighten to 13–20 ft. lbs. (18–26 Nm).

16. Install the pressure and return lines on the rack and pinion assembly. Connect the pressure and return hoses to the lines.

17. Position the tie rod end studs in the knuckles and install the nuts. Tighten the nuts to 43–58 ft. lbs. (59–78 Nm) and install new cotter pins.

18. Install the splash shield and the wheel and tire assemblies. Lower the vehicle and bleed the power steering system.

Front Strut

REMOVAL & INSTALLATION

1. Raise and safely support the vehicle. Remove the wheel and tire assembly.

2. Support the lower control arm with a jack.

3. Remove the clip attaching the brake hose to the strut and disconnect the hose from the strut.

4. Remove the strut-to-knuckle attaching bolts and nuts.

5. Working in the engine compartment, remove the 4 attaching nuts from the strut tower and remove the strut assembly from the vehicle.

6. Remove the rubber cap from the upper mounting block. Loosen the upper attaching nut, but do not remove it.

7. Install a suitable spring compressor and compress the coil spring.

8. Remove the upper attaching nut and slowly relieve the tension on the coil spring, using the spring compressor. When the spring is no longer under tension, remove the spring compressor.

9. Remove the upper mounting block, upper spring seat, spring seat, coil spring, bump stopper and ring rubber from the strut.

To install:

10. Secure the strut in a vise equipped with protective jaw covers, so the strut will not be damaged.

11. Apply a suitable rubber grease to the ring rubber and install it on the bump stopper. Install the bump stopper on the strut.

12. Attach the spring compressor to the coil spring and compress the spring.

13. Install the compressed spring on the strut and install the spring seat.

14. Install the upper spring seat. The flat of the strut rod must fit correctly into the upper spring seat.

15. Install the upper mounting block. Install and loosely tighten the upper attaching nut.

16. Remove the spring compressor. Be sure the spring is properly seated in the upper and lower spring seats.

17. Secure the upper spring seat in a vise and tighten the upper attaching nut to 47–59 ft. lbs. (64–80 Nm). Install the rubber cap on the upper mounting block.

18. Install the strut assembly in the strut tower, making sure the white mark on the upper mounting block is in the front-inside direction. Install the attaching nuts and tighten to 34–46 ft. lbs. (47–62 Nm).

19. Install the strut to the knuckle and tighten the attaching bolts and nuts to 69–86 ft. lbs. (93–117 Nm).

20. Position the brake hose on the strut

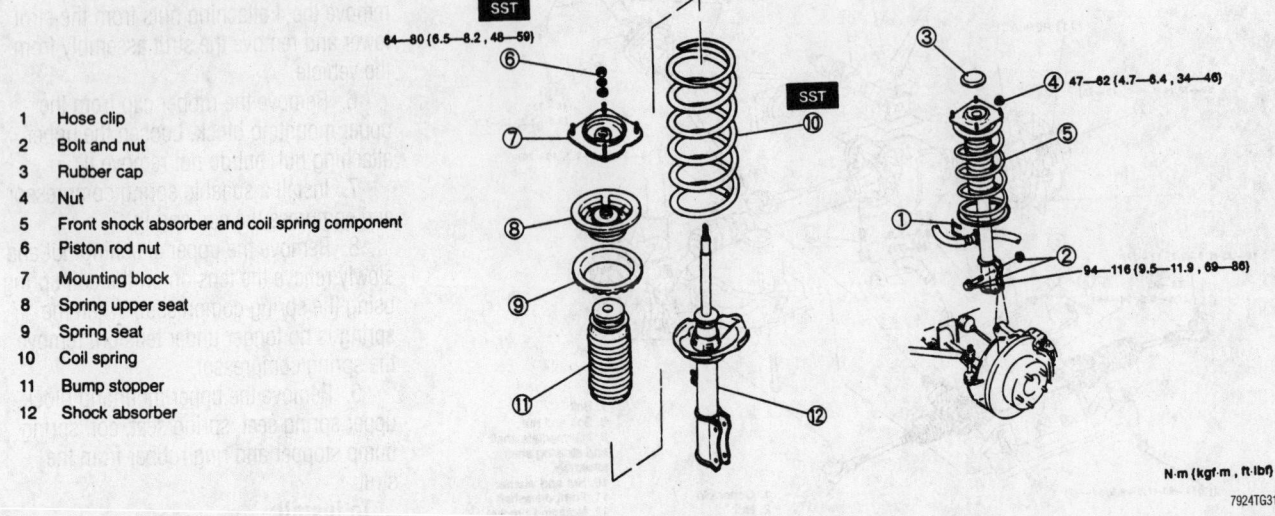

1 Hose clip
2 Bolt and nut
3 Rubber cap
4 Nut
5 Front shock absorber and coil spring component
6 Piston rod nut
7 Mounting block
8 Spring upper seat
9 Spring seat
10 Coil spring
11 Bump stopper
12 Shock absorber

Exploded view of the strut assembly

and install the clip. Remove the jack from under the lower control arm.

21. Install the wheel and tire assembly and lower the vehicle. Check the front end alignment.

Rear Shock Absorber/ Coil Spring

REMOVAL & INSTALLATION

1. Raise and safely support the vehicle. Remove the splash shield.

2. Remove the stabilizer bar.

3. If equipped, remove the nut and disconnect the height sensor from the rear axle.

4. Remove the bolt attaching the parking brake cable bracket.

5. Support the rear axle housing with a jack. Raise the jack slightly to take the load off the shock absorbers.

6. Remove the attaching bolts and nuts and disconnect the shock absorbers from the lower axle housing.

7. Slowly lower the axle housing until the spring tension is relieved. Remove the coil springs.

8. Remove the spring seats and bump stopper, if equipped.

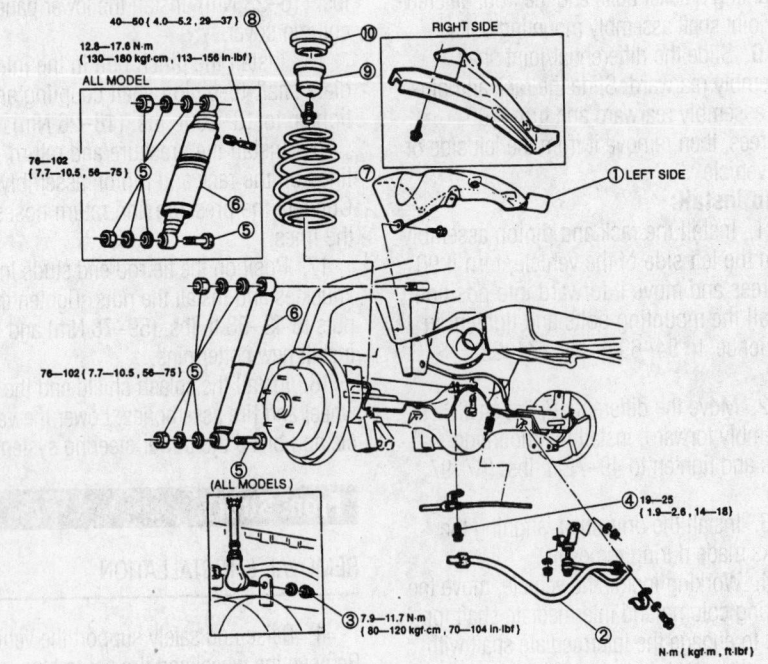

1 Splash shield
2 Stabilizer
3 Nut (ALL model)
4 Bolt
5 Bolt, nut, and washer
6 Shock absorber
7 Coil spring
8 Bolt
9 Bump stopper
10 Spring seat

Exploded view of the rear coil spring, shock absorber and related component mounting

To install:

9. Install the upper and lower spring seats and the bump stopper, if removed.

10. Install the coil springs, making sure the larger diameter coil is toward the axle housing.

11. Raise the axle housing enough to connect the shock absorbers. Install the attaching bolts and nuts and tighten to 56–75 ft. lbs. (76–102 Nm). Remove the jack.

12. Install the bolt attaching the parking brake cable bracket and the nut attaching the height sensor.

13. Install the stabilizer bar. Tighten the link bolt nut until 0.28 in. (7mm) of thread is exposed at the top of the link bolt. Do not tighten the stabilizer bar bushing bracket bolts at this time.

14. Lower the vehicle. With the vehicle unladen, tighten the stabilizer bar bushing bracket bolts to 26–37 ft. lbs. (35–50 Nm).

15. Connect the height sensor, if disconnected.

16. Install the splash shield.

Lower Ball Joint

REMOVAL & INSTALLATION

On 2WD vehicles, the lower ball joints are pressed into the lower control arm. The ball joints cannot be removed from the lower control arms and in the event of the defective ball joint, the lower control arm and ball joint must be replaced as an assembly.

1. On 4WD vehicles, raise and safely support the vehicle. Remove the wheel and tire assembly.

2. Disconnect the sway bar from the lower control arm.

3. Remove the cotter pin and nut from the lower ball joint stud. Separate the ball joint from the knuckle.

4. Remove the 2 upper bolts, one through-bolt and remove the ball joint.

To install:

5. Install the ball joint to the lower control arm.

6. Install the ball stud into the steering knuckle. Tighten the nut to 116–137 ft. lbs. (157–186 Nm) and install the cotter pin.

7. Connect the sway bar to the lower control arm.

8. Install the wheel and tire assembly and lower the vehicle.

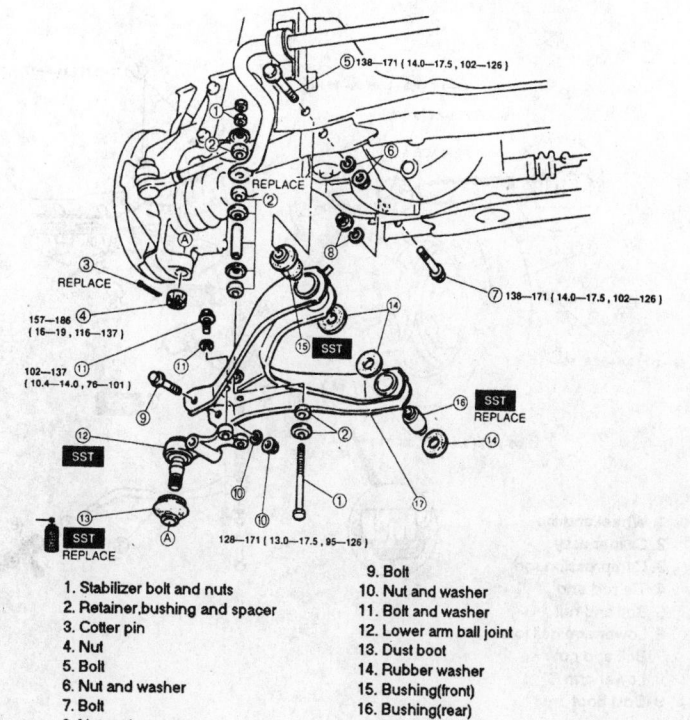

1. Stabilizer bolt and nuts	9. Bolt
2. Retainer,bushing and spacer	10. Nut and washer
3. Cotter pin	11. Bolt and washer
4. Nut	12. Lower arm ball joint
5. Bolt	13. Dust boot
6. Nut and washer	14. Rubber washer
7. Bolt	15. Bushing(front)
8. Nut and washer	16. Bushing(rear)
	17. Lower arm

N·m (kgf·m , ft·lbf)

Exploded view of the ball joint and lower control arm assembly—4WD vehicles

Lower Control Arm

REMOVAL & INSTALLATION

2WD Models

1. Raise and safely support the vehicle. Remove the wheel and tire assembly.

2. Remove the brake caliper and support it aside with mechanics wire, do not let it hang by the brake hose.

3. Remove the nuts, bolts, spacer, washers and bushings and remove the compression rod from the lower control arm and chassis and disconnect the stabilizer bar from the lower control arm.

4. Remove the cotter pin and nut and separate the tie rod end from the knuckle.

5. Remove the bolts and nuts and disconnect the strut from the knuckle.

6. Remove the cotter pin and nut from the lower ball joint stud and separate the lower ball joint from the knuckle.

7. Remove the mounting bolt and nut and remove the lower control arm from the vehicle.

To install:

8. Position the lower control arm to the chassis and install the bolt and nut, but do not tighten at this time.

9. Install the knuckle to the lower control arm. Tighten the lower ball joint stud nut to 87–115 ft. lbs. (118–156 Nm) and install a new cotter pin.

10. Connect the strut to the knuckle and tighten the attaching bolts and nuts to 69–86 ft. lbs. (94–116 Nm).

11. Connect the tie rod end to the knuckle. Tighten the tie rod end stud nut to 43–58 ft. lbs. (59–78 Nm) and install a new cotter pin.

12. Install the compression rod to the lower control arm and chassis. Tighten the compression rod-to-lower control arm mounting bolts to 76–93 ft. lbs. (103–126 Nm) and the compression rod bushing-to-chassis bolts to 61–76 ft. lbs. (83–103 Nm). Install the compression rod nut but do not tighten at this time.

➡The left-hand compression rod nut has left-hand threads.

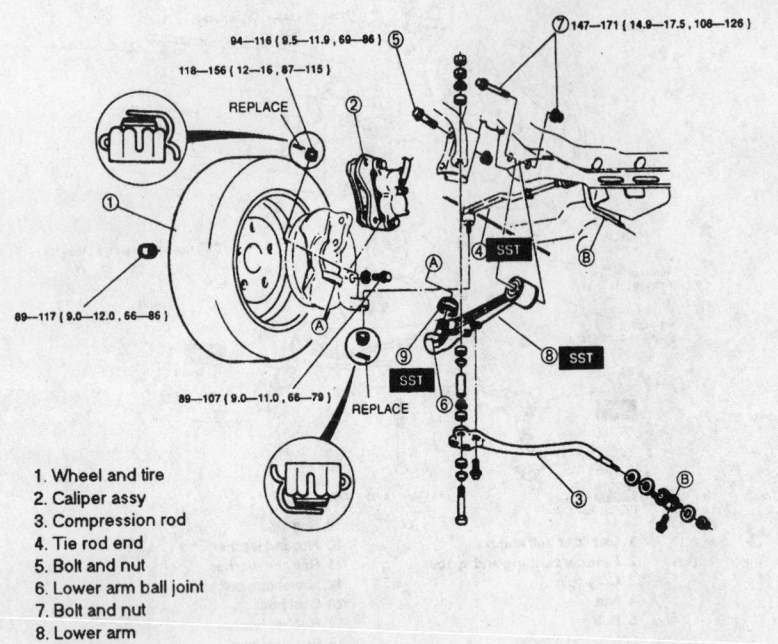

94—116 (9.5—11.9 , 69—86)

118—156 (12—16 , 87—115)

REPLACE

7 147—171 (14.9—17.5 , 108—126)

SST

A

B

89—117 (9.0—12.0 , 66—86)

A

SST

89—107 (9.0—11.0 , 66—79)

SST

REPLACE

B

N·m (kgf·m , ft·lbf)

7924TG34

1. Wheel and tire
2. Caliper assy
3. Compression rod
4. Tie rod end
5. Bolt and nut
6. Lower arm ball joint
7. Bolt and nut
8. Lower arm
9. Dust boot

Exploded view of the ball joint and lower control arm assembly—2WD vehicles

13. Connect the stabilizer bar to the control arm with the bolt, washers, bushings, spacer and nuts. Tighten the nuts so 0.24 in. (6mm) of thread is exposed at the end of the bolt.

14. Install the caliper and the wheel and tire assembly. Lower the vehicle.

15. With the vehicle unloaded, tighten the lower control arm-to-chassis bolt and nut to 94–108 ft. lbs. (146–172 Nm). Tighten the compression rod nut to 108–126 ft. lbs. (147–171 Nm).

16. Lower the vehicle.

17. Check the front end alignment.

Wheel Bearing

ADJUSTMENT

Front

1. Raise and support the vehicle safely. Remove the tire and wheel assembly.

2. Remove and properly support the caliper assembly.

3. Position a dial indicator gauge against the dust cap. Push and pull the disc brake rotor or brake drum in and out in the axial direction and measure the end-play of the wheel bearing.

4. End-play should not exceed 0.002 in. (0.05mm).

5. If end-play is excessive, check the hub nut torque or replace the bearing.

Rear

1. Raise and support the vehicle safely. Remove the tire and wheel assembly.

2. Remove the brake drum.

3. Position a dial indicator gauge against the axle shaft. Push and pull the axle shaft by hand in and out in the axial direction and measure the end-play of the wheel bearing.

4. End-play should not exceed 0.0224 in. (0.57mm).

5. If end-play is excessive, replace the bearing.

REMOVAL & INSTALLATION

Front

2WD MODELS

1. Raise and safely support the vehicle.
2. Remove the wheel assembly.
3. Remove the hub dust cap.
4. Remove the locknut.
5. Remove the brake caliper.
6. Remove the disc plate.

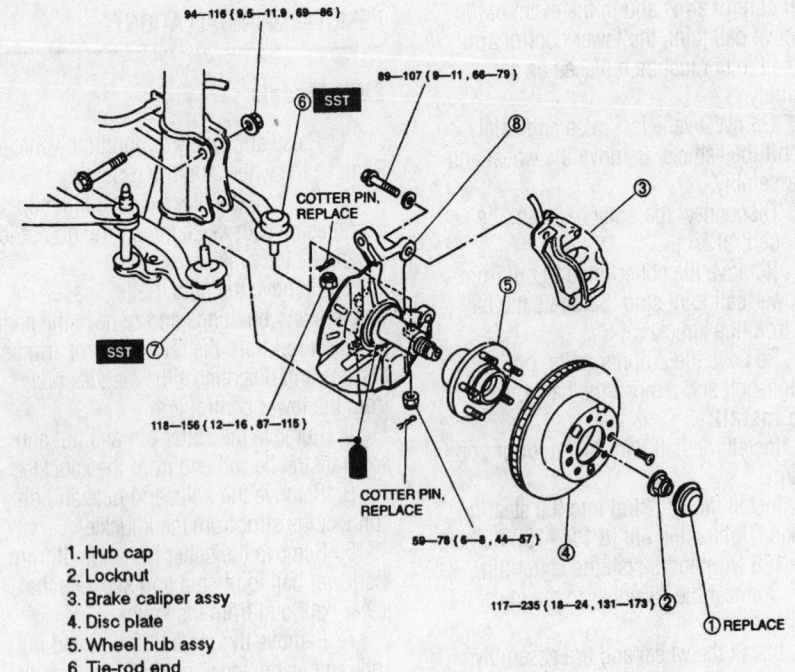

94—116 (9.5—11.9 , 69—86)

89—107 (8—11 , 66—79)

6 SST

8

COTTER PIN. REPLACE

SST

3

5

SST 7

118—156 (12—16 , 87—115)

COTTER PIN. REPLACE

59—78 (6—8 , 44—57)

4

117—235 (18—24 , 131—173)

2

1 REPLACE

1. Hub cap
2. Locknut
3. Brake caliper assy
4. Disc plate
5. Wheel hub assy
6. Tie-rod end
7. Lower arm
8. Knuckle spindle and dust cover

N·m (kgf·m , ft·lbf)

7924TG35

Exploded view of the front hub and related components—2WD vehicles

7. Remove the hub assembly.

8. Disconnect the tie-rod end from the knuckle/spindle.

9. Disconnect the lower arm.

10. Remove the knuckle/spindle assembly.

11. Remove the wheel bearings from the hub assembly, if needed.

To install:

12. Install wheel bearings to the hub assembly, if removed.

13. Install the knuckle/spindle assembly. Tighten the strut mounting nut to 69–86 ft. lbs. (94–116 Nm).

14. Install the lower arm, ball joint to the knuckle/spindle assembly. Tighten the ball joint nut to 87–115 ft. lbs. (118–156 Nm).

15. Connect the tie-rod end, tighten the nut to 44–57 ft. lbs. (59–78 Nm).

16. Install the hub assembly.

17. Install the disc plate.

18. Install the brake caliper, tighten the mounting bolts to 66–79 ft. lbs. (89–107 Nm).

19. Install the locknut, tighten to 131–173 ft. lbs. (117–235 Nm).

20. Install the hub dust cap.

21. Install the wheel assembly.

22. Lower the vehicle.

4WD MODELS

1. Raise and safely support the vehicle.

2. Remove the wheel assembly.

3. Remove the locknut.

4. Remove the brake caliper.

5. Remove the disc plate retaining screw(s).

6. Disconnect the tie-rod end from the knuckle.

7. Disconnect the lower ball joint.

8. Remove the disc plate.

9. Remove the ball joint mounting nuts and bolts.

10. Remove the knuckle, wheel hub and dust plate as an assembly.

11. Remove the wheel bearings from the hub assembly, if needed.

To install:

12. Install wheel bearings to the hub assembly, if removed.

13. Install the knuckle assembly. Tighten the strut mounting nut to 69–86 ft. lbs. (94–116 Nm).

14. Install the ball joint mounting nuts and bolts. Tighten the upper mounting bolts to 76–101 ft. lbs. (102–137 Nm). Tighten the through-bolt nut to 95–106 ft. lbs. (128–171 Nm).

15. Replace the disc plate.

16. Install ball joint to the knuckle assembly. Tighten the ball joint nut to 116–137 ft. lbs. (157–186 Nm).

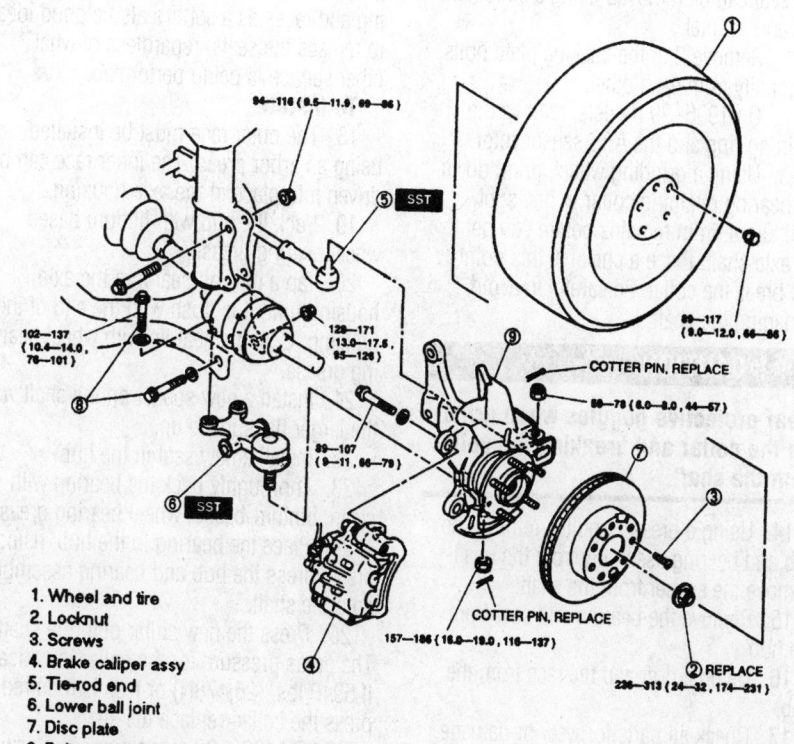

1. Wheel and tire
2. Locknut
3. Screws
4. Brake caliper assy
5. Tie-rod end
6. Lower ball joint
7. Disc plate
8. Bolts, washers and nuts
9. Knuckle, wheel hub and dust cover

N·m (kgf·m , ft-lbf)

7924TG36

Exploded view of the front hub and related components—4WD vehicles

17. Connect the tie-rod end, tighten the nut to 44–57 ft. lbs. (59–78 Nm).

18. Install the brake caliper, tighten the mounting bolts to 66–79 ft. lbs. (89–107 Nm).

19. Install the disc plate retaining nut.

20. Install the locknut, tighten to 174–231 ft. lbs. (236–313 Nm).

21. Install the wheel assembly.

22. Lower the vehicle.

Rear

1. Raise and support the rear end on jackstands.

2. Remove the wheel(s).

3. Remove the brake caliper assembly.

4. Remove the disc/drum plate.

5. Remove the parking brake shoe assembly.

6. Disconnect the parking brake cable.

7. Disconnect and cap the brake line(s).

8. Remove the dust cover and rear axle assembly by using a slide hammer.

9. Slide the axle shaft from the axle housing. Be careful to avoid damaging the oil seal with the shaft.

10. If the seal in the axle housing is damaged in any way, it must be replaced.

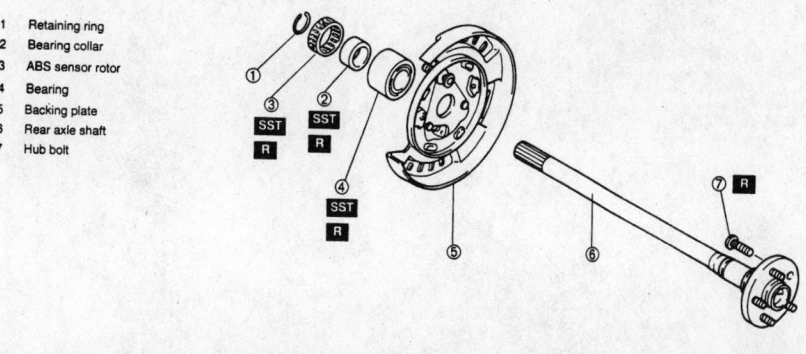

1 Retaining ring
2 Bearing collar
3 ABS sensor rotor
4 Bearing
5 Backing plate
6 Rear axle shaft
7 Hub bolt

7924TG37

Rear axle and bearing components—1996–99 shown

The seal can be removed using a slide hammer and adapter.

11. Remove 2 of the backing plate bolts, diagonally from each other.

12. On 1996–99 models, remove the retaining ring and the ABS sensor rotor.

13. Using a grinding wheel, grind down the bearing retaining collar in one spot, until about 5mm remains before you get to the axle shaft. Place a chisel at this point and break the collar. Be careful to avoid damaging the shaft.

✷ CAUTION

Wear protective goggles when grinding the collar and breaking the collar from the shaft.

14. Using a press or puller, remove the hub and bearing assembly from the shaft. Remove the spacer from the shaft.

15. Remove the bearing and seal from the hub.

16. Using a drift, tap the race from the hub.

17. Check all parts for wear or damage. If either race is to be replaced, both must be replaced. The race in the axle housing can be removed with a slide hammer and adapter. It's a good idea to replace the bearing and races as a set. It's also a good idea to replace the seals, regardless of what other service is being performed.

To install:

18. The outer race must be installed using an arbor press. The inner race can be driven into place in the axle housing.

19. Pack the hub with lithium based wheel bearing grease.

20. Tap a new oil seal into the axle housing until it is flush with the end of the housing. Coat the seal lip with wheel bearing grease.

21. Install a new spacer on the shaft with the larger flat surface up.

22. Install a new seal in the hub.

23. Thoroughly pack the bearing with clean, lithium-based, wheel bearing grease.

24. Place the bearing in the hub, using a press, press the hub and bearing assembly onto the shaft.

25. Press the new collar onto the shaft. The press pressure for the collar is critical. If 5940 lbs. (26,478N) or less is required to press the collar, replace the shaft.

26. On 1996–99 models, press a new ABS speed sensor rotor to the axle shaft and install a new retainer.

27. Install one shaft in the housing being very careful to avoid damaging the inner seal.

28. If only on shaft was being serviced, the other must now be removed to check bearing play on the serviced axle. If both shafts were removed, leave the other one out for now.

29. Tighten the backing plate bolts on the one installed axle to 80 ft. lbs. (109 Nm).

30. Mount a dial indicator on the backing plate, with the pointer resting on the axle shaft flange. Check the axial play. Standard bearing play should be 0.0224 in. (0.57mm).

31. If play is not within specifications, shims are available for correcting it.

32. Install the other shaft and tighten the backing plate bolts. Check the play as on the first shaft.

33. Install the brake line and parking brake cable.

34. Install the parking brake shoe assembly.

35. Install the brake disc/drum.

36. Install the caliper assembly.

37. Install the wheel(s).

38. Lower the vehicle.

MERCEDES-BENZ

ML320

29

ENGINE REPAIR

➡Disconnecting the negative battery cable on some vehicles may interfere with the functions of the on board computer systems and may require the computer to undergo a relearning process, once the negative battery cable is reconnected.

Ignition Timing

ADJUSTMENT

This vehicle's ignition timing is controlled by the engine control computer and is not adjustable.

Engine Assembly

REMOVAL & INSTALLATION

1. Verify that the rear engine lifting eyes are correct. The left lifting eye is marked with a star and code number 04 and the right lifting eye is marked with a star and code number 02.

2. Disconnect the negative battery cable.

3. Remove the engine undercover.

❄❄ CAUTION

Never open, service or drain the radiator or cooling system when hot; serious burns can occur from the steam and hot coolant. Also, when draining engine coolant, keep in mind that cats and dogs are attracted to ethylene glycol antifreeze and could drink any that is left in an uncovered container or in puddles on the ground. This will prove fatal in sufficient quantities. Always drain coolant into a sealable container. Coolant should be reused unless it is contaminated or is several years old.

4. Drain and recycle the engine coolant and oil.

❄❄ CAUTION

The EPA warns that prolonged contact with used engine oil may cause a number of skin disorders, including cancer! You should make every effort to minimize your exposure to used

engine oil. Protective gloves should be worn when changing the oil. Wash your hands and any other exposed skin areas as soon as possible after exposure to used engine oil. Soap and water, or waterless hand cleaner should be used.

5. Remove the engine cooling fan and clutch, then the fan shroud.

➡The fan clutch is equipped with right-hand thread.

6. Place a guard plate behind the radiator/condenser to protect it from damage during removal and installation.

7. Remove the air cleaner housing.

8. Remove the resonance pipe and body.

9. Disconnect the coolant hoses from the water pump and thermostat housing.

10. Remove the coolant expansion tank.

11. Disconnect the transmission dipstick tube from the cylinder head cover.

12. Disconnect the brake booster vacuum hose from the rear of the intake manifold.

13. Drain the power steering fluid from the pump.

14. Label and disconnect the hoses from the P/S pump, then plug the openings.

15. Disconnect the vacuum hose at the purge control valve.

16. Properly relieve the fuel system pressure.

17. Disconnect the fuel pipe.

18. Detach the heater hose form the rear of the cylinder head.

19. Disconnect the engine wiring harness.

20. Lock the automatic belt tensioner by rotating the tensioner counterclockwise until a 5mm drift or pin fits through the tensioner, then remove the serpentine belt.

21. Unbolt the A/C compressor and position it aside, leaving the hoses attached.

22. Disconnect the exhaust system from the manifolds.

23. Remove the cable for the park lock interlock.

24. Matchmark the torque converter-to-ring gear, then remove the torque converter bolts.

25. Disconnect the left and right oxygen sensors.

26. Unbolt the starter.

27. Remove the engine-to-transmission mounting bolts, be sure to leave the two upper bolts attached.

28. Unbolt the motor mounts from the front suspension support.

29. Disconnect the generator wiring harness.

30. Detach the A/C compressor electrical connector.

31. Disconnect the engine ground cable at the power steering pump.

32. Attach the engine hoist to the lifting eyes, then raise the engine and support the transmission.

➡ **Be sure that the engine does not touch the body at the rear.**

33. Remove the two remaining engine-to-transmission mounting bolts.

34. Slowly, pull the engine out to the front and lift out.

To install:

35. Slowly lower the engine into the vehicle.

36. Mate the engine to the transmission and install the upper two mounting bolts.

37. Attach the engine ground cable to the power steering pump.

38. Install the motor mount bolts and tighten to 26 ft. lbs. (35 Nm).

39. Remove the engine hoist from the lifting eyes.

40. Install the remaining engine-to-transmission mounting bolts.

41. Attach the A/C compressor electrical connector.

42. Connect the generator wiring harness.

43. Install the starter.

44. Connect the left and right oxygen sensor.

45. Align the torque converter-to-ring gear matchmarks and tighten the bolts to 31 ft. lbs. (42 Nm).

46. Connect the cable for the park lock interlock.

47. Connect the exhaust system to the manifolds and tighten the mounting nuts to 15 ft. lbs. (20 Nm).

48. Install the A/C compressor.

49. Install the serpentine belt and remove the locking pin.

50. Connect the engine wiring harness.

51. Connect the coolant hoses to the water pump, cylinder head and thermostat housing.

52. Connect the fuel pipe.

53. Connect the vacuum hose to the purge control valve.

54. Connect the P/S hoses to the pump and fill the reservoir.

55. Connect the brake booster vacuum hose to the intake manifold.

56. Attach the transmission dipstick tube to the cylinder head cover.

57. Install the coolant expansion tank.

58. Install the resonance pipe and body, then the air cleaner housing.

59. Remove the radiator/condenser guard plate.

60. Install the fan shroud and fan.

➡ **The fan clutch is equipped with right-hand thread.**

61. Install the engine undercover.

62. Fill the engine with coolant and oil.

63. Connect the negative battery cable.

64. Read fault memory, encode the radio and normalize the power windows.

Water Pump

REMOVAL & INSTALLATION

❄❄ CAUTION

Never open, service or drain the radiator or cooling system when hot; serious burns can occur from the steam and hot coolant. Also, when draining engine coolant, keep in mind that cats and dogs are attracted to ethylene glycol antifreeze and could drink any that is left in an uncovered container or in puddles on the ground. This will prove fatal in sufficient quantities. Always drain coolant into a sealable container. Coolant should be reused unless it is contaminated or is several years old.

1. Disconnect the negative battery cable.

2. Remove the engine cooling fan and clutch, then the fan shroud.

➡ **The fan clutch is equipped with right-hand thread.**

3. Drain and recycle the engine coolant.

4. Remove the engine cover.

5. Lock the automatic belt tensioner by rotating the tensioner counterclockwise until a 5mm drift or pin fits through the tensioner, then remove the serpentine belt.

6. Disconnect the coolant hoses from the water pump.

7. Remove the belt pulley.

8. Remove the water pump mounting bolts, then the water pump.

9. Clean and dry the gasket mating surface for the water pump.

To install:

10. Install the water pump and gasket, and tighten the M6 bolts to 88 inch lbs. (10 Nm) and the M8 bolts to 177 inch lbs. (20 Nm).

11. Install the water pump belt pulley and tighten the mounting bolts to 88 inch lbs. (10 Nm).

12. Connect the coolant hoses to the water pump.

13. Install the serpentine belt and remove the locking pin.

14. Install the engine cover.

15. Install the fan shroud and fan.

➡ **The fan clutch is equipped with right-hand thread.**

16. Fill the engine with coolant.

17. Connect the negative battery cable.

18. Read fault memory, encode the radio and normalize the power windows.

19. Start the vehicle and check for leaks.

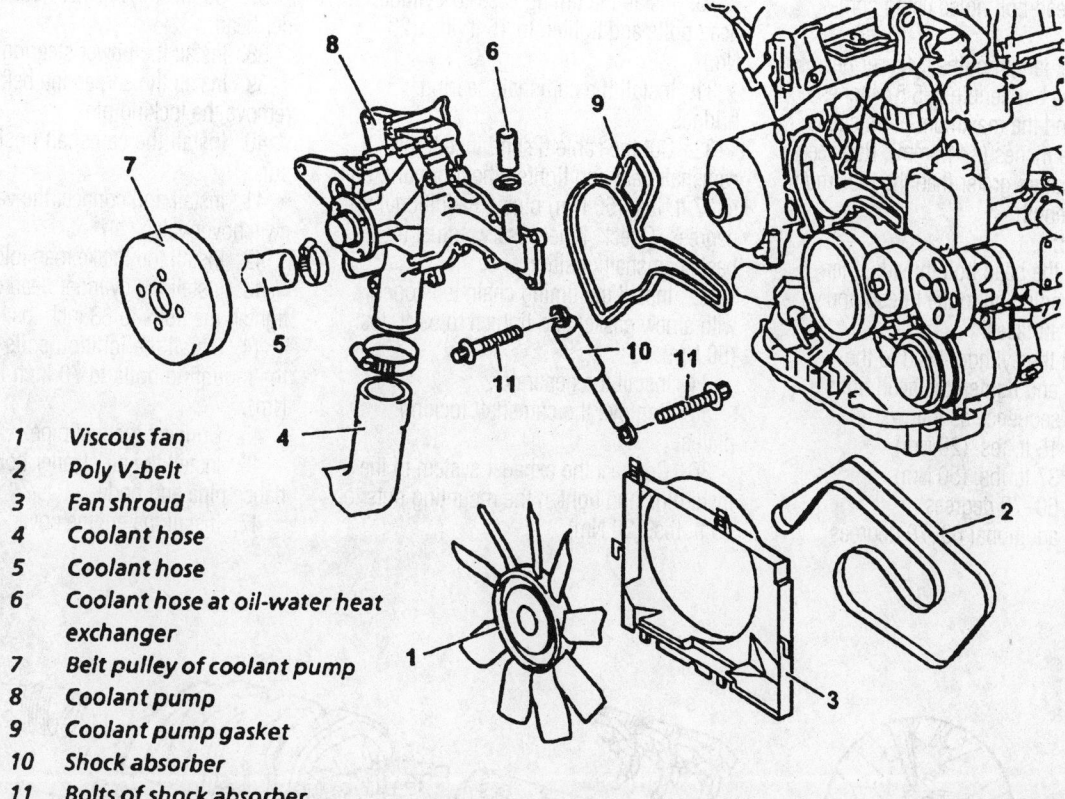

1   Viscous fan
2   Poly V-belt
3   Fan shroud
4   Coolant hose
5   Coolant hose
6   Coolant hose at oil-water heat exchanger
7   Belt pulley of coolant pump
8   Coolant pump
9   Coolant pump gasket
10   Shock absorber
11   Bolts of shock absorber

7924Z101

Exploded view of the water pump mounting and related components

Cylinder Head

REMOVAL & INSTALLATION

1. Disconnect the negative battery cable.

✳✳ CAUTION

Never open, service or drain the radiator or cooling system when hot; serious burns can occur from the steam and hot coolant. Also, when draining engine coolant, keep in mind that cats and dogs are attracted to ethylene glycol antifreeze and could drink any that is left in an uncovered container or in puddles on the ground. This will prove fatal in sufficient quantities. Always drain coolant into a sealable container. Coolant should be reused unless it is contaminated or is several years old.

2. Drain and recycle the engine coolant.
3. Remove the engine cooling fan and clutch, then the fan shroud.

➡**The fan clutch is equipped with right-hand thread.**

4. Place a guard plate behind the radiator/condenser to protect it from damage during removal and installation.
5. Remove the engine cover.
6. Remove the air cleaner housing, resonance pipe and body.
7. Properly relieve the fuel system pressure.
8. Disconnect the fuel line.
9. Remove the ignition coils.
10. Remove the cylinder head covers.

➡**The intake manifold system must not be disassembled.**

11. Remove the intake manifold.
12. Label and remove the vacuum switchover valve.
13. Remove the camshaft position sensor.
14. Lock the automatic belt tensioner by rotating the tensioner counterclockwise until a 5mm drift or pin fits through the tensioner, then remove the serpentine belt.
15. Remove the power steering pump and position it aside leaving the hoses attached.
16. Disconnect the heater hose at the firewall.
17. Detach the exhaust system from the exhaust manifolds.

18. Rotate the engine clockwise to position the crankshaft 40 degrees after top dead center.

✳✳ WARNING

Engine must not be rotated backwards.

19. Lock the camshafts using the Camshaft Locking tools 112 589 00 32 00 and 112 589 01 32 00, or their equivalent.
20. Remove the generator, then the timing chain tensioner.
21. Unbolt the camshaft gears and attach them to the chain with a cable tie.
22. Remove the camshaft bearing bridges.
23. Remove the timing case-to-cylinder head bolts.
24. Loosen and remove the cylinder head bolts in stages following the illustrated sequence.
25. Lift the cylinder head off the engine block.
26. Remove all gasket material from the sealing surfaces of the cylinder head and engine block. Be careful not to gouge or scratch the surface of the aluminum head. Be sure the cylinder head locating dowels are positioned in the engine block. Clean

and dry the head bolt holes using compressed air.

27. Inspect length of the cylinder head bolt shaft, new bolt length is 5.57 inches (141.5mm) and the maximum permissible length is 5.69 inches (144.5mm). Replace bolts that measure grater than the maximum permissible length.

To install:

28. Clean the head bolt threads, then apply clean engine oil to the thread and head contact surfaces.

29. Install the cylinder head to the engine block and tighten the head bolts according to sequence as follows:

- Step 1. 15 ft. lbs. (20 Nm)
- Step 2. 37 ft. lbs. (50 Nm)
- Step 3. 60–70 degrees
- Step 4. additional 60–70 degrees

30. Install the timing case-to-cylinder head bolts and tighten to 15 ft. lbs. (20 Nm).

31. Install the camshaft bearing bridges.

32. Cut the cable tie and install the camshaft gear and tighten the mounting bolt to 37 ft. lbs. (50 Nm) plus an additional 90 degrees. Check, if necessary, adjust the basic camshaft position.

33. Install the timing chain tensioner with a new gasket and tighten to 59 ft. lbs. (80 Nm).

34. Install the generator.

35. Remove the camshaft locking plates.

36. Connect the exhaust system to the manifolds and tighten the mounting nuts to 15 ft. lbs. (20 Nm).

37. Connect the heater hose to the cylinder head.

38. Install the power steering pump.

39. Install the serpentine belt and remove the locking pin.

40. Install the camshaft position sensor.

41. Install and connect the vacuum switchover valve.

42. Install the intake manifold.

43. Install the cylinder head covers and tighten the bolts to 88 inch lbs. (10 Nm).

44. Install the ignition coils and tighten the mounting bolts to 70 inch lbs. (8 Nm).

45. Connect the fuel pipe.

46. Install the air cleaner housing, resonance pipe and body.

47. Install the engine cover.

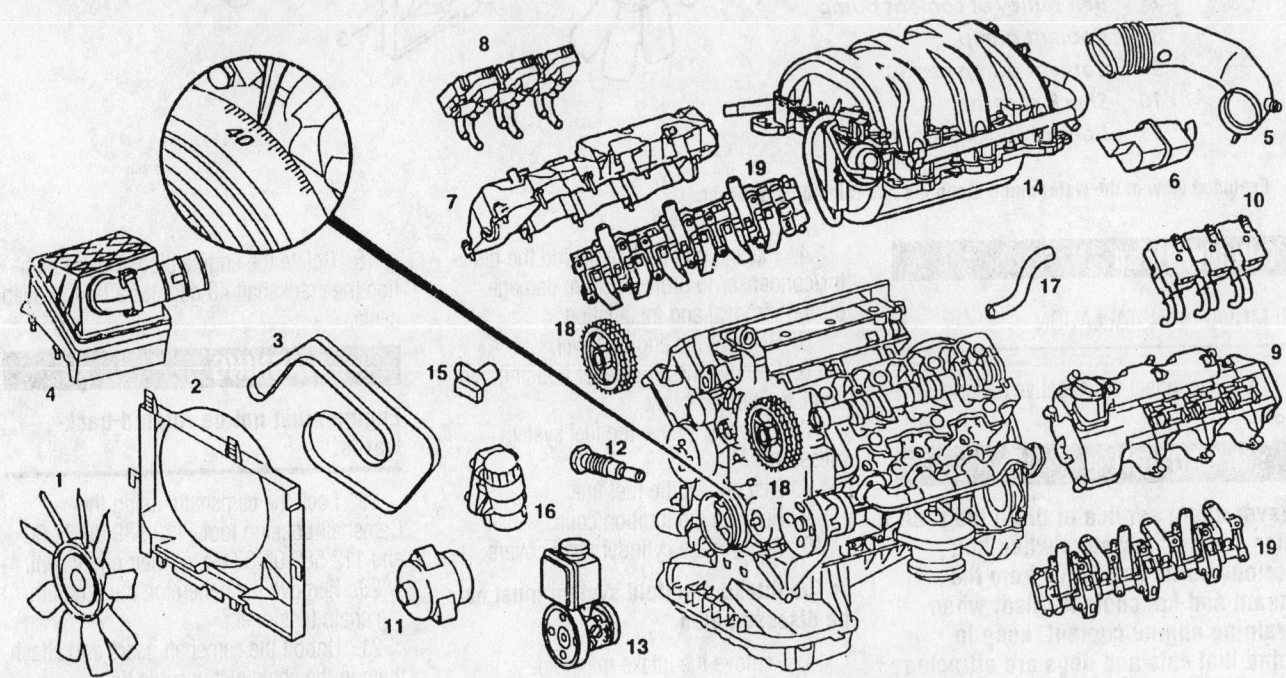

1	Viscous fan	11	Generator
2	Fan shroud	12	Chain tensioner
3	Poly V-belt	13	Power steering pump with reservoir
4	Air cleaner housing with HFM-SFI	14	Intake manifold
5	Resonance pipe	15	Camshaft position sensor
6	Resonance body	16	Oil filter housing
7	Right cylinder head cover	17	Heating hose
8	Right ignition coils	18	Camshaft gears
9	Left cylinder head cover	19	Camshaft bearing bridges
10	Left ignition coils		

Exploded view of the cylinder head accessory components

7924Z102

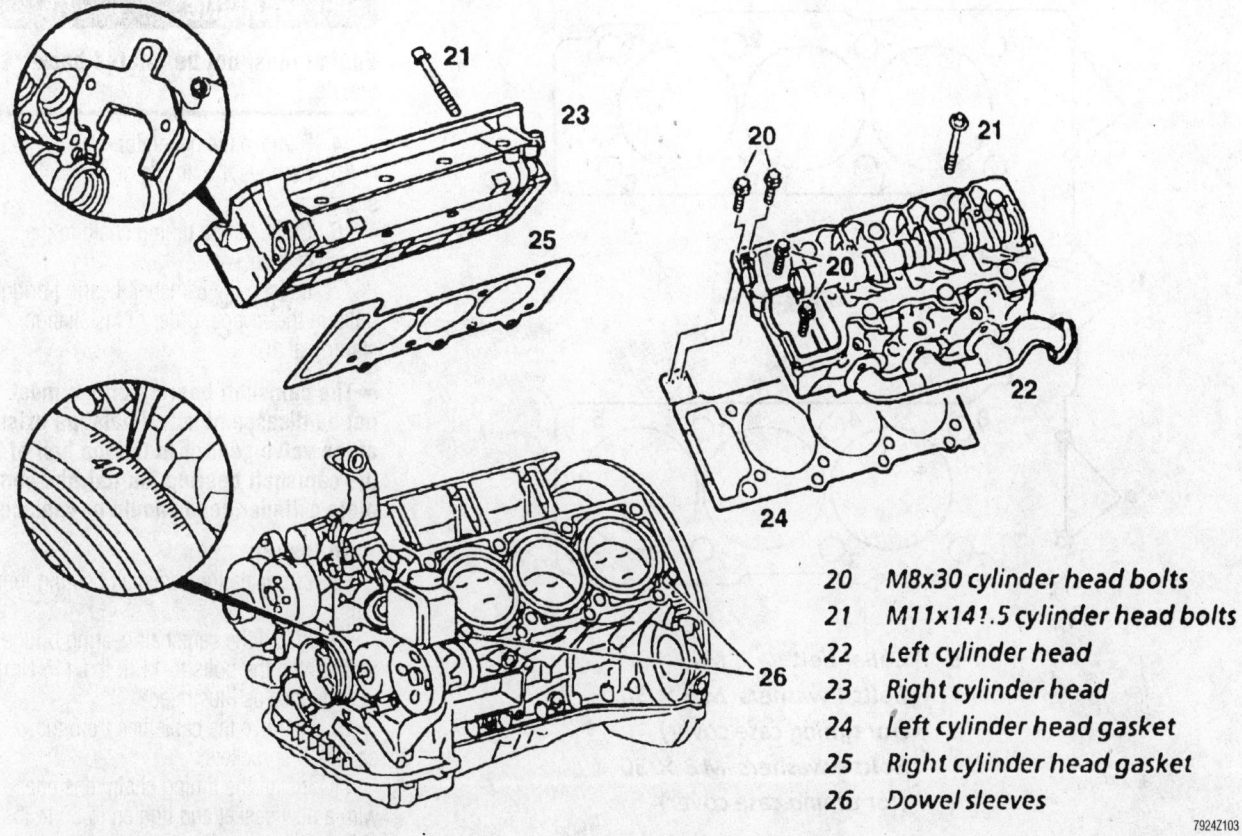

20 M8x30 cylinder head bolts
21 M11x141.5 cylinder head bolts
22 Left cylinder head
23 Right cylinder head
24 Left cylinder head gasket
25 Right cylinder head gasket
26 Dowel sleeves

Exploded view of the cylinder head removal

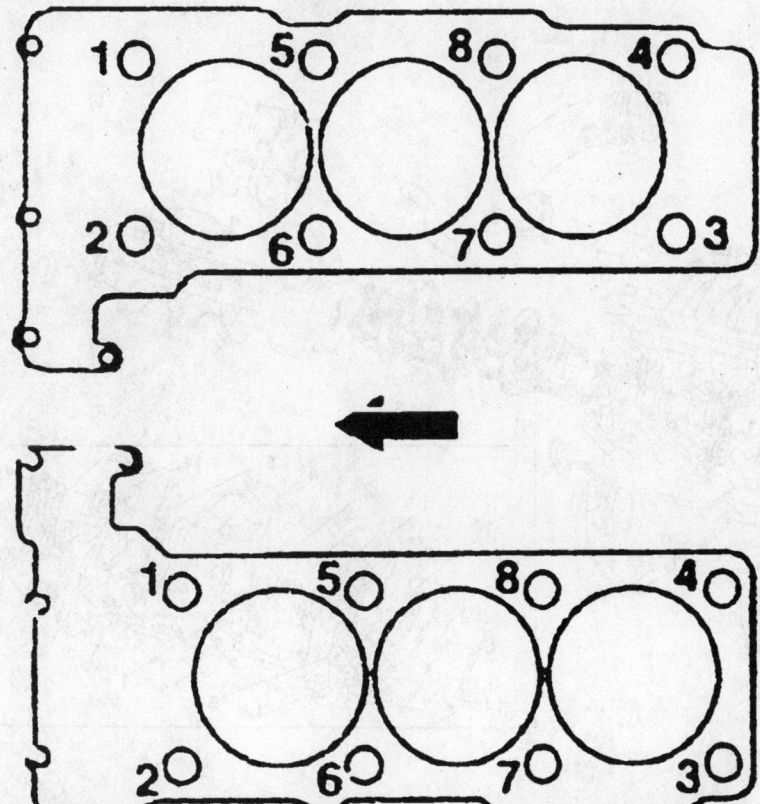

Cylinder head bolt removal sequence

48. Remove the guard plate from the radiator/condenser.

49. Install the fan shroud, then the cooling fan.

➡**The fan clutch is equipped with right-hand thread.**

50. Fill the engine with coolant.

51. Connect the negative battery cable.

52. Read fault memory, encode the radio and normalize the power windows.

53. Start the vehicle and check for leaks.

Rocker Arms/Shafts

The rocker arm/shaft is part of the camshaft bearing cap assembly and is called the camshaft bearing bridge. This procedure is for removing and installing the camshaft bearing bridge.

REMOVAL & INSTALLATION

1. Disconnect the negative battery cable.

2. Remove the cylinder head cover.

3. Rotate the engine clockwise to position the crankshaft 40 degrees after top dead center.

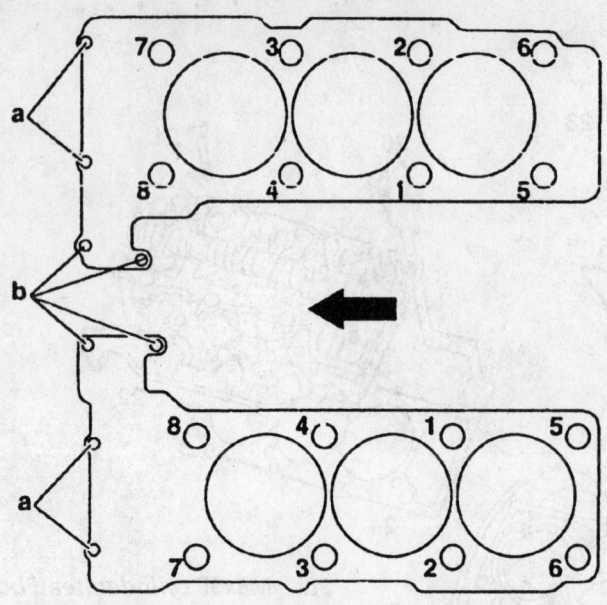

1-8	Collar bolt	M11
a	Bolts + washers M 8 × 50 (for timing case cover)	
b	Bolts + washers M 8 × 30 (for timing case cover)	

7924Z105

Cylinder head tightening sequence

✳✳ WARNING

Engine must not be rotated backwards.

---

4. Remove the generator.
5. Remove the timing chain tensioner.
6. Cable tie the timing chain to the camshaft sprocket.
7. Loosen the camshaft bearing bridge bolts in the reverse order of installation, starting at 16.

➡ **The camshaft bearing bridge must not be disassembled. If damage exists at the valve gear or at the top half of the camshaft bearing journal, the complete cylinder head should be replaced.**

To install:
8. Lubricate the camshaft bearing journals.
9. Install the camshaft bearing bridge and tighten the bolts to 11 ft. lbs. (15 Nm) in sequence as illustrated.
10. Remove the cable ties from the camshaft sprockets.
11. Install the timing chain tensioner with a new gasket and tighten to 59 ft. lbs. (80 Nm).

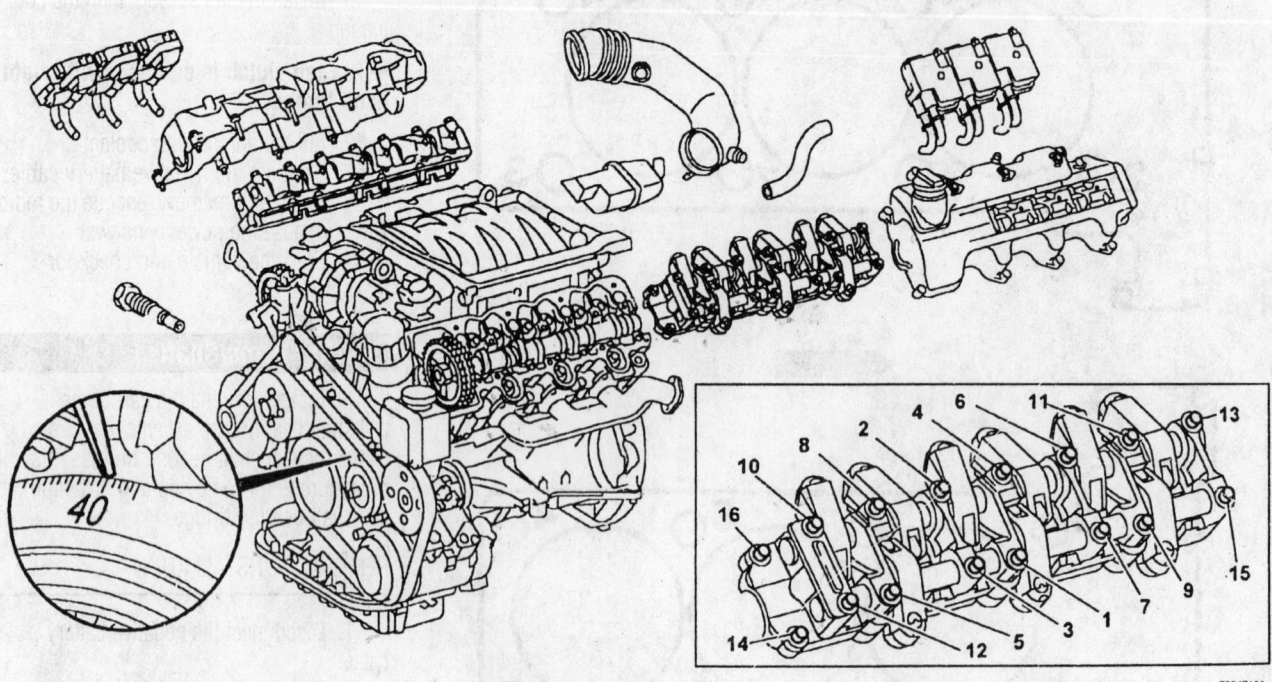

7924Z106

Exploded view of the camshaft bearing bridge removal components

12. Install the cylinder head covers to 88 inch lbs. (10 Nm).

13. Connect the negative battery cable.

14. Read fault memory, encode the radio and normalize the power windows.

15. Start the vehicle and check for leaks.

Intake Manifold

REMOVAL & INSTALLATION

1. Disconnect the negative battery cable.

2. Remove the cylinder head cover.

3. Remove the hot film mass air flow sensor with the intake pipe.

4. Properly relieve the fuel system pressure.

5. Remove the fuel rail with the injectors.

6. Label and disconnect the vacuum lines from the intake manifold.

7. Label and detach any electrical connections to the intake manifold.

8. Disconnect the EGR valve.

9. Remove the combination valve, then the intake manifold mounting bolts.

10. Remove the intake manifold and gaskets.

11. Place clean shop rags into the intake passages to prevent dirt from entering. Clean the gasket mating surfaces.

To install:

12. Install the new gaskets and verify the secondary air injection passage opening in the gasket.

13. Remove the shop rags from the intake passages.

14. Install the intake manifold to the engine and tighten the mounting bolts to 15 ft. lbs. (20 Nm).

15. Install the combination valve and tighten the bolts to 15 ft. lbs. (20 Nm).

16. Connect the EGR valve.

17. Attach the electrical connections to the intake manifold.

18. Connect the vacuum lines to the manifold.

19. Install the fuel rail with the injectors.

20. Install the mass air flow sensor with the air intake pipe.

21. Install the cylinder head cover and tighten the bolts to 88 inch lbs. (10 Nm).

22. Connect the negative battery cable.

23. Read fault memory, encode the radio and normalize the power windows.

24. Start the vehicle and check for leaks.

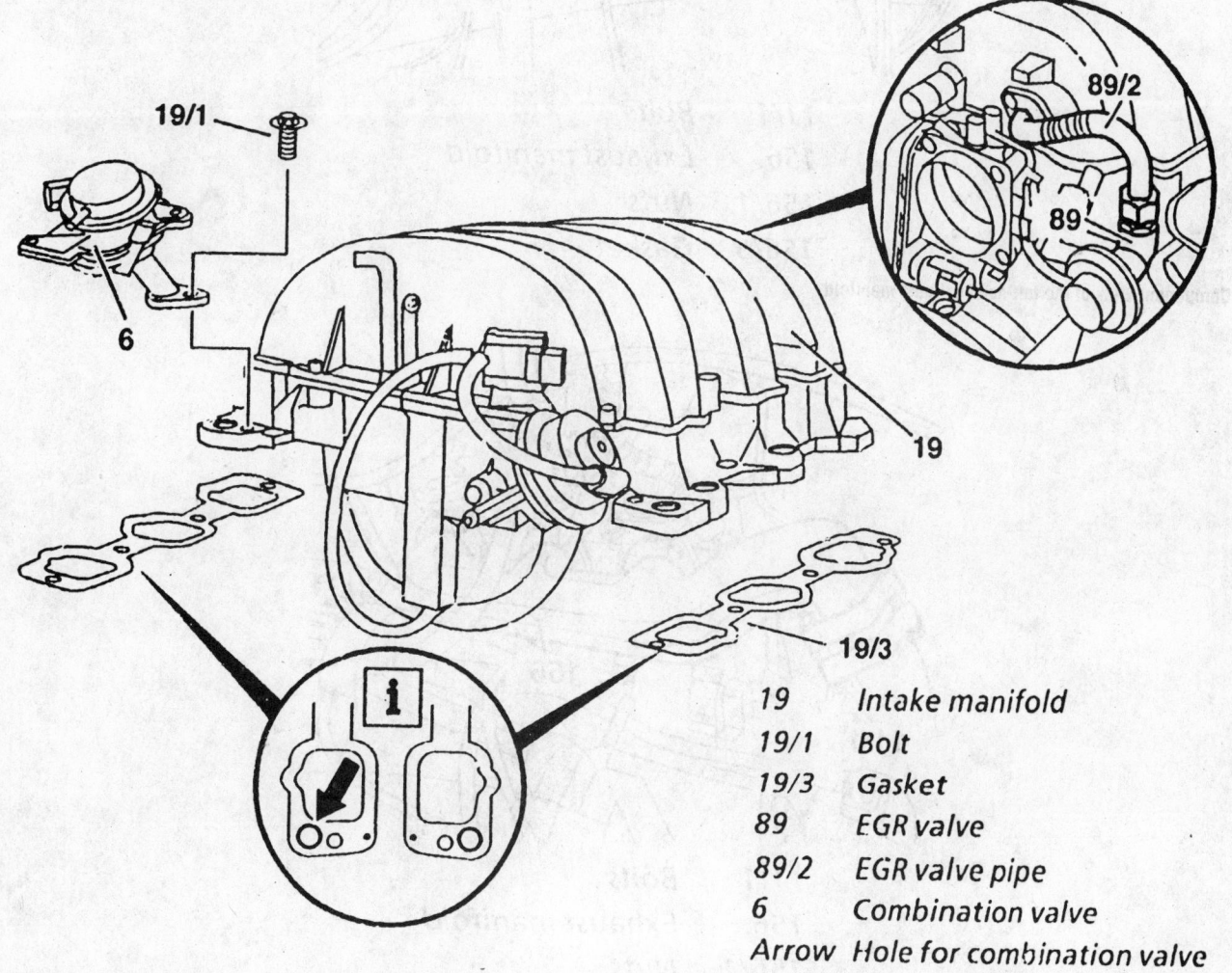

19	Intake manifold
19/1	Bolt
19/3	Gasket
89	EGR valve
89/2	EGR valve pipe
6	Combination valve
Arrow	Hole for combination valve

Exploded view of the intake manifold and related components

7924Z107

Exhaust Manifold

REMOVAL & INSTALLATION

1. Disconnect the negative battery cable.
2. Raise and safely support the front of the vehicle and remove the front wheels.
3. Remove the plastic inner fender liners.
4. Remove the exhaust manifold heat shields.

➡**If the bolts are difficult to remove or the threads show signs of damage, replace the rivet nuts in the manifold.**

5. Remove the exhaust system-to-manifold flanged connection mounting bolts.
6. Remove the front exhaust pipe at the transmission exhaust bracket.
7. Remove the exhaust manifold-to-cylinder head mounting bolts, then remove the manifolds.

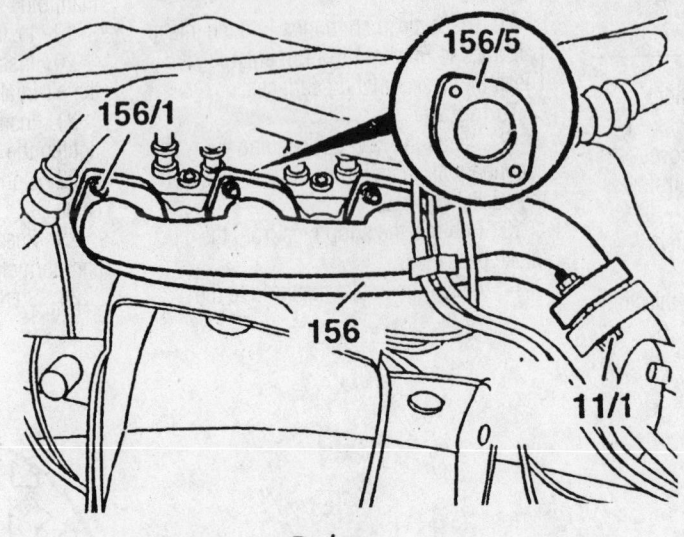

11/1	Bolts
156	Exhaust manifold
156/1	Nuts
156/5	Gasket

79242108

Component view of the left-hand exhaust manifold

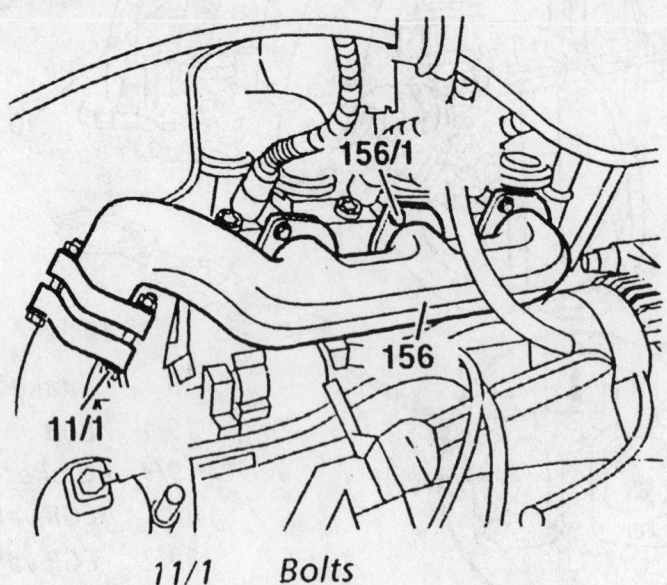

11/1	Bolts
156	Exhaust manifold
156/1	Nuts
156/5	Gasket

79242109

Component view of the right-hand exhaust manifold

8. Clean the gasket mating surfaces.

To install:

9. Install the exhaust manifold using new gaskets and nuts. Tighten the nuts to 12 ft. lbs. (16 Nm).

10. Connect the front exhaust pipe to the transmission exhaust bracket.

11. Connect the exhaust system to the manifolds and tighten the mounting bolts to 15 ft. lbs. (20 Nm).

12. Install the exhaust manifold heat shields.

13. Install the plastic inner fender liners.

14. Install the front wheels, lower the vehicle and tighten the lug bolts to 110 ft. lbs. (150 Nm).

15. Connect the negative battery cable.

16. Read fault memory, encode the radio and normalize the power windows.

17. Start the vehicle and check for leaks.

Camshaft

REMOVAL & INSTALLATION

1. Disconnect the negative battery cable.

2. Remove the cylinder head cover.

3. Rotate the engine clockwise to position the crankshaft 40 degrees After Top Dead Center (ATDC).

❄❄ WARNING

Engine must not be rotated backwards.

4. Remove the generator.

5. Remove the timing chain tensioner.

6. Cable tie the timing chain to the camshaft sprocket.

7. Remove the camshaft position sensor.

8. Lock the camshafts using the Camshaft Locking tools 112 589 00 32 00 and 112 589 01 32 00, or their equivalent.

9. Unbolt the camshaft gears.

10. Remove the camshaft bearing bridge.

11. Carefully remove the camshaft from the cylinder head.

To install:

➡**Be sure to install the correct camshaft for the corresponding cylinder head.**

12. Apply clean engine oil to the camshaft contact surfaces, then install the camshaft.

13. Install the camshaft bearing bridge.

➡**The camshafts can be rotated 40 degrees after top dead center of the No. 1 cylinder without the valves touching the pistons.**

14. Position the camshaft so that the groove points centered towards the contact surface of the cylinder head cover, then attach the camshaft fixing plate. Repeat this step for the other camshaft.

15. Install the camshaft sprockets and tighten the attaching bolt to 37 ft. lbs. (50 Nm) plus 90–100 °.

16. Remove the camshaft locking tools and the cable ties from the timing chain.

17. Install the camshaft position sensor and tighten the mounting bolt to 70 inch lbs. (8 Nm).

18. Install the timing chain tensioner with a new gasket and tighten to 59 ft. lbs. (80 Nm).

19. Install the cylinder head cover.

20. Connect the negative battery cable.

21. Read fault memory, encode the radio and normalize the power windows.

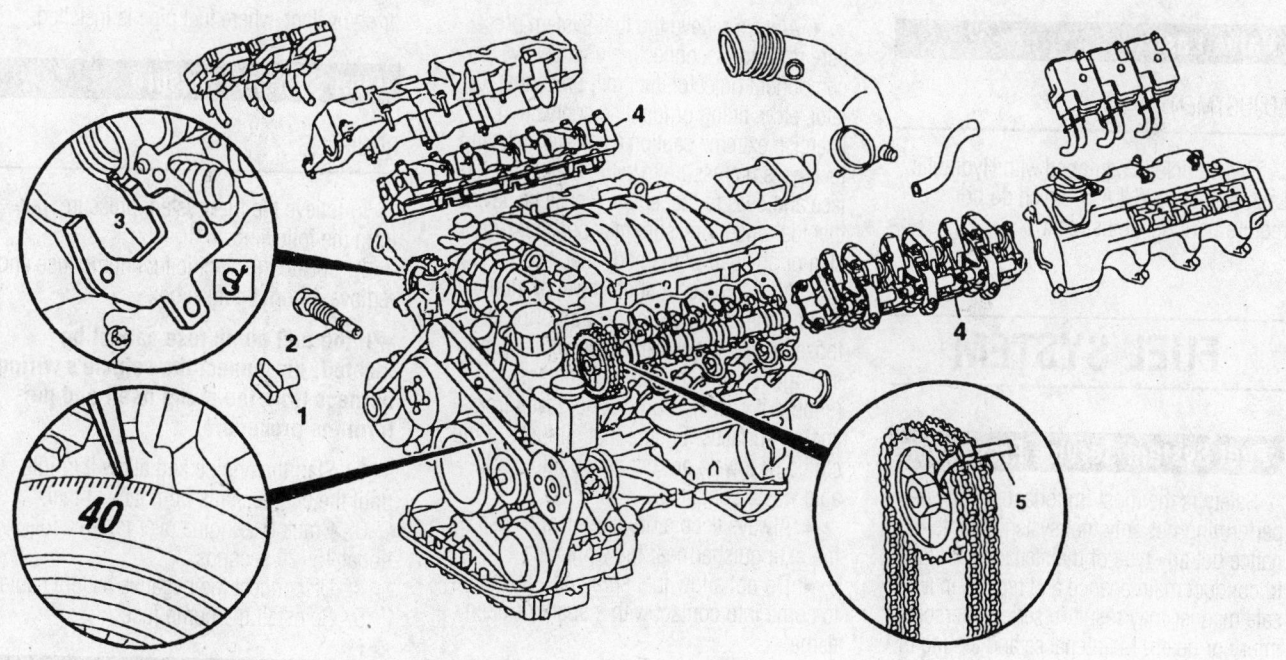

1	Camshaft Hall sensor	3	Fixing plate of right camshaft 40° after first ignition TDC
2	Chain tensioner	4	Camshaft bearing bridge
		5	Cable strap

7924Z110

Exploded view of the camshaft removal and related components

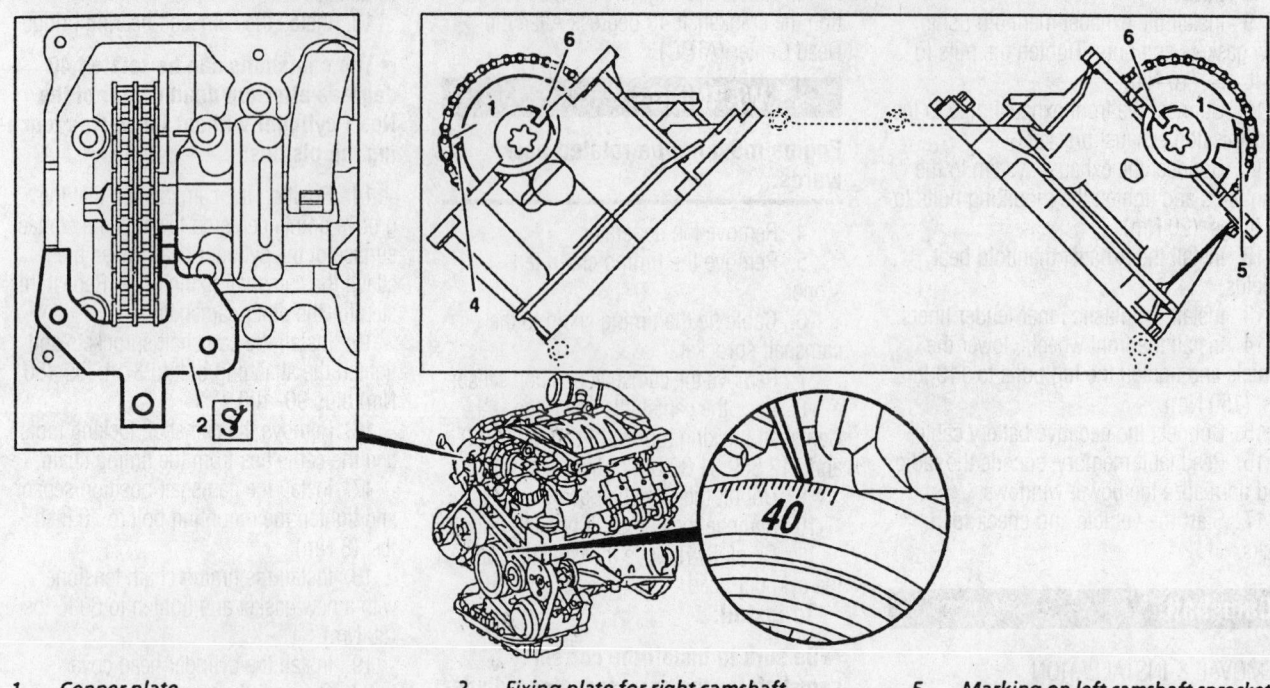

1. Copper plate
2. Fixing plate for right camshaft
4. Marking on right camshaft sprocket
5. Marking on left camshaft sprocket
6. Groove in camshaft

7924Z111

Be sure the camshafts are at their basic positions before installing the cam sprockets

Valve Lash

ADJUSTMENT

This vehicle is equipped with Hydraulic Lash Adjuster's (HLA's), which do not require periodic adjustment.

FUEL SYSTEM

Fuel System Service Precautions

Safety is the most important factor when performing not only fuel system maintenance but any type of maintenance. Failure to conduct maintenance and repairs in a safe manner may result in serious personal injury or death. Maintenance and testing of the vehicle's fuel system components can be accomplished safely and effectively by adhering to the following rules and guidelines.

• To avoid the possibility of fire and personal injury, always disconnect the negative battery cable unless the repair or test procedure requires that battery voltage be applied.

• Always relieve the fuel system pressure prior to disconnecting any fuel system component (injector, fuel rail, pressure regulator, etc.), fitting or fuel line connection. Exercise extreme caution whenever relieving fuel system pressure to avoid exposing skin, face and eyes to fuel spray. Please be advised that fuel under pressure may penetrate the skin or any part of the body that it contacts.

• Always place a shop towel or cloth around the fitting or connection prior to loosening to absorb any excess fuel due to spillage. Ensure that all fuel spillage (should it occur) is quickly removed from engine surfaces. Ensure that all fuel soaked cloths or towels are deposited into a suitable waste container.

• Always keep a dry chemical (Class B) fire extinguisher near the work area.

• Do not allow fuel spray or fuel vapors to come into contact with a spark or open flame.

• Always use a back-up wrench when loosening and tightening fuel line connection fittings. This will prevent unnecessary stress and torsion to fuel line piping. Always follow the proper torque specifications.

• Always replace worn fuel fitting O-rings with new ones. Do not substitute fuel hose or equivalent, where fuel pipe is installed.

Fuel System Pressure

RELIEVING

To relieve the fuel system pressure, perform the following:

1. Locate the electric fuel pump fuse and remove it from the fuse box.

➡ **If the fuel pump fuse cannot be located, disconnect the vehicle's wiring harness from the pump itself and perform the procedure.**

2. Start the engine and allow it to idle until the engine stalls from lack of fuel.
3. Crank the engine over for an additional 15–20 seconds.
4. Disconnect the negative battery cable.
5. Reinstall the pump fuse.

Fuel Filter

REMOVAL & INSTALLATION

1. Disconnect the negative battery cable.
2. Relieve the pressure in the fuel

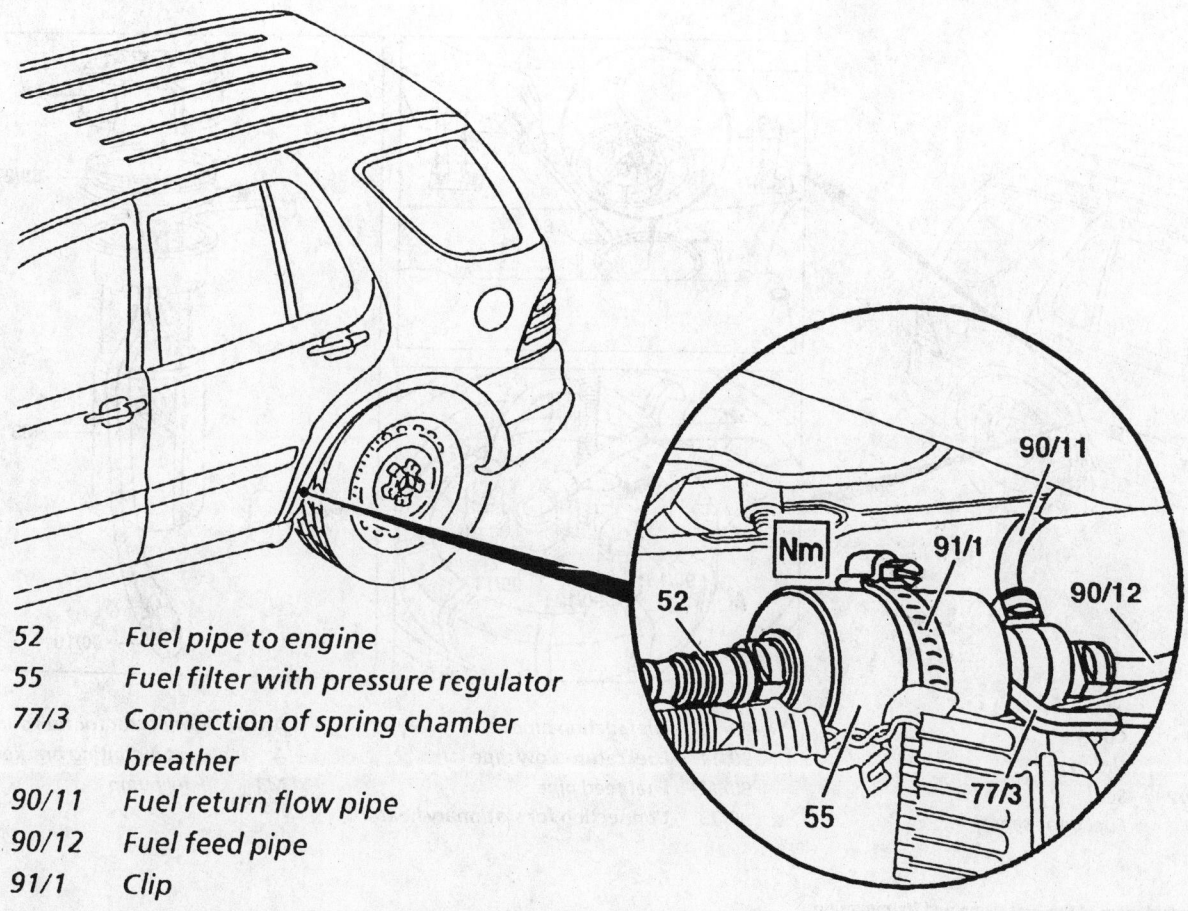

52	Fuel pipe to engine
55	Fuel filter with pressure regulator
77/3	Connection of spring chamber breather
90/11	Fuel return flow pipe
90/12	Fuel feed pipe
91/1	Clip

View of the fuel filter mounting location and component identification

tank by opening, then tightening the filler cap.

3. Raise and safely support the rear of the vehicle and remove the left rear wheel.

4. Remove the plastic inner fender liner.

➡**The fuel lines must not be kinked.**

5. Compress the locking catches, then remove the fuel feed and return pipes from the fuel filter.

6. Detach the breather hose from the filter assembly.

7. Loosen the filter securing clip, then remove the filter/pressure regulator assembly to the front.

To install:

8. Insert the new fuel filter/pressure regulator into the housing and tighten the securing clip to 27 inch lbs. (3 Nm).

9. Attach the breather hose to the filter assembly.

10. Connect the fuel line to the filter assembly.

11. Install the plastic inner fender liner.

12. Install the left rear wheel, lower the vehicle and tighten the lug bolts to 110 ft. lbs. (150 Nm).

13. Connect the negative battery cable.

14. Read fault memory, encode the radio and normalize the power windows.

15. Start the vehicle and check for leaks.

Fuel Pump

REMOVAL & INSTALLATION

1. Disconnect the negative battery cable.

2. Empty the fuel tank into a suitable container.

3. Raise the left rear seat approximately 20 inches (50 cm).

4. Lift the carpeting to gain access to the fuel pump cover, then remove the cover.

5. Detach the fuel pump electrical connector.

➡**Be sure not to kink the fuel pipes.**

6. Unclip the supply and return fuel pipes.

7. Unscrew the union nut mounting the fuel pump-to-the tank.

8. Slowly remove the fuel pump from the tank.

To install:

➡**Lightly oil the fuel pump sealing O-ring to simplify the installation.**

9. Install the fuel pump into the tank using a new union nut and O-ring.

10. Tighten the union nut to 50 ft. lbs. (65 Nm).

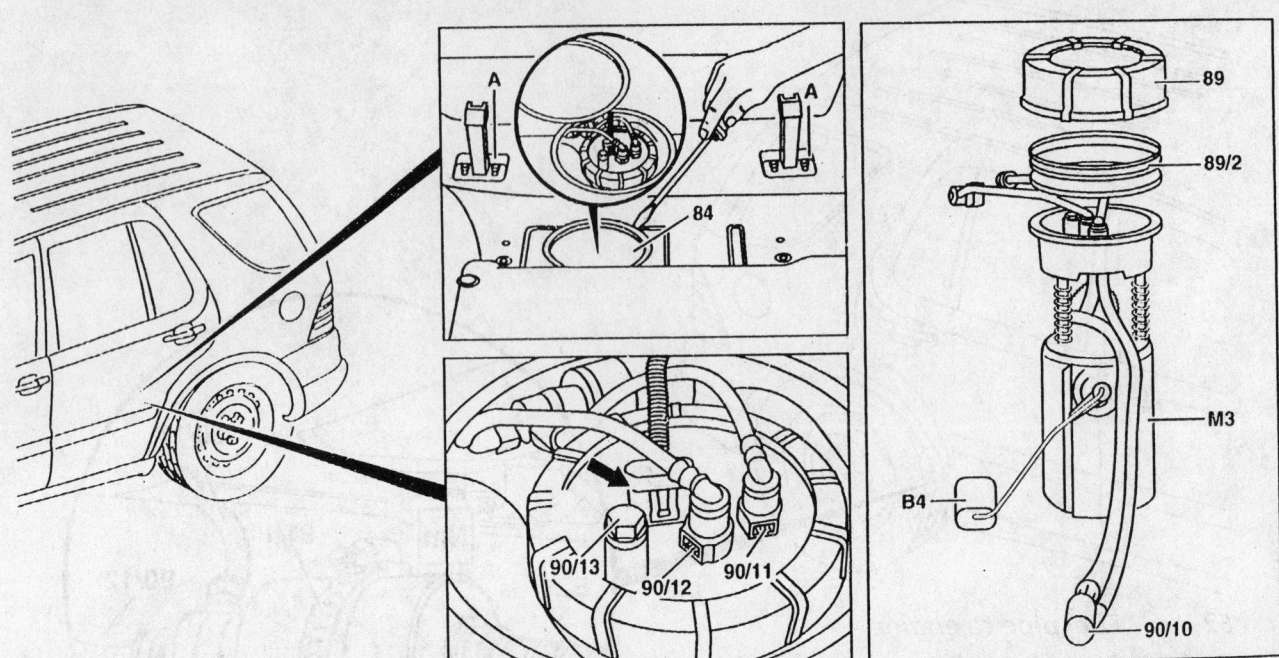

84	Cover	90/10	Fuel suction pipe	Arrow	Connector for fuel pump
89	Union nut	90/11	Fuel return flow pipe	A	Seat mounting brackets
89/2	Seal	90/12	Fuel feed pipe	M3	Fuel pump
B4	Fuel gage sensor	90/13	Connection for stationary heater		

7924Z113

Exploded view of the fuel pump and its mounting

11. Connect the supply and return lines to the fuel pump.

12. Attach the fuel pump electrical connector.

13. Install the fuel pump access cover and the rear seat.

14. Fill the fuel tank.

15. Connect the negative battery cable.

16. Read fault memory, encode the radio and normalize the power windows.

17. Start the vehicle and check for leaks.

ENGINE REPAIR

➡Disconnecting the negative battery cable on some vehicles may interfere with the functions of the on board computer systems and may require the computer to undergo a relearning process, once the negative battery cable is reconnected.

Distributor

REMOVAL

1. Disconnect the negative battery cable.
2. Disconnect the distributor pick-up lead wires and vacuum hose(s), if equipped.
3. Unfasten the distributor cap retaining clips or screws and lift off the distributor cap with all ignition wires connected. Remove the coil wire if necessary.
4. Matchmark the rotor to the distributor housing and the distributor housing to the engine.

➡Do not crank the engine during this procedure. If the engine is cranked, the matchmark must be disregarded.

5. Remove the retaining nut and remove the distributor from the engine.

INSTALLATION

Timing Not Disturbed

1. Install a new distributor housing O-ring.
2. Install the distributor in the engine so the rotor is aligned with the matchmark on the housing and the housing is aligned with the matchmark on the engine. Be sure the distributor is fully seated and the distributor shaft is fully engaged.
3. Install the retaining nut finger-tight only. Connect the vacuum hose(s), if removed.
4. Connect the distributor pick-up electrical harness.
5. Install the distributor cap and secure.
6. Connect the negative battery cable.
7. Check the ignition timing and adjust as required. Tighten the retaining nut.

Timing Disturbed

1. Install a new distributor housing O-ring.
2. Rotate the engine so No. 1 piston is on TDC of compression stroke and the tim-

ing mark on the vibration damper is aligned with **T** on the timing indicator.
3. Install the distributor so the rotor is aligned with the No. 1 ignition wire on the distributor cap. Take note that the distributor shaft is fully engaged and the housing is fully seated.

➡Some distributor caps may contain runners inside the cap. If so, be sure the rotor is pointing to where the No. 1 runner originates inside the cap and not where the No. 1 ignition wire plugs into the cap.

4. Install the retaining nut finger-tight only. Connect the vacuum hose(s), if removed.
5. Connect the distributor electrical harness.
6. Install the distributor cap and secure.
7. Connect the negative battery cable.
8. Adjust the ignition timing and tighten the retaining nut.

Ignition Timing

ADJUSTMENT

Montero and Montero Sport

3.0L 12 VALVE ENGINE

Before attempting to adjust the ignition timing, be sure of the following:
- The engine should be at normal operating temperature.
- The lights and all accessories should be OFF.
- If equipped with an automatic transmission, the transmission should be in **P** or **N**.

1. Insert a paper clip into the one-pin connector between the primary side of the ignition coil and the noise filter. The connector should not be disconnected.
2. Connect a primary voltage detection tachometer to the paper clip.

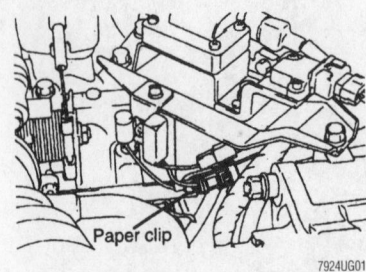

7924UG01

Insert a paper clip into the one-pin connector between the primary side of the ignition coil and the noise filter—Montero with 3.0L engines

➡Do not use the scan tool. When the scan tool is connected to the data link connector, the ignition timing will be unchanged, instead of reverting to the base ignition timing.

3. Connect the timing light and run the engine at idle speed.
4. Verify that the idle speed is 600–800 rpm.
5. Turn the ignition switch **OFF** and disconnect the brown waterproof female connector from the ignition timing adjustment connector.
6. Use a jumper wire to ground the ignition timing adjustment terminal.

➡Grounding this terminal sets the engine to base ignition timing.

7. Start the engine and run it at idle. Check the base timing; it should be 3–7° BTDC.
8. If the base timing is out of specification, loosen the hold-down nut and turn the distributor. Turning the distributor clockwise will advance the timing and counter-clockwise will retard the ignition timing. After the base timing is set to specifications, tighten the distributor hold-down nut and recheck the base timing.
9. Turn the ignition switch **OFF** and disconnect the jumper wire from the ignition timing adjustment connector.
10. Start the engine and allow it to idle. The ignition timing at idle should be approximately 15° BTDC.

➡Ignition timing under computer control can vary as much as 7° up or down, even under normal operating conditions. The ignition timing is automatically advanced by about 5°–15° BTDC at higher altitudes.

11. Remove all of the test equipment.

3.0L 24 VALVE AND 3.5L ENGINES

Before attempting to adjust the ignition timing, be sure of the following:
- The engine should be at normal operating temperature.
- The lights and all accessories should be OFF.
- If equipped with an automatic transmission, the transmission should be in **P** or **N**.

1. Insert a paper clip into the engine speed detection connector (blue), then connect a tachometer to the paper clip.

➡Do not use the scan tool. When the scan tool is connected to the data link connector, the ignition timing will be unchanged, instead of reverting to the base ignition timing.

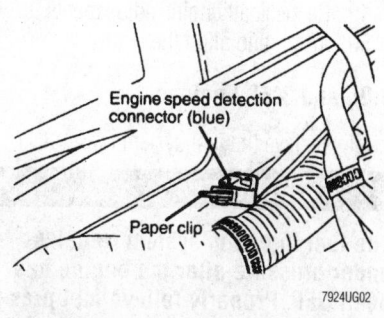

Insert a paper clip into the blue engine speed detection connector—3.0L 24 valve and 3.5L engine

2. Connect the timing light and run the engine at idle speed.

3. Idle speed should be 600–800 rpm.

➡**The reading on the tachometer indicates ⅓ of the actual engine speed. The actual engine speed is really 3 times the tachometer reading.**

4. Turn the ignition switch **OFF** and disconnect the waterproof female connector from the ignition timing adjustment connector (brown).

5. Use a jumper wire to ground the ignition timing adjustment terminal.

➡**Grounding this terminal sets the engine to the base ignition timing.**

6. Start the engine and run it at idle speed. Check the base timing, it should be 2–8° BTDC.

➡**The ignition timing is controlled by the ECM and is not adjustable. The ECM determines the timing based on input from the crankshaft position sensor.**

7. Turn the ignition switch **OFF** and disconnect the jumper wire from the ignition timing adjustment connector.

8. Start the engine and allow it to idle.

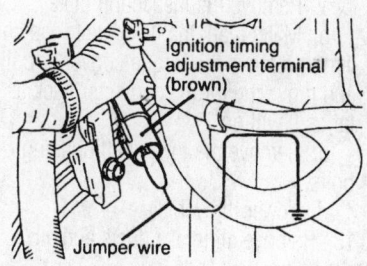

Use a jumper wire to ground the ignition timing adjustment terminal—3.0L 24 valve and 3.5L engine

---

The ignition timing at idle should be approximately 15° BTDC.

➡**Ignition timing under computer control can vary by as much as 7°, even under normal operating conditions. The ignition timing is automatically advanced by about 5° from 15° BTDC at higher altitudes.**

9. Turn the ignition switch **OFF** and remove all the test equipment.

Mighty Max

1. Firmly apply the parking brake and block the drive wheels.

2. Start the engine and allow to operate until normal operating temperature is reached.

3. Turn all lights and accessories **OFF**.

4. Place the transmission lever in **N**.

5. Without disconnecting the connector, insert a paper clip into the tachometer terminal.

   a. Connect the red lead of a tachometer to the paper clip and connect the black lead to a ground.

   b. Set the idle speed to specifications.

6. Turn the engine **OFF**.

7. Remove the water-proof cover from the ignition timing adjusting connector.

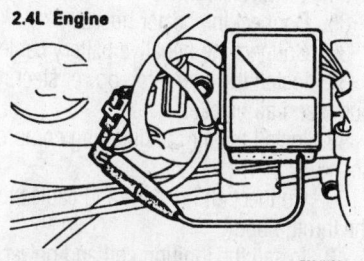

Connect the tachometer to the location shown—Mighty Max with the 2.4L and 3.0L engines

Connect a jumper wire from the ignition timing adjusting terminal to ground—Mighty Max with the 2.4L and 3.0L engines

---

Connect a jumper wire from the ignition timing adjusting terminal to ground.

8. Connect a conventional timing light to No. 1 cylinder spark plug wire. Start the engine and allow to idle.

9. Aim the timing light at the timing scale.

10. Basic ignition timing should be 3–7° BTDC.

11. Loosen the distributor nut to allow for distributor rotation.

12. Turn the distributor in the proper direction until the specified timing is reached. Tighten the retainer nut and recheck the timing.

13. Turn the engine **OFF**.

14. Remove the jumper wire from the ignition timing adjusting terminal and install the water-proof cover.

15. Start the engine and check the actual ignition timing. This reading should be 8° BTDC for the 2.4L engine or 15° BTDC for the 3.0L engine.

➡**The actual timing may fluctuate according to the control mode of the engine control unit; this is a normal condition.**

16. Turn the engine **OFF** and remove all test equipment.

Engine Assembly

REMOVAL & INSTALLATION

2.4L Engine

✳✳ CAUTION

The fuel injection system remains under pressure after the engine has been OFF. Properly relieve fuel pressure before disconnecting any fuel lines. Failure to do so may result in fire or personal injury.

1. Relieve the fuel system pressure.

2. Disconnect the negative battery cable.

✳✳ CAUTION

Wait at least 90 seconds after the negative battery cable is disconnected to prevent possible deployment of the air bag.

3. Matchmark and remove the hood.

4. Remove the oil dipstick.

5. Raise the vehicle and support safely.

6. Remove the engine undercover.

7. Drain the engine oil and coolant. Remove the lower radiator hose.

8. Remove the starter.

9. Remove the exhaust pipe from the exhaust manifold.

10. If equipped with 4WD, remove transfer case from vehicle.

11. If equipped with a manual transmission, remove the transmission and all related parts.

12. If equipped with an automatic transmission and 2WD:

   a. Remove the inspection plate.

   b. Matchmark the flexplate to the converter; remove the torque converter bolts and move the torque converter back as far as it will go.

   c. Remove the lower bell housing bolts.

   d. Lower the vehicle.

13. Remove all intake ducts and air intake hoses.

14. Disconnect all linkages and cables from the throttle body.

15. Cover the fuel line connections with a clean shop rag and disconnect and plug the fuel lines.

16. If equipped with air conditioning, unbolt the air conditioning compressor from the engine and position it aside. It is not necessary to remove the lines from the compressor.

17. Remove the radiator and shroud. Remove the fan and all related parts.

18. Disconnect the heater hoses.

19. Unbolt the power steering pump from its brackets and position it to the side. Do not remove the hoses from the pump.

20. Remove the alternator. Remove the ignition coil and power transistor assembly, if equipped.

21. Label and disconnect all remaining electrical connectors, vacuum hoses and check for any other items preventing engine removal.

22. Attach an engine removal device to the engine support eyes on the engine.

23. If equipped with an automatic transmission, support the transmission with a floor jack or equivalent. Remove the remaining bell housing bolts.

24. Remove the engine mount nuts and remove the engine from the vehicle.

To install:

25. Lower the engine into position and install the engine mount nuts. Tighten the nuts to 14–22 ft. lbs. (30–40 Nm).

26. Install the upper bell housing bolts.

27. Remove the engine removal device and the transmission support. Install the oil dipstick.

28. Raise the vehicle and support safely. Install the remaining bell housing bolts.

29. Install transfer case, if equipped.

30. If equipped with a manual transmission, install the transmission and all related parts.

31. If equipped with an automatic transmission, align the torque converter and flexplate and install the bolts.

32. Install the inspection plate and starter motor.

33. Install the exhaust pipe to the exhaust manifold using new gaskets. Install the lower radiator hose.

34. Lower the vehicle.

35. Connect the heater hoses.

36. Connect the negative battery cable.

37. Install the alternator, power steering pump and all brackets.

38. Install the air conditioning compressor.

39. Connect all linkages and cables to the throttle body.

40. Install the ignition coil and power transistor assembly.

41. Connect all electrical connectors and vacuum hoses.

42. Install the fan and all related parts. Adjust all belt tensions, as required.

43. Install the radiator, shroud and upper hose.

44. Install the air cleaner assembly, ducts and air intake hose.

45. Fill the engine with the specified amount of oil and fill the radiator with coolant.

46. Connect the negative battery cable.

47. Connect the jumper wire from the fuel pump activation terminal to the positive battery post. Inspect the system for leaks.

48. Check the automatic transmission fluid level, if equipped.

49. Recheck all engine adjustments.

50. Install and align the hood.

3.0L and 3.5L Engines

1. Relieve the fuel system pressure.

2. Disconnect the negative battery cable.

3. Matchmark and remove the hood.

4. Remove the oil dipstick.

5. Raise the vehicle and support safely.

6. Remove the engine undercover.

7. Drain the engine oil and coolant. Remove the lower radiator hose.

8. Remove the starter.

9. Remove the exhaust pipe from the exhaust manifolds.

10. If equipped with 4WD, remove transfer case from vehicle.

11. If equipped with a manual transmission, remove the transmission and all related parts.

12. If equipped with an automatic transmission and 2WD:

   a. Remove the inspection plate.

   b. Matchmark the flexplate to the converter; remove the torque converter bolts and move the torque converter back as far as it will go.

   c. Remove the lower bell housing bolts.

   d. Lower the vehicle.

13. Remove all intake ducts and air intake hoses.

14. Disconnect all linkages and cables from the throttle body.

15. Cover the fuel line connections with a clean shop rag and disconnect and plug the fuel lines.

16. If equipped with air conditioning, unbolt the air conditioning compressor from the engine and position it aside. It is not necessary to remove the lines from the compressor.

17. Remove the radiator and shroud. Remove the fan and all related parts.

18. Disconnect the heater hoses.

19. Unbolt the power steering pump from its brackets and position it to the side. Do not remove the hoses from the pump.

20. Remove the alternator.

21. If equipped, remove the ignition coil and power transistor assembly.

22. Label and disconnect all remaining electrical connectors, vacuum hoses and check for any other items preventing engine removal.

23. Attach an engine removal device to the engine support eyes on the engine.

24. If equipped with an automatic transmission, support the transmission with a floor jack or equivalent. Remove the remaining bell housing bolts.

25. Remove the engine mount nuts and remove the engine from the vehicle.

To install:

26. Lower the engine into position and install the engine mount nuts. Tighten the nuts to 20 ft. lbs. (27 Nm), on the Montero, tighten to 33 ft. lbs. (44 Nm).

27. Install the upper bell housing bolts.

28. Remove the engine removal device and the transmission support. Install the oil dipstick.

29. Raise the vehicle and support safely. Install the remaining bell housing bolts.

30. Install transfer case, if equipped.

31. If equipped with a manual transmission, install the transmission and all related parts.

32. If equipped with an automatic transmission, align the torque converter and flexplate and install the bolts.

33. Install the inspection plate and starter motor.

34. Install the exhaust pipe to the exhaust manifolds using new gaskets. Install the lower radiator hose.

35. Lower the vehicle.

36. Connect the heater hoses.

37. Connect the negative battery cable.

38. Install the alternator, power steering pump and all brackets.

39. Install the air conditioning compressor.

40. Connect all linkages and cables to the carburetor or throttle body.

41. Install the ignition coil and power transistor assembly, if equipped.

42. Connect all electrical connectors and vacuum hoses.

43. Install the fan and all related parts. Adjust all belt tensions, as required.

44. Install the radiator, shroud and upper hose.

45. Install the air cleaner assembly, ducts and air intake hose.

46. Fill the engine with the specified amount of oil and fill the radiator with coolant.

47. Connect the negative battery cable.

48. Connect the jumper wire from the fuel pump activation terminal to the positive battery post. Inspect the system for leaks.

49. Check the automatic transmission fluid level, if equipped.

50. Recheck all engine adjustments.

51. Install and align the hood.

Water Pump

REMOVAL & INSTALLATION

1. If necessary, properly release the fuel pressure.

2. Disconnect the negative battery cable.

> ❋❋ **CAUTION**
>
> **Wait at least 90 seconds after the negative battery cable is disconnected to prevent possible deployment of the air bag.**

3. Drain the cooling system.

> ❋❋ **CAUTION**
>
> **The fuel injection system remains under pressure after the engine has been OFF. Properly relieve fuel pressure before disconnecting any fuel lines. Failure to do so may result in fire or personal injury.**

4. Remove the upper radiator shroud.

5. Remove all accessory belts. Remove the air conditioning compressor tensioner pulley, if equipped.

6. Remove the cooling fan and clutch assembly and remove the water pump pulley.

7. Disconnect the radiator hose from the water pump.

8. Remove the crankshaft pulley(s).

9. Remove the timing belt covers. If the same timing belt will be reused, mark the direction of the timing belt's rotation, for installation in the same direction. Be sure the engine is positioned so the No. 1 cylinder is at the TDC of its compression stroke and the sprockets timing marks are aligned with the engine's timing mark indicators. Remove the timing belt.

10. The water pump bolts are different lengths, note their positions before removing. Remove the water pump mounting bolts and remove the pump from the block and the water pipe connection. Remove the O-ring from the water pipe connection.

To install:

11. Clean and dry the mating surfaces of the block and water pump. Install a new O-ring to the water pipe connection. Coat the new O-ring with water to aid in installation.

12. Install the water pump with a new gasket to the block and tighten the bolts to:

- 2.4L engine: 14 ft. lbs. (19 Nm)
- 3.0L 12 valve and 3.5L engines: 17 ft. lbs. (23 Nm)
- 3.0L 24 valve engine: 12–14 ft. lbs. (17–20 Nm)

13. Tighten the alternator bracket bolt to 17 ft. lbs. (23 Nm).

14. Install the timing belt(s) and covers.

15. Install the crankshaft pulley(s).

16. Connect the radiator hose to the water pump.

17. Install the water pump pulley. Install the cooling fan and clutch assembly.

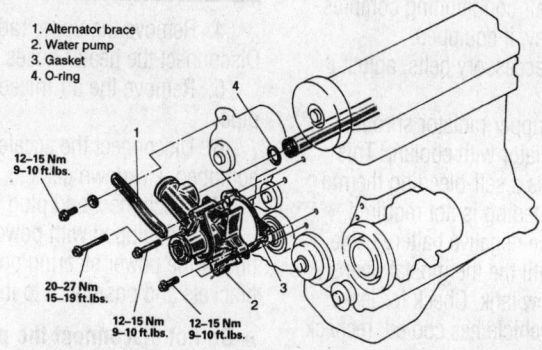

1. Alternator brace
2. Water pump
3. Gasket
4. O-ring

12–15 Nm
9–10 ft. lbs.

20–27 Nm
15–19 ft. lbs.

12–15 Nm
9–10 ft. lbs.

12–15 Nm
9–10 ft. lbs.

Water pump and related components—2.4L engine

7924UG06

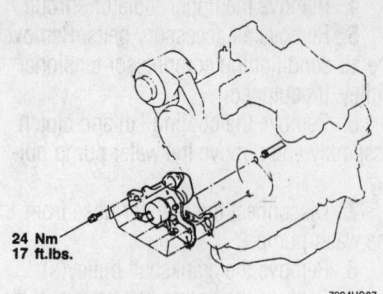

24 Nm
17 ft.lbs.

7924UG07

Water pump mounting—3.0L 12 valve engine

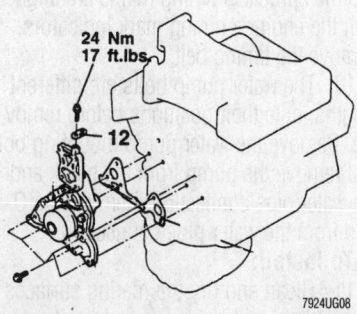

24 Nm
17 ft.lbs.

12

7924UG08

Water pump mounting—3.0L 24 valve engine

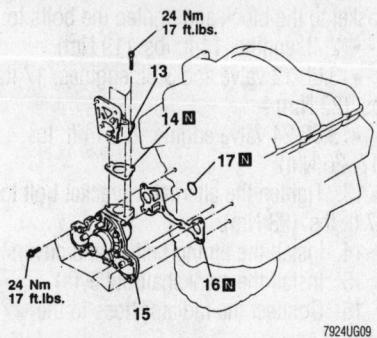

24 Nm
17 ft.lbs.

13

14

17

16

24 Nm
17 ft.lbs.

15

7924UG09

Water pump and related components—3.5L engine

18. Install the air conditioning compressor tensioner pulley, if equipped.

19. Install the accessory belts, adjust if necessary.

20. Install the upper radiator shroud.

21. Fill the radiator with coolant. This cooling system has a self-bleeding thermostat, so system bleeding is not required.

22. Connect the negative battery cable, run the vehicle until the thermostat opens and fill the overflow tank. Check for leaks.

23. Once the vehicle has cooled, recheck the coolant level.

Cylinder Head

REMOVAL & INSTALLATION

2.4L Engine

1. Disconnect negative battery cable.

✳✳ CAUTION

Some models covered by this manual may be equipped with a Supplemental Restraint System (SRS), which uses an air bag. Whenever working near any of the SRS components, such as the impact sensors, the air bag module, steering column and instrument panel, disable the SRS.

2. Properly relieve the fuel system pressure.

✳✳ CAUTION

The fuel injection system remains under pressure after the engine has been OFF. Properly relieve fuel pressure before disconnecting any fuel lines. Failure to do so may result in fire or personal injury.

3. Drain the cooling system.

✳✳ CAUTION

Never open, service or drain the radiator or cooling system when hot; serious burns can occur from the steam and hot coolant. Also, when draining engine coolant, keep in mind that cats and dogs are attracted to ethylene glycol antifreeze and could drink any that is left in an uncovered container or in puddles on the ground. This will prove fatal in sufficient quantities. Always drain coolant into a sealable container. Coolant should be reused unless it is contaminated or is several years old.

4. Remove the upper radiator hose. Disconnect the heater hoses.

5. Remove the air intake hose and pipe.

6. Disconnect the accelerator, and if equipped, kickdown cable. .

7. Disconnect and plug the fuel lines.

8. If equipped with power steering, unbolt the power steering pump from its brackets and position it to the side.

➡**Do not disconnect the power steering lines.**

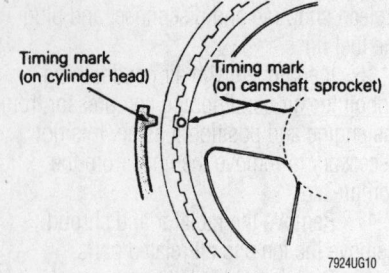

Timing mark (on cylinder head)

Timing mark (on camshaft sprocket)

7924UG10

Align the camshaft sprocket timing mark with the timing mark on the cylinder head—2.4L engines

9. Remove the timing belt upper cover and valve cover.

10. Rotate the crankshaft clockwise and align the timing mark on the camshaft sprocket with the timing mark on the cylinder head.

11. Remove the camshaft bolt.

12. Remove the sprocket from the camshaft (with the timing belt attached) and allow it to rest on the lower cover. Secure the belt to the sprocket so that they do not become disengaged.

✳✳ WARNING

Do not rotate the crankshaft after the camshaft sprocket is removed from the camshaft. Secure the sprocket and timing belt so there is no slack in the belt. Be sure the sprocket does not become disengaged from the timing belt. If the engine is disturbed or the timing belt moved, the camshaft timing will have to be reset.

13. Label and disconnect the spark plug wires from the spark plugs. Remove the distributor cap and wires.

14. Mark the position of the rotor and distributor housing in relation to the cylinder head and remove the distributor.

15. Label and disconnect all vacuum lines, hoses and wiring connectors from the manifolds and cylinder head.

16. Raise the vehicle and support safely.

17. Remove the exhaust pipe from the exhaust manifold.

18. Lower the vehicle.

19. Remove cylinder head bolts, starting from the outside and working inward. Remove the cylinder head from the engine.

20. If necessary, remove the intake and exhaust manifolds from the cylinder head.

21. Clean the cylinder head gasket mating surfaces.

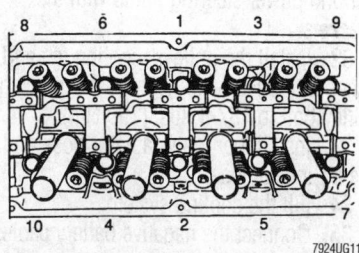

Tighten the cylinder head bolts according to the sequence shown for proper cylinder sealing—2.4L engines

To install:

22. If removed, install the intake and exhaust manifolds to the cylinder head.

23. Install a new head gasket to the block and position the cylinder head assembly with all head bolts and washers. Tighten the bolts in sequence, in 2 or 3 steps, to 72 ft. lbs. (100 Nm).

24. Install the camshaft sprocket to the camshaft and tighten the bolt to 72 ft. lbs. (100 Nm).

25. Install the distributor, aligning the marks made during removal. Install the distributor cap and spark plug wires.

26. Install the power steering pump and adjust the belt tension.

27. Connect the heater hoses and upper radiator hose.

28. Install the valve cover and upper timing belt cover.

29. Connect the fuel lines. Connect the accelerator, and if equipped, kickdown cable.

30. Connect all remaining wiring connectors, vacuum lines and hoses in their proper locations.

31. Install the air intake pipe and hose.

32. Refill the cooling system.

33. Connect the negative battery cable.

34. Start the engine and check for leaks. Check the ignition timing.

3.0L Engines

12 VALVE HEADS

✳✳ CAUTION

The fuel injection system remains under pressure after the engine has been OFF. Properly relieve fuel pressure before disconnecting any fuel lines. Failure to do so may result in fire or personal injury.

1. Relieve the fuel system pressure.
2. Disconnect the negative battery cable.

✳✳ CAUTION

Work must be started after 90 seconds from the time the ignition switch is turned to the LOCK position and the negative battery cable is disconnected.

3. Drain the cooling system. Remove the upper radiator hose.

4. Remove the air intake hose.

5. Remove the accessory drive belts, fan and pulleys.

6. Remove the air conditioning compressor, power steering pump and mounting brackets and position them to the side, without disconnecting the lines.

7. If removing the right cylinder head, disconnect the wiring and remove the alternator cover, alternator and alternator stay.

8. Remove the timing belt covers.

9. Remove the timing belt as follows:

   a. Rotate the crankshaft and bring the piston in No. 1 cylinder to TDC on the compression stroke. Align the camshaft and crankshaft sprocket timing marks.

   b. Mark the timing belt in the direction of rotation for reinstallation purposes.

   c. Loosen the timing belt tensioner bolt and turn the tensioner counterclockwise. Remove the timing belt.

✳✳ WARNING

Do not rotate the crankshaft or camshaft sprockets after the timing belt has been removed.

10. Label and disconnect the spark plug wires from the spark plugs.

11. If removing the left cylinder head, remove the distributor cap. Mark the position of the rotor and the distributor housing in relation to the cylinder head, then remove the distributor.

12. Remove the valve cover.

13. Remove the EGR pipe and the air intake plenum stays.

14. Disconnect the accelerator, if equipped, throttle control cable. Disconnect and plug the fuel lines.

15. Label and disconnect the wiring connectors, vacuum lines and hoses from the air intake plenum, intake manifold and cylinder head.

16. Remove the air intake plenum and intake manifold.

17. Remove the exhaust manifold.

18. If removing the right cylinder head, remove the dipstick tube.

19. If necessary, remove the camshaft sprocket bolt and camshaft sprocket.

Remove the alternator bracket and/or timing belt rear cover.

20. Remove the cylinder head bolts starting from the outside and working inward.

21. Remove the cylinder head from the engine.

22. Clean the gasket mounting surfaces.

To install:

23. Install the new cylinder head gasket over the dowels on the engine block, with the identification mark at front top.

24. Install the cylinder head on the engine and tighten the cylinder head bolts in sequence using 3 even steps, to 70 ft. lbs. (95 Nm).

25. If removed, install the timing belt rear cover and/or alternator bracket. Tighten the alternator bracket bolts in the proper sequence.

26. If removed, install the dipstick tube using a new O-ring coated with clean engine oil.

27. Install the exhaust manifold.

28. Install the intake manifold and air intake plenum.

29. Install the EGR pipe and air intake plenum stays.

30. Connect the fuel lines, accelerator, if equipped, throttle control cable, wiring connectors, vacuum lines and hoses.

31. Install the valve cover.

32. Install the distributor, if removed, aligning the marks made during removal. Install the distributor cap and spark plug wires.

33. Be sure the camshaft and crankshaft sprocket timing marks are aligned.

34. Turn the timing belt tensioner to the extreme counterclockwise position and temporarily tighten the bolt.

35. Install the timing belt in the original rotation direction. Loosen the timing belt tensioner bolt and allow the spring force of the tensioner to tension the belt.

36. Turn the crankshaft 2 turns in the normal direction of rotation and check the timing mark alignment.

37. If the timing is correct, tighten the tensioner bolt to 21 ft. lbs. (30 Nm). If the timing is incorrect, repeat the belt installation procedure.

38. Install the timing belt covers.

39. If removed, install the alternator, alternator cover and alternator stay.

40. Install the air conditioning compressor and power steering pump with the brackets.

41. Install the pulleys, cooling fan and accessory drive belts. Tighten the crankshaft pulley bolt to 137 ft. lbs. (190 Nm).

42. Install the upper radiator hose and the air intake hose.

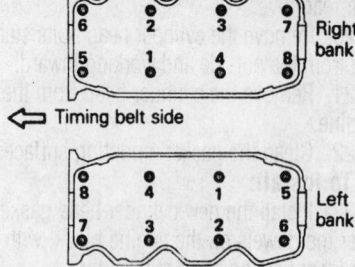

Right bank

Left bank

← Timing belt side

7924UG12

Tighten the cylinder head bolts in the order shown, in 2 or 3 passes—3.0L engines

43. Fill the cooling system.
44. Connect the negative battery cable.
45. Start the engine and check for leaks. Check the ignition timing.

24 VALVE HEADS

1. Relieve the fuel system pressure.
2. Disconnect the negative battery cable.

✳✳ CAUTION

Never open, service or drain the radiator or cooling system when hot; serious burns can occur from the steam and hot coolant. Also, when draining engine coolant, keep in mind that cats and dogs are attracted to ethylene glycol antifreeze and could drink any that is left in an uncovered container or in puddles on the ground. This will prove fatal in sufficient quantities. Always drain coolant into a sealable container. Coolant should be reused unless it is contaminated or is several years old.

3. Drain the cooling system. Remove the upper radiator hose.
4. Remove the accessory drive belts, fan and pulleys.
5. Remove the air conditioning compressor, power steering pump and mounting brackets and position them to the side, without disconnecting the lines.
6. Remove the timing belt covers.
7. Remove the timing belt as follows:
   a. Rotate the crankshaft and bring the piston in No. 1 cylinder to TDC on the compression stroke. Align the camshaft and crankshaft sprocket timing marks.
   b. Mark the timing belt in the direction of rotation for reinstallation purposes.
   c. Loosen the timing belt tensioner bolt and turn the tensioner counterclockwise. Remove the timing belt.

✳✳ WARNING

Do not rotate the crankshaft or camshaft sprockets after the timing belt has been removed.

8. Label and disconnect the spark plug wires from the spark plugs.
9. Remove the valve cover.
10. Disconnect and plug the fuel lines.
11. Label and disconnect the wiring connectors, vacuum lines and hoses from the air intake plenum, intake manifold and cylinder head.
12. Remove the air intake plenum and intake manifold.
13. Remove the exhaust manifold.
14. If necessary, remove the camshaft sprocket bolt and camshaft sprocket. Remove the alternator bracket and/or timing belt rear cover.
15. Remove the cylinder head bolts starting from the outside and working inward.
16. Remove the cylinder head from the engine.
17. Clean the gasket mounting surfaces.

To install:

18. Install the new cylinder head gasket over the dowels on the engine block, with the identification mark at front top.
19. Install the cylinder head on the engine and tighten the cylinder head bolts in sequence using 3 even steps, to 70 ft. lbs. (95 Nm).
20. Install the exhaust manifold.
21. Install the intake manifold and air intake plenum.
22. Connect the fuel lines, accelerator, if equipped, throttle control cable, wiring connectors, vacuum lines and hoses.
23. Install the valve cover.
24. Be sure the camshaft and crankshaft sprocket timing marks are aligned.
25. Turn the timing belt tensioner to the extreme counterclockwise position and temporarily tighten the bolt.
26. Install the timing belt in the original rotation direction. Loosen the timing belt tensioner bolt and allow the spring force of the tensioner to tension the belt.
27. Turn the crankshaft 2 turns in the normal direction of rotation and check the timing mark alignment.
28. If the timing is correct, tighten the tensioner bolt to 21 ft. lbs. (30 Nm). If the timing is incorrect, repeat the belt installation procedure.
29. Install the timing belt covers.
30. If removed, install the alternator, alternator cover and alternator stay.
31. Install the air conditioning compres-

sor and power steering pump with the brackets.
32. Install the pulleys, cooling fan and accessory drive belts. Tighten the crankshaft pulley bolt to 137 ft. lbs. (190 Nm).
33. Install the upper radiator hose and the air intake hose.
34. Fill the cooling system.
35. Connect the negative battery cable.
36. Start the engine and check for leaks. Check the ignition timing.

3.5L Engine

✳✳ CAUTION

The fuel injection system remains under pressure after the engine has been OFF. Properly relieve fuel pressure before disconnecting any fuel lines. Failure to do so may result in fire or personal injury.

1. Relieve fuel system pressure. Disconnect the negative battery cable.

✳✳ CAUTION

Work must be started after 90 seconds from the time the ignition switch is turned to the LOCK position and the negative battery cable is disconnected.

2. Drain the cooling system.
3. Remove the air intake hoses.
4. Remove air intake plenum and intake manifold.
5. Remove the exhaust manifold.
6. Remove the timing belt.
7. Remove the breather hose.
8. Remove the spark plug cable center cover and remove the spark plug cables.
9. Remove the valve cover.
10. To remove the intake camshaft sprocket, hold the camshaft with a wrench on the hexagon near the end of the camshaft and remove the sprocket bolt.
11. Remove the rear timing belt cover.
12. Remove the ignition coil.
13. Disconnect all water hoses from the thermostat housing and remove the housing.
14. Disconnect the water inlet from the front head and discard O-ring.
15. Loosen the cylinder head mounting bolts in three steps, starting from the outside and working inward. Lift off the cylinder head assembly and remove the head gasket.

To install:

16. Thoroughly clean and dry the mating surfaces of the head and block. Check the

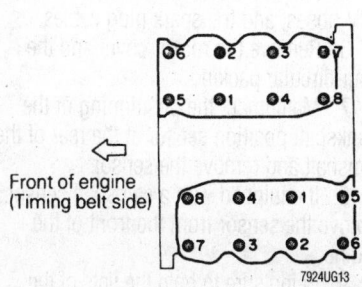

Be sure to tighten the cylinder head bolts in the correct sequence—3.5L engines

cylinder head for cracks, damage or engine coolant leakage. Remove scale, sealing compound and carbon. Clean oil passages thoroughly. Check the head for flatness. End to end, the head should be within 0.0012 in., normally with 0.008 in. the maximum allowed out of true. The total thickness allowed to be removed from the head and block is 0.008 in. maximum.

17. Place a new head gasket on the cylinder block with the identification marks in the front top (upward) position. Do not use sealer on the gasket.

18. Carefully install the cylinder head on the block. Be sure the head bolt washers are installed with the chamfered edge upward. Using three even steps, tighten the head bolts in sequence, to 76–83 ft. lbs. (105–115 Nm) for non-turbocharged cold engine or 87–94 ft. lbs. (120–130 Nm) for turbocharged cold engine.

19. Install new O-ring and connect the water inlet to the front head.

20. Replace the gaskets and install the thermostat housing and connect the hoses.

21. Install the ignition coil and center rear timing belt cover.

22. Install the intake camshaft sprocket. Use hex flange on camshaft to secure and tighten the retaining bolt to 65 ft. lbs. (90 Nm).

23. Apply sealer to the lower edges of the half-round portions of the belt-side of the new gasket and install the valve cover. Tighten the bolts to 7 ft. lbs. (10 Nm).

24. Connect the spark plug cables and install the center cover.

25. Install the breather hose.

26. Install the timing belt and all related items.

27. Using all new gaskets, install the intake manifold, air intake plenum, turbocharger and exhaust manifold, following the proper torque sequences.

28. Install the air intake hoses.

29. Change the engine oil and oil filter.

30. Fill the system with coolant.

31. Connect the negative battery cable, run the vehicle until the thermostat opens, fill the radiator completely.

32. Adjust the accelerator cable. Check and adjust the idle speed and ignition timing.

33. Once the vehicle has cooled, recheck the coolant level.

Rocker Arms/Shafts

REMOVAL & INSTALLATION

2.4L and 1995 3.0L Engines

1. Install the special clips (MB 998443–01) to hold the auto-adjusters in place. Carefully loosen the retaining bolts at each bearing cap. Keep light downward pressure on the rockers to hold the camshaft in place.

2. When all the bolts are loose, remove them from the front (pulley end) to the rear while keeping the same light pressure on the assembly. Have an assistant hold the camshaft firmly in place as you remove the rocker assembly. The tension of the timing belt will try to pop the camshaft out of its journals; this must NOT be allowed to hap-

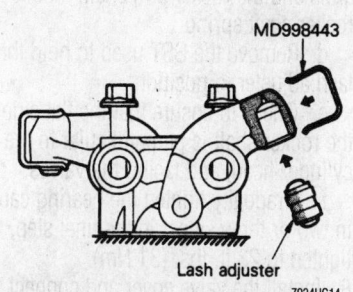

Insert the Lash Adjuster Holder tool MD998443 or equivalent to prevent the lash adjuster from falling out—2.4L and 3.0L engines

pen. If this should happen, you'll need to reinstall the camshaft and timing belt. Instructions for each are found later in this section.

3. As soon as the rocker assembly is clear of the vehicle, remove the rearmost bearing cap and install it in its original position on the cam. Tighten the bolts just snug enough to hold it in place. If, during the inspection process, this bearing cap must be removed for cleaning or replacement, install the No. 3 bearing cap before removing the No. 4 cap.

➡**Take care during removal that no oil, grease or dirt comes into contact with the timing belt.**

To install:

4. Disassemble the rocker assembly, checking each component for wear, scoring, or plugged oil passages. Check the roller for correct and smooth rotation. Inspect the inner diameter of each rocker for any scoring or enlargement. If wear is found inside the rocker, replace it and inspect the shaft for damage.

5. Before reassembly, coat the contact faces liberally with clean motor oil. Observe the numbers on the bearing caps so that they are replaced in the correct location. Reassemble the rocker shafts into the front bearing cap so that the notches face outward. As you continue to assemble the springs, rockers and bearing caps, remember that the arrows on the bearing caps must point in the same direction as the arrow on the head. This is particularly important if the rockers have been removed from both heads.

6. When the rocker arm assembly is complete (except for the rear bearing cap, which is holding the camshaft down), have your assistant hold the camshaft while the rear cap is removed. Assemble the last cap onto the rocker assembly and carefully fit the assembly onto the head. Remember that

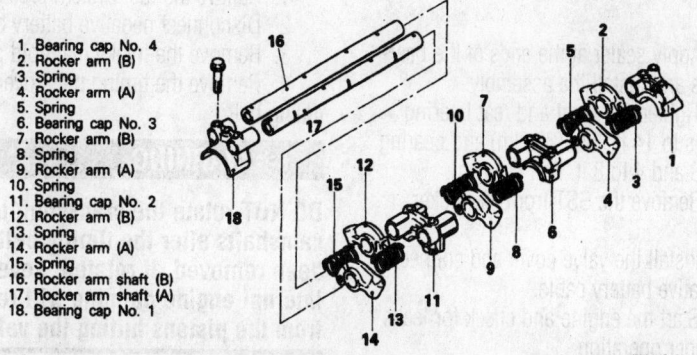

1. Bearing cap No. 4
2. Rocker arm (B)
3. Spring
4. Rocker arm (A)
5. Spring
6. Bearing cap No. 3
7. Rocker arm (B)
8. Spring
9. Rocker arm (A)
10. Spring
11. Bearing cap No. 2
12. Rocker arm (A)
13. Spring
14. Rocker arm (A)
15. Spring
16. Rocker arm shaft (B)
17. Rocker arm shaft (A)
18. Bearing cap No. 1

Exploded view of the rocker arms and shafts—3.0L engine

the camshaft must be held in place during this exchange.

7. Install the retaining bolts finger-tight, making sure each rocker is correctly aligned with the camshaft and valve stem. Once the rockers are securely in place, the special retaining clips may be removed.

8. Starting at the center bearing cap and working outward in two passes, tighten the bearing cap bolts to 15 ft. lbs. (20 Nm).

9. Apply sealer to the correct locations and reinstall the valve cover.

1996–99 3.0L Engines

12 VALVE HEADS

1. Disconnect the negative battery cable.

✳✳ CAUTION

Wait at least 90 seconds from the time the ignition switch is turned to the LOCK position and the negative battery cable is disconnected to allow the air bag time to discharge.

2. Remove the valve cover.

3. Install auto lash adjuster retainers SST MD998443–01 or equivalent, on the rocker arms.

4. Remove the camshaft bearing caps, but do not remove the bolts from the caps. Remove the rocker arms, rocker shafts and bearing caps, as an assembly.

To install:

5. Lubricate the camshaft journals and camshaft with clean engine oil.

6. Align the camshaft bearing caps with the arrow mark (depending on cylinder numbers) and in numerical order.

➡ **The arrow marks on the left rocker shaft bearing caps face in same direction as the arrow marked on the left cylinder head which is away from the timing belt. The arrows on the right cylinder head and the right rocker bearing caps face toward the timing belt.**

7. Apply sealer at the ends of the bearing caps and install the assembly.

8. Tighten the front and rear bearing cap bolts to 14 ft. lbs. (20 Nm) and bearing caps 2,3 and 4 to 8 ft. lbs. (11 Nm).

9. Remove the SST from the rocker arms.

10. Install the valve cover and connect the negative battery cable.

11. Start the engine and check for leaks and proper operation.

24 VALVE HEADS

1. Disconnect the negative battery cable.

✳✳ CAUTION

Work must be started after 90 seconds from the time the ignition switch is turned to the LOCK position and the negative battery cable is disconnected.

2. Remove the valve cover.

3. Install auto lash adjuster retainers SST MD998443 or equivalent, on the rocker arms.

4. Remove the rocker arms, rocker shafts and bearing caps, as an assembly.

5. Inspect the bearing journals on the camshaft and the cylinder head.

To install:

6. Lubricate the camshaft journals and camshaft with clean engine oil.

7. Install the rocker arms, rocker arm shaft and the rocker shaft spring.

a. Temporarily tighten the rocker shaft with the bolts at the condition that all the intake valve rocker arms do not push the valves.

b. Insert the rocker shaft spring from above and mount it at right angles to the plug guide.

c. Before installing the exhaust rocker arms and the rocker arm shaft, mount the rocker shaft spring.

d. Remove the SST used to hold the lash adjuster in position.

e. Check to ensure that the flat side of the rocker shaft is perpendicular to the cylinder head, and facing the valves.

f. Gradually tighten the bearing caps in two or three steps. In the final step, tighten to 23 ft. lbs. (31 Nm).

8. Install the valve cover and connect the negative battery cable.

9. Start the engine and check for leaks and proper operation.

3.5L Engine

1. Relieve the fuel system pressure.

2. Disconnect negative battery cable.

3. Remove the intake manifold plenum.

4. Remove the timing belt cover and the timing belt.

✳✳ WARNING

DO NOT rotate the crankshaft or camshafts after the timing belt has been removed. If rotated, severe internal engine damage will result from the pistons hitting the valves.

5. Remove the center cover, breather, PCV hoses, and the spark plug cables.

6. Remove the rocker cover and the semi-circular packing.

7. Matchmark the positioning of the crankshaft position sensor at the rear of the camshaft and remove the sensor.

8. If equipped with a camshaft sensor, remove the sensor from the front of the engine.

9. Being sure to hold the flats of the camshaft, loosen the camshaft sprocket bolts.

10. Noting the positioning and location of the sprockets, remove the sprockets from the camshafts.

➡ **Be sure to note the positioning of the knock pin at the end of the camshafts for reinstallation purposes.**

➡ **Be sure to keep the valvetrain components labeled and in proper order for reassembly.**

11. Loosen the bearing cap bolts in 2–3 steps. Label and remove all camshaft bearing caps.

➡ **If the bearing caps are difficult to remove, use a plastic hammer to gently tap the components.**

12. Mark the components and remove the intake and the exhaust camshafts.

13. Remove the rocker arms and the lash adjusters. Be sure to note the location of the valvetrain components for reinstallation purposes.

14. Check the camshaft journals for wear or damage. Check the cam lobes for damage. Also, check the cylinder head oil holes for clogging.

To install:

➡ **Lubricate the valvetrain components with clean engine oil.**

15. Bleed and install the lash adjusters to the to the original bores in the cylinder head.

16. Install the rocker arms to the cylinder head.

17. Lubricate the camshafts with clean engine oil and position the camshafts on the cylinder head.

✳✳ WARNING

Be sure to properly position the knock pins of the camshaft to prevent valve to piston interference.

➡ **Do not confuse the intake camshaft with the exhaust camshaft. The intake camshaft on the Montero has a P** stamped on the hexagon. The exhaust camshaft on the Montero has a K stamped on the hexagon.

→Install the bearing caps according to the identification mark and cap number. Bearing caps No. 2, 3 and are marked as such. The caps also are marked I for intake or E for exhaust.

18. Install the bearing caps. Tighten the caps in sequence and in 2 or 3 steps. Caps 2, 3 and 4 have a front mark. Install with the mark aligned with the front mark on the cylinder head. Tighten the retaining bolts for caps No. 2, 3 and 4 to 8 ft. lbs. (11 Nm) and tighten the retaining bolts for the front and rear caps to 14 ft. lbs. (20 Nm).

19. Apply a coating of engine oil to the oil seals and install the oil seals to the front and rear of the camshafts.

20. Holding the flats of the camshaft, install and tighten the sprocket bolts to 65 ft. lbs. (90 Nm).

21. If removed, install the camshaft position sensor and tighten the mounting bolts to 78 inch lbs. (9 Nm).

22. Aligning the matchmark, install the crankshaft position sensor at the rear of the camshaft and tighten the mounting nut to 7 ft. lbs. (12 Nm).

23. Align the marks on the camshaft and crankshaft sprockets. Install the timing belt assembly.

24. Install the rocker cover and the semi-circular packing.

25. Install the intake manifold plenum.

26. Install the spark plug cables, center cover, breather and PCV hoses.

27. Connect the negative battery cable and check for leaks.

Intake Manifold

REMOVAL & INSTALLATION

2.4L Engine

1. Relieve the fuel pressure.

> ❋❋ **CAUTION**
>
> **The fuel injection system remains under pressure after the engine has been OFF. Properly relieve fuel pressure before disconnecting any fuel lines. Failure to do so may result in fire or personal injury.**

2. Disconnect the negative battery cable.

> ❋❋ **CAUTION**
>
> **Wait at least 90 seconds after the negative battery cable is disconnected to prevent possible deployment of the air bag.**

3. Drain the engine coolant. Disconnect the upper radiator hose from the thermostat housing.

4. Remove the air intake hoses, breather hose and the air intake pipe.

5. Disconnect all wires, hoses and linkages to the throttle body.

6. Remove the ignition coil.

7. Disconnect the brake booster hose and vacuum hose cluster from the air intake plenum.

8. Unbolt the air intake plenum from the intake manifold and remove the plenum from the engine.

9. Cover the fuel line with a clean shop rag and disconnect the fuel lines from the fuel rail. Keep the line covered or plugged.

10. Remove the fuel rail assembly with injectors intact.

11. Disconnect the heater hose from the manifold.

12. Disconnect the wires to the engine coolant switches.

13. Matchmark the rotor to the housing and the housing to the cylinder head and remove the distributor.

14. Unbolt the intake manifold from the cylinder head and remove from the engine.

15. Clean and dry the mating surfaces of the manifold and cylinder head.

To install:

16. Using a new gasket, install the intake manifold to the head. Starting from the middle and working outward, tighten the retaining nuts to 11–14 ft. lbs. (15–20 Nm).

17. Connect the wires to the engine coolant switches. Install the distributor with matchmarks aligned.

18. Install the fuel rail assembly to the manifold and connect the fuel line using a new O-ring.

19. Connect the heater hose to the manifold.

20. Install the air intake plenum with a new gasket. Tighten the retaining bolts to 12 ft. lbs. (16 Nm).

21. Connect the vacuum hoses cluster, brake booster hose and all wires, hoses and linkages to the throttle body.

22. Install the ignition coil.

23. Install the air intake pipe and hoses.

24. Connect the upper radiator hose.

25. Fill the radiator with coolant.

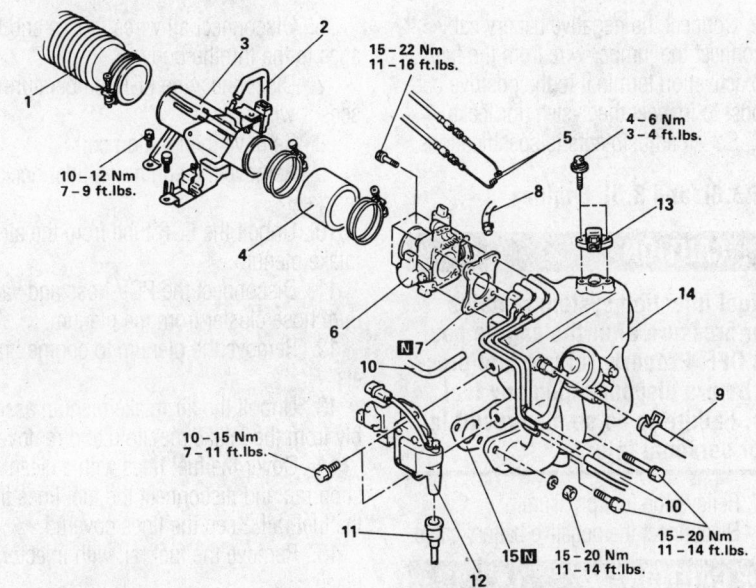

1. Air intake hose
2. Breather hose
3. Air intake pipe
4. Air hose
5. Accelerator cable and kick down plate
6. Throttle body
7. Gasket
8. Water hose
9. Brake booster vacuum hose
10. Vacuum hose connection
11. High tension cable
12. Ignition coil
13. Manifold difference pressure sensor <1996 models>
14. Intake manifold plenum assembly
15. Intake manifold plenum gasket

7924UG39

Exploded view of the intake plenum chamber and related components—1995–96 2.4L engine shown

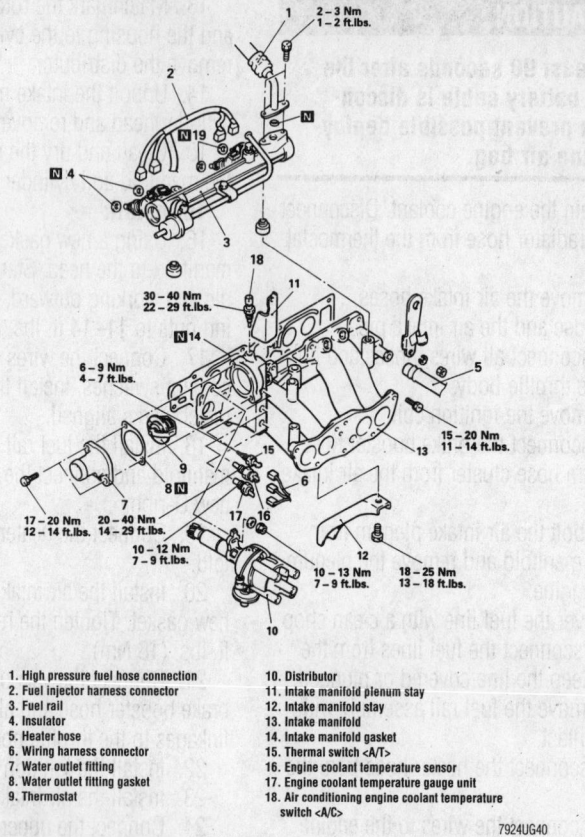

2 – 3 Nm
1 – 2 ft.lbs.

19

4

30 – 40 Nm
22 – 29 ft.lbs.

14

6 – 9 Nm
4 – 7 ft.lbs.

15 – 20 Nm
11 – 14 ft.lbs.

17 – 20 Nm
12 – 14 ft.lbs.

20 – 40 Nm
14 – 29 ft.lbs.

10 – 12 Nm
7 – 9 ft.lbs.

10 – 13 Nm
7 – 9 ft.lbs.

18 – 25 Nm
13 – 18 ft.lbs.

1. High pressure fuel hose connection
2. Fuel injector harness connector
3. Fuel rail
4. Insulator
5. Heater hose
6. Wiring harness connector
7. Water outlet fitting
8. Water outlet fitting gasket
9. Thermostat
10. Distributor
11. Intake manifold plenum stay
12. Intake manifold stay
13. Intake manifold
14. Intake manifold gasket
15. Thermal switch <A/T>
16. Engine coolant temperature sensor
17. Engine coolant temperature gauge unit
18. Air conditioning engine coolant temperature switch <A/C>

7924UG40

Exploded view of the intake manifold and related components—1995–96 2.4L engine shown

26. Connect the negative battery cable and connect the jumper wire from the fuel pump activation terminal to the positive battery post to inspect the system for leaks.

27. Set all adjustments to specifications.

1995 3.0L and 3.5L Engines

> ※ **CAUTION**
>
> **The fuel injection system remains under pressure after the engine has been OFF. Properly relieve fuel pressure before disconnecting any fuel lines. Failure to do so may result in fire or personal injury.**

1. Relieve the fuel pressure.
2. Disconnect the negative battery cable.

> ※ **CAUTION**
>
> **Never open, service or drain the radiator or cooling system when hot; serious burns can occur from the steam and hot coolant.**

3. Drain the engine coolant.
4. Disconnect the upper radiator hose from the thermostat housing.
5. Remove the air intake hose from the throttle body.

6. Disconnect all wires, hoses and linkages to the throttle body.
7. Disconnect the EGR temperature sensor wire.
8. Remove the ignition coil.
9. Remove the engine oil filler neck bracket.
10. Unbolt the EGR tube from the air intake plenum.
11. Disconnect the PCV hose and vacuum hose cluster from the plenum.
12. Remove the plenum to engine brackets.
13. Unbolt the air intake plenum assembly from the intake manifold and remove.
14. Cover the fuel lines with a clean shop rag and disconnect the fuel lines from the fuel rail. Keep the lines covered.
15. Remove the fuel rail with injectors in place.
16. Disconnect the bypass hose and the upper radiator hose from the thermostat housing. Disconnect the wires to the coolant temperature switches.
17. Remove the intake manifold retaining nuts and remove the manifold from the cylinder heads.
18. Remove the gaskets and thoroughly clean and dry the mating surfaces of the manifold and heads.

To install:

19. Position the manifold over the studs and install the retaining nuts. Tighten to 10 ft. lbs. (14 Nm) for the 3.0L 12 valve and the 3.5L engines and 16 ft. lbs. (21 Nm) for the 3.0L 24 valve engine. Start from the center and work out toward the corners.
20. Connect the hoses and connect the wires to the coolant switches.
21. Install the fuel rail assembly and connect the fuel hoses.
22. Using a new gasket, install the air intake plenum to the intake manifold. Tighten the nuts and bolts to 12 ft. lbs. (16 Nm). Install the plenum to engine brackets.
23. Connect the PCV hose and vacuum hose cluster to the plenum.
24. Connect the EGR tube.
25. Install the engine oil filler neck bracket.
26. Install the ignition coil assembly.
27. Connect the EGR temperature sensor wire.
28. Connect all wires, hoses and linkages to the throttle body.
29. Install the air intake hose to the throttle body.
30. Connect the upper radiator hose to the thermostat housing.
31. Refill the radiator with coolant.
32. Connect the negative battery cable.
33. Connect the jumper wire from the fuel pump activation terminal to the positive battery post; check the system for fuel leaks.
34. Set all adjustments to specifications.

1996–99 3.0L Engine

1. Relieve the fuel pressure.

> ※ **CAUTION**
>
> **The fuel injection system remains under pressure after the engine has been OFF. Properly relieve fuel pressure before disconnecting any fuel lines. Failure to do so may result in fire or personal injury.**

2. Disconnect the negative battery cable.

> ※ **CAUTION**
>
> **Work must be started after 90 seconds from the time the ignition switch is turned to the LOCK position and the negative battery cable is disconnected.**

3. Drain the engine coolant.
4. Remove the air intake hose from the throttle body.

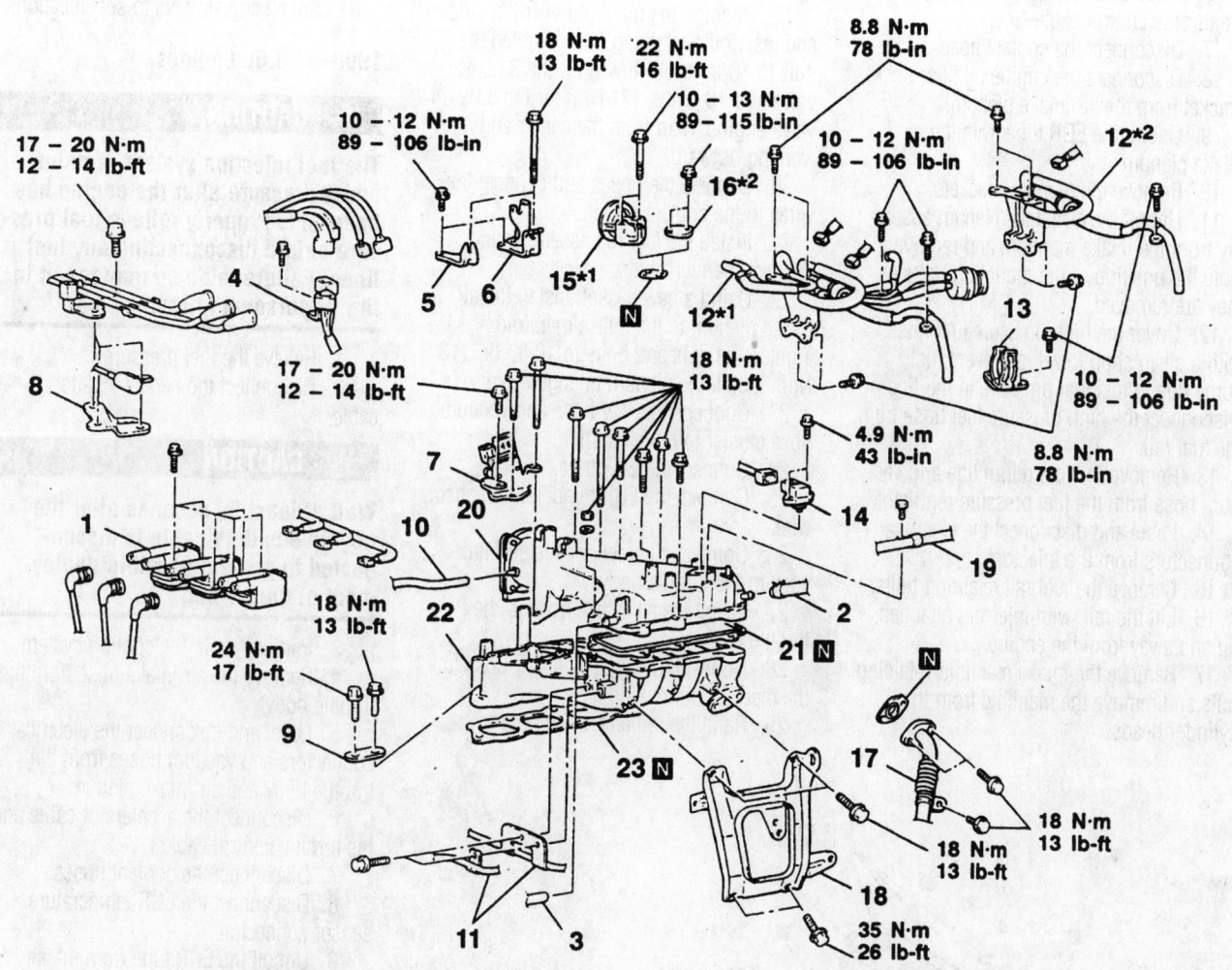

1. IGNITION COILS
2. BRAKE BOOSTER VACUUM HOSE CONNECTION
3. PCV HOSE CONNECTION
4. CRANKSHAFT POSITION SENSOR AND CAM POSITION SENSOR CONNECTOR
5. ACCELERATOR CABLE BRACKET <M/T>
6. THROTTLE CABLE BRACKET <A/T>
7. IGNITION POWER TRANSISTOR
8. WATER OUTLET FITTING BRACKET
9. WATER PUMP STAY
10. VACUUM HOSE CONNECTION
11. FUEL PIPE CONNECTION
12. SOLENOID VALVE AND VACUUM HOSE ASSEMBLY

13. VCV BRACKET
14. MDP SENSOR
15. EGR VALVE
16. COVER
17. EGR PIPE CONNECTION
18. INTAKE MANIFOLD PLENUM STAY
19. THROTTLE CABLE CONNECTION
20. AIR INTAKE FITTING
21. AIR INTAKE FITTING GASKET
22. UPPER INTAKE MANIFOLD
23. INTAKE MANIFOLD PLENUM GASKET

NOTE
*¹: Vehicles for Federal
*²: Vehicles for California

7924UG41

Exploded view of the upper intake manifold and related components—1997 3.0L 24valve engine shown

5. Label and disconnect the electrical connectors and vacuum hoses from the throttle body and air intake plenum.

6. Disconnect the accelerator cable and the throttle control cable.

7. Disconnect the coolant hoses.

8. Disconnect the engine oil filler neck bracket from the air intake plenum.

9. Unbolt the EGR tube from the air intake plenum.

10. Remove the plenum brackets.

11. Unbolt the air intake plenum assembly from the intake manifold and remove. Note the position of the mounting bolts as they are removed.

12. Cover the high pressure fuel hose with a clean shop towel to prevent fuel spray due to residual pressure in the line. Disconnect the high pressure fuel hose from the fuel rail.

13. Remove the fuel return line and vacuum hose from the fuel pressure regulator.

14. Label and disconnect the electrical connectors from the injectors.

15. Remove the fuel rail retaining bolts.

16. Lift the rail, with injectors attached, up and away from the engine.

17. Remove the intake manifold retaining nuts and remove the manifold from the cylinder heads.

18. Remove the gaskets and thoroughly clean and dry the mating surfaces of the manifold and heads.

To install:

19. Position the manifold over the studs and install the retaining nuts. Tighten the nuts to 10 ft. lbs. (14 Nm) for the 3.0L 12 valve and 16 ft. lbs. (21 Nm) for the 3.0L 24 valve engine. Start from the center and working outward.

20. Connect the hoses and connect the wires to the coolant switches.

21. Install the fuel rail assembly and connect the fuel hoses.

22. Using a new gasket, install the air intake plenum to the intake manifold. Tighten the nuts and bolts to 10 ft. lbs. (13 Nm). Install the plenum to engine brackets.

23. Connect the PCV hose and vacuum hose cluster to the plenum.

24. Connect the EGR tube.

25. Connect the EGR temperature sensor wire.

26. Connect all wires, hoses and linkages to the throttle body.

27. Install the air intake hose to the throttle body.

28. Connect the upper radiator hose to the thermostat housing.

29. Refill the radiator with coolant.

30. Connect the negative battery cable.

31. Connect the jumper wire from the fuel pump activation terminal to the positive battery post; check the system for fuel leaks.

32. Set all adjustments to specifications.

1996–99 3.5L Engines

✳✳ CAUTION

The fuel injection system remains under pressure after the engine has been OFF. Properly relieve fuel pressure before disconnecting any fuel lines. Failure to do so may result in fire or personal injury.

1. Relieve the fuel pressure.

2. Disconnect the negative battery cable.

✳✳ CAUTION

Wait at least 90 seconds after the negative battery cable is disconnected to prevent possible deployment of the air bag.

3. Partially drain the cooling system.

4. Remove the air intake hose from the throttle body.

5. Label and disconnect the electrical connectors and vacuum hoses from the throttle body and air intake plenum.

6. Disconnect the accelerator cable and the throttle control cable.

7. Disconnect the coolant hoses.

8. Disconnect the EGR temperature sensor connector.

9. Unbolt the EGR tube from the air intake plenum.

10. Remove the intake manifold plenum cover.

11. Remove the intake manifold plenum stay brackets.

12. Unbolt the air intake plenum assembly from the intake manifold and remove. Note the position of the mounting bolts as they are removed.

13. Remove the induction control valve assembly.

14. Cover the high pressure fuel hose with a clean shop towel to prevent fuel spray due to residual pressure in the line. Disconnect the high pressure fuel hose from the fuel rail.

15. Remove the fuel return line and vacuum hose from the fuel pressure regulator.

16. Label and disconnect the electrical connectors from the injectors.

17. Remove the fuel rail retaining bolts.

18. Lift the rail, with injectors attached, up and away from the engine.

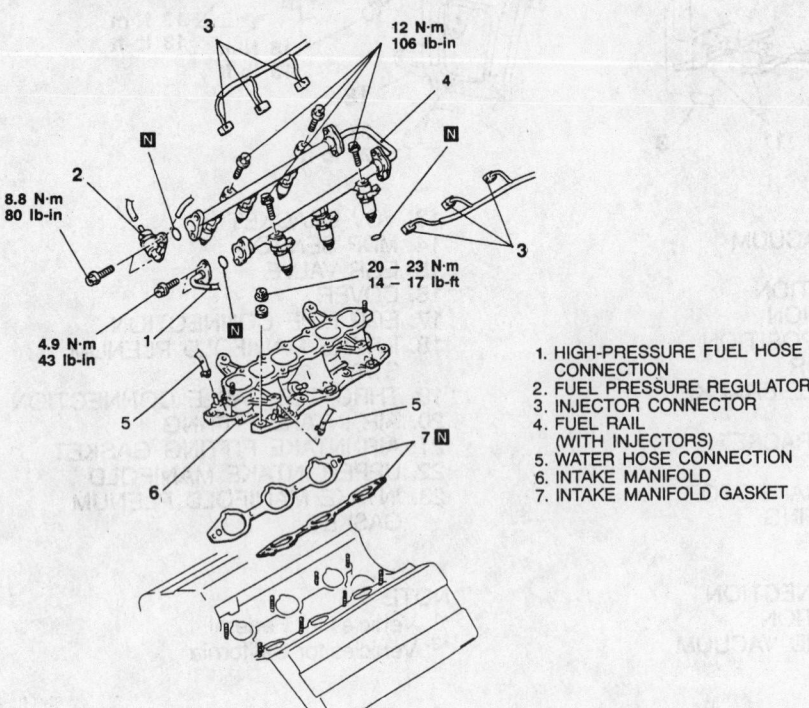

12 N·m
106 lb-in

8.8 N·m
80 lb-in

20 – 23 N·m
14 – 17 lb-ft

4.9 N·m
43 lb-in

1. HIGH-PRESSURE FUEL HOSE CONNECTION
2. FUEL PRESSURE REGULATOR
3. INJECTOR CONNECTOR
4. FUEL RAIL (WITH INJECTORS)
5. WATER HOSE CONNECTION
6. INTAKE MANIFOLD
7. INTAKE MANIFOLD GASKET

7924UG42

Exploded view of the lower intake manifold and related components—1997 3.0L 24valve engine shown

19. Remove the intake manifold retaining nuts and remove the manifold from the cylinder heads.

20. Remove the gaskets and thoroughly clean and dry the mating surfaces of the manifold and heads.

To install:

21. Position the manifold over the studs and install the retaining nuts. Tighten the nuts to 10 ft. lbs. (14 Nm). Start from the center and work outward.

22. Connect the hoses and connect the wires to the coolant switches.

23. Install the fuel rail assembly and connect the fuel hoses.

24. Install the induction control valve assembly and tighten to 6 ft. lbs. (9 Nm).

25. Using a new gasket, install the air intake plenum to the intake manifold. Tighten the nuts and bolts to 13 ft. lbs. (18 Nm). Install the plenum to engine brackets.

26. Connect the PCV hose and vacuum hose cluster to the plenum.

27. Connect the EGR tube.

28. Connect the EGR temperature sensor wire.

29. Connect all wires, hoses and linkages to the throttle body.

30. Install the air intake hose to the throttle body.

31. Connect the upper radiator hose to the thermostat housing.

32. Refill the radiator with coolant.

33. Connect the negative battery cable.

34. Connect the jumper wire from the fuel pump activation terminal to the positive battery post; check the system for fuel leaks.

35. Set all adjustments to specifications.

Exhaust Manifold

REMOVAL & INSTALLATION

2.4L Engine

1. Disconnect the negative battery cable.

❊❊ CAUTION

Wait at least 90 seconds after the negative battery cable is disconnected to prevent possible deployment of the air bag.

2. Remove the heat cowl from the exhaust manifold.

3. If equipped, remove the aspirator valve assembly.

4. Raise and safely support the vehicle.

5. Disconnect the exhaust pipe from the manifold. Lower the vehicle.

6. If equipped, disconnect the oxygen sensor connector and ground cable.

7. Remove the manifold mounting nuts and remove the manifold and gasket from the engine.

To install:

8. Install the exhaust manifold and tighten the mounting nuts to 13 ft. lbs. (18 Nm), starting from the middle and working outward.

9. If removed, connect the oxygen sensor connector and ground cable.

10. Raise and safely support the vehicle.

11. Connect the exhaust pipe to the manifold. Lower the vehicle.

12. If removed, install the aspirator valve assembly.

13. Install the heat cowl to the exhaust manifold.

14. Connect the negative battery cable.

15. Start the engine and check for exhaust leaks.

3.0L 12 Valve Engines

1995 VEHICLES

1. Disconnect the negative battery cable. Raise and safely support the vehicle.

2. Disconnect the oxygen sensor.

3. Disconnect the exhaust pipe from the exhaust manifolds.

4. If removing the left manifold:

a. Disconnect the EGR tube.

b. Remove the air intake plenum bracket.

c. Remove the bracket from the exhaust manifold.

5. If removing the right manifold:

a. Remove the alternator bracket.

b. Remove the engine hanger.

c. Remove the air duct.

6. Remove the heat shield.

7. Remove the manifold to cylinder head nuts and remove the manifolds.

8. Clean the gasket mounting surfaces. Inspect the manifolds for cracks, flatness and/or damage.

To install:

9. Install the new gasket and manifold. Tighten the manifold-to-cylinder head nuts to 14 ft. lbs. (19 Nm).

➡The numbers 2–4–6 on the gasket indicate that the gasket should be installed on the left cylinder head. The numbers 1–3–5 indicate that the gasket should be installed on the right cylinder head.

10. Connect the heat shield.

11. Connect the EGR tube to the manifold, exhaust manifold bracket and air intake plenum bracket, if removed.

12. Install the alternator bracket, engine hanger and air duct, if removed.

13. Connect the exhaust pipe to the exhaust manifolds.

14. Connect the oxygen sensor.

15. Lower the vehicle and connect the negative battery cable.

16. Start the engine and check for exhaust leaks.

1996–99 VEHICLES

1. Disconnect the negative battery cable. Raise and safely support the vehicle.

2. Remove the oil dipstick and the guide.

3. Remove the right heat protector, the engine hanger and the right exhaust manifold and gasket.

4. Remove the left heat protector, the bracket and the exhaust manifold and gasket.

5. Remove the water hoses, the heater pipe and gasket, the water pipe and O-ring and the water inlet pipe.

To install:

6. Install the water inlet pipe, the water pipe and gasket, the heater pipe and gasket and the water hoses.

7. Install the new gasket and manifold. Tighten the exhaust manifold head nuts to 14 ft. lbs. (19 Nm).

➡The numbers 2–4–6 on the gasket indicate that the gasket should be installed on the left cylinder head. The numbers 1–3–5 indicate that the gasket should be installed on the right cylinder head.

8. Connect the heat shield.

9. Install the bracket and the engine hanger.

10. Install the oil dipstick and guide with a new O-ring.

11. Lower the vehicle and connect the negative battery cable.

12. Start the engine and check for exhaust leaks.

3.0L 24 Valve and 3.5L Engines

1995 VEHICLES

1. Disconnect the battery cables and remove the battery and tray.

2. Raise and safely support the vehicle.

3. Remove the air duct and the air cleaner cover.

4. Disconnect the exhaust pipe from the exhaust manifolds.

5. Remove the heat shield.

6. Remove the EGR tube.

7. Remove the manifold to cylinder head nuts and remove the manifolds.

8. Clean the gasket mounting surfaces. Inspect the manifolds for cracks, flatness and/or damage.

To install:

9. Install the new gasket and manifold. Tighten the manifold head nuts to 22 ft. lbs. (29 Nm).

10. Connect the heat shield.

11. Connect the EGR tube to the manifold and the exhaust manifold bracket.

12. Install the air duct.

13. Connect the exhaust pipe to the exhaust manifolds.

14. Lower the vehicle, install the battery and connect the cables.

15. Start the engine and check for exhaust leaks.

1996–99 VEHICLES

1. Disconnect the negative battery cable.

2. Raise and safely support the vehicle.

3. Disconnect the exhaust pipe from the exhaust manifolds.

4. Remove the oil dipstick guide and O-ring.

5. Remove the heat shields.

6. Remove the exhaust manifolds.

7. Clean the gasket mounting surfaces. Inspect the manifolds for cracks, flatness and/or damage.

To install:

8. Install the new gasket and manifold. Tighten the manifold head nuts to:
- 3.0L: 22 ft. lbs. (29 Nm)
- 3.5L: 33 ft. lbs. (45 Nm)

9. Connect the heat shield.

10. Connect the exhaust pipe to the exhaust manifolds.

11. Install the oil dipstick guide with a new O-ring.

12. Lower the vehicle, and connect the negative battery cable.

13. Start the engine and check for exhaust leaks.

Front Crankshaft Seal

REMOVAL & INSTALLATION

2.4L Engines

1. Disconnect the negative battery cable.

⁂ CAUTION

Wait at least 90 seconds after the negative battery cable is disconnected to prevent possible deployment of the air bag.

⁂ CAUTION

Never open, service or drain the radiator or cooling system when hot; serious burns can occur from the steam and hot coolant.

2. Remove the radiator and cooling fan.

3. Remove the accessory drive belts and the air conditioner tension pulley if equipped.

4. Remove the crankshaft pulley and the timing belt front covers.

5. Reinstall the crankshaft pulley bolt and use it to rotate the engine clockwise until the timing marks are aligned.

6. Remove the timing belt and crankshaft sprocket.

7. Remove the inner timing belt and the inner crankshaft sprocket.

8. Carefully pry out the crankshaft seal without scratching the crankshaft.

To install:

9. Place the seal guide over the crankshaft and coat it with engine oil.

10. Slide the seal over the guide until it touches the front case assembly, then use the appropriate seal driver to install the seal. Remove the seal guide.

11. Install the inner crankshaft sprocket and the inner timing belt. Adjust the belt tension.

12. Install the flange and the timing belt sprocket on the crankshaft.

13. Install the main timing belt. Adjust the belt tension.

14. Install the timing belt front covers and the crankshaft pulleys.

15. Install the air conditioner belt tension pulley if equipped.

16. Install the accessory drive belts. Adjust the belts to specifications.

17. Install the cooling fan, radiator and the fan shroud.

18. Fill the cooling system to the proper level.

19. Connect the negative battery cable, start the engine and check for leaks.

3.0L and 3.5L Engines

1995 VEHICLES

1. Disconnect the negative battery cable.

2. Remove the accessory drive belts.

3. Remove the crankshaft pulley.

4. Remove the timing belt covers and the timing belt.

5. Remove the crankshaft sprocket.

6. Pry out the oil seal with a flat prying tool, being careful not to damage the crankshaft.

To install:

7. Coat the lip of the new seal with oil and install the seal using the proper seal driver.

8. Install the crankshaft sprocket and the timing belt.

9. Install the timing belt covers.

10. Install the crankshaft pulley.

11. Install the accessory drive belts.

12. Connect the negative battery cable.

13. Start the engine and check for proper operation.

1996–99 VEHICLES

1. Disconnect the negative battery cable.

⁂ CAUTION

Wait at least 90 seconds after the negative battery cable is disconnected to prevent possible deployment of the air bag.

2. Remove the accessory drive belts.

3. Remove the crankshaft pulley.

4. Remove the timing belt covers and the timing belt.

5. Remove the crankshaft sprocket.

6. Cut out a portion in the crankshaft oil seal lip and pry out the oil seal with a flat prying tool, being careful not to damage the crankshaft.

To install:

7. Coat the lip of the new seal with oil and install the seal using the proper seal driver.

8. Install the crankshaft sprocket and the timing belt.

9. Install the timing belt covers.

10. Install the crankshaft pulley.

11. Install the accessory drive belts.

12. Connect the negative battery cable.

13. Start the engine and check for proper operation.

Camshaft and Valve Lifters

REMOVAL & INSTALLATION

2.4L Engine

1. Disconnect the negative battery cable.

⁂ CAUTION

Wait at least 90 seconds after the negative battery cable is disconnected to prevent possible deployment of the air bag.

2. Remove the valve cover and the upper timing belt cover.

3. Rotate the crankshaft clockwise and align the camshaft sprocket timing mark with the timing mark on the cylinder head.

4. Matchmark the rotor to the distributor housing and remove the distributor.

5. Remove the camshaft sprocket retaining bolt. Pull the sprocket from the camshaft, with the timing belt attached, and place it on top of the timing belt front lower cover. Secure the sprocket in place so the belt remains taut.

※※ WARNING

Do not rotate the crankshaft after the sprocket is removed from the camshaft. Be sure there is no slack in the timing belt. Be sure the timing belt does not disengage from the sprocket. If the crankshaft is rotated or the timing belt position is disturbed, timing belt and sprocket alignment will have to be set.

6. Remove the camshaft cap bolts evenly and gradually.

7. Install auto lash adjuster retainer MD998443 or equivalent, to each rocker arm to hold the auto lash adjusters in place.

8. Remove the caps, shafts, rocker arms and bolts together as an assembly.

9. Remove the camshaft with the front seal from the engine.

To install:

10. Install a new roll pin to the camshaft.

11. Lubricate the camshaft and install with the front seal in place.

12. Install the camshaft so the hole in the sprocket will line up with the roll pin.

13. Install the caps, shafts and arms assembly. Tighten the camshaft bearing cap bolts to 85 inch lbs. (10 Nm), in the following order:

    a. No. 3, No. 2, No. 4, front cap, rear cap.

    b. Repeat the sequence increasing the torque to 14 lbs. (20 Nm).

14. Install the sprocket to the camshaft, engaging the roll pin. Tighten the bolt to 65 ft. lbs. (90 Nm).

15. Install the distributor, aligning the mark made during removal.

16. Install the valve cover and upper timing belt cover. Connect the negative battery cable.

17. Start the engine and check for leaks. Check the ignition timing.

3.0L Engines

1995 12 VALVE ENGINES

1. Disconnect the negative battery cable.

2. Remove the valve cover.

3. Remove the timing belt and remove the sprocket from the camshaft.

4. Install auto lash adjuster retainers MD998443 or equivalent, on the rocker arms.

5. If removing the left side camshaft, remove the distributor and the distributor extension.

6. Remove the camshaft bearing caps, but do not remove the bolts from the caps. Remove the rocker arms, rocker shafts and bearing caps, as an assembly.

7. Remove the camshaft from the cylinder head.

8. Inspect the bearing journals on the camshaft, cylinder head and bearing caps.

To install:

9. Lubricate the camshaft journals and camshaft with clean engine oil and install the camshaft in the cylinder head.

10. Align the camshaft bearing caps with the arrow mark (depending on cylinder numbers) and in numerical order.

➡The arrow marks on the left rocker shaft bearing caps face in same direction as the arrow marks on the left cylinder head which is away from the timing belt. The arrows on the right cylinder head and the right rocker bearing caps face toward the timing belt.

11. Apply sealer at the ends of the bearing caps and install the assembly.

12. Tighten the bearing cap bolts to 85 inch lbs. (10 Nm), in the following sequence: No. 3, No. 2, No. 1 and No. 4.

13. Repeat the sequence increasing the torque to 175 inch lbs. (18 Nm).

14. Install the distributor, if removed.

15. Install the sprockets, timing belt and timing belt cover.

16. Install the valve cover and connect the negative battery cable.

17. Start the engine and check for leaks and proper operation.

1996–99 12 VALVE ENGINES

1. Disconnect the negative battery cable.

※※ CAUTION

Work must be started after 90 seconds from the time the ignition switch is turned to the LOCK position and the negative battery cable is disconnected.

2. Remove the valve cover.

3. Remove the timing belt and remove the sprocket from the camshaft.

4. Install auto lash adjuster retainers SST MD998443–01 or equivalent, on the rocker arms.

5. If removing the left side camshaft, remove the distributor and the distributor extension.

6. Remove the camshaft bearing caps, but do not remove the bolts from the caps. Remove the rocker arms, rocker shafts and bearing caps, as an assembly.

7. Remove the camshaft from the cylinder head.

8. Inspect the bearing journals on the camshaft, cylinder head and bearing caps.

To install:

9. Lubricate the camshaft journals and camshaft with clean engine oil and install the camshaft in the cylinder head.

10. Align the camshaft bearing caps with the arrow mark (depending on cylinder numbers) and in numerical order.

➡The arrow marks on the left rocker shaft bearing caps face in same direction as the arrow marked on the left cylinder head which is away from the timing belt. The arrows on the right cylinder head and the right rocker bearing caps face toward the timing belt.

11. Apply sealer at the ends of the bearing caps and install the assembly.

12. Tighten the front and rear bearing cap bolts to 14 ft. lbs. (20 Nm) and bearing caps 2,3 and 4 to 8 ft. lbs. (11 Nm).

13. Remove the SST from the rocker arms.

14. If removed, install the distributor.

15. Install the sprockets, timing belt and the timing belt cover.

16. Install the valve cover and connect the negative battery cable.

17. Start the engine and check for leaks and proper operation.

24 VALVE ENGINES

1. Disconnect the negative battery cable.

※※ CAUTION

Work must be started after 90 seconds from the time the ignition switch is turned to the LOCK position and the negative battery cable is disconnected.

2. Remove the valve cover.

3. Remove the timing belt and remove the sprocket from the camshaft.

4. Install auto lash adjuster retainers SST MD998443 or equivalent, on the rocker arms.

5. If removing the left side camshaft, remove the distributor and the distributor extension.

6. Remove the rocker arms, rocker shafts and bearing caps, as an assembly.

7. Remove the camshaft from the cylinder head.

8. Inspect the bearing journals on the camshaft and the cylinder head.

To install:

9. Lubricate the camshaft journals and camshaft with clean engine oil and install the camshaft in the cylinder head.

10. Install the rocker arms, rocker arm shaft and the rocker shaft spring.

   a. Temporarily tighten the rocker shaft with the bolts at the condition that all the intake valve rocker arms do not push the valves.

   b. Insert the rocker shaft spring from above and mount it at right angles to the plug guide.

   c. Before installing the exhaust rocker arms and the rocker arm shaft, mount the rocker shaft spring.

   d. Remove the SST used to hold the lash adjuster in position.

   e. Check to ensure that the flat side of the rocker shaft is perpendicular to the cylinder head, and facing the valves.

   f. Gradually tighten the bearing caps in two or three steps. In the final step tighten to 23 ft. lbs. (31 Nm).

11. If removed, install the distributor.

12. Install the sprockets, timing belt and timing belt cover.

13. Install the valve cover and connect the negative battery cable.

14. Start the engine and check for leaks and proper operation.

3.5L Engine

❄❄ CAUTION

Observe all applicable safety precautions when working around fuel. Whenever servicing the fuel system, always work in a well ventilated area. Do not allow fuel spray or vapors to come in contact with a spark or open flame. Keep a dry chemical fire extinguisher near the work area. Always keep fuel in a container specifically designed for fuel storage; also, always properly seal fuel containers to avoid the possibility of fire or explosion.

1. Relieve the fuel system pressure.
2. Disconnect negative battery cable.
3. Remove the intake manifold plenum.
4. Remove the timing belt cover and the timing belt.

❄❄ WARNING

DO NOT rotate the crankshaft or camshafts after the timing belt has been removed. If rotated, severe internal engine damage will result from the pistons hitting the valves.

5. Remove the center cover, breather, PCV hoses, and the spark plug cables.

6. Remove the rocker cover and the semi-circular packing.

7. Matchmark the positioning of the crankshaft position sensor at the rear of the camshaft and remove the sensor.

8. If equipped with a camshaft sensor, remove the sensor from the front of the engine.

9. Being sure to hold the flats of the camshaft, loosen the camshaft sprocket bolts.

10. Noting the positioning and location of the sprockets, remove the sprockets from the camshafts.

➡**Be sure to note the positioning of the knock pin at the end of the camshafts for reinstallation purposes.**

➡**Be sure to keep the valvetrain components labeled and in proper order for reassembly.**

11. Loosen the bearing cap bolts in 2–3 steps. Label and remove all camshaft bearing caps.

➡**If the bearing caps are difficult to remove, use a plastic hammer to gently tap the components.**

12. Mark the components and remove the intake and the exhaust camshafts.

13. Remove the rocker arms and the lash adjusters. Be sure to note the location of the valvetrain components for reinstallation purposes.

14. Check the camshaft journals for wear or damage. Check the cam lobes for damage. Also, check the cylinder head oil holes for clogging.

To install:

➡**Lubricate the valvetrain components with clean engine oil.**

15. Bleed and install the lash adjusters to the to the original bores in the cylinder head.

16. Install the rocker arms to the cylinder head.

17. Lubricate the camshafts with clean engine oil and position the camshafts on the cylinder head.

❄❄ WARNING

Be sure to properly position the knock pins of the camshaft to prevent valve to piston interference.

➡**Do not confuse the intake camshaft with the exhaust camshaft. The intake camshaft has a P stamped on the hexagon. The exhaust camshaft has a K stamped on the hexagon.**

18. Install the bearing caps. Tighten the caps in sequence and in 2 or 3 steps. Caps 2, 3 and 4 have a front mark. Install with the mark aligned with the front mark on the cylinder head. Tighten the retaining bolts for caps No. 2, 3 and 4 to 8 ft. lbs. (11 Nm) and tighten the retaining bolts for the front and rear caps to 14 ft. lbs. (20 Nm).

➡**Install the bearing caps according to the identification mark and cap number. Bearing caps No. 2, 3 and are marked as such. The caps also are marked I for intake or E for exhaust.**

19. Apply a coating of engine oil to the oil seals and install the oil seals to the front and rear of the camshafts.

20. Holding the flats of the camshaft, install and tighten the sprocket bolts to 65 ft. lbs. (90 Nm).

21. If removed, install the camshaft position sensor and tighten the mounting bolts to 78 inch lbs. (9 Nm).

22. Aligning the matchmark, install the crankshaft position sensor at the rear of the camshaft and tighten the mounting nut to 7 ft. lbs. (12 Nm).

23. Align the marks on the camshaft and crankshaft sprockets. Install the timing belt assembly.

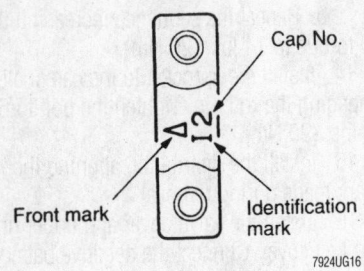

The camshaft bearing caps have identification marks on them—3.5L engine

24. Install the rocker cover and the semi-circular packing.

25. Install the intake manifold plenum.

26. Install the spark plug cables, center cover, breather and PCV hoses.

27. Connect the negative battery cable and check for leaks.

Oil Pan

REMOVAL & INSTALLATION

1. Disconnect the negative battery cable.

2. Raise the vehicle and support it safely.

3. Remove the skid plate and the engine undercover.

4. Place a suitable container into position. Remove the oil drain plug and drain the engine oil.

5. If necessary, remove the front exhaust pipe.

6. Remove the oil pan.

To install:

7. Before installing, thoroughly clean the oil pan and cylinder block mating surfaces.

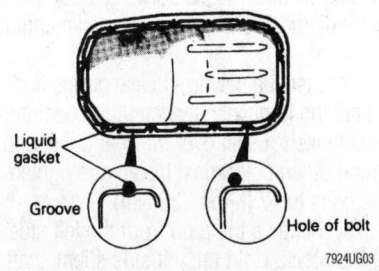

Apply a bead of sealant around the oil pan flange as shown—all engines are similar

8. Apply liquid gasket around the surface of the oil pan.

➡**Assemble the oil pan to the cylinder block within 15 minutes after applying the liquid gasket.**

9. Install the oil pan and tighten the oil pan retaining bolts to:

- 3.0L 12 valve engines: 29 ft. lbs. (39 Nm)

- 3.0L 24 valve and 3.5L: 7 engines-9 ft. lbs. (10–12 Nm)

- 3.5L engine: 4–6 ft. lbs. (6–8 Nm)

10. If removed, install the front exhaust pipe.

11. Install the skid plate and the engine undercover.

12. Fill with engine oil.

13. Connect the negative battery cable.

14. Start the vehicle and check for leaks.

Oil Pump

REMOVAL & INSTALLATION

2.4L Engine

1. Disconnect the negative battery cable.

❋❋ CAUTION

Wait at least 90 seconds after the negative battery cable is disconnected to prevent possible deployment of the air bag.

2. Remove the timing belt covers, timing belts and sprockets.

3. Raise the vehicle and support safely.

4. Drain the oil and remove the oil filter.

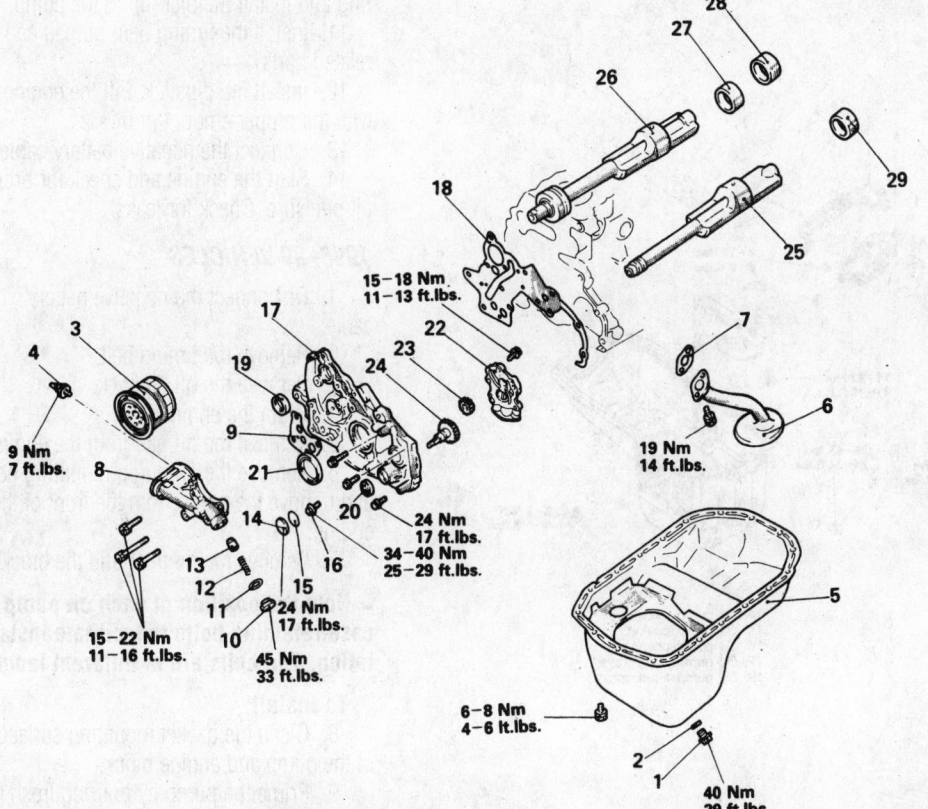

1. Drain plug
2. Gasket
3. Oil filter
4. Oil pressure switch
5. Oil pan
6. Oil screen
7. Gasket
8. Oil filter bracket
9. Gasket
10. Relief plug
11. Gasket
12. Relief spring
13. Relief plunger
14. Plug cap
15. O-ring
16. Driven gear bolt
17. Front case
18. Gasket
19. Oil seal
20. Oil seal
21. Crankshaft front oil seal
22. Oil pump cover
23. Oil pump driven gear
24. Oil pump drive gear
25. Left silent shaft
26. Right silent shaft
27. Silent shaft front bearing
28. Right silent shaft rear bearing
29. Left silent shaft rear bearing

Exploded view of the oil pump, oil pan and related components—2.4L engine shown-other engines are similar

5. Remove the oil pan and gasket. Remove the oil pump pick-up and gasket.

6. Remove the oil pressure relief plunger plug and gasket. Remove the spring and plunger from the oil filter bracket.

7. Remove the 4 bracket mounting bolts and remove the oil filter mount and gasket.

8. Using special tool MD998162, remove the plug cap and gasket that covers the oil pump driven gear shaft. This is located on the right side of the front case, just above the protruding drive gear shaft.

9. Using a long socket, remove the retaining bolt from the oil pump driven gear located behind the plug removed earlier.

10. Remove the front case mounting bolts and remove the case from the block.

11. Remove the case gasket from the block.

To install:

12. Prime the pump by pouring fresh oil into the pump intake and turning the drive-shaft until oil comes out of the pressure port. Repeat this a few times until no air bubbles are present. Replace all seals on the case assembly.

13. Install a special seal guide to the crankshaft, MD998285 or equivalent so the

smaller diameter faces outward. Coat the outer diameter of the seal with clean engine oil.

14. Install a new front case gasket and install the front case by carefully positioning the crankshaft seal over the seal guide and lining up all bolt holes. Install and tighten the bolts to 17 ft. lbs. (23 Nm).

15. Remove the plug from the left side of the block. Hold the left side silent shaft by inserting a tool in the plug hole and tighten the driven gear bolt to 26 ft. lbs. (35 Nm). Using a new O-ring, install the plug cover.

16. Install the oil filter mounting bracket gasket. Install the mounting bracket and bolts tightening the oil filter mounting bracket bolts to 12 ft. lbs. (16 Nm).

17. Clean or replace the oil pick-up screen and install with a new gasket.

18. Install the oil pan using a new gasket.

19. Install the timing sprockets, belts and covers.

20. Fill the engine with the proper amount of engine oil.

21. Connect the negative battery cable and check for proper oil pressure and leaks.

3.0L and 3.5L Engines

1995 VEHICLES

1. Disconnect the negative battery cable.

2. Remove the dipstick.

3. Raise the vehicle and support safely.

4. Remove the timing belt.

5. Drain the engine oil and remove the oil pan from the engine. Remove the oil pick-up.

6. Remove the oil pump mounting bolts and remove the pump from the front of the engine.

➡Note the position of each oil pump case retaining bolts to facilitate installation. The bolts are of different length.

To install:

7. Clean the gasket mounting surfaces of the pump and engine block.

8. Prime the pump by pouring fresh oil into the inlet and turning the rotors or by packing pump with petroleum jelly. Using a new gasket, install the oil pump on the engine and tighten all bolts to 10 ft. lbs. (14 Nm).

9. Install the balancer and crankshaft sprockets.

10. Clean out the oil pick-up or replace as required. Replace the oil pick-up gasket ring and install the pick-up to the pump.

11. Install the timing belt, oil pan and all related parts.

12. Install the dipstick. Fill the engine with the proper amount of oil.

13. Connect the negative battery cable.

14. Start the engine and check for proper oil pressure. Check for leaks.

1996–99 VEHICLES

1. Disconnect the negative battery cable.

2. Remove the timing belt.

3. Remove the oil dipstick.

4. Drain the engine oil.

5. Remove the oil pan from the engine.

6. Remove the oil pump mounting bolts and remove the pump from the front of the engine.

7. Remove the oil filter and the bracket.

➡Note the position of each oil pump case retaining bolts to facilitate installation. The bolts are of different length.

To install:

8. Clean the gasket mounting surfaces of the pump and engine block.

9. Prime the pump by pouring fresh oil into the inlet and turning the rotors or by packing pump with petroleum jelly. Using a

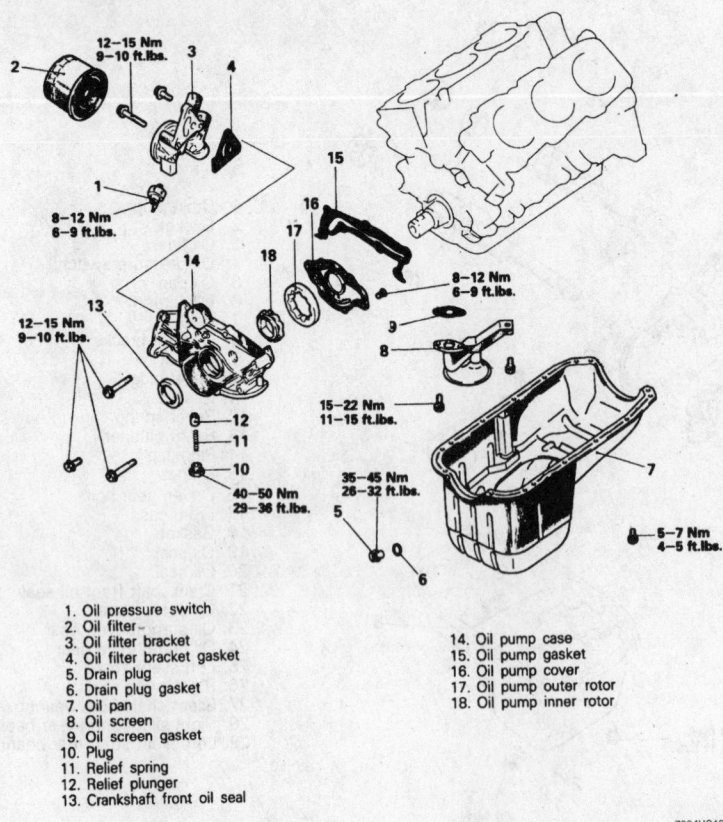

1. Oil pressure switch
2. Oil filter
3. Oil filter bracket
4. Oil filter bracket gasket
5. Drain plug
6. Drain plug gasket
7. Oil pan
8. Oil screen
9. Oil screen gasket
10. Plug
11. Relief spring
12. Relief plunger
13. Crankshaft front oil seal
14. Oil pump case
15. Oil pump gasket
16. Oil pump cover
17. Oil pump outer rotor
18. Oil pump inner rotor

7924UG19

Exploded view of the oil pump, oil pan and related components—3.0L and 3.5L engines

new gasket, install the oil pump on the engine and tighten all bolts to 10 ft. lbs. (14 Nm).

10. Clean out the oil pick-up or replace as required. Replace the oil pick-up gasket ring and install the pick-up to the pump.

11. Install the oil filter and the bracket.

12. Install the oil pan.

13. Install the timing belt.

14. Install the dipstick. Fill the engine with the proper amount of oil.

15. Connect the negative battery cable.

16. Start the engine and check for proper oil pressure. Check for leaks.

Rear Main Seal

REMOVAL & INSTALLATION

2.4L Engine

1. Remove the transmission and clutch assembly, if so equipped.

2. Matchmark and remove the flywheel or the driveplate and the adapter plate, if so equipped.

3. Unbolt and remove the lower bell housing cover from the rear of the engine. Remove the rear plate from the upper portion of the rear of the block.

4. The lower surface of the oil seal casing presses against the oil pan gasket (or seal) at the rear of the pan.

➡**You may want to loosen the oil pan bolts slightly at the rear to make it easier to separate the two surfaces. If the gasket is damaged, the oil pan will have to be removed and the gasket replaced.**

5. On engines employing sealant around the oil pan, unbolt and remove the pan. Clean the pan and block surfaces of all traces of sealant.

6. Remove the oil seal retainer bolts, then pull the case straight off the rear of the crankshaft.

7. Remove the seal retainer from the case and pry out the seal. Take care not to gouge or damage the metal surrounding the seal. Inspect the sealing surface at the rear of the crankshaft. If a deep groove is worn into the surface, the crankshaft will have to be replaced. Lubricate the sealing surface with clean engine oil.

To install:

8. Using a seal installer of the correct size, install the new seal into the bore of rear oil seal case. Make certain that the flat side of the seal will face outward when the case is installed on the engine. The inside

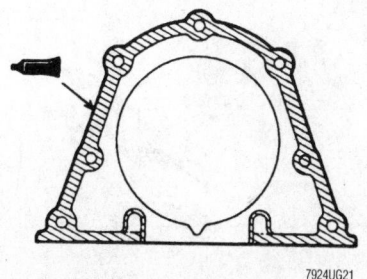

7924UG21

Apply a bead of RTV sealant to the seal retainer before installation—2.4L engine

of the seal must be flush with the inside surface of the seal case.

9. Apply RTV sealant to the retainer as shown and install it with the small hole located at the bottom. Tighten the bolts to 95 inch lbs. (11 Nm).

10. Reinstall the oil pan if it was removed, making certain the sealant is correctly applied. If the pan was only loosened, retighten the bolts to the correct torque. If the oil was drained, install the correct amount of motor oil.

11. Install the rear plate and bell housing cover.

12. Observing the matchmarks made earlier, install the flywheel or drive plate.

13. Install the transmission and related components as necessary.

3.0L and 3.5L Engines

1. Remove the transmission and clutch assembly, if so equipped.

2. Matchmark and remove the flywheel or driveplate and adapter plate. For the 3.0L engines, use the Mitsubishi tools (MB990767–01 and MIT308239) to hold the crankshaft and flywheel stationary while loosening the flywheel bolts. For the 3.5L engines, use the Mitsubishi tool (MD998781) to hold the flywheel in position.

3. Remove the rear oil seal.

   a. Cut out a portion in the crankshaft oil seal lip.

   b. Cover the tip of a small prytool with a cloth and apply it to the cutout in the oil seal to pry the oil seal out.

✳✳ CAUTION

Take care not to damage the crankshaft and oil seal case.

To install:

4. Inspect the sealing surface at the rear of the crankshaft. If a deep groove is worn

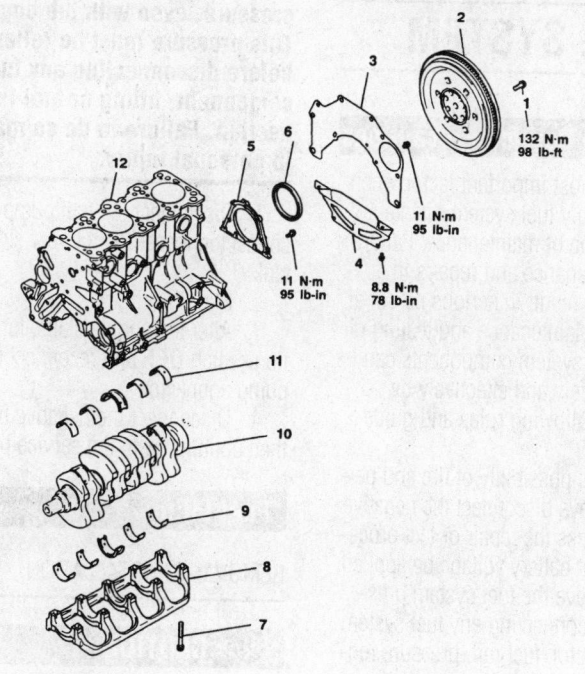

2 — 132 N·m / 98 lb-ft
— 11 N·m / 95 lb-in
— 8.8 N·m / 78 lb-in
— 11 N·m / 95 lb-in

1. FLYWHEEL BOLT
2. FLYWHEEL
3. REAR PLATE
4. BELL HOUSING COVER
5. OIL SEAL CASE
6. OIL SEAL
7. BEARING CAP BOLT
8. BEARING CAP
9. CRANKSHAFT BEARING (LOWER)
10. CRANKSHAFT
11. CRANKSHAFT BEARING (UPPER)
12. CYLINDER BLOCK

7924UG20

Exploded view of the crankshaft, rear main seal, and related components—2.4L engines

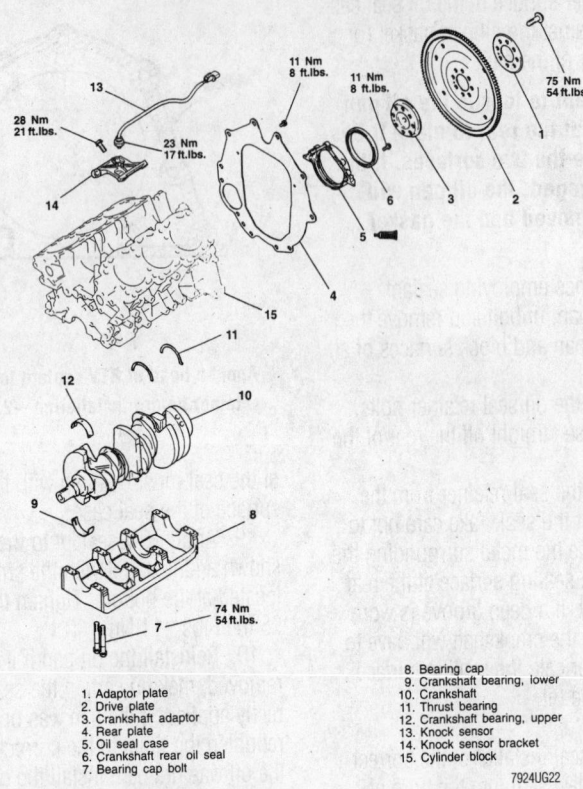

1. Adaptor plate
2. Drive plate
3. Crankshaft adaptor
4. Rear plate
5. Oil seal case
6. Crankshaft rear oil seal
7. Bearing cap bolt
8. Bearing cap
9. Crankshaft bearing, lower
10. Crankshaft
11. Thrust bearing
12. Crankshaft bearing, upper
13. Knock sensor
14. Knock sensor bracket
15. Cylinder block

7924UG22

Exploded view of the crankshaft, rear main seal and related components—3.5L engine shown, 3.0L engines is similar

into the surface, the crankshaft will have to be replaced. Coat the sealing lip of the seal with fresh, clean engine oil. Press the new seal into the case with a seal installing tool. The seal must be pressed in square until it bottoms in the case. It is necessary to use the proper tool (MD998718–01) to fit the seal into place.

5. Install the rear plate.

6. Install the transmission mounting plate.

7. Install the flywheel or drive plate and adapter.

8. Install the transmission and related components as necessary.

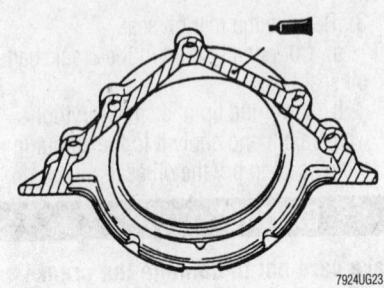

7924UG23

If removed, apply a bead of RTV sealant to the seal retainer before installation—2.4L engine

FUEL SYSTEM

Fuel System Service Precautions

Safety is the most important factor when performing not only fuel system maintenance but any type of maintenance. Failure to conduct maintenance and repairs in a safe manner may result in serious personal injury or death. Maintenance and testing of the vehicle's fuel system components can be accomplished safely and effectively by adhering to the following rules and guidelines.

• To avoid the possibility of fire and personal injury, always disconnect the negative battery cable unless the repair or test procedure requires that battery voltage be applied.

• Always relieve the fuel system pressure prior to disconnecting any fuel system component (injector, fuel rail, pressure regulator, etc.), fitting or fuel line connection. Exercise extreme caution whenever relieving fuel system pressure to avoid exposing skin, face and eyes to fuel spray. Please be advised that fuel under pressure may penetrate the skin or any part of the body that it contacts.

• Always place a shop towel or cloth around the fitting or connection prior to loosening to absorb any excess fuel due to spillage. Ensure that all fuel spillage (should it occur) is quickly removed from engine surfaces. Ensure that all fuel soaked cloths or towels are deposited into a suitable waste container.

• Always keep a dry chemical (Class B) fire extinguisher near the work area.

• Do not allow fuel spray or fuel vapors to come into contact with a spark or open flame.

• Always use a back-up wrench when loosening and tightening fuel line connection fittings. This will prevent unnecessary stress and torsion to fuel line piping. Always follow the proper torque specifications.

• Always replace worn fuel fitting O-rings with new. Do not substitute fuel hose or equivalent, where fuel pipe is installed.

Fuel System Pressure

RELIEVING

✳✳ CAUTION

The fuel system is under constant pressure, even with the engine off. This pressure must be relieved before disconnecting any fuel system component, fitting or fuel line connection. Failure to do so may result in personal injury.

1. Disconnect the fuel pump electrical connector, located at the rear side of the fuel tank.

2. Start the engine.

3. After the engine stalls, turn the ignition switch **OFF** and reconnect the fuel pump connector.

4. Disconnect the negative battery cable, then continue with the service procedure.

Fuel Filter

REMOVAL & INSTALLATION

✳✳ CAUTION

The fuel injection system remains under pressure after the engine has been OFF. Properly relieve fuel pressure before disconnecting any fuel lines. Failure to do so may result in fire or personal injury.

✳✳ CAUTION

Do not allow fuel spray or fuel vapors to come in contact with a spark or open flame. Keep a dry chemical fire extinguisher nearby. Never store fuel in an open container due to risk of fire or explosion.

1. Relieve the fuel system pressure.
2. Disconnect the negative battery cable.
3. Remove the fuel filter protector if equipped.
4. Using a back-up wrench disconnect the fuel line(s) from the filter. If the filter uses a push-on type connector, press the retainer to release the connection.
5. Remove the filter from the mounting bracket.

To install:

6. Install the fuel filter to the mounting bracket in the proper direction.
7. Connect the fuel lines to the filter. Use a back-up wrench to hold the fuel filter. Tighten the banjo bolt(s) to 18–25 ft. lbs. (25–35 Nm) or the line fitting to 27 ft. lbs. (36 Nm).
8. Install the fuel filter protector if equipped.

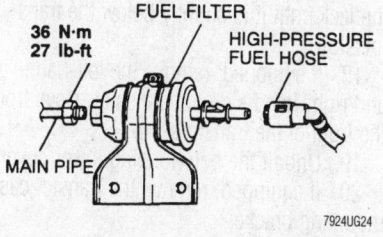

Fuel filter removal—Montero Sport shown

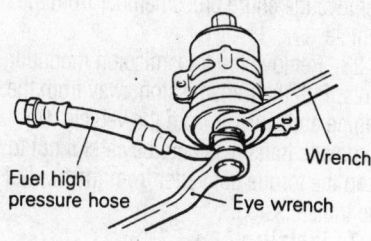

Always use a back-up wrench when removing or installing fuel lines to the filter

9. Connect the negative battery cable.
10. Start the engine and check for leaks.

Fuel Pump

REMOVAL & INSTALLATION

Montero

1. Disconnect the negative battery cable.

✳✳ CAUTION

The fuel injection system remains under pressure after the engine has been OFF. Properly relieve fuel pressure before disconnecting any fuel lines. Failure to do so may result in fire or personal injury.

2. Relieve the fuel system pressure.

➡The manufacturer recommends draining of the fuel tank.

✳✳ CAUTION

Do not allow fuel spray or fuel vapors to come in contact with a spark or open flame. Keep a dry chemical fire extinguisher nearby. Never store fuel in an open container due to risk of fire or explosion.

3. Remove the rear floor carpeting.
4. Remove the fuel pump cover.
5. Disconnect the fuel pump connector and the fuel hoses.
6. Remove the fuel pump assembly.

To install:

7. Install the fuel pump assembly into the fuel tank.

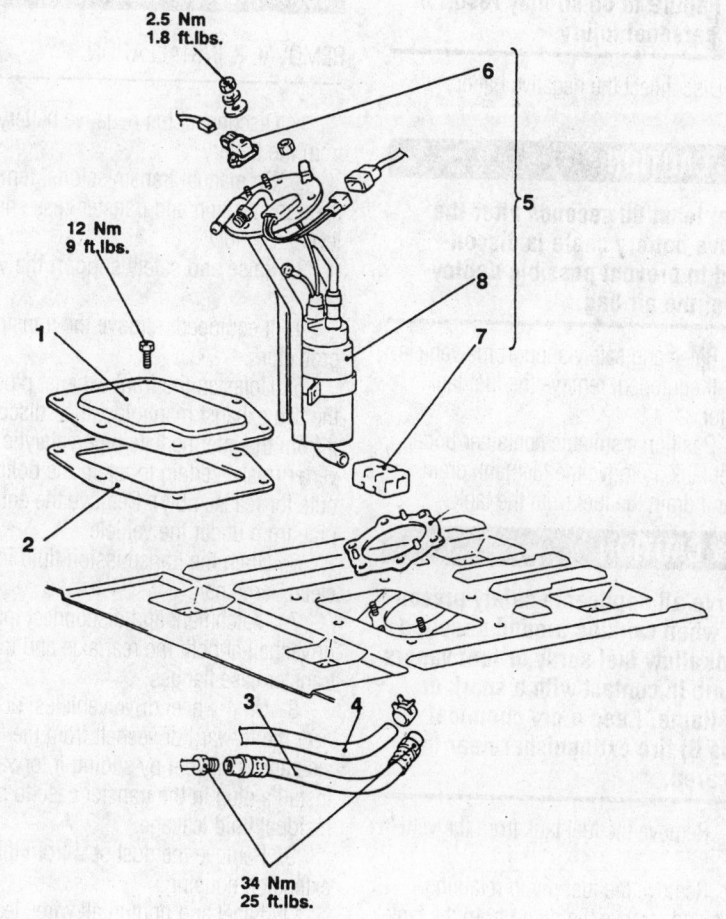

1. Floor cover
2. Packing
3. High-pressure fuel hose
4. Fuel return hose connection
5. Fuel pump and filter assembly
6. Fuel tank differential pressure sensor
7. Filter
8. Fuel pump assembly

The fuel pump on the Montero is removed through the rear floor pan

8. Tighten the nuts to 24 inch lbs. (2.5 Nm).

9. Connect the fuel lines and the fuel pump connector.

10. Install the fuel pump cover and tighten the bolts to 9 ft. lbs. (12 Nm).

11. Install the rear floor carpeting.

12. If drained, refill the fuel tank.

13. Connect the negative battery cable.

14. Start the vehicle; check for leaks and proper operation.

Mighty Max and Montero Sport

1. Properly relieve the fuel system pressure.

✳✳ CAUTION

The fuel injection system remains under pressure after the engine has been OFF. Properly relieve fuel pressure before disconnecting any fuel lines. Failure to do so may result in fire or personal injury.

2. Disconnect the negative battery cable.

✳✳ CAUTION

Wait at least 90 seconds after the negative battery cable is disconnected to prevent possible deployment of the air bag.

3. Raise and safely support the vehicle.

4. If equipped, remove the fuel tank protector.

5. Position a suitable container under the fuel tank. Remove the fuel tank drain plug and drain the fuel from the tank.

✳✳ CAUTION

Observe all applicable safety precautions when working around gasoline. Do not allow fuel spray or fuel vapors to come in contact with a spark or open flame. Keep a dry chemical (Class B) fire extinguisher near the work area.

6. Remove the fuel tank from the vehicle.

7. Remove the fuel pump retaining screws and remove the pump from the tank.

To install:

8. Clean the seal area of the tank. Install a new gasket.

9. Install the fuel pump in the same position as originally installed.

10. Install the fuel pump retaining screws and tighten to 1.8 ft. lbs. (2.5 Nm).

11. Install the fuel tank and tighten to 18–22 ft. lbs. (25–30 Nm).

12. Install the fuel tank drain plug and the fuel tank protector, if equipped.

13. Lower the vehicle. Refill the fuel tank and install the cap.

14. Connect the negative battery cable.

15. Connect a jumper wire from the fuel pump check connector on the back of the fuse block to the positive battery terminal and operate the fuel pump. Check for leaks.

DRIVE TRAIN

Transmission Assembly

REMOVAL & INSTALLATION

1. Disconnect the negative battery cable from the battery.

2. On manual transmissions, remove the transmission and transfer case shift lever assembly.

3. Raise and safely support the vehicle.

4. If equipped, remove the transfer case protector.

5. Unfasten the front exhaust pipe from the two exhaust manifolds, then disconnect it from the intermediate pipe/catalytic converter (make certain to retain the bolts and nuts for reassembly). Remove the entire pipe from under the vehicle.

6. Drain the transmission fluid into a clean, large pan.

7. Matchmark and disconnect the rear driveshaft at both the rear axle and the transfer case flanges.

8. On 4-wheel drive vehicles, disconnect the forward driveshaft from the front axle and remove it by sliding it forward. Install a plug in the transfer case to prevent residual fluid leakage.

9. Remove the dust seal from the rear extension housing.

10. Label and unplug all wires leading to the transmission, some are: Ground cables, 4WD indicator light switch connector, pulse generator connector, oxygen sensor connec-

tor, back-up light switch connector, HI/LO detection switch connector, center differential lock detection switch connection. There may be others depending on the particular year, model and engine the vehicle came equipped with.

11. Unscrew the retaining ring holding the speedometer in the transmission. Pull the speedometer cable out of the transmission and label it.

12. On manual transmissions, remove the clutch cylinder heat protector. Remove the retaining bolts, then the clutch release cylinder (with the clutch hose connected to it) from the transmission. Suspend it from the body by using a piece of wire or a similarly safe method.

13. Unbolt the starter motor from the front of the transmission and remove it. Remove the heat shield, if so equipped.

14. On automatic transmissions, remove the bolts attaching the torque converter to the flexplate, remove the dipstick, disconnect the fluid cooling lines and the shift linkage at the transmission.

15. Place a floor jack and a block of wood below the engine oil pan.

16. Lift the floor jack under the engine just until the weight of the engine is taken onto the jack—the engine should only barely be lifted by the jack.

17. Use a transmission jack or second floor jack to place under the transmission. Don't support the transmission yet, only lift the jack until it is slightly below the transmission.

18. If equipped, remove the left-hand and right-hand side transmission stays from the front of the transmission.

19. Unbolt the bell housing lower cover.

20. If equipped, remove the transfer case mounting bracket.

21. Lift the floor jack up until the transmission is being slightly supported by it.

22. Remove the two transmission-to-crossmember bolts. Lift the jack about ¼ in. (6mm) off of the crossmember support. Remove the entire crossmember from the vehicle.

23. Remove the transmission mounting blots. Pull the transmission away from the engine and lower it from the vehicle. On automatic transmissions, be careful not to drop the torque converter from the front of the transmission.

To install:

24. Lift the transmission and transfer assembly close to position with the floor jack.

25. On the engine side, there are two centering locations. Be sure that the transmission mounting bolt holes are aligned with them before mounting the transmission and transfer assembly to the engine. Lowering the rear of the engine SLIGHTLY may help align the two assemblies.

26. Slide the transmission assembly onto the engine making sure the aligning areas stay aligned.

27. Install and tighten the bolts to 54 ft. lbs. (75 Nm).

28. Lift the transmission/transfer assembly with the floor jack. Since the engine is now attached to the transmission, it also will rise slightly—adjust its jack to keep only slight support.

29. Hold the crossmember in place and secure with the mounting bolts. Tighten the mounting bolts to 47 ft. lbs. (65 Nm).

30. Lower the transmission and transfer case assembly onto the crossmember. Install the two crossmember-to-transmission bolts and tighten them to 13–18 ft. lbs. (18–25 Nm) on Pick-ups and to 29 ft. lbs. (39 Nm) on Montero.

31. If equipped, install the transfer mounting bracket back onto the transfer case.

32. Install the bell housing lower cover, then mount the left-hand and right-hand side transmission stays.

33. Mount the starter motor and clutch release cylinder (and heat shield, if so equipped) to the transmission.

34. Insert the speedometer cable into the transmission and secure it there with the retaining ring.

35. Plug all of the electrical harness connectors back together.

36. On automatic transmissions, install the flexplate-to-torque converter bolts. Tighten the bolts to 25–30 ft. lbs. (35–42 Nm).

37. On automatic transmissions, install the dipstick tube and connect the shift linkage and fluid cooler lines. Tighten the line fittings to 32 ft. lbs. (44 Nm).

38. Tap the dust seal guard back onto the rear extension housing with a rubber or plastic mallet.

39. Slide the front driveshaft into the transfer case, then attach it to the front differential. Install the rear driveshaft also. Make certain that the matchmarks line up.

40. Fill the transmission and transfer case with oil.

41. Install the front exhaust pipe to the catalytic converter and the exhaust manifolds. Use new gaskets if it was equipped with them. Be sure that the wiring does not lay on or very close to the exhaust pipes or exhaust manifolds, if they do, reroute them away from the exhaust components.

42. If equipped, install the transfer case protector.

43. Lower the vehicle to the ground. On manual transmissions, install the transmission and transfer shift lever assembly.

44. Connect the negative battery cable to the battery.

45. Start the vehicle and check for any leaks.

Clutch

ADJUSTMENT

Mighty Max

➡**The following procedure is for cable operated clutches only.**

1. Measure the clutch pedal height from the face of the pedal pad to the floorboard. This distance should be 6.5–6.7 in. (166–171mm).

2. If the pedal height is not correct, turn the pedal stopper bolt or clutch switch until the pedal height is adjusted to its standard value.

3. Measure the clutch pedal free-play. The standard value should be 0.8–1.4 in. (20–35mm).

4. If the clutch pedal free-play is not within the standard value, adjust it as follows:

   a. Pull the clutch cable lightly at the toe board and turn adjusting nut until the adjusting jut clearance is adjusted to 0.12–0.16 in. (3–4mm).

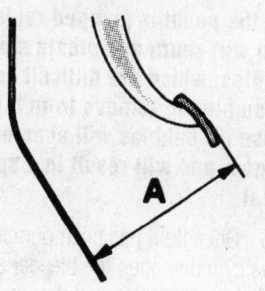

7924UG27

Adjust the pedal height "A" to the specified dimension—cable operated clutches only

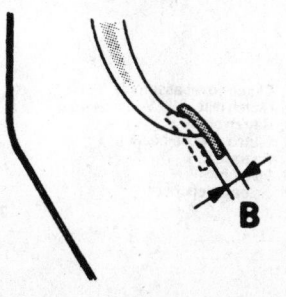

7924UG28

Adjust the pedal free-play "B" to the specified dimension—cable operated clutches only

   b. After making the adjustment, depress the clutch pedal several times and check that the clutch pedal play is with the standard value.

5. Measure the distance between the clutch pedal and floorboard when the clutch is disengaged. The standard value should be 2.4 in. (60mm) or more.

REMOVAL & INSTALLATION

1. Disconnect the negative battery cable.

2. Remove the transmission assembly and if equipped, the transfer assembly from the vehicle.

3. Insert a suitable tool in the flywheel pilot bearing hole to keep the clutch disc from falling off. Loosen the clutch cover retainer bolts gradually in a crisscross fashion. Remove the clutch cover and disc.

4. Check the release bearing for scorching, damage or strange noise. Replace, if necessary.

5. Inspect the flywheel surface for heat cracks or scoring. Reface or replace the flywheel as required.

   To install:

6. Apply high temperature grease to the clutch disc splines, input shaft, contact points of the release fork and inside diameter of the release bearing.

➡**Do not allow oil or grease to contact the clutch facing and pressure plate.**

7. Align the clutch disc to the flywheel, using a suitable tool.

➡**When installing the clutch disc, be sure that the surface having the manufacturer's stamped mark is on the pressure plate side.**

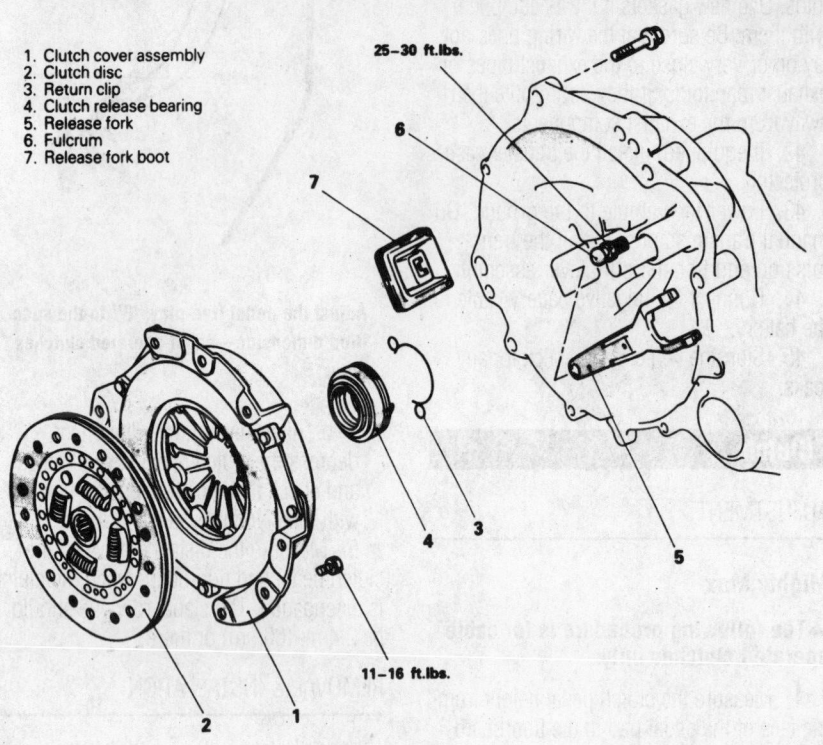

1. Clutch cover assembly
2. Clutch disc
3. Return clip
4. Clutch release bearing
5. Release fork
6. Fulcrum
7. Release fork boot

25–30 ft.lbs.

11–16 ft.lbs.

7924UG29

Exploded view of the typical clutch assembly components

8. Install the clutch cover with the dowel pin holes in alignment with the dowel pins in the flywheel and tighten the bolts gradually in a crisscross fashion. Tighten the bolts to 14 ft. lbs. (19 Nm).

9. Install the transmission assembly and if equipped, the transfer assembly to the vehicle.

10. Connect the negative battery cable. Road test the vehicle for proper operation.

Hydraulic Clutch System

BLEEDING

※ WARNING

When bleeding, keep the facial area well away from the slave cylinder and protect all painted surfaces from fluid contact. Brake fluid will damage painted surfaces and could cause physical injury.

1. Fill the clutch master cylinder with fresh DOT 3 brake fluid.

2. Have a helper sit in the vehicle.

3. Raise the vehicle and support it safely.

4. Remove the bleeder screw cap.

5. If the system is empty, the most efficient way to get fluid down to the cylinder is to loosen the bleeder about ½–¾ turn; place a finger firmly over the bleeder and have the helper pump the brakes slowly until fluid pressure is felt at the bleeder. Once fluid is at the bleeder, close before the pedal is released.

➡**If the pedal is pumped rapidly, the fluid will churn and create small air bubbles, which are difficult and time consuming to remove from the system. These air bubbles will eventually congregate and will result in a spongy pedal.**

6. Once fluid has been pumped to the slave cylinder, open the bleeder screw, have the helper depress the clutch pedal, lock the bleeder and have the helper release the pedal. Wait 15 seconds and repeat the procedure (including the 15 second wait) until

no air bubbles flow from the bleeder. Remember to close the bleeder before the pedal is released. If the bleeder is left open when the pedal is released, air will be induced into the system.

7. If a helper is not available, connect a small hose to the bleeder, submerge the other end in a clean container of fresh brake fluid placed in a position that is visible from the driver's seat. Pump the pedal until no air comes out of the tube.

Transfer Case Assembly

REMOVAL & INSTALLATION

Montero and Montero Sport

The transfer case is removed from the vehicle along with the transmission. Refer to the transmission removal and installation procedure for information.

Mighty Max

1. Disconnect the negative battery cable.

2. Raise and safely support the vehicle.

3. Remove the transmission and transfer case as an assembly.

4. Remove the plug from right side of the transfer case, under the control housing.

5. Remove the select spring and plunger from the housing bore.

6. On automatic transmission equipped vehicles, remove the control lever housing assembly, cover and gasket.

7. On manual transmission equipped vehicles, remove the spring pin that retains the shift changer to the control shaft using a pin punch.

8. Remove the transfer case to transmission retaining nuts and separate the transfer case from the transmission.

To install:

9. Align and seat transfer case on transmission.

10. Install and tighten the attaching nuts to 30 ft. lbs. (42 Nm).

11. On automatic transmission equipped vehicles, install the control lever housing assembly, cover and gasket.

12. On manual transmission equipped vehicles, install the shift changer to the control shaft and install a new roll pin.

13. Install the select spring, plunger and plug into the housing bore.

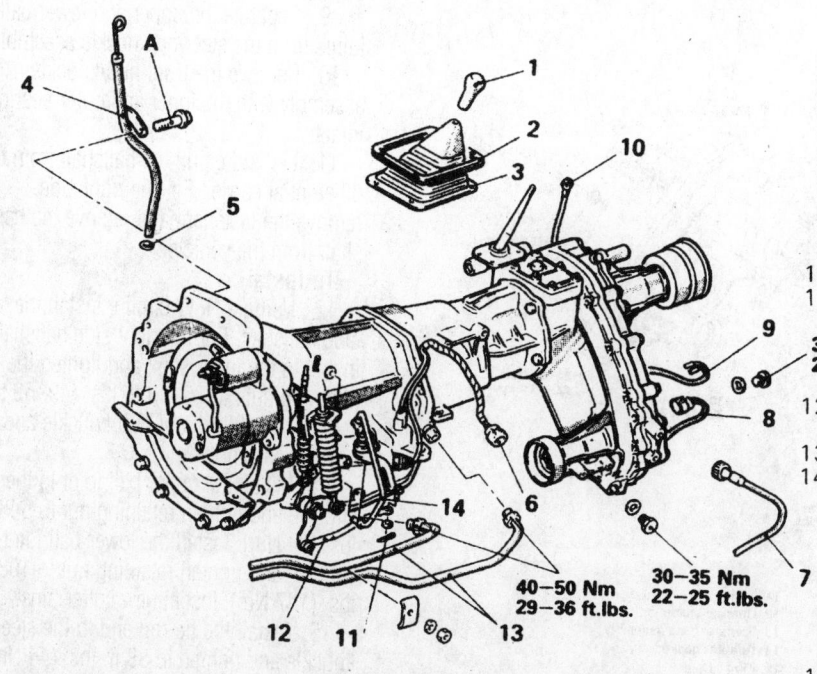

1. Transfer shift lever knob
2. Dust cover retaining plate
3. Control lever cover
4. Oil filler tube
5. O-ring
6. Transmission harness connector
7. Speedometer cable
8. Pulse generator connector
9. 4WD indicator light switch connector
10. Ground cable
11. Cotter pin

30—35 Nm
22—25 ft.lbs.

12. Transmission control rod (Transmission side)
13. Automatic transmission cooler tube
14. Transmission throttle lever (Bell crank bracket side)

40—50 Nm
29—36 ft.lbs.

30—35 Nm
22—25 ft.lbs.

	Nm	ft.lbs.	O.D.×Length mm (in.)	Bolt identification
A	43 – 55	31 – 40	7 10×50 (.4×2.0)	7 D×L
B	43 – 55	31 – 40	7 10×70 (.4×2.8)	
C	30 – 42	21 – 30	7 10×16 (.4×.6)	

15. Exhaust pipe mounting bracket
16. Bell housing cover
17. Special bolt
18. Transfer roll stopper
19. Transfer mounting bracket
20. Bolt
21. No. 2 crossmember
22. Starter motor
23. Bell crank bracket assembly
24. Transmission and transfer assembly

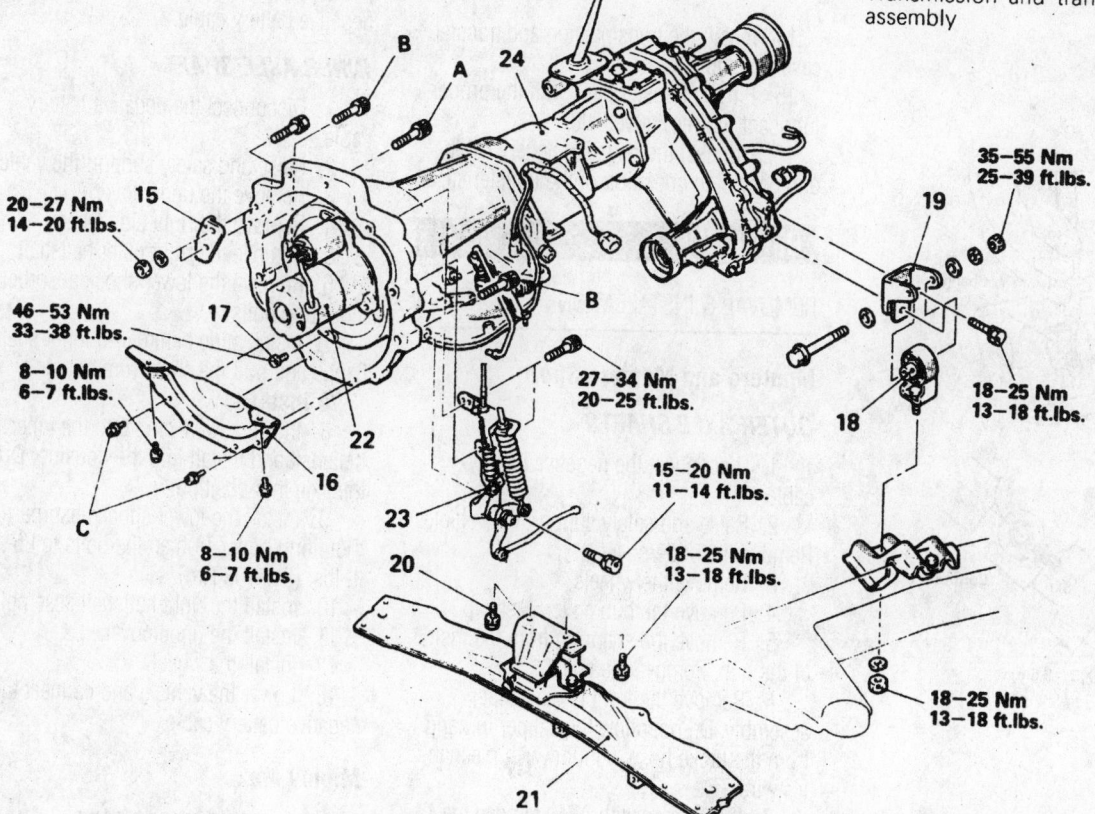

20—27 Nm
14—20 ft.lbs.

46—53 Nm
33—38 ft.lbs.

8—10 Nm
6—7 ft.lbs.

8—10 Nm
6—7 ft.lbs.

35—55 Nm
25—39 ft.lbs.

27—34 Nm
20—25 ft.lbs.

18—25 Nm
13—18 ft.lbs.

15—20 Nm
11—14 ft.lbs.

18—25 Nm
13—18 ft.lbs.

18—25 Nm
13—18 ft.lbs.

Transfer case assembly removal and installation—Mighty Max with automatic transmissions

7924UG30

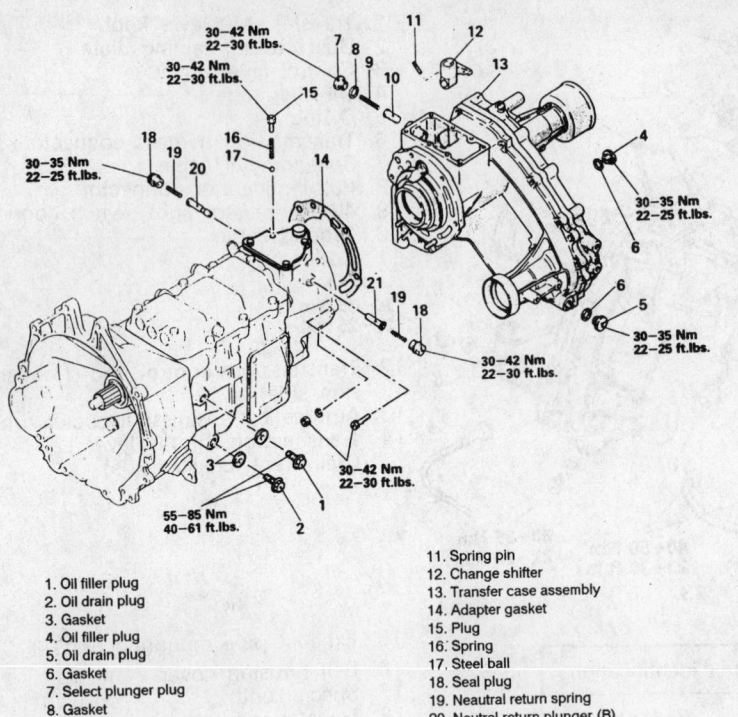

1. Oil filler plug
2. Oil drain plug
3. Gasket
4. Oil filler plug
5. Oil drain plug
6. Gasket
7. Select plunger plug
8. Gasket
9. Select spring
10. Select plunger

11. Spring pin
12. Change shifter
13. Transfer case assembly
14. Adapter gasket
15. Plug
16. Spring
17. Steel ball
18. Seal plug
19. Neautral return spring
20. Neutral return plunger (B)
21. Neutral return plunger (A)

7924UG31

Transfer case assembly removal and installation—Mighty Max with manual transmission and 3.0L (VIN H) engine

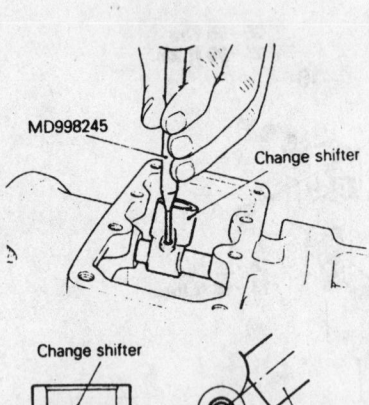

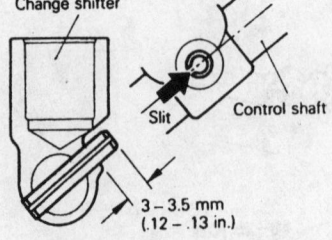

7924UG32

Spring pin removal and installation— Mighty Max

14. Install the transmission and transfer case assembly.

15. Fill the transfer case with the proper lubricant. Lower the vehicle.

16. Connect the negative battery cable and check the operation of the transfer case.

Halfshaft

REMOVAL & INSTALLATION

Montero and Montero Sport

OUTER AXLE SHAFTS

1. Disconnect the negative battery cable.

2. Raise and safely support the vehicle. Remove the undercover.

3. Remove the wheels.

4. Remove the hub cover dust cap.

5. Remove the snapring from the inside of the hub. Remove the shim.

6. Remove the front brake caliper assembly. Do not allow the caliper to hang from the brake hose, support with mechanics wire.

7. If equipped with ABS, remove the speed sensor.

8. Separate the tie rod from the steering knuckle assembly.

9. Separate the upper and lower ball joints from the steering knuckle assembly.

10. Remove the front hub/knuckle assembly with the inner and outer bearings intact.

11. Left side, pull the halfshaft from the differential carrier. For the right side, remove the fasteners and remove the halfshaft from the vehicle.

To install:

12. Using a NEW circlip, install the left side halfshaft. For the right side halfshaft, install to the inner shaft and tighten the retaining nuts to 36–43 ft. lbs. (49–59 Nm).

13. Install the front hub/knuckle and bearing assembly.

14. Install the upper ball joint to the knuckle and tighten retaining nut to 54 ft. lbs. (74 Nm). Install the lower ball joint to knuckle and tighten retaining nut to 108 ft. lbs. (147 Nm). Install new cotter pins.

15. Install the tie rod end to the steering knuckle and tighten to 33 ft. lbs. (44 Nm). Install new cotter pin.

16. Install the speed sensor, if removed.

17. Install the front brake assembly.

18. Install the shim and snapring to the axle shaft. Install the front hub dust cover.

19. Install the wheels and the undercover.

20. Lower the vehicle and connect the negative battery cable.

INNER AXLE SHAFT

1. Disconnect the negative battery cable.

2. Raise and safely support the vehicle.

3. Remove the undercover.

4. Remove the right side wheel.

5. Remove the right outer halfshaft.

6. Remove the lower shock absorber mounting bolts.

7. Install slide hammer to inner shaft flange and pull the shaft from housing.

To install:

8. Install a NEW circlip to the inner halfshaft and install into the housing. Drive the axle into position.

9. Install the lower shock absorber mounting bolts. Tighten the bolts to 65–76 ft. lbs. (88–103 Nm).

10. Install the right halfshaft assembly.

11. Install the undercover.

12. Install the wheel.

13. Lower the vehicle and connect the negative battery cable.

Mighty Max

OUTER RIGHT HALFSHAFT

1. Place the free-wheeling hub in the free condition by placing the transfer lever

in the **2H** position and moving in reverse for about 6 or 7 feet.

2. Disconnect negative battery cable.

3. Raise and safely support the vehicle. Remove the skid plate, if equipped.

4. Remove the tire and wheel assembly.

5. Remove the hub cover with the use of an oil filter wrench. Install a protective cloth between the wrench and the cover to avoid damage to the cover.

6. Remove the snapring from the inside of the hub. Remove the shim.

7. Remove the front brake caliper and brake pads from the vehicle. Do not allow the caliper to hang from the brake hose, support with mechanics wire.

8. Separate the tie rod from the steering knuckle.

9. Separate the upper and lower ball joints from the steering knuckle.

10. Remove the front hub/knuckle assembly with the inner and outer bearings intact.

11. Remove the halfshaft to axle housing retaining nuts and remove the halfshaft from the vehicle.

To install:

12. Install the halfshaft and the retaining nuts and tighten to 43 ft. lbs. (60 Nm).

13. Install the front hub/knuckle and bearing assembly.

14. Install the upper ball joint to the knuckle and tighten retaining nut to 130 ft. lbs. (180 Nm). Install the lower ball joint to knuckle and tighten retaining nut to 65 ft. lbs. (90 Nm). Install new cotter pins.

15. Install the tie rod end to the steering knuckle and tighten to 33 ft. lbs. (45 Nm). Install new cotter pin.

16. Install the shim and snapring to the axle shaft. Install the front hub cover.

17. Install front brake caliper assembly.

18. Install the tire and wheel assembly. Install skid plate, if removed.

19. Lower the vehicle and connect the negative battery cable.

INNER RIGHT AXLE SHAFT

1. Disconnect negative battery cable.
2. Raise and safely support the vehicle.
3. Remove the skid plate, if equipped.
4. Remove the tire and wheel assembly.
5. Remove the right outer halfshaft.
6. Remove the lower shock absorber mounting bolts.
7. Install slide hammer to inner shaft flange and pull from housing. Press the bearing from the axle, as required.

To install:

8. Press new bearing and seal on axle, as required. Install new circlip to inner half-shaft and install into housing. Drive the axle into position.

9. Install the lower shock absorber mounting bolts.

10. Install the right outer halfshaft and related parts.

11. Install the skid plate, if equipped.

12. Install the tire and wheel assembly.

13. Lower the vehicle and connect the negative battery cable.

LEFT HALFSHAFT

1. Place the free-wheeling hub in the free condition by placing the transfer lever in the **2H** position and moving in reverse for about 6 or 7 feet.

2. Disconnect negative battery cable.

3. Raise and safely support the vehicle. Remove the skid plate, if equipped.

4. Remove the tire and wheel assembly.

5. Remove the hub cover with the use of an oil filter wrench. Install a protective cloth between the wrench and the cover to avoid damage to the cover.

6. Remove the snapring from the inside of the hub. Remove the shim.

7. Remove the front brake caliper and brake pads from the vehicle. Do not allow the caliper to hang from the brake hose, support with mechanics wire.

8. Separate the tie rod from the steering knuckle.

9. Separate the upper and lower ball joints from the steering knuckle.

10. Remove the front hub/knuckle assembly with the inner and outer bearings intact.

11. Pull the left halfshaft out from the differential carrier. Use care not to damage the oil seal with the splines of the shaft.

To install:

12. Replace circlip on the end of the shaft. Install the halfshaft into the front differential case and drive into position using a plastic hammer.

13. Install the front hub/knuckle and bearing assembly.

14. Install the upper ball joint to the knuckle and tighten retaining nut to 130 ft. lbs. (180 Nm). Install the lower ball joint to knuckle and tighten retaining nut to 65 ft. lbs. (90 Nm). Install new cotter pins.

15. Install the tie rod end to the steering knuckle and tighten to 33 ft. lbs. (45 Nm). Install new cotter pin.

16. Install the shim and snapring to the axle shaft. Install the front hub cover.

17. Install front brake caliper assembly.

18. Install the tire and wheel assembly. Install skid plate, if removed.

19. Lower the vehicle and connect the negative battery cable.

Locking Hubs

REMOVAL & INSTALLATION

1. Place the free-wheeling hub in the free condition. To do this, shift the transfer shift lever to the 2H position, then move the vehicle 4–7 ft. (1–2 m) backwards.

2. Remove the front wheels of the vehicle, and support the raised front end of the vehicle with jackstands.

3. Remove the automatic free-wheeling hub cover. When the cover cannot be loosened by hand, use an oil filter wrench with a protective cloth in between not to damage the cover. Remove the O-ring from the underside of the hub cover. A new O-ring will be needed upon reassembly.

4. Using a pair of snapring pliers, remove the snapring from the driveshaft.

5. Slide the washer (shim) off of the driveshaft.

6. Unscrew the 6 retaining bolts from the automatic free-wheeling hub assembly and slide the assembly off of the driveshaft.

7. Remove the rear housing C-clip, which is easily removed by pushing the brake in and using a small prytool to pry it out.

8. Remove brake (B), brake (A) and the brake spring from the housing.

9. Use a small prytool to remove the housing snapring.

10. Using the special tool MB990811–01, lightly push (with a press) the drive gear in and remove the retainer (B) C-ring. Since the return spring relaxes approximately 1.5 in. (40mm) the stroke of the press should be set to more than 1.5 in. (40mm).

✳✳ CAUTION

Use a protective cover so as not damage the cover attaching surface of the housing before setting it on the press table. Be sure that the pressing force does not exceed 44.1 lbs. (200 N).

11. Slowly reduce the pressure of the press until the return spring relaxes fully.

Removal steps

1. Cover
Adjustment of drive shaft end play
2. Snap ring
3. Shim
4. Front brake assembly

5. Bolts
6. Automatic free-wheeling hub assembly
7. Shim
8. Lock washer
9. Lock nut
10. Front hub assembly

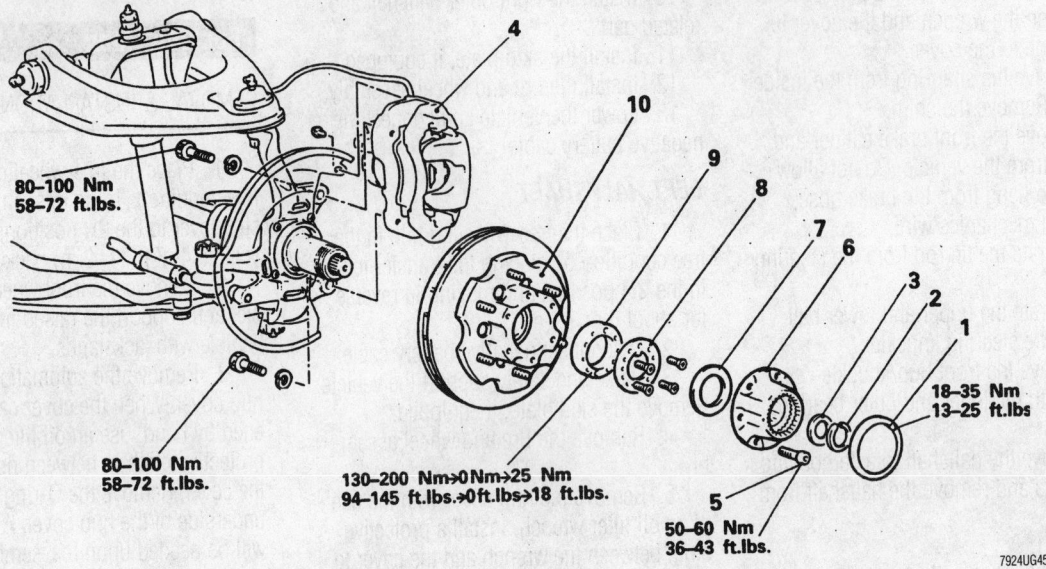

80–100 Nm
58–72 ft.lbs.

80–100 Nm
58–72 ft.lbs.

130–200 Nm→0 Nm→25 Nm
94–145 ft.lbs.→0 ft.lbs.→18 ft.lbs.

50–60 Nm
36–43 ft.lbs.

18–35 Nm
13–25 ft.lbs

7924UG45B

Axle hub and free-wheeling hub assemblies' removal and installation—automatic-change 4WD systems

Disassembly steps

1. Cover
2. O-ring
3. Housing
4. Housing C ring
5. Brake (B)
6. Brake (A)
7. Brake spring

8. Housing snap ring
9. Retainer (B) C ring
10. Drive gear assembly
11. Slide gear assembly
12. Return spring
13. Retainer (B)
14. Retainer bearing

15. Drive gear snap ring
16. Retainer (A)
17. Drive gear
18. Slide gear C ring
19. Cam
20. Spring holder
21. Shift spring
22. Slide gear

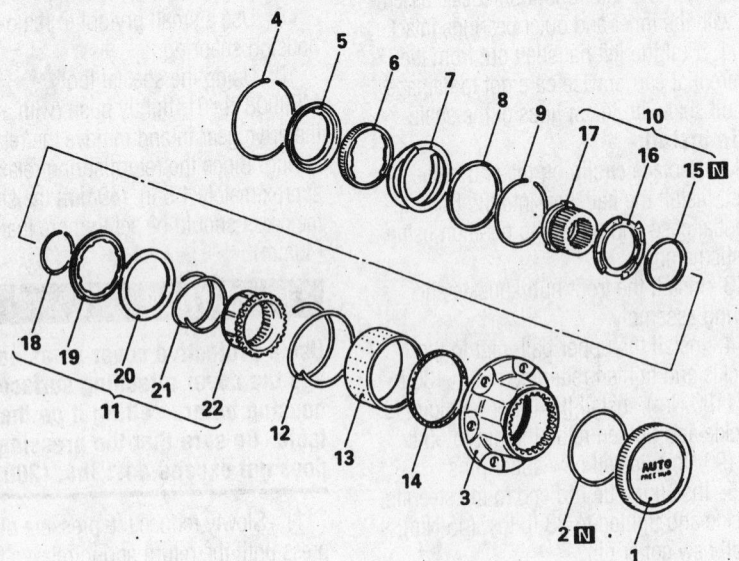

7924UG46

Automatic free-wheeling hub assembly disassembly and assembly components

Remove the drive gear assembly, the slide gear assembly and the return spring.

> ※※ **CAUTION**

When the pressure of the press is removed, be sure that the retainer (A) is not caught by retainer (B).

12. Remove the retainer (B) and the retainer bearing.

13. Remove the drive gear snapring from the drive gear assembly. Separate the retainer (A) and the drive gear.

➡**When the drive gear snapring is removed, be sure to replace it with a new one.**

14. To remove the slide gear C-ring in the slide gear assembly, push the cam in and remove the slide gear C-ring with the spring compressed. Remove the cam, spring holder, shift spring and the slide gear from the slide gear assembly.

To install:

15. Assemble the slide gear, the shift spring, the spring holder and the cam to make the slide gear assembly. Install a new C-ring to the slide gear assembly.

16. Assemble the drive gear and the retainer (A) to form the drive gear assembly.

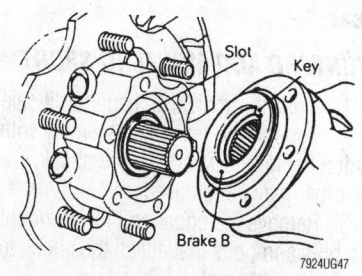

When assembling the free-wheeling hub assembly onto the spindle, line the key and the key slot up—automatic free-wheeling 4WD systems

Install a new drive gear snapring to the drive gear assembly.

17. Grease the following components with multipurpose grease SAE J310, NLGI No. 2:

   a. Retainer (B)—pack the grooves of the retainer (B) with grease.

   b. Brake (B)—pack the grooves of brake (B) with grease.

18. Install the components to the hub housing in the following order: retainer bearing, retainer (B), return spring, slide gear assembly, drive gear assembly, new retainer (B) C-ring, new housing snapring, brake spring, brake (A), brake (B), new housing C-ring. When installing the return spring, install it with the smaller coil diameter side toward the cam.

19. Using a spring scale, measure the front hub turning resistance without the free-wheeling hub housing installed yet. Note the amount of force needed to turn the hub assembly on the vehicle.

20. Apply a coating of 3M® ART Part No. 8661, or No. 8663, or the equivalent sealant, equally all around and without missing any spots, to the free-wheeling hub body assembly and the front hub contact surfaces.

✳✳ CAUTION

Be sure that there is no excess sealant on the hub outside surface.

21. Align the key of the brake (B) and the keyway of the knuckle spindle, then loosely install the automatic free-wheeling hub assembly to the spindle.

22. Check that the hub proper and the automatic free-wheeling hub assembly are brought into intimate contact when the assembly is forced lightly against the hub proper. If not, turn the hub until intimate contact is achieved.

23. Tighten the free-wheeling hub mounting bolts to 36–43 ft. lbs. (50–60 Nm). Use the spring scale to measure the front hub turning resistance again. Subtract the value measured from before from that now measured to find the turning resistance of the free-wheeling hub. The limit of resistance is 8.7 inch lbs. (1 Nm) or 3.1 lbs. (14 N) on the spring scale. If the free-wheeling hub turning resistance exceeds the limit, disassemble and reassemble the free-wheeling hub again.

24. Install the shim and the snapring to the driveshaft end. Screw the cover onto the hub assembly to 13–25 ft. lbs. (18–35 Nm).

25. Install the wheels to the vehicle. Lower the vehicle to the ground, then finish tightening the wheel lug nuts.

STEERING AND SUSPENSION

Air Bag

✳✳ CAUTION

Some vehicles are equipped with an air bag system, also known as the Supplemental Inflatable Restraint (SIR) or Supplemental Restraint System (SRS). The system must be disabled before performing service on or around system components, steering column, instrument panel components, wiring and sensors. Failure to follow safety and disabling procedures could result in accidental air bag deployment, possible personal injury and unnecessary system repairs.

PRECAUTIONS

Several precautions must be observed when handling the inflator module to avoid accidental deployment and possible personal injury.

• Never carry the inflator module by the wires or connector on the underside of the module.

• When carrying a live inflator module, hold securely with both hands, and ensure that the bag and trim cover are pointed away.

• Place the inflator module on a bench or other surface with the bag and trim cover facing up.

• With the inflator module on the bench, never place anything on or close to the module which may be thrown in the event of an accidental deployment.

DISARMING

To avoid personal injury when working on vehicles equipped with an air bag, the negative battery cable must be disconnected and at least 60 seconds must elapse before working on the system. Failure to do so may result in deployment of the air bag.

Recirculating Ball Power Steering Gear

REMOVAL & INSTALLATION

1. On vehicles equipped with a supplemental restraint system (SRS), turn the front wheel to the straight ahead position and remove the ignition key to prevent the steering wheel from turning.

2. Disconnect the negative battery cable.

3. Drain the power steering fluid.

4. Raise and support the vehicle.

5. Remove the pinch bolt securing the steering shaft to the steering gear.

6. Using the proper tools separate the Pitman arm from the relay rod.

7. Disconnect the fluid lines from the steering gear. Tag them if necessary for installation.

8. Remove the mounting bolts securing the gear to the frame rail and remove the steering gear.

To install:

9. Install the steering gear on the frame rail. Tighten the mounting nuts to 40–47 ft. lbs. (54–64 Nm).

10. Using new O-rings, connect the fluid lines to the steering gear. Tighten the fittings to 11 ft. lbs. (15 Nm).

11. Install the relay rod to the Pitman arm. Tighten the nut to 33 ft. lbs. (44 Nm).

12. Position the steering shaft on the steering gear and install the pinch bolt. Tighten the bolt to 13 ft. lbs. (18 Nm).

13. Connect the negative battery cable.

14. Refill and bleed the power steering system.

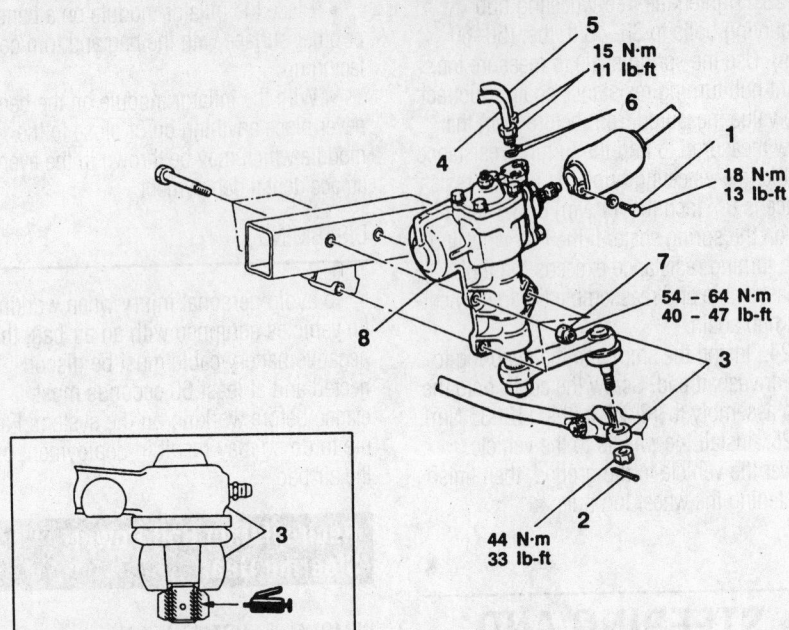

1. CONNECTING BOLT FOR STEERING GEAR BOX AND STEERING SHAFT
2. COTTER PIN
3. CONNECTION FOR PITMAN ARM AND RELAY ROD
4. PRESSURE TUBE
5. RETURN TUBE
6. O-RING
7. SELF-LOCKING NUT
8. POWER STEERING GEAR BOX

7924UG33

Exploded view of a typical power steering gear mounting

Shock Absorber

REMOVAL & INSTALLATION

Front

1. Raise the vehicle just enough to gain access to the upper shock nut.
2. Remove the upper shock mounting nut, washer and bushing.
3. Fully raise the vehicle and support it safely.
4. Remove the lower mounting bolts and shock assembly.
   To install:

➡ **If the shock absorber has a white paint mark on the lower end, be sure the mark faces the outside of the vehicle when installed.**

5. Install the shock and tighten the lower nut to:
   - Mighty Max: 7–10 ft. lbs. (9–14 Nm)
   - Montero: 65–76 ft. lbs. (88–103 Nm)
6. Tighten the upper nut to 9–13 ft. lbs. (12–18 Nm) for the vehicle and 11 ft. lbs. (15 Nm) for the Montero.
7. Test drive the vehicle and check the alignment.

Rear

MONTERO AND MONTERO SPORT

1. Raise the vehicle and support it safely.
2. Support the rear axle assembly with a hydraulic floor jack, so that the shock absorber may be removed.
3. Remove the upper and lower mounting nuts and bolts that attach the shock to the frame and bracket.
4. Remove the shock from the vehicle.
   To install:
5. Install the shock.
6. Tighten the lower bolt to 181 ft. lbs. (245 Nm)
7. Tighten the upper nut to 11 ft. lbs. (15 Nm).
8. Remove the floor jack from under the axle assembly.
9. Lower the vehicle.

MIGHTY MAX

1. Raise the vehicle and support it safely.
2. Remove the upper and lower mounting nuts and bolts that attach the shock to the frame and bracket.
3. Remove the shock from the vehicle.
   To install:
4. Install the shock.

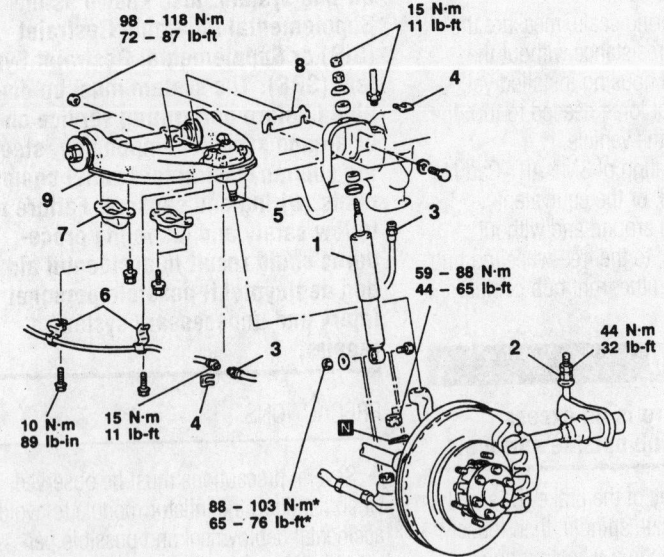

1. SHOCK ABSORBER
   ● BUMP STOPPER AND BUMP STOPPER BRACKET CLEARANCE ADJUSTMENT
2. REAR ANCHOR ARM ADJUSTING NUT
3. BRAKE HOSE CONNECTION
4. HOSE CLIP
5. UPPER ARM BALL JOINT CONNECTION
6. SPEED SENSOR BRACKET <VEHICLES WITH ABS>
7. REBOUND STOPPER
8. SHIMS
9. UPPER ARM
10. UPPER ARM BALL JOINT ASSEMBLY

Caution
*: Indicates parts which should be temporarily tightened, and then fully tightened with the vehicle on the ground in an unladen condition.

7924UG34

Common shock absorber and upper control arm components

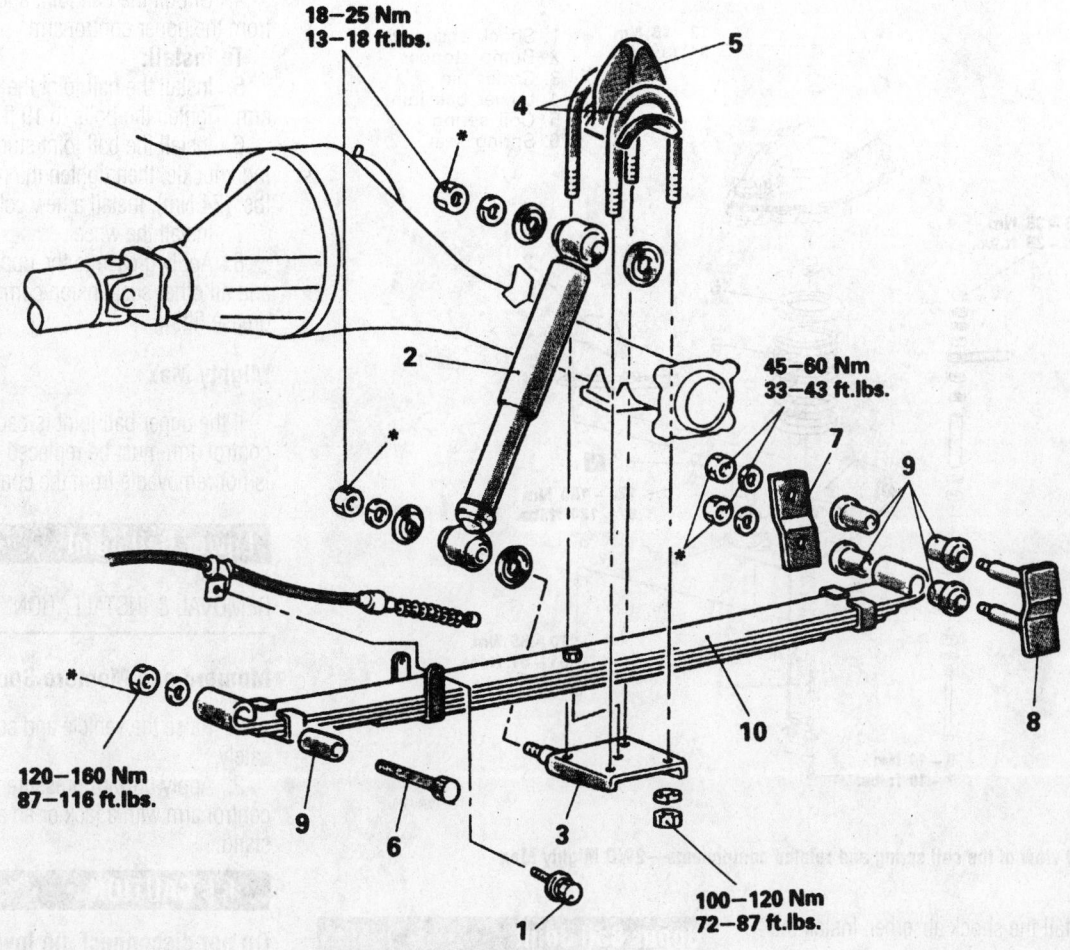

18—25 Nm
13—18 ft.lbs.

45—60 Nm
33—43 ft.lbs.

120—160 Nm
87—116 ft.lbs.

100—120 Nm
72—87 ft.lbs.

7924UG38

1. Parking brake cable attaching bolt
2. Shock absorber
3. U-bolt seat
4. U-bolts
5. Bump stopper
6. Bolt
7. Shackle plate
8. Shackle assembly
9. Rubber bushings
10. Rear spring

Rear leaf spring suspension components—Mighty Max

5. Tighten the upper and lower nuts to 18 ft. lbs. (25 Nm).
6. Lower the vehicle.

Coil Spring

REMOVAL & INSTALLATION

Front

1. Raise and support the vehicle safely.
2. Remove the shock absorber.

3. Disconnect the stabilizer bar from the lower control arm.
4. Install a suitable spring compressor and compress the spring.
5. Remove the cotter pin and lower ball joint nut.
6. Release the lower ball joint taper using a suitable tool.
7. Remove the tool and the ball stud from the control arm. Release the compressor tool from the coil spring.
8. Pull the arm down and remove the

spring with the rubber isolation pad from the vehicle.
To install:
9. Install the spring with the rubber isolator. Install the compressor tool and compress spring so the lower ball joint can be inserted through the knuckle.
10. Tighten the lower ball joint nut to 87–130 ft. lbs. (120–180 Nm). Install a new cotter pin. Remove the spring compressor.
11. Connect the sway bar to the lower control arm, if equipped.

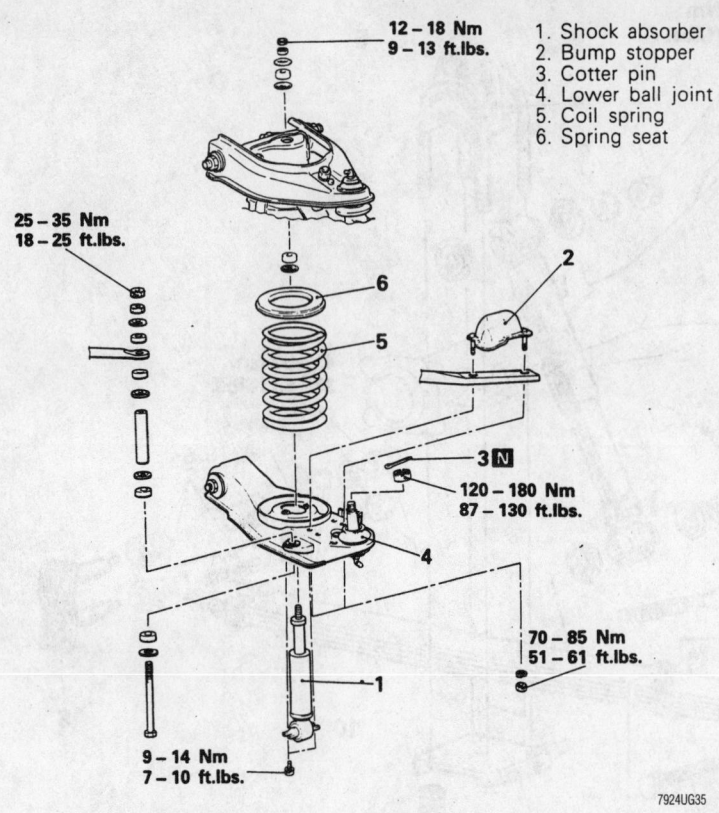

12 – 18 Nm
9 – 13 ft.lbs.

1. Shock absorber
2. Bump stopper
3. Cotter pin
4. Lower ball joint
5. Coil spring
6. Spring seat

25 – 35 Nm
18 – 25 ft.lbs.

120 – 180 Nm
87 – 130 ft.lbs.

70 – 85 Nm
51 – 61 ft.lbs.

9 – 14 Nm
7 – 10 ft.lbs.

7924UG35

Exploded view of the coil spring and related components—2WD Mighty Max

12. Install the shock absorber. Install the wheel and tire assembly.

Rear

MONTERO

1. Raise the vehicle and support it safely.
2. Remove the parking brake cable attaching bolt.
3. Using the proper equipment, support the weight of the axle.
4. Remove the bolt that attaches the lateral rod to the body.
5. Remove the lower shock mounting bolts.
6. Lower the axle and remove the coil springs with their seats.
To install:
7. Install the coil to the lower axle.
8. Install the lower shock mounting bolts and tighten to 181 ft. lbs. (245 Nm).
9. Support the axle and install the lateral rod bolt. Tighten to 170 ft. lbs. (230 Nm).
10. Attach the parking brake cable.
11. Test drive the vehicle and check the alignment.

Upper Ball Joint

REMOVAL & INSTALLATION

Montero and Montero Sport

1. Raise and safely support the vehicle.
2. Remove the front wheel.
3. Support the lower control arm, then disconnect the upper ball joint from the steering knuckle.

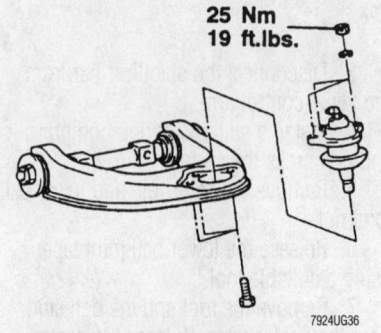

25 Nm
19 ft.lbs.

7924UG36

Exploded view of the upper ball joint and related components—Montero shown

4. Unbolt the ball joint and remove it from the upper control arm.
To install:
5. Install the ball joint the upper control arm. Tighten the bolts to 19 ft. lbs. (25 Nm)
6. Install the ball joint stud in the steering knuckle, then tighten the nut to 54 ft. lbs. (74 Nm). Install a new cotter pin.
7. Install the wheel.
8. Apply grease to the upper ball joint and all other suspension components with a grease fitting.

Mighty Max

If the upper ball joint is bad, the upper control arm must be replaced. The ball joint is not removable from the control arm.

Lower Ball Joint

REMOVAL & INSTALLATION

Montero and Montero Sport

1. Raise the vehicle and support it safely.
2. Apply upward pressure to the lower control arm with a jack or an adjustable stand.

✳✳ CAUTION

Do not disconnect the lower ball joint stud from the steering knuckle unless the lower control arm has a stand or a jack under it.

3. Remove the ball joint stud nut and disconnect the stud from the steering knuckle.
4. Remove the ball joint retaining nuts and bolts and remove the ball joint from the arm.
To install:
5. Install the lower ball joint to the control arm. Tighten the ball joint retaining nuts and bolts to 60 ft. lbs. (81 Nm).
6. Tighten the ball stud nut to 108 ft. lbs. (147 Nm). Install a new cotter pin.
7. Lubricate the ball joint with a grease gun.
8. If equipped with a torsion bar, adjust the riding height.
9. Check and adjust the alignment if necessary.

Mighty Max

1. Raise the vehicle and support it safely.

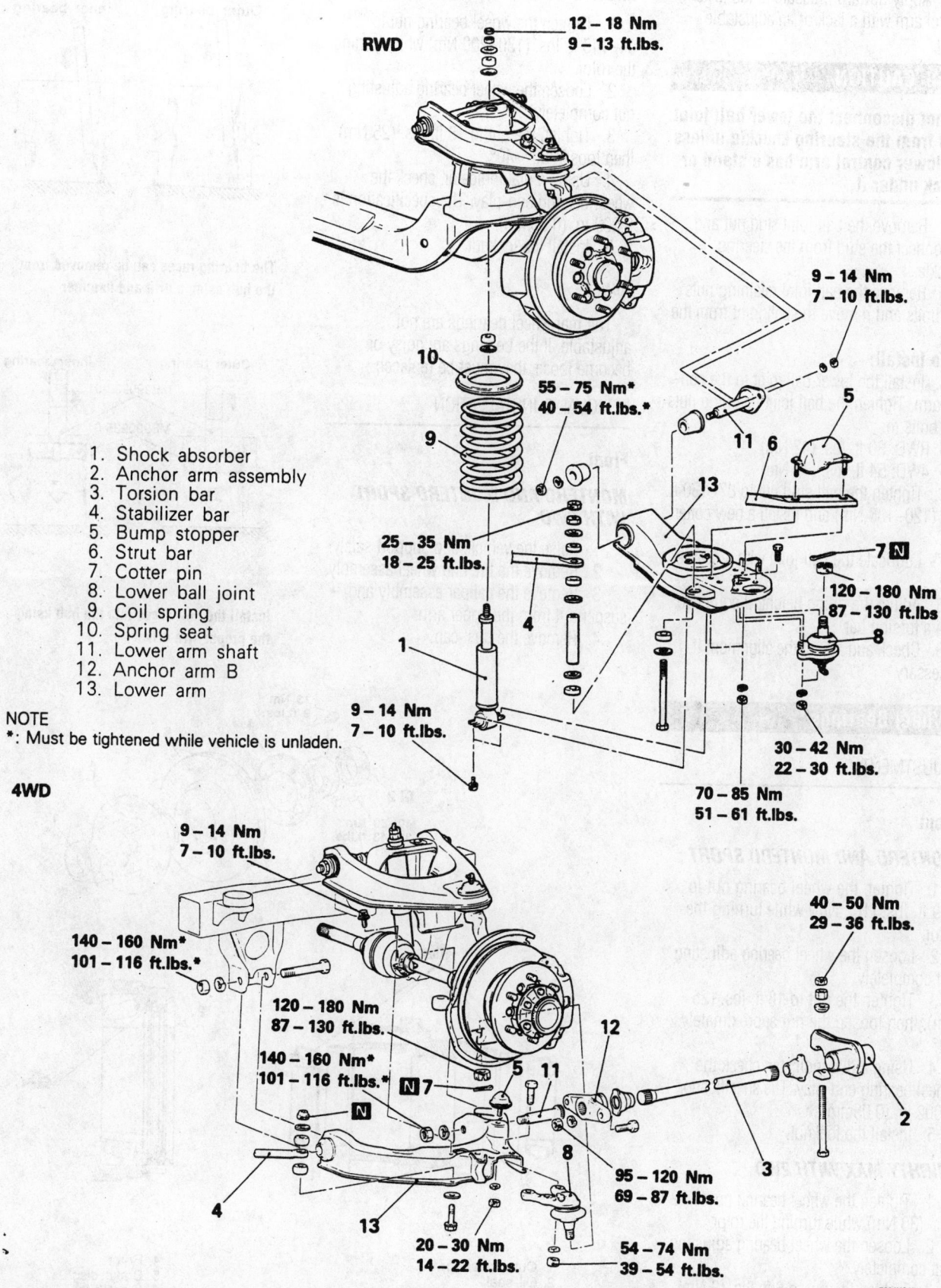

RWD

12 – 18 Nm
9 – 13 ft.lbs.

9 – 14 Nm
7 – 10 ft.lbs.

10

55 – 75 Nm*
40 – 54 ft.lbs.*

9

13

5

1. Shock absorber
2. Anchor arm assembly
3. Torsion bar
4. Stabilizer bar
5. Bump stopper
6. Strut bar
7. Cotter pin
8. Lower ball joint
9. Coil spring
10. Spring seat
11. Lower arm shaft
12. Anchor arm B
13. Lower arm

25 – 35 Nm
18 – 25 ft.lbs.

4

11   6

7

120 – 180 Nm
87 – 130 ft.lbs

8

1

9 – 14 Nm
7 – 10 ft.lbs.

30 – 42 Nm
22 – 30 ft.lbs.

NOTE
*: Must be tightened while vehicle is unladen.

70 – 85 Nm
51 – 61 ft.lbs.

4WD

9 – 14 Nm
7 – 10 ft.lbs.

140 – 160 Nm*
101 – 116 ft.lbs.*

120 – 180 Nm
87 – 130 ft.lbs.

40 – 50 Nm
29 – 36 ft.lbs.

140 – 160 Nm*
101 – 116 ft.lbs.*   7

12

5   11

2

95 – 120 Nm
69 – 87 ft.lbs.

3

13

8

20 – 30 Nm
14 – 22 ft.lbs.

54 – 74 Nm
39 – 54 ft.lbs.

7924UG37

Exploded view of the front suspension—Mighty Max shown

2. Apply upward pressure to the lower control arm with a jack or an adjustable stand.

❋❋ CAUTION

Do not disconnect the lower ball joint stud from the steering knuckle unless the lower control arm has a stand or a jack under it.

3. Remove the ball joint stud nut and disconnect the stud from the steering knuckle.

4. Remove the ball joint retaining nuts and bolts and remove the ball joint from the arm.

To install:

5. Install the lower ball joint to the control arm. Tighten the ball joint retaining nuts and bolts to:
- RWD: 30 ft. lbs. (42 Nm)
- 4WD: 54 ft. lbs. (74 Nm)

6. Tighten the ball stud nut to 87–130 ft. lbs. (120–180 Nm) and install a new cotter pin.

7. Lubricate the ball joint with a grease gun.

8. Adjust the riding height, if equipped with a torsion bar.

9. Check and adjust the alignment if necessary.

Wheel Bearings

ADJUSTMENT

Front

MONTERO AND MONTERO SPORT

1. Tighten the wheel bearing nut to 119 ft. lbs. (162 Nm) while turning the rotor.

2. Loosen the wheel bearing adjusting nut completely.

3. Tighten the nut to 18 ft. lbs. (25 Nm), then loosen the nut approximately 30°.

4. Using a dial indicator, check the wheel bearing end-play. The specification is 0.002 in. (0.05mm).

5. Install the locknut.

MIGHTY MAX WITH 2WD

1. Tighten the wheel bearing nut to 22 ft. lbs. (30 Nm) while turning the rotor.

2. Loosen the wheel bearing adjusting nut completely.

3. Tighten the nut to 6 ft. lbs. (8 Nm).

4. Check the wheel bearing end-play. The specification is 0.001–0.003 in.

5. Install the nut lock and cotter pin.

MIGHTY MAX WITH 4WD

1. Tighten the wheel bearing nut to 94–145 ft. lbs. (130–200 Nm) while turning the rotor.

2. Loosen the wheel bearing adjusting nut completely.

3. Tighten the nut to 18 ft. lbs. (25 Nm), then loosen 30°–40°.

4. Using a dial indicator, check the wheel bearing end-play. The specification is 0.002 in. (0.05mm).

5. Install the locknut.

Rear

The rear wheel bearings are not adjustable. If the bearings are noisy or become loose, they must be replaced.

REMOVAL & INSTALLATION

Front

MONTERO AND MONTERO SPORT WITH 2WD

1. Raise the vehicle and support safely.
2. Remove the tire and wheel assembly.
3. Remove the caliper assembly and suspend it from the upper arm.
4. Remove the dust cap.

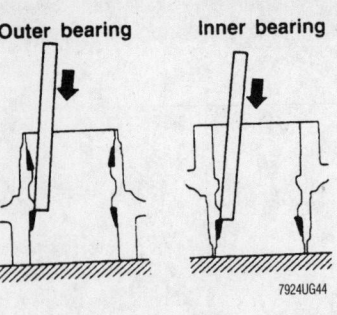

The bearing races can be removed from the hub using a drift and hammer

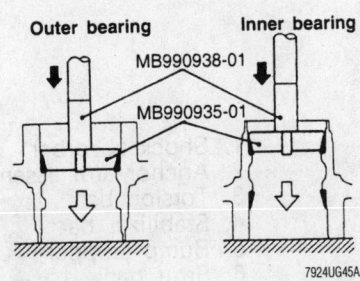

Install the new races into the hub using the proper size driver

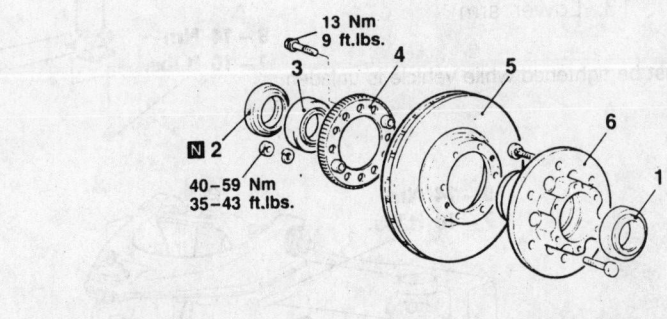

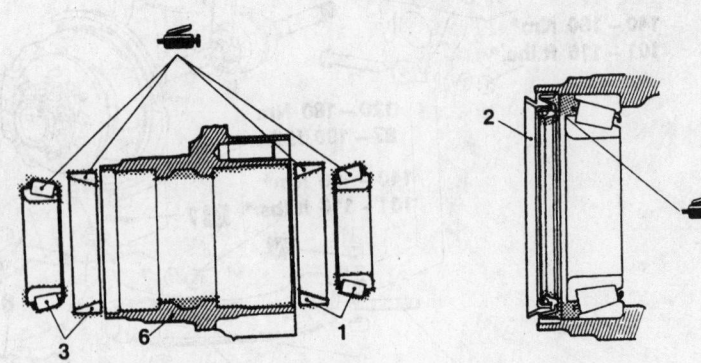

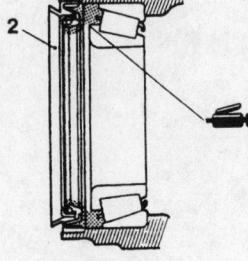

1. Outer bearing
2. Oil seal
3. Inner bearing

4. Rotor
5. Brake disc
6. Front hub

7924UG43

Exploded view of the hub and wheel bearing assembly—Montero

5. Remove the cotter pin, castellated nut lock, wheel bearing nut and washer from the spindle.

6. Remove the outer wheel bearing.

7. Remove the hub and rotor as an assembly.

8. Remove the grease seal and inner wheel bearing.

9. If required, press the inner and outer bearing outer races from the hub assembly.

10. If replacement of the hub is necessary, matchmark the brake disc with the hub, then separate the hub from the disc.

To install:

11. If removed, assemble the brake disc to the hub, while aligning the matchmarks. Tighten the mounting bolts to 34–38 ft. lbs. (47–52 Nm).

12. If removed, press-fit the inner and outer bearing outer races into the hub assembly.

13. Lubricate the seal lip and inside surface of the front hub with MP grease. Repack and install the inner wheel bearing. Install a new grease seal.

14. Install the hub assembly to the spindle.

15. Lubricate and install the outer wheel bearing, washer and nut. When the bearing preload is properly set, install the nut lock and a new cotter pin.

16. Install the grease cap.

17. Install the brake pads and caliper.

18. Install the tire and wheel assembly.

MONTERO AND MONTERO SPORT WITH 4WD

1. Raise the vehicle and support safely.

2. Remove the tire and wheel assembly.

3. If equipped with free-wheeling hub, place the free-wheeling hub in the free condition.

➡ **The free condition can be obtained by shifting the transfer shift lever to the 2H position, then moving in reverse for approximately 3.3–6.5 ft.**

   a. Remove the hub cover.

   b. Remove the snapring from the driveshaft.

   c. Remove the bolts and remove the automatic free-wheeling hub.

4. Remove the caliper assembly and suspend it from the upper arm.

5. Remove the lockwasher and locknut.

6. Remove the hub and rotor as an assembly from the knuckle together with the inner and outer bearings.

7. Remove the outer bearing, grease seal and inner wheel bearing.

8. If required, press the inner and outer bearing outer races from the hub assembly.

9. If replacement of the hub is necessary, matchmark the brake disc with the hub, then separate the hub from the disc.

To install:

10. If removed, assemble the brake disc to the hub, while aligning the matchmarks.

11. If removed, press-fit the inner and outer bearing outer races into the hub assembly.

12. Lubricate the seal lip and inside surface of the front hub with MP grease. Repack and install the inner wheel bearing. Install a new grease seal.

13. Install the hub assembly to the spindle.

14. Lubricate and install the outer wheel bearing and locknut. When the bearing preload is properly set, install the lockwasher.

15. If equipped with free-wheeling hub:

   a. Apply a coating of semi-drying sealant to the free-wheeling hub body and front hub contact surfaces.

   b. Align the key of the brake "B" and the keyway of the knuckle spindle and loosely install the automatic free-wheeling hub assembly. Tighten the mounting bolts to 36–43 ft. lbs. (50–60 Nm).

16. Install the wheel and tire assembly. Lower the vehicle.

MIGHTY MAX WITH 2WD

1. Raise the vehicle and support safely.

2. Remove the tire and wheel assembly.

3. Remove the caliper assembly and suspend it from the upper arm.

4. Remove the dust cap.

5. Remove the cotter pin, castellated nut lock, wheel bearing nut and washer from the spindle.

6. Remove the outer wheel bearing.

7. Remove the hub and rotor as an assembly.

8. Remove the grease seal and inner wheel bearing.

9. If required, press the inner and outer bearing outer races from the hub assembly.

10. If replacement of the hub is necessary, matchmark the brake disc with the hub, then separate the hub from the disc.

To install:

11. If removed, assemble the brake disc to the hub, while aligning the matchmarks. Tighten the mounting bolts to 58–72 ft. lbs. (80–100 Nm).

12. If removed, press-fit the inner and outer bearing outer races into the hub assembly.

13. Lubricate the seal lip and inside surface of the front hub with MP grease.

Repack and install the inner wheel bearing. Install a new grease seal.

14. Install the hub assembly to the spindle.

15. Lubricate and install the outer wheel bearing, washer and nut. When the bearing preload is properly set, install the nut lock and a new cotter pin.

16. Install the grease cap.

17. Install the brake pads and caliper.

18. Install the tire and wheel assembly.

MIGHTY MAX WITH 4WD

1. Raise the vehicle and support safely.

2. Remove the tire and wheel assembly.

3. If equipped with free-wheeling hub, place the free-wheeling hub in the free condition.

➡ **The free condition can be obtained by shifting the transfer shift lever to the 2H position, then moving in reverse for approximately 3.3–6.5 ft.**

   a. Remove the hub cover.

   b. Remove the snapring from the driveshaft.

   c. Remove the bolts and remove the automatic free-wheeling hub.

4. Remove the caliper assembly and suspend it from the upper arm.

5. Remove the lockwasher and locknut.

6. Remove the hub and rotor as an assembly from the knuckle together with the inner and outer bearings.

7. Remove the outer bearing, grease seal and inner wheel bearing.

8. If required, press the inner and outer bearing outer races from the hub assembly.

9. If replacement of the hub is necessary, matchmark the brake disc with the hub, then separate the hub from the disc.

To install:

10. If removed, assemble the brake disc to the hub, while aligning the matchmarks.

11. If removed, press-fit the inner and outer bearing outer races into the hub assembly.

12. Lubricate the seal lip and inside surface of the front hub with MP grease. Repack and install the inner wheel bearing. Install a new grease seal.

13. Install the hub assembly to the spindle.

14. Lubricate and install the outer wheel bearing and locknut. When the bearing preload is properly set, install the lockwasher.

15. If equipped with free-wheeling hub:

   a. Apply a coating of semi-drying sealant to the free-wheeling hub body and front hub contact surfaces.

   b. Align the key of the brake "B" and

the keyway of the knuckle spindle and loosely install the automatic free-wheeling hub assembly. Tighten the mounting bolts to 36–43 ft. lbs. (50–60 Nm).

16. Install the wheel and tire assembly. Lower the vehicle.

Rear

1. Loosen the wheel lug nuts only 1/2 a turn.

2. Raise and support the vehicle on jackstands.

3. Remove the wheel(s) from the vehicle.

4. Loosen the bleeder valve on the right rear caliper and drain the brake fluid into a container. Disconnect the rear brake caliper brake hose from the hard line on the frame.

5. Pull the rear brake caliper off of the rear disc and remove from the vehicle.

6. Pull the rear disc off of the rear axle.

7. Remove the parking cable attaching bolt and disconnect the parking brake cable end from the brake assembly. Remove the parking brake assembly from the end of the axle.

8. On vehicles with the ABS system, remove the speed sensor.

9. Pull the rear axle shaft out of the axle

housing. If the rear axle shaft is difficult to remove, use a slide hammer (impact puller) to remove it.

✲✲ WARNING

Do not damage the oil seal during removal.

10. Remove the snapring from the inside end of the axle shaft. Remove one retainer bolt from the backing plate with a plastic mallet. Apply gummed cloth tape around the edge of the bearing case for protection. Fix the axle shaft in a vise or with a similar method. Using a grinder, grind down the

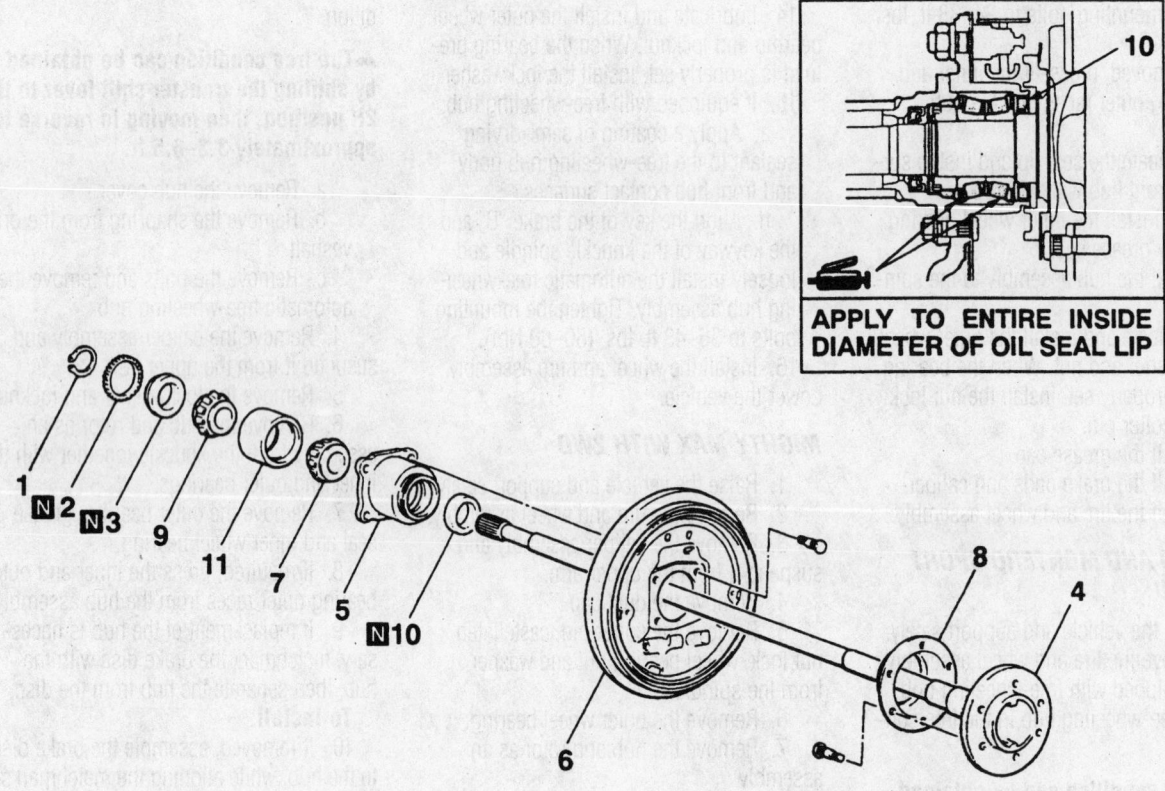

APPLY TO ENTIRE INSIDE DIAMETER OF OIL SEAL LIP

DISASSEMBLY STEPS

1. SNAP RING
2. ABS ROTOR
3. RETAINER
4. AXLE SHAFT
5. BEARING CASE
6. BACKING PLATE
7. OUTER BEARING INNER RACE
8. DUST COVER
9. INNER BEARING INNER RACE
10. OIL SEAL
11. BEARING OUTER RACE

ASSEMBLY STEPS

11. BEARING OUTER RACE
9. INNER BEARING INNER RACE
7. OUTER BEARING INNER RACE
10. OIL SEAL
8. DUST COVER
6. BACKING PLATE
5. BEARING CASE
4. AXLE SHAFT
3. RETAINER
2. ABS ROTOR
1. SNAP RING

7924UG48

Exploded view of the typical rear axle shaft, bearings and races

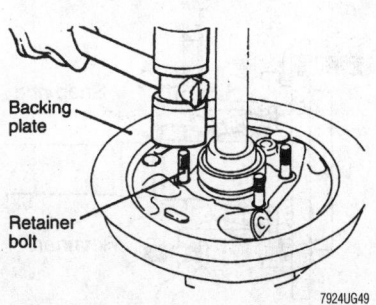

Remove one of the rear axle studs before attempting to grind down the retainer

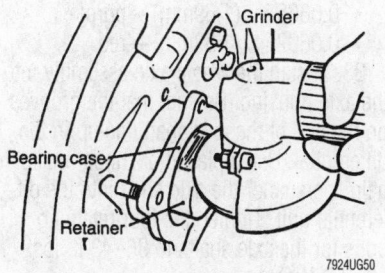

Using a grinder, grind the retainer, on one side, down to 1–2mm (0.04–0.08 in.) thickness

retainer flat, on one side, until the thickness of the retainer is only 0.04–0.08 in. (1–2mm). That is that the retainer is ground down toward the axle shaft, not toward the flange. Cut, with a chisel, the place where the retainer ring has been shaven down and remove the retainer.

❋❋ CAUTION

Be careful not to damage the bearing case and the axle shaft.

➥**Only the retainer ring is to be ground down—NOT the axle shaft, the axle flange, the bearing or any other component.**

11. Grind the plate of special tool MB990861 with a grinder (see illustration) so that there will be no interference between the plate and the bearing case. While adjusting the height of the hanger, secure the washers, plate and nuts in order so that the processed plate is as shown in the illustration.

➥**The washers are used to eliminate the difference in height of the bearing case so that the plate and the bearing case are parallel.**

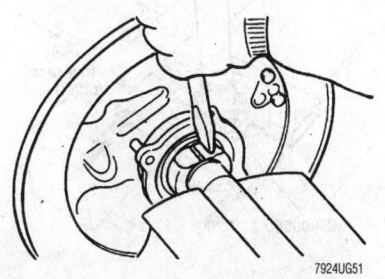

On the ground down spot on the rear axle bearing retainer, use a chisel to split the retainer, then remove it

Place the end of the bolt against the center of the axle shaft, then tighten the nuts to remove the axle shaft from the bearing case assembly.

➥**The hanger and plate must be placed so that they are parallel.**

12. Remove the bearing inner race and the bearing outer race. To remove the races, install the tool MB990560 and use a press to remove the bearing race from the axle shaft.

13. Remove the oil seal and the dust cover (vehicles without ABS).

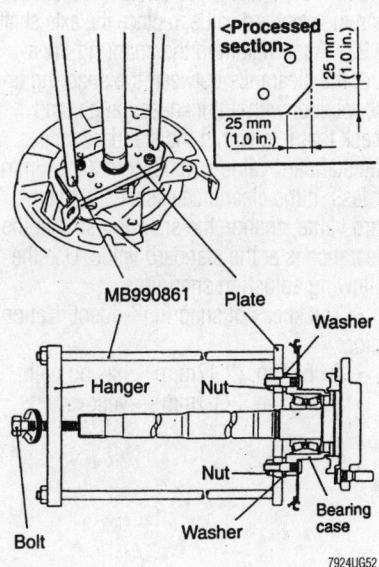

Use the special tool (MB990861) to remove the rear axle shaft from the bearing case

14. On vehicles without ABS, insert an iron plate of approximately 0.04 in. (1mm) thickness between the rotor assembly and the axle shaft, then use a press to remove the rotor assembly.

❋❋ WARNING

In order not to bend the rotor assembly plate, place the support in contact with the axle shaft when using the press.

15. Remove the axle shaft from the remaining bearings and components.

16. Remove the backing plate.

17. Reinstall the bearing inner race that was removed previously, then use the tool MB990799–01 and press to remove the bearing outer race.

18. Remove the bearing case.

19. Remove the O-ring from the end of the axle housing tube.

20. Remove the oil seal from the end of the rear axle housing using the tool MB990211–01 (slide hammer with a hooked end), if necessary.

21. Check the dust cover for deformation or damage. Check the oil seal for damage. Check the inner and outer bearings for seizure, discoloration and rough raceway surface. Check the axle shaft for cracks, wear and damage. For there are any of these indications, replace the part with a new one. The retainer, the bearing inner (inner and outer) and outer races and the oil seal need to be replaced with new components upon reassembly. After all of this work, it is probably a good idea to replace the bearings and the axle housing tube oil seals.

To install:

22. Drive the new oil seal into the end of the rear axle housing using the tools MB990932–01 and MB990938–01, if necessary.

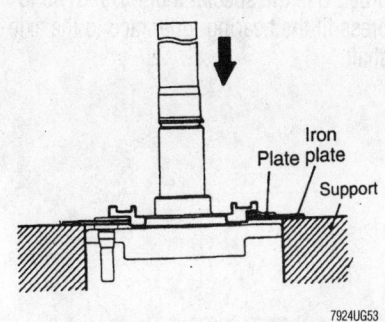

Use a iron plate and supports to remove the rotor assembly

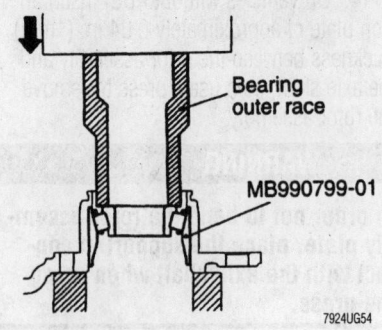

Use the tool (MB990799–01) to install and remove the rear axle bearing races

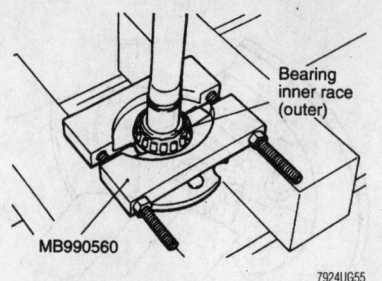

Use the MB990560 tool to hold the bearing inner race (outer), then use a plastic hammer to drive the axle out of the race—do not let the axle fall onto a hard floor

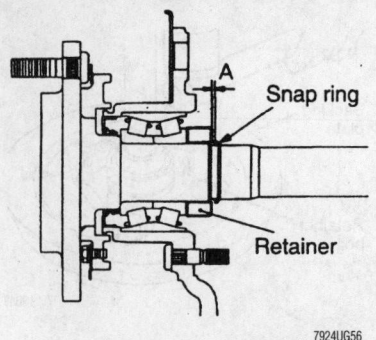

Measure the clearance (A) between the snapring and the retainer edge

23. Install the new O-ring into the axle tube.

24. Apply multipurpose grease to the external surface of the bearing out race. Press-fit the bearing outer race into the bearing case by using the toll MB990890–01.

25. Install the speed sensor bracket to the back of the backing plate.

26. Install the rotor assembly to the axle shaft by press-fitting (plastic mallet will also work) it on using the special tool MB991388. Slide the backing plate onto the axle shaft.

27. Install the dust cover to the backing plate if the vehicle is equipped with ABS.

28. Install the bearing inner race (outer) to the bearing case. Install the oil seal to the front end of the bearing case. To do this, apply multipurpose grease to the outside of the oil seal. Use the special tools MB990936–01 and MB990938–01 to press-fit the oil seal until it is flush with the end of the bearing case. Apply multipurpose grease to the lip of the oil seal.

29. Pass the axle shaft through the bearing inner race, the bearing case and the second bearing inner race in that order. Use the special tool MB990799 to press fit the bearing inner race to the axle shaft.

✳✳ WARNING

Both bearing inner race sets should be press-fitted together. The left and right lengths of the axle shaft are different in vehicles with rear differential locks. The right side is longer, be careful when installing.

30. Use the tool MB990799–01 to press-fit the retainer onto the axle shaft, while checking that the press-fitting force is at the following values:

a. Initial press-fitting force is 11,023 lbs. (50,000 N) or more.

b. Final press-fitting force is 22,046–24,251 lbs. (100,000–110,000 N).

31. If the initial press-fitting force is less than the standard value, replace the axle shaft.

32. After installing the snapring, measure the clearance between the snapring and the retainer with a thickness gauge, and check that it is within the standard values. The standard value is 0.0065 in. (0.166mm) or less. If the clearance exceeds the standard value, change the snapring so that the clearance is at the standard value. Use the following adjusting snaprings:

• Thickness of snapring—identification color.

• 0.0854 in. (2.17mm)—has no color.
• 0.0791 in. (2.01mm)—yellow.

• 0.0728 in. (1.85mm)—blue.
• 0.0665 in. (1.69mm)—purple.
• 0.0602 in. (1.53mm)—red.

33. Install the whole axle assembly into the axle housing. Be sure that the grooves on the end of the axle shaft line up in the differential. Use a plastic or rubber mallet to help "persuade" the axle shaft into the differential unit. Tighten the four retaining bolts for the axle shafts to 36–43 ft. lbs. (49–59 Nm).

34. Attach the speed sensor and install the various parking brake assembly components to the axle flange.

35. Attach the parking brake cable to the parking brake assembly, then secure it in place with the cable bracket.

36. Slide the brake rotor onto the axle shaft, then install the brake caliper. Tighten the brake caliper bolts to 65 ft. lbs. (88 Nm).

37. Tighten the brake hose to the frame brake line. Tighten the flare nut to 11 ft. lbs. (15 Nm).

38. Install the wheels and tighten the lug nuts as tight as possible with the vehicle not on the ground.

39. Bleed the brake system.

40. Lower the vehicle until the wheels are touching the ground, then finish tightening the lug nuts. Lower the vehicle the rest of the way to the ground.

41. Road test the vehicle and check for leaks.

ENGINE REPAIR

➡ **Disconnecting the negative battery cable on some vehicles may interfere with the functions of the on board computer systems and may require the computer to undergo a relearning process, once the negative battery cable is reconnected.**

Distributor

REMOVAL

1. Disconnect the negative battery cable.
2. Disconnect and remove the distributor cap with the plug wires attached.
3. Using a piece of chalk, make alignment marks on the distributor-to-engine and rotor-to-distributor locations; the alignment marks are used for reinstallation.
4. Disconnect the distributor electrical harness connector.
5. Remove the distributor hold-down bolt(s) and lift the distributor assembly from the engine.

INSTALLATION

1. If the engine was undisturbed, install the distributor, align the matchmarks and reverse the removal procedures. Start the engine; check and/or adjust the timing.
2. If the crankshaft was turned, the engine disturbed in any manner (while the distributor was removed), or the alignment marks were not drawn, perform the following procedures:

   a. Remove the No. 1 cylinder spark plug.

   b. Turn the crankshaft until the No. 1 piston is positioned on the Top Dead Center (TDC) of the compression stroke.

➡ **To determine the TDC of the compression stroke, place your thumb over the spark plug hole and feel the air being forced from the cylinder. Stop turning the crankshaft when the timing marks, used to time the engine, are aligned.**

   c. Oil the distributor housing-to-cylinder block surface.

   d. Install the distributor so the rotor points toward the No. 1 spark plug terminal tower of the distributor cap (when installed).

   e. When the distributor shaft has reached the bottom of the hole, move the rotor back and forth slightly until the driv-

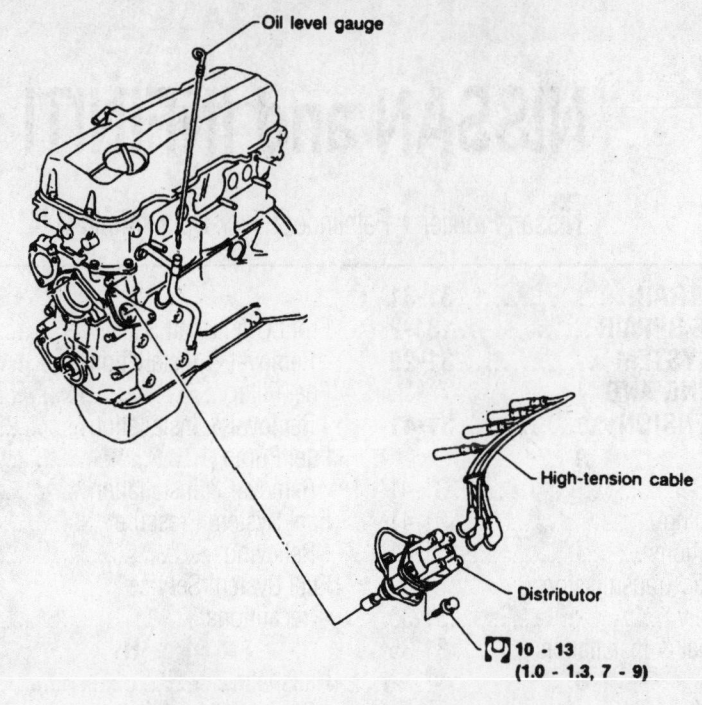

: N•m (kg-m, ft-lb)

: Apply liquid gasket.

7924VG01

Exploded view of the distributor assembly and related components—2.4L engine

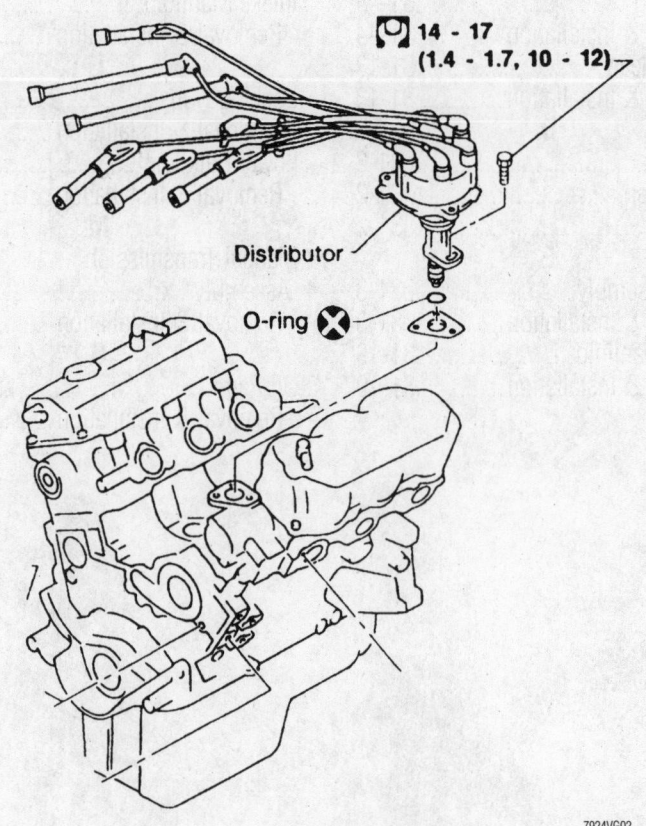

7924VG02

Exploded view of the distributor assembly and related components—3.0L and 3.3L engines

ing lug on the end of the distributor shaft enters the slots cut in the end of the oil pump shaft and the distributor assembly slides down into place.

3. Install the distributor cap.

4. Tighten the distributor hold-down bolt(s).

5. Connect the negative battery cable.

6. Start the engine, check and/or adjust the ignition timing.

Ignition Timing

ADJUSTMENT

1. Visually inspect:
   a. The air cleaner for clogging
   b. Check the hoses/ducts for leaks
   c. Check the EGR valve operation
   d. Check all the electrical connectors
   e. Check the gaskets
   f. Check the throttle valve and throttle sensor operation.

2. Locate the timing marks on the crankshaft pulley and the front of the engine.

3. Clean the timing marks.

4. Using chalk or white paint, color the mark on the crankshaft pulley and the mark on the scale which will indicate the correct timing when aligned with the notch on the crankshaft pulley.

5. Attach a tachometer to the engine.

6. Attach a timing light to the engine, to number one cylinder ignition wire.

7. Check to be sure all of the wires clear the fan, then, start the engine and allow it to reach normal operating temperatures.

8. Block the front wheels and set the parking brake. Shift the transmission into **NEUTRAL** ; do not stand in front of the vehicle when making adjustments.

9. Perform the following procedures:
   a. Race the engine at 2000 rpm for

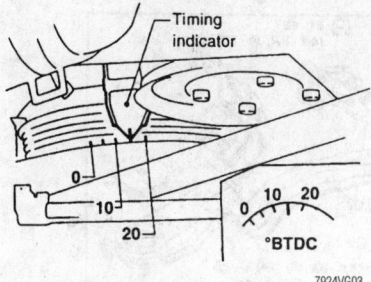

Timing marks—1995–97 2.4L, 3.0L and 3.3L engines

7924VG03

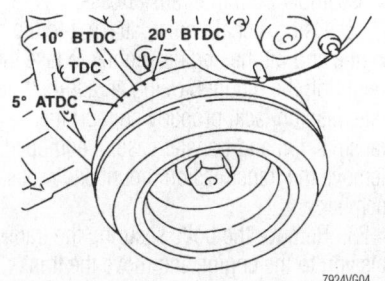

Timing indicator—1998–99 2.4L engine

7924VG04

about two minutes under a no-load condition; be sure all of the accessories are turned off.

   b. Perform on-board engine diagnostics and repair any fault code.

   c. Race the engine 2–3 times under no-load, then run the engine for one minute at idle.

   d. Stop the engine and disconnect the throttle position sensor.

   e. Race the engine at 2000 rpm for about two minutes under a no-load condition; be sure all of the accessories are turned off.

   f. Run the engine at idle speed.

10. Aim the timing light at the timing marks. If the marks on the pulley and the engine are aligned when the light flashes, the timing is correct. Turn the engine **OFF** and remove the tachometer and the timing light. If the marks are not in alignment, proceed with the following steps.

11. Turn the engine **OFF**.

12. Loosen the bolts that secure the distributor just enough so it can be turned.

13. Start the engine. Keep the wires of the timing light clear of the cooling fan.

14. With the timing light aimed at the pulley and the marks on the engine, turn the distributor for the proper adjustment. The ignition timing specification is 13–17° BTDC on the 3.3L engine; 8–12° BTDC on the 1995–97 2.4L engine and 18–22° BTDC on the 1998–99 2.4L engine.

15. Race the engine 2–3 times under no-load, then run the engine it at idle.

16. Aim the timing light at the timing marks. If the marks on the pulley and the engine are aligned when the light flashes, the timing is correct.

17. Tighten the bolt that secures the distributor and recheck the timing.

18. Turn the engine **OFF**, then remove the tachometer and the timing light.

19. Connect the throttle position sensor.

REMOVAL & INSTALLATION

2.4L Engine

✳✳ CAUTION

The fuel injection system remains under pressure after the engine has been turned OFF. Properly relieve fuel pressure before disconnecting any fuel lines. Failure to do so may result in fire or personal injury.

1. Relieve the fuel system pressure.

2. Disconnect the battery cables, then remove the battery.

3. Matchmark the location of the hood hinges and remove the hood.

4. Remove the air cleaner assembly.

5. Wrap a shop rag around the fuel filter outlet and disconnect the hose. Disconnect the fuel return hose.

6. Raise and safely support the vehicle. If equipped, remove the splash pan from under the engine.

7. Drain the engine oil and the cooling system, including the block drains. Dispose of the old fluids properly.

8. Remove the upper and lower radiator hoses.

9. Remove the radiator.

10. If equipped with air conditioning, loosen the belt tension and remove the belt. Disconnect the wiring, remove the compressor and secure it out of the way. Do not disconnect the pressure hoses.

11. If equipped with power steering, remove the drive belt and the power steering pump and secure it out of the way. Do not disconnect the pressure hoses.

12. Label and disconnect all wiring and vacuum hoses.

13. Disconnect the heater hoses from the engine and disconnect the throttle cable.

14. Remove the starter.

15. Matchmark the driveshaft flange at the rear pinion flange and remove the driveshaft. Plug the extension housing opening to prevent the oil from draining out.

16. If equipped with 4WD, matchmark both front driveshaft flanges so the shaft can be installed in the same position. Remove the front driveshaft.

17. Disconnect the exhaust pipe from the manifold and from the catalytic converter and remove the pipe.

18. Disconnect the speedometer cable and the wiring from the transmission.

19. If equipped an automatic transmission perform the following:

a. Disconnect the selector lever and throttle cables from the transmission.

b. Remove the dipstick tube and disconnect the cooler lines.

c. Remove the torque converter housing dust cover. Matchmark the converter with the driveplate for reassembly; these are balanced together at the factory. Remove the torque converter-to-driveplate (flywheel) bolts. Use a wrench on the crankshaft pulley bolt to rotate the crankshaft to expose the hidden torque converter bolts.

20. If equipped with a manual transmission perform the following:

a. Remove the shifter knob and boot and remove the snapring to lift the shift lever out of the transmission. Stuff a rag in the opening to keep dirt out of the transmission.

b. Without disconnecting the hydraulic hose, remove the clutch slave cylinder from the transmission and secure it aside.

21. If equipped with 4WD perform the following:

a. Remove the torsion bars and the second crossmember

b. Remove the transfer case shift lever assembly from the transfer case.

22. Using a chain hoist, attach it to the engine and lift the engine slightly to take the weight off the mounts. Using an appropriate transmission jack, properly support the transmission and transfer case, if equipped. Remove the transmission mount and crossmember.

23. Remove the bolts securing the transmission to the engine and move the transmission back from the engine. If equipped with an automatic transmission, secure the torque converter to the transmission. Lower the transmission from the vehicle.

➡**When removing the engine mounts, do not loosen the 4 mount cover nuts. The mount is fluid filled and will not function properly if the fluid leaks out.**

24. Check to be sure all wires and hoses have been disconnected. Remove the front engine mount bolts and carefully lift the engine out.

To install:

25. Carefully guide the engine into place and start the mount bolts. Tighten the bolts temporarily.

26. If equipped with a manual transmission, perform the following steps:

a. Lightly grease the input shaft splines. On 4WD, apply a silicone sealant to the engine block and rear plate to seal the engine to the transmission.

b. Fit the transmission into place and start the bolts attaching the engine to the transmission. Be sure the input shaft fits properly into the clutch disc and pilot bearing.

c. Tighten the bolts to specification.

27. If equipped with an automatic transmission perform the following:

a. Use a dial indicator to check the driveplate run-out while turning the crankshaft. Maximum allowable run-out is 0.020 in. (0.5mm); if beyond specifications, replace the driveplate.

b. Measure and adjust how far the torque converter is recessed into the transmission housing. The distance between the front mounting surface of the transmission and the torque converter-to-driveplate bolt boss should be at least 1.024 in. (26mm).

c. Install the transmission and start the bolts that attach the transmission to the engine. Tighten the transmission attaching bolts to specification.

28. If equipped with an automatic transmission, align the matchmarks on the drive-

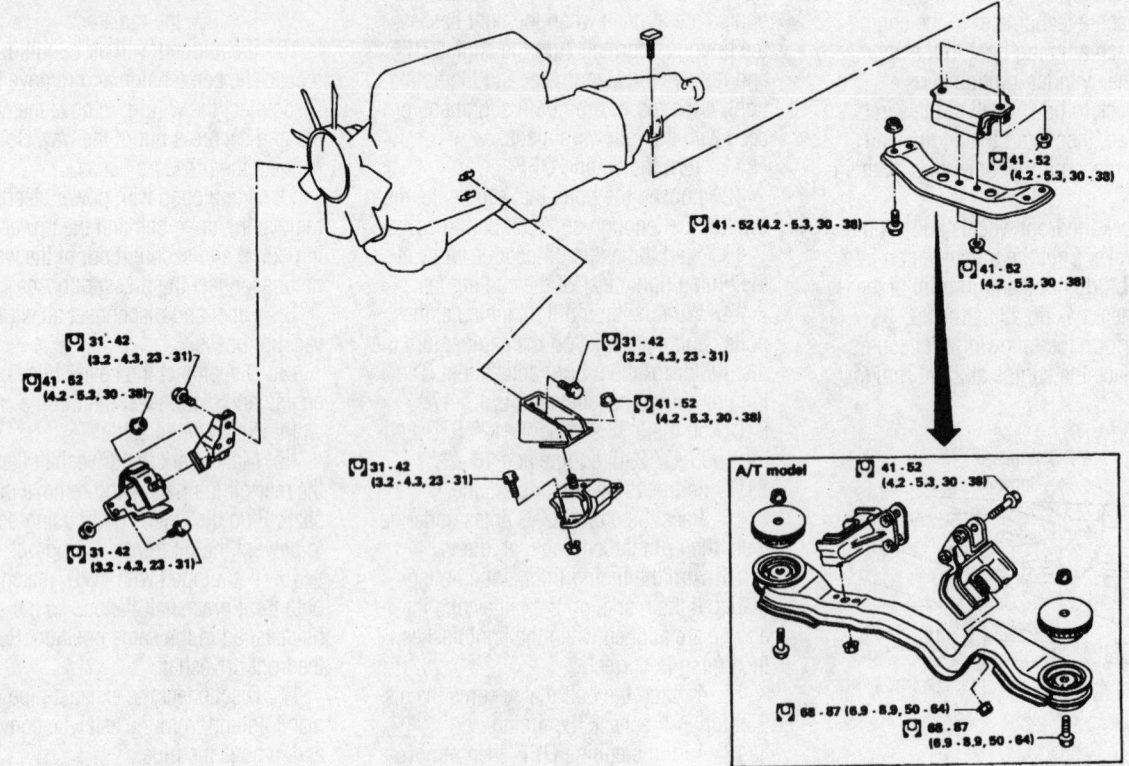

The engine is secured to the frame at three locations as shown—2.4L engine

7924VG05

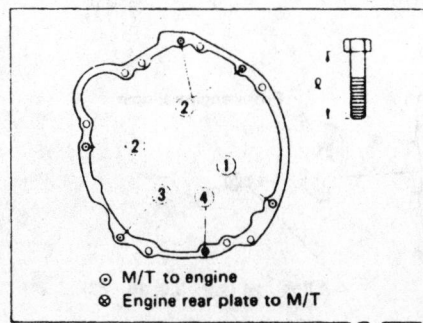

Bolt No.	Tightening torque N·m (kg-m, ft-lb)	ℓ mm (in)
1	39 - 49 (4.0 - 5.0, 29 - 36)	65 (2.56)
2	39 - 49 (4.0 - 5.0, 29 - 36)	60 (2.36)
3*	19 - 25 (1.9 - 2.5, 14 - 18)	25 (0.98)
4	19 - 25 (1.9 - 2.5, 14 - 18)	16 (0.63)

*: With nut

Manual transmission bolt specifications—2.4L engine

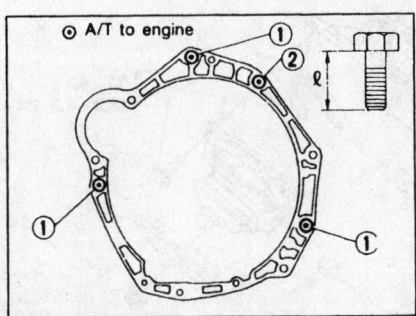

Tightening torque N·m (kg-m, ft-lb)	Bolt length "ℓ" mm (in)
① 39 - 49 (4.0 - 5.0, 29 - 36)	45 (1.77)
② 39 - 49 (4.0 - 5.0, 29 - 36)	40 (1.57)

Automatic transmission bolt specifications—2.4L engine

plate and torque converter, install the bolts and tighten to 33–43 ft. lbs. (44–59 Nm). Turn the crankshaft after tightening the bolts to be sure there is no binding at the drive-plate.

29. Install the transmission crossmember and the transmission mount.

30. Loosen the engine mount bolts, tighten the transmission mount bolts, then the engine mounts.

31. If equipped with a manual transmission, perform the following steps:

a. Install the clutch slave cylinder and tighten the bolts to 22–30 ft. lbs. (30–40 Nm).

b. Remove the rag from the transmission and install the shifter, then install the snapring to hold the shifter in position.

c. Install the shifter boot and knob.

32. If equipped with an automatic transmission perform the following:

a. Install the tighten converter housing dust cover.

b. Install the dipstick tube and connect the cooler lines.

c. Connect the selector lever and throttle cables to the transmission.

33. If the torsion bars were removed, install them in their original location. Be sure

the splines are in their original position and set the adjustment to its original position.

34. When installing the driveshafts, be sure to align the matchmarks. Tighten the bolts on the front driveshaft (4WD) to 29–33 ft. lbs. (39–44 Nm). If equipped with a one-piece driveshaft, tighten the flange bolts to 58–65 ft. lbs. (78–88 Nm). If equipped with a two-piece driveshaft, tighten the flange bolts to 29–33 ft. lbs. (39–44 Nm) and tighten the center bearing bracket bolts to 12–16 ft. lbs. (16–22 Nm).

35. When installing the exhaust pipe, use new gaskets and tighten the flange bolts to 20–27 ft. lbs. (26–36 Nm). Tighten the bolts attaching the pipe to the catalytic converter to 23–31 ft. lbs. (31–42 Nm).

36. Connect the speedometer cable. Connect any electrical leads to the transmission.

37. Install the starter and connect the starter leads.

38. Connect the throttle cable and the heater hoses to the engine.

39. Connect the wiring and hoses.

40. Install the power steering pump and its bracket, if equipped; be sure to connect the ground strap to the bracket.

41. Install the air conditioning compressor and drive belt.

42. Adjust the drive belts.

43. Install the radiator and shroud. If equipped with an automatic transmission, unplug oil cooler lines and connect them to the radiator.

44. Install the upper and lower radiator hoses.

45. Install the splash pan, if equipped.

46. Connect the fuel return hose and the hose to the filter outlet.

47. Install the air cleaner assembly.

48. Fill the transmission, if equipped, the transfer case.

49. Fill the engine with fresh oil.

50. Install and adjust the hood. Install the undercover.

51. Connect the negative battery cable, start the engine, allow it to reach normal operating temperatures and check for leaks.

3.0L Engine

※ CAUTION

The fuel injection system remains under pressure after the engine has been turned OFF. Properly relieve fuel pressure before disconnecting any fuel lines. Failure to do so may result in fire or personal injury.

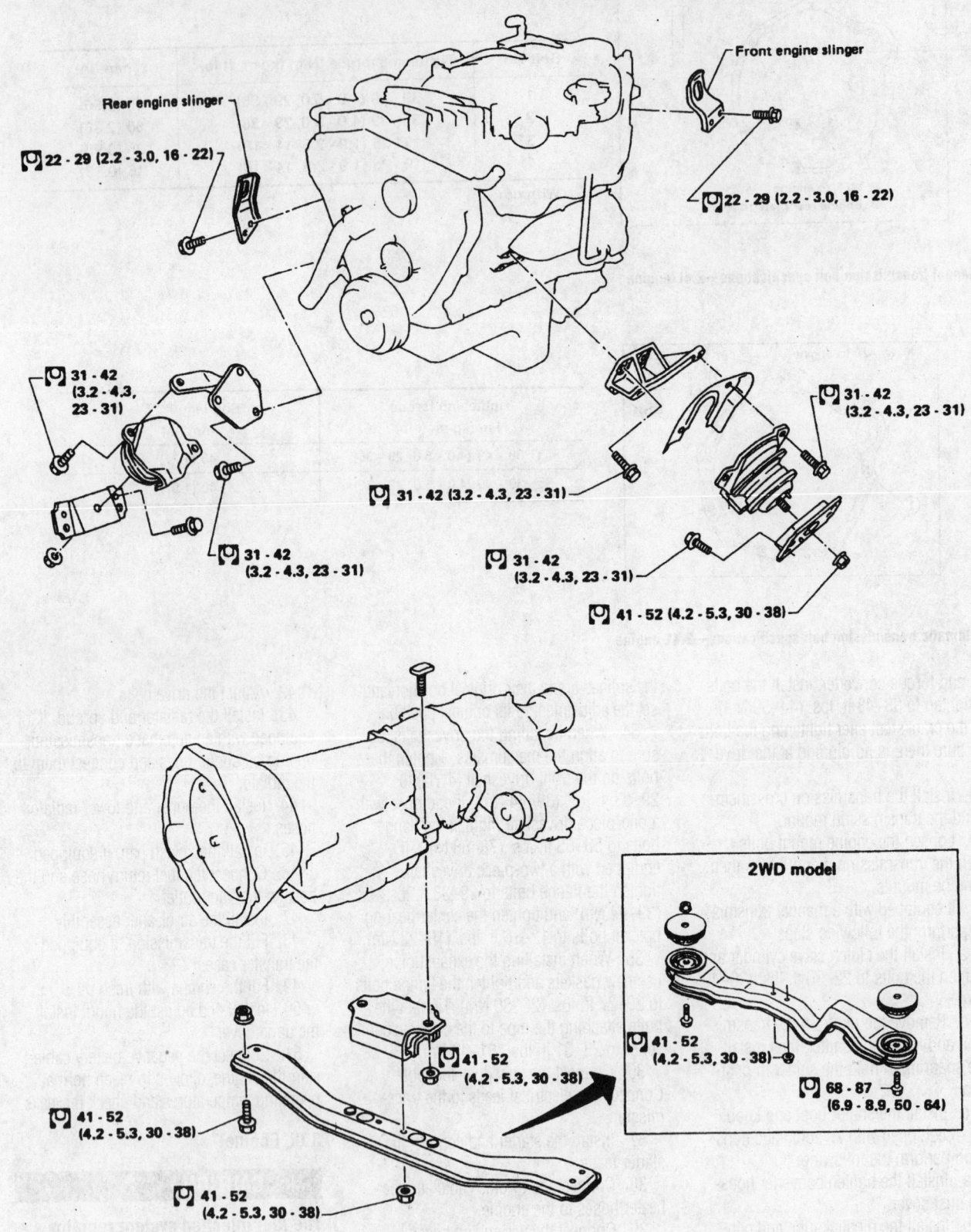

Front engine slinger

Rear engine slinger

22 - 29 (2.2 - 3.0, 16 - 22)

22 - 29 (2.2 - 3.0, 16 - 22)

31 - 42 (3.2 - 4.3, 23 - 31)

31 - 42 (3.2 - 4.3, 23 - 31)

31 - 42 (3.2 - 4.3, 23 - 31)

31 - 42 (3.2 - 4.3, 23 - 31)

31 - 42 (3.2 - 4.3, 23 - 31)

41 - 52 (4.2 - 5.3, 30 - 38)

41 - 52 (4.2 - 5.3, 30 - 38)

41 - 52 (4.2 - 5.3, 30 - 38)

41 - 52 (4.2 - 5.3, 30 - 38)

2WD model

41 - 52 (4.2 - 5.3, 30 - 38)

68 - 87 (6.9 - 8.9, 50 - 64)

: N·m (kg-m, ft-lb)

7924VG08

Engine mounts and related components—3.0L engine

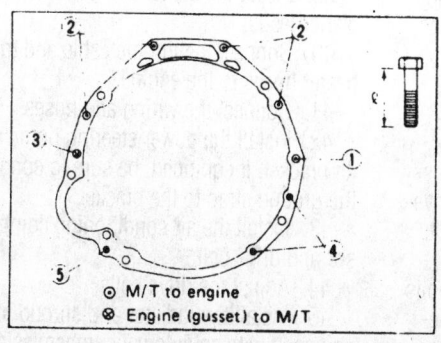

Bolt No.	Tightening torque N·m (kg·m, ft-lb)	ℓ mm (in)
1	39 - 49 (4.0 - 5.0, 29 - 36)	65 (2.56)
2	39 - 49 (4.0 - 5.0, 29 - 36)	60 (2.36)
3	29 - 39 (3.0 - 4.0, 22 - 29)	55 (2.17)
4	29 - 39 (3.0 - 4.0, 22 - 29)	30 (1.18)
5	29 - 39 (3.0 - 4.0, 22 - 29)	25 (0.98)

Manual transmission bolt specifications—3.0L engine

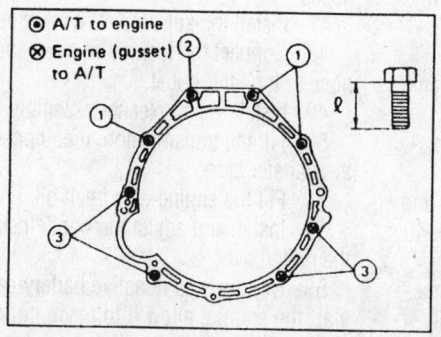

Bolt No	Tightening torque N·m (kg-m, ft-lb)	Bolt length ''ℓ'' mm (in)
1	39 - 49 (4.0 - 5.0, 29 - 36)	45 (1.77)
2	39 - 49 (4.0 - 5.0, 29 - 36)	50 (1.97)
3	29 - 39 (3.0 - 4.0, 22 - 29)	25 (0.98)
Gusset to engine	29 - 39 (3.0 - 4.0, 22 - 29)	20 (0.79)

Automatic transmission bolt specifications—3.0L engine

1. Relieve the fuel system pressure.
2. Disconnect the battery cables and remove the battery.
3. Matchmark the location of the hood hinges and remove the hood.
4. Remove the air cleaner assembly.
5. Wrap a shop rag around the fuel filter outlet and disconnect the hose. Disconnect the fuel return hose.
6. Raise and safely support the vehicle. If equipped, remove the splash pan from under the engine.

✳✳ CAUTION

The EPA warns that prolonged contact with used engine oil may cause a number of skin disorders, including cancer! You should make every effort to minimize your exposure to used engine oil. Protective gloves should be worn when changing the oil. Wash your hands and any other exposed skin areas as soon as possible after exposure to used engine oil. Soap and water, or waterless hand cleaner should be used.

7. Drain the engine oil and the cooling system, including the block drains. Dispose of old fluids properly.

8. Remove the upper and lower radiator hoses.
9. Remove the radiator.
10. If equipped with air conditioning, loosen the belt tension and remove the belt. Disconnect the wiring, remove the compressor and secure it out of the way. Do not disconnect the pressure hoses.
11. If equipped with power steering, remove the drive belt and the power steering pump and secure it out of the way. Do not disconnect the pressure hoses.
12. Label and disconnect all wiring and vacuum hoses.
13. Disconnect the heater hoses from the engine and disconnect the throttle cable.
14. Remove the starter.
15. Matchmark the driveshaft flange at the rear pinion flange and remove the driveshaft. Plug the extension housing opening to prevent the oil from draining out.
16. If equipped with 4WD, matchmark both front driveshaft flanges so the shaft can be installed in the same position. Remove the front driveshaft.
17. Disconnect the exhaust pipe from the manifolds and from the catalytic converter and remove the pipe.
18. Disconnect the speedometer cable and the wiring from the transmission.

19. Remove the bracket securing the transmission to the engine.
20. If equipped an automatic transmission perform the following:
   a. Disconnect the selector lever and throttle cables from the transmission.
   b. Remove the dipstick tube and disconnect the cooler lines.
   c. Remove the torque converter housing dust cover. Matchmark the converter with the driveplate for reassembly; these are balanced together at the factory. Remove the torque converter-to-driveplate (flywheel) bolts. Use a wrench on the crankshaft pulley bolt to rotate the crankshaft to expose the hidden torque converter bolts.
21. If equipped with a manual transmission perform the following:
   a. Remove the shifter knob and boot and remove the snapring to lift the shift lever out of the transmission. Stuff a rag in the opening to keep dirt out of the transmission.
   b. Without disconnecting the hydraulic hose, remove the clutch slave cylinder from the transmission and secure it aside.
22. If equipped with 4WD perform the following:

a. Remove the torsion bars and the second crossmember

b. Remove the transfer case shift lever assembly from the transfer case.

23. Using a chain hoist, attach it to the engine and lift the engine slightly to take the weight off the mounts. Using an appropriate transmission jack, properly support the transmission and transfer case, if equipped. Remove the transmission mount and crossmember.

24. Remove the bolts securing the transmission to the engine and move the transmission back from the engine. If equipped with an automatic transmission, secure the torque converter to the transmission. Lower the transmission from the vehicle.

➡**When removing the engine mounts, do not loosen the 4 mount cover nuts. The mount is fluid filled and will not function properly if the fluid leaks out.**

25. Check to be sure all wires and hoses have been disconnected. Remove the front engine mount bolts and carefully lift the engine out.

To install:

26. Carefully guide the engine into place and start the mount bolts. Tighten the bolts temporarily.

27. If equipped with a manual transmission, perform the following steps:

a. Lightly grease the input shaft splines. On 4WD, apply a silicone sealant to the engine block and rear plate to seal the engine to the transmission.

b. Fit the transmission into place and start the bolts attaching the engine to the transmission. Be sure the input shaft fits properly into the clutch disc and pilot bearing.

c. Tighten the 2.36 in. (60mm) and 2.56 in. (65mm) transmission bolts to 29–36 ft. lbs. (39–49 Nm).

d. Tighten the remaining bolts to 22–29 ft. lbs. (29–39 Nm).

28. If equipped with an automatic transmission perform the following:

a. Use a dial indicator to check the driveplate run-out while turning the crankshaft. Maximum allowable run-out is 0.020 in. (0.5mm); if beyond specifications, replace the driveplate.

b. Measure and adjust how far the torque converter is recessed into the transmission housing. The distance between the front mounting surface of the transmission and the torque converter-to-driveplate bolt boss should be at least 1.024 in. (26mm).

c. Install the transmission. Install the transmission attaching bolts and tighten the 1.77 in. (45mm) and the 1.97 in. (50mm) bolts to 29–36 ft. lbs. (39–49 Nm), and tighten the 0.98 in. (25mm) bolts to 22–29 ft. lbs. (29–39 Nm).

29. If equipped with an automatic transmission, align the matchmarks on the driveplate and torque converter, install the bolts and tighten to 33–43 ft. lbs. (44–59 Nm). Turn the crankshaft after tightening the bolts to be sure there is no binding at the driveplate.

30. Install the bracket securing the transmission to the engine and tighten the attaching bolts to 22–29 ft. lbs. (29–39 Nm).

31. Install the transmission crossmember and the transmission mount.

32. Loosen the engine mount bolts, tighten the transmission mount bolts, then the engine mounts.

33. If equipped with a manual transmission, perform the following steps:

a. Install the clutch slave cylinder and tighten the bolts to 22–30 ft. lbs. (30–40 Nm).

b. Remove the rag from the transmission and install the shifter, then install the snapring to hold the shifter in position.

c. Install the shifter boot and knob.

34. If equipped with an automatic transmission perform the following:

a. Install the torque converter housing dust cover.

b. Install the dipstick tube and connect the cooler lines.

c. Connect the selector lever and throttle cables to the transmission.

35. If the torsion bars were removed, install them in their original location. Be sure the splines are in their original position and set the adjustment to its original position.

36. When installing the driveshafts, be sure to align the matchmarks. Tighten the bolts on the front driveshaft (4WD) to 29–33 ft. lbs. (39–44 Nm). On the rear driveshaft, tighten the bolts to 58–65 ft. lbs. (78–88 Nm). If equipped with a center bearing bracket tighten the bolts to 12–16 ft. lbs. (16–22 Nm).

37. When installing the exhaust pipe, use new gaskets and tighten the flange bolts to 20–27 ft. lbs. (26–36 Nm). Tighten the bolts attaching the pipe to the catalytic converter to 32–41 ft. lbs. (43–55 Nm).

38. Connect the speedometer cable. Connect any electrical leads to the transmission.

39. Install the starter and connect the starter leads.

40. Connect the throttle cable and the heater hoses to the engine.

41. Connect the wiring and hoses.

42. Install the power steering pump and its bracket, if equipped; be sure to connect the ground strap to the bracket.

43. Install the air conditioning compressor and drive belt.

44. Adjust the drive belts.

45. Install the radiator and shroud. If equipped with an automatic transmission, unplug oil cooler lines and connect them to the radiator.

46. Install the upper and lower radiator hoses.

47. Install the splash pan, if equipped.

48. Connect the fuel return hose and the hose to the filter outlet.

49. Install the air cleaner assembly.

50. Fill the transmission, if equipped, the transfer case.

51. Fill the engine with fresh oil.

52. Install and adjust the hood. Install the undercover.

53. Connect the negative battery cable, start the engine, allow it to reach normal operating temperatures and check for leaks.

3.3L Engine

❄❄ CAUTION

The fuel injection system remains under pressure after the engine has been turned OFF. Properly relieve fuel pressure before disconnecting any fuel lines. Failure to do so may result in fire or personal injury.

1. Relieve the fuel system pressure.

2. Disconnect the battery cables and remove the battery.

3. Matchmark the location of the hood hinges and remove the hood.

4. Remove the air cleaner assembly.

5. Raise and safely support the vehicle. If equipped, remove the splash pan from under the engine.

6. Drain the engine oil and the cooling system, including the block drains. Dispose of old fluids properly.

7. Remove the upper and lower radiator hoses.

8. Remove the radiator, shroud and cooling fan.

9. If equipped with air conditioning, loosen the belt tension and remove the belt. Disconnect the wiring, remove the compres-

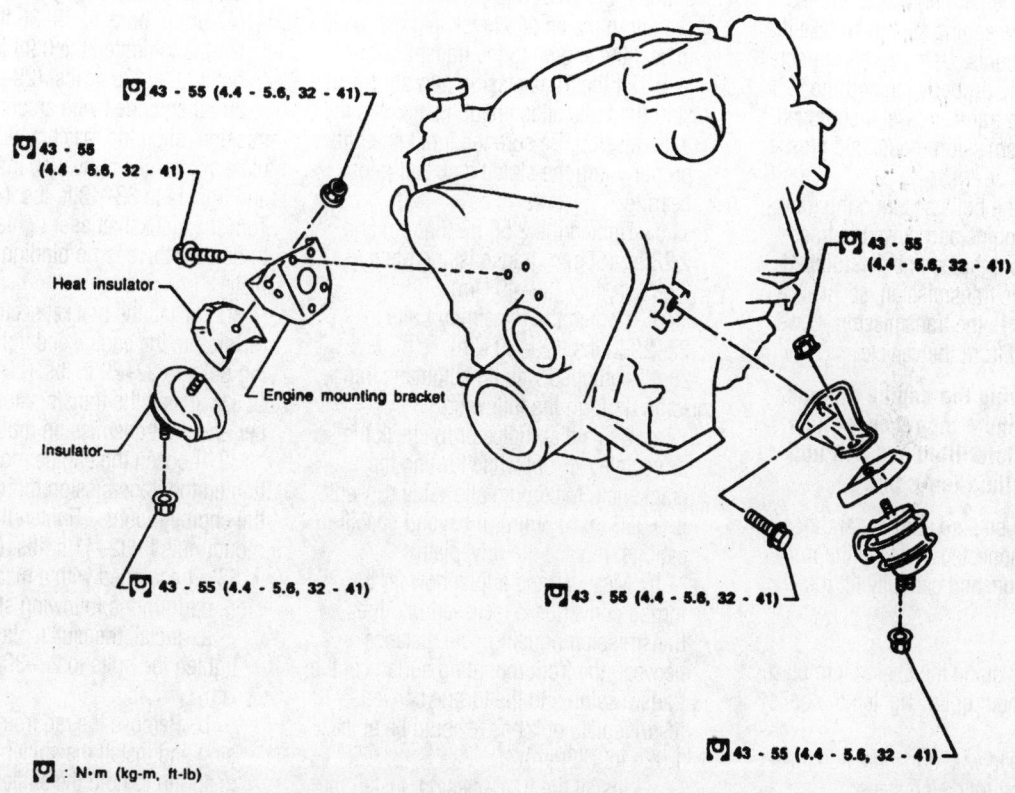

43 - 55 (4.4 - 5.6, 32 - 41)

43 - 55 (4.4 - 5.6, 32 - 41)

Heat insulator

Engine mounting bracket

Insulator

43 - 55 (4.4 - 5.6, 32 - 41)

43 - 55 (4.4 - 5.6, 32 - 41)

43 - 55 (4.4 - 5.6, 32 - 41)

43 - 55 (4.4 - 5.6, 32 - 41)

: N·m (kg-m, ft-lb)

7924VG11

Engine mounts and related components—3.3L engine

sor and secure it out of the way. Do not disconnect the pressure hoses.

10. If equipped with power steering, remove the drive belt and the power steering pump and secure it out of the way. Do not disconnect the pressure hoses.

11. Label and disconnect all wiring and vacuum hoses.

12. Disconnect the heater hoses from the engine and disconnect the throttle cable.

13. Remove the starter.

14. Matchmark the driveshaft flange at the rear pinion flange and remove the driveshaft. Plug the extension housing opening to prevent the oil from draining out.

15. If equipped with 4WD, matchmark both front driveshaft flanges so the shaft can be installed in the same position. Remove the front driveshaft.

16. Disconnect the exhaust pipe from the manifolds and from the catalytic converter and remove the pipe.

17. Disconnect the speedometer cable and the wiring from the transmission.

18. Remove the bracket securing the transmission to the engine.

19. If equipped with an automatic transmission perform the following:

a. Disconnect the selector lever and throttle cables from the transmission.

b. Remove the dipstick tube and disconnect the cooler lines.

c. Remove the torque converter housing dust cover. Matchmark the converter with the driveplate for reassembly; these are balanced together at the factory. Remove the torque converter-to-driveplate (flywheel) bolts. Use a wrench on the crankshaft pulley bolt to rotate the crankshaft to expose the hidden torque converter bolts.

20. If equipped with a manual transmission perform the following:

a. Remove the shifter knob and boot and remove the snapring to lift the shift

lever out of the transmission. Stuff a rag in the opening to keep dirt out of the transmission.

b. Without disconnecting the hydraulic hose, remove the clutch slave cylinder from the transmission and secure it aside.

21. If equipped with 4WD perform the following:

a. Remove the torsion bars and the second crossmember.

b. Remove the transfer case shift lever assembly from the transfer case.

22. Attach engine slingers to the engine.

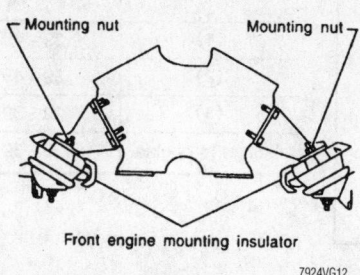

Mounting nut   Mounting nut

Front engine mounting insulator

7924VG12

Engine mounting nuts—3.3L engine

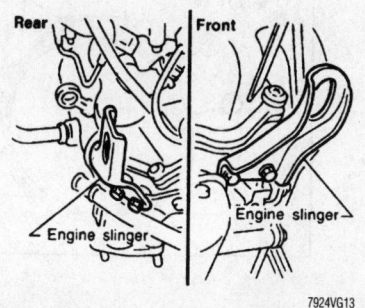

Rear | Front

Engine slinger

Engine slinger

7924VG13

Engine slingers installed—3.3L engine

23. Using a chain hoist, attach it to the engine and lift the engine slightly to take the weight off the mounts. Using an appropriate transmission jack, properly support the transmission and transfer case, if equipped. Remove the transmission mount and cross-member.

24. Remove the bolts securing the transmission to the engine and move the transmission back from the engine. If equipped with an automatic transmission, secure the torque converter to the transmission. Lower the transmission from the vehicle.

➡️**When removing the engine mounts, do not loosen the 4 mount cover nuts. The mount is fluid filled and will not function if the fluid leaks out.**

25. Check to be sure all wires and hoses have been disconnected. Remove the front engine mount nuts and carefully lift the engine out.

To install:

26. Carefully guide the engine into place and start the mount bolts. Tighten the bolts temporarily.

27. If equipped with a manual transmission, perform the following steps:

a. Lightly grease the input shaft splines. On 4WD, apply a silicone sealant to the engine block and rear plate to seal the engine to the transmission.

b. Fit the transmission into place and start the bolts attaching the engine to the transmission. Be sure the input shaft fits properly into the clutch disc and pilot bearing.

c. Tighten the 2.56 in. (65mm) and 2.28 in. (58mm) transmission bolts to 29–36 ft. lbs. (39–49 Nm).

d. Tighten the remaining bolts to 22–29 ft. lbs. (29–39 Nm).

28. If equipped with an automatic transmission perform the following:

a. Use a dial indicator to check the driveplate run-out while turning the crankshaft. Maximum allowable run-out is 0.020 in. (0.5mm); if beyond specifications, replace the driveplate.

b. Measure and adjust how far the torque converter is recessed into the transmission housing. The distance between the front mounting surface of the transmission and the torque converter-to-driveplate bolt boss should be at least 1.024 in. (26mm).

c. Install the transmission. Install the transmission attaching bolts and tighten the 2.283 in. (58mm) and the 1.87 in. (47.5mm) bolts to 29–36 ft. lbs. (39–49 Nm), and tighten the 0.98 in. (25mm) bolts to 22–29 ft. lbs. (29–39 Nm).

29. If equipped with an automatic transmission, align the matchmarks on the driveplate and torque converter, install the bolts and tighten to 33–43 ft. lbs. (44–59 Nm). Turn the crankshaft after tightening the bolts to be sure there is no binding at the driveplate.

30. Install the bracket securing the transmission to the engine and tighten the attaching bolts to 22–29 ft. lbs. (29–39 Nm).

31. Install the transmission crossmember and the transmission mount.

32. Loosen the engine mount nuts, tighten the transmission mount bolts, then the engine mounts. Tighten the engine mount nuts to 32–41 ft. lbs. (43–55 Nm).

33. If equipped with a manual transmission, perform the following steps:

a. Install the clutch slave cylinder and tighten the bolts to 22–30 ft. lbs. (30–40 Nm).

b. Remove the rag from the transmission and install the shifter, then install the snapring to hold the shifter in position.

c. Install the shifter boot and knob.

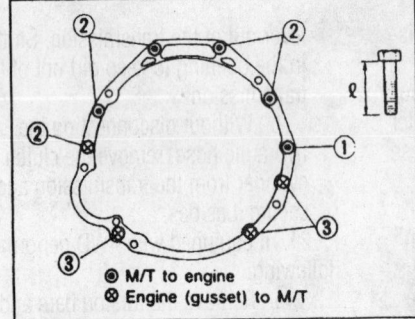

Bolt No.	Tightening torque N·m (kg-m, ft-lb)	ℓ mm (in)
①	39 - 49 (4.0 - 5.0, 29 - 36)	65 (2.56)
②	39 - 49 (4.0 - 5.0, 29 - 36)	58 (2.28)
③	29 - 39 (3.0 - 4.0, 22 - 29)	25 (0.98)
Gusset to engine	29 - 39 (3.0 - 4.0, 22 - 29)	20 (0.79)

⊙ M/T to engine
⊗ Engine (gusset) to M/T

7924VG14

Manual transmission bolt specifications—3.3L engine with 2WD

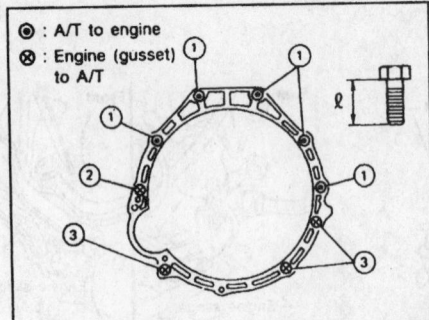

⊙ : A/T to engine
⊗ : Engine (gusset) to A/T

Bolt No.	Tightening torque N·m (kg-m, ft-lb)	Bolt length "ℓ" mm (in)
①	39 - 49 (4.0 - 5.0, 29 - 36)	47.5 (1.870)
②	39 - 49 (4.0 - 5.0, 29 - 36)	58.0 (2.283)
③	29 - 39 (3.0 - 4.0, 22 - 29)	25.0 (0.984)
Gusset to engine	29 - 39 (3.0 - 4.0, 22 - 29)	20.0 (0.787)

7924VG15

Automatic transmission bolt specifications—3.3L engine with 2WD

34. If equipped with an automatic transmission perform the following:

a. Install the torque converter housing dust cover.

b. Install the dipstick tube and connect the cooler lines.

c. Connect the selector lever and throttle cables to the transmission.

35. If the torsion bars were removed, install them in their original location. Be sure the splines are in their original position and set the adjustment to its original position.

36. When installing the driveshafts, be sure to align the matchmarks. Tighten the bolts on the front driveshaft (4WD) to 29–33 ft. lbs. (39–44 Nm). On the rear driveshaft, tighten the bolts to 58–65 ft. lbs. (78–88 Nm). If equipped with a center bearing bracket tighten the bolts to 12–16 ft. lbs. (16–22 Nm).

37. When installing the exhaust pipe, use new gaskets and tighten the flange bolts to 20–27 ft. lbs. (26–36 Nm). Tighten the bolts attaching the pipe to the catalytic converter to 32–41 ft. lbs. (43–55 Nm).

38. Connect the speedometer cable. Connect any electrical leads to the transmission.

39. Install the starter and connect the starter leads.

40. Connect the throttle cable and the heater hoses to the engine.

41. Connect the wiring and hoses.

42. Install the power steering pump and its bracket, if equipped; be sure to connect the ground strap to the bracket.

43. Install the air conditioning compressor and drive belt.

44. Adjust the drive belts.

45. Install the radiator and shroud. If equipped with an automatic transmission, unplug oil cooler lines and connect them to the radiator.

46. Install the upper and lower radiator hoses.

47. Install the splash pan, if equipped.

48. Connect the fuel return hose and the hose to the filter outlet.

49. Install the air cleaner assembly.

50. Fill the transmission, if equipped, the transfer case.

51. Fill the engine with fresh oil and fill the cooling system with the proper coolant/water mixture.

52. Install and adjust the hood. Install the undercover.

53. Connect the negative battery cable, start the engine, allow it to reach normal operating temperatures and check for leaks. Make all the necessary adjustments.

Water Pump

REMOVAL & INSTALLATION

2.4L Engine

1. Disconnect the negative battery cable.

2. Drain the cooling system and engine block, using the block drain.

3. Remove the upper radiator hose to provide working room and remove the drive belt(s) from the pulleys.

4. Remove the retaining screws, and lift the fan shroud from the engine.

5. While holding the pulley, remove the nuts retaining the fan and pulley to the water pump.

6. Remove the mounting bolts and pull the water pump from the engine.

➡ **The mounting bolts are different sizes and must be reinstalled in the correct location, therefore it is a good idea to arrange the bolts so that they can be easily identified during installation.**

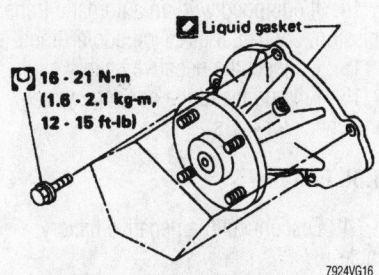

Water pump assembly—2.4L engine

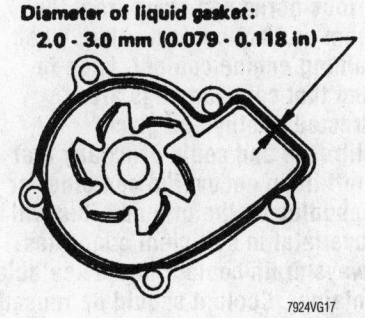

Be sure to apply the liquid gasket to the pump assembly before installation—2.4L engine

To install:

7. Be sure all gasket surfaces are clean and properly apply silicone sealer to the pump. Install the pump to the engine and tighten the bolts to 12–15 ft. lbs. (16–21 Nm).

8. Install the fan clutch, fan, and pulley and tighten the nuts or bolts to 5–6 ft. lbs. (7–8 Nm).

9. Install the fan shroud and drive belt(s).

10. Connect the upper hose, then fill and bleed the cooling system.

11. Connect the negative battery cable.

12. Start the engine to check for leaks.

3.0L Engine

1. Disconnect the negative battery cable.

2. Drain the coolant from the radiator and the drain plugs on both sides of the engine block.

3. Remove the radiator hoses, on automatic transmission, disconnect and plug the fluid cooling lines.

4. Remove the lower section of the fan shroud and remove the screws to lift the shroud from the engine. Remove the bracket bolts and lift the radiator out of the vehicle.

5. Remove all the accessory drive belts.

6. Hold the pulley and remove the nuts to remove the fan and pulley from the water pump.

7. Remove the timing belt covers.

8. Remove the bolts to remove the water pump from the engine.

➡ **Water pump mounting bolts are different sizes and must be reinstalled in their original locations.**

To install:

9. Be sure all gasket surfaces are clean and use a new gasket or silicone sealer when installing the pump to the engine. Tighten the bolts to 15 ft. lbs. (21 Nm).

10. Install the timing belt covers. On 4WD models, be sure the sealing surfaces are clean and carefully install the rubber seal when installing the cover. The timing belt must be properly protected from dirt and oil.

11. Install the pulley, fan clutch, and the fan.

12. Install the accessory drive belts and adjust the tension.

13. Install the radiator and fan shroud; connect the cooling system hoses.

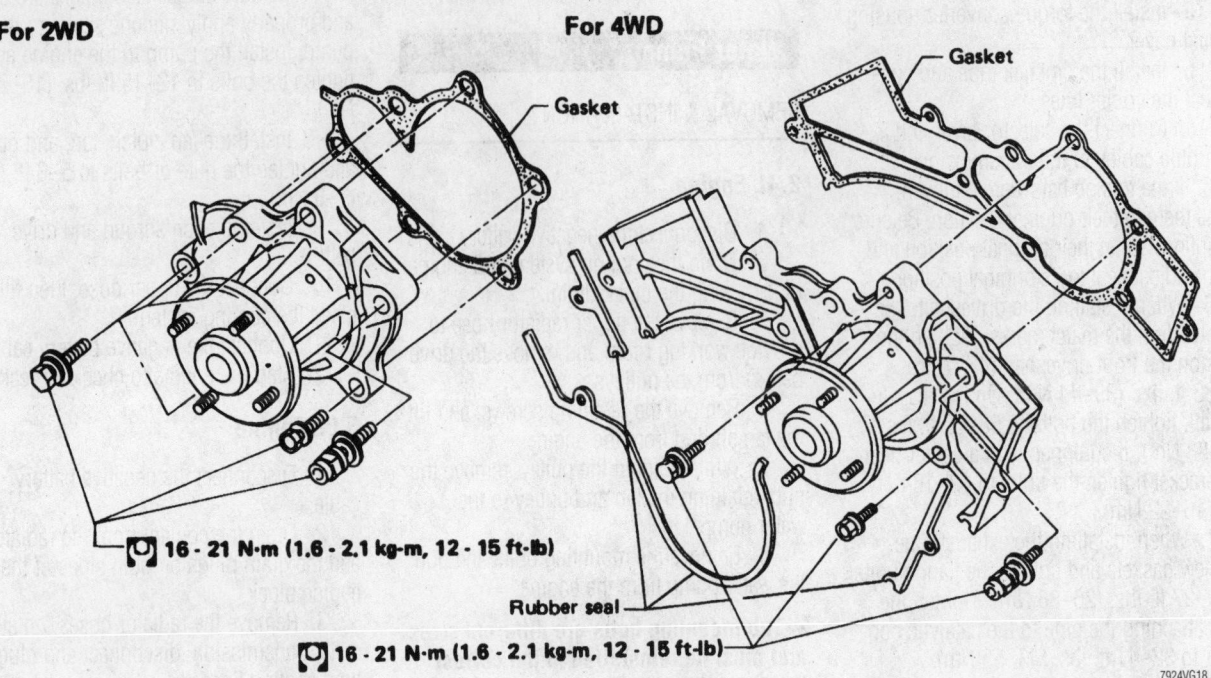

For 2WD

Gasket

For 4WD

Gasket

⊙ **16 - 21 N·m (1.6 - 2.1 kg-m, 12 - 15 ft-lb)**

Rubber seal

⊙ **16 - 21 N·m (1.6 - 2.1 kg-m, 12 - 15 ft-lb)**

7924VG18

Exploded view of the water pump assembly—3.0L engine

RIGHT SIDE:

DRAIN PLUG

LEFT SIDE:

DRAIN PLUG

7924VG19

Remove the plugs to drain the coolant from the engine block—6-cylinder engines

14. If equipped with an automatic transmission, connect the A/T oil cooler lines.

15. Connect the negative battery cable.

16. Fill and bleed the cooling system and check for leaks.

3.3L Engine

1. Disconnect the negative battery cable.

❋❋ CAUTION

Never open, service or drain the radiator or cooling system when hot; serious burns can occur from the steam and hot coolant. Also, when draining engine coolant, keep in mind that cats and dogs are attracted to ethylene glycol antifreeze and could drink any that is left in an uncovered container or in puddles on the ground. This will prove fatal in sufficient quantities. Always drain coolant into a sealable container. Coolant should be reused unless it is contaminated or is several years old.

2. Drain the coolant from the radiator and the drain plugs on both sides of the engine block.

3. Remove the upper and lower radiator hoses.

4. Remove the fan shroud.

5. Remove the drive belts.

6. Remove the cooling fan and the water pump pulley.

7. Remove the crankshaft pulley.

8. Remove the upper and lower timing belt covers.

➡**Water pump mounting bolts are different sizes and must be reinstalled in their original locations.**

9. Remove the water pump. Don't let the engine coolant get on the timing belt.

To install:

10. Clean the gasket mating surfaces on the water pump and engine block.

11. Using a new gasket, install the water pump. Tighten the mounting bolts to 12–15 ft. lbs. (16–21 Nm).

12. Install the timing belt covers.

13. Install the crankshaft pulley.

[◘] 16 - 21 (1.6 - 2.1, 12 - 15)

Gasket ⊗

Rubber seal ⊗

Water pump

[◘] 16 - 21 (1.6 - 2.1, 12 - 15)

[◘] 16 - 21 (1.6 - 2.1, 12 - 15)

Rubber seal ⊗

[◘] : N·m (kg-m, ft-lb)

7924VG20

Exploded view of the water pump assembly—3.3L engine

14. Install the water pump pulley and the cooling fan.

15. Install the drive belts.

16. Install the fan shroud and the radiator hoses.

17. Connect the negative battery cable.

18. Refill the engine with coolant and bleed the system. Check for leaks.

Cylinder Head

REMOVAL & INSTALLATION

2.4L Engine

✳✳ CAUTION

The fuel injection system remains under pressure after the engine has been turned OFF. Properly relieve fuel pressure before disconnecting any fuel lines. Failure to do so may result in fire or personal injury.

➡**After completing this procedure, allow the rocker cover-to-cylinder head rubber plugs to dry for 30 minutes before starting the engine. This will allow the liquid gasket sealer to cure properly.**

1. Release the fuel system pressure.

2. Disconnect the negative battery cable.

3. Drain the cooling system into a sealable container.

4. Remove the air cleaner assembly.

5. Remove the power steering drive belt, power steering pump, idler pulley, and the power steering pump brackets.

6. Tag and disconnect all vacuum hoses, water hoses, fuel tubes, and wiring harnesses necessary to gain access to cylinder head.

7. Detach the accelerator bracket. If necessary mark the position and remove the accelerator cable wire end from the throttle drum.

8. Tag and disconnect the high tension wires from the spark plugs on the exhaust side of the engine.

9. Disconnect the oxygen sensor electrical connector, then, remove the exhaust manifold cover.

10. Disconnect the EGR tube from the exhaust manifold.

11. Remove the exhaust manifold mounting nuts and bolts, then remove the manifold from the cylinder head. Discard the gaskets.

12. Remove the intake manifold.

13. Remove the rocker cover. If cover sticks to the cylinder head, tap it with a rubber hammer. Be careful not to strike the rocker arms when removing the rocker arm cover.

14. Remove the spark plugs to protect them from damage.

15. Set No. 1 cylinder piston at TDC on its compression stroke. The No. 1 will be at TDC when the timing pointer is aligned with the red timing mark on the crankshaft pulley.

16. Mark the relationship of the camshaft sprocket to the timing chain with paint or chalk. If this is done, it will not be necessary to locate the factory timing marks. Before removing the camshaft sprocket, it will be necessary to wedge the chain in place so that it will not fall down into the front cover, using special tool KV10105800 or equivalent.

17. Remove the camshaft sprocket bolt and carefully remove the camshaft sprocket.

18. Remove the bolts securing the cylinder head to the front cover assembly.

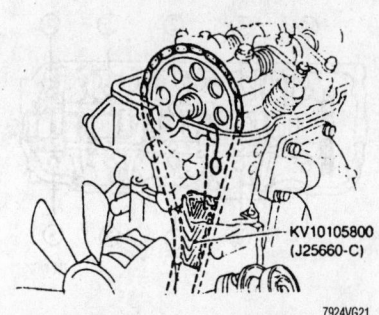

KV10105800
(J25660-C)

7924VG21

Wedge the chain in place so that it will not fall down into the front cover

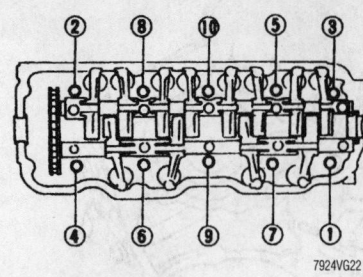

7924VG22

Remove the cylinder head bolts in the sequence shown to prevent damaging the head—2.4L engine

19. Remove the cylinder head bolts in the correct sequence. Lift the cylinder head off the engine block. It may be necessary to tap the head lightly with a rubber mallet to loosen it.

➡The cylinder head bolts should be loosened in two or three steps, in the correct order to prevent head warpage or cracking.

To install:

20. Thoroughly clean the cylinder block and head surfaces and check both for warpage.

21. Fit the new head gasket. Don't use sealant. Be sure that no open valves are in the way of raised pistons, and **never** rotate the crankshaft or camshaft separately because of possible damage which might occur to the valves and/or pistons.

22. Confirm that the No. 1 piston is at TDC on its compression stroke as follows: Align timing mark with the red (0 degree) mark on the crankshaft pulley. Be sure the distributor rotor head is set at No. 1 on the distributor cap. Confirm that the knock pin on the camshaft is set at the top position.

23. Install the cylinder head and tighten the head bolts in sequence using the following 5 step procedure:

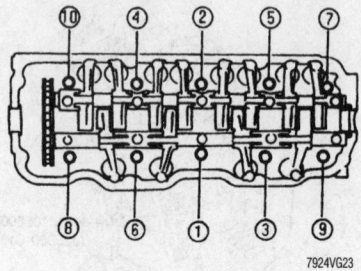

7924VG23

Tighten the cylinder head bolts in the proper sequence to help prevent leaks—2.4L engine

a. Tighten all the bolts to 22 ft. lbs. (29 Nm).
b. Tighten all the bolts to 58 ft. lbs. (78 Nm).
c. Loosen all the bolts completely.
d. Tighten all the bolts to 22 ft. lbs. (29 Nm).
e. Tighten all the bolts to 54–61 ft. lbs. (74–83 Nm), or if an angle wrench is used, turn all bolts 80–85 degrees clockwise.

✳✳ WARNING

Do not rotate crankshaft and camshaft separately, or valves will hit the tops of the pistons.

24. Install the bolts securing the cylinder head to the front cover assembly.

25. Position the timing chain on the camshaft sprocket by aligning each matchmark. Install the camshaft sprocket to the camshaft, then remove the wedge from the timing chain.

26. Hold the camshaft sprocket stationary, and tighten the sprocket bolt to 87–116 ft. lbs. (118–157 Nm).

27. Install the spark plugs and tighten them to 14–22 ft. lbs. (20–29 Nm).

28. Install the intake manifold with new gaskets.

29. Install the exhaust manifold onto the engine with new gaskets. Tighten the nuts/bolts working from the center out, tighten the nuts/bolts to 12–15 ft. lbs. (16–21 Nm).

30. Install the EGR tube to the exhaust manifold.

31. Install the exhaust manifold cover and tighten the bolts to 3–4 ft. lbs. (4–5 Nm), then connect the oxygen sensor electrical connector.

32. Apply liquid gasket to the rubber plugs and install the rubber plugs in the correct location in the cylinder head. The seating surface of the rubber plugs must be clean and dry. The rubber plugs should be installed within 5 minutes of the sealant application. After the sealant is applied and the rubber plugs are in place, rock the plugs back and forth a few times to distribute the sealant evenly. Wipe the excess sealant from the cylinder head with a clean rag.

33. Check each valve lifter in its free position by forcefully pushing it with your finger. If the lifter moves more than 0.04 in. (1mm) air may be in it.

34. Position the rocker cover and install two of the attaching bolts on opposite sides of the cover. Tighten the two bolts to 2 ft. lbs. (3 Nm).

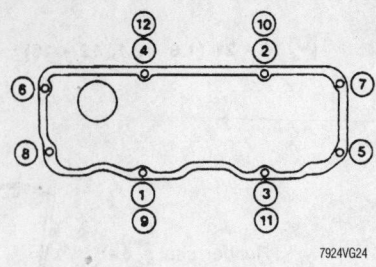

7924VG24

Tighten the rocker cover bolts in the sequence shown—2.4L engine

35. Tighten the rocker cover bolts in the proper sequence to 5–8 ft. lbs. (7–11 Nm).

36. Attach the accelerator bracket and cable.

37. Connect all the vacuum hoses, water hoses, fuel tubes, and electrical connections that were removed to gain access to cylinder head.

38. Install the spark plugs wires in the correct location.

39. Install the power steering bracket and tighten the bracket attaching bolts to 16–22 ft. lbs. (22–29 Nm).

40. Install the idler pulley and the power steering pump.

41. Install and adjust the drive belts.

42. Drain the engine oil into a sealable container, then refill the engine with fresh oil.

43. Fill and bleed the cooling system.

44. Install the air cleaner assembly.

45. Connect the negative battery cable.

46. Run the engine at 1000 rpm with no load for approximately 20 minutes to bleed the air from the valve lifters. If the lifters continue to make noise they need to be replaced.

47. Check the engine for any leaks.

48. Check and/or adjust the ignition timing.

3.0L and 3.3L Engines

1. Release the fuel pressure.

✳✳ CAUTION

The fuel injection system remains under pressure after the engine has been turned OFF. Properly relieve fuel pressure before disconnecting any fuel lines. Failure to do so may result in fire or personal injury.

2. Disconnect the negative battery cable.

3. Set No. 1 cylinder to TDC.

➡️**Do not rotate either the crankshaft or camshaft from this point onward, or the valves could be bent by hitting the pistons.**

4. Drain the coolant from the engine.
5. Remove the air duct hose.
6. Tag and separate the ASCD and accelerator control wire from the intake manifold collector.

7. Remove the intake collector and intake manifold assembly.
8. Remove the timing belt covers and the timing belt.
9. Remove the camshaft pulleys and the rear timing belt cover.
10. Tag, then disconnect the ignition wires from the spark plugs. Disconnect the ignition wire from the ignition coil.
11. Match mark the distributor position,

then remove the distributor attaching bolt and the distributor.
12. Remove the harness clamp from the right side cylinder head cover.
13. Unbolt the forward exhaust pipe at the manifold and move it out of the way.
14. Remove the drive belts and the A/C compressor and alternator. Remove the five mounting bolts, then remove the compressor bracket.

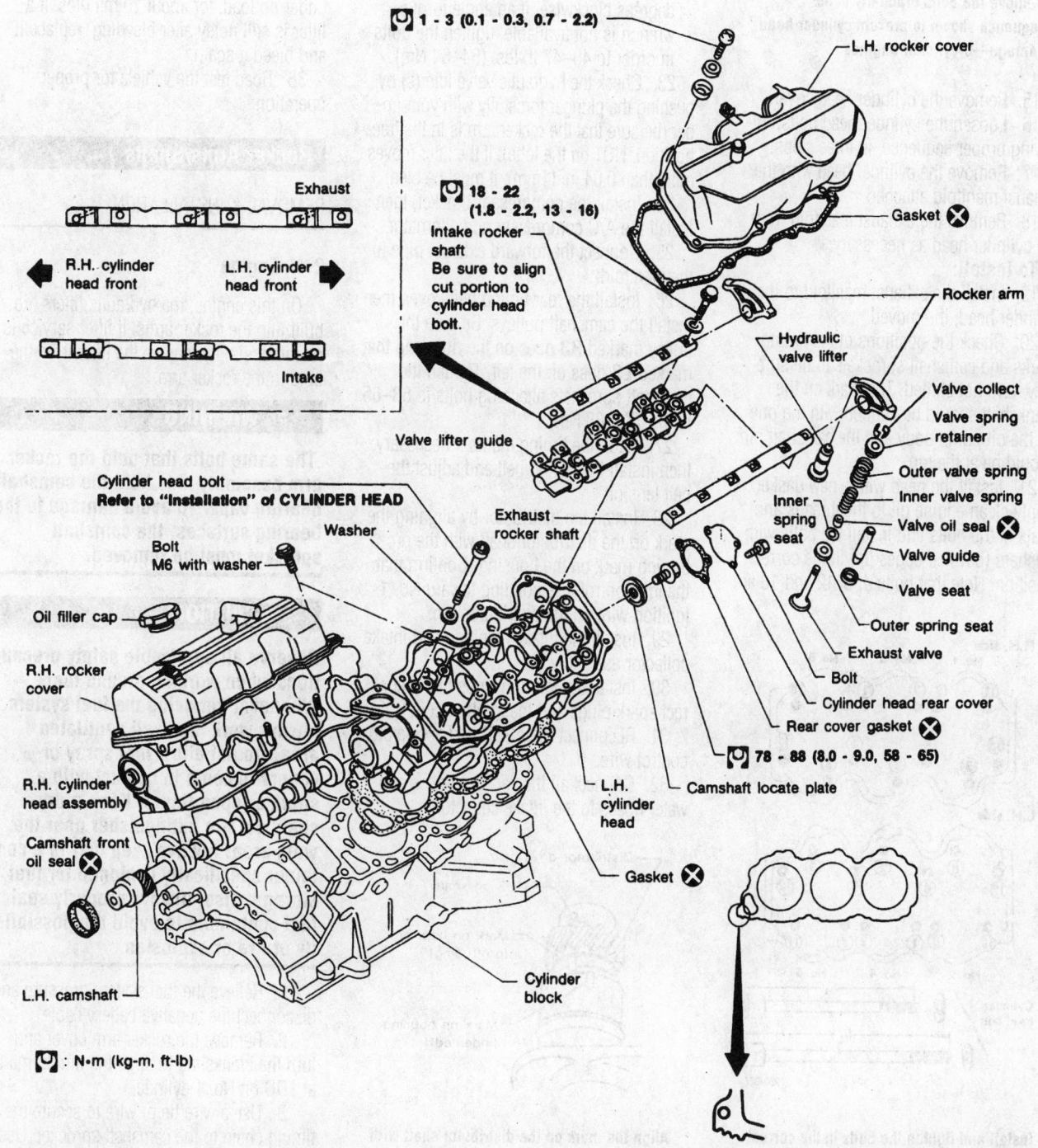

Exploded view of the cylinder head assembly—6-cylinder engines

7924VG25

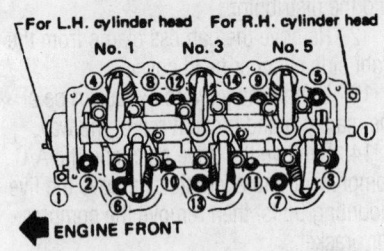

Remove the bolts gradually in the sequence shown to prevent cylinder head damage—6-cylinder engines

15. Remove the cylinder head covers.

16. Loosen the cylinder head bolts, following proper sequence, in three steps.

17. Remove the cylinder head with the exhaust manifold attached.

18. Remove the exhaust manifold from the cylinder head as necessary.

To install:

19. Install the exhaust manifold to the cylinder head, if removed.

20. Check the positions of the timing marks and camshaft sprockets to be sure they have not shifted. The mark on the crankshaft should be aligned with the one on the oil pump body and the camshaft pin should be at the top.

21. Install the head with a new gasket. Apply clean engine oil to the threads and seats of the bolts and install the bolts with washers (beveled edges up) in the correct position. Note that bolts 4, 5, 12 and 13 are

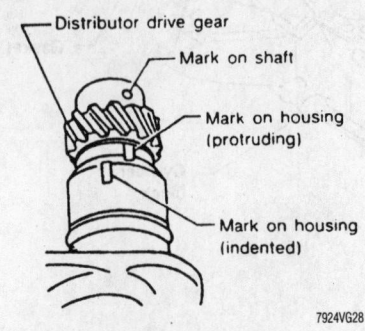

Install and tighten the bolts in the correct sequence to ensure cylinder sealing—6-cylinder engines

longer than the others, 5.00 in. (127mm). Other bolts are 4.17 in. (10.6cm).

22. Tighten the bolts in the proper sequence, in the following stages:

a. Tighten all bolts, in order, to 22 ft. lbs. (29 Nm)

b. Tighten all bolts, in order, to 43 ft. lbs. (59 Nm)

c. Loosen all bolts completely.

d. Tighten all bolts, in order, to 22 ft. lbs. (29 Nm)

e. Turn all bolts, in order, 60–65 degrees clockwise. If an angle torque wrench is not available, tighten the bolts in order to 40–47 ft. lbs. (54–64 Nm).

23. Check the hydraulic valve lifter(s) by pushing the plunger forcefully with your finger (be sure that the rocker arm is in the free position, NOT on the lobe). If the lifter moves more than 0.04 in. (1mm), it must be bled.

24. Install the compressor bracket, then install the A/C compressor and alternator.

25. Connect the forward exhaust pipe to the manifold.

26. Install the rear timing belt cover, then install the camshaft pulleys. Be sure the pulley marked **R3** goes on the right and that marked **L3** goes on the left. Tighten the camshaft sprockets attaching bolts to 58–65 ft. lbs. (78–88 Nm).

27. Align the timing marks if necessary, then install the timing belt and adjust the belt tension.

28. Install the distributor by aligning the mark on the distributor shaft with the protruding mark on the housing. Confirm that the ignition rotor is pointing toward No. 1 ignition wire on the distributor cap.

29. Install the intake manifold and intake collector assembly.

30. Install the ignition wires to the correct spark plugs and the ignition coil.

31. Reconnect the ASCD and accelerator control wire.

32. Connect all the vacuum hoses and water hoses to the intake collector.

Align the mark on the distributor shaft with the protruding mark on the housing — 6-cylinder engines

33. Install the air duct hose.

34. Drain the engine oil into a sealable container, then refill the engine with new oil.

35. Connect the negative battery cable.

36. Refill the cooling system. Start the engine, then check the engine timing. After the engine reaches the normal operating temperature, check for the correct coolant level.

37. If the hydraulic valve lifter(s) needed bleeding, run the engine at about 1000 rpm, under no load, for about 10 minutes. If a lifter is still noisy after bleeding, replace it and bleed it again.

38. Road test the vehicle for proper operation.

Rocker Arms/Shafts

REMOVAL & INSTALLATION

2.4L Engine

On this engine, the hydraulic lifters are built into the rocker arms. If lifter service is required, simply remove the lifter from the bore in the rocker arm.

❋❋ WARNING

The same bolts that hold the rocker arm assembly also hold the camshaft bearing caps. To avoid damage to the bearing surfaces, the camshaft sprocket must be removed.

❋❋ CAUTION

Observe all applicable safety precautions when working around fuel. Whenever servicing the fuel system, always work in a well ventilated area. Do not allow fuel spray or vapors to come in contact with a spark or open flame. Keep a dry chemical fire extinguisher near the work area. Always keep fuel in a container specifically designed for fuel storage; also, always properly seal fuel containers to avoid the possibility of fire or explosion.

1. Relieve the fuel system pressure and disconnect the negative battery cable.

2. Remove the rocker arm cover and turn the crankshaft to align the timing marks at TDC on No. 1 cylinder.

3. Use a wire tie or wire to secure the timing chain to the camshaft sprocket. Use special tool KV10105800 to support the timing chain.

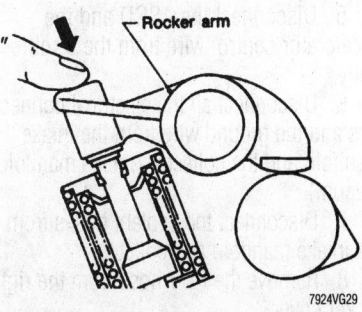

Press down on the valve side of the rocker are to check hydraulic lifter (lash adjuster) operation—2.4L engine

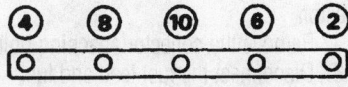

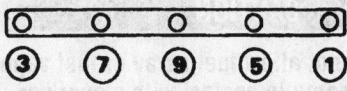

Loosen the rocker arm shaft bolts according to the sequence shown—2.4L engine

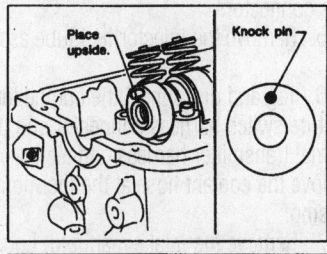

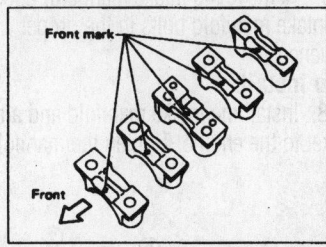

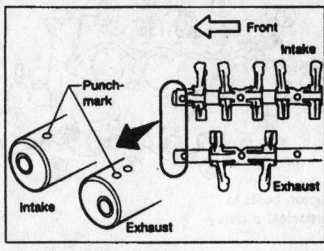

Rocker arm shaft identification marks—2.4L engine

4. Hold the camshaft sprocket to loosen the bolt and remove the sprocket. Secure the sprocket so the chain does not fall off the crankshaft sprocket.

5. Loosen each rocker shaft bolt 1 turn at a time in the proper sequence to prevent bending the shafts.

6. When all the bolts are loose, remove the rocker arm shafts with the bolts still in the shafts. This will hold the assembly together.

7. If the rocker arms are to be removed from the shafts, mark them so they can be returned into their original position. Remove the bolts from the shaft assembly and remove the parts.

➡ **Do not allow the arms to lay on their side or they will become air bound. Keep the rocker arms upright or lay them in a pan of new engine oil.**

To install:

8. Lubricate the shafts with engine oil and assemble them with the punch marks facing up. Use the bolts to hold the assembly together. Be sure the camshaft and the bearing surfaces are in good condition and lubricate with engine oil. Be sure the pin on the camshaft sprocket end is up.

➡ **Punch marks on the front of each shaft that tell which shaft is for the intake side and which is for the exhaust side. This is important for correct rocker arm oiling.**

9. Install the rocker arm shafts and tighten the bolts in the proper sequence 1 turn at a time to draw the shafts down evenly against the valve springs without bending the shafts. Tighten the bolts in reverse order of the loosening sequence, tighten bolts from the inside out. Tighten the bolts to 27–30 ft. lbs. (37–41 Nm).

10. Install the camshaft sprocket and remove the tie securing the chain. Install the sprocket bolt but don't tighten it yet. Rotate the crankshaft 2 full turns to be sure the

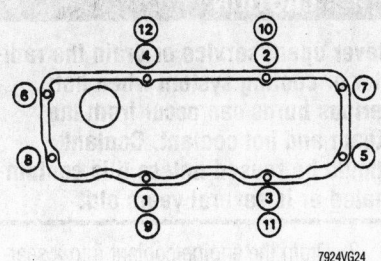

Tighten the rocker cover bolts in the sequence shown to prevent oil leaks

timing marks line up. When the valve timing is correct, tighten the sprocket bolt to 87–116 ft. lbs. (118–157 Nm).

11. Use a silicone sealer on the rubber end plugs and install the rocker arm cover with a new gasket. Install two attaching bolts at opposite sides of the rocker cover, then tighten them to 2 ft. lbs. (3 Nm).

12. Tighten the rocker cover bolts in the proper sequence to 5–8 ft. lbs. (7–11 Nm).

13. Connect the negative battery cable.

14. When the engine is first started, the hydraulic valve lifters may be noisy. Run the engine for 20 minutes at about 1000 rpm. If the noise has not subsided, the lifter will probably never pump up and must be replaced.

3.0L and 3.3L Engines

1. Disconnect the negative battery cable.

2. Align the timing marks to bring No. 1 cylinder to TDC.

3. Remove the valve covers.

4. Loosen the rocker shaft bolts in two or three stages, then remove the rocker shafts with rocker arms, as an assembly.

5. Separate the rocker arms from the shaft.

➡ **When separating the rocker arms from the rocker arm shafts, be sure to keep the parts in order for reinstallation purposes.**

6. Check the rocker arms and the shafts for damage. If necessary, replace the damaged components.

7. Attach a wire to the top of the lifters so that they will not drop from the lifter guide. Carefully remove the lifter guide and lifters from the cylinder head. Put an identification mark on the lifters to avoid mixing them up if they are removed from the guide and to be reused. If the lifters are damaged replace them as necessary.

To install:

8. Install new lifters if replacing them or install the old lifters to their original locations.

�★ CAUTION

When installing the rocker arm shafts, be certain that they are installed in their original positions.

9. Slide the rocker arms onto the shafts in their proper positions.

10. Be sure that cylinder No. 1 is at TDC.

11. Install the left cylinder head valve lifter guide assembly and remove the wire

from the lifters. Install the rocker arm shaft assemblies and attaching bolts, coat the bolt threads and seat surfaces before installing them. Tighten the bolts gradually in three steps to 13–16 ft. lbs. (18–22 Nm).

12. Rotate the crankshaft clockwise 180°, to bring cylinder No. 4 to TDC. Install the right cylinder head valve lifter guide assembly and remove the wire from the lifters. Install the rocker arm shaft assemblies and attaching bolts, coat the bolt threads and seat surfaces before installing them. Tighten the bolts gradually in three steps to 13–16 ft. lbs. (18–22 Nm).

13. Install the rocker covers with new gaskets, tighten the rocker cover bolts to 9–26 inch lbs. (1–3 Nm).

14. Install the valve cover.

15. Connect the negative battery cable.

Intake Manifold

REMOVAL & INSTALLATION

2.4L Engine

> ❈ **CAUTION**
>
> **The fuel injection system remains under pressure after the engine has been turned OFF. Properly relieve fuel pressure before disconnecting any fuel lines. Failure to do so may result in fire or personal injury.**

1. Following the proper procedure, relieve the fuel system pressure.

> ❈ **CAUTION**
>
> **Never open, service or drain the radiator or cooling system when hot; serious burns can occur from the steam and hot coolant. Coolant should be reused unless it is contaminated or is several years old.**

2. Drain the coolant into a sealable container.

3. Disconnect the negative battery cable.

4. Remove the air cleaner and disconnect the hoses.

5. Disconnect the cooling system hoses from the manifold.

6. Remove the throttle cable and disconnect the fuel feed and return fuel lines. Plug the fuel lines to prevent spilling fuel.

7. Tag, then disconnect the electrical connectors from the throttle body and intake manifold.

8. Remove the EGR and PCV tubes from the rear of the intake manifold.

9. Remove the intake manifold stay (support bracket).

10. Unbolt and remove the intake manifold. Remove the manifold with the fuel injectors, EGR valve, and the throttle body still attached.

To install:

11. Clean the gasket mounting surfaces, then install the intake manifold on the engine with a new intake manifold gasket. Tighten the mounting bolts following the proper sequence to 12–15 ft. lbs. (16–21 Nm). The bolts should be tightened from the center towards the ends.

12. Install the intake manifold stay.

13. Connect the EGR and PCV tubes to the rear of the intake manifold.

14. Connect the electrical connections to the throttle body and intake manifold.

15. Connect the fuel feed and return lines.

16. Connect the throttle cable to the throttle body.

17. Connect the cooling system hoses to the intake manifold.

18. Install the air cleaner and the air cleaner hoses.

19. Fill and bleed the air from the cooling system.

20. Connect the negative battery cable.

21. Start the engine and check for leaks.

3.0L and 3.3L Engines

> ❈ **CAUTION**
>
> **The fuel injection system remains under pressure after the engine has been turned OFF. Properly relieve fuel pressure before disconnecting any fuel lines. Failure to do so may result in fire or personal injury.**

1. Release the fuel system pressure and disconnect the battery cables.

> ❈ **CAUTION**
>
> **Never open, service or drain the radiator or cooling system when hot; serious burns can occur from the steam and hot coolant. Coolant should be reused unless it is contaminated or is several years old.**

2. Drain the engine coolant into a sealable container.

3. Remove the air duct hose.

4. Tag, then disconnect the spark plug wires from the spark plugs.

5. Disconnect the ASCD and the accelerator control wire from the throttle body.

6. Disconnect all the electrical connectors and the ground wire from the intake manifold and the collector (intake manifold plenum).

7. Disconnect the coolant hoses from the intake manifold and collector.

8. Remove the PCV hose from the right rocker cover.

9. Tag, then disconnect the vacuum hoses for the canister, power brake booster, and the fuel pressure regulator.

10. Disconnect the purge hose from the canister.

11. Disconnect the EGR tube from the collector.

12. Remove the collector attaching bolts.

13. Disconnect the fuel feed and fuel return lines from the injector fuel tube assembly.

> ❈ **CAUTION**
>
> **Do not allow fuel spray or fuel vapors to come in contact with a spark or open flame. Keep a dry chemical fire extinguisher nearby. Never store fuel in an open container due to risk of fire or explosion.**

14. Disconnect all the fuel injector harness connectors.

15. Remove the injector fuel tube assembly.

16. Tag and disconnect the engine temperature switch harness connector and the thermal transmitter harness connector. Remove the coolant hose at the thermostat housing.

17. Remove the intake manifold. Loosen the intake manifold bolts in the proper sequence.

To install:

18. Install the intake manifold and a new gasket to the engine. Tighten the manifold

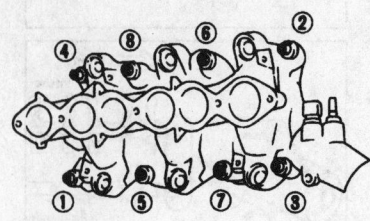

Loosen bolts in numerical order.

7924VG32

Remove the bolts in the sequence shown to prevent warping or breaking the manifold—V6 engines

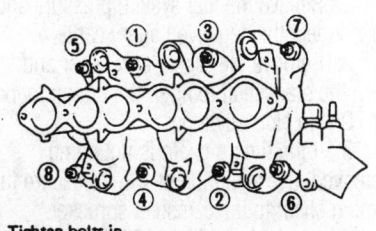

Tighten bolts in
numerical order.

7924VG33

**Be sure to tighten the bolts according to
the sequence shown—V6 engines**

bolts and nuts in two stages until reaching a
total torque of 12–14 ft. lbs. (16–20 Nm) on
all bolts and 17–20 ft. lbs. (24–27 Nm) on
all nuts. On 1996 models, tighten the nuts
and bolts in sequence to 2.2–3.6 ft. lbs.
(3–5 Nm), then tighten them to 13–16 ft.
lbs. (18–22 Nm).

19. Connect the engine temperature
switch harness connector and the thermal
transmitter harness connector. Install the
coolant line at the thermostat housing.

20. Install the injector fuel tube assembly.

21. Connect all fuel injector harness
connectors.

22. Connect the fuel feed and fuel
return lines from the injector fuel tube
assembly.

23. Install the collector with new gaskets. Tighten collector attaching bolts in two
stages, to 13–16 ft. lbs. (18–22 Nm).

24. Connect the EGR tube to the collector.

25. Connect the purge hose to the canister.

26. Connect the vacuum hoses for the
canister, power brake booster, and the fuel
pressure regulator.

27. Install the PCV hose to the right
rocker cover.

28. Connect the coolant hoses to the
intake manifold and collector.

29. Connect the electrical connectors
and the ground wire to the intake manifold
and the collector.

30. Connect and adjust the ASCD and
the accelerator control wire to the throttle
body.

31. Connect the spark plug wires to the
correct spark plugs.

32. Install the air duct hose.

33. Fill and bleed the air from the cooling system.

34. Connect the battery cables.

35. Check the fluid levels, start the
engine, and check for leaks.

Exhaust Manifold

REMOVAL & INSTALLATION

2.4L Engine

1. Disconnect the negative battery
cable.

2. Remove the air cleaner assembly.
Remove the exhaust manifold heat shield.

3. Tag and disconnect the high tension
wires from the spark plugs on the exhaust
side of the engine.

4. Disconnect the air induction and/
or the EGR tubes from the exhaust manifold.

5. Disconnect the oxygen sensor electrical connector.

6. Disconnect the front exhaust pipe
from the exhaust manifold.

➡**Soak the exhaust pipe retaining bolts
with penetrating oil if necessary to
loosen them.**

7. Remove the exhaust manifold
mounting nuts, then remove the manifold
from the cylinder head.
   To install:

8. Using a gasket scraper, clean the
gasket mounting surfaces.

9. Install the manifold onto the engine
with new Gaskets. Tighten the nuts/bolts
working from the center out, tighten the
exhaust manifold nuts/bolts to 12–15 ft.
lbs. (16–21 Nm).

10. Install the air induction and/or the
EGR tubes to the exhaust manifold.

11. Connect the oxygen sensor electrical
connector.

12. Connect the exhaust pipe to the
manifold.

13. Connect spark plug wires, the air
cleaner, and any related hoses.

14. Connect the negative battery
cable.

15. Start engine and check for exhaust
leaks.

3.0L Engine

1. Disconnect the negative battery
cable.

2. Remove the exhaust manifold sub-
cover and manifold cover.

3. Remove the EGR tube from the right
side exhaust manifold.

4. Remove the exhaust manifold stay.

5. Disconnect the left side exhaust
manifold at the exhaust pipe by removing
retaining nuts, then disconnect the right
side manifold from the connecting pipe.

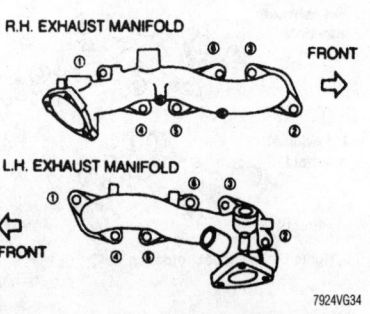

**Remove bolts for each manifold in the
order shown—3.0L engine**

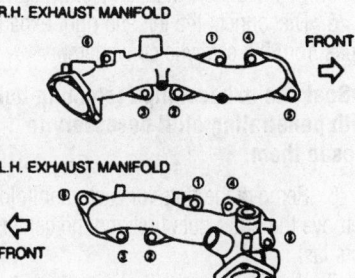

**Tighten the exhaust manifold bolts as
shown to ensure proper sealing—3.0L
engine**

➡**Soak the exhaust pipe retaining bolts
with penetrating oil if necessary to
loosen them.**

6. Remove bolts for each manifold in
the order shown.
   To install:

7. Clean all gasket surfaces. Install new
gaskets.

8. Install the manifold to the engine,
tightening the mounting bolts alternately, in
two stages, in sequence. Tighten the left
side bolts to 13–16 ft. lbs. (18–22 Nm);
tighten the right side bolts to 16–20 ft. lbs.
(22–27 Nm).

9. Connect the exhaust pipe and the
connecting pipe. Be careful not to break
these bolts.

10. Install the exhaust manifold stay and
the EGR tube to the right side manifold.

11. Install the exhaust manifold covers.

12. Connect the negative battery cable.

13. Start the engine and check for
exhaust leaks.

3.3L Engine

1. Disconnect the negative battery cable.

2. Remove the exhaust manifold cover.

3. Remove the EGR tube from the left
side exhaust manifold.

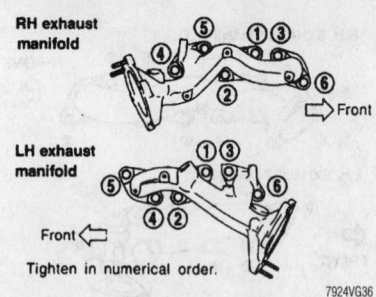

RH exhaust manifold

LH exhaust manifold

Front

Tighten in numerical order.

7924VG36

Tighten the exhaust manifold bolts in the sequence shown to prevent any possible leaks—3.3L engine

4. Remove the exhaust manifold stay.

5. Disconnect the left and right exhaust pipes from the catalytic converters.

➡**Soak the exhaust pipe retaining bolts with penetrating oil if necessary to loosen them.**

6. Remove the nuts for each manifold. Remove the outer nuts first and the center nuts last.

7. Remove the manifold with the catalytic converter attached.

To install:

8. Clean all gasket surfaces. Install new gaskets.

9. Install the manifold and converter to the engine, tightening the mounting nuts in two stages, in sequence. Tighten the nuts to 21–25 ft. lbs. (28–33 Nm).

10. Connect the exhaust pipe to the converter. Be careful not break these studs.

11. Install the EGR tube to the left side manifold. Tighten the tube nut to 29–36 ft. lbs. (39–49 Nm).

12. Install the exhaust manifold covers.

13. Connect the negative battery cable.

14. Start the engine and check for exhaust leaks.

Front Crankshaft Seal

REMOVAL & INSTALLATION

➡**The front crankshaft seal procedures only apply to engines equipped with timing belts. For timing chain or gear engines, please refer to the applicable procedure later in this section.**

3.0L and 3.3L Engines

➡**The front oil seal is a part of the oil pump body.**

1. Disconnect the negative battery cable.

2. Remove the timing belt and the crankshaft sprocket.

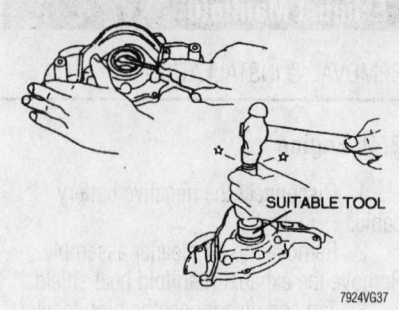

SUITABLE TOOL

7924VG37

Removing and installing the front crankshaft oil seal—6-cylinder engines

3. Remove the oil pump assembly.

4. Remove the oil seal from the oil pump body using a prytool. Be careful not to damage the oil pump body during seal removal.

To install:

5. Apply clean engine oil to the new oil seal. Install the seal using proper size driver.

6. Install the oil pump assembly to the engine.

7. Install the remaining components in reverse order of removal.

8. Connect the negative battery cable.

Camshaft and Valve Lifters

REMOVAL & INSTALLATION

2.4L Engine

1995–97 MODELS

The same bolts that hold the rocker arm assembly also hold the camshaft bearing caps. The hydraulic lifters are built into the rocker arms. If the rocker arm shafts are disassembled, do not allow the arms to lie on their side or they will become air-bound. Keep the rocker arms upright or lay them in a pan of new engine oil.

✳✳ CAUTION

Observe all applicable safety precautions when working around fuel. Whenever servicing the fuel system, always work in a well ventilated area. Do not allow fuel spray or vapors to come in contact with a spark or open flame. Keep a dry chemical fire extinguisher near the work area. Always keep fuel in a container specifically designed for fuel storage; also, always properly seal fuel containers to avoid the possibility of fire or explosion.

1. Relieve the fuel system pressure and disconnect the negative battery cable.

2. Remove the rocker arm cover and turn the crankshaft to align the timing marks at TDC on No. 1 cylinder.

3. If the timing chain is not being removed, use a wire tie or wire to secure the timing chain to the camshaft sprocket. Use special tool KV10105800 to hold the timing chain in position.

4. Hold the camshaft sprocket to loosen the bolt and remove the sprocket. Secure the sprocket so the chain does not fall off the crankshaft sprocket.

5. Loosen each rocker shaft bolt 1 turn at a time to prevent bending the shafts.

6. When all the bolts are loose, remove the rocker arm shafts with the bolts still in the shafts. This will hold the assembly together.

7. If they are not already identified, mark the bearing caps so they can be installed in their original position facing the same direction. Lift the caps off and lift the camshaft out.

To install:

8. Inspect the camshaft and the bearings:

a. Be sure the camshaft and the bearing surfaces are in good condition.

b. Install the bearing caps without the camshaft, tighten the rocker arm shaft bolts to specification and measure the inside diameter of the bearing circle.

c. Measure the diameter of the camshaft bearings.

d. The difference between the measurements is the camshaft journal clearance; it should be no more than 0.0047 in. (0.12mm)

e. Install the camshaft without the rocker arms and tighten the bolts to specification. The camshaft end-play should be no more than 0.008 in. (0.2mm).

9. Lubricate the camshaft with engine oil and set it in place. Be sure the pin on the sprocket end is up.

10. Install the rocker arm shafts and tighten the bolts in the proper sequence 1 turn at a time to draw the shafts down evenly against the valve springs without bending the shafts. Tighten the bolts to 27–30 ft. lbs. (37–41 Nm).

11. Install the camshaft sprocket and remove the tie securing the chain. Install the sprocket bolt but don't tighten it yet. Rotate the crankshaft 2 full turns to be sure the timing marks line up. When the valve timing is correct, tighten the sprocket bolt to 87–116 ft. lbs. (118–157 Nm).

12. Use a silicone sealer on the rubber end plugs and install the rocker arm cover with a new gasket. Install two of the attach-

ing bolts and tighten the two bolts to 2 ft. lbs. (3 Nm).

13. Tighten the rocker cover bolts in the proper sequence to 5–8 ft. lbs. (7–11 Nm).

14. Drain the engine oil and refill the engine with fresh oil.

15. Connect the negative battery cable. Run the engine and check for proper operation.

16. When the engine is first started, the hydraulic valve lifters may be noisy. Run the engine for 10–20 minutes at about 1000 rpm. If the noise has not subsided, the lifter will probably never pump up and must be replaced.

1998–99 MODELS

The lower timing chain is installed in the same manor as the timing chain on the 1995–97 2.4L engine.

1. Disconnect the negative battery cable.

✳✳ CAUTION

Never open, service or drain the radiator or cooling system when hot; serious burns can occur from the steam and hot coolant. Coolant should be reused unless it is contaminated or is several years old.

2. Drain the engine coolant.

✳✳ CAUTION

Observe all applicable safety precautions when working around fuel. Whenever servicing the fuel system, always work in a well ventilated area. Do not allow fuel spray or vapors to come in contact with a spark or open flame. Keep a dry chemical fire extinguisher near the work area. Always keep fuel in a container specifically designed for fuel storage; also, always properly seal fuel containers to avoid the possibility of fire or explosion.

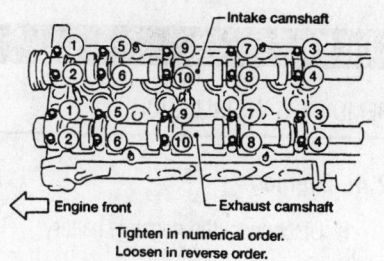

Tighten in numerical order.
Loosen in reverse order.

7924VG51

Tighten the bearing cap bolts in the correct sequence—1998–99 2.4L (DOHC) engine

3. Disconnect any vacuum hoses, fuel lines and wires that may interfere with the removal of the exhaust manifold and timing chain cover.

4. Remove the exhaust manifold heat shield and the front exhaust pipe.

5. Remove the exhaust manifold.

6. Remove the air intake duct, cooling fan with coupling and the radiator shroud.

7. Unplug the fuel injector connectors and remove the fuel rail with injectors.

8. Position No. 1 piston at TDC, then remove the distributor.

9. Remove the rocker arm cover.

10. Mark the position of the chain in relation to the sprocket with paint.

11. Remove the rubber plugs from the sprocket cover.

12. Remove the camshaft sprockets.

13. Position the chain out of the way and remove the camshaft bearing cap bolts in the sequence shown. Keep the bearing caps in order, they must be installed in their original positions.

14. Remove the camshafts.

To install:

15. Lubricate the camshafts and position them on the cylinder head.

16. Install the bearing caps. Tighten the

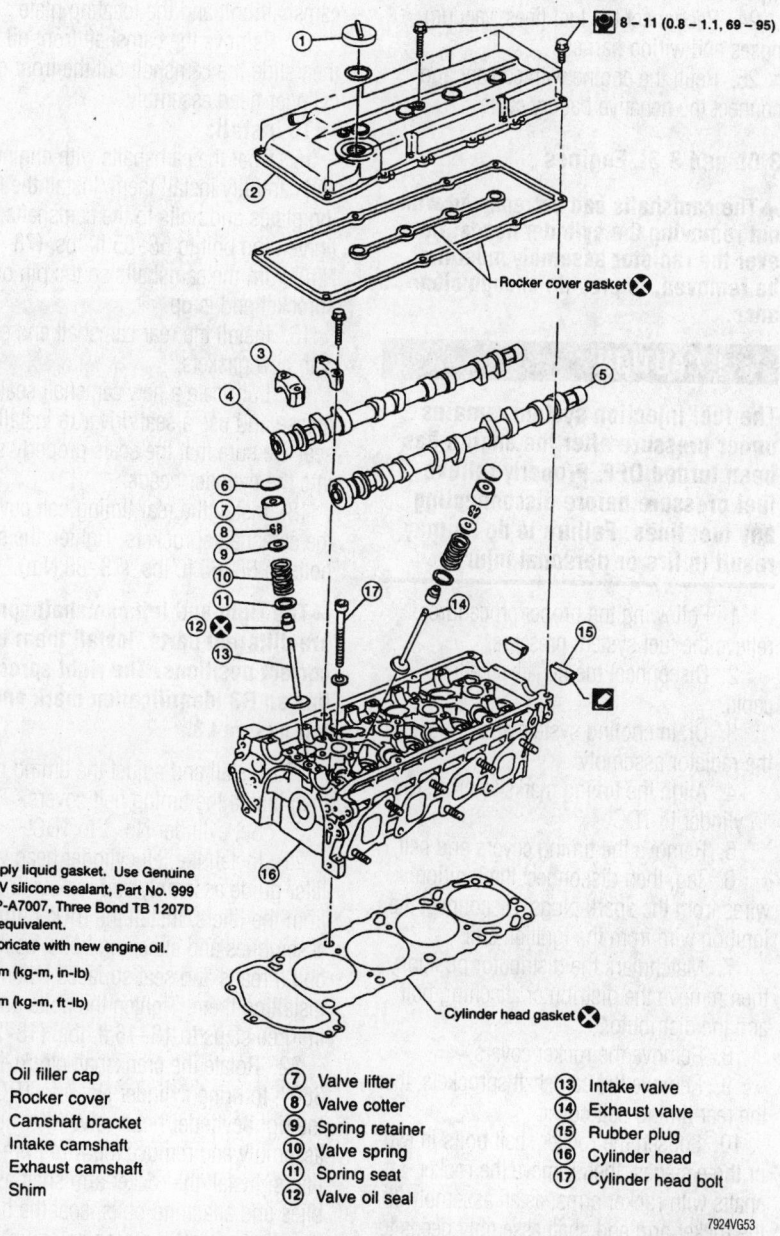

8 – 11 (0.8 – 1.1, 69 – 95)

Rocker cover gasket ✖

Apply liquid gasket. Use Genuine RTV silicone sealant, Part No. 999 MP-A7007, Three Bond TB 1207D or equivalent.

Lubricate with new engine oil.

: N·m (kg-m, in-lb)

: N·m (kg-m, ft-lb)

① Oil filler cap	⑦ Valve lifter	⑬ Intake valve
② Rocker cover	⑧ Valve cotter	⑭ Exhaust valve
③ Camshaft bracket	⑨ Spring retainer	⑮ Rubber plug
④ Intake camshaft	⑩ Valve spring	⑯ Cylinder head
⑤ Exhaust camshaft	⑪ Spring seat	⑰ Cylinder head bolt
⑥ Shim	⑫ Valve oil seal	

7924VG53

Exploded view of the camshafts and related components—1998–99 2.4L (DOHC) engine

bolts in two steps in the correct sequence first to 17 inch lbs. (2 Nm), then to 80–104 inch lbs. (9–12 Nm).

17. Install the sprockets on the camshafts. Tighten the bolts to 123–130 ft. lbs. (167–177 Nm).

18. Apply RTV sealant to the rubber plugs and install them on the sprocket cover flush with the surface of the cylinder head.

19. Install the rocker arm cover.

20. Install the distributor.

21. Install the fuel rail/injector assembly and connect them to the harness.

22. Install the radiator shroud, fan with coupling and air intake duct.

23. Install the exhaust manifold, exhaust pipe and heat shield.

24. Reconnect the fuel lines, vacuum hoses and wiring harness.

25. Refill the engine with coolant and connect the negative battery cable.

3.0L and 3.3L Engines

➡ **The camshafts can be removed without removing the cylinder heads, however the radiator assembly must first be removed, to provide enough clearance.**

❋❋ CAUTION

The fuel injection system remains under pressure after the engine has been turned OFF. Properly relieve fuel pressure before disconnecting any fuel lines. Failure to do so may result in fire or personal injury.

1. Following the proper procedures, relieve the fuel system pressure.

2. Disconnect the negative battery cable.

3. Drain cooling system and remove the radiator assembly.

4. Align the timing marks to bring No. 1 cylinder to TDC.

5. Remove the timing covers and belt.

6. Tag, then disconnect the ignition wires from the spark plugs. Disconnect the ignition wire from the ignition coil.

7. Matchmark the distributor position, then remove the distributor attaching bolt and the distributor.

8. Remove the rocker covers.

9. Remove the camshaft sprockets, then the rear timing belt cover.

10. Loosen the rocker shaft bolts in two or three stages, then remove the rocker shafts with rocker arms, as an assembly. If the rocker arm and shaft assembly needs to be disassembled for service, note the location of the components as they are

removed. The rocker arms must be installed in the same position if reused.

11. Attach a wire to the top of the lifters so that they will not drop from the lifter guide. Carefully remove the lifter guide and lifters from the cylinder head. Put an identification mark on the lifters to avoid mixing them up if they are removed from the guide and are being reused.

12. Measure the camshaft end-play, it should be 0.0012–0.0024 in. (0.03–0.06mm). If the end-play is out of specification, the locate plate will have to be replaced with a plate of the correct thickness.

13. At the rear of the cylinder head, remove the cylinder head rear cover, the camshaft bolt and the locating plate.

14. Remove the camshaft front oil seal, then slide the camshaft out the front of the cylinder head assembly.

To install:

15. Coat the camshafts with engine oil and carefully install them. Install the locating plates and bolts to the camshafts, tighten the bolt to 58–65 ft. lbs. (78–88 Nm). Turn the camshafts so the pin on the sprocket end is up.

16. Install the rear camshaft end covers with new gaskets.

17. Lubricate a new camshaft seal with grease and use a seal driver to install the seal. Be sure that the seals properly seat into the cylinder heads.

18. Install the rear timing belt cover and the camshaft sprockets. Tighten the sprocket bolts to 58–65 ft. lbs. (78–88 Nm).

➡ **The right and left camshaft sprockets are different parts. Install them in their correct positions. The right sprocket has an R3 identification mark and the left has an L3.**

19. Install and adjust the timing belt, then install the timing belt covers.

20. Set cylinder No. 1 to TDC.

21. Install the left cylinder head valve lifter guide assembly and remove the wire from the lifters. Install the rocker arm shaft assemblies and attaching bolts, coat the bolt threads and seat surfaces before installing them. Tighten the bolts gradually in three steps to 13–16 ft. lbs. (18–22 Nm).

22. Rotate the crankshaft clockwise 360°, to bring cylinder No. 4 to TDC. Install the right cylinder head valve lifter guide assembly and remove the wire from the lifters. Install the rocker arm shaft assemblies and attaching bolts, coat the bolt threads and seat surfaces before installing them. Tighten the bolts gradually in three steps to 13–16 ft. lbs. (18–22 Nm).

23. Install the rocker covers with new gaskets, tighten the rocker cover bolts to 1–2 ft. lbs. (1–3 Nm).

24. Set cylinder No. 1 to TDC by rotating the engine 360° and aligning the timing marks. Install the distributor assembly and connect the wires to their proper locations. Do not tighten the distributor attaching bolt until the timing has been checked and adjusted as necessary.

25. Install the radiator assembly, then refill and bleed the cooling system.

26. Connect the negative battery cable and run the engine for 10–20 minutes, at about 1000 rpm, to pump up the lifter assemblies.

➡ **If the hydraulic valve lifters are still noisy, replace them and bleed the air from them.**

VALVE LIFTER BLEEDING

➡ **The hydraulic valve lifters are installed in the rocker arms. The rocker arms should be stored standing up or in a bath of new engine oil when removed from the engine.**

Bleed the air from the hydraulic valve lifters (lash adjusters) by running the engine under no load at 1,000 rpm for 10 minutes (6-cylinder engines) or 20 minutes (4-cylinder engine). If any valve lifters are still noisy, replace them.

Valve Lash

ADJUSTMENT

All of the engines in this section utilize hydraulic valve lifters (lash adjusters) that do not require periodic adjustment. If a valve remains noisy after the bleeding procedure, replace the valve lifter. If the valve is still noisy, inspect the camshaft, rocker arm and valve for excessive wear. Replace all worn components.

Oil Pan

REMOVAL & INSTALLATION

2.4L Engine

1. Disconnect the negative battery cable.

2. Raise the front of the vehicle and support it with safety stands.

3. Drain the engine oil into a suitable container.

Prolonged contact with used engine oil may cause a number of skin disorders, including cancer. Make every effort to minimize exposure to used engine oil.

4. Remove the front stabilizer bar mounting nuts and bolts from the side member.

5. Remove the nuts from the engine mounts and raise the engine.

6. Loosen the oil pan bolts in the proper sequence. Insert a seal cutter tool between the cylinder block and the oil pan and tap it around the circumference of the pan with a hammer. Remove the oil pan. Pull it out from the front side.

➡Be careful not to drive the seal cutter into the oil pump or rear oil seal retainer as you will damage the aluminum mating surface.

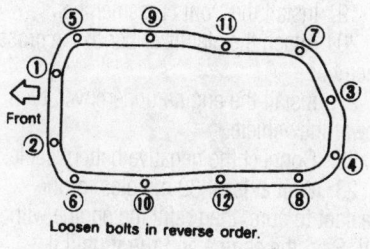

Loosen bolts in reverse order.

7924VG38

Remove the oil pan bolts in the correct sequence—2.4L engine

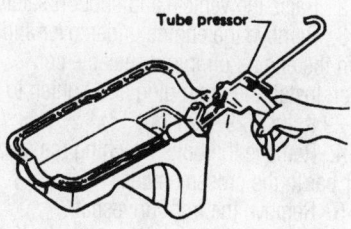

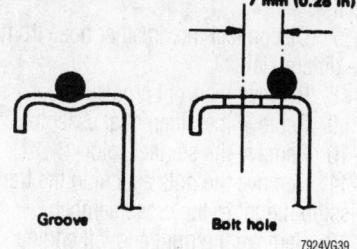

7924VG39

Apply sealant to the oil pan as shown— 2.4L engine-other engines similar

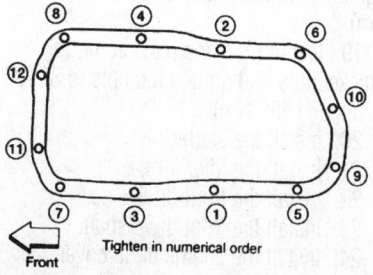

Tighten in numerical order

Front

7924VG40

Oil pan bolt installation sequence—2.4L engine

To install:

7. Remove all traces of gasket material from the pan and block mating surfaces.

8. Apply a continuous bead of sealant 0.138–0.177 in. (3.5–4.5mm) to the oil pan mating surface. Be sure to trace the sealant bead to the inside of the bolt holes where there is no groove.

9. Install the pan within five minutes and tighten all bolts in the proper sequence. Tighten the bolts to 5–6 ft. lbs. (7–8 Nm).

10. Lower the engine and install the engine mount nuts. Tighten the nuts to 30–38 ft. lbs. (41–52 Nm).

11. Connect the stabilizer bar.

12. Lower the vehicle.

13. Connect the negative battery cable.

14. Wait at least 30 minutes, then refill the engine with oil. Run the engine until it reaches normal operating temperature, then check for leaks.

3.0L Engine

The EPA warns that prolonged contact with used engine oil may cause a number of skin disorders, including cancer! You should make every effort to minimize your exposure to used engine oil. Protective gloves should be worn when changing the oil. Wash your hands and any other exposed skin areas as soon as possible after exposure to used engine oil. Soap and water, or waterless hand cleaner, should be used.

2WD VEHICLES

1. Disconnect the negative battery cable.

2. Raise the vehicle and support safely.

3. Remove the engine undercover and drain the engine oil into a sealable con-

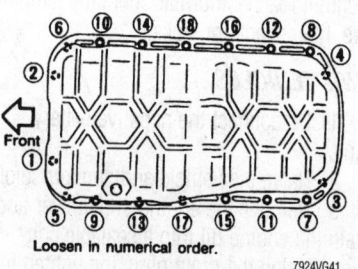

Loosen in numerical order.

7924VG41

Remove the oil pan bolts in the correct sequence to prevent warping it—3.0L engine with 2WD

tainer. Install the drain plug and tighten to 22–29 ft. lbs. (29–39 Nm).

4. Remove the bolts attaching the stabilizer bar to the crossmember.

5. Remove the front crossmember.

6. Remove the idler arm.

7. Remove the starter motor.

8. Remove the bracket securing the engine to the transmission.

9. Remove the oil pan mounting bolts in the proper sequence. Insert a seal cutter tool between the cylinder block and the oil pan and tap it around the circumference of the pan with a hammer. Remove the oil pan.

➡Be careful not to drive the seal cutter into the oil pump or rear oil seal retainer as you will damage the aluminum mating surface.

To install:

10. Remove all traces of gasket material from the pan and block mating surfaces.

11. Apply sealant to the oil pump and oil seal retainer gasket.

12. Apply a continuous bead of sealant 0.138–0.177 (3.5–4.5mm) to the oil pan mating surface. Be sure to trace the sealant bead to the inside of the bolt holes where there is no groove.

13. Install the pan within five minutes of sealant application and tighten all bolts in the reverse order of removal. Tighten the bolts to 5 ft. lbs. (7 Nm).

14. Install the bracket securing the engine to the transmission.

15. Install the starter.

16. Install the idler arm.

17. Install the front crossmember.

18. Attach the stabilizer bar to the crossmember.

19. Install the engine undercover, then lower the vehicle.

20. Connect the negative battery cable.

21. Wait at least 30 minutes, then refill the engine with oil. Start the engine and run

it until it reaches normal operating temperature, then check for leaks.

4WD VEHICLES

1. Disconnect the negative battery cable.

2. Raise the vehicle and support safely.

3. Remove the engine undercover and drain the engine oil into a sealable container. Install the drain plug and tighten to 22–29 ft. lbs. (29–39 Nm).

4. Remove the front driveshaft.

5. Remove the front drive axle from the vehicle.

6. Remove the idler arm.

7. Remove the starter motor.

8. Remove the transmission mount bracket nuts.

9. Remove the bolts attaching the engine mount.

10. Remove the bracket securing the engine to the transmission.

11. Attach an engine hoist and raise the engine slightly.

✵✵ WARNING

It may be necessary to disconnect the exhaust to avoid damaging it when raising the engine. When lifting the engine be careful not to contact any adjacent parts, especially the accelerator wire casing end, brake tubes and the brake master cylinder.

12. Remove the oil pan mounting bolts in the proper sequence. Insert a seal cutter tool between the cylinder block and the oil pan and tap it around the circumference of the pan with a hammer. Remove the oil pan.

➡Be careful not to drive the seal cutter into the oil pump or rear oil seal retainer as you will damage the aluminum mating surface.

To install:

13. Remove all traces of gasket material from the pan and block mating surfaces.

14. Apply sealant to the oil pump and oil seal retainer gasket.

15. Apply a continuous bead of sealant 0.138–0.177 in. (3.5–4.5mm) to the oil pan mating surface. Be sure to trace the sealant bead to the inside of the bolt holes where there is no groove.

16. Install the pan within five minutes of sealant application and tighten all bolts in the reverse order of removal. Tighten the bolts to 5 ft. lbs. (7 Nm).

17. Install the bracket securing the engine to the transmission.

18. Install the engine mount bolts and

tighten the bolts to 23–31 ft. lbs. (31–42 Nm).

19. Install the transmission mount bracket nuts and tighten the nuts to 30–38 ft. lbs. (41–52 Nm).

20. Install the starter.

21. Install the idler arm.

22. Install the front drive axle.

23. Install the front drive shaft.

24. Install the engine undercover, then lower the vehicle.

25. Connect the negative battery cable.

26. Wait at least 30 minutes, then refill the engine with oil. Start the engine and run it until it reaches normal operating temperature, then check for leaks.

3.3L Engine

2WD MODELS

1. Disconnect the negative battery cable.

2. Raise the vehicle and support safely.

3. Remove the engine undercover and drain the engine oil into a sealable container. Install the drain plug and tighten to 22–29 ft. lbs. (29–39 Nm).

4. Remove the bolts attaching the stabilizer bar to the crossmember.

5. Remove the front crossmember.

6. Remove the starter motor.

7. Remove the nuts attaching the transmission mount to the crossmember.

8. Remove the right and left engine mounting bolts and nuts.

9. Remove the right and left power steering mounting brackets.

10. Raise and safely support the engine. Disconnect the front exhaust pipes if needed.

11. Remove the oil pan mounting bolts in the proper sequence. Insert a seal cutter tool between the cylinder block and the oil pan and tap it around the circumference of the pan with a hammer. Remove the oil pan.

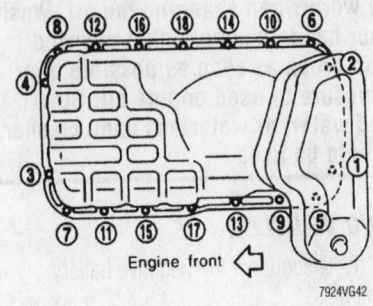

Oil pan bolt removal sequence—3.3L engine with 2WD

➡Be careful not to drive the seal cutter into the oil pump or rear oil seal retainer as you will damage the aluminum mating surface.

To install:

12. Remove all traces of gasket material from the pan and cylinder block mating surfaces.

13. Apply sealant to the oil pump and oil seal retainer gasket.

14. Apply a continuous bead of sealant 0.138–0.177 in. (3.5–4.5mm) to the oil pan mating surface. Be sure to trace the sealant bead to the inside of the bolt holes where there is no groove.

15. Install the pan within five minutes of sealant application and tighten all bolts in the reverse order of removal. Tighten the bolts to 5 ft. lbs. (7 Nm).

16. Lower the engine and install the mounting bolts and nuts. Connect the front exhaust pipes if removed.

17. Install the power steering mounting brackets.

18. Install the starter.

19. Install the front crossmember.

20. Attach the stabilizer bar to the crossmember.

21. Install the engine undercover, then lower the vehicle.

22. Connect the negative battery cable.

23. Wait at least 30 minutes for the sealant to cure, then refill the engine with oil. Start the engine and run it until it reaches normal operating temperature, then check for leaks.

4WD MODELS

1. Disconnect the negative battery cable.

2. Raise the vehicle and support safely.

3. Remove the engine undercover and drain the engine oil into a sealable container. Install the drain plug and tighten to 22–29 ft. lbs. (29–39 Nm).

4. Remove the bolts attaching the stabilizer bar to the crossmember.

5. Remove the front driveshaft.

6. Remove the front axle shafts (half shafts).

7. Disconnect the breather hose from the differential.

8. Remove the front crossmember.

9. Remove the differential assembly.

10. Remove the starter motor.

11. Remove the nuts attaching the transmission mount to the crossmember.

12. Remove the right and left engine mounting bolts and nuts.

13. Remove the right and left power steering mounting brackets.

14. Raise and safely support the engine. Disconnect the front exhaust pipes if needed.

15. Remove the oil pan mounting bolts in the proper sequence. Insert a seal cutter tool between the cylinder block and the oil pan and tap it around the circumference of the pan with a hammer. Remove the oil pan.

➡ **Be careful not to drive the seal cutter into the oil pump or rear oil seal retainer as you will damage the aluminum mating surface.**

To install:

16. Remove all traces of gasket material from the pan and cylinder block mating surfaces.

17. Apply sealant to the oil pump and oil seal retainer gasket.

18. Apply a continuous bead of sealant 0.138–0.177 (3.5–4.5mm) to the oil pan mating surface. Be sure to trace the sealant bead to the inside of the bolt holes where there is no groove.

19. Install the pan within five minutes of sealant application and tighten all bolts in the reverse order of removal. Tighten the bolts to 5 ft. lbs. (7 Nm).

20. Lower the engine and install the mounting bolts and nuts. Connect the front exhaust pipes if removed.

21. Install the power steering mounting brackets.

22. Install the starter.

23. Install the differential assembly mounting bracket.

24. Install the differential assembly.

25. Connect the differential breather hose.

26. Install the front half shafts.

27. Install the front crossmember.

28. Install the driveshaft.

29. Attach the stabilizer bar to the crossmember.

30. Install the engine undercover, then lower the vehicle.

31. Connect the negative battery cable.

32. Wait at least 30 minutes for the sealant to cure, then refill the engine with oil. Start the engine and run it until normal operating temperature, then check for leaks.

Oil Pump

REMOVAL & INSTALLATION

2.4L Engine

The oil pump is mounted externally on the engine, eliminating the need to remove the oil pan in order to remove the oil pump.

1. Turn the engine to TDC. Matchmark and remove the distributor.

2. Disconnect the negative battery cable.

3. Drain the engine oil.

4. Remove the front stabilizer bar.

5. Remove the splash shield.

6. Loosen the mounting bolts and remove the oil pump body with the drive spindle assembly.

To install:

7. If the crankshaft has been moved, turn the crankshaft so that the No. 1 piston is at TDC of the compression stroke before installing the oil pump in the engine.

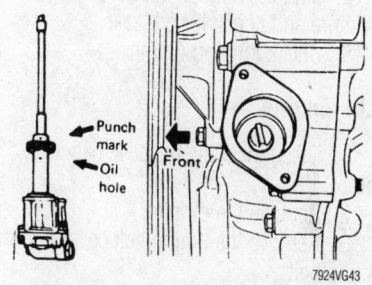

Punch mark
Oil hole
Front
7924VG43

Align the punch mark with the oil hole before oil pump installation—2.4L engine

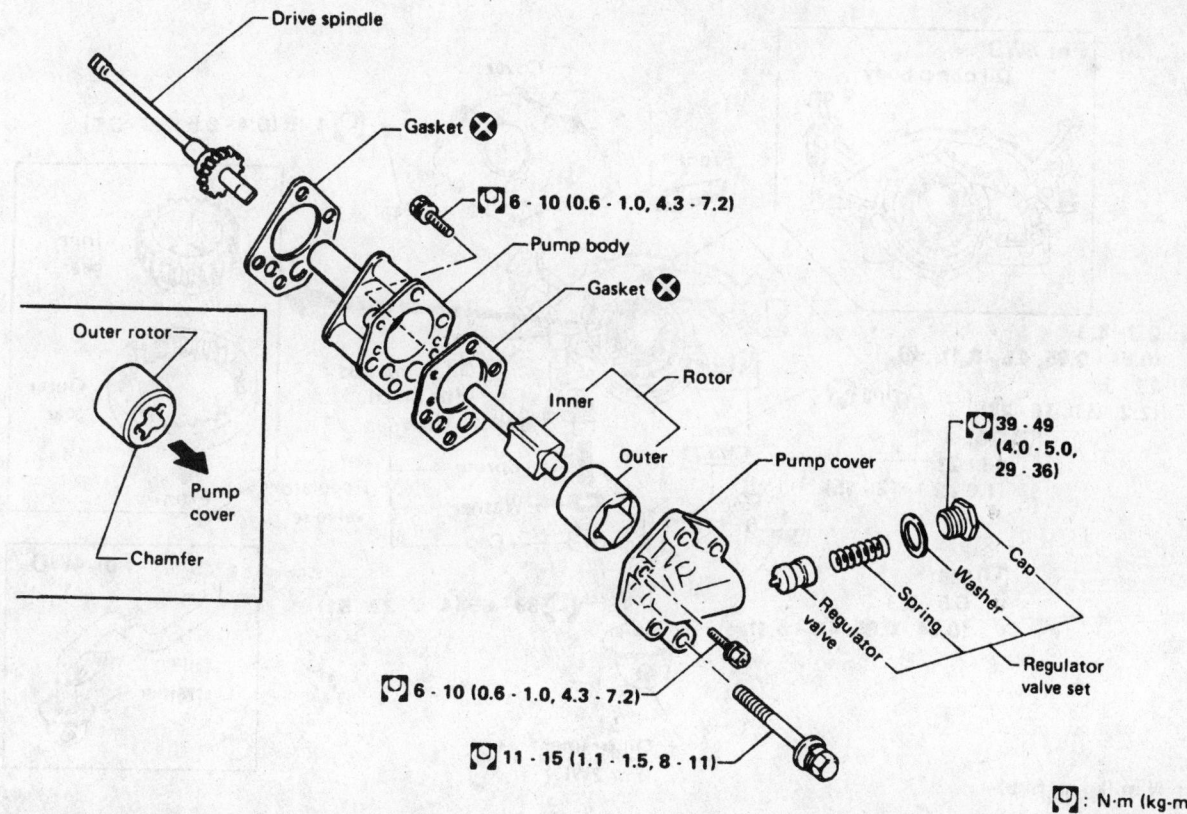

- Drive spindle
- Gasket ✕
- 6 - 10 (0.6 - 1.0, 4.3 - 7.2)
- Pump body
- Gasket ✕
- Outer rotor
- Pump cover
- Chamfer
- Inner
- Rotor
- Outer
- Pump cover
- Regulator valve
- Spring
- Washer
- Cap
- 39 - 49 (4.0 - 5.0, 29 - 36)
- Regulator valve set
- 6 - 10 (0.6 - 1.0, 4.3 - 7.2)
- 11 - 15 (1.1 - 1.5, 8 - 11)
- : N·m (kg-m, ft-lb)
- 7924VG44

Exploded view of oil pump assembly—2.4L engine

8. Fill the pump housing with engine oil, then align the punch mark on the spindle with the hole in the oil pump.

9. With a new gasket placed over the drive spindle, install the oil pump and drive spindle assembly so that the projection on the top of the drive spindle is located in the 11 o'clock position.

10. Align matchmarks and install the distributor with the metal tip of the rotor pointing toward the No. 1 spark plug tower, of the distributor cap.

11. Install the splash shield and front stabilizer bar.

12. Refill the engine with oil.

13. Connect the negative battery cable.

14. Start the engine, ensure proper oil pressure and check for oil leaks.

15. Check ignition timing.

3.0L Engine

1. Disconnect the negative battery cable.

2. Remove the oil pan.

3. Remove the timing belt cover and the timing belt.

4. Remove the crankshaft timing sprocket (it may be necessary to use a puller) and the timing belt plate.

5. Remove the oil pump strainer and pick-up tube from the oil pump.

6. Loosen the oil pump retaining bolts, then remove the oil pump.

To install:

7. Use new gaskets and install a new oil seal. Tighten the 6mm bolts to 4–6 ft. lbs. (6–8 Nm) and the 8mm bolts to 16–22 ft. lbs. (22–39 Nm).

➡ **Before installing the oil pump, be sure to pack the pump's cavity with petroleum jelly, then be sure the O-ring is properly fitted.**

8. Connect the oil strainer and pick-up tube to the pump body.

9. Clean gasket surfaces and install the oil pan.

10. Install the timing belt plate and the crankshaft sprocket.

11. Install the timing belt and front cover.

12. Connect the negative battery cable.

13. Refill the engine with oil, start the engine, and check for any leaks.

3.3L Engine

1. Disconnect the negative battery cable.

2. Drain the engine oil and the coolant

from the radiator. Save the coolant so it can be reused.

3. Remove the air duct between the mass air flow sensor and the throttle body.

4. Remove the cooling fan.

5. Remove the upper and lower radiator hoses and the fan shroud.

6. Remove the drive belts.

7. Remove the crankshaft pulley.

8. Remove the upper and lower timing belt covers.

9. Remove the oil pan.

10. Remove the oil strainer (pick-up).

11. Remove the oil pump mounting bolts and the oil pump.

To install:

12. Replace the oil seal in the pump.

13. Using a new gasket, install the oil pump. Tighten the mounting bolts to 12–15 ft. lbs. (16–21 Nm).

14. Install the oil strainer and the oil pan.

15. Install the timing belt covers and the crankshaft pulley.

16. Install the drive belts.

17. Install the fan shroud and the radiator hoses.

18. Install the cooling fan.

19. Install the air duct between the throttle body and mass air flow sensor.

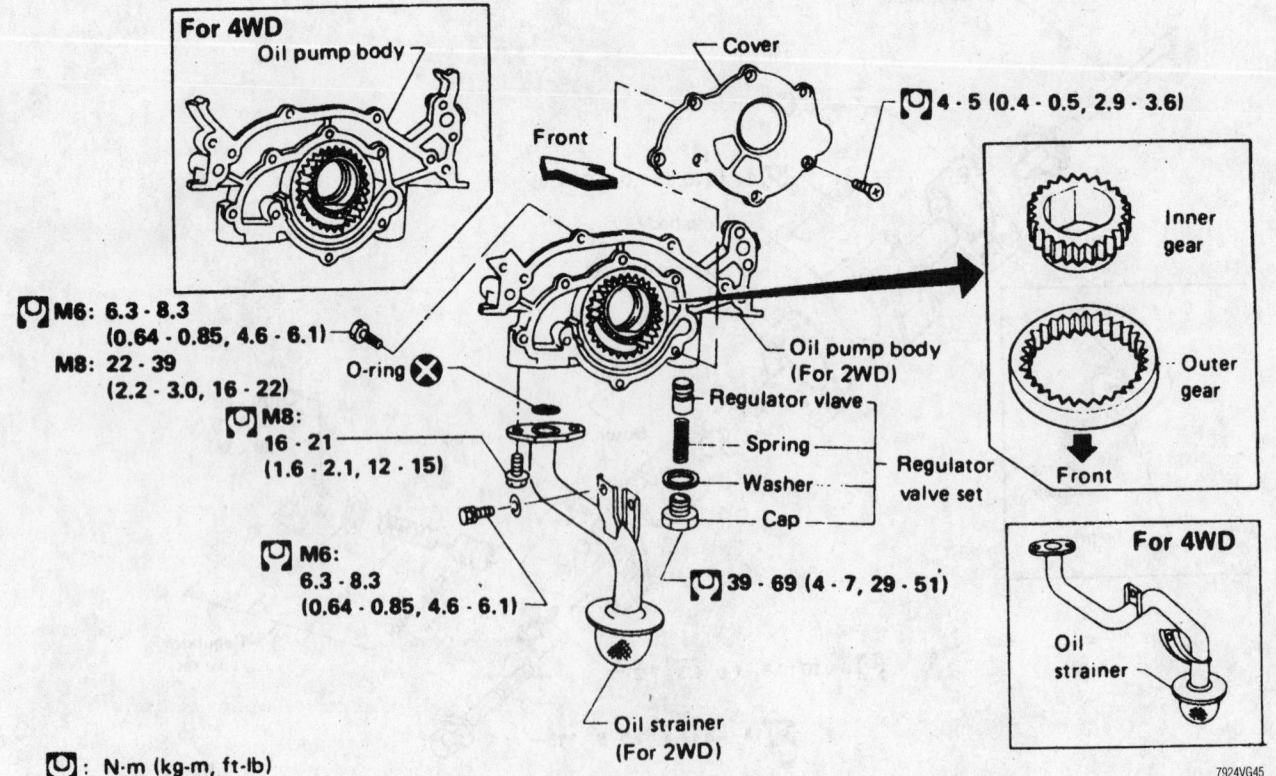

Exploded view of oil pump assembly—3.0L engine

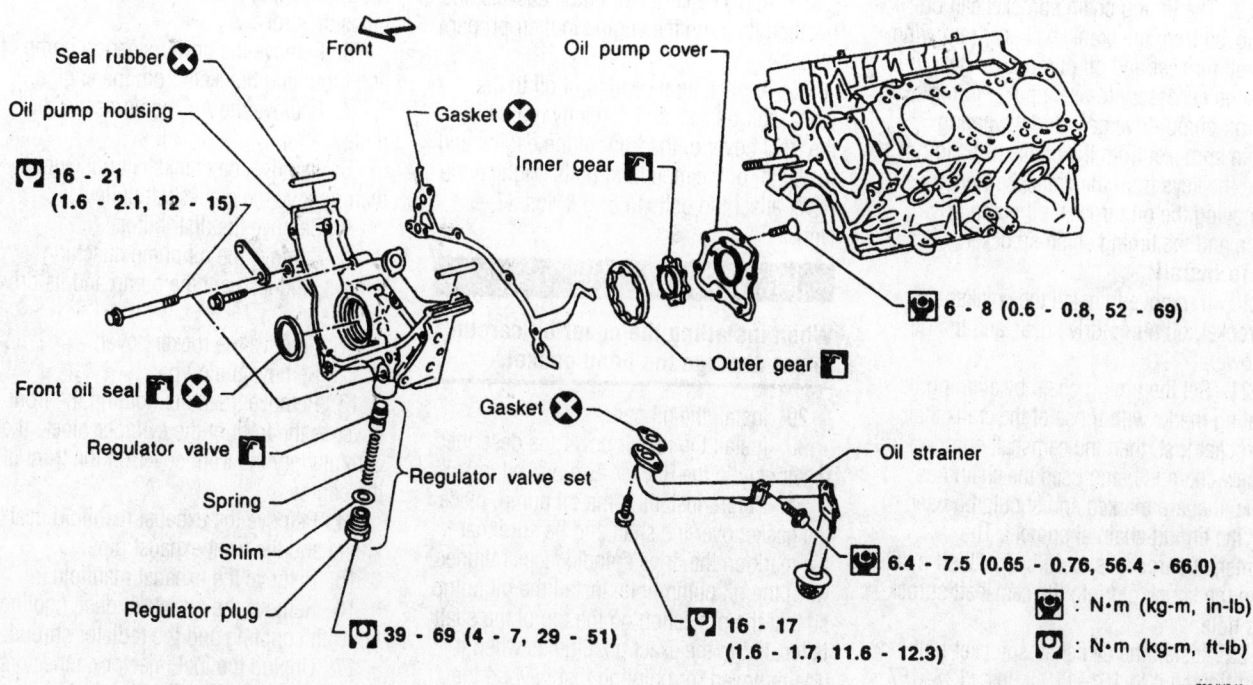

Oil pump assembly exploded view—3.3L engine

20. Refill the radiator with the coolant that was removed unless the coolant is being changed.

21. Fill the engine with the proper amount of new engine oil.

22. Connect the negative battery cable, start the engine and check for leaks.

Rear Main Seal

REMOVAL & INSTALLATION

1. Remove the transmission.
2. Remove the flywheel/flexplate.
3. Remove the seal retainer, then remove the seal from the retainer.

To install:

4. Install the new seal in the retainer.

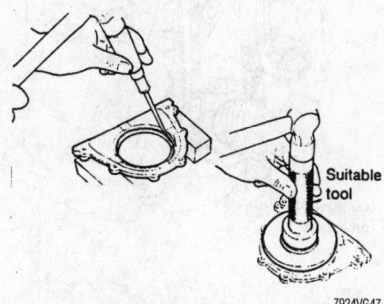

Tap out the old seal from the back of the retainer and install the new seal from the front

5. Apply clean engine oil to the lip of the seal.

6. Using a new gasket, install the seal retainer to the engine.

7. Install the flexplate/flywheel.

8. Install the transmission.

Timing Chain, Sprockets, Front Cover and Seal

REMOVAL & INSTALLATION

2.4L Engine

1995–97 MODELS

1. Disconnect the negative battery cable from the battery.

2. Drain the cooling system into a sealable container, then remove the radiator together with the upper and lower radiator hoses.

3. Remove the cooling fan.
4. Remove the drive belts.
5. Remove all of the spark plugs, then set the No. 1 cylinder to TDC of its compression stroke.
6. Remove the power steering pump and mounting brackets from the engine.
7. Remove the A/C compressor idler pulley.
8. Remove the crankshaft pulley bolt, then remove the crankshaft pulley.
9. Remove the distributor.
10. Remove the oil pump attaching

screws, and take out the pump and its drive spindle.

11. Remove the rocker cover.
12. Remove the oil pan.
13. Remove the bolts holding the front cover to the front of the cylinder block, then carefully pry the front cover off the front of the engine.
14. With the No. 1 piston at TDC, the timing marks on the camshaft sprocket and the timing chain should be visible. If the marks on the chain and sprocket are not visible matchmark the chain and the sprocket with paint, if the chain is going to be reused.
15. With the timing marks on the camshaft sprocket located, locate the timing marks on the crankshaft sprocket. If the marks on the chain and sprocket are not visible matchmark the chain and the sprocket with paint, if the chain is going to be reused.
16. Carefully remove the tensioner, the tensioner is spring loaded.
17. Remove the timing chain guides.
18. Remove the camshaft sprocket attaching bolt, then carefully remove the sprocket and timing chain.

✷✷ WARNING

After removing the timing chain, do not turn the crankshaft or camshaft separately or the pistons will hit the valves.

19. The timing chain sprocket can be removed from the crankshaft after removing the oil thrower and oil pump drive gear. It may be necessary to use a puller to remove the oil pump drive gear and the timing chain sprocket from the crankshaft. Do not lose the keys from the crankshaft when removing the oil thrower, oil pump drive gear, and the timing chain sprocket.

To install:

20. If removed, install the crankshaft sprocket, oil pump drive gear, and the oil thrower.

21. Set the timing chain by aligning its mating marks with those of the crankshaft sprocket first, then the camshaft sprocket. If a new chain is being used the chain has links that are marked and should be used for the timing chain alignment. The camshaft sprocket should be installed by fitting the knock pin into the camshaft sprockets hole.

22. Install the camshaft sprocket bolt and tighten it to 101–116 ft. lbs. (137–157 Nm).

23. Install the chain guide and tighten the bolts to 9–14 ft. lbs. (13–19 Nm).

24. Install the tensioner guide and tighten the bolt to 9–14 ft. lbs. (13–19 Nm).

25. Carefully install the tensioner and tighten the mounting bolts to 5–6 ft. lbs. (7–8 Nm).

26. Install and lightly oil the new front seal.

27. Apply sealer to all of the gaskets and position them on the engine in their proper places.

28. Apply a light coating of oil to the crankshaft oil seal and carefully mount in the front cover to the front of the engine and install all of the mounting bolts. Tighten the cover attaching bolts to 5–6 ft. lbs. (7–8 Nm).

✳✳ WARNING

When installing the cover be careful to not damage the head gasket.

29. Install the oil pan.

30. Install the rocker cover (as described previously in the Rocker Arm procedure).

31. Before installing the oil pump, place the gasket over the shaft and be sure that the mark on the drive spindle faces (aligned with) the oil pump hole. Install the oil pump so that the projection on the top of the shaft is located in the exact position as when it was removed (or pointing just beyond the 11 o'clock and 5 o'clock positions when the piston in the No. 1 cylinder is placed at TDC on the compression stroke, if the engine was disturbed since disassembly). Tighten the oil pump attaching screws to 8–10 ft. lbs. (11–15 Nm).

32. Install the distributor in the correct position.

33. Install the crankshaft pulley and bolt. Tighten the bolt to 87–116 ft. lbs. (118–157 Nm).

34. Install the A/C compressor idler pulley.

35. Install the power steering mounting brackets and the pump.

36. Install the fan pulley and cooling fan.

37. Install and adjust the drive belts.

38. Install the spark plugs and tighten them to 14–22 ft. lbs. (20–29 Nm).

39. Install the radiator; reconnect the upper and lower radiator hoses.

40. Fill and bleed the air from the cooling system.

41. Connect the negative battery cable.

42. Start the engine, check ignition timing, and check for leaks.

1998–99 2.4L MODELS

1. Disconnect the negative battery cable from the battery.

2. Drain the cooling system into a sealable container, then remove the radiator together with the upper and lower radiator hoses.

3. Remove the cooling fan.

4. Remove the drive belts.

5. Remove all of the spark plugs, then set the No. 1 cylinder to TDC of its compression stroke.

6. Remove the power steering pump and mounting brackets from the engine.

7. Remove the A/C compressor idler pulley.

8. Remove the crankshaft pulley bolt, then remove the crankshaft pulley.

9. Remove the distributor.

10. Remove the oil pump attaching screws, and take out the pump and its drive spindle.

11. Remove the rocker cover.

12. Remove the oil pan.

13. Remove the bolts holding the front cover to the front of the cylinder block, then carefully pry the front cover off the front of the engine.

14. Remove the exhaust manifold heat shield and the front exhaust pipe.

15. Remove the exhaust manifold.

16. Remove the air intake duct, cooling fan with coupling and the radiator shroud.

17. Unplug the fuel injector connectors and remove the fuel rail with injectors.

18. Position No. 1 piston at TDC, then remove the distributor.

19. Remove the rocker arm cover.

20. Mark the position of the chain in relation to the sprocket with paint.

21. Remove the camshaft sprockets.

22. Remove the camshaft sprocket cover.

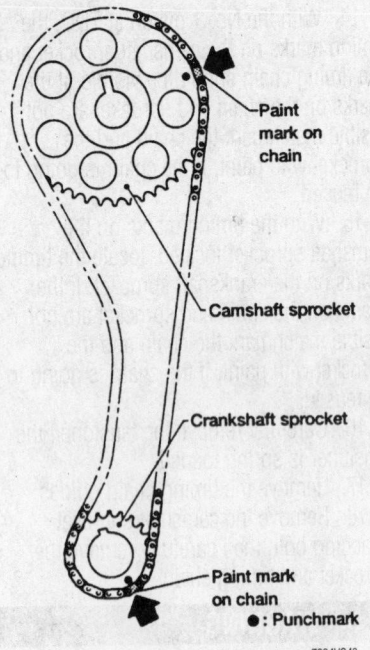

Align the paint marks on the chain with the marks on the sprockets—1995–97 2.4L engine

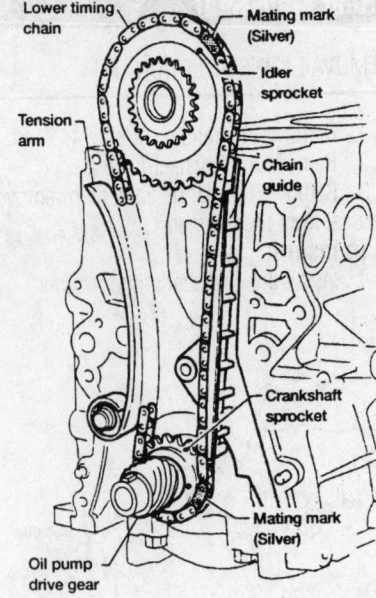

Align the silver links on the chain with the marks on the sprockets—1998–99 2.4L engine

23. Push in the chain tensioner and install a suitable pin through the tensioner to retain it.

24. Paint a mark on the chain and sprocket before removing the chain.

25. Remove the upper timing chain from the idler sprocket.

26. Drain the engine oil and remove the oil pan.

27. Remove the oil strainer.

28. Remove the drive belts and the A/C compressor idler pulley.

29. Remove the crankshaft pulley, oil pump and front cover.

30. Push in the chain tensioner and insert a suitable pin through the hole to retain the tensioner.

31. Remove the chain tensioner arm and lower chain guide.

32. Remove the chain from the sprocket.

33. If necessary, remove the oil thrower, oil pump drive gear and sprocket from the crankshaft.

To install:

34. If removed, install the sprocket, oil pump drive gear and oil thrower on the crankshaft.

35. Align the silver link on the chain with the mark on the sprocket and install the chain on the sprockets.

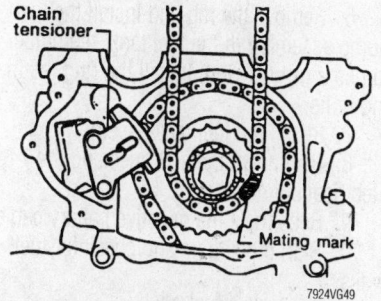

Align the matchmark and position the upper chain on the idler sprocket—1998–99 2.4L engine

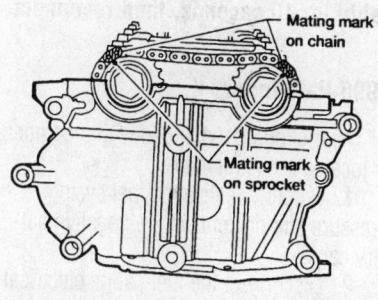

Align the paint marks on the upper chain with the marks on the camshaft sprockets—1998–99 2.4L engine

36. Install the chain tension arm and lower chain guide.

37. Remove the pin from the tensioner.

38. Install the front cover, as follows:

   a. Apply sealer to all of the gaskets and position them on the engine in their proper places.

   b. Apply a light coating of oil to the crankshaft oil seal and carefully mount in the front cover to the front of the engine and install all of the mounting bolts. Tighten the cover attaching bolts to 5–6 ft. lbs. (7–8 Nm).

✳✳ WARNING

When installing the cover be careful to not damage the head gasket.

39. Install the oil pan.

40. Install the rocker cover (as described previously in the Rocker Arm procedure).

41. Before installing the oil pump, place the gasket over the shaft and be sure that the mark on the drive spindle faces (aligned with) the oil pump hole. Install the oil pump so that the projection on the top of the shaft is located in the exact position as when it was removed (or pointing just beyond the 11 o'clock and 5 o'clock positions when the piston in the No. 1 cylinder is placed at TDC on the compression stroke, if the engine was disturbed since disassembly). Tighten the oil pump attaching screws to 8–10 ft. lbs. (11–15 Nm).

42. Install the distributor in the correct position.

43. Install the crankshaft pulley and bolt. Tighten the bolt to 87–116 ft. lbs. (118–157 Nm).

44. Install the A/C compressor idler pulley.

45. Install the crankshaft pulley, oil strainer and pan, A/C compressor idler pulley and oil pump.

46. Align the matchmark and position the upper chain on the idler sprocket.

47. Remove the pin from the tensioner.

48. Apply silicone sealant to the camshaft sprocket cover and install the cover. Tighten the vertical (large) bolts to 12–14 ft. lbs. (16–19 Nm) and the horizontal (small) bolts to 56–66 inch lbs. (6.4–7.5 Nm).

49. Align the matchmarks and install the chain on the camshaft sprockets.

50. Install the sprockets on the camshafts. Tighten the bolts to 123–130 ft. lbs. (167–177 Nm).

51. Install the rocker arm cover.

52. Install the distributor.

53. Install the fuel rail/injector assembly and connect them to the harness.

54. Install the radiator shroud, fan with coupling and air intake duct.

55. Install the exhaust manifold, exhaust pipe and heat shield.

56. Reconnect the fuel lines, vacuum hoses and wiring harness.

57. Refill the engine with coolant and connect the negative battery cable.

FUEL SYSTEM

Fuel System Service Precautions

Safety is the most important factor when performing not only fuel system maintenance but any type of maintenance. Failure to conduct maintenance and repairs in a safe manner may result in serious personal injury or death. Maintenance and testing of the vehicle's fuel system components can be accomplished safely and effectively by adhering to the following rules and guidelines.

• To avoid the possibility of fire and personal injury, always disconnect the negative battery cable unless the repair or test procedure requires that battery voltage be applied.

• Always relieve the fuel system pressure prior to disconnecting any fuel system component (injector, fuel rail, pressure regulator, etc.), fitting or fuel line connection. Exercise extreme caution whenever relieving fuel system pressure to avoid exposing skin, face and eyes to fuel spray. Please be advised that fuel under pressure may penetrate the skin or any part of the body that it contacts.

• Always place a shop towel or cloth around the fitting or connection prior to loosening to absorb any excess fuel due to spillage. Ensure that all fuel spillage (should it occur) is quickly removed from engine surfaces. Ensure that all fuel soaked cloths or towels are deposited into a suitable waste container.

• Always keep a dry chemical (Class B) fire extinguisher near the work area.

• Do not allow fuel spray or fuel vapors to come into contact with a spark or open flame.

• Always use a back-up wrench when loosening and tightening fuel line connection fittings. This will prevent unnecessary stress and torsion to fuel line piping. Always follow the proper torque specifications.

• Always replace worn fuel fitting O-rings with new. Do not substitute fuel hose or equivalent, where fuel pipe is installed.

Fuel System Pressure

RELIEVING

1. Remove the fuel pump fuse from the panel in the vehicle.
2. Start the engine and let it run until it stalls.
3. After the engine stalls, try to restart it; if the engine will not start, the fuel pressure has been released.
4. Turn the ignition switch **OFF**. Reinstall the fuel pump fuse.

➡**Do not crank the engine or turn the ignition switch ON after the fuel pump fuse has been reinstalled, or the fuel pressure will be re-established.**

Fuel Filter

REMOVAL & INSTALLATION

✳✳ CAUTION

The fuel injection system remains under pressure after the engine has been turned OFF. Properly relieve fuel pressure before disconnecting any fuel lines. Failure to do so may result in fire or personal injury.

1. Relieve the fuel system pressure.
2. Locate the fuel filter.
3. Loosen the hose clamps at the fuel inlet and outlet lines and slide each line off the filter nipples.
4. Remove the fuel filter from the mounting bracket.
   To install:
5. Mount the filter in the mounting bracket and secure.

➡**Always use a high pressure-type fuel filter.**

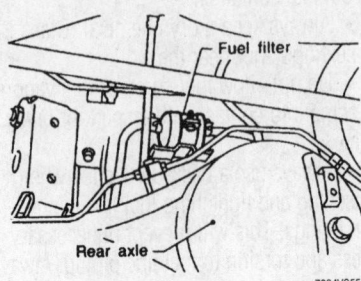

The fuel filter is located near the rear axle—Pathfinder and QX4 with the 3.3L engine

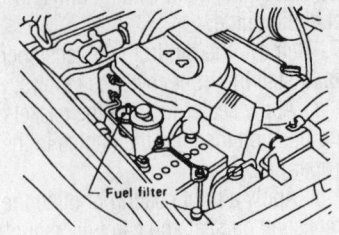

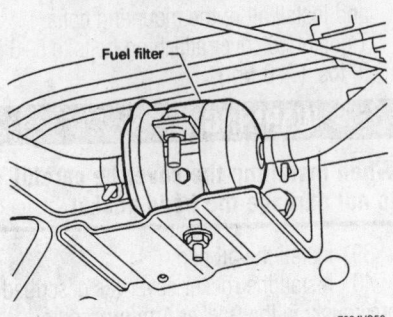

The fuel filter is located in the engine compartment—1995–97 Pick-up vehicles with the 2.4L engine

6. Connect the fuel line hoses and secure them with new hose clamps.
7. Start the engine and check for leaks.

Fuel Pump

REMOVAL & INSTALLATION

Pick-Up and Frontier

✳✳ CAUTION

The fuel injection system remains under pressure after the engine has been turned OFF. Properly relieve fuel pressure before disconnecting any fuel lines. Failure to do so may result in fire or personal injury.

The fuel pump, equipped with a damper, is located in the fuel tank.

1. Properly relieve the fuel system pressure and disconnect the negative battery cable.
2. Disconnect the fuel gauge electrical connector.
3. Disconnect the fuel outlet and the return hoses.
4. Remove the fuel tank assembly.
5. Remove the ring retaining bolts and the O-ring, then lift the fuel pump assembly from the fuel tank. Plug the opening with a clean rag to prevent dirt from entering the system.

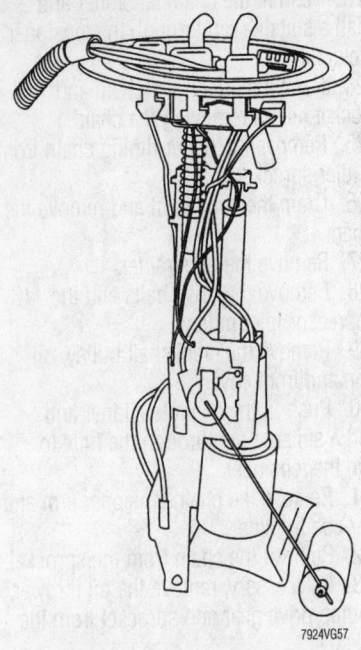

Fuel pump assembly—1998 Pick-up vehicle shown

➡**When removing or installing the fuel pump assembly, be careful not to damage or deform it and always install a new O-ring.**

To install:
6. Remove the rag and install the fuel pump assembly in the fuel tank; remember to use a **new** O-ring. Install the ring retaining bolts.
7. Install the fuel tank assembly.
8. Reconnect the fuel lines and the electrical connections
9. Reconnect the negative battery cable.
10. Start the engine and check for fuel leaks.

➡**On some models, the Check Engine Light will stay on after installation is completed. The memory code in the control unit must be erased. To erase the code, disconnect the battery cable for 10 seconds, then reconnect it.**

1995 Pathfinder

The fuel pump, equipped with a damper, is located in the fuel tank.

1. Properly relieve the fuel system pressure and disconnect the negative battery cable.
2. Disconnect the fuel gauge electrical connector and remove the fuel tank inspection cover.

3. Disconnect the fuel outlet and the return hoses.

4. Support and lower the fuel tank. Only lower the tank enough to remove the fuel pump.

5. Remove the ring retaining bolts and the O-ring, then lift the fuel pump assembly from the fuel tank. Plug the opening with a clean rag to prevent dirt from entering the system.

➡When removing or installing the fuel pump assembly, be careful not to damage or deform it and always install a new O-ring.

To install:

6. Remove the rag and install the fuel pump assembly in the fuel tank; remember to use a **new** O-ring.

7. Install the ring retaining bolts.

8. Connect the fuel lines and the electrical connections.

9. Raise tank up and secure it in vehicle.

10. Install the inspection cover.

11. Connect the negative battery cable.

12. Start engine and check for fuel leaks.

➡On some models, the Check Engine Light will stay on after installation is completed. The memory code in the control unit must be erased. To erase the code, disconnect the negative battery cable for 10 seconds, then reconnect it.

1996–99 Pathfinder and QX4

1. Relieve the fuel system pressure.

2. Disconnect the negative battery cable.

3. Remove the inspection hole cover located behind the rear seat.

4. Disconnect the harness connectors and the fuel tubes from the upper plate of the fuel gauge.

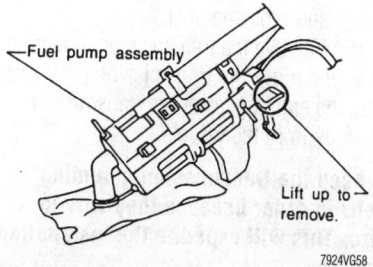

Remove the fuel pump with bracket while lifting the pawl of the pump bracket upward —1996–99 Pathfinder and QX4

5. Remove the fuel gauge retainer and fuel gauge.

6. Remove the fuel pump with bracket while lifting the pawl of the pump bracket upward.

7. Separate the pump from the sender if necessary.

To install:

8. Connect the pump to the sender if removed.

9. Install the fuel pump assembly in the tank.

10. Install the fuel gauge retainer.

11. Connect the harness connector and the fuel tubes.

12. Install the inspection hole cover.

13. Connect the negative battery cable and check for leaks.

DRIVE TRAIN

Manual Transmission Assembly

REMOVAL & INSTALLATION

Except 1996–99 Pathfinder

2WD MODELS

1. Disconnect the negative battery cable.

2. Remove the shifter knob and boot, then remove the snapring to lift the shift lever out of the transmission. Stuff a rag in the opening to keep dirt out of the transmission.

3. Raise and safely support the vehicle. If equipped, remove the splash pan or skid plate.

4. If equipped with a V6 engine, disconnect the exhaust pipe from the manifolds and from the catalytic converter. Remove the pipe.

5. Remove the starter and drain the oil from the transmission.

6. If equipped, remove the crankshaft

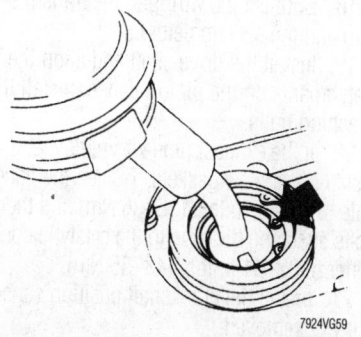

Remove the snapring to release the shifter from the transmission—2WD models

position sensor from the upper side of the transmission.

7. Matchmark the driveshaft flange at the rear pinion flange and remove the driveshaft. Plug the extension housing opening to prevent dirt from getting in.

8. Disconnect the speedometer cable and the wiring from the transmission.

9. Without disconnecting the hydraulic hose, remove the clutch slave cylinder from the transmission and secure it aside.

10. Support the engine by placing a jack under the oil pan, do not place the jack under the drain plug.

11. Using an appropriate transmission jack, properly support the transmission and remove the transmission mount and crossmember.

12. Remove the transmission mounting bolts and move the transmission back away from the engine. Lower the transmission carefully from the vehicle.

➡Keep the transmission mounting bolts in order because they vary in size. This will expedite the installation process.

To install:

13. Lightly grease the input shaft splines. Fit the transmission into place and start all the transmission mounting bolts. Be sure the input shaft fits properly into the clutch disc and pilot bearing. Tighten the transaxle bolts to specification.

14. Install the crossmember and tighten the crossmember mounting bolts to 30–38 ft. lbs. (41–52 Nm).

15. Install the transmission mount and tighten the nuts to 30–38 ft. lbs. (41–52 Nm).

16. Remove the transmission jack from the transmission and the jack from under the oil pan.

17. Install the clutch slave cylinder and tighten the mounting bolts to 22–30 ft. lbs. (30–40 Nm).

18. Connect the wiring to the transmission and the speedometer cable.

19. Install the drive shaft and align the matchmarks on the pinion flange. Install the attaching bolts.

20. If the exhaust pipe was removed, install it with new gaskets. Tighten the flange nuts to 20–27 ft. lbs. (26–36 Nm) and the bolts attaching the pipe to the catalytic converter to 32–41 ft. lbs. (43–55 Nm).

21. Install the crankshaft position sensor if it was removed.

22. Install the starter.

23. If removed, install the splash pan or skid plate.

24. Refill the transmission with oil and lower the vehicle.

25. Install the shift lever into the transmission and install the snapring. Install the shifter boot and the shifter knob.

26. Connect the negative battery cable.

4WD MODELS

1. Disconnect the negative battery cable.

2. Remove the shifter knob and boot and remove the snapring to lift the shift lever out of the transmission. Stuff a rag in the opening to keep dirt out of the transmission.

3. Raise and safely support the vehicle. If equipped, remove the splash pan or skid plate.

4. Remove the starter.

5. Remove the crankshaft position sensor from the upper side of the transmission.

6. Drain the oil from the transmission and transfer case.

7. If equipped with a V6 engine, disconnect the exhaust pipe from the manifolds and from the catalytic converter, then remove the pipe.

8. Matchmark the rear driveshaft flange at the rear pinion flange and remove the driveshaft. Plug the extension housing opening to prevent dirt from getting in.

9. Matchmark both front driveshaft flanges so the shaft can be installed in the same position. Remove the driveshaft.

10. Disconnect the speedometer cable from the transfer case and the wiring from the transmission.

11. Without disconnecting the hydraulic hose, remove the clutch slave cylinder from the transmission and secure it aside.

12. Remove the torsion bars from the vehicle.

13. Remove the transfer case shift lever.

14. Support the engine by placing a jack under the oil pan, do not place the jack under the drain plug.

15. Using an appropriate transmission jack, properly support the transmission and

transfer case, then remove the transmission mount and crossmember.

16. Remove the transmission mounting bolts and move the transmission back away from the engine. Lower the transmission carefully from the vehicle.

➡ **Keep the transmission mounting bolts in order because they vary in size. This will expedite the installation process.**

To install:

➡ **Apply a silicone sealant to the engine block or rear plate to seal the engine to the transmission.**

17. Lightly grease the input shaft splines, then fit the transmission into place and start all the transmission mounting bolts. Be sure the input shaft fits properly into the clutch disc and pilot bearing. Tighten the transaxle bolts to specification.

18. Install the crossmember and tighten the crossmember mounting bolts to 30–38 ft. lbs. (41–52 Nm).

19. Install the rear transmission mount bolts, then tighten the mount bolts to 30–38 ft. lbs. (41–52 Nm).

20. Remove the transmission jack from the transmission and the jack from under the oil pan.

21. Install the transfer case shift lever.

22. Install the torsion bars in their original locations. Be sure the splines are in their original position and set the adjustment to its original position.

23. Install the clutch slave cylinder and tighten the mounting bolts to 22–30 ft. lbs. (30–40 Nm).

24. Connect the wiring to the transmission and the speedometer cable to the transfer case.

25. Install the driveshafts, be sure to align the matchmarks.

26. If the exhaust pipe was removed, install it with new gaskets. Tighten the flange nuts to 20–27 ft. lbs. (26–36 Nm) and the bolts attaching the pipe to the catalytic converter to 32–41 ft. lbs. (43–55 Nm).

27. Install the crankshaft position sensor if it was removed.

28. Install the starter.

29. Refill the transmission and transfer case.

30. If removed, install the splash pan or skid plate, then lower the vehicle.

31. Install the shift lever into the transmission and install the snapring. Install the shifter boot and the shifter knob.

32. Connect the negative battery cable.

1996–99 Pathfinder and All QX4

2WD MODELS

1. Disconnect the negative battery cable.

2. Remove the shifter knob and boot, then remove the snapring to lift the shift lever out of the transmission. Place a rag in the opening to keep dirt out of the transmission.

3. Raise and safely support the vehicle. If equipped, remove the splash pan or skid plate.

4. Remove the crankshaft position sensor from the upper side of the transmission.

5. Matchmark the driveshaft flange at the rear pinion flange and remove the driveshaft. Install a plug to the extension housing opening.

6. Without disconnecting the hydraulic hose, remove the clutch slave cylinder from the transmission and secure it aside.

7. Remove the exhaust tube mounting and brackets.

8. Disconnect the vehicle speed sensor, back-up lamp and neutral position switch harness connectors.

9. Remove the starter motor assembly.

10. Support the engine by placing a jack with a block of wood under the oil pan. Do not place the jack under the drain plug.

11. Using an appropriate transmission jack, properly support the transmission and remove the transmission mount and crossmember.

12. Remove the transmission mounting bolts and move the transmission back away from the engine. Lower the transmission carefully from the vehicle.

➡ **Keep the transmission mounting bolts in order because they vary in size. This will expedite the installation process.**

To install:

13. Lightly grease the input shaft splines. Fit the transmission into place and start all the transmission mounting bolts. Be sure the input shaft fits properly into the clutch disc and pilot bearing. Tighten the transaxle bolts as follows:

   a. Bolts No. 1 and No. 2 tighten to 29–39 ft. lbs. (39–49 Nm).

   b. Bolts No. 3 tighten to 22–29 ft. lbs. (29–39 Nm).

   c. Gusset-to-engine bolts tighten to 22–29 ft. lbs. (29–39 Nm).

14. Install the crossmember and tighten the crossmember-to-frame mounting bolts to 57–77 ft. lbs. (77–105 Nm).

15. Install the transmission mount and tighten the nuts to 32–41 ft. lbs. (43–55 Nm).

◖ : N·m (kg-m, in-lb)

◗ : N·m (kg-m, ft-lb)

▦ 1 : Fill multi-purpose grease up.

▦ 2 : Apply multi-purpose grease.

*1 : Securely bend pawls during assembly.
     Be careful not to damage boot.

*2 : Do not touch boot with a sharp-pointed
     or a hard tool as it breaks easily.

Control knob

Transfer control lever

Pin

Bush

Snap ring ⊗

Bush

Snap ring ⊗

Boot band *1 ⊗

Boot *2 ⊗

Boot band *1 ⊗

Spring

Snap ring ⊗

▦ 1

Control lever (holder assembly)

▦ 2

Control lever bracket

◗ 16 - 21 (1.6 - 2.1, 12 - 15)

◖ 8 - 11
(0.8 - 1.1, 69 - 95)

Guide plate

Outer shift lever

Washer

Ball joint linkage

A

Control lever
(holder assembly)

◗ 15.7 - 20.6
(1.6 - 2.1,
11.6 - 15.2)

Ball joint linkage

Outer shift lever

Exploded view of the transfer case shifter lever and related components—1996-99 Pathfinder with 4WD

7924VG60

16. Remove the transmission jack from the transmission and the jack from under the oil pan.

17. Install the starter motor assembly.

18. Connect the vehicle speed sensor, back-up lamp and neutral position switch harness connectors.

19. Install the exhaust tube mounting and brackets.

20. Install the clutch slave cylinder and tighten the mounting bolts to 22–30 ft. lbs. (30–40 Nm).

21. Install the drive shaft and align the matchmarks on the pinion flange. Install the attaching bolts.

22. Install the crankshaft position sensor.

23. If removed, install the splash pan or skid plate.

24. Refill the transmission with oil and lower the vehicle.

25. Install the shift lever into the transmission and install the snapring. Install the shifter boot and the shifter knob.

26. Connect the negative battery cable and road test the vehicle.

4WD MODELS

1. Disconnect the negative battery cable.

2. Remove the shifter knob and boot, then remove the snapring to lift the shift lever out of the transmission. Place a rag in the opening to keep dirt out of the transmission.

3. Remove the control knob from the transfer case control lever.

4. Raise and safely support the vehicle. If equipped, remove the splash pan or skid plate.

5. Remove the crankshaft position sensor from the upper side of the transmission.

6. Matchmark the driveshaft flange at the front and rear pinion flanges and remove the driveshafts. Install a plug to the extension housing opening of the transmission.

7. Without disconnecting the hydraulic hose, remove the clutch slave cylinder from the transmission and secure it aside.

8. Remove the exhaust system front and rear pipes.

9. Disconnect the vehicle speed sensor, back-up lamp 4WD switch and neutral position switch harness connectors.

10. Remove the starter motor assembly.

11. Disconnect the transfer control lever linkage and remove the control lever mounting bracket.

12. Remove the control lever from the vehicle.

13. Support the engine by placing a jack with a block of wood under the oil pan. Do not place the jack under the drain plug.

14. Using an appropriate transmission jack, properly support the transmission/transfer case and remove the transmission/transfer case mount and crossmember.

15. Remove the transmission/transfer case mounting bolts and move the unit back away from the engine. Lower the transmission/transfer case carefully from the vehicle.

➡**Keep the transmission/transfer case mounting bolts in order because they vary in size. This will expedite the installation process.**

To install:
➡**Apply sealant to the rear of the engine block prior to the installation of the transmission/transfer case.**

16. Lightly grease the input shaft splines. Fit the transmission/transfer case into place and start all the transmission mounting bolts. Be sure the input shaft fits properly into the clutch disc and pilot bearing. Tighten the transaxle bolts as follows:

   a. Bolts No. 1 and No. 2 to 29–39 ft. lbs. (39–49 Nm).

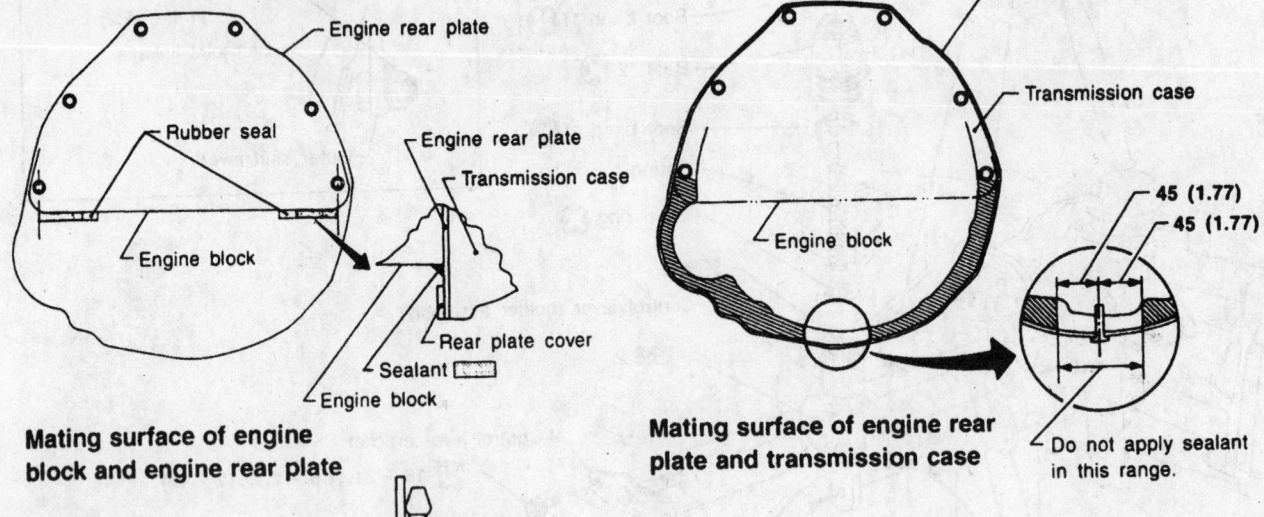

Mating surface of engine block and engine rear plate

Mating surface of engine rear plate and transmission case

45 (1.77)
45 (1.77)

Do not apply sealant in this range.

▨ : Apply recommended sealant (Nissan genuine part: KP510-00150) or equivalent.

▧ : Apply recommended sealant (Nissan genuine part: KP610-00250) or equivalent.

Unit: mm (in)

7924VG61

Apply sealant to the indicated areas on the rear of the engine block—3.3L engine

b. Bolt No. 3 to 22–29 ft. lbs. (29–39 Nm).

c. Gusset-to-engine bolts to 22–29 ft. lbs. (29–39 Nm).

17. Install the crossmember and tighten the crossmember-to-frame mounting bolts to 57–77 ft. lbs. (77–105 Nm).

18. Install the transmission mount and tighten the nuts to 32–41 ft. lbs. (43–55 Nm).

19. Remove the transmission jack from the transmission and the jack from under the oil pan.

20. Install the transfer case control lever and tighten the lever mounting bracket to 12–15 ft. lbs. (16–20 Nm).

21. Install the transfer case control lever linkage and tighten the mounting nuts to 12–15 ft. lbs. (16–20 Nm).

22. Install the starter motor assembly.

23. Connect the vehicle speed sensor, back-up lamp, 4WD switch and neutral position switch harness connectors.

24. Install and connect the front and rear exhaust pipes and the mounting brackets.

25. Install the clutch slave cylinder and tighten the mounting bolts to 22–30 ft. lbs. (30–40 Nm).

26. Install the driveshafts and align the matchmarks on the pinion flanges. Install the attaching bolts.

27. Install the crankshaft position sensor.

28. If removed, install the splash pan or skid plate.

29. Refill the transmission with oil and lower the vehicle.

30. Install the shift lever into the transmission and install the snapring. Install the shifter boot and the shifter knob.

31. Install the transfer case shifter knob.

32. Connect the negative battery cable and road test the vehicle.

Automatic Transmission Assembly

REMOVAL & INSTALLATION

Except 1996–99 Pathfinder

2WD VEHICLES

1. Disconnect the negative battery cable.

2. Raise and safely support the vehicle. If equipped, remove the splash pan or skid plate.

3. If equipped with a V6 engine, disconnect the exhaust pipe from the manifolds and from the catalytic converter, then remove the pipe.

4. Remove the dipstick tube and disconnect the oil cooler lines from the transmission.

5. Matchmark the driveshaft flange at the rear pinion flange and remove the driveshaft. Plug the extension housing opening to prevent fluid from leaking out.

6. Disconnect the speedometer cable and the wiring from the transmission.

7. Disconnect the selector lever and throttle cables from the transmission.

8. Remove the starter.

9. If equipped with a V6, remove the bracket securing the transmission to the engine.

10. Remove the torque converter housing dust cover. Matchmark the torque converter with the driveplate for reassembly; these are balanced together at the factory. Remove the torque converter-to-driveplate (flywheel) bolts. Use a wrench on the crankshaft pulley bolt to rotate the crankshaft to expose the hidden torque converter bolts.

11. Using an appropriate transmission jack, properly support the transmission and remove the transmission mount and crossmember.

12. Remove the bolts attaching the transmission to the engine and move the transmission back away from the engine. Secure the torque converter to the transmission to prevent it from dropping.

13. Tilt, then lower the transmission carefully from the vehicle.

To install:

14. Before installing the transmission, perform the following checks:

a. Use a dial indicator to check the driveplate run-out while turning the crankshaft. Maximum allowable run-out is 0.020 in. (0.5mm); if beyond specification, replace the driveplate.

b. Measure and adjust how far the torque converter is recessed into the transmission housing. The distance between the front mounting surface of the transmission and the torque converter-to-driveplate bolt boss should be 1.024 in. (26mm).

15. Install the transmission and the bolts attaching the transmission to the engine. If equipped with a 4 cylinder engine, tighten the transmission attaching bolts to 29–36 ft. lbs. (39–49 Nm). If equipped with a V6 engine, tighten the 1.77 in. (45mm) and the 1.97 in. (50mm) bolts to 29–36 ft. lbs. (39–49 Nm), and tighten the 0.98 in. (25mm) bolts to 22–29 ft. lbs. (29–39 Nm).

16. Install the transmission crossmember and tighten the attaching bolts to 50–64 ft. lbs. (68–87 Nm).

17. If equipped with a 2.4L engine, install the transmission mounts. Tighten the bolts to 30–38 ft. lbs. (41–52 Nm) and the nuts to 50–64 ft. lbs. (68–87 Nm).

18. If equipped with a 3.0L engine, install the transmission mount and tighten the nuts to 30–38 ft. lbs. (41–52 Nm).

19. Remove the transmission jack from the vehicle.

20. Align the matchmarks on the driveplate and torque converter, then install the bolts. Tighten the bolts to 33–43 ft. lbs. (44–59 Nm). Turn the crankshaft after tightening the bolts to be sure there is no binding at the driveplate.

21. If equipped with a 3.0L engine, install the bracket securing the transmission to the engine and tighten the attaching bolts to 22–29 ft. lbs. (29–39 Nm).

22. Install the starter.

23. Connect the selector lever and throttle cables to the transmission.

24. Connect the wiring to the transmission and the speedometer cable.

25. Install the driveshaft and align the matchmarks on the pinion flange. Install the attaching bolts.

26. Connect the oil cooler lines to the transmission and install the dipstick tube.

27. If the exhaust pipe was removed, install it with new gaskets. Tighten the flange nuts to 20–27 ft. lbs. (26–36 Nm) and the bolts attaching the pipe to the catalytic converter to 32–41 ft. lbs. (43–55 Nm).

28. If removed, install the splash pan or skid plate.

29. Lower the vehicle.

30. Connect the negative battery cable.

31. Refill the transmission with fluid and adjust as required.

4WD VEHICLES

1. Disconnect the negative battery cable.

2. Raise and safely support the vehicle. If equipped, remove the splash pan or skid plate.

3. If equipped with a 3.0L engine, disconnect the exhaust pipe from the manifolds and from the catalytic converter, then remove the pipe.

4. Remove the dipstick tube and disconnect the oil cooler lines from the transmission.

5. Matchmark the front and rear driveshaft flanges remove the driveshafts.

6. Remove the transfer case shift linkage.

7. Remove the torsion bars and the second crossmember.

8. Disconnect the speedometer cable from the transfer case and the wiring from the transmission.

9. Disconnect the selector cable and throttle cables from the transmission.

10. Remove the starter.

11. If equipped with a 3.0L engine, remove the bracket securing the transmission to the engine.

12. Remove the torque converter housing dust cover. Matchmark the torque converter with the driveplate for reassembly; these are balanced together at the factory. Remove the torque converter-to-driveplate (flywheel) bolts. Use a wrench on the crankshaft pulley bolt to rotate the crankshaft to expose the hidden torque converter bolts.

13. Using an appropriate transmission jack, properly support the transmission and transfer case and remove the transmission mount and crossmember.

14. Remove the bolts attaching the transmission to the engine and remove the transmission and transfer case from the vehicle.

To install:

15. Before installing the transmission, perform the following checks:

   a. Use a dial indicator to check the driveplate run-out while turning the crankshaft. Maximum allowable run-out is 0.020 in. (0.5mm); if beyond specification, replace the driveplate.

   b. Measure and adjust how far the torque converter is recessed into the transmission housing. The distance between the front mounting surface of the transmission and the torque converter-to-driveplate bolt boss should be at least 1.024 in. (26mm).

16. Install the transmission and the bolts attaching the transmission to the engine. If equipped with a 2.4L engine, tighten the transmission attaching bolts to 29–36 ft. lbs. (39–49 Nm). If equipped with a 3.0L engine, tighten the 1.77 in. (45mm) and the 1.97 in. (50mm) bolts to 29–36 ft. lbs. (39–49 Nm), and tighten the 0.98 in. (25mm) bolts to 22–29 ft. lbs. (29–39 Nm).

17. Install the transmission crossmember. If equipped with a 2.4L engine, tighten the bolts to 50–64 ft. lbs. (68–87 Nm). If equipped with a 3.0L engine, tighten the attaching bolts to 30–38 ft. lbs. (41–52 Nm).

18. If equipped with a 2.4L engine, install the transmission mounts. Tighten the bolts to 30–38 ft. lbs. (41–52 Nm) and the nuts to 50–64 ft. lbs. (68–87 Nm).

19. If equipped with a 3.0L engine, install the transmission mount and tighten the nuts to 30–38 ft. lbs. (41–52 Nm).

20. Remove the transmission jack from the vehicle.

21. Align the matchmarks on the driveplate and torque converter, then install the bolts. Tighten the bolts to 33–43 ft. lbs.

(44–59 Nm). Turn the crankshaft after tightening the bolts to be sure there is no binding at the driveplate.

22. If equipped with a 3.0L engine, install the bracket securing the transmission to the engine and tighten the attaching bolts to 22–29 ft. lbs. (29–39 Nm).

23. Connect the selector cable and throttle cables to the transmission.

24. Connect the speedometer cable to the transfer case and the wiring to the transmission.

25. Install the second crossmember into the vehicle.

26. Install the torsion bars to their original location. Be sure the splines are in their original position and set the adjustment to its original position.

27. Install the transfer case shift lever.

28. Install the front and rear driveshafts, making sure to align the matchmarks.

29. Connect the oil cooler lines to the transmission and install the dipstick tube.

30. If the exhaust pipe was removed, install it with new gaskets. Tighten the flange nuts to 20–27 ft. lbs. (26–36 Nm) and the bolts attaching the pipe to the catalytic converter to 32–41 ft. lbs. (43–55 Nm).

31. If removed, install the splash pan or skid plate.

32. Lower the vehicle.

33. Connect the negative battery cable.

34. Refill the transmission with fluid and adjust as required.

1996–99 Pathfinder

2WD VEHICLES

1. Disconnect the negative battery cable.

2. Raise and safely support the vehicle. If equipped, remove the splash pan or skid plate.

3. Remove the dipstick tube and disconnect the oil cooler lines from the transmission. Be sure to plug the fluid line openings.

4. Matchmark the driveshaft flange at the rear pinion flange and remove the driveshaft. Plug the extension housing opening to prevent fluid from leaking out.

5. Disconnect the speedometer cable and the wiring from the transmission.

6. Disconnect the selector lever and throttle cables from the transmission.

7. Remove the starter.

8. Remove the torque converter housing access cover.

9. Matchmark the torque converter with the driveplate for reassembly; these are balanced together at the factory.

10. Remove the torque converter-to-driveplate (flywheel) bolts. Use a wrench on the crankshaft pulley bolt to rotate the crankshaft to expose the hidden torque converter bolts.

11. Using an appropriate transmission jack, properly support the transmission and remove the transmission mount and crossmember.

➡ **The bolts that secure the transmission are different in length. Be sure to note the proper positioning for reinstallation.**

12. Remove the bolts attaching the transmission to the engine and move the transmission back away from the engine.

✳✳ CAUTION

Secure the torque converter to the transmission to prevent it from falling out.

13. Tilt, then lower the transmission carefully from the vehicle.

To install:

14. Before installing the transmission, perform the following checks:

   a. Use a dial indicator to check the driveplate run-out while turning the crankshaft. Maximum allowable run-out is 0.020 in. (0.5mm); if beyond specification, replace the driveplate.

   b. Measure and adjust how far the torque converter is recessed into the transmission housing. The distance between the front mounting surface of the transmission and the torque converter-to-driveplate bolt boss should be 1.024 in. (26mm).

15. Install the transmission and the bolts attaching the transmission to the engine.

16. Tighten the bolts as follows:

   a. Tighten bolts No. 1 and No. 2 to 29–36 ft. lbs. (39–49 Nm).

   b. Tighten bolts No. 3 to 22–29 ft. lbs. (29–39 Nm).

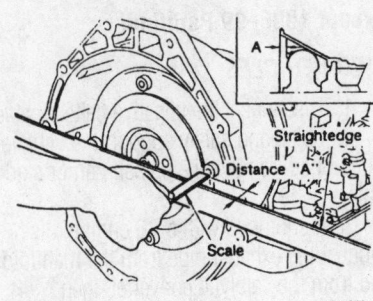

Measuring the depth of the torque converter to be sure it is fully seated— 1996–99 Pathfinder with 2WD

7924VG62

c. Tighten the gusset-to-engine bolts to 22–29 ft. lbs. (29–39 Nm).

17. Install the transmission crossmember and tighten the crossmember-to-frame attaching bolts to 57–77 ft. lbs. (77–105 Nm). Tighten the nuts that secure the transmission-to-crossmember to 32–41 ft. lbs. (43–55 Nm).

18. Remove the transmission jack from the vehicle.

19. Align the matchmarks on the driveplate and torque converter, then install the bolts. Tighten the bolts to 33–43 ft. lbs. (44–59 Nm). Turn the crankshaft after tightening the bolts to be sure there is no binding at the driveplate.

20. Install the torque converter access cover.

21. Install the starter assembly.

22. Connect the selector lever and throttle cables to the transmission.

23. Connect the wiring to the transmission and the speedometer cable.

24. Install the driveshaft and align the matchmarks on the pinion flange. Install the attaching bolts.

25. Connect the oil cooler lines to the transmission and install the dipstick tube.

26. If removed, install the splash pan or skid plate.

27. Lower the vehicle.

28. Connect the negative battery cable.

29. Refill the transmission with fluid and road test.

4WD VEHICLES

1. Disconnect the negative battery cable.

2. Raise and safely support the vehicle. If equipped, remove the splash pan or skid plate.

3. Disconnect and remove the front and rear exhaust pipe from the vehicle.

4. Remove the dipstick tube and disconnect the oil cooler lines from the transmission. Be sure to plug the fluid line openings.

5. Matchmark the driveshaft flange at the rear pinion flange and remove the driveshaft. Plug the extension housing opening to prevent fluid from leaking out.

6. Matchmark the driveshaft flange at the front pinion flange and remove the driveshaft.

7. Disconnect the control linkage from the transfer case.

8. Disconnect the speedometer cable and the wiring from the transmission.

9. Disconnect the selector lever and throttle cables from the transmission.

10. Remove the starter.

11. Remove the torque converter housing access cover.

12. Matchmark the torque converter with the driveplate for reassembly; these are balanced together at the factory.

13. Remove the torque converter-to-driveplate (flywheel) bolts. Use a wrench on the crankshaft pulley bolt to rotate the crankshaft to expose the hidden torque converter bolts.

14. Using an appropriate transmission jack, properly support the transmission and transfer case.

15. Remove the transmission mount and crossmember.

➡**The bolts that secure the transmission are different in length. Be sure to note the proper positioning for reinstallation.**

16. Remove the bolts attaching the transmission and transfer case to the engine and move the transmission back away from the engine.

❋❋ CAUTION

Secure the torque converter to the transmission to prevent it from falling out.

17. Tilt, then lower the transmission and transfer case carefully from the vehicle.

To install:

18. Before installing the transmission, perform the following checks:

a. Use a dial indicator to check the driveplate run-out while turning the crankshaft. Maximum allowable run-out is 0.020 in. (0.5mm); if beyond specification, replace the driveplate.

b. Measure and adjust how far the torque converter is recessed into the transmission housing. The distance between the front mounting surface of the transmission and the torque converter-to-driveplate bolt boss should be 1.024 in. (26mm).

19. Install the transmission and the bolts attaching the transmission to the engine.

20. Tighten the bolts as follows:

a. Tighten bolts No. 1 and No. 2 to 29–36 ft. lbs. (39–49 Nm).

b. Tighten bolts No. 3 to 22–29 ft. lbs. (29–39 Nm).

c. Tighten the gusset-to-engine bolts to 22–29 ft. lbs. (29–39 Nm).

21. Install the transmission crossmember and tighten the crossmember-to-frame attaching bolts to 57–77 ft. lbs. (77–105 Nm). Tighten the nuts that secure the transmission-to-crossmember to 32–41 ft. lbs. (43–55 Nm).

22. Remove the transmission jack from the vehicle.

23. Align the matchmarks on the driveplate and torque converter, then install the bolts. Tighten the bolts to 33–43 ft. lbs. (44–59 Nm). Turn the crankshaft after tightening the bolts to be sure there is no binding at the driveplate.

24. Install the torque converter access cover.

25. Install the starter assembly.

26. Connect the selector lever and throttle cables to the transmission.

27. Connect the wiring to the transmission and the speedometer cable.

28. Connect the control linkage to the transfer case.

29. Install the driveshafts and align the matchmarks on the pinion flanges. Install the attaching bolts.

30. Connect the oil cooler lines to the transmission and install the dipstick tube.

31. Using new gaskets, install the exhaust pipes.

32. If removed, install the splash pan or skid plate.

33. Lower the vehicle.

34. Connect the negative battery cable.

35. Refill the transmission with fluid and road test.

Clutch

REMOVAL & INSTALLATION

1. Disconnect the negative battery cable; raise and safely support the vehicle.

2. Remove the transmission.

3. Using a piece of chalk or a center punch, paint or punch mark the relationship of the clutch assembly to the flywheel so it can be reassembled in the same position from which it is removed.

4. Using a clutch aligning tool, insert it into the clutch disc hub.

5. Loosen the bolts attaching the clutch cover to the flywheel, a turn at a time in an alternating sequence, until the spring tension is relieved to avoid distorting or bending the clutch cover. Remove the clutch assembly.

6. Inspect the flywheel for scoring, roughness or signs of overheating. Light scoring may be cleaned up with emery cloth, but any deep grooves or overheating (blue marks) warrant replacement or resurfacing of the flywheel. If the clutch facings or flywheel are oily, inspect the transmission front cover oil seal, the pilot bushing and engine rear seals, etc. for leakage; replace any leaking seals before replacing the clutch.

7. If the crankshaft pilot bushing is worn, replace it. Install it using a soft hammer. The factory supplied part does not have to be

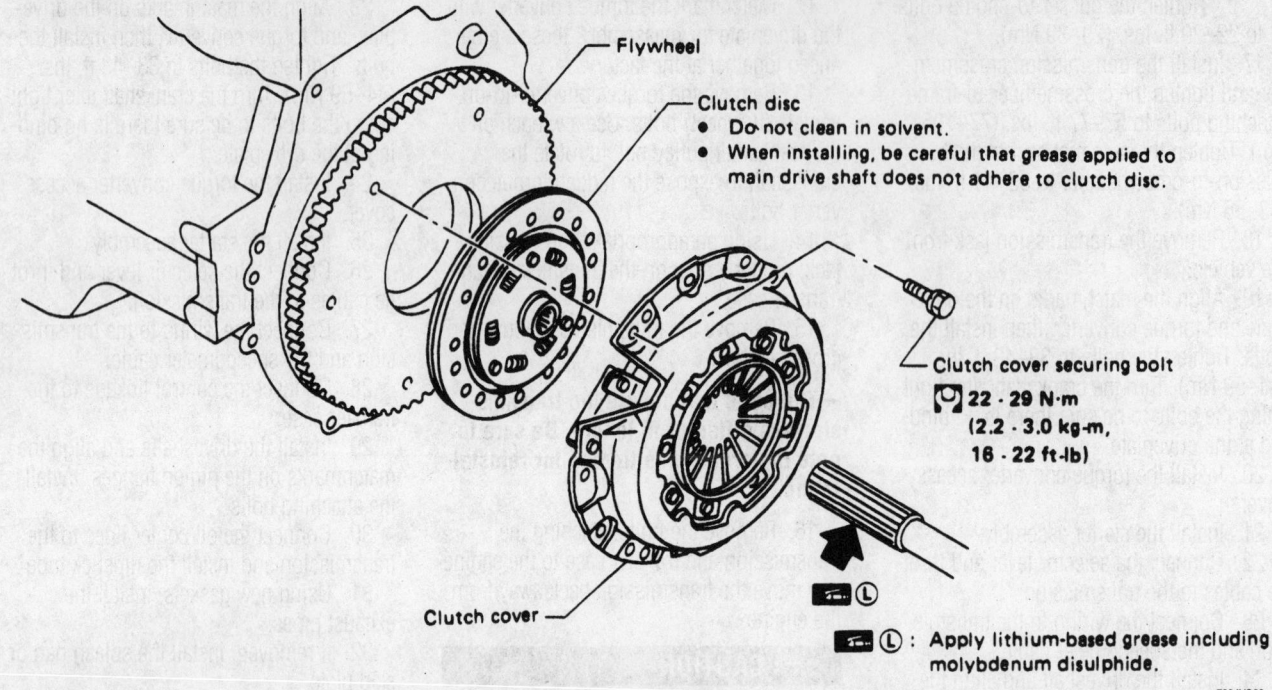

Exploded view of the pressure plate and clutch disc and related components—all models

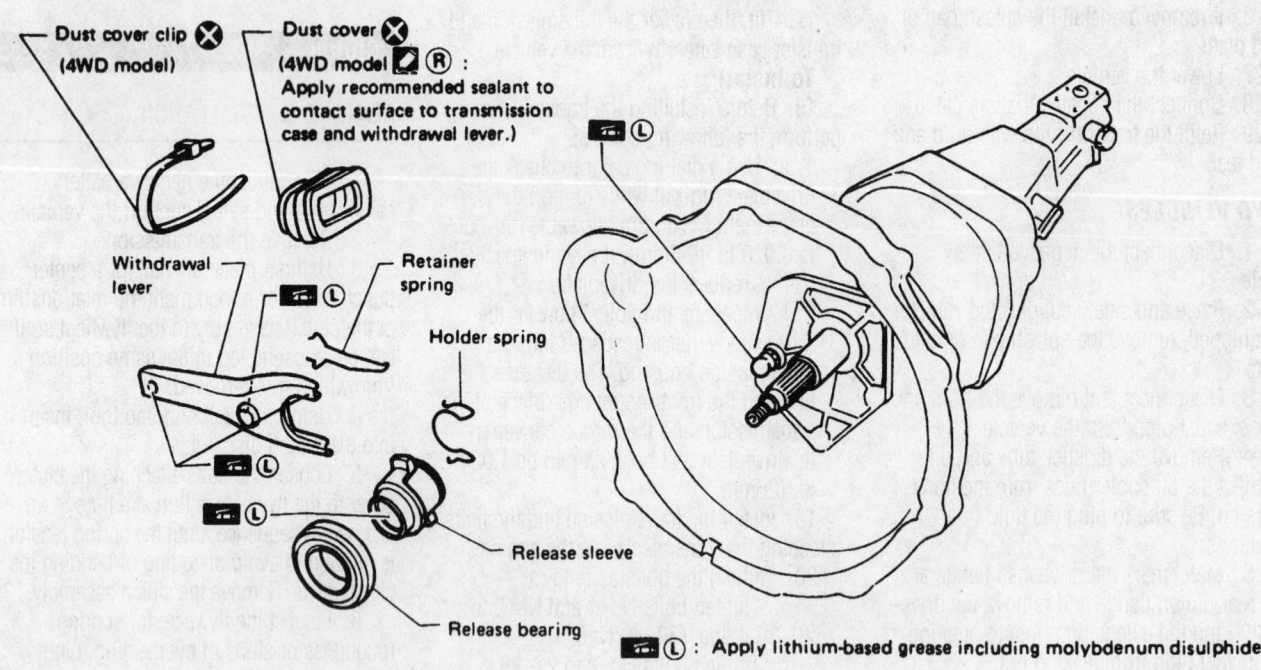

Clutch release mechanism exploded view—all models

oiled, but check the procedure if using an aftermarket part. Inspect the clutch cover for wear or scoring and replace it, if necessary.

➡**The pressure plate and spring cannot be disassembled; replace the clutch cover as an assembly.**

To install:

8. Inspect the clutch release bearing. If it is rough or noisy, it should be replaced. The bearing can be removed from the sleeve with a puller; this requires a press to install the new bearing. After installation, coat the sleeve groove, the release lever contact sur-

faces, the pivot pin/sleeve and the release bearing-to-transmission/transaxle contact surfaces with a light coat of grease. Be careful not to use too much grease, which will run at high temperatures and get onto the clutch facings. Reinstall the release bearing on the lever.

9. Apply a thin coat of grease to the pressure plate wire ring, diaphragm spring, clutch cover grooves and the pressure plate drive bosses.

10. Apply a thin coat of Lubriplate® to the splines in the driven plate. Slide the clutch disc onto the splines and move it back and forth several times. Remove the disc and wipe off the excess lubricant. Be very careful not to get any grease on the clutch facings.

11. Assemble the clutch cover and the clutch plate on the clutch alignment arbor.

12. To complete the installation, align the clutch assembly and flywheel alignment marks and install the bolts. Tighten the bolts 1 or 2 turns at a time in a crisscross pattern to avoid distorting the cover. Tighten the bolts to 16–22 ft. lbs. (22–29 Nm).

13. Install the transmission and adjust the pedal height as necessary.

Hydraulic Clutch System

BLEEDING

1. Check and refill the clutch fluid reservoir to the full mark. During the bleeding process, continue to check and replenish the reservoir to prevent the fluid level from getting lower than ½ full.

2. Connect a clear vinyl hose to the bleeder screw on the slave cylinder. Immerse the other end of the hose in a clear jar ½ filled with brake fluid.

3. Have an assistant pump the clutch pedal several times and hold it down. Loosen the bleeder screw slowly.

4. Tighten the bleeder screw and release the clutch pedal gradually. Repeat this operation until the air bubbles disappear from the brake fluid being expelled out through the bleeder screw.

5. When the air is completely removed, securely tighten the bleeder screw and replace the dust cap.

6. Check and refill the master cylinder reservoir as necessary.

7. Depress the clutch pedal several times to check the operation of the clutch and check for leaks.

Transfer Case Assembly

REMOVAL & INSTALLATION

1. Disconnect the negative battery cable.

2. Raise and safely support the vehicle. If equipped, remove the splash pan or skid plate.

3. Remove the starter. Drain the oil from both the transmission and the transfer case.

4. Matchmark the driveshaft flange at the rear differential pinion flange and at both front driveshaft flanges. Remove both driveshafts.

5. Disconnect the selector lever assembly from the transfer case.

6. The torsion bars must be removed:

   a. Working under the vehicle, measure and record the length of the threads on the torsion bar adjustment.

   b. At the front of the bar, pull the boot back and matchmark the bar to the mounting plate. The spline on the bar must be reinstalled in the same position on the plate.

   c. Remove the locknut and adjustment nut and remove the 3 nuts at the mounting plate to remove each bar. Mark the bars left and right side for proper installation.

7. Using an appropriate transmission jack, properly support the transmission and remove the transmission mount and crossmember.

8. Remove the transfer case-to-transmission bolts and move the unit back away from the transmission.

➡**Mounting bolts are different lengths on manual transmission models.**

To install:

9. Clean the mating surfaces and apply a bead of silicone sealant to the transfer case mounting flange.

10. Carefully fit the case into place and start all the mounting bolts. Tighten the bolts to 23–30 ft. lbs. (31–41 Nm).

11. Install the crossmember and tighten the bolts to 58 ft. lbs. (78 Nm). Install the mount bolts and tighten to 38 ft. lbs. (52 Nm).

12. Install the driveshafts and be sure to align the matchmarks:

   a. On the front driveshaft, tighten the bolts to 33 ft. lbs. (44 Nm).

   b. On two-piece rear driveshafts, tighten the flange bolts to 33 ft. lbs. (44 Nm) and the center bearing bracket bolts to 16 ft. lbs. (22 Nm).

   c. On single-piece rear driveshafts, tighten the flange bolts to 65 ft. lbs. (88 Nm).

13. Install the selector lever assembly.

14. Install the torsion bars in their original location. Be sure the splines are in their original position and set the adjustment to its original position.

15. Install the remaining components and fill the pathfarce case and transmission with oil. Check and adjust front suspension height.

Halfshaft

REMOVAL & INSTALLATION

Except 1996–99 Pathfinder

1. Raise and safely support the front of the vehicle.

2. Remove the wheel.

3. Remove the bolts attaching the axle shaft to the differential while the brake pedal is being depressed.

4. Remove the free running hub assembly with the brake pedal depressed.

5. Remove the brake caliper assembly without disconnecting the hydraulic brake line. Support or hang the brake caliper with a wire to avoid damaging the hose.

6. Remove the brake rotor.

7. Disconnect the tie rod ball joint from the steering knuckle.

8. Support the lower link with a jack and remove the nuts attaching the lower ball joint to the lower link.

9. Remove the upper ball joint attaching bolts.

10. Remove the shock absorber lower attaching bolt.

11. Cover the axle shaft boot with a suitable protector, then remove the axle shaft with the knuckle still attached.

12. Separate the axle shaft from the knuckle by lightly tapping it with a rubber mallet.

To install:

13. Install the bearing spacer onto the axle shaft, making sure that the bearing spacer is facing in the proper direction, then install the axle shaft into the knuckle.

14. Install the axle shaft and steering knuckle assembly.

15. Connect the shock absorber and tighten the bolt on to 87–108 ft. lbs. (118–147 Nm).

16. Connect the upper ball joint and tighten the bolts to 12–15 ft. lbs. (16–21 Nm).

17. Connect the lower ball joint to the lower link and tighten the nuts to 35–45 ft. lbs. (47–61 Nm).

18. Connect the tie rod ball joint to the steering knuckle.

19. Install the brake rotor and caliper.

20. Temporarily install a new snapring on the axle shaft at the same thickness as it was before removal, then measure the axial end-play of the axle shaft with a dial gauge. The axial end-play should be 0.1–0.3mm. Select another snapring if not within specifications.

21. Install the hub, then connect the axle shaft to the differential and tighten the bolts to 25–33 ft. lbs. (34–44 Nm).

22. Install the wheel, then lower the vehicle.

1996–99 Pathfinder

1. Raise and safely support the front of the vehicle.
2. Remove the wheel.

❊❊❊ WARNING

Before removing the axle shaft, disconnect the ABS wheel sensor and move it out of the way. Failure to do so may result in damage to the sensor wires, which would render the sensor inoperative.

3. Remove the brake caliper assembly without disconnecting the hydraulic brake line. Support or hang the brake caliper with a wire to avoid damaging the hose.
4. Remove the hub cap and snapring.
5. Remove the bolts attaching the axle shaft to the final drive.
6. Remove the lower control arm (transverse link) fixing nut and bolts.
7. Separate the axle shaft from the knuckle by lightly tapping it with a copper hammer.

➡ **Cover the CV-boots with a towel to avoid damage when removing the axle shaft.**

To install:

8. Apply multi-purpose grease to the opening of the knuckle.
9. Install a thrust washer onto the end of the axle shaft. Be sure that the thrust washer is facing in the proper direction, then apply multi-purpose grease.
10. Insert the wheel side end of the axle shaft into the knuckle. Then, align and position the other end of the shaft with the final drive.
11. Install the transverse link fixing nuts and bolts.
12. Install the bolts attaching the axle shaft to the final drive.
13. Adjust the shaft's axial end-play by selecting a suitable snapring as follows:
    a. Temporarily install a new snapring (of the same thickness which was previously removed) on the end of the axle shaft.

➡ **Do not reuse the old snapring.**

    b. Attach a dial gauge to the end of the axle shaft.
    c. Measure the axial end-play. If it is not 0.0177 in. (0.45mm) or less, select a thicker snapring.
14. Install the hub cap.
15. Install the brake caliper and connect the ABS wheel sensor.
16. Install the wheel, then lower the vehicle.

Locking Hubs

REMOVAL & INSTALLATION

Auto-Lock Free Running Hub

1. Set the auto-lock free running hub to the free position.
2. Remove the six bolts and the hub assembly.
3. Remove the snap-ring, spindle washer and thrust washer.

To install:
4. Lubricate the axle splines with bearing grease.
5. Install the thrust and spindle washers, then the snap-ring.
6. Install the hub assembly. Tighten the bolts to 18–25 ft. lbs. (25–34 Nm).

Manual Locking Free Running Hub

1. Set the manual-lock free running hub to the free position.
2. Remove the six bolts and the hub assembly.

To install:
3. Lubricate the axle splines with bearing grease.
4. Install the hub assembly. Tighten the bolts to 18–25 ft. lbs. (25–34 Nm).

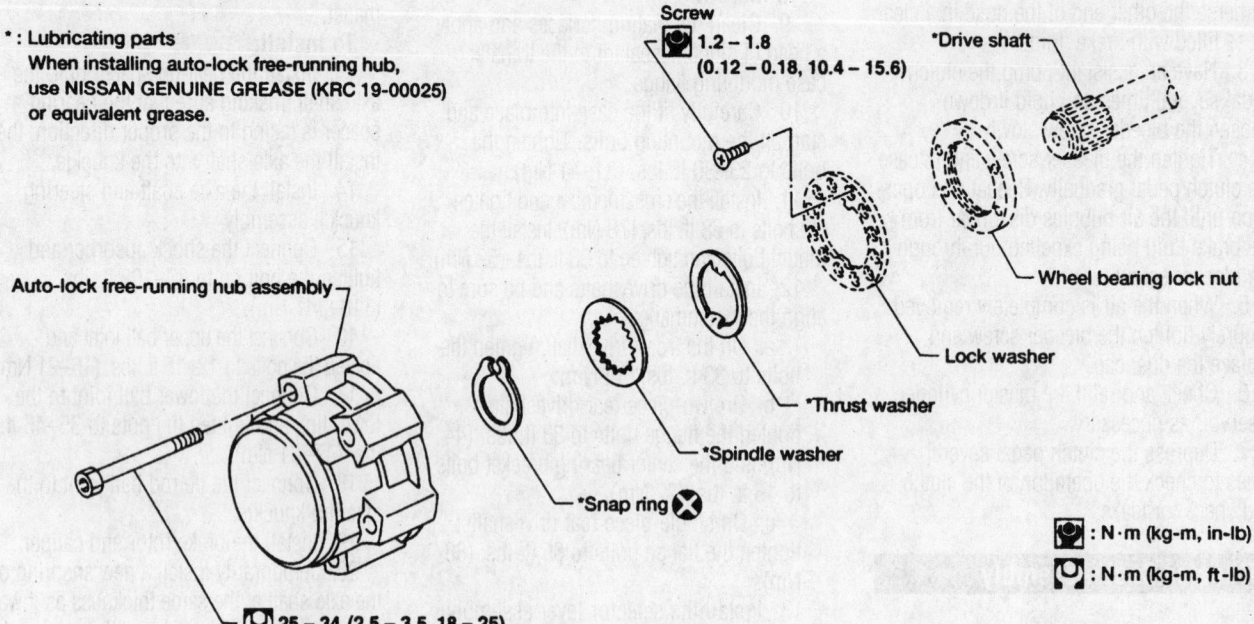

* : Lubricating parts
When installing auto-lock free-running hub, use NISSAN GENUINE GREASE (KRC 19-00025) or equivalent grease.

Screw
1.2 – 1.8
(0.12 – 0.18, 10.4 – 15.6)

*Drive shaft

Auto-lock free-running hub assembly

Wheel bearing lock nut

Lock washer

*Thrust washer

*Spindle washer

*Snap ring ⊗

25 – 34 (2.5 – 3.5, 18 – 25)

: N·m (kg-m, in-lb)

: N·m (kg-m, ft-lb)

7924VG73

Exploded view of the Auto-lock free running hub assembly—Pathfinder and QX4 shown

SEC. 400
*: Lubricating parts
When installing manual-lock free-running hub,
use multi-purpose grease.

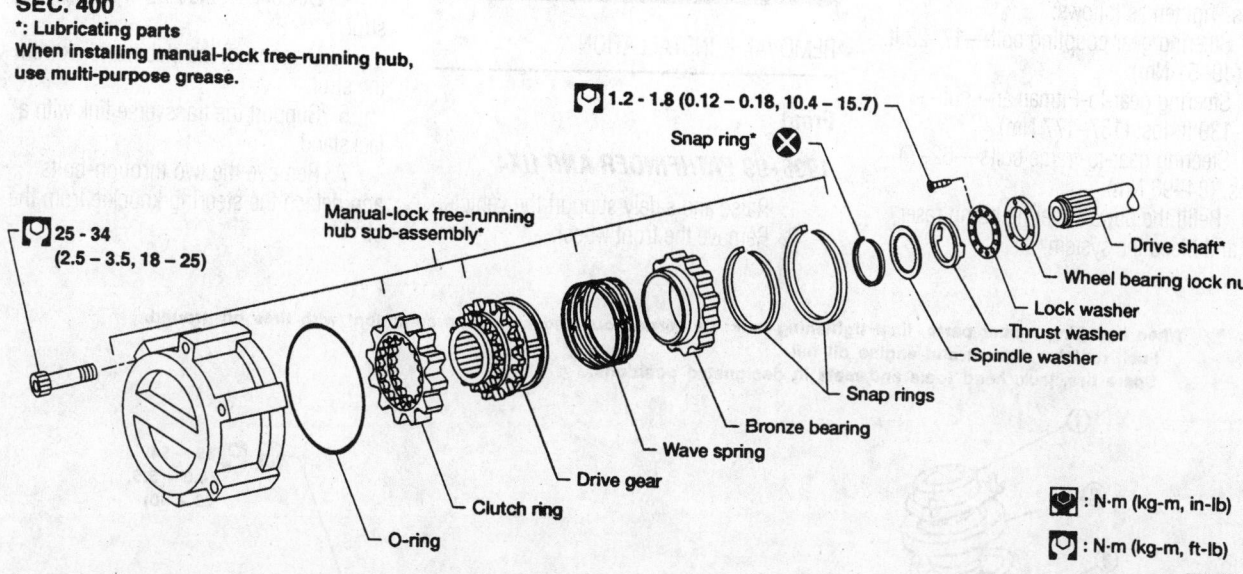

Exploded view of the Manual-lock free running hub assembly—Pathfinder and QX4 shown

STEERING AND SUSPENSION

Air Bag

☀ CAUTION

Some vehicles are equipped with an air bag system, also known as the Supplemental Inflatable Restraint (SIR) or Supplemental Restraint System (SRS). The system must be disabled before performing service on or around system components, steering column, instrument panel components, wiring and sensors. Failure to follow safety and disabling procedures could result in accidental air bag deployment, possible personal injury and unnecessary system repairs.

PRECAUTIONS

Several precautions must be observed when handling the inflator module to avoid accidental deployment and possible personal injury.

• Never carry the inflator module by the wires or connector on the underside of the module.

• When carrying a live inflator module, hold securely with both hands, and ensure that the bag and trim cover are pointed away.

• Place the inflator module on a bench or other surface with the bag and trim cover facing up.

• With the inflator module on the bench, never place anything on or close to the module which may be thrown in the event of an accidental deployment.

DISARMING

☀ CAUTION

To avoid rendering the SRS inoperative, which could lead to personal injury or death in the event of a severe frontal collision, extreme caution must be taken when servicing the electrical related systems.

➡**All SRS electrical wiring harnesses and connectors are covered with YELLOW outer insulation. Do not use electrical test equipment on any circuit related to the SRS (air bag) sensors. When installing SRS components, always install with the arrow marks facing the front of the vehicle.**

To disarm the **SRS** system turn the ignition switch to the **OFF** position. Then, disconnect both battery cables starting with the negative cable first and wait at least 10 minutes after the cables are disconnected. Be sure to insulate the battery terminal ends.

To rearm the **SRS** system, turn the ignition switch to the **OFF** position. Connect

both battery cables starting with the positive cable first.

➡The SRS or air bag system is equipped with a self-diagnostic operation. After turning the ignition key to the ON or START position, the AIR BAG warning lamp will illuminate for 7 seconds. After 7 seconds, the AIR BAG lamp will extinguish if no malfunction is detected. If the AIR BAG lamp does not extinguish after 7 seconds, check the SRS self diagnostic system for a malfunction.

Recirculating Ball Power Steering Gear

REMOVAL & INSTALLATION

1. Raise and safely support the vehicle.
2. Remove the wormshaft-to-rubber coupling bolt.
3. Matchmark the Pitman arm and sector shaft, with the wheels in a straight-ahead position, remove the Pitman arm-to-sector shaft nut.
4. Disconnect the fluid lines from the gear, then cap the lines and openings in the gear.
5. Using the steering gear arm puller, press the gear arm from the steering knuckle.
6. Remove the steering gear-to-chassis bolts and the steering gear from the vehicle.

7. To install, reverse the removal procedures. Tighten as follows:
- Steering gear coupling bolt—17–22 ft. lbs. (49–51 Nm)
- Steering gear-to-Pitman arm nut—101–130 ft. lbs. (137–177 Nm)
- Steering gear-to-frame bolts—62–71 ft. lbs. (84–96 Nm)

8. Refill the power steering pump reservoir and bleed the system.

Strut

REMOVAL & INSTALLATION

Front

1996–99 PATHFINDER AND QX4

1. Raise and safely support the vehicle.
2. Remove the front wheel.
3. Detach the brake tube from the strut.
4. Disconnect the ABS wiring from the strut.
5. Disconnect the stabilizer link from the strut.
6. Support the transverse link with a jackstand.
7. Remove the two through-bolts and detach the steering knuckle from the strut.

When installing rubber parts, final tightening must be carried out under unladen condition* with tires on ground.
Fuel, radiator coolant and engine oil full.
Spare tire, jack, hand tools and mats in designated positions.

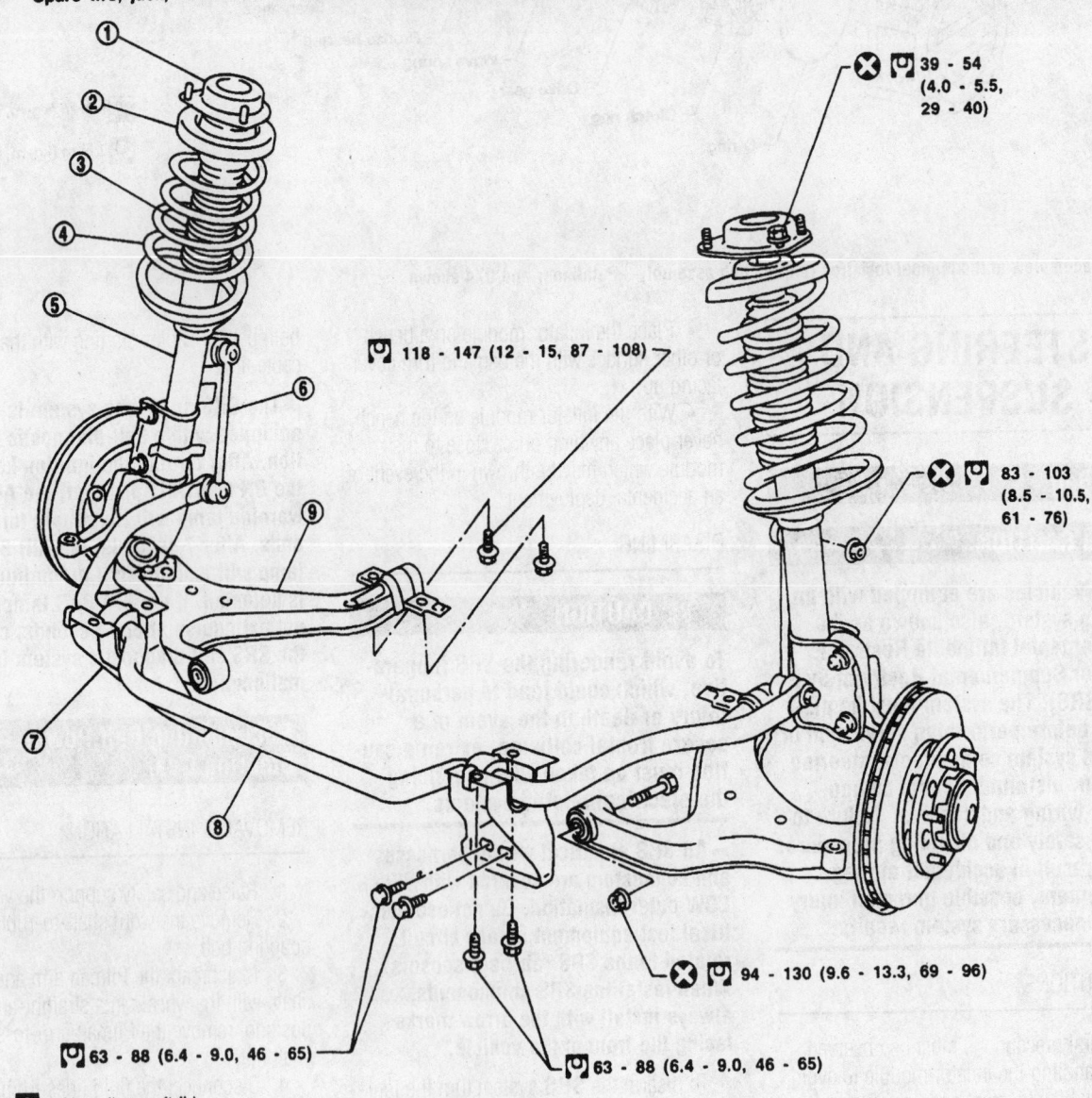

39 - 54 (4.0 - 5.5, 29 - 40)

118 - 147 (12 - 15, 87 - 108)

83 - 103 (8.5 - 10.5, 61 - 76)

94 - 130 (9.6 - 13.3, 69 - 96)

63 - 88 (6.4 - 9.0, 46 - 65)

63 - 88 (6.4 - 9.0, 46 - 65)

: N•m (kg-m, ft-lb)

1. Strut mounting insulator
2. Spring upper seat
3. Bound bumper
4. Coil spring
5. Strut assembly
6. Stabilizer connecting rod
7. Bracket
8. Stabilizer bar
9. Transverse link

7924VG66

Exploded view of the front suspension—1996–99 2WD Pathfinder; 4WD similar

➡Note the positioning of the strut alignment (cutout) mark for reassembly purposes.

8. Support the strut and remove the three upper attaching nuts. Remove the strut from the vehicle.

✳✳ CAUTION

Never loosen the center spring retaining nut until the coil spring is compressed, or serious injury or vehicle damage may occur.

9. Place the strut assembly in a vise with the special holding tool (part # ST35652000) or in a spring compressor.
10. Loosen but do not remove the piston rod locknut.
11. Compress the spring with the spring compressor, then remove the piston rod locknut.

➡Before removing the strut from the coil spring, note the positioning of the strut in relationship to the coil spring for reassembly.

12. Remove the strut mounting insulator bracket, strut mounting bearing and upper spring seat.
13. Remove the strut, leaving the coil spring compressed.
14. Remove the piston boot and rebound bumper from the strut.
To install:
15. Install the rebound bumper and the boot to the strut piston.
16. Install the strut into the coil spring, be sure the strut and spring are properly positioned.
17. Install the upper spring seat, strut mounting bearing, and the strut mounting insulator bracket. Be sure that the cutout on the upper spring seat is facing the inside of the vehicle.
18. Install the piston rod locknut, then remove the spring compressor.
19. Tighten the piston rod locknut to 30–39 ft. lbs. (41–53 Nm).

➡When installing the strut, be sure to position the alignment mark toward the inside of the vehicle.

20. Position the strut to the vehicle and install the 3 upper attaching nuts. Tighten the upper mounting nuts to 29–40 ft. lbs. (39–54 Nm).
21. Connect the steering knuckle to the strut and tighten mounting nuts of the

mounting bolts to 111–122 ft. lbs. (151–165 Nm).
22. Connect the stabilizer link to the strut and tighten the new mounting nut to 61–76 ft. lbs. (83–103 Nm).
23. Connect the brake tube to the strut and connect the ABS wiring to the strut.
24. Bleed the brake system and install the wheel.
25. Perform a front end alignment.

Shock Absorber

REMOVAL & INSTALLATION

Front

EXCEPT 1996–99 PATHFINDER AND QX4

1. Raise and safely support the vehicle. Remove the wheel assembly.
2. While holding the upper stem of the shock absorber, remove the shock absorber attaching nut, washer and rubber bushing.
3. Remove the bolt attaching the shock absorber to the lower control arm and remove the shock absorber from the vehicle.
To install:
4. Install the shock with a new bushing on the upper shock mounting stud.
5. With the shocks upper stud positioned in the chassis, install the lower attaching bolt. Tighten the lower bolt on 2WD vehicles to 65 ft. lbs. (88 Nm). Tighten the lower bolt on 4WD vehicles to 108 ft. lbs. (147 Nm).
6. Install the bushing, washer and attaching nut to the shock stud. Tighten the nut to 16 ft. lbs. (22 Nm).
7. Install the front wheels and lower the vehicle.

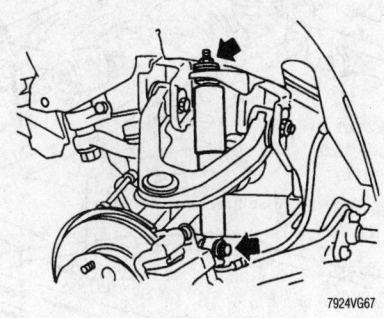

Front shock absorber mounting points

Rear

PATHFINDER AND QX4

➡For rear shock absorber replacement the vehicle chassis and axle weight must be supported separately, requiring the use of two separate lifting devices.

1. Raise and properly support vehicle. Remove both rear wheels.
2. If equipped, disconnect the electrical connector from the shock absorber.
3. Working on one side at a time, jack one side of the rear axle and remove upper and lower attaching nuts. While supporting the rear, remove the shock absorber.
To install:
4. Align shock and install both attaching nuts. Do not tighten nuts completely until the full weight of the vehicle is on the ground.
5. If equipped, connect the electrical connector to the shock absorber.
6. Install wheels, remove the jack from under the rear axle and lower the vehicle.
7. For 1994–95 models, tighten attaching nuts to 22–30 ft. lbs. (30–40 Nm), or for 1996 models tighten the nuts to 36–49 ft. lbs. (49–67 Nm).

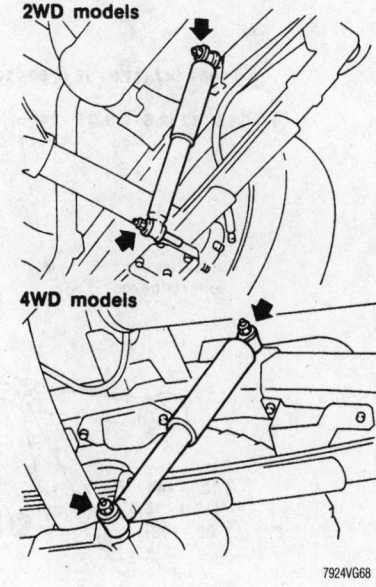

Rear shock absorber mounting points— Pick-up shown

PICK-UP AND FRONTIER

➡For rear shock absorber replacement the vehicle chassis and axle weight must be supported separately, requiring the use of two separate lifting devices.

1. Raise and properly support vehicle. Remove both rear wheels.

2. Working on one side at a time, jack one side of the rear axle and remove upper and lower attaching nuts. While supporting the rear, remove the shock absorber.

To install:

3. Align the shock and install both attaching nuts. Do not tighten the bolts completely until the full weight of the vehicle is on the ground.

4. Install wheels, remove the jack from under the rear axle, and lower vehicle.

5. Tighten attaching nuts to 22–30 ft. lbs. (30–40 Nm) on all models.

Coil Spring

REMOVAL & INSTALLATION

Front

1996–99 PATHFINDER AND QX4

Refer to the strut removal and installation procedure for information on removing the coil spring.

Rear

The spring removal procedure for the 2WD and 4WD models is the same for both vehicles.

➡The coil spring is a load bearing component, therefore the vehicle chassis and axle weight must be supported separately, requiring the use of two separate lifting devices.

1. Raise the vehicle and support it safely.

2. Using the proper equipment, support the weight of the rear axle.

3. Disconnect the Panhard rod at the axle assembly and secure it to the body.

➡Working on one side at a time, complete the following steps for removal and installation of the coil spring.

4. Remove the nut that attaches the shock to the lower mounting bracket.

➡Mark the spring installation direction for reinstallation.

5. Lower the axle assembly until the spring and upper insulator can be removed. Remove the coil spring and lower insulator.

❊❊ WARNING

Do not stretch the brake hose or parking brake cable.

To install:

➡Check the spring for identification marks and properly align.

6. Position the insulator and install the spring.

7. Raise the axle and loosely attach the lower shock absorber mounting nut.

8. Loosely connect the Panhard rod to the mount.

9. Remove the rear axle support and lower the vehicle.

10. With the suspension supporting the weight of the vehicle, tighten the lower shock retaining nut. On 1995 vehicles, tighten the nut to 22–30 ft. lbs. (30–40 Nm), or on 1996 vehicles, tighten the nut to 49–65 ft. lbs. (67–88 Nm).

Upper Ball Joint

REMOVAL & INSTALLATION

Except 1996–99 Pathfinder and QX4

1. Raise and safely support the vehicle.
2. Remove the wheel/tire assembly.
3. Place a floor jack under the steering knuckle and support it.
4. Remove and discard the cotter pin from the ball joint stud, then loosen the nut. Using the ball joint removal tool, press the

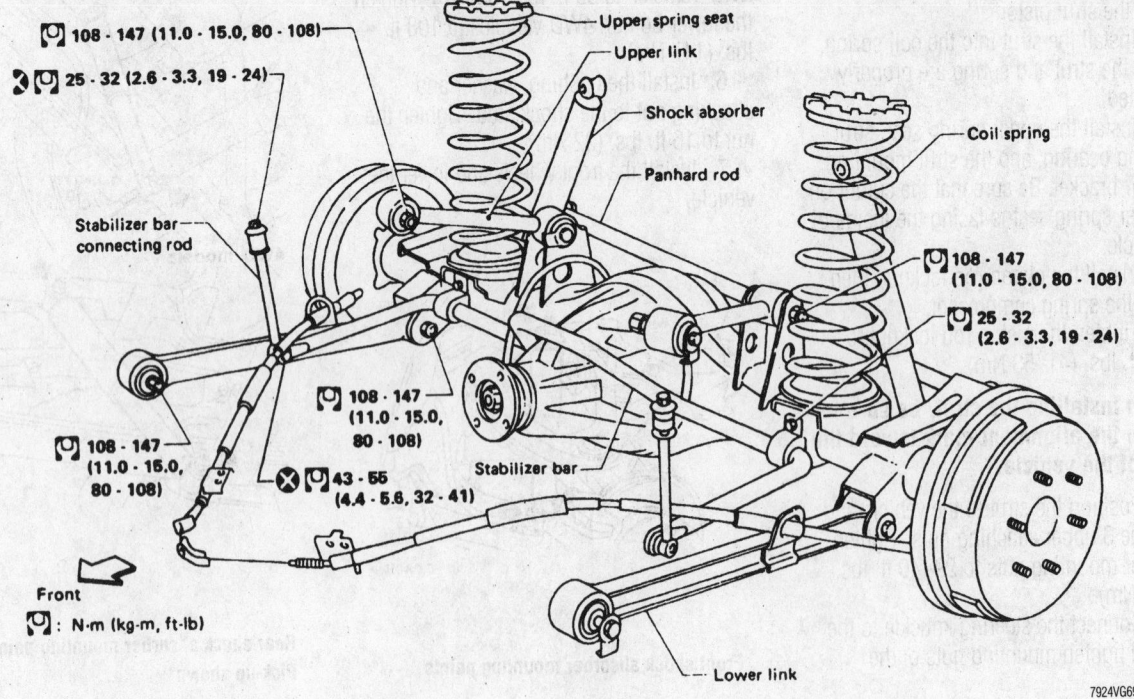

Rear suspension component identification—Pathfinder

upper ball joint from the steering knuckle. Remove the upper ball joint nut.

5. Remove the bolts attaching the upper ball joint to the upper control arm and remove the ball joint from the vehicle.

To install:

6. Install a new ball joint and tighten the bolts attaching the ball joint to the upper control arm to 12–17 ft. lbs. (16–23 Nm).

7. Install the ball joint into the steering knuckle and tighten the nut to 58–108 ft. lbs. (78–147 Nm). Install a new cotter pin to secure the nut.

8. Remove the jack from under the steering knuckle.

9. Install the tire and wheel assembly, then lower the vehicle.

10. Check and/or adjust the ride height and the front end alignment.

Lower Ball Joint

REMOVAL & INSTALLATION

Except 1996–99 Pathfinder and QX4

➡The lower ball joint on 2WD models is integral with the lower control arm. They are removed and replaced as an assembly.

1. Raise and safely support the front of the vehicle under the frame rails.

2. Remove the front wheels.

3. Make matching marks on the anchor arm crossmember when loosening the adjusting nut until there is no tension on the torsion bar. Remove the torsion bar assembly.

4. Unbolt the shock absorber from the lower arm.

5. Remove the ball joint nut.

6. Using a ball joint separator, disconnect the ball joint from the knuckle.

7. Unbolt the ball joint from the lower arm.

To install:

8. Install the ball joint to the lower arm and tighten the nuts to 45 ft. lbs. (61 Nm).

9. Press the ball stud into the knuckle and tighten the nut to 87–141 ft. lbs. (118–191 Nm). Be sure you use a new cotter pin!

10. Connect the lower end of the shock absorber.

11. Install the torsion bar to the lower arm and align the matchmarks made during disassembly.

12. Install the wheels and lower the vehicle.

13. Check the front end alignment.

1996–99 Pathfinder and QX4

1. Raise and safely support the front of the vehicle under the frame rails.

2. Remove the front wheels.

3. Remove the cotter pin from the lower ball joint castle nut.

4. Remove the ball joint nut.

5. Using a ball joint separator, disconnect the ball joint from the knuckle.

6. Unbolt the ball joint from the lower arm and remove the ball joint.

To install:

7. Install the ball joint to the lower control arm and tighten the nuts and bolts to 76–94 ft. lbs. (103–127 Nm).

8. Position the ball joint stud into the knuckle and tighten the nut to 87–123 ft. lbs. (118–167 Nm). Be sure to install a new cotter pin.

9. Install the wheels and lower the vehicle.

10. Check the front end alignment.

Wheel Bearings

ADJUSTMENT

Front

2WD MODELS

➡Adjust the wheel bearing after the bearing has been replaced or the front axle has been reassembled.

1. Raise the front wheel(s) off the ground and support the vehicle safely.

2. Remove the front wheel(s).

3. Using a suitable tool, spread the brake pads to reduce the drag on the brake rotor.

4. Remove the bearing cap and the cotter pin.

5. Tighten the locknut to 25–29 ft. lbs. (34–39 Nm).

6. Turn the hub assembly several times in both directions to seat the bearings.

7. Again tighten the locknut to 25–29 ft. lbs. (34–39 Nm). Turn the locknut counterclockwise 45°. Install the locking cap and a new cotter pin. If the cotter pin cannot be installed, loosen the nut no more than 15° and install the cotter pin.

8. Measure the turning tighten using a spring scale hooked to a hub bolt. Turning torque should be 2.2–6.4 lbs. (9.8–28.4 N) with a new grease seal or 2.2–5.3 lbs. (9.8–23.5 N) with a used grease seal.

9. Repeat the procedure until the correct specification is achieved.

10. Install the wheel(s) and lower the vehicle to the floor.

✳✳ CAUTION

Pump the brakes to reposition the pads against the rotor before moving the vehicle.

4WD MODELS

➡Adjust the wheel bearing after the bearing has been replaced or the front axle has been reassembled.

1. Raise the front wheel(s) off the ground and support the vehicle safely.

2. Remove the front wheel(s).

3. Using a suitable tool, spread the brake pads to reduce the drag on the brake rotor.

4. Remove the bearing cap and the cotter pin.

5. Tighten the locknut with a torque wrench and special tool KV40105400, or equivalent, to 58–72 ft. lbs. (78–98 Nm).

6. Turn the hub assembly several times in both directions to seat the bearings.

7. Loosen the locknut completely.

8. Tighten the locknut with special tool to 0.4–1.1 ft. lbs. (0.5–1.5 Nm).

9. Turn the hub assembly several times in both directions to seat the bearings.

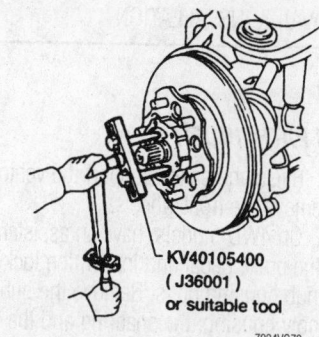

KV40105400
(J36001)
or suitable tool

7924VG70

Tighten the locknut with the special tool— front wheel bearings on 4WD models

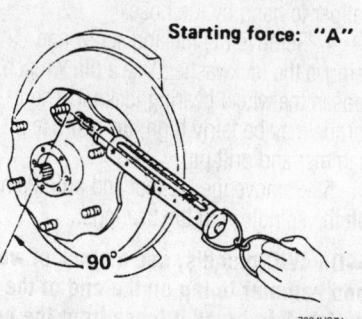

Starting force: "A"

90°

7924VG71

Use a spring scale to measuring the starting force "A"—front wheel bearings on 4WD models

10. Again tighten the locknut with special tool to 0.4–1.1 ft. lbs. (0.5–1.5 Nm).

11. Measure the starting force A using a spring scale hooked to a hub bolt and record the reading.

12. Install the lockwasher by tightening the locknut 15–30°. Turn the hub assembly several times in both directions to seat the bearings.

13. Measure the starting force with a spring scale after the lockwasher has been installed. Record the measurement and call his measurement B.

14. Calculate wheel bearing preload C using the equation C=B-A. Preload C should be 1.59–4.72 lbs. (7.06–20.99 N).

15. Repeat the procedure until the correct specification is achieved.

16. Install the wheel(s) and lower the vehicle to the floor.

✳✳ CAUTION

Pump the brakes to reposition the pads against the rotor before moving the vehicle.

Rear

No adjustment is possible on the rear wheel (axle) bearings. If they become loose or make noise, they must be replaced.

REMOVAL & INSTALLATION

Front

1995 PATHFINDER

1. Raise and safely support the vehicle and remove the front wheels.

2. On 4WD models, have an assistant hold the brake pedal and loosen the locking front hub housing bolts. Remove the hub assembly housing, the snapring and the hub assembly.

3. Without disconnecting the hydraulic line, remove the brake caliper and hang it from the body with wire. Do not allow the caliper to hang by the hose.

4. Remove the locking screw and remove the lockwasher. Use a pin wrench to loosen the wheel bearing locknut. The torque may be fairly high, **do not** use a hammer and drift pin.

5. Remove the locknut and pull the hub off the spindle with the bearings.

➡**On 4WD models, use a block of wood and hammer to tap on the end of the halfshaft to break it loose from the hub spline.**

6. Pry the inner grease seal out to remove the inner bearing. Discard the seal.

7. Clean and inspect the wheel bearings and replace if worn or heat damaged.

➡**Always replace wheel bearings and races together as sets.**

8. Using a hammer and punch, drive the bearing races from the hub.

To install:

9. Carefully install the new inner races using the proper size driver, making sure they are fully seated in the hub.

10. Pack the bearings with new grease and pack grease into the hub. Install the inner bearing and press a new inner seal into the hub.

11. Slip the hub assembly onto the spindle and install the outer bearing. Install the halfshaft and grease the locknut. Thread the locknut into place.

12. To set the bearing pre-load, perform the following steps:

a. Use a pin wrench to tighten the locknut to 58–72 ft. lbs. (78–98 Nm). Turn the hub in both directions several times while tightening the nut.

b. Loosen the locknut, then tighten again to 13 inch lbs. (1.5 Nm).

c. Turn the hub several times and check the nut torque again.

d. Install the lockwasher. When installing the screw, be sure the locknut turns no more than 30 degrees in either direction.

e. When bearing pre-load is properly set, there will be no end-play in the hub and it will require no more than 4.7 lbs. of pull at the wheel stud to turn the hub.

13. On 4WD models, install the locking hub and tighten the hub bolts to 25 ft. lbs. (34 Nm).

14. Install the brake caliper, disc brake pads, and wheel assembly.

15. Lower the vehicle and pump the brakes until the pedal is firm.

1996–99 PATHFINDER AND QX4

1. Raise and safely support the vehicle and remove the front wheels.

2. Remove the ABS sensor from the steering knuckle/spindle.

3. Without disconnecting the hydraulic line, remove the brake caliper and hang it from the body with wire. Do not allow the caliper to hang by the hose.

4. Remove the hub cap using a suitable tool.

5. On 4WD models, remove the snapring and O-ring.

6. On 4WD models, remove the nuts that secure the drive flange and remove the flange assembly.

7. Remove the screw that secures the lockwasher and remove the lockwasher.

8. Use a pin wrench to loosen the wheel bearing locknut. The torque may be fairly high, **do not** use a hammer and drift pin.

9. Remove the locknut and pull the hub off the spindle with the bearings.

➡**On 4WD models, use a block of wood and hammer to tap on the end of the halfshaft to break it loose from the hub spline.**

10. Pry the inner grease seal out to remove the inner bearing. Discard the seal.

11. Clean and inspect the wheel bearings and replace if worn or heat damaged.

➡**Always replace wheel bearings and races together as sets.**

12. Using a hammer and punch, drive the bearing races from the hub.

To install:

13. Carefully install the new inner races using the proper size driver, making sure they are fully seated in the hub.

14. Pack the bearings with new grease and pack grease into the hub. Install the inner bearing and press a new inner seal into the hub.

15. Slip the hub assembly onto the spindle and install the outer bearing.

16. Install the wheel bearing locknut. Use a special tool such as KV40105400 to tighten the locknut. The torque may be fairly high, **do not** use a hammer and drift pin.

17. To set the bearing pre-load, perform the following steps:

a. Use a pin wrench to tighten the locknut to 58–72 ft. lbs. (78–98 Nm). Turn the hub in both directions several times while tightening the nut.

b. Loosen the locknut, then tighten again to 4.3–13 inch lbs. (0.5–1.5 Nm).

c. Turn the hub several times and check the nut torque again.

d. Install the lockwasher. When installing the screw, be sure the locknut turns no more than 30 degrees in either direction.

e. Tighten the lockwasher mounting screw to 11–15 inch lbs. (1.2–1.8 Nm).

f. When bearing pre-load is properly set, there will be no end-play in the hub and it will require no more than 4.7 lbs. of pull at the wheel stud to turn the hub.

18. On 4WD models, install the drive flange and tighten the mounting nuts to 18–26 ft. lbs. (25–35 Nm). Be sure to install new O-rings and pack the flange groves with grease. Also apply grease to the O-rings.

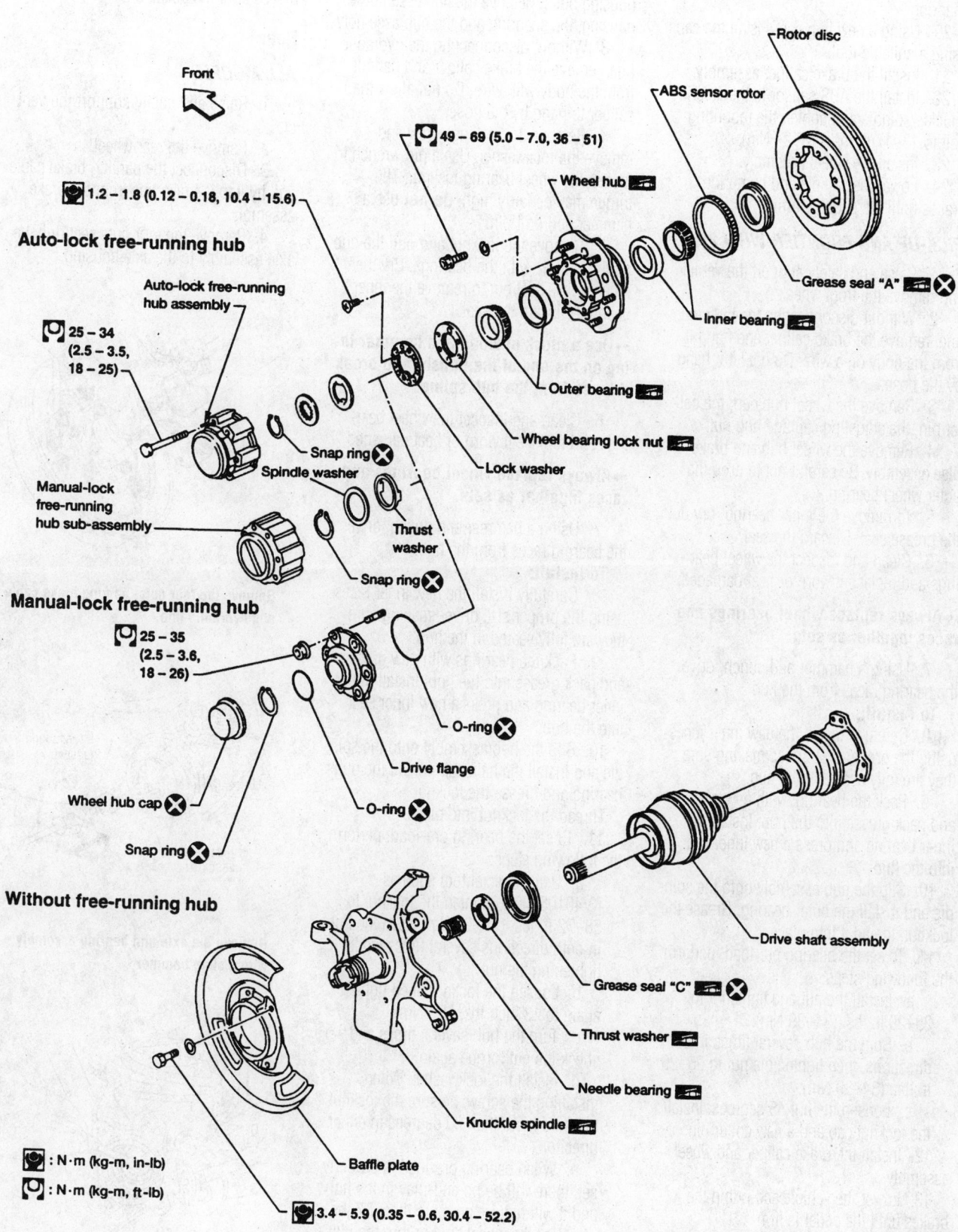

Front

🔩 49 – 69 (5.0 – 7.0, 36 – 51)

Rotor disc

ABS sensor rotor

Wheel hub 🔧

🔩 1.2 – 1.8 (0.12 – 0.18, 10.4 – 15.6)

Auto-lock free-running hub

Auto-lock free-running hub assembly

🔩 25 – 34 (2.5 – 3.5, 18 – 25)

Inner bearing 🔧

Grease seal "A" 🔧 ⊗

Outer bearing 🔧

Wheel bearing lock nut 🔧

Lock washer

Snap ring ⊗

Spindle washer

Thrust washer

Snap ring ⊗

Manual-lock free-running hub sub-assembly

Manual-lock free-running hub

🔩 25 – 35 (2.5 – 3.6, 18 – 26)

O-ring ⊗

Drive flange

Wheel hub cap ⊗

O-ring ⊗

Snap ring ⊗

Drive shaft assembly

Without free-running hub

Grease seal "C" 🔧 ⊗

Thrust washer 🔧

Needle bearing 🔧

Knuckle spindle 🔧

🔧 : N·m (kg-m, in-lb)

🔩 : N·m (kg-m, ft-lb)

Baffle plate

🔩 3.4 – 5.9 (0.35 – 0.6, 30.4 – 52.2)

7924VG72

Exploded view of the wheel bearing assembly on 4WD vehicles—1996–99 Pathfinder and QX4

19. On 4WD models, install the snapring.

20. Using a new hub cap, install the cap using a suitable tool.

21. Install the brake caliper assembly.

22. Install the ABS sensor to the steering knuckle/spindle and tighten the mounting bolt to 13–17 ft. lbs. (18–24 Nm).

23. Install the wheel assembly.

24. Lower the vehicle and pump the brakes until the pedal is firm.

PICK-UP AND FRONTIER WITH 2WD

1. Raise and safely support the vehicle and remove the front wheels.

2. Without disconnecting the hydraulic line, remove the brake caliper and hang it from the body on a wire. Do not let it hang by the hose.

3. Remove the wheel hub cup, the cotter pin, the adjusting cap and hub nut.

4. Remove the wheel hub and brake disc assembly. Be careful not to drop the outer wheel bearing.

5. To remove the inner bearing, pry out the grease seal. Discard the seal.

6. Clean and inspect the wheel bearings and replace if worn or heat damaged.

➡**Always replace wheel bearings and races together as sets.**

7. Using a hammer and punch, drive the bearing races from the hub.

To install:

8. Carefully install the new inner races using the proper size driver, making sure they are fully seated in the hub.

9. Pack the bearings with new grease and pack grease into the hub. Install the inner bearing and press a new inner seal into the hub.

10. Slip the hub assembly onto the spindle and install the outer bearing. Grease the locknut, thread it into place.

11. To set the bearing pre-load, perform the following steps:

a. Install the nut and tighten it to 25–29 ft. lbs. (34–39 Nm).

b. Spin the hub several times in both directions, then tighten the nut to 25–29 ft. lbs. (34–39 Nm).

c. Loosen the nut 45 degrees. Install the locknut cap and a new cotter pin.

12. Install the brake caliper and wheel assembly.

13. Lower the vehicle and pump the brakes until the pedal is firm.

PICK-UP AND FRONTIER WITH 4WD

1. Raise and safely support the vehicle and remove the front wheels.

2. Have an assistant hold the brake pedal and loosen the locking front hub housing bolts. Remove the hub assembly housing, the snapring and the hub assembly.

3. Without disconnecting the hydraulic line, remove the brake caliper and hang it from the body with wire. Do not allow the caliper to hang by the hose.

4. Remove the locking screw and remove the lockwasher. Use a pin wrench to loosen the wheel bearing locknut. The torque may be fairly high, **do not** use a hammer and drift pin.

5. Remove the locknut and pull the hub off the spindle with the bearings. Pry the inner grease seal out to remove the inner bearing. Discard the seal.

➡**Use a block of wood and hammer to tap on the end of the halfshaft to break it loose from the hub spline.**

6. Clean and inspect the wheel bearings and replace if worn or heat damaged.

➡**Always replace wheel bearings and races together as sets.**

7. Using a hammer and punch, drive the bearing races from the hub.

To install:

8. Carefully install the new inner races using the proper size driver, making sure they are fully seated in the hub.

9. Pack the bearings with new grease and pack grease into the hub. Install the inner bearing and press a new inner seal into the hub.

10. Slip the hub assembly onto the spindle and install the halfshaft. Install the outer bearing and grease the locknut. Thread the locknut into place.

11. To set the bearing pre-load, perform the following steps:

a. Use a special tool such as KV40105400 to tighten the locknut to 58–72 ft. lbs. (78–98 Nm). Turn the hub in both directions several times while tightening the nut.

b. Loosen the locknut, then tighten again to 13 inch lbs. (1.5 Nm).

c. Turn the hub several times and check the nut torque again.

d. Install the lockwasher. When installing the screw, be sure the locknut turns no more than 30 degrees in either direction.

e. When bearing pre-load is properly set, there will be no end-play in the hub and it will require no more than 4.7 lbs. of pull at the wheel stud to turn the hub.

12. Install the locking hub and tighten the hub bolts to 18–25 ft. lbs. (25–34 Nm).

13. Install the brake caliper, brake pads, and wheel assembly.

14. Lower the vehicle and pump the brakes until the pedal is firm.

Rear

ALL MODELS

1. Raise and safely support the vehicle.

2. Remove the rear wheel.

3. Disconnect the parking brake cable and hydraulic brake line from the brake assembly.

4. Remove the four nuts securing the axle assembly to the axle housing.

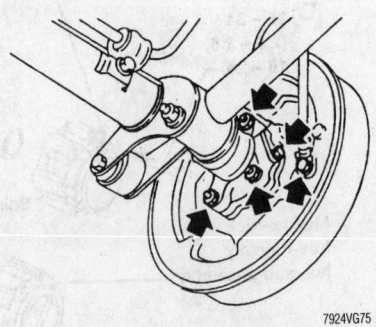

Remove the four nuts, parking brake cable and hydraulic line

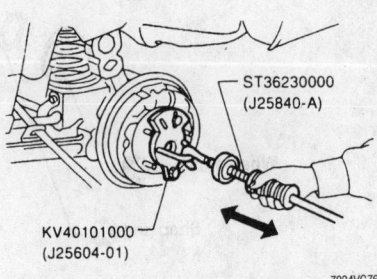

Remove the axle and bearing assembly with a slide hammer

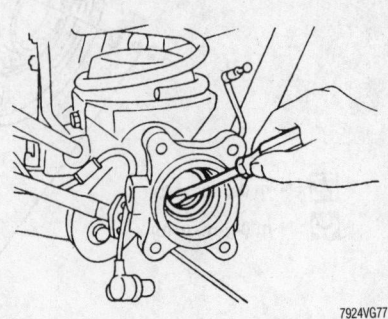

Remove the seal from the axle housing

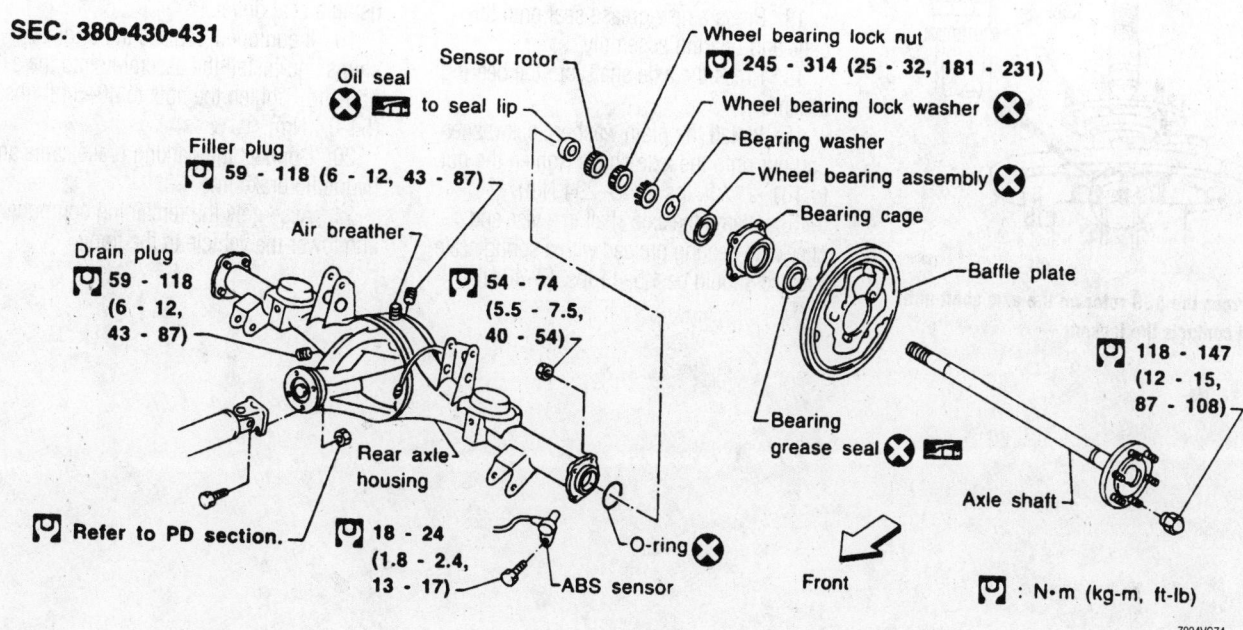

SEC. 380•430•431

Exploded view of the rear axle bearing and related components—4-WD QX4 shown, others are similar

5. Using a slide hammer, pull the axle assembly out of the housing.

6. Using a puller with the proper adapters, remove the ABS rotor.

7. Unbend the lockwasher and remove the bearing nut.

8. Using a puller with the proper adapters, press the axle shaft out of the bearing assembly.

9. Remove the grease seal with a hammer and drift.

10. Press the bearing assembly out of the bearing cage.

11. Remove the old seal from the axle housing with a suitable tool.

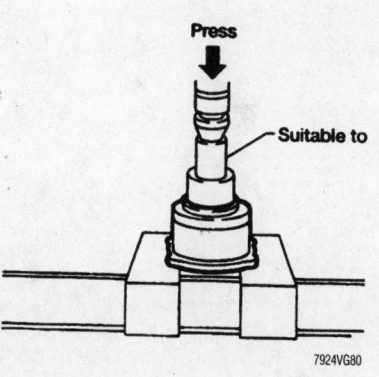

Remove the wheel bearing from the cage

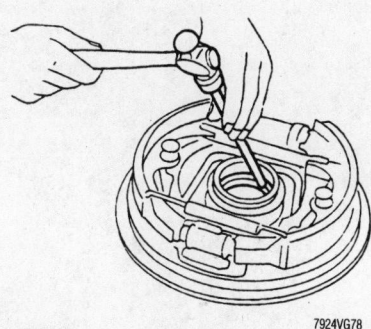

Remove the seal from the bearing cage with a hammer and drift

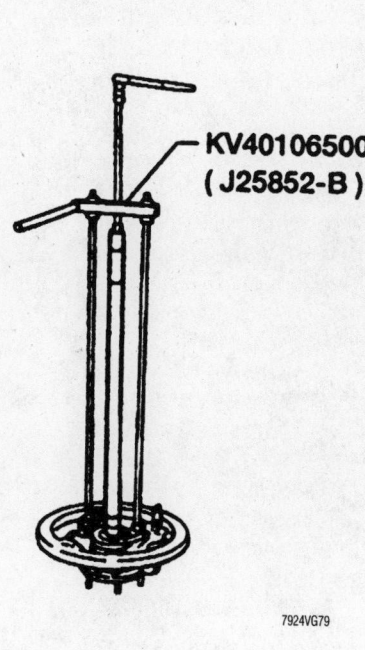

Press the axle shaft out of the bearing assembly

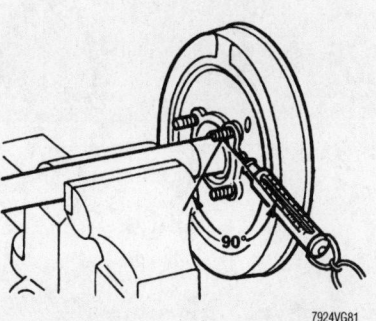

Measure the bearing preload using a spring scale

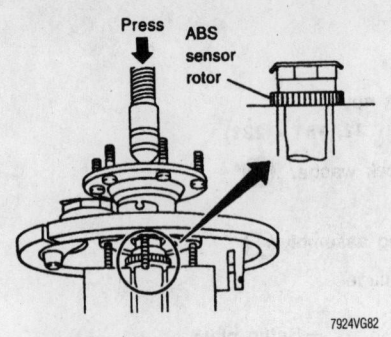

Press the ABS rotor on the axle shaft until it contacts the locknut

To install:

12. Press a new bearing into the cage until it bottoms.

13. Press a new grease seal onto the cage and bearing assembly.

14. Press the axle shaft into the bearing assembly.

15. Install the plain washer, lockwasher and nut onto the axle shaft. Tighten the nut to 181–217 ft. lbs. (245–294 Nm).

16. Clamp the axle shaft in a vise and check the bearing preload with a spring scale. Preload should be 1.5–11 lbs. (7–48 N).

17. Press the ABS rotor on the axle shaft until it contacts the locknut.

18. Install a new seal in the axle housing using a seal driver.

19. If equipped, replace the original shims and install the assembly into the axle housing. Tighten the nuts to 40–54 ft. lbs. (54–74 Nm).

20. Connect the parking brake cable and hydraulic brake line.

21. Assembly the remaining components and lower the vehicle to the floor.

NISSAN and MERCURY

Nissan-Quest • Mercury-Villager

32

ENGINE REPAIR

➡ **Disconnecting the negative battery cable on some vehicles may interfere with the functions of the on board computer systems and may require the computer to undergo a relearning process, once the negative battery cable is reconnected.**

Distributor

REMOVAL & INSTALLATION

These vehicles use a camshaft-driven distributor. The Camshaft Position Sensor (CMP) is built into the distributor. The CMP sensor monitors engine speed (rpm) and piston position and sends signals to the Powertrain Control Module (PCM). The 3.0L engine uses a power transistor, resistor, condenser and ignition coil mounted separately from the distributor. As the power transistor grounds the primary circuit, the inductive charge built up in the secondary circuit sends a spark from the ignition coil to the distributor. The distributor rotor and cap, then send a spark to each spark plug. The spark advance and retard functions are controlled by the PCM. If engine knocking occurs, the Knock Sensor detects the condition and a signal is set to the PCM. The PCM retards the ignition timing to prevent engine knocking. The base ignition timing is programmed in the anti-knocking zone, if the recommended fuel is used. Therefore, the knock sensor system does not operate under normal driving conditions.

Engine firing order is 1–2–3–4–5–6. Number one cylinder is on the rear bank, forward cylinder.

1. Disconnect the negative battery cable.
2. Remove the distributor cover.

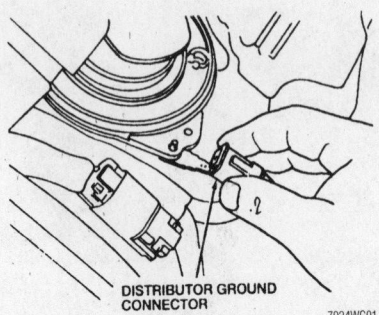

DISTRIBUTOR GROUND CONNECTOR

Disengage the distributor ground connector when removing the distributor

3. Loosen the three distributor cap screws and set aside the distributor cap with all the wires still attached.
4. Disengage the distributor ground connector from the tab on the distributor housing.
5. Disengage the distributor harness electrical connector and remove it from its bracket.

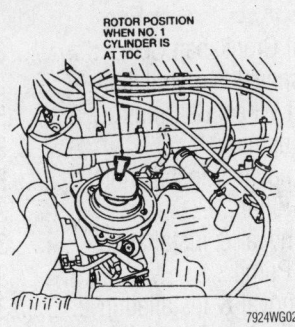

ROTOR POSITION WHEN NO. 1 CYLINDER IS AT TDC

Note the position of the rotor when the No. 1 piston is at TDC on the compression stroke

6. Rotate the crankshaft until the No. 1 piston is at TDC on the compression stroke. Check to be sure that the timing mark (may be yellow) on the crankshaft pulley and the timing pointer on the front cover are aligned.
7. Note the relation of the distributor rotor to the engine. Make a mark on a nearby engine component to assist with installation.
8. Remove the distributor hold-down bolt and lift up the distributor with the base gasket.
9. Note that if the rotor is being removed, a small retainer bolt on the side must first be removed.

To install:

10. Verify that the crankshaft is still at TDC No. 1 cylinder, compression stroke (firing position).
11. Install a new distributor base gasket. Lower the distributor into the engine. Note that during installation, the distributor rotor will tend to turn as the gears engage and mesh. Be sure that the distributor rotor aligns with the mark made on the engine component during removal. Install the hold-down bolt only finger-tight.
12. Install the distributor electrical connector and the ground connector.
13. Install the distributor cap and wires. Be sure the cap is seated properly. Install the cover.
14. Connect the negative battery cable.
15. Start the engine and allow to warm to operating temperature. Adjust the ignition

timing using the recommended procedure. After all adjustments are made, tighten the distributor hold-down bolt to 10–12 ft. lbs. (14–17 Nm).

Ignition Timing

ADJUSTMENT

1. Apply the parking brake and be sure that the vehicle is in PARK.
2. Start and run the engine until it reaches normal operating temperature.
3. Be sure the throttle is not touching the fast idle cam and the engine speed is below 1,000 rpm.
4. Turn off all electrical loads.
5. Run the engine at about 2000 rpm for 2 minutes under no-load.
6. Check for trouble codes and make necessary repairs if needed.
7. Turn **OFF** the engine and disconnect the Throttle Position Sensor (TPS).
8. Start the engine
9. Rev the engine 2 or 3 times to between 2,000–3,000 rpm and return the engine to idle speed.

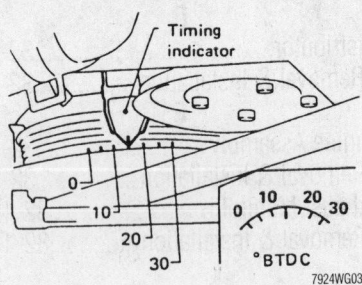

Timing indicator

°BTDC

Adjust the timing so the pointer on the engine indicates 15 degrees before top dead center (three notches from TDC) on the crankshaft pulley

10. Connect a timing light to the No. 1 cylinder spark plug wire at the distributor end and check the ignition timing. Be sure that the timing pointer is pointing to the 15 degrees BTDC mark on the crankshaft pulley.

➡ **Each notch on the crankshaft pulley represents 5 degrees.**

11. If the timing is not within the specification, loosen the distributor mounting bolt and adjust the distributor until the timing is at the proper specification.
12. Tighten the distributor mounting bolt to 10–12 ft. lbs., (14–17 Nm).
13. Stop the engine and connect the TPS.

Engine Assembly

REMOVAL & INSTALLATION

The engine is removed with the transaxle attached. The engine and transaxle are lowered from the vehicle as an assembly.

✻✻ CAUTION

Observe all applicable safety precautions when working around fuel. Whenever servicing the fuel system, always work in a well ventilated area. Do not allow fuel spray or vapors to come in contact with a spark or open flame. Keep a dry chemical fire extinguisher near the work area. Always keep fuel in a container specifically designed for fuel storage; also, always properly seal fuel containers to avoid the possibility of fire or explosion.

1. Relieve the fuel system pressure using the recommended procedure.
2. Disconnect the negative battery cable.

✻✻ CAUTION

Never open, service or drain the radiator or cooling system when hot; serious burns can occur from the steam and hot coolant. Also, when draining engine coolant, keep in mind that cats and dogs are attracted to ethylene glycol antifreeze and could drink any that is left in an uncovered container or in puddles on the ground. This will prove fatal in sufficient quantities. Always drain coolant into a sealable container. Coolant should be reused unless it is contaminated or is several years old.

3. Drain and properly contain the coolant from the cooling system.
4. Remove the air intake tube and resonator.

✻✻ CAUTION

The EPA warns that prolonged contact with used engine oil may cause a number of skin disorders, including cancer! You should make every effort to minimize your exposure to used engine oil. Protective gloves should be worn when changing the oil. Wash your hands and any other exposed skin areas as soon as possible after exposure to used engine oil. Soap and water, or waterless hand cleaner should be used.

5. Drain the engine oil.
6. Disconnect the radiator overflow hose from the radiator filler neck and remove the reservoir.
7. Label and disengage all electrical connectors from the engine and transmission.
8. Disconnect and plug the fuel tubes.
9. Disconnect the vacuum hoses from the evaporative emission canister.
10. Disconnect the main engine wiring harness from the crankcase vent tube brackets.
11. Remove the two ground connections from the upper intake manifold.
12. Disconnect the throttle cable.
13. Remove the upper and lower radiator hoses.
14. If equipped with A/C, remove the two upper A/C compressor bolts.
15. Disconnect the heater hoses.
16. Disconnect the brake booster hose.
17. Remove the ground cable from the oil filler tube.
18. If equipped with A/C, remove the A/C drive belt.
19. Raise and safely support the vehicle on jackstands
20. Remove the front wheels.
21. Remove the inner and outer engine and transmission splash shields.
22. Remove the shift cable nut from the Transmission Range (TR) switch.
23. Using a screwdriver release the shift cable locking pin from the shift cable bracket.
24. Remove the accessory drive belts.
25. Remove the power steering pump. Access the front power steering pump-to-bracket bolts by inserting a socket through the pulley holes.
26. If equipped with A/C, remove the two lower compressor bolts and position the compressor aside without disconnecting the refrigerant lines.
27. Remove the exhaust inlet pipe.
28. Remove the front halfshafts using the recommended procedure.
29. Disconnect and label the oil cooler tubes.
30. Disconnect the transaxle cooler lines.
31. Disconnect the transaxle ground strap.
32. Place a suitable jack or lift under the engine transaxle assembly.
33. Remove the three front transaxle mount bolts.
34. Remove the three transaxle rear mount nuts.
35. Remove the two rear refrigerant/heater pipes hold-down bracket bolts.
36. Remove the four transverse member (crossmember) bolts and the transverse member.

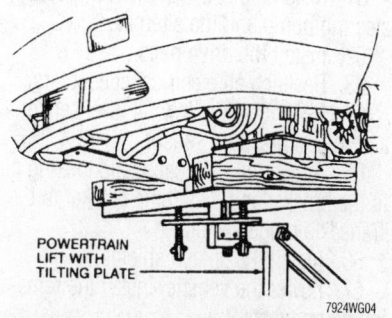

POWERTRAIN LIFT WITH TILTING PLATE

7924WG04

Carefully lower the engine/transaxle assembly from the vehicle

37. Lower the assembly from the vehicle.
38. Remove the upper transaxle to engine bolts.
39. Remove the bolts from both transaxle braces.
40. Remove the lower transaxle bolt.
41. Remove the torque converter bolts.
42. Separate the transaxle from the engine.

To install:

43. Assemble the transaxle to the engine. Use care to see that the alignment dowels are properly positioned.
44. Install the lower transaxle to engine bolt and tighten to 22–30 ft. lbs. (30–40 Nm).
45. Install the torque converter bolts and tighten to 33–43 ft. lbs. (44–59 Nm).
46. Install the two transaxle braces.
47. Tighten the transaxle brace bolts to 22–30 ft. lbs. (30–40 Nm.
48. Install the exhaust bracket on to the transaxle and tighten to 22–30 ft. lbs. (30–40 Nm.
49. Install the upper transaxle-to-engine bolts. Tighten the bolts to 29–36 ft. lbs. (39–49 Nm).
50. Raise the assembly into the vehicle and install the four transverse bolts.
51. Tighten the transverse bolts to 58–65 ft. lbs. (78–88 Nm).
52. Install the two rear A/C heater brackets to the transaxle.
53. Install the three rear transaxle support bracket nuts and tighten to 32–42 ft. lbs. (43–55 Nm).
54. Install the three front transaxle mount bolts and tighten to 30–38 ft. lbs. (41–52 Nm).
55. Remove the engine lift.
56. Reconnect the transaxle ground strap.
57. Reconnect the transaxle cooler lines.
58. Connect the oil cooler lines to the proper connections.
59. Install both halfshafts. Install all of the remaining components.
60. Install the exhaust inlet pipe.

61. Install the A/C compressor, power steering pump and the alternator.

62. Install the drive belts.

63. Reattach all electrical connectors.

64. Reconnect the low oil level sensor and the oil pressure sensor.

65. Attach the shift cable and locking pin to the bracket and install the bracket to the Transaxle range switch.

66. Install the splash shields.

67. Lower the vehicle. Install the remaining components.

68. Fill the crankcase with the correct amount of oil.

69. Fill the cooling system.

70. Reconnect the negative battery cable.

71. Bleed the cooling system.

72. Start the engine.

73. Check for leaks and proper operation.

Water Pump

REMOVAL & INSTALLATION

1. Drain the cooling system.

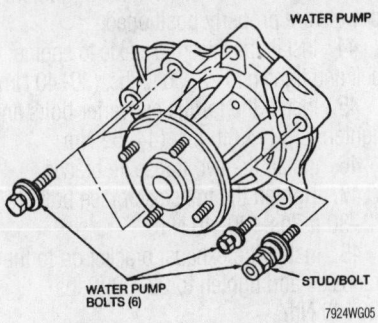

Water pump mounting. Note the location of the stud/bolt

❊❊ CAUTION

Never open, service or drain the radiator or cooling system when hot; serious burns can occur from the steam and hot coolant. Also, when draining engine coolant, keep in mind that cats and dogs are attracted to ethylene glycol antifreeze and could drink any that is left in an uncovered container or in puddles on the ground. This will prove fatal in sufficient quantities. Always drain coolant into a sealable container. Coolant should be reused unless it is contaminated or is several years old.

2. Disconnect the negative battery cable.

3. Remove the alternator drive belt, the water pump and power steering pump drive belt and the A/C compressor drive belt (if equipped).

4. Use a strap wrench to hold the water pump pulley while removing the four water pump pulley bolts.

5. Remove the water pump pulley from the water pump.

6. Remove the crankshaft pulley using the following procedure.

   a. Raise and safely support the vehicle on jackstands

   b. Remove the five right side inner engine and transmission splash shield bolts and two screws and remove the inner engine and transmission shield.

   c. Remove the four right side outer engine and transmission splash shield bolts and two screws and remove the right side outer engine and transmission splash shields.

   d. Use a strap wrench to hold the crankshaft pulley while removing the crankshaft pulley bolt.

   e. Use a crankshaft damper remover to draw the crankshaft pulley off the front of the crankshaft.

7. Remove the five lower engine front cover bolts and take of the front cover.

8. Remove the six water pump bolts. Make note of the locations of the bolts since one should be a stud/bolt and must be returned to its original location. Remove the water pump.

To install:

9. Clean all parts well. The bolt threads should be cleaned of any old sealer or corrosion. Be sure the mating surfaces between the water pump and the engine block are cleaned of any old sealant. Apply a continuous bead of gasket maker type sealer approximately ⅛ inch wide onto the water pump and position the water pump on the engine block.

10. Install the six water pump bolts. Refer to any notes made at removal so the bolts can be returned to their original locations. Do not over-tighten the water pump bolts. Tighten the water pump bolts evenly to 12–15 ft. lbs. (16–21 Nm).

11. Position the water pump pulley on the water pump and install the four pulley bolts. Use a strap wrench to hold the pulley as the bolts are tightened to 12–15 ft. lbs. (16–21 Nm).

12. Install the front engine cover and the five lower front cover bolts. Tighten to 27–44 inch lbs. (3–5 Nm).

13. Install the crankshaft pulley using the following procedure.

   a. Install the crankshaft pulley and pulley bolt.

   b. Hold the pulley with a strap wrench. Tighten the crankshaft pulley bolt to 90–98 ft. lbs. (123–132 Nm).

   c. Install the inner and outer engine and transmission splash shields.

14. Install and adjust the drive belts.

15. Connect the negative battery cable.

16. Refill the cooling system.

17. Start the engine, bleed the cooling system and verify no leaks.

Cylinder Head

REMOVAL & INSTALLATION

The 3.0L SOHC V6 engine is a free-wheeling engine design with an in-head camshaft containing 12 valves, one each for intake and exhaust per cylinder.

The factory specifies that the cylinder head bolts ARE NOT to be reused. Obtain the proper replacement parts before beginning this procedure. Check carefully that all bolts are removed before attempting to remove a cylinder head. A tab, part of the head, contains one lightly tightened head bolt that is external to the valve cover. Do not overlook this "hidden" bolt or the head will be damaged.

❊❊ CAUTION

The fuel injection system remains under pressure, even after the engine has been turned OFF. The fuel system pressure must be relieved before disconnecting any fuel lines. Failure to do so may result in fire and/or personal injury.

1. Properly relieve the fuel system pressure using the recommended procedure.

2. Drain the cooling system.

3. Disconnect the negative battery cable.

4. Remove the air intake tube.

5. Remove the timing belt using the recommended procedure.

6. Remove the upper intake manifold (plenum) using the recommended intake manifold procedure.

7. Tag each spark plug wire for identification. This saves time at assembly. Disconnect the spark plug wires and ignition coil to distributor high tension wires.

8. Remove the distributor using the recommended procedure.

9. Tag for identification and remove the following electrical connectors:

    a. Exhaust Gas Recirculation Control (EGRC) solenoid

    b. Mass Air Flow (MAF) sensor

    c. Vehicle Speed Sensor (VSS)

    d. Valve body wiring harness

    e. Transmission Range (TR) switch

    f. A/C clutch pulley

    g. Power transistor

    h. A/C cut-off switch

    i. Fuel injectors

    j. Ignition coil

    k. Water temperature indicator sender unit.

    l. Engine Coolant Temperature (ECT) sensor.

10. Remove the main wiring harness bracket from the water hose connection.

11. Remove the fuel tube bracket bolt from the EGR valve bracket.

12. Remove the four Allen-head fuel injection supply manifold (fuel rail) and position the fuel rail and injectors aside.

13. Disconnect the upper heater water hose.

14. Using two steps, remove the four Allen-head intake manifold to cylinder head

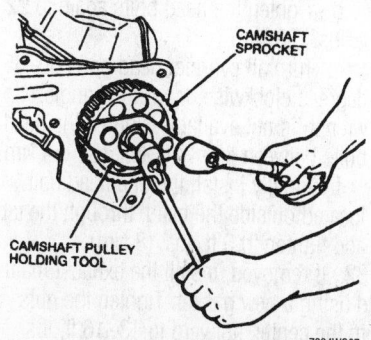

Hold the camshaft sprocket while removing the sprocket retaining bolt

bolts and the four nuts. Work from the outer fasteners, inward. Remove the lower intake manifold from the vehicle. Discard the gaskets.

15. If removing the front cylinder head, remove the nine valve cover screws, take off the cover and discard the gasket.

16. Use Camshaft Pulley Holding Tool T92P-6312-aH or equivalent to hold the camshaft sprocket while removing the sprocket retaining bolt. Remove the camshaft sprocket and the four seal plate bolts and seal plate.

17. If removing the front cylinder head, remove the oil level indicator (dipstick) bolts and take off the dipstick tube bracket.

18. Raise and safely support the vehicle on jackstands

19. If removing the front cylinder head, remove the three exhaust manifold to inlet pipe nuts. Remove and discard the gasket.

20. Remove the front exhaust manifold to mounting bracket bolt.

21. If removing the front cylinder head, remove the two lower A/C compressor bolts, if equipped.

22. Lower the vehicle.

23. If removing the front cylinder head, and if equipped with A/C, loosen the two upper A/C compressor bolts and move the A/C compressor out of he way. Secure with wire. Note that the upper A/C compressor bolts are too long to remove from the compressor. Remove them once the compressor is moved aside.

24. If removing the front cylinder head, remove the two upper alternator regulator mounting bracket bolts. Remove the two coolant crossover tube bracket bolts.

25. If removing the rear cylinder head, remove the six rear exhaust manifold nuts working from the center, outward. Remove the manifold and discard the gasket. Remove the nine rear valve cover screws, lift off the valve cover and discard the gasket.

26. The cylinder head bolts must be removed in sequence, working from the outside, inward. Please note that there is a "tab" on one end of each cylinder head. This tab contains a head bolt. Be sure that the first bolt in the removal sequence is this bolt outside the cylinder head. This bolt is easily forgotten or overlooked and the tab can be broken off if the cylinder head is moved prior to the bolt being removed. Also note that the head bolts are not to be reused. They must be replaced with Original Equipment Manufacturer (OEM) parts. Loosen the head bolts in sequence, in two steps. Remove the head bolt washers and discard the head bolts. Remove the front cylinder head along with the front exhaust manifold and discard the gasket.

To install:

27. Clean all parts well. With the intake and exhaust valves in place to protect the valve seats, remove deposits from the combustion chambers and valve heads with a scraper and wire brush. Be careful not to damage the head gasket surface. After the valves are removed, clean the guide bores. Use a suitable solvent to remove dirt, grease and other deposits. Clean all head bolt holes. Remove all deposits from the valves with a fine wire brush.

28. Inspect the cylinder head for damage, cracks and leakage of water and oil. If

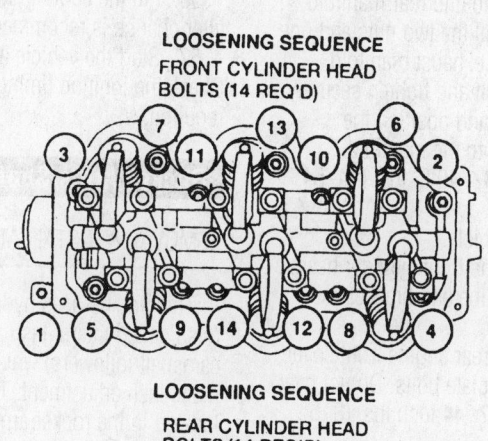

LOOSENING SEQUENCE FRONT CYLINDER HEAD BOLTS (14 REQ'D)

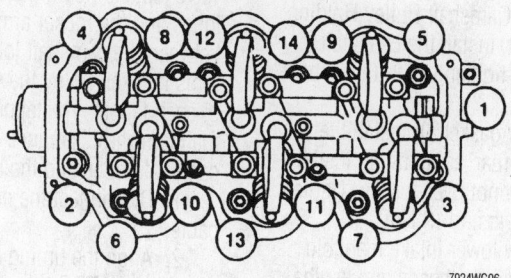

LOOSENING SEQUENCE REAR CYLINDER HEAD BOLTS (14 REQ'D)

Remove the cylinder head bolts and nuts in the proper sequence to avoid warping the head. Don't forget the hidden bolt outside of the valve cover area

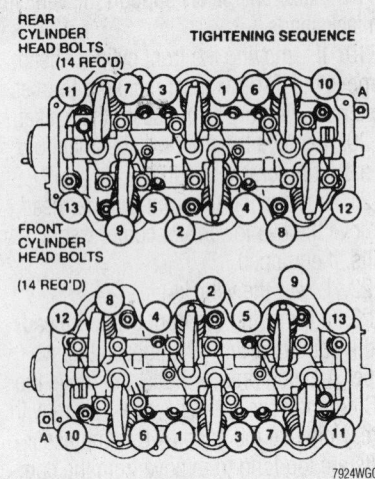

REAR CYLINDER HEAD BOLTS (14 REQ'D)

TIGHTENING SEQUENCE

FRONT CYLINDER HEAD BOLTS

(14 REQ'D)

7924WG08

Follow the tightening sequence and the steps carefully to achieve a good head gasket seal. The "hidden" bolt is indicated as "A"

necessary, replace the head. Check the head gasket surface for burrs and nicks. If the head is cracked, it must be replaced.

29. Damaged spark plug threads can be repaired with commercially-available thread inserts. A properly installed insert will be flush to 0.03937 inch (1mm) below the spark plug counterbore seat.

30. Using a straightedge and a feeler gauge, check the head for flatness. Measure lengthwise and across the head. Maximum distortion is 0.004 inch (0.10mm). If the head distortion exceeds this specification, the head should be resurfaced. Use care. The overall height of the cylinder head must not be reduced too much. The cylinder head must be replaced if the head height is not within 4.205–4.220 inches (106.8–107.2mm) tall.

31. Position a new head gasket and either the front or rear cylinder head on the block. Examine the head bolt washers. Note that the washers have a chamfer or bevel on one side. The beveled side should face "up" when installed. Examine the new replacement head bolts. There are different lengths. The head bolts in positions 4, 5, 12 and 13 are 5.00 inches (127mm) long and the rest are 4.17 inches (106mm) long. Be sure the new cylinder head bolts are installed in the correct positions. Tighten the new head bolts in the following sequence:

    a. Tighten all head bolts to 22 ft. lbs. (29 Nm).

    b. Tighten all head bolts to 43 ft. lbs. (59 Nm).

    c. Loosen all of the bolts completely using the loosening sequence (loosening from the center outwards).

    d. Tighten the head bolts again to 22 ft. lbs. (29 Nm).

    e. Turn all cylinder head bolts 60–65 degrees clockwise. If a torque angle wrench is not available, tighten the head bolts between 40–47 ft. lbs. (54–64 Nm).

    f. Finally install the one head bolt located outside the head, through the tab and tighten to 6 ft. lbs. (8 Nm).

32. If removed, install the exhaust manifold using a new gasket. Tighten the nuts from the center, outward to 13–16 ft. lbs. (18–22 Nm).

33. If installing the front cylinder head, position the coolant crossover tube on the bracket on the cylinder head and install the bolt. Install the two upper alternator regulator mounting bracket bolts and tighten securely.

34. If installing the front cylinder head and if equipped with A/C, position the A/C compressor on the alternator regulator mounting bracket and install the upper compressor bolts. Tighten to 33–44 ft. lbs. (45–60 Nm).

35. Raise and safely support the vehicle.

36. If installing the front cylinder head and if equipped with A/C, install the two lower compressor bolts. Tighten to 33–44 ft. lbs. (45–60 Nm).

37. Install a new gasket between the front exhaust manifold and rear manifold crossover tube, install the two nuts and one bolt. Install the front exhaust manifold mounting bracket bolt and tighten securely. Install a new gasket and position the exhaust inlet pipe onto the manifold. Tighten the nuts to 32–40 ft. lbs. (44–54 Nm).

38. Lower the vehicle.

39. If installing the front cylinder head, install the dipstick tube and bracket and secure.

40. Position the rear engine front cover install the four seal plate bolts. Do not over-torque. Tighten to 27–44 inch lbs. (3–5 Nm).

41. Using the Camshaft Pulley Holding Tool or equivalent, install the camshaft sprocket bolt and tighten to 58–65 ft. lbs. (78–88 Nm).

42. Using new gasket(s), install the valve cover(s), front or rear, as required. Install the nine bolts. Do not over-torque. Tighten to just 9–26 inch lbs. (1–3 Nm).

43. Install new lower intake manifold gaskets on the cylinder heads and lay the manifold in lace. Install the four Allen-head bolts and four manifold nuts. Tighten in sequence, working from the center, outward as follows:

    a. Step 1: Tighten to 27–44 inch lbs. (3–5 Nm).

    b. Step 2: Tighten to 12–14 ft. lbs. (16–20 Nm).

    c. Step 3: Tighten again to 12–14 ft. lbs. (16–20 Nm).

44. Connect the upper water hose.

45. Position the fuel injectors into their respective ports and install the fuel rail bolts. Tighten evenly to 17–20 ft. lbs. (24–27 Nm).

46. Connect the water bypass hose to the intake manifold.

47. Bolt the fuel tube bracket back onto the EGR valve bracket.

48. Attach the main wiring harness bracket to the intake manifold water hose connection.

49. Connect the electrical connectors removed at disassembly.

50. Install the distributor using the recommended procedure.

51. Connect the spark plug wires using the identification made at removal.

52. Install the upper intake manifold (plenum) using the recommended intake manifold procedure.

53. Install the timing belt using the recommended procedure.

54. Install the air cleaner and intake tubes as required.

55. Connect the negative battery cable.

56. Fill the cooling system. An oil and filter change is recommended.

57. Start the vehicle and check for leaks. Check the ignition timing and adjust as required.

Rocker Arms/Shafts

REMOVAL & INSTALLATION

This engine uses hydraulic valve tappets (also called lifters, lash adjusters or camshaft followers) which provide automatic lash adjustment. The valve tappets are located in the rocker arm shaft support. The valve tappets ride between the camshaft lobes and the rocker arms. Any clearance between the camshaft lobes and the rocker arms is taken up by the hydraulic extension of each tappet. The tappets are designed to maintain zero clearance between the camshaft lobes and the rocker arms.

1. Disconnect the negative battery cable.

2. Align the timing marks to bring No. 1 cylinder to TDC compression stroke (firing position).

3. Remove the valve cover(s) as required.

4. Loop a length of mechanic's wire around the tops of the tappets to hold them in place when the rocker arm and shaft assembly is removed.

✳✳ CAUTION

Loosening the rocker arm shaft bolts in one step may distort or break the rocker arm shaft.

5. Loosen the rocker shaft bolts in two or three steps. Mark the location of the rocker arms to ease installation. Remove the rocker shafts with rocker arms, as an assembly.

6. Separate the rocker arms from the shaft.

➡**When separating the rocker arms from the rocker arm shafts, be sure to keep the parts in order for reinstallation, if any of the parts are to be reused.**

7. Check the rocker arms and the shafts for damage. If necessary, replace the damaged components.

To install:

8. Clean all parts well. Any worn parts should be replaced.

9. Attach a loop of mechanic's wire to the top of the lifters so that they will not drop from the lifter guide during assembly. Carefully remove the lifter guide and lifters from the cylinder head. Put an identification mark on the lifters to avoid mixing them up if they are removed from the guide and to be reused. If the lifters are damaged replace them as necessary.

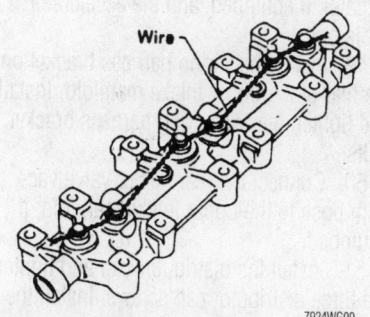

Wire

7924WG09

Wire the lifters on top of the guide so they won't fall out when the guide is removed from the head

10. Install new lifters if replacing them or install the old lifters to their original locations. Coat all parts with clean engine oil as they are assembled.

✳✳ CAUTION

When installing the rocker arm shafts, be certain that they are installed in their original positions.

11. Slide the rocker arms onto the shafts in their proper positions.

12. Be sure that cylinder No. 1 is still at TDC.

13. Install the rocker arm/shaft and lifter guide assembly. When seated, remove the wire that held the lifters in place. Coat the bolt threads and seat surfaces with clean engine oil before installing them. Tighten the bolts gradually in three steps to 13–16 ft. lbs. (18–22 Nm).

14. Rotate the crankshaft clockwise 180 degrees (½ turn) to bring cylinder No. 4 to TDC. Install the right cylinder head rocker arm/shaft and lifter guide assembly. Coat the bolt threads and seat surfaces with clean engine oil before installing them. Tighten the bolts gradually in three steps to 13–16 ft. lbs. (18–22 Nm).

15. Install the rocker covers with new gaskets. Tighten the rocker cover bolts to 12–24 inch lbs. (1–3 Nm).

16. Connect the negative battery cable.

17. An oil and filter change is recommended.

18. Test run the engine and verify lack of valve noise.

Intake Manifold

REMOVAL & INSTALLATION

This Multi-port Fuel Injection (MFI) system is classified as a multi-point, pulse time, mass airflow fuel injection system. This system supplies the engine with the air/fuel mixture necessary for combustion. An air induction system and fuel injection system work in conjunction with the Electronic Engine Control (EEC) system which consists of various sensors, switches and a Powertrain Control Module (PCM). The PCM uses the signals that it receives from the various sensors and switches to compute fuel injector timing and pulse width.

The intake manifold is a two-piece design. The upper half, usually called the plenum, mounts the throttle body, control cables, fast idle solenoid, idle air control solenoid and emission connections. The lower manifold bolts directly to the engine and carries the fuel injectors.

Air enters the system through the air cleaner intake tube. The air, then flows through the dry element air cleaner and is metered by the Mass Air Flow (MAF) sensor. The metered air passes through the air cleaner to intake manifold tube and enters the throttle body. From the throttle body, the air passes through the upper intake manifold to the lower intake manifold where it is mixed with fuel for combustion. To reduce intake noise, three engine air intake resonators are part of the system. These three components absorb air flow pulsations as air is drawn into the system.

Upper Intake Manifold (Plenum)

✳✳ CAUTION

Observe all applicable safety precautions when working around fuel. Whenever servicing the fuel system, always work in a well ventilated area. Do not allow fuel spray or vapors to come in contact with a spark or open flame. Keep a dry chemical fire extinguisher near the work area. Always keep fuel in a container specifically designed for fuel storage; also, always properly seal fuel containers to avoid the possibility of fire or explosion.

1. Relieve the fuel system pressure.

2. Disconnect the negative battery cable.

✳✳ CAUTION

Never open, service or drain the radiator or cooling system when hot; serious burns can occur from the steam and hot coolant. Also, when draining engine coolant, keep in mind that cats and dogs are attracted to ethylene glycol antifreeze and could drink any that is left in an uncovered container or in puddles on the ground. This will prove fatal in sufficient quantities. Always drain coolant into a sealable container. Coolant should be reused unless it is contaminated or is several years old.

3. Drain the cooling system.

4. Remove the air cleaner intake tube and resonator. Use the following procedure.

   a. Remove the battery.

   b. Remove the air cleaner intake tube bracket bolt.

c. Separate the air cleaner intake tube from the engine air cleaner.

d. Remove and separate the air cleaner intake tube from the engine air intake resonator No. 1.

5. Disconnect the idle switch electrical connector and the Throttle Position (TP) sensor. Both connectors are located near the throttle body opening. Slide the TP sensor electrical wiring out of the bracket.

6. Remove the coolant hose from the throttle body.

7. Disconnect the Exhaust Gas Recirculation Control (EGRC) solenoid electrical connector.

8. Disconnect the Evaporative Emission canister vacuum hose from the throttle body.

9. Remove the fuel pressure regulator-to-upper intake manifold vacuum hose from the upper intake manifold.

10. Tag and disconnect the No. 1, 3 and 5 cylinder distributor to spark plug wires from the spark plugs, then remove the spark plug wire boots from between the upper intake manifold runners. Remove the two spark plug wire bracket bolts and spark plug wire brackets from the upper intake manifold.

11. Remove the distributor dust cover. Loosen the three distributor cap screws. Note that the screws are not removable from the distributor cap. Remove the cap from the distributor. Position the cap and plug wires aside to gain access to the main engine wiring harness.

12. Disconnect the rear heater valve vacuum hose from the side of the upper intake manifold, if equipped.

13. Remove the two wiring harness bracket bolts and remove the wiring harness bracket from the upper intake manifold and set aside.

14. Disconnect the speed control actuator from the throttle lever, if equipped. Disconnect the accelerator cable from the throttle lever.

15. Remove the speed control actuator from the hanging bracket, if equipped.

16. Remove the accelerator cable from the bracket attached to the top of the upper intake manifold.

17. Loosen the accelerator cable locknut and remove the accelerator cable from the bracket on top of the upper intake manifold.

18. If equipped, loosen the speed control actuator locknut, then remove the speed control actuator from the bracket on the top of the upper intake manifold and position aside.

19. Disconnect the brake booster hose from the power brake booster check valve.

20. Remove the Exhaust Gas Recirculation (EGR) valve-to-Back Pressure Transducer (BPT) valve tube from the EGR valve. Remove the EGR valve to manifold tube from the EGR valve.

21. Disengage the Fast Idle Control solenoid electrical connector.

22. Disconnect the main engine wiring harness clips from the breather tube brackets. Remove the two ground wire bolts from the upper intake manifold and position aside the ground wires and the harness.

23. Remove the two front breather tube bracket-to-upper intake manifold bolts. Disconnect the front breather hose from the front valve cover.

24. Remove the fuel tube mounting bracket bolt.

25. Disengage the EGR temperature sensor electrical connector.

26. Disengage the Bypass Air (BPA) valve electrical connector.

27. Disengage the Idle Air Control (IAC) solenoid electrical connector.

28. Disconnect the Positive Crankcase Ventilation (PCV) hose from the upper intake manifold.

29. Remove the five Allen-head upper intake manifold to lower intake manifold bolts. Separate the upper intake manifold from the lower intake manifold and discard the gasket.

To install:

30. Clean all parts well. Use care working with light alloy parts, not to scratch or damage the gasket sealing surface or any threaded openings. Be sure all traces of old gasket and or sealer are removed.

31. Lay a new gasket on the lower intake manifold carefully aligning all openings. Set the upper intake manifold into place and install the five Allen-head bolts. Tighten the upper manifold bolts evenly, working from the center, outward, to 13–16 ft. lbs. (18–22 Nm). Do not over-tighten.

32. Connect the PCV hose to the upper intake manifold.

33. Connect the IAC solenoid electrical connector.

34. Connect the BPA valve electrical connector.

35. Connect the EGR temperature sensor electrical connector.

36. Connect the coolant hose to the top heater core pipe.

37. Install the fuel tube mounting bracket.

38. Connect the front breather hose to the front valve cover.

39. Position the front breather tube onto the upper intake manifold and install the two front breather tube bracket-to-upper intake manifold bolts. Tighten the front breather tube bracket-to-upper manifold bolts.

40. Position the ground wires and main wiring harness into place. Install and tighten the ground wire bolts. Be sure the connection is clean and tight.

41. Connect the main engine wiring clips on the breather tube brackets.

42. Connect the FIC solenoid electrical connector.

43. Install the EGR valve to manifold tube. Connect the EGR valve to BPT valve tube to the EGR valve.

44. Connect the brake booster hose to the power brake booster check valve.

45. Install the speed control actuator, if equipped, and the accelerator cable to the bracket on the top of the upper intake manifold.

46. Connect the speed control actuator on the throttle lever, if equipped.

47. Connect the accelerator cable on the throttle lever.

48. Tighten the nut on the speed control actuator, if equipped, and the accelerator cable nut.

49. Install the wiring harness bracket on the rear of the upper intake manifold. Install and tighten the two wiring harness bracket bolts.

50. Connect the rear heater valve vacuum hose to the upper intake manifold, if equipped.

51. Install the distributor cap and tighten the three distributor cap screws. Install the distributor cap dust cover.

52. Install the spark plug wire bracket on the upper intake manifold and install the bracket bolts. Install the spark plug wire boots in between the upper intake manifold runners and connect the No. 1, 3 and 5

cylinder spark plug wires onto the spark plugs as tagged during removal.

53. Connect the fuel pressure regulator vacuum hose to the intake manifold. Connect the EVAP canister hose to the throttle body.

54. Connect the EGRC solenoid electrical connector.

55. Connect the coolant hose to the throttle body.

56. Slide the TP sensor electrical wiring into the bracket and connect the TP sensor connector. Connect the idle switch connector.

57. Install the air cleaner intake tube and engine air intake resonator.

58. Fill the cooling system.

59. Connect the negative battery cable.

60. Start the engine and check for leaks and for proper engine operation.

Lower Intake Manifold

❊❊❊ CAUTION

The fuel injection system remains under pressure, even after the engine has been turned OFF. The fuel system pressure must be relieved before disconnecting any fuel lines. Failure to do so may result in fire and/or personal injury.

1. Relieve the fuel system pressure.

2. Disconnect the negative battery cable.

3. Remove the upper intake manifold (plenum) using the previous procedure.

4. Disengage the six fuel injector electrical connectors.

5. Disconnect the rear breather hose from the rear valve cover.

6. Remove the crossover breather tube bracket bolt and the crossover breather tube from the upper engine front cover.

7. Remove the four fuel injection supply manifold (fuel rail) bolts, then position the supply manifold and the insulators out of the way.

8. Disengage the Engine Coolant Temperature (ECT) sensor electrical connector.

9. Disengage the water temperature indicator sender unit electrical connector.

10. Remove the upper radiator hose from the water hose connection. Remove the upper radiator hose bracket bolt.

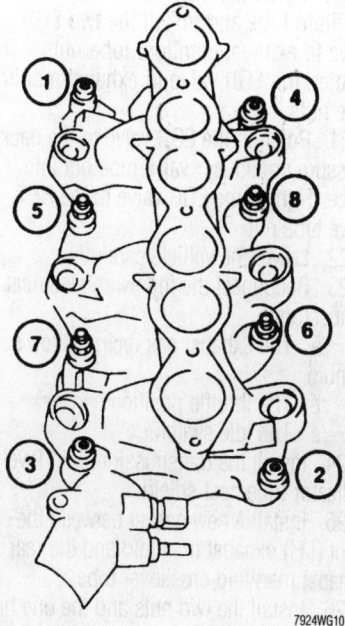

LOOSENING SEQUENCE

7924WG10

Gradually loosen the lower manifold bolts in two steps using the proper sequence

11. Remove the water bypass hose from the water hose connection.

12. Remove the bolt securing the water hose connection to the upper engine front cover.

13. Remove the heater hose from the top heater core pipe.

14. Remove the four Allen-head lower intake manifold bolts and the four manifold nuts, in two steps, using the proper sequence, working from the outside fasteners, inward.

15. Remove the lower intake manifold washers. Lift the lower intake manifold from the engine block. Discard the gaskets.

To install:

16. Clean all parts well. Use care working with light alloy parts, not to scratch or damage the gasket sealing surface or any threaded openings. Be sure all traces of old gasket and or sealer are removed.

17. Lay a new lower intake manifold gaskets on the cylinder heads and position the lower intake manifold on the engine block. Install the lower intake manifold washers

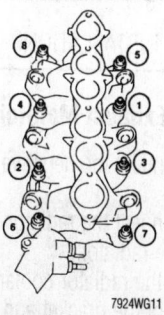

TIGHTENING SEQUENCE

7924WG11

Tighten the bolts and nuts in three steps using this sequence

and the Allen-head bolts and the manifold to cylinder head nuts.

18. Working from the center outward, tighten the lower intake manifold bolts and nuts in three steps using following sequence:

   a. Tighten the lower intake manifold to engine block bolts to 26–44 inch lbs. (3–5 Nm).

   b. Tighten the lower intake manifold to engine block bolts to 12–14 ft. lbs. (16–20 Nm) and the nuts to 17–20 ft. lbs. (24–27 Nm).

   c. Again, tighten the lower intake manifold to engine block bolts to 12–14 ft. lbs. (16–20 Nm) and the nuts to 17–20 ft. lbs. (24–27 Nm).

19. Install the heater hose to the top heater core pipe.

20. Install the bolt securing the water hose connection to the upper engine front cover.

21. Install the water bypass tube or hose onto the water hose connection.

22. Install the upper radiator hose bracket and bolt. Connect the upper radiator hose onto the water hose connec-tion.

23. Connect the water temperature indicator sender unit connector. Connect the Engine Coolant Temperature sensor wire.

24. Position the fuel rail with the insulators in place and install the four bolts.

25. Install the breather tube assembly and connect the rear breather to the rear valve cover. Install the crossover breather tube and the crossover tube bracket bolt.

26. Connect the six fuel injector electrical connectors.

27. Install the upper intake manifold (plenum assembly) using the recommended procedure given earlier.

Exhaust Manifold

REMOVAL & INSTALLATION

Rear (RH) Exhaust Manifold

1. Disconnect the negative battery cable.

2. Disconnect the radiator overflow hose from the radiator.

3. Slide the radiator coolant-recovery reservoir off of the bracket and remove the reservoir.

4. Remove the air cleaner intake tube and the engine air intake resonator.

5. Remove the six rear (RH) exhaust manifold crossover tube heat-shield bolts and remove the heat shields.

6. Remove the two nuts and the one bolt securing the rear (RH) exhaust manifold tube to the front (LH) exhaust manifold. Discard the gasket.

7. Remove the transmission fluid level indicator tube heat shield.

8. Disengage the following electrical connectors:

   a. The idle switch.

   b. The throttle position sensor.

   c. The exhaust gas recirculation control solenoid.

9. Raise and safely support the vehicle.

10. Remove the exhaust gas recirculation valve to back-pressure transducer valve tube nut and position it out of the way.

11. Remove the two EGR valve to exhaust manifold tube nuts and remove the EGR valve to exhaust manifold tube.

12. Remove the six rear exhaust manifold nuts in the reverse order of the tightening sequence.

13. Safely lower the vehicle, remove the exhaust manifold and discard the exhaust manifold gasket.

To install:

14. Raise and safely support the vehicle on jackstands

15. Be sure that both the exhaust manifold and the cylinder head mating surfaces are clean of any old gasket material.

16. Position the rear (RH) exhaust manifold gasket onto the exhaust manifold mounting studs.

17. Lower the vehicle safely.

18. Place the rear (RH) exhaust manifold onto the studs.

19. Safely raise the vehicle and install the six rear (RH) exhaust manifold nuts. Tighten the nuts in sequence to 13–16 ft. lbs. (18–22 Nm).

20. Install the EGR valve to exhaust manifold tube and install the two EGR valve to exhaust manifold tube nuts. Tighten the EGR valve to exhaust manifold tube nuts.

21. Position the EGR valve to the back-pressure transducer valve tube nut into place. Tighten the EGR valve to the BPT valve tube nut.

22. Lower the vehicle carefully.

23. Reconnect the following electrical connectors:

   a. The exhaust gas recirculation solenoid.

   b. The throttle position sensor.

   c. The idle switch.

24. Install the transmission fluid level indicator tube heat shield.

25. Install a new gasket between the front (LH) exhaust manifold and the rear exhaust manifold crossover tube.

26. Install the two nuts and the one bolt securing the rear (RH) exhaust manifold crossover tube to the front (LH) exhaust-manifold. Tighten the rear exhaust manifold crossover tube-to-front (LH) exhaust manifold nuts and bolt.

27. Reinstall the rear (RH) exhaust manifold crossover tube heat shield with the six mounting bolts.

28. Tighten the rear (RH) exhaust manifold crossover tube bolts.

29. Install the air cleaner intake tube and the engine air intake resonator.

30. Install the radiator coolant recovery reservoir and reconnect the radiator overflow hose to the radiator.

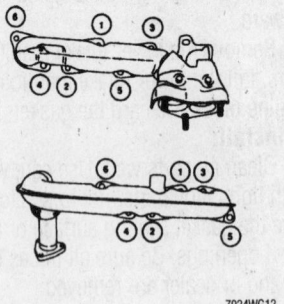

7924WG12

To avoid warping the exhaust manifolds, use this sequence when tightening the bolts

31. Reconnect the negative battery cable, start the engine and check for leaks and proper operation.

Front (LH) Exhaust Manifold

1. Disconnect the negative battery cable and wait at least 90 seconds before performing any work. This allows time for the SRS or air bag system to deplete its back up energy supply.

2. Remove the two nuts and the one bolt securing the front (LH) exhaust manifold to the rear (RH) exhaust manifold crossover tube. Discard the gasket.

3. Remove the transmission fluid level indicator tube heat shield.

4. Loosen the six front (LH) exhaust manifold nuts in two steps in the reverse order of the tightening sequence. Do not remove the three lower front (LH) exhaust manifold nuts.

5. Remove the front (LH) exhaust manifold-to-mounting bracket bolt.

6. Raise and safely support the vehicle on jackstands

7. Disengage the heated oxygen sensor electrical connector.

8. Remove the three front (LH) exhaust manifold-to-inlet pipe nuts.

9. Remove the exhaust system flex tube bracket bolt.

10. Remove the LH inner engine and transmission splash shield bolts and screws and remove the LH inner engine and transmission splash shield.

11. Remove the three lower exhaust manifold nuts.

12. Remove the front (LH) exhaust manifold and discard the exhaust manifold gasket.

To install:

13. Be sure that both the exhaust manifold and the cylinder head mating surfaces are clean of any old gasket material.

14. Position a new front exhaust manifold gasket in place and install the front (LH) exhaust manifold. Install the three lower exhaust manifold mounting nuts. Do not tighten the nuts at this time.

15. Install the LH inner engine and transmission splash shield with their mounting bolts and screws.

16. Reinstall the exhaust system flex tube bracket bolt.

17. Install the three exhaust manifold-to-exhaust inlet pipe nuts.

18. Reconnect the heated oxygen sensor electrical connector.

19. Lower the vehicle.

20. Install the front (LH) exhaust manifold-to-mounting bracket bolt.

21. Install the three upper exhaust manifold mounting bolts and tighten all six exhaust manifold mounting bolts in sequence. Tighten the bolts to 13–16 ft. lbs. (18–22 Nm).

22. Install the transmission fluid level indicator tube heat shield.

23. Install the two nuts and the one bolt securing the front (LH) exhaust manifold to the rear (RH) exhaust manifold crossover tube.

24. Reconnect the negative battery cable, start the engine, check for leaks and road test for proper operation.

Front Crankshaft Seal

REMOVAL & INSTALLATION

➡ **The front oil seal is a part of the oil pump body.**

1. Disconnect the negative battery cable.

2. Remove the timing belt and the crankshaft sprocket.

3. Remove the oil pump assembly.

4. Remove the oil seal from the oil pump body using a prytool. Be careful not to damage the oil pump body during seal removal.

To install:

5. Apply clean engine oil to the new oil seal. Install the seal using the proper size driver.

6. Install the oil pump assembly to the engine.

7. Install the remaining components in reverse order of removal.

8. Connect the negative battery cable.

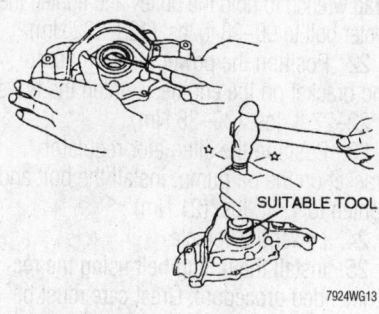

SUITABLE TOOL

7924WG13

Removing and installing the front crankshaft oil seal

Camshaft And Valve Lifters

REMOVAL & INSTALLATION

1. Disconnect the negative battery cable.

2. Drain the coolant from the cooling system and the engine. Coolant from the engine can be drained by removing the drain plug on the cylinder block.

3. Remove the timing belt assembly.

4. Remove the collector assembly and intake manifold.

5. Remove the cylinder head from the engine.

6. With cylinder head mounted on a suitable workbench, remove the rocker shafts with rocker arms. Bolts should be loosened in 2–3 steps.

7. Remove hydraulic valve lifters and lifter guide.

8. Hold hydraulic valve lifters with wire so they will not drop from lifter guide.

9. Remove the camshaft front oil seal and slide camshaft out the front of the cylinder head assembly.

To install:

10. Install camshaft, locate plate, cylinder head rear cover and a new front oil seal. Set camshaft knock pin at 12:00 o'clock position. Install cylinder head with new gasket to engine.

11. Install valve lifter guide assembly. Assemble valve lifters in their original position. After installing them in the correct location remove the wire holding them in lifter guide.

12. Install rocker shafts in correct position with rocker arms. Tighten bolts in 2–3 stages to 13–16 ft. lbs. Before tightening, be sure to set camshaft lobe in a position where it is not opening the valve. You can set each cylinder one at a time or follow the procedure below (timing belt must be installed in the correct position):

   a. Set No. 1 piston at TDC on its compression stroke and tighten rocker shaft bolts for No. 2, No. 4 and No. 6 cylinders.

   b. Set No. 4 piston at TDC on its compression stroke and tighten rocker shaft bolts for No. 1, No. 3 and No. 5 cylinders.

   c. Torque specification for the rocker shaft retaining bolts is 13–16 ft. lbs.

13. Install the intake manifold and collector assembly.

14. Install the timing belt cover and camshaft sprocket. The left and right camshaft sprockets are different parts. Install the correct sprocket in the correct position.

15. Install the timing belt assembly, fill coolant and set engine timing to specifications.

Valve Lash

Adjustment

The 3.0L (SOHC) engine uses hydraulic valve lifters that automatically adjust the valve lash. No periodic adjustment is needed.

Oil Pan

REMOVAL & INSTALLATION

1. Disconnect the negative battery cable.

2. Raise and safely support the vehicle on jackstands

3. Drain the engine oil. When all the oil has been drained, install the drain plug and tighten it to 22–25 ft. lbs. (29–39 Nm).

4. The front and rear engine mounts must be disconnected. This means the engine must be safely supported. Place an underbody type screw-jack or equivalent under the crankshaft pulley. Cover the end of the jack with a pad made of shop rags to prevent damage to the crankshaft pulley. Take up just enough of the engine's weight to allow the engine mounts to be disconnected.

5. Remove the front engine mount (support) insulator through-bolt. Remove the rear engine mount (support) through-bolt.

6. Remove the two rear refrigerant/heater pipe hold down bracket bolts.

7. Remove the four crossmember (also called a transverse member) bolts and remove the crossmember.

8. Remove the exhaust inlet pipe.

9. Remove the four rear transaxle-to-engine brace bolts and the five front transaxle-to-engine brace bolts.

10. Remove the front transaxle-to-engine brace.

11. Disengage the low oil level sensor electrical connector.

12. Remove the 18 oil pan bolts in the reverse order of the tightening sequence, working from the outside, towards the center bolts.

13. Remove the pan and discard the seals.

To install:

14. Clean all parts well. Be sure that all old sealing material is removed from the oil pan and engine mating surfaces.

15. Install new oil pan seals. Apply Loctite® Ultra Gray 599 Silicone Sealer or equivalent to the ends of the oil pan seals.

16. Apply a bead of Loctite® Ultra Gray 599 Silicone Sealer or equivalent to the oil pan gasket rail inboard of the bolt holes.

17. Install the oil pan on the engine block. Tighten the 18 oil pan bolts in sequence, working from the inside, towards the outer bolts. Do not overtighten. Tighten to 62–70 inch lbs. (7–8 Nm).

18. Connect the low oil level sensor electrical connector.

19. Install the front and rear transaxle braces. Tighten all bolts to 22–30 ft. lbs. (30–40 Nm).

20. Install the exhaust inlet pipe.

21. Install the crossmember (also called transverse member) and tighten the bolts 58–65 ft. lbs. (78–88 Nm).

22. Install both engine support through-bolts and tighten to 58–65 ft. lbs. (78–88 Nm).

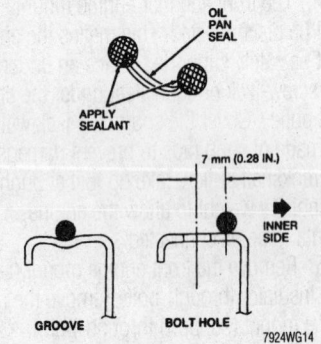

Apply RTV silicone sealer to the seal ends and to the oil pan gasket rail

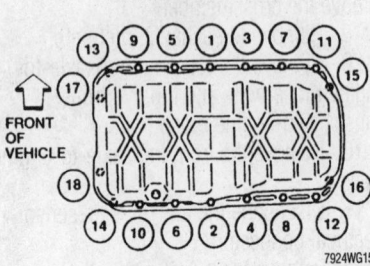

Tighten the 18 oil pan bolts in sequence, working from the inside, towards the outer bolts

23. Remove the support jack from under the crankshaft pulley.

24. Lower the vehicle.

25. Fill the engine with the specified engine oil to the required level.

26. Connect the negative battery cable. Start the engine and check for leaks.

Oil Pump

REMOVAL & INSTALLATION

The engine lubrication system on the 3.0L engine uses a positive-displacement, crankshaft driven oil pump that draws oil from the oil pan, through the oil pump screen cover and tube. The oil pump, then sends the oil to the oil filter where it is directed to the engine block and the cylinder heads. The oil, then drains back into the pan. The oil filter adapter is mounted on the oil pump. A full-flow oil filter is mounted to the oil filter adapter. The oil filter adapter uses an oil pressure relief valve that allows engine oil pressure to bypass a clogged or blocked filter and continue to lubricate the engine main oil gallery. This condition is not allowable for long periods of time because the engine oil is not being filtered. Under the recommended normal engine operating conditions, all of the engine oil will pass through the oil filter.

1. Disconnect the negative battery cable.

2. Remove the alternator belt, water pump and power steering belt and the A/C compressor belt, if equipped.

3. Remove the timing belt using the recommended procedure.

4. Raise and safely support the vehicle on jackstands

5. Drain the engine oil. After the oil has drained, install the drain plug and tighten it to 22–29 ft. lbs. (29–39 Nm).

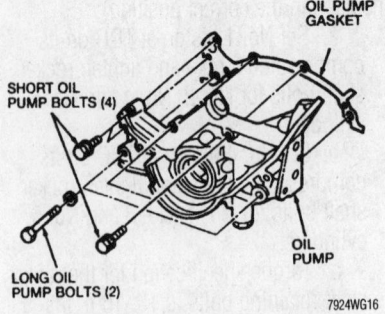

The oil pump is mounted on the front of the engine and driven by the crankshaft

6. Remove the alternator regulator mounting bracket-to-oil pump bolt and position the bracket out of the way.

7. Remove the power steering pump bracket bolts, remove the bracket and position the power steering pump aside.

8. Remove the crankshaft pulley.

9. Remove the outer timing belt guide, the crankshaft drive sprocket and the inner timing belt guide.

10. Disengage the oil pressure sensor electrical connector.

11. Remove the oil filter adapter bolts and remove the oil filter/adapter assembly. Discard the two oil filter adapter O-rings.

12. Remove the oil pan using the recommended procedure.

13. Remove the two oil pump screen and pick-up tube assembly bolts and remove the tube assembly. Discard the oil pump pick-up tube O-rings.

14. Remove the six oil pump bolts and remove the pump assembly from the engine. Discard the oil pump gasket.

To install:

15. With all pump and engine block sealing surfaces clean, install the oil pump using a new gasket. Tighten the long bolts to 108–144 inch lbs. (12–16 Nm) and the short bolts to 38–45 inch lbs. (6–7 Nm).

16. Install a new O-ring on the oil pump pick-up tube and install the tube/screen assembly. Tighten the two bolts to 144–180 inch lbs. (16–21 Nm).

17. Install the oil pan using the recommended procedure.

18. Install two new O-rings to the oil filter adapter and fit to the oil pump. Tighten the three oil filter adapter bolts to 144–180 inch lbs. (16–21 Nm).

19. Connect the oil pressure sensor electrical connector.

20. Install the inner timing belt guide, the crankshaft drive sprocket and the outer timing belt guide.

21. Install the crankshaft pulley. Use a strap wrench to hold the pulley and tighten the center bolt to 90–98 ft. lbs. (123–132 Nm).

22. Position the power steering pump and bracket on the engine. Tighten the bolts to 22–27 ft. lbs. (30–36 Nm).

23. Position the alternator regulator bracket on the oil pump, install the bolt and tighten to 15 ft. lbs. (21 Nm).

24. Lower the vehicle.

25. Install the timing belt using the recommended procedure. Great care must be exercised to properly align all timing marks.

26. Install and adjust all drive belts using the recommended adjustment procedure.

27. Fill the engine with the specified engine oil to the required level.

28. Connect the battery ground cable. Start the engine and check for leaks. Oil pressure can be verified by installing a mechanical oil pressure gauge in the pressure sensor outlet, just above the oil filter rim. Oil pressure with the engine running at no load should be approximately 17 psi at idle and 57–70 psi at 3200 rpm. Install the oil pressure sensor with sealant.

Rear Main Seal

REMOVAL & INSTALLATION

1. Disconnect the negative battery cable.
2. Remove the transaxle from the vehicle.
3. Remove the flexplate from the crankshaft.
4. Remove the rear oil seal retainer.

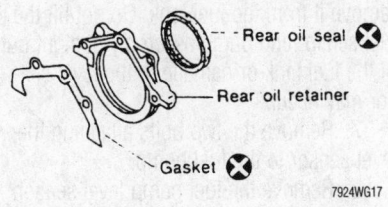

- Rear oil seal ⊗
- Rear oil retainer
- Gasket ⊗

7924WG17

Exploded view of the oil seal, retainer and gasket

❋❋ WARNING

Do not scratch the seal bore of the oil seal retainer when removing the oil seal.

5. Remove the oil seal from the seal retainer.

To install:

6. Apply clean engine oil to the lip and outer surface of the new seal to aid during installation.

7. Install the seal in the retainer using a suitable seal driver.

8. Using a new gasket install the retainer on the engine. Tighten the bolts to 52–61 inch lbs. (6–7 Nm).

9. Install the flexplate. Tighten the bolts to 61–69 ft. lbs. (83–93 Nm).

10. Install the transaxle and remaining components.

FUEL SYSTEM

Fuel System Service Precautions

Safety is the most important factor when performing not only fuel system maintenance but any type of maintenance. Failure to conduct maintenance and repairs in a safe manner may result in serious personal injury or death. Maintenance and testing of the vehicle's fuel system components can be accomplished safely and effectively by adhering to the following rules and guidelines.

- To avoid the possibility of fire and personal injury, always disconnect the negative battery cable unless the repair or test procedure requires that battery voltage be applied.

- Always relieve the fuel system pressure prior to disconnecting any fuel system component (injector, fuel rail, pressure regulator, etc.), fitting or fuel line connection. Exercise extreme caution whenever relieving fuel system pressure to avoid exposing skin, face and eyes to fuel spray. Please be advised that fuel under pressure may penetrate the skin or any part of the body that it contacts.

- Always place a shop towel or cloth around the fitting or connection prior to loosening to absorb any excess fuel due to spillage. Ensure that all fuel spillage (should it occur) is quickly removed from engine surfaces. Ensure that all fuel soaked cloths or towels are deposited into a suitable waste container.

- Always keep a dry chemical (Class B) fire extinguisher near the work area.

- Do not allow fuel spray or fuel vapors to come into contact with a spark or open flame.

- Always use a back-up wrench when loosening and tightening fuel line connection fittings. This will prevent unnecessary stress and torsion to fuel line piping. Always follow the proper torque specifications.

- Always replace worn fuel fitting O-rings with new. Do not substitute fuel hose or equivalent, where fuel pipe is installed.

Fuel System Pressure

RELIEVING

❋❋ CAUTION

The fuel injection system remains under pressure, even after the engine has been turned OFF. The fuel system pressure must be relieved before disconnecting any fuel lines. Failure to do so may result in fire and/or personal injury.

Relieve the fuel system pressure using the following procedure.

1. Remove the left side engine compartment relay panel cover.

2. Locate and remove the fuel pump relay from the relay panel.

3. Start the engine.

4. Allow the engine to run until it stalls from fuel starvation. After the engine stalls, crank the engine over two more times to ensure all pressure has been released.

5. Turn the ignition switch to the **OFF** position and install the fuel pump relay.

6. Most service work that follows fuel pressure relief also requires that the negative battery cable (ground) be disconnected before service work begins. This also prevents accidental fuel pump energizing that could repressurize the system.

Fuel Filter

REMOVAL & INSTALLATION

Clean fuel is key to a trouble-free fuel injection system. These vehicle use several filters. A replaceable in-line filter is the primary fuel filter. This is a high-pressure in-line type that provides extremely fine filtration to protect the small metering orifices of the fuel injection nozzles. The filter is of one-piece construction which cannot be cleaned. If the fuel filter becomes clogged or restricted, it should be replaced.

A low-pressure nylon filter is inside the fuel tank attached to the fuel pump/level sensor. It protects the fuel pump from larger contaminates in the tank. This strainer also allows the passage of small quantities of water which may accumulate in the fuel tank. The design and placement of this filter is such that if this filter becomes clogged, the fuel tank must be removed to service the filter. This provides the opportunity to clean the inside of the fuel tank.

A fuel injector screen is located at the top of each fuel injector on the engine and is not serviceable. If the fuel injector screen becomes clogged, the entire fuel injector must be replaced.

To replace the in-line fuel filter, use the following procedure.

✳✳ CAUTION

The fuel injection system remains under pressure after the engine has been turned OFF. The fuel system pressure must be relieved before disconnecting any fuel lines. Failure to do so may result in fire and/or personal injury.

1. Relieve the fuel system pressure using the recommended procedure.
2. Disconnect the negative battery cable.
3. Raise and safely support the vehicle on jackstands
4. Remove the fuel hose clamps.
5. Disconnect and plug the hoses to prevent leakage.
6. Remove the fuel filter from the bracket.

To install:

7. Install the fuel filter into the bracket with the arrow facing up, in the direction of the fuel travel to the engine.
8. Reconnect the fuel hoses.
9. Install and tighten the hose clamps. Verify that the clamps are properly tightened. System operating pressure is approximately 36 psi (248 kPa) and fuel will leak is connections are not properly made.
10. Lower the vehicle.
11. Reconnect the negative battery cable.
12. Check for leaks.

Fuel Pump

REMOVAL & INSTALLATION

These vehicles are equipped with Sequential Multi-port Fuel Injection (SFI) and an electric fuel pump. The fuel pump system consists of: the fuel pump and bracket assembly, an Inertia Fuel Shutoff (IFS) switch, a fuel level sensor and sensor filter, an in-line fuel filter, an evapora-

tive emission valve, a fuel pump relay, a fuel pressure regulator, the fuel supply manifold (fuel rail) and the fuel injectors. Note that the fuel pressure regulator maintains the fuel system pressure at 36–38 psi (248–262 kPa) with the engine running.

The SFI system has a fuel pump relay located in the left side engine compartment relay panel. The fuel pump relay is controlled by the Powertrain Control Module (PCM). The fuel pump is, in turn, controlled by the fuel pump relay. When the ignition switch is turned to the **ON** position, the PCM sends a signal to the fuel pump relay to close its contacts. The fuel pump is energized for five seconds prior to vehicle start-up. When the ignition switch is turned to the **OFF** position, the PCM sends a signal to the fuel pump relay to open its contacts to shut the fuel pump off.

This system uses an Inertia Fuel Shutoff (IFS) switch. Be aware of this when troubleshooting a No-Start or No-Fuel Pressure/Fuel Pump Inoperative complaint. The IFS switch is located behind the left side cowl trim panel below the hood release handle. In the event of a collision, the electric contacts in the IFS switch open. When the contacts open, the fuel delivery circuit is open, disabling the fuel system. The engine may continue to run for several seconds after the IFS switch is opened, due to fuel pressure remaining in the fuel supply line. Once the fuel pressure has been released, the engine will stall. To reset the IFS switch, depress the reset button on the IFS switch.

The fuel tank must be removed from the vehicle to service the fuel pump.

✳✳ CAUTION

The fuel injection system remains under pressure, even after the engine has been turned OFF. The fuel system pressure must be relieved before disconnecting any fuel lines. Failure to do so may result in fire and/or personal injury.

1. Relieve the fuel system pressure using the following procedure.
   a. Remove the left side engine compartment relay panel cover.
   b. Locate and remove the fuel pump relay from the relay panel.
   c. Start the engine.

d. Allow the engine to run until it stalls from fuel starvation. After the engine stalls, crank the engine over two more times to ensure all pressure has been released.
   e. Turn the ignition switch to the **OFF** position and install the fuel pump relay.
2. After relieving the fuel system pressure, disconnect the negative battery cable.
3. Raise and safely support the vehicle on jackstands
4. Remove the fuel tank using the recommended draining and tank removal procedure.

✳✳ CAUTION

Observe all applicable safety precautions when working around fuel. Do not allow fuel spray or fuel vapors to come in contact with a spark or open flame. Keep a dry chemical (Class B) fire extinguisher near the work area. Never drain store fuel in an open container due to the possibility of fire or explosion.

5. Remove the six fuel pump bolts.
6. Lift the fuel pump out of the fuel tank. Use care. The fuel level sensor and fuel pump and bracket must be tipped to remove it from the fuel tank. Do not lift the fuel sensor and pump assembly straight out of the fuel tank or damage to the level sensor may occur.
7. Remove the two bolts attaching the level sensor to the fuel pump.
8. Remove the fuel pump level sensor and the gasket.
9. Discard the gasket.
10. Remove the fuel pump from the bracket.

To install:

11. Position the fuel level sensor on the fuel pump and bracket and install the two bolts.
12. Install a new level sensor gasket. Carefully install the level sensor and pump assembly.
13. Install the six fuel pump bolts. Do not over-tighten the bolts. Tighten the bolts to just 18 to 21 inch lbs. (2 to 3 Nm).
14. Install the fuel tank.
15. Lower the vehicle. Refill the fuel tank as required.
16. Connect the negative battery cable. Verify that the fuel pump relay has been properly installed. Start the engine and check for proper operation.

DRIVE TRAIN

Automatic Transaxle Assembly

REMOVAL & INSTALLATION

1. Disconnect the negative battery cable.

2. Remove the starter motor from the vehicle.

3. Disengage the Park/Neutral Position (PNP, also called the Transmission Range or TR) switch electrical connector from the transaxle and disengage the 2 electrical connectors secured on the electrical connector retaining brackets.

4. Remove the 2 retaining brackets for the electrical connectors.

5. Remove the upper and lower mounting bolts for the oil level indicator (dipstick) and remove the oil level indicator.

6. Remove the upper transaxle-to-engine mounting bolts.

7. Raise and safely support the vehicle on jackstands.

8. Remove the left side inner engine and transaxle splash shield.

9. Remove the right and left front wheel halfshaft assemblies using the recommended procedure.

10. Place an oil pan under the transaxle and drain the transaxle fluid from the unit.

11. Remove the catalytic converter inlet pipe-to-exhaust bracket retaining bolt.

12. Place a suitable jack or lift under the transaxle assembly.

13. Remove the 2 exhaust bracket-to-transaxle assembly nuts. Remove the exhaust bracket.

14. Remove the 3 rear transaxle-to-engine brace mounting bolts. Remove the rear transaxle-to-engine brace.

15. Remove the 4 front transaxle-to-engine brace mounting bolts. Remove the front transaxle-to-engine brace.

16. Remove the transaxle-to-engine mounting bolt. Remove the 2 exhaust system bracket-to-transaxle studs and torque converter inspection plate.

17. Remove the shift cable nut and shift cable from the Manual Lever Position (MLP) switch.

18. Using a screwdriver release the shift cable locking pin from the shift cable bracket and remove the shift cable from the shift cable bracket.

19. Disengage the wiring harness electrical connector to the valve body.

20. Disconnect the transaxle ground strap.

21. Disengage the electrical connector to the MLP switch.

22. Remove the Vehicle Speed Sensor (VSS) and VSS hold-down bracket from the transaxle.

23. Disconnect the transaxle breather tube.

24. Using a socket tool to prevent the crankshaft from rotating, remove the 4 torque converter-to-flywheel mounting bolts.

25. Secure the transaxle assembly to the transaxle jack.

26. Remove the 3 rear transaxle assembly support insulator nuts and the rear insulator through-bolt from the rear support insulator. Remove the rear support insulator from the rear transaxle support bracket.

27. Remove the 3 front transaxle assembly support insulator bolts and the front insulator through-bolt from the front support insulator.

28. Using a pair of pliers, slide the hose clamps for the transaxle oil cooler tube hose away from the transaxle.

29. Disconnect and plug the transaxle oil cooler tubes from the transaxle assembly.

30. Carefully separate the transaxle assembly from the engine assembly. Lower the assembly from the vehicle.

To install:

31. Be sure that the transaxle is secured firmly to the transaxle jack.

32. Carefully raise the transaxle into the vehicle and align the transaxle to the engine assembly, making sure that the alignment dowels are positioned properly.

33. Install the lower transaxle-to-engine bolt and tighten to 22–30 ft. lbs. (30–40 Nm).

34. Install the rear transaxle support insulator and tighten the 3 insulator mounting nuts to 32–41 ft. lbs. (43–55 Nm).

35. Install and tighten the rear transaxle support insulator through-bolt to 32–41 ft. lbs. (43–55 Nm).

36. Install the front transaxle support insulator and tighten the 3 insulator mounting nuts to 30–38 ft. lbs. (41–52 Nm).

37. Install and tighten the front transaxle support insulator through-bolt to 47–54 ft. lbs. (64–74 Nm).

38. Using a socket on the crankshaft pulley bolt to prevent the crankshaft from turning, install the 4 torque converter-to-flywheel bolts and tighten them to 38 ft. lbs. (51 Nm). Remove the socket.

39. Remove the transaxle jack.

40. Install the transaxle oil cooler tubes along with new hose clamps.

41. Install the transaxle breather tube.

42. Install the VSS into the transaxle case and install the hold-down bracket. Tighten the VSS hold-down bracket bolt to 44–61 inch lbs. (5–7 Nm).

43. Install the shift cable into the shift cable bracket and snap the cable locking pin into position. Align the cable and bracket, then tighten the cable nut securely.

44. Reconnect the transaxle ground strap.

45. Install the torque converter inspection plate and exhaust bracket-to-transaxle studs.

46. Install the front and rear transaxle-to-engine braces. Tighten the brace bolts to 22–30 ft. lbs. (30–40 Nm).

47. Install the exhaust bracket on to the transaxle and tighten to 27 ft. lbs. (35 Nm).

48. Install and tighten the catalytic converter inlet pipe-to-exhaust bracket bolt to 32 ft. lbs. (43 Nm).

49. Reconnect the electrical connectors to the MLP switch and the valve body wiring harness.

50. Install the right and left halfshaft assemblies.

51. Install the inner engine and transaxle splash shield.

52. Lower the vehicle.

53. Install the upper transaxle to engine bolts.

54. Tighten the bolts to 33 ft. lbs. (44 Nm).

55. Install the transaxle oil level indicator and tighten the upper and lower retaining bolts.

56. Install the 2 electrical connector retaining brackets and tighten the retaining bolts. Reconnect the electrical connectors on the retaining brackets.

57. Reconnect the electrical connector to the PNP switch.

58. Install the starter motor. Reconnect the starter motor wiring.

59. Reconnect the negative battery cable. Refill the transaxle with the correct amount and type of clean transaxle fluid.

60. Start the engine.

61. Check for leaks and proper operation.

1 Rear Transaxle Support Insulator Through Bolt
2 Rear Transaxle Support Bracket Brace
3 Rear Transaxle Support Bracket Bolt (3 Req'd)
4 Rear Transaxle Support Bracket
5 Rear Transaxle Support Insulator Through Bolt Nut
6 Rear Transaxle Support Insulator
7 Rear Transaxle Support Insulator Bracket Bolts (4 Req'd)
8 Rear Transaxle Support Insulator Bracket
9 Rear Transaxle Support Insulator Nut (3 Req'd)
10 Front Transaxle Support Insulator Through Bolt Nut
11 Front Transaxle Support Bracket
12 Front Transaxle Support Insulator Through Bolt
13 Front Transaxle Support Insulator Bolt (3 Req'd)
14 Front Transaxle Support Insulator
15 Front Engine Support Insulator Through Bolt
16 Front Engine Support Bracket Bolt
17 Front Engine Support Bracket
18 Front Engine Support Insulator
19 Engine Insulator Mounting Bolt Nut (4 Req'd)
20 Transverse Member
21 Transverse Member Nuts (4 Req'd)
22 Transverse Member Bolts (4 Req'd)
23 Engine Insulator Mounting Bolt, Front (2 Req'd)
24 Engine Insulator Mounting Bolt, Rear (2 Req'd)
25 Rear Engine Support Insulator Through Bolt
26 Rear Engine Support Bracket Bolt (2 Req'd)
27 Rear Engine Support Bracket
28 Rear Engine Support Insulator Through Bolt Nut
29 Rear Transaxle Support Insulator
30 Front Engine Support Insulator Through Bolt Nut
A Tighten to 43-55 N·m (32-41 Lb-Ft)
B Tighten to 41-52 N·m (30-38 Lb-Ft)
C Tighten to 64-74 N·m (47-54 Lb-Ft)
D Tighten to 78-88 N·m (58-65 Lb-Ft)

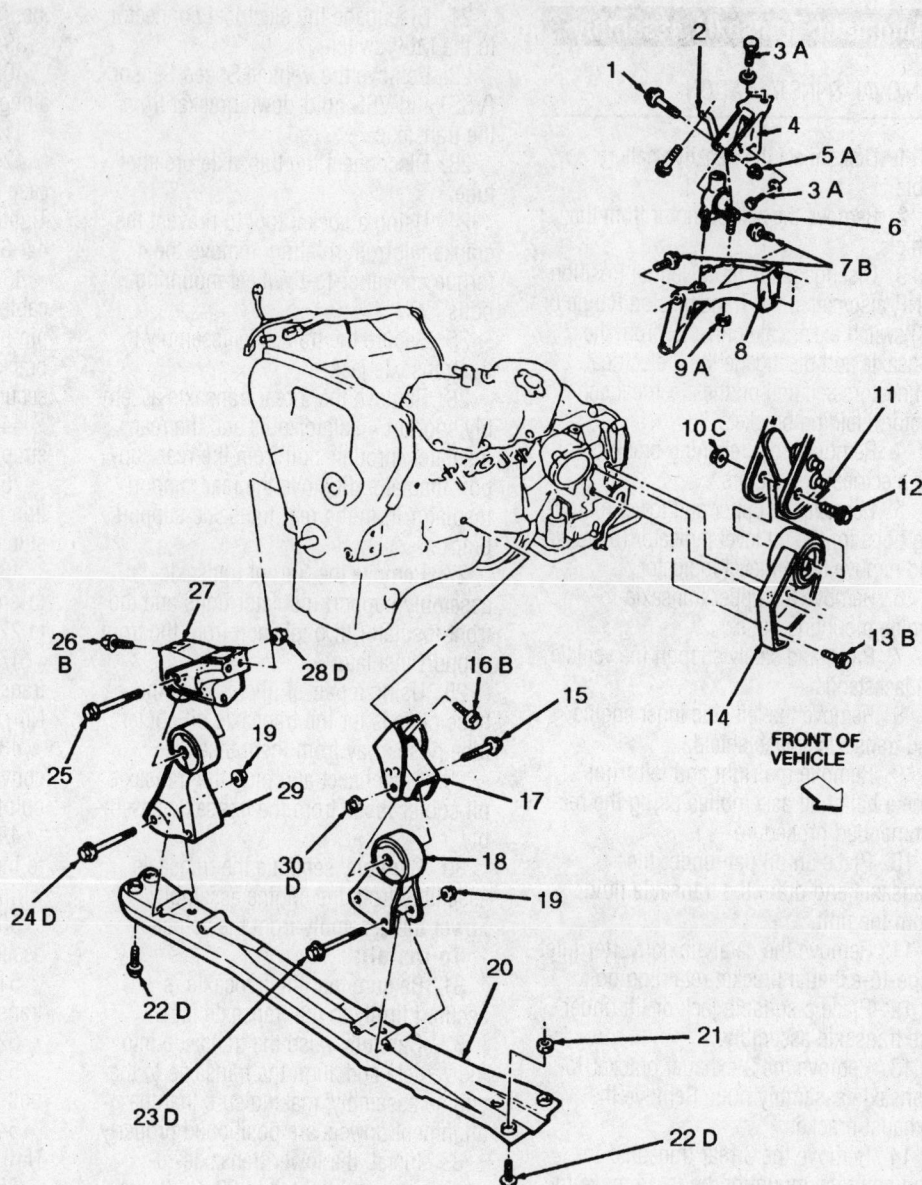

Exploded view of the engine and transaxle mounting

7924WG24

Halfshaft

REMOVAL & INSTALLATION

The front wheel driveshaft and joints, also known as halfshafts, are the mechanical links that transfer engine power from the transaxle and differential to the front wheels. At the transaxle end, the halfshaft joints are splined to the differential side gears. Disengagement of the left side halfshaft from the differential side gears is prevented by an expanding spring steel circlip. During installation, the circlip compresses around the shaft as it enters the gear. Once through the differential side gear, the circlip expands into a counterbore machined in the back of the differential side gear. The right side halfshaft and joint is secured by a front axle bearing.

The wheel ends of the halfshafts and joints are splined to the wheel hubs which are supported on one-piece wheel bearings. Disengagement of the halfshaft from the wheel hub is prevented by the front axle hub retainers, nut and washers, secured with a cotter pin.

Backlash between the wheel hub and halfshaft is eliminated by the splines. The wheel hub splines are machined straight while the halfshaft joint splines are machined with a slight helical cut. The difference in splines provides a tight backlash-free coupling without the removal and installation problems associated with an interference fit.

Halfshafts and their Constant Velocity Joints (CV-Joints) are precision made and should be handled with care. Do not angle a halfshaft CV-Joint more than about 20 degrees or the joint can be damaged. Never allow a halfshaft to hang if it is disconnected from the transaxle or the hub. Wire the halfshaft to a convenient underbody component. Never strike a halfshaft or CV-Joint component with a metal hammer. Do not drop an assembled halfshaft since internal components may be damaged or the protective boot can be cut allowing the special grease to run out of the joint and dirt and water into the joint, shortening the joint's service life.

1. Raise and safely support the vehicle on jackstands

2. Remove the wheel and tire assembly.

3. If necessary, remove the six right side inner fender splash shield screws and remove the right side inner splash shield. Remove the seven right side or seven left side engine and transmission splash shield screws. Remove the splash shields.

4. Remove and discard the cotter pin.

5. Remove the nut retainer, the nut and the hub retainer washers from the front hub assembly.

6. Remove and discard the cotter pin from the lower ball joint nut. Loosen the ball joint nut until it contacts the halfshaft joint. Strike the front wheel knuckle with a soft-faced hammer while pulling down on the lower control arm until the lower ball joint separates from the knuckle. There should now be enough clearance to remove the ball joint nut.

7. Disconnect the sway bar (also called the stabilizer bar) from the lower control arm at the sway bar link nut.

8. Carefully pry down on the lower control arm to separate the ball joint stud completely from the knuckle.

9. Use a prybar to separate the sway bar link from the lower control arm.

10. Separate the halfshaft and CV-Joint from the wheel hub.

11. Position a drain pan under the transaxle since some fluid may run out when the inner joint is disengaged from the transaxle.

12. A prybar is used to separate the inner CV-Joint from the transaxle. Use great care that the prybar does not damage the transaxle

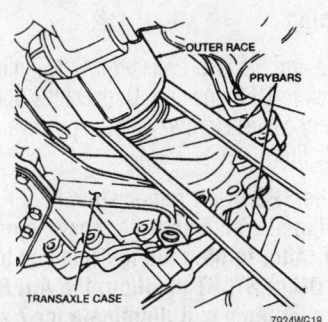

Removing the left side halfshaft by gently prying with two prybars to unseat the circlip

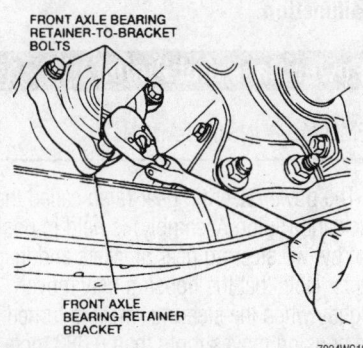

Right side halfshaft bearing retainer bracket

case, differential oil seal, outer race or boot. If removing the left side halfshaft, position prybars on both sides of the outer race, between the outer race and the transaxle case. Gently pry outward to unseat the circlip.

13. When removing the right side halfshaft, it is not be necessary to remove the halfshaft bearing retainer bracket from the cylinder block. Remove the three bearing retainer bolts and pull the right side halfshaft CV-Joint with the bearing retainer from the differential side gear.

14. Support the halfshafts and remove them from the vehicle. Use care not to damage the boots. Place the halfshafts on a flat, protected work area.

To install:

✳✳ CAUTION

Do not reuse the circlip used on the left side halfshaft.

15. To prevent over-expanding the circlip, install the circlip carefully, starting one end in the shaft groove, then working the circlip over the CV-Joint splined end. Always use a new circlip. No circlip is used on the right side halfshaft.

16. Inspect the CV-Joint boots. If service is required, replace the CV-Joint boots.

17. Inspect the differential oil seals. If damaged, the factory recommends using a hook-type puller and slide hammer arrangement to remove the seals. A seal driver is used to install the replacement differential oil seals.

18. If installing the left side halfshaft and CV-Joint assembly, position the CV-Joint so the splines are aligned with the differential side gear splines, then push the halfshaft joint into the differential case. As the circlip locks into the differential side gear groove, a click will be felt.

19. If installing the right side halfshaft and CV-Joint assembly, simply push the CV-Joint into the differential side gear. Position the bearing retainer onto the bearing retainer bracket which should still be on the cylinder block. Install the three bolts and tighten to 8–14 ft. lbs. (13–19 Nm).

20. Position the halfshaft through the wheel hub.

21. Insert the lower ball joint stud partially through the wheel knuckle and start the nut on the stud.

22. Push the lower ball joint completely into the knuckle and tighten the lower ball joint stud nut to 52–63 ft. lbs. (71–86 Nm). Secure the nut with a new cotter pin.

23. Install the sway bar link to the lower control arm and tighten the link nut to 12–16 ft. lbs. (16–22 Nm).

24. Install the front wheel outer bearing retainer, washer and axle nut. Tighten the hub nut to 174–231 ft. lbs. (235–314 Nm). Install the nut retainer and secure with a new cotter pin.

25. If removed, install the right side inner fender splash shield and tighten the screws securely. Position the engine and transmission splash shields, and tighten the screws securely.

26. Install the wheel and tire assembly. Tighten the lug nuts to 72–87 ft. lbs. (98–118 Nm).

27. Lower the vehicle.

28. Check the transaxle fluid level.

29. Road test the vehicle to verify correct operation and no noise or vibration.

STEERING AND SUSPENSION

Air Bag

✴✴ CAUTION

The Supplemental Inflatable Restraint (SIR) or Supplemental Restraint System (SRS) must be disabled before performing service on or around system components, steering column, instrument panel components, wiring and sensors. Failure to follow safety and disabling procedures could result in accidental air bag deployment, possible personal injury and unnecessary system repairs.

PRECAUTIONS

Several precautions must be observed when handling the inflator module to avoid accidental deployment and possible personal injury.

• Never carry the inflator module by the wires or connector on the underside of the module.

• When carrying a live inflator module, hold securely with both hands, and ensure that the bag and trim cover are pointed away.

• Place the inflator module on a bench or other surface with the bag and trim cover facing up.

• With the inflator module on the bench, never place anything on or close to the module which may be thrown in the event of an accidental deployment.

DISARMING

✴✴ CAUTION

To avoid rendering the Supplemental Restraint System (SRS) inoperative, which could lead to personal injury or death in the event of a severe frontal collision, extreme caution must be taken when servicing the electrical related systems.

➡**All SRS electrical wiring harnesses and connectors are covered with YELLOW outer insulation. Do not use electrical test equipment on any circuit related to the SRS (air bag) sensors. When installing SRS components, always install with the arrow marks facing the front of the vehicle.**

Disarming

To disarm the SRS system turn the ignition switch to the **OFF** position. Then, disconnect the both battery cables starting with the negative cable first and wait at least 10 minutes after the cables are disconnected. Be sure to insulate the battery terminal ends.

Arming

To arm the SRS system turn the ignition switch to **OFF** position. Connect the both battery cables starting with the positive cable first.

➡**The SRS or air bag system is equipped with a self-diagnostic operation. After turning the ignition key to the ON or START position, the AIR BAG warning lamp will illuminate for 7 seconds. After 7 seconds, the AIR BAG lamp will extinguish if no malfunction is detected. If the AIR BAG lamp does not extinguish after 7 seconds, check the SRS self diagnostic system for a malfunction.**

Power Rack and Pinion

REMOVAL & INSTALLATION

The power steering gear (also called the Rack and Pinion Assembly) is held in position by two steering gear brackets and insulators. Note that the housing may move slightly when the steering wheel is turned. If the housing moves more than 0.080 inch (2mm), replace the steering gear insulators. If one or both of the brackets move, check

the torque of the bracket bolts. The correct torque for these bolts is 54–72 ft. lbs. (73–97 Nm).

1. Place a drain pan under the steering rack.

2. Remove the brake master cylinder remote reservoir bracket screws. Position the reservoir out of the way and secure with wire.

3. Remove the junction block/high pressure line from the steering rack. Position the junction block and line out of the way.

4. Raise and safely support the vehicle.

5. Remove both front wheels and tires.

6. Remove the front sway bar (stabilizer bar).

7. Disconnect the tie rod ends from the steering knuckles using a press-type tool to avoid damage to the tie rod ends.

8. Pull back the steering ball stud dust seal and have an assistant turn the steering rack tie rod until the clamp bolt on the lower steering column is accessible.

9. Remove the lower steering column shaft clamp bolt.

10. Remove the power steering fluid return hose and position out of the way.

11. Remove the five steering rack clamp bracket bolts.

12. Lower the steering rack from the vehicle.

To install:

13. Carefully slide the steering gear rack and pinion assembly in place from the left side of the vehicle. Position the input shaft so it is just below the lower steering column shaft clamp.

14. Raise the steering gear until the plastic aligning tab on the input shaft enters the clamp bolt gap on the lower column shaft. Do not install the clamp bolt yet.

15. Examine the steering gear brackets. They should be marked UP with arrows pointing to one end of the bracket. Be sure the brackets are installed correctly. Tighten

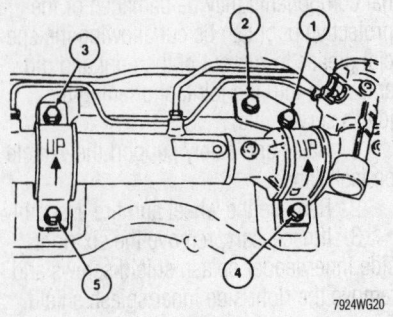

Torque the power steering rack mounting bolts in the sequence shown

the five steering gear bracket bolts to 54–72 ft. lbs. (73–97Nm) in sequence, working counterclockwise from the number one bolt (upper right side).

16. Connect the fluid return line to the steering gear.

17. Install the steering column shaft clamp bolt. Tighten the bolt to 17–22 ft lbs. (24–29 Nm). Install the dust cover.

18. Connect the tie rod ends.

19. Install the stabilizer bar.

20. Install the front tire and wheel assemblies. Torque the lug nuts to 72–87 ft. lbs. (98–118 Nm).

21. Lower the vehicle.

22. Install the junction block. Torque the high pressure line to 11–18 ft. lbs. (15–25 Nm).

23. Install the brake master cylinder reservoir.

24. Bleed the power steering system.

25. Check for leaks and proper operation.

MacPherson Strut

REMOVAL & INSTALLATION

1. Disconnect the negative battery cable.

2. Matchmark the front strut upper mounting bracket and the chassis strut tower.

3. Raise and safely support the vehicle on jackstands.

4. Remove the front wheel.

5. If equipped, remove the two front brake anti-lock sensor cable bracket bolts and position the anti-lock sensor cable out of the way.

6. Detach the brake tube from the strut.

7. Support the control arm.

8. Matchmark the knuckle to the strut so it can installed in the same position. This is important for the camber angle of the front wheel.

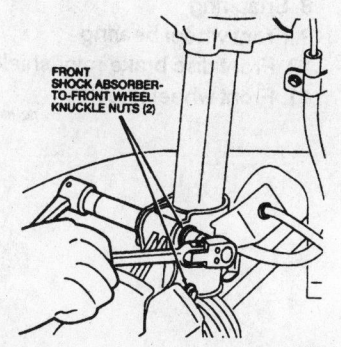

FRONT SHOCK ABSORBER-TO-FRONT WHEEL KNUCKLE NUTS (2)

7924WG21

The strut is attached to the knuckle with two large bolts

9. Remove the strut-to-steering knuckle bolts.

10. Support the strut and remove the 3 upper strut-to-chassis nuts. Remove the strut from the vehicle.

❊❊ WARNING

Never loosen the strut center nut until the spring is compressed or serious injury or vehicle damage may occur.

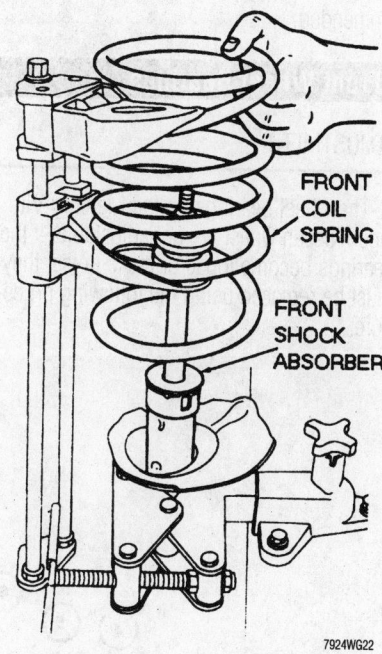

FRONT COIL SPRING

FRONT SHOCK ABSORBER

7924WG22

Compress the coil spring in a good spring compressor

11. Place the strut and coil spring assembly in a suitable vise and remove the strut nut cover.

12. Slightly loosen, but **do not** remove the front strut nut.

If desired, use the following steps to remove the coil spring from the strut.

13. Using an approved coil spring compressor, compress the coil spring.

14. Remove the strut assembly top nut.

15. Remove the following components from the strut assembly:

   a. The upper mounting bracket.

   b. The strut bearing.

   c. The bearing seat.

   d. The upper coil spring seat and dust boot.

   e. The coil spring.

16. Slowly release the tension of the coil spring compressor and remove the coil spring from the compressor tool.

17. Remove the coil spring insulator and slide the jounce bumper off of the strut assembly.

To install:

18. Slide the jounce bumper onto the strut assembly and install the coil spring insulator.

19. Carefully compress the coil spring with an approved coil spring compressor.

20. Reinstall the following components to the strut assembly.

   a. The coil spring.

➡**Install the coil spring to the strut assembly with the end of the spring in the lower coil spring seat indentation.**

   b. The upper coil spring seat and dust boot.

   c. The bearing seat and the bearing.

   d. The upper mounting bracket.

21. Install and tighten the strut assembly nut and tighten the nut to 43–58 ft. lbs. (59–78 Nm).

22. Install the strut assembly onto the vehicle and tighten the following:

- Strut-to-body nuts: 29–40 ft. lbs. (39–54 Nm)
- Strut-to-knuckle bolts: 89–91 ft. lbs. (113–123 Nm)

23. Reattach the brake tube to the strut assembly.

24. Install and tighten the two front brake anti-lock sensor cable bracket bolts.

25. Reinstall the tire and wheel assembly.

26. Connect the negative battery cable and the adjustable strut electrical connectors, if equipped.

27. Check and/or adjust the wheel alignment.

Shock Absorber

REMOVAL & INSTALLATION

1. Raise and safely support the vehicle.

2. Support the rear axle and slightly lower the vehicle enough to lessen tension on the shock absorber.

3. Remove the lower shock absorber retaining nut and washer.

4. Disconnect the lower end of the shock absorber from the mounting stud.

5. Remove the shock absorber upper end retaining nut and washer.

6. Remove the shock absorber from the vehicle.

To install:

7. Install the shock absorber onto the upper and lower mounting studs of the vehicle.

8. Install the washers and retaining nuts. Tighten the upper and lower retaining nuts to 22–30 ft. lbs. (30–41 Nm).

9. Lower the vehicle.

Lower Ball Joints

REMOVAL & INSTALLATION

To check if ball joint replacement is required, raise and safely support the vehicle clear of the floor and try to rock the wheel up and down. If any play is felt, have an assistant rock the wheel while observing the front suspension lower arm ball joint at the bottom of the steering knuckle. If any movement is seen, the ball joint should be replaced. If not, any wheel play indicates wheel bearing wear.

1. Raise and safely support the vehicle on jackstands

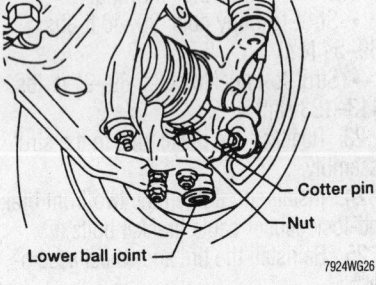

Loosen the nut on the lower ball joint stud

2. Remove the tire and wheel.

3. Remove and discard the ball joint cotter pin. Loosen the ball joint attaching nut from the steering knuckle. Because of tight clearance, the nut likely cannot be removed until the ball joint stud is loosened and lower slightly.

4. Strike the front knuckle with a hammer while pulling down on the lower control arm. There should now be enough clearance to allow removal of the ball joint stud nut. Separate the ball joint from the steering knuckle.

5. Remove the three bolts attaching the ball joint to the control arm.

6. Remove the ball joint from the control arm.

To install:

7. Install the ball joint to the control arm and install the attaching bolts.

8. Tighten the bolts to 56–80 ft. lbs. (76–109 Nm).

9. Install the ball joint into the steering knuckle, just enough to get the nut started on the stud. Then, push the ball joint stud fully in place. Tighten the nut to 52–63 ft. lbs. (71–86 Nm). Secure the nut with a new cotter pin.

10. Install the tire and wheel.

11. Lower the vehicle.

12. A front end alignment check is recommended.

Front Wheel Bearings

ADJUSTMENT

The wheel bearings on the Mercury Villager/Nissan Quest are not adjustable. If the bearings become loose or make noise, they must be replaced using the following procedure.

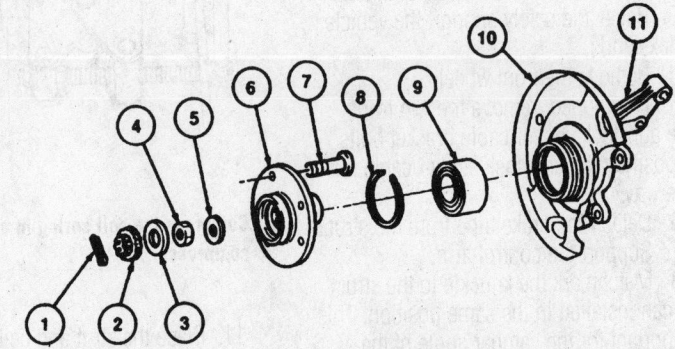

1. Cotter pin
2. Nut retainer
3. Insulator
4. Front axle wheel hub retainer
5. Front wheel outer bearing retainer washer
6. Wheel hub
7. Wheel hub bolt
8. Snap ring
9. Front wheel bearing
10. Front disc brake rotor shield
11. Front wheel knuckle

Exploded view of the knuckle, hub and bearing

REMOVAL & INSTALLATION

The front wheel knuckles transmit steering input, pivoting on the lower control arm ball joints and upper front strut bearing, house driveline components and support the disc brake calipers.

The front wheel hubs and the front steering knuckles can be replaced independently. The front hub/knuckle assemblies can also be disassembled in order to replace the knuckles, hubs, wheel hub bolts, front wheel bearings and front disc brake rotor shield. The front wheels attach to the wheel hub and the brake rotor. The wheel hub and front disc brake rotor are supported by a one-piece front wheel bearing pressed into the knuckle.

1. Raise and safely support the vehicle on jackstands

2. Remove the wheel and tire.

3. Remove the brake caliper assembly. DO NOT disconnect the brake hose. Hang the caliper on a piece of wire from a near by support such as the strut.

4. Remove the brake rotor.

5. Remove and discard the cotter pin from the end of the outboard CV-Joint stub shaft. Remove the hub nut retainer, washer and the hub nut. There should be another washer under the hub nut that acts as a front wheel bearing outer bearing retainer.

6. Disengage the lower ball joint stud from the steering knuckle using the following procedure.

   a. Remove and discard the cotter pin from the front lower ball joint.

   b. Loosen the lower ball joint nut until it contacts the front halfshaft joint.

   c. Strike the front knuckle with a hammer while pulling down on the lower control arm until the ball joint stud separates from the knuckle.

   d. Remove the ball joint nut.

   e. Disengage the lower ball joint stud from the steering knuckle.

7. Disengage the outer tie rod end stud from the steering knuckle using the following procedure.

   a. Remove and discard the cotter pin from the outer tie rod end stud.

   b. Remove the outer tie rod end retaining nut.

   c. Use a tie rod end puller to carefully press the tie rod end from the steering knuckle.

8. Remove the front ABS sensor bolt.

9. Remove the two front strut-to-front knuckle nuts and remove the two bolts. Disengage the strut from the steering knuckle.

10. Use a 2-jaw puller to separate the front halfshaft outboard CV-Joint stub shaft from the knuckle/bearing assembly.

11. Remove the front wheel hub, knuckle and wheel bearing assembly from the vehicle.

12. If the knuckle is being replaced with a service part, changeover the steer-ing stop bolt and jam nut from the old knuckle to the replacement part.

13. To remove the front wheel bearing, jig up a puller to bear against the front wheel bearing inner race and pull the race from the hub/knuckle assembly.

14. Use a shop press to press out damaged wheel studs and also to press out the outer bearing race.

15. Use a shop press to press out the inner bearing race.

To install:

16. If the front wheel bearings were removed, assemble the ABS sensing ring, if removed and the disc brake dust shield under the steering knuckle. Use a shop press to push in new front wheel bearing inner and outer races. Support the knuckle and press the front wheel bearing into the knuckle and install the snap ring retainer. Support the bearing assemblies and press

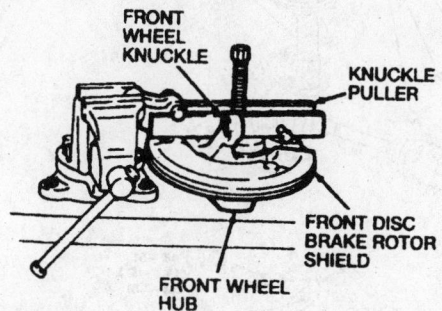

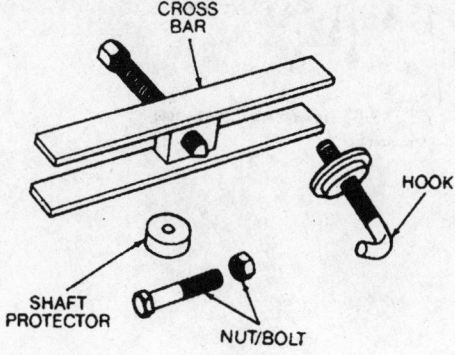

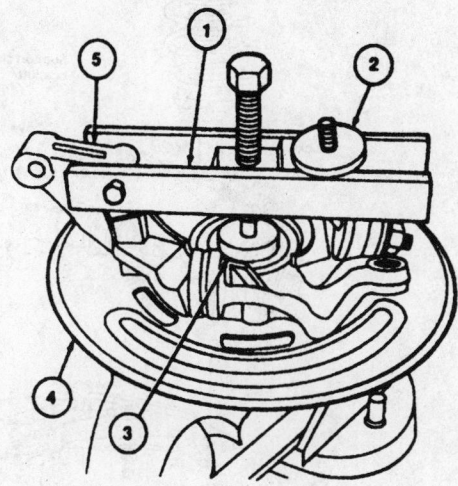

1. Knuckle puller
2. Knuckle puller adapter
3. Step plate adapter
4. Front disc brake rotor shield
5. Front wheel knuckle

7924WG27

Example of a puller set up to bear against the front wheel bearing inner race

the hub onto the knuckle and wheel bearing assembly.

17. Install the hub, knuckle and bearings as an assembly. Position the assembly on the halfshaft outer CV-Joint stub axle end. Guide the knuckle into the front strut and install the two knuckle-to-strut bolts and nuts. Tighten the nuts to 83–91 ft. lbs. (113–123 Nm).

18. Install the ABS sensor bolt. Do not overtighten. Tighten to just 16–21 inch lbs. (1.8– 2.4 Nm).

19. Install the outer tie rod end to the steering knuckle. Tighten the nut to 22–29 ft. lbs. (29–39 Nm). If the cotter pin holes

do not align, tighten the nut slightly until they do. Never loosen the nut to align the holes. Secure the nut with a new cotter pin.

20. Start the lower ball joint stud to the steering knuckle and partially install the nut, then push the ball joint stud fully in place. Tighten the ball joint stud nut to 52– 63 ft. lbs. (71– 86 Nm). Secure the nut with a new cotter pin.

21. Install the front wheel outer bearing retaining washer and the hub retainer nut. Tighten to 174–231 ft. lbs. (235–314 Nm). Install the nut retainer, insulator and a new cotter pin.

22. Install the front brake rotor and install the disc brake caliper.

23. If removed, install the steering stop bolt.

24. Install the tire and wheel assembly. Tighten the lug nuts to 72 to 87 ft. lbs. (98 to 118 Nm).

25. Lower the vehicle. Pump the brake pedal slowly to seat the front brake pads. Do not move the vehicle until a firm pedal is obtained.

26. A front end alignment is recommended.

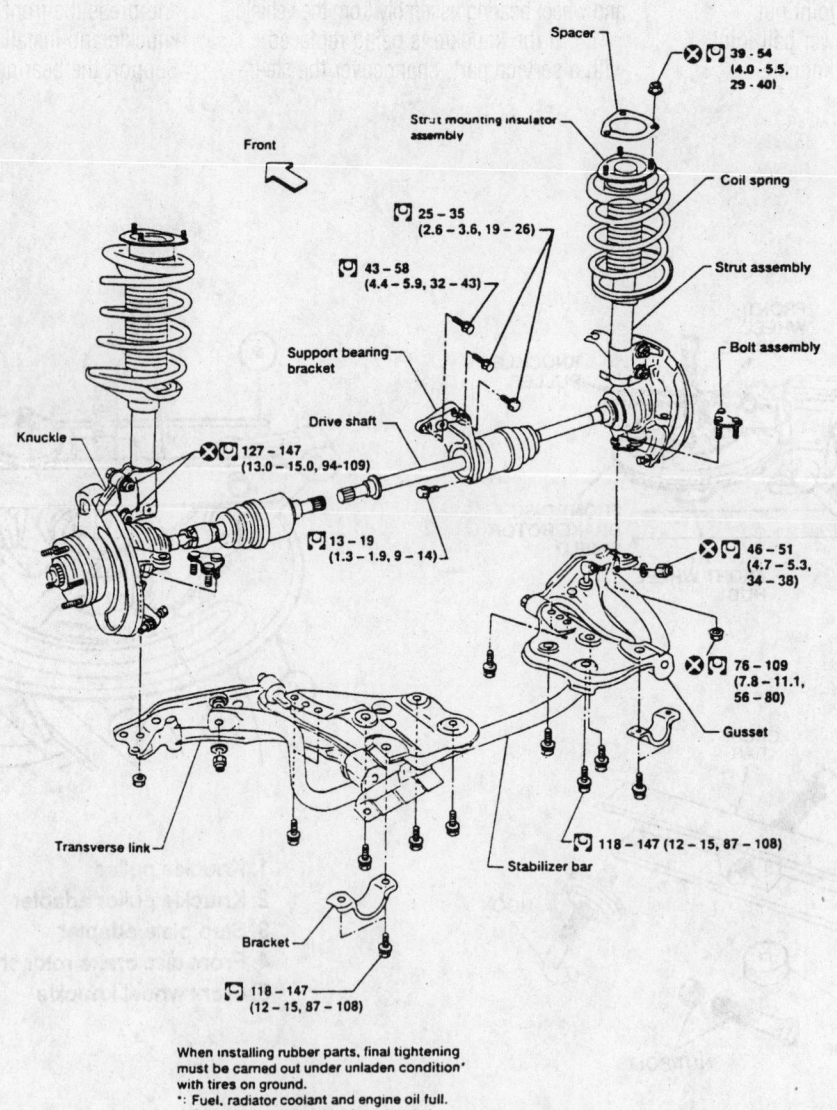

Exploded view of the front suspension and drive axles

ENGINE REPAIR

➡**Disconnecting the negative battery cable on some vehicles may interfere with the functions of the on board computer systems and may require the computer to undergo a relearning process, once the negative battery cable is reconnected.**

Distributor

The Forester model is equipped with a distiburtorless ignition system.

Ignition Timing

ADJUSTMENT

The ignition timing is controlled by the engine control computer and is not adjustable. To check the ignition timing proceed as follows:

1. Warm up the engine, then turn the ignition **OFF**.
2. Connect a timing light to the No. 1 spark plug wire according to the manufacture's directions.
3. Start engine, with the vehicle at idle check the timing.
4. The timing should be 7–23° Before Top Dead Center (BTDC) at 700 RPM.
5. If the timing is not correct, there could be a problem in the ignition control system.

Engine Assembly

REMOVAL & INSTALLATION

2.5L Engine

✳✳ CAUTION

The fuel injection system remains under pressure after the engine has been turned OFF. Properly relieve fuel pressure before disconnecting any fuel lines. Failure to do so may result in fire or personal injury.

1. Relieve the fuel system pressure.
2. Disconnect the battery cables, negative first, then positive. Remove the battery from the vehicle.

3. Remove the engine undercover.
4. Drain the engine oil and coolant into suitable containers.
5. Disconnect the radiator hoses and fan motor harness, then remove the radiator.
6. If equipped with A/C, discharge the system using an approved recovery/recycling machine. Disconnect and cap the lines from the compressor.
7. Remove the air intake duct.
8. Remove the air cleaner element and upper cover.
9. Remove the evaporator canister and bracket.
10. Unfasten the following electrical connectors and cables:
- O$_2$sensor
- Engine ground terminal
- Crank angle sensor connector
- Cam angle sensor connector
- Knock sensor connector
- Alternator connector and terminal
- A/C compressor connectors, if equipped
- Accelerator cable
- Cruise control cable, if equipped
11. Disconnect the following hoses:
- Brake booster hose
- Heater inlet and outlet hoses
12. Remove the alternator drive belt.
13. Disconnect the wires from the spark plugs on the left side of the engine.
14. Remove the power steering pump line bracket.
15. Remove the power steering pump, leaving the lines connected and position it aside.
16. Raise and support the engine safely.
17. Remove the exhaust Y-pipe.
18. Remove the lower starter nuts.
19. Remove the lower engine-to-transmission nuts.
20. Remove the front engine mount-to-crossmember nuts.
21. Lower the vehicle.
22. Remove the starter.
23. If equipped with an automatic transmission, perform the following:
    a. Remove the torque converter service hole plug.
    b. Matchmark the torque converter-to-driveplate
    c. Rotate the engine to remove the torque converter-to-driveplate bolts as they become accessible.

24. If equipped with a manual transmission, remove the flywheel cover.
25. Remove the pitching stopper.
26. Disconnect the fuel delivery, return and evaporation hoses.
27. Support the engine with a suitable lifting device attached to the engine lifting eyes.
28. Slightly raise the engine.
29. Raise the transmission with a floor jack.
30. If equipped with a manual transmission, pull the engine forward then up and out of the vehicle to clear the transmission mainshaft.
31. If equipped with an automatic transmission, pull the engine forward then up and out of the vehicle.

To install:

32. If equipped with a manual transmission, apply a small amount of grease to the splines of the mainshaft.
33. Position the engine in the engine compartment and align it with the transmission.
34. Install the engine and tighten the upper bolts to 34–40 ft. lbs. (44–54 Nm).
35. Remove the lifting device and floor jack.
36. Install the pitching stopper and tighten the bolts to the following specifications:
- Body side—49 ft. lbs. (67 Nm)
- Bracket side—40 ft. lbs. (54 Nm)
37. If equipped with a manual transmission, install the flywheel cover.
38. If equipped with an automatic transmission, perform the following:
    a. Align the matchmarks, install the torque converter-to-driveplate bolts while rotating the engine, and tighten to 20 ft. lbs. (26 Nm).
    b. Install the service hole cover.

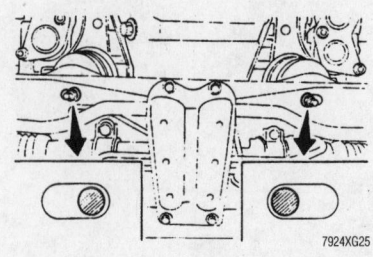

7924XG25

Be sure to tighten the front cushion rubber mounting bolts in the innermost elliptical hole in the front crossmember

39. Install the evaporator canister and bracket.

40. Install the power steering pump. Tighten the retainer bolts to 22–36 ft. lbs. (29–47 Nm).

41. Install and tension the drive belt.

42. Install the starter. Tighten the bolts to 34–40 ft. lbs. (44–52 Nm).

43. Raise and support the vehicle safely.

44. Install the lower engine-to-transmission nuts and tighten them to 34–40 ft. lbs. (44–52 Nm).

45. Install the lower engine mounting nuts. Tighten them to 61 ft. lbs. (83 Nm) in the inner most elliptical hole in the front crossmember so the clearance is 0.16–0.24 in. (4–6mm).

46. Install the exhaust Y-pipe with new gaskets and nuts.

47. Connect the following hoses:
- Brake booster hose
- Heater inlet and outlet hoses

48. Attach the following cables:
- Accelerator cable
- Cruise control cable, if equipped

49. Fasten the following electrical connectors:
- Engine harness connectors
- O₂sensor
- Engine ground terminal
- Crank angle sensor connector
- Cam angle sensor connector
- Knock sensor connector
- Alternator connector and terminal
- A/C compressor connectors, if equipped

50. Install the air cleaner element and cover.

51. If equipped, connect the A/C lines with new O-rings and tighten the bolts to 23 ft. lbs. (31 Nm).

52. Install the radiator.

53. Install the engine undercover.

54. Check the engine oil and add if necessary.

55. Install the battery.

56. Fill and bleed the cooling system.

57. Charge the A/C system using an approved recovery/recycling machine.

58. If equipped, check the automatic transmission fluid level and add Dexron II® if necessary.

59. Start the engine and allow it to reach normal operating temperature. Check for leaks.

Water Pump

REMOVAL & INSTALLATION

2.5L Engine

1. Disconnect the negative battery cable.

2. Raise and safely support the vehicle and remove the engine undercover.

3. Drain the coolant into a suitable container.

4. Disconnect the radiator outlet hose.

5. Remove the radiator fan motor assembly.

6. Remove the accessory drive belts.

7. Remove the timing belt, tensioner and camshaft angle sensor. Refer to the procedure in this section for removal and installation steps.

8. Remove the left side camshaft pulley(s) and left side rear timing belt cover. Remove the tensioner bracket.

9. Disconnect the radiator hose and heater hose from the water pump.

10. Remove the water pump retainer bolts.

11. Remove the water pump.

To install:

12. Clean the gasket mating surfaces thoroughly. Always use new gaskets during installation.

13. Install the water pump and tighten the bolts, in sequence, to 7–10 ft. lbs. (10–14 Nm). After tightening the bolts once, retighten to the same specification again.

14. Inspect the radiator hoses for deterioration and replace as necessary. Connect the radiator hose and heater hose to the water pump.

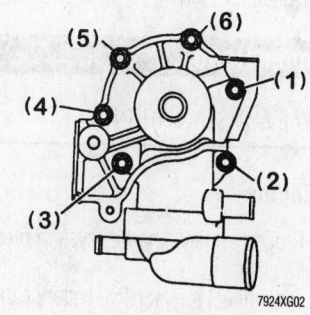

Water pump bolt tightening sequence— 2.5L engine

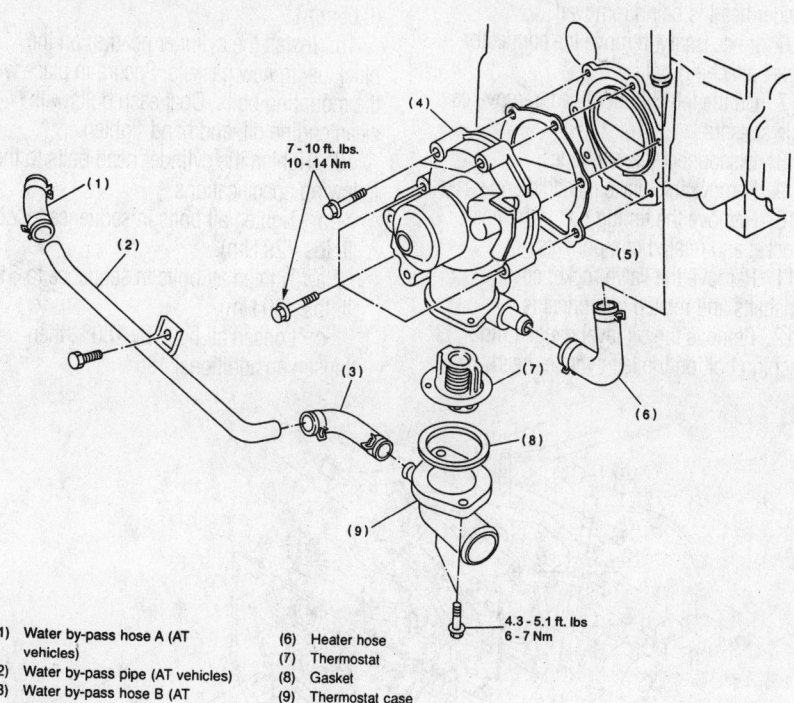

(1) Water by-pass hose A (AT vehicles)
(2) Water by-pass pipe (AT vehicles)
(3) Water by-pass hose B (AT vehicles)
(4) Water pump ASSY
(5) Gasket
(6) Heater hose
(7) Thermostat
(8) Gasket
(9) Thermostat case

Exploded view of the water pump mounting and related components—2.5L engine

15. Install the left side rear timing belt cover, left side camshaft pulley(s) and tensioner bracket.

16. Install the camshaft angle sensor, tensioner and timing belt.

17. Install the accessory drive belts.

18. Install the radiator fan motor assembly.

19. Install the radiator outlet hose.

20. Install the engine undercover.

21. Fill the system with coolant.

22. Connect the negative battery cable.

23. Start the engine and allow it to reach operating temperature.

24. Check for leaks.

Cylinder Head

REMOVAL & INSTALLATION

2.5L Engine

1. Properly relieve the fuel system pressure.

2. Disconnect the negative battery cable.

3. Remove the engine undercover.

4. Remove the V-belt.

5. Remove the engine accessories and brackets from the side of the engine whose cylinder head is being removed.

6. If necessary, remove the connector bracket attaching bolt.

7. On the left cylinder head, remove cam angle sensors.

8. Disconnect the fuel pipes.

9. Remove the intake manifold and gasket.

10. Remove the timing belt, camshaft sprocket and related components.

11. Remove the valve rocker cover, camshafts and related components.

12. Remove the oil level gauge guide attaching bolt on the left cylinder head.

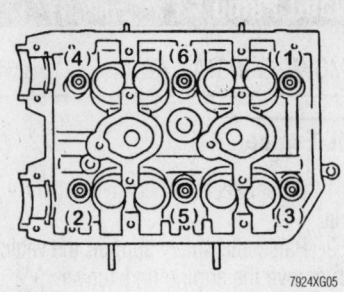

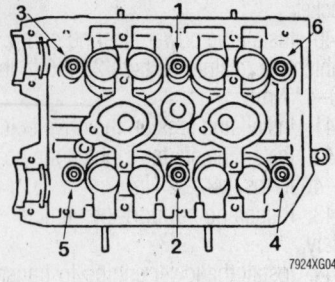

Cylinder head bolt loosening sequence— 2.5L engine

Cylinder head bolt tightening sequence— 2.5L engine

13. Remove the cylinder head bolts in the proper sequence. Leave bolts 1 and 3 installed loosely to prevent the cylinder head from falling.

14. Separate the cylinder head from the block, Use a plastic-faced hammer, if needed.

15. Remove bolts 1 and 3. Remove the cylinder head and gasket.

16. Clean all gasket material from both mating surfaces.

To install:

17. Inspect the cylinder head for warpage. Warpage should not exceed 0.0020 in. (0.05mm).

18. Install the cylinder head(s) on the block using new gaskets. Secure in place with the mounting bolts. Coat each bolts with clean engine oil, and hand-tighten.

19. Tighten the cylinder head bolts to the following specifications:

   a. Tighten all bolts in sequence to 22 ft. lbs. (29 Nm).

   b. Tighten all bolts in sequence to 51 ft. lbs. (69 Nm).

   c. Loosen all bolts by 180°, then loosen an additional 180°.

   d. Tighten bolts 1 and 2 to 25 ft. lbs. (24 Nm).

   e. Tighten bolts 3, 4, 5 and 6 to 11 ft. lbs. (15 Nm).

   f. Tighten all bolts in sequence by 80–90°.

   g. Tighten all bolts in sequence an additional 80–90°.

➡ **Do not exceed 180° total tightening.**

20. Install the oil level gauge guide attaching bolt on the left cylinder head.

21. Install the valve rocker cover, camshafts and related components.

22. Install the timing belt, camshaft sprocket and related components.

23. Install the intake manifold and tighten bolts to 21–25 ft. lbs. (28–34 Nm)

24. Connect the fuel delivery pipes.

25. On the left cylinder head, install cam angle sensors.

26. Install the connector bracket attaching bolt.

27. Connect the spark plug wires.

28. As necessary, install the engine accessories and brackets, then install and tension the drive belts.

29. Connect the negative battery cable. Start the engine and allow it to reach operating temperature. Check for leaks.

Intake Manifold

REMOVAL & INSTALLATION

2.5L Engine

✳✳ CAUTION

The fuel injection system remains under pressure after the engine has been turnedOFF. Properly relieve fuel pressure before disconnecting any fuel lines. Failure to do so may result in fire or personal injury.

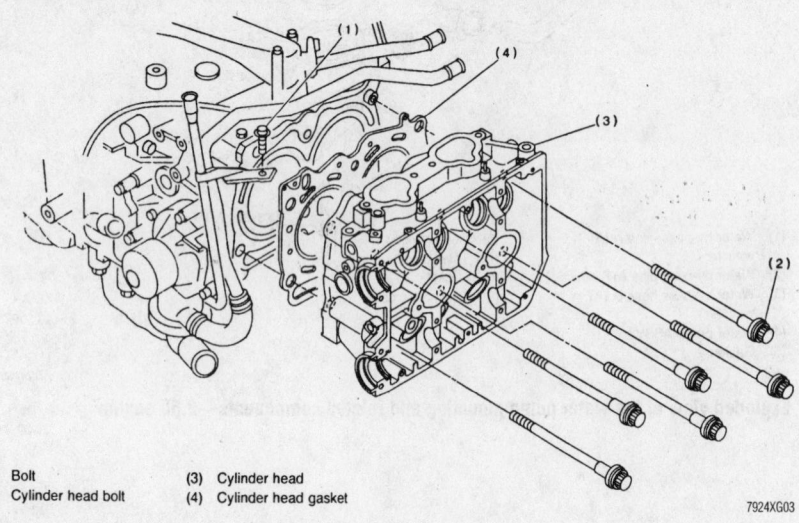

(1) Bolt
(2) Cylinder head bolt
(3) Cylinder head
(4) Cylinder head gasket

Exploded view of the cylinder head mounting—2.5L engine

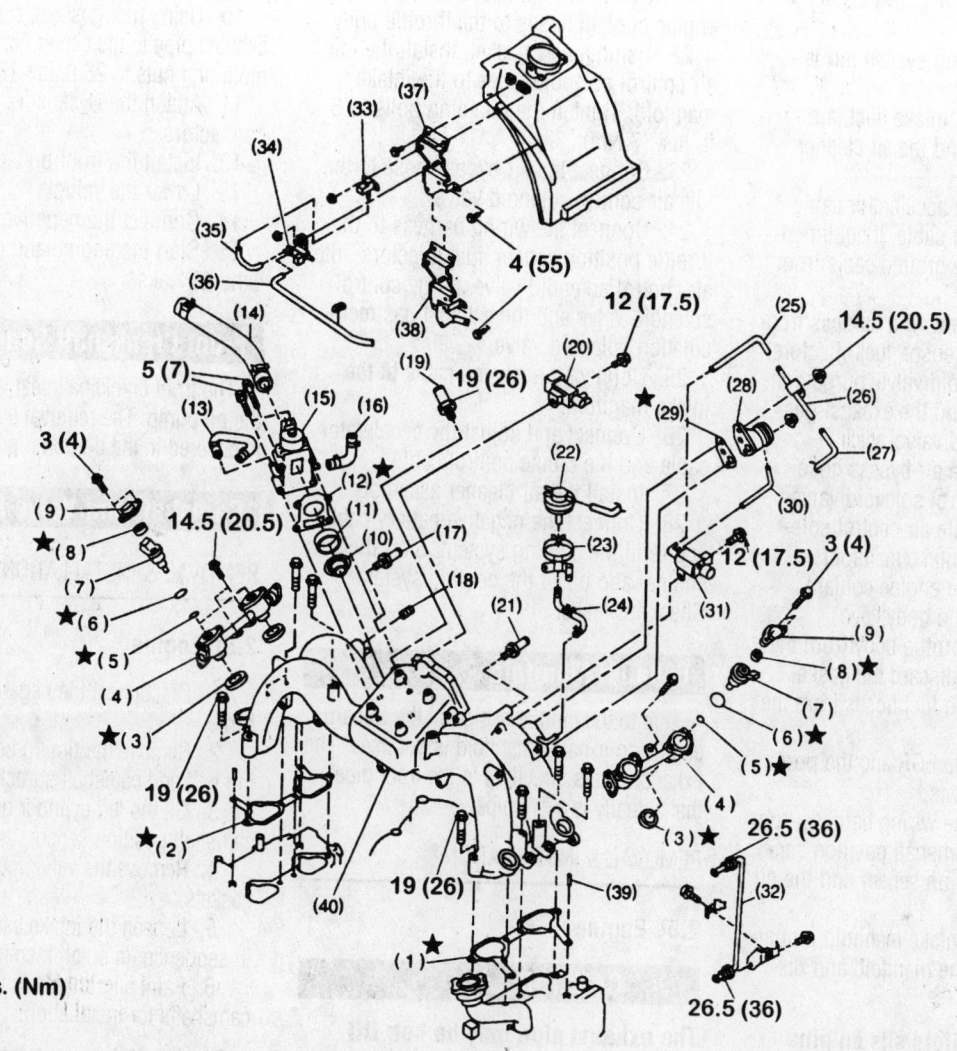

ft. lbs. (Nm)

(1) Intake manifold gasket LH
(2) Intake manifold gasket RH
(3) Fuel injector pipe insulator
(4) Fuel injector pipe
(5) O-ring A
(6) O-ring B
(7) Fuel injector
(8) Insulator
(9) Fuel injector cap
(10) Plate
(11) Sealing
(12) Gasket
(13) Engine coolant hose B
(14) Air by-pass hose
(15) Idle air control solenoid valve
(16) Engine coolant hose A

(17) Nipple (Equipped cruise control model)
(18) Plug
(19) PCV valve
(20) Purge control solenoid valve
(21) Nipple
(22) BPT
(23) BPT holder bracket
(24) Back pressure hose
(25) EGR vacuum hose A
(26) EGR vacuum pipe
(27) EGR vacuum hose C
(28) EGR valve
(29) Gasket
(30) EGR vacuum hose B
(31) EGR solenoid valve
(32) EGR pipe

(33) Pressure sensor
(34) Pressure sources switching solenoid valve
(35) Vacuum hose A
(36) Vacuum hose B
(37) Bracket (Except Canada spec. vehicles)
(38) Bracket (For Canada spec. vehicles)
(39) Collar
(40) Intake manifold

Exploded view of the intake manifold—2.5L engine

7924XG06

1. Properly relieve the fuel system pressure.

2. Disconnect the negative battery cable.

3. Drain the cooling system into a suitable container.

4. Remove the air intake duct, air cleaner upper cover and the air cleaner element.

5. Disconnect the accelerator cable and the cruise control cable, if equipped.

6. Disconnect the ground cable from the intake manifold.

7. Disconnect the wiring harness from the throttle position sensor, fuel injectors, idle air control solenoid valve, purge control solenoid valve, and the exhaust gas recirculation solenoid valve.

8. Disconnect the air bypass hose from the idle air control solenoid valve.

9. Remove the idle air control solenoid valve from the intake manifold.

10. Disconnect the engine coolant hoses from the throttle body.

11. Remove the throttle body from the intake manifold and discard the gasket.

12. Disconnect the fuel hoses from the fuel pipes.

13. Disconnect the EGR and the purge control solenoid valves.

14. Disconnect the wiring harness from the knock sensor, camshaft position sensor, crankshaft position sensor and the oil pressure switch.

15. Remove the intake manifold mounting bolts. Remove the manifold and discard the gaskets.

➡️**The intake manifold sits on pins that protrude from the cylinder heads. Be sure the pins remain in the cylinder heads.**

To install:

16. Using new gaskets, install the manifold to the engine. Tighten the mounting bolts to 19 ft. lbs. (26 Nm).

17. Connect the wiring to the knock sensor, camshaft position sensor, crankshaft position sensor and the oil pressure switch.

18. Connect the EGR and the purge control solenoid valves.

19. Connect the fuel hoses to the fuel pipes. Be sure to secure the hoses with new clamps.

20. Using new gaskets, install the throttle body to the intake manifold.

Tighten the retaining bolts to 16 ft. lbs. (22 Nm).

21. Using new clamps, connect the engine coolant hoses to the throttle body.

22. Using a new gasket, install the idle air control solenoid valve to the intake manifold. Tighten the retaining bolts to 5 ft. lbs. (7 Nm).

23. Connect the air bypass hose to the idle air control solenoid valve.

24. Connect the wiring harness to the throttle position sensor, fuel injectors, idle air control solenoid valve, purge control solenoid valve and the exhaust gas recirculation solenoid valve.

25. Connect the ground cable to the intake manifold.

26. Connect and adjust the accelerator cable and the cruise control cable.

27. Install the air cleaner assembly.

28. Connect the negative battery cable and refill the cooling system. Start the engine, and bleed the cooling system. Check for leaks.

Exhaust Manifold

Due to the unique design of the Subaru engine an exhaust manifold is not used. The exhaust enters directly into the front pipe that actually is a "Y" pipe.

REMOVAL & INSTALLATION

2.5L Engine

✴️ CAUTION

The exhaust pipe may be hot; DO NOT perform any work until the system has completely cooled.

1. Disconnect the negative battery cable.

2. Remove the front under cover.

3. Disconnect the O_2 sensors electrical connectors.

4. Remove the bolts securing the exhaust manifold covers and remove the covers.

5. Remove the front pipe-to-front catalytic converter mounting nuts.

6. Remove the nuts that secure the exhaust pipe to the cylinder head and remove the exhaust pipe.

7. Discard the gaskets.

To install:

8. Clean all gasket surfaces completely.

9. Install the exhaust pipe to the cylinder head using new gaskets. Tighten the mounting nuts to 22 ft. lbs. (30 Nm).

10. Using new gaskets, connect the exhaust pipe to the center pipe. Tighten the mounting nuts to 26 ft. lbs. (35 Nm).

11. Attach the O_2 sensors electrical connectors.

12. Install the front under cover.

13. Lower the vehicle.

14. Connect the negative battery cable.

15. Start the engine and check for exhaust leaks.

Front Crankshaft Seal

The front crankshaft seal is mounted in the oil pump. The removal and installation is covered in the oil pump procedure.

Camshaft and Valve Lifters

REMOVAL & INSTALLATION

2.5L Engine

1. Disconnect the negative battery cable.

2. Remove the timing belt covers, timing belt and camshaft sprockets.

3. On the left cylinder head, remove the camshaft position sensor.

4. Remove the valve rocker covers and gaskets.

5. Loosen the intake camshaft cap bolts in sequence, in small increments.

6. Paint alignment marks on the camshafts for installation.

➡️**Be sure to keep the intake and exhaust bearing caps and camshafts in proper order for reassembly. Also note the positioning and location of the camshafts for reinstallation.**

7. Remove the intake camshaft bearing caps, then remove the camshaft.

➡️**When removing the exhaust camshaft, the valve lifter and valve shim may fall out. Have an assistant help keep the lifters and shims in the proper order.**

8. Loosen the exhaust camshaft cap bolts in sequence, in small increments.

9. Remove the exhaust camshaft bearing caps, then remove the camshaft.

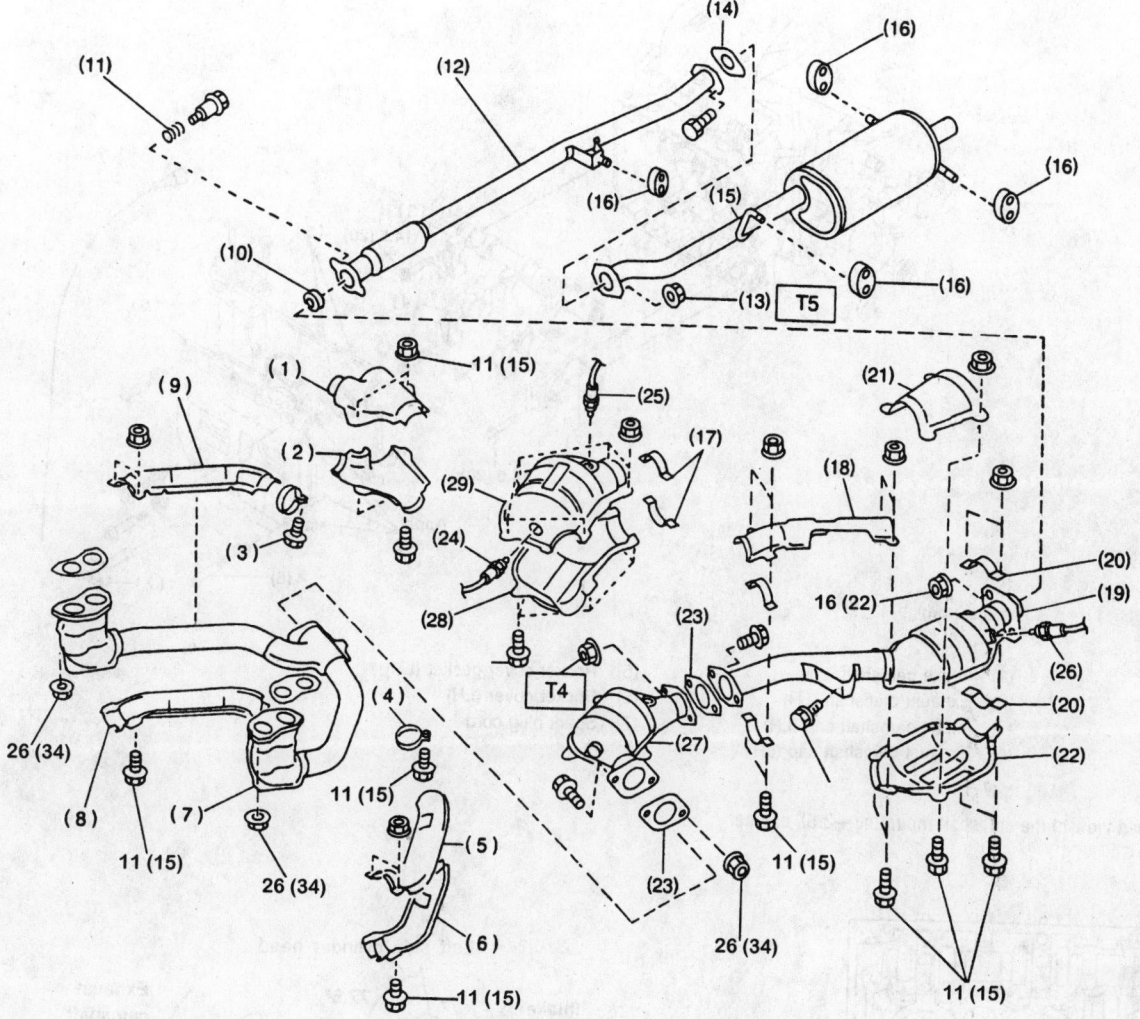

ft. lbs. (Nm)

(1) Upper front exhaust pipe cover CTR
(2) Lower front exhaust pipe cover CTR
(3) Band RH
(4) Band LH
(5) Upper front exhaust pipe cover LH
(6) Lower front exhaust pipe cover LH
(7) Front exhaust pipe
(8) Lower front exhaust pipe cover RH
(9) Upper front exhaust pipe cover RH
(10) Gasket
(11) Spring

(12) Rear exhaust pipe
(13) Self-locking nut
(14) Gasket
(15) Muffler
(16) Cushion rubber
(17) Clamp
(18) Upper center exhaust pipe cover
(19) Center exhaust pipe
(20) Clamp B
(21) Upper rear catalytic converter cover
(22) Lower rear catalytic converter cover
(23) Gasket
(24) Front oxygen sensor
(25) Rear oxygen sensor (California spec. vehicles)

(26) Rear oxygen sensor (Except California spec. vehicles)
(27) Front catalytic converter
(28) Lower front catalytic converter cover
(29) Upper front catalytic converter cover

Exploded view of the exhaust manifold and related components

7924XG07

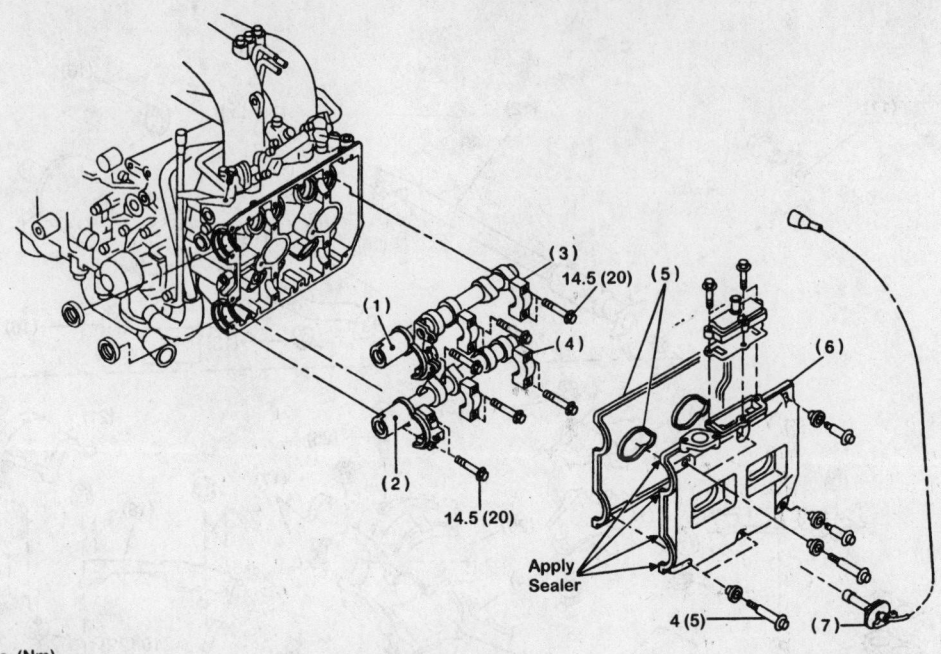

ft. lbs. (Nm)

(1) Intake camshaft (LH)
(2) Exhaust camshaft (LH)
(3) Intake camshaft cap (LH)
(4) Exhaust camshaft cap (LH)

(5) Rocker cover gasket (LH)
(6) Rocker cover (LH)
(7) Spark plug cord

7924XG08

Exploded view of the camshaft mounting—2.5L engine

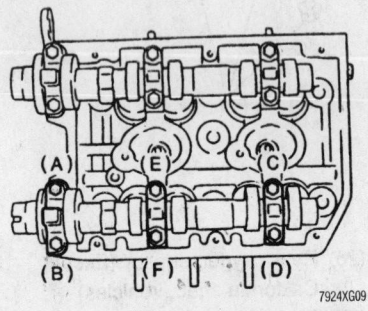

7924XG09

Camshaft bearing cap removal sequence—exhaust camshaft shown, intake camshaft is the same

To install:

➡**Lubricate the camshaft bearings prior to camshaft installation.**

10. Install the camshafts so the base circle (non-lobe portion) of the camshafts are in contact with the lash adjusters. This will position the lobes of the camshafts away from the valves.

➡**The left camshaft will need to be rotated for timing belt alignment.**

11. Apply Three Bond® 1215 or the equivalent fluid packing to the front bearing cap mating surfaces, then install the bearing

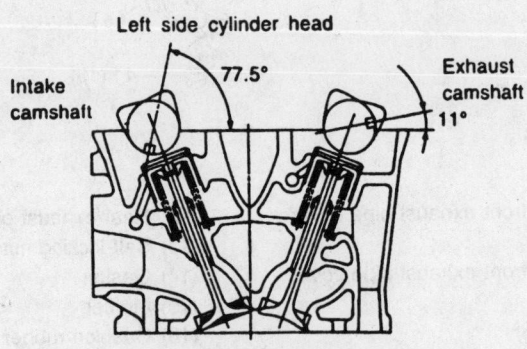

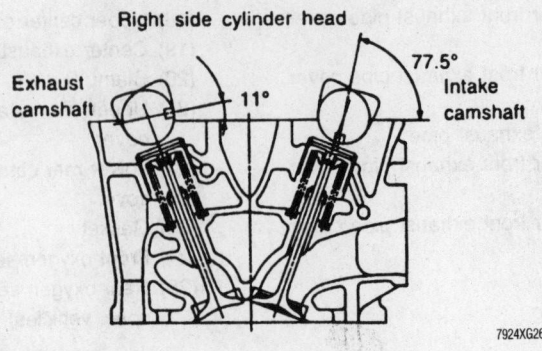

7924XG26

Cut away view of the camshaft positioning

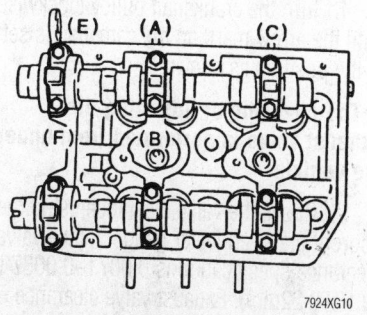

Camshaft bearing cap tightening sequence—intake camshaft shown, exhaust camshaft the same

caps. Tighten the caps in sequence in two progressive steps to 14.5 ft. lbs. (20 Nm).

12. Install new oil seals in the cylinder heads using a suitable seal installation tool.

✳✳ WARNING

Only rotate camshafts the specified amount. If the camshafts are rotated beyond the specified amount, the valves will contact each other and cause severe internal damage.

13. For correct timing belt alignment, rotate the left-hand intake camshaft 80 degrees clockwise and the left-hand exhaust camshaft 45 degrees counterclockwise.

14. If new camshafts were installed, check and adjust the valve lash.

15. Using a new gasket, install the rocker covers. Be sure to apply liquid sealant to the front edges of the gasket at the camshaft opening.

16. If removed, install the camshaft position sensor.

17. Install the camshaft sprockets and tighten the retaining bolts to 58 ft. lbs. (78 Nm). Be sure to secure the sprockets when tightening the bolts.

18. Check the timing sprockets for proper alignment and install the timing belt.

19. Connect the negative battery cable.

20. Check the fluid levels and start the engine.

21. Allow the engine to reach operating temperature and check for leaks.

Valve Lash

ADJUSTMENT

➡**The valve adjustment should be performed while the engine is cold. A Shim Replacer Kit 498187100 or equivalent will be needed to perform the valve adjustment.**

1. Disconnect the negative battery cable.
2. For the right cylinder head perform the following:

   a. Remove the upper and center timing belt cover mounting bolts.

   b. Raise and safely support the vehicle, and remove the engine undercover.

   c. Remove the timing belt cover lower mounting bolts, then the covers.

   d. Lower the vehicle.

   e. Disconnect the Mass Air Flow (MAF) sensor, and remove the air intake duct with the air cleaner assembly.

   f. Disconnect the blow-by hose.

   g. Label and disconnect the spark plug wires.

   h. Place a drip tray under the vehicle, and remove the rocker cover.

3. For the left cylinder head perform the following:

   a. Remove the battery and tray.

   b. Disconnect and remove the washer tank.

   c. Label and disconnect the spark plug wires.

   d. Place a drip tray under the vehicle, and remove the rocker cover.

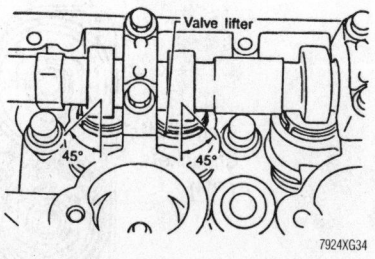

Position the lifter notch as shown, to remove the shim

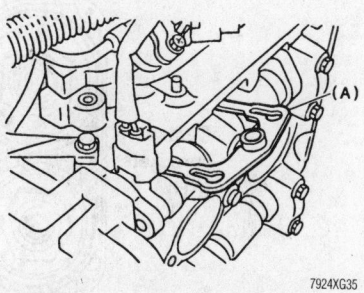

First install part "A" of the tool to the camshaft . . .

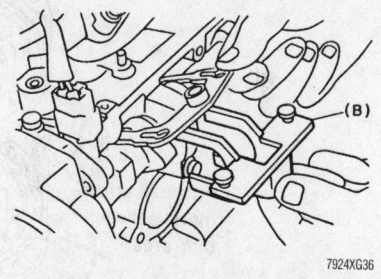

. . . then part "B" under part "A" as shown

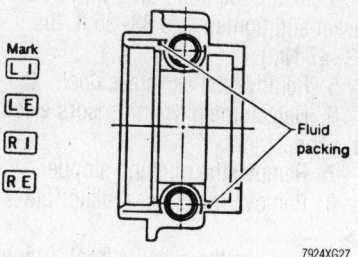

Apply the fluid packing as shown to prevent oil leaks

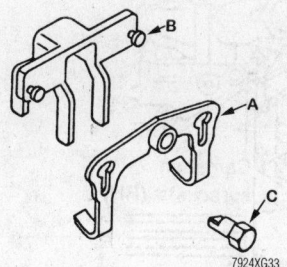

Exploded view of the shim replacer kit

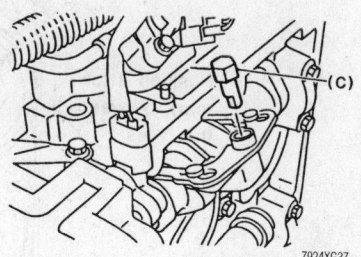

Turn part "C" until the adjusting shim can be removed

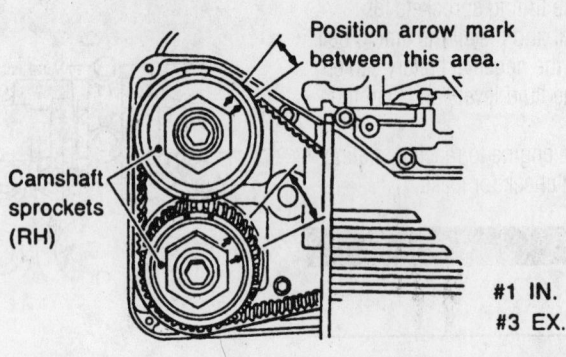

Position the camshaft as shown to adjust No. 1 intake valve and No. 3 exhaust valve

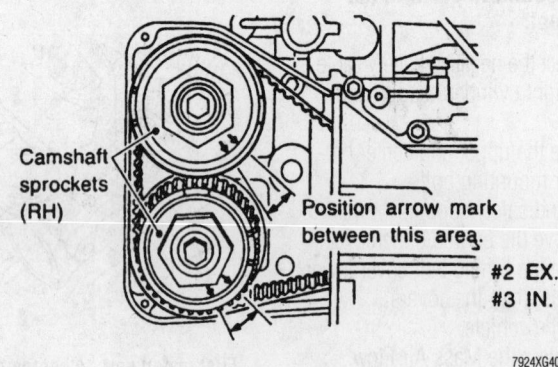

Position the camshaft as shown to adjust No. 3 intake valve and No. 2 exhaust valve

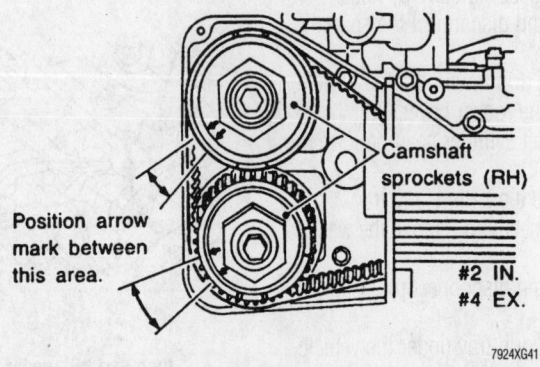

Position the camshaft as shown to adjust No. 2 intake valve and No. 4 exhaust valve

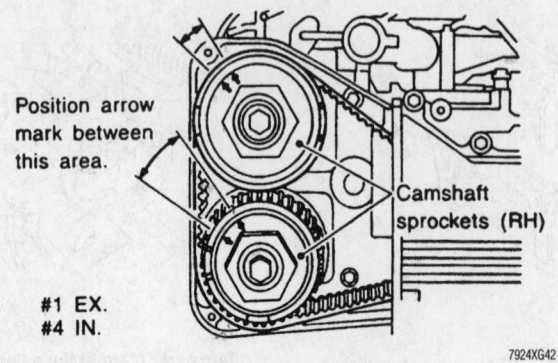

Position the camshaft as shown to adjust No. 4 intake valve and No. 1 exhaust valve

4. Turn the crankshaft pulley clockwise until the arrow mark on the camshaft is set to the position as shown.

➡**The checking or adjusting the exhaust valve is performed from under the vehicle.**

5. Check the valve clearance using the appropriate sized feeler gauge. Intake valve clearance specification is 0.0071–0.0087 in. (0.18–0.22mm). Exhaust valve clearance specification is 0.0090–0.0106 in. (0.23–0.27mm). If any valve needs adjustment, perform the following:

a. Rotate the notch of the lifter outward by 45°.

b. Install part "A" of the Replacer on to the camshaft.

c. Install part "B" of the Replacer as shown.

d. Install part "C" and turn until part "B" pushes the lifter away.

e. Insert tweezers into the notch of the valve lifter, and take out the shim.

f. Measure the thickness of the shim, then using the chart select and install a new shim and recheck the clearance.

6. Remove the adjusting tools.

7. Install the rocker covers.

8. Connect the spark plug wires.

9. Install the components removed to access the rocker covers.

10. Check the engine oil level.

11. Install the battery tray and battery.

Oil Pan

REMOVAL & INSTALLATION

2.5L Engine

1. Raise and support the vehicle safely.

2. Remove the engine undercover.

3. Drain the oil from the engine into a suitable container.

4. Install the drain plug with a new gasket and tighten it to 33–36 ft. lbs. (43–47 Nm).

5. Remove the air intake duct.

6. Detach the oxygen sensors electrical connectors.

7. Remove the pitching stopper.

8. Remove the upper radiator brackets.

9. Remove the exhaust front Y-pipe.

10. Remove the nuts which secure the front engine mounts to the front crossmember.

Intake valve (mm): S = (V + T) − 0.20
Exhaust valve (mm): S = (V + T) − 0.25

S: Shim thickness to be used
V: Measured valve clearance
T: Shim thickness required

Part No.	Thickness mm (in)	Part No.	Thickness mm (in)
13218AC230	2.22 (0.0874)	13218AC480	2.52 (0.0992)
13218AE000	2.23 (0.0878)	13218AC490	2.53 (0.0996)
13218AC240	2.24 (0.0882)	13218AC500	2.54 (0.1000)
13218AE010	2.25 (0.0886)	13218AC510	2.55 (0.1004)
13218AC250	2.26 (0.0890)	13218AC520	2.56 (0.1008)
13218AE020	2.27 (0.0894)	13218AC530	2.57 (0.1012)
13218AC260	2.28 (0.0898)	13218AC540	2.58 (0.1016)
13218AE030	2.29 (0.0902)	13218AC550	2.59 (0.1020)
13218AC270	2.30 (0.0906)	13218AC560	2.60 (0.1024)
13218AE040	2.31 (0.0909)	13218AC570	2.61 (0.1028)
13218AC280	2.32 (0.0913)	13218AC580	2.62 (0.1031)
13218AC290	2.33 (0.0917)	13218AC590	2.63 (0.1035)
13218AC300	2.34 (0.0921)	13218AC600	2.64 (0.1039)
13218AC310	2.35 (0.0925)	13218AC610	2.65 (0.1043)
13218AC320	2.36 (0.0929)	13218AC620	2.66 (0.1047)
13218AC330	2.37 (0.0933)	13218AC630	2.67 (0.1051)
13218AC340	2.38 (0.0937)	13218AC640	2.68 (0.1055)
13218AC350	2.39 (0.0941)	13218AC650	2.69 (0.1059)
13218AC360	2.40 (0.0945)	13218AC660	2.70 (0.1063)
13218AC370	2.41 (0.0949)	13218AE050	2.71 (0.1067)
13218AC380	2.42 (0.0953)	13218AC670	2.72 (0.1071)
13218AC390	2.43 (0.0957)	13218AE060	2.73 (0.1075)
13218AC400	2.44 (0.0961)	13218AC680	2.74 (0.1079)
13218AC410	2.45 (0.0965)	13218AE070	2.75 (0.1083)
13218AC420	2.46 (0.0969)	13218AC690	2.76 (0.1087)
13218AC430	2.47 (0.0972)	13218AE080	2.77 (0.1091)
13218AC440	2.48 (0.0976)	13218AC700	2.78 (0.1094)
13218AC450	2.49 (0.0980)	13218AE090	2.79 (0.1098)
13218AC460	2.50 (0.0984)	13218AC710	2.80 (0.1102)
13218AC470	2.51 (0.0988)	13218AE100	2.81 (0.1106)

7924XG38

Valve adjusting shim chart

T1: 3.6 ft. lbs. (5Nm)   T5: 32.5 ft. lbs. (44Nm)
T2: 3.6 ft. lbs. (5Nm)
T3: 4.7 ft. lbs. (6.4Nm)
T4: 7 ft. lbs. (10Nm)

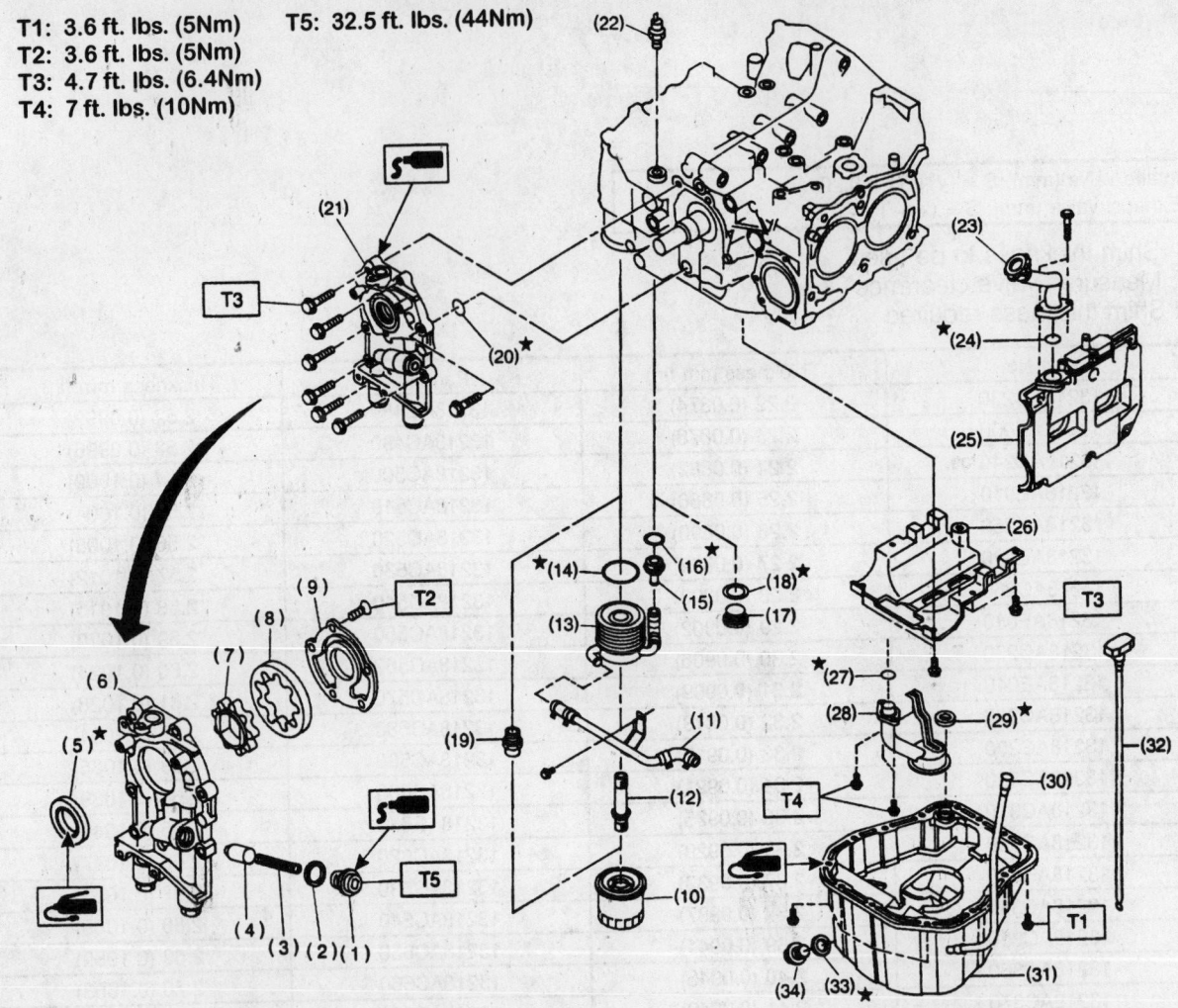

(1) Plug
(2) Washer
(3) Relief valve spring
(4) Relief valve
(5) Oil seal
(6) Oil pump case
(7) Inner rotor
(8) Outer rotor
(9) Oil pump cover
(10) Oil filter
(11) Oil cooler pipe and hose ASSY
    (AT vehicles)
(12) Connector (AT vehicles)
(13) Oil cooler (AT vehicles)
(14) O-ring (AT vehicles)

(15) Nipple (AT vehicles)
(16) Gasket (AT vehicles)
(17) Oil cooler connector (MT vehicles)
(18) Gasket (MT vehicles)
(19) Oil filter connector (MT vehicles)
(20) O-ring
(21) Oil pump ASSY
(22) Oil pressure switch
(23) Oil filler duct
(24) O-ring
(25) Cylinder head cover
(26) Baffle plate
(27) O-ring
(28) Oil strainer

(29) Gasket
(30) Oil level gauge guide
(31) Oil pan
(32) Oil level gauge
(33) Metal gasket
(34) Drain plug

7924XG28

Oil pan and lubrication components—2.5L engine

11. Support the engine with a suitable lifting device.

12. Lift up the engine slightly.

13. Remove the bolts that secure the oil pan.

14. While supporting the oil pan. use a rubber mallet and tap the oil pan to free it from the engine. Be sure to support the oil pan.

15. Clean all gasket material from both mating surfaces.

To install:

16. Apply a continuous bead of sealer to a new oil pan gasket.

17. Install the oil pan assembly. Tighten the bolts to 3–4 ft. lbs. (4–5 Nm).

18. Lower the engine onto the front crossmember.

19. Install the front engine mount nuts and tighten to 61 ft. lbs. (83 Nm).

20. Remove the engine lifting device.

21. Install the front Y-pipe with new gaskets. Tighten the nuts that secure the pipe to the engine to 23 ft. lbs. (30 Nm).

22. Attach the oxygen sensors electrical connectors.

23. Install the pitching stopper and tighten the bolts as follows:

- Front bolt—40 ft. lbs. (54 Nm)
- Rear bolt—49 ft. lbs. (67 Nm)

24. Install the upper radiator brackets.

25. Install the air intake duct.

26. Install the engine undercover.

27. Fill the engine to the proper level with the recommended oil and run the engine. Check for leaks.

Oil Pump

REMOVAL & INSTALLATION

2.5L Engine

1. Disconnect the negative battery cable.

2. Remove the engine undercover.

3. Drain the cooling system into a suitable container.

4. Raise and support the vehicle safely.

5. Drain the engine oil into a separate container.

6. Remove the belt covers, timing belt and related parts.

7. Remove the belt tensioner bracket.

8. Remove the engine coolant pipe.

9. Remove the water pump assembly.

10. Remove the oil pump mounting bolts and carefully pry the pump from the engine block.

※ WARNING

Use extreme care not to damage the engine block or the oil pump during removal of the pump.

To install:

11. Install a new front seal to the oil pump.

12. Apply a continuous bead sealant to the mating surfaces of the oil pump.

13. Install a new O-ring to the oil pump.

14. Install the oil pump and tighten the bolts to 56 inch lbs. (6.4 Nm).

15. Install the water pump.

16. Install the engine coolant pipe.

17. Install the belt tensioner bracket, timing belt and the covers.

18. Fill the engine to the proper level with the recommended oil and coolant.

19. Install the engine undercover.

20. Connect the negative battery cable.

21. Fill and bleed the cooling system.

Rear Main Seal

REMOVAL & INSTALLATION

1. Remove the engine from the vehicle.

2. Remove the clutch assembly/flywheel (MT). If equipped with an AT, remove the torque converter flexplate from the crankshaft.

3. Using a seal removal tool, pry the oil seal from the housing.

To install:

4. Utilizing the appropriate seal installer, install the new oil seal and press it into the flywheel housing using the appropriate driver.

5. Install the flywheel/flexplate and tighten the bolts to 53 ft. lbs. (72 Nm).

6. Install the engine into the vehicle.

FUEL SYSTEM

Fuel System Service Precautions

Safety is the most important factor when performing not only fuel system maintenance but any type of maintenance. Failure to conduct maintenance and repairs in a safe manner may result in serious personal injury or death. Maintenance and testing of

the vehicle's fuel system components can be accomplished safely and effectively by adhering to the following rules and guidelines.

- To avoid the possibility of fire and personal injury, always disconnect the negative battery cable unless the repair or test procedure requires that battery voltage be applied.

- Always relieve the fuel system pressure prior to disconnecting any fuel system component (injector, fuel rail, pressure regulator, etc.), fitting or fuel line connection. Exercise extreme caution whenever relieving fuel system pressure to avoid exposing skin, face and eyes to fuel spray. Please be advised that fuel under pressure may penetrate the skin or any part of the body that it contacts.

- Always place a shop towel or cloth around the fitting or connection prior to loosening to absorb any excess fuel due to spillage. Ensure that all fuel spillage (should it occur) is quickly removed from engine surfaces. Ensure that all fuel soaked cloths or towels are deposited into a suitable waste container.

- Always keep a dry chemical (Class B) fire extinguisher near the work area.

- Do not allow fuel spray or fuel vapors to come into contact with a spark or open flame.

- Always use a back-up wrench when loosening and tightening fuel line connection fittings. This will prevent unnecessary stress and torsion to fuel line piping. Always follow the proper torque specifications.

- Always replace worn fuel fitting O-rings with new. Do not substitute fuel hose or equivalent, where fuel pipe is installed.

Fuel System Pressure

RELIEVING

➡**This procedure must be performed prior to servicing any component of the fuel injection system.**

1. Disconnect the fuel pump harness at the fuel pump, under the rear seat access panel.

2. Crank the engine for 5 seconds or more to relieve the fuel pressure. If the engine starts during this time, allow it to run until it stalls.

3. Connect the fuel pump harness.

Fuel Filter

REMOVAL & INSTALLATION

1. Locate the fuel filter in the engine compartment on the left inside fender.
2. Properly relieve the fuel system pressure.
3. Disconnect the negative battery cable.
4. Disconnect the fuel delivery hoses from the fuel filter.
5. Remove the fuel filter from its holder.

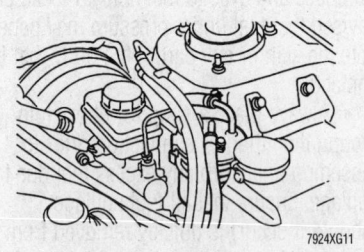

View of fuel filter mounting

To install:

6. Install the fuel filter into its mounting bracket.
7. Connect the fuel delivery hoses and tighten the hose clamps.
8. Connect the negative battery cable.

Fuel Pump

REMOVAL & INSTALLATION

1. Properly relieve the fuel system pressure.
2. Disconnect the negative battery cable.
3. Clean any debris away from the fuel pump mounting to prevent it from entering the tank.
4. Remove the fuel delivery and return hoses.
5. Remove the fuel pump mounting nuts.
6. Carefully lift the fuel pump out of the fuel tank.

To install:

7. Replace the sealing gaskets for the fuel pump.

8. Install the pump into the tank, and tighten the mounting nuts in sequence to 3.3 ft. lbs. (4.4 Nm).
9. Connect the fuel delivery and return hoses.
10. Attach the fuel pump electrical connector.
11. Tighten the fuel filler cap.
12. Connect the negative battery cable.
13. Start the vehicle. Check for leaks.
14. Install the fuel pump access cover.

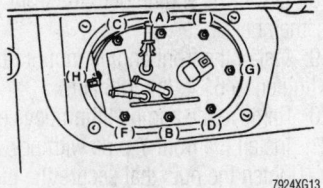

Fuel pump mounting nut tightening sequence

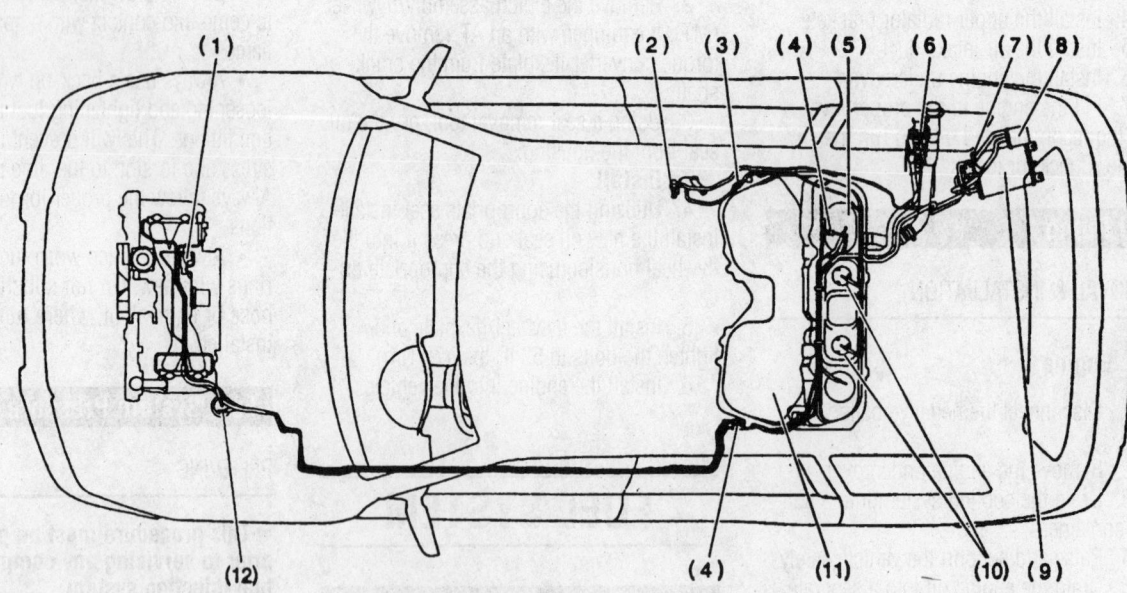

(1)	Purge control solenoid valve	(5)	Fuel pump	(9)	Canister
(2)	Roll over valve	(6)	Fuel tank pressure sensor	(10)	Fuel cut valve
(3)	Pressure control solenoid	(7)	Vent control solenoid valve	(11)	Fuel tank
(4)	Quick connector	(8)	Air filter	(12)	Fuel filter

Fuel system component locations

DRIVE TRAIN

Transmission Assembly

REMOVAL & INSTALLATION

1. Open the hood fully, and support it with the prop-rod.

2. Remove the engine undercover.

3. Disconnect the negative battery cable.

4. Remove the air intake and chamber, then the camber stays.

5. Label and detach the following connectors:
   - Front and rear oxygen sensor connectors
   - Transmission harness connector (AT vehicles)
   - Transmission ground terminal
   - Neutral position switch connector (MT vehicles)
   - Back-up light switch connector (MT vehicles)
   - Two vehicle speed sensors

6. Label and detach the starter wires, then remove the starter.

7. Remove the pitching stopper.

8. On AT vehicles, remove timing hole inspection plug matchmark the torque con-verter-to-driveplate, then remove the 4 bolts which hold torque converter to driveplate, then remove the ATF level gauge.

9. On MT vehicles, disconnect and remove the clutch slave cylinder.

10. Install Engine Support Assembly ST 41099AA020 or equivalent. (Also available as part no. 927670000)

11. Remove the bolt securing the right upper side of the transmission to the engine.

12. Raise and support the vehicle safely.

13. Remove the front Y-pipe.

14. Remove the center exhaust pipe and the heat shield cover.

15. Remove the hanger bracket from the right side of the transmission.

16. On AT vehicles, drain the ATF and the front differential, then disconnect the transmission cooler lines.

17. Matchmark and remove the rear driveshaft from the vehicle. Plug the opening at the rear of extension housing to prevent oil from flowing out.

18. On MT vehicles, remove the spring and disconnect the shifter stay and rod from the transmission.

19. On AT vehicles, disconnect the gear shift cable from the transmission select lever.

20. Remove swaybar from transverse link.

21. Remove parking brake cable bracket from transverse link and bolt holding transverse link to crossmember on each side. Lower the transverse link.

22. Remove spring pin and separate halfshaft from transmission on each side.

➡ **Use a small punch to remove spring pin. Discard old spring pin and always install a new pin.**

23. Disconnect the halfshaft from transmission on each side. Be sure to remove axle shaft from transmission by pushing the rear of tire outward.

24. Remove the remaining engine to transmission mounting nuts.

25. Place a transmission jack under the transmission. Always support transmission case with a transmission jack.

➡ **Do not place jack under the transmission oil pan, otherwise the oil pan may be damaged.**

26. Remove the rear transmission crossmember, then remove the transmission.

To install:

27. Install the transmission to the engine.

28. Install the rear transmission crossmember and tighten the nuts/bolts to specification.

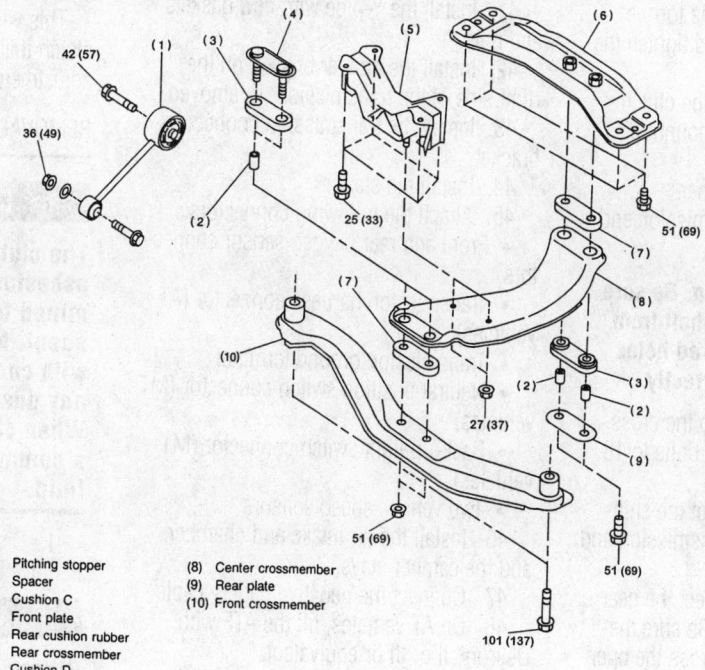

(1)	Pitching stopper
(2)	Spacer
(3)	Cushion C
(4)	Front plate
(5)	Rear cushion rubber
(6)	Rear crossmember
(7)	Cushion D
(8)	Center crossmember
(9)	Rear plate
(10)	Front crossmember

ft. lbs. (Nm)

Exploded view of the manual transmission mounting

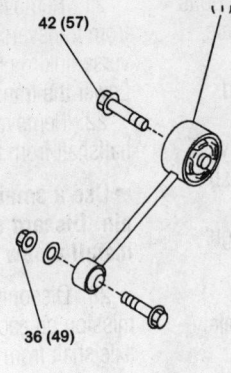
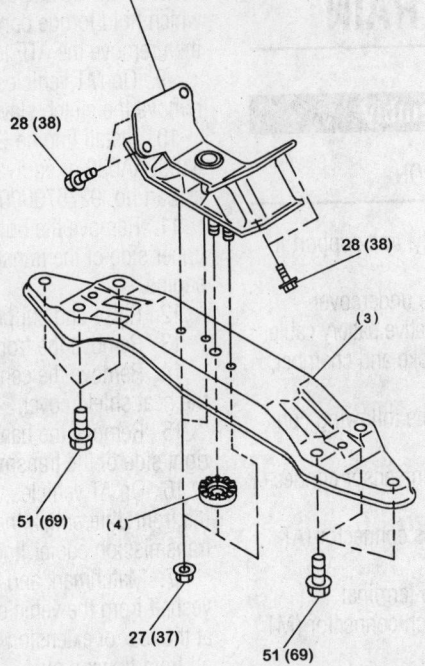

(1) Pitching stopper
(2) Rear cushion rubber
(3) Crossmember
(4) Stopper

7924XG15

Exploded view of the automatic transmission mounting

29. Remove the transmission jack.
30. Install the transmission-to-engine mounting nuts/bolts and tighten to 37 ft. lbs. (50 Nm).
31. On AT vehicles, install the torque converter-to-driveplate bolts and tighten the bolts to 18 ft. lbs. (25 Nm).
32. On MT vehicles, install the clutch slave cylinder and tighten the mounting bolts to 27 ft. lbs. (37 Nm).
33. Install the pitching stopper.
34. Install halfshaft to transmission and install spring pin into place.

➡**Always use new spring pin. Be sure to align the axle shaft and shaft from the transmission at chamfered holes and install shaft splines correctly.**

35. Connect the sway bar to the crossmember and tighten the clamp bolts to 18 ft. lbs. (25 Nm).
36. On MT vehicles, connect the shift control rod and stay to the transmission and install the spring.
37. On AT vehicles, reconnect the gear shift cable to the select lever. Be sure the lever operates smoothly all across the operating range.
38. On AT vehicles, install the ATF level gauge guide, and connect the ATF cooler lines.

39. Install the driveshaft and tighten the bolts to 23 ft. lbs. (31 Nm).
40. Install the heat shield cover, if removed.
41. Install the Y-pipe with new gaskets and nuts.
42. Install the hanger bracket on the right side of the transmission, if removed.
43. Install the transmission connectors bracket.
44. Install the starter.
45. Attach the following connectors:
• Front and rear oxygen sensor connectors
• Transmission harness connector (AT vehicles)
• Transmission ground terminal
• Neutral position switch connector (MT vehicles)
• Back-up light switch connector (MT vehicles)
• Two vehicle speed sensors
46. Install the air intake and chamber, and the camber stays.
47. Connect the negative battery cable.
48. On AT vehicles, fill the ATF with Dexron® II or III or equivalent.
49. On MT vehicles, check and fill the transmission with 75W-90 gear oil.
50. Road test the vehicle.

Clutch

ADJUSTMENT

This vehicle is equipped with a hydraulic clutch that is self-adjusting, therefore no adjustment is possible or necessary.

REMOVAL & INSTALLATION

❊❊ CAUTION

The clutch driven disc may contain asbestos, which has been determined to be a cancer causing agent. Never clean clutch surfaces with compressed air! Avoid inhaling any dust from any clutch surface! When cleaning clutch surfaces, use a commercially available brake fluid.

1. Disconnect the negative battery cable. Remove the transmission.

❊❊ WARNING

Removing the bolts on one side of the pressure plate will warp the pressure plate, rendering it useless.

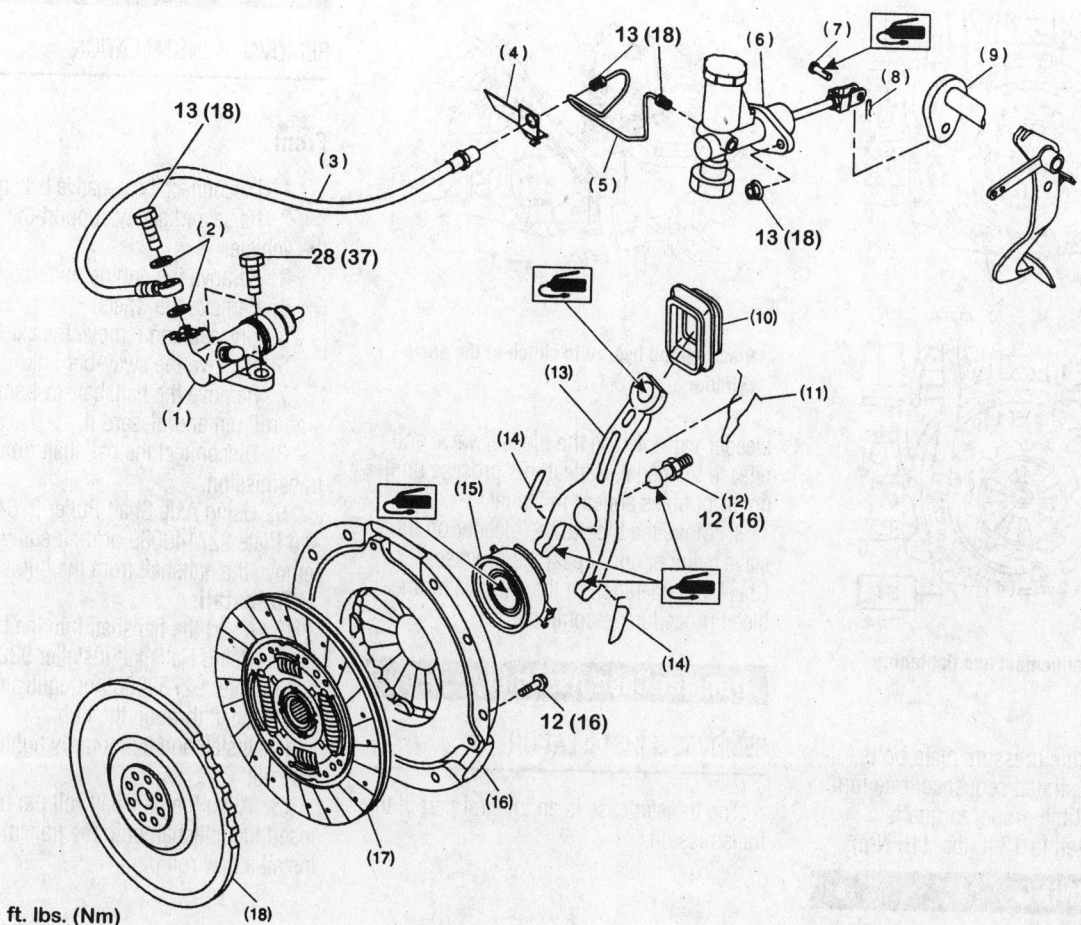

ft. lbs. (Nm)

(1) Operating cylinder
(2) Washer
(3) Clutch hose
(4) Bracket
(5) Pipe
(6) Master cylinder ASSY
(7) Clevis pin
(8) Snap pin
(9) Lever
(10) Clutch release lever sealing
(11) Retainer spring
(12) Pivot
(13) Release lever
(14) Clip
(15) Release bearing
(16) Clutch cover
(17) Clutch disc
(18) Flywheel

7924XG17

Exploded view of clutch system

2. Gradually unscrew the six bolts (6mm) which hold the pressure plate assembly on the flywheel. Loosen the bolts only one turn at a time, working around the pressure plate. Do not unscrew all the bolts on one side at one time.

3. When all of the bolts have been removed, remove the clutch plate and disc.

✳✳ WARNING

Do not get oil or grease on the clutch facing.

4. Remove the two retaining springs and remove the throwout bearing and the release fork.

➡**Do not disassemble either the clutch cover or disc. Inspect the parts for wear or damage and replace any parts as necessary. Replace the clutch disc if there is any oil or grease on the facing. Do not wash or attempt to lubricate the throwout bearing, because it is sealed and permanently lubricated. If it requires replacement, the bearing may be**

removed and a new one installed in the holder by means of a press.

To install:
5. Fit the release fork boot on the front of the transmission housing. Install the release fork.

6. Insert the throwout bearing assembly and secure it with the two springs. Coat the inside diameter of the throwout bearing and the release lever contact points with grease.

7. Insert a clutch alignment tool through the clutch cover and disc, then insert the end of the tool into the needle bearing.

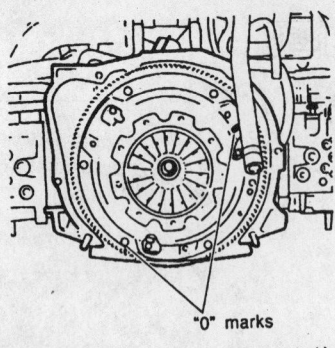

"0" marks

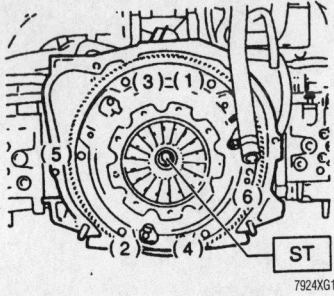

Clutch cover alignment and tightening sequence

8. Tighten the pressure plate bolts following the illustrated sequence, one turn at a time, until the proper torque is reached. Tighten to 12 ft. lbs. (16 Nm).

❋❋ WARNING

When installing the clutch pressure plate assembly, be sure that the O marks on the flywheel and the clutch pressure plate assembly are at least 120° apart. These marks indicate the direction of residual unbalance. Also, be sure that the clutch disc is installed properly, noting the FRONT and REAR markings.

9. Install the transmission.

Hydraulic Clutch System

BLEEDING

➡To properly bleed the system, it must be bled at the slave cylinder and at the damper. Each of these has an air bleeder on it.

1. Connect a vinyl tube to the air bleeder on the damper and put the other end in a jar with clean clutch fluid.

➡Do not let the fluid level fall too low in the master cylinder. Do not release the pedal with the bleeder open.

2. With the help of an assistant depressing the clutch pedal, slowly open the

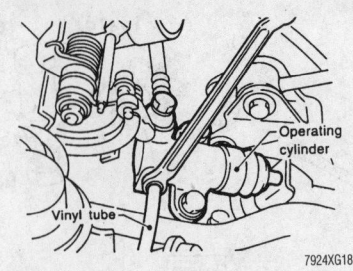

Bleeding the hydraulic clutch at the slave cylinder

bleeder valve. Close the bleeder valve and release the pedal. Repeat this process until no air bubbles appear in the jar.

3. Move the tube to the bleeder on the slave cylinder and repeat the process. Check the operation of the clutch after the bleed procedure is complete.

Transfer Case Assembly

REMOVAL & INSTALLATION

The transfer case is an integral part of the transmission.

Halfshafts

REMOVAL & INSTALLATION

Front

1. Disconnect the negative battery cable.
2. Raise and safely support the front of the vehicle.
3. Remove the engine undercover.
4. Remove the wheel.
5. Unstake and remove the axle nut.
6. Remove the sway-bar link.
7. Remove the halfshaft-to-transmission roll pin and discard it.
8. Disconnect the halfshaft from the transmission.
9. Using Axle Shaft Puller 926470000 and Plate 927140000 or their equivalents, remove the halfshaft from the hub.

To install:

10. Insert the halfshaft into the hub.
11. Using Halfshaft Installer 922431000 and Adapter 927390000 or equivalent, pull the halfshaft through the hub.
12. Install and temporarily tighten a new axle nut.
13. Align the halfshaft roll pin hole and insert the halfshaft onto the transmission. Install a new roll pin.

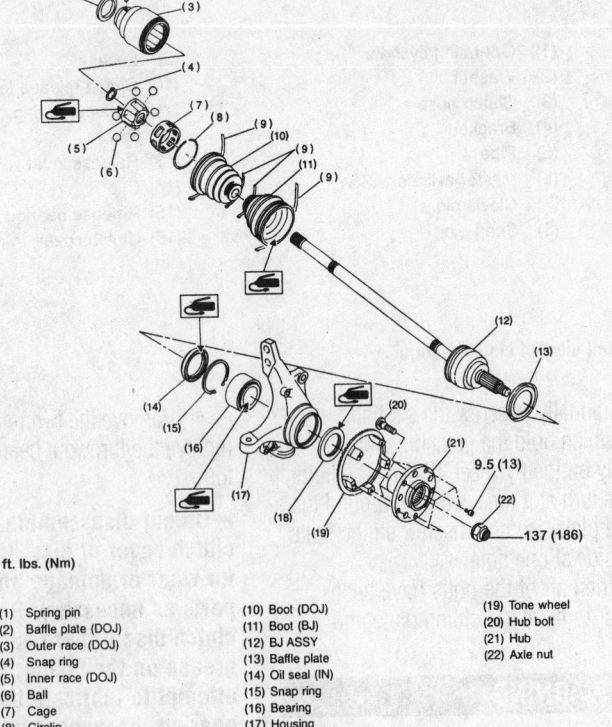

ft. lbs. (Nm)

(1) Spring pin	(10) Boot (DOJ)
(2) Baffle plate (DOJ)	(11) Boot (BJ)
(3) Outer race (DOJ)	(12) BJ ASSY
(4) Snap ring	(13) Baffle plate
(5) Inner race (DOJ)	(14) Oil seal (IN)
(6) Ball	(15) Snap ring
(7) Cage	(16) Bearing
(8) Circlip	(17) Housing
(9) Boot band	(18) Oil seal (OUT)

(19) Tone wheel
(20) Hub bolt
(21) Hub
(22) Axle nut

Exploded view of the front halfshaft—manual transmission

14. Connect the lower control arm to the crossmember and tighten a new self-locking nut to 83 ft. lbs. (113 Nm).

15. Install the sway bar link.

16. Tighten the new axle nut to 197 ft. lbs. (186 Nm) and stake the nut.

17. Install the engine undercover.

18. Install the wheel.

19. Lower the vehicle.

20. Connect the negative battery cable.

Rear

1. Disconnect the negative battery cable.

2. Raise and support the vehicle safely.

3. Unstake and remove the axle nut.

4. Remove the sway bar bracket.

5. Remove the lower control arm-to-rear housing bolt and nut.

6. Remove the trailing link assembly-to-rear housing bolt and nut.

7. Remove the halfshaft-to-differential roll pin.

8. Disconnect the halfshaft from the differential.

9. Using Axle Shaft Puller 926470000 and Plate 927140000 or their equivalents, remove the halfshaft from the hub.

To install:

10. Insert the halfshaft into the hub.

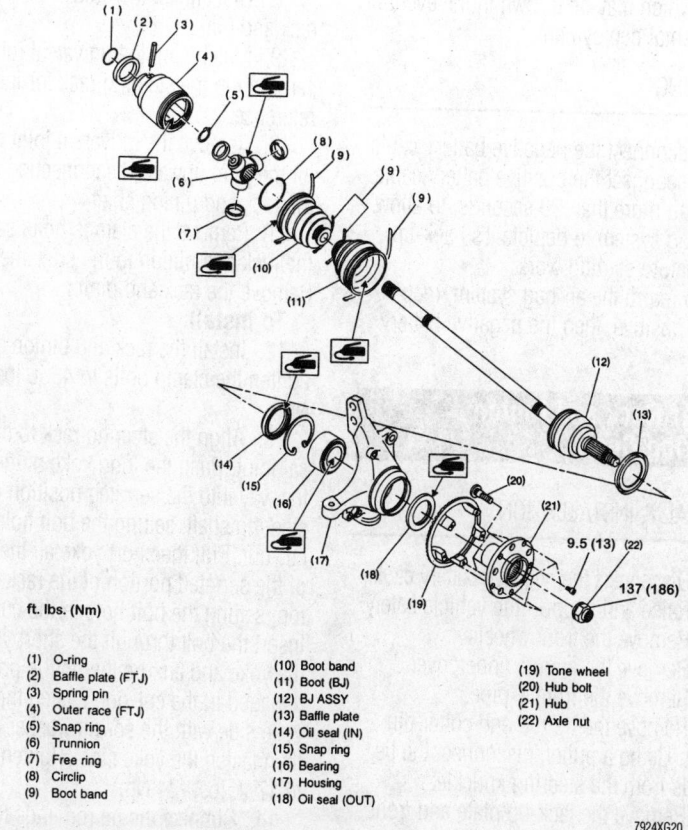

ft. lbs. (Nm)

(1) O-ring	(10) Boot band
(2) Baffle plate (FTJ)	(11) Boot (BJ)
(3) Spring pin	(12) BJ ASSY
(4) Outer race (FTJ)	(13) Baffle plate
(5) Snap ring	(14) Oil seal (IN)
(6) Trunnion	(15) Snap ring
(7) Free ring	(16) Bearing
(8) Circlip	(17) Housing
(9) Boot band	(18) Oil seal (OUT)

(19) Tone wheel
(20) Hub bolt
(21) Hub
(22) Axle nut

7924XG20

Exploded view of the front halfshaft—automatic transmission

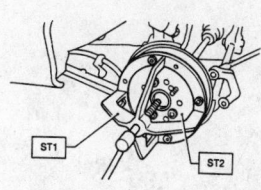

ST1 926470000 AXLE SHAFT PULLER
ST2 927140000 PLATE

7924XG29

Be sure not to damage the threads when removing the front or rear halfshafts

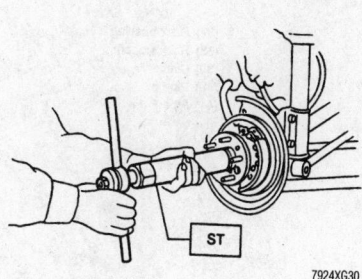

ST

7924XG30

To avoid using a hammer when installing the halfshafts, use the proper tools as shown

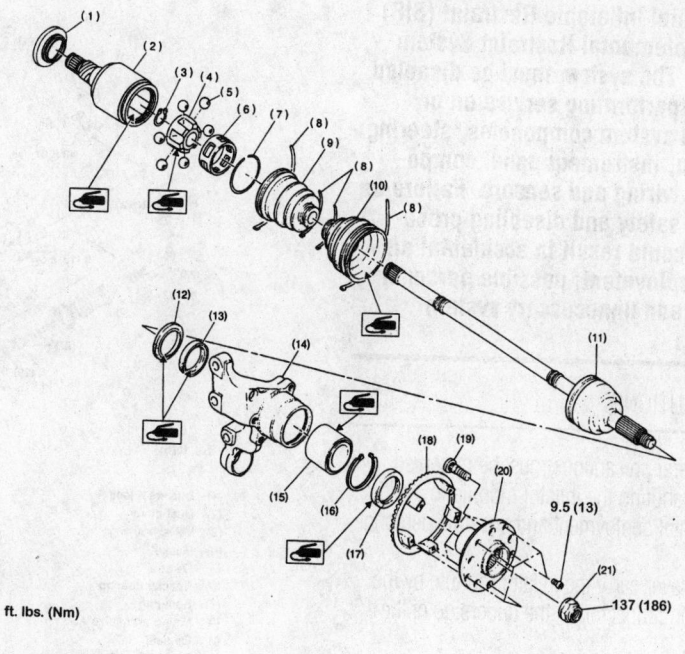

ft. lbs. (Nm)

(1) Baffle plate (DOJ)	(10) Boot (BJ)
(2) Outer race (DOJ)	(11) BJ ASSY
(3) Snap ring	(12) Oil seal (IN. No. 2)
(4) Inner race	(13) Oil seal (IN. No. 3)
(5) Ball	(14) Housing
(6) Cage	(15) Bearing
(7) Circlip	(16) Snap ring
(8) Boot band	(17) Oil seal (OUT)
(9) Boot (DOJ)	(18) Tone wheel

(19) Hub bolt
(20) Hub
(21) Axle nut

Exploded view of the rear halfshaft

7924XG21

11. Using Halfshaft Installer 922431000 and Adapter 927390000 or equivalent, pull the halfshaft into place.

12. Install and temporarily tighten a new axle nut.

13. Align the halfshaft-to-differential roll pin holes and slide the halfshaft onto the splines. Install a new roll pin.

14. Connect the trailing link assembly to the rear housing. Install the bolt and new nut, then tighten them to 108 ft. lbs. (147 Nm).

15. Install the sway bar bracket.

16. Tighten the new axle nut to 137 ft. lbs. (186 Nm).

17. Install the wheel.

18. Lower the vehicle.

19. Connect the negative battery cable.

STEERING AND SUSPENSION

Air Bag

❋❋ CAUTION

All vehicles are equipped with an air bag system, also known as the Supplemental Inflatable Restraint (SIR) or Supplemental Restraint System (SRS). The system must be disabled before performing service on or around system components, steering column, instrument panel components, wiring and sensors. Failure to follow safety and disabling procedures could result in accidental air bag deployment, possible personal injury and unnecessary system repairs.

PRECAUTIONS

Several precautions must be observed when handling the inflator module to avoid accidental deployment and possible personal injury.

• Never carry the inflator module by the wires or connector on the underside of the module.

• When carrying a live inflator module, hold securely with both hands, and ensure that the bag and trim cover are pointed away.

• Place the inflator module on a bench or other surface with the bag and trim cover facing up.

• With the inflator module on the bench, never place anything on or close to the

module which may be thrown in the event of an accidental deployment.

DISARMING

1. Disconnect the negative battery cable.

2. Disconnect the positive battery cable.

3. Wait more than 20 seconds, to allow the air bag system to deplete it's back-up power, before starting work.

4. To rearm the air bag system, reconnect the positive, then the negative battery cables.

Power Rack and Pinion Steering Gear

REMOVAL & INSTALLATION

1. Disconnect the negative battery cable.

2. Raise and support the vehicle safely.

3. Remove the front wheels.

4. Remove the engine undercover.

5. Remove the front Y-pipe.

6. Remove the tie rod end cotter pin and nut. Using a puller, disconnect the tie rod ends from the steering knuckle.

7. Remove the jack-up plate and front sway bar.

8. Disconnect the fluid lines from the rack and pinion.

9. Matchmark the universal joint to the serration in the steering rack for installation reference.

10. Remove the universal joint bolts and lift the joint upward disconnecting it from the rack and pinion shaft.

11. Remove the clamps bolts securing the rack and pinion to the crossmember. Remove the rack and pinion.

To install:

12. Install the rack and pinion and tighten the clamp bolts to 43 ft. lbs. (59 Nm).

13. Align the steering rack to the universal joint. Push the long yoke of the joint all the way into the serrated position of the steering shaft, setting the bolt hole in the cut-out. Pull the short yoke all the way out of the serrated portion of the rack and pinion, setting the bolt hole in the cut-out. Insert the bolt through the short yoke. Pull the yoke and ensure the bolt is properly engaged in the cut-out. Fasten the short yoke side with the spring washer and bolt, then fasten the yoke side. Tighten the bolts to 17 ft. lbs. (24 Nm).

14. Connect the tie rod ends to the steering knuckle.

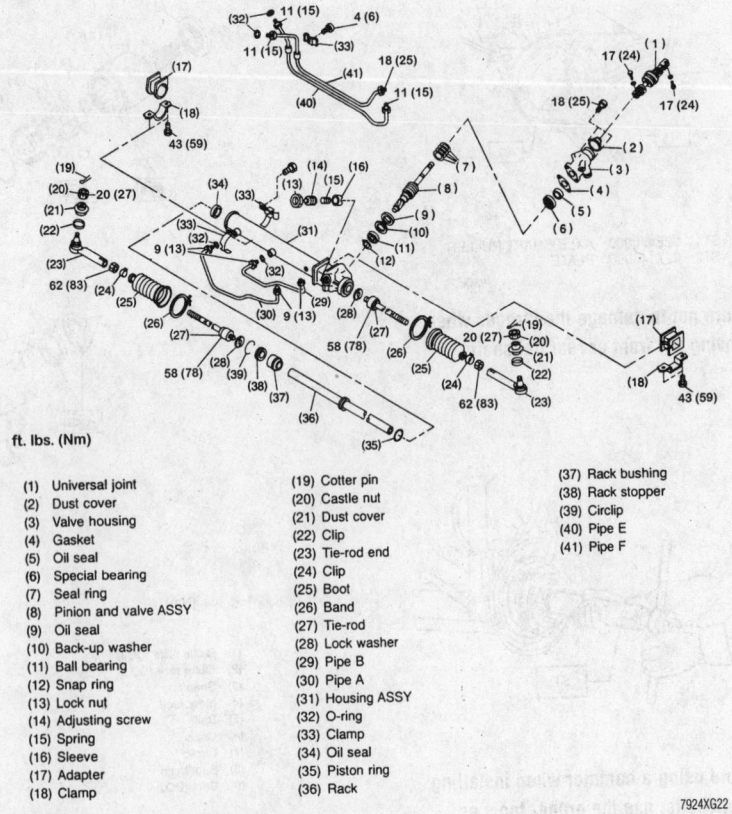

ft. lbs. (Nm)

(1) Universal joint	(19) Cotter pin
(2) Dust cover	(20) Castle nut
(3) Valve housing	(21) Dust cover
(4) Gasket	(22) Clip
(5) Oil seal	(23) Tie-rod end
(6) Special bearing	(24) Clip
(7) Seal ring	(25) Boot
(8) Pinion and valve ASSY	(26) Band
(9) Oil seal	(27) Tie-rod
(10) Back-up washer	(28) Lock washer
(11) Ball bearing	(29) Pipe B
(12) Snap ring	(30) Pipe A
(13) Lock nut	(31) Housing ASSY
(14) Adjusting screw	(32) O-ring
(15) Spring	(33) Clamp
(16) Sleeve	(34) Oil seal
(17) Adapter	(35) Piston ring
(18) Clamp	(36) Rack

(37) Rack bushing
(38) Rack stopper
(39) Circlip
(40) Pipe E
(41) Pipe F

Exploded view of the rack and pinion steering gear

7924XG22

15. Install the sway bar and jack-up plate.

16. Install the Y-pipe with new gaskets and nuts.

17. Install the engine undercover.

18. Install the wheels.

19. Lower the vehicle.

20. Fill and bleed the steering system.

Strut

REMOVAL & INSTALLATION

Front

> ❈❈ **CAUTION**
>
> **Do not remove the large nut on top of the strut assembly unless the coil spring is properly compressed with a suitable spring compressor.**

1. Disconnect the negative battery cable.

2. Raise and support the vehicle safely.

3. Remove the front wheel assembly.

4. Disconnect the ABS sensor, if equipped.

5. Remove the caliper, leaving the line connected and suspend it out of the way with a piece of wire or string.

6. Remove the clip attaching the brake line to the strut housing.

7. Matchmark the camber adjustment bolt to the strut housing as reference for installation.

8. If equipped with ABS, remove the bolt securing the sensor harness.

9. Remove the two bolts and nuts securing the strut to the steering knuckle. Notice that the shaft of the top bolt is not round. This bolt is used for camber adjustment, and most always be installed in the top hole.

10. Remove the three nuts securing the strut to the body in the engine compartment.

11. Remove the strut and coil spring assembly from the vehicle.

To install:

12. Install the strut assembly into vehicle.

13. Install the upper strut retainer nuts, and tighten the nuts to 15 ft. lbs. (20 Nm).

14. If equipped, install the ABS sensor harness, and tighten the bolt to 15 ft. lbs. (20 Nm).

15. Install the lower strut nuts and bolts. Ensure the alignment adjustment bolt is installed in the top mounting hole. Tighten the nuts, while securing the bolts to 112 ft. lbs. (152 Nm).

16. Install the caliper.

17. Attach the brake line to the strut and install the clip.

18. Install the front wheel.

19. Lower the vehicle to the floor.

20. Connect the negative battery cable.

21. Check and adjust the front end alignment.

Rear

> ❈❈ **CAUTION**
>
> **Do not remove the large nut on top of the strut assembly unless the coil spring is properly retained with a spring compressor.**

1. Remove the strut mount cap located at the right rear interior quarter trim.

2. Raise and support the vehicle safely.

3. Remove the wheel and tire assembly.

4. Remove the brake hose clip.

5. Remove the union bolt from the brake caliper. Move the brake hose out of the way.

6. Remove the lower nuts and bolts securing the strut to the rear wheel housing.

7. From inside the vehicle, loosen and remove the retainer nuts securing the strut bearing cap to the strut tower.

8. Lower and remove the strut from the vehicle.

To install:

9. Install the strut on to the vehicle, making sure to position the strut with the "4WD" mark on the strut mount facing the outside of the vehicle. Refer to the illustration if needed. Install the retainer nuts, and tighten to 15 ft. lbs. (20 Nm).

10. Connect the strut to the rear wheel knuckle assembly, using the retainer nuts and bolts, and tighten the bolts to 145 ft. lbs. (196 Nm).

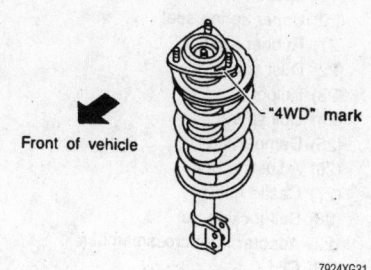

Front of vehicle — "4WD" mark

7924XG31

Position the upper strut bearing as shown

11. Install the brake union bolt, and tighten to 13 ft. lbs. (18 Nm).

12. Insert the brake hose clip.

13. Bleed the brakes.

14. Install the wheel.

15. Lower the vehicle.

16. Install the strut mount cap.

Coil Spring

REMOVAL & INSTALLATION

Front

> ❈❈ **CAUTION**
>
> **Do not remove the large nut on top of the strut assembly unless the coil spring is properly compressed with a suitable spring compressor.**

1. Remove the strut assembly from the vehicle.

2. Place the strut assembly in a vise with a holding tool and install a spring compressor.

3. Compress the spring slightly.

4. Loosen but do not remove the bearing cap locknut.

5. Unload the spring seat using the spring compressor, then remove the locknut.

6. Remove the strut bearing cap, mounting insulator bracket and upper spring seat.

7. Remove the coil assembly, leaving the spring compressed.

8. Remove the strut boot and rebound bumper from the strut. Inspect and replace if worn.

9. Remove the strut retainer nut using a suitable wrench. Remove the strut insert from the assembly.

To install:

10. Install the strut into the chamber, and install the retainer nut. Tighten the nut snugly.

11. Install the rebound bumper and the boot to the strut piston rod.

12. Install the coil spring on the strut assembly. Ensure the spring is properly positioned on the lower bracket.

13. Install the upper spring seat, mounting insulator and bearing cap. Ensure the upper spring seat is facing the proper direction.

14. Install the locknut, and tighten to 36–43 ft. lbs. (47–56 Nm).

15. Loosen and remove the spring compressor from the coil spring.

16. Install the strut to the vehicle.

Rear

> ✳✳ **CAUTION**
>
> **Do not remove the large nut on top of the strut assembly unless the coil spring is properly retained with a spring compressor.**

1. Remove the strut assembly from the vehicle and secure in a soft jawed vise.

2. Compress the coil spring with a spring compressor until the upper spring seat can be turned by hand.

3. Remove the self-locking nut on the top of the strut assembly, then remove the upper spring seat.

4. Remove the coil spring and compressor. If the spring is being replaced, slowly release the spring from the compressor and compress the new coil spring.

To install:

5. Place the proper end of the coil spring on the lower spring seat on the strut.

6. Install the insulator, upper spring seat and strut mount on the strut piston. Install a new self-locking nut. Tighten the nut to 43 ft. lbs. (59 Nm).

7. Slowly release the spring compressor.

8. Install the strut on to the vehicle.

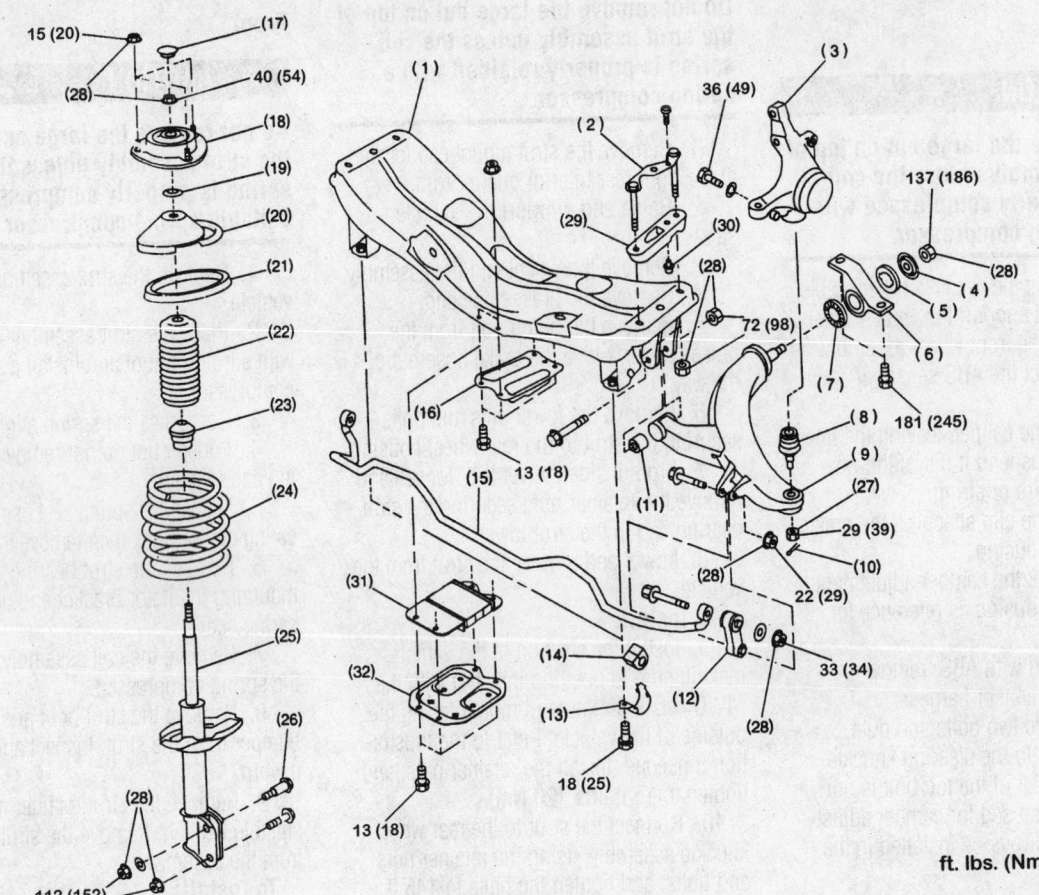

ft. lbs. (Nm)

(1) Front crossmember
(2) Bolt ASSY
(3) Housing
(4) Washer
(5) Stopper rubber (Rear)
(6) Rear bushing
(7) Stopper rubber (Front)
(8) Ball joint
(9) Transverse link
(10) Cotter pin
(11) Front bushing
(12) Stabilizer link
(13) Clamp
(14) Bushing
(15) Stabilizer
(16) Jack-up plate (Except MT model)

(17) Dust seal
(18) Strut mount
(19) Spacer
(20) Upper spring seat
(21) Rubber seat
(22) Dust cover
(23) Helper
(24) Coil spring
(25) Damper strut
(26) Adjusting bolt
(27) Castle nut
(28) Self-locking nut
(29) Adapter front crossmember
(30) Clip
(31) Dynamic damper (MT model)
(32) Jack-up plate (MT model)

7924XG23

Exploded view of the front suspension

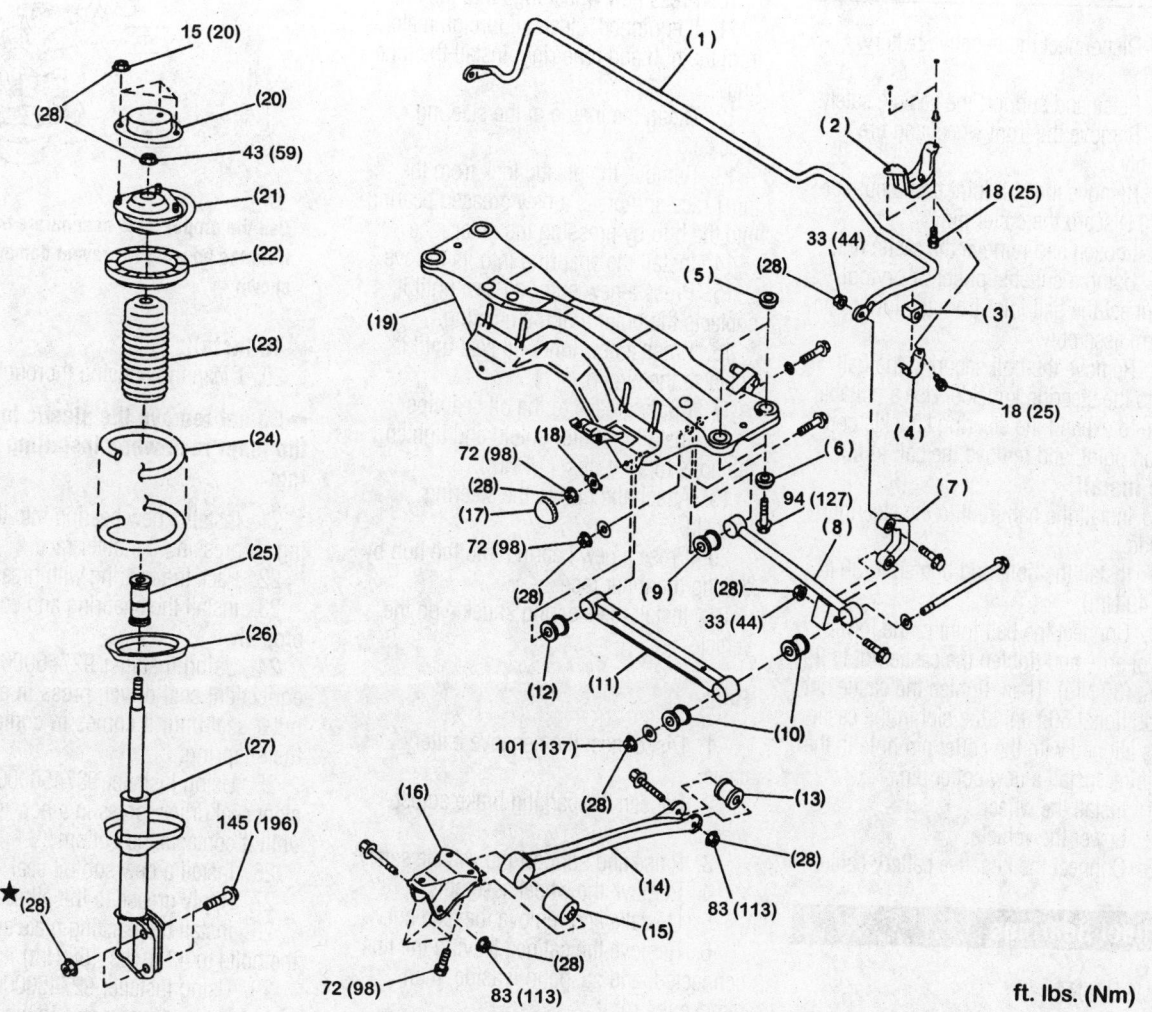

ft. lbs. (Nm)

(1) Stabilizer
(2) Stabilizer bracket
(3) Stabilizer bushing
(4) Clamp
(5) Floating bushing
(6) Stopper
(7) Stabilizer link
(8) Rear lateral link
(9) Bushing (C)
(10) Bushing (A)
(11) Front lateral link
(12) Bushing (B)
(13) Trailing link rear bushing
(14) Trailing link

(15) Trailing link front bushing
(16) Trailing link bracket
(17) Cap (Protection)
(18) Washer
(19) Rear crossmember
(20) Strut mount cap
(21) Strut mount
(22) Rubber seat upper
(23) Dust cover
(24) Coil spring
(25) Helper
(26) Rubber seat lower
(27) Damper strut
(28) Self-locking nut

7924XG24

Exploded view of the rear suspension

Lower Ball Joint

REMOVAL & INSTALLATION

1. Disconnect the negative battery cable.
2. Raise and support the vehicle safely.
3. Remove the front wheel and tire assembly.
4. Remove the ball joint castle nut cotter pin. Discard the cotter pin.
5. Loosen and remove the castle nut.
6. Using a suitable puller or prytool, disconnect the ball joint from the lower control arm assembly.
7. Remove the bolt securing the ball joint to the steering knuckle. Use a suitable wedge to expand the steering knuckle connection point, and remove the ball joint.

To install:
8. Install the ball joint to the steering knuckle.
9. Install the bolt, and tighten to 36 ft. lbs. (49 Nm).
10. Connect the ball joint to the lower control arm, and tighten the castle nut to 29 ft. lbs. (39 Nm). Then, tighten the castle nut an additional 60° until the slot in the castle nut is aligned with the cotter pin hole in the ball joint. Install a new cotter pin.
11. Install the wheel.
12. Lower the vehicle.
13. Connect the negative battery cable.

Wheel Bearings

ADJUSTMENT

The wheel bearings are not adjustable.

REMOVAL & INSTALLATION

Front

1. Remove the steering knuckle assembly from the vehicle.
2. Position the steering knuckle in a soft-jawed vise.
3. Press the hub from the steering knuckle. If the inner bearing race remains in the hub, press it out.
4. Remove the rotor shield.
5. Remove the inner and outer seals.
6. Remove the snapring from the steering knuckle.
7. Press the inner bearing race to remove the outer bearing.
8. If equipped with ABS, remove the tone ring.
9. Press the wheel lugs from the hub.

➡ **To prevent deforming the hub, do not hammer the lugs out.**

To install:
10. Press new wheel lugs into the hub.
11. If equipped, clean all foreign material from the hub and tone ring. Install the tone ring.
12. Clean the inside of the steering knuckle.
13. Remove the plastic lock from the inner race and press a new greased bearing into the hub by pressing the outer race.
14. Install the snapring into its groove.
15. Press a new outer oil seal until it contacts the bottom of the housing.
16. Press a new inner oil seal until it contacts the circlip.
17. Apply grease to the oil seal lips.
18. Install the rotor shield and tighten the bolts to 10 ft. lbs. (14 Nm).
19. Attach the hub to the steering knuckle.
20. Press a new bearing into the hub by driving the inner race.
21. Install the steering knuckle on the vehicle.

Rear

1. Disconnect the negative battery cable.
2. Loosen the parking brake adjustment.
3. Raise and support the vehicle safely.
4. Remove the wheel assembly.
5. Unstake and remove the axle nut.
6. Remove the caliper, leaving the line connected, and suspend it aside, then remove the rotor.
7. Disconnect the parking brake cable.
8. Remove the sway bar clamp.
9. Remove the bolt securing the lateral link to the housing.
10. Remove the bolts securing the trailing link to the housing.
11. Remove the halfshaft.
12. Remove the bolts securing the strut to the housing.
13. If equipped with ABS, remove the speed sensor from the backing plate.
14. Remove the housing assembly.
15. Using Hub Stand 92708000 and Hub Remover 927420000 or equivalent, remove the hub from the rear housing.
16. Remove the backing plate from the housing.
17. Remove the outer, inner and sub oil seals.
18. Remove the snapring.
19. Remove the bearing by pressing the inner race.

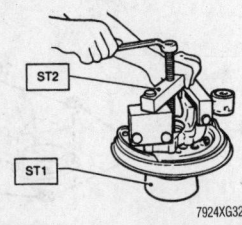

7924XG32

Use the proper tools to separate the hub from the housing to prevent damage—as shown

To install:
20. Clean the housing thoroughly.

➡ **Do not remove the plastic lock from the inner race when installing the bearing.**

21. Install a new bearing into the housing by pressing the outer race.
22. Pack the bearing with grease.
23. Install the snapring and ensure it fits properly.
24. Using Installer 927460000 or equivalent seal driver, press in a new outer seal until it comes in contact with the snapring.
25. Using Installer 927450000 or equivalent seal driver, press in a new inner seal until it contacts the bottom.
26. Install a new sub oil seal.
27. Apply grease to the oil seal lip.
28. Install the backing plate and tighten the bolts to 43 ft. lbs. (58 Nm).
29. Using Installer 927450000 or equivalent bearing driver, press in the hub into the housing.
30. Connect the housing to the strut and tighten the bolts to 119 ft. lbs. (162 Nm).
31. If equipped with ABS, install the speed sensor.
32. Install the halfshaft.
33. Connect the trailing link to the housing and tighten the bolt and new nut to 94 ft. lbs. (127 Nm).
34. Connect the lateral link to the housing and tighten the bolt and new nut to 116 ft. lbs. (157 Nm).
35. Install the sway bar clamp.
36. Connect the parking brake cable.
37. Install the rear brake assembly.
38. Install a new axle nut and tighten it to 152 ft. lbs. (206 Nm). Stake the nut.
39. Install the wheel.
40. Lower the vehicle.
41. Adjust the parking brake cable.
42. Connect the negative battery cable.

TOYOTA

T-100 • Tacoma • 4Runner

34

ENGINE REPAIR

➡ Disconnecting the negative battery cable on some vehicles may interfere with the functions of the on board computer systems and may require the computer to undergo a relearning process, once the negative battery cable is reconnected.

Distributor

REMOVAL

1. Disconnect the negative battery cable.

✳✳ WARNING

The air bag system is equipped with a back-up power source. To avoid possible air bag deployment, do not start working on the vehicle until 90 seconds has elapse from the time the ignition switch is turned OFF and the negative battery terminal is disconnected.

2. Unplug the distributor connectors.
3. Remove the distributor cap without disconnecting the secondary leads and position it aside.
4. Matchmark the rotor with the distributor housing and housing with the cylinder block.
5. Remove the distributor hold-down bolt and pull the distributor from the cylinder block.

INSTALLATION

Timing Not Disturbed

1. Install a new O-ring to the distributor and lubricate it with engine oil.
2. Install the distributor into the cylinder block, while aligning the matchmarks made during removal. Install the distributor hold-down bolt.
3. Install the distributor cap and attach the distributor connector.
4. Connect the negative battery cable. Start the engine and allow normal operating temperature to be reached.
5. Check and if necessary, adjust the ignition timing.

Timing Disturbed

1. Install a new O-ring to the distributor and lubricate it with engine oil.

2. Remove the No. 1 cylinder spark plug.
   a. Place a finger or compression gauge over the spark plug hole.
   b. Turn the crankshaft until compression starts to build up. Continue turning the crankshaft until the crankshaft pulley groove align with the timing mark **0** of the timing chain cover.
3. If necessary, remove the valve cover.
   a. Check that the timing marks with 1 and 2 dots are aligned on the camshaft sub-gears.
   b. If not, turn the crankshaft 1 revolution (360 degrees) and align the crankshaft pulley groove with the timing mark **0** of the timing chain cover.
4. Align the protrusion of the distributor housing with the groove on the driven gear.
5. Insert the distributor into the cylinder block. Install the distributor hold-down bolt.
6. Install the valve cover.
7. Install the distributor cap and attach the distributor connector.
8. Connect the negative battery cable. Start the engine and allow normal operating temperature to be reached.
9. Check and if necessary, adjust the ignition timing.

Ignition Timing

ADJUSTMENT

The 1997 4Runner/Tacoma with 2.7L (3RZ-FE), all 3.4L (5VZ-FE), and all 1998–99 engines use a distributorless ignition system referred to as Direct Ignition System (DIS). It is a fixed ignition timing system which means that basic ignition timing cannot be adjusted. All spark advance is permanently set by the Powertrain Control Module (PCM).

2.4L (2RZ-FE) Engine

➡ The ignition timing is not adjustable, but can be checked.

1. Warm the engine to normal operating temperature.
2. Attach a hand-held tester to the DLC3 connector under the dashboard on the drivers side.
3. Jumper terminals T_{E1} and E_1 of the DLC1.
4. Check the idle speed.
5. Aim the timing light at the timing indicator and check the ignition timing. Timing should be between 3–7 degrees BTDC at idle.

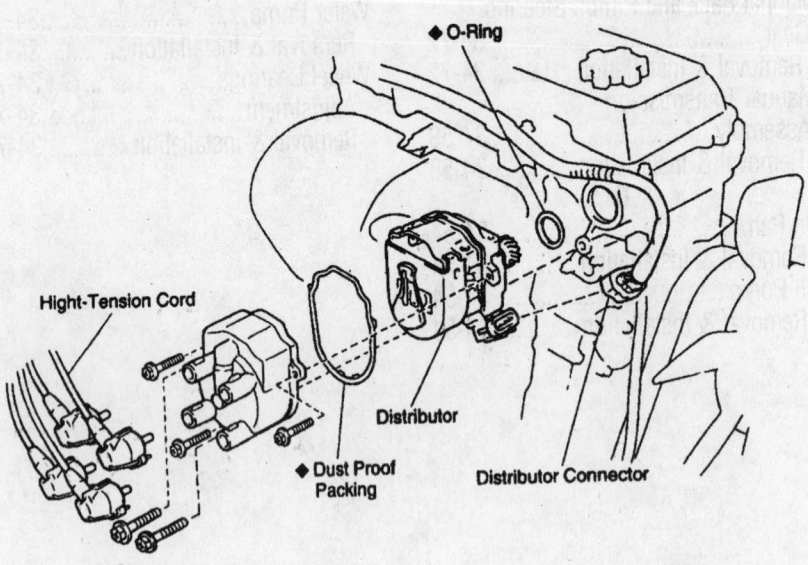

◆ O-Ring

Hight-Tension Cord

Distributor

Distributor Connector

◆ Dust Proof Packing

◆ Non-reusable part

7924YG99

Exploded view of the distributor and related components—Tacoma and T-100 With 2.4L (2RZ-FE) and 2.7L (3RZ-FE) engines

6. For a further check on ignition timing, disconnect the hand-held tester from the DLC3 and disconnect the jumper wire from the DLC1.

7. Point the timing light at the crankshaft pulley and read the timing. Timing should be between 7–18 degrees BTDC at idle.

8. Remove timing light from the engine.

2.4L (22R-E) and 3.0L (3VZ-E) Engines

1. Warm the engine to normal operating temperature.

2. Connect a tachometer and timing light to the engine.

3. Jumper terminals T_{E1} and E_1 of the DLC1.

4. Check the idle speed.

5. Aim the timing light at the timing indicator and check the ignition timing. Timing should be 5° BTDC for the 22R-E engine and 10° BTDC for the 3VZ-E engine at idle.

6. If adjustment is necessary, loosen the distributor hold-down bolt and adjust by turning. Tighten the hold-down bolt and recheck the timing.

7. Remove the jumper connector. Check that the ignition timing advances.

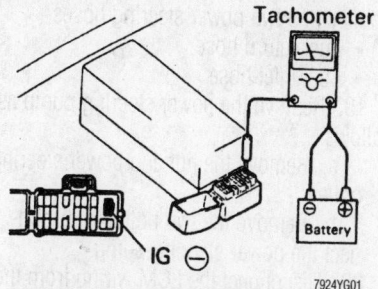

Connect a tachometer to the engine as shown—2.4L (22R-E) and 3.0L (3VZ-E) engines

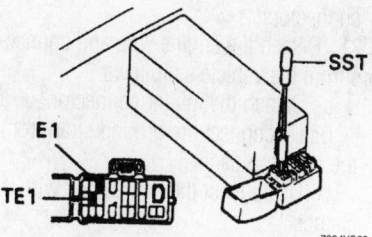

Jumper terminals of the DLC1 as shown—2.4L (22R-E) and 3.0L (3VZ-E) engines

8. Disconnect the tachometer and timing light from the engine.

Engine Assembly

REMOVAL & INSTALLATION

2.4L (22R-E) Engine

1. Remove the hood.
2. Release the fuel system pressure.

> **✼✼ CAUTION**
>
> **The fuel injection system remains under pressure after the engine has been turned OFF. Properly relieve fuel pressure before disconnecting any fuel lines. Failure to do so may result in fire or personal injury.**

3. Disconnect the battery and remove it from the vehicle.

> **✼✼ CAUTION**
>
> **Wait 90 seconds from the time the key is turned to LOCK and the negative battery cable is disconnected to begin work. This allows the SRS capacitor to discharge and prevent deployment of the air bag(s).**

4. Raise and support the vehicle safely.
5. Remove the engine under covers.
6. Drain the engine coolant from the radiator and the cylinder block.
7. Drain the engine oil.
8. Remove the air cleaner assembly.
9. Lower the vehicle and remove the radiator.
10. Remove the power steering belt.
11. If equipped, remove the air conditioning belt.
12. Remove the alternator belt, the fluid coupling, and the fan pulley.
13. Detach the following wires and connectors:
   a. Ground strap form the left fender apron
   b. Alternator connector and wire
   c. Igniter connector
   d. High tension cord for the ignition coil
   e. Distributor wire from the igniter
   f. Ground strap from the engine rear side
   g. ECM connectors
   h. Manual transmission: starter relay connector
   i. Check connector
   j. Air conditioning: compressor connector

14. Disconnect the following hoses:
   a. Power steering air hoses from the gas filter and the air pipe
   b. Brake booster hose
   c. Cruise: Cruise control vacuum hose
   d. Charcoal canister hose from the canister
   e. Fuel inlet and return lines.
15. Disconnect the following cables:
   a. Accelerator cable
   b. Automatic transmission: throttle cable
   c. Cruise: cruise control cable
16. If equipped with power steering, remove the pump from the bracket.
17. Disconnect the ground strap from the power steering pump bracket.
18. If equipped with air conditioning, remove the compressor from the bracket.
19. Disconnect the ground straps from the engine rear side and the right side.
20. Manual transmission: Remove the shift lever(s) from the inside of the vehicle.
21. Remove the rear driveshaft.
22. 2-Wheel Drive automatic transmission: Disconnect the manual shift linkage from the Park Neutral Position (PNP) switch.
23. 4-Wheel Drive automatic transmission: Disconnect the transfer shift linkage.
24. Disconnect the speedometer cable.
25. 4-Wheel Drive: Remove the transfer undercover.
26. Remove the stabilizer bar.
27. 4-Wheel Drive: Remove the front driveshaft.
28. Detach the oxygen sensor connector.
29. Manual transmission: Remove the clutch release cylinder with the bracket from the transmission.
30. 4-Wheel Drive: Remove the No. 1 front floor heat insulator and brake tube heat insulator.
31. 2-Wheel Drive: Remove the engine rear mounting and the bracket.
32. 4-Wheel Drive: Remove the No. 2 frame crossmember from the side frame.
33. Attach the engine hoist chain to the lift brackets to the engine.
34. Remove the mounting nuts and bolts. Lift the engine/transmission out of the vehicle slowly. Be sure that the engine/transmission is clear of all wiring and hoses.
35. Remove the engine with the transmission from the vehicle.
36. Remove the transmission from the engine.
   a. Automatic transmission: remove the automatic transmission oil cooler pipes.

b. Remove the starter.

c. Remove the two stiffener plates and the exhaust pipe bracket from the engine.

d. Remove the transmission from the engine.

37. For manual transmission: remove the clutch cover and the disc.

To install:

38. If removed, install the clutch disc and the cover to the flywheel.

39. Connect the transmission to the engine.

40. Lower the engine with the transmission into the engine compartment.

41. 4-Wheel Drive: Place a jack under the transmission. Use a wooden block between the jack and the transmission pan.

42. Jack the transmission up and put it onto the member.

43. Install the engine mount to the frame bracket.

44. Install the engine mount bolts on each side of the engine and remove the chain hoist.

45. For 2-wheel drive models, install the engine rear mount and bracket. Tighten the bolts holding the rear engine mount bracket to the support member to 9 ft. lbs. (13 Nm).

46. Lower the transmission and rest it on the extension housing. Install the rear engine mounting bracket to the mount and tighten the bolts to 19 ft. lbs. (25 Nm).

47. For 4-wheel drive models, install the No. 2 frame crossmember to the side frame and tighten the bolts to 70 ft. lbs. (95 Nm).

48. Lower the transmission and the transfer and secure the transmission mount to the crossmember. Tighten the bolts to 9 ft. lbs. (13 Nm).

49. For 4-wheel drive models, install the brake tube heat insulator and the No. 1 front floor heat insulator.

50. Manual transmission: Install the clutch release cylinder with the bracket to the transmission. Tighten the bracket to 29 ft. lbs. (39 Nm) and the release cylinder to 9 ft. lbs. (12 Nm).

51. Install the front exhaust pipe. Connect the oxygen sensor.

52. For 4-wheel drive models, install the front driveshaft, stabilizer bar, transfer case undercover, and connect the transfer shift linkage.

53. Connect the speedometer cable.

54. 2-Wheel Drive: Connect the manual shift linkage to the PNP switch.

55. Install the rear driveshaft.

56. Manual transmission: Install the shift lever(s).

57. Connect the ground straps to the engine rear side and the right side.

58. If removed, install the compressor to the bracket.

59. Connect the ground strap for the power steering pump bracket.

60. If removed, install the power steering pump with the bracket.

61. Reconnect all cables, hoses and wires:

62. Install the fan pulley, belt guide, fluid coupling, and the alternator drive belt.

63. If removed, install the air conditioning belt.

64. Install the power steering belt. Tighten the power steering pump pulley locknut to 32 ft. lbs. (43 Nm).

65. Install the radiator.

66. Install the air cleaner case and the intake air connector.

67. Fill with engine oil and coolant. If drained, refill the transmission with the appropriate fluid.

68. Install the battery and connect the cables.

69. Install the hood.

70. Start the engine, check for leaks, and perform all engine adjustments.

71. Install the engine undercover.

72. Road test the vehicle.

73. Recheck the coolant, engine oil, and transmission fluid levels.

2.4L (2RZ-FE) Engine

TACOMA

1. Properly relieve the fuel system pressure.

2. Turn the ignition switch **OFF.** Disconnect the battery cables; negative cable first.

WARNING

The air bag system is equipped with a back-up power source. To avoid possible air bag deployment, do not start working on the vehicle until 90 seconds has elapsed from the time the ignition switch is turned OFF and the negative battery terminal is disconnected.

3. Matchmark the hood hinges and remove the hood.

4. Remove the engine undercover.

5. Drain the engine oil, transmission oil and cooling system.

6. Remove the radiator as outlined in this section.

7. If equipped with power steering, loosen the lockbolt and adjusting bolt to the idler pulley and remove the drive belt.

8. If equipped with air conditioning, loosen the idler pulley nut and adjusting bolt. Remove the air conditioning drive belt from the engine.

9. Remove the alternator drive belt, fan (with fan clutch), water pump pulley, and fan shroud as outlined in this section.

10. If equipped with a manual transaxle, disconnect the accelerator cable from the throttle body.

11. If equipped with a automatic transaxle, disconnect the accelerator and throttle cables from the throttle body.

12. If equipped with cruise control, remove the actuator cover and disconnect the cruise control cable from the actuator.

13. Remove the air cleaner cap, MAF meter and resonator. Remove the air cleaner case.

14. Remove the intake air connector.

15. If equipped with air conditioning, disconnect the air conditioning compressor and bracket as outlined in this section.

16. Disconnect the alternator wires from the alternator.

17. Disconnect the heater hoses at the cowl panel.

18. Disconnect the following hoses:

- Brake booster vacuum hose
- EVAP hose
- If equipped with 4-Wheel Drive with Automatic Disconnecting Differential (ADD), disconnect the vacuum hose.
- If equipped with power steering, disconnect the two power steering hoses
- Fuel return hose
- Fuel inlet hose

19. Remove the power steering pump as follows:

a. Remove the nut and power steering pulley.

b. Remove the two bolt and disconnect the power steering pump.

20. Disconnect the ECM wiring from the ECM as follows:

a. Remove the four screws to the right front door scuff plate. Remove the scuff plate from the vehicle.

b. Remove the cowl panel side trim by removing the clip.

c. Detach the four ECM electrical connectors.

21. Detach the engine wire and connectors from the vehicle as follows:

a. Detach the igniter connector.

b. Disconnect the ground strap from the cowl top panel.

c. Disconnect the two engine wire clamps.

d. Remove the nuts holding the engine wire retainer to the cowl panel and pull out the engine wire from the vehicle cabin.

22. Disconnect the front exhaust pipe

from the exhaust manifold and catalytic converter.

23. If equipped with manual transmission, remove the shift lever assembly as follows:

    a. Remove the shift lever knob.

    b. Remove the four screws and shift lever boot.

    c. Remove the six bolts, shift lever assembly and baffle.

24. Remove the driveshaft from the vehicle.

25. Disconnect the speedometer cable from the transmission.

26. If equipped with manual transmission, remove the clutch release cylinder.

27. If equipped with automatic transmission, remove the cross-shaft.

28. Disconnect the wires at the starter.

29. Position a jack and wooden block under the transmission and remove the rear engine mounting bracket.

30. Attach a suitable engine hoist to the engine hangers.

31. Remove the nuts and bolts from the engine mounts.

➡**Be sure the engine/transmission assembly is clear of all wiring and hoses.**

32. Carefully lift the engine/transmission assembly out of the vehicle.

To install:

33. Attach the engine hoist to the engine hangers. Carefully lower the engine/transmission assembly into the vehicle. Keep the engine level, while aligning the engine mounts.

34. Install the engine mount fasteners, but do not fully tighten them.

35. Position a jack and wooden block under the transmission and install the rear engine mounting bracket. Tighten the bolts to the frame to 19 ft. lbs. (58 Nm) and the bolts to the mount to 13 ft. lbs. (18 Nm).

36. Remove the jack and engine hoist. Tighten the engine mounts to 28 ft. lbs. (38 Nm).

37. Connect the starter wires to the starter.

38. If equipped with manual transmission, install the clutch release cylinder.

39. If equipped with automatic transmission, install the cross-shaft.

40. Connect the speedometer to the transmission.

41. Install the driveshaft.

42. If equipped with manual transmission, install the shift lever assembly as follows:

    a. Install the baffle and shift lever assembly with the six bolts.

    b. Install the shift lever boot with the four screws.

    c. Install the shift lever knob.

43. Install the front exhaust pipe to the exhaust manifold.

44. Attach all wires and connectors.

    a. Install the cowl side trim and clip.

    b. Install the front door scuff plate and four screws.

45. Connect the alternator wires to the alternator.

46. Install the power steering pump as follows:

    a. Connect the power steering pump to the bracket with the two bolts. Tighten the bolts to 43 ft. lbs. (58 Nm).

    b. Install the power steering pulley with the nut. Tighten the nut to 32 ft. lbs. (43 Nm).

47. Connect all hoses previously removed.

48. If equipped with air conditioning, install the compressor.

49. Install the intake air connector. Tighten the two bolts to 13 ft. lbs. (18 Nm).

50. Install the water pump pulley, fan shroud, fan (with fan clutch), and alternator drive belt.

51. If equipped with air conditioning, install and adjust the drive belt.

52. Install and adjust the power steering drive belt.

53. If equipped with manual transmission, connect the accelerator cable to the throttle body.

54. If equipped with automatic transmission, connect the accelerator and throttle cables to the throttle body.

55. Install the air cleaner case.

56. Install the MAF meter, resonator and air cleaner cap.

57. Install the radiator to the vehicle with the tabs on the supports through the radiator service holes. Install the four bolts and tighten the bolts to 9 ft. lbs. (13 Nm).

58. Connect the lower radiator hose to the radiator.

59. If equipped with automatic transmission, connect the oil cooler hoses to the radiator.

60. Install the No. 2 fan shroud.

61. Connect the radiator reservoir hose to the radiator.

62. Connect the upper radiator hose to the radiator.

63. If removed, connect the air pipe with the two bolts.

64. Install the radiator grille to the vehicle with the 11 clips.

65. Install the two fillers.

66. Install the clearance lights to the grille with the four bolts and two clips.

67. Fill the engine oil, engine coolant, and transmission oil.

68. Connect the negative and positive cables to the battery.

69. Start the engine and check for leaks.

70. Check ignition timing.

71. Install the engine undercover.

72. Install the hood.

73. Road test the vehicle and check all fluids.

2.7L (3RZ-FE) Engine

T-100

1. Properly relieve the fuel system pressure.

2. Turn the ignition switch **OFF**. Disconnect the battery cables; negative cable first.

✳✳ WARNING

The air bag system is equipped with a back-up power source. To avoid possible air bag deployment, do not start working on the vehicle until 90 seconds has elapse from the time the ignition switch is turned OFF and the negative battery terminal is disconnected.

3. Matchmark the hood hinges and remove the hood.

4. Remove the battery and battery tray.

5. Drain the engine oil, transmission oil and cooling system.

6. Remove the expansion tank.

7. Remove the radiator as outlined in this section.

8. Remove the air cleaner cap, MAF meter and resonator. Remove the air cleaner case.

9. If equipped with a manual transaxle, disconnect the accelerator cable from the throttle body.

10. If equipped with a automatic transaxle, disconnect the accelerator and throttle cables from the throttle body.

11. Remove the intake air connector.

12. If equipped with air conditioning, disconnect the air conditioning compressor and bracket.

13. Disconnect the following hoses:
- Brake booster vacuum hose
- EVAP hose
- Two power steering hoses
- Fuel return hose
- Fuel inlet hose

14. Remove the power steering pump as follows:

    a. Remove the nut and power steering pulley.

b. Remove the two bolt and disconnect the power steering pump.

15. Disconnect the alternator wires from the alternator.

16. Disconnect the ECM wiring from the ECM as follows:

a. Remove the four screws to the right front door scuff plate. Remove the scuff plate from the vehicle.

b. Remove the cowl panel side trim by removing the clip.

c. Detach the four ECM electrical connectors.

17. Detach the engine wire and connectors from the vehicle as follows:

a. Disconnect the igniter.

b. Disconnect the ground strap from the cowl top panel.

c. Disconnect the four engine wire clamps.

d. Pull out the engine wire from the vehicle cabin.

18. If equipped with manual transmission, remove the shift lever assembly as follows:

a. Remove the shift lever knob.

b. Remove the four screws and shift lever boot.

c. Remove the six bolts, shift lever assembly and baffle.

19. Remove the sway bar as follows:

a. Remove the nuts and cushions holding the sway bar to the lower control arms.

b. Remove the sway bar bolts and brackets and remove the sway bar from the suspension.

20. Remove the driveshaft from the vehicle.

21. Disconnect the speedometer cable from the transmission.

22. Disconnect the front exhaust pipe from the exhaust manifold and catalytic converter.

23. If equipped with manual transmission, remove the clutch release cylinder.

24. If equipped with automatic transmission, remove the cross-shaft.

25. Disconnect the wires at the starter.

26. Position a jack and wooden block under the transmission and remove the rear engine mounting bracket.

27. Attach a suitable engine hoist to the engine hangers.

28. Remove the nuts and bolts from the engine mounts.

➡**Be sure the engine/transmission assembly is clear of all wiring and hoses.**

29. Carefully lift the engine/transmission assembly out of the vehicle.

To install:

30. Attach the engine hoist to the engine hangers. Carefully lower the engine/transmission assembly into the vehicle. Keep the engine level, while aligning the engine mounts.

31. Install the engine mount fasteners, but do not fully tighten.

32. Position a jack and wooden block under the transmission and install the rear engine mounting bracket. Tighten the bolts to the frame to 42 ft. lbs. (58 Nm) and the bolts to the mount to 13 ft. lbs. (18 Nm).

33. Remove the jack and engine hoist. Tighten the engine mounts to 28 ft. lbs. (38 Nm).

34. Connect the starter wires to the starter.

35. If equipped with manual transmission, install the clutch release cylinder.

36. If equipped with automatic transmission, install the cross-shaft.

37. Install the front exhaust pipe to the exhaust manifold.

38. Connect the speedometer to the transmission.

39. Install the driveshaft.

40. Install the sway bar as follows:

a. Place the sway bar in position and install both sway bar bushings and brackets to the frame.

b. Tighten the sway bar mounting bolts to 22 ft. lbs. (930 Nm).

c. Connect the sway bar to the lower control arms with the brackets and cushions. Tighten the nuts to 9 ft. lbs. (13 Nm).

41. If equipped with manual transmission, install the shift lever assembly as follows:

a. Install the baffle and shift lever assembly with the six bolts.

b. Install the shift lever boot with the four screws.

c. Install the shift lever knob.

42. Reconnect all engine wires.

43. Install the cowl side trim and clip.

a. Install the front door scuff plate and four screws.

44. Install the power steering pump as follows:

a. Connect the power steering pump to the bracket with the two bolts. Tighten the bolts to 43 ft. lbs. (58 Nm).

b. Install the power steering pulley with the nut. Tighten the nut to 32 ft. lbs. (43 Nm).

45. Connect all hoses previously removed.

46. Connect the engine heater hoses at the cowl panel.

47. If equipped with air conditioning, install the compressor as follows:

a. Install the air conditioning compressor bracket with the four bolts. Tighten the bolts to 32 ft. lbs. (44 Nm).

b. Connect the air conditioning compressor to the bracket with the four bolts. Tighten the bolts to 18 ft. lbs. (25 Nm).

48. Install the intake air connector. Tighten the two bolts to 13 ft. lbs. (18 Nm).

49. If equipped with manual transmission, connect the accelerator cable to the throttle body.

50. If equipped with automatic transmission, connect the accelerator and throttle cables to the throttle body.

51. Install the air cleaner case.

52. Install the MAF meter, resonator and air cleaner cap.

53. Install the radiator as outlined.

54. If equipped with air conditioning, install and adjust the air conditioning compressor drive belt.

55. Install and adjust the power steering drive belt.

56. Connect the radiator reservoir hose to the radiator.

57. Connect the upper radiator hose to the radiator.

58. Install the radiator grille to the vehicle with the 11 clips and four screws.

59. Install the clearance lights to the grille with the four bolts to each light.

60. Install the battery and battery clamp.

61. Install the radiator expansion tank.

62. Fill the engine oil, engine coolant, and transmission oil.

63. Connect the negative and positive cables to the battery.

64. Start the engine and check for leaks.

65. Check ignition timing.

66. Install the engine undercover.

67. Install the hood.

68. Road test the vehicle and check all fluids.

TACOMA

1. Properly relieve the fuel system pressure.

2. Turn the ignition switch **OFF**. Disconnect the battery cables; negative cable first.

✲✲ WARNING

The air bag system is equipped with a back-up power source. To avoid possible air bag deployment, do not start working on the vehicle until 90 seconds has elapsed from the time the ignition switch is turned OFF and the negative battery terminal is disconnected.

3. Matchmark the hood hinges and remove the hood.

4. Remove the engine undercover.

5. Drain the engine oil, transmission oil and cooling system.

6. Remove the radiator as outlined in this section.

7. If equipped with power steering, loosen the lockbolt and adjusting bolt to the idler pulley and remove the drive belt.

8. If equipped with air conditioning, loosen the idler pulley nut and adjusting bolt. Remove the air conditioning drive belt from the engine.

9. Remove the alternator drive belt, fan (with fan clutch), water pump pulley, and fan shroud.

10. If equipped with a manual transaxle, disconnect the accelerator cable from the throttle body.

11. If equipped with a automatic transaxle, disconnect the accelerator and throttle cables from the throttle body.

12. If equipped with cruise control, remove the actuator cover and disconnect the cruise control cable from the actuator.

13. Remove the air cleaner cap, MAF meter and resonator. Remove the air cleaner case.

14. Remove the intake air connector.

15. If equipped with air conditioning, disconnect the air conditioning compressor and bracket.

16. Disconnect the alternator wires from the alternator.

17. Disconnect the heater hoses at the cowl panel.

18. Disconnect the following hoses:
• Brake booster vacuum hose
• EVAP hose
• If equipped with 4-Wheel Drive with Automatic Disconnecting Differential (ADD), disconnect the vacuum hose.
• If equipped with power steering, disconnect the two power steering hoses
• Fuel return hose
• Fuel inlet hose

19. Remove the power steering pump as follows:

a. Remove the nut and power steering pulley.

b. Remove the two bolt and disconnect the power steering pump.

20. Disconnect the ECM wiring from the ECM as follows:

a. Remove the four screws to the right front door scuff plate. Remove the scuff plate from the vehicle.

b. Remove the cowl panel side trim by removing the clip.

c. Detach the four ECM electrical connectors.

21. Detach the engine wire and connectors from the vehicle as follows:

a. Disconnect the igniter.

b. Disconnect the ground strap from the cowl top panel.

c. Disconnect the two engine wire clamps.

d. Remove the nuts holding the engine wire retainer to the cowl panel and pull out the engine wire from the vehicle cabin.

22. Disconnect the front exhaust pipe from the exhaust manifold and catalytic converter.

23. If equipped with manual transmission, remove the shift lever assembly as follows:

a. Remove the shift lever knob.

b. Remove the four screws and shift lever boot.

c. Remove the six bolts, shift lever assembly and baffle.

24. Remove the driveshaft from the vehicle.

25. Disconnect the speedometer cable from the transmission.

26. If equipped with manual transmission, remove the clutch release cylinder.

27. If equipped with automatic transmission, remove the cross-shaft.

28. Disconnect the wires at the starter.

29. Position a jack and wooden block under the transmission and remove the rear engine mounting bracket.

30. Attach a suitable engine hoist to the engine hangers.

31. Remove the nuts and bolts from the engine mounts.

➡**Be sure the engine/transmission assembly is clear of all wiring and hoses.**

32. Carefully lift the engine/transmission assembly out of the vehicle.

To install:

33. Attach the engine hoist to the engine hangers. Carefully lower the engine/trans-

mission assembly into the vehicle. Keep the engine level, while aligning the engine mounts.

34. Install the engine mount fasteners, but do not fully tighten them.

35. Position a jack and wooden block under the transmission and install the rear engine mounting bracket. Tighten the bolts to the frame to 19 ft. lbs. (58 Nm) and the bolts to the mount to 13 ft. lbs. (18 Nm).

36. Remove the jack and engine hoist. Tighten the engine mounts to 28 ft. lbs. (38 Nm).

37. Connect the starter wires to the starter.

38. If equipped with manual transmission, install the clutch release cylinder.

39. If equipped with automatic transmission, install the cross-shaft.

40. Connect the speedometer to the transmission.

41. Install the driveshaft.

42. If equipped with manual transmission, install the shift lever assembly as follows:

a. Install the baffle and shift lever assembly with the six bolts.

b. Install the shift lever boot with the four screws.

c. Install the shift lever knob.

43. Install the front exhaust pipe to the exhaust manifold.

44. Connect all engine wires.

a. Install the cowl side trim and clip.

b. Install the front door scuff plate and four screws.

45. Connect the alternator wires to the alternator.

46. Install the power steering pump as follows:

a. Connect the power steering pump to the bracket with the two bolts. Tighten the bolts to 43 ft. lbs. (58 Nm).

b. Install the power steering pulley with the nut. Tighten the nut to 32 ft. lbs. (43 Nm).

47. Connect all hoses previously removed.

48. Connect the engine heater hoses at the cowl panel.

49. If equipped with air conditioning, install the compressor.

50. Install the intake air connector. Tighten the two bolts to 13 ft. lbs. (18 Nm).

51. Install the water pump pulley, fan shroud, fan (with fan clutch), and alternator drive belt as follows:

a. Place the fan (with the fan clutch), water pump pulley and fan shroud in position.

b. Install the water pump pulley mounting nuts but do not tighten the nuts at this time.

c. Install the alternator drive belt to the engine.

d. Stretch the alternator belt tight and tighten the fan nuts to 16 ft. lbs. (21 Nm).

e. Adjust the drive belt for the alternator.

52. If equipped with air conditioning, install and adjust the drive belt.

53. Install and adjust the power steering drive belt.

54. If equipped with manual transmission, connect the accelerator cable to the throttle body.

55. If equipped with automatic transmission, connect the accelerator and throttle cables to the throttle body.

56. Install the air cleaner case.

57. Install the MAF meter, resonator and air cleaner cap.

58. Install the radiator to the vehicle with the tabs on the supports through the radiator service holes. Install the four bolts and tighten the bolts to 9 ft. lbs. (12.5 Nm).

59. Connect the lower radiator hose to the radiator.

60. If equipped with automatic transmission, connect the oil cooler hoses to the radiator.

61. Install the No. 2 fan shroud.

62. Connect the radiator reservoir hose to the radiator.

63. Connect the upper radiator hose to the radiator.

64. If removed, connect the air pipe with the two bolts.

65. Install the radiator grille to the vehicle with the 11 clips.

66. Install the two fillers.

67. Install the clearance lights to the grille with the four bolts and two clips.

68. Fill the engine oil, engine coolant, and transmission oil.

69. Connect the negative and positive cables to the battery.

70. Start the engine and check for leaks.

71. Check ignition timing.

72. Install the engine undercover.

73. Install the hood.

74. Road test the vehicle and check all fluids.

1996–99 4RUNNER (2-WHEEL DRIVE)

1. Properly relieve the fuel system pressure.

2. Disconnect the negative battery cable.

✳✳ CAUTION

Wait 90 seconds from the time the key is turned to LOCK and the negative battery cable is disconnected to begin work. This allows the SRS capacitor to discharge and prevent deployment of the air bag(s).

✳✳ CAUTION

The fuel injection system remains under pressure after the engine has been turned OFF. Properly relieve fuel pressure before disconnecting any fuel lines. Failure to do so may result in fire or personal injury.

3. Raise and safely support the vehicle.

4. Remove the engine undercover.

5. Drain the engine coolant.

6. Drain the engine oil and the transmission oil.

7. Remove the hood.

8. Remove the radiator.

9. Remove the drive belt for the alternator and the water pump pulley.

10. Disconnect the accelerator cable from the throttle body.

11. If equipped with cruise control, remove the actuator cover and disconnect the cruise control cable from the actuator.

12. Remove the air cleaner assembly.

a. Detach the IAT and the MAF meter connectors.

b. Disconnect the three wire clamps and the engine wire.

c. Loosen the air cleaner hose clamp.

d. Remove the three bolts and the MAF meter, resonator and the air cleaner assembly.

13. If equipped with air conditioning, disconnect the air conditioning compressor.

14. Detach the alternator connector.

15. Disconnect the heater hoses.

16. Disconnect the following hoses:

• Brake booster vacuum hose

• EVAP hose

• Two air hoses for the power steering idle-up

• Fuel return hose

• Fuel inlet hose

17. Remove the power steering pump from the engine.

18. Disconnect the engine wire from the cabin.

a. Remove the glove box door.

b. Lower the finish No. 2 panel.

c. Detach the four ECM connectors.

d. Detach the two cassette connectors and the two wire clamps from the lower finish panel.

e. Disconnect the igniter.

f. Disconnect the ground strap from the cowl top panel.

g. Disconnect the two engine wire clamps.

h. Remove the two nuts holding the engine wire retainer to the cowl panel and pull out the engine wire from the cabin.

19. Disconnect the heated oxygen sensor and remove the front exhaust pipe.

20. Manual transmission: Remove the shift lever assembly.

a. Remove the shift lever knob.

b. Remove the four screws and the shift lever boot.

c. Remove the six bolts, the shift lever assembly and baffle.

21. Remove the driveshaft.

22. Disconnect the speedometer cable.

23. Manual transmission: Remove the clutch release cylinder.

24. Automatic transmission: Remove the cross-shaft.

25. Disconnect the starter wire.

26. Place a jack under the transmission and remove the engine rear mounting bracket.

27. Install a rear engine hanger in the correct direction.

28. Attach the engine hoist chain to the two engine hangers.

29. Remove the four bolts and nuts holding the engine front mounting insulators to the frame.

➡**Be sure that the engine is clear of all wiring and hoses.**

30. Lift the engine and the transmission assembly onto the stand.

31. Separate the engine from the transmission.

Be sure to support the engine before removing the right and left engine mounts—4Runner (2-Wheel Drive) with 2.7L (3RZ-FE) engine

To install:

32. Install the transmission to the engine.

33. Attach a chain hoist to the engine hangers.

34. Lower the engine and transmission assembly into the engine compartment.

35. Keep the engine level and align the right and left mounting and body mountings.

36. Attach the right and left mounting insulators to the body mountings and temporarily install the bolts and nuts.

37. Jack up and put the transmission onto the frame.

38. Remove the chain hoist.

39. Remove the bolt and the rear engine hanger.

40. Install the engine rear mounting bracket and tighten to:
- Bolt A: 13 ft. lbs. (19 Nm)
- Bolt B: 19 ft. lbs. (26 Nm)

41. Tighten the left and right engine mounting insulator bolts and nuts to 28 ft. lbs. (38 Nm).

42. Connect the starter wire.

43. Manual transmission: Install the clutch release cylinder. Tighten the clutch line bolt to 29 ft. lbs. (39 Nm) and the clutch release cylinder bolts to 9 ft. lbs. (13 Nm).

44. Automatic transmission: Install the cross-shaft and tighten the bolt to 29 ft. lbs. (39 Nm) and the nut to 13 ft. lbs. (18 Nm).

45. Connect the speedometer cable.

46. Install the driveshaft.

47. Manual transmission: Install the shift lever assembly.

48. Install the front exhaust pipe.

a. Install the new gaskets and the front exhaust pipe assembly and tighten the three new nuts to 46 ft. lbs. (62 Nm).

b. Install the support bracket and tighten the bolts to 29 ft. lbs. (39 Nm).

c. Connect the three-way catalytic converter with a new gasket to the tail pipe and tighten to 29 ft. lbs. (39 Nm).

d. Connect the heated oxygen sensor.

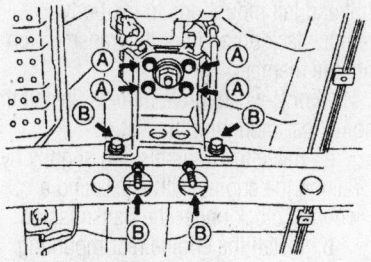

7924YG04

Bolt tightening pattern for the engine rear mounting bracket—4Runner (2-Wheel Drive) with 2.7L (3RZ-FE) engine

49. Connect the engine wire to the cabin.

50. Install the power steering pump.

51. Connect all hoses previously removed.

52. Connect the alternator wire.

53. If removed, install the air conditioning compressor and tighten the bolts to 18 ft. lbs. (25 Nm).

54. Install the intake air connector and tighten the bolts to 13 ft. lbs. (18 Nm).

55. Install the air cleaner assembly.

56. Connect the throttle cable to the throttle body.

57. If disconnected, connect the cruise control cable to the actuator and install the actuator cover.

58. Install the drive belt for the alternator and the water pump pulley.

59. Install the radiator.

60. Refill the engine oil, coolant, and transmission oil.

61. Connect the negative battery cable, start the engine, and check for leaks.

62. Check the ignition timing.

63. Install the engine undercover.

64. Install the hood.

65. Road test the vehicle and recheck the fluid levels.

1996–99 4RUNNER (4-WHEEL DRIVE)

⁕⁕ CAUTION

The fuel injection system remains under pressure after the engine has been turned OFF. Properly relieve fuel pressure before disconnecting any fuel lines. Failure to do so may result in fire or personal injury.

1. Properly relieve the fuel system pressure.

2. Disconnect the negative battery cable.

⁕⁕ CAUTION

Wait 90 seconds from the time the key is turned to LOCK and the negative battery cable is disconnected to begin work. This allows the SRS capacitor to discharge and prevent deployment of the air bag(s).

3. Raise and safely support the vehicle.

4. Remove the transmission.

5. Remove the engine undercover.

6. Drain the engine coolant.

7. Drain the engine oil.

8. Remove the hood.

9. Remove the radiator.

10. Remove the drive belt for the alternator and the water pump pulley.

11. Disconnect the accelerator cable from the throttle body.

12. If equipped with cruise control, remove the actuator cover and disconnect the cruise control cable from the actuator.

13. Remove the air cleaner assembly.

14. Remove the intake air connector.

15. If equipped with air conditioning, remove the air conditioning compressor.

16. Detach the alternator connector.

17. Disconnect the heater hoses.

18. Disconnect the following hoses:
- Brake booster vacuum hose
- EVAP hose
- Two air hoses for the power steering idle-up
- With Automatic Disconnecting Differential (ADD)—Vacuum hose
- Fuel return hose
- Fuel inlet hose

19. Remove the power steering pump from the engine.

20. Disconnect the engine wire from the cabin.

a. Remove the glove box door.

b. Lower the finish No. 2 panel.

c. Detach the ECM connectors.

d. Detach the two cassette connectors and the two wire clamps from the lower finish panel.

e. Variable Switching Valve (VSV) connector for the EVAP and clamp.

f. Disconnect the igniter.

g. Disconnect the ground strap from the cowl top panel.

h. Disconnect the two engine wire clamps.

i. Remove the two nuts holding the engine wire retainer to the cowl panel and pull out the engine wire from the cabin.

21. Install a rear engine hanger in the correct direction.

22. Attach the engine hoist chain to the two engine hangers.

23. Remove the bolts and nuts holding the engine front mounting insulators to the frame.

24. Lift the engine out of the vehicle slowly and carefully. Be sure that the engine is clear of all wiring and hoses.

25. Remove the engine from the vehicle.

To install:

26. Attach a chain hoist to the engine hangers.

27. Lower the engine assembly into the engine compartment.

28. Keep the engine level and align the right and left mounting and body mountings.

29. Attach the right and left mounting insulators to the body mountings and temporarily install the bolts and nuts.

30. Remove the chain hoist.

31. Remove the bolt and the rear engine hanger.

32. Tighten the right and left engine mounting insulator bolts and nuts to 28 ft. lbs. (38 Nm).

33. Connect the engine wire to the cabin.

34. Connect the alternator wire.

35. Install the power steering pump.

36. Connect all hoses:

37. If removed, install the air conditioning compressor and tighten the bolts to 18 ft. lbs. (25 Nm).

38. Install the intake air connector and tighten the bolts to 13 ft. lbs. (18 Nm).

39. Install the air cleaner assembly.

40. Connect the throttle cable to the throttle body.

41. If removed, install the cruise control cable to the actuator and install the actuator cover.

42. Install the radiator.

43. Install the hood.

44. Fill with engine with oil and the coolant system with coolant.

45. Install the transmission.

46. Install the engine undercover.

47. Connect the negative battery cable.

48. Fill the transmission fluid.

49. Check the ignition timing.

50. Test drive the vehicle and check for leaks.

51. Recheck fluid levels.

3.0L (3VZ-E) Engine

1995 4RUNNER

1. Remove the hood.
2. Relieve the fuel system pressure.

> ❋❋ **CAUTION**
>
> **The fuel injection system remains under pressure after the engine has been turned OFF. Properly relieve fuel pressure before disconnecting any fuel lines. Failure to do so may result in fire or personal injury.**

3. Disconnect the battery and remove it from the vehicle.

> ❋❋ **CAUTION**
>
> **Wait 90 seconds from the time the key is turned to LOCK and the negative battery cable is disconnected to begin work. This allows the SRS capacitor to discharge and prevent deployment of the air bag(s).**

4. Raise and safely support the vehicle.
5. Remove the engine under covers.

6. Drain the engine coolant.

7. Drain the engine oil.

8. Lower the vehicle and remove the air cleaner and hose.

9. Disconnect the hoses, fan shrouds, and remove the radiator. On the automatic transmission, disconnect the oil cooler hoses.

10. For manual transmission models, disconnect the clutch release cylinder hose.

11. Remove the power steering drive belt and pump pulley.

12. Disconnect the power steering pump from the engine.

13. Remove the cooling fan.

14. Remove the alternator belt.

15. Detach the following straps, wires, and connectors:

a. Ground strap from the left fender apron

b. Generator connector and wire

c. Igniter connector

d. Oil pressure sender gauge connector

e. Ground strap from the engine rear side

f. ECM connectors

g. Variable Switching Valve (VSV) connectors

h. Air conditioning compressor connector

i. Manual transmission: Starter relay connector

j. Solenoid resister connector

k. Data Link Connector No. 1 (DLC 1)

l. With Automatic Disconnecting Differential (ADD): ADD switch connector

16. Disconnect the following hoses:

a. Power steering air hoses from the gas filter and air pipe

b. Brake booster hose

c. Cruise control: Cruise control vacuum hose

d. Charcoal canister hose from the canister

e. Variable Switching Valve (VSV) vacuum hoses

17. Disconnect the following cables:

a. Accelerator cable

b. For automatic transmission models, remove the throttle cable

c. Cruise: Cruise control cable

18. Disconnect the heater hoses.

19. Disconnect the fuel inlet and the outlet hoses.

20. Disconnect the air conditioning compressor from the engine.

21. Disconnect the heated oxygen sensor and remove the front exhaust pipe.

22. Manual transmission: Remove the shift levers.

23. Remove the rear driveshaft.

24. 4-Wheel Drive: Remove the front driveshaft.

25. 2-Wheel Drive automatic transmission: Disconnect the manual shift linkage.

26. 4-Wheel Drive automatic transmission: Disconnect the transfer case shift linkage.

27. Detach the speedometer connector.

28. 4-Wheel Drive: Remove the transfer undercover and the stabilizer bar.

29. On 4-Wheel Drive vehicles, remove the No. 1 front floor heat insulator and the brake tube heat insulator.

30. 2-Wheel Drive: Remove the engine rear mounting bracket.

31. 4-Wheel Drive: Remove the No. 2 frame crossmember:

a. Remove the four bolts holding the engine rear mounting insulator to the frame crossmember.

b. Raise the transmission slightly with a jack.

c. Remove the eight bolts holding the frame crossmember to the side frame. Remove the frame crossmember.

32. Attach the engine chain hoist to the engine hangers.

33. Remove the four bolts holding the right and left engine mounting insulators to the body mountings.

34. Lift the engine and transmission assembly out of the vehicle slowly and carefully. Be sure that the engine is clear of all wiring, hoses, and cables.

35. Remove the transmission from the engine.

36. Manual transmission: Remove the clutch cover and the disc.

To install:

37. If removed, install the clutch cover and the disc.

38. Install the transmission to the engine.

39. Attach the engine hoist to the engine hangers and slowly lower the engine and transmission assembly into the engine compartment.

40. Keep the engine level and align the right and left mountings to the body mounts. Jack up and put the transmission onto the member.

41. For 2-wheel drive models, install the engine rear mounting bracket.

a. Raise the transmission slightly by raising the engine with a jack and a wooden block under the transmission.

b. Install the engine rear mounting bracket to the support member and tighten to 19 ft. lbs. (25 Nm).

c. Lower the transmission and rest it on the extension housing.

d. Install the mounting bracket to the

mounting insulator. Tighten the bolts to 9 ft. lbs. (13 Nm).

42. For 4-wheel drive models, install the No. 2 frame crossmember.

    a. Raise the transmission slightly with a jack.

    b. Install the frame crossmember to the side frame. Tighten the bolts to 70 ft. lbs. (95 Nm).

    c. Lower the transmission and the transfer.

    d. Install the frame crossmember to the engine rear mounting insulator and tighten the bolts to 9 ft. lbs. (13 Nm).

43. Tighten the left and right engine mounting insulator bolts to 27 ft. lbs. (37 Nm).

44. For 4-wheel drive models, install the No. 1 front floor and brake tube heat insulator.

45. For 4-wheel drive models, install the stabilizer bar and the transfer undercover.

46. Attach the speedometer connector.

47. For 2-wheel drive models equipped with automatic transmissions, connect the manual shift linkage.

48. For 4-wheel drive models equipped with automatic transmissions, connect the transfer shift linkage.

49. For 4-wheel drive models, install the front driveshaft.

50. Install the rear driveshaft.

51. Manual transmission: Install the shift levers.

52. Install the front exhaust pipe to the exhaust manifold and tighten to 46 ft. lbs. (62 Nm). Tighten the bolts connecting the exhaust to the catalytic converter to 29 ft. lbs. (39 Nm).

53. Connect the heated oxygen sensor.

54. Manual transmission: Connect the clutch release cylinder hose.

55. Install the air conditioning compressor. Install the remaining components.

56. Fill the engine with coolant and with oil. If drained, refill the transmission with the appropriate fluid.

57. Start the engine and check for leaks.

58. Perform all engine adjustments.

59. Install the engine undercover.

60. Install the hood.

61. Road test the vehicle.

62. Recheck the engine coolant, oil, and transmission fluid levels.

3.4L (5VZ-FE) Engine

1995 TACOMA AND 1995–99 T-100—2-WHEEL DRIVE

1. Properly relieve the fuel system pressure.

2. Remove the hood.

3. Disconnect the battery and remove it from the vehicle.

✳✳ CAUTION

Work must be started after 90 seconds from the time the ignition switch is turned to the LOCK position and the negative battery cable is disconnected.

4. Raise and safely support the vehicle.

5. Remove the engine under covers.

6. Drain the engine coolant.

7. Drain the engine oil.

8. Remove the radiator from the vehicle.

9. Remove the power steering (PS) drive belt as follows:

    a. Stretch the belt and loosen the fan pulley mounting nuts.

    b. Loosen the lockbolt, pivot bolt and adjusting bolt and remove the drive belt from the engine.

10. If equipped with air conditioning, remove the air conditioning drive belt by loosening the idle pulley nut and adjusting bolt.

11. Loosen the lockbolt, pivot bolt and adjusting bolt and the alternator drive belt.

12. Remove the fan with the fluid coupling and fan pulleys.

13. Disconnect the PS pump from the engine and set aside. Do not disconnect the lines from the pump.

14. If equipped with air conditioning, disconnect the compressor from the engine and set aside. Do not disconnect the lines from the compressor.

15. Remove the air cleaner cap, MAF meter and resonator.

16. Remove the air cleaner case and filter.

17. Disconnect the following cables:

• If equipped with cruise control, disconnect the actuator cable with the bracket.

• Accelerator cable

• With automatic transmission, Throttle cable

18. Disconnect the heater hoses.

19. Disconnect the following hoses:

• Brake booster vacuum hose

• EVAP hose

• Fuel return hose

• Fuel inlet hose

20. Detach the starter wire and connectors as follows:

    a. Remove the ground strap by removing the bolt.

    b. Remove the nuts and disconnect the positive cable from the battery.

    c. Detach the three starter wire clamps and connector.

21. Detach the alternator connector and wire.

22. Detach the engine wire and connectors as follows:

    a. Remove the four screws to the right front door scuff plate. Remove the scuff plate from the vehicle.

    b. Remove the cowl panel side trim by removing the clip.

    c. Disconnect the ECM.

    d. Detach the two connectors from the cowl wire.

    e. Disconnect the igniter.

    f. Disconnect the ground strap.

    g. Disconnect the six engine wire clamps.

    h. Pull out the engine wire from the cabin.

23. If equipped with manual transmission, remove the shift lever assembly as follows:

    a. Remove the shift lever knob.

    b. Remove the four screws and the shift lever boot.

    c. Remove the shift lever assembly and gasket by removing the six bolts.

24. Remove the stabilizer bar.

25. Remove the driveshaft from the transmission.

26. Disconnect the speedometer cable.

27. Remove the front exhaust pipe.

28. If equipped with a manual transmission, remove the clutch release cylinder.

29. If equipped with an automatic transmission, remove the cross-shaft.

30. Place a jack under the transmission.

31. Remove the transmission rear mounting bracket by removing the eight bolts.

32. If equipped with air conditioning, remove the bolt and disconnect the air conditioning compressor wire clamp.

33. If necessary, install a No. 2 engine hanger with two bolts. Tighten the two bolts to 30 ft. lbs. (40 Nm).

34. Attach the engine hoist chain to the two engine hangers.

35. Remove the four bolts and nuts holding the engine front mounting insulators to the frame.

36. Lift the engine and transmission out of the vehicle.

To install:

37. Install the engine assembly to the vehicle. Attach the engine mounts to the body mountings. Install the bolts and nuts but do not tighten at this time.

38. Remove the engine chain hoist the No. 2 engine hanger.

39. If equipped with air conditioning, connect the air conditioning wire with the bolt.

40. Raise the transmission slightly and install the transmission mounting bracket. Tighten the bolts to the frame to 43 ft. lbs. (58 Nm) and the bolts to the mounting insulator to 13 ft. lbs. (18 Nm).

41. Tighten the engine mounting nuts and bolts to 28 ft. lbs. (38 Nm).

42. If equipped with an automatic transmission, install the cross-shaft.

43. If equipped with a manual transmission, install the clutch release cylinder. Tighten the bolts to 9 ft. lbs. (12 Nm).

44. Install the front exhaust pipe.

45. Connect the speedometer cable.

46. Install the driveshaft.

47. Install the stabilizer bar.

48. Install the shift lever assembly as follows:

a. Install a new gasket and shift lever assembly with the six bolts.

b. Install the shift lever boot with the four screws.

c. Install the shift lever knob.

49. Connect all engine wires, hoses and cables.

50. Install the air cleaner case and air filter.

51. Install the MAF meter, resonator and air cleaner cap.

52. If equipped, install the air conditioning compressor. Install the remaining components.

53. Fill the engine with oil.

54. Fill the engine and radiator with coolant.

55. Install the engine undercover.

56. Start the engine and check for leaks.

1995 TACOMA AND 1995–99 T-100— 4-WHEEL DRIVE

1. Properly relieve the fuel system pressure.

2. Remove the transmission from the vehicle.

3. Remove the hood.

4. Disconnect the battery and remove it from the vehicle.

❈❈ CAUTION

Work must be started after 90 seconds from the time the ignition switch is turned to the LOCK position and the negative battery cable is disconnected.

5. Raise and safely support the vehicle.

6. Remove the engine under covers.

7. Drain the engine coolant.

8. Drain the engine oil.

9. Remove the radiator from the vehicle.

10. Remove the power steering (PS) drive belt as follows:

a. Stretch the belt and loosen the fan pulley mounting nuts.

b. Loosen the lockbolt, pivot bolt and adjusting bolt and remove the drive belt from the engine.

11. If equipped with air conditioning, remove the air conditioning drive belt by loosening the idle pulley nut and adjusting bolt.

12. Loosen the lockbolt, pivot bolt and adjusting bolt and the alternator drive belt.

13. Remove the fan with the fluid coupling and fan pulleys.

14. Disconnect the PS pump from the engine and set aside. Do not disconnect the lines from the pump.

15. If equipped with air conditioning, disconnect the compressor from the engine and set aside. Do not disconnect the lines from the compressor.

16. Remove the air cleaner cap, MAF meter and resonator.

17. Remove the air cleaner case and filter.

18. Disconnect the following cables:

• If equipped with cruise control, disconnect the actuator cable with the bracket.

• Accelerator cable

• With automatic transmission, Throttle cable

19. Disconnect the heater hoses.

20. Disconnect the following hoses:

• Brake booster vacuum hose

• EVAP hose

• Automatic Disconnecting Differential (ADD) vacuum hose

• Fuel return hose

• Fuel inlet hose

21. Detach the starter wire and connectors as follows:

a. Remove the ground strap by removing the bolt.

b. Remove the nuts and disconnect the positive cable from the battery.

c. Detach the three starter wire clamps and connector.

d. Detach the Automatic Disconnecting Differential (ADD) indicator switch connector.

22. Detach the alternator connector and wire.

23. Detach the engine wire and connectors as follows:

a. Remove the four screws to the right front door scuff plate. Remove the scuff plate from the vehicle.

b. Remove the cowl panel side trim by removing the clip.

c. Disconnect the ECM.

d. Detach the two connectors from the cowl wire.

e. Detach the igniter connector.

f. Disconnect the ground strap.

g. Disconnect the six engine wire clamps.

h. Pull out the engine wire from the cabin.

24. If equipped with air conditioning, remove the bolt and disconnect the air conditioning compressor wire clamp.

25. If necessary, install a No. 2 engine hanger with two bolts. Tighten the two bolts to 30 ft. lbs. (40 Nm).

26. Attach the engine hoist chain to the two engine hangers.

27. Remove the four bolts and nuts holding the engine front mounting insulators to the frame.

28. Lift the engine out of the vehicle.

To install:

29. Install the engine assembly to the vehicle. Attach the engine mounts to the body mountings. Install the bolts and nuts but do not tighten at this time.

30. Remove the engine chain hoist the No. 2 engine hanger.

31. If equipped with air conditioning, connect the air conditioning wire with the bolt.

32. Tighten the engine mounting nuts and bolts to 28 ft. lbs. (38 Nm).

33. Connect all engine wires, hoses and cables.

34. Install the air cleaner case and air filter.

35. Install the MAF meter, resonator and air cleaner cap.

36. If equipped, install the air conditioning compressor.

37. Install the fan with the fluid coupling and fan pulleys. Tighten the nuts to 48 inch lbs. (5.4 Nm).

38. Install the alternator drive belt.

39. If equipped, install and adjust the air conditioning drive belt.

40. Install the PS pump, pump pulley and the drive belt.

41. Install the radiator to the vehicle.

42. Fill the engine with oil.

43. Fill the engine and radiator with coolant.

44. Install the hood.

45. Install the engine undercover.

46. Install the transmission to the vehicle.

47. Start the engine and check for leaks.

1996–99 4RUNNER (2-WHEEL DRIVE)

1. Properly relieve the fuel system pressure.

2. Remove the hood.

3. Disconnect the battery and remove it from the vehicle.

✳✳ CAUTION

Wait 90 seconds from the time the key is turned to LOCK and the negative battery cable is disconnected to begin work. This allows the SRS capacitor to discharge and prevent deployment of the air bag(s).

4. Raise and safely support the vehicle.
5. Remove the engine under covers.
6. Drain the engine coolant.
7. Drain the engine oil.
8. Remove the radiator from the vehicle.
9. Remove the fan with the fluid coupling and fan pulleys.
10. Remove the air cleaner cap, MAF meter, and the resonator.
11. Remove the air cleaner case and filter.
12. Disconnect the heater hoses.
13. Disconnect the following hoses:
- Brake booster vacuum hose
- EVAP hose
- Fuel return hose
- Fuel inlet hose
14. Detach the starter wire and connectors as follows:
   a. Remove the ground strap by removing the bolt.
   b. Detach the three starter wire clamps and connector.
15. Detach the alternator connector and wire.
16. Disconnect the engine wire from the cabin.
   a. Remove the glove box door.
   b. Lower the finish No. 2 panel.
   c. Detach the four ECM connectors.
   d. Detach the two cassette connectors and the two wire clamps from the lower finish panel.
   e. Disconnect the engine wire clamp.
   f. Disconnect the following:
- Igniter connector
- Ground strap
- Variable Switching Valve (VSV) connector for the EVAP
- Vapor pressure sensor connector and clamp
- Vapor connector for the vapor pressure sensor and clamp
   g. Remove the bolt and wire bracket.
   h. Remove the two nuts holding the engine wire retainer to the cowl panel and pull out the engine wire from the cabin.
17. Remove the driveshaft from the transmission.

18. Disconnect the speedometer cable.
19. Remove the front exhaust pipe.
20. Remove the nut and the control cable.
21. Place a jack under the transmission.
22. Remove the transmission rear mounting bracket by removing the eight bolts.
23. If equipped with air conditioning, remove the bolt and disconnect the air conditioning compressor wire clamp.
24. If necessary, install a No. 2 engine hanger with two bolts. Tighten the two bolts to 30 ft. lbs. (40 Nm).
25. Attach the engine hoist chain to the two engine hangers.
26. Remove the four bolts and nuts holding the engine front mounting insulators to the frame.
27. Lift the engine and transmission out of the vehicle.

To install:

28. Install the engine assembly to the vehicle. Attach the engine mounts to the body mountings. Install the bolts and nuts but do not tighten at this time.
29. Remove the engine chain hoist the No. 2 engine hanger.
30. If equipped with air conditioning, connect the air conditioning wire with the bolt.
31. Raise the transmission slightly and install the transmission mounting bracket. Tighten the bolts to the frame to 43 ft. lbs. (58 Nm) and the bolts to the mounting insulator to 13 ft. lbs. (18 Nm).
32. Tighten the engine mounting nuts and bolts to 28 ft. lbs. (38 Nm).
33. Install the control cable.
34. Install the front exhaust pipe.
35. Connect the speedometer cable.
36. Install the driveshaft.
37. Connect all engine wire, hoses and cables.
38. Install the fan with the fluid coupling and fan pulleys. Tighten the nuts to 48 inch lbs. (5.4 Nm).
39. Install the air cleaner case and air filter.
40. Install the MAF meter, resonator, and the air cleaner cap.
41. Install the radiator to the vehicle.
42. Fill the engine with oil.
43. Fill the engine and radiator with coolant.
44. Install the engine undercover.
45. Install the battery and connect the cables.
46. Start the engine and check for leaks.
47. Make any necessary adjustments, install the hood, and road test the vehicle.

1996–97 4RUNNER (4-WHEEL DRIVE)

1. Remove the transmission from the vehicle.
2. Remove the hood.
3. Release the fuel system pressure.

✳✳ CAUTION

The fuel injection system remains under pressure after the engine has been turned OFF. Properly relieve fuel pressure before disconnecting any fuel lines. Failure to do so may result in fire or personal injury.

4. Disconnect the battery and remove it from the vehicle.

✳✳ CAUTION

Wait 90 seconds from the time the key is turned to LOCK and the negative battery cable is disconnected to begin work. This allows the SRS capacitor to discharge and prevent deployment of the air bag(s).

5. Raise and safely support the vehicle.
6. Remove the engine under covers.
7. Drain the engine coolant.
8. Drain the engine oil.
9. Remove the radiator from the vehicle.
10. Remove the fan with the fluid coupling and fan pulleys.
11. Remove the air cleaner cap, MAF meter, and the resonator.
12. Disconnect the heater hoses.
13. Disconnect the following hoses:
- Brake booster vacuum hose
- EVAP hose
- Automatic Disconnecting Differential (ADD) vacuum hose
- Fuel return hose
- Fuel inlet hose
14. Detach the starter wire and connectors as follows:
   a. Remove the ground strap by removing the bolt.
   b. Detach the three starter wire clamps and connector.
15. Detach the alternator connector and wire.
16. Disconnect the engine wire from the cabin.
   a. Remove the glove box door.
   b. Lower the finish No. 2 panel.
   c. Detach the four ECM connectors.
   d. Detach the two cassette connectors and the two wire clamps from the lower finish panel.
   e. Disconnect the engine wire clamp.
   f. Disconnect the following:

- Igniter connector
- Ground strap
- Variable Switching Valve (VSV) connector for the EVAP
- Vapor pressure sensor connector and clamp
- Vapor connector for the vapor pressure sensor and clamp

g. Remove the bolt and wire bracket.

h. Remove the two nuts holding the engine wire retainer to the cowl panel and pull out the engine wire from the cabin.

17. If equipped with air conditioning, remove the bolt and disconnect the air conditioning compressor wire clamp.

18. If necessary, install a No. 2 engine hanger with two bolts. Tighten the two bolts to 30 ft. lbs. (40 Nm).

19. Attach the engine hoist chain to the two engine hangers.

20. Remove the four bolts and nuts holding the engine front mounting insulators to the frame.

21. Lift the engine out of the vehicle.

To install:

22. Install the engine assembly to the vehicle. Attach the engine mounts to the body mountings. Install the bolts and nuts but do not tighten at this time.

23. Remove the engine chain hoist the No. 2 engine hanger.

24. If equipped with air conditioning, connect the air conditioning wire with the bolt.

25. Tighten the engine mounting nuts and bolts to 28 ft. lbs. (38 Nm).

26. Connect the engine wire to the cabin.

a. Push the engine wire through the cowl panel.

b. Install the bolt and wire bracket.

c. Connect the engine wire clamp.

d. Connect all wires, hoses and cables.

27. Install the fan with the fluid coupling and fan pulleys. Tighten the nuts to 48 inch lbs. (5.4 Nm).

28. Install the air cleaner case and air filter.

29. Install the MAF meter, resonator, and the air cleaner cap.

30. Install the radiator to the vehicle.

31. Fill the engine with oil.

32. Fill the engine and radiator with coolant.

33. Install the transmission and refill it with transmission oil.

34. Install the engine undercover.

35. Install the battery and connect the battery cables.

36. Start the engine, make any necessary adjustments, install the hood, and check for leaks.

Water Pump

REMOVAL & INSTALLATION

2.4L (22R-E) Engine

1. Disconnect the negative battery cable.

2. Drain the cooling system.

3. If equipped with an air conditioning compressor or power steering pump drive belts, it may be necessary to loosen the adjusting bolt, remove the drive belt(s) and move the component(s) out of the way.

4. Remove the fluid coupling with the fan and water pump pulley.

5. Remove the water pump.

6. Clean the gasket mounting surfaces.

To install:

7. Install the replacement water pump using a new gasket.

8. Install the water pump pulley and fluid coupling with the fan.

9. Install the removed engine drive belts.

10. Fill the cooling system.

11. Connect the negative battery cable. Start the engine and check for leaks.

12. Bleed the cooling system.

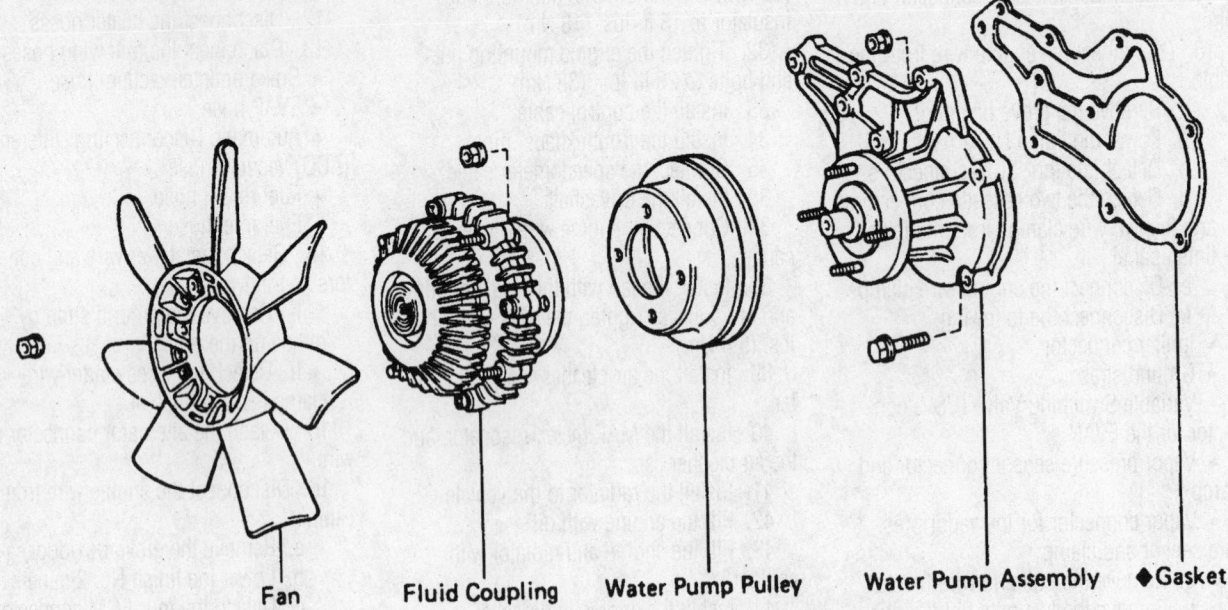

Fan   **Fluid Coupling**   **Water Pump Pulley**   **Water Pump Assembly**   ◆**Gasket**

◆ **Non-reusable part**

Exploded view of the water pump—2.4L (22R-E) engine

7924YG05

3.0L (3VZ-E) and 3.4L (5VZ-FE) Engines

4RUNNER

1. Disconnect the negative battery cable.
2. Drain the cooling system.
3. Remove the timing belt.
4. Remove the thermostat.
5. Disconnect the No. 2 oil cooler hose from the water pump.
6. Remove the water pump by removing the bolts.
7. Thoroughly clean the mating surfaces.

To install:

8. Apply sealant (PN 08826–00100 or equivalent) to the water pump. Parts must be assembled within five minutes of application. Otherwise the material must be removed and reapplied.
9. Install the water pump and tighten the bolts to 14 ft. lbs. (20 Nm).
10. Connect the No. 2 oil cooler hose.
11. Install the thermostat.
12. Install the timing belt.
13. Connect the negative battery cable.
14. Fill the cooling system.
15. Start the engine and check for leaks.

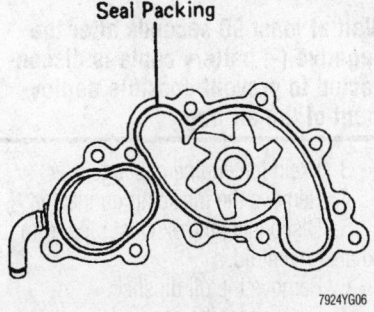

Water pump with seal packing—1995 4Runner with 3.0L (3VZ-E) engine

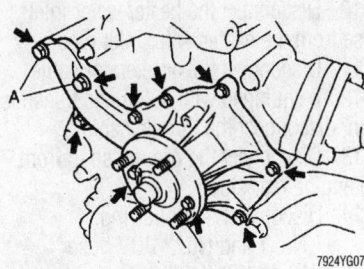

Water pump mounting bolt locations— 2.7L (3RZ-FE) engine

2.4L (2RZ-FE) and 2.7L (3RZ-FE) Engines

T-100, TACOMA AND 4RUNNER

1. Disconnect the negative battery cable from the battery.
2. Remove the engine undercover.
3. Drain the cooling system.
4. For the California vehicles with 3RZ-FE engine, remove the two bolts and disconnect the air pipe.
5. Disconnect the upper radiator hose from the radiator.
6. Remove the oil dipstick guide by removing the bolt.
7. If equipped with power steering, remove the power steering drive belt by loosening the lockbolt and adjusting bolt to the idler pulley.
8. Remove the No. 2 fan shroud by removing the two clips.
9. Remove the No. 1 fan shroud by removing the four bolts.
10. If equipped with air conditioning, loosen the idler pulley nut and adjusting bolt. Remove the air conditioning drive belt from the engine.
11. Remove the alternator drive belt, fan (with fan clutch), water pump pulley, and the fan shroud as follows:
    a. Stretch the belt and loosen the water pump pulley mounting nuts.
    b. Loosen the lock, pivot, and the adjusting bolts for the alternator and remove the alternator drive belt from the engine.
    c. Remove the four water pump pulley mounting nuts.
    d. Remove the fan (with fan clutch) and the water pump pulley.
12. Remove the 10 bolts and remove the water pump and gasket from the engine.

To install:

13. Clean all surfaces and apply a thin layer of liquid sealant to a new gasket.
14. Place the gasket and water pump into position. Tighten the 14mm head bolts **A** to 18 ft. lbs. (25 Nm) and the 12mm head bolts to 78 inch lbs. (9 Nm).
15. Install the water pump pulley, fan shroud, fan (with fan clutch), and the alternator drive belt as follows:
    a. Place the fan (with the fan clutch), water pump pulley, and the fan shroud in position.
    b. Install the water pump pulley mounting nuts but do not tighten the nuts at this time.
    c. Install the alternator drive belt to the engine.
    d. Stretch the alternator belt tight and tighten the fan nuts to 16 ft. lbs. (21 Nm).

e. Adjust the drive belt for the alternator.
16. If equipped with air conditioning, install and adjust the drive belt.
17. Install the No. 1 fan shroud by installing the four bolts.
18. Install the No. 2 fan shroud with the two clips.
19. Install and adjust the power steering drive belt.
20. Install the oil dipstick guide with the bolt.
21. Connect the upper radiator hose to the radiator.
22. If removed, connect the air pipe with the two bolts.
23. Fill and bleed the cooling system.
24. Connect the negative battery cable to the battery.
25. Start the engine and check for leaks.
26. Install the engine undercover.

3.4L (5VZ-FE) Engine

T-100 AND TACOMA

1. Disconnect the negative battery cable.
2. Raise and safely support the vehicle.
3. Remove the engine undercover.
4. Drain the engine coolant.
5. Disconnect the upper radiator hose from the engine.
6. Remove the power steering drive belt as follows:
    a. Stretch the belt and loosen the fan pulley mounting nuts.
    b. Loosen the lockbolt, pivot bolt, and the adjusting bolt and remove the drive belt from the engine.
7. Remove the air conditioning drive belt by loosening the idler pulley nut and adjusting bolt.
8. Loosen the lockbolt, pivot bolt, and the adjusting bolt. Remove the alternator drive belt.
9. Remove the No. 2 fan shroud by removing the two clips.
10. Remove the fan with the fluid coupling and fan pulleys.
11. Disconnect the power steering pump from the engine and set aside. Do not disconnect the lines from the pump.
12. If equipped with air conditioning, disconnect the compressor from the engine and set aside. Do not disconnect the lines from the compressor.
13. If equipped with air conditioning, disconnect the air conditioning bracket.
14. Remove the No. 2 timing belt cover as follows:
    a. Detach the camshaft position sen-

sor connector from the No. 2 timing belt cover.

b. Disconnect the three spark plug wire clamps from the No. 2 timing belt cover.

c. Remove the six bolts and remove the timing belt cover.

15. Remove the fan bracket as follows:

a. Remove the power steering adjusting strut by removing the nut.

b. Remove the fan bracket by removing the bolt and nut.

16. Set the No. 1 cylinder at TDC of the compression stroke.

a. Turn the crankshaft pulley and align its groove with the timing mark **0** of the No. 1 timing belt cover.

b. Check that the timing marks of the camshaft timing pulleys and the No. 3 timing belt cover are aligned. If not, turn the crankshaft pulley one revolution (360°).

➡ **If re-using the timing belt, be sure that you can still read the installation marks. If not, place new installation marks on the timing belt to match the timing marks of the camshaft timing pulleys.**

17. Remove the timing belt tensioner by alternately loosening the two bolts.

18. Remove the camshaft timing pulleys.

a. Using Variable Wrench Set No. 09960–10010 or equivalent, remove the pulley bolt, the timing pulley, and the knock pin. Remove the two timing pulleys with the timing belt.

19. Remove the thermostat.

20. Disconnect the No. 2 oil cooler hose from the water pump.

21. Remove the water pump by removing the seven bolts.

22. Thoroughly clean the mating surfaces.

To install:

23. Apply sealant (PN 08826–00100 or equivalent) to the water pump. Parts must be assembled within five minutes of appli-

cation. Otherwise the material must be removed and reapplied.

24. Install the water pump. Tighten the bolts to 14 ft. lbs. (20 Nm).

25. Connect the No. 2 oil cooler hose.

26. Install the thermostat.

27. Install the left camshaft timing pulley. Tighten the pulley bolt to 81 ft. lbs. (110 Nm).

28. Set the No. 1 cylinder to TDC of the compression stroke.

29. Connect the timing belt to the left camshaft timing pulley. Check that the installation mark on the timing belt is aligned with the end of the No. 1 timing belt cover.

a. Using Variable Pin Wrench Set 09960–01000 or equivalent, slightly turn the left camshaft timing pulley clockwise. Align the installation mark on the timing belt with the timing mark of the camshaft timing pulley, and hang the timing belt on the left camshaft timing pulley.

b. Align the timing marks of the left camshaft pulley and the No. 3 timing belt cover.

c. Check that the timing belt has tension between the crankshaft timing pulley and the left camshaft timing pulley.

30. Install the right camshaft timing pulley and the timing belt.

31. Set the timing belt tensioner as follows:

a. Using a press, slowly press in the pushrod using 220–2,205 lbs. (981–9,807 N) of force.

b. Align the holes of the pushrod and housing, pass a 1.5mm hexagon wrench through the holes to keep the setting position of the pushrod.

c. Release the press and install the dust boot to the tensioner.

32. Install the timing belt tensioner and alternately tighten the bolts to 20 ft. lbs. (28 Nm). Using pliers, remove the 1.5mm hexagon wrench from the belt tensioner.

33. Check the valve timing.

a. Slowly turn the crankshaft pulley two revolutions from the TDC to TDC. Always turn the crankshaft pulley clockwise.

b. Check that each pulley aligns with the timing marks. If the timing marks do not align, remove the timing belt and reinstall it.

34. Install the fan bracket with the bolt and nut. Install the remaining components.

35. Fill with engine coolant.

36. Connect the negative battery cable.

37. Start the engine and check for leaks.

Cylinder Head

REMOVAL & INSTALLATION

2.4L (22R-E) Engine

➡ **The rocker arms and rocker arm shaft are secured by the cylinder head bolts. The cylinder head, camshaft, and valvetrain should be removed as an assembly, then disassembled off the vehicle.**

1. Release the fuel system pressure.
2. Disconnect the negative battery cable.

※※ CAUTION

Wait at least 90 seconds after the negative (-) battery cable is disconnected to prevent possible deployment of the air bag.

3. Drain the engine coolant.
4. Remove the intake air connector.
5. Disconnect the exhaust pipe from the exhaust manifold.
6. Remove the oil dipstick.
7. Disconnect the spark plug wires from the spark plugs. Make note of the proper firing order for installation.
8. Remove the distributor and the spark plugs.
9. Remove the radiator inlet hose.
10. Disconnect the heater water inlet hose from the heater water inlet pipe.
11. Disconnect the accelerator cable.
12. If equipped with automatic transmission, disconnect the throttle cable.
13. Disconnect the ground strap from the engine rear side.
14. Disconnect the following:

a. No. 1 and No. 2 PCV hose

b. Brake booster hose

c. With power steering, the air control valve hoses

d. With air conditioning, the Variable

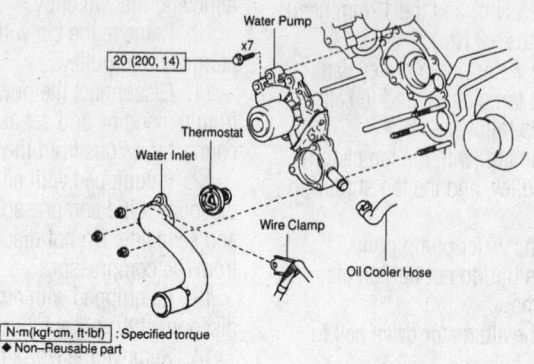

Exploded view of the water pump mounting—3.4L (5VZ-FE) engine

7924YG08

Switching Valve (Variable Switching Valve (VSV) hoses

e. EVAP hose

f. EGR vacuum modulator and EGR valve hoses

g. Fuel pressure up hose

h. PAIR valve hose

i. Pressure regulator hose

j. Vacuum hoses from the throttle body

k. No. 2 and No. 3 water bypass hoses from the throttle body

l. With oil cooler, disconnect the No. 1 oil cooler hose from the intake manifold

m. Without oil cooler, disconnect the No. 1 water bypass hose from the intake manifold

15. Remove the EGR vacuum modulator.

16. Disconnect the following wires:

a. Cold start injector wire

b. Throttle position wire

c. California only, EGR gas temperature sensor wire

17. Remove the air intake chamber with the throttle body.

18. Disconnect the fuel return hose.

19. Disconnect the following wires:

a. Knock sensor wire

b. Oil pressure sender gauge wire

c. Starter wire from terminal 50.

d. Transmission wires

e. With air conditioning, the compressor wires

f. Fuel injector wires

g. Engine coolant temperature sender gauge wire

h. With automatic transmission, the OD temperature switch wire

i. Igniter wire

j. Variable Switching Valve (VSV) wires

k. Cold start injector time switch wire

l. ECT sensor wire

20. Disconnect the fuel hose from the delivery pipe.

✳✳ CAUTION

The fuel injection system remains under pressure after the engine has been turned OFF. Properly relieve fuel pressure before disconnecting any fuel lines. Failure to do so may result in fire or personal injury.

21. Disconnect the bypass hose from the intake manifold.

22. If equipped with power steering, remove the drive belt.

23. Remove the power steering bracket.

24. Remove the nuts, grommets, cylinder head cover, and the gasket.

25. Set the No. 1 cylinder to TDC of the

compression stroke. Place matchmarks on the sprocket and the chain. Remove the camshaft sprocket bolt.

26. Remove the distributor drive gear and the camshaft thrust plate.

27. Remove the camshaft sprocket.

28. Remove the timing chain cover bolt.

✳✳ WARNING

The timing chain cover bolt must be removed prior to loosening any of the cylinder head / rocker arm shaft bolts.

29. Remove the rocker arm assembly by removing the cylinder head bolts in the reverse order of tightening sequence. Loosen the bolts in two to three stages.

30. Remove the cylinder head rear cover.

31. Remove the cylinder head.

32. Remove the rocker shaft assembly, camshaft caps, and the camshaft.

33. To remove the rocker arms, remove the screws from the rocker shafts and slide the rocker shaft stands, rocker arms, and springs from the rocker shafts.

➡ **Keep all parts in order as they are removed. All parts must be reinstalled in their original position.**

34. Remove the intake manifold, exhaust manifold, the right engine hanger, the left engine hanger, the EGR valve, and the ground strap connector.

To install:

35. Place the camshaft in the cylinder head and install the bearing caps in the numbered order from the front with the arrows pointed toward the front. Tighten the bolts to 14 ft. lbs. (20 Nm). Turn the camshaft to position the dowel at the top.

36. Install the cylinder head rear cover.

37. Install the left engine hanger and ground strap.

38. Install the right engine hanger.

39. Install the exhaust manifold. Tighten the nuts to 33 ft. lbs. (44 Nm). Install the heat insulator and tighten the bolts to 14 ft. lbs. (19 Nm).

40. Install the EGR valve.

41. Install the intake manifold. Tighten the nuts and bolts to 14 ft. lbs. (19 Nm).

42. Apply seal packing and to the cylinder block. Only apply sealant at the two areas where the timing chain cover meets the cylinder block.

43. Install the cylinder head on the cylinder block. Be sure to align the dowel pins on the block with the cylinder head.

44. Assemble the rocker arms, springs, and the rocker shaft stands in the same order that they were removed.

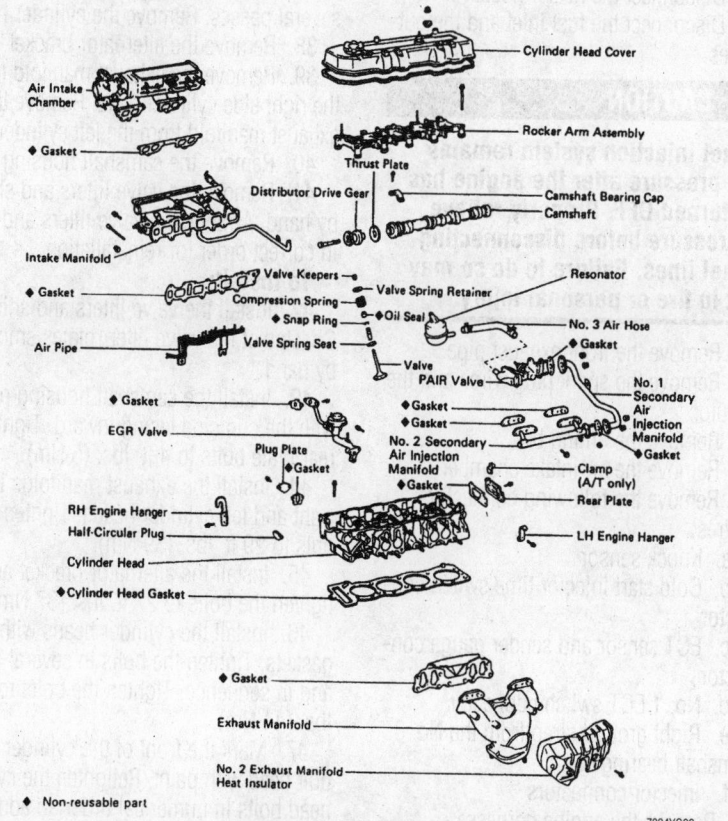

Exploded view of the cylinder head and related components—2.4L (22R-E) engine

Air Intake Chamber
Cylinder Head Cover
◆ Gasket
Rocker Arm Assembly
Thrust Plate
Distributor Drive Gear
Camshaft Bearing Cap
Camshaft
Intake Manifold
Valve Keepers
Resonator
◆ Gasket
Compression Spring
Valve Spring Retainer
◆ Snap Ring
◆ Oil Seal
No. 3 Air Hose
Air Pipe
Valve Spring Seat
◆ Gasket
Valve
PAIR Valve
No. 1 Secondary Air Injection Manifold
◆ Gasket
◆ Gasket
◆ Gasket
EGR Valve
Plug Plate
No. 2 Secondary Air Injection Manifold
◆ Gasket
◆ Gasket
Clamp (A/T only)
Rear Plate
RH Engine Hanger
Half-Circular Plug
LH Engine Hanger
Cylinder Head
◆ Cylinder Head Gasket
◆ Gasket
Exhaust Manifold
No. 2 Exhaust Manifold Heat Insulator
◆ Non-reusable part

7924YG09

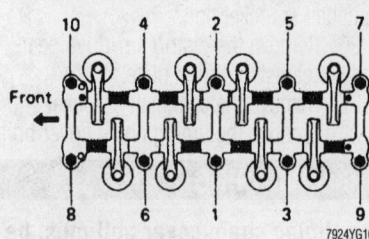

Cylinder head bolt tightening sequence—2.4L (22R-E) engine

45. Place the rocker arm assembly over the dowels on the cylinder head.

46. Install and tighten the head bolts. Following the proper sequence, tighten the bolts in three stages to 58 ft. lbs. (78 Nm).

47. Align matchmarks made during removal and install the camshaft sprocket and chain.

48. Install the chain cover bolt and tighten it to 9 ft. lbs. (13 Nm).

49. Install the distributor drive gear and camshaft thrust plate. Tighten the bolt to 58 ft. lbs. (78 Nm).

50. Check and adjust valve clearance.

51. Install the cylinder head cover with the grommets and the four nuts. Tighten the nuts to 4 ft. lbs. (5 Nm).

52. If removed, install the power steering bracket and tighten the bolts to 33 ft. lbs. (44 Nm). Install the drive belt and adjust the belt tension. Install the remaining components.

53. Check the engine oil level and add as necessary.

54. Fill the engine coolant.

55. Connect the negative battery cable.

56. Start the engine, check for leaks, and bleed the cooling system.

57. Perform engine adjustments (ignition timing, etc.).

58. Road test the vehicle for proper operation.

3.0L (3VZ-E) Engine

1. Disconnect the negative battery cable.

✳✳ CAUTION

Wait 90 seconds from the time the key is turned to LOCK and the negative battery cable is disconnected to begin work. This allows the SRS capacitor to discharge and prevent deployment of the air bag(s).

2. Relieve the fuel system pressure.

3. Drain the cooling system and the engine oil.

4. Disconnect the air cleaner and the hose.

5. Remove the radiator.

6. On manual transmissions, disconnect the clutch release cylinder hose.

7. Remove the power steering belt and remove the pump.

8. Remove the air conditioning belt.

9. Remove the cooling fan.

10. Remove the alternator belt.

11. Tag and detach all of the wires and connectors that interfere in the cylinder head removal.

12. Disconnect the following hoses:

a. Power steering air hoses from the gas filter and the air pipe

b. Brake booster hose

c. Cruise control vacuum hose, if equipped

d. Charcoal canister hose from the canister

e. Variable Switching Valve (VSV) vacuum hoses

13. Disconnect the following cables:

a. Accelerator cable

b. Automatic transmission: Throttle cable

c. W/Cruise: Cruise control cable

14. Disconnect the heater hoses.

15. Disconnect the fuel inlet and the outlet hoses.

✳✳ CAUTION

The fuel injection system remains under pressure after the engine has been turned OFF. Properly relieve fuel pressure before disconnecting any fuel lines. Failure to do so may result in fire or personal injury.

16. Remove the front exhaust pipe.

17. Remove the spark plug wires and the distributor.

18. Remove the timing belt.

19. Remove the air intake chamber.

20. Remove the following connectors and wires:

a. Knock sensor

b. Cold start injector time switch connector

c. ECT sensor and sender gauge connector

d. No. 1 ECT switch connector

e. Right ground strap from the No. 3 camshaft bearing cap

f. Injector connectors

21. Remove the engine harness.

22. Remove the union bolts, then remove the No. 2 and 3 fuel pipes.

23. Remove the No. 4 timing belt cover. Remove the No. 2 idler pulley and the No. 3 timing belt cover.

24. Remove the Variable Switching Valve (VSV) bracket and the Variable Switching Valve (VSV) from the PAIR reed valve.

25. Remove the PAIR reed valve and the No. 1 air injection manifold.

26. Remove the fuel delivery pipes with their injectors.

27. Remove the water bypass outlet.

28. Remove the intake manifold.

29. Remove the knock sensor wire.

30. Remove the exhaust crossover pipe.

31. Disconnect the water bypass pipe from the right cylinder head.

32. Remove the alternator.

33. Remove the oil dipstick guide and the dipstick.

34. Remove the No. 2 engine hanger from the left side cylinder head.

35. Remove the cylinder head covers.

36. Remove the camshaft bearing cap bolts in reverse order of the tightening sequence. Remove the camshafts.

➡ **Arrange the bearing caps in correct order for installation.**

37. Remove the cylinder head bolts in the reverse order of tightening sequence, in several passes. Remove the cylinder heads.

38. Remove the alternator bracket.

39. Remove the exhaust manifold from the right side cylinder head. Remove the exhaust manifold from the left cylinder head.

40. Remove the camshaft housing plugs.

41. Remove the valve lifters and shims by hand. Arrange the valve lifters and shims in correct order for reinstallation.

To install:

42. Install the valve lifters and shims. Check that the valve lifter rotates smoothly by hand.

43. Install the camshaft housing plugs with the cup side facing inward. Tighten the rear plate bolts to 4 ft. lbs. (5 Nm).

44. Install the exhaust manifolds to the right and left cylinder heads. Tighten the nuts to 29 ft. lbs. (39 Nm).

45. Install the alternator bracket and tighten the bolts to 27 ft. lbs. (37 Nm).

46. Install the cylinder heads with new gaskets. Tighten the bolts in several passes and in sequence. Tighten the bolts to 33 ft. lbs. (44 Nm).

47. Mark the front of the cylinder head bolt head with paint. Retighten the cylinder head bolts in numerical order an additional 90°.

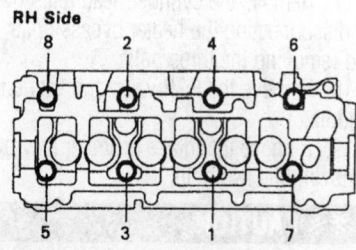

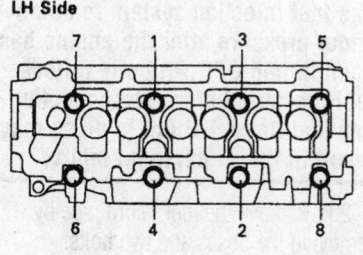

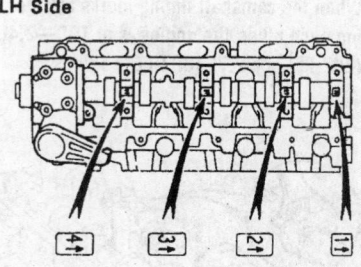

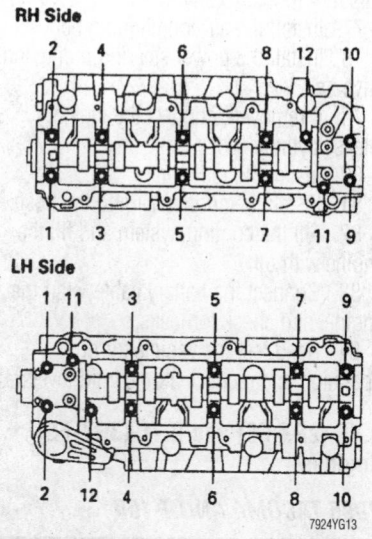

Cylinder head bolt tightening sequence—3.0L (3VZ-E) engine

Position bearing caps with arrows pointing toward the front (RH side) or rear (LH side)—3.0L (3VZ-E) engine

Tighten bearing cap bolts in numerical sequence shown—3.0L (3VZ-E) engine

48. Retighten the cylinder head bolts by an additional 90°. Check that the painted mark is now facing rearward.

❋❋ WARNING

Do not attempt to combine steps 49 and 50. The correct sequence is 33 ft. lbs. (42 Nm) plus 90° plus an additional 90°. It is very important that these instructions are followed precisely to avoid damage to the cylinder head.

49. Install the cylinder head six pointed head bolt to each head and tighten to 30 ft. lbs. (41 Nm).

50. Install the camshafts. Install the bearing caps in their proper locations and tighten the bearing cap bolts to 12 ft. lbs. (16 Nm).

51. Check and adjust valve clearance, if necessary.

52. Install the cylinder head covers and tighten to 4 ft. lbs. (5.4 Nm).

53. Install the water bypass pipe to the right side cylinder head.

54. Install the No. 2 engine hanger and tighten the bolts to 30 ft. lbs. (40 Nm).

55. Install the oil dipstick guide and the

dipstick. Tighten the holding bolt to 27 ft. lbs. (37 Nm).

56. Install the alternator.

57. Install the exhaust crossover pipe and tighten the bolts to 29 ft. lbs. (39 Nm).

58. Install the knock sensor wire.

59. Install the intake manifold with new gaskets and tighten the mounting bolts to 13 ft. lbs. (18 Nm).

60. Install the water bypass outlet and tighten the 2 bolts to 13 ft. lbs. (18 Nm). Connect the No. 3 water bypass hose to the No. 1 water bypass pipe.

61. Install the fuel delivery pipes and injectors. Tighten the delivery pipe holding nuts to 9 ft. lbs. (13 Nm).

62. Install the PAIR reed valve and the No. 1 injection manifold. Tighten the bolts to 27 ft. lbs. (37 Nm) and the nuts to 22 ft. lbs. (29 Nm).

63. Install the VSV bracket and the VSV to the PAIR reed valve.

64. Install the No. 2 idler pulley. Tighten the bolts to 13 ft. lbs. (18 Nm).

65. Install the No. 3 and No. 4 timing belt covers and tighten the bolts to 74 inch lbs. (8.3 Nm).

66. Install the fuel pipes and tighten the

union bolts to 25 ft. lbs. (34 Nm). Connect the vacuum hose to the TVV.

67. Install the engine harness. Attach the following connectors and straps:

a. Injector connectors

b. Right ground strap from the No. 3 camshaft bearing cap

c. No. 1 ECT switch connector

d. ECT sensor and sender gauge connector

e. Cold start injector time switch connector

f. Knock sensor

68. Install the air intake chamber and tighten the nuts and bolts to 13 ft. lbs. (18 Nm).

69. Install the timing belt.

70. Install the distributor and the spark plug wires.

71. Install the front exhaust pipe. Tighten the nuts holding the pipe to the left exhaust manifold to 46 ft. lbs. (62 Nm) and the bolts holding the pipe to the catalytic converter to 29 ft. lbs. (39 Nm).

72. Connect the fuel inlet and outlet hoses.

73. Connect the heater hoses.

74. Attach the cables, straps, wiring, connectors and hose removed prior.

75. Install the alternator drive belt.
76. Install the cooling fan and tighten the nuts to 4 ft. lbs. (5.9 Nm).
77. Install the air conditioning belt.
78. Install the power steering pump and drive belt.
79. If removed, connect the clutch release cylinder hose.
80. Install the radiator.
81. Install the air cleaner and the hose.
82. Fill the cooling system and fill the engine with oil.
83. Connect the battery cable, start the engine, and check for leaks.
84. Road test the vehicle for proper operation and recheck all the fluid levels.

2.4L (2RZ-FE) and 2.7L (3RZ-FE) Engines

1995 TACOMA AND T-100

❊❊ CAUTION

The fuel injection system remains under pressure after the engine has been turned OFF. Properly relieve fuel pressure before disconnecting any fuel lines. Failure to do so may result in fire or personal injury.

1. Properly relieve the fuel system pressure.
2. Disconnect the negative battery cable.
3. Drain the engine coolant.
4. Remove the air cleaner cap, MAF meter, and the resonator.
5. If equipped with a manual transaxle, disconnect the accelerator cable from the throttle body.
6. If equipped with a automatic transaxle, disconnect the accelerator and throttle cables from the throttle body.
7. Remove the intake air connector.
8. If equipped with air conditioning, remove the air conditioning idle-up valve.
9. Remove the power steering drive belt, idler pulley, pump and bracket.
10. Remove the No. 1 and No. 2 PCV hoses.
11. Remove the distributor connector, hold-down bolts, and the distributor.
12. Remove the water housing.
13. Remove the throttle body.
14. Detach the following connectors:
   • If equipped with air conditioning, disconnect the air conditioning compressor.
   • Disconnect the oil pressure sensor.
   • Disconnect the ECT sensor.
   • Disconnect the EGR gas temperature sensor.
   • Disconnect the EGR Variable Switching Valve (VSV).

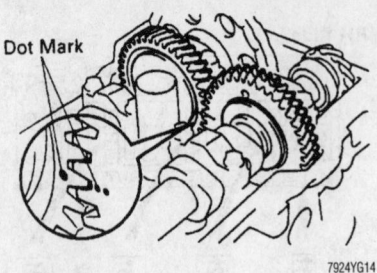

When the camshaft timing marks are facing each other, the engine is at TDC—2.4L (2RZ-FE) and 2.7L (3RZ-FE) engines

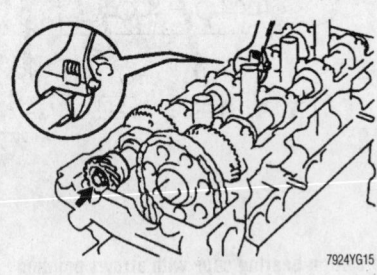

Hold the camshaft at the hexagon wrench head portion—2.4L (2RZ-FE) and 2.7L (3RZ-FE) engines

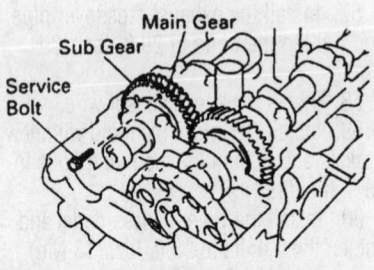

Secure the exhaust camshaft sub-gear to the main gear with a service bolt—2.4L (2RZ-FE) and 2.7L (3RZ-FE) engines

15. Disconnect the engine wire as follows:
   a. Remove the two bolts and disconnect the engine wire from the intake chamber.
   b. Disconnect the five engine wire clamps and engine wire.
   c. Detach the following connectors:
   • Knock sensor connector
   • Crankshaft position sensor connector
   • Fuel pressure control Variable Switching Valve (VSV) connector
   d. Disconnect the DLC1 from the bracket.
   e. Disconnect the two engine wire clamps.

f. Remove the bolt and disconnect the engine wire from the engine.
16. Disconnect the fuel injectors.
17. Remove the cylinder head rear cover by disconnecting the heater bypass hose and removing the three bolts.
18. Remove the EGR valve and vacuum modulator.
19. Remove the intake chamber stay by removing the two bolts.

❊❊ CAUTION

The fuel injection system remains under pressure after the engine has been turned OFF. Properly relieve fuel pressure before disconnecting any fuel lines. Failure to do so may result in fire or personal injury.

20. Remove the fuel return pipe by removing the hoses and two bolts.
21. Remove the intake chamber assembly.
22. Remove the fuel inlet tube by removing the union bolts.
23. Remove the delivery pipe and injectors.

➡ **Be careful not to drop the injectors when removing the delivery pipe.**

24. Remove the intake manifold by removing the three bolts and two nuts.
25. Disconnect the front exhaust pipe from the exhaust manifold.
26. Remove the heat insulator by removing the two bolts and two nuts.
27. Remove the exhaust manifold and gasket by removing the six nuts.
28. Remove the No. 1 and No. 2 engine hangers.
29. Remove the cylinder head cover by removing the ten bolts.
30. Remove the spark plug wires and plugs from the engine.
31. Set No. 1 cylinder to TDC compression stroke. The groove on the crankshaft pulley should align with the **0** mark on the timing chain cover and the timing marks (one and two dots) of the camshaft gears should form a straight line in respect to the cylinder head surface. If not, turn the crankshaft 1 revolution (360 degrees).
32. Remove the chain tensioner and gasket by removing the two nuts.
33. Remove the camshaft timing gear.

➡ **Since the thrust clearance of the camshaft is small, the camshaft must be kept level while it is being removed. If the camshaft is not kept lever, the portion of the cylinder head receiving the shaft thrust may crack or be damaged, causing the camshaft to seize or break.**

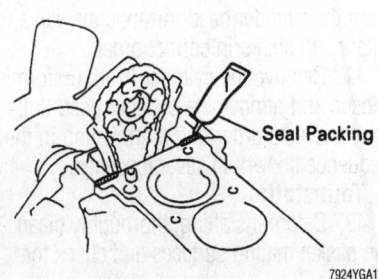

7924YGA1

To ensure proper sealing of the head gasket, apply sealant as shown—2.4L (2RZ-FE) and 2.7L (3RZ-FE) engines

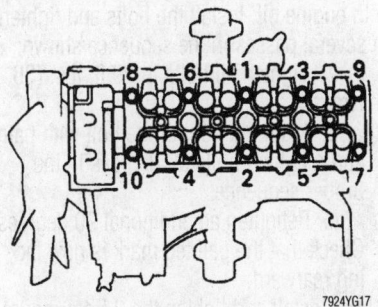

7924YG17

Uniformly tighten the cylinder head bolts in the sequence shown—2.4L (2RZ-FE) and 2.7L (3RZ-FE) engines

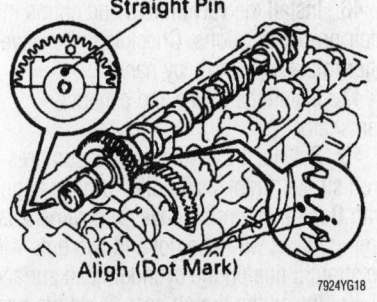

7924YG18

Engage both camshaft gears while matching the timing marks as shown—2.4L (2RZ-FE) and 2.7L (3RZ-FE) engines

34. Remove exhaust and intake camshafts.

35. Remove the 2 bolts in the front of the head before the other head bolts are removed. Uniformly loosen and remove the remaining head bolts, in the reverse order of the tightening several passes, in the sequence shown.

36. Lift the cylinder head from the block and place the head on wooden blocks on a bench.

To install:

37. Before installing, thoroughly clean the gasket mating surfaces and check for warpage.

38. Apply sealant (PN 08826–00080 or equivalent) to the 2 locations, as shown. Place a new head gasket on the block and install the cylinder head.

39. Lightly coat the cylinder head bolts with engine oil. Install the bolts and tighten in several passes in the sequence shown:

    a. Tighten all bolts to 29 ft. lbs. (39 Nm)

    b. Mark the front of the bolt with paint and retighten bolts 90 degrees in the proper sequence.

    c. Retighten an additional 90 degrees. Check that the painted mark is now facing rearward.

40. Install and tighten the 2 front mounting bolts to 15 ft. lbs. (21 Nm).

➡**If any of the bolts break, deform or do not meet the tighten specification, replace them.**

41. Install the intake and exhaust camshafts.

42. Set No. 1 cylinder to TDC compression stroke: Crankshaft pulley groove align with **0** mark on timing cover and camshafts timing marks with one dot and two dots will be straight line on the cylinder head surface.

43. Install the timing gear. Place the gear over the straight pin of the intake camshaft.

44. Install the chain tensioner, using a new gasket (mark toward the front).

45. Check and adjust the valve clearance. Intake valve clearance is 0.006–0.010 inch (0.15–0.25mm) and exhaust valve clearance is 0.010–0.014 inch (0.25–0.35mm).

46. Recheck the engine for proper valve timing. Check and adjust the valve clearance.

47. Install the spark plugs and the semi-circular plug.

48. Recheck the engine for proper valve timing. Install the valve cover and engine hangers. Tighten the engine hanger bolts to 30 ft. lbs. (42 Nm).

49. Install the exhaust manifold and gasket to the engine and install the six nuts. Tighten the nuts to 36 ft. lbs. (49 Nm).

50. Install the heat insulator with the two bolts and two nuts. Tighten the bolts and nuts to 48 inch lbs. (5.5 Nm).

51. Install the front exhaust pipe to the exhaust manifold.

52. Install the intake manifold using a new gasket. Tighten the bolts and nuts to 22 ft. lbs. (29 Nm).

53. Install the injectors and delivery pipe. Tighten the bolts holding the delivery pipe to 15 ft. lbs. (21 Nm).

54. Install the fuel tube with four new gaskets. Tighten the union bolts to 22 ft. lbs. (29 Nm).

55. Install the air intake chamber.

56. Install the fuel return pipe by installing the two bolts and hoses.

57. Install the intake chamber stay by installing the two bolts. Tighten the bolts to 14 ft. lbs. (20 Nm).

58. Install the EGR valve, vacuum modulator and any other hoses associated with the units.

59. Install the cylinder head rear cover with a new gasket. Tighten the three bolts to 10 ft. lbs. (14 Nm).

60. Connect the heater water bypass pipe.

61. Attach the injector connectors.

62. Connect the engine wire to the engine as follows:

    a. Connect the engine wire to the intake manifold with the bolt.

    b. Connect the two engine wire clamps.

    c. Connect the DLC1 to the bracket.

    d. Attach the following connectors:
- Fuel pressure control Variable Switching Valve (VSV) connector
- Knock sensor connector
- Crankshaft position sensor connector

    e. Connect the five engine wire clamps.

    f. Connect the engine wire to the intake chamber with the two bolts.

63. Attach the connectors removed prior.

64. Install the throttle body.

65. Install the water outlet with a new gasket. Install the two bolts and tighten the bolts to 14 ft. lbs. (20 Nm). Attach the ECT sender gauge connector and radiator inlet hose.

66. Install the distributor.

67. Install the No. 1 and No. 2 PCV hoses.

68. Install the power steering pump bracket and pump.

69. Install the power steering drive belt and idler pulley.

70. If equipped with air conditioning, install the air conditioning idle up valve.

71. Install the air intake connector by installing the two bolts, hose clamp, and two air hoses.

72. If equipped with a manual transaxle, connect the accelerator cable to the throttle body.

73. If equipped with a automatic transaxle, connect the throttle and accelerator cables to the throttle body.

74. Install the MAF meter, resonator, and the air cleaner cap.

75. Refill the cooling system. Drain and refill the engine oil, if required.

76. Connect the negative battery cable. Start the engine and check for leaks.

77. Check the ignition timing. Road test the vehicle for proper operation.
78. Recheck all fluid levels.

1996–99 4RUNNER

1. Release the fuel system pressure.
2. Disconnect the negative battery cable.
3. Drain the engine coolant.
4. Remove the air cleaner cap, MAF meter and resonator.
5. If equipped with a manual transmission, disconnect the accelerator cable from the throttle body.
6. If equipped with a automatic transmission, disconnect the accelerator and throttle cables from the throttle body.
7. If equipped with cruise control, disconnect the cruise control cable from the actuator.
8. Remove air cleaner cap, MAF meter and resonator assembly.
9. Remove the intake air connector. Disconnect the following:
   a. Air hose for IAC
   b. Vacuum sensing hose
   c. Wire clamp for the engine wire
10. Remove the oil dipstick guide.
11. Remove the power steering belt.
12. Remove the power steering pulley, pump, and bracket.
13. Remove the PCV hoses.
14. Remove the distributor.
15. Disconnect the spark plug wires from the spark plugs.
16. Disconnect the engine wire clamps and engine wire.
17. Detach the following connectors:
   • If equipped with air conditioning, disconnect the air conditioning compressor
   • Disconnect the oil pressure sensor
   • Disconnect the ECT sensor connector
   • Engine coolant temperature sender gauge connector
   • EGR gas temperature sensor connector
   • Variable Switching Valve (VSV) connector for the EGR
   • Two vacuum hose from the Variable Switching Valve (VSV) for the EGR
   • Ground strap from the cowl top panel
   • Engine wire from the air intake chamber
   • Throttle position sensor connector
   • IAC valve connector
   • Crankshaft position sensor connector
   • Knock sensor connector
   • DLC1 from the bracket
   • Engine wire clamp
18. Remove the EGR pipe.
19. Remove the intake chamber stay.
20. Remove the air intake chamber assembly.

21. Disconnect the following hoses:
   a. EVAP hose from the throttle body
   b. Brake booster vacuum hose from the union
   c. Water bypass hose from the water bypass pipe
   d. Water bypass hose from the cylinder head rear cover
22. Detach the injector connectors.

※※ CAUTION

The fuel injection system remains under pressure, even after the engine has been turned OFF. The fuel system pressure must be released before disconnecting any fuel lines. Failure to do so may result in fire and/or personal injury.

23. Remove the fuel inlet pipe.
24. Disconnect the hoses and remove the fuel return pipe.
25. Remove the delivery pipe and injectors.
26. Remove the intake manifold.
27. Remove the front exhaust pipe.
28. Remove the exhaust manifold and gasket.
29. Remove the water outlet.
30. Remove the cylinder head rear cover.
31. Remove the spark plugs.
32. Remove the front engine hanger.
33. Remove the engine wire brackets.
34. Remove the cylinder head cover.
35. Set No. 1 cylinder to TDC compression stroke. The groove on the crankshaft pulley should align with the **0** mark on the timing chain cover and the timing marks (one and two dots) of the camshaft gears should form a straight line in respect to the cylinder head surface. If not, turn the crankshaft 1 revolution (360 degrees).
36. Remove the chain tensioner and gasket.
37. Remove the camshaft timing gear.
38. Remove exhaust camshafts.
39. To remove the intake camshaft, uniformly loosen and remove the bearing cap bolts in the reverse order of the tightening in several passes, in the sequence shown. Remove the bearing caps and camshaft. Make a note of the bearing cap positions for proper installation.

➡**If the camshaft is not being lifted out straight and level, reinstall the No. 3 bearing cap with the two bolts. Then, alternately loosen and remove the two bearing cap bolts with the camshaft gear pulled up.**

40. Remove the valve lifters and shims from the cylinder head. Arrange the valve lifters and shims in correct order.
41. Remove the cylinder head, uniformly loosen and remove the cylinder head bolts in the reverse order of the tightening in the sequence shown, in several passes.
To install:
42. Before installing, thoroughly clean the gasket mating surfaces and check for warpage.
43. Apply sealant (PN 08826–00080 or equivalent) to the 2 locations, as shown. Place a new head gasket on the block and install the cylinder head.
44. Lightly coat the cylinder head bolts with engine oil. Install the bolts and tighten in several passes in the sequence shown:
   a. Tighten all bolts to 29 ft. lbs. (39 Nm)
   b. Mark the front of the bolt with paint and retighten bolts 90 degrees in the proper sequence.
   c. Retighten an additional 90 degrees. Check that the painted mark is now facing rearward.
45. Install and tighten the 2 front mounting bolts to 15 ft. lbs. (21 Nm).

➡**If any of the bolts break, deform or do not meet the torque specification, replace them.**

46. Install the valve lifters and shims in their proper locations. Check that the valve lifter rotates smoothly by hand.
47. Install the intake and exhaust camshafts.
48. Set No. 1 cylinder to TDC compression stroke: Crankshaft pulley groove align with **0** mark on timing cover and camshafts timing marks with one dot and two dots will be straight line on the cylinder head surface.
49. Install the timing gear. Place the gear over the straight pin of the intake camshaft.
   a. Hold the intake camshaft with a wrench. Install and tighten the bolt to 54 ft. lbs. (74 Nm).
   b. Hold the exhaust camshaft and install the bolt and distributor gear. Tighten the bolt to 34 ft. lbs. (46 Nm).
50. Install the chain tensioner, using a new gasket (mark toward the front).
51. Recheck the engine for proper valve timing. Check and adjust the valve clearance.
52. Install the spark plugs.
53. Install the semi-circular plug.
54. Recheck the engine for proper valve timing.
55. Install the cylinder head cover with a new gasket.

56. Install the engine wire brackets.

57. Install the front engine hanger and tighten the mounting bolts to 30 ft. lbs. (42 Nm).

58. Install the cylinder head rear cover. Tighten the bolts to 10 ft. lbs. (13.5 Nm).

59. Install the water outlet with a new gasket. Tighten the bolts to 14 ft. lbs. (20 Nm). Connect the upper radiator hose.

60. Install the exhaust manifold. Tighten the bolts to 36 ft. lbs. (49 Nm). Install the remaining components.

61. Fill the engine and radiator with engine coolant.

· 62. Connect the negative battery cable. Start the engine and check for leaks.

63. Check the ignition timing. Road test the vehicle for proper operation.

64. Recheck all fluid levels.

3.4L (5VZ-FE) Engine

1995–99 T-100 AND TACOMA

1. Disconnect the negative battery cable.

✳✳ CAUTION

Wait 90 seconds from the time the key is turned to LOCK and the negative battery cable is disconnected to begin work. This allows the SRS capacitor to discharge and prevent deployment of the air bag(s).

2. Relieve the fuel system pressure.
3. Remove the engine undercover.
4. Drain the cooling system.
5. Remove the front exhaust pipe.
6. Disconnect the air cleaner cap, MAF meter, and the resonator.
7. Disconnect the following cables:
• If equipped with cruise control, disconnect the actuator cable with the bracket.
• Accelerator cable
• With automatic transmission, Throttle cable
8. Disconnect the heater hose.
9. Disconnect the upper radiator hose from the engine.
10. Remove the power steering drive belt.
11. Remove the air conditioning drive belt by loosening the idle pulley nut and adjusting bolt.
12. Loosen the lockbolt, pivot bolt and adjusting bolt and the alternator drive belt.
13. Remove the No. 2 fan shroud by removing the two clips.
14. Remove the fan with the fluid coupling and fan pulleys.
15. Disconnect the power steering pump from the engine and set aside. Do not disconnect the lines from the pump.

16. If equipped with air conditioning, disconnect the compressor from the engine and set aside. Do not disconnect the lines from the compressor.

17. If equipped with air conditioning, disconnect the air conditioning bracket.

18. Remove the spark plug wires with the ignition coils.

19. Remove the spark plugs.

20. Remove the No. 2 timing belt cover.

21. Remove the fan bracket as follows:
   a. Remove the power steering adjusting strut by removing the nut.
   b. Remove the fan bracket by removing the bolt and nut.

22. Set the No. 1 cylinder at TDC of the compression stroke.
   a. Turn the crankshaft pulley and align its groove with the timing mark **0** of the No. 1 timing belt cover.
   b. Check that the timing marks of the camshaft timing pulleys and the No. 3 timing belt cover are aligned. If not, turn the crankshaft pulley one revolution (360°).

➡**If re-using the timing belt, be sure that you can still read the installation marks. If not, place new installation marks on the timing belt to match the timing marks of the camshaft timing pulleys.**

23. Remove the timing belt tensioner by alternately loosening the two bolts.

24. Remove the camshaft timing pulleys.
   a. Using Variable Pin Wrench Set tool 09960–10010 or equivalent, remove the pulley bolt, the timing pulley and the knock pin. Remove the two timing pulleys with the timing belt.

25. Remove the bolt and the No. 2 idler pulley.

26. Remove the alternator from the engine.

27. If equipped with an EGR valve, remove the nuts and remove the EGR pipe and two gaskets.

28. Remove the intake chamber stay as follows:
   a. Remove the oil filler tube and No. 1 throttle cable clamp by removing the bolt and two nuts.
   b. Remove the intake chamber stay by removing the two bolts.

29. Remove the following connectors:
• Variable Switching Valve (VSV) connector for the fuel pressure control.
• Disconnect the throttle position sensor
• IAC valve connector
• If equipped with an EGR valve, disconnect the EGR gas temperature
• If equipped with an EGR valve, discon-

nect the Variable Switching Valve (VSV) connector for the EGR valve

30. Disconnect the following hoses:
• Disconnect the PCV hoses.
• Disconnect the water bypass hoses.
• Detach the air assist hose from the intake air connector.
• Two vacuum sensing hoses from the Variable Switching Valve (VSV)
• EVAP hose
• Air hose from the power steering
• If equipped with air conditioning, disconnect the air hose from the air conditioning idle up valve.

31. Remove the four bolts, two nuts and remove the air intake chamber assembly from the engine.

32. Remove the intake air connector.

33. Disconnect the engine wire from the intake manifold as follows:
   a. Detach the following connectors:
• Oil pressure sensor connector
• Crankshaft position sensor connector
• Six injector connectors
• ECT sender gauge connector
• ECT sensor connector
• Knock sensor connector
• Camshaft position sensor connector
   b. Disconnect the three engine wire clamps.
   c. Remove the three bolts and disconnect the engine wire from the cylinder head.

34. Remove the camshaft position sensor.

35. Remove the No. 3 (rear) timing belt cover by removing the six bolts.

36. Remove the fuel pressure regulator.

✳✳ CAUTION

The fuel injection system remains under pressure after the engine has been turned OFF. Properly relieve fuel pressure before disconnecting any fuel lines. Failure to do so may result in fire or personal injury.

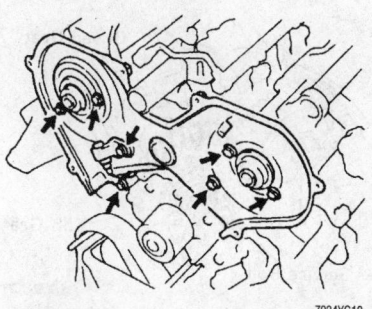

7924YG19

Rear timing belt cover bolt locations— 3.4L (5VZ-FE) engine

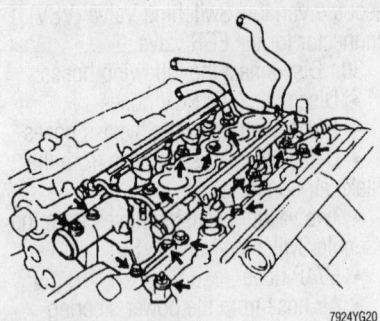

Intake manifold bolts and nuts locations—T-100 and Tacoma with 3.4L (5VZ-FE) engine

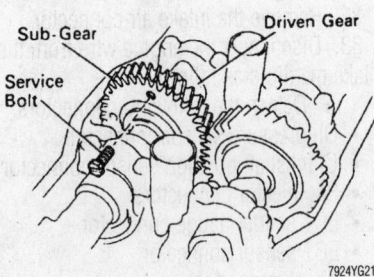

Drive gear service bolt (right side)—3.4L (5VZ-FE) engine

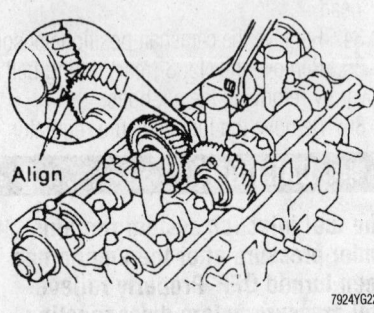

Aligning the timing mark (1 dot mark) of the left camshafts—3.4L (5VZ-FE) engine

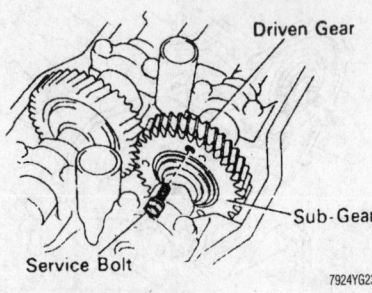

Drive gear service bolt (left side)—3.4L (5VZ-FE) engine

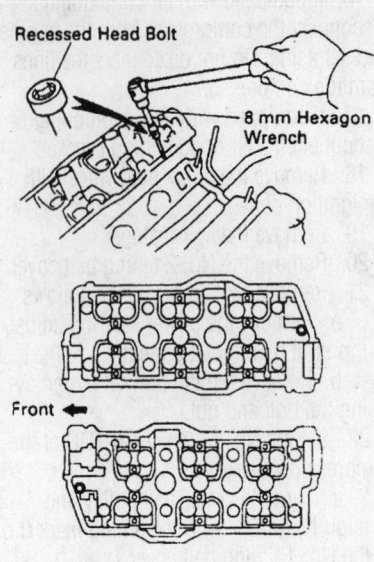

Cylinder head recessed bolts—3.4L (5VZ-FE) engine

37. Remove the intake manifold assembly.

38. Remove the power steering pump bracket.

39. Remove the oil dipstick and guide.

40. Remove the exhaust crossover pipe and gaskets by removing the six nuts.

41. Remove the left-hand exhaust manifold by removing the heat insulator and six nuts for the exhaust manifold.

42. Remove the right-hand exhaust manifold by removing the heat insulator and six nuts for the exhaust manifold.

43. Remove the eight bolts, seal washers, cylinder head cover and gasket. Remove both cylinder head covers.

44. Remove the semi circular plugs.

45. Remove the right exhaust camshafts.

46. Remove the right-hand intake camshaft.

47. Remove the left exhaust camshafts.

48. Remove the left-hand intake camshaft.

49. Remove the valve lifters and shims from the cylinder head. Arrange the valve lifters and shims in correct order.

50. Remove the cylinder heads as follows:

 a. Remove the bolt and disconnect the ground strap.

 b. Using a 8mm hexagon wrench,

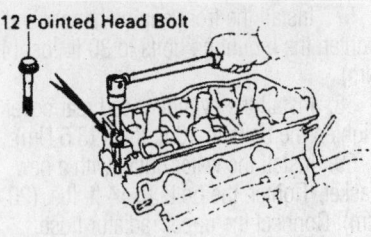

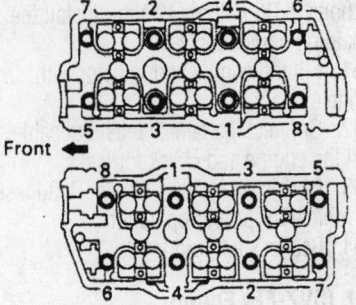

Cylinder head bolt installation sequence—3.4L (5VZ-FE) engine

remove the cylinder head (recessed head) bolt on each cylinder head, then repeat for the other side.

 c. Uniformly loosen and remove the eight cylinder head (12 pointed head) bolts on each cylinder head. Loosen the bolts in several passes and in the reverse order of the tightening sequence shown.

 d. Remove the 16 cylinder head bolts and plate washers.

 e. Lift the cylinder head from the dowels on the cylinder block.

To install:

51. Clean all surfaces.

52. Place two new cylinder head gaskets in position on the cylinder block.

53. Place the two cylinder heads on the dowels of the cylinder block.

54. Apply a light coat of engine oil on the threads and under the heads of the cylinder head bolts.

55. Install and uniformly tighten the cylinder head bolts on each cylinder as follows:

 a. In several passes and in the sequence shown, tighten the cylinder bolts to 25 ft. lbs. (34 Nm).

 b. Mark the front of the cylinder head bolt with paint.

 c. Retighten the cylinder head bolts by 90° in order.

 d. Check that the painted mark is now at a 90° angle to the front.

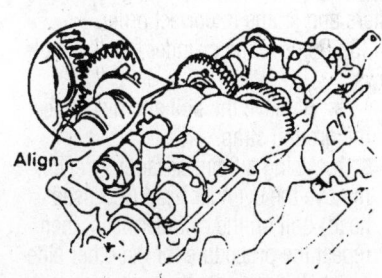

7924YG26

Aligning the right camshafts for installation—3.4L (5VZ-FE) engine

56. Install the recessed head cylinder head bolts as follows:

   a. Apply a light coat of engine oil on the threads and under the heads of the cylinder head bolts.

   b. Using a 8mm hexagon wrench, install the cylinder head bolt on each cylinder head, then repeat for the other side, as shown. Tighten the bolts to 13 ft. lbs. (18 Nm).

   c. Install the bolt and connect the ground strap.

57. Install the valve lifters and shims. Check that the valve lifter rotates smoothly by hand.

58. Following proper procedures, install the camshafts.

59. Check and adjust the valve clearance.

60. Install the semi circular plugs.

61. Install the cylinder head covers. Uniformly tighten the bolts in several passes to 53 inch lbs. (6 Nm).

62. Install the exhaust manifolds with new gaskets. Tighten the nuts to 30 ft. lbs. (40 Nm).

63. Install the exhaust manifold heat insulators with the nuts. Tighten the nuts to 71 inch lbs. (8 Nm).

64. Install the exhaust crossover pipe and tighten the nuts to 33 ft. lbs. (45 Nm).

65. Install the alternator bracket and tighten to 14 ft. lbs. (18 Nm).

66. Install the oil dipstick and guide using a new O-ring.

67. Install the power steering bracket and tighten to 14 ft. lbs. (18 Nm).

68. Install two new gaskets and the intake manifold assembly. Install the four plate washers, eight bolts and four nuts. Tighten the bolts and nuts to 13 ft. lbs. (18 Nm).

69. Install the intake manifold stay with the two bolts. Tighten the bolts to 14 ft. lbs. (18 Nm).

70. Connect the fuel inlet hose.

71. Install the fuel pressure regulator.

72. Install the No. 3 timing belt cover with the six bolts. Tighten the bolts to 80 inch lbs. (9 Nm).

73. Install the camshaft position sensor and tighten to 71 inch lbs. (8 Nm).

74. Connect the engine wire to the intake manifold in the reverse order of removal.

75. Install the intake air connector as follows:

   a. Install the intake manifold to the engine by installing the three bolts and two nuts. Tighten the bolts and nuts to 14 ft. lbs. (19 Nm).

   b. Connect the DLC1 to the bracket on the intake manifold.

   c. Connect the ground strap to the intake manifold by installing the bolt.

   d. Attach the brake booster vacuum hose to the intake air connector.

   e. Connect the two fuel return hoses.

   f. Connect the engine wire to the intake manifold by installing the bolt.

   g. If equipped with air conditioning, connect the idle up valve connector.

76. Install the air intake chamber assembly to the engine by installing the four bolts and two nuts. Tighten the bolts and nuts to 14 ft. lbs. (18.5 Nm).

77. Connect the hoses removed prior.

78. Attach the following connectors:

• Variable Switching Valve (VSV) connector for the fuel pressure control.

• Connect the throttle position sensor connector.

• IAC valve connector

• If equipped with an EGR valve, connect the EGR gas temperature connector

• If equipped with an EGR valve, connect the Variable Switching Valve (VSV) connector for the EGR valve

79. Install the intake chamber stay.

80. Install two new gaskets and the EGR pipe with the nuts. Tighten the clamp nuts to 71 inch lbs. (8 Nm) and the EGR pipe nuts to 14 ft. lbs. (18 Nm).

81. Install the alternator put do not tighten the bolts and nuts at this time.

82. Install the No. 2 timing belt idler with the bolt. Tighten the bolt to 30 ft. lbs. (40 Nm). Check that the pulley bracket moves smoothly.

83. Install the left camshaft timing pulley.

84. Set the No. 1 cylinder to TDC of the compression stroke.

   a. Turn the crankshaft pulley, and align its groove with the timing mark **0** of the No. 1 timing belt cover.

   b. Turn the camshaft, align the knock pin hole of the camshaft with the timing mark of the No. 3 timing belt cover.

   c. Turn the camshaft timing pulley, align the timing marks of the camshaft timing pulley and the No. 3 timing belt cover.

85. Connect the timing belt to the left camshaft timing pulley. Check that the installation mark on the timing belt is aligned with the end of the No. 1 timing belt cover.

   a. Using Variable Pin Wrench Set 09960–01000 or equivalent, slightly turn the left camshaft timing pulley clockwise. Align the installation mark on the timing belt with the timing mark of the camshaft timing pulley, and hang the timing belt on the left camshaft timing pulley.

   b. Align the timing marks of the left camshaft pulley and the No. 3 timing belt cover.

   c. Check that the timing belt has tension between the crankshaft timing pulley and the left camshaft timing pulley.

86. Install the right camshaft timing pulley and the timing belt as follows:

   a. Align the installation mark on the timing belt with the timing mark of the right camshaft timing pulley, and hang the timing belt on the right camshaft timing pulley with the flange side facing inward.

   b. Slide the right camshaft timing pulley on the camshaft. Align the timing marks on the right camshaft timing pulley and the No. 3 timing belt cover.

   c. Align the knock pin hole of the camshaft with the knock pin groove of the pulley and install the knock pin. Install the bolt and tighten to 81 ft. lbs. (110 Nm).

87. Set the timing belt tensioner as follows:

   a. Using a press, slowly press in the pushrod using 220–2,205 lbs. (981–9,807 N) of force.

   b. align the holes of the pushrod and housing, pass a 1.5mm hexagon wrench through the holes to keep the setting position of the pushrod.

   c. Release the press and install the dust boot to the tensioner.

88. Install the timing belt tensioner and alternately tighten the bolts to 20 ft. lbs. (28 Nm). Using pliers, remove the 1.5mm hexagon wrench from the belt tensioner.

89. Check the valve timing. Install the remaining components.

90. Fill the radiator with engine coolant.

91. Connect the negative battery cable to the battery.

92. Start the engine and check for leaks.

93. Check the ignition timing.

94. Install the engine undercover.
95. Road test the vehicle.
96. Recheck all fluid levels.

1996–99 4RUNNER

1. Disconnect the negative battery cable.

> ✵✵ **CAUTION**

Wait at least 90 seconds from the time the ignition switch is turned to the LOCK position and the negative (-) battery cable is disconnected before starting the repair procedure.

2. Relieve the fuel system pressure.
3. Remove the engine undercover.
4. Drain the cooling system.
5. Remove the front exhaust pipe.
6. Disconnect the air cleaner cap, MAF meter, and the resonator.
7. Disconnect the following cables:
• If equipped with cruise control, disconnect the actuator cable with the bracket.
• Accelerator cable
• With automatic transmission, Throttle cable
8. Disconnect the heater hose.

> ✵✵ **CAUTION**

The fuel injection system remains under pressure after the engine has been turned OFF. Properly relieve fuel pressure before disconnecting any fuel lines. Failure to do so may result in fire or personal injury.

9. Disconnect the following hoses:
• Brake booster vacuum hose
• EVAP hose
• 4-Wheel Drive: Automatic Disconnecting Differential (ADD) vacuum hose
• Fuel inlet and fuel return hose
10. Remove the spark plug wires with the ignition coils.
11. Remove the spark plugs.
12. Remove the intake chamber stay.
13. Remove the No. 2 timing belt cover.
14. Remove the air intake chamber assembly.
15. Remove the following connectors and hoses:
• Throttle position sensor connector
• IAC valve connector
• PCV hoses
• Water bypass hoses
• Air assist hose from the throttle body
16. Remove the intake air connector.
17. Disconnect the engine wire protector.
  a. Disconnect the six injector connectors.

  b. Disconnect the ECT sensor and sender gauge connectors.
  c. Disconnect the engine wire protector from the cylinder head.
18. Remove the fuel pressure regulator.
19. Remove the intake manifold assembly.
20. Set the No. 1 cylinder at TDC of the compression stroke.
  a. Turn the crankshaft pulley and align its groove with the timing mark **O** of the No. 1 timing belt cover.
  b. Check that the timing marks of the camshaft timing pulleys and the No. 3 timing belt cover are aligned. If not, turn the crankshaft pulley one revolution (360°).

➡ **If re-using the timing belt, be sure that you can still read the installation marks. If not, place new installation marks on the timing belt to match the timing marks of the camshaft timing pulleys.**

21. Remove the timing belt tensioner by alternately loosening the two bolts.
22. Remove the timing belt.
23. Remove the camshaft timing pulleys.
  a. Using Variable Pin Wrench Set 09960–10010 or equivalent, remove the pulley bolt, the timing pulley and the knock pin. Remove the two timing pulleys with the timing belt.
24. Remove the bolt and the No. 2 idler pulley.
25. Remove the camshaft position sensor.
26. Remove the No. 3 timing belt cover.
27. Remove the alternator from the engine.
28. Remove the alternator bracket.
29. Disconnect the power steering pump from the engine and set aside. Do not disconnect the lines from the pump.
30. Remove the exhaust crossover pipe and gaskets by removing the six nuts.
31. Remove the left-hand exhaust manifold by removing the heat insulator and six nuts for the exhaust manifold.
32. Remove the right-hand exhaust manifold by removing the heat insulator and six nuts for the exhaust manifold.
33. Remove the eight bolts, seal washers, cylinder head cover and gasket. Remove both cylinder head covers.
34. Remove the semi circular plugs.
35. Remove the right exhaust and intake camshafts.
36. Remove the left exhaust and intake camshafts.
37. Remove the valve lifters and shims

from the cylinder head. Arrange the valve lifters and shims in correct order.
38. Remove the cylinder heads as follows:
  a. Remove the bolt and disconnect the ground strap.
  b. Using a 8mm hexagon wrench, remove the cylinder head (recessed head) bolt on the cylinder head, then repeat the procedure for the other side.
  c. Uniformly loosen and remove the eight cylinder head (12 pointed head) bolts on each cylinder head. Loosen the bolts in several passes and in the reverse order of the tightening sequence shown.
  d. Remove the 16 cylinder head bolts and plate washers.
  e. Lift the cylinder head from the dowels on the cylinder block.

To install:
39. Clean all surfaces.
40. Place two new cylinder head gaskets in position on the cylinder block.
41. Place the two cylinder heads on the dowels of the cylinder block.
42. Apply a light coat of engine oil on the threads and under the heads of the cylinder head bolts.
43. Install and uniformly tighten the cylinder head bolts on each cylinder as follows:
  a. In several passes and in the sequence shown, tighten the cylinder bolts to 25 ft. lbs. (34 Nm).
  b. Mark the front of the cylinder head bolt with paint.
  c. Retighten the cylinder head bolts by 90° in order.
  d. Check that the painted mark is now at a 90° angle to the front.
44. Install the recessed head cylinder head bolts as follows:
  a. Apply a light coat of engine oil on the threads and under the heads of the cylinder head bolts.
  b. Using a 8mm hexagon wrench, install the cylinder head bolt on each cylinder head, then repeat for the other side, as shown. Tighten the bolts to 13 ft. lbs. (18 Nm).
  c. Install the bolt and connect the ground strap.
45. Install the valve lifters and shims. Check that the valve lifter rotates smoothly by hand.
46. Install the right intake and exhaust camshafts.
47. Install the left intake and exhaust camshaft.
48. Check and adjust the valve clearance.

49. Install the semi circular plugs.

50. Install the cylinder head covers. Uniformly tighten the bolts in several passes to 53 inch lbs. (6 Nm).

51. Install the exhaust manifolds with new gaskets. Tighten the nuts to 30 ft. lbs. (40 Nm).

52. Install the exhaust manifold heat insulators with the nuts. Tighten the nuts to 71 inch lbs. (8 Nm).

53. Install the exhaust crossover pipe and tighten the nuts to 33 ft. lbs. (45 Nm).

54. Install the power steering pump.

55. Install the alternator bracket and tighten to 14 ft. lbs. (18 Nm).

56. Install the alternator.

57. Install the No. 3 timing belt cover with the six bolts. Tighten the bolts to 80 inch lbs. (9 Nm).

58. Install the camshaft position sensor and tighten to 71 inch lbs. (8 Nm).

59. Install the timing belt.

60. Install the No. 2 timing belt idler with the bolt. Tighten the bolt to 30 ft. lbs. (40 Nm). Check that the pulley bracket moves smoothly.

61. Install the left camshaft timing pulley.

62. Set the No. 1 cylinder to TDC of the compression stroke. Connect the timing belt to the left camshaft timing pulley. Check that the installation mark on the timing belt is aligned with the end of the No. 1 timing belt cover. Install the right camshaft timing pulley and the timing belt. Set the timing belt tensioner.

63. Install the timing belt tensioner and alternately tighten the bolts to 20 ft. lbs. (28 Nm). Using pliers, remove the 1.5mm hexagon wrench from the belt tensioner.

64. Check the valve timing.

65. Install two new gaskets and the intake manifold assembly. Install the four plate washers, eight bolts and four nuts. Tighten the bolts and nuts to 13 ft. lbs. (18 Nm).

66. Install the intake manifold stay with the two bolts. Tighten the bolts to 14 ft. lbs. (18 Nm).

67. Install the fuel pressure regulator.

68. Connect the engine wire to the intake manifold as follows:

   a. Install the engine wire to the cylinder head by installing the three bolts.

   b. Connect the three engine wire clamps.

   c. Attach the following connectors:

- Six injector connectors
- ECT sender gauge connector
- ECT sensor connector

69. Install the intake air connector.

70. Install the air intake chamber assembly to the engine by installing the four bolts and two nuts. Tighten the bolts and nuts to 13 ft. lbs. (18 Nm).

71. Install the intake chamber stay.

72. Install the No. 2 timing belt cover. Tighten the bolts to 80 inch lbs. (9 Nm).

73. Connect the following:

- Connect the PCV hoses.
- Connect the water bypass hoses.
- Connect the air assist hose to the throttle body.
- IAC valve connector.
- Throttle position sensor connector.
- The camshaft position sensor connector to the No. 2 timing belt cover.
- The three spark plug wire clamps.

74. Connect the following hoses:

- Brake booster vacuum hose
- EVAP hose
- 4-Wheel Drive: Automatic Disconnecting Differential (ADD) vacuum hose
- Fuel inlet and fuel return hose

75. Install the oil dipstick and guide using a new O-ring.

76. Install the spark plugs.

77. Install the spark plug wires with the ignition coils.

78. Install the alternator drive belt.

79. Install the drive belt.

80. Connect the heater hose.

81. Connect the following cables:

- If equipped with cruise control, connect the actuator cable with the bracket.
- Accelerator cable
- With automatic transmission, connect the throttle cable

82. Install the MAF meter, resonator, and the air cleaner cap.

83. Install the front exhaust pipe.

84. Fill the radiator with engine coolant.

85. Connect the negative battery cable to the battery.

86. Start the engine and check for leaks.

87. Check the ignition timing.

88. Install the engine undercover.

89. Road test the vehicle.

90. Recheck all fluid levels.

Rocker Arms/Shafts

REMOVAL & INSTALLATION

2.4L (22R-E) Engine

➡ Only the 2.4L (22R-E) engine uses rocker arms. All other engines operate the valves through direct action of the camshaft.

➡ The rocker arms and rocker arm shaft are secured by the cylinder head bolts. The cylinder head, camshaft, and valvetrain should be removed as an assembly, then disassembled off the vehicle.

1. Release the fuel system pressure.
2. Disconnect the negative battery cable.

✳ CAUTION

Wait at least 90 seconds after the negative (-) battery cable is disconnected to prevent possible deployment of the air bag.

3. Drain the engine coolant.
4. Following proper procedures, remove all cylinder head components necessary for removal.
5. Remove the rocker arm assembly by removing the cylinder head bolts in sequence. Loosen the bolts in two to three stages in the reverse order of the tightening sequence.
6. Remove the cylinder head rear cover.
7. Remove the cylinder head.
8. Remove the rocker shaft assembly, camshaft caps, and the camshaft.

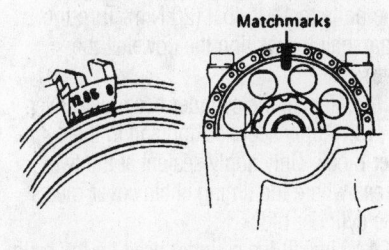

Matchmark the camshaft sprocket to the timing chain—2.4L (22R-E) engine

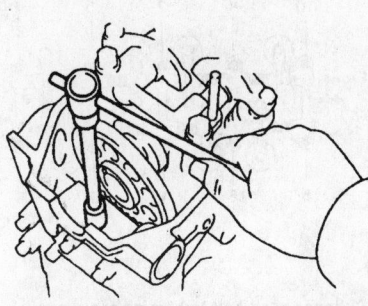

Removing the timing chain cover bolt—2.4L (22R-E) engine

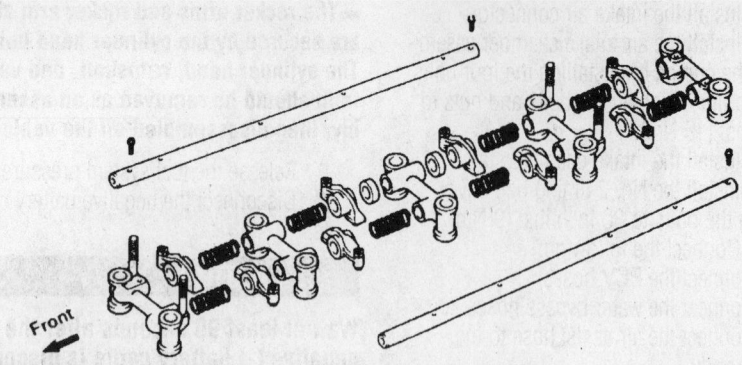

Exploded view of the rocker shaft assembly—2.4L (22R-E) engine

9. To remove the rocker arms, remove the screws from the rocker shafts and slide the rocker shaft stands, rocker arms, and springs from the rocker shafts.

➡**Keep all parts in order as they are removed. All parts must be reinstalled in their original position.**

10. Remove the intake manifold, exhaust manifold, the right engine hanger, the left engine hanger, the EGR valve, and the ground strap connector.

To install:

11. Place the camshaft in the cylinder head and install the bearing caps in the numbered order from the front with the arrows pointed toward the front. Tighten the bolts to 14 ft. lbs. (20 Nm). Turn the camshaft to position the dowel at the top.

12. Install the cylinder head rear cover.

13. Apply seal packing and to the cylinder block. Only apply sealant at the two areas where the timing chain cover meets the cylinder block.

14. Install the cylinder head on the cylinder block. Be sure to align the dowel pins on the block with the cylinder head.

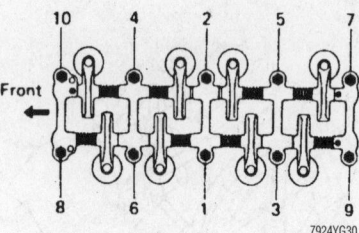

Cylinder head bolt tightening sequence— 2.4L (22R-E) engine

15. Assemble the rocker arms, springs, and the rocker shaft stands in the same order that they were removed.

16. Place the rocker arm assembly over the dowels on the cylinder head.

17. Install and tighten the head bolts. Following the proper sequence, tighten the bolts in three stages to 58 ft. lbs. (78 Nm).

18. Align matchmarks made during removal and install the camshaft sprocket and chain.

19. Install the chain cover bolt and tighten it to 9 ft. lbs. (13 Nm).

20. Install the distributor drive gear and camshaft thrust plate. Tighten the bolt to 58 ft. lbs. (78 Nm).

21. Check and adjust valve clearance.

22. Install the cylinder head cover with the grommets and the four nuts. Tighten the nuts to 4 ft. lbs. (5 Nm). Install the remaining components.

23. Check the engine oil level and add as necessary.

24. Fill the engine coolant.

25. Connect the negative battery cable.

26. Start the engine, check for leaks, and bleed the cooling system.

27. Perform engine adjustments (ignition timing, etc.).

28. Road test the vehicle for proper operation.

Intake Manifold

REMOVAL & INSTALLATION

2.4L (22R-E) Engine

1995 MODELS

1. Drain the cooling system.
2. Disconnect the negative battery cable.

Work must be started after 90 seconds from the time the ignition switch is turned to the LOCK position and the negative (-) battery cable is disconnected.

3. Relieve the fuel pressure.

The fuel injection system remains under pressure after the engine has been turned OFF. Properly relieve fuel pressure before disconnecting any fuel lines. Failure to do so may result in fire or personal injury.

4. Remove the intake air connector.
5. Remove the distributor cap and spark plug wires.
6. Remove the radiator inlet hose.
7. Remove the heater water inlet hose from the heater water inlet pipe.
8. Disconnect the accelerator cable.
9. If equipped with an automatic transmission, disconnect the throttle cable from the bracket and the clamp.
10. Disconnect the ground strap from the rear side of the engine.
11. Disconnect the No. 1 and No. 2 PCV hoses.
12. Tag and disconnect the all hoses in the way of removing the intake manifold.
13. Remove the EGR vacuum modulator.
14. Disconnect the following wires:
   a. Cold start injector wire
   b. Throttle position wire
   c. California only: EGR gas temperature sensor wire
15. Remove the chamber with the throttle body.
16. Disconnect the fuel return hose.

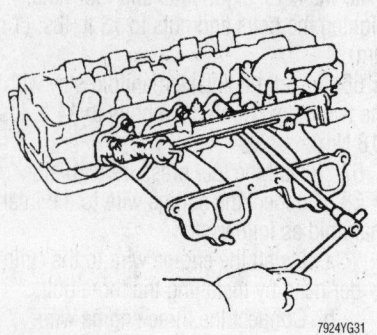

Removing the intake manifold—1995 2.4L (22R-E) engine

17. Tag and disconnect any wiring associated in the removal of the intake manifold.

18. Disconnect the fuel hose from the delivery pipe.

19. Disconnect the bypass hose from the intake manifold.

20. If equipped with power steering, remove the power steering belt.

21. Disconnect the power steering bracket from the cylinder head.

22. Remove the intake manifold with the delivery pipe.

   a. Remove the heater inlet pipe from the cylinder head.

   b. Remove the No. 1 air pipe.

   c. Remove the intake manifold together with the delivery pipe, injectors and the heater water inlet pipe.

23. Remove the EGR valve from the intake manifold.

To install:

24. Install the EGR valve to the intake manifold.

25. Install the intake manifold with a new gasket and tighten the nuts and bolts to 14 ft. lbs. (19 Nm).

   a. Install the No. 1 air pipe.

   b. Install the heater inlet pipe to the cylinder head.

26. Connect the power steering bracket to the cylinder head. Tighten the bolts to 33 ft. lbs. (44 Nm).

27. Install the power steering belt and adjust the belt to the proper tension.

28. Connect the bypass hose to the intake manifold.

29. Connect the fuel hose to the delivery pipe and tighten the bolt to 33 ft. lbs. (44 Nm).

30. Connect the tagged wires removed prior.

31. Connect the fuel return line.

32. Install the chamber with the throttle body.

33. Connect the following wires:

   a. California models only, EGR gas temperature sensor wire.

   b. Throttle position wire.

   c. Cold start injector wire.

34. Install the EGR vacuum modulator.

35. Connect the hoses removed prior.

36. Connect the No. 1 and No. 2 PCV hoses.

37. If removed, connect the throttle cable to the bracket and the clamp.

38. Connect the accelerator cable.

39. Connect the ground strap at the rear side of the engine.

40. Connect the heater water inlet hose to the heater water inlet pipe.

41. Connect the radiator inlet pipe.

42. Install the distributor cap and wires.

43. Install the intake air connector.

44. Connect the negative battery cable.

45. Refill and bleed the cooling system.

46. Start the engine and check for leaks.

2.4L (2RZ-FE) Engine

1. Relieve the fuel system pressure.

2. Disconnect the negative battery cable.

3. Drain the engine coolant.

4. Remove the air cleaner cap, MAF meter, and the resonator.

5. If equipped with a manual transaxle, disconnect the accelerator cable from the throttle body.

6. If equipped with a automatic transaxle, disconnect the accelerator and throttle cables from the throttle body.

7. Remove the intake air connector.

8. If equipped with air conditioning, remove the air conditioning idle-up valve.

9. Remove the No. 1 and No. 2 PCV hoses.

10. Remove the spark plug wires from the spark plugs.

11. Remove the throttle body.

12. Disconnect the following connectors:

   • If equipped with air conditioning, disconnect the air conditioning compressor connector.

   • Disconnect the oil pressure sensor connector.

   • Disconnect the ECT sensor connector.

   • Disconnect the EGR gas temperature sensor connector.

   • Disconnect the EGR Variable Switching Valve (VSV) connector.

13. Disconnect the engine wire as follows:

   a. Remove the two bolts and disconnect the engine wire from the intake chamber.

   b. Disconnect the five engine wire clamps and engine wire.

   c. Disconnect the following connectors:

   • Knock sensor connector

   • Crankshaft position sensor connector

   • Fuel pressure control Variable Switching Valve (VSV) connector

   d. Disconnect the DLC1 from the bracket.

   e. Disconnect the two engine wire clamps.

   f. Remove the bolt and disconnect the engine wire from the engine.

14. Disconnect the fuel injectors.

15. Remove the EGR valve and vacuum modulator.

16. Remove the intake chamber stay by removing the two bolts.

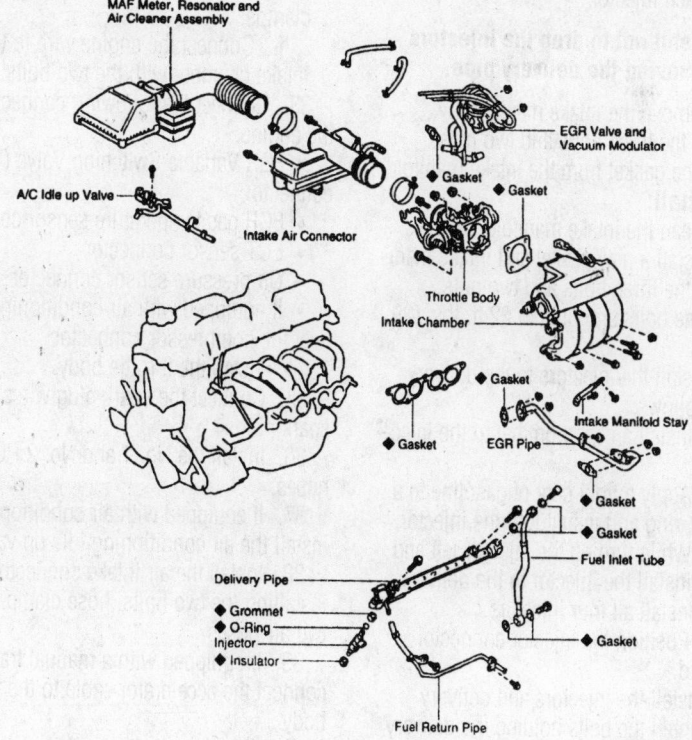

◆ Non-reusable part

Exploded view of the intake manifold assembly—2.4L (2RZ-FE) and 2.7L (3RZ-FE) engines

17. Remove the fuel return pipe by removing the hoses and two bolts.

18. Remove the intake chamber as follows:

a. Disconnect the vacuum hose from the gas filter.

b. Disconnect the brake booster vacuum hose from the intake chamber.

c. Remove the three bolts, two nuts, air intake chamber and gasket.

✳✳ CAUTION

The fuel injection system remains under pressure after the engine has been turned OFF. Properly relieve fuel pressure before disconnecting any fuel lines. Failure to do so may result in fire or personal injury.

19. Remove the fuel inlet tube by removing the union bolts.

20. Remove the delivery pipe and injectors as follows:

a. Disconnect the vacuum hose from the fuel pressure regulator.

b. Remove the two bolts and delivery pipe together with the four injectors.

c. Remove the four insulators from the four spacers.

d. Pull out the four injectors from the delivery pipe.

e. Remove the O-ring and grommet from each injector.

→Be careful not to drop the injectors when removing the delivery pipe.

21. Remove the intake manifold by removing the three bolts and two nuts. Remove the gasket from the intake manifold.

To install:

22. Clean the intake manifold surfaces.

23. Install a new gasket and intake manifold with the three bolts and two nuts. Tighten the bolts and nuts to 22 ft. lbs. (29 Nm).

24. Install the injectors to the delivery pipe as follows:

a. Install a new grommet to the injector.

b. Apply a light coat of gasoline to a new O-ring and install it to the injector.

c. While turning the injector left and right, install the injector to the delivery pipe. Install all four injectors.

d. Position the injector connector upward.

25. Install the injectors and delivery pipe. Tighten the bolts holding the delivery pipe to 15 ft. lbs. (21 Nm). Check that the injectors rotate smoothly.

26. Install the fuel tube with four new gaskets. Tighten the union bolts to 22 ft. lbs. (29 Nm).

27. Install the air intake chamber as follows:

a. Install a new gasket and the air intake chamber with the three bolts and two nuts. Tighten the bolts and nuts to 15 ft. lbs. (21 Nm).

b. Connect the vacuum hose to the gas filter.

c. Connect the brake booster vacuum hose to the intake chamber.

28. Install the fuel return pipe by installing the two bolts and hoses.

29. Install the intake chamber stay by installing the two bolts. Tighten the bolts to 14 ft. lbs. (20 Nm).

30. Install the EGR valve and vacuum modulator.

31. Connect the injector connectors.

32. Connect the engine wire to the engine as follows:

a. Connect the engine wire to the intake manifold with the bolt.

b. Connect the two engine wire clamps.

c. Connect the DLC1 to the bracket.

d. Connect the following connectors:

• Fuel pressure control Variable Switching Valve (VSV) connector

• Knock sensor connector

• Crankshaft position sensor connector

e. Connect the five engine wire clamps.

f. Connect the engine wire to the intake chamber with the two bolts.

33. Connect the following connectors to the engine:

• EGR Variable Switching Valve (VSV) connector

• EGR gas temperature sensor connector

• ECT sensor connector

• Oil pressure sensor connector

• If equipped with air conditioning, connect the compressor connector

34. Install the throttle body.

35. Connect the spark plug wires to the spark plugs.

36. Install the No. 1 and No. 2 PCV hoses.

37. If equipped with air conditioning, install the air conditioning idle up valve.

38. Install the air intake connector by installing the two bolts, hose clamp, and two air hoses.

39. If equipped with a manual transaxle, connect the accelerator cable to the throttle body.

40. If equipped with a automatic transaxle, connect the throttle and accelerator cables to the throttle body.

41. Install the MAF meter, resonator, and the air cleaner cap.

42. Refill the cooling system.

43. Connect the negative battery cable. Start the engine and check for leaks.

44. Check the ignition timing. Road test the vehicle for proper operation.

45. Recheck all fluid levels.

2.7L (3RZ-FE) Engine

T-100 AND TACOMA

1. Relieve the fuel system pressure.

2. Disconnect the negative battery cable.

3. Drain the engine coolant.

4. Remove the air cleaner cap, MAF meter, and the resonator.

5. If equipped with a manual transaxle, disconnect the accelerator cable from the throttle body.

6. If equipped with a automatic transaxle, disconnect the accelerator and throttle cables from the throttle body.

7. Remove the intake air connector.

8. If equipped with air conditioning, remove the air conditioning idle-up valve.

9. Remove the No. 1 and No. 2 PCV hoses.

10. Remove the spark plug wires from the spark plugs.

11. Remove the throttle body.

12. Disconnect the following connectors:

• If equipped with air conditioning, disconnect the air conditioning compressor connector.

• Disconnect the oil pressure sensor connector.

• Disconnect the ECT sensor connector.

• Disconnect the EGR gas temperature sensor connector.

• Disconnect the EGR Variable Switching Valve (VSV) connector.

13. Disconnect the engine wire as follows:

a. Remove the two bolts and disconnect the engine wire from the intake chamber.

b. Disconnect the five engine wire clamps and engine wire.

c. Disconnect the following connectors:

• Knock sensor connector

• Crankshaft position sensor connector

• Fuel pressure control Variable Switching Valve (VSV) connector

d. Disconnect the DLC1 from the bracket.

e. Disconnect the two engine wire clamps.

f. Remove the bolt and disconnect the engine wire from the engine.

14. Disconnect the fuel injectors.

15. Remove the EGR valve and vacuum modulator.

16. Remove the intake chamber stay by removing the two bolts.

17. Remove the fuel return pipe by removing the hoses and two bolts.

18. Remove the intake chamber as follows:

a. Disconnect the vacuum hose from the gas filter.

b. Disconnect the brake booster vacuum hose from the intake chamber.

c. Remove the three bolts, two nuts, air intake chamber and gasket.

※※ CAUTION

The fuel injection system remains under pressure after the engine has been turned OFF. Properly relieve fuel pressure before disconnecting any fuel lines. Failure to do so may result in fire or personal injury.

19. Remove the fuel inlet tube by removing the union bolts.

20. Remove the delivery pipe and injectors as follows:

a. Disconnect the vacuum hose from the fuel pressure regulator.

b. Remove the two bolts and delivery pipe together with the four injectors.

c. Remove the four insulators from the four spacers.

d. Pull out the four injectors from the delivery pipe.

e. Remove the O-ring and grommet from each injector.

➡ **Be careful not to drop the injectors when removing the delivery pipe.**

21. Remove the intake manifold by removing the three bolts and two nuts. Remove the gasket from the intake manifold.

To install:

22. Clean the intake manifold surfaces.

23. Install a new gasket and intake manifold with the three bolts and two nuts. Tighten the bolts and nuts to 22 ft. lbs. (29 Nm).

24. Install the injectors to the delivery pipe as follows:

a. Install a new grommet to the injector.

b. Apply a light coat of gasoline to a new O-ring and install it to the injector.

c. While turning the injector left and right, install the injector to the delivery pipe. Install all four injectors.

d. Position the injector connector upward.

25. Install the injectors and delivery pipe. Tighten the bolts holding the delivery

pipe to 15 ft. lbs. (21 Nm). Check that the injectors rotate smoothly.

26. Install the fuel tube with four new gaskets. Tighten the union bolts to 22 ft. lbs. (29 Nm).

27. Install the air intake chamber as follows:

a. Install a new gasket and the air intake chamber with the three bolts and two nuts. Tighten the bolts and nuts to 15 ft. lbs. (21 Nm).

b. Connect the vacuum hose to the gas filter.

c. Connect the brake booster vacuum hose to the intake chamber.

28. Install the fuel return pipe by installing the two bolts and hoses.

29. Install the intake chamber stay by installing the two bolts. Tighten the bolts to 14 ft. lbs. (20 Nm).

30. Install the EGR valve and vacuum modulator as follows:

a. Install a new gasket, EGR valve and vacuum modulator with the bolt and two nuts. Tighten the bolt to 74 inch lbs. (9 Nm) and the nuts to 14 ft. lbs. (19 Nm).

b. Connect the following hoses:

• Two vacuum hoses to the EGR Variable Switching Valve (VSV)

• Water bypass hose to the water bypass pipe

c. Install two new gaskets and EGR pipe with the bolt and four nuts. Tighten the bolts and nuts as follows:

• Bolt to 14 ft. lbs. (18 Nm)

• Nuts to intake manifold to 14 ft. lbs. (19 Nm)

• Nuts to cylinder head to 15 ft. lbs. (20 Nm).

31. Connect the injector connectors.

32. Connect the engine wire to the engine as follows:

a. Connect the engine wire to the intake manifold with the bolt.

b. Connect the two engine wire clamps.

c. Connect the DLC1 to the bracket.

d. Connect the following connectors:

• Fuel pressure control Variable Switching Valve (VSV) connector

• Knock sensor connector

• Crankshaft position sensor connector

e. Connect the five engine wire clamps.

f. Connect the engine wire to the intake chamber with the two bolts.

33. Connect the following connectors to the engine:

• EGR Variable Switching Valve (VSV) connector

• EGR gas temperature sensor connector

• ECT sensor connector

• Oil pressure sensor connector

• If equipped with air conditioning, connect the compressor connector

34. Install the throttle body.

35. Connect the spark plug wires to the spark plugs.

36. Install the No. 1 and No. 2 PCV hoses.

37. If equipped with air conditioning, install the air conditioning idle up valve.

38. Install the air intake connector by installing the two bolts, hose clamp, and two air hoses.

39. If equipped with a manual transaxle, connect the accelerator cable to the throttle body.

40. If equipped with a automatic transaxle, connect the throttle and accelerator cables to the throttle body.

41. Install the MAF meter, resonator, and the air cleaner cap.

42. Refill the cooling system.

43. Connect the negative battery cable. Start the engine and check for leaks.

44. Check the ignition timing. Road test the vehicle for proper operation.

45. Recheck all fluid levels.

1996–99 4RUNNER

1. Disconnect the negative battery cable.

※※ CAUTION

Wait 90 seconds from the time the key is turned to LOCK and the negative battery cable is disconnected to begin work. This allows the SRS capacitor to discharge and prevent deployment of the air bag(s).

2. Release the fuel system pressure.

3. Drain the engine coolant.

4. Remove the air cleaner cap, MAF meter, and the resonator.

5. If equipped with a manual transmission, disconnect the accelerator cable from the throttle body.

6. If equipped with a automatic transmission, disconnect the accelerator and throttle cables from the throttle body.

7. If equipped with cruise control, disconnect the cruise control cable from the actuator.

8. Remove the intake air connector. Disconnect the following:

a. Air hose for IAC

b. Vacuum sensing hose

c. Wire clamp for the engine wire

9. Remove the PCV hoses.

10. Disconnect the engine wire clamps and engine wire.

11. Disconnect the following connectors:

• If equipped with air conditioning, disconnect the air conditioning compressor connector

• Disconnect the oil pressure sensor connector

• Disconnect the ECT sensor connector

• Engine coolant temperature sender gauge connector

• EGR gas temperature sensor connector

• Variable Switching Valve (VSV) connector for the EGR

• Two vacuum hoses from the Variable Switching Valve (VSV) for the EGR

• Ground strap from the cowl top panel

• Engine wire from the air intake chamber

• Throttle position sensor connector

• IAC valve connector

• Crankshaft position sensor connector

• Knock sensor connector

• DLC1 from the bracket

• Engine wire clamp

12. Remove the EGR pipe.

13. Remove the intake chamber stay.

14. Remove the air intake chamber assembly.

15. Disconnect the following hoses:

a. EVAP hose from the throttle body

b. Brake booster vacuum hose from the union

c. Water bypass hose from the water bypass pipe

d. Water bypass hose from the cylinder head rear cover

16. Disconnect the injector connectors.

❋❋ CAUTION

The fuel injection system remains under pressure after the engine has been turned OFF. Properly relieve fuel pressure before disconnecting any fuel lines. Failure to do so may result in fire or personal injury.

17. Remove the fuel inlet pipe.

18. Disconnect the hoses and remove the fuel return pipe.

19. Remove the delivery pipe and injectors.

a. Remove the two bolts and the delivery pipe together with the four injectors.

b. Remove the four insulators form the four spacers.

c. Pull out the four injectors from the delivery pipe.

d. Remove the O-ring and grommets from each injector.

e. Carefully pry out the four spacers.

20. Remove the intake manifold.

To install:

21. Install the intake manifold and tighten the bolts to 22 ft. lbs. (29 Nm).

22. Install the injectors and the delivery pipe.

23. Install the fuel return pipe.

24. Install the fuel inlet pipe with a new gasket and tighten the bolts to 22 ft. lbs. (29 Nm).

25. Connect the injector connectors.

26. Install the air intake chamber assembly. Tighten the bolts to 15 ft. lbs. (21 Nm). Connect the following hoses:

a. EVAP hose to the throttle body

b. Brake booster vacuum hose to union

c. Water bypass hose to water bypass pipe

d. Water bypass hose to cylinder head rear cover

27. Install the air intake chamber stay and tighten the bolts to 15 ft. lbs. (20 Nm).

28. Install the EGR pipe. Tighten the nuts and bolts to:

• Bolt: 13 ft. lbs. (18 Nm)

• Nut A: 14 ft. lbs. (19 Nm)

• Nut B: 15 ft. lbs. (20 Nm)

29. Connect the engine wire.

• If disconnected, connect the air conditioning compressor connector

• Connect the oil pressure sensor connector

• Connect the ECT sensor connector

• Engine coolant temperature sender gauge connector

• EGR gas temperature sensor connector

• Variable Switching Valve (VSV) connector for the EGR

• Two vacuum hose to the Variable Switching Valve (VSV) for the EGR

• Ground strap to the cowl top panel

• Engine wire to the air intake chamber

• Throttle position sensor connector

• IAC valve connector

• Crankshaft position sensor connector

• Knock sensor connector

• DLC1 to the bracket

• Engine wire clamp

30. Install the PCV hoses.

31. Install the intake air connector and tighten the bolts to 13 ft. lbs. (18 Nm).

32. If equipped with cruise control, connect the cruise control cable to the actuator.

33. If equipped with a manual transmission, connect the accelerator cable to the throttle body.

34. If equipped with a automatic trans-

mission, connect the accelerator and throttle cables to the throttle body.

35. Fill the engine and radiator with engine coolant.

36. Install the air cleaner cap, MAF meter, and the resonator assembly.

37. Connect the negative battery cable, start the engine, and check for leaks.

38. Road test the vehicle for proper operation.

39. Recheck all fluid levels.

3.0L Engine

1995 MODELS

1. Relieve the fuel system pressure.

❋❋ CAUTION

The fuel injection system remains under pressure after the engine has been turned OFF. Properly relieve fuel pressure before disconnecting any fuel lines. Failure to do so may result in fire or personal injury.

2. Disconnect the negative battery cable. See the air bag warning!

3. Drain the cooling system.

4. Disconnect the air intake hose from both the air cleaner assembly on one end and the air intake chamber on the other.

5. If equipped with manual transmission, disconnect the clutch release cylinder hose.

6. Disconnect the following straps, wires, and connectors:

a. The ground strap from the LH fender apron.

b. Alternator connector and wire.

c. Igniter connector.

d. Oil pressure sender gauge connector.

e. The ground strap from the rear of the engine.

f. The ECM and Variable Switching Valve (VSV) connectors.

g. The air conditioning compressor connector.

h. The Data Link Connector (DLC) No. 1.

i. With manual transmission: the starter relay and solenoid resistor connectors.

j. If equipped with Automatic Disconnecting Differential (ADD), unplug the ADD switch connector.

7. Disconnect the following hoses:

a. With power steering: the air hoses.

b. Power brake booster hose.

c. With Cruise: cruise control vacuum hose.

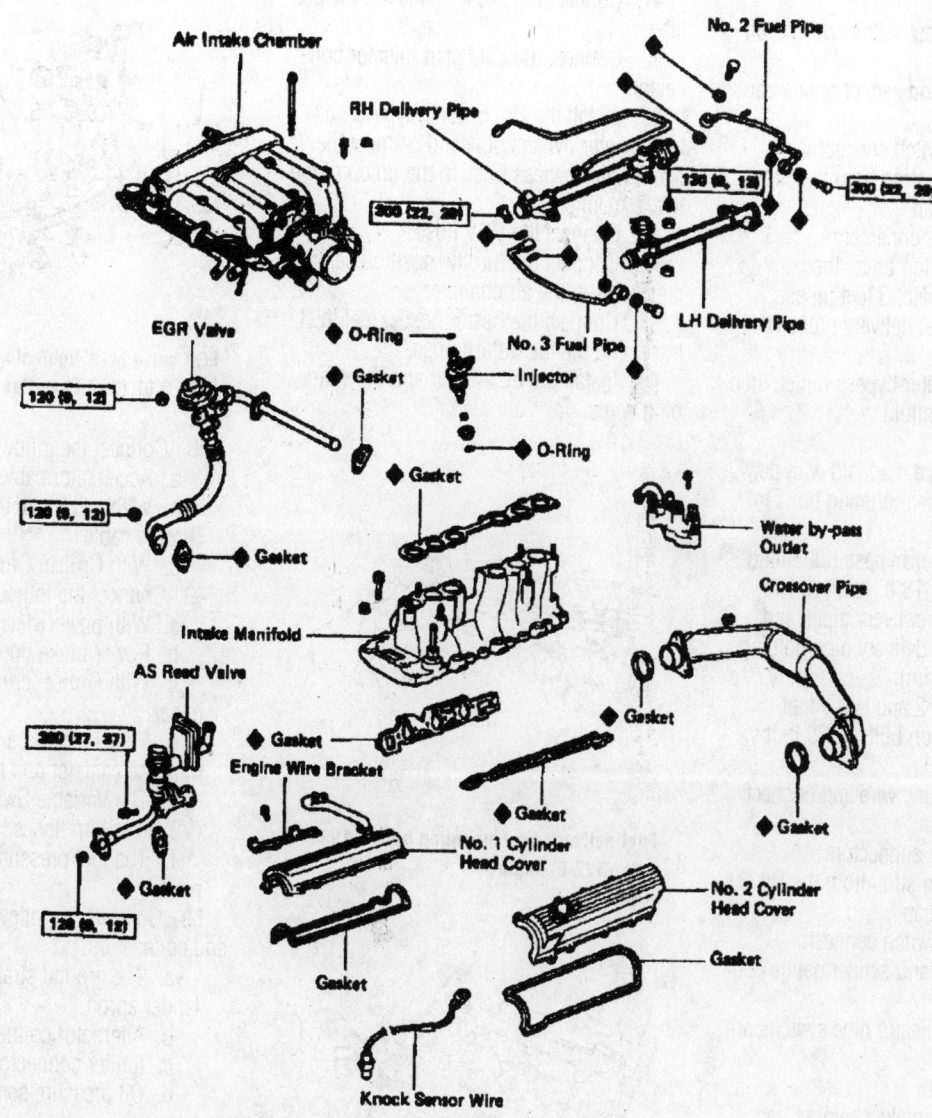

Air Intake Chamber

No. 2 Fuel Pipe

RH Delivery Pipe

130 (9, 13)

300 (22, 29)

LH Delivery Pipe

EGR Valve

O-Ring

No. 3 Fuel Pipe

Gasket

Injector

120 (9, 12)

O-Ring

120 (9, 12)

Gasket

Gasket

Water by-pass Outlet

Crossover Pipe

Intake Manifold

AS Reed Valve

Gasket

300 (27, 37)

Gasket

Engine Wire Bracket

Gasket

Gasket

No. 1 Cylinder Head Cover

No. 2 Cylinder Head Cover

Gasket

120 (9, 12)

Gasket

Gasket

Knock Sensor Wire

kg-cm (ft-lb, N-m) : Specified torque

◆ Non-reusable part

7924YG33

Exploded view of the intake manifold and related components—1995 with 3.0L (3VZ-E) engine

   d. The charcoal canister vacuum hose from the canister and the throttle body.
   e. The Variable Switching Valve (VSV) vacuum hoses.
   f. The fuel pressure regulator vacuum hose.
  8. Disconnect the following cables:
   a. Accelerator cable.
   b. With automatic transmission: Throttle cable.
   c. With Cruise: Cruise control cable.
  9. Remove the spark plug wires and the distributor.
  10. Disconnect the heater hoses, the fuel inlet hose, and the fuel return hose.

  11. Disconnect the throttle position sensor connector at the air chamber. Disconnect the PCV hose at the union.
  12. Disconnect the No. 4 water bypass hose at the manifold. Remove the No. 5 bypass hose at the water bypass pipe.
  13. Disconnect the cold start injector connector and remove the vacuum hose from the gas filter.
  14. Remove the union bolt and gaskets, then remove the cold start injector tube.
  15. Disconnect the EGR gas temperature sensor connector and the EGR vacuum hoses from the air pipe and the vacuum modulator.

  16. Remove the intake chamber stay and the throttle cable bracket.
  17. Remove the No. 1 engine hanger.
  18. Remove the power steering pump and bracket.
  19. Remove the EGR valve.
  20. Disconnect the No. 1 air hose at the PAIR reed valve.
  21. Disconnect the vacuum hoses from the air pipes.
  22. Remove the accelerator cable bracket.
  23. Remove the air intake chamber, then remove the engine wire. Disconnect the following connectors:

a. Knock sensor

b. Cold start injector time switch connector

c. ECT sensor and sender gauge connector

d. No. 1 ECT switch connector

e. Right ground strap from the No. 3 camshaft bearing cap

f. Fuel injector connectors

24. Remove the union bolts, then remove the No. 2 and No. 3 fuel pipes.

25. Remove the fuel delivery pipes with their injectors.

26. Remove the water bypass outlet, then remove the intake manifold.

To install:

27. Install the intake manifold with new gaskets and tighten the mounting bolts to 13 ft. lbs. (18 Nm).

28. Install the water bypass outlet and tighten the 2 bolts to 13 ft. lbs. (18 Nm).

29. Install the fuel delivery pipes and injectors. Tighten the delivery pipe holding nuts to 9 ft. lbs. (13 Nm).

30. Install the No. 2 and No. 3 fuel pipes; tighten the union bolts to 25 ft. lbs. (34 Nm).

31. Install the engine wire and connect the following:

a. Fuel injector connectors

b. Right ground strap from the No. 3 camshaft bearing cap

c. No. 1 ECT switch connector

d. ECT sensor and sender gauge connector

e. Cold start injector time switch connector

f. Knock sensor

32. Install the air intake chamber and tighten the nuts and bolts to 13 ft. lbs. (18 Nm).

33. Install the accelerator cable bracket.

34. Connect the vacuum hoses to the air pipes.

35. Connect the No. 1 air hose to the reed valve.

36. Install the EGR valve and pipes assembly, the air intake chamber stay, and the throttle cable bracket. Tighten bolt A to 22 ft. lbs. (29 Nm) and bolt B to 13 ft. lbs. (18 Nm).

37. Install the power steering pump and bracket.

38. Install the No. 1 engine hanger. Tighten the bolt to 30 ft. lbs. (40 Nm).

39. Connect the EGR temperature sensor connector and the EGR vacuum hoses to the air pipe and the vacuum modulator.

40. Connect the cold start injector tube with new gaskets and the union bolt. Tighten to 11 ft. lbs. (15 Nm).

41. Connect the vacuum hose to the gas filter.

42. Connect the cold start injector connector.

43. Install the No. 5 water bypass hose to the water bypass pipe and connect the No. 4 water bypass hose to the union of the intake manifold.

44. Connect the PCV hose.

45. Connect the throttle position sensor connector to the air chamber.

46. Connect the heater hoses, fuel inlet hose, and the fuel return hose.

47. Install the distributor and the spark plug wires.

Fuel delivery pipe mounting bolts—1995 3.0L (3VZ-E) engine

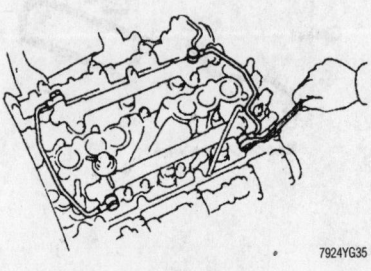

Location of the No. 2 and No. 3 fuel pipes—1995 3.0L (3VZ-E) engine

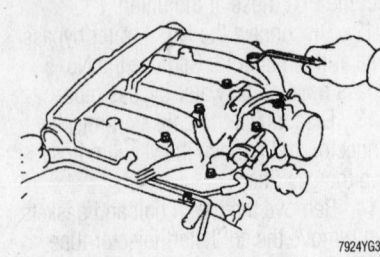

Air intake chamber mounting bolts—1995 3.0L (3VZ-E) engine

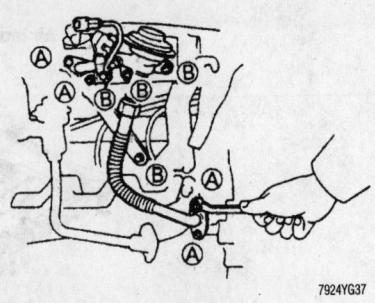

EGR valve bolts tightening sequence—1995 3.0L (3VZ-E) engine

48. Connect the following cables:

a. Accelerator cable.

b. With automatic transmission: Throttle cable.

c. With Cruise: Cruise control cable.

49. Connect the following hoses:

a. With power steering: the air hoses.

b. Power brake booster hose.

c. With Cruise: cruise control vacuum hose.

d. The charcoal canister vacuum hose from the canister and the throttle body.

e. The Variable Switching Valve (VSV) vacuum hoses.

f. The fuel pressure regulator vacuum hose.

50. Connect the following straps, wires, and connectors:

a. The ground strap from the LH fender apron.

b. Alternator connector and wire.

c. Igniter connector.

d. Oil pressure sender gauge connector.

e. The ground strap from the rear of the engine.

f. The ECM and Variable Switching Valve (VSV) connectors.

g. The air conditioning compressor connector.

h. The Data Link Connector (DLC) No. 1.

i. With manual transmission: the starter relay and solenoid resistor connectors.

j. With Automatic Disconnecting Differential (ADD): the ADD switch connector.

51. If removed, connect the clutch release cylinder hose.

52. Connect the air intake hose to both the air cleaner assembly on one end and the air intake chamber on the other end.

53. Refill the engine with coolant.

54. Connect the negative battery cable.

55. Start the engine, check for leaks, check the ignition timing, and bleed the cooling system.

56. Perform any necessary adjustments and road test the vehicle.

3.4 (5VZ-FE) Engine

1996–97 4RUNNER

1. Disconnect the negative battery cable.

> ※※ **CAUTION**
>
> **Wait 90 seconds from the time the key is turned to LOCK and the negative battery cable is disconnected to begin work. This allows the SRS capacitor to discharge and prevent deployment of the air bag(s).**

2. Relieve the fuel system pressure.
3. Remove the engine undercover.
4. Drain the cooling system.
5. Disconnect the air cleaner cap, MAF meter, and the resonator.
6. Disconnect the following cables:
   - If equipped with cruise control, disconnect the actuator cable with the bracket.
   - Accelerator cable
   - If equipped with automatic transmission, remove the throttle cable
7. Disconnect the heater hose.
8. Disconnect the following hoses:
   - Brake booster vacuum hose
   - EVAP hose
   - 4-Wheel Drive: Automatic Disconnecting Differential (ADD) vacuum hose
   - Fuel inlet and fuel return hose

> ※※ **CAUTION**
>
> **The fuel injection system remains under pressure after the engine has been turned OFF. Properly relieve fuel pressure before disconnecting any fuel lines. Failure to do so may result in fire or personal injury.**

9. Remove the spark plug wires with the ignition coils.
10. Remove the intake chamber stay.
11. Remove the No. 2 timing belt cover.
12. Remove the air intake chamber assembly.
13. Remove the following connectors and hoses:
    - Throttle position sensor connector
    - IAC valve connector
    - PCV hoses
    - Water bypass hoses
    - Air assist hose from the throttle body
14. Remove the intake air connector.
    a. Remove the bolt and disconnect the engine wire.

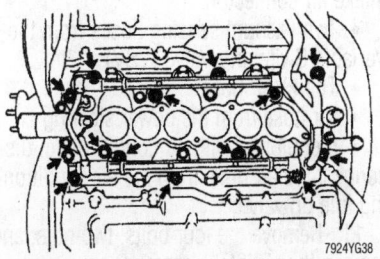

Intake manifold bolts and nuts—4Runner with 3.4 (5VZ-FE) engine

b. Disconnect the fuel return hose.
c. Disconnect the vacuum hose from the fuel pressure regulator.
d. Disconnect the ground strap from the intake air connector.
e. Disconnect the DLC1 from the bracket.

15. Disconnect the engine wire protector.
    a. Disconnect the six injector connectors.
    b. Disconnect the ECT sensor and sender gauge connectors.
    c. Disconnect the engine wire protector from the cylinder head.
16. Remove the fuel pressure regulator.
17. Remove the intake manifold assembly.
    a. Remove the intake manifold stay.
    b. Remove the eight bolts, four nuts, four plate washers, the intake manifold, delivery pipes and the injectors assembly with the gaskets.

To install:

18. Install two new gaskets and the intake manifold assembly. Install the four plate washers, eight bolts and four nuts. Tighten the bolts and nuts to 13 ft. lbs. (18 Nm).

19. Install the intake manifold stay with the two bolts. Tighten the bolts to 14 ft. lbs. (18 Nm).

20. Install the fuel pressure regulator.
21. Connect the engine wire to the intake manifold as follows:
    a. Install the engine wire to the cylinder head by installing the three bolts.
    b. Connect the three engine wire clamps.
    c. Connect the following connectors:
    - Six injector connectors
    - ECT sender gauge connector
    - ECT sensor connector
22. Install the intake air connector as follows:
    a. Install the intake manifold to the engine by installing the three bolts and

two nuts. Tighten the bolts and nuts to 14 ft. lbs. (18.5 Nm).
    b. Connect the DLC1 to the bracket on the intake manifold.
    c. Connect the ground strap to the intake manifold by installing the bolt.
    d. Connect the brake booster vacuum hose to the intake air connector.
    e. Connect the two fuel return hoses.
    f. Connect the engine wire to the intake manifold by installing the bolt.
23. Install the air intake chamber assembly to the engine by installing the four bolts and two nuts. Tighten the bolts and nuts to 14 ft. lbs. (18.5 Nm).
24. Install the intake chamber stay as follows:
    a. Install the intake chamber stay and install the two bolts. Tighten the bolts to 30 ft. lbs. (40 Nm)
    b. Install a new O-ring to the oil filler tube.
    c. Push in the oil filler tube end into the tube hole in the oil pan.
    d. Install the oil filler tube and No. 1 throttle cable clamp and install the bolt and two nuts.
25. Install the No. 2 timing belt cover and tighten the bolts to 80 inch lbs. (9 Nm).
26. Connect the following:
    - Connect the PCV hoses.
    - Connect the water bypass hoses.
    - Connect the air assist hose to the throttle body.
    - IAC valve connector.
    - Throttle position sensor connector.
27. Connect the following hoses:
    - Brake booster vacuum hose
    - EVAP hose
    - 4-Wheel Drive: Automatic Disconnecting Differential (ADD) vacuum hose
    - Fuel inlet and fuel return hose
28. Connect the three clamps for the spark plug wires to the No. 2 timing belt cover.
29. Connect the camshaft position sensor connector to the No. 2 timing belt cover.
30. Install the spark plug wires with the ignition coils.
31. Connect the heater hose.
32. Connect the following cables:
    - If equipped with cruise control, connect the actuator cable with the bracket.
    - Accelerator cable
    - If equipped with automatic transmission, connect the throttle cable
33. Install the MAF meter, resonator, and the air cleaner cap.
34. Fill the radiator with engine coolant.
35. Connect the negative battery cable to the battery.
36. Start the engine and check for leaks.

37. Install the engine undercover.
38. Road test the vehicle.
39. Recheck all fluid levels.

1995–97 T-100 AND TACOMA

1. Disconnect the negative battery cable.

❈❈ CAUTION

Wait 90 seconds from the time the key is turned to LOCK and the negative battery cable is disconnected to begin work. This allows the SRS capacitor to discharge and prevent deployment of the air bag(s).

2. Relieve the fuel system pressure.

❈❈ CAUTION

The fuel injection system remains under pressure after the engine has been turned OFF. Properly relieve fuel pressure before disconnecting any fuel lines. Failure to do so may result in fire or personal injury.

3. Drain the engine coolant.
4. Disconnect the spark plug wires from the spark plugs.
5. Remove the air cleaner cap, MAF meter, and the resonator.
6. Disconnect the following cables:
• If equipped with cruise control, disconnect the actuator cable with the bracket.
• Accelerator cable
• If equipped with automatic transmission, disconnect the throttle cable
7. If equipped with an EGR valve, remove the nuts and remove the EGR pipe and two gaskets.
8. Remove the intake chamber stay as follows:
   a. Remove the oil filler tube and No. 1 throttle cable clamp by removing the bolt and two nuts.
   b. Remove the intake chamber stay by removing the two bolts.
9. Remove the following connectors:
• Variable Switching Valve (VSV) connector for the fuel pressure control.
• Disconnect the throttle position sensor connector.
• IAC valve connector
• If equipped with an EGR valve, disconnect the EGR gas temperature connector
• If equipped with an EGR valve, disconnect the Variable Switching Valve (VSV) connector for the EGR valve
10. Disconnect the following hoses:
• Disconnect the PCV hoses.
• Disconnect the water bypass hoses.

• Disconnect the air assist hose from the intake air connector.
• Two vacuum sensing hoses from the Variable Switching Valve (VSV)
• The EVAP hose
• Air hose from the power steering
• If equipped with air conditioning, disconnect the air hose from the air conditioning idle up valve.
11. Remove the four bolts, two nuts, and remove the air intake chamber assembly from the engine.
12. Remove the intake air connector as follows:
   a. Disconnect the engine wire from the intake air connector by removing the bolt.
   b. Disconnect the two fuel return hoses.
   c. Disconnect the brake booster vacuum hose from the intake air connector.
   d. Remove the bolt and disconnect the ground strap from the intake air connector.
   e. Disconnect the DLC1 from the bracket of the intake air connector.
   f. If equipped with air conditioning, disconnect idle up valve connector.
   g. Remove the intake air connector from the engine by removing the three bolts and two nuts.
13. Disconnect the upper radiator hose from the engine.
14. Disconnect the engine wire from the intake manifold as follows:
   a. Disconnect the following connectors:
• Oil pressure sensor connector
• Crankshaft position sensor connector
• Six injector connectors
• ECT sender gauge connector
• ECT sensor connector
• Knock sensor connector
• Camshaft position sensor connector
   b. Disconnect the three engine wire clamps.
   c. Remove the three bolts and disconnect the engine wire from the cylinder head.
15. Remove the fuel pressure regulator.
16. Disconnect the heater hose.
17. Remove the camshaft position sensor.
18. Remove the intake manifold assembly as follows:
   a. Disconnect the fuel inlet hose.
   b. Remove the two bolts and the intake manifold stay.
   c. Remove the eight bolts, four nuts, four plate washers, and the intake manifold assembly.

To install:
19. Clean all surfaces.
20. Install two new gaskets and the intake manifold assembly. Install the four plate washers, eight bolts and four nuts. Tighten the bolts and nuts to 13 ft. lbs. (18 Nm).
21. Install the intake manifold stay with the two bolts. Tighten the bolts to 13 ft. lbs. (18 Nm).
22. Connect the fuel inlet hose.
23. Install the camshaft position sensor and tighten to 71 inch lbs. (8 Nm).
24. Connect the engine wire to the intake manifold as follows:
   a. Install the engine wire to the cylinder head by installing the three bolts.
   b. Connect the three engine wire clamps.
   c. Connect the following connectors:
• Oil pressure sensor connector
• Crankshaft position sensor connector
• Six injector connectors
• ECT sender gauge connector
• ECT sensor connector
• Knock sensor connector
• Camshaft position sensor connector
25. Install the heater hose.
26. Install the intake air connector as follows:
   a. Install the intake manifold to the engine by installing the three bolts and two nuts. Tighten the bolts and nuts to 14 ft. lbs. (18.5 Nm).
   b. Connect the DLC1 to the bracket on the intake manifold.
   c. Connect the ground strap to the intake manifold by installing the bolt.
   d. Connect the brake booster vacuum hose to the intake air connector.
   e. Connect the two fuel return hoses.
   f. Connect the engine wire to the intake manifold by installing the bolt.
   g. If equipped with air conditioning, connect the idle up valve connector.
27. Install the air intake chamber assembly to the engine by installing the four bolts and two nuts. Tighten the bolts and nuts to 14 ft. lbs. (18.5 Nm).
28. Connect the following hoses:
• Connect the PCV hoses.
• Connect the water bypass hoses.
• Connect the air assist hose to the intake manifold.
• Two vacuum sensing hoses to the Variable Switching Valve (VSV)
• The EVAP hose
• Air hose to the power steering
• If equipped with air conditioning, connect the air hose to the air conditioning idle up valve.

29. Connect the following connectors:
- Variable Switching Valve (VSV) connector for the fuel pressure control.
- Connect the throttle position sensor connector.
- IAC valve connector
- If equipped with an EGR valve, connect the EGR gas temperature connector
- If equipped with an EGR valve, connect the Variable Switching Valve (VSV) connector for the EGR valve

30. Install the intake chamber stay as follows:

a. Install the intake chamber stay and install the two bolts. Tighten the bolts to 30 ft. lbs. (40 Nm)

b. Install a new O-ring to the oil filler tube.

c. Push in the oil filler tube end into the tube hole in the oil pan.

d. Install the oil filler tube and No. 1 throttle cable clamp and install the bolt and two nuts.

31. Install two new gaskets and the EGR pipe with the nuts. Tighten the clamp nuts to 71 inch lbs. (8 Nm) and the EGR pipe nuts to 14 ft. lbs. (18 Nm).

32. Install the fuel pressure regulator.

33. Connect the three clamps for the spark plug wires to the No. 2 timing belt cover.

34. Connect the camshaft position sensor connector to the No. 2 timing belt cover.

35. Connect the upper radiator hose.

36. Fill with engine coolant.

37. Connect the spark plug wires to the spark plugs.

38. Connect the following cables:
- If equipped with cruise control, connect the actuator cable with the bracket.
- Accelerator cable
- If equipped with automatic transmission, connect the throttle cable

39. Install the air cleaner hose.

40. Fill the radiator with engine coolant.

41. Connect the negative battery cable to the battery.

42. Start the engine and check for leaks.

Exhaust Manifold

REMOVAL & INSTALLATION

2.4L (22R-E) Engine

1. Disconnect the negative battery cable.

2. Raise and safely support the vehicle.

3. Working from under the vehicle, remove the two bolts holding the front exhaust pipe to the mounting bracket.

4. Remove the three nuts and disconnect the exhaust pipe.

5. Disconnect the main oxygen sensor and the sub oxygen sensor connectors.

6. Remove the three bolts and the exhaust manifold heat insulator.

7. Remove the eight nuts, the exhaust manifold, if equipped, the secondary air injection manifold.

To install:

8. Install a new gasket and the exhaust manifold. Uniformly tighten the nuts in several passes. Tighten the nuts to 33 ft. lbs. (44 Nm).

9. Install the exhaust manifold heat insulator with the three bolts and tighten to 14 ft. lbs. (19 Nm).

10. Connect the main oxygen and the sub oxygen sensor connectors.

11. Install the front exhaust pipe with a new gasket to the exhaust manifold with the three nuts.

12. Secure the front exhaust pipe to the exhaust pipe clamp.

13. Lower the vehicle safely and connect the negative battery cable.

14. Start the engine and be sure that there are no exhaust leaks.

2.4L (2RZ-FE) and 2.7L (3RZ-FE) Engines

1. Raise and safely support the vehicle.

2. Disconnect the front exhaust pipe from the exhaust manifold as follows:

a. Loosen the clamp bolt and disconnect the clamp from the support bracket.

b. Remove the support bracket by removing the two bolts.

c. Disconnect the three nuts, front exhaust pipe, and the gaskets from the exhaust manifold.

3. Lower the vehicle.

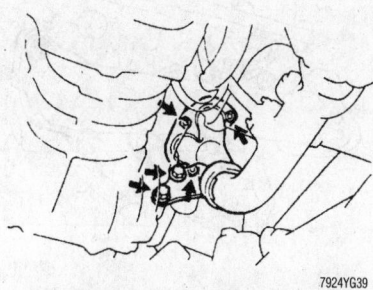

7924YG39

Front exhaust pipe to exhaust manifold nut and bolt locations—2.4L (2RZ-FE) and 2.7L (3RZ-FE) engine

7924YG40

Exhaust manifold nuts—2.4L (2RZ-FE) and 2.7L (3RZ-FE) engine

4. Remove the heat insulator by removing the two bolts and two nuts.

5. Remove the exhaust manifold and gasket by removing the six nuts.

To install:

6. Install the exhaust manifold and gasket to the engine and install the six nuts. Tighten the nuts to 36 ft. lbs. (49 Nm).

7. Install the heat insulator with the two bolts and two nuts. Tighten the bolts and nuts to 48 inch lbs. (5.5 Nm).

8. Raise and safely support the vehicle.

9. Install the front exhaust pipe to the exhaust manifold as follows:

a. Install the two gaskets and the front exhaust pipe assembly to the exhaust manifold. Install the three nuts and tighten the nuts to 46 ft. lbs. (62 Nm).

b. Install the support bracket with the two bolts. Tighten the brackets to 29 ft. lbs. (39 Nm).

c. Connect the clamp and tighten the clamp bolt. Tighten the bolt to 14 ft. lbs. (19 Nm).

10. Lower the vehicle and start the engine.

11. Check for exhaust leaks.

3.0L (3VZ-E) Engine

1. Disconnect the negative battery cable.

2. Raise and safely support the vehicle.

3. Working from under the vehicle, disconnect the heated oxygen sensor connector.

4. Loosen the pipe clamp bolt.

5. Remove the two bolts and the pipe bracket.

6. Remove the three nuts and disconnect the exhaust pipe from the exhaust manifold. Remove the gasket.

7. Remove the two bolts, the joint retainer, the exhaust pipe and the gasket from the catalytic converter.

8. Remove the six nuts, two gaskets, and the exhaust crossover pipe.

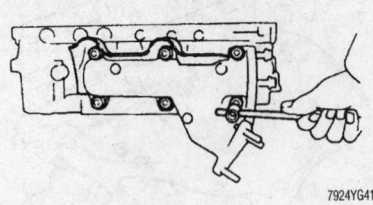

Left side exhaust manifold—3.0L (3VZ-E) engine

7924YG41

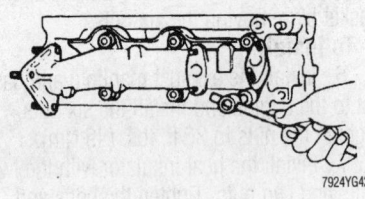

Right side exhaust manifold—3.0L (3VZ-E) engine

7924YG42

9. Remove the bolts, nuts, and the exhaust manifold heat insulators. Remove the six nuts, the left and the right exhaust manifolds, and the gaskets.

To install:

10. Install the new gaskets and the left and the right exhaust manifolds. Uniformly tighten the nuts in several passes. Tighten the nuts to 29 ft. lbs. (39 Nm).

11. Install the exhaust crossover pipe with two new gaskets and six nuts. Tighten the nuts to 29 ft. lbs. (39 Nm).

12. Install the left and right exhaust manifold heat insulators with the nuts and bolts.

13. Connect the exhaust pipe to the left exhaust manifold with a new gasket and tighten the three new nuts to 46 ft. lbs. (62 Nm).

14. Connect the exhaust pipe to the catalytic converter with a new gasket and tighten the two bolts to 29 ft. lbs. (39 Nm).

15. Connect the heated oxygen sensor connector.

16. Secure the front exhaust pipe to the exhaust pipe clamp.

17. Lower the vehicle safely and connect the negative battery cable.

18. Start the engine and be sure that there are no exhaust leaks.

3.4L (5VZ-FE) Engine

1996–99 4RUNNER, T-100 AND TACOMA

> ❄❄ **CAUTION**
>
> **Make certain all surfaces are cool to the touch before beginning work.**

1. Disconnect the exhaust crossover pipe from the exhaust manifold by removing the three nuts.

2. On the left manifold, if equipped with an EGR valve, remove the nuts and disconnect the EGR pipe from the exhaust manifold.

3. Remove the exhaust manifold heat insulator by removing the three nuts.

4. Remove the exhaust manifold by removing the six nuts.

Exhaust crossover pipe mounting nut locations—3.4L (5VZ-FE) engine

7924YG43

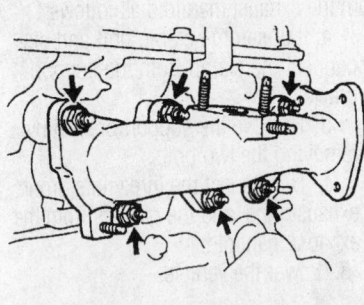

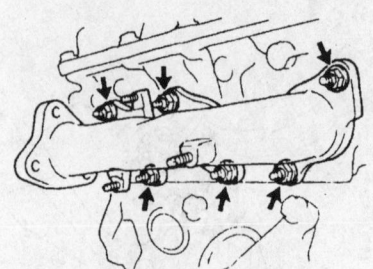

Exhaust manifold nuts—3.4L (5VZ-FE) engine

7924YG44

To install:

5. Install the exhaust manifold with a new gasket. Tighten the six nuts to 30 ft. lbs. (40 Nm).

6. Install the exhaust heat insulator by installing the three nuts. Tighten the nuts to 71 inch lbs. (8 Nm).

7. If removed, with an EGR valve, connect the EGR pipe to the exhaust manifold. Tighten the nuts to the manifold to 14 ft. lbs. (18 Nm). Tighten the clamp nuts to 71 inch lbs. (8 Nm).

8. Connect the crossover pipe to the exhaust manifold with the three bolts and a new gasket. Tighten the nuts to 33 ft. lbs. (45 Nm).

Front Crankshaft Seal

REMOVAL & INSTALLATION

➡ **This procedure is for timing belt equipped engines only. For engines which utilize timing chains or gears, please refer to the chain or gear procedure.**

3.0L (3VZ-E) and 3.4L (5VZ-FE) Engines

➡ **There are 2 methods to replace the oil seal, which are as follows:**

OIL PUMP BODY INSTALLED

1. Disconnect the negative battery cable.

> ❄❄ **CAUTION**
>
> **Wait 90 seconds from the time the key is turned to LOCK and the negative battery cable is disconnected to begin work. This allows the SRS capacitor to discharge and prevent deployment of the air bag(s).**

2. On the 3VZ-E engines, remove the engine undercover (4-Wheel Drive vehicles).

3. Remove the timing belt and crankshaft pulley.

4. Using a knife, cut off the oil seal lip.

5. Using a suitable tool, pry out the oil seal. Be careful not to damage the crankshaft.

To install:

6. Apply Multi-Purpose (MP) grease to the new oil seal lip.

7. Using Seal Driver tool 09309–37010, or equivalent, and a mallet, tap in the new oil seal until its surface is flush with the oil pump case edge.

8. Install the crankshaft pulley and the timing belt.

9. If removed, install the engine undercover.

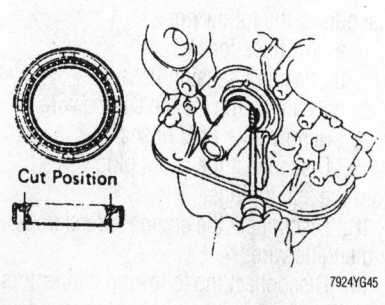

Cutting the oil seal lip—3.0L (3VZ-E) engines

10. Connect the negative battery cable.

OIL PUMP BODY REMOVED

1. Carefully pry out the seal using a suitable tool.

2. Apply Multi-Purpose (MP) grease to the new oil seal lip.

3. Using Seal Driver tool 09309–37010, or equivalent, drive the new seal into place.

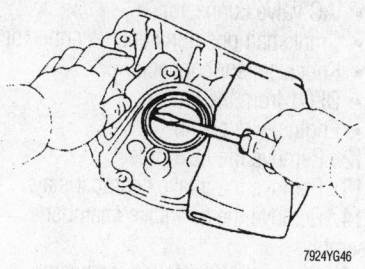

Oil seal removal—3.0L (3VZ-E) engines

Camshaft and Valve Lifters

REMOVAL & INSTALLATION

2.4L (22R-E) Engine

1. Perform the Cylinder Head Removal procedure (for your engine) far enough to gain access to the camshaft bearing cap bolts. If you are going to remove the head anyway, remove the cam after removing the cylinder head.

2. Remove the bearing cap bolts. Remove the caps. Keep them in order, or mark them.

3. Measure the bearing oil clearance by placing a piece of Plastigage® on each journal. Replace the caps and tighten their bolts to 13–16 ft. lbs. (18–22 Nm).

4. Remove the caps and measure each piece of Plastigage®. If the clearance is greater than 2 in. (0.1mm), replace the head and cam.

5. Lift the camshaft out of the head.

To install:

6. Coat all of the camshaft bearing journals with engine oil.

7. Lay the camshaft in the head.

8. Install the bearing caps in numerical order with their arrows pointing forward (toward the front of the engine).

9. Install the cap bolts and tighten them in three passes and in the correct order, to 13–16 ft. lbs. (18–22 Nm).

10. Complete the cylinder head installation procedure and/or valve rocker installation.

2.4L (2RZ-FE) Engine

1. Remove the timing chain from the engine.

2. Remove the exhaust camshaft by bringing the service bolt hole of the driven sub-gear upwards. Turn the hexagon wrench head portion of the exhaust camshaft with a wrench.

3. Secure the exhaust camshaft sub-gear to the main gear with a service bolt. The thread diameter should be 0.23 in. (6mm) with a thread pitch of 0.04 in. (1.0mm) and a bolt length of 0.63–0.79 in. (16–20mm).

➡**When you remove the camshaft, be sure that the torsional spring force of the sub-gear has been eliminated by the above operation.**

4. Uniformly loosen and remove the exhaust bearing cap bolts (10 of them), in several passes. Use the reverse order of the tightening sequence. Remove the 5 bearing caps and the camshaft. Do the same for the intake camshafts.

➡**If the camshaft is not being lifted out straight and level, reinstall the No. 3 cap with the 2 bolts. Alternately loosen, then remove the bearing cap bolts with the camshaft pulled up. Do not pry on or force the camshaft.**

5. Inspect the camshafts for runout. Inspect the cam lobes and journals. The bearings are part of the cam and should be inspected for flaking or scoring. If the bearings are damaged, replace the caps and the cylinder head as a set. The camshaft journal oil and thrust clearances should be checked.

To install:

➡**When installing the camshafts; since the thrust clearance of the shafts is**

small, the cam must be kept level while it is being installed. If it is not kept level, the portion of the head receiving the shaft thrust may crack or be damaged. This can cause the camshaft to seize or break.

6. Install the intake camshaft as follows:

a. Apply multi-purpose grease to the thrust portion of the intake camshaft.

b. Position the intake camshaft with the pin facing upward.

c. Install the bearing caps in their proper locations. Apply a light coat of engine oil to the threads and install the cap bolts. Uniformly tighten the cap bolts in the sequence shown to 12 ft. lbs. (16 Nm).

7. Install the exhaust camshaft as follows:

a. Apply engine oil to the thrust portion of the intake camshaft.

b. Engage the exhaust camshaft gear to the intake camshaft gear by matching the timing marks (one and two dots) on each other.

c. Roll down the exhaust camshaft onto the bearing journals while engaging the gears with each other. Install the bearing caps in their proper locations.

d. Apply a light coat of engine oil to the threads and install the cap bolts. Uniformly tighten the cap bolts in the sequence shown to 12 ft. lbs. (16 Nm).

e. Remove the service bolt from the driven sub-gear. Check that the intake and exhaust camshafts turn smoothly.

8. Set No. 1 cylinder to TDC compression stroke. The crankshaft pulley groove aligns with the **0** mark on timing cover and camshaft timing marks with one dot and two dots will be in a straight line on the cylinder head surface.

9. Install the timing gear. Place the

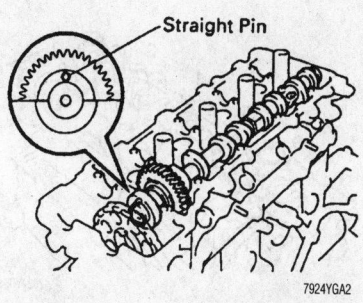

Install the camshaft with the pin facing upwards—2.4L (2RZ-FE) engine

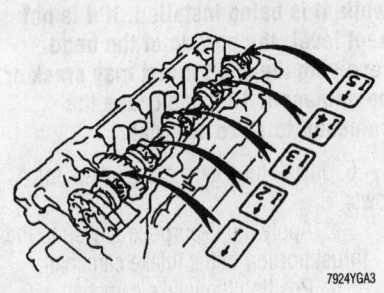

Intake camshaft bearing cap locations— 2.4L (2RZ-FE) engine

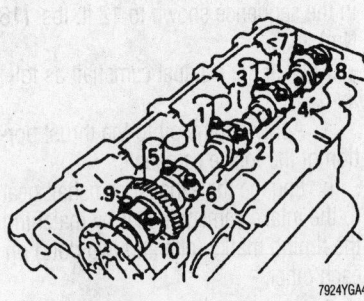

Tighten the intake bearing caps following this order—2.4L (2RZ-FE) engine

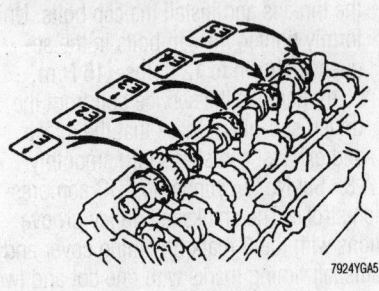

Position the exhaust camshaft bearing caps as shown—2.4L (2RZ-FE) engine

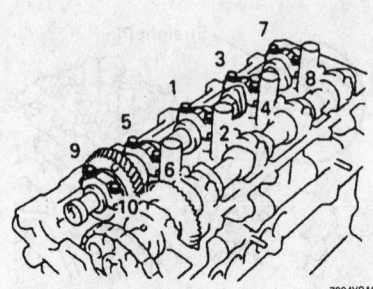

Tighten the exhaust bearing cap bolts in this order—2.4L (2RZ-FE) engine

gear over the straight pin of the intake camshaft.

   a. Hold the intake camshaft with a wrench. Install and tighten the bolt to 54 ft. lbs. (74 Nm).

   b. Hold the exhaust camshaft and install the bolt and distributor gear. Tighten the bolt to 34 ft. lbs. (46 Nm).

10. Install the chain tensioner, using a new gasket (mark toward the front) as follows:

   a. Release the ratchet pawl, fully push in the plunger and apply the hook to the pin so that the plunger cannot spring out.

   b. Turn the crankshaft pulley clockwise to provide some slack for the chain on the tensioner side.

   c. Push the tensioner by hand until it touches the head installation surface, then install the 2 nuts. Tighten the nuts to 13 ft. lbs. (18 Nm). Check that the hook of the tensioner is not released.

   d. Turn the crankshaft to the left so that the hook of the chain tensioner is released from the pin of the plunger, allowing the plunger to spring out and the slipper to be pushed into the chain.

11. Check and adjust the valve clearance. Intake valve clearance is 0.006–0.010 inch (0.15–0.25mm) and exhaust valve clearance is 0.010–0.014 inch (0.25–0.35mm).

12. Recheck the engine for proper valve timing. Check and adjust the valve clearance.

13. Install the spark plugs and the semicircular plug.

14. Recheck the engine for proper valve timing. Install the valve cover and engine hangers. Tighten the engine hanger bolts to 30 ft. lbs. (42 Nm).

15. Reinstall all other parts from the timing chain removal. Fill any fluids, start the engine, top off the fluids.

2.7L (3RZ-FE) Engine

1. Disconnect the negative battery cable.
2. Drain the engine coolant.
3. Remove the air cleaner cap, MAF meter, and the resonator.
4. If equipped with a manual transmission, disconnect the accelerator cable from the throttle body.
5. If equipped with a automatic transmission, disconnect the accelerator and throttle cables from the throttle body.
6. If equipped with cruise control, disconnect the cruise control cable from the actuator.

7. Remove the intake air connector. Disconnect the following:

   a. Air hose for IAC
   b. Vacuum sensing hose
   c. Wire clamp for the engine wire

8. Remove the PCV hoses.
9. Disconnect the spark plug wires from the spark plugs.
10. Disconnect the engine wire clamps and engine wire.
11. Disconnect the following connectors:

• If equipped with air conditioning, disconnect the air conditioning compressor connector
• Disconnect the oil pressure sensor connector
• Disconnect the ECT sensor connector
• Engine coolant temperature sender gauge connector
• EGR gas temperature sensor connector
• Variable Switching Valve (VSV) connector for the EGR
• Two vacuum hoses from the Variable Switching Valve (VSV) for the EGR
• Ground strap from the cowl top panel
• Engine wire from the air intake chamber
• Throttle position sensor connector
• IAC valve connector
• Crankshaft position sensor connector
• Knock sensor connector
• DLC1 from the bracket
• Engine wire clamp

12. Remove the EGR pipe.
13. Remove the intake chamber stay.
14. Remove the air intake chamber assembly.
15. Disconnect the following hoses:

   a. EVAP hose from the throttle body
   b. Brake booster vacuum hose from the union
   c. Water bypass hose from the water bypass pipe

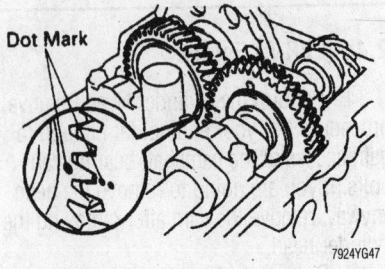

Camshafts TDC/compression timing marks. Marks with 1 and 2 dots will be in straight line on cylinder head surface— 2.7L (3RZ-FE) engine

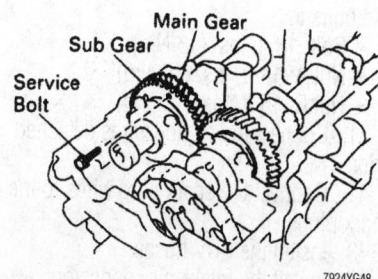

Secure the exhaust camshaft sub-gear to the main gear with a service bolt—2.7L (3RZ-FE) engine

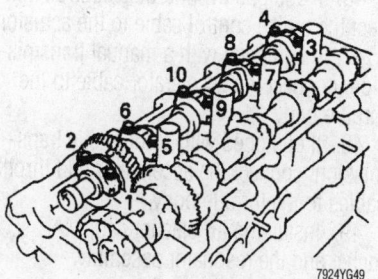

Loosen and remove the exhaust camshaft bearing cap bolts in the sequence shown—2.7L (3RZ-FE) engine

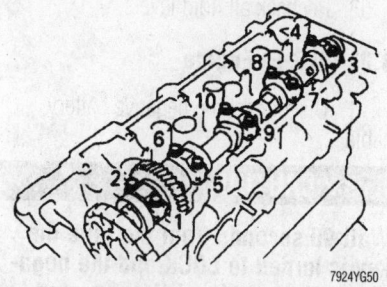

Loosen and remove the intake camshaft bearing cap bolts in the sequence shown—2.7L (3RZ-FE) engine

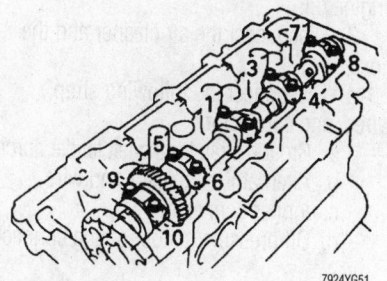

Tighten the intake camshaft bearing cap bolts in the sequence shown—2.7L (3RZ-FE) engine

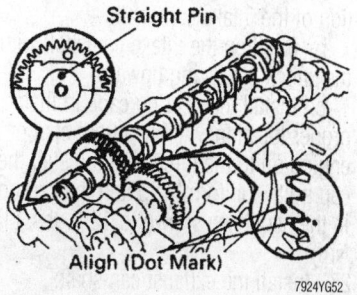

Engage both camshaft gears while matching the timing marks as shown—2.7L (3RZ-FE) engine

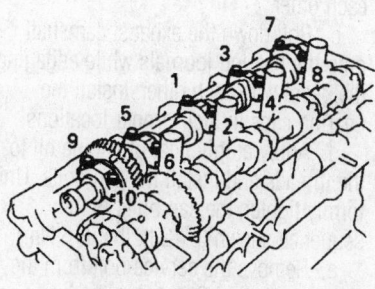

Tighten the exhaust camshaft bearing cap bolts in the sequence shown—2.7L (3RZ-FE) engine

d. Water bypass hose from the cylinder head rear cover

16. Remove the front engine hanger.
17. Remove the engine wire brackets.
18. Remove the cylinder head cover.
19. Set No. 1 cylinder to TDC compression stroke. The groove on the crankshaft pulley should align with the **0** mark on the timing chain cover and the timing marks (one and two dots) of the camshaft gears should form a straight line in respect to the cylinder head surface. If not, turn the crankshaft 1 revolution (360 degrees).
20. Remove the chain tensioner and gasket.
21. Remove the camshaft timing gear as follows:

a. Remove the 2 semi-circular plugs.

b. Place matchmarks on the camshaft timing gear and No. 1 timing chain.

c. Hold the hexagon head portion of the exhaust camshaft with a wrench and remove the fastener and distributor gear.

d. Hold the hexagon head portion of the intake camshaft with a wrench and remove the bolt.

e. Remove the camshaft timing gear and chain from the intake camshaft and leave on the slipper and damper.

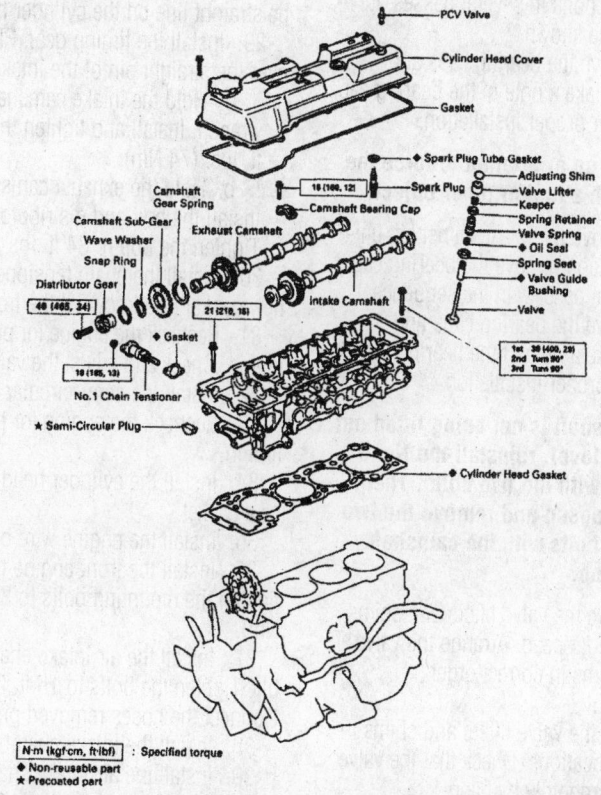

Exploded view of the cylinder head components—2.7L (3RZ-FE) engine

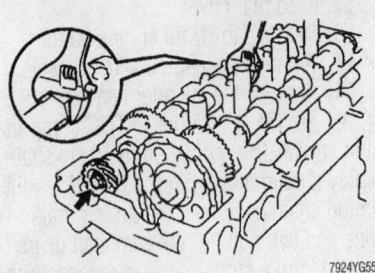

7924YG55

Using a wrench to hold the camshaft— 2.7L (3RZ-FE) engine

22. Remove exhaust camshafts:

a. Bring the service bolt hole of the driven sub-gear upward by turning the hexagon head portion of the exhaust camshaft with a wrench.

b. Secure the exhaust camshaft sub-gear to the driven gear with a service bolt (6mm diameter, 0.63–0.79 inches in length and 1.0mm in thread pitch).

➡**When removing the camshaft, be sure that the torsional spring force of the sub-gear has been eliminated by the above operation.**

c. Uniformly loosen and remove the bearing cap bolts in several passes, in the sequence shown.

d. Remove the bearing caps and camshaft. Make a note of the bearing cap positions for proper installation.

➡**Do not pry on or attempt to force the camshaft with a tool or other object.**

23. To remove the intake camshaft, uniformly loosen and remove the bearing cap bolts in several passes, in the sequence shown. Remove the bearing caps and camshaft. Make a note of the bearing cap positions for proper installation.

➡**If the camshaft is not being lifted out straight and level, reinstall the No. 3 bearing cap with the two bolts. Then, alternately loosen and remove the two bearing cap bolts with the camshaft gear pulled up.**

24. Remove the valve lifters and shims from the cylinder head. Arrange the valve lifters and shims in correct order.

To install:

25. Install the valve lifters and shims in their proper locations. Check that the valve lifter rotates smoothly by hand.

26. Install the intake camshaft:

a. Apply engine oil to the thrust portion of the intake camshaft.

b. Position the intake camshaft with the knock pin facing upward.

c. Install the bearing caps in their proper locations. Apply a light coat of engine oil to the threads and install the cap bolts. Uniformly tighten the cap bolts in the sequence shown to 12 ft. lbs. (16 Nm).

27. Install the exhaust camshaft:

a. Apply engine oil to the thrust portion of the intake camshaft.

b. Engage the exhaust camshaft gear to the intake camshaft gear by matching the timing marks (one and two dots) on each other.

c. Roll down the exhaust camshaft onto the bearing journals while engaging the gears with each other. Install the bearing caps in their proper locations.

d. Apply a light coat of engine oil to the threads and install the cap bolts. Uniformly tighten the cap bolts in the sequence shown to 12 ft. lbs. (16 Nm).

e. Remove the service bolt from the driven sub-gear. Check that the intake and exhaust camshafts turns smoothly.

28. Set No. 1 cylinder to TDC compression stroke: Crankshaft pulley groove align with **O** mark on timing cover and camshafts timing marks with one dot and two dots will be straight line on the cylinder head surface.

29. Install the timing gear. Place the gear over the straight pin of the intake camshaft.

a. Hold the intake camshaft with a wrench. Install and tighten the bolt to 54 ft. lbs. (74 Nm).

b. Hold the exhaust camshaft and install the bolt and distributor gear. Tighten the bolt to 34 ft. lbs. (46 Nm).

30. Install the chain tensioner, using a new gasket (mark toward the front).

31. Recheck the engine for proper valve timing. Check and adjust the valve clearance.

32. Install the semi-circular plug.

33. Recheck the engine for proper valve timing.

34. Install the cylinder head cover with a new gasket.

35. Install the engine wire brackets.

36. Install the front engine hanger and tighten the mounting bolts to 30 ft. lbs. (42 Nm).

37. Install the air intake chamber assembly. Tighten the bolts to 15 ft. lbs. (21 Nm). Connect the hoses removed prior.

38. Install the intake chamber stay.

39. Install the air intake chamber stay and tighten the bolts to 15 ft. lbs. (20 Nm).

40. Install the EGR pipe. Tighten the nuts and bolts to:
- Bolt: 13 ft. lbs. (18 Nm)
- Nut A: 14 ft. lbs. (19 Nm)
- Nut B: 15 ft. lbs. (20 Nm)

41. Connect the engine wires detached prior.

42. Connect the spark plug wires to the spark plugs.

43. Install the PCV hoses.

44. Install the intake air connector. Tighten the bolts to 13 ft. lbs. (18 Nm).

45. Connect the following:

a. Air hose for the IAC.

b. Vacuum sensing hose.

c. Wire clamp for the engine wire.

46. If equipped with cruise control, connect the cruise control cable to the actuator.

47. If equipped with a manual transmission, connect the accelerator cable to the throttle body.

48. If equipped with a automatic transmission, connect the accelerator and throttle cables to the throttle body.

49. Install the air cleaner cap, MAF meter, and the resonator assembly.

50. Fill the engine and radiator with engine coolant.

51. Connect the negative battery cable. Start the engine and check for leaks.

52. Check the ignition timing. Road test the vehicle for proper operation.

53. Recheck all fluid levels.

3.0L (3VZ-E) Engine

1. Disconnect the negative battery cable.

✳✳ CAUTION

Wait 90 seconds from the time the key is turned to LOCK and the negative battery cable is disconnected to begin work. This allows the SRS capacitor to discharge and prevent deployment of the air bag(s).

2. Drain the cooling system and the engine oil.

3. Disconnect the air cleaner and the hose.

4. Disconnect the following strap, wires, and connectors:

a. Ground strap for the left fender apron

b. Alternator connector and wire

c. Igniter connector

d. Oil pressure sender gauge connector

e. Ground strap from the engine rear side

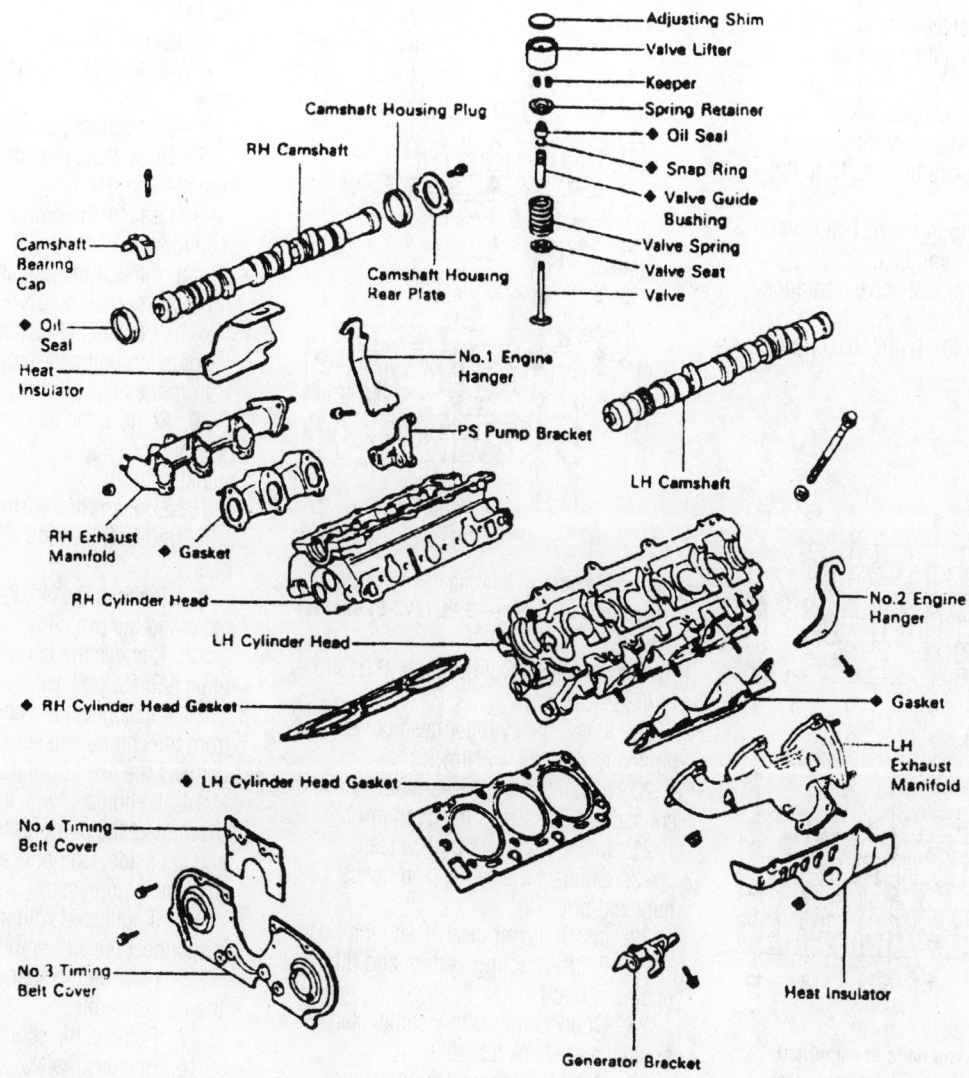

Adjusting Shim
Valve Lifter
Keeper
Spring Retainer
◆ Oil Seal
◆ Snap Ring
◆ Valve Guide Bushing
Valve Spring
Valve Seat
Valve

Camshaft Housing Plug
RH Camshaft
Camshaft Bearing Cap
◆ Oil Seal
Heat Insulator
Camshaft Housing Rear Plate
No.1 Engine Hanger
PS Pump Bracket
LH Camshaft
RH Exhaust Manifold
◆ Gasket
RH Cylinder Head
LH Cylinder Head
No.2 Engine Hanger
◆ RH Cylinder Head Gasket
◆ Gasket
LH Exhaust Manifold
◆ LH Cylinder Head Gasket
No.4 Timing Belt Cover
No.3 Timing Belt Cover
Heat Insulator
Generator Bracket

◆ Non-reusable part

7924YG56

Exploded view of the camshaft and related components—3.0L (3VZ-E) engine

f. ECM connectors

g. Variable Switching Valve (VSV) connectors

h. Air conditioning compressor connector.

i. Manual transmission: Starter relay connector

j. Solenoid resister connector

k. W/ Automatic Disconnecting Differential (ADD): ADD switch connector

5. Disconnect the following hoses:

a. Power steering air hoses from the gas filter and the air pipe

b. Brake booster hose

c. W/Cruise: Cruise control vacuum hose

d. Charcoal canister hose from the canister

e. Variable Switching Valve (VSV) vacuum hoses

6. Disconnect the following cables:

a. Accelerator cable

b. Automatic transmission: Throttle cable

c. W/Cruise: Cruise control cable

7. Remove the spark plug wires.

8. Remove the timing belt.

9. Remove the following connectors and wires:

a. Knock sensor

b. Cold start injector time switch connector

c. ECT sensor and sender gauge connector

d. No. 1 ECT switch connector

e. Right ground strap from the No. 3 camshaft bearing cap

f. Injector connectors

10. Remove the cylinder head covers.

11. Remove the No. 4 timing belt cover.

12. Remove the camshaft bearing cap bolts in sequence. Remove the camshafts.

➡**Arrange the bearing caps in correct order for installation.**

13. Remove the valve lifters and shims by hand. Arrange the valve lifters and shims in correct order for reinstallation.

To install:

14. Install the valve lifters and shims. Check that the valve lifter rotates smoothly by hand.

15. Install the camshafts. Install the bearing caps in their proper locations and tighten the bearing cap bolts to 12 ft. lbs. (16 Nm).

16. Install the No. 4 timing belt cover.

17. Install the timing belt.

18. Check and adjust valve clearance to:

    a. Intake (Cold): 0.007–0.011 in. (0.18–0.28mm)

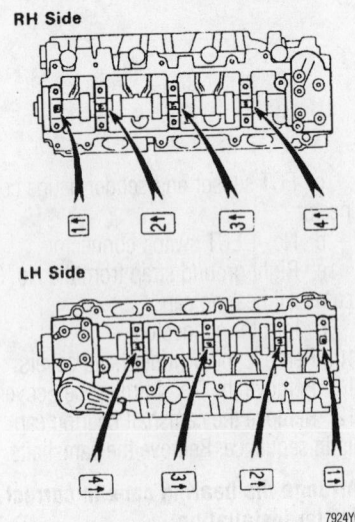

Tighten bearing cap bolts in numerical sequence shown—3.0L (3VZ-E) engine

Position bearing caps with arrows pointing toward the front (RH side) or rear (LH side)—3.0L (3VZ-E) engine

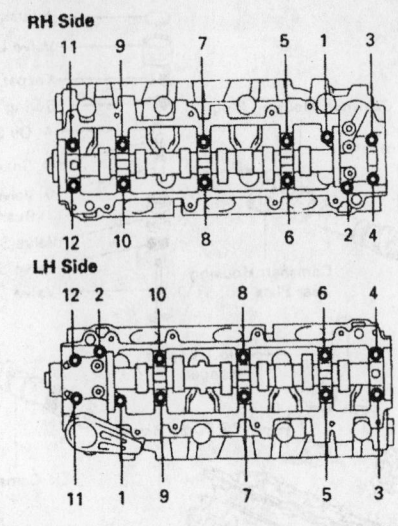

Loosen camshaft bearing cap bolts in sequence as shown—3.0L (3VZ-E) engine

    b. Exhaust (Cold): 0.009–0.013 in. (0.22–0.32mm)

19. Install the cylinder head covers and tighten to 4 ft. lbs. (5 Nm).

20. Install the engine harness. Attach the connectors and straps removed prior.

21. Install the spark plug wires.

22. Connect the cables and hoses removed prior.

23. Install the air cleaner and the hose.

24. Fill the cooling system and fill the engine with oil.

25. Connect the battery cable, start the engine, and check for leaks.

26. Road test the vehicle for proper operation and recheck all the fluid levels.

3.4L (5VZ-FE) Engine

1995–99 T-100 AND TACOMA

1. Disconnect the negative battery cable.

> ### ❊❊❊ CAUTION
>
> **Wait 90 seconds from the time the key is turned to LOCK and the negative battery cable is disconnected to begin work. This allows the SRS capacitor to discharge and prevent deployment of the air bag(s).**

2. Remove the engine undercover.

3. Drain the cooling system.

4. Disconnect the air cleaner cap, MAF meter, and the resonator.

5. Disconnect the following cables:

• If equipped with cruise control, disconnect the actuator cable with the bracket.

    • Accelerator cable

    • With automatic transmission, Throttle cable

6. Disconnect the heater hose.

7. Disconnect the upper radiator hose from the engine.

8. Remove the power steering drive belt as follows:

    a. Stretch the belt and loosen the fan pulley mounting nuts.

    b. Loosen the lockbolt, pivot bolt and adjusting bolt and remove the drive belt from the engine.

9. Remove the air conditioning drive belt by loosening the idle pulley nut and adjusting bolt.

10. Loosen the lockbolt, pivot bolt and adjusting bolt and the alternator drive belt.

11. Remove the No. 2 fan shroud by removing the two clips.

12. Remove the fan with the fluid coupling and fan pulleys.

13. Disconnect the power steering pump from the engine and set aside. Do not disconnect the lines from the pump.

14. If equipped with air conditioning, disconnect the compressor from the engine and set aside. Do not disconnect the lines from the compressor.

15. If equipped with air conditioning, disconnect the air conditioning bracket.

16. Remove the spark plug wires with the ignition coils.

17. Remove the spark plugs.

18. Remove the No. 2 timing belt cover as follows:

    a. Disconnect the camshaft position sensor connector from the No. 2 timing belt cover.

    b. Disconnect the three spark plug wire clamps from the No. 2 timing belt cover.

    c. Remove the six bolt and remove the timing belt cover.

19. Remove the fan bracket as follows:

    a. Remove the power steering adjusting strut by removing the nut.

    b. Remove the fan bracket by removing the bolt and nut.

20. Set the No. 1 cylinder at TDC of the compression stroke.

    a. Turn the crankshaft pulley and align its groove with the timing mark **0** of the No. 1 timing belt cover.

    b. Check that the timing marks of the camshaft timing pulleys and the No. 3 timing belt cover are aligned. If not, turn the crankshaft pulley one revolution (360°).

➡**If re-using the timing belt, be sure that you can still read the installation marks. If not, place new installation marks on the timing belt to match the timing marks of the camshaft timing pulleys.**

21. Remove the timing belt tensioner by alternately loosening the two bolts.
22. Remove the camshaft timing pulleys.

    a. Using Variable Pin Wrench Set 09960–10010 or equivalent, remove the pulley bolt, the timing pulley and the knock pin. Remove the two timing pulleys with the timing belt.

23. Remove the bolt and the No. 2 idler pulley.
24. Remove the alternator from the engine.
25. Disconnect the following hoses:
- Disconnect the PCV hoses.
- Disconnect the water bypass hoses.
- Disconnect the air assist hose from the intake air connector.
- Two vacuum sensing hoses from the Variable Switching Valve (VSV)
- EVAP hose
- Air hose from the power steering
- If equipped with air conditioning, disconnect the air hose from the air conditioning idle up valve.
26. Remove the four bolts, two nuts and remove the air intake chamber assembly from the engine.
27. Remove the intake air connector.
28. Remove the camshaft position sensor.
29. Remove the No. 3 (rear) timing belt cover by removing the six bolts.
30. Remove the eight bolts, seal washers, cylinder head cover and gasket. Remove both cylinder head covers.
31. Remove the semi circular plugs.
32. Remove the right exhaust camshafts as follows:

    a. Bring the service bolt hole of the driven sub-gear upward by turning the hexagon head portion of the exhaust camshaft with a wrench.

    b. Align the timing mark (2 dot marks) of the camshaft drive and driven gears by turning the camshaft with a wrench.

    c. Secure the exhaust camshaft sub-gear to the driven gear with a service bolt (6mm diameter, 16–20mm bolt length and 1.0mm in thread pitch).

➡**When removing the camshaft, be sure that the torsional spring force of the sub-gear has been eliminated by the above operation.**

    d. Uniformly loosen and remove the bearing cap bolts in several passes, in the sequence shown.

    e. Remove the bearing caps and camshaft. Make a note of the bearing cap positions for proper installation.

➡**Do not pry on or attempt to force the camshaft with a tool or other object.**

33. Remove the right-hand intake camshaft as follows:

    a. Uniformly loosen and remove the bearing cap bolts in several passes, in the sequence shown.

    b. Remove the bearing caps, oil seal and camshaft. Make a note of the bearing cap positions for proper installation.

34. Remove the left exhaust camshafts as follows:

    a. Align the timing mark (1 dot mark) of the camshaft drive and driven gears by turning the camshaft with a wrench.

    b. Secure the exhaust camshaft sub-gear to the driven gear with a service bolt (6mm diameter, 16–20mm bolt length and 1.0mm in thread pitch).

➡**When removing the camshaft, be sure that the torsional spring force of the sub-gear has been eliminated by the above operation.**

    c. Uniformly loosen and remove the bearing cap bolts in several passes, in the sequence shown.

    d. Remove the bearing caps and camshaft. Make a note of the bearing cap positions for proper installation.

➡**Do not pry on or attempt to force the camshaft with a tool or other object.**

35. Remove the left-hand intake camshaft as follows:

    a. Uniformly loosen and remove the bearing cap bolts in several passes, in the sequence shown.

    b. Remove the bearing caps, oil seal and camshaft. Make a note of the bearing cap positions for proper installation.

36. Remove the valve lifters and shims from the cylinder head. Arrange the valve lifters and shims in correct order.

To install:

37. Clean all surfaces.
38. Install the valve lifters and shims.

Check that the valve lifter rotates smoothly by hand.

39. Install the right intake camshaft as follows:

    a. Apply engine oil to the thrust portion of the intake camshaft.

    b. Position the intake camshaft at 90° angle of the timing mark (2 dot marks) on the cylinder head.

    c. Install the bearing caps in their proper locations. Apply a light coat of engine oil to the threads and install the cap bolts.

    d. Apply a light coat of engine oil on the threads and under the heads of the bearing cap bolts.

    e. Uniformly tighten the cap bolts in the sequence shown to 12 ft. lbs. (16 Nm).

40. Install the right exhaust camshaft:

    a. Apply engine oil to the thrust portion of the intake camshaft.

    b. Align the timing marks (2 dot marks) of the camshaft drive and driven gears.

    c. Roll down the exhaust camshaft onto the bearing journals while engaging the gears with each other. Install the bearing caps in their proper locations.

    d. Apply a light coat of engine oil to the threads and install the cap bolts.

    e. Apply a light coat of engine oil on the threads and under the heads of the bearing cap bolts.

    f. Uniformly tighten the cap bolts in the sequence shown to 12 ft. lbs. (16 Nm).

    g. Remove the service bolt from the driven sub-gear. Check that the intake and exhaust camshafts turns smoothly.

    h. Align the timing marks (2 dot mark) of the camshaft drive and driven gears by turning the camshaft with a wrench.

41. Install the left intake camshaft as follows:

    a. Apply engine oil to the thrust portion of the intake camshaft.

    b. Position the intake camshaft at 90° angle of the timing mark (1 dot marks) on the cylinder head.

    c. Install the bearing caps in their proper locations. Apply a light coat of engine oil to the threads and install the cap bolts.

    d. Apply a light coat of engine oil on the threads and under the heads of the bearing cap bolts.

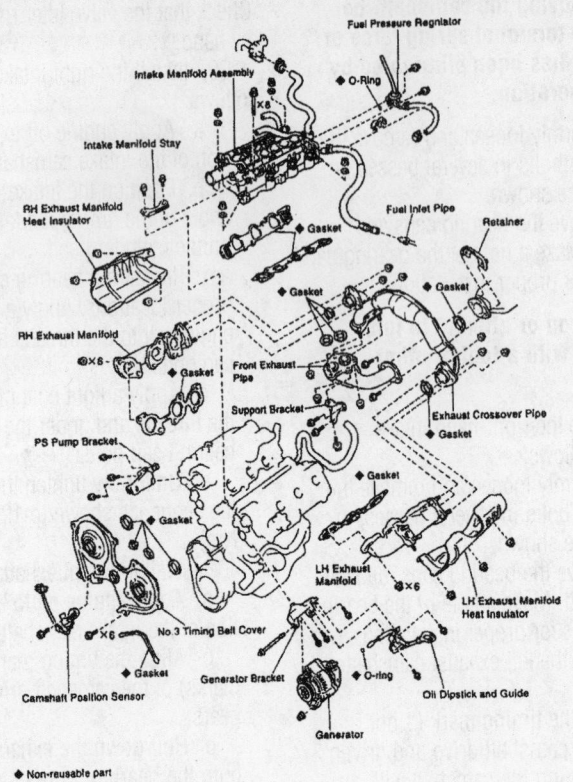

◆ Non-reusable part

7924YG60

Exploded view of the cylinder head component assembly—T-100 and Tacoma with 3.4L (5VZ-FE) engine

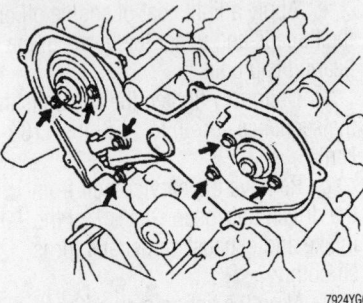

7924YG61

Rear timing belt cover mounting bolt locations—all models with 3.4L (5VZ-FE) engine

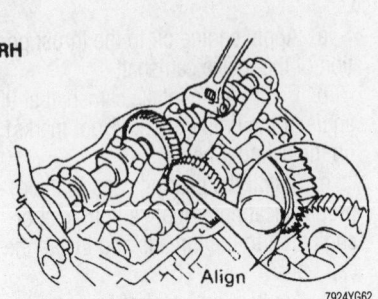

7924YG62

Aligning the timing marks (2 dot marks) of the right camshafts—all models with 3.4L (5VZ-FE) engine

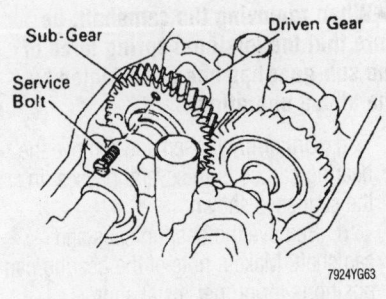

7924YG63

Drive gear service bolt (right side)—all models with 3.4L (5VZ-FE) engine

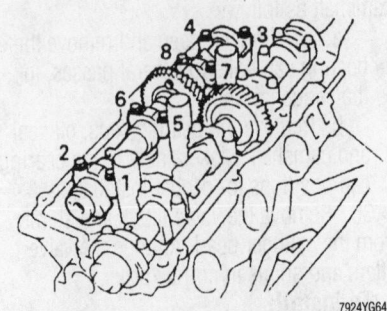

7924YG64

Right exhaust camshaft bolts removal sequence—all models with 3.4L (5VZ-FE) engine

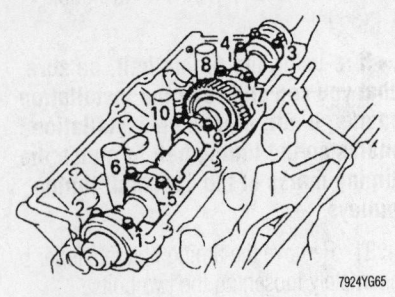

7924YG65

Right intake camshaft bolts removal sequence—all models with 3.4L (5VZ-FE) engine

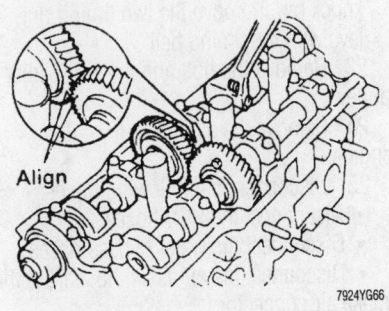

7924YG66

Aligning the timing mark (1 dot mark) of the left camshafts—all models with 3.4L (5VZ-FE) engine

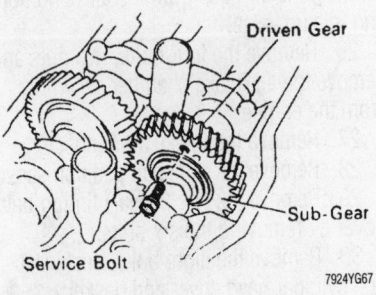

7924YG67

Drive gear service bolt (left side)—all models with 3.4L (5VZ-FE) engine

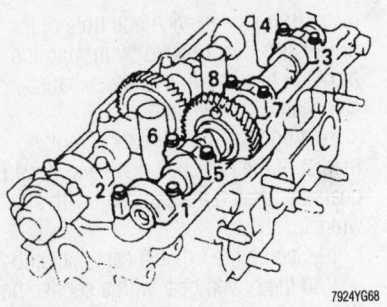

7924YG68

Left exhaust camshaft bolts removal sequence—all models with 3.4L (5VZ-FE) engine

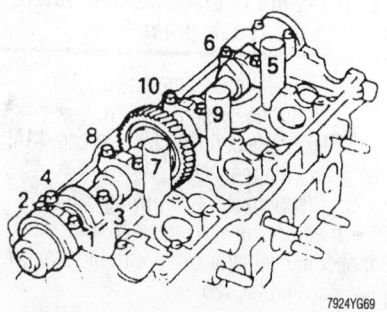

Left intake camshaft bolts removal sequence—all models with 3.4L (5VZ-FE) engine

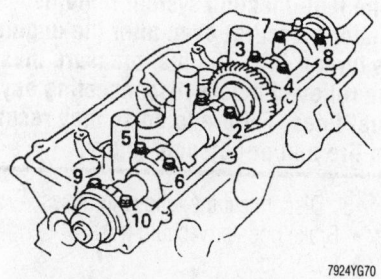

Right intake camshaft tightening sequence—all models with 3.4L (5VZ-FE) engine

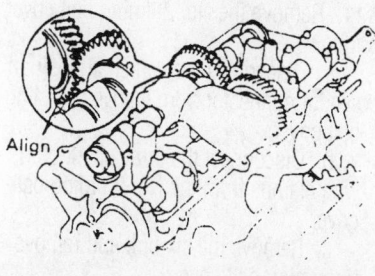

Aligning the right camshafts for installation—all models with 3.4L (5VZ-FE) engine

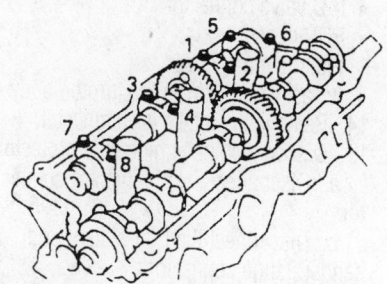

Right exhaust camshaft bolts tightening sequence—all models with 3.4L (5VZ-FE) engine

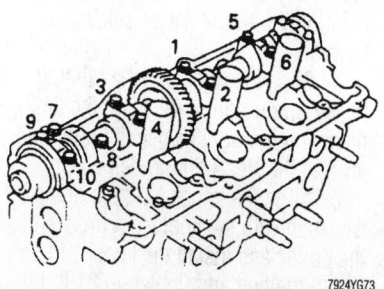

Left intake camshaft bolts tightening sequence—all models with 3.4L (5VZ-FE) engine

Left exhaust camshaft bolts tightening sequence—all models with 3.4L (5VZ-FE) engine

e. Uniformly tighten the cap bolts in the sequence shown to 12 ft. lbs. (16 Nm).

42. Install the left exhaust camshaft:

a. Apply engine oil to the thrust portion of the intake camshaft.

b. Align the timing marks (1 dot marks) of the camshaft drive and driven gears.

c. Roll down the exhaust camshaft onto the bearing journals while engaging the gears with each other. Install the bearing caps in their proper locations.

d. Apply a light coat of engine oil to the threads and install the cap bolts.

e. Apply a light coat of engine oil on the threads and under the heads of the bearing cap bolts.

f. Uniformly tighten the cap bolts in the sequence shown to 12 ft. lbs. (16 Nm).

g. Remove the service bolt.

43. Check and adjust the valve clearance.

44. Install the semi circular plugs.

45. Install the cylinder head covers and uniformly tighten the bolts in several passes to 53 inch lbs. (6 Nm).

46. If removed, install the alternator

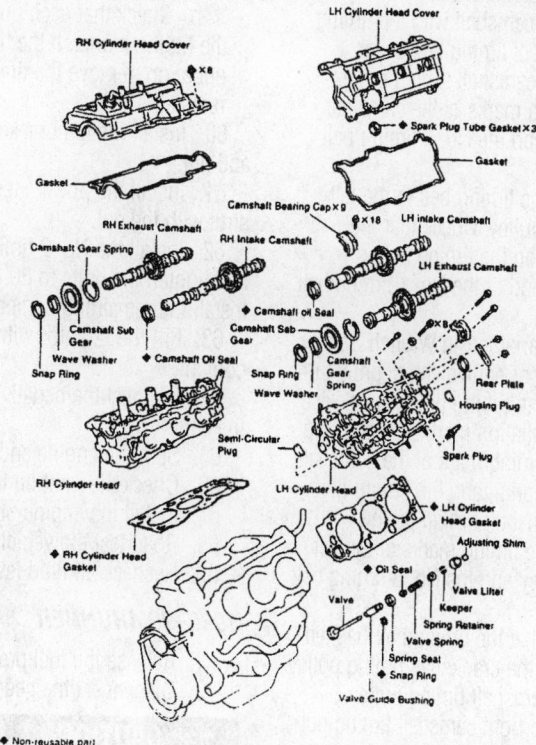

Exploded view of the cylinder head component assembly—all models with 3.4L (5VZ-FE) engine

bracket and tighten the bolts to 14 ft. lbs. (18 Nm).

47. Install the No. 3 timing belt cover with the six bolts. Tighten the bolts to 80 inch lbs. (9 Nm).

48. Install the camshaft position sensor and tighten to 71 inch lbs. (8 Nm).

49. Install the intake air connector.

50. Connect the hoses removed prior.

51. Install the alternator put do not tighten the bolts and nuts at this time.

52. Install the No. 2 timing belt idler with the bolt. Tighten the bolt to 30 ft. lbs. (40 Nm). Check that the pulley bracket moves smoothly.

53. Install the left camshaft timing pulley.

 a. Install the knock pin to the camshaft.

 b. Align the knock pin hose of the camshaft with the knock pin groove of the timing pulley.

 c. Slide the timing pulley on the camshaft with the flange side facing outward. Tighten the pulley bolt to 81 ft. lbs. (110 Nm).

54. Set the No. 1 cylinder to TDC of the compression stroke.

 a. Turn the crankshaft pulley, and align its groove with the timing mark **0** of the No. 1 timing belt cover.

 b. Turn the camshaft, align the knock pin hole of the camshaft with the timing mark of the No. 3 timing belt cover.

 c. Turn the camshaft timing pulley, align the timing marks of the camshaft timing pulley and the No. 3 timing belt cover.

55. Connect the timing belt to the left camshaft timing pulley. Check that the installation mark on the timing belt is aligned with the end of the No. 1 timing belt cover.

 a. Using Variable Pin Wrench Set 09960–01000 or equivalent, slightly turn the left camshaft timing pulley clockwise. Align the installation mark on the timing belt with the timing mark of the camshaft timing pulley, and hang the timing belt on the left camshaft timing pulley.

 b. Align the timing marks of the left camshaft pulley and the No. 3 timing belt cover.

 c. Check that the timing belt has tension between the crankshaft timing pulley and the left camshaft timing pulley.

56. Install the right camshaft timing pulley and the timing belt as follows:

 a. Align the installation mark on the timing belt with the timing mark of the right camshaft timing pulley, and hang the timing belt on the right camshaft timing pulley with the flange side facing inward.

 b. Slide the right camshaft timing pulley on the camshaft. Align the timing marks on the right camshaft timing pulley and the No. 3 timing belt cover.

 c. Align the knock pin hole of the camshaft with the knock pin groove of the pulley and install the knock pin. Install the bolt and tighten to 81 ft. lbs. (110 Nm).

57. Set the timing belt tensioner as follows:

 a. Using a press, slowly press in the pushrod using 220–2,205 lbs. (981–9,807 N) of force.

 b. Align the holes of the pushrod and housing, pass a 1.5mm hexagon wrench through the holes to keep the setting position of the pushrod.

 c. Release the press and install the dust boot to the tensioner.

58. Install the timing belt tensioner and alternately tighten the bolts to 20 ft. lbs. (28 Nm). Using pliers, remove the 1.5mm hexagon wrench from the belt tensioner.

59. Check the valve timing.

 a. Slowly turn the crankshaft pulley two revolutions from the TDC to TDC. Always turn the crankshaft pulley clockwise.

 b. Check that each pulley aligns with the timing marks. If the timing marks do not align, remove the timing belt and reinstall it.

60. Install the fan bracket with the bolt and nut.

61. Install the power steering adjusting strut with the nut.

62. Install the No. 2 timing belt cover and tighten the bolts to 80 inch lbs. (9 Nm). Install the remaining components.

63. Fill the radiator with engine coolant.

64. Connect the negative battery cable to the battery.

65. Start the engine and check for leaks.

66. Check the ignition timing.

67. Install the engine undercover.

68. Road test the vehicle.

69. Recheck all fluid levels.

1996–99 4RUNNER

1. Release the fuel pressure.

2. Disconnect the negative battery cable.

⁂ CAUTION

Wait at least 90 seconds from the time the ignition switch is turned to the LOCK position and the negative (-)

battery cable is disconnected before starting the repair procedure.

3. Remove the engine undercover.

4. Drain the cooling system.

5. Disconnect the air cleaner cap, MAF meter, and the resonator.

6. Disconnect the following cables:

• If equipped with cruise control, disconnect the actuator cable with the bracket.

• Accelerator cable

• With automatic transmission, Throttle cable

7. Disconnect the heater hose.

⁂ CAUTION

The fuel injection system remains under pressure, even after the engine is turned OFF. The fuel pressure must be relieved before disconnecting any fuel lines. Failure to do so may result in fire and/or personal injury.

8. Disconnect the following hoses:

• Brake booster vacuum hose

• EVAP hose

• 4-Wheel Drive: Automatic Disconnecting Differential (ADD) vacuum hose

• Fuel inlet and fuel return hose

9. Remove the spark plug wires with the ignition coils.

10. Remove the intake chamber stay.

11. Remove the No. 2 timing belt cover as follows:

 a. Disconnect the camshaft position sensor connector from the No. 2 timing belt cover.

 b. Disconnect the three spark plug wire clamps from the No. 2 timing belt cover.

 c. Remove the six bolt and remove the timing belt cover.

12. Remove the air intake chamber assembly.

13. Remove the following connectors and hoses:

• Throttle position sensor connector

• IAC valve connector

• PCV hoses

• Water bypass hoses

• Air assist hose from the throttle body

14. Remove the intake air connector.

15. Disconnect the engine wire protector.

 a. Disconnect the six injector connectors.

 b. Disconnect the ECT sensor and sender gauge connectors.

 c. Disconnect the engine wire protector from the cylinder head.

16. Set the No. 1 cylinder at TDC of the compression stroke.

a. Turn the crankshaft pulley and align its groove with the timing mark **0** of the No. 1 timing belt cover.

b. Check that the timing marks of the camshaft timing pulleys and the No. 3 timing belt cover are aligned. If not, turn the crankshaft pulley one revolution (360°).

➡**If re-using the timing belt, be sure that you can still read the installation marks. If not, place new installation marks on the timing belt to match the timing marks of the camshaft timing pulleys.**

17. Remove the timing belt tensioner by alternately loosening the two bolts.

18. Remove the timing belt.

19. Remove the camshaft timing pulleys.

a. Using Variable Pin Wrench Set 09960–10010 or equivalent, remove the pulley bolt, the timing pulley and the knock pin. Remove the two timing pulleys.

20. Remove the bolt and the No. 2 idler pulley.

21. Remove the camshaft position sensor.

22. Remove the timing belt cover.

23. Remove the eight bolts, seal washers, cylinder head cover and gasket. Remove both cylinder head covers.

24. Remove the semi circular plugs.

25. Remove the right exhaust camshafts as follows:

a. Bring the service bolt hole of the driven sub-gear upward by turning the hexagon head portion of the exhaust camshaft with a wrench.

b. Align the timing mark (2 dot marks) of the camshaft drive and driven gears by turning the camshaft with a wrench.

c. Secure the exhaust camshaft sub-gear to the driven gear with a service bolt (6mm diameter, 16–20mm bolt length and 1.0mm in thread pitch).

➡**When removing the camshaft, be sure that the torsional spring force of the sub-gear has been eliminated by the above operation.**

d. Uniformly loosen and remove the bearing cap bolts in several passes, in the sequence shown.

e. Remove the bearing caps and camshaft. Make a note of the bearing cap positions for proper installation.

➡**Do not pry on or attempt to force the camshaft with a tool or other object.**

26. Remove the right-hand intake camshaft as follows:

a. Uniformly loosen and remove the bearing cap bolts in several passes, in the sequence shown.

b. Remove the bearing caps, oil seal and camshaft. Make a note of the bearing cap positions for proper installation.

27. Remove the left exhaust camshafts as follows:

a. Align the timing mark (1 dot mark) of the camshaft drive and driven gears by turning the camshaft with a wrench.

b. Secure the exhaust camshaft sub-gear to the driven gear with a service bolt (6mm diameter, 16–20mm bolt length and 1.0mm in thread pitch).

➡**When removing the camshaft, be sure that the torsional spring force of the sub-gear has been eliminated by the above operation.**

c. Uniformly loosen and remove the bearing cap bolts in several passes, in the sequence shown.

d. Remove the bearing caps and camshaft. Make a note of the bearing cap positions for proper installation.

➡**Do not pry on or attempt to force the camshaft with a tool or other object.**

28. Remove the left-hand intake camshaft as follows:

a. Uniformly loosen and remove the bearing cap bolts in several passes, in the sequence shown.

b. Remove the bearing caps, oil seal and camshaft. Make a note of the bearing cap positions for proper installation.

29. Remove the valve lifters and shims from the cylinder head. Arrange the valve lifters and shims in correct order.

To install:

30. Install the valve lifters and shims. Check that the valve lifter rotates smoothly by hand.

31. Install the right intake camshaft as follows:

a. Apply engine oil to the thrust portion of the intake camshaft.

b. Position the intake camshaft at 90° angle of the timing mark (2 dot marks) on the cylinder head.

c. Install the bearing caps in their proper locations. Apply a light coat of engine oil to the threads and install the cap bolts.

d. Apply a light coat of engine oil on

the threads and under the heads of the bearing cap bolts.

e. Uniformly tighten the cap bolts in the sequence shown to 12 ft. lbs. (16 Nm).

32. Install the right exhaust camshaft:

a. Apply engine oil to the thrust portion of the intake camshaft.

b. Align the timing marks (2 dot marks) of the camshaft drive and driven gears.

c. Roll down the exhaust camshaft onto the bearing journals while engaging the gears with each other. Install the bearing caps in their proper locations.

d. Apply a light coat of engine oil to the threads and install the cap bolts.

e. Apply a light coat of engine oil on the threads and under the heads of the bearing cap bolts.

f. Uniformly tighten the cap bolts in the sequence shown to 12 ft. lbs. (16 Nm).

g. Remove the service bolt from the driven sub-gear. Check that the intake and exhaust camshafts turns smoothly.

h. Align the timing marks (2 dot mark) of the camshaft drive and driven gears by turning the camshaft with a wrench.

33. Install the left intake camshaft as follows:

a. Apply engine oil to the thrust portion of the intake camshaft.

b. Position the intake camshaft at 90° angle of the timing mark (1 dot marks) on the cylinder head.

c. Install the bearing caps in their proper locations. Apply a light coat of engine oil to the threads and install the cap bolts.

d. Apply a light coat of engine oil on the threads and under the heads of the bearing cap bolts.

e. Uniformly tighten the cap bolts in the sequence shown to 12 ft. lbs. (16 Nm).

34. Install the left exhaust camshaft:

a. Apply engine oil to the thrust portion of the intake camshaft.

b. Align the timing marks (1 dot marks) of the camshaft drive and driven gears.

c. Roll down the exhaust camshaft onto the bearing journals while engaging the gears with each other. Install the bearing caps in their proper locations.

d. Apply a light coat of engine oil to the threads and install the cap bolts.

e. Apply a light coat of engine oil on

the threads and under the heads of the bearing cap bolts.

f. Uniformly tighten the cap bolts in the sequence shown to 12 ft. lbs. (16 Nm).

g. Remove the service bolt.

35. Check and adjust the valve clearance.

36. Install the semi circular plugs.

37. Install the cylinder head covers. Uniformly tighten the bolts in several passes to 53 inch lbs. (6 Nm).

38. Install the No. 3 timing belt cover with the six bolts. Tighten the bolts to 80 inch lbs. (9 Nm).

39. Install the camshaft position sensor and tighten to 71 inch lbs. (8 Nm).

40. Install the No. 2 timing belt idler with the bolt. Tighten the bolt to 30 ft. lbs. (40 Nm). Check that the pulley bracket moves smoothly.

41. Install the left camshaft timing pulley.

a. Install the knock pin to the camshaft.

b. Align the knock pin hose of the camshaft with the knock pin groove of the timing pulley.

c. Slide the timing pulley on the camshaft with the flange side facing outward. Tighten the pulley bolt to 81 ft. lbs. (110 Nm).

42. Set the No. 1 cylinder to TDC of the compression stroke.

a. Turn the crankshaft pulley, and align its groove with the timing mark **0** of the No. 1 timing belt cover.

b. Turn the camshaft, align the knock pin hole of the camshaft with the timing mark of the No. 3 timing belt cover.

c. Turn the camshaft timing pulley, align the timing marks of the camshaft timing pulley and the No. 3 timing belt cover.

43. Connect the timing belt to the left camshaft timing pulley. Check that the installation mark on the timing belt is aligned with the end of the No. 1 timing belt cover.

44. Install the right camshaft timing pulley and the timing belt.

45. Set the timing belt tensioner as follows:

a. Using a press, slowly press in the pushrod using 220–2,205 lbs. (981–9,807 N) of force.

b. align the holes of the pushrod and housing, pass a 1.5mm hexagon wrench through the holes to keep the setting position of the pushrod.

c. Release the press and install the dust boot to the tensioner.

46. Install the timing belt tensioner and alternately tighten the bolts to 20 ft. lbs. (28 Nm). Using pliers, remove the 1.5mm hexagon wrench from the belt tensioner.

47. Check the valve timing.

a. Slowly turn the crankshaft pulley two revolutions from the TDC to TDC. Always turn the crankshaft pulley clockwise.

b. Check that each pulley aligns with the timing marks. If the timing marks do not align, remove the timing belt and reinstall it.

48. Connect the engine wire to the intake manifold as follows:

a. Install the engine wire to the cylinder head by installing the three bolts.

b. Connect the three engine wire clamps.

c. Connect the following connectors:
- Six injector connectors
- ECT sender gauge connector
- ECT sensor connector

49. Install the intake air connector.

50. Install the air intake chamber assembly to the engine by installing the four bolts and two nuts. Tighten the bolts and nuts to 13 ft. lbs. (18 Nm).

51. Install the intake chamber stay as follows:

a. Install the intake chamber stay and install the two bolts. Tighten the bolts to 30 ft. lbs. (40 Nm)

b. Install a new O-ring to the oil filler tube.

c. Push in the oil filler tube end into the tube hole in the oil pan.

d. Install the oil filler tube and No. 1 throttle cable clamp and install the bolt and two nuts.

52. Install the No. 2 timing belt cover. Tighten the bolts to 80 inch lbs. (9 Nm). Install the remaining components.

53. Fill the radiator with engine coolant.

54. Connect the negative battery cable to the battery.

55. Start the engine and check for leaks.

56. Check the ignition timing.

57. Install the engine undercover.

58. Road test the vehicle.

59. Recheck all fluid levels.

Valve Clearance

ADJUSTMENT

2.2L (22R-E) Engine

1. Start the engine and warm it up to normal operating temperature.

2. Turn the engine **OFF**. Remove the air cleaner and housing, along with the hot air and cold air intake ducts.

> ❋❋ **CAUTION**
>
> **Components will be hot. The engine head, block and radiator will be very hot.**

3. Remove any other hoses, cables, or wires attached to the valve cover. The valve cover (or cylinder head cover) is the domed steel item with the oil filler in it.

4. Remove the small nuts holding the valve cover, then lift the cover off. Retrieve the rubber gasket and put it aside; it can be reused if not damaged or crushed out of shape. Beware of hot oil dripping from the inside of the cover.

5. Use a large wrench on the crankshaft pulley bolt to turn the engine clockwise until the timing mark on the pulley to **0** on the scale. Turning the engine will be easier if the spark plugs are removed, but this is not required.

> ❋❋ **CAUTION**
>
> **Do not attempt to align the engine by using the ignition switch to turn the engine. Doing will splash hot oil onto everything in the area, including you.**

6. Check that the rockers on No.1 cylinder are loose and the rockers on No.4 are under tension. (No.1 is closest to the radiator; No. 4 is closest to the firewall.) If this is true, the engine is aligned with No.1 piston at top dead center. If it is not true, turn the engine one full revolution clockwise and realign the timing mark at zero; recheck the rockers.

7. Adjust the clearance 0.008 in. (0.20mm) intake and 0.012 in. (0.30mm) exhaust. Insert a feeler gauge and check for proper clearance between the top of the valve stem and the bottom of the rocker

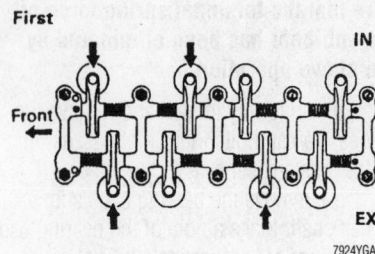

7924YGA7

Adjust the clearance of half the valves, do the arrowed ones first—2.4L (22R-E) engine

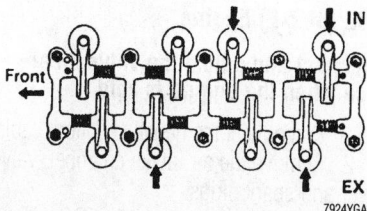

Second

Front

IN

EX

7924YGA8

Adjust the second set of valves the same as the first—2.4L (22R-E) engine

arm on the No. 1 intake valve. To adjust, loosen the locknut on the end of the rocker arm and turn the adjusting screw until the clearance is correct. Tighten the locknut and recheck the clearance; there should be a slight drag felt when the feeler gauge is pulled through the gap. Repeat the procedure for No 1 exhaust, No 2 intake and No. 3 exhaust.

8. Turn the crankshaft pulley one full rotation clockwise until the marks align at **0** and for the remaining valves.

9. Clean the valve cover thoroughly with a lint-free rag. Wipe any oil off the cylinder head edges in the area of the valve cover gasket.

10. Fit the gasket into the valve cover, making sure it is not crimped or twisted. If the half-moon rubber plugs came out of the valve cover, clean them and apply sealant to the part of the plug contacting the valve cover; install the half-moon plug.

➡**The use of sealants on the valve cover gasket is not recommended.**

11. Install the valve cover onto the head. Make certain is squarely seated and not pinching any adjacent wires or cables.

12. Install the valve cover retaining nuts. Tighten them to 43–60 inch lbs. (5–7 Nm) This is little more than finger-tight; over-tightening will deform the cover and cause leaks.

13. Connect the lines, hoses and cables which were removed for access. Make certain electrical and ignition wires are firmly held by their clips or brackets.

14. Install the air cleaner with the hoses and duct work.

15. If still in place, remove the wrench and socket from the crankshaft pulley.

2.4L (2RZ-FE) Engine

1. Disconnect the negative battery cable.
2. Remove the air intake connector.

3. Remove the PCV hoses.
4. Disconnect the spark plug wires.
5. Disconnect the four engine wire clamps and the engine wire.
6. Disconnect the following connectors:
   a. With air conditioning: air conditioning compressor connector
   b. Oil pressure sensor connector
   c. ECT sensor connector
   d. Distributor connector
7. Remove the cylinder head cover.
8. Set the No. 1 cylinder to TDC of the compression stroke.
   a. Turn the crankshaft pulley clockwise and align its groove with the O mark on the timing chain cover.
   b. Check that the timing marks (one and two dots) of the camshaft drive and driven gears are in a straight line on the cylinder head surface. If not, turn the crankshaft one revolution (360°) and align the marks.
9. Inspect the valve clearance.
   a. Measure the clearance between the valve lifter and the camshaft. Measure the first and second intake and the first and third exhaust valves.
   b. Turn the crankshaft pulley one revolution (360°) and align the marks as above. Measure the third and fourth

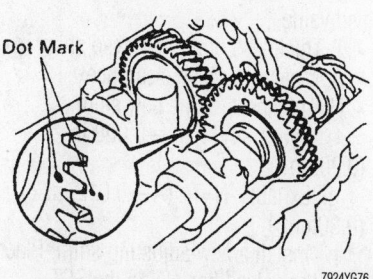

Dot Mark

7924YG76

Aligning the timing marks—Tacoma with 2.4L (2RZ-FE) and 2.7L (3RZ-FE) engine

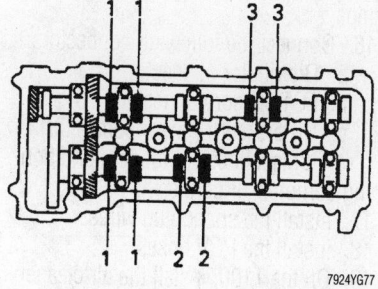

7924YG77

First valve adjustment—2.4L (2RZ-FE) and 2.7L (3RZ-FE) engines

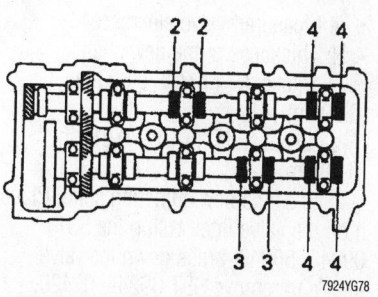

7924YG78

Second valve adjustment—2.4L (2RZ-FE) and 2.7L (3RZ-FE) engines

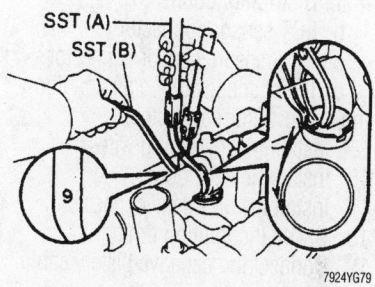

SST (A)

SST (B)

7924YG79

Removing adjusting shim using the special tools shown above—2.4L (2RZ-FE) and 2.7L (3RZ-FE) engines

intake and the second and fourth exhaust valves.

10. Valve clearance cold should be:
   • Intake: 0.006–0.010 in. (0.15–0.25mm)
   • Exhaust: 0.010–0.014 in. (0.25–0.35mm)

11. Adjust the valve clearance by using adjusting shims.
   a. Turn the equipment driveshaft so that the cam lobe for the valve to be adjusted faces up.
   b. Using SST 09248–55040 or equivalent, press down the valve lifter and place SST 09248–05420 or equivalent, between the camshaft and the valve lifter. Remove SST 09248–55040.
   c. Remove the adjusting shim with a small flat prying tool and a magnetic finger.
   d. Determine the replacement adjusting shim size according to the following formula, or use the adjusting shim charts.
   e. Using a micrometer, measure the thickness of the removed shim. Calculate the thickness of a new shim so that the valve clearance comes within the specified value.
   • T: Thickness of the removed shim

- A: Measured valve clearance
- N: Thickness of the new shim

  f. Intake: N=T+ (A—0.008 in. (0.20mm))

  g. Exhaust: N=T+ (A—0.012 in. (0.30mm))

  h. Install a new adjusting shim. Place it on the valve lifter. Using the SST 09248–55040, press down the valve lifter and remove SST 09248–05420.

  i. Recheck the valve clearance.

12. Reinstall the cylinder head cover.

13. Reconnect the engine wire and clamps.

14. Connect the following connectors:
  a. Distributor connector
  b. ECT sensor connector
  c. Oil pressure sensor connector
  d. If disconnected, the air conditioning compressor connector

15. Install the spark plug wires.

16. Install the PCV hoses.

17. Install the air intake connector.

18. Check the ignition timing.

19. Connect the negative battery cable.

2.7L Engine (3RZ-FE) Engine

TACOMA, T100 AND 1996–99 4RUNNER

1. Disconnect the negative battery cable.

2. Drain the engine coolant.

3. On the Tacoma and 4Runner, remove the intake air connector.

4. On the T100, remove the air cleaner cap, MAF meter, and the resonator.

5. Remove the PCV hoses.

6. Disconnect the spark plug wires.

7. Disconnect the engine wire clamps and the engine wire.

8. Disconnect the following connectors:
  a. With air conditioning: air conditioning compressor connector
  b. Oil pressure sensor connector
  c. ECT sensor connector
  d. Distributor connector (Tacoma and 4Runner)

9. Remove the cylinder head cover.

10. Set the No. 1 cylinder to TDC of the compression stroke.
  a. Turn the crankshaft pulley clockwise and align its groove with the **0** mark on the timing chain cover.
  b. Check that the timing marks (one and two dots) of the camshaft drive and driven gears are in a straight line on the cylinder head surface. If not, turn the crankshaft one revolution (360°) and align the marks.

11. Inspect the valve clearance.

  a. Measure the clearance between the valve lifter and the camshaft. Measure the first and second intake and the first and third exhaust valves.

  b. Turn the crankshaft pulley one revolution (360°) and align the marks as above. Measure the third and fourth intake and the second and fourth exhaust valves.

12. Valve clearance cold should be:
- Intake: 0.006–0.010 in. (0.15–0.25mm)
- Exhaust: 0.010–0.014 in. (0.25–0.35mm)

13. Adjust the valve clearance by using adjusting shims.

  a. Turn the camshaft so that the cam lobe for the valve to be adjusted faces up.

  b. Using SST 09248–55040 or equivalent, press down the valve lifter and place SST 09248–05420 or equivalent, between the camshaft and the valve lifter. Remove SST 09248–55040.

  c. Remove the adjusting shim with a small flat prying tool and a magnetic finger.

  d. Determine the replacement adjusting shim size according to the following formula, or use the adjusting shim charts.

  e. Using a micrometer, measure the thickness of the removed shim. Calculate the thickness of a new shim so that the valve clearance comes within the specified value.

- T: Thickness of the removed shim
- A: Measured valve clearance
- N: Thickness of the new shim

  f. Intake: N=T+ (A—0.008 in. (0.20mm))

  g. Exhaust: N=T+ (A—0.012 in. (0.30mm))

  h. Install a new adjusting shim. Place it on the valve lifter. Using the SST 09248–55040, press down the valve lifter and remove SST 09248–05420.

  i. Recheck the valve clearance.

14. Reinstall the cylinder head cover.

15. Reconnect the engine wire and clamps.

16. Connect the following connectors:
  a. Distributor connector
  b. ECT sensor connector
  c. Oil pressure sensor connector
  d. If disconnected, the air conditioning compressor connector

17. Install the spark plug wires.

18. Install the PCV hoses.

19. On the T100, install the air cleaner cap, MAF meter, and the resonator.

20. On the 4Runner and Tacoma, install the intake air connector.

21. Refill with engine coolant.

22. Check the ignition timing.

23. Connect the negative battery cable.

3.0L (3VZ-E) Engine

→**Inspect and adjust the valve clearance when the engine is cold.**

1. Disconnect the negative battery cable.

2. Remove the air intake chamber, valve cover and spark plugs.

3. Set the No. 1 cylinder to TDC/compression. Rotate the crankshaft until the groove on the crankshaft pulley align with "0" on the No. 1 timing belt cover. Check that the lifters on the No. 1 cylinder are loose and valve lifters on the No. 4 are tight.

4. Measure the clearance between the valve lifter and camshaft, using a thickness gauge. Record the clearance; it will be used later to determine the required shim. Valve clearance cold should be:
- Intake: 0.007–0.011 in. (0.18–0.28mm)
- Exhaust: 0.009–0.013 in. (0.22–0.32mm)

  a. Measure the clearance of No. 6 (intake) and No. 2 (exhaust) valves.

  b. Turn the crankshaft pulley ⅓ revolution (120 degrees). Measure the clearance of No. 1 (intake) and No. 3 (exhaust) valves.

  c. Turn the crankshaft pulley ⅓ revolution (120 degrees). Measure the clearance of No. 2 (intake) and No. 4 (exhaust) valves.

  d. Turn the crankshaft pulley ⅓ revolution (120 degrees). Measure the clearance of No. 3 (intake) and No. 5 (exhaust) valves.

  e. Turn the crankshaft pulley ⅓ revolution (120 degrees). Measure the clearance of No. 4 (intake) and No. 6 (exhaust) valves.

  f. Turn the crankshaft pulley ⅓ revolution (120 degrees). Measure the clearance of No. 5 (intake) and No. 1 (exhaust) valves.

5. To adjust:
  a. Turn the crankshaft to position the lobe of the camshaft on the adjusting valve upward.
  b. Using SST (A), press down the valve lifter and place SST (B) between the camshaft and valve lifter flange. Remove SST (A).
  c. Remove the adjusting shim using a small screw driver and magnet.

6. Determine the replacement shim size using the following formula:
  a. Measure the thickness of the shim removed, using a micrometer.
  b. Calculate the thickness of the

replacement shim by adding the measured clearance (recorded earlier), plus the thickness of the shim removed.

c. Select a new shim with a thickness as close as possible to the calculated value.

7. Install the new adjusting shim.

8. Install the spark plugs, valve cover and air intake chamber.

9. Connect the negative battery cable.

10. Check and adjust the ignition timing and idle speed, as required.

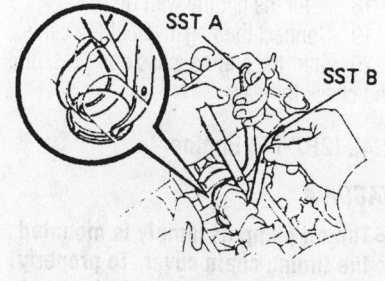

Remove adjusting shim using SST 09248–55020 (09248–05011, 09248–05021)—3.0L (3VZ-E) engine

3.4L (5VZ-FE) Engine

1. Disconnect the negative battery cable.
2. Drain the engine coolant.
3. Remove the air intake connector.
4. Remove the cylinder head cover.
5. Set the No. 1 cylinder to TDC of the compression stroke.

a. Turn the crankshaft pulley clockwise and align its groove with the **0** mark on the timing chain cover.

b. Check that the timing marks (one and two dots) of the camshaft drive and driven gears are in a straight line on the cylinder head surface. If not, turn the crankshaft one revolution (360°) and align the marks.

6. Inspect the valve clearance.

a. Measure the clearance between the valve lifter and the camshaft. Measure the first intake and the third exhaust valves on the right head and the sixth intake and the second exhaust valves on the left head.

b. Turn the crankshaft ⅔ of a revolution (240°) and adjust the third intake and the fifth exhaust valves on the right head and the second intake and the fourth exhaust valves on the left head.

c. Turn the crankshaft ⅔ of a revolution (240°) and adjust the fifth intake and the first exhaust valves on the right head

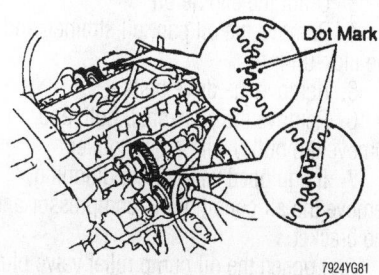

Aligning the timing marks—3.4L (5VZ-FE) engine

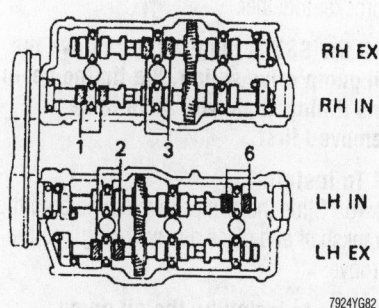

First valve adjustment—3.4L (5VZ-FE) engine

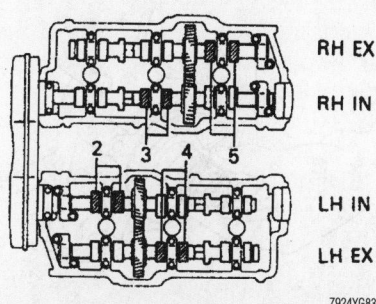

Second valve adjustment—3.4L (5VZ-FE) engine

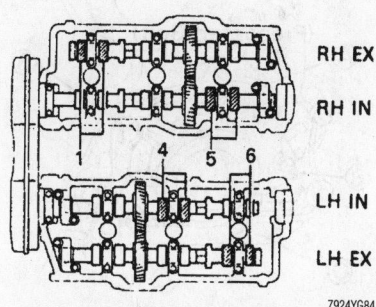

Third valve adjustment—3.4L (5VZ-FE) engine

and the fourth intake and the sixth exhaust valves on the left head.

7. Valve clearance cold should be:
- Intake: 0.006–0.009 in. (0.13–0.23mm)
- Exhaust: 0.011–0.014 in. (0.27–0.37mm)

8. Adjust the valve clearance by using adjusting shims.

a. Turn the equipment camshaft so that the cam lobe for the valve to be adjusted faces up.

b. Turn the valve lifter so that the notches are perpendicular to the camshaft.

c. Using SST 09248–55040 or equivalent, press down the valve lifter and place SST 09248–05420 or equivalent, between the camshaft and the valve lifter. Remove SST 09248–55040.

d. Remove the adjusting shim with a small flat prying tool and a magnetic finger.

e. Determine the replacement adjusting shim size according to the following formula, or use the adjusting shim charts.

f. Using a micrometer, measure the thickness of the removed shim. Calculate the thickness of a new shim so that the valve clearance comes within the specified value.

- T: Thickness of the removed shim
- A: Measured valve clearance
- N: Thickness of the new shim

g. Intake: N=T+ (A—0.007 in. (0.18mm))

h. Exhaust: N=T+ (A—0.013 in. (0.32mm))

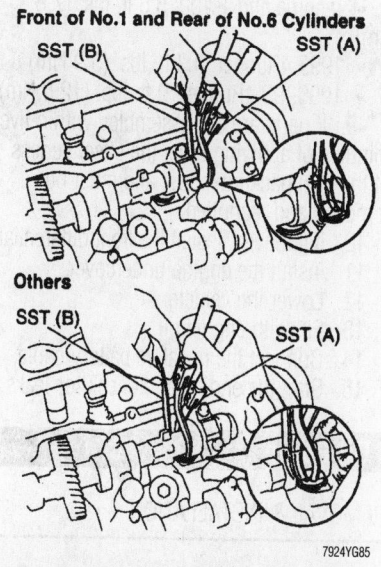

Front of No.1 and Rear of No.6 Cylinders

Others

Removing the adjusting shim—3.4L (5VZ-FE) engine

i. Install a new adjusting shim. Place it on the valve lifter. Using the SST 09248–55040, press down the valve lifter and remove SST 09248–05420.

j. Recheck the valve clearance.

9. Reinstall the cylinder head cover.

10. Install the intake air connector.

11. Refill with engine coolant.

12. Connect the negative battery cable.

13. Start the engine and check for leaks.

Oil Pan

REMOVAL & INSTALLATION

1. Disconnect the negative battery cable.

✳✳ CAUTION

Wait 90 seconds from the time the key is turned to LOCK and the negative battery cable is disconnected to begin work. This allows the SRS capacitor to discharge and prevent deployment of the air bag(s).

2. Raise and safely support the vehicle.

3. Drain the engine oil.

4. Remove the engine undercover.

5. If equipped with 4-Wheel Drive, remove the front differential.

6. Remove the oil pan by removing the bolts and nuts.

7. Using SST 09032–00100 or equivalent and a brass bar, separate the oil pan from the cylinder block.

To install:

8. Apply seal packing to the oil pan and install the pan to the cylinder block. Tighten the nuts and bolts to:

• Tacoma and T-100: 5.6 ft. lbs. (7.6 Nm)

• 1995 4Runner: 4.3 ft. lbs. (5.9 Nm)

• 1996–99 4Runner: 9 ft. lbs. (12.5 Nm)

9. If parts are not assembled within five minutes of applying time, the effectiveness of the seal packing is lost and must be removed and reapplied.

10. If removed, install the front differential.

11. Install the engine undercover.

12. Lower the vehicle.

13. Fill with engine oil.

14. Connect the negative battery cable.

15. Start the engine and check for leaks.

Oil Pump

REMOVAL & INSTALLATION

2.4L (22R-E) Engine

1. Disconnect the negative battery cable.

2. Raise and safely support the vehicle.

3. Drain the engine oil.

4. Remove the oil pan, oil strainer, and the pick-up tube.

5. Remove the drive belts.

6. Remove the crankshaft bolt, and remove the pulley with a gear puller.

7. If equipped with air conditioning, remove the air conditioning compressor and the bracket.

8. Loosen the oil pump relief valve plug and remove the oil pump assembly.

9. Remove the oil pump drive spline. If the drive spline cannot be removed by hand, use SST 09213–36020, or equivalent, and remove the spline and the crankshaft timing sprocket together.

➡**If the SST is required to remove the oil pump drive spline, the timing cover and timing chain will have to be removed first.**

To install:

10. Slide the pump drive spline onto the crankshaft and place a new O-ring into the groove.

➡**Prior to installing the oil pump, lubricate the pump drive and driven gears with clean engine oil.**

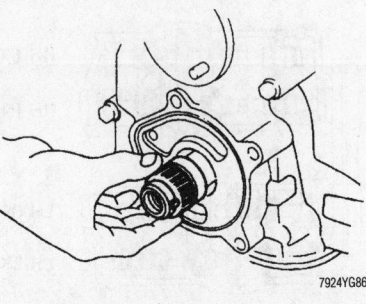

Be sure the oil pump drive splines are not damaged—1995 with 2.4L (22R-E) engine

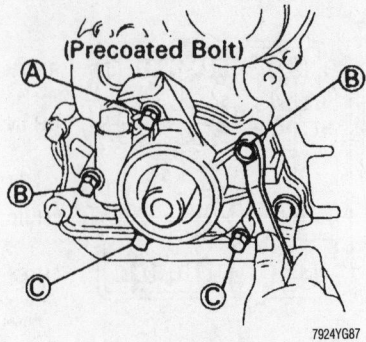

Oil pump bolt tightening diagram—1995 with 2.4L (22R-E) engine

11. Install the pump assembly. Tighten the bolts to:

a. A: 18 ft. lbs. (25 Nm)

b. B: 14 ft. lbs. (19 Nm)

c. C: 9 ft. lbs. (13 Nm)

12. Tighten the relief valve plug to 27 ft. lbs. (37 Nm).

13. Install the crankshaft pulley and tighten the bolt to 116 ft. lbs. (157 Nm).

14. If removed, install the air conditioning compressor and the bracket.

15. Install and adjust the drive belts.

16. Install the oil strainer.

17. Install the oil pan and tighten the nuts and bolts to 3.8 ft. lbs. (5.9 Nm).

18. Refill the engine with oil.

19. Connect the negative battery cable.

20. Start the engine, check oil pressure, and check for leaks.

2.4L (2RZ-FE) Engine

TACOMA

➡**The oil pump assembly is mounted in the timing chain cover. To properly service the oil pump, the timing chain cover should be removed from the cylinder block.**

1. Disconnect the negative battery cable.

2. Drain the oil and the cooling system.

3. Raise and safely support the vehicle.

4. Remove the engine undercover.

5. If equipped with 4-Wheel Drive, remove the front differential and halfshaft assembly.

6. Disconnect the upper radiator hose from the radiator.

7. Remove the oil dipstick guide by removing the bolt.

8. If equipped with power steering, remove the power steering drive belt by loosening the lockbolt and adjusting bolt.

9. Remove the No. 2 fan shroud by removing the two clips.

10. Remove the No. 1 fan shroud by removing the four bolts.

11. If equipped with air conditioning, loosen the idler pulley nut and adjusting bolt and remove the drive belt from the engine.

12. Remove the alternator drive belt, fan (with fan clutch), water pump pulley, and the fan shroud.

13. Remove the cylinder head from the engine.

14. If equipped with air conditioning, disconnect the air conditioning compressor and bracket.

15. Remove the alternator, adjusting bar and bracket.

16. Remove the crankshaft position sensor by removing the two bolts.

17. If equipped with 2-Wheel Drive, remove the stiffener plates by removing the eight bolts.

18. Remove the flywheel housing under-cover and dust seal.

19. Remove the oil pan by removing the 16 mounting bolts and 2 nuts.

➡**Be careful not to damage the flanges of the oil pan and cylinder block.**

20. Remove the two bolts, two nuts, oil strainer, and gasket.

21. Remove the crankshaft pulley.

22. Remove the timing chain cover.

23. Disassemble the oil pump from the front cover by removing the nine screws, pump cover, drive rotor, driven rotor and O-ring.

24. Remove the relief valve as follows:

a. Using snapring pliers, remove the snapring for the relief valve.

b. Remove the retainer, spring(s) and relief valve from the front cover.

To install:

25. Install the relief valve as follows:

a. Install the relief valve, spring(s) and retainer to the valve cover.

b. Using snapring pliers, install the snapring to hold the relief valve.

26. Install the drive and driven rotors as follows:

a. Place the drive and driven rotors into the pump body.

b. Place a new O-ring to the pump body.

c. Install the pump cover with the nine screws.

27. Install the remaining components in the reverse order of removal.

28. Lower the vehicle.

29. Fill the cooling system and fill the engine with oil.

30. Connect the negative battery cable.

31. Start the engine and check for leaks.

32. Adjust ignition timing. Road test the vehicle for proper operation.

33. Recheck all fluid levels.

2.7L (3RZ-FE) Engine

1996–99 4RUNNER

1. Disconnect the negative battery cable.

✸✸ CAUTION

Wait 90 seconds from the time the key is turned to LOCK and the nega-tive battery cable is disconnected to begin work. This allows the SRS capacitor to discharge and prevent deployment of the air bag(s).

2. Remove the cylinder head assembly.

3. Remove the water inlet and housing.

4. Remove the timing chain cover.

5. Remove the nine screws and separate the oil pump from the timing chain cover.

To install:

6. Install the oil pump assembly to the timing chain cover and tighten the nine screws.

7. Install the timing chain cover.

8. Install the two water bypass pipe mounting nuts and tighten to 14 ft. lbs. (20 Nm).

9. Install the cylinder head assembly.

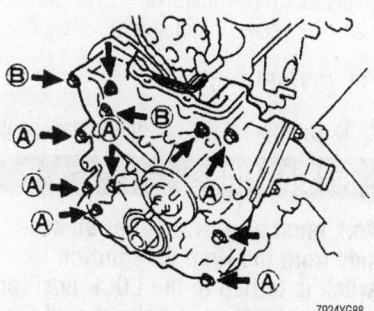

Timing cover bolt pattern—1996–99 4Runner and T-100 with 2.7L (3RZ-FE) engine

T-100

➡**The oil pump assembly is mounted in the timing chain cover. To properly service the oil pump, the timing chain cover should be removed from the cylinder block.**

1. Disconnect the negative battery cable.

2. Drain the oil and cooling system.

3. Remove the cylinder head from the engine.

4. Raise and safely support the vehicle.

5. Remove the engine undercover by removing the four bolts.

6. If equipped with air conditioning, loosen the idler pulley nut and adjusting bolt and remove the drive belt from the engine.

7. Remove the alternator drive belt, fan (with fan clutch), water pump pulley, and fan shroud.

8. If equipped with air conditioning, remove the air conditioning compressor and bracket. Set the compressor aside with the lines attached.

9. Remove the alternator, adjusting bar, and bracket.

10. Remove the crankshaft position sen-sor by removing the two bolts.

11. Remove the oil pan by removing the 16 mounting bolts and 2 nuts.

➡**Be careful not to damage the flanges of the oil pan and cylinder block.**

12. Remove the two bolts, two nuts, oil strainer and gasket.

13. Remove the crankshaft pulley.

14. Remove the timing chain cover.

15. Disassemble the oil pump from the front cover by removing the nine screws, pump cover, drive rotor, driven rotor and O-ring.

16. Remove the relief valve as follows:

a. Using snapring pliers, remove the snapring for the relief valve.

b. Remove the retainer, spring and relief valve from the front cover.

To install:

17. Install the relief valve as follows:

a. Install the relief valve, spring and retainer to the valve cover.

b. Using a snapring pliers, install the snapring to hold the relief valve.

18. Install the drive and driven rotors as follows:

a. Place the drive and driven rotors into the pump body.

b. Place a new O-ring to the pump body.

c. Install the pump cover with the nine screws.

19. Install the timing chain cover.

20. Install the 2 rear timing chain cover mounting bolts and water bypass pipe mounting nuts. Tighten to 13 ft. lbs. (18 Nm).

21. Install the remaining components in the reverse order of the removal sequence.

22. Install the cylinder head.

23. Lower the vehicle.

24. Fill the cooling system. Fill the engine with oil.

25. Connect the negative battery cable.

26. Start the engine and check for leaks.

27. Adjust ignition timing. Road test the vehicle for proper operation.

28. Recheck all fluid levels.

TACOMA

➡**The oil pump assembly is mounted in the timing chain cover. To properly service the oil pump, the timing chain cover should be removed from the cylinder block.**

1. Disconnect the negative battery cable.

2. Drain the oil and the cooling system.

3. Raise and safely support the vehicle.

4. Remove the engine undercover.

5. If equipped with 4-Wheel Drive, remove the front differential and halfshaft assembly.

6. For the California vehicles with 3RZ-FE engine, remove the two bolts and disconnect the air pipe.

7. Disconnect the upper radiator hose from the radiator.

8. Remove the oil dipstick guide by removing the bolt.

9. If equipped with power steering, remove the power steering drive belt by loosening the lockbolt and adjusting bolt.

10. Remove the No. 2 fan shroud by removing the two clips.

11. Remove the No. 1 fan shroud by removing the four bolts.

12. If equipped with air conditioning, loosen the idler pulley nut and adjusting bolt and remove the drive belt from the engine.

13. Remove the alternator drive belt, fan (with fan clutch), water pump pulley, and the fan shroud.

14. Remove the cylinder head from the engine.

15. If equipped with air conditioning, remove the air conditioning compressor and bracket. Set the compressor aside with the lines attached.

16. Remove the alternator, adjusting bar and bracket.

17. Remove the crankshaft position sensor by removing the two bolts.

18. If equipped with 2-Wheel Drive, remove the stiffener plates by removing the eight bolts.

19. Remove the flywheel housing undercover and dust seal.

20. Remove the oil pan by removing the 16 mounting bolts and 2 nuts.

➡**Be careful not to damage the flanges of the oil pan and cylinder block.**

21. Remove the two bolts, two nuts, oil strainer, and gasket.

22. Remove the crankshaft pulley.

23. Remove the timing chain cover.

24. Disassemble the oil pump from the front cover by removing the nine screws, pump cover, drive rotor, driven rotor and O-ring.

25. Remove the relief valve as follows:

　a. Using snapring pliers, remove the snapring for the relief valve.

　b. Remove the retainer, spring(s) and relief valve from the front cover.

To install:

26. Install the relief valve as follows:

　a. Install the relief valve, spring(s) and retainer to the valve cover.

　b. Using snapring pliers, install the snapring to hold the relief valve.

27. Install the drive and driven rotors as follows:

　a. Place the drive and driven rotors into the pump body.

　b. Place a new O-ring to the pump body.

　c. Install the pump cover with the nine screws.

28. Install the remaining components in the reverse order of removal.

29. Fill the cooling system and fill the engine with oil.

30. Connect the negative battery cable.

31. Start the engine and check for leaks.

32. Adjust ignition timing. Road test the vehicle for proper operation.

33. Recheck all fluid levels.

3.0L (3VZ-E) Engine

1. Disconnect the negative battery cable.

✱✱ CAUTION

Work must be started after 90 seconds from the time the ignition switch is turned to the LOCK position and the negative (-) battery cable is disconnected.

2. Remove the engine undercover.

3. If equipped with 4-Wheel Drive, remove the front differential.

4. Remove the timing belt and crankshaft pulley.

5. Remove the oil pan, strainer, and the baffle plate. Be careful not to damage the baffle plate flange.

6. Remove the oil pump body.

7. Remove the O-ring from the cylinder block.

To install:

8. Apply seal packing (PN 08826–00080 or equivalent) to the oil pump. Place a new O-ring into the groove of the cylinder block.

9. Install the oil pump to the crankshaft with the spline teeth of the drive rotor engaged with the large teeth of the crankshaft. Tighten the oil pump bolts to 14 ft. lbs. (20 Nm).

10. Apply seal packing to the baffle plate before installation.

11. Install the oil strainer and tighten the bolts to 5 ft. lbs. (6.9 Nm).

12. Apply seal packing to the oil pan before installation, tighten the nuts and bolts to 4.3 ft. lbs. (5.9 Nm).

13. Install the timing belt.

14. Install the crankshaft pulley and tighten the bolt to 181 ft. lbs. (245 Nm).

15. If removed, install the differential.

16. Install the engine undercover.

17. Fill with engine oil.

18. Connect the negative battery cable.

19. Start the engine and check for leaks.

3.4L (5VZ-FE) Engine

1. Disconnect the negative battery cable.

✱✱ CAUTION

Wait 90 seconds from the time the key is turned to LOCK and the negative battery cable is disconnected to begin work. This allows the SRS capacitor to discharge and prevent deployment of the air bag(s).

2. Remove the engine undercover.

3. Remove the crankshaft timing pulley.

4. If equipped with 4-Wheel Drive, remove the front differential.

5. Drain the engine oil from the engine.

6. Remove the timing belt and crankshaft gear.

7. If equipped with automatic transmission, remove the oil cooler tube and clamp.

8. Remove the stiffener plate.

9. Remove the flywheel housing undercover and dust cover.

10. Remove the rear end cover and dust cover.

11. Disconnect the starter wire clamp.

12. Remove the crankshaft position sensor.

13. Remove the oil pan by removing the 15 bolts and 4 nuts. Be careful not to damage the baffle plate flange.

14. Remove the oil strainer by removing the bolt and three nuts.

15. Remove the oil baffle plate by removing the nut and two bolts.

16. Remove the oil pump body by removing the eight bolts.

17. Remove the O-ring from the cylinder block.

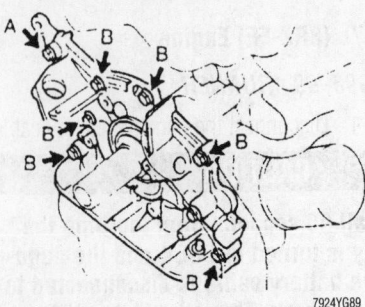

Oil pump bolt identification—4Runner, Tacoma and T-100 with 3.4L (5VZ-FE) engine

To install:

18. Apply seal packing (PN 08826–00080 or equivalent) to the oil pump. Place a new O-ring into the groove of the cylinder block.

19. Install the oil pump to the crankshaft with the spline teeth of the drive rotor engaged with the large teeth of the crankshaft. Tighten the oil pump bolts to:

- Bolt A to 15 ft. lbs. (20 Nm)
- Bolt B to 31 ft. lbs. (42 Nm)

20. Install the crankshaft position sensor.

21. Install the oil pan baffle plate.

22. Install the oil strainer with a new gasket and tighten the bolts to 13 ft. lbs. (18 Nm).

23. Install the remaining components in the reverse order of removal.

24. Fill with engine oil.

25. Connect the negative battery cable.

26. Start the engine and check for leaks.

FUEL SYSTEM

Fuel System Service Precautions

Safety is the most important factor when performing not only fuel system maintenance but any type of maintenance. Failure to conduct maintenance and repairs in a safe manner may result in serious personal injury or death. Maintenance and testing of the vehicle's fuel system components can be accomplished safely and effectively by adhering to the following rules and guidelines.

- To avoid the possibility of fire and personal injury, always disconnect the negative battery cable unless the repair or test procedure requires that battery voltage be applied.
- Always relieve the fuel system pressure prior to disconnecting any fuel system component (injector, fuel rail, pressure regulator, etc.), fitting or fuel line connection. Exercise extreme caution whenever relieving fuel system pressure to avoid exposing skin, face and eyes to fuel spray. Please be advised that fuel under pressure may penetrate the skin or any part of the body that it contacts.

- Always place a shop towel or cloth around the fitting or connection prior to loosening to absorb any excess fuel due to spillage. Ensure that all fuel spillage (should it occur) is quickly removed from engine surfaces. Ensure that all fuel soaked cloths or towels are deposited into a suitable waste container.

- Always keep a dry chemical (Class B) fire extinguisher near the work area.

- Do not allow fuel spray or fuel vapors to come into contact with a spark or open flame.

- Always use a back-up wrench when loosening and tightening fuel line connection fittings. This will prevent unnecessary stress and torsion to fuel line piping. Always follow the proper torque specifications.

- Always replace worn fuel fitting O-rings with new. Do not substitute fuel hose or equivalent, where fuel pipe is installed.

Fuel System Pressure

RELIEVING

❊❊ CAUTION

The fuel injection system remains under pressure after the engine has been turned OFF. Properly relieve fuel pressure before disconnecting any fuel lines. Failure to do so may result in fire or personal injury.

1. Disconnect the negative battery terminal.

2. Place a catch-pan under the joint to be disconnected. A large quantity of fuel may be released when the joint is opened.

3. Wear eye or full face protection.

4. Place a shop towel over the area and slowly loosen the joint using a wrench of the correct size. Use a back-up wrench if needed.

5. Allow the fuel left in the line to bleed off slowly before fully disconnecting the joint.

6. Plug the opened lines immediately to prevent fuel spillage or the entry of dirt.

7. Dispose of the released fuel properly.

8. After connecting fuel lines, connect the negative battery cable and start the engine.

9. Check for leaks and repair as needed.

Fuel Filter

REMOVAL & INSTALLATION

1. Relieve the fuel system pressure.

❊❊ CAUTION

The fuel injection system remains under pressure after the engine has been turned OFF. Properly relieve fuel pressure before disconnecting any fuel lines. Failure to do so may result in fire or personal injury.

2. Disconnect the negative battery cable.

➡ **The fuel filter is located in the engine compartment, at the inlet line to the fuel rail.**

3. Disconnect and plug the inlet and outlet lines from the filter.

4. Remove the fuel filter retaining bolts and remove the filter from the vehicle.

5. Remove the fuel filter bracket from the fuel filter.

To install:

6. Install the fuel filter bracket to the fuel filter.

7. Install the fuel filter and tighten the two bolts to 14 ft. lbs. (20 Nm).

8. Use two new gaskets and tighten the union bolts to 22 ft. lbs. (30 Nm).

9. Connect the negative battery cable.

10. Start the engine and check for leaks.

Fuel Pump

REMOVAL & INSTALLATION

1. Relieve the fuel pressure.

Exploded view of the fuel delivery components—2.4L (2RZ-FE) and 2.7L (3RZ-FE) engines

7924YG90

Reference (2WD)

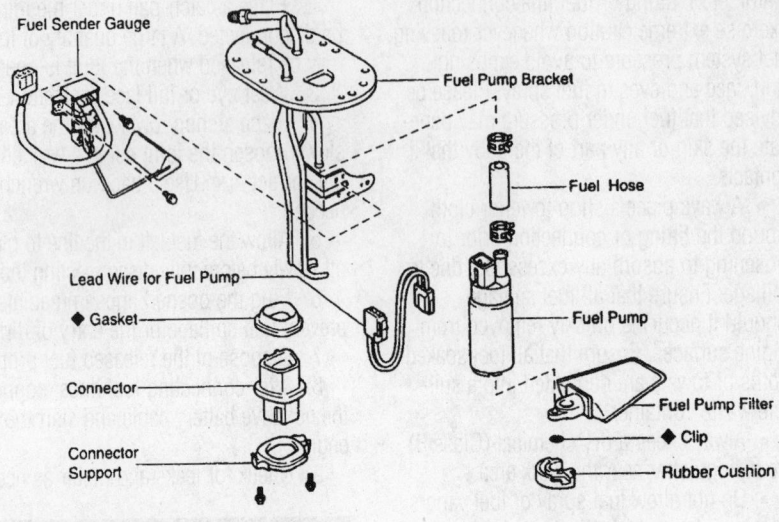

Exploded view of the fuel pump assembly and related components—Tacoma shown

---

✳✳ CAUTION

The fuel injection system remains under pressure after the engine has been turned OFF. Properly relieve fuel pressure before disconnecting any fuel lines. Failure to do so may result in fire or personal injury.

2. Disconnect the negative battery cable from the battery.

✳✳ CAUTION

Work must be started after 90 seconds from the time the ignition switch is turned to the LOCK position and the negative (-) battery cable is disconnected.

3. Drain the fuel from the fuel tank.

✳✳ CAUTION

Do not allow fuel spray or fuel vapors to come in contact with a spark or open flame. Keep a dry chemical fire extinguisher nearby. Never store fuel in an open container due to risk of fire or explosion.

4. Remove the fuel tank from the vehicle.

5. Disconnect the fuel pump connector from the clamp.

6. Remove the access plate bolts, then pull out the fuel pump assembly from the fuel tank.

7. Remove the gasket(s) from the pump bracket.

8. Disconnect the fuel pump connector.

9. Pull the bracket from the lower side of the fuel pump and remove the fuel pump from the fuel hose.

10. Remove the rubber cushion, the clip, and the fuel filter from the bottom of the fuel pump.

To install:

11. Install the fuel pump filter to the fuel pump with a new clip.

12. Install the fuel pump to the fuel pump bracket.

13. Connect the fuel hose to the outlet port of the fuel pump.

14. Connect the fuel pump connector.

15. Install the fuel pump assembly with a new gasket(s). Tighten the bolts to 31 inch lbs. (4 Nm).

16. Connect the fuel pump connector to the clamp.

17. Install the fuel tank and connect all electrical and fuel connections.

18. Connect the negative battery cable.

19. Refill the fuel tank and check for leaks.

DRIVE TRAIN

Manual Transmission Assembly

REMOVAL & INSTALLATION

1995 4Runner

2-WHEEL DRIVE MODELS

1. Disconnect the negative battery cable.

2. Remove the fan shroud bolts.

3. Remove the transmission shift lever from the inside of the vehicle.

   a. Remove the screws holding the shift lever boot retainer and remove the shift lever boot.

   b. Cover the shift lever cap with a cloth. Pressing down on the shift lever cap, rotate it counterclockwise to remove it.

4. Raise and safely support the vehicle. Drain the transmission fluid.

5. Disconnect the driveshaft.

6. Disconnect the back-up light switch and the vehicle speed sensor.

7. Remove the exhaust pipe clamp and the exhaust pipe.

8. Remove the clutch release cylinder. Do not disconnect the clutch line.

9. Remove the stabilizer bar bracket set bolts.

10. Remove the frame auxiliary cross-member.

11. Using a transmission jack, support the transmission.

12. Remove the four bolts from the engine rear mounting.

13. Remove the mounting bracket.

14. Remove the engine rear mounting from the transmission.

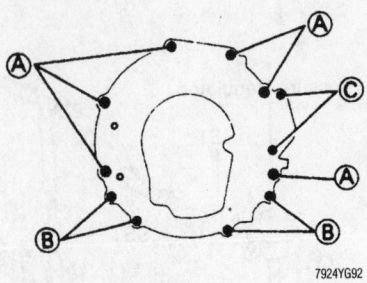

Transmission, stiffener and starter bolt locations—1995 4Runner

15. Place a piece of wood between the engine oil pan and the front crossmember.

16. Lower the transmission.

17. Disconnect the wiring at the starter. Remove the starter mounting bolts and lower the starter out of the vehicle.

18. Remove the transmission bolts, draw the transmission rearward and down, away from the engine.

To install:

19. Raise the transmission into position under the vehicle.

    a. Install the extension housing between the member and the floor, then slide the transmission forward. Align the input shaft spline with the clutch disc, and install the transmission to the engine.

20. Install and tighten the transmission, stiffener, and the starter bolts to:

    a. Transmission bolt: 53 ft. lbs. (72 Nm)

    b. Stiffener plate bolt: 27 ft. lbs. (37 Nm)

    c. Starter bolt: 29 ft. lbs. (39 Nm)

21. Install the engine rear mounting and the bracket. Tighten the rear mounting bolts to 19 ft. lbs. (25 Nm).

    a. Raise the transmission slightly by raising the engine with a jack and a wooden block under the transmission.

    b. Install the engine rear mounting bracket to the support member and tighten the four bolts to 43 ft. lbs. (59 Nm).

    c. Lower the transmission and rest it on the extension housing.

    d. Install the bracket to the mounting and tighten the four bolts to 22 ft. lbs. (29 Nm).

22. Remove the piece of wood from the front crossmember.

23. Install the exhaust pipe, bracket, and the clamp.

    a. Install the exhaust pipe and a new gasket to the exhaust manifold. Tighten the three new nuts to 46 ft. lbs. (62 Nm).

    b. Connect the exhaust pipe to the catalytic converter front side with a new gasket. Tighten the two bolts to 29 ft. lbs. (39 Nm).

    c. Install the pipe bracket to the clutch housing and tighten the two bolts to 29 ft. lbs. (39 Nm).

    d. Install and tighten the exhaust pipe clamp set bolt to 14 ft. lbs. (19 Nm).

24. Install the clutch release cylinder and tighten the two bolts to 9 ft. lbs. (12 Nm).

25. Install the stabilizer bar bracket set bolts and tighten to 22 ft. lbs. (29 Nm).

26. Install the frame auxiliary crossmember and tighten the bolts to 70 ft. lbs. (95 Nm).

27. Connect the vehicle speed sensor and the back-up light switch connector.

28. Connect the driveshaft.

29. Lower the vehicle.

30. Refill the transmission to the correct level.

31. Install the shift lever.

    a. Apply MP grease to the shift lever.

    b. Align the groove of the shift lever cap and the pin part of the case cover.

    c. Cover the shift lever cap with a cloth. Pressing down on the shift lever cap, rotate it clockwise to install.

    d. Install the shift lever boot and retainer with the screws.

32. Install the fan shroud set bolts.

33. Connect the negative battery cable. Start the engine and check for leaks.

34. Road test the vehicle for proper operation. Recheck all fluid levels.

4-WHEEL DRIVE MODELS

1. Disconnect the negative battery cable.

2. Remove fan shroud set bolts.

3. Remove the heater hose clamp.

4. Remove the transmission shift lever from the inside of the vehicle.

    a. Remove the screws holding the shift lever boot retainer and remove the shift lever boot.

    b. Cover the shift lever cap with a cloth. Pressing down on the shift lever cap, rotate it counterclockwise to remove it.

5. Remove the transfer shift lever from the inside of the vehicle.

    a. Using pliers, remove the snapring and pull out the shift lever from the transfer.

6. Raise the vehicle and drain the transmission and the transfer oil.

7. Remove the driveshaft dust cover sub-assembly.

8. Disconnect the driveshaft.

9. Disconnect the vehicle speed sensor, back-up light switch connector, and the transfer indicator switch connector.

10. Remove the exhaust pipe bracket.

11. Remove the two bolts holding the clutch release cylinder and lay it along side the engine. Do not disconnect the clutch line.

12. Remove the front differential assembly.

13. Remove the stabilizer bar bracket set bolts.

14. Remove the No. 2 frame crossmember from the side frame.

    a. Remove the four bolts from the engine rear mounting.

    b. Using a transmission jack, support the transmission.

    c. Remove the eight bolts and the No. 2 frame crossmember from the side frame.

15. Lower the transmission with the transfer case.

16. Remove the starter.

17. Remove the exhaust pipe bracket and the stiffener plate bolts.

18. Remove the remaining transmission bolts.

19. Remove the transmission with the transfer toward the rear of the vehicle.

20. Remove the four bolts and the engine rear mounting from the transfer.

21. For the regular cab with the planetary gear type transfer, remove the dynamic damper.

22. Remove the driveshaft upper dust cover and the transfer from the transmission.

    a. Remove the dust cover bolt from the bracket.

    b. Remove the transfer adapter rear mounting bolts.

    c. Pull the transfer straight up and remove it from the transmission.

➡**Take care not to damage the adapter rear oil seal with the transfer input gear spline.**

To install:

23. Install the transfer and driveshaft upper dust cover to the transmission with a new gasket.

    a. Shift the two shift fork shafts to the high-four position.

    b. Apply MP grease to the adapter oil seal.

    c. Place a new gasket to the transfer adapter. Install the transfer to the transmission. Be careful not to damage the oil seal by the input gear spline when installing the transfer.

    d. Install and tighten the bolts with the driveshaft upper dust cover to 27 ft. lbs. (37 Nm).

    e. Install the dust cover bolt to the bracket. Tighten to 17 ft. lbs. (23 Nm).

24. Install the engine rear mounting and tighten the four bolts to 19 ft. lbs. (25 Nm).

25. If removed, install the dynamic damper and tighten to 27 ft. lbs. (37 Nm).

26. Place the transmission with the transfer at the installation position.

    a. Support the transmission with a jack. Align the input shaft spline with the clutch disc, and push the transmission with the transfer fully into position.

27. Install and tighten the transmission, stiffener and the starter bolts. Tighten to:

a. Transmission bolt: 53 ft. lbs. (72 Nm)

b. Stiffener plate bolt: 27 ft. lbs. (37 Nm)

c. Starter bolt: 29 ft. lbs. (39 Nm)

28. Install the No. 2 frame crossmember.

a. Raise the transmission slightly with a jack.

b. Install the No. 2 frame crossmember to the side frame and tighten the eight bolts to 70 ft. lbs. (95 Nm).

c. Lower the transmission and transfer.

d. Install the four mounting bolts to the engine rear mounting and tighten the bolts to 9 ft. lbs. (13 Nm).

29. Install the stabilizer bar bracket bolts and tighten to 22 ft. lbs. (29 Nm).

30. Install the front differential assembly.

a. Install and tighten the three bolts holding the differential carrier cover to the frame to 108 ft. lbs. (147 Nm) and the others to 123 ft. lbs. (167 Nm).

31. Install a new gasket and the exhaust pipe to the exhaust manifold. Tighten the three new nuts to 46 ft. lbs. (62 Nm).

32. Install the exhaust pipe bracket to the clutch housing and tighten the two bolts to 29 ft. lbs. (39 Nm).

33. Install the exhaust pipe clamp set bolt to 14 ft. lbs. (19 Nm).

34. Install the exhaust pipe and a new gasket to the catalytic converter front side. Tighten the two bolts to 29 ft. lbs. (39 Nm).

35. Install the clutch release cylinder and tighten the two bolts to 9 ft. lbs. (12 Nm).

36. Install the driveshaft dust cover sub-assembly.

a. Install the cover and tighten the bolt **A** to 27 ft. lbs. (36 Nm) and bolt **B** to 17 ft. lbs. (23 Nm).

37. Connect the vehicle speed sensor, the back-up light switch connector and the transfer indicator switch connector.

38. Connect the driveshaft.

39. Refill the transmission and transfer case to the correct level with the appropriate fluid.

40. Lower the vehicle.

41. Apply MP grease to the transfer shift lever and install the lever. Using pliers, install the snapring.

42. Install the transmission shift lever.

a. Apply MP grease to the transmission shift lever.

b. Align the groove of the shift lever cap and the pin par of the case cover. Cover the shift lever cap with a cloth. Pressing down on the shift lever cap, rotate it clockwise to install.

43. Install the heater hose clamp.

44. Install the fan shroud set bolts.

45. Connect the negative battery cable. Start the engine and check for leaks.

46. Road test the vehicle for proper operation. Recheck all fluid levels.

1996–99 4Runner

W59 AND R150 TRANSMISSIONS

1. Disconnect the negative battery cable from the battery.

2. Remove the transmission shift lever from the inside of the vehicle.

a. Remove the four screws and front console box.

b. Remove the screws holding the shift lever boot retainer and remove the shift lever boot.

c. Cover the shift lever cap with a cloth. Pressing down on the shift lever cap, rotate it counterclockwise to remove it.

3. On 4-wheel drive models, remove the transfer shift lever from the inside of the vehicle.

a. Using snapring pliers, remove the snapring and pull out the shift lever from the transfer case.

4. Raise the vehicle and drain the transmission and the transfer oil.

5. Remove the No. 1 and No. 2 engine undercover.

6. Disconnect the front and rear driveshafts.

7. Disconnect the vehicle speed sensor, back-up light switch, and the 4-Wheel Drive position switch connectors.

8. If equipped with ABS and/or Differential lock: Disconnect the L4 position switch connector.

9. Remove the two bolts holding the clutch release cylinder and lay it along side the engine. Do not disconnect the clutch line.

10. Remove the exhaust pipe bracket.

11. Remove the rear end plate by removing the nuts and two bolts.

12. Remove the crossmember.

a. Support the transmission rear side.

b. Remove the four bolts from the engine rear mounting.

c. Remove the four bolts, nuts and the crossmember.

13. Remove the four bolts and the engine rear mounting from the transfer case.

14. Using a transmission jack, support the transmission.

15. Disconnect the wiring and the connector and remove the starter.

16. Disconnect the wiring harness from the transmission.

17. Remove the transmission mounting bolts from the engine and lower the transmission with the transfer case (on 4WD models) down and to the rear.

18. If equipped, remove the transfer case from the transmission.

To install:

19. If applicable, install the transfer case to the transmission. Tighten the bolts to 17 ft. lbs. (24 Nm). Be careful not to damage the oil seal by the input gear spline when installing the transfer.

20. Place the transmission with the transfer case (4WD models) at the installation position.

21. Support the transmission with a jack. Align the input shaft spline with the clutch disc, and push the transmission with the transfer case fully into position.

22. Install the engine to transmission bolts and tighten the bolts to 53 ft. lbs. (72 Nm)

23. Install the starter by installing the two bolts and connecting the electrical wiring. Tighten the two bolts to 29 ft. lbs. (39 Nm).

24. Install the engine rear mounting and tighten the four bolts to 48 ft. lbs. (65 Nm).

25. Install the crossmember.

a. Raise the transmission slightly with a jack.

b. Install the crossmember and tighten the four bolts to 48 ft. lbs. (65 Nm).

c. Lower the transmission and transfer.

d. Install the four mounting bolts to the engine rear mounting and tighten the bolts to 14 ft. lbs. (19 Nm).

26. Install the rear end plate by installing the four bolts and nuts. Tighten the bolts to 27 ft. lbs. (37 Nm).

27. Install the front exhaust pipe. Tighten the bracket bolts to 52 ft. lbs. (71 Nm), the exhaust pipe-to-catalytic converter bolts to 35 ft. lbs. (48 Nm) and the support bracket bolts to 14 ft. lbs. (19 Nm).

28. Install the clutch release cylinder and tighten the two bolts to 9 ft. lbs. (12 Nm).

29. If disconnected, connect the L4 position switch connector.

30. Connect the vehicle speed sensor, back-up light switch, and the 4-Wheel Drive position switch connectors.

31. Connect the front and rear driveshafts.

32. Install the No. 1 and No. 2 engine under covers.

33. Refill the transmission to the correct level.

34. Apply MP grease to the transfer shift lever and install the lever. Using snapring pliers, install the snapring.

35. Install the transmission shift lever.

a. Apply MP grease to the transmission shift lever.

b. Align the groove of the shift lever cap and the pin par of the case cover. Cover the shift lever cap with a cloth. Pressing down on the shift lever cap, rotate it clockwise to install.

c. Install shift lever boot retainer with the four screws.

d. Install the front console box with the four screws.

36. Connect the negative battery cable, start the engine, and check for leaks.

37. Road test the vehicle for proper operation and recheck all the fluid levels.

T-100 with Model R150 and R150F Transmission

2-WHEEL DRIVE MODELS

➡ **The transmission is removed with the engine.**

1. Turn the ignition switch **OFF**. Disconnect the battery cables; negative cable first.

✳✳ WARNING

The air bag system is equipped with a back-up power source. To avoid possible air bag deployment, do not start working on the vehicle until 90 seconds has elapsed from the time the ignition switch is turned OFF and the negative battery terminal is disconnected.

2. Matchmark the hood hinges and remove the hood.

3. Remove the battery and battery tray.

4. Drain the engine oil, transmission oil and cooling system.

5. Remove the expansion tank.

6. Remove the radiator.

7. Remove the air cleaner cap, MAF meter and resonator. Remove the air cleaner case.

8. If equipped with a manual transaxle, disconnect the accelerator cable from the throttle body.

9. If equipped with a automatic transaxle, disconnect the accelerator and throttle cables from the throttle body.

10. Remove the intake air connector.

11. If equipped with air conditioning, remove the air conditioning compressor and bracket. Position the compressor aside with the A/C lines attached.

12. Disconnect the heater hoses at the cowl panel.

13. Disconnect the following hoses:
- Brake booster vacuum hose

- EVAP hose
- Two power steering hoses
- Fuel return hose
- Fuel inlet hose

14. Remove the power steering pump.

15. Disconnect the alternator wires from the alternator.

16. Disconnect the ECM wiring from the ECM.

17. Disconnect the engine wire and connectors from the vehicle as follows:

a. Disconnect the igniter connector.

b. Disconnect the ground strap from the cowl top panel.

c. Disconnect the four engine wire clamps.

d. Pull out the engine wire from the vehicle cabin.

18. If equipped with manual transmission, remove the shift lever assembly as follows:

a. Remove the shift lever knob.

b. Remove the four screws and shift lever boot.

c. Remove the six bolts, shift lever assembly and baffle.

19. Remove the sway bar.

20. Remove the driveshaft from the vehicle.

21. Disconnect the speedometer cable from the transmission.

22. Disconnect the front exhaust pipe from the exhaust manifold and catalytic converter.

23. If equipped with manual transmission, remove the clutch release cylinder.

24. If equipped with automatic transmission, remove the cross-shaft.

25. Disconnect the wires at the starter.

26. Position a jack and wooden block under the transmission and remove the rear engine mounting bracket.

27. Attach a suitable engine hoist to the engine hangers.

28. Remove the nuts and bolts from the engine mounts.

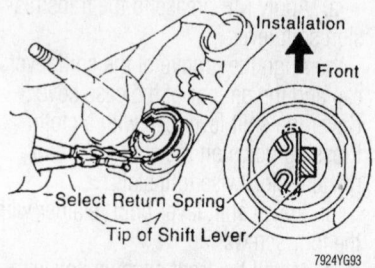

Removing the transfer shift lever— Tacoma, 4Runner and T-100 with R150 and R150F transmissions

➡ **Be sure the engine/transmission assembly is clear of all wiring and hoses.**

29. Carefully lift the engine/transmission assembly out of the vehicle.

30. Safely support the engine/transmission assembly.

31. Remove the rear end plate by removing the nuts and four bolts.

32. Remove the starter by removing the two bolts.

33. Remove the transmission from the engine by removing the six mounting bolts from the engine.

34. Pull out the transmission toward the rear.

35. Remove the transmission mount by removing the four bolts.

To install:

36. Install the transmission mount by installing the four bolts. Tighten the bolts to 18 ft. lbs. (25 Nm).

37. Connect the transmission to the engine by installing the six bolts. Tighten the bolts to 53 ft. lbs. (72 Nm).

38. Install the starter and tighten the two bolts to 29 ft. lbs. (39 Nm).

39. Install the rear end plate and tighten the four nuts and bolts to 27 ft. lbs. (37 Nm).

40. Attach the engine hoist to the engine hangers. Carefully lower the engine/transmission assembly into the vehicle. Keep the engine level, while aligning the engine mounts.

41. Install the engine mount fasteners, but do not fully tighten.

42. Position a jack and wooden block under the transmission and install the rear engine mounting bracket. Tighten the bolts to the frame to 42 ft. lbs. (58 Nm) and the bolts to the mount to 13 ft. lbs. (18 Nm).

43. Remove the jack and engine hoist. Tighten the engine mounts to 28 ft. lbs. (38 Nm).

44. Connect the starter wires to the starter.

45. If equipped with manual transmission, install the clutch release cylinder.

46. If equipped with automatic transmission, install the cross-shaft.

47. Install the front exhaust pipe to the exhaust manifold. Tighten the exhaust pipe-to-manifold bolts to 46 ft. lbs. (62 Nm), the support bracket bolts and the exhaust pipe-to-catalytic converter bolts to 29 ft. lbs. (39 Nm), and the exhaust pipe clamp nuts to 14 ft. lbs. (19 Nm).

48. Install the remaining components in the reverse order of removal.

49. Fill the engine oil, engine coolant, and transmission oil.

50. Connect the negative and positive cables to the battery.

51. Start the engine and check for leaks.

52. Install the engine undercover.

53. Install the hood.

54. Road test the vehicle and check all fluids.

4-WHEEL DRIVE MODELS

1. Disconnect the negative battery cable from the battery.

2. Remove the transmission shift lever from the inside of the vehicle.

   a. Remove the four screws and front console box.

   b. Remove the screws holding the shift lever boot retainer and remove the shift lever boot.

   c. Cover the shift lever cap with a cloth. Pressing down on the shift lever cap, rotate it counterclockwise to remove it.

3. Remove the transfer shift lever from the inside of the vehicle.

   a. Using snapring pliers, remove the snapring and pull out the shift lever from the transfer case.

4. Raise the vehicle and drain the transmission and the transfer oil.

5. Disconnect the driveshafts from the vehicle.

6. Disconnect the speedometer cable, back-up light switch connector and the transfer indicator switch connector.

7. Remove the two bolts holding the clutch release cylinder and lay it along side the engine. Do not disconnect the clutch line.

8. Remove the exhaust pipe bracket.

9. Remove the starter by disconnecting the starter wires and removing the two bolts.

10. Remove the stiffener plate by removing the four bolts.

11. Remove the rear end plate by removing the nuts and two bolts.

12. Remove the stabilizer bar from the suspension.

13. Using a transmission jack, support the transmission.

14. Remove the frame crossmember from the side frame as follows:

   a. Remove the four bolts from the engine rear mounting.

   b. Remove the eight bolts and the frame crossmember from the side frame.

15. Remove the six transmission mounting bolts from the engine.

16. Disconnect the three wire clamps from the transmission.

17. Remove the transmission with the transfer toward the rear of the vehicle.

18. Remove the four bolts and the engine rear mounting from the transfer.

19. Remove the transfer adapter rear mounting bolts.

20. Pull the transfer straight up and remove it from the transmission.

To install:

21. Apply MP grease to the adapter oil seal and shift the two shift fork shafts to the high 4 position.

22. Install the transfer to the transmission. Install the bolts and tighten the bolts to 27 ft. lbs. (37 Nm). Be careful not to damage the oil seal by the input gear spline when installing the transfer.

23. Install the engine rear mounting and tighten the four bolts to 18 ft. lbs. (25 Nm).

24. Place the transmission with the transfer at the installation position.

25. Support the transmission with a jack. Align the input shaft spline with the clutch disc, and push the transmission with the transfer fully into position.

26. Install the engine to transmission bolts. Tighten the bolts to 53 ft. lbs. (72 Nm)

27. Install the No. 2 frame crossmember.

   a. Raise the transmission slightly with a jack.

   b. Install the No. 2 frame crossmember to the side frame and tighten the eight bolts to 70 ft. lbs. (95 Nm).

   c. Lower the transmission and transfer.

   d. Install the four mounting bolts to the engine rear mounting and tighten the bolts to 9 ft. lbs. (13 Nm).

28. Install the stabilizer bar.

29. Install the rear end plate by installing the two bolts and nuts. Tighten to 27 ft. lbs. (37 Nm).

30. Install the stiffener plate with the four bolts. Tighten the bolts to 27 ft. lbs. (37 Nm).

31. Install the remaining components in the reverse order of removal.

32. Install the transmission shift lever.

   a. Apply MP grease to the transmission shift lever.

   b. Align the groove of the shift lever cap and the pin par of the case cover. Cover the shift lever cap with a cloth. Pressing down on the shift lever cap, rotate it clockwise to install.

   c. Install shift lever boot retainer with the four screws.

   d. Install the front console box with the four screws.

33. Connect the negative battery cable. Start the engine and check for leaks.

34. Road test the vehicle for proper operation. Recheck all fluid levels.

1995–99 T-100 and 2-Wheel Drive Tacoma

1. Remove the transmission with the engine.

2. Remove the left and right side stiffener plates.

3. Remove the rear end plate.

4. Disconnect the connector and remove the starter.

5. Place a stand under the transmission.

6. Remove the transmission mounting bolts and pull the transmission toward the rear.

7. Remove the rear engine mount by removing the four bolts.

To install:

8. Install the rear engine mount by installing the four bolts. Tighten the bolts to 48 ft. lbs. (65 Nm).

9. Install the transmission to the engine.

   a. Align the input shaft spline with the clutch disc and install the transmission to the engine. Tighten the three bolts to 53 ft. lbs. (72 Nm).

10. Install the starter and tighten the bolts to 29 ft. lbs. (39 Nm). Connect the connectors.

11. Install the rear end plate and tighten the bolts to 27 ft. lbs. (37 Nm).

12. Install the left and right side stiffener plates.

13. Install the transmission with the engine assembly.

4-Wheel Drive Tacoma

1. Disconnect the negative battery cable from the battery.

2. Remove the transmission shift lever from the inside of the vehicle.

   a. Remove the four screws and front console box.

   b. Remove the screws holding the shift lever boot retainer and remove the shift lever boot.

   c. Cover the shift lever cap with a cloth. Pressing down on the shift lever cap, rotate it counterclockwise to remove it.

3. Remove the transfer shift lever from the inside of the vehicle.

   a. Using snapring pliers, remove the snapring and pull out the shift lever from the transfer case.

4. Raise the vehicle and drain the transmission and the transfer oil.

5. Disconnect the front and rear drive-shafts.

6. Disconnect the speedometer cable and the back-up light switch connector.

7. On the Standard cab: disconnect the 4-Wheel Drive position switch connector.

8. On the Extra cab: Disconnect the L4 position switch connector.

9. Remove the two bolts holding the clutch release cylinder and lay it along side the engine. Do not disconnect the clutch line.

10. Remove the exhaust pipe bracket.

11. Remove the starter by disconnecting the starter wires and removing the two bolts.

12. Remove the rear end plate by removing the nuts and two bolts.

13. Remove the crossmember.

   a. Support the transmission rear side.

   b. Remove the four bolts from the engine rear mounting.

   c. Disconnect the O-ring and remove the four bolts, nuts and the crossmember.

14. Using a transmission jack, support the transmission.

15. Remove the six transmission mounting bolts from the engine.

16. Disconnect the three wire clamps from the transmission.

17. Remove the transmission with the transfer toward the rear of the vehicle.

18. Remove the four bolts and the engine rear mounting from the transfer.

19. Remove the transfer adapter rear mounting bolts.

20. Pull the transfer straight up and remove it from the transmission.

To install:

21. Apply MP grease to the adapter oil seal and shift the two shift fork shafts to the high 4 position.

22. Install the transfer to the transmission. Install the bolts and tighten the bolts to 17 ft. lbs. (24 Nm). Be careful not to damage the oil seal by the input gear spline when installing the transfer.

23. Place the transmission with the transfer at the installation position.

24. Support the transmission with a jack. Align the input shaft spline with the clutch disc, and push the transmission with the transfer fully into position.

25. Install the engine to transmission bolts. Tighten the bolts to 53 ft. lbs. (72 Nm)

26. Install the engine rear mounting and tighten the four bolts to 48 ft. lbs. (65 Nm).

27. Install the crossmember.

   a. Raise the transmission slightly with a jack.

   b. Install the crossmember and tighten the four bolts to 48 ft. lbs. (65 Nm).

   c. Lower the transmission and transfer.

   d. Install the four mounting bolts to the engine rear mounting and tighten the bolts to 14 ft. lbs. (19 Nm).

28. Install the rear end plate by installing the four bolts and nuts. Tighten to 13 ft. lbs. (18 Nm) on R150 and R150F transmissions, or to 27 ft. lbs. (37 Nm) on W59 transmissions.

29. Install the starter by installing the two bolts and connecting the electrical wiring. Tighten the two bolts to 29 ft. lbs. (39 Nm).

30. Install the front exhaust pipe. Tighten the exhaust pipe-to-manifold bolts to 46 ft. lbs. (62 Nm), the exhaust bracket bolts to 33 ft. lbs. (44 Nm) and the exhaust pipe-to-catalytic converter bolts to 35 ft. lbs. (48 Nm).

31. Install the clutch release cylinder and tighten the two bolts to 9 ft. lbs. (12 Nm).

32. Connect the L4 position switch connector on the extra cab and the 4-Wheel Drive position switch connector on the standard cab.

33. Connect the vehicle speed sensor and the back-up light switch connector.

34. Connect the front and rear drive-shafts.

35. Lower the vehicle.

36. Refill the transmission to the correct level.

37. Apply MP grease to the transfer shift lever and install the lever. Using pliers, install the snapring.

38. Install the transmission shift lever.

   a. Apply MP grease to the transmission shift lever.

   b. Align the groove of the shift lever cap and the pin par of the case cover. Cover the shift lever cap with a cloth. Pressing down on the shift lever cap, rotate it clockwise to install.

   c. Install shift lever boot retainer with the four screws.

   d. Install the front console box with the four screws.

39. Connect the negative battery cable. Start the engine and check for leaks.

40. Road test the vehicle for proper operation. Recheck all fluid levels.

Clutch Assembly

REMOVAL & INSTALLATION

1. Disconnect the negative battery cable from the battery.

2. Remove the transmission assembly from the vehicle.

3. Matchmark the clutch cover to the flywheel.

4. At the clutch cover, loosen each bolt one turn until spring tension is released.

5. Remove the set bolts to the clutch cover and pull off the clutch cover with the clutch disc.

6. Remove the retaining clip and withdraw the release bearing.

7. Remove the release fork and boot assembly.

To install:

8. Using a suitable clutch disc alignment tool, install the clutch disc onto the flywheel.

9. Position the clutch cover onto the flywheel and if reusing the old pressure plate, align the matchmarks.

10. Install the clutch cover retaining bolts. Tighten the bolts in a crisscross pattern to 14 ft. lbs. (19 Nm).

11. Lubricate the release fork pivot and contact points, the release bearing, bearing hub, and input shaft spline surfaces with a suitable molybdenum disulfide lithium based or multi-purpose grease.

➡**Be careful not to apply too much grease or else it will get on the clutch disc and cause it to slip or grab.**

12. Install the boot, release fork, hub, and the bearing assemblies.

13. Install the transmission to the vehicle.

14. Connect the negative battery cable to the battery.

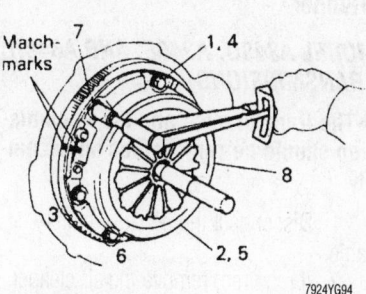

Bolt tightening sequence for the clutch cover—all engines

Hydraulic Clutch System

BLEEDING

➡**If any maintenance on the clutch system was performed or the system is suspected of containing air, bleed the system. Use care; brake fluid will remove the paint from any surface. If the brake fluid spills onto any painted surface, wash it off immediately with soap and water.**

1. Fill the clutch reservoir with brake fluid. Check the reservoir level frequently and add fluid as needed.
2. Connect one end of a vinyl tube to the bleeder plug on the slave cylinder and submerge the other end into a clear container half-filled with brake fluid.
3. Slowly pump the clutch pedal several times.
4. Have an assistant hold the clutch pedal down and loosen the bleeder plug until fluid and/or air starts to run out of the bleeder plug. Close the bleeder plug while the pedal is held to the floor.

➡**Do not allow the pedal to rise back up while the bleeder is still open. If this happens, it will allow air to enter the slave cylinder and cause the clutch system not to work properly.**

5. Repeat Steps 2 and 3 until all the air bubbles are removed from the system.
6. Tighten the bleeder plug when all the air is gone.
7. Refill the master cylinder to the proper level as required.
8. Check the system for leaks.

Automatic Transmission Assembly

REMOVAL & INSTALLATION

4Runner

MODEL A340D, A340E, AND A340H TRANSMISSIONS

➡**The transfer case and the transmission should be removed as an assembly.**

1. Disconnect the negative battery cable.
2. If required, remove the air cleaner assembly.
3. Disconnect the transmission throttle cable from the throttle body.
4. Raise and safely support the vehicle.
5. Remove the engine undercover.

6. Drain the transmission and transfer case (if applicable) fluid.
7. Disconnect the wiring connectors from the transmission and transfer case (if applicable).
8. Disconnect the starter wiring at the starter. Remove the mounting bolts and the starter from the engine.
9. Make matchmarks on the front and rear driveshaft flanges and the differential pinion flanges. These marks must be aligned during installation.
10. Unbolt the front and rear driveshaft flanges. If the vehicle has a two piece driveshaft, remove the center bearing bracket bolts. Remove the driveshaft from the vehicle.
11. Disconnect the speedometer cable.
12. Remove the front exhaust pipe and the bracket.
13. Disconnect the transmission oil cooler lines at the transmission.
14. Disconnect the oil cooler lines bracket and remove the transmission oil filler tube, as required.
15. Support the transmission, using a jack with a wooden block placed between the jack and the transmission pan. Raise the transmission, just enough to take the weight off of the rear mount.
16. Remove the rear engine mount with the bracket, the rear crossmember, and the transfer case (if applicable) undercover.
17. Remove the dynamic damper (Regular Cab only) and the No. 2 cross-shaft bracket.
18. Place a wooden block(s) between the engine oil pan and the front frame crossmember.
19. Slowly, lower the transmission until the engine rests on the wooden block(s).
20. Remove the torque converter cover at the rear of the engine in order to gain access to the converter bolts.
21. Rotate the crankshaft to access the bolts through the service holes and remove the torque converter bolts.
22. Remove the stiffener plates from the transmission.
23. Disconnect the shift control rod and the transfer case shift lever.
24. For the A340H transmissions, perform the following:
    a. Remove the cross-shaft and the No. 2 shifting rod.
    b. Remove the front stabilizer bar.
    c. Support the front differential with a jack and remove the differential mounting bolts.
    d. Slowly lower the front differential

so there is enough clearance to remove the transmission and transfer case (if applicable). If enough clearance can't be obtained, remove the differential from the vehicle.
25. Remove the stabilizer bar, if equipped, the auxiliary frame crossmember.
26. For A340D transmissions, obtain a bolt of the same dimensions as the torque converter bolts. Cut the head off of the bolt and hacksaw a slot in the bolt opposite the threaded end. Thread the guide pin into one of the torque converter bolt holes. The guide pin will help keep the converter with the transmission.

➡**This modified bolt is used as a guide pin. Two guides pins are needed to properly install the transmission.**

27. Remove the transmission bolts, then carefully move the transmission rearward by prying on the dowel pins through the service hole.
28. Pull the transmission rearward and lower it out of the vehicle.

➡**Be careful not to drop the torque converter.**

29. Separate the transfer case from the transmission
To install:
30. Connect the transfer case to the transmission.
31. Apply a coat of multi-purpose grease to the torque converter stub shaft and the corresponding pilot hole in the flexplate.
32. Install the torque converter into the front of the transmission. Push inward on the torque converter while rotating it to completely couple the torque converter to the transmission.
33. To be sure the converter is properly installed, measure the distance between the torque converter mounting lugs and the front mounting face of the transmission. The proper distance is 0.71 in. (18mm) for the A340H transmission, or 0.79 in. (20mm) for the A340D, A340E and A340F transmissions.
34. For A340D transmissions, install guide pins into 2 opposite mounting lugs of the torque converter.
35. Raise the transmission to the engine and align the transmission with the dowel pins.
36. Install and tighten the transmission mounting bolts to 47 ft. lbs. (63 Nm).
37. Rotate the crankshaft and install the torque converter mounting bolts. Evenly, tighten the converter mounting bolts to 30 ft. lbs. (41 Nm) for the A340H, A3430D and

A340E transmissions, or to 20 ft. lbs. (27 Nm) for the A340F transmission.

38. Install the torque converter access cover.

39. Raise the transmission slightly and remove the wood block(s) from under the engine oil pan.

40. Install the transmission crossmember. Tighten the crossmember bolts to 70 ft. lbs. (95 Nm).

41. Install the rear mount and the mounting bracket. Tighten the bracket mounting bolts to 43 ft. lbs. (58 Nm) and tighten the bracket to the rear mount bolts to 9 ft. lbs. (13 Nm).

42. Lower the transmission onto the crossmember and install the transmission to mount bolts. Tighten the bolts to 18 ft. lbs. (25 Nm).

43. Remove the wooden blocks from between the frame and the engine. Remove the support from under the transmission.

44. For the A340H transmission, install the front differential. Tighten the two rear mounting bolts to 123 ft. lbs. (167 Nm). Tighten the front mounting through-bolt to 108 ft. lbs. (147 Nm).

➡**If the differential oil was drained, refill it at this time.**

45. Connect the shift control rod and the transfer case shift lever.

46. Install the front stabilizer bar, if applicable.

47. Install the cross-shaft and the No. 2 shifting rod, if applicable.

48. Install the stiffener plates and tighten the bolts to 27 ft. lbs. (37 Nm).

49. If equipped, install the transfer case undercover and the dynamic damper. Tighten the dynamic damper mounting bolts to 27 ft. lbs. (37 Nm).

50. Install the No. 2 cross-shaft bracket.

51. Install the oil filler tube and the oil cooler pipe bracket.

52. Connect the oil cooler lines to the transmission and tighten the fittings to 25 ft. lbs. (34 Nm).

53. Install the front exhaust pipe and the support bracket.

54. Connect the speedometer cable.

55. Align the matchmarks of the front and rear driveshaft flanges and the differential pinion flanges and tighten the bolts to 54 ft. lbs. (74 Nm).

56. Install the starter and connect the wiring.

57. Connect the wiring connectors to the transmission and the transfer case (if applicable).

58. Install the engine undercover.

59. Lower the vehicle.

60. Install and adjust the transmission throttle cable.

61. If removed, install the air cleaner assembly.

62. Refill the transmission and the transfer case (if applicable).

63. Connect the negative battery cable.

64. Start the engine and check for leaks.

65. Road test the vehicle for proper operation.

66. Recheck all fluid levels.

MODEL A340F TRANSMISSION

1. Disconnect the negative battery cable.

2. Disconnect the throttle cable from the engine compartment.

3. Remove the ATF level gauge.

4. 3RZ-FE engine: Remove the oil filler pipe upper side bolt.

5. 5VZ-FE engine: Remove the oil filler pipe.

6. Remove the transmission shift lever assembly and transfer shift lever.

  a. Remove the rear console upper panel and disconnect the connectors.

  b. Pull off the heater control knobs.

  c. Remove the center cluster finish panel and disconnect the connectors.

  d. Without 2–4 selector: Remove the transfer shift lever knob.

  e. With 2–4 selector: Remove the bolt and disconnect the transfer shift lever knob.

  f. Remove the front console upper panel.

  g. If equipped, disconnect the 2–4 selector connector and remove the transfer shift lever knob.

  h. Remove the shift control rod.

  i. Disconnect the connector and remove the eight screws and the transmission shift lever assembly.

  j. Using pliers, remove the snapring and pull out the shift lever from the transfer.

7. Remove the engine undercover.

8. Remove the front and rear driveshafts.

9. Remove the exhaust pipe.

10. 3RZ-FE engine: Remove the oil filler pipe.

11. Disconnect the following connectors from the transmission:

- No. 2 vehicle speed sensor connector
- Solenoid connector
- ATF temperature sensor connector
- Park/neutral position switch connector

12. Disconnect the following connectors from the transfer:

- No. 1 vehicle speed sensor connector (3RZ-FE)

- Transfer neutral position switch connector
- Transfer L4 position switch connector
- Transfer 4-Wheel Drive position switch connector
- Actuator connector (w/ 2–4 selector only)

13. Separate the wiring harness from the transmission and the transfer.

14. Disconnect the two oil cooler pipes.

15. Remove the rear end plate and torque converter clutch mounting bolt.

16. Support the transmission with a jack stand and remove the engine rear mounting bolts.

17. Remove the four bolts and the crossmember.

18. Disconnect the starter wire and remove the starter.

19. Remove the transmission.

To install:

20. Install the transmission in the vehicle. Tighten the bolts to 53 ft. lbs. (71 Nm).

21. Install the starter and tighten the bolts to 29 ft. lbs. (39 Nm). Connect the wiring for the starter.

22. Install the crossmember and tighten the four bolts to 48 ft. lbs. (65 Nm). Install the four engine rear mounting bolts and tighten to 14 ft. lbs. (19 Nm).

23. Install the clutch converter bolts by installing the green colored bolt before the other five. Tighten them to 30 ft. lbs. (41 Nm).

24. Install the rear end plate and tighten the bolts to 13 ft. lbs. (18 Nm).

25. Install the two oil cooler pipes and tighten to 25 ft. lbs. (34 Nm).

26. Install the three bolts for the oil cooler pipe clamps and tighten to:

- 10mm head bolt: 4 ft. lbs. (5 Nm)
- 12mm head bolt: 9 ft. lbs. (12 Nm)

27. Connect the harness from the transmission and the transfer.

28. Install the remaining components in the reverse order of removal.

29. Fill the transmission and transfer case with transmission fluid.

30. Connect the throttle cable to the cable clamps in the engine compartment.

31. Connect the negative battery cable.

T-100

MODEL A340E TRANSMISSION WITH 2.7L (3RZ-FE) ENGINE

1. Turn the ignition switch **OFF**. Disconnect the battery cables; negative cable first.

The air bag system is equipped with a back-up power source. To avoid possible air bag deployment, do not start working on the vehicle until 90 seconds has elapse from the time the ignition switch is turned OFF and the negative battery terminal is disconnected.

2. Matchmark the hood hinges and remove the hood.

3. Remove the battery and battery tray.

4. Drain the engine oil, transmission oil, and the cooling system.

5. Remove the expansion tank.

6. Remove the radiator.

7. Remove the air cleaner cap, MAF meter, and the resonator. Remove the air cleaner case.

8. Disconnect the accelerator and throttle cables from the throttle body.

9. Remove the intake air connector.

10. If equipped with air conditioning, remove the air conditioning compressor and bracket. Position the compressor aside with the A/C lines attached.

11. Disconnect the heater hoses at the cowl panel.

12. Disconnect the following hoses:
- Brake booster vacuum hose
- EVAP hose
- Two power steering hoses
- Fuel return hose
- Fuel inlet hose

13. Remove the power steering pump.

14. Disconnect the alternator wires from the alternator.

15. Disconnect the ECM wiring from the ECM.

16. Disconnect the engine wire and connectors from the vehicle as follows:

a. Disconnect the igniter connector.

b. Disconnect the ground strap from the cowl top panel.

c. Disconnect the four engine wire clamps.

d. Pull out the engine wire from the vehicle cabin.

17. Remove the sway bar.

18. Remove the driveshaft from the vehicle.

19. Disconnect the speedometer cable from the transmission.

20. Disconnect the front exhaust pipe from the exhaust manifold and catalytic converter.

21. Remove the cross-shaft as follows:

a. Remove the clip and disconnect the No. 2 gear shifting rod.

b. Remove the nut, washer, four bolts, and the cross-shaft.

22. Disconnect the wires at the starter.

23. Position a jack and wooden block under the transmission and remove the rear engine mounting bracket.

24. Attach a suitable engine hoist to the engine hangers.

25. Remove the nuts and bolts from the engine mounts.

➡**Be sure the engine/transmission assembly is clear of all wiring and hoses.**

26. Be sure all connectors, wires, and hoses are disconnected and away from the engine and transmission.

27. Carefully lift the engine/transmission assembly out of the vehicle.

28. Remove the rear end plate by removing the two nuts and four bolts.

29. Turn the crankshaft to gain access to the torque converter bolts.

30. Remove the torque converter bolts.

31. Remove the starter by removing the two bolts.

32. Remove the three mounting bolts from the engine.

33. Pull out the transmission toward the rear.

To install:

34. Connect the engine to the transmission. Install the three bolts and tighten the bolts to 53 ft. lbs. (71 Nm).

35. Install the torque converter bolts to the torque converter. Tighten the bolts to 30 ft. lbs. (41 Nm).

36. Install the rear end plate and tighten the nuts and bolts to 27 ft. lbs. (37 Nm).

37. Connect the starter and bolts. Tighten the bolts to 29 ft. lbs. (39 Nm).

38. Attach the engine hoist to the engine hangers. Carefully lower the engine/transmission assembly into the vehicle. Keep the engine level, while aligning the engine mounts.

39. Install the engine mount fasteners, but do not fully tighten.

40. Position a jack and wooden block under the transmission and install the rear engine mounting bracket. Tighten the bolts to the frame to 42 ft. lbs. (58 Nm) and the bolts to the mount to 13 ft. lbs. (18 Nm).

41. Remove the jack and engine hoist. Tighten the engine mounts to 28 ft. lbs. (38 Nm).

42. Connect the starter wires to the starter.

43. Install the cross-shaft.

44. Install the remaining components in the reverse order of removal. Be sure to tighten the exhaust pipe-to-exhaust manifold nuts to 46 ft. lbs. (62 Nm), the exhaust pipe support bracket bolts to 29 ft. lbs. (39 Nm), the exhaust pipe clamp nuts to 14 ft. lbs. (19 Nm), the exhaust pipe-to-catalytic converter bolts to 29 ft. lbs. (39 Nm), the sway bar mounting bolts to 22 ft. lbs. (30 Nm), the power steering pump-to-bracket bolts to 43 ft. lbs. (58 Nm), the A/C compressor bracket-to-engine bolts to 32 ft. lbs. (44 Nm), and the A/C compressor-to-bracket bolts to 18 ft. lbs. (25 Nm).

45. Fill the engine oil, engine coolant, and the transmission oil.

46. Connect the negative and positive cables to the battery.

47. Start the engine and check for leaks.

48. Check the ignition timing.

49. Install the engine undercover.

50. Install the hood.

51. Road test the vehicle and check all fluids.

MODEL A340E TRANSMISSION WITH 3.4L (5VZ-FE) ENGINE

1. Remove the hood.

2. Disconnect the battery and remove it from the vehicle.

Work must be started after 90 seconds from the time the ignition switch is turned to the LOCK position and the negative (-) battery cable is disconnected.

3. Raise and safely support the vehicle.

4. Remove the engine under covers.

5. Drain the engine coolant.

6. Drain the engine oil.

7. Remove the radiator from the vehicle.

8. Remove the power steering (PS) drive belt.

9. If equipped with air conditioning, remove the air conditioning drive belt by loosening the idler pulley nut and adjusting bolt.

10. Loosen the lockbolt, pivot bolt, and the adjusting bolt. Remove the alternator drive belt.

11. Remove the fan with the fluid coupling and fan pulleys.

12. Disconnect the PS pump from the engine and set aside. Do not disconnect the lines from the pump.

13. If equipped with air conditioning, disconnect the compressor from the engine and set aside. Do not disconnect the lines from the compressor.

14. Remove the air cleaner cap, MAF meter, and the resonator.

15. Remove the air cleaner case and filter.

16. Disconnect the following cables:
- If equipped with cruise control, disconnect the actuator cable with the bracket.
- Accelerator cable
- Throttle cable

17. Disconnect the heater hoses.

18. Disconnect the following hoses:
- Brake booster vacuum hose
- EVAP hose
- Fuel return hose
- Fuel inlet hose

19. Disconnect the starter wire and connectors.

20. Disconnect the alternator connector and wire.

21. Disconnect the engine wire and connectors as follows:

a. Remove the four screws to the right front door scuff plate. Remove the scuff plate from the vehicle.

b. Remove the cowl panel side trim by removing the clip.

c. Disconnect the ECM electrical connectors.

d. Disconnect the two connectors from the cowl wire.

e. Disconnect the igniter connector.

f. Disconnect the ground strap.

g. Disconnect the six engine wire clamps.

h. Pull out the engine wire from the cabin.

22. Remove the stabilizer bar.

23. Remove the driveshaft from the transmission.

24. Disconnect the speedometer cable.

25. Remove the front exhaust pipe.

26. Remove the cross-shaft.

27. Place a jack under the transmission.

28. Remove the transmission rear mounting bracket by removing the eight bolts.

29. If equipped with air conditioning, remove the bolt and disconnect the air conditioning compressor wire clamp.

30. If necessary, install a No. 2 engine hanger with two bolts. Tighten the two bolts to 30 ft. lbs. (40 Nm).

31. Attach the engine hoist chain to the two engine hangers.

32. Remove the four bolts and nuts holding the engine front mounting insulators to the frame.

33. Lift the engine and transmission out of the vehicle.

34. Remove the starter by removing the two bolts.

35. Remove the transmission six mounting bolts from the engine.

36. If equipped, remove the rear end plate by removing the four bolts and four nuts.

37. Turn the crankshaft to gain access to the torque converter bolts.

38. Remove the torque converter bolts.

39. Pull out the transmission toward the rear.

40. Remove the three mounting bolt from the engine.

41. Pull out the transmission toward the rear.

To install:

42. Connect the engine to the transmission. Install the three bolts and tighten the bolts to 53 ft. lbs. (71 Nm).

43. Install the torque converter bolts to the torque converter. Tighten the bolts to 30 ft. lbs. (41 Nm).

44. Install the rear end plate and tighten the nuts and bolts to 27 ft. lbs. (37 Nm).

45. Connect the starter and bolts. Tighten the bolts to 29 ft. lbs. (39 Nm).

46. Install the transmission to the engine and install the six mounting bolts. Tighten the bolts to 53 ft. lbs. (71 Nm).

47. Install the starter and install the two bolts. Tighten the bolts to 29 ft. lbs. (39 Nm).

48. Install the engine assembly to the vehicle. Attach the engine mounts to the body mounts. Install the bolts and nuts but do not tighten at this time.

49. Remove the engine chain hoist the No. 2 engine hanger.

50. If equipped with air conditioning, connect the air conditioning wire with the bolt.

51. Raise the transmission slightly and install the transmission mounting bracket. Tighten the bolts to the frame to 43 ft. lbs. (58 Nm) and the bolts to the mounting insulator to 13 ft. lbs. (18 Nm).

52. Tighten the engine mounting nuts and bolts to 28 ft. lbs. (38 Nm).

53. Install the cross-shaft.

54. Install the remaining components in the reverse order of removal. Make certain to tighten the exhaust pipe-to-exhaust manifold bolts to 46 ft. lbs. (62 Nm), the exhaust pipe support bracket bolts to 33 ft. lbs. (44 Nm), the exhaust pipe-to-catalytic converter bolts to 35 ft. lbs. (48 Nm) and the cooling fan-to-fluid clutch nuts to 48 inch lbs. (5.4 Nm).

55. Fill the engine with oil and fill the transmission with fluid.

56. Fill the engine and radiator with coolant.

57. Install the engine undercover.

58. Install the battery and connect the cables.

59. Start the engine and check for leaks.

60. Install the hood.

MODEL A340F TRANSMISSION

1. Disconnect the negative battery cable.

2. Disconnect the transmission throttle cable and clamp from the throttle body.

3. Raise and safely support the vehicle.

4. Remove the engine undercover.

5. Drain the transmission fluid.

6. Remove the transfer shift lever front the inside of the vehicle as follows:

a. Remove the shift lever knob.

b. Remove the four screws and the boot.

c. Using pliers, remove the snapring, and pull out the shift lever from the transfer case.

7. Remove the transmission oil filler tube.

8. Remove the front and rear driveshafts.

9. Remove the front exhaust pipe.

10. Disconnect the speedometer cable.

11. Disconnect the No. 2 vehicle speed sensor connector.

12. Disconnect the solenoid connector by removing the electrical connector and bolt.

13. Disconnect the transfer case neutral position switch.

14. Disconnect the transfer case L4 position switch.

15. Remove the clip and disconnect the No. 2 gear shifting rod.

16. Remove the nut, four bolts, and the cross-shaft.

17. Disconnect the starter wires.

18. Remove the oil cooler pipe by removing the bolts and clamps.

19. Disconnect the ATF temperature sensor connector.

20. Disconnect the park/neutral position switch connector.

21. Remove the starter from the engine by removing the two bolts.

22. Remove the stiffener plate and rear end plate by removing the eight bolts.

23. Remove the sway bar.

24. Support the transmission, using a jack with a wooden block placed between the jack and the transmission pan. Raise the transmission just enough to take the weight off of the rear mount.

25. Remove the rear engine mounting bracket.

26. Remove the rear support member by removing the eight bolts.

27. Rotate the crankshaft to access the torque converter bolts. Remove the six bolts from the torque converter.

28. Disconnect or remove any component that will get in the way of removing the transmission.

29. Remove the transmission bolts, then carefully move the transmission rearward.

30. Pull the transmission rearward and lower it out of the vehicle.

To install:

31. Raise the transmission into place and install the bolts. Tighten the transmission bolts to 53 ft. lbs. (71 Nm).

32. Install the torque converter bolts and tighten the bolts to 30 ft. lbs. (41 Nm).

33. Install the rear support member and tighten the eight bolts to 70 ft. lbs. (97 Nm).

34. Install the rear mounting bracket and install the four bolts. Tighten the bolts to 13 ft. lbs. (18 Nm).

35. Install the dynamic damper with the two bolts. Tighten the bolts to 44 ft. lbs. (61 Nm).

36. Remove the jack supporting the transmission.

37. Install the sway bar.

38. Install the stiffener plate and rear end plate and tighten the bolts to 27 ft. lbs. (37 Nm).

39. Install the starter and tighten the bolts to 29 ft. lbs. (39 Nm).

40. Connect the park/neutral position switch connector.

41. Connect the ATF temperature sensor connector.

42. Install the oil cooler pipes and clamps. Tighten the two oil cooler pipes to 25 ft. lbs. (34 Nm).

43. Connect the starter wires.

44. Install the cross-shaft with the four bolts, washer, and nut. Tighten the nut to 9 ft. lbs. (13 Nm) and the bolts as follows:
- Transmission side: 9 ft. lbs. (13 Nm)
- Frame side: 21 ft. lbs. (28 Nm)

45. The balance of installation is the reverse of the removal procedure.

46. Fill the transmission with the proper fluid.

47. Connect the negative battery cable to the battery.

48. Road test the vehicle and check for leaks.

49. Check all fluids.

Tacoma

MODEL A340F TRANSMISSION

1. Remove the ATF level gauge.
2. Remove the engine undercover.
3. Drain the transmission fluid.
4. Disconnect the throttle cable.
5. Remove the No. 1 fan shroud.
6. Remove the transmission shift lever assembly and the transfer shift lever.

a. Remove the two bolts and the four screws and remove the rear console box.

b. Remove the front console box with the transfer shift lever knob.

c. Disconnect the connectors.

d. Remove the nut and washer and disconnect the shift control rod.

e. Disconnect the connector and remove the eight screws and the transmission shift lever assembly.

f. Using snapring pliers, remove the snapring and pull out the shift lever from the transfer.

7. Remove the oil filler pipe with the O-ring.

8. Remove the front and rear driveshaft.

9. Remove the exhaust pipe.

10. Disconnect the speedometer cable.

11. Disconnect the No. 2 vehicle speed sensor connector.

12. Disconnect the solenoid connector.

13. Disconnect the transfer neutral position switch connector.

14. Disconnect the transfer L4 position switch connector.

15. Disconnect the transfer indicator switch.

16. Disconnect the oil cooler pipe.

a. Remove the three bolts and clamps.

b. Loosen the two union nuts and disconnect the two oil cooler pipes.

17. Disconnect the ATF temperature sensor connector.

18. Disconnect the park/neutral position switch connector.

19. Disconnect the connector and remove the starter.

20. Remove the four stabilizer bar bracket mounting bolts.

21. Remove the torque converter clutch mounting bolt.

a. Remove the nuts and bolts and the flywheel housing undercover.

b. While turning the crankshaft to gain access, remove the six bolts.

22. Remove the front differential rear mounting cushion.

a. Using a hexagon wrench, remove the nut.

b. Lift up the front differential. Be careful not to touch the torque converter clutch housing and the front differential companion flange.

c. Remove the two rear mounting cushion mounting bolts.

23. Remove the crossmember.

a. Support the transmission rear side.

b. Remove the four engine rear mounting bolts.

c. Supporting the transmission with a jack, remove the four nuts, bolts and the crossmember.

24. Lower the transmission rear side, separate the wiring harness and remove the bolts and the transmission.

To install:

25. Install the transmission to the engine and install the transmission to engine bolts. Tighten the bolts to 53 ft. lbs. (71 Nm).

26. Install the crossmember and tighten the bolts to 48 ft. lbs. (65 Nm).

27. Install the engine rear mounting bolts and tighten to 14 ft. lbs. (19 Nm).

28. Install the front differential rear mounting cushion and tighten the nut to 64 ft. lbs. (41 Nm).

29. Install the torque converter clutch mounting bolt. First install the green colored bolt, then the five others. Tighten to 30 ft. lbs. (41 Nm).

30. Install the flywheel housing undercover and tighten to:
- 3.4L (5VZ-FE): 13 ft. lbs. (18 Nm)
- 2.7L (3RZ-FE): 27 ft. lbs. (37 Nm)

31. Install the four stabilizer bar bracket mounting bolts and tighten to 19 ft. lbs. (25 Nm).

32. Install the starter and tighten the bolts to 29 ft. lbs. (39 Nm). Connect the connector and terminal.

33. Install the remaining components in the reverse order of removal.

34. Install the ATF level gauge.

35. Fill and check the fluid level.

36. Test drive and check for proper shifting.

1995 MODEL A340E TRANSMISSION

1. Remove the hood.
2. Disconnect the battery and remove it from the vehicle.

✳✳ CAUTION

Work must be started after 90 seconds from the time the ignition switch is turned to the LOCK position and the negative battery cable is disconnected.

3. Raise and safely support the vehicle.
4. Remove the engine under covers.
5. Drain the engine coolant.
6. Drain the engine oil.
7. Remove the radiator from the vehicle.
8. Remove the power steering drive belt as follows:

a. Stretch the belt and loosen the fan pulley mounting nuts.

b. Loosen the lockbolt, pivot bolt and adjusting bolt and remove the drive belt from the engine.

9. If equipped with air conditioning, remove the air conditioning drive belt by

loosening the idle pulley nut and adjusting bolt.

10. Loosen the lockbolt, pivot bolt, and the adjusting bolt and the alternator drive belt.

11. Remove the fan with the fluid coupling and fan pulleys.

12. Disconnect the power steering pump from the engine and set aside. Do not disconnect the lines from the pump.

13. If equipped with air conditioning, disconnect the compressor from the engine and set aside. Do not disconnect the lines from the compressor.

14. Remove the air cleaner cap, MAF meter, and the resonator.

15. Remove the air cleaner case and filter.

16. Disconnect the following cables:
- If equipped with cruise control, disconnect the actuator cable with the bracket.
- Accelerator cable
- With automatic transmission, Throttle cable

17. Disconnect the heater hoses.

18. Disconnect the following hoses:
- Brake booster vacuum hose
- Automatic Disconnecting Differential (ADD) vacuum hose
- EVAP hose
- Fuel return hose
- Fuel inlet hose

19. Disconnect the starter wire and connectors as follows:

a. Remove the ground strap by removing the bolt.

b. Remove the nuts and disconnect the positive cable from the battery.

c. Disconnect the three starter wire clamps and connector.

20. Disconnect the alternator connector and wire.

21. Disconnect the engine wire and connectors as follows:

a. Remove the four screws to the right front door scuff plate. Remove the scuff plate from the vehicle.

b. Remove the cowl panel side trim by removing the clip.

c. Disconnect the ECM electrical connectors.

d. Disconnect the two connectors from the cowl wire.

e. Disconnect the igniter connector.

f. Disconnect the ground strap.

g. Disconnect the six engine wire clamps.

h. Pull out the engine wiring harness from the cabin.

22. Remove the stabilizer bar.

23. Remove the driveshaft from the transmission.

24. Disconnect the speedometer cable.

25. Remove the front exhaust pipe.

26. Remove the cross-shaft.

27. Place a jack under the transmission.

28. Remove the transmission rear mounting bracket by removing the eight bolts.

29. If equipped with air conditioning, remove the bolt and disconnect the air conditioning compressor wire clamp.

30. If necessary, install a No. 2 engine hanger with two bolts. Tighten the two bolts to 30 ft. lbs. (40 Nm).

31. Attach the engine hoist chain to the two engine hangers.

32. Remove the four bolts and nuts holding the engine front mounting insulators to the frame.

33. Lift the engine and transmission out of the vehicle.

34. Place the engine and transmission on a stand.

35. Remove the ATF level gauge.

36. Remove the transmission oil filler pipe.

37. Loosen the two oil cooler pipe union nuts.

38. Remove the three bolts, three clamps, and the two oil cooler pipes.

39. Disconnect the park/neutral position switch, solenoid connector and No. 2 vehicle speed sensor.

40. Separate the wiring harness from the transmission.

41. Remove the flywheel housing undercover by removing the four bolts.

42. Turn the crankshaft to gain access to the torque converter bolts. Remove the torque converter bolts.

43. Remove the starter mounting bolts and the starter.

44. Remove the six transmission mounting bolts from the engine.

45. Pull out the transmission toward the rear.

To install:

46. Install the transmission to the engine and install the transmission to engine bolts. Tighten the bolts to 53 ft. lbs. (71 Nm).

47. Install the starter with the two bolts. Tighten the bolts to 29 ft. lbs. (39 Nm).

48. Install the green colored torque converter bolt, then install the other five bolts. Tighten the torque converter bolts to 30 ft. lbs. (41 Nm).

49. Install the flywheel housing cover by installing the four bolts. Tighten the bolts to 13 ft. lbs. (18 Nm).

50. Install the wiring harness to the transmission.

51. Connect the No. 2 vehicle speed sensor, solenoid connector, and the park/neutral position switch.

52. Install the oil cooler pipes.

53. Install the oil filler pipe.

54. Install the ATF gauge.

55. Install the engine assembly to the vehicle. Attach the engine mounts to the body mounts. Install the bolts and nuts but do not tighten at this time.

56. Remove the engine chain hoist the No. 2 engine hanger.

57. If equipped with air conditioning, connect the air conditioning wire with the bolt.

58. Raise the transmission slightly and install the transmission mounting bracket. Tighten the bolts to the frame to 43 ft. lbs. (58 Nm) and the bolts to the mounting insulator to 13 ft. lbs. (18 Nm).

59. Tighten the engine mounting nuts and bolts to 28 ft. lbs. (38 Nm).

60. The balance of installation is the reverse of the removal procedure. Make certain to tighten the exhaust pipe-to-exhaust manifold bolts to 46 ft. lbs. (62 Nm), the exhaust pipe support bracket bolts to 33 ft. lbs. (44 Nm), the exhaust pipe-to-catalytic converter bolts to 35 ft. lbs. (48 Nm) and the cooling fan-to-fluid clutch nuts to 48 inch lbs. (5.4 Nm).

61. Fill the engine with oil.

62. Fill the engine and radiator with coolant.

63. Install the engine undercover.

64. Start the engine and check for leaks.

1996–99 MODEL A340D AND A340E TRANSMISSIONS

1. Remove the transmission with the engine.

2. Place the engine/transmission assembly on a stand.

3. Remove the bolts, two stiffener plates and rear end plate.

4. Turn the crankshaft to gain access to the torque converter bolts. Remove the torque converter bolts.

5. Remove the starter by removing the two bolts.

6. Remove the transmission mounting bolts from the engine.

7. Pull out the transmission toward the rear and separate the engine from the transmission.

To install:

8. Connect the transmission to the engine and install the bolts. Tighten the bolts to 53 ft. lbs. (71 Nm).

9. Install the starter with the two bolts. Tighten the bolts to 29 ft. lbs. (39 Nm).

10. Install the torque converter bolts and tighten to 30 ft. lbs. (41 Nm).

11. Install the stiffener plate and rear end

plate with the bolts. Tighten the bolts to 27 ft. lbs. (37 Nm).

12. Connect the starter wires to the starter.

13. Install the transmission with the engine.

Transfer Case Assembly

REMOVAL & INSTALLATION

1. Disconnect the negative battery cable.
2. Raise and safely support the vehicle.
3. Drain the transmission and the transfer case.
4. Remove the transfer case with the transmission.
5. If equipped with an automatic transmission, disconnect the breather hose from the transfer upper cover and the transmission control retainer.
6. Remove the rear engine mounting.
7. Remove the dynamic damper.
8. Remove the driveshaft upper dust cover and the transfer from the transmissions follows:
   a. Remove the dust cover bolt from the bracket.
   b. Remove the transfer adapter rear mounting bolts.
   c. Pull the transfer straight up and remove it from the transmission. Be careful not to damage the adapter rear oil seal with the transfer input gear spline.

To install:

9. Install the transfer and the driveshaft upper dust cover to the transmission with a new gasket as follows:
   a. Shift the two shift fork shafts to the high four position.
   b. Apply MP grease to the adapter oil seal.
   c. Place a new gasket to the transfer adapter.
   d. Install the transfer to the transmission. Take care not to damage the oil seal by the input gear spline.
   e. Install the transfer adapter rear mounting bolts and tighten the bolts to 27 ft. lbs. (37 Nm).
   f. Install the dust cover bolt to the bracket. Tighten the bolt to 17 ft. lbs. (23 Nm).
10. Install the engine rear mounting and tighten the bolts to 19 ft. lbs. (25 Nm).
11. Install the dynamic damper and tighten the bolts to 27 ft. lbs. (37 Nm).
12. If equipped with an automatic transmission, install the breather hose.
13. Install the transfer case with the transmission to the engine.

14. Fill the transmission and the transfer case with oil.
15. Test drive the vehicle and check the abnormal noise and smooth operation.
16. Recheck the fluid levels.

Halfshaft

REMOVAL & INSTALLATION

1995 4Runner and All T-100

1. Raise and safely support the vehicle.
2. Remove the wheel(s) and tire assembly.
3. While having an assistant hold the brake pedal, remove the six nuts holding the halfshaft to the differential.
4. If equipped with a free wheeling hub, remove the free-wheel hub as follows:
   a. Set the control handle to FREE.
   b. Remove the cover bolts and pull off the cover.
   c. Remove the center bolt with washer.
   d. Remove the mounting nuts and washer to the hub body.
   e. Using a brass bar and hammer, tap on the bolts head and remove the cone washer.

   f. Pull off the free wheel hub body and gasket.
5. If equipped without a free wheeling hub, remove the flange for the axle hub as follows:
   a. Remove the grease cap from the flange.
   b. Remove the bolt from the flange.
   c. Remove the six mounting nuts to the flange.
   d. Using a brass bar and hammer, tap on the bolts head and remove the six cone washers.
   e. Install the two bolts to the flange. Tighten the bolts to remove the flange.
   f. Remove the gasket for the flange.
6. Using a snapring expander, remove the snapring from the end of the halfshaft, then remove the spacer.
7. Remove the halfshaft from the differential, then pull the halfshaft from the steering knuckle.

➡It may be necessary to tap the end of the halfshaft with a rubber hammer.

To install:

8. Install the halfshaft to the steering knuckle and differential.
9. Install the six nuts to the differential but do not tighten the nuts at this time.

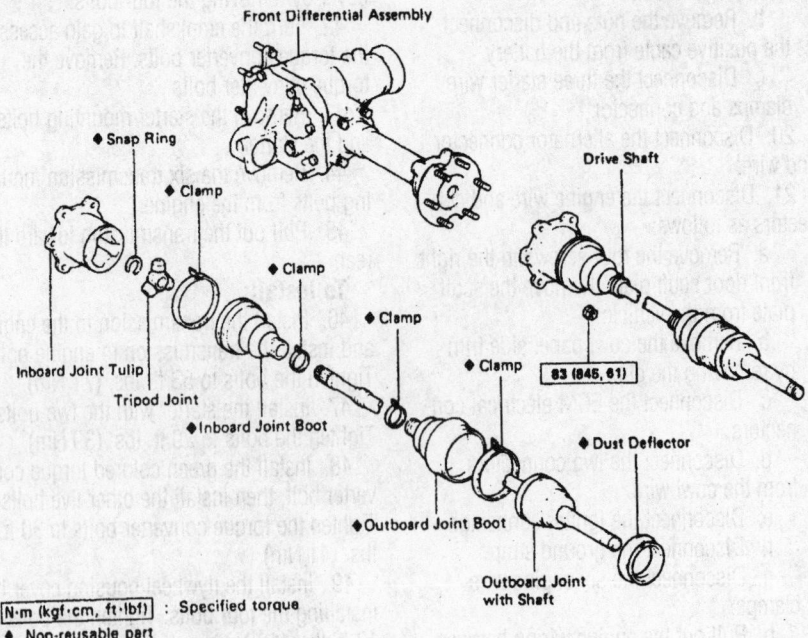

N·m (kgf·cm, ft·lbf) : Specified torque
◆ Non-reusable part

83 (845, 61)

Front Differential Assembly
Drive Shaft
◆ Snap Ring
◆ Clamp
◆ Clamp
◆ Clamp
◆ Clamp
Inboard Joint Tulip
Tripod Joint
◆ Inboard Joint Boot
◆ Outboard Joint Boot
◆ Dust Deflector
◆ Outboard Joint with Shaft

7924YG95

Exploded view of the halfshaft components—1995 4Runner and All T-100 models

10. Install the spacer and using a snapring expander, install a new snapring to the halfshaft.

11. If equipped without a free wheeling hub, install the flange as follows:

a. Place a new gasket in position on the axle hub.

b. Install the flange to the axle hub.

c. Install the six cone washers, plate washers and nuts.

d. Install the six nuts. Tighten the nuts to 23 ft. lbs. (31 Nm).

e. Install the bolt and tighten the bolt to 13 ft. lbs. (18 Nm).

f. Install the grease cap.

12. If equipped with a free wheeling hub, install the hub as follows:

a. Place a new gasket in position on the front axle hub.

b. Install the free wheeling hub body with the six cone washers and nuts. Tighten the nuts to 23 ft. lbs. (31 Nm).

c. Install the bolt with the washer. Tighten the bolt to 13 ft. lbs. (18 Nm).

d. Apply multi purpose grease to the inner hub splines.

e. Set the control handle and clutch to the FREE position.

f. Place a new gasket in position on the cover.

g. Install the cover to the hub body with the follower pawl tabs aligned with the non-toothed portions of the hub body.

h. Install the mounting bolts and tighten the cover bolts to 7 ft. lbs. (10 Nm).

13. With an assistant holding down the brake pedal, tighten the six nuts holding the halfshaft to the differential. Tighten the nuts to 61 ft. lbs. (83 Nm).

14. Install the front wheel(s) and lower the vehicle.

1996–99 4Runner

1. Raise and safely support the vehicle.

2. Remove the front wheel.

3. Drain the differential oil from the differential.

4. Remove the half shaft locknut.

a. Remove the grease cap.

b. Remove the cotter pin and the lock cap. While applying the brakes, remove the locknut.

5. Using a brass bar and a hammer, disconnect the halfshaft.

6. Disconnect the lower control arm.

7. Push the steering knuckle outward and remove the halfshaft.

8. Remove the snapring from the inboard shaft.

To install:

9. Install the snapring to the inboard shaft.

10. Install the halfshaft and install the steering knuckle.

11. Connect the lower control arm and tighten the nut to 105 ft. lbs. (142 Nm).

12. Connect the halfshaft.

a. Set the snapring opening side facing downward.

b. Using SST 09631–10030 and a hammer, strike the inboard joint into the differential.

c. Check that the halfshaft cannot be pulled out by hand.

13. While applying the brakes install the locknut. Tighten to 174 ft. lbs. (235 Nm).

14. Install the grease cap.

15. Fill the differential with oil.

16. Install the front wheel.

Tacoma

1. Raise and safely support the vehicle.

2. Drain the differential oil from the differential.

3. If not equipped with a free-wheeling hub, disconnect the halfshaft from the steering knuckle as follows:

a. Using a screwdriver, remove the grease cap.

b. Remove the cotter pin and lock cap from the halfshaft.

c. While having an assistant apply the brakes, remove the locknut from the halfshaft.

4. If equipped with free-wheeling hub, remove the free wheel hub as follows:

a. Set the control handle to FREE.

b. Remove the cover bolts and pull off the cover.

c. Remove the center bolt with washer.

d. Remove the mounting nuts and washer to the hub body.

e. Using a brass bar and hammer, tap on the bolts head and remove the cone washer.

f. Pull off the free wheel hub body and gasket.

g. Using a snapring expander, remove the snapring from the end of the halfshaft.

5. Using a brass bar and hammer, disconnect the halfshaft from the differential.

6. Remove the cotter pin and nut from the lower ball joint.

7. Using SST 09628–62011 or equivalent, disconnect the lower control arm from the lower ball joint.

8. After disconnecting the ball joint

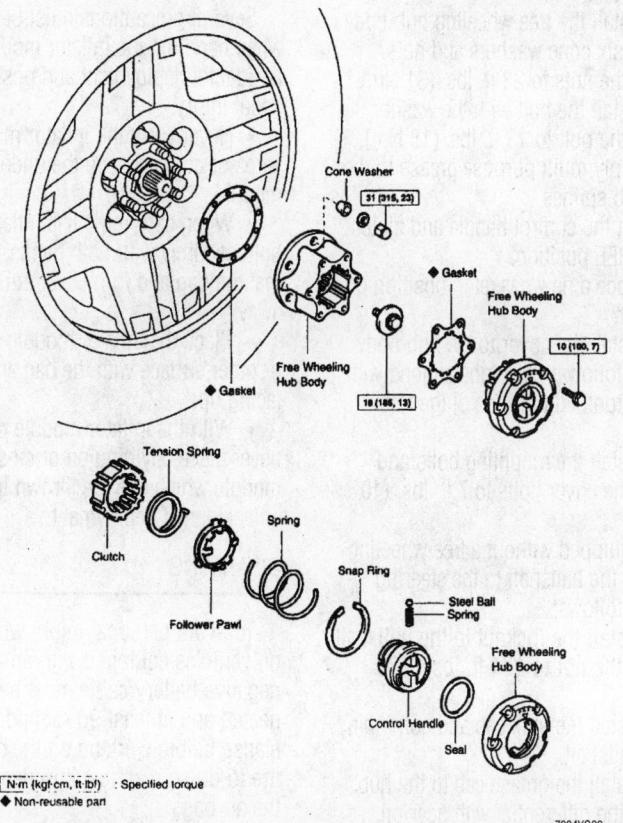

Exploded view of the free wheeling hub assembly—Tacoma model shown

7924YG96

from the lower control arm, push the steering knuckle outwards and remove the halfshaft from the steering knuckle and vehicle.

➡**If it is difficult to remove the halfshaft from the steering knuckle, use a rubber hammer and tap the halfshaft from the steering knuckle.**

9. Remove the snapring from the inboard shaft.

To install:

10. Install a new snapring to the inboard shaft.

11. Connect the halfshaft to the steering knuckle.

12. Push the steering knuckle inwards and at the same time, push the halfshaft into the differential with the snapring opening facing downward. Be sure the halfshaft is fully installed to the differential by checking that it cannot be pulled out by hand.

13. Connect the lower control arm to the lower ball joint and install the nut. Tighten the nut to 112 ft. lbs. (152 Nm). Install a new cotter pin.

14. If equipped with a free wheeling hub, install the hub as follows:

　a. Install the spacer and using a snapring expander, install the snapring to the halfshaft.

　b. Place a new gasket in position on the front axle hub.

　c. Install the free wheeling hub body with the six cone washers and nuts. Tighten the nuts to 23 ft. lbs. (31 Nm).

　d. Install the bolt with the washer. Tighten the bolt to 13 ft. lbs. (18 Nm).

　e. Apply multi purpose grease to the inner hub splines.

　f. Set the control handle and clutch to the FREE position.

　g. Place a new gasket in position on the cover.

　h. Install the cover to the hub body with the follower pawl tabs aligned with the non-toothed portions of the hub body.

　i. Install the mounting bolts and tighten the cover bolts to 7 ft. lbs. (10 Nm).

15. If equipped without a free wheeling hub, install the halfshaft to the steering knuckle as follows:

　a. Install the locknut to the halfshaft. Tighten the nut to 174 ft. lbs. (235 Nm).

　b. Install the lock cap and cotter pin to the halfshaft.

　c. Install the grease cab to the hub.

16. Fill the differential with gear oil.

17. Install the wheels and lower the vehicle.

STEERING AND SUSPENSION

Air Bag

❋❋ CAUTION

Some vehicles are equipped with an air bag system, also known as the Supplemental Inflatable Restraint (SIR) or Supplemental Restraint System (SRS). The system must be disabled before performing service on or around system components, steering column, instrument panel components, wiring and sensors. Failure to follow safety and disabling procedures could result in accidental air bag deployment, possible personal injury and unnecessary system repairs.

PRECAUTIONS

Several precautions must be observed when handling the inflator module to avoid accidental deployment and possible personal injury.

• Never carry the inflator module by the wires or connector on the underside of the module.

• When carrying a live inflator module, hold securely with both hands, and ensure that the bag and trim cover are pointed away.

• Place the inflator module on a bench or other surface with the bag and trim cover facing up.

• With the inflator module on the bench, never place anything on or close to the module which may be thrown in the event of an accidental deployment.

DISARMING

To avoid personal injury when working on vehicles equipped with an air bag, the negative battery cable must be disconnected and at least 90 seconds must elapse before working on the system. Failure to do so may result in deployment of the air bag.

Manual Rack and Pinion Steering Gear

REMOVAL & INSTALLATION

Tacoma

1. Disconnect the negative battery cable.

❋❋ CAUTION

Work must be started after 90 seconds from the time the ignition switch is turned to the LOCK position and the negative (-) battery cable is disconnected.

2. Raise and safely support the vehicle.

3. Disconnect the right and left tie rod ends from the knuckle.

4. Disconnect the intermediate No. 2 shaft from the steering rack.

5. Remove the manual steering rack.

To install:

6. Install the manual steering rack and tighten the bolts to 148 ft. lbs. (201 Nm).

7. Connect the intermediate No. 2 shaft to the steering rack.

8. Connect the right and left tie rod ends to the steering knuckle. Tighten the castle nuts to 53 ft. lbs. (72 Nm) and install new cotter pins.

9. Connect the negative battery cable.

10. Check the steering wheel center point.

11. Bleed the power steering system.

12. Check the front wheel alignment. Tighten the tie rod end locknuts to 67 ft. lbs. (90 Nm)

Power Rack and Pinion Steering Gear

REMOVAL & INSTALLATION

T-100

1. Disconnect the negative battery cable.

❋❋ CAUTION

Work must be started after 90 seconds from the time the ignition switch is turned to the LOCK position and the negative (-) battery cable is disconnected.

2. Raise and safely support the vehicle.

3. Disconnect the right and left tie rod ends from the knuckle.

4. Matchmark and disconnect the intermediate shaft from the steering rack.

5. Using SST 09631–22020, or equivalent, remove the pressure feed and the return tubes.

6. Remove the mount bracket and the grommet from the power steering rack assembly.

7. Remove the power steering rack and pinion.

To install:

8. Install the power steering rack and pinion. Tighten the mounting bolts to 65 ft. lbs. (88 Nm).

9. Install the grommet and the mount bracket to the gear assembly. Tighten the bolts to 65 ft. lbs. (88 Nm).

10. Install a new O-ring and install the pressure feed and return tubes. Tighten the line fittings to 14 ft. lbs. (19 Nm).

11. Align the matchmarks and connect the intermediate shaft to the steering rack.

➡**If installing a new rack assembly, be sure the steering wheel and the rack are centered.**

12. Connect the right and left tie rod ends and tighten nuts to 67 ft. lbs. (90 Nm). Install new cotter pins.

13. Connect the negative battery cable.

14. Check the steering wheel center point.

15. Check the fluid level and bleed the power steering system.

16. Check the front wheel alignment.

Tacoma

1. Disconnect the negative battery cable.

✳✳ CAUTION

Work must be started after 90 seconds from the time the ignition switch is turned to the LOCK position and the negative (-) battery cable is disconnected.

2. Raise and safely support the vehicle.

3. Disconnect the right and left tie rod ends from the knuckle.

4. Disconnect the intermediate No. 2 shaft from the steering rack.

5. Using SST 09631–22020 or equivalent, remove the pressure feed and the return tubes.

6. Remove the power steering rack.

To install:

7. Install the power steering rack and tighten the bolts to 148 ft. lbs. (201 Nm) for

2-wheel drive models, or to the following values for 4-wheel drive models:

• Rack assembly bolt—123 ft. lbs. (167 Nm)

• Rack assembly nut—141 ft. lbs. (191 Nm)

• Bracket nut and bolt—123 ft. lbs. (167 Nm)

8. Install a new O-ring and install the pressure feed tube and tighten to 33 ft. lbs. (45 Nm). Install the return tube and tighten to 36 ft. lbs. (49 Nm) for 2-wheel drive models, or to 29 ft. lbs. (40 Nm) for 4-wheel drive models.

9. Connect the intermediate No. 2 shaft to the steering rack.

10. Connect the right and left tie rod ends to the steering knuckle. Tighten the castle nuts to specification and install new cotter pins.

11. Connect the negative battery cable.

12. Check the steering wheel center point.

13. Bleed the power steering system.

14. Check the front wheel alignment. Tighten the tie rod end locknuts to 67 ft. lbs. (90 Nm)

Strut

REMOVAL & INSTALLATION

Front

1996–99 4RUNNER

1. Raise and safely support the front of the vehicle.

2. Remove the front wheel.

3. Disconnect the strut from the lower control arm by removing the bolt.

4. Remove the three nuts and the strut assembly.

To install:

5. Install the strut assembly and tighten the three nuts to 47 ft. lbs. (64 Nm).

6. Install the lower bolt to hold the strut assembly to the lower control arm. Tighten the bolt to 101 ft. lbs. (135 Nm).

7. Install the wheels and lower the vehicle.

8. Check the vehicle alignment.

4-WHEEL DRIVE TACOMA

1. Raise and safely support the vehicle. Place the jack stands under the frame of the vehicle.

2. Remove the nut and bolt holding the strut to the lower control arm.

3. Remove the three nuts holding the strut to the strut tower.

4. Remove the strut from the vehicle.

✳✳ WARNING

Never loosen the center nut on the strut unless a spring compressor is installed. Serious injury or vehicle damage may result.

To install:

5. Install the strut to the vehicle.

6. Install the three nuts to hold the strut to the strut tower. Tighten the nuts to 47 ft. lbs. (64 Nm).

7. Install the lower bolt and nut to hold the strut to the lower control arm. Tighten the nut to 101 ft. lbs. (135 Nm).

8. Install the front wheels and lower the vehicle.

9. Check the vehicle alignment.

Shock Absorber

REMOVAL & INSTALLATION

Front

T-100, TACOMA, AND 1995 4RUNNER 2-WHEEL DRIVE MODELS

1. Raise and safely support the front of the vehicle.

2. Remove the front wheel.

3. Disconnect the shock absorber from the lower control arm by removing the two bolts.

4. Remove the nut, retainers, and the cushion from the top of the shock absorber.

5. Remove the shock absorber from the vehicle.

6. Remove the retainers and cushion from the shock absorber.

To install:

7. Install the retainers and cushion to the shock absorber.

8. Install the shock absorber to the vehicle

9. Install the retainers, cushion and nut to the top of the shock absorber. Tighten the nut to 18 ft. lbs. (25 Nm).

10. Install the lower bolts to hold the shock absorber to the lower control arm. Tighten the bolts to 13 ft. lbs. (19 Nm).

11. Install the wheels and lower the vehicle.

T-100, TACOMA, AND 1995 4RUNNER 4-WHEEL DRIVE MODELS

1. Raise and safely support the front of the vehicle.

2. Remove the front wheel.

3. Disconnect the shock absorber from the lower control arm by removing the nut, washer, and the through-bolt.

4. Remove the nut, retainers, and cushion from the top of the shock absorber.

5. Remove the shock absorber from the vehicle.

6. Remove the retainers and cushion from the shock absorber.

To install:

7. Install the retainers and cushion to the shock absorber.

8. Install the shock absorber to the vehicle

9. Install the retainers, cushion, and the nut to the top of the shock absorber. Tighten the nut to 18 ft. lbs. (25 Nm).

10. Install the lower through-bolt, washer, and the nut to hold the shock absorber to the lower control arm. Tighten the bolt to 101 ft. lbs. (137 Nm).

11. Install the wheels and lower the vehicle.

Rear

1. Raise and safely support the frame with stands.

2. Support the axle housing with a floor jack.

3. Remove the wheel and tire assemblies.

4. Lower the floor jack to take tension off of the spring.

5. Disconnect the shock absorber from the rear axle housing by removing the bolt.

6. Remove the nut, retainers, and the cushions holding the shock absorber to the frame.

7. Remove the shock absorber from the vehicle with the washers and bushings.

To install:

8. Install the shock absorber to the frame with the washers and bushings.

9. Tighten the nut to hold the shock absorber to the frame to the following values:

- 1995 4Runner models—18 ft. lbs. (25 Nm)
- 1996–97 4Runner models—14 ft. lbs. (20 Nm)
- 1995–97 T-100 models—19 ft. lbs. (25 Nm)
- Tacoma models with 2WD—19 ft. lbs. (25 Nm)
- Tacoma models with 4WD—53 ft. lbs. (72 Nm)

10. Connect the shock absorber to the rear axle housing and tighten the bolt to the following specifications:

- 1995–97 4Runner models—47 ft. lbs. (64 Nm)
- 1995–97 T-100 models—19 ft. lbs. (25 Nm)
- Tacoma models with 2WD—19 ft. lbs. (25 Nm)
- Tacoma models with 4WD—53 ft. lbs. (72 Nm)

11. Install the wheels.

12. Lower the vehicle to the ground.

Coil Spring

REMOVAL & INSTALLATION

Front

1996–99 4RUNNER

1. Raise and safely support the vehicle.

2. Remove the strut assembly.

3. Using SST 09727–30030 or equivalent, compress the coil spring until there is clearance on both ends.

4. Remove the support center nut.

⁕⁕ CAUTION

Do not remove the center nut without compressing the spring. Failure to do so may cause property damage or personal injury.

5. Remove the two retainers, cushion, suspension support and the coil spring.

To install:

6. Compress the coil spring and install to the strut.

7. Fit the lower end of the coil spring into the gap of the spring seat of the strut.

8. Install the suspension support.

   a. Install the two retainers, suspension support and the cushion to the rod.

   b. Temporarily tighten the support center nut.

   c. Align the suspension support with the strut lower bushing.

   d. Face the lower end of the coil spring to the outside of the vehicle.

   e. Remove the spring compressor and tighten the center nut to 18 ft. lbs. (25 Nm).

9. Install the shock absorber.

TACOMA 2-WHEEL DRIVE MODELS

1. Raise and safely support the front of the vehicle. Place the jack stands under the frame of the vehicle.

2. Remove the shock absorber from the suspension by removing the two bottom bolts and top nut.

3. Using SST 09727–22011, or equivalent (spring compressor), compress the coil spring.

⁕⁕ CAUTION

The proper tools for this procedure must be used. The spring is under high pressure and can cause serious injury if not properly removed and installed.

4. Remove the nut and disconnect the sway bar link from the lower control arm.

5. Remove the two sway bar bracket bolts on the side of the suspension that the lower control arm is being removed. This will allow access to the lower control arm through-bolt.

6. Support the steering knuckle and upper control arm.

⁕⁕ WARNING

Before removing the lower control arm, be sure the coil spring is safely compressed.

7. Remove the cotter pin and nut from the lower ball joint.

8. Using SST 09628–62011, or equivalent, disconnect the lower ball joint from the lower control arm.

9. Remove the nut from the lower control arm set bolt.

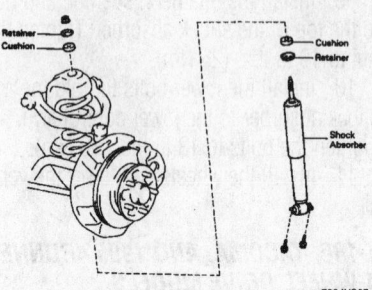

Shock absorber component assembly (2-Wheel Drive)—Tacoma model shown

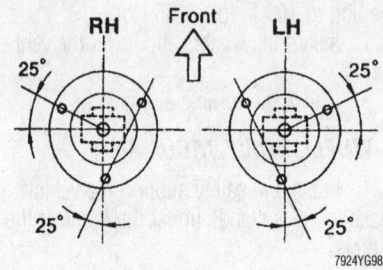

Aligning the strut support to the strut's lower bushing—4Runner

10. Remove the nut from the strut bar front set bolt.

11. Pull out the two bolts and remove the lower control arm and strut bar as an assembly. When the lower control arm is removed, set the coil spring aside.

To install:

12. Place the end of the coil spring in contact with the lower control arm seat.

13. Install the lower control arm, spring, and strut arm to the suspension. Install the strut arm bolt and lower control arm bolt.

14. Install the nuts for the strut arm bolt and lower control arm bolt. Do not tighten the bolts at this time.

15. Connect the lower control arm to the lower ball joint. Install the nut and tighten the nut to 80 ft. lbs. (110 Nm). Install a new cotter pin.

16. Remove the support from the upper control arm and steering knuckle.

17. Connect the sway bar bracket to the suspension and install the two bolts. Tighten the bolts to 22 ft. lbs. (29 Nm).

18. Connect the sway bar link to the lower control arm and install the nut. Tighten the nut to 29 ft. lbs. (39 Nm).

19. Making sure the coil spring is in its correct position, slowly remove the spring compressor from the coil.

20. Install the shock absorber. Tighten the top nut to 18 ft. lbs. (25 Nm) and the bottom two bolts to 29 ft. lbs. (39 Nm).

21. Install the wheel and lower the vehicle.

22. Stabilize the suspension by pushing up and down on the vehicle.

23. Tighten the strut bar nut and bolt to 221 ft. lbs. (300 Nm) and the lower control arm bolt and nut to 148 ft. lbs. (200 Nm).

24. Check the front wheel alignment.

TACOMA 4-WHEEL DRIVE MODELS

1. Raise and safely support the vehicle. Place the jack stands under the frame of the vehicle.

2. Remove the nut and bolt holding the strut to the lower control arm.

3. Remove the three nuts holding the strut to the strut tower.

4. Remove the strut from the vehicle.

✴✴ CAUTION

The proper tools for this procedure must be used. The spring is under high pressure and can cause serious injury if not properly removed and installed.

5. Using SST 09727–30030, or equivalent, compress the coil spring until there is a clearance on both ends.

6. Remove the strut center nut.

7. Remove the suspension support and coil spring.

8. Remove the insulator from the suspension support.

To install:

9. Install the insulator to the suspension support.

➡**Match the bolt of the suspension support with the cut out part of the insulator.**

10. Using the coil spring compressor, compress the coil spring and install the coil spring to the strut.

➡**Fit the lower end of the coil spring into the gap of the spring seat of the strut.**

11. Install the suspension support to the strut rod.

12. Temporarily tighten a new suspension support center nut.

13. Position the suspension support so that a line drawn between the two bolts would be parallel to the direction of the lower bushing.

14. Remove the compressor from the spring.

15. Tighten the strut center nut to 22 ft. lbs. (29 Nm).

16. Install the strut to the vehicle.

17. Install the three nuts to hold the strut to the strut tower. Tighten the nuts to 47 ft. lbs. (64 Nm).

18. Install the lower bolt and nut to hold the strut to the lower control arm. Tighten the nut to 101 ft. lbs. (135 Nm).

19. Install the front wheels and lower the vehicle.

Rear

1995 4RUNNER

✴✴ CAUTION

The spring on the axle carrier is under high pressure and can cause serious injury if not properly removed and installed.

1. Raise and safely support the vehicle at the frame.

2. Support the axle housing with a floor jack.

3. Remove the wheel and tire assembly.

4. Remove the shock absorber to axle housing bolt. Disconnect both shock absorbers from the axle housing.

5. Disconnect the sway bar brackets from the axle housing.

6. Remove the two bolts and disconnect the shackle bracket from the lateral control rod.

7. Remove the nut, washer, and the bolt from the frame and disconnect the lateral control rod.

8. If necessary, disconnect the brake hose as follows:

   a. Disconnect the brake line from the brake hose at the body bracket.

   b. Remove the clip and disconnect the brake hose from the body bracket.

9. Lower the floor jack, then remove the coil spring(s) and the insulators.

To install:

10. Place the coil spring insulators into place.

11. Place the springs into position and raise the axle housing. Be sure to fit the lower end of the coil spring into the gap of the spring seat on the lower control arm.

12. Raise the rear axle housing.

13. Connect the lateral control rod to the frame with the bolt, washer and nut. Install the bolt from the front of the vehicle but do not tighten at this time.

14. Connect the shackle bracket to the lateral control rod with the two bolts.

15. Connect the shock absorber to the lower control arm and install the nut. Tighten the nut to 47 ft. lbs. (64 Nm).

16. If necessary, install the brake hose as follows:

   a. Connect the brake hose to the bracket and install the clip.

   b. Connect the brake tube to the brake hose and tighten the tube.

17. Install the sway bar brackets to the rear axle housing and tighten the bolts to 9 ft. lbs. (13 Nm).

18. Bleed the brake system.

19. Install the wheels and lower the vehicle.

20. Stabilize the suspension and tighten the lateral control rod nut to 101 ft. lbs. (137 Nm).

1996–99 4RUNNER

1. Raise and safely support the vehicle.

2. Remove the wheel assemblies.

3. Support the axle housing with a floor jack.

4. Remove the brake drum from the axle housing.

5. Disconnect the parking brake cable from the brake shoe.

6. Remove the bolt and disconnect the parking brake cable from the axle housing.

7. Place matchmarks on the flanges for the driveshaft and differential.

8. Remove the four bolt and nuts and disconnect the driveshaft from the differential.

9. Disconnect the brake hose line from the brake hose.

10. Remove the clip holding the brake hose to the brake bracket and disconnect the brake hose from the body.

11. If equipped with ABS, remove the ABS wiring harness bracket.

12. Disconnect the shock absorbers from the axle housing.

13. Remove the nuts and bolts to the lateral control rod and remove the control rod from the suspension.

✳✳ CAUTION

The spring on the axle carrier is under high pressure and can cause serious injury if not properly removed and installed.

14. Slowly lower the rear axle housing and remove the coil spring.

To install:

15. Place the springs into position and raise the axle housing. Be sure to fit the lower end of the coil spring into the gap of the spring seat on the lower control arm.

16. Install the lateral control rod to the suspension and tighten the bolts and nuts to 64 ft. lbs. (86 Nm).

17. Connect the shock absorbers to the axle housing. Tighten the bolt to 47 ft. lbs. (64 Nm).

18. If equipped, install the ABS wiring harness bracket.

19. Connect the brake hose to the bracket and install the clip.

20. Connect the brake line to the brake hose and tighten the tube.

21. Connect the parking brake cable bracket to the axle housing and tighten to 9 ft. lbs. (13 Nm).

22. Connect the parking brake cable to the brake shoes.

23. Align the matchmarks and connect the driveshaft to the differential. Install the bolts and nuts and tighten them to 54 ft. lbs. (73 Nm).

24. Fill the differential with the proper amount and type of oil.

25. Install the brake drum.

26. Bleed the brake system.

27. Install the wheel assemblies.

28. Lower the vehicle and bounce the vehicle several times to stabilize the suspension.

29. Tighten the lower control arm to 107 ft. lbs. (145 Nm).

Upper Ball Joint

REMOVAL & INSTALLATION

1995 4Runner and All T-100

2-WHEEL DRIVE MODELS

1. Raise and safely support the vehicle. Place the jack stands under the frame of the vehicle.

2. Remove the wheel and tire assembly.

3. Support the lower control arm with a floor jack.

4. Remove the brake caliper and support it out of the way with a wire.

5. Remove the cotter pin and nut from the upper ball joint.

6. Using a ball joint removal tool, separate the ball joint from the knuckle arm.

7. Remove the four nuts, washers and bolts holding the ball joint to the upper control arm.

8. Remove the ball joint from the upper control arm.

To install:

9. Install the ball joint to the upper control arm. Install the bolts, washers, and nuts to hold the upper ball joint to the upper control arm. Tighten the ball joint-to-upper control arm bolts and ball joint-to-steering knuckle castle nuts to the following specifications:

   a. On 1995 Pick-up and 4Runner, tighten the ball joint-to-upper control arm bolts to 23 ft. lbs. (31 Nm). Tighten the ball joint-to-steering knuckle castle nut to 80 ft. lbs. (108 Nm) and install a new cotter pin.

   b. On T-100, tighten the ball joint-to-upper control arm bolts 23 ft. lbs. (31 Nm) for 2-Wheel Drive vehicles or 25 ft. lbs. (33 Nm) for 4-Wheel Drive vehicles. Tighten the ball joint-to-control arm bolts to 80 ft. lbs. (108 Nm) and install a new cotter pin.

➡**Be sure to grease the ball joints before moving the vehicle.**

10. Install the wheel and lower the vehicle.

4-WHEEL DRIVE MODELS

1. Raise and safely support the vehicle. Place the jack stands under the frame of the vehicle.

2. Remove the wheel and tire assembly.

3. Support the lower control arm with a floor jack.

4. Remove the steering knuckle.

5. Remove the four nuts and disconnect the upper ball joint from the upper control arm.

To install:

6. Install the upper ball joint to the upper control arm and install the four nuts. Tighten the nuts to 25 ft. lbs. (33 Nm).

7. Install the steering knuckle.

8. Install the wheel and tire assembly.

9. Lower the vehicle.

1996–99 4Runner

1. Raise and safely support the vehicle.

2. Remove the front wheels.

3. Remove the strut assembly.

4. Remove the grease cap.

5. 4-Wheel Drive: Disconnect the half-shaft.

   a. Remove the cotter pin and lock cap.

   b. While applying the brakes, remove the locknut.

6. With ABS: Remove the ABS speed sensor and wiring harness clamp from the steering knuckle.

7. Remove the brake line bracket from the steering knuckle.

8. Remove the front brake caliper and the rotor.

9. Remove the four bolts and disconnect the lower ball joint.

10. Remove the steering knuckle with the axle hub.

    a. Remove the cotter pin and loosen the nut.

    b. Using SST 09950–40010 or equivalent, disconnect the steering knuckle from the upper control arm.

    c. Remove the nut and the steering knuckle.

11. Remove the upper ball joint.

    a. Remove the wire and the boot.

    b. Remove the snapring.

    c. Using SST 09950–40010 or equivalent, and a deep socket wrench, remove the upper ball joint.

To install:

12. Install the upper ball joint.

    a. Install a new ball joint with a new snapring.

    b. Install a new boot and fix it with a new wire.

13. Install the steering knuckle with the axle hub to the upper control arm. Tighten the nut to 80 ft. lbs. (108 Nm). Install a new cotter pin.

14. Install the lower ball joint and tighten the four bolts to 59 ft. lbs. (80 Nm).

15. Install the rotor and the caliper. Tighten the caliper bolts to 90 ft. lbs. (123 Nm).

16. Install the brake line bracket to the steering knuckle and tighten to 21 ft. lbs. (28 Nm).

17. If removed, install the ABS speed sensor and wiring harness clamp to the steering knuckle. Tighten the bolts to 6 ft. lbs. (8 Nm).

18. If disconnected, install the halfshaft and tighten the nut to 174 ft. lbs. (235 Nm).

19. Install the grease cap.

20. Install the strut assembly.

21. Install the front wheel.

22. Lower the vehicle and check the alignment.

Tacoma

2-WHEEL DRIVE MODELS

1. Raise and safely support the front of the vehicle. Place jackstands under the frame of the vehicle.

2. Remove the wheel.

3. Support the lower control arm with a floor jack.

4. Disconnect the ABS speed sensor wire from the upper control arm.

5. Remove the two bolts and camber adjusting shims from the upper control arm.

➡**Before removing the shims from the upper control arm, make a note of each shim size and position. It is important to replace the shims into their original position.**

6. Remove the cotter pin and nut from the upper control arm. Using SST 09628–62011 or equivalent, disconnect the upper ball joint from the steering knuckle.

7. Remove the upper control arm from the vehicle.

8. Remove the four nuts and bolts connecting the upper control arm to the upper ball joint. Disconnect the upper control arm from the upper ball joint.

To install:

9. Connect the upper control arm to a new ball joint and install the four bolts and nuts. Tighten the nuts to 29 ft. lbs. (39 Nm).

➡**Be sure to grease the ball joint.**

10. Install the upper control arm to the vehicle.

11. Install the camber adjusting shims and two bolts to the upper control arm. Tighten the two bolts to 94 ft. lbs. (130 Nm).

12. Connect the upper ball joint to steering knuckle. Install the ball joint nut and tighten the nut to 80 ft. lbs. (110 Nm).

13. Connect the ABS speed sensor wire to the upper control arm. Install and tighten the ABS bolt to 71 inch lbs. (8 Nm).

14. Install the wheels and lower the vehicle.

15. Check the wheel alignment.

4-WHEEL DRIVE MODELS

1. Raise and safely support the vehicle.

2. Remove the wheel and tire assembly.

3. Remove the strut from the suspension.

4. If not equipped with a FREE wheeling hub, disconnect the halfshaft from the steering knuckle as follows:

   a. Using a screwdriver, remove the grease cap.

   b. Remove the cotter pin and lock cap from the halfshaft.

   c. While having an assistant apply the brakes, remove the locknut from the halfshaft.

5. If equipped with FREE wheeling hub, remove the free wheel hub as follows:

   a. Set the control handle to FREE.

   b. Remove the cover bolts and pull off the cover.

   c. Remove the center bolt with washer.

   d. Remove the mounting nuts and washer to the hub body.

   e. Using a brass bar and hammer, tap on the bolts head and remove the cone washer.

   f. Pull off the free wheel hub body and gasket.

   g. Using a snapring expander, remove the snapring and spacer from the end of the halfshaft.

6. If equipped with ABS brakes, disconnect the ABS speed sensor from the steering knuckle.

7. Disconnect the brake hose from the steering knuckle by removing the bolt.

8. Remove the two brake caliper support bracket bolts and wire the caliper to the side. Do not allow the caliper to hang from the brake hose.

9. Remove the rotor from the steering knuckle.

10. Remove the four bolts and disconnect the lower ball joint from the steering knuckle.

11. Remove the cotter pin and nut to the upper control arm.

12. Using SST 09950–40010 or equivalent, disconnect the steering knuckle from the upper control arm.

13. Remove the steering knuckle from the vehicle.

➡**If it is difficult to remove the halfshaft from the steering knuckle, use a rubber hammer and tap the halfshaft from the steering knuckle.**

14. Remove the wire and boot from the upper ball joint.

15. Using a snapring expander, remove the snapring from the ball joint.

16. Using SST 09950–40010 (puller set) or equivalent and a deep socket wrench, remove the upper ball joint from the steering knuckle.

To install:

17. Using SST 09309–37010 or equivalent and a socket wrench, press in a new upper ball joint.

18. Using a snapring expander, install a new snapring.

19. Install a new boot and hold it down with a new piece of wire.

➡**Be sure to grease the ball joint.**

20. Install the steering knuckle to the halfshaft.

21. Connect the steering knuckle to the lower ball joint by installing the four bolts. Do not tighten the bolts at this time.

22. Push down on the upper control arm and connect the upper ball joint to the steering knuckle.

23. Install the upper ball joint nut. Tighten the nut to 80 ft. lbs. (105 Nm). Install a new cotter pin.

24. Tighten the lower ball joint to steering knuckle bolts to 59 ft. lbs. (80 Nm).

25. Install the brake rotor.

26. Connect the caliper support bracket to the steering knuckle and install the two bolts. Tighten the bolts to 90 ft. lbs. (123 Nm).

27. Connect the brake hose clamp to the steering knuckle by installing the bolt. Tighten the bolt to 13 ft. lbs. (18 Nm).

28. If equipped with ABS brakes, connect the ABS speed sensor and wiring harness to the steering knuckle.

29. Install the spacer and using a snapring expander, install the snapring to the halfshaft.

30. If equipped with a free wheeling hub, install the hub as follows:

   a. Place a new gasket in position on the front axle hub.

   b. Install the free wheeling hub body with the six cone washers and nuts. Tighten the nuts to 23 ft. lbs. (31 Nm).

   c. Install the bolt with the washer. Tighten the bolt to 13 ft. lbs. (18 Nm).

   d. Apply multi purpose grease to the inner hub splines.

   e. Set the control handle and clutch to the FREE position.

   f. Place a new gasket in position on the cover.

   g. Install the cover to the hub body

with the follower pawl tabs aligned with the non-toothed portions of the hub body.

h. Install the mounting bolts and tighten the cover bolts to 7 ft. lbs. (10 Nm).

31. If equipped without a free wheeling hub, install the halfshaft to the steering knuckle as follows:

a. Install the locknut to the halfshaft. Tighten the nut to 174 ft. lbs. (235 Nm).

b. Install the lockcap and cotter pin to the halfshaft.

c. Install the grease cab to the hub.

32. Install the strut to the vehicle. Tighten the nut holding the strut to the lower control arm to 101 ft. lbs. (135 Nm). Tighten the upper three nuts to 47 ft. lbs. (64 Nm)

33. Install the front wheels and lower the vehicle.

34. Check the wheel alignment.

Lower Ball Joint

REMOVAL & INSTALLATION

1995 4Runner and All T-100

2-WHEEL DRIVE MODELS

1. Raise and safely support the vehicle.

2. Remove the wheel and tire assembly.

3. Support the lower control arm with a floor jack.

4. Remove the cotter pin and nut from the lower ball joint.

5. Using a ball joint removal tool, separate the lower ball joint from the steering knuckle.

6. Remove the ball joint from the lower control arm by removing the three mounting nuts, washers, and bolts.

7. Remove the ball joint from the vehicle.

To install:

8. Install the lower ball joint to the vehicle.

9. Install the lower ball joint bolts, washers, and nuts, then install the ball joint to the steering knuckle. Tighten the fasteners to the following specifications:

a. 1995 4Runner and 1995–97 T-100, tighten the ball joint-to-control arm nuts to 55 ft. lbs. (75 Nm). Tighten the ball joint-to-steering knuckle bolts to 105 ft. lbs. (142 Nm). Install a new cotter pin.

➡**Be sure to grease the ball joints before moving the vehicle.**

10. Install the wheel and lower the vehicle.

4-WHEEL DRIVE MODELS

1. Raise and safely support the vehicle. Place the jack stands under the frame of the vehicle.

2. Remove the wheel and tire assembly.

3. Support the lower control arm with a floor jack.

4. Remove the steering knuckle.

5. Remove the four nuts, washers and bolts and disconnect the lower ball joint from the lower control arm.

To install:

6. Install the lower ball joint to the lower control arm and install the four bolts, washers and nuts. Tighten the nuts to 25 ft. lbs. (33 Nm).

7. Install the steering knuckle.

8. Install the wheel and tire assembly.

9. Lower the vehicle.

1996–99 4Runner

1. Raise and safely support the vehicle. Place the jack stands under the frame of the vehicle.

2. Remove the wheel and tire assembly.

3. Disconnect the tie rod end.

a. Loosen the four bolts.

b. Remove the cotter pin and nut from the tie rod end.

c. Using SST 09610–20012 or equivalent, disconnect the tie rod end from the steering knuckle.

4. Remove the lower ball joint.

a. Remove the cotter pin and the nut from the lower ball joint.

b. Using a ball joint separator, disconnect the lower ball joint from the lower suspension arm.

c. Remove the four bolts. While lifting the upper suspension arm and the steering knuckle, remove the lower ball joint.

To install:

5. Install the lower ball joint to the lower control arm. Tighten the four bolts to 59 ft. lbs. (80 Nm).

6. Install the nut and tighten to 105 ft. lbs. (142 Nm).

7. Connect the tie rod end to the steering knuckle and tighten the nut to 66 ft. lbs. (90 Nm).

8. Install the wheel and tire assembly.

Tacoma

2-WHEEL DRIVE MODELS

1. Raise and safely support the vehicle. Place the jacks under the frame.

2. Remove the wheel from the vehicle.

3. Support the lower control with a floor jack.

Do not remove the floor jack from the lower control arm until all components are replaced. The coil spring is under high pressure and if not removed properly, the spring could cause damage and/or injury.

4. Loosen the two lower ball joint set bolts.

5. Remove the cotter pin and nut from the tie rod.

6. Using SST 09628–62011 or equivalent, disconnect the tie rod from the ball joint bracket.

7. Remove the cotter pin and nut from the lower ball joint.

8. Using SST 09628–62011 or equivalent, disconnect the lower ball joint from the lower control arm.

9. Remove the two lower ball joint set bolts and remove the ball joint from the suspension.

To install:

10. Install the lower ball joint to the steering knuckle and lower control arm.

11. Install the two lower ball joint set bolts. Do not tighten the bolts at this time.

12. Install the lower ball joint nut to hold the lower ball joint to the lower control arm. Tighten the nut to 80 ft. lbs. (110 Nm). Install a new cotter pin to the lower ball joint.

13. Connect the tie rod end to the ball joint bracket. Install the nut to hold the tie rod end to the lower ball joint bracket. Tighten the nut to 53 ft. lbs. (72 Nm). Install a new cotter pin to the tie rod end.

14. Tighten the two lower ball joint set bolts to 116 ft. lbs. (160 Nm).

15. Remove the floor jack from the lower control arm.

16. Install the wheel and lower the vehicle.

17. Check the wheel alignment.

4-WHEEL DRIVE MODELS

1. Raise and safely support the vehicle. Place the jacks under the frame.

2. Remove the wheel from the vehicle.

3. Loosen the four lower ball joint set bolts.

4. Remove the cotter pin and nut from the tie rod.

5. Using SST 09610–20012 or equivalent, disconnect the tie rod from the ball joint bracket.

6. Remove the cotter pin and nut from the lower ball joint.

7. Using SST 09628–62011 or equiva-

lent, disconnect the lower ball joint from the lower control arm.

8. Remove the four lower ball joint set bolts.

9. While lifting the upper control arm and steering knuckle, remove the ball joint from the suspension.

To install:

10. Install the lower ball joint to the steering knuckle and lower control arm.

11. Install the four lower ball joint set bolts. Do not tighten the bolts at this time.

12. Install the lower ball joint nut to hold the lower ball joint to the lower control arm. Tighten the nut to 112 ft. lbs. (152 Nm). Install a new cotter pin to the lower ball joint.

13. Connect the tie rod end to the ball joint bracket. Install the nut to hold the tie rod end to the lower ball joint bracket. Tighten the nut to 67 ft. lbs. (90 Nm). Install a new cotter pin to the tie rod end.

14. Tighten the two lower ball joint set bolts to 83 ft. lbs. (113 Nm).

15. Remove the floor jack from the lower control arm.

16. Install the wheel and lower the vehicle.

17. Check the wheel alignment.

Wheel Bearings

ADJUSTMENT

Front

1996–97 4RUNNER AND 4-WHEEL DRIVE TACOMA

The wheel bearing is not adjustable and must be replaced if a problem is found.

EXCEPT 1996–97 4RUNNER AND 4-WHEEL DRIVE TACOMA

The front wheel bearing adjustment is included in the front wheel bearing removal and installation procedure.

REMOVAL & INSTALLATION

Front

1995 4RUNNER, PICK-UP AND ALL T-100 2-WHEEL DRIVE MODELS

1. Raise and safely support the vehicle.
2. Remove the wheel and tire.
3. Remove the disc brake caliper support bracket by removing the two bolts and wire the caliper aside.
4. Remove the cap, cotter pin, lock cap, and the nut from the spindle.

5. Remove the hub and disc together with the outer bearing and thrust washer.

6. Using a small prybar, pry the grease seal from the disc/hub assembly, then remove the inner bearing from the assembly.

7. Using a shop cloth, wipe the grease from inside the disc/hub assembly.

8. Using a brass drift, drive the outer bearing races from each side of the disc/hub assembly.

9. Place matchmarks on the disc and axle hub.

10. Remove the six bolts and separate the disc and axle hub.

11. Using solvent, clean all of the parts.

To install:

12. Align the marks on the disc and axle hub. Install the six bolts and tighten the bolts to 47 ft. lbs. (64 Nm).

13. Using a bearing installation tool, drive the outer races into the disc/hub assembly until they seat against the shoulder.

14. Using multi-purpose grease, coat the area between the races and pack the bearings.

15. Place the inner bearing into the rear of the disc/hub assembly. Using a bearing installation tool, drive a new grease seal into the rear of the disc/hub assembly until it is flush with the housing.

16. Install the disc/hub assembly onto the axle shaft, the outer bearing, the thrust washer and the adjusting nut.

17. To adjust the bearing preload, perform the following:

   a. Tighten the adjusting nut to 26 ft. lbs. (35 Nm).

   b. Turn the disc/hub assembly 2–3 times, from the left to the right.

   c. Loosen the adjusting nut until it can be turned by hand.

   d. Attach a spring tension gauge to 1 lug on the hub assembly. Pull on the gauge and measure the frictional force. Frictional force should be 1.1–31. ft. lbs. (5.0–14.0 N).

   e. Adjust the preload by tightening the nut.

18. Measure the hub axial play. The limit is 0.0020 in. (0.05mm).

19. Install the locknut, cotter pin, and the grease cap.

20. Install the disc brake caliper. Install the wheel and tire assembly.

21. Lower the vehicle.

1995 4RUNNER AND ALL T-100 4-WHEEL DRIVE MODELS

1. Raise and safely support the vehicle.

2. Remove the wheel and tire assembly.

3. If equipped with ABS brakes, disconnect the ABS speed sensor from the steering knuckle.

4. Remove the two brake caliper support bracket bolts and wire the caliper to the side. Do not allow the caliper to hang from the brake hose.

5. Remove the free-wheel hub:

   a. Set the control handle to FREE.

   b. Remove the cover bolts and pull off the cover.

   c. Remove the center bolt with washer.

   d. Remove the mounting nuts and washer to the hub body.

   e. Using a brass bar and hammer, tap on the bolts head and remove the cone washer.

   f. Pull off the free wheel hub body and gasket.

6. If equipped without a free wheeling hub, remove the flange for the axle hub as follows:

   a. Remove the grease cap from the flange.

   b. Remove the bolt from the flange.

   c. Remove the six mounting nuts to the flange.

   d. Using a brass bar and hammer, tap on the bolts head and remove the six cone washers.

   e. Install two bolts to the flange. Tighten the bolts to remove the flange.

   f. Remove the gasket for the flange.

7. Using a screwdriver, release the taps on the lockwasher.

8. Using SST 09607–60020 or equivalent, remove the locknut.

9. Remove the lockwasher and adjusting nut.

10. Remove the claw washer.

11. Remove the axle hub and rotor as an assembly. Remove the outer bearing with the hub and disc.

12. Remove the oil seal and inner bearing, using a suitable puller.

13. If replacing bearing outer race, drive out the outer bearing race using a brass bar and hammer.

To install:

14. If removed, drive in a new bearing outer race using a suitable installer.

15. Pack the bearings with MP grease. Coat the inside of the hub and cap with MP grease.

16. Install the inner bearing and oil seal. Coat the oil seal with MP grease.

17. Install the hub on the spindle. Install the outer bearing and claw washer.

18. Adjust the preload:

a. Tighten the bearing adjusting nut to 43 ft. lbs. (59 Nm).

b. Turn the hub right and left 2 or 3 times and retighten.

c. Loosen the nut until it can be turned by hand.

d. Retighten the nut to 18 ft. lbs. (25 Nm).

e. Check the bearing preload with a spring scale. The preload should be 6.4–12.6 ft. lbs. (28–56 N).

19. Install the lockwasher and nut. Tighten to 35 ft. lbs. (47 Nm)

20. Check that there is no bearing end-play.

21. Secure the locknut by bending one lockwasher tooth inward and another outward.

22. If equipped without a free wheeling hub, install the flange as follows:

a. Place a new gasket in position on the axle hub.

b. Install the flange to the axle hub.

c. Install the six cone washers, plate washers and nuts.

d. Install the six nuts. Tighten the nuts to 23 ft. lbs. (31 Nm)

e. Install the bolt and tighten the bolt to 13 ft. lbs. (18 Nm).

f. Install the grease cap.

23. If equipped with a free wheeling hub, install the hub as follows:

a. Place a new gasket in position on the front axle hub.

b. Install the free wheeling hub body with the six cone washers and nuts. Tighten the nuts to 23 ft. lbs. (31 Nm).

c. Install the bolt with the washer. Tighten the bolt to 13 ft. lbs. (18 Nm).

d. Apply multi purpose grease to the inner hub splines.

e. Set the control handle and clutch to the FREE position.

f. Place a new gasket in position on the cover.

g. Install the cover to the hub body with the follower pawl tabs aligned with the non-toothed portions of the hub body.

h. Install the mounting bolts and tighten the cover bolts to 7 ft. lbs. (10 Nm).

24. Install the brake caliper support bracket to the steering knuckle and tighten the two bolts to 90 ft. lbs. (123 Nm).

25. Clean the threads of the bolts and steering knuckle.

26. Apply sealant to the bolt threads and connect the knuckle arm with the brake line bracket to the steering knuckle. Tighten the bolts to 135 ft. lbs. (183 Nm).

27. If equipped with ABS brakes, connect the ABS speed sensor to the steering knuckle.

28. Install the front wheel and lower the vehicle.

29. Check the ABS speed sensor signal.

1996–99 4RUNNER

1. Raise and safely support the vehicle.
2. Remove the front wheels.
3. Remove the shock absorber.
4. Remove the grease cap.
5. 4-Wheel Drive: Disconnect the half-shaft.

a. Remove the cotter pin and lock cap.

b. While applying the brakes, remove the locknut.

6. With ABS: Remove the ABS speed sensor and wiring harness clamp from the steering knuckle.

7. Remove the brake line bracket from the steering knuckle.

8. Remove the front brake caliper and the rotor.

9. Remove the four bolts and disconnect the lower ball joint.

10. Remove the steering knuckle with the axle hub.

a. Remove the cotter pin and loosen the nut.

b. Using SST 09950–40010 or equivalent, disconnect the steering knuckle.

11. Clamp the axle hub in a soft jaw vise.

12. 2-Wheel Drive: remove the grease cap.

13. 4-Wheel Drive: remove the inside oil seal.

14. Remove the four bolts and shift the brake dust cover towards the hub side. Using SST 09710–30021 or equivalent, remove the axle hub from the steering knuckle.

15. Remove the bearing spacer and ABS speed sensor rotor/spacer.

16. Using a flat prying tool, remove the oil seal (outside) from the steering knuckle.

17. Remove the bearing from the steering knuckle.

a. Remove the snapring.

b. Using SST 09950–60020 and 09950–70010 or equivalent, and a press, remove the bearing from the steering knuckle.

To install:

18. Install a new bearing.

a. Using SST 09527–17011 and 09950–60020 or equivalent, and a press, install a new bearing to the steering knuckle.

b. Install a new snapring.

19. Using SST 09223–15030 or equivalent, and a plastic hammer, install a new outside oil seal.

a. Coat MP grease to the oil seal lip.

20. Install the brake dust cover to the steering knuckle with the four bolts and tighten to 13 ft. lbs. (18 Nm).

21. Using a press, install the axle hub to the steering knuckle.

22. Install the ABS speed sensor rotor/spacer. Be careful not to scratch the serration of the speed sensor rotor.

23. Using a press, install the bearing spacer.

24. If removed, install the grease cap.

25. If removed, install a new inside oil seal. Using SST 09527–17011 or equivalent, and a plastic hammer, strike the circumference evenly.

26. Install the steering knuckle with the axle hub. Tighten the nut to 80 ft. lbs. (108 Nm). Install a new cotter pin.

27. Install the lower ball joint and tighten the four bolts to 59 ft. lbs. (80 Nm).

28. Install the rotor and the caliper. Tighten the caliper bolts to 90 ft. lbs. (123 Nm).

29. Install the brake line bracket to the steering knuckle and tighten to 21 ft. lbs. (28 Nm).

30. If removed, install the ABS speed sensor and wiring harness clamp to the steering knuckle. Tighten the bolts to 6 ft. lbs. (8 Nm).

31. If disconnected, install the halfshaft and tighten the nut to 174 ft. lbs. (235 Nm).

32. Install the grease cap.

33. Install the shock absorber.

34. Install the front wheel.

35. Connect the negative battery cable.

TACOMA 2-WHEEL DRIVE MODELS

1. Raise and safely support the front of the vehicle. Place the jackstands under the frame of the vehicle.

2. Remove the brake caliper support bracket by removing the two bolts. Support the brake caliper with a piece of wire. Do not allow the caliper to hang from the brake hose.

3. Remove the cotter pin, lock cap, nut, and the claw washer from the axle hub and disc.

4. Remove the axle hub with the disc from the steering knuckle. Do not drop the outer bearing when removing the hub.

5. Using a suitable tool, remove the inner oil seal.

6. Remove the inner bearing.

7. Using SST 09527–17011, or equivalent, a brass bar, and a hammer, drive out the bearing outer races.

8. If it is necessary to separate the hub and rotor, place matchmarks on the hub and

rotor. Remove the five bolts and remove the hub from the rotor.

To install:

9. Using SST 09527–17011 and a press, install new bearing races.

10. Connect the hub to the rotor and install the five bolts. Tighten the five bolts to 47 ft. lbs. (64 Nm).

11. Clean all parts.

12. Repack the bearings with multi purpose grease and apply the same grease to the outer bearings.

13. Install the inner bearing and seal to the hub. Coat the inner seal with multi purpose grease.

14. Install the outer bearing to the hub.

15. Install the hub to the steering knuckle.

16. Install the claw washer and nut to holding the axle hub to the steering knuckle.

17. To adjust the bearing preload, perform the following:

a. Tighten the adjusting nut to 25 ft. lbs. (34 Nm).

b. Turn the disc/hub assembly 2–3 times, from the left to the right.

c. Loosen the adjusting nut until it can be turned by hand.

d. Attach a spring tension gauge to 1 lug on the hub assembly. Pull on the gauge and measure the frictional force, which should be 1.3–4.0 ft. lbs. (6.0–18.0 N).

e. Adjust the preload by tightening the nut.

18. Measure the hub axial play. Limit 0.0020 in. (0.05mm).

19. Install the locknut, cotter pin, and the grease cap.

20. Install the disc brake caliper by installing the two bolts. Tighten the two bolts to 80 ft. lbs. (108 Nm).

21. Install the wheel and tire assembly.

22. Lower the vehicle.

TACOMA 4-WHEEL DRIVE MODELS

1. Raise and safely support the vehicle.

2. Remove the wheel and tire assembly.

3. If not equipped with a FREE wheeling hub, disconnect the halfshaft from the steering knuckle as follows:

a. Remove the grease cap.

b. Remove the cotter pin and lock cap from the halfshaft.

c. While having an assistant apply the brakes, remove the locknut from the halfshaft.

4. If equipped with FREE wheeling hub, remove the free wheel hub as follows:

a. Set the control handle to FREE.

b. Remove the cover bolts and pull off the cover.

c. Remove the center bolt with washer.

d. Remove the mounting nuts and washer to the hub body.

e. Using a brass bar and hammer, tap on the bolts head and remove the cone washer.

f. Pull off the free wheel hub body and gasket.

g. Using a snapring expander, remove the snapring from the end of the half-shaft.

5. If equipped with ABS brakes, disconnect the ABS speed sensor from the steering knuckle.

6. Disconnect the brake hose from the steering knuckle by removing the bolt.

7. Remove the two brake caliper support bracket bolts and wire the caliper to the side. Do not allow the caliper to hang from the brake hose.

8. Remove the rotor from the steering knuckle.

9. Remove the four bolts and disconnect the lower ball joint from the steering knuckle.

10. Remove the cotter pin and nut to the upper control arm.

11. Using SST 09950–40010 or equivalent, disconnect the steering knuckle from the upper control arm.

12. Remove the steering knuckle from the vehicle.

➡**If it is difficult to remove the half-shaft from the steering knuckle, use a rubber hammer and tap the halfshaft from the steering knuckle.**

13. Place the axle hub in a soft jaw vise.

14. Using a prytool, remove the inside oil seal.

15. If equipped with free wheeling hubs, remove the locknut and ABS speed sensor rotor/spacer as follows:

a. Using a hammer and chisel, loosen the staked part of the locknut.

b. Using SST 09318–12010 or equivalent, remove the locknut from the hub.

c. Remove the ABS speed sensor rotor/spacer. Take care not to scratch the serration of the speed sensor rotor.

16. Remove the axle hub from the steering knuckle as follows:

a. Remove the four bolts for the backing plate and shift the plate towards the hub side.

b. Using the proper tools, press the axle hub from the steering knuckle.

c. If the vehicle is not equipped with free wheeling hubs, remove the bearing spacer and ABS speed sensor rotor/spacer.

17. Using a prytool, remove the outside oil seal from the steering knuckle.

18. Using a snapring pliers, remove the snapring from the hub.

19. Using SST 09608–35014 or equivalent and a press, remove the bearing from the steering knuckle.

To install:

20. Using SST 09527–17011 and a press, install a new bearing to the steering knuckle.

21. Using a snapring pliers, install the snapring to the hub.

22. Using SST 09223–15030, 09527–17011 and a plastic hammer, install a new outside oil seal. Coat the multi purpose grease to the oil seal lip.

23. Turn the backing plate back into place and install the four bolts. Tighten the bolts to 13 ft. lbs. (18 Nm).

24. Using SST 09649–17010 or equivalent and a press, install the axle hub to the steering knuckle.

25. Install the ABS speed sensor rotor/spacer.

26. If equipped with free wheeling hubs, install a new locknut to the hub and tighten the nut to 203 ft. lbs. (274 Nm). Stake the nut with a chisel and hammer.

27. If the vehicle is not equipped with free wheeling hubs, install the bearing spacer with SST 09950–60010 and a press.

28. Using SST 09527–17011 or equivalent and a plastic hammer, install a new inside oil seal. Coat the multi purpose grease to the oil seal lip.

29. Install the steering knuckle to the halfshaft.

30. Connect the steering knuckle to the lower ball joint by installing the four bolts. Do not tighten the bolts at this time.

31. Push down on the upper control arm and connect the upper ball joint to the steering knuckle.

32. Install the upper ball joint nut. Tighten the nut to 80 ft. lbs. (105 Nm). Install a new cotter pin.

33. Tighten the lower ball joint to steering knuckle bolts to 59 ft. lbs. (80 Nm).

34. Install the brake rotor.

35. Connect the caliper support bracket to the steering knuckle and install the two bolts. Tighten the bolts to 90 ft. lbs. (123 Nm).

36. Connect the brake hose clamp to the steering knuckle by installing the bolt. Tighten the bolt to 13 ft. lbs. (18 Nm).

37. If equipped with ABS brakes, connect the ABS speed sensor and wiring harness to the steering knuckle.

38. Install the spacer and using a snapring expander, install the snapring to the halfshaft.

39. If equipped with a free wheeling hub, install the hub as follows:

a. Place a new gasket in position on the front axle hub.

b. Install the free wheeling hub body with the six cone washers and nuts. Tighten the nuts to 23 ft. lbs. (31 Nm).

c. Install the bolt with the washer. Tighten the bolt to 13 ft. lbs. (18 Nm).

d. Apply multi purpose grease to the inner hub splines.

e. Set the control handle and clutch to the FREE position.

f. Place a new gasket in position on the cover.

g. Install the cover to the hub body with the follower pawl tabs aligned with the non-toothed portions of the hub body.

h. Install the mounting bolts and tighten the cover bolts to 7 ft. lbs. (10 Nm).

40. If equipped without a free wheeling hub, install the halfshaft to the steering knuckle as follows:

a. Install the locknut to the halfshaft. Tighten the nut to 174 ft. lbs. (235 Nm).

b. Install the lock cap and cotter pin to the halfshaft.

c. Install the grease cab to the hub.

41. Install the strut to the vehicle. Tighten the nut holding the strut to the lower control arm to 101 ft. lbs. (135 Nm). Tighten the upper three nuts to 47 ft. lbs. (64 Nm)

42. Install the front wheels and lower the vehicle.

TOYOTA and LEXUS

35

Lexus-Rx300 • **Toyota-** Previa • RAV4 • Sienna

ENGINE REPAIR

➡ **Disconnecting the negative battery cable on some vehicles may interfere with the functions of the on board computer systems and may require the computer to undergo a relearning process, once the negative battery cable is reconnected.**

Distributor

The Sienna, RX 300 and the 1998–99 RAV4 models are equipped with a distributorless ignition.

REMOVAL

2.0L (VIN P) Engine
1995–97 MODELS

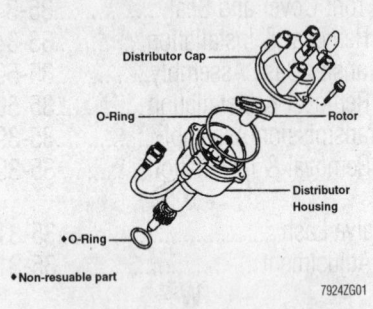

Non-resuable part   7924ZG01

Exploded view of the distributor assembly—2.0L (VIN P) engine

1. Disconnect the negative battery cable.

✳✳ CAUTION

Wait 90 seconds from the time the key is turned to LOCK and the negative battery cable is disconnected to begin work. This allows the SRS capacitor to discharge and prevent deployment of the air bag(s).

2. Remove the air cleaner cap assembly.
3. Detach the electrical connector to the distributor.
4. Remove the high tension cable from the coil.
5. Mark and remove the spark plug wires from the distributor. Remove the distributor cap bolts, then remove the cap.
6. Matchmark the rotor to the distributor housing and the distributor housing to the engine block. This will aid in correct positioning of the distributor during installation.

7. Remove the distributor hold-down clamp bolt and the distributor from the engine.

INSTALLATION

Timing Not Disturbed

1. Install a new O-ring to the distributor and lubricate it with engine oil.
2. Insert the distributor into the engine block by aligning the matchmarks made during removal.
3. Engage the distributor drive with the slit in the intake camshaft.
4. Install the distributor hold-down clamp, the cap, the high tension wire, the spark plug wires and the electrical connector.
5. Install the air cleaner cap assembly.
6. Connect the negative battery cable.
7. Start the engine. Check and adjust the ignition timing.

Timing Disturbed

1. Install a new O-ring to the distributor and lubricate it with engine oil.
2. Remove the No. 1 cylinder spark plug.
   a. Place a finger or compression gauge over the spark plug hole.
   b. Turn the crankshaft until compression starts to build up. Continue turning the crankshaft until the crankshaft pulley groove align with the timing mark **0** of the timing chain. The position of the slit of the intake camshaft should be as shown.
3. On the distributor, align the cutout of the coupling with the line on the housing.

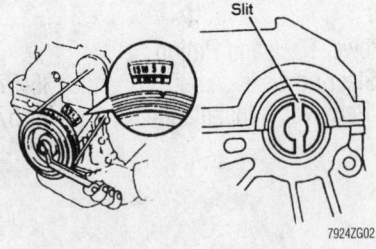

Slit

7924ZG02

Crankshaft TDC mark and intake camshaft slit position—2.0L (VIN P) engine

4. Insert the distributor, aligning the center of the flange with that of the bolt hole on the cylinder head.
5. Tighten the hold-down bolt to 14 ft. lbs. (19 Nm).
6. Install the distributor cap.
7. Install the spark plug wires and the electrical connector.

8. Install the No. 1 cylinder spark plug. Install the high tension wire on the coil.
9. Connect the negative battery cable.
10. Start the engine and inspect the timing.

REMOVAL

2.4L (VIN A and T) Engines

1. On the 2TZ-FZE engine, remove the exhaust pipe heat insulator.
2. Disconnect the negative battery cable. Wait at least 90 seconds prior to working on the vehicle on models equipped with and airbag.
3. Label and disconnect the spark plug wires.
4. Disconnect the distributor wiring and ventilation hoses.
5. Remove the cap and packing.

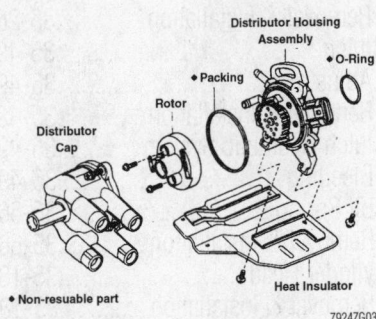

Distributor Housing Assembly

♦ Packing

♦ O-Ring

Rotor

Distributor Cap

Heat Insulator

Non-resuable part   7924ZG03

Exploded view of the distributor assembly—2.4L (VIN A and T) Engines

6. Set the No. 1 cylinder to TDC of the compression stroke. Install the service bolt and nut into the equipment driveshaft to turn the crankshaft pulley until the timing mark is aligned with the 0 mark on the timing chain cover.
7. Turn the crankshaft one turn if the rotor is not facing No. 1 spark plug wire.

➡ **Check that the rotor direction is as shown, if not, turn the drive pulley one complete revolution.**

8. Place markings on the distributor housing, and rotor positions. Remove the two hold-down bolts and pull the distributor out of the engine.

INSTALLATION

Timing Not Disturbed

1. Install a new O-ring to the distributor and lubricate with engine oil if it has not recently been replaced.

2. Insert the distributor, aligning the center of the distributor housing flange with the bolt hole on the cylinder head.

3. Engage the distributor drive with the oil pump drive shaft.

4. Install the distributor hold-down clamp, the cap, the high tension wire, the primary wire or the electrical connector and the vacuum line(s).

5. Install the spark plugs cables.

6. Connect the negative battery cable.

7. Reset all digital components such as the radio.

Timing Disturbed

1. If the engine was disturbed while the distributor was removed, continue as follows:

2. Turn the drive pulley clockwise, and position the slit of the exhaust camshaft as shown in the illustration.

➡ **Be sure the slit in the exhaust camshaft is in the proper position.**

3. Remove the service bolt and nut.

4. Align the cut out portion of the coupling with the groove on the housing.

5. Install the distributor and align the center of the flange with the bolt hole on the cylinder head.

6. Install the hold-down bolt loosely.

7. Install the seal packing, distributor cap, air hoses and connect the wiring.

Ignition Timing

ADJUSTMENT

Ignition timing is controlled by the ECM and is not adjustable.

Engine Assembly

REMOVAL & INSTALLATION

RAV4

✶✶ CAUTION

The fuel injection system remains under pressure after the engine has been turned OFF. Properly relieve fuel pressure before disconnecting any fuel lines. Failure to do so may result in fire or personal injury.

1. Relieve the fuel system pressure.

2. Disconnect the negative battery cable.

✶✶ CAUTION

On models with an air bag, wait at least 90 seconds from the time that the ignition switch is turned to the LOCK position and the battery is disconnected before performing any further work.

3. Remove the battery.

4. Remove the hood.

5. Remove the engine undercover, then drain the engine coolant and oil.

6. Drain the transaxle assembly.

7. Remove the air cleaner and case.

8. Disconnect the accelerator cable from the throttle body, bracket and clamps.

9. Disconnect the engine wire from the No. 2 relay block as follows:

a. Disconnect the No. 2 relay block from the body by removing the two bolts.

b. Remove the upper cover to the relay block.

c. Detach the connectors.

d. Remove the engine wire by removing the two nuts.

10. Remove the charcoal canister.

11. Remove the alternator as follows:

a. Disconnect the electrical wiring from the alternator.

b. Loosen the adjusting lockbolt and the pivot bolt.

c. If equipped with A/C, loosen the adjusting bolt to relieve tension on the drive belt.

d. Remove the drive belt. It may be necessary to remove other belts for access.

e. Remove the pivot bolt first, support the alternator and remove the adjusting lockbolt.

f. Remove the alternator from the vehicle.

12. Remove the upper and lower radiator hoses.

13. Remove the water inlet from the engine by removing the two nuts.

14. Disconnect the heater hoses.

✶✶ CAUTION

Fuel injection systems remain under pressure after the engine has been turned OFF. Properly relieve fuel pressure before disconnecting any fuel lines. Failure to do so may result in fire or personal injury.

15. Place a rag under the fuel inlet hose and disconnect the hose.

16. For vehicles with M/T, remove the starter by disconnecting the electrical connectors and two bolts.

17. Disconnect the ground cable from the transaxle by removing the bolt.

18. For vehicles with M/T, remove the clutch release cylinder from the transaxle.

19. Disconnect the transaxle control cables (two cable for M/T and one for A/T) from the transaxle.

20. For vehicles with A/T disconnect the transaxle cable from the front suspension crossmember and engine mounting center member by removing the two bolts.

21. For vehicles with A/T or 4WD with M/T, disconnect the transaxle oil cooler hoses.

22. Detach the following:

- Vapor pressure sensor connector
- Igniter connector
- Ignition coil connector
- Noise filter connector
- Ignition coil wire
- MAP sensor connector
- MAP sensor vacuum hose from the gas filter on the intake manifold
- Brake booster hose from the intake manifold
- If equipped with 4WD M/T, detach the differential lock control solenoid connector
- Ground strap from cowl

23. Detach the engine wire from the passenger compartment as follows:

a. Remove the right-hand scuff plate.

b. Remove the right-hand side trim.

c. Remove the right-hand carpet center cover.

d. Detach the two ECM connectors.

e. Detach the two connector from the connectors on the bracket.

f. Remove the connector from the No. 4 junction block.

g. Detach the wire clamp from the bracket.

h. Pull out the engine wire from the passenger compartment.

24. Remove the front exhaust pipe as follows:

a. Using a 14mm. deep socket wrench, remove the three nuts and the gasket to disconnect the front exhaust pipe from the exhaust manifold.

b. Remove the two bolts and two nuts holding the front exhaust pipe to the catalytic converter.

c. Remove the front exhaust pipe and two gaskets.

25. Disconnect the compressor from the engine and suspend the compressor securely. It is not necessary to remove the A/C compressor lines in order to remove the engine.

26. If equipped with 4WD, remove the drive shaft.

27. Remove the halfshaft from the vehicle.

28. Remove the sway bar.

29. Remove the front suspension crossmember assembly as follows:

   a. Remove the two center member set nuts holding the center member to the middle of the crossmember.

   b. Remove the two rack and pinion assembly set bolts and nuts from the crossmember. Securely suspend the steering gear assembly.

   c. Disconnect the catalytic converter with pipe from the ring.

   d. Support the suspension crossmember with a jack.

   e. Remove the six bolts from the suspension crossmember.

   f. Remove the suspension crossmember with the lower suspension arms.

30. Remove the engine mounting center member as follows:

   a. Remove the two bolts holding the center member to the front engine mounting insulator.

   b. Remove the two bolts holding the center member to the body and remove the center member.

31. Disconnect the power steering pump from the engine as follows:

   a. Disconnect the two vacuum hoses from the steering pump.

   b. Remove the adjusting bolt for the power steering unit. Loosen the pivot bolt to the power steering pump and remove the drive belt. Use Torque Wrench Adapter tool 09249–63010 or equivalent, and a deep socket to loosen the pivot bolt.

   c. Remove the power steering pump from the engine by removing the three bracket bolts.

➡ **It is not necessary to disconnect the power steering lines from the pump.**

32. Install an engine hanger to the engine.

33. Attach the engine sling device to the engine hangers.

34. Disconnect the left-hand engine mounting bracket from the mounting insulator by removing the two nuts and two bolts.

35. Detach the ground connector next to the right-hand engine mount.

36. Disconnect the right-hand engine mounting bracket from the mounting insulator by removing the bolt and two nuts.

37. Lower the engine and transaxle out of the vehicle slowly and carefully. At the same time, raise the vehicle to gain clearance to the remove the engine.

38. Place the assembly on a stand and separate the engine from the transaxle.

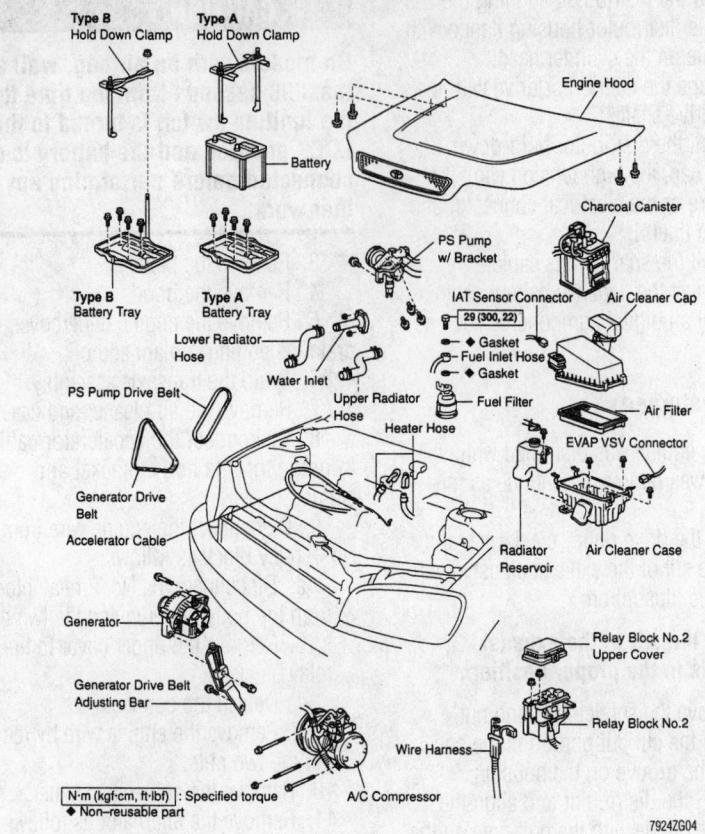

Exploded view of the engine accessory removal components—RAV4 model

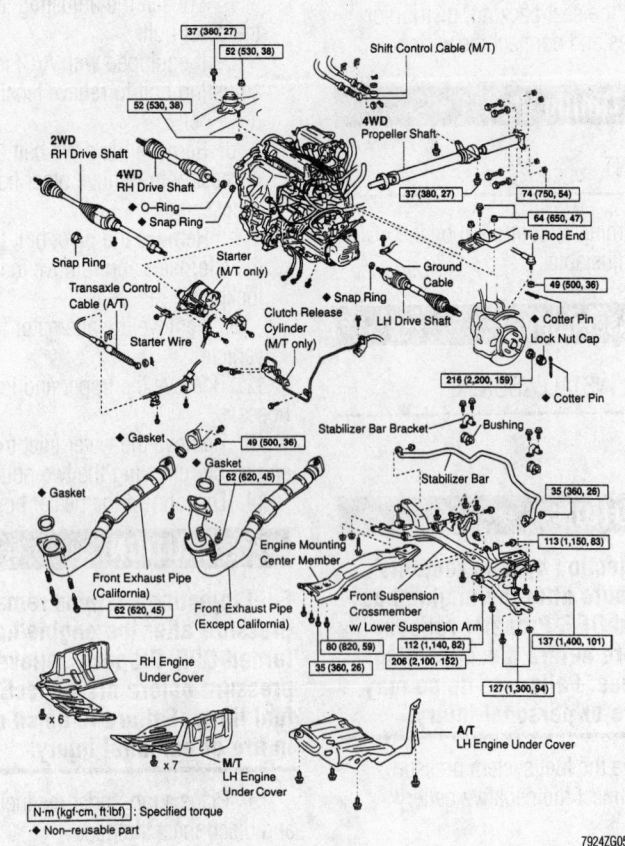

Exploded view of the engine removal—RAV4 model

To install:

39. Install the engine and transaxle in the vehicle.

40. Attach the left-hand engine mounting bracket to the mounting insulator and install the two nuts and two bolts. Tighten the bolts and nuts to 47 ft. lbs. (64 Nm).

41. Install the bolt and two nuts to hold the right-hand engine mounting bracket to the mounting insulator. Tighten the bolt to 27 ft. lbs. (37 Nm) and the two nuts to 38 ft. lbs. (52 Nm).

42. Attach the ground connector next to the right-hand engine mount.

43. Remove the engine sling and hanger.

44. Install the power steering pump as follows:

   a. Install the pump with the bracket. Install the three bolts and tighten the bolts to 32 ft. lbs. (43 Nm).

   b. Install the pivot bolt and tighten the bolt to 32 ft. lbs. (43 Nm). Tighten the adjusting bolt to 29 ft. lbs. (39 Nm).

   c. Install the drive belt and adjust the tension.

   d. Connect the two air hoses to the power steering pump.

45. Install the engine mounting center member to the body and install the four bolts. Do not tighten the bolts at this time.

46. Install the front crossmember as follows:

   a. Raise the suspension crossmember with the lower control arms. Install the two bolts to hold the crossmember to the vehicle. Tighten the bolts to 152 ft. lbs. (206 Nm).

   b. Connect the rack and pinion and install the two bolts and nuts. Tighten to 83 ft. lbs. (113 Nm).

   c. Connect the center member to the crossmember and install the two set nuts. Tighten the nuts to 82 ft. lbs. (112 Nm).

   d. Tighten the lower control arm rear brackets to 101 ft. lbs. (137 Nm).

   e. Tighten the two bolts to hold the engine mounting center member to the front engine mounting insulator. Tighten the bolts to 59 ft. lbs. (80 Nm).

   f. Tighten the two bolts to hold the engine mounting center member to the body. Tighten to 26 ft. lbs. (35 Nm).

47. Install the sway bar.

48. Install the halfshafts.

49. If equipped with 4WD, install the driveshaft.

50. Install the A/C compressor with the two bolts. stud bolt and nut. Tighten as follows:

- Stud bolt to 34 ft. lbs. (47 Nm)
- Bolt to 27 ft. lbs. (37 Nm)
- Nut to 20 ft. lbs. (27 Nm)

51. Attach the A/C compressor connector.

52. Install the front exhaust pipe with new gaskets. Tighten the three front nuts to 46 ft. lbs. (62 Nm) and the two bolts to 35 ft. lbs. (48 Nm).

53. Attach the engine wire to the passenger compartment as follows:

   a. Push in the engine wire through the cowl panel.

   b. Install the wire clamp to the bracket.

   c. Attach the two ECM connectors.

   d. Attach the two connector to the connectors on the bracket.

   e. Connect the No. 4 junction block.

   f. Install the right-hand floor carpet center cover.

   g. Install the right-hand cowl side trim

   h. Install the right-hand scuff plate

54. Attach the following:

- Vapor pressure sensor connector
- Igniter connector
- Ignition coil connector
- Noise filter connector
- Ignition coil wire
- MAP sensor connector
- MAP sensor vacuum hose to the gas filter on the intake manifold
- Brake booster hose to the intake manifold
- If equipped with 4WD M/T, attach the differential lock control solenoid connector
- Ground strap from cowl

55. For A/T and 4WD with M/T, connect the transaxle oil cooler hoses.

56. Connect the transaxle control cable(s) to the transaxle.

57. For A/T, install the transaxle control cable to the front crossmember and engine mounting center member.

58. For M/T, install the clutch release cylinder. Tighten the two bolts to 9 ft. lbs. (12 Nm).

59. Connect the ground cable to the transaxle.

60. For M/T, install the starter.

61. Using new gaskets, connect the fuel inlet hose to the fuel filter. Tighten the union bolt to 22 ft. lbs. (29 Nm).

62. Connect the heater hoses.

63. Connect the water inlet to the engine with the two nuts. Tighten the nuts to 78 inch lbs. (8.8 Nm).

64. Install the radiator upper and lower hoses.

65. Install the alternator to the engine.

66. Install the charcoal canister.

67. Connect the engine wire to the No. 2 relay box as follows:

   a. Connect the engine wire to the No. 2 relay block with the two nuts.

   b. Attach the connector.

   c. Install the upper cover.

   d. Connect the No. 2 relay block to the body with the two bolts.

68. Install the accelerator cable to the throttle body, cable bracket and clamps.

69. Install the air cleaner case and cap.

70. Install the battery.

71. Fill the transaxle with oil.

72. Fill the engine with oil.

73. Fill the engine coolant.

74. Connect the battery cables.

75. Start the engine and check for leaks.

76. Check the front wheel alignment.

77. Install the engine undercovers.

78. Install the hood.

79. Recheck all fluid levels.

Previa

> **✵✵ CAUTION**
>
> **The fuel injection system remains under pressure after the engine has been turned OFF. Properly relieve fuel pressure before disconnecting any fuel lines. Failure to do so may result in fire or personal injury.**

1. Relieve the fuel system pressure.

2. Disconnect the negative battery cable. Wait at least 90 seconds to proceed working on the vehicle if equipped with an airbag.

3. Drain the engine coolant and oil.

4. Raise the vehicle and support safely. Remove the engine undercovers.

5. Drain the engine oil and cooling system.

6. On 4WD vehicles, disconnect the front driveshaft.

7. Remove the rear driveshaft.

8. Remove the air duct.

9. Matchmark and disconnect the equipment (separated accessory drive system) driveshaft from the crankshaft pulley.

10. Disconnect the A/T shift cable.

11. Remove the air intake duct.

12. Disconnect the ground strap from the left-hand front engine mounting.

13. Disconnect the starter wire.

14. Disconnect the following hoses:

- No. 4 radiator hose from the water inlet
- No. 1 radiator hose from the water inlet
- Heater hose from the water pump
- Oil auto feeder hose from the No. 1 oil return pipe
- Disconnect the A/C idle up air hose from the union under the intake manifold
- Disconnect the power steering idle up air hose from the union under the intake manifold

• Disconnect the water bypass hose from the floor pipe

• Disconnect the brake booster hose from the floor pipe

• Disconnect the two vacuum hoses for the fuel pressure control Vacuum Switching Valve (VSV) from the engine wire and vacuum transmitting pipe on the throttle body

• Air hose for the distributor ventilation from the water bypass pipe under the intake manifold

• Disconnect the vacuum hose for the EVAP from the charcoal canister

15. If equipped with automatic transmission, disconnect the shift cable.

16. Remove the intake pipe hose.

17. Disconnect the accelerator cable from the throttle body.

18. Disengage the Vacuum Switching Valve (VSV) connector for the fuel pressure control.

19. Disconnect the engine wire from the engine left side as follows:

a. Disengage the igniter connector.

b. Disconnect the two Electronic Control Module (ECM) harnesses.

c. Disconnect the four harnesses from the cowl wire on the front floor panel.

d. Separate the engine wire from the front floor panel by removing the bolt and detaching the three clamps.

e. Pull out the engine wire from the front floor panel hose.

20. Remove the A/T oil dipstick.

21. Disconnect the fuel inlet and return hoses.

22. Remove the front exhaust pipe.

23. Remove the exhaust pipe heat insulator and ground strap by removing the four bolts.

24. Disconnect the two A/T oil cooler hoses.

25. Remove the ignition coil and disconnect the ground strap for the engine.

26. Disconnect the condenser wiring.

27. Disconnect the four clamps and engine wire.

28. Disconnect the A/T and park/neutral position switch wiring.

29. Remove the engine with the transmission as follows:

a. Support the engine and transmission with a supporting device.

b. Lower the vehicle while supporting the engine and transmission with the engine lifter.

c. Remove the two bolts, two nuts and two plate washers holding the right and left engine mountings to the engine front support member.

d. Remove the four through-bolts, four plate washers and four nuts holding the rear mounting to the No. 2 rear engine mounting bracket.

e. Be sure the engine and transmission are clear of all wiring, hoses and cables.

f. Lower the engine and transmission to the floor.

To install:

30. Install the engine in the reverse order of removal while paying close attention to the following.

31. Raise the engine and install the two bolts, two plate washers and two nuts to hold the right and left engine front mountings to the engine front support member. Tighten the bolts and nuts to 27 ft. lbs. (37 Nm).

32. Install the two through-bolts, four washers and four nuts to hold the engine rear mounting to the No. 2 rear engine mounting bracket. Tighten the bolts and nuts to 31 ft. lbs. (42 Nm).

33. Install the rear driveshaft. If equipped with 4WD, install the front driveshaft. Install a new oil filter and fill the engine with oil. Fill the engine with engine coolant. Install the engine undercovers. Start the engine and check for leaks.

Sienna and RX300

1. Matchmark the hood position and remove the hood.

2. Remove the wiper and blade assembly.

3. Remove the top cowl seal and panel.

4. Label and disconnect the window washer hoses from the ventilator louvers.

5. Remove the left and right ventilator louvers.

6. Remove the heater air duct.

> **※ CAUTION**
>
> **The fuel injection system remains under pressure after the engine has been turned OFF. Properly relieve fuel pressure before disconnecting any fuel lines. Failure to do so may result in fire or personal injury.**

7. Properly relieve the fuel system pressure.

8. Disconnect the negative, then the positive battery cable. Remove the battery and tray.

9. Drain and recycle the engine coolant.

10. Drain the engine oil into a suitable container.

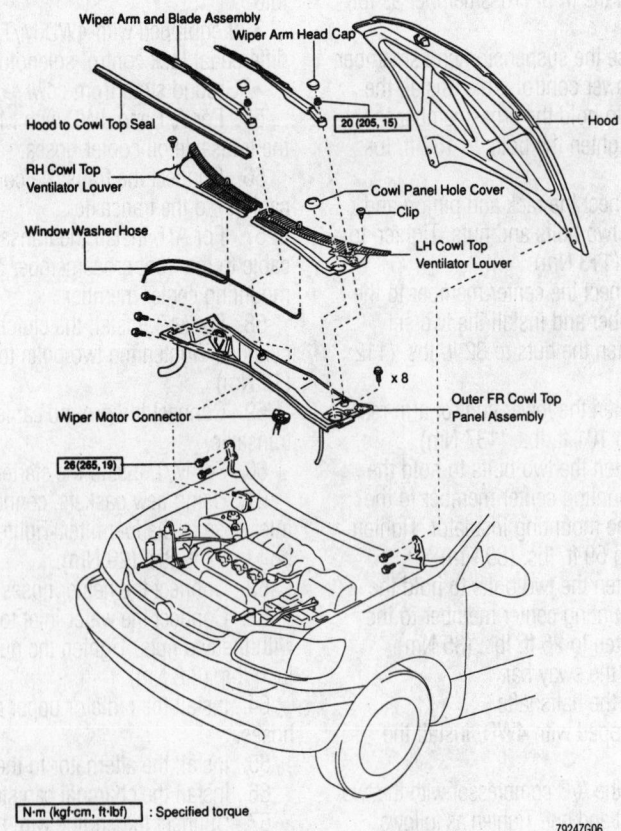

N·m (kgf·cm, ft·lbf) : Specified torque

7924ZG06

Exploded view of the top cowl removal and related components—Sienna model

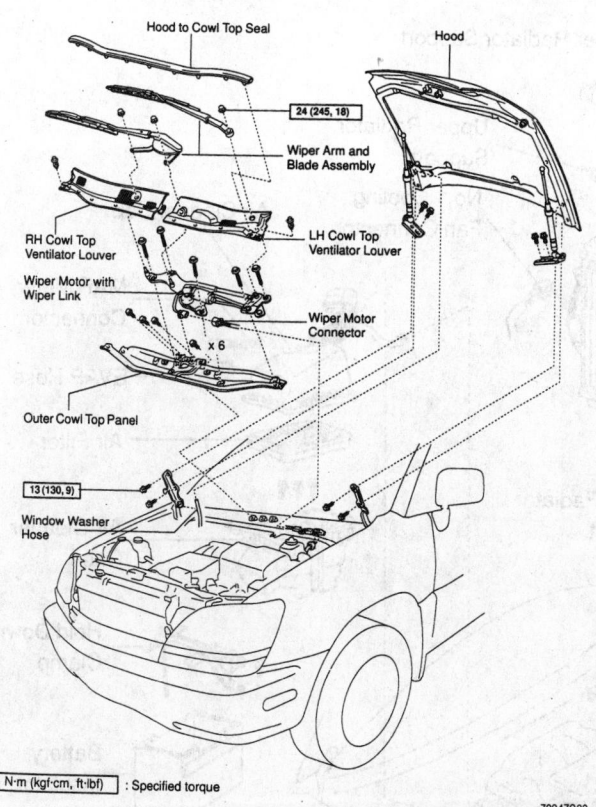

Hood to Cowl Top Seal

24 (245, 18)

Hood

Wiper Arm and Blade Assembly

RH Cowl Top Ventilator Louver

LH Cowl Top Ventilator Louver

Wiper Motor with Wiper Link

Wiper Motor Connector

x 6

Outer Cowl Top Panel

13 (130, 9)

Window Washer Hose

N·m (kgf·cm, ft·lbf) : Specified torque

7924ZG83

Exploded view of the top cowl removal and related components—RX300 model

11. Remove the intake air cleaner and case assembly.

12. If equipped, remove the cruise control actuator.

13. On RX300 models, remove the upper suspension brace.

14. Disconnect the upper and lower radiator hoses, then remove the radiator.

15. Disconnect and plug the A/T oil cooler lines.

16. Label and detach any connectors, hoses and sensors that would interfere with engine removal.

17. Disconnect the ECM engine wiring harness from inside the glove box, and pull the harness into the engine compartment.

➡**Only a MVAC-trained, EPA-certified, automotive technician should service the A/C system or its components.**

18. Disconnect the compressor from the engine and suspend the compressor securely. It may be necessary to remove the

A/C compressor lines in order to remove the engine.

19. Disconnect the A/T shifter cable from the transaxle.

20. Disconnect the header pipes from the exhaust manifolds.

21. Remove the left and right fender apron seals.

22. Remove the halfshafts.

23. On 4WD RX300 models, remove the front driveshaft.

24. Remove the stabilizer links and the steering intermediate shaft.

25. Disconnect the power steering pump.

26. Remove the engine undercover.

27. Install a engine hanger to the engine.

28. Attach the engine sling device to the engine hangers, as shown.

29. Remove the right-hand motor mount and moving control rod.

30. Remove the front suspension lower braces.

31. Lower the engine, transaxle and front suspension member as an assembly from the vehicle.

To install:

32. Raise the engine, transaxle and front suspension member as an assembly into the vehicle.

33. Install the front suspension lower braces, and tighten the fasteners as follows:

- Bolt A: 134 ft. lbs. (181 Nm)
- Bolt B: 24 ft. lbs. (32 Nm)
- Nut C: 27 ft. lbs. (36 Nm)

34. Install the moving control rod and tighten the bolts to 47 ft. lbs. (64 Nm).

35. Install the right-hand motor mount and tighten the bolts to 23 ft. lbs. (32 Nm).

36. Remove the engine sling device from the engine hangers.

37. Install the engine undercover.

38. Connect the power steering pump hoses.

39. Connect the stabilizer links and the steering intermediate shaft.

40. On 4WD RX300 models, install the front driveshaft.

41. Install the halfshafts.

42. Install the left and right fender apron seals.

43. Connect the header pipes to the exhaust manifolds.

44. Connect the A/T shifter cable to the transaxle.

➡**Only a MVAC-trained, EPA-certified, automotive technician should service the A/C system or its components.**

45. Connect the A/C compressor to the engine.

46. Push the wiring harness into the glove box and connect to the ECM.

47. Attach any connectors, hoses and sensors that were removed.

48. Connect the A/T oil cooler lines.

49. Connect the upper and lower radiator hoses, and fit the radiator.

50. On RX300 models, install the front upper suspension brace, and tighten the mounting nuts to 59 ft. lbs. (80 Nm).

51. If removed, install the cruise control actuator.

52. Install the intake air cleaner and case assembly.

53. Fill the engine oil to proper level.

54. Fill the engine with coolant.

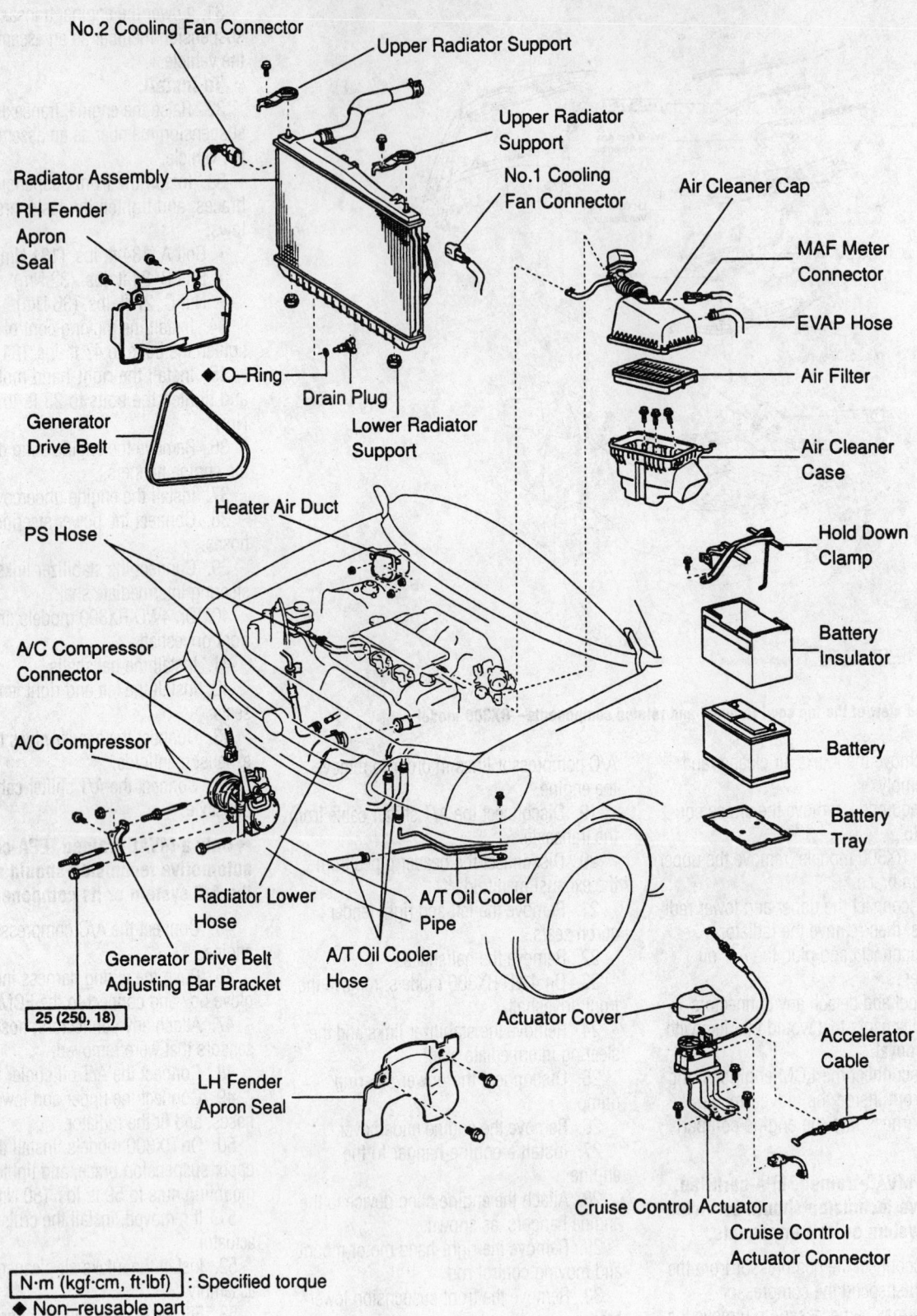

No.2 Cooling Fan Connector

Upper Radiator Support

Upper Radiator Support

No.1 Cooling Fan Connector

Air Cleaner Cap

MAF Meter Connector

Radiator Assembly

RH Fender Apron Seal

EVAP Hose

Air Filter

Air Cleaner Case

◆ O–Ring

Drain Plug

Lower Radiator Support

Generator Drive Belt

Heater Air Duct

PS Hose

Hold Down Clamp

Battery Insulator

A/C Compressor Connector

Battery

A/C Compressor

Battery Tray

Radiator Lower Hose

A/T Oil Cooler Pipe

Generator Drive Belt Adjusting Bar Bracket

A/T Oil Cooler Hose

| 25 (250, 18) |

Actuator Cover

Accelerator Cable

LH Fender Apron Seal

Cruise Control Actuator

Cruise Control Actuator Connector

| N·m (kgf·cm, ft·lbf) | : Specified torque
◆ Non–reusable part

7924ZG07

Exploded view of engine pre-removal components—Sienna model

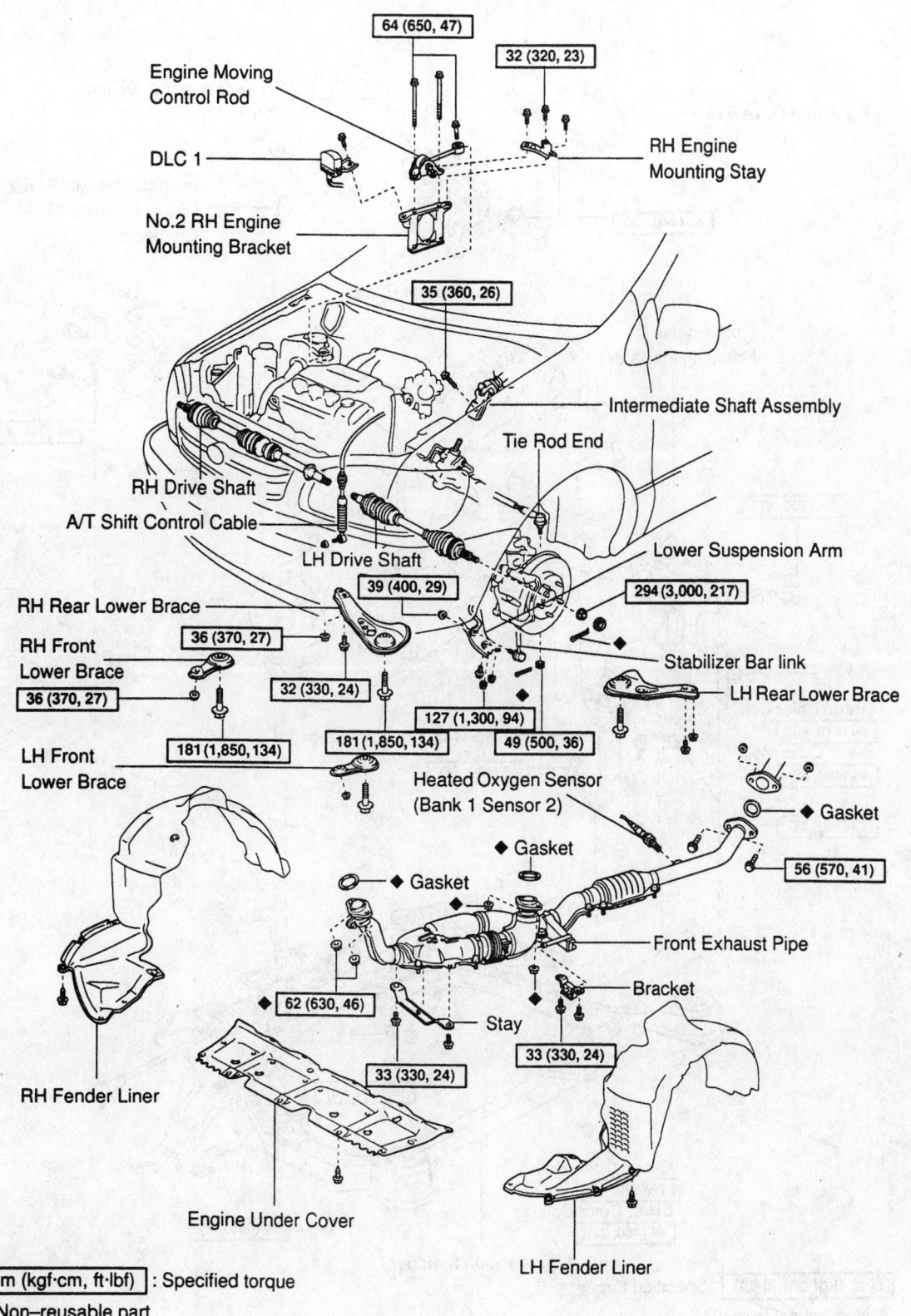

64 (650, 47)

32 (320, 23)

Engine Moving
Control Rod

DLC 1

RH Engine
Mounting Stay

No.2 RH Engine
Mounting Bracket

35 (360, 26)

Intermediate Shaft Assembly

Tie Rod End

RH Drive Shaft

A/T Shift Control Cable

LH Drive Shaft

Lower Suspension Arm

RH Rear Lower Brace

39 (400, 29)

294 (3,000, 217)

RH Front
Lower Brace

36 (370, 27)

Stabilizer Bar link

36 (370, 27)

32 (330, 24)

LH Rear Lower Brace

LH Front
Lower Brace

181 (1,850, 134)

181 (1,850, 134)

127 (1,300, 94)

49 (500, 36)

Heated Oxygen Sensor
(Bank 1 Sensor 2)

◆ Gasket

56 (570, 41)

◆ Gasket

◆ Gasket

RH Fender Liner

◆ 62 (630, 46)

Stay

Front Exhaust Pipe

Bracket

33 (330, 24)

33 (330, 24)

Engine Under Cover

LH Fender Liner

N·m (kgf·cm, ft·lbf) : Specified torque

◆ Non–reusable part

7924ZG08

Exploded view of engine removal and installation tightening specifications of the related components—Sienna model

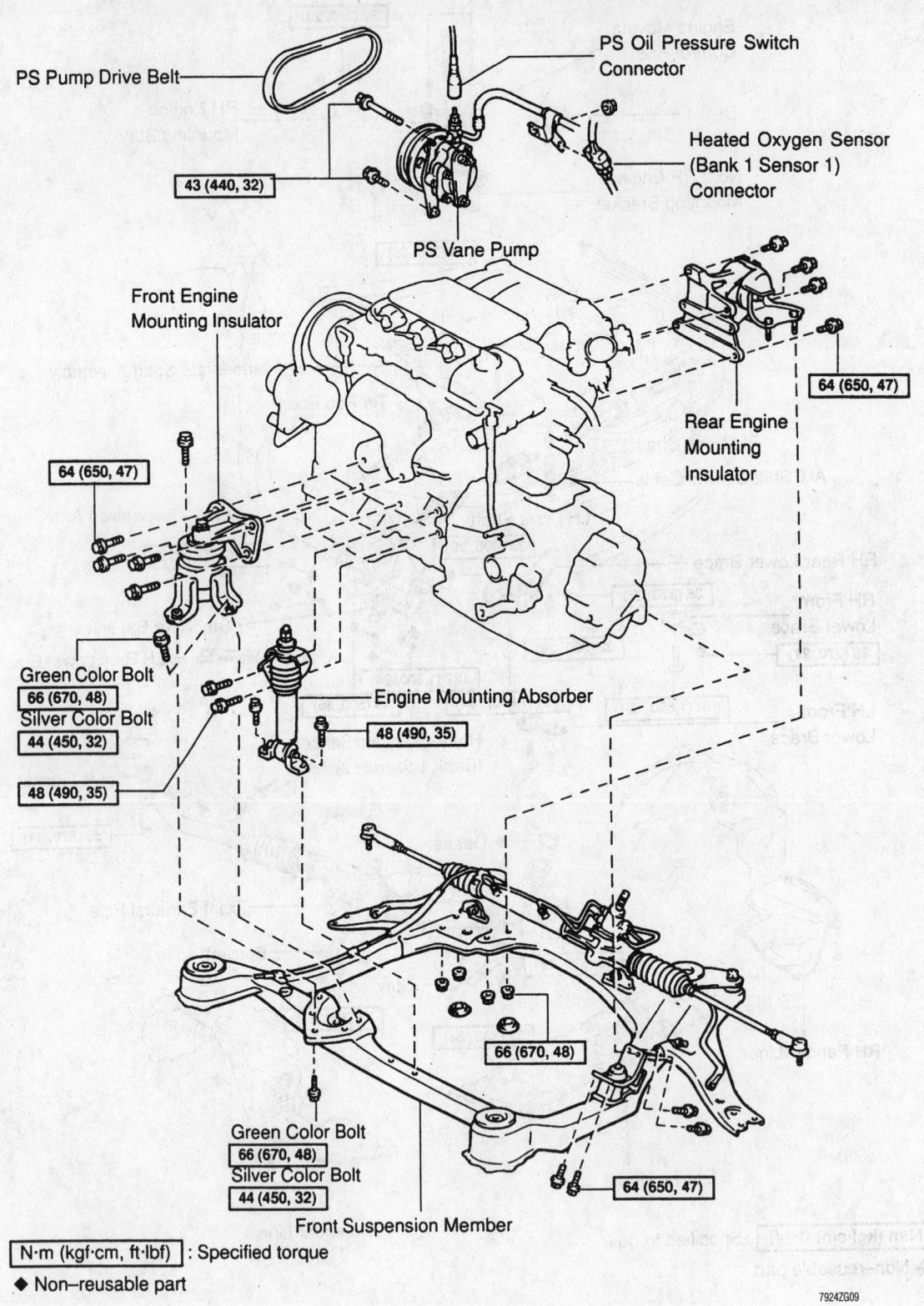

PS Pump Drive Belt

PS Oil Pressure Switch
Connector

43 (440, 32)

PS Vane Pump

Heated Oxygen Sensor
(Bank 1 Sensor 1)
Connector

Front Engine
Mounting Insulator

Rear Engine
Mounting
Insulator

64 (650, 47)

64 (650, 47)

Green Color Bolt
66 (670, 48)
Silver Color Bolt
44 (450, 32)

48 (490, 35)

Engine Mounting Absorber

48 (490, 35)

66 (670, 48)

Green Color Bolt
66 (670, 48)
Silver Color Bolt
44 (450, 32)

64 (650, 47)

Front Suspension Member

N·m (kgf·cm, ft·lbf) : Specified torque

◆ Non–reusable part

7924ZG09

Exploded view of the suspension component removal and installation for engine removal—Sienna model

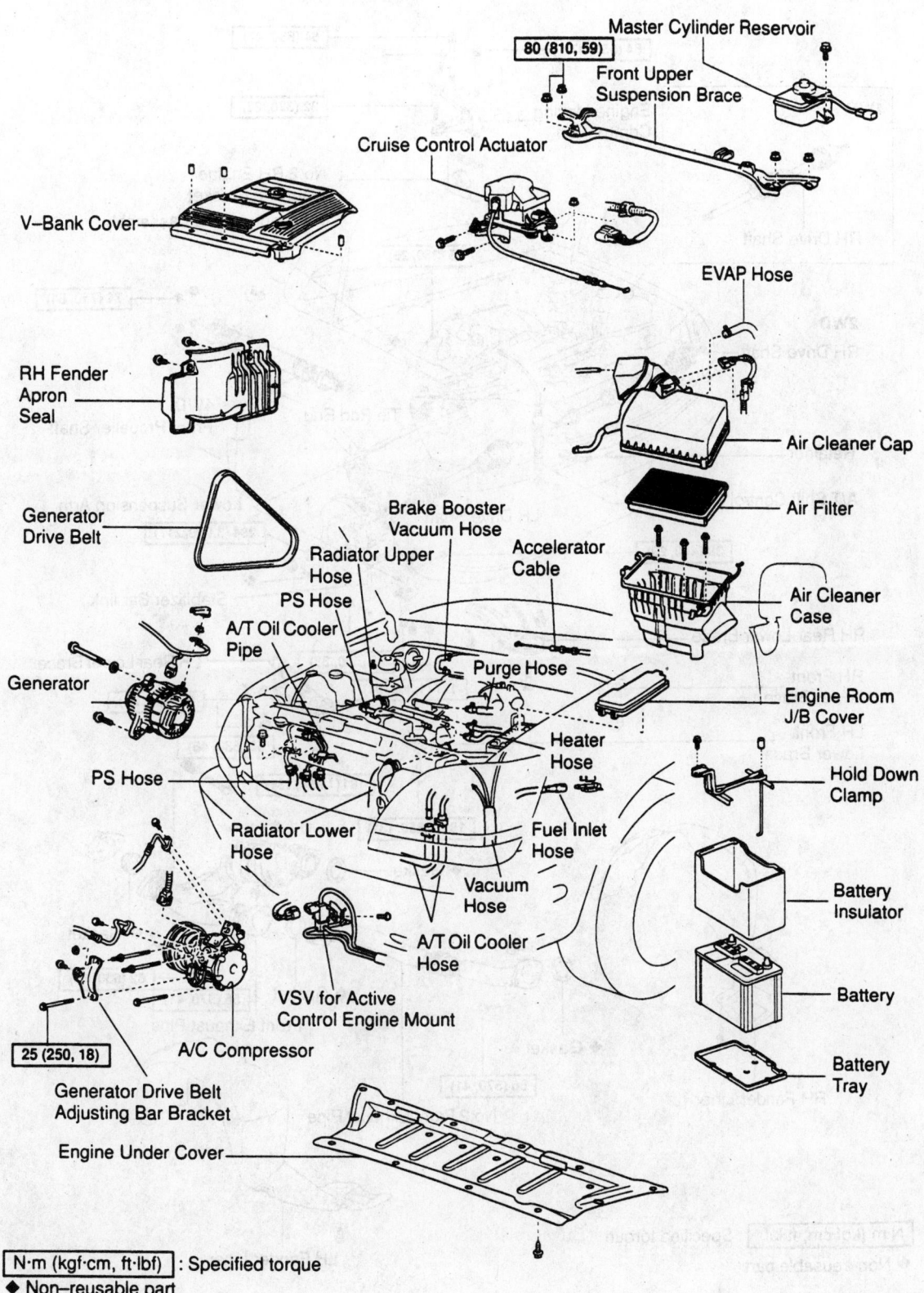

Master Cylinder Reservoir

80 (810, 59)

Front Upper Suspension Brace

Cruise Control Actuator

V–Bank Cover

EVAP Hose

RH Fender Apron Seal

Air Cleaner Cap

Air Filter

Generator Drive Belt

Brake Booster Vacuum Hose

Accelerator Cable

Air Cleaner Case

Radiator Upper Hose

PS Hose

A/T Oil Cooler Pipe

Purge Hose

Generator

Heater Hose

Engine Room J/B Cover

PS Hose

Hold Down Clamp

Radiator Lower Hose

Fuel Inlet Hose

Vacuum Hose

Battery Insulator

A/T Oil Cooler Hose

Battery

VSV for Active Control Engine Mount

25 (250, 18)

A/C Compressor

Battery Tray

Generator Drive Belt Adjusting Bar Bracket

Engine Under Cover

N·m (kgf·cm, ft·lbf) : Specified torque
◆ Non–reusable part

7924ZG84

Exploded view of engine pre-removal components—RX300 model

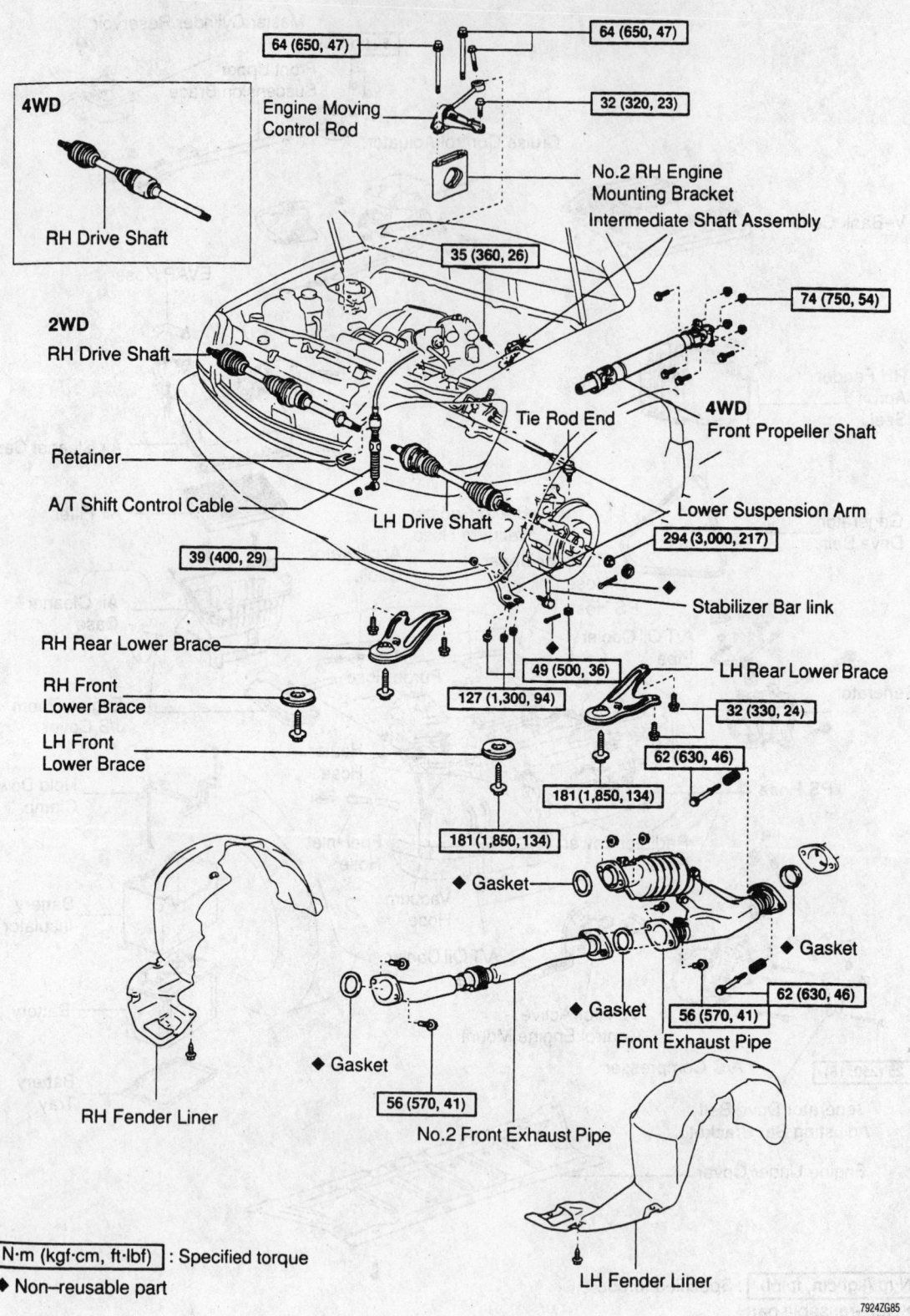

4WD

RH Drive Shaft

64 (650, 47)

64 (650, 47)

Engine Moving
Control Rod

32 (320, 23)

No.2 RH Engine
Mounting Bracket

Intermediate Shaft Assembly

35 (360, 26)

74 (750, 54)

2WD

RH Drive Shaft

Tie Rod End

4WD
Front Propeller Shaft

Retainer

A/T Shift Control Cable

LH Drive Shaft

Lower Suspension Arm

294 (3,000, 217)

39 (400, 29)

Stabilizer Bar link

RH Rear Lower Brace

49 (500, 36)

LH Rear Lower Brace

RH Front
Lower Brace

127 (1,300, 94)

32 (330, 24)

LH Front
Lower Brace

62 (630, 46)

181 (1,850, 134)

181 (1,850, 134)

◆ Gasket

◆ Gasket

62 (630, 46)

◆ Gasket

56 (570, 41)

Front Exhaust Pipe

◆ Gasket

RH Fender Liner

56 (570, 41)

No.2 Front Exhaust Pipe

◆ Gasket

LH Fender Liner

N·m (kgf·cm, ft·lbf) : Specified torque

◆ Non–reusable part

7924ZG85

Exploded view of engine removal and installation tightening specifications of the related components—RX300 model

2WD

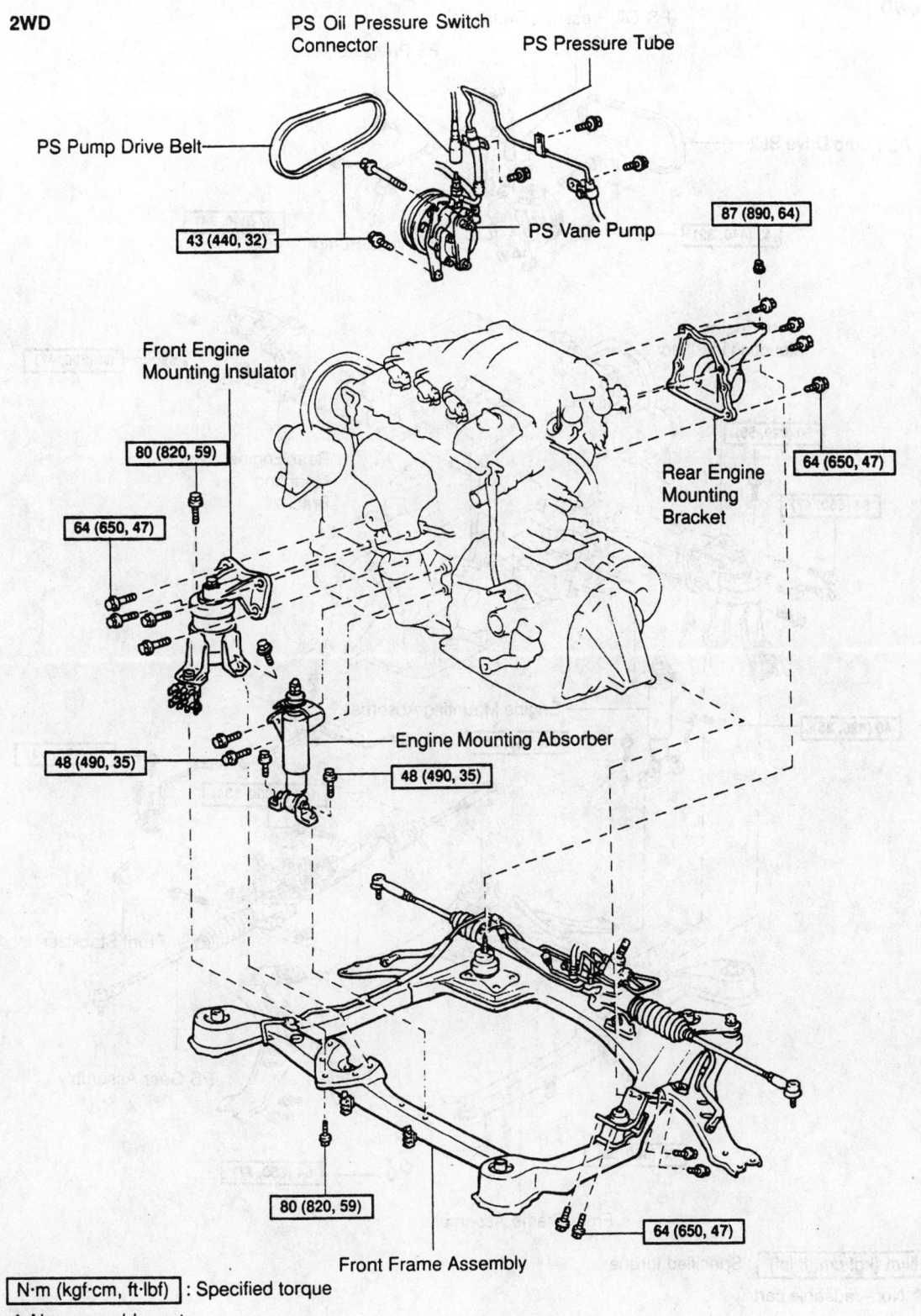

PS Oil Pressure Switch
Connector

PS Pressure Tube

PS Pump Drive Belt

PS Vane Pump

87 (890, 64)

43 (440, 32)

Front Engine
Mounting Insulator

Rear Engine
Mounting
Bracket

80 (820, 59)

64 (650, 47)

64 (650, 47)

Engine Mounting Absorber

48 (490, 35)

48 (490, 35)

80 (820, 59)

64 (650, 47)

Front Frame Assembly

N·m (kgf·cm, ft·lbf) : Specified torque

◆ Non–reusable part

7924ZG86

Exploded view of the suspension component removal and installation for engine removal—RX300 with 2WD model

4WD

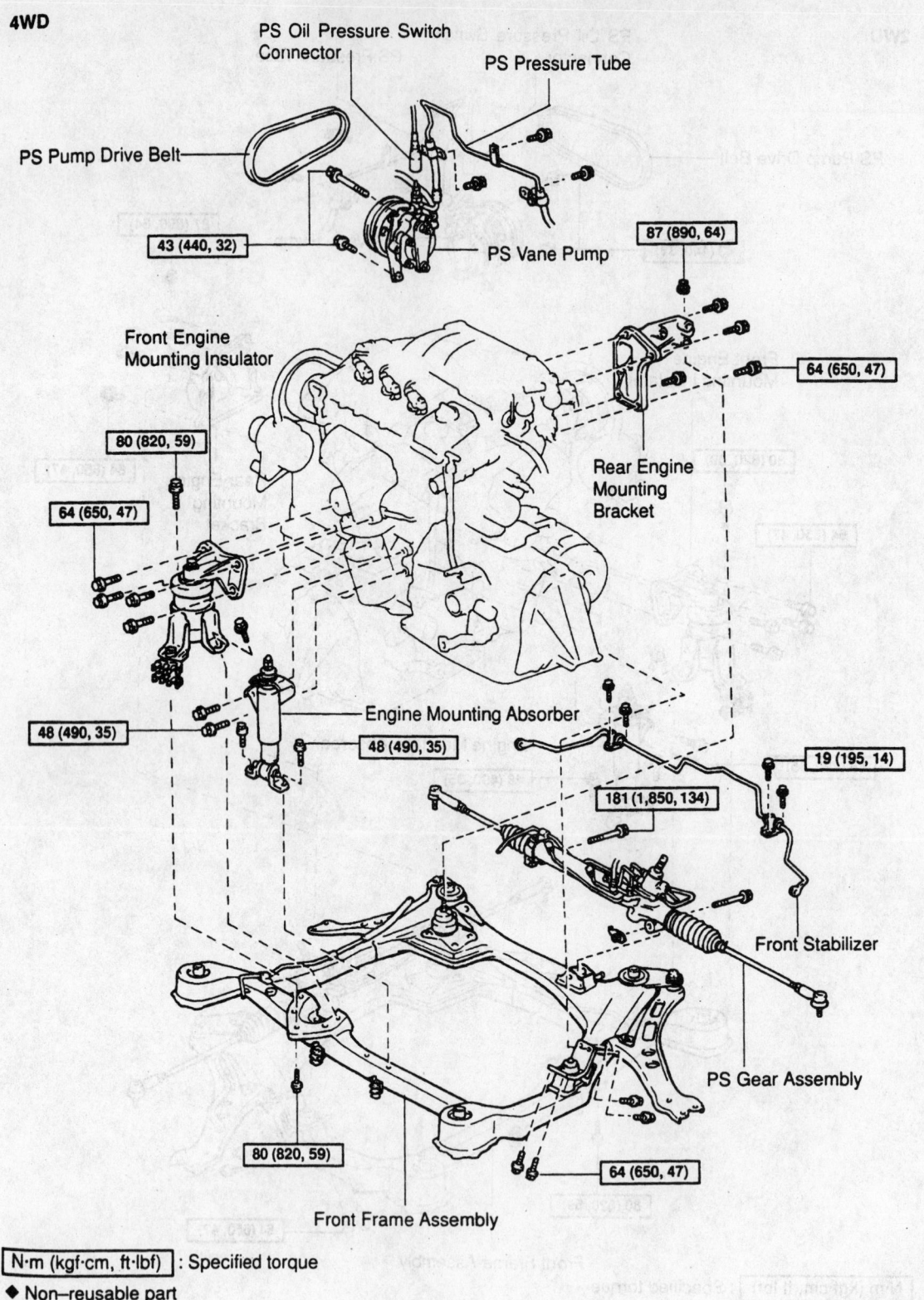

PS Oil Pressure Switch Connector

PS Pressure Tube

PS Pump Drive Belt

43 (440, 32)

87 (890, 64)

PS Vane Pump

Front Engine Mounting Insulator

64 (650, 47)

80 (820, 59)

64 (650, 47)

Rear Engine Mounting Bracket

Engine Mounting Absorber

48 (490, 35)

48 (490, 35)

19 (195, 14)

181 (1,850, 134)

Front Stabilizer

PS Gear Assembly

80 (820, 59)

64 (650, 47)

Front Frame Assembly

N·m (kgf·cm, ft·lbf) : Specified torque

◆ Non—reusable part

79242G87

Exploded view of the suspension component removal and installation for engine removal—RX300 with 4WD model

55. Install the battery tray and battery.

56. Connect the positive, then the negative battery cable.

57. Install the heater air duct.

58. Install the left and right ventilator louvers.

59. Connect the window washer hoses from the ventilator louvers.

60. Install the top cowl seal and panel.

61. Install the wiper and blade assembly.

62. Align the matchmarks and install the hood.

63. Install a new oil filter and fill the engine with oil. Fill the engine with engine coolant. Install the engine undercovers. Start the engine and check for leaks.

Water Pump

REMOVAL & INSTALLATION

RAV4

1. Disconnect the negative battery cable.

✱✱ CAUTION

Wait 90 seconds from the time the key is turned to LOCK and the negative battery cable is disconnected to begin work. This allows the SRS capacitor to discharge and prevent deployment of the air bag(s).

2. Remove the right-hand engine undercover.

3. Drain the engine coolant from the radiator and engine.

4. Remove the timing belt.

5. Disconnect the lower radiator hose from the water inlet.

6. Remove the timing belt tension spring and the No. 2 idler pulley.

7. Disconnect the crankshaft position sensor connector clamp.

8. Remove the alternator drive belt adjusting bar.

9. Remove the two nuts holding the water pump to the water bypass pipe.

10. Remove the three bolts in the sequence shown.

11. Disconnect the water pump cover from the water bypass pipe and remove the water pump and water pump cover assembly.

12. Remove the gasket and two O-rings from the water pump and water bypass pipe.

13. Remove the water pump from the water pump cover by removing the three bolts, water pump and gasket.

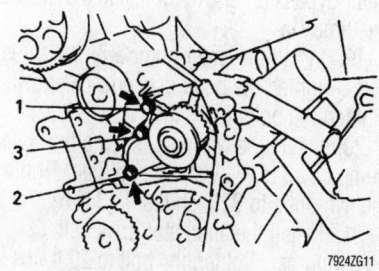

Loosening sequence for the water pump bolts—RAV4 model

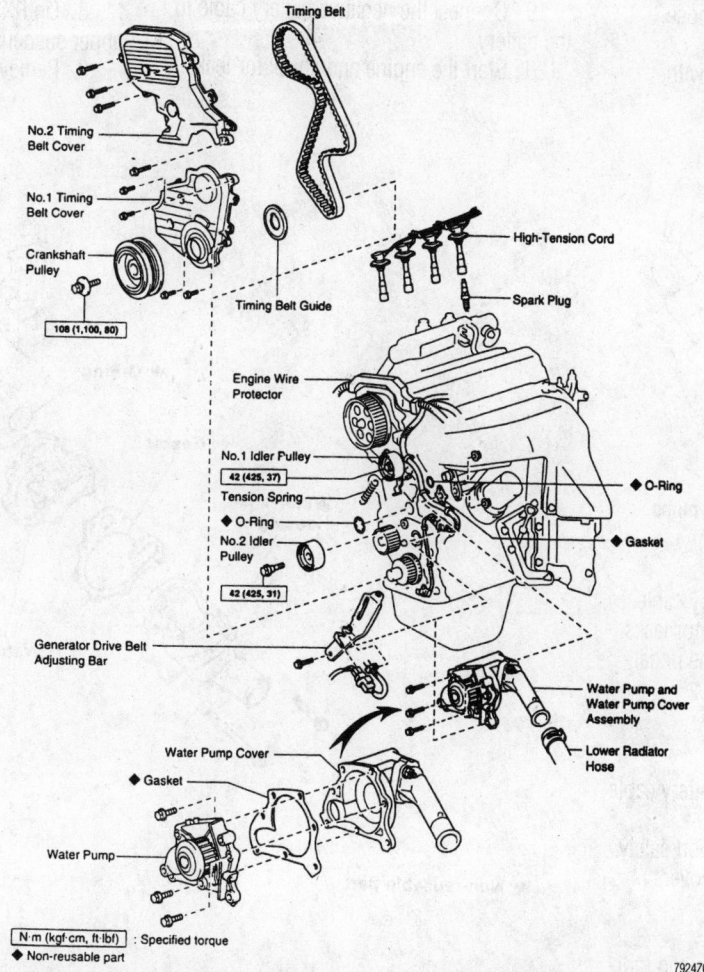

Exploded view of the water pump and related components—RAV4 model

To install:

14. Using a new gasket, install the water pump to the water pump cover. Install the three bolts and tighten the bolts to 78 inch lbs. (8.8 Nm).

15. Install a new O-ring and gasket to the water pump cover.

16. Install a new O-ring to the water bypass pipe.

17. Apply soapy water to the O-ring on the water bypass pipe.

18. Connect the water pump cover to the water bypass pipe. Do not install the nuts at this time.

19. Install the water pump with the three bolts. Tighten the bolts in sequence shown. Tighten the bolts to 78 inch lbs. (8.8 Nm).

20. Install the two nuts holding the water pump cover to the water pump pipe. Tighten the two bolts to 82 inch lbs. (9.3 Nm).

21. Install the alternator drive belt adjusting bar. Tighten the bolt to 20 ft. lbs. (27 Nm).

22. Attach the crankshaft position sensor connector clamp.

23. Install the No. 2 idler pulley and timing belt tension spring.

24. Connect the lower radiator hose.

25. Install the timing belt.

26. Fill the engine and radiator with engine coolant.

Tightening sequence for the water pump bolts—RAV4 model

27. Connect the negative battery cable.

28. Start the engine and check for leaks.

29. Install the right-hand engine undercover.

Previa

1. Disconnect the negative battery cable from the battery.

2. Raise the vehicle and support safely.

3. Remove the engine undercovers.

4. Drain the engine coolant.

5. Drain the engine oil.

6. Disconnect the heater hose and radiator outlet hoses.

7. Remove the oil filter bracket.

8. Disconnect the water hose from the water pump.

9. Remove the water pump retaining bolts and pump from the timing cover.

10. Remove the O-ring from the water pump.

11. Remove the water pump from the housing by removing the two bolts.

To install:

12. Install the water pump with a new gasket and tighten the bolts to 14 ft. lbs. (20 Nm).

13. Install the water pump to the timing cover and install the bolts. Tighten the bolts for the water pump as follows:

- Bolt A: 14 ft. lbs. (20 Nm)
- Bolt B: 21 ft. lbs. (28 Nm)

14. Connect the water hose to the water pump.

15. Install the oil filter bracket to the engine using a new O-ring.

16. Connect the heater hose and radiator outlet hose.

17. Fill the engine with oil.

18. Fill the engine and radiator with coolant.

19. Connect the negative battery cable to the battery.

20. Start the engine and check for leaks.

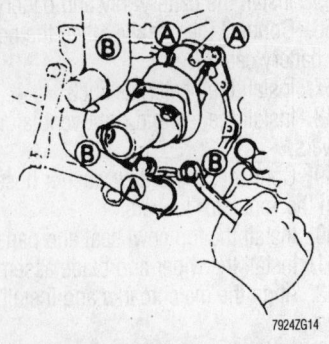

Tightening the water pump bolts—Previa model

Sienna and RX300

1. Disconnect the negative battery cable.

2. Drain the engine coolant.

3. Remove the wiper and blade assembly.

4. Remove the top cowl seal and panel.

5. Label and disconnect the window washer hoses from the ventilator louvers.

6. Remove the left and right ventilator louvers.

7. Remove the heater air duct.

8. On RX300 models, remove the front upper suspension brace.

9. Remove the timing belt.

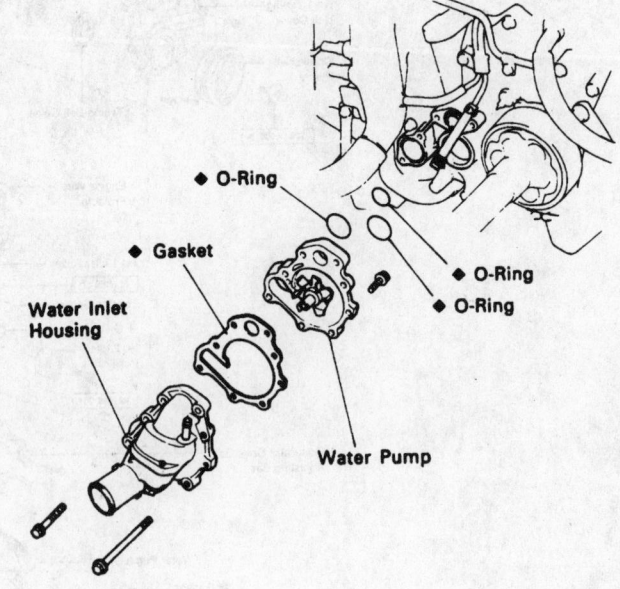

◆ **Non-reusable part**

Exploded view of the water pump components—Previa model

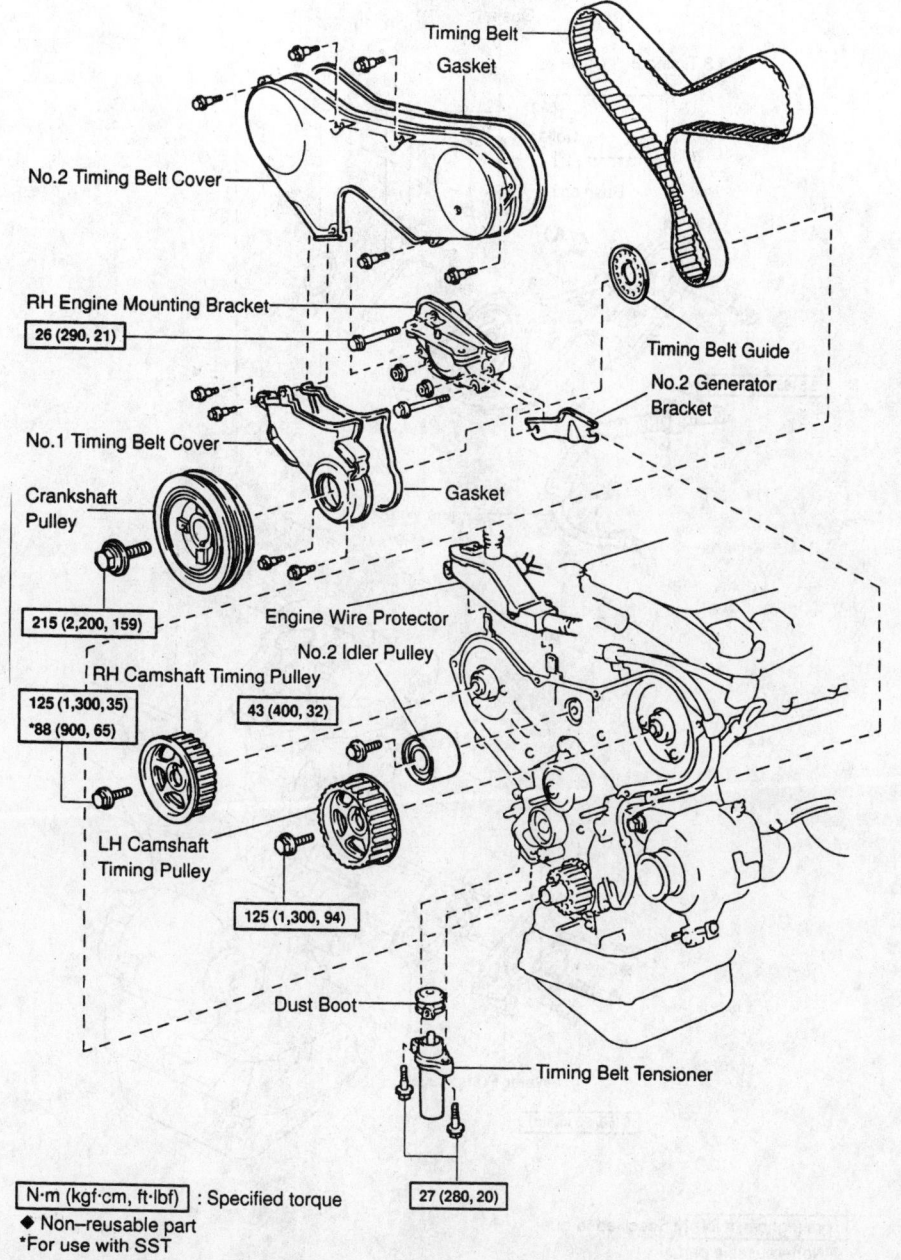

Timing Belt

Gasket

No.2 Timing Belt Cover

Timing Belt Guide

No.2 Generator Bracket

RH Engine Mounting Bracket

26 (290, 21)

No.1 Timing Belt Cover

Gasket

Crankshaft Pulley

215 (2,200, 159)

Engine Wire Protector

No.2 Idler Pulley

RH Camshaft Timing Pulley

125 (1,300, 35)
*88 (900, 65)

43 (400, 32)

LH Camshaft Timing Pulley

125 (1,300, 94)

Dust Boot

Timing Belt Tensioner

N·m (kgf·cm, ft·lbf) : Specified torque
◆ Non–reusable part
*For use with SST

27 (280, 20)

7924ZG15

Exploded view of the components to gain access to the water pump—Sienna and RX300 models

10. Mark the left and right camshaft pulleys with a touch of paint. Using a spanner wrench, remove the bolts to the right and left camshaft pulleys. Separate the pulleys from the engine. Be sure not to mix up the pulleys.

11. Remove the No. 2 idler pulley by removing the bolt.

12. Disconnect the three clamps and engine wire from the rear timing belt cover.

13. Remove the six bolts holding the No. 3 timing belt cover to the engine block.

14. Remove the bolts and nuts to extract the water pump.

15. Raise the engine slightly and remove the water pump and the gasket from the engine.

To install:

16. Check that the water pump turns smoothly. Also check the air hole for coolant leakage.

17. Using a new gasket, apply liquid sealer to the gasket, water pump and engine block.

18. Install the gasket and pump to the engine and install the four bolts and two nuts. Tighten the nuts and bolts to 53 inch lbs. (6 Nm).

19. Install the rear timing belt cover and tighten the six bolts to 74 inch lbs. (9 Nm).

20. Connect the engine wire with the three clamps to the rear timing belt cover.

21. Install the No. 2 idler pulley with the bolt. Tighten the bolt to 32 ft. lbs. (43 Nm). After tightening the bolt, be sure the idler pulley moves smoothly.

22. With the flange side **outward** , install the right-hand camshaft pulley to the engine. Be sure to align the knock pin hole on the camshaft pulley with the knock pin on the camshaft. Using the same tools as removal, tighten the camshaft bolt to 65 ft. lbs. (88 Nm).

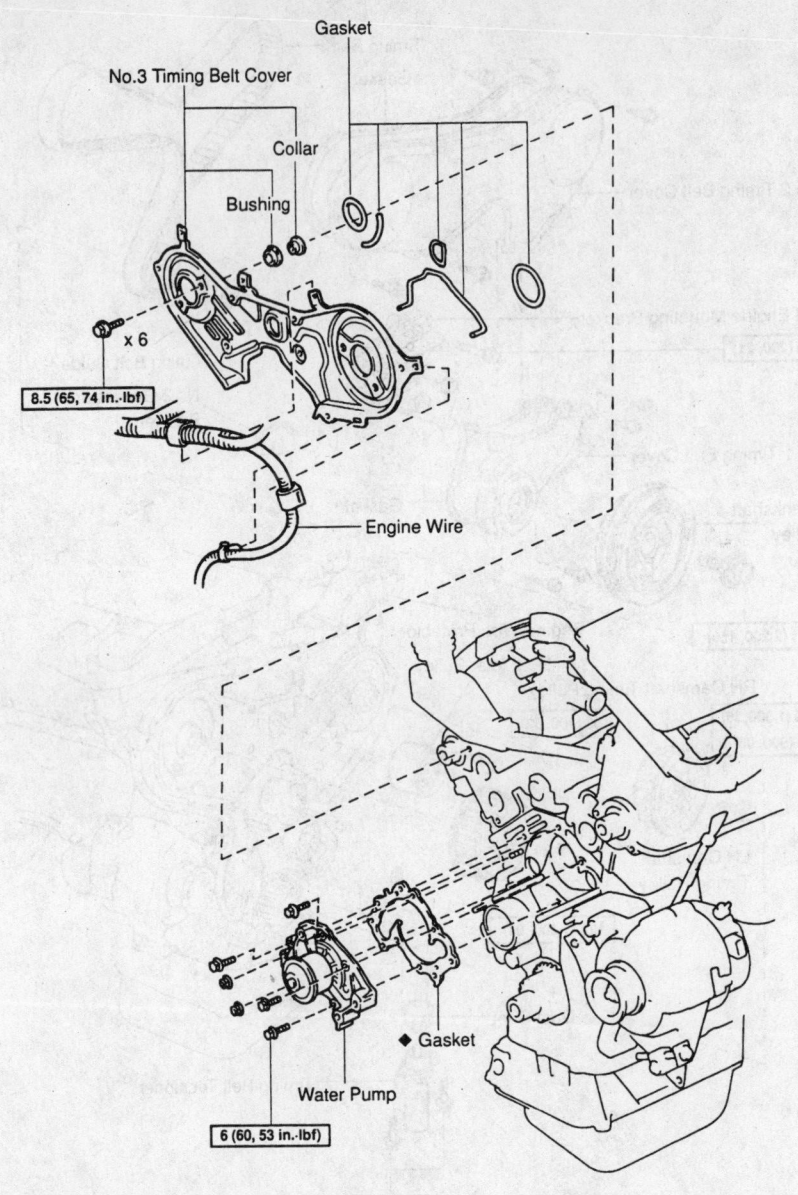

N·m (kgf·cm, ft·lbf) : Specified torque
◆ Non–reusable part

7924ZG16

Exploded view of the water pump and related components—Sienna and RX300 models

23. With the flange side **inward**, install the left-hand camshaft pulley to the engine. Be sure to align the knock pin hole on the camshaft pulley with the knock pin on the camshaft. Using the same tools as removal, tighten the camshaft bolt to 94 ft. lbs. (125 Nm).

24. Install the timing belt to the engine.

25. On RX300 models, install the front upper suspension brace and tighten the mounting nuts to 59 ft. lbs. (80 Nm).

26. Fill the engine coolant.

27. Install the heater air duct.

28. Install the left and right ventilator louvers.

29. Connect the window washer hoses from the ventilator louvers.

30. Install the top cowl seal and panel.

31. Install the wiper and blade assembly.

32. Connect the negative battery cable to the battery and start the engine.

33. Top off the engine coolant and check for leaks.

Cylinder Head

REMOVAL & INSTALLATION

RAV4

✳✳ CAUTION

The fuel injection system remains under pressure after the engine has been turned OFF. Properly relieve fuel pressure before disconnecting

any fuel lines. Failure to do so may result in fire or personal injury.

1. Release the fuel system pressure.
2. Disconnect the negative battery cable.

✳✳ CAUTION

Wait 90 seconds from the time the key is turned to LOCK and the negative battery cable is disconnected to begin work. This allows the SRS capacitor to discharge and prevent deployment of the air bag(s).

3. Remove the right-hand engine undercover.
4. Drain the engine coolant into a suitable container.
5. See the procedure under Camshaft Removal and Installation and remove the camshafts.
6. Uniformly loosen and remove the cylinder head bolts in several passes and in the reverse of the removal sequence. Lift the cylinder head with the intake manifold from the cylinder block. Disengage the cylinder head from the block dowel pins.
7. Remove the air hose from the intake manifold. Remove the two bolts and the air tube.
8. Remove the six bolts, two nuts, and the intake manifold and gasket.
9. Disconnect the air hose from the cylinder head port and remove the air hose.
10. Remove the fuel delivery pipe and the injectors.
11. Remove the oil pressure switch.

To install:

12. Install the oil pressure switch.
13. Install the fuel injectors and the delivery pipe.
14. Install the air hose to the cylinder head port.
15. Install the intake manifold with new gaskets. Install the six bolts and two nuts, and tighten the intake manifold to 14 ft. lbs. (19 Nm)
16. Install the air tube with the two bolts and connect the air hose to the intake manifold.
17. Clean the gasket mating surfaces using care not to damage the aluminum components, replace the gasket, then lower the cylinder head onto the engine. Be sure the dowel pins are aligned and no hoses or wires are between the head and cylinder block.
18. The cylinder head bolts are tightened in two progressive steps. Apply a light coat of engine oil to the cylinder head bolts. Uniformly tighten the 10 cylinder head bolts in

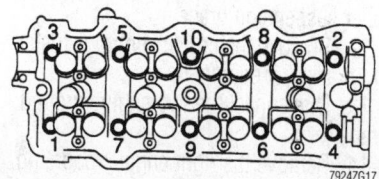

Cylinder head bolts installation sequence—RAV4 model

several passes and in sequence. The torque for the head bolts is 36 ft. lbs. Mark the front of the cylinder head bolt with paint. Retighten the cylinder head bolts by 90° in sequence. Retighten an additional 90° and be sure that the paint mark is now positioned toward the rear.

19. Install the intake and exhaust camshafts and remaining parts as described in the Camshaft Removal and Installation procedure.
20. Connect the negative battery cable, fill the engine with coolant, start the engine, warm up, and check for leaks. Bleed the cooling system and top off coolant as necessary.
21. Install the right-hand engine undercover, check ignition timing, and road test the vehicle for proper operation.

Previa

➡**In order remove the cylinder head, the engine and transmission must be pulled from the vehicle.**

1. Disconnect the negative battery cable. Wait at least 90 seconds when working on vehicles equipped with an airbag.
2. Relieve the fuel system pressure.

✳✳ CAUTION

The fuel injection system remains under pressure after the engine has been turned OFF. Properly relieve fuel pressure before disconnecting any fuel lines. Failure to do so may result in fire or personal injury.

3. Remove the engine/transmission assembly from the vehicle.
4. Label and remove the engine wiring from the engine and move it aside.
5. Unbolt and remove the No. 2 head cover.
6. Remove the distributor.
7. Remove the EGR valve.

8. Remove the union bolts and gaskets from the delivery pipe and cold start injector. Unbolt and remove the pressure regulator with the hose from the delivery pipe. Unbolt and extract the delivery pipe.
9. Disconnect the hose and unbolt, then extract the water outlet from the engine.
10. Remove the PCV hose.
11. Remove the fuel delivery pipe with insulators.
12. Remove the intake manifold.
13. Remove the right side engine mounting.
14. Remove the exhaust manifold and heat insulator.
15. Remove the No. 1 oil return pipe.
16. Remove the no. 1 cylinder head cover and half moons.
17. Remove the camshafts.
18. Remove the 2 bolts in front of the head before the other head bolts are removed.
19. Using a 12 sided socket wrench, remove the 10 cylinder head retaining bolts in the proper sequence.
20. Remove the cylinder head from the block as follows:
   a. Remove the two bolts in front of the head before the other head bolts are removed. The two bolts are located in front of the timing chain.
   b. Uniformly remove the 10 head bolts in the reverse of the torque sequence.
   c. Lift the cylinder head from the dowels on the cylinder block.
   d. Remove the cylinder head gasket.

To install:

➡**Always clean and inspect the cylinder head and mating surface for cracks and flatness when removed.**

21. Clean the gasket mating surfaces and check for warpage.
22. Apply seal packing to two locations on the cylinder block as shown.
23. Install the head gasket and install the cylinder head.
24. If the camshaft timing gear was removed, align the matchmarks placed on the timing gear and chain during removal.
25. Place the cylinder head in position on the gasket.
26. Oil the bolts and using the proper sequence, tighten the bolts in three steps.
   a. Uniformly tighten the head bolts to 29 ft. lbs. (39 Nm).
   b. Mark the front of the cylinder head bolt with paint.
   c. Retighten the cylinder head bolts 90° in the numerical order as shown.

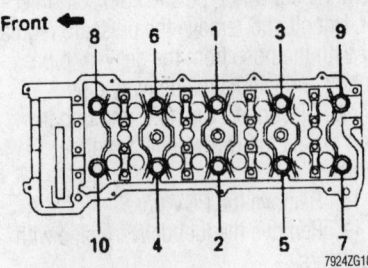

Cylinder head bolt tightening sequence—Previa model

d. Check that the painted mark is now facing sideward.

e. Retighten the cylinder head bolts an additional 90°.

f. Check that the painted mark is now facing rearward.

27. Install and tighten the two front mounting bolts to 15 ft. lbs. (21 Nm).

28. Install the camshafts.

29. Install all remaining components.

30. Install the engine/transmission assembly into the vehicle.

31. Fill the cooling system and fill the engine with oil.

32. Connect the battery cable, start the engine, and check for leaks.

33. Road test the vehicle for proper operation and recheck all fluid levels.

Sienna and RX300

> ❋❋ **CAUTION**
>
> **The fuel injection system remains under pressure after the engine has been turned OFF. Properly relieve fuel pressure before disconnecting any fuel lines. Failure to do so may result in fire or personal injury.**

1. Remove the wiper and blade assembly.

2. Remove the top cowl seal and panel.

3. Label and disconnect the window washer hoses from the ventilator louvers.

4. Remove the left and right ventilator louvers.

5. Remove the heater air duct.

6. Relieve the fuel pressure.

7. Turn the ignition key to the **OFF** position. Disconnect the negative battery cable. Wait at least 90 seconds from the time the negative battery was disconnected to start work.

8. Drain the cooling system.

9. Disconnect the accelerator cable and the throttle cable on vehicles equipped with an automatic transaxle.

10. Remove the air cleaner cover, air flow meter, and the air duct.

11. On RX300 models, remove the front upper suspension brace.

12. Remove the cruise control actuator and bracket, if equipped.

13. Disconnect the two engine ground straps.

14. Remove the right engine mounting support.

15. Disconnect the radiator hoses.

16. Disconnect the two heater hoses.

17. Disconnect and plug the fuel feed and return lines from the fuel rail assembly.

18. Disconnect and plug the pressure hose from the hydraulic motor.

19. Remove the V-bank cover.

20. Disconnect the following vacuum hoses:

- Fuel pressure control VSV
- Fuel pressure regulator
- Cylinder head rear plate
- Intake air control valve VSV
- EGR vacuum modulator
- EGR valve

21. Disconnect the following wiring and hoses:

- Intake air control valve
- Fuel pressure regulator
- EGR VSV

22. Remove the two nuts and the emission control valve set.

23. Disconnect the following hoses;

- Brake booster vacuum hose
- PCV hose
- Intake air control valve vacuum hose

24. Remove the data link connector from the mounting bracket.

25. Remove the two ground straps from the intake chamber.

26. Remove the hydraulic motor pressure hose from the intake chamber.

27. Remove the right oxygen sensor connector from the power steering pressure tube.

28. Remove the two nuts and the power steering pressure tube from the intake chamber.

29. Disconnect the two power steering air hoses.

30. Remove the engine hanger and the intake chamber support.

31. Remove the EGR pipe and gaskets.

32. Disconnect the following wiring;

- Throttle position sensor connector
- IAC valve connector
- EGR gas temperature connector
- A/C idle up connector

33. Disconnect the following vacuum hoses:

- Two vacuum hoses from the TVV
- Vacuum hose from the cylinder head rear plate

- Vacuum hose from the charcoal canister

34. Disconnect the air assist hose and the two water bypass hoses.

35. Remove the air intake chamber.

36. Disconnect the left engine wiring harness and position it out of the way.

37. Remove the wiring harness from the rear of the engine.

38. Disconnect the right engine wiring harness and position it out of the way.

39. Remove the ignition coils and lay them aside in the exact positions where they are placed in the heads.

40. Remove the timing belt.

41. Remove the camshaft pulleys and the timing belt rear cover.

42. Remove the cylinder head rear plate.

43. Remove the water inlet pipe.

44. Remove the air assist hose and vacuum hose.

45. Remove the intake manifold and fuel rail assembly.

46. Remove the water outlet.

47. Remove the EGR pipe from the right exhaust manifold.

48. Remove the front exhaust pipe and exhaust manifolds.

49. Remove the dipstick assembly and the power steering pump bracket.

50. Remove the valve covers and the camshaft position sensor.

51. Remove the camshafts following the proper sequences and procedures.

52. Be sure the engine is at or near ambient temperature and remove the two (one on each head) 8mm recessed hex bolts. Loosen and remove the eight head bolts evenly, in three passes, in the reverse order of the installation sequence. Carefully lift the head from the engine; if it is necessary to pry the head loose, take great care not to damage the mating surfaces. Place the head on wood blocks in a clean work area.

➡**If the cylinder head bolts are loosened out of sequence, warpage or cracking could result.**

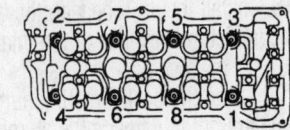

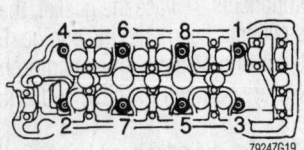

Cylinder head bolt loosening sequence—Sienna and RX300 models

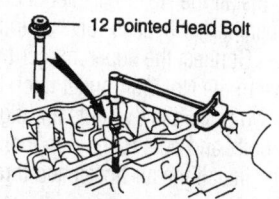

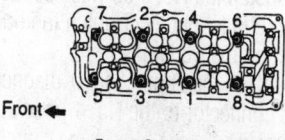

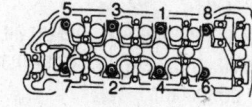

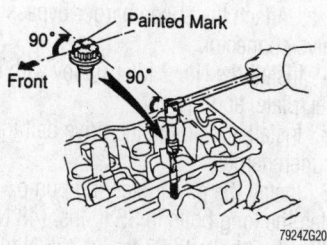

Cylinder head bolt tightening sequence—Sienna and RX300 models

53. Remove the cylinder head gasket. With a gasket scraper, carefully remove all the old gasket material from the cylinder head and engine block surfaces.

To install:

54. Place the new cylinder head gasket onto the cylinder block. Place the cylinder head onto the gasket.

55. Coat the threads of the eight cylinder head bolts (12-sided) with clean engine oil and install the bolts into the cylinder head. Uniformly tighten the bolts in sequence in three steps to 40 ft. lbs. (54 Nm), using the proper sequence. If any bolt does not meet the torque, replace it.

56. Mark the forward edge of each bolt with paint, then tighten each bolt, in proper sequence, an additional 90°. Check that each painted mark is now at a 90° angle to the front. The paint mark should have been applied to the bolt in the 9 O'clock position and should now be in the 12 O'clock position.

57. Coat the threads of the two remaining 8mm bolts with engine oil and install them. Tighten to 13 ft. lbs. (18 Nm).

58. Install the camshafts following the proper sequences and procedures.

59. Check and adjust the valves.

60. Apply sealant to the cylinder heads where the camshaft supports meet the cylinder heads.

61. Use new gaskets and install the cylinder head covers.

62. Install the dipstick and power steering pump bracket.

63. Install the exhaust manifolds. Tighten the nuts to 36 ft. lbs. (49 Nm).

64. Install the EGR pipe to the right exhaust manifold.

65. Install the water outlet.

66. Install the intake manifold and the fuel rail assembly. Tighten the intake manifold nuts and bolts to 11 ft. lbs. (15 Nm).

67. Install the air assist hose and the two water bypass hoses.

68. Install the water inlet pipe and the cylinder head rear plate.

69. Install the timing belt rear cover and the camshaft pulleys.

70. Install the timing belt.

71. Install the spark plugs and the ignition coils.

72. Install the right engine wiring harness.

73. Install the wiring harness to the rear of the engine.

74. Install the left engine wiring harness.

75. Attach and secure the air intake chamber.

76. Use new gaskets and install the EGR pipe.

77. Connect the following vacuum hoses:
- The two TVV vacuum hoses.
- The vacuum hose to the rear cylinder head plate.
- Charcoal canister vacuum hose.

78. Connect the following electrical wiring:
- Throttle position sensor connector.
- IAC valve connector.
- EGR gas temperature connector.
- A/C idle up connector.

79. Install the engine hanger and the intake chamber support.

80. Connect the two power steering air hoses.

81. Install the power steering pressure tube to the intake chamber.

82. Install the oxygen sensor connector to the pressure tube.

83. Install the two ground straps to the intake chamber.

84. Install the data link connector to the bracket.

85. Connect the following hoses:
- Power brake booster vacuum hose.
- PCV hose.
- IAC valve vacuum hose.

86. Install the emission control valve set and related vacuum hoses and connectors.

87. Install the V-bank cover.

88. Connect the pressure hose to the hydraulic motor.

89. Connect the fuel lines to the fuel rail assembly.

90. Connect the heater and radiator hoses.

91. Install the right engine mounting support.

92. Connect the two engine ground straps.

93. If removed, install the upper front suspension brace and tighten the mounting nuts to 59 ft. lbs. (80 Nm).

94. Install the cruise control actuator and bracket.

95. Install the air cleaner, air flow meter, and air duct assembly.

96. Connect the accelerator cable and the throttle cable on vehicles equipped with an automatic transaxle.

97. Fill the cooling system to the proper level with coolant.

98. Connect the negative battery cable.

99. Install the heater air duct.

100. Install the left and right ventilator louvers.

101. Connect the window washer hoses from the ventilator louvers.

102. Install the top cowl seal and panel.

103. Install the wiper and blade assembly.

104. Start the engine and check for leaks. Bleed the air from the cooling system.

105. Road test the vehicle and check for unusual noise, shock, slippage, correct shift points and smooth operation.

106. Recheck the coolant and engine oil levels.

Supercharger

REMOVAL & INSTALLATION

Previa with 2.4L (VIN A) Engine

1. Disconnect the negative battery cable from the battery. Wait at least 90 seconds after the cable has been removed to perform any work on models equipped with an airbag.

2. Drain the engine coolant from the radiator.

3. Remove the air duct.

4. Remove the engine coolant reservoir tank and bracket.

5. Remove the air damper case.

6. Remove the supercharger blower.

7. Disconnect the power steering reservoir.

8. Remove the radiator.

9. Remove the throttle body.

10. Remove the alternator/power steering drive belt.

11. Remove the power steering pump.

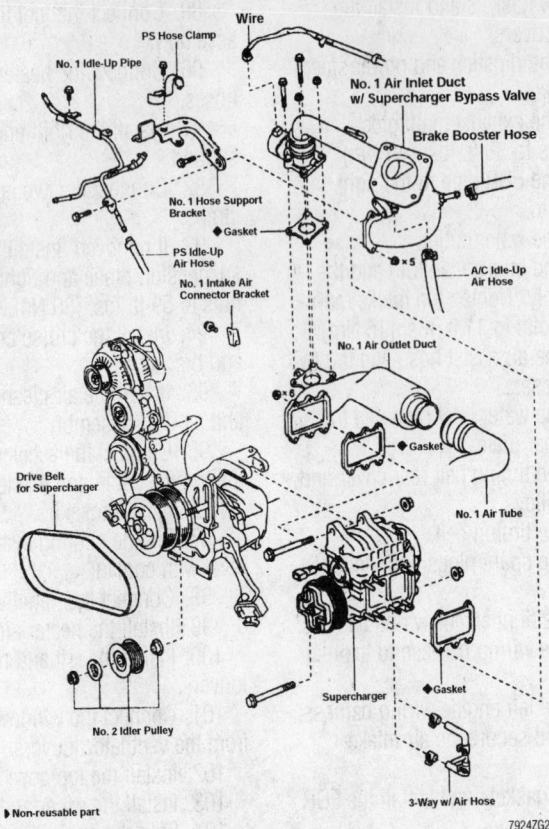

Exploded view of the supercharger and related components

12. Remove the drive belt for the super-charger.

13. Remove the No. 2 idler pulley by extracting the nut, plate, and the spacer.

14. Remove the No. 1 air inlet duct with the supercharger bypass valve as follows:

    a. Disconnect the supercharger bypass valve harness.

    b. Disconnect the brake booster hose.

    c. Disconnect the A/C idle up air hose.

    d. Disengage the supercharger magnetic clutch connector.

    e. Disengage the supercharger magnetic clutch connector from the No.1 hose support bracket.

    f. Remove the air hoses and three way.

    g. Remove the two bolts and two nuts holding the supercharger bypass valve to the No. 1 air outlet duct.

    h. Remove the five nuts and the No. 1 air inlet duct with the supercharger bypass valve.

    i. Remove the supercharger bypass valve and No. 1 air inlet duct gaskets.

15. Pull off the No. 1 idle up pipe by removing the bolt and air hose.

16. Disconnect the No. 1 air tube.

17. Remove the No. 1 intake air connector bracket by removing the two bolts.

18. Remove the supercharger as follows:

    a. Remove the two bolts and two nuts holding the supercharger to the equipment drive housing.

    b. Remove the six nuts holding the supercharger to the No. 1 air outlet duct.

    c. Separate the supercharger and No. 1 air outlet duct and remove the gasket.

    d. Remove the supercharger and No. 1 air outlet duct from the vehicle.

To install

19. Install the supercharger as follows:

    a. Install the supercharger and No. 1 air outlet duct to the vehicle.

    b. Connect the supercharger and No. 1 air outlet duct with a new gasket.

    c. Install the six nuts to hold the supercharger to the No. 1 air outlet duct. Tighten the nuts to 82 inch lbs. (9 Nm).

    d. Install the two bolts and two nuts to hold the supercharger to the equipment drive housing. Tighten the bolts and nuts to 27 ft. lbs. (37 Nm).

20. Install the No. 1 intake air connector bracket with the two bolts. Tighten the bolts to 13 ft. lbs. (18 Nm).

21. Connect the No.1 air tube.

22. Install the No. 1 idle up pipe by connecting the air hose and installing the bolt. Tighten the bolts to 69 inch lbs. (7.5 Nm).

23. Install the No. 1 air inlet duct with the supercharger bypass valve as follows:

    a. Connect the supercharger bypass valve to the No. 1 air outlet duct. Install the two bolts and two nuts and tighten the bolts and nuts to 48 inch lbs. (5 Nm).

    b. Install the air hoses and three way valve.

    c. Install the No. 1 air hose support bracket and two bolts. Tighten the bolts to 69 inch lbs. (8 Nm).

    d. Attach the supercharger magnetic clutch connector to the No. 1 hose support bracket.

    e. Connect the A/C idle up air hose.

    f. Attach the supercharger magnetic clutch connector.

    g. Connect the brake booster hose.

    h. Attach the supercharger bypass valve connector.

24. Install the No. 2 idler pulley with the spacer, plate, and the nut.

25. Install and adjust the drive belt for the supercharger.

26. Install the power steering pump. Tighten the long bolts to 35 ft. lbs. (48 Nm) and the short bolts to 27 ft. lbs. (36 Nm).

27. Install and adjust the alternator/power steering drive belts.

28. Install the throttle body.

29. Install the radiator. Tighten the radiator bolts to 13 ft. lbs. (18 Nm). Attach all connectors, hoses and shrouds.

30. Connect the power steering reservoir with the two bolts. Tighten the bolts to 9 ft. lbs. (13 Nm).

31. Install the supercharger blower.

32. Install the air damper case.

33. Install the engine coolant reservoir and bracket.

34. Install the air duct.

35. Fill the radiator with engine coolant.

36. Connect the battery cable to the battery. Reset any electronic components such as the radio.

37. Check all fluids.

Intake Manifold

REMOVAL & INSTALLATION

RAV4

✺✺ CAUTION

The fuel injection system remains under pressure after the engine has been turned OFF. Properly relieve fuel pressure before disconnecting any fuel lines. Failure to do so may result in fire or personal injury.

1. Properly relieve the fuel system pressure.

2. Disconnect the negative battery cable to the battery.

3. Remove the air cleaner assembly.

4. Remove the throttle body from the intake manifold.

5. Disconnect the engine wire from the intake manifold:

   a. Detach the four injector connectors.

   b. Disconnect the two engine wire clamps from the wire brackets on the intake manifold.

   c. Disconnect the engine wire protector from the right-hand side of the intake manifold by removing the bolt.

   d. Disconnect the clamp of the engine wire from the wire clamp.

6. Remove the EGR valve, EGR pipe and modulator as follows:

   a. Disconnect the two vacuum hoses from the VSV for the EGR.

   b. Disconnect the vacuum modulator from the clamp on the intake manifold.

   c. Loosen the cylinder head side of the EGR pipe union nut.

   d. Remove the two nuts, the EGR valve, pipe assembly and gasket. Remove the vacuum modulator.

7. Disconnect the following hoses:

• Fuel filter vacuum sensor hose on the intake manifold

• Brake booster vacuum hose from the intake manifold

• Ground strap from the intake manifold

8. Remove the intake manifold stay by removing the two bolts.

9. If equipped with A/T, disconnect the control cable from the clamp on the rear side of the intake manifold.

10. Disconnect the air hose from the intake manifold.

11. Remove the air tube from the intake manifold by removing the two bolts.

12. Remove the six bolts and two nuts from the intake manifold.

13. Remove the intake manifold from the vehicle.

To install:

14. Install the intake manifold to the engine and install the six bolts and two nuts. Tighten the bolts and nuts to 14 ft. lbs. (19 Nm).

15. Install the air tube with the two bolts.

16. Connect the air hose to the intake manifold.

17. If equipped with A/T, connect the control cable to the clamp on the rear side of the intake manifold.

18. Install the intake manifold stay with the two bolts. Tighten the bolts to 31 ft. lbs. (42 Nm).

19. Connect the following hoses:

• Ground strap to the intake manifold

• Brake booster vacuum hose to the intake manifold

• Fuel filter vacuum sensor hose to the intake manifold

20. Install the EGR valve, EGR pipe and the vacuum modulator as follows:

   a. Install the vacuum modulator.

   b. Install the EGR valve and pipe. Tighten the two nuts to 9 ft. lbs. (13 Nm) and the union nut to 43 ft. lbs. (59 Nm).

   c. Connect all the vacuum hoses.

21. Connect the engine wire and injectors.

➡ **The No. 1 and No. 3 injector connectors are brown, and the No. 2 and No. 4 injector connectors are gray.**

22. Install the throttle body to the intake manifold.

23. Install the air cleaner assembly.

24. Connect the negative battery cable to the battery.

Previa

✳✳ CAUTION

The fuel injection system remains under pressure after the engine has been turned OFF. Properly relieve fuel pressure before disconnecting any fuel lines. Failure to do so may result in fire or personal injury.

1. Properly relieve the fuel system pressure.

2. Disconnect the negative battery cable. Wait at least 90 seconds before performing any work after the cable has been disconnected on models with an airbag.

3. Remove the air intake connector.

4. Disconnect and tag all wires, harnesses, coolant and vacuum hoses from the intake manifold.

5. Disconnect the shift and accelerator cables.

6. Remove the fuel pipes.

7. Remove the distributor and EGR valve.

8. Remove the PCV hose.

9. Remove the water outlet, bypass pipe and gasket from the manifold.

10. Disconnect the water hose form the water pump and remove the bolt holding the water bypass pipe and timing chain case.

11. Remove the intake manifold stays.

12. Remove the two nuts and four bolts and extract the intake manifold with gasket. Remove the cylinder block insulators.

To install:

13. Position the cylinder block insulator on the cylinder head.

14. Position a new gasket on the cylinder head and install the intake manifold with the two nuts and four bolts. Tighten the fasteners to 15 ft. lbs. (21 Nm).

15. Install the intake manifold stays and secure to 27 ft. lbs. (37 Nm) to the block and 13 ft. lbs. (18 Nm) to the intake manifold.

16. Install the bolt holding the water bypass pipe and timing chain case and tighten to 13 ft. lbs. (18 Nm). Connect the water hose to the water pump.

17. Install the delivery pipe, water outlet and EGR valve.

18. Install the distributor.

19. Attach all wires, connectors, coolant and vacuum hoses from the intake manifold.

20. Connect the shift and accelerator cables.

21. Install the air intake connector.

22. Fill and bleed the cooling system.

23. Connect the negative battery cable. Start the engine and check for leaks.

Sienna and RX300

1. Remove the wiper and blade assembly.

2. Remove the top cowl seal and panel.

3. Label and disconnect the window washer hoses from the ventilator louvers.

4. Remove the left and right ventilator louvers.

5. Remove the heater air duct.

6. On RX300 models, remove the front upper suspension brace.

✳✳ CAUTION

The fuel injection system remains under pressure after the engine has been turned OFF. Properly relieve fuel pressure before disconnecting any fuel lines. Failure to do so may result in fire or personal injury.

7. Properly relieve the fuel system pressure.

8. Remove the battery and battery tray.

9. Drain and recycle the engine coolant.

10. Disconnect the accelerator cable on automatic transaxles. Disconnect the throttle cable.

11. Remove the air cleaner cap assembly. Disconnect and label any wiring or hoses interfering with removal.

12. Remove the right side engine mounting stay.

13. Disconnect the radiator and heater hoses in the way of the intake manifold removal.

14. Remove the V-bank cover.

15. Disconnect all the vacuum hose and wiring for the emission control valve set and remove.

16. Unbolt and remove the air intake chamber. Label all wiring and hoses. Discard the old gasket.

17. Remove the EGR pipe discard the old gaskets.

18. Unbolt and remove the hydraulic motor pressure hose from the air intake chamber.

19. Disconnect, unbolt and label the engine wiring harnesses from the left side, right side, rear and No. 3 timing belt cover.

20. Disconnect the front exhaust pipe and remove if necessary.

21. Remove the timing belt, camshaft timing pulleys, No. 2 idler pulley and No. 3 timing belt cover. Unbolt and remove the cylinder head rear plate.

22. Remove the two bolts, nuts and plate washers with the intake manifold. The delivery pipes with injectors will be attached to the manifold. Remove any other fuel related components such as the No. 2 fuel pipe and pulsation damper if needed.

23. Separate the delivery pipes from the intake manifold.

24. Clean and inspect the intake manifold mating surfaces. Scrape all old gasket martial off.

To install:

25. Install the delivery pipes with injectors to the intake manifold. Be sure to place four spacers in position on the manifold. Temporarily install four bolts to retain the delivery pipes to the manifold. Inspect the injectors for smooth rotation. Once the injectors are seated properly, tighten the delivery pipes retaining bolts to 7 ft. lbs. (10 Nm).

26. Install the No. 2 fuel pipe with union bolts and gaskets. Tighten the bolts to 24 ft. lbs. (32 Nm).

27. Install the No. 1 fuel pipe with pulsation damper, four new gaskets and the bolt, tighten the damper to 35 ft. lbs. (32 Nm) and the bolt to 11 ft. lbs. (15 Nm).

28. Install the fuel pressure regulator if removed.

29. Attach the intake manifold and tighten the nine retaining bolts and two nuts in a crisscross pattern to 11 ft. lbs. (15 Nm). Be sure the gasket is in place properly prior to tightening.

30. Retighten the water outlet mounting bolts and nuts to 11 ft. lbs. (15 Nm) if loosened. Install the air assist hose and water inlet pipe. Place a new O-ring on the end of the water inlet pipe, apply a small amount of soapy water and tighten to 14 ft. lbs. (20 Nm).

31. Attach the ground strap. Attach any vacuum hoses removed to the air intake chamber and vacuum tank.

32. Install any remaining components, remember to use all new gaskets. Tighten the air intake chamber bolts and nuts to 32 ft. lbs. (43 Nm), the EGR pipe nuts to 9 ft. lbs. (12 Nm), the emission control valve set to 69 inch lbs. (8 Nm).

33. Install the air cleaner assembly. Connect the heater hoses. Install the battery and tray.

34. Connect the throttle cable with bracket onto the throttle body.

35. On vehicles equipped with automatic transaxle, connect the accelerator cable and adjust it.

36. On RX300 models, install the front upper suspension brace and tighten the mounting nuts to 59 ft. lbs. (80 Nm).

37. Fill the cooling system to the proper level and connect the negative battery cable.

38. Install the heater air duct.

39. Install the left and right ventilator louvers.

40. Connect the window washer hoses from the ventilator louvers.

41. Install the top cowl seal and panel.

42. Install the wiper and blade assembly.

43. Start the engine and inspect for leaks.

Exhaust Manifold

REMOVAL & INSTALLATION

RAV4

1. Disconnect the negative battery cable.

✷✷ CAUTION

Wait 90 seconds from the time the key is turned to LOCK and the negative battery cable is disconnected to begin work. This allows the SRS capacitor to discharge and prevent deployment of the air bag(s).

2. Raise and safely support the vehicle.

3. Using a 14mm. deep socket wrench, remove the three nuts and the gasket to disconnect the front exhaust pipe from the exhaust manifold.

4. Detach the main oxygen sensor and the sub oxygen sensor connectors.

5. Remove the six bolts and the upper manifold heat insulator.

6. Remove the two bolts holding the right-hand exhaust manifold stay to the cylinder block.

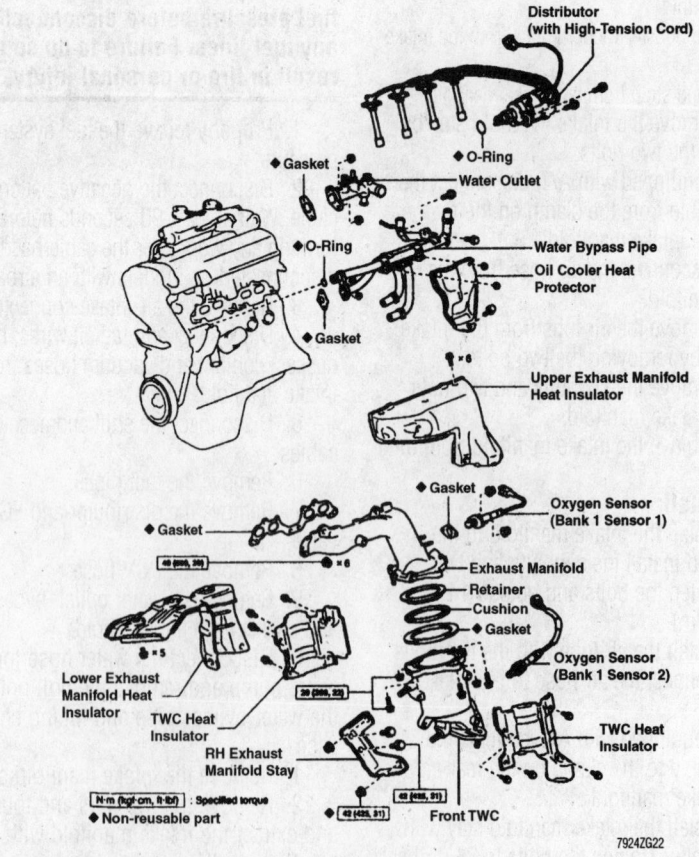

Exploded view of the exhaust manifold and components—RAV4

7. Remove the six nuts, the exhaust manifold, and the three-way catalytic converter assembly.

8. Separate the exhaust manifold and front catalytic converter.

To install:

9. Install the catalytic converter to the exhaust manifold. Tighten the three bolts and two nuts to 22 ft. lbs. (29 Nm).

10. Install a new gasket, the exhaust manifold, and the front TWC assembly with the six nuts. Uniformly tighten the nuts in several passes. Tighten the nuts to 36 ft. lbs. (49 Nm).

11. Install the right-hand manifold stay with the two bolts and tighten to 31 ft. lbs. (42 Nm).

12. Install the manifold upper heat insulator with the six bolts and connect the main oxygen and the sub oxygen sensor connectors.

13. Install the front exhaust pipe with a new gasket to the TWC. Install the three nuts using a 14mm. deep socket wrench. Tighten the nuts to 46 ft. lbs. (62 Nm).

14. Lower the vehicle safely and connect the negative battery cable.

15. Start the engine and be sure that there are no exhaust leaks.

Previa

1. Disconnect the negative battery cable.

❄ CAUTION

Wait 90 seconds from the time the key is turned to LOCK and the negative battery cable is disconnected to begin work. This allows the SRS capacitor to discharge and prevent deployment of the air bag(s).

2. Raise and safely support the vehicle.

3. Disconnect the front exhaust pipe from the exhaust manifold by removing the three nuts.

4. Remove the five nuts and remove the exhaust manifold and gasket.

To install:

5. Install the exhaust manifold with a new gasket. Install and tighten the exhaust manifold nuts to 36 Nm (41 Nm).

6. Connect the front exhaust pipe to the manifold by installing new gaskets and three nuts. Tighten the nuts to 46 ft. lbs. (62 Nm).

7. Lower the vehicle and connect the negative battery cable.

8. Start the engine and check for leaks.

Sienna and RX300

FRONT MANIFOLD

➡**Removing the oil filter helps gain access to a lower bolt in the front exhaust manifold.**

1. Disconnect the negative battery cable from the battery.

2. Raise and safely support the vehicle.

3. Remove the engine undercovers from the vehicle.

4. From below the engine, disconnect the front exhaust pipe from the exhaust manifolds by removing the nuts.

➡**Check for access to some of the manifold lower bolts, if so remove any possible.**

5. Lower the vehicle to access the upper manifold bolts.

6. Disconnect and remove the heated oxygen sensor.

7. Remove the exhaust manifold stay by removing the bolt and nut.

8. Remove the remaining nuts to the exhaust manifold and separate the exhaust manifold from the engine.

To install:

9. Using a new gasket, install the exhaust manifold to the engine and install the six nuts. Uniformly tighten, then tighten the bolts to 36 ft. lbs. (49 Nm).

10. Install the exhaust manifold stay and install the bolt and nut. Tighten the bolt and nut to 15 ft. lbs. (20 Nm).

11. Attach the heated oxygen sensor to the exhaust manifold.

12. Raise the vehicle and safely support.

13. Attach the front exhaust pipe to the exhaust manifold. Use a new gasket and tighten the two nuts to 46 ft. lbs. (62 Nm).

14. Install the engine undercovers to the vehicle.

15. Lower the vehicle.

16. Connect the negative battery cable to the battery.

REAR MANIFOLD

1. Disconnect the negative battery cable from the battery.

2. Raise and safely support the vehicle.

3. Remove the engine undercovers from the vehicle.

4. From below the engine, disconnect the front exhaust pipe from both exhaust manifolds.

5. Remove the EGR pipe from the rear exhaust manifold by removing the four nuts.

6. Disconnect the heated oxygen sensor wiring to the right exhaust manifold.

7. Remove the exhaust manifold stay by removing the bolt and nut.

8. Remove the six nuts to the exhaust manifold and separate the exhaust manifold from the engine.

To install:

9. Using a new gasket, install the exhaust manifold to the engine and install the six nuts. Uniformly tighten, then tighten the bolts to 36 ft. lbs. (49 Nm).

10. Install the exhaust manifold stay and install the bolt and nut. Tighten the bolt and nut to 15 ft. lbs. (20 Nm).

11. Attach the heated oxygen sensor wiring to the exhaust manifold.

12. Using new gaskets, install the EGR pipe to the exhaust manifold and the engine. Tighten the four nuts to 9 ft. lbs. (12 Nm).

13. Attach the front exhaust pipe to the exhaust manifold. Use a new gasket and tighten the two nuts to 46 ft. lbs. (62 Nm).

14. Install the engine undercovers to the vehicle.

15. Lower the vehicle.

16. Connect the negative battery cable to the battery.

Front Crankshaft Seal

REMOVAL & INSTALLATION

RAV4

❄ CAUTION

On models with an air bag, wait at least 90 seconds from the time that the ignition switch is turned to the LOCK position and the battery is disconnected before performing any further work.

➡**The front oil seal can be removed from the engine without removing the oil pump.**

1. Disconnect the negative battery cable from the battery.

2. Remove the timing belt covers and the timing belt from the engine.

3. Using SST tool 09950–50010 or equivalent crankshaft gear puller, remove the front crankshaft gear from the crankshaft. Be sure not to damage any part of the crankshaft.

4. Using a knife, cut off the oil seal lip.

5. Using a suitable tool, pry out the oil seal. Wrap the edge of the tool with a rag or tape to prevent damaging the crankshaft. Be careful not to damage the crankshaft.

To install:

6. Using a new seal, apply a thin layer of liquid sealer to the outside of the seal.

7. Apply multi purpose grease to the new oil seal lip.

8. Using SST tool 09226–00010 or equivalent oil seal installer and a hammer, tap in the oil seal until its surface is flush with the oil pump body edge.

9. Install the timing belt and the timing belt covers.

10. Install all other components, then connect the negative battery cable to the battery.

11. Start the engine and check for leaks.

Sienna and RX300

1. Remove the engine coolant reservoir tank and the alternator belt.

2. Remove the right front wheel and the splash shield.

3. Remove the power steering pump drive belt by loosening the two bolts.

4. Detach the two ground wire connectors.

5. Remove the right engine mounting stay.

6. Remove the engine moving control rod and the No. 2 right engine mounting bracket.

➡**To extract the engine bracket and control rod, you will need to raise the engine slightly.**

7. Remove the No. 2 alternator bracket.

8. Using a prybar and wrench or Crankshaft Pulley Holding tool 09213–54015 and Flange Holding tool 09330–00021 or their equivalents, remove the crankshaft pulley bolt.

9. Using a puller, remove the crankshaft pulley.

10. Remove the No. 1 timing belt cover by removing four bolts.

11. Remove the No. 2 timing belt cover as follows:

   a. Remove the bolt and disconnect the engine wire protector from the No. 3 (rear) timing belt cover.

   b. Disconnect the engine wire protector clamp from the No. 3 timing belt cover.

   c. Remove the five bolts from the No. 2 timing belt cover.

   d. Remove the No. 2 cover from the engine.

To install:

12. Apply the new gasket to the No. 2 timing belt cover. Install it evenly to the part of the belt cover shaded black. After installation, press down on it so that the adhesive sticks to the belt cover firmly.

13. Install the No. 2 timing belt cover with the five bolts. Tighten the bolts to 74 inch lbs. (8 Nm).

14. Install the engine wire protector clamp to the No. 3 timing belt cover.

15. Install the engine wire protector to the No. 3 timing belt cover with the bolt.

16. Apply new gasket the way you did on the No. 2 cover on the No. 1 timing cover. Secure the No. 1 timing belt cover with four bolts. Tighten the bolts to 74 inch lbs. (8 Nm).

17. Install the crankshaft pulley. Tighten the bolt to 159 ft. lbs. (215 Nm).

18. Install the No. 2 alternator bracket. Tighten the nut to 21 ft. lbs. (28 Nm). Do not tighten the pivot bolt at this time.

19. Install the No. 2 right engine mounting bracket and the moving control rod.

20. Install the right engine mounting stay.

21. Attach the two ground wire connectors.

22. Install and adjust the drive belts.

23. Install the coolant reservoir.

24. Install the right front splash shield and wheel assembly.

25. Connect the negative battery cable.

26. Start the vehicle and check for any leaks.

27. Recheck the ignition timing.

Camshaft and Valve Lifters

REMOVAL & INSTALLATION

RAV4 and Previa

1. Disconnect the negative battery cable.

2. Remove the cylinder head cover and the upper timing belt cover.

3. Rotate the crankshaft to set the engine at TDC/compression for the No. 1 cylinder.

➡**Due to the small thrust clearance on both the intake and exhaust camshafts, the camshafts must be kept level during removal. If the camshafts are removed without being kept level, the camshaft may be caught in the cylinder head causing the head to break or the camshaft to seize.**

4. Remove the camshaft timing sprocket and the timing belt.

5. Set the knock pin of the intake camshaft at 10–45° BTDC of camshaft angle. This angle will help to lift the exhaust camshaft level and evenly by pushing No. 2 and No. 4 cylinder camshaft lobes of the exhaust camshaft toward their valve lifters.

6. Secure the exhaust camshaft sub-gear to the main gear using a service bolt. The manufacturer recommends a bolt 0.63–0.79 in. (16–20mm) long with a thread diameter of 6mm and a 1mm thread pitch. When removing the exhaust camshaft be sure that the torsional spring force of the sub-gear has been eliminated.

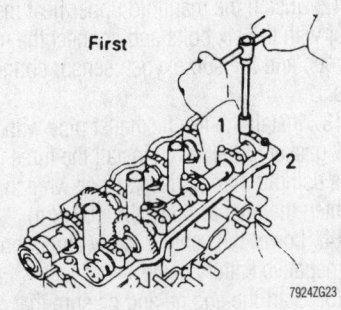

Exhaust camshaft bolt removal—Step 1— 2.0L (VIN P) and 2.4L (VIN A and T) Engine

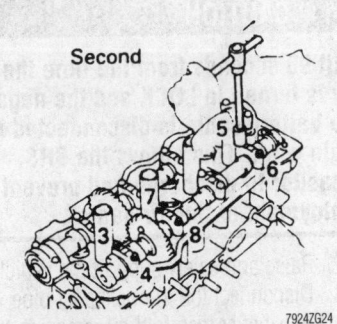

Exhaust camshaft bolt removal—Step 2— 2.0L (VIN P) and 2.4L (VIN A and T) Engine

7. Remove the No. 1 and No. 2 rear bearing cap bolts and remove the cap. Uniformly loosen and remove bearing cap bolts No. 3 to No. 8 in several passes and in the proper sequence. Do not remove bearing cap bolts No. 9 and 10 at this time. Remove the No. 1, 2, and 4 bearing caps.

8. Alternately loosen and remove bearing cap bolts No. 9 and 10. As these bolts are loosened check to see that the camshaft is being lifted out straight and level.

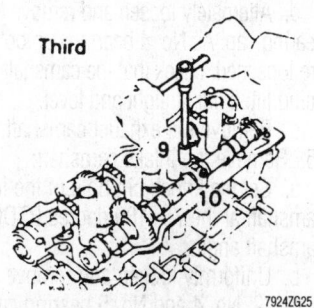

Exhaust camshaft bolt removal—Step 3—2.0L (VIN P) and 2.4L (VIN A and T) Engine

➡**If the camshaft is not lifting out straight and level retighten No. 9 and 10 bearing cap bolts. Reverse the order of Steps 5 through 7 and reset the intake camshaft knock pin to 10–45 degrees BTDC and repeat Steps 5 through 7 again. Do not attempt to pry the camshaft from its mounting.**

9. Remove the No. 3 bearing cap and exhaust camshaft from the engine.

10. Set the knock pin of the intake camshaft at 80–115° BTDC of camshaft angle. This angle will help to lift the intake camshaft level and evenly by pushing No. 1 and No. 3 cylinder camshaft lobes of the intake camshaft toward their valve lifters.

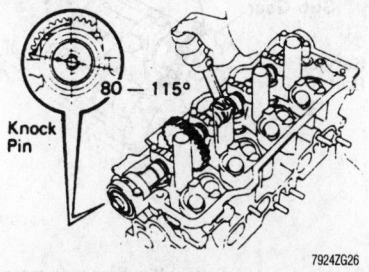

Intake camshaft knock pin alignment—2.0L (VIN P) and 2.4L (VIN A and T) Engine

11. Remove the No. 1 and No. 2 front bearing cap bolts and remove the front bearing cap and oil seal. If the cap will not come apart easily, leave it in place without the bolts.

12. Uniformly loosen and remove bearing cap bolts No. 3 to No. 8 in several phases and in the proper sequence. Do not remove bearing cap bolts No. 9 and 10 at this time. Remove No. 1, 3, and 4 bearing caps.

13. Alternately loosen and remove bearing cap bolts No. 9 and 10. As these bolts are loosened and after breaking the adhesion on the front bearing cap, check to see that the camshaft is being lifted out straight and level.

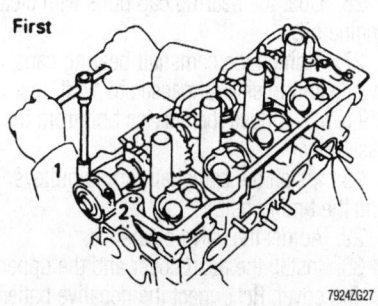

Intake camshaft bolt removal—Step 1—2.0L (VIN P) and 2.4L (VIN A and T) Engine

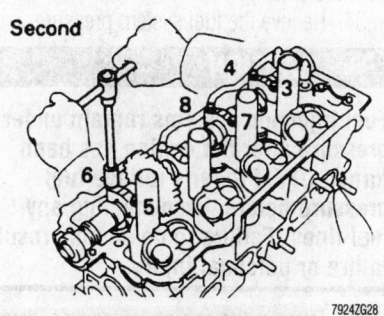

Intake camshaft bolt removal—Step 2—2.0L (VIN P) and 2.4L (VIN A and T) Engine

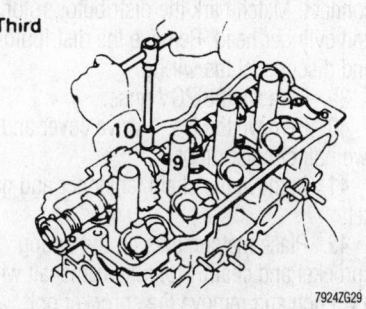

Intake camshaft bolt removal—Step 3—2.0L (VIN P) and 2.4L (VIN A and T) Engine

➡**If the camshaft is not lifting out straight and level retighten No. 9 and 10 bearing cap bolts. Reverse steps 10 through 12, than start over from Step 10. Do not attempt to pry the camshaft from its mounting.**

14. Remove the No. 2 bearing cap with the intake camshaft from the engine.

15. Remove the valve adjusting shims from the engine. Be sure to replace the shims to their original location.

To install:

16. Install the valve adjusting shims to the engine.

17. Before installing the intake camshaft, apply multi-purpose grease to the thrust portion of the camshaft.

18. Position the camshaft at 80–115° BTDC of camshaft angle on the cylinder head.

19. Apply sealant to the front bearing cap.

20. Coat the bearing cap bolts with clean engine oil.

21. Tighten the camshaft bearing caps evenly and in several passes to 14 ft. lbs. (19 Nm) in the proper sequence.

22. Set the knock pin of the camshaft at 10–45° BTDC of camshaft angle.

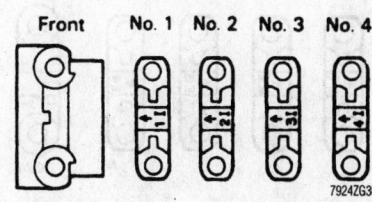

Intake camshaft bearing cap positioning—2.0L (VIN P) and 2.4L (VIN A and T) Engine

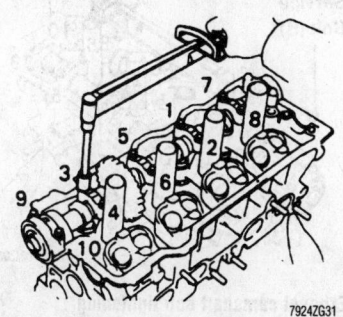

Intake camshaft bolt tightening sequence—2.0L (VIN P) and 2.4L (VIN A and T) Engine

23. Apply multipurpose grease to the thrust portion of the camshaft.

24. Position the exhaust camshaft gear with the intake camshaft gear so that the timing marks are in alignment with one another. Be sure to use the proper alignment marks on the gears. Do not use the assembly reference marks.

25. Turn the intake camshaft clockwise or counterclockwise little by little until the exhaust camshaft sits in the bearing journals evenly without rocking the camshaft on the bearing journals.

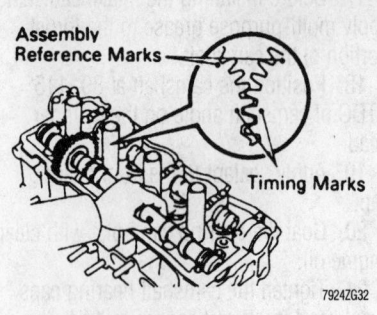

Camshaft timing mark alignment—2.0L (VIN P) and 2.4L (VIN A and T) Engine

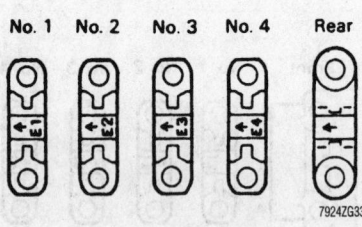

Exhaust camshaft bearing cap positioning—2.0L (VIN P) and 2.4L (VIN A and T) Engine

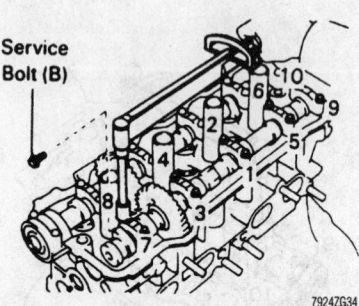

Exhaust camshaft bolt tightening sequence—2.0L (VIN P) and 2.4L (VIN A and T) Engine

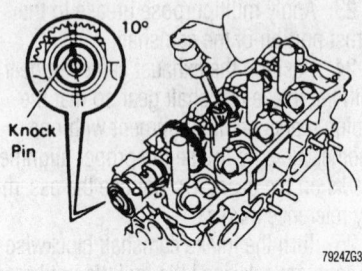

Exhaust camshaft knock pin alignment—2.0L (VIN P) and 2.4L (VIN A and T) Engine

26. Coat the bearing cap bolts with clean engine oil.

27. Tighten the camshaft bearing caps evenly and in several passes to 14 ft. lbs. (19 Nm). Remove the service bolt from the assembly.

28. Install the camshaft timing pulleys and the timing belt.

29. Adjust the valve clearance.

30. Install the head cover and the upper timing cover. Reconnect the negative battery cable.

31. Start the engine and check for leaks.

32. Check and adjust the ignition timing.

33. Disconnect the negative battery cable from the battery.

34. Relieve the fuel system pressure.

✳✳ CAUTION

Fuel injection systems remain under pressure after the engine has been turned OFF. Properly relieve fuel pressure before disconnecting any fuel lines. Failure to do so may result in fire or personal injury.

35. Remove the engine/transaxle assembly from the vehicle.

36. Remove the engine wiring from the engine and move it aside.

37. Remove the No. 2 valve cover.

38. Mark the spark plug wires and disconnect. Matchmark the distributor, rotor and cylinder head. Remove the distributor and disconnect the wiring.

39. Remove the PCV hose.

40. Remove the No. 1 valve cover and two half circular plugs.

41. Remove the chain tensioner and gasket.

42. Place matchmarks on the timing sprocket and chain. Hold the camshaft with a wrench and remove the sprocket bolt.

43. Remove the No. 6 camshaft bearing cap.

44. Remove exhaust camshaft:

   a. Set the knock pin hole of the exhaust camshaft at the 5–30 degree BTDC of camshaft angle. Hint: The above angle helps to lift the exhaust camshaft level and evenly by pushing No. 2 and No. 4 cylinder cam lobes of the exhaust camshaft to their valve lifters.

   b. Secure the exhaust camshaft sub-gear to main gear with a service bolt.

   c. Uniformly loosen and remove No. 1, No. 2, No. 3 and No. 5 bearing caps in several passes in the proper sequence.

➡ **Do not remove No. 4 bearing cap bolt at this stage.**

   d. Alternately loosen and remove No. 4 bearing cap. As No. 4 bearing cap bolts are loosened, check that the camshaft is being lifted out straight and level.

   e. Remove the exhaust camshaft.

45. Remove the intake camshaft:

   a. Set the knock pin hole of the intake camshaft at the 75–100 degree BTDC of camshaft angle.

   b. Uniformly loosen and remove No. 1, No. 2, No. 4 and No. 5 bearing caps in several passes in the proper sequence.

➡ **Do not remove No. 3 bearing cap bolt at this stage.**

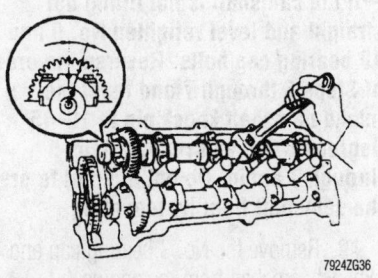

Set knock pin hole of exhaust camshaft at 5–30 degrees BTDC of camshaft angle—2.0L (VIN P) and 2.4L (VIN A and T) Engine

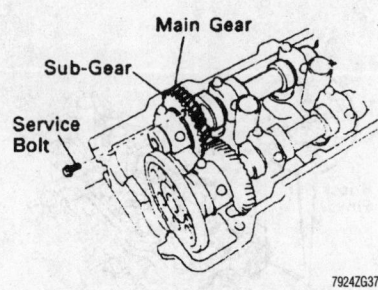

Secure exhaust camshaft sub-gear to main gear with service bolt—2.0L (VIN P) and 2.4L (VIN A and T) Engine

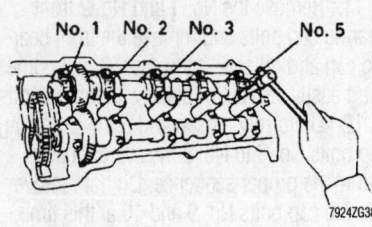

Remove No. 1, 2, 3 and 5 bearing caps (exhaust camshaft) in sequence shown—2.0L (VIN P) and 2.4L (VIN A and T) Engine

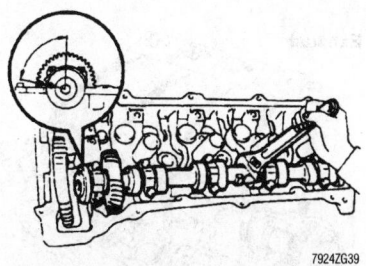

Set knock pin of intake camshaft at 75–100 degrees BTDC of camshaft angle—2.0L (VIN P) and 2.4L (VIN A and T) Engine

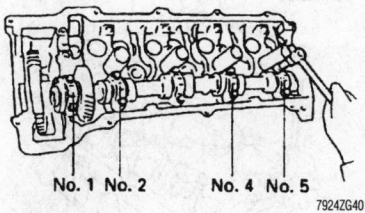

No. 1 No. 2   No. 4 No. 5

Remove No. 1, 2, 4 and 5 bearing caps (intake camshaft) in sequence shown— 2.0L (VIN P) and 2.4L (VIN A and T) Engine

c. Alternately loosen and remove No. 3 bearing cap. As No. 3 bearing cap bolts are loosened, check that the camshaft is being lifted out straight and level.

d. Remove the intake camshaft.

46. Remove the valve lifters from the engine. Keep the shims and the lifters together. The lifters and the shims must be reinstalled in their original location.

To install:

➡**If any of the bolts break, deform or do not meet the torque specification replace them.**

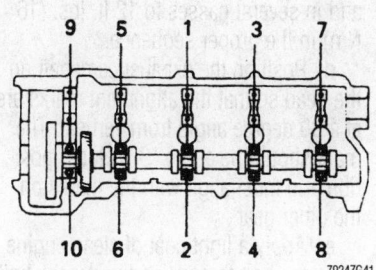

Camshaft bearing cap bolts tightening sequence—2.0L (VIN P) and 2.4L (VIN A and T) Engines

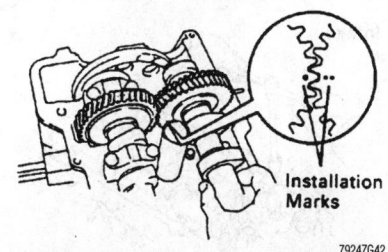

Installation Marks

Engage the exhaust and intake camshafts by matching the installation marks to each gear—2.0L (VIN P) and 2.4L (VIN A and T) Engine

47. Install the valve lifters and shims.
48. Install the intake camshaft:

a. Apply MP grease to the thrust portion of the intake camshaft.

b. Place the intake camshaft at 75–100 degrees BTDC. Install the bearing caps with the marking arrows facing forward. Uniformly tighten the bearing cap bolts in several passes in the proper sequence to 12 ft. lbs. (16 Nm).

49. Install the exhaust camshaft:

a. Set the knock pin of the intake camshaft at 5–30 degrees BTDC of camshaft angle.

b. Apply MP grease to the thrust portion of the exhaust camshaft.

c. Engage the exhaust camshaft gear to the intake camshaft gear by matching the installation marks (timing marks).

d. Roll down the exhaust camshaft onto the bearing journals while engaging the gears with each other. Be sure the exhaust and intake camshaft gear alignment marks are facing each other. The one gear has 2 dots and the other has one dot.

e. Install the bearing caps with the marking arrows facing forward.

f. Uniformly tighten the bearing cap bolts in several passes in the proper sequence to 12 ft. lbs. (16 Nm).

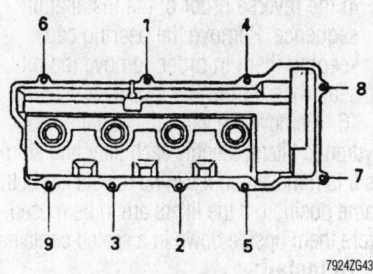

Cylinder head cover tightening sequence—2.0L (VIN P) and 2.4L (VIN A and T) Engine

50. Apply sealer to the bottom of No. 6 bearing cap and install. Tighten the cap to 12 ft. lbs. (16 Nm).

51. Install the camshaft sprocket and chain. Tighten the bolt to 54 ft. lbs. (74 Nm).

52. Release the chain tensioner ratchet pawl. Fully push in the plunger and apply the hook to the pin so the plunger can not spring out and install the tensioner. Tighten the bolts to 15 ft. lbs. (21 Nm).

53. Set the tensioner: Turn the crankshaft to the left so the hook of the tensioner is released from the pin. If it does not spring out, press the slipper into the tensioner to release the hook.

54. Adjust the valve clearance.

55. Apply sealant (P/N 08826–00080 or equivalent) to the cylinder head. Install the two half-circular plugs to the cylinder head and install the valve cover. Tighten the bolts to 69 inch lbs. (7.8 Nm).

56. Install the PCV hose.
57. Install the distributor.
58. Install the No. 2 cylinder head cover.

59. Install the engine wire and attach all connectors.

60. Install the engine/transaxle assembly into the vehicle.

61. Fill the cooling system. Fill the engine with oil.

62. Connect the battery cable. Start the engine and check for leaks.

63. Road test the vehicle for proper operation. Recheck all fluid levels.

Sienna and RX300

1. Remove the timing belt and idler pulley.

2. Remove the camshaft timing pulleys.

3. Remove the cylinder head covers.

➡**The thrust clearance on both the intake and exhaust camshafts is very small; the camshafts must be kept level during removal. If the camshafts are removed without being kept level, the camshaft may be caught in the cylinder head, causing the head to break or the camshaft to seize.**

4. To remove the exhaust and intake camshafts from the right side cylinder head:

a. Turn the camshaft with a wrench until the two pointed marks drive and driven gears are aligned. (The right camshaft gears have two marks apiece; the left side camshaft gears have one mark each.)

b. Secure the exhaust camshaft sub-gear to the main gear using a service bolt. A bolt 0.63–0.79 in. (16–20mm) long with a 6mm thread diameter and a 1mm pitch is recommended. When removing the exhaust camshaft be sure the sub-gear is not loaded; all the force must be eliminated.

c. Uniformly loosen and remove the exhaust camshaft bearing cap bolts in several passes and in the proper sequence. Remove the eight bearing cap bolts and remove the caps, keeping them in the correct order.

d. Remove the exhaust camshaft from the engine.

e. Uniformly loosen and remove the 10 bearing cap bolts in several passes, in the proper sequence. Remove the bearing caps, keeping them in order, remove the oil seal, then lift out the intake camshaft.

5. To remove the exhaust and intake camshafts from the left side cylinder head:

a. Turn the camshaft with a wrench until the pointed marks on the drive and driven gears are aligned. (The right camshaft gears have two marks apiece; the left side camshaft gears have one mark each.)

Intake

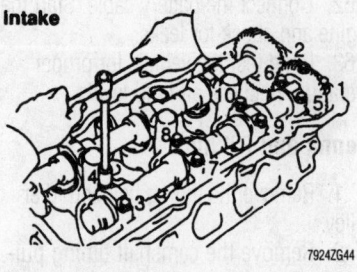

Right intake camshaft bearing cap bolt loosening sequence—Sienna and RX300 models

Exhaust

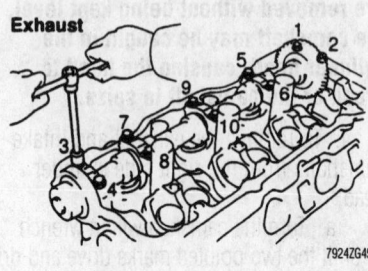

Right side exhaust camshaft bearing cap bolt loosening sequence—Sienna and RX300 models

Intake

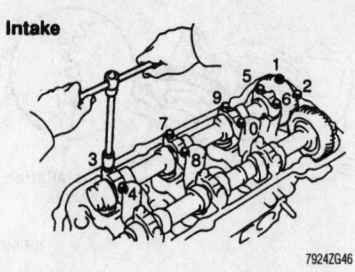

Left intake camshaft bearing cap bolt loosening sequence—Sienna and RX300 models

Exhaust

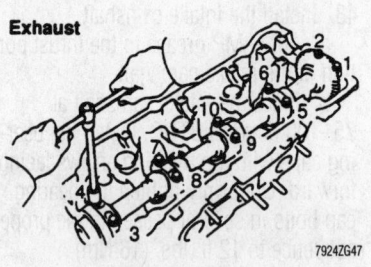

Left side exhaust camshaft bearing cap bolt loosening sequence—Sienna and RX300 models

b. Secure the exhaust camshaft sub-gear to the main gear using a service bolt. A bolt 0.63–0.79 inch (16–20mm) long with a 6mm thread diameter and a 1mm pitch is recommended. When removing the exhaust camshaft be sure the sub-gear is not loaded; all the force must be eliminated.

c. Uniformly loosen and remove the exhaust camshaft bearing cap bolts in several passes and in the proper sequence. Remove the eight bearing cap bolts and remove the caps. Keep the caps in the correct order.

d. Remove the exhaust camshaft from the engine.

e. Uniformly loosen and remove the 10 bearing cap bolts in several passes, in the reverse order of the installation sequence. Remove the bearing caps, keeping them in order, remove the oil seal, then lift out the intake camshaft.

6. Remove the valve lifter shims and hydraulic lifters. Identify each lifter and shim as it is removed so it can be reinstalled in the same position. If the lifters are to be reused, store them upside down in a sealed container.

To install:

7. Install the valve lifters into their original positions and install the shims. Check valve clearance and replace the shims as necessary.

Exhaust

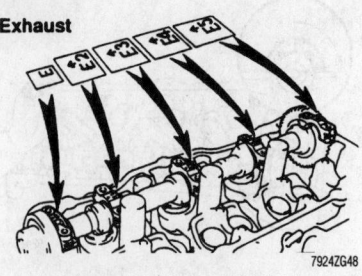

Right exhaust bearing caps must be placed in their proper locations—Sienna and RX300 models

Exhaust

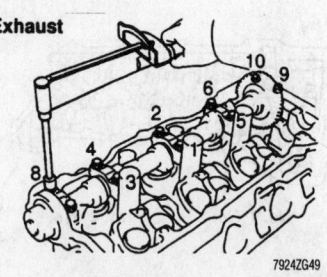

Right exhaust camshaft bearing cap bolt tightening sequence—Sienna and RX300 models

8. When reinstalling, remember that the camshafts must be handled carefully and kept straight and level to avoid damage.

9. Before installing the camshafts in either cylinder head, apply multi-purpose grease to each camshaft.

10. To install the right camshafts:

a. Position the intake camshaft on the head so that the alignment marks are at a 90 degree angle from vertical. The mark should be at the "3 O'clock" position.

b. Apply sealant to the No. 1 bearing cap.

c. Apply a light coat of clean engine oil to the bolt threads and under the bolt head. Install the bearing caps to their proper position. Tighten the bolts evenly and in several passes to 12 ft. lbs. (16 Nm) in the proper sequence.

d. Position the exhaust camshaft on the head so that the alignment marks are at a 90 degree angle from vertical. The mark should be at the "9 O'clock" position and must align with the marks on the other gear.

e. Apply a light coat of clean engine oil to the bolt threads and under the bolt head. Install the bearing caps to their proper position. Tighten the bolts evenly and in several passes to 12 ft. lbs. (16 Nm) in the proper sequence.

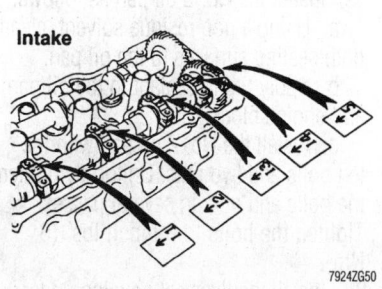

7924ZG50

Right intake bearing caps must be placed in their proper locations—Sienna and RX300 models

f. Remove the service bolt.

11. To install the left camshafts:

a. Position the intake camshaft on the head so that the alignment mark is at a 90 degree angle from vertical. The mark should be at the "9 O'clock" position.

b. Apply sealant to the No. 1 bearing cap.

c. Apply a light coat of clean engine oil to the bolt threads and under the bolt head. Install the bearing caps to their proper position. Tighten the bolts evenly and in several passes to 12 ft. lbs. (16 Nm) in the proper sequence.

d. Position the exhaust camshaft on the head so that the alignment marks are at a 90 degree angle from vertical. The mark should be at the "3 O'clock" position and must align with the marks on the other gear.

e. Apply a light coat of clean engine oil to the bolt threads and under the bolt head. Install the bearing caps to their proper position. Tighten the bolts

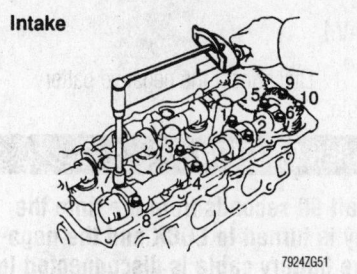

7924ZG51

Right intake camshaft bearing cap bolt tightening sequence—Sienna and RX300 models

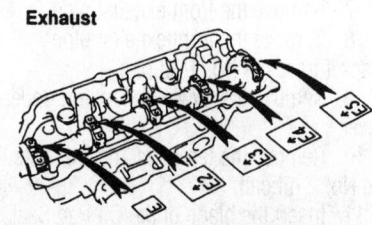

7924ZG52

Exhaust

Exhaust

Left exhaust bearing caps locations and bolt tightening sequence—Sienna and RX300 models

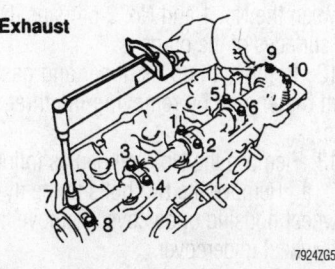

Intake

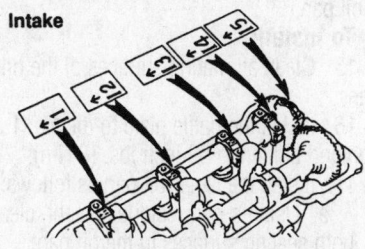

Intake

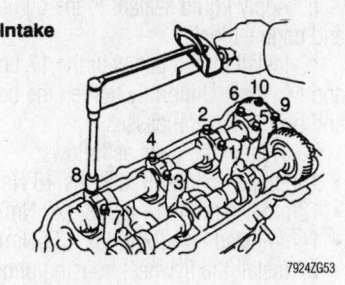

7924ZG53

Left intake camshaft bearing cap locations and bolt tightening sequence—Sienna and RX300 models

evenly and in several passes to 12 ft. lbs. (16 Nm) in the proper sequence.

f. Remove the service bolt.

12. Apply multi-purpose grease to new camshaft oil seals. Install the seals.

13. Install the No. 3 (rear) timing belt cover.

14. Install the camshaft timing gears.

15. Install the idler pulley, timing belt and covers.

16. Check and adjust the valve clearance.

17. Install the cylinder head (valve) covers.

18. Start the engine. Check the ignition timing.

19. Test drive the vehicle.

20. Check all fluid levels.

Valve Lash

ADJUSTMENT

These vehicles are equipped with Hydraulic Lash Adjusters (HLA), no adjustment is necessary.

Oil Pan

REMOVAL & INSTALLATION

RAV4

1. Raise and safely support the vehicle.

2. Remove the right-hand engine undercover.

3. Drain the oil and remove the dipstick.

4. Disconnect the front exhaust pipe.

5. Remove the stiffener plate from the engine by removing the two (M/T) or three (A/T) bolts.

6. Remove the two nuts and 17 bolts from the oil pan.

7. Separate the oil pan from the cylinder block.

8. Remove the oil pan from the vehicle.

To install:

9. Clean all gasket surfaces completely.

10. Apply a thin bead of sealer to the oil pan mounting surfaces.

➡**Avoid applying too much sealant to the oil pan.**

11. Place the oil pan against the block and install the bolts and nuts. Tighten the nuts and bolts to 4 ft. lbs. (5.4 Nm)

12. Install the stiffener plate and tighten the mounting bolts to 27 ft. lbs. (37 Nm).

13. Install the front exhaust pipe.

14. Fill the engine with oil to the proper level.

15. Lower the vehicle.

16. Start the engine and check for leaks. Recheck the engine oil level.

17. Install the right engine cover.

Previa

➡**This engine has two oil pans. If the crankshaft is going to be serviced, the side crankcase pan has to be removed. If the oil pump sump is going to be serviced, the bottom oil pan has to be removed.**

1. Disconnect the negative battery cable. Wait at least 90 seconds after the negative battery cable is disconnected before performing work on models equipped with an airbag.

2. Drain the engine oil.

3. Remove the oil level sensor and gasket. Be careful not to drop the sensor when removing.

4. Remove the 14 bolts and two nuts. Carefully pry the pan from the engine, being careful not to damage the flange.

To install:

5. Before installing, thoroughly clean the gasket mating surfaces. Apply gasket sealer 08826–00080 or equivalent, to the pan and assembly within 5 minutes.

6. Install the pan and tighten the bolts and nuts to 48 inch lbs. (5 Nm).

7. Install the gasket, oil sensor and tighten to 9 ft. lbs. (13 Nm).

8. Install the remaining components.

9. Refill the engine with oil.

10. Connect the negative battery cable. Start the engine and check for leaks.

Sienna and RX300

1. Turn the ignition key to the **OFF** position. Disconnect the negative battery cable. Wait at least 90 seconds from the time the negative battery was disconnected to start work.

2. Raise and safely support the front of the vehicle.

3. Remove the right front wheel.

4. Remove the fender apron seal.

5. Remove the engine undercover.

6. Drain the engine oil from the engine.

7. Remove the front exhaust pipe.

8. Remove the front exhaust pipe bracket from the No. 1 oil pan.

9. Remove the flywheel housing undercover.

10. Remove the ten bolts and two nuts to the No. 2 oil pan.

11. Insert the blade of the Oil Pan Seal Cutting tool 09032–00100 or equivalent between the No. 1 and No. 2 oil pans. Clean the surfaces of the oil pans.

12. Remove the oil strainer and gasket from the engine by removing the three nuts.

13. Remove the No.1 oil pan as follows:

a. Remove the two bolts to the flywheel housing undercover. Remove the flywheel undercover.

b. Remove the 17 bolts and two nuts to the No. 1 oil pan. Make a note of the position of the each bolt. When replacing the bolts into the oil pan, place each bolt in the position from which it was removed.

c. Remove the oil pan by prying the portions between the cylinder block and the oil pan. Be careful not to damage the contact surfaces.

14. Remove the baffle plate from the No. 1 oil pan.

To install:

15. Clean all mating surfaces of the oil pans.

16. Install the baffle plate to the No. 1 oil pan and tighten to 69 inch lbs. (8 Nm).

17. Install the No. 1 oil pan as follows:

a. Using a non residue solvent, clean both sealing surfaces to the oil pan.

b. Apply liquid sealant to the oil pan and engine block.

c. Install the oil pan with the 17 bolts and two nuts. Uniformly tighten the bolts and nuts in several passes.

d. Tighten the bolts as follows:

• 10mm head bolt-69 inch lbs. (8 Nm)
• 12mm head bolt-14 ft. lbs. (20 Nm)
• 14mm head bolt-27 ft. lbs. (37 Nm)

e. Install the flywheel housing undercover with the two bolts. Tighten the bolts to 69 inch lbs. (8 Nm).

18. Install the oil strainer with the three nuts. Tighten the nuts to 69 inch lbs. (8 Nm).

19. Install the No. 2 oil pan as follows:

a. Using a non residue solvent, clean both sealing surfaces to the oil pan.

b. Apply liquid sealant to the oil pan and engine block.

c. Install the No. 2 oil pan with the ten bolts and two nuts. Uniformly tighten the bolts and nuts in several passes. Tighten the bolts to 69 inch lbs. (8 Nm).

20. Install the flywheel housing undercover.

21. Install the front exhaust pipe bracket to the No. 1 oil pan. Tighten the bolts to 15 ft. lbs. (21 Nm).

22. Install the front exhaust pipe as follows:

a. Temporarily install the three new gaskets and the front exhaust pipe with the two bolts and six nuts.

b. Tighten the four nuts holding the exhaust manifolds to the front exhaust pipe. Tighten the four nuts to 46 ft. lbs. (62 Nm).

c. Tighten the two bolts and two nuts holding the front exhaust pipe to the center exhaust pipe. Tighten the bolts and nuts to 41 ft. lbs. (56 Nm).

d. Install the bracket with the two bolts and tighten to 14 ft. lbs. (19 Nm).

e. Install the support stay with the two bolts and tighten to 22 ft. lbs. (29 Nm).

23. Install the engine undercover.

24. Install the right fender apron seal.

25. Install the right front wheel and lower the vehicle.

26. Fill the engine with oil.

27. Start the engine and check for leaks.

Oil Pump

REMOVAL & INSTALLATION

RAV4

1. Disconnect the negative battery cable.

✱✱ CAUTION

Wait 90 seconds from the time the key is turned to LOCK and the negative battery cable is disconnected to begin work. This allows the SRS capacitor to discharge and prevent deployment of the air bag(s).

2. Remove the hood.

3. Raise and safely support the vehicle.

4. Remove the right-hand engine undercover.

5. Drain the engine oil.

6. Remove the front exhaust pipe.

7. Remove the rear end stiffener plate.

8. Remove the oil dipstick.

9. Remove the 17 bolts and two nuts from the oil pan.

10. Insert the blade of the Oil Pan Seal Cutting tool 09032–00100 or equivalent between the oil pan and the cylinder block, and cut off the applied sealer and remove the oil pan.

➡ Do not use the tool for the oil pump body side and rear oil seal retainer.

11. Remove the bolts, nuts, oil strainer and gasket.

12. Carefully suspend the engine with a sling device or equivalent.

13. Remove the timing belt.

14. Remove the No. 2 idler pulley and crankshaft timing pulley.

15. Using the Variable Pin Wrench Set 09960–10010 or equivalent, remove the nut and pulley to the oil pump.

16. Remove the crankshaft position sensor.

17. Remove the 12 mounting bolts, the oil pump, and the gasket.

To install:

18. Install a new gasket and the oil pump with the 12 bolts. Tighten the bolts to 82 inch lbs. (9 Nm).

➡ **The long bolts are 1.38 in. and all the others are 0.98 in.**

19. Install the crankshaft position sensor.

20. Install the oil pump pulley and install the nut. Tighten the nut to 18 ft. lbs. (24 Nm).

21. Install the crankshaft timing pulley and No. 2 idler pulley.

22. Install the timing belt.

23. Remove the engine sling.

24. Install the oil strainer with a new gasket, bolt, and nuts. Tighten to 48 inch lbs. (5 Nm).

25. Remove any old sealant from the oil pan flange and thoroughly clean both sealing surfaces.

26. Apply a 3–5mm bead of sealant to the oil pan flange.

➡ **The pan must be installed within five minutes of sealant application or the procedure will have to be repeated.**

27. Install the oil pan with the 17 bolts and two nuts. Uniformly tighten the bolts and nuts in several passes. Tighten the bolts and nuts to 48 inch lbs. (5 Nm) and install the dipstick.

28. Install the rear end stiffener plate and tighten the bolts to 27 ft. lbs. (37 Nm).

29. Install the front exhaust pipe.

30. Lower the vehicle and fill the engine with oil.

✸✸ WARNING

Be sure to prime the oil pump prior to initial engine start-up or engine damage may occur because of low oil pressure.

31. Connect the negative battery cable, start the engine, and check for leaks.

32. Recheck the engine oil level and install the hood.

33. Install the right-hand engine undercover.

Previa

➡ **The oil pump is part of the timing chain case on all Previa models. In order to do the oil pump, the engine/transmission assembly must be removed from the vehicle and the cylinder head removed.**

1. Disconnect the negative battery cable. Wait at least 90 seconds after the negative battery cable is disconnected before performing work on models equipped with an airbag.

2. Disconnect the equipment driveshaft from the crankshaft pulley.

3. Remove the crankshaft pulley.

4. Remove the oil pump cover screws and cover. Remove the O-ring.

5. Remove the timing chain case as follows:

   a. Remove the three bolts from the rear of the timing chain cover.

   b. Remove the 12 bolts and two nuts from the front of the chain cover.

➡ **Beware of the three bolts in the cover that are not to be removed, refer to the illustration.**

c. Using a plastic faced hammer, tap the chain case and remove the timing chain case and two gaskets.

To install:

6. Install the timing chain case as follows:

   a. Clean the gasket surface for the timing chain case.

   b. Install two new gaskets over the dowels.

   c. Slide on the chain case over the dowels.

   d. Install the bolts and nuts and tighten the bolts as follows:

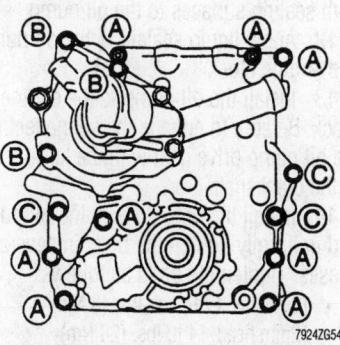

Tighten the bolts A, B and C to specifications

- A to 14 ft. lbs. (20 Nm)
- B to 21 ft. lbs. (28 Nm)
- C to 32 ft. lbs. (44 Nm)

   e. Install and tighten the three chain case bolts (rear) to 13 ft. lbs. (18 Nm).

7. Place a new O-ring into the groove of the timing chain case.

8. Install the oil pump cover. Tighten the screws to 8 ft. lbs. (10 Nm).

9. Install the crankshaft pulley and equipment driveshaft.

10. Connect the negative battery cable. Road test the vehicle for proper operation.

Sienna and RX300

1. Remove the oil pan.

2. Remove the crankshaft position sensor by removing the connector and bolt.

3. On the oil pump, remove the nine bolts. Make a note of the position of the each bolt. When replacing the bolts into the oil pump body, place each bolt in the position from which it was removed.

4. Remove the oil pump body by prying between the oil pump and main bearing cap.

5. Remove the O-ring from the cylinder block.

6. Remove the plug, gasket, spring, and relief valve from the oil pump body.

7. Remove the nine screws, pump body cover, drive, and driven rotors.

To install:

8. Install the driven rotors, drive, pump body cover, then install the nine screws.

9. Install the oil pump relief valve, spring, gasket, and the plug to the oil pump body.

10. Place a new O-ring on the cylinder block.

11. Using a non residue solvent, clean both sealing surfaces to the oil pump.

12. Apply liquid sealant to the oil pump and engine block.

13. Install the oil pump to the engine block. Be sure to engage the spline teeth of the oil pump drive gear with the large teeth of the crankshaft.

14. Install the nine bolts to the oil pump and uniformly tighten the bolts in several passes. Tighten the bolts as follows:
- 10mm head-69 inch lbs. (8 Nm)
- 12mm head-14 ft. lbs. (20 Nm)

15. Install the crankshaft position sensor and install the bolt. Tighten the bolt to 69 inch lbs. (8 Nm).

16. Install the baffle plate to the No. oil pan and tighten to 69 inch lbs. (8 Nm).

17. Install the No. 1 oil pan, oil strainer and No. 2 oil pan.

18. Refill the engine with oil.

19. Start the engine and inspect for leaks.

20. Recheck the engine oil level.

Rear Main Seal

REMOVAL & INSTALLATION

If the rear oil seal retainer is not installed to the block, use a taped ended screwdriver and hammer to remove the oil seal. Apply multi-purpose grease to the new oil seal lip. Using a seal driver, tap the seal into place. Be careful not to install it slantwise.

If the rear oil seal retainer is installed on the cylinder block, using a knife, cut off the lip of the seal. Using a taped ended prytool, pry the old seal out of the retainer. inspect the oil seal lip contacting surface of the crankshaft for cracks or damage. Apply multipurpose grease to the new oil seal, then tap the seal in place with a seal installer. Be careful not to install the seal slantwise.

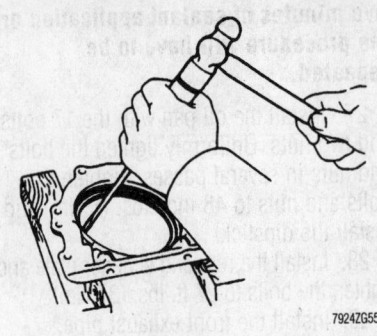

Carefully tap the old seal from the retainer

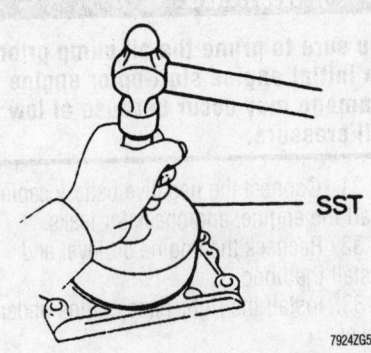

Use the proper sized driver to seat the seal

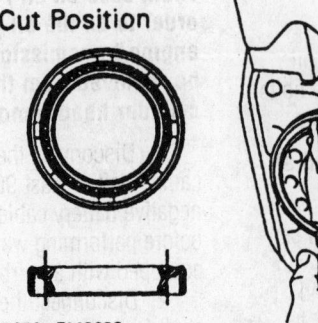

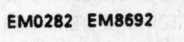

EM0282 EM8692

Cut off the oil seal lip, then pry the seal out of the retaining plate

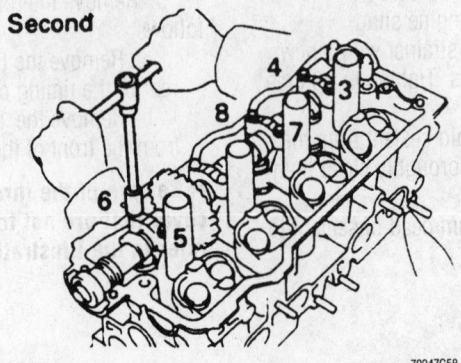

Tap a new seal into place

Timing Chain, Sprockets, Front Cover and Seal

REMOVAL & INSTALLATION

Previa

The engine/transmission assembly must to be extracted from the vehicle and the cylinder head removed before performing work on the timing chain.

1. Remove the engine from the vehicle.

2. Separate the engine and transmission.

3. Remove the cylinder head from the engine block.

4. Remove the crankshaft pulley as follows:

a. Using a holding device, secure the crankshaft. Loosen the pulley bolt.

b. Remove the tool and pulley bolt.

c. Using a puller, extract the crankshaft pulley.

5. Loosen the left engine mounting bolts and the left-hand mounting stay, then extract the assembly from the engine.

6. Remove the oil pressure switch.

7. Remove the No. 1 engine oil dipstick.

8. Unbolt and remove the No. 2 engine hanger.

9. Unbolt and extract the ventilation case, discard the gasket.

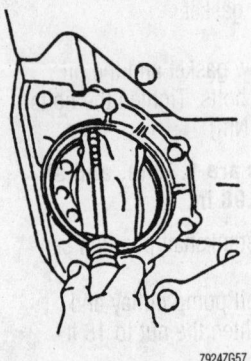

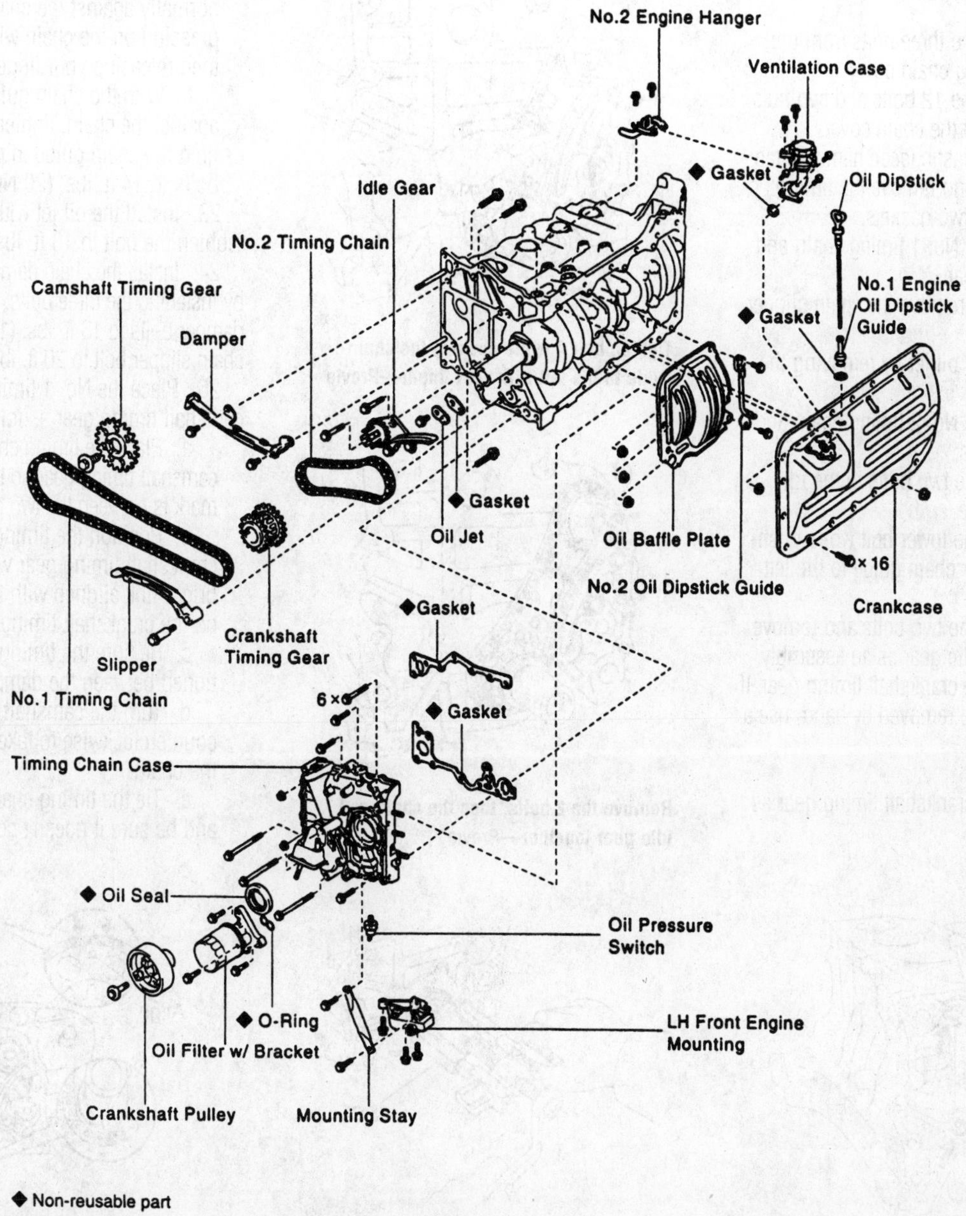

No.2 Engine Hanger

Ventilation Case

Idle Gear

◆ Gasket

Oil Dipstick

No.2 Timing Chain

◆ Gasket

No.1 Engine
Oil Dipstick
Guide

Camshaft Timing Gear

Damper

Gasket

Oil Jet

Oil Baffle Plate

No.2 Oil Dipstick Guide

× 16

Crankcase

Crankshaft
Timing Gear

◆ Gasket

Slipper

No.1 Timing Chain

6 × 6

◆ Gasket

Timing Chain Case

◆ Oil Seal

Oil Pressure
Switch

◆ O-Ring

Oil Filter w/ Bracket

LH Front Engine
Mounting

Crankshaft Pulley

Mounting Stay

◆ Non-reusable part

7924ZG90

Exploded view of the timing chain and gear component locations—Previa

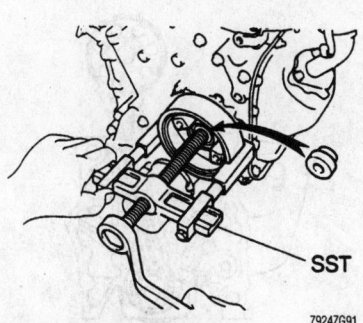

SST

7924ZG91

**Place a puller on the end of the crankshaft
pulley and remove it—Previa**

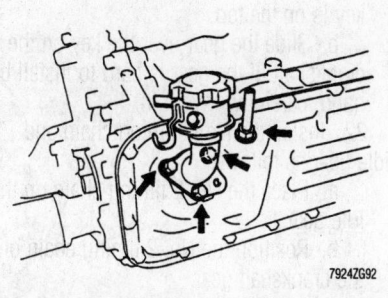

7924ZG92

**Unbolt and remove the ventilation case
and No. 1 oil dipstick guide—Previa**

10. Remove the No. 1 oil dipstick guide
and gasket.

11. Remove the crankshaft position sensor if equipped.

12. Remove the crankcase as follows:

a. Remove the 16 bolts and 2 nuts on
the case.

b. Using a prytool (09032–00100),
and a brass bar, separate the crankcase
from the cylinder block.

13. Remove the No. 2 oil dipstick guide
and oil baffle plate by removing the two
bolts and three nuts.

14. Unbolt and remove the oil filter
bracket with the oil filter. Remove the O-ring
from the timing chain case.

15. Remove the timing chain case as follows:

a. Remove the three bolts from the rear of the timing chain cover.

b. Remove the 12 bolts and two nuts from the front of the chain cover.

c. Using a plastic faced hammer, tap the chain case and remove the timing chain case and two gaskets.

16. Remove the No. 1 timing chain and camshaft timing gear.

17. Unbolt and remove the chain slipper and damper.

18. Remove the oil jet by removing the bolt.

19. Remove the No. 2 timing chain and idle gear as follows:

a. Loosen the two bolts to the idle gear chain guide.

b. Tighten the lower bolt while pushing the idle gear chain guide to the left with your finger.

c. Remove the two bolts and remove the chain and idle gear as an assembly.

20. Remove the crankshaft timing gear. If the gear can not be removed by hand, use a puller to extract it.

To install:

21. Install the crankshaft timing gear as follows:

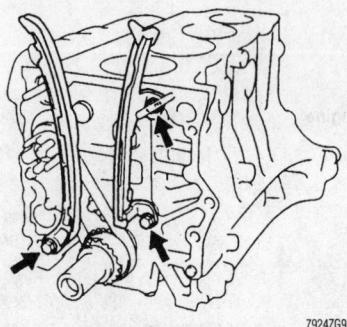

Unbolt the chain slipper and damper from the timing area—Previa

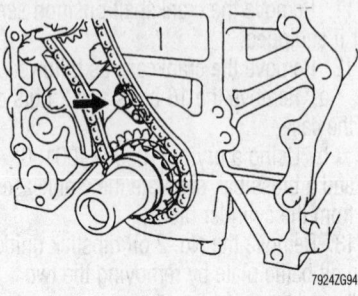

The oil jet is retained with one bolt in the center of the No. 2 timing chain—Previa

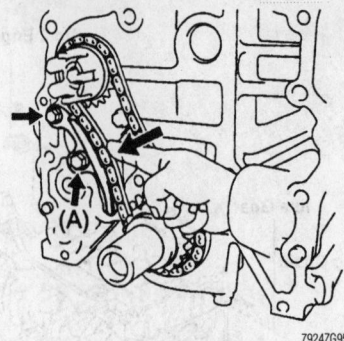

Tighten bolt A while pushing the chain guide to the left with your finger—Previa

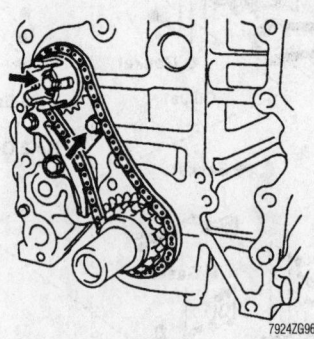

Remove the 2 bolts, then the chain and idle gear together—Previa

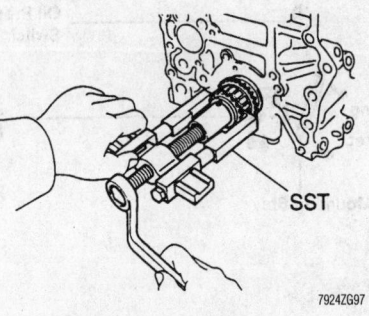

Using a puller to extract the crankshaft timing gear—Previa

a. Turn the crankshaft until the shaft key is on the top.

b. Slide the gear over the key on the crankshaft. If the gear is hard to install by hand, carefully drive it in.

22. Install the No. 2 timing chain and idle gear as follows:

a. Place the No. 2 timing chain on the idle gear.

b. Position the No. 2 timing chain on the crankshaft gear.

c. Install and tighten the two bolts to 14 ft. lbs. (20 Nm).

d. Loosen the lower bolt so that the chain guide presses against the chain.

e. Check that the spring is operating normally against the chain guide by pressing on the chain with your finger, then releasing your finger.

f. With the chain guide pressing against the chain, tighten the bolts to hold the chain guide in place. Tighten the bolts to 14 ft. lbs. (20 Nm).

23. Install the oil jet with a new gasket. Tighten the bolt to 13 ft. lbs. (18 Nm).

24. Install the chain damper and slipper by installing the three bolts. Tighten the chain damper bolts to 13 ft. lbs. (18 Nm) and the chain slipper bolt to 20 ft. lbs. (27 Nm).

25. Place the No. 1 timing chain and camshaft timing gear as follows:

a. Place the timing chain on the camshaft timing gear so that the timing mark is between the two bright chain links.

b. Position the timing chain on the crankshaft timing gear with the single bright link aligned with the timing mark on the crankshaft timing gear.

c. Be sure the timing chain is positioned between the damper and slipper.

d. Turn the camshaft timing gear counterclockwise to take the slack out of the chain.

e. Tie the timing chain with a cord and be sure it doesn't come loose.

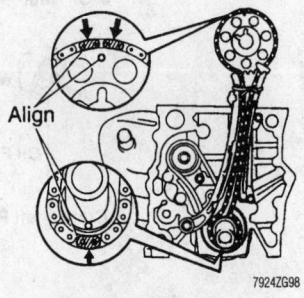

Align the timing chain on the camshaft gear so that the timing mark is between the two bright links—Previa

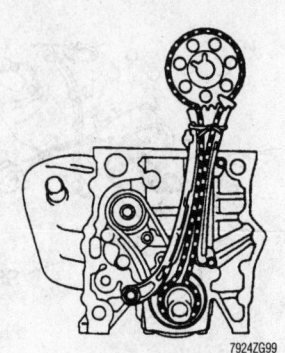

Tie the timing chain together as shown—Previa

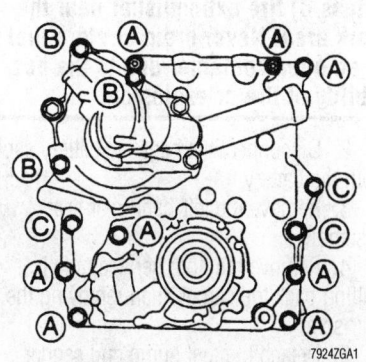

Timing chain cover bolt locations—Previa

26. Install the timing chain case as follows:

a. Clean the gasket surface for the timing chain case.

b. Install two new gaskets over the dowels.

c. Slide on the chain case over the dowels.

d. Install the bolts and nuts and tighten the bolts as follows:
- A to 14 ft. lbs. (20 Nm)
- B to 21 ft. lbs. (28 Nm)
- C to 32 ft. lbs. (44 Nm)

e. Install and tighten the three chain case bolts (rear) to 13 ft. lbs. (18 Nm).

27. Using a new O-ring, Install the oil filter bracket with the oil filter. Tighten the three bolts to 14 ft. lbs. (20 Nm).

28. Install the baffle plate and the No. 2 oil dipstick guide. Tighten the three baffle plate nuts to 43 inch lbs. (5 Nm) and the two No. 2 oil dipstick guide bolts to 13 ft. lbs. (18 Nm).

29. Install the crankcase to the engine. Tighten the 16 bolts and two nuts to 9 ft. lbs. (12 Nm).

30. Install the ventilation case with a new gasket. Tighten the three bolts to 69 inch lbs. (8 Nm).

31. Install the No. 1 oil dipstick guide with a new gasket. Tighten the nut to 22 ft. lbs. (29 Nm).

32. Install the No. 2 engine hanger with the four bolts. Tighten the bolts to the cylinder head to 27 ft. lbs. (37 Nm) and the bolts to the ventilation side to 69 inch lbs. (8 Nm).

33. Install the engine oil dipstick.

34. Install the oil pressure switch and tighten to 11 ft. lbs. (15 Nm).

35. Install the left-hand engine mounting and stay. Tighten the bolts for the mounting to 30 ft. lbs. (41 Nm) and the stay bolts to 27 ft. lbs. (37 Nm).

36. Install the crankshaft pulley as follows:

a. Install the crankshaft pulley to the crankshaft with the spline teeth of the crankshaft pulley engaged with the large teeth of the oil pump.

b. Rotate the crankshaft pulley to the left and right and check that the key groove of the crankshaft pulley correctly fits the crankshaft key.

c. Install the crankshaft pulley bolt.

d. Using the holding device, 09213–58012 and 09330–00021, or equivalent, tighten the bolt to 192 ft. lbs. (260 Nm).

37. Remove the cord from the timing chain.

38. Install the cylinder head.

39. Connect the engine and transmission.

40. Install the engine and transmission to the vehicle.

41. Top off all fluid levels. Test drive the vehicle.

FUEL SYSTEM

Fuel System Service Precautions

Safety is the most important factor when performing not only fuel system maintenance but any type of maintenance. Failure to conduct maintenance and repairs in a safe manner may result in serious personal injury or death. Maintenance and testing of the vehicle's fuel system components can be accomplished safely and effectively by adhering to the following rules and guidelines.

- To avoid the possibility of fire and personal injury, always disconnect the negative battery cable unless the repair or test procedure requires that battery voltage be applied.

- Always relieve the fuel system pressure prior to disconnecting any fuel system component (injector, fuel rail, pressure regulator, etc.), fitting or fuel line connection. Exercise extreme caution whenever relieving fuel system pressure to avoid exposing skin, face and eyes to fuel spray. Please be advised that fuel under pressure may penetrate the skin or any part of the body that it contacts.

- Always place a shop towel or cloth around the fitting or connection prior to loosening to absorb any excess fuel due to spillage. Ensure that all fuel spillage (should it occur) is quickly removed from engine surfaces. Ensure that all fuel soaked cloths or towels are deposited into a suitable waste container.

- Always keep a dry chemical (Class B) fire extinguisher near the work area.

- Do not allow fuel spray or fuel vapors to come into contact with a spark or open flame.

- Always use a back-up wrench when loosening and tightening fuel line connection fittings. This will prevent unnecessary stress and torsion to fuel line piping. Always follow the proper torque specifications.

- Always replace worn fuel fitting O-rings with new. Do not substitute fuel hose or equivalent, where fuel pipe is installed.

Fuel System Pressure

RELIEVING

✳✳ CAUTION

Fuel injection systems remain under pressure after the engine has been turned OFF. Properly relieve fuel pressure before disconnecting any fuel lines. Failure to do so may result in fire or personal injury.

1. Disconnect the negative battery terminal. Wait at least 90 seconds prior to working on models equipped with an airbag.

✳✳ CAUTION

Work must be started after 90 seconds from the time the ignition switch is turned to the LOCK position and the negative battery cable has been disconnected. The SRS is equipped with a back-up power source so that if work is started within 90 seconds of disconnecting the negative battery cable, the SRS may deploy. When the negative terminal cable is disconnected from the battery, memory of the clock and radio will be canceled. Before you start working, make a note of the contents memorized by the audio memory system. When you have finished working, reset the audio systems and adjust the clock. Never use a back-up power supply from outside the vehicle.

2. Place a catch-pan under the joint to be disconnected. A large quantity of fuel may be released when the joint is opened.

➡**Wear eye or full face protection.**

3. Place a shop towel over the area and slowly loosen the joint using a wrench of the correct size. Use a back-up wrench if needed.

4. Allow the fuel left in the line to bleed off slowly before fully disconnecting the joint.

5. Plug the opened lines immediately to prevent fuel spillage or the entry of dirt.

6. Dispose of the released fuel properly.

7. After adjoining fuel lines, connect the negative battery cable and start the engine.

8. Check for leaks and repair as needed.

Fuel Filter

REMOVAL & INSTALLATION

RAV4

✷✷ CAUTION

On models with an air bag, wait at least 90 seconds from the time that the ignition switch is turned to the LOCK position and the battery is disconnected before performing any further work.

1. Properly release fuel system pressure and disconnect the negative battery cable.

2. Unbolt the retaining screws and remove the protective shield for the fuel filter.

3. Place a pan under the delivery pipe to catch the dripping fuel and slowly loosen the union bolt or flare nut to bleed off the fuel pressure.

4. Drain the remaining fuel.

5. Disconnect and plug the inlet and outlet lines.

6. Unbolt and remove the fuel filter from the vehicle.

To install:

7. Coat the flare nut, union nut and bolt threads with engine oil.

8. Hand-tighten the inlet line to the fuel filter.

➡**When tightening the fuel line bolts to the fuel filter, use a torque wrench. The tightening torque is very important, as under or over tightening may cause fuel leakage. Insure that there is no fuel line interference and that there is sufficient clearance between it and any other parts.**

9. Install the fuel filter, then tighten the inlet bolts to 22 ft. lbs. (30 Nm).

10. Reconnect the delivery pipe using new gaskets, then tighten the union bolt to 22 ft. lbs. (30 Nm).

11. Run the engine for a few minutes and check for any fuel leaks.

12. Install the protective shield.

Except RAV4

1. Disconnect the negative battery cable.

2. Relieve the fuel system pressure.

✷✷ CAUTION

Fuel injection systems remain under pressure after the engine has been turned OFF. Properly relieve fuel pressure before disconnecting any fuel lines. Failure to do so may result in fire or personal injury.

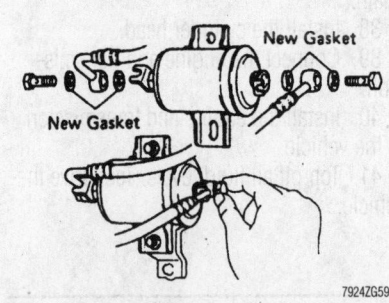

Exploded view of the fuel filter—except RAV4

➡**The fuel filter is located in the engine compartment, at the inlet line to the fuel rail.**

3. Disconnect and plug the inlet and outlet lines from the filter.

4. Remove the fuel filter retaining bolts and remove the filter.

To install:

5. Install the fuel filter.

6. Use new O-rings and tighten the lines to 22 ft. lbs. (29 Nm).

7. Connect the negative battery cable.

8. Start the engine and check for leaks.

Fuel Pump

REMOVAL & INSTALLATION

RAV4 and RX300

1. Relieve the fuel system pressure.

✷✷ CAUTION

Observe all applicable safety precautions when working around fuel. Do not allow fuel spray or fuel vapors to come into contact with a spark or open flame. Keep a dry chemical

(Class B) fire extinguisher near the work area. Never drain or store fuel in an open container due to the possibility of fire or explosion.

2. Disconnect the negative battery cable from the battery.

3. Remove the left-hand rear seat assembly.

4. Remove the floor service hole by pulling back the carpet, then removing the four screws.

5. Detach the fuel pump and sender gauge connector.

➡**Loosen the fuel cap to relieve any fuel pressure within the tank.**

6. Remove the union bolt and two gaskets to the fuel pipe. Disconnect the outlet pipe from the fuel pump.

7. Disconnect the return vent hose from the fuel pump.

8. Remove the eight bolts to the fuel pump and remove the pump assembly from the tank.

To install:

9. Install the fuel pump to the fuel tank and install the eight bolts. Tighten the bolts to 31 inch lbs. (3.5 Nm).

10. Connect the return vent hose to the fuel pump.

11. Connect the outlet pipe to the fuel pump. Using new gaskets, tighten the union bolts to 22 ft. lbs. (29 Nm).

12. Attach the fuel pump and sender gauge connector.

13. Install the floor hole cover with the four screws. Replace the carpet to its original position.

14. Install the left rear seat assembly.

15. Connect the negative battery cable to the battery.

16. Tighten the fuel cap and start the vehicle. Check for leaks.

Previa and Sienna

✷✷ CAUTION

Fuel injection systems remain under pressure after the engine has been turned OFF. Properly relieve fuel pressure before disconnecting any fuel lines. Failure to do so may result in fire or personal injury.

1. Relieve the fuel pressure.

2. Disconnect the negative battery cable. Wait at least 90 seconds before proceeding on models with an airbag.

⁜ CAUTION

Work must be started after 90 seconds from the time the ignition switch is turned to the LOCK position and the negative battery cable has been disconnected. The SRS is equipped with a back-up power source so that if work is started within 90 seconds of disconnecting the negative battery cable, the SRS may deploy. When the negative terminal cable is disconnected from the battery, memory of the clock and radio will be canceled. Before you start working, make a note of the contents memorized by the audio memory system. When you have finished working, reset the audio systems and adjust the clock. Never use a back-up power supply from outside the vehicle.

3. Drain the fuel from the gas tank.

4. Remove the fuel tank from the vehicle.
5. Remove the access plate bolts, then pull out the fuel pump assembly.
6. Disconnect the electrical wires from the fuel pump. Pull the bracket from the lower side of the fuel pump and remove the fuel pump from the fuel hose.
7. Remove the rubber cushion, the clip and the fuel filter from the bottom of the fuel pump.

To install:

8. Install the fuel pump filter to the fuel pump with a new clip.
9. Install the fuel pump to the fuel pump bracket and use new gaskets.
10. Connect the fuel hose to the outlet port of the fuel pump.
11. Install the fuel pump bracket. Tighten the bolts to 26 inch lbs. (3 Nm).
12. Install the fuel tank and connect all electrical and fuel harness.
13. Connect the negative battery cable.
14. Refill the fuel tank and check for leaks.

DRIVE TRAIN

Transmission Assembly

REMOVAL & INSTALLATION

Manual

RAV4 WITH 2-WHEEL DRIVE

1. Disconnect the negative battery cable.
2. Remove the air cleaner case assembly with hose.
3. Remove the engine coolant reservoir tank.
4. Remove the set nut of the engine wire clamp.
5. Remove the starter by removing the electrical connectors and the two set bolts.
6. Remove the clutch release cylinder as follows:

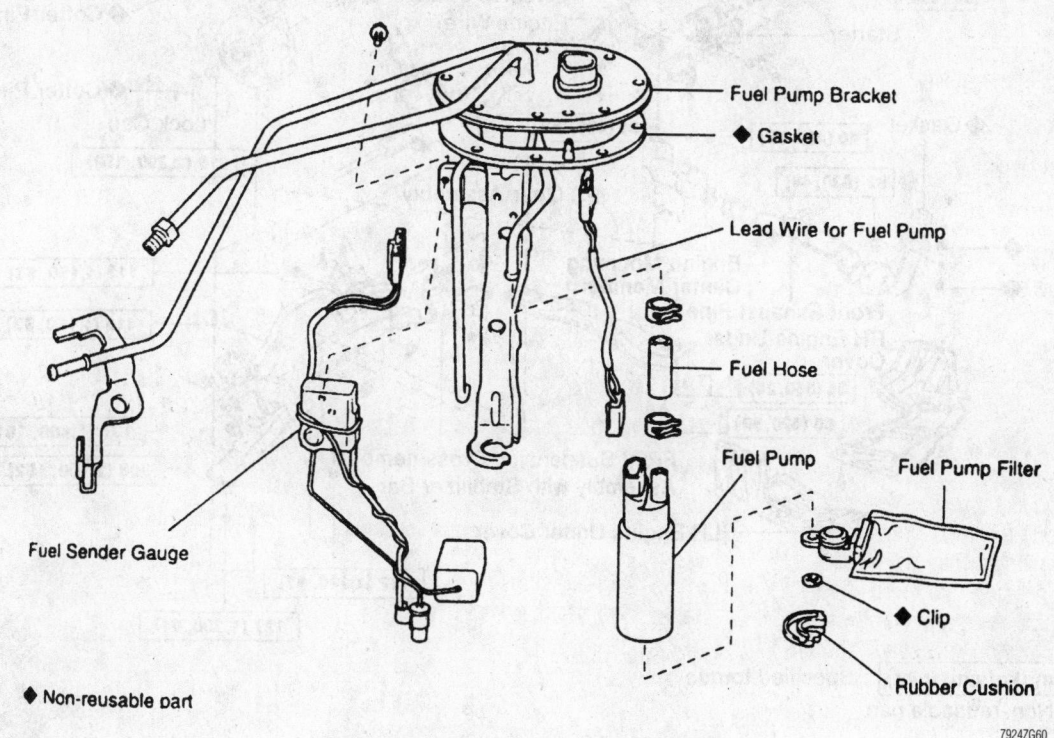

◆ Non-reusable part

Fuel Pump Bracket
◆ Gasket
Lead Wire for Fuel Pump
Fuel Hose
Fuel Pump
Fuel Pump Filter
Fuel Sender Gauge
◆ Clip
Rubber Cushion

7924ZG60

Exploded view of the fuel pump, bracket and related components—all vehicles similar

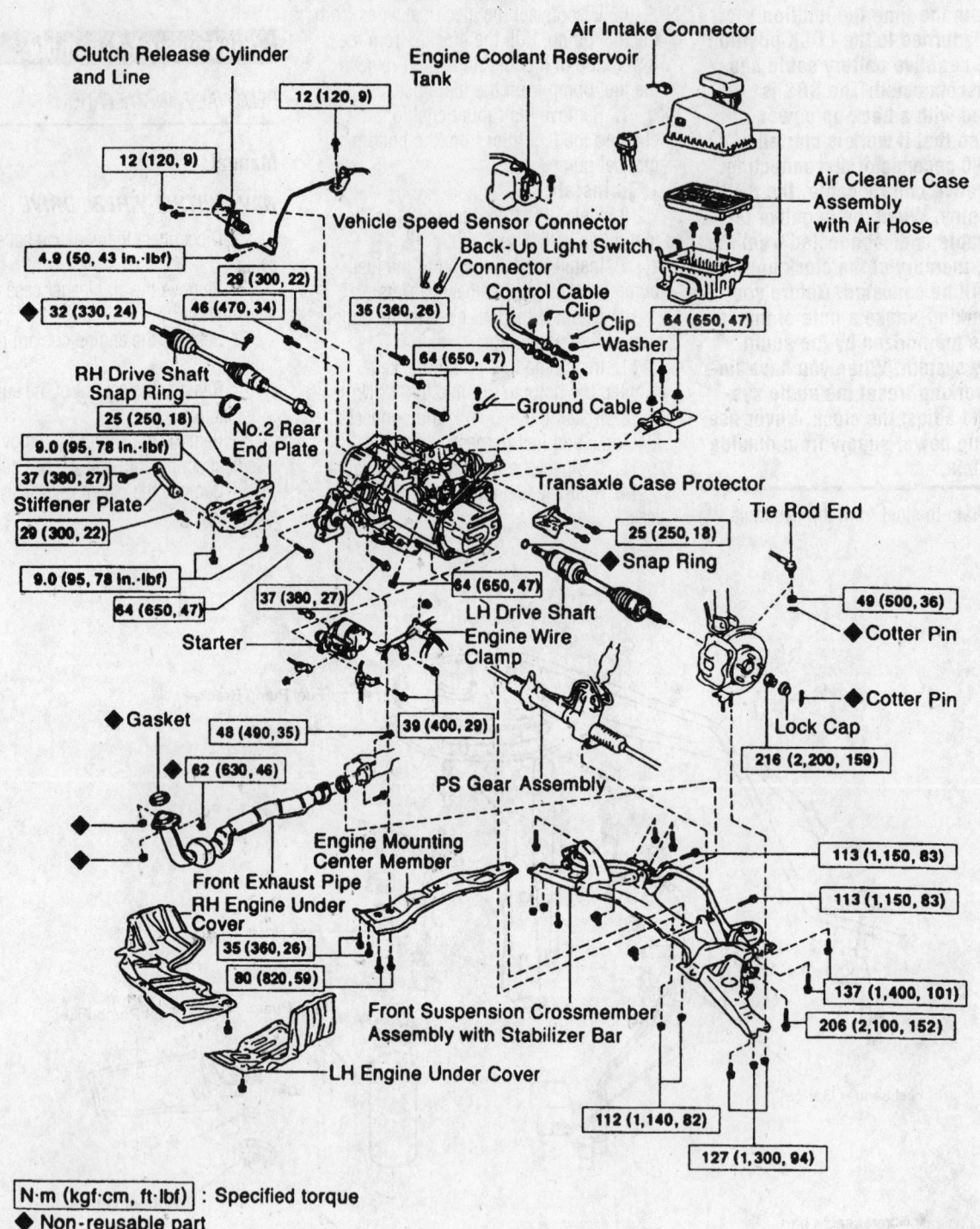

Clutch Release Cylinder
and Line

Engine Coolant Reservoir
Tank

Air Intake Connector

12 (120, 9)

12 (120, 9)

Air Cleaner Case
Assembly
with Air Hose

4.9 (50, 43 in.·lbf)

Vehicle Speed Sensor Connector

Back-Up Light Switch
Connector

29 (300, 22)

Control Cable
Clip

46 (470, 34)

35 (360, 26)

Clip
Washer

◆ 32 (330, 24)

64 (650, 47)

64 (650, 47)

RH Drive Shaft
Snap Ring

Ground Cable

25 (250, 18)

No.2 Rear
End Plate

Transaxle Case Protector

Tie Rod End

9.0 (95, 78 in.·lbf)

37 (360, 27)

25 (250, 18)

Stiffener Plate

◆ Snap Ring

49 (500, 36)

29 (300, 22)

◆ Cotter Pin

9.0 (95, 78 in.·lbf)

LH Drive Shaft

◆ Cotter Pin

37 (360, 27)

64 (650, 47)

Engine Wire
Clamp

Lock Cap

64 (650, 47)

216 (2,200, 159)

Starter

◆ Gasket

48 (490, 35)

39 (400, 29)

62 (630, 46)

PS Gear Assembly

Engine Mounting
Center Member

113 (1,150, 83)

113 (1,150, 83)

Front Exhaust Pipe
RH Engine Under
Cover

35 (360, 26)

80 (820, 59)

137 (1,400, 101)

Front Suspension Crossmember
Assembly with Stabilizer Bar

206 (2,100, 152)

LH Engine Under Cover

112 (1,140, 82)

127 (1,300, 94)

N·m (kgf·cm, ft·lbf) : Specified torque

◆ Non-reusable part

7924ZG61

Transaxle exploded view—2WD RAV4

a. Remove the set bolts holding the clutch line bracket to the transaxle.

b. Remove the two bolts, release cylinder and line.

7. Disconnect the ground cable from the transaxle by removing the bolt.

8. Detach the vehicle speed sensor and back up light switch connector.

9. Disconnect the control cable by removing the four clips and washers.

10. Remove the four upper side transaxle bolts connecting the transaxle to the engine.

11. Remove the bolt and two nuts holding the left mounting insulator to the vehicle.

12. Install a engine support to the engine.

13. Support rack and pinion to the engine support fixture with a rope.

14. Raise and safely support the front of the vehicle.

15. Remove the front wheels.

16. Remove the left and right-hand engine undercovers.

17. Drain the transaxle oil.

18. Remove the left and right halfshafts.

19. Remove the front exhaust pipe as follows:

a. Remove the three nuts and gasket from the exhaust manifold.

b. Remove the two bolts holding the exhaust pipe to the center exhaust pipe.

c. Disconnect the exhaust pipe from the vehicle.

20. Remove the front suspension crossmember assembly with the sway bar as follows:

a. Support the front suspension crossmember with a jack.

b. Disconnect the ring from the center exhaust pipe.

c. Remove the two set bolts and nuts of the power steering rack and pinion assembly.

d. Remove the suspension cross-member assembly with the sway bar by removing the two nuts and six bolts.

21. Remove the engine mounting center member by removing the four bolts.

22. Jack up the transaxle slightly.

23. Disconnect the left mounting bracket from the mounting insulator by removing the set bolt.

24. Remove the stiffener plate, No. 2 rear end plate and transaxle lower side mounting bolt.

25. Lower the engine left side and remove the transaxle from the engine.

26. Remove the transaxle case protector by removing the two bolts.

To install:

27. Install the transaxle case protector with the two bolts. Tighten the bolts to 18 ft. lbs. (25 Nm).

28. Install the transaxle to the engine.

29. Install the No. 2 rear end plate and transaxle bolts. Tighten the bolts as follows:

- Bolt C to 22 ft. lbs. (29 Nm)
- Bolt D to 34 ft. lbs. (46 Nm)
- Bolt E to 18 ft. lbs. (25 Nm)
- Bolt F to 78 inch lbs. (9.0 Nm)

30. Install the stiffener plate with the two bolts. Tighten the bolts to 27 ft. lbs. (37 Nm).

31. From underneath the vehicle, connect the engine left mounting insulator to the left mounting bracket. Tighten the bolt to 47 ft. lbs. (64 Nm).

32. Install the engine mounting center member with the four bolts. Tighten the bolts to the radiator support to 26 ft. lbs. (35 Nm) and the mounting insulator to 59 ft. lbs. (80 Nm).

33. Install the front suspension cross-member with the sway bar as follows:

a. Install the sway bar and suspension crossmember and install the six bolts and two nuts. Tighten the bolts as follows:

- Bolt A to vehicle: 152 ft. lbs. (206 Nm)
- Bolt B to lower control arm bracket: (101 ft. lbs. (137 Nm)
- Bolt C to rear mounting bracket: 82 ft. lbs. (112 Nm)

b. Connect the rack and pinion to the crossmember with the two bolts and two nuts. Tighten to 83 ft. lbs. (113 Nm).

c. Connect the ring for the center exhaust pipe.

34. Install the front exhaust pipe as follow:

a. Install the pipe with new gaskets.

b. Connect the front pipe to the center exhaust pipe with the two bolts. Tighten the bolts to 35 ft. lbs. (48 Nm).

c. Install the three nuts to hold the front exhaust pipe to the exhaust manifold. Tighten the three nuts to 46 ft. lbs. (62 Nm).

35. Install the left and right halfshafts.

36. Install the front wheels and lower the vehicle.

37. Install the set bolt and two nuts for the engine left mounting insulator. Tighten to 47 ft. lbs. (64 Nm).

38. Remove the engine support fixture.

39. Install the four transaxle upper side mounting bolts. Tighten bolt A to 47 ft. lbs. (64 Nm) and bolt B to 26 ft. lbs. (35 Nm).

40. Connect the ground cable with the clips and washers.

41. Attach the vehicle speed sensor and back up light switch connector.

42. Connect the ground cable to the transaxle and install the bolt.

43. Install the clutch release cylinder and line.

44. Install the starter and tighten the two bolts to 29 ft. lbs. (39 Nm). Install the electrical connectors.

45. Install engine wire clamp with the nut.

46. Install the engine coolant reservoir tank.

47. Install the air cleaner case assembly with the air hose.

48. Fill the transaxle with fluid. Check all fluids.

49. Connect the negative battery cable to the battery.

RAV4 WITH 4-WHEEL DRIVE

1. Remove the transaxle with the engine.

2. Remove the transaxle case protector by removing the two bolts.

3. Remove the starter by removing the two bolts.

4. Remove the transfer vacuum actuator bracket by removing the four bolts.

5. Remove the transfer vacuum actuator assembly as follows:

a. Disconnect the four solenoid hoses from the transfer vacuum actuator assembly.

b. Remove the transfer vacuum actuator assembly by removing the two bolts.

6. Remove the right transfer stiffener plate by removing the five bolts.

7. Remove the center transfer stiffener plate by removing the three bolts.

8. Remove the stiffener plate by removing the two bolts.

9. Remove the transaxle from the engine by removing the nine transaxle mounting bolts.

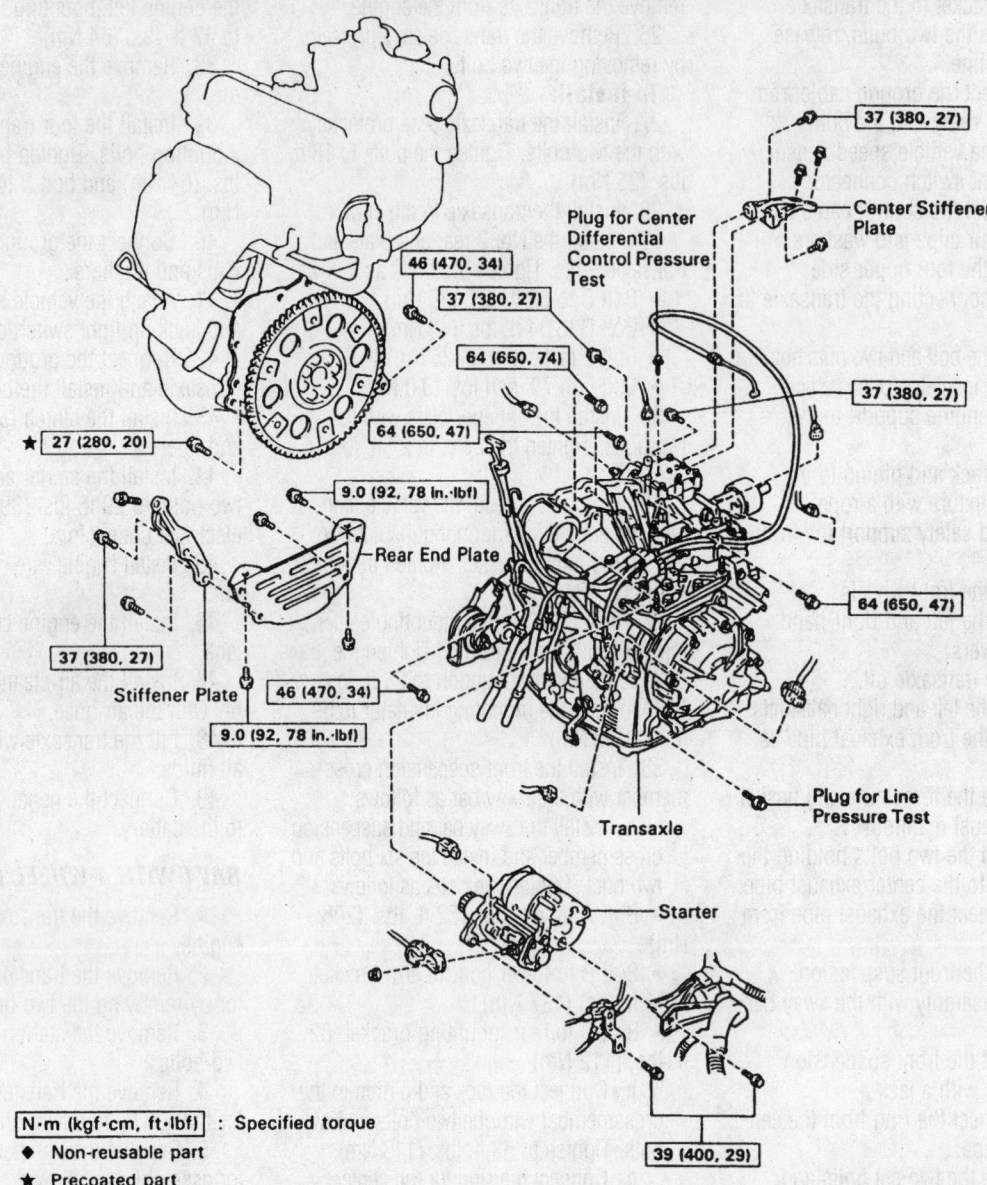

Plug for Center Differential Control Pressure Test

Center Stiffener Plate

37 (380, 27)

37 (380, 27)

46 (470, 34)

37 (380, 27)

64 (650, 74)

64 (650, 47)

27 (280, 20)

9.0 (92, 78 In.·lbf)

Rear End Plate

37 (380, 27)

Stiffener Plate

46 (470, 34)

9.0 (92, 78 In.·lbf)

64 (650, 47)

Plug for Line Pressure Test

Transaxle

Starter

39 (400, 29)

N·m (kgf·cm, ft·lbf) : Specified torque
◆ Non-reusable part
★ Precoated part

7924ZG62a&b

Transaxle exploded view—4WD RAV4

To install:

10. Connect the transaxle to the engine and install the nine bolts. Tighten the bolts as follows:

- Bolt A: 47 ft. lbs. (64 Nm)
- Bolt B: 26 ft. lbs. (35 Nm)
- Bolt C: 22 ft. lbs. (29 Nm)
- Bolt D: 34 ft. lbs. (46 Nm)
- Bolt E 18 ft. lbs. (25 Nm)
- Bolt F 78 inch lbs. (9.0 Nm)

11. Install the stiffener plate with the two bolts. Tighten the bolts to 27 ft. lbs. (37 Nm).

12. Install the center transfer stiffener plate with the three bolts. Tighten the bolts to 27 ft. lbs. (37 Nm).

13. Install the right transfer stiffener plate with the five bolts. Tighten the bolts to 27 ft. lbs. (37 Nm).

14. Install the transfer vacuum actuator assembly as follows:

a. Install the transfer vacuum actuator assembly with the two bolts. Tighten the bolts to 27 ft. lbs. 937 Nm).

b. Connect the four solenoid hoses to the transfer vacuum actuator assembly.

15. Install the transfer vacuum actuator bracket with the four bolts. Tighten the bracket bolts to 27 ft. lbs. (37 Nm).

16. Install the starter with the two bolts. Tighten the bolts to 29 ft. lbs. (39 Nm).

17. Install the transaxle case protector with the two bolts, Tighten the two bolts to 18 ft. lbs. (25 Nm).

18. Install the transaxle with engine assembly.

PREVIA

1. Disconnect the negative battery cable. Wait at least 90 seconds to perform any work on models equipped with air bags.

2. Drain the transmission fluid.

3. Raise the vehicle and support safely.

4. Remove the starter motor. Match-mark the driveshafts-to-flange and remove the front (4WD) and rear driveshafts.

5. Remove the clutch release cylinder, hose and bracket.

6. Remove the exhaust pipe bracket which is retained by four bolts.

7. Disconnect the control cables/bracket and speed sensor wiring.

8. Remove the engine-to-transmission stiffener plate.

9. Place a suitable transmission jack under the transmission.

10. Remove the engine rear mounting bolts and raise the rear side of the engine.

11. Remove the engine-to-transmission bolts, pull the transmission toward the rear and extract.

To install:

12. Align the input shaft with the clutch disc and push the transmission fully into position.

13. Install the transmission bolts and tighten to 53 ft. lbs. (72 Nm).

14. Install the rear engine mounts and stiffener plate. Tighten the bolts to 27 ft. lbs. (37 Nm).

15. Connect the speed sensor and control cables.

16. Install the exhaust pipe bracket and tighten to 37 ft. lbs. (51 Nm).

17. Install the clutch release cylinder, starter and driveshafts. Tighten the starter to 41 ft. lbs. (56 Nm) and driveshaft bolts to 20 ft. lbs. (25 Nm).

18. Lower the vehicle.

19. Connect the battery cable and refill with transmission fluid.

Automatic

RAV4 WITH 4-WHEEL DRIVE

1. Disconnect the negative battery cable from the battery.

2. Remove the engine and transaxle assembly from the vehicle.

3. Remove the starter by removing the two bolts.

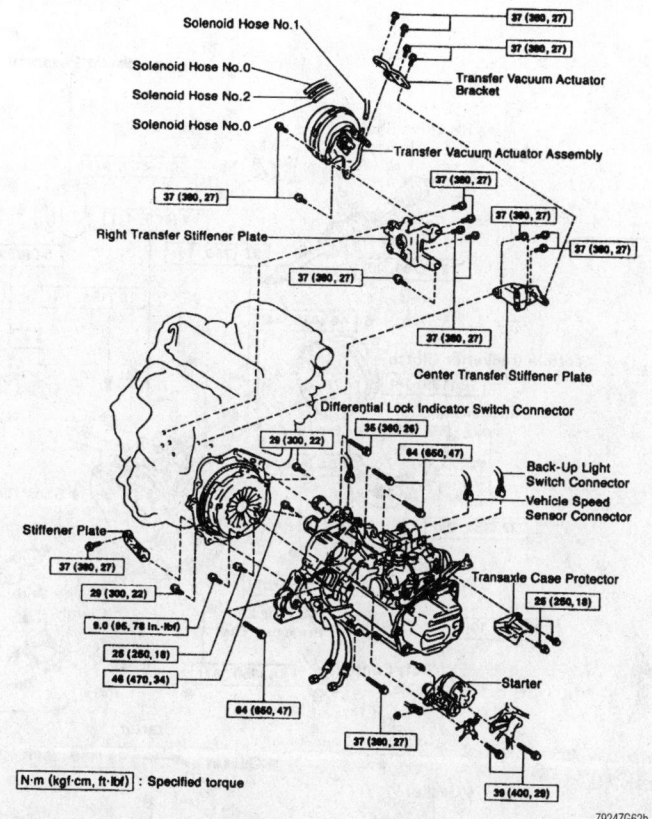

Transaxle exploded view—4WD RAV4

4. Remove the stiffener plate by removing the three bolts.

5. Remove the rear end plate by removing the four bolts.

6. Remove the six torque converter clutch mounting bolts.

7. Detach the connectors and wiring harness from the transaxle.

8. Remove the center stiffener plate by removing the four bolts.

9. Remove the transaxle with the transfer assembly as follows:

a. Remove the two bolts.

b. Remove the five transaxle mounting bolts.

c. Separate the transaxle assembly from the engine.

To install:

10. Install the transaxle to the engine.

11. Install the five transaxle mounting bolts and tighten the bolts as follows:

• 14mm head bolts to 47 ft. lbs. (64 Nm)

• 12mm head bolts to 34 ft. lbs. (46 Nm)

12. Install the two bolts and tighten the bolts to 27 ft. lbs. (37 Nm).

13. Install the center stiffener plate with the four bolts and tighten the bolts to 27 ft. lbs. (37 Nm).

14. Attach the connectors and the wiring harness to the transaxle.

15. Install the torque converter clutch mounting bolts. Install and tighten all the bolts evenly. Tighten each bolt to 20 ft. lbs. (27 Nm).

➡ **Coat the threads of the bolts with an approved locking compound.**

16. Install the rear end plate with the four bolts. Tighten the bolts to 80 inch lbs. (9.0 Nm).

17. Install the stiffener plate with the three bolts. Tighten the bolts to 27 ft. lbs. (37 Nm).

18. Install the starter with the two bolts. Tighten the bolts to 29 ft. lbs. (39 Nm).

19. Install the engine and transaxle assembly to the vehicle.

20. Connect the negative battery cable to the battery.

RAV4 WITH 2-WHEEL DRIVE

1. Disconnect the negative battery cable from the battery.

2. Disconnect the throttle cable.

3. Remove the engine coolant reservoir tank.

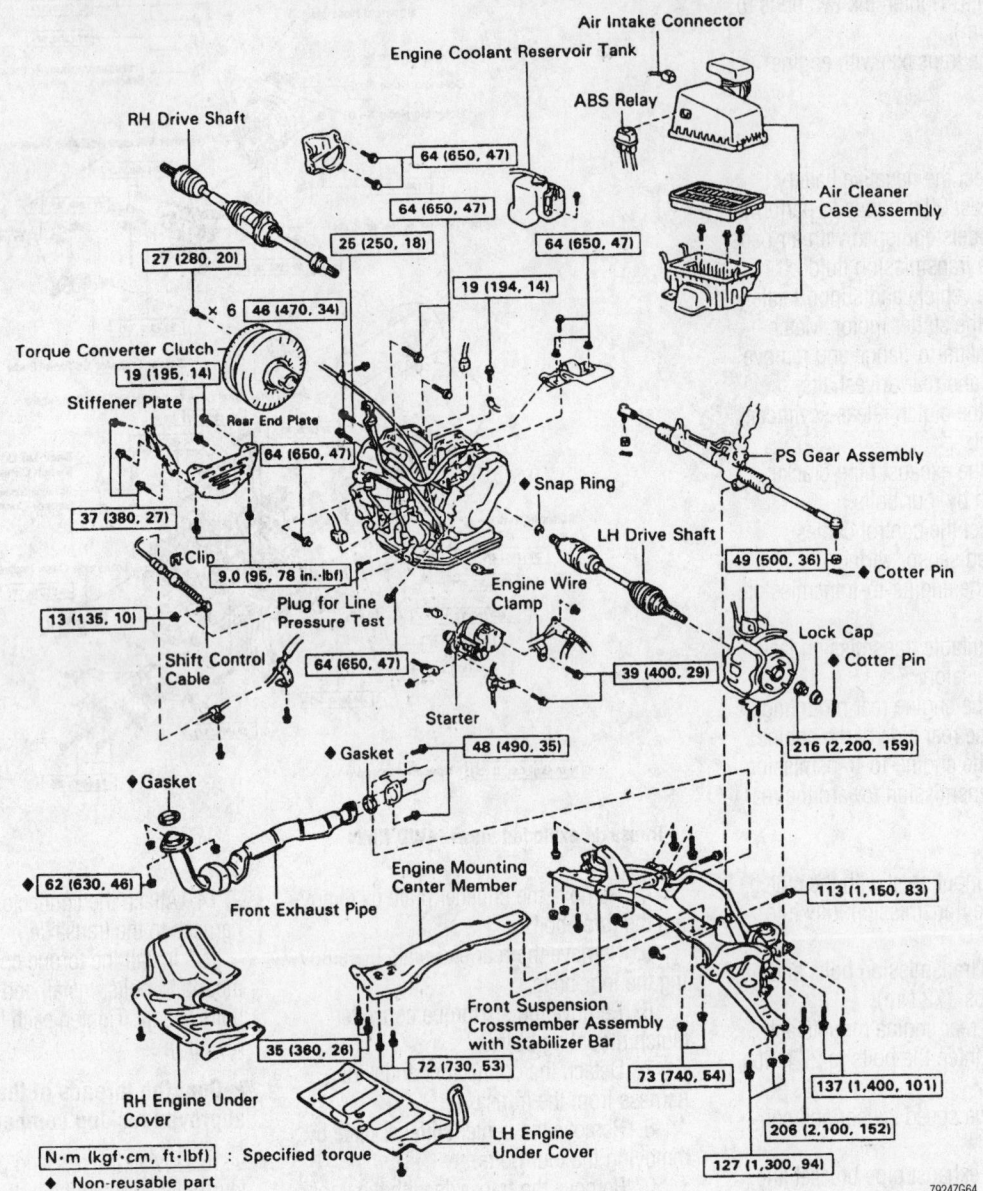

Air Intake Connector
Engine Coolant Reservoir Tank
ABS Relay
Air Cleaner Case Assembly

RH Drive Shaft

64 (650, 47)
64 (650, 47)
26 (250, 18)
64 (650, 47)
27 (280, 20)
19 (194, 14)

× 6   46 (470, 34)

Torque Converter Clutch
19 (195, 14)
Stiffener Plate
Rear End Plate
64 (650, 47)
37 (380, 27)
Clip
9.0 (95, 78 in.·lbf)
13 (135, 10)
Plug for Line Pressure Test
Shift Control Cable
64 (650, 47)

♦ Snap Ring

PS Gear Assembly

LH Drive Shaft
49 (500, 36)   ♦ Cotter Pin

Engine Wire Clamp
Lock Cap
♦ Cotter Pin
39 (400, 29)
216 (2,200, 159)

Starter
48 (490, 35)
♦ Gasket

♦ Gasket
62 (630, 46)
Front Exhaust Pipe

Engine Mounting Center Member
113 (1,150, 83)

Front Suspension Crossmember Assembly with Stabilizer Bar
35 (360, 26)
72 (730, 53)
73 (740, 54)
137 (1,400, 101)
RH Engine Under Cover
LH Engine Under Cover
206 (2,100, 152)
127 (1,300, 94)

N·m (kgf·cm, ft·lbf) : Specified torque
♦ Non-reusable part

7924ZG64

Transaxle exploded view—2WD RAV4

4. Remove the air cleaner assembly.

5. Remove the ground cable from the transaxle by removing the bolt.

6. Remove the set nut of the engine wire clamp.

7. Remove the starter from the engine as follows:

a. Detach the connector and nut from the starter.

b. Remove the two bolts and disconnect the engine wire.

c. Remove the starter from the engine.

8. Remove the three upper side transaxle mounting bolts.

9. Install an engine support fixture.

10. Remove the two bolts and two nuts from the left engine mounting.

11. Remove the engine undercovers.

12. Drain the fluid from the transaxle.

13. Remove the left and right half-shafts.

14. Remove the front exhaust pipe as follows:

a. Remove the two bolts and gasket holding the front exhaust pipe to the center exhaust pipe.

b. Remove the three nuts and gasket holding the front exhaust pipe to the exhaust manifold.

c. Remove the exhaust manifold from the vehicle.

15. Disconnect the shift control cable from the transaxle and frame as follows:

a. Remove the nut from the control shaft lever.

b. Remove the clip and disconnect the control cable from the transaxle.

c. Remove the two bolts holding the shift control cable to the center member and crossmember.

16. Detach the following connectors:
- Shift solenoid valve connector
- Park/neutral position switch connector
- Vehicle speed sensor connector

17. Disconnect the oil cooler hoses from the transaxle.

18. Disconnect the rack and pinion from the crossmember by removing the two nuts and two bolts. Support the rack and pinion.

19. Support the suspension crossmember with a floor jack. Remove the ten bolts and two nuts to the crossmember and center member. Lower the crossmember from the vehicle with the sway bar.

20. Remove the stiffener plate by removing the three bolts.

21. Remove the rear end plate by removing the four bolts.

22. Remove the six torque converter bolts.

23. Raise the transaxle slightly and remove the two rear side transaxle mounting bolts.

24. Lower the transaxle and remove the transaxle from the vehicle.

To install:

25. Install the transaxle to the vehicle and raise the transaxle slightly.

26. Install the two rear side transaxle mounting bolts. Tighten the top bolt to 18 ft. lbs. (25 Nm) and the lower bolt to 34 ft. lbs. (46 Nm).

27. Install the torque converter bolts and tighten the bolts to 20 ft. lbs. (27 Nm).

➡**First install the gray bolt, then install the five black bolts.**

28. Install the rear end plate with the four bolts. Tighten the bolts to the engine to 78 inch lbs. (9.0 Nm) and the transaxle bolts to 14 ft. lbs. (19 Nm).

29. Install the stiffener plate with the three bolts. Tighten the bolts to 27 ft. lbs. (37 Nm).

30. Install the front suspension crossmember and center member with the sway bar. Install the ten bolts and two nuts and tighten the bolts as follows:

- Bolt A: 152 ft. lbs. (206 Nm)
- Bolt B: 101 ft. lbs. (137 Nm)
- Bolt C to 26 ft. lbs. (35 Nm)
- Bolt D to 53 ft. lbs. (72 Nm)
- Nut to 54 ft. lbs. (73 Nm)

31. Connect the rack and pinion to the crossmember by installing the two bolts and two nuts. Tighten the nuts to 83 ft. lbs. (113 Nm).

32. Connect the oil cooler hoses with the two clips.

33. Attach the following connectors:
- Shift solenoid valve connector
- Park/neutral switch connector
- Vehicle speed sensor connector

34. Connect the shift control cable. Tighten the nut to hold the control cable to the transaxle to 10 ft. lbs. (13 Nm).

35. Install the front exhaust pipe as follows:

a. Using new gaskets, install the front exhaust pipe.

b. Connect the front exhaust pipe to the exhaust manifold with the three nuts. Tighten the nuts to 46 ft. lbs. (62 Nm).

c. Connect the front exhaust pipe to the center exhaust pipe with the two bolts. Tighten the bolts to 35 ft. lbs. (48 Nm).

36. Install the left and right halfshafts.

37. Install the engine undercovers.

38. Install the two bolts and two nuts for the left engine mount. Tighten the mount bolts and nuts to 47 ft. lbs. (64 Nm).

39. Remove the engine fixture.

40. Install the three upper side transaxle mounting bolts. Tighten the three bolts to 47 ft. lbs. (64 Nm).

41. Install the starter with the two bolts. Tighten the bolts to 29 ft. lbs. (39 Nm). Connect the engine wire and install the starter wire with the nut.

42. Install the set nut of the engine wire clamp.

43. Install the ground cable to the transaxle with the bolt. Tighten the bolt to 14 ft. lbs. (19 Nm).

44. Install the air cleaner assembly.

45. Install the engine coolant reservoir tank.

46. Connect the throttle cable.

47. Check all fluids and install the negative battery cable to the battery.

PREVIA WITH 4-SPEED

1. Disconnect the negative battery terminal from the battery. Wait at least 90 seconds after the battery cable is disconnected before working on the vehicle if equipped with air bags.

2. If required, remove the air cleaner assembly.

3. Disconnect the transmission throttle cable from the throttle body.

4. Raise and safely support the vehicle.

5. Drain the transmission fluid.

6. Disconnect the wiring harness for the neutral start switch and the back-up light switch. If equipped, disconnect the solenoid (overdrive) switch wiring at the same location.

7. If equipped, disconnect the oil level gauge.

8. Disconnect the starter wiring at the starter. Remove the mounting bolts and the starter from the engine.

9. Make matchmarks on the rear driveshaft flange and the differential pinion flange. These marks must be aligned during installation.

10. Unbolt the rear driveshaft flange. If the vehicle has a two piece driveshaft, remove the center bearing bracket-to-frame bolts. Remove the driveshaft from the vehicle.

11. Disconnect the speedometer cable (tie it aside). Disconnect the shift linkage from the transmission.

12. Disconnect the transmission oil cooler lines at the transmission.

13. Disconnect the exhaust pipe clamp and remove the oil filler tube, as required.

14. Support the transmission, using a jack with a wooden block placed between the jack and the transmission pan. Raise the transmission, just enough to take the weight off of the rear mount.

15. Remove the rear engine mount with the bracket and the engine undercover, to gain access to the engine crankshaft pulley.

16. Remove the stiffener plates, if equipped.

17. Place a wooden block (or blocks) between the engine oil pan and the front frame crossmember.

18. Slowly, lower the transmission until the engine rests on the wooden block.

19. Remove the rubber plug(s) from the service holes located at the rear of the engine in order to gain access to the torque converter bolts.

20. Rotate the crankshaft (to remove the torque converter bolts) to access the bolts through the service holes.

21. Obtain a bolt of the same dimensions as the torque converter bolts. Cut the head off of the bolt and hacksaw a slot in the bolt opposite the threaded end.

➡**This modified bolt is used as a guide pin. Two guides pins are needed to properly install the transmission.**

22. Thread the guide pin into one of the torque converter bolt holes. The guide pin will help keep the converter with the transmission.

23. Remove the stiffener plates from the transmission.

24. Remove the transmission-to-engine bolts, then carefully move the transmission rearward by prying on the guide pin through the service hole.

25. Pull the transmission rearward and lower it (front end down) out of the vehicle.

26. Installation is the reverse of removal. Please note the following important steps.

27. Apply a coat of Multi-purpose grease to the torque converter stub shaft and the corresponding pilot hole in the flexplate.

28. Install the torque converter into the front of the transmission. Push inward on the torque converter while rotating it to completely couple the torque converter to the transmission.

29. To be sure the converter is properly installed, measure the distance between the torque converter mounting lugs and the front mounting face of the transmission. The proper distance is 1995 models: 1.250 inch (31.75mm) and on 1996–99 models: 0.079 inch (20mm).

30. Install guide pins into two opposite mounting lugs of the torque converter.

31. Raise the transmission to the engine, align the transmission with the engine alignment dowels and position the converter guide pins into the mounting holes of the flexplate.

32. Install and tighten the transmission-to-engine mounting bolts. Tighten the bolts to specifications.

33. Remove the converter guide pins and install the converter mounting bolts. Rotate the crankshaft as necessary to gain access to the guide pins and bolts through the service holes. Evenly, tighten the converter mounting bolts to specifications. Install the rubber plugs into the access holes.

34. Install the remaining components by reversing the removal procedure.

35. Adjust the transmission throttle cable.

36. Refill the transmission.

37. Connect the negative battery cable. Start the engine and check for leaks.

38. Road test the vehicle for proper operation.

39. Recheck all fluid levels.

PREVIA WITH 3-SPEED

1. Disconnect the negative battery terminal from the battery. Wait at least 90 seconds after the battery cable is disconnected before working on the vehicle if equipped with air bags.

2. Remove the ATF level gauge.

3. Loosen the nut and disconnect the throttle cable.

4. Remove the equipment driveshaft.

5. Raise and safely support the vehicle.

6. Remove the filler pipe. Remove the driveshaft.

7. Disconnect the control cable.

8. Disconnect the No. 1 and No. 2 vehicle speed sensor harnesses. Disconnect the starter solenoid wiring.

9. Disengage the neutral safety switch harness.

10. Remove the bolt and disconnect the oil pipe clamp. Disconnect the oil cooler pipes. On 4WD disconnect the A/T fluid temperature switch harness.

11. Remove the starter.

12. Using a jack, support the transmission. Remove the stiffener plate.

13. Remove the torque converter clutch cover and turn the crankshaft to gain access and remove the six bolts.

14. Remove the exhaust pipe bracket. Remove the rear mounting bolts.

15. Remove the transmission mounting bolts, then detach the wiring harness, and lower the transmission.

16. Installation is the reverse of removal. Please note the following important steps.

17. Connect wiring harness and install the transmission. Tighten the bolts to specifications.

18. Install the remaining components and tighten to specifications. Install the equipment driveshaft.

19. Connect the throttle cable and tighten.

20. Install the ATF level gauge.

21. Check the shift lever position.

22. Check the fluid level and fill if necessary.

23. Connect the negative battery cable.

SIENNA AND RX300

✳✳ CAUTION

Fuel injection systems remain under pressure even after the engine has been turned OFF. The fuel system pressure must be relieved before disconnecting any fuel lines. Failure to do so may result in fire and/or personal injury.

1. Matchmark the hood position and remove the hood.

2. Remove the wiper and blade assembly.

3. Remove the top cowl seal and panel.

4. Label and disconnect the window washer hoses from the ventilator louvers.

5. Remove the left and right ventilator louvers.

6. Remove the heater air duct.

7. Disconnect the negative, then the positive battery cable. Remove the battery and tray.

8. Disconnect the throttle cable.

9. On RX300 models, remove the front upper suspension brace.

10. If equipped, remove the cruise control actuator with its bracket.

11. Remove the starter and the shift control cable.

12. On RX300 models with 4WD, remove the driveshaft.

13. Disconnect the body-to-engine ground strap.

14. Detach the park/neutral position switch, solenoid, and ATF temperature connectors.

15. Remove the five upper transaxle-to-engine mounting bolts.

16. Remove the front tire and wheel assembly, then the engine undercover.

17. Remove the halfshafts.

18. Disconnect and remove the front exhaust pipe.

19. Remove the stabilizer bar.

20. Remove the two steering gear mounting bolts and support it in the vehicle.

21. Disconnect the shift control cable from its bracket.

22. Unbolt the power steering pipe and the oil cooler clamps from the frame.

23. Remove the two left-side transaxle mounting nuts.

24. Remove the rear-side engine mounting nuts, then the engine shock absorber mounting bolts.

25. Remove the three front-side engine mounting bolts.

26. Attach an engine sling to the engine hangers in order to support the engine weight.

27. Remove the front frame mounting bolts, then the frame.

28. Disconnect the transaxle oil cooler lines.

29. Support the transaxle with a transmission jack.

30. Remove the torque converter access cover, then remove the six converter mounting bolts.

31. Remove the three lower transaxle-to-engine mounting bolts, then separate the engine from the transaxle and remove.

To install:

32. Raise the transaxle into the vehicle and mate it to the engine.

33. Install the three lower transaxle-to-engine mounting bolts and tighten to the illustrated value.

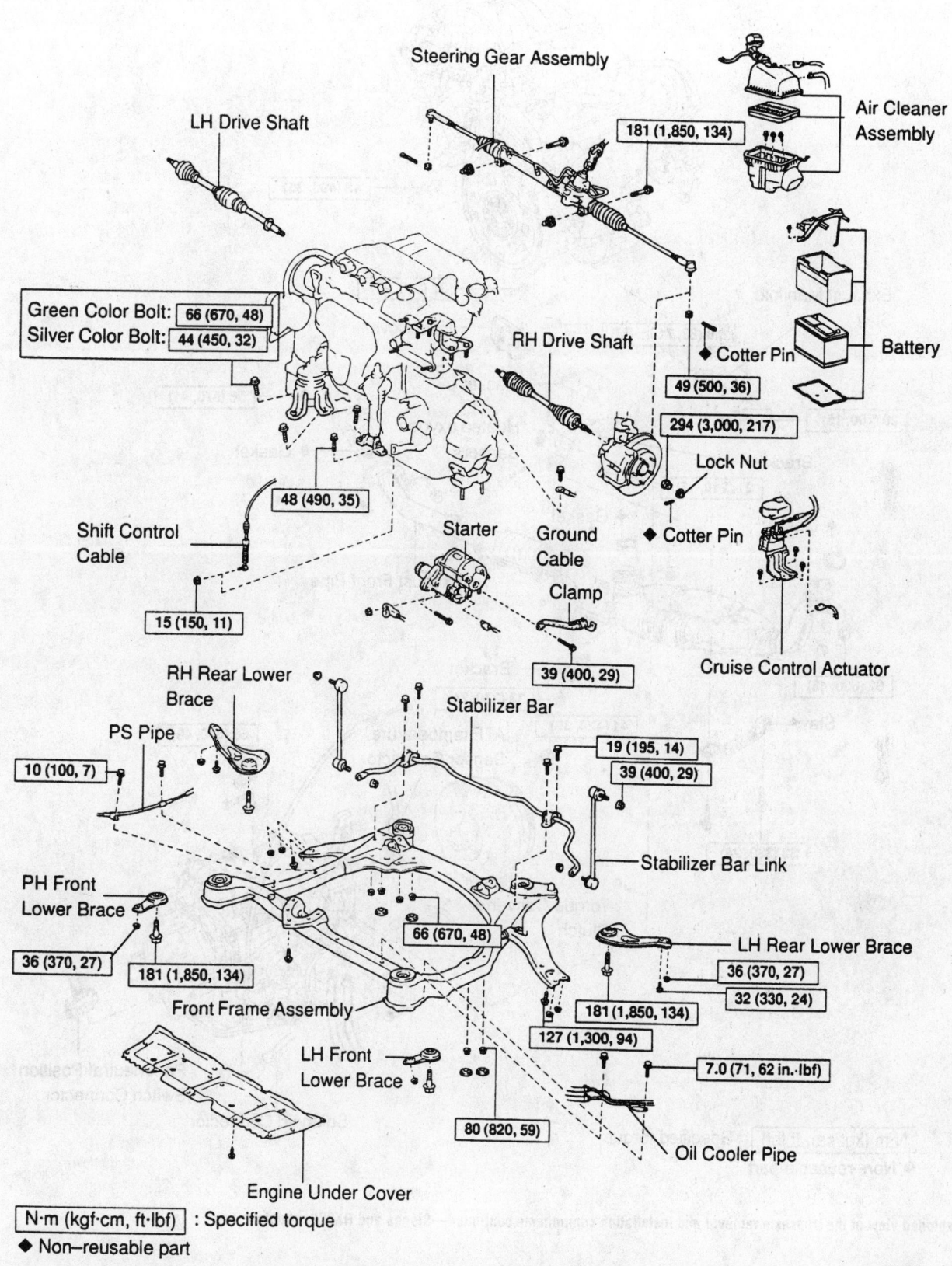

Steering Gear Assembly

LH Drive Shaft

Air Cleaner Assembly

181 (1,850, 134)

Green Color Bolt: 66 (670, 48)
Silver Color Bolt: 44 (450, 32)

RH Drive Shaft

Cotter Pin

Battery

49 (500, 36)

294 (3,000, 217)

Lock Nut

48 (490, 35)

Shift Control Cable

Starter

Ground Cable

Cotter Pin

Clamp

Cruise Control Actuator

15 (150, 11)

39 (400, 29)

RH Rear Lower Brace

Stabilizer Bar

19 (195, 14)

39 (400, 29)

PS Pipe

10 (100, 7)

Stabilizer Bar Link

PH Front Lower Brace

LH Rear Lower Brace

36 (370, 27)

66 (670, 48)

36 (370, 27)

181 (1,850, 134)

32 (330, 24)

Front Frame Assembly

181 (1,850, 134)

127 (1,300, 94)

LH Front Lower Brace

7.0 (71, 62 in.·lbf)

80 (820, 59)

Oil Cooler Pipe

Engine Under Cover

N·m (kgf·cm, ft·lbf) : Specified torque

◆ Non–reusable part

7924ZG65

Exploded view of the transaxle removal and installation components—Sienna and RX300 models

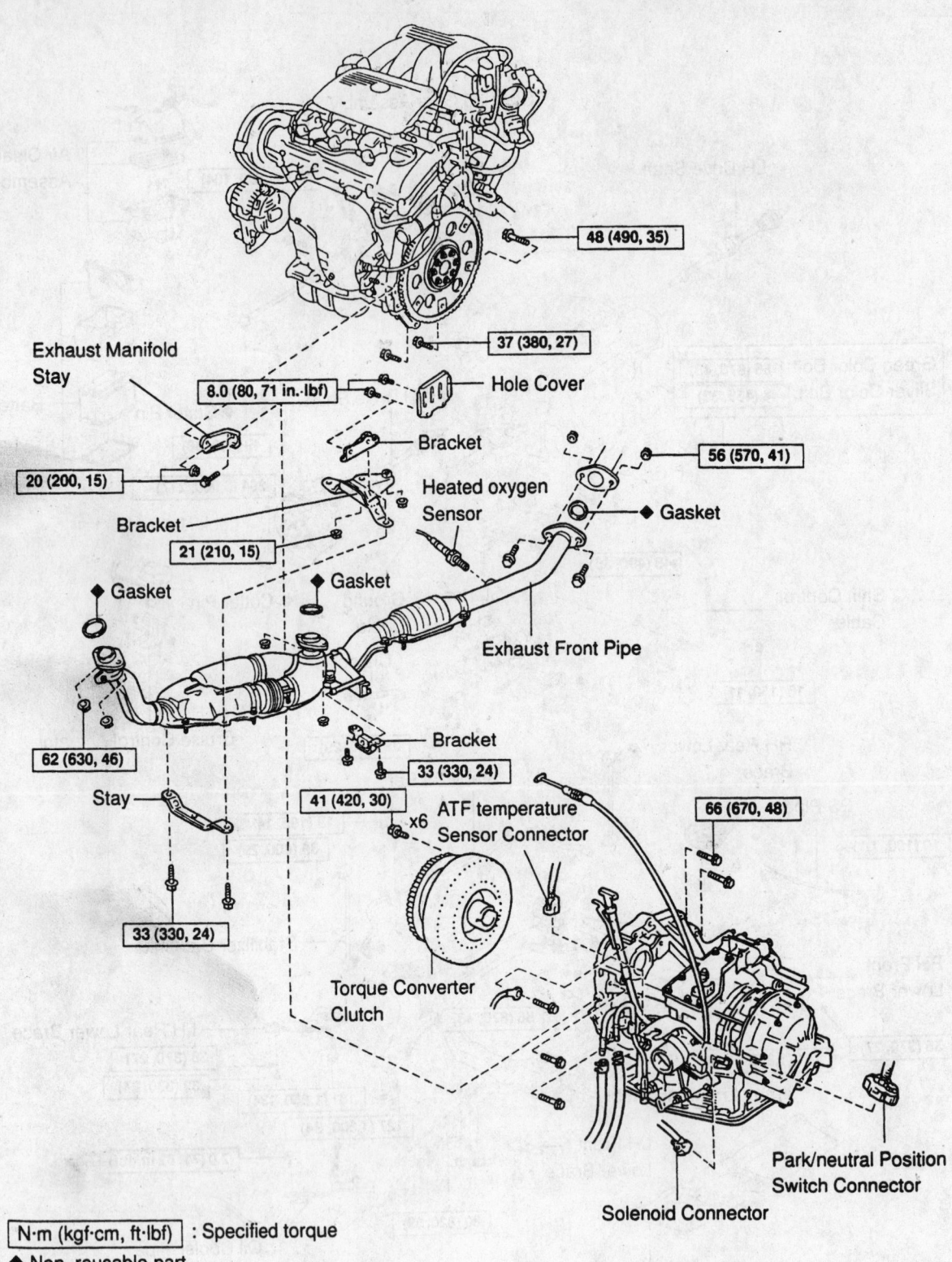

48 (490, 35)

37 (380, 27)

Exhaust Manifold Stay

8.0 (80, 71 in.·lbf)

Hole Cover

Bracket

56 (570, 41)

Heated oxygen Sensor

20 (200, 15)

◆ Gasket

Bracket

21 (210, 15)

◆ Gasket

◆ Gasket

Exhaust Front Pipe

62 (630, 46)

Bracket

33 (330, 24)

Stay

41 (420, 30)

ATF temperature Sensor Connector

x6

66 (670, 48)

33 (330, 24)

Torque Converter Clutch

Park/neutral Position Switch Connector

Solenoid Connector

N·m (kgf·cm, ft·lbf) : Specified torque

◆ Non–reusable part

7924ZG66

Exploded view of the transaxle removal and installation components continued—Sienna and RX300 models

34. Install the torque converter-to-flexplate bolts, starting with the black bolt, then the other five.

35. The completion of the installation is the reverse of the removal referring to the illustrations for the tightening specifications.

Clutch

ADJUSTMENT

Free Play

RAV4

1. Check that the pedal height is correct. Pedal height from the floor panel should be: 6.889–7.283 in. (175–185mm)

2. If necessary to adjust the pedal height, loosen the locknut and turn the stopper bolt until the height is correct. Tighten the locknut.

3. Push in on the pedal until the beginning of the clutch resistance is felt. Free-play should be 0.197–0.591 in. (5–15mm).

4. Gently push on the pedal until the resistance begins to increase a little. Pushrod play at the pedal top should be 0.039–0.197 in. (1–5mm).

5. If necessary to adjust the pedal free-play and the pushrod play,

   a. Loosen the locknut and turn the push the rod until the free-play and pushrod play are correct.

   b. Tighten the locknut.

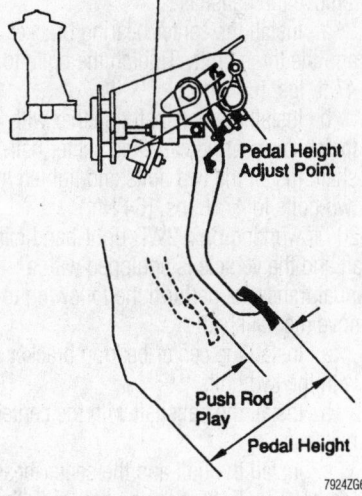

Clutch pedal height measurement location—RAV4

REMOVAL & INSTALLATION

✳✳ CAUTION

To avoid personal injury and accidental deployment of the air bag, work must be started after about 90 seconds or longer from the time the ignition switch is turned to the LOCK position and the battery cable is disconnected from the battery.

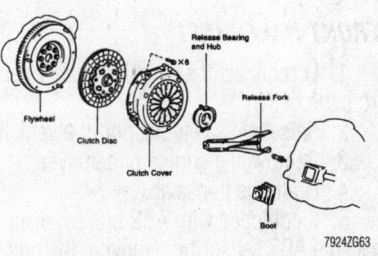

Clutch component assembly—Exploded view

1. Disconnect the negative battery cable from the battery.

2. Remove the transaxle assembly from the vehicle.

3. Matchmark the clutch cover to the flywheel.

4. Remove the clutch pressure plate retaining bolts in small amounts and in a crisscross pattern to relieve the clutch disc spring tension.

5. At the clutch cover, loosen each bolt one turn until spring tension is released.

6. On RAV4, remove the set bolts to the clutch cover and pull off the clutch cover with the clutch disc.

7. On Previa, remove the clutch cover-to-flywheel bolts. Remove the clutch cover and the clutch disc.

8. If the clutch release bearing is to be replaced, perform the following:

   a. Remove the bearing retaining clip(s), the bearing and hub.

   b. Remove the release fork and the boot.

   c. The bearing is press fitted to the hub.

   d. Clean all parts and lightly grease the input shaft splines and all of the contact points.

   e. Install the bearing/hub assembly, the fork, the boot and the retaining clip(s) in their original locations.

To install:

9. Inspect the flywheel surface for cracks, heat scoring (blue marks) and warpage. Replace or resurface the flywheel, if any damage is present.

➡Before installing any new parts, be sure they are clean. During installation, do not get grease or oil on any of the components, as this will shorten clutch life considerably.

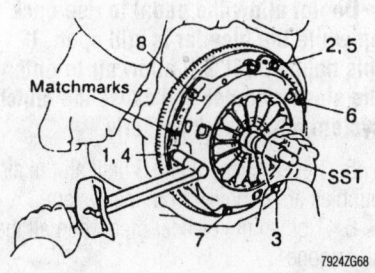

Torque sequence for the clutch cover

10. Using a clutch alignment tool, position the clutch disc against the flywheel. The raised center section of the disc faces the transaxle.

11. Position the clutch cover onto the flywheel and align the matchmarks.

12. Install the clutch cover retaining bolts. Tighten the bolts in a crisscross pattern to 14 ft. lbs. (19 Nm).

13. Lubricate the release fork pivot and contact points, release bearing, bearing hub and input shaft spline surfaces with a suitable molybdenum disulfide lithium based or multi-purpose grease.

14. Install the boot, release fork, hub, and the bearing assemblies.

15. Install the transaxle to the vehicle.

16. Connect the negative battery cable to the battery.

Clutch Hydraulic System

BLEEDING

➡If any maintenance on the clutch system was performed or the system is suspected of containing air, bleed the system. Use care; brake fluid will remove the paint from any surface. If the brake fluid spills onto any painted surface, wash it off immediately with soap and water.

1. Fill the clutch reservoir with brake fluid. Check the reservoir level frequently and add fluid as needed.

2. Connect one end of a vinyl tube to the bleeder plug on the slave cylinder and submerge the other end into a clear container half-filled with brake fluid.

3. Slowly pump the clutch pedal several times.

4. Have an assistant hold the clutch pedal down and loosen the bleeder plug until fluid and/or air starts to run out of the bleeder plug. Close the bleeder plug while the pedal is held to the floor.

➡**Do not allow the pedal to rise back up while the bleeder is still open. If this happens, it will allow air to enter the slave cylinder and cause the clutch system not to work properly.**

5. Repeat Steps 2 and 3 until all the air bubbles are removed from the system.

6. Tighten the bleeder plug when all the air is gone.

7. Refill the master cylinder to the proper level as required.

8. Check the system for leaks.

Transfer Case Assembly

REMOVAL & INSTALLATION

Previa

1. Remove the transmission and transfer case assembly from the vehicle.

➡**Have a pan ready to catch any spilt fluid.**

2. Remove the bolt attaching the speedometer cable and set the cable aside.

3. If necessary, disconnect and remove the No. 1 and No. 2 speed sensors.

4. If necessary, remove the speedometer driven gear.

5. Unbolt the transfer from the transmission.

6. If necessary, remove the transfer adapter. Remove the six bolts and the adapter. It may be necessary to tap the adapter with a plastic hammer to loosen it. Discard the gasket.

To install:

7. If removed, install the transfer adapter with a new gasket. Apply sealant such as Three bond® 1344 or LOCTITE® 242 or equivalent. Tighten the bolts to 25 ft. lbs. (34 Nm).

8. Attach the transfer case to the transmission. Tighten the bolts to 27–37 ft. lbs. (36–50 Nm).

9. Attach the speedometer cable and tighten the bolt to 33 ft. lbs. (45 Nm).

10. If removed, install the speedometer gear and the No. 1 and No. 2 speed sensors.

11. Replace any necessary seals or gaskets. Install the transfer/transmission assembly into the vehicle.

Halfshaft

REMOVAL & INSTALLATION

RAV4

FRONT HALFSHAFT

1. Disconnect the negative battery cable from the battery.

2. Raise and safely support the vehicle.

3. Remove the engine undercover.

4. Drain the transaxle.

5. If equipped with ABS brakes, disconnect the ABS sensor by removing the bolt.

6. Remove the cotter pin, lock cap, and the locknut holding the halfshaft to the steering knuckle.

7. Disconnect the tie rod ends from the steering knuckle.

8. Disconnect the sway bar link from the lower control arm.

9. Disconnect the lower ball joint from the lower control arm.

10. Using a plastic hammer, disconnect the halfshaft from the axle hub.

11. If working on a 2WD right-hand halfshaft and the vehicle is equipped with a manual transaxle, perform the following to remove the halfshaft:

   a. Using a screwdriver and hammer, remove the snapring from the center bearing bracket.

   b. Remove the bolt and the center bearing bracket.

   c. Remove the halfshaft with the center halfshaft.

   d. Remove the two bolts and the center bearing bracket.

12. If working on a 2WD right-hand halfshaft and the vehicle is equipped with an automatic transaxle, perform the following to remove the halfshaft:

   a. Remove the two bolts of the center bearing bracket and pull out the halfshaft together with the center bearing case and center halfshaft.

   b. Remove the three bolts and the center bearing bracket.

13. If working on a 2WD left-hand, perform the following:

   a. Using a brass bar and hammer, remove the halfshaft from the transaxle.

   b. Remove the snapring from the transaxle.

14. If working of a 4WD right-hand halfshaft, perform the following:

   a. Using a brass bar and hammer, remove the halfshaft.

   b. Remove the snapring from the transaxle.

   c. Using a screwdriver, remove the O-ring.

15. If working on a 4WD left-hand side, perform the following:

   a. Remove the air cleaner from the vehicle.

   b. Remove the transaxle case protector.

   c. Using a hub wrench, pry the halfshaft out.

   d. Remove the snapring.

To install:

16. If working on a 4WD left-hand side, perform the following:

   a. Install the snapring.

   b. Install the halfshaft to the transaxle.

   c. Install the transaxle case protector.

   d. Install the air cleaner to the vehicle.

17. If working of a 4WD right-hand halfshaft, perform the following:

   a. Install the snapring from the transaxle.

   b. Install a new O-ring.

   c. Install the halfshaft to the transaxle.

18. If working on a 2WD left-hand, perform the following:

   a. Install the snapring.

   b. Install the halfshaft to the transaxle.

19. If working on a 2WD right-hand halfshaft and the vehicle is equipped with an automatic transaxle, perform the following to remove the halfshaft:

   a. Install the center bearing bracket and the three bolts. Tighten the bolts to 47 ft. lbs. (64 Nm).

   b. Install the halfshaft together with the center bearing case and center halfshaft. Install the two bolts and tighten the two bolts to 47 ft. lbs. (64 Nm).

20. If working on a 2WD right-hand halfshaft and the vehicle is equipped with a manual transaxle, perform the following to remove the halfshaft:

   a. Install the center bearing bracket with the two bolts.

   b. Install the halfshaft with the center halfshaft.

   c. Install the bolt and the center bearing bracket. Tighten the bolt to 24 ft. lbs. (32 Nm).

   d. Install the snapring to the center bearing bracket.

21. Connect the halfshaft to the axle hub.

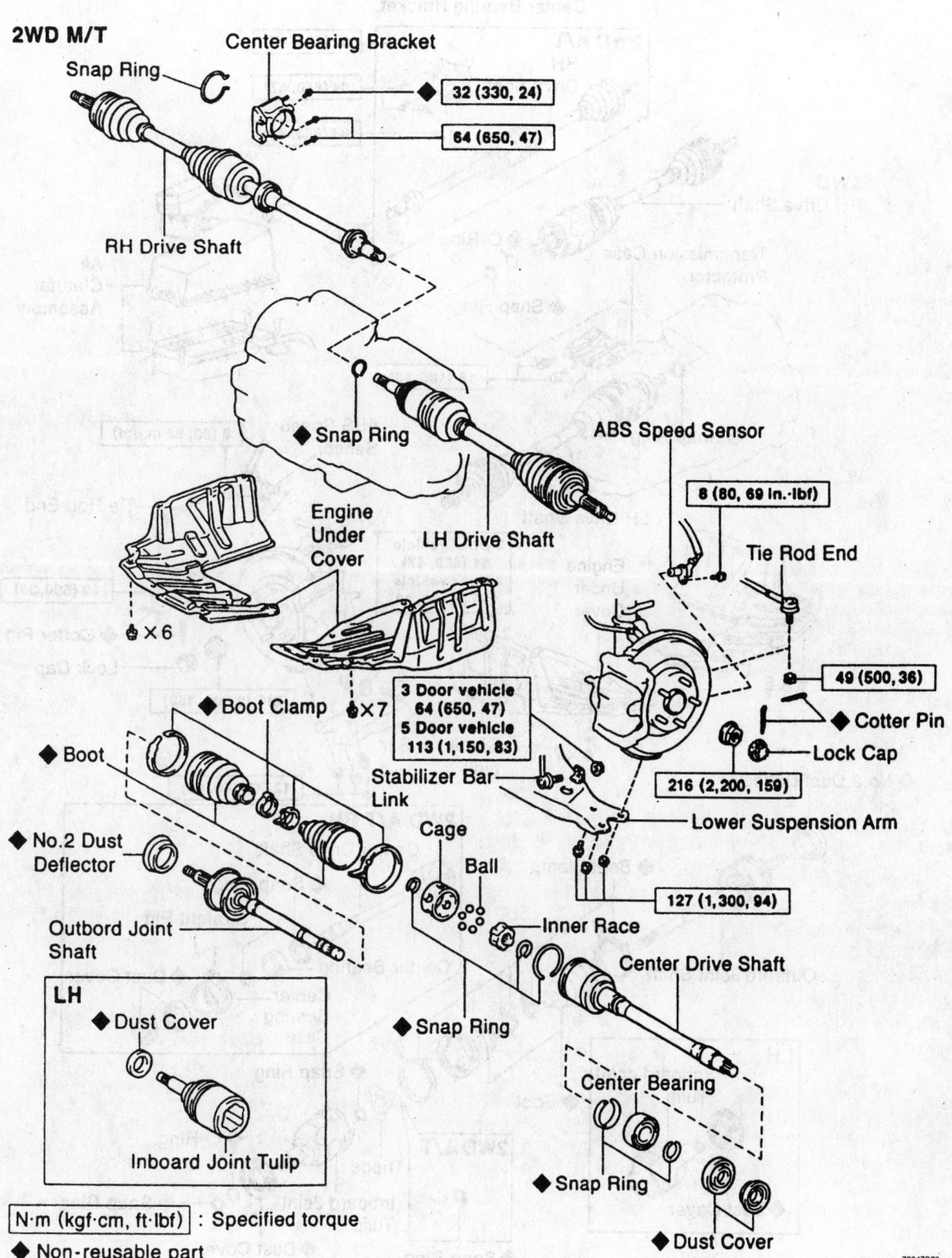

2WD M/T

Snap Ring

Center Bearing Bracket

32 (330, 24)

64 (650, 47)

RH Drive Shaft

◆ Snap Ring

Engine Under Cover

LH Drive Shaft

ABS Speed Sensor

8 (80, 69 in.·lbf)

Tie Rod End

◆×6

49 (500, 36)

◆ Cotter Pin

Lock Cap

216 (2,200, 159)

◆ Boot Clamp

◆×7

3 Door vehicle 64 (650, 47)
5 Door vehicle 113 (1,150, 83)

Stabilizer Bar Link

◆ Boot

Cage

Ball

Lower Suspension Arm

◆ No.2 Dust Deflector

127 (1,300, 94)

Inner Race

Outbord Joint Shaft

Center Drive Shaft

LH

◆ Dust Cover

◆ Snap Ring

Center Bearing

Inboard Joint Tulip

◆ Snap Ring

◆ Dust Cover

N·m (kgf·cm, ft·lbf) : Specified torque

◆ Non-reusable part

7924ZG70

Front halfshaft exploded view (2WD with M/T only)—RAV4

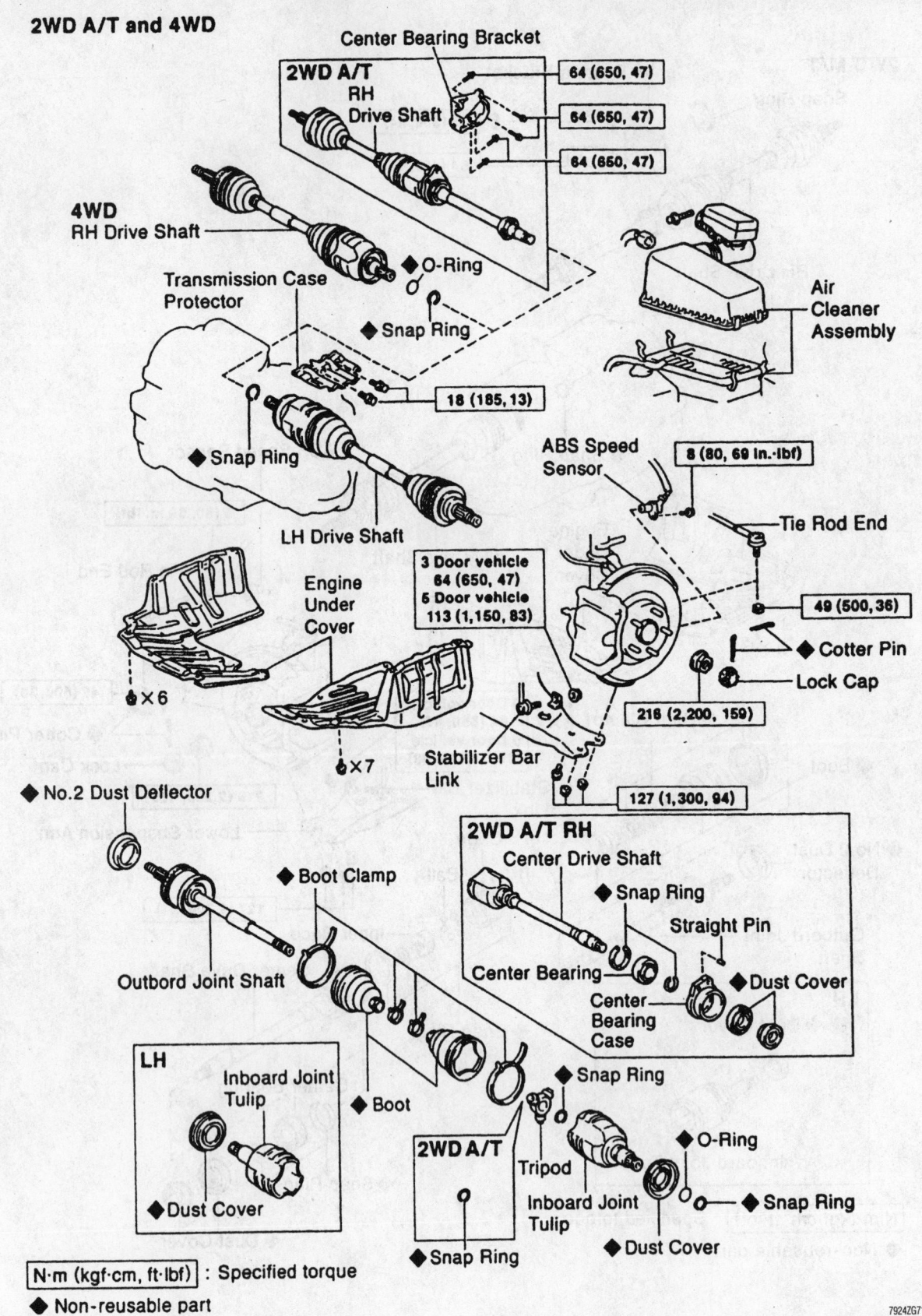

2WD A/T and 4WD

Center Bearing Bracket

2WD A/T RH Drive Shaft

64 (650, 47)
64 (650, 47)
64 (650, 47)

4WD RH Drive Shaft

◆ O-Ring

◆ Snap Ring

Transmission Case Protector

Air Cleaner Assembly

18 (185, 13)

◆ Snap Ring

LH Drive Shaft

ABS Speed Sensor

8 (80, 69 In.-lbf)

Tie Rod End

Engine Under Cover

3 Door vehicle
64 (650, 47)
5 Door vehicle
113 (1,150, 83)

49 (500, 36)

◆ Cotter Pin

Lock Cap

×6

216 (2,200, 159)

×7

Stabilizer Bar Link

127 (1,300, 94)

◆ No.2 Dust Deflector

Boot Clamp

2WD A/T RH

Center Drive Shaft

◆ Snap Ring

Straight Pin

Center Bearing

◆ Dust Cover

Center Bearing Case

Outbord Joint Shaft

LH

Inboard Joint Tulip

◆ Boot

◆ Snap Ring

◆ Dust Cover

2WD A/T

Tripod

◆ O-Ring

◆ Snap Ring

Inboard Joint Tulip

◆ Dust Cover

◆ Snap Ring

N·m (kgf·cm, ft·lbf) : Specified torque

◆ Non-reusable part

7924ZG71

Front halfshaft exploded view (2WD with A/T and 4WD)—RAV4

22. Connect the lower ball joint to the lower control arm. Tighten the bolt and two nuts to 94 ft. lbs. (127 Nm).

23. Connect the sway bar link to the lower control arm and tighten the nut as follows:

- 3 door vehicles to 47 ft. lbs. (64 Nm).
- 5 door vehicles to 83 ft. lbs. (113 Nm).

24. Connect the tie rod end to the steering knuckle. Tighten the nut to 36 ft. lbs. (49 Nm) and install a new cotter pin.

25. Install the cotter pin, lock cap, and the locknut to hold the halfshaft to the axle hub. Tighten the nut to 159 ft. lbs. (216 Nm).

26. If equipped with ABS, install the ABS speed sensor with the bolt.

27. Fill the transaxle with gear oil (M/T) or ATF (A/T).

28. Install the engine undercover.

29. Install the wheels and lower the vehicle.

30. Connect the negative battery cable to the battery.

31. Check the ABS sensor signal.

REAR HALFSHAFT

1. Disconnect the negative battery cable from the battery.

2. Raise and safely support the rear of the vehicle.

3. Remove the rear wheels.

4. If equipped with ABS brakes, remove the ABS speed sensor from the axle assembly by removing the bolt.

5. Remove the cotter pin, lock cap, and the nut holding the halfshaft to the axle carrier.

6. Place matchmarks on the halfshaft and side gear shaft.

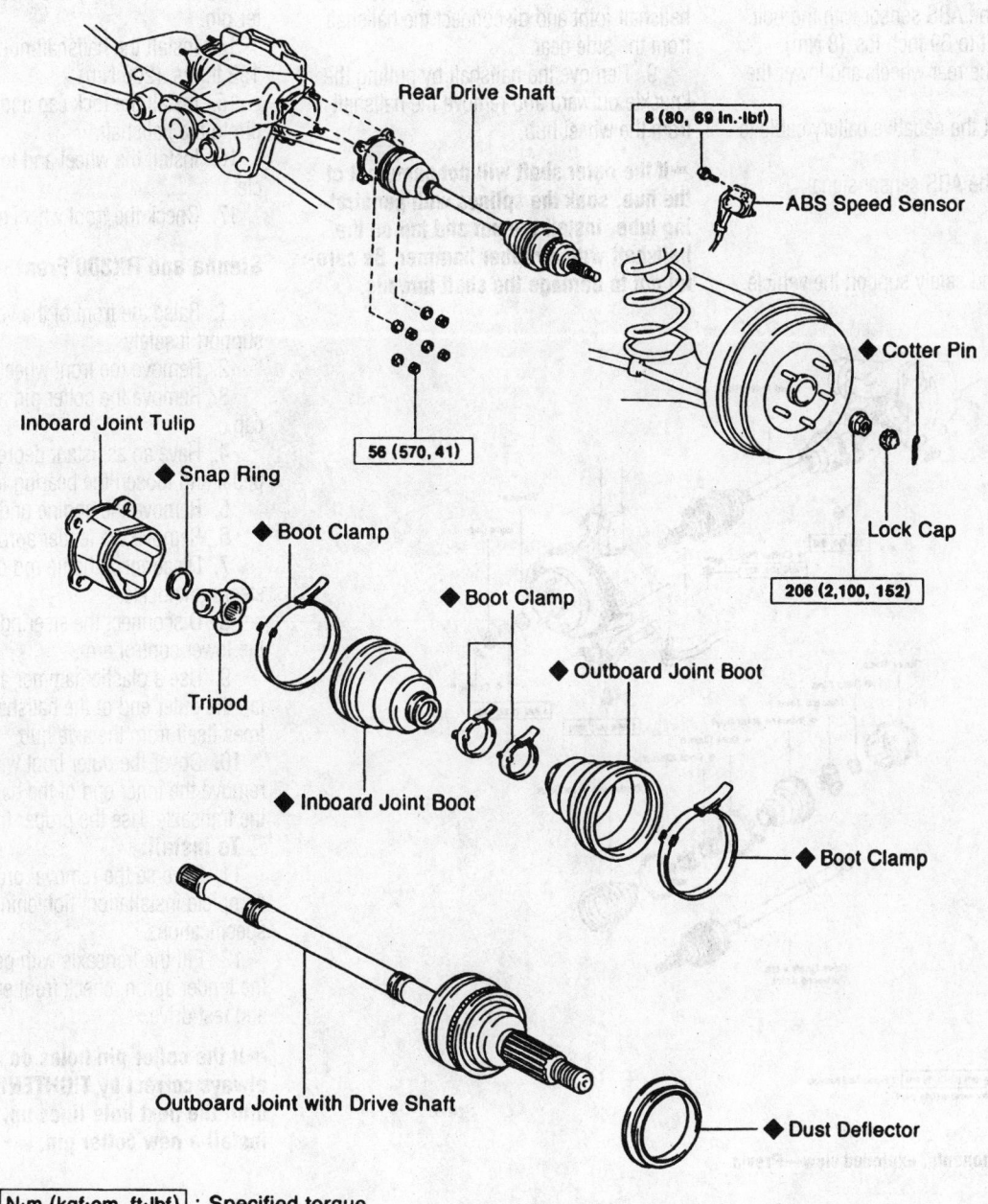

Rear Drive Shaft

8 (80, 69 in.·lbf)

ABS Speed Sensor

56 (570, 41)

Inboard Joint Tulip

◆ Snap Ring

◆ Boot Clamp

◆ Boot Clamp

◆ Outboard Joint Boot

Tripod

◆ Inboard Joint Boot

◆ Boot Clamp

Cotter Pin

Lock Cap

206 (2,100, 152)

Outboard Joint with Drive Shaft

◆ Dust Deflector

N·m (kgf·cm, ft·lbf) : Specified torque

◆ Non-reusable part

7924ZG72

Rear halfshaft removal and installation (4WD only)—RAV4

7. Disconnect the halfshaft from the differential side gear shaft by removing the four nuts and washers.

8. Using a plastic hammer, disconnect the halfshaft from the axle carrier.

To install:

9. Install the halfshaft to the axle carrier.

10. Aligning the marks, connect the halfshaft to the differential side gear shaft with the four nuts. Tighten the nuts to 41 ft. lbs. (56 Nm).

11. Install the nut, lock cap, and the cotter pin to hold the halfshaft to the axle carrier. Tighten the nut to 152 ft. lbs. (206 Nm).

12. Install the ABS sensor with the bolt. Tighten the bolt to 69 inch lbs. (8 Nm).

13. Install the rear wheels and lower the vehicle.

14. Connect the negative battery cable to the battery.

15. Check the ABS sensor signal.

Previa

1. Raise and safely support the vehicle.

2. Remove the wheel and tire assembly.

3. Remove the cotter pin and lock cap from the halfshaft.

4. Remove the locknut from the halfshaft.

5. Remove the cotter pin and locknut to the tie rod end and disconnect the tie rod end from the knuckle.

6. Remove the lower ball joint from the steering knuckle by removing the two mounting bolts.

7. Place matchmarks on the halfshaft and side gear.

8. Remove the six bolts from the inner halfshaft joint and disconnect the halfshaft from the side gear.

9. Remove the halfshaft by pulling the knuckle outward and remove the halfshaft from the wheel hub.

➡ **If the outer shaft will not come out of the hub, soak the splines with penetrating lube, install the nut and tap on the halfshaft with a rubber hammer. Be careful not to damage the shaft threads.**

To install:

➡ **Coat the halfshaft splines with anti-seize compound to prevent spline seizure. This will help for future halfshaft removal.**

10. Connect the halfshaft to the steering knuckle.

11. Connect the inner halfshaft joint to the side gear by installing the six bolts. Tighten the bolts to 51 ft. lbs. (61 Nm).

12. Connect the lower ball joint and tighten the bolts to 94 ft. lbs. (127 Nm).

13. Install the tie rod end, tighten the nut to 36 ft. lbs. (49 Nm) and install a new cotter pin.

14. Install the halfshaft nut and tighten to 152 ft. lbs. (206 Nm).

15. Install the lock cap and a new cotter pin to the halfshaft.

16. Install the wheel and lower the vehicle.

17. Check the front wheel alignment.

Sienna and RX300 Front

1. Raise the front of the vehicle and support it safely.

2. Remove the front wheels.

3. Remove the cotter pin and locknut cap.

4. Have an assistant depress the brake pedal and loosen the bearing locknut.

5. Remove the engine undercover.

6. Remove the fender apron seal.

7. Disconnect the tie rod end from the steering knuckle.

8. Disconnect the steering knuckle from the lower control arm.

9. Use a plastic hammer and carefully tap the outer end of the halfshaft until it frees itself from the axle hub.

10. Cover the outer boot with a rag, then remove the inner end of the halfshaft from the transaxle. Use the proper tools.

To install:

11. Reverse the removal procedures to complete installation, tightening fasteners to specifications.

12. Fill the transaxle with gear oil, install the fender apron, check front end alignment and test drive.

➡ **If the cotter pin holes do not line up, always correct by TIGHTENING the nut until the next hole lines up. Then, install a new cotter pin.**

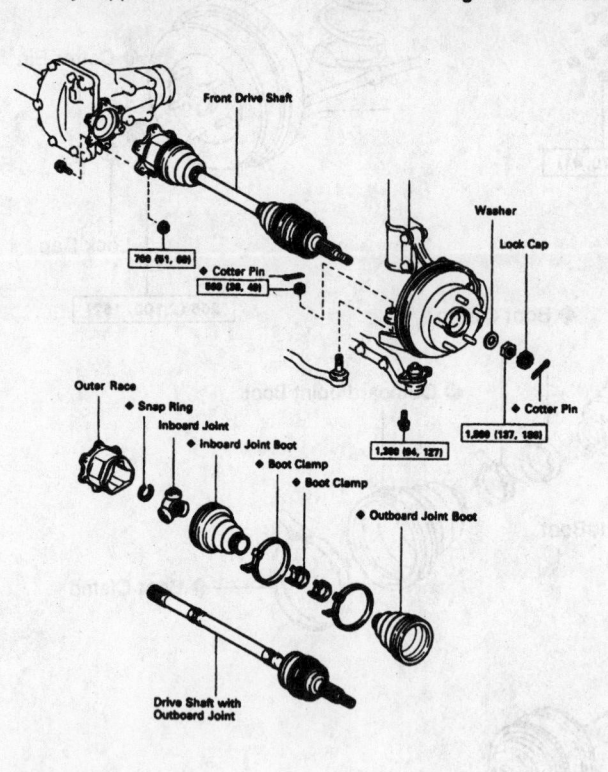

Front Drive Shaft

Washer

Lock Cap

◆ Cotter Pin
700 (51, 69)

◆ Cotter Pin
660 (36, 49)

◆ Cotter Pin
1,500 (127, 180)

1,300 (94, 127)

Outer Race

◆ Snap Ring

Inboard Joint

◆ Inboard Joint Boot

◆ Boot Clamp

◆ Boot Clamp

◆ Outboard Joint Boot

Drive Shaft with
Outboard Joint

kg-cm (ft-lb, N·m) : Specified torque
◆ Non-reusable part

7924ZG69

Halfshaft components, exploded view—Previa

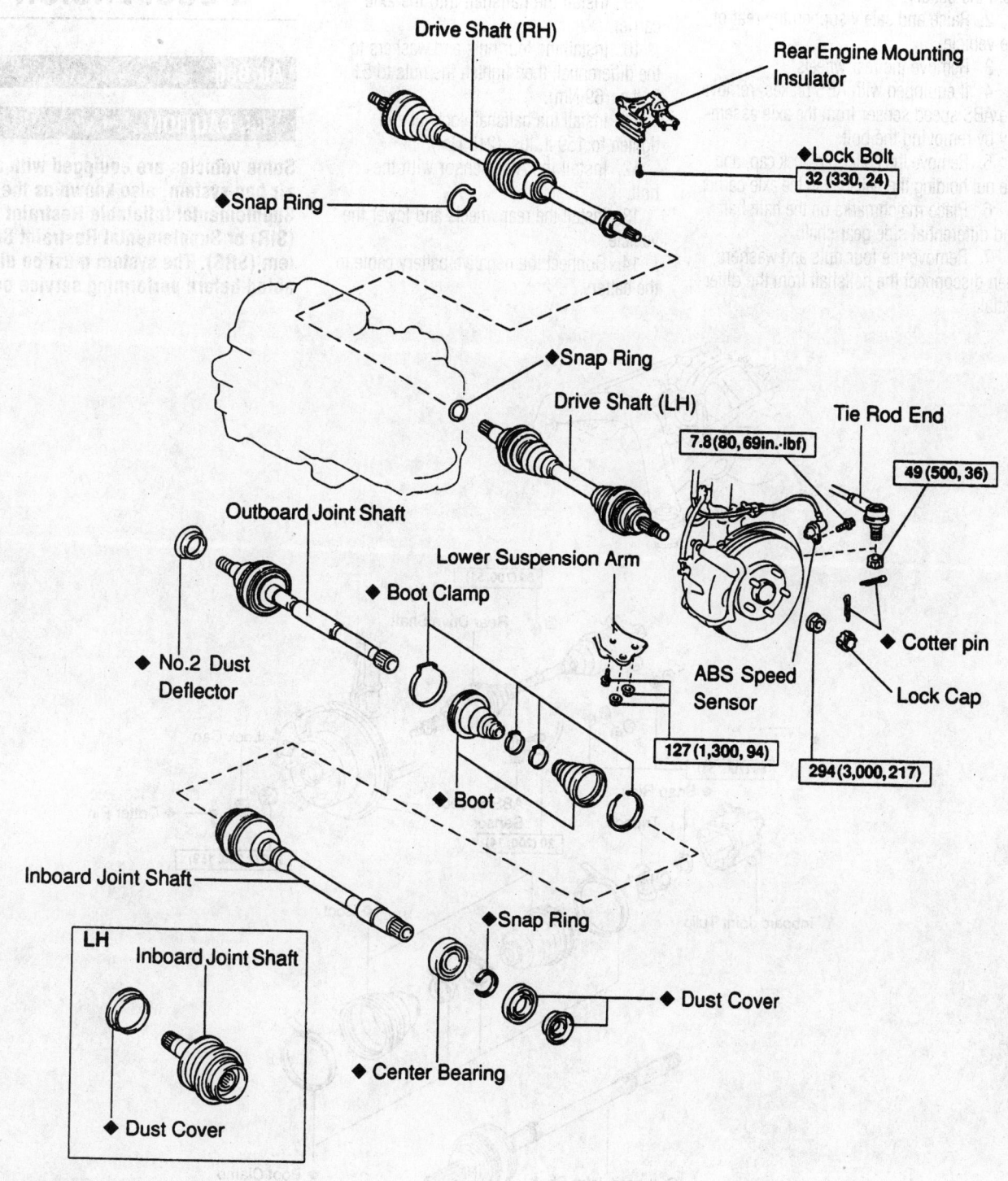

Drive Shaft (RH)

Rear Engine Mounting Insulator

◆Lock Bolt
32 (330, 24)

◆Snap Ring

◆Snap Ring

Drive Shaft (LH)

Tie Rod End

7.8(80, 69in.·lbf)

49 (500, 36)

Outboard Joint Shaft

Lower Suspension Arm

◆ Boot Clamp

◆ No.2 Dust Deflector

ABS Speed Sensor

◆ Cotter pin

Lock Cap

◆ Boot

127(1,300, 94)

294(3,000,217)

Inboard Joint Shaft

◆Snap Ring

◆ Dust Cover

LH
Inboard Joint Shaft

◆ Dust Cover

◆ Center Bearing

N·m (kgf·cm, ft·lbf) : Specified torque

◆ Non–reusable part

7924ZG73

Exploded view of halfshaft—Sienna and RX300

RX300 Rear

1. Disconnect the negative battery cable from the battery.
2. Raise and safely support the rear of the vehicle.
3. Remove the rear wheels.
4. If equipped with ABS brakes, remove the ABS speed sensor from the axle assembly by removing the bolt.
5. Remove the cotter pin, lock cap, and the nut holding the halfshaft to the axle carrier.
6. Place matchmarks on the halfshaft and differential side gear shaft.
7. Remove the four nuts and washers, then disconnect the halfshaft from the differential.

8. Remove the halfshaft from the axle carrier.

To install:

9. Install the halfshaft into the axle carrier.
10. Install the four nuts and washers to the differential, then tighten the nuts to 51 ft. lbs. (69 Nm).
11. Install the halfshaft locknut and tighten to 159 ft. lbs. (216 Nm).
12. Install the ABS sensor with the bolt.
13. Install the rear wheels and lower the vehicle.
14. Connect the negative battery cable to the battery.

STEERING & SUSPENSION

Air Bag

✳✳ CAUTION

Some vehicles are equipped with an air bag system, also known as the Supplemental Inflatable Restraint (SIR) or Supplemental Restraint System (SRS). The system must be disabled before performing service on

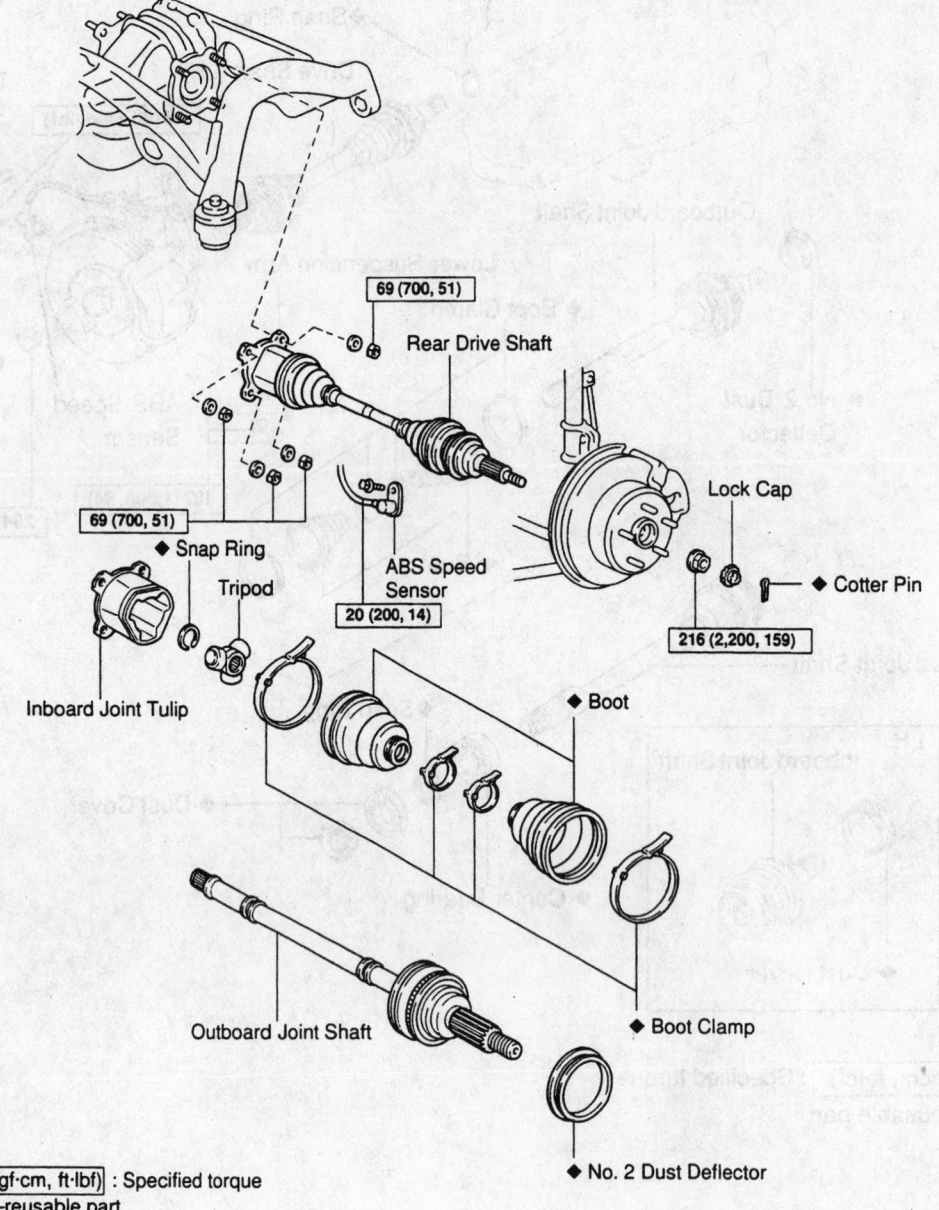

N·m (kgf·cm, ft·lbf) : Specified torque
◆ Non–reusable part

Exploded view of the rear halfshaft—RX300 model with 4WD

7924ZG88

or around system components, steering column, instrument panel components, wiring and sensors. Failure to follow safety and disabling procedures could result in accidental air bag deployment, possible personal injury and unnecessary system repairs.

PRECAUTIONS

Several precautions must be observed when handling the inflator module to avoid accidental deployment and possible personal injury.

• Never carry the inflator module by the wires or connector on the underside of the module.

• When carrying a live inflator module, hold securely with both hands, and ensure that the bag and trim cover are pointed away.

• Place the inflator module on a bench or other surface with the bag and trim cover facing up.

• With the inflator module on the bench, never place anything on or close to the module which may be thrown in the event of an accidental deployment.

DISARMING

To avoid personal injury when working on vehicles equipped with an air bag, the negative battery cable must be disconnected and at least 90 seconds must elapse before working on the system. Failure to do so may result in deployment of the air bag.

Power Rack and Pinion Steering Gear

REMOVAL & INSTALLATION

RAV4

1. Disconnect the negative battery cable from the battery.

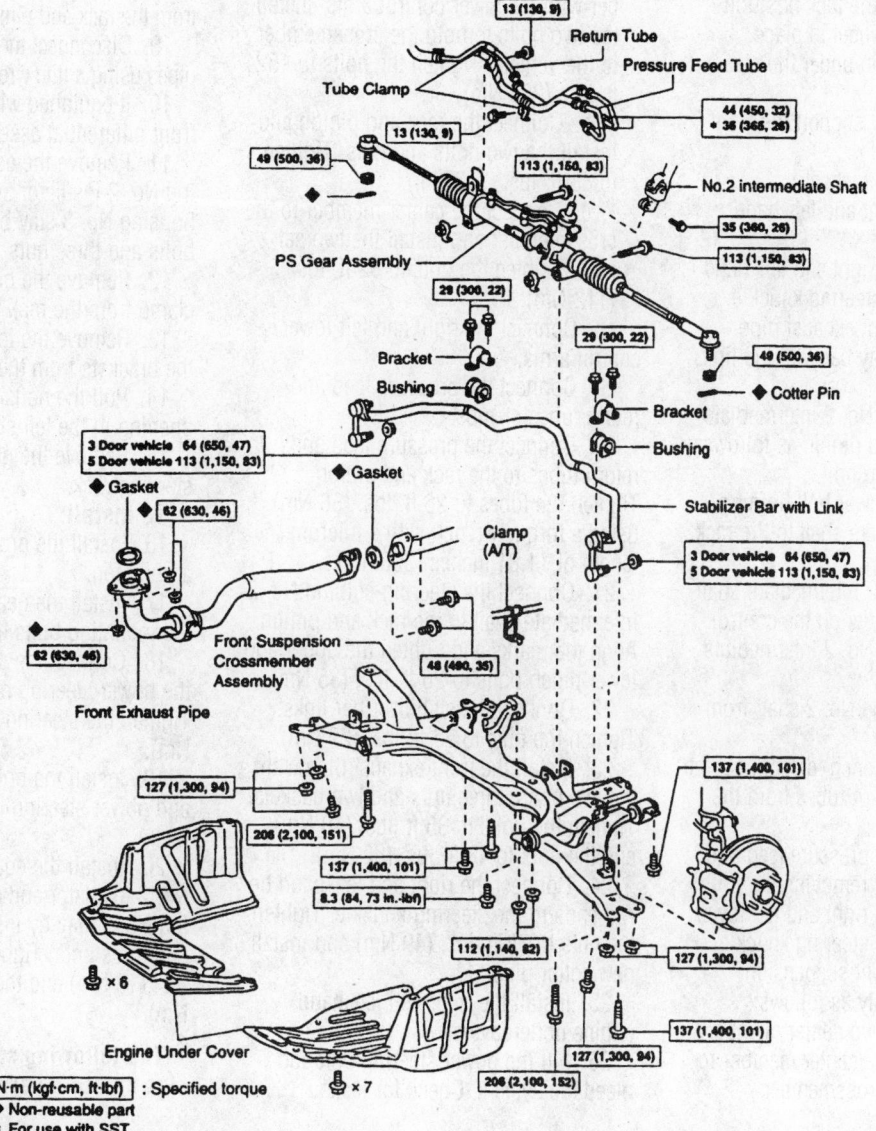

Rack and pinion exploded view—RAV4

✳✳ CAUTION

To avoid personal injury when working on air bag equipped vehicles, work must be started after 90 seconds or longer from the time the ignition switch is turned to the LOCK position and the negative battery terminal is disconnected. If the air bag system is disconnected with the ignition switch at the ON or ACC, diagnostic codes will be set. When removing the air bag, take care not to pull the air bag wiring harness. When carrying the wheel pad, carry it with the upper surface facing away. When storing it, keep the upper surface of the pad facing upward.

2. Turn the key to the lock position and lock the steering wheel in place.

3. Place a drain pan under the steering rack.

4. Raise and safely support the front of the vehicle.

5. Remove the front wheels.

6. Remove the right and left-hand engine undercovers.

7. Disconnect the right and left-hand tie rod ends from the steering knuckle.

8. Remove the front exhaust pipe.

9. Remove the sway bar with the links from the vehicle.

10. Disconnect the No. 2 intermediate shaft from the rack and pinion as follows:

    a. Loosen the top bolt.

    b. Remove the lower bolt holding the No. 2 intermediate shaft to the rack and pinion.

    c. Shift the No. 2 intermediate shaft and place matchmarks on the control valve shaft and the No. 2 intermediate shaft.

    d. Disconnect the No. 2 shaft from the rack and pinion.

11. Using a line wrench, disconnect the pressure feed and return tubes from the rack and pinion.

12. Disconnect the pressure feed and return tube clamps by removing the bolt.

13. Disconnect the right and left lower control arms from the steering knuckle.

14. Remove the front suspension crossmember assembly as follows:

    a. Remove the two center member set nuts holding the center member to the middle of the crossmember.

    b. Remove the two rack and pinion assembly set bolts and nuts from the crossmember. Securely suspend the steering gear assembly.

    c. Support the suspension crossmember with a jack.

    d. Remove the two bolts from the suspension crossmember.

    e. Remove the suspension crossmember with the lower suspension arms.

15. Remove the rack and pinion from the vehicle.

To install:

16. Install the rack and pinion to the vehicle.

17. Install the crossmember to the vehicle as follows:

    a. Raise the suspension crossmember with the lower control arms. Install the two bolts to hold the crossmember to the vehicle. Tighten the bolts to 152 ft. lbs. (206 Nm).

    b. Connect the rack and pinion and install the two bolts and nuts. Tighten to 83 ft. lbs. (113 Nm).

    c. Connect the center member to the crossmember and install the two set nuts. Tighten the nuts to 82 ft. lbs. (112 Nm).

18. Connect the right and left lower control arms.

19. Connect the pressure feed and return tubes clamps.

20. Connect the pressure feed and return tubes to the rack and pinion. Tighten the tubes to 26 ft. lbs. (36 Nm) using a torque wrench with a fulcrum length of 11.81 inches (300mm).

21. Connect the steering column No. 2 intermediate shaft to the rack and pinion. Align the marks and tighten the upper and lower pinch bolts to 26 ft. lbs. (35 Nm).

22. Connect the stabilizer bar links. Tighten the nuts to 22 ft. lbs. (29 Nm).

23. Install the front exhaust pipe with the two bolts, three nuts and two gaskets. Tighten the bolts to 35 ft. lbs. (48 Nm) and the nuts to 46 ft. lbs. (62 Nm).

24. Connect the right and left-hand tie rod ends to the steering knuckle. Tighten the nuts to 36 ft. lbs. (49 Nm) and install new cotter pins.

25. Install the right and left-hand engine undercovers.

26. Fill the power steering unit and bleed the system. Check for leaks.

27. Install the front wheels and lower the vehicle. Check the front wheel alignment.

Previa

1. Remove the battery from the vehicle.

2. Raise the vehicle and support safely.

3. Remove the front wheels.

4. Remove the engine undercovers.

5. Remove the cotter pins and the nuts to the tie rod ends.

6. Using a separator tool (SST 0961112010 or equivalent), disconnect the tie rod ends from the steering knuckle.

7. Place matchmarks on the universal joint and steering column shaft.

8. Remove the lower bolt and loosen the upper bolt to the sliding yoke. Slide the yoke upward to disconnect the sliding yoke from the rack and pinion.

9. Disconnect the pressure and return pipes using a line wrench.

10. If equipped with 4WD, remove the front differential assembly.

11. Remove the equipment drive housing No. 2 insulator and equipment drive housing No. 3 stay by removing the two bolts and three nuts.

12. Remove the bolt and disconnect the clamp from the rack housing.

13. Remove the four bracket bolts and the brackets from the steering rack.

14. Pull the housing out through the opening in the left side of the body.

15. Remove the grommets from the steering rack.

To install:

16. Install the grommets to the rack and pinion.

17. Install the gear housing and tighten the mounting bolts to 70 ft. lbs. (95 Nm).

18. Connect the power steering lines to the power steering rack and pinion. Tighten the steering lines to 27 ft. lbs. (36 Nm).

19. Install the bolt to hold the clamp and power steering lines to the steering rack.

20. Install the equipment drive housing No. 2 insulator and equipment drive housing No. 3 stay by installing the two bolts and three nuts. Tighten the bolts to 13 ft. lbs. (18 Nm) and the nuts to 18 ft. lbs. (25 Nm).

➡**The following steps are for all vehicles unless otherwise specified.**

21. Install the lower bolt to the sliding yoke. Tighten the upper and lower bolts to 26 ft. lbs. (35 Nm).

22. 4WD only: install the front differential assembly.

23. Connect the tie rod ends to the steering knuckle and tighten the nuts to 36 ft. lbs. (49 Nm). Install a new cotter pin.

24. Install the engine undercovers.

25. Install the battery to the vehicle.

26. Install the front wheels and lower the vehicle.

27. Perform a front end alignment.

Sienna and RX300

1. Disconnect the negative battery cable. Wait at least 90 seconds before working on the vehicle to allow the SRS system to disarm.

2. Remove the right and left side fender apron seals.

3. Disconnect the right and left tie rod ends.

4. Place matchmarks on the intermediate shaft and remove the pinch bolt, then side the shaft out from under the vehicle.

5. Disconnect the power steering line clamp.

6. Disconnect the pressure and feed lines.

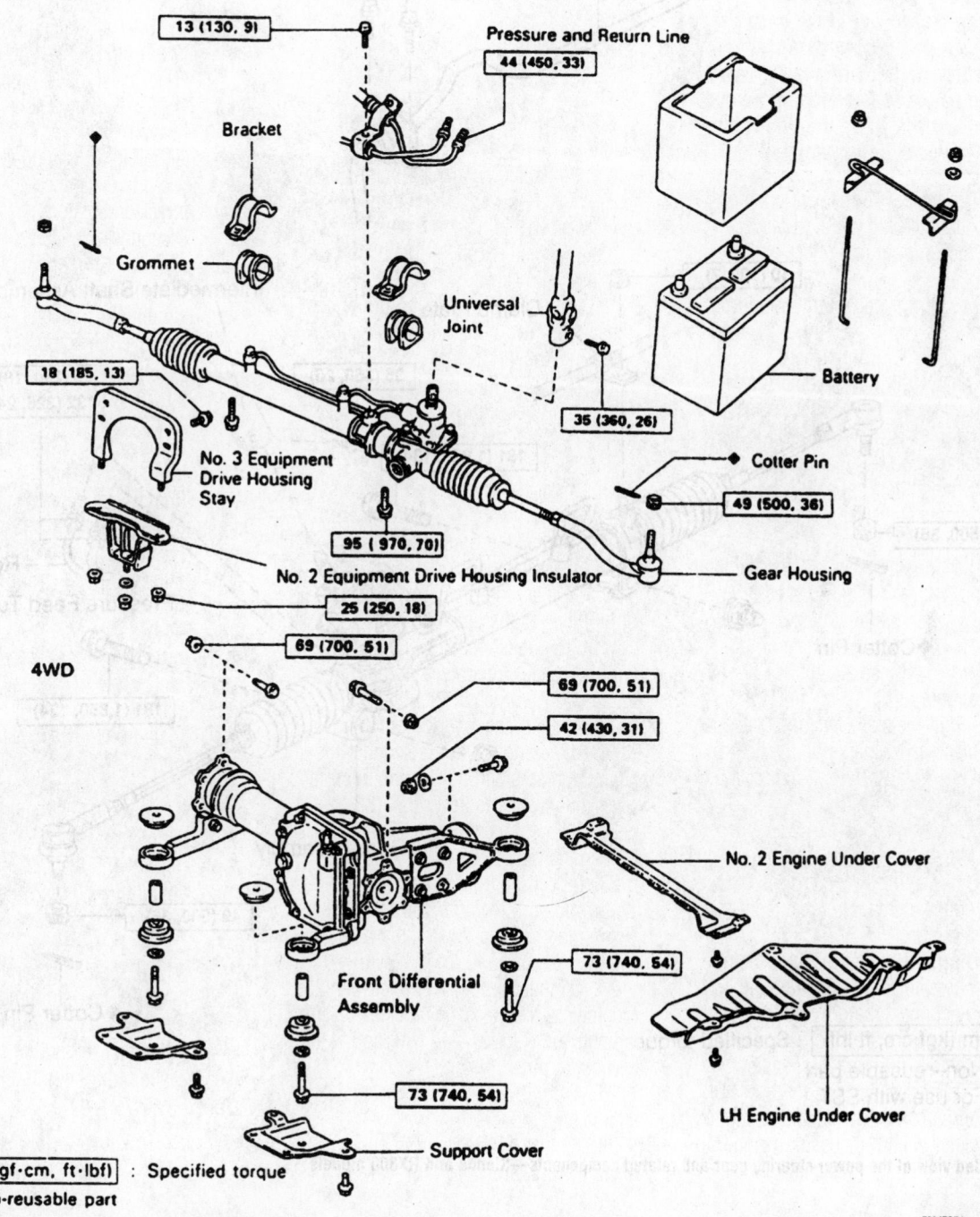

Rack and pinion removal and installation exploded view—Previa

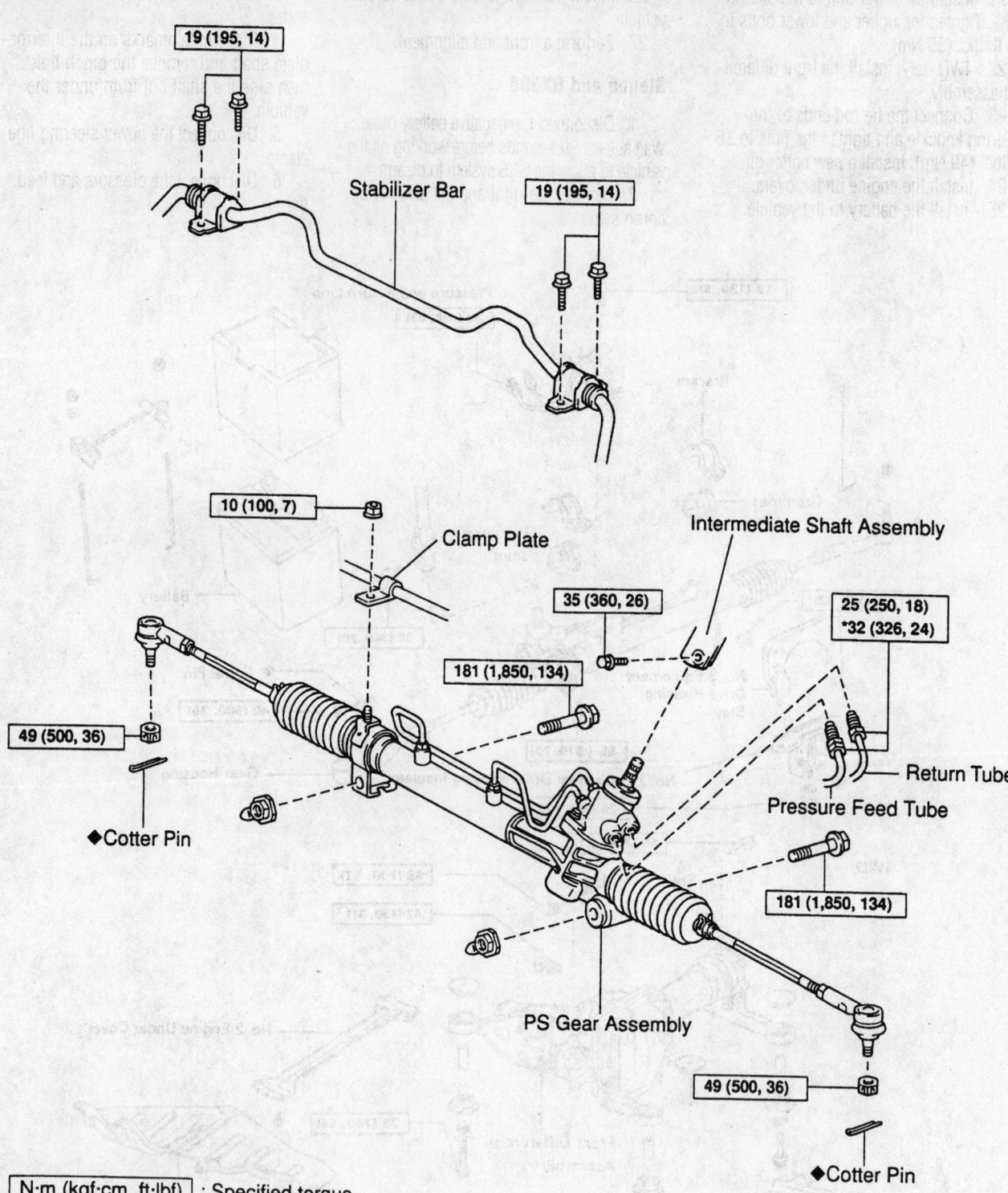

19 (195, 14)

Stabilizer Bar

19 (195, 14)

10 (100, 7)

Clamp Plate

Intermediate Shaft Assembly

35 (360, 26)

25 (250, 18)
*32 (326, 24)

181 (1,850, 134)

Return Tube

Pressure Feed Tube

49 (500, 36)

◆Cotter Pin

181 (1,850, 134)

PS Gear Assembly

49 (500, 36)

◆Cotter Pin

N·m (kgf·cm, ft·lbf) : Specified torque
◆ Non–reusable part
* For use with SST

7924ZG76

Exploded view of the power steering gear and related components—Sienna and RX300 models

7. Unbolt the stabilizer bar, but do not remove it.

8. Remove the heated oxygen sensor.

9. Lift up the stabilizer bar and remove the two gear assembly set bolts and nuts. Remove the gear assembly from the left side of the vehicle.

To install:

10. Install the gear assembly from the left side of the vehicle. Be careful not to damage the power steering lines. Lift the stabilizer bar and install the set bolts. Tighten the two set bolts and nuts to 134 ft. lbs. (181 Nm).

11. Install the heated oxygen sensor.

12. Position the stabilizer bar and tighten the bolt to 14 ft. lbs. (19 Nm) and the nut to 29 ft. lbs. (39 Nm).

13. Attach the pressure and feed return lines. Tighten them to 18 ft. lbs. (25 Nm).

14. Connect the clamp to the lines and tighten the nut to 7 ft. lbs. (10 Nm).

15. Connect the intermediate shaft. Be sure to align the matchmarks on the joint and main shaft. Tighten to 26 ft. lbs. (35 Nm).

16. Connect the tie rod ends.

17. Install the fender apron seals and securely tighten the bolts.

18. Remove the steering wheel pad.

19. Remove the steering wheel.

20. Position the front wheels facing straight ahead. Do this with the front of the vehicle on jackstands.

21. Center the spiral cable.

22. Install the steering wheel at the straight ahead position. Temporarily tighten the wheel set nut. Attach the wiring.

23. Bleed the power steering system.

24. Check the steering wheel center point. Tighten the steering nut to 26 ft. lbs. (35 Nm).

25. Check the front wheel alignment.

Strut

REMOVAL & INSTALLATION

RAV4

The RAV4 is equipped with front strut and rear shock absorber type suspension arrangement.

1. Disconnect the negative battery cable at the battery

2. Unfasten the lug nuts and remove the wheel.

3. Raise and support the vehicle safely.

➡**Do not support the weight of the vehicle on the suspension arm.**

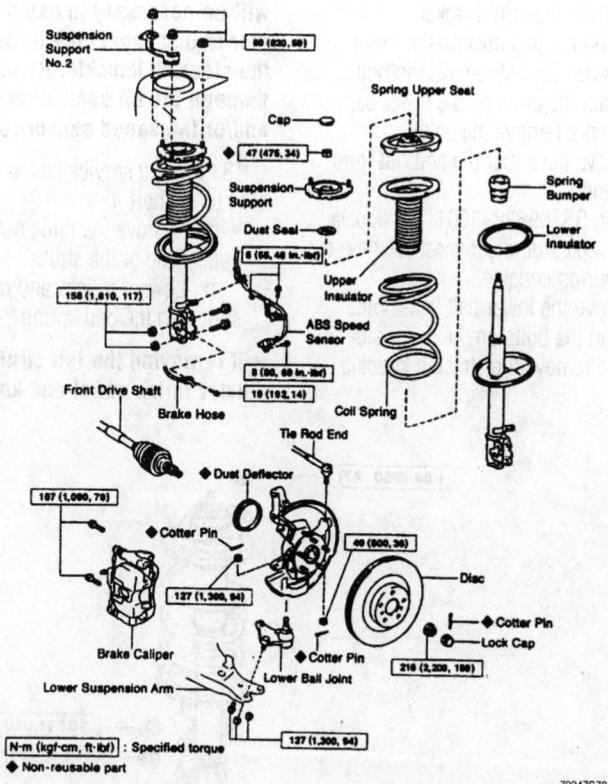

Strut assembly exploded view—RAV4

4. Remove the bolt and disconnect the brake hose from the strut.

5. If the vehicle is equipped with ABS brakes, remove the bolt holding the ABS electrical connection to the strut.

➡**It is not necessary to disconnect the brake hose from the brake caliper.**

6. Disconnect the strut from the steering knuckle by removing the bolts and nuts.

7. Remove the suspension support bracket at the top of the strut tower.

8. Remove the strut with coil spring from the vehicle.

To install:

9. Place the suspension support bracket at the top of the strut tower.

10. Align the strut to the strut tower bolt holes and secure the strut to the strut tower with the three nuts. Tighten the nuts to 59 ft lbs. (80 Nm).

11. Connect the steering knuckle to the strut lower bracket.

12. Install the two bolts and tighten the nuts to the bolts. Tighten the nuts to 117 ft. lbs. (158 Nm).

13. Connect the ABS electrical connector to the strut using the bolt. Tighten the bolt to 48 inch lbs. (5.4 Nm).

14. Install the brake line bolt to secure the brake line to the strut. Tighten the brake line bolt to 14 ft. lbs. (19 Nm).

15. If the brake lines were opened, add brake fluid and bleed the brake system.

16. Install the wheel and lower the vehicle.

17. Connect the negative battery cable to the battery.

18. Perform a front wheel alignment.

Previa

The Previa is equipped with front strut and rear shock absorber type suspension arrangement.

1. Raise and safely support the vehicle.

2. Remove the wheel and tire assembly.

3. If equipped with 4WD, remove the driveshaft locknut:

   a. Remove the cotter pin and lock cap.

   b. While applying the brake, remove the locknut.

   c. Remove the washer.

4. Disconnect the sway bar link from the strut by removing the nut.

5. If equipped with ABS, remove the speed sensor.

6. Using a line wrench, disconnect the brake line at the strut.

7. Remove the clips and disconnect the brake hose from the strut bracket.

8. Loosen the two nuts on the lower side of the strut. Do not remove the bolts.

9. Loosen the bolts on the lower ball joint, but do not remove the bolts.

10. Remove the cotter pin and nut from the tie rod end.

11. Using SST09628–10011, or equivalent tie rod separator, disconnect the tie rod from the steering knuckle.

12. Remove the lower ball joint bolts, two nuts, and the bolts on the lower side of the strut and remove the steering knuckle.

➡ If equipped with four wheel drive, it will be necessary to use a rubber hammer to disconnect the halfshaft from the steering knuckle. Be careful not to damage the oil seal, driveshaft boot, and/or the speed sensor rotor.

13. Place a service jack underneath the strut to support it.

   a. Remove the three nuts on the upper side of the strut.

   b. Lower the jack and remove the strut with the coil spring.

➡ If removing the left strut, remove the cluster finish panel and knee panel to gain access to the three upper strut bolts. If removing the right strut, remove the glove compartment door.

To install:

14. Place the strut with coil spring into position and support it with a jack. Install the three upper side nuts and tighten to 47 ft. lbs. (64 Nm). Install the knee panel and cluster finish panel if installing the left strut and the glove compartment if installing the right strut.

15. Install the steering knuckle and temporarily install the bolts to hold the steering knuckle to the strut.

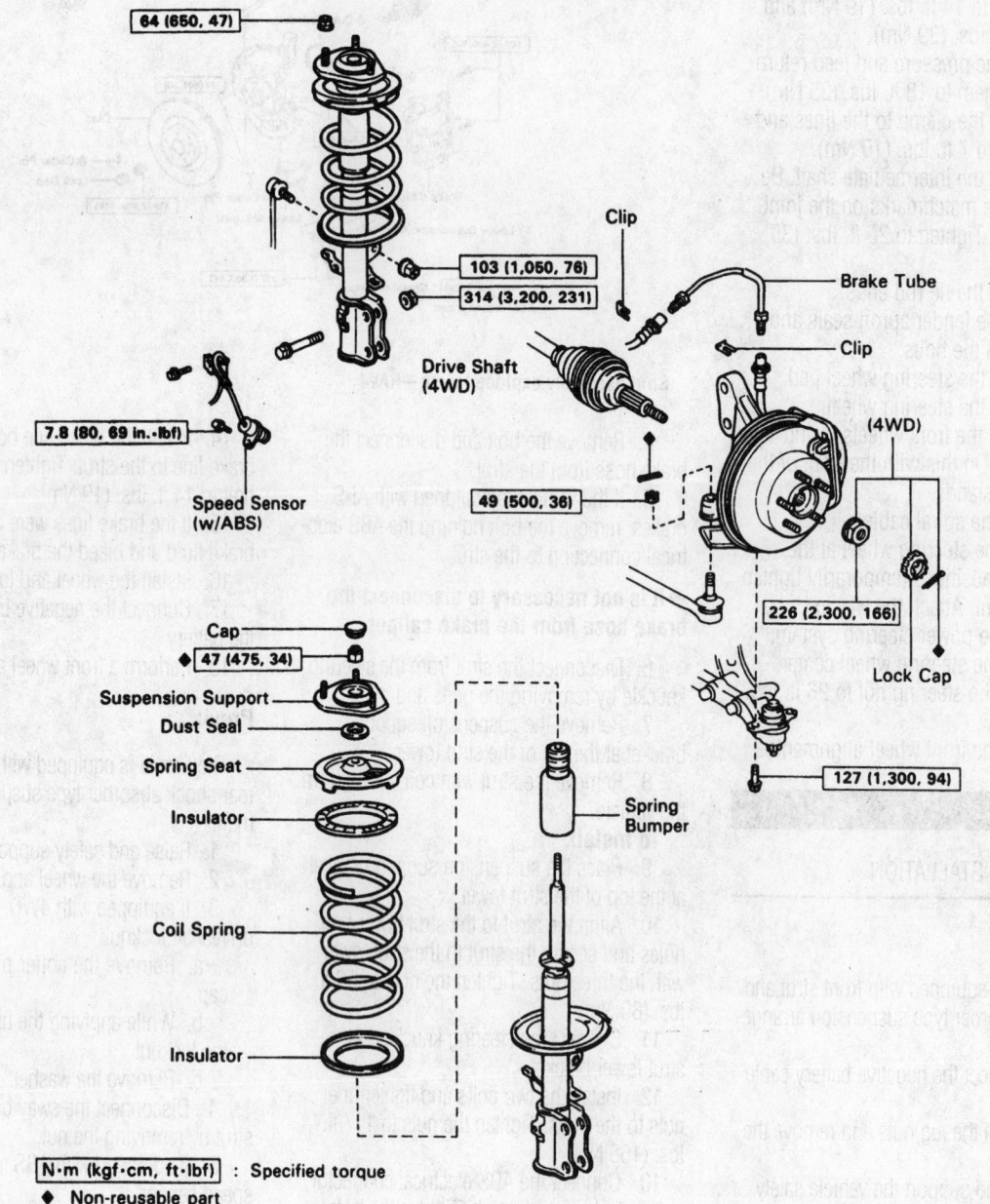

- 64 (650, 47)
- 103 (1,050, 76)
- 314 (3,200, 231)
- Clip
- Brake Tube
- Clip
- Drive Shaft (4WD)
- 7.8 (80, 69 in.·lbf)
- Speed Sensor (w/ABS)
- 49 (500, 36)
- (4WD)
- 226 (2,300, 166)
- Lock Cap
- Cap
- ◆ 47 (475, 34)
- Suspension Support
- Dust Seal
- Spring Seat
- Insulator
- Coil Spring
- Insulator
- Spring Bumper
- 127 (1,300, 94)

N·m (kgf·cm, ft·lbf) : Specified torque
◆ Non-reusable part

Front strut exploded view—Previa

7924ZG77

16. Connect the tie rod end to the steering knuckle. Tighten the nut to 36 ft. lbs. (49 Nm). Install a new cotter pin.

17. Tighten the nuts on the lower side of the strut to 231 ft. lbs. (314 Nm) and the ball joint bolts to 94 ft. lbs. (127 Nm).

18. Secure the brake hose to the strut and connect the brake line to the brake hose.

19. If equipped with ABS, install the speed sensor.

20. Connect the stabilizer bar link to the strut. Tighten the nut to 76 ft. lbs. (103 Nm).

21. Bleed the brake system and check for leaks.

22. If equipped with 4WD, install the driveshaft locknut. Tighten the nut to 152 ft. lbs. (206 Nm). Install the lock cap and a new cotter pin.

23. Install the wheel and tire assembly.

24. Check the front wheel alignment.

Sienna

The Sienna is equipped with front struts and rear shock absorbers.

REAR

1. Raise and support the vehicle safely.

➡ Do not support the weight of the vehicle on the suspension arm; the arm will deform under its weight.

2. Unfasten the lug nuts and remove the wheel.

3. Remove the brake hose and the ABS speed sensor wire from the strut.

4. Disconnect the sway bar link from the strut.

5. On Sienna model, remove the outer front cowl top panel.

6. Matchmark on the strut lower bracket and camber adjust cam, if equipped. Remove the two bolts and nuts which attach

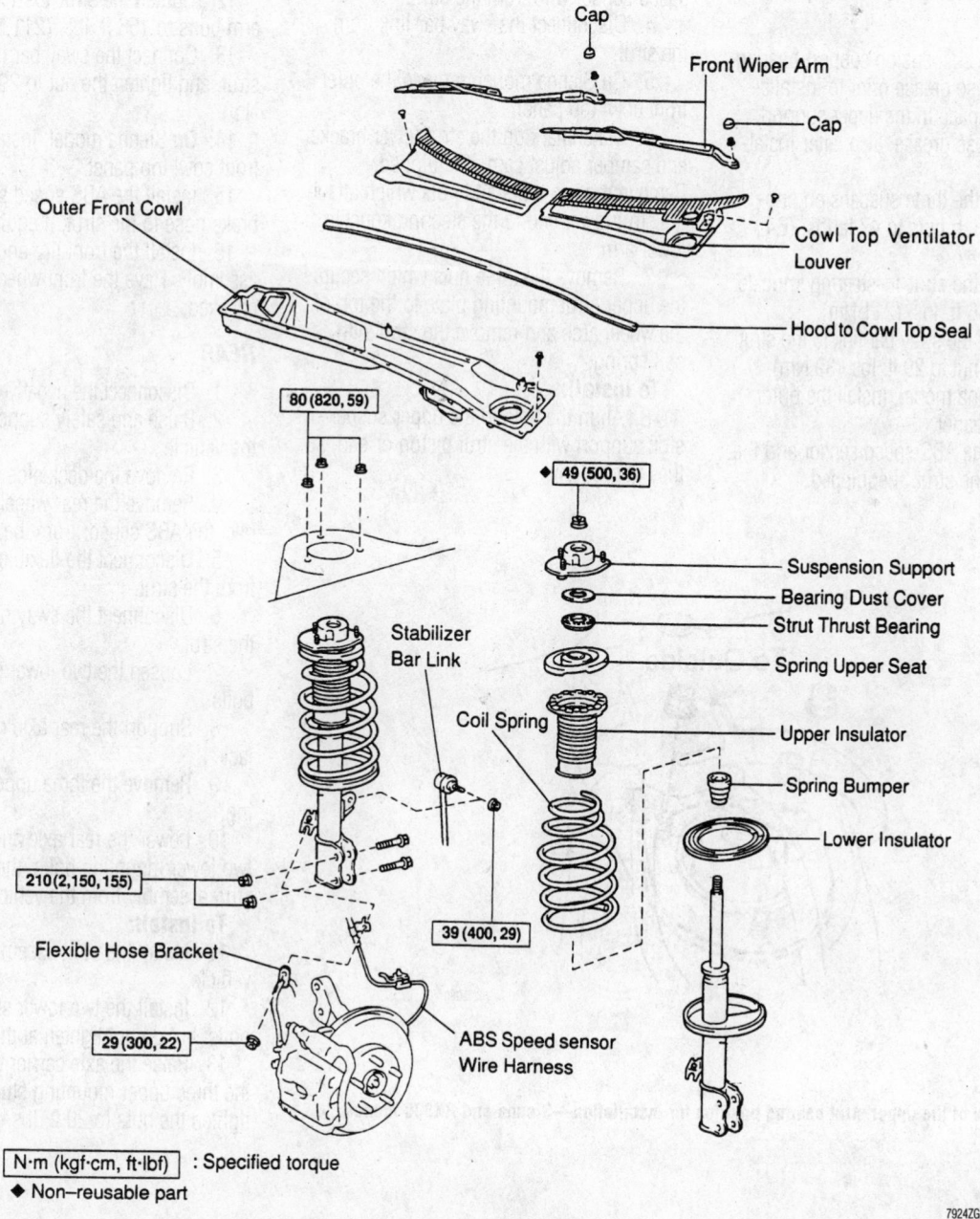

Cap
Front Wiper Arm
Cap
Cowl Top Ventilator Louver
Hood to Cowl Top Seal
Outer Front Cowl
80 (820, 59)
49 (500, 36)
Suspension Support
Bearing Dust Cover
Strut Thrust Bearing
Spring Upper Seat
Stabilizer Bar Link
Coil Spring
Upper Insulator
Spring Bumper
Lower Insulator
210 (2,150, 155)
39 (400, 29)
Flexible Hose Bracket
29 (300, 22)
ABS Speed sensor Wire Harness

N·m (kgf·cm, ft·lbf) : Specified torque
◆ Non–reusable part

7924ZG79

View of the front strut assembly and related components—Sienna and RX300 models

the strut lower end to the steering knuckle lower arm.

7. Remove the three nuts which secure the upper strut mounting plate to the top of the wheel arch and remove the strut with coil spring.

To install:

8. Align the hole in the upper suspension support with the strut piston or end, so they fit properly.

9. Always use a new nut and nylon washer on the strut piston rod end when securing it to the upper suspension support. Tighten the nut to 29–40 ft. lbs. (39–54 Nm).

➡ **Do not use an impact wrench to tighten the nut.**

10. Coat the suspension support bearing with multipurpose grease prior to installation. Pack the space in the upper support with multipurpose grease, also, after installation.

11. Tighten the three suspension support-to-wheel arch nuts to 47 ft. lbs. (64 Nm).

12. Tighten the strut-to-steering knuckle arm bolts to 156 ft. lbs. (211 Nm).

13. Connect the sway bar link to the strut, and tighten the nut to 29 ft. lbs. (39 Nm).

14. On Sienna model, install the outer front cowl top panel.

15. Install the ABS speed sensor and the brake hose to the strut, if equipped.

16. Install the front tire and wheel assembly. Have the front wheel alignment checked.

RX300

FRONT

1. Raise and support the vehicle safely.

➡ **Do not support the weight of the vehicle on the suspension arm; the arm will deform under its weight.**

2. Unfasten the lug nuts and remove the wheel.

3. Remove the brake hose and the ABS speed sensor wire from the strut.

4. Disconnect the sway bar link from the strut.

5. On Sienna model, remove the outer front cowl top panel.

6. Matchmark on the strut lower bracket and camber adjust cam, if equipped. Remove the two bolts and nuts which attach the strut lower end to the steering knuckle lower arm.

7. Remove the three nuts which secure the upper strut mounting plate to the top of the wheel arch and remove the strut with coil spring.

To install:

8. Align the hole in the upper suspension support with the strut piston or end, so they fit properly.

9. Always use a new nut and nylon washer on the strut piston rod end when securing it to the upper suspension support. Tighten the nut to 29–40 ft. lbs. (39–54 Nm).

➡ **Do not use an impact wrench to tighten the nut.**

10. Coat the suspension support bearing with multipurpose grease prior to installation. Pack the space in the upper support with multipurpose grease, also, after installation.

11. Tighten the three suspension support-to-wheel arch nuts to 47 ft. lbs. (64 Nm).

12. Tighten the strut-to-steering knuckle arm bolts to 156 ft. lbs. (211 Nm).

13. Connect the sway bar link to the strut, and tighten the nut to 29 ft. lbs. (39 Nm).

14. On Sienna model, install the outer front cowl top panel.

15. Install the ABS speed sensor and the brake hose to the strut, if equipped.

16. Install the front tire and wheel assembly. Have the front wheel alignment checked.

REAR

1. Disconnect the negative battery cable.

2. Raise and safely support the rear of the vehicle.

3. Remove the deck side cover.

4. Remove the rear wheels, and disconnect the ABS sensor from the strut bracket.

5. Disconnect the flexible brake hose from the strut.

6. Disconnect the sway bar link from the strut.

7. Loosen the two lower strut mounting bolts.

8. Support the rear axle carrier with a jack.

9. Remove the three upper strut mounting.

10. Lower the rear axle and remove the two lower mounting bolts, then remove the strut assembly from the vehicle.

To install:

11. Install the strut assembly into the vehicle.

12. Install the two lower strut mounting bolts, but do not tighten at this time.

13. Raise the axle carrier while aligning the three upper mounting studs, then tighten the nuts to 29 ft. lbs. (39 Nm).

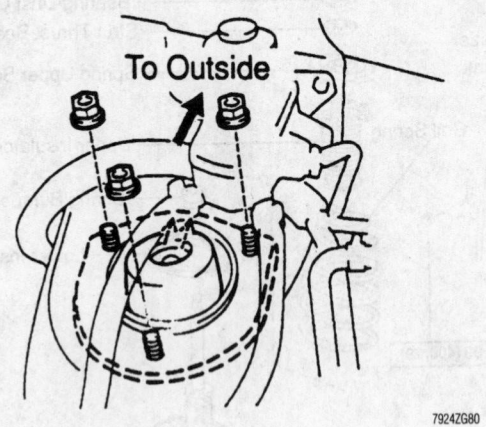

To Outside

7924ZG80

Cutaway view of the upper strut bearing position for installation—Sienna and RX300 models

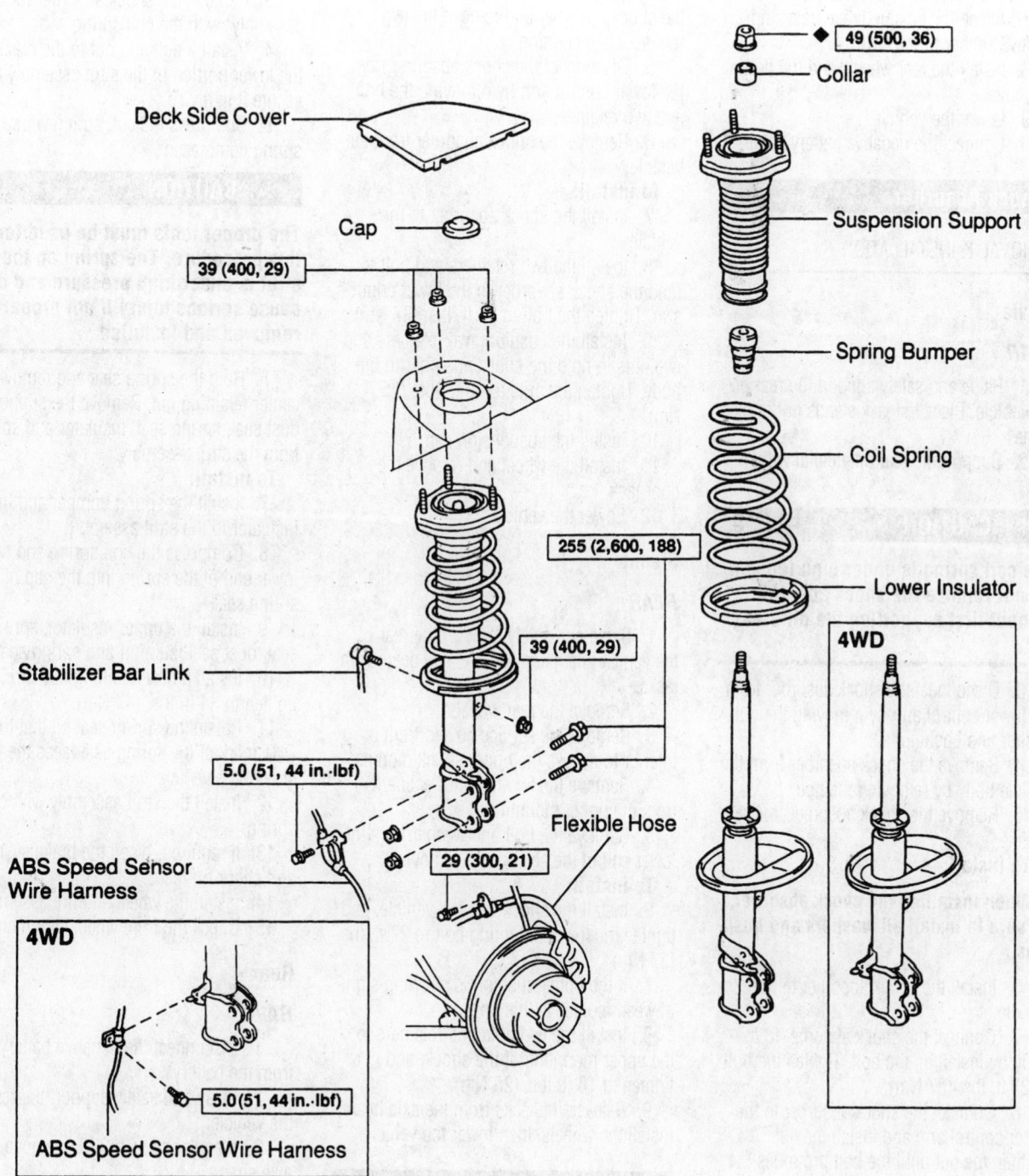

49 (500, 36) ◆

Collar

Suspension Support

Spring Bumper

Coil Spring

255 (2,600, 188)

Lower Insulator

4WD

Deck Side Cover

Cap

39 (400, 29)

39 (400, 29)

Stabilizer Bar Link

5.0 (51, 44 in.·lbf)

ABS Speed Sensor
Wire Harness

Flexible Hose

29 (300, 21)

4WD

5.0 (51, 44 in.·lbf)

ABS Speed Sensor Wire Harness

N·m (kgf·cm, ft·lbf) : Specified torque

◆ Non–reusable part

7924ZG89

Exploded view of the rear strut assembly—RX300

14. Lower the axle carrier.

15. Tighten the two lower mounting bolts to 188 ft. lbs. (255 Nm).

16. Connect the sway bar link and tighten the nut to 29 ft. lbs. (39 Nm).

17. Attach the flexible brake hose and the ABS sensor to the strut.

18. Install the rear wheels and the deck side cover.

19. Lower the vehicle.

20. Connect the negative battery cable.

Shock Absorber

REMOVAL & INSTALLATION

Previa

REAR

1. Raise and safely support the rear of the vehicle. Place the jack stands under the frame.

2. Support the rear differential with a jack.

✳✳ CAUTION

The coil spring is under high tension. Do not remove the shock absorber without first supporting the differential.

3. Disconnect the shock absorber from the lower control arm by removing the nut, washer and bushing.

4. Remove the shock absorber from the vehicle body by removing the bolt.

5. Remove the shock absorber from the vehicle.

To install:

➡**When installing the shock absorber, be sure to install all washers and bushings.**

6. Install the shock absorber to the vehicle.

7. Connect the shock absorber to the body by installing the bolt. Tighten the bolt to 27 ft. lbs. (37 Nm)

8. Connect the shock absorber to the lower control arm and install the nut. Tighten the nut until the bolt protrudes 0.0059 inch (1.5mm) or more.

9. With the shock absorbers installed, remove the jacks from the vehicle.

10. Lower the vehicle to the ground.

RAV4

REAR

1. Raise and safely support the rear of the vehicle.

2. Remove the rear wheel and support the No. 1 control arm with a floor jack.

3. Remove the suspension cap from inside the vehicle.

4. Remove the two nuts from the top of the shock absorber and remove the two retainers and cushion.

5. Disconnect the shock absorber from the lower control arm by removing the bolt and two retainers.

6. Remove the shock absorber from the vehicle.

To install:

7. Install the shock absorber to the vehicle.

8. Install the two retainers and bolt to hold the shock absorber to the lower control arm. Tighten the bolt to 27 ft. lbs. (37 Nm).

9. Install the cushion, two retainers and two nuts to hold the shock absorber to the body. Tighten the nuts to 18 ft. lbs. (25 Nm).

10. Install the suspension cap.

11. Install the wheel and remove the floor jack.

12. Lower the vehicle.

Sienna

REAR

1. Raise and safely support the rear of the vehicle, then support the axle beam with jacks.

2. Remove the rear wheels.

3. Remove the service covers from the interior to access the upper shock mounts.

4. Remove the two nut and retainers from the upper mounting.

5. Remove the bolt and washer from the lower end of the shock, and remove.

To install:

6. Install the shock into the vehicle, and tighten the lower mounting bolt to 27 ft. lbs. (37 Nm).

7. If the upper cushion is showing signs of wear, replace at this time.

8. Install the two nuts and retainers to the upper mounting of the shock, and tighten to 18 ft. lbs. (25 Nm).

9. Release the jacks from the axle beam, install the wheels, then lower the vehicle.

Coil Spring

REMOVAL & INSTALLATION

Front

ALL MODELS

1. Raise and safely support the vehicle.

2. Remove the wheel and tire assembly.

➡**If equipped, be careful not to damage the oil seal, driveshaft boot and/or speed sensor rotor when removing the steering knuckle.**

3. Remove the shock absorber (strut assembly) with the coil spring.

4. Install a bolt and nut to the bracket at the lower portion of the strut assembly and secure it in a vice.

5. Compress the coil spring with a spring compressor.

✳✳ CAUTION

The proper tools must be used for this procedure. The spring on the strut is under high pressure and can cause serious injury if not properly removed and installed.

6. Hold the spring seat and remove the center retaining nut. Remove the support, dust seal, spring seat, insulator and spring from the strut assembly.

To install:

7. Install the spring bumper and lower insulator to the strut assembly.

8. Compress the coil spring and fit the lower end of the spring into the gap of the spring seat.

9. Install the upper insulator, spring seat, dust seal, support and spring seat.

10. Install a new retaining nut and tighten to 34 ft. lbs. (47 Nm).

11. Rotate the spring seat so that the OUT mark of the spring seat faces the outside of the vehicle.

12. Install the strut assembly with coil spring.

13. If required, bleed the brake system and check for leaks.

14. Install the wheel and tire assembly.

15. Check the front wheel alignment.

Rear

RAV4

1. Disconnect the negative battery cable from the battery.

2. Raise and safely support the rear of the vehicle.

3. If equipped with 2WD, remove the axle shaft.

4. If equipped with 4WD, remove the halfshaft.

5. Remove the brake drum.

6. Remove the two brake line clamp bolts.

7. Remove the parking brake cable clamp bolt.

8. If equipped with ABS brakes, remove the ABS speed sensor and wiring harness.

9. Remove the rear axle hub with the brake by removing the four bolts. Support the hub securely.

10. Support the control arm with a floor jack. Disconnect the shock absorber from the control arm by removing the bolt.

➡**The control arm must be supported before removing the bolt for the shock absorber. Leave the floor jack under the control arm. Later, the floor jack will be lowered to remove the coil spring.**

11. Remove the cotter pins and nuts holding the lower and upper suspension arms to the control arm.

12. Using SST 09628–62011 or equivalent, disconnect the upper and lower control arms from the control arm.

13. Remove the coil spring and control arm as follows:

a. Place matchmarks on the toe adjust cam and body.

b. Loosen the bolt and lower the control arm to remove the coil spring and upper insulator.

c. Remove the bolt, toe-adjust cam, two attachments, nut and control arm.

14. Remove the bolt and spring bumper.

To install:

15. Install the spring bumper and bolt. Tighten the bolt to 9 ft. lbs. (13 Nm).

16. Install the control arm, two attachments, toe-adjust cam, bolt, and nut. Do not tighten the bolt at this time.

17. Install the spring and upper insulator and raise the control arm with a floor jack.

18. Install the upper and lower suspension arms to the control arm. Install and tighten the nuts to 76 ft. lbs. (103 Nm) and install new cotter pins.

19. Connect the shock absorber to the control arm. Tighten the bolt to 27 ft. lbs. (37 Nm).

20. Install the rear axle hub with the brake. Install the four bolts and tighten the bolts to 59 ft. lbs. (80 Nm).

21. If equipped with ABS brakes, install the ABS speed sensor and wiring harness. Tighten the ABS speed sensor to 69 inch lbs. (8 Nm) and the wiring harness to 9 ft. lbs. (13 Nm).

22. Install the parking brake cable clamp bolt. Tighten the bolt to 14 ft. lbs. (19 Nm).

23. Install the two brake line cable clamp bolts. Tighten the bracket bolt to 13 ft. lbs. (18 Nm) and the clamp bolt to 9 ft. lbs. (13 Nm).

24. Install the brake drum.

25. If equipped with 4WD, install the rear halfshaft.

26. If equipped with 2WD, install the axle shaft.

27. Install the rear wheel.

28. Lower the rear of the vehicle and stabilize the suspension. Align the matchmarks to the toe-adjust cam and tighten the bolt to 98 ft. lbs. (132 Nm).

29. Check the wheel alignment.

30. Connect the negative battery cable to the battery.

PREVIA

✳✳ CAUTION

The spring on the axle carrier is under high pressure and can cause serious injury if not properly removed and installed.

1. Raise and safely support the vehicle at the frame.

2. Support the axle housing with a floor jack.

3. Remove the wheel and tire assembly.

4. Remove the shock absorber to axle housing bolt. Disconnect both shock absorbers from the axle housing.

5. Disconnect the lateral control arm from axle housing by removing the nut.

6. Disconnect the LSPV spring from the lower control arm by removing the nut.

7. Disconnect the brake line from the brake hose at the body bracket.

8. Remove the clip and disconnect the brake hose from the body bracket.

9. If equipped with ABS brakes, remove the ABS wiring harness bracket.

10. Disconnect the parking brake cable from the lower control arm.

11. Lower the floor jack, then remove the coil spring(s) and the insulators.

To install:

12. Place the springs into position and raise the axle housing. Be sure to fit the lower end of the coil spring into the gap of the spring seat on the lower control arm.

13. Raise the rear axle housing.

14. Connect the parking brake cable to the lower control arm.

15. Install the lateral control rod to the suspension and tighten the bolts and nuts as follows:

• Body side bolt to 156 ft. lbs. (211 Nm).

• Axle housing side to 43 ft. lbs. (59 Nm).

16. Connect the shock absorber to the lower control arm and install the nut.

17. Connect the LSPV spring to the lower control arm and install the nut. Tighten the nut to 9 ft. lbs. (13 Nm).

18. If equipped, install the ABS wiring harness bracket.

19. Connect the brake hose to the bracket and install the clip.

20. Connect the brake tube to the brake hose and tighten the tube.

21. Bleed the brake system.

22. Install the wheels and lower the vehicle.

SIENNA

1. Remove the shocks.

2. Lower the jacks that are supporting the axle beam.

3. Remove the coil springs.

To install:

4. Install the coil springs.

5. Using jacks, raise the axle beam enough to apply tension on the springs.

6. Install the shocks.

Lower Ball Joint

REMOVAL & INSTALLATION

RAV4

✳✳ CAUTION

The Supplemental Inflatable Restraint (SIR) system must be disarmed before removing the ball joint. Failure to do so may cause accidental deployment of the air bag, resulting in unnecessary SIR system repairs and/or personal injury.

1. Disconnect the negative battery cable.

2. Raise the front of the vehicle and support it safely.

3. Remove the front wheel(s).

4. Remove the steering knuckle with the axle hub, from the vehicle.

5. Pry the dust deflector from the knuckle.

6. Remove the cotter pin and the nut from the ball joint stud.

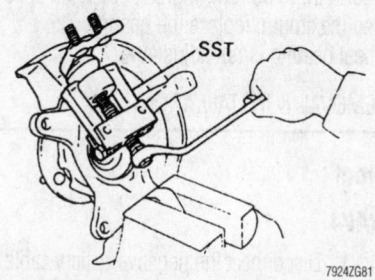

Use a two jaw puller to remove the lower ball joint—RAV4

7. Using the appropriate two jaw puller, remove the lower ball joint from the steering knuckle.

To install:

8. Install the lower ball joint onto the steering knuckle and tighten nut to 94 ft. lbs. (127 Nm). Install new cotter pin.

9. Align the hole in the dust deflector with the ABS speed sensor. Using the appropriate driver, install a new dust deflector.

10. Install the steering knuckle and hub onto the vehicle.

11. Install the front wheel(s).

12. Connect the negative battery cable.

Except RAV4

1. Raise and safely support the vehicle.

2. Remove the wheel and tire assembly.

3. Remove the steering knuckle with the axle hub, from the vehicle.

4. Pry the dust deflector from the knuckle.

5. Remove the cotter pin and nut from the ball joint.

6. Disconnect the ball joint from the steering knuckle by removing the two bolts.

7. Using a suitable ball joint separator tool (SST 09628–62011), remove the lower ball joint.

To install:

8. Install the lower ball joint with the nut and bolts. Tighten the nut to 76 ft. lbs. (103 Nm) and the two bolts to 94 ft. lbs. (127 Nm).

9. Install a new cotter pin to the ball joint.

10. Install the wheel and tire assembly and lower the vehicle.

Wheel Bearings

ADJUSTMENT

Front and Rear

Check the bearing play in the axial direction and also check the axle hub deviation. The maximum play for both checks should be 0.0020 in. (0.05mm). If greater than the specified maximum, replace the bearing. The wheel bearing is not adjustable.

REMOVAL & INSTALLATION

Front

RAV4

1. Disconnect the negative battery cable.

2. Raise the vehicle and support safely.

3. Remove the front wheels.

4. Remove the cotter pin and lock cap from the end of the halfshaft.

5. While applying the front brakes, remove the halfshaft locknut.

6. Remove the brake caliper and use a wire to support it out of the way.

➡**Never allow the caliper to hang freely from the brake hose.**

7. Matchmark the rotor to the hub and remove the rotor.

8. If equipped with ABS brakes, remove the ABS speed sensor from the steering knuckle.

9. Loosen the nuts on the lower end of the strut.

10. Disconnect and separate the tie rod end from the steering knuckle.

11. Disconnect the lower control arm from the ball joint by removing the bolt and two nuts.

12. Remove the halfshaft from the axle hub. Secure the shaft out of the way using a wire. Be careful not to damage the shaft boot or ABS sensor rotor.

13. Remove the two nuts on the lower end of the strut and remove the steering knuckle.

14. Clamp the steering knuckle in a vise with soft jaws to protect the knuckle.

15. Carefully pry the dust deflector from the hub.

16. Remove the ball joint from the steering knuckle.

17. Using slide hammer, remove the hub from the knuckle.

18. Using press and arbor tool, remove the inner race from the hub.

19. Remove the four bolts to the dust cover, then remove dust cover.

20. Using Seal Removal tool SST 09308–00010 or equivalent, remove the inner oil seal.

21. Using Seal Removal tool SST 09308–00010 or equivalent, remove the outer oil seal.

22. Using snapring pliers, remove the snapring.

23. Take the inner race (removed from hub) and install it on the outside of the bearing.

24. Using a bearing driver, drive the bearing from the steering knuckle.

To install:

25. Clean bearing seating surfaces with a clean, dry rag.

26. Using a press and Bearing Installer tool SST 09608–32010 or equivalent, install the bearing into the knuckle.

27. Install the snapring.

28. Install the dust cover. Tighten the four bolts to 74 inch lbs. (8.3 Nm).

29. Using a seal driver, install a new outer oil seal. Apply multi-purpose grease to the oil seal lip.

30. Press the hub into the steering knuckle.

31. Using a seal driver, install a new inner oil seal. Apply multi-purpose grease to the oil seal lip.

32. Install the lower ball joint to the steering knuckle. Tighten the nut to 94 ft. lbs. (127 Nm) and install a new cotter pin.

33. Align the hole in the dust deflector and the hole for the ABS speed sensor and install the dust deflector.

34. Position the knuckle to the lower strut and install the bolts.

35. Install the lower ball joint to the lower arm. Tighten the bolts to 94 ft. lbs. (127 Nm).

36. Connect the tie rod end to the steering knuckle. Tighten the nut to 36 ft. lbs. (49 Nm).

37. Install the halfshaft the hub and knuckle.

38. Install and tighten the nuts on the lower strut to 117 ft. lbs. (158 Nm).

39. Install the ABS speed sensor. Tighten the mounting bolt to 69 inch lbs. (7.8 Nm).

40. Align the matchmark and install the rotor on the hub. Install the brake caliper. Tighten the mounting bolts to 79 ft. lbs. (107 Nm).

41. Have a helper apply the brakes and install the axle locknut. Tighten the nut to 159 ft. lbs. (216 Nm). Install the lock cap and a new cotter pin.

42. Install the wheel.

43. Turn the wheel by hand, verify that the wheel turns without noise and without binding.

44. Lower the vehicle.

45. Connect the negative battery cable to the battery and check the signal from the ABS sensor.

PREVIA

1. Raise and safely support the vehicle.

2. Remove the wheel and tire assembly.

3. Remove the steering knuckle from the vehicle.

4. On 2WD vehicles, pry off the grease cap using a small prybar.

   a. Using a chisel and hammer, release the nut caulking.

   b. Remove the locknut from the hub.

   c. Remove the spacer (w/o ABS) or speed sensor rotor (w/ABS).

5. Using a suitable puller (SST 09520–00031), remove the wheel hub from the knuckle.

6. Remove the bearing from the hub using a press and suitable bearing separator tool.

7. Remove the oil seal from the axle hub.

8. Remove the backing plate by removing the three bolts.

9. If equipped with 4WD, remove the dust deflector and oil seal.

10. Remove the bearing snapring from the knuckle.

11. Press the inner bearing from the knuckle.

To install:

12. Using a press and arbor tool 09608–10010 or equivalent, press the inner bearing into the steering knuckle.

13. Install the snapring.

14. Install the outer bearing.

15. Install the outer oil seal. The seal should be flush with the end surface of the steering knuckle.

16. Install the dust deflector with the three bolts.

17. Press the axle hub onto the steering knuckle.

18. On 2WD vehicles, install the spacer (w/o ABS) or speed sensor rotor (w/ABS).

a. Install a new nut to the hub and tighten to 147 ft. lbs. (199 Nm). Caulk the nut.

b. Install the grease cap.

19. On 4WD vehicles, install a new inner oil seal. Install the dust deflector.

20. Install the steering knuckle assembly onto the vehicle.

21. Install the tie rod end to the steering knuckle.

22. Install the wheel. Lower the vehicle.

23. Check the vehicle's front end alignment.

SIENNA AND RX300

1. Raise the vehicle and support safely. Remove the front wheels and the fender apron seal.

2. Check the bearing backlash and axle hub deviation.

a. Remove the two brake caliper set bolts.

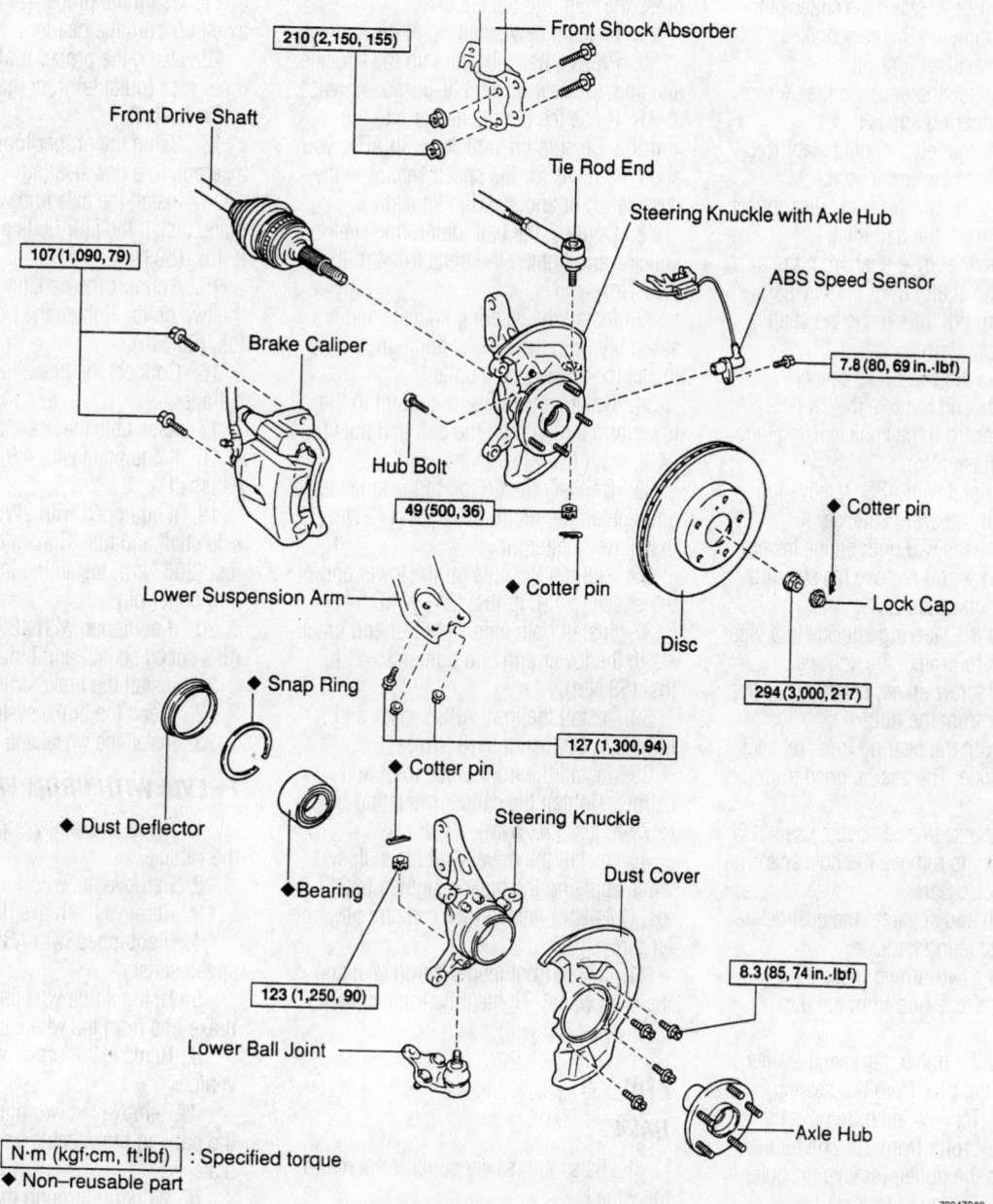

Exploded view of the front hub, bearing and steering knuckle assembly—Sienna

7924ZG82

b. Hang the caliper using stiff wire on the shock absorber assembly.

c. Remove the rotor.

d. Place a dial indicator near the center of the axle hub and check the backlash in the bearing shaft direction.

e. Backlash maximum should read 0.0020 inch (0.05mm). If the specification is greater than this, replace the bearing.

f. Using the dial indicator, check the deviation at the surface of the axle hub outside and hub bolt. Maximum is 0.0020 inch (0.05mm). If greater than specified, replace the axle hub.

3. Install the rotor and caliper assembly. Remove the cotter pin and lock cap off the center hub nut. Discard the cotter pin.

4. While applying the front brakes, remove the drive shaft locknut.

5. Disconnect and separate the tie rod end from the steering knuckle.

6. Remove the left and right stabilizer end brackets from the lower arms.

7. Remove the two nuts and disconnect the lower arm from the ball joint.

8. Remove the drive shaft from the axle hub. Secure the shaft out of the way using wire. Be careful not to damage the shaft boot or ABS sensor rotor.

9. Remove the two brake caliper mounting bolts and remove the caliper. Support caliper from the vehicle using wire. Remove the brake rotor.

10. If equipped with ABS, remove the sensor from the steering knuckle.

11. Remove the two nuts on the lower end of the shock and remove the steering knuckle and hub assembly.

12. Clamp the steering knuckle in a vise with soft jaws to protect the knuckle.

13. Using screw driver, carefully pry the dust deflector from the hub.

14. Drive out the bearing inner oil seal from the knuckle. The seal is pried from the knuckle bore.

15. After the seal is removed, use snapring pliers to remove the hole snapring from the knuckle bore.

16. Unbolt and separate the dust deflector from the steering knuckle.

17. Using a two-armed mechanical puller, pull the axle hub from the dust deflector.

18. Using the puller, remove the inner (inside) bearing race from the bearing.

19. Using Torx® wrench, remove the sensor control rotor from the axle hub.

20. Using the puller, remove the outer bearing race. Set the outer race aside.

21. Remove the outer bearing seal in the same manner as the inner seal.

22. Take the inner (outside) race and install it inside the bearing.

23. With a piece of brass stock, tap the bearing from the steering knuckle.

To install:

24. Clean all the oil seal and bearing seating surfaces with a clean, dry rag.

25. Using Bearing Driver SST No. 09608–32010 or equivalent and a press, drive the bearing into the bore.

26. Turn and insert the side lip of the new outer oil seal into the factory tool and drive the seal into the steering knuckle.

27. Attach the brake disc cover to the steering knuckle with the bolts.

28. Apply multi-purpose grease between the oil seal lip, oil seal and bearing and press the hub into the knuckle.

29. Install a new snapring in the knuckle.

30. Press a new oil seal into the knuckle and coat the seal with multi-purpose grease.

31. Press the dust deflector into the knuckle. Be sure on vehicles with ABS, you align the holes for the speed sensor in the dust deflector and steering knuckle.

32. Connect the ball joint to the steering knuckle and tighten the bolts to 94 ft. lbs. (127 Nm).

33. Install the steering knuckle and hub assembly onto the vehicle and temporarily install the lower shock bolts.

34. Connect the lower ball joint to the lower arm and tighten the bolt and nuts to 94 ft. lbs. (127 Nm).

35. Connect the tie rod to the knuckle and tighten the nut to 36 ft. lbs. (49 Nm). Install new cotter pin.

36. Tighten the nuts on the lower end of the shock to 156 ft. lbs. (211 Nm).

37. Install both side stabilizer end brackets to the lower arm and tighten to 43 ft. lbs. (58 Nm).

38. Install the front ABS sensor and tighten to 69 inch lbs. (8 Nm).

39. Install the front brake rotor and caliper. Tighten the caliper mounting bolts to 79 ft. lbs. (107 Nm).

40. Install the drive shaft locknut, and while applying the brakes, tighten to 217 ft. lbs. (294 Nm). Install lock cap and new cotter pin.

41. Install front fender apron seal and the front wheel. Tighten the front wheel to 76 ft. lbs. (103 Nm).

Rear

RAV4

1. Raise and safely support the rear of the vehicle.

2. Remove the wheel.

3. Remove the brake drum.

4. If equipped with ABS brakes, remove the bolt and the ABS speed sensor.

5. If equipped with 2WD, remove the cotter pin and lock cap to the axle shaft. Remove the nut to the axle shaft and remove the shaft from the hub.

6. If equipped with 4WD, remove the rear halfshaft from the hub.

7. Disassembly the rear brake components.

8. Disconnect the brake line from the wheel cylinder.

9. Disconnect the parking brake cable from the backing plate by removing the two bolts.

10. Remove the rear axle hub with the backing plate by removing the four bolts.

11. Using the proper tools, press out the axle hub from the bearing.

12. Using the proper tools, remove the inner race (outside) from the axle hub.

To install:

13. Using the proper tools, install the axle hub to a new bearing.

14. Install the axle hub with the backing plate. Install the four bolts and tighten to 59 ft. lbs. (80 Nm).

15. Connect the parking brake cable with the two bolts. Tighten the bolts to 69 inch lbs. (80 Nm).

16. Connect the brake line to the wheel cylinder.

17. Assemble the brake assembly.

18. If equipped with 4WD, install the halfshaft.

19. If equipped with 2WD, install the axle shaft and nut. Tighten the nut to 152 ft. lbs. (206 Nm). Install the lock cap and a new cotter pin.

20. If equipped with ABS, connect the ABS speed sensor and bolt.

21. Install the brake drum.

22. Bleed the brake system.

23. Install the wheel and lower the vehicle.

PREVIA WITH DRUM BRAKES

1. Raise and safely support the rear of the vehicle.

2. Remove the wheel and tire assembly.

3. Remove the brake drum.

4. If equipped with ABS, remove the speed sensor.

5. Using a line wrench, disconnect the brake line from the wheel cylinder.

6. Remove the brake shoes from the vehicle.

7. Remove the two bolts and remove the parking brake cable from the backing plate.

8. Working through the hole in the axle flange, remove the four backing plate mounting nuts.

9. Using SST 09520–00031 or equivalent (slide hammer puller), pull the axle shaft from the housing.

10. Remove the backing plate.

11. Remove the end gasket to the axle housing.

12. If equipped with ABS, press the seal and speed sensor rotor from the axle shaft.

13. Using a grinder, grind down the inner bearing retainer on the axle shaft. Using a chisel and a hammer, cut off the retainer and remove it from the shaft.

✲✲ WARNING

When removing the bearing, be careful not to damage the axle shaft.

14. Using a press, press the bearing from the axle shaft.

15. Remove the bearing outer retainer.

To install:

16. Install the bearing outer retainer to the axle shaft.

17. Using SST 09506–30012 or equivalent bearing driver and a press, install a new bearing.

18. Heat the new inner retainer to approximately 302° F (150° C) in an oil bath. Using a suitable installer and a press, install the inner retainer to the axle shaft while the retainer is still hot.

➡**Face the non-beveled side of the inner retainer toward the bearing.**

19. If equipped with ABS, carefully install the speed sensor rotor. Using a suitable installer and a press, install the new oil seal.

20. Apply liquid sealant on a new axle housing gasket and install the end gasket on the rear axle housing.

21. Install the backing plate.

22. Using a suitable tool, install the rear axle shaft.

➡**Be careful not to damage the oil seal and speed sensor rotor (w/ABS).**

23. Install the backing plate mounting nuts. Tighten the nuts to 59 ft. lbs. (59 Nm).

24. Install the parking brake cable, brake shoes, and the drum.

25. Connect the brake line to the wheel cylinder.

26. Bleed the brake system.

27. Install the rear wheel and lower the vehicle.

28. Road test the vehicle for proper operation.

PREVIA WITH DISC BRAKES

1. Raise and safely support the rear of the vehicle.

2. Remove the wheel and tire assembly.

3. If equipped with ABS brakes, remove the speed sensor.

4. Using a line wrench, disconnect the brake line from the brake hose.

5. Disconnect the brake hose from the axle bracket by removing the clip.

6. Remove the brake caliper support by removing the two bolts.

7. Remove the disc and parking brake shoes.

8. Remove the parking brake cable.

9. Remove the backing plate by removing the four mounting nuts.

10. Using SST 09520–00031 or equivalent slide hammer puller, pull the axle shaft from the housing.

11. Remove the axle housing end gasket.

12. Remove the four bolts and disconnect the backing plate from the axle shaft.

13. If equipped with ABS, press the seal and speed sensor rotor from the axle shaft.

14. Using a grinder, grind down the inner bearing retainer on the axle shaft. Using a chisel and a hammer, cut off the retainer and remove it from the shaft.

✲✲ WARNING

When removing the bearing, be careful not to damage the axle shaft.

15. Using a press, press the bearing from the axle shaft.

16. Remove the bearing outer retainer.

To install:

17. Place a new retainer gasket and bearing on the backing plate. Using a socket wrench and hammer, install the four bolts for the backing plate.

18. Install the backing plate to the axle shaft.

19. Using SST 09506–30012 or equivalent bearing driver and a press, install a new bearing.

20. Heat the new inner retainer to approximately 302° F (150° C) in an oil bath. Using a suitable installer and a press, install the inner retainer to the axle shaft while the retainer is still hot.

➡**Face the non-beveled side of the inner retainer toward the bearing.**

21. If equipped with ABS, carefully install the speed sensor rotor. Using a suitable installer and a press, install the new oil seal.

22. Apply liquid sealant on a new axle housing gasket and install the end gasket on the rear axle housing.

23. Using a suitable tool, install the rear axle shaft.

➡**Be careful not to damage the oil seal and speed sensor rotor (w/ABS).**

24. Install the backing plate mounting nuts. Tighten the nuts to 59 ft. lbs. (59 Nm).

25. Install the parking brake cable and parking brake shoes.

26. Install the rotor.

27. Install the brake caliper support to the vehicle and install the two bolts. Tighten the bolts to 65 ft. lbs. (88 Nm).

28. Install the brake hose to the axle bracket and install the clip.

29. Connect the brake line to the brake hose.

30. If equipped with ABS brakes, install the speed sensor.

31. Install the wheels to the vehicle.

32. Lower the vehicle.

SIENNA AND RX300 WITH 2WD

1. Raise and safely support the rear of the vehicle.

2. Remove the rear wheel.

3. If equipped with drum brakes, remove the drum.

4. If equipped with disk brakes, disconnect the flexible brake hose from the rear strut assembly, remove the brake caliper and support it from the vehicle using wire, then remove the brake rotor.

5. Detach the ABS speed sensor connector.

6. Remove the four nuts securing the rear axle hub assembly, then withdrawal the hub.

To install:

7. Install the new hub assembly and tighten the nuts to 59 ft. lbs. (80 Nm).

8. Connect the ABS speed sensor.

9. If equipped with disk brakes, install the brake rotor, install the brake caliper and tighten the mounting bolts to 34 ft. lbs. (47 Nm), then connect the flexible brake hose to the rear strut assembly and tighten the mounting bolt to 21 ft. lbs. (29 Nm).

10. If equipped with drum brakes, install the brake drum.

11. Install the wheels to the vehicle.

12. Lower the vehicle and test drive.

RX300 WITH 4WD

1. Raise and safely support the rear of the vehicle.

2. Remove the rear wheel.

3. Remove the cotter pin, lockcap, then the halfshaft locknut.

4. Disconnect the flexible brake hose from the rear strut assembly, remove the brake caliper and support it from the vehicle using wire, then remove the brake rotor.

5. Detach the ABS speed sensor connector.

6. Disassemble the parking brake assembly, and disconnect the parking brake cable.

7. Loosen the two lower strut mounting bolts.

8. Disconnect the strut rod rear bolt and nut, then separate it from the axle carrier.

9. Disconnect the No. 1 and No. 2 lower suspension arms.

10. Remove the two lower strut mounting bolts, then the rear axle hub with the carrier.

11. Using a slide hammer, remove the axle hub assembly from the axle carrier.

12. Using a press and the appropriate bearing driver, remove the inner race from the axle carrier.

13. Remove the four backing plate mounting bolts, then the backing plate.

14. Remove the inner and outer oil seals utilizing an appropriate seal remover.

15. Remove the snapring from the axle carrier.

16. Using a press and the appropriate bearing driver, remove the wheel bearing from the axle carrier.

To install:

17. Press the new bearing into place.

18. Using snapring pliers, install a new snapring.

19. Install a new outer oil seal, using a hammer and an appropriate seal driver.

Apply MP grease to the lip of the oil seal.

20. Install the backing plate and tighten the mounting bolts to 53 ft. lbs. (72 Nm).

21. Press the axle hub onto the axle carrier.

22. Install a new inner oil seal, using a hammer and an appropriate seal driver. Apply MP grease to the lip of the oil seal.

23. The completion of installation is the reverse of the removal procedure noting the following items:

• Tighten the No. 1 and No. 2 lower suspension arms to 131 ft. lbs. (177 Nm)

• Tighten the strut rod mounting to 91 ft. lbs. (123 Nm)

• Tighten the two lower strut mounting bolts to 188 ft. lbs. (255 Nm)

• Tighten the halfshaft locknut to 159 ft. lbs. (216 Nm)

24. Install the rear wheels and lower the vehicle.